BIOLOGICAL TREATMENT OF SOLID WASTE

Enhancing Sustainability

Edited by

Elena Cristina Rada, PhD

APPLE ACADEMIC PRESS

Apple Academic Press Inc.
3333 Mistwell Crescent
Oakville, ON L6L 0A2
Canada

Apple Academic Press Inc.
9 Spinnaker Way
Waretown, NJ 08758
USA

©2016 by Apple Academic Press, Inc.

Exclusive worldwide distribution by CRC Press, a member of Taylor & Francis Group

No claim to original U.S. Government works

Printed in the United States of America on acid-free paper

International Standard Book Number-13: 978-1-77188-279-8 (Hardcover)

International Standard Book Number-13: 978-1-77188-280-4 (ebook)

This book contains information obtained from authentic and highly regarded sources. Reprinted material is quoted with permission and sources are indicated. Copyright for individual articles remains with the authors as indicated. A wide variety of references are listed. Reasonable efforts have been made to publish reliable data and information, but the authors, editors, and the publisher cannot assume responsibility for the validity of all materials or the consequences of their use. The authors, editors, and the publisher have attempted to trace the copyright holders of all material reproduced in this publication and apologize to copyright holders if permission to publish in this form has not been obtained. If any copyright material has not been acknowledged, please write and let us know so we may rectify in any future reprint.

Trademark Notice: Registered trademark of products or corporate names are used only for explanation and identification without intent to infringe.

Library and Archives Canada Cataloguing in Publication

Biological treatment of solid waste : enhancing sustainability / edited by Elena Cristina Rada, PhD.

Includes bibliographical references and index.
ISBN 978-1-77188-279-8 (bound)
1. Sewage--Purification--Biological treatment. 2. Soil microbiology. 3. Compost. I. Rada, Elena Cristina, author, editor

TD755.B52 2015 628.3'5 C2015-903021-8

Library of Congress Cataloging-in-Publication Data

Biological treatment of solid waste : enhancing sustainability / Elena Cristina Rada, PhD, editor.

pages cm
Includes bibliographical references and index.
ISBN 978-1-77188-279-8 (alk. paper)
1. Sewage--Purification--Biological treatment. 2. Soil microbiology. 3. Compost. I. Rada, Elena Cristina, editor.

TD755.B4635 2015 628.3'5--dc23 2015016061

Apple Academic Press also publishes its books in a variety of electronic formats. Some content that appears in print may not be available in electronic format. For information about Apple Academic Press products, visit our website at **www.appleacademicpress.com** and the CRC Press website at **www.crcpress.com**

About the Editor

ELENA CRISTINA RADA, PhD

Elena Cristina Rada, PhD, earned her master's degree in Environmental Engineering from the Politehnica University of Bucharest, Romania; she received a PhD in Environmental Engineering and a second PhD in Power Engineering from the University of Trento, Italy, and the Politehnica University of Bucharest. Her post-doc work was in Sanitary Engineering from the University of Trento, Italy. She has been a professor in the Municipal Solid Waste master's program at Politehnica University of Bucharest, and has served on the organizing committees of "Energy Valorization of Sewage Sludge," an international conference held in Rovereto, Italy, and Venice 2010, an International Waste Working Group international conference. She also teaches seminars in the bachelor, master, and doctorate modules in the University of Trento and Padua and Politehnica University of Bucharest and has managed university funds at national and international level. Dr. Rada is a reviewer of international journals, a speaker at many international conferences, and the author or co-author of about a hundred research papers. Her research interests are bio-mechanical municipal solid waste treatments, biological techniques for biomass characterization, environmental and energy balances regarding municipal solid waste, indoor and outdoor pollution (prevention and remediation) and health, and innovative remediation techniques for contaminated sites and streams.

Contents

Acknowledgment and How to Cite ix
List of Contributors xi
Introduction xv

Part I: Microbial Technologies

1. **Management Options of Food Waste: A Review** 3
F. Girotto, L. Alibardi, and R. Cossu

2. **Effect of Increasing Total Solids Contents on Anaerobic Digestion of Food Waste under Mesophilic Conditions: Performance and Microbial Characteristics Analysis** 23
Jing Yi, Bin Dong, Jingwei Jin, and Xiaohu Dai

3. **Microbial Anaerobic Digestion (Bio-Digesters) as an Approach to the Decontamination of Animal Wastes in Pollution Control and the Generation of Renewable Energy** 47
Christy E. Manyi-Loh, Sampson N. Mamphweli, Edson L. Meyer, Anthony I. Okoh, Golden Makaka, and Michael Simon

4. **New Steady-State Microbial Community Compositions and Process Performances in Biogas Reactors Induced by Temperature Disturbances** 83
Gang Luo, Davide De Francisci, Panagiotis G. Kougias, Treu Laura, Xinyu Zhu, and Irini Angelidaki

Part II: Composting

5. **Composting of Organic Fraction of Municipal Solid Waste: A Pilot Plant in Maxixe District, Mozambique** 107
C. Collivignarelli, A. Perteghella, and M. Vacchari

6. **Changes in Bacterial and Fungal Communities across Compost Recipes, Preparation Methods, and Composting Times** 119
Deborah A. Neher, Thomas R. Weicht, Scott T. Bates, Jonathan W. Leff, and Noah Fierer

7. **Effects of Bulking Agents, Load Size or Starter Cultures in Kitchen-Waste Composting... 145**
Norazlin Abdullah, Nyuk Ling Chin, Mohd Noriznan Mokhtar, and Farah Saleena Taip

8. **Microbial Diversity of Vermicompost Bacteria that Exhibit Useful Agricultural Traits and Waste Management Potential 169**
Jayakumar Pathma and Natarajan Sakthivel

Part III: Biodrying

9. **Criteria for Assessing the Viability of a Small Scale MSW Bio-Drying Plant Aimed at RDF Production for Local Use 219**
E. C. Rada and M. Ragazzi

10. **Technical and Economic Efficiency of Utilization of Biogas from Animal Waste for Energy Generation 235**
E. Minciuc, R. Patrascu, M. Norisor, and D. Tutica

11. **Potential of Bio-Drying Applied to Exhausted Grape Marc ... 255**
Elena Cristina Rada and Marco Ragazzi

Author Notes 271

Index.. 275

Acknowledgment and How to Cite

The editor and publisher thank each of the authors who contributed to this book. The chapters in this book were previously published elsewhere. To cite the work contained in this book and to view the individual permissions, please refer to the citation at the beginning of each chapter. Each chapter was carefully selected by the editor; the result is a book that looks at the sustainable treatment of solid waste from a variety of perspectives. The chapters included are broken into three sections, which describe the following topics:

- When considering the biologic treatment of solid wastes, microbial technologies are the foundation for all other considerations, while at the same time they also offer unique treatments of their own. The articles in chapters 1 through 4 were chosen to specifically examine the contributions and implications of various forms of microbial technology.
- Composting is one of the most ancient of all waste management systems, dating at least as far back as the first century of the Common Era. Today's composting treatments offer important solutions to solid waste problems. The articles in chapters 5 through 8 were selected to represent investigations into important aspects of this technology, including vermicomposting.
- Biodrying, the process by which biodegradable waste is rapidly heated through initial stages of composting to remove moisture from a waste stream, thus reducing its overall weight, is the next step in our considerations. Biodrying reduces the moisture content of waste without the need for supplementary fossil fuels, and with minimal electricity consumption—and it can take as few as 8 days to dry waste in this manner, which in turn reduces the cost of solid waste management. Biodrying is not yet a perfect technology, however, and it calls for ongoing research and development. The articles chosen for chapters 9 to 11 represent a sample of this most recent ongoing research.

List of Contributors

Norazlin Abdullah
Department of Process and Food Engineering, Faculty of Engineering, Universiti Putra Malaysia, 43400 UPM Serdang, Selangor, Malaysia

L. Alibardi
Department of Civil, Environmental and Architectural Engineering, University of Padova - Via Marzolo 9, 35131,Padova, Italy

Irini Angelidaki
Department of Environmental Engineering, Technical University of Denmark, Copenhagen, DK-2800, Kgs Lyngby, Denmark

Scott T. Bates
Department of Plant Pathology, University of Minnesota, Minneapolis, Minnesota, United States of America

Nyuk Ling Chin
Department of Process and Food Engineering, Faculty of Engineering, Universiti Putra Malaysia, 43400 UPM Serdang, Selangor, Malaysia

C. Collivignarelli
DICATAM, University of Brescia, via Branze 43, 25123 Brescia, Italy

R. Cossu
Department of Industrial Engineering, University of Padova - Via Marzolo 9, 35131, Padova, Italy

Xiaohu Dai
National Engineering Research Center for Urban Pollution Control, College of Environmental Science and Engineering, Tongji University, Shanghai, People's Republic of China

Davide De Francisci
Department of Environmental Engineering, Technical University of Denmark, Copenhagen, DK-2800, Kgs Lyngby, Denmark

Bin Dong
National Engineering Research Center for Urban Pollution Control, College of Environmental Science and Engineering, Tongji University, Shanghai, People's Republic of China

Noah Fierer
Cooperative Institute for Research in Environmental Sciences, University of Colorado, Boulder, Colorado, United States of America and Department of Ecology and Evolutionary Biology, University of Colorado, Boulder, Colorado, United States of America

F. Girotto
Department of Civil, Environmental and Architectural Engineering, University of Padova - Via Marzolo 9, 35131,Padova, Italy

Jingwei Jin
National Engineering Research Center for Urban Pollution Control, College of Environmental Science and Engineering, Tongji University, Shanghai, People's Republic of China

Panagiotis G. Kougias
Department of Environmental Engineering, Technical University of Denmark, Copenhagen, DK-2800, Kgs Lyngby, Denmark

Treu Laura
Department of Environmental Engineering, Technical University of Denmark, Copenhagen, DK-2800, Kgs Lyngby, Denmark

Jonathan W. Leff
Cooperative Institute for Research in Environmental Sciences, University of Colorado, Boulder, Colorado, United States of America

Gang Luo
Shanghai Key Laboratory of Atmospheric Particle Pollution and Prevention (LAP3), Department of Environmental Science and Engineering, Fudan University, Shanghai, 200433, China

Golden Makaka
Department of Physics, University of Fort Hare, Alice Campus, Alice 5700, Eastern Cape Province, South Africa

Sampson N. Mamphweli
Fort Hare Institute of Technology, University of Fort Hare, Alice Campus, Alice 5700, Eastern Cape Province, South Africa

Christy E. Manyi-Loh
Fort Hare Institute of Technology, University of Fort Hare, Alice Campus, Alice 5700, Eastern Cape Province, South Africa and Applied and Environmental Microbiology Research Group (AEMREG), Department of Biochemistry and Microbiology, University of Fort Hare, Alice Campus, Alice 5700, Eastern Cape Province, South Africa

Edson L. Meyer
Fort Hare Institute of Technology, University of Fort Hare, Alice Campus, Alice 5700, Eastern Cape Province, South Africa

E. Minciuc
University Politehnica of Bucharest, Romania

Mohd Noriznan Mokhtar
Department of Process and Food Engineering, Faculty of Engineering, Universiti Putra Malaysia, 43400 UPM Serdang, Selangor, Malaysia

Deborah A. Neher
Department of Plant and Soil Science, University of Vermont, Burlington, Vermont, United States of America

M. Norisor
Power Engineering Faculty, University Politehnica of Bucharest

Anthony I. Okoh
Applied and Environmental Microbiology Research Group (AEMREG), Department of Biochemistry and Microbiology, University of Fort Hare, Alice Campus, Alice 5700, Eastern Cape Province, South Africa

Jayakumar Pathma
Department of Biotechnology School of Life Sciences, Pondicherry University, Kalapet, Puducherry, 605014, India

R. Patrascu
Faculty of Power Engineering, University Politehnica of Bucharest, 060032 Bucharest, Romania

A. Perteghella
DICATAM, University of Brescia, via Branze 43, 25123 Brescia, Italy

Elena Cristina Rada
Civil and Environmental Department, University of Trento, via Mesiano 77, 38100, Trento, Italy

Marco Ragazzi
Civil and Environmental Department, University of Trento, via Mesiano 77, 38100, Trento, Italy

Natarajan Sakthivel
Department of Biotechnology School of Life Sciences, Pondicherry University, Kalapet, Puducherry, 605014, India

Michael Simon
Fort Hare Institute of Technology, University of Fort Hare, Alice Campus, Alice 5700, Eastern Cape Province, South Africa

Farah Saleena Taip
Department of Process and Food Engineering, Faculty of Engineering, Universiti Putra Malaysia, 43400 UPM Serdang, Selangor, Malaysia

D. Tutica
Academy of Romanian Scientists, Romania

M. Vacchari
DICATAM, University of Brescia, via Branze 43, 25123 Brescia, Italy

Thomas R. Weicht
Department of Plant and Soil Science, University of Vermont, Burlington, Vermont, United States of America

Jing Yi
National Engineering Research Center for Urban Pollution Control, College of Environmental Science and Engineering, Tongji University, Shanghai, People's Republic of China

Xinyu Zhu
Department of Environmental Engineering, Technical University of Denmark, Copenhagen, DK-2800, Kgs Lyngby, Denmark

Introduction

Throughout most of history, humans had very little problem deciding what to do with left-over substances and objects for which they had no use. Waste material produced during pre-modern times was mainly ashes, vegetable refuse, and human and animal waste. These could all be released back into earth with minimum environmental impact. Objects made from metal—tools and jewelry, for the most part—were expected to last a lifetime, and then were passed on through generations.

As dense population centers grew, however, waste disposal became a problem, one that was exacerbated by industrialization. Solid wastes were often dumped into waterways, and diseases such as cholera spread through the polluted water. By the mid-nineteenth century, the problem had become so severe that it became a public debate that led to the first sanitation legislation.

Incinerators were the most common means of waste disposal through the rest of the nineteenth century and into the twentieth century. In the twentieth century, landfills became another option. Both forms of waste management had serious potential for polluting the land, air, and water around urban waste management plants.

Today, the development of sustainable solutions for waste management is one of society's great challenges. In this book, we will discuss biological treatment methods, including microbial technologies, composting, and biodrying.

Food waste is a part of the solid waste challenge. In chapter 1, Girotto and colleagues discuss several possibilities that are currently available for food waste. They offer a management hierarchy that begins with prevention and reduction of food waste. The next step is salvaging food for the poor from food waste, followed by reusing food waste to feed animals. Next in line are industrial uses (biofuels and plastic production, for example). Once these most sustainable options have been exhausted, composting is next in line. Incineration and landfill should be the last option considered—and hopefully, having passed through all the other layers of the hierarchy, the solid waste that reaches the end will be greatly reduced. The benefits of this are obvious.

Microbes play a significant role in two levels of this hierarchy: industrial uses and composting. Recently, numerous studies have analyzed microbial communities for their potential to play a part in sustainable management of solid waste. Pyrosequencing has gained increasing attention as next-generation technology for studying microbial diversity. In chapter 2, Jing Yi and colleagues apply this to various environmental samples, such as source waste, membrane filtration systems, and soil, in order to compare the microbial community structures in anaerobic digestion of food waste at different organic loading rates.

Overall, anaerobic digestion reduces biomass wastes and mitigates a wide spectrum of solid wastes' impact on the environment. It improves sanitation, helps to limit air and water pollution, and reduces greenhouse gas emissions. It also provides a high-quality nutrient-rich fertilizer and can yield energy in the form of biogas. Particularly in developing countries, biogas offers important solutions to societal demands. It can be used as fuel for cooking, lighting, and heating; it reduces the demand for wood and charcoal for cooking, which helps preserve forests and other natural environments, and it produces less indoor pollution when used than wood and charcoal do. In Western countries, biogas is converted to electricity and heat.

In chapter 3, Manyi-Loh and her colleagues offer a comprehensive description of anaerobic digestion as a means to resolve or mitigate the current dangers from animal waste disposal. They emphasize certain types of biodigesters (microbial communities), and conclude that anaerobic digestion of animal manure is a strong option for either safely reusing wastes or else transforming them into valuable materials and energy. The decomposition process reduces wastes' oxygen demand, destroys pathogenic microbes that pose health risks, destroys volatile fatty acids, and reduces greenhouse gas emissions. Ultimately, as already mentioned, it generates biogas and high-quality nutrient-rich fertilizer.

Based on these considerations, the objective of Gang Luo and colleagues in chapter 4 was to understand the role of stochastic factors and disturbance in the steady-state microbial community and its function in biogas reactors. They used three replicate biogas reactors to treat cattle manure in order to determine whether similar microbial communities would be achieved at steady states if the reactors were operated under the same conditions. Temperature stability is an essential factor for the biogas process, but biogas reactors are subject to temperature fluctuations due to mechanical errors and breakdowns. Gang Luo and colleagues therefore intentionally fluctuated the temperature in the three reactors in order to determine how much the temperature disturbance would alter the steady-state microbial community. They also monitored the reactors' performances as to biogas production, pH, and total volatile fatty acids. They found that similar steady-state process performance and microbial community profiles were achieved in the three reactors, suggesting that stochastic factors have a minor role in shaping the profile of the microbial community composition and activity in biogas reactor. Temperature disturbance, however, played a very important role in the microbial community composition as well as process performance. After temperature disturbances, all three reactors had increased methane yields and decreased volatile fatty acids concentrations. The authors also observed new steady-state microbial community profiles in all the biogas reactors after a temperature disturbance.

In chapter 5, we turn to composting, a form of solid waste management that also offers the opportunity to produce fertilizer for farming. Authors Collivignarelli and colleagues compare two compost heaps and discuss temperature trends and physical-chemical analyses. Their results were intended for practical application to a full-scale plant that treats 6 tons of organic waste per month. Their test on the pilot plant revealed that heap overturning frequencies can influence the composting process in terms of duration, temperature, and final product quality. More frequent heap overturning (every 5 days) avoided possible temperature peaks. Just as temperature fluctuations can have a negative impact on bioreactors, as discussed in chapter 4, they can also impede the microbial communities in compost. Overturning can also reduce the biological process duration and therefore also reduce the material's treatment time in the compost plant, allowing the plant to handle a larger quantity of daily waste. The authors found that too much overturning, however, can cause an excessive volatilization of organic parameters. Even then, the quality of compost produced was good, since there was no heavy metal pollution. Collivignarelli and colleagues conclude that the pilot tests' results represented a good starting point for the full-scale composting plant, while at the same time they called for ongoing monitoring and adjustments in order to achieve good quality compost production

Microbial communities in compost are abundant and diverse, influenced by both recipe and post-thermophilic treatment. Chapter 6, by Neher and her colleagues, is a comprehensive assessment of the bacteria and fungi associated with compost and the influence of both composting recipe and process on the structure of microbial communities. Of particular interest is their analysis of the ways in which communities change through time when compost is produced on a commercial-scale. They note that composition starts similarly after the thermophilic phase, but then shifts dynamically through time. Economic considerations encourage commercial composters to speed up the composting process, which has contributed to a focus on the effectiveness of the thermophilic phase. The authors note, however, that the curing phase offers a substrate and climate conducive for microbial recolonization, which can be accomplished either by inoculating post-thermophilic compost or preparing a palatable substrate that provides a competitive advantage for microbial colonization, slow-release fertility, and plant growth. Neher and her colleagues call for future research building on their results, in order to determine which recipe and post-thermophilic phase will best promote the agricultural goals of weed management, disease suppression, and plant growth promotion.

Composting has various practical requirements. One of these is a bulking agent. In chapter 7, Abdullah and colleagues study the effects of two common bulking media: newspaper and onion peels. They found that onion peels were more suitable; at a smaller waste load, compost maturity was attained more quickly with the onion peels than with the newspaper.

Vermicomposting is a cost-effective and eco-friendly waste management technology that has many advantages over traditional thermophilic composting. Vermicomposts are excellent sources of biofertilizers that improve the physiochemical and biological properties of agricultural soil. Vermicomposting amplifies the diversity and population of beneficial microbial communities. In chapter 8, Pathma and Sakthivel conclude that vermicomposting is beneficial in numerous ways, including soil reclamation, soil fertility, plant growth, and control of pathogens, pests, and nematodes. They point to this method as one of the most promising for sustainable agriculture.

Last, we consider biodrying. In chapter 9, we outline the advantages of this technology. These include reducing the cost per unit of thermal energy, as well as reducing the global impact of fossil fuels. We recommend that those nations complying with the Kyoto protocol should consider biodrying technology as a means to achieve their targets.

In Chapter 10, Minciuc and his colleagues focus on biological degradation of organic matter of animal origin aimed to energy exploitation. They discuss the formation of combustible gases due to the biological degradation. This is a process that takes place naturally on our planet. It is how natural gas was formed, and now we have the option of replicating that process. The authors present two solutions in order to use the animal waste for biogas production and its utilization for energy generation: electricity is used either in the plant or is sold to the company managing the national grid. They present also an economic balance and conclude that within the context of sustainable development, where economic, energy, and environmental issues must all be addressed simultaneously, the proposed technology depends on the input material that influences the biogas production.

Finally, in chapter 11, we examine the ways in which biodrying exhausted grape marc can be an improved alternative to thermal drying. Thermal drying generally requires an integrated and centralized plant of thermal pretreatment and combustion. Grape marc biodrying could open an option for decentralized pretreatment of this organic substrate before a centralized combustion.

Our world can no longer afford to consider waste as something that can be discarded with no regard for future use. Instead, if addressed correctly through policy and practice, solid waste can become a valuable resource. With rational and consistent waste management practices come opportunities to reap a range of benefits. Those benefits include economic, social, environmental factors. Biological treatment of solid wastes can help build a better world for future generations, one that has a healthier economy, a healthier society, and a healthier environment

Elena Cristina Rada

Food waste can be defined as material intended for human consumption that is instead discharged, lost, degraded or contaminated. In Chapter 1, Girotto and colleagues propose several solutions for the proper management of this food waste. Such solutions can be prioritised similarly to the waste management hierarchy. First steps are minimisation and use to feed poor. Food waste can then be used in industrial processes for the production of biofuels or biopolymers. Further steps foresee the recovery of nutrients and the fixation of carbon by composting. Final and less desirable solutions are incineration and landfilling.

The total solids content of feedstocks affects the performances of anaerobic digestion and the change of total solids content will lead the change of microbial morphology in systems. In order to increase the efficiency of anaerobic digestion, it is necessary to understand the role of the total solids content on the behavior of the microbial communities involved in anaerobic digestion of organic matter from wet to dry technology. In Chapter 2, Yi and colleagues compared the performances of mesophilic anaerobic digestion of food waste with different total solids contents from 5% to 20% and investigated the microbial communities in reactors were using 454 pyrosequencing technology. Three stable anaerobic digestion processes were achieved for food waste biodegradation and methane generation. Better performances mainly including volatile solids reduction and methane yield were obtained in the reactors with higher total solids content. Pyrosequencing results revealed significant shifts in bacterial community with increasing total solids contents. The proportion of phylum *Chloroflexi* decreased obviously with increasing total solids contents while other functional bacteria showed increasing trend. *Methanosarcina* absolutely dominated in archaeal communities in three reactors and the relative abundance of this group showed increasing trend with increasing total solids contents. These results revealed the effects of the total solids content on the performance parameters and the behavior of the microbial communities involved in the anaerobic digestion of food waste from wet to dry technologies.

With an ever increasing population rate; a vast array of biomass wastes rich in organic and inorganic nutrients as well as pathogenic microorganisms will result from the diversified human, industrial and agricultural activities. In Chapter 3, Manyi-Loh and colleagues posit anaerobic digestion as one of the best ways to properly handle and manage these wastes. Animal wastes have been recognized as suitable substrates for anaerobic digestion process, a natural biological process in which complex organic materials are broken down into simpler molecules in the absence of oxygen by the concerted activities of four sets of metabolically linked microorganisms. This process occurs in an airtight chamber (biodigester) via four stages represented by hydrolytic, acidogenic, acetogenic and methanogenic microorganisms. The microbial population and structure can be identified by the combined use of culture-based, microscopic and molecular techniques. Overall, the process is affected by bio-digester design, operational factors and manure characteristics. The purpose of anaerobic digestion is the production of a renewable energy source (biogas) and an odor free nutrient-rich fertilizer. Conversely, if animal wastes are accidentally found in the environment, it can cause a drastic chain of environmental and public health complications.

The microbial community in a biogas reactor greatly influences the process performance. However, only the effects of deterministic factors (such as temperature and hydraulic retention time (HRT)) on the microbial community and performance have been investigated in biogas reactors. Little is known about the manner in which stochastic factors (for example, stochastic birth, death, colonization, and extinction) and disturbance affect the stable-state microbial community and reactor performances. In Chapter 4, Luo and colleagues ran three replicate biogas reactors treating cattle manure to examine the role of stochastic factors and disturbance in shaping microbial communities. In the triplicate biogas reactors with the same inoculum and operational conditions, similar process performances and microbial community profiles were observed under steady-state conditions. This indicated that stochastic factors had a minor role in shaping the profile of the microbial community composition and activity in biogas reactors. On the contrary, temperature disturbance was found to play an important role in the microbial community composition as well as process performance for biogas reactors. Although three different temperature disturbances were applied to each biogas reactor, the increased methane yields (around 10% higher) and decreased volatile fatty acids (VFAs) concentrations at steady state were found in all three reactors after the temperature disturbances. After the temperature disturbance, the biogas reactors were brought back to the original operational conditions; however, new steady-state microbial community profiles were observed in all the biogas reactors. The present study demonstrated that temperature disturbance, but not stochastic factors, played an important role in shaping the profile of the microbial community composition and activity in biogas reactors. New steady-state microbial community profiles and reactor performances were observed in all the biogas reactors after the temperature disturbance.

Organic solid waste and the current lack of sound treatment for its valorisation are aspects of high priority for municipal decision makers in developing countries. Despite the remarkable use of composting processes in the field in developing countries, the availability of databases and information like those of developed countries is still lacking in literature. Thus, this lack represents constrains towards the sustainable implementation of the composting. The research and monitoring activities conducted on Maxixe composting pilot plant were carried out in order to obtain valuable and useful data and information for their further adoption on the full scale composting plant. Chapter 5, by Collivignarelli and colleagues, shows how different overturning frequencies influence the composting process of two heaps with the same initial characteristics, comparing also chemical analyses of the final compost. Moreover this paper is aimed to spread out results from field work and contributing at the same time to fill the current gap of real field data concerning composting process in developing countries.

Compost production is a critical component of organic waste handling, and compost applications to soil are increasingly important to crop production. However, we know surprisingly little about the microbial communities involved in the composting process and the factors shaping compost microbial dynamics. In Chapter 6, Neher and colleagues used high-throughput sequencing approaches to assess the diversity and composition of both bacterial and fungal communities in compost produced at a commercial-scale. Bacterial and fungal communities responded to both compost recipe and composting method. Specifically, bacterial communities in manure and hay recipes contained greater relative abundances of Firmicutes than hardwood recipes with hay recipes containing relatively more Actinobacteria and Gemmatimonadetes. In contrast, hardwood recipes contained a large relative abundance of Acidobacteria and Chloroflexi.

Fungal communities of compost from a mixture of dairy manure and silage-based bedding were distinguished by a greater relative abundance of Pezizomycetes and Microascales. Hay recipes uniquely contained abundant Epicoccum, Thermomyces, Eurotium, Arthrobotrys, and Myriococcum. Hardwood recipes contained relatively abundant Sordariomycetes. Holding recipe constant, there were significantly different bacterial and fungal communities when the composting process was managed by windrow, aerated static pile, or vermicompost. Temporal dynamics of the composting process followed known patterns of degradative succession in herbivore manure. The initial community was dominated by Phycomycetes, followed by Ascomycota and finally Basidiomycota. Zygomycota were associated more with manure-silage and hay than hardwood composts. Most commercial composters focus on the thermophilic phase as an economic means to insure sanitation of compost from pathogens. However, the community succeeding the thermophilic phase begs further investigation to determine how the microbial dynamics observed here can be best managed to generate compost with the desired properties.

To prevent the interruption of the carbon cycle by the disposal of waste to landfills, organic kitchen waste requires proper treatment such as composting to reduce its uncontrolled degradation on disposal sites and subsequent greenhouse gases, odour emissions and nutrient losses. Chapter 7, by Abdullah and colleagues, investigated the effects of bulking agent, newspaper and onion peels, composting waste load sizes of 2 and 6 kg, or the use of starter culture on kitchen-waste composting consisting of nitrogen-riched substrates, vegetable scraps and fish processing waste in an in-vessel system. The optimised formulation of kitchen waste mixture was used for a 30-day composting study, where the temperature profiles were recorded and the carbon-to-nitrogen ratios were measured as an indication of compost maturity. The kitchen-waste composting process was conducted in parallel in two fabricated kitchen waste composters. It was found that the onion peels were more suitable in producing matured compost where the carbon-to-nitrogen ratio reduced to 10 within 16 days of composting. A smaller kitchen waste load size of 2 kg gave a shorter composting time by half when compared to the 6 kg. The use of a microbial cocktail consisting seven types of bacteria and eight types of fungi isolated from soils as a starter culture for this kitchen-waste composting did not show advantages in accelerating the composting process. The results suggest that the in-vessel kitchen-waste composting can be efficient with a minimal load of about 2 kg using onion peels without additional starter culture.

Vermicomposting is a non-thermophilic, boioxidative process that involves earthworms and associated microbes. This biological organic waste decomposition process yields the biofertilizer namely the vermicompost. Vermicompost is a finely divided, peat like material with high porosity, good aeration, drainage, water holding capacity, microbial activity, excellent nutrient status and buffering capacity thereby resulting the required physiochemical characters congenial for soil fertility and plant growth. Vermicompost enhances soil biodiversity by promoting the beneficial microbes which inturn enhances plant growth directly by production of plant growth-regulating hormones and enzymes and indirectly by controlling plant pathogens, nematodes and other pests, thereby enhancing plant health and minimizing the yield loss. Due to its innate biological, biochemical and physiochemical properties, Pathma and Sakthivel argue in Chapter 8 that vermicompost may be used to promote sustainable agriculture and also for the safe management of agricultural, industrial, domestic and hospital wastes which may otherwise pose serious threat to life and environment.

In the frame of municipal solid waste management, one of the available options is based on the generation of refuse derived fuel (RDF) for industrial use. To this end, in Chapter 9 Rada and Raguzzi propose bio-drying for a decentralized management of waste: a few satellite plants could generate RDF for a centralized use. This decentralization could be based on a small scale plant for local RDF exploitation. To this concern in the sector there is not yet an approach useful for assessing the viability of bio-drying when a factory requiring heat wants to generate exactly the amount of RDF to be used in its own dedicated RDF burner. In the present paper some criteria for assessing the viability of this strategy at a very small scale are presented. Co-generation is not taken into account because of the unfavourable scale. The method takes into account many aspects: characteristics of the factory, of the waste, of the RDF that can be generated and also of the local economy.

In Chapter 11, Rada and Raguzzi carried out an experimentation to study the behavior of exhausted grape marc during the bio-drying process. This process was chosen as an alternative to the typical grape marc thermal drying approach. The aim was to reduce the moisture level thanks to the biological exothermal reactions, and to increase the energy content in the bio-dried grape marc. The target was the generation of a product interesting for energy options. For the development of the research, a biological pilot reactor and a respirometric apparatus were used. Results demonstrated that bio-drying can decrease the water content saving the original energy content. The final material could be assumed like a Solid Recovered Fuel, class 5:1:1 with a very low potential rate of microbial self heating.

PART I

MICROBIAL TECHNOLOGIES

CHAPTER 1

Management Options of Food Waste: A Review

F. GIROTTO, L. ALIBARDI, AND R. COSSU

1.1 INTRODUCTION

The management of food waste is currently a growing problem in the waste management field and the search for sustainable solutions represents a challenge not only for this specific sector but also for the agricultural and industrial sectors and requires the involvement of retailers and consumers.

The Food and Agriculture Organization of the United Nations (FAO) provides a definition for food waste. It is the food loss occurring during the retail and final consumption stages due to the behaviour of retailers and consumers, where the term food loss itself measures the decrease in edible food mass (excluding inedible parts and seeds) throughout the part of the supply chain that specifically leads to edible food for human consumption, that is, loss at the production, postharvest and processing stages (FAO, 1983).

Girotto F, Alibardi L, and Cossu R. "Management Options of Food Waste: A Review" SUM 2014, Second Symposium on Urban Mining Bergamo, Italy; 19 – 21 May 2014. *CISA Publisher (2014). Used with permission from the publisher.*

A deep analysis of a whole Food Supply Chain (FSC) system can highlight the fact that the production of waste material (organic waste or food waste) regards all sectors involved in the production, distribution and consumption of food.

In the definition provided by FAO and also in those from other sources (Smill, 2004; Stuart, 2009; Gustavsson et al., 2011), the concept of food waste is wide and regards not only the production of food waste at household level but also the decrease in edible food mass throughout the human FSC. In this food waste definition therefore, all those materials that are edible food but that are wasted for several reasons, are considered.

A FSC starts with food production from the agricultural sector where both framing and husbandry produce waste or subproducts that can be both organic waste (i.e. cornstolk, manure) or food waste (i.e. low quality fruits or vegetable, damaged productions left in the field, good products or co-products but with low or no commercial value). The edible (or potentially edible) materials are considered food waste. The further food processing and manufacturing industrial sector can produce food losses and food waste during the entire production phase for several reasons. Damage during transport or non-appropriate transport systems, problems during storage, losses during processing or contaminations, inappropriate packaging, are some examples of food waste production in the agro-industrial sector. The retailing system and markets also produce food waste, due to problems in conservation or handling, lack in cooling/cold storage. The production of food waste from the final consumer is due to over- or non appropriate purchasing, bad storage conditions, over-preparation, portioning and cooking as well as confusion between the terms "best before" or "use by" dates. (Papargyropoulou et al., 2014). The generation of food waste at household level is influenced by several interconnected factors, the main ones being socio-demographic characters of household, consumption behaviour and food patterns (Glanz and Schneider, 2009).

Food waste production has impacts at environmental, social and economical levels. From an environmental point of view, food waste contributes to Green House Gas (GHG) emissions during final disposal in landfills (uncontrolled methane release) and during activities associated to food production, processing, manufacturing, transportation, storage and distribution. Other environmental impacts associated to food waste are natural resource depletion in terms of soil, nutrients, water and energy, disruption of biogenic cycles due to intensive agricultural activities and all other characteristics impacts in any step of the FSC. Social impacts of food waste can be ascribed at ethical and moral dimension within the general concept of global food security. Economical impacts are due to the costs related to food westage and their effects on farmers and consumers incomes (Lipinski et al., 2013; Papargyropoulou et al., 2014).

This paper reviews the management options reported and discussed in the scientific literature with the final aim of providing a clear and complete vision on the magnitude of food waste production and the possibilities currently available for food waste management and their state of the art.

1.2 PRODUCTION OF FOOD WASTE

The Food and Agriculture Organization of the United Nations (FAO, 2004) estimates that 32% of all food produced in the world was lost or wasted in 2009. While 870 million people are reported to be chronically undernourished, about 1.3 billion tons/year, that is one third of the food produced for human consumption, is wasted globally (Kojima and Ishikawa, 2013). In Europe, food waste production is estimated at about 90 million tons annually (EC, 2013), in the United States it is estimated around 27% of the industrially produced food (Garcia, 2011) and in Japan it is about 21 million tons in 2010 (Kojima and Ishikawa, 2013). Quested et al., (2013) reported a production of food waste at household level of 160 kg per year in United Kingdom (UK), representing 12% of the food and drink entering a home and 30% of the general waste stream from UK household. The order of magnitude of food waste production is consistent and is not limited to developed countries. Gustavson at al. (2011) reported data on the food waste production around the World and results are summarised in Table 1.

Food waste production in industrialised countries has similar order of magnitudes as in developing countries (DCs) (EC, 2014). Nevertheless industrialized and developing countries differ substantially as reported in Table 1. In the latter, more than 40% of food losses occur at the postharvest and processing stages, while in the former, about 40% of losses occur at the retail and consumer levels and, on a per-capita basis, much more food is wasted in the industrialized world than in developing countries (Gustavsson et al., 2011).

TABLE 1: Average annual food waste production per person in different areas of the World (Modified from Gustavson et al., 2011)

Area	Food waste production (kg/person/year)		
	Total	Production and retail stage	Consumer's stage
Europe	280	190	90
North America and Oceania	295	185	110
Industrialized Asia	240	160	80
sub-Saharan Africa	160	155	5
North Africa, West and Central Asia	215	180	35
South and Southeast Asia	125	110	15
Latin America	225	200	25

The causes of food losses and waste in low-income countries are mainly connected to financial, managerial and technical limitations in harvesting techniques, storage and cooling facilities in difficult climatic conditions, infrastructure, packaging and marketing systems. Given that many smallholder farmers in DCs live on the margins of food insecurity, a reduction in food losses could have an immediate and significant impact on their livelihoods. The food supply chains in DCs need to be strengthened, encouraging small farmers to organize, diversify and upscale their production and marketing. Investments in infrastructure, transportation, food industries and packaging industries are also required. Both the public and private sectors have a role to play in achieving this.

The causes of food losses and waste in medium/high-income countries mainly relate to consumer behaviour as well as to a lack of coordination between different actors in the supply chain. Farmer-buyer sales agreements may contribute to quantities of farm crops being wasted. Food can be wasted due to quality standards, which reject food items not perfect in shape or appearance. At the consumer level, insufficient purchase planning and expiring "best before dates" also cause large amounts of waste, in combination with the careless attitude of those consumers who can afford to waste food. Food waste in industrialized countries can be reduced by raising awareness among food industries, retailers and consumers. This inevitably also means that huge amounts of the resources used in food production are used in vain, and that the greenhouse gas emissions caused by production of food that gets lost or wasted are also emissions in vain.

According to Venkat (2011), at least 123 million metric tonnes of CO_2 emissions are added to the atmosphere each year from the production, transport and disposal of the uneaten food. This translates to over 13% of all food-related emissions in the US and about 1.5% of total US emissions, and most of these emissions come from the production stage. The cost estimation indicated that consumers and businesses waste nearly $200 billion worth of raw food commodities annually (Venkat, 2011). Including in the evaluation other food commodities that are produced and wasted in smaller quantities, together with other emission sources such as additional processing, packaging and cooking, the overall cost and climate change impact of food waste would be higher than the reported preliminary calculation. This should stand for an additional incentive to push the food industry to reduce food waste production in order to gain benefits on both the financial and environmental fronts. A much larger share of the food waste burden rests with consumers, counting household food waste and all of the plate loss away from home. Since consumers do not have the tools and systems to manage their inventories and food preparation, putting a dent in this part of the waste will require a combination of education, the availability of more optimal portion sizes away from home, innovative food packaging/preservation techniques. Cutting household waste food would be helpful to go one step further towards reducing food-related emissions besides representing an efficient way for families to save money.

1.3 FOOD WASTE MANAGEMENT HIERARCHY

Similarly to the Waste Management Hierarchy introduced in Europe in 1989 (European Parliament Council, 1989), based on a hierarchy of solutions of five distinct steps (waste prevention, reuse, recovery and recycling of materials, energy recovery and safe landfilling of residues) and often graphically represented by a reverse triangle (Cossu, 2009), the Environmental Protection Agency (EPA, 2014) defined the following hierarchy concept in relation to food waste management: source reduction, feed hungry people, feed animals, industrial uses, composting, incineration or landfilling. A graphical representation is reported in Figure 1, as a reverse triangle, where the larger is the bar of the step, the more viable is the management option.

The first step in the reduction of food waste production starts from the issue of the undesirable food surplus, the prevention of over-production and over-supply of food beyond human nutritional needs (Papargyropoulou et al., 2014).

The first step of the hierarchy also involves food waste reduction throughout the entire FSC. This focus has to be different from country to country as highlighted by the work of Gustavsson et al. (2011). In developed Countries, food waste prevention should focus on the consumer's behaviours at household level while in developing Countries it should focus more on the retail and distribution system. Very close to food waste prevention are the issues of food security and the utilisation of food surplus to satisfy the nutritional needs of poor people.

The next steps of the hierarchy are the utilisation of food waste to feed animals and in industrial sectors. Within industrial utilisation, several options can be described varying from the use of food waste for energy production by anaerobic digestion (bio-hydrogen or bio-methane productions) to the production of specific chemical compounds as precursors for plastic materials production, chemical or pharmaceutical applications.

The composting step has the main aim of nutrients recovery and carbon sequestration by humic substances formation. The last, less desirable step is represented by landfilling or incineration. It is well known that biodegradable organic material represents the main source of adverse environmental impacts and risks in traditional landfilling (odours, fires, VOC's, groundwater contamination by leachate, global climate changes, etc.) (i.a. Manfredi et al., 2010; Thomsen et al., 2012; Beylot et al., 2013) while thermal treatment, even though providing energy recovery, is limited by the low heating values of organic waste (Nelles et al., 2010). For these reasons, such options represent the less desired ones.

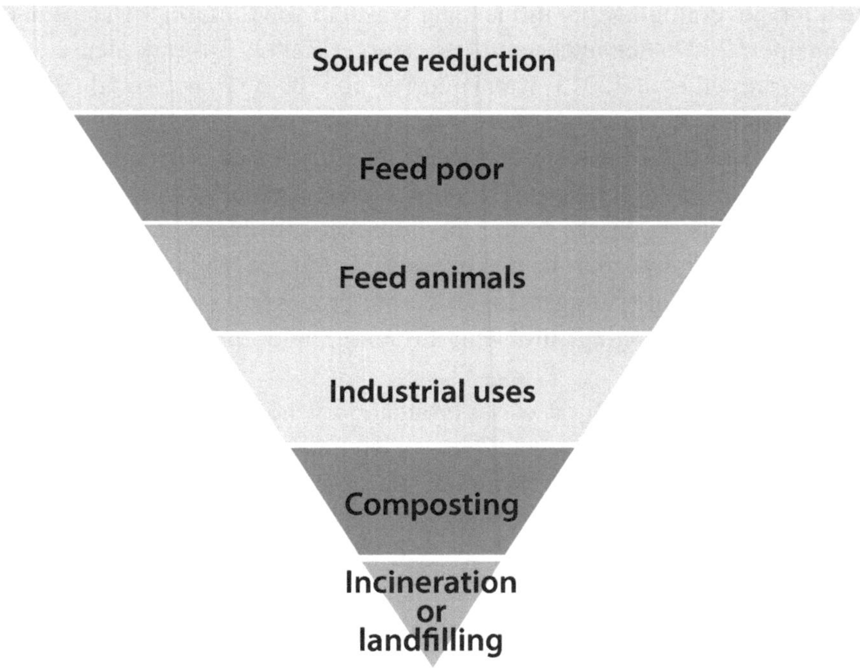

FIGURE 1: Food waste management hierarchy pyramid (EPA, 2014).

1.3.1 PREVENTION

One way of dealing with food waste is the reduction of its production. This attitude has been promoted by campaigns from advisory and environmental groups and by concentrated media attention on the subject. Several papers have analysed the behaviour of companies and citizens in developed countries at different levels (household, restaurant, retail) to assess the governing factors influencing waste of food products (Glanz and Schneider, 2009; Schneider and Lebersorger, 2009; Silvennoinen et al., 2012; Quested et al., 2013; Katajajuuri et al., 2014; Garrone et al., 2014; Graham-Rowe et al., 2014; Mena et al., 2014).

To prevent food waste production, different types of actions carried out at several levels are required. The retail system can cause food waste production in various phases of the food product distribution and purchase: damages during transport or non appropriate transport systems, problems during intermediate storages, losses during processing or contaminations, inappropriate packaging, problems in conservation or handling, lack in cooling/cold storage.

A huge impact on food waste production is held by the behaviour of consumers at household level. Over- or non appropriate purchasing, bad storage conditions, over-preparation, portioning and cooking as well as confusion between the terms "best before" or "use by" dates are some of the main factors affecting food loss. This behaviour is influenced by several interconnected factors, the main ones being socio-demographic characters of household, consumption behaviour and food patterns. The barriers to surpass to reach food loss minimisation at household level can also involve emotional and psychological aspects. One example is the wish to be "good provider" in terms of healthy and abundant food for the family. Lack in food may produce a sense of inappropriate ability to take care of the needs of the family thus pushing the purchase of more items even though not necessary ("bread every day in the table" or "never an empty fridge"). Another example is the avoidance of frequent trips to shops, thus pushing people to buy more food products to prevent or avoid inconveniences. A general lack in awareness of the amount of food waste generated at household level can have a strong impact on food waste production, due to the fact that small quantities thrown away a bit at a time with other waste does not provide the proper order of magnitude of the problem to consumers (Graham-Rowe et al., 2014).

1.3.2 FEEDING THE POOR

Organizations can donate safe and healthy food to food banks or food rescue organizations. This simple effort would lead to a double achievement: the reduction of food sent to landfills and the feeding of those in need. In the US, for example, the "*Emerson Good Samaritan Food Donation Act*" was created to encourage food donation to non-profits by minimizing liability (Culver, 2013). Donating wholesome and edible food can also claim tax benefits (EPA, 2014).

More and more widespread is garbage picking: the practice of sifting through commercial or residential waste to find items that have been discarded by their owners, but may prove useful to the garbage picker. Garbage picking may take place in dumpsters or in landfills. Nowadays, unfortunately, because of the globally critical economical situation, many people find themselves compelled to be "dumpster divers" also for food as well as clothing, furniture, etc. Dumpster diving is practiced differently in developed countries and developing countries. In many developing countries, food is rarely thrown away unless it is rotten. As already stated, in fact, the highest percentage of food loss occurs at the postharvest and processing stages. In developed nations, like the United States, where 40–50% of food is wasted, the trash contains a lot more food to gather (Harrison, 2004). In regions where people practice dumpster diving, food waste is reduced. However, it can pose a health risk to these people and there may also be questions of legality.

1.3.3 FEEDING ANIMALS

The feeding of food scraps to animals is, historically, the most common way of dealing with household food waste. From the end of the 19th century through the middle of the 20th Century, many municipalities collected food waste (called "garbage" as opposed to "trash") separately. This was typically disinfected by steaming and fed to pigs, either on private farms or in municipal piggeries. Now feeding scraps to worms that produce soil as a by-product is also a widespread practise called vermicomposting.

1.3.4 INDUSTRIAL USES

1.3.4.1 BIOFUEL PRODUCTION

Currently more and more effort is invested in the fourth step of the food waste management hierarchy. The definition of effective and stable ways to obtain biofuel and bio-plastic from waste food represents in fact one of the most interesting goals of this century. This possibility could demonstrate benefits from an environmental point of view due to the reduction of methane gas emissions from landfills and the preservation of natural resources such as coal and fossil fuels, from a social point of view because there would be no food vs. fuel competition to obtain bioethanol or biodiesel, and from an economical point of view thanks to costs saving connected to surplus food production and specific investments to grow no food crops dedicated to biofuel or bioplastic production.

The industrial concept behind food waste utilisation is the biorefinery. Similarly to the way that oil refineries convert petroleum into fuels and ingredients for hundreds of consumer products, biorefineries convert corn, sugar cane, and other plant-based material into a range of ingredients for bio-based fuels and other products.

Improving energy security and mitigating climate changes are among the most important bioenergy drivers in most countries. Concerns about the negative environmental effects due to the utilization of fossil fuels (GHG emissions into atmosphere and global warming effects), rising prices of crude oil and increasing demand for transportation fuel, are the major constraints in the economic development of many nations that could stimulate investment, research and industrial application of the biorefinery concept. Of course, much attention has to be paid in order to develop the right techniques and strategies to obtain the desired substances without implying the increase of the prices of edible products and without causing land use competition between food and biofuels. Biofuels and bio-based materials will only be beneficial if they are produced in a sustainable way with both biodiversity and "food vs. fuel"-debate in mind (Nigam and Singh, 2011; Refaat, 2012).

Biodiesel can be defined as fatty acid alkyl esters (methyl/ethyl esters) of short-chain alcohols and long-chain fatty acids derived from natural biological lipid sources like vegetable oils or animal fats, which have had their viscosity reduced using a process called transesterification and can be used in conventional diesel engines and distributed through existing fuel infrastructure. Any fatty acid source may be used to prepare biodiesel (Refaat, 2012). Thus, any animal or plant lipid should be a ready substrate for the production of biodiesel. However the use of edible vegetable oils and animal fats for biodiesel production has recently been of great concern because they compete with food materials while the use of nonedible vegetable oils for biodiesel production is also questionable because the growing of crops for fuel wastes land, water, and energy resources vital for the production of food for human consumption. The conclusion that can be derived is that the use of waste oil for the production of biodiesel is the most realistic and effective. The new process technologies developed during recent years have made it possible to produce biodiesel from recycled frying oils comparable in quality to that of virgin vegetable oil biodiesel: both are composed of methyl esters of fatty acids and, therefore, they have very similar properties and potential in reducing pollutant emissions from the engine. Anyhow there is an imperative need to improve the existing biodiesel production methods from both economic and environmental viewpoints, and to investigate alternative and innovative production processes. The identification of some key parameters (acid value and FFA content, moisture content, viscosity, and fatty acid profile of the used oil) is a prerequisite for determining the viability of the vegetable oil transesterification process and, therefore, is essential for identifying the right processes to achieve the best results with respect to yield and purity of the produced biodiesel. In most cases, a simple pretreatment (removal by filtration of solid particles and esterification process to reduce the content of FFAs) is enough for subsequent transesterification. These results are expected to encourage the public and private sectors to improve the collection and recycling of used cooking oil to produce biodiesel. Using this fuel would not only avoid generating more waste, but also would provide a more eco-friendly fuel which cuts in half the greenhouse gases when compared to standard diesel fuels (Refaat, 2012).

First-generation bioethanol can be derived from renewable sources of virgin feedstock; typically starch and sugar crops such as corn, wheat, or sugarcane. Indeed most of the feedstocks used for first generation biofuel production are food crops. For this reason biofuel expansion can compete with food production directly (food crops diverted for biofuel production) and indirectly (competition for land and agricultural labor) (Gasparatos et al., 2011). These barriers can be partly overcome by the utilization of lignocellulosic materials for the production of the so-called second-generation bioethanol. One potential advantage for cellulosic ethanol technologies is that they can avoid direct competition for crops used in the food supply chain as it is not edible but this option should be limited to cases where actual and sustainable surplus of crops occurs or where crop wastes and wood wastes are available as a feedstock. (Timilsina and Shrestha, 2011; Pirozzi et al., 2012; Refaat, 2012). Cellulosic ethanol has a number of potential benefits over corn grain ethanol. Cellulosic ethanol is projected to be much more cost-effective, environmentally beneficial, and have a greater energy output to input ratio than grain ethanol. Cellulosic ethanol production, in particular, can result in a fuel with a net energy yield that is close to CO_2 neutral (Refaat, 2012). Although the cost of biomass is low, releasing fermentable sugars from these materials remains challenging.

Butanol can be obtained from food waste by fermentation processes using *Clostridium acetobutylicum* bacteria. This organism has a number of unique properties, including the ability to use variety of starchy substances and to produce much better yields of acetone and butanol than did Fernbach's original culture (Stoeberl et al., 2011). Butanol as fuel or blending component has some advantages compared to ethanol, for example a lower vapour pressure, improved combustion efficiency, higher energy density and it can be dissolved with vegetable oils in any ratio reducing their viscosity. Data on butanol production indicate that a potential of 0.3 g of butanol from 1 g carbohydrates from waste whey being a substrate characterised by high lactose content (Stoeberl et al., 2011).

Anaerobic digestion for biogas production (methane rich gas) is a well-established technology perfectly suitable for food waste management. In this framework, interest for anaerobic digestion (AD) has been continuously growing in the last decades, being more and more frequently promoted by national programmes for energy production from renewable resources. AD processes are also considered to be the best option for the biological production of hydrogen, the latter being recognized as one of the most interesting and promising biofuels (Guo et al., 2010; Ozkan et al., 2010; De Gioannis et al., 2013).

Numerous investigators demonstrated that if fermentation of biodegradable organic substrates is appropriately operated in a two-staged mode, separation of the acidogenic and methanogenic phases can be accomplished (i.a. De Giannis et al., 2013). Production of hydrogen by means of anaerobic digestion processes, combined with methane production (Kapdan and Kargi, 2006) may therefore represent an interesting solution for synergising sustainable management of food waste with renewable energy production.

A number of potentially suitable residual substrates have been evaluated for biohydrogen generation potential through dark fermentation. Among these, food waste may represent relatively inexpensive and suitable sources of biodegradable organic matter for H_2 production, mainly due to their high carbohydrate content and wide availability (i.a. De Gioannis et al., 2013).

1.3.4.2 BIOPRODUCTS PRODUCTION

The current frontier in the bioprocessing of organic materials lies in the biorefinery concept where organic waste is considered as a feedstock for the biological production of high value commodities. There is particular interest in the production of metabolites as renewable and biodegradable substitutes for petrochemical products. These metabolites include: lactate for the production of polylactate, a plastic constituent; polyhydroxyalkanotes, particularly polyhydroxybutyrate, which are natural storage polymer of many bacterial species with properties similar to polyethylene and polypropylene and harvestable from mixed cultures fed with organic wastes; succinate, a valuable and flexible precusor for pharmaceutical, plastic and detergent production, fermentable from carbohydrate rich wastes by selected bacterial species (Clarke and Alibardi, 2010).

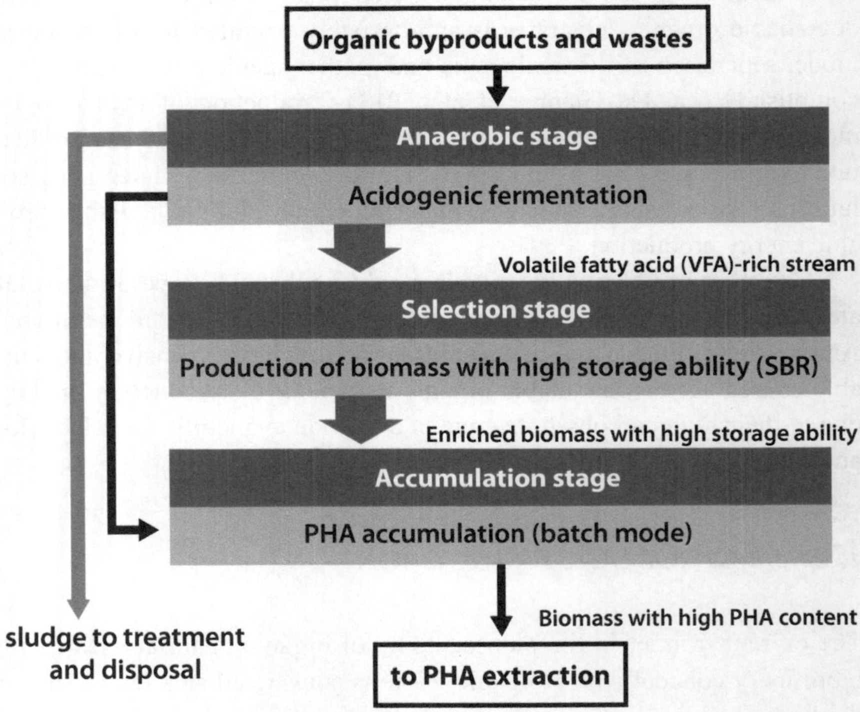

FIGURE 2: Schematic representation of a three-stage sequential process for Polyhydroxyalkanoate (PHA) production from waste/surplus-based feedstocks by mixed cultures using sequencing batch reactor (SBR) to carry out culture selection (from Reis et al., 2011).

The challenge of finite fossil resources has been addressed by academic and industrial research with the development of polymers based on renewable resources (besides the previously mentioned biofuels). The corresponding monomers are accessible either through fermentation of carbohydrate feedstocks by microbes, often genetically modified, or by chemical processing of plant oils (Fuessl et al., 2012). As for the biofuel production from virgin feedstocks, there is much debate about manufacturing bioplastics from natural materials and whether they are a negative impact to human food supply. 2.65 kg of corn are required to produce 1 kg of polylactic acid (PLA), the most common commercially compostable plastic. Since 270 million tonnes of plastic are made every year, replacing conventional plastic with corn-derived polylactic acid would remove 715 million tonnes of corn from the world's food supply (Harris, 2013). In this context, the opportunity to use waste food for bio-plastics seems a well-working option.

At present, among the various types of starch-based biodegradable plastics (such as PLA and PVA-polyvinylacetate), polyhydroxyalkanoates (PHA) is one of the most promising.

Polyhydroxyalkanoates (PHAs) are linear polyesters of hydroxyacids (hydroxyalkanoate monomers) synthesized by a wide variety of bacteria through bacterial fermentation (Poirier, 2003; Reis et al., 2011). Its strength and toughness are good, it is completely resistant to moisture and it has very low oxygen permeability. Therefore PHA can be used for bottles and water resistant film production (Van Wegen et al., 1998; Poirier, 2003). The simplest type of PHA is polyhydroxybutyrate (PHB). Pure PHB possesses several undesirable mechanical properties, so the copolymer poly(3-hydroxybutyrate-co-3-hydroxyvalerate) (PHBV) is generally used. Commercially-produced PHBV are marketed under the name "BIOPOL" (Van Wegen et al., 1998).

PHAs represent a large group of bacterial polyesters, which can include over 150 different hydroxyacids. The majority of bacteria synthesizing PHAs can be broadly subdivided into two groups. One group produces short-chain-length PHAs (SCL-PHAs) with monomers ranging from 3 to 5 carbons in length, while a distinct group synthesizes medium-chain-length PHAs (MCL-PHAs) with monomers from 6 to 16 carbons (Poirier, 2003). PHAs accumulate in bacteria cytoplasm as a high molecular weight polymer forming intracellular granules of 0.2–0.7mm in diameter. Typically, PHAs accumulate to a significant proportion of the cell dry weight when bacteria are grown in a media that is limited in a nutrient essential for growth (typically nitrogen or phosphorus), but with an abundant supply of carbon (for example glucose). Under these conditions, bacteria convert the extracellular carbon into an intracellular storage form, namely PHA. When the limiting nutrient is resupplied, intracellular PHA is degraded and the resulting carbon is used for growth (Poirier, 2003; Reis et al., 2011).

The main limitation in using bacterial PHAs as a source of biodegradable polymers is their production cost. In particular the average cost is by far the most significant contributor to overall PHB price, approximately two and a quarter times greater than the capital cost of equipment (Van Wegen et al., 1998). Using agro-industrial food waste as substrate instead of virgin feedstock of refined sugar such as glucose, sucrose and corn steep liquor could, again, represent the turning point. Sugarcane and beet molasses, cheese whey effluents, plant oils, swine waste liquor, vegetable and fruit wastes, effluents of palm oil mill, olive oil mill, paper mill, pull mill and hydrolysates of starch (e.g., corn and tapioca), cellulose and hemicellulose are all excellent alternatives characterized by a high organic fraction (Reis et al., 2011).

A three stage biotechnological process was proposed by Reis and coworkers (Reis et al., 2011) demostrating good potential for PHA production from waste/surplus-based feedstocks using enriched mixed cultures. The basic concept, reported in Figure 2, was based on an initial acidogenic fermentation phase of the feedstock, a second phase of selection and production of PHA-storing bacterial biomass under dynamic feeding, and the last phase where PHA was accumulated in batch conditions.

It is important to underline the main role played by the initial acidogenic fermentation stage that is needed to overcome the weak point of the process represented by the fact that most waste and surplus feedstocks contain several organic compounds that are not equally suitable for PHA production. Carbohydrates in fact are not stored by mixed cultures as PHA, but rather as glycogen. Thus, the acidogenic fermentation is an essential stage to increase the potential of producing PHA by mixed cultures from surplus-based feedstocks. Carbohydrates and other compounds are, in this way, transformed into VFA which are readily convertible into PHA (Reis et al., 2011).

Interconnection of biotechnological processes for the co-production of bio-fuels and bio-products represents therefore a key strategy to maximise the food waste utilisation and the potential income of the entire bioprocess chain.

1.5 CONCLUSIONS

The development of sustainable solutions for food waste management represents one of the main challenges for the current society. Sustainable solutions will be capable of exploiting the precious resources represented by food waste to reach social, economical and environmental benefits.

Several possibilities are currently available: reuse to feed animals, production of biofuels, exploitation to obtain monomers to be used for plastic production.

The benefits of recycling food waste are clear: bio-energy recovery, lower GHG emissions and lower space requirements for landfilling.

Data and analysis to assess the potential of food waste prevention are still required. The actions should interest all the sectors of food supply chains, from producers to retailers and consumers.

REFERENCES

1. Beylot A., Villeneuve J., Bellenfant G. (2013). Life Cycle Assessment of landfill biogas management: Sensitivity to diffuse and combustion air emissions. Waste Manage. 33, 401-411.
2. Cossu R., (2009). From triangles to cycles. Waste Manage. 29, 2915–2917
3. Culver D. (2013). Feed the hungry, reduce food waste by promoting and publicizing the obscure Bill Emerson Good Samaritan Act of 1996. Evergreen Digest: A Journal of Progress for the Rest of Us.
4. De Gioannis G., Muntoni A., Polettini A., Pomi R. (2013). A review of dark fermentative hydrogen production from biodegradable municipal waste fractions. Waste Manage. 33, 1345-1361.
5. EC (2013). European Commission - Stop food waste. http://ec.europa.eu/food/food/sustainability/index_en.htm
6. EPA (2014). http://www.epa.gov/foodrecovery/

7. FAO (1983). Food loss prevention in perishable crops.
8. FAO (2014). Food and Agriculture Organisation of United Nations, http://fa.org/
9. Fuessl A., Yamamoto M. and Schneller A. (2012). Opportunities in bi-based building blocks for polycondensates and vinyl polymers., Elsevier.
10. Garcia N. (2011). Food Waste Recycled as Biofuel. Stanford University.
11. Garrone P., Melacini M., Perego A. (2014). Opening the black box of food waste reduction. Food policy, 46, 129-139.
12. Glanz R., Schneider F. (2009). Causes of food waste generation in household. Proceedings Sardinia 2009, Twelfth International Waste Management and Landfill Symposium. S. Margherita di Pula, Cagliari. CISA Publisher, Italy.
13. Graham-Rowe E., Jessop D.C., Sparks P. (2014) Identifying motivations and barriers to minimising household food waste. Resources, Conservation and Recycling, 84, 15-23.
14. Guo X.M., Trably E., Latrille E., Carrere H., Steyer J-P. (2010). Hydrogen production from agricultural waste by dark fermentation: a review. Int. J. Hydrogen Energy 35, 10660-10673.
15. Gustavson J., Cederberg C., Sonesson U., van Otterdijk R and Meybeck A. (2011). Global food losses and food waste. Extent, causes and prevention. Rome.
16. Harris W. (2013). How long does it take for plastics to biodegrade?. How Stuff Works.
17. Harrison J. (2004). U.S. Wastes Half of Food Produced. UA News, Arizona.
18. Kapdan I.K., Kargi F. (2006). Bio-hydrogen production from waste materials. Enzyme Microb. Technol. 38, 569-582.
19. Katajajuuri J.M., Silvennoinen K., Hartikainen H., Heikkila L., Reinikainen A. (2014). Food waste in the Finnish food chain. Journal of Cleaner Production, http://dx.doi.org/10.1016/j.jclepro.2013.12.057.
20. Kojima R. and Ishikawa M. (2013). Prevention and Recycling of Food Wastes in Japan: Policies and Achievements. Resilent cities, Kobe University, Japan.
21. Lipinski B, Hanson C., Lomax J., Kitinoja L., Waite R. and Searchinger T. (2013). Reducing food loss and waste. Installment 2 of Creating a Sustainable Food Future, World Resources Institute, Washington DC.
22. Manfredi S., Tonini D., Christensen T.H. (2010). Contribution of individual waste fractions to the environmental impacts from landfilling of municipal solid waste. Waste Manage. 30, 433-440.
23. Mena C., Terry L.A., Williams A., Ellram L. (2014). Causes of waste across multi-tier supply networks: Cases in the UK food sector. Int. J. Production Economics, http://dx.doi.org/10.1016/j.ijpe.2014.03.012.
24. Nelles M., Arena U., Bilitewski B. (2010). Thermal waste treatment – An essential component of a sustainable waste treatment system. Waste Manage. 30, 1159-1160.
25. Nigam P.S., Singh A. (2011). Production of liquid biofuels from renewable resources, Elsevier.
26. Ozkan L., Erguder T.H., Demirer G.N. (2010). Investigation of the effect of culture type on biological hydrogen production from sugar industry wastes. Waste Manage. 30, 792-798.

27. Papargyropoulou E., Lozano R., Steinberger J., Wright N., bin Ujang Z. (2014) The food waste hierarchy as a framework for the management of food surplus and food waste. Journal of Cleaner Production, doi: 10.1016/j.jclepro.2014.04.020.
28. Pirozzi D., Ausiello A. and Yousuf A. (2012). Exploitation of lignocellulosic materials for the production of II generation biodiesel. Proceesing Venice 2012, Fourth International Symposium on Energy from Biomass and Waste. Cini Foundation, Venice Italy. CISA Publisher, Italy.
29. Poirier Y. (2003). Biopolimers, Elsevier.
30. Quested T.E., Marsh E., Stunell D., Parry A.D. (2013). Paghetti soup: The complex world of food waste behaviours. Resource Conservation and Recycling, 79, 43-51.
31. Rafaat A. (2012). Biofuels from Waste Materials, Elsevier.
32. Reis M., Albuquerque M., Villano M., Majone M. (2011). Mixed Culture Processes for Polyhydroxyalkanoate Production from Agro-Industrial Surplus/Wastes as Feedstocks, Elsevier.
33. Schneider F., Lebersorger S. (2009) Household attitudes and behaviour towards wasting food - A case study. Proceedings Sardinia 2009, Twelfth International Waste Management and Landfill Symposium. S. Margherita di Pula, Cagliari. CISA Publisher, Italy.
34. Silvennoinen K., Katajajuuri J.M., Hartikainen H., Jalkanen L., Koivupuro H.K., Reinikainen A. (2012). Food waste volume and composition in the finnish supply chain: special focus on food service sector. Proceesing Venice 2012, Fourth International Symposium on Energy from Biomass and Waste. Cini Foundation, Venice Italy. CISA Publisher, Italy.
35. Smill V. (2004). Improving efficiency and reducing waste in our food system. Environmental Science, 1, 17-26.
36. Stoeberla M., Werkmeistera R., Faulstichb M., Russa W. (2011). Biobutanol from food wastes – fermentative production, use as biofuel an the influence on the emissions. Procedia Food Waste, 1, 1868-1974.
37. Stuart T. (2009). Waste. Uncovering the global food scandal. London: Penguin.
38. Timilsina G.R., Shrestha A. (2011). How much hope should we have for biofuels?, Elsevier.
39. Thomsen N.I., Milosevic N., Bjerg P.L. (2012). Application of a contaminant mass balance method at an old landfill to assess the impact on water resources. Waste Manage. 32, 2406-2417.
40. Van Wegen R. J., Ling Y.and Middelberg A. P. J. (1998). Industrial production of polyhydroxyalkanoates using escherichia coli: an economic analysis. Trans IChemE, 76, 417-426.
41. Venkat K. (2011). The anatomy of food waste. Environmental Leader, http://www.environmentalleader.com/2011/09/28/the-anatomy-of-food-waste/

CHAPTER 2

Effect of Increasing Total Solids Contents on Anaerobic Digestion of Food Waste under Mesophilic Conditions: Performance and Microbial Characteristics Analysis

JING YI, BIN DONG, JINGWEI JIN, AND XIAOHU DAI

2.1 INTRODUCTION

Food waste (FW), usually from residential, commercial establishments, institutional and industrial sources, is generated at an ever-increasing rate (higher than 10% every year) with the rapid population growth and rising living standards in China [1]. It seems to be a good idea to reuse this favorable feedstock for energy recovery and municipal solid waste (MSW) reduction because FW contains high moisture and biodegradable organics and accounts for 40–50% of the weight of MSW. Anaerobic digestion (AD) is the most attractive and cost-effective technology for treating sorted organic fraction of MSW, especially food wastes [2]. Various AD

Effect of Increasing Total Solids Contents on Anaerobic Digestion of Food Waste under Mesophilic Conditions: Performance and Microbial Characteristics Analysis. © *Yi J, Dong B, Jin J, Dai X.* PLoS ONE *9,7 (2014). http://journals.plos.org/plosone/article?id=10.1371/journal.pone.0102548. Licensed under Creative Commons International License, http://creativecommons.org/licenses/by/4.0/.*

processes have been widely developed in many countries for the treatment of FW.

So far, three main types of AD technologies have been developed according to the total solids (TS) content of feedstocks: conventional wet (≤10% TS), semi-dry (10–20% TS) and modern dry (≥20% TS) processes. Dry anaerobic digestion, so called "high-solids" technology, has become attractive and was applied widely because it requires smaller reactor volume, lower energy requirements for heating, less material handling, and so on [2]–[4]. The TS content of solid waste influences anaerobic digestion performance, especially biogas and methane production efficiency [5]. Previous reports have investigated that role of TS content on AD performance in order to determine conditions for optimum gas production. Abbassi-Guendouz et al., showed that the total methane production decreased with TS contents increasing from 10% to 25% in batch anaerobic digestion of cardboard under mesophilic conditions [6]. The results obtained by Duan et al., showed that high-solids system could reach much higher volumetric methane production rate compared with low-solids system at the same solid retention time (SRT) in mesophilic anaerobic reactors treating sewage sludge [3]. Forster-Carneiro et al., showed that the biogas and methane production decreased with the total solids contents increasing from 20% to 30% in dry batch anaerobic digestion of food waste [2].

Anaerobic digestion is a multi-stage biochemical process in which the complex organic materials undergo hydrolysis, acidogenesis, and methanogenesis in series and each metabolic stage is functioned by different types of microorganisms [4]. They are present in a mixed culture but differ in their nutritional and pH requirement, growth kinetics, and their ability to tolerate environment stresses [7]. Characterization of microbial community structures in anaerobic digesters has been attractive from the point of review of engineering because understanding of microbial behavior can provides valuable information to optimize fermentation process to favor efficient breakdown of wastes [8]. However, the available literature is mainly about performance and corresponding the structure and dynamic of microbial community in either thermophilic or mesophilic anaerobic digestion of food waste, or only simply about performance comparisons. The AD performances at steady state and the comprehensive characterizations of microbial community in anaerobic digestion of FW with different

TS contents (wet, semi-dry and dry) were not compared in parallel. In order to increase the efficiency of anaerobic digestion of FW, it is necessary to understand the role of the TS contents on the behavior of the microbial community structure involved in the anaerobic digestion of degradation from wet to dry technology.

Recently, various molecular microbial ecology tools have been applied in numerous studies to analyze microbial communities in different anaerobic digesters and their influences on the efficiency and stability of AD processes [9]. Pyrosequencing, as a next generation sequencing technology, has gained increasing attention as a novel tool for studying the microbial diversity [4]. Recently, this technology has been widely and successfully used to characterize the microbial community structures in various environmental samples, such as source waster [10], membrane filtration systems [11], soil [12]. Meanwhile, the microbial community structures were compared by this technology in anaerobic digestion of food waste at different organic loading rates (OLRs) [4].

Hence, the aim of this study was to conduct a comprehensive comparison of the microbial community structure using 454 high throughput pyrosequencing technology and related these microbial findings to their respective performances of mesophilic anaerobic digesters treating FW with different TS contents ranging from 5%–20%. It was expected that the reported work herein will reveal the role of the TS content on the behavior of the microbial community structure to increasing TS contents and hence to effective guide high solids anaerobic digestion of FW and to optimize the operational conditions for high anaerobic digestion efficiency.

2.2 MATERIALS AND METHODS

2.2.1 SUBSTRATES AND INOCULUMS

FW used in this study was collected every 30 days from a dining room at Tongji University in Shanghai. After removing bones, shells, and other indigestible materials, the FW was finely smashed using an electrical crusher and sufficient mixed and stored at 4°C. The TS of the FW ranged from 26% to 28% (w/w) and volatile solid (VS) accounted for 92%–95% of

TS. The mesophilic seed sludge was obtained from a full-scale anaerobic digester at Bailonggang municipal wastewater treatment plant (WWTP) (Shanghai, China). It had TS of 4.1% (w/w) and VS of 52.3% of TS. The main characteristics (average data plus standard deviations in duplicate tests) of substrates and inoculums are listed in Table 1. The collected FW was heated to 35°C before daily feeding.

TABLE 1: Characteristics of the substrates and inoculums.

Parameters	FW[a] 1 (days 1–30)	FW 2 (days 31–60)	FW 3 (days 61–90)	FW 4 (days 91–120)	Inoculums
TS[b] (%, w/w)	26.5 ± 0.6	27. 8 ± 1.1	27.3 ± 1.2	26.8 ± 1.2	4.1 ± 0.1
VS[c]/TS (%)	94.7 ± 3.9	92.2 ± 3.7	93.4 ± 4.6	93.9 ± 4.2	52.3 ± 2.4
pH	4.72 ± 0.21	4.64 ± 0.11	4.79 ± 0.24	4.87 ± 0.23	7.9 ± 0.3
C/N (w/w)	13.4 ± 0.6	14.2 ± 0.7	13.9 ± 0.4	13.6 ± 0.6	–
TAN[d] (mg/L)	538 ± 24	546 ± 19	534 ± 25	543 ± 19	299 ± 13

–Not determined. [a]FW: food waste. [b]TS: total solids. [c]VS: volatile solids. [d]TAN: total ammonia nitrogen

2.2.2 REACTORS AND OPERATION

Three identical reactors (numbered R1, R1 and R3), with liquid working volume of 6.0 L, were equipped with helix-type stirrers to provide sufficient mixing for substrates. The rotation speed was set at a rate of 60 rpm (rotations per minute) with 9 min stirring and 1 min break, continuously. Daily feeding was carried out by pushing semi-fluid substrate through the feeding piston. Since the digestate of FW in each reactor was completely fluid, daily draw-off was easily carried out by opening the discharge valve.

On the first day of the experiments, 6.0 L seed sludge was added to each reactor, which was operated semi-continuously (once-a-day draw-off and feeding) under single phase mesophilic conditions (35°C). The reactors were purged with N_2 for 10 min in order to provide anaerobic conditions. During the start-up period, the OLR was increased stepwise with high-solids FW before the TS content of the substrate in each reactor did not reach its designed TS level. Once the TS of the substrate in each reactor approached its designed level, the feeding FW was diluted to its designed TS level (5%, 15% and 20%, respectively) with de-ionized water before feeding. Each reactor was operated for five SRTs at 20 days SRT. For a full understading of the microbial community structures in anaerobic fermentation reactors with different TS contents, the anaerobically digested FW samples were taken on Day 100 when the systems could be deemed to have reached their steady state operation (determined by constant methane yield and VS reduction) after running for more than 3 months. The fermentation substrate samples in the reactors were taken every three days during the operation period of the fifth SRT for reactor performance analysis.

2.2.3 DNA EXTRACTION, PCR AND PYROSEQUENCING

To analyze the bacterial and archaeal communities in mesophilic anaerobic digesters with different feeding TS levels, 0.5 g of sample in reactor operated for 100 d was used for DNA extraction using a Fast DNA Spin Kit (QBIOgene, Carlsbad, CA, USA) following the manufacturer's instructions. For each sample, two independent PCR reactions were conducted using the primer pairs of 27F (5′-AGAGTTTGATCCTGGCTCAG-3′) and 533R (5′-TTACCGCGGCTGCTGGCAC-3′) for bacteria and 344F (5′-ACGGGGYGCAGCAGGCGCGA-3′) and 915R (5′-GTGCTCCCCCGCCAATTCCT-3′) for archaea [4]. To achieve the sample multiplexing during pyrosequencing, barcodes were incorporated in the 5′end of reverse primers 553R and 915R. All PCR reactions were carried out in a 25 uL mixture containing 0.5 uL of each primer at 30 $mmolL^{-1}$, 1.5 uL of template DNA (10 ng), and 22.5 uL of Platinum PCR SuperMix (Invitrogen, Shanghai, China). The PCR amplification program contained an initial denature at 95°C for 5 min, followed by 25 cycles of denaturing at 95°C for 30 s, annealing at 55°C for 30 s, and extension at 72°C for 30 s, followed by a final extension at 72°C for 5 min. The thermal cycling for archaea was similar to that for bacteria except that the annealing temperature was 57°C. After amplification, the PCR products were purified and quantified, and an equal amount of the PCR product was combined in a single tube to be run on a Roche GS FLX 454 Pyrosequencing machine at Majorbio Bio-Pharm Technology Co., Ltd., Shanghai, China.

2.2.4 ANALYSIS OF PYROSEQUENCING-DERIVED DATA

After sequencing completed, all sequence reads were quality checked using Mothur software [13]. Raw sequence reads were filtered before subsequence analyses to minimize the effect of random sequencing errors. The sequence reads that did not contain the correct primer sequence after the initial quality check (primer sequences were subsequently removed), were shorter than 200 bp, contained one or more ambiguous base(s), or checked as chimeric artifact were eliminated. Finally, the high-quality sequences after filtering were assigned to samples according to barcodes. Sequences were aligned in according with SILVA alignment [14]. Mothur was also used to conduct rarefaction curve, abundance base coverage estimator (ACE), richness (Chao), Shannon diversity, Simpson diversity indices and Good's coverage analysis, assign sequences to operational taxonomic units (OTUs, 97% similarity) using furthest neighbor approach. For taxonomy-based analysis, the SILVA database project (http://www.arb-silva.de) was used as a repository for aligned rRNA sequences. The sequences have been deposited into the NCBI short read archive (SRA) under the accession number SRX484115 for bacteria and SRX485028 for archaea.

2.2.5 ANALYTICAL METHODS

Volumes of produced biogas were measured by wet gas meters every day. The methane content of the biogas was measured by a gas chromatograph (GC) (Agilent Technologies 6890N, CA, USA) with a thermal conductivity detector equipped with Hayseq Q mesh and Molsieve 5A columns. For the analysis of volatile fatty acid (VFA), the fermentation mixtures withdrawn from digesters were centrifuged at 10, 000 ×g for 10 min, and then the supernatants were immediately filtered through 0.45 um cellulose nitrate membrane fiber paper. The filtrate was collected in a 1.5 ml gas chromatography (GC) vial and acidified by formic acid to adjust the pH to approximately 2.0, and then analyzed using a gas chromatograph (GC, Agilent 7820) with a flame ionization detector (FID) and equipped with a 52 CB column (30 m×0.32 mm×0.25 mm). The concentration of total VFA was calculated as the sum of the measured acetic, propionic, n-butyric iso-butyric, n-valeric, and iso-valeric acids. Metrohm 774 pH-meter was used in all pH measurements. The TS, VS, total alkalinity (TA) and total ammonia-nitrogen (TAN) were measured according to Standard Methods [15]. Free ammonia-nitrogen (FAN) was calculated in the same way as described by Østergaard [16]. The degradation or removal level based on VS (i.e., VS reduction) was calculated by the same formula as reported previously [3]. All experimental analyses were performed in triplicate. The data on performances of each reactor were expressed as mean±standard deviation of the samples.

2.3 RESULTS AND DISCUSSION

2.3.1 EFFECT OF TS CONTENT ON ANAEROBIC DIGESTION PERFORMANCE

Table 2 summarizes the values of the main parameters indicating system stability (pH, VFA, TA) and potential inhibitory chemicals (TAN and FAN) for three reactors operated at different TS contents, and the performance data were the average values of the last five samples during the operation period of the fifth SRT after the system reached steady state (determined by constant methane yield and VS reduction).

For each semi-continuously experiment with a good anaerobic digestion performance (between 5% and 20% TS), there was no accumulation of VFA and low pH. The concentration values of VFA showed increasing trend with increasing TS contents. Under mesophilic semi-dry anaerobic digestion of sorted organic fraction of municipal solid waste (OFMSW), Li et al., also observed an increasing trend of the VFA concentrations with TS contents increasing (for TS contents of 11.0%, 13.5% and 16.0%), the maximum VFA value was 4.2 g L^{-1}, 6.8 g L^{-1} and 22.4 g L^{-1}, respectively [17]. In this study, higher VFA concentrations were obtained in the reactors with higher TS contents, which could be explained by the fact that more organic matter was hydrolyzed and transformed to VFA in the reactors. High VFA levels and almost steady VS reduction (Table 3) in reactors indicated that the acidogenic activity was not influenced significantly. In addition, the reactor stability was maintained and the digestion occurred normally because a constant pH was maintained for each reactor. The average pH value was about 7.39, 7.68 and 7.82 at 5%, 15% and 20% TS, respectively. These pH values were within the permissible range for AD 6.5–8.5 but not with the optimal range 6.8–7.4 [18]. As we all know, the increase of VFA concentration contributes to the decrease of pH. However, low pH value was not observed in R3 in which the VFA concentration was highest. It could be explained by the fact that high buffering capacity was observed in high-solids anaerobic system at TS 20%, for which the total alkalinity value of 13.8 g $CaCO_3$/L was detected.

TABLE 2: Summary of performance parameters on system stability and inhibition in three reactors.

reactor	SRT[a]	OLR[b] (Kg VS m^{-3} d^{-1})	pH	TA[c] (g/L)	TAN[d] (g/L)	FAN[e] (mg/L)	VFA[f] (g/L)	
							Total	Acetic
R1 (5%)	20	2.35	7.39 ± 0.08	3.8 ± 0.1	0.40 ± 0.01	11 ± 0.4	0.12 ± 0.01	0.11 ± 0.01
R2 (15%)	20	7.01	7.68 ± 0.06	10.9 ± 0.3	1.31 ± 0.15	66 ± 2.5	0.53 ± 0.02	0.43 ± 0.01
R3 (20%)	20	9.41	7.82 ± 0.09	13.8 ± 0.2	1.92 ± 0.04	163 ± 8.0	0.94 ± 0.01	0.64 ± 0.02

[a]SRT: solid retention time. [b]OLR: organic loading rate. [c]TA: total alkalinity. [d]TAN: total ammonia nitrogen. [e]FAN: free ammonia nitrogen. [f]VFA: volatile fatty acid.

It was known that ammonia nitrogen concentration (especially free ammonia concentration) was an important factor influencing the stability of anaerobic digestion system. The TAN and FAN concentrations in three reactors at steady state were also observed. They showed a similar trend to that of above parameters with increasing TS contents. However, the maximum FAN value was just 163 mg/L. It has been reported that the FAN at concentrations above 200–1100 mg/L can inhibit the anaerobic system [19]. Therefore, the effect of FAN concentration on the system stability was probably negligible for the three reactors with TS contents ranged from 5% to 20%.

Biogas generation and methane efficiency of different reactors are shown in Table 3. The average daily cumulative biogas (based on added VS) of R1-5%, R2-15% and R3-20% accounted to 700, 760 and 870 ml and 370, 410 and 480 ml methane content, respectively. Hence, both of biogas production and methane content showed increasing trend with increasing TS contents. This result was in contrast with a previous work [2], in which the reactors with smaller TS contents showed higher biogas production and methane percentage in the batch anaerobic digestion of FW. It was suggested that the increasing of feeding TS contents lower than 20% has positive effect on the methane production. A maximum methane content of 55.1% in R3 agreed with the previous study on anaerobic digestion of FW [1], but was lower than in another reference [20], which was probably due to the differences in substrate composition. In addition, it could also be observed that reactors with higher TS contents showed higher volumetric biogas and methane production rate. It is well known that FW is a high degradable substrate for anaerobic digestion [21]. For reactors R1-R3 at a fixed 20 days SRT, increased feeding TS content of FW meant higher applied OLR and larger proportion of easily degradable substrate for microorganisms, which results in higher volumetric biogas yield and methane production rate. As showed in Table 3, higher VS reduction was observed in the anaerobic digesters with higher TS contents. The reasons for this important result obtained were investigated from the microbiology aspect in the following chapters. The specific biogas and methane product rate based on removed VS increased slightly. The highest specific biogas production rate determined on removed VS was 1.01 L gVS^{-1} removed in R3, which was higher than corresponding data obtained in a previous study [1].

TABLE 3: Performance parameters of three reactors with different total solids contents.

Reactor	SRT (d)	OLR (Kg VS m^{-3} d^{-1}	Y_{biogas} [a] (LBiogas gVS^{-1}_{added})	CH_4 (%)	$Y_{methane}$ [b] (L CH_4 g VS^{-1}_{added})	VS_r [c] (%)	SBP[d] (L Biogas $gVS^{-1}_{removed}$)	SMP[e] (LCH_4 $gVS^{-1}_{removed}$)	BP[f] (Biogas $L^{-1}d^{-1}$)	MP[g] (LCH_4 L^{-1} d^{-1})
R1 (5%)	20	2.35	0.70 ± 0.02	52.5 ± 2.1	0.37 ± 0.01	80.1 ± 2.4	0.88 ± 0.02	0.46 ± 0.01	1.65 ± 0.06	0.87 ± 0.03
R2 (15%)	20	7.01	0.76 ± 0.01	54.2 ± 2.7	0.41 ± 0.01	82.4 ± 2.2	0.92 ± 0.05	0.50 ± 0.01	5.36 ± 0.2	2.90 ± 0.07
R3 (20%)	20	9.41	0.87 ± 0.02	55.1 ± 2.6	0.48 ± 0.01	85.6 ± 2.6	1.01 ± 0.04	0.56 ± 0.02	8.21 ± 0.34	4.52 ± 0.05

[a]*Y_{biogas}: biogas yield.* [b]*$Y_{methane}$: methane yield.* [c]*VS_r: VS reduction.* [d]*SBP: specific biogas production rate based on removed VS.* [e]*SMP: specific methane production rate based on removed VS.* [f]*BP: volumetric biogas production rate.* [g]*MP: volumetric methane production rate.*

2.3.2 OVERALL ANALYSIS OF PYROSEQUENCING

The latest developed 454 high-throughput pyrosequencing that can generate huge amounts of DNA reads is widely employed to investigate the bacterial and archaeal community structures and dynamics in various environmental samples. To investigate the compositions of microbial populations involved in the fermentative reactors with different TS contents, a total of 9571, 7769 and 5598 trimmed bacterial 16S rRNA gene sequences and 5245, 4654 and 4432 trimmed archaeal 16S rRNA gene sequences were recovered from samples R1, R2 and R3 (Table S1), respectively. The sequences were grouped into OTUs at a distance level of 3% to estimate the phylogenetic diversities of microbial communities. The OTUs number identified by bacterial sequences in R1 was the largest among three samples. The bacterial community richness levels can also reflected using ACE, Chao, Shannon and Simpson diversity indices (Table S1), which also revealed that the R1-5% had the highest bacterial diversity among three samples. However, the number of archaeal OTUs in R2 was the largest. The rarefaction curves of three samples generated at 3% cutoff for bacterial and archaeal communities are shown in the Figure S1 (Supporting information), demonstrating clearly that the bacterial community richness of R1 and the archaeal community richness of R2 was the highest among these samples, respectively. However, none of the curves approached a plateau, suggesting that this sequencing depth was still not enough to cover the whole microbial diversity and further sequencing would have resulted in more OTUs for each sample. Pyrosequencing analysis of environmental samples can obtain much more sequences and OTUs than conventional cloning and sequencing methods [11], [12]. The bacterial (or archaeal) PCR amplicons from anaerobic digester were grouped into only 238–514 (or 8–26) OTUs according to the clone library in a previous publication [22]. To the authors's knowledge, this was the first study using pyrosequencing technology to characterize the microbial communities in anaerobic digesters with different TS contents. It can be found that compared with traditional clone library, 454 high-throughput pyrosequencing could be a powerful tool to elucidate the microbial community structures and diversities in anaerobic reactors treating food waste with different TS contents.

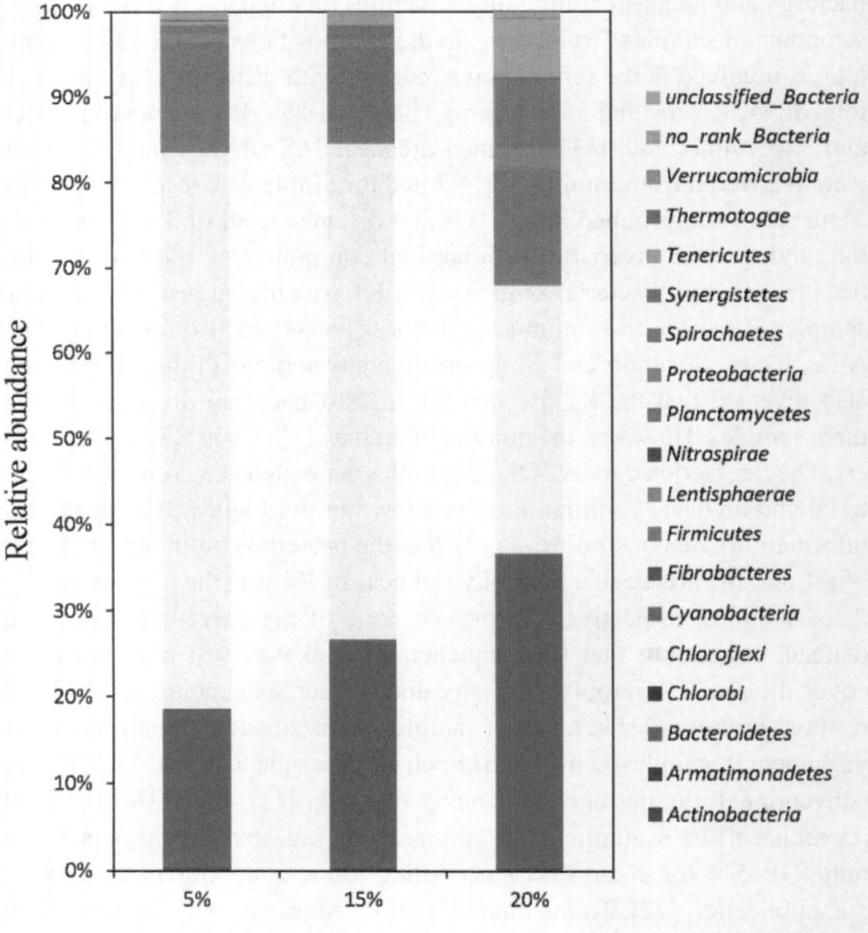

FIGURE 1: Taxonomic compositions of bacterial communities at phyla level in each sample retrieved from pyrosequencing.

TABLE 4: Taxonomic composition of bacterial communities at the genus level for the sequences retrieved from each sample.

Phylum	Genus	Relative abundance		
		5%	15%	20%
	Bacteroidales	0.43%	0.40%	1.54%
	Bacteriodes	0.54%	0.27%	0.82%
	Barnesiella	2.83%	0.08%	0.00%
	Marinilabiaceae	0.37%	0.22%	0.61%
Bacteriodetes	*Parabacteroides*	0.18%	0.06%	0.34%
	Petrimonas	0.45%	0.37%	1.11%
	Proteiniphilum	1.15%	2.59%	4.29%
	Rikenellaceae	11.16%	21.70%	26.58%
	Sphingobacteriales	0.48%	0.17%	0.52%
Chloroflexi	*Anaerolineaceae*	64.99%	58.03%	31.37%
	Anaerobranca	0.11%	0.36%	1.43%
	Christensenellaceae	0.30%	0.14%	0.18%
	Clostridiales	1.73%	4.00%	3.88%
	Erysipelotrichaceae	7.17%	1.85%	2.41%
Firmicutes	*Fastidiosipila*	0.30%	0.79%	1.09%
	Gelria	0.85%	0.85%	0.45%
	Lachnospiraceae	0.13%	0.72%	0.39%
	Lutispora	0.09%	0.06%	0.36%
	Ruminococcaceae	1.11%	0.75%	0.86%
Proteobacteria	*Novosphingobium*	0.21%	0.14%	0.77%
	Rhizobiales	0.07%	0.06%	0.36%
Spirochaetes	*Spirochaeta*	0.28%	0.26%	0.73%
	Spirochaetes	0.46%	1.08%	6.98%
Tenericutes	*Acholeplasma*	0.40%	1.12%	6.75%
	Minor group	4.21%	3.93%	6.20%

2.3.3 EFFECT OF TOTAL SOLIDS CONTENT ON FUNCTIONAL BACTERIAL POPULATIONS INVOLVED IN FOOD WASTE HYDROLYSIS AND ACIDIFICATION

Large numbers of bacterial populations are involved in the hydrolysis and acidification processes of anaerobic fermentation for food waste. The distribution of sequences at the phylum level in each sample is shown in Figure 1. There are seven phyla with relative abundance of higher than 0.5% in at least one sample. From the phylum assignment results, it can be seen that most bacterial sequences in the anaerobic digester treating food waste were distributed among three major phyla: *Chloroflexi, Bacteroidetes* and *Firmicutes*, the total relative abundances of them accounted for 96.13%, 95.61% and 81.35% in R1, R2 and R3, respectively, along with other phyla at minor predominance. Similarly, *Bacteroidetes, Firmicutes, Chloroflexi, Synergistetes,* and *Actinobacteria* were reported to the major populations at phylum level in the mesophilic anaerobic digester treating food waste [4]. The dominance of *Bacteroidetes, Firmicutes* and *Chloroflexi* was also found in other previous studies [7], [23]. In addition, R3 with 20% feeding TS content had high relative abundance of *Spirochaetes* (8.09%), *Tenericutes* (6.86%) and *Proteobacteria* (2.16%).

Although most bacteria in reactors were affiliated to these dominant phyla, the relative abundances of these phyla in each reactor were different and each digester had its own characteristic bacterial community composition. The proportion of phylum *Chloroflexi* in each reactor was the highest in this study. This was in good accordance with previous reports that *Choroflexi* populations were abundant in anaerobic digesters, as determined by membrane hybridization [7], FISH [23] and 16S rRNA gene clone analysis [23], [24]. Rivière et al., also found large proportions (25–45%) of *Chloroflexi* sequences in municipal WWTP sludge samples [22]. An important trend is the small proportion of *Choroflexi* at the highest TS content: 31% for the 20% TS, compared to 58% with the 15% TS and 65% at the 5% TS. The proliferation of *Choroflexi* (formerly known as Green Nonsulfur Bacteria), a well known scavenger biomass-derived organic carbon such as soluble microbial products (SMP), supports a greater influence of difficult-to-biodegrade organic materials from the input substrates and from endogenous dacay of the anaerobic biomass [22], [25].

For R1-R3 at a fixed SRT, increased feeding TS of FW meant higher applied OLR and larger amount of easily degradable substrate per unit volume for microorganisms, which resulted in a smaller relative abundance of phylum *Choroflexi*.

On the other hand, the *Bacteroidetes* population was enriched in the reactors with higher TS contents (from 18.2% at the 5% TS to 26.40% at the 15% TS and 36.33% at the 20% TS). The phylum *Bacteroidetes* are proteolytic bacteria and were probably involved in the degradation of various proteins used for anaerobic digestion studies [22], [25]. The majority of proteolytic microorganisms are able to metabolize amino acids to produce VFA such as acetate, propionate and succinate and NH_3 [22]. Interestingly, their selective enrichment at high TS contents seems to be in consistent with the observation of high protein-input rate and VFA production in the reactors with higher TS contents (Table 2). This result indicated the importance of the *Bacteroidetes* performing protein hydrolysis. However, the changing trend of relative abundance of the phylum *Firmicutes* was not obvious with increasing TS contents. The average value of *Firmicutes* proportion was 12% in three reactors. *Firmicutes* are well-known to be acetogenic and syntrophic bacteria that can degrade VFA, such as butyrate and its analogs. The prevalence of organisms belonging to *Firmicutes* suggested that these products are readily available due to the prior fermentation of these simple VFA and played a critical role in anaerobic digestion of FW, especially on the production of acetic acid, an essential step for methane production by acetoclastic methanogenic microorganisms. In addition, the relative abundances of other phyla including *Proteobacteria, Spirochaetes* and *Tenericutes* obviously increased with the feeding TS contents increasing. It has been suggested that they might play important roles in the degradation of FW. *Proteobacteria* are also involved in the first step of the degradation of organic wastes and they are important consumers of propionate, butyrate, and acetate [23]. *Spirochaetes* are reported to ferment carbohydrates or amino acids into, mainly, acetate, H_2 and CO_2 [8] and *Tenericutes* was found to be related with lignin utilization [26].

In order to further compare the difference of bacterial communities in anaerobic digesters with different feeding TS contents, it is preferable to deconstruct the sequencing date at the subdivision level. Therefore, the relative abundance of each genus in three samples was calculated. The sequence distributions at genus level in each sample are shown in Table 4. A total of 17 genera were detected among which 7 genera with relative abundance of higher than 0.5% in at least one sample were screened as the abundant genera. Other genera were grouped into the minors. As mentioned in the previous section, lower proportions of population from the phylum *Choroflexi* were markedly detected in the reactors with higher TS contents. All sequences classified to phylum *Choroflexi* in three reactors were assigned to genus *Anaerolineaceae* (Figure 1 and Table 4) and class *Anaerolineae* at class level (previous known as "subphylum I" [24]) (Table S2), and the relative abundance of genus *Anaerolineaceae* decreased with increasing TS contents. Because all the characterized species of the class *Anaerolineae* are anaerobic bacteria that decompose carbohydrates via fermentation [27], the genus *Anaerolineaceae* seemed to be involved in carbohydrate decomposition in anaerobic digestion of FW. Similarly, in the previous studies, it was found that all the *Choroflexi* sequences obtained from the up-flow anaerobic sludge blanket reactors treating various food-processing and high-strength organic wastewaters belong to the class *Anaerolineae* [27] and *Anaerolineaceae* group was dominant in phylum *Choroflexi* with its maximum proportion of 8.9% at the 58 days in mesophilic anaerobic digestion of FW [4].

Concerning *Bacteroidetes*, another very abundant phylum which increased with the increasing TS contents, the subdivisions at genus level were multiple and many genera were mainly presented in three anaerobic reactors. *Rikenellaceae* spp. and *Proteiniphilum* spp. were the mostly major genera within this dominant phylum and the changing trends of the relative abundances of these two genera were the same as that of the *Bacteroidetes*. *Rikenellaceae* spp. showed a remarkable proportion from 11% to 27%. The genus *Rikenellaceae* could utilize lactate as substrate in the fermentation processes, and acetate and propionate are the main end-products [28]. *Proteiniphilum*, a relatively new genus showed an unusual ability to grow well at 20–45°C and pH 6.0–9.7. The strains were proteolytic and yeast extract, peptone and l-arginine could be used as carbon and energy sources. Acetic acid and NH_3 were produced after utilizing these substrates [29]. The predominance of *Proteiniphilum* was also obtained in other anaerobic digesters by using a meta-analysis approach [9]. Other genera in this phylum with individual proportion higher or lower than 0.5% might also have played important roles in FW degradation. Regarding to *Firmicutes*, the generic distributions were also distinct with genera *Clostridiales* and *Erysipelotrichaceae* as the main groups in three anaerobic reactors (Table 4). The latter was especially notable in sample R1-5% with relative abundance of 7.13%. Moreover, the proportion of the reigning genera *Spirochaetes* within the abundant phylum *Spirochaetes* and *Acholeplasma* within the *Tenericutes* increased obviously with TS contents increasing. From the analyses made above, it can be seen that the changing patterns of main microbial population abundances were closely related to the performance variations with TS contents increasing, especially for VS reduction. The increasing degradation of organic matter to precursors for methanogenesis was jointly accomplished by the compatible collaborations of these microorganisms which played their respective roles in one of several trophic levels including hydrolysis, fermentation and acetogenesis.

2.3.4 EFFECT OF TOTAL SOLIDS CONTENT ON FUNCTIONAL ARCHAEAL POPULATIONS INVOLVED IN FOOD WASTE METHANOGENESIS

The diversities of archaeal populations in three anaerobic digesters were also revealed by high-throughput pyrosequencing target 16S rRNA gene segments. All species richness estimators including ACE, Chao Shannon and Simpson indices are shown in Table S1. The Good's coverage estimated at least 97% coverage at a similarity of 97%, indicating good coverage of archaeal community. Two hydrogen-utilizing methanogenic groups, *Methanobacteriales* and *Methanomicrobiales*, and acetoclastic methanogenic order *Methanosarcinales* were detected in three reactors. The sum relative abundances of these three methanogenic groups accounted for 99.64%, 99.19% and 99.62% of total archaeal sequences in R1, R2, R3, respectively. However, *Methanococcales* was not detected in any DNA samples in this study (Figure 2). This result was in accordance with previous work characterizing the microbial community shifts in anaerobic digestion of secondary sludge [30]. The relative abundances of sequencing data were also analyzed more specifically at genus level (Table. S3). It was showed that the phylogenetic diversity of methanogens was much lower than that of the bacterial community due partly to the inherent phylogenetic low diversity of methanogens.

As shown in Table S3, there was no large gap in terms of methanogens diversity and distinct discrimination in the taxonomic compositions at genus level. Most of methanogens were assigned to the genus *Methanosarcina* (accounting for 84.4%, 89.5% and 90.9% of total archaeal sequences in R1, R2, R3, respectively), indicating that acetoclastic methanogens played important roles in anaerobic digestion of FW and acetoclastic methanogenesis was the principal pathway of methane production. The low-solids anaerobic digester R1 was secondly dominated by hydrogenotrophic *Methanoculleus* while another hydrogenotrophic methanogens *Methanomicrobiales* was the second most detected group in anaerobic digesters R2 and R3.

Methanosarcina, a typical member of acetoclastic methanogens, have been often reported as the dominant methanogens in AD [31]. The ability of genus *Methanosarcina* having high growth rates and forming irregular cell clumps makes them more tolerant to changing in pH and high concentrations of toxic ionic agents [32]. The genus *Methanosarcina* produce methane from acetate, although some species are more versatile and can also utilize H_2/CO_2, methylated amines and methanol. In addition, *Methanosarcina* spp. are able to use both the acetoclastic and the hydrogenotroph methanogenesis pathways, making them more tolerant to specific inhibitors of the acetoclastic pathway compared to *Methanosaeta* spp. Therefore, anaerobic digester dominantly based on *Methanosaricna* spp. could potentially achieve stable methanogenesis [33], as their special morphological characteristics and flexibility in metabolism.

Besides, the changing patterns of the proportions of three major genera with TS contents increasing were different. The relative proportion of the genus *Methanosaricna* slightly increased from 84% to 90.9% with the TS content increased from 5% in R1 to 20% in R3. On the basis of stable operation, increased feeding TS contents of FW meant higher applied OLR and more VS for microorganisms, which resulted in higher VFA concentrations. In this study, it was observed in Table 2 that the acetate concentration increased with increasing TS contents. It is suggested that higher acetate concentrations would favor the growth of *Methanosarcina* [33].

Therefore, higher concentrations of VFA (especially acetate) and, by extension, at higher OLR caused by the anaerobic systems with higher TS contents induced the selective proliferation of *Methanosarcina*.

The relative abundance of genus *Methanoculleus* obvious decreased from 7.63 to 2.91% with increasing TS contents, indicating that hydrogenotrophic methanogenesis by *Methanoculleus* contributed less to the methane production in high-solids AD than it did in low-solids AD. It has been reported that *Methanoculleus* methanogens had been widely distributed with large proportion in various thermophilic ananerobic digesters [34] and their population ratio seems to be affected by HRT, OLR, or the concentration of VFA. In this study, similar result was obtained that the dominance of *Methanoculleus* declined in the mesophilic anaerobic digesters with TS content increasing resulting in the increase of OLR and the concentration of VFA. Summarily, the changing of microbial communities in mesophilic anaerobic digestion of FW was responsible for the different performances of the reactors with the increasing TS contents. The results obtained in this study expand our knowledge about the role of the TS content on the behavior of the microbial community structure involved in the anaerobic digestion degradation of solids, from low-solids to high-solids technology, and hence to provide valuable information to optimize fermentation process to favor efficient breakdown of food waste.

2.4 CONCLUSIONS

Three stable processes were achieved for AD of food waste with TS contents increasing from 5% to 20%. Better performances, mainly including VS reduction and methane yield and significant shifts in bacterial community, were obtained with the increasing TS contents. The relative abundance of phylum *Chloroflexi* decreased while other functional bacteria increased. The genus *Methanosarcina* absolutely dominated in archaeal communities in three reactors and the relative abundance of this group showed increasing trend with TS contents increasing. These results revealed the effect of the TS content on the performance parameters and the behavior of the microbial community involved in the AD of food waste from wet to dry technologies.

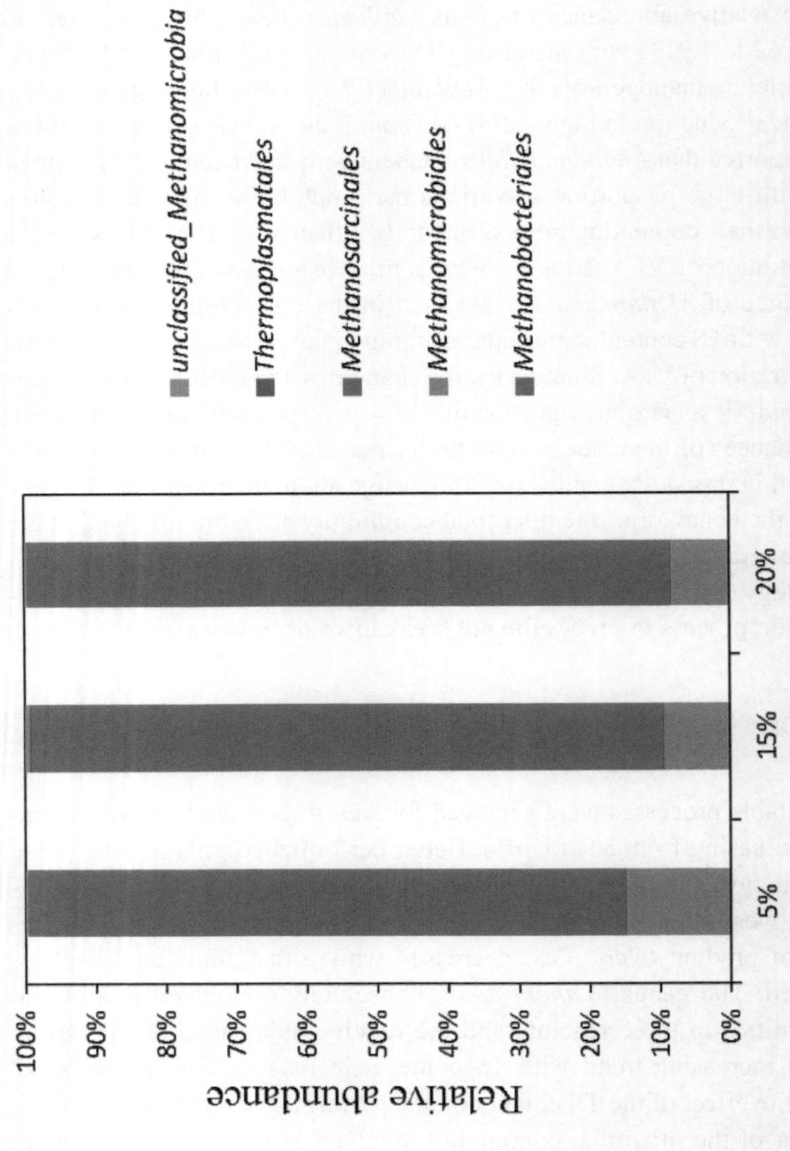

FIGURE 2: Taxonomic compositions of methanogens at order level in each sample retrieved from pyrosequencing.

REFERENCES

1. Dai X, Duan N, Dong B, Dai L (2013) High-solids anaerobic co-digestion of sewage sludge and food waste in comparison with mono digestions: Stability and performance. Waste Manage 33: 308–316. doi: 10.1016/j.wasman.2012.10.018
2. Forster-Carneiro T, Pérez M, Romero L (2008) Influence of total solid and inoculum contents on performance of anaerobic reactors treating food waste. Bioresour Technol 99: 6994–7002. doi: 10.1016/j.biortech.2008.01.018
3. Duan N, Dong B, Wu B, Dai X (2012) High-solid anaerobic digestion of sewage sludge under mesophilic conditions: feasibility study. Bioresour Technol 104: 150–156. doi: 10.1016/j.biortech.2011.10.090
4. Guo X, Wang C, Sun F, Zhu W, Wu W (2014) A comparison of microbial characteristics between the thermophilic and mesophilic anaerobic digesters exposed to elevated food waste loadings. Bioresour Technol 152: 420–428. doi: 10.1016/j.biortech.2013.11.012
5. Pavan P, Battistoni P, Mata-Alvarez J (2000) Performance of thermophilic semi-dry anaerobic digestion process changing the feed biodegradability. Water Sci Technol 41: 75–81.
6. Abbassi-Guendouz A, Brockmann D, Trably E, Dumas C, Delgenès JP, et al. (2012) Total solids content drives high solid anaerobic digestion via mass transfer limitation. Bioresour Technol 111: 55–61. doi: 10.1016/j.biortech.2012.01.174
7. Chouari R, Le PD, Daegelen P, Ginestet P, Weissenbach J, et al. (2005) Novel predominant archaeal and bacterial groups revealed by molecular analysis of an anaerobic sludge digester. Environ Microbiol 7: 1104–1115. doi: 10.1111/j.1462-2920.2005.00795.x
8. Fernández A, Huang S, Seston S, Xing J, Hickey R, et al. (1999) How stable is stable? Function versus community composition. Applied Environ Microb 65: 3697–3704.
9. Nelson MC, Morrison M, Yu Z (2011) A meta-analysis of the microbial diversity observed in anaerobic digesters. Bioresour Technol 102: 3730–3739. doi: 10.1016/j.biortech.2010.11.119
10. Pinto AJ, Xi C, Raskin L (2012) Bacterial community structure in the drinking water microbiome is governed by filtration processes. Environ Sci Technol 46: 8851–8859. doi: 10.1021/es302042t
11. Kwon S, Moon E, Kim TS, Hong S, Park HD (2011) Pyrosequencing demonstrated complex microbial communities in a membrane filtration system for a drinking water treatment plant. Microbes Environ 26: 149–155. doi: 10.1264/jsme2.me10205
12. Roesch L, Fulthorpe RR, Riva A, Casella G, Hadwin AK, et al. (2007) Pyrosequencing enumerates and contrasts soil microbial diversity. ISME J 1: 283–290. doi: 10.1038/ismej.2007.53
13. Schloss PD, Westcott SL, Ryabin T, Hall JR, Hartmann M, et al. (2009) Introducing mothur: open-source, platform-independent, community-supported software for describing and comparing microbial communities. Appl Environ Microb 75: 7537–7541. doi: 10.1128/aem.01541-09

14. Quast C, Pruesse E, Yilmaz P, Gerken J, Schweer T, et al. (2013) The SILVA ribosomal RNA gene database project: improved data processing and web-based tools. Nucleic acids res 41: 590–596. doi: 10.1093/nar/gks1219
15. APHA (American Public Health Association) (1995) Standard Methods for the Examination of Water and Wastewater, 19th ed, Washington, DC, USA.
16. Østergaard N (1985) Biogasproduktion i det thermofile temperaturinterval: Kemiteknik, Teknologisk Institut.
17. Li R, Chen S, Li X (2010) Biogas production from anaerobic co-digestion of food waste with dairy manure in a two-phase digestion system. Appl Environ Microb 160: 643–654. doi: 10.1007/s12010-009-8533-z
18. Malina J, Pohland JF, Frederick G (1992) Design of anaerobic processes for the treatment of industrial and municipal wastes: CRC Press 7.
19. Hansen KH, Angelidaki I, Ahring BK (1998) Anaerobic digestion of swine manure: inhibition by ammonia. Water Res 32: 5–12. doi: 10.1016/s0043-1354(97)00201-7
20. Lou XF, Nair J, Ho G (2012) Field performance of small scale anaerobic digesters treating food waste. Energy Sustain Dev 16: 509–514. doi: 10.1016/j.esd.2012.06.004
21. Heo NH, Park SC, Kang H (2004) Effects of mixture ratio and hydraulic retention time on single-stage anaerobic co-digestion of food waste and waste activated sludge. J Environ Sci Heal A 39: 1739–1756. doi: 10.1081/ese-120037874
22. Rivière D, Desvignes V, Pelletier E, Chaussonnerie S, Guermazi S, et al. (2009) Towards the definition of a core of microorganisms involved in anaerobic digestion of sludge. ISME J 3: 700–714. doi: 10.1038/ismej.2009.2
23. Ariesyady HD, Ito T, Okabe S (2007) Functional bacterial and archaeal community structures of major trophic groups in a full-scale anaerobic sludge digester. Water Res 4: 1554–1568. doi: 10.1016/j.watres.2006.12.036
24. Yamada T, Sekiguchi Y (2009) Cultivation of uncultured *Chloroflexi* subphyla: significance and ecophysiology of formerly uncultured *Chloroflexi* 'subphylum I' with natural and biotechnological relevance. Microbes Environ 24: 205–216. doi: 10.1264/jsme2.me09151s
25. Kindaichi T, Ito T, Okabe S (2004) Ecophysiological interaction between nitrifying bacteria and heterotrophic bacteria in autotrophic nitrifying biofilms as determined by microautoradiography-fluorescence in situ hybridization. Appl Environ Microb 70: 1641–1650. doi: 10.1128/aem.70.3.1641-1650.2004
26. Boucias DG, Cai Y, Sun Y, Lietze VU, Sen R, et al. (2013) The hindgut lumen prokaryotic microbiota of the termite Reticulitermes flavipes and its responses to dietary lignocellulose composition. Mol EcoL 22: 1836–1853. doi: 10.1111/mec.12230
27. Narihiro T, Terada T, Kikuchi K, Iguchi A, Ikeda M, et al. (2008) Comparative analysis of bacterial and archaeal communities in methanogenic sludge granules from upflow anaerobic sludge blanket reactors treating various food-processing, high-strength organic wastewaters. Microbes Environ 24: 88–98. doi: 10.1264/jsme2.me08561
28. Su Y, Li B, Zhu WY (2012) Fecal microbiota of piglets prefer utilizing dl-lactate mixture as compared to d-lactate and l-lactate in vitro. Anaerobe 19: 27–33. doi: 10.1016/j.anaerobe.2012.11.006

29. Chen S, Dong X (2005) Proteiniphilum acetatigenes gen nov, sp nov, from a UASB reactor treating brewery wastewater. Int J Syst Evol Micr 55: 2257–2261. doi: 10.1099/ijs.0.63807-0
30. Shin SG, Lee S, Lee C, Hwang K, Hwang S (2010) Qualitative and quantitative assessment of microbial community in batch anaerobic digestion of secondary sludge. Bioresour Technol 101: 9461–9470. doi: 10.1016/j.biortech.2010.07.081
31. Demirel B, Scherer P (2008) The roles of acetotrophic and hydrogenotrophic methanogens during anaerobic conversion of biomass to methane: a review. Rev Environ Sci Biotechnol 7: 173–190. doi: 10.1007/s11157-008-9131-1
32. Conklin A, Stensel HD, Ferguson J (2006) Growth kinetics and competition between *Methanosarcina* and Methanosaeta in mesophilic anaerobic digestion. Water Environ Res 78: 486–496. doi: 10.2175/106143006x95393
33. Vrieze JD, Hennebel T, Boon N, Verstraete W (2012) *Methanosarcina*: the rediscovered methanogen for heavy duty biomethanation. Bioresour Technol 112: 1–9. doi: 10.1016/j.biortech.2012.02.079
34. Bourque JS, Guiot S, Tartakovsky B (2008) Methane production in an UASB reactor operated under periodic mesophilic-thermohilic conditions. Biotechnol Bioeng 100: 1115–1121. doi: 10.1002/bit.21850

There are several supplemental files that are not available in this version of the article. To view this additional information, please use the citation on the first page of this chapter.

CHAPTER 3

Microbial Anaerobic Digestion (Bio-Digesters) as an Approach to the Decontamination of Animal Wastes in Pollution Control and the Generation of Renewable Energy

CHRISTY E. MANYI-LOH, SAMPSON N. MAMPHWELI, EDSON L. MEYER, ANTHONY I. OKOH, GOLDEN MAKAKA, AND MICHAEL SIMON

3.1 INTRODUCTION

Biomass encompasses materials derived from plants, animals, humans as well as their wastes. In addition, food processing, agro-industrial and industrial wastes are other sources of biomass wastes, so also is the microbial population metabolically active and cultivable plus metabolically active but non-cultivable cells existing within these wastes. Depending on the characteristics of these wastes, they can be converted into energy/and or fuel by combustion, gasification, co-firing with other fuels and ultimately by anaerobic digestion [1].

Microbial Anaerobic Digestion (Bio-Digesters) as an Approach to the Decontamination of Animal Wastes in Pollution Control and the Generation of Renewable Energy © *Manyi-Loh CE, Mamphweli SN, Meyer EL, Okoh AI, Makaka G, and Simon M.* International Journal of Environmental Research and Public Health ***10****,9 (2013). doi:10.3390/ijerph10094390. Licensed under a Creative Commons Attribution 3.0 Unported License, http://creativecommons.org/licenses/by/3.0/.*

So far, the conventional sources of energy that have provided power for developing and maintaining the technologically advanced modern world are the fossil fuels including coal, oil and natural gas. Yet, fossil resources are finite and their continued recovery and use appreciably impact our environment and affect the global climate due to the emission of greenhouse gases. Moreover, shortening of oil and gas are becoming imminent and to prepare for a transition to more sustainable sources of energy, viable alternatives for conservation, supplementation and replacement must be explored [2]. In this regard, biomass materials have been viewed as a way to expand energy supply, help mitigate growing dependence on fossil fuels and alleviate environmental and health hazards emanating as side effects from the use of fossil resources in many developing and developed countries [3].

Anaerobic digestion of biomass wastes could have a huge impact on renewable energy requirements. It is best suited to convert organic wastes from agriculture, livestock, industries, municipalities and other human activities into energy and fertilizer. It has become popular in developing countries such as China, India and Nepal; however, in South Africa, biogas digesters are principally constructed and installed in the Western and Kwa-Zulu Natal provinces of the country [4]. Owing to the important roles demonstrated by rumen microorganisms in anaerobic digestion [3], animal manures have been established as suitable sources of biogas production in Africa although, they are co-digested with energy crops in Denmark and Germany [5,6]. Co-digestion refers to the simultaneous anaerobic digestion of multiple organic wastes in one digester. This principle enhances methane yield due to positive synergisms established in the digestion medium, bacterial diversities in different wastes and the supply of missing nutrients by the co-substrates [7].

Furthermore, these wastes obtained from different animals vary in chemical composition and physical forms as a result of principal differences in the digestive physiology of the various species, the composition and form of diet, the stage of growth of the animal and lastly the management system of waste collection and storage [8]. Moreover, Sakar et al. [9] and St-Pierre and Wright [10] stated that a large proportion of the agricultural sector in both developing and developed countries is involved with poultry and livestock farming resulting in huge quantities of animal manure and other wastes from livestock operations which merit public, environmental and social concerns. Consequently numerous digesters are designed and installed on farms for the proper management of these wastes [11].

Overall, anaerobic digestion reduces biomass wastes and mitigates a wide spectrum of environmental undesirables, it improves sanitation, helps in air and water pollution control and reduces greenhouse gas emissions. Also, it provides a high-quality nutrient-rich fertilizer and yield energy in the form of biogas. The uses of biogas vary greatly from developing to developed countries. In Africa, biogas generated can be used as fuel for cooking, lighting and heating; it reduces the demand for wood and charcoal for cooking therefore helps preserve forested areas and natural vegetation, and can also help alleviate a very serious health problem due to poor indoor air quality associated with wood and charcoal used for cooking [12,13].

In Western countries (e.g., Germany & America), biogas is converted to electricity and heat for on-farm purposes by combined heat and power units after removing water and sulphur from its mixture [14]. Alternatively, it is upgraded to bio-methane constituting 95–99% methane wherein it opens up more utilization opportunities. Bio-methane is fed into the gas grid and used as power, transportation fuel and for heating [15].

Against this background, this paper appraises insights on environmental and public health implications arising from improper disposal of animal wastes and a comprehensive description of anaerobic digestion of these animal wastes as a means of resolving the ills; with emphasis on types of bio-digesters, microbial communities engaged in the process and techniques for their identification as well as factors affecting the digestion process.

3.2 ENVIRONMENTAL AND PUBLIC HEALTH IMPLICATIONS OF ANIMAL MANURE

Wastes from agricultural animals (poultry and livestock) often contain high concentrations of human pathogens, spilled feed, bedding material, fur, process-generated wastewater, undigested feed residues, feces and urine therefore must be effectively managed to minimize environmental and public health risks. However, the type and pathogenic microbial load depend on the type of the waste and its composition [16]. Figure 1 shows an overview of different contaminants in animal wastes and plausible implications. The following contaminants including pathogens (bacteria, viruses and protozoa), nutrients (phosphorus, nitrogen and sulphur), heavy metals (zinc and copper), veterinary pharmaceuticals (antibiotics) and naturally excreted hormones are present in animal manure [17].

3.2.1 SOURCES OF CONTAMINANTS IN ANIMAL MANURE

The intestinal tract of human and animals have been found to be the major sources of *Salmonella* and *Escherichia coli* in nature [18], which could be shed in feces. These pathogens may persist for days to weeks to months depending on the type of pathogen, the medium and the environmental conditions. Approximately 1% to 3% of all domestic animals are infected with *Salmonella*e [12,19]. Furthermore, other non bacterial pathogens that may be present with fecal material include protozoa (*Cyptosporidium* and *Giardia*) and viruses (Swine Hepatitis E- virus). The management and disposal of animal wastes harboring such pathogens can increase the risk of infections and diseases that threatens human health if these wastes are not properly treated and contained [20].

Antibiotics are routinely used in animal farming to prevent the spread of diseases or treat infected animals and simply added to animal feed to promote/increase growth. Further compounding the problem is the fact that the misuse or overuse of antibiotics could speed up the development of resistance or increase resistance of the microbial population present [17] due to the fact that resistance genes may be transferred between the microbial communities present. The use of any one antibiotic can yield resistance to multiple antibiotics. However, it is devastating that most antibiotics are designed to be quickly excreted from the treated organisms. Thus, they are commonly found in animal wastes.

Consequently, microorganisms in animal manures are thought to affect human health via multiple pathways either through direct or indirect contact with food, water, air or anywhere manure goes [21]. Further transmission of pathogens off-farm from farm workers to family members is also possible. It has been noted that most human-acquired infections result from these resistant strains [22]. Antimicrobial resistance of microorganisms is a local, national and global challenge; therefore the quest to identify the sources of these antibiotic resistant microbes and to seek for means to stop the spread of their resistance genes or not to select for further resistant strains is imperative.

Furthermore, animal manure may cause environmental pollution of water bodies as it is described to be rich in nutrients. Seppälä et al. [23] noted that these wastes harbor both micro and macronutrients including zinc and copper. These metals (Zn & Cu) are micronutrients found in animal manure originating from feed, supplements, antibiotics and water consumed by the animals [9].

3.2.2 ADVERSE EFFECTS OF ANIMAL WASTES ON THE ENVIRONMENT AND HUMANS

Taking into consideration the concentration of contaminants in animal wastes, it does have the potential to pollute land, water and air if containment and treatment do not adequately manage it. Haulage of these contaminants in animal wastes is dependent on the chemical characteristics, soil properties, climatic conditions and crop management practices. It is most probable that rain may wash/flush these wastes into streams, rivers or may cause waste to seep through the soil into underground springs and wells that humans use for sanitation and domestic purposes [24].

From the environmental point of view, excessive nutrients (especially phosphorus and nitrogen) in conjunction with elevated levels of biological oxygen demand (BOD) and chemical oxygen demand (COD) in these water bodies can contribute to algal blooms and cyanobacterial growth thus presenting serious socioeconomic hazards [25]. As a long term effect, they may cause shifts in phytoplankton community structure from desirable species to noxious species by means of holding back the growth of desirable aquatic species. Antibiotics in the soil may affect the natural ecosystem functions such as soil microbial activity and bacterial denitrification [17].

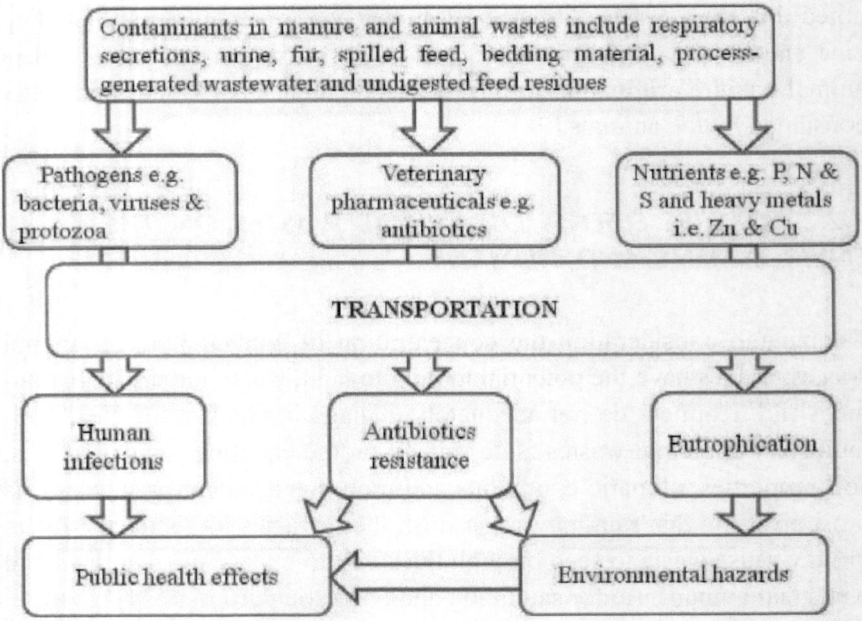

FIGURE 1: Environmental and public health implications of animal manure.

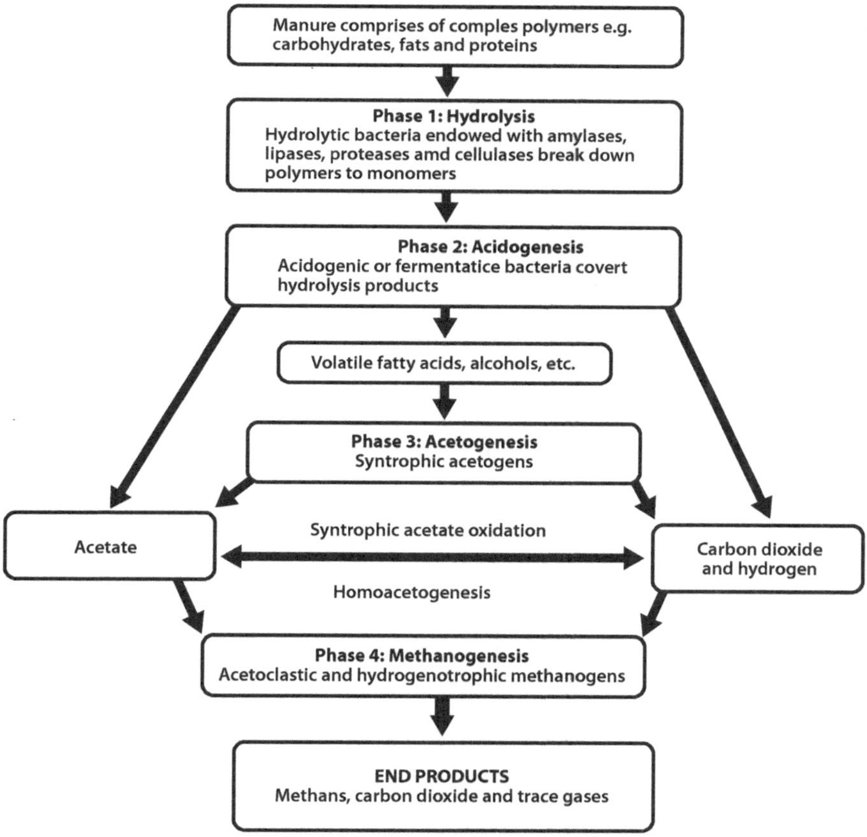

FIGURE 2: A schematic presentation of anaerobic digestion process.

Of profound public health effect is the reality that waterborne pathogens can be accidentally ingested during recreational activities or drinking water that is contaminated with animal feces [26]. This might result in acute gastrointestinal upset, e.g., nausea, diarrhea and vomiting. Also, contact with affected surface water during recreational activities can cause skin, ear or eye infection. Cyanobacteria (blue green algae) in surface water can produce neuro-toxins and hepatotoxins which are known to cause acute and chronic health complications [27]. However, the disease conditions in susceptible individuals including the very young, the elderly, pregnant women, and immunocompromised may be more severe, chronic or at times fatal [26]. In addition, antibiotic resistance interferes with antimicrobial chemotherapy causing treatment failure of some infections thereby presenting life threatening situations. Consequently, anaerobic digestion of animal manure in bio-digesters can substantially help to address the above-mentioned troubles arising from the improper or careless disposal of these wastes.

3.3 ANAEROBIC DIGESTION OF ANIMAL WASTES IN BIO-DIGESTERS

Growth and intensification of livestock operations often result to great quantities of manure that have to be properly managed. Even when stored, manure generates and releases methane (a greenhouse gas) into the atmosphere [10]. Moreover, anaerobic degradation has usually taken place in the lower digestive tract of animals and then continues in the manure piles resulting in malodorous compounds. These malodorous compounds originate from the incomplete breakdown of organic matter in manure by anaerobic microbes under uncontrolled environmental conditions [28].

Farm-based anaerobic digester presents as an alternative to the proper management of these wastes [29]. There are thousands of on-farm digester plants worldwide including Blue Spruce Farm, Green Mountain Diary, Chaput Family Farm, Cantabria Diary Plant, Buttermilk Hall Farm, Bulcote Farm, Minnesota mid-sized Diary Farm and a host of others [10,30]. However, Lutge and Standish [31] noted that very few on-farm anaerobic digesters are available in South Africa. Nevertheless, the type of digester used is being influenced by the characteristics of manure collected which is dependent on the animal's diet and on-farm practices [32]. All the same, these farms employ anaerobic digestion which is a consistent technology for their waste management/treatment.

Taking into consideration the aforementioned differences in livestock practices existing between the farms, the essential step of liquid-solid separation of manure mixture may be performed before or after anaerobic treatment [33]. In addition, the manure collected may be mixed with milk house waste for anaerobic digestion [34]. Generally, manure is collected with or without milk house waste and slurry prepared by adding water to it. The slurry is pumped to the separator for screening, separating the mixture into liquid and solid fractions. Subsequently, the screened liquid fraction is fed into the digester whilst the solid fraction could be dewatered and redistributed to areas lacking nutrients, used as bedding and or composted to serve as an additional source of more carbon and nitrogen [9]. In addition, the digested liquid fraction can be processed to obtain concentrated fertilizers or post-treated to obtain clean water for recycling and irrigation purposes [30,32].

Overall, during anaerobic digestion complex polymers in animal wastes are catabolized through a series of steps by complex consortia of microorganisms in the digester to ultimately yield methane and carbon dioxide [35]. Basically, this process can be divided into four phases: hydrolysis, acidogenesis, acetogenesis and methanogenesis in which hydrolytic, fermentative bacteria, acetogens and methanogens play distinct roles, respectively, as shown in Figure 2.

During hydrolysis, complex polymers like carbohydrates, proteins and fats are being degraded into sugars, amino acids and long chain fatty acids respectively. This breaking down process occurs primarily through the activity of extracellular enzymes (lipases, proteases, cellulases & amylases) secreted by hydrolytic bacteria attached to a polymeric substrate [36].

Later, fermentative or acidogenic bacteria transform hydrolysis products into acetic acid and intermediate compounds, such as ethanol, lactic acid, short chain fatty acids (C3–C6), hydrogen and carbon dioxide [35]. Acetate, carbon dioxide, formate, methylamines, methyl sulphide, acetone and methanol produced in this phase can be directly utilized for methanogenesis. Consequently, the other intermediary products from acidogenesis are converted to acetate, formate or CO_2 & H_2 by syntrophic acetogens in a bid to maximize methane production.

As a final point, methane is produced during methanogenesis by methanogens in two ways: either through cleavage of acetic acid molecules to produce methane and carbon dioxide or reduction of carbon dioxide with hydrogen by acetotrophic and hydrogenotrophic methanogens, respectively [37]. The biogas generated constitutes mainly methane (50–75%), CO_2 (25–45%) and traces of other gases like CO, H_2S, NH_3, O_2, water vapor (Table 1).

TABLE 1: Various constituents of biogas generated from the anaerobic digestion process; its average composition adopted from de Graaf and Fendler, [38].

Component	Symbol	Percentage
Methane	CH_4	50–75
Carbon dioxide	CO_2	25–45
Hydrogen	H_2	1–2
Ammonia	NH_3	<1
Water vapor	H_2O	2–7
Oxygen	O_2	<2
Hydrogen sulphide	H_2S	<1

3.3.1. MICROBIAL COMMUNITIES INVOLVED IN ANAEROBIC DIGESTION OF ANIMAL MANURE AND METHODS OF THEIR IDENTIFICATION

The specific microbes and their metabolic activities during anaerobic digestion depend on the chemical composition of the feedstock/waste, environmental factors and digester operating conditions [39,40]. Four sets of microorganisms are involved and these groups of microorganisms are tightly attached metabolically whereby the early stages of digestion yield reduced intermediate products that are utilized by acetogens and methanogens [37]. However, the interrelationship between the acetogens and methanogens is highly complex. These microorganisms are classified as anaerobes therefore oxygen poses a threat via the disruption of metabolic pathways causing oxidation of cellular factors that normally occur in reduced form. Contrarily, in recent times, it has been documented that several methanogens adapt to oxygen due to the presence of genes that synthesize enzymes (e.g., catalase and superoxide dismutase) within their genomes, which serve a role in defense against oxygen toxicity [41]. Several authors have reported the high tolerance of methanogens including *Methanobacterium thermoautotrophicum, Methanobrevibacter arboriphilus* and *Methanosarcina barkerii* to oxygen and dessication [42,43].

More elaborately, Anderson et al. [44] also noted that after dessication process, *M. barkeri* had innate capability to survive extended periods of exposure to air and lethal temperatures owing to the synthesis of thick outer cell layers composed of extracellular polysaccharide (EPS); added to the accumulation of cyclic 2,3-diphosphoglycerate (a novel metabolite which may be used to stabilize proteins at elevated temperatures). In addition, the membrane lipids of archael species have glycerol molecules bound by ether linkages to branched isoprene hydrocarbon molecules causing the organisms to adjust to such extreme temperatures [45]. On the whole, microbial community within a digester system can be grouped into acidogens, syntrophic acetogens and methanogens [46].

3.3.1.1 ACIDOGENS

It has been documented that the bacterial species active in the polymer hydrolysis phase are also active during the acidogenic phase. Hence, the hydrolytic and acidogenic bacteria are sometimes referred to as fermentative bacteria. They can be either facultative anaerobic bacteria (i.e., can survive under both aerobic and anaerobic conditions) or strict anaerobes. The family *Enterobacteriaceae* or enteric bacteria (a group of bacteria that inhabit the intestine of humans and other animals) are active fermenters and are among the organisms responsible for the first step in the bioconversion of carbohydrates to CH_4 [18].

In addition, Blumer-Schuette et al. [47] and Wirth et al. [48] documented the relevance to biomass deconstruction of the following microorganisms: *Caldicellulosiruptor saccharolyticus, Thermotoga maritima, Clostridium thermocellum, Anaerocellum thermophilum, Escherichia coli, Clostridium kluyveri, Bacillus cereus, Ruminococcus albus* etc. Other known groups of anaerobic cellulose-degrading bacteria are found in the following genera; *Aminobacterium, Psychrobacter, Anaerococcus, Bacteroides, Acetivibrio, Butyrivibrio, Halocella, Spirochaeta, Caldicellulosiruptor* and *Cellulomonas* (a facultative anaerobe of the phylum *Actinobacteria*) [49,50].

3.3.1.2 SYNTROPHIC ACETOGENS

Syntrophic acetogens, e.g., *Syntrophobacter wolinii, Syntrophomonas wolfei* and *Smithella* sp. are responsible for the syntrophic metabolism of alcohols, short chain fatty acids (C3–C6), some amino acids and aromatic compounds to yield methanogenesis substrates [51]. The conversion of the above-mentioned substrates to yield methanogenesis products is thermodynamically unfavorable but it becomes favorable with the presence of a syntrophic partner (hydrogenotrohphs) [51,52].

However, accumulation of volatile fatty acids results in decrease in pH, increase acidification, destroy methanogens activity and leads to failure of digester ultimately [7]. While the syntrophic acetogens are converting intermediary metabolites to acetate and other methanogenesis substrates, homoacetogens also produce acetate from the reduction of carbon dioxide with hydrogen via the acetyl Co-A pathway [53].

Overall, methane is produced during methanogenesis by acetotrophic, hydrogenotrophic and methylotrophic pathways. In regards to acetate degradation to yield methane via the acetoclastic pathway, specific methanogens from the order *Methanosarcinales* are responsible. Contrarily, a group of acetate-oxidizing bacteria occur in syntrophic relationships with hydrogenotrophic methanogens wherein they oxidize acetate to form methane in association with the latter microorganisms.

These bacteria are equally referred to as syntrophic acetogens and include both mesophilic and thermophilic bacteria (Table 2). Syntrophic acetate-oxidizing bacteria are involved in the reversed reductive acetogenesis [54] and can be identified by the combinatorial use of flux measurement and transcriptional profiling of formyltetrahydrofolate synthetase (FTHFS) gene, an ecological biomarker engaged in reductive acetogenesis [52].

TABLE 2: Syntrophic acetate-oxidizing bacteria in association with hydrogenotrophic methanogens.

Acetate-oxidizing bacteria	Microbial description	Hydrogenotrophic methanogens	References
AOR	Anaerobic, rod-shaped, gram positive, non-spore forming and thermophilic (60 °C)	*Methanobacterium* sp. strain THF	Lee and Zinder [55]
Clostridium ultunense	Anaerobic, spore-forming, rod-shaped, gram negative and mesophilic (37 °C)	*Methanoculleus* sp. strain MAB1	Schnürer et al. [56]
Thermacetogenium phaeum	Anaerobic, rod-shaped, gram negative but with gram positive cell wall structure and thermophilic (between 55 and 58 °C)	*Methanothermobacter thermoautotrophicus* TM	Hattori et al. [57]
Thermotoga lettingae	Anaerobic, rod-shaped, non-spore forming, mobile, gram negative and thermophilic (65 °C)	*Methanothermobacter thermoautotrophicus* or *Thermodesulfovibrio yellowstonii*	Balk et al. [58]
Syntrophaceticus schinkii	Anaerobic, spore-forming, variable cell shape, gram variable and mesophilic (between 25 and 40 °C)	*Methanoculleus* sp. strain MAB1	Westerholm et al. [59]

3.3.1.3 METHANOGENS (ARCHAEA)

Methanogens are found in a wide range of anaerobic habitats including freshwater and marine habitat, sewage digesters, the digestive tracts of herbivores, mammals and wood and humus feeding insects etc. [60,61,62,63]. They belong to the domain *Archaea* and they occupy a key position in the anaerobic digestion process because it is in this last step of the process where the valuable methane is produced [45]. During an unstable anaerobic digestion process in a poorly performing anaerobic digester, the methanogenic populations are especially sensitive to acidity (pH), concentrations of volatile fatty acids, and free ammonia and ammonium ions in the digesting substrate [64].

In addition, methanogens are classified into six orders i.e., *Methanobacteriales, Methanococcales, Methanomicrobiales, Methanosarcinales, Methanocellales* and *Methanopyrales* [65]. Members of the order *Metha-*

nosarcinales utilize acetate which has for a long time been known as the major precursor for more than 70% of methane produced in most engineered anaerobic digester [66]. In other words, previous knowledge assumed that in totality, two-thirds of methane are obtained from the acetoclastic methanogenesis pathway and one-third from hydrogenotrophic methanogenesis pathway according to anaerobic digestion model 1 (ADM1) [66]. This is misleading since reports were based on anaerobic digestion of waste water and sewage sludge with very low organic content.

Clearly, biomethanation involves a complex community of specialized microorganisms that depend on each other either for substrate supply or metabolism of end products in order to favor their metabolic activity. Moreover, microbial species require specific combination of physical and chemical conditions viz temperature, pH, salinity besides substrate availability to thrive. Therefore, microbial species obtained from different environment associated with different physical and chemical factors tend to vary even though they could perform anaerobic digestion through the same stages of hydrolysis, acidogenesis, acetogenesis and methanogenesis [10]. However, this might influence the biogas yield obtained under different physical, chemical and substrate conditions.

Nowadays, results obtained from several studies conducted by different authors have contradicted the assumptions of ADM1. Krakat et al. [67,68] demonstrated that a higher temperature of 60 °C and a drastic reduction in hydraulic retention time resulted in dominance of hydrogenotrophs among the microbial communities with a corresponding increase in methane production in thermophilic and mesophilic biogas fermentors respectively, digesting energy crops. Similarly, Klocke et al. [69,70] reported varying substrate utilization during methanogenesis within the biogas plant as depicted from the dominance fraction of hydrogenotrophic methanogens relative to acetoclastic counterparts.

Furthermore, members of *Methanosarcinales* are described as acetoclastic and comprise of two families *Methanosarcinaceae* and *Methanosaetaceae*. However, these two families of acetoclastic methanogens differ in their physiology, biokinetics and growth environment with respect to the acetate concentration [2]. Conclusively, the interactions of the different groups of anaerobic microorganisms are incredibly complicated, and the effective performance of the biological process strongly depends on the balance of these relationships [71].

3.3.2 TECHNIQUES FOR IDENTIFYING MICROORGANISMS INVOLVED IN THE ANAEROBIC DIGESTION PROCESS

St-Pierre and Wright [10] mentioned that the microbial communities within an anaerobic digester treating animal manure are not fully characterized. However, due to the diversity of microorganisms in the system, a variety of methodological approaches are required for a detailed analysis of community structure in a bid to unravel the complex antagonistic and synergistic effects between microbial communities in order to eventually improve process stability and efficiency of biogas formation [72,73]. This can be achieved by the combined use of traditional culture-based, microscopic and molecular techniques.

Conversely, culture-based techniques such as plate counts, membrane filtration and most probable number (MPN) have an inherent limitation because only the viable population will grow to produce colonies under specific growth conditions whereas others that are important in the original sample do not proliferate [74]. However, these traditional culturing methods employed with environmental samples also underestimate the total number of microorganisms due to the selective nature of the media used, the lack of detection of active but non cultivable (ABNC) microbes and failure to count microbes that are present as aggregates or associated with particles. Likewise, it is impossible to obtain pure cultures of most microorganisms in natural environment due to the complex syntrophic and symbiotic relationships that are abundant in nature [75].

Contrarily, direct microscopic methods e.g., DAPI epifluorescence microscopy allows the direct observation and total enumeration of viable and non-viable microorganisms in the feedstock [72]. Specifically, the identification and enumeration of methanogenic microbes can be achieved by epi-fluorescence microscopy, a technique based on their unique fluorescent pigment, factor F_{420} [76]. The coenzyme F_{420} shows autofluorescence (blue-green color) of methanogenic cells when excited by UV light. Hence, this autofluorescence serves as a diagnostic tool used to count autofluorescent methanogens [77]. Results from this technique can be combined with those from molecular methods for a better overview of microbial population and structure in an anaerobic digester [78].

Molecular techniques targeting particularly 16S rRNA genes (the only RNA component of 30S ribosomal subunit), can also be employed as conventional methods for identification of microbial community in a digester. These methods include cloning, fluorescent in situ hybridization (FISH), denaturing gradient gel electrophoresis (DGGE), single strand conformation polymorphism (SSCP), restriction fragment length polymorphism (RFLP), quantitative real-time PCR (qPCR) and DNA sequencing (Sanger and next generation sequencing methods). Each of these methods has its advantages and shortcomings that have been deliberated and presented elsewhere [74,79].

Most previous studies on anaerobic digestion of biomass materials were based on the construction of 16S rRNA clone libraries and subsequent sequencing (Sanger method) of individual clones [69]. The resulting sequences corresponded to different taxa that were employed in the phylogenetic classification thereby revealing the structure of the underlying community. In addition to knowledge on phylogenetic diversity, the microbial population present in an anaerobic digester can be quantified using FISH and qRT-PCR as well as uncultured microbes will be uncovered by the use of molecular methods based on 16S rRNA [80]. However, the analysis of 16S rRNA gene does not cover the whole complexity of the environment due to sequencing of limited number of clones as well as low cloning efficiencies [81].

In addition to the methods relying on 16S rRNA for the detection of microorganisms in an anaerobic digester, other methods are available incorporating other remarkable genes which could serve as a diagnostic tool. The hydrazine oxidoreductase genes (*hzo* gene) can be employed in phylogenetic diversity and functional analysis of anaerobic ammonium-oxidizing bacterium (anammox) in a community or an environment. The hzo genes are specific for the identification of anammox bacteria and can be used as a functional marker to give activity-based information regarding the anammox bacteria in a community [82]. Ozgun et al. [83] identified and quantified anammox bacteria from waste water treatment plant by adopting a FISH-based approach with the PCR primers designed for the amplification of *hzo* specific genes.

Furthermore, hydrazine synthase (hzs A) also represents a unique phylogenetic marker for anammox bacteria hence; it can be used for identification purpose. Moreover, Harhangi et al. [84] determined the presence and biodiversity of anammox bacteria from samples of wastewater treatment systems, fresh water and marine environments as well as anammox enriched cultures by the development of PCR primer set targeting a subunit of hydrazine synthase. The tested primers successfully retrieved hzs A gene sequences covering all known anammox genera thus indicating that the use of 16S rRNA gene does not directly relate to the physiology of the targeted microbes and also the primer set currently available does not give tangible information pertaining to the diversity of the anammox species.

Also, methanogens can be distinguished from other microbes by virtue of their cell wall components, unique membrane lipids, 16S rRNA gene sequences as well as key enzymes that are involved in methanogenesis. The key enzymes encompass specific coenzyme and cofactors such as F_{420}, methanopterin and coenzyme M that are engage in methanogenesis [85]. These enzymes are strongly conserved and found only among methanogens. Consequently, methanogens can be specifically identified by targeting genes that code these peculiar enzymes.

Of great significance is the methyl Coenzyme M reductase (an enzyme complex) that catalyzes the reduction of the methyl group bound to coenzyme M and is encoded by the gene *mcr A*. It has been documented that methanogenic biodiversity displayed upon utilization of *mcr A* is similar to that revealed by 16S rRNA therefore validating the application of the former gene for identification [86,87]. However, a prominent drawback linked to the incorporation of *mcr A* gene is that its available sequences are very limited in the database therefore primers designed based on these sequences may lead to inefficient amplification and an adequate representation of the methanogenic community will not be delineated [85].

Interestingly, next generation sequencing methods (NGS) are emerging as a robust technology that creates abundant data and has caused a fundamental shift in molecular biology. These new methods involve whole genome sequencing; function-driven metagenomics and high throughput sequencing [85] that have the potential to demonstrate new insight into the entire genome of the microbial environment resulting in the rapid characterization of targeted sequences at less cost compared to the first-generation sequencing method, "traditional Sanger method" [88].

Sequencing of the whole genome is often aimed at obtaining information about the complete set of genes in any particular genome [89]. Genome sequencing opens up unexpected opportunities with information that cannot be obtained with conventional microbiological methods. For instance, Leahy and colleagues [90] unravelled the presence of vaccine and chemogenomic targets in their attempt to assess the sequencing of the M1 genome (first ever identified genome of rumen archael species) of *M. ruminatium.* Attwood et al. [61] equally noted the presence of a phage sequence residing within the genome of *M. ruminatium* and postulated that enzymes from the phage could serve as potential methanogen control agents. However, knowledge of the complete genetic makeup of an organism is not sufficient as information pertaining to their function is missing thus indicating that the biochemical function of each gene product is crucial.

On the other hand, metagenomics or community genomics or environmental genomics is the analysis of genomic DNA of the whole microbial communities directly in their natural environments by extracting or isolating the total DNA from an environmental sample [91]. Specifically, function-driven metagenomics is a powerful tool that has the potential to identify entire new classes of genes of new or known functions thus providing information on the metabolic activities of all members of a microbial community, including even those that were uncultivable and undefined previously [91].

In addition, Gilbert and Dupont [92] further described that metagenomics embodies two aspects; the environmental single-gene surveys in which single targets are amplified and the PCR- amplified products are sequenced and secondly, random shotgun studies of all environmental genes where the total DNA is isolated and sequenced; giving a profile of all the genes occurring in the community. However, the extent of community coverage depends on the depth of sequencing.

With the advent of high throughput sequencing technologies such as Roche 454, Illumina/Solexa, Applied Biosystems/SOLiD and Helicos BioSciences, more sequence data are obtained than has ever been possible with the traditional Sanger sequencing method [10]. These improved sequencing methodologies do not require cloning of the DNA before sequencing thereby; they bypass one of the main biases in environmental sampling. Consequently, bioinformatics strategies and tools can be used for the analysis of the huge data obtained [85].

In conclusion, conventional molecular tools based on 16S rRNA (viz PCR, qRT-PCR, FISH, DGGE, cloning and sequencing of gene library) are still of relevance since they provide an initial selection of collected samples prior to comprehensive analyses by next generation sequencing technologies [85]. Moreover, from clone libraries of 16S rRNA, usually precise sequences are assigned as a result of larger sequence length relative to the short read lengths obtained from high throughput sequencing technologies [50]. Therefore, it is very promising integrating conventional molecular methods, epi-fluorescence microscopy and high throughput sequencing technologies in describing the microbial structure of a biogas plant. Combining these methods will reveal the entire complexity of the microbial communities in an anaerobic digester as well as the physiology and function/activity of the community. Also, previously unidentified microbes with culture-based techniques will be delineated and their function within the community will be recognized [91].

3.3.3 TYPES OF BIO-DIGESTERS FOR TREATING ANIMAL MANURE

A biogas digester consists of one or more airtight reservoirs (chambers) into which animal manure or a mixture of manure and co-substrate is placed, either in batches or by continuous feed [93]. These biogas generating systems could be categorized on the basis of the number of reactors used into single (one) stage or multi (two) stage and on the mode of feeding into continuous and batch feeding systems [12].

In single stage processes, the three stages of anaerobic process occur in one reactor; however, the growth rate of fermentative bacteria is faster than that of acetogenic and methanogenic bacteria [33]. Consequently, acids accumulate; the pH falls and the growth of methanogenic bacteria is inhibited due to increase organic loading rate and inappropriate other process parameters. Whereas multi-stage processes make use of two or more reactors that separate the acetogenesis and methanogenesis stages in space and allows the establishment of operational conditions that reduce the start time and microbiota specialization in each reactor, thereby allowing the most desirable products at each stage to be obtained [35,69].

In a batch experimental set up, the digester is loaded with the feedstock at the beginning of the reaction and the product discharged at the end of each cycle whereas in continuous feeding, the organic material is continuously charged and discharged [12].

Livestock operational practices differ between individuals and they influence the characteristics of the manure obtained which in turn determines the choice of the digester [28]. Manure can be collected either by scraping with an automated device or flushing with water [32]. Ideally, scraped manure can be digested by a complete mix digester (e.g., continuously stirred tank reactor, CSTR) and a plug flow digester whereas flushed manure warrants the use of covered lagoons and anaerobic fixed film digesters [31,94].

Traditionally, animal manure is often flushed, pretreated by means of mechanical screening, sedimentation or both in a bid to achieve two separate fractions of liquid and sludge; with the liquid portion pushed into covered lagoons for storage and anaerobic treatment [28]. However, the anaerobic digestion process in the lagoon is affected by climatic conditions (temperature) as well as the water table on the site especially since liquid can seep into underground spring and streams [95]. In recent times, with the quest for retaining active microbial population in the digester in order to improve process stability and control; biodigesters are designed with active microbial populations attached to inert supports as biofilms or form aggregates or granules [9,94]. Sakar et al. [9] revealed that up flow anaerobic sludge blanket is most suitable for the anaerobic digestion of poultry wastes.

More elaborately, conventional digesters used for anaerobic treatment of animal manure are CSTR and plug-flow reactors with appreciable holding capacity though with long HRT compared to fixed film digesters [11]. On the other hand, anaerobic fixed film reactors have the potential of retaining microbial mass (as biofilms) on support materials and also reduce the retention time for anaerobic digestion to several hours or a few days [94]. However, Lutge and Standish [31] mentioned that South Africa has the potential of implementing the use of CSTR and covered lagoons for on-site animal manure treatment.

3.3.4 FACTORS INFLUENCING ANAEROBIC DIGESTION OF ANIMAL MANURE

Generally, factors affecting the performance of an anaerobic digester include operational factors (pH, temperature, organic loading rate (OLR)/ hydraulic retention time (HRT), free ammonia concentration), substrate characteristics/biodegradability and biodigester design [96]. However, Wilkie [28] reported that temperature, biodegradability, OLR and HRT have great impact on the anaerobic digestion of animal manure. Notwithstanding, other factors should not be overlooked.

3.3.4.1 TEMPERATURE

Based on temperature, anaerobic microorganisms can be categorized into psychrophiles (<20 °C), mesophiles (25–37 °C) and thermophiles (55–65 °C) [97]. Some methanogenic species exhibit a preference of extreme heat (90–100 °C) thus, are classified as hyperthermophilic methanogens [45]. Examples are *Methanocaldococcus jannaschii* and *Methanococcus vulcanius* [62]. Temperature can be considered as the most important environmental factor influencing the growth of microbes. Albeit, each microorganism has a certain temperature range within which it can grow and multiply. When temperature is increased within a certain range, the chemical and enzymatic reactions increase at a faster rate and growth increases [98].

However, above optimum temperature, key chemical reactions in the different metabolic pathways being catalyzed by enzymes cannot occur because enzymes are irreversibly destroyed since they are protein like in nature. Enzymes are crucial to metabolism because they allow organisms to drive desirable reactions that require energy and will not occur by themselves, by coupling them to spontaneous reactions that release energy. Consequently, the growth rate of microbes will equally stop [99]. However, different microbial species respond differently to abrupt changes in temperature. Moreover, temperature does not only influence the rate of metabolism of the microorganisms but also affect other process parameters such as OLR and ammonia concentration [29,100].

Generally, anaerobic digestion of biomass wastes could occur both at mesophilic (25–37 °C) and thermopilic (55–65 °C) temperature ranges. However, the ratio of free ammonia to total ammonium ion is higher at thermophilic temperature ranges. Consequently, animal wastes (containing nitrogen and ammonia compounds) are digested at mesophilic temperature range (25–37 °C) in a bid to avoid ammonia mediated inhibition of methanogenesis [29]. Moreover, thermophilic treatment requires high energy thereby may reduce the net energy obtained from the overall digestion process [100]. In spite of the abovementioned drawbacks of thermophilic fermentation; it undoubtedly causes significant destruction of pathogens and weed seeds and also causes higher metabolic rate resulting in higher methane yield [100,101].

3.3.4.2 PH AND ALKALINITY

In regards to anaerobic digestion, it is more appropriate to discuss pH alongside alkalinity since the latter can be used to control pH thus buffering the acidity of the system derived from acidogenesis phase [102]. Therefore, the amount of alkalinity present in an anaerobic digester represents the buffering capacity.

The pH range of anaerobic digestion normally occurs near neutral pH range and it is dependent on the OLR (which depends on reactor type) and the buffering capacity of the substrate. Livestock wastes (rich in ammonia and nitrogen compounds) such as cow, swine and poultry manure have high buffering capacity as they produce alkalinity when degraded upon by microorganisms [103]. However, anaerobic digestion of these wastes is often maintained at higher pH values of 7.6 [29]. An increase in OLR with a corresponding decrease in HRT can result to accumulation of volatile fatty acids which causes a drop in pH due to increased acidity of the digesting medium [104]. However, in instances where the pH has to be adjusted, several chemicals such as sodium hydroxide, potassium hydrogen carbonate, sodium carbonate, calcium carbonate, calcium hydroxide etc. can be added for alkalinity supplementation [40].

3.3.4.3 AMMONIA CONCENTRATION

Anaerobic digestion of urea- and protein-rich wastes such as animal wastes is often faced with the challenge of high levels of free ammonia due to their high organic nitrogen concentration which upon biological degradation results in high concentration of total ammonium ion plus free ammonia [105]. The quantity of ammonia produced during the digestion process is attributed to substrate concentration of nitrogen, reactor loading, C/N ratio, buffering capacity and temperature. In aqueous solution, inorganic ammonia nitrogen exists in two principal forms; ammonium ion (NH_4^+) and unionized ammonia or free ammonia (NH_3) in a pH dependent equilibrium state. Ammonia toxicity is influenced by the operating pH and temperature [29].

An increase in pH will cause increase ammonia toxicity of the system since a greater part of the total ammonia nitrogen will be free ammonia, the form that has been recognized as a toxic agent [106]. On the other hand, reduction in pH to a level within the optimum pH necessary for growth of the microorganisms will help to counteract free ammonia concentration. However, process instability provoked by ammonia toxicity often results in increased level of volatile fatty acids with a corresponding decrease in methane yield [64].

In addition, Strik et al. [107] noted that high ammonia concentration led to poor biogas quality requiring treatment, decreased COD removal efficiency, decreased biogas generation and malodor, besides process inhibition. High free ammonia content has usually been associated with unstable process performance and increased risk of process failure as a result of its inhibitory effect on methanogens (specifically acetate-utilizing methanogens). Therefore, in the presence of elevated levels of ammonia in a fermentor, a shift occurs in the biomethanation process from acetoclastic methanogenesis (performed by acetate-utilizing methanogens) to syntrophic acetate oxidation conducted by syntrophic acetogens in collaboration with hydrogenotrophs [64]. Moreover, El-Mashad et al. [100]

revealed that ammonia toxicity does not only affect the acetoclastic methanogens but also hydrolysis and acidification processes.

Furthermore, chemical equilibriums especially of free ammonia concentration at a fixed total ammonium concentration can be affected by the operating temperature (i.e., mesophilic and thermophilic temperature ranges) of the digester system. Even though the temperature is pivotal in the thermodynamics and kinetics of microbial reactions in methanogenesis, challenges are encountered during treatment at thermophilic temperatures (i.e., 55–65 °C) of ammonium-, urea- and protein-rich biomass materials owing to a high level of free ammonia [108]. Garcia et al. [29] noted that at higher temperatures the ratio of free ammonia to total ammonium was much higher; consequently affecting methane generation due to free ammonia inhibition. Nevertheless, an increase in temperature within the mesophilic range relieved the digester system of ammonia toxicity. Therefore, anaerobic digestion of animal manure at mesophilic temperature offers better process stability and performance of the digester system than at thermophilic temperatures [101].

It has been observed that co-digesting animal wastes with carbon-rich co-substrates will help to prevent both volatile fatty acid and ammonia mediated inhibition [103]. However, the inhibitory ammonia threshold concentration is not standardized because of the conflicting results obtained from different studies conducted under different environmental conditions with different substrates and inocula in conjunction with the complex nature of the anaerobic digestion process and acclimation periods [106]. The way microbes tend to adapt to increased ammonia level is dependent on the rate of ammonia formation which is influenced by the OLR and HRT.

3.3.4.4 HYDRAULIC RETENTION TIME AND ORGANIC LOADING RATE

HRT is the average period of time that the substrate resides in the anaerobic digester and OLR describes the amount of organic matter expressed in g COD/L or g TS/L or g VS/L added to the digester per reactor volume and unit time. HRT is inversely proportional to OLR and both are very useful parameters that contribute knowledge on design and performance of the reactor [109].

Biological decomposition of animal manure is affected greatly by its retention time in the reactor [110]. Retention time is determined by solid content of manure, temperature as well as the type of reactor used for treatment [11]. More elaborately, CSTR and plug flow reactors for animal manure treatment require retention time of 20–30 days whereas fixed film reactors usually have a shorter retention time of several hours to a few days [101]. However, covered lagoons require a longer retention time of 60 days [28]. In addition, HRT also affects the quality of effluent in terms of microbial load, nutrient content as well as the methane yield. Umaña et al. [94] investigated the influence of HRT on anaerobic fixed bed reactor and noted that the quality of effluent and methane yield increased due to increase in HRT.

On the other hand, OLR is dependent on temperature and HRT. An abrupt increase in OLR causes system failure attributed to decreased COD removal efficiency, methane production rate and pH [111]. More elaborately, a higher OLR beyond the optimum capacity elevates the rate of production of intermediary products (fatty acids) by hydrolytic and acidogenic bacteria. Subsequently, these fatty acids would accumulate due to the slow rate of their consumption by methanogens thus; pH will drop thereby inhibiting methanogenic activity [112]. Furthermore, Rincòn et al. [113] documented that a higher OLR influenced the bacterial community within the digester system; with the genus *Clostridium* being predominant at low OLR and the classes and phyla; *Gammaproteobacteria, Deferribacteres, Actinobacteria* and *Bacteroidetes* respectively, predominated at high OLR.

3.3.4.5 SUBSTRATE CHARACTERISTICS AND HEAVY METALS

The constituents of manure directly determine the biogas yield and the level of biochemical reactions that would take place within the digester system [28]. The composition of the manure will depend on the livestock operations which includes the diet and the handling/storage procedure of the wastes [8]. Evidently, for the proper functioning and continuous reproduction of microbes implicated in the anaerobic digestion process, there is a need for available sources of energy; carbon for the synthesis of new cellular materials, inorganic elements such as nitrogen, phosphorus, potassium, sulfur, calcium and magnesium as well as organic nutrients [114]. As a consequence, the physical and chemical characteristics including the moisture content, total solids content, volatile solids content, phosphorus, nitrogen and carbon content of the feedstock must be evaluated before commencement of the digestion process [40].

Volatile solids of manure are a very critical parameter as it consists of the biodegradable portion which includes carbohydrates, fats and proteins and the refractory portion which cannot be anaerobically digested described as lignocellulosic [28]. The term biodegradability of manure is indicated by biogas or methane yield and percentage of solids (total or volatile solids) that are destroyed in the anaerobic digestion process [115].

Microorganisms require a trace amount of some metals (nickel, cobalt, copper, iron, zinc, molybdenum etc.) for optimum growth and performance. Matseh [116] noted that these trace elements are usually known as stimulatory micro-nutrients and do occur in coenzymes and cofactors. The stimulatory effects potentiated by these metals on biogas process performance are linked to increased methane production, substrate utilization and reactor stability. However, there are wide ranges in the quantity of these metals that are needed in order to become stimulatory; this has been ascribed to differences in pH, OLR, HRT, substrate characteristics and the complex chemical and biological processes monitoring trace metal bioavailability [117].

Moreover, the stimulatory effect varies between the different types of trace metals. This is affirmed by the work of Pobeheim et al. [118] which recorded an increase in methane yield upon addition of a well-defined trace element solution composed of Co, Ni and Mo. Whereas a higher decrease in methane generation and process stability was noted due to the elimination of Ni from the solution. Results further revealed that 0.4–2 μM concentration of Co caused a 10% increase in methane production but the addition of Mo exhibited no profound effect on methane production.

Interestingly, animal manure has been reported to contain a good level of both macro and micronutrients (trace elements) [23]. However, process failure caused by trace element deficiency has been demonstrated during anaerobic digestion of single substrates such as maize silage. Some mono substrates (e.g., maize silage, potato etc.) or even food wastes cannot provide both the micro and macro nutrients essential for the growth of anaerobic microbes that are present in the anaerobic digestion process [119]. Therefore, they have to be supplemented with these nutrients before commencement of the digestion process. Better still, they can be co-digested with animal manure such that animal manure provides good buffering capacity and required nutrients whilst the energy crop provides increases in the energy yield of the process [23,119].

A deficiency of these metals can cause shifts in microbial community structure; Gustavsson et al. [117] documented a shift in microbial community structure from *Methanosarcinales* dominance during a stable process performance with Co and Ni supplementation to *Methanomicrobiales* dominance at both Co and Ni deficiency. However, too high a concentration of these heavy metals would lead to toxicity of the system thereby hampering the biological process via the enzymes involved by interfering with their function and structure. Apparently, they may substitute for naturally occurring metals in an enzyme prosthetic group or by binding to the SH group of the enzyme [106].

3.3.4.6 MIXING

Of great value in the anaerobic digestion of animal manure is the extent of contact between the incoming animal manure and a viable bacterial population; this is a function of mixing in the reactor [110]. The benefits of mixing digester contents during anaerobic process have been documented by several authors and include: it prevents scum formation inside the digester, ensures uniform distribution of microorganisms and substrate throughout the mixture and intensifies contact between them, prevents stratification within the digester therefore enables uniform distribution of heat throughout the mixture and lastly helps to release gas from the mixture [30,120].

According to Ghanimeh et al. [121], stirring can result in reduction of particle sizes of the substrate as anaerobic digestion progresses. However, what is unclear about the aspect of mixing is the intensity and the duration of mixing considering the fact that different modes (mechanical mixers and recirculation pumps) could be used [110]. In the characterization of manure, total and volatile solids are very paramount because there is a certain limit above which the manure will no longer be a slurry hence posing problems of mixing and pumping operations [28]. As a result, Rico et al. [30] mentioned that low total solids added to long HRT minimizes the need of mixing in anaerobic digestion of dairy manure.

3.4 CONCLUSIONS

Anaerobic digestion of animal manure is looked upon as a strong option in safely reusing wastes or transforming them into valuable materials and energy. This decomposition process that occurs within a confinement contributes to pollution control as it presents with the following benefits; it reduces biological oxygen demand (BOD) and chemical oxygen demand (COD) of wastes; it destroys pathogenic microbes reducing the microbial load to a level which could be safely handled by humans with limited health risks [16] and it destroys volatile fatty acids and many malodorous compounds present in the feedstock and reduces the emission of greenhouse gases. Ultimately, it generates biogas and high quality nutrient-rich fertilizer from animal manure thus upholds the concept of waste to wealth in enhancing sustainability of development [3].

REFERENCES

1. Federal Energy Management Program, Biomass Energy-Focus on Wood Waste. In Biomass and Alternative Methane Fuels; BAMF Fact Sheet: Oak Ridge, TN, USA, 2004.
2. Wilkie, A.C. Biomethane from Biomass, Biowaste and Biofuels. In Bionergy; Wall, J.D., Harwood, C.S., Deamin, A.L., Eds.; ASM Press: Washingston, DC, USA, 2008; pp. 195–215.
3. Uzodinma, E.O.; Ofoefule, A.U.; Eze, J.I.; Mbaeyi, I.; Onwuka, N.D. Effect of some organic wastes on the biogas yield from carbonated soft drink sludge. Sci. Res. Essays 2008, 3, 401–405.
4. Mukumba, P.; Makaka, G.; Mamphweli, S.; Simon, M.; Meyer, E. An insight into the status of biogas digesters technologies in South Africa with reference to the Eastern Cape Province. Fort Hare Pap. 2012, 19, 5–29.
5. Karakashev, D.; Batstone, D.J.; Angelidaki, I. Influence of environmental conditions on methanogenic compositions in anaerobic biogas reactors. Appl. Environ. Microbiol. 2005, 71, 331–338.

6. Kröber, M.; Bekel, T.; Diaz, N.N.; Goesmann, A.; Jaenicke, S.; Krause, L.; Miller, D.; Runte, K.J.; Viehöver, P.; Pühler, A.; et al. Phylognetic characterization of a biogas plant microbial community integrating clone library 16S rDNA sequences and metagenome sequence data obtained by 454-pyrosequencing. J. Biotechnol. 2009, 142, 38–49.
7. Li, J.; Jha, A.K.; He, J.; Ban, Q.; Chang, S.; Wang, P. Assessment of the effects of dry anaerobic co-digestion of cow dung with waste water sludge on biogas yield and biodegradability. Int. J. Phys. Sci. 2011, 6, 3679–3688.
8. Anunputtikul, W.; Rodtong, S. Laboratory Scale Experiments for Biogas Production from Cassava Tubers. In Proceedings of the Joint International Conference on "Sustainable Energy and Environment (SEE)", Hua Hin, Thailand, 1–3 December 2004; 3-017(0). pp. 238–243.
9. Sakar, S.; Yetilmezsoy, K.; Kocak, E. Anaerobic digestion technology in poultry and livestock waste treatment. Waste Manag. Res. 2009, 27, 3–18.
10. St-Pierre, B.; Wright, A.D.G. Metagenomic analysis of methanogen populations in three full- scale mesophilic anaerobic manure digesters operated on dairy farms in Vermont, USA. Bioresour. Technol. 2013, 138, 277–284.
11. Wilkie, A.C. Anaerobic Digestion: Holistic Bioprocessing of Animal Manure. In Proceedings of the Animal Residuals Management Conference, Alexandria, VA, USA, 14–18 October 2000; pp. 1–12.
12. Brown, V.J. Biogas a bright idea for Africa. Environ. Health Perspect. 2006, 114, A300–A303.
13. Liu, F.H.; Wang, S.B.; Zhang, J.S.; Zhang, J.; Yan, X.; Zhou, H.K.; Zhao, G.P.; Zhou, Z.H. The structure of the bacterial and archaeal community in a biogas digester as revealed by denaturing gradient gel electrophoresis and 16S rDNA sequencing analysis. J. Appl. Microbiol. 2009, 106, 952–966.
14. Mauky, E. Biogas Use. Technologies and Trends in Germany; DBFZ, Federal Ministry of Economics and Technology, Eclareon: Berlin, Germany, 2009; pp. 1–28.
15. Rechberger, P. Biogas Markets and Opportunities—A European Review; Anaerobic Digestion in Ireland, Tullamore, AEBIOM: Brussels, Belgium, 2009.
16. Jenkins, S.R.; Armstrong, C.W.; Monti, M.M. Health Effects of Biosolids Applied to Land: Available Scientific Evidence. Virginia Department of Health, 2007. Available online: http://www.vdh.virginia.gov/epidemiology/DEE/documents/biosolids.pdf (accessed on 6 July 2013).
17. Burkholder, J.; Libra, B.; Weyer, P.; Heathcote, S.; Kolpin, D.; Thorne, P.S.; Wichman, M. Impacts of waste from concentrated animal feeding operations on water quality. Environ. Health Perspect. 2007, 115, 308–312.
18. Carbone, S.R.; da Silva, F.M.; Tavares, C.R.G.; Dias Filho, B.P. Bacterial population of a two-phase anaerobic digestion process treating effluent of cassava starch factory. Environ. Technol. 2002, 23, 591–597.
19. Litchfield, J.H. *Salmonella* Food Poisoning. In Safety of Food, 2nd ed.; Graham, C.W., Ed.; AVI Publishing: Westport, CT, USA, 1980; pp. 120–122.
20. Eriksson, O.; Reich, M.C.; Frostell, B.; Bjorklund, A.; Assefa, G.; Sundqvist, J.-O.; Granath, J.; Baky, A.; Thyselius, L. Municipal solid waste management from a systems perspective. J. Clean. Prod. 2005, 13, 241–252.

21. Willey, J.M.; Sherwood, L.M.; Woolverton, C. Microbial Interactions. In Prescott's Microbiology, 8th ed.; McGraw-Hill Companies Inc.: New York, USA, 2011; pp. 713–728.
22. Nyachoti, C.M.; Omogbenigun, F.O.; Rademacher, M.; Blank, G. Performance responses and indicators of gastrointestinal health in early-weaned pigs fed low-protein amino acid-supplemented diets. J. Anim. Sci. 2006, 84, 125–134.
23. Seppälä, M.; Pyykkönen, V.; Väisänen, A.; Rintala, J. Biomethane production from maize and liquid cow manure-effect of the share of maize, post methanation potential and digestate characteristics. Fuel 2013, 107, 209–216.
24. Nwanta, J.A.; Onunkwo, J.; Ezenduka, E. Analysis of Nsukka metropolitan abattoir solid waste and its bacterial contents in south eastern Nigeria: Public health implication. Arch. Environ. Occup. Health. 2010, 65, 21–26.
25. Ribaudo, M.; Gollehon, N.; Ailley, M.; Kaplan, J.; Johansson, R.; Agapoff, J.; Christenan, L.; Breneman, V.; Peters, M. Manure Management for Water Quality: Costs to Animal Feeding Operations of Applying Manure Nutrients to Land. In Agricultural Economic Report; United States Department of Agriculture: Washington, DC, USA, 2003; No. AER-824; p. 97.
26. Health Care Canada, Guidelines for Canadian Drinking Water Quality: Technical Guideline Document-Bacterial Waterborne Pathogens-Current and Emerging Organisms of Concern. Water Quality and Health Bureau, Healthy Environment and Consumer Safety Branch, Health Canada: Ottawa, ON, Canada, 2006; pp. 1–34.
27. Rapala, J.; Lahti, K.; Rasanen, L.A.; Esala, A.L.; Niemela, S.I.; Sivonen, K. Endotoxins associated with cyanobacteria and their removal during drinking water treatment. Water Res. 2002, 36, 2627–2635.
28. Wilkie, A.C. Anaerobic Digestion of Dairy Manure: Design and Process Considerations. In Dairy Manure Management: Treatment, Handling and Community Relations; Natural Resource, Agriculture, and Engineering Service, Cornell University: Ithaca, NY, USA, 2005; NRAES-176; pp. 301–312.
29. Garcia, M.L.; Angenent, L.T. Interactions between temperature and ammonia in mesophilic digesters for animal waste treatment. Water Res. 2009, 43, 2373–2382.
30. Rico, C.; Rico, J.L.; Muňoz, N.; Gòmez, B.; Tejero, I. Effect of mixing on biogas production during mesophilic anaerobic digestion of screened dairy manure in a pilot plant. Eng. Life Sci. 2011, 11, 476–481.
31. Lutge, B.; Standish, B. Assessing the potential for electricity generation from animal waste biogas on South African farms. Agrekon: Agric. Econ. Res. Policy Pract. S. Afr. 2013, 52, 1–24.
32. Burke, D.A. Dairy Waste Anaerobic Digestion Handbook: Options for Recovering Beneficial Products from Animal Manure; Environmental Energy Company: Olympia, WA, USA, 2001; pp. 1–51. Available online: http://www.makingenergy.com (accessed on 6 July 2013).
33. Tucker, M.F. Farm digesters for small dairies in Vermont. BioCycle 2008, 49, 44.
34. Goodrich, P.R.P.E. Anaerobic Digester Systems for Mid-Sized Dairy Farms; The Minnesota Project: St. Paul, MN, USA, 2005; pp. 1–46.
35. Lozano, C.J.S.; Mendoza, M.V.; de Arango, M.C.; Monroy, E.F.C. Microbiological characterization and specific methanogenic activity of anaerobe sludges used in urban solid waste treatment. Waste Manag. 2009, 29, 704–711.

36. Song, H.; Clarke, W.P.; Blackall, L.L. Concurrent microscopic observations and activity measurements of cellulose hydrolyzing and methanogenic populations during the batch anaerobic digestion of crystalline cellulose. Biotechnol. Bioeng. 2005, 91, 369–378.
37. Franke-Whittle, I.H.; Goberna, M.; Pfister, V.; Insam, H. Design and development of the anaerochip microarray for investigation of methanogenic communities. J. Microbiol. Methods 2009, 79, 279–288.
38. De Graaf, D.; Fendler, R. Biogas Production in Germany; Federal Environment Agency. Dessau Rosslau, Baltic Sea Region Programme. Dessau Rosslau, Baltic Sea Region Programme: Dessau-Rosslau, Germany, 2010; pp. 1–24.
39. Cha, G.-C.; Chung, H.-K.; Kim, D.-J. Characteristics of temperature change on the substrate degradation and bacterial population in one-phase and two-phase anaerobic digestion. Environ. Eng. Res. 2001, 6, 99–108.
40. Demirel, B.; Scherer, P. The roles of acetotrophic and hydrogenotrophic methanogens during anaerobic conversion of biomass to methane: A review. Rev. Environ. Sci. Biotechnol. 2008, 7, 173–190.
41. Brioukhanov, A.L.; Netrusov, A.I.; Eggen, R.I.L. The catalase and superoxide dismutase genes are transcriptionally up-regulated upon oxidative stress in the strictly anaerobic archaeon Methanosarcina barkeri. Microbiology 2006, 152, 1671–1677.
42. Keiner, A.; Leisinger, T. Oxygen sensitivity of methanogenic bacteria. Syst. Appl. Microbiol. 1983, 4, 305–312.
43. Fetzer, S.; Bak, F.; Conrad, R. Sensitivity of methanogenic bacteria from paddy soil to oxygen and desiccation. FEMS Microbiol. Ecol. 1993, 12, 107–115.
44. Anderson, K.L.; Apolinario, E.E.; Sowers, K.R. Desiccation as a long-term survival mechanism for the archaeon Methanosarcina barkeri. Appl. Environ. Microbiol. 2012, 78, 1473–1479.
45. Barber, R.D.; Ferry, J.G. Methanogenesis. Encyclopedia for Life; Nature Publishing Group: New York, NY, USA, 2001; pp. 1–8. Available online: http://www.els.net (accessed on 5 April 2013).
46. McInerney, M.J.; Sieber, J.R.; Gunsalus, R.P. Syntrophy in anaerobic global carbon cycles. Curr. Opin. Biotechnol. 2009, 20, 623–632.
47. Blumer-Schuette, S.E.; Kataeva, I.; Westpheling, J.; Adams, M.W.W.; Kelly, R.M. Extremely thermophilic microorganisms for biomass conversion: status and prospects. Curr. Opin. Biotechnol. 2008, 19, 210–217.
48. Wirth, R.; Kovács, E.; Maròti, G.; Bagi, Z.; Rakhely, G.; Kovács, K.L. Characterization of a biogas—Producing microbial community by short-read next generation DNA sequencing. Biotechnol. Biofuels 2012, 5, 41.
49. Burrell, P.C.; O'Sullivan, C.; Song, H.; Clarke, W.P.; Black-all, L.L. The identification, detection and spatial resolution of Clostridium populations responsible for cellulose degradation in a methanogenic landfill leachate bioreactor. Appl. Environ. Microbiol. 2004, 70, 2414–2419.
50. Li, A.; Chu, Y.; Wang, X.; Ren, L.; Yu, J.; Liu, X.; Yan, J.; Zhang, L.; Wu, S.; Li, S. A pyrosequencing-based metagenomic study of methane-producing microbial community in solid-state biogas reactor. Biotechnol. Biofuels 2013, 6, 3.
51. McInerney, M.J.; Struchtemeyer, C.G.; Sieber, J.; Mouttaki, H.; Stams, A.J.M.; Schnink, B.; Rohlin, L.; Gunsalus, R.P. Physiology, Ecology, Phylogeny, and Ge-

nomics of Microorganisms Capable of Syntrophic Metabolism. Ann. N. Y. Acad. Sci. 2008, 1125, 58–72.
52. Hori, T.; Sasaki, D.; Haruta, S.; Shigematsu, T.; Ueno, Y.; Ishii, M.; Igarashi, Y. Detection of active, potentially acetate-oxidizing syntrophs in an anaerobic digester by flux measurement and formyltetrahydrofolate synthetase expression profiling. Microbiology 2011, 157, 1980–1989.
53. Siriwongrungson, V.; Zeng, R.J.; Angelidaki, I. Homoacetogenesis as the alternative pathway for H2 sink during thermophilic anaerobic degradation of butyrate under suppressed methanogenesis. Water Res. 2007, 41, 4202–4210.
54. Hattori, S.; Galushko, A.S.; Kamagata, Y.; Schink, B. Operation of the CO dehydrogenase/acetyl coenzyme A pathway in both acetate oxidation and formation by the syntrophically acetate oxidizing bacterium Thermacetogenium phaeum. J. Bacteriol. 2005, 187, 3471–3476.
55. Lee, M.J.; Zinder, S.H. Isolation and characterization of a thermophilic bacterium which oxidizes acetate in syntrophic association with a methanogen and which grows acetogenically on H2-CO2. Appl. Environ. Microbiol. 1988, 52, 124–129.
56. Schnürer, A.; Schink, B.; Svensson, B.H. Clostridium ultunense sp. nov., a mesophilic bacterium oxidizing acetate in syntrophic association with a hydrogenotrophic methanogen bacterium. Int. J. Syst. Bacteriol. 1996, 46, 1145–1152.
57. Hattori, S.; Kamagata, Y.; Hanada, S.; Shuon, H. Thermacetogenium phaeum gen. nov., sp. nov., a strictly anaerobic thermophilic, syntrophic acetate-oxidizing bacterium. Int. J. Syst. Evol. Microbiol. 2000, 50, 1601–1609.
58. Balk, M.; Weijma, J.; Stams, A.J.M. Thermotoga lettingae sp. nov., a novel thermophilic, methanol-degrading bacterium isolated from a thermophilic anaerobic reactor. Int. J. Syst. Evol. Microbiol. 2002, 52, 1361–1368.
59. Westerholm, M.; Roos, S.; Schnürer, A. Syntrophaceticus schinkii gen. nov., sp. nov., an anaerobic syntrophic acetate-oxidizing bacterium isolated from a mesophilic anaerobic filter. FEMS Microbiol. Lett. 2010, 309, 100–104.
60. Zhu, W.; Reich, C.I.; Olsen, G.J.; Giometti, C.S.; Yates, J.R. Shotgun proteomics of Methanococcus jannaschii and insights into methanogenesis. J. Proteome Res. 2004, 3, 538–548.
61. Attwood, G.T.; Kelly, W.J.; Altermann, E.H.; Leahy, S.C. Analysis of the Methanobrevibacter ruminantium draft genome: Understanding methanogen biology to inhibit their action in the rumen. Aust. J. Exp. Agric. 2007, 48, 83–88.
62. Ver Eecke, H.C.; Butterfield, D.A.; Huber, J.A.; Lilley, M.D.; Olson, E.J.; Roe, K.K.; Evans, L.J.; Merkel, A.Y.; Cantin, H.V.; Holden, J.F. Hydrogen limited growth of hyperthermophilic methanogens at deep-sea hydrothermal vent. Proc. Natl. Acad. Sci. USA 2012, 109, 13674–13679.
63. Brune, A. Methanogenesis in the Digestive Tracts of Insects. In Handbook of Hydrocarbon and Lipid Microbiology; Timmis, K.W., Ed.; Springer-Verlag: Berlin/Herdelberg, Germany, 2010; pp. 707–728.
64. Westerholm, M.; Levén, L.; Schnürer, A. Bioaugmentation of syntrophic acetate-oxidizing culture in biogas reactors exposed to increasing levels of ammonia. Appl. Environ. Microbiol. 2012, 78, 7619–7625.
65. De Macario, E.C. Taxonomy of Methanogens. In Bergey's Manual of Systematic Bacteriology, 2nd ed.; Springer: New York, NY, USA, 2008.

66. Batstone, D.J.; Keller, J.; Angelidaki, I.; Kalyuzhnyi, S.V.; Pavlostathis, S.G.; Rozzi, A.; Sanders, W.T.M.; Siegrist, H.; Vavilin, V.A. The IWA anaerobic digestion model No.1 (ADM1). Water Sci. Technol. 2002, 45, 65–73.
67. Krakat, N.; Westphal, A.; Schmidt, S.; Scherer, P. Anaerobic digestion of renewable biomass: Thermophilic temperature governs methanogen population dynamics. Appl. Environ. Microbiol. 2010, 76, 1842–1850.
68. Krakat, N.; Schmidt, S.; Scherer, P. Mesophilic fermentation of renewable biomass: Does hydraulic retention time regulate methanogen diversity. Appl. Environ. Microbiol 2010, 76, 6322–6326.
69. Klocke, M.; Mähnert, P.; Mundt, K.; Souidi, K.; Linke, B. Microbial community analysis of a biogas-producing completely stirred tank reactor fed continuously with fodder beet silage as mono-substrate. Syst. Appl. Microbiol. 2007, 30, 139–151.
70. Klocke, M.; Nettmann, E.; Bergmann, I.; Mundt, K.; Souidi, K.; Mumme, J.; Linke, B. Characterization of the methanogenic archaea within two-phase biogas reactor systems operated with plant biomass. Syst. Appl. Microbiol. 2008, 31, 190–205.
71. Amani, T.; Nosrati, M.; Sreekrishnan, T.R. Anaerobic digestion from the viewpoint of microbiological, chemical, and operational aspects: A review. Environ. Rev. 2010, 18, 255–278.
72. Solera, R.; Romero, L.I.; Sales, D. Determination of the microbial population in thermophilic anaerobic reactor: Comparative analysis by different counting methods. Anaerobe 2001, 7, 79–86.
73. Ziganshin, A.M.; Schmidt, T.; Scholwin, F.; Il'inskaya, O.N.; Harms, H.; Kleinsteuber, S. Bacteria and Archaea involved in anaerobic digestion of distillers grains with solubles. Appl. Microbiol. Biotechnol. 2011, 89, 2039–2052.
74. Rivière, D.; Desvignes, V.; Pelletier, E.; Chaussonnerie, S.; Guermazi, S.; Weissenbach, J.; Li, T.; Camacho, P.; Sghir, A. Towards the definition of a core of microorganisms involved in anaerobic digestion of sludge. Int. Soc. Microb. Ecol. 2009, 3, 700–714.
75. Scully, C.; Collins, G.; O'Flaherty, V. Assessment of anaerobic wastewater treatment failure using terminal restriction fragment length polymorphism analysis. J. Appl. Microbiol. 2005, 99, 1463–1471.
76. Kataoka, N.; Tokiwa, Y.; Takeda, K. Improved technique for identification and enumeration of methanogenic bacterial colonies on roll tubes by epifluorescence microscopy. Appl. Environ. Microbiol. 1991, 57, 3671–3673.
77. Singh, L.S.; Mazumder, P.B. Differential approaches for studying methanogens: Methods, analysis and prospects. Assam Univ. J. Sci. Technol. 2010, 6, 123–128.
78. Schlüter, A.; Bekel, T.; Diaz, N.N.; Dondrup, M.; Eichenlaub, R.; Gartemann, K-H.; Krahn, I.; Krause, L.; Krömeke, H.; Kruse, O.; et al. The metagenome of a biogas-producing microbial community of a production-scale biogas plant fermenter analyzed by the 454-pyrosequencing technology. J. Biotechnol. 2008, 136, 77–90.
79. Lee, C.; Kim, J.; Shin, S.G.; Hwang, S. Monitoring bacterial and archaeal community shifts in a mesophilic anaerobic batch reactor treating a high-strength organic wastewater. FEMS Microbiol. Ecol. 2008, 65, 544–554.
80. Cirne, D.G.; Lehtomäki, A.; Björnsson, L.; Blackall, L.L. Hydrolysis and microbial community analyses in two-stage anaerobic digestion of energy crops. J. Appl. Microbiol. 2006, 103, 516–527.

81. Jaenicke, S.; Ander, C.; Bekel, T.; Bisdorf, R.; Dröge, M.; Gartemann, K.-H.; Jüneman, S.; Kaiser, O.; Krause, L.; Tille, F.; et al. Comparative and joint analyses of two-metagenomic dataset from a biogas fermenter obtained by 454-pyrosequencing. PLoS One 2011, 6, e14519.
82. Li, M.; Cao, H.; Hong, Y.-G.; Gu, J.-D. Seasonal dynamics of anammox bacteria in estuarial sediment Mai Po nature reserve revealed by analyzing the 16S rRNA and hydrazine oxidoreductase (hzo) genes. Microbes Environ. 2011, 26, 15–22.
83. Ozgun, D.; Basak, S.; Cinar, O. Current Molecular Biologic Techniques for Anaerobic Ammonium Oxidizing (Anammox) Bacteria. In Proceedings of the Sixteenth International Water Technology Conference, IWTC, Istanbul, Turkey, 7–10 May 2012; pp. 1–15.
84. Harhangi, H.R.; Roy, M.L.; Alen, T.V.; Hu, B.-L.; Groen, J.; Kartal, B.; Tringe, S.G.; Quan, Z.-X.; Jetten, M.S.M.; den Camp, H.J.M.O. Hydrazine synthase, a unique phylomarker with which to study the presence and biodiversity of anammox bacteria. Appl. Environ. Microbiol. 2012, 78, 752–758.
85. Zhou, M.; McAllister, T.A.; Guan, L.L. Molecular identification of rumen methanogen: Technologies, advances and prospects. Anim. Feed Sci. Technol. 2011, 166–167, 76–86.
86. Lutton, P.E.; Wayne, J.M.; Sharp, R.J.; Riley, P.W. The mcrA gene as an alternative to 16S rRNA in the phylogenetic analysis of methanogen population in landfills. Microbiology 2002, 148, 3521–3530.
87. Denman, S.E.; Tomkins, N.W.; McSweeney, C.S. Quantitation and diversity analysis of ruminal methanogenic populations in response to the antimethanogenic compound bromochloromethane. FEMS Microbiol. Ecol. 2006, 62, 313–322.
88. Jiang, B.; Song, K.; Ren, J.; Deng, M.; Sun, F.; Zhang, X. Comparison of metagenomic samples using sequence signatures. BMC Genomics 2012, 13, 730.
89. Mashhadi, Z. Identification and Characterization of the Enzymes Involved in Biosynthesis of FAD and Tetrahydromethanopterin in Methanococcus jannaschii. Ph.D. Thesis, Virginia Polytechnic Institute and State University (Virginia Tech), Blacksburg, VA , USA, In: Doctor of philosophy Biochemistry.. 30 June 2010; pp. 1–136.
90. Leahy, S.C.; Kelly, W.J.; Altermann, E.; Ronimus, R.S.; Yeoman, C.J.; Pacheco, D.M.; Li, D.; Kong, Z.; McTavish, S.; Sang, C.; et al. The genome sequence of the rumen methanogen Methanobrevibacter ruminantium reveals new possibilities for controlling ruminant methane emission. PLoS One 2010, 5, e8926.
91. Handelsman, J. Metagenomics: Application of genomics to uncultured microorganisms. Microbiol. Mol. Biol. Rev. 2004, 68, 669–685.
92. Gilbert, J.A.; Dupont, C.L. Microbial metagenomics: Beyond the genome. Annu. Rev. Mar. Sci. 2004, 3, 347–371.
93. Ilaboya, I.R.; Assekhame, F.F.; Ezugwu, M.O.; Erameh, A.A.; Omofuma, F.E. Studies on biogas generation from agricultural wastes; analysis of the effects of alkaline on gas generation. World Appl. Sci. J. 2010, 9, 537–545.
94. Umaňa, O.; Nikolaeva, S.; Sanchez, E.; Borja, R.; Raposo, F. Treatment of screened dairy manure by upflow anaerobic fixed bed reactors packed with waste tyre rubber and a combination of waste tyre rubber and Zeolite: Effect of the hydraulic retention time. Bioresour. Technol. 2008, 99, 7412–7417.

95. Balsam, J.; Ryan, D. Anaerobic Digestion of Animal Wastes: Factors to Consider; ATTRA: Butte, MT, USA, 2006; pp. 1–10.
96. Cioabla, A.E.; Lonel, L.; Dumitrel, G.-A.; Popescu, F. Comparative study on factors affecting anaerobic digestion of agricultural vegetal residues. Biotechnol. Biofuels 2012, 5, 39.
97. Choorit, W.; Wisarnwan, P. Effect of temperature on the anaerobic digestion of palm oil mill effluent. Electron. J. Biotechnol. 2007, 10, 376–385.
98. Saleh, M.M.A.; Mahmood, U.F. Anaerobic Digestion Technology for Industrial Waste Water Treatment. In Proceedings of the Eighth International Water Technology Conference, IWTC, Alexandria, Egypt, 26–28 March 2004; pp. 817–833.
99. Rittmann, B.E.; McCarty, P.L. Environmental Biotechnology: Principles and Applications; McGraw-Hill Book Co.: New York, NY, USA, 2001; p. 768.
100. El-Mashad, H.M.; Zeeman, G.; van Loon, W.K.P.; Bot, G.P.A.; Lettinga, G. Effect of temperature and temperature fluctuation on thermophilic anaerobic digestion of cattle manure. Bioresour. Technol. 2004, 95, 191–201.
101. Campos, E.; Palatsi, J.; Flotats, X. Co-Digestion of Pig Slurry and Organic Wastes from Food Industry. In Proceedings of the 2th International Symposium on Anaerobic Digestion of Solid Waste, Barcelona, Junio, Spain, 15–18 June 1999; pp. 192–195.
102. Gerardi, M.H. The Microbiology of Anaerobic Digesters; John Wiley and Sons Inc.: Hoboken, NJ, USA, 2003; pp. 91–118.
103. Molinuevo-Salces, B.; García-González, M.C.; González-Fernández, C.; Cuetos, M.J.; Morán, A.; Gòmez, X. Anaerobic co-digestion of livestock wastes with vegetable processing wastes: A statistical analysis. Bioresour. Technol. 2010, 101, 9479–9485.
104. Veeken, A.; Kalyuzhnyi, S.; Scharff, H.; Hamelers, B. Effect of pH and VFA on hydrolysis of organic solid waste. J. Environ. Eng. 2000, 126, 1076–1081.
105. El Hadj, T.B.; Astals, S.; Galí, A.; Mace, S.; Mata-Álvarez, J. Ammonia influence in anaerobic digestion of OFMSM. Water Sci. Technol. 2009, 59, 1153–1158.
106. Chen, Y.; Cheng, J.J.; Creamer, K.S. Inhibition of anaerobic digestion process. A review. bioresour. Technol. 2008, 99, 4044–4064.
107. Strik, D.P.B.T.B.; Domnanovich, A.M.; Holubar, P. A pH-based control of ammonia in biogas during anaerobic digestion of artificial pig manure and maize silage. Process Biochem. 2006, 41, 1235–1238.
108. Angelidaki, I.; Ahring, B.K. Thermophilic anaerobic digestion of livestock waste: The effect of ammonia. Appl. Microbiol. Biotechnol. 1993, 38, 560–564.
109. Bolzonella, D.; Pavan, P.; Battistoni, P.; Cecchi, F. Mesophilic anaerobic digestion of waste activated sludge: Influence of solid retention time in the wastewater treatment process. Process Biochem. 2005, 40, 1453–1460.
110. Karim, K.; Hoffmann, R.; Klasson, K.T.; Al-Dahhan, M.H. Anaerobic digestion of animal waste: Effects of mode of mixing. Water Res. 2009, 39, 3597–3606.
111. Babaee, A.; Shayegan, J. Effects of Organic Loading Rates (OLR) on Production of Methane from Anaerobic Digestion of Vegetable Waste. In Proceedings of the World Renewable Energy Congress, Linköping, Sweden, 8–13 May 2011; pp. 411–417.
112. Rincòn, B.; Travieso, L.; Sanchez, E.; Martín, M.L.A.; Martín, A.; Raposo, F.; Borja, R. The effect of organic loading rate on the anaerobic digestion of two-phase olive

mill solid residue derived from fruits with low ripening index. J. Chem. Technol. Biotechnol. 2007, 82, 259–266.
113. Rincòn, B.; Borja, R.; González, J.M.; Portillo, M.C.; Sáiz-Jiménez, C. Influence of organic loading rate and hydraulic retention time on the performance, stability, and microbial communities of one-stage anaerobic digestion of two-phase olive mill solid residue. Biochem. Eng. J. 2008, 40, 253–261.
114. Tchobanoglous, G.; Burton, F.L.; Stensel, H.D. Wastewater Engineering: Treatment and Reuse, 4th ed.; Metcalf and Eddy, Inc., Tata Mcgraw-Hill Publishing Company Ltd.: New Delhi, India, 2003.
115. Zhang, R.; El-Mashad, H.M.; Hartman, K.; Wang, F.; Liu, G.; Choate, C.; Gamble, P. Characterization of food waste as feedstock for anaerobic digestion. Bioresour. Technol. 2007, 98, 929–935.
116. Matseh, I. Effect of Ni and Co as trace elements on digestion performance and biogas produced from the fermentation of palm oil mill effluent. Int. J. Waste Resour. 2012, 2, 16–19.
117. Gustavsson, J. Cobalt and Nickel Bioavailability for Biogas Formation. Ph.D. Thesis, Department of Thematic studies, University of Linköping, Linköping, Sweden, Water and Environmental Studies. 19 January 2012; pp. 1–64.
118. Pobeheim, H.; Munk, B.; Johansson, J.; Guebitz, G.M. Influence of trace elements on methane formation froma synthetic model substrate for maize silage. Bioresour. Technol. 2010, 101, 836–839.
119. Facchin, V.; Cavinato, C.; Fatone, F.; Pavan, P.; Cecchi, F.; Bolzonella, D. Effect of trace element supplementation on the mesophilic anaerobic digestion of food wastes in batch trials. The influence of inoculum origin. Biochem. Eng. J. 2013, 70, 71–77.
120. Rojas, C.; Fang, S.; Uhlenhut, F.; Borchert, A.; Stein, I.; Schlaak, M. Stirring and biomass starter influences the anaerobic digestion of different substrates for biogas production. Eng. Life Sci. 2010, 10, 339–347.
121. Ghanimeh, S.; El Fadel, M.; Saikaly, P. Mixing effect on thermophilic anaerobic digestion of source-sorted organic fraction of municipal solid waste. Bioresour. Technol. 2012, 117, 63–71.

CHAPTER 4

New Steady-State Microbial Community Compositions and Process Performances in Biogas Reactors Induced by Temperature Disturbances

GANG LUO, DAVIDE DE FRANCISCI, PANAGIOTIS G. KOUGIAS, TREU LAURA, XINYU ZHU, AND IRINI ANGELIDAKI

4.1 BACKGROUND

Anaerobic digestion (AD) is widely used in the treatment of organic wastes to achieve reduction of the wastes with simultaneous production of biogas [1]. The production of biogas via AD is a complex process, involving many different microbial species [2]. The complex organic compounds are first hydrolyzed into oligomers and monomers, and then further converted to acetate, CO_2, and H_2 by various fermenting bacteria. The methanogenesis is the final step to convert acetate, CO_2, and H_2 to CH_4 by methanogenic archaea. The syntrophic relationship between bacteria and archaea is essential for the stability of the biogas process [3].

New Steady-State Microbial Community Compositions and Process Performances in Biogas Reactors Induced by Temperature Disturbances. © *Luo G, De Francisci D, Kougias PG, Laura T, Zhu X, and Angelidaki I.* Biotechnology for Biofuels *8,3 (2015), doi:10.1186/s13068-014-0182-y. Licensed under a Creative Commons Attribution 4.0 International License, http://creativecommons.org/licenses/by/4.0/.*

It is crucial to understand the processes and factors controlling the microbial community composition in biogas reactors. The mechanisms of community assembly and the critical factors shaping species composition and structure remain controversial in ecology [4]-[6]. The traditional niche-based theory supports the idea that the community is shaped mainly by deterministic factors such as competition and niche differentiation, and thereby asserts that community composition should converge toward a single pattern under similar environmental conditions [7]. In contrast to niche-based theory, neutral theory assumes that many natural community patterns can be generated under similar environmental conditions by stochastic factors considering birth, death, dispersal, and speciation and disregards the differences between species at the same trophic level [8]. In addition, disturbance was also shown to play an important role in the community assembly since the disturbance could kill or damage certain species and promote the growth of other species that are resistant to the disturbance [9],[10].

Microbial community compositions in biogas reactors have been studied for several decades, and deterministic factors including temperature, hydraulic retention time (HRT), and substrate type have been demonstrated to play an important role in shaping microbial communities [11],[12]. However, based on the neutral theory, stochastic factors may be important in shaping the highly diverse microbial communities in biogas reactors. Up to now, it is still unknown whether there are different microbial community patterns under the same environmental conditions in biogas reactors if stochastic factors are determining the microbial communities. Although different disturbances (such as temperature and organic loading) on the biogas process have been evaluated before, most of the studies focused only on the reactor performances, and the effect of disturbance on the community assembly was not documented [13]-[16]. It is still unknown whether the disturbance in the biogas reactors would lead to different steady-state microbial community compositions and reactor performances.

Traditional molecular technologies for microbial community analysis (for example, polymerase chain reaction-denaturing gradient gel electrophoresis, terminal restriction fragment length polymorphism, and cloning) can only identify the most abundant microorganisms in the microbial community. The less abundant yet functionally important microorganisms cannot be detected [17]. Therefore, using traditional molecular technologies, it is difficult to study variations of less abundant microorganisms in different samples. With the newly developed sequencing technologies, it is possible to define microbial community composition with a high sequencing depth. The Ion Torrent Personal Genome Machine (PGM) (Life Technologies), launched in early 2011, has the highest throughput compared with 454 GS Junior (Roche) and Miseq (Illumina), thus making the high-throughput sequencing cost effective and time saving [18].

Based on the above considerations, the objective of this study was to understand the role of stochastic factors (based on neutral theory) and disturbance in the steady-state microbial community assembly and functions in biogas reactors. We ran three replicate biogas reactors treating cattle manure to first determine whether similar microbial communities would be achieved at steady states where the reactors were operated under the same conditions. In most modern biogas plants one attempts to keep a constant temperature, as temperature stability is of utmost importance for the biogas process. Nevertheless, biogas reactors may be subjected to undesired temperature fluctuations due to various technical problems such as heat exchanger or pump failures or fouling in temperature sensors [16],[19]. Therefore, temperature disturbance was introduced to the three reactors in order to determine if and to what extent the temperature disturbance would alter the steady-state microbial community. In addition, the reactor performances including biogas production, pH, and total volatile fatty acids (VFAs) were monitored. The microbial community composition was analyzed by Ion Torrent sequencing of 16 s rRNA gene amplicons.

4.2 RESULTS AND DISCUSSION

4.2.1 REACTOR PERFORMANCE

The monitoring profiles for methane yield, pH, and total VFAs in the three reactors are shown in Figure 1, and the overall performances of the reactors at steady state are summarized in Table 1. The initial higher methane yield was due to the methane production from the inoculum, since there are still organics in the inoculum which can be digested. After around 30 days' operation, the methane yields were relatively stable. The steady-state methane yields for the three reactors were not significantly different ($P<0.05$): 194±7.3, 189±14.5, and 195±6.9 mL/g volatile solids (VS) for reactors A, B, and C, respectively. Both the pH values (around 7.5) and the total VFAs concentrations were also similar for all three reactors. Acetate was the dominant VFAs, as seen in Figure 2. The above results indicated that the replicate reactors (A, B, and C) did not present obvious differences in their performances.

TABLE 1: Summary of the reactor performances at steady state before (phase I, 0 to 50 days) and after (phase II, 61 to 112 days) temperature disturbance

Parameter	A_b	A_a	B_b	B_a	C_b	C_a
Methane yield (mL CH_4/gVS)	194 ± 7.3	220±17.5	189±14.5	213±11.7	195±6.9	214±13.1
pH	7.50± 0.06	7.64 ± 0.02	7.51±0.04	7.66±0.01	7.52±0.04	7.60±0.02
Total VFAs (mM)	25.5± 3.9	3.3±0.9	24.9±1.7	5.3±0.6	21.1±1.9	7.8±1.2

Subscript b means before temperature disturbance, subscript a means after temperature disturbance.

From day 50 (phase II), the temperatures of the reactors were changed (A 25°C, B 45°C, C 55°C) from the original temperature of 37°C. A sharp decrease in methane yields was observed in all the reactors, together with a decrease in pH and an increase in total VFAs. After 10 days at the new temperatures, the total VFAs increased to around 60 mM for reactor A, and to around 90 mM for both reactors B and C, which clearly indicates that the increase of temperature had a more profound effect on the stability of the reactors. There are several reasons leading to the higher VFAs accumulation when the temperatures were increased. It could be due to faster adaptation of acidogenic bacteria or to a greater temperature span of acidogens at higher temperatures compared to methanogens, as they generally grow more slowly and have a narrower temperature span [20],[21]. Among the total VFAs, acetate was still the dominant component, although propionate was also accumulated, which is consistent with previous reports that propionate is easy to accumulate when the biogas reactor is disturbed [22],[23].

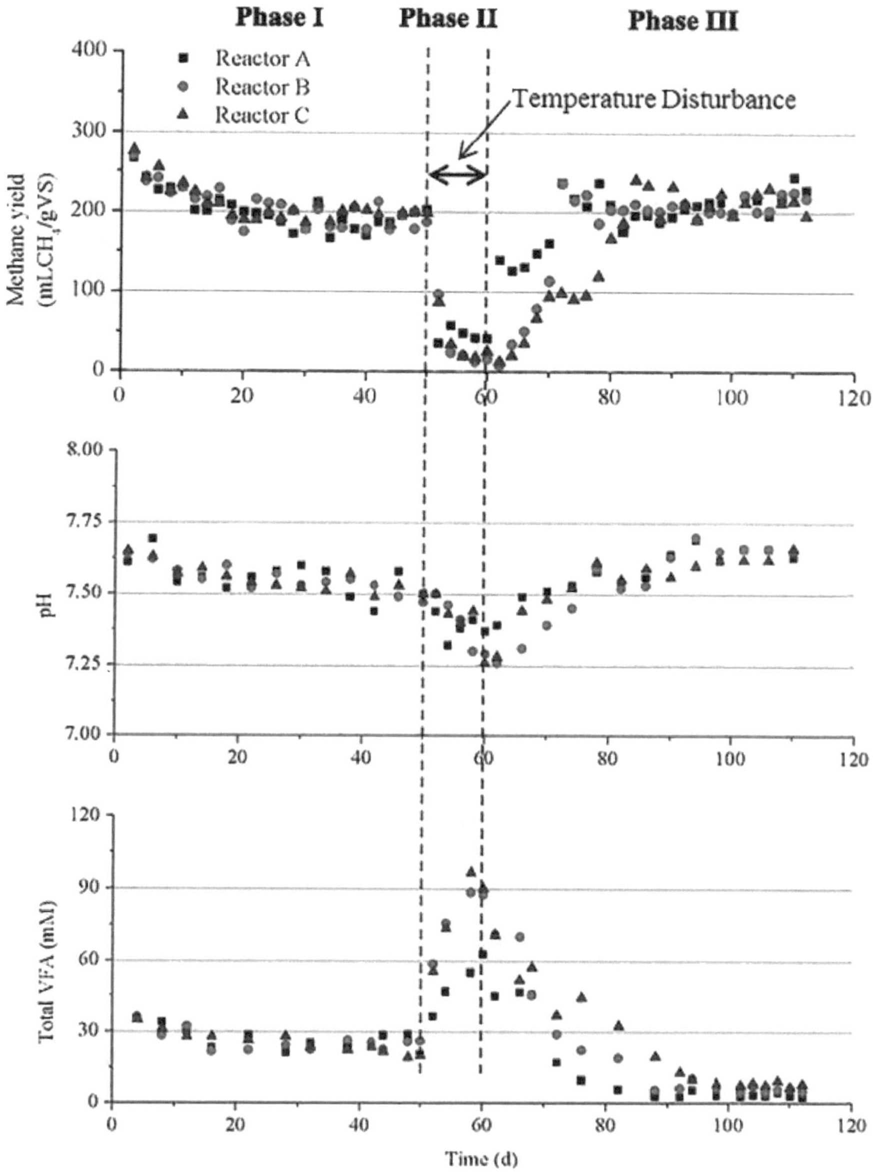

FIGURE 1: Reactor performances for the whole operational period.

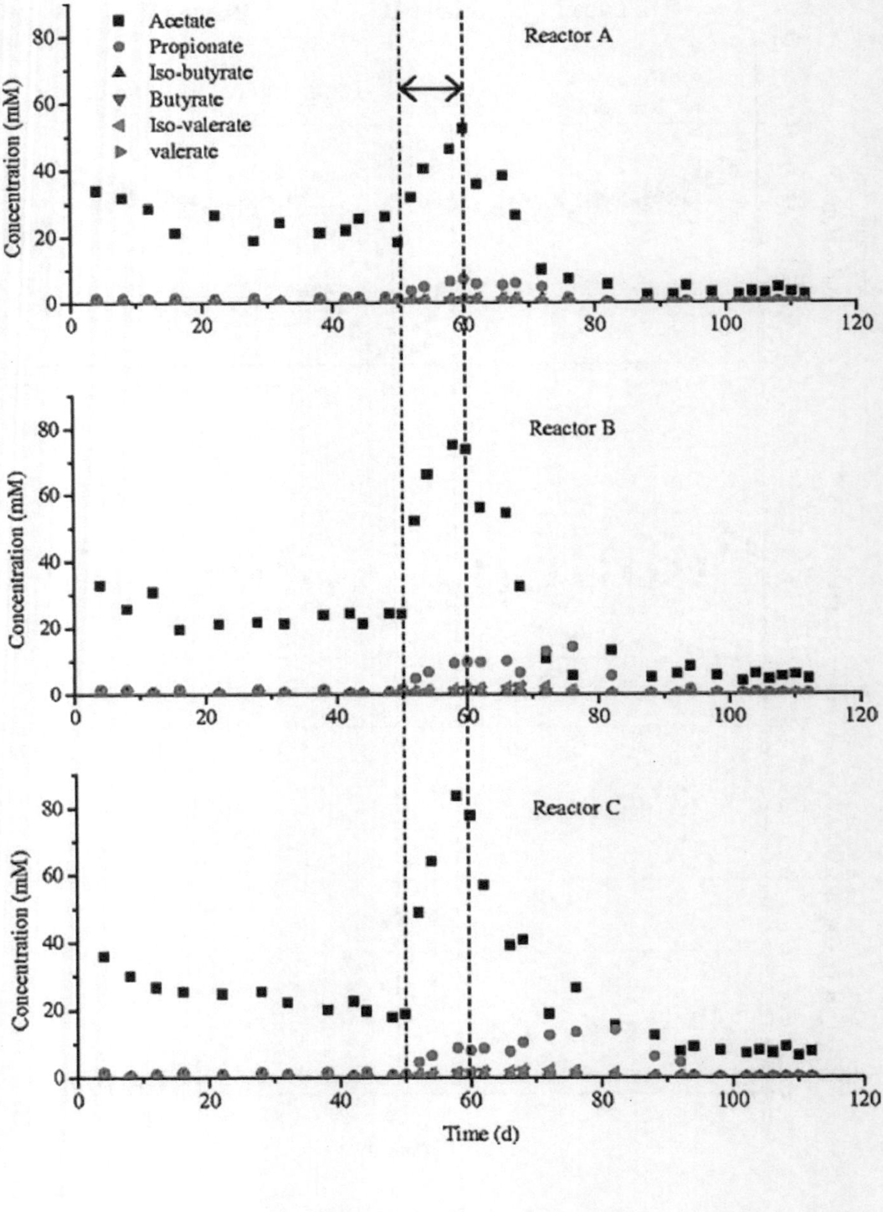

FIGURE 2: Individual VFAs changes for the whole operational period.

From day 60 (phase III), the operating temperatures of all the reactors were changed back to 37°C. The methane yield increased immediately for reactor A. However, a slower recovery of the methane yield was observed for reactors B and C, which might indicate that the higher temperature disturbances (45°C and 55°C) had a more negative impact on the stability of the biogas reactors. In particular, reactor C took around 10 days before the methane yield increased to a similar level to that of the steady-state level in phase I. The fast increase of methane yield in reactor A was in good agreement with the fast decrease of total VFAs. During the steady states of phase III, the methane yields for reactors A, B, and C were 220±17.5, 213±11.7, and 214±13.1 mL/gVS, respectively, and there were no significant differences ($P<0.05$) among the reactors. However, the methane yields were all significantly higher (around 10%) than those in phase I. The results indicated that the temperature disturbances affected the performances of all the reactors, although the operational conditions were the same for all the reactors in both phases I and III. The lower VFAs concentrations in phase III compared with those in phase I could explain the increased methane yields. However, the decreased VFAs concentration only accounted for 10% or less of the increased methane yield. This indicates that the increased methane yield could also be related with the increased hydrolysis of the solid part in cattle manure, which would lead to higher methane production. The lower VFAs concentrations in phase III for all the three reactors also resulted in the relatively higher pH (around 7.6).

4.2.2 MICROBIAL COMMUNITY ANALYSIS

The numbers of sequences after quality filtration from different samples are shown in Additional file 1: Table S1. The average sequence lengths were around 273 bp for all the samples. The high-quality sequences were assigned to taxonomic classifications by the Ribosomal Database Project (RDP) classifier. Since the primers used in the present study were universal primers, sequences belonging to both bacteria and archaea were obtained at the same time [24],[25]. The phylogenetic classification of sequences assigned to bacteria from all the samples is summarized in Figure 3. Samples A1, B1, and C1 had similar distributions of the sequences at the phylum, class, and genus levels. *Firmicutes, Bacteroidetes*, and *Proteobacteria* were dominant at the phylum level, and their dominance in biogas reactors was in accordance with other studies [26],[27]. *Clostridia, Bacteroidia*, and *Gammaproteobacteria* were dominant at the class level. However, a considerable amount of the sequences (around 50%) were unclassified at the genus level, which could be due to some new microorganisms that have not yet been identified. The high percentages of unclassified sequences at the genus level were also found in previous studies [1],[28]. The temperature disturbances had different effects on the shift of bacterial communities in reactors A, B, and C. The decrease of temperature from 37°C to 25°C in reactor A resulted in an increased abundance of *Proteobacteria* (A2), and the reason might be that some of the classes (for example, *Gammaproteobacteria*) belonging to *Proteobacteria* can grow well at lower temperatures [29]. In reactor B, the increase of temperature from 37°C to 45°C led to an increased relative abundance of the unclassified sequences at the phylum level (B2). The increased relative abundance of *Firmicutes* (C2) was observed with a further increase of temperature from 37°C to 55°C in reactor C. The relative abundance of class *Clostridia*, belonging to *Firmicutes*, was enriched in sample C2, which could be due to their spore-forming character and gradual adaptation to thermophilic conditions [26],[30]. The bacterial communities (A3, B3, and C3, sampled on day 60) continued to change after the temperatures in all the reactors returned to 37°C, which was consistent with the unstable reactor performances (Figure 1, day 60). Samples A4, B4, and C4 were obtained during the steady states of phase III, and they had similar distributions at the phylum, class, and genus levels. However, compared with A1, B1, and C1, the relative abundance of *Bacteroidetes* in A4, B4, and C4 increased and that of *Proteobacteria* decreased. Differences at the class and genus levels were also observed between A1, B1, C1 and A4, B4, C4. The above results showed that although the biogas reactors, before and after the temperature disturbance, were run under exactly the same operational conditions, the bacterial communities did not return to the original bacterial composition. New steady-state bacterial community compositions, distinct from the original, were established after the temperature disturbances.

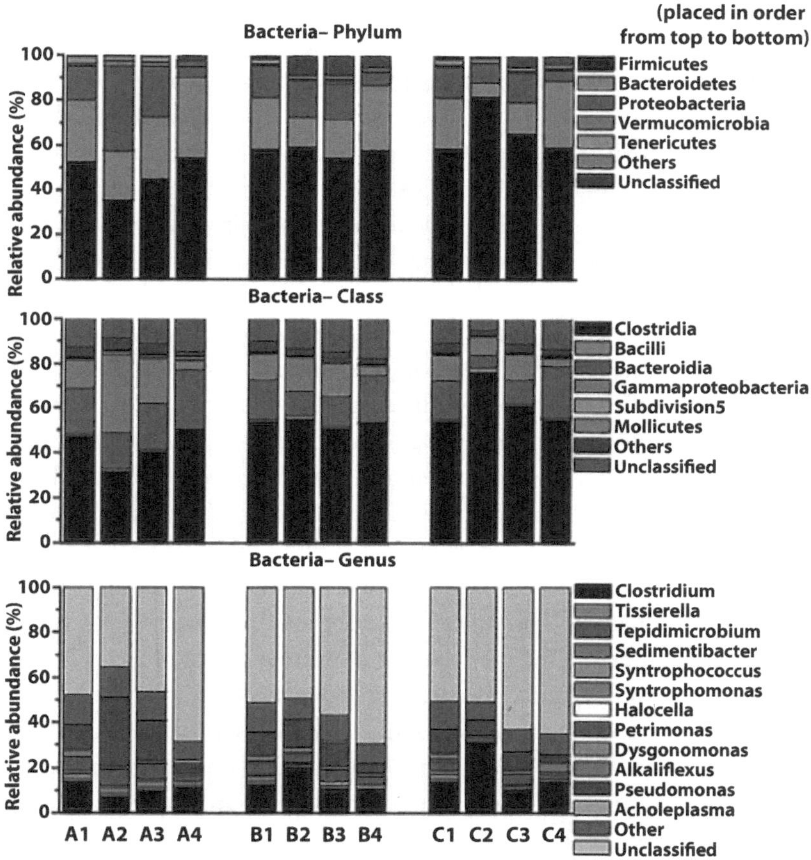

FIGURE 3: Taxonomic classification of the bacteria communities. Phyla, classes, and genera making up less than 1% of total composition in all the samples were classified as Others.

The phylogenetic classification of sequences assigned to archaea from all the samples is summarized in Figure 4. The archaea mediating hydrogenotrophic and aceticlastic methanogenesis were found mainly within four orders (*Methanobacteriales*, *Methanococcales*, *Methanomicrobiales*, and *Methanosarcinales*) [31]. Therefore, only order- and genus-level classifications are shown in Figure 4. Samples A1, B1, and C1 were all dominated by *Methanomicrobiales* and *Methanobacteriales* in phase I, which belonged to hydrogenotrophic methanogens. Similar distributions of samples A1, B1, and C1 at the genus level were also observed, and the dominant genera were *Methanoculleus*, *Methanocorpusculum*, *Methanobrevibacter*, and *Methanobacterium*. An increase of *Methanobacteriales* was found in all the reactors after temperature disturbance, which may indicate that the archaea belonging to this order were more resistant to the temperature disturbance (both downwards and upwards). It has been reported that the most frequently observed hydrogen utilizers are members of *Methanobacteriales*, present in both manure and sewage sludge digesters [31]. Further study is needed in order to understand why *Methanobacteriales* were more resistant to temperature changes than other methanogens. In reactor C, the temperature increase resulted in the increased relative abundance of order *Methanosarcinales*, which are mainly aceticlastic methanogens. Since the methane production during temperature shock was significantly reduced, the changes of archaeal communities during the temperature disturbance might be due to the different decay rates of the archaea rather than the different growth rates of the archaea. After the temperature was changed back to 37°C, *Methanosarcinales* became dominant in all the reactors (A3, B3, and C3). At the steady states of phase III, *Methanomicrobiales*, *Methanosarcinales*, and *Methanobacteriales* were all dominant in reactors A, B, and C. The genus *Methanosarcina*, belonging to *Methanosarcinales*, mainly mediates aceticlastic methanogenesis, and therefore it is expected that the methanogenic pathway was changed before and after temperature disturbance. The dominance of *Methanosarcina* might be related to the better reactor performances in phase III compared with phase I. It has been reported that the dominance of hydrogenotrophic methanogenesis is always related to extreme conditions such as high ammonia or acetate concentration [32]-[34]. It is possible that the higher acetate concentration in phase I in all the reactors induced the dominance of archaea belonging to hydrogenotrophic methanogens.

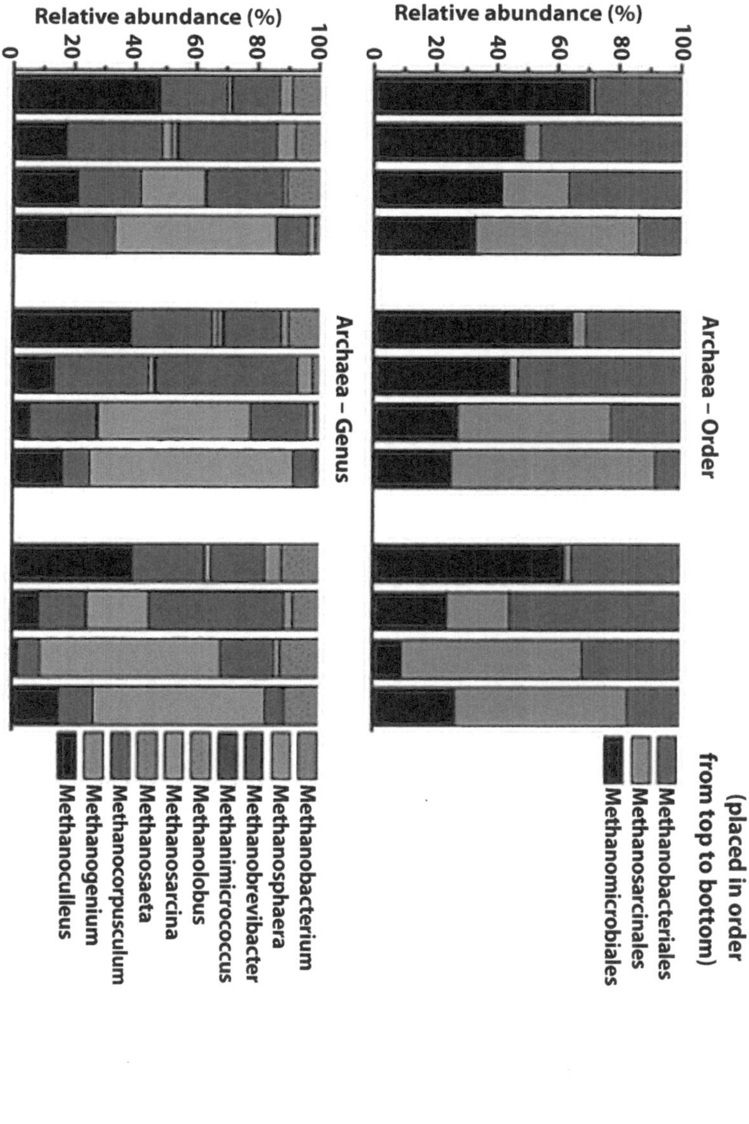

FIGURE 4: Taxonomic classification of the archaea communities.

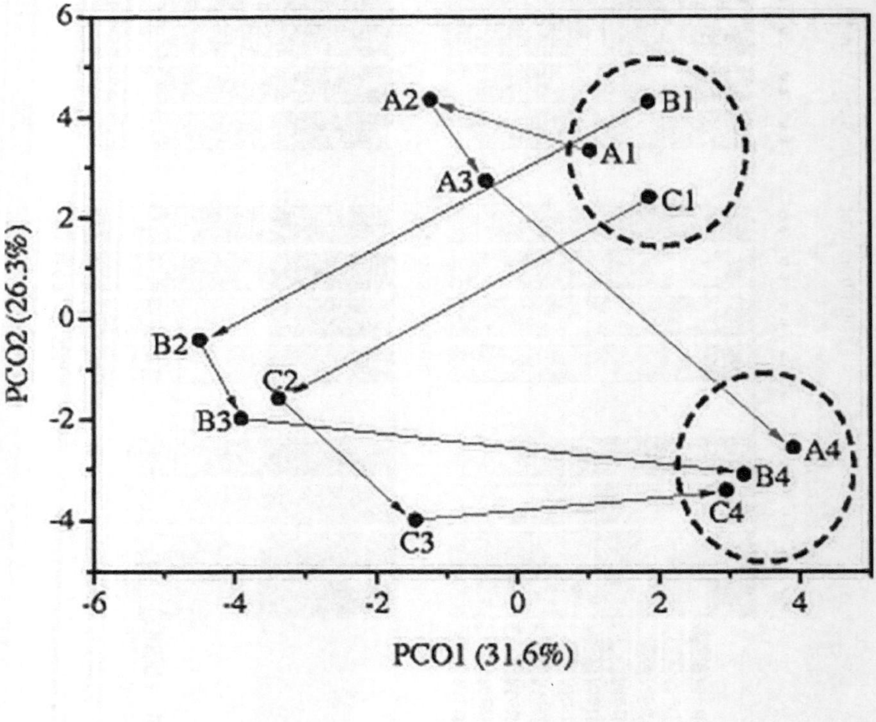

FIGURE 5: PCoA of all the samples.

The differences between the microbial communities of different samples were further assessed by principal coordinates analysis (PCoA) and hierarchical cluster analysis. The results from PCoA are shown in Figure 5. Principal components 1 and 2 explained 31.6% and 26.3% of the total community variations, respectively. A1, B1, and C1 were clustered together, and were well separated from the group of A4, B4, and C4. The results were consistent with the taxonomic analysis that steady-state microbial community compositions were changed after temperature disturbances in reactors A, B, and C, which could also be used to explain the different steady-state reactor performances in phase I and phase III, as discussed in the preceding section. A2 was close to A1, while B2 and C2 were far away from B1 and C1, which indicated that the increase of temperature to 45°C or 55°C in the biogas reactor induced significant changes in the microbial communities. A3, B3, and C3 were closer to A4, B4, and C4, which suggested that the microbial communities gradually changed to the new steady states after the temperature changed back to 37°C. The hierarchical cluster analysis (Additional file 1: Figure S1) also clustered A1, B1, and C1 as one group and A4, B4, and C4 as one group, which further supported the cluster results from PCoA.

4.2.3 MICROBIAL COMMUNITY ASSEMBLY AND FUNCTION IN ANAEROBIC DIGESTION OF CATTLE MANURE

The results in the present study clearly showed that replicate biogas reactors treating cattle manure had similar steady-state reactor performances under the same environmental conditions. The replicate biogas reactors also had similar microbial community compositions based on RDP classification, PCoA, and hierarchical cluster analysis. However, Zhou et al. found that, under the same environmental conditions, both bacterial community compositions and functions in replicate microbial electrolysis cells were different, and they proposed that stochastic assembly played a dominant role in determining not only community structure but also ecosystem functions [35]. Based on our results, the microbial community compositions and functions in anaerobic digestion of cattle manure were not obviously affected by stochastic assembly. It seems that stochastic assembly played different roles in different ecosystems. A recent publication, which found that community history affects the predictability of microbial ecosystem development, might explain the differences between our results and Zhou's results. Pagaling et al.[36] demonstrated that the final community composition and function are unpredictable when the source communities (inoculum) colonize a novel environment, while the community development is more reproducible when source communities are pre-conditioned to their new habitat. In our study, the inoculum was obtained from a biogas reactor, which was adapted to the anaerobic condition for methane production. Therefore, the steady-state microbial community composition and reactor performance were reproducible in the replicate biogas reactors. However, all the reactors in Zhou's study were inoculated with wastewater not pre-conditioned in the microbial electrolysis cell reactors, which might have been the reason for the variation of microbial community compositions and functions observed. Further investigation is imperative to elucidate whether other inocula sources (sewage sludge, soil, and others), which are not derived from anaerobic reactors, would lead to different stable-state microbial communities and reactor performances in biogas reactors under the same operational conditions.

Although the effect of temperature disturbance on biogas production has been studied before [13],[16], the previous studies only focused on the reactor performances. The rapid recovery of biogas production after the temperature disturbance was also previously observed [13], but in the cited study the biogas production reached the same level as that prior to the disturbance. We are the first to report the increased biogas production after temperature disturbance. The difference in biogas production recovery levels could be due to the fact that the substrate was already efficiently degraded before temperature shock in Chae's study [13]. Our results suggested that temperature disturbance could be used as a strategy to stimulate biogas production in the biogas reactor, where the biogas production was not so efficient (possibly due to lower HRT). In a previous study, Nielsen et al. investigated the effects of disturbance of oleate on the performance of a biogas reactor treating cattle and pig manure, and they also found that a lower VFAs concentration along with a higher methane production were achieved in the biogas reactor after the disturbance of oleate

[15]. However, they could not link the microbial community composition with the reactor performance due to the lack of microbial analysis. Our results combined with the above literature suggest that the idea that different types (temperature, oleate) of disturbances might have similar stimulation effects on the biogas process, and the effects of other disturbances (ammonia, organic loading shock, and others) on the community assembly and functioning of the biogas reactor deserve to be further investigated. Although in the present study, three different temperature disturbances were investigated, the establishment of new steady-state microbial community compositions in all the reactors after temperature disturbance was observed. More importantly, the three new steady-state microbial community compositions were clustered together and were clearly distinguished from the steady-state microbial community compositions before the temperature disturbances. A comparison of steady-state microbial community compositions before and after temperature disturbance showed that the temperature disturbance played an important role in the microbial community assembly and ecosystem function during the anaerobic digestion of cattle manure.

4.3 CONCLUSIONS

The present study showed that similar steady-state process performances and microbial community profiles were achieved in three biogas reactors with the same inoculum and operational conditions, which suggested a minor role of stochastic factors in shaping the profile of the microbial community composition and activity in biogas reactor. Instead, temperature disturbance played an important role in the microbial community composition as well as process performance in biogas reactors. Increased methane yields (around 10% higher) and decreased VFAs concentrations at steady states were found in all three reactors after the temperature disturbances, although three different temperature disturbances were applied to each biogas reactor. New steady-state microbial community profiles were also observed in all the biogas reactors after the temperature disturbance.

4.4 MATERIALS AND METHODS

4.4.1 INOCULUM AND SUBSTRATE

The inoculum (total solids (TS) 41.5 ± 1.4 g/L, VS 30.2 ± 1.2 g/L) used in this study was obtained from a mesophilic full-scale biogas plant (Hashoj biogas plant, Denmark) co-digesting cattle manure, pig manure, intestinal content from pig abattoirs, and fat and flotation sludge from pig abattoirs, fish and food processing industries, and so on. After collection, the inoculum was stored in an incubator with the temperature controlled at 37°C. The substrate used in the present study was cattle manure, and its characteristics were as follows: pH 7.5 ± 0.1, TS 36.6 ± 1.2 g/L, VS 25.2 ± 1.4 g/L, total nitrogen 2.2 ± 0.3 g/L.

4.4.2 REACTOR OPERATION

Three identical 1000-mL continuously stirred tank reactors (A, B, and C) with working volumes of 700 mL were used. The reactors were mixed by magnetic stirrer at a stirring speed of 150 rpm, and the produced biogas was collected by gas bags. All the reactors were put into an incubator with the temperature controlled at 37°C. The HRT of the reactors was 14 d. Initially, all the reactors were filled with the above-mentioned inoculum (600 mL) and cattle manure (100 mL). After one week's digestion, all the reactors were fed daily with cattle manure. Once steady states were achieved in all the reactors, the temperatures in the reactors were changed to 25°C, 45°C, and 55°C for reactors A, B, and C by putting them into different incubators with the corresponding temperatures, while the daily feeding with cattle manure continued. The temperatures were changed back to 37°C after 10 days' operation period by putting all the reactors back into the initial incubator with the temperature controlled at 37°C, and all the reactors were operated with daily feeding at 37°C, until new steady states were achieved. A steady state in this study was defined as a stable biogas production with a daily variation of lower than 10% for at least 10 days.

4.4.3 MICROBIAL ANALYSIS

Four samples were collected from the mixture in each reactor for microbial analysis. Samples A1, B1, and C1 were collected during the steady states of reactors A, B, and C before the temperature disturbance (day 50). Samples A2, B2, and C2 were collected 10 days after the temperatures were changed to 25°C, 45°C, and 55°C for reactors A, B, and C, respectively (day 60). Samples A3, B3, and C3 were collected 10 days after the temperatures of all the reactors were changed back to 37°C (day 70). Samples A4, B4, and C4 were collected during the steady states of reactors A, B, and C after the temperature disturbance (day 112).

The total genomic DNA of the collected samples was extracted using QIAamp DNA Stool Mini Kit (51504, QIAGEN, Valencia, CA) according to the manufacturer's instructions. The extracted DNA was amplified with the universal primers 515f and 806r [24],[25]. The forward primer was 5'-GTGCCAGCMGCCGCGGTAA-3', and the reverse primer was 5'-GGACTACHVGGGTWTCTAAT-3'. PCR conditions were set according to a previous study [24]. The PCR products were purified using the QIAquick spin columns (QIAGEN) to remove the excess primer dimers and dNTPs, and the concentration of PCR amplicons was measured by NanoDrop spectrophotometer [37]. The samples were then sent out for barcoded libraries preparation and sequencing on an Ion Torrent PGM system with 316 chip using the Ion Sequencing 400 bp Kit (all from Life Technologies) according to the standard protocol (Ion Xpress™ Plus gDNA and Amplicon Library Preparation, Life Technologies). The low-quality sequences without exact matches to the forward and reverse primers, with length shorter than 100 bp, and containing any ambiguous base calls, were removed from the raw sequencing data by RDP tools [38]. Chimeras were removed from the data by using the Find Chimeras web tool. The data were submitted to MG-RAST (ID 4556097.3-4556107.3). The sequences were phylogenetically assigned to taxonomic classifications by the RDP classifier with a confidence threshold of 50%. The PCoA and hierarchical cluster analysis were conducted by MG-RAST [39].

4.4.4 ANALYTICAL METHODS

TS, VS, and total nitrogen were analyzed according to the American Public Health Association (APHA) [40]. The concentrations of acetate, propionate, isobutyrate, butyrate, iso-valerate, and valerate were determined by a gas chromatograph (Hewlett Packard, HP5890 series II) equipped with a flame ionization detector and HP FFAP column (30 m × 0.53 mm × 1.0 μm). CH_4 was analyzed using a gas chromatograph with a thermal conductivity detector (GC-TCD) fitted with a parallel column of 1.1 m × 3/16 Molsieve 137 and 0.7 m × 1/4 Chromosorb 108. Detailed information about the operating conditions for the GC was given previously [41]. An analysis of variance was used to test the significance of results, and $P<0.05$ was considered to be statistically significant. The methane yield was calculated based on the daily methane production (mL_{CH4}/d) divided by daily VS feeding (g_{VS}/d).

REFERENCES

1. Luo G, Wang W, Angelidaki I: Anaerobic digestion for simultaneous sewage sludge treatment and CO biomethanation: process performance and microbial ecology. Environ Sci Technol 2013, 47:10685-93.
2. Luo G, Angelidaki I: Co-digestion of manure and whey for in situ biogas upgrading by the addition of H2: process performance and microbial insights. Appl Microbiol Biotechnol 2013, 97:1373-81.
3. Werner JJ, Knights D, Garcia ML, Scalfone NB, Smith S, Yarasheski K, et al.: Bacterial community structures are unique and resilient in full-scale bioenergy systems. Proc Natl Acad Sci U S A 2011, 108:4158-63.
4. Chase JM: Stochastic community assembly causes higher biodiversity in more productive environments. Science 2010, 328:1388-91.
5. Chase JM: Drought mediates the importance of stochastic community assembly. Proc Natl Acad Sci U S A 2007, 104:17430-4.
6. Ofiteru ID, Lunn M, Curtis TP, Wells GF, Criddle CS, Francis CA, et al.: Combined niche and neutral effects in a microbial wastewater treatment community. Proc Natl Acad Sci U S A 2010, 107:15345-50.
7. Fargione J, Brown CS, Tilman D: Community assembly and invasion: an experimental test of neutral versus niche processes. Proc Natl Acad Sci U S A 2003, 100:8916-20.
8. Chave J: Neutral theory and community ecology. Ecol Lett 2004, 7:241-53.

9. Jiang L, Patel SN: Community assembly in the presence of disturbance: A microcosm experiment. Ecology 2008, 89:1931-40.
10. Trexler JC, Loftus WF, Perry S: Disturbance frequency and community structure in a twenty-five year intervention study. Oecologia 2005, 145:140-52.
11. Rincón B, Borja R, González JM, Portillo MC, Sáiz-Jiménez C: Influence of organic loading rate and hydraulic retention time on the performance, stability and microbial communities of one-stage anaerobic digestion of two-phase olive mill solid residue. Biochem Eng J 2008, 40:253-61.
12. Nielsen HB, Mladenovska Z, Westermann P, Ahring BK: Comparison of two-stage thermophilic (68°C/55°C) anaerobic digestion with one-stage thermophilic (55°C) digestion of cattle manure. Biotechnol Bioeng 2004, 86:291-300.
13. Chae KJ, Jang A, Yim SK, Kim IS: The effects of digestion temperature and temperature shock on the biogas yields from the mesophilic anaerobic digestion of swine manure. Bioresour Technol 2008, 99:1-6.
14. Boe K, Angelidaki I: Serial CSTR digester configuration for improving biogas production from manure. Water Res 2009, 43:166-72.
15. Nielsen HB, Ahring BK: Responses of the biogas process to pulses of oleate in reactors treating mixtures of cattle and pig manure. Biotechnol Bioeng 2006, 95:96-105.
16. Ahn JH, Forster CF: The effect of temperature variations on the performance of mesophilic and thermophilic anaerobic filters treating a simulated papermill wastewater. Process Biochem 2002, 37:589-94.
17. Ye L, Zhang T: Pathogenic bacteria in sewage treatment plants as revealed by 454 pyrosequencing. Environ Sci Technol 2011, 45:7173-9.
18. Loman NJ, Misra RV, Dallman TJ, Constantinidou C, Gharbia SE, Wain J, et al.: Performance comparison of benchtop high-throughput sequencing platforms. Nat Biotechnol 2012, 30:434-9.
19. Lescure JP, Delannoy B, Verrier D, Albagnac G. Consequence of a thermal accident on the microbial activity of an industrial anbaerobic filter. Proceedings of the 5th international conference on anaerobic digestion 1988:211–214.
20. Yu HQ, Fang HHP: Acidogenesis of gelatin-rich wastewater in an upflow anaerobic reactor: influence of pH and temperature. Water Res 2003, 37:55-66.
21. Demirel B, Scherer P: The roles of acetotrophic and hydrogenotrophic methanogens during anaerobic conversion of biomass to methane: a review. Rev Environ Sci Biotechnol 2008, 7:173-90.
22. Ma JX, Carballa M, Van de Caveye P, Verstraete W: Enhanced propionic acid degradation (EPAD) system: Proof of principle and feasibility. Water Res 2009, 43:3239-48.
23. Boe K, Batstone DJ, Steyer JP, Angelidaki I: State indicators for monitoring the anaerobic digestion process. Water Res 2010, 44:5973-80.
24. Bates ST, Berg-Lyons D, Caporaso JG, Walters WA, Knight R, Fierer N: Examining the global distribution of dominant archaeal populations in soil. ISME J 2011, 5:908-17.
25. Caporaso JG, Lauber CL, Walters WA, Berg-Lyons D, Lozupone CA, Turnbaugh PJ, et al.: Global patterns of 16S rRNA diversity at a depth of millions of sequences per sample. Proc Natl Acad Sci U S A 2011, 108:4516-22.

26. Sundberg C, Al-Soud WA, Larsson M, Alm E, Yekta SS, Svensson BH, et al.: 454 pyrosequencing analyses of bacterial and archaeal richness in 21 full-scale biogas digesters. FEMS Microbiol Ecol 2013, 85:612-26.
27. Riviere D, Desvignes V, Pelletier E, Chaussonnerie S, Guermazi S, Weissenbach J, et al.: Towards the definition of a core of microorganisms involved in anaerobic digestion of sludge. ISME J 2009, 3:700-14.
28. Lu L, Xing DF, Ren NQ: Pyrosequencing reveals highly diverse microbial communities in microbial electrolysis cells involved in enhanced H-2 production from waste activated sludge. Water Res 2012, 46:2425-34.
29. Debowski M, Korzeniewska E, Filipkowska Z, Zielinski M, Kwiatkowski R: Possibility of hydrogen production during cheese whey fermentation process by different strains of psychrophilic bacteria. Int J Hydrog Energy 2014, 39:1972-8.
30. Lay JJ: Biohydrogen generation by mesophilic anaerobic fermentation of microcrystalline cellulose. Biotechnol Bioeng 2001, 74:280-7.
31. Karakashev D, Batstone DJ, Angelidaki I: Influence of environmental conditions on methanogenic compositions in anaerobic biogas reactors. Appl Environ Microbiol 2005, 71:331-8.
32. Hao LP, Lu F, He PJ, Li L, Shao LM: Predominant contribution of syntrophic acetate oxidation to thermophilic methane formation at high acetate concentrations. Environ Sci Technol 2011, 45:508-13.
33. Fotidis IA, Karakashev D, Kotsopoulos TA, Martzopoulos GG, Angelidaki I: Effect of ammonium and acetate on methanogenic pathway and methanogenic community composition. FEMS Microbiol Ecol 2013, 83:38-48.
34. De Francisci D, Kougias PG, Treu L, Campanaro S, Angelidaki I. Microbial diversity and dynamicity of biogas reactors due to radical changes of feedstock composition. Bioresour Technol. 2015;176:56-64, doi: http://dx.doi.org/10.1016/j.biortech.2014.10.126.
35. Zhou JZ, Liu WZ, Deng Y, Jiang YH, Xue K, He ZL, et al. Stochastic Assembly Leads to Alternative Communities with Distinct Functions in a Bioreactor Microbial Community. Mbio 2013;4:e00584.
36. Pagaling E, Strathdee F, Spears BM, Cates ME, Allen RJ, Free A: Community history affects the predictability of microbial ecosystem development. ISME J 2014, 8:19-30.
37. Zhang H, Banaszak JE, Parameswaran P, Alder J, Krajmalnik-Brown R, Rittmann BE: Focused-Pulsed sludge pre-treatment increases the bacterial diversity and relative abundance of acetoclastic methanogens in a full-scale anaerobic digester. Water Res 2009, 43:4517-26.
38. Cole JR, Wang Q, Cardenas E, Fish J, Chai B, Farris RJ, et al.: The ribosomal database project: improved alignments and new tools for rRNA analysis. Nucleic Acids Res 2009, 37:D141-5.
39. Meyer F, Paarmann D, D'Souza M, Olson R, Glass EM, Kubal M, et al. The metagenomics RAST server - a public resource for the automatic phylogenetic and functional analysis of metagenomes. BMC Bioinformatics 2008;9:386.
40. APHA: Standard methods for the examination of water and wastewater, 19 th ed. American Public Health Association, New York, USA; 1995.

41. Luo G, Talebnia F, Karakashev D, Xie L, Zhou Q, Angelidaki I: Enhanced bioenergy recovery from rapeseed plant in a biorefinery concept. Bioresour Technol 2010, 102:1310-3.

There is one supplemental file that is not available in this version of the article. To view this additional information, please use the citation on the first page of this chapter.

PART II

COMPOSTING

CHAPTER 5

Composting of Organic Fraction of Municipal Solid Waste: A Pilot Plant in Maxixe District, Mozambique

C. COLLIVIGNARELLI, A. PERTEGHELLA, AND M. VACCARI

5.1 INTRODUCTION

Maxixe is the largest city and economic capital of the province of Inhambane, Mozambique, with a population of 121,097 inhabitants. Maxixe town is divided in urban, sub-urban and rural area for 20.3%, 35.6% and 44.1% respectively. The absence of proper waste management (Figure 1) in Maxixe District (Mozambique) is a common issue that has to be faced, like in many other African countries (Hoornweg and Bhada-Tata, 2012). Usually waste are abandoned along the streets in the center of the city or are often burned directly inside the municipal street container in order to empty them, reducing the waste volume. As stated by Cointreau (2006), lack of proper waste management, especially concerning the organic frac-

Collivignarelli C, Perteghella A, and M. Vacchari M. "Composting of Organic Fraction of Municipal Solid Waste: A Pilot Plant in Maxixe District, Mozambique." Venice 2014, Fifth International Symposium on Energy from Biomass and Waste, San Servolo, Venice, Italy; 17 - 20 November 2014. *CISA Publisher (2014). Used with permission from the publisher.*

tion of municipal solid waste (OFMSW), causes health risks for the population due to proliferation of disease carrying vectors (rats, mosquitoes, flies, birds, etc.), direct contact with waste (mainly for children and waste pickers), air pollution through indiscriminate burning, blocking of drains and flooding, water and soil pollution and greenhouse gas emissions.

According to Fernando and Carmo Lima (2012), the OFMSW of Maxixe district and Chambone, Rumbana and Malalane neighborhoods is approximately 50% of the total amount of produced waste, as showed in Table 1; it is coherent with the organic waste percentage of low, middle and upper income countries (Table 2). Even the percentage of paper, plastic, glass and metal of Maxixe district are comparable with data showed in Table 2.

TABLE 1: Composition of MSW in Maxixe district and in 3 others Maxixe neighborhoods (Fernando and Carmo Lima, 2012)

Monitored Location	Organic (%)	Paper (%)	Plastic (%)	Glass (%)	Metal (%)	Other (%)
Maxixe district	48	4	2	4	3	39
Chambone neighborhood	40	7	4	4	5	41
Rumbana neighborhood	58	3	2	4	2	32
Malalane neighborhood	46	2	1	3	3	45
Average	48	4	2	3	3	39

TABLE 2: Composition of MSW related to population income levels (Hoornweg and Bhada-Tata 2012)

Income Level	Organic (%)	Paper (%)	Plastic (%)	Glass (%)	Metal (%)	Other (%)
Low Income	64	5	8	3	3	17
Low Middle Income	59	9	12	3	2	15
Upper Middle income	54	14	11	5	3	13
High Income	28	31	11	7	6	17

FIGURE 1: Inadequate waste management practices in Maxixe District.

FIGURE 2: Pilot plant wood boxes.

As pointed out by Table 1, Maxixe district needs to treat the organic fraction properly, thus the composting process seems to be an appropriate solution in order to avoid environmental and sanitary problems. The composting process (Mansoor er al., 2004) allows transforming the organic matter into compost, valuable product that improves the soil quality (Gajalakshmi and Abbasi, 2008) from a physical-chemical point of view, enhancing at the same time the agronomic production and food security (Edmondson et al., 2014; Mdluli et al., 2013). The composting technology is widely used in developing countries, and it's possible to distinguish 3 main different scales of technology management: i) household composting process, ii) decentralized/community composting plants, iii) centralized composting plants. Each plant can treat the OFMSW with 2 different types of aerobic process, the windrow type (Rytz, 2001) and the vessel/bin/barrel type (Zurbrügg et al., 2004). Nevertheless scientific literature presents a big gap concerning data and information of composting plants running in low and middle income countries. This entails difficulties when decision makers want to implement the composting solution, with low possibilities to take as example other field experiences.

CeLIM, an Italian NGO, has implemented a project, which is still going on, with the aim to reduce the environmental and sanitary problems, caused by inadequate waste management, in Maxixe district. To reach the project aim, a lot of awareness campaigns were carried out for the population, and new street containers were introduced in order to increase the volume of waste collection and reduce the uncontrolled dumping along the streets. Moreover, a little centralized composting plant was built, near Maxixe dump in order to recover the organic fraction of municipal solid waste and produce compost to reduce the environmental burden, reduce the chemical fertilizer use and increase the soil quality towards sustainable agriculture and food security. CeTAmb LAB (Research laboratory on Appropriate Technologies for Environmental Management in resource-limited Countries, University of Brescia) has been involved for the study and monitoring of the pilot plant in order to assess its treatment capacity and increase its operational management.

This paper reports the results based on a comparison of two heaps realized at the composting pilot plant. The authors have compared and discussed the temperature trend and the physical-chemical analyses carried out on the final compost. The obtained results will be applied and test on the full scale plant, designed to treat 6 tons of organic waste per month.

5.2 MATERIALS AND METHODS

5.2.1 PILOT AREA

The composting pilot plant was built before the full scale plant, close to the Maxixe open dump, in order to represent approximately the same characteristics of the full scale plant, especially considering 2 working hypotheses: a duration equal to 100 days for the biological degradation process and an heaps overturning with frequency equal to 10 days. Ten 1 m3 wood boxes (Figure 2) were employed to guarantee the respect of the predetermined hypothesis. Different tests were performed on the pilot plant in order to understand what kind of organic waste mix is more suitable for the initial heap composition and its process, as well as to monitor the trend temperature and the amount of water required during the aerobic process.

5.2.2 EXPERIMENTAL ACTIVITY

Two heaps, named A and B, were considered to investigate how different heaps overturning frequencies influence the temperatures trend, the degradation process duration, and the quality of the final product. As shown in Table 3, the heaps were composed with the same organic substrate percentages and with the same initial mass.

TABLE 3: Initial amount of waste and waste composition of the analyzed heaps.

	Unit	Heap A	Heap B
Initial mass of the heap	kg	282.5	267.4
Garden pruning	%	49	47
Food waste	%	23	24
Coconut sawdust	%	17	18
Compost residue from sieving	%	11	11

Heap A and heap B were overturned every 10 and 5 days respectively. All the operations as heap formation, grinding, heap overturning, watering and sieving were manually performed. An operator collected initial heaps mass, heaps temperature, environmental temperature and heaps overturning and he has reported them on a monitoring form according to the date in which measurement and operations were performed.

The equipment used for the measurement was: i) 1 m long temperature probe (for heap temperature), ii) standard electronic thermometer (for environmental temperature), iii) mechanical balance (for initial and final heap mass). The heap temperature was measered putting the temperature probe inside the heap from the top layer, considering 3 measuring points: one in the heap core, one on the left and one on the right of the core point.

Due to difficulties in finding a chemical laboratory close to Maxixe district, the compost samples of heap A and B were analyzed in an Italian chemical laboratory, according to the Italian procedures and regulation laws (D. Lgs. 75/2010).

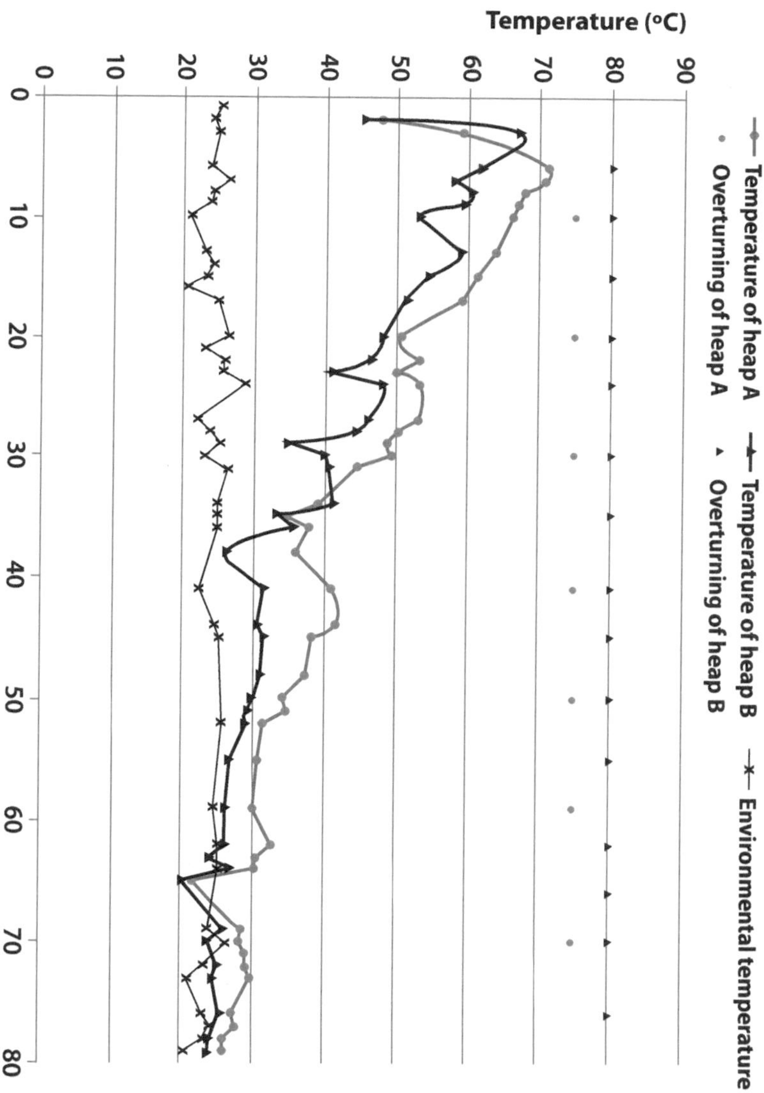

FIGURE 3: Temperature trend and frequency overturning: comparison between heap A and B.

5.3 RESULTS AND DISCUSSION

Heap A and B were removed from the wood box after 79 days of biological treatment. Subsequently the final product was spread out on the concrete basement in order to reduce the water content by natural drying. Then the compost was sieved and the final amount of saleable compost was 123.3 kg and 111.4 kg for heap A and B respectively. The compost production ratio, that are the kilograms of final compost produced per each kilograms of organic waste launched at the process, was 44% and 42% for heap A and B, respectively. Moreover, the water amount dosed during the process, necessary to control heaps temperature and moisture, was approximately 0.013 L/kg for both the heaps. Apparently the amount of water used seems to be adequate because no apparent problems raised during the process, as confirmed by the temperature trend (Figure 3).

Nevertheless it has not been possible to make a comparison with other scientific data because there are not similar data in scientific literature, and at the same time this parameter strongly depends on the process management and the climate.

5.3.1 TEMPERATURE TREND

Figure 3 shows the core temperature trend with time of the heap A and B, according with different overturning frequencies. Both of the temperature heaps increase rapidly in the first days, reaching the peak up to 70 °C within 1 week from the test launching. After reaching the peak, the temperature decreased gradually and constantly to the environmental temperature that stated out the end of the process after 80 days approximately.

According to Figure 3, heap A and B had the same temperature trend, but since for all the process duration the temperature trend of heap B was lower than the temperature trend of heap A, heap B reached the environmental temperature about a week before the heap A. A more frequent heap overturning (every 5 days) allows increasing the oxygen supply intensifying the biological degradation rate and reducing the duration process. Moreover it allows supervising the temperature, avoiding temperature peaks higher than 82°C that could severely impede the microbial community (Gajalakshmi and Abbasi, 2008).

TABLE 4: Characteristics of the final compost produced in the two experimental runs.

Parameters	Units	Heap A	Heap B	Italian law quality parameters
Organic carbon	% DM	23.96	16.47*	> 20
Total nitrogen	% DM	0.71	0.42	-
Organic nitrogen	5% N tot	87.35	90.64	> 80% tot. N
Humic and fulvic acids	% DM	4.11*	2.44*	> 7
C/N	-	33.74*	39.21*	< 25
pH	-	8.6*	8.71*	6-8.5
Moisture	% t.q.	38.08	22.91	< 50
Dry matter	% t.q.	61.92	77.09	> 50
Cu	mg/kg DM	9.13	8.09	< 230
Zn	mg/kg DM	80.19	59.59	< 500
Cd	mg/kg DM	0.55	0.12	< 1.5
Ni	mg/kg DM	5.81	8.68	< 100
Hg	mg/kg DM	0.16	0.15	< 1.5
Cr^{IV}	mg/kg DM	0.33	Below Detection Limit	< 0.5
Salmonella	MPN/g	Absence	Absence	No presence
Escherichia Coli	CFU/g	Absence	Absence	< 1000
Germination Index	%	94.07	72.02	> 60

* *Not in accordance with Italian law quality parameters set by Legislative Decree 2010/75*

5.3.2 CHEMICAL ANALYSES

Table 4 point out the results of the physical-chemical analyses carried out on two compost samples taken from the heaps A and B. All the results comply with Italian law parameters, except the organic carbon and humic and fulvic acids that are lower while carbon-nitrogen ratio (C/N) and pH are higher than the prescribed value. Globally the chemical analyses show good compost quality for both the heaps. All the measured parameters respect the Italian law limit, except for organic carbon, humic and fulvic acids, carbon-nitrogen ratio (C/N) and pH. Concerning the organic carbon measured in the heap B, the lower value was probably due to the low content of organic carbon in the initial organic waste, but at the same time due to the more frequent overturning that increased the organic carbon volatilization rate. Moreover the low content of organic nitrogen, for both the analyzed samples, determines a compost C/N ratio a little bit higher than the Italian limit. As concerns humic and fulvic acids, the low values measured in both of the heaps are probably due to the incomplete process of organic carbon conversion into humic and fulvic acids, that at the same time could depend on the organic carbon concentration in the initial used substrates. The low humic and fulvic acid concentrations contribute to have a basic pH of the produced compost. All the analyzed heavy metals (Cu, Zn, CD, Ni, Hg, CrVI) are below the Italian standard, and the good germination test results confirm the absence of uncontrolled heavy metal pollution.

Moreover the mean temperature reached by heap A and B in the first 20 days (between 50-70 °C) was fundamental for the pathogenic bacteria control (Gajalakshmi and Abbasi, 2008), indeed the microbiological analyses confirm the complete deactivation of *Salmonella* and *Escherichia coli* (Table 4), guaranteeing the microbiological safety of the use of that compost in agriculture.

5.4 CONCLUSIONS

The tests carried out on the composting pilot plant show how two different heap overturning frequencies can influence the composting process in term of duration, temperature and final product quality. As pointed out by heap B results, a more frequent heap overturning (every 5 days) allows to reduce the temperature trend, avoiding possible temperature peaks that could severely impede the microbial community, reduce the biological process duration (increasing the oxygen supply increase the biological degradation ratio) and therefore the treatment time of the material in the processing plant, allowing an increasing of the daily quantity of treated waste. At the same time an intensive heap overturning can cause an excessive volatilization of organic parameters as shown by heap B concerning the organic carbon and humic and fulvic acids that are below the Italian law limits. Anyway, even if the organic carbon (heap B), humic and fulvic acids (heap A and B), C/N ratio (heap A and B) and pH (heap A and B) do not respect the Italian standard limit, the quality of compost produced was good because there was not heavy metal pollution and at the same time the germination test shown good results.

The achieved results from the pilot tests represent a good starting point for the start-up of the full scale composting plant, nevertheless they need to be applied at the full scale plant in order to verify and correct them properly according to the composting process requirements for a good quality compost production.

REFERENCES

1. Cointreau S. (2006). Occupational and Environmental Health Issues of Solid Waste Management. Available at http://siteresources.worldbank.org/INTUSWM/Resources/up-2.pdf. The World Bank.
2. Edmondson J. L., Davies Z. G., Gaston K. J., Leake J. R. (2014).Urban cultivation in allotments maintains soil qualities adversely affected by conventional agriculture. J Appl Ecol, article in press.
3. D. Lgs. 29 aprile 2010, n.75. "Riordino e revisione della disciplina in materia di fertilizzanti, a norma dell'articolo 13 della legge 7 luglio 2009, n. 88". Pubblicato nella Gazzetta Ufficiale n. 121 del 26 maggio 2010.
4. Fernando A., Carmo Lima S. (2012). Caracterizacao dos residuos solidos urbanos do municipio de Maxixe/Mozambique. Caminhos de geografia. ISSN 1678-6343. Available at http://www.seer.ufu.br/index.php/caminhosdegeografia
5. Gajalakshmi S., Abbasi S. A. (2008). Solid Waste Management by Composting: State of the Art, Critical Reviews in Environmental Science and Technology. Available at http://dx.doi.org/10.1080/10643380701413633
6. Mansoor A., Harper M., Pervez A., Rouse J., Drescher S., Zurbrügg C., (2004). Sustainable Composting. Case studies and guidelines for developing countries. Available at http://r4d.dfid.gov.uk/pdf/outputs/R8063.pdf. WEDC, Loughborough University, England.
7. Mdluli F., Thamaga-Chitja J., Schmidt S. (2013). Appraisal of Hygiene Indicators and Farming Practices in the Production of Leafy Vegetables by Organic Small-Scale Farmers in uMbumbulu (Rural KwaZulu-Natal, South Africa). Int J Environ Res Public Health, vol. 10, 4323-4338.

8. Rytz I. (2001). Assessment of a decentralised composting scheme in Dhaka, Bangladesh. Technical, operational, organizational and financial aspects. Available at http://www.eawag.ch/forschung/sandec/publikationen/swm/dl/Rytz_2001.pdf
9. Zurbrügg C., Drescher S., Patel A., Sharatchandra H. C. (2004). Decentralised composting of urban waste – an overview of community and private initiatives in Indian cities. Waste Manage vol. 24, 655–662.
10. Zurbrügg C., Gfrerer M., Ashadi H., Brenner W., Küper D. (2012). Determinants of sustainability in solid waste management – The Gianyar Waste Recovery Project in Indonesia. Waste Manage vol. 32, 2126-2133.

CHAPTER 6

Changes in Bacterial and Fungal Communities across Compost Recipes, Preparation Methods, and Composting Times

DEBORAH A. NEHER, THOMAS R. WEICHT, SCOTT T. BATES, JONATHAN W. LEFF, AND NOAH FIERER

6.1 INTRODUCTION

Municipalities, industry and agricultural farms are generating substantial amounts of organic wastes. These wastes not only strain landfill space, but also pose serious threats to the environment. Compostable materials (including paper, food wastes, and grass clippings) comprised 62% (155 million tons) of this waste stream in the US [1]. Composting represents an important solution for a more sustainable management of organic waste. Not only does composting remove waste, it can effectively convert the waste into a nutrient-rich organic amendment for a variety of agricultural, horticultural or landscaping applications. Countries, states, and municipalities are increasingly enacting legislation and regulation to

Changes in Bacterial and Fungal Communities across Compost Recipes, Preparation Methods, and Composting Times. © *Neher DA, Weicht TR, Bates ST, Leff JW, and Fierer N.* PLoS ONE ***8**,11 (2013). http://journals.plos.org/plosone/article?id=10.1371/journal.pone.0079512. Licensed under a Creative Commons Attribution 3.0 Unported License, http://creativecommons.org/licenses/by/3.0/.*

promote the diversion of organics from solid waste disposal facilities to recycling and composting.

Composting is a controlled aerobic process that degrades organic waste to stable material, with the resident microbial community mediating the biodegradation and conversion processes. There are three distinct successional phases driving chemical and microbial changes through time, phases that are determined primarily by changes in temperature [2]: mesophilic phase (moderate temperatures rising to ~45°C), thermophilic phase (high temperatures peaking at ~70°C), curing phase (cooling to ambient temperature). Compost recipes can vary widely. For example, carbon sources can include straw, paper, woodchips, sawdust, or bark; whereas nitrogen sources can include animal manures, sewage sludge, and/or municipal solid waste. Large-scale commercial composting requires a high-temperature phase designed to facilitate the removal of human and plant pathogens. The primary types of commercial composting methods are windrow, aerated static pile, and vermicomposting. Although vermiculture does not inherently include a thermophilic phase [3], [4], material can be pre-composted through aerated static piles and windrow to remove substances toxic to earthworms, inactivate plant seeds and remove human and plant pathogens. Generally, we have a poor understanding of the biological dynamics that occur during the composting process and there are no current regulations or guidelines that define desirable microbiological properties of compost.

Microbial communities found in compost have not been well characterized and relatively few studies describe both their bacterial and fungal diversity even though both groups are likely important mediators of the composting process. Most studies have utilized culture-based methods [2], [4], [5], [6], [7] that are known to only capture a small portion of the microbial diversity found in environmental samples [8]. While a few studies have employed cultivation-independent approaches, a comprehensive perspective on compost microbial dynamics is still lacking because these studies have focused on only one of the three compost phases [9], [10], [11], [12], [13], [14] or have used fingerprinting techniques that offer limited taxonomic resolution [15], [16], [17], [18], [19], [20], [21]. The few available high-throughput sequencing studies have focused primarily on bacteria [22], [23], [24], [25] even though fungi likely play an important role in the compost process and may enhance the quality of compost. Overall, we still have a limited understanding of how microbial communities are influenced by different composting recipes or methods, or the phases of heating and cooling that occur during the compost process.

This study represents a comprehensive assessment of both bacteria and fungi associated with compost using high-throughput sequencing on the Illumina MiSeq platform. We investigated the influence of composting recipe and process on the structure of microbial communities, and how communities change through time when compost is produced on a commercial-scale.

6.2 METHODS

6.2.1 FIELD SITE AND SAMPLING

To insure validity and applicability of results, all compost recipes were produced at a commercial compost production facility at Highfields Center for Composting (HCC; Hardwick, Vermont). All recipes contained 25% manure-silage from the same source, and all composting processes included a thermophilic phase. For all experiments, a sample is defined as a composite of 10 subsamples collected at random depths from a given pile that are mixed to be representative of a pile. Samples were frozen at −20°C until processed.

6.2.2 EXPERIMENT 1: COMPOST RECIPES

Three recipes contained manure from the same source with varying carbon sources were prepared and composted using the aerated static pile process. Recipe one was 100% manure-silage from a typical Vermont dairy barn, with a C:N ratio of 17:1, as a control. Recipe two used hay as a carbon source, and was mixed in a 3:1 ratio (volume basis) with manure/silage resulting in a C:N ratio of 23:1. Recipe three used hardwood as a carbon source mixed in a 5:5:3 ratio of manure/silage:hardwood bark: softwood shavings resulting in a C:N ratio of 34:1. During the early curing phase, compost was delivered to two farm locations where it continued to cure for three months prior to sampling. In total, eight samples were taken from each of the three recipes for a total of 24 samples. Hereafter, the recipes will be referred to as 'manure-silage', 'hay', and 'hardwood', respectively.

6.2.3 EXPERIMENT 2: COMPOSTING PROCESS

A standard commercial recipe was used in each of the three composting processes: windrow, aerated static pile, vermicompost. This recipe was comprised of 20% food residuals, 10–15% 2.5 cm woody material (e.g., hardwood bark and mixed wood chips), 10% hay, up to 5% shredded paper, up to 2% dry sawdust or shavings, and 50–60% mixed livestock manures (e.g., horse, cow, heifer, calf) mixed with various bedding materials (e.g., straw and hay). Four replicate samples were taken per treatment, for a total of 12 samples. Total genomic DNA was extracted from two subsamples for each of two piles per composting process. Windrow, aerated static pile, and vermicompost samples were collected after curing, i.e., 9, 6, and 7 months of composting, respectively.

6.2.3.1 WINDROW

Windrow involves placing a mixture of organic waste materials into long, narrow piles on a composting pad which are turned frequently [3]. Piles were mixed with a bucket loader and were capped with manure/bedding to meet Vermont and National Organic Standards (NOS) Board regulations (www.ams.usda.gov/nop) and the piles were managed to maintain a temperature between 55–77°C for a minimum of 15 days and turned with a bucket loader a minimum of five times to ensure all materials have been subjected to the minimum temperature requirements.

6.2.3.2 AERATED STATIC PILE (ASP)

ASP systems force air throughout the pile and does not require turning once the pile has been formed, thus allowing for larger piles to be produced [3]. Piles were mixed with a bucket loader, placed in a three-sided ASP bay, and capped with manure/bedding. Each ASP pile was built to a height of 2 to 2.5 m after settling. Initial piles were 57–76 m^3 when placed in ASP bays. Piles were aerated in place with an in-floor air delivery system that uses a 20 cm layer of wood chips between the duct-work and pile to evenly distribute air. Blower fans were managed with speed control and timers, to meet 'Process to Further Reduce Pathogens' (PFRP) requirements as dictated by Vermont and NOS regulations. Briefly, piles were aerated in the ASP system for 3–6 weeks so the pile attained temperatures of 55°C for a minimum of three days. Piles were then re-stacked on composting pads, and turned regularly to continue composting.

6.2.3.3 VERMICOMPOST

Vermicompost is a mesophilic process that employs earthworms to stabilize organic residues [4]. Material entering the Continuous-Flow Worm Reactor was taken from piles on the composting pad, after they have been through the ASP procedure (outlined above), and re-stacked on the pads. Material had already met PFRP, and was generally four to six weeks old. Fresh material (0.76 m^3) was spread out weekly in a 3.8 cm layer on top of the bed, where it was allowed to continue to decompose, and be consumed by earthworms (*Eisenia fetida*). The bed was 1.52 m wide, 12.19 m long, and 0.6 m deep. The worm bed was housed in an indoor, heated room, and the compost temperature was 21–27°C. The compost remained in the worm bin for 60 to 90 days, the time it took for the fresh compost to be decomposed and move downward and out of the bottom of the bed.

6.2.4 EXPERIMENT 3: CHANGES IN MICROBIAL COMMUNITIES DURING THE COMPOSTING PROCESS

A standard commercial recipe of HCC was composted through windrow piles, aerated static pile, and vermicompost. Samples were collected on 20 September 2012 and represent various ages throughout the thermophilic and curing phases of the compost process for windrow, aerated static pile, and vermicomposting. Duplicate samples were analyzed at each time point for each composting method yielding a total of 24 samples.

6.2.5 DNA EXTRACTION, PCR AMPLIFICATION, SEQUENCING

Genomic DNA was extracted using the MoBio PowerSoil™ kit (MoBio, Carlsbad, CA, USA) according to the manufacturer's instructions following the method described in Lauber et al. [26]. PCR amplification of the 16S rRNA gene (for bacteria and archaea) or the internal transcribed spacer region (ITS1) of the nuclear ribosomal RNA gene (for fungi) followed the approach described in Fierer et al. [27]. Briefly, each sample was amplified in triplicate, and amplicons were composited together in equimolar concentrations prior to sequencing. PCR reactions contained 13 μL PCR-grade water, 10 μL 5 Prime Hot Master Mix, 0.5 μL each of the forward and reverse primers (10 μM final concentration), and 1.0 μL genomic DNA (diluted 1:10 with PCR-grade water). Reactions were held at 94°C for 3 min to denature the DNA, with amplification proceeding for 35 cycles at 94°C for 45 s, 50°C for 60 s, and 72°C for 90 s; a final extension of 10 min at 72°C was added to ensure complete amplification. For the bacterial and archaeal analyses, the PCR primers (515f/806r) targeted the V4 region of the 16S rRNA gene [28]. For the fungal analyses, we used PCR primers (ITS1-F/ITS2) to amplify the ITS1 spacer [29]. Both primer pairs contained 12-bp barcodes unique to each sample and the appropriate adapters to permit sequencing on the Illumina MiSeq platform [28], [30].

6.2.6 DATA ANALYSIS

Quality filtering, assignment of sequences to samples based on their barcodes, and clustering of sequences into operational taxonomic units (OTUs) was done following the standard QIIME pipeline [31] with sequence data quality-filtered as described previously. OTUs were determined using an open reference-based approach that implements reference-based clustering followed by de novo clustering using the UCLUST algorithm [32]. Clustering was conducted at the 97% similarity level using pre-clustered versions of the October 2012 Greengenes database (for 16S rRNA) [33] and November 2012 UNITE database (fungal ITS gene) [34] for the sequence reference set. Fungal ITS sequence processing followed the procedure outlined in McGuire et al. [30]. Sequences were assigned to taxonomic groups using the RDP classifier [35]. To keep sequencing depth consistent across all samples, the sequence data were rarified by randomly subsampling 2,000 and 100 reads per sample before downstream analyses of the 16S and fungal ITS rRNA datasets, respectively. Amplicon sequences were deposited in the public EMBL-EBI database (http://www.ebi.ac.uk/) and may be accessed using the accession numbers, 'ERP003625' and 'ERP003626' for the 16S and ITS sequences, respectively.

6.2.7 STATISTICAL ANALYSIS

We used the PRIMER v.6 software package (PRIMER-E, Plymouth, WA, USA) [36] for the calculation of pair-wise differences in community composition (Bray-Curtis distances) and the subsequent analyses of the pair-wise dissimilarity matrices via principal coordinate analysis and permutational multivariate analysis of variance (PERMANOVA). We used PERMANOVA to assess the effects of compost type and process type on the composition of the bacterial and fungal community compositions. For compost type, we included the farm identity as a random factor in our model to account for variation between farms. Differences in the relative abundance of specific taxa across recipe and methods were determined using multiple Kruskal-Wallis tests in R [37] and applying false discovery rate corrections to p-values to account for the multiple comparisons. Tests were only performed for the more abundant taxa (those with median relative abundances greater than 1.0% in any of the recipes or processes).

6.3 RESULTS

We obtained a total of 799,030 and 35,280 150-bp quality-filtered 16S and fungal ITS rRNA gene sequences across all samples, respectively. The number of 16S rRNA sequences obtained per sample varied from 770 to

13,390 (median = 5,398), and the number of fungal ITS sequences per sample varied from 51 to 663 (median = 215).

TABLE 1: Mean ± 1 SD (n = 8) of dominant bacterial phyla and sub-phyla, expressed as percentage of sequences in cured manure, hay and hardwood compost recipes.

Taxon	Manure	Hay	Hardwood
*Acidobacteria****	1.2 ± 0.7	0.7 ± 0.1	7.4 ± 2.7
Actinobacteria*	4.9 ± 2.0	9.9 ± 3.8	6.8 ± 2.0
Bacteroidetes*	27.0 ± 4.8	27.7 ± 2.7	21.0 ± 3.6
*Chloroflexi****	5.5 ± 3.9	2.3 ± 0.6	11.8 ± 8.5
Firmicutes**	5.2 ± 2.7	6.8 ± 1.6	1.5 ± 0.5
Gemmatimonadetes**	1.9 ± 0.8	4.4 ± 0.6	2.9 ± 0.8
Planctomycetes$^{n.s.}$	2.6 ± 1.4	1.8 ± 0.3	3.0 ± 0.7
α-Proteobacteria$^{n.s}$	7.8 ± 1.6	6.2 ± 2.4	6.3 ± 1.7
β-Proteobacteria$^{n.s}$	7.0 ± 3.7	4.1 ± 0.7	4.0 ± 1.9
δ-Proteobacteria$^{n.s}$	7.0 ± 1.7	7.0 ± 2.3	7.2 ± 1.5
γ-Proteobacteria*	12.5 ± 5.9	13.5 ± 1.5	9.2 ± 2.8
Verrucomicrobia***	1.7 ± 1.0	1.5 ± 0.3	4.6 ± 1.1

False Discovery Rate (FDR) p-value from Kruskal-Wallis Test. $^{n.s}$: $p_{FDR} > 0.05$. *: $0.01 < p_{FDR} < 0.05$. **: $0.01 < p_{FDR} < 0.05$. ***: $0.001 < p_{FDR} < 0.01$.

6.3.1 COMPOST RECIPES

All pair-wise comparisons of recipes contained unique communities of bacteria ($p_{(perm)} = 0.001$)and fungi ($p_{(perm)} = 0.001$). There was far more variability in bacterial (Table S1) and fungal community composition between compost recipes than across replicate samples collected from the same recipe (Figure 1).

Proteobacteria and *Bacteroidetes* were the most abundant bacterial phyla among compost recipes (Table 1). Within *Proteobacteria*, the γ-proteobacteria was more abundant than α-proteobacteria, β-proteobacteria, and δ-proteobacteria. *Archaea* were rare (median 1.2%

of reads for all samples), comprised mostly of *Crenarchaeota* and *Euryarchaeota*. Similarly, manure and hay contained greater relative abundances of *Firmicutes* than hardwood. Uniquely, hay contained more *Actinobacteria* and *Gemmatimonadetes*. In contrast, hardwood contained a greater relative abundance of *Acidobacteria* and *Chloroflexi*.

Ascomycota was the most abundant fungal phylum among recipes, about nearly two-fold more abundant than the *Basidiomycota* (Table 2). *Basidiomycota* were represented by *Agaricomycetes* and two undefined taxonomic classes. Manure-silage was distinguished by containing the greatest abundance of *Pezizomycetes* (including *Ascobolus*) and *Microascales*. Uniquely, hay contained greater abundances of *Epicoccum, Thermomyces, Eurotium, Arthrobotrys*, and *Myriococcum* (Table 2). Manure-silage and hay, but not hardwood, contained *Chytridomycota* and *Zygomycota* with hardwood containing greater abundances of *Sordariomycetes*.

6.3.2 COMPOST PROCESS

Holding recipe constant, bacterial (p(perm) = 0.002, Fig. 2A) and fungal (p(perm) = 0.003, Figure 2B) communities varied according to the type of managed compost process (Table S2). The compost prepared by ASP and windrow harbored bacterial and fungal communities that were more similar to one another than to the vermicompost-treated compost (Figure 2).

Bacterial communities in the windrow method were characterized by having a greater relative abundance of *Chloroflexi* and *Chlorobi* (Table 3). In contrast, the product of the vermicompost process contained relatively abundant *Bacteroidetes, γ-Proteobacteria*, and *Verrucomicrobia*. Fungal communities of windrow were dominated by *Sordariomycetes, Acremonium,* and an unclassified group in the *Basidiomycota* (Table 4). ASP was distinguished by a greater relative abundance of an unclassified family of *Pezizales*. Vermicompost contained the greatest relative abundance of *Arthrobotrys, Microascaceae, Zopfiella, Agaricomycetes,* and *Mortierella.*

TABLE 2: Mean ± 1 SD of fungal genera, expressed as percentage of sequences classified to phylum level in cured manure, hay, and hardwood compost recipes.

					n =8	n = 7	n = 8
Phylum	Class	Order	Family	Genus	Manure	Hay	Hardwood
Ascomycota[ns]					**61.5 ± 16.4**	**55.9 ± 14.7**	**69.4 ± 22.6**
	Dothideomycetes*	Pleosporales*	Pleosporaceae**	*Epicoccum*[a**]	0 ± 0	2.1 ± 1.5	0.3 ± 0.8
	Eurotiomycetes**	Eurotiales**	Unknown[b,ns]	*Thermomyces*[ns]	0 ± 0	2.7 ± 3.7	0 ± 0
		Eurotiales**	Trichocomaceae**	*Eurotium***	0 ± 0	2.0 ± 2.0	0 ± 0
	Orbiliomycetes*	Orbiliales*	Orbiliaceae*	*Arthrobotrys*[c*]	0.4 ± 0.9	6.1 ± 7.0	0 ± 0
	Pezizomycetes*	Pezizales*	Ascobolaceae*	*Ascobolus**	9.3 ± 12.3	2.4 ± 3.7	0 ± 0
		Unknown*	Unknown*	Unknown*	3.2 ± 3.1	1.1 ± 1.3	0 ± 0
	Sordariomycetes*	Hypocreales*	Unknown**	*Acremonium**	0 ± 0	1.4 ± 1.1	0 ± 0
		Microascales**	Microascacae**	*Pseudallescheria*[ns]	2.7 ± 3.7	0.2 ± 0.4	0 ± 0
		Microascales**	Microascacae**	*Scedosporium*[d*]	0.2 ± 0.5	7.7 ± 16.1	0 ± 0
		Microascales**	Microascacae**	Unknown**	14.1 ± 12.9	1.8 ± 2.5	0 ± 0
		Microascales**	Unknown**	Unknown**	7.7 ± 7.4	0.5 ± 1.2	0 ± 0
		Unknown**	Unknown**	Unknown**	14.9 ± 10.6	7.0 ± 4.5	54.2 ± 16.7
		Sordariales*	Lasiosphaeriaceae[ns]	*Zopfiella*[ns]	0.3 ± 0.9	1.5 ± 2.8	3.4 ± 5.4
		Sordariales*	Lasiosphaeriaceae[ns]	Unknown**	0 ± 0	0 ± 0	2.3 ± 1.8
		Sordariales*	Unknown**	Unknown**	0 ± 0	0.2 ± 0.4	2.8 ± 2.6
	Unknown[ns]	Unknown[ns]	Unknown[ns]	Unknown[ns]	4.9 ± 3.4	5.3 ± 2.6	5.4 ± 4.4
Basidiomycota[ns]					**29.0 ± 15.8**	**35.3 ± 16.7**	**30.3 ± 22.9**
	Agaricomycetes*	Agaricales[ns]	Psathyrellaceae[ns]	*Coprinellus*[ns]	0.3 ± 0.9	0 ± 0	6.6 ± 9.6

TABLE 2: ***Cont.***

Phylum	Class	Order	Family	Genus	n =8	n = 7	n = 8
					Manure	Hay	Hardwood
		Agaricales[ns]	Psathyrellaceae[ns]	*Coprinus*[ns]	3.2 ± 6.9	0 ± 0	1.0 ± 2.1
		Agaricales[ns]	Psathyrellaceae[ns]	Unknown*	3.4 ± 3.9	1.5 ± 2.1	0 ± 0
		Corticiales[ns]	Corticiaceae[ns]	Unknown[ns]	0 ± 0	1.3 ± 1.4	2.5 ± 3.7
		Unknown**	Unknown**	*Myriococcum*[e]**	0 ± 0	9.8 ± 8.7	0 ± 0
		Unknown[ns]	Unknown[ns]	Unknown[ns]	0.7 ± 1.0	18.0 ± 15.5	18.4 ± 24.6
	Unknown**	Unknown**	Unknown**	Unknown**	16.8 ± 11.0	0 ± 0	0 ± 0
	Unknown*	Unknown*	Unknown*	Unknown*	3.2 ± 1.6	2.8 ± 2.8	0.8 ± 1.3
Chytridiomycota*					**6.7 ± 6.7**	**5.0 ± 5.6**	**0 ± 0**
	Chytridiomycetes*	Spizellomycetales[ns]	Spizellomycetaceae[ns]	*Gaertneriomyces*[ns]	0.3 ± 0.8	1.6 ± 4.2	0 ± 0
		Unknown*	Unknown*	Unknown*	4.8 ± 6.6	3.4 ± 3.3	0 ± 0
Zygomycota*					**2.9 ± 2.2**	**3.8 ± 3.2**	**0.3 ± 0.8**
	Incertae_sedis*	Harpellales*	Legeriomycetaceae*	*Smittium*[f]	1.3 ± 1.5	0.3 ± 0.6	0 ± 0
		Mortierellales[ns]	Mortierellaceae[ns]	*Mortierella*[ns]	1.1 ± 1.6	1.1 ± 2.9	0 ± 0
		Mucorales[ns]	Mucoraceae[ns]	*Mucor*[ns]	0 ± 0	2.1 ± 2.9	0 ± 0

False Discovery Rate (FDR) p-values from Kruskal-Wallis test, [ns]: $P_{FDR} > 0.05$, *: $0.01 < p_{FDR} < 0.05$, **: $0.001 < p_{FDR} < 0.01$. [a]: *rank order of Epicoccum species abundance: E. sp_CHTAM7, E. sp_TMS_201 l.* [b]*represents a) a sequence from an un described taxon, b) from an environmental sequence were the organism was not identified, or c) a sequence matches a described species that is not represented in the reference database.* [c] *: rank order of Arthrobotrys species abundance: A. amerospora>A. flagrans >A. oligospora.* [d]*: rank order of Scedosporium species abundance: S. prolificans>S. aurantiacum>S. apiospermum.* [e]*: dominant species: Myriococcum th ermophilum.* [f]*: rank order of Smittium species abu ndance: Smittium sp.> S. orthocladii.*

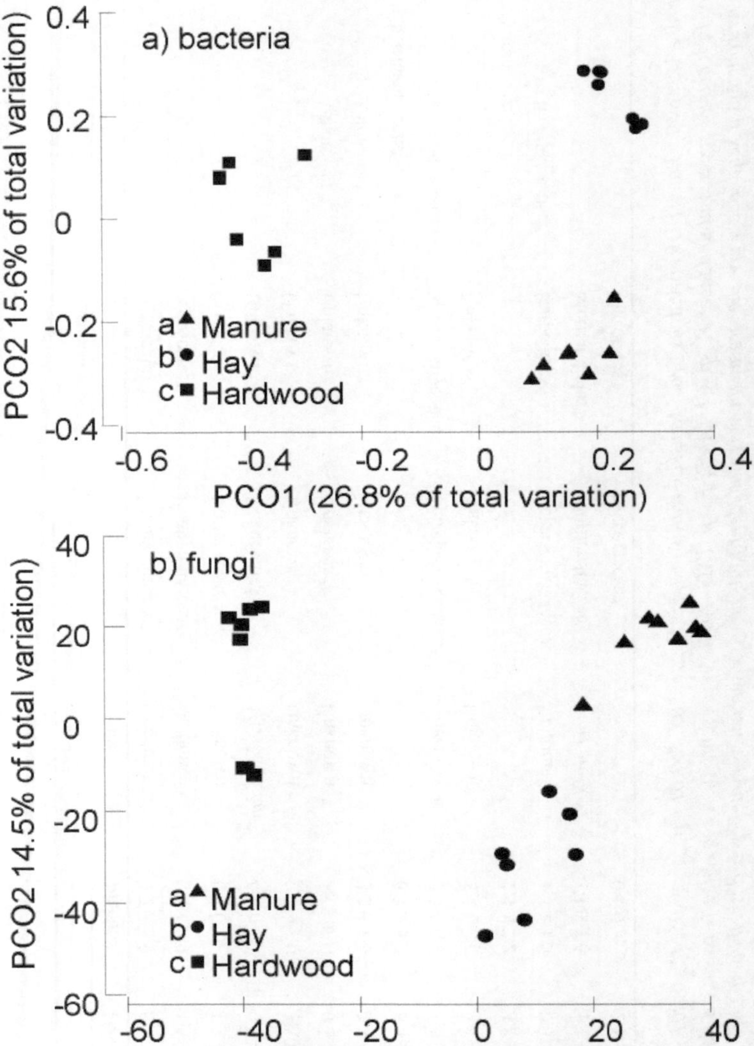

FIGURE 1: Principal coordinates analysis biplot for a) bacterial and b) fungal communities in three compost recipes (triangle: manure-silage, circle: hay, square: hardwood), n = 7 per treatment. Permutational multivariate analysis of variance indicated that differences between communities were highly significant ($p \leq 0.001$). Contrasting superscripts indicate that treatments are significantly different ($p \leq 0.05$).

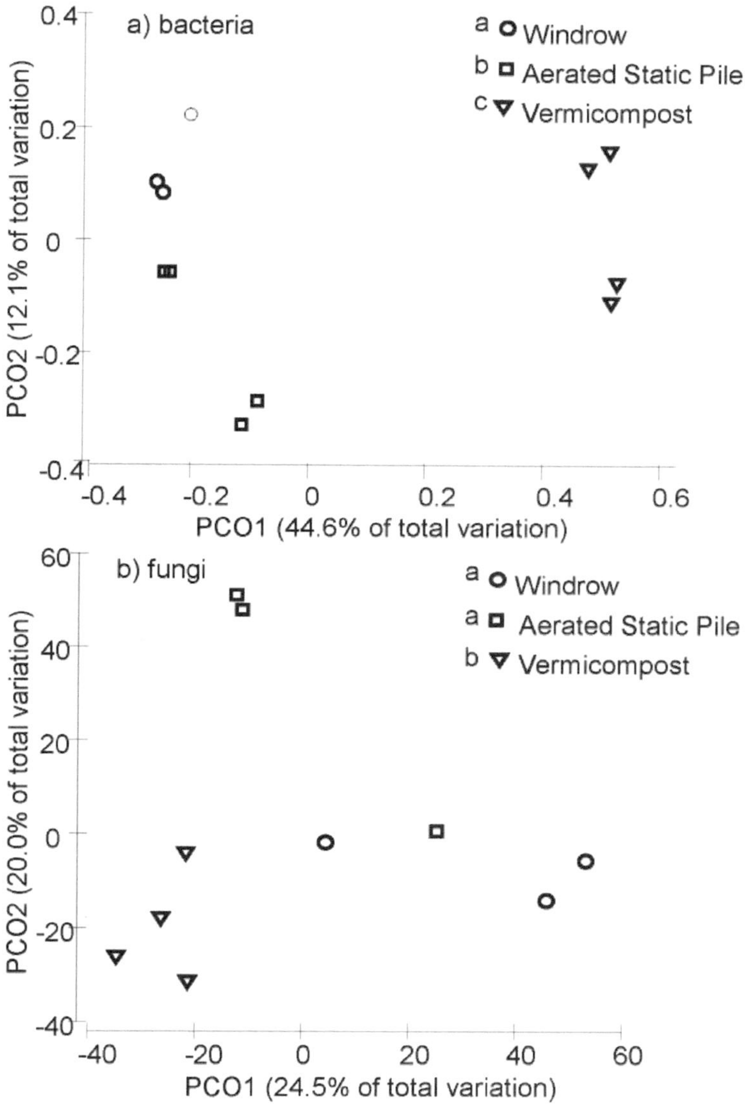

FIGURE 2: Principal coordinates analysis biplot for a) bacterial and b) fungal communities in the end product of three compost processes (circle: windrow, square: aerated static pile, inverted triangle: vermicompost), n = 4 per process. Permutational multivariate analysis of variance indicated that differences between communities were highly significant ($p \leq 0.001$). Contrasting superscripts indicate that treatments are significantly different ($p \leq 0.05$).

a) Bacteria

Bacteria	n =2	n =2	n =2	n =2	n =2	n =2	n =2	n =0	n =2	n =2	n =2	n =2	n =2
	Windrow					Aerated Static Pile					Vermicompost		
Taxa / Time	155	191	239	281	323	22	56	155	169	183	53	143	203
Bacteroidetes	24.4	25.4	14.7	20.5	22.3	2.0	12.9	.	26.6	23.6	10.1	29.7	29.2
Chloroflexi	16.6	21.6	21.1	23.1	16.9	2.3	1.7	.	10.6	17.6	1.4	1.4	3.4
γ Proteobacteria	11.5	10.4	11.2	8.7	9.9	4.6	27.8	.	15.9	9.9	44.3	17.2	15.9
δ-proteobacteria	10.7	7.0	8.3	7.8	6.7	2.5	3.6	.	9.6	7.7	1.5	11.5	7.2
Acidobacteria	2.0	3.0	5.5	5.4	4.9	0.1	0.5	.	1.9	4.5	0.0	1.7	4.3
Planctomycetes	2.3	2.7	3.9	2.7	4.8	0.5	0.9	.	2.7	3.6	0.3	3.2	4.2
δ-Proteobacteria	4.5	4.1	4.5	3.2	4.6	3.7	3.5	.	6.4	3.0	5.7	6.2	8.2
β-Proteobacteria	2.3	2.9	3.9	4.2	4.6	2.1	6.6	.	3.5	3.7	3.9	3.8	5.4
Firmicutes	3.0	3.1	2.6	2.1	2.5	62.1	22.6	.	2.3	2.4	22.8	1.1	2.1
Actinobacteria	3.8	2.1	1.4	1.4	2.2	7.5	3.1	.	1.9	3.2	1.3	2.0	2.6
Verrucomicrobia	3.0	1.5	2.6	1.2	1.7	0.2	0.6	.	2.8	2.2	0.3	5.0	3.5
Gemmatimonadetes	2.6	3.1	4.2	2.1	1.6	3.9	2.5	.	1.7	2.8	0.5	1.2	2.1

b) Fungi (Ascomycota)

				n =2	n =0	n =2	n =1	n =2	n =1	n =2	n =2	n =2	n =1	n =0	n =2	n =2
				Windrow					Aerated Static Pile					Vermicompost		
class	order	family	genus	155	191	239	281	323	22	56	155	169	183	53	143	203
Dothideomycetes	Capnodiales	Mycosphaerellaceae	*Cladosporium*	9.3	.	45.1	38.3	58.2	1.0	2.3	1.1	3.3	50.0	.	7.0	3.3
	Incertae_sedis[b]	Eremomycetaceae	*Arthrographis*	0.0	.	0.0	0.0	0.0	2.0	10.5	55.7	54.3	0.0	.	0.0	1.7
Eurotiomycetes	Eurotiales	Incertae_sedis	*Thermomyces*	30.0	.	1.2	9.0	1.3	0.0	2.3	12.6	14.6	0.0	.	22.2	8.7
	Eurotiales	Trichocomaceae	*Penicillium*	13.2	.	8.7	15.6	3.8	5.0	3.8	10.2	13.9	4.7	.	6.8	3.8
	Eurotiales	Trichocomaceae	*Talaromyces*	4.2	.	0.0	0.0	0.0	72.0	0.8	0.0	1.3	0.0	.	0.9	0.0
	Eurotiales	unknown	unknown	0.6	.	1.8	3.3	5.1	3.0	25.2	1.7	1.3	6.3	.	0.9	11.2
	unknown	unknown	unknown	4.2	.	33.3	0.0	0.0	0.0	0.0	0.0	0.0	0.0	.	0.0	8.1
Lecanoromycetes	unknown	unknown	unknown	0.0	.	0.6	0.0	0.0	0.0	0.0	0.0	2.0	0.0	.	16.8	14.9
Leotiomycetes	Helotiales	Incertae sedis	*Scytalidium*	16.2	.	0.0	2.7	1.3	0.0	2.4	2.3	0.7	0.0	.	0.0	0.6
	unknown	unknown	unknown	0.0	.	0.0	0.0	1.3	1.0	3.0	0.0	0.0	0.0	.	0.0	19.7
Orbiliomycetes	Orbiliales	Orbiliaceae	*Arthrobotrys*	2.9	.	0.0	2.6	1.3	0.0	1.6	3.4	2.6	4.7	.	3.0	1.8
Pezizomycetes	Pezizales	Ascobolaceae	*Ascobolus*	0.0	.	0.0	0.0	0.0	1.0	10.6	3.4	0.7	0.0	.	0.0	1.1
	Pezizales	Pezizaceae	unknown	0.0	.	0.0	0.0	0.0	0.0	0.0	0.0	0.0	7.8	.	16.4	0.0
	Pezizales	unknown	unknown	3.2	.	0.6	15.6	2.5	0.0	1.5	0.0	0.0	6.3	.	0.0	2.5
	unknown	unknown	unknown	0.0	.	0.0	0.0	2.5	2.0	4.9	3.4	0.0	0.0	.	0.0	0.0
Saccharomycetes	Saccharomycetales	Dipodascaceae	*Galactomyces*	0.0	.	0.0	0.0	0.0	0.0	0.0	0.0	0.0	0.0	.	12.6	1.9
Sordariomycetes	Hypocreales	Incertae_sedis	*Acremonium*[c]	0.0	.	0.0	1.6	16.5	0.0	0.0	0.0	0.0	0.0	.	0.0	0.0
	Hypocreales	unknown	unknown	0.6	.	1.2	0.8	0.0	0.0	5.7	0.0	0.0	0.0	.	3.0	0.0
	Microascales	Microascaceae	*Graphium*	0.0	.	1.3	0.0	0.0	0.0	8.5	0.0	0.0	1.6	.	0.0	0.0
	Microascales	Microascaceae	*Pseudallescheria*	0.0	.	0.0	0.0	1.3	0.0	1.6	0.6	0.0	6.3	.	1.0	0.7
	unknown	unknown	unknown	0.0	.	0.6	0.0	0.0	1.0	5.7	0.6	0.7	0.0	.	1.0	1.3

[a] represents a) a sequence from an described taxon, b) from an environmental sequence were the organism was not identified, or c) a sequence matches a described species that is not represented in the reference database

[b] from a known genus, family, but the phylogeny is not resolved

[c]: rank order of *Acremonium* abundance: *A.* sp_ATT126 > *A. alcalophilum* > *A. chrysogenum* > *A. rutilum*

FIGURE 3: Heat map illustrating changes in A) bacterial and B) fungal composition through time for the same recipe composted by three processes: windrow, aerated static pile or vermicompost. All fungi illustrated are *ascomycota*. Time is expressed as days of decomposition. The thermophillic phase occurred prior to sampling in windrow, days 22–56 for aerated static pile, and day 53 for vermicompost. Units illustrated as mean percentages of total sequences (bacteria) and percentage of taxa classified to phylum (fungi). Dots represent missing samples.

TABLE 3: Mean ± 1 SD (n = 4) of total sequences classified as bacteria in a common recipe processed by windrow, aerated static pile or vermicompost.

	Windrow	Aerated Static Pile	Vermicompost
Bacteroidetes*	21.5 ± 4.6	16.3 ± 0.7	29.4 ± 10.5
Chlorobi*	2.9 ± 1.1	1.0 ± 0.2	0.2 ± 1.1
Chloroflexi*	19.8 ± 3.8	8.0 ± 1.2	2.4 ± 7.3
γ-Proteobacteria^	10.3 ± 1.3	14.6 ± 1.2	16.5 ± 9.7
Verrucombicrobia^	2.0 ± 0.8	1.5 ± 1.3	4.2 ± 1.2

*Values are expressed as percentages. *: $p \leq 0.05$ false discovery rate (adjusted) from KW and unadjusted P-values. ^: $p \leq 0.05$ for unadjusted P-value, but ≤ 0.1 for false discovery rate (adjusted).*

6.3.3 DYNAMICS OF COMPOSTING PROCESS

The temporal shifts in bacterial and fungal communities during the composting process were influenced primarily by whether the curing phase was managed as windrow, ASP or vermicompost (Figure 3). For the bacteria, *Bacteroidetes* varied by compost process through time (Figure 3A). *Bacteroidetes* were abundant following the thermophilic phase, but declined and subsequently increased in relative abundance as composting progressed in the windrow process but steadily increased through time in the ASP and vermicompost processes. The thermophilic phase of ASP and vermicompost was dominated by *Firmicutes*. The relative abundance of *γ-Proteobacteria* increased soon after the thermophilic phase and declined through time for all composting processes but to different extents. *Chloroflexi* were relatively abundant at the end of the thermophilic phase of ASP, fluctuated in abundance through time in windrow, and were uncommon in vermicompost.

Fungal community composition varied through time and in different trajectories for the three composting processes (Figure 3B). In windrow, *Thermomyces* and *Scytalidium* were the dominant fungi after the thermophilic phase and declined through time whereas *Cladosporium* increased through time (Table 3b). *Acremonium* was second to *Cladosporium* in dominance within the finished compost. *Talaromyces* dominated the thermophilic phase of ASP while *Arthrographis* and *Cladosporium* predominate later in the ASP process. Vermicompost had approximately equal abundances of T*hermomyces, Pezizaceae, Galactomyces*, and *Lecanoromycetes* with similar abundances of *Lecanoromycetes* remaining in finished compost, accompanied by *Eurotiales* and *Leotimycetes*.

6.4 DISCUSSION

As the most comprehensive assessment of compost bacterial and fungal communities conducted to date, this work provides unique insight into microbial dynamics across different compost recipes, preparation techniques, and through time as compost cures. The types of bacterial and fungal taxa found in compost with the high-throughput sequencing methods employed here are similar to previous studies that have used other approaches to describe microbial communities in composts of similar feedstock and/or process as in this study [2], [7], [10], [12], [23], [25], [38]. For example, we found large numbers of sequences for *Bacteroidetes, Proteobacteria, Acremonium*, *Ascobolus*, and *Mortierella*, taxa that have been commonly associated with compost. Furthermore, *Aspergillus, Penicillium, Mucor,* and *Alternaria* were present as common saprophytic fungi on food wastes [2], [4], [5].

6.4.1 COMPOST RECIPE

There are distinct types of microbial communities in finished compost products that originate from different source materials. We found taxa similar to those reported for compost and its starting ingredients of manure, hay, and hardwood. For example, temporal dynamics of the composting process followed known patterns of ecological succession in herbivore manure [39]. The initial community was dominated by *Phycomycetes,* mostly *Mucorales,* such as *Mucor* and *Mortierella,* followed by ascomycota such as *Ascobolus* and *Chaetomium* spp., and finally basidiomycota such as *Coprinus* and *Stropharia* spp. [39]. Although manure- silage contained the greatest volume of manure, the other recipes also contained manure which explains why taxa commonly associated with animal feces (e.g., *Bacteroidetes, Firmicutes, γ-Proteobacteria, Chaetomium, Coprinis,* and *Ascobolus*) were found in all recipes [2], [19], [40]. *Zygomycota* in the *Harpelles, Mortierellales,* and *Mucorales* were associated more with manure-silage and hay than hardwood composts.

Fungi associated with tree bark were more commonly associated with hardwood compost, e.g., *Sordariomycetes* and *Agaricomycetes*. All recipes contained *Zopfiella* which typically arrives later in succession [41]. *Trichoderma, Alternaria,* and *Aspergillus* were not dominant on finished hardwood compost. To our knowledge, this is the first report of hardwood compost containing relatively abundant bacterial taxa within the *Acidobacteria* and *Verrucomicrobia* phyla, or of hay being a favorable habitat for *Actinobacteria* and *Gemmatimonadetes. Acidobacteria, Gemmatimonadetes,* and *Chloroflexi* have all been reported in waste water and sludge [42], [43], [44], but this is the first study to note their importance in the compost process. These taxa are notoriously difficult to culture and the ecological attributes of many members of these groups are not well-known.

6.4.2 COMPOST PROCESS

Different methods of composting after the thermophilic phase affect the dynamics and resulting composition of bacterial and fungal communities (Figure 2). We expected *Actinobacteria* and *Firmicutes* from windrow and ASP, and more *Chloroflexi, Acidobacteria, Bacteroidetes,* and *Gemmatimonadetes* in vermicompost [10]. The dominance of *Ascomycota* in both windrow and vermicompost processes has been documented in both culture-based [4] and 454-pyrosequencing [38] studies. However, in contrast to reports based on culturable fungi, we did not find *Fusarium* species to be a magnitude of order greater in windrow than vermicompost, or for *Trichoderma* species (anamorph: *Hypocrea, Hypocreaceae*) to be present exclusively in windrow [4].

Vermicompost had substantially different microbial communities when compared to those from ASP and windrow processes. These differences may be driven, in part, by differences in temperature regimes. Although vermicompost can be exclusively mesophilic, we inserted a thermophilic stage (ASP) prior to vermicomposting. We observed a greater diversity of bacteria in vermicompost than windrow and ASP (Figure 4). Our results support earlier reports that earthworms promote growth of bacteria [45] including *Bacteroidetes, Verrucomicrobia, Firmicutes*, and *Proteobacteria* [46]. There was also a trend for greater diversity of fungi in vermicompost and a relatively high abundance of fungi including *Mortierella* and *Arthrobotrys* supporting earlier reports [4], [47]. *Arthrobotrys* is a nematode-trapping fungus and been reported to be associated with earthworm inhabited soils [48]. In addition to the partially decomposed compost, regions of the digestive tract of earthworms are colonized by distinct communities of bacteria and fungi that may contribute to the overall microbial community of the compost product [49].

6.4.3 DYNAMICS OF COMPOSTING PROCESS

We know the thermophilic phase is important to reduce pathogen loads during composting (www.ams.usda.gov/nop). In contrast, the post-thermophilic phase is often ignored. Although recipe was held constant, microbial community composition diversified rapidly through time depending on whether the compost process was managed by windrow, ASP or vermicompost. To our knowledge, this is the first report of temporal dynamics in ASP or vermicompost.

Most of what we know about degradative succession in compost is from windrow studies. Relatively few taxa dominate during the thermophilic phase. For example, we observed that *Bacteroidetes* dominated at the end of thermophilic phase for windrow, supporting previous work [19]. In contrast, *Firmicutes* and γ-*Proteobacteria* were dominant after the thermophilic phase in ASP and vermicompost, respectively. All three of these bacterial phyla contain thermophiles [2]. Furthermore, relative abundance of γ-*Proteobacteria, Firmicutes* and *Actinobactera* are reported as indicators of disease suppression [50].

The fungi we observed at the end of thermophilic phase were similar to those reported previously from compost. For example, *Dothideomycetes* and *Eurotiales* were both abundant after the thermophilic phase [2], [15]. In contrast to Anastasi et al. [4], we found *Taloromyces* in compost made by all three processes rather than just vermicompost. Other fungi that were relatively abundant after the thermophilic phase have been reported for composts containing ingredients similar to our study. For example, *Cladosporium* has been reported previously on compost based on cattle manure [2]. *Thermomyces* and *Penicillium* have both been isolated from composts containing hardwood bark and manure [2].

Near the end of the composting process, a different and more complex community develops that includes chytrids, protists, *Ascomycota*, and *Stamenopiles* [15]. In windrow, *Chloroflexi* and γ-*Proteobacteria* decreased in relative abundance during the cooling and curing phases and were surpassed in abundance by *Bacteroidetes* in finished composts. This supports earlier reports that *Bacteroidetes* are relatively abundant, and more abundant than γ-*Proteobacteria*, in finished compost [10], [16]. In contrast, *Chloroflexi* decreased and abundances of α-*Proteobacteria* were half those of γ-*Proteobacteria* in the final product [23]. *Actinobacteria* become more abundant during the curing phase [51], [52], [53], [54]. *Actinobacteria* are less likely to be found in vermicompost with the more active microbial communities promoted by earthworms [47].

Numerous mesophilic fungi proliferate during the cooling and curing phases [50]. Many of the fungi we found are known to be widespread saprophytes on soil and dead plant tissue [55]. Similar to Anastasi et al. [4], *Acremonium* and *Cladosporium* occurred in both windrow and vermicompost, and *Scytalidium* was more abundant in windrow than vermicompost. It is no surprise that *Arthrographis* and *Galactomyces* occur in compost given their ability to produce cellulolytic enzymes [56], [57], however, this is the first report for composts made by ASP and vermicompost processes, specifically.

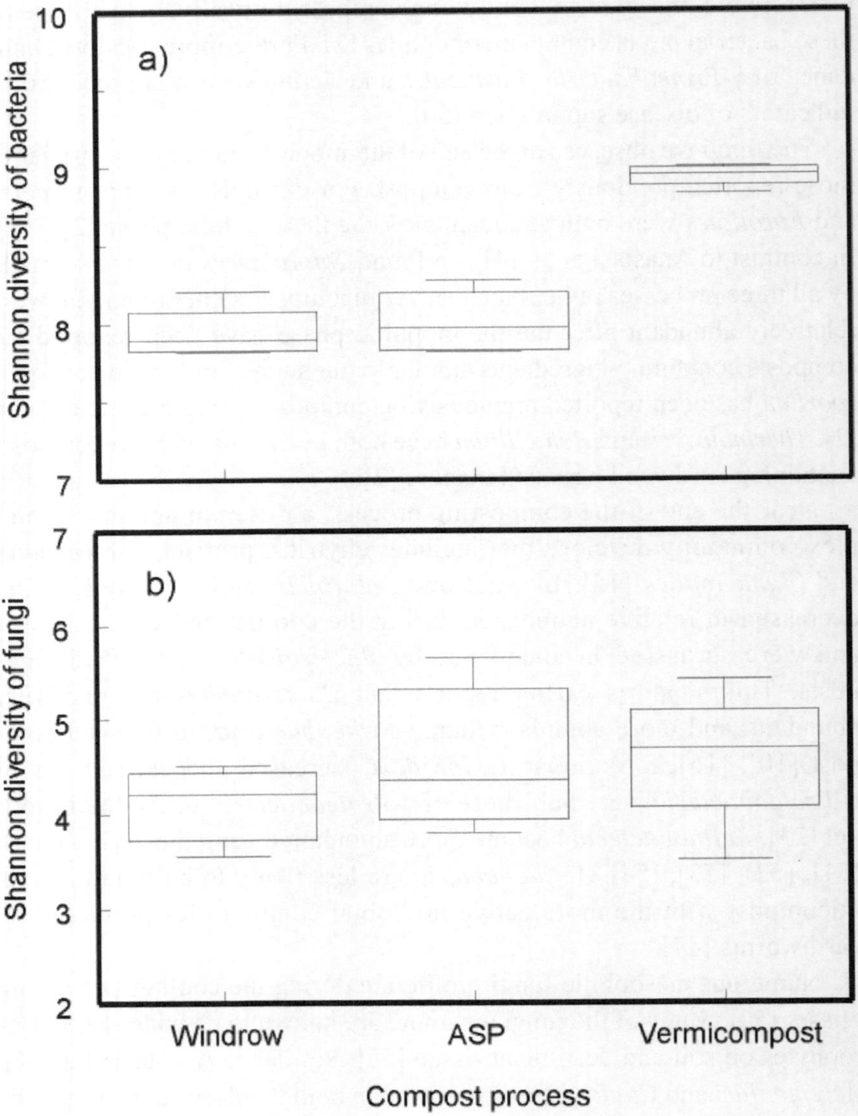

FIGURE 4: Shannon diversity of a) bacteria and b) fungal communities within a standardized recipe finished by windrow, aerated static pile (ASP) or vermicompost. Shannon diversity is computed as $H' = -\Sigma(p_i \ln p_i)$ where p represents the proportion of taxon i in the community. Box-whisker plots are illustrated.

6.5 CONCLUSION

Microbial communities are abundant and diverse in compost. Communities are organized and influenced by recipe, and post-thermophilic treatment. Composition starts similarly after thermophilic phase and shifts dynamically through time. Economic considerations have driven commercial composters to expedite the composting process. As a result, the focus has been on the effectiveness of the thermophilic phase. The curing phase offers a substrate and climate conducive for microbial recolonization which can be accomplished either by inoculating post-thermophilic compost or preparing a palatable substrate that provides a competitive advantage for colonization by bacteria and fungi that offer biological control, slow-release fertility, and/or promote plant growth. Future research can build on the microbial results presented here to determine which recipe and post-thermophilic phase are best to achieve desired agricultural goals of weed management, disease suppression, and plant growth promotion.

REFERENCES

1. US-EPA (2011) Municipal solid waste generation, recycling, and disposal in the United States Tables and Figures for 2010. Available: http://www.epa.gov/osw/nonhaz/municipal/msw99.htm. Accessed May 10, 2013.
2. Ryckeboer J, Mergaert J, Vaes K, Klammer S, de Clercq D, et al. (2003) A survey of bacteria and fungi occurring during composting and self-heating processes. Ann Microbiol 53: 349–410.
3. Ahmad R, Jilani G, Arshad M, Zahir ZA, Khalid A (2007) Bio-conversion of organic wastes for their recycling in agriculture: an overview of perspectives and prospects. Ann Microbiol 57: 471–479. doi: 10.1007/bf03175343
4. Anastasi A, Varese GC, Marchisio VF (2005) Isolation and identification of fungal communities in compost and vermicompost. Mycologia 97: 33–44. doi: 10.3852/mycologia.97.1.33
5. Ashraf R, Shahid F, Ali TA (2007) Association of fungi, bacteria and actinomycetes with different composts. Pak J Bot 39: 2141–2141.
6. Franke-Whittle IH, Knapp BA, Fuchs J, Kaufmann R, Insam H (2009) Application of COMPOCHIP microarray to investigate the bacterial communities of different composts. Microbial Ecol 57: 510–521. doi: 10.1007/s00248-008-9435-2
7. Green SJ, Michel FC, Hadar Y, Minz D (2004) Similarity of bacterial communities in sawdust- and straw-amended cow manure composts. FEMS Microbiol Lett 233: 115–123. doi: 10.1016/j.femsle.2004.01.049

8. Hill GT, Mitkowski NA, Aldrich-Wolfe L, Emele LR, Jurkonie DD, et al. (2000) Methods for assessing the composition and diversity of soil microbial communities. Appl Soil Ecol 15: 25–36. doi: 10.1016/s0929-1393(00)00069-x
9. Dees PM, Ghiorse WC (2001) Microbial diversity in hot synthetic compost as revealed by PCR-amplified rRNA sequences from cultivated isolates and extracted DNA. FEMS Microbiol Ecol 35: 207–216. doi: 10.1111/j.1574-6941.2001.tb00805.x
10. Fracchia L, Dohrmann A, Martinotti M, Tebbe C (2006) Bacterial diversity in a finished compost and vermicompost: differences revealed by cultivation-independent analyses of PCR-amplified 16S rRNA genes. Appl Microbiol Biot 71: 942–952. doi: 10.1007/s00253-005-0228-y
11. Hansgate AM, Schloss PD, Hay AG, Walker LP (2005) Molecular characterization of fungal, community dynamics in the initial stages of composting. FEMS Microbiol Ecol 51: 209–214. doi: 10.1016/j.femsec.2004.08.009
12. Guo Y, Zhang J, Deng C, Zhu N (2012) Spatial heterogeneity of bacteria: Evidence from hot composts by culture-independent analysis. Asian Austral J Anim 25: 1045–1054. doi: 10.5713/ajas.2011.11341
13. Nakasaki K, Nag K, Karita S (2005) Microbial succession associated with organic matter decomposition during thermophilic composting of organic waste. Waste Manage Res 23: 48–56. doi: 10.1177/0734242x05049771
14. Peters S, Koschinsky S, Schwieger F, Tebbe CC (2000) Succession of microbial communities during hot composting as detected by PCR–single-strand-conformation polymorphism-based genetic profiles of small-subunit rRNA genes. Appl Environ Microbiol 66: 930–936. doi: 10.1128/aem.66.3.930-936.2000
15. Bonito G, Isikhuemhen OS, Vilgalys R (2010) Identification of fungi associated with municipal compost using DNA-based techniques. Biorsource Technol 101: 1021–1027. doi: 10.1016/j.biortech.2009.08.109
16. Danon M, Franke-Whittle IH, Insam H, Chen Y, Hadar Y (2008) Molecular analysis of bacterial community succession during prolonged compost curing. FEMS Microbiol Ecol 65: 133–144. doi: 10.1111/j.1574-6941.2008.00506.x
17. Hultman J, Vasara T, Partanen P, Kurola J, Kontro MH, et al. (2010) Determination of fungal succession during municipal solid waste composting using a cloning-based analysis. J Appl Microbiol 108: 472–487. doi: 10.1111/j.1365-2672.2009.04439.x
18. Klamer M, Baath E (1998) Microbial community dynamics during composting of straw material studied using phospholipid fatty acid analysis. FEMS Microbiol Ecol 27: 9–20. doi: 10.1111/j.1574-6941.1998.tb00521.x
19. Sasaki H, Nonaka J, Otawa K, Kitazume O, Asano R, et al. (2009) Analysis of the structure of the bacterial community in the livestock manure-based composting process. Asian Austral J Anim 22: 113–118.
20. Tiquia SM (2005) Microbial community dynamics in manure composts based on 16S and 18S rDNA T-RFLP profiles. Environ Technol 26: 1101–1114. doi: 10.1080/09593332608618482
21. Viera B, Madayanti VE, Aryantha F, Akhmaloka INP (2012) Succession of eukaryotic communities during traditional composting of domestic waste based on PCR-DGGE analysis. J Pure Appl Microbiol 6: 525–536.

22. Alfreider A, Peters S, Tebbe CC, Rangger A, Insam H (2002) Microbial community dynamics during composting of organic matter as determined by 16S ribosomal DNA analysis. Compost Sci Util 10: 303–312. doi: 10.1080/1065657x.2002.10702094
23. de Gannes VD, Eudoxie G, Hickey WJ (2013a) Prokaryotic successions and diversity in composts as revealed by 454-pyrosequencing. Bioresource Technol 133: 573–580. doi: 10.1016/j.biortech.2013.01.138
24. Fernandez-Gomez MJ, Nogales R, Insam H, Robero E, Goberna M (2012) Use of DGGE and COMPOCHIP for investigating bacterial communities of various vermicomposts produced from different wastes under dissimilar conditions. Sci Total Environ 414: 664–671. doi: 10.1016/j.scitotenv.2011.11.045
25. Partanen P, Hultman J, Paulin L, Auvinen P, Romantschuk M (2010) Bacterial diversity at different stages of the composting process. BMC Microbiol 10: 94 Available: http://www.biomedcentral.com/1471-2180/10/94. Accessed 25 January 2013.
26. Lauber CL, Hamady M, Knight R, Fierer N (2009) Pyrosequencing-based assessment of soil pH as a predictor of soil bacterial community structure at the continental scale. Appl Environ Microbiol 75: 5111–5120. doi: 10.1128/aem.00335-09
27. Fierer N, Leff JW, Adams BJ, Nielsen UN, Bates ST, et al. (2012) Cross-biome metagenomic analysis of soil microbial communities and their functional attributes. PNAS doi 10.1073pnas.1215210110/pnas.1215210110.
28. Caporaso JG, Lauber CL, Walters WA, Berg-Lyons D, Lozupone CA, et al. (2012) Ultra-high-throughput microbial community analysis on the Illumina HiSeq and MiSeq platforms. The ISME J 6: 1621–1624. doi: 10.1038/ismej.2012.8
29. Gardes M, Bruns TD (1993) ITS primers with enhanced specificity for basidiomycetes – application to the identification of mcorrhizae and rusts. Mol Ecol 2: 113–118. doi: 10.1111/j.1365-294x.1993.tb00005.x
30. McGuire KL, Payne SG, Palmer MI, Gillikin CM, Keefe D, et al. (2013) Digging the New York City skyline: Soil fungal communities in green roofs and city parks. PLoS One 8: e58020. doi: 10.1371/journal.pone.0058020
31. Caporaso JG, Kuczynski J, Stombaugh J, Bittinger K, Bushman FD, et al. (2010) QIIME allows integration and analysis of high-throughput community sequencing data. Nat Methods 7: 335–336. doi: 10.1038/nmeth.f.303
32. Edgar RC (2010) Search and clustering orders of magnitude faster than BLAST. Bioinformatics 26: 2460–2461. doi: 10.1093/bioinformatics/btq461
33. McDonald D, Price MN, Goodrich J, Nawrocki EP, DeSantis TZ, et al. (2011) An improved Greengenes taxonomy with explicit ranks for ecological and evolutionary analyses of bacteria and archaea. ISME J 6: 610–618. doi: 10.1038/ismej.2011.139
34. Bates S, Ahrendt S, Bik H, Bruns T, Caporaso J, Cole J, et al. (2013) Meeting Report: Fungal ITS Workshop (October 2012). Standards In Genomic Sciences 8 doi:10.4056/sigs.3737409.
35. Wang Q, Garrity GM, Tiedje JM, Cole JR (2007) Naive Bayesian classifier for rapid assignment of rRNA sequences into the new bacterial taxonomy. Appl Environ Microbiol 73: 5261–5267. doi: 10.1128/aem.00062-07
36. Clarke KR, Gorley RN (2001) Plymouth Routines in Multivariate Ecological Research (Primer) Version 5: User Manual/Tutorial. Primer-E Ltd, Plymouth, UK.

37. R Core Team (2012) R: A language and environment for statistical computing. R Foundation for Statistical Computing, Vienna, Austria. Available: http://www.R-project.org/. Accessed 7 January 2013.
38. de Gannes VD, Eudoxie G, Hickey WJ (2013b) Insights into fungal communities in composts revealed by 454-pyrosequencing: implications for human health and safety. Front Microbiol doi:10.3389/fmicb.2013.00164.
39. Hudson HJ (1968) The ecology of fungi on plant remains above the soil. New Phytol 67: 837–874. doi: 10.1111/j.1469-8137.1968.tb06399.x
40. Richardson MJ (2001) Diversity and occurrence of coprophilous fungi. Mycol Res 105: 387–402. doi: 10.1017/s0953756201003884
41. Pillinger JM, Cooper JA, Ridge I, Barrett PRF (1992) Barley straw as an inhibitor of algal growth III: the role of fungal decomposition. J Appl Phycol 4: 353–355. doi: 10.1007/bf02185793
42. Bjornsson L, Hugenholtz P, Tyson GW, Blackall LL (2002) Filamentous *Chloroflexi* (green non-sulfur bacteria) are abundant in wastewater treatment processes with biological nutrient removal. Microbiol-Sgm 148: 2309–2318.
43. Quaiser A, Ochsenreiter T, Lanz C, Schuster SC, Treusch AH, et al. (2003) *Acidobacteria* form a coherent but highly diverse group within the bacterial domain: evidence from environmental genomics. Mol Microbiol 50: 563–575. doi: 10.1046/j.1365-2958.2003.03707.x
44. Zhang HJ, Sekiguchi Y, Hanada S, Hugenholtz P, Kim H, et al. (2003) Gemmatimonas aurantiaca gen. nov., sp. nov., a gram-negative, aerobic, polyphosphate-accumulating micro-organism, the first cultured representative of the new bacterial phylum Gemmatimonadetes phyl. nov. Int J Syst Evol Microbiol 53: 1155–1163. doi: 10.1099/ijs.0.02520-0
45. Yakushev AV, Bubnov IA, Semenov AM (2011) Estimation of the effects of earthworms and initial substrates on the bacterial community in vermicomposts. Eurasian Soil Sci 44: 1117–1124. doi: 10.1134/s1064229311100164
46. Bernard L, Chapuis-Lardy L, Razafimbelo T, Razafindrakoto M, Pablo AL, et al. (2012) Endogeic earthworms shape bacterial functional communities and affect organic matter mineralization in a tropical soil. ISME J 6: 213–222. doi: 10.1038/ismej.2011.87
47. Lazcano C, Gomez-Brandon M, Dominguez J (2008) Comparison of the effectiveness of composting and vermicomposting for the biological stabilization of cattle manure. Chemosphere 72: 1013–1019. doi: 10.1016/j.chemosphere.2008.04.016
48. Sukhjeet K, Kaul VK (2006) Evaluation of substrates for growth of nematophagous fungus, Arthrobotrys oligospora. Ann Plant Prot Sci 14: 456–458.
49. Byzov BA, Nechitaylo TY, Burmazhkin BK, Kurakov AV, Golyshin PN, et al. (2009) Culturable microorganisms from the earthworm digestive tract. Microbiology 78: 360–368. doi: 10.1134/s0026261709030151
50. Hadar Y, Papadopoulou KK (2012) Suppressive composts: microbial ecology links between abiotic environments and healthy plants. Annu Rev Phytopathol 50: 133–153. doi: 10.1146/annurev-phyto-081211-172914
51. Cahyani VR, Matsuya K, Asakawa S, Kimura M (2003) Succession and phylogenetic composition of bacterial communities responsible for the composting process

of rice straw estimated by PCR-DGGE analysis. Soil Sci Plant Nutr 49: 619–630. doi: 10.1080/00380768.2003.10410052

52. Steger K, Jarvis Å, Vasara T, Romantschuk M, Sundh I (2007) Effects of differing temperature management on development of Actinobacteria populations during composting. Res Microbiol 158: 617–624. doi: 10.1016/j.resmic.2007.05.006
53. Tang JC, Kanamori T, Inoue Y, Yasuta T, Yoshida S, et al. (2004) Changes in the microbial community structure during thermophilic composting of manure as detected by the quinone profile method. Process Biochem 39: 1999–2006. doi: 10.1016/j.procbio.2003.09.029
54. Xiao Y, Zeng GM, Yang ZH, Ma YH, Huang C, et al. (2011) Effects of continuous thermophilic composting (CTC) on bacterial community in the active composting process. Microb Ecol 62: 599–608. doi: 10.1007/s00248-011-9882-z
55. Kirk PM, Cannon PF, Minter DW, Stalpers JA (2008) Ainsworth & Bisby's Dictionary of the Fungi, Tenth Edition. Wallingsford: CABI. 784 p.
56. Eida MF, Nagaoka T, Wasaki J, Kouno K (2011) Evaluation of cellulolytic and hemicellulolytic abilities of fungi isolated from coffee residue and sawdust composts. Microbes Environ 28: 220–227. doi: 10.1264/jsme2.me10210
57. Okeke BC, Obi SKC (1993) Production of cellulolytic and xylanolytic enzymes by an *Arthrographis* species. World J Microb Biot 9: 345–349. doi: 10.1007/bf00383077

There is one table and several supplemental files that are not available in this version of the article. To view this additional information, please use the citation on the first page of this chapter.

CHAPTER 7

Effects of Bulking Agents, Load Size or Starter Cultures in Kitchen-Waste Composting

NORAZLIN ABDULLAH, NYUK LING CHIN,
MOHD NORIZNAN MOKHTAR, AND FARAH SALEENA TAIP

7.1 INTRODUCTION

Disposal of kitchen waste, which contains about 80% of moisture to the landfills, causes various problems like easy putrefaction, offensive odour and pollution of ground and surface water by leachate (Rogoshewski et al. 1983; Wang et al. 2001). Due to interruption of the carbon cycle by disposal of waste to landfills, organic kitchen waste requires proper composting system to reduce its uncontrolled degradation on disposal sites and subsequent greenhouse gases, odour and nutrient emissions (Luostarinen and Rintala 2007). In addition, kitchen waste may be wasted if it is just dumped to landfills as it will break up naturally and never be used directly again, having its nutritious matter lost within the waste. Therefore, a

Effects of Bulking Agents, Load Size or Starter Cultures in Kitchen-Waste Compostin © Abdullah N, Chin NL, Mokhtar MN, and Taip FS. International Journal of Recycling of Organic Waste in Agriculture *2,3 (2013), doi:10.1186/2251-7715-2-3. Licensed under Creative Commons Attribution 2.0 Generic License, http://creativecommons.org/licenses/by/2.0/.*

kitchen waste composter was designed to allow the composting process to be conducted in a consistent manner.

While people give attention to recycled inorganic wastes such as plastics, glass and metals, kitchen waste which is rich in organic material and possesses more than 90% of biodegradability can be easily recycled into compost (Veeken and Hamelers 1999). Composting of kitchen waste can be an effective method to reduce waste in landfills while helping conserve the environment. As kitchen waste is produced everyday and everywhere from processed and unprocessed food for human consumption, its composition is quite variable. An optimised kitchen waste formulation and composition involving the use of bulking materials, waste load size and presence of microbes are important in ensuring the commencement of an effective composting process (Fang et al. 2001; Ishii and Takii 2003; Cekmecelioglu et al. 2005; Stabnikova et al. 2005; Cayuela et al. 2006; Chang and Hsu 2008).

Recognised bulking agents used for composting include sawdust (Sundberg and Jönsson 2005; Kalamdhad et al. 2008), rice hulls and chips of tree cuttings (Chikae et al. 2006), horticultural waste compost (Stabnikova et al. 2005) and mulch hay and wood shavings (Cekmecelioglu et al. 2005). Newspaper contains 9% of moisture content (MC) (Wayman et al. 1992) and 94% of volatile solids with lignin content of the volatile solids ranging from 16% to 22% (Sun and Cheng 2002), while the onion peels contain about 54% of volatile solids (Lubberding et al. 1988) with 1.5% dry weight of lignin (Suutarinen et al. 2003). Both can be suitable bulking materials in the composting of kitchen waste as they contain high carbon contents (Abdullah and Chin 2010). Newspaper is distributed throughout the land, in practically every house and building. The Malaysian Audit Bureau of Circulations reports that newspaper circulations have increased from 2005 to 2009, where the total of publications for the last 6 months of 2009 was 4,100,486 and 576,663 for West and East Malaysia, respectively. Onions are one of the major vegetables consumed in Malaysia (Kamil et al. 2010) and onion peels are the most common waste disposed in almost every kitchen. Among the three important constituents of plant cell wall material, the cellulose, lignin and hemicellulose, lignin is particularly difficult to biodegrade and reduces the bioavailability of the other cell wall constituents (Naik et al. 2012). The domination of different indigenous microorganism population at various stages of composting plays a distinct role in degrading lignin (Raut et al. 2008; Belyaeva and Haynes 2009; Huang et al. 2010).

Waste load size influences the temperature development during a composting process (Cayuela et al. 2006). It affects the achievement of the thermophilic phase where it is an essential stage for sanitation and killing all pathogens. Small heap sizes may not produce enough heat to reach the thermophilic phase, while too large heap sizes may prevent air passage into the centre which will paralyse the composting process (Cayuela et al. 2006). The windrow system requires a large heap to achieve the thermophilic stage. This fact is supported by Cekmecelioglu et al. (2005) who performed a composting process using about 11-m long, 2.5-m wide and 1.2-m high heap using conventional layering and mixing methods. Elsewhere, Sellami et al. (2008) have developed cone-shaped heap of 2-m high with a 3-m wide base for co-composting of exhausted olive cake, poultry manure and sesame bark. The other composting method using the in-vessel system requires a closed container, which is also known as a reactor to perform the composting process in small volume. This system can be placed inside or outside the building, as long as the substrates are protected from environmental effects and the process is under controlled conditions. Ishii and Takii (2003) found that smaller stacks are better for food-waste composting. A smaller scale of food waste progresses faster as food waste contains much more easily degradable compounds, and it takes a shorter time to turn into compost.

The application of starter culture has been found advantageous in composting especially for cellulosic waste, which is difficult to be broken down into smaller pieces (Volchatova et al. 2002). Its use is not a new practice for composting, where Wang et al. (2003) added a starter bacterial culture of *Bacillus thermoamylovorans* SW25 in composting mixtures of anaerobic dewatered sewage sludge and vegetable food waste. Although Fang et al. (2001) suggested that starter culture might assist in the composting process, Stabnikova et al. (2005) found that starter cultures are useless in food waste aerobic thermophilic bioconversion as sufficient air and thermophilic temperature ensure a quick composting process. The starter culture governs the process by reducing the accumulation of actinomycetes and fungi, which discharge allergenic spores into the air, and lowering the risk of harmful microorganism growth in the end product (Wang et al. 2003). Previous researchers have used *B. thermoamylovorans*

SW25 (Wang et al. 2003; Stabnikova et al. 2005), cultures of three *Bacillus* species (such as *B. brevis, B. coagulans* and *B. licheniformis* (Fang et al. 2001)), and a specially developed association of microorganism (Volchatova et al. 2002) in their composting studies.

The objectives of this work are to study the effects of two common bulking media, newspaper or onion peels with compost load sizes of 2 and 6 kg and the use of starter culture in kitchen-waste composting consisting of vegetable scraps and fish processing waste using an optimised formulation using the in-vessel system.

TABLE 1: Percentage (%) of moisture, carbon and nitrogen content, and CN ratio (Abdullah and Chin 2010)

Substrates	Moisture content (%)	Carbon content (%)	Nitrogen content (%)	CN ratio
Vegetable scraps	93.32±0.33	37.87±1.14	2.94±0.14	12.87±0.24
Fish processing waste	75.83±0.43	42.60±1.73	8.14±1.24	5.30±0.61
Newspaper	6.50±0.32	50.56±0.02	0.18±0.01	288.15±13.36
Onion peels	11.69±0.22	50.27±0.63	0.83±0.01	60.91±1.82

Percentage and CN ratio are in mean and standard deviation.

7.2 METHODS

7.2.1 COMPOSTING MATERIAL PREPARATION

The main composting substrates were vegetable scraps from spinach (*Spinacia oleracea* L.) and fish processing waste of Indian mackerels (*Rastrelliger kanagurta* C.). Newspaper and onion peels were chosen as they are easily available bulking agents, can act as moisture adjuster due to their low moisture content and have high cellulose content, which can be a good source of carbon. The characteristics of both waste and bulking agents in terms of its percentage of moisture, carbon and nitrogen contents, and the values of the carbon-to-nitrogen (CN) ratio have been measured in an earlier study (Abdullah and Chin 2010) and are given in Table 1.

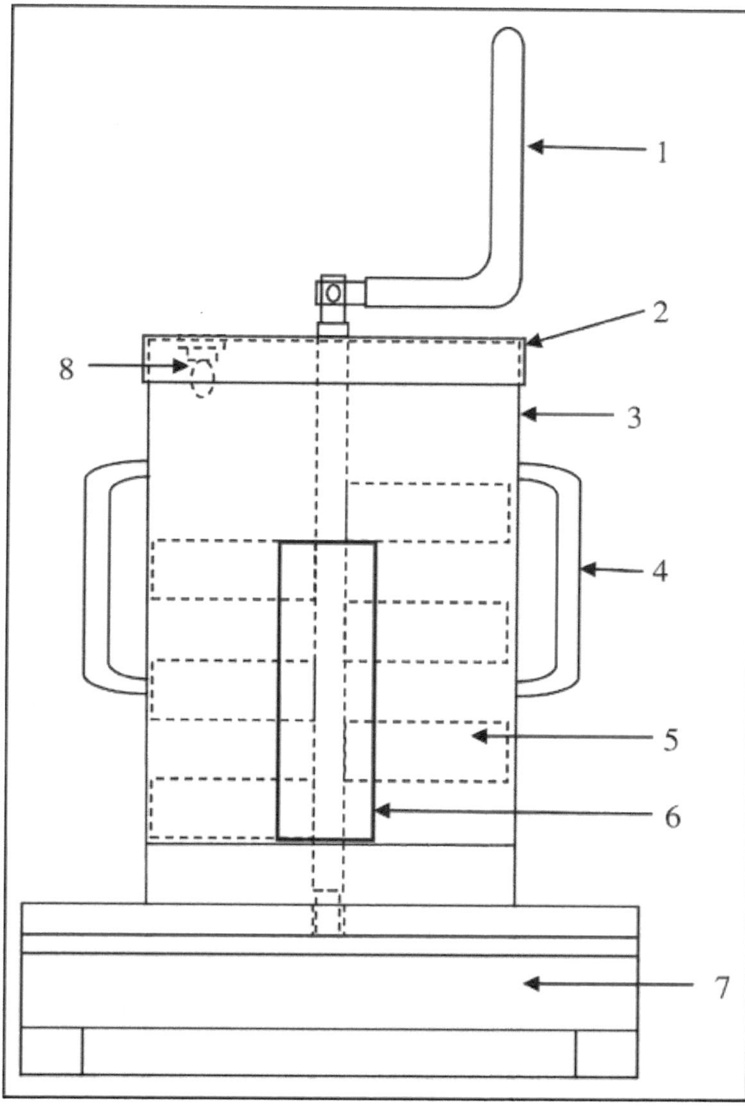

FIGURE 1: Schematic diagram of the designed kitchen waste composter. 1=Turner's handle, 2=perforated cover, 3=body, 4=handle, 5=blade, 6=window, 7=tray, 8=light.

7.2.2 KITCHEN WASTE COMPOSTER DESIGN

As kitchen waste which contains very high moisture is difficult to be self-heated up to 45°C, two self-designed kitchen waste composters which included features for temperature and moisture content control were fabricated to allow parallel experiments to be conducted consistently (Figure 1). The kitchen composter consists of three parts, which are the perforated cover, the body and a collector of water and end product. All parts were made from stainless steel 304, except for its transparent window made of polycarbonate that is a heat-resistant plastic for convenient viewing of the waste inside the composter. The external part of body and the perforated cover were insulated using cloth to minimise heat loss. A 25-W bulb was attached to the perforated cover to provide heat internally. The opening at the top allows feeding of fresh waste, and the small holes in the perforated cover allow air movement to maintain an aerobic process. The long L-shaped handle on the top is for manual turning of the mixer blade to aerate and homogenise the sample. The blades were slanted 5° in opposite directions to keep a good blending by moving the mixture up and down. The two side handles at the body of the composter aid the rotation of the body up to 360° when collecting end products. A collector tray was inserted below the body to collect the leachate water or end product. Between the body and the collector, there was a medium hole to collect the end product and many tiny holes to allow excessive water draining.

7.2.3 COMPOSTING AND SAMPLING

The optimised mixture formulation for kitchen-waste composting was referred from Abdullah and Chin (2010) following the fixed CN ratio of 30 and MC of 60%. Table2 shows the amounts of substrate compositions for investigations on bulking agents, load size and use of starter culture. All substrates were weighed using a scale (SSB12001, Mettler Toledo, Switzerland), then mixed manually in a basin before loading into the composter. The composting was conducted simultaneously using two composters for each experiment. The first experiment was the comparison of the effectiveness of the bulking media, i.e. between the newspaper and onion peels at 2-kg load size. After finding that the onion peels were more effective in composting, they were used at two load sizes, 2 and 6 kg, before further investigation on the effect of presence of starter cultures using 6 kg of kitchen waste. Each compost cycle was for 30-days, and the composters were placed in a laboratory at 27±5°C. The composts were also turned and mixed in the composters each time after sampling to prevent shortage of oxygen pore required for aerobic composting. The pipe water sprayed onto the composting substrates inside the composters to maintain an MC of 60% was left overnight before use to reduce its chlorine content. Samples were taken on days 0, 1, 4, 8, 10, 16, 23 and 30 of the composting period. The bulb in the composter was switched on to help the composting material to achieve the thermophilic temperature of 45°C and until the compost temperature stabilised towards the end of the composting process where the CN ratio reduced to less than 15 or the compost reached the matured stage.

TABLE 2: Amount of substrate compositions used in each experiment

Substrates	Amount (g)	
	Composter 1	Composter 2
Bulking agents	Newspaper	Onion peels
Vegetable scraps	971	880
Fish processing waste	355	395
Newspaper	675	0
Onion peels	0	725
Load size	2 kg	6 kg
Vegetable scraps	880	2,641
Fish processing waste	395	1,185
Onion peels	725	2,174
Starter culture	With	Without (control)
Vegetable scraps	2,641	2,641
Fish processing waste	1,185	1,185
Onion peels	2,174	2,174

For composting with added starter culture, a solution of bacteria from *Aeromonas* sp., *Azotobacter* sp., *Bacillus* sp., *Clostridium* sp., *Pseudomonas* sp., *Thermomonaspora* sp. and *Trichurus* sp., and eight plates of fungi from *Aspergillus* sp., *Cellulomonas* sp., *Chaetomium* sp., *Coprinus* sp., *Microbispora* sp., *Penicillium* sp., *Thermoactinomyces* sp. and *Trichoderma* sp. were provided by the Biotechnology Research Centre, Malaysian Agricultural Research and Development Institute. This microbe cocktail was formulated after considering the functions of each microorganism isolated from soil. The *Bacillus* sp., *Clostridium* sp., *Pseudomonas* sp., *Aspergillus* sp., *Cellulomonas* sp., *Chaetomium* sp., *Penicillium* sp. and *Trichoderma* sp. play important roles in degrading cellulose and other carbohydrates (Bhatt and Kausadikar 2010). Proteins are degraded to individual amino acids mainly by fungi, *Actinomycetes, Bacillus* sp., *Clostridium* sp. and *Pseudomonas* sp. (Bhatt and Kausadikar 2010). Bacteria like the *Pseudomonas* sp., *Clostridium* sp. and *Bacillus* sp. and fungi like *Trichoderma* sp., *Penicillium* sp. and *Aspergillus* sp. help degrade toxic chemicals and pesticides into non-toxic substances, thus minimising any damage caused by harmful chemicals to the ecosystem (Bhatt and Kausadikar 2010). Both bacteria and fungi were mixed using a kitchen blender (PB-323 T, Pensonic, Malaysia) before being sprayed onto the waste in composter 1 on day 1 of composting. This is to allow the waste inside the kitchen waste composter to stabilise and so that the starter culture could easily adapt to the surroundings inside the kitchen waste composter and perform good composting activities.

7.2.4 ANALYTICAL METHODS

During the 30-days of kitchen-waste composting, the progression of temperature and MC (Abdullah and Chin 2010) were determined. The temperature was measured at 50% depth from the surface of compost materials at four different positions. In assessing the maturity level of the compost, the CN ratio was determined. The content of the volatile solids was calculated after measuring the ash content using the dry ashing method to obtain the total organic carbon (TOC) content (Abdullah and Chin 2010). The total nitrogen content was measured using the micro-Kjeldahl method

(Mohee et al. 2008; Unmar and Mohee 2008), where the digested sample was distilled with 45% sodium hydroxide and 2% boric acid, and titrated with 0.05 N sulphuric acid until neutral; the same procedure was repeated for the blank sample. The CN ratio was calculated by dividing the TOC content with the total nitrogen content. All the analyses were conducted on samples collected using a standard sampling method, except for temperature measurements which were performed in situ.

7.2.5 STATISTICAL ANALYSES

All samples were analysed in triplicates, except for the temperature measurements, which were repeated four times. The averages and standard deviation of the means as the error bars were calculated using Microsoft Excel 2007 (Vista Edition, Microsoft Corporation, USA). Statistical analyses were made using statistical analysis software (SAS 9.2, SAS Institute, Inc., USA). The data were subjected to one-way analysis of variance (ANOVA), generalised linear model (GLM) and Duncan's multiple range tests (DMRT) at alpha level, $\alpha=0.05$. ANOVA was implemented to all parameters, except for those data with unbalanced sample size and missing data using the GLM statistical analyses. Where significant differences were obtained giving $p<0.05$, individual means were tested using the DMRT to compare the significant difference between the two treatments means, i.e. newspaper blend with the onion peels blend, 2 kg with 6 kg of load size and starter culture with control.

7.3 RESULTS AND DISCUSSION

In general, the MC of the compost increased with time due to vaporisation and the trapped moisture loss inside the composter. The values were still acceptable as long as the air supply in the compost mass is sufficient in keeping the microbes alive (Unmar and Mohee 2008). The pH of the waste mixture for all treatments increased up to 9 from acidic conditions through the 30-days of composting, while the electrical conductivity (EC) increased and was in the range of 3 to 6 dS/m. The EC was low at the thermophilic temperature because a high amount of nutrients that were ingested by the microorganisms made the nutrients insoluble in water and consequently produced a low salinity. With colour measurements, the blend with onion peels became black more rapidly than the blend with newspaper ($p<0.0001$), showing the colour of matured and stable compost. The waste compost colour changes to black in the 2- and 6-kg load sizes, and both mixtures with and without starter culture were not significantly different with $p=0.2031$ and $p=0.8383$, respectively. There were no clear trends in the all microbial numbers due to a diverse microbial community in the organic waste, which comply with the reports by Sundberg et al. (2004) who believed that the temperatures affected the active microbial numbers.

7.3.1 TEMPERATURE PROFILES

Figure 2 illustrates the temperature profiles showing typical phases of a composting process for investigations using bulking agents of newspaper and onion peels, investigations of load sizes at 2 and 6 kg, and use of starter culture during a 30-day kitchen-waste composting process. The temperature profiles displayed quite similar patterns for all composting batches where it increased at the early stage before decreasing gradually. For the first 24 h in Figure 2a, the compost with newspaper was self-heated from 26.7°C to about 39.2°C, while the blend with the onion peels was self-heated from 30.3°C to about 44.9°C. The temperature change in onion peels blend during composting followed a typical pattern displayed by many other composting system, i.e. organic fraction of municipal solid wastes, raw sludge, anaerobically digested sludge, animal by-products from slaughterhouses and partially hydrolysed hair from the leather industry (Pagans et al. 2006), co-composting of exhausted olive-cake with poultry manure and sesame shells (Sellami et al. 2008), green leaves, green branches, grass, brown branches and brown leaves Unmar and Mohee (2008), and co-composting of green tea waste and rice bran (Khan et al. 2009). The newspapers, however, did not assist the compost to achieve thermophilic temperature of 45°C. The temperature of the compost with the onion peels rose to 51.3°C on day 4 and dropped sharply to 44.1°C on day 7. It stayed in the thermophilic phase for about 3 days. The temperature decrease is probably due to the excessive loss of compost volume, and when the compost moves into the cool phase, it produces stable compost. The average temperature of the two composts with different bulking agents, newspaper and onion peels was significantly different at $p<0.0001$ when compared with ambient temperatures. The DMRT resulted greater mean of temperature in compost with onion peels, followed by compost with newspaper and ambient temperature.

Both compost load sizes at 2 and 6 kg using onion peels as shown in Figure 2b present a quite similar temperature trend with compost using food waste, manure and bulking agent mixture by Cekmecelioglu et al. (2005). The composting started with mesophilic temperature, continued to thermophilic temperature and then drop to ambient temperature. During the first 24 h of composting, the temperature of the waste rose 17.5°C for the 2-kg load and 8°C for the 6 kg. The 6-kg load contained more carbon to be degraded compared to the 2 kg; thus, it was observed that the 2-kg load has entered the mesophilic stage 10 days earlier than the 6 kg, on day 6 and 16, respectively. The temperature fluctuations observed during the composting process were because of the turning and watering activities. The 6-kg load size has a higher mean of temperature followed by 2-kg load size and ambient temperature.

Figure 2c shows that the temperatures of the composts with added starter culture and control increased quickly to thermophilic temperatures of 53°C and 51.1°C, respectively, on day 3 of composting due to the compost load size. The presence of an active microbial community also helps start the degradation process immediately and reduces the particle size (Sundberg and Jönsson 2005). The thermophilic phase of both composts lasted for a long period of 17 days, probably due to the availability of easily degradable organic matter, energetic compounds (like protein) and large amounts of organic nitrogen in the waste (Pagans et al. 2006). With a large amount of waste, the high thermophilic temperatures worked on breaking down the proteins, fats and complex carbohydrates like cellulose and hemicelluloses. The compost with added starter culture achieved the highest temperature of 54.3°C on day 8, which is slightly higher and one day earlier than the control (53.8°C on day 9). The temperature of the mixtures began to decline on day 8 for mixture with starter culture and day 9 for control, probably due to convective loss (Palmisano et al. 1993) and a higher amount of readily degradable carbon. At the end of the process, heat was released progressively, causing the temperature for both mixtures to decrease and tended to meet the ambient temperature. This tendency was found with composting fish offal in reactors (Laos et al. 2002) and composting of green tea waste and rice bran (Khan et al. 2009), which implied that the rapidly degradable organic matter had been reduced (Sundberg and Jönsson 2005). The average temperatures of the compost with added starter culture and the control were not significantly different from the DMRT results.

7.3.2 DEGRADATION RATE PROGRESSION

Figure 3 illustrates the decreasing trends of volatile solids content during the 30-day composting period. All three figures were comparable with the findings of Cekmecelioglu et al. (2005) in terms of volatile solids content. Unmar and Mohee (2008) considered that the volatile solids are a good indicator of how biological degradation occurred over time. Figure 3a shows that the initial volatile solids contents for both blends (onion peels and newspaper) were the same at 85%. The final volatile solids in compost with onion peels then reduced to 60.9% compared to the newspaper at 76.1%, indicating that onion peel compost had a higher degradation as it contained higher amounts of biodegradable matter. The calculated percentage of volatile solids loss for blend with newspaper was only 9.98% compared to 28.3% for blend with onion peels. The volatile solids of both blends were significantly different at $p=0.0004$, with onion peels having lower means compared to the newspaper. Figure 3b presents that the percentage of volatile solids loss for the 2 kg was higher than that of the 6 kg, at 24.5% and 21.4%, respectively. These findings were lower than those of Cekmecelioglu et al. (2005) who discovered that the percent volatile solids loss of food waste in various windrow systems was in the range of 32.6% to 52.6%. The means of volatile solids content of the two load sizes were significantly different at $p=0.0427$, where the 2-kg load size contained lower volatile solids than the 6 kg. With the same volatile solids contents of 87%, the compost with added starter culture attained a higher volatile solids loss of 22.5% compared to the control at 21.7% (Figure 3c). The starter culture could have been involved in degrading the materials slightly faster, although the average values between the two types of composts were not significantly different at $p=0.4511$. This is because the starter culture serves as an inoculum, increasing bacterial density and perhaps reducing lag phase duration.

7.3.3 COMPOST MATURITY EVALUATION

The CN ratio is an accurate indication of compost maturity (Gray et al. 1971; Chanyasak and Kubota 1981; Jimenez and Garcia 1989). In these kitchen-waste composting studies, the TOC content in the compost material reduced arbitrarily higher than the reduction of total nitrogen contents resulting in a corresponding reduction of CN ratio as composting proceeds. The carbon provides the primary energy source for microbial metabolism, and nitrogen is critical for microbial population growth (Stoffella and Kahn 2001).

Figure 4a shows that the TOC content of the onion peel blend decreased to 28%, while the newspaper blend was at 10% over the 30-day period. The reduction of TOC is mainly because of the mineralisation process (Grigatti et al. 2004), a process where microbes employ organic matter and leave behind inorganic substances such as minerals, carbon dioxide and water. The blend with onion peels was prone to a higher mineralisation than the blend with newspaper because newspaper contains cellulose fibres sheathed in lignin, which is a compound found in wood with highly resistant to biological degradation (Trautmann et al. 1996). In addition, the thermophilic bacteria ($p<0.0001$) and the thermophilic fungi ($p=0.0002$) were governing the blend with onion peels, as they help accelerate the composting. In the blend with newspaper, the mesophilic fungi ($p<0.0001$) was dominating the 30-day period of composting, which indicates that the mineralisation is low and takes a longer time for composting. The nitrogen content for the newspaper blend fluctuated at the range of 0.8% to 1.3%, while the onion peels increased to 4.2%. Both average carbon and nitrogen contents were significantly different with $p=0.0004$ and $p<0.0001$, respectively, with DMRT results showing lower carbon content and higher nitrogen content for the blend with onion peels. The resulting CN ratio in Figure 4c shows that the blend with newspaper underwent a slow composting process with its CN ratio value reduced to 37.04 compared to the blend with onion peels with a CN ratio of 8.15 after 30 days. Benito et al. (2006) stated that a CN ratio between 14 and 24 would be ideal for ready-to-use compost. For these reasons, the final products compost with onion peels were ready to be used on day 8 due to its CN ratio of 16.2, but the blend with newspaper needed more time for decomposition activity. The means of CN ratio for both blends were significantly different ($p<0.0001$), with the onion peel blend having lower means of CN ratio than the newspaper. The initial increase in CN ratio of the newspaper blend could be due to the incomplete degradation process, and it did not reach thermophilic temperatures.

For load size studies, the 2-kg compost had lower TOC content and higher nitrogen content than the 6 kg (Figure 5a,b). The CN ratio of the 2-kg load increased slightly after 24 h before dropping rapidly to 10.94 on day 10 as compared to the 6 kg which required 30 days to reach a CN ratio of 10.15 (Figure 5c). The 2-kg load size decomposed more quickly compared to the 6-kg load size. The final CN ratio of the 2- and 6-kg loads was 9.22 and 10.15, respectively. They were considered matured and suitable to be applied to soil based on the study of Shiralipour et al. (1992) who have identified that if the CN ratio is less than 30, it will help crop production with its microbial mobilisation. Molnar and Bartha (1988) believed that a stable and high quality compost has a CN ratio in the range of 15 to 17. Hence, it was suggested that the end product of the 2-kg load size can be collected on day 8, half the time of the 6-kg load size on day 16. The means of the TOC content for the 2- and 6-kg load sizes were significantly different with $p=0.0427$. The means of the total nitrogen content and CN ratio between the 2- and 6-kg load sizes, however, were not significantly different at $p=0.0533$ and $p=0.0701$, respectively.

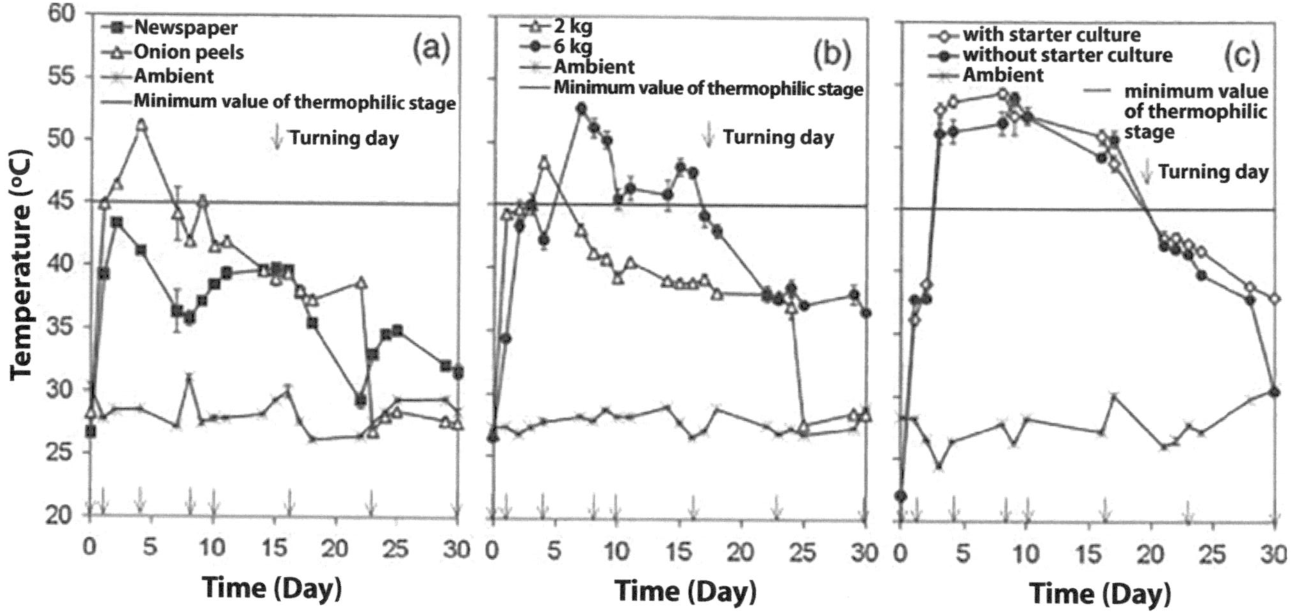

FIGURE 2: Temperature profiles of the compost. (a) With blends of newspaper or onion peels at 2-kg load size, (b) at 2- or 6-kg load size both with onion peels, and (c) with or without starter culture both at 6-kg load size using onion peels.

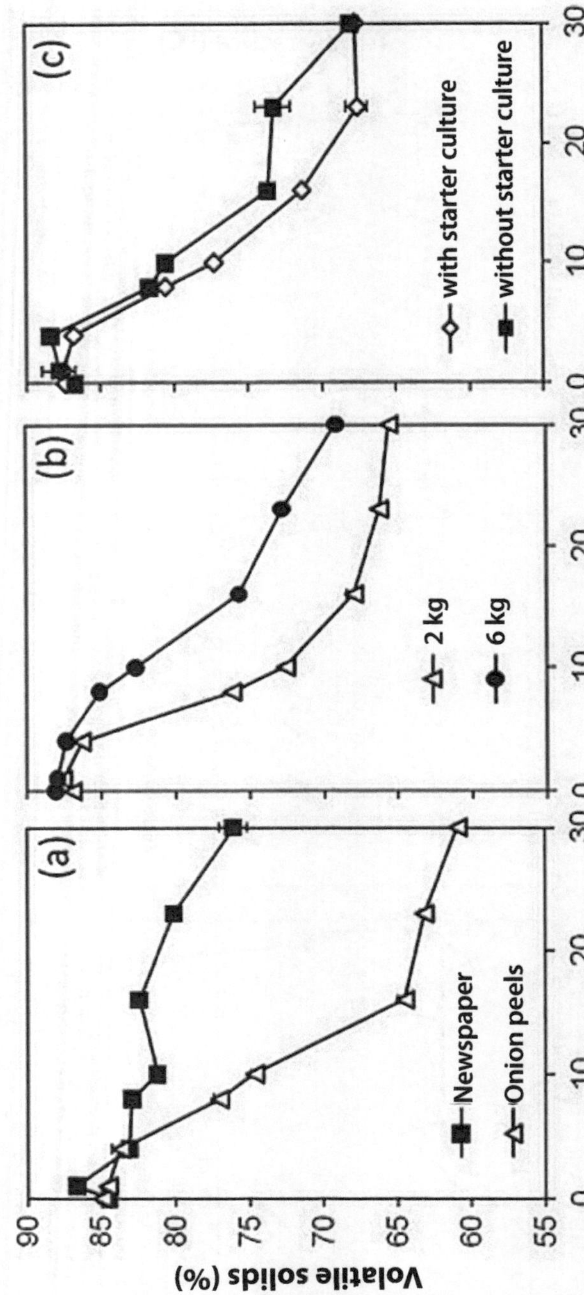

FIGURE 3: Volatile solids of compost. (a) With blends of newspaper or onion peels at 2-kg load size, (b) at 2- or 6-kg load size both with onion peels, and (c) with or without starter culture both at 6-kg load size using onion peels.

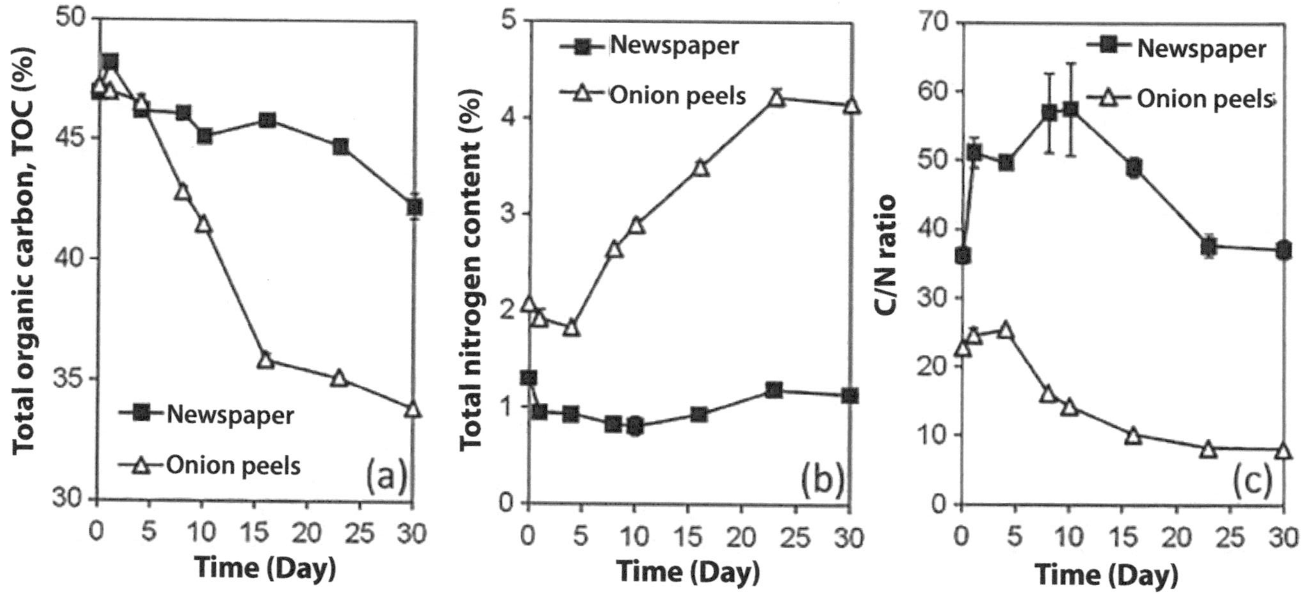

FIGURE 4: Composting using different blends of bulking media, newspaper or onion peels at 2-kg load size. (a) TOC, (b) nitrogen content and (c) CN ratio profiles.

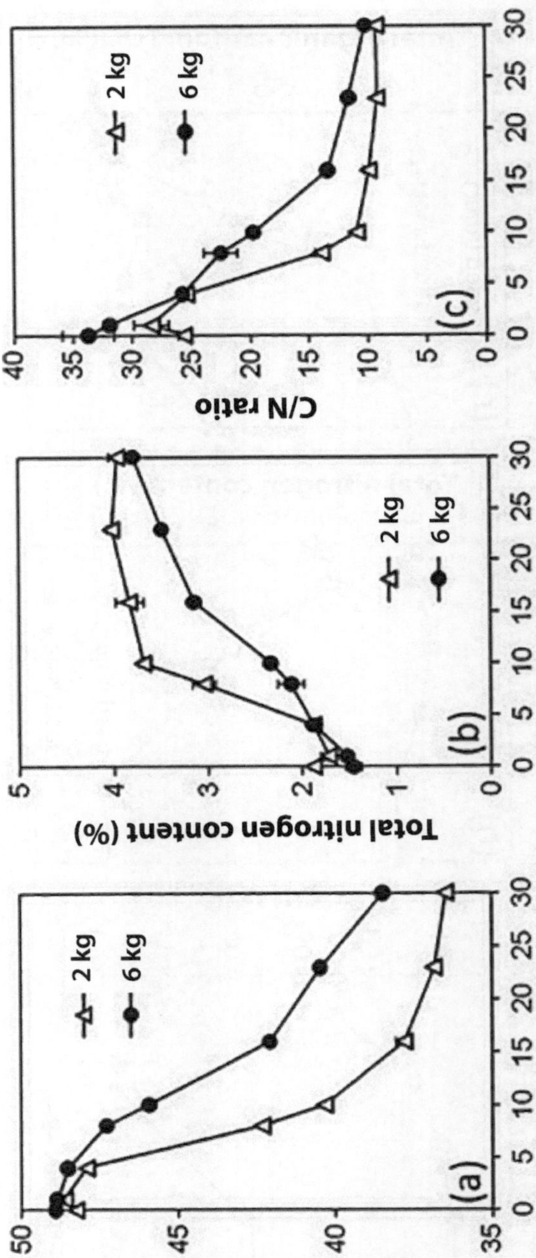

FIGURE 5: Composting using different load size of 2 or 6 kg with onion peels. (a) TOC, (b) nitrogen content and (c) CN ratio profiles.

Figure 6 shows that the mixture with added starter culture and control have no significant difference in terms of TOC content ($p = 0.4512$), total nitrogen content ($p = 0.9183$) and CN ratio ($p = 0.8158$). The starter culture has just a slight influence on the reduction of TOC due to the presence of microbial growth, although thermophilic temperatures do increase the level of carbonaceous material (Ginnivan et al. 1981). The microbial growth caused a carbon substrate limited condition even though the thermophilic temperatures could increase the level of carbon. The compost with added starter culture had lower nitrogen contents before day 10 and higher nitrogen contents after day 10 when compared to the control. The presence of microbes could have accelerated the decomposition activities in the beginning of the composting process. At the early composting stage, the mesophilic bacteria dominated the control up to 9 log colony-forming unit (CFU)/g, while the thermophilic bacteria dominated the mixture with added starter culture up to 13 log CFU/g. Both the TOC and nitrogen contents of the compost with added culture and control reached similar values at the end of composting. The decrease in CN ratio generally is explained by the bigger loss of TOC to produce carbon dioxide and smaller increase of nitrogen content as the decomposition progressed. The decreasing pattern of the CN ratio and TOC, and the increasing pattern of nitrogen content were similar to the findings of Goyal et al. (2005) on sugarcane trash, cattle dung, press mud, poultry waste and water hyacinth composting. The slight nitrogen losses of the control on day 10 which caused the slight CN ratio increase was due to nitrogen losses in the form of ammonia. This interpretation is consistent with Goyal et al. (2005) who found that nitrogen losses during ammonia volatilisation caused the increase of the CN ratio. The compost with added starter culture and control both possessed quite similar CN ratio of 9.33 and 9.35, respectively, which are less than 12 and in the range of acceptable degree of maturation (Bernal et al. 1998; Paredes et al. 2005). The optimum CN ratio of 10 has been suggested by Pöpel and Ohnmacht (1972) for the complete oxidation of the waste by aerobic bacteria.

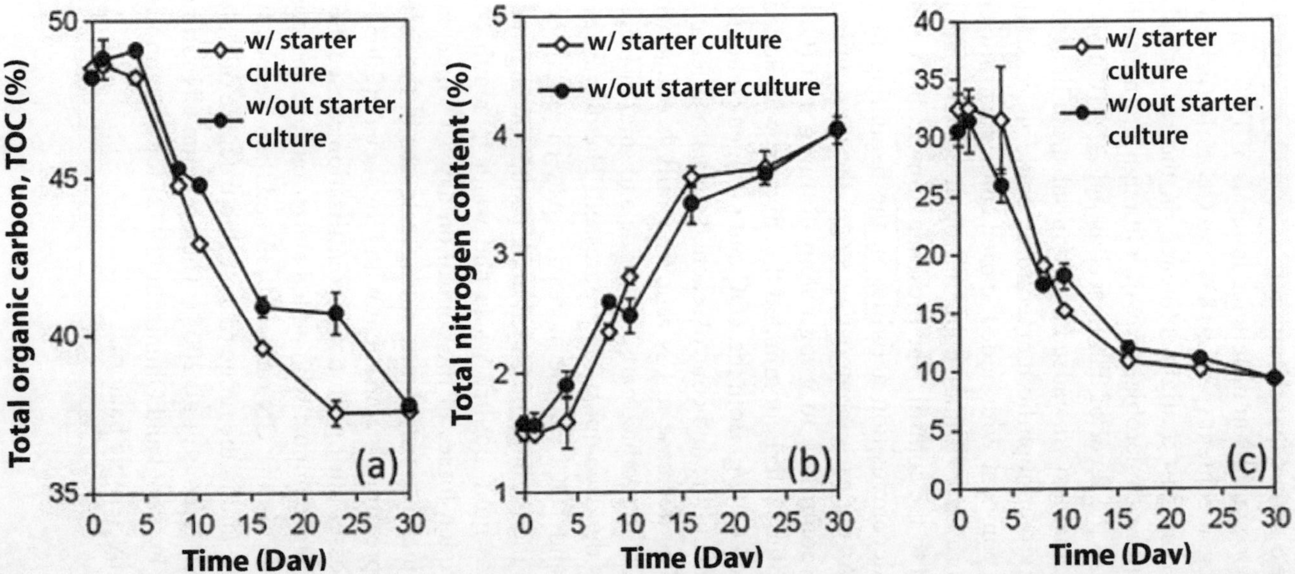

FIGURE 6: Composting using microbial treatments at 6-kg load size using onion peels. (a) TOC, (b) nitrogen content and (c) CN ratio profiles.

7.4 CONCLUSIONS

The onion peels were more suitable; at a smaller waste load, compost maturity with CN ratio below 10 was attained more quickly. The larger compost load size plays a vital role to achieve higher temperature. However, the lack of oxygen may have influenced the loss of volatile solids, and the lack of ventilation certainly influenced the loss of nitrogen by ammonia volatilisation when the maximum volume for the reactor, which is 6-kg load size, was used. No apparent differences were found in compost with added starter culture. The kitchen-waste composting process is said to require no additional microbes as accelerant, although it did help to achieve a slightly higher thermophilic temperature during the early stage of composting.

REFERENCES

1. Abdullah N, Chin NL (2010) Simplex-centroid mixture formulation for optimised composting of kitchen waste. Bioresour Technol 101(21):8205-8210
2. Belyaeva ON, Haynes RJ (2009) Chemical, microbial and physical properties of manufactured soils produced by co-composting municipal green waste with coal fly ash. Bioresour Technol 100(21):5203-5209
3. Benito M, Masaguer A, Moliner A, De Antonio R (2006) Chemical and physical properties of pruning waste compost and their seasonal variability. Bioresour Technol 97:2071-2076
4. Bernal PM, Navarro AF, Sánchez-Monedero MA, Roig A, Cegarra J (1998) Influence of sewage sludge compost stability and maturity on carbon and nitrogen mineralization in soil. Soil Biol Biochem 30(3):305-313
5. Bhatt R, Kausadikar H (2010) Definition of soil microbiology and soil in view of microbiology. http://www.scribd.com/doc/27429437/Soil-Microbiology . Accessed 18 Feb 2012
6. Cayuela ML, Sánchez-Monedero MA, Roig A (2006) Evaluation of two different aeration systems for composting two-phase olive mill wastes. Process Biochem 41(3):616-623
7. Cekmecelioglu D, Demirci A, Graves RE, Davitt NH (2005) Applicability of optimised in-vessel food waste composting for windrow systems. Biosyst Eng 91(4):479-486
8. Chang JI, Hsu TE (2008) Effects of compositions on food waste composting. Bioresour Technol 99(17):8068-8074
9. Chanyasak V, Kubota H (1981) Carbon/organic nitrogen ratio in water extract as measure of compost degradation. J Ferment Technol 59(3):215-219

10. Chikae M, Ikeda R, Kerman K, Morita Y, Tamiya E (2006) Estimation of maturity of compost from food wastes and agro-residues by multiple regression analysis. Bioresour Technol 97(16):1979-1985
11. Fang M, Wong MH, Wong JWC (2001) Digestion activity of thermophilic bacteria isolated from ash-amended sewage sludge compost. Water Air Soil Pollut 126(1–2):1-12
12. Ginnivan MJ, Woods JL, O'Callaghan JR (1981) Thermophilic aerobic treatment of pig slurry. J Agric Eng Res 26(6):455-466
13. Goyal S, Dhull SK, Kapoor KK (2005) Chemical and biological changes during composting of different organic wastes and assessment of compost maturity. Bioresour Technol 96(14):1584-1591
14. Gray KR, Sherman K, Biddlestone AJ (1971) A review of composting. Part I. Microbiology and biochemistry. Process Biochem 6:32-36
15. Grigatti M, Ciavatta C, Gessa C (2004) Evolution of organic matter from sewage sludge and garden trimming during composting. Bioresour Technol 91(2):163-169
16. Huang DL, Zeng GM, Feng CL, Hu S, Lai C, Zhao MH, Su FF, Tang L, Liu HL (2010) Changes of microbial population structure related to lignin degradation during lignocellulosic waste composting. Bioresour Technol 101(11):4062-4067
17. Ishii K, Takii S (2003) Comparison of microbial communities in four different composting processes as evaluated by denaturing gradient gel electrophoresis analysis. J Appl Microb 95:109-119
18. Jimenez EI, Garcia VP (1989) Evaluation of city refuse compost maturity: a review. Biol Wastes 27:115-142
19. Kalamdhad AS, Pasha M, Kazmi AA (2008) Stability evaluation of compost by respiration techniques in a rotary drum composter. Res Cons Recycl 52(5):829-834
20. Kamil NK, Alwi SA, Singh M (2010) Malaysia. Malaysia. http://www.readbag.com/avrdc-pdf-dynamics-malaysia . Accessed 18 Feb 2012
21. Khan MAI, Ueno K, Horimoto S, Komai F, Tanaka K, Ono Y (2009) Physicochemical, including spectroscopic, and biological analyses during composting of green tea waste and rice bran. Biol Fert Soils 45(3):305-313
22. Laos F, Mazzarino MJ, Walter I, Roselli L, Satti P, Moyano S (2002) Composting of fish offal and biosolids in northwestern Patagonia. Bioresour Technol 81(3):179-186
23. Lubberding HJ, Gijzen HJ, Heck M, Vogels GD (1988) Anaerobic digestion of onion waste by means of rumen microorganisms. Biol Wastes 25(1):61-67
24. Luostarinen S, Rintala J (2007) Anaerobic on-site treatment of kitchen waste in combination with black water in UASB-septic tanks at low temperatures. Bioresour Technol 98:1734-1740
25. Mohee R, Driver M-FB, Sobratee N (2008) Transformation of spent broiler litter from exogenous matter to compost in a sub-tropical context. Bioresour Technol 99(1):128-136
26. Molnar L, Bartha I (1988) High solids anaerobic fermentation for biogas and compost production. Biomass 16(3):173-182
27. Naik VN, Sharma DD, Kumar PMP, Yadav RD (2012) Efficacy of ligno-cellulolytic fungi on recycling sericultural wastes. Acta Biologica Indica 1(1):47-50

28. Pagans E, Barrena R, Font X, Sánchez A (2006) Ammonia emissions from the composting of different organic wastes. Dependency on process temperature. Chemosphere 62(9):1534-1542
29. Palmisano AC, Maruscik DA, Ritchie CJ, Schwab BS, Harper SR, Rapoport RA (1993) A novel bioreactor simulating composting of municipal solid waste. J Microbiol Meth 18(2):99-112
30. Paredes C, Cegarra J, Bernal MP, Roig A (2005) Influence of olive mill wastewater in composting and impact of the compost on a Swiss chard crop and soil properties. Environ Int 31(2):305-312
31. Pöpel F, Ohnmacht C (1972) Thermophilic bacterial oxidation of highly concentrated substrates. Water Res 6(7):807-815
32. Raut MP, Prince William SPM, Bhattacharyya JK, Chakrabarti T, Devotta S (2008) Microbial dynamics and enzyme activities during rapid composting of municipal solid waste—a compost maturity analysis perspective. Bioresour Technol 99(14):6512-6519
33. Rogoshewski P, Bryson H, Wagner K (1983) Remedial action technology for waste disposal sites. Park Ridge, New Jersey: Noyes Data Corporation.
34. Sellami F, Jarboui R, Hachicha S, Medhioub K, Ammar E (2008) Co-composting of oil exhausted olive-cake, poultry manure and industrial residues of agro-food activity for soil amendment. Bioresour Technol 99(5):1177-1188
35. Shiralipour A, McConnell DB, Smith WH (1992) Physical and chemical properties of soils as affected by municipal solid waste compost application. Biomass Bioenergy 3(3–4):261-266
36. Stabnikova O, Ding H-B, Tay J-H, Wang J-Y (2005) Biotechnology for aerobic conversion of food waste into organic fertilizer. Waste Manage Res 23:39-47
37. Stoffella PJ, Kahn BA (2001) Compost utilization in horticultural cropping systems. Boca Raton: Lewis Publishers.
38. Sun Y, Cheng J (2002) Hydrolysis of lignocellulosic materials for ethanol production: a review. Bioresour Technol 83(1):1-11
39. Sundberg C, Jönsson H (2005) Process inhibition due to organic acids in fed-batch composting of food waste—influence of starting culture. Biodegradation 16:205-213
40. Sundberg C, Smårs S, Jönsson H (2004) Low pH as an inhibiting factor in the transition from mesophilic to thermophilic phase in composting. Bioresour Technol 95(2):145-150
41. Suutarinen M, Mustranta A, Autio K, Salmenkallio-Marttila M, Ahvenainen R, Buchert J (2003) The potential of enzymatic peeling of vegetables. J Sci Food Agric 83(15):1556-1564
42. Trautmann NM, Richard T, Krasny ME (1996) Cornell composting: science and engineering. Compost chemistry. http://compost.css.cornell.edu/chemistry.html . Accessed 18 Feb 2012
43. Unmar G, Mohee R (2008) Assessing the effect of biodegradable and degradable plastics on the composting of green wastes and compost quality. Bioresour Technol 99(15):6738-6744
44. Veeken A, Hamelers B (1999) Effect of temperature on hydrolysis rates of selected biowaste components. Bioresour Technol 69:249-254

45. Volchatova IV, Belovezhets LA, Medvedeva SA (2002) Microbiological and biochemical investigation of succession in lignin-containing compost piles. Microbiology 71(4):467-470
46. Wang JY, Stabnikova O, Ivanov V, Tay STL, Tay JH (2003) Intensive aerobic bioconversion of sewage sludge and food waste into fertiliser. Waste Manage Res 21(5):405-415
47. Wang Q, Yamabe K, Narita J, Morishita M, Ohsumi Y, Kusano K, Shirai Y, Ogawa HI (2001) Suppression of growth of putrefactive and food poisoning bacteria by lactic acid fermentation of kitchen waste. Process Biochem 37:351-357
48. Wayman M, Chen S, Doan K (1992) Bioconversion of waste paper to ethanol. Process Biochem 27(4):239-245

CHAPTER 8

Microbial Diversity of Vermicompost Bacteria that Exhibit Useful Agricultural Traits and Waste Management Potential

JAYAKUMAR PATHMA AND NATARAJAN SAKTHIVEL

8.1 INTRODUCTION

Soil, is the soul of infinite life that promotes diverse microflora. Soil bacteria viz., *Bacillus*, *Pseudomonas* and *Streptomyces* etc., are prolific producers of secondary metabolites which act against numerous co-existing phytopathogeic fungi and human pathogenic bacteria (Pathma et al. 2011b). Earthworms are popularly known as the "farmer's friend" or "nature's plowman". Earthworm influences microbial community, physical and chemical properties of soil. They breakdown large soil particles and leaf litter and thereby increase the availability of organic matter for microbial degradation and transforms organic wastes into valuable vermicomposts by grinding and digesting them with the help of aerobic and anaerobic microbes (Maboeta and Van Rens-

Microbial Diversity of Vermicompost Bacteria that Exhibit Useful Agricultural Traits and Waste Management Potential. © Pathma J and Sakthivel N. SpringerPlus *1,26 (2012), doi:10.1186/2193-1801-1-26. Licensed under Creative Commons Attribution 2.0 Generic License, http://creativecommons.org/licenses/by/2.0/.*

burg 2003). Earthworms activity is found to enhance the beneficial microflora and suppress harmful pathogenic microbes. Soil wormcasts are rich source of micro and macro-nutrients, and microbial enzymes (Lavelle and Martin 1992). Vermicomposting is an efficient nutrient recycling process that involves harnessing earthworms as versatile natural bioreactors for organic matter decomposition. Due to richness in nutrient availability and microbial activity vermicomposts increase soil fertility, enhance plant growth and suppress the population of plant pathogens and pests. This review paper describes the bacterial biodiversity and nutrient status of vermicomposts and their importance in agriculture and waste management.

8.1.1 EARTHWORMS

Earthworms are capable of transforming garbage into 'gold'. Charles Darwin described earthworms as the 'unheralded soldiers of mankind', and Aristotle called them as the 'intestine of earth', as they could digest a wide variety of organic materials (Darwin and Seward 1903; Martin 1976). Soil volume, microflora and fauna influenced by earthworms have been termed as "drilosphere" and the soil volume includes the external structures produced by earthworms such as surface and below ground casts, burrows, middens, diapause chambers as well as the earthworms body surface and internal gut associated structures in contact with the soil (Lavelle et al. 1989; Brown et al. 2000). Earthworms play an essential role in carbon turnover, soil formation, participates in cellulose degradation and humus accumulation. Earthworm activity profoundly affects the physical, chemical and biological properties of soil. Earthworms are voracious feeders of organic wastes and they utilize only a small portion of these wastes for their growth and excrete a large proportion of wastes consumed in a half digested form (Edwards and Lofty 1977; Kale and Bano 1986; Jambhekar 1992). Earthworms intestine contains a wide range of microorganisms, enzymes and hormones which aid in rapid decomposition of half-digested material transforming them into vermicompost in a short time (neary 4–8 weeks) (Ghosh et al. 1999; Nagavallemma et al. 2004) compared to traditional composting process which takes the advantage of microbes alone and thereby requires a prolonged period (nearly 20 weeks) for compost production (Bernal et al. 1998; Sánchez-Monedero et al. 2001). As the organic matter passes through the gizzard of the earthworm it is grounded into a fine powder after which the digestive enzymes, microorganisms and other fermenting substances act on them further aiding their breakdown within the gut, and finally passes out in the form of "casts" which are later acted upon by earthworm gut associated microbes converting them into mature product, the "vermicomposts" (Dominguez and Edwards 2004).

Earthworms, grouped under phylum annelida are long, narrow, cylindrical, bilaterally symmetrical, segmented soil dwelling invertebrates with a glistening dark brown body covered with delicate cuticle. They are hermaphrodites and weigh over 1,400–1,500 mg after 8–10 weeks. Their body contains 65% protein (70–80% high quality 'lysine rich protein' on a dry weight basis), 14% fats, 14% carbohydrates, and 3% ash. Their life span varies between 3–7 years depending upon the species and ecological situation. The gut of earthworm is a straight tube starting from mouth followed by a muscular pharynx, oesophagus, thin walled crop, muscular gizzard, foregut, midgut, hindgut, associated digestive glands, and ending with anus. The gut consisted of mucus containing protein and polysaccharides, organic and mineral matter, amino acids and microbial symbionts viz., bacteria, protozoa and microfungi. The increased organic carbon, total organic carbon and nitrogen and moisture content in the earthworm gut provide an optimal environment for the activation of dormant microbes and germination of endospores etc. A wide array of digestive enzymes such as amylase, cellulase, protease, lipase, chitinase and urease were reported from earthworm's alimentary canal. The gut microbes were found to be responsible for the cellulase and mannose activities (Munnoli et al. 2010). Earthworms comminutes the substrate, thereby increases the surface area for microbial degradation constituting to the active phase of vermicomposting. As this crushed organic matter passes through the gut it get mixed up with the gut associated microbes and the digestive enzymes and finally leaves the gut in partially digested form as "casts" after which the microbes takes up the process of decomposition contributing to the maturation phase (Lazcano et al. 2008).

Association of earthworms with microbes is found to be complex. Certain groups of microbes were found to be a part of earthworm's diet which is evidenced by the destruction of certain microbes as they pass through the earthworms digestive system. Few yeasts, protozoa and certain groups of fungi such as *Fusarium oxysporum*, *Alternaria solani*, and microfungi were digested by the earthworms, *Drawida calebi*, *Lumbricus terrestris* and *Eisenia foetida*. *Bacillus cereus* var *mycoides* were reported to decrease during gut passage while *Escherichia coli* and *Serratia marcessens* were completely eliminated during passage through earthworm gut (Edwards and Fletcher, 1988).

TABLE 1: Ecological categories and niches of earthworms and their characteristic features and beneficial traits

Species	Ecological category	Ecological niche	Characteristic features	Beneficial trait
Eisenia foetida,	Epigeics	Superficial soil layers, leaf litter, compost	Smaller in size, body uniformly pigmented, active gizzard, short life cycle, high reproduction rate and regeneration, tolerant to disturbance, phytophagous	Efficient bio-degraders and nutrient releasers, efficient compost producers, aids in litter comminution and early decomposition
Lumbricus rubellus,				
L. castaneus,				
L. festivus,				
Eiseniella tetraedra,				
Bimastus minusculus,				
B. eiseni,				
Dendrodrilus rubidus,				
Dendrobaena veneta,				
D. octaedra				
Aporrectodea caliginosa,	Endogeics	Topsoil or subsoil	Small to large sized worms, weakly pigmented, life cycle of medium duration, moderately tolerant to disturbance, geophagous	Brings about pronounced changes in soil physical structure, can efficiently utilize energy from poor soils hence can be used for soil improvements

TABLE 1: *Cont.*

Species	Ecological category	Ecological niche	Characteristic features	Beneficial trait
A. trapezoides,				
A. rosea,				
Millsonia anomala,				
Octolasion cyaneum,	Polyhumic endogeic	Top soil (A1)	Small size, unpigmented, forms horizontal burrows, rich soil feeder	
O. lacteum,				
Pontoscolex corethrurus,	Meso-humic endogeic	A and B horizon	Medium size, unpigmented, forms extensive horizontal burrows, bulk (A1) soil feeder	
Allolobophora chlorotica,				
Aminthas sp.	Oligo-humic endogeic	B and C horizon	Very large in size, unpigmented, forms extensive horizontal burrows, feeds on poor, deep soils	
L. terrestris,	Anecics	Permanent deep burrows in soil	Large in size, dorsally pigmented, forms extensive, deep, vertical permanent burrows, low reproductive rate, sensitive to disturbance, phytogeophagous, nocturnal	Forms vertical burrows affecting air-water relationship and movement from deep layers to surface helps in efficient mixing of nutrients
L. polyphemus,				
A. longa				

Earthworms are classified into epigeic, anecic and endogeic species based on definite ecological and trophic functions (Brown 1995; Bhatnagar and Palta 1996) (Table 1). Epigeic earthworms are smaller in size, with uniformly pigmented body, short life cycle, high reproduction rate and regeneration. They dwell in superficial soil surface within litters, feeds on the surface litter and mineralize them. They are phytophagous and rarely ingest soil. They contain an active gizzard which aids in rapid conversion of organic matter into vermicomposts. In addition epigeic earthworms are efficient bio-degraders and nutrient releasers, tolerant to disturbances, aids in litter comminution and early decomposition and hence can be efficiently used for vermicomposting. Epigeic earthworms includes *Eisenia foetida, Lumbricus rubellus, L. castaneus, L. festivus, Eiseniella tetraedra, Bimastus minusculus, B. eiseni, Dendrodrilus rubidus, Dendrobaena veneta, D. octaedra.* Endogeics earthworms are small to large sized worms, with weakly pigmented body, life cycle of medium duration, moderately tolerant to disturbance, forms extensive horizontal burrows and they are geophagous feeding on particulate organic matter and soil. They bring about pronounced changes in soil physical structure and can efficiently utilize energy from poor soils, hence can be used for soil improvements. Endogeics include *Aporrectodea caliginosa, A. trapezoides, A. rosea, Millsonia anomala, Octolasion cyaneum, O. lacteum, Pontoscolex corethrurus, Allolobophora chlorotica* and *Aminthas* sp. They are further classified into polyhumic endogeic which are small sized, rich soil feeding earthworms, dwelling in top soil (A1); mesohumic endogeic which are medium sized worms, dwelling in A and B horizon, feeding on bulk (A1) soil; and oligohumic endogeic which are very large worms, dwelling in B and C horizons, feeding on poor, deep soil. Aneceics are larger, dorsally pigmented worms, with low reproductive rate, sensitive to disturbance, nocturnal, phytogeophagous, bury the surface litter, forms middens and extensive, deep, permanent vertical burrows, and live in them. Formation of vertical burrows affects airwater relationship and movement from deep layers to surface helps in efficient mixing of nutrients. *Lumbricus terrestris, L. polyphemus* and *Aporrectodea longa* are examples of aneceics earthworms (Kooch and Jalilvand 2008). Epigeics and aneceics are harnessed largely for vermicomposting (Asha et al. 2008). Epigeics namely *Eisenia foetida* (Hartenstein et al. 1979), *Eudrilus eugeniae* (Kale and Bano 1988), *Perionyx excavatus* (Sinha et al. 2002; Suthar and Singh 2008) and *Eisenia anderi* (Munnoli et al. 2010) have been used in converting organic wastes into vermicompost. These surface dwellers capable of working on litter layers converting them into manure are of no significant value in modifying the soil structure. In contrast, anecics such as *Lampito mauritii* are efficient creators of an effective drilosphere as well as excellent compost producers (Ismail 1997). Earthworms thus act as natural bio-reactors, altering the nature of the organic waste by fragmenting them.

Earthworm activity engineers the soil by forming extensive burrows which loosen the soil and makes it porous. These pores improve aeration, water absorption, drainage and easy root penetration. Soil aggregates formed by earthworms and associated microbes, in the casts and burrow walls play an indispensible role in soil air ecosystem. These aggregates are mineral granules bonded in a way to resist erosion and to avoid soil compaction both in wet and dry condition. Earthworms speed up soil reclamation and make them productive by restoring beneficial microflora (Nakamura 1996). Thus degraded unproductive soils and land degraded by mining could be engineered physically, chemically and biologically and made productive by earthworms. Hence earthworms are termed as ecosystem engineers (Brown et al. 2000; Munnoli et al. 2010).

8.1.2 VERMICOMPOSTING

Vermicomposting is a non-thermophilic biological oxidation process in which organic material are converted into vermicompost which is a peat like material, exhibiting high porosity, aeration, drainage, water holding capacity and rich microbial activities (Edwards 1998; Atiyeh et al. 2000b; Arancon et al. 2004a), through the interactions between earthworms and associated microbes. Vermiculture is a cost-effective tool for environmentally sound waste management (Banu et al. 2001; Asha et al. 2008). Earthworms are the crucial drivers of the process, as they aerate, condition and fragment the substrate and thereby drastically alter the microbial activity and their biodegradation potential (Fracchia et al. 2006; Lazcano et al. 2008). Several enzymes, intestinal mucus and antibiotics in earthworm's intestinal tract play an important role in the breakdown of organic macromolecules. Biodegradable organic wastes such as crop residues, municipal, hospital and industrial wastes pose major problems in disposal and treatment. Release of unprocessed animal manures into agricultural fields contaminates ground water causing public health risk. Vermicomposting is the best alternative to conventional composting and differs from it in several ways (Gandhi et al. 1997). Vermicomposting hastens the decomposition process by 2–5 times, thereby quickens the conversion of wastes into valuable biofertilizer and produces much more homogenous materials compared to thermophilic composting (Bhatnagar and Palta 1996; Atiyeh et al. 2000a). Distinct differences exist between the microbial communities found in vermicomposts and composts and hence the nature of the microbial processes is quite different in vermicomposting and composting (Subler et al. 1998). The active phase of composting is the thermophilic stage characterized by thermophilic bacterial community where intensive decomposition takes place followed by a mesophilic maturation phase (Lazcano et al. 2008; Vivas et al. 2009). Vermicomposting is a mesophilic process characterized by mesophilic bacteria and fungi (Benitez et al. 1999). Vermicomposting comprises of an active stage during which earthworms and associated microbes jointly process the substrate and the maturation phase that involves the action of associated microbes and occurs once the worm's moves to the fresher layers of undigested waste or when the product is removed from the vermireacter. The duration of the active phase depends on the species and density of the earthworms involved (Ndegwa et al. 2000; Lazcano et al. 2008; Aira et al. 2011). A wide range of oganic wastes viz., horticultural residues from processed potatoes (Edwards 1988); mushroom wastes (Edwards 1988; Tajbakhsh et al. 2008); horse wastes (Hartenstein et al. 1979; Edwards et al. 1998); pig wastes (Chan and Griffiths 1988; Reeh 1992); brewery wastes (Butt 1993); sericulture wastes (Gunathilagraj and Ravignanam 1996); municipal sewage sludge (Mitchell et al. 1980; Dominguez et al. 2000); agricultural residues (Bansal and Kapoor 2000); weeds (Gajalakshmi et al. 2001); cattle dung (Gunadi et al. 2002); industrial refuse such as paper wastes (Butt 1993; Elvira et al. 1995; Gajalakshmi et al. 2002); sludge from paper mills and dairy plants (Elvira et al. 1997; Banu et al. 2001); domestic kitchen wastes (Sinha et al. 2002); urban residues and animal wastes (Edwards et al. 1985; Edwards 1988) can be vermicomposted (Sharma et al. 2005).

Effects of vermicomposting on pH, electrical conductivity (EC), C:N ratio and other nutrients have been documented. Earthworm activity reduced pH and C:N ratio in manure (Gandhi et al. 1997; Atiyeh et al. 2000b). Chemical analysis showed vermicompost had a lower pH, EC, organic carbon (OC) (Nardi et al. 1983; Albanell et al. 1988; Mitchell 1997),

C:N ratio (Riffaldi and Levi-Minzi 1983; Albanell et al. 1988), nitrogen and potassium and higher amounts of total phosphorous and micronutrients compared to the parent material (Hashemimajd et al. 2004). Slightly decreased pH values of vermicompost compared to traditional compost might be attributed due to mineralization of N and P, microbial decomposition of organic materials into intermediate organic acids, fulvic acids, humic acids (Lazcano et al. 2008; Albanell et al. 1988; Chan and Griffiths 1988; Subler et al. 1998) and concomitant production of CO_2 (Elvira et al. 1998; Garg et al. 2006). Vermicomposting of paper mill and dairy sludge resulted in 1.2–1.7 fold loss of organic carbon as CO_2 (Elvira et al. 1998). In contrast to the parent material used, vermicomposts contain higher humic acid substances (Albanell et al. 1988). Humic acid substances occur naturally in mature animal manure, sewage sludge or paper-mill sludge, but vermicomposting drastically increases the rate of production and their amount from 40–60 percent compared to traditional composting. The enhancement in humification processes is by fragmentation and size reduction of organic matter, increased microbial activity within earthworm intestine and soil aeration by earthworm feeding and movement (Dominguez and Edwards, 2004). EC indicates the salinity of the organic amendment. Minor production of soluble metabolites such as ammonium and precipitation of dissolved salts during vermicomposting lead to lower EC values. Compared to the parent material used, vermicomposts contain less soluble salts and greater cation exchange capacity (Holtzclaw and Sposito 1979; Albanell et al. 1988). C:N ratio is an indicator of the degree of decomposition. During the process of biooxidation, CO_2 and N is lost and loss of N takes place at a comparatively lower rate. Comparison of compost and vermicompost showed that vermicompost had significantly less C:N ratios as they underwent intense decomposition (Lazcano et al. 2008).

Vermicomposting of cow manure using earthworm species *E. andrei* (Atiyeh et al. 2000b) and *E. foetida* (Hand et al. 1988) favored nitrification, resulting in the rapid conversion of ammonium-nitrogen to nitrate-nitrogen. Vermicomposting increased the concentration of nitrate-nitrogen to 28 fold after 17 weeks, while in conventional compost there was only 3-fold increase (Subler et al. 1998; Atiyeh et al. 2000a). Increase in ash concentration during vermicomposting suggests that vermicomposting accelerates the rate of mineralization (Albanell et al. 1988). Mineralization is the process in which the chemical compounds in the organic matter decompose or oxidise into forms that could be easily assimilated by the plants. Increase in ash content increases the rate of mineralization. Ash is an alkaline substance which hinders the formation of H_2S as well as improves the availibility of O_2 and thereby renders composts odorless. Thus vermicomposting increases the ash content and accelerates the rate of mineralization which is essential to make nutrients available to plants. The observed increase of total phosphorous (TP) in vermicompost is probably due to mineralization and mobilization of phosphorus resulting from the enhanced phosphatase activity by microorganisms in the gut epithelium of the earthworms (Zhang et al. 2000; Garg et al. 2006). Vermicomposts showed a significant increase in exchangeable Ca^{2+}, Mg^{2+} and K^{+} compared to fresh sludge indicating the conversion of nutrients to plant-available forms during passage through the earthworm gut (Garg et al. 2006; Yasir et al. 2009a). Vermicomposts contain higher nutrient concentrations, but less likely to produce salinity, than composts. Additionally, vermicomposts possess outstanding biological properties and have microbial populations significantly larger and more diverse compared to conventional composts (Edwards 1998). Soil supplemented with vermicompost showed better plant growth compared to soil treated with inorganic fertilizers or cattle manure (Kalembasa 1996; Subler et al. 1998).

8.1.3 DIVERSITY OF BACTERIA ASSOCIATED WITH EARTHWORMS

Earthworm's ability to increase plant nutrient availability is likely to be dependent on the activity of earthworm gut microflora. Earthworms indirectly influence the dynamics of soil chemical processes, by comminuting the litter and affecting the activity of the soil micro-flora (Petersen and Luxton 1982; Lee 1985; Edwards and Bohlen 1996). Interactions between earthworms and microorganisms seem to be complex. Earthworms ingest plant growth-promoting rhizospheric bacteria such as *Pseudomonas*, *Rhizobium, Bacillus, Azosprillium, Azotobacter,* etc. along with rhizospheric soil, and they might get activated or increased due to the ideal micro-environment of the gut. Therefore earthworm activity increases the population of plant growth-promoting rhizobacteria (PGPR) (Sinha et al. 2010). This specific group of bacteria stimulates plant growth directly by solubilization of nutrients (Ayyadurai et al. 2007; Ravindra et al. 2008), production of growth hormone, 1-aminocyclopropane-1-carboxylate (ACC) deaminase (Correa et al. 2004), nitrogen fixation (Han et al. 2005), and indirectly by suppressing fungal pathogens. Antibiotics, fluorescent pigments, siderophores and fungal cell-wall degrading enzymes namely chitinases and glucanases (Han et al. 2005; Sunish et al. 2005; Ravindra et al. 2008; Jha et al. 2009; Pathma et al. 2010; Pathma et al. 2011a, b) produced by bacteria mediate the fungal growth-suppression. Earthworms are reported to have association with such free living soil bacteria and constitute the drilosphere (Ismail 1995). Earthworm microbes mineralize the organic matter and also facilitate the chelation of metal ions (Pizl and Novokova 1993; Canellas et al. 2002). Gut of earthworms *L. terrestris, Allolobophora caliginosa* and *Allolobophora terrestris* were reported to contain higher number of aerobes compared to soil (Parle 1963). Earthworms increased the number of microorganisms in soil as much as five times (Edwards and Lofty 1977) and the number of bacteria and 'actinomycetes' contained in the ingested material increased upto 1,000 fold while passing through their gut (Edwards and Fletcher 1988). Similar increase was observed in plate counts of total bacteria, proteolytic bacteria and actinomycetes by passage through earthworms gut (Parle 1963; Daniel and Anderson 1992; Pedersen and Hendriksen 1993; Devliegher and Verstraete 1995). Similarly microbial biomass either decreased (Bohlen and Edwards 1995; Devliegher and Verstraete 1995), or increased (Scheu 1992) or remained unchanged (Daniel and Anderson 1992) after passage through the earthworm gut. An oxalate-degrading bacterium *Pseudomonas oxalaticus* was isolated from intestine of *Pheretima* species (Khambata and Bhat 1953) and an actinomycete *Streptomyces lipmanii* was identified in the gut of *Eisenia lucens* (Contreras 1980). Scanning electron micrographs provided evidence for endogenous microflora in guts of earthworms, *L. terrestris* and *Octolasion cyaneum* (Jolly et al. 1993). Gut of *E. foetida* contained various anaerobic N_2-fixing bacteria such as *Clostridium butyricum, C. beijerinckii* and *C. paraputrificum* (Citernesi et al. 1977). Alimentary canal of *Lumbricus rubellus* and *Octolasium lacteum* were found to contain more numbers of aerobes and anerobes (Karsten and Drake 1995) and culturable denitrifiers

(Karsten and Drake 1997). List of vermicompost bacteria and their beneficial traits is presented in Table 2.

TABLE 2: Biodiversity of vermicompost bacteria and their beneficial traits

Vermicompost earthworm	Names of bacteria	Beneficial traits	References
Pheretima sp.	*Pseudomonas oxalaticus*	Oxalate degradation	Khambata and Bhat, 1953
Unspecified	*Rhizobium trifolii*	Nitrogen fixation and growth of leguminous plants	Buckalew et al. 1982
Lumbricus rubellus	*R. japonicum, P. putida*	Plant growth promotion	Madsen and Alexander 1982
L. terrestris	*Bradyrhizobium japonicum*	Improved distribution of nodules on soybean roots	Rouelle, 1983
Aporrectodea trapezoids,	*P. corrugata* 214OR	Suppress *Gaeumannomyces* graminis var. Tritd in wheat	Doube et al. 1994
A. rosea			
A. trapezoids,	*R. meliloti* L5-30R	Increased root nodulation and nitrogen fixation in legumes	Stephens et al. 1994b
Microscolex dubius			
Eisenia foetida	*Bacillus* spp., *B. megaterium, B. pumilus, B. subtilis*	Antimicrobial activity against Enterococcus faecalisDSM 2570, Staphylococcus aureus DSM 1104	Vaz-Moreira et al. 2008
L. terrestris	Fluorescent pseudomonads,	Suppress Fusarium oxysporum f. sp. asparagi and F. proliferatum in asparagus, Verticillium dahlia in eggplant and *F. oxysporum* f. sp. lycopersici Race 1 in tomato	Elmer, 2009
Filamentous actinomycetes			

TABLE 2: *Cont.*

Vermicompost earthworm	Names of bacteria	Beneficial traits	References
Eudrilus sp.	Free-living N_2 fixers, *Azospirillum, Azotobacter, Autotrophic Nitrosomonas, Nitrobacter,* Ammonifying bacteria, Phosphate solubilizers, Fluorescent pseudomonads	Plant growth promotion by nitrification, phosphate solubilisation and plant disease suppression	Gopal et al. 2009
E. foetida	*Proteobacteria, Bacteroidetes,*	Antifungal activity against Colletotrichum coccodes, *R. solani, P. ultimum, P. capsici* and *F. moliniforme*	Yasir et al. 2009a
	Verrucomicrobia, Actinobacteria,		
	Firmicutes		
Unspecified	*Eseniicola composti* YC06271T	Antagonistic activity against *F. moniliforme*	Yasir et al. 2009b

Earthworms harbor 'nitrogen-fixing' and 'decomposer microbes' in their gut and excrete them along with nutrients in their excreta (Singleton et al. 2003). Earthworms stimulate and accelerate microbial activities by increasing the population of soil microorganisms (Binet et al. 1998), microbial numbers and biomass (Edwards and Bohlen 1996), by improving aeration through burrowing actions. Vermicomposting modified the original microbial community of the waste in a diverse way. Actinobacteria and Gammaproteobacteria were abundant in vermicompost, while conventional compost contained more Alphaproteobacteria and Bacteriodetes, the bacterial phylogenetic groups typical of non-cured compost (Vivas et al. 2009). Total bacterial counts exceeded 10^{-10}/ g of vermicompost and it included nitrobacter, azotobacter, rhizobium, phosphate solubilizers and actinomycetes (Suhane 2007). Molecular and culture-dependent analyses of bacterial community of vermicompost showed the presence of α-Proteobacteria, β-Proteobacteria, γ-Proteobacteria, Actinobacteria, Planctomycetes, Firmicutes and Bacteroidetes (Yasir et al. 2009a). Several findings showed considerable increase in total viable counts of actinomycetes and bacteria in the worm treated compost (Parthasarathi and Ranganathan 1998; Haritha Devi et al. 2009). The increase of microbial population may be due to the congenial condition for the growth of microbes within the digestive tract of earthworm and by the ingestion of nutrient rich organic wastes which provide energy and also act as a substrate for the growth of microorganisms (Tiwari et al. 1989). The differences in microbial species, numbers and activity between the earthworm alimentary canal or burrow and bulk soil indirectly support the hypothesis that the bacterial community structures of these habitats are different from those of the soil. Specific phylogenetic groups of bacteria such as *Aeromonas hydrophila* in *E. foetida* (Toyota and Kimura 2000), fluorescent pseudomonads in *L. terrestris* (Devliegher and Verstraete 1997), and Actinobacteria in *L. rubellus* (Kristufek et al. 1993) have been found in higher numbers in earthworm guts, casts, or burrows.

Enzymatic activity characterization and quantification has a direct correlation with type and population of microbes and reflects the dynamics of the composting process in terms of the decomposition of organic matter and nitrogen transformations and provide information about the maturity of the compost (Tiquia 2005). Wormcasts contain higher activities of cellulase, amylase, invertase, protease, peroxidase, urease, phosphatase and dehydrogenase (Sharpley and Syers 1976; Edwards and Bohlen 1996). Dehydrogenase is an intracellular enzyme related to the oxidative phosphorylation process (Trevors 1984) and is an indicator of microbial activity in soil and other biological ecosystems (Garcia et al. 1997). The maximum enzyme activities (cellulase, amylase, invertase, protease and urease) were observed during 21–35 days in vermicomposting and on 42–49 days in conventional composting. Also, microbial numbers and their extracellular enzyme profiles were more abundant in vermicompost produced from fruitpulp, vegetable waste, groundnut husk and cowdung compared to the normal compost of the same parental origin (Haritha Devi et al. 2009). *Pseudomonas*, *Paenibacillus*, *Azoarcus, Burkholderia, Spiroplasm, Acaligenes*, and *Acidobacterium,* the potential degraders of several categories of organics are seen associated with the earthworm's intestine and vermicasts (Singleton et al. 2003). Firmicutes viz., *Bacillus benzoevorans, B. cereus, B. licheniformis, B. megaterium, B. pumilus, B. subti-*

lis, B. macroides; Actinobacteria namely *Cellulosimicrobium cellulans, Microbacterium spp., M. oxydans*; Proteobacteria such as *Pseudomonas* spp., *P. libaniensis*; ungrouped genotypes *Sphingomonas* sp., *Kocuria palustris* and yeasts namely *Geotrichum* spp. and *Williopsis californica* were reported from vermicomposts (Vaz-Moreira et al. 2008). Pinel et al. (2008) reported the presence of a novel nephridial symbiont, *Verminephrobacter eiseniae* from *E. foetida. Ochrobactrum* sp., *Massilia* sp., *Leifsonia* sp. and bacteria belonging to families Aeromonadaceae, Comamonadaceae, Enterobacteriaceae, Flavobacteriaceae, Moraxellaceae, Pseudomonadaceae, Sphingobacteriaceae, Actinobacteria and Microbacteriaceae were reported to occur in earthworms alimentary canal (Byzov et al. 2009). The microbial flora of earthworm gut and cast are potentially active and can digest a wide range of organic materials and polysaccharides including cellulose, sugars, chitin, lignin, starch and polylactic acids Zhang et al. (2000; Aira et al. 2007; Vivas et al. 2009). Single-strand conformation polymorphism (SSCP) profiles on the diversity of eight bacterial groups viz., Alphaproteobacteria, Betaproteobacteria, Bacteroidetes, Gammaproteobacteria, Deltaproteobacteria, Verrucomicrobia, Planctomycetes, and Firmicutes from fresh soil, gut, and casts of the earthworms *L. terrestris* and *Aporrectodea caliginosa* showed the presence of Bacteroidetes, Alphaproteobacteria, Betaproteobacteria and representatives of classes Flavobacteria, Sphingobacteria (Bacteroidetes) and *Pseudomonas* spp. in the worm casts in addition to unclassified Sphingomonadaceae (Alphaproteobacteria) and *Alcaligenes* spp. (Betaproteobacteria) (Nechitaylo et al. 2010).

8.1.4 ROLE OF VERMICOMPOST IN SOIL FERTILITY

Vermicomposts can significantly influence the growth and productivity of plants (Kale et al. 1992; Kalembasa 1996; Edwards 1988; Sinha et al. 2009) due to their micro and macro elements, vitamins, enzymes and hormones (Makulec 2002). Vermicomposts contain nutrients such as nitrates, exchangeable phosphorus, soluble potassium, calcium, and magnesium in plant available forms (Orozco et al. 1996; Edwards 1998) and have large particular surface area that provides many microsites for microbial activity and for the strong retention of nutrients (Shi-wei and Fu-zhen 1991).

Uptake of nitrogen (N), phosphorus (P), potassium (K) and magnesium (Mg) by rice (*Oryza sativa*) plant was highest when fertilizer was applied in combination with vermicompost (Jadhav et al. 1997). N uptake by ridge gourd (*Luffa acutangula*) was higher when the fertilizer mix contained 50% vermicompost (Sreenivas et al. 2000). Apart from providing mineralogical nutrients, vermicomposts also contribute to the biological fertility by adding beneficial microbes to soil. Mucus, excreted through the earthworm's digestive canal, stimulates antagonism and competition between diverse microbial populations resulting in the production of some antibiotics and hormone-like biochemicals, boosting plant growth (Edwards and Bohlen 1996). In addition, mucus accelerates and enhances decomposition of organic matter composing stabilized humic substances which embody water-soluble phytohormonal elements (Edwards and Arancon 2004) and plant-available nutrients at high levels (Atiyeh et al. 2000c). Adding vermicasts to soil improves soil structure, fertility, plant growth and suppresses diseases caused by soil-borne plant pathogens, increasing crop yield (Chaoui et al. 2002; Scheuerell et al. 2005; Singh et al. 2008). Kale (1995) reported the nutrient status of vermicomposts with organic carbon 9.15-17.98%, total nitrogen 0.5-1.5%, available phosphorus 0.1-0.3%, available potassium 0.15%, calcium and magnesium 22.70-70 mg/100 g, copper 2–9.3 ppm, zinc 5.7-11.5 ppm and available sulphur 128–548 ppm.

Effects of a variety of vermicomposts on a wide array of field crops (Chan and Griffiths 1988; Arancon et al. 2004b), vegetable plants (Edwards and Burrows 1988; Wilson and Carlile 1989;Subler et al. 1998; Atiyeh et al. 2000b), ornamental and flowering plants (Edwards and Burrows 1988; Atiyeh et al. 2000c) under greenhouse and field conditions have been documented. Vermicomposts are used as alternative potting media due to their low-cost, excellent nutrient status and physiochemical characters. Considerable improvements in plant growth recorded after amending soils with vermicomposts have been attributed to the physico-chemical and biological properties of vermicomposts.

Vermicompost addition favorably affects soil pH, microbial population and soil enzyme activities (Maheswarappa et al. 1999) and also reduces the proportion of water-soluble chemical, which cause possible environmental contamination (Mitchell and Edwards 1997). Vermicompost addition increases the macropore space ranging from 50–500 μm, resulting in improved air-water relationship in the soil, favourably affecting plant growth (Marinari et al. 2000). Evaluation of various organic and inorganic amendments on growth of raspberry proves that vermicompost has beneficial buffering capability and ameliorate the damage caused by excess of nutrients which may otherwise cause phytotoxicity (Subler et al. 1998). Thus, vermicompost acts a soil conditioner (Albanell et al. 1988) and a slow-release fertilizer (Atiyeh et al. 2000a). During vermicomposting the heavy metals forms complex, aggregates with humic acids and other polymerized organic fractions resulting in lower availability of heavy metals to the plant, which are otherwise phytotoxic (Dominguez and Edwards 2004). Soil amended with vermicompost produced better quality fruits and vegetables with less content of heavy metals or nitrate, than soil fertilized with mineral fertilizers (Kolodziej and Kostecka 1994).

8.1.5 ROLE OF VERMICOMPOST BACTERIA IN BIOMEDICAL WASTE MANAGEMENT

The importance of sewage sludge, biosolids and biomedical waste management by safe, cheap and easy methods need no further emphasis. All these wastes are infectious and have to be disinfected before being disposed into the environment. Biosolids also contain an array of pathogenic microorganisms (Hassen et al. 2001). Biocomposting of wastes bring about biological transformation and stabilization of organic matter and effectively reduces potential risks of pathogens (Burge et al. 1987; Gliotti et al. 1997; Masciandaro et al. 2000). Vermicomposting does not involves a thermophilic phase which might increase the risk of using this technology for management of infectious wastes, but surprisingly vermicomposting resulted into a noticeable reduction in the pathogen indicators such as fecal coliform, Salmonella sp., enteric virus and helminth ova in the biosolids (Eastman 1999; Sidhu et al. 2001). Vermicomposting of biosolids resulted in reduction of faecal coliforms and *Salmonella* sp. from 39,000 MPN/g to 0 MPN/g and < 3 MPN to < 1MPN/g respectively (Dominguez and Edwards 2004). Vermicomposting of municipal sewage sludge with *L. mauritii* eliminated *Salmonella* and *Escherichia* sp., and the earthworm gut analysis also proved that *Salmonella* sp. ranging $15–17 \times 10^3$ CFU/g and *Escherichia* sp. ranging 10–14 × 10^2 CFU/g were completely eliminated in the gut after 70 days of vermicomposting period (Ganesh Kumar and Sekaran 2005). Activities by earthworms on sludge reduced levels of pathogens and odors of putrefaction and accelerated sludge stabilization (Mitchell 1978; Brown and Mitchell 1981; Hartenstein 1983). The reduction or removal of these enteric bacterial populations at the end of vermicomposting period, correlates with the findings that earthworm's diet include microorganisms and earthworms ability to selectively digest them (Bohlen and Edwards 1995; Edwards and Bohlen 1996). Apart from solid waste management, earthworms are also used in sewage water treatment. Earthworms promote the growth of 'beneficial decomposer bacteria' in wastewater and acts as aerators, grinders, crushers, chemical degraders, and biological stimulators (Dash 1978; Sinha et al. 2002). Earthworms also granulate the clay particles and increase the hydraulic conductivity and natural aeration and further grind the silt and sand particles and increase the total specific surface area and thereby enhance adsorption of the organic and inorganic matter from the wastewater. In addition, earthworms body acts as a 'biofilter' and remove the biological oxygen demand (BOD), chemical oxygen demand (COD), total dissolved solids (TDS) and total suspended solids (TSS) from wastewater by 90%, 80–90%, 90–92% and 90–95% respectively by 'ingestion' and biodegradation of organic wastes, heavy metals, and solids from wastewater and by their 'absorption' through body walls (Sinha et al. 2008).

Reports reveal that vermicomposting converts the infected biomedical waste containing various pathogens viz., *Staphylococcus aureus, Proteus vulgaris, Pseudomonas pyocyaneae* and *Escherichia coli* to an innocuous waste containing commensals like *Citrobactor freundii* and aerobic spore bearing microorganism usually found in the soil and alimentary canal of earthworms (Umesh et al. 2006). Vermicomposting plays a vital role for safe management of biomedical wastes and solid wastes generated from wastewater treatment plants and its bioconversion into valuable composts free from enteric bacterial populations. Depending on the earthworm species, vermicomposting was known to reduce the level of different pathogens such as *Salmonella enteriditis, Escherichia coli,* total and faecal coliforms, helminth ova and human viruses in different types of waste. Direct means of reduction in these microbial numbers during gut passage might be due to the digestive enzymes and mechanical grinding, while indirect means of pathogen removal might be due to promotion of aerobic conditions which could bring down the load of coliforms (Monroy et al. 2009; Edwards 2011; Aira et al. 2011).

8.1.6 ROLE OF VERMICOMPOST IN PLANT GROWTH PROMOTION

Use of vermicomposts as biofertilizers has been increasing recently due to its extraordinary nutrient status, and enhanced microbial and antagonistic activity. Vermicompost produced from different parent material such as food waste, cattle manure, pig manure, etc., when used as a media supplement, enhanced seedling growth and development, and increased productivity of a wide variety of crops (Edwards and Burrows 1988; Wilson and Carlile 1989; Buckerfield and Webster 1998; Edwards 1998; Subler et al. 1998; Atiyeh et al. 2000c). Vermicompost addition to soil-less bedding plant media enhanced germination, growth, flowering and fruiting of a wide range of green house vegetables and ornamentals (Atiyeh et al. 2000a, b, c), marigolds (Atiyeh et al. 2001), pepper (Arancon et al. 2003a), strawberries (Arancon et al. 2004b) and petunias (Chamani et al. 2008). Vermicompost application in the ratio of 20:1 resulted in a significant and consistent increase in plant growth in both field and greenhouse conditions (Edwards et al. 2004), thus providing a substantial evidence that biological growth promoting factors play a key role in seed germination and plant growth (Edwards and Burrows 1988; Edwards 1998). Investigations revealed that plant hormones and plant-growth regulating substances (PGRs) such as auxins, gibberellins, cytokinins, ethylene and abscisic acid are produced by microorganisms (Barea et al. 1976; Arshad and Frankenberger 1993).

Several researchers have documented the presence of plant growth regulators such as auxins, gibberellins, cytokinins of microbial origin (Krishnamoorthy and Vajranabhiah 1986; Grappelli et al. 1987; Tomati et al. 1988; Muscolo et al. 1999) and humic acids (Senesi et al. X1992; Masciandaro et al. 1997; Atiyeh et al. 2002) in vermicompost in appreciable quantities. Cytokinins produced by *Bacillus* and *Arthrobacter* spp.

in soils increase the vigour of seedlings (Inbal and Feldman 1982; Jagnow 1987). Microbially produced gibberellins influence plant growth and development (Mahmoud et al. 1984; Arshad and Frankenberger 1993) and auxins produced by *Azospirillum brasilense* affects the growth of plants belonging to paoceae (Barbieri et al. 1988). Extensive investigations on the biological activities of humic substances showed that they also posses plant growth stimulating property (Chen and Aviad 1990). Humic substances increased the dry matter yields of corn and oat seedlings (Lee and Bartlett 1976; Albuzio et al. 1994); number and length of tobacco roots (Mylonas and Mccants 1980); dry weights of roots, shoots and number of nodules of groundnut, soyabean and clover plants (Tan and Tantiwiramanond 1983) and vegetative growth of chicory plants (Valdrighi et al. 1996) and induced root and shoot formation in plant tissue culture (Goenadi and Sudharama 1995). High levels of humus have been reported from vermicomposts originating from food wastes, animal manure, sewage and paper mill sludges (Atiyeh et al. 2002; Canellas et al. 2002; Arancon et al. 2003c). The humic and fulvic acid in the humus dissolves insoluble minerals in the organic matter and makes them readily available to plants and in addition they also help plants to overcome stress and stimulates plant growth (Sinha et al. 2010). Studies on biological activities of vermicompost derived humic substances, revealed that they had similar growth-promoting hormonal effect (Dell'Agnola and Nardi 1987; Nardi et al. 1988; Muscolo et al. 1993). The humic materials extracted from vermicomposts have been reported to produce auxin-like cell growth and nitrate metabolism in carrots (*Daucus carota*) (Muscolo et al. 1996). Humates obtained from pig manure vermicompost increased growth of tomato (Atiyeh et al. 2002) and those obtained from cattle, food and paper waste vermicompost increased the growth of strawberries and peppers (Arancon et al. 2003a).

Earthworms produce plant growth regulators (Gavrilov 1963). Since earthworms increase the microbial activity by several folds they are considered as important agents which enhance the production of plant growth regulators (Nielson 1965; Graff and Makeschin 1980; Dell'Agnola and Nardi 1987; Grappelli et al. 1987; Tomati et al. 1987, 1988; Edwards and Burrows 1988; Nardi et al. 1988; Edwards 1998). Plant growth stimulating substances of microbial origin were isolated from tissues of *Aporrectodea longa, L. terrestris* and *Dendrobaena rubidus* and indole like substances were detected from the tissue extracts of *A. caliginosa, L. rubellus* and *E. foetida* which increased the growth of peas (Nielson 1965) and dry matter production of rye grass (Graff and Makeschin 1980). A. trapezoids aided in the dispersal of *Rhizobium* through soil resulting in increased root colonization and nodulation of leguminous plants (Bernard et al. 1994). Use of earthworm casts in plant propagation promoted root initiation, increased root numbers and biomass. The hormone-like effect produced by earthworm casts on plant metabolism, growth and development causing dwarfing, stimulation of rooting, internode elongation and precociousness of flowering was attributed to the fact of presence of microbial metabolites (Tomati et al. 1987; Edwards 1998). Earthworm casts stimulated growth of ornamental plants and carpophore formation in *Agaricus bisporus* when used as casing layer in mushroom cultivation (Tomati et al. 1987). Aqueous extracts of vermicompost produced growth comparable to the use of hormones such as auxins, gibberellins and cytokinins on Petunia, Begonia and Coleus, providing solid evidence that vermicompost is a rich source of plant growth regulating substances (Grappelli et al. 1987; Tomati et al. 1987, 1988). Addition of vermicompost at very low levels to the growth media dramatically increased the growth of hardy ornamentals *Chamaecyparis lawsonian, Elaeagnus pungens, Pyracantha* spp., *Viburnum bodnantense, Cotoneaster conspicus* and *Cupressocyparis leylandi.* Cucumber (Hahn and Bopp 1968), dwarf maize (Sembdner et al. 1976) and coleus bioassays (Edwards et al. 2004) evidenced that vermicompost contained appreciable amounts of cytokinins, gibberellins and auxins respectively. Maize seedlings dipped in vermicompost water showed marked difference in plumule length compared to normal water indicating that plant growth promoting hormones are present in vermicompost (Nagavallemma et al. 2004). Comparative studies on the impact of vermiwash and urea solution on seed germination, root and shoot length in *Cyamopsis tertagonoloba* proved that vermiwash contained hormone like substances (Suthar 2010). High performance liquid chromatography (HPLC) and gas chromatography-mass spectroscopy (GC-MS) analyses of aqueous extracts of cattle waste derived vermicompost showed presence of significant amounts indole-acetic-acid (IAA), gibberellins and cytokinins (Edwards et al. 2004).

Earthworm gut associated microbes enrich vermicomposts with highly water-soluble and light-sensitive plant growth hormones, which gets absorbed on humic acid substances in vermicompost making them extremely stable and helps them persist longer in soils thereby influencing plant growth (Atiyeh et al. 2002; Arancon et al. 2003c). This is confirmed by presence of exchangeable auxin group in the macrostructure of humic acid extract from vermicompost (Canellas et al. 2002). Apart from the rich nutritional status and ready nutrient availability, presence of humic acids and plant growth regulating substances makes vermicompost a biofertilizer which increases germination, growth, flowering and fruiting in a wide range of crops. Vermicompost substitution in a relatively small proportion (10–20%) to the potting mixture increased dry matter production and tomato growth significantly (Subler et al. 1998). Soil amended with 20% vermicompost was more suitable for tomato seedling production (Valenzuela et al. 1997). Similarly vermicompost addition upto 50% in the medium resulted in enhanced growth of *Chamaecyparis lawsoniana* (Lawson's Cypress), *Juniperus communis* (Juniper) and *Elaeagnus pungens* (Silverberry) rooted liners (Bachman and Edgar Davice 2000).

Vermicompost application increased plant spread (10.7%), leaf area (23.1%), dry matter (20.7%) and increased total strawberry fruit yield (32.7%) (Singh et al. 2008). Substitution of vermicompost drastically reduced the incidence of physiological disorders like albinism (16.1–4.5%), fruit malformation (11.5–4.0%) and occurrence of grey mould (10.4–2.1%) in strawberry indicating its significance in reducing nutrient-related disorders and Botrytis rot, thereby increasing the marketable fruit yield upto 58.6% with better quality parameters. Fruit harvested from plant receiving vermicompost were firmer, had higher total soluble solids (TSS), ascorbic acid content and attractive colour. All these parameters appeared to be dose dependent and best results were achieved at 7.5 t ha^{-1} (Singh et al. 2008). Vermicompost application showed significant increase in germination percent (93%), growth and yield of mung bean (*Vigna radiata*) compared to the control (Karmegam et al. 1999). Similarly, the fresh and dry matter yields of cowpea (*Vigna unguiculata*) were higher in soil amended with vermicompost than with biodigested slurry, (Karmegam and Daniel 2000). Combined application of vermicompost with N fertilizer gave higher dry matter (16.2 g $plant^{-1}$) and grain yield (3.6 t ha^{-1}) of wheat (*Triticum aestivum*) and higher dry matter yield (0.66 g $plant^{-1}$) of the following coriander (*Coriandrum sativum*) crop in wheat-coriander cropping system (Desai et al. 1999). Vermicompost application produced herbage yields of coriander cultivars comparable to those obtained with chemical fertilizers (Vadiraj et al. 1998). Yield of pea (*Pisum sativum*) increased with the application of vermicompost (10 t ha-1) along with recommended NPK (Meena et al. 2007). Vermicompost application to sorghum (*Sorghum bicolor*) (Patil and Sheelavantar 2000), sunflower (*Helianthus annuus*) (Devi et al. 1998), tomato (*Lycopersicon esculentum*) (Nagavallemma et al. 2004), eggplant (*Solanum melangona*) (Guerrero and Guerrero, 2006), okra (*Abelmoschus esculentus*) (Gupta et al. 2008), hyacinth bean (*Lablab purpureas*) (Karmegam and Daniel 2008), grapes (Buckerfield and Webster 1998) and cherry (Webster 2005) showed a positive result. Vermicompost amendment at the rate of 10 t ha^{-1} along with 50% of recommended dose of NPK fertilizer increased the number and fresh weight of flowers per plant, flower diameter and yield, while at the rate of 15 t ha^{-1} along with 50% of recommended dose of NPK increased vase life of *Chrysanthemum chinensis* (Nethra et al. 1999). Red Clover and cucumber grown in soil amended with vermicompost showed an increase in mineral contents viz., Ca, Mg, Cu, Mn and Zn in their shoot tissues (Sainz et al. 1998). Vermicomposted cow manure stimulated the growth of lettuce and tomato plants while the unprocessed parent material did not (Atiyeh et al. 2000b). Similarly, vermicomposted duck wastes resulted in better growth of tomatoes, lettuce, and peppers than the unprocessed wastes (Wilson and Carlile 1989). The enhancement in plant growth might be attributed to the fact that processed waste had improved physicochemical characteristics and nutrients, in forms readily available to the plant as well as the presence of plant growth promoting and antagonistic disease suppressing beneficial bacteria.

8.1.7 ROLE OF VERMICOMPOST IN PLANT DISEASE MANAGEMENT

8.1.7.1 PLANT PATHOGEN CONTROL

Soils with low organic matter and microbial activity are conducive to plant root diseases (Stone et al. 2004) and addition of organic amendments can effectively suppress plant disease (Raguchander et al. 1998; Blok et al.

2000; Lazarovits et al. 2000). Several researchers reported the disease suppressive properties of thermophilic compost (Hoitink et al. 1997; Goldstein 1998; Pitt et al. 1998) on a wide range of phytopathogens viz., *Rhizoctonia* (Kuter et al. 1983), *Phytopthora* (Hoitink and Kuter 1986; Pitt et al. 1998), *Plasmidiophora brassicae* and *Gaeumannomyces graminis* (Pitt et al. 1998) and *Fusarium* (Kannangowa et al. 2000; Cotxarrera et al. 2002). Microbial antagonism might be one of the possible reasons for disease suppression as organic amendments enhances the microbial population and diversity. Traditional thermophilic composts promote only selected microbes while non-thermophilic vermicomposts are rich sources of microbial diversity and activity and harbour a wide variety of antagonistic bacteria thus acts as effective biocontrol agents aiding in suppression of diseases caused by soil-borne phytopathogenic fungi (Chaoui et al. 2002; Scheuerell et al. 2005; Singh et al. 2008).

Earthworm feeding reduces the survival of plant pathogens such as *Fusarium* sp. and *Verticillium dahliae* (Yeates 1981; Moody et al. 1996) and increases the densities of antagonistic fluorescent pseudomonads and filamentous actinomycetes while population densities of *Bacilli* and *Trichoderma* spp. remains unaltered (Elmer 2009). Earthworm activities reduce root diseases of cereals caused by *Rhizoctonia* (Doube et al. 1994). It has been proved that earthworms decreased the incidence of field diseases of clover, grains, and grapes incited by *Rhizoctonia* spp. (Stephens et al. 1994a; Stephens and Davoren 1997) and *Gaeumannomyces* spp. (Clapperton et al. 2001). Earthworms *Aporrectodea trapezoides* and *Aporrectodea rosea* act as vectors of *Pseudomonas* corrugata 214OR, a biocontrol agent for wheat take-all caused by *G. graminis* var. *tritd* (Doube et al. 1994). Greenhouse studies on augmentation of pathogen infested soils with *L. terrestris* showed a significant reduction of disease caused by *Fusarium oxysporum* f. sp. *asparagi* and *F. proliferatum* on susceptible cultivars of asparagus (*Asparagus officinalis*), *Verticillium dahliae* on eggplant (*Solanum melongena*) and *F. oxysporum* f. sp. *lycopersici* race 1 on tomato. Plant weights increased by 60-80% and disease severity reduced by 50-70% when soils were augmented with earthworms. Incorporation of soil with vermicompost effectively suppressed *R. solani* in wheat (Stephens et al. 1993), *Phytophthora nicotianae* (Nakamura 1996; Szczech 1999; Szczech and Smolinska 2001) and *Fusarium* in tomatoes (Nakamura

1996; Szczech 1999), *Plasmodiophora brassicae* in tomatoes and cabbage (Nakamura 1996), *Pythium* and *Rhizoctonia* (root rot) in cucumber and radish (Simsek Ersahin et al. 2009), *Botrytis cineria* (Singh et al. 2008) and *Verticillium* (Chaoui et al. 2002) in strawberry and *Sphaerotheca fulginae* in grapes (Edwards et al. 2004). Vermicompost application drastically reduced the incidence of 'Powdery Mildew', 'Color Rot' and 'Yellow Vein Mosaic' in Lady's finger (*Abelmoschus esculentus*) (Agarwal et al. 2010). Substitution of vermicompost in the growth media reduced the fungal diseases caused by *R. solani, P. drechsleri* and *F. oxysporum* in gerbera (Rodriguez et al. 2000). Amendment of vermicompost at low rates (10-30%) in horticulture bedding media resulted in significant suppression of *Pythium* and *Rhizoctonia* under green house conditions (Edwards et al. 2004). Research findings proved that vermicompost when added to container media significantly reduced the infection of tomato plants by *P. nicotianae* var. *nicotianae* and *F. oxysporum* sp. *lycopersici* (Szcech et al. 1993; Szczech 1999). Club-rot of cabbage caused by *P. brassicae* was inhibited by dipping cabbage roots into a mixture of clay and vermicompost (Szcech et al. 1993). Potato plants treated with vermicompost were less susceptible to *P. infestans* than plants treated with inorganic fertilizers (Kostecka et al. 1996a). Aqueous extracts of vermicompost inhibited mycelial growth of *B. cineria, Sclerotinia sclerotiorum, Corticium rolfsii, R. solani* and *F. oxysporum* (Nakasone et al. 1999), effectively controlled powdery mildew of barley (Weltzien 1989) and affected the development of powdery mildews on balsam (*Impatiens balsamina*) and pea (*Pisum sativum*) caused by *Erysiphe cichoracearum* and *E. pisi*, respectively in field conditions (Singh et al. 2003).

8.1.7.2 MECHANISMS THAT MEDIATE PATHOGEN SUPPRESSION

Two possible mechanisms of pathogen suppression have been described, one depends on systemic plant resistance and the other is mediated by microbial competition, antibiosis and hyperparasitism (Hoitink and Grebus 1997). The microbially mediated suppression is again classified into two mechanisms viz., 'general suppression' where a wide range of microbes suppress the pathogens such as *Pythium* and *Phytopthora* (Chen et al. 1987) and 'specific suppression' where a narrow range of organisms facilitates suppression, for instance disease caused by *Rhizoctonia* (Hoitink et al. 1997). The disease suppressive effect of vermicompost against fusarium wilt of tomato clearly depicted that fungus inhibition was purely biotic and no chemical factors played any role, since the experiments with heat-sterilized vermicompost failed to control the disease (Szczech 1999). Experiments on suppression of damping-off caused by *R. solani*, in vermicompost amended nurseries of white pumpkin proved that vermicompost suppressed the disease in a dosage and temperature dependent manner (Rivera et al. 2004). Earthworm castings are rich in nutrients (Lunt and Jacobson 1944; Parle 1963) and calcium humate, a binding agent (Edwards 1998) that reduces desiccation of individual castings and favors the incubation and proliferation of beneficial microbes, such as *Trichoderma* spp. (Tiunov and Scheu 2000), *Pseudomonas* spp. (Schmidt et al. 1997), and mycorrhizal spores (Gange 1993; Doube et al. 1995). Earthworm activity increased the communities of Gram-negative bacteria (Clapperton et al. 2001; Elmer 2009). Vermicompost associated chitinolytic bacterial communities viz., *Nocardioides oleivorans*, several species of *Streptomyces* and *Staphylococcus epidermidis* showed inhibitory effects against plant phytopathogens such as, *R. solani, Colletotrichum coccodes, Pythium ultimum, P. capsici* and *Fusarium moniliforme* (Yasir et al. 2009a).

8.1.8 ROLE OF VERMICOMPOST IN ARTHROPOD PEST CONTROL

Addition of organic amendments helped in suppression of various insect pests such as European corn borer (Phelan et al. 1996), other corn insect pests (Biradar et al. 1998), aphids and scale insects (Culliney and Pimentel 1986; Costello and Altiei 1995; Huelsman et al. 2000) and brinjal shoot and fruit borer (Sudhakar et al. 1998). Several reports also evidenced that vermicompost addition decreased the incidence of *Spodoptera litura, Helicoverpa armigera*, leaf miner (*Apoaerema modicella)*, jassids (*Empoasca kerri*), aphids (*Aphis craccivora*) and spider mites on groundnuts (Rao et al. 2001; Rao 2002, 2003) and psyllids (*Heteropsylla cubana*) on a tropical leguminous tree (*Leucaena leucocephala*) (Biradar et al. 1998).

Vermicompost amendment decreased the incidence of sucking pests under field conditions (Ramesh 2000) and suppressed the damage caused by of two-spotted spider mite (*Tetranychus* spp.), aphid (*Myzus persicae*) (Edwards et al. 2007) and mealy bug (*Pseudococcus* spp.) under green house conditions (Arancon et al. 2007). Vermicompost substitution to soil less plant growth medium MetroMix 360 (MM360) at a rate less then 50% reduced the damage caused by infestation of pepper seedlings by *M. persicae* and *Pseudococcus* spp. and tomato seedlings by *Pseudococcus* spp., cabbage seedlings by *M. persicae* and cabbage white caterpillars (*Pieris brassicae* L.) (Arancon et al. 2005). Greenhouse cage experiments conducted on tomatoes and cucumber seedlings infested with *M. persicae*, citrus mealybug (*Planococcus citri*), two spotted spider mite (*Tetranychus urticae*); striped cucumber beetles (*Acalymna vittatum*) attacking cucumbers and tobacco hornworms (*Manduca sexta*) attacking tomatoes proved that treatment of infested plants with aqueous extracts of vermicompost suppressed pest establishment, and their rates of reproduction. Vermicompost teas at higher dose also brought about pest mortality (Edwards et al. 2010b). Suppression of aphid population gains importance since they are key vectors in transmission of plant viruses. Addition of solid vermicompost reduced damage by *A. vittatum* and spotted cucumber beetles (*Diabotrica undecimpunctata*) on cucumbers and larval hornworms (*Manduca quinquemaculata*) on tomatoes in both greenhouse and field experiments (Yardim et al. 2006). Combined application of vermicompost and vermiwash spray to chilli (*Capiscum annum*) significantly reduced the incidence of 'Thrips' (*Scirtothrips dorsalis*) and 'Mites' (*Polyphagotarsonemus latus*) (Saumaya et al. 2007).

8.1.8.1 MECHANISMS THAT MEDIATE PEST CONTROL

Plants grown in inorganic fertilizers are more prone to pest attack than those grown on organic fertilizers (Culliney and Pimentel 1986; Yardim and Edwards 2003; Phelan 2004). Inorganic nitrogen fertilization improves the nutritional quality and palatability of the host plants, inhibits the raise of secondary metabolite concentrations (Fragoyiannis et al. 2001; Herms 2002), enhances the fecundity of insects dieting on them, attracts more individuals for oviposition (Bentz et al. 1995) and increases the population growth rates of insects (Culliney and Pimentel 1986; Jannsson and Smilowitz 1986). Though organic fertilizer has an enhanced nutritional composition they release nutrients at a slower rate (Patriquin et al. 1995) hence plants grown with organic fertilizers possess decreased N levels (Steffen et al. 1995) and have higher phenol content (Asami et al. 2003) resulting in resistance of these plants to pest attack. Similarly vermicomposts exhibit a slow, balanced nutritional release pattern, particularly in release of plant available N, soluble K, exchangeable Ca, Mg and P (Edwards and Fletcher 1988; Edwards 1998). Vermicomposts are rich in humic acid and phenolic compounds. Phenolic compounds act as feeding deterrents and hence significantly affect pest attacks (Kurowska et al. 1990; Summers and Felton 1994; QiTian 2004; Hawida et al. 2007; Koul 2008; Mahanil et al. 2008; Bhonwong et al. 2009). Soil containing earthworms contained polychlorinated phenols and their metabolites (Knuutinen et al. 1990). An endogenous phenoloxidase present in *L. rubellus* bioactivate compounds to form toxic phenols viz., p-nitrophenol (Park et al. 1996). Monomeric phenols could be absorbed by humic acids in the gut of earthworms (Vinken et al. 2005). Uptake of soluble phenolic compounds from vermicompost, by the plant tissues makes them unpalatable thereby affecting pest rates of reproduction and survival (Edwards et al. 2010a; Edwards et al. 2010b).

8.1.9 ROLE OF VERMICOMPOST IN NEMATODE CONTROL

It has been well documented that addition of organic amendments decreases the populations of plant parasitic nematodes (Addabdo 1995; Sipes et al. 1999; Akhtar and Malik 2000). Vermicompost amendments appreciably suppress plant parasitic nematodes under field conditions (Arancon et al. 2003b). Vermicomposts also suppressed the attack of Meloidogyne incognita on tobacco, pepper, strawberry and tomato (Swathi et al. 1998; Edwards et al. 2007; Arancon et al. 2002; Morra et al. 1998) and decreased the numbers of galls and egg masses of *Meloidogyne javanica* (Ribeiro et al. 1998).

8.1.9.1 MECHANISMS THAT MEDIATE NEMATODE CONTROL

There are several feasible mechanisms that attribute to the suppression of plant parasitic nematodes by vermicompost application and it involves both biotic and abiotic factors. Organic matter addition to the soil stimulates the population of bacterial and fungal antagonists of nematodes (e.g., *Pasteuria penetrans*, *Pseudomonas* spp. and chitinolytic bacteria, *Trichoderma* spp.), and other typical nematode predators including nematophagous mites viz., *Hypoaspis calcuttaensis* (Bilgrami 1996), Collembola and other arthropods which selectively feeds on plant parasitic nematodes. (Thoden et al. 2011). Vermicompost amendment promoted fungi capable of trapping nematode and destroying nematode cysts (Kerry 1988) and increased the population of plant growth-promoting rhizobacteria which produce enzymes toxic to plant parasitic nematodes (Siddiqui and Mahmood 1999). Vermicompost addition to soils planted with tomatoes, peppers, strawberry and grapes showed a significant reduction of plant parasitic nematodes and increased the population of fungivorous and bacterivorous nematodes compared to inorganic fertilizer treated plots (Arancon et al. 2002). In addition, few abiotic factors viz., nematicidal compounds such as hydrogen sulphide, ammonia, nitrates, and organic acids released during vermicomposting, as well as low C/N ratios of the compost cause direct adverse effects while changes in soil physiochemical characterists viz., bulk density, porosity, water holding capacity, pH, EC, CEC and nutrition posses indirect adverse effects on plant parasitic nematodes (Rodriguez-Kabana 1986; Thoden et al. 2011).

8.2 CONCLUSION

Vermicomposting is a cost-effective and eco-friendly waste management technology which takes the privilege of both earthworms and the associated microbes and has many advantages over traditional thermophilic composting. Vermicomposts are excellent sources of biofertilizers and their addition improves the physiochemical and biological properties of agricultural soil. Vermicomposting amplifies the diversity and population of beneficial microbial communities. Although there are some reports indicating that few harmful microbes such as spores of *Pythium* and *Fusarium* are dispersed by earthworms (Edwards and Fletcher 1988), the presence and amplification of antagonistic disease-suppressing and other plant growth-promoting beneficial bacteria during vermicomposting out weigh these harmful effects (Edwards and Fletcher 1988; Gammack et al. 1992; Brown 1995). Vermicomposts with excellent physio-chemical properties and buffering ability, fortified with all nutrients in plant available forms, antagonistic and plant growth-promoting bacteria are fantabulous organic amendments that act as a panacea for soil reclamation, enhancement of soil fertility, plant growth, and control of pathogens, pests and nematodes for sustainable agriculture.

REFERENCES

1. Addabdo TD (1995) The nematicidal effect of organic amendments: a review of the literature 1982–1994. Nematol Mediterranea 23:299-305
2. Aira M, Monroy F, Dominguez J (2007) Earthworms strongly modify microbial biomass and activity triggering enzymatic activities during vermicomposting independently of the application rates of pig slurry. Sci Total Environ 385:252-261
3. Aira M, Gómez-Brandón M, González-Porto P, Domínguez J (2011) Selective reduction of the pathogenic load of cow manure in an industrial-scale continuous-feeding vermireactor. Bioresource Technol 102:9633-9637
4. Agarwal S, Sinha RK, Sharma J, et al. (2010) Ver-miculture for sustainable horticulture: Agronomic impact studies of earthworms, cow dung compost and vermi-compost vis-à-vis chemical fertilizers on growth and yield of lady's finger (Abelmoschus esculentus). In: Sinha RK (ed) Special Issue on 'Vermiculture Technology', International Journal of Environmental Engineering, Inderscience Publishers, Geneva, Switzerland.
5. Akhtar M, Malik A (2000) Role of organic amendments and soil organisms in the biological control of plant parasitic nematodes: a review. Bioresour Technol 74:35-47
6. Albanell E, Plaixats J, Cabrero T (1988) Chemical changes during vermicomposting (Eisenia fetida) of sheep manure mixed with cotton industrial wastes. Biol Fertil Soils 6:266-269
7. Albuzio A, Concheri G, Nardi S, Dell'Agnola G (1994) Effect of humic fractions of different molecular size on the development of oat seedlings grown in varied nutritional conditions. In: Senesi N, Miano TM (eds) Humic substances in the Global Environment and Implications on Human Health, Elsevier, Amsterdam, Netherlands. pp 199-204

8. Arancon NQ, Edwards CA, Atiyeh R, Metzger JD (2004) Effects of vermicomposts produced from food waste on the growth and yields of greenhouse peppers. Bioresour Technol 93:139-144
9. Arancon NQ, Edwards CA, Bierman P, Metzger JD, Lee S, Welch C (2003) Effects of vermicomposts to tomatoes and peppers grown in the field and strawberries under high plastic tunnels. Pedobiologia 47:731-735
10. Arancon NQ, Edwards CA, Bierman P, Welch C, Metzger JD (2004) The influence of vermicompost applications to strawberries: Part 1. Effects on growth and yield. Bioresour Technol 93:145-153
11. Arancon NQ, Edwards CA, Lee S (2002) Management of plant parasitic nematode populations by use of vermicomposts. Proceedings Brighton Crop Protection Conference – Pests and Diseases. 705-716
12. Arancon NQ, Edwards CA, Yardim EN, Oliver TJ, Byrne RJ, Keeney G (2007) Suppression of two-spotted spider mite (Tetranychus urticae), mealy bug (Pseudococcus sp) and aphid (Myzus persicae) populations and damage by vermicomposts. Crop Prot 26:29-39
13. Arancon NQ, Galvis PA, Edwards CA (2005) Suppression of insect pest populations and damage to plants by vermicomposts. Bioresour Technol 96:1137-1142
14. Arancon NQ, Galvis P, Edwards CA, Yardim E (2003) The trophic diversity of nematode communities in soils treated with vermicomposts. Pedobiologia 47:736-740
15. Arancon NQ, Lee S, Edwards CA, Atiyeh RM (2003) Effects of humic acids and aqueous extracts derived from cattle, food and paper-waste vermicomposts on growth of greenhouse plants. Pedobiologia 47:744-781
16. Arshad M, Frankenberger WT Jr (1993) Microbial production of plant growth regulators. In: Metting FB Jr (ed) Soil Microbial Ecology: Applications in Agricultural and Environmental Management, Marcell Dekker, New York. pp 307-347
17. Asami DK, Hang YJ, Barnett DM, Mitchell AE (2003) Comparison of the total phenolic and ascorbic acid content of freeze-dried and air-dried marionberry, strawberry and corn grown using conventional organic and sustainable agricultural practices. J Agric Food Chem 51:1237-1241
18. Asha A, Tripathi AK, Soni P (2008) Vermicomposting: A Better Option for Organic Solid Waste Management. J Hum Ecol 24:59-64
19. Atiyeh RM, Arancon NQ, Edwards CA, Metzger JD (2001) The influence of earthworm-processed pig manure on the growth and productivity of marigolds. Bioresour Technol 81:103-108
20. Atiyeh RM, Dominguez J, Subler S, Edwards CA (2000) Changes in biochemical properties of cow manure during processing by earthworms (Eisenia andrei, Bouché) and the effects on seedling growth. Pedobiologia 44:709-724
21. Atiyeh RM, Lee S, Edwards CA, Arancon NQ, Metzger JD (2002) The influence of humic acids derived from earthworm-processed organic wastes on plant growth. Bioresour Technol 84:7-14
22. Atiyeh RM, Subler S, Edwards CA, Bachman G, Metzger JD, Shuster W (2000) Effects of vermicomposts and composts on plant growth in horticulture container media and soil. Pedobiologia 44:579-590

23. Atiyeh RM, Arancon NQ, Edwards CA, Metzger JD (2000) Influence of earthworm-processed pig manure on the growth and yield of green house tomatoes. Bioresour Technol 75:175-180
24. Ayyadurai N, Ravindra Naik P, Sakthivel N (2007) Functional characterization of antagonistic fluorescent pseudomonads associated with rhizospheric soil of rice (Oryza sativa L.). J Microbiol Biotechnol 17:919-927
25. Bachman GR, Edgar Davice W (2000) Growth of magnolia virginiana liners in vermicompost-amended media. Proceeding of SNA Research Conference, Southern Nursery Association, Atlanta. pp 65-67
26. Bansal S, Kapoor KK (2000) Vermicomposting of crop residues and cattle dung with Eisenia foetida. Bioresour Technol 73:95-98
27. Banu JR, Logakanthi S, Vijayalakshmi GS (2001) Biomanagement of paper mill sludge using an indigenous (Lampito mauritii) and two exotic (Eudrilus eugineae and Eisenia foetida) earthworms. J Environ Biol 22:181-185
28. Barbieri P, Bernardi A, Galli E, Zanetti G (1988) Effects of inoculation with different strains of Azospirillum brasilense on wheat roots development. In: Klingmüller W (ed) Azospirillum IV, Genetics, Physiology, Ecology, SpringerVerlag, Berlin. pp 181-188
29. Barea JM, Navarro E, Montana E (1976) Production of plant growth regulators by rhizosphere phosphate solubilizing bacteria. J Appl Bacteriol 40:129-134
30. Benitez E, Nogales R, Elvira C, Masciandaro G, Ceccanti B (1999) Enzymes activities as indicators of the stabilization of sewage sludges composting by Eisenia foetida. Bioresour Technol 67:297-303
31. Bentz JA, Reeves J, Barbosa P, Francis B (1995) Nitrogen fertilizer effect on selection, acceptance and suitability of Euphorbia pulcherrima (Euphorbiaceae) as a host plant to Bemisia tabaci (Homoptera: Aleyrodidae). Environ Entomol 24:40-45
32. Bernal MP, Faredes C, Sanchez-Monedero MA, Cegarra J (1998) Maturity and stability parameters of composts prepared with a, wide range of organic wastes. Bioresour Technol 63:91-99
33. Bernard MD, Peter MS, Christopher WD, Maarten HR (1994) Interactions between earthworms, beneficial soil microorganisms and root pathogens. Appl Soil Ecol 1:3-10
34. Bhatnagar RK, Palta RK (1996) Earthworm-Vermiculture and Vermicomposting. Kalyani Publishers, New Delhi.
35. Bhonwong A, Stout MJ, Attajarusit J, Tantasawat P (2009) Defensive role of tomato polyphenoloxidases against cotton bollworm (Helicoverpa armigera) and beet army worm (Spodoptera exigua). J Chem Ecol 35:28-38
36. Bilgrami AL (1996) Evaluation of the predation abilities of the mite Hypoaspis calcuttaensis, predaceous on plant and soil nematodes. Fund Appl Nematol 20:96-98
37. Binet F, Fayolle L, Pussard M (1998) Significance of earthworms in stimulating soil microbial activity. Biol Fertil Soils 27:79-84
38. Biradar AP, Sunita ND, Teggel RG, Devaradavadgi SB (1998) Effect of vermicompost on the incidence of subabul psyllid. Insect- Environ 4:55-56
39. Blok WJ, Lamers JG, Termoshuizen AJ, Bollen GJ (2000) Control of soil-borne plant pathogens by incorporating fresh organic amendments followed by tarping. Phytopathology 90:253-259

40. Bohlen PJ, Edwards CA (1995) Earthworm effects on N dynamics and soil respiration in microcosms receiving organic and inorganic nutrients. Soil Biol Biochem 27:341-348
41. Brown BA, Mitchell MJ (1981) Role of the earthworm, Eisenia fetida, in affecting survival of Salmonella enteritidis ser. Typhimurium. Pedobiologia 22:434-438
42. Brown GG (1995) How do earthworms affect microfloral and faunal community diversity? Plant Soil 170:209-231
43. Brown GG, Barois I, Lavelle P (2000) Regulation of soil organic matter dynamics and microbial activity in the drilosphere and the role of interactions with other edaphic functional domains. Eur J Soil Biol 36:177-198
44. Buckalew DW, Riley RK, Yoder WA, Vail WJ (1982) Invertebrates as vectors of endomycorrhizal fungi and Rhizobium upon surface mine soils. West Virginia Acad Sci Proc 54:1
45. Buckerfield JC, Webster KA (1998) Worm-worked waste boosts grape yields: prospects for vermicompost use in vineyards. Australian and New Zealand Wine Industry Journal 13:73-76
46. Burge WD, Enkiri NK, Hussong D (1987) Salmonella regrowth in compost as influenced by substrate. Microbial Ecol 14:243-253
47. Butt KR (1993) Utilization of solid paper mill sludge and spent brewery yeast as a feed for soil dwelling earthworms. Bioresour Technol 44:105-107
48. Byzov BA, Nechitaylo TY, Bumazhkin BK, Kurakov AV, Golyshin PN, Zvyagintsev DG (2009) Culturable microorganisms from the earthworm digestive tract. Microbiology 78:360-368
49. Canellas LP, Olivares FL, Okorokova FAR (2002) Humic acids isolated from earthworm compost enhance root elongation, lateral root emergence and plasma membrane H+− ATPase activity in maize roots. Plant Physiol 130:1951-1957
50. Chamani E, Joyce DC, Reihanytabar A (2008) Vermicompost Effects on the Growth and Flowering of Petunia hybrida 'Dream Neon Rose'. American-Eurasian J Agric And Environ Sci 3:506-512
51. Chan LPS, Griffiths DA (1988) The vermicomposting of pretreated pig manure. Biol Wastes 24:57-69
52. Chaoui H, Edwards CA, Brickner M, Lee S, Arancon N (2002) Suppression of the plant diseases, Pythium (damping off), Rhizoctonia (root rot) and Verticillum (wilt) by vermicomposts. Proceedings of Brighton Crop Protection Conference – Pests and Diseases II(8B-3):711-716
53. Chen W, Hoitink HA, Schmitthenner AF, Touvinen O (1987) The role of microbial activity in suppression of damping off caused by Pythium ultimum. Phytopathology 78:314-322
54. Chen Y, Aviad T (1990) Effects of humic substances on plant growth. In: MacCarthy P, Clapp CE, Malcolm RL, Bloom PR (eds) Humic Substances in Soil and Crop Sciences, Selected Reading ASA and SSSA, Madison. pp 161-186
55. Citernesi U, Neglia R, Seritti A, Lepidi AA, Filippi C, Bagnoli G, Nuti MP, Galluzzi R (1977) Nitrogen fixation in the gastro-enteric cavity of soil animals. Soil Biol Biochem 9:71-72

56. Clapperton MJ, Lee NO, Binet F, Conner RL (2001) Earthworms indirectly reduce the effect of take-all (Gaeumannomyces graminis var. tritici) on soft white spring wheat (Triticium aestivum cv. Fielder). Soil Biol Biochem 33:1531-1538
57. Contreras E (1980) Studies on the intestinal actinomycete flora of Eisenia lucens (Annelida, Oligochaeta). Pedobiologia 20:411-416
58. Correa JD, Barrios ML, Galdona RP (2004) Screening for plant growth promoting rhizobacteria in Chamaecytisus proliferus (tagasaste), a forage tree-shrub legume endemic to the Canary Islands. Plant Soil 266:75-84
59. Costello MJ, Altiei MA (1995) Abundance, growth rate and parasitism of Brevicoryne brassicae and Myzus persicae (Homoptera: Aphididae) on broccoli grown in living mulches. Agric Ecosyst Environ 52:187-196
60. Cotxarrera L, Trillas-Gayl MI, Steinberg C, Alabouvette C (2002) Use of sewage sludge compost and Trichoderma asperellum isolates to suppress Fusarium wilt of tomato. Soil Biol Biochem 34:467-476
61. Culliney TW, Pimentel D (1986) Ecological effects of organic agricultural practices on insect populations. Agric Ecosyst Environ 15:253-256
62. Daniel O, Anderson JM (1992) Microbial biomass and activity in contrasting soil material after passage through the gut of earthworm Lumbricus rubellus Hoffmeister. Soil Biol Biochem 24:465-470
63. Darwin F, Seward AC (1903) More letters of Charles Darwin. In: John M (ed) A record of his work in series of hitherto unpublished letters, London. p 508
64. Dash MC (1978) Role of earthworms in the decomposer system. In: Singh JS, Gopal B (eds) Glimpses of ecology, India International Scientific Publication, New Delhi. pp 399-406
65. Dell'Agnola G, Nardi S (1987) Hormone-like effect and enhanced nitrate uptake induced by depolyconder humic fractions obtained from Allolobophora rosea and A. caliginosa faeces. Biol Fertil Soils 4:111-118
66. Desai VR, Sabale RN, Raundal PV (1999) Integrated nitrogen management in wheat-coriander cropping system. J Maharasthra Agric Univ 24:273-275
67. Devi D, Agarwal SK, Dayal D (1998) Response of sunflower (Helianthus annuus) to organic manures and fertilizers. Indian J Agron 43:469-473
68. Devliegher W, Verstraete W (1995) Lumbricus terrestris in a soil core experiment: nutrient-enrichment processes (NEP) and gut-associated processes (GAP) and their effect on microbial biomass and microbial activity. Soil Biol Biochem 27:1573-1580
69. Devliegher W, Verstraete W (1997) Microorganisms and soil physicochemical conditions in the drilosphere of Lumbricus terrestris. Soil Biol Biochem 29:1721-1729
70. Dominguez J, Edwards CA, Webster M (2000) Vermicomposting of sewage sludge: effects of bulking materials on the growth and reproduction of the earthworm Eisenia andrei. Pedobiologia 44:24-32
71. Dominguez J, Edwards CA (2004) Vermicomposting organic wastes: A review. In: Shakir Hanna SH, Mikhail WZA (eds) Soil Zoology for sustainable Development in the 21st century, Cairo. pp 369-395
72. Doube BM, Ryder MH, Davoren CW, Meyer T (1995) Earthworms: a down under delivery system service for biocontrol agents of root disease. Acta Zool Fennica 196:219-223

73. Doube BM, Stephens PM, Davorena CW, Ryderb MH (1994) Interactions between earthworms, beneficial soil microorganisms and root pathogens. Appl Soil Ecol 1:3-10
74. Eastman BR (1999) Achieving pathogen stabilization using vermicomposting. BioCycle 40:62-64
75. Edwards CA (1988) Breakdown of animal, vegetable and industrial organic wastes by earthworms. In: Edwards CA, Neuhauser EF (eds) Earthworms in Waste and Environmental Management SPB, The Hague, Netherlands. pp 21-31
76. Edwards CA (1998) The use of earthworms in the breakdown and management of organic wastes. In: Edwards CA (ed) Earthworm Ecology, CRC Press, Boca Raton. pp 327-354
77. Edwards CA (2011) Human pathogen reduction during vermicomposting. In: Edwards CA, Arancon NQ, Sherman R (eds) Vermiculture technology: earthworms, organic wastes and environmental management, CRC Press, Boca Raton. pp 249-261
78. Edwards CA, Arancon NQ, Bennett MV, Askar A, Keeney G, Little B (2010) Suppression of green peach aphid (Myzus persicae) (Sulz.), citrus mealybug (Planococcus citri) (Risso), and two spotted spider mite (Tetranychus urticae) (Koch.) attacks on tomatoes and cucumbers by aqueous extracts from vermicomposts. Crop Prot 29:80-93
79. Edwards CA, Arancon NQ, Bennett MV, Askar A, Keeney G (2010) Effect of aqueous extracts from vermicomposts on attacks by cucumber beetles (Acalymna vittatum) (Fabr.) on cucumbers and tobacco hornworm (Manduca sexta) (L.) on tomatoes. Pedobiologia 53:141-148
80. Edwards CA, Arancon NQ, Emerson E, Pulliam R (2007) Supressing plant parasitic nematodes and arthropod pests with vermicompost teas. Biocycle. 38-39
81. Edwards CA, Arancon NQ (2004) Vermicomposts suppress plant pest and disease attacks. BioCycle 45:51-53
82. Edwards CA, Bohlen PJ (1996) Biology and Ecology of earthworms. Chapman and Hall, London. p 426
83. Edwards CA, Burrows I, Fletcher KE, Jones BA (1985) The use of earthworms for composting farm wastes. In: Gasser JKR (ed) Composting Agricultural and Other Wastes, Elsevier, London. pp 229-241
84. Edwards CA, Burrows I (1988) The potential of earthworm composts as plant growth media. In: Edwards CA, Neuhauser E (eds) Earthworms in Waste and Environmental Management, SPB Academic Press, The Hague. pp 21-32
85. Edwards CA, Dominguez J, Arancon NQ (2004) The influence of vermicomposts on pest and diseases. In: Shakir Hanna SH, Mikhail WZA (eds) Soil Zoology for Sustainable Development in the 21st centuary, Cairo. pp 397-418
86. Edwards CA, Dominguez J, Neuhauser EF (1998) Growth and reproduction of Perionyx excavatus (Perr.) (Megascolecidae) as factors in organic waste management. Biol Fertil Soils 27:155-161
87. Edwards CA, Fletcher KE (1988) Interaction between earthworms and microorganisms in organic matter breakdown. Agric Ecosyst Environ 20:235-249
88. Edwards CA, Lofty R (1977) The Biology of Earthworms. Chapmann and Hall, London.

89. Elmer WH (2009) Influence of earthworm activity on soil microbes and soilborne diseases of vegetables. Plant Dis 93:175-179
90. Elvira C, Dominguez J, Sampedro L, Mato S (1995) Vermicomposting for the pulp industry. Biocycle 36:62-63
91. Elvira C, Sampedro L, Benítez E, Nogales R (1998) Vermicomposting of sludges from paper mill and dairy industries with Eisenia andrei: a pilot-scale study. Bioresour Technol 63:205-211
92. Elvira C, Sampedro L, Dominguez J, Mato S (1997) Vermicomposting of wastewater sludge from paper-pulp industry with nitrogen rich materials. Soil Biol Biochem 9:759-762
93. Fracchia L, Dohrmann AB, Martinotti MG, Tebbe CC (2006) Bacterial diversity in a finished compost and vermicompost: differences revealed by cultivation independent analyses of PCR-amplified 16S rRNA genes. Appl Microbiol Biotechnol 71:942-952
94. Fragoyiannis DA, McKinlay RG, D'Mello JPF (2001) Interactions of aphids herbivory and nitrogen availability on the total foliar glycoalkoloid content of potato plants. J Chem Ecol 27:1749-1762
95. Gajalakshmi S, Ramasamy EV, Abbasi SA (2001) Assessment of sustainable vermiconversion of water hyacinth at different reactor efficiencies employing Eudrilus engeniae Kingburg. Bioresour Technol 80:131-135
96. Gajalakshmi S, Ramasamy EV, Abbasi SA (2002) Vermicomposting of paper waste with the anecic earthworm Lampito mauritii Kingburg. Indian J Chem Technol 9:306-311
97. Gammack SM, Paterson E, Kemp JS, Cresser MS, Killham K (1992) Factors affecting movement of microorganisms in soils. In: Stotzky G, Bolla LM (eds) Soil Biochemistry, 7, Marcel Dekker, New York. pp 263-305
98. Gandhi M, Sangwan V, Kapoor KK, Dilbaghi N (1997) Composting of household wastes with and without earthworms. Environ Ecol 15:432-434
99. Ganesh kumar A, Sekaran G (2005) Enteric pathogen modification by anaecic earthworm, Lampito Mauritii. J Appl Sci Environ Mgt 9:15-17
100. Gange AC (1993) Translocation of mycorrhizal fungi by earthworms during early succession. Soil Biol Biochem 25:1021-1026
101. Garcia C, Hernandez T, Costa F (1997) Potential use of dehydrogenase activity as an index of microbial activity in degraded soils. Commun Soil Sci Plant Anal 28:123-134
102. Garg P, Gupta A, Satya S (2006) Vermicomposting of different types of waste using Eisenia foetida: a comparative study. Bioresour Technol 97:391-395
103. Gavrilov K (1963) Earthworms, producers of biologically active substances. Zh Obshch Biol 24:149-154
104. Ghosh M, Chattopadhyay GN, Baral K (1999) Transformation of phosphorus during vermicomposting. Bioresour Technol 69:149-154
105. Gliotti C, Giusquiani PL, Businelli D, Machioni A (1997) Composition changes of dissolved organic matter in a soil amended with municipal waste compost. Soil Sci 162:919-926
106. Goenadi DH, Sudharama IM (1995) Shoot initiation by humic acids of selected tropical crops grown in tissue culture. Plant Cell Rep 15:59-62

107. Goldstein J (1998) Compost suppresses diseases in the lab and fields. BioCycle 39:62-64
108. Gopal M, Gupta A, Sunil E, Thomas VG (2009) Amplification of plant beneficial microbial communities during conversion of coconut leaf substrate to vermicompost by Eudrilus sp. Curr Microbiol 59:15-20
109. Graff O, Makeschin F (1980) Beeinlussung des Ertrags von Weidelgrass (Lolium muttiflorum) Ausscheidungen von Regenwurmen dreier verschiedener Arten. Pedobiologia 20:176-180
110. Grappelli A, Galli E, Tomati U (1987) Earthworm casting effect on Agaricus bisporus fructification. Agrochimica 2:457-462
111. Guerrero RD, Guerrero LA (2006) Response of eggplant (Solanum melongena) grown in plastic containers to vermicompost and chemical fertilizer. Asia Life Sciences 15:199-204
112. Gunadi B, Blount C, Edward CA (2002) The growth and fecundity of Eisenia foetida (Savigny) in cattle solids pre-composted for different periods. Pedobiologia 46:15-23
113. Gunathilagraj K, Ravignanam T (1996) Vermicomposting of sericulture wastes. Madras Agric J 83:455-457
114. Gupta AK, Pankaj PK, Upadhyava V (2008) Effect of vermicompost, farm yard manure, biofertilizer and chemical fertilizers (N, P, K) on growth, yield and quality of lady's finger (Abelmoschus esculentus). Pollution Research 27:65-68
115. Hahn H, Bopp M (1968) A cytokinin test with high specificity. Planta 83:115-118
116. Han J, Sun L, Dong X, Cai Z, Yang H, Wang Y, Song W (2005) Characterization of a novel plant growth-promoting bacteria strain Delftia tsuruhatensis HR4 both as a diazotroph and a potential biocontrol agent against various pathogens. Syst Appl Microbiol 28:66-76
117. Hand P, Hayes WA, Frankland JC, Satchell JE (1988) Vermicomposting of cow slurry. Pedobiologia 31:199-209
118. Haritha Devi S, Vijayalakshmi K, Pavana Jyotsna K, Shaheen SK, Jyothi K, Surekha Rani M (2009) Comparative assessment in enzyme activities and microbial populations during normal and vermicomposting. J Environ Biol 30:1013-1017
119. Hartenstein R, Neuhauser EF, Kaplan DL (1979) Reproductive potential of the earthworm Eisenia foetida. Oecologia 43:329-340
120. Hartenstein R (1983) Assimilation by earthworm Eisenia fetida. In: Satchell JE (ed) Earthworm ecology. From Darwin to vermiculture, Chapman and Hall, London. pp 297-308
121. Hashemimajd K, Kalbasi M, Golchin A, Shariatmadari H (2004) Comparison of vermicompost and composts as potting media for growth of tomatoes. J Plant Nutr 27:1107-1123
122. Hassen A, Belguith K, Jedidi N, Cherif A, Cherif M, Boudabous A (2001) Microbial characterization during composting of municipal solid waste. Bioresour Technol 80:217-225
123. Hawida S, Kapari L, Ossipov V, Ramtala MJ, Ruuhola T, Haukioja E (2007) Foliar phenolics are differently associated with Epirrita autmnata growth and immuno competence. J Chem Ecol 33:1013-1023

124. Herms DA (2002) Effects of fertilization on insect resistance of woody ornamental plants. Environ Entomol 31:923-933
125. Hoitink HA, Kuter GA (1986) Efects of composts in growth media on soil-bome pathogens. In: Chen Y, Avnimelech Y (eds) The role of organic matter in modern agriculture, Martinus Nijhoff Publishers, Dordrecht. pp 289-306
126. Hoitink HA, Stone AG, Han DY (1997) Suppression of plant diseases by compost. Hort Sci 32:184-187
127. Hoitink HA, Grebus ME (1997) Composts and Control of Plant Diseases. In: Hayes MHB, Wilson WS (eds) Humic Substances Peats and Sludges Health and Environmental Aspects, Royal Society of Chemistry, Cambridge. pp 359-366
128. Holtzclaw KM, Sposito G (1979) Analytical properties of the soluble metal-complexing fractions in sludge-soil mixtures. IV. Determination of carboxyl groups in fulvic acid. Soil Sci Soc Am J 43:318-323
129. Huelsman MF, Edwards CA, Lawrence JL, Clarke-Harris DO (2000) A study of the effect of soil nitrogen levels on the incidence of insect pests and predators in Jamaican sweet potato (Ipomoea batatus) and Callaloo (Amaranthus). Proc Brighton Pest Control Conference: Pests and Diseases 8D–13:895-900
130. Inbal E, Feldman M (1982) The response of a hormonal mutant of common wheat to bacteria of the Azospirillium. Israel J Bot 31:257-263
131. Ismail SA (1995) Earthworms in soil fertility management. In: Thampan PK (ed) Organic Agriculture, pp. 77-100
132. Ismail SA (1997) Vermicology: The biology of Earthworms. Orient Longman Limited, Chennai.
133. Jadhav AD, Talashilkar SC, Pawar AG (1997) Influence of the conjunctive use of FYM, vermicompost andurea on growth and nutrient uptake in rice. J Maharashtra Agric Univ 22:249-250
134. Jagnow G (1987) Inoculation of cereal crops and forage grasses with nitrogen-fixing rhizosphere bacteria: a possible cause of success and failure with regard to yield response - a review. Z Pflanzenernaehr Dueng Bodenkde 150:361-368
135. Jambhekar H (1992) Use of earthworm as a potential source of decompose organic wastes. Proc Nat Sem Org Fmg, Coimbatore. pp 52-53
136. Jannsson RK, Smilowitz Z (1986) Influence of nitrogen on population parameters of potato insects: abundance, population growth and within-plant distribution of the green peach aphid, Myzus persicae (Homoptera: Aphididae). Environ Entomol 15:49-55
137. Jha BK, Gandhi Pragash M, Cletus J, Raman G, Sakthivel N (2009) Simultaneous phosphate solubilization potential and antifungal activity of new fluorescent pseudomonad strains, Pseudomonas aeruginosa, P. plecoglossicida and P. mosselii. W J Microbiol Biotech 25:573-581
138. Jolly JM, Lappin-Scott HM, Anderson JM, Clegg CD (1993) Scanning electron microscopy of the gut microflora of two earthworms: Lumbricus terrestris and Octolasion cyaneum. Microbial Ecol 26:235-245
139. Kale RD, Bano K (1986) Field Trials with vermicompost (vee comp. E. 8. UAS) on organic fertilizers. In: Dass MC, Senapati BK, Mishra PC (eds) Proceedings of the national seminar on organic waste utilization, Sri Artatrana Ront, Burla. pp 151-157

140. Kale RD, Bano K (1988) Earthworm cultivation and culturing techniques for the production of vee COMP83E UAS. Mysore J Agric Sci 2:339-344
141. Kale RD, Mallesh BC, Bano K, Bagyaray DJ (1992) Influence of vermicompost application on the available macronutrients and selected microbial populations in paddy field. Soil Biol Biochem 24:1317-1320
142. Kale RD (1995) Vermicomposting has a bright scope. Indian Silk 34:6-9
143. Kalembasa D (1996) The influence of vermicomposts on yield and chemical composition of tomato. Zesz Probl Post Nauk Roln 437:249-252
144. Kannangowa T, Utkhede RS, Paul JW, Punja ZK (2000) Effect of mesophilic and thermophilic composts on suppression of Fusarium root and stem rot of greenhouse cucumber. Canad J Microbiol 46:1021-1022
145. Karmegam N, Alagermalai K, Daniel T (1999) Effect of vermicompost on the growth and yield of greengram (Phaseolus aureus Rob.). Trop Agric 76:143-146
146. Karmegam N, Daniel T (2000) Effect of biodigested slurry and vermicompost on the growth and yield of cowpea (Vigna unguiculata (L.). Environ Ecol 18:367-370
147. Karmegam N, Daniel T (2008) Effect of vermi-compost and chemical fertilizer on growth and yield of Hyacinth Bean (Lablab purpureas). Dynamic Soil, Dynamic Plant, Global Science Books 2:77-81
148. Karsten GR, Drake HL (1995) Comparative assessment of the aerobic and anaerobic microfloras of earthworm guts and forest soils. Appl Environ Microbiol 61:1039-1044
149. Karsten GR, Drake HL (1997) Denitrifying bacteria in the earthworm gastrointestinal tract and in vivo emission of nitrous oxide (N2O) by earthworms. Appl Environ Microbiol 63:1878-1882
150. Kerry B (1988) Fungal parasites of cyst nematodes. In: Edwards CA, Stinner BR, Stinner D, Rabatin S (eds) Biological Interactions in Soil, Elsevier, Amsterdam. pp 293-306
151. Khambata SR, Bhat JV (1953) Studies on a new oxalate-decomposing bacterium, Pseudomonas oxalaticus. J Bacteriol 66:505-507
152. Knuutinen J, Palm H, Hakala H, Haimi J, Huhta V, Salminen J (1990) Polychlorinated phenols and their metabolites in soil and earthworms of a saw mill environment. Chemosphere 20:609-623
153. Kolodziej M, Kostecka J (1994) Some qualitative features of the cucumbers and carrots cultivated on the vermicompost. Zeszyty Naukowe Akademii Rolniczej W Krakowie 292:89-94
154. Kooch Y, Jalilvand H (2008) Earthworm as ecosystem engineers and the most important detritivors in forst soils. Pak J Boil Sci 11:819-825
155. Kostecka J, Blazej JB, Kolodziej M (1996a) Investigations on application of vermicompost in potatoes farming in second year of experiment. Zeszyty Naukowe Akademii Rolniczej W Krakowie 310:69-77
156. Koul O (2008) Phytochemicals and insect control: an antifeedant approach. Crit Rev Plant Sci 27:1-24
157. Krishnamoorthy RV, Vajranabhiah SN (1986) Biological activity of earthworm casts: An assessment of plant growth promoter levels in casts. Proc Indian Acad Sci (Anim Sci) 95:341-35

158. Kristufek V, Ravasz K, Pizl V (1993) Actinomycete communities in earthworm guts and surrounding soil. Pedobiologia 37:379-384
159. Kurowska A, Gora J, Kalemba D (1990) Effects of plant phenols on insects. Pol Wiad Chem 44:399-409
160. Kuter GA, Nelson GB, Hoitink HA, Madden LV (1983) Fungal population in container media amended with composted hardwood bark suppressive and conductive to Rhizoctonia damping-off. Phytopathology 73:1450-1456
161. Lavelle E, Barois I, Martin A, Zaidi Z, Schaefer R (1989) Management of earthworm populations in agro-ecosystems: A possible way to maintain soil quality? In: Clarholm M, Bergstrom L (eds) Ecology of Arable Land: Perspectives and Challenges, Kluwer Academic Publishers, London. pp 109-122
162. Lavelle P, Martin A (1992) Small-scale and large-scale effects of endogeic earthworms on soil organic matter dynamics in soils of the humid tropics. Soil Biol Biochem 12:1491-1498
163. Lazarovits G, Tenuta M, Conn KL, Gullino ML, Katan J, Matta A (2000) Utilization of high nitrogen and swine manure amendments for control of soil-bome diseases: efficacy and mode of action. Acta Hortic 5:559-564
164. Lazcano C, Gomez-Brandon M, Dominguez J (2008) Comparison of the effectiveness of composting and vermicomposting for the biological stabilization of cattle manure. Chemosphere 72:1013-1019
165. Lee KE (1985) Earthworms: Their Ecology and Relationships with Soils and Land Use. Academic Press, Sydney.
166. Lee YS, Bartlett RJ (1976) Stimulation of plant growth by humic substances. Soil Sci Soc Am J 40:876-879
167. Lunt HA, Jacobson HGM (1944) The chemical composition of earthworm casts. Soil Sci 58:367-375
168. Maboeta MS, Van Rensburg L (2003) Vermicomposting of industrially produced wood chips and sewage sludge utilizing Eisenia foetida. Ecotoxicol Environ Saf 56:265-270
169. Madsen EL, Alexander M (1982) Transport of Rhizobium and Pseudomonas through soil. Soil Sci Soc Am J 46:557-560
170. Mahanil S, Attajarusit J, Stout MJ, Thipayong P (2008) Over expression of tomato phenol oxidase increases resistance to the common cutworm. Plant Sci 174:456-466
171. Maheswarappa HP, Nanjappa HV, Hegde MR (1999) Influence of organic manures on yield of arrowroot, soil physico-chemical and biological properties when grown as intercrop in coconut garden. Ann Agr Res 20:318-323
172. Mahmoud SA, Ramadan Z, Thabet EM, Khater T (1984) Production of plant growth promoting substance rhizosphere organisms. Zentrbl Mikrobiol 139:227-232
173. Makulec G (2002) The role of Lumbricus rubellus Hoffm. In determining biotic and abiotic properties of peat soils. Pol J Ecol 50:301-339
174. Marinari S, Masciandaro G, Ceccanti B, Grego S (2000) Influence of organic and mineral fertilisers on soil biological and physical properties. Bioresour Technol 72:9-17
175. Martin JP (1976) Darwin on earthworms: the formation of vegetable moulds. Bookworm Publishing, Ontario.

176. Masciandaro G, Ceccanti B, Gracia C (1997) Soil agro-ecological management: fertigation and vermicompost treatments. Bioresour Technol 59:199-206
177. Masciandaro G, Ceccanti B, Gracia C (2000) `In situ' vermicomposting of biological sludges and impacts on soil quality. Soil Biol Biochem 32:1015-1024
178. Meena RN, Singh Y, Singh SP, Singh JP, Singh K (2007) Effect of sources and level of organic manure on yield, quality and economics of garden pea (Pisum sativam L.) in eastern uttar pradesh. Vegetable Science 34:60-63
179. Mitchell A, Edwards CA (1997) The production of vermicompost using Eisenia fetida from cattle manure. Soil Biol Biochem 29:3-4
180. Mitchell A (1997) Production of Eisenia fetida and vermicompost from feed-lot cattle manure. Soil Biol Biochem 29:763-766
181. Mitchell MJ, Hornor SG, Abrams BI (1980) Decomposition of sewage sludge in drying beds and the potential role of the earthworm, Eisenia foetida. J Environ Qual 9:373-378
182. Mitchell MJ (1978) Role of invertebrates and microorganisms in sludge deomposition. In: Hartenstein R (ed) Utilisation of soil organisms in sludge management. natural technology information services, Springfield, Virginia. pp 35-50
183. Monroy F, Aira M, Domínguez J (2009) Reduction of total coliform numbers during vermicomposting is caused by short-term direct effects of earthworms on microorganisms and depends on the dose of application of pig slurry. Sci Tot Environ 407:5411-5416
184. Moody SA, Piearce TG, Dighton J (1996) Fate of some fungal spores associated with wheat straw decomposition on passage through the guts of Lumbricus terrestris and Aporrectodea longa. Soil Biol Biochem 28:533-537
185. Morra L, Palumbo AD, Bilotto M, Ovieno P, Ptcascia S (1998) Soil solarization: organic fertilization grafting contribute to build an integrated production system in a tomato-zucchini sequence. Colture-Protte 27:63-70
186. Munnoli PM, Da Silva JAT, Saroj B (2010) Dynamics of the soil-earthworm-plant relationship: a review. Dynamic soil, dynamic plant. 1-21
187. Muscolo A, Bovalo F, Gionfriddo F, Nardi S (1999) Earthworm humic matter produces auxin-like effect on Daucus carota cell growth and nitrate metabolism. Soil Biol Biochem 31:1303-1311
188. Muscolo A, Felici M, Concheri G, Nardi S (1993) Effect of earthworm humic substances on esterase and peroxidase activity during growth of leaf explants of Nicotiana plumbaginifolia. Biol Fertil Soils 15:127-131
189. Muscolo A, Panuccio MR, Abenavoli MR, Concheri G, Nardi S (1996) Effect of molecular complexity acidity of earthworm faeces humic fractions on glutamale dehydrogenase, glutamine synthetase, and phosphoenolpyruvate carboxylase in Daucus carota II cells. Biol Fertil Soils 22:83-88
190. Mylonas VA, Mccants CB (1980) Effects of humic and fulvic acids on growth of tobacco. I. Root initiation and elongation. Plant Soil 54:485-490
191. Nagavallemma KP, Wani SP, Stephane L, Padmaja VV, Vineela C, Babu Rao M, Sahrawat KL (2004) Vermicomposting: Recycling wastes into valuable organic fertilizer. Global Theme on Agrecosystems Report no.8. Patancheru 502324. International Crops Research Institute for the Semi-Arid Tropics, Andhra Pradesh. p 20

192. Nakamura Y (1996) Interactions between earthworms and microorganisms in biological control of plant root pathogens. Farming Jpn 30:37-43
193. Nakasone AK, Bettiol W, de Souza RM (1999) The effect of water extracts of organic matter on plant pathogens. Summa Phytopathology 25:330-335
194. Nardi S, Arnoldi G, Dell'Agnola G (1988) Release of hormone-like activities from Alloborophora rosea and Alloborophora caliginosa feces. J Soil Sci 68:563-657
195. Nardi S, Dell'Agnola G, Nuti PM (1983) Humus production from farmyard wastes by vermicomposting. Proc. Int. Symp. On Agricultural and Environmental Prospects in Earthworm Farming. Rome. pp 87-94
196. Ndegwa PM, Thompson SA, Das KC (2000) Effects of stocking density and feeding rate on vermicomposting of biosolids. Bioresour Technol 71:5-12
197. Nechitaylo TY, Yakimov MM, Godinho M, Timmis KN, Belogolova E, Byzov BA, et al. (2010) Effect of the earthworms Lumbricus terrestris and Aporrectodea caliginosa on bacterial diversity in soil. Microbial Ecol 59:574-587
198. Nethra NN, Jayaprasad KV, Kale RD (1999) China aster (Callistephus chinensis (L)) cultivation using vermicompost as organic amendment. Crop Research, Hisar 17:209-215
199. Nielson RL (1965) Presence of plant growth substances in earthworms demonstrated by paper chromatography and the Went pea test. Nature 208:1113-1114
200. Orozco FH, Cegarra J, Trujillo LM, Roig A (1996) Vermicomposting of coffee pulp using the earthworm Eisenia fetida: effects on C and N contents and the availability of nutrients. Biol Fertil Soils 22:162-166
201. Park SR, Cho EJ, Yu KH, Kim YS, Suh JJ, Chang CS (1996) Endogenous phenoloxidase from an earthworm Lumbricus rubellus. Tongmul Hakoehi 39:36-46
202. Parle JN (1963) A Microbiological Study of Earthworm Casts. J Gen Microbiol 31:13-22
203. Parthasarathi K, Ranganathan LS (1998) Pressmud vermicast are hot spots of fungi and bacteria. Ecol Environ Cons 4:81-86
204. Pathma J, Ayyadurai N, Sakthivel N (2010) Assessment of Genetic and Functional Relationship of Antagonistic Fluorescent Pseudomonads of Rice Rhizosphere by Repetitive Sequence, Protein Coding Sequence and Functional Gene Analyses. J Microbiol 48:715-727
205. Pathma J, Kamaraj Kennedy R, Sakthivel N (2011a) Mechanisms of fluorescent pseudomonads that mediate biological control of phytopathogens and plant growth promotion of crop plants. In: Maheswari DK (ed) Bacteria in Agrobiology: Plant Growth Responses, SpringerVerlag, Berlin. pp 77-105
206. Pathma J, Rahul GR, Kamaraj Kennedy R, Subashri R, Sakthivel N (2011) Secondary metabolite production by bacterial antagonists. Journal of Biological Control 25:165-181
207. Patil SL, Sheelavantar MN (2000) Effect of moisture conservation practices, organic sources and nitrogen levels on yield, water use and root development of rabi sorghum (Sorghum bicolor (L.)) in the vertisols of semiarid tropics. Ann Agric Res 21:32-36
208. Patriquin DG, Baines D, Abboud A (1995) Diseases, pests and soil fertility. In: Cook HF, Lee HC (eds) Soil Management in Sustainable Agriculture, Wye College Press, Wye. pp 161-174

209. Pedersen JC, Hendriksen NB (1993) Effect of passage through the intestinal tract of detritivore earthworms (Lumbricus spp.) on the number of selected gram-negative and total bacteria. Biol Fertil Soils 16:227-232
210. Petersen H, Luxton MA (1982) A comparative analysis of soil fauna populations and their role in decomposition process. Oikos 39:287-388
211. Phelan PL, Norris KH, Mason JF (1996) Soil management history and host preference by Ostrinia nubilatis: evidence for plant mineral balance mediating insect-plant interactions. Environ Entom 25:1329-1336
212. Phelan PL (2004) Connecting below-ground and above-ground food webs: the role of organic matter in biological buffering. In: Magadoff F, Well RR (eds) Soil Organic Matter in sustainable agriculture, CRC Press, Boca Raton. pp 199-226
213. Pinel N, Davidson SK, Stahl DA (2008) Verminephrobacter eiseniae gen. nov., sp. nov., a nephridial symbiont of the earthworm Eisenia foetida (Savigny). Int J Syst Evol Microbiol 58:2147-2157
214. Pitt D, Tilston EL, Groenhof AC, Szmidt RA (1998) Recycled organic materials (ROM) in the control of plant disease. Acta Hortic 469:391-403
215. Pizl V, Novokova A (1993) Interactions between microfungi and Eisenia andrei (Oligochaeta) during cattle manure vermicomposting. Pedobiologia 47:895-899
216. QiTian S (2004) Research on prevention and elimination of agricultural pests by ginkgo phenols phenolic acids. Chem Ind For Prod 24:83
217. Raguchander T, Rajappan K, Samiyappan R (1998) Influence of biocontrol agents and organic amendments on soybean root rot. Int J Trop Agri 16:247-252
218. Ramesh P (2000) Effects of vermicomposts and vermicornposting on damage by sucking pests to ground nut (Arachis hypogea). Indian J Agri Sci 70:334
219. Rao KR, Rao PA, Rao KT (2001) Influence of fertilizers and manures on the population of coccinellid beetles and spiders in groundnut ecosystem. Ann Plant Protect Sci 9:43-46
220. Rao KR (2002) Induce host plant resistance in the management sucking pests of groundnut. Ann Plant Protect Sci 10:45-50
221. Rao KR (2003) Influence of host plant nutrition on the incidence of Spodoptera litura and Helicoverpa armigera on groundnuts. Indian J Entomol 65:386-392
222. Ravindra NP, Raman G, Badri Narayanan K, Sakthivel N (2008) Assessment of genetic and functional diversity of phosphate solubilizing fluorescent pseudomonads isolated from rhizospheric soil. BMC Microbiol 8:230
223. Reeh U (1992) Influence of population densities on growth and reproduction of the earthworm Eisenia andrei on pig manure. Soil Biol Biochem 24:1327-1331
224. Ribeiro CF, Mizobutsi EH, Silva DG, Pereira JCR, Zambolim L (1998) Control of Meloidognye javanica on lettuce with organic amendments. Fitopatol Brasileira 23:42-44
225. Riffaldi R, Levi-Minzi R (1983) Osservazioni preliminari sul ruolo dell Eisenia foetida nell'umificazione del letame. Agrochimica 27:271-274
226. Rivera AMC, Wright ER, López MV, Fabrizio MC (2004) Temperature and dosage dependent suppression of damping-off caused by Rhizoctonia solani in vermicompost amended nurseries of white pumpkin. Phyton 53:131-136

227. Rodriguez JA, Zavaleta E, Sanchez P, Gonzalez H (2000) The effect of vermicomposts on plant nutrition, yield and incidence of root and crown rot of gerbera (Gerbera jamesonii H. Bolus). Fitopatol 35:66-79
228. Rodriguez-Kabana R (1986) Organic and inorganic amendments to soil as nematode suppressants. J Nematol 18:129-135
229. Rouelle J (1983) Introduction of an amoeba and Rhizobium Japonicum into the gut of Eisenia fetida (Sav.) and Lumbricus terrestris L. In: Satchell JE (ed) Earthworm Ecology: From Darwin to Vermiculture, Chapman and Hall, New York. pp 375-381
230. Sánchez-Monedero MA, Roig A, Paredes C, Bernal MP (2001) Nitrogen transformation during organic waste composting by the Rutgers system and its effects on ph, EC and maturity of the composting mixtures. Bioresour Technol 78:301-308
231. Sainz MJ, Taboada-Castro MT, Vilariño A (1998) Growth, mineral nutrition and mycorrhizal colonization of red clover and cucumber plants grown in a soil amended with composted urban wastes. Plant Soil 205:85-92
232. Saumaya G, Giraddi RS, Patil RH (2007) Utility of vermiwash for the management of thrips and mites on chilli (Capiscum annum) amended with soil organics. Karnataka J Agric Sci 20:657-659
233. Scheu S (1992) Automated measurement of the respiratory response of soil microcompartments: active microbial biomass in earthworm faeces. Soil Biol Biochem 24:1113-1118
234. Scheuerell SJ, Sullivan DM, Mahaffee WF (2005) Suppression of seedling damping-off caused by Pythium ultimum, and Rhizoctonia solani in container media amended with a diverse range of Pacific Northwest compost sources. Phytopathology 95:306-315
235. Schmidt O, Doubre BM, Ryder MH, Killman K (1997) Population dynamics of Pseudomonas corrugata 2140R LUX8 in earthworm food and in earthworm cast. Soil Biol Biochem 29:523-528
236. Sembdner G, Borgman E, Schneider G, Liebisch HW, Miersch O, Adam G, Lischewski M, Schieber K (1976) Biological activity of some conjugated gibberellins. Planta 132:249-257
237. Senesi N, Saiz-Jimenez C, Miano TM (1992) Spectroscopic characterization of metal-humic acid-like complexes of earthworm-composted organic wastes. Sci Total Environ 117–118:111-120
238. Sharma S, Pradhan K, Satya S, Vasudevan P (2005) Potentiality of earthworms for waste management and in other uses - A Review. The Journal of American Science 1:4-16
239. Sharpley AN, Syers JK (1976) Potential role of earthworm casts for the phosphorous enrichment of runoff waters. Soil Biol Biochem 8:341-346
240. Shi-wei Z, Fu-zhen H (1991) The nitrogen uptake efficiency from 15N labeled chemical fertilizer in the presence of earthworm manure (cast). In: Veeresh GK, Rajagopal D, Viraktamath CA (eds) Advances in Management and Conservation of Soil Fauna, Oxford and IBH publishing Co, New Delhi. pp 539-542
241. Siddiqui ZA, Mahmood I (1999) Role of bacteria in the management of plant parasitic nematodes: a review. Bioresour Technol 69:167-179

242. Sidhu J, Gibbs RA, Ho GE, Unkovich I (2001) The role of indigenous microorganisms in suppression of Salmonella regrowth in composted biosolids. Water Res 35:913-920
243. Simsek Ersahin Y, Haktanir K, Yanar Y (2009) Vermicompost suppresses Rhizoctonia solani Kühn in cucumber seedlings. J Plant Dis Protect 9:15-17
244. Singh R, Sharma RR, Kumar S, Gupta RK, Patil RT (2008) Vermicompost substitution influences growth, physiological disorders, fruit yield and quality of strawberry (Fragaria x ananassa Duch.). Bioresour Technol 99:8507-8511
245. Singh UP, Maurya S, Singh DP (2003) Antifungal activity and induced resistance in pea by aqueous extract of vermicompost and for control of powdery mildew of pea and balsam. J Plant Dis Protect 110:544-553
246. Singleton DR, Hendrixb PF, Colemanb DC, Whitmana WB (2003) Identification of uncultured bacteria tightly associated with the intestine of the earthworm Lumbricus rubellus (Lumbricidae; Oligochaeta). Soil Biol Biochem 35:1547-1555
247. Sinha RK, Agarwal S, Chauhan K, Valani D (2010) The wonders of earthworms and its vermicompost in farm production: Charles Darwin's 'friends of farmers', with potential to replace destructive chemical fertilizers from agriculture. Agricultural sciences 1:76-94
248. Sinha RK, Bharambe G, Chaudhari U (2008) Sewage treatment by vermifiltration with synchronous treatment of sludge by earthworms: a low-cost sustainable technology over conventional systems with potential for decentralization. The Environmentalist 28:409-420
249. Sinha RK, Heart S, Agarwal S, Asadi R, Carretero E (2002) Vermiculture technology for environmental management: study of the action of the earthworms Eisenia foetida, Eudrilus euginae and Perionyx excavatus on biodegradation of some community wastes in India and Australia. The Environmentalist 22:261-268
250. Sinha RK, Herat S, Valani D, Chauhan K (2009) Vermiculture and sustainable agriculture. Am-Euras J Agric and Environ Sci, IDOSI Publication 5:1-55
251. Sipes BS, Arakaki AS, Schmitt DP, Hamasaki RT (1999) Root-knot nematode management in tropical cropping systems with organic products. J Sustain Agr 15:69-76
252. Sreenivas C, Muralidhar S, Rao MS (2000) Vermicompost, a viable component of IPNSS in nitrogen nutrition of ridge gourd. Ann Agr Res 21:108-113
253. Steffen KL, Dan MS, Harper JK, Fleischer SJ, Mkhize SS, Grenoble DW, MacNab AA, Fager K (1995) Evaluation of the initial season for implementation of four tomato production systems. J Am Soc Hort Sci 120:148-156
254. Stephens PM, Davoren CW, Doube BM, Ryder MH (1993) Reduced superiority of Rhizoctonia solani disease on wheat seedlings associated with the presence of the earthworm Aporrectodea trapezoids. Soil Biol Biochem 11:1477-1484
255. Stephens PM, Davoren CW, Ryder MH, Doube BM, Correll RL (1994) Field evidence for reduced severity of Rhizoctonia bare-patch disease of wheat, due to the presence of the earthworms Aporrectodea rosea and Aporrectodea trapezoides. Soil Biol Biochem 26:1495-1500
256. Stephens PM, Davoren CW, Ryder MH, Doube BM (1994) Influence of the earthworm Aporrectodea trapezoides (Lumbricidae) on the colonization of alfalfa (Medicago sativa L.) roots by Rhizobium melilotti strain LS-30R and the survival of L5-30R in soil. Biol Fertil Soils 18:63-70

257. Stephens PM, Davoren CW (1997) Influence of the earthworms Aporrectodea trapezoides and A. rosea on the disease severity of Rhizoctonia solani on subterranean cloves and ryegrass. Soil Biol Biochem 29:511-516
258. Stone AG, Scheurell SJ, Darby HM (2004) Suppression of soilborne diseases in field agricultural systems: organic matter management, cover cropping and other cultural practices. In: Magdoff F, Weil (eds) Soil Organic Matter in Sustainable Agriculture, CRC Press LLC, Boca Raton. pp 131-177
259. Subler S, Edwards CA, Metzger PJ (1998) Comparing vermicomposts and composts. Biocycle 39:63-66
260. Sudhakar K, Punnaiah KC, Krishnayya PV (1998) Influence of organic and inorganic fertilizers and certain insecticides on the incidence of shoot and fruit borer, Leucinodes orbonalis Guen, infesting brinjal. J Entomol Res 22:283-286
261. Suhane RK (2007) Vermicompost. Publication of Rajendra Agriculture University, Pusa. 88
262. Summers G, Felton GW (1994) Prooxidation effects of phenolic acids on the generalist herbivore Helicoverpa zea: potential mode of action of phenolic compounds on plant anti-herbivory chemistry. Insect Biochem Mol Biol 24:943-953
263. Sunish KR, Ayyadurai N, Pandiaraja P, Reddy AV, Venkateshwarlu Y, Prakash O, Sakthivel N (2005) Characterization of antifungal metabolite produced by a new strain Pseudomonas aeruginosa PuPa3 that exhibits broad-spectrum antifungal activity and biofertility traits. J Appl Microbiol 98:145-154
264. Suthar S (2010) Evidence of plant hormone like sub-stances in vermiwash: An ecologically safe option of synthetic chemicals for sustainable farming. J Ecol Eng 36:1089-1092
265. Suthar S, Singh S (2008) Vermicomposting of domestic waste by using two epigeic earthworms (Perionyx excavatus and Perionyx sansibaricus). Int J Evniron Sci and Technol 5:99-106
266. Swathi P, Rao KT, Rao PA (1998) Studies on control of root-knot nematode Meloidogyne incognita in tobacco miniseries. Tobacco Res 1:26-30
267. Szcech M, Rondomanski W, Brzeski MW, Smolinska U, Kotowski JF (1993) Suppressive effect of a commercial earthworm compost on some root infecting pathogens of cabbage and tomato. Biol Agric and Hortic 10:47-52
268. Szczech M, Smolinska U (2001) Comparison of suppressiveness of vermicomposts produced from animal manures and sewage sludge against Phytophthora nicotianae Breda de Haan var. nicotiannae. J Phytopathology 149:77-82
269. Szczech MM (1999) Suppressiveness of vermicomposts against fusarium wilt of tomato. J Phytopathology 147:155-161
270. Tajbakhsh J, Abdoli MA, Mohammadi Goltapeh E, Alahdadi I, Malakouti MJ (2008) Trend of physico-chemical properties change in recycling spent mushroom compost through vermicomposting by epigeic earthworms Eisenia foetida and E. andrei. J Agric Technol 4:185-198
271. Tan KH, Tantiwiramanond D (1983) Effect of humic acids on nodulation and dry matter production of soybean, peanut, and clover. Soil Sci Soc Am J 47:1121-1124
272. Thoden TC, Korthals GW, Termorshuizen (2011) Organic amendments and their influences on plant-parasitic and free living nematodes: a promosing method for nematode management. Nematology 13:133-153

273. Tiquia SM (2005) Microbiological parameters as indicators of compost maturity. J Appl Microbiol 99:816-828
274. Tiunov AV, Scheu S (2000) Microfungal communities in soil litter and casts of Lumbricus terrestris (Lumbricidae): a laboratory experiment. Appl Soil Ecol 14:17-26
275. Tiwari SC, Tiwari BK, Mishra RR (1989) Microbial populations, enzyme activities and nitrogen, phosphorous, potassium enrichment in earthworm casts and in the surrounding soil of pine apple plantation. Biol Fertil Soils 8:178-182
276. Tomati U, Grapppelli A, Galli E (1987) The presence of growth regulators in earthworm worked waste. In: Bonvicini Paglioi AM, Omodeo P (eds) On Earthworms. Proceedings of International Symposium on Earthworms, Selected Symposia and Monographs, Union Zoologica Italian, 2, Modena, Mucchi. pp 423-435
277. Tomati U, Grapppelli A, Galli E (1988) The hormone-like effect of earthworm casts on plant growth. Biol Fertil Soils 5:288-294
278. Toyota K, Kimura M (2000) Microbial community indigenous to the earthworm Eisenia foetida. Biol Fertil Soils 31:187-190
279. Trevors JT (1984) Dehydrogenase activity in soil. A comparison between the INT and TTC assay. Soil Biol Biochem 16:673-674
280. Umesh B, Mathur LK, Verma JN, Srivastava (2006) Effects of vermicomposting on microbiological flora of infected biomedical waste. ISHWM Journal 5:28-33
281. Vadiraj BA, Siddagangaiah D, Potty SN (1998) Response of coriander (Coriandrum sativum L.) cultivars to graded levels of vermicompost. J Spices Aromatic Crops 7:141-143
282. Valdrighi MM, Pera A, Agnolucci M, Frassinetti S, Lunardi D, Vallini G (1996) Effects of compost-derived humic acids on vegetable biomass production and microbial growth within a plant (Cichorium intybus) soil system: a comparative study. Agric Ecosyst Environ 58:133-144
283. Valenzuela O, Gluadia Y, Gallardo S (1997) Use of vermicompost as a growing medium for tomato seedlings (cv. Pltense). Revista Cientifica Agropecuaria 1:15-21
284. Vaz-Moreira I, Maria E, Silva CM, Manaia Olga C, Nunes (2008) Diversity of Bacterial Isolates from Commercial and Homemade Composts. Microbial Ecol 55:714-722
285. Vinken R, Schaeffer A, Ji R (2005) Abiotic association of soil-borne monomeric phenols with humic acids. Org Geochem 36:583-593
286. Vivas A, Moreno B, Garcia-Rodriguez S, Benitez E (2009) Assessing the impact of composting and vermicomposting on bacterial community size and structure, and functional diversity of an olive-mill waste. Bioresour Technol 100:1319-1326
287. Webster KA (2005) Vermicompost increases yield of cherries for three years after a single application. EcoResearch, South Australia.
288. Weltzien HC (1989) Some effects of composted organic materials on plant health. Agric Ecosyst Environ 27:439-446
289. Wilson DP, Carlile WR (1989) Plant growth in potting media containing worm-worked duck waste. Acta Hortic 238:205-220
290. Yardim EN, Arancon NQ, Edwards CA, OliverTJ BRJ (2006) Suppression of tomato hornworm (Manduca quinquemaculata) and cucumber beetles (Acalymma vittatum and Diabotrica undecimpunctata) populations and damage by vermicomposts. Pedobiologia 50:23-29

291. Yardim EN, Edwards CA (2003) Effects of organic and synthetic fertilizer sources on pest and predatory insects associated with tomatoes. Phytoparasitica 31:324-329
292. Yasir M, Aslam Z, Kim SW, Lee SW, Jeon CO, Chung YR (2009) Bacterial community composition and chitinase gene diversity of vermicompost with antifungal activity. Bioresour Technol 100:4396-4403
293. Yasir M, Aslam Z, Song GC, Jeon CO, Chung YR (2009b) Eiseniicola composti gen. nov., sp. nov., with antifungal activity against plant pathogenic fungi. Int J Sys Evol Microbiol 60:268
294. Yeates GW (1981) Soil nematode populations depressed in the presence of earthworms. Pedobiologiaogia 22:191-202
295. Zhang BG, Li GT, Shen TS, Wang JK, Sun Z (2000) Changes in microbial biomass C, N, and P and enzyme activities in soil incubated with the earthworms Metaphire guillelmi or Eisenia foetida. Soil Biol Biochem 32:2055-2062

PART III

BIODRYING

CHAPTER 9

Criteria for Assessing the Viablility of a Small Scale MSW Bio-Drying Plant Aimed to RDF Production for Local Use

E. C. RADA AND M. RAGAZZI

9.1 INTRODUCTION

One of the leading principles for EU countries is the concept of a waste management where in the first place there is the waste prevention, in the second one the most desirable options, recycling and energy generation, and in the last one the disposal of waste with no recovery of either materials and/or energy. The technologies who attracted considerable interest in the latest years are the Mechanical Biological Treatment (MBT) and the Biological Mechanical Treatment (BMT) (Muller, 2009; Tintner et al., 2009; Ragazzi and Rada 2009a). A part of the output from these processes, named generally Refuse Derived Fuel (RDF), can have industrial uses, for instance as a substitute of coal (Ragazzi and Rada, 2009b; Rada et al., 2005b; Gleis 2009).

Rada EC and Ragazzi M. "Criteria for Assessing the Viablility of a Small Scale MSW Bio-Drying Plant Aimed to RDF Production for Local Use." SARDINA 2009, Twelfth International Waste Management and Landfill Symposium, S. Margherita di Pula, Cagliari, Italy; 5 - 9 October 2009. CISA Publisher (2009). Used with permission from the publisher.

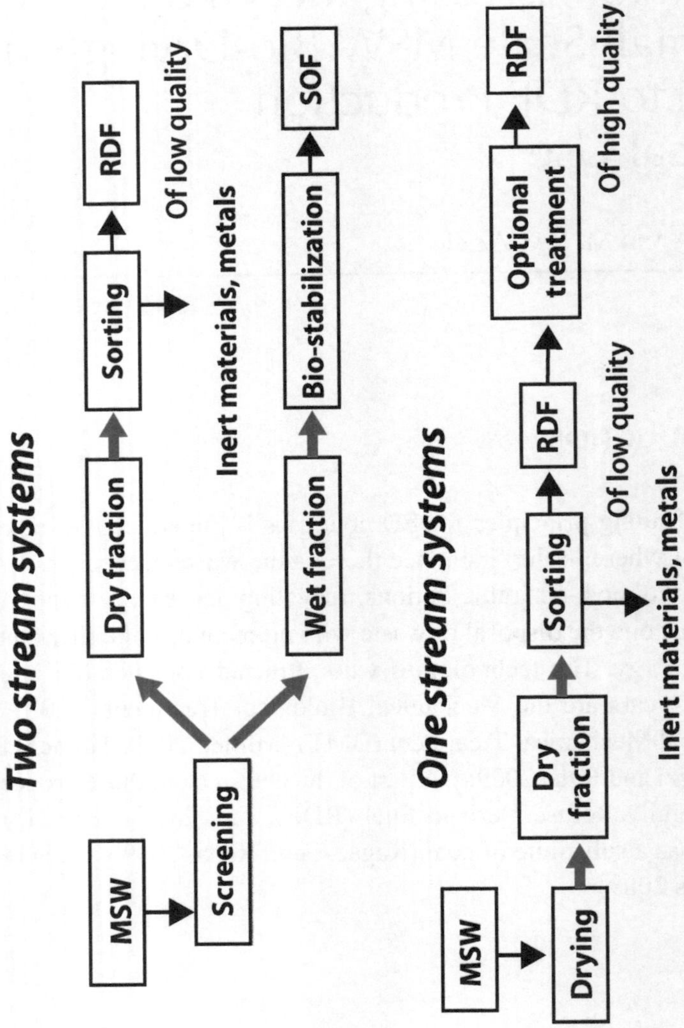

FIGURE 1: Two concepts of MBT plants

In the frame of Municipal Solid Waste (MSW) management, one of the topics more discussed is the treatment of the residual fraction that must be handled even when a good selective collection is performed. This fraction is named residual MSW (RMSW) and, from the theoretical point of view, should be composed exclusively by non-recyclable materials. In real scale the separation at the source cannot be perfect, thus RMSW has a variable composition, but generally one of the available options for its handling can be based on the generation of RDF. To this concern a typical strategy is co-combustion in cement works. In this case the partial substitution of coal follows the principles of the Kyoto protocol as a part of RDF is biomass.

Among the options for RDF generation, the two schemes that are often adopted are presented in Figure 1. One is based on the concept of two-streams treatment of MSW: an initial sieving separates MSW into an oversieve aimed to RDF generation and into an undersieve aimed to a biological treatment. The undersieve is generally too contaminated for allowing a generation of a good compost. This is the critical point of this strategy.

In the field of aerobic processes a distinction must be made between bio-stabilization and bio-drying (Rada et al., 2006a):

- Bio-stabilization: treatment of aerobic bioconversion generally applied to the fine fraction of residual MSW (under-sieve from mechanical selection, etc.) with the aim of generating a stabilized organic fraction (SOF) suitable to be landfilled or used for capping operations during the cultivation of the sanitary landfill;
- Bio-drying: treatment of aerobic bioconversion applied to residual MSW. After an inert separation, the final product is aimed to the use as RDF.

In the last two decades another approach has been developed: the one-stream treatment. In this case the main initial stage is the biological one, where a bio-drying process is performed. All the waste entering into the plant is introduced into the aerobic reactor with the target of decreasing the MSW water content, thanks to the exothermal reactions of a biochemical oxidation. The bio-dried material is easily refined in order to extract recyclable materials (glass, metals and inert). The obtained product after refinement could be classified as RDF, but in some cases, a post-sieving is required in order to separate the fine fractions (with a low energy content) and to obtain a high quality RDF (RDF_HQ). This post-sieving seems to be compulsory when a very high Lower Heating Value (LHV) must be guaranteed for the generated RDF.

Bio-drying is generally proposed for a decentralized management of MSW: a few satellite plants could generate RDF for a centralized use. This decentralization could be based on a small scale plant for local RDF exploitation. To this concern in the MSW sector there is not yet an approach useful for assessing the viability of bio-drying when a factory requiring heat wants to generate exactly the amount of RDF to be used in its own dedicated RDF burner.

In the present paper some criteria for assessing viability of this strategy (bio-drying plant coupled with a strictly dedicated RDF burner) are presented.

9.2 METHOD

9.2.1 DEFINITION OF THE CRITERIA

The criteria to be used for the analysis of the viability of bio-drying at small scale for industrial uses aimed to heat generation can be related to the following concepts:

- heat needed from the factory;
- costs for the conventional fuel to be substituted;
- residual MSW characteristics in the area around the factory;
- expected bio-drying performances referred to the available residual MSW in the area;
- assessment of RDF characteristics after post-refining;
- assessment of the amount of RDF needed for generating the required heat;
- assessment of MSW needed as input of the system to be implemented;
- assessment of the population involved for guaranteeing the MSW availability;
- assessment of the capital cost for MSW treatment;
- assessment of operating costs for MSW treatment;
- assessment of the capital cost for the RDF burner serviced to the factory;
- assessment of operating costs for the RDF burner;
- analysis of thermally integrated solutions (in case of variable energy demand);

- analysis of waste integrated solutions (in case of availability of special waste too);
- economic comparison with conventional fuels adoption.

9.2.2 CRITICAL ASPECTS OF THE CRITERIA

9.2.2.1 HEAT NEEDED FROM THE FACTORY

The method presented in this paper concerns the small scale applications of RDF, thus the target is basically related to heat generation (electricity generation needs a larger scale as the efficiency of the most conventional option for co-generation, that is a steam cycle, is very low at small scale). As a consequence it is important to know the amount of $MWh_{th}\ y^{-1}$ necessary for the plant where the substitution of conventional fuels by RDF must be analysed. It is clear that the matter is not only the stream of heat on yealy basis as, looking at a deeper detail this option, the characteristics of the heat generated by RDF must be compatible with the characteristics of the heat necessary for the plant. In this paper we suppose that the need of heat can be satisfied by the generation of off-gas (from RDF burning) at high temperature.

9.2.2.2 COSTS FOR THE CONVENTIONAL FUEL

When some calculations must be performed with long-time scenarios a big trouble is related to the cost of conventional fuel in a period that is even longer than 10 years. Suppose to decide to construct a RDF based strategy. The timing for the implementation of the system could be 2 years (indeed a preliminary step for desining and authorisation must be taken into account). The lasting of the RDF plants for the new strategy will be 15 years. Thus the cost-benefits analysis must take into account the dynamics of costs for conventional fuels for the next 17 years. Concerning the trend for conventional fuels it is clear that this parameter is not steady and can fluctuate significantly in the short time. The trend in the long time, more related to the calculations interesting for this paper, can be anyway assumed growing.

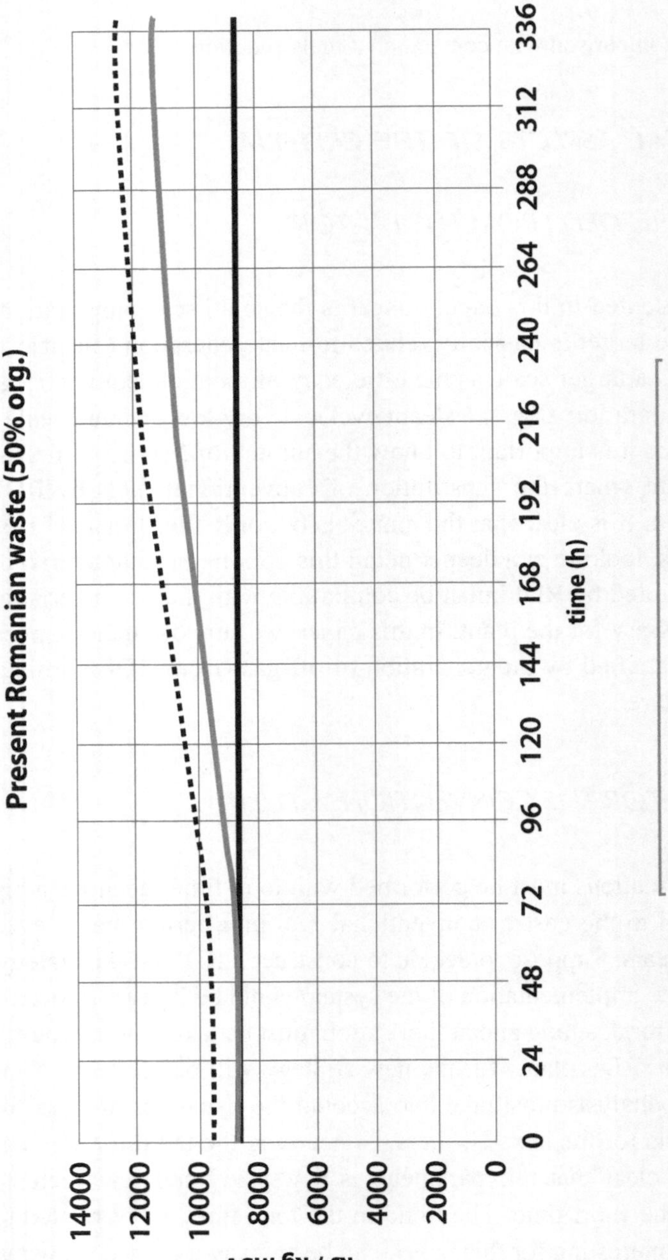

FIGURE 2: LHV dynamics during the bio-drying process.

9.2.2.3 RESIDUAL MSW CHARACTERISTICS

This is another critical aspect of the calculations as the MSW management can change dramatically if there is not a clear plan in the region where the plant is located. In particular, the strategy and the efficiency of selective collection can strongly modify the characteristics and the amount of MSW available in the area. More in general the economic development can increase the amount of MSW generated on per-capita base and can increase the energy content of MSW because of the increase of packaging. In the same time, the activation of more and more efficient strategies of selective collection can decrease the amount of MSW available for the studied strategy. In order to limit this problem it is important to set a strong agreement with the local authorities responsible for the management of MSW. Particular attention must be paied if the bio-drying process is proposed in an area where a high efficiency of organic fraction selective collection is performed or planned: in this case the organic fraction in the residual MSW can drop down to values that are not compatible with bio-drying (Rada et al., 2009a).

9.2.2.4 BIO-DRYING PERFORMANCES

An example of the Lower Heating Value (LHV) dynamics of RMSW, bio-dried material and RDF during the bio-drying treatment is reported in Figure 2, (Rada et al., 2008). In the reported example (referred to an area of Romania) the biodried material and LHV increase of RDF after two weeks is respectively around 30% and 44%. This is not an energy increase because it must be taken into account that the available mass of fuel is lower after the process.

The process allows "concentrating" the initial energy with a contemporary consumption of electrical energy. Generally the energy available at the end of the process is about 3% lower than the initial one (Rada 2005a), apart from the electricity needs that changes depending on the adopted technology. On large scale this two aspects are a weakness of the strategy if the final target were a burning a bio-dried material: you would pay for having less energy, but if the pre-treatment of residul MSW is aimed to the generation of a "product" (RDF) that open towards new options, this contradiction is no longer present. The international literature (Rada et al., 2006b; Rada et al., 2007a,b; Shao et al., 2009; Scotti 2009; Rada et al., 2009b; Grosso et al., 2009) make it available all the parameters needed to assess the behaviour of bio-drying referred to a residual MSW with a known composition.

9.2.2.5 RDF CHARACTERISTICS

From the theoretical point of view, a first calculation of the characteristics of RDF can be made assuming RDF as a refined bio-dried material: a simplified refining can be performed by separating metals, glass and inert. Depending on the initial value of LHV for the treated waste, the result of this simplified separation could be an RDF with a LHV higher or lower than 15 MJ kg^{-1} that could be assumed as a threshold for a viable result if RDF has to be used in a simplified burner (Rada 2005a). In case there is a need of increasing the LHV of RDF, an additional post-refining is necessary but a clear consequence is an increase in the amount of residues to be landfilled.

9.2.2.6 RDF AMOUNT

The amount of RDF needed is related to the amount of heat to be guaranteed at the factory. There are some details to be taken into account: the generation of RDF is not 365 days per year as at least a couple of weeks must be planned for maintenance problems. Also the use of RDF is not performed 365 days per year as the industrial process of most of the factories is based on a stop of at least a couple of weeks per year. This last stop could be synchronised with the first stop, but a correct design of the system should take into account also unfavourable situations. As a consequence there is the need of a (small) storage area for managing correctly this aspect.

9.2.2.7 MSW COLLECTION AREA

The extension of the area that must supply the system thanks to the collection of residual MSW can be easily assessed. It is compulsory that a clear MSW management plan must be previously adopted in order to allow generating reliable scenarios for the present and for the future. Talking about small scale of RDF generation, the area interested to the initiative is expected to be small. That means that the presence of industrial district close to a region with many small towns could couple towns and plants.

9.2.2.8 POPULATION INVOLVED

This parameter is a consequence of the above ones. The initiative does not affect the behaviour of the population of the area. The criteria of collection and the communication to the population will be the same of a conventional approach. The public communication will be simpler than the conventional strategy (centralized direct combustion of residual MSW) as the capacity of the RDF burner serviced to a factory could be very small.

9.2.2.9 CAPITAL COSTS FOR BMT

The capital costs for the construction of a BMT plant do not vary a lot from country to country as the main machines are constructed in few regions and thus the capital costs cannot be strongly affected by a different economical scenario (cheaper manpower). Some differences of capital costs could be anyway associated to a different approach in term of environmental impact minimisation. The process air treatment line can be performed in different ways with different costs (Rada 2005a): the most simple approach is related to the adoption of a bio-filter, while the most expensive solution is based on a thermal regenerative oxidation. In this last case there are also strong differences in term of operating costs because of a consume of methane.

9.2.2.10 OPERATING COSTS FOR BMT

If we consider the cost for manpower in the centre of Europe and we compare it with some areas of the east of Europe we can find differences even of 5 times. That strongly affect the economical balance as the cost for generating one unit of energy as RDF is very different from country to country. Of course the operating costs for bio-drying are only a part of the total costs. Taking into account the reduced capacity of BMT serviced to a specific factory, it could be interesting to integrate the personnel of BMT in the personnel of the factory as the limited amount of waste to be managed and the automation of the management of the process allow adopting a part-time.

9.2.2.11 CAPITAL COSTS FOR RDF BURNER

The capital costs for the construction of a RDF burner do not vary a lot changing country: also in this case the main machines are constructed in few regions and thus the capital costs cannot be strongly affected by a different economical scenario.

9.2.2.12 OPERATING COSTS FOR RDF BURNER

The different scenario of salaries for the workers can affect this item of cost. If the burner is constructed integrated in a factory, some operating costs could decrease thanks to an integrated management of the machines.

9.2.2.13 THERMALLY INTEGRATED SOLUTIONS

Suppose to have an existing factory with an existing natural gas burner. In case of fluctuations in the heat demand, a solution could be based on a steady generation of heat by burning RDF in a new burner and adopting the existing natural gas burner to follow the fluctuations of demand. In this case the integration simplify the management of the plant and guarantee a lower environmental impact as RDF burned by a steady combustion is less impacting. The steady combustion in also guaranteed by the fact that RDF has steady characteristics compared with the original characteristics of waste.

9.2.2.14 WASTE INTEGRATED SOLUTIONS

Until now the scenario has been discussed as based only on the explotation of MSW. In reality, if the area of interest is characterised by a significant generation of special waste with combustible characteristics, the integration of MSW with special waste could guarantee a better economical balance: special waste in some industrial districts could be found with an interesting energy content (wood, packaging, cardboard, etc.).

9.2.2.15 ECONOMIC COMPARISON ANALYSIS

The overall economical analysis can be developed according to the conventional cost-benefits approaches. Some advantages can be taken in countries where special funds (as the European Union structural funds) are available for a co-financing of the plants. Of course, that is not money freely received: some limitations in co-financing are to be taken into account. For instance it is possible to have a support for the BMT plant as it services the population of a region and, may be, the co-financing of a private burning of RDF could not be viable. Additional limitations could be related to the tariff to be applied to the population.

9.2.3 DEFINITION OF A CASE-STUDY

In some regions of the east of Europe there are many industrial districts under construction. The case-study developed in this paper refers to the case of an area with no selective collection (today), with a present generation of 1 kg of MSW per capita, with a LHV equal to 8 MJ kg^{-1}, with many small towns in the surroundings, with an uncorrect MSW management (today). The factory to be considered uses natual gas for generating hot off gas for its industrial process, with a power of 1 MW_{th} and no fluctuation of energy requierement. The manpower is cheap compared to the European averge (300 Euro $month^{-1}$ as a base salary).

TABLE 1: Parameter and results of the analyzed case-study.

parameter	value	units	comments
$power_{needed}$	1	MW_{th}	heat
operation	8000	$h\ y^{-1}$	
$energy_{needed}$	8000	$MWh_{th}\ y^{-1}$	heat
$fuel_{to\ substitute}$	0.45	Euro Nm^{-3}	natural gas
$LHV_{of\ methane}$	9.3	$kWh_{th}\ Nm^{-3}$	
$gas_{required}$	860215.05	$Nm^3\ y^{-1}$	
LHV_{RMSW}	8	$MJ\ kg^{-1}$	
$RMSW_{production}$	1	$kg\ inh^{-1}\ d^{-1}$	
$bio\text{-}drying_{loss}$	28%	-	weight loss
$bio\text{-}drying_{energy\ loss}$	3%	-	overall loss
Post-secycling	28%	-	weight loss
LHV_{RDF}	15	$MJ\ kg^{-1}$	viable as is
$ratio_{RDF/MSW}$	52%	-	
RDF_{needed}	1920	$t\ y^{-1}$	
MSW_{needed}	3692.3	$t\ y^{-1}$	
$people_{involved}$	10116	inh.	
$methane_{cost}$	387096.77	Euro y^{-1}	
$MSW_{present\ cost}$	30	Euro t^{-1}_{MSW}	Savings
equivalence cost	74.85	Euro t^{-1}_{MSW}	RMSW tariff

9.3 RESULTS AND DISCUSSION

In the Table 1 some details and the results of the case-study are presented. It must be pointed out that the approach has been developed "factory oriented", in the sense that the final result is the tariff of equilibrium that must be taken into account for veryfing if the cost for treating MSW according to this strategy is competitive with the cost of the present strategy. In the discussed case-study the present strategy is based on the use of natural gas. In practice, the tariff of equilibrium is the one that allow generating a unit of heat from MSW with the same cost of the generation of a unit of heat from natural gas (in different scenarios the conventional fuel could be coal, etc.).

The volume of the bio-container can be easily assessed taking into account the lasting of bio-drying (see Figure 2) and the density of residual MSW: 0.2 t m^{-3}. If only one factory activates this strategy, a second paralle bio-container must be constructed in order to guarantee the acceptance of MSW when the first reactor is full and operating. The cost for thermal energy generation can be lower in case of RDF based strategy if we take into account that the equivalence cost is reasonable: even if, at very small scale, we can increase the influence of personnel costs per unit of waste treated, the tariff seems to be viable. This approach is even more interesting where structural funds can be activated. The substitution of methane gives a higher local impact into the environment for some parameters as the quality of fuel is different. Anyway the local impact can be kept low also thanks to the small stream of waste locally combusted. On the contrary the global impact is lower (a part of RDF is biomass-like).

9.5 CONCLUSIONS

Results show that in areas where the cost for workers is low, the cost for a unit of thermal energy can be lower in case of RDF based strategy and where methane can be substituted. Anyway the method can be adopted to a variety of case-studies. This approach is more interesting for countries where European Union structural funds can be used to support the re-organization of the MSW management. In this case the advantage is higher as these funds can solve the problem of finding financing for the initial investment for covering capital costs, at least for a part of the installations (bio-drying). It is clear that the substitution of methane gives a higher local impact into the environment for some parameters, but the local impact can be anyway kept under acceptable levels. On the contrary the global impact is lower thanks the substitution of methane by RDF. Thus, this approach seems to be more interesting in countries where a target from the Kyoto protocol must be complied with.

REFERENCES

1. Gleis Markus (2009). Technische Lösungen zur Abfallentsorgung in Europa -Strategien und Erfahrungen aus 20 Jahren, Proceedings of Tagung :Gesellschaft, Umwelt und Gesundheit, Ecocenter Bolzano.
2. Grosso M., Rigamonti L., Malpei F (2009). I trattamenti biologici all'interno dei sistemi di gestione integrata dei rifiuti: bilanci energetici ed ambientali, In Compost ed energia da biorifiuti: Vismara, Grossu, Centemero, editore Dario Flaccovio, 63-90
3. Muller Wolfgang (2009). Mechanical Biological Treatment and its role in Europe, Proceedings of Waste–to-Resources 2009, 3rd International Symposium MBT&MRF, Cuvillier Verlag publischer, 1-10.
4. Rada E. C. (2005a). Municipal Solid Waste bio-drying before energy generation, PhD Thesis, University of Trento & Politehnica University of Bucharest.
5. Rada E. C., Franzinelli A., Taiss M., Ragazzi M., Panaitescu V., Apostol T. (2007a). Lower Heating Value dynamics during municipal solid waste bio-drying, Environmental Technology, vol. 28, 463-469.
6. Rada E. C., Istrate I. A., Ragazzi M. (2009b). Trends in the management of the Residual Municipal Solid Waste, Environmental Technology, Vol. 30, No. 7, 651–661.
7. Rada E. C., Ragazzi M., Apostol T. (2008). Role of Refuse Derived Fuel in the Romanian industrial sector after the entrance in EU, Proceesings of Fourth International Conference on Waste Management and the Environment, edited by Zamorano M, 89-96.
8. Rada E. C., Ragazzi M., Apostol T. (2009a). MSW bio-drying in different countries, Proceedings of Waste–to-Resources 2009, 3rd International Symposium MBT&MRF, Cuvillier Verlag publischer, 571-576.
9. Rada E. C., Ragazzi M., Panaitescu V., Apostol T (2006a). MSW bio-drying and bio-stabilization : an experimental comparison, Proceedings of ISWA 2005, International Conference: Towards integrated urban solid waste management system, CD version.
10. Rada E. C., Ragazzi M., Panaitescu V., Apostol T (2005b). Energy from waste: the role of bio-drying, Stintific Buletin, seria C: Electrical Engineering, vol 67, nr. 2, 69-76.
11. Rada E. C., Ragazzi M., Panaitescu V., Apostol T. (2006b). Experimental characterization of Municipal Solid waste bio-drying, Proceedings of Third International Conference on Waste Management and the Environment, Efited by Popov V.,Vol. 92, 295 – 302.
12. Rada E.C., Franzinelli A, Ragazzi M., Panaitescu V., Apostol T (2007b). Modelling of PCDD/F release from MSW bio-drying, Chemosphere, vol. 68, 1669-1674.

13. Ragazzi M, Rada E.C. (2009a). Tecnologie per la bioessiccazione, In Compost ed energia da biorifiuti: Vismara, Grossu, Centemero, Dario Flaccovio editore 307-330.
14. Ragazzi M., Rada E. C. (2009b). MSW bio-drying eco-balance and Kyoto protoco, Proceedings of Waste–to-Resources 2009, 3rd International Symposium MBT&MRF, Cuvillier Verlag publischer, 373-380.
15. Scotti Sergio (2009). Recyclable materials recovery after biological treatment of the residual fraction: quality improuvement and contribution to landfill diversion targets, 3rd International Symposium MBT&MRF, Cuvillier Verlag publischer, 381-388.
16. Shao Li-Ming, Zhang Dong-Qing, He Pin-Jin (2009). Effect of bio-drying on sorting and combustion performances of municipal solid waste, Proceedings of Waste–to-Resources 2009, 3rd International Symposium MBT&MRF, Cuvillier Verlag publischer, 389-399.
17. Tintner J., Smidt E., Meissl K., Binner E (2009). How in MBT-technology in 20 years?, , Proceedings of Waste–to-Resources 2009, 3rd International Symposium MBT&MRF, Cuvillier Verlag publischer, 111-118.

CHAPTER 10

Technical and Economic Efficiency of Utilization of Biogas from Animal Waste for Energy Generation

E. MINCIUC, R. PATRASCU, M. NORISOR, AND D. TUTICA

10.1 THE OPPORTUNITY OF BIOGAS UTILIZATION

Biogas is a combustible gas that is similar to natural gas or to propane-butane gas. The biogas (the biologic gas) traditionally can be obtained after anaerobe fermentation of organic matter.

Due to the global efforts for reduction of emissions of Green House Gases (GHG) and for replacing fossil fuels with renewable energy sources, in the last decade the interest towards the biogas has considerably increased. In addition, there should be mentioned that biogas generation represents an attractive solution for organic waste treatment with great advantages for the environment. In this context biogas and thus energy

Minciuc E, Patrascu R, Norisor M, and Tutica D. "Technical and Economic Efficiency of Utilization of Biogas from Animal Waste for Energy Generation." Printed with permission from the authors.

generation from animal and agricultural wastes represents a priority and needs state support, [1].

The formation of combustible gases due to the process of degradation of organic matter in an environment with molecular oxygen is a process that takes place naturally on our planet. This is the way the underground sediments were created, i.e. natural gas, due to degradation of plants and animals from the oceans and seas of pre-historic times. Identically, on the Earth surface there are formed combustible gases anywhere where there is organic matter that are in anaerobic conditions, [2].

The calorific value of biogas depends on non-combustible elements (N, CO_2); their presence reduces the energy quality of biogas. Lower Calorific Value of biogas varies between 20 and 24 MJ/m^3, [3].

TABLE 1: General characteristics of biogas, [4]

Composition	55-70 % methane 30-45 % carbon dioxide Other gases
Energy content	6.0 – 6.5 kWh/ m^3
Fuel equivalent	0.6- 0.65 l oil/m^3 biogas
Explosion limits	6-12 % biogas in the air
Ignition temperature	650 – 750 °C
Critical pressure	75-89 bar
Critical temperature	82.5 °C
Density	1.2 kg/ m^3
Odor	Rotten eggs (treated biogas has almost no odor)
Molar mass	16,.043 kg/mole

10.2 THE DOMAIN OF VARIATION OF CALORIFIC VALUE

Generally speaking, the biogas is a mixture of gases—methane, carbon dioxide, nitrogen, carbon monoxide, etc. The higher the methane weight in the biogas the higher is the energy quality of the fuel. Carbon dioxide is a neutral gas, it is not combustible, but the carbon monoxide is a combustible gas. The composition of generated biogas depends very much on the type of biomass used as raw material and on the technology used for biogas production, [1].

Typical composition of biogas, [5, 6, 7]:

- 40-75 % Methane, CH_4;
- 25-60 % Carbon dioxide, CO_2;
- 0-10 % Nitrogen, N_2;
- 0-1 % Hydrogen, H_2;
- 0-3 % Hydrogen sulphide, H_2S.

TABLE 2: Energy value of biogas compared to other fuels

Fuel	Unit	Calorific value kcal/unit	Equivalent of unit for 1 m^3 of biogas
Biogas with cu 60 % methane, 0 °C, 1 bar	m^3	5130	1
Fresh wood	kg	1300-1800	3.95-2.85
Dried wood	kg	1800-2200	2.85-2.34
Lignite	kg	1800-3800	2.85-1.35
Coal dust briquettes	kg	4000-6800	1.28-0.76
Oil fuel	kg	9400-9500	0.55-0.54
Diesel	kg	10000-11000	0.51-0.47
Natural gas methane	m^3	8500	0.60
Liquefied petroleum gas	m^3	22000	0.23

10.3 SOURCES FOR BIOGAS GENERATION

Biogas can be generated from any organic matter of vegetal or animal origin, including wastes. The raw material used for biogas generation can be of different origin, [1]:

Biomass from agriculture:

- Different herbage, lucerne.
- Wastes (tree leafs, roots, straws, etc.).
- Plants (sunflower, sugar beet, etc.) and their silage.

Livestock products:

- Animal manure.

Municipal organic wastes:

- Water and sewage sludge.
- Municipal solid organic wastes.

Organic wastes of industrial origin:

- Food wastes (from restaurants, canteen, etc.).
- Wastes and residues from agro-food industry.
- Wastes from production processes of food industry.
- Fats.

The biogas that is generated using the process of anaerobic digestion is cheap and is a renewable energy source. In the combustion process of biogas a neutral CO_2 is generated. The anaerobic digestion also offers the possibility of treatment and recycling of different types of wastes, such as agricultural wastes, sewage waters and sludge, industrial waste water, etc. using a sustainable and environmentally friendly solution, [8].

10.4 CASE STUDY: ANALYSIS OF EFFICIENCY OF UTILIZATION OF BIOGAS FROM A PIG FARM IN A COGENERATION PLANT

10.4.1 DESCRIPTION OF THE ANALYZED CASE STUDY

Technical and economic analysis has as starting point the fact that animal waste treatment is compulsory and, thus, it is desired to find an advantageous solution for energy valorization of generated biogas.

The reference solution, for technical and economic evaluation of analyzed biogas generation solution it is considered that the agricultural wastes are treated in specialized tanks and generated biogas is flame burned. For covering energy demands of the farm electricity is bought from the national power grid, and heat is generated locally in a natural gas fuelled boiler.

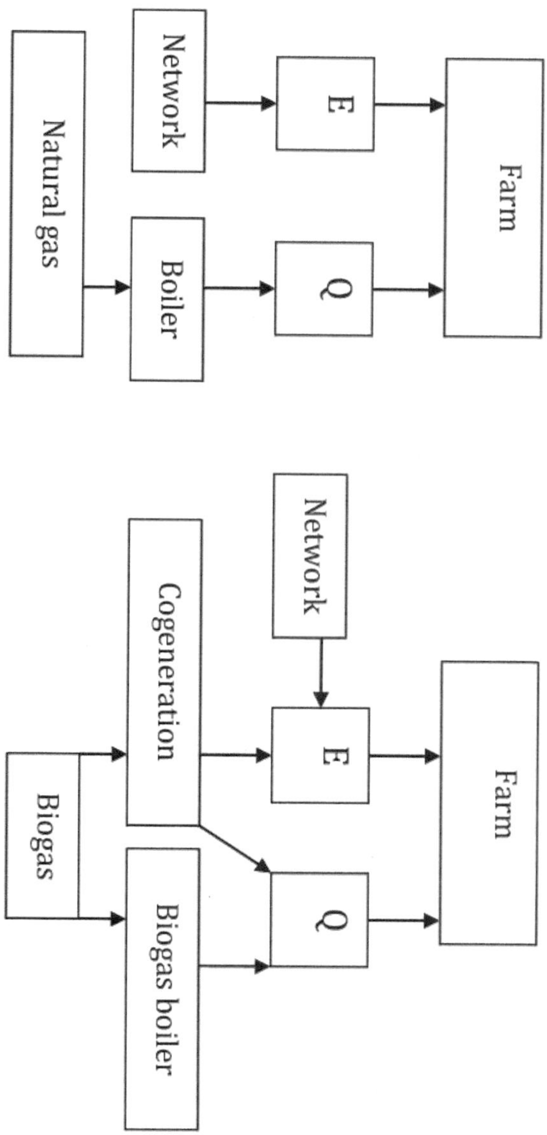

FIGURE 1: Reference solution (a) and biogas cogeneration solution (b) Notations used: E – electricity (consumed at the farm) and Q – heat (consumed at the farm)

The analyzed cogeneration solution involves installation at the farm of a anaerobic digester, a storage tank, a biogas treatment plant and an internal combustion engine that operates on biogas. When the internal combustion engine is not operated the biogas is combusted in a boiler and the surplus is flame burned.

10.4.2 DETERMINATION OF DAILY BIOGAS ENERGY POTENTIAL

The waste quantity that comes from an animal species depends on many factors, such as: animal category, applied keeping system (indoor, outdoor, etc.), existence or non-existence of bedding, type of used bedding, floor type (with or without grate), the way of watering the animals and cleaning the floor, type of forage used to feed animals, which leads to different water consumption, the way of feeding the animals, [1].

The daily biogas flow is:

$$D = SV \cdot ef \cdot d_{SV} \cdot N \cdot \varepsilon \ [Nm^3/zi] \quad (1)$$

where:

- N: number of animals
- SV: quantity of volatile solids [kg/day animal]
- ef: efficiency
- ε: efficiency of wastes collection
- d_{sv}: flow of destroyed SV [Nm^3/kgSV]

The calorific value of biogas is:

$$PCI_b = c_i^{CH4} \cdot p_{CH4} \ [Wh/Nm^3] \quad (2)$$

where:

- p_{CH4}: specific calorific value [Wh/%CH_4]
- c_i^{CH4}: methane concentration [%]

Energy generated daily through anaerobic digestion of animal wastes is:

$$B = D \cdot PCI_b \text{ [Wh/zi]} \quad (3)$$

10.4.3 DETERMINATION OF DAILY ENERGY CONSUMPTIONS

The estimated values of maximum specific energy consumptions are:

- Electricity specific consumption e= 6.95 [We/cap]
- Heat specific consumption q= 16.65 [Wt/cap]

Daily electricity quantity consumed during winter days:

$$E = e \cdot N \text{ [kWe/day]} \quad (4)$$

Electricity consumption for spring-autumn days has been considered as being equal to 0.7 from the winter days consumption, and for summer days electricity consumption has been considered as being equal to 0.4 from winter days consumption.

Daily heat quantity

$$Q = N \cdot q \text{ [kWT/day]} \quad (5)$$

The heat consumption during spring-autumn days has been considered as being equal to 0.5 from the consumption of winter days, and for summer days has been considered equal to 0.2 from the consumption of winter days.

The number of days taken into consideration for the calculus is as follows:

- Winter days : 150 days/year
- Spring-autumn days: 125 days/year
- Summer days: 90 days/year

The quantities of electricity and heat that are consumed annually at the farm have been determined through multiplication of daily consumption with typical number of days.

There are taken the following hypotheses:

- The own electricity consumption of the farm has been estimated as being 60 % from the total consumption; the rest is needed for biogas generation and utilization.
- The own heat consumption of the farm has been estimated as being 50 % from the total consumption; the rest is needed for biogas generation and utilization.

10.4.4 EQUIPMENT SIZING FOR BIOGAS VALORIZATION

The volume of the digester needed for anaerobic fermentation of wastes is:

$$V = \lambda \cdot v \cdot N \cdot \varepsilon \text{ [m}^3\text{]} \quad (6)$$

where:

- λ: coefficient of over-sizing the digester
- v: specific volume [m^3/animal]

The size of the internal combustion engine and boiler that use biogas is determined taking into account the daily energy content of produced biogas.

The biogas utilization module has the following priority phases:

1. Combustion in internal combustion engine
2. Combustion in boiler

3. Storage
4. Flame burning

The installed electric power of the internal combustion engine is:

$$P_{coge} = B \cdot \eta_e / 24 \text{ [We]} \quad (7)$$

where:

- η_e: electric efficiency of the cogeneration unit

During the periods when the engine is not operated biogas shall be used in boiler. The maximum capacity of the boiler is:

$$q_{caz} = B \cdot \eta_{caz}/24 \text{ [Wt]} \quad (8)$$

where:

- η_{caz}: boiler's efficiency

A basic condition is to use biogas without reselling the excess of electricity (case A).

The biogas consumption in the cogeneration unit is:

$$b_{mot} = P_{el} / \eta_{el} \text{ [W]} \quad (9)$$

The biogas consumption in the boiler is:

$$b_{caz} = q_{caz} / \eta_{caz} \text{ [W]} \quad (10)$$

The annual electricity generation in cogeneration unit is:

$$E_{an} = P_{el} \cdot T_{an,e} \text{ [kWh/year]} \tag{11}$$

where:

- $T_{an,e}$: number of operating hours during the year at maximum load

The annual heat generation in the boiler is:

$$Q_{caz,an} = Q_{caz} \cdot T_{an,caz} \text{ [kWh/year]} \tag{12}$$

10.4.5 TECHNICAL AND ECONOMIC ANALYSIS

From economic point of view there have been analyzed 2 different solutions:

1. An integrated plant: For this solution it has been considered that wastes are treated and produced biogas is used for covering a part farms own needs. Produced electricity is consumed on-site. The electricity peak demand is covered from the national power grid. In this case the company shall not obtain any green certificates.
2. Separate plant: For this solution it has been considered installation of a CHP plant and a biogas boiler, which will produce heat for farm and digesters, and whole amount of electricity shall be sold into the national power grid, thus obtaining green certificates.

The analysis of economic efficiency has been performed based on technical and economic criteria using discounted values:

1. Net Present Value, NPV: is the most comprehensive and observable indicator of economic efficiency. It includes all economic efforts and effects that have been realized during the project lifetime.
2. Internal Rate of Return, IRR: it reflects the project's capacity to generate profit, offering clues about acceptability of loans. IRR represents

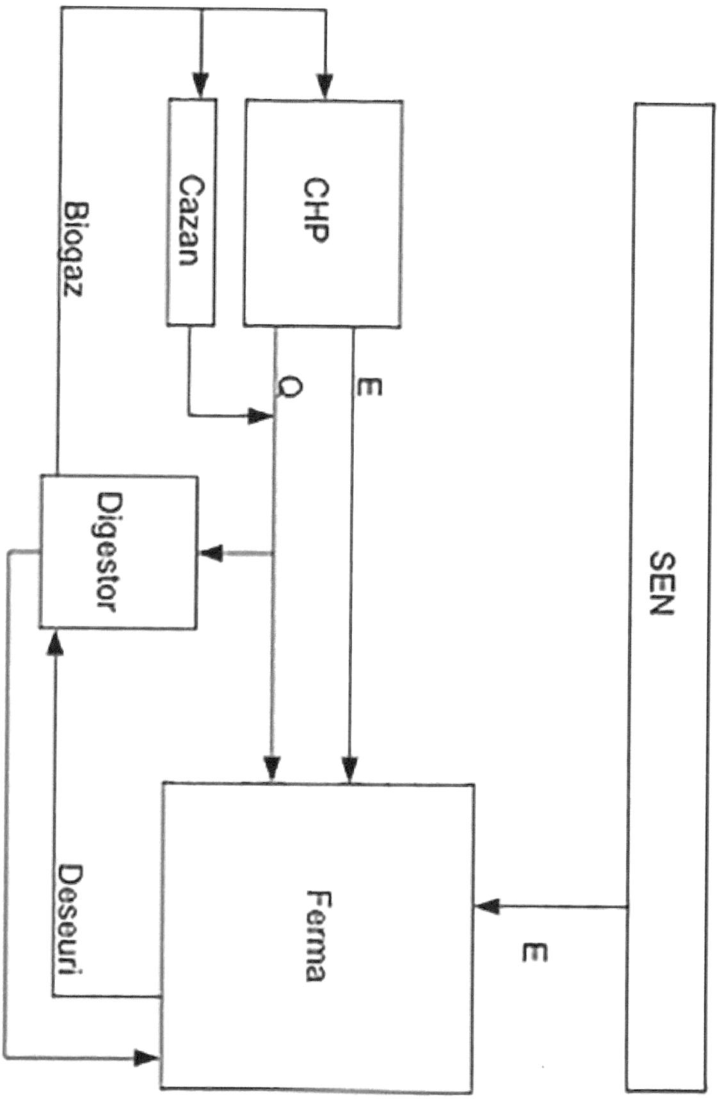

FIGURE 2: General layout of the integrated plant

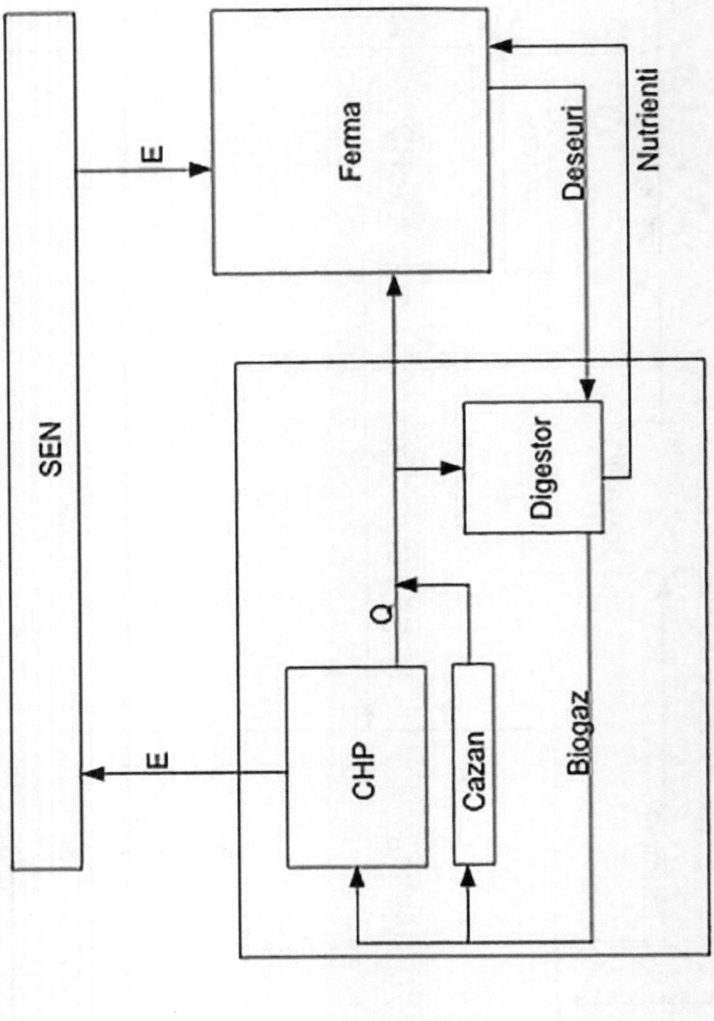

FIGURE 3: General layout of separate plant

the value of the discount rate for which NPV is zero. For the case IRR is greater than discount rate the project is considered to be feasible.

3. The Discount Payback Period, DPP: it represents the period of investment return using discount values.

For determination of the above mentioned criteria there have been considered the following elements:

Investment Estimation

There have been considered investments in the following equipment: digester, CHP unit, biogas boiler, and biogas storage tank and filtration equipment.

The investment in the digester is calculated by multiplying the specific investment for digester, i_{dig} with digester volume:

$$I_{dig} = i_{dig} V \ [€] \quad (13)$$

The investment in the cogeneration unit is estimated using specific investment and installed capacity:

$$I_{cog} = i_{cog} P_{cog} \ [€] \quad (14)$$

The investment in the biogas boiler is estimated using specific investment:

$$I_{caz} = i_{caz} q_{caz} \ [€] \quad (15)$$

For the filtration equipment the investment depends on the maximal biogas flow that is consumed by internal combustion engine:

$$I_{filt} = i_{filt}\, b_{mot} \ [€] \tag{16}$$

The investment in the biogas storage tank is calculated using specific investment and the tank volume.

$$I_{stoc} = {}_{istoc} V_{stoc} \ [€] \tag{17}$$

The total investment for the biogas plant shall be the sum of all investments in the different equipment needed for the process:

$$I_{total} = I_{dig} + I_{cog} + I_{caz} + I_{filt} + I_{stoc} \ [€] \tag{18}$$

Estimation of the Maintenance Costs

These costs are estimated taking into consideration specific maintenance costs for different equipment.

Estimation of Expenses with Fuel and Electricity

For the cogeneration solution the expenses with fuel include expenses for natural gas purchase needed for start up the cogeneration unit that then generates heat for heating up the digester till biogas is generated; these expenses are negligible during the summer time.

For the case (A) there is a need for a certain amount of electricity bought from the national power grid for covering up all farm's needs. For increasing economic efficiency of the cogeneration unit there should be used first all electricity generated by it and then to buy the difference from the grid.

The annual sum allocated for these expenses shall the electricity quantity purchased from the national power grid multiplied by electricity tariff.

For the case (B) it is necessary to buy all the amount of electricity from the national power grid for covering farm's needs and for biogas generation; it is due to the fact that all electricity generated by CHP unit is sold into the national power grid with the purpose of getting green certificates.

Estimation of Revenues

1. For the case of the integrated plant the annual revenues of the project represent the savings due to power and heat generation on-site using biogas.

Annual savings due to electricity generation on-site are:

$$VA_{el} = t_{el}\, E_{an} \quad (22)$$

Annual savings due to heat generation on-site using biogas are:

$$VA_{t} = t_{t}\, (E_{t,cog} + E_{t,caz}) \quad (23)$$

The total annual revenues/savings shall be the sum of the previous two:

$$VA_{tot} = VA_{el} + VA_{t} \quad (24)$$

2. For the case of separate plant the annual revenues represent the incomes from electricity selling into the grid and from selling green certificates.

10.4.6 HYPOTHESES USED FOR THE ANALYSIS

For the technical and economic analysis there have been used the following hypotheses:

- There has been considered a farm with 10000 animals.
- The daily biogas flow at the farm is considered to be constant.
- The discount rate is 10%.
- The number of days used for annual energy consumption calculation have been considered as follows:

- Winter days: 150
- Spring-autumn days: 125
- Summer days: 90
- The capacity of the storage tank shall ensure the stocking of biogas generation for 6 hours. The biogas storage equipment shall allow a uniform biogas flow and high biogas quality.
- The installed capacity of the boiler shall be chosen so the boiler can burn as much biogas as possible during the periods when the internal combustion engine is not operated, but without producing more heat than necessary for the farm.
- Biogas valorization generated through waste treatment involves biogas treatment. There shall be installed biogas treatment equipment that are sized depending in the flow of treated biogas.

10.5 RESULTS OF THE ANALYSIS FOR A PIG FARM

The results of the technical and economic analysis for a 10000 pig farm are shown in table 3.

TABLE 3: Results of technical and economic analysis for a pig farm

Item		Unit	Value case A	Value case B
Number of animals		-	10000	
Efficiency of waste collection		%	95	
Specific daily flow of biogas		Nm^3/day animal	0.12	
Average methane concentration		%	58	
Lower calorific value of biogas		kWh/m^3	6.09	
Daily biogas flow		Nm^3/day	1117.2	
Daily energy potential of biogas		kWh/day	6803.75	
Hourly biogas energy		kW	283	
Digester volume		m^3	1596	
Engine electric installed capacity		kWe	112.5	
Engine heat installed capacity		kWt	143.8	
Boiler installed capacity		kWt	250	
Engine operation time	winter	h/year	2225	4170
	spring-autumn	h/year	1300	1750
	summer	h/year	500	500

TABLE 3: *Cont.*

Item		Unit	Value case A	Value case B
Boiler operation time	winter	h/year	1120	0
	spring-autumn	h/year	250	0
	summer	h/year	0	0
Annual quantity of generated electricity		MWhe	452.812	722.250
Annual quantity of heat generated in engine		MWht	578.795	923.196
Annual quantity of heat generated in boiler		MWht	342.500	0
Biogas surplus flame burned annually		m^3		
Annual amount of electricity bought from grid		MWhe	3.83	0
Average tariff of bought electricity		€/MWhe		70
Heat tariff in network		€/MWht	40	-
Tariff for selling heat to farm (B)		€/MWht	-	30
Tariff for selling electricity to grid		€/MWhe	-	40
Minimum price of green certificate		€/CV	-	28.8
Digester investment		Thousands €		164.39
CHP unit investment		Thousands €		90.90
Boiler investment		Thousands €		23.75
Filtration equipment investment		Thousands €		6.14
Storage tank investment		Thousands €		33.00
Works		Thousands €		40.00
Cost for capital repair discounted (B, year 7)		Thousands €	-	12
Total investment		Thousands €	358.18	370.18
Annual maintenance costs (engine+boiler)		Thousands €	4.65	5.78
Annual costs with electricity purchase		Thousands €	0.267	-
Annual cost with electricity for own needs (B)		Thousands €	-	12.68
Total annual expenses		Thousands €	4.92	18.46
Annual savings due to electricity		Thousands €	19.02	-
Annual savings due to heat		Thousands €	18.43	-
Annual revenues due to green certificates		Thousands €	-	62.48
Annual revenues due to selling of electricity		Thousands €	-	28.89
Annual revenues due to selling heat to farm			-	13.85
Total annual revenues		Thousands €	37.44	105.14
Net annual revenue		Thousands €	32.52	86.68
Discount rate		%		10
Project lifetime		years		10

TABLE 3: *Cont.*

Item	Unit	Value case A	Value case B
Total discounted costs	Thousands €	388.43	483.7
Total discounted revenues	Thousands €	230.28	646.61
NPV	Thousands €	<0	162.91
IRR	%	-	19 %
DPP	years	-	2.3
Grant need for IRR 12 %	%	50	-

10.6 CONCLUSIONS

The biomass and waste utilization can be justified especially due to ecological considerations, due to the need of waste treatment.

For increasing biogas production there can be combined different types of wastes/raw materials. There can be mixed different animal manure and also different vegetal substances.

Within the context of sustainable development, where the energy and environmental issues should be solved simultaneously, biogas production and its utilization as an energy source is of great interest due to many advantages, such as, [9]:

1. Energy generation.
2. Waste treatment.
3. Stabilization of animal manure.
4. Reduction of pathogen loads.

The feasibility of implementation of biogas equipment, generally, depends on many key factors:

- Biogas production scale.
- Availability of energy crops and their specific costs.
- Waste production at analyzed farm.
- Policies and regulations regarding waste treatment.
- The value of bio-fertilizers.
- Possibility of local use of heat.

- Reasonable costs for investments.
- Availability of equipment and buildings that can be integrated in the biogas plant concept.
- Availability of grants.

The results of the analyzed cases lead to the following conclusions:

- The heat demand of agricultural farms can be integrally covered by biogas generated through waste treatment.
- With no financial facilities a CHP biogas project for agricultural farms can hardly be feasible from the economic point of view.
- In situation with green certificates case B is the best one.
- For the case when green certificates award for the next 10 years is not certain it is preferable the solution with financial grants.

The performed analysis established a set of criteria that should be analyzed when implementing such a solution:

- The biogas unit should be placed near to raw materials to minimize the costs.
- The biogas unit should be placed near to town so the heat can be sold.
- More than 70 % of raw material should come from own sources or there should contracts for supply for at least 10 years.
- The raw material flow should quiet stable.
- The utilization of secondary product, digestate, can significantly increase economic feasibility of the project.

REFERENCES

1. Valentin Arion, Olga Şveţ şi Constantin Borosan, Producerea biogazului din deşeuri animaliere, Universitatea Tehnică a Moldovei, Chişinău 2013
2. Tudor Ambros, Surse regenerabile de energie , editura „Tehnica –info",1999
3. Adrian Badea, Horia Necula, Surse regeerabile de energie, Agir Bucuresti 2013
4. Ing.Adrian Eugen Cioabă ,Contribuţii teoretice şi experimentale privind poducerea de biogaz din deşeuri de biomasă, teză de doctorat 2009
5. McKendry Peter, Energy production from biomass:Conversion technology, Bioresource Technology, 2001
6. http://www.nikolicivasilie.ro/lucrari-stiintifice/Biogaz%20curs.pdf ;
7. Ioan Bitir-Istrate,Eduard Minciuc, Valorificarea biogazului pentru producera energiei electrice si termice, Cartea Univrsitară,2003
8. http://www.gazetadeagricultura.info/eco-bio/565-energie-regenerabila/14414-tehnologii-actuale-de-obtinere-a-biogazului.html
9. http://www.probiopol.de/10-Care-sunt-componentele-une.45.0.html?&L=1

10. Augustin Ofiţeru, Mihai Adamescu, Florian Bodescu,Biogazul ghid practic, 2008
11. Ioan Paunescu,Gigel Paraschiv, Managementul mediului si obtinerea biogazului în fermele suinicole, Facultatea de Ingineria Sistemelor Biotehnice - UPB
12. Valentin Arion, Olga Şveţ şi Constantin Borosan,Utilizarea biogazului la producerea căldurii şi electricităţii, Chişinău, 2013
13. V. Arion, dr.hab., prof.univ., C. Gherman, T. Tutunaru ,Fezabilitatea economico-financiară a producerii energiei electrice şi termice la mini-cet prin valorificarea biogazului , Universitatea Tehnicăa Moldovei,2009
14. Dominik Rutz,Utilizarea durabilă a energiei termice a instalaţiilor de biogas, 2012 WIP Renewable Energies, Munchen, Germania, 2012
15. Adrian Eugen Cioabla, Ioana Ionel, Biogazul, energie pentru viitor, editura Politehnica , 2011
16. Victor Emil Lucian ,Surse alternative de energie, Bucuresti, matrix Rom 2011
17. Ing. Nicolae Lontiş, Cercetări teoretice şi experimentale privind cogenerarea cu motot M.A.I., funcţionînd cu biocombustibil, teză de doctorat, 2008
18. Robert Pecsi,Utilizarea biogazului ca sursă de energie electrică, Facultatea de Instalatii, UTCB
19. Dr Ralf Utermohlen ,Măsuri, metode şi soluţii pentru poligenerarea industrială a Biogazului în România. (AGIMUS) Germania , martie 2009
20. Screening the rise of fermentable wastes & market prices for energy and waste treatment in Romania, Research report carried out by SC Project Developer (ProDev), Romania, December 2008
21. Michael Kuttner, dr. Sigrid Kusch, Achim Kaiser, dominik Durrie, Economic modelling of anaerobic digestion/ biogas installations in a range of rural scenarios in cornwall, prepared byinternational biogas and bioenergycentre of competence, 21 august 2008

CHAPTER 11

Potential of Bio-Drying Applied to Exhausted Grape Marc

ELENA CRISTINA RADA AND MARCO RAGAZZI

11.1 INTRODUCTION

The wine industry produces every years in the world about 270 millions of hectolitres of wine, of which about the half in European Union (EU), and in particular in Spain, Italy and France; Italy and France compete each year for being the main world wine producer.

The central wine regulations in the EU is entitled *Council Regulation on the common organization of the market in wine,* of 29 April 2008 and was supplemented by *Council Regulation (EC) No 491/2009* of 25 May 2009 amending Regulation (EC) No 1234/2007 that establish a common organization of agricultural markets and specific provisions for certain agricultural products. From the regulatory point of view the wastewater generated from the sector must be treated according to the principals available for the management of this kind of discharge.

This article is an original article. Reprinted with the authors' permission.

The Province of Trento, Italy, where the present research was performed, generates about 800,000 hectolitres of wine per year. To produce one liter of wine about 1.3 kg of grape are utilized, and about the 20% of this grape represents a 'waste' of the wine industry (1). The exhausted grape marc (GM) or spent GM, is the material that results after fermentation and distillation of the grapes in the wine companies and has a high organic content (2). This exhausted grape marc contains generally about 72% skins, 20-22% seeds and 7-8% stalks (3). The exhausted GM used for the experimentation is presented in Figure 1.

Grape marc has traditionally been used to produce pomace brandy (such as grappa), grape-seed oil, polymers and seldom as supplement in animal feed (4,5,6). Today, it is mostly used as fertilizer or for producing renewable energy. However, in 2011, Devesa-Rey presented a study regarding the possibility of valorization of winery waste vs. the costs of not recycling (2). The valorization scenarios didn't take into account the potential role of GM bio-drying. This lack confirms the originality of the contents of the present paper, where an alternative technique to treat exhausted grape marc is presented for energy purposes. The chosen process is an aerobic biological-mechanical treatment: the bio-drying one. The aim of this process is the exploitation of the exothermic reactions for the evaporation of the highest part of the water present in the treated material with the lowest conversion of organic carbon.

The experimental results of a pilot scale bio-drying process applied to exhausted grape marc are presented and discussed in this paper. An important advantage of the use of bio-drying is that the final product, like Solid Recovered Fuel (SRF), is considered biomass opening to the incentives market for renewable energy.

An important parameter for the SRF characterization is the potential rate of microbial self heating. This rate can be determined by the real dynamic respiration index (RDRI = average value of the respiration indexes representing 24 h showing the highest aerobic microactivity). Until 2010 in Italy this parameter has been requested only for the stabilized organic fraction and not for fuels (7). With the transposition of the technical norm CEN/TR 15590 in the decree 205/2010, this RDRI parameter became a key parameter for SRF (8).

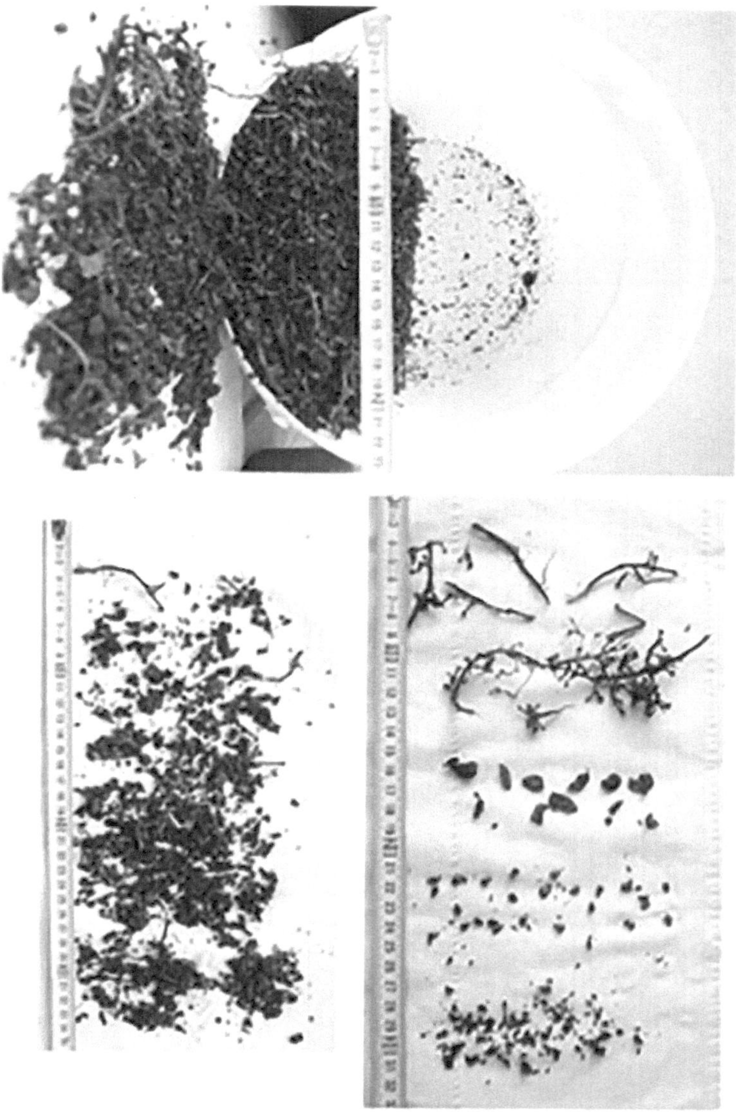

FIGURE 1: Exhausted grape marc: seeds, skins and stalks

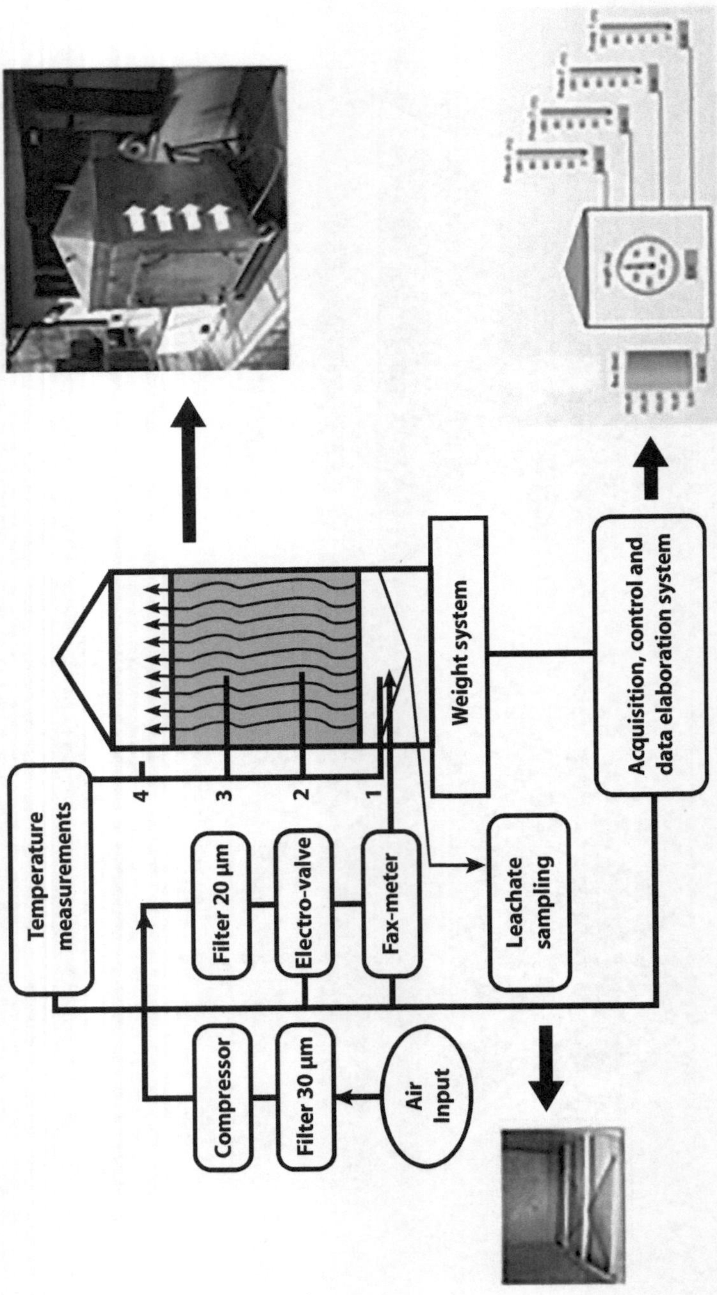

FIGURE 2: Biological reactor

Bio-drying of exhausted grape marc can be an alternative to thermal drying generally performed in the sector by an integrated plant of thermal-pre-treatment and combustion. This integration has the disadvantage of requiring a centralization of the pre-treatment. Grape marc bio-drying could open to a decentralized pre-treatment of this organic substrate before a centralized combustion.

11.2 DESIGN AND METHODS

In order to develop the bio-drying process, a bioreactor of 1m^3 was used at University of Trento (9,10). When performing a run, the adopted reactor (Figure 2) was placed on an electronic balance for monitoring the mass loss during the bio-drying process. The process air was sent into the reactor through a steel diffuser, placed at the bottom. The air crossed upward the matter from the lower part, activating the biological reactions and went out of the biological reactor from the upper part, extracting a part of the water content of the material. The leachate was discharged continuously in order to avoid alterations of the dynamics of weight loss. For monitoring the temperature during the bio-drying process, four temperature probes were placed inside the reactor, one at the air outlet/inlet and two at intermediate levels. All these equipments were connected to a data acquisition system developed for a good management of the process.

A bio-chemical model based on an energy and mass balance was used for a correct interpretation of the experimental data reported in this paper (9). In particular the model reconstructs the composition of the bio-dried material through a process balance taking into account the mass, the bio-chemical reactions stoichiometry and the available energy, resulting at the end a determined mathematical system that has as input the quantity of water, carbon, hydrogen, oxygen and nitrogen contained in the initial mass of exhausted GM and as output the amount of water, carbon, hydrogen, oxygen and nitrogen consumed/removed during the bio-drying process. The model also describes the dynamics of the calorific value during the bio-drying process. Additionally the model gives the dynamics of the following parameters of interest:

- internal energy consumption (overall available Lower Heating Value, LHV);
- volatile solids reduction (assessed from an energy balance);
- net moisture extraction (taking into account the role of hydrogen);
- highest theoretical NH_3 emission factor (supposing that 100% of nitrogen is converted into ammonia).

The aeration criteria were adopted in order to optimize the temperature of the biological bio-drying process taking into account the microbial activity needs: the highest value was limited at 65°C.

For the experimental run, that lasted 30 days, about 96 kg of exhausted grape marc were used. The height of the material in the bio-drying pilot plant was about 22 cm. The moisture content of the exhausted grape marc and of the bio-dried one was determined using a stove at 105°C for 24 hours.

For the volatile and total solids (VS and TS) content and also for the elemental composition a provincial laboratory was used. Carbon, Hydrogen, Oxygen, Sulfur and Nitrogen were determined taking into account the UNI 9903-6:1992 norm, and for the Copper content the EPA 3050B and EPA 6010C were used. For the Chlorine and Mercury content needed for discussing the role of bio-drying towards SRF generation, the determination was made taking into account the ISTISAN'99 and EPA 50505 or/and EPA 9056A.

Generally the lasting of the bio-drying runs is about 2 weeks, when adopted for municipal solid waste (11). For a better understanding of the potential of the bio-drying process applied to exhausted grape marc:

- lasting of was doubled;
- data from respirometry tests regarding the oxygen consumption of the exhausted grape marc before and after the bio-drying process, were generated;
- Chemical Oxygen Demand, COD, was determined in specific cases.

The dynamic respirometer AIR NL (Respirometric Index Analyzer with Not Limiting oxygen) developed by the University of Trento was used in order to have data about the Oxygen Uptake Rate of the organic matrix (12). Inside of this instrument a controlled aeration in no limiting oxygen conditions is made. The Respirometric Index (IR= potential rate of microbial self heating) calculation starts from the decrease of the oxygen concentration curve in the interstitial air of the waste sample, due to bacterial activity. This rate can be determined by the real dynamic respiration index (IR_{24} = RDRI = average value of the respiration indexes representing 24 h showing the highest aerobic microactivity).

11.3 RESULTS

In Table 1 data regarding VS and TS of the exhausted grape marc before bio-drying are presented. Table 2, instead, presents data regarding the ratios VS/TS and TS/GM, the moisture of the initial exhausted GM and the moisture of the bio-dried GM. These values are similar with the ones available in the literature (13). The chlorine and mercury contents were measured: the values for the used exhausted GM sample resulted Cl < 50 mg/kg and Hg = 0.0022 mg/kg respectively.

TABLE 1: Elemental composition of initial exhausted grape marc [%TS and % VS]

	Carbon	Hydrogen	Oxygen	Nitrogen	Sulfur	Ash
% of TS	54.94	5.38	32.08	2.09	0.21	4.85
% of VS	57.74	6.13	33.72	2.20	0.22	-

TABLE 2: VS/TS, TS/GM ratios and moisture and copper content

	kg_{VS} / kg_{TS}	kg_{TS} / kg_{GM}	kg_{H2O} / kg_{GM}	mg_{Cu}/$kg_{bio\text{-}driedGM}$
Initial exhausted GM	93.06%	34.94%	65.06%	-
Bio-dried GM	81.38%	47.08%	52.95%	130

Using the cited bio-chemical model the dynamics of the main parameters for characterizing the exhausted grape marc bio-drying were developed (9). Data concerning temperature dynamic during the bio-drying run are reported in Figure 3.

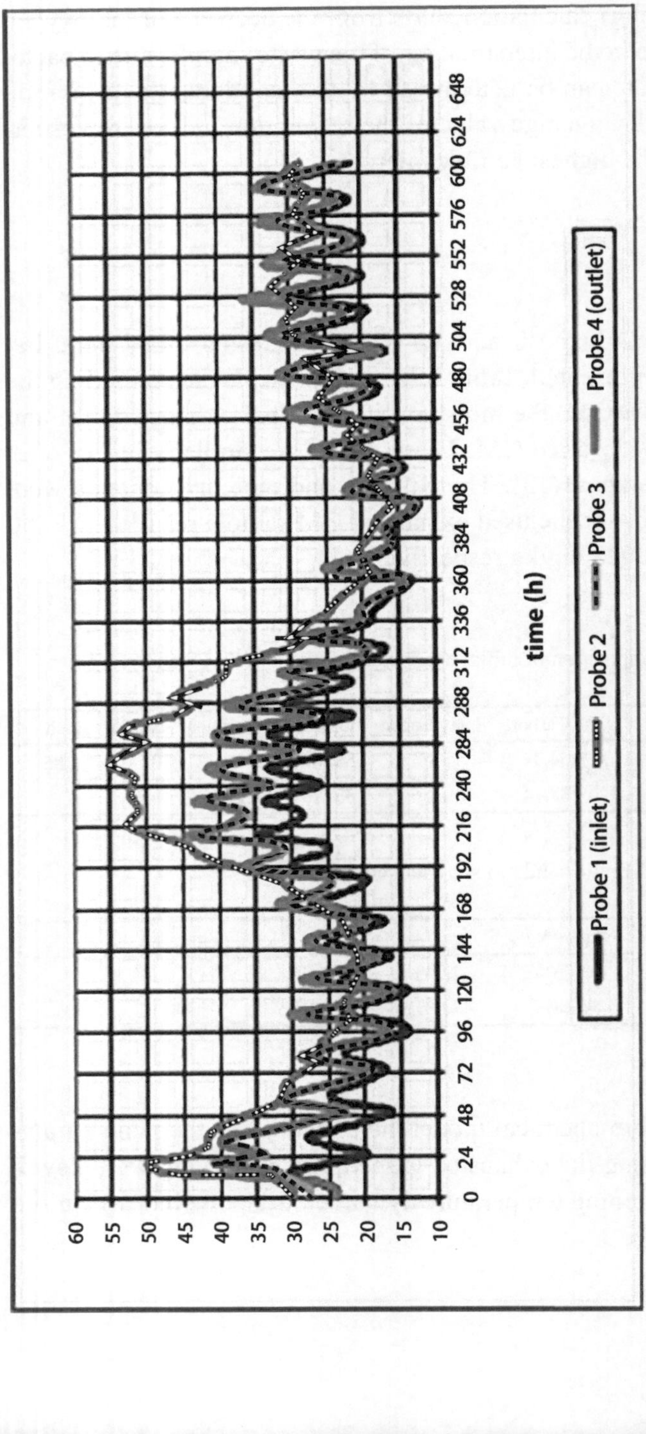

FIGURE 3: Temperature dynamics during the bio-drying run

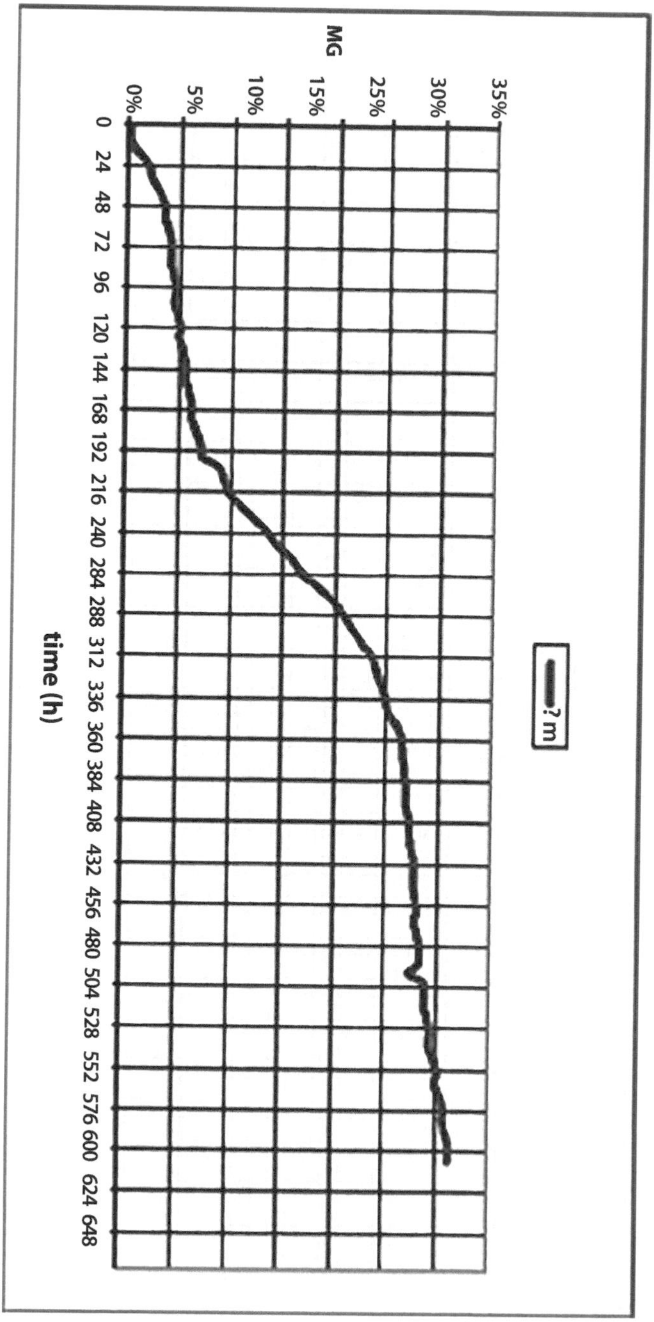

FIGURE 4: Mass loss dynamics during the bio-drying run

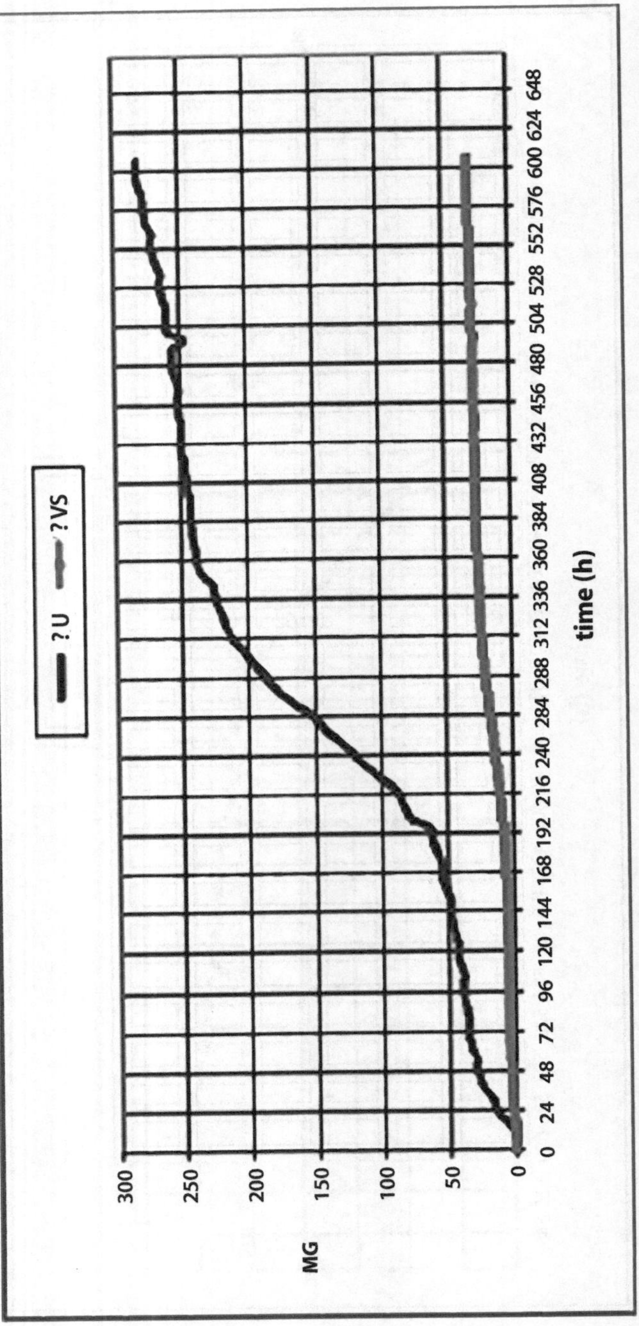

FIGURE 5: Moisture and volatile solids dynamics (ΔU and ΔVS) during the bio-drying run

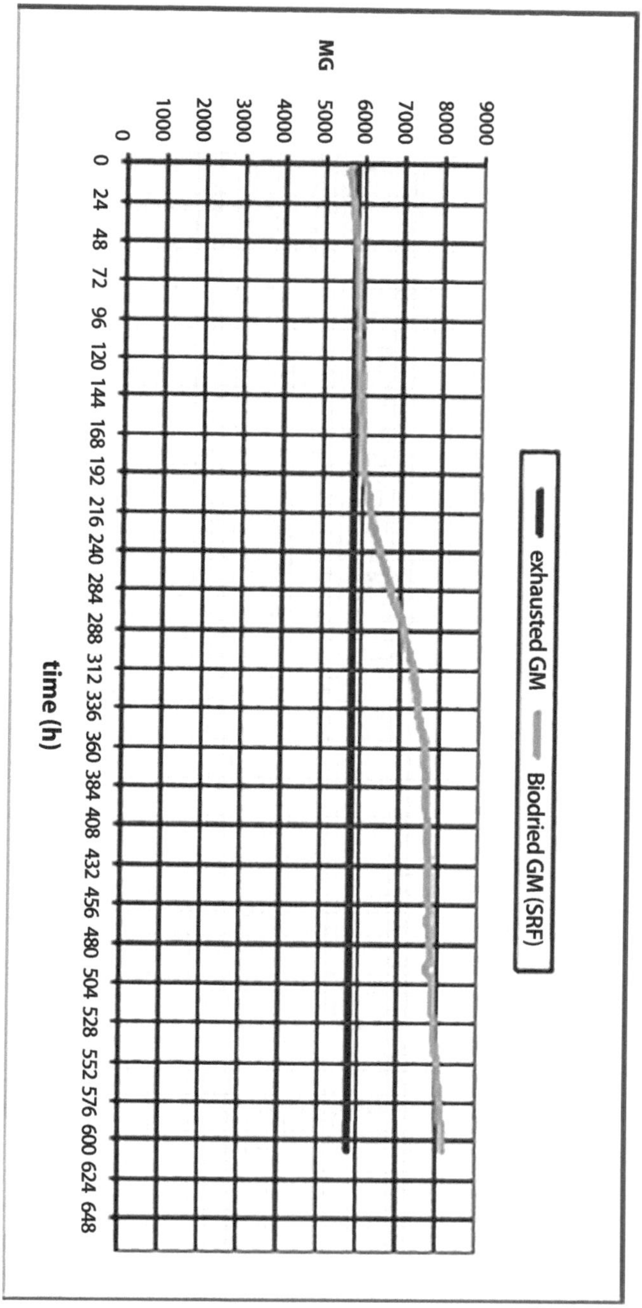

FIGURE 6: LHV dynamics during the bio-drying run

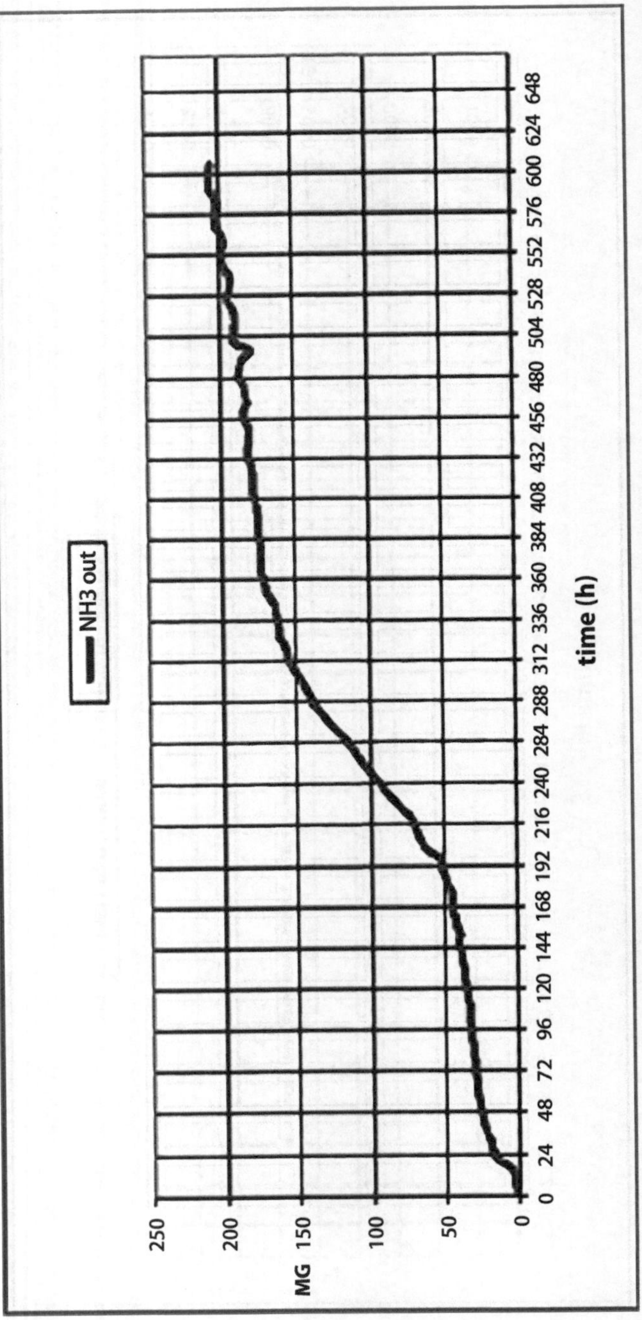

FIGURE 7: Ammonia dynamics during the bio-drying run

Mass loss during the bio-drying run is presented in Figure 4. The mass loss depended strongly on the quantity of putrescible materials present in the input material.

The VS and net water loss were assessed by the quoted model and are presented in Figure 5 (9). The water loss (ΔU) was assessed taking into account the role of hydrogen during the bio-oxidation that characterizes the bio-drying process. VS consumption was minimized thanks to the avoidance of water addition. By this way the presence of carbon was maximized, giving interesting results in term of specific Lower Heating Value (LHV) increase as it can be seen in Figure 6.

The emission factors of ammonia were assessed by modeling and referred to a theoretical case where all the nitrogen present in the treated material is converted into ammonia and emitted through the process air (9). The results are presented in Figure 7.

For a better understanding of the exhausted grape marc use, COD values were generated. To this concern, a part of a sample was dried at 105°C for 24 hours, shredded and homogenized. The COD of the dried exhausted GM resulted 1,961 g_{O2}/kg_{DM}. Another part of the sample was washed and centrifuged at 4000 rpm for 5 minutes. The COD of the washed exhausted GM resulted 1,624 g_{O2}/kg_{DM}. The COD of the liquid phase was determined too. Its value was 27.8 gO_2/l.

The potential rate of microbial self heating of initial and bio-dried exhausted GM were also determined. The obtained results (IR_{24}) were 750 and 500 $mgO_2\ kg^{-1}TS\ h^{-1}$ respectively.

11.4 DISCUSSION

Grape marc, even if exhausted showed a residual activity that supported a quick increase of temperature. Indeed, this phenomenum is confirmed by the fact that at the beginning the temperature in the core of the exhausted grape marc was about 24°C, after only one hour increased and after half day arrived to about 52°C.

Even if the lasting of the process was doubled the results after the first two weeks were suitable instead the results for the last 2 weeks were less important, the process slowed down significantly:

- the temperatures in the last 2 weeks didn't overcome 40°C; on the contrary in the first 2 weeks they reached 55°C (Figure 3);
- the mass, VS and water losses in the last 2 weeks didn't exceed 9%, 9 $g_{(C+H+O+N)}/kg_{GM}$ and 61 g_{H2O}/kg_{GM} respectively; on the contrary in the first 2 weeks they arrived to about 23%, 23 $g_{(C+H+O+N)}/kg_{GM}$ and 221 g_{H2O}/kg_{GM} respectively (Figure 4 and 5);
- the LHV increase in the last 2 weeks didn't exceed 15%; on the contrary in the first 2 weeks it reached 27% (Figure 6);
- NH_3 dynamics (Figure 7) shows a potential emission factor into the atmosphere of about 200 mg/kg_{GM} during the whole period; taking into account that the specific flow rate resulted about 10 Nm^3/kg_{GM}, an potential average concentration could be around 20 mg/ Nm^3. This value make it necessary a treatment of the process air.
- Cu content (Table 2) of the bio-dried exhausted GM resulted moderate, in agreement with the content of the original material pointed out in the literature (14). This content is lower than the limit set in the EU for the agriculture use of sludge (15).
- the high concentration of COD in the liquid phase (27.8 g_{COD}/l) can demonstrate that the process of bio-drying is water limited: an addition of water could reactivate the biological activity (16). Anyway the final respirometric index value is typical of a good stabilization.

Concerning the overall process and the obtained results , the optimal lasting for the exhausted grape marc bio-drying could be around 2 weeks.

The measured data from Table 1 and 2 and the modeled data from Figure 5 demonstrate that the used model is reliable in terms of water balance: the final moisture was 52.9% measured versus 53.0% modeled.

The possibility to use the bio-dried material as a SRF for energy proposed was also taken into account in this paper. The top class is expected for Cl (< 50 mg/kg) and Hg (0.022 mg/kg) whilst the final moisture content after bio-drying, being relatively low cannot allow a final LHV typical of the best SRF. This LHV is about 7 $MJ/kg_{biodriedGM}$ after 2 week and respectively 8 $MJ/kg_{biodriedGM}$ after one month. The values are similar to the ones of green wood.

Taking into account all these data, the final product is like an SRF:5,1,1 that is similar with the one obtained from the initial exhausted grape marc: SRF: 5,1,1 (17); the potential microbial self heating for the final product is very low, and low for the initial exhausted GM. Both SRFs-like can be used in a cement factory, but the SRF from bio-drying can be easily stored as its stability is decreased thanks to the moisture limiting content (18).

REFERENCES

1. ISTAT [Internet]. Roma: The Italian National Institute for Statistics, [cited 2012]. Available from http://www.istat.it/it/
2. Devesa-Rey R., Vecino X., Varela-Alende J.L., et al. Valorization of winery waste vs. the costs of not recycling. Waste Manag 2011;31:2327–2335.
3. Fiori L., Florio L. Gasification and combustion of grape marc: comparison among different scenarios. Waste Biomass Valor 2010;2:191-200.
4. Thorngate J.H., Singleton V.L. Localization of procyanidins in grape seeds. Am. J. Enol. Vitic. 1994;45:259–262.
5. Karleskind A. Sources et monographies des principaux corps gras. In: Manuel des Corps Gras. Paris: Lavoisier; 1992. p. 140–144.
6. Maugenet J. Evaluation of the by-products of wine distilleries. II. Possibility of recovery of proteins in the vinasse of wine distilleries. C. R. Seances Acad. Agric. Fr 1973;59:481–487.
7. UNI/TS 11184: 2006 Waste and refuse derived fuels - Determination of biological stability by dynamic respirometric index, ICS : [13.030.01] [75.160.10], October 2006.
8. CEN/TR 15590:2007 Determination of potential rate of microbial self heating using the real dynamic respiration index, ICS : [75.160.10], October 2007.
9. Rada E.C., Franzinelli A., Taiss M., et al. Lower Heating Value dynamics during municipal solid waste bio-drying. Environ. Technol. 2007;28:463-69.
10. Ragazzi M., Rada E.C., Antolini D. Material and energy recovery in integrated waste management systems: An innovative approach for the characterization of the gaseous emissions from residual MSW bio-drying. Waste Manag 2011;31:2085-91.
11. Rada E.C., Venturi M., Ragazzi M., et al. Bio-drying role in changeable scenarios of Romanian MSW management. Waste Biomass Valor 2010;1:271-279.
12. Andreottola G., Dallago L., Ragazzi M. Dynamic respirometric tests for assessing the biological activity of waste, Proceedings of the Tenth International Waste Management and Landfill Symposium; 2005 Oct. 3-7; S. Margherita di Pula (CA), Italy.
13. Ciuta M.S., Marculescu C., Dinca C., Badea A. Primary characterization of wine making and oil refining industry wastes. Sci B - Electr Eng 2011;73:307-320.
14. La Pera L., Dugon G., Rando R., et al. Statistical study of the influence of fungicide treatments (mancozeb, zoxamide and copper oxychloride) on heavy metal concentrations in Sicilian red wine. Food Addit And Contam 2008;25:302-312.
15. 86/278/EEC:1998 Council Directive of on the protection of the environment, and in particular of the soil, when sewage sludge is used in agriculture, 12 June 1986.
16. Rada E.C., Ragazzi M., Panaitescu V., Apostol T. Bio-drying or bio-stabilization process? Sci B Sci B - Electr Eng 2005;67:51-60.
17. UNI 9903-1:2004. Non mineral refuse derived fuels - Specifications and classification, ICS : [75.160.10], March 2004.
18. UNI CEN/TR 15508:2008. Key properties on solid recovered fuels to be used for establishing a classification system, ICS : [75.160.10], October 2006.

Author Notes

CHAPTER 2

Data Availability

The authors confirm that all data underlying the findings are fully available without restriction. All sequences have been deposited into the NCBI short read archive (SRA) under the accession number SRX484115 for bacteria and SRX485028 for archaea.

Funding

This research has been supported financially by National Key Technologies R&D Program of China (2010BAC67B04) and the key projects of National Water Pollution Control and Management of China (2011ZX07316-004). The funders had no role in study design, data collection and analysis, decision to publish, or preparation of the manuscript.

Competing Interests

The authors have declared that no competing interests exist.

Author Contributions

Conceived and designed the experiments: XD. Performed the experiments: JY. Analyzed the data: BD. Contributed reagents/materials/analysis tools: JJ. Contributed to the writing of the manuscript: JY.

CHAPTER 3

Acknowledgments

The authors wholeheartedly appreciate the financial support offered by Claude Leon Foundation, South Africa.

Conflicts of Interest

The authors declare no conflict of interest.

CHAPTER 4

Competing Interests

The authors declare that they have no competing interests.

Author Contributions

GL and IA designed the experiment. GL and XYZ carried out the experiment. DDF, PGK, and TL participated in the molecular analysis. GL performed the bioinformatics analysis and drafted the manuscript. All the authors have read and approved the final manuscript.

Acknowledgments

This study was funded by the Yangfan project from the Science and Technology Commission of Shanghai Municipality (14YF1400400), the National Natural Science Foundation of China (51408133), and The Danish Council for Independent Research (12-126632) and Strategic Research (12-132654).

CHAPTER 5

Acknowledgments

The authors are grateful to CeLIM NGO for logistical and financial support and to all the people and local bodies for supporting the local activities, in particular CeLIM local staff and Maxixe municipality. This work was part of the "Urban and peri-urban environmental protection: a project for Maxixe Municipality", founded by Comune di Milano and Peppino Vismara foundation.

CHAPTER 6

Funding

DAN received USDA Hatch VT-HO1609MS funding and NF received funding from the National Science Foundation to support this research. The funders had no role in study design, data collection and analysis, decision to publish, or preparation of the manuscript.

Competing Interests

The authors have declared that no competing interests exist.

Acknowledgments

The authors thank Tom Gilbert (HCC), James McSweeney (HCC), and Alexander Utevsky (HCC) for formulating recipes and managing variations of the thermophilic process at a realistic commercial scale. We thank Jessica Henley and Chris Lauber for technical assistance with the laboratory analyses.

Author Contributions

Conceived and designed the experiments: DAN TRW NF. Performed the experiments: DAN TRW. Analyzed the data: DAN JWL. Contributed reagents/materials/analysis tools: DAN NF STB. Wrote the paper: DAN TRW STB JWF NF.

CHAPTER 7

Acknowledgments

The authors wish to acknowledge the R&D collaborative work of UPM-O3 Solutions (2008 to 2010) and Dr. Chon Seng Tan from the Biotechnology Research Centre, Malaysian Agricultural Research and Development Institute, for their assistance and valuable advice.

CHAPTER 8

Competing Interests

The authors declare that they have no competing interests.

Author Contributions

JP: Collected and reviewed the literature and drafted the manuscript. NS: Formulated the objectives, provided guidance and improved the quality of the manuscript. Both authors read and approved the final manuscript.

Acknowledgments
The financial support from Department of Biotechnology (DBT), New Delhi, India, and Department of Science and Technology (DST), New Delhi, India, through Fund for Improvement of Science and Technology Infrastructure in Higher Educational Institutions (FIST), is gratefully acknowledged.

CHAPTER 11

Author Contributions:
Both the authors contributed to conception and design, to acquisition of data and to analysis and interpretation of data; both the authors drafted the article and revisited it critically for important intellectual content; and both the authors gave final approval of the version to be published.

Index

A

abattoir, 76, 98
acetate, 37–38, 40–41, 55–56, 58–60, 69, 78, 83, 86, 92, 100, 102
acetogens, 55–58, 69
  syntrophic acetogens, 56–58, 69
acetogenesis, 39, 55, 58, 60, 66
acidogens, 57, 86
acidogenesis, 24, 55, 60, 68, 101
Acremonium, 127–128, 134, 137
Actinobacteria, xxii, 36, 58, 71, 126–127, 135, 137, 143, 181–183
active but non cultivable (ABNC), 61
aerated static pile (ASP), xxiii, 120–123, 127, 131, 133–138
agriculture, xviii–xx, xxiv, 3–5, 14, 20, 48–49, 76, 80–81, 111, 116–117, 119, 139, 145, 152, 169–170, 175–176, 197–199, 202–203, 206, 210–211, 213–214, 236–238, 253, 255, 268–269, 273
agro-industrial, 4, 18, 21, 47
alcohols, 13, 58
algae, 54
Alternaria, 134–135, 172
amino acids, 37, 55, 58, 152, 171
ammonia (NH_3), 26, 29–30, 37–38, 44, 56, 59, 67–70, 76, 78, 81, 92, 97, 163, 165, 167, 197, 260, 266–268
  ammonia concentration, 30, 67–70
ammonia-nitrogen, 29
  free ammonia-nitrogen (FAN), 29–31
  total ammonia-nitrogen (TAN), 29–30, 188, 214, 244, 273
amplification, 27, 62–63, 124, 198, 205
anaerobic, xvi, xx–xxi, 8, 14–15, 23–27, 29–31, 33, 35–41, 43–45, 47–49, 51, 53–63, 65–83, 95–97, 100–102, 147, 166, 169, 179, 207, 236, 238, 240–242, 254
anaerobic digestion (AD), xvi, xx–xxi, 8, 14–15, 23–25, 27, 29–31, 33, 35, 37–41, 43–45, 47–49, 51, 53–57, 59–63, 65–77, 79, 81–83, 95, 97, 100–102, 166, 206, 209, 238, 241, 254
  conventional wet, 24
  modern dry, 24
  semi-dry, 24–25, 29, 43
Anaerolineae, 38
animal, xv–xvi, xxi, 8, 11, 13, 19, 47–57, 59, 61, 63, 65–77, 79, 81, 120, 135, 154, 175–177, 188, 201, 203, 214, 235–243, 245, 247, 249–253, 256
antibiotic, 50–51, 54, 175, 179, 184
  antibiotic resistance, 54
Archaea, 27–28, 59, 79, 83, 89, 92, 124, 126, 141, 271
Ascobolus, 127–128, 134–135
Aspergillus, 134–135, 152
autofluorescence, 61

B

Bacteroidetes, 36–38, 71, 90, 126–127, 133–137, 181, 183
bio-digesters, xxi, 47, 49, 54, 65

bio-drying, xxiv–xxv, 219, 221–229, 231–233, 255–257, 259–269
bio-filter, 227
bio-hydrogen, 8, 20
bio-methane, 8, 49
bio-stabilization, 221, 232, 269
bioavailability, 72, 82, 146
biodegradability, 8, 15, 17–19, 23, 43, 67, 72, 75, 146, 156, 167, 175
biodiesel, 12–13, 21
biodiversity, xxiv, 13, 63, 80, 100, 170, 180
bioethanol, 14
biofuels, xv, xx, 12, 14–15, 17, 19–21, 74, 77, 81, 83
biogas, xvi–xvii, xix, xxi–xxii, 14, 19, 24, 28, 31–32, 44, 48–49, 56, 60–61, 65, 69, 71–72, 74–86, 89–90, 95–98, 100–102, 166, 235–245, 247–254
  biogas boiler, 244, 247
  biogas production, xvii, xix, 14, 31–32, 44, 48, 75–77, 82, 85, 96, 98, 101, 237, 252
  biogas reactors, xvii, xxi–xxii, 74, 77–79, 83–86, 89–90, 95–97, 102
biogenic cycles, 5
bioinformatics, 65, 102, 141, 272
Biological Mechanical Treatment (BMT), 219, 227–229
biological oxygen demand (BOD), 51, 74, 186
biomass, xvi, xx, 14, 18, 21, 36, 45, 47–49, 57, 62, 68, 70, 74, 77, 79, 82, 101, 107, 166–167, 179, 181, 189, 198, 202, 212, 215–216, 221, 231, 237, 252–253, 256, 269
biooxidation, 177
bioprocessing, 15, 75
biorefinery, 12, 15, 103
buffering, xxiv, 30, 68–69, 73, 185, 198, 211
bulking, xviii, xxiii, 145–151, 153–155, 157, 159, 161, 163, 165, 167, 202
burning, 108, 223, 225, 228–229, 243
butanol, 14

C

carbohydrates, 14–15, 17–18, 37–38, 55, 57, 72, 152, 155, 171
carbon cycle, xxiii, 145
carbon dioxide (CO_2), 7, 14, 37, 40, 55–56, 58, 78, 83, 157, 163, 177, 236–238, 276
chemical oxygen demand (COD), 51, 69–71, 74, 186, 260, 267–268
chlorine (Cl), 101, 141, 151, 166, 260–261, 268
Chloroflexi, xx, xxii, 35–36, 41, 44, 126–127, 133, 135, 137, 142
cleavage, 56
climate change, 7–8, 12
clone, 33, 36, 62, 65, 75, 84, 140
Clostridium acetobutylicum, 14
co-digestion, 43–44, 48, 75, 81, 100
coal, 12, 48, 165, 219, 221, 231, 237
cogeneration, 238–240, 243, 247–248
combustion, xix, 14, 19, 47, 221, 227, 229, 233, 238, 240, 242–243, 247, 250, 259, 269
competition, 12, 14, 45, 84, 184, 193
compost, xvii–xviii, xxii–xxiii, 110–112, 114–117, 119–121, 123, 125–128, 130–131, 133–137, 139–141, 146–148, 151–160, 163, 165–168, 170, 172, 175, 177, 181–182, 192, 197–198, 201–202, 204–207, 212, 214–215, 221, 232–233
  compost mass, 153
  compost maturity, xviii, xxiii, 157, 165–167, 215

compost process, 121, 123, 127, 133, 135–136
compost volume, 155
biocomposting, 185
vermicompost, xviii–xix, xxiii–xxiv, 12, 120, 122–124, 127, 131, 133–140, 142, 169–171, 173–211, 213–216
composters, xviii, xxiii, 139, 146, 149–153, 166
composting, xv–xviii, xx, xxii–xxiv, 8, 105, 107, 109–111, 113, 115–124, 133–137, 139–143, 145–148, 150–158, 161–167, 170, 175–177, 182, 197, 200, 203–205, 208, 212, 215
commercial composting, 120
conservation, 4, 10, 20–21, 48, 210, 212
consumer, 3–8, 10–12, 19, 37, 76
consumer behavior, 6
cooking, xvi, 4, 7, 10, 13, 49
cooking oil, 13
copper (Cu), 50–51, 72, 115–116, 184, 55–56, 58, 157, 163, 236–237, 260–261, 268–269, 276
corn, 12, 14, 17–18, 188, 194, 199
cost, xviii–xix, 5, 7, 12, 14, 18, 23, 64, 76, 85, 175, 184, 197, 213, 222–223, 227–231, 248, 251–253, 256, 269
capital cost, 18, 222, 227–228, 231
maintenance costs, 248, 251
operating costs, 222, 227–228
production cost, 18
cyanobacteria, 54, 76

D

decomposition, xvi, xxiv, 38, 71, 74, 133, 140, 142, 158, 163, 170–172, 174, 176–177, 182, 184, 209, 211
degradation, xix, xxiii, 25, 29, 37, 39, 41, 54, 58, 69, 77–78, 101, 111–112, 114, 116, 145, 155–158, 165–166, 169–171, 180, 236
denaturing gradient gel electrophoresis (DGGE), 62, 65, 75, 84, 140–141, 143, 166
diet, 48, 55, 71, 171, 186
digester, xxi, 24–28, 31, 33, 36–41, 43–44, 47–49, 54–62, 65–68, 70–71, 73–76, 78, 81, 92, 101–102, 240, 242, 244, 247–248, 250–251
discount payback period (DPP), 247, 252
disease, xviii, 54, 108, 137, 139, 181, 191–192, 194, 198, 202–203, 211, 213–214
disease suppression, xviii, 137, 139, 181, 192
diversity, xvi, xviii, xxii, 25, 28, 33, 39–40, 43, 48, 61–63, 79–80, 101–102, 120, 136, 138, 140–142, 169, 171, 173, 175, 177–179, 181, 183, 185, 187, 189, 191–193, 195, 197, 199, 201, 203–205, 207, 209–211, 213, 215–216
DNA, 27, 33, 39, 62, 64–65, 77, 99, 122, 124, 140–141
DNA sequencing, 62, 77
next generation sequencing methods (NGS), 62–63
drains, 108
dumping, 111

E

earthworms, xxiv, 120, 123, 136–137, 142, 169–186, 188–189, 192, 194, 196–216
anecic, 173, 204
epigeic, 173–174, 214

ecology, 25, 77, 84, 100–101, 142, 199–200, 202–203, 205, 208, 212
economic, xviii–xix, xxiii, 4–5, 11–13, 19–21, 75–76, 107, 139, 209, 223, 225, 227–229, 235, 238, 244, 248–250, 253–254
    economic efficiency, viii, 235, 244, 248
edible, 3–4, 11–14
education, 7
Eisenia fetida, 123, 198, 201, 205, 209–210, 212
electrical conductivity (EC), xvi, 49, 76, 153, 176–177, 197, 212, 219, 255
electricity, 223, 225, 238–239, 241–244, 248–249, 251
    electricity consumption, 241–242
    electricity peak demand, 244
electrolysis, 95–96, 102
energy security, 12
Enterobacteriaceae, 57, 183
equilibrium, 69–70, 230–231
Escherichia coli, 21, 50, 58, 115–116, 172, 185–186
esterification, 13
ethanol, 14, 55, 167–168
    cellulosic ethanol, 14
ethical, 5
extracellular polysaccharide (EPS), 57

F

farms, xvii, 5–6, 12, 48–51, 54–55, 75–76, 117, 119, 121, 125, 169, 203, 205, 207, 210, 213–214, 238–240, 242, 244, 248–253
fatty acids, xvi–xvii, xxi, 13, 28, 30, 55, 58–59, 68–71, 74, 85, 140
    fatty acid alkyl esters, 13
feedstock, 14–15, 18, 23, 56, 61, 66, 72, 74, 82, 102, 134
fermentation, 14–15, 17–18, 20, 24, 27–28, 36–39, 41, 68, 79, 82, 102, 166, 168, 235, 242, 256
    acidogenic fermentation, 18
    bacterial fermentation, 17
fertilizer, xvi–xvii, xxi, 48–49, 55, 74, 111, 167, 178, 184–185, 190–191, 193, 195–198, 200, 202, 205–207, 209, 211–214, 216, 252, 256
filtration, xvi, 13, 25, 43, 61, 89, 247, 251
Firmicutes, xxii, 35–37, 39, 90, 126–127, 133, 135–137, 181–183
flame ionization detector (FID), 29, 100
flooding, 108
fluorescent in situ hybridization (FISH), xxiii, 36, 62, 65, 98, 102, 148, 151, 156, 166
food
    Food and Agriculture Organization (FAO), 3, 5, 20
    food donation, 11
    food insecurity, 6
    food mass, 3–4
    food processing, 4, 47, 98
    food security, 5, 8, 110–111
    food supply chain (FSC), 4–6, 8, 14, 19
    food waste (FW), xv–xvi, xx, 3–15, 17–21, 23–27, 29, 31, 33, 35–41, 43–45, 82, 112, 147, 155–156, 165, 167–168, 187, 199
formic acid, 28
formyltetrahydrofolate synthetase (FTHFS), 58, 78
fossil fuels, xix, 12, 48, 235
fungi, xviii, xxiii, 120–121, 124, 126, 133–137, 139–140, 142–143, 147, 152, 157, 166, 169, 172, 176, 192, 197, 201, 204, 210, 216

G

Gammaproteobacteria, 71, 90, 181, 183
garbage picker, 11
gas chromatograph (GC), 28–29, 100, 139, 189, 216
gas chromatography-mass spectroscopy (GC-MS), 189
gasification, 47, 269
genus, 35, 37–41, 71, 90, 92, 128
Giardia, 50
global warming, 12
grape marc (GM), xix, xxiv–xxv, 141, 143, 166, 255–257, 259–261, 263, 265, 267–269
greenhouse gas (GHG), xvi, xxiii, 4, 7, 12, 14, 19, 48–49, 54, 74, 108, 145, 235

H

health hazards, 48
heat, xvi, xxiv–xxv, 8, 24, 49, 67, 73, 85, 121, 133, 139, 147, 150, 156, 194, 222–223, 225–226, 228, 230–232, 238–239, 241–242, 244, 248–253, 256, 260–261, 267–269
heating value, 222, 225, 232, 260, 267, 269
heavy metal, xviii, 50, 71, 73, 116–117, 185–186, 269
high performance liquid chromatography (HPLC), 189
high-throughput sequencing, xxii, 85, 101, 120–121, 134
hormones, xxiv, 50, 170, 179, 183–184, 187, 189, 202, 210, 214–215
household, 4–5, 7–8, 10–11, 20–21, 110, 204
humic, 8, 115–117, 177, 184–185, 187–188, 190, 196, 198–199, 201–202, 204, 206, 208–209, 212, 214–215
humic acids, 177, 185, 187, 190, 196, 199, 201, 204, 212, 214–215
hydraulic retention time (HRT), xxi, 41, 44, 60, 66–68, 70–72, 74, 79–80, 82, 84, 96, 98, 101
hydrazine oxidoreductase genes (*hzo*), 62, 80
hydrogen, 8, 15, 19–20, 39, 55–56, 58, 68, 78, 92, 102, 197, 237, 259–261, 267
hydrolysis, 24, 36–37, 39, 55, 57, 60, 70, 79, 81, 89, 167

I

incineration, xv, xx, 8
indole-acetic-acid (IAA), 189
infrastructure, 6, 13, 274
inoculum, xxi, 25–26, 43, 82, 85, 96–98, 157
internal rate of return (IRR), 244, 252
investment, 12, 231, 247–248, 251

K

Kyoto Protocol, xix, 221, 232

L

lactose, 14
lag phase, 157
landfill, xv, xx, xxiii, 4, 7–8, 11–12, 19–21, 77, 80, 119, 145–146, 219, 221, 233, 269
leachate, 8, 77, 145, 150, 259
library, 33, 62, 65, 75, 99
lignin, 37, 146–147, 157, 166, 168, 183
lignocellulosic, 14, 21, 72, 166–167

litter, 166, 169, 172, 174, 178, 215
livestock, 48–49, 54–55, 66, 68, 71, 75, 81, 122, 140, 238

M

manure, xvi–xvii, xxi–xxiii, 4, 44, 48–52, 54–56, 61, 65–68, 70–76, 80–81, 85, 89, 92, 95–98, 100–101, 120–123, 126–128, 130, 135, 137, 139–140, 142–143, 147, 154–155, 167, 174–178, 187–188, 191, 198–202, 205, 208–209, 211–212, 214, 238, 252
  cattle manure, xvii, xxi, 81, 85, 89, 95, 97–98, 101, 137, 142, 178, 187, 208–209, 211
  dairy manure, xxiii, 44, 74, 76, 80
marketing, 6
Mechanical Biological Treatment (MBT), 219, 221, 232–233
membrane, xvi, 25, 28, 36, 43, 57, 61, 63, 201
  membrane filtration, xvi, 25, 43, 61
mesophilic, xx, 23–27, 29, 36, 38, 41, 43, 45, 58–60, 68, 70, 75–76, 78–79, 81–82, 98, 101–102, 120, 123, 136–137, 155, 157, 163, 167, 176, 207
methane (CH_4), xvii, xx–xxi, 4, 8, 12, 14–15, 24, 27–29, 31–32, 37, 40–41, 45, 48–49, 54–60, 68–72, 74, 77, 80–83, 85–86, 89, 92, 96–97, 100–102, 227, 230–231, 236–237, 240–241, 250, 276, 280
  methane production, 15, 24, 31–32, 37, 40–41, 45, 56, 60, 71–72, 85, 89, 92, 96, 100
  methane yield, xvii, xx–xxi, 27, 29, 32, 41, 48, 68–69, 71–72, 85–86, 89, 97, 100
Methanobacteriales, 39, 59, 92
Methanobacterium, 57, 59, 92
Methanococcales, 39, 59, 92
Methanoculleus, 40–41, 59, 92
methanogens, 40–42, 45, 55–60, 62–64, 67, 69–71, 75, 77–80, 86, 92, 101–102
methanogenesis, 24, 39–41, 55–56, 58, 60, 63, 66, 68–70, 77–78, 83, 92
Methanomicrobiales, 39–40, 59, 73, 92
Methanosarcinales, 39, 58–60, 73, 92
microbial communities, xvi–xviii, xx–xxii, 24–25, 27, 33, 39, 41, 43, 45, 49–50, 56–57, 60–62, 64–65, 73, 75, 77, 79, 82–85, 87, 89, 91, 93, 95–97, 101–103, 115–116, 120–121, 123, 134, 136–137, 139–141, 143, 154–155, 166, 169, 176, 181, 198, 205, 215
microbial diversity, xvi, 25, 33, 43, 102, 120, 140, 169, 171, 173, 175, 177, 179, 181, 183, 185, 187, 189, 191–193, 195, 197, 199, 201, 203, 205, 207, 209, 211, 213, 215
mixing, 26, 73–74, 76, 81–82, 147, 173–174
moisture, xxv, 13, 17, 23, 72, 114–115, 145–146, 148, 150, 153, 171, 210, 260–261, 264, 268
money, 7, 229
Mortierella, 127, 134–136
most probable number (MPN), 61, 115, 185
mucus, 171, 175, 184
municipal solid waste (MSW), xxiv, 20, 23, 29, 75, 82, 107–109, 111, 113, 115, 117, 120, 139–140, 167, 205, 219, 221–223, 225–227, 229–233, 260, 269

N

nematodes, xix, xxiv, 136, 196–200, 203, 207, 212–214, 216

net present value (NPV), 244, 251
newspaper, xviii, xxiii, 146, 148, 151, 153–161
nutrient, xvi, xx–xxi, xxiii–xxiv, 5, 8, 18, 48–51, 55, 71–74, 76, 119, 142, 145, 154, 170, 172–174, 176, 178–179, 181–185, 187, 190–191, 194, 196, 198, 201–202, 206, 210

O

odor, xxi, xxiii, 8, 145, 186, 236
oleate, 96–97, 101
onion peels, xviii, xxiii–xxiv, 146, 148, 151, 153–162, 164–165
organic loading rate (OLR), xvi, 25–26, 30–32, 37, 40–41, 65, 67–68, 70–72, 81–82, 101
oxygen supply, 114, 116

P

packaging, 4, 6–7, 10, 225, 229
paper mill, 18, 177, 188, 200–201, 204
pathogens, xix, xxiii–xxiv, 49–51, 54, 68, 76, 120, 123, 136, 147, 170, 179, 184–187, 191–193, 198, 200, 203–206, 210, 214, 252
Penicillium, 134, 137, 152
permeability, 17
Personal Genome Machine (PGM), 85, 99
pH, xvii, 24, 26, 28–30, 38, 40, 58–60, 65, 67–69, 71–72, 80–82, 85–86, 89, 98, 101, 115–117, 141, 153, 167, 176–177, 184, 197, 212
phytoplankton, 51
pipeline, 124
plant growth, xviii–xix, xxiv, 139, 170, 178–181, 184–185, 187–191, 195, 197–203, 205, 207–208, 210, 215
  plant growth-promoting rhizobacteria (PGPR), 179, 197
plant metabolism, 189
pollution, xvi, xviii, 47, 49, 51, 74, 108, 116–117, 145, 205, 271
  air pollution, 108
polyhydroxyalkanoates (PHAs), 16–19, 21
polylactic acid (PLA), 17
population, xv, xviii, xx–xxi, 23, 37–39, 41, 47, 50, 61–62, 66, 73, 75, 77, 79–80, 107–108, 111, 146, 157, 166, 170, 178, 181–182, 184, 192, 195–197, 206, 208, 211–212, 222, 227, 229
  population growth, 23, 157, 196, 206
postharvest, 3, 5, 11
poultry, 48–49, 66, 68, 75, 147, 154, 163, 167
power grid, 238, 244, 248
pretreatment, xix, 13
principal coordinates analysis (PCoA), 94–95, 99, 130–131
processing, xxiii, 3–5, 7, 10–11, 17, 38, 44, 47, 81, 98, 116, 125, 148, 151, 199
production, xv, xvii–xxii, xxiv, 3–8, 10, 12–21, 24, 31–32, 37, 40–41, 44–45, 48, 56, 60, 71–72, 75–77, 79, 81–83, 85, 89, 92, 96, 98, 100–102, 110, 114, 117, 121, 143, 158, 166–167, 171, 177, 179, 184, 188–190, 199–200, 207–210, 213–215, 219, 237–238, 252–253
Proteobacteria, 35–37, 90, 126–127, 133–137, 181, 183
Pseudomonas, 152, 169, 178–180, 182–183, 186, 192, 194, 197, 206–208, 212, 214
public health, xxi, 44, 47, 49–50, 52, 54, 76, 100, 102, 117, 175

purity, 13
pyrosequencing, xvi, xx, 25, 27–28, 33–34, 39, 42–43, 75, 77, 79–80, 101–102, 135, 141–142

R

reactor, xvii, xx–xxii, xxv, 16, 24, 26–27, 29–33, 36–39, 41, 43–45, 65–66, 68–74, 77–80, 83–87, 89–90, 92, 95–99, 101–102, 123, 147, 156, 165, 175, 204, 221, 231, 258–259
recovery, xx, 7–8, 19, 23, 48, 89, 96, 103, 118, 219, 233, 269
  energy recovery, 7–8, 19, 23, 269
recycling, 7, 13, 19–21, 55, 120, 139, 145, 166, 170, 209, 214, 219, 238, 256, 269
Refuse Derived Fuel (RDF), xxiv, 219, 221–223, 225–229, 231–232
restriction fragment length polymorphism (RFLP), 62, 79, 84, 140
retailers, 3, 7, 19
reuse, 7, 19, 23, 82
Ribosomal Database Project (RDP), 89, 95, 99, 102, 125

S

salinity, 60, 154, 177–178
Salmonella, 50, 75, 115–116, 185–186, 201, 213
sanitation, xv–xvi, xxiii, 49, 51, 147
sawdust, 112, 120, 122, 139, 143, 146
scum, 73
sewage, 24, 43, 59–60, 92, 96, 100–101, 120, 147, 165–166, 168, 176–177, 185–186, 188, 200, 202, 208–209, 213–214, 238, 269
Shannon diversity, 28, 138
Simpson diversity, 28, 33
single strand conformation polymorphism (SSCP), 62, 183
sludge, 24, 26, 36, 38–39, 43–45, 60, 66, 74–75, 79, 81, 92, 96, 98, 100, 102, 120, 135, 147, 154, 165–166, 168, 176–178, 185–186, 200–202, 204, 206, 208–209, 213–214, 238, 268–269
social, xix, 4–5, 12, 19, 49
socio-demographic, 4, 10
socioeconomic, 51
soil, xvi, xviii–xix, xxii, xxiv, 5, 12, 25, 43, 51, 77, 96, 101, 108, 110–111, 117, 137, 140–143, 152, 158, 165–167, 169–175, 177–179, 181–187, 189–192, 195–204, 206–216, 269
  soil fertility, xix, xxiv, 170, 183, 198, 206, 210
  soil structure, 174, 184
solid recovered fuel (SRF), xxv, 256, 260, 268
solid retention time (SRT), 24, 27, 29–32, 37, 81
soluble microbial products (SMP), 36
Spirochaetes, 35–37, 39
standard deviation, 26, 29, 148, 153
stochastic assembly, 95, 102
stochastic factors, xvi–xvii, xxi–xxii, 84–85, 97
stoichiometry, 259
storage, 4, 6, 10, 15, 18, 48, 66, 71, 226, 240, 242, 247–248, 250–251
straw, 120, 122, 139–140, 142–143, 209
Streptomyces, 169, 179, 194
substrate, xviii–xix, xxi, xxiii, 13–15, 18, 25–27, 31, 36–38, 48, 55, 58–60, 65, 67–73, 77, 79, 82, 84, 96, 98, 112, 116, 139, 142, 147–148, 150–151, 163, 167, 171, 175–176, 182, 201, 205, 259
sugar cane, 12

supply chain, 3–4, 6, 14, 21

T

tax, 11
taxonomic, 28, 34–35, 40, 42, 78, 89, 91–92, 95, 99, 120, 124–125, 127, 141
temperature, xvii, xxi–xxiii, 27, 60, 66–71, 76–77, 79, 81, 83–86, 89–90, 92, 95–99, 101, 111–112, 114, 116, 120, 122–123, 136, 143, 147, 150–156, 159, 165, 167, 194, 211, 223, 236, 259–262, 267
  ambient temperature, 120, 155–156
  temperature disturbance, xvii, xxi–xxii, 85–86, 90, 92, 96–97, 99
  temperature fluctuations, xvii, 85, 155
Tenericutes, 35–37, 39
thermophilic, xviii, xxiii–xxiv, 24, 41, 43, 58–60, 68, 70, 77–79, 81–82, 90, 101–102, 120–121, 123, 133–137, 139–140, 143, 147, 151, 154–155, 157–158, 163, 165–167, 175–176, 185, 192, 197, 207, 273
total dissolved solids (TDS), 186
total organic carbon (TOC), 152–153, 157–158, 161–164, 171
total solids (TS), xx, 23–27, 29–33, 35–41, 43, 45, 70, 72, 74, 98, 100, 216, 260–261, 269
total suspended solids (TSS), 186, 190
transesterification, 13
transport, 4, 7, 10, 208
transportation, 4, 6, 12, 49
Trichoderma, 135–136, 152, 192, 194, 197, 202

V

valorization, 238, 242, 250, 256, 269
vegetable oil, 13–14
volatile fatty acid (VFA), xvi–xvii, xxi, 19, 28–30, 37, 40–41, 58–59, 68–70, 74, 81, 85–86, 88–89, 96–97
volatile solid (VS), 12–13, 25–27, 29–32, 39–41, 70, 86, 98, 100, 256, 260–261, 267–269

W

waste
  agro-industrial, 4, 18, 21, 47
  food waste, xv–xvi, xx, 3–15, 17–21, 23–27, 29, 31, 33, 35–37, 39, 41, 43–45, 82, 112, 147, 155–156, 165, 167–168, 187, 199
  industrial, xv–xvi, xx, xxiv, 3–4, 8, 12, 17–18, 21, 23, 44, 47, 81, 101, 167, 175–176, 198, 203, 219, 222, 226–227, 229–230, 232, 238
  kitchen waste, xxiii, 145–146, 149–152, 165–166, 168
  milk house waste, 55
  waste management, xv, xvii–xx, xxiv, 3, 5, 7–9, 12, 14, 19–21, 55, 75, 107, 109, 111, 117–118, 169–170, 175, 185–186, 197, 199, 203, 212, 219, 232, 269
  waste management hierarchy, xx, 7, 9, 12
  waste prevention, 7–8, 19, 219
  waste volume, 21, 107
wastewater treatment plant (WWTP), 26, 36
water, xv–xvi, xxiv–xxv, 5, 13, 17, 21, 27, 43–45, 49–51, 54–56, 60, 62–63, 66, 75–76, 78–82, 101–102, 108, 111, 114, 124, 135, 145, 150–151, 154, 157, 163, 165–167, 173, 175, 184–186, 189, 197, 204, 210, 213, 221, 238, 240, 256, 259, 267–268, 271
  surface water, 54, 145

weed, xviii, 68, 139, 176
wheat, 14, 180, 190, 192, 200, 202, 206, 209, 213
wine, 201, 255–256, 269

Z

zinc (Zn), 50–51, 72, 115–116, 184, 191

AF828879

CUSTOM'S FUTURE

Although customary international law has long been an important source of rights and obligations in international relations, there has been extensive debate in recent years about whether this body of law is equipped to address complex modern problems such as climate change, international terrorism, nuclear proliferation, global financial instability, and cyber warfare. In addition, there is growing uncertainty about how, precisely, international and domestic courts should identify rules of customary international law.

Custom's Future seeks to address this uncertainty by providing a better understanding of how customary international law has developed over time, the way in which it is applied in practice, and the challenges that it faces going forward. Reflecting an interdisciplinary mix of historical, empirical, economic, philosophical, and doctrinal analysis, and containing chapters by leading international law experts, it will be of use to lawyers, judges, and researchers alike.

Curtis A. Bradley is the William Van Alstyne Professor of Law at Duke Law School specializing in international law and U.S. foreign relations law.

Custom's Future

INTERNATIONAL LAW IN A CHANGING WORLD

Edited by

CURTIS A. BRADLEY

Duke University School of Law

CAMBRIDGE
UNIVERSITY PRESS

CAMBRIDGE
UNIVERSITY PRESS

32 Avenue of the Americas, New York, NY 10013-2473, USA

Cambridge University Press is part of the University of Cambridge.

It furthers the University's mission by disseminating knowledge in the pursuit of education, learning, and research at the highest international levels of excellence.

www.cambridge.org
Information on this title: www.cambridge.org/9781107443105

© Cambridge University Press 2016

This publication is in copyright. Subject to statutory exception and to the provisions of relevant collective licensing agreements, no reproduction of any part may take place without the written permission of Cambridge University Press.

First published 2016

Printed in the United States of America

A catalog record for this publication is available from the British Library.

Library of Congress Cataloging in Publication Data
Names: Bradley, Curtis A., editor.
Title: Custom's future : international law in a changing world / edited by Curtis A. Bradley.
Description: New York : Cambridge University Press, 2016. |
Includes bibliographical references and index.
Identifiers: LCCN 2015043520| ISBN 9781107082670 (hardback) |
ISBN 9781107443105 (paperback)
Subjects: LCSH: Customary law, International. | BISAC: LAW / International.
Classification: LCC KZ1277.C87 2016 | DDC 340.5–dc23
LC record available at http://lccn.loc.gov/2015043520

ISBN 978-1-107-08267-0 Hardback
ISBN 978-1-107-44310-5 Paperback

Cambridge University Press has no responsibility for the persistence or accuracy of URLS for external or third-party Internet Web sites referred to in this publication and does not guarantee that any content on such Web sites is, or will remain, accurate or appropriate.

DAMAGED

To Kathy, my beloved wife, for her patience and support during this project.

Contents

Notes on Contributors — *page* ix

Introduction: Custom's Future — 1
Curtis A. Bradley

1 **Custom's Past** — 11
Emily Kadens

2 **Customary International Law Adjudication as Common Law Adjudication** — 34
Curtis A. Bradley

3 **Customary International Law as a Dynamic Process** — 62
Brian D. Lepard

4 **Custom, *Jus Cogens*, and Human Rights** — 95
John Tasioulas

5 **Customary International Law: How Do Courts Do It?** — 117
Stephen J. Choi and Mitu Gulati

6 **Custom's Method and Process: Lessons from Humanitarian Law** — 148
Monica Hakimi

7 **The Growing Obsolescence of Customary International Law** — 172
Joel P. Trachtman

8 **The Strange Vitality of Custom in the International Protection of Contracts, Property, and Commerce** — 205
Chin Leng Lim

9 **The Decline of Customary International Law as a Source of International Criminal Law** 230
Larissa van den Herik

10 **Customary International Law and Public Goods** 253
Niels Petersen

11 **Reinvigorating Customary International Law** 275
Andrew T. Guzman and Jerome Hsiang

12 **The Evolution of Codification: A Principal-Agent Theory of the International Law Commission's Influence** 305
Laurence R. Helfer and Timothy Meyer

13 **Custom and Informal International Lawmaking** 332
Jan Wouters and Linda Hamid

14 **Custom's Bright Future: The Continuing Importance of Customary International Law** 360
Omri Sender and Michael Wood

Bibliography 371
Index 375

Notes on Contributors

Curtis A. Bradley is the William Van Alstyne Professor at Duke Law School in Durham, North Carolina. Professor Bradley has been teaching and writing about international law and U.S. foreign relations law for more than twenty years. His books include *International Law in the U.S. Legal System* (Oxford 2d ed. 2015) and *Foreign Relations Law: Cases and Materials* (with Jack Goldsmith) (Wolters Kluwer 5th ed. 2014). He has served as Counselor on International Law in the Legal Adviser's Office of the U.S. State Department and is currently a reporter for the American Law Institute's restatement (fourth) project on U.S. foreign relations law.

Stephen J. Choi is the Murray and Kathleen Bring Professor at NYU Law School in New York. His research interests focus on the theoretical and empirical analysis of corporations and capital markets as well as on judicial performance. He is the coauthor of *Securities Regulation: Cases and Analysis* (West 4th ed. 2015).

Mitu Gulati is on the faculty of Duke Law School in Durham, North Carolina. He teaches and writes in a number of areas, including contract law, corporate law, and international law, and he often uses empirical methods to study legal problems. His recent research has included a focus on the historical evolution of contract provisions in cross-border debt instruments (particularly those involving sovereigns). He is the coauthor, with Curtis Bradley, of "Withdrawing from International Custom," published in 2010 in the *Yale Law Journal*.

Andrew T. Guzman is the Dean and Carl Mason Franklin Chair in Law and Political Science at USC Gould School of Law. He is the author of *How International Law Works* (Oxford 2008); *Overheated: The Human Cost of Climate Change* (Oxford 2013); and *International Trade Law* (with Joost Pauwelyn) (Wolters Kluwer 2d ed. 2012). Dean Guzman's writings include work on public international law, private international law, international trade law, and international investment law.

Monica Hakimi is a Professor of Law at the University of Michigan Law School. Her research focuses on the informal ways in which international law adapts to contemporary challenges, particularly in the contexts of human rights law, humanitarian law, and the law governing the use of armed force. Between 2002 and 2006, Professor Hakimi served as an attorney-adviser in the Legal Adviser's Office at the U.S. Department of State.

Linda Hamid is a Junior Researcher at the Leuven Centre for Global Governance Studies – Institute for International Law, University of Leuven. She holds a bachelor degree in law (first class honors) from Babes-Bolyai University (Romania) and an LLM in public international law (cum laude) from Utrecht University (the Netherlands). After graduating, Linda gained practical legal experience in various international courts and tribunals, including the International Criminal Court, the European Court of Human Rights, and the Special Tribunal for Lebanon. In addition, she has worked as a teaching assistant in public international law at Babes-Bolyai University and has recently been admitted to practice law in the Cluj Bar (Romania).

Laurence R. Helfer is the Harry R. Chadwick, Sr. Professor of Law and Codirector of the Center for International and Comparative Law at Duke University in Durham, North Carolina. He is also a Permanent Visiting Professor at iCourts: The Center of Excellence for International Courts at the University of Copenhagen, which awarded him an honorary doctorate in 2014. Professor Helfer's research interests include international human rights, treaty design, international adjudication, and the interdisciplinary analysis of international law and institutions. He has authored more than seventy publications on these and other topics. He is a member of the Board of Editors of the *American Journal of International Law*.

Larissa van den Herik is Professor of Public International Law and Vice Dean at the Grotius Centre for International Legal Studies of Leiden Law School. Her areas of expertise include international criminal law and peace and security, particularly counterterrorism and UN sanctions. She is the Chair of the ILA Study Group on UN Sanctions and International Law and Vice Chair of the Government's Advisory Committee on Public International Law Issues.

Jerome Hsiang is a JSD candidate at the University of California, Berkeley School of Law. He writes at the intersection of international law and international relations, focusing on the development of nonconsensual transnational rulemaking and its impact on state behavior. Mr. Hsiang is also a partner at Shiang Law Firm, PC in Los Angeles.

Emily Kadens is a professor of law at the Northwestern University School of Law, where she teaches both commercial law and legal history. Prior to attending law school, she received a PhD in medieval history. Her scholarship on premodern European legal history has been recognized by such diverse groups as the American Society for Legal History and the National Conference of Bankruptcy Judges. She has also been awarded research fellowships at the Library of Congress and the George Washington University School of Law.

Brian D. Lepard is the Harold W. Conroy Distinguished Professor of Law at the University of Nebraska College of Law in Lincoln. He has written numerous books and articles on international law, comparative law, human rights, world religions, ethics, and tax law, including *Customary International Law: A New Theory with Practical Applications*, published by Cambridge University Press in 2010. He also serves as Director of the law college's LLM in Global Legal Practice Program for international students and non-U.S. lawyers.

Chin Leng Lim is Professor of Law at Hong Kong University, the Lionel A. Sheridan Visiting Professor at the National University of Singapore, and a barrister at Keating Chambers, London. He once served as State Counsel, International Affairs to Singapore and as a UN lawyer.

Timothy Meyer is a Professor of Law and the Enterprise Scholar at Vanderbilt University Law School in Nashville. He teaches and writes on a wide range of international law issues, including international economic law, international environmental law, the codification of customary international law, and international energy governance. Prior to entering the academy, Professor Meyer served as an attorney-adviser in the Office of the Legal Adviser, U.S. Department of State.

Niels Petersen is Professor of Public Law, International Law, and EU Law at the University of Münster. His main areas of research are international law, comparative constitutional law, and constitutional theory, and he has published several articles on different issues of customary international law and sources of international law.

Omri Sender is counsel for the World Bank and a consultant in public international law. Since 2012, he has assisted the Special Rapporteur of the UN International Law Commission on the topic "Identification of customary international law." A graduate of Tel Aviv University (LLB, LLM) and New York University (LLM), he is currently pursuing a doctoral degree in international law (SJD) at the George Washington University Law School.

John Tasioulas is Director of the Yeoh Tiong Lay Centre for Politics, Philosophy, and Law at the Dickson Poon School of Law, King's College London. He has published extensively in the fields of moral, legal, and political philosophy, with a recent focus on human rights and international law. He is the coeditor of *The Philosophy of International Law* (Oxford 2010).

Joel P. Trachtman is Professor of International Law at The Fletcher School of Law and Diplomacy. His recent books include *The Future of International Law: Global Government*, (Cambridge 2013); *The Tools of Argument* (Createspace 2013); *The International Law of Economic Migration: Toward the Fourth Freedom* (Upjohn Institute 2009); *Ruling the World: Constitutionalism, International Law, and Global Governance* (Cambridge 2009); *Developing Countries in the WTO Legal System* (Oxford 2009); and *The Economic Structure of International Law* (Harvard 2008). Professor Trachtman has served as a member of the boards of the American Journal of International law, the European Journal of International Law, the Journal of International Economic Law, the Cambridge Review of International Affairs, and the Singapore Yearbook of International Law.

Sir Michael Wood is a member of the International Law Commission and its Special Rapporteur for "Identification of Customary International Law." He is a Senior Fellow of the Lauterpacht Centre for International Law, University of Cambridge, and a barrister at 20 Essex Street, London, where he practices in the field of public international law, including before international courts and tribunals. He was Legal Adviser to the UK's Foreign and Commonwealth Office between 1999 and 2006, having joined as an Assistant Legal Adviser in 1970.

Jan Wouters is Jean Monnet Chair *Ad Personam* EU and Global Governance and founding Director of the Institute for International Law and of the Leuven Centre for Global Governance Studies at the University of Leuven (KU Leuven). He is a Member of the Royal Academy of Belgium for Sciences and Arts, is President of the United Nations Association Flanders Belgium, and practices law as *Of Counsel* at Linklaters, Brussels. He has published widely on international, EU, corporate and financial law. His recent books include *The European Union and Multilateral Governance* (Palgrave Macmillan 2012) (coeditor); *The EU's Role in Global Governance* (Oxford 2013) (coeditor); and *The Law of EU External Relations: Cases, Materials, and Commentary* (2013) (coauthor).

Introduction

Custom's Future

Curtis A. Bradley

Customary international law (CIL) is, along with treaties, one of the two principal sources of international law. Before the proliferation of treaty making in the twentieth century, CIL was the predominant source, regulating issues such as diplomatic immunity, rights at sea, and the conduct of war. In recent years, however, there have been increasing questions about CIL's continued relevance. Most areas of international law traditionally regulated by CIL are now also regulated by treaties. Treaties, moreover, have potential advantages as a source of law as compared with CIL: they are typically written, making their content easier to determine; they are expressly negotiated and ratified, making them more consensual; and they can be crafted with more precision, making them a potentially better vehicle for addressing the complexity of modern problems.

CIL is also subject to a variety of conceptual and evidentiary uncertainties. The conventional view today is that CIL arises out of state practice that is followed out of a sense of legal obligation.[1] Agreement on this "two-element" definition of CIL, however, has the potential to obscure a lack of agreement over issues such as what constitutes state practice; how much state practice is enough; and what materials demonstrate a sense of legal obligation (also known as *opinio juris*). There is also little agreement about the extent to which treaties can serve as evidence of CIL, and about the role that non-state actors, including international institutions, can play in generating or confirming rules of CIL. In addition, the increasing resort by nations to "soft law" has raised new questions about the relationship between nonbinding norms and the development of CIL, and even about whether there should be a

[1] *See, e.g.*, Restatement (Third) of the Foreign Relations Law of the United States § 102(2) (1987) ("Customary international law results from a general and consistent practice of states followed by them from a sense of legal obligation."); North Sea Continental Shelf Cases (Fed. Rep. Ger./Denmark), 1969 I.C.J. 4, 44 (Feb. 20) ("Not only must the acts concerned amount to a settled practice, but they must also be such, or be carried out in such a way, as to be evidence of a belief that this practice is rendered obligatory by the existence of a rule of law requiring it.").

sharp distinction between nonbinding norms and CIL in an international system that relies heavily on informal methods of enforcement.

More generally, the two-element definition of CIL has itself faced substantial criticism. Part of the concern is normative: by tethering CIL so heavily to what states do, CIL is potentially too regressive, leaving little room for aspirational legal development. A different concern is empirical: there is suspicion that institutional actors may voice the two-element definition but do not actually follow it in practice. The two-element definition also implicates deeper jurisprudential questions about CIL as a source of law. One famous such question is known as the "chronological paradox": if a rule of CIL can arise only after states are following a practice out of a sense of legal obligation, it is not clear how this sense of legal obligation develops in the first place. While it might be tempting to dismiss this paradox as a mere pet concern of legal philosophers, it may provide further reason to doubt that the standard account of CIL accurately describes how this body of law operates in practice.

For these and other reasons, scholars have been attempting to gain a better understanding of the nature of CIL and how it operates in the international system. This scholarly focus has produced a substantial body of work that reflects a willingness to reconsider standard accounts of CIL. Some scholars have questioned the continued vitality of CIL as a source of international law.[2] Others have questioned whether CIL operates as a genuine constraint on state behavior.[3] Still others have proposed new ways of thinking about CIL in an effort either to revitalize it or better explain it.[4] Not confined to doctrinal legal analysis, some of this modern scholarship encompasses interdisciplinary insights from empirical research, economic theory, philosophical inquiry, and historical study.

Efforts to better explain CIL extend beyond the academy. In 2000, the International Law Association adopted an extensive "statement of principles" concerning the formation of CIL, after years of work by a special committee.[5] Although rightly regarded as a major contribution, the statement provoked controversy by

[2] *See, e.g.*, J. Patrick Kelly, "The Twilight of Customary International Law," 40 VA. J. INT'L L. 449 (2000).

[3] *See, e.g.*, JACK L. GOLDSMITH & ERIC A. POSNER, *THE LIMITS OF INTERNATIONAL LAW* ch. 1 (2006).

[4] *See, e.g.*, Anthea E. Roberts, "Traditional and Modern Approaches to Customary International Law: A Reconciliation," 95 AM. J. INT'L L. 757 (2001); Andrew T. Guzman, "Saving Customary International Law," 27 *MICH. J. INT'L L.* 115 (2005); Pierre-Hugues Verdier & Erik Voeten, "Precedent, Compliance, and Change in Customary International Law: An Explanatory Theory," 108 AM. J. INT'L L. 389 (2014).

[5] *See* International Law Association, London Conference, Committee on Formation of Customary (General) International Law, *Statement of Principles Applicable to the Formation of General Customary International Law* (2000), adopted at the sixty-ninth Conference of the International Law Association in London, on July 29, 2000.

de-emphasizing the *opinio juris* requirement for CIL. Debate about the methodology for determining CIL rules also emerged in the wake of a widely discussed study in 2005 by the International Committee of the Red Cross on customary international humanitarian law.[6] Then, in 2012, the UN's International Law Commission initiated an important new project on CIL. First entitled "Formation and Evidence of Customary International Law," and then re-titled "Identification of Customary International Law," the project seeks to explain how to identify rules of customary international law and their content in order to assist nonspecialists in international law. This project has already generated substantial commentary and likely will continue to do so for years to come.

In the meantime, new books about CIL continue to appear.[7] CIL, moreover, is invoked with some frequency by international adjudicatory institutions such as the International Court of Justice (ICJ), with an especially noteworthy example being the ICJ's extensive consideration of CIL as it concerns sovereign immunity in the *Germany v. Italy* decision in 2012.[8] In addition, debates about CIL were triggered in 2013, when Great Britain claimed that there was a CIL exception to the UN Charter's limitations on the use of military force for certain instances of "humanitarian intervention."[9] CIL, in short, seems to be receiving more attention than ever just as claims about its growing obsolescence have intensified. It is therefore an ideal time to be considering the future of this body of international law.

The roots of this particular book can be traced to collaborations with my colleague and close friend, Mitu Gulati. In 2011–12, we managed a school-wide project at Duke Law School on "Custom and Law." With the generous support of our dean, David Levi, we organized a series of workshops and conferences on this crosscutting topic, including a symposium that produced a diverse set of articles by our faculty in the *Duke Law Journal*.[10] Mitu and I had previously published an article on CIL in

[6] *See* JEAN-MARIE HENCKAERTS & LOUISE DOSWALD-BECK, *CUSTOMARY INTERNATIONAL HUMANITARIAN LAW: VOLUME I: RULES* (2005); *see also, e.g.*, John B. Bellinger, III & William J. Haynes II, "A US Government Response to the International Committee of the Red Cross Study Customary International Humanitarian Law," 89 *INT'L REV. RED CROSS* 443 (2007); Jean-Marie Henckaerts, "Customary International Humanitarian Law: A Response to US Comments," 89 *INT'L REV. RED CROSS* 473 (2007).

[7] *See, e.g.*, NOORA ARAJARVI, *THE CHANGING NATURE OF CUSTOMARY INTERNATIONAL LAW: METHODS OF INTERPRETING THE CONCEPT OF CUSTOM IN INTERNATIONAL CRIMINAL TRIBUNALS* (2014); BRIAN D. LEPARD, *CUSTOMARY INTERNATIONAL LAW: A NEW THEORY WITH PRACTICAL APPLICATIONS* (2010); MICHAEL P. SCHARF, *CUSTOMARY INTERNATIONAL LAW IN TIMES OF FUNDAMENTAL CHANGE: RECOGNIZING GROTIAN MOMENTS* (2013).

[8] *See* Jurisdictional Immunities of the State (Ger. v. It.), 2012 I.C.J. 99 (Feb. 3).

[9] *See* Chemical Weapon Use by Syrian Regime – UK Government Legal Position (Aug. 29, 2013), at https://www.gov.uk/government/publications/chemical-weapon-use-by-syrian-regime-uk-government-legal-position/chemical-weapon-use-by-syrian-regime-uk-government-legal-position-html-version.

[10] *See* volume 62, issue 3 of the *Duke Law Journal*, at http://dlj.law.duke.edu/archive/volume-62-number-3-december-2012-special-symposium-issue-on-custom-and-law/.

the *Yale Law Journal*,[11] and Duke's international law journal published a collection of commentary that was focused on that article.[12] Important commentary was also published in the *Yale Law Journal Online*.[13] To continue exploring the rich theoretical and practical issues surrounding CIL, I organized a conference in Geneva in the summer of 2012, as part of the Duke-Geneva Institute in Transnational Law, on "The Role of *Opinio Juris* in Customary International Law." A number of the authors in the present volume participated in that conference.

Professor Ingrid Wuerth at Vanderbilt Law School was also a participant in the Geneva conference, and she persuaded me that the ideas generated at the conference should be developed into a book, with a focus on CIL more generally rather than just on *opinio juris*. She and I recruited a number of additional authors for this volume, for the sake of methodological and subject matter variety. Unlike with some edited volumes, we made a conscious effort to promote a dialogue among the authors, so that the volume would be a collaborative enterprise. To foster that dialogue, we held a conference at Duke Law School in October 2014, at which the authors met to discuss their draft chapters. We also encouraged the authors to reach out to each other after the conference with comments, and the two of us offered editorial feedback to each of the authors. Unfortunately, Ingrid was unable to continue serving as coeditor of the book project, but her initial work on the project was invaluable, and this volume benefited greatly from it.

It was originally envisioned that the book would be entitled *Custom in Crisis*. At the conference at Duke, however, it became apparent that this title would not encompass the full range of perspectives on CIL represented in the volume and might sound more alarmist than was intended. After consulting with the authors, I decided to re-title the book *Custom's Future* to better capture the diversity of views reflected in the chapters, and also to highlight the book's effort to contribute to the study of CIL's role going forward. My hope is that, with the benefit of the differing accounts presented in this book, readers will be in a better position to form their own judgments about the direction that CIL will, and should, take in the future.

[11] *See* Curtis A. Bradley & Mitu Gulati, "Withdrawing from International Custom," 120 *YALE L.J.* 202 (2010).

[12] *See* volume 21, issue 1 of the *Duke Journal of Comparative & International Law* (2010), with contributions from David Bederman, Rachel Brewster, Samuel Estreicher, Laurence Helfer, Barbara Koremenos & Allison Nau, Dino Kritsiotis, C. L. Lim & Olufemi Elias, Christiana Ochoa, Anthea Roberts, Paul Stephan, Edward Swaine, and Joel Trachtman, at http://scholarship.law.duke.edu/djcil/vol21/iss1/. *See also* Curtis A. Bradley & Mitu Gulati, "Customary International Law and Withdrawal Rights in an Age of Treaties," 21 *DUKE J. COMP. & INT'L L.* 1 (2010).

[13] *See* forum for volume 120, *Yale Law Journal Online*, with contributions from Lea Brilmayer & Isaias Yemane Tesfalidet, William Dodge, David Luban, and Carlos Vazquez. *See also* Curtis A. Bradley & Mitu Gulati, "Mandatory Versus Default Rules: How Can Customary International Law Be Improved?" 120 *YALE L.J. ONLINE* 421 (2011).

The book begins with a chapter by Emily Kadens on "Custom's Past." As Professor Kadens reminds us, many of the difficulties surrounding the use of custom as a source of law are longstanding. Indeed, she suggests that "publicists' arguments today look little different from those of medieval jurists, and the decisions of the International Court of Justice rather resemble those of premodern courts."[14] In particular, Professor Kadens notes "the insistence on a definition of custom that may describe no phenomenon that truly existed in the real world of communities governing themselves bottom-up without enacted law."[15] She also provocatively contends that "when we assert the existence of a custom of international law we essentially perform an act of legislation or invention."[16]

Continuing this emphasis on the legislative aspect of custom identification, my own chapter follows on "Customary International Law Adjudication as Common Law Adjudication." This chapter contends that the application of CIL by an international adjudicator "is best understood in terms similar to judicial development of the common law: that is, an approach whereby adjudicators look to past practice but necessarily make choices about how to describe it, which baselines to apply in evaluating it, and whether and when to extend it to new situations."[17] Understanding the adjudication of CIL in this way, the chapter further argues, "avoids many of the difficulties surrounding the standard view of CIL."[18]

Under a common law account, CIL can potentially be more forward-looking and progressive than under the standard, two-element view. A similar effort to conceptualize CIL in such terms characterizes Brian Lepard's chapter on "Customary International Law as a Dynamic Process." Continuing themes from his 2010 book, Professor Lepard contends that CIL is "in a jurisprudential crisis," and he argues that the way to resolve this crisis is to conceive of CIL not as a static form of law "embedded" in international practice, but rather "as a dynamic method of lawmaking."[19] In particular, he suggests that the *opinio juris* element of CIL should be reconceptualized "as a belief by states generally that it is desirable now or in the near future to have an authoritative legal principle or norm prescribing, permitting, or prohibiting certain conduct, apart from treaty obligations."[20] Such an approach, Professor Lepard explains, "permit[s] new rules to be recognized quickly to solve urgent problems."[21]

14 Emily Kadens, "Custom's Past" (in this volume), at 11.
15 *Id.* at 12.
16 *Id.* at 33.
17 Curtis A. Bradley, "Customary International Law Adjudication as Common Law Adjudication" (in this volume), at 34.
18 *Id.*
19 Brian D. Lepard, "Customary International Law as a Dynamic Process" (in this volume), at 63, 64.
20 *Id.* at 63.
21 *Id.* at 93.

John Tasioulas, like Professor Lepard, thinks that the traditional account of CIL needs revision. In his chapter on "Custom, *Jus Cogens*, and Human Rights," Professor Tasioulas sketches a "moral judgment-based account" of CIL, which he contends "makes best sense of what custom as a source of international law has become in recent decades."[22] As he explains, under this account, "*opinio juris* involves the judgment that a norm is already a part of customary international law and that (compliance with) it is morally justified; or that, as a moral matter, it should be established as law through the process of general state practice and *opinio juris*; or else some mixture of these two attitudes."[23] Professor Tasioulas considers implications of this account for international human rights law in particular, including for *jus cogens* norms.

In contrast with Professor Tasioulas's chapter, the focus of the chapter by Stephen Choi and Mitu Gulati is empirical rather than normative. In "Customary International Law: How Do Courts Do It?," Professors Choi and Gulati study what the International Court of Justice and other international tribunals actually cite in support of their determinations regarding CIL. Compiling a dataset of 175 different CIL determinations, they find that "the type of evidence that looks to be the most important for determinations of CIL is the international treaty."[24] They also find that "three other forms of evidence that are largely aspirational in character also play a big role in CIL determinations," namely "UN resolutions, other UN material (committee reports, conference reports, etc.), and domestic statutes."[25] By contrast, they find relatively little citation to traditional forms of state practice. They conclude that "the question that jump-started our inquiry – how do judges apply the CIL definition that seems both analytically impossible to apply and normatively unattractive – has a simple answer: they ignore it."[26]

That the method for determining CIL may be somewhat elusive is not necessarily a vice, as Monica Hakimi points out in her chapter, "Custom's Method and Process: Lessons from Humanitarian Law." Echoing some of the earlier chapters, Professor Hakimi contends that "CIL finding is deeply entangled with CIL making," and she describes the CIL process as "chaotic, unstructured, and politically charged."[27] Focusing on international humanitarian law, she suggests two conclusions: "First, nonstate actors who are charged with finding CIL can be

22 John Tasioulas, "Custom, Jus Cogens, and Human Rights" (in this volume), at 115.
23 *Id.* at 97.
24 Stephen J. Choi & Mitu Gulati, "Customary International Law: How Do Courts Do It?" (in this volume), at 132.
25 *Id.* at 133.
26 *Id.* at 147.
27 Monica Hakimi, "Custom's Method and Process: Lessons from Humanitarian Law" (in this volume), at 149.

extremely influential in making CIL.... Second, no particular method for finding CIL is capable of disciplining global actors or imposing order on CIL, because the process for making CIL is so heavily *un*disciplined and *dis*ordered."[28] This does not mean that "anything goes" in determining CIL, she explains. Rather, "It means that the limits on CIL must – and do – come from *within* the CIL process."[29]

Many of the remaining chapters in the book reflect more generally on the future of CIL. In "The Growing Obsolescence of Customary International Law," Joel Trachtman has doubts about CIL's continued usefulness in addressing the world's problems. His chapter discusses in detail a number of CIL's limitations and notes the comparative advantages of treaties and other "legislated law." While acknowledging that CIL has certain benefits that make it useful in particular situations, Professor Trachtman contends that "in general these benefits seem to be outweighed by the related detriments."[30] He also conducts an extensive study of the overlap between CIL and existing treaties, finding that "there are few asserted rules of CIL that have not been included in conventions."[31] Professor Trachtman suggests, among other things, that "states and international organizations should focus their international legal analytical resources on legislated law."[32]

Chin Lim contests some of Professor Trachtman's assessment in "The Strange Vitality of Custom in the International Protection of Contracts, Property, and Commerce." In particular, Professor Lim "defend[s] custom's continued importance in international economic law, particularly in the interpretation of investment treaty clauses, that is, in the protection of property and contracts."[33] He also questions some of CIL's purported limitations, arguing that "(i) custom is at least as susceptible as treaties to detailed customization in design, (ii) like treaties, customary lawmaking can also be formed through complex tradeoffs ... and, finally, (iii) custom's lack of predictability is overstated."[34] In addition, Professor Lim contends that "increased treaty lawmaking and the increased institutionalization of international law create a greater, not lesser, role for custom in modern international economic law and regulation."[35] Instead of the displacement of CIL through treaties, he finds that in

28 *Id.*
29 *Id.* at 171.
30 Joel P. Trachtman, "The Growing Obsolescence of Customary International Law" (in this volume), at 174.
31 *Id.* at 194.
32 *Id.* at 204.
33 C. L. Lim, "The Strange Vitality of Custom in the International Protection of Contracts, Property, and Commerce" (in this volume), at 205.
34 *Id.* at 206.
35 *Id.*

the area of international economic law, CIL and treaties have become intertwined, such that "custom's fate is bound to that of treaty law."[36]

The next chapter, by Larissa van den Herik, also focuses on a particular area of international law, in this case international criminal law. In "The Decline of Customary International Law as a Source of International Criminal Law," Professor van den Herik explains that, although CIL has played a significant role in adjudication in the ad hoc criminal tribunals for former Yugoslavia and Rwanda, the establishment of the International Criminal Court has led to a decline in CIL's role in international criminal law. As she explains, "The codification movement and the gradual maturation of [international criminal law] made resort to CIL less necessary in a practical sense."[37] In addition, she notes how reliance on CIL for international criminal adjudication presents particular difficulties, both with respect to legality concerns as well as methodological issues relating to the nature of the relevant state practice.

Shifting to another subject area focus, Niels Petersen considers CIL's ability to address global public goods problems, such as climate change. In "Customary International Law and Public Goods," Professor Petersen argues that "economic theory and experimental evidence show that the establishment of public goods through decentralized cooperation is difficult."[38] He concludes, therefore, that CIL, which is a form of decentralized lawmaking, has limited potential in this context. Even some of the proposed "modern" approaches to CIL are unlikely to be especially useful in this context, suggests Professor Petersen. A more promising avenue for protecting global public goods, he argues, is through initially unilateral actions by states to extend their authority.

A more optimistic picture of CIL's continued importance is presented in "Reinvigorating Customary International Law," by Andrew Guzman and Jerome Hsiang. The authors emphasize in particular that CIL is "a tool that can promote cooperation in situations where consent-based rule making proves impractical."[39] Although they acknowledge that "CIL in its current form is not entirely coherent conceptually," they believe that it can be "reinvigorated for the cooperative benefit of modern states."[40] They put forth a "rational choice" understanding of how CIL works, pursuant to which "states are incentivized to act in certain ways by reputational consequences attached to compliance with international law."[41] Under this account,

[36] *Id.* at 229.

[37] Larissa van den Herik, "The Decline of Customary International Law as a Source of International Criminal Law" (in this volume), at 251.

[38] Niels Petersen, "Customary International Law and Public Goods" (in this volume), at 273–74.

[39] Andrew T. Guzman & Jerome Hsiang, "Reinvigorating Customary International Law" (in this volume), at 275–76.

[40] *Id.* at 276.

[41] *Id.* at 293.

they argue, the *opinio juris* element of CIL should be understood as "stand[ing] for the subjective beliefs of observing states regarding the legality of a rule," and proof of state practice should not be formally necessary as a separate element.[42] As they explain, "To the extent that observing states believe there exists a legal obligation, the potential violator faces the reputational consequences of violating CIL."[43]

A social science perspective also characterizes the next chapter by Laurence Helfer and Timothy Meyer. In "The Evolution of Codification: A Principal-Agent Theory of the International Law Commission's Influence," the authors consider a shift by the UN's International Law Commission in recent years away from codification efforts to "principles, conclusions, and draft articles that it does not recommend be turned into treaties."[44] Using principal-agent theory, the authors explain the shift by positing that "increasing political gridlock in the [UN] General Assembly ... has led the Commission to modify the form of its work to preserve its influence in shaping the evolution of international law."[45] Professors Helfer and Meyer also contend that "the shift away from draft treaties increases the salience of the methodology that the ILC uses to enhance its influence when the traditional constraint of UNGA review is unavailable due to gridlock."[46] The authors predict that the ILC will "select and, relatively consistently, adhere to a methodological approach aimed at attracting the support of the audience it hopes to persuade."[47] Viewed this way, they suggest that the ILC's current project on the identification of CIL "promises to limit, and by so limiting also expand, the ILC's importance in a post-treaty world."[48]

Jan Wouters and Linda Hamid similarly consider CIL against the backdrop of a shift away from codification, but their focus is on CIL's relationship to "informal international lawmaking," which includes, but is not limited to, soft law. This informal international lawmaking has expanded in recent years, they explain, at the same time that there has been a slowdown in traditional international lawmaking. Instead of replacing traditional international law, they contend that informal international lawmaking "exists alongside it."[49] Nevertheless, Professor Wouters and Ms. Hamid point out that informal international lawmaking offers an alternative to CIL, while avoiding some of CIL's conceptual, methodological, and normative limitations. Ultimately, they conclude that "there is a domain for both informal law, as well

42 *Id.* at 291.
43 *Id.* at 302.
44 Laurence R. Helfer & Timothy Meyer, "The Evolution of Codification: A Principal-Agent Theory of the International Law Commission's Influence" (in this volume), at 305.
45 *Id.*
46 *Id.* at 306.
47 *Id.* at 327.
48 *Id.* at 329.
49 Jan Wouters & Linda Hamid, "Custom and Informal International Lawmaking" (in this volume), at 341.

as custom," but that "in an age of increasing informality, custom's domain is quite limited."[50] At the same time, they note some ways in which informal international lawmaking can contribute to the development of CIL, and they wonder whether soft-law instruments might "emerge as the 'new' custom of the twenty-first-century global governance, gradually replacing state-centred notions of custom."[51]

The book concludes with a chapter entitled "Custom's Bright Future: The Continuing Importance of Customary International Law." In this chapter, Sir Michael Wood, the special rapporteur for the International Law Commission's project on CIL, and Omri Sender, who has since 2012 been working with Sir Michael closely on the project, deny that CIL is facing a crisis and instead contend that it "is now more necessary and important than ever."[52] The recent work of the International Law Commission on CIL, the authors suggest, shows that there is consensus on many issues surrounding the identification of CIL. The authors also observe that "the theoretical torment that accompanies custom in the books simply does not impede it in action."[53] The authors make clear, however, that they do not expect the Commission's work to resolve all of the theoretical issues surrounding CIL. "That is not to be regretted," they note, since "those engaged in the practice of law may benefit much from such debates, as did members of the [Commission]," and they graciously observe that "the present volume, too, contains very valuable contributions to the doctrinal discussions."[54]

As indicated by these brief descriptions of the chapters, the book contains a range of perspectives and methodological approaches. A number of recurring themes emerge: that CIL faces new challenges in a world in which treaty making and soft law seem better suited than CIL for addressing modern problems; CIL may not operate the same way across different institutional contexts; it is important to probe doctrinal categories in order to better understand how CIL actually operates in practice; the identification and application of CIL is in part a creative exercise and is affected by baseline assumptions and prescriptive judgments; and that, along with its virtues as a source of law, CIL has structural and normative limitations. In addition to shedding light on these and other issues as they relate to the current state of CIL, many of the chapters offer insights into how this traditional body of international law might be revitalized going forward. In that sense, this book aims to make its own contribution to custom's future.

[50] *Id.* at 350.
[51] *Id.* at 357.
[52] Omri Sender & Michael Wood, "Custom's Bright Future: The Continuing Importance of Customary International Law" (in this volume), at 369.
[53] *Id.* at 365.
[54] *Id.* at 369.

1

Custom's Past

Emily Kadens

The chapters in this volume put on display the difficulties lawyers, judges, and lawmakers continue to have making sense of custom as a source of law. These debates, whether about the lack of a workable definition, the instability of norms, the democracy deficit, the lack of fidelity to the strictly descriptive – rather than normative – nature of custom, or the unpredictability of courts, are not new. Despite the efforts of lawyers over the centuries to "solve" the problem of custom, publicists' arguments today look little different from those of the medieval jurists, and the decisions of the International Court of Justice rather resemble those of premodern courts. As David Bederman pointed out not long ago in his book, *Custom as a Source of Law*, the discussion of custom has not advanced very far in all this time, and the uncertainties have come no closer to being resolved.[1]

This chapter argues that the modern publicists' problems with custom grow out of the efforts of the medieval jurists to fit custom into the hierarchy of law. Trained in formal law, lawyers and judges expect all legally binding rules to have the characteristics of rules found in statute books and judicial opinions. This sort of lawyerly bias has its origin in the twelfth century, when the European tradition of formal legal study began. But custom had a prelegal existence,[2] and in this "natural" state it did not fit the mold of enacted law. Natural custom was fluid, uncertain, equitable, and communitarian – features of a system of social regulation that lawyers no longer equate with law. Instead, for nearly 900 years, jurists and judges have been trying to force custom to look like what they have been trained to believe law is, and for nearly 900 years they have failed. Natural custom might, in certain circumstances, have functioned as law, but it did not function like law.

1 David J. Bederman, *Custom as a Source of Law* 135 (2010).

2 Dominique Bauer, "Custom in Canon Law and the Expansion of Legal Reality," in *Custom: The Development and Use of a Legal Concept in the Middle Ages* 107, 107 (Per Anderson & Mia Münster-Swendsen eds., 2009).

At the root of the lawyers' failure to fit the square peg of custom into the round hole of law lay the insistence on a definition of custom that may describe no phenomenon that truly existed in the real world of communities governing themselves bottom-up without enacted law. The definition Western legal systems inherited from the Romans and bequeathed upon the rest of the world is a legal fiction. Unable to correctly articulate the nonlegal reality of custom, the definition created a law-like phantom. Little wonder, then, that it has given rise to so many problems.

This chapter traces the historical development of the concept of custom from the twelfth to the seventeenth centuries. The first part of the chapter examines the anthropological phenomenon of custom to argue that its prelegal reality does not correspond to the definition of custom imposed by jurists. This background helps to explain the trouble jurists had and continue to have with forcing custom into the straitjacket of law. The second part considers how medieval jurists co-opted the concept of custom and refashioned it into something that looked and functioned like enacted law but that no longer looked and functioned like custom in its original habitat. The third part picks up the story of the legal concept of custom at the historical moment of the creation of the idea of customary international law in the seventeenth century. Having largely lost its law-creative role in private law, custom was reborn as a catchall default for jurists looking for ways to legitimize legal rules they could not fit into any other category.

PART ONE: THE EVOLUTION OF CUSTOM

The problem of custom begins with a definition that was not meant to define. François Gény's commonly accepted definition of custom as repeat action plus *opinio juris sive necessitatis* is the product of a long historical evolution that ultimately traces back to the texts of the Roman law as they were interpreted by the jurists of the twelfth and thirteenth centuries.[3] From the beginning, the definition included both subjective and objective factors. While jurists of various eras have used different words and analogies to describe the subjective factor, the essential idea of the community's sense that a certain act was legally mandatory remained consistent.

3 1 François Gény, *Méthode d'interprétation et sources en droit privé (Method of Interpretation and Sources in Private Law)* 360 (Paris, F. Pichon et Durand-Auzias, 2d ed. 1919). Gény did not invent this definition. As he credits in his footnotes, it goes back at least to Christian Friedrich von Glück, who used the term *opinio obligationis* ("sense of obligation") in his multivolume commentary on the Roman law in the late eighteenth century. 1 Christian Friedrich Glück, *Versuch einer ausführlichen Erläuterung der Pandecten (Attempt at a Detailed Explanation of the Pandects)* 448 (Erlangen, Johann Jacob Palm 1790); *see also* Carl Christian Wilhelm Klötzer, *Versuch eines Beytrags zur Revision der Theorie vom Gewohnheitsrecht (Attempt at a Contribution to the Theory of Customary Law)* 190 (Jena, J.M. Mauke und Söhne 1813) (writing about *opinio juris et obligationis*).

Gény's explanation that the subjective element consists "in the sentiment that [the people involved in the conduct] act by virtue of an unexpressed rule imposed on themselves as a rule of positive law"[4] does not differ greatly from the definition of custom by Bartolus, the leading jurist of the fourteenth century, who wrote, "Custom is a certain law, instituted by repeated acts, which is acknowledged as law."[5]

The medieval jurists' understanding of custom started with several fragments in the Digest of Justinian, a sixth-century compilation of excerpts from the writings of the classical Roman jurists that, beginning in the twelfth century, became the basis for the renewed study of law in continental Europe. The most basic of the texts for these jurists was book 1, title 3, law 32, which established the defining elements of custom: repeat behavior over time and the tacit willingness of the people of the community to abide by the behavior.[6] A custom that met these two criteria had a force like that of enacted law (*pro lege*).[7]

A recent article, however, has pointed out that Julian, the second-century Roman jurist who authored this Digest fragment, was not writing with the intent to create a definition of custom.[8] He had been presented with the problem of what rule one should follow in the absence of a written statute. His answer: do what people in the community usually do under this or similar situations.[9] In fact, as far as the evidence shows, the Romans had no particular interest in custom, never developed a theory of custom, and never wrote treatises about it.[10] Ever pragmatic, the Roman jurists simply observed that what people did was worth following, at least when it did not conflict with an enacted law.[11]

In the twelfth century, however, the medieval jurists, newly rediscovering the Roman law, did not focus on the context of Julian's discussion. They took his text as a pure definition. And it was a definition of great use to them, for, in contrast to their Roman predecessors, the medieval jurists were intensely interested in custom.

4 GÉNY, *supra* note 3, at 363 ("dans le sentiment qu'ils agissent, en vertu d'une règle non exprimée, s'imposant à eux comme règle de droit objectif").

5 BARTOLUS, *IN PRIMAM DIGESTI VETERIS PARTEM COMMENTARIA* (*COMMENTARY ON THE FIRST PART OF THE OLD DIGEST*) 19r. (Turin, Nicholaus Beuilaquam, 1574) (*repetitio ad* Dig. 1.3.32, § 6) ("Consuetudo est ius quoddam morib[us] institutum, q[uod] pro [lege] suscipitur").

6 1 *THE DIGEST OF JUSTINIAN* 13 (Alan Watson ed., 1985) (at Dig. 1.3.32).

7 *Id.* ("Inueterata consuetudo pro lege non immerito custoditur.")

8 Caroline Humfress, "Law and Custom Under Rome," in *LAW, CUSTOM, AND JUSTICE IN LATE ANTIQUITY AND THE EARLY MIDDLE AGES* 23, 26–29 (Alice Rio ed., 2011).

9 1 *DIGEST OF JUSTINIAN*, *supra* note 6, at 13 (at Dig. 1.3.32). With reference to the acts of the community, the text in full reads, "What ought to be held to in those cases where we have no applicable written law is the practice established by customs and usage. And if this is in some way deficient, we should hold to what is most nearly analogical to and entailed by such a practice. If even this is obscure, then we ought to apply law as it is in use in the City of Rome."

10 Humfress, *supra* note 8, at 25.

11 *See also* 2 *CORPUS IURIS CIVILIS: CODEX IUSTINIANUS* 362 (Paul Krueger ed., 1906) (Cod. 8.52[53].1 and 8.52[53].3).

They lived in a world in which the local custom of towns and villages, parishes, and bishoprics was the predominate form of law.[12]

François Gény astutely described the moment at which the twelfth-century jurists turned their attention to custom:

> For a long time, custom had [authority]. Not only in the age when it reigned more or less alone, but all the time since that it dominated as a source of positive law, people barely concerned themselves, or discussed, or justified the creative power of custom. It was invoked and submitted to as an ineluctable fact.... But, from the day when the written law occupied an increasingly important place alongside customary law, people took up the notion of scrutinizing custom's qualifications and believed it necessary to verify its authority.[13]

Until the Roman law resurfaced, people did not think to analyze and define custom. It simply existed as a fact of life. And then the written law intruded, trained legal experts appeared, and custom had to be assigned legal relevance and meaning. In doing this, the jurists changed custom's essence and turned it into law.

Medieval lawyers assumed that the texts of the Digest were both authoritative and correct, and as a consequence, they did not explore the possibility that the custom they saw around them demanded a definition different from the one in the Digest. Instead, they tried to fit what they lived into the definition they had inherited from a quite different society with a quite different attitude toward local norms. Their unquestioning adoption of the Roman law's accidental definition thus paved the way for centuries of confusion.

According to the Roman law, "Every definition in civil law is dangerous; for it is rare for the possibility not to exist of its being overthrown."[14] The definition of custom is no exception. In a community in which something that we might call custom controlled behavior, "repeat behavior plus a sense of being bound" is an overly optimistic description of what really happens. Instead, the historical and

[12] JOHN GILISSEN, *LA COUTUME* 41 (1982) ("Le droit medieval est presqu'exclusivement coutumier; autrement dit, la coutume est la principale, et même à certaines époques, dans certaines regions, pour certains domains, l'unique source du droit."); R. C. Van Caenegem, "Aantekeningen bij het middeleeuwse gewoonterecht" ("Notes on Medieval Customary Law"), 64 *TIJDSCHRIFT VOOR RECHTSGESCHIEDENIS (LEGAL HIST. REV.)* 97, 97 (1996) (Neth.) ("De gewoonte was in de Middeleeuwen de belangrijkste bron van het recht.").

[13] 1 GÉNY, *supra* note 3, at 334–35:

> Pendant longtemps, il en dut être ainsi de la coutume. Non seulement à l'époque où elle régnait à peu près seule, mais tout le temps encoure qu'elle domina comme source de droit positif, on ne songea guère, ni à discuter, ni à justifier, la puissance créatrice de la coutume. On l'invoquait et on la subissait comme un fait inéluctable... Mais, du jour où le droit écrit occupa une place importante, et de plus en plus large, à coté du droit coutumier, on s'avisa de scruter les titres de celui-ci, on crut nécessaire d'en vérifier l'autorité.

[14] 4 *THE DIGEST OF JUSTINIAN*, *supra* note 6, at 969 (at Dig. 50.17.202).

anthropological evidence suggests that custom, before the medieval lawyers got a hold of it, functioned as a relatively flexible and malleable set of social norms ostensibly held in the memory of the community, but in many instances more likely invented or reinvented on the spot to solve a current problem. This sort of custom does not have to produce formal rules of decision, nor does it have to be consistently stated or applied across individual instances. Thus, it is not particularly law like.

Philological evidence suggests that the very concept of custom as a set of rules arising from the community's behavior is not necessarily natural. In Romance languages, the word for, and presumably the notion of, custom derived from Latin and the legal practices of the Roman Empire. By contrast, philologists have shown that the Germanic word for custom was introduced only as an early medieval loanword from the Latin via the canon law.[15] This suggests that in the sorts of small, tightly knit, highly stable communities in which social control can function without positive lawmaking, "the way we do things" does not even need to be identified or articulated as a set of rules – "our custom" – to be effective.

This lack of a sense of rules informs the problem many scholars in this volume and elsewhere have pointed out in the so-called chronological paradox.[16] According to the subjective-objective definition of custom, for repeat behavior to become binding custom it must at a certain point move from an "is" to an "ought." Obviously, however, the first person to behave in a certain way cannot believe themselves to be bound; the second person cannot believe themselves required to act in this way because the first person was bound; and so on. Like the modern commentators who "either ignore the paradox or treat it as if it were a mere rhetorical question that need not be answered,"[17] the medieval jurists noted and admitted the insolubility of the problem.[18]

Viewing custom not through the lens of law as a search for formal rules but rather on its own terms, however, reveals this to be a false problem. In a customary community, individuals do an act in mimicry of others simply because they believe, rightly or wrongly, that everyone else is doing it. The community does not need to pinpoint the moment when this belief in the mandatory nature of the act arises because individuals do not define their actions in terms of legal obligation but rather

[15] Klaus von See, *Altnordische Rechtswörter. Philologische Studien zur Rechtsauffassung und Rechtsgesinnung der Germanen (Old Norse Legal Words. Philological Studies on the Legal Conception and the Legal Mindset of the Germanic Peoples)* 92–96, 100–02, 249–55 (1964); Helle Vogt, "Secundum consuetudinem et leges patrie": "The Concept of Law and Custom in Thirteenth Century Denmark," in *Custom: The Development and Use of a Legal Concept in the Middle Ages* 67, 68 (Per Anderson & Mia Münster-Swendsen eds., 2009).

[16] *See, e.g.*, Alan Watson, *The Evolution of Western Private Law* 93–94 (expanded ed. 2001).

[17] Curtis A. Bradley, "Customary International Law Adjudication as Common Law Adjudication" (in this volume).

[18] Bartolus, *supra* note 5, at 19v. (*repetitio ad* Dig. 1.3.32, § 15).

in terms of perceived social obligation: we in this community act this way because it is the way we act and have always acted, not because we must act this way.

Part of the lawyerly insistence on custom as a legal rule, with a beginning in time and in the minds of the community members, may arise from the articulation of a norm or custom in the form of a rule statement ("the custom is …"). This habit tends to persuade us that the description stands for a static, generally followed, known, and articulable rule of decision. Yet this image of stability, definition, and awareness may in many instances be misleading. Trained as we are to assume the fixity of laws, we make the mistake of equating the single frame of a custom at one moment in time based on one version of the behavior in question for the ever-changing movie that is custom in practice. As anthropologists studying customary societies have discovered, assuming that a description of a custom represents a stable rule that, when codified, will accurately reflect the community's understanding of the custom imposes a lawyerly concept of legal rules upon a prelegal form of social organization.[19]

In fact, the signal characteristic of the prelegal form of social organization known as custom is its malleability, and that malleability derives from two sources: the fallibility of memory and the allowable variability of behavior. In a society without written records, the past must be held in memory. A past fact is held in memory, however, only so long as the knowledge of it is useful. A memory remains useful only insofar as it can be adapted and manipulated to serve present purposes. In other words, the memory does not preserve an objective historical truth. As the English legal historian Michael Clanchy has pointed out, in illiterate societies "historical facts in the remembered past are under constant pressure from the needs of the present. The memory adapts them, transposes them, or eliminates them altogether, when they are no longer useful."[20] Memory stores the "fact" of a custom in a way that will serve the argument being made in the present.[21] Unless objected to, the new version of the historical fact becomes the new truth.[22] When it ceases to be useful, a memory, and any custom attached to it, is

[19] Martin Chanock, *Law, Custom and Social Order. The Colonial Experience in Malawi and Zambia* 54–55, 59, 62 (1998); Erica Harper, *Customary Justice: From Program Design to Impact Evaluation* 17–19 (2011) ("The governing principles and rules [of custom] are not static but are constantly evolving in response to cultural interactions, socio-economic and demographic shifts, political processes and environmental change.… 'To view customary practices as 'law' is essentially a Western-centric approach,' which does not facilitate a sound understanding of customary justice.").

[20] M. T. Clanchy, "Remembering the Past and the Good Old Law," 55 *History* 165, 166 (1970).

[21] David Ibbetson, "Custom in Medieval Law," in *The Nature of Customary Law* 151, 174–75 (Amanda Perreau-Saussine & James Bernard Murphy eds., 2007).

[22] J. A. Barnes, "History in a Changing Society," 11 *Rhodes-Livingstone J.* 1, 5–6 (1951). *See also* Walter Ullmann, "Bartolus on Customary Law," 52 *Jurid. Rev.* 265, 277 (1940) (discussing Bartolus's view that silence in the face of a judicial decision not in conformity with the custom "presume[d] consent to this new decision").

simply forgotten.[23] The memory, and the custom attached to it, disappear from both the community's present and its past.[24]

The memory of a past behavior is complicated in another way. Individuals engaged in a behavior may have no reason to identify it as a custom, and therefore as obligatory upon them. When asked later, perhaps years later, to describe their actions in order to give evidence in a dispute over whether a particular behavior is mandatory as a binding custom, the witnesses may have to manufacture a memory. While they could relate what they in fact did, they might also describe what they imagine, in light of the current dispute, they did, what they assume their audience wants to hear that they did, what they believe other people did, or what they think they, and others, should have done.[25]

Even when the behavior at the root of a supposed custom is current, and thus does not require an act of memory, individuals' exercise of the behavior could vary without inciting notice or sanction. A person's belief about the custom and how he should perform it would be formed by what he knew of how he and the others he knew had performed. But people could perform the act in somewhat different ways, and none might have had a reason to discover the full spectrum of behavior covered by the norm.[26] For example, in the Middle Ages, maritime custom had the rule that in the case of a shipwreck, merchants only had to pay for freightage.[27] While at a

[23] 1 MARC BLOCH, *FEUDAL SOCIETY* 114 (L. A. Manyon trans., 1961); 1 FRITZ KERN, *KINGSHIP AND LAW IN THE MIDDLE AGES* 179 (S. B. Chrimes trans., 1956) (customary law "quietly passes over obsolete laws, which sink into oblivion, and die peacefully").

[24] Jack Goody and Ian Watt, "The Consequences of Literacy," 5 *COMP. STUD. SOC'Y & HIST.* 304, 307 (1963) ("Whatever parts of [the tradition] have ceased to be of contemporary relevance are likely to be eliminated by the process of forgetting").

[25] SALLY FALK MOORE, *SOCIAL FACTS AND FABRICATIONS: "CUSTOMARY" LAW ON KILIMANJARO, 1880–1980*, at 194 (1986); Simon Roberts, "The Recording of Customary Law: Some Problems of Method," in 1 *FOLK LAW: ESSAYS IN THE THEORY AND PRACTICE OF* LEX NON SCRIPTA 331, 333 (Alison Dundes Renteln & Alan Dundes eds., 1994) (concerning ascertaining African custom from witnesses, "there is the risk of distortion on the part of the informant: he may tell you what he thinks you would like the answer to be; what he would like the answer to be; or, what the answer might have been in the past"). *See also* CHANOCK, *supra* note 19, at 17:

> Claimed custom is sometimes simply ideological, but it is often pragmatic, a claim put forward in a form in which it is likely to be successful. In circumstances of conflict and change where there is an unbridgeable gap between social ideals and the actual ways in which life can be lived, custom, or customary law, cannot be a rule which emerges from, is descriptive of and which governs practice or social system. In this changing world claims about custom were competitive rather than descriptive.

[26] *See* Richard Craswell, "Do Trade Customs Exist?" in *THE JURISPRUDENTIAL FOUNDATIONS OF CORPORATE AND COMMERCIAL LAW* 118, 127 (Jody S. Kraus & Steven D. Walt eds., 2000) (discussing how different people could frame the same custom).

[27] Edda Frankot, "The Practice of Maritime Law in the Town Courts of Fifteenth-Century Northern Europe," in *THE DYNAMICS OF ECONOMIC CULTURES IN THE NORTH SEA- AND BALTIC REGION IN THE LATE MIDDLE AGES AND EARLY MODERN PERIOD* 136, 144 (Hanno Brand & Leos Müller eds., 2007).

high level of generality, this custom seems to have been largely uniform; in fact, the specific practice varied greatly because its application depended on the point in the voyage at which the wreck occurred. A shipwreck early might mean paying only half freight, while a shipwreck late in the voyage might mean full freight, except where it meant paying freight *pro rata* according to the distance traveled.[28] The merchant who only knew about shipwrecks early in a voyage might believe that the custom was half freight, while one who knew only about shipwrecks late in the voyage might think it was full freight or *pro rata*. Unless a dispute arose or an accusation of nonconformity was levied, each individual could have had reason to believe that the actions he observed and engaged in were conforming and represented the totality of the custom. And even then his behaviors or beliefs might change over time without his consciously articulating it.[29]

Such a situation of diversity within an overarching assumption of conformity could exist for a long time in a customary society. No one would think that his own behavior violated the custom or that his behavior constituted only a part of the custom until a dispute arose. But where a custom was common and well accepted, in other words, where it was indeed universally acknowledged to be the custom and whatever variations that existed were tolerated, disputes might rarely, if ever, arise. As Philippe de Beaumanoir, the author of a massive thirteenth-century collection of northern French customs, wrote, some customs are "so clear that I have never seen [them] challenged."[30] This does not necessarily mean that everyone agreed on the exact nature of the perhaps unarticulated custom. It meant only that they did not punish variant but still acceptable performance of the custom.

And even when disputes did arise, courts or elders did not necessarily need to apply rules of decision to resolve them. In the sorts of small, stable, and close-knit communities governed by the mimicking and social policing of behavior, dispute resolution often meant finding equitable solutions that the community was willing to accept.[31] This might help explain why it is rather unusual in the historical

[28] *Id.* at 144.

[29] Van Caenegem, *supra* note 12, at 106–07.

[30] THE COUTUMES DE BEAUVAISIS OF PHILIPPE DE BEAUMANOIR 249, ¶ 683 (F.R.P. Akehurst trans., Univ. of Pa. Press 1992) (1283).

[31] WATSON, *supra* note 16, at 97–98. *See also* Carol Bohmer, "Community Values, Domestic Tranquility, and Customary Law in Upper Volta," 16 *J. MODERN AFRICAN STUD.* 295, 307 (1978):

> [The judges in an African court] were not looking ahead to any final decision, but rather were involved in discussing the issues which hopefully would lead eventually to some outcome acceptable to all parties. This may be a reflection of a broader social trait in African culture where the concept of "winner takes all" is difficult to understand and even abhorrent. Making a judgement was viewed as a last resort, to be engaged in only when all attempts at accommodation failed. Cases were frequently postponed to avoid having to make an either/or decision. The judges used the force of public opinion in the form of the participant audience to help them achieve their accommodation. This use of public opinion, and response to it, was in

evidence to find customs stated as rules of decision to justify a judicial decision in premodern legal records and why anthropologists found customary courts that simply did not apply the supposed "customary law" to resolve disputes allegedly arising from violations of the custom.[32] Part of the problem, as ethnographers studying Indonesian adat law discovered, is that even when customs are articulated, they are often stated more as maxims or standards than as rules. They thus leave a great deal of room for the community's sense of fairness or equity to come into play in regulating each individual set of facts.[33]

Claiming a custom is a rhetorical move more complex both socially and legally than alleging a law written in the statute books or approved by a court.[34] Customs, based on behavior, are conjured into existence as a "rule" only when disputants need to do so to claim primacy in a dispute. Prior to being recorded in writing (something that raises its own problems),[35] these customs did not exist in suspended animation hovering immanent in the air. To claim a custom meant creating it, calling it into life, even inventing it (sometimes fraudulently) if necessary.[36] It also meant articulating one specific description of the asserted behavior and obtaining community agreement to that particular articulation. To the extent that a community was not aware that it had a customary way of doing something until that behavior was denominated as such in a dispute, to the extent that the claimed custom selected among a number of possible variant but potentially conforming behaviors, and to the extent that the claimed behavior may or may not have existed at all in the way asserted, stating a custom as a positive rule of decision is itself an act of deliberate lawmaking.[37]

direct contrast to that western view of jurisprudence which has the judge making an objective decision on the case before him, unswayed by public opinion.

32 Bohmer, *supra* note 31, at 298, 301–02 (discussing evidence that judges neither knew nor cared what the customary law was, and when they did claim to know the custom they might disagree on what it was); Lyndan Warner, "Customary Law? Roman Law? Sixteenth-Century Lawyers' Pleadings Before the Parlement de Paris," in *Europa und seine Regionen: 2000 Jahre Rechtsgeschichte* [*Europe and Its Regions: 2000 Years of Legal History*] (Andreas Bauer & Karl H. L. Welker eds., 2007) 253, 255 (giving example of a sixteenth-century lawyer discussing how the Parlement of Paris would ignore custom and rule based on fairness).

33 Chanock, *supra* note 19, at 63.

34 *Id.* at 17.

35 James J. Fentress & Chris Wickham, *Social Memory: New Perspectives on the Past* 9 (1992) ("Writing not only freezes memory, but freezes memory in textual forms which evolve in ways quite unlike those of oral memory.").

36 Emily Kadens, "The Myth of the Customary Law Merchant," 90 *Tex. L. Rev.* 1153, 1189–90 (2012) (concerning the invention of insurance customs in sixteenth-century London); Jürgen Weitzel, "Gewohnheitsrecht und fränkisch-deutsches Gerichtsverfahren" ["Customary Law and Frankish-German Legal Proceedings"], in *Gewohnheitsrecht und Rechtsgewohnheiten im Mittelalter* [*Customary Law and Legal Customs in the Middle Ages*] 67, 79 (Gerhard Dilcher, et al. eds., 1992).

37 Chanock, *supra* note 19, at 62–63. *Cf.* Bradley, *supra* note 17 (discussing creative aspects to the identification of customary international law).

Before the sort of natural custom described thus far became juridicized by trained lawyers and judges, custom was a system for controlling behavior that did not concern itself overmuch with legal rules versus social practices, true memories versus pragmatically invented ones, or the niceties of what, exactly, a certain behavior required, or when – or whether – it became binding. Instead, custom allowed for the equitable development of social norms and tolerated a certain degree of variance as long as peace and cooperation were maintained. Custom, in other words, did not create formal rules of decision that looked or functioned like enacted law.[38]

If this description of custom is accurate, then customs discovered in international law, if they are truly customs and not something else merely called custom, should resemble the customs of the prelegal communities of medieval Europe and of the communities studied by anthropologists. They should be malleable, equitable, inconsistently-applied, and subject to change, extinction, and invention. The repeat behavior plus sense of being bound formula could only accurately depict such custom if the custom so defined were not expected to describe a fixed legal rule of decision. And yet that is precisely what lawyers since the twelfth century have asked the definition to do.

PART TWO: THE DEVOLUTION OF CUSTOM

The reintroduction of the Roman law in the twelfth century changed both the identity and the historical course of custom.[39] Many medieval students who went to study law at one of the new law schools took with them the notion of a society governed by oral social rules held in memory and guided by the need of small, tightly knit, stable communities to resolve disputes amicably and equitably. This custom was the possession of the people; they helped form it, and they helped adjudicate it. The custom depended on the interpretation of the facts of individual cases more than it did upon the articulation or creation of rules of decision.

Studying the law must inevitably have altered the way these students saw custom. They learned to view the law as an authoritative body of written rules. This law no longer offered the option of forgetfulness. The rules existed apart from the community calling them into being by acts of memory or discretion. This law was the possession of a new, more restricted community: the lawyers. The technical terms and techniques of interpretation required to use it meant that the law no longer belonged to the community at large.

38 Weitzel, *supra* note 36, at 73–74, 77–78.

39 Laurent Mayali, "La Coutume dans la doctrine romaniste au Moyen Age" ["Custom in Romanist Doctrine in the Middle Ages"], in *LA COUTUME*, 2, 11, 15 (Recueils de la Société Jean Bodin 52, 1990).

The trained lawyer's challenge became placing custom into a legal framework that made sense in his new, Roman-law-inflected worldview and that mediated between his book-learned understanding of law and his life experience of custom. Once medieval jurists had learned to think about law as a system of rules established by the approved (written) sources of law, the first step in dealing with custom had to have been figuring out where in the hierarchy of sources it fell. And indeed, in their early writings, the twelfth-century jurists initially appear to have taken a particular interest in the relationship between custom and enacted law (*lex*), asking whether and when custom could abrogate statute.[40] Of course, this would have been a significant issue in a world in which, until the rediscovery of the Roman law, custom – whether in the anthropological sense or in the sense of "our (local) law" – had been the predominate form of law. But the inquiry also had the effect of pulling custom out of the world of community social regulation and placing it into the sphere of formal law.[41]

Once the hierarchy had been established, the jurists turned to cataloguing custom's characteristics. It no longer sufficed to describe custom simply as "the way we do things." If custom was to be a source of formal law, it needed to provide lawyers with elements that could be evaluated and proved in court. The Roman law provided the basis for such elements. It described custom as: (1) some behavior (*moribus constitutum*)[42] that was (2) frequently and tenaciously followed (*frequenter ... servata sunt* and *servata tenaciter*),[43] and (3) observed over an indeterminate long period (*inveterata* or *per annos plurimos observata*), and which was accompanied by (4) the tacit agreement of the community to observe the behavior (*tacita civium conventio ... servantur*).[44]

By the early thirteenth century at the latest, the jurists had realized that the definition they built from the Roman law elements required them to work out three characteristics of custom: time, acts, and assent to be bound.[45] Their efforts created

40 André Gouron, "Coutume Contre Loi Chez les Premiers Glosseteurs" ["Custom Versus Law According to the First Glossators"], *in* RENAISSANCE DU POUVOIR LÉGISLATIF ET GENÈSE DE L'ÉTAT [RENAISSANCE OF LEGISLATIVE POWER AND GENESIS OF THE STATE] 117 (André Gouron & Albert Rigaudière eds., 1988); GUSTAVUS HAENEL, DISSENSIONES DOMINORUM [DISAGREEMENTS OF THE MASTERS] 151–53, 585 (Scientia Verlag 1964) (1834); PLACENTINUS, SUMMA CODICIS [SUMMA ON THE CODEX] 416 (Mainz, Ivo Schoeffer 1536) (Turin, Bottega D'Erasmo, repr. 1962); 4 FRIEDRICH CARL VON SAVIGNY, GESCHICHTE DES RÖMISCHEN RECHTS IM MITTELALTER [HISTORY OF THE ROMAN LAW IN THE MIDDLE AGES], 459 (2d ed., J. C. B. Moher 1850) (Irnerian gloss).

41 Mayali, *supra* note 39, at 12, 18.

42 1 DIGEST OF JUSTINIAN, *supra* note 6, at 13 (Dig 1.3.32).

43 2 CODEX IUSTINIANUS, *supra* note 11, at 362 (Cod 8.52(53).1 and 8.52(53).3).

44 1 DIGEST OF JUSTINIAN, *supra* note 6, at 13–14 (Dig 1.3.32 and 1.3.35).

45 PETRUS REBUFFUS, "Tractatus de Consuetudine, Usu, et Stylo, in Judiciis Valde Frequens et Utilis" ["Treatise on Custom, Practice, Style, Exceedingly Frequent and Useful in Trials"], in CONSTITUTIONES REGIAS COMMENTARIUS [COMMENTARIES ON ROYAL LEGISLATION] 475 para. 54

insoluble problems both theoretically and, when they tried to apply the definition to ascertain actual customs in court, practically. They debated for centuries how tacit consent was to be determined, how long the behavior had to have continued, and how many times it needed to be repeated to qualify as customary. These discussions are enlightening because their similarity to today's scholarly arguments demonstrate how longstanding have been the difficulties with the definition of custom.

The jurists' debate over time is particularly telling. In a society that believes that custom is "what we have always done," the members of the community cannot reconcile "remembering" a time before the custom with considering themselves bound to abide by it. This is true even if the custom was, in fact, of relatively recent vintage.[46] Assigning it a longstanding history, whether true or not, increased the custom's authority.[47] The earliest jurists to write about custom adopted this idea: for a behavior to be a custom it had to have existed since time immemorial.[48] Later generations, generations presumably ever-more deeply imbued with the Roman law mindset, began to put a defined time limit of ten, thirty, or forty years on custom. The limits depended upon the jurist, the particular custom claimed, and whether the applicable law was Roman or canon.[49]

At least one medieval jurist even tried out the idea of instant custom. Jacques de Révigny, a late thirteenth-century French law professor, argued that a custom could be created by a single, grand act, as long as the community tacitly assented to the act.[50] However, a later jurist called this notion of instant custom *ficta et intitulata*, which seems to have had the sense of "contrived and so-called."[51]

Nothing in law required the change to a defined time horizon. In the English common law, the pretense of pleading that a local custom (at least notionally) has existed since time immemorial (eventually statutorily defined as since 1189) continued into modern times.[52] But the common law was also less antagonistic toward custom than the civil law. Ever since the Middle Ages, English lawyers spoke

(Amsterdam, Johannes Schipper 1668) (citing Baldus on the four characteristics of custom, the fourth being *ratio*).

46 Clanchy, *supra* note 20, at 172.

47 1 KERN, *supra* note 23, at 151, 156–57.

48 PLACENTINUS, *supra* note 40, at 416.

49 AZO, SUMMA AZONIS [SUMMA OF AZO] 874 (Venice, Franciscus Bindonus 1566).

50 L. WAELKENS, *LA THÉORIE DE LA COUTUME CHEZ JACQUES DE RÉVIGNY [JACQUES DE RÉVIGNY'S THEORY OF CUSTOM]* 485–86 (1984) (*repetitio ad* Dig. 1.3.32, § 2). *See also* CYNUS PISTORIENSIS, *LECTURA SUPER CODICE [READING ON THE CODEX]* 524r. (Frankfurt am Main, Feyerabendt 1578) (Rome, Il Cigno Galileo Galilei, repr. 1998) (*repetitio ad* Cod. 8.52(53).2, § 21) (allowing that, "rarely," a single act could demonstrate the consent of the community).

51 1 JOSEPHUS MASCARDUS, *DE PROBATIONIBUS [ON EVIDENCE]* 420 (Frankfurt am Main, Erasmus Kempfferus 1619) (conclusio 424, para. 49).

52 F. A. Greer, "Custom in the Common Law," 9 *LAW Q. REV.* 153, 156–57 (1893).

of the common law as the custom of the realm.[53] In other words, they thought of the law enunciated by judges as often coming from custom. This custom had an independent value and prestige. It did not exist as the disfavored stepchild to a private law codified in an authoritative text.

The Roman law-trained jurists, in assigning minimum time requirements to custom, made it more law-like. While a custom with a defined lifespan might theoretically have been easier to prove, witnesses had to clear a higher bar of actually demonstrating that the alleged custom had been followed consistently for the requisite time. Vague assertions that the community had always done things in a certain way would not suffice.[54] More importantly for the development of the lawyers' conception of custom, the latter became a source of law with a beginning. Like enacted law, it was made, not found, and its authority rested on the assent of present actors rather than on an assumed tradition.

Rules about the frequency of acts went through a different evolution of perspective over time. In a society living by custom, repetition needed to be perceived, or at least asserted and accepted, rather than quantified. But asking the abstract question, "Is this the way we do things?" is not the same as asking, "How many times have we done this thing?" The medieval jurists focused on the latter question. They took two initial approaches to the Roman law text describing custom as occurring *frequenter*.[55] Most thirteenth-century jurists demanded specific quantification, deciding that a custom could be proved by two acts within ten years.[56] Such jurists asked the lawyer's question of proof: how can a court know that a custom has been formed?

But not all of their contemporaries agreed. Jacobus Balduinus (d. 1225), an Italian romanist, argued against the "twice make a custom" maxim, saying that "this certainly cannot be, because custom is said to be an habitual practice, but can two acts or three be said to be an habitual practice? Certainly not, because lawmakers disdain what happens once or twice, and rights are adjusted to that which happens frequently."[57] Balduinus viewed the question of repeat behavior in a different light. Rather than as a factor to be proved, he appears to have focused on the community forming a norm through repeat behavior.

Although a minority voice in his time, Balduinus's position essentially triumphed in the end, although for a different purpose. Later commentators continued to

53 *Id.* at 158–59; Frederic L. Cheyette, "*Custom, Case Law, and Medieval 'Constitutionalism'*: *A Re-Examination*," 78 POL. SCI. Q. 362, 377–78 (1963).

54 CYNUS, *supra* note 50, at 524r. (*repetitio ad* Cod. 8.52(53).2, § 21).

55 Cod 8.52(53).1.

56 1 ODOFREDUS, LECTURA SUPER DIGESTO VETERI [READING ON THE OLD DIGEST] 16r. (Lyon, Petrus Compater & Blasius Guido 1550) (reprint Bologna, Forni 1967) (*repetitio ad* Dig 1.3.32, § 13)

57 *Id.* (*repetitio ad* Dig 1.3.32, § 14) ("certe hoc esse non potest. quia [con]suetudo d[icitu]r vsus co[n]suetus: sed potest ne dici vsus co[n]suetus binus vsus vel trin[us] vsus certe no[n]: q[uia] que semel aut bis accidunt despiciunt legislatores & ad ea que freque[n]ter accidunt iura adaptantur").

repeat the two-acts rule,[58] but by the fourteenth century they had begun to require sufficient acts to give form to the tacit assent requirement. Their question ceased to be "How many times has the community done this" and became instead "Has the community done this enough times to show that it believes itself to be obligated to do it?"[59] This was not quite Balduinus's approach. He focused on the regularity of the behavior only to show the usage. The later jurists focused on the regularity to draw the distinction between a mere usage and a legally binding custom. They thus treated the definitional elements as a formal evidentiary matter: to show that the subjective part of the definition had been met litigants would have to show that the objective part had been met. This gave the repeated behavior a different significance than it had had when asserting a custom had been a matter of assuming the community's agreement to rhetoric about "what we have always done."[60]

Counting acts as a means of measuring assent permitted the jurists to provide courts with a workable solution to the problem of ascertaining the tacit consent to be bound, even if it did not solve the theoretical problem of distinguishing the subjective and objective elements of custom. Other solutions had simply frustrated the jurists, as they do modern scholars. For example, when confronted with the hypothetical about a village in which for 100 years every man had done the pilgrimage to Santiago de Compostella, some jurists argued that mere repeated behavior, even over a long period, did not prove consent to be bound. The men, they reasoned, may have done the pilgrimage without having felt compelled to do it. These jurists proposed that tacit assent to be bound could be identified only when someone did an act in opposition to the custom and the community sanctioned him for it.[61] Others, however, pointed out that the long-standing absence of any oppositional acts better demonstrated the community's consent.[62]

Lay people might have laughed at the jurists' struggles to resolve this issue. For them the question of whether the pilgrimage was a custom would not have been so difficult to answer – either the community agreed it was "what we do" or it did not. For the lawyers, on the other hand, an attitude of "we know it when we (want to) see it" did not suffice. They needed to find a metric to determine each part of their definition of custom. Not only did such metrics satisfy their concept of law as a hierarchical set of determinate rules, but more practically they also allowed the jurists to provide a mechanism for courts to ascertain when a custom existed and to distinguish between custom as law and mere nonbinding common practice.

58 REBUFFUS, *supra* note 45, at 474 paras. 40–41.

59 BARTOLUS, *supra* note 5, at 19r. (*repetitio ad* Dig 1.3.32, § 10) ("tacitus co[n]sensus, quod colligit[ur] ex vsu, & moribus"); Ullmann, *supra* note 22, at 276, 279 (citing Bartolus *repetitio ad* Cod 8.52.2, § 12: "populus non videtur consensisse, nisi frequenter illum actum exerceat").

60 Weitzel, *supra* note 36, at 71.

61 WAELKENS, *supra* note 50, at 488 (*repetitio ad* Dig 1.3.32, § 4).

62 *Id.* at 488–89 (*repetitio ad* Dig. 1.3.32, § 3).

Of course, when parties actually litigated about a custom, they by definition disagreed about its meaning or existence, and so did their witnesses. To establish consent under such circumstances, the jurists turned to voting theory: a repeat usage became a binding custom when the relevant (adult male) majority of the community had tacitly accepted that they were bound by it.[63] Majority rule allowed for the reality of variation in the performance of the custom. It also conveniently permitted judges to resolve disputes without actually having to find evidence that the whole community in fact abided by the custom that one party alleged. When witnesses disagreed about the exact custom, judges could, and did, simply count heads.[64]

Ultimately, the jurists worked out the implications of the definition of custom in order to provide litigants and judges a means by which they could allege and ascertain a custom in court. Unfortunately, custom did not lend itself to easy proof.[65] As the claimed remembrances of a community concerning its past behavior, it did not present a court with a clean set of facts. Custom in its natural habitat was behavior: under-determined, perhaps unarticulated or incompletely articulated, but still (sometimes) performed by the community. The behavior could allow for some variation within nonidentical but nonetheless conforming acts, and those variations could flourish as long as no disputes arose that required the custom to be expressed as a formal rule. Thus, when called upon to give evidence of the custom, witnesses could in good faith offer various answers. Contemporary lawyers understood this difficulty. According to the sixteenth-century French jurist Petrus Rebuffus, the fifteenth-century jurist "Panormitanus says that custom is difficult to prove. Many believe they can prove it, but they are wrong. Hostiensis [a thirteenth-century jurist] and others say that it is almost impossible to prove custom because sometimes it is black, and sometimes it is white."[66] The French lawyer Charles Dumoulin added

63 Ullmann, *supra* note 22, at 270–73.

64 2 *Les Olim ou registres des arrêts [The Olim or Registers of Decisions]* 678–81 (J.-C. Beugnot ed., Paris, Imprimerie Royale 1842) (appeal concerning "notorious" custom in which witnesses for each side testified to different rules and majority prevailed); Alain Wijffels, "Business Relations Between Merchants in Sixteenth-Century Belgian Practice-Orientated Civil Law Literature," in *From lex mercatoria to Commercial Law* 255, 270 (Vito Piergiovanni ed., 2005) (majority prevailed in dispute about custom concerning thief in the chain of title).

65 *See also* James Q. Whitman, "Why Did the Revolutionary Lawyers Confuse Custom and Reason?" 58 *U. Chi. L. Rev.* 1321, 1332–33 (1991).

66 Rebuffus, *supra* note 45, at 480 (art. I, gl. 3) ("Panor[mitanus] dicit difficillimum esse probare consuetudinem. Et multi credunt probare illam, sed aberrant. Host[iensis] et alii dicunt esse quasi impossibile probare consuetudinem, quia modò alba, modò nigra.") (Internal citations omitted). *See also* John Gilissen, *"La Preuve de la Coutume dans l'Ancien Droit Belge"* [*"Proof of Custom in Old Belgian Law"*], in *Hommage au Professeur Paul Bonenfant* 563, 567 (1965) (quoting author of fourteenth-century customary law manual that the proof of custom is difficult and dangerous).

in the sixteenth century that not only is custom not black and white, it is also "multi-colored even in the same town."[67]

Undaunted, the jurists tried to tame this uncertainty by applying the rules of evidence to the definition of custom. The person alleging the custom had to prove its existence.[68] This was normally done through witness testimony, and required at least two credible witnesses testifying to the performance of the customary act for the requisite number of years: ten to forty or longer depending upon the type of custom and whether the issue fell under the civil or canon law.[69]

The rules went into great detail about this witness testimony. Testimony that "such was the custom" was considered "silly," despite the fact that in a customary society without lawyers this was probably exactly the sort of testimony that had been allowed.[70] The witnesses had to testify to affirmative acts about which they had actual knowledge, and they had to agree on the details.[71] The testimony had to satisfactorily prove that the acts had been done repeatedly, with the knowledge of many people, from which could be assumed the community's tacit consent to the custom.[72] Finally, the witnesses had to testify to the community's actual awareness that the act was required of them as a custom.[73]

While this may give the impression of objectivity and certainty, the process contained two elements that at each step removed the found custom further from its natural state and turned it more into formal law. First, the witnesses or an inquest summoned with the purpose of stating the custom as a rule of decision had to articulate it for the use of the court,[74] and to do this they may have had to select among the constellation of variant but conforming acts making up the customary behavior in practice in order to express it as a single, definite rule. Thus, once a custom was proved in court and applied to a certain set of facts, it acquired a more rule-like counterpart to go along with its fluid behavioral original.[75]

67 CHARLES DUMOULIN, "Oratio de Concordia et Unione Consuetudinum Franciae" ["Oration on the Concord and Union of the Customs of France"], in 2 *OPERA OMNIA* 1771 (Paris, Officina Nivelliana 1612) ("modò alba, modò nigra, modò versicolor erat etiam in eodem municipio").

68 HOSTIENSIS, *SUMMA AUREA* [*GOLDEN SUMMA*] 20v. (Lyon, n.p. 1568) (*De consuetudine* § 9).

69 1 MASCARDUS, *supra* note 51, at 417–20 (collecting the disputes and received wisdom of three centuries of jurists on the issue of proof of custom).

70 *Id.* at 421 (*conclusio* 424, § 63) ("Immo testimonium talis testis ita deponentis, videlicet quod ita est consuetudo, est fatuum.").

71 *Id.* at 417–19.

72 *Id.* at 420 (*conclusio* 424, § 52).

73 *Id.* at 418, 420 (*conclusio* 424, §§ 18, 52–53); ALEXANDER IMOLENSIS, *LIBER SECUNDUS CONSILIORUM ALEXANDRI IMOLENSIS* [*SECOND BOOK OF THE OPINIONS OF COUNSEL OF ALEXANDER IMOLENSIS*] 94r. (Lyon, Thomas Bertheau 1544) (*consilium* 136, § 18).

74 M. GUILLEMARD, *L'ENQUÊTE CIVILE EN BOURGOGNE SPÉCIALEMENT À L'ÉPOQUE DES DERNIERS DUCS* [*THE CIVIL INQUEST IN BURGUNDY IN PARTICULAR DURING THE EPOQUE OF THE LAST DUKES*] 33 (1906).

75 The late fifteenth-century German jurist Ulrich Zasius (1461–1536) described this phenomenon. He said that custom had judicial and extrajudicial sides: "Judicial is when two similar judicial sentences

Second, the evidentiary process permitted judges to exercise a great deal of discretion. As mentioned previously, they could decide to count heads rather than investigate claims, especially if the judges had no other way of ascertaining the truth about an unrecorded custom with which they were unfamiliar. They could also hear testimony and decide to multiply its effect. For instance, in discussing how a judge should evaluate testimony about a custom, the early fourteenth-century Italian jurist, Cynus da Pistoia, explained that where witnesses had testified that an alleged custom of a village of 1,000 people had been adjudged in court with 100 people present, and again with 200 people present, and a third time with 300 people present, then the custom had been proved if the witnesses say that 10 years have elapsed since these events.[76] But if the witnesses testified to earlier adjudication but they did not know how many people were present,

> The Doctors say that it is for the Judge to estimate how many could have been present. For example, assume that witnesses had seen that [the alleged custom] had been adjudged four times in the usual manner and rarely does it fail that during trials fewer than 20 men are present. The Judge can estimate that at each trial 20 men were present, and that four times 20 men had consented [to the adjudged custom]. If they are the greater part of the populace, the custom can be proved by this.[77]

By contrast, judges could hear testimony and decide to reduce its impact by finding that the witnesses had not obtained their knowledge from a sufficiently diverse number of sources and events.[78]

Indeed, the ability of judges and parties to manipulate custom is exactly the complaint made by the anonymous author of a polemic about the practice of determining insurance customs in sixteenth-century London. He described a system in which parties produced competing petitions signed by merchants and alleging opposing customs. The judges or arbitrators decided between the petitions, often

concerning some matter have been produced [as evidence of the custom], that is, when [the custom] has been adjudicated at least twice in similar cases in adversarial proceedings. Extrajudicial is that which has been introduced by the longstanding practice of the populace and which has existed for at least ten years." "UDALRICUS ZASIUS," IN *PRIMAM DIGESTORUM* PARTEM *PARATITLA [PARATITLES ON THE FIRST PART OF THE DIGEST]* 8 (Basel, Michael Ising 1539) ("Iudicialis, quando duae sententiae co[n]formes super aliquo negocio productae sunt, id est, quando ad minus bis iudicatam est in causis similibus in iudicio co[n]tradictorio. Extraiudicialis, quae per diuturnu[m] usum populi indicitur, ad quod ad minus decem anni exigu[n]tur.") (internal citations omitted). *See also* 1 ODOFREDUS, *supra* note 56, at 16r. (*repetitio ad* Dig. 1.3.32, § 16) (expressing a similar sentiment several centuries earlier).

76 CYNUS, *supra* note 50, at 524r. (*repetitio ad* Cod. 8.52(53).2, § 22).

77 *Id.* ("Dicunt Docto[res] quod est in ipso Iudice aestimare, quot possunt fuisse praesentes. Verbi gratia: Pone, quod quater vidit testis ita esse iudicatum: modò de consuetudine vel raròdeficit, quando in iudicio sint benè 20. homines fuerunt praesentes, & ita quater 20. homines consentientes erant. & si illi sunt maior pars populi, consuetudo potest ex hoc argui.").

78 ALEXANDER IMOLENSIS, *supra* note 73, at 94r. (*consilium* 136, § 20).

simply by considering the number and identity of the signatories. The resultant ruling became the custom although the alleged practice "was never or but a small time before recorded," and although, had the signatories been examined individually, "they wil[l] be founde of divers opinions" about the existence of the practice.[79]

Even mechanisms that kept the power to ascertain custom in the hands of the community did not completely answer these objections. In 1270, the king of France promulgated a statute (*ordonnance*) establishing an official procedure for proving customs in royal courts.[80] This so-called *enquête par turbe*, or inquest by jury, required that

> a number of wise men, above suspicion, be called together. The custom is put to them orally by one among them after they have gathered, and it is given to them in writing. The custom having been set forth, they are to swear that they will honestly speak and recount that which they know and believe and have seen to have been done concerning that custom. The oath having been taken, they shall take themselves to another place, and they shall deliberate, and they shall record those deliberations, and they shall relate amongst themselves what they have seen of that custom, and who, in what case, and in what place, and if it had been adjudged and under what circumstances. And all of this should be written down and given to the court sealed under the seals of the members of the inquest, and they shall all speak for themselves and relate the case as a group.[81]

The evidentiary requirements of the enquête par turbe appear to have answered the lawyers' desire to turn "the way we do things" into an express legal rule while acknowledging that the only way to discover this was to allow the community to demonstrate its understanding of their long-standing behavior.

And yet litigants disliked the procedure. Not only was it expensive, but people also doubted that the turbiers told the truth. In the fourteenth century, Jehan Boutillier, author of an important French customal, "warn[ed] that relying upon the proof of private customs 'was highly dangerous', as inquests were unpredictable and records

[79] British Library, Add. MS 48020, 348v.

[80] L. Waelkens, L'Origine de l'Enquête par Turbe ["The Origin of the Enquête par Turbe"], 53 *Tijdschrift voor Rechtsgeschiedenis [Legal Hist. Rev.]* 337, 338–39 (1985).

[81] 1 Actes du Parlement de Paris: 1254–1299, 242, no 2547 B (Edgard Boutaric ed., Paris, Henri Plon 1863):

> Vocabuntur plures sapientes carentes suspicione. Ipsis vocatis, proponetur eis consuetudo per os unius ex ipsis, et dabitur eis in scripto. Qua proposita iurabunt quod ipsi dicent et fideliter referent illud quod sciunt et credunt et viderunt usitari super illa consuetudine. Quo juramento prestito, trahent se ad partem et deliberabunt et referent deliberationem illam, et dicent inter quos viderunt illam consuetudinem, et qui in quo casu et quo loco, et si fuerit judicatum, et de circunstanciis; et omnia redigentur in scriptis et mittantur in curiam clausa sub sigillis inquisitorum, et reddent omnes causam dicti sui, eciam in turba.

rare."[82] The French even had a proverb: "Fol est qui se met enqueste, car qui mieux abreuve, mieux prevue" – "He is crazy who puts himself upon an inquest, because [the turbiers] who drink the best, prove the best."[83] In other words, the turbiers could be bribed.

Furthermore, the written records of the results of enquêtes par turbe (and its parallel procedure in Germany) became a sort of code of custom. Once decided, the custom determined through the inquest became a statute. This makes sense if one views custom through the lens of enacted law: once established, the law remains the same over time. But it does not make sense to view natural custom this way, because that sort of custom changed constantly as the community wanted it to change to meet present needs. Yet in the process of defining it, surrounding it with rules, and creating a procedure for it, the jurists had altered what custom was. It now looked, and functioned, as something the lawyers could understand as law.

In addition, the enquête par turbe procedure raises difficult questions that remain equally valid in international law today. If the custom were well known and universally followed, it would not be put to the inquest at all but would be judicially noticed by the judge as a "notorious" custom.[84] But if a custom were only put to turbiers if it were not well known and universally followed, how could it be called a custom? And how could the turbiers claim, under oath, that the custom was something everyone knew when apparently it was not? In other words, the inquest might often have occurred when the litigants, and presumably the community, as well, genuinely disagreed about the claimed custom.[85] Consequently, turbiers could honestly differ in their interpretations and experiences. Did the turbiers select one version, perhaps even an idealized version, of the alleged behavior? This could help explain the distrust of the inquest proceedings. The uncertainty surrounding custom – whether because of honest differences of understanding or because of the ease with which claims of custom could be manipulated – could result in self-serving assertions.

Perhaps unsurprisingly, given the potential for fraud and the interest of lawyers and judges in proving custom in a manner befitting their formal procedures, the trajectory of the enquête par turbe led in the sixteenth century to using the lawyers themselves as turbiers. The commissioner of the inquest would summon ten well-known lawyers, who would give their responses to the question seriatim.

[82] Esther Cohen, *The Crossroads of Justice: Law and Culture in Late Medieval France* 40 (1993).

[83] Guillemard, *supra* note 74, at 35.

[84] Hippolyte Pissard, "Essai sur la connaissance et la prevue des coutumes en justice, dans l'ancien droit français et dans le système romano-canonique" ["Essay on the Knowledge and Proof of Customs at Justice in the Ancient French Law and in the Romano-Canonical System"] 20 (Paris, A. Rousseau 1910).

[85] Watson, *supra* note 16, at 97 (quoting the Ordonnance de Montil-les-Tours of April 1453 by Charles VII of France: "It often happens that in one single region, the parties rely on contrary customs, and sometimes customs are silent and vary at will.").

The answers were recorded in writing and signed by the turbiers.[86] These responses became the statement of the customary law.

By this time, in any event, custom was in rapid decline as a living source of law. Where cities grew and the sovereigns of nation-states consolidated their power, custom withered as a meaningful source of law. Urbanization required a steady stream of new rules and regulations that could solve immediate problems immediately. As a community became larger and less stable, the effectiveness of equitable decision making decreased, and with it the ability of anthropological custom to regulate behavior.[87] On the national scale, powerful kings sought to centralize courts and lawmaking. Custom was a hyper-local phenomenon, varying greatly over even small distances because it arose only when people came into contact frequently enough to mimic each other's behavior. The kings broke the power of this local law by ordering the codification of customs and by creating appellate courts that gave preference to written law over oral custom.[88] By the late seventeenth century, even the enquête par turbe had been abolished in France, and courts were ordered to look to the codifications of custom instead.[89]

PART THREE: REVOLUTION IN CUSTOM

The formal law had won. Over the course of the early modern period, custom gradually lost any significant law-creative function. Yet at precisely this moment, custom experienced a rebirth of sorts as a catch-all category for law, or claims about law, which jurists could fit nowhere else.

Using custom to label miscellaneous law was not new. Medieval lay people called many things custom that did not meet the Roman law definition. When towns began to write down their own law in the High Middle Ages, they called their compilations of local oral customs, legislated aldermanic regulations, important public contracts, judicial opinions, and charters granted by the prince their "custom."[90] In the context of the invention of customary international law, however, the pragmatic borrowing of the concept of custom to identify otherwise ill-defined law takes on added relevance.

86 GUILLEMARD, *supra* note 74, at 35.
87 *See* CHANOCK, *supra* note 19, at 57.
88 John P. Dawson, "The Codification of the French Customs," 38 *MICH. L. REV.* 765 (1940).
89 GUILLEMARD, *supra* note 74, at 36.
90 John Gilissen, "Loi et Coutume: Quelques Aspects de l'Interpénétration des Sources du Droit dans l'Ancien Droit Belge" ["Law and Custom: Some Aspects of the Interpenetration of the Sources of Law in the Ancient Belgian Law"], 21 *TIJDSCHRIFT VOOR RECHTSGESCHIEDENIS [LEGAL HIST. REV.]* 257, 278 (1953).

Readers of this volume know that in his 1613 treatise, *Tractatus de Legibus ac Deo Legislatore*, Francisco Suárez would be the first to equate the law of nations with custom.[91] Suárez arrived at this conclusion by process of elimination. The law of nations could not be national, enacted law because national law governed countries *intra se*. But the law of nations governed countries *inter se*.[92] And the law of nations could not be natural law, because natural law was immutable. The law of nations, however, could change.[93] In addition, the law of nations was unwritten, and only custom could take the form of unwritten law.[94]

Despite the trouble Suárez had in his book explaining how the law of nations fit the characteristics of custom, his theory became influential, and he remains one of the intellectual godfathers of modern customary international law. But Suárez's use of custom to fill in a gap in the hierarchy of law occurred in a context that is not as well known. Custom, it turns out, was the *de rigueur* source of law for seventeenth-century legal thinkers making generally propagandistic claims about the authority and antiquity of a certain class of law that did not have some other obvious and legitimating source.

We have two excellent examples in the history of English law. The participants in the struggle between the king and parliament in the early seventeenth century emphasized the origin of the common law in the common custom of the realm.[95] While modern legal historians do not accept this story,[96] it was a powerful source of legitimacy and stability for the English law. At the same moment, the merchant and statesman Gerard Malynes advocated the notion of an ancient customary law merchant shared by all nations but not established by national law.[97] This, too, was a claim that would have a long-lasting influence, but which is, nonetheless, also inaccurate.[98]

91 2 Francisco Suárez, *Selections from Three Works of Francisco Suárez* 351 (Gwladys L. Williams, Ammi Brown, & John Waldron eds. & trans., 1944).

92 *Id.* at 345 (Bk. 2, ch. 19, § 5).

93 *Id.* at 342 (Bk. 2, ch. 19, § 2).

94 *Id.* at 345 (Bk. 2, ch. 19, § 6).

95 J. G. A. Pocock, *The Ancient Constitution and the Feudal Law: A Study of English Historical Thought in the Seventeenth Century* 31–33, 36 (reissue 1987).

96 6 John Baker, *Oxford History of the Laws of England 1483–1558*, at 471–72 (John Baker ed., 2003); J. H. Baker, *The Law's Two Bodies* 67–70, 81–82, 161–69 (2001).

97 1 Gerard Malynes, *Consuetudo, vel Lex Mercatoria, or The Ancient Law-Merchant* (London, Adam Islip 1662) (at the second and third pages of the unpaginated dedicatory epistle "To the Courteous Reader") ("Customary Law approved by the authoritie of all Kingdomes and commonweales, and not a Law established by the Soveraigntie of any Prince, either in the first foundation or by continuance of time.") (italics omitted).

98 *See* Emily Kadens, "The Medieval Law Merchant: The Tyranny of a Construct," *J. Legal Analysis* (forthcoming 2015) (providing bibliography of the historical works criticizing the historical story of a customary law merchant).

Custom was similarly useful to the seventeenth-century Dutch jurist, Hugo Grotius, in explaining why Dutch law recognized contracts not recognized in the Roman law. Rather than admit to the influence of the medieval Catholic canon law, the Protestant Grotius looked to the supposed ancient Germanic custom of accepting all good faith contracts "which proceed from any reasonable causes, whatever be the forms of words employed."[99]

The fact that custom was a reputable, convenient, and effectively untestable source to claim for law that did not fit anywhere else did not make such claims true. Nor did it mean that contemporaries all accepted such claims uncritically. In 1699, for example, a German legal dissertation scoffed at the notion of a customary international law (*ius gentium*). "They wished," the author wrote,

> to imagine the *ius gentium* to be a species of human law, but being at a loss, sought refuge in customary law and thought, for example, that certain nations had uniformly introduced by uniform acts the inviolability of ambassadors, and thus the rest [of the nations] should be bound to the same law because custom is a type of unwritten law.[100]

All of these claims about the common law, the law merchant, ancient Germanic law, and international law were rhetorical, using custom to score a legal point in much the same way as did litigants in medieval – or modern African customary – courts.[101] In the customary societies, such claims of custom were known to be pragmatic. We have, perhaps, lost such healthy skepticism today.

CONCLUSION

It is that skepticism of earlier jurists that could stand as the lesson of history. If we understand the natural state of custom as well as the limited ability of the historical

[99] 1 Hugo Grotius, *The Jurisprudence of Holland* 305 (Robert Warden Lee trans., 1953) (III.i.52).

[100] Christian Thomasius & Petrus Herff, *Dissertatio Inauguralis Iuridica, Sistens Coniecturas de Iure Consuetudinis et Observantiae [Inaugural Dissertation in Law, Putting Conjectures on Customary Law and Observance]* (Halle Magdeburg, Christiani Ludovici Sympheri 1740) (originally 1699) ("Quamuis enim deprehensi fuerint viri quidam eruditi, qui, dum Ius Gentium Iuris humani speciem effingere voluerunt, & iis aqua haesit, in iure consuetudinario refugium quaesiuerunt, atque putarunt v.g. quod gentes quaedam vniformiter legatorum inviolabilitatem vniformibus actibus introduxerint, & sic reliqui obligati sint ad idem ius, cum consuetudo sit species iuris non scripti.").

[101] *Cf.* Stephen J. Choi & Mitu Gulati, "Customary International Law: How Do Courts Do It?" (in this volume) ("The question that jump-started our inquiry – how do judges apply the CIL definition that seems both analytically impossible to apply and normatively unattractive – has a simple answer: they ignore it."); Monica Hakimi, "Custom's Method and Process: Lessons from Humanitarian Law" (in this volume) ("To make CIL, disparate actors advance and respond to one another's legal claims, as they pursue their own interests.").

definition of custom to capture that original character, then we are empowered to decide how we wish to treat claims of international custom. We can accept that true custom is messy and not neatly and accurately tied up in the bow of a legal rule. In this case, we can stop trying to force custom into the narrow bounds of formal law and look for an alternative way to categorize and make rules about the practices we witness or hope to encourage. Or we can accept that applying the definition of custom means creating a construct. If we do so, we can also acknowledge that when we assert the existence of a custom of international law we essentially perform an act of legislation or invention. We do not merely perform a neutral act of articulating a preexisting custom.

2

Customary International Law Adjudication as Common Law Adjudication

Curtis A. Bradley

The standard view today of customary international law (CIL) is that it arises from the widespread and consistent practice of states followed out of a sense of legal obligation. Although commonly recited, this account is plagued by evidentiary, normative, and conceptual difficulties, and it has been subjected to increasing criticism in recent years. This chapter suggests that these difficulties stem in part from the effort to formulate one conception of CIL that applies across all institutional contexts. This chapter posits a particular account of CIL, considered from the perspective of international adjudication. The application of CIL by an international adjudicator, this chapter suggests, is best understood in terms similar to the judicial development of the common law: that is, as an approach whereby adjudicators look to past practice but necessarily make choices about how to describe it, which baselines to apply in evaluating it, and whether and when to extend or analogize it to new situations. These choices, moreover, are shaped by assessments of both state preferences and the likely consequences of adopting a particular formulation of the relevant international rule, as well as moral and ethical considerations. Unlike the standard view of CIL, this common law account recognizes a significant element of judgment and creativity in determining the content of CIL. Understanding the adjudication of CIL in this way, the chapter contends, avoids many of the difficulties surrounding the standard view of CIL.

PROBLEMS WITH THE STANDARD VIEW OF CIL

As noted, the generally accepted view among international lawyers and scholars today is that CIL arises from the consistent practices of states that are followed out of a sense of legal obligation,[1] an account that this chapter will refer to as the "standard

[1] *See, e.g.*, RESTATEMENT (THIRD) OF THE FOREIGN RELATIONS LAW OF THE UNITED STATES § 102(2) (1987); SEAN D. MURPHY, *PRINCIPLES OF INTERNATIONAL LAW* 92–93 (2d ed. 2012); 1 *OPPENHEIM'S INTERNATIONAL LAW* § 2 (Robert Jennings & Arthur Watts eds., 9th ed. 1996).

view." Under this standard view, CIL is conceived of as having two components: an objective, state practice component, and a subjective, sense of legal obligation component. The second component is sometimes referred to by the Latin phrase *opinio juris sive necessitatis*, which translates as "a belief that something is required by law or necessity," although commentators and courts often shorten the phrase simply to *opinio juris*. This account of CIL has been endorsed by international tribunals, including the International Court of Justice (ICJ),[2] and it is often recited by representatives of states.[3]

Despite its general acceptance, the standard view of CIL suffers from a variety of difficulties. Some of these difficulties are evidentiary. While it is accepted that one must show state practice and *opinio juris* to establish a rule of CIL, there is no consensus about what evidence is sufficient to support either of these two elements. It is not clear, for example, how much state practice is required, or how longstanding it needs to be. Nor is it clear how the beliefs of nation-states about the content of CIL are to be determined when, as is often the case, the states fail to articulate a position. In addition, although adjudicators frequently point to treaties as evidence of CIL, the extent to which treaties constitute valid evidence of CIL is mysterious and controversial.[4] The growth in recent years of nonbinding instruments in lieu of treaties has further complicated the evidentiary issues surrounding CIL.[5]

Other difficulties are practical. There are more than 190 nations in the world today, and, for many issues, it is challenging, if not impossible, for interpreters to determine the practices and beliefs of these nations. Sometimes, this is because language and other barriers make it difficult to access the relevant materials. More often, it is because there simply is no relevant evidence in most of the countries. Even though the standard conception of CIL purports to require widespread and consistent state practice, for contested issues there is often little state practice directly on point, and what practice there is (or at least the practice that is actually considered) arises from a small number of states.[6] Indeed, actions by states often become points of controversy precisely because they involve new situations not

2 *See, e.g.*, Jurisdictional Immunities of the State (*Ger. v. It.*), 2012 I.C.J. 99, 122 (Feb. 3); North Sea Continental Shelf Cases (Fed. Rep. Ger./Denmark), 1969 I.C.J. 4, 44 (Feb. 20).

3 *See* Special Rapporteur, *Second Report on Identification of Customary International Law*, 9–10, Int'l L. Comm'n, U.N. Doc. A/CN.4/672 (May 22, 2014).

4 *See* Curtis A. Bradley & Mitu Gulati, "Withdrawing from International Custom," 120 *Yale L.J.* 202, 213 (2010); *see also* R. R. Baxter, "Multilateral Treaties as Evidence of Customary International Law," 41 *Brit. Y.B. Int'l L.* 275 (1965–66); Curtis A. Bradley & Mitu Gulati, "Customary International Law and Withdrawal Rights in an Age of Treaties," 21 *Duke J. Comp. & Int'l L.* 1 (2010).

5 *See, generally*, Jan Wouters & Linda Hamid, "Custom and Informal International Lawmaking" (in this volume).

6 *See* J. Patrick Kelly, "The Twilight of Customary International Law," 40 *Va. J. Int'l L.* 449, 472–73 (2000).

specifically covered by past practice, and yet the standard conception of CIL seems to have little to offer in that scenario.[7]

These practical difficulties may in turn help explain an empirical difficulty. While adjudicatory institutions often recite the standard conception of CIL, they do not actually seem to follow it. The ICJ, for example, often cites relatively little state practice in support of its claims about the content of CIL.[8] International criminal tribunals, too, seem to find rules of CIL through means other than the standard account.[9] And the decisions of these international adjudicatory institutions are in turn cited themselves as evidence of the content of CIL.[10]

7 For similar observations, see, for example, DANIEL BODANSKY, *THE ART AND CRAFT OF INTERNATIONAL ENVIRONMENTAL LAW* 198 (2010) ("Although writers claim to adhere to the traditional account of customary international law, they do not base their assertions about customary international law on systematic surveys of state practice."); and Niels Petersen, "Customary Law Without Custom?: Rules, Principles, and the Role of State Practice in International Norm Creation," 23 *AM. U. INT'L L. REV.* 275, 277 (2008) ("A survey of customary international law is often highly selective and takes into account only major powers and the most affected states."). Even in its *Jurisdictional Immunities of the State* decision, see *supra* note 2, which had an unusually extensive discussion of state practice; the ICJ looked to legislation in only ten states and judicial decisions from only about a dozen states.

8 *See, e.g.*, GLEIDER I. HERNANDEZ, *THE INTERNATIONAL COURT OF JUSTICE AND THE JUDICIAL FUNCTION* 91 (2014) ("Despite the Court's doctrinal insistence on State practice and *opinio juris*, it in fact rarely refers to those elements in its judgments."); Richard H. Geiger, "Customary International Law in the Jurisprudence of the International Court of Justice: A Critical Appraisal," in *FROM BILATERALISM TO COMMUNITY INTEREST: ESSAYS IN HONOUR OF BRUNO SIMMA* 673, 692 (Ulrich Fastenrath et al. eds., 2011) ("In general the [ICJ] does not follow its self-proclaimed method of finding customary international law."); Stefan Talmon, "Determining Customary International Law: The ICJ's Methodology Between Induction, Deduction and Assertion," 26 *EUR. J. INT'L L.* 417, 441 (2015) ("In a majority of cases the [ICJ] has not examined, whatever it may say on the matter, the practice and *opinio juris* of States, but has simply asserted the rules which it applies."). *See also* Stephen Choi & Mitu Gulati, "Customary International Law: How Do Courts Do It?" (in this volume).

9 *See, e.g.*, William Schabas, "Customary Law or 'Judge Made' Law: Judicial Creativity at the UN Criminal Tribunals," in *THE LEGAL REGIME OF THE INTERNATIONAL CRIMINAL COURT* 100 (Jose Doria et al. eds., 2009) ("Overall, customary international law [as applied by the international criminal tribunals] mainly seems to provide a convenient license for judicial lawmaking, a process similar in many respects to the creation of judge-made rules of the English common law."). This may be less true, however, in the International Criminal Court than in other international criminal tribunals. *See* Larissa van den Herik, "The Decline of Customary International Law as a Source of International Criminal Law" (in this volume).

10 *See, e.g.*, Alberto Alvarez-Jimenez, "Methods for the Identification of Customary International Law in the International Court of Justice's Jurisprudence: 2000–2009," 60 *INT'L & COMP. L.Q.* 681, 685 (2011) ("When it intervenes, the [ICJ] remains as a prominent actor within the process of creation and evolution of customary international rules."); Roozbeh (Rudy) B. Baker, "Customary International Law in the 21st Century: Old Challenges and New Debates," 21 *EUR. J. INT'L L.* 173, 186 (2010) (noting the phenomenon of "commentators at times seizing on the jurisprudence of the ICTY and ICTR as evidence of the formation of new customary international law"); Christian J. Tams, "The ICJ as a 'Law-Formative Agency': Summary and Synthesis," in *THE DEVELOPMENT OF INTERNATIONAL LAW BY THE INTERNATIONAL COURT OF JUSTICE* 377, 378 (2013) ("ICJ pronouncements are generally considered to be authoritative pronouncements on the law").

There are also serious and growing questions about the usefulness of CIL. CIL seems structurally unable to address many of the world's most pressing issues, such as (to name a few examples) climate change, international financial stability, nuclear proliferation, terrorism, and cyber warfare. For problems such as these, a requirement that a substantial number of nations already act in accordance with the desired norm, out of a sense that such behavior is legally required, seems to require the impossible.[11] Relatedly, the proliferation of multilateral treaties has raised new questions about the need for CIL as a distinct source of international law. Most of the major issue areas that were historically regulated by CIL are now regulated, to one degree or another, by treaties.[12] Treaties have a variety of advantages over CIL, in that they provide more direct evidence of what states want (since they are the product of express negotiation), can provide for greater specificity (since they are typically in writing), and can establish institutional mechanisms to promote monitoring, adjudication, and enforcement of the norms. In addition to multilateral treaties, nations are also increasingly resorting to the creation of "soft law" – that is, nonbinding international agreements. As with treaties, but unlike with CIL, soft-law agreements are typically committed to writing and involve express negotiation. The possible result of these developments, as Joel Trachtman has observed, may be "the increasing marginalization of custom."[13]

Finally, there are conceptual difficulties with the standard conception of CIL. Many of these difficulties relate to the concept of *opinio juris*. As Professor Hugh Thirlway memorably noted: "The precise definition of the *opinio juris*, the psychological element in the formation of custom, the philosophers' stone which transmutes the inert mass of accumulated usage into the gold of binding legal rules, has probably caused more academic controversy than all the actual contested claims made by States on the basis of alleged custom, put together."[14] Some scholars

[11] These issues may also require a level of specificity and administrative structure that CIL cannot provide. For an observation along these lines more than a half-century ago, see WOLFGANG FRIEDMANN, *THE CHANGING STRUCTURE OF INTERNATIONAL LAW* 122 (1964) ("Custom is an unsuitable vehicle for international 'welfare' or 'co-operative' law. The latter demands positive regulation of economic, social, cultural and administrative matters, a regulation that can only be effective by specific formulation and enactment."). *See also* Niels Petersen, "Customary International Law and Public Goods" (in this volume).

[12] *See* Joel Trachtman, "The Growing Obsolescence of Customary International Law" (in this volume).

[13] Joel P. Trachtman, "Persistent Objectors, Cooperation, and the Utility of Customary International Law," 21 *DUKE J. COMP. & INT'L L.* 221, 232 (2010); *see also* Joel P. Trachtman, "Reports of the Death of Treaty Are Premature, but Customary International Law May Have Outlived Its Usefulness," AJIL UNBOUND (Apr. 29, 2014) at http://www.asil.org/blogs/reports-death-treaty-are-premature-customary-international-law-may-have-outlived ("CIL is increasingly ill-fitted to respond to the needs for [an] international law of cooperation.").

[14] H. W. A. THIRLWAY, *INTERNATIONAL CUSTOMARY LAW AND CODIFICATION* 47 (1972); *see also, e.g.*, Jörg Kammerhofer, "Uncertainty in the Formal Sources of International Law: Customary International

question whether it is even possible to find *opinio juris* in the pure sense that seems to be contemplated by the standard view.[15]

A related conceptual problem is that, even though one of the purported virtues of CIL is that it can evolve in response to changing circumstances, the standard view does not seem well structured to allow for such evolution. The principal account of how it occurs is that individual nations are supposed to violate rules of CIL and then hope that other nations will acquiesce in the violation.[16] Since the violating state by definition would not be acting with *opinio juris*,[17] this account appears to require a sufficiently large number of violations to cause the CIL rule to collapse due to lack of supportive state practice.[18] Even assuming this is realistic, however, basing a central feature of a legal regime on widespread violations of the law is normatively questionable.[19]

These difficulties have made CIL ripe for criticism and rethinking. Some commentators have questioned CIL's continued usefulness and legitimacy.[20] Others have criticized CIL for being insufficiently consensual and, relatedly, undemocratic.[21] Still others have argued that CIL is structured in such a way that

Law and Some of its Problems," 15 *Eur. J. Int'l L.* 523, 532 (2004) ("[*Opinio Juris*] is the most disputed, least comprehended component of the workings of customary international law.").

15 *See, e.g.*, Kelly, *supra* note 6, at 475 ("There is no methodology that has the capacity to determine whether states have, in fact, accepted a norm as law.").

16 *See, e.g.*, Jonathan I. Charney, "The Persistent Objector Rule and the Development of Customary International Law," 56 *Brit. Y.B. Int'l L.* 1, 21 (1985) ("Nations forge new law by breaking existing law, thereby leading the way for other nations to follow."); Anthony D'Amato, "The President and International Law: A Missing Dimension," 81 *Am. J. Int'l L.* 375, 377 (1987) ("Existing customary law, then, contains the seeds of its own violation; otherwise it could never change itself."); Michael J. Glennon, "How International Rules Die," 93 *Geo. L.J.* 939, 957 (2005) ("Customary international law is thought to be altered by acts that initially constitute violations of old rules; that is how it changes.").

17 *See* Int'l Law Ass'n, Final Report of the Committee, *Statement of Principles Applicable to the Formation of General Customary International Law*, at 33 (2000) (hereinafter ILA Report) ("It is hard to see how a State, if properly advised, could entertain the belief that its conduct is permitted [or required] by existing law when that conduct is, by definition, a departure from it.").

18 *See* Glennon, *supra* note 16; Kammerhofer, *supra* note 14, at 530–31; *cf.* Karol Wolfke, *Custom in Present International Law* 65 (2d ed. 1993) ("Much less practice and manifestations of its acceptance as law are needed for the arising of a custom in a relatively virgin, that is, unregulated section of international intercourse, then where a change or termination of an old, well-settled customary rule is at stake.").

19 *See, e.g.*, G. J. H. van Hoof, *Rethinking the Sources of International Law* 99 (1983) ("It must be quite an extraordinary system of law which incorporates as its main, if not the only, vehicle for change the violation of its own provisions."); Michael Akehurst, "Custom as a Source of International Law," 47 *Brit. Y.B. Int'l L.* 1, 8 (1974–75) ("There is no doubt that customary rules can be changed in this way, but the process is hardly one to be recommended by anyone who wishes to strengthen the rule of law in international relations.").

20 *See, e.g.*, Kelly, *supra* note 6.

21 *See, e.g.*, John O. McGinnis & Ilya Somin, "Should International Law Be Part of Our Law?" 59 *Stan. L. Rev.* 1175 (2007).

it is unlikely to produce efficient rules for the international system.[22] As Andrew Guzman has observed, CIL is "under attack from all sides."[23]

These criticisms have prompted various efforts to reconceptualize CIL in a manner that would put it on a sounder empirical or normative footing. Some scholars have suggested a focus primarily on state practice rather than *opinio juris*,[24] whereas others have suggested precisely the opposite.[25] Some of the scholars who emphasize the state practice element seek to limit CIL to "old-style" CIL induced from longstanding patterns of behavior rather than "new-style" CIL deduced from materials such as international resolutions, treaties, advocacy by NGOs, and academic opinion.[26] Other scholars, by contrast, have sought to reconcile and unify these two types of CIL reasoning.[27] Still others have proposed explanations of CIL that are grounded in state interests and rational decision making rather than in legal doctrine, while reaching different conclusions about whether and to what extent CIL is likely to affect state behavior.[28]

[22] *See* Eugene Kontorovich, "Inefficient Customs in International Law," 48 *Wm. & Mary L. Rev.* 859, 889–94 (2006).

[23] Andrew T. Guzman, "Saving Customary International Law," 27 *Mich. J. Int'l L.* 115, 116 (2005); *see also* David J. Bederman, "Acquiescence, Objection and the Death of Customary International Law," 21 *Duke J. Comp. & Int'l L.* 31, 43 (2010) (noting that this is "a time when customary international law is coming under attack by both extreme positivists [who suggest that its processes are illegitimate and non-transparent] and those of a naturalist bent [who regard CIL as merely pandering to State interests]"); George Norman & Joel P. Trachtman, "The Customary International Law Game," 99 *Am. J. Int'l L.* 541, 541 (2005) ("[CIL] is under attack as behaviorally epiphenomenal and doctrinally incoherent.").

[24] *See, e.g.*, Maurice H. Mendelson, "The Formation of Customary International Law," 272 *Recueil des Cours* 155 (1998).

[25] Most notably, claims relating to CIL concerning human rights have tended to de-emphasize state practice in favor of other materials. *See* Curtis A. Bradley & Jack L. Goldsmith, "Customary International Law as Federal Common Law: A Critique of the Modern Position," 110 *Harv. L. Rev.* 815, 838–42 (1997) (describing this "new CIL"). For a critique of this phenomenon, see Bruno Simma & Philip Alston, "The Sources of Human Rights Law: Custom, Jus Cogens, and General Principles," 12 *Aust'l Y.B. Int'l L.* 82 (1989).

[26] *See, e.g.*, Samuel Estreicher, "A Post-Formation Right of Withdrawal from Customary International Law?: Some Cautionary Notes," 21 *Duke J. Comp. & Int'l L.* 57 (2010); Edward T. Swaine, "Bespoke Custom," 21 *Duke J. Comp. & Int'l L.* 207 (2010).

[27] *See, e.g.*, Frederic L. Kirgis, Jr., "Custom on a Sliding Scale," 81 *Am. J. Int'l L.* 146 (1987); Anthea E. Roberts, "Traditional and Modern Approaches to Customary International Law: A Reconciliation," 95 *Am. J. Int'l L.* 757 (2001).

[28] *See, e.g.*, Jack L. Goldsmith & Eric A. Posner, *The Limits of International Law* 43 (2005) (arguing that CIL "is not an exogenous influence on state behavior"); Guzman, *supra* note 23, at 130 ("As a matter of theory, then, CIL may be effective."); Pierre-Hugues Verdier & Erik Voeten, "Precedent, Compliance, and Change in Customary International Law: An Explanatory Theory," 108 *Am. J. Int'l L.* 389, 390 (2014) ("A state may comply [with CIL] because it knows its decision to defect creates a precedent that may undermine a cooperative norm it values.").

THE CHRONOLOGICAL PARADOX

One of the conceptual difficulties surrounding the standard view of CIL, which is referred to in the literature as the "chronological paradox," merits special attention. The paradox is as follows: If state practices do not become binding as CIL until the states involved act out of a sense of legal obligation, how do the states develop that sense of legal obligation in the first place? In other words, the standard conception of CIL paradoxically seems to require that CIL develop before it can develop, in which case we would never have CIL. As one scholar describes the paradox, "Nothing can be a source of new customary international law if *opinio juris* requires that any action must be in accordance with the existing law."[29]

Some commentators who articulate the standard view of CIL either ignore the paradox or treat it as if it were a mere rhetorical question that need not be answered. The *Restatement (Third) of the Foreign Relations Law of the United States*, for example, simply notes:

> There have been philosophical debates about the very basis of the definition: how can practice build law? Most troublesome conceptually has been the circularity in the suggestion that law is built by practice based on a sense of legal obligation: how, it is asked, can there be a sense of legal obligation before the law from which the legal obligation derives has matured? Such conceptual difficulties, however, have not prevented acceptance of customary law essentially as here defined.[30]

The special rapporteur for the International Law Commission's project on the identification of customary international law has similarly taken note of the

[29] Olufemi Elias, "The Nature of the Subjective Element in Customary International Law," 44 *Int'l & Comp. L.Q.* 501, 504 (1995). *See also, e.g.*, David J. Bederman, *Custom as a Source of Law* 20 (2010) ("Taken to its logical extreme, no custom can ever be created unilaterally because every 'first-mover' was acting erroneously as to the 'binding' character or 'sense of obligation' of the practice."); Anthony A. D'Amato, The *Concept of Custom in International Law* 53 (1971) ("But if custom *creates* law, how can a component of custom require that that the creative acts be in accordance with some *prior* right or obligation in international law?"); Raphael M. Walden, "The Subjective Element in the Formation of Customary International Law," 12 *Isr. L. Rev.* 344, 363 (1977) ("It is not possible consistently to maintain at one and the same time both that custom is creative of new law and not declaratory of existing law, and also that it always requires to be accompanied by a belief that the conduct in question is already law."); Dissenting Opinion of Judge Manfred Lachs, in North Sea Continental Shelf Cases, 1969 I.C.J. Rep. 3, 231 (Feb. 20) ("To postulate that all States, even those which initiate a given practice, believe themselves to be acting under a legal obligation is to resort to a fiction – and in fact to deny the possibility of developing such rules.").

[30] Restatement (Third) of the Foreign Relations Law of the United States, *supra* note 1, § 102, reporters' note 2. *See also, e.g.*, Hugh Thirlway, "The Sources of International Law as a Consensual Bond," in *International Law* 125 (Malcolm Evans, ed., 2d ed. 2006) ("The process by which customary rules change and develop thus presents theoretical difficulties; but it is a process which does occur.").

paradox (along with other conceptual problems associated with the *opinio juris* concept) and simply observed that "the theoretical torment which may accompany [the concept of *opinio juris*] in the books has rarely impeded its application in practice."[31]

The paradox, however, is not merely of academic interest: it suggests that the standard view of CIL, although often recited, *does not correctly account for how CIL is actually discerned and applied in practice.* Therefore, as John Tasioulas has observed, for legitimacy and other reasons, the paradox must be confronted "head-on."[32]

A variety of theories have been proposed to resolve the chronological paradox, while retaining the standard view of CIL, but none of them seems satisfactory.[33] The most common theory for how CIL develops despite the requirement of *opinio juris* is that states initially make a mistake and believe themselves to be under a legal obligation even though they are not.[34] There are at least two problems with this "mistake theory." First, according to most accounts of CIL, in order for CIL to arise, a large number of states need to believe that a practice is legally binding. But, unless there is some reason to think that states (many of which are sophisticated actors) regularly make mistakes and, moreover, make the same mistakes en masse, CIL would rarely, if ever, develop and evolve.[35] Second, if one assumes that the subset of states that makes mistakes is the least sophisticated, that means that changes in the law would be driven by the least sophisticated actors, which hardly seems like a recipe for effective and desirable law making.

[31] Wood, *supra* note 3, at 48.

[32] John Tasioulas, "Customary International Law and the Quest for Global Justice," in *The Nature of Customary Law: Legal, Historical and Philosophical Perspectives* 322 (Amanda Perreau-Saussine & James Bernard Murphy eds., 2007); *see also* Martti Koskenniemi, *From Apology to Utopia: The Structure of International Legal Argument* 420 (2005) (orig. 1989) (noting that the paradox "is not merely an academic's pet dilemma").

[33] *See, e.g.*, Hilary Charlesworth, "Customary International Law and the Nicaragua Case," 11 *Aust'l Y.B. Int'l L.* 1, 9 (1984–87) (noting that the "paradox of the traditional theory of customary international law has never been persuasively resolved"); Josef L. Kunz, "The Nature of Customary International Law," 47 *Am. J. Int'l L.* 662, 667 (1953) ("There is here, certainly, a challenging theoretical problem which, as far as this writer can see, has not yet found a satisfactory solution.").

[34] *See, e.g.*, van Hoof, *supra* note 19, at 99.

[35] *See, e.g.*, Michael Byers, *Custom, Power and the Power of Rules* 131 (1999) ("This [mistake] approach is unsatisfactory because it is inconceivable that an entire legal process ... could be based on a persistent misconception."); Michael Akehurst, "Custom as a Source of International Law," 47 *Brit. Y.B. Int'l L.* 1, 32 (1974–75) ("It is stretching credulity to suggest that all the many rules of customary law existing today originated on the basis of such mistakes."); Hiroshi Taki, "*Opinio Juris* and the Formation of Customary International Law: A Theoretical Analysis," 51 *German Y.B. Int'l L.* 447, 449 (2008) ("[I]t is strange and unreasonable to consider that a mistake by the acting individuals in relation to the existence of the law is indispensable in order to establish customary international law.").

Another theory that has been suggested could be called the "fiat theory." The idea is that one or a small number of states purposely assert the existence of a customary rule, even though they do not believe that it is required by preexisting custom, and other states accept the claim and act accordingly. To the extent that other states accept the claim out of a mistaken belief that it correctly reflects preexisting custom, this theory tends to merge into the mistake theory, and it has similar problems. Another possibility is that states accept the claim because they are coerced into doing so by virtue of the power of the states asserting the claim. (Imagine Britain in the nineteenth century, for example, imposing CIL rules of admiralty law when it had the most powerful navy.) While such a scenario seems realistic, it does not genuinely involve practice followed out of a sense of legal obligation. Moreover, a legal system in which a small number of states in effect impose law on the rest of the international community, although perhaps an accurate description of how CIL worked in the past, may be even more normatively problematic than the mistake theory.

In addition to the mistake and fiat theories, there are a variety of other theories about how a custom that is not initially perceived as legally binding might come to be understood that way, after passing through certain stages of development.[36] The actual mechanism of this transformation, however, is not specified, so the chronological paradox is left unresolved.[37] Other efforts to resolve the paradox involve alterations to the standard view of CIL. For example, Hans Kelsen argued – contrary to the standard view – that it need not be established that states perceive that a practice is *legally* obligatory, as long as they perceive that it is *normatively* required.[38] To take a much more recent example, Pierre-Hugues Verdier and Erik Voeten have argued that the chronological paradox is avoided if the *opinio juris* element of CIL is understood as merely requiring a claim about the legal status of a norm rather than a subjective belief that it has that legal status.[39] As with those efforts, this chapter contends that, in order to resolve the chronological paradox, it is necessary to revisit the standard view of CIL.

36 *See, e.g.*, VAN HOOF, *supra* note 19, at 92–93; Herman Meijers, "How Is International Law Made? The Stages of Growth of International Law and the Use of its Customary Rules," 9 *NETH. Y.B. INT'L L.* 1 (1978).

37 *See* KOSKENNIEMI, *supra* note 32, at 421 (noting that the gradual ripening theory "brings in nothing to solve the problem" because "we are still unable to reveal how the transformation from a political *opinio necessitatis* into a legally motivated *opinio juris* was possible"); HUGH THIRLWAY, *THE SOURCES OF INTERNATIONAL LAW* 77 (2014) ("The riddle remains, however, as to how this transition occurs: if for it to occur requires the presence of a justified belief that it has already occurred, it can in fact never occur.").

38 *See* HANS KELSEN, *PRINCIPLES OF INTERNATIONAL LAW* 307 (1952) ("They must believe that they apply a norm, but they need not believe that it is a legal norm which they apply.").

39 Verdier & Voeten, *supra* note 28, at 418.

REFERENCE TO CUSTOM IN THE ICJ STATUTE

Before considering an alternative account of CIL, it is important to understand that the standard conception of CIL was not always the prevailing view, and, in fact, does not precisely accord with the reference to custom in the ICJ's governing statute. Article 38(1) of the ICJ Statute lists the sources of law to be applied by the ICJ, and it includes in the list "international custom, as evidence of a general practice accepted as law." Instead of saying that CIL is based on practices and *opinio juris*, Article 38(1) says that international custom is *evidence* of practices and *opinio juris*. Moreover, the phrasing seems to entail only one component – "international custom" – rather than two components as under the standard view. Some commentators ignore these differences and simply assert that the phrasing articulates the standard view.[40] Other commentators acknowledge the differences but attribute them to poor drafting.[41]

In fact, Article 38(1) does not articulate the standard view, and it is not the product of poor drafting. Rather, it reflects a different conception of CIL than the one that is now widely recited, including by the ICJ itself. To understand this point, it is necessary to trace the origins of the *opinio juris* component of the standard view of CIL.

A number of scholars have traced the *opinio juris* component to a French jurist, Francois Geny.[42] In a treatise first published in 1899, entitled *Methode d'Interpretation et Sources en Droit Prive Positif* (*Method of Interpretation and Sources of Positive Private Law*), Geny attempted to distinguish between legally binding custom and mere social usage, and for the former he suggested that one look for "a feeling among the persons who practice it that they act on [the] basis of an unexpressed rule which

40 *See, e.g.*, LORI F. DAMROSCH ET AL., *INTERNATIONAL LAW: CASES AND MATERIALS* 59 (5th ed. 2009) (contending that "the order of words [in Article 38] makes little difference"); DAVID J. BEDERMAN, *INTERNATIONAL LAW FRAMEWORKS* 15 (2001) ("There are two key elements in the formation of a customary international law rule. They are elegantly and succinctly expressed in Article 38 of the ICJ Statute.").

41 *See, e.g.*, JAMES CRAWFORD, *BROWNLIE'S PRINCIPLES OF PUBLIC INTERNATIONAL LAW* 23 (8th ed. 2012) (describing the phrasing in Article 38 as "prima facie defective"); VAN HOOF, *supra* note 19, at 87 (noting that "this much criticized formulation has probably contributed to the confusion surrounding custom"); Kunz, *supra* note 33, at 664 (referring to the "extremely bad drafting of article 38(1)" of the ICJ Statute).

42 *See, e.g.*, ANTHONY A. D'AMATO, *THE CONCEPT OF CUSTOM IN INTERNATIONAL LAW* 49 (1971); BEDERMAN, *CUSTOM AS A SOURCE OF LAW*, *supra* note 29, at 142; V. D. DEGAN, *SOURCES OF INTERNATIONAL LAW* 144 (1997); MALCOLM N. SHAW, *INTERNATIONAL LAW* 53 (7th ed. 2014); *see also* Peter E. Benson, notes and comments, "Francois Geny's Doctrine of Customary Law," 20 *CAN. Y.B. INT'L L.* 267 (1982) ("Historians single out Francois Geny as the writer whose work crowned the nineteenth century's analysis of custom and whose ideas were directly incorporated into our approach towards international custom."). Some writers, however, connect the standard view of CIL either to the German scholar Franz von Liszt (who was a cousin of the famous composer), or the Swiss scholar Alphonse Rivier. *See, e.g.*, STEPHEN C. NEFF, *JUSTICE AMONG NATIONS: A HISTORY*

is binding for them as a rule of law."[43] Although Geny was writing about domestic private law, the subjective element of his formulation is similar to the *opinio juris* requirement under what is now the standard view of CIL.

Although suggestions are sometimes made that *opinio juris* has deeper intellectual roots, David Bederman's book-length survey of the role of custom suggests otherwise. As he explained: "Contemporary public international law's doctrine of *opinio juris* bears no real resemblance to antecedents in Roman law, canon law, the *ius commune*, or English common law.... It sits as an oddity in the law of nations, and, perhaps is the only thing that distinguishes the exceptional character of CIL."[44] For example, some commentators have suggested that the *opinio juris* concept finds support in Blackstone's treatise on the laws of England,[45] but in fact, Blackstone merely argued that customs needed to have a mandatory rather than discretionary character in order to qualify as law, not that they had to be shown to be followed out of a sense of legal obligation.[46]

Importantly, the phrasing of the ICJ statute appears to have intellectual roots that are distinct from the roots of the modern concept of *opinio juris*. The language of Article 38(1) was carried over from what had been Article 38 of the Statute of the Permanent Court of International Justice (PCIJ). The initial drafting of the PCIJ statute was delegated to an advisory committee of jurists working in 1920, which was chaired by Baron Descamps of Belgium. (The committee consisted of delegates from Japan, Spain, Brazil, Belgium, Norway, France, the Netherlands, England, Italy, and the United States.) The committee specifically rejected a proposal by Descamps that would have referred to "international custom, being practice between nations accepted by them as law."[47] In other words, the committee rejected a description of CIL that would, like the standard view today, have set forth a separate *opinio juris* requirement.[48]

OF *INTERNATIONAL LAW* 253 (2014) ("Alphonse Rivier has been credited with being the first to clearly articulate this twofold picture of custom (albeit without the use of the actual expression '*opinio juris*').").

43 FRANCOIS GENY, *METHODE D'INTERPRETATION ET SOURCES EN DROIT PRIVE POSITIF* 248 (2d ed. 1954) (transl. Louisiana State Law Institute 1963).

44 BEDERMAN, *CUSTOM AS A SOURCE OF LAW*, *supra* note 29, at 173; *see also* Gerald J. Postema, "Custom in International Law: A Normative Practice Account," in *THE NATURE OF CUSTOMARY LAW*, *supra* note 32, at 279, 280 (noting that the "additive understanding" of CIL, whereby *opinio juris* is required in addition to practice, "is of relatively recent vintage"). This is not to suggest, of course, that there were not earlier accounts of customary law. *See* Emily Kadens, "Custom's Past" (in this volume).

45 *See, e.g.*, CLIVE PARRY, *THE SOURCES AND EVIDENCES OF INTERNATIONAL LAW* 61 (1965).

46 *See* 1 WILLIAM BLACKSTONE, *COMMENTARIES ON THE LAWS OF ENGLAND* 78 (1765) (arguing that customs "must be (when established) *compulsory*; and not let to the option of every man, whether he will use them or no").

47 *See* Permanent Court of International Justice, Advisory Committee of Jurists, Procés-Verbaux of the Proceedings of the Committee, June 16–July 24 1920, at 306; *see also* BRIAN D. LEPARD, *CUSTOMARY INTERNATIONAL LAW: A NEW THEORY WITH PRACTICAL APPLICATIONS* 129 (2010).

48 *See also* Alain Pellet, Article 38, in *THE STATUTE OF THE INTERNATIONAL COURT OF JUSTICE: A COMMENTARY* 749 (Andreas Zimmerman et al. eds., 2d ed. 2012) (noting that "the

The phrasing of the PCIJ statute's reference to custom was apparently influenced by the "historical school" of jurisprudence of the nineteenth century, which is generally traced to (among others) the German jurist Friedrich Carl von Savigny.[49] This school hypothesized that customary law emanated from a collective spirit or will of the people – a "Volksgeist."[50] Importantly, Francois Geny – whose writings are said to be the intellectual genesis of *opinio juris* for CIL – expressly disagreed with the Volksgeist concept.[51]

The Volksgeist concept has a natural law character that is probably not viable in today's more positivistic legal culture.[52] It also became associated to some extent with Nazi ideology.[53] Not surprisingly, therefore, it has generally been rejected in the post–World War II custom literature.[54] The Volksgeist approach did, however, have one significant advantage over what is now the standard view: it had an answer to the famous chronological paradox that has plagued the standard view of CIL.

The approach of the historical school to CIL avoided the chronological paradox by hypothesizing that custom was evidence of something deeper and preexisting. Under that approach, the development of a customary obligation did not depend on nations following a practice out of a sense of legal obligation. Rather, the obligation would exist, and the custom would then arise to reflect it. Savigny specifically recognized the need to avoid the chronological paradox, in critiquing the mistake theory of *opinio juris*:

> According to [this idea], the first act must have been induced by the necessitatis opinio; consequently if it rests upon an error it must not be of any account at all as to the origination of customary law. The same is true of the second act which

Committee of Jurists in 1920 clearly did not have in mind a splitting-up of the definition of custom into two distinct elements").

49 *See* KELSEN, *supra* note 38, at 308–09; *see also* G. M. DANILENKO, *LAW-MAKING IN THE INTERNATIONAL COMMUNITY* 77 (1993) (noting that the wording of the PCIJ Statute's reference to custom "might also reflect the continued influence of natural law theories which regard custom not as a law-creating fact but as evidence of an already existing law").

50 *See* ILA Report, *supra* note 17, at 32; *see also* Walden, *supra* note 29, at 358 ("As originally formulated by the adherents of the historical school, the theory of *opinio juris* was intimately connected with their doctrine which saw law as an expression of the *Volksgeist*."). French legal theorists developed a somewhat analogous idea that custom was evidence of an underlying social solidarity (*solidarite sociale*). *See* KELSEN, *supra* note 38, at 309.

51 *See* GENY, *METHODE D'INTERPRETATION*, *supra* note 43, at 248; *see also* Walden, *supra* note 29, at 358–59 (noting that the modern doctrine of *opinio juris* is not connected to the Volksgeist concept).

52 *See* KELSEN, *supra* note 38, at 310–11; KOSKENNIEMI, *supra* note 32, at 415–16.

53 *See, e.g.*, C. K. ALLEN, *LAW IN THE MAKING* 86 (6th ed. 1958) ("Without disrespect for their scholarly genius, it is difficult not to feel that unconsciously (for they could hardly guess what would be built upon the foundations which they laid) Savigny and his followers were National Socialists before the National Socialists."); CHRISTOPHER ROEDERER & DARREL MOELLENDORF, *JURISPRUDENCE* 127 (2004; reprinted 2007) (noting that the historical school "appealed to the apologistic lawyers of Nazi Germany").

54 *See* Kunz, *supra* note 33, at 663–64; Alan Watson, "An Approach to Customary Law," 1984 *ILL. L. REV.* 561, 566. *See also* Robert E. Rodes, Jr., "On the Historical School of Jurisprudence," 49 *AM.*

now becomes the first and of the third and every following one. The formation of a customary law is hence, unless one of those conditions is given up, wholly impossible.[55]

The approach of the historical school avoided this problem, Savigny explained, because "the rule of law was merely manifested by the custom, not generated by it; consequently in the first demonstrable act the *necessitatis opinio*, free of all error, might and must have been present."[56]

There is another important respect in which some of the earlier theories of CIL differed from today's standard view. Instead of requiring that states believe that a practice was already legally required, some of the earlier theories hypothesized that it was sufficient if states believed that a legal rule in support of the practice was *necessary*. Alphonse Rivier, for example, wrote that "the custom or the usage of the nations is the manifestation of the international juridical consciousness operated by the facts that are continuously repeated *with the sense of their necessity*."[57] Such an approach appears to be consistent with the broader phrase *opinio juris sive necessitatis*, the latter portion of which is often omitted in modern discussions of CIL.[58]

In sum, the standard, two-part conception of CIL has not always been the dominant view, and it is not fully reflected in the ICJ's governing statute. A key reason why some of the earlier approaches died away was the intellectual shift toward positivism. Instead of hypothesizing that custom is a reflection of some underlying spirit, will, or consciousness, the standard view today attempts to ground CIL in the actual practices and beliefs of states. As recounted, however, such an inductive conception of CIL has difficulty resolving a host of evidentiary, normative, and conceptual problems associated with deriving law in this fashion.

Other efforts to articulate an inductive account of CIL have also failed to overcome these problems. Long before the International Law Commission's current project on customary international law, there was Manley Hudson's effort in the Commission in the mid-twentieth century. In 1947, the UN General Assembly established the Commission and approved its governing statute. Article 24 of the statute directed the Commission to "consider ways and means for making the evidence of customary

J. Juris. 165 (2004) (presenting a more sympathetic account of the historical school but noting that "everyone is polite to [the historical school], and no one explicitly disowns it, but no one really takes it seriously").

55 1 Friedrich Carl von Savigny, *System of the Modern Roman Law* 141 (William Holloway transl., 1867).

56 *Id.*

57 1 Alphonse Rivier, *Principes du Droit des Gens* 35 (1896).

58 *See* Thirlway, *supra* note 37, at 78 ("The phrase in its entirety signifies that it is or may be sufficient if there is an *opinio* to the effect that the action [or refraining from it, as the case may be] is required as being, in some sense, necessary.").

international law more readily available, such as the collection and publication of documents concerning State practice and of the decisions of national and international courts on questions of international law, and shall make a report to the General Assembly on this matter." Hudson, who had served as a judge on the Permanent Court of International Justice and was the first U.S. member of the International Law Commission, was subsequently asked to prepare a working paper on customary international law and to submit it to the Commission.[59]

Based on his canvassing of an array of international law treatises from around the world, Hudson found that there seemed to be general agreement that, in order for a CIL principle or rule to emerge, the following four elements were required:

(a) concordant practice by a number of States with reference to a type of situation falling within the domain of international relations;
(b) continuation or repetition of the practice over a considerable period of time;
(c) conception that the practice is required by, or consistent with, prevailing international law; and
(d) general acquiescence in the practice by other States.[60]

The report then proceeded to discuss the accessibility of various types of evidence of CIL – treaties, decisions of international tribunals, decisions of national courts, national legislation, diplomatic correspondence, and opinions of national legal advisers – and it suggested various measures to make such evidence more readily available.

Hudson's report generated substantial discussion within the Commission.[61] At the outset, Hudson observed that, in his view, Article 38(b) of the ICJ Statute "was not very happily worded," and he said that he had "felt it would be useful to try to set down what should be understood by 'customary international law.'"[62] He subsequently noted that "the authors of article 38 of the statute of the International

[59] Article 18 of the ILC statute provides that "the Commission shall survey the whole field of international law with a view to selecting topics for codification, having in mind existing drafts whether governmental or not." In a 1949 "Survey of International Law in Relation to the Work of Codification of the International Law Commission," Hersch Lauterpacht had expressed the view that "it is doubtful whether any useful purposes would be served by attempts to make [Article 38 of the ICJ Statute] more specific, as, for instance, by defining the conditions of the creation and of the continued validity of international custom." U.N. Secretary-General, *Survey of Int'l Law: In Relation to the Work of Codification of the ILC, Preparatory Work within the Purview of Article 18, Paragraph 1, of the Statute of the ILC, Memorandum Submitted by the Secretary-General*, ¶ 33, U.N. Doc. A/CN.4/1/Rev. 1 (Feb. 10, 1949).

[60] Manley O. Hudson, "Article 24 of the Statute of the International Law Commission," [1950] 2 Y.B. INT'L L. COMM'N 24, 26, UN Doc. A/CN.4/16 (1950).

[61] For an account of the discussions, see WOLFKE, *supra* note 18, at 29–36.

[62] *Summary Records of the 40th Meeting*, [1950] 1 Y.B. INT'L L. COMM'N 4, U.N. Doc. A/CN.4/Ser.A/ (1950).

Court, and of article 24 of the statute of the Commission, had no very clear idea as to what constituted international custom."[63]

There were a number of questions from Commission members about precisely what was meant by "acquiescence" in Hudson's proposed element (d). In addition, Hudson's proposed requirement of a "considerable period of time" in element (b), which Hudson admitted he found "difficult to define," generated questions and disagreement. James Brierly observed, for example, that he was "not convinced that sub-heading (b) was necessary."[64] "What was required," said Brierly, "was *opinio juris necessitatis*," and he noted that "it had been known for this to arise at a moment's notice."[65] Even sharper disagreements emerged concerning the requirement in element (c) that there be a "conception that the practice is required by, or consistent with, prevailing international law." In particular, several Commission members objected to element (c) on the ground that CIL would not be able to generate new law if states had to have a conception that the practice was already required by or consistent with international law.[66] These members, in other words, were concerned about the chronological paradox surrounding the concept of *opinio juris*.[67]

In subsequent discussions, the Commission became skeptical about the feasibility and desirability of trying to specify the elements of CIL. Brierly, for example, "wondered whether it was desirable for the Commission to embark on a question of doctrine," and he "felt it would be difficult to find a formula on which all members of the Commission could agree."[68] The Commission ultimately voted not to include Hudson's proposed elements of CIL in its report.

COMMON LAW ACCOUNT OF CIL ADJUDICATION

A fundamental problem with much of the theorizing about CIL, this chapter contends, is that it fails to identify which institutional context it has in mind. Instead,

[63] *Id.* at 6. Years earlier, Hudson had noted in a book that "the customary part of international law is in an unsatisfactory condition." MANLEY O. HUDSON, *PROGRESS IN INTERNATIONAL ORGANIZATION* 83 (1932).

[64] Summary Records, *supra* note 62, at 5.

[65] *Id.*

[66] *See id.* at 6 (Ypes) (pointing out that "if, as was stated in sub-heading (c), a custom must be consistent with international law, it ceased to be a source for that law"); *id.* (Scelle) (noting that "it was somewhat contradictory to state on the one hand that custom is the basis of law, and on the other that it must be consistent with law"); *id.* (Ypes) (expressing the view that "custom could depart from prevailing international law; otherwise it had no raison d'être"); *id.* (Amado) ("Custom was the primordial source of law; but to judge from sub-heading (c) it had ceased to be so."). *See also* BEDERMAN, *CUSTOM AS A SOURCE OF LAW*, *supra* note 29, at 143 ("If Hudson was expecting quick approval of his gloss on Article 38(1)(b), he was sadly disappointed.").

[67] For a critique of each of Hudson's proposed four elements of CIL, see Anthony A. D'Amato, "Wanted: A Comprehensive Theory of Custom in International Law," 4 *TEX. INT'L L. F.* 28 (1968).

[68] Summary Records, *supra* note 62, at 275.

the discussion proceeds as if CIL existed in the abstract without any particular human entity to interpret and apply it. In fact, customary practices do not operate in a self-liquidating manner to produce law. Merely looking out into the world to see what nations have done and said does not itself reveal rules of international law. Among other things, there will always be questions about the appropriate level of generality at which to describe the practices.[69] Moreover, disputes tend to arise only when the norms are unclear for some reason. As will be seen, once a decision maker is hypothesized, it becomes easier to gain traction on some of the difficulties surrounding CIL.

The remainder of this chapter sketches an alternate account of CIL that addresses the chronological paradox as well as other well-known problems with the standard view of CIL. Under this "common law" account, adjudicators look to past practice, just as adjudicators look to past decisions when developing the common law.[70] But, as with common law adjudication, judges necessarily make choices about how to describe the practice, which baselines to apply in evaluating it, and whether and when to extend or analogize it to new situations.[71] Furthermore, the choices that they make are informed by judgments about the perceived preferences of the regulated states, the likely consequences of adopting a particular formulation of the relevant international rule, and moral and ethical considerations. In other words,

[69] *See, e.g.*, Richard Craswell, "Do Trade Customs Exist?", in THE JURISPRUDENTIAL FOUNDATIONS OF CORPORATE AND COMMERCIAL LAW 118, 122 (Jody S. Kraus & Steven D. Walt eds., 2000) ("Any history of prior decisions will always underdetermine the possible patterns that might be ascribed to that history."); Frederick Schauer, "Pitfalls in the Interpretation of Customary Law," in THE NATURE OF CUSTOMARY LAW, *supra* note 32, at 13, 21 ("The relevant custom for some decision does not simply leap out and grab the interpreter, but rather is selected by that interpreter from among the entire field of that jurisdiction's normative customs."). *Cf.* N. C. H. Dunbar, "The Myth of Customary International Law," 1983 AUST. Y.B. INT'L L. 1, 8 ("The myth is in assuming that universal state practice *ipso facto* creates law. Law can only be created by legislation or by the judgment of a court, or, in the case of international law, by a treaty.").

[70] This common law account of CIL adjudication is distinct from CIL's status within domestic common law, an issue that has been the subject of substantial debate in the United States. *Compare, e.g.*, Bradley & Goldsmith, *supra* note 25, *with* Harold Hongju Koh, "Is International Law Really State Law?," 111 HARV. L. REV. 1824 (1998).

[71] For a discussion of these and other modes of reasoning associated with common law adjudication, see MELVIN ARON EISENBERG, THE NATURE OF THE COMMON LAW ch. 6 (1988). There are of course varying accounts of what precisely is entailed in common law adjudication, and how the common law tradition has been understood at various times. The account presented here is consistent with modern, post–Legal Realist understandings of the common law process, in which "law is made, developed, and changed in the context of real litigated controversies." FREDERICK SCHAUER, *THINKING LIKE A LAWYER: A NEW INTRODUCTION TO LEGAL REASONING* 111 (2009); *see also, e.g.*, *Sosa v. Alvarez-Machain*, 542 U.S. 692, 725 (2004) ("Now, however, in most cases where a court is asked to state or formulate a common law principle in a new context, there is a general understanding that the law is not so much found or discovered as it is either made or created.").

under a common law account, adjudicators are understood as *developing* CIL rather than merely finding it.

Because CIL adjudication involves an element of lawmaking as well as law identification, it is not captured by the standard, two-part definition of CIL, which treats CIL as stemming directly from state practice and *opinio juris*. The paradigm adjudicatory institution considered here is the ICJ, although the analysis potentially applies to other institutions as well and might also apply (probably with modifications) to domestic courts applying CIL.[72] That said, under this account one should expect to see variation among tribunals in how CIL is identified and applied because different tribunals will presumably face different levels of constraint on their common law-type discretion.[73]

This common law account of customary international law adjudication helps shed light on a number of the problems surrounding the standard account of CIL.

Principles and Rules

The common law account helps makes sense of the difference between general principles and specific rules, a distinction often glossed over under the standard view of CIL. Sometimes – probably often – there is a generally agreed-upon principle but no widespread practice concerning the specific issue, and the decision maker must decide whether the principle applies to that specific issue. To the extent that commentators maintain that there is CIL relating to protection of the environment, for example, they generally describe it in terms of principles rather than specific rules.[74] In that situation, the standard view of CIL seems to offer no help, because there is no state practice directly on point, let alone state practice followed out of a sense of legal obligation. Under a common law approach, by contrast, a decision maker would be free to apply the principle to the new set of facts. (The possibility that a similar result could be accomplished by utilizing the "general principles"

[72] *Cf.* Anthea Roberts, "Comparative International Law?: The Role of National Courts in Creating and Enforcing International Law," 60 INT'L & COMP. L.Q. 57, 60 (2011) (challenging "the frequent presumption that domestic courts should necessarily or only function as international actors that seek to impartially enforce international law instead of national actors seeking to create and shape international norms").

[73] The other chapters in this volume appear to confirm this expectation. *See also* Harlan Grant Cohen, "International Law's Erie Moment," 34 *MICH. J. INT'L L.* 249, 270 (2013) ("The development of 'custom' via adjudication is not merely a continuation of that rule's development, but its transformation into something new.").

[74] *See, e.g.*, Pierre-Marie Dupuy, "Formation of Customary International Law and General Principles," in *THE OXFORD HANDBOOK OF INTERNATIONAL ENVIRONMENTAL LAW* 450, 452–53 (Daniel Bodansky, Jutta Brunee, and Ellen Hay eds., 2008); Catherine Redgwell, "International Environmental Law," in *INTERNATIONAL LAW*, *supra* note 30, at 657, 663–65.

category of the ICJ's jurisdiction rather than the custom category is considered at the end of this chapter.)

A good example of the distinction between general principles and specific rules is the famous *Schooner Exchange* decision by the U.S. Supreme Court that is said to be the fount of the international law of sovereign immunity.[75] In that decision, the Court had to decide whether to accord sovereign immunity in an admiralty case to a French warship. There was little state practice directly on point, so the Court explained that it would be "necessary to rely much on general principles and on a train of reasoning founded on cases in some degree analogous to this."[76] The Court proceeded to invoke the general principle in CIL concerning the "perfect equality and absolute independence of sovereigns," and to invoke by analogy various CIL doctrines, such as the CIL governing head of state immunity and the CIL governing diplomatic immunity.[77] Despite the lack of direct evidence of state practice, the Court concluded that there was "a principle of public law that national ships of war entering the port of a friendly power open for their reception are to be considered as exempted by the consent of that power from its jurisdiction."[78] The Court was, in other words, *both applying and developing CIL at the same time*.[79] This is a typical feature of CIL adjudication, this chapter contends, but the standard account of CIL does not have an explanation for it.

Another prominent example is the PCIJ's decision in the *Lotus* case.[80] There, the court held that Turkey did not violate international law in regulating the conduct of a French citizen on board a French vessel, conduct that was alleged to have caused a collision on the high seas resulting in the deaths of Turkish nationals. The majority reasoned that "restrictions upon the independence of States cannot ... be presumed," and it found no affirmative basis for concluding that Turkey had violated an established rule of international law.[81] Its decision contains no significant review of either state practice or *opinio juris*. Rather, the court noted that it was aware of no relevant decisions of international tribunals and that the decisions of national courts that had been cited to it "sometimes support one view and sometimes the other."[82] Because it appeared that national jurisprudence was divided, the court said it need not consider such jurisprudence further because "it is hardly possible

75 11 U.S. (7 Cranch) 116 (1812).

76 *Id.* at 136.

77 *Id.* at 137.

78 *Id.* at 145. *Cf. Jurisdictional Immunities of the State*, *supra* note 2, ¶ 57 ("Exceptions to the immunity of the State represent a departure from the principle of sovereign equality.").

79 *Cf.* Monica Hakimi, "Custom's Method and Process: Lessons from Humanitarian Law" (in this volume), at 149 (arguing that "CIL finding is deeply entangled with CIL making").

80 The Case of the S.S. Lotus (*Fr. v. Turk.*), 1927 P.C.I.J. (ser. A) No. 10 (Sept. 7).

81 *Id.* at 18.

82 *Id.* at 28.

to see in it an indication of the existence of the restrictive rule of international law."[83] Most of the dissenting opinions also failed to focus on evidence of state practice or *opinio juris*; rather, they referred to propositions such as "the spirit of international law,"[84] the "consensus omnium" of the international community,[85] and the "principle" of territoriality.[86] The two partial exceptions are the dissents of Judge Moore, which reviewed a number of domestic cases, and Judge Altamira, who reviewed the legislation of a number of states. Of particular relevance to this chapter, Altamira concluded his dissent by acknowledging that when international tribunals adjudicate issues of CIL

> there are moments in time in which the rule, implicitly discernible, has not as yet taken shape in the eyes of the world, but is so forcibly suggested by precedents that it would be rendering good service to the cause of justice and law to assist its appearance in a form in which it will have all the force rightly belonging to rules of positive law appertaining to that category.[87]

Finally, consider the Truman Proclamation in 1945, asserting that the United States had the right to control the resources of its continental shelf beneath the high seas. Imagine if an international adjudicator were asked to decide the legality of the proclamation shortly after it was issued. The adjudicator would need to decide how to reconcile two competing principles: freedom of the seas, and territorial control. There would not have been any state practice specifically on point. Of course, there was a practice of inaction: states had not previously asserted exclusive control over the resources in their continental shelves (although they had at various times made other claims concerning the seabeds of the continental shelves). But that was likely due to the fact that they had no technological ability to access those resources. As in *The Lotus*, much would have therefore turned on the court's choice of baseline: would the United States be required to show an established practice in support of its claim, or would a challenger be required to show an established practice disallowing it? The key point is that the standard view does not help resolve such a case.[88]

83 *Id.* at 29.
84 *Id.* at 34 (Loder).
85 *Id.* at 43–44 (Weiss).
86 *See id.* at 52–53 (Lord Finlay), 59 (Nyholm).
87 *Id.* at 106–07 (Altamira).
88 For another example, see *Case Concerning the Continental Shelf (Libya v. Malta)*, 1985 I.C.J. 4, 24 (June 3) (applying a distance principle developed for the exclusive economic zone to the context of the continental shelf "for juridical and practical reasons"). *See also* Jorg Kirchner, "Thoughts About a Methodology of Customary International Law," 43 Austrian J. Pub. Int'l L. 215, 236 (1992) (noting that the ICJ "manages to extend the scope of customary rules by means of deduction, analogy and the shift of the burden of proof," an approach "which does not wait for the formation of new rules through state practice").

Evidentiary Issues

The common law account also helps address a variety of evidentiary uncertainties that surround the standard account of CIL. First, this account avoids artificially excluding state practices from consideration merely because they might be followed for reasons other than a sense of legal obligation. For example, nations presumably follow many customs out of self-interest (and, indeed, one would assume that customs develop because they are generally in the interest of the participants), yet it is unclear why behavior motivated by self-interest should not count when discerning CIL.[89] Similarly, nations may follow a custom out of a sense of morality (abolishing the slave trade in the nineteenth century, for example), but, again, it is unclear why that should count against the custom qualifying as CIL, since the law and morality presumably overlap. Finally, many customs described today as CIL are probably followed out of bureaucratic habit rather than a conscious sense of legal obligation, and yet such a state of mind (even though not *opinio juris*) probably is desirable for international law compliance.[90]

There has also been much debate in the literature about whether verbal acts by states can be considered a form of state practice. Those who object to such classification worry that these acts will end up being "double counted" as both practice and evidence of *opinio juris*.[91] Under the common law account, verbal acts would certainly be considered, and the classification debate is unimportant, as is the worry about double counting.[92] To be sure, verbal actions should be considered with caution, since they might simply be "cheap talk" as opposed to the expression of a genuine preference. But this concern relates to the weight to be given to the evidence, not its relevance.

A common law account further helps explain why treaties that do not purport to codify CIL are nevertheless relevant to the identification of rules of CIL, a question that has generated significant debate and uncertainty. Although parties to a treaty presumably have a sense of legal obligation *to the treaty*, it is not clear why treaties show that nations are acting out of a sense of legal obligation

[89] *See* George Norman & Joel P. Trachtman, "The Customary International Law Game," 99 AM. J. INT'L L. 541, 571 (2005) (noting that "legal obligation and self-interest are not mutually exclusive categories").

[90] *See generally* ABRAM CHAYES & ANTONIA HANDLER CHAYES, *THE NEW SOVEREIGNTY: COMPLIANCE WITH INTERNATIONAL REGULATORY AGREEMENTS* (1995).

[91] For discussion of this concern about double counting, see Special Rapporteur, *Third Report on the Identification of Customary International Law*, Int'l L. Comm'n, U.N. Doc. A/CN.4/682, at 10 (Mar. 27, 2015).

[92] *See, e.g.*, Maurice Mendelson, "The International Court of Justice and the Sources of International Law," in *FIFTY YEARS OF THE INTERNATIONAL COURT OF JUSTICE* 63, 87 (Vaughan Lowe & Malgosia Fitzmaurice eds., 1996).

under CIL.[93] Nevertheless, treaties are among the most common evidence cited by courts in support of *opinio juris*. Indeed, in many cases, including many cases decided by the ICJ, they are essentially the *only* evidence cited.[94] The explanation, according to the account provided in this chapter, is that adjudicators use treaties as evidence of state preferences, and, relatedly, as evidence of social and moral desirability. In many cases, of course, a treaty will reveal only a preference for binding consenting parties, but in some cases a widely ratified treaty might reveal a preference for a universal, community-wide rule. For this and other evidentiary issues, a common law approach unifies the "traditional" inductive CIL with the "modern" deductive CIL.[95] Relatedly, it explains why adjudicators seem to use a sliding scale, requiring more evidence of practice when there is less evidence of *opinio juris*, and vice-versa.[96]

Consequentialist Considerations

A common law approach also helps explain why decision makers often take into account consequentialist considerations, such as considerations of efficiency and international comity, when discerning rules of CIL. To take one of many examples, in the *Arrest Warrant Case*, the ICJ relied heavily on consequentialist considerations concerning the ability of foreign ministers to carry out duties on behalf of their states.[97] As Professor Kirgis has noted, "A reasonable rule is always more likely to be found reflective of state practice and/or the *opinio juris* than is an unreasonable (for example, a highly restrictive

93 *Cf.* Case Concerning Ahmadou Sadio Diallo (*Guinea v. Dem. Rep. Congo*), 2007 I.C.J. 582, 615 (May 24):

> The fact invoked by Guinea that various international agreements, such as agreements for the promotion and protection of foreign investments and the Washington Convention, have established special legal régimes governing investment protection, or that provisions in this regard are commonly included in contracts entered into directly between States and foreign investors, is not sufficient to show that there has been a change in the customary rules of diplomatic protection; it could equally show the contrary.

See also North Sea Continental Shelf Cases (Federal Republic of Germany/Denmark), 1969 I.C.J. Rep. 3, 44 (Feb. 20) ("From [the action of states that were parties to a treaty setting forth the equidistance principle] no inference could legitimately be drawn as to the existence of a rule of customary international law in favour of the equidistance principle.").

94 *See* Choi & Gulati, *supra* note 8. This is not a particularly new phenomenon. *See, e.g.*, R. R. Baxter, "Multilateral Treaties as Evidence of Customary International Law," 41 *Brit. Y.B. Int'l L.* 275, 275 (1965–66) ("Both multilateral and bilateral treaties are not infrequently cited as evidence of the state of customary international law.").

95 *Cf.* Anthea E. Roberts, "Traditional and Modern Approaches to Customary International Law: A Reconciliation," 95 *Am. J. Int'l L.* 757 (2001).

96 *See* Frederic L. Kirgis, Jr., "Custom on a Sliding Scale," 81 *Am. J. Int'l L.* 146 (1987); *see also, e.g., Prosecutor v. Kupreskic*, Case No. IT-95-16-T, Judgment, ¶ 527 (Jan. 14, 2000) (noting that "principles of international humanitarian law may emerge through a customary process under the pressure of the demands of humanity or the dictates of public conscience, even where State practice is scant or inconsistent").

97 *See Case Concerning the Arrest Warrant of 11 April 2000*, 2002 I.C.J. 3, 21–22 (Feb. 14).

or inflexible) rule."[98] Giving weight to these considerations is entirely consistent with a common law approach, pursuant to which adjudicators attempt to articulate rules in a manner that promotes social welfare.[99]

Another good example of this phenomenon is the famous *Trail Smelter* arbitration between the United States and Canada in 1941, in which the tribunal held that "no state has the right to use or permit the use of its territory in such a manner as to cause injury by fumes in or to the territory of another or the properties or persons therein, when the case is of serious consequence and the injury is established by clear and convincing evidence."[100] The tribunal did not base this conclusion on any specific state practice or *opinio juris*. The tribunal noted that "no case of air pollution dealt with by an international tribunal has been brought to the attention of the Tribunal nor does the Tribunal know of any such case."[101] Nor could the tribunal find any international decisions concerning water pollution. Instead, the tribunal invoked a "general principle" that states have a duty to protect other states from injurious acts emanating from their territories, and it analogized to decisions of the U.S. Supreme Court concerning cross-border pollution issues between U.S. states.[102]

The possibility that international adjudicators consider social desirability when discerning purported rules of CIL helps address the critique that, under the standard view, CIL is unlikely to result in efficient rules. As Eugene Kontorovich has argued, factors such as the large and heterogeneous nature of the world community, the frequent lack of widespread past interactions concerning issues in dispute, and the difficulty of legally exiting from the rules cast doubt on the likelihood that CIL can be expected to be efficient.[103] But this critique assumes that CIL rules are discerned primarily inductively, as maintained by the standard view. If such rules are partly created in adjudication, as argued here, there will be more opportunity for adjudicators to make them efficient.[104]

[98] Kirgis, *supra* note 93, at 149.

[99] *Cf.* Eyal Benvenisti, *Sharing Transboundary Resources: International Law and Optimal Resource Use* 203 (2002) (arguing that the international community accepts a role along these lines for the ICJ in certain types of cases). *See also* Ted L. Stein, "The Approach of the Different Drummer: The Principle of the Persistent Objector in International Law," 26 *Harv. Int'l L.J.* 457 (1985) ("Correspondingly, *opinio juris* is no longer seen as a consciousness that matures slowly over time (and finally imparts obligatory force to a practice once motivated by habit, convenience, or moral sentiment), but instead as a conviction that instantaneously attaches to a rule believed to be socially necessary or desirable.").

[100] Trail Smelter Case (United States/Canada), III Reports of International Arbitral Awards 1965 (Mar. 11, 1941)

[101] *Id.* at 1963.

[102] *Id.* at 1963–64.

[103] *See* Kontorovich, *supra* note 22.

[104] *See* Eyal Benvenisti, "Customary International Law as a Judicial Tool for Promoting Efficiency," *in The Impact of International Law in International Cooperation* (Eyal Benvenisti & Moshe Hirsch eds., 2004).

Changes to CIL

A common law approach also does a better job than the standard view of explaining how CIL changes. As previously discussed, the only mechanism for change specified under the standard account is through violations of the norm, and these violations need to be either widespread or widely accepted in order to effectuate change. This process for change seems both difficult to achieve in practice and normatively questionable. Under a common law approach, by contrast, an adjudicator can recognize that the deviation, along with other evidence (such as evidence of changed circumstances), shows that the international community no longer prefers the prior rule, or, relatedly, that the old rule is no longer working well.[105]

Under this approach, one can imagine, for example, the recognition of an erosion of sovereign or individual official immunity based on changed normative commitments reflected in treaties and other materials. Indeed, consider a shift in immunity that has already taken place: the shift during the twentieth century from the absolute theory of sovereign immunity to the restrictive theory.[106] If an international adjudicator had been called upon to consider the validity of this shift in its early stages, the standard view of CIL suggests that the adjudicator should have held that it was illegal – a decision that could have prevented the further development of CIL on this subject. The common law account, by contrast, allows for a greater ability to recognize and permit evolutionary changes to CIL.

The Chronological Paradox and First Violators

Conceiving of customary international law adjudication as akin to common law adjudication also helps resolve the chronological paradox. Under the common law account, the recognition of a CIL rule does not require proof that states are already following a practice out of a sense of legal obligation before a CIL rule can be recognized. Rather, a CIL rule can be recognized when it is evident – from state practices, statements, and other evidence – that the rule is something that the relevant community of states wishes to have as a binding norm going forward and that it is socially and morally desirable.[107] To be sure, the past practices of states are still highly

[105] *Cf.* Schauer, *supra* note 71, at 105 ("It is characteristic of common-law method that judges have the power to change the rules in the very act of applying them, typically in the context of a hitherto unforeseen situation in which the existing rule would produce a poor outcome.").

[106] *See* Gamal Moursi Badr, *State Immunity: An Analytical and Prognostic View* 115–28 (1984).

[107] For a somewhat similar perspective, see Lepard, *supra* note 47, at 97–98 (arguing that "*opinio juris* be interpreted as a requirement that *states generally believe that it is desirable now or in the near future to have an authoritative legal principle or rule prescribing, permitting, or prohibiting certain state conduct*") (emphasis in original); *see also* Brian D. Lepard, "Customary International Law as a Dynamic Process" (in this volume); Eric A. Posner & Alan O. Sykes, *Economic Foundations*

relevant under this account, because such practices are often the best evidence of state preferences. But, because it need not be shown that nations are already following a practice out of a legal obligation, the chronological paradox is avoided.

For similar reasons, the common law account is able to explain how CIL can be applied against the first violator of a norm. If a practice is uniform and has no deviations, there may be no opportunity for a sense of legal obligation to develop. When the first deviation occurs, however, the issue of whether the practice is legally binding suddenly becomes relevant. Under the standard view of CIL, the past practice might not count, since it was not necessarily being followed out of a sense of legal obligation. Under a common law approach, by contrast, the past practice would be relevant in assessing whether states prefer to have a binding rule governing the issue, and an adjudicator would be open to finding that the first deviation violates CIL.[108]

This account also confirms the argument, made most famously by Bin Cheng, that it is possible for there to be "instant custom."[109] Traditionally, it was thought that the creation of CIL tended to be a longstanding process of action and reaction among states. The modern view, however, is that CIL can be created quickly, and the ICJ has confirmed this possibility.[110] Commentators note that technological changes in communication, along with international fora such as the United Nations, allow for much more extensive and rapid dialogue concerning what the law is and should be. Although this possibility is recognized by a number of commentators, it is not clear how it can be reconciled with the standard view of CIL. It is consistent with the common law account, however, because there will be situations in which it will

OF INTERNATIONAL LAW 57 (2013) ("When trying to determine whether a norm of customary international law exists, we are really trying to figure out what is optimal and sustainable between states."); Jonathan I. Charney, "Universal International Law," 83 AM. J. INT'L L. 529, 538 (1993) ("Certainly, those searching for norms seek to determine whether states that have expressed interest in a matter have reached consensus on establishing a corresponding norm as law."); Walden, *The Subjective Element*, *supra* note 50, at 97 ("What is involved may be, not a belief that the practice is *already* legally binding, but a claim that it *ought* to be legally binding.").

108 *Cf. Hilton v. Guyot*, 159 U.S. 113, 163 (1895):

> When, as is the case here, there is no written law upon the subject, the duty still rests upon the judicial tribunals of ascertaining and declaring what the law is, whenever it becomes necessary to do so in order to determine the rights of parties to suits regularly brought before them. In doing this, the courts must obtain such aid as they can from judicial decisions, from the works of jurists and commentators, and from the acts and usages of civilized nations.

109 *See, e.g.*, Bin Cheng, "United Nations Resolutions on Outer Space: 'Instant' International Customary Law?" 5 INDIAN J. INT'L L. 23 (1965).

110 *See* North Sea Continental Shelf (F.R.G./Den.; F.R.G./Neth.), 1969 I.C.J. 3, 74 (Feb. 20) (noting that "the passage of only a short period of time is not necessarily, or of itself, a bar to the formation of a new rule of customary international law on the basis of what was originally a purely conventional rule"); *see also* WOLFKE, *supra* note 18, at 59 ("At present . . . an international custom can arise even in a very short time.").

be possible to have a good sense of state preferences even before there has been enough time for specific state practices to develop.

An oft-cited example of instant CIL is the Truman Proclamation in 1945. Within a few years after this Proclamation, it was clear that many states accepted the claim that states had exclusive sovereignty over the resources in their continental shelves, even though this might seem to contradict the traditional principle of freedom of the seas. If CIL developed this rapidly, however, it was not because there was already extensive state practice involving exclusive mineral extraction on the shelves, something that was not even technologically possible for most states. Instead, it was because numerous state declarations indicated that the relevant states preferred a CIL rule along the lines of the one advocated by Truman.[111] As Michael Scharf has discussed in his book on "Grotian Moments," there are a number of modern examples like this one in which CIL is viewed as developing rapidly despite the absence of extensive state practice.[112]

Judicial Legislation?

It might be objected that the common law account envisions a quasi-legislative role for adjudicators, a role that naturally raises questions about the extent to which adjudicators have been properly charged to act in this fashion. Critics are likely to contend that adjudicators should be limited to applying the law as it already exists (*lex lata*) rather than the law as it should be (*lex ferenda*).

Questions of judicial authority to develop the law are common in a domestic legal system, but they are more pronounced in the international system that (for the most part) lacks an agreed-upon central judiciary. A partial response to this concern is that "judicial legislation" is inherently more constrained than ordinary legislation – by, for example, the contours of the cases that come before the courts, the expectation that decisions will be supported by a certain type of public reasoning, and the heavy reliance of most international adjudicatory institutions on voluntary compliance.[113]

In any event, a significant check on judicial creativity in this context is the ability of states to withdraw (prospectively) from many international adjudicatory institutions.[114]

[111] *See* SUZETTE V. SUAREZ, *THE OUTER LIMITS OF THE CONTINENTAL SHELF: LEGAL ASPECTS OF THEIR ESTABLISHMENT* ch. 8 (2008).

[112] *See* MICHAEL P. SCHARF, *CUSTOMARY INTERNATIONAL LAW IN TIMES OF FUNDAMENTAL CHANGE: RECOGNIZING GROTIAN MOMENTS* (2013).

[113] *Cf.* ALLEN, *supra* note 53, at 149 (arguing that it is "misleading ... to call [the judicial application of custom], without qualification, 'judicial legislation'" and that "what applies to the adoption and application of Common Law rules ... apply with equal force – neither more nor less – to the adoption and application of custom").

[114] In 1986, the United States withdrew from the ICJ's compulsory jurisdiction in response to the ICJ's decision to exercise jurisdiction in a case brought by Nicaragua concerning U.S. covert and military

Moreover, under the persistent objector doctrine, when a rule of CIL is first recognized, it should not be applied to states that have clearly disagreed with the rule. Although not part of existing doctrine, it might also make sense to allow for subsequent withdrawal rights for certain types of CIL rules.[115] In any event, states can override CIL rules as between themselves by treaty, as long as the rules do not have the status of *jus cogens* norms. More generally, literature on international courts suggests that, even when they act creatively, these courts are constrained in a variety of ways.[116] This is true regardless of whether the court is conceived of as an agent of states, a trustee of states, or a hybrid of the two.[117]

The concern about judicial lawmaking can also be reduced by limiting the formal precedential effect of international adjudicatory decisions. For example, it may make sense to give greater emphasis to the fact that ICJ decisions are technically binding only on the parties and thus should not automatically be treated as the last word by the international community on the content of CIL.[118] The modern "fragmentation" of international law and institutions, while posing potential challenges to international law's coherence and legitimacy, may make this easier by allowing for greater variation in judicial claims about the content of CIL.[119]

Another reason to allow for such variation in claims about CIL, and to limit the precedential effect of determinations of CIL, is that there is inherent tension between centralized adjudication and the maintenance of a system of customary law.[120] As

activities in that country. *See* United States: Department of State Letter and Statement Concerning Termination of Acceptance of ICJ Compulsory Jurisdiction, 24 I.L.M. 1742 (1985). In 2005, after a series of ICJ rulings against the United States concerning the Vienna Convention on Consular Relations, the United States withdrew from an optional protocol that allowed the ICJ to exercise jurisdiction in such cases. *See* Charles Lane, "U.S. Quits Pact Used in Capital Cases," WASH. POST, Mar. 10, 2005.

115 *See* Bradley & Gulati, *Withdrawing from International Custom*, *supra* note 4; Guzman, *supra* note 23, at 169–71.

116 *See, e.g.*, Tom Ginsburg, "Political Constraints on International Courts," in THE OXFORD HANDBOOK OF INTERNATIONAL ADJUDICATION (Cesare P. R. Romano, Karen J. Alter, and Chrisanthi Avgerou eds., 2013).

117 *See, e.g.*, Karen J. Alter, "Agents or Trustees? International Courts in Their Political Context," 14 EUR. J. INT'L REL. 33 (2008); *see also* Laurence R. Helfer & Anne-Marie Slaughter, "Why States Create International Tribunals: A Response to Professors Posner and Yoo," 93 CAL. L. REV. 899, 942–54 (2005).

118 *See* UN CHARTER, art. 94(1) (requiring that each member of the United Nations comply with any ICJ decision "to which it is a party").

119 *See, e.g.*, Paul B. Stephan, "Disaggregating Customary International Law," 21 DUKE J. COMP. & INT'L L. 191, 195–96 (2010). A much-discussed example is the different standards adopted by the ICJ and the International Criminal Tribunal for the former Yugoslavia concerning state responsibility. *See id.* at 196. For additional discussion of the phenomenon of fragmentation, see Martii Koskenniemi & Paivi Leino, "Fragmentation of International Law? Postmodern Anxieties," 15 LEIDEN J. INT'L L. 553 (2002).

120 *See, e.g.*, Suzanne Katzenstein, "International Adjudication and Custom Breaking by Domestic Courts," 62 DUKE L.J. 671 (2012).

previously noted, one of the purported virtues of custom is that it can continue to evolve in response to changing conditions. Judicial decisions concerning its content, however, have the potential to stifle such evolution if states end up coordinating around the decision. This potential is exacerbated by a backward-looking approach to CIL, as called for by the standard view, which envisions that changes to CIL require violations of the norm. In the *Jurisdictional Immunities of the State* case, for example, the ICJ concluded that there was no *jus cogens* exception to state sovereign immunity. If states feel constrained by that decision, or other adjudicatory institutions feel obliged to follow it, evolution that might otherwise have occurred toward a *jus cogens* exception may be stifled.[121]

General Principles Instead of Custom?

Some commentators have acknowledged that when international adjudicators purport to identify rules of CIL they often do not seem to be applying the standard, two-component view, but these commentators contend that this phenomenon should be labeled as the application of "general principles" rather than the application of CIL.[122] The objection to using the label "customary international law" appears to be at least partly linguistic: in the absence of state practice in support of a CIL rule, it seems strange to refer to the rule as being "customary." As Robert Jennings critically observed, "Most of what we perversely insist on calling customary international law is not only *not* customary law: it does not even faintly resemble a customary law."[123] Moreover, the ICJ statute, these commentators point out, specifically distinguishes between the application of international custom and the application of general principles.[124]

It is certainly true that the standard view of CIL has difficulty capturing the way in which international adjudicators identify and apply general principles when deciding cases, even though adjudicators often purport to be doing this under the general umbrella of CIL. It is unclear, however, how much would be gained by simply shifting the account of this phenomenon to a different label. The same concerns about judicial lawmaking, for example, would continue to exist. Moreover, the general principles category referred to in the ICJ statute has traditionally been

121 This point suggests, for example, that the European Court of Human Rights went too far in concluding, in *Jones v. United Kingdom*, that the ICJ's analysis of the CIL of sovereign immunity in *Jurisdictional Immunities of the State* "must be considered by this Court as authoritative." Case of *Jones and Others v. The United Kingdom* ¶ 198 (Judgment) (Jan. 14, 2014).

122 *See, e.g.*, BODANSKY, *supra* note 7, at 199–203; Simma & Alston, *supra* note 25, at 102; Petersen, *supra* note 7.

123 Robert Y. Jennings, "The Identification of International Law," in *INTERNATIONAL LAW: TEACHING AND PRACTICE* 3, 5 (Bin Cheng ed., 1982).

124 *See* ICJ STATUTE, art. 38(1)(c).

thought to be primarily a reference to gap-filling rules derived from common features of domestic legal systems (such as rules concerning remedies and defenses), rather than primary rules of international conduct, so it is not clear how well suited it is for taking on a broader role.[125]

In any event, the deficiencies with the standard view of CIL do not primarily concern difficulties in articulating general principles, such as "freedom on the seas," "territorial integrity," or "equality of states." Instead, the difficulties come in attempting to reconcile potentially competing principles and in applying the principles to specific fact patterns.[126] This is what international adjudication typically requires, but the standard view of CIL often fails to describe how judges go about it. Simply positing a separate category of general principles that do not depend on the standard view of CIL does not advance the analysis concerning the application of such principles. This chapter contends that a common law account, by contrast, does help to advance the analysis.

CONCLUSION

As noted at the outset of this chapter, there are increasing doubts about the usefulness and viability of CIL as a source of international law, making this an especially appropriate time to rethink some of the assumptions concerning the operation of CIL in theory and practice. In this regard, it is worth emphasizing that the claim here is not only that a common law approach is a better account of how CIL adjudication *should* work. It is also a claim that it is a better description of how it already *does* work in institutions such as the ICJ. These institutions may recite the standard view of CIL, but they do not actually follow it in practice.

125 *See, e.g.*, Louis Henkin, *International Law: Politics and Values* 40 (1995) ("It is a source available only as necessary for interstitial use, to fill out what international law requires but has not recognized as customary law because it has not yet been invoked often and widely enough, and it is too cumbersome for the system to negotiate by multilateral treaty."); Oscar Schachter, *International Law in Theory and Practice* 52 (1991) ("The international cases show such use [of the general principles category] in a limited degree, nearly always as a supplement to fill in gaps left by the primary sources of treaty and custom."). The intended scope of the general principles category is unclear from the PCIJ drafting history. *See* Giorgo Gaja, "General Principles of Law," in *IV Max Planck Encyclopedia of Public International Law* 370–78 (Rudiger Wolfrum ed., 2012). For a controversial argument in favor of a much broader role for general principles, see the Separate Opinion of Judge Cancado Trindade, in *Case Concerning Pulp Mills on the River Uruguay (Arg. v. Uru.)*, 2010 I.C.J. 14 (April 20).

126 *See* Hersch Lauterpacht, *The Development of International Law by the International Court* 370 (1958) ("While in most fields of international law there is agreement as to broad principle, there is almost invariably a pronounced degree of divergence with regard to the application of specific rules.").

3

Customary International Law as a Dynamic Process

Brian D. Lepard

Today there is great debate about just how long customary international law takes to form, and how it can be changed. Ranged on one side are those "traditionalists" who maintain that customary law requires both (1) consistent state practice, and (2) *opinio juris sive necessitatis*, a belief among states that a customary practice is legally binding.[1] Moreover, these traditionalists argue that both elements must persist over some extended period of time. That is, state practice must be longstanding, and even the *opinio juris* must be well grounded and consistent through time. This long gestation period, in turn, gives customary norms permanence and rootedness. This same quality of rootedness can make customary law norms difficult to change. That is because for a norm to change, both elements must be modified, and this modification, too, ought to take some time under the traditional view.

According to the traditional view, customary international law is like a giant ocean liner. It takes a long time to get up to cruising speed, and once it is headed in a particular direction, much effort is required to cause it to change course. Furthermore, the traditional view also is not merely a jurisprudential one about what characteristics customary international law "has." It also incorporates a normative dimension, and traditionalists argue that there are good reasons for making customary international law difficult to create, and difficult to modify.

Ranged against this traditional view is an army of new approaches to customary international law, all of which view it as a more dynamic process and as more susceptible to change. According to some of these theories, a new consistent state practice can arise very quickly; no particular duration of the practice is required to establish a corresponding new norm of customary international law. Likewise, *opinio juris* can be formed in an "instant," or at least very quickly. And some theories minimize or dispense with either the state practice requirement or the *opinio juris*

[1] See, for example, the formulation in North Sea Continental Shelf Cases (*F.R.G. v. Denmark; F.R.G. v. Neth.*), 1969 I.C.J. Rep. 3, 44, para. 77 (Feb. 20), discussed presently.

requirement, thus making it even easier for customary norms to be created, or changed, since only one element needs changing.

This clash of approaches and theories has left customary law in a jurisprudential crisis. We might dismiss this as just another academic conundrum, of little interest to practitioners, except that customary international law is assuming enormous importance practically in a wide variety of fields. The traditional theory as well as new theories are appearing with increasing frequency in judicial opinions, and therefore must be taken into account by ministries of foreign affairs and legal advisers to governments. While, as Joel Trachtman points out in his chapter, treaties have proliferated and occupied more legal "terrain" that used to be covered only by customary law,[2] treaties only bind states that have ratified them. Many states are not bound by particular treaty norms. Moreover, nonstate actors are not bound by them. This is why customary law plays a key role in the mandates and decision making of international criminal tribunals like the International Criminal Court (ICC) and the criminal tribunals for the former Yugoslavia and Rwanda.[3] Furthermore, treaties have many "gaps" that can be filled by customary law. And treaties must be interpreted in a broader legal context, with customary law often providing that context. For all these reasons, it is critical to resolve the crisis in customary law.

This chapter argues that customary international law is, and ought to be, conceived of as a dynamic method of lawmaking. It also argues that the essence of customary international law is *opinio juris*, and that state practice is best viewed as evidence of *opinio juris*. In particular, the chapter contends that *opinio juris* should be reconceptualized as a belief by states generally that it is desirable now or in the near future to have an authoritative legal principle or norm prescribing, permitting, or prohibiting certain conduct, apart from treaty obligations.[4] Their beliefs can and should be ascertained through examination of a wide range of evidence, including the text of treaties, statements by states about their views (including the significance of the treaties they enter into), the provisions of national legislation, and national judicial decisions, among others. Moreover, state beliefs ought to be evaluated in the context of certain fundamental ethical principles that states themselves have endorsed. These perspectives mean that a customary norm can emerge fairly quickly, and be changed fairly quickly, if there is sufficient evidence of such a belief of states in the desirability of creating or modifying an authoritative legal norm,

2 *See* Joel P. Trachtman, "The Growing Obsolescence of Customary International Law" (in this volume).

3 However, some scholars have argued that the role of customary international law in the decision making of international criminal tribunals is now declining in favor of "codification" of international crimes, as exemplified by certain provisions of the ICC Statute. *See* Larissa van den Herik, "The Decline of Customary International Law as a Source of International Criminal Law" (in this volume).

4 *See* Brian D. Lepard, *Customary International Law: A New Theory with Practical Applications* 8 (2010).

and particularly where the change promotes the realization of fundamental ethical principles.

In the following sections, the chapter explores the traditional view and its justifications and weaknesses, modern approaches and their benefits and shortcomings, and the proposed new perspective on the dynamic quality of customary international law. It also explains how this perspective reinforces, but is also distinct from, some of the intriguing views offered by other contributors to this volume.

CUSTOMARY INTERNATIONAL LAW AS "EMBEDDED" LAW

The traditional view of customary international law is that it evolves over a long period of time, and thus becomes "embedded" in the society of states. Under this perspective, there is a static quality to customary international law. It is rooted in interstate society, and serves, indeed, as a kind of legal base or foundation for a network of international legal rules. Moreover, not only is it difficult to change, but normatively, viewed through this lens, it *should* be difficult to change. Without this quality of rootedness, of permanence, the argument goes, customary international law would be like shifting sands, and any legal edifice constructed on it runs the risk of toppling over.

Historical Evolution of the Traditional View

It is evident that certain patterns of behavior by states developed over time. These "customs" were transmitted from state to state, and from generation to generation of state leaders. But were these customs law? Judges and lawyers eventually arrived at a view that custom becomes international law when there is "*opinio juris sive necessitatis*" – a belief by those states subject to the rule that it *is* a legal rule. This led to the traditional bipartite definition of customary international law as a consistent practice among states accompanied by *opinio juris*.

In short, as relations among nations grew in size and complexity, it was natural that states would develop certain practices and accept them as legally binding in order to achieve a variety of goals, including the facilitation of trade and the maintenance of peace, and when war occurred, the minimization of its harmful effects. These customary norms supplemented those developed by contract in the form of treaties. Of course, treaties themselves could also give rise to customary norms that took on a life of their own apart from the treaties. As Hans Kelsen has famously pointed out, even the law of treaties began as customary rules, and therefore customary international law is the foundation of the international legal order.[5]

[5] *See* HANS KELSEN, *GENERAL THEORY OF LAW AND STATE* 369 (Anders Wedberg trans., 1945).

The bipartite definition of customary international law just described did not develop in a vacuum. As Emily Kadens helpfully explores in her chapter, Roman lawyers as well as medieval European jurists were quite familiar with the concept of customary law and endorsed some form of the two-element definition.[6] It is also notable, as she underscores, that going back to these early conceptions, customary law has always sat uneasily alongside written law, because it is "fluid, uncertain, equitable, and communitarian."

Growing out of these early precedents, customary local or national law has long been applied in common law countries, at least in discrete types of cases. William Blackstone's famous *Commentaries on the Laws of England* established various criteria for the recognition of customary law by common law courts.[7] Moreover, a number of countries with a civil code permit a judge to decide a case by resort to customary law as a fallback method if there is no governing written law.[8] Some codes explicitly adopt the bipartite definition. For example, the Louisiana Civil Code affirms in Article 3 that "custom results from practice repeated *for a long time* and generally accepted as having acquired the force of law."[9]

In both common law and civil law systems, there traditionally has been a requirement of longstanding practice. Thus, English common law requires that a custom be "immemorial." Indeed, one of Blackstone's requirements for customary law is that the custom "have been used so long, that the memory of man runneth not to the contrary."[10] In addition, some of the civil codes referred to here, such as the Louisiana Code, specify that a practice must be of long duration.

The treatise writers of international law, and governments, too, eventually adopted these ideas about customary law drawn from national law and elevated them to the level of international law. Not surprisingly, the jurists who began to codify international law, and especially the members of the Advisory Committee of Jurists who drafted Article 38 of the Statute of the Permanent Court of International Justice (PCIJ) in 1920, which became verbatim Article 38 of the Statute of the International Court of Justice (ICJ), drew on their knowledge of the operation of customary law at the domestic level in articulating its longstanding function at the international level.

6 *See* Emily Kadens, "Custom's Past" (in this volume).

7 For a discussion of his criteria and their application by modern courts, see David Callies, "How Custom Becomes Law in England," in THE ROLE OF CUSTOMARY LAW IN SUSTAINABLE DEVELOPMENT 158 (Peter Ørebech et al. eds., 2005).

8 *See, e.g.*, Swiss Civil Code, art. 1, para. 2, *available at* http://www.admin.ch/ch/e/rs/2/210.en.pdf.

9 Louisiana Civil Code, art. 3, Acts 1987, No. 124, §1, *available at* http://www.legis.la.gov/legis/Law.aspx?d=110037 (emphasis added).

10 WILLIAM BLACKSTONE, COMMENTARIES ON THE LAWS OF ENGLAND 76–77 (1st ed.), *quoted in* Callies, "How Custom Becomes Law in England," at 166. On the requirement of immemoriality, see generally *id.* at 166–70.

Thus, Article 38(1)(b) allows the ICJ to apply, in addition to treaties, "international custom, as evidence of a general practice accepted as law."[11]

International courts, like their domestic counterparts, eventually formulated the well-known bipartite definition of customary international law, which has been expressed on a number of occasions by the ICJ.[12] Notably, the ICJ has insisted that a practice must be "settled" before it can become law.[13] Many, if not most, scholars have also emphasized the traditional requirements of both state practice and *opinio juris*. They adopt the view that state practice is essential to the formation of a customary rule. For example, Sir Michael Wood, in his second report presented in 2014 to the International Law Commission as special rapporteur, concluded that "to determine the existence of a rule of customary international law and its content, it is necessary to ascertain whether there is a general practice accepted as law."[14] Moreover, most publicists still insist that practice be of long duration. However, they take the view that no particular length of time is required for the formation of customary law.[15]

Even if no particular duration is necessary, under the traditional view, customary international law should require a fairly long period of gestation to emerge. And similarly, once a rule becomes entrenched, there ought to be a fairly long following period of contrary practice to change or overturn it. In this connection, legal scholar Karol Wolfke has affirmed that more practice and greater uniformity of practice are required to terminate "an old, well-settled customary rule" than to create a new one.[16]

Two initial points are worth noting here about the *opinio juris* element of the traditional view of customary international law. First, even if a very long period of practice relating to some issue exists among states, *opinio juris* is an essential requirement for the formation of a customary rule. Why? Because a custom, even an ancient one, is not by itself a rule. As a pattern of behavior, rather, it could be described as "consistent with" a variety of incipient or potential rules. To give but one example, a pattern of states not arresting ambassadors of other states could be equally consistent with a rule forbidding any arrests of ambassadors, a rule only

[11] I.C.J. STATUTE, art. 38(1)(b).

[12] *See, e.g.*, Continental Shelf Case (*Libya v. Malta*), 1985 I.C.J. Rep. 13, 29, para. 27 (June 3) (stating that the substance of customary international law must be "looked for primarily in the actual practice and *opinio juris* of States").

[13] *See* North Sea Continental Shelf Cases, 1969 I.C.J. Rep. 3, 44, para. 77; Jurisdictional Immunities of the State (*Germany v. Italy*), 2012 I.C.J. Rep. 99, 122, para. 55 (Feb. 3) (observing that "the existence of a rule of customary international law requires that there be 'a settled practice' together with *opinio juris*").

[14] Michael Wood, Special Rapporteur of the International Law Commission, Draft Conclusion 3, *in Second Report on Identification of Customary International Law*, U.N. Doc. A/CN.4/672 (2014), at 65.

[15] *See, e.g.*, Draft Conclusion 9, para. 3; in *id* at 67 ("Provided that the practice is sufficiently general and consistent, no particular duration is required.").

[16] KAROL WOLFKE, *CUSTOM IN PRESENT INTERNATIONAL LAW* 65 (2d rev. ed. 1993).

allowing arrests of ambassadors for certain crimes (which no ambassadors happen to commit), a rule requiring states to give "due consideration" to the sanctity of ambassadors and embassies, or a rule allowing states freely to arrest ambassadors (which states choose not to take advantage of in the interests of promoting good diplomatic relations).

Second, although some notion of *opinio juris* is essential, the traditional definition of *opinio juris* manifests a "paradox."[17] How can a belief by states that a custom *already* reflects a legal rule be considered a precondition for recognition of a *new* legal rule? Certainly the definition could work well enough for existing legal rules. But it is wholly unsatisfactory for explaining or justifying the creation of new customary law. This is because before the custom becomes a legal rule it is not a legal rule. Yet in order for it to become a legal rule, the participants must erroneously believe that it already is one. This chapter will propose a solution to this paradox.

Advantages of the Traditional View

A number of arguments can be made in support of the traditional view that international customary norms ought to be difficult to create, and difficult to change, some of which were just touched upon. First, clear, longstanding, and resilient rules can solve interstate coordination problems very well, so long as the nature of those problems does not change. Many issue areas governed by international law can be viewed as coordination dilemmas, such as rules on maritime navigation (i.e., ships approaching head-on at sea must pass on the right)[18] and problems involving delimitation of the continental shelf. The same is true for norms designed to solve prisoners' dilemmas, where every state has a self-interested preference for cheating. A "hard," entrenched rule may be necessary to combat these incentives and prevent all states from winding up with their least-favored outcomes. Many issues addressed by international law could be reasonably perceived as prisoners' dilemmas, where a clear, stable rule enforced by sanctions is desirable to counteract the incentives for defection and avoid the worst results for states.

Furthermore, the traditional view of customary international law can help ensure that societal rules enforcing minimum moral rules of social conduct are enduring and cannot easily be overturned. Just as the prohibition of murder, which originated

[17] On this paradox, see, among other sources, David J. Bederman, *Custom as a Source of Law* 20, 149 (2010).

[18] This rule is now codified as Rule 14(a) of the International Regulations for Preventing Collisions at Sea, *in* Convention on the International Regulations for Preventing Collisions at Sea (1972), Annex, 1050 U.N.T.S. 17, entered into force 15 July 1977 (providing that "when two power-driven vessels are meeting on reciprocal or nearly reciprocal courses so as to involve risk of collision, each shall alter her course to starboard so that each shall pass on the port side of the other").

as a customary rule in societies before statutory criminal enactments,[19] ought to be entrenched because of its morally compelling character and not made susceptible to easy change through contrary practice, so also norms regarding basic human rights that are recognized at the international level should be difficult to modify. For example, there are good normative reasons for treating the prohibition of torture as an embedded norm and not allowing it to be changed easily, including by recent practices of "enhanced interrogation" by Western and other powers.

Indeed, many norms of customary international law qualify as peremptory norms (*jus cogens*) in large part (or exclusively) because of their compelling moral character. *Jus cogens* norms enjoy a privileged status. Even states that persistently object to these rules cannot exempt themselves from their reach.[20] And these rules are not easily susceptible to change, by design. They cannot, for example, be modified simply by treaty; indeed, any treaty that conflicts with them is considered entirely void.[21] In the words of Article 53 of the 1969 Vienna Convention on the Law of Treaties, "A peremptory norm of general international law is a norm accepted and recognized by the international community of States as a whole as a norm from which no derogation is permitted *and which can be modified only by a subsequent norm of general international law having the same character*."[22] Of course, many human rights norms are now properly regarded as *jus cogens*, and therefore have an entrenched character. They are "super norms" that can only be changed by other "super norms."[23]

At the same time, law is not always the best means of dealing with international problems generally. There are many other means, including voluntary persuasion of states to behave in a certain way. Setting high barriers to the formation of customary rules in the first place, according to this perspective, rightly favors "non-law" over law in influencing behavior. It may be desirable to allow members of the society of states as much freedom as possible, unrestrained by legal obligations unless absolutely necessary. Given that state sovereignty itself is a fundamental and well-recognized norm of customary international law, it is important not to burden states with "too much law." Certainly, the traditional view of customary international law helps restrain its reach. It also allows a wider sphere of operation for "soft law," which could encompass legal norms (including customary norms) that impose only persuasive obligations rather than binding ones, as well as norms that are not legal

[19] On the development of customary laws against homicide, see BEDERMAN, *supra* note 17, at 13–14.

[20] *See* LEPARD, *supra* note 4, at 250–52.

[21] *See* Vienna Convention on the Law of Treaties, art. 53.

[22] *Id.* (emphasis added).

[23] For a discussion of *jus cogens* norms and their relationship to moral values, see LEPARD, *supra* note 4, at 243–60. On the recognition of some *jus cogens* norms as a "form of natural law 'super-custom,'" see BEDERMAN, *supra* note 17, at 159.

in character at all. Many commentators have argued that soft law norms can provide unique benefits because of their ability to encourage desirable state behavior and cooperation without the burdens of strict legal obligation. One example of an area of law where soft law norms play an important role is the law of outer space.[24]

A related merit of the traditional view of customary international law is that its restraints on the creation of new law help to make customary international law more representative of the will of states, and in this sense, more "democratic." Insistence on widespread and enduring state practice and *opinio juris* ensures that asserted customary norms are not merely the whims and wishes of international judges who are pursuing their personal policy agendas. Given that states always have the option of entering into treaties with their explicit consent, according to this perspective it is desirable to limit the scope of customary law, to which states typically consent either not as explicitly or not at all. A number of academic commentators have criticized modern views of customary international law for being "undemocratic" in this way.[25]

By insisting on the recognition only of rules that develop over a long time, the traditional view also has the benefit of making it more likely that states know what the law is and are not surprised by novel assertions about customary international law. This is arguably fairer to states. Where international courts are applying international criminal law to individual defendants, it is also critical that the law respect the fundamental principle of *nullem crimen sine lege*, holding that one cannot be punished for an act that was not a crime when the act occurred. For example, the ICTY has generally been careful to insist on clear evidence of state practice before convicting defendants for violations of customary international criminal law. Theodor Meron has defended the state practice requirement for this reason.[26] Larissa van den Herik also refers to the problem of legality in her chapter.[27]

Disadvantages of the Traditional View

Despite these apparent advantages of the traditional view of customary international law, it also possesses its share of weaknesses. Here again, for example, the problem of interpretation arises. *Opinio juris* is essential to identify the rule that states believe exists (or should exist) and is consistent with a pattern of practice. Moreover, the

[24] For a study of soft law norms involving outer space, see the essays collected in SOFT LAW IN OUTER SPACE: THE FUNCTION OF NON-BINDING NORMS IN INTERNATIONAL SPACE LAW (Irmgard Marboe ed., 2012).

[25] *See, e.g.*, J. Patrick Kelly, "The Twilight of Customary International Law," 40 VA. J. INT'L L. 449, 518–23 (2000).

[26] *See* Theodor Meron, "Editorial Comment: Revival of Customary Humanitarian Law," 99 AM. J. INT'L L. 817, 821–34 (2005).

[27] *See* van den Herik, *supra* note 3.

well-accepted formulation of customary law again raises the paradox of *opinio juris*. It should be noted just how difficult it can be to legitimately recognize a new customary international law rule under the traditional definition of custom plus *opinio juris* as belief in the existing legal character of a rule. While some commentators tend to minimize the importance of the paradox of *opinio juris* in practice,[28] the doctrinal formulation nevertheless can be a barrier to judges or other decision makers finding that a new customary international law norm has been created.

The traditional view can also make it difficult for customary international law to adapt to new global problems. The pace of technological developments, such as those related to computing, the Internet (including cyber security and data privacy), trade, and advanced weaponry, can make longstanding international law rules obsolete in the blink of an eye. States need a mechanism to allow them to create rules quickly to solve these new problems, and without having to resort to the often laborious and time-consuming process of multilateral treaty drafting. Similarly, the increasing paralysis of certain international bodies, such as the UN Security Council, makes it desirable to allow customary international law to evolve quickly to fill these normative voids.

Thus, entrenched customary international law norms designed to solve coordination problems of a prior era may no longer work when the fundamental nature of the problem has changed. For example, businesses routinely trade across international borders, raising challenging problems of coordination among the world's many national taxing authorities that require new legal rules.[29] Likewise, situations that previously were not prisoners' dilemma situations internationally may evolve into them. A simple example involves pollution. Centuries ago, a customary practice may well have developed according to which every state bore none of the cost of externalities of the pollution caused by its inhabitants to inhabitants of other states. However, this situation quickly evolved into a prisoners' dilemma given the increase in the number of polluters and the broad extent of transboundary harm, requiring new rules to prevent defection and prevent worst outcomes. Accordingly, customary rules such as the "good neighbor" principle and the "polluter pays" principle were developed in response to this prisoners' dilemma.[30]

28 *See, e.g.*, ANTHONY A. D'AMATO, *THE CONCEPT OF CUSTOM IN INTERNATIONAL LAW* 73–74 (1971) (referring to the paradox of *opinio juris* as "harmless" in the case of existing customary norms, but acknowledging problems with the paradox in the recognition of new norms).

29 One particular coordination issue, involving transfer pricing, is discussed in LEPARD, *supra* note 4, at 285–305.

30 On the development of these principles in customary international law, see Catherine Redgwell, "International Environmental Law," in *INTERNATIONAL LAW* 687, 695 (Malcolm D. Evans ed., 3d ed. 2010).

The traditional theory can also impede recognition of new customary international law rules consistent with more progressive trends in moral thinking. For example, prior to the adoption of the UN Charter in 1945 and the Universal Declaration of Human Rights in 1948, customary international law apparently permitted a state to mistreat its citizens as it wished, with a few potential exceptions such as for crimes against humanity. Had this centuries-old rule been treated as sacrosanct and embedded, it might well have taken another century to modify, even in the face of a plethora of human rights instruments like the Charter and the Universal Declaration.

In this connection, while the ICJ has at times seemed too eager to embrace modern views of customary international law and to recognize new norms based primarily on changing moral perspectives, at other times it has staunchly upheld longstanding rules of customary international law even though they run counter to ethical principles found in contemporary international law. For example, in the 2002 *Arrest Warrant Case*,[31] the court stated that under a longstanding rule of customary international law ministers of foreign affairs enjoy absolute immunity from criminal prosecution in other states for all official acts, including those constituting war crimes and crimes against humanity, even after leaving office.[32] The court explicitly discounted principles in the Rome Statute of the International Criminal Court and other treaties establishing international criminal tribunals that allow national courts to exercise jurisdiction over persons suspected of having committed war crimes and crimes against humanity, including ministers of foreign affairs, in its assessment of customary international law. The court reasoned that "jurisdiction does not imply absence of immunity, while absence of immunity does not imply jurisdiction."[33] This kind of approach, which is supported by the traditional view, may be too backward-looking and conservative, ignoring important trends in state practice and views evidenced by the aforementioned treaties.

Moreover, while sovereignty is generally a value worthy of protection, the resistance of the traditional view to recognizing new limitations on state discretion can have, as just noted, deleterious effects on the realization of competing moral values, such as respect for human rights and protection of the environment. Thus, "non-law" is not always to be preferred to "law." There are good reasons to allow customary international law to change and grow quickly, but in a measured way, to prevent serious affronts to fundamental moral values under the banner of state "sovereignty."

31 Arrest Warrant of 11 April 2000 (*Democratic Republic of the Congo v. Belgium*), 2002 I.C.J. Rep. 3 (Feb. 14).

32 *See id.* at 24–26, paras. 58–61.

33 *See id.* at 24, para. 59.

The traditional view of customary international law may also not be so "democratic." Customary international law is often the product of the behavior and attitudes of the most powerful states. Indeed, critics have pointed out that many norms recognized as customary law under the traditional definition are simply the policies favored by Western powers.[34] The practices of less-powerful states typically have been ignored or discounted in the assessment of state practice, and likewise their views on the legality of a practice have often been given short shrift.

It may also not be true that the traditional view of customary international law makes it easier for states to know their obligations. There are still many uncertainties lurking in the concepts of longstanding consistent practice and *opinio juris*. States may have to undertake extensive studies to ascertain whether a practice is widespread and longstanding or *opinio juris* is similarly of long duration. And they may not know just when the magic time period required to achieve a "settled" state practice has been traversed. In other words, the advertised objective certainties of the traditional view may often be illusory.

Finally, the static, and even backward-looking, quality of the traditional bipartite formulation lends itself to a narrow conception of the legitimate range and sphere of operation of customary international law. The higher the bar that is set for the two requirements of practice and *opinio juris*, the more difficult customary international law is to find in the first place. Furthermore, there would appear to be a need for a sufficiently high quantity of discordant practice and contrary *opinio juris* to change an entrenched rule. Any discordant state practice would first, by necessity, be labeled a "violation" rather than treated as an "experiment" in favor a revised norm. Similarly, if *opinio juris* about the existing legal character of a norm must be widespread and convincing in the first place for the norm to be recognized, then discordant views expressed afterward would be regarded with suspicion and contrary action in accordance with these views would be regarded as violations of the rule. This is the so-called "first mover" problem.[35] It could take a rather significant mass of contrary opinion to force the rule to be revisited.

CUSTOMARY INTERNATIONAL LAW AS ADAPTABLE

In an effort to address some of these disadvantages of the traditional view of customary international law, scholars and some judges have proposed a number of alternatives.

34 *See, e.g.*, J. Patrick Kelly, "Customary International Law in Historical Context: The Exercise of Power Without General Acceptance," in *Reexamining Customary International Law* (Brian D. Lepard ed., forthcoming 2016) (affirming that the history of customary international law "suggests that to a large degree publicists and powerful nations ignored inconvenient state practice and generated customary international law norms based on prior assumed values or perceived self-interest irrespective of the general acceptance of that norm").

35 *See* Bederman, *supra* note 17, at 149.

According to these modern views of customary international law, customary law should be flexible and relatively easy to modify.[36] Proponents of these new views have justified them based on the acceleration of development of new technologies, and new shared moral sensibilities, that require innovative rules to achieve coordination or solve prisoners' dilemmas, ensure respect for basic moral values, or even avoid the destruction or disintegration of states.

Survey of Modern Views

Some of these views maintain adherence to the traditional twofold requirements for customary law of state practice and *opinio juris*, but argue that it may not take a long period of practice for a customary norm to emerge, or that *opinio juris* similarly need not be longstanding. In this vein, legal scholar Michael Scharf has proposed that some rules of customary international law can arise quickly based on new *opinio juris*, and with less state practice, in what he calls a "Grotian Moment."[37] As he explains:

> The Grotian Moment concept illuminates how and why customary international law can sometimes develop with surprising rapidity and limited state practice. The concept reflects the reality that in periods of fundamental change, whether by technological advances, the commission of new forms of crimes against humanity, or the development of new means of warfare or terrorism, rapidly developing customary international law may be necessary to keep up with the pace of developments.[38]

Regarding the traditional requirement of a "settled" practice, some observers believe that in the case of some norms we cannot wait for a significant time for substantial and nearly universal state practice to accrete. Thus, for example, the advent of nuclear weapons that could be launched from space necessitated the rapid establishment of a rule prohibiting such weapons in space and binding all space-faring nations.[39] Another example of customary norms that have developed quickly involves the continental shelf. It is generally accepted that rules on jurisdiction of states over the shelf evolved rapidly after the Truman Proclamation of 1945.[40]

36 On so-called modern theories of customary international law, see, for example, Anthea Elizabeth Roberts, "Traditional and Modern Approaches to Customary International Law: A Reconciliation," 95 AM. J. INT'L L. 757 (2001).

37 *See* MICHAEL P. SCHARF, *CUSTOMARY INTERNATIONAL LAW IN TIMES OF FUNDAMENTAL CHANGE: RECOGNIZING GROTIAN MOMENTS* (2013).

38 *Id.* at 8.

39 On the rapid development of customary international space law as a "Grotian Moment," see *id.* at 123–37.

40 *See id.* at 107–122.

Similarly, advocates of modern views have argued that *opinio juris* can likewise develop rapidly. They observe that since the creation of the UN in 1945, its organs, and especially the General Assembly, can provide a centralized forum for the expression of views of states about the existence or desirability of particular legal rules. Thus, General Assembly resolutions can, under certain circumstances, serve as a kind of "shortcut" in evidencing *opinio juris*. According to many commentators, it is no longer necessary to pore over diplomatic documents and statements of governments, state by state, and establish that these many documents, over some length of time, evidence a view that particular rules are law. Rather, a single General Assembly resolution, or a series of them, can provide the same level of evidence of government views in "one fell swoop." Even the ICJ has emphasized the ability of General Assembly resolutions to rally and encapsulate *opinio juris*. For example, in its judgment in the 1986 *Nicaragua Case*, the Court affirmed that

> *opinio juris* may, though with all due caution, be deduced from, *inter alia*, the attitude of the Parties and the attitude of States towards certain General Assembly resolutions.... The effect of consent to the text of such resolutions ... may be understood as an acceptance of the validity of the rule or set of rules declared by the resolution by themselves. The principle of non-use of force, for example, may thus be regarded as a principle of customary international law.[41]

Some commentators have further allowed for the rapid development of customary norms by discounting either one of the two elements. Some have taken an "agnostic" position about which is the more important, arguing, instead, that increased evidence of consistent state practice can compensate for little evidence of *opinio juris*, or conversely that significant evidence of *opinio juris* can compensate for a paucity of evidence of consistent state practice. This view is represented by Frederick Kirgis's famous "sliding scale" theory.[42] This kind of approach can allow for the speedier recognition of customary law norms to the extent that it minimizes the need to establish longstanding practice or *opinio juris*, as the case may be.

Of course, other commentators have systematically diminished the importance of a particular element. Their single-minded focus on one element can permit the quicker recognition of customary law rules. For example, the International Law Association has taken the position that evidence of *opinio juris* is not essential to establish a customary law norm.[43] That may allow a norm involving state practices

[41] Military and Paramilitary Activities in and against Nicaragua (*Nicar. v. U.S.*), 1986 I.C.J. Rep. 14, 99–100, para. 188 (June 27) [hereinafter *Nicaragua Case*].

[42] *See* Frederick L. Kirgis, Jr., "Custom on a Sliding Scale," 81 AM. J. INT'L L. 146 (1987).

[43] *See* International Law Association, London Conference (2000), Committee on the Formation of Customary (General) International Law, *Final Report of the Committee: Statement of Principles Applicable to the Formation of General Customary International Law*, sect. 1, Commentary, para. (b)

that accumulate rapidly to be recognized even though there is little evidence of *opinio juris* in favor of the norm. Conversely, some scholars have argued that the essence of customary law is *opinio juris*, and that state practice is either entirely unnecessary to prove, or that it at least serves as desirable, but not essential, evidence of *opinio juris*.[44] John Tasioulas, among others, appears to adopt this view in his contribution to this volume.[45] Such a view means that customary norms can be recognized as soon as there is sufficient evidence of *opinio juris*, and without waiting for concordant state practice to accumulate.[46] Moreover, this evidence of *opinio juris* can itself appear rapidly, perhaps in the form of a single General Assembly resolution. Bin Cheng espoused this view, arguing that customary law could be created instantaneously with the appropriate evidence of *opinio juris*. He declared: "There is no reason why an *opinio juris communis* may not grow up in a very short period of time among all or simply some Members of the United Nations with the result that a new rule of international customary law comes into being among them."[47]

All of these approaches have found favor in various judicial opinions, including those issued by the ICJ. For example, the ICJ, while doctrinally adhering to the definition of customary international law as arising from consistent practice and *opinio juris*, has in dictum recognized that practices need not be of long duration. In the 1969 *North Sea Continental Shelf Cases* it affirmed:

> Although the passage of *only a short period of time* is not necessarily, or of itself, a bar to the formation of a new rule of customary international law ... an indispensable requirement would be that within the period in question, *short though it might be*, State practice, including that of states whose interests are specially affected, should have been both extensive and virtually uniform in the sense of the provision invoked.[48]

In the 1986 *Nicaragua Case*, the ICJ ruled that a state has no right under customary international law to participate in collective military action based on a right of collective self-defense in response to an opposing military activity falling short of an armed attack.[49] In doing so, it relied primarily on Articles 2(4) and 51

(4) (affirming that "it is not *usually* necessary to demonstrate the existence of the subjective element [*opinio juris*] before a customary rule can be said to have come into being") (emphasis in original).

44 *See, e.g.*, Bin Cheng, "United Nations Resolutions on Outer Space: 'Instant' International Customary Law?" in INTERNATIONAL LAW: TEACHING AND PRACTICE 237 (Bin Cheng ed., 1982); ANDREW T. GUZMAN, HOW INTERNATIONAL LAW WORKS: A RATIONAL CHOICE THEORY 200 (2008).

45 *See* John Tasioulas, "Custom, *Jus Cogens*, and Human Rights" (in this volume).

46 *See also, e.g.*, GUZMAN, *supra* note 44, at 200 ("If CIL requires only *opinio juris*, then customary rules can change as quickly as *opinio juris* changes").

47 Cheng, "United Nations Resolutions on Outer Space," at 252.

48 North Sea Continental Shelf Cases, 1969 I.C.J. Rep. 3, 43, para. 74 (emphasis added).

49 *See Nicaragua Case*, 1986 I.C.J. Rep. 14, 99–103, paras. 188–93; 110–11, paras. 210–11.

of the UN Charter and two UN General Assembly resolutions,[50] The decision was widely criticized by commentators for focusing solely on *opinio juris*, as evidenced by the UN Charter and the UN resolutions, and ignoring state practice – including a long history of the use of forceful "reprisals" against low-scale military and terrorist activities not rising to the level of an armed attack.[51] The court evidently took the position that the UN Charter and the UN resolutions had demonstrated a clear *opinio juris* that changed the prior practice, thus resulting in the relatively rapid formation of a new, more prohibitive, customary law rule.

Similarly, some international criminal tribunals have invoked customary law primarily based on treaties and resolutions as evidence of *opinio juris*, and without imposing strict requirements for a showing of longstanding and consistent state practice. For example, in the case of *Prosecutor v. Kupreškić*, decided in 2000, the trial chamber of the International Criminal Tribunal for the former Yugoslavia (ICTY) found that there was a customary law prohibition of reprisals against civilians, despite a paucity of state practice.[52] Earlier, in its 1995 decision in *Tadic*,[53] the ICTY had likewise expanded the scope of customary international humanitarian law relating to noninternational armed conflicts based primarily on *opinio juris* (and moral considerations). As both Larissa van den Herik and Monica Hakimi point out in their chapters for this volume, while these decisions have been widely criticized, they underscore a trend in judicial decision making toward focusing on *opinio juris* rather than practice.[54]

Some theorists of international law have, moreover, argued that the customary lawmaking process – whether based on state practice, *opinio juris*, or both – should be opened up to non-state actors to take account of the important role played by these actors in international affairs. For example, Jordan Paust has affirmed that, "contrary to false myth perpetrated in the early twentieth century, the subjective element of customary international law (*i.e.*, *opinio juris* or expectations that something is legally appropriate or required) is to be gathered from patterns of generally shared legal expectation among humankind, not merely among official State elites."[55] Tasioulas in his chapter similarly argues that the practices and *opinio juris* of non-state actors

[50] *See* Declaration on Principles of Law Concerning Friendly Relations and Co-Operation Among States in Accordance with the Charter of the United Nations, G.A. Res. 2625 (XXV) (1970); Definition of Aggression, G.A. Res. 3314 (XXIX) (1974).

[51] For a representative critique of the opinion, see Anthony D'Amato, "Trashing Customary International Law," 81 AM. J. INT'L L. 101 (1987).

[52] *See Prosecutor v. Kupreškić*, 14 January 2000, IT-95-16-T, para. 527, available at http://www.refworld.org/docid/40276c634.html.

[53] *See Prosecutor v. Tadic*, 2 October 1995, Case No. IT-94-1-1, Decision on Defence Motion for Interlocutory Appeal on Jurisdiction.

[54] *See* van den Herik, *supra* note 3; Monica Hakimi, "Custom's Method and Process: Lessons from Humanitarian Law" (in this volume).

[55] JORDAN J. PAUST, *INTERNATIONAL LAW AS LAW OF THE UNITED STATES* 4 (2d ed. 2003).

should be taken into account where appropriate.[56] And Hakimi notes that customary international lawmaking in the field of international humanitarian law includes claims by a wide variety of non-state actors with their own normative agenda, and that these claims may ultimately influence the evolution of customary law norms.[57]

Disadvantages of Modern Views

There is no doubt that these modern theories introduce flexibility into the recognition of customary international law. They all allow it to adapt to changing circumstances far more quickly than application of the traditional model. Treaties can take years, if not decades, to negotiate; they often fail to "keep up" with the needs of the time. In the meantime, states can be bereft of legal rules to guide their behavior. Modern views of customary law allow it to be created and be modified rapidly to fill this void. For example, in the area of space law, states rushed to adopt a rule that space can only be used for "peaceful purposes"; arguably, this rule became part of customary law in a short time frame, despite the fact that only a few states had the capacity to send objects into orbit.[58]

On the other hand, all of the modern views have certain weaknesses. Most importantly, they can lead to uncertainty about the existence and content of particular norms of customary law. Without certain safeguards, they could result in violations of the principle of *nullem crimen sine lege* in the application of international criminal law. Furthermore, the new theories can make it more difficult to separate legal norms from moral norms – or law as it is (*lex lata*) from law as it ought to be (*lex ferenda*). There can be legitimate concerns that these theories, while allowing the "law" to change more easily, mask moral or political agendas on the part of those scholars or practitioners who promote them. They can result in what legal scholar Fernando Tesón has called "fake custom."[59] In effect, these theories can lead to the claim that new norms are customary law even though the norms constitute nothing more than a legal "wish list" on the part of the proponents (which can include certain governments).

Moreover, some of the specific new views about how customary law can evolve exhibit their own particular vulnerabilities. The view according to which both elements of consistent practice and *opinio juris* are required, but can be generated in a short period of time, may seem like the most benign of the modifications of

56 *See* Tasioulas, *supra* note 45.

57 *See* Hakimi, *supra* note 54.

58 *See, e.g.*, Frans G. von der Dunk, "Customary International Law and Outer Space," in REEXAMINING CUSTOMARY INTERNATIONAL LAW (Brian D. Lepard ed., forthcoming 2016).

59 *See* Fernando R. Tesón, "Fake Custom," in REEXAMINING CUSTOMARY INTERNATIONAL LAW (Brian D. Lepard ed., forthcoming 2016).

the traditional view and the one most likely to maintain its advantages. However, this view suffers from the fact that it continues to perpetuate the paradox of *opinio juris*. It also has the potential to allow modest changes in practice and rather thin evidence of *opinio juris* to be used to declare the existence of a new customary norm. It is noteworthy that the ICJ has cautioned against the easy extrapolation of customary law from prior treaty norms, affirming that while treaty norms might evidence *opinio juris* and help create customary law over time, "this result is not lightly to be regarded as having been attained."[60]

Furthermore, the modified traditional view, by requiring at least some changes in state practice, can still decline to recognize new norms that win wide, if not universal, support among states, simply because their practice has not yet "caught up" with these norms that they clearly endorse. Human rights norms offer a prime example of this problem. States may use treaties or declarations to articulate new human rights norms they intend to be universally legally binding, thus serving as clear evidence of *opinio juris*. At the same time, practice may continue to lag, and may not even have changed at all immediately prior to or after adoption of the treaty or declarations evidencing the *opinio juris*. Thus, even the more flexible two-element view shares some of the change-inhibiting features of the traditional view.

The view under which only consistent state practice is required for a new norm to form, without evidence of *opinio juris*, runs into a number of hurdles. Most importantly, as emphasized above, state practice always needs to be interpreted. Practice itself is not a norm, and any given practice may be consistent with a variety of norms, many of them contradictory. To illustrate, let us return to the prohibition of torture. It appears to be the case, especially in the post–9/11 world, that many states have sometimes employed torture. At the same time, they have laws against it and it is prohibited in international human rights instruments. And there are many punishments inflicted that fall short of torture. So which practice "counts"? It is not easy to say. One might look at the widespread practice of torture and declare that a new norm has evolved allowing states to use it in extreme circumstances, particularly against suspected terrorists. Or one might infer a rule that torture is always allowed if states deem it useful. Or one might characterize the widespread use of torture as simply the rampant violation of an absolute rule against it. We need some concept of *opinio juris* to tell us which rule is most defensible.

Another failing of the "state practice only" school of thought is that it can make existing practice-based norms particularly difficult to change – not easier to change. Why? Because if one does not take *opinio juris* into account, any modification of an existing widespread practice could be viewed as a violation of that practice. This is not necessarily the case if we factor *opinio juris* into the customary law equation.

60 North Sea Continental Shelf Cases, 1969 I.C.J. Rep. 3, 41, para. 71.

That is because the new practice might well be endorsed by various evidences of *opinio juris*. This endorsement would mean the practice could be viewed as in conformity with a new or revised norm, rather than simply a violation of the existing norm.

On the other hand, we cannot take the position that there is really no such thing as a violation because in apparently violating an existing customary norm a state is always making a "bid" for a new norm.[61] We have to evaluate that apparent violation in a wider context, and ask such questions as how the state itself views its own conduct and how other states react to it.[62] These views are evidence of *opinio juris*. The key point is that we cannot evaluate the significance of the new practice without reference to *opinio juris*. In short, while "state practice only" theories superficially appear to allow for the more dynamic evolution of customary law, they can lead to confusion about the content of new norms because of the need to interpret practice. And they can actually impede the formation of new or revised norms.

Theories that emphasize *opinio juris* and downplay consistent practice would appear to cure these defects. To the extent *opinio juris* clearly states a rule, there is no problem interpreting practice. And similarly, if ample evidence of *opinio juris* endorses a new practice, the practice does not have to first overcome the challenge of being labeled a violation of a preexisting customary norm. That is to say, new *opinio juris* could precede new state practice – contrary to the traditional view that *opinio juris* can only endorse a preexisting practice.[63] This would seem to allow for much greater flexibility in the evolution of customary international law.

On the other hand, "*opinio juris* only theories" exhibit their own unique weaknesses. First, if they adopt the traditional definition of *opinio juris*, they are marred by the paradox of *opinio juris* just described. This paradox can impede recognition of new norms if courts take the traditional definition of *opinio juris* seriously.

Another potential drawback of these views is that they can hinder the formation of new norms through changing state practices if they insist on relying on "old" evidence of *opinio juris*. That is, in some cases new norms are created through the way states behave, which can be strong evidence of a new *opinio juris*, while for various reasons states may be slow or reluctant to endorse rhetorically a revised rule consistent with their new behavior. One example is the concept of humanitarian intervention – the right of one or more states to intervene militarily to protect victims of mass atrocities in a third state where the third state is unwilling or unable

[61] Anthony D'Amato has argued in this vein that "an 'illegal' act by a state contains the seeds of a new legality." D'AMATO, *supra* note 28, at 97.

[62] *See* LEPARD, *supra* note 4, at 278.

[63] On this point, see *id.* at 277.

to protect them or is itself the perpetrator of the atrocities. In the last twenty-five years the global community has witnessed a number of potential examples of such intervention, including in Serbia and Kosovo in 1999 by NATO forces without Security Council approval, in Libya in 2011 with the blessing of the UN Security Council, and in 2014 and 2015 in Syria and Iraq against the Islamic State without Security Council endorsement. Could this pattern of interventions have already resulted in a new customary rule? This is at least plausible, even though UN member states have been manifestly reluctant to articulate any such new norm that would carve out an exception to the rules in Articles 2(4), 39–42, and 51 of the UN Charter. These rules provide that state uses of force on the territory of another state are permissible only in self-defense against an armed attack or as part of Security Council-authorized action. Although the UN General Assembly has accepted the idea of a "responsibility to protect" victims of mass atrocities,[64] it has stopped well short of endorsing any general doctrine conferring a unilateral right of humanitarian intervention, and few states have lent their support to such a doctrine, as well. Without prejudging the issue, here, then, is an example where practice may well "lead the charge" toward creation of a new customary law norm, while formal evidence of *opinio juris* lags.

Perhaps the biggest problem with *opinio juris*-focused theories is that they risk treating *lex ferenda* as *lex lata*. They are particularly prone to "wishful thinking." Without the confirmation of consistent state practice, there is at least the possibility that states do not actually endorse the norm as a legal norm that should or does bind them; it may be merely aspirational.

The weaknesses of theories that emphasize either state practice or *opinio juris* also besmirch the "sliding scale" theory propounded by Kirgis and others, for similar reasons. Such a theory invites confusion about how customary norms evolve. For example, just what is "ample" state practice that can compensate for "thin" *opinio juris*? This is not clear. Moreover, state practice of one form or another (including abstentions from acting) can always be characterized as "widespread." So just when is strong evidence of *opinio juris* required?

Conversely, it is not always the case that little evidence of consistent practice need be demonstrated if there is strong *opinio juris*. For example, one could argue that there is strong *opinio juris*, based on a number of General Assembly resolutions, but especially Article 24 of the Universal Declaration of Human Rights, in favor of a human right to take holidays with pay and a binding obligation on the part of all states to require that workers enjoy such a right.[65] But the practice on this point

64 *See* United Nations World Summit Outcome Resolution, G.A. Res. 60/1 (2005), paras. 138–39.

65 Article 24 of the Universal Declaration asserts: "Everyone has the right to rest and leisure, including reasonable limitation of working hours and periodic holidays with pay." Universal Declaration of Human Rights, G.A. Res. 217A(III) (1948), art. 24.

is conflicting; some states require paid holidays, while others do not. If we turn our back on this unclear record of state practice, we may be missing important information about states' true views concerning recognition, as customary law, of the norm endorsed by the *opinio juris*.

As to theories that introduce a role for the practice or opinions of non-state actors, there is no doubt that these actors – whether political groups, armed opposition groups, nongovernmental organizations, or intergovernmental organizations, among others – are playing a much more important and multifaceted role in international affairs. The question is whether their practices or views contribute to international law. As a social construct, international law is understood as a law created *by states*. To accept a coequal role in customary law formation or change by non-state actors would be to transform international law into some other kind of law. The better view is that an approach to customary international law must be state centered, as the traditional view presupposes,[66] while acknowledging that there can be important influences of non-state actors on the practices and beliefs of states that can affect the evolution of customary international law.[67] As noted, Monica Hakimi appears to endorse this latter perspective in her contribution.

CUSTOMARY INTERNATIONAL LAW AS A DYNAMIC PROCESS

This chapter now proposes a new approach that sees customary law as a dynamic process and attempts to integrate the advantages of both the traditional and modern views. This approach begins with the proposition that customary international law is, in essence, an informal method of lawmaking among states. We saw that customary international law arises over time as states come to believe that certain norms are desirable and act in conformity with those norms. Sometimes articulation of the norm precedes the behavior, but more commonly there is a coincidence of behavior that in time results in more conscious recognition of a norm requiring or permitting it.

A New Approach to Opinio Juris

Here again, however, we encounter the problem of the paradox of *opinio juris*. The traditional definition works well enough for existing norms, but is indeed problematic in justifying the recognition of new norms if courts apply it rigorously.

66 In keeping with this view, Michael Wood concludes that "the requirement, as an element of customary international law, of a general practice means that it is *primarily the practice of States* that contributes to the creation, or expression, of rules of customary international law." Michael Wood, Draft Conclusion 5, in *Second Report*, *supra* note 14, at 66 (emphasis added).

67 *See* LEPARD, *supra* note 4, at 185–87.

Accordingly, consistent with the views of certain other scholars, the chapter proposes that in the case of new customary international law norms, *opinio juris* be defined as a general belief shared by states that *it is desirable, now or in the near future, to have an authoritative legal rule prescribing, proscribing, or permitting certain conduct.* In other words, the focus is on beliefs about the desirability of a new rule rather than beliefs that a particular rule already exists. The conception proposed here is one in which states are constantly evaluating what rules should govern their relations and behavior outside of contractual obligations formed through treaties. Accordingly, states' beliefs about what the law *should* be can help the law change.

Moreover, the chapter maintains that in ascertaining state beliefs, decision makers must take into account particular "fundamental ethical principles" that have been recognized by states themselves in a variety of modern-day instruments. Fundamental ethical principles are defined for this purpose as principles identified in these instruments, including the UN Charter and the Universal Declaration, that are in turn rationally related to a preeminent ethical principle of "unity in diversity." This principle of unity of diversity affirms that "all states and individuals form part of global communities of states and human beings that ethically should be united at the same time that they take pride in their fundamental autonomy and diversity of culture, ethnic origin, religion, and belief."[68]

For example, the Universal Declaration supports the concept of unity in diversity by referring in its preamble to "recognition of the inherent dignity and of the equal and inalienable rights of *all members of the human family*."[69] The declaration also endorses respect for individual and cultural diversity, protecting freedom of belief, freedom of expression, freedom of association, and freedom to participate in the cultural life of one's community.[70] In short, the declaration promotes as a core value "unity in diversity." A number of principles merit the status of "fundamental ethical principles" that are logically related to this principle of unity in diversity. These include principles of human dignity and human rights, significant state autonomy, a trust theory of government, limited state sovereignty, the right to freedom of moral choice, punishment of criminals, open-minded consultation, the existence of a global community of states that promotes fundamental ethical principles, and the duty of states to honor treaties.[71] It should be emphasized that these are ethical principles, *not* norms of customary international law, although they may be relevant in determining whether or not particular norms of customary law should be recognized.

68 *Id.* at 8.
69 Universal Declaration, preamble (emphasis added).
70 *See id.*, arts. 18–20, art. 27, para. 1.
71 *See* LEPARD, *supra* note 4, at 82–92.

This reformulation of *opinio juris* has a number of merits. Most importantly, it gives *opinio juris* a dynamic quality, allowing explicitly for the recognition of new norms and the revision or termination of existing ones, without any false beliefs on the part of states. Even if the practical impact on judicial or government decision making of the current conception of *opinio juris* is difficult to gauge, there is no doubt that at the margins a requirement that states believe a norm already to be the law can be a disincentive to the recognition of new or modified law. This new concept of *opinio juris* removes this barrier to dynamism in the evolution of customary law. Other scholars have similarly suggested that the traditional definition of *opinio juris* needs to be modified along the lines suggested here. For example, in his chapter in this volume, Curtis Bradley argues that a rule of customary international law "can be recognized when it is evident – from state practices, statements, and other evidence – that the rule is something that the relevant community of states *wishes to have as a binding norm going forward* and that it is socially and morally desirable."[72]

There is another critical element to the theory proposed here – namely, that *opinio juris*, rather than state practice, is at the center of customary law, and that consistent state practice is evidence of *opinio juris*, but not an essential requirement in its own right for every type of norm. Indeed, one element of this theory is that it distinguishes different types of norms designed to solve different types of problems, rather than adopting a "one size fits all" approach. It draws distinctions regarding the amount of consistent practice that should be demanded as evidence of *opinio juris* based on these different problem types. For example, it distinguishes norms designed to solve coordination problems (in which case practice is normally very important evidence of *opinio juris*, since coordination without coordinated practice is impossible to achieve) from norms designed to uphold fundamental human rights (in which case practice is less important evidence of *opinio juris* because any practice of respecting human rights furthers the moral goals of the norms).[73] That said, in most cases lawyers and jurists would be hard pressed to conclude that there is sufficient *opinio juris* (as redefined) in favor of a putative norm in the absence of any state practice in support of it.

This perspective obviously resonates with some of the *opinio juris* only theories described, but it is different from them in a number of important respects, as discussed next. Most importantly, it sets a high bar for finding the requisite *opinio juris* in favor of a new or revised customary norm, one carefully formulated to distinguish *lex lata* from *lex ferenda*.

[72] Curtis A. Bradley, "Customary International Law Adjudication as Common Law Adjudication" (in this volume) (emphasis added).

[73] *See, e.g.*, LEPARD, *supra* note 4, at 122–26.

Importance and Role of Opinio Juris

Before further exploring this conception of customary law as a dynamic process, some further explanation of the emphasis on *opinio juris* is necessary. At first blush, it seems to fly in the face of the normal understanding of "customary" international law, which apparently originates with "customs." To take the "custom" out of customary international law would seem to convert it into a different animal altogether – at best, to "general principles of law" described in Article 38(1)(c) of the ICJ Statute, or at worst, to an indeterminate category of norms based on wishful thinking, but detached from the actual practices of states.

As a matter of historical fact, customary law has typically been born out of customs among peoples and among states. However, these customs did not become *law* unless and until those peoples or states *recognized them as binding* and *articulated a norm that explained and justified them*. That process of societal recognition of the norm became the "tipping point" – the critical and essential factor – that led to the recognition of a custom as something more than a mere coincidence of behavior, and indeed as the expression of a legal norm. If our focus is on identifying legal norms, then, this critical belief, or *opinio juris*, must be viewed as the most important component of customary law.

Of course, this focus on *opinio juris* is totally consistent with the traditional bipartite definition of customary law. It does not by itself negate the relevance of consistent practice. As just pointed out, the fact is that historically much of customary law did originate with widespread local or international practice. One reason is that many foundational norms within a local or national society or within the global community of states are coordination norms that depend on consistent state practice to establish a desired convention that solves the coordination problem. These include, for example, many norms involving international transportation and trade.

However, as the society of states has developed, it has moved beyond simple coordination norms and begun to address a series of more complex problems, including protection of the environment and fundamental human rights, among many others. These problems do not involve simple coordination dilemmas; they may involve prisoners' dilemmas and they have a strong moral content. And like many morality-based norms, their demands almost by definition will exceed current practice. That is to say, one feature of a moral norm is that its very purpose is to require behavior that is not motivated by self-interest alone and that asks more of states than what they are already practicing. To require consistent state practice *prior to* recognition of these norms as binding law might well prevent them from ever being recognized as law.

Some observers might say this is a good thing; after all, such moral norms not supported by practice are the classic category of *lex ferenda* – norms that ought to be the law, but are not yet the law. And to recognize them as law before states "put their money where their mouth is" degrades the very concept of law. That is no doubt a legitimate concern. The problem is that taken to its logical conclusion this critique could prevent the formation of virtually all moral-based norms at the global level. To return to the example of torture, if almost every state tortures some of the time, how could a customary law prohibition of torture ever be recognized under a view demanding a widespread and consistent state practice of *not* torturing?

Moreover, the drafters of Article 38 of the Statute of the PCIJ appear to have been well aware of these difficulties. Indeed, the text itself of Article 38(1)(b) supports the evidentiary role for practice suggested here. First, that text plainly refers to custom as "evidence" of a "general practice accepted as law." The word "evidence" appears explicitly in the Statute. It is instructive to note, furthermore, that an early draft of Article 38(1)(b) prepared by Baron Descamps of Belgium, which became the basis of the final version, did not contain the word "evidence." The drafters consciously added this word to the final version, suggesting its importance.[74] Furthermore, while the clause also refers to a "general practice accepted as law," thus appearing to require a "general practice," the clear import of this phrase, alongside the word "custom," is to emphasize the need for *opinio juris* – acceptance as law. And as Curtis Bradley underscores in his chapter, the drafters may well have been influenced by the "historical school" of jurisprudence propounded by Friedrich Carl von Savigny, according to which custom was evidence of a deeper and preexisting norm or obligation.

Naturally, the drafters had in mind that normally there would be some consistent practice. However, the language they adopted provides strong support for the conclusion that they viewed the "heart" of customary law to be acceptance of a norm as law (*opinio juris*) and believed that the primary function of a custom, or general practice, was to evidence this view. While not a model of clarity, the language they chose was an advance over earlier, more simplistic formulations of the "state practice plus *opinio juris*" concept, and one intended to clarify the evidentiary role of practice. It is also worth underscoring that the drafters were not strict "positivists"; they consciously adopted the language in Article 38(1)(c) referring to "general principles of law" that could exist without the need for practice. This at least opens the door to a more flexible interpretation of the language they agreed upon in Article 38(1)(b), as proposed here. It also implies the possibility of referring to ethical principles that were widely accepted at the time as "general principles of law," including the principle, for example, of "good faith."[75]

74 *See id.* at 129.

75 On good faith as a general principle of law recognized by the ICJ, see JAMES CRAWFORD, *BROWNLIE'S PRINCIPLES OF PUBLIC INTERNATIONAL LAW* 36 (8th ed. 2012).

Several other points are important to note about the evidentiary role of state practice. This role is implied in actual decisions of the ICJ. Thus, in the *North Sea Continental Shelf Cases*, the Court stated that a practice must be "*evidence* of a belief that this practice is rendered obligatory by the existence of a rule of law requiring it. The need for such a belief, i.e., the existence of a subjective element, is implicit in the very notion of the *opinio juris sive necessitatis*."[76] In many cases, like the *Nicaragua Case*, the Court has emphasized *opinio juris* without any serious inquiry into state practice, apparently treating the latter as only one source of evidence of the former.

Furthermore, both international and national courts in general, when considering the existence of a customary norm, pay far more attention to *opinio juris* than to state practice, as documented empirically by Stephen Choi and Mitu Gulati's chapter in this volume.[77] Moreover, Choi and Gulati's preliminary findings suggest that courts tend to apply something like the normative version of *opinio juris* advocated here. As they report, courts analyzing customary law "are generally engaged in a forward-looking or aspirational exercise." In other words, regardless of the traditional doctrine, what courts are doing in practice is treating *opinio juris* as the core of customary law, and state practice as important, but not always essential, evidence of that *opinio juris*. For all these reasons, other scholars, such as Andrew Guzman, have similarly urged that state practice "is best considered as evidence of *opinio juris*."[78]

Finally, it should be emphasized that this view is simply an *interpretation* of the traditional doctrine long propounded by jurists and scholars; it does not dispense wholesale with the concept of either state practice or *opinio juris*. It represents a refinement of the conventional view informed by judicial and practical experience with applying it, and that better accords with what courts actually do than does the old doctrine. All international legal doctrines have been revised, and should be susceptible to revision and refinement, over time, better to serve the needs of states; that is an undeniable process that has kept international law relevant over the centuries. To take but one example, the doctrine of absolute state sovereignty has, over the last 100 years, gradually been refined so that sovereignty, while important, is no longer unqualified. One might even go so far as to argue that the very science of jurisprudence involves constant reexamination and refinement of existing legal doctrine.

76 North Sea Continental Shelf Cases, 1969 I.C.J. Rep. 3, 44, para. 77 (emphasis added).

77 Stephen J. Choi & Mitu Gulati, "Customary International Law: How Do Courts Do It?" (in this volume).

78 GUZMAN, *supra* note 44, at 200. In their contribution to this volume, Guzman and Hsiang adopt the even stronger view that "state practice is irrelevant.... The only place for state practice is as evidence that states hold some kind of belief about a rule." Andrew T. Guzman & Jerome Hsiang, "Reinvigorating Customary International Law" (in this volume).

In short, the view advanced here is that the society of states is now grappling with such complex problems, many involving moral considerations, that the old method of customary lawmaking in which widespread practice always precedes *opinio juris* cannot be viewed as a requirement for every type of customary norm. Rather, customs are best viewed as evidence of *opinio juris*, the weight of which will depend on the nature of the problem states are trying to solve. And at the end of the day, a customary norm is created by the sincere belief by the generality of states that the norm ought to be instituted as an authoritative legal norm now or in the near future.

Preserving the Benefits of the Traditional View

The proposed reformulation of *opinio juris*, and conception of state practice as evidence of *opinio juris*, incorporate many of the benefits of the traditional view of customary law just explored. Most importantly, the traditional view gives customary law a rootedness that allows state expectations to converge around norms and puts states on fair notice about what is expected of them under those norms. The strict requirements proposed here for *opinio juris* are intended to fulfill similar objectives. For example, the definition of *opinio juris* looks to the beliefs of states, not those of scholars, nongovernmental organizations, or judges. Thus, the focus is on what states themselves believe should be the rules and not on the wishful thinking of others. This feature of the proposed definition is similar to that of the traditional view.

Second, the definition requires that states "generally" believe that a given rule is desirable. This means that there must be a minimum of majority support among states for a rule to be created or changed. This is concordant with the traditional requirement of "consistent" or "widespread" practice and *opinio juris* among states. This requirement prevents a minority of states from changing an established rule (unless they ultimately win over a majority). However, it is appropriate to "weight" states' views based on a variety of factors, including the extent to which they take into account views of their citizens as part of their policymaking.[79]

Third, the definition requires that states generally believe that a rule would be desirable to implement "now or in the near future" – not at some distant time. This requirement is intended to help distinguish *lex lata* from *lex ferenda*. States must believe that they should be subject to the rule now or soon; that is, they must be willing to abide by it in the present. This is an important qualification that may eliminate many aspirational norms from recognition as new customary law.

Fourth, states must believe that it is desirable to implement an "authoritative" legal rule. This means they believe that it is appropriate to limit their own decision making in some way. The authoritative character of the rule might be binding (and

[79] *See* LEPARD, *supra* note 4, at 155–56.

thus preempt states' own consideration of how to act), or it might be persuasive in nature (requiring states to give great weight to the norm in their decision making). In either event, no customary rule is created under this test if states merely believe that "it is desirable for states to act in a particular way." To give an example, no customary rule on limitation of greenhouse gas emissions can arise under this test for *opinio juris* simply because states believe it is desirable for them to take measures to reduce these emissions. Rather, they must believe it is desirable to constrain their own decision making and force themselves to either limit emissions or give great weight to the limitation of emissions in their policymaking. Again, this sets a high barrier to recognition of new or revised customary law norms.

Finally, the test requires that states believe it is desirable to implement an authoritative "legal" rule – not a moral, social, or political one. That is an important qualification. It means that states must believe that there should be some legal remedy for states, individuals, or other persons who are victims of violations of the rule. In many cases, states well might endorse a rule as a moral one (as is the case with many rules or principles upheld in UN General Assembly resolutions), but not as a legal one.

All of these requirements serve as a "check" on what might be called "reckless" lawmaking under the guise of customary law. All of them introduce a key element of objectivity into what otherwise might be a wildly subjective enterprise. Taken together, they should help prevent abuses of this new interpretation of the two-part test for customary law. All of them also help ensure that customary law norms recognized by the definition have a quality of stability, while allowing for change. And customary law norms cannot change without the concurrence of the generality of states. In this sense, it honors states' legitimate expectations and does not thrust upon them rules not of their own making.

Moreover, while at first glance this definition may appear very subjective compared to, for example, a mechanical evaluation of state practice, thus placing states in a situation of uncertainty about the rules that bind them, in fact it is "fairer" to states than the traditional definition of customary international law. Why? Because the strict requirements just referred to, and the insistence of the definition on clear evidence of *opinio juris*, mean that states can more easily identify norms that so qualify. Indeed, the relegation of state practice to an evidentiary role is a benefit to states in this regard. They can rely, in general, on what are generally public and easily acceptable sources of evidence of *opinio juris*, including UN General Assembly resolutions, treaties, and public declarations of other states. Of course, they must also consult state practice, but they do not face the hurdle of having to "prove" the existence of some undefined quantum of practice as an essential element of recognizing a customary rule binding on them, as under the traditional view.

Avoiding the Disadvantages of the Traditional View

At the same time, the proposed theory avoids the pitfalls of the traditional view. Most importantly, it views customary law as a dynamic process, and gives states the opportunity to change existing customary law rather quickly if they view it as so desirable. States do not have to wait for decades or centuries to demonstrate some longstanding practice and *opinio juris* before they can benefit from a new rule. A new rule can emerge simply from their views about the desirability of that rule – but only with all of the safeguards just described.

The approach also takes into account the context of a particular problem area as perceived by states, rather than applying a blanket doctrine bluntly to all issues. It demands that we ask whether states reasonably perceive an issue to constitute a simple coordination problem, or a prisoners' dilemma, or one in which moral values are directly affected. It makes certain presumptions about states' views concerning the desirability of a legal norm based on the nature of the problem. For example, in the case of a prisoners' dilemma, it presumes that states desire a legal norm to prevent defectors, but only if they can be assured of adequate enforcement, and only if there is a high degree of consensus in favor of the proposed legal norm.

To take one example, the issue of climate change could reasonably be perceived as a prisoners' dilemma, as noted by Niels Petersen in his chapter.[80] This might mean that states believe a legal rule regulating emissions is desirable apart from a treaty, but only if there is adequate enforcement. Without enforcement and supervision (such as that provided by the Kyoto Protocol), states may not in fact want binding limitations on emissions. All evidence of state views must be considered, including views expressed during negotiations on the Kyoto Protocol and on its implementation.

Of course, in this respect, the theory proposed in this chapter shares the virtues of the modern theories described here. It allows customary international law to respond to new technologies and solve new problems. It focuses on and strengthens the key advantage of customary law compared to treaties – namely, its flexibility and adaptability, not to mention its ability to bind all states other than states qualifying as persistent objectors.

Avoiding the Disadvantages of Other Modern Views

At the same time, the proposed theory seeks to remedy some of the deficiencies in modern views. The most promising modern view might be the one according to which both consistent state practice and *opinio juris* are required for a customary

80 *See* Niels Petersen, "Customary International Law and Public Goods" (in this volume).

law rule to emerge or change, but both can arise over a much shorter period of time than was conventionally believed. This would certainly remedy the problem with the traditional view impeding the rapid development of norms. The difficulties with this modern view, however, include its insistence that there must be consistent state practice in every case, and its reliance on the traditional definition of *opinio juris*, which is overly broad and paradoxical. The proposed approach avoids these obstacles.

As already explained, an exclusive focus on state practice can lead to confusion about customary law norms, since practice must always be interpreted. And it can also impede the development of new norms by insisting that a change of practice must precede recognition of those norms. This is definitely not a requirement under the theory proposed here. New norms can be created simply through states' beliefs that they should be recognized, before practice changes. Of course, the *opinio juris* only theories share this benefit, too, and the dynamic approach advanced here might well be described as such a theory. There are a variety of ways, however, in which it is distinct from those approaches and can help overcome some of their weaknesses.

First, the dynamic approach avoids the paradox of the traditional view of *opinio juris*, which is usually adopted by proponents of these *opinio juris*-focused theories. The traditional definition can act as a brake on the recognition of new norms that states strongly desire to see implemented. A number of authors contributing to this volume allude to problems with the traditional definition of *opinio juris*, requiring states to act out of a belief that an international norm *already* binds them. For example, Larissa van den Herik believes that there can be no legitimate *opinio juris* supporting customary international criminal law because states do not adopt national criminal laws "with the belief that this is mandated by an international rule."[81] That may well be true, and it shows a problem with the traditional definition of *opinio juris*. On the other hand, consistent with the revised understanding of *opinio juris* proposed here, it is very possible that states believe that concepts in their national criminal laws *ought to be* the law internationally and bind other states as well as themselves.

Second, the proposed theory does not regard state practice as irrelevant to the determination of customary international law. Quite the contrary. In the case of most kinds of norms it will be very important evidence of state views. Thus, in this regard, the proposed theory is not an *opinio juris* only theory. This continued attention to state practice allows the theory to find a new *opinio juris* in cases where approaches that rely on traditional evidence of *opinio juris* might find it lacking.

For example, again without prejudging the issue, it is possible that the recent practice of humanitarian intervention, at least with some kind of direct or indirect

[81] *See* van den Herik, *supra* note 3.

endorsement by United Nations organs, might be good evidence of a new *opinio juris* allowing for such intervention under limited circumstances. This might be the case even though longstanding documents such as the UN Charter would appear to evidence an *opinio juris* against it. In other words, what states "really" believe should be the law may be better evidenced by their actions than their words. Indeed, it is important to examine all evidence of state views beyond just formal written documents such as the UN Charter or General Assembly resolutions, including the "action" of states tacitly accepting unilateral interventions by other states without protest, or even if they do protest, their "action" of expressing approval of the underlying humanitarian goals of the interventions.

Third, the definition of *opinio juris* proposed here contains many safeguards, as already explained. Existing *opinio juris*-based approaches may not share these protections. They may allow aspirational norms endorsed in UN General Assembly resolutions to be recognized as customary law even where there is clear evidence that states supporting the resolutions had no belief that the norms in them should be treated as authoritative legal rules now or in the near future. Thus, the theory prevents recognition of "fake custom" and rules that are mere wishes on the part of their advocates. At the same time, it maintains a distinction between customary international law and general principles of law, based primarily on the degree of generality of the rule in question and whether it establishes binding or persuasive obligations.[82]

Fourth, existing *opinio juris*-based approaches typically do not include an explicit and specific ethical test, as does the theory proposed here, although they may refer to ethics generally as a factor in finding *opinio juris*. This is the case with the approach proposed by John Tasioulas, both in his contribution to this volume and elsewhere.[83] By contrast, the view proposed in this chapter insists that an evaluation of state beliefs be made in the context of what this chapter has called "fundamental ethical principles" – allowing for the possibility that it is appropriate *not* to find an *opinio juris* in favor of a norm if it would directly contravene fundamental ethical principles. Thus, for example, the many resolutions adopted by the UN Human Rights Council calling for criminal laws prohibiting the "defamation of religion" should be discounted as evidence of *opinio juris* because they would violate the essential human right to freedom of expression.[84]

[82] For a fuller discussion of the relationship between customary international law and general principles of law, see LEPARD, *supra* note 4, at 162–68.

[83] *See, e.g.*, John Tasioulas, "Customary International Law and the Quest for Global Justice," in *THE NATURE OF CUSTOMARY LAW* 307, 310 (Amanda Perreau-Saussine & James Bernard Murphy eds., 2007).

[84] On the defamation of religion resolutions, see Brian D. Lepard, "Parochial Restraints on Religious Liberty," in *PAROCHIALISM, COSMOPOLITANISM AND THE FOUNDATIONS OF INTERNATIONAL LAW* 225, 231–32, 245–46 (M. N. S. Sellers ed., 2012).

In addition, the approach proposed here differs from that proposed by Tasioulas in that it continues to focus on the *views of states* about what norms should be law in the context of ethical principles endorsed by states themselves. Tasioulas argues that in evaluating *opinio juris* we should make reference to ethical principles that are objectively determined, apart from the views of states. After all, states might choose to endorse as "ethical" certain principles that are, according to some external yardstick, quite immoral (like an "ethical" norm of absolute state sovereignty that could allow for mass human rights violations). While Tasioulas' proposal might lead to recognition of a kind of objective "moral law," this would not be customary international law as historically understood. International law is itself a social construct; it is ultimately dependent on the views of states. Of course, states may opt to incorporate some kind of moral law into international law, as they arguably have done with respect to both customary law and general principles. Moreover, part of the proposed test for a "fundamental ethical principle" is the existence of a rational and objective relationship with the foundational principle of unity in diversity. Nevertheless, if we are to ascertain "international law," it is important to take into account in some way the attitudes of states – and even the concept of unity in diversity has been endorsed by them.

Furthermore, the theory advanced here addresses some of the deficiencies of sliding scale theories. It offers a coherent explanation of why the core of customary international law is *opinio juris*, and when and why state practice should be considered important evidence of *opinio juris*. It distinguishes among different categories of norms, rather than lumping them all together. More importantly, it better addresses the needs of states by asking important questions about why they believe a norm is desirable and whether or not they believe it ought to be instituted as an authoritative legal norm.

Finally, as noted, some scholars, such as Curtis Bradley in this volume, have courageously argued that we ought to formulate a doctrine of customary law "from scratch" based on what adjudicators actually do. Bradley proposes a "common law" model of customary international law finding, under which adjudicators' choices about how to interpret practice and *opinio juris*, and identify a customary law rule, "are shaped by assessments of state preferences as well as social and moral considerations."[85] Bradley's reasons for such an innovative proposal resonate with some of the arguments made here for a reinterpretation both of the *opinio juris* test and the state practice requirement. So how is the theory presented here different? One way is that it proposes an ethical background system for evaluating state beliefs about the desirability of norms and attempts to specify relevant ethical principles. While Bradley's approach allows for "moral considerations" to be taken into account

85 Bradley, *supra* note 72.

by an adjudicator, it does not give much detail about how the adjudicator should find, specify, or rank those moral considerations. Of course, identifying relevant ethical principles does not make the inquiry into the customary status of a norm easy, but it sharpens and focuses that inquiry.

Furthermore, the theory presented here entails an interpretation of existing customary law doctrine, thus preserving its character as *customary* law. It does not dispense with state practice altogether, instead viewing it as evidence of state beliefs about what the law is or should be. It maintains that we should not discard the bipartite definition altogether in favor of some wholly new model of customary law because the *opinio juris* concept is a critical one in the formation of law, and because state practice is, most often, a very important source of evidence about states' beliefs.

CONCLUSION

In short, the proposed approach does not merely "allow for" change in customary international law as a kind of safety valve, and only after major and enduring shifts in practice and *opinio juris*, as does the traditional view. Rather, it fundamentally conceptualizes customary international law as a dynamic process. It sees states as engaged in a constant dialogue about the rules that should govern their relations and behavior apart from treaties. Parts of this "dialogue" are nonverbal, and take the form of practice, which is why state practice should be viewed as evidence of *opinio juris*. But of course this dialogue also involves verbal exchanges. In every case, it is critical to evaluate the content of states' communications to determine their views about the desirability of recognizing a new norm, or modification of an existing norm, as an authoritative legal rule for all states now or in the near future. This involves a rigorous inquiry, we have seen, and high hurdles must be cleared to find that the requisite *opinio juris* exists.

Nevertheless, this conception of an ongoing dialogue can allow for a new dynamism in customary international law. It can permit new rules to be recognized quickly to solve urgent problems, a common goal of all the modern views described here. The proposed approach is more inclusive, recognizing a place for the views of all states, including the less powerful, and for opinions expressed in more representative bodies, such as the UN General Assembly. It thus acknowledges that customary international law can, and should, no longer be made just by the "great powers."

The theory recognizes, too, that states are engaged in dialogue and discussion of a wide range of issues, in multiple forums, and many types of state organs participate in these discussions and generate "practice."[86] Moreover, the fact that

[86] On the appropriate weight to be given to the views of organs of executive, legislative, and judicial branches, see LEPARD, *supra* note 4, at 171–80.

states participate in an ongoing dialogue against the background of fundamental ethical principles that they have generally accepted – and the fact that the proposed approach explicitly takes these principles into account – mean that customary law can progress more easily in an ethical direction.

The proposed approach can, it is hoped, rescue customary international law from the current crisis in which it finds itself. In the world of medicine, every medical crisis must ultimately be resolved: either the patient lives, or the patient dies. It is critical to apply the appropriate remedy – adopting a view of customary international law as a dynamic process – to ensure its survival and relevance.

4

Custom, *Jus Cogens*, and Human Rights

John Tasioulas

Immanuel Kant notoriously declared that it was a "scandal of philosophy" that it had not yet furnished us with a convincing proof of the existence of an external world. International lawyers have their equivalent occupational scandal: the failure to achieve clarity or consensus on the nature of customary international law. Custom, after all, is arguably the most fundamental source of international law, at least insofar as treaty law is itself embedded within a customary framework. This framework includes various principles bearing on the interpretation of treaties and arguably also the grundnorm of treaty law, *pacta sunt servanda*. Indeed, the international lawyer's scandal goes deeper. All of us, philosophers or not, standardly proceed on the basis that a world external to our senses exists. By contrast, assertions about customary international law are largely confined to international lawyers, although their being taken seriously occasionally has real practical consequences for others.

It is not enough to respond to this state of affairs with a knee-jerk pragmatism: the shop-worn thesis that customary international law works well enough "in practice" and so requires no explication "in theory." After all, this simply presupposes that we already know what customary international law is, and merely shifts attention to whether it "works." In any case, it is doubtful that anything can satisfactorily "work" in a discursive and legitimacy-claiming practice if its very nature remains stubbornly opaque or conceptually problematic. Equally, we should not be put off by the skeptical dogma that all of our moral-political ideas are infected with contradictions at their very core, so that the search for an explanation that makes good sense of them is doomed from the outset. Even the alluring consolations of intellectual resignation need to be earned by argument rather than mere fiat.

In this chapter, by drawing on, clarifying, and extending previous work, I try to sketch the argument that the pragmatists and skeptics take to be either unnecessary or impossible. I offer a moral judgment-based account (MJA) of customary international law, one that challenges the orthodox idea that there is a deep connection between custom and consent, and I mobilize the ensuing account in relation to human rights norms in particular.

THE NECESSITY FOR MORAL JUDGMENT

There is a perfectly intelligible sense in which all law, including customary international law, derives from what might loosely be called "practice." It is the product of what states and other agents actually do or refrain from doing, where this importantly includes the performance of speech acts that give expression to their objectives and beliefs. To this extent, *opinio juris*, as a factor in the genesis of customary international law, should not be contrasted with "practice," as if it denominated some occult phenomenon unfolding behind the scenes of ordinary human activity. This is not just for the quite general metaphysical reason that, as Wittgenstein put it, "an 'inner process' stands in need of outward criteria." It follows more directly from the public and intentional – the *positive* or *posited* – character of legislative activity. Law is paradigmatically created through publicly accessible acts that are undertaken precisely *as* law creating. Hence, all law is practice based in this wide sense.

Nonetheless, in seeking to make good sense of the orthodox understanding of customary international law reflected in Article 38(1) of the Statute of the International Court of Justice – according to which two elements, general state practice and *opinio juris*, bear on its formation – we can regard both ingredients as forms of practice, or two aspects under which practice may be interpreted. In determining whether a norm exists as a matter of international custom, we begin by giving its putative content (for example, about the length of the territorial sea, the requirements of diplomatic protection, the immunity of sovereign states from certain forms of intervention, and so on). This content will specify some pattern of (state) conduct[1] to which a normative modality (obligatory, impermissible, permissible, etc.) is assigned under certain conditions. In order to determine whether this so far merely notional legal norm, call it X, exists as a matter of customary international law, we must address the following two questions:[2]

(1) State practice: Is there evidence that states generally conform their conduct to X, engaging in behavior consistent with the normative content of X? For example, if X is an obligation-imposing norm, is it the case that states generally do, or refrain from doing, what X enjoins them to do, or refrain from doing?

[1] I use "conduct" here in a broad sense to encompass both acts and omissions.

[2] For the sake of convenience, I consider only the activities of states as bearing on the practice and *opinio juris* relevant to custom. I later revisit the issue as to *opinio juris*.

(2) *Opinio juris*: Is there evidence that states adopt one or other of the following attitudes to X:

[OJ1] the creation of an international legal rule according to which the specified pattern of behavior has the normative significance attached to it by X is morally justified, and such a legal rule should be created by means of a process that involves general state practice consistent with X and a moral endorsement by states of X's establishment as a legal rule, or

[OJ2] X is already a norm of customary international law, that is, it exists as a matter of general state practice and *opinio juris* (i.e., OJ1), and the status of X as a legal norm (or compliance with it as such) is morally justified.

In short, *opinio juris* involves the judgment that a norm is already part of customary international law and that (compliance with) it is morally justified (OJ2); or that, as a moral matter, it should be established as law through the process of general state practice and *opinio juris* (OJ1); or else some mixture of these two attitudes.

Let me elaborate on the elements of this understanding of general state practice and *opinio juris*. The first thing to notice is that it involves a narrower interpretation of "state practice" than that which is sometimes deployed. On the view just outlined, state practice consists in the behavior of states insofar as it is in conformity with the putative norm. A positive showing of state practice depends on evidence of general state conformity with what the supposed norm stipulates as obligatory, impermissible, permissible, and so forth. Positive state practice, therefore, is redeemable in the hard currency of actual conformity to the norm. Forms of state behavior that evidence some kind of belief regarding the existence or otherwise of the norm, but which do not relate to conformity with it, do not fall within the category of state practice. Instead, they will bear on the separate matter of *opinio juris*.

One advantage of this way of distinguishing the two elements of custom is that it marks the distinctive significance of whether states actually generally conform to a supposed norm as opposed to other things they may do in relation to that norm, such as expressing their approval of it. This is broadly the significance of putting your money where your mouth is: of actually *conforming* ("state practice") to the (putative) legal norm that you *avow* ("*opinio juris*") to be morally justified. It therefore avoids the unorthodox claim that state practice can amount to nothing more than evidence of *opinio juris*.[3] Another advantage is that it prevents undue "double counting," whereby systematically one and the same course of conduct, for example, diplomatic correspondence, votes on resolutions by international

3 This is the view advanced in BRIAN D. LEPARD, CUSTOMARY INTERNATIONAL LAW: A NEW THEORY WITH PRACTICAL APPLICATIONS ch. 8 (2010).

organizations, and so forth, is treated as both state practice and *opinio juris*.[4] Nonetheless, this framing of the distinction allows that state practice, interpreted against a suitable background, can be evidence of *opinio juris*.[5] But it avoids the wholesale conflation of the two ingredients entailed by an expansive interpretation of state practice.

Turn now to the disjunctive interpretation of *opinio juris*. It is not fanciful to say that this interpretation is already literally foreshadowed by its full Latin tag: *opinio juris sive necessitatis* ("an opinion of law or necessity"). Firstly, what is at issue is an *opinion* or judgment about what is or ought to be the case, rather than the mere expression of a desire or a preference. Second, the content of that judgment relates *either* to what the law is and whether, as a moral matter, it may be complied with *or else* what it ought to be (and, on either alternative, a moral "necessity"). Let me expand on both of these points.

What is central to both variants of *opinio juris*, OJ1 and OJ2, is an imputed attitude at the core of which is a *judgment* that a norm is, or would be if established, *morally justified* as a norm of customary international law. The judgment is one about moral justification because only this species of justification is adequate to the task of upholding the claim to legitimacy inherent in law, that is, its claim to impose obligations of obedience on its purported subjects.[6] Only a justification grounded in moral standards, as opposed to mere considerations of self-interest, for example, can vindicate the claim of the law to be morally binding on its subjects. This understanding of *opinio juris*, as reflecting a moral judgment, should be contrasted with two other understandings, the one unduly broad, the other unduly narrow.

The overly broad view characterizes *opinio juris* in terms of state preferences.[7] Preferences are a subject's positive attitudes toward some particular outcome, which typically reflect what the subject takes to be reasons. These reasons may differ greatly in kind, from reasons of pleasure or self-interest, at one extreme, to moral reasons at the other. But a preference, thus broadly understood, does not necessarily purport

4 The failure to avoid double counting mars the analysis in Michael Wood, Special Rapporteur, International Law Commission, *Second Report on Identification of Customary International Law* 9–11 (May 2014), *at* http://legal.un.org/ilc/sessions/66/a_cn4_672(advance).pdf. On the one hand, the report affirms the necessity for both elements, general practice and *opinio juris* (or, "acceptance as law") in order for a customary international legal norm to emerge (Draft conclusion 3). On the other hand, it gives an account of general practice that seems to subsume, as a component part, *opinio juris*, at least insofar as the manifestations of the former seem to encompass most, if not all, of those of the latter (compare Draft conclusion 7[2] with Draft conclusion 11[2]). The double counting entailed by the second aspect of the report largely undercuts its insistence on the necessity for the presence of *both* state practice and *opinio juris* in the case of each norm of customary international law.

5 *See, e.g.*, Delimitation of the Maritime Boundary in the Gulf of Maine Area (*Can. v. U.S.*), I.C.J. Rep. 246, 299, para. 111 (ICJ Oct. 12, 1984) (Judgment).

6 Joseph Raz, "Hart on Moral Rights and Legal Duties," 4 Oxford J. Leg. Stud. 123 (1984).

7 There are intimations of this view in Curtis A. Bradley, "Customary International Law Adjudication as Common Law Adjudication" (in this volume).

to identify a consideration that is even in principle capable of justifying the claim to legitimacy (moral bindingness) inherent in law. Notice, in addition, that we can often intelligibly speak of a discrepancy between what a state would prefer the law to be and what it publicly judges that it should be as a moral matter. Its self-interested preference (e.g., as a powerful, or land-locked, or culturally homogeneous state) may point in one direction, but its assessment of the moral merits regarding the content of international law may point in the opposite direction. But it is only the latter that counts as *opinio juris*. All this is compatible with two observations. First, that the moral judgments made by states will often be skewed by considerations of self-interest or mere preference. Indeed, they may sometimes be insincerely avowed. This is simply a pitfall to which all moral judgment is prey. Second, that even when not so skewed, the relevant moral judgment may be one to the effect that permitting states to pursue their preferences or self-interest in various ways is justifiable. In other words, state preferences do have a potentially substantial role to play in the formation of customary international law, but only as regulated by background moral judgments regarding their suitability to do so. It is these background judgments, not the preferences, that are the core of *opinio juris*.

If the preference-based interpretation of *opinio juris* fails in virtue of being overly broad, another much more familiar interpretation is unduly narrow. The latter usually takes *opinio juris* to be an attitude accompanying state practice, one according to which the state acts out of a "sense of obligation" when engaging in the relevant pattern of conduct. So, for example, in the first ICJ case to invoke the notion of *opinio juris*, it is incorrectly described as a matter of states feeling "legally compelled to ... [perform the relevant act] by reason of a rule of customary international law obliging them to do so."[8] And most recently, in the Second Report on Identification of Customary International Law, the International Law Commission's special rapporteur glossed *opinio juris* as a matter of general practice being "accepted as law." This, in turn, was understood to mean that "the practice in question must be accompanied by a sense of legal obligation."[9] This is an apt characterization of *opinio juris* for the specific case in which a state is complying with an already existing international norm that it takes to impose an obligation on itself. But it fails to embrace two other cases. First, if the obligation is understood as an already existing legal obligation, this analysis does not properly capture *opinio juris* that gives rise to a new norm (i.e., the type covered by OJ1). Second, even in the case of an existing customary norm, it does not cover situations where the

8 North Sea Continental Shelf (Judgment), 1969 I.C.J. Rep. 3, 44–45, para. 78 (ICJ Feb. 20, 1969). The centrality of a sense of obligation is echoed in many other ICJ judgments. *See, e.g.*, Colombian-Peruvian asylum case, 1950 I.C.J. Reports, 266, at 286 (ICJ Nov. 20, 1950) (Judgment) ("a feeling of legal obligation").

9 Draft Conclusion 10 (1), 51.

norm is thought by the state to confer a *right* or a *liberty* upon it to engage in the specified pattern of conduct.[10] The characterization of *opinio juris* I have given, by contrast, accommodates all of these normative modalities. What it requires is that the relevant norm, whether it is taken by the state to impose a duty or confer a right or a liberty on it, is judged by the state to be morally justified in doing so. It therefore does not clamp *opinio juris* to one specific normative modality, that of obligation, even though many norms of customary international law will of course be obligation imposing.

Now, an advantage of this disjunctive specification of *opinio* juris – as OJ1 or OJ2 – is that it defuses the so-called paradox of custom, according to which the creation of new customary international law is inescapably premised on *error* or *deception* on the part of states.[11] Specifically, it is premised on the mistaken belief, or pretended mistaken belief, that a norm that is not already part of customary international law actually possesses this status. Although some question the practical significance of this paradox, I have argued elsewhere that it tarnishes the legitimacy of customary international law. This is because there is a transparency constraint on any form of law making to the effect that its successful operation must not necessarily depend on mistaken beliefs (or pretended such beliefs) on the part of the agents that create the law as to what it is they are doing. Political legitimacy demands that exercises of political power, including acts of law making, must be publicly assessable in terms of standards that appropriately bear on political decision making. A corollary of their being assessable in this way is a transparency requirement, according to which exercises of political power must be sincerely and accurately presented and defended by their agents as the acts that they are.

On the disjunctive analysis of *opinio juris* we do not need to assume that the generation of a new customary norm requires the existence of widespread error or deception as to the existing state of the law. This is because the relevant kind of *opinio juris* may be of the type OJ1. Once the customary norm has come into existence, however, its continued existence across time can be sustained by *opinio*

[10] The ICJ seemed to hold that a unilateral declaration of independence could be such a liberty: "it is entirely possible for a particular act – such as a unilateral declaration of independence – not to be in violation of international law without necessarily constituting the exercise of a right conferred by it." Accordance with International Law of the Unilateral Declaration of Independent in respect of Kosovo, 2010 I.C.J. Rep., 425–26, para. 56 (ICJ, 2010).

[11] For a fuller elaboration of this argument, see J. Tasioulas, "Opinio Juris and the Genesis of Custom: A Solution to the 'Paradox,'" 26 AUST'L Y.B. INT'L L. 199 (2007). In interpreting *opinio juris* as "practice accepted as law," the ILC's Special Rapporteur Second Report rather optimistically, and without argument, suggests that "use of this term from the [ICJ's] Statute goes a large way towards overcoming the *opinio juris* paradox" (50–51). This makes it sound as if the paradox stems from the Latin expression and will be eradicated by the use of plain English. On the contrary, the "practice accepted as law" formulation accentuates that very paradox, making it clear that a practice has to be accepted as already legal in order to *become* legal.

juris of the second sort, OJ2. But there is no "paradox" involved in a legal norm's *continued* existence depending in part on the fact of its being taken to already exist.

Another, more widely credited threat to the legitimacy of customary international law is the accusation that it is hopelessly indeterminate; in particular, that there is no determinate account of the role and relations of state practice and *opinio juris* in the generation of customary international law. Of course, any such skeptical conclusion has to be earned by argument, in the same way as an anti-skeptical account must pay its way, and cannot be presumed true by default. I have offered a general interpretative framework for responding to this skeptical challenge in previous work. This work takes as its focus the conception of customary international law that emerged in ICJ decisions such as the *Nicaragua* and *Nuclear Weapons* cases (a conception that by now has achieved the status of orthodoxy in judicial practice, even if not in the estimation of traditionalist authors with positivist inclinations).[12] I shall not rehearse this interpretative framework here, save to point out one implication that does not follow from it and three that do. The implication that the MJA does *not* have is that of ensuring that all valid customary laws are also just or ethically sound. Instead, their formation relies on *opinio juris*, which embodies moral judgments that may or may not be correct. Nor does the moral judgment required in the interpretative process of adducing customary norms guarantee that any valid norm will be free of ethical flaws. By contrast, the following three implications genuinely do flow from the MJA:

(1) Although state practice and *opinio juris* are, as a conceptual matter, independent variables in the formation and persistence of customary international law, the paradigm case is that in which general state practice and widespread *opinio juris* are both present. Indeed, especially in the case of OJ2, the existence of *opinio juris* is what in part explains the state practice. Practice, after all, is the natural product of *opinio juris*, the practical manifestation of the value judgment that the latter embodies.

(2) In appropriate cases, state practice and *opinio juris* can be traded off against each other within the interpretative framework for assessing the formation of customary norms. In particular, customary norms can come into being despite the absence of much supporting general state practice, or at the extreme, even in the teeth of considerable countervailing practice.[13] This is because a

[12] John Tasioulas, "In Defence of Relative Normativity: Communitarian Values and the Nicaragua Case," 16 Oxford J. Leg. Stud. 85 (1996); John Tasioulas, "Customary International Law and the Quest for Global Justice," in The Nature of Customary Law: Legal, Philosophical and Historical Perspectives (Amanda Perreau-Saussine & James B. Murphy eds., 2007).

[13] The report of the ILC's special rapporteur is ambiguous on this issue. Some formulations suggest that both elements, state practice and *opinio juris*, must be present in the case of each and every norm of customary international law. Other formulations suggest only the weaker thesis that state practice and

dearth of state practice can be compensated for by high levels of *opinio juris*, especially if there is a strong moral case for the norm in question. That case must typically be constructed around those values that are especially salient for the legitimacy of international law, such as peaceful co-existence, human rights, environmental protection, and so forth. Notice, however, that this evaluative dimension to the formation of customary international law requires judgments about objective values; it cannot be reduced to a set of further facts concerning what states or other actors *believe* to be important values (this point is developed further in the next section).

(3) Whereas evidence of *opinio juris* can establish a customary norm in the absence of supporting general state practice, the reverse position very seldom, if ever, obtains. To the extent that it does so, it will probably involve cases in which *opinio juris* is primarily inferred from general state practice. It follows that *opinio juris* is always necessary for the formation of customary international law, even if sometimes its existence is inferred primarily from a pattern of general state practice.

These features of the MJA, especially feature (2), bring in their train a number of benefits. Chief among them are the following:

(a) it allows new, potentially universally binding law to come into existence (or to do so more rapidly than would otherwise be the case) in areas in which it is needed, but where there is much contrary state practice, for example, human rights norms and the laws of war, or in cases where state practice has not yet had an opportunity to develop, for example, the law of outer space;
(b) it enables the law to be changed through large-scale shifts in *opinio juris*, thereby avoiding the legitimacy-undermining idea that the only way to reform existing customary international law is through a vast program orchestrating its persistent violation;[14]
(c) by construing *opinio juris* as an ingredient independent of state practice, it potentially enables the *opinio juris* of non-state actors, such as organs of the United Nations, international organizations and tribunals, nongovernmental organizations, expert academic opinion, and so forth to be taken into account where this is appropriate in terms of enhancing the legitimacy of international

opinio juris are relevant to the existence of any given customary international norm, without making the presence of the former necessary in each case. The effect of this ambiguity is ameliorated by the fact that, as previously noted, on the report's account state practice and *opinio juris* massively overlap.

14 Cf. the idea of reform through illegality proposed independently in ALLEN E. BUCHANAN, *JUSTICE, LEGITIMACY, AND SELF-DETERMINATION: MORAL FOUNDATIONS FOR INTERNATIONAL LAW* ch. 11 (2004); and Robert Goodin, "Toward an International Rule of Law: Distinguishing International Law-Breakers from Would-Be Law-Makers," 9 *J. ETHICS* 225 (2005).

law. Indeed, by a further extension, it allows us to take into account the appropriate practice of non-state actors, especially in relation to putative norms that regulate the activities of such actors, such as the various organs of the United Nations.

IS *OPINIO JURIS* AN EXPRESSION OF CONSENT?

Does *opinio juris* operate as a means of expressing consent to be bound by a putative norm, a consent perhaps conditional on a sufficient number of other states similarly consenting, and which is at least a necessary condition for the bindingness of any norm vis-à-vis a given state? Nothing in the MJA implies the correctness of such a voluntarist interpretation of *opinio juris*. Still, we might ask whether voluntarism fits the existing practice of customary international law and, if not, whether there are compelling reasons to revise the latter along voluntarist lines.

A modicum of reflection shows that the voluntarist interpretation fails to fit the communitarian character of the institution of customary international law, as ordinarily conceived.[15] If there is adequate state practice and *opinio juris* among the community of states (and other relevant actors), the resultant customary norm may bind a state that did not consent to it. Hence, a binding customary norm arises from a convergence of collective practice and a consensus of community opinion, not from the aggregation one by one of discrete episodes of consent. Now, it might be replied that the so-called persistent objector rule testifies to the voluntarist basis, in the last resort, of the existing customary regime.[16] But that rule, even supposing it exists, which is far from self-evident, is heavily qualified in two ways. First, it does not apply to customary norms whose emergence predated the coming into existence of a given state. Such a state will therefore be bound by the norm even though it had no opportunity to object persistently during its formation. Second, the persistent objector rule does not in any case apply to the body of *jus cogens* norms, which is a category of customary norms that are opposable against all states independently of their individual volition (see the discussion of *jus cogens* norms in the next section).[17]

[15] My remarks here are confined to customary international law. Voluntarism is obviously a much better fit for the institution of treaty law. For a helpful discussion of the non-voluntarist character of customary international law, see Andrew T. Guzman, "Against Consent," 52 VA. J. INT'L. L. 747, 775–77 (2012).

[16] Fisheries 1951 I.C.J. Rep. 1167 at 131 (ICJ, 1951).

[17] The failure to distinguish consent from consensus is ubiquitous, even among critics of voluntarism. See, for example, Ronald Dworkin's claim that the main sources of international law, including customary international law and *jus cogens* norms, are conceived in existing doctrine as consent based, see Ronald Dworkin, "A New Philosophy of International Law," 41 PHIL. & PUB. AFF. 5 (2013). But on Dworkin's account talk of consent in describing existing doctrines alternates with the quite different talk of "recognition": "If enough states to constitute 'the international community of States' have

The voluntarist picture is also incompatible with another, perhaps less obvious, feature of existing doctrine. This is the significance that is attached to the *opinio juris* of actors that are not themselves within the scope of the norm in question. This may be because the content of the norm does not extend to them, as in the case of the *opinio juris* of an international organ, such as the Security Council, regarding a matter that binds states. More speculatively perhaps, there is also the scenario in which a state's *opinio juris* is taken into account regarding a norm that on its face regulates the former's conduct, but where the state in question claims to be exempt from the norm's operation whilst supporting its imposition on other states. This is one manifestation of the phenomenon of exceptionalism. For example, a multilateral treaty may be adduced as evidence of *opinio juris* despite the fact that some states' acceptance of the treaty is qualified by hefty reservations.

Of course, a sufficiently hardheaded voluntarist will be undeterred by the infelicities of doctrinal fit previously identified. He may well respond that the regime of customary law should be purged of its non-voluntarist elements. And the reason for this drastic revisionism is the goal of better aligning *legality* – what we recognize as valid norms of customary international law – and *legitimacy* – the moral bindingness of such law on its putative subjects. In other words, the voluntarist insists that it is at least a necessary condition of a legal norm's bindingness with respect to a given state that that state has consented to it. Therefore, consent should be a condition on the validity of a legal norm, since it is inherent in law to claim legitimacy. Yet although it is widely credited, the consent-based account is deeply problematic in the context of both municipal and international law.

To focus on the international case, why should the bindingness of international norms that serve valuable goals not otherwise achievable hinge on whether they have been accepted by states whose rulers do not themselves govern with the consent of their own subjects? Or who, more to the point, flout the basic rights of their subjects? And even in the case where the government of a state rules by consent and respects basic rights, why should it be able to opt out of an international regime that promises to achieve great benefits for all, preventing it from achieving those benefits or significantly raising the cost of its doing so? There is no need to expand on the anti-voluntarist case here. Voluntarism has been subject to heavy general criticism that applies to both the municipal and international spheres.[18] Still,

recognized fundamental rules as preemptory and nonnegotiable, then these rules are preemptory and nonnegotiable for the whole international community." *Id.* at 6. The failure to disentangle consent from a consensus of recognition has downstream implications for Dworkin's argument, since it means that he defends his "more basic principle," *id.* at 9, for identifying international law only against the contrast afforded by a voluntarist model that has a poor foothold in existing practice.

18 For a rejection of consent-based accounts of international law's legitimacy, see Allen E. Buchanan, "The Legitimacy of International Law," in THE PHILOSOPHY OF INTERNATIONAL LAW (Samantha

we should recognize that the voluntarist is motivated by a laudable concern: to align processes of law-formation more closely with the conditions of legitimacy. His error consists in misidentifying those conditions.

But this leaves the MJA facing a challenge: what is the understanding of legitimacy that undergirds its account of custom? In other writings, I have extended Joseph Raz's service conception of legitimate authority to the case of international law. According to this conception, a body of law will be binding on its putative subjects to the extent that they are more likely to comply with the independent reasons that apply to them by treating it as binding than by not doing so. On this view, it is objective values, and the reasons they generate, and not subjective volition, that is the ultimate touchstone of legitimacy.[19] Most importantly, objective values are not to be confused with the value beliefs actually held by states; instead, they are the values that states *should* recognize and conform to, whether they actually do so or not. Unfortunately, some who superficially appear to adopt an approach akin to the MJA falter at this crucial hurdle, making the operative values ultimately a matter of value judgments that command widespread assent in international society at any given time.[20] But nothing is a genuine value just in virtue of its being believed to be so, although there may be *independent* normative reasons for according a certain significance to whether an ethical belief is widespread or not, for example, it may in some cases bear on the appropriate objects of toleration within an international legal regime, or on the morally salient issue of the latter's efficacy in securing conformity among its putative subjects.

Besson & John Tasioulas eds., 2010). For a rejection of consent as a general theory of legitimate authority, see Joseph Raz, *The Morality of Freedom* Part I (OUP, 1986).

19 John Tasioulas, "The Legitimacy of International Law," in *The Philosophy of International Law* (Samantha Besson & John Tasioulas eds., 2010); and John Tasioulas, "Human Rights, Legitimacy, and International Law," 58 *Am. J. Juris.* 1 (2013).

20 See, for example, the approaches to customary international law of Roberts and Lepard, which make the moral attractiveness of a customary norm ultimately a function of the extent to which it is endorsed by states rather than a matter of objective truth. Thus, Roberts appeals to "commonly held subjective views about actions that are right or wrong, which a representative majority of states has recognized in treaties and declarations," Anthea E. Roberts, "Traditional and Modern Approaches to Customary International Law," 95 *Am. J. Int'l L.* 757, 760 (2001). Similarly, Lepard invokes "fundamental ethical principles" that are "the product of the views of states" (Lepard, *supra* note 3, at 374–75). But a simple insistence on Hume's Law – the idea that no "ought" can be derived exclusively from statements about what is the case, including, what opinions are generally adhered to – is enough to expose the fatal flaw in such views. It is also worth noting that such consensus-based views often involve highly optimistic assumptions about the content of the principles regarding which there is a global consensus, leaving the impression that they are surreptitiously driven by undefended ethical commitments that are not consensus based. None of this is intended to deny the existence of objective moral reasons for giving some kinds of significance to whether or not a moral proposition is widely accepted by states, but such acceptance is not ultimately determinative of whether or not a given proposition is correct.

At an abstract level, the affinity of the MJA with the service conception of legitimacy should be evident.[21] Both make objective values, the sources of corresponding objective reasons, central in their respective inquiries into law-formation and legitimacy. However, something further can be said about why *opinio juris* can have the role I have assigned it (especially feature [2], in the previous section) under a service conception. Rather than having the import for legitimacy suggested by the consent theory, the existence of widespread *opinio juris* is capable of enhancing the legitimacy of a norm in at least three ways under the service conception.

First, the consensus of view it embodies among many and diverse states, as well as other relevant global actors, can constitute a distillation of collective *wisdom*, one that outstrips the knowledge that a state acting on its independent assssment could reliably bring to bear in decision making. A state may therefore be more likely to conform with reason by aligning itself with custom than by resorting to its own judgment on the matter. Second, it may indicate that the norm in question is more likely to be *efficacious* in securing compliance among relevant actors, and hence to achieve its goals, than one not backed by communal consensus. This is an especially salient feature when the legitimacy of the norm in question turns on its capacity to provide a solution to coordination problems. In such cases, of course, the presence of supporting state practice will be especially helpful in confirming the efficacy of the relevant norm. This is part of the distinctive significance of *compliance* (state practice) with a putative norm, as opposed to mere *endorsement* of it (*opinio juris*) that was previously noted.[22] Finally, taking into account the views of multiple states, as well as other global actors, ministers to the value of political *participation* by independent, self-determining states, groups, and individuals in the communal processes of international governance. Among the reasons that states have are reasons to advance widespread participation in international lawmaking

[21] It may be objected that the service conception of authority justifies a "hard" positivist account of law incompatible with the MJA, insofar as the latter makes moral judgments necessary to the discovery of customary international law. *See, e.g.*, Joseph Raz, "Authority, Law, and Morality," in ETHICS IN THE PUBLIC DOMAIN (1994). Answering this large objection would take us too far afield, so I simply note two points. First, a subscriber to hard positivism could nonetheless accept the MJA as an account not of customary international law, but rather of how courts should decide cases involving customary international law, on the assumption that they have a responsibility to develop such law in appropriate ways. Second, even if we stick fast to MJA as an account of the grounds of customary international law, I would join those who question whether the service conception of authority, properly understood, necessitates the embrace of Raz's brand of hard positivism.

[22] The first consideration may justify according greater weight in the process of customary norm formation, other things being equal, to the practice and *opinio juris* of states whose history reveals a more steadfast commitment to relevant values of global justice. The second consideration justifies according greater weight to the practice and *opinio* of more powerful and influential states, or to those states especially affected by the putative norm. Both considerations make the existence of *opinio* and practice among states that are *representative* of salient global differences – cultural, religious, economic, military, etc. – a matter of significance.

processes by affected political communities and individuals. In short, although the voluntarist is right to link insistence on *opinio juris* to considerations of legitimacy, they tell an overly simplistic story about how that link is forged.

HUMAN RIGHTS AS CUSTOM AND *JUS COGENS*

Let us turn now to the implications of the MJA for international human rights law. How do we determine which, if any, human rights norms have acquired the status of customary international law? Potentially, this will be a subset of all the norms of human rights law insofar as there may exist some human rights norms, for example, based on treaties, that do not have customary status. Having isolated the customary international law of human rights, we proceed to identify those customary human rights norms that belong to the elite category of *jus cogens*. On this view, the *jus cogens* norms of human rights are a subset of the customary international law of human rights. More generally, it is a presupposition of this view that *jus cogens* norms are a special category of customary norms.[23] Their distinctive character, within the general class of customary norms, consists mainly in their possession of the following features: (a) they legally bind all states (and other relevant international agents) without exception; (b) their binding character for any given state is independent of whether that state has accepted, or failed to object to, the norm in question. In particular, the supposed "persistent objector rule" for evading a law's opposability is inapplicable to *jus cogens* norms. So, for example, South Africa's supposed persistent objections to norms prohibiting racial discrimination and apartheid were legally nugatory; (c) noncompliance with a norm of this class cannot be legally justified, except perhaps insofar as this is permitted by another norm that also possesses *jus cogens* standing. As a specific implication of (c), any treaty agreement to depart from a *jus cogens* norm is null and void – certainly the offending provision itself and arguably the entire treaty of which it is a part. Of course, it does not follow from the fact that *jus cogens* norms are non-derogable in this way that they are not subject to limitations on how states and others may seek to enforce them or respond to their violation, for example, by means of rules establishing jurisdiction with respect to

[23] A presupposition that seems to be widely accepted, including by the ICJ, e.g., in the *Nicaragua Case* 1986 I.C.J. Rep. 14, 100, ¶ 190 (ICJ, 1986), which held that the prohibition of the use of force is 'not only a principle of customary international law but also a fundamental or cardinal principle of such law' and in *Obligation to Extradite or Prosecute* case 2012 I.C.J. Rep 457, ¶ 99 (ICJ, 2012), which held that the prohibition of torture is part of customary law and has become a peremptory norm (*jus cogens*), and is "grounded in a widespread international practice and on the *opinio juris* of States." However, some have argued that *jus cogens* norms can be grounded in general principles of law. *See, e.g.*, Bruno Simma and Philip Alston, "The Sources of Human Rights Law: Custom, Jus Cogens, and General Principles," 12 AUST'L Y.B. INT'L L. 82, 102–06 (1988–89). I shall set aside this possibility in what follows.

their violation.[24] In particular, it is worth observing that although *jus cogens* norms impose obligations on all states (obligations *omnium*) it does not logically follow that all states have a right to the enforcement of these obligations (obligations *erga omnes*).[25] In short, *jus cogens* norms are an elite class of norms that seek to protect what the international community regards as particularly important values. This perceived importance is marked by the formal characteristics of being judged properly to be, or to become, legal norms that are (a) universal, (b) peremptory, and (c) non-derogable (in the sense of being indefeasible in relation to non-*jus cogens* norms. It may be that, in addition, some *jus cogens* norms are indefeasible *tout court*).

If the *jus cogens* human rights norms are a subset of the customary human rights norms, how are they to be identified? The answer to this question involves at least two dimensions. On the first dimension, the *jus cogens* character of a norm will be indicated by the distinctive content of the *opinio juris* associated with it. In particular, that *opinio juris* will contain a moral judgment, explicit or implicit, to the effect that the norm in question either already properly has or may properly be accorded the three features (a)–(c) enumerated in the preceding paragraph. So far this is predominantly a matter of fit with social facts: a successful argument for the existence of a customary norm must adequately exemplify a *dimension of fit* with such facts. But the interpretive process of identifying *jus cogens* norms, like the process of identifying customary norms generally, operates against a background grasp of the sorts of norms that are appropriately accorded the relevant legal status. This demands some view, on the part of the interpreter, as to which norms may justifiably be ascribed features (a)–(c) given background evaluative considerations bearing on international order, and given also the prospect of realizing them through the medium of law at a given juncture of history. The *dimension of justification* will implicate judgments as to whether the norm in question is an eligible object of international concern, such that it ought to find expression in international legal norms, and whether it has the kind of importance sufficient to justify attributing to it features (a)–(c). But even then a further judgment will need to be made, one more pragmatic in character, about the overall effect of recognizing such a legal norm at the international legal system's present stage of evolution.

[24] See, for example, 2006 *Armed Activities Case (D.R.C. v Rwanda)*, ¶ 64, 125. In *Jurisdictional Immunities of the State* (Germany v. Italy), Judgment, No.143 at 83–5, 96 it was decided that there is no exception to state immunity for violations of norms of *jus cogens*.

[25] It has been argued that observations in the ICJ's *Barcelona Traction* decision suggest that all *jus cogens* norms are also *erga omnes*, see Erika de Wet, "*Jus Cogens* and Obligations Erga Omnes," in THE OXFORD HANDBOOK OF INTERNATIONAL HUMAN RIGHTS LAW, 554–5 (2013). This view seems neither plausible nor desirable, see LEPARD, *supra* note 3, at 343.

It is evident that both of the dimensions of fit and justification described here may have the effect of generating changing interpretative outcomes over time. This accounts for the dynamic quality of *jus cogens* doctrine, with new norms coming into being with the passage of time, while existing norms change shape or even lapse from this exalted status. Hence, there is no implication here of the heavy-duty "natural law" thesis that *jus cogens* doctrine picks out a set of immutable norms that form part of the international legal system from its very inception, its unvarying constitutional structure, as it were. Irrespective of whether or not any such norms exist, they are not synonymous with the idea of *jus cogens*. For all the elevated ethical status it connotes, the latter is a form of *positive* law, and hence cannot be grounded exclusively in moral reasoning.

Consider, however, a challenge to the thesis that *jus cogens* norms are a subset of customary norms. Antonio Cassese has championed the view that *jus cogens* norms constitute the constitutional groundwork of the international legal order, and that certain basic human rights norms figure prominently in this category. However, he denies that the *jus cogens* character of a norm presupposes its status as a customary norm. This is because the former sort of norm may exist despite lacking the backing in *opinio juris* and state practice required for the latter. Instead, *jus cogens* norms are *sui generis* with respect to customary international law, rather than a species of it, and can be adduced provided that a representative majority of states evince "acceptance" of them as such norms:

> [W]hat is required [for *jus cogens* status] is the *acceptance* by the *majority* of states provided that such majority includes states which are representative of the various political and geographic areas of the world. We can conclude that, unlike the customary process, the two elements of *usus* and *opinion juris* are not required. It may suffice for the majority of members of the world community in some way to evince their "acceptance" of a customary rule as having the rank of a peremptory norm. Such "acceptance" does not necessarily involve actual conduct, or a positive assertion; it may involve an express or tacit manifestation of will, which can take the form of a statement or declaration, or acquiescence in statements by other international legal subjects or in recommendations or declarations by intergovernmental organizations or in decisions by judicial bodies. No consistent practice of states and other international legal subjects (*usus*) is necessary.[26]

Cassese's position has the curious upshot that, *ceteris paribus*, the bar for qualification as a customary norm (state practice and *opinio juris*) is, in a significant

[26] Antonio Cassese, "For an Enhanced Role of *Jus Cogens*," in *Realizing Utopia: The Future of International Law*, 165 (Antonio Cassese ed., 2012). For an apparently similar view, see Bruno Simma, "From Bilateralism to Community Interest," 250 *Recueil des Cours* 217, 292 (199).

respect, set considerably higher than that for *jus cogens* status. And this is so notwithstanding the fact that the normative consequences of *jus cogens* status are far more significant than those of customary status. Now, the strain in this position emerges when Cassese himself plausibly rejects as "illogical" one writer's view[27] that it may be possible to establish the customary nature of a rule *after* finding that it is a *jus cogens* norm. But just this possibility seems to be countenanced by Cassese's own view in making it possible for a norm to be *jus cogens* without first establishing it as a customary norm. If, over time, the "acceptance" of that norm by a representative majority of states comes to be supplemented by general state practice, it is difficult to see why it should not eventually be affirmed as customary in nature. The air of "illogicality," I suggest, stems from the assumption – common to Cassese and the position he is criticizing – that *jus cogens* norms are not a subset of customary norms. This is the real reason why it seems decidedly odd to say that a *jus cogens* norm can become customary law at some point subsequent to its initial emergence.

A more logical view, I think, is to maintain the customary character of *jus cogens* norms while adopting a more flexible understanding of the process of customary norm formation than the one assumed by Cassese. On the MJA, it is indeed possible for a customary norm to emerge precisely through something like the process of "acceptance" he describes, without the support of widespread and consistent state practice. There is no need, in that case, to countenance a troubling bifurcation of customary international law and *jus cogens* doctrine.

Assuming the MJA is correct, general state practice is relevant, but not always necessary, in the formation of a *jus cogens* norm. Now, some have argued that there are three distinctive problems in establishing state practice in favor of a human rights norm, especially one that is supposed to be *jus cogens* in character: (1) Many human rights norms are essentially prohibitions, requiring that states refrain from certain forms of conduct. But there seems to be a special difficulty in adducing evidence of supportive "negative state practice"; (2) human rights norms essentially concern a state's treatment of its own subjects, rather than some inherently interstate matter, yet custom is all about norms hammered out by states through a give-and-take adjustment of competing interests in the process of inter-state contact; (3) it has been argued that state practice supporting a *jus cogens* norm, such as the norm prohibiting genocide, requires evidence of "failed attempts to contract out of" the application of the *jus cogens* norm. This would paradigmatically consist in some authoritative body treating as void a treaty that permitted or recognized the

[27] Cassesse, *supra* note 26, at 164 n.17.

possibility of doing what the *jus cogens* norm prohibits. Of course, such evidence is extremely thin on the ground.[28]

All three of these supposed difficulties, however, are either overdrawn or false. Problem (1) exaggerates the extent to which human rights impose negative duties as opposed to positive duties to undertake a course of action. Moreover, the imposition of negative duties is hardly unique to human rights law, but also extends to many other customary norms, including the cornerstone norms prohibiting the use of force and intervention against sovereign states. Finally, the objection exaggerates the difficulty of adducing evidence of compliance with a prohibition. One can plausibly infer relevant state practice, for example, in situations in which a state had a strong interest in engaging in the prohibited conduct but refrained to do so, and cited the relevant norm by way of justification. One can also adduce cases in which states engage in positive action, such as compensation, economic sanctions, or humanitarian intervention, that they claim is justified on the basis that some other state violated the negative duty imposed by the customary norm. In other words, the "state practice" relevant to the existence of a norm is not only compliance with the primary duty it imposes, but also acting in accordance with the normative implications of noncompliance with those primary duties, such as duties to compensate or rights of intervention.

Problem (2) also misfires. It takes a commonplace but contingent feature of much general state practice (its emergence via process of interstate activity) and without obvious warrant converts it into a necessary characteristic. In any case, human rights law does not exclusively concern a state's treatment of its own citizens, for example, in cases of the extradition of foreign nationals to states whose criminal justice systems flout basic human rights norms. And even when it does so, an interstate dimension bearing on state practice is not invariably absent. Consider, for example, the appeal to gross human rights violations to justify economic sanctions or engagement in forms of intervention that are otherwise incompatible with state sovereignty.

Problem (3), if genuine, would render state practice in favor of a *jus cogens* norm of human rights extremely scarce.[29] It would be most unambiguously confined to

[28] *See* Hugh Thirlway, The Sources of International Law 156 (2014) ("To support a customary rule not merely forbidding genocide, but also providing that the obligation not to commit it is a matter of *jus* cogens, the practice invoked would have to consist of, or include, one or more failed attempts to contract out of its application, by treaty or informal agreement – 'failed' in the sense that it was accepted, or authoritatively declared, that the norm over-rode the agreement."). Thirlway also appears to give credence to versions of the first two objections.

[29] But not completely nonexistent. For example, in *Aloeboetoe v Surinam*, 15 Inter-Am. Ct. HR (ser C.) (1993), the Inter-American Court objected to the invocation of a treaty that included a clause on the return of runaway slaves.

the situation in which states made a treaty with the explicit purpose of departing from human rights norms, and some authoritative body invalidated the treaty on the grounds that the norms in question are *jus cogens*.[30] However, this is manifestly an overly narrow construction of what such practice might involve. As noted previously, the distinctiveness of *jus cogens* norms does not simply consist in the fact that they are non-derogable by means of a treaty, but rather in the three broader characteristics, (a)–(c), specified. Now, the further point is that the finding of state practice need not be made independently of registering the presence of *opinio juris*. Hence, state practice supporting a norm can be adduced in cases where states justify their action or inaction, whether in compliance with the norm or in response to its violation by others, on the basis of the judgment embodied in an *opinio juris* that attributes to the norm characteristics (a)–(c).

In short, general state practice is not a *necessary* condition in order to establish a *jus cogens* norm, and, insofar as it is *relevant*, the prospects of adducing it are nothing like as bleak as has been claimed by sponsors of the three difficulties we have just surveyed.

WHICH HUMAN RIGHTS ARE *JUS COGENS*?

In response to the question of which customary human rights norms belong to the category of *jus cogens*, one answer is that necessarily all of them do, since it is of the essence of human rights that to affirm their existence is to believe that they are universally binding quite independently of consent.[31] Or, put another way, in

[30] As Andrea Biondi has aptly put it:

> It is an irony of sorts that we know the most about the effects of a violation of *jus cogens* in the area in which they are least likely to be relevant: the law of treaties. It is indeed highly unlikely that two or more states would make a treaty to commit an act of genocide or to subject certain individuals to torture. "Human Rights and the Magic of *Jus Cogens*," 19 *Eur. J. Int'l L.* 491, 495–96 (2008).

However, the extent of this unlikelihood should not be exaggerated. As Gerry Neuman (personal communication) has pointed out to me, the conflict between extradition treaties and the non-refoulement obligation (assuming the latter is *jus cogens*) is a very real one, as shown by the activities of the Human Rights Committee relating to the refoulement of "extremists" under the Minsk Convention and the Shanghai Cooperation Agreement.

[31] There is more than a whiff of this view in Judge Tanaka's dissent in *South West Africa Cases*: "If we can introduce in the international law field a category of law, namely *jus cogens* … a kind of imperative law which constitutes the contrast to the *jus dispositivum*, capable of being changed by way of agreement between States, surely the law concerning the protection of human rights may be considered to belong to the *jus cogens*." 1966 I.C.J. Rep. 250, 298 (1966). *See also Kadi v. Council of European Union*, Judgment of the Court of First Instance of the European Communities (Second Chamber, Extended Composition), Case T-315/01 ¶ 226–31 (Sept. 21, 2005).

the case of human rights norms, the requisite *opinio juris* for customary status is indistinguishable from that for *jus cogens* status. But this argument purchases its sweeping conclusion too cheaply, since it fails to distinguish between moral human rights norms and legal human rights norms. Commitment to moral human rights norms does indeed involve the judgment that they are morally binding on all states regardless of whether they have accepted or failed to object to them. By contrast, there is nothing incoherent in a state believing that a particular human rights norm is or should be part of customary international law without it also judging, or being committed to the judgment, that it is or should be a norm of *jus cogens*. Moreover, it seems undesirable, from the point of view of human rights morality itself, to adopt the position that customary human rights law more or less automatically has *jus cogens* status. By collapsing these two categories, the predictable effect is to inhibit unduly the growth of the customary international law of human rights. This is because states will be less prepared to affirm human rights as customary if the upshot is that they will thereby also achieve *jus cogens* status. This would deprive us of the real advantages of enshrining certain human rights in customary international law, even if they lack *jus cogens* status.

Arguments such as the one we have just rejected, which seek to establish on conceptual or similar grounds a shortcut to the conclusion that customary human rights norms inherently possess *jus cogens* status, seem overly ambitious. Instead, an independent argument is needed to assess the *jus cogens* status of any given human right in customary international law. The cumulative effect of such piecemeal inquiry would be to arrive at one of the following conclusions:

(a) All customary human rights norms are *jus cogens*.
(b) Only some, but not all, customary human rights norms are *jus cogens*.
(c) No customary human rights norms are *jus cogens*.

In the scholarship on this topic, (b) is the leading contender, but there is also some support for (a) and perhaps to a lesser extent for (c), although the latter often tends to be a specific implication of a wholesale skepticism about the category of *jus cogens* as such. However, it is worth identifying a further possibility which strikes me as highly plausible, but which has tended to go unremarked:

(d) Regarding some customary human rights norms that are *jus cogens*, it is only some aspects of their normative content, and not the whole of it, that possess this status.

In other words, the *jus cogens* status of any given customary human right, such as the *jus cogens* status of the general class of customary human rights, is not an all-or-nothing matter. Just as some human rights can be *jus cogens* and not others, so, too, some aspects of a given human right's normative content can have *jus cogens* status, while other aspects of its content are merely customary. I return to (d) after making some remarks about (b).

A useful starting point in the quest for human rights that are *jus cogens* is the list of human rights that the *Restatement (Third) of the Foreign Relations Law of the United States* indicates are violations of *customary law*: "(a) genocide, (b) slavery or slave trade, (c) murder or causing the disappearance of individuals, (d) torture or other cruel, inhuman, or degrading treatment or punishment, (e) prolonged arbitrary detention, (f) systematic racial discrimination, or (g) a consistent pattern of gross violations of internationally recognized human rights."[32] Of course, the rights prohibiting these wrongs can be properly upheld as *jus cogens* only if there is sufficient evidence of *opinio juris* of the requisite variety, together with any supporting state practice. As the *Restatement* acknowledges, it is an open question whether or not other human rights norms have achieved customary status in addition to those set out in (a)–(f).[33]

There is another way in which the *Restatement* is helpful to us, and that is its observation, made in the Reporters' Notes, that "Nonderogability in emergency and *jus cogens* are different principles, responding to different concerns, and they are not necessarily congruent."[34] Non-derogability in an emergency is certainly not sufficient for *jus cogens* status. After all, as noted previously, *jus cogens* norms are distinguished by the three aspects of universality, peremptoriness, and non-derogability. Even a standard customary or treaty norm can be non-derogable in this sense. Nor does non-derogability in an emergency seem to be necessary; instead, the non-derogability characteristic of *jus cogens* is to be understood in a broader way – as the idea that there is no justification for noncompliance except perhaps by reference to other *jus cogens* norms – which does not obviously entail the more specific idea of non-derogability in an emergency. This more stringent form of non-derogability may be characteristic of some *jus cogens* human rights norms, such as the right not to be tortured, but not all of them. Imposing it sets a needlessly high bar for *jus cogens* status.

Return now to proposition (d). This alerts us to the possibility that some, but not all, aspects of a given human right in customary international law may have acquired *jus cogens* status. No decisive reason exists for making assessments of the *jus cogens* standing of individual human rights on an all-or-nothing basis. There may be requisite support for the *jus cogens* character of some elements of a given norm,

[32] RESTATEMENT (THIRD) OF THE FOREIGN RELATIONS LAW OF THE UNITED STATES § 702 (1987).

[33] *See id.* § 702, cmt. a. Specific reference is made to the right not to be subject to systematic religious discrimination, the right to own and not be arbitrarily deprived of property, and the right against gender discrimination, as norms that have already achieved, or are on the verge of achieving, customary status. *Id.*, cmts, j–l.

[34] *Id.*, rptrs. note 11. *See also* LEPARD, *supra* note 3, at 338–39.

but not for others. So, for example a right to religious freedom may form part of the corpus of *jus cogens* norms. However, it is not necessary for someone defending this position to argue that all of the normative requirements we properly associate with that right in customary law possess *jus cogens* status. So, for example, persecution on religious grounds may be a violation of *jus cogens*, whereas restrictions on individuals' freedom to proselytize or to change their religion, although prohibited by customary international law, lack the requisite basis in *opinio juris* and state practice to be classified as *jus cogens*. Similarly, within the general right to freedom of expression, it may be possible to isolate some core content that possesses *jus cogens* status, such as a right to criticize government officials without fear of punishment, even if other customary law elements lack this status.

At this point, it might be objected that states should not be permitted to pick and choose *jus cogens* status *within* a customary norm of human rights. Is it not disingenuous, the objection goes, for a state to accept as *jus cogens* some part of a human right, but not another? On this view, international law should not connive at assaults on the integrity of human rights norms through the fragmentation of their legal status. But this is a difficult line to maintain once we have admitted that states can be selective with respect to *jus cogens* status within the general class of customary human rights norms. Presumably, they are permitted to do so in order properly to reflect their judgments that some of these norms are better suited for the legal status of *jus cogens* than others. But once we have made that concession, it is difficult to comprehend why similar judgments of relative fittingness for *jus cogens* status should not be permitted *within* a given customary norm of human rights. After all, it is not as if there are no good reasons, relating to such matters as relative moral importance or the likelihood of attracting a widespread consensus, for drawing such distinctions within the content of a given human right. Indeed, it most likely enhances the ability of vital demands of global justice to ascend to *jus cogens* status if we permit this kind of splitting off of human rights norms than if we rigidly insist on an all-or-nothing approach.

CONCLUSION

In this chapter I have sketched the contours of an MJA of customary international law, one that I believe makes best sense of what custom as a source of international law has become in recent decades. It is a moral judgment-based account along at least two dimensions. First, *opinio juris* is construed as centrally involving some form of moral judgment. On the disjunctive view elaborated here, it is either a moral judgment that a given norm is appropriately law or else that it may be appropriately established as law. Second, moral judgment is called for when an interpreter of law seeks to determine whether state practice and *opinio juris* suffice to establish the existence of a putative customary norm. I have argued that, in this interpretative process, it is possible for a customary norm to be established predominantly on the basis of widespread *opinio juris*, despite the absence of supporting state practice. Often it is an overly inclusive conception of "state practice," one that overlaps substantially with *opinio juris*, that obscures this possibility in the analyses offered by more conservative commentators.

Beyond defending the MJA, I have also argued that human rights norms of *jus cogens* status are a subclass of customary norms of human rights. *Jus cogens* norms are those customary norms whose grounding *opinio juris* and state practice establishes them as universal, peremptory, and non-derogable norms. I rejected both the thesis that human rights customary norms are automatically *jus cogens* and the contrasting thesis that there are near-insuperable barriers to a human rights norm achieving either customary or *jus cogens* standing. Instead, I argued that a subset of customary human rights norms are *jus cogens* in status and, moreover, that sometimes only a subset of the obligations associated with such a right is *jus cogens*. I have not, however, offered more than a general framework and some conjectures regarding the normative question of which human rights norms *ought to be* elevated to *jus cogens* status in international law. This is a topic that must be left for another time, and one which I am inclined to believe is best addressed in a piecemeal fashion.

5

Customary International Law

How Do Courts Do It?

Stephen J. Choi and Mitu Gulati

INTRODUCTION: THE ODIOUS DEBT DEBATE

Around 2003, there was a surge in interest among international debt scholars in the question of what evidence courts look to in determining customary international law (CIL). The new post-Saddam Iraqi government had inherited an economy saddled with roughly $140 billion in Saddam-era debt. Since portions of that debt had almost certainly been incurred to fund things such as weapons to put down insurrections and palaces for the dictator, it is unsurprising that the post-Saddam Iraqi government bridled at the prospect of having to pay it. A key legal question, therefore, was whether international law excused a new government from paying some or all of the debts of a prior regime that had been, to use the technical term, "odious." Or, did the international law governing governmental succession to debts contain an exception for odious debts?

There are two primary sources of international law, treaties and customary law. On the treaty front, the drafters of one international treaty, the 1983 Vienna Convention on the Succession of States, had discussed the inclusion of an odious debts exception to the general rules of state and governmental succession. But the idea failed to garner support beyond a handful of states.[1] The next question, therefore, was whether an odious debt exception existed as a matter of custom. The treatises on international debt and state succession were clear on the rule: governments had to pay the debts of their predecessor governments regardless of matters such as political differences.[2] Many of the treatises, however, were from a half-century or more prior and the norms of international law had changed dramatically during that

1 *See* STEPHANIA BONILLA, *ODIOUS DEBT: LAW AND ECONOMICS PERSPECTIVES* 11 (2010).

2 *See* Lee C. Buchheit, Mitu Gulati, & Robert B. Thompson, "The Dilemma of Odious Debts," 56 *DUKE L.J.* 1201 (2004) (describing the prior literature on the laws of state and governmental succession).

post–World War II period. Scholars, therefore, took on the question of whether it could be argued that, over time, a customary exception of odious debts had evolved.

Even a rudimentary sense of history told these scholars that there had been occasions when governments had refused to pay debts of prior regimes. In the early 1900s, the communist governments in Russia and China had famously refused to pay the debts of the prior monarchies. In the United States in the 1800s, a number of post-Reconstruction southern governments had explicitly repudiated the debts of their predecessors on the grounds that those prior regimes had been illegitimate. More recently, there were multiple examples of popular uprisings where there had been consideration of challenging the legitimacy of contracts entered into by prior oppressive regimes. Among those were: Iran after the Shah, South Africa after the end of apartheid, Haiti after Baby Doc, the Philippines after Marcos, and Argentina after the Dirty War. But was there evidence from these overthrows of oppressive regimes that could be added up to make a claim that a legal defense had evolved, so much so that Iraq could assert it in court or arbitration against creditor claims brought by those who had made loans to Saddam's government?[3]

The textbooks and treatises generally set forth a two-part test that is to be applied to the historical evidence to determine whether CIL can be said to have emerged. Sean Murphy articulates the standard view when he writes that CIL exists when there is evidence of: "(1) a relatively uniform and consistent state practice regarding a particular matter; and (2) a belief among states that such practice is legally compelled."[4] Similarly, the *Restatement (Third) of the Foreign Relations Law of the United States*, states that CIL "results from a general and consistent practice of states followed by them from a sense of legal obligation."[5]

To return to the search for an odious debt doctrine a decade ago, while the authorities seemed to be in agreement on the need for significant evidence of state practice and state belief (also referred to as *opinio juris*), the terms they used to describe the quantum of state practice and belief required were all rather vague. They were terms such as "uniform and consistent," "settled," "extensive," or "virtually

3 As it turned out, the primary forum where the Iraqi debts were resolved was neither a court nor an arbitration, but the Paris Club. The Paris Club is a setting where the rich debtor countries gather in Paris, under the auspices of the French Ministry of Finance, to determine, collectively, how much relief to give poorer nations that have defaulted on their debts to the rich nations who are members of the club. The issue of whether there was legal defense against debt claims on the ground of odiousness was relevant here, though, as well.

4 Sean D. Murphy, *Principles of International Law* 92–93 (2d ed. 2012).

5 *Restatement (Third) of the Foreign Relations Law of the United States*, § 102 (2) (1987). Much quoted in the literature are statements in a set of classic ICJ cases. Those include the ICJ's *North Sea Continental Shelf Cases*, stating that for a practice to amount to CIL, the practice must be "settled" (1969, ICJ Rep. 3, ¶ 77) and the ICJ's *Gulf of Maine* case, stating that the presence of *opinio juris* can be tested based on the analysis of a "sufficiently extensive and convincing practice" (1984, ICJ Rep. 246, ¶ 111).

uniform." No one provided a translation of what those dictates meant in terms of how much practice or belief there needed to be, either in terms of the number of states engaged in the practice or belief or for how long they needed to have been engaging in it.[6] Nor did the treatises explain how the practices or beliefs of the various states were to be weighed in making the ultimate determination. Were, for example, the views and practices of states that had historically been more powerful, such as Great Britain or the United States, to be given more weight than those of the weaker and smaller states such as Iceland or Malta? Or what about giving differential weights to the more populous states such as India and China over a sparsely populated but geographically large state such as Argentina? Or, for that matter, what constituted state belief? Was it the beliefs of the population of a state or the senior government officials, and, if the latter, what types of officials?

For those working on building an argument in favor of the odious debt exception in the period 2003–04, it was frustrating that the authoritative materials provided little sense of what type or how much evidence we needed. Nevertheless, the standard doctrinal analysis seemed clear on a couple of things. First, there needed to be an empirical analysis – the doctrine was satisfied only if a certain quantum of evidence was obtained (we did not know how much, but the definition suggested that a significant amount was needed). Working from the doctrinal definition alone, one might think that, at a minimum, the practice probably would need to have been engaged in or explicitly accepted consistently over a significant period of time by at least a majority of all the states in existence.[7]

[6] *E.g.*, Curtis F.J. Doebbler, *Introduction to International Human Rights Law* 33 (2006) ("Just how large a number of states are necessary to create a rule of customary international law is not settled."). For a wonderfully vague attempt to translate the definition of CIL into the quantum of evidence required, see Herman Meijers, "How Is International Law Made?" 9 *Netherlands Y.B. Int'l L.* 3, 5 (1978).

[7] In the course of discussing this question in class at the time, one of us asked a set of second-year law students to translate the standard two-part definition of CIL into rough numbers. The table below breaks down the responses of roughly eighty students at Duke Law School and the University of Virginia Law School in 2004 to the question to them of "How Many States Does it Take?"

Fraction of states who need to engage in the practice for it to become CIL	Fraction of students who specified this percentage
0%–10%	0%
10%–20%	0%
20%–30%	0%
30%–40%	0%
40%–50%	0%
50%–60%	12%

(continued)

Based on what was probably a similar reading of the standard rule for when CIL could be said to have evolved, multiple scholars from around the globe spent months, if not years, searching the historical evidence on the odious debt exception. No one was able to come up with more than a half dozen or so of instances where the evidence even arguably fit the doctrinal demands – that is, cases of governments not only repudiating debts of prior odious governments, but seeming to do so with a belief that they were legally entitled to because of the odiousness.[8]

In terms of the evidence that was unearthed, there were two historical occurrences cited by almost every commentator writing on the topic. The first piece drew from the Spanish-American treaty negotiations of 1898. In the aftermath of the Spanish-American War, when the United States was negotiating the removal of Cuba from Spanish control, the question arose as to responsibility for Spanish debts that had been secured by Cuban revenues. U.S. negotiators argued that Cuba did not bear responsibility for those debts because the loan proceeds had been used in a manner contrary to Cuban interests – indeed, that they had been used, in part, to counter the Cuban opposition to the Spanish. The Spanish delegation rejected the argument. But something resembling an odiousness-type argument appeared to have been made.

The second piece of evidence was typically drawn from an arbitral proceeding in 1923 between Costa Rica and the Royal Bank of Canada. The Royal Bank of Canada had made loans, purportedly to the Costa Rican state, but that seemed to go directly into the hands of the then-dictator, Fredrico Tinoco. The arbitrator, Chief Justice Taft of the U.S. Supreme Court, ruled in favor of excusing the post-Tinoco Costa Rican government on the ground that the loans were full of "irregularities." Translated, it should have been clear to the lenders that the loans in question were for Tinoco's personal purposes rather than for the Costa Rican state. This rationale was not quite an application of the odiousness doctrine, but it is in the ballpark in its excusal of a debt obligation based on the purpose of the debt not for the Costa Rican state.

Fraction of states who need to engage in the practice for it to become CIL	Fraction of students who specified this percentage
60%–70%	0%
70%–80%	40%
80%–90%	0%
90%–100%	48%

8 *See, e.g.*, Al Brophy et al., "*Applied Legal History: Demystifying the Doctrine of Odious Debts*," 11 THEORETICAL INQUIRIES IN L. 247 (2010).

Once one got beyond these two pieces of evidence, scholars varied in what they pointed to as possible support. Many cited to the Chinese, Russian, and U.S. state repudiations mentioned previously, the preliminary drafts of the 1983 Vienna Convention, and the writings of an obscure Russian scholar who some writers suggested had been an adviser to the Czar in the early 1900s (turned out he was not). A smaller set of scholars added in claims made by Iran after the fall of the Shah and perhaps preliminary steps taken in the Philippines after the fall of Marcos. Thus, except for perhaps the most ardent proponents of the doctrine, there were no more than about six or seven instances of state repudiations of debt where there had been contemporaneous statements of state officials asserting something close to an odiousness rationale for their actions.[9] There was also the failed attempt at a treaty exception in Vienna and the writings of the obscure Russian scholar. In sum, there were maybe a dozen or so weak pieces of evidence, nothing close to the majority or supermajority approval level that the standard understanding of CIL seemed to require.[10]

In the end, Iraq ended up restructuring its debts without explicitly invoking the odious debts doctrine.[11] The United Nations Security Council passed a directive immunizing Iraq's oil revenues from seizure by creditors; and that protection from suit enabled Iraq to obtain significant relief from its creditors.[12]

Despite the resolution of Iraq's debt, the evidentiary question of CIL poses a continuing puzzle. The standard doctrine for finding CIL, on its face, seemed impossible to satisfy in the odious debt doctrine area. Taken seriously, CIL could

[9] There was almost never any evidence that the statements were backed up by a belief on the part of the officials making the statements that international law supported such an odiousness defense to creditor claims.

[10] *See* Robert E. Scott & Paul B. Stephan, *The Limits of Leviathan: Contract Theory and the Enforcement of International Law* 207 (2005) (describing the evidence as "scant"); Buchheit et al., *supra* note 2; Anna Gelpern: "What Iraq and Argentina Might Learn From Each Other," 6 *Chi. J. Int'l L.* 391 (2005) (similar); Christoph G. Paulus, "The Concept of 'Odious Debts': A Historical Survey," Duke Law School Working Paper (2007) (available at http://papers.ssrn.com/sol3/papers.cfm?abstract_id=1077688) (similar). *But see* August Reinisch, "A History of the Doctrine of Odious Debts: Serving Individual/Bilateral or Community Interests," in *From Bilateralism to Community Interests: Essays in Honor of Judge Bruno Simma* 1239 (Ulrich Fastenrath et al., eds. 2011) (getting to his conclusion by largely ignoring the dictates of the standard definition of CIL and looking instead at more general legal principles).

[11] There were mutterings, including in the Iraqi legislature, about how Saddam's debts were illegitimate and should not have to be paid. There was no meaningful discussion of the international law issues, however, as best we know, in the legislature. That discussion took place primarily in academic and policy papers and ultimately did no more than provide some of the atmospherics for the debt negotiations between Iraq and its creditors. Jai Damle, "The Odious Debt Doctrine After Iraq," 70 *L. & Contemp. Prob.* 139 (2007).

[12] *See* Lee C. Buchheit, Mitu Gulati, & Ignacio Tirado, "The Problem of Holdout Creditors in Eurozone Sovereign Debt Restructurings," *Butterworths J. of Int'l Banking & Fin. L.* 191 (2013).

never emerge with respect to odious debt because of the lack of long-term adoption of such a doctrine by a majority of countries that consider the application of the doctrine part of their legal obligation. Yet, in other areas, courts find that CIL emerges quite regularly. That disparity then set up an intriguing set of research questions: Is the evidence equally as sparse in other areas, and, if so, how were judges periodically finding that it was nonetheless satisfied in determining the presence of CIL? How much evidence did it really take to persuade a court? Was there some other kind of analysis that judges were using in the cases where they were finding CIL?

After a brief summary of literature examining CIL evolution, the paper moves to the data and describes results from 175 CIL determinations from the International Court of Justice (ICJ) and a variety of other international tribunals.

BACKGROUND LITERATURE: THE THREE CIL PUZZLES

It turns out that the scholars struggling with applying the standard doctrine of CIL evolution in the Iraqi debt context were not alone in their frustration. Numerous scholars within the international law community have both expressed frustration with the standard doctrine and observed, at least casually, that courts (and especially international courts) do not appear to be following the formal doctrinal requirements.[13] Three of the frustrations expressed by the scholars stand out.

Implausible and Inefficient

The textbooks tell us that CIL is formed only when there is evidence of a consistent and established practice among states *and* when states engage in that practice out of a sense of legal obligation. But, there is a circularity in the two parts of this rule; for a certain set of acts to become legal, the rule requires that states believe that those acts are legal before they are legal. As a practical matter, how is it possible for a multitude of states to be engaging in this practice for a long period of time without realizing that what they think is lawful is in fact not legal under existing international law?[14] It is

[13] *See, e.g.*, ALAN BOYLE & CHRISTINE CHINKIN, *THE MAKING OF INTERNATIONAL LAW* 279 (2007) ("The [ICJ] has not been consistent in applying its own criteria for the determination of customary international law."); Rudolf H. Geiger, "Customary International Law in the Jurisprudence of the International Court of Justice: A Critical Appraisal," in *FROM BILATERALISM TO COMMUNITY INTEREST: ESSAYS IN HONOR OF JUDGE BRUNO SIMMA* 673, 692 (Ulrich Fastenrath, et. al. eds., 2011) ("In general the Court does not follow its self-proclaimed method of finding customary international law."); Monica Hakimi, "Custom's Method and Process: Lessons from Humanitarian Law" (in this volume) (similar); A. Mark Weisburd, "The International Court of Justice and the Concept of State Practice," 31 *U. PA. J. INT'L L.* 295 (2009) (similar).

[14] One way around this problem is to conceptualize CIL as a gap-filling doctrine that judges create in situations where there is no law, but law is needed to resolve the dispute at hand. This is not,

not, unless one either ignores the strict dictates of the doctrine or believes that a large fraction of states are systematically mistaken about the law over long periods of time. This is an implausible scenario, many scholars have concluded.[15] But even assuming that scenario, which states would be willing to follow the rules evolving out of such an evolutionary pattern? As Curtis Bradley asks in his contribution to this volume: Would states really want the law governing their practices to arise solely out of the mistaken beliefs of their brethren? In other words, the evolution of law would be decided by those states least knowledgeable about the law (and, therefore, most likely to make systematic mistakes over the long term).[16] Another implausible scenario.

Superhuman Research Skills

There are roughly 200 states in the world. Even the most expansive reading of the doctrine of CIL evolution surely requires affirmative evidence of practice of at least a quarter of this number of states, and that, too, over a long period of time. At the end of a couple of years of research on odious debts, the conclusion that many scholars reached was that it was impossible to collect the kind of evidence that the definition required for even a small subset of these states. Not only was the type of evidence being requested unlikely to exist as a logical matter as discussed (over fifty states engaging in a practice for multiple decades, all the while believing that they were acting consistent with the law, even though they were wrong about the law?), but it was impossible to collect, as a practical matter, unless one somehow assembled an extraordinary team of anthropologists, economists, historians, political scientists, and lawyers who would then be able to spend decades excavating the historical record.[17] Again, implausible.

however, the standard conception of CIL. For discussions of the circularity problem, see, for example, Curtis A. Bradley, "Customary International Law Adjudication as Common Law Adjudication" (in this volume); Olufemi Elias, "The Nature of the Subjective Element in Customary International Law," 44 *Int'l & Comp. L.Q.* 501, 504 (1995); Raphael M. Walden, "The Subjective Element in the Formation of Customary International Law," 12 *Isr. L. Rev.* 344, 363 (1977).

15 *See, e.g.*, G. J. H. Van Hoof, *Rethinking the Sources of International Law* 99 (1983) ("It must be quite an extraordinary system of law which incorporates as its main, if not the only, vehicle for change the violation of its own provisions."); Michael Akehurst, "Custom as a Source of International Law," 47 *Brit. Y.B. Int'l L.* 1, 8 (1974–75) ("There is no doubt that customary rules can be changed in this way, but the process is hardly one to be recommended by anyone who wishes to strengthen the rule of law in international relations.").

16 *See* Bradley, *supra* note 14. John Tasioulas, for his part, points out that this lack of logical structure also undermines the legitimacy of any law that might be created. *See* John Tasioulas, "Custom, *Jus Cogens*, and Human Rights" (in this volume).

17 For observations along similar lines, see Stefan Talmon, "Determining Customary International Law: The ICJ's Methodology Between Induction, Deduction and Assertion," Bonn Research Papers on Public International Law, Paper # 4/2014 (2014 draft).

Aspirational CIL

Last is the puzzle of what is sometimes called "modern" or "soft" (as opposed to "traditional" or "hard") CIL.[18] This type of CIL is aspirational. Often, there is little evidence of past practice to support it. If anything, on matters such as torture, environmental harms, nuclear proliferation, and international financial coordination, where claims of modern CIL are sometimes made, either state practice or *opinio juris*, if seriously examined, would probably cut in the opposite direction.[19] Yet such aspirational CIL doctrines are periodically asserted to have arisen, even though the standard definition does not appear to allow for exceptions. Indeed, as commentators have pointed out, there is often little that is customary about this modern CIL.[20] Relatedly, there is the matter of international treaties sometimes being used as evidence to support the existence of CIL. The use of treaties as evidence poses a puzzle because many treaties are likely entered into by states because there is an absence of law (hence the need to enter into a treaty to provide law).[21] In other words, treaties may be at best evidence of aspirational CIL. The exception here might be treaties that were entered into to clarify the law. That said, even here there is a puzzle because the traditional definition requires that the customary practice in question have been well settled and widely understood for a long period of time. That kind of customary practice hardly needs clarification.

Despite the unhappiness with the traditional CIL definition and the implausibility of its application that many have expressed, the standard treatises seem to be willing to paper over these problems and assume that they are somehow overcome in practice. Bradley's paper for this volume quotes an illustrative passage from the *Restatement on Foreign Relations Law* that basically says, "Yes, there are problems,

[18] *E.g.*, Niels Peterson, "Customary International Law and Public Goods" (in this volume); Anthea E. Roberts, "Traditional and Modern Approaches to Customary International Law: A Reconciliation," 95 *Am. J. Int'l L.* 757 (2001); Anja Seibert-Fohr, "Modern Concepts of Customary International Law as a Manifestation of a Value-Based International Order," in *Unity and Diversity in International Law* 257–83 (Andreas Zimmerman & Reiner Hoffman eds. 2014).

[19] *E.g.*, Melissa Boudreau & Mitu Gulati, "The IMF's Imperiled Priority," *Duke J. of Const. L. & Pol'y* (forthcoming 2015).

[20] *E.g.*, Daniel Bodansky, "Customary (And not so Customary) International Environmental Law," 3 *Ind. J. Global Legal Stud.* 105, 118 (1995); Niels Petersen, "Customary Law Without Custom? Rules, Principles, and the Role of State Practice in International Norm Creation," 23 *Am. U. Int'l. L. Rev.* 275, 303–05 (2008); Bruna Simma & Philip Alston, "The Sources of Human Rights Law: Custom, Jus Cogens, and General Principles," 12 *Aust. Yearbook Int'l L.* 82, 89–90 (1988–89).

[21] *See* Bradley, *supra* note 14. For a discussion from a half-century ago of the debate over whether treaties should be a source of evidence in CIL determinations, see Anthony D'Amato, "Treaties as a Source of General Rules of International Law," 3 *Harv. Int'l L. J.* 1 (1962).

but they are surmounted in practice because CIL does, after all, evolve."[22] But if the doctrine is (a) impossible to apply and (b) it is implausible to expect that anyone would want to apply it, then *how* is it being applied?[23]

In terms of the literature, there appears to be little in the way of empirical examinations of court determinations of CIL. This chapter is an attempt to take a preliminary step in that direction and ask: What do judges do when they are faced with a doctrine that is both impossible to apply and would produce normatively unattractive results if it could be applied? The data collection process is as yet ongoing, but our initial conclusions are the foregoing. First, in terms of what the judges are *not* doing: Judges, while often paying homage to the formal definition of CIL, do not try to follow it with any meaningful degree of fidelity. Second, in terms of what they *are* doing instead (and this is necessarily more speculative): They are engaging in what is essentially a form of pragmatic common law decision making in trying to figure out when it is that a new rule of CIL needs to be announced.

DATA AND MEASURES

How Do Courts Find CIL?

Our question is simple. We want to know how courts apply the dictates regarding the requirements for finding that a norm has evolved into CIL.

Our starting assumption was that courts probably do look carefully at evidence of state practice – that is, at prong one of the traditional definition of CIL (we were wrong). What we were most interested in was how courts tackled the analysis of prong two, the *opinio juris* bit of the equation. In theory, the best approach would have been to code the various court decisions for the evidence that courts examine on the second prong alone. Unfortunately, we quickly realized when we began our coding that the court decisions were not going to cooperate. Courts in this area, it turns out, do not neatly separate out the evidence that they look at in terms of saying X piece of evidence helped persuade them on prong one and Y piece of evidence helped persuade them on prong two. Instead, they tend to bundle all the evidence into a single discussion and then assert whether the two-prong test is satisfied (and sometimes they do not even mention the two-prong test). Given that reality, our inquiry reports on the evidence that

[22] *See* Bradley, *supra* note 14. Not all scholars find as much fault with doctrinal requirements for CIL evolution. *E.g.*, Joseph W. Dellapenna, "The Customary International Law of Transboundary Fresh Waters," 1 *Int'l J. Global Env'tl Issues* 264, 267 (2001) ("Despite obvious difficulties in determining the obvious content of customary international law, the system has been rather remarkably successful.").

[23] *See also* Tasioulas, *supra* note 16 (expressing frustration with the "it works in practice and therefore requires no explication in theory" position).

courts look to when confronting CIL, and then tries to back out the question of what pieces are likely to have impacted the *opinio juris* determinations.

Which Courts?

Historically, the most important court in terms of determinations of customary international law has been the International Court of Justice.[24] It is the highest court in the hierarchy of international courts and the principal legal organ of the United Nations.[25] This is the court whose determinations are cited most often by scholars and other courts as the key authority in terms of what CIL is and how it should be determined. We therefore collected all of the judgments of the ICJ that could arguably be said to have made determinations of CIL. We supplemented those cases with the judgments of the ICJ's predecessor court, the Permanent Court of International Justice (PCIJ). In addition, we examined a subset of decisions from the numerous other international tribunals, such as the International Tribunal on the Law of the Sea. We did not have the resources to collect the decisions of every one of these secondary international tribunals. Hence, we tried to limit ourselves to the most important decisions. We did that by looking to all cases from these tribunals that were cited in the 2013 report of the International Law Commission's (ILC) study group on CIL.[26] Finally, we considered a random sample of roughly fifty individual opinions authored by judges in the decisions mentioned, to see whether judges writing separately were applying the standard definition differently.

To obtain our observations, we searched each database of cases for all of the judicial opinions that discussed any one of the following terms: "custom," "international law," "customary international law," "*opinio juris*," "practice," or "law of nations." As noted, we supplemented the list of cases here with the cases mentioned in the 2013 draft of the ILC's report on CIL.[27] Once the cases had been identified, we examined

[24] Alberto Alvarez-Jimenez, "Methods for the Identification of Customary International Law in the International Court of Justice's Jurisprudence: 2000–2009," 60 *Int'l & Comp. L.Q.* 681, 685 (2011) ("When it intervenes, the [ICJ] remains as a prominent actor within the process of creation and evolution of customary international rules."); Christian J. Tams, "The ICJ as a 'Law-Formative Agency': Summary and Synthesis," in *The Development of International Law by the International Court of Justice* 377, 378 (2013) ("ICJ pronouncements are generally considered to be authoritative pronouncements on the law.").

[25] Guide to Archives of International Institutions, available at http://www.unesco.org/archives/sio/Eng/presentation_print.php?idOrg=1016.

[26] The ILC consists of prominent international lawyers who are elected by the UN General Assembly and are charged with developing and codifying international law. For details, see "Origin and Background of the Development and Codification of International Law, International Law Commission, available at http://legal.un.org/ilc/ilcintro.htm.

[27] Because of our primary interest in a question about sovereign debt, we also added any international tribunal cases that we could find that tackled the topic (there were only a handful).

the portions of the opinions discussing CIL and coded each determination of CIL for the types of evidence that were discussed as either rising to the level of CIL or not. Within a single case that might have multiple CIL issues in it, we coded each issue separately. If there were dissenting or concurring opinions that were doing independent analyses of the CIL question, we coded those separately, as well. The goal was to examine a wide array of judicial determinations so as to try and understand how judges and courts got around the intractability problem in the traditional definition of CIL.

We encountered a few threshold issues. The first was whether to code only opinions that concluded that there was CIL or also those that concluded that CIL was not present. The argument in favor of limiting the data to cases affirmatively finding CIL is that since we are interested in examining the type and amount of evidence courts think is adequate to find CIL, we should limit our analysis to those instances where CIL was found. The counterargument is that the types of disputes that end up in court – particularly at the appeals level or at the level at which states are willing to go to an international tribunal – will necessarily be those where the legal question is a close one. Otherwise, as the research on the economics of litigation teaches us, the cases would have settled. This approach is also relevant if one believes that – consistent with a large body of literature on judicial behavior – factors other than simply the quantum of evidence can make a difference as to the outcome in close cases.[28] We accordingly include both cases that find and do not find CIL (although we test the difference between such cases).

The second threshold issue was whether to code only cases from one type of court (the ICJ, for example) or include a variety of types of courts. On the matter of including multiple courts, including domestic courts, the argument in favor of limiting the analysis to international tribunals such as the ICJ is that domestic courts are often going to be biased toward domestic interests and that therefore an examination of their analyses is less valuable.[29] For us, it would have been interesting to be able to examine a wide range of domestic tribunals, but resource constraints again kept us focused on the international tribunals.

Which Variables?

In choosing the variables to code for, we began with the standard rule (extensive practice plus *opinio juris*) and set forth the types of evidence that the definition

[28] *Cf.* Eric A. Posner & Miguel F.P. Figueiredo, "Is the International Court of Justice Biased?" 34 *J. Legal Stud.* 599 (2005).

[29] Along these lines, see Eyal Benvenisti, "Judicial Misgivings Regarding the Application of International Law: An Analysis of the Attitudes of National Courts," 4 *Eur. J. Int'l L.* 159 (1993).

likely contemplated. Next, we did an initial round of coding of roughly a dozen ICJ cases to look at what types of evidence they were citing as support for their inquiries. The combination of those two steps gave us eleven evidentiary variables to code for. The variables fit roughly into four categories. By looking at the kinds of evidence that courts use in applying the standard CIL rule of evolution, we hope to draw inferences about how they are tackling the problems posed by the rule.

First, there are the variables that capture the use by courts of direct evidence supporting the standard CIL definition ("Direct Evidence"). Second are the variables that capture the degree of delegation by courts of the task of collecting and analyzing the evidence to third parties, such as treatise writers or academics ("Delegation Evidence"). Third are the variables capturing something that is not at all traditional CIL, but that we will call "aspirational" CIL ("Aspirational Evidence"). Finally, there are the variables that capture what one might call the procedural elements in the CIL calculus ("Procedural Evidence").

Direct Evidence Supporting the Standard Definition

State acts: Part one of the standard definition of CIL tells us to look for widespread and settled evidence of state practice. The first variable we coded for, therefore, was evidence of actions by states. Specifically, assertions along the lines of, "On X date, State Y took Z action."

Statements by state officials: Under the traditional definition of CIL, states not only have to act, but they also have to believe that their actions are consistent with international law. That is, there needs to be *opinio juris*. Courts might cite to evidence of statements by state officials indicating their understanding of certain legal obligations. There might also be evidence of what other states believe at the time.

Evidence of Delegation to Intermediaries

Treatises: Our expectation was that academic treatises would play an important role in assisting courts with their determinations of CIL. Assuming that treatise writers seek the fame that comes with being cited by courts as authoritative, there is an incentive for these writers to build reputations for credibly reporting and analyzing the historical evidence that points to or away from a certain set of practices having evolved toward CIL status.

Academic articles: The rationale for academic articles constituting evidence of CIL is the same as with treatises.

Reports of international committees: Interest groups often have distinct perspectives on how law should evolve. They therefore have an incentive to try and influence the way in which CIL evolves by putting together groups of eminent lawyers and scholars to do the kind of background research work that courts are unable to do. Examples of the types of bodies that we code for under this variable include the International Law Association, expert committees put together by the Red Cross, and various bar association committees. We code both citations to committee reports and conference reports under this variable.

Note, however, that we are making substantial assumptions regarding the content of some of these types of materials. There will certainly be academic articles and committee reports that are largely normative, which would perhaps better fit under the next category of evidence.

Evidence of Aspirational CIL

Treaties: The problem with looking solely at the statements of state officials as evidence of their understanding of the law is that those officials might be acting strategically in making those statements. In this context, treaties might provide more credible evidence of state beliefs. After all, a state that enters into a treaty is subjecting itself to legal liability. Here, however, we encounter an issue under the traditional definition of CIL: Treaties are good evidence of what states want the law to be, but they are not necessarily good evidence of what the law is. After all, the need for a treaty will often arise because of the absence of law, not when it is widespread and well established.

UN resolutions: UN resolutions are relevant to CIL determinations in the same manner that treaties are: They are indications of what states consider to be important issues for the international community. However, like treaties, they are also problematic as a source of evidence of CIL since it is unlikely that there would be a need for a UN resolution on topics where there was already longstanding and widespread agreement on a certain principle of international law. Plus, there is the problem of cheap talk, since nations may say things in the UN and not necessarily intend to align their practice with the statements; something made easier given that UN resolutions are not legally binding. Along those lines, it is hard to see the presence of a UN resolution as affirmative evidence of an existing CIL rule.

U.N. conference and committee reports: The different organs of the UN are constantly producing a vast amount of material on issues of international concern, such as: the proliferation of nuclear weapons, labor rights, human trafficking, environmental standards, and odious debts. The level of support from the various nations for these UN studies or conference reports can range from none at all to widespread support. At most, this type of evidence suggests that there are issues of concern. Individual states may support or even fund UN conferences and reports, but that does not tell us whether the state supports CIL on the matter. The state may simply be interested in studying the matter or seeking to placate some minority domestic constituency.

Domestic statutes: Domestic statutes may indicate individual state preferences as to the future of international law. Or, even if those preferences vis-à-vis international law have not formed, perhaps courts could read an aspiration in that direction into them. The link is not necessarily obvious, however. The fact that a state has local legislation on an issue affecting its citizens does not mean that it takes the same view regarding the rights or obligations of citizens of other countries. Both contemporary and historical practice are rife with examples of how states, particularly the rich ones, have very different conceptions of what their own citizens are entitled to and what outsiders to their nations are entitled to. Would one infer, for example, from the domestic statutes in northern European nations that there is CIL mandating that a basic level of social support be provided in all nations (and that all other nations are responsible for)? Probably not.[30] Objections aside, we code for the number of statutes cited because statutes do seem be cited as support of CIL.

Evidence on Procedural or Other Background Matters

Case Law – international tribunal cases: Prior cases are not, in and of themselves, evidence of either state behavior or *opinio juris*. That is, unless the argument is made that courts are state actors. And they may be in some very rare instances, such as where the state has delegated certain tasks to the courts.[31] But, for the most part, courts in most jurisdictions and in most contexts are avowedly not seeking to reflect either the views of the executive or those of the general public. Prior court decisions might get cited, however, because they set precedent in terms of how CIL cases are to be decided or what kinds of evidence had been considered relevant in a prior dispute. For that reason, we code them as a procedural form of evidence. We have been asked whether it is possible that these prior cases are being cited because the prior case had done an extensive evaluation of the evidence of state practices and *opinio juris*. For that reason, our rule of thumb while coding was to look for any indication that this was the reason for the citation (and if there was, we would go to the prior case to count the evidence). But absent any such indication, we did

[30] The exception here might be statutes dealing with matters such as the immunities of foreign states within a domestic jurisdiction. These statutes, unlike, say, a domestic statute specifying the limits on punishment of minors, are perhaps properly viewed as state practice vis-à-vis another state. In future work on this topic, we plan to try and separate out the citations to statutes by the types of statutes. The numbers of statutes cited overall, however, are small enough that the basic results should remain unchanged.

[31] In such cases, we would expect the court to explain the context to us. We did not find any such explanations backing up court citations in our data, however.

not assume that the citation to the prior case was a proxy for some other kind of evidence.

Case law – domestic cases: The rationale for using municipal court decisions as evidence is the same as that for international tribunal cases.

We recognize that others might disagree with our categorizations of the variables. For that reason, we report the results on all the variables individually.

In addition to the foregoing variables, we also coded for agreements among the parties regarding what the applicable CIL was ("Parties' Agreement"). We had not expected this to be a relevant form of evidence, but came across it in a handful of cases.

Our unit of analysis is an individual CIL determination. For each case, we look at each opinion in the case, including majority, concurring, and dissenting opinions. Within each opinion, we search for CIL determinations, treating each separate CIL determination as a separate observation. For each CIL observation, we coded the number of unique evidentiary items that were cited. If the treaty creating the International Monetary Fund was cited, we counted that as 1 in the treaty column. If, in addition, the treaty implementing the European Stabilization Mechanism was also cited, our number of treaties cited went up to two. We did not count the number of times the same piece of material was cited, however; just whether it was cited at least once.

ANALYSIS

The Total Counts

Figure 5.1 provides a breakdown of the full dataset, in terms of the types of evidence described (for simplicity, we have combined some of them). This project began with the inquiry of how much and what kind of evidence was enough to satisfy the traditional CIL definition. For example, how many statements from state officials were being discussed as relevant in each judicial determination with respect to CIL?

Our dataset consists of 175 different CIL determinations by international tribunals. Each determination is an observation in our analysis. When we began this project to inquire into how courts determined *opinio juris*, we expected to primarily find courts citing evidence of various government officials making statements as to what they believed CIL in a particular case to be. This, after all, would be classic *opinio juris*, if one could somehow solve the problem of determining whether the statements were genuine or strategic. What we see in the data, however, is that there are only 32 citations to statements by government officials in our total 175 determinations.

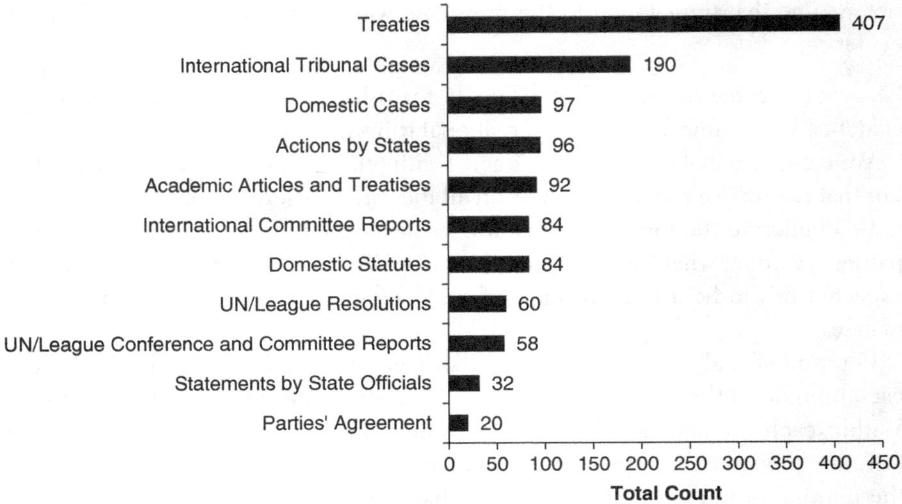

FIGURE 5.1. Materials cited in finding customary international law (N = 175).

Courts examine statements of state officials only rarely, inconsistent with the view that courts follow the classic *opinio juris* method of determining CIL.

Instead, the type of evidence that looks to be the most important for determinations of CIL is the international treaty. In the 175 CIL determinations, there were 407 citations to treaties. Compared with our other types of evidence, treaties are the most important type of evidence in CIL determinations. After citations to treaties come citations to prior cases (domestic and international) – what we expect are largely citations explaining procedural context; that is, how prior determinations of similar issues might have been made. As noted earlier, the fact that treaties are cited at all as evidence of CIL is puzzling if one expects to see that the traditional definition is being followed. Treaties, after all, are often entered into because there is no existing law and nations need some law. Or, more charitably, perhaps the existing law is unclear, and the treaty is adding clarity and providing specifics – but even here, the fact that clarification was needed suggests that the legal norm was not well established (that is, the opposite of what one needs to show to satisfy the CIL definition). Yet here, we find that not only are treaties cited, but they are the dominant type of material cited in CIL determinations.

Although most treaties are forward-looking, and the traditional definition of CIL asks for backward-looking evidence, there are exceptions to the objection to treaties as evidence for CIL. In the case of some treaties, for example, the drafters of the treaty might purport to be doing nothing more than codifying established and consistent state practice. In those cases, however, one would expect the court in question to specify that it was citing a treaty because that treaty represented a codification of past state practice and *opinio juris*.[32] And this does happen. But it happens rarely. To examine this question, we took a random subset of fifty of our coded CIL determinations and examined the question of whether there was any indication on the part of the judge that the treaty being cited was meant to codify existing practice. We found such indications in fewer than 20 percent of the determinations. If one plausibly assumes that the remainder of the treaties were cited as evidence of what states wanted as their laws, then it strikes us as reasonable to conclude that the dominant form of evidence being cited is forward-looking or aspirational.[33]

And it is not just treaties that receive a high amount of attention from courts in their CIL determinations. Treaties dominate, but three other forms of evidence that are largely aspirational in character also play a big role in CIL determinations. Those are UN resolutions, other UN material (committee reports, conference reports, etc.), and domestic statutes. Added together, these materials constitute another significant portion of the material cited in CIL determinations. Treaties by themselves, as Figure 5.1 shows, have 407 citations in the 175 CIL determinations. Add in the UN materials, UN committee reports, and state statutes and we have more than 600 citations to this aspirational or forward-looking material.[34] By contrast, there are only 128 cites to material that might directly support the traditional CIL definition, including cites to statements by state officials and actions by states. Even the 128 citation number is misleading. There were fifty citations to actions by states in one CIL determination alone. Removing this one outlier leaves only seventy-eight citations to materials that directly support the traditional CIL definition.

32 For example, take the following language from a U.S. domestic case applying CIL, *Khulumani v. Barclay Nat. Bank Ltd.*, 504 F.3d 254, 271 (2d Cir. 2007): "Moreover, other courts, international bodies, and scholars have recognized that the principles set out in the London Charter and applied by the International Military Tribunal are significant not only because they have garnered broad acceptance, but also because they were viewed as reflecting and crystallizing preexisting customary international law."

33 Even if the drafters of the treaty did assert that they were codifying past state practice and *opinio juris*, it is not clear that a court, if it were following the traditional definition of CIL, should be able to assume that what was said by states during the codification process was true. Among the other complications involved in using treaties as evidence of widespread and settled state practices is that most treaties have only a subset of states that agree to them, as other states may not have even been invited to participate. For a discussion of some of these complications, see, for example, Gary L. Scott & Craig L. Carr, "Multilateral Treaties and the Formation of Customary International Law," 25 *Denv. J. Int'l L. & Pol'y* 71 (1996); and Arthur M. Weisburd, "Customary International Law: The Problem of Treaties," 21 *Vand. J. Transnat'l L.* 1 (1988).

34 As noted earlier, it may be the case that some state statutes – such as those governing sovereign immunity – should be viewed as state practice vis-à-vis other states, whereas others – such as those governing the punishment of minors – are more domestic (and, if cited in the CIL context, aspirational). Our basic point, however, does not change if one takes state statutes out of the equation.

Despite the numbers reported here, one might nevertheless wonder whether courts, while they are not looking to direct evidence supporting the traditional definition, are delegating the job of collecting the evidence and analyzing it to academic scholars. After all, the role that academic scholars played for a long time (and still do in some jurisdictions) was to collect evidence and synthesize law. From the citation count data in Figure 5.1, we do see that there are a number of cites (ninety-two) to academic articles and treatises. But that is still just roughly one cite to an article or treatise every second CIL determination, not enough to support arguments about widespread *opinio juris*.

To recap, then, based on Figure 5.1, it looks like courts are neither pursuing direct evidence fitting the traditional definition nor are they delegating that task to academics and treatise writers. The majority of the action in CIL cases is occurring on the aspirational or forward-looking front; little is occurring on the traditional or backward-looking front. To borrow from Bradley's chapter for this volume, it looks as if courts are engaged in a type of common law adjudication – forward-looking adjudication focused on solving global problems as opposed to some kind of backward-looking historical aggregation of evidence (something that courts probably would not have the expertise to do, even if they wanted to).[35]

Dealing with Outliers

Most of the determinations that we examined cited to only a handful of materials. This creates a potential issue with outliers if there is a subset of cases where the judges cite a lot of materials, while the vast majority of cases have very few citations. For example, imagine that there is one opinion that cites to seventy-five treaties and every other determination cites only to one or two treaties. With a limited dataset, such as the one we use, that single opinion with the seventy-five treaties would skew our results.

To correct the problem, we estimate counts of materials being cited, in individual cases, by giving a maximum score of 1 if a particular type of material is cited. So, if a particular CIL determination cites to five different treaties, we code that as a 1 (as opposed to a 5). What we get with this calculation is the fraction of CIL determinations that use treaties or direct evidence of state acts and so forth. We end up undercounting the influence of individual variables here, but we correct for the problem of outliers. Figure 5.2 reports these modified results.

35 *See* Bradley, *supra* note 14. The type of backward-looking analysis that the CIL definition calls for, if taken seriously, would require courts to be experts in history, economics, political science, and anthropology, in addition to the expertise in law that we assume they have. One does not have to read more than a couple of these CIL opinions to realize that these courts are not demonstrating those types of expertise or borrowing it from the arguments of the parties. Rather, they are doing what pragmatic courts do.

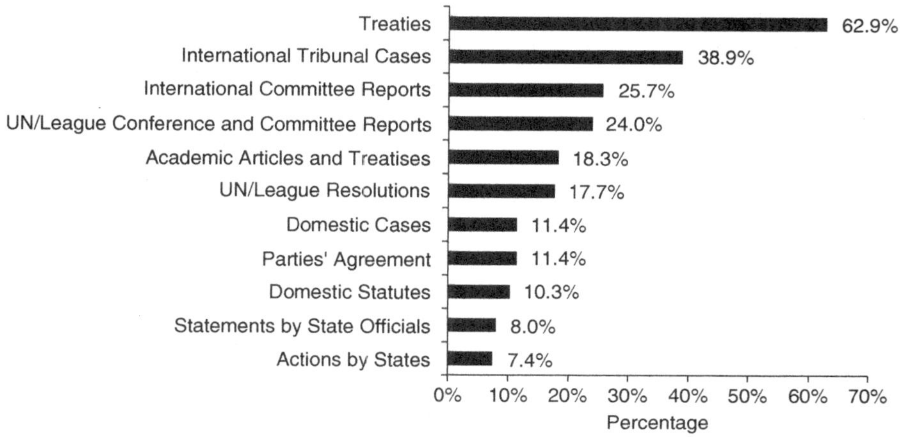

FIGURE 5.2. Materials cited at least once in finding customary international law as a percentage of all CIL determinations (N = 175)

The primary result remains: treaties constitute the dominant form of evidence used in CIL determinations. Note that 62.9 percent of the determinations use at least one treaty as support for their analysis (and we know from Figure 5.1 that they probably use more than one). By contrast, statements of state officials are used in only a small fraction of cases (8.0 percent), as are actions by states (7.4 percent). The other forms of aspirational evidence, as Figure 5.2 demonstrates, also show up in relatively high fractions of cases as compared to statements of state officials. UN or League of Nations conference reports and committee reports show up in 24.0 percent of the determinations. UN or League of Nations resolutions also show up in 17.7 percent of the determinations, and other international committee materials show up in 25.7 percent of the determinations. The point that emerges from Figure 5.2, therefore, is essentially the same as in Figure 5.1: it is the forward-looking or aspirational types of evidence that dominate the determination of CIL. Judges do look, to some extent, to evidence of past practices. But for the most part, the inference we draw from the data is that judges seem to be trying to figure out not what legal norms among states were in the past and whether they believed they were law, but rather what states generally believe *should* be law for their collective.

Parsing the Subsets

Explicit Mention of the Two-Part Test

We have coded the cases using a generous definition of when the courts might be determining a CIL rule. It might be argued that some of these determinations that we have coded are no more than exercises in clarification of the CIL rule and do not call for the courts to analyze evidence. It strikes us that this is not a particularly meaningful distinction, since clarification of the CIL rule is essentially a form of determining whether there is CIL.[36] Still, perhaps cases in which the judge explicitly says that she is applying the two-part definition are ones that the judge is more serious about illustrating how the definition applies.

To address this objection, we coded each of our determinations for whether the two-part test was explicitly invoked. Table 5.1 reports on the comparison of the types of evidence used in the subset of determinations invoking the two-part test compared with the types used in the remainder of cases. We report here the percentage of times a type of evidence show up at least once in a determination (and not the raw numbers of times the type of evidence shows up).[37]

We find significant differences between determinations that explicitly invoke the two-part test and those that do not. As a percentage, determinations that invoke the two-part test tend to cite more to statements by state officials, UN/League Resolutions, and International Tribunal Cases. In terms of our key variable, the use of treaties is higher in the determinations invoking the two-part test (80.7 percent of the determinations) than in the other determinations (59.0 percent). This difference is significant at the 5 percent confidence level. The difference, however, goes in the direction of supporting our primary finding because if we conjecture that cases where the two-part test is invoked are ones in which the courts are actively determining new CIL, we find that treaties are an even more important part of the calculus.

Cases That Find CIL

Another possible objection to the numbers in Figures 5.1 and 5.2 is that those numbers are from all CIL determinations, including where CIL is found and where CIL is rejected. Perhaps, as one colleague argued to us, treaties are only being cited in cases where CIL claims are being rejected? If we are interested in what types of evidence are used in determining whether CIL is present, this colleague argued, we should have focused only on the cases where the tribunal found CIL (as opposed to looking also at the determinations where CIL was not found).

To answer the objection, we coded all of our determinations for whether CIL was found or not. Table 5.2 reports the results of comparing the determinations where CIL was found versus those where it was not.

A couple of things are worth noting in Table 5.2. First, CIL is found in roughly half the determinations and not found in the other half. That suggests that these

36 *See* Hakimi, *supra* note 13.

37 We do this for all three tables. The basic results hold true if one looks at the raw number breakdowns, as well. We choose to report the percentage numbers though, so as to be able to also report the significance levels (which would not make sense in the case of the raw numbers).

TABLE 5.1. *CIL determinations with explicit mention of the two-part test*

	Two-part test not mentioned		Two-part test mentioned		
Variable	(n = 144)	**Percent**	(n = 31)	**Percent**	**p-value**
Actions by states	10	6.9%	3	9.7%	0.599
Statements by state officials	7	4.9%	7	22.6%	0.001
Direct evidence	**16**	**11.1%**	**9**	**29.0%**	**0.010**
Academic articles and treatises	27	18.8%	5	16.1%	0.732
International committee reports	35	24.3%	10	32.3%	0.358
Delegation	**52**	**36.1%**	**12**	**38.7%**	**0.787**
Treaties	85	59.0%	25	80.7%	0.024
UN/League resolutions	19	13.2%	12	38.7%	0.001
UN/League conference and committee reports	34	23.6%	8	25.8%	0.795
Domestic statutes	14	9.7%	4	12.9%	0.597
Aspirational	**99**	**68.8%**	**27**	**87.1%**	**0.039**
International tribunal cases	51	35.4%	17	54.8%	0.044
Domestic cases	15	10.4%	5	16.1%	0.365
Procedural	**56**	**38.9%**	**17**	**54.8%**	**0.104**
Parties' agreement	**16**	**11.1%**	**4**	**12.9%**	**0.776**

Direct Evidence is equal to 1 if Actions by states or Statements by state officials is equal to 1 and 0 otherwise. Delegation is equal to 1 if Academic (articles or treatises) or International committee reports is equal to 1 and 0 otherwise. Aspirational is equal to 1 if Treaties, UN/League resolutions, UN/League conference and committee reports, or Domestic statutes is equal to 1 and 0 otherwise. Procedural is equal to 1 if International tribunal cases or Domestic cases is equal to 1 and 0 otherwise

cases, when they are brought to an international tribunal, stand in equipoise – where the litigants have roughly equal chances of winning. Importantly for our purposes, however, we see that the use of treaties as evidence is greater in the cases where CIL is found (68.1 percent of determinations use treaty evidence) as compared to where CIL is not found (56.8 percent). Unlike the prior parsing, the difference is significant at only the 12.3 percent level, beyond conventional levels of significance. Whether CIL is found or not, courts tend to cite to treaties. More generally, note from Table 5.2 that opinions that find CIL tend to cite more to aspirational evidence.

TABLE 5.2. *CIL determinations categorized by whether CIL was found*

	CIL not found		CIL found		
Variable	(n = 81)	**Percent**	(n = 94)	**Percent**	**p-value**
Actions by states	6	7.4%	7	7.5%	0.992
Statements by state officials	7	8.6%	7	7.5%	0.771
Direct evidence	**12**	**14.8%**	**13**	**13.8%**	**0.854**
Academic articles and treatises	14	17.3%	18	19.2%	0.750
International committee reports	17	21.0%	28	29.8%	0.184
Delegation	**25**	**30.9%**	**39**	**41.5%**	**0.147**
Treaties	46	56.8%	64	68.1%	0.123
UN/League resolutions	9	11.1%	22	23.4%	0.034
UN/League conference and committee reports	19	23.5%	23	24.5%	0.876
Domestic statutes	8	9.9%	10	10.6%	0.869
Aspirational	**53**	**65.4%**	**73**	**77.7%**	**0.073**
International tribunal cases	25	30.9%	43	45.7%	0.044
Domestic cases	12	14.8%	8	8.5%	0.191
Procedural	**29**	**35.8%**	**44**	**46.8%**	**0.143**
Parties' agreement	**3**	**3.7%**	**17**	**18.1%**	**0.003**

Direct Evidence is equal to 1 if Actions by states or Statements by state officials is equal to 1 and 0 otherwise. Delegation is equal to 1 if Academic (articles or treatises) or International committee reports is equal to 1 and 0 otherwise. Aspirational is equal to 1 if Treaties, UN/League resolutions, UN/League conference and committee reports, or Domestic statutes is equal to 1 and 0 otherwise. Procedural is equal to 1 if International tribunal cases or Domestic cases is equal to 1 and 0 otherwise

Opinion Characteristics

Given the domination of treaties, and more generally, aspirational evidence in CIL determinations, we look at whether characteristics specific to the particular CIL determination matter for the use of different types of evidentiary sources.

Individual Rights

Some scholars have suggested that there are two different categories of CIL, roughly speaking: traditional CIL and modern CIL.[38] Traditional CIL emphasizes the first

[38] *E.g.*, Roberts, *supra* note18.

prong of the textbook definition (state practice) while modern CIL emphasizes the second prong (*opinio juris*). Traditional CIL is made up of the rules that have long governed interstate relationships, such as the rules governing diplomatic immunities, the conduct of war, and the regulation of ships in international waters. Here, traditional practices are arguably more relevant in determining CIL. Then there is modern CIL, which governs matters that fit more into the category of individual rights, such as prohibitions on states committing genocide. In these individual-rights cases, the argument is sometimes made that evidence of *opinio juris* is more important than state practice – largely because longstanding state practices are often inconsistent with the rights being sought.

For purposes of this inquiry, we have divided cases into those involving collective interests (more traditional CIL) versus individual rights (modern CIL). We define individual rights determinations as those CIL determinations where an individual's rights against a state are at stake as opposed to a determination where the rules governing interstate relations are involved. The paradigmatic individual rights cases, of course, are in the human rights area. However we also found individual rights cases in areas involving citizenship, immigration, sovereign immunity, commercial interests (debt), and terrorism.[39] Table 5.3 reports summary statistic comparisons in the use of different types of CIL evidence for collective interests compared with individual rights determinations.

Table 5.3 shows evidence of courts looking to different types of evidence in rights cases versus those involving intergovernmental interaction issues. We see that treaties are cited significantly more often in rights-type cases (81.4 percent) than in those involving collective interests including intergovernmental relations (56.8 percent). More generally, Aspirational evidence is cited more in Individual rights determinations. This finding is consistent with what has generally been hypothesized by scholars, regarding the more "modern" or "forward-looking" forms of CIL. Interestingly, we also find Delegation evidence, including both academic articles and treatises and international committee reports, are cited more in individual rights determinations. Again, this finding is consistent with the conjecture of some scholars that academic articles and international committees – particularly in the individual rights area – often try to push new and progressive conceptions of CIL in their work.[40]

39 Our coding here was necessarily subjective, and there is often an argument that can be made as to why an individual rights case is really an interstate one.

40 *See* Ed Whelan, "Harold Koh's Transnationalism," *Nat'l. Rev.*, April 6, 2009 (discussing Curtis A. Bradley & Jack L. Goldsmith, "Customary International Law as Federal Common Law: A Critique of the Modern Position," 110 *Harv. L. Rev.* 815 [1997]), at http://www.nationalreview.com/bench-memos/50307/harold-kohs-transnationalism-mdash-customary-international-law/ed-whelan.

TABLE 5.3. *CIL determinations broken down by subject type (collective interest v. individual rights)*

	Collective interest		Individual rights		
Variable	(n = 132)	**Percent**	(n = 43)	**Percent**	**p-value**
Actions by states	13	9.9%	0	0.0%	0.032
Statements by state officials	9	6.8%	5	11.6%	0.313
Direct evidence	**20**	**15.2%**	**5**	**11.6%**	**0.569**
Academic articles and treatises	20	15.2%	12	27.9%	0.060
International committee reports	28	21.2%	17	39.5%	0.017
Delegation	**44**	**33.3%**	**20**	**46.5%**	**0.121**
Treaties	75	56.8%	35	81.4%	0.004
UN/League resolutions	23	17.4%	8	18.6%	0.860
UN/League conference and committee reports	27	20.5%	15	34.9%	0.054
Domestic statutes	12	9.1%	6	14.0%	0.362
Aspirational	**87**	**65.9%**	**39**	**90.7%**	**0.002**
International tribunal cases	44	33.3%	24	55.8%	0.009
Domestic cases	13	9.9%	7	16.3%	0.250
Procedural	**48**	**36.4%**	**25**	**58.1%**	**0.012**
Parties' agreement	**18**	**13.6%**	**2**	**4.7%**	**0.108**

Direct Evidence is equal to 1 if Actions by states or Statements by state officials is equal to 1 and 0 otherwise. Delegation is equal to 1 if Academic (articles or treatises) or International committee reports is equal to 1 and 0 otherwise. Aspirational is equal to 1 if Treaties, UN/League resolutions, UN/League conference and committee reports, or Domestic statutes is equal to 1 and 0 otherwise. Procedural is equal to 1 if International tribunal cases or Domestic cases is equal to 1 and 0 otherwise

Dissenting Opinions

Judges who author a dissenting opinion may face different considerations in writing the opinion as opposed to judges who author a majority or concurring opinion. The author of a dissent may seek to invoke others to support their minority view. We expect that a dissenting opinion will tend to cite to secondary sources, including academic articles and treatises. We therefore expect more reliance on Delegation evidence in a dissent. We also expect those who author a dissent do so because they have strong views about the application of CIL – whether the view is that CIL should be applied or not. These strong views may in turn lead the author of a dissent

TABLE 5.4. *Types of opinions*

	Majority		Dissent		
Variable	(n = 114)	Percent	(n = 40)	Percent	p-value
Actions by states	7	6.1%	5	12.5%	0.197
Statements by state officials	7	6.1%	5	12.5%	0.197
Direct evidence	**14**	**12.3%**	**0**	**0.2%**	**0.233**
Academic articles and treatises	11	9.7%	16	40.0%	0.000
International committee reports	27	23.7%	10	25.0%	0.867
Delegation	**31**	**27.2%**	**22**	**55.0%**	**0.001**
Treaties	72	63.2%	24	60.0%	0.723
UN/League resolutions	17	14.9%	7	17.5%	0.698
UN/League conference and committee reports	20	17.5%	14	35.0%	0.022
Domestic statutes	10	8.8%	3	7.5%	0.803
Aspirational	**79**	**69.3%**	**29**	**72.5%**	**0.706**
International tribunal cases	50	43.9%	13	32.5%	0.209
Domestic cases	13	11.4%	6	15.0%	0.552
Procedural	**53**	**46.5%**	**15**	**37.5%**	**0.328**
Parties' agreement	**17**	**14.9%**	**2**	**5.0%**	**0.101**

Direct Evidence is equal to 1 if Actions by states or Statements by state officials is equal to 1 and 0 otherwise. Delegation is equal to 1 if Academic (articles or treatises) or International committee reports is equal to 1 and 0 otherwise. Aspirational is equal to 1 if Treaties, UN/League resolutions, UN/League conference and committee reports, or Domestic statutes is equal to 1 and 0 otherwise. Procedural is equal to 1 if International tribunal cases or Domestic cases is equal to 1 and 0 otherwise

to cite more to Aspirational evidence that support these views. Table 5.4 reports summary statistic comparisons between majority opinions on the one hand and dissenting opinions on the other hand. For this comparison, we omit concurrences. We later include concurrences in our multivariate tests.

From Table 5.4, note that there generally are not significant differences for the different subcategories of evidence between Majority and Dissenting opinions. Dissents tend to cite more to academic articles and treatises (difference significant at the 1 percent level). Consequently, the Delegation category of evidence is used in Dissents more than in Majority opinions (difference significant at the 1 percent level). Contrary to our expectation, we do not find that Dissents cite more to Aspirational evidence.

Post–World War II

A key factor that determines what types of evidence judges cite in CIL determinations is the time period during which the determination was made. Scholars have suggested that World War II forms an important dividing line in international law, in that international organizations such as the UN moved to center stage and individual rights (and particularly human rights) became increasingly important. In addition, the sheer number of nations has increased dramatically in the post-war period, with the fall of the colonial empires – a fact that translates into more international treaties among subsets of nations. Putting those elements together, we accordingly expect that Post–World War II opinions will tend to cite more to Delegation, Aspirational, and Procedural evidence.

As reported in Table 5.5, there is a significant increase in citations to International committee reports (Delegation evidence), treaties and UN/League conference and committee reports (Aspirational evidence), and to International tribunal cases (Procedural evidence) in the post–World War II period. Unlike in these other categories, there is no increase in citations to Direct evidence in the Post– compared with Pre–World War II period. Unlike in the other categories, there likely was not an increase in the availability of Direct evidence in the Post–World War II period.

Multivariate Analysis

In order to explore these differences between Individual rights, Dissent, and Post–World War II in affecting the types of evidence used in CIL determinations in greater depth, we use the incidence of each of the four types of evidentiary sources in a series of logit models. Our first logit model, using Direct evidence as the dependent variable, is estimated at the level of specific CIL determinations with errors clustered by opinion as follows:

$$\begin{aligned}\text{Prob(Direct evidence)}_i = \alpha &+ \beta_{1i}\text{Individual rights}_i \\ &+ \beta_{2i}\text{Dissent}_i + \beta_{3i}\text{Concurrence}_i \\ &+ \beta_{4i}\text{Post–World War II}_i + \varepsilon_i\end{aligned}$$

We estimate the model using errors clustered on specific opinions to control for the possible presence of multiple CIL determinations in the same opinion. For independent variables, we include indicator variables for the CIL issue characteristics, including whether the CIL issue involves individual rights (Individual rights), whether the CIL issue is determined in a dissenting opinion (Dissent), whether the CIL issue is determined in a concurring opinion (Concurrence), and whether the CIL issue is decided after World War II (Post–World War II).

TABLE 5.5. *Pre– Versus Post–World War II*

	Pre–World War II		Post–World War II		
Variable	(n = 17)	**Percent**	(n = 158)	**Percent**	**p-value**
Actions by states	1	5.9%	12	7.6%	0.798
Statements by state officials	0	0.0%	14	8.9%	0.201
Direct evidence	**1**	**5.9%**	**24**	**15.2%**	**0.300**
Academic articles and treatises	2	11.8%	30	19.0%	0.464
International committee reports	0	0.0%	45	28.5%	0.011
Delegation	**2**	**11.8%**	**62**	**39.2%**	**0.025**
Treaties	5	29.4%	105	66.5%	0.003
UN/League resolutions	1	5.9%	30	19.0%	0.179
UN/League conference and committee reports	0	0.0%	42	26.6%	0.015
Domestic statutes	1	5.9%	17	10.8%	0.529
Aspirational	**5**	**29.4%**	**121**	**76.6%**	**0.000**
International tribunal cases	1	5.9%	67	42.4%	0.003
Domestic cases	1	5.9%	19	12.0%	0.449
Procedural	**2**	**11.8%**	**71**	**44.9%**	**0.008**
Parties' agreement	**1**	**5.9%**	**19**	**12.0%**	**0.449**

Direct Evidence is equal to 1 if Actions by states or Statements by state officials is equal to 1 and 0 otherwise. Delegation is equal to 1 if Academic (articles or treatises) or International committee reports is equal to 1 and 0 otherwise. Aspirational is equal to 1 if Treaties, UN/League resolutions, UN/League conference and committee reports, or Domestic statutes is equal to 1 and 0 otherwise. Procedural is equal to 1 if International tribunal cases or Domestic cases is equal to 1 and 0 otherwise

Model 1 of Table 5.6 reports the results. We reestimate the same model using the use of delegated sources of evidence (Delegation) as the dependent variable and report the results in Model 2. We reestimate the model using the use of aspirational sources of evidence (Aspirational) as the dependent variable and report the results in Model 3. We reestimate the model using the use of procedural and other matters as sources of evidence (Procedural) as the dependent variable and report the results in Model 3.

As reported in Model 1 of Table 5.6, none of our case characteristics correlate with the use of Direct evidence. Regardless of whether the subject matter involves Individual rights, the opinion contains more than one determination with respect to CIL, the opinion is a dissenting opinion, or the time period of the decision

TABLE 5.6. *Full sample*

	Model 1	Model 2	Model 3	Model 4
	Direct evidence	Delegation	Aspirational	Procedural and other matters
Individual rights	−0.503	0.418	1.445**	0.678+
	(−0.92)	(0.99)	(2.96)	(1.67)
	[−0.053]	[0.097]	[0.222]	[0.166]
Dissent	0.684	1.291**	0.211	−0.390
	(1.11)	(2.78)	(0.43)	(−0.89)
	[0.091]	[0.306]	[0.039]	[−0.091]
Concurrence	0.0207	1.054*	0.967	−1.058+
	(0.04)	(2.04)	(1.32)	(−1.82)
	[0.002]	[0.254]	[0.148]	[−0.224]
Post–World War II	1.291	1.624+	1.692*	1.711*
	(1.08)	(1.83)	(2.36)	(2.09)
	[0.104]	[0.274]	[0.385]	[0.317]
Constant	−3.063*	−2.621**	−0.952	−1.890*
	(−2.55)	(−2.96)	(−1.42)	(−2.35)
N	175	175	175	175
Log likelihood	0.026	0.088	0.120	0.070
r2_p	−69.87	−104.8	−91.29	−110.5

z statistics in parentheses; + $p < 0.10$, $^{*}p < 0.05$, $^{**}p < 0.01$. Errors clustered by opinion. Marginal probability measured at the mean of all the other independent variables in brackets

is Post–World War II, judges making a CIL determination are not more or less likely to cite to our indicia of direct evidence. Perhaps because of the high cost of obtaining direct evidence, both for multiple countries and across time, judges simply do not undertake such an analysis regardless of the characteristics of the specific case.

Judges are more willing to delegate the task of finding evidence on CIL in certain situations. Model 2 of Table 5.6 reports that the coefficients on Dissent and Post–World War II are positive and significant at the 1 percent and 10 percent levels, respectively. Judges authoring a dissenting opinion on a CIL issue correspond to a 30.6 percentage point increase in the probability that the judges will cite to an academic article or treatise or international committee report. This is consistent with a dissenting judge seeking additional authority to buttress their minority position. The positive coefficient on Post–World War II indicates that judges are more likely to cite to these secondary sources after World War II, perhaps due to an increasing supply of such sources. Post–World War II cases correspond to a 27.4 percentage point increase in the probability that a court will turn to Delegation evidence. Note also that the coefficient on Concurrence is positive and significant at the 5 percent level. As with those authoring a dissenting opinion, those authoring a concurring opinion are also more likely compared with majority opinions to cite to an academic article or treatise or international committee report (delegation type evidence). While a concurrence generally is aligned with the majority opinion, the concurrence typically is not identical to the majority opinion. The author of the concurrence may therefore value additional authority to support their position.

Judges are more willing to use Aspirational sources of CIL evidence for Individual rights cases. Model 3 of Table 5.6 reports that the coefficient on Individual rights is positive and significant at the 1 percent level. An Individual rights case corresponds to a 22.2 percentage point increase in the probability that a court will use an Aspirational source of evidence. As we have conjectured, because the development of individual rights in CIL does not build upon existing understanding of international law but instead seeks to expand the law, courts applying CIL in an individual rights case are more likely to turn to Aspirational sources of evidence. In addition, the coefficient on Post–World War II is positive and significant at the 5 percent level. Post–World War II cases correspond to a 38.5 percentage point increase in the probability that a court will use an Aspirational source of evidence.

Lastly, judges are more likely to use Procedural evidence after World War II. As reported in Model 4 of Table 5.6, the coefficient on Post–World War II is positive and significant at the 5 percent level. Post–World War II cases correspond to a 31.7 percentage point increase in the probability that a court will turn to a Procedural source of evidence. The coefficient on Individual rights is positive and significant at the 10 percent level. Judges authoring a CIL determination dealing with individual rights correspond to a 16.6 percentage point increase in the probability that there will be a cite to a Procedural source of evidence. The coefficient on Concurrence is negative and significant at the 10 percent level. Judges authoring a concurring opinion correspond to a 22.4 percentage point decrease in the probability that there will be a cite to a Procedural source of evidence, possibly because the concurrence will rely on many of the procedural citations in the majority opinion. In contrast, we find no evidence that a Dissent corresponds to a change in the likelihood that a court will refer to procedural evidence.

Our analysis focuses on CIL determinations by international tribunals. Not all international tribunals operate in the same manner or have the same influence,

TABLE 5.7. *ICJ and PCIJ only*

	Model 1	Model 2	Model 3	Model 4
	Direct evidence	Delegation	Aspirational	Procedural and other matters
Individual rights	−0.251	0.560	1.186*	0.892*
	(−0.43)	(1.13)	(2.40)	(2.01)
	[−0.030]	[0.125]	[0.194]	[0.207]
Dissent	0.790	1.123*	0.137	−0.0413
	(1.13)	(2.11)	(0.25)	(−0.08)
	[0.114]	[0.258]	[0.026]	[−0.009]
Concurrence	0.117	1.448**	1.314	−0.841
	(0.21)	(2.58)	(1.41)	(−1.34)
	[0.015]	[0.341]	[0.194]	[−0.163]
Post–World War II	1.255	2.037*	1.812*	2.028+
	(0.99)	(1.99)	(2.27)	(1.80)
	[0.111]	[0.290]	[0.414]	[0.304]
Constant	−3.022*	−3.222**	−1.067	−2.623*
	(−2.35)	(−3.20)	(−1.43)	(−2.34)
N	136	136	136	136
Log likelihood	0.028	0.107	0.135	0.087
r2_p	−56.85	−77.71	−70.53	−80.63

z statistics in parentheses; + $p < 0.10$, * $p < 0.05$, ** $p < 0.01$. Errors clustered by opinion. Marginal probability measured at the mean of all the other independent variables in brackets

however.[41] To address this issue, we reestimate the models of Table 5.6 only for CIL determinations by the ICJ or PCIJ. Table 5.7 reports our results.

Note from Table 5.7 that we obtain largely the same results for our multivariate models when we limit our sample to ICJ or PCIJ CIL determinations as in Table 5.6 (the full sample). The only difference is that the coefficient on Concurrence in Model 4 in Table 5.7 is not significantly different from zero.

CONCLUSION

The results are fairly consistent, no matter how we parse them. The data suggest that international courts do not come anywhere close to engaging in the type of

[41] For example, see the discussion of the International Criminal Court and the WTO in Larissa van den Herik, "The Decline of Customary International Law as a Source of International Criminal Law" (in this volume).

analysis the officially stated two-part rule for the evolution of CIL sets up. Instead, as best we can tell, courts analyzing CIL – whether they find it or not – are generally engaged in a forward-looking or aspirational exercise. We draw this inference from the fact that, by far, the dominant form of evidence used in determinations of CIL is that gleaned from international treaties. Our assumption is that treaties are (a) a strong indicator of the rules that countries want to govern their future relationships with each other, and (b) a sign that such rules are not already well established in existing custom. We also find that other forms of aspirational evidence, such as UN resolutions and international committee reports, are often used as evidence in analyses of the existence or not of CIL. These materials tend to generally focus on questions of what types of new rules would serve the world community in the future, rather than analyses of what rules have been utilized in the past. Under this rubric, the question that jump-started our inquiry – how do judges apply the CIL definition that seems both analytically impossible to apply and normatively unattractive – has a simple answer: they ignore it.

What are the judges doing instead? The optimistic story of what they actually do in the face of these intractable legal requirements is a kind of a common-law type analysis that looks to what is best for global welfare. But there are other stories that might get revealed with more detailed analyses – maybe the judges are finding CIL when it benefits their home nations and allies and not finding it in situations where they do not get a benefit? We do not have enough information as yet to tell.

Before ending, we should emphasize the preliminary nature of our inquiry and tentativeness of our conclusions. Our primary finding is that the type of evidence used most often in CIL determinations is the international treaty. But there is much more that can be done to unpack that finding. In particular, the data on treaties could be coded for the type of treaty. Treaties, after all, vary considerably in terms of factors such as the number of nations who have signed up to them, the types of nations that they are among, the withdrawal rights that they allow, whether they purport to be codifying existing state practice, and so on. Understanding the implications of our finding regarding the dominance of treaties as a form of evidence requires assessing the type of treaty being cited and when it is that certain types of treaties are cited. And then there is the question of whether the use of treaties in determining CIL differs across types of courts – not only the different types of international tribunals, but also domestic courts. Last but not least is the question of whether this preference for looking to treaties is being driven by the lawyers or the judges (a difficult matter to distinguish, we suspect). We leave these questions for future research.

6

Custom's Method and Process

Lessons from Humanitarian Law

Monica Hakimi

A central question in the literature on customary international law (CIL) goes to method: what is the proper method for "finding" CIL – that is, for determining that particular norms qualify as CIL?[1] The traditional method is to identify a widespread state practice, plus evidence that states believe that the practice reflects the law (*opinio juris*). That method has long been criticized as incoherent, unworkable, and out of touch with modern sensibilities.[2] Thus, much of the CIL literature addresses its perceived problems. The principal goals of this literature are to help resolve whether norms that are claimed to be CIL are *really* CIL, and thus to reduce the volatility and susceptibility to abuse in CIL.

I argue in this chapter that the method for finding CIL might be so elusive because the question itself is misconceived. The question of how to find CIL presupposes that finding CIL is an objective exercise and somehow removed from the process for making CIL. This process is notoriously undisciplined and politically charged. To make CIL, disparate actors advance and respond to one another's legal claims,

1 *See, e.g.*, Int'l Law Comm'n, *First Report on Formation and Evidence of Customary International Law* (by Michael Wood, Special Rapporteur), ¶ 22, UN Doc. A/CN.4/663 (May 17, 2013) [hereinafter *Wood's First Report*] (reviewing literature on method and asserting that the goal of the International Law Commission's project is to "provide guidance on how to identify a rule of customary international law"); International Law Association [ILA], *Statement of Principles Applicable to the Formation of General Customary International Law*, at 4 (2000) [hereinafter "*ILA Report*"] ("[I]t was felt that what would be most useful was ... some practical guidance" for finding CIL.).

2 *See, e.g.*, Robert Y. Jennings, "The Identification of International Law," in INTERNATIONAL LAW: TEACHING AND PRACTICE 3, 4–6 (Bin Cheng ed., 1982) (describing the method as incoherent and full of "inner mysteries"); LOUIS HENKIN, INTERNATIONAL LAW: POLITICS AND VALUES 29 (1995) ("The definition is easy to state but not easy to interpret and apply, and it continues to raise difficult questions, some 'operational,' some conceptual-jurisprudential."); Jonathan I. Charney, "Universal International Law," 87 AM. J. INT'L L. 529, 543–45 (1993) (arguing that the traditional method is inapt, given the increased number and diversity of states and the expanded subjects of international regulation).

as they pursue their own interests.[3] The methodological question assumes that CIL finding is distinct – that actors who find CIL do not advance their own agendas but rather assess the evidence objectively and thus that their decisions help settle CIL and weed out invalid claims. I use the recent rise of CIL in international humanitarian law to show that these assumptions are flawed. CIL finding is deeply entangled with CIL making. The two exercises operate in much the same way and through the same process, so they share similar limitations.

My argument has two practical implications. First, nonstate actors who are charged with finding CIL can be extremely influential in making CIL. Some of these actors play a much larger role in the formation of CIL than the literature now recognizes. Second, no particular method for finding CIL is capable of disciplining global actors or imposing order on CIL, because the process for making CIL is so heavily *un*disciplined and *dis*ordered.

THE QUESTION OF METHOD

The frustrations with finding CIL stem directly from the nature of the CIL process. This process is chaotic, unstructured, and politically charged. The participants make and respond to competing claims on the law as they advance their own agendas. Because the process lacks any structure, these claims and counterclaims can take multiple forms and appear in varied arenas. Some CIL claims are advanced collectively, such as through international organizations.[4] Others appear in national documents or press briefings. Still others are communicated nonverbally, as states act and react in concrete cases. In short, the process produces vast amounts of disconnected raw data – disparate claims and counterclaims on an issue.

The methodological question asks how to identify the claims that pass successfully through that process and emerge as CIL. For all the attention that that question receives, its answer is still only rudimentary and shifty. Consider three methodological guideposts that will again be relevant in the discussion on international humanitarian law. First, there is broad agreement that *states* generate almost all of the relevant input for finding CIL.[5] The extent to which other kinds of actors participate in CIL's

[3] *See* Myers S. McDougal, Editorial Comment, "The Hydrogen Bomb Tests and the International Law of the Sea," 49 AM. J. INT'L L. 356, 357 (1955) (describing the process as one of "continuous interaction … in which the decision-makers … unilaterally put forward claims … and in which other decision-makers … weigh and appraise these competing claims … and ultimately accept or reject them").

[4] *See* Charney, *supra* note 2, at 543–50.

[5] Int'l Law Comm'n, *Second Report on Identification of Customary International Law* (by Michael Wood, Special Rapporteur), Draft Conclusion 5, UN Doc. A/CN.4/672 (May 22, 2014) [hereinafter "*Wood's Second Report*"] ("[I]t is primarily the practice of States that contributes to the creation, or expression, of rules of customary international law."); Christiana Ochoa, "The Individual and Customary International Law Formation," 48 VA. J. INT'L L. 119, 135 (2007) ("What is remarkable in

formation is unclear. For example, the dominant position in the CIL literature is that international courts and tribunals do not help make CIL.[6] Judicial decisions do not qualify as state practice and are, at best, only indirect evidence of *opinio juris*. Yet as a practical matter, global actors rely on these decisions to determine whether particular norms qualify as CIL – that is, to find CIL.[7] Separately, state-comprised bodies of international organizations, such as the UN General Assembly, are known to participate in the formation of CIL. Questions remain on when these bodies participate,[8] and on the extent to which bodies that are *not* comprised of states also participate.[9]

Second, a CIL finding will be uncontroversial if the data clearly show a widespread state practice plus strong evidence of *opinio juris*. But where – as is usually the case – the data are inconsistent or ambiguous, the independence and relative weight of each of those elements is contested.[10] Some argue that CIL may be found on the basis of the practice alone, without much evidence of *opinio juris*.[11] Others argue the opposite.[12] Still others contend that a bounty of one element can compensate for a dearth of the other.[13] And those who accept that both elements might be necessary

this literature is that virtually all of it has accepted the core premise that only states can form CIL."); Yoram Dinstein, "The Interaction Between Customary International Law and Treaties," 322 *Recueil des Cours* 243, 267 (2006) ("[States] have a quasi-monopoly over the formation of custom.").

6 *See, e.g.*, *ILA Report*, *supra* note 1, at 18 ("Although international courts and tribunals ultimately derive their authority from States, it is not appropriate to regard their decisions as a form of State practice."); Dinstein, *supra* note 5, at 317 ("[I]nternational judicial bodies – not being the organs of any single State – never contribute as such to the practice of States.").

7 *See, e.g.*, *Prosecutor v. Akayesu*, Case No. ICTR-96-4-T, Judgment, ¶ 495 (Sept. 2 1998) (citing an ICJ decision to find that the Genocide Convention reflects CIL); Panel Report, *Korea – Measures Affecting Government Procurement*, ¶ 7.123, WT/DS163/R (May 1, 2000) ("Since this article has been derived largely from the case law of the ... PCIJ and the ICJ, there can be little doubt that it presently represents customary international law.").

8 Legality of the Threat or Use of Nuclear Weapons, Advisory Opinion, 1996 I.C.J. 226, ¶ 70 (July 8) ("General Assembly resolutions ... can, in certain circumstances, provide evidence important for establishing the existence of a rule or the emergence of an *opinio juris*.").

9 Reservations to Convention on Prevention and Punishment of Crime of Genocide, Advisory Opinion, 1951 I.C.J. 15, 22 (May 28) (considering the UN Secretary General's statement but suggesting that other evidence was weightier); *Wood's Second Report*, *supra* note 5, at ¶ 43 (asserting that secretariats can produce relevant data but that such data should be given less weight than the output of bodies that are comprised of states).

10 *Wood's Second Report*, *supra* note 5, at ¶ 3(a) ("[I]t was recognized that the two elements may sometimes be 'closely entangled,' and that the relative weight to be given to each may vary according to the circumstances.").

11 *See, e.g.*, Anthony D'Amato, Comment, "Customary International Law: A Reformulation," 4 *Int'l Legal Theory* 1 (1998); Maurice H. Mendelson, "The Formation of Customary International Law," 272 *Recueil des Cours* 155, 250, 289 (1998).

12 *See, e.g.*, Andrew T. Guzman, "Saving Customary International Law," 27 *Mich. J. Int'l L.* 115, 153 (2005); Rudolf Bernhardt, "Custom and Treaty in the Law of the Sea," 205 *Recueil des Cours* 247, 266 (1987).

13 *See* Frederic L. Kirgis, Jr., "Custom on a Sliding Scale," 81 *Am. J. Int'l L.* 146 (1987); *ILA Report*, *supra* note 1, at 40.

still debate whether the same evidence may be "double counted," once as state practice and then again as *opinio juris*.[14]

Third, treaties and CIL are known to interact, but the question of how they interact in particular contexts – that is, what a treaty provision signifies for CIL and vice versa – remains open-ended. The International Law Association's high-profile study on CIL summarized the conventional wisdom as follows: a treaty provision might be irrelevant to the formation of CIL, it might reflect preexisting CIL, or it might help create CIL.[15] Although the International Law Association did not say so, most scholars accept that CIL can also, in rare cases, *displace* a contrary treaty provision.[16] The upshot is that a treaty provision cannot be presumed to have any one of these effects; the provision must be assessed in light of the other evidence for and against the CIL claim.[17]

All of this imprecision in the method for finding CIL breeds uncertainty in CIL and is thought to weaken it as a source of law.[18] A CIL finding might well turn on how a particular decisionmaker interprets the raw data; whether she considers only data generated by states or also other kinds of data; whether she focuses more on the operational practice or on the *opinio juris*; and what she makes of any treaty provision on the issue. As the methodological variance expands, so too does the range of plausible interpretations of the raw data – and the indeterminacy on the CIL status and substantive content of the putative norm.

Many scholars argue that the indeterminacy undercuts CIL's efficacy in shaping state behavior.[19] They argue that, because CIL allows so much discretion at the

14 *ILA Report*, *supra* note 1, at 7.

15 *ILA Report*, *supra* note 1, at 43–48; *see also Wood's First Report*, *supra* note 1, at ¶ 34 ("[I]t is generally recognized that treaties may be reflective of pre-existing rules of customary international law; generate new rules and serve as evidence of their existence; or ... have a crystallizing effect for emerging rules of customary international law.").

16 *See infra* note 108 and accompanying text.

17 *ILA Report*, *supra* note 1, at 43–48; *see also* Curtis A. Bradley & Mitu Gulati, "Withdrawing from International Custom," 120 *Yale L.J.* 202, 213 (2010) ("[A]lthough treaty and CIL obligations frequently overlap, there is debate over whether and to what extent treaties can serve as evidence of CIL.").

18 *See, e.g.*, Anthony A. D'Amato, *The Concept of Custom in International Law* 6 (1971) ("When the rules for finding the rules of law are themselves vague or ambiguous, law becomes unpredictable.").

19 *See, e.g.*, Andrew T. Guzman & Timothy L. Meyer, "Customary International Law in the 21st Century," in *Progress in International Law* 197, 197 (Russell A. Miller & Rebecca M. Bratspies eds., 2008) ("This lack of a defining procedure creates considerable confusion as to the content of CIL and its relevance to state behavior."); K. Wolfke, "Some Persistent Controversies Regarding Customary International Law," 24 *Neth. Y.B. Int'l L.* 1, 15 (1993) (noting "the frequently expressed doubts about the present usefulness of customary international law" and asserting that, "[c]onsidering the complexity, imprecision and relative slowness of international custom as a law-generating instrument, such doubts seem to be fully justified").

point of application, it invites states to make CIL claims that evade regulation.[20] These claims can be difficult to refute because the CIL process is, at bottom, decentralized: no actor is formally designated to settle CIL or, in most cases, to apply a putative norm to specific facts. Moreover, even if CIL can effectively regulate states on some issues, its fluidity and lack of structure might make it particularly ill-suited for addressing complex modern problems, like climate change and nuclear nonproliferation. Such problems typically demand detailed regulation, tradeoffs among differently situated states, and some institutional oversight.[21]

Separately, CIL's indeterminacy is said to undercut its legitimacy as law. The argument here is that, because CIL is so malleable, it is easily infected by factors, such as raw power, that ought to be exogenous to the law.[22] For example, Jean d'Aspremont recently proclaimed that many CIL norms are so "*dangerously* indeterminate" that "their authority is gravely enfeebled."[23] Fernando Tesón characterizes as deeply problematic what he calls "fake custom" – "pieces of advocacy disguised as law."[24] The idea that CIL is easily abused or opportunistically invoked is sprinkled throughout the CIL literature.[25]

20 This kind of argument is commonly made not just about CIL but about ambiguity in international law more generally. *See, e.g.*, JEAN D'ASPREMONT, *FORMALISM AND THE SOURCES OF INTERNATIONAL LAW* 29–30 (2011) (asserting that "uncertainty regarding the existence of international legal rules prevents them from providing for meaningful commands" or "generating any change in the behavior of its addressees"); Andrew T. Guzman, "A Compliance-Based Theory of International Law," 90 *CALIF. L. REV.* 1823, 1863 (2002) ("As the uncertainty of an obligation increases, the reputational cost from a violation decreases."); Thomas M. Franck, "Legitimacy in the International System," 82 *AM. J. INT'L L.* 705, 714 (1988) ("Indeterminate normative standards not only make it harder to know what conformity is expected, but also make it easier to justify noncompliance.").

21 *See* Niels Peterson, "Customary International Law and Public Goods" (in this volume); Joel Trachtman, "The Growing Obsolescence of Customary International Law" (in this volume).

22 *See e.g.*, J. Patrick Kelly, "The Twilight of Customary International Law," 40 *VA. J. INT'L L.* 449, 451 (2000) (asserting that, because CIL is indeterminate, "it cannot function as a legitimate source of substantive legal norms"); Philip R. Trimble, "A Revisionist View of Customary International Law," 33 *UCLA L. REV.* 665, 716–31 (1986). Like the efficacy argument, this legitimacy argument is made not just about CIL but about ambiguous international legal norms more generally. *See, e.g.*, Thomas M. Franck, "The Power of Legitimacy and the Legitimacy of Power: International Law in an Age of Power Disequilibrium," 100 *AM. J. INT'L L.* 88, 93 (2006) (arguing that the most important element of a norm's legitimacy is its determinacy); Martti Koskenniemi, "Solidarity Measures: State Responsibility as a New International Order?" 72 *BRIT. Y.B. INT'L L.* 337, 341 (2002) ("[V]ague clauses would give too much room for political abuse.").

23 D'ASPREMONT, *supra* note 20, at 164 (emphasis added).

24 Fernando R. Tesón, "Fake Custom," in *CUSTOMARY LAW* (Brian Lepard ed., forthcoming 2015), *available at* http://ssrn.com/abstract=2527523.

25 ALAN BOYLE & CHRISTINE CHINKIN, *THE MAKING OF INTERNATIONAL LAW* 21 (2007) ("Uncertainties about the existence and content of rules of customary law allow opportunistic claims."); Pierre-Hugues Verdier & Erik Voeten, "Precedent, Compliance, and Change in Customary International Law: An Explanatory Theory," 108 *AM. J. INT'L L.* 389, 413 (2014) ("[S]tates often appear to assess state practice opportunistically in the context of specific disputes or when arguing for a desired rule.").

These criticisms of CIL are largely what motivate the persistent focus on method: if the method for finding CIL were more refined, states and other global actors would have less discretion to characterize as CIL their own, parochial policy preferences – and the uncertainty and susceptibility to abuse that are the hallmarks of CIL might be reduced. Indeed, the International Law Commission has explained its current project on CIL in precisely these terms:

> The view was expressed that the ambiguities surrounding the identification of customary international law had given rise to legal uncertainty and instability, as well as opportunistic or bad faith arguments regarding the existence of a rule of customary international law. The proposed effort to clarify the process by which a rule of customary international law is identified was thus generally welcomed.[26]

Of course, those who focus on method recognize that CIL finding involves some interpretive discretion. Still, the goal is to limit that discretion by making the exercise more methodologically disciplined.

In short, although different methods vary in the relevance and weight that they attribute to particular data, they all present the same detached question: do the data support this or that CIL claim? The methods also have similar goals: to help resolve whether norms that are claimed to be CIL are really CIL and thus to reduce the volatility and political excess in CIL. And they rest on a set of common assumptions. They assume that CIL finding is an objective exercise – and thus that it stands apart from and can impose order on the process for making CIL. The recent experience in international humanitarian law (IHL), the law that governs the conduct of hostilities in wartime, suggests that these assumptions do not hold. CIL finding is entangled with CIL making and part of the same chaotic process.

THE DEMAND FOR CUSTOMARY INTERNATIONAL HUMANITARIAN LAW

Customary IHL presents a tidy case for studying the methodological question because the recent shift in this area has been so drastic and undeniable. From the mid-nineteenth century until the end of the twentieth century, IHL was dominated by treaties. There are now dozens of IHL treaties, the principal ones being the four Geneva Conventions of 1949 and their two Additional Protocols of 1977.[27] Customary

[26] Rep. of the Int'l Law Comm'n, *56th Sess.*, May 6–June 7, July 8–Aug. 9, 2013, ¶ 75, UN Doc. A/68/10.

[27] Geneva Convention (I) for the Amelioration of the Condition of the Wounded and Sick in Armed Forces in the Field, Aug. 12, 1949, 6 U.S.T. 3114, 75 U.N.T.S. 31 [hereinafter *Geneva Convention I*]; Geneva Convention (II) for the Amelioration of the Condition of Wounded, Sick and Shipwrecked Members of Armed Forces at Sea, Aug. 12, 1949, 6 U.S.T. 3217, 75 U.N.T.S. 85 [hereinafter *Geneva Convention II*]; Geneva Convention (III) Relative to the Treatment of Prisoners of War, Aug. 12, 1949, 6 U.S.T. 3316, 75 U.N.T.S. 135 [hereinafter *Geneva Convention III*]; Geneva Convention (IV) Relative to the Protection of Civilian Persons in Time of War, Aug. 12, 1949, 6 U.S.T. 3516, 75 U.N.T.S 287

IHL was always understood to operate alongside these treaties but was thought to do little independent work. Well into the 1990s, the dominant position in the literature was that customary IHL, at best, reflected the most minimal provisions of the IHL treaties.[28] Customary IHL has since flourished; it is now widely understood to *exceed* the scope of the IHL treaties.

That shift in customary IHL resulted from a concerted effort to "fix" two perceived deficiencies in the treaties' scope of application.[29] First, the treaties bind only state parties. This is not an issue for the Geneva Conventions because they are universally ratified. But dozens of states, including militarily active ones, have either declined to ratify or attached reservations to the Additional Protocols. Most significantly, the United States and Israel refused to ratify the Additional Protocols because of fundamental disagreements with certain provisions. Several other states, including states that participate in the North Atlantic Treaty Organization, ratified the Additional Protocols with reservations.[30] These nonparties and reserving parties create gaps in the treaty regime.

Second, IHL developed primarily to regulate interstate conflicts ("international conflicts"), so the treaties under-regulate most conflicts involving armed nonstate groups ("non-international conflicts").[31] Of the hundreds of provisions in the Geneva Conventions, only Article 3 – which is common to all four Conventions – applies in non-international conflicts. Common Article 3 establishes minimal protections for people in wartime.[32] Additional Protocol II also applies in non-international conflicts, but this protocol is much more rudimentary than the first, which applies in international conflicts. Moreover, Additional Protocol II applies by its terms only in extreme cases: when an armed nonstate group operates under responsible command,

[hereinafter *Geneva Convention IV*]; Protocol Additional to the Geneva Conventions of 12 August 1949, and Relating to the Protection of Victims of International Armed Conflicts, June 8, 1977, 1125 U.N.T.S. 3 [hereinafter *Additional Protocol I*]; Protocol Additional to the Geneva Conventions of 12 August 1949, and relating to the Protection of Victims of Non-International Armed Conflicts, June 8, 1977, 1125 U.N.T.S. 609 [hereinafter *Additional Protocol II*].

28 *See, e.g.*, Final Report of the Commission of Experts Established Pursuant to Security Council Resolution 780 (1992), ¶ 52, UN Doc. S/1994/674 Annex (May 27, 1994) ("It is probable that common article 3 would be viewed as a statement of customary international law, but unlikely that the other instruments would be so viewed."); Sandesh Sivakumaran, "Re-Envisaging the International Law of Internal Armed Conflict," 22 *Eur. J. Int'l L.* 219, 228 (2011) ("Until the 1990s, the view that there were more than simply a handful of customary rules applicable in internal armed conflict was never seriously entertained, and identifying even those rules proved rather problematic.").

29 *See* 1 Jean-Marie Henckaert et al., Int'l Comm. Red Cross, *Customary International Humanitarian Law* xxxiv (2005) [hereinafter *ICRC Study*].

30 *See* Julie Gaudreau, "The Reservations to the Protocols Additional to the Geneva Conventions for the Protection of War Victims," 849 *Int'l Rev. Red Cross* 143 (2003).

31 IHL now classifies as "international" one set of conflicts against nonstate groups: conflicts involving national liberation movements. *See Additional Protocol I*, *supra* note 27, art. 1.

32 *Geneva Conventions I, II, III, IV*, *supra* note 27, art. 3.

controls part of a state's territory, and carries out "sustained and concerted military operations."[33] As late as 1977, then, states collectively decided that international law would lightly regulate non-international conflicts.

That decision became untenable with the maturation of international human rights law. Human rights law's initial impact on IHL was not doctrinal. Although both regimes aim to protect people from harm, each developed for a different context. IHL was designed primarily for interstate wars, while human rights law focused on a state's everyday relations with its own people. The two regimes might reasonably establish different rules for each context. Rather, human rights law's initial impact on IHL was conceptual. By declaring that international law *is* concerned with how states treat nonstate actors, human rights law painted as outmoded the minimal regulation of non-international conflicts under IHL.[34]

Using CIL to correct perceived deficiencies in the treaties was an ambitious goal. States had deliberately limited the treaties' scope of application and had not shown an appetite for altering that arrangement. Moreover, the extent to which states were exceeding, let alone meeting, their treaty obligations – so as to create parallel substantive norms that apply more broadly as a matter of CIL – was, to say the least, unclear. Compliance with IHL has long been a problem. Nevertheless, the customary IHL project has been extremely successful. Although global actors still quibble about the content or CIL status of particular IHL norms, the prevailing view is now that customary IHL applies in all conflicts and is modeled not after the minimal treaty provisions for non-international conflicts but after the much more robust provisions for *international* conflicts.[35]

THE PROCESS FOR CUSTOMARY INTERNATIONAL HUMANITARIAN LAW

The experience with customary IHL exposes a conceptual flaw in the literature on method. Asking how to find CIL assumes that CIL finding is a distinct thing – specifically, that it is meaningfully distinguishable from CIL making. Studying the rise of customary IHL shows that, in fact, CIL finding looks and operates much like CIL making. Both exercises involve constructing the preexisting data to make a prescriptive claim. Both are susceptible to opportunism and abuse – claims that push hard to change or undercut expectations, even as they pretend to reflect expectations. And both depend for their relevance on their reception within the

[33] *Additional Protocol II, supra* note 27, art. 1.

[34] *See, e.g., Prosecutor v. Tadic*, Case No. IT-94-1-I, Decision on the Defence Motion for Interlocutory Appeal on Jurisdiction, ¶ 97 (Int'l Crim. Trib. for the Former Yugoslavia Oct. 2, 1995) ("A State-sovereignty-oriented approach has been gradually supplanted by a human-being-oriented approach.").

[35] *See infra* notes 51–52, 57–74, and accompanying text.

chaotic CIL process. The fact that the two exercises overlap so heavily calls into question the assumption that CIL finding is objective and disciplined, even as CIL making clearly is not.

Law Finding as Lawmaking

Decisions that purport to find CIL look a lot like decisions that aim to make CIL. The CIL process produces a mass of disconnected, ambiguous, and inevitably contradictory data. "Finding" CIL entails sifting through and advancing a prescriptive claim on the data – a claim that the data should be interpreted as supporting this or that CIL norm.[36] Such decisions are usually articulated as if they describe an objective phenomenon: "CIL is X." In fact, the decisions construct the data and push CIL in a particular direction.[37] Decisions that aim to make CIL use very similar moves. They usually assert that CIL is X in an effort to shift CIL toward X.

Indeed, because the CIL process is so unstructured, it lacks the formal controls that might inhibit the participants from pushing hard for particular norms – making the opportunistic claims that methods for finding CIL are supposed to weed out. Such claims are an ordinary part of the normative contestation that occurs during the process for making CIL.[38] Actors that participate in CIL making, such as states or nongovernmental advocates, might benefit from treating their own contested claims as law. Their legal decisions thus might contain very tendentious CIL claims. Actors that are specifically charged with *finding* CIL are presumed to stand apart from that process and to assess the evidence objectively. But their decisions sometimes also contain opportunistic or tendentious CIL claims.

Consider the 1995 *Tadic* decision of the International Criminal Tribunal for the Former Yugoslavia (ICTY).[39] *Tadic* determined that, in addition to replicating Common Article 3 and "the core of Additional Protocol II,"[40] customary IHL prohibits certain tactics that the treaties proscribe only in international conflicts.[41]

[36] *See, e.g.*, STARKE'S INTERNATIONAL LAW 36 (I.A. Shearer ed., 11th ed. 1994) ("The difficulties involved in extracting a customary rule or principle of international law from the mass of heterogeneous documentation of state practice … are not to be minimised.").

[37] *Cf.* Curtis A. Bradley, "Customary International Law Adjudication as Common Law Adjudication" (in this volume) (arguing that courts "find" CIL by constructing and evaluating the past practice, assessing state preferences, and accounting for social or moral considerations).

[38] *See* DAVID J. BEDERMAN, *THE SPIRIT OF INTERNATIONAL LAW* 57 (2002) (describing the process as one of "struggle and resistance," and as a " 'marketplace in which states affirmatively (and self-consciously) 'bid' and 'barter' and 'trade' in new rules of conduct"); D'AMATO, *supra* note 18, at 266 ("[C]ustom represents a type of structured legal argument that has recurred in many claim-conflict situations.").

[39] *Tadic, supra* note 34.

[40] *Id.* at ¶ 98.

[41] *Id.* at ¶ 119.

This position was extremely expansive for the time.[42] The overwhelmingly dominant view was that customary IHL at best reflected the minimal treaty provisions for non-international conflicts.[43] Although *Tadic* purported only to find CIL, it pushed hard to create CIL. It drove customary IHL toward the treaty law for international conflicts.[44]

The literature already recognizes that CIL finding bleeds into CIL making. But even scholars who underscore this point go on to ask the methodological question – and thus to assume that the two exercises are, in some way, distinct.[45] For example, Maurice Mendelson explains in his Hague Course that CIL creation and CIL application often occur simultaneously.[46] Still, Mendelson insists that global actors need a method for finding CIL.[47] His course then works through the methodological question. Likewise, Anthea Roberts recognizes that CIL finding involves some level of CIL making and that the two exercises can look alike.[48] However, Roberts's stated purpose is to offer a new method for finding CIL. Brian Lepard's chapter in this volume is similar. Lepard argues for conceiving of CIL as a dynamic lawmaking process. Nevertheless, Lepard proposes a method for finding CIL. He argues that CIL should be found by sifting through a broad range of evidence and identifying the position that states generally believe to be desirable.[49]

42 *See, e.g.*, Christopher Greenwood, "International Humanitarian Law and the *Tadic* Case," 7 EUR. J. INT'L L. 265, 278 (1996) ("[T]he Appeals Chamber has gone further than other bodies by determining that there are rules applicable to internal armed conflicts which are not based upon either common Article 3 or Additional Protocol II.").

43 *See supra* note 28 and accompanying text.

44 To be clear, *Tadic* recognized certain differences between the customary IHL that applies in all conflicts and the treaty law that applies in international conflicts. *Tadic* explained that "only a number of rules and principles governing international armed conflicts have gradually been extended to apply to internal conflicts." *Tadic*, *supra* note 34, at ¶ 126. But as Sandesh Sivakumaran has explained, "the consistent approach of the ICTY" has been to "analogize to the law of international armed conflict." Sivakumaran, *supra* note 28, at 230. Given the preexisting view of customary IHL, *Tadic*'s push toward the treaty law for international conflicts was significant.

45 *See also, e.g.*, KAROL WOLFKE, *CUSTOM IN PRESENT INTERNATIONAL LAW* 53–54 (2d rev. ed. 1993) (recognizing that CIL "formation and ascertainment, are of course closely interdependent," but then insisting that "the formation of a custom and the ascertaining of custom or customary rule are two different notions"); Nicholas Greenwood Onuf, "Global Law-Making and Legal Thought," in *LAW-MAKING IN THE GLOBAL COMMUNITY* 1, 14–19 (Nicholas Greenwood Onuf ed., 1982) (asserting that "law-making and law-using tend to be aspects of the same phenomenon," but then addressing the question of how to find CIL).

46 Mendelson, *supra* note 11, at 176 ("[I]t is not always possible or helpful to label one part [of the process] 'creation' and another 'application.'").

47 *Id.* (explaining that decision makers must be able to determine "whether, *at the moment the appreciation is being made*, the practice has matured into a rule of law").

48 Anthea Elizabeth Roberts, "Traditional and Modern Approaches to Customary International Law: A Reconciliation," 95 AM. J. INT'L L. 757, 761, 763 (2001).

49 Brian D. Lepard, "Customary International Law as a Dynamic Process" (in this volume).

Perhaps the distinction between CIL making and CIL finding is intended to capture the idea that CIL decisions vary in the extent to which they reflect, rather than challenge, preexisting expectations. Decisions might be said to shift from CIL making toward CIL finding as their prior support increases. Yet that distinction elides both the fluidity of the CIL process and the prescriptive effect of later-in-time decisions. Because the CIL process is always ongoing, the meaning and relevance of a CIL norm fluctuate as different actors engage with it over time.[50] Any decision relating to the norm – even a decision that might reasonably be characterized as a "finding" – has some prescriptive effect. As I elaborate below, *Tadic* would have made a prescriptive move even if it had more accurately reflected expectations. Adopting the then-dominant position on customary IHL would have helped solidify that position as law. This prescriptive effect might seem trivial if the position were already widely accepted as CIL. But in that event, any *finding* that the position was CIL would also be trivial. Decisions that purport to find CIL are most relevant when the CIL norm is uncertain or contested – and thus when it benefits from more prescription.

Law Finding, Lawmaking, and the CIL Process

Decisions that purport to find CIL and decisions that push to create CIL do not only look alike and make similar moves. They also operate in similar ways in the CIL process. This process is highly unstructured; it lacks a central authority for settling the law or a designated forum for ratifying or challenging particular decisions. All CIL decisions – decisions that aptly reflect preexisting expectations and decisions that push hard to alter expectations – feed into the same, chaotic process. In this process, a decision's relevance and impact turn not necessarily on the decision's prior level of support but on its reception among other actors going forward. To the extent that other actors seize on a decision and treat its claim as law, the claim grows stronger and solidifies as CIL. To the extent that they instead reject or continue to debate the claim, it remains vulnerable and unstable. As a practical matter, this means that norms that are claimed to be CIL vary considerably in their strength. Norms that are widely accepted and treated as law tend to be strong. Norms with shallower or narrower support are weaker; they might be treated as law only by some actors, in particular venues, or in certain respects. Decisions that purport to find CIL do not weed out invalid claims or stabilize CIL because they operate like other prescriptive claims do; they become part of the lawmaking mix.

Consider again the *Tadic* decision. *Tadic*'s position on customary IHL is now widely accepted not because it accurately reflected the customary IHL at the

[50] *See* W. Michael Reisman, "International Incidents: Introduction to a New Genre in the Study of International Law," 10 Yale J. Int'l L. 1, 4–5 (1984); Mendelson, *supra* note 11, at 175.

time – it did not – but because of how the lawmaking process unfolded. States essentially adopted that position in the Rome Statute of the International Criminal Court.[51] Like *Tadic*, the Rome Statute maintains a distinction between international and non-international conflicts. The statute's list of war crimes is more abbreviated for non-international conflicts.[52] But like *Tadic*, the statute identifies as war crimes much conduct that the IHL treaties do not specifically proscribe in non-international conflicts. States effectively converged on *Tadic*'s claim in the Rome Statute.

Tadic would have fed into the same process even if it had come out the other way. Adopting the then-dominant position on customary IHL would have helped strengthen that position and weaken any competing claims on the law. These competing claims would have been more difficult to sustain because their pretense of simply reflecting reality would have been less plausible; they would have looked like unilateral efforts to change the law. Still, solidified CIL norms *can* deteriorate. The then-dominant position would have remained viable only so long as it sustained a base of support.[53] States might have agitated against that position by, for example, behaving in ways that reflected their discontent, openly promoting a change in the law, or simply adopting the Rome Statute.

Most CIL decisions are unlike *Tadic* in that they are not rapidly and collectively endorsed. Their relevance is diffuse and drawn out. The customary IHL study that the International Committee of the Red Cross (ICRC) published in 2005 is illustrative.[54] This study is much more ambitious than *Tadic* or the Rome Statute; it addresses almost all of IHL, not just the most serious violations that qualify as war crimes and trigger the jurisdiction of international criminal courts.[55] The ICRC purported to find 161 rules of customary IHL that largely replicate the treaty law

[51] Rome Statute of the International Criminal Court, art. 8(2)(a)–8(2)(b), July 17, 1998, 2187 U.N.T.S. 90 [hereinafter *Rome Statute*]; *cf.* Jean-Marie Henckaerts, "Customary International Humanitarian Law: A Response to U.S. Comments," 89 *Int'l Rev. Red Cross* 473, 486 (2007) [hereinafter Henckaerts, *A Response to US Comments*] ("[E]very humanitarian law treaty adopted since 1996 has been made applicable to both international and non-international conflicts."). On this development, see Allison Marston Danner, "When Courts Make Law: How the International Criminal Tribunals Recast the Laws of War," 59 *Vand. L. Rev.* 1, 25–37 (2006); Claus Kress, "War Crimes Committed in Non-International Armed Conflict and the Emerging System of International Criminal Justice," 30 *Isr. Y.B. Hum. Rts.* 103, 104–07 (2000).

[52] *Compare Rome Statute*, *supra* note 51, art. 8(2)(a)-8(2)(b), *with id.* art. 8(2)(c), 8(2)(e).

[53] *See, e.g.*, Malcolm N. Shaw, *International Law* 89 (6th ed. 2008) ("Generally, where states are seen to acquiesce in the behaviour of other states without protesting against them, the assumption must be that such behaviour is accepted as legitimate."); Thomas Franck, *The Power of Legitimacy Among Nations* 24 (1990) (arguing that noncompliance undermines a norm's "compliance pull," making it less likely that others will comply going forward); Michael J. Glennon, "How International Rules Die," 93 *Geo. L.J.* 939, 960 (2005) ("[A]t some point state practice that is inconsistent with a norm is simply too thick to justify the conclusion that states really accept the norm as obligatory.").

[54] *ICRC Study*, *supra* note 29.

[55] The study excludes from its scope a few discrete aspects of IHL. *See id.* at xxxvi–xxxvii.

for international conflicts but apply more broadly: to states that are not parties or have attached reservations to the Additional Protocols, and in non-international conflicts.[56] As the ICRC proudly proclaimed, "State practice … has led to the creation of rules parallel to those in Additional Protocol I, but applicable as customary law to non-international armed conflicts."[57]

No matter to what extent the study reflected prior expectations – and there is good reason to believe that some of its conclusions were heavily prescriptive[58] – it is now part of the CIL process and input for further lawmaking. The ICRC invokes and applies its own CIL positions when it educates people about IHL and tries to influence warring parties to comply.[59] It thus helps create the expectation that its positions are law. Moreover, some of these positions have clearly resonated with other actors and helped shape customary IHL. For instance, since the study was published, several states have incorporated into their national laws or military manuals specific conclusions from the study. A few states have even replicated the study's precise language.[60] Similarly, international and national courts have cited the study as evidence of CIL.[61] Courts most often cite the study for rules, such as the

56 The study recognizes that several rules (rules 3, 4, 41, 49, 51, 106–08, 114, 130, and 145–47) might apply only in international conflicts. Most of these rules concern combatant status or occupations. In addition, the study formulates a handful of rules (rules 124, 126, 128–29) differently for international than for non-international conflicts. *See* Malcom MacLauren & Felix Schwendimann, "An Exercise in the Development of International Law: The New ICRC Study on Customary International Humanitarian Law," 6 German L.J. 1217, 1228–29 (2005).

57 *ICRC Study*, *supra* note 29, at xxxv.

58 The study was strongly criticized on methodological grounds. *See, e.g.*, John B. Bellinger, III & William J. Haynes, II, "A U.S. Government Response to the International Committee of the Red Cross Study Customary International Humanitarian Law," 89 *Int'l Rev.* Red Cross 443 (2007) (raising a litany of methodological objections); Daniel Bethlehem, "The Methodological Framework of the Study," in Perspectives on the *ICRC* Study on Customary International Humanitarian Law 3, 10 (Elizabeth Wilmshurst & Susan Breau eds., 2007) (arguing that the practice was too often equivocal or sparse to support the claimed CIL norms); Yoram Dinstein, "The ICRC Customary International Humanitarian Law Study," 36 *Isr. Y.B. Hum. Rts.* 1 (2006) (questioning the weight that the study attributed to different kinds of data); George H. Aldrich, "Customary International Humanitarian Law – An Interpretation on Behalf of the International Committee of the Red Cross," 76 *Brit. Y.B. Int'l L.* 503, 507 (2005) (asserting that the study's "credibility crumbles" as a result of its overly expansive claims). For examples of the ICRC's responses, see Jean-Marie Henckaerts, "Customary International Humanitarian Law – A Rejoinder to Judge Aldrich," 76 *Brit. Y.B. Int'l L.* 525 (2006); and Henckaerts, "A Response to US Comments," *supra* note 51.

59 On the ICRC's work, see Steven R. Ratner, "Law Promotion Beyond Law Talk: The Red Cross, Persuasion, and the Laws of War," 22 *Eur. J. Int'l L.* 459 (2011).

60 *See, e.g.*, ICRC, *Practice Relating to Rule* 54, Customary IHL Database, retrieved from https://www.icrc.org/customary-ihl/eng/docs/v2_rul_rule54 (last visited Sept. 18, 2014) (Spain); ICRC, *Practice Relating to Rule* 120, Customary IHL Database, retrieved from https://www.icrc.org/customary-ihl/eng/docs/v2_rul_rule120 (last visited Sept. 18, 2014) (Guinea and Mexico).

61 *See, e.g.*, *Prosecutor v. Halilovic*, Case No. IT-01-48-T, Judgment, ¶ 90 (Int'l Crim. Trib. for the Former Yugoslavia Nov. 16, 2005) (on command responsibility); *Janowiec v. Russia*, App. 55508/07, 58 Eur.

ones that define and prohibit attacks on civilian objects, that were already widely regarded as CIL before the study was published.[62] The study has a reinforcing effect on these rules. As global actors repeatedly endorse a CIL claim, it grows stronger and more difficult to refute. It solidifies as CIL.

The reactions to other positions in the study have been more mixed, and the normative contestation that is endemic to CIL remains apparent. Consider the study's core claim that customary IHL is modeled after the treaty law for international conflicts. This claim is now widely accepted at a high level of abstraction.[63] But the claim's relevance and meaning continue to be unstable. IHL experts who take the claim for granted still disagree on how to translate into the customary IHL for non-international conflicts the treaty law that was designed for international conflicts.[64] This disagreement is evident, for example, in the controversy surrounding the ICRC's 2009 *Interpretive Guidance on the Notion of Direct Participation in Hostilities*.[65] The *Interpretive Guidance* was meant to help clarify an issue that lies at the heart of IHL: when may someone be targeted for attack? IHL treaties answer the question for international conflicts: members of state armed forces generally identify themselves as such, comprise the fighting class, and are targetable.[66] The rule for civilians is the reverse: civilians are generally *not* targetable because they are presumed not to participate in the fighting. Under Additional Protocol I, civilians are targetable only if they directly participate in hostilities.[67] Those treaty rules translate poorly to non-international conflicts in which the combatant-civilian distinction breaks down.[68] The *Interpretive Guidance* sought to resolve the targeting

H.R. Rep. 30 ¶ 157 (2014) (on duty to investigate war crimes); *Hamdan v. Rumsfeld*, 548 U.S. 557, 632 (2006) (on trying people before a "regularly constituted court"); *Al-Sirri v. Sec. State for Home Dept.*, [2013] 1 A.C. 745, 787 (U.K.) (on attacking peacekeeping personnel); HCJ 3799/02, *Adalah Legal Center v. IDF Central Commander* [2005] (2) Isr. 206, ¶ 20 (Isr.) (on advance warnings of attacks that might affect civilians).

62 *Compare, e.g., ICRC Study, supra* note 29, at 29–32, *with Prosecutor v. Kordić and Čerkez*, Motion to Dismiss for Lack of Jurisdiction, No. IT_95-14/2-PT, ¶ 15–31 (Mar. 2, 1999); *Prosecutor v. Hadžihasanović & Kubura*, Case No. IT-01-47-AR73.3, Decision on the Joint Defence, ¶ 30 (Int'l Crim. Trib. for the Former Yugoslavia Mar. 11, 2005); *and Prosecutor v. Garda*, Case No. ICC-02/05-02/09, Decision on the Confirmation of Charges, n. 130 (Feb. 8, 2010).

63 *See* Sandesh Sivakumaran, *The Law of Non-International Armed Conflict* 56–57, 61 (2012).

64 For a discussion and critique of this method, see Monica Hakimi, "A Functional Approach to Targeting and Detention," 110 *Mich. L. Rev.* 1365 (2012).

65 Nils Melzer, ICRC, *Interpretive Guidance on the Notion of Direct Participation in Hostilities Under International Humanitarian Law* (2009).

66 On the people who qualify as combatants, see *Geneva Convention III, supra* note 27, art. 4; *and Additional Protocol I, supra* note 27, arts. 1(4), 43–44. On the point that members of state armed forces must identify themselves as such, see *Geneva Convention III, supra* note 27, art. 4; *and Additional Protocol I, supra* note 27, art. 44(3). On the well-accepted exceptions to this targeting rule, see *Additional Protocol I, supra* note 27, art. 41.

67 *Additional Protocol I, supra* note 27, art. 51(3).

68 *See* Hakimi, *supra* note 64, at 1377–79, 1398–03; *see also* Laurie Blank & Amos Guiora, "Teaching an Old Dog New Tricks: Operationalizing the Law of Armed Conflict in New Warfare," 1 *Harv.*

rules that apply, mostly as a matter of customary IHL, in these conflicts. Its conclusions were heavily criticized.[69]

Separately, the ICRC's core claim is under pressure from actors who question the extent to which customary IHL, as modeled after the treaty law for international conflicts, is even the appropriate regulatory framework for modern conflicts. Now more than ever, domestic courts, human rights institutions, and civil society groups appraise situations involving armed nonstate groups. These decisionmakers are not especially invested in "pure" IHL, and they regularly invoke and apply hybrids that mix IHL with norms from other sources.[70] For example, the Israeli Supreme Court has used domestic law to modify the norms that, in its view, would have applied as a matter of customary IHL.[71] The European Court of Human Rights has repeatedly applied human rights law, instead of IHL, to situations that are also covered by IHL.[72] And civil society groups regularly invoke their preferred versions of IHL to condemn conduct that is consistent with more mainstream positions on IHL.[73] IHL purists might insist that these moves do not affect IHL because they are

NAT'L SECURITY J. 45, 53 (2010) (asserting that "new warfare" is "characterized by fighting in highly populated areas with a blurring of the lines between military forces and civilian persons and objects").

69 For an analysis of the areas of agreement and disagreement, see Hakimi, *supra* note 64, at 1377–79, 1398–03. For evidence of the disagreement, see, for example, W. Hays Parks, "Part IX of the ICRC 'Direct Participation in Hostilities' Study: No Mandate, No Expertise, and Legally Incorrect," 42 *N.Y.U. J. INT'L L. & POL.* 769 (2010); Michael N. Schmitt, "The Interpretive Guidance on the Notion of Direct Participation in Hostilities: A Critical Analysis," 1 *HARV. NAT'L SECURITY J.* 5 (2010); Kenneth Watkin, "Opportunity Lost: Organized Armed Groups and the ICRC 'Direct Participation in Hostilities' Interpretive Guidance," 42 *N.Y.U. J. INT'L L. & POL.* 641 (2010); Bill Boothby, "'And for Such Time as': The Time Dimension to Direct Participation in Hostilities," 42 *N.Y.U. J. INT'L L. & POL.* 741 (2010).

70 The ICRC study incorporates human rights law to the extent that the norms in human rights law are analogous to the ones in IHL. However, the study does not try to resolve any discrepancies between the two regimes. *See ICRC Study*, *supra* note 29, at xxxvi–xxxvii.

71 HCJ 769/02 Pub. Comm. Against Torture in *Isr. v. Israel*, 2006(2) Isr. Rep. 459, 490–95 [2005] (Isr.).

72 *See, e.g.*, *Isayeva v. Russia*, App. No. 57950/00, 41 Eur. H.R. Rep. 791 (2005); *Al-Skeini v. United Kingdom*, App. No. 55721/07, Eur. Ct. H.R. (2011). The court's approach in these cases might be justified on the ground that the court's mandate is to apply human rights law, not IHL. This justification is incomplete. Other bodies with similar mandates have occasionally applied IHL. *See, e.g.*, *Bámaca Velásquez v. Guatemala*, Case 11.129, 2001 Inter-Am. Ct. H.R. (ser. C) No. 70, at 321 (Nov. 25, 2000); Human Rights Comm., *Concluding Observations: Croatia*, ¶ 9, UN Doc. CCPR/C/79/Add.15 (Dec. 28, 1992).

73 *Compare, e.g.*, *Geneva Convention IV*, *supra* note 27, art. 19 ("The protection to which civilian hospitals are entitled shall not cease unless they are used to commit, outside their humanitarian duties, acts harmful to the enemy."); *and* HUMAN RIGHTS WATCH, *OFF TARGET: THE CONDUCT OF THE WAR AND CIVILIAN CASUALTIES IN IRAQ* 73 (2003) ("By using hospitals as military headquarters, Iraqi forces turned them into military objectives."); *with* ANTHONY H. CORDESMAN, *THE "GAZA WAR": A STRATEGIC ANALYSIS* 65–66 (2009) (claiming that Hamas used a hospital as a military base during the 2008–2009 Gaza conflict); *and* DAN MAGEN, *"ILL MORALS": GRAVE VIOLATIONS OF THE RIGHT TO HEALTH DURING THE ISRAELI ASSAULT ON GAZA* 41–44 (Ran Yaron ed., Saul Vardi et al. trans., 2009) (condemning Israel's attacks on hospitals in Gaza in 2008–2009).

external to the regime or plainly opportunistic. But the moves unquestionably shape expectations on whether and how IHL applies – either as a matter of treaty law or as a matter of customary law that is modeled after the treaties. The ICRC's "finding" that customary IHL is modeled after the treaty law for international conflicts has not suppressed these competing positions or stabilized the law in this area.[74]

In short, although the ICTY and ICRC purported only to find CIL, they actually helped make CIL. They advanced particular CIL claims. These claims fed back into the CIL process, as other prescriptive claims do. And like other claims, their viability has turned on how the process has unfolded. Some claims already had support and have grown stronger. Other claims were tendentious when they were made but have since gained legal traction. Still others are contested, which means that the putative norms will continue to be invoked and applied inconsistently, depending on who is acting and in which forum.[75] This is how the CIL process works. It works this way no matter to what extent a particular decision aims to reflect or to challenge preexisting expectations.

IMPLICATIONS

The argument that CIL finding is not meaningfully distinguishable from CIL making has two important practical implications. First, nonstate actors who are charged with finding CIL can be extremely influential in making CIL. Second, no method for finding CIL is likely to discipline these actors or help settle CIL, so long as the process for making CIL remains as it is. I address each of these implications in turn.

Participants in the Process

Because CIL finding and CIL making are intertwined, nonstate actors who are charged with finding CIL sometimes play a significant role in making CIL. The dominant view in the literature – which all but the most esoteric methods for finding

74 Indeed, even the ICRC acknowledges that "the study should not be seen as the final word on customary IHL because it cannot be exhaustive and because the formation of CIL is an ongoing process." Henckaerts, *A Response to U.S. Comments*, *supra* note 51.

75 *Cf.* W. Michael Reisman, "The Democratization of Contemporary International Law-Making Processes and the Differentiation of Their Application," in DEVELOPMENTS OF INTERNATIONAL LAW IN TREATY MAKING 15, 28–29 (Rüdiger Wolfrum & Volker Röben eds., 2005) [hereinafter Reisman, *Democratization*]:

> Because the question of whether international law will be effective in a particular dispute will increasingly depend upon the arena or forum in which the dispute is heard, scholarly and practitioner statements of what the law is, which are provided to clients to assist them in their planning, will increasingly have to be qualified by reference to *where* a potential dispute in the future may be initially characterized in terms of law and *where* those characterizations will thereafter be put to political use.

CIL reinforce – is that *states* drive the formation of CIL.[76] That view rests, again, on the distinction between CIL finding and CIL making. Nonstate actors are known to find CIL but are said to play a marginal or subsidiary role in making CIL. The experience with customary IHL suggests otherwise. The ICTY and ICRC have not just participated in making CIL; they have had a significant impact.

The conventional view is incompatible with the very nature of the CIL process. Because the process is unstructured, it lacks the formal mechanisms that might exclude certain participants. Anyone can make a claim on CIL. Of course, some actors are more influential than others; some push harder, command a broader audience, or are more effective in promoting their views.[77] Yet there is no reason to believe that states alone are capable of exercising outsized influence. Other kinds of actors might be influential in part because they are *not* states – and thus are perceived not to be advancing particular, nationalistic agendas. International courts and tribunals, high-level officials of intergovernmental bodies, and expert nongovernmental groups often have an authority, at least in the eyes of some relevant audiences, that individual states lack.[78] These actors can be extremely influential in shaping expectations about CIL. In the face of ambiguity or competition – "CIL is X," "no, CIL is Y" – the claim that appears authoritative tends to be sticky.[79] Competing claims might still be advanced, but these claims become harder to sustain. Their pretense of describing CIL becomes less credible once an actor with some authority has said otherwise.

The literature already recognizes that two nonstate institutions – the International Court of Justice (ICJ) and the International Law Commission – sometimes shape CIL. A recent volume on the ICJ concludes that its "decisions are accorded 'a truly astonishing deference'" in international law.[80] Several CIL claims that were

76 *See* sources cited at *supra* note 5.

77 North Sea Continental Shelf (Ger., Den., Neth.), Judgment, 1969 I.C.J. 3, 43 (Feb. 20) (explaining that states that are especially invested in an issue play a larger role in creating the relevant CIL than do states that are disengaged); Michael Byers, "Introduction: Power, Obligation, and Customary International Law," 11 *Duke J. Comp. & Int'l L.* 81, 84 (2001) (explaining that powerful states can play an outsized role).

78 *See* Harlan Grant Cohen, "Finding International Law, Part II: Our Fragmenting Legal Community," 44 *N.Y.U. J. Int'l L. & Pol.* 1049 (2012); W. Michael Reisman, "International Law-Making: A Process of Communication," 75 *Am. Soc'y Int'l L. Proc.* 101, 110 (1981).

79 *Cf.* Fernando Lusa Bordin, "Reflections of Customary International Law: The Authority of Codification Conventions and the ILC Draft Articles in International Law," 63 *Int'l & Comp. L.Q.* 535, 547 (2014) ("[I]t is the inherent uncertainty by which customary international law is characterized that accounts for the influence that non-legislative [i.e., non-binding] codifications exercise."); Andrew T. Guzman & Timothy L. Meyer, "International Common Law: The Soft Law of International Tribunals," 9 *Chi. J. Int'l L.* 515, 516–17 (2009) (characterizing as "soft law" the decisions of institutions that, although lacking in formal authority, shape expectations about the law).

80 Christian J. Tams, "The ICJ as a 'Law-Formative Agency': Summary and Synthesis," in *The Development of International Law by the International Court of Justice* 377, 379 (Christian J. Tams & James Sloan eds., 2013).

controversial when the ICJ endorsed them have since "come to be accepted."[81] Similarly, the International Law Commission is known to engage in some lawmaking as it codifies CIL.[82] Its prescriptive moves can be significant. For instance, the provisions on countermeasures in the Draft Articles on State Responsibility were extremely controversial during the commission's work on that project.[83] Now that the commission has adopted those provisions, they are regularly cited as CIL.[84]

And yet, the conventional view is still that nonstate actors participate in CIL making only at the margins or by "finding" what states do or believe. Much of the literature posits that nonstate actors have, at best, a peripheral or subsidiary role in CIL making.[85] A common refrain is that they participate only to the extent that states delegate lawmaking authority to them or subsequently endorse their CIL positions. For example, Allison Danner has argued that *Tadic*'s effect on CIL is best explained by states' tacit delegation of lawmaking authority to the ICTY.[86] Anthea Roberts and Sandesh Sivakumaran likewise cite *Tadic* for the proposition that "[a]ny role that state-empowered bodies play in law creation is thus dependent on initial state consent and at least some level of ongoing state consent."[87]

[81] *Id.; see also, e.g.*, Antonio Cassese, "The International Court of Justice: It is High Time to Restyle the Respected Old Lady," in *Realizing Utopia: The Future of International Law* 239, 240 (Antonio Cassese ed., 2012) ("Once the ICJ has stated that a legal standard is part of customary international law, few would seriously challenge such a finding."); Mahnoush H. Arsanjani, "The United Nations and International Law-Making," 362 *Recueil des Cours* 11, 29, 38 (2012) ("[T]he International Court has been involved … on occasion, in making international law, by declaring a particular practice as having become 'customary international law.'").

[82] Boyle & Chinkin, *supra* note 25, at 174; Bordin, *supra* note 79, at 542–46.

[83] *See* Koskenniemi, *supra* note 22, at 340.

[84] *See, e.g.*, *Archer Daniels Midland Co. v. The United Mexican States*, ICSID Case No. ARB(AF)/04/05, Award, ¶¶ 125–80 (Nov. 21, 2007), *retrieved from* http://icsid.worldbank.org/ICSID/Index.jsp (search "Archer Daniels"); *Corn Prods. Int'l v. The United Mexican States*, ICSID Case No., ARB(AF)/04/01, Decision on Responsibility, ¶¶ 144–92 (Jan. 15, 2008), *retrieved from* http://icsid.worldbank.org/ICSID/Index.jsp (search "corn products").

[85] *See, e.g.*, Ochoa, *supra* note 5, at 144 ("Even if (or when) the individual is formally recognized as a legitimate participant in the creation of CIL, the state will continue to hold this position of prominence."); Mendelson, *supra* note 11, at 203 (asserting that nonstate actors participate only indirectly in the formation of CIL, by influencing other actors); *cf.* Reisman, *Democratization*, *supra* note 75, at 19–24 (underscoring that nonstate actors regularly participate in international lawmaking but noting that treaty law and CIL are still largely made by states). As discussed, the literature does recognize that the ICJ can play a significant role in CIL making. But many scholars who recognize this point still try to link the ICJ's role to states. *See, e.g.*, Eyal Benvenisti, "Customary International Law as a Judicial Tool for Promoting Efficiency," in *The Impact of International Law on International Cooperation* 85, 86–87 (Eyal Benvenisti & Moshe Hirsch eds., 2004) (arguing that the ICJ "has, in fact, the authority to invent the custom," as "a sort of trustee acting in the best interests of the states and the global community," and that "States accept and welcome such leaps.").

[86] Danner, *supra* note 51, at 47–48.

[87] Anthea Roberts & Sandesh Sivakumaran, "Lawmaking by Nonstate Actors: Engaging Armed Groups in the Creation of International Humanitarian Law," 37 *Yale J. Int'l L.* 107, 117–18 (2012).

Those accounts are unconvincing. Any delegation of lawmaking authority to the ICTY or ICRC was, at best, implicit and incomplete. Danner herself recognizes that, when the ICTY was created, several states asserted that the ICTY was only to apply existing CIL, not to help make CIL.[88] In addition, states had not committed to accepting the ICTY's positions. True, *Tadic*'s CIL positions *were* soon endorsed by states, but the same cannot be said of the ICRC's. The ICRC study is very much a nongovernmental product. States did not participate in drafting the study, and most states responded with silence when the study was released.[89] Nevertheless, the study has clearly been shaping customary IHL.

Of course, states continue to play a role in this process. A CIL claim that many states reject will have little legal traction, no matter whether that claim is advanced by a nonstate actor, such as the ICRC, or by a state. The point is that nonstate actors can participate in the formation of CIL just as – and sometimes more effectively than – individual states can. Where a CIL norm is uncertain or contested, a nonstate actor's "finding" can tip the scale toward one side or the other. Indeed, the principal threat to the ICRC's core claim on customary IHL appears to be coming not from states – which are still equivocal when engaging with IHL – but from other nonstate actors. Recall that human rights institutions and civil society groups are resisting the idea that customary IHL, as modeled after the treaty law for international conflicts, is the appropriate regulatory framework for many modern conflicts.

Method as Discipline

One response might be to criticize the ICTY and ICRC for exceeding their mandates and trying to make CIL. Sure, all CIL finding involves some level of CIL making. But those who are charged with finding CIL are supposed to be objective and disciplined; their prescriptive discretion should be constrained.[90] And, the

88 Danner, *supra* note 51, at 21 ("Many delegates asserted that the Tribunals could not create new law."); *see also* UN Secretary-General, *Rep. of the Secretary-General Pursuant to Paragraph 2 of Security Council Resolution 808*, ¶ 34, UN Doc. S/RES/808 (May 3, 1993), *reprinted in* 32 ILM 1163 (asserting that the ICTY was to apply only norms that were "beyond any doubt part of customary law").

89 The International Conference of the Red Cross and Red Crescent – which consists of states – did ask the ICRC to prepare a report on customary IHL. ICRC, *From Law to Action: Report on the Follow-Up to the International Conference for the Protection of War Victims*, 26th International Conference of the Red Cross and Red Crescent Res. 1 (Dec 3–7, 1995). But this request cannot reasonably be characterized as a delegation of CIL making authority or an endorsement of whatever substantive positions the ICRC would take. States did not themselves work on the study. And to the extent that governmental experts were consulted, they were consulted in their personal capacities and only once the study was very far along. *See* Iain Scobbie, "The Approach to Customary International Law in the Study," in PERSPECTIVES ON THE *ICRC* STUDY ON CUSTOMARY INTERNATIONAL HUMANITARIAN LAW, *supra* note 58, 15, 16–17.

90 *Cf.* D'ASPREMONT, *supra* note 20, at 168 ("[T]he idea that courts and tribunals are expressly endowed with custom-making power is not yet accepted."); Mendelson, *supra* note 11, at 178 ("[T]he role of the

logic goes, their discretion *might* be constrained if the proper method for finding CIL were refined. Indeed, this logic motivates much of the focus on method in the CIL literature: identifying the proper method for finding CIL should impose some discipline on and help settle CIL. However, global actors are unlikely to be methodologically disciplined, unless they believe that using an accepted method matters – for example, that it enhances the efficacy or legitimacy of their decisions. The evidence that they do is weak. After all, their decisions feed into the same chaotic process no matter whether they are or are not methodologically restrained.

The proper method for finding CIL is especially contested where, as in IHL, the relevant norms relate to human dignity or security. States commonly endorse these norms discursively but deviate from them in the operational practice. The methodological question asks how to assess those inconsistencies – specifically, how to account for the abhorrent physical practice. That practice arguably prevents the norms from attaining the status of CIL. Many international lawyers thus defend the so-called modern method for finding CIL.[91] This method generally downplays the unsavory physical practice and emphasizes instead the positive verbal pronouncements.[92] Different variants of the modern method give the physical practice more or less weight.

The variant that is now within the mainstream of legal thinking – but not by any means universally accepted[93] – is the one that the ICJ endorsed in *Military and Paramilitary Activities in and against Nicaragua*.[94] The *Nicaragua* court made two key methodological moves. First, it treated verbal acts as state practice.[95] Verbal acts that support a norm thus might counterbalance the physical acts that undercut the norm. Second, the court discounted the physical acts that states themselves did not own.[96] Bad physical conduct might not detract from a CIL finding if states condemn or try to justify that conduct by reference to the norm.

judge is to be an impartial third party, and so far as possible to take an objective photograph, as it were, of the state of the law, frozen at a particular moment in time.").

91 *See e.g.*, Henckaerts, *A Response to U.S. comments*, *supra* note 51, at 478–80 (defending this approach in the ICRC study on the ground that the alternative is to permit "violators [to] dictate the law or stand in the way of rules emerging").

92 *See* Roberts, *supra* note 48, at 758–60.

93 *See, e.g.*, Bruno Simma & Philip Alston, "The Sources of Human Rights Law: Custom, *Jus Cogens*, and General Principles," 1988–89 *Austl. Y.B. Int'l L.* 82; Prosper Weil, "Towards Relative Normativity in International Law?" 77 *Am. J. Int'l L.* 413, 434–35 (1983).

94 Military and Paramilitary Activities in and Against Nicaragua (*Nicar. v. U.S.*), 1986 I.C.J. 14, ¶¶ 183–207 (June 27). For scholars who embrace this method, see, for example, Theodor Meron, *Human Rights and Humanitarian Norms as Customary Law* 94 (1989); and Charney, *supra* note 2, at 545–46.

95 *Nicaragua*, *supra* note 94, ¶¶ 183–207; *see also ILA Report*, *supra* note 1, at n.32 (citing *Nicaragua* for the same).

96 *Nicaragua*, *supra* note 94, ¶ 186.

As the *Nicaragua* court articulated its method, it licensed CIL finders not to ignore the bad practice but rather to weigh that practice against the data that support a putative norm. Yet when the court applied its method, it appeared not even to account for the bad practice. *Nicaragua* relied entirely on verbal evidence to support its CIL claims.[97] That approach to method is part of a broader pattern: the ICJ does not consistently use the method that it itself propounds.[98] The court regularly finds CIL on the basis of scant evidentiary support or without conducting the kind of exhaustive review that its method seems to require.[99] Still, the court *does* focus on method; once it articulates a particular method, it arguably limits its discretion to "find" the CIL norms that it prefers. At the very least, *Nicaragua* needed solid verbal evidence to offset the unsavory physical practice.

The ICTY and ICRC decisions are striking, however, in that they expressly endorse very unconventional methods.[100] *Tadic* plainly pushed past *Nicaragua*. Whereas *Nicaragua* insisted that the operational practice matters, *Tadic* did not.[101] *Tadic* offered "a word of caution" about even trying to consider that practice:

> [I]t is difficult, if not impossible, to pinpoint ... whether [troops] in fact comply with, or disregard, certain standards of behavior. This examination is rendered extremely difficult by the fact that not only is access to the theater of military operations normally refused to independent observers ... but information on the actual conduct of hostilities is withheld by the parties to the conflict; what is worse,

97 *See, e.g.*, John Tasioulas, "In Defense of Relative Normativity: Communiatarian Values and the Nicaragua Case," 16 *Oxford J. Legal Stud.* 85, 97 (1996).

98 *See, e.g.*, Boyle & Chinkin, *supra* note 25, at 279 ("[T]he [ICJ] has not been consistent in applying its own criteria for the determination of customary international law."); Rudolf H. Geiger, "Customary International Law in the Jurisprudence of the International Court of Justice: A Critical Appraisal," in *From Bilateralism to Community Interest: Essays in Honour of Judge Bruno Simma* 673, 692 (Ulrich Fastenrath, et. al. eds., 2011) ("In general the Court does not follow its self-proclaimed method of finding customary international law.").

99 *See, e.g.*, Alain Pellet, "Shaping the Future of International Law: The Role of the World Court in Lawmaking," in *Looking to the Future: Essays on International Law in Honor of W. Michael Reisman* 1065, 1076 (Mahnoush H. Arsanjani et. al., eds., 2011) (asserting that the International Court of Justice has a "marked tendency to assert the existence of a customary rule more than to prove it"); *see also* Stephen J. Choi & Mitu Gulati, "Customary International Law: How do Courts Do It?" (in this volume) (showing that international courts generally ignore the traditional method and instead "find" CIL by looking to treaties).

100 Judge Meron has asserted that the ICTY and ICRC used fairly rigorous methods for finding CIL. *See* Meron, "The Revival of Customary Humanitarian Law," 99 *Am. J. Int'l L.* 817, 817 (2005). However, Meron calls these methods rigorous against a background understanding that the proper method is generally quite relaxed and flexible. He endorses a method that allows for considerable prescriptive discretion – and that many others reject.

101 *Tadic*, *supra* note 34, ¶ 99; *see also* Theodor Meron, "The Continuing Role of Custom in the Formation of International Humanitarian Law," 90 *Am. J. Int'l L.* 238, 239–40 (1996) (explaining that the tribunal used verbal acts to "compensate for scarcity of supporting practice").

> often recourse is had to misinformation with a view to misleading the enemy as well as public opinion and foreign Governments.[102]

The tribunal here is not using an accepted method for finding CIL. It does not even pretend to be methodologically disciplined.

The ICRC's approach was different. The ICRC at least claimed that the operational practice mattered.[103] Yet like the court in *Nicaragua*, the ICRC appeared to pay little, if any, attention to that practice.[104] The data that it collected to support its CIL claims are overwhelmingly verbal in form.[105] Further, after claiming that the operational practice mattered, the ICRC invoked an extremely expansive method that rendered that practice irrelevant. The ICRC asserted that certain norms qualify as CIL, notwithstanding the extensive practice to the contrary and the absence of any *opinio juris*: "It appears that international courts and tribunals on occasion conclude that a rule of customary international law exists when the rule is a desirable one … for the protection of the human person, *provided that there is no important contrary* opinio juris."[106]

However one assesses the ICTY's and ICRC's methods in the abstract, they become much more dubious as applied to IHL. Neither institution seriously grappled with the IHL treaties that contradicted its CIL claims. Several states expressly rejected the treaty provisions that the ICTY or ICRC characterized as CIL.[107] And the states that negotiated the treaties deliberately did not extend to non-international conflicts the panoply of norms that govern international conflicts. The literature on method suggests that CIL can supersede a contrary treaty arrangement only if the treaty parties so intend; the evidentiary burden is thought to be substantial.[108] The ICTY and ICRC did not

[102] *Tadic*, *supra* note 34, ¶ 99.

[103] *ICRC Study*, *supra* note 29, at xxxviii, xliiv–xliv

[104] Elizabeth Wilmshurst, "Conclusions," in PERSPECTIVES ON THE ICRC STUDY ON CUSTOMARY INTERNATIONAL HUMANITARIAN LAW, *supra* note 58, 401, 402–03 (summarizing the views of several commentators by concluding that "the authors of the Study have sometimes adopted an approach which is less conservative than is claimed" and on occasion "fairly relaxed").

[105] *See generally* ICRC, *Practice*, CUSTOMARY IHL DATABASE, retrieved from https://www.icrc.org/customary-ihl/eng/docs/v2 (last visited Sept. 23, 2014).

[106] *ICRC Study*, *supra* note 29, at xlviii (emphasis added).

[107] The ICRC study claimed to pay particular attention to the practice of states that are not party to the Additional Protocols. *ICRC Study*, *supra* note 29, at l. But the weight that the ICRC gave to this practice is unclear. The ICRC also took into account the practice of states that *are* party and "the fact that, at the time of this writing, Additional Protocol I has been ratified by 162 States and Additional Protocol II by 157 States." *Id.*

[108] *See, e.g.*, RESTATEMENT (THIRD) FOREIGN RELATIONS LAW OF THE UNITED STATES § 102, n. 4 (1987) ("Provisions in international agreements are superseded by principles of customary law that develop subsequently, where the parties to the agreement so intend."); Michael Akehurst, "The Hierarchy of the Sources of International Law," 47 *BRIT. Y.B. INT'L L.* 273, 276 (1977) ("[S]ubsequent custom can terminate a treaty only when there is clear evidence that that is what the parties intend.").

even try to satisfy that burden. The ICRC, in particular, was criticized for trying to use CIL to circumvent the limitations in the IHL treaties.[109]

Surely, the ICTY and ICRC realized that their methods were outside the mainstream. They presumably used these methods because more conventional ones would have complicated their efforts to achieve their desired results. Their mandates to find CIL did not constrain them from openly invoking and using dubious methods to try to make CIL. And their methodological indiscipline appears to have had little, if any, effect on how their decisions were received. Refining the method for finding CIL is unlikely to alter this incentive structure. Rather, actors who have a normative agenda are likely to continue advancing heavily prescriptive claims, even when they are charged only with finding CIL. These claims will sometimes be effective.

CONCLUSION

The question of how to find CIL assumes that finding CIL is an objective exercise that stands apart from the messy process for making CIL. I have used the experience with customary IHL to cast doubt on that assumption. My argument here should not be entirely surprising. The scholarly literature already recognizes that CIL finding entails some degree of CIL making; that actors who purport to find CIL commonly help shape CIL; and that these actors routinely disregard their own professed CIL-finding methods. What is surprising is that, even as the literature recognizes these points, it insists that CIL finding is different from CIL making and can impose some order on the CIL process. This persistent focus on method reflects, at bottom, a deep discomfort with the volatility and political excess of that process. Global actors sometimes invoke and treat as CIL norms that are, in fact, novel or contested. Methods for finding CIL are, I have argued, unlikely to discipline these actors or to reduce the contestation and uncertainty in CIL.

These methods might do other work. At the very least, the methodological guideposts provide a shared language that disparate actors use to advance varied

[109] *See, e.g.*, Kenneth Anderson, *My Initial Reactions to the ICRC Customary International Humanitarian Law Study*, Kenneth Anderson's Law of War and Just Theory Blog, retrieved from http://kennethandersonlawofwar.blogspot.com/2005/11/my-initial-reactions-to-icrc-customary.html (accessed Nov. 14, 2015) ("In very important, very disturbing ways, the Study represents a sort of customary law 'end run' around the failure of Protocol I to achieve near universal or customary status itself."); Scobbie, *supra* note 89, at 34 ("Can States be expected to accept as customary that which they have rejected as a conventional obligation?"); Bethlehem, *supra* note 58, at 8 ("Particularly when heavy reliance is placed on treaties to which a number of States are not parties, initiatives to derive customary rules may be seen as an attempt to circumvent the requirement of express consent necessary for a State to be bound by the treaty-based rule."); Aldrich, *supra* note 58, at 505–06 ("[S]ubstantive and public dissent by at least some [non-parties] to various provisions of the Protocol makes it impossible to suggest that the Protocol … should be considered to represent as a whole a codification of customary international humanitarian law.").

international norms. Moreover, any individual actor might craft and use a method that helps achieve its own goals. For example, the U.S. Supreme Court now uses a method in the human rights context that is intended to restrict the CIL claims that U.S. courts assess or endorse.[110] A nongovernmental advocate might use a very loose method to pressure states to change their behavior. The point is that no one method for finding CIL is likely to help settle or impose order on CIL.

This does not mean that "anything goes" in CIL. It means that the limits on CIL must – and do – come from *within* the CIL process. Actors might refrain from advancing controversial CIL claims for all sorts of reasons that are internal to that process. A controversial claim might be less stable and therefore less appealing to advance than is a widely accepted compromise. The controversial claim might detract from an actor's credibility with particular audiences. Or the claim might be challenged or ignored. Indeed, this is probably the strongest constraint on CIL: for a CIL claim to be effective, it must actually resonate with other actors. Others must be willing to treat the claim as law.

[110] *See Sosa v. Alvarez-Machain*, 542 U.S. 692, 725 (2004) ("[W]e think courts should require any claim based on the present-day law of nations to rest on a norm of international character accepted by the civilized world and defined with a specificity comparable to the features of the 18th-century paradigms we have recognized.").

7

The Growing Obsolescence of Customary International Law

Joel P. Trachtman

Consider the great modern challenges to international society: global environmental protection, international public health, cybersecurity, financial crisis, and liberalization of movement of goods, services, and people. As will be explained in detail, it is unimaginable that customary international law (CIL) can provide effective mechanisms by which to address these challenges. Consider also some of the longer-term challenges: suppressing and regulating war, protecting human rights, and reducing poverty. While CIL has played an important role in regulating war and protecting human rights, this chapter argues that it has now done about all that it can do in those areas. It is time to focus the world's attention on more effective mechanisms by which to address all of these challenges. These more effective mechanisms will include treaty-based hard law and soft law.

CIL has not been around forever, and it responded to particular historical circumstances. While CIL had roots in Roman law,[1] and while an early mention of a variant of modern CIL was contained in Francisco Suarez's *De Legibus ac cleo legislatore*, in 1612, Hugo Grotius is considered to have articulated the modern version of CIL later in the seventeenth century.[2] CIL doctrine has changed over time since then, and there is no reason to think that it will not change in the future, and even that it will not be superseded by other mechanisms for international cooperation.

The demands made of international law up until the latter part of the twentieth century were quite different from those in view today. The law of coexistence, in Wolfgang Friedmann's dichotomy, provided the basis for diplomatic interaction that allowed for informal – nonlegal – give and take. As Friedmann explained in his

[1] *See* Raphael M. Walden, "The Subjective Element in the Formation of Customary International Law," 12 *Isr. L. Rev.* 344 (1977).

[2] Hugo Grotius, *De Juri Belli Ac Pacis Libri Tres [The Law of War and Peace in Three Books]*, book 1, chapter 1, section 12 (F. Kelsey trans. 1925) (original 1625).

seminal 1964 work, *The Changing Structure of International Law*, "The principal preoccupation of the classical international law, as formulated by Grotius and the other founders, was the formalization, and the establishment of generally acceptable rules of conduct in international diplomacy."[3] Note that these were formal rules designed to facilitate informal interaction. They formalized the process of diplomacy, not the substance of international cooperation. The law of coexistence was largely established by CIL. In the law of coexistence, states were equal, their interests were the same, and the rules did not need to change much to face new problems. CIL served these purposes reasonably well.[4]

Friedmann's "law of cooperation," on the other hand, addresses all of the substantive commitments that states need to make in order to manage collective goods and externalities. It includes international law relating to the full range of governmental action, including trade, environment, finance, investment, public health, intellectual property, and telecommunications. With the rise of globalization, and the rise of the regulatory state, there are increasing demands for the law of cooperation.[5] The law of cooperation will increasingly be contractual and legislative: it will be made by treaty and by non-unanimous voting in international organizations. (I will refer to these types of law as "legislated law," as distinct from customary international law, which is generally not self-consciously "legislated.") Friedmann found that custom is "an unsuitable vehicle for international 'welfare' or 'co-operative' law. The latter demands positive regulation of economic, social, cultural, and administrative matters, a regulation that can only be effective by specific formulation and enactment."[6]

CIL has several important limitations: (i) it cannot be made in a coordinated manner in advance of events; (ii) it cannot be made with sufficient detail; (iii) it cannot be made with sufficiently heterogeneous reciprocity between states; (iv) it cannot be made with specifically designed organizational support; (v) it is generally not subject to national parliamentary control; (vi) it purports to bind states that did not consent but failed to object to its formation; and (vii) it provides excessive space for auto-interpretation by states, or for sometimes insufficiently disciplined interpretation by judges. Legislated law can perform better on all of these dimensions. The chapter reviews each of these limitations in detail, explaining the advantages of legislated law.

Of course, there are some dimensions of CIL that may be understood as attractive. First, it is informal and, to some degree, decentralized in its production. This may

3 Wolfgang Friedmann, *The Changing Structure of International Law* 5 (1964).

4 For an argument that it does little more, see J. Patrick Kelly, "The Twilight of Customary International Law," 40 Va. J. Int'l L. 449 (2000).

5 *See* Joel P. Trachtman, *The Future of International Law: Global Government* (2013).

6 Friedmann, *supra* note 3, at 122.

allow it to be more responsive to emerging social contexts. Second, some of the above "glitches" can be understood as "features." In some circumstances, lack of national parliamentary control, the ability to bind states without their actual consent, and the ability to bind later-created states may be useful. The chapter considers the features of CIL that may make it superior to legislated law in particular contexts. While in general these benefits seem to be outweighed by the related detriments, in particular circumstances the benefits may outweigh the detriments.

The chapter, supplemented by the Appendix,[7] further shows that CIL is largely superseded by treaty law and other codification; that is, only in very few circumstances are there CIL rules of law that have not been reflected in treaty or other codification. This serves as evidence that states have already responded to some of CIL's limitations. Because treaty obligations only bind adherents, and treaties do not apply universally as does most CIL, CIL will not be useless anytime soon. Nevertheless, the extent to which CIL has been superseded is striking.

FUNCTIONAL LIMITATIONS OF CIL

This chapter operates from a contractual perspective on international law. Under this perspective, international law is entered into by states in order to advance the interests of the states as expressed through the states' political systems. This contractual perspective recognizes that it is generally beneficial if states are able to make the types of commitments that they desire, with the types of effects they desire. Exceptions arise if nonparticipating third states are harmed, or if other external individuals or groups are harmed in a way that is not sufficiently taken into account.

So, the inevitable question about CIL is in what way is CIL the efficient system for making particular types of commitments with particular effects. It will be seen that the functional limitations of CIL are that (i) it cannot be made in a tightly coordinated manner in advance of events, (ii) it cannot be made with sufficient detail, (iii) it cannot be made with sufficiently heterogeneous reciprocity between states, and (iv) it cannot be made with custom-designed organizational support.

Coordination

To understand the functional limitations of CIL, it is useful to consider the relative utility of CIL and treaties to address international public goods cooperation problems. In particular, the fact that CIL cannot be made in a tightly coordinated manner may prevent it from being made at all in an important set of international cooperation circumstances.

7 Online appendix available at: www.cambridge.org/9781107443105

Reduction of carbon emissions to address global warming is an example of a public good: consumption of the benefits of carbon reduction is (i) non-excludible, and (ii) non-rival.[8] Because carbon reduction is a public good, without intervention it is likely to be undersupplied. Each state is best off continuing to pollute while other states reduce emissions. Thus, the underlying game has the structure of a prisoner's dilemma, in which each state's best strategy is to pollute. This strategy is individually rational, but collectively irrational. For simplicity, consider a two-person prisoner's dilemma game.

For simplicity's sake, assume full compliance with applicable international law, consistent with the requirements of conventional doctrine. While obviously counterfactual in a number of real-world contexts, this assumption allows us to model a two-stage game in which states are assumed capable of making binding commitments, and draw insights from that model.

The stages that are to be analyzed include (i) acceptance of the rule ("adherence"), and (ii) compliance with the rule ("compliance"). Under this two-stage game, adapted from Barrett,[9] the players choose in stage i whether to accept the CIL rule or treaty obligation (to adhere), or to take persistent objector (PO) status or decline to enter the treaty (decline to adhere).[10] In stage ii, adherents and non-adherents choose whether to comply or violate, although in accordance with our assumption, for any state that has adhered to the rule there is no choice as to whether to comply, provided that there are other adherents who are in a position to enforce the initial adherent's obligations. I assume that any international legal rule is binding, that each player knows what happened at the prior stage, and that each player anticipates its choices at the second stage when determining what to do at the first stage – whether to adhere or decline to adhere. By this process of backward induction, we can determine whether states would adhere to the rule or not. Following from our assumption that international law is binding, the only interesting question is adherence.

The following analysis is adapted from Black, Levi, and de Meza.[11] Assume that the payoff to each state from compliance with the hypothetical carbon reduction rule is rb-c, where r is the number of states that comply, b is the benefit produced

[8] For an incisive analysis, *see* Niels Petersen, "Customary International Law and Public Goods" (in this volume).

[9] Scott Barrett, *Environment and Statecraft: The Strategy of Environmental Treaty-Making*, chapter 7 (2003).

[10] Of course, under CIL, adherence would often take the form of compliance. This does not affect the approach, assuming that *future* compliance has a sufficiently large value to overcome the incentives to simply take advantage of the first state's initial compliance. States can also signal adherence through statements or other behavior short of actual compliance.

[11] Jane Black, Maurice D. Levi, & David de Meza, "Creating a Good Atmosphere: Minimum Participation for Tackling the 'Greenhouse Effect,'" 60 *Economica* 281 (1993).

by each complying state's compliance, and c is the cost to each complying state of compliance. Note that in this payoff structure the aggregate benefits rise in proportion to the number of states that comply. The payoff to a noncomplying state simply equals rb: under the public-good assumption of non-excludability, they get the benefit of compliance by others without the cost of their own compliance. Assume that b is less than c. Otherwise, there would be no need for a rule of international law: each state would by definition benefit from its own compliance in an amount greater than the cost of compliance.

For example, with two players that both comply, and assuming $b = 4$ and $c = 6$, the payoff to each state from compliance is 2. If neither complies, then the payoff to each is 0. If one complies while the other violates, the complying state player gets a payoff of –2, while the noncomplying state gets a payoff of 4. This is a prisoner's dilemma.

Recall that we are assuming that any rule of international law is strictly binding, so that it always results in compliance. We know from the fact that the underlying situation has the structure of a prisoner's dilemma that a non-adherent, not being bound, will play "violate" at stage ii – this is the dominant solution for a non-adherent. In this two-player game, assume that one state adheres to the rule in stage i. Since it is the only adherent, (assume that) the rule either never comes into being or cannot be enforced against it, and so it plays "violate" in stage ii, as it anticipates that the other state will also play "violate" in stage ii: it understands the other state's dominant solution, and in fact it is acting out its own dominant solution. The outcome is that both play violate: the same type of inefficient equilibrium that we expect to result from a prisoner's dilemma.

Anticipating the inefficient solution to the prisoner's dilemma game, both parties examine their choices at stage i. If one of the parties (A) adheres to the CIL rule at stage 1, the other party (*B*) faces the following choice. If *B* declines to adhere, then A will play violate in stage ii, as discussed in the prior paragraph. *B* anticipates that it will receive a payoff of 0 if both parties violate. On the other hand, if *B* adheres, irrevocably binding itself to comply, A will be required to comply, securing a payoff of 1 for *B* (as well as for A). So, in this setting, *B* will adhere. We might understand A's adherence as an offer to contract, which *B* may accept by adherence. Adherence is a (weakly) dominant solution for both players in stage i.

As suggested by Barrett, this works well for a two-person game, and the two-person prisoner's dilemma when transformed into a cooperative game (in which binding agreement is possible) is easily resolved. This is intuitive.

But we are interested in plurilateral and multilateral international law. When this two-person model is extended to multiple persons, whether states will adhere to a multilateral legal rule will depend on the structure of the payoffs. In a game with n players, it may well be that a benefit is created through adherence by a

TABLE 7.1. *A public goods game prisoner's dilemma*

	Number of states that cooperate					
	0	1	2	3	4	5
Payoff for noncooperation	0	0	4	8	12	16
Payoff for cooperation	0	−2	2	6	10	14

coalition that is less than n. So reversion to Nash (non-adherence, and consequent noncompliance) may not be attractive to that smaller group: they may be induced to adhere and to comply even if other states do not. Furthermore, under our assumption of non-excludability, the non-adhering states still receive the benefit of the performance by the adhering states.

Perhaps counterintuitively, the possibility of a benefit created by compliance of a subset of states may cause other states to seek to "free ride" once a benefit is created. Using the same formula provided, we recall that the payoff to an adhering (and therefore complying) state is $rb\text{-}c$, while the payoff to a non-adhering state simply equals rb. We assume that the payoff from noncompliance, assuming all others fail to comply, is 0. Therefore, k states will adhere if $kb\text{-}c \geq 0$. Therefore, if the number of states $k \geq c/b$, then these states will adhere. Using the values of $c = 6$ and $b = 4$, if the number of states is greater than or equal to 1.5, they will adhere. So, in this example, adherence by two states results (as stated previously) in a payoff of 1, and since $1 > 0$, they will adhere. However, once the number of signatories reaches this level, other states will have no incentive to adhere – they will have an incentive to free ride. Adherence will result in costs of 6 incurred by the marginal adhering state, with incremental benefits to that state of only 4.

The matrix in Table 7.1 depicts this circumstance of a prisoner's dilemma involving five states.[12] For every state that complies, a pure public good is produced, giving all states a benefit of 4, with a cost to the complying state alone of 6. Each state can play either of two strategies: cooperation or noncooperation. The dominant strategy is for each state to free ride and play noncooperation, because each of the payoffs for noncooperation in the top row is greater than the payoff for cooperation in the bottom row. The Nash equilibrium is an outcome where no state cooperates: at each stage, noncooperation yields a greater benefit than cooperation. This is an inefficient equilibrium.

What happens when we introduce binding international law? In a two-stage game with the possibility of adherence to a binding rule of international law, two will be expected to adhere under these assumptions – that is, in equilibrium, the number

[12] Adapted from Todd M. Sandler, *Treaties: Strategic Considerations*, 2008 U. ILL. L. REV. 155 (2008).

of adherents is two. This is because the payoff to non-adherents is greater than the payoff to adherents once two have adhered. For example, if three states adhere, the payoff is 6 to each adherent but 12 to each non-adherent (non-adherents do not bear the cost).

Barrett shows that under these assumptions the *gains from cooperation* increase with *b* (the number of states), and decrease with *c* (the cost of compliance). This is intuitive. However, less intuitive is the fact that the *equilibrium number of state adherents* increases with *c*, and decreases with *b*. This means that the equilibrium number of states will tend to be small when the gain from cooperation is large, and large when the gain from cooperation is small. Furthermore, in equilibrium, non-adherents can free ride and get a higher payoff than adherents. Of course, this assumes an isolated adherence game, without the ability to subject states to scrutiny or punishment for "unilateralism," in the sense of failure to join a plurilateral regime. Much depends on the values of the costs and benefits to each state.

The core problem in this public-goods problem is one of free riding: some states may receive the benefit of compliance by others, without incurring any costs themselves. Indeed, each incremental state would prefer to free ride, if it could ensure that enough other states would adhere. If, on the other hand, states that fail to comply can be excluded from sharing the benefits, the strategic challenge becomes smaller, although not trivial.

To be clear, if states can be excluded from benefiting from the cooperation of others, all states should then be willing to adhere. By adhering they would, under this condition, achieve a greater payoff than they would receive by failing to adhere. Even if the cooperation problem is characterized by a public-goods-type set of payoffs, there may be a relatively easy solution given the assumption of binding international law. Under the public-goods-type payoffs, assuming binding international law, the decision to adhere to the (binding) international law rule, or to decline to do so, would have the characteristics of a "chicken" game, as described in Table 7.2. The following is a bilateral illustration for simplicity, but this illustration is readily generalized to multiple player circumstances.

Under these payoffs, each state's best outcome is to abstain from agreement, while others form a stable coalition that will generate the relevant public good. The second-best outcome is to adhere while others adhere. The worst outcome is if no state adheres. In this chicken game, neither player has a dominant strategy. In this particular case, there is no unique efficient equilibrium. Each player has two Nash equilibria: object when the other adheres, and adhere when the other objects.

However, both players wish to avoid the circumstance where they each play object, and the even split in the northwest quadrant seems intuitively attractive, although it is unstable. In order to achieve it, they should each commit to adhere. They may do so through a number of mechanisms. The simplest (in the treaty

TABLE 7.2. *A chicken game*

		B	
		Adhere	Object
A	Adhere	3,3	2,4
	Object	4,2	0,0

Source: Adapted from STEVEN J. BRAMS & D. MARC KILGOUR, GAME THEORY AND NATIONAL SECURITY 41 (1998).

context) is a signing conference where each state signs a treaty simultaneously.[13] Only slightly more complex is a specification of a minimum number of adherents prior to the entry into force of the treaty.

But, and this is the main point, in the CIL context, it is difficult to say how the chicken game can be avoided. Perhaps the PO status can be used: states go about complying, but if one state claims PO status, others can avoid formation of the incipient rule, or simply claim PO status themselves. Or perhaps the requirement of "general practice" addresses this issue: if enough states fail to adopt the practice, it does not become a rule.

These international legal circumstances are comparable to the chicken game, but instead of two wild teenagers hurtling toward a cliff, in the treaty context we have sophisticated diplomats sitting eyeball to eyeball. While there may still be incentives to try to avoid contributing, and these incentives may sometimes hold sway, the diplomatic context takes place in a simultaneous and broadly linked setting, where unilateralism may be criticized and may be subject to punishment.

In the CIL setting, it may be that the benefits of simultaneity are lost, but states might still anticipate the unraveling of cooperation, and veer away from defection. The problem is that the first state is unlikely to adopt a compliant practice, because it will recognize that the dominant behavior of other states, after observing the initial compliant practice, is to free ride.

In order to form custom, states must act consistently, and do so with *opinio juris*. Unless there is an agreement to act simultaneously, or unless the circumstances occasioning action occur simultaneously, some states will have to act in advance of other states. Assuming that it is costly to act while others do not, this involves an act of faith on the part of first movers. Under the prisoner's dilemma assumptions described, such an act of faith would be in vain: others would have an incentive to defect while the first mover complied.

[13] It seems reasonable here to elide the distinction between signature and adherence, as signature brings certain obligations, including obligations to seek ratification.

Some have argued that United Nations General Assembly resolutions may form the basis for "instantaneous customary international law," and Michael Scharf has suggested that there are "Grotian moments" in which CIL can be made in an accelerated fashion.[14] It is possible that the requirements for formation of CIL could change to overcome the coordination problem highlighted in this subsection. However, such accelerated formation may raise other problems, and there are several other limitations of CIL, described presently. It is difficult to see why treaties would not always dominate CIL as a tool of cooperation in connection with this type of problem.

Normatively, we would first hope that the international legal system contains appropriate structures to allow states the opportunity to form efficient agreements, without limitations due to limited forms of agreement. Based on a contract analogy, international law is valuable because it allows states to contract. We must also recognize that not all international legal contexts are the same. There is international law that responds to coordination problems and international law that responds to cooperation problems. And there is a range of different types of cooperation problems.

Some argue that global warming may be addressed through nonlegal, informal cooperation. Assuming that global warming has the public good characteristic described, perhaps they are suggesting that a CIL-type equilibrium (one without legal force) may emerge that would induce carbon reduction. The analysis does not exclude that possibility, but indicates that it would be significantly easier for an efficient equilibrium to emerge under treaty. This is especially true if the treaty negotiations allow linkage to diverse international obligations, or if the treaty negotiations context facilitates side payments, as discussed next.

Insufficient Precision

Modern cooperation problems in environment, trade, finance, cyberspace, public health, and other important areas are addressed in extraordinarily complex terms, with very detailed conditions of application of rules. This complexity is observed in many areas. Examples include international trade law and international food safety law requirements for national food safety regulation; the complex reporting, monitoring, and substantive requirements of international health law; the very detailed financial organization capital requirements under (soft) international financial rules; and the complex arrangements for mutual deference under bilateral

[14] Michael P. Scharf, *Customary International Law in Times of Fundamental Change: Recognizing Grotian Moments* (2013); Bin Cheng, "United Nations Resolutions on Outer Space: "Instant" International Customary Law?" 5 *Indian J. Int'l L.* 23 (1965).

tax treaties. It is difficult to imagine these varied and complex arrangements arising from and becoming empirically identifiable under a CIL regime. There are so many details and contingencies, conditions, and exceptions that it would be difficult to identify the custom. "We should bear in mind that customary law is, by nature, often general, vague, and imprecise."[15]

For example, in the *Pulp Mills on the River Uruguay* case, the International Court of Justice (ICJ) found to exist "a practice, which in recent years has gained so much acceptance among States that it may now be considered a requirement under general international law to undertake an environmental impact assessment where there is a risk that the proposed industrial activity may have a significant adverse impact in a transboundary context, in particular, on a shared resource."[16] However, the court also observed that this customary international law rule did not "specify the scope and content of an environmental impact assessment."[17] CIL can set broad standards, but does not seem to be able to generate detailed specific rules.

Yet most of the concerns of modern international law require detailed specific responses. One reason is that international law increasingly reaches "behind the border," restricting national regulatory autonomy. In order to ensure that these restrictions are not overbroad, unnecessarily restricting policy space, or under-inclusive, international law is formulated so as to apply only upon the satisfaction of complex and precise conditions.

Law generally is used in society to set rules and responsibilities for subsequent action. CIL only begins to cover subsequent action after it has emerged as law, with the requisite practice and *opinio juris*. Law established *ex ante* can be specified with greater or lesser specificity. But there is a wide range of legal situations in which greater specificity would be superior to lesser specificity. The law and economics-based literature of "rules and standards" seeks to guide cost-benefit analysis in determining when greater or lesser specificity is appropriate.

In the rules versus standards literature, a law is a "rule" *to the extent that* it is specified in advance of the conduct to which it is applied.[18] Thus, a law against littering is a rule to the extent that "littering" is well defined. Must there be an intent not to pick up the discarded item; are organic or readily biodegradable substances covered; is littering on private property covered; is the distribution of leaflets by air

[15] MARK E. VILLIGER, CUSTOMARY INTERNATIONAL LAW AND TREATIES: A MANUAL ON THE THEORY AND PRACTICE OF THE INTERRELATION OF SOURCES 103 (2d ed. 1997).

[16] Pulp Mills on the River Uruguay (*Arg. v. Uru.*), 2010 I.C.J. ¶ 204 (Apr. 20), *retrieved from* http://www.icj-cij.org/docket/files/135/15877.pdf.

[17] *Id.* at ¶ 205.

[18] For an introduction to the rules versus standards discussion in law and economics, *see* Louis Kaplow, "General Characteristics of Rules," in ENCYCLOPEDIA OF LAW AND ECONOMICS (B. Bouckaert & G. De Geest eds., 1998); Louis Kaplow, "Rules Versus Standards: An Economic Analysis," 42 DUKE L.J. 557 (1992). *See, also* Cass R. Sunstein, "Problems with Rules," 83 CAL. L. REV. 955 (1995).

covered? Any lawyer knows that there are always questions to ask, so that every law is incompletely specified in advance, and therefore incompletely a rule.

A standard, on the other hand, is a law that is farther toward the more general end of the spectrum, in relative terms. It establishes general guidance to both the person governed and the person charged with applying the law, but does not specify in detail in advance the conduct required or proscribed. Incompleteness of specification may not simply be a result of conservation of resources. It may be a more explicitly political decision to either agree to disagree for the moment, to avoid the political price that may arise from immediate hard decisions, or to cloak the hard decisions in the false inevitability of judicial interpretation.

Rules are more expensive to develop than standards, *ex ante*, because rules entail greater specification costs, including drafting costs and negotiation costs. In order to reach agreement on specification – in order to legislate specifically – there may be greater political costs.[19] On the other hand, standards often involve greater *ex post* costs. These *ex post* costs include the costs of *ex post* decision making after the circumstances calling for a decision arise, but also include (a) the costs of defection by counterparties under regimes of auto-interpretation, and/or (b) the agency costs associated with delegation to a judicial, administrative, or other *ex post* decision maker. The *ex post* decision maker may shirk its duties or may be subject to conflicts of interest, or may simply fail to decide in a way that achieves the best interests of the collective principals. These agency costs and defection costs associated with incomplete contracts can be substantial.

Rules are generally thought to provide greater predictability. There are two moments at which to consider predictability. First, is the ability of persons subject to the law to be able to plan and conform their conduct *ex ante*, sometimes known as "primary predictability."[20] The second moment in which predictability is important is *ex post*, after the relevant conduct has taken place. Where the parties can predict the outcome of dispute resolution – where they can predict the tribunal's determination of their respective rights and duties – they will spend less money on litigation. This type of predictability is secondary predictability. Both types of predictability can reduce costs.

While rules appear to provide primary and secondary predictability, tribunals may construct exceptions in order to achieve goals they think worthy, and thereby reduce predictability. It may be difficult to constrain the ability of tribunals to do this.

19 *See* Gillian K. Hadfield, "Weighing the Value of Vagueness: An Economic Perspective on Precision in the Law," 82 *CAL. L. Rev.* 541, 550 (1994), *citing* Linda R. Cohen & Roger G. Noll, "How to Vote, Whether to Vote: Strategies for Voting and Abstaining on Congressional Role Calls," 13 *POL. BEHAV.* 97 (1991).

20 For this use of the terms "primary predictability" and "secondary predictability," *see* William F. Baxter, "Choice of Law and the Federal System," 16 *STAN. L. REV.* 1, 3 (1963).

Furthermore, some degree of uncertainty – of unpredictability – may enhance the ability of the parties to bargain to a lower-cost solution. Thus, simple predictability is not the only measure of a legal norm; rather, we must also be concerned with the ability of the legal norm to provide satisfactory outcomes.

As we consider the relative allocative efficiency of potential outcomes, we must recognize that there is a temporal distinction between rules and standards. Standards may be used earlier in the development of a field of law, before sufficient experience to form a basis for more complete specification is acquired. Indeed, one possibility is that CIL serves as a precursor for treaty. In many areas of law, courts develop a jurisprudence that forms the basis for codification – or even rejection – by legislatures. With this in mind, legislatures may set standards at an early point in time, and determine to establish rules at a later point in time after some experience is gained.[21] A common law system, or one that otherwise relies on precedent, may be understood as a hybrid standards-rules system, in which standards evolve into rules through experience-based precedent.

Kaplow points out that where instances of the relevant behavior are more frequent, economies of scale will indicate that rules become relatively more efficient. For circumstances that arise only infrequently, it is more difficult to justify promulgation of specific rules. In addition, rules provide compliance benefits: they are cheaper to obey, because the cost of determining the required behavior is lower. Rules are also cheaper to apply by a court: the court must only determine the facts and compare them to the rule.

Johnston analyzes rules and standards from a strategic perspective, finding that, under a standard, bargaining may under certain circumstances yield immediate efficient agreement.[22] Johnston qualifies the "standard supposition in the law and economics literature … that private bargaining between [two parties] over the allocation of [a] legal entitlement is most likely to be efficient if the entitlement is clearly defined and assigned ex ante according to a rule, rather than made contingent upon a judge's ex post balancing of relative value and harm."[23] That standard supposition would often argue for use of more specific treaty rules, as opposed to more general CIL standards. Johnston suggests this supposition may be incorrect in particular circumstances:[24] "When the parties bargain over the entitlement when there is private information about value and harm, bargaining may be more efficient under a blurry balancing test than under a certain rule."[25] This is because under a

[21] *See* Kaplow, "General Characteristics of Rules," *supra* note 18, at 10.
[22] Jason Scott Johnston, "Bargaining under Rules versus Standards," 11 *J. L. Econ. & Org.* 256 (1995).
[23] *Id.* (citations omitted).
[24] *See also* Carol Rose, "Crystals and Mud in Property Law," 40 *Stan. L. Rev.* 577 (1988); Joel P. Trachtman, "Externalities and Extraterritoriality," in *Jagdeep Bhandari & Alan O. Sykes, Economic Dimensions of International Law* 642 (1998).
[25] Johnston, *supra* note 22, at 257.

certain rule, the holder of the entitlement will have incentives to "hold out" and decline to provide information about the value to him of the entitlement. Under a standard, where presumably it cannot be known with certainty *ex ante* who owns the entitlement, the person not possessing the entitlement may credibly threaten to take it, providing incentives for the other person to bargain. This possible benefit of less-precise CIL depends on the existence of important private information, and the ability of one of the states involved to "hold out" for an excessive compensation if it had a more clear, presumably treaty-based, entitlement.

It is not possible to consider the costs and benefits of rules and standards separately from the strategic considerations that would cause countries to select a rule as opposed to a standard. Where adjudication of an international legal rule is unlikely, which is the case for most international legal rules, specific rules may result in greater compliance, which might be understood as a benefit. This is the role of the concept of "precision" advanced in the "legalization" literature on international law.[26] The greater specificity possible under treaty, as opposed to CIL, may have compliance benefits: to the extent that the legal rule is made clear through greater specificity, then it will be harder for states to claim falsely that they are in compliance under a regime of auto-interpretation. Reputational consequences will be promoted by greater specificity. States will have greater incentives to comply in the first place. Under a regime of adjudication, states will also have greater incentives to comply, and tribunals will be able to adjudicate more swiftly and surely.

Less-clear rules would also have the effect of permitting powerful states greater flexibility to make claims about the effect of the rules, and to force less-powerful states to accept those claims. In this connection, it is also important to note that, under a regime of adjudication, standards have the side-effect of assigning discretion to tribunals charged with applying the standard. These very tribunals may be empowered by the use of standards as opposed to rules, and may therefore promote the use of CIL, as compared to treaties that have greater potential for specification.

Heterogenous Reciprocity

Most issues addressed by the body of CIL that developed during its classical period (up through the nineteenth century) can be characterized as symmetric: you protect my diplomats, and I will protect yours. We can call this type of exchange "identical performance reciprocity." This is Friedmann's law of coordination, and it is easier to make, and to enforce, if everyone has symmetric interests.

[26] *See* Kenneth W. Abbott, Robert O. Keohane, Andrew Moravcsik, Anne-Marie Slaughter, & Duncan Snidal, "The Concept of Legalization," 54 *Int'l Org.* 401 (2000).

However, identical performance reciprocity leaves no room for diverse preferences or comparative advantage. In more modern times, with the development of the need for an international law of cooperation, it makes perfect sense in many areas for states to make reciprocal commitments that involve different performances by each state (let us call this "diverse performance reciprocity"). States have varying preferences, and varying comparative advantage in regulation. For example, you reduce your carbon emissions and I will reduce my tariff on computers from 30 percent to 10 percent in exchange for State B reducing its tariff on bananas from 25 percent to 20percent. Or perhaps State A wants State B to provide greater protection to intellectual property rights. Both are types of diverse-performance reciprocity. As compared with identical-performance reciprocity, diverse-performance reciprocity responds to the fact that different states have different situations, different capabilities, and different preferences. In addition, common but differentiated responsibilities may be based on differences in wealth.

CIL seems unable efficiently to encompass diverse performance reciprocity. Therefore, under CIL, the set of possible international legal transactions is excessively constrained: only in-kind barter is possible, not a market of diverse exchange. There have been examples of diplomatic conferences in which package deals of newly "agreed" CIL have been developed,[27] but this is an unusual occurrence, and would be limited to topics within those addressed at the diplomatic conference.

Diverse-performance reciprocity is akin to a side payment, in which a kind of package deal is constructed that might not be possible without assembling a set of concessions that will sufficiently reciprocate. "With all side payments prohibited, there is no assurance that collective action will be taken in the most productive way."[28] So, CIL is a blunt instrument, indeed.

On the other hand, it might be argued that specific reciprocity in forming a particular arrangement is unnecessary – that Robert Keohane's concept of diffuse reciprocity (which might be understood as an extreme form of diverse-performance reciprocity) can induce transactions even where a particular transaction is one-sided, because the loser will anticipate future transactions in which reciprocity will be evened out.[29] This requires a relatively long horizon and a high level of trust. And perhaps it is true that for smaller transactions, where the amount at stake if the counter-parties defect is small, diffuse reciprocity would be satisfactory. However, if the amount at stake is sufficient, it would be more likely that states would anticipate noncompliance by their partner, and decline to comply themselves. Cooperation may thus unravel.

[27] *See* BEN CHIGARA, *LEGITIMACY DEFICIT IN CUSTOM: A DECONSTRUCTIONIST CRITIQUE* 34 (2001).

[28] JAMES M. BUCHANAN & GORDON TULLOCK, *THE CALCULUS OF CONSENT* 153 (1999).

[29] Robert Keohane, "Reciprocity in International Relations," 40:1 *INT'L ORGS.* 1 (1986).

A related problem to that of insufficient heterogeneous reciprocity is the tendency of CIL to be universal. The drive to universality limits the ability to establish more narrowly tailored reciprocity among sub-universal groups of states.

Customized Organizational Support

CIL itself cannot create organizations, or specify particular organizational features. Rather, unless organizational features are established by treaty, CIL rules must accept the default international organizational system. CIL rules cannot be matched with appropriate organizational features *through the operation of CIL*.

Consider the World Trade Organization's (WTO's) dispute settlement system or the Law of the Sea Tribunal or the legislative mechanism of the European Union as examples of customized organizational features intended to deal with specified legal rules. It is possible to create organizations by treaty that can service specified CIL rules, so this does not mean that CIL rules cannot have these customized organizational features. But increasingly complex international legal rules will be matched with specialized organizational structures, and it will be impossible for efficient CIL rules to develop these types of specialized structures without treaty negotiation. It would be more efficient to design the organization at the same time at which, and in the same context in which, the rules are designed, because there are synergies and interdependencies between these pursuits.

LEGITIMACY LIMITATIONS AND POLITICAL ECONOMY OF CIL

CIL gives international law a bad name. The concerns are well known,[30] so I will be brief. First, CIL is not dependent on explicit consent. Second, it generally is not dependent on parliamentary approval or other domestic ratification. Importantly, as more and more topics of international cooperation interest relate to issues of domestic regulation – so-called behind-the-border issues – it seems less and less legitimate to make decisions without opportunities for parliamentary approval. Third, its evidentiary basis has become so loose that it provides excessive discretion for states to claim violations by others or defenses for themselves.[31] Furthermore, where CIL is to be applied by a court, it provides excessive discretion to the court.

It is worth noting at this juncture that the aspects of CIL that are seen as the most valuable – the ability to bind states that neither consent nor object, and even the ability to evade the need for parliamentary approval – are problematic from the standpoint of legitimacy or democratic control.

[30] *See* Kelly, *supra* note 4.
[31] *See id.*

Lack of Consensual Basis

CIL can be made without the express consent of all states, and comes into effect against all states that do not persistently object. Is this a feature or a glitch? The argument that it is a feature is that we increasingly need international legislative capacity, and unanimity is a very high transaction-cost way to make new rules.[32] As discussed presently, this capacity is one of the reasons that CIL may have some utility for the near future – at least until other legislative capacity is developed.

It might initially appear that a requirement of explicit consent is perfectly democratic and legitimate, insofar as new rules cannot be made unless all consent. However, in the context of the law of cooperation, a rule of consensus is better understood as broadly undemocratic, allowing the tyranny of the minority to reign, insofar as the smallest minority can block collective action. The rising importance of global collective goods shows the strategic inadequacy of the consent rule. Of course, under formal requirements of unanimity, informal coercion or logrolling can reduce the inhibitive effect of a rule of unanimity.

So, why is the implicit consent requirement a glitch? It is a glitch compared to the alternative: an explicit and transparent system of majority voting that would not sneak rules past inattentive states.

Lack of National Democratic Control

CIL has another, more important, class of legitimacy problems, and this occurs not within the international legal system itself, but within national legal systems. Because of the way that most states are structured internally, and because most actions that would constitute state practice, acquiescence, or *opinio juris*, originate or are taken through the executive, the CIL process privileges executives at the expense of parliamentary democracy. This may not have been much of a problem when international law was largely concerned with more constrained diplomatic issues that did not engage domestic public policy, but now that more and more "behind-the-border" issues are addressed through international law, democratic accountability becomes essential.

It is also possible that certain interests within each state will benefit disproportionately from the use of CIL. These interests may include certain lobbies that may be able to achieve their international legal goals with less democratic accountability through the use of CIL. Executives in general are empowered by CIL, and thus would tend to support CIL. Another group that is empowered by CIL

[32] *See* Andrew T. Guzman, "Against Consent," 52 VA. J. INT'L L. 747 (2012); Nico Krisch, "The Decay of Consent: International Law in the Age of Global Public Goods," 108 AM. J. INT'L L. (2014).

is international law academics. Their teaching, studies, and advice become more important, and better compensated, when CIL is more important. This is somewhat different from treaty, which is generally subject to more objective determination and accessibility. CIL tends to provide reduced access for nonexperts and reduced accountability. Kelly states it well:

> The CIL articulated in judicial opinions and treatises remains the domain of experts familiar with and able to selectively manipulate a vast body of legal materials and contradictory legal theory. A knowledge of CIL requires detailed study of [ICJ] decisions and those of its League of Nations predecessor, the Permanent Court of International Justice, a willingness to examine old and venerable treatises, and familiarity with difficult to obtain materials, such as international arbitral findings and individual state practices. This has become the work of a highly specialized group of experts, not the residue of customary norms understood and accepted by members of a society.[33]

A final group that is empowered by CIL is international judges, whether at the ICJ or in other tribunals, such as international investment arbitration tribunals. As discussed in more detail next, their discretion in determining the existence and terms of CIL is a source of power to them, and also limits their accountability.

Excessive Discretion

The rule of recognition in CIL is insufficiently delineated, as other chapters in this volume argue.[34] The problems are well known: how much practice, what counts as practice, how powerful an *opinio juris*, and what counts as *opinio juris*? This lack of precision results from loose doctrine relating to evidence of requisite state practice and *opinio juris*, failure to codify existing CIL, and lack of mandatory and definitive adjudication which might otherwise lead to a more precise jurisprudence.

The discretion afforded by the lack of precision in CIL is significant, and raises serious concerns regarding accountability and legitimacy. Of course, some of these issues of discretion could be addressed by a stronger set of doctrinal and professional constraints, and in some areas characterized by mandatory dispute settlement, such as international trade law and international investment law, we see some movement in this direction. In other areas, discourse is less constrained.[35]

[33] Kelly, *supra* note 4, at 475.

[34] *See* Curtis Bradley, "Customary International Law Adjudication as Common Law Adjudication" (in this volume).

[35] For an analysis of existing constraints in the Security Council, as well as a proposal for greater constraint, *see* Ian Johnstone, "Security Council Deliberations: The Power of the Better Argument," 14 EUR. J. INT'L L. 437 (2003).

This problem of excessive discretion accorded to international legal judges might be seen as a feature, not a glitch. It could be that these judges would make good decisions, and respond to international cooperation needs in ways that states themselves cannot, due to a variety of bargaining and political economy problems.[36] Indeed, we have seen the European Court of Justice do just this during the period of Eurosclerosis, and perhaps other courts have heroically filled gaps in legislative capacity at particular times. However, international adjudication has limited jurisdiction, and there are grave limitations of authority, substantive capacity, and legitimacy that make such creative adjudication more a glitch than a feature.

WHEN IS CIL STILL USEFUL?

Customary international law has five major types of advantages. First, it is formed in a decentralized and socially immanent way. Second, it allows formation of international law without explicit communication or diplomatic conferences. Third, it can come into effect without unanimity – it is established by a process in which a sufficient number of states act in accordance with the rule, and have *opinio juris* with respect to the rule, while those that persistently object are not subject to the rule. Fourth, it can come into effect without ratification. Fifth, it can bind subsequently established states. In this section, I will compare CIL with treaty by reference to these advantages, and suggest why these advantages are less valuable today than they were during the prime of CIL's life.

Note that it is possible for a treaty to provide for consensus-based decision making, but treaties themselves may not otherwise be established by consensus, but instead require unanimity: no state is a party that does not specifically consent. However, the practical difference between consensus and unanimity is not necessarily great – it is only that states may be bound by silence under custom, whereas they cannot generally be bound by silence under treaty (under some international organization treaty regimes, rules may bind states unless they "opt out").

Decentralized Formation and Modification

Because it has its roots in practice, CIL can in theory be initiated by a single state that decides to propose, through its conduct, a new or different rule of law. In this sense, it is unconstrained by bureaucracy or procedural requirements, and it can respond quickly to new social demands.

[36] For an argument along these lines, *see* Eyal Benvenisti, "Customary International Law as a Judicial Tool for Promoting Efficiency," in THE IMPACT OF INTERNATIONAL LAW ON INTERNATIONAL COOPERATION (Eyal Benvenisti & Moshe Hirsh eds., 2004).

During the early growth of CIL, then-existing technology available made communication and meeting costly. In addition, prior to globalization and the rise of the regulatory state, there were fewer incentives to meet or communicate to make international law. Given the declining costs of communications and transportation, and the increasing benefits of cooperation, as well as the increasing complexity of international legal cooperation, the ability to make law without communication or conferences is of little benefit. Also, recall the discussion of international public goods cooperation problems, which can be resolved by simultaneous commitment, but not by sequential commitment: that discussion showed that it is not possible to make appropriately responsive law without communications and diplomatic conferences.

Formation of International Law without Unanimity

It is often efficient to be able to produce international legal rules without unanimity. While CIL is initially attractive because it has this capacity, its dynamic does not fit well with modern legislative needs. First, CIL formation without unanimity requires that the non-consenting state be ignorant. Under modern communications and information dissemination, such ignorance is decreasingly likely. Moreover, the resulting rule of law would suffer from a lack of legitimate process. Finally, this lack of legitimate process would often disadvantage less advanced countries.

Giovanni Maggi and Massimo Morelli show that a key parameter in determining whether to choose a rule of majority voting versus a rule of unanimity, focusing on the likelihood of compliance, is the governments' discount factors, reflecting their patience.[37] As Maggi and Morelli explain, their model posits that "the voting rule is chosen ex ante, under a veil of ignorance about future issues. Thus, the optimal voting rule maximizes the ex ante expected utility of the representative member subject to self-enforcement constraint: a government must have incentive to comply with the collective decision even if it happens to disagree with it."[38] One important finding of the Maggi and Morelli work is that if external enforcement *is* available, the ex ante efficient rule is typically some type of majority voting.[39]

Conversely, Maggi and Morelli also show that where an international agreement must be self-enforcing (there is no external enforcement), an organizational rule of

[37] Giovanni Maggi & Massimo Morelli, "Self-Enforcing Voting in International Organizations," 96 AM. ECON. REV. 1137, 1138 (2006); *see also* George Norman & Joel Trachtman, "The Customary International Law Game," 99 AM. J. INT'L L. 541 (2005).

[38] Maggi & Morelli, *supra* note 37, at 1138.

[39] *See generally* Philippe Aghion & Patrick Bolton, "Incomplete Social Contracts," 1 J. EURO. ECON. ASSN. 38 (2003); Joel Guttman, "Unanimity and Majority Rule: The Calculus of Consent Reconsidered," 14 EUR. J. POLIT. ECON. 189 (1998).

unanimity may be efficient under certain circumstances. In order to do so, they use a model in which unanimity eliminates the possibility for defection. Their model involves voting over just two options: against or in favor of a collective action that can provide benefits only if all participants contribute. Given the binary nature of voting in this model, a voter can never gain from voting cynically: if they vote for the provision of the public good, but then do not contribute, they will obtain no benefits. However, in many other policy circumstances, states may have incentives to vote cynically, so that unanimity would not provide compliance benefits. CIL seems to eliminate the possibility of cynical "voting," because at least enough states to form sufficient practice would have to comply prior to formation.

In addition, even without cynicism, unanimity can result from "package deals" under circumstances in which a state agreeing to the package would prefer not to comply with one or more components of the package.[40] CIL would seem to be immune to this problem, also, because, as discussed, it lacks significant capacity to produce package deals.

For these two reasons, CIL may have compliance advantages as a system for non-unanimous production of international legal rules. However, these compliance advantages can easily be replicated in the institutional design of treaties. For example, in connection with the formation of the World Trade Organization, the stability of the package deals established there were preserved by specifically providing for mandatory dispute settlement, combined with the possibility for retaliation in diverse fields – so-called cross-retaliation. Thus, if India were to defect in protecting intellectual property rights, the United States could retaliate by blocking imports of Indian fabrics. Diverse performance reciprocity can support compliance.

Furthermore, circumstances may change such that a domestic political equilibrium that supported entry into the relevant obligation no longer exists to support compliance with the obligation. Consent would tend to imply compliance in circumstances only (i) where consent is known to be based on domestic political support for the behavior that constitutes compliance (even then the domestic political equilibrium may change so as to undermine the tendency to comply); and/or (ii) where the giving of consent activates a lobby that believes that promises should be kept. So, consent is not sufficient for compliance, but nor is it necessary for compliance, because, as I have argued, diverse performance reciprocity may induce compliance in a way that makes the larger relationship self-enforcing while the narrower rule is not. Indeed, Maggi and Morelli suggest that the availability of transfers expands the range of discount factors for which a rule of majority voting is sustainable. However, I argue presently that a rule of unanimity, as suggested by Maggi and Morelli, would not necessarily solve enforcement problems, because

40 Maggi and Morelli recognize this issue. *See* Maggi & Morelli, *supra* note 37, at 1151.

unanimity does not necessarily imply subsequent compliance. So there may be fewer circumstances in which a rule of unanimity would be attractive.

In addition, with high discount factors – greater valuation of future payoffs from cooperation – there are smaller incentives for defection, and less need for what Maggi and Morelli determine are the compliance benefits of a rule of unanimity. Greater likelihood of repeated play and greater frequency of interaction – increasing the amount at stake over a shorter time – also promote compliance. But note that the frequency of interaction parameter need not be limited to interaction *within* a particular organization or issue area.[41]

Thus, their model "predicts that a non-unanimous rule is more likely to be adopted in organizations where governments are more stable, and in 'busier' organizations."[42] They also find that greater correlation in the preferences among member states increases the likelihood of a non-unanimous voting rule. Furthermore, a non-unanimous rule may be efficient where there is external enforcement.

But perhaps even more importantly, the assumption that most rules of international law or treaties must be self-enforcing may be too strong, as well. The common understanding of international law in the economics literature assumes that for international law to be effective, it must have the characteristics of a self-enforcing contract.[43] The tools of analysis used by economics for particular legal rules or regimes is thus noncooperative game theory, where it is assumed to be impossible to enter into exogenously binding contracts. Rather, the internal dynamics of the rule or regime must be structured so that it is endogenously binding – so that it has internal dynamics that will result in compliance.

Much of this analytical perspective is factually dependent on the definition of the scope of the game being played. For example, if the game is isolated as the "reduction of tariffs on bananas game," or even as the broader "trade liberalization game," then perhaps the requirement for self-enforcing contracts is appropriate. If, instead, we understand the trade liberalization game as part of a broader "general international law game," then it is not true that the trade liberalization component must itself be self-enforcing. Rather, the question is whether the general international law game is self-enforcing. Similarly, in domestic society, we do not need to ask whether the domestic "environmental protection game" is self-enforcing, because it is embedded in a broader legal and constitutional system. The domestic system starts out equally anarchic to the international system.[44]

[41] *See generally* Norman & Trachtman, *supra* note 37 (explaining the role of multisector interaction).

[42] Maggi & Morelli, *supra* note 37, at 1138.

[43] *See, generally* Giovanni Maggi, "The Role of Multilateral Institutions in International Trade Cooperation," 89 AM. ECON. REV. 190 (1999).

[44] *See, generally* Jack Goldsmith & Daryl Levinson, "Law for States: International Law, Constitutional Law, Public Law," 122 *HARV. L. REV.* 1791 (2009).

It is true that in anarchy, we must search for self-enforcing agreements. But similar to the fundamental basis for domestic law, the self-enforcing character of international law must be assessed looking at the totality of the international law relationships between states, rather than by isolating individual relationships. There is a much greater possibility for reciprocity, and for what we might call a self-enforcing system, when we examine the entire set of relationships, and examine them across time, than when we examine a single obligation.

The economists' perspective, expecting individual legal rules or organizations to be self-enforcing when considered separately and in the short term, is both ignorant of the networked power of law and impotent to address long-term, highly asset-specific, and asymmetric, cooperation problems. As to both domestic law and international law, we might say that each individual thread of obligation may be weak or strong – if it is weak, it will be unable to sustain cooperation. However, once a variety of threads are woven together in a "fabric" of society, the fabric can be considerably stronger than any individual thread.

To conclude on this point, the general perspective is that there will be many circumstances in which the ability to form international law without unanimity is beneficial. However, CIL does not have a monopoly on this ability: treaty regimes can specify a rule of majority voting for secondary law. Furthermore, CIL doctrine does not provide a rule of majority voting, but merely a special rule of consensus: if it is formed, those who fail to object to it are generally thought to be bound. So this possible advantage of CIL is insignificant.

Avoiding Domestic Ratification

This feature is the converse of a glitch described earlier. There will be circumstances in which executives will prefer not to engage in a process of domestic ratification of international legal rules, and so they might prefer the CIL process to the treaty process. However, this capacity can only be understood as a feature in circumstances where the domestic political structure is stuck in an inefficient equilibrium in which it cannot either ratify or determine to allow to come into force without ratification in appropriate circumstances.

Binding Subsequently Created States

With respect to international cooperation problems that can only be solved if all states are bound, including "weakest link" public goods in which the public good is only successfully provided if all parties comply, it is important to bind all parties. Also, there are circumstances in which each incremental complying state adds significantly to the quality of the public good, so that there are important benefits from binding all states. Therefore, to the extent that CIL binds subsequently created states, this feature may assist in solving certain types of cooperation problems. On the other hand, a rule that successor states succeed to treaty obligations would have a similar effect in the context of treaty.

MUCH EXISTING CIL IS ALREADY SUPERSEDED

Recognizing the limitations of CIL, the world community has already moved away from it, and has largely superseded it with treaties. As Anthony D'Amato suggested in 1971, "The growth of treaty regulation of international behavior has so accelerated with the passage of time that today there are few areas of international law untouched by the complex network of international treaty law."[45]

Michael Wood, in the *First Report on Formation and Evidence of Customary International Law to the International Law Commission*, has made the following statement (citations omitted):

> It is sometimes suggested that treaties are now a more important source of international law than customary international law. Such generalizations are neither particularly illuminating nor necessarily accurate. Even in fields where there are widely accepted "codification" conventions, the rules of customary international law continue to govern questions not regulated by the conventions and continue to apply in relations with and between non-parties. Rules of customary international law may also fill possible lacunae in treaties, and assist in their interpretation. An international court may also decide that it may apply customary international law where a particular treaty cannot be applied because of limits on its jurisdiction.[46]

However, none of the facts cited is inconsistent with the claim that Wood purports to refute. It is true that CIL retains relevance where it provides rules that have not been included in codifying conventions, and it is true that the relevant CIL continues to bind states that have not joined such conventions. But this observation only supports a claim of continued relevance of CIL; it does not refute the claims of *declining relevance* of CIL relative to treaty.

As explained presently, the evidence shows that there are few asserted rules of CIL that have not been included in conventions. While CIL may indeed "fill lacunae in treaties," or more simply provide rules that treaty parties did not see fit to include in the treaties, the analysis shows that those lacunae are limited in scope. Again, the observation that CIL can fill lacunae by no means refutes claims of declining relevance of CIL *relative to* treaty. Finally, the *Nicaragua* case-type context in which

[45] Anthony D'Amato, The Concept of Custom in International Law 105 (1971).

[46] Special Rapporteur, *First Rep. on Formation and Evidence of Customary International Law*, Int'l Law Comm'n, UN Doc. A/CN.4/663 (May 17, 2013) (by Michael Wood).

an international court can apply CIL that parallels a treaty, when that treaty is not included in the relevant submission to jurisdiction, is another very limited area of CIL relevance. That case seems to rely more on exploiting inartful submissions to jurisdiction than on a substantive characteristic of CIL.

Other work has suggested that there is some evidence in the past decade or two of a decline in the number of new *multilateral* treaties, as well as in a few states' practice in connection with *sub-multilateral* treaties.[47] There are a number of problems in drawing any conclusions regarding a decline in salience of treaties from these observations:

- First, findings of decline are extremely sensitive to the period chosen for observation. If we were to examine a time series describing the establishment of new international law rules over the past 100 years, focusing on the *ratio* between use of CIL and use of treaty, I suspect we would find a consistent and substantial move from CIL to treaty. It is in any event too early to tell whether recent trends in multilateral treaties are a pause in a broader trend of increase, or a new trend.
- Second, the data provided with respect to all countries seems to count only multilateral treaties. While there is data regarding a few countries' practice relating to sub-multilateral treaties, this does not tell us anything about general practice.
- Third, and even more trenchantly, these claims focus on the number of new treaties. If you are serious about measuring the influence of treaty, the better question to ask is about the stock, rather than the flow, of treaties, compared to the stock of CIL, and how that *ratio* has changed over time. While, as suggested, it is by no means clear that the flow of treaties has declined over a meaningful period, or for structural as opposed to temporary reasons, there is simply no doubt that the stock of treaties is growing. As the Appendix shows, there are few CIL rules that are not incorporated in treaty. But a moment's thought will show that there are many treaty rules that are not incorporated in CIL. Furthermore, even if it is true that the flow is reducing, this would be consistent with the possibility that major issues have been addressed early, with fewer issues remaining to be addressed. Deceleration may be perfectly natural, and consistent with increasing utility of treaty.
- Fourth, the number of treaties is not necessarily a good indicator of the number of rules, or of their importance in international society. The subject matter addressed in treaties in modern times often involves significant complication, and many rules. Again, if you are serious about measuring the influence of

47 Joost Pauwelyn, Ramses A. Wessel, & Jan Wouters, "When Structures Become Shackles: Stagnation and Dynamics in International Lawmaking," *Eur. J. Int'l L.* (forthcoming 2015).

treaties, the focus should be on the number and importance of *rules* included in treaties. The unit of analysis should be rules: it would be foolish to compare CIL rules with treaties, rather than with treaty rules.
- Fifth, decline in absolute numbers does not necessarily indicate a decline in relative importance. Indeed, the same work that suggests decline in treaty also suggests decline in CIL, both in favor of soft law.[48]

Greater statistical nuance will be needed in order to address these concerns, and in order to be sure that observed trends are not artifacts of a particular state's politics, or of a particular change in practice.

The principal concern here, however, is with identifying the extent to which CIL rules are replicated by treaty law rules. With one caveat, this might indicate a reduced relevance for CIL, suggesting that the world community is moving toward treaty as its favored means of legislation. The caveat is that it might be argued that the world's process of making treaty often begins with CIL. Indeed, this may be a correct characterization of many circumstances in the past,[49] such as international humanitarian law, and of some cases today. Treaties that build on CIL can add precision and institutional support that are generally not available in CIL. However, for the reasons outlined earlier, it is difficult to expect CIL to respond effectively to modern international cooperation needs.

One example is in the area of international investment law. CIL arguably provided two critical protections for foreign direct investment: (i) disciplines on expropriation, and (ii) fair and equitable treatment. Bilateral investment treaties, where applicable, have gone a long way toward concretizing and clarifying each of these protections. It is true that the bilateral investment treaties built on CIL, but it is also true that they added a far greater degree of precision, as well as institutional support in the form of investor-state dispute settlement, which were not available under CIL. In turn, investor-state dispute settlement has added more precision. They also added a number of additional obligations that did not exist in CIL, such as requirements of market access, protection of ability to make transfers, and restrictions on performance requirements.

It might also be argued that treaty is a way of making CIL, and there is an extensive literature supporting this idea.[50] However, a treaty cannot by itself make CIL.[51] Rather, the only role a treaty may play is as evidence of state practice. Treaty is

48 *Id.*

49 VILLIGER, *supra* note 16.

50 *See, e.g.*, Richard R. Baxter, "Multilateral Treaties as Evidence of Customary International Law," 41 BRIT. Y.B. INT'L L. 275 (1965).

51 *See* Ahmadou Sadio Diallo (*Republic of Guinea v. Democratic Republic of the Congo*), ¶ 90 (Judgment) (ICJ May 24, 2007); North Sea Continental Shelf Cases (Federal Republic of Germany/Denmark), 1969 I.C.J. Rep. 3, 44 (ICJ Feb. 20, 1969) (Judgment).

ambivalent as evidence of *opinio juris* as to CIL, because the fact of making the treaty suggests that the signatories considered that there was no preexisting rule of CIL. In the *Nicaragua* case, the ICJ referred to subsequent resolutions expressing *opinio juris* in order to find that treaty rules prohibiting the use of force in international relations had attained CIL status.[52] Furthermore, again, for the reasons recounted earlier, it would seem strange intentionally to use treaty to make CIL, except if there is a need to bind states without their consent. That use seems devious, and must be of declining utility under modern capabilities for dissemination of information.

Is CIL still an important independent source of international legal rules? In order to determine the extent to which there is important CIL that has not been transposed to treaty, I engaged in the following analysis. First, I developed a general taxonomy of CIL rules. I did so by examining an arbitrarily selected, but highly regarded, group of secondary sources on international law to identify these rules. Where International Law Commission statements found rules of CIL, I simply accepted those rules as part of the taxonomy. I did not attempt to identify the specifics of any rules included, but merely to name the basic rule asserted. So, the taxonomy operates at a fairly general level, failing to take account of detailed components of rules. I did not examine state practice or *opinio juris* in order to confirm or invalidate the rules asserted. My goal was to construct a rough inventory of colorable rules, not to write a treatise.

Second, I examined the general extent to which the rules identified were incorporated in treaty law, or codified by the International Law Commission (ILC). This was easiest to do with well-known multilateral treaties, and much more difficult to do with treaties among smaller groups of states. For non-multilateral treaties, I focused on broad groups of treaties, such as regimes of bilateral investment treaties and bilateral tax treaties, recognizing that there is significant diversity among these treaties.

As you can see by examining the Appendix,[53] much of CIL has been codified in the form of treaty or ILC codification. To a remarkable extent, due to the growth of treaty law incorporating the same norms, if CIL were abolished today, most of the international legal system would remain intact. Although I cannot prove that the rise of treaty rules has resulted from the factors discussed here, those factors are a good set of reasons why states might have determined to produce treaties that include, but are not limited to, the rules that exist in CIL.

The taxonomy of rules of CIL provided in the Appendix identifies a non-exhaustive number of rules of CIL in all major areas of international law. The study is organized

[52] Military and Paramilitary Activities in and against Nicaragua (*Nicar. v. U.S.*), 1986 I.C.J. ¶ 188 (June 27) (Jurisdiction and Admissibility).

[53] Online appendix available at: www.cambridge.org/9781107443105

around these major areas of international law: (i) fundamentals of international law; (ii) use of force; (iii) law of treaties; (iv) international environmental law; (v) international human rights law; (vi) nationality, aliens, and refugees; (vii) international humanitarian law; (viii) succession of states; (ix) international economic law; (x) privileges and immunities; (xi) state responsibility; and (xii) law of the sea.

A first caveat must be mentioned here: the taxonomy does not claim to be an exhaustive list of all existing rules of CIL, but rather represents a comprehensive enumeration of rules of CIL. The attempt to draft an exhaustive list of *all* existing rules of CIL would not only be an enterprise of years or even decades of research, but, as it would turn out at the end of the research, a highly illusionary enterprise. It would be an impossible task because the customary status of many ostensible rules of CIL remains heavily contested by states and scholars. The complex structure of rules of CIL – *opinio juris*, state practice, and the required link between these two – inevitably keeps the field of contestation open.

Therefore, the taxonomy of CIL provided in the Appendix offers not an exhaustive but a comprehensive list of (less-controversial) existing rules of CIL. The purpose of the taxonomy is, after all, not to confirm what rules exist, or what they say, but to determine the extent to which rules of CIL are codified in international treaties or ILC codifications. In this regard, a wide random sample of CIL rules, rather than an exhaustive list, is all that I need.

The general methodology adopted for the identification of rules of CIL is based on the consultation of five secondary sources on general public international law. The five treatises are all written by leading scholars in the field, considered as highly authoritative and frequently quoted:

(i) Malcolm Shaw, *International Law* (6th ed.);
(ii) James Crawford, *Brownlie's Principles of International Law* (8th ed.);
(iii) *Restatement Third of the Foreign Relations Law of the United States*;
(iv) *Oppenheim's International Law* (9th ed.); and
(v) *The Max Planck Encyclopedia of Public International Law*.

The identification of specific rules of CIL was conducted as follows. A specific rule is regarded as a rule of CIL if at least three of the treatises mentioned consider the rule as part of CIL and the remaining two treatises remain silent. In instances where a treatise does not explicitly refer to a rule as CIL, but the context or textual framing strongly suggest that it is part of CIL, it was assumed that the author simply did not consider explicit reference necessary.

Furthermore, where the ILC explicitly considers a rule of international law part of CIL, the view of the ILC is accepted as authoritative. In these instances, the position of the ILC was adopted and only minor additional analysis conducted.

The secondary sources consulted are general treatises on international law. Nevertheless, they reflect professional specializations of their authors. The edition of *Brownlie's Principles of International Law* referenced, for instance, reflects the professional background of its editor, James Crawford, as ILC special rapporteur on the law of state responsibility and deals only marginally with international environmental law. In other areas, such as international humanitarian law, the five secondary sources only provide an overview of the substantive aspects of the respective fields. The statements on CIL in these fields, that is, international humanitarian law or international law on treaties, are either too broad, identify only a few rules of CIL, or are entirely absent. Therefore, identification of existing CIL in these areas required the consultation of more specific treatises. For instance, the identification of customary international humanitarian law relies on the authoritative study done by the International Committee of the Red Cross and the precise identification of customary rules in the law of treaties relies on three authoritative commentaries on the law of treaties.[54] In areas where the taxonomy adopts an approach that deviates from the general methodology, specific sections provide detailed reasons and explanations.

The codification of rules of international law in treaties is important because it reveals how states view specific rules. First, on a rather abstract level, codification reveals that states consider the regulation of a specific field as important enough to engage in the arduous and time-consuming process of treaty negotiation. Second, on a more specific level, reservations and statements of interpretation on specific rules of a treaty provide insight into the specific scope and meaning of a rule.

As the taxonomy reveals, the vast majority of rules of CIL are codified in international treaties or codification instruments. Of (the randomly even number of) 300 identified rules of CIL, only 13 (4.33 percent) have not been either incorporated in treaties or codified. Even if we remove ILC codifications from the data, only three additional rules, plus the rules relating to state responsibility (twenty-seven), are not incorporated in treaty. Thus, the aggregate of all CIL rules not incorporated in treaty are 43/300 (14.33 percent). This is still not a major role for CIL.

It is not disputed that in a few specific areas CIL retains importance. First, CIL is still relevant in areas where a rule of CIL is either more precise than a codified rule or where codification is absent. In international humanitarian law, for instance, the provisions of CIL remain significant in the regulation of non-international armed conflicts. The rules of customary international humanitarian law go beyond the protection offered by Additional Protocol II and are thus binding even for states that

[54] These three secondary sources are: Anthony Aust, *Modern Treaty Law and Practice* (2d ed. 2013); Mark E. Villiger, *Commentary on the 1969 Vienna Convention on the Law of Treaties* (2009); *The Vienna Conventions on the Law of Treaties* (Olivier Corten & Pierre Klein eds., 2011).

are not parties to Additional Protocol II.[55] Furthermore, the nationality requirement of article four of the Fourth Geneva Convention regarding the definition of protected persons has been broadened by the ICTY to a "substantial relations" doctrine so as to include persons who technically do not meet the nationality requirements of article four.[56] Similarly, the precise extent of combatant status is only marginally covered by treaty law. Sassoli points out, however, that the precise customary status of the requirements of retaining combatant status is difficult to determine.[57] The taxonomy has been written with this caveat in mind. But, at the same time, it must be noted that the jurisprudence of the European Court of Human Rights develops standards and rules in international humanitarian law significantly further by relying on the extra-territorial application of specific provisions of the European Convention of Human Rights.

Furthermore, while Chin Lim's contribution to this volume sees creative and extensive use of CIL in investment treaty arbitration,[58] I see increasing displacement of CIL by investment treaties, including especially in the current U.S. practice, where the international law minimum standard is referenced, but specifically defined. The CIL and treaty, as Lim argues, are indeed intertwined, but the treaty selectively incorporates and adapts CIL in a sphere in which, a mere fifty years ago, there were no treaties. It is a story of remarkable growth and dominance of treaty. Furthermore, in order to resolve difficult issues in investment arbitration jurisprudence, greater treaty refinement will be necessary in the future.

Second, CIL remains important in areas where it is codified by an international treaty or instrument, but the instrument is not legally binding or only few states are parties to the treaty. With regard to the former, the ILC Articles on State Responsibility (2001) are not binding qua codification, but as CIL. The ILC articles portray pretty well the reasons why states might abstain from treaty codification. First, the extremely long and institutionalized codification process – the first public codification attempts started at the Hague conference in 1930 under the League of Nations – resulted in clear and very uncontroversial rules on state responsibility. The articles were annexed to General Assembly (GA) Resolution 56/83 and commended "to the attention of Governments without prejudice to … their future adoption or other appropriate action."[59] Subsequent GA resolutions deferred a decision on the

55 Jean-Marie Haenckerts, "Study on Customary International Humanitarian Law: A Contribution to the Understanding and Respect for the Rule of Law in Armed Conflict," 857 *Int'l Rev. Red Cross* 175, 189 (2005).

56 *Prosecutor v. Delalic*, Case No. IT-96-21-A, Judgment of the Appeals Chamber (Int'l Crim. Trib. for the Former Yugoslavia Feb. 20, 2001).

57 Marco Sassòli, *Max Planck Encyclopedia of Public Int'l L.* 13 (Rüdiger Wolfrum ed., 2014).

58 Chin Leng Lim, "The Strange Vitality of Custom in the International Protection of Contracts, Property, and Commerce" (in this volume).

59 GA Res 56/83, ¶ 3 UN GAOR, 56th Sess., Supp. No. 10, U.N. Doc. A/56/10 and Corr.1 (Dec. 12, 2001). The articles annexed to the resolution were adopted without a vote.

final status of the draft articles.[60] Since the draft articles had been annexed to the adopted GA resolution, international and domestic courts have extensively relied upon them. Because the rules are reasonably clear and codified, there might simply be no reason to convene a costly treaty conference in which states would propose new rules to be incorporated. The statement of the United Kingdom representative put it succinctly: "It is difficult to see what would be gained by the adoption of a convention.... The draft articles are already providing their worth and are entering the fabric of international law through state practice, decisions of courts and tribunals and writing."[61] In this context treaty law does not present advantages and there seems to be no need for the arduous process of double codification. Interestingly, one might conjecture that in this relatively technical field, a combination of custom and the authoritative statement provided by the ILC articles does all the work that is needed.

The rules on state succession provide an example of low participation in a treaty regime in force or not yet entered into force. Only twenty-two states are currently parties to the Vienna Convention on the Succession of States in Respect of Treaties (1978),[62] and the Vienna Convention on Succession of States in Respect of States Property, Archives and Debts has not yet entered into force.[63] However, issues on state succession are mainly resolved in bilateral negotiations between the predecessor and the successor state and between creditor and debtor states, respectively. Furthermore, practice on state succession varies greatly and only a few principles can be said to have achieved customary law status. For other principles there is not more than mere evolving state practice.[64] The low support for the treaty regimes on state succession reflects the few rules of CIL in this legal field.

For some, prescriptive jurisdiction might be seen as an area of important CIL, without significant coverage in treaty. Yet, most of the treatises consulted to produce the taxonomy did not mention these rules as part of CIL, as the treatises either point to varying state practice and *opinio juris* or do not mention rules of jurisdiction to prescribe as part of CIL. Furthermore, national differences on prescriptive jurisdiction make the identification of an exact scope, and thus a precise CIL rule, difficult.

60 GA Res 59/35, UN GAOR, 59th Sess., Supp. No. 35, UN Doc. A/59/35 (Oct. 8, 2004); GA Res 62/61, UN GAOR, 62nd Sess., Supp. No. 12, UN Doc. A/62/61 (Dec. 6, 2007); GA Res 65/19, UN GAOR, 65th Sess., Supp. No. 75, UN Doc. A/65/463 (Dec. 6, 2010).

61 Alain Pellet, "The ILC's Articles on State Responsibility," in *The Law of International Responsibility* 87 (James Crawford et al. eds., 2010).

62 Vienna Convention on the Succession of States in Respect of Treaties, Aug. 23, 1978, 1946 U.N.T.S. 3.

63 Chapter III Privileges and Immunities, Diplomatic and Consular Relations, Etc., United Nations Treaty Collection, retrieved from https://treaties.un.org/pages/ViewDetails.aspx?src=TREATY&mtdsg_no=III-12&chapter=3&lang=en (last visited July 23, 2014).

64 Anthony Aust, *Handbook of International Law* 235 (2010).

IMPLICATIONS AND CONCLUSION

An aphorism attributed to the film producer Samuel Goldwyn is that "a verbal contract is not worth the paper it is printed on." CIL is somewhat equivalent to an unwritten contract made in part by noncontractual statements and in part by actions. One imagines that the reason for Goldwyn's criticism of verbal contracts is that they are difficult to prove, and this criticism seems aptly applied to CIL. But an additional, more nuanced, interpretation of Goldwyn's aphorism is that verbal contracts are also insufficiently detailed in a modern world of complex contracting. This insufficient detail leads to excessively incomplete contracts, which cannot be applied reliably.

For the reasons set forth in this paper, CIL cannot respond effectively to the great modern challenges confronting international society: global environmental protection, international public health, cybersecurity, financial crisis, and liberalization of movement of goods, services, and people. Nor can it address the enduring challenges of suppressing and regulating war, protecting human rights, and reducing poverty.

Treaties provide the ability to coordinate in advance of events, specify obligations in detail and with nuance, provide for diverse performance reciprocity allowing efficient transactions in authority among heterogeneous states, and provide for customized institutional support, including dispute settlement. Treaties carry greater legitimacy because they are transparent, they do not bind states without their actual, as opposed to constructive, consent, and they are subject to whatever domestic ratification process particular states decide. Even the purported advantages of CIL can be recreated in a more deliberate and nuanced way in the treaty context. Furthermore, in a constitutional moment, treaty can be constructed with transnational features like binding dispute settlement or majority voting, in a way that carries the legitimacy of general consent.

It is true that soft law also has a role in replacing CIL.[65] In a sense, this is the story of the ILC Articles on State Responsibility, and we can extend this observation to much of international financial cooperation, for example, under the Basle process. However, there are a number of features of law that are not available with soft law, so the rise of soft law is simply another part of the increasing variety of types of international rules. In a sense, all law is social order, labeled law. States may be expected to move from non-international law equilibrium behavior to international law where the latter either makes equilibrium possible that would not otherwise be

[65] *See, e.g.*, Nico Krisch, "The Decay of Consent: International Law in the Age of Global Public Goods," 108 AM. J. INT'L L. (2014).

possible, or enables that equilibrium to be achieved more efficiently than through other means.

The international legal system must become more complex and variegated in order to take on new responsibilities.[66] The Grotian innovation, essentially inventing CIL in the seventeenth century, was a sound one, and CIL was an appropriate and durable institutional response to the needs of the international system from the seventeenth to the twentieth centuries. But nothing lasts forever, and the types of "law of coexistence" problems that CIL addressed well are of a fundamentally different nature than the "law of cooperation" problems that characterize the modern international agenda. The origin of all law is in custom. In the village or tribal setting, customary law is established through social practice within a given community and accepted as obligatory by its members.[67] This process is analogous to the CIL process. While custom has enduring roles, in both informal and formal settings, purposive or legislated law seems to become more important as societies become more industrialized and complex.[68]

Sir Henry Maine provided an early description of the legal development of societies, painting with a broad brush. In the earliest days of humankind, law was not a statement or enunciation of principles, but rather determinations ex post facto, that is, the pronunciation of right and wrong after certain facts had occurred. At a later stage, the monopoly of knowing the law was vested in aristocracies who referred to the customs of a particular tribe, ethnicity, or society, claiming that they knew what the customs were. This role may be similar to that of international lawyers today. Subsequently, the era of codes followed, with the Twelve Tables as perhaps the earliest and best known example.

These developments, of course, did not occur in a historically linear way. Nevertheless, Maine's account of legal development is based on a linear logic of progressive codification of laws. According to this account, every human society, as it develops from its premodern existence to modern society, develops its laws from custom to statute, unwritten to written form, status to contract, instinct to knowledge.[69]

66 *See* Joel P. Trachtman, *The Future of International Law: Global Government* (Cambridge 2013).

67 T. W. Bennett, "Comparative Law and African Customary Law," in *The Oxford Handbook of Comparative Law*, (Mathias Reimann & Reinhard Zimmermann [eds.] Oxford 2006).

68 *See* Emily Kadens, "Custom's Past" (in this volume) (arguing that "in the sorts of small, tight-knit, highly stable communities in which social control can function without positive lawmaking, 'the way we do things' does not even need to be identified or articulated as a set of rules – 'our custom' – to be effective").

69 Sir Henry Maine, *Ancient Law* xiv (5th ed. 1873). While this story maps well with the story I articulate here about the diminishing extent to which CIL can respond to the needs of modern international society, it is also true that the colonial experience, and the postcolonial construction of

What are the implications of the analysis presented in this chapter? First and foremost, states and international organizations should focus their international legal analytical resources on legislated law. Instead of endlessly arguing about whether a particular CIL rule exists, or still exists, or what it is – trying to establish responsibilities for others without exchange – we should focus on transactions that are desirable to both sides. Not only will these rules have a better pedigree, but they will also generally have a greater chance for compliance: a rule that is created in reciprocity is more likely to be characterized by continuing reciprocity.[70]

For NGOs or other advocates, instead of trying to charge states with our preferred environmental or human rights law rule by divining CIL, we should advocate and negotiate specific legal obligations. For both states and citizens, instead of hoping that we can sneak or bootstrap a desired CIL rule past a target state or group of states, we should engage with those states.

For academics, we should focus our analysis on the politically immanent, interdisciplinary, work of developing proposed rules that are administratively workable and effective, and that achieve actual social goals. International lawyers should focus less on advocacy seeking to establish or contradict purported rules of CIL, and more on assisting international processes of treaty making.

This will lead to the professionalization of the international law academy. For too long, the international law academy has seen itself as a priesthood with superior normative judgment and special capabilities of reading mysterious entrails to determine what the law is, whether it comes from the gods or from constructed custom. If this is the vocation of international law, there is no reason to engage with scientists, economists, generals, or politicians. Indeed, there is no need for knowledge about what people want, or how best to achieve it. The law is the law.

If, on the other hand, we recognize that international law does not simply develop autonomously in ways that only international lawyers can see and judge, then we will have the beginning of a useful, responsible profession. This responsible profession can then work pragmatically with other experts, and can itself provide international legal expertise to governments and international organizations in order to assist in achieving the goals of peace, justice, environmental protection, and prosperity that international law, and the world, has set for itself.

states in traditional societies, had much to do with the growth of law, and in particular of legislated law, in these societies. Furthermore, there are important reasons why the international system is different from municipal societies.

70 *See* Joel P. Trachtman, "International Law and Domestic Political Coalitions: The Grand Theory of Compliance with International Law," 11 *CHI. J. INT'L L.* (2010).

8

The Strange Vitality of Custom in the International Protection of Contracts, Property, and Commerce

Chin Leng Lim

Large parts of the international economic order remain governed by "soft laws," with the more notable exception of international trade and investment regulation.[1] But because the trade and investment areas have been the result of treaty lawmaking, the role of custom has become concealed under a patchwork of treaty rules. Out of sight, custom is not only out of mind, but its utility is scorned. In other areas, such as the immunity of states from commercial claims, a combination of treaty, custom, and national laws has filled the gap in public international law regulation. By shifting attention from lawmaking to international economic law's "adjudicative aspects," this chapter intends to, first, defend custom's continued importance in international economic law, particularly in the arbitral interpretation of investment treaty clauses, that is, in the protection of property and contracts.[2] In contrast, custom's potential role in respect of trade-related treaty clauses is still largely unrealized.[3] The chapter proceeds to discuss how questions about foreign sovereign immunity to commercial claims has been the result of minute domestic lawmaking, which in turn fashions detailed customary rules that are fleshed out in even greater detail by domestic courts. These illustrations from the trade, investment, and commercial spheres illustrate some of the ways in which custom continues to be useful, whether or not custom is as fully utilized – as in the trade area – as it should be.

[1] This is also true in the field of international monetary cooperation, see C. L. Lim, "'The Law Works Itself Pure': The Fragmented Disciplines of Trade and Monetary Cooperation, and the Chinese Currency Problem," in *International Economic Law after the Global Crisis: A Tale of Fragmented Disciplines* (C. L. Lim & Bryan Mercurio eds., 2015).

[2] There is some comfort to be drawn from the fact that the chapter does not tread entirely new ground. International investment lawyers especially have paid particularly close attention to the fact that the investment law field may not be reducible to the workings of investment treaties, viewed in clinical isolation from general international law. *See, especially*, José E. Alvarez, "A BIT on Custom," 42 *NYU J. Int'l L & Pol'y* 17, 76 (2009) (drawing on Andreas Lowenfeld's earlier work).

[3] *See* Tomer Broude, "Nothing to Declare: The Silence of Custom in International Trade Law" (manuscript, on file).

Second, this chapter responds to current criticisms of custom's supposed deficiencies in making new international law, particularly in global economic regulation.[4] It argues, against the grain of the current literature, that (i) custom is at least as susceptible as treaties to detailed customization in design, (ii) like treaties, customary lawmaking can also be formed through complex tradeoffs (and that custom is not simply achieved through "in-kind" reciprocity), and, finally, (iii) custom's lack of predictability is overstated. My contention is that custom is more useful than we might think in making new laws because the same diplomatic, organizational, and institutionalized methods that are used today to make and remake treaty laws also create custom and that these methods address many of the weaknesses that are usually attributed, in the abstract, to customary lawmaking. As with the "adjudicative" context mentioned earlier in which tribunals resort to custom in order to interpret and apply treaties, which this chapter calls the "weak intertwination" of custom and treaty, an equally complex interrelationship between treaty and custom in the lawmaking or "legislative" realm bears witness to the ways in which modern methods of multilateral treaty making can help to address some of the weaknesses found in customary lawmaking, including its lack of careful advance negotiation, lack of detail, democratic deficits, and legitimacy problems. This chapter calls the intertwination of customary lawmaking in modern, multilateral treaty and similar negotiations "strong intertwination" for want of a better term, in order to distinguish it from the application of customary rules by tribunals in the interpretation of treaty rules.

In short, increased treaty lawmaking and the increased institutionalization of international law create a greater, not lesser, role for custom in modern international economic law and regulation. As the chapter will explain, modern developments have led to, among other things: investment treaties that incorporate customary rules; tribunals that resort to custom, even where there is no such express treaty mention or incorporation; a "systemic-integrationist" need on the part of tribunals to mark out the boundaries of treaty regimes from previous customary international law rules; and the organizationally co-embedded and intertwined nature of treaty and customary international lawmaking.

SOME TRADITIONAL NORMS IN INTERNATIONAL ECONOMIC LAW, AND THEIR INTERRELATIONSHIPS WITH MODERN TREATY RULES

Norms of nondiscrimination form a golden thread that runs through the fabric of international economic law, and they have long been present in both the

[4] *See, e.g.*, Joel Trachtman, "Reports of the Death of Treaty Are Premature, but Customary International Law May Have Outlived Its Usefulness," *AJIL Unbound* (Apr. 29, 2014), *retrieved from* http://www.asil.org/blogs/reports-death-treaty-are-premature-customary-international-law-may-have-outlived; Joel P. Trachtman, "The Growing Obsolescence of Customary International Law" (in this volume).

international trade and investment law fields. Lord McNair,[5] for example, cited the British law officers for how they had favored the unconditional "most favored nation" (MFN) clause in commercial treaties, while John Jackson once described the MFN norm as Europe's "common commercial law" in the nineteenth century.[6] But these were commercial treaty clauses, and in no other field was the replacement of customary law by treaty rules – in this case, bilateral treaty rules – more celebrated than in the international investment law field.

In international investment law, a customary minimum standard of treatment (MST) norm had originated in nineteenth-century classical writings and the early twentieth-century jurisprudence. That norm became fiercely contested during the latter half of the twentieth century and in the twenty-first century, precisely because states and their legal advisors knew custom's role in international economic law to be of critical and functional importance. Beginning with Germany's 1959 bilateral investment treaty (BIT) with Pakistan, European nations, subsequently the United States, and finally the Asian states, too, turned to BITs because debates over the content of customary international law protection of foreign investment had descended into deep contestation throughout the 1960s through the 1980s. It became easy to forget the lurking, constant presence of custom, which is now beginning to reassert itself in new and surprising ways.

In more recent times, in *S.D. Myers v. Government of Canada*, for example,[7] the North American Free Trade Agreement (NAFTA) tribunal in that case famously cited Dr. F. A. Mann's view that the fair and equitable treatment (FET) standard is in fact the most basic expression of the nondiscrimination rule in the international investment law field, and that in that field at least it even underpins both the ideas of MFN treatment and National Treatment.[8] Some readers may recall that Mann himself believed this rule to be, unlike in the trade or commercial-treaty sphere, *customary* in nature. Mann had claimed that a plethora of bilateral treaties containing key rules such as the FET and full protection and security (FPS) rules had themselves formed customary international law.[9] He pointed to the similarity in the nature, content, and structure of the various examples of BIT clauses, based among other things on the blueprint of the 1959 Abs/Shawcross Draft.[10] In an early defense of the interrelationship between treaty and customary international

5 Lord McNair, *Law of Treaties* 278 (1961).

6 John Jackson, *World Trade and the Law of GATT* 250 (1969) (citing Rist).

7 40 I.L.M. 1048, paras. 265–66 (2001).

8 F. A. Mann, "British Treaties for the Promotion and Protection of Investments," 52 *Brit. Y.B. Int'l L.* 241, 243 (1981).

9 Mann, *supra* note 8, at 249. For a recent articulation of this view, *see* Patrick Dumberry, "Are BITs Representing the 'New' Customary International Law in International Investment Law?" 29 *Penn. St. Int'l L. Rev.* 675 (2010).

10 *See* Rudolph Dolzer & Christoph Schreuer, *Principles of International Investment Law* 8 (2d ed. 2012).

economic law rules, states, according to Mann, cannot seriously claim in good faith that their widespread participation in treaties containing a particular rule actually denies that rule's existence as custom.[11]

Today, the interrelationship of treaty and custom in the field of international investment law is a point of increasing, not decreasing, importance. We see a recurring need for treaty interpretation to turn to preexisting and constantly evolving customary rules. In fact, there are various ways by which the interpretation of a treaty rule might have to refer to the customary MST norm in international investment law, and there is now an extensive literature on the subject. This debate has a real, practical significance for the international law protection of property and contractual rights.[12] Even critics of Dr. Mann admit that "far from being obsolete, the question of the content of customary international law remains of fundamental importance" in international investment law.[13]

INTERTWINED CUSTOMARY AND TREATY PROTECTION OF PROPERTY AND CONTRACTUAL RIGHTS: THE PRACTICE OF INVESTMENT TRIBUNALS

Disputes over the standard of customary protection of alien property had formed the basis of the modern international law of state responsibility, and of international investment law. But if there is any field in which custom has since been widely considered to have fallen into "obsolescence," to use Joel Trachtman's term, it is in the field of the protection of foreign-owned property.

The modern story of controversy over custom's role in property protection is so well known that it needs, if at all, to be mentioned only briefly. The Russian Revolution led to uncompensated expropriations, and Mexico underwent a revolution and adopted the Latin American Calvo doctrine. These upheavals foreshadowed a century of intense challenges to the classical, Western view of the international law protection of alien property. Initially, the effects were slow to be

[11] Another example of the "intertwining" of treaty and custom was *Texaco v. Libya*, in which the arbitrator famously took the view that state participation in the Convention on the Settlement of Investment Disputes between States and Nationals of Other States (ICSID Convention) had contributed to a minimum international customary standard of compensation for expropriation. *Texaco Overseas Petroleum et al v. Libyan Arab Republic*, 17 I.L.M. 1, para. 89 (1978). Doctrine also tells us that a treaty may have codified preexisting custom, crystallized an emerging customary rule, or even influenced the subsequent adoption of a new customary rule. *See* North Sea Continental Shelf Cases (*FRG v. Denmark*, *FRG v. Netherlands*), 1969 ICJ Rep. 3, para. 71; *Nicaragua v. U.S.* (Merits), 1986 ICJ Rep. 14, paras. 175–78.

[12] *See, e.g.*, Margaret Clare Ryan, "*Glamis Gold, Ltd. v. The United States* and the Fair and Equitable Treatment Standard," 56 McGill L.J. 919 (2011).

[13] *See* Dumberry, *supra* note 9, at 700 (giving the examples of tribunals filling in treaty gaps in applying the customary law defense of necessity, customary rules on calculating damages, and the customary international law rules of state responsibility).

felt in the handling of international claims. In arbitral and judicial practice, the award in the 1922 *Norwegian Shipowners' Arbitration*[14] had continued to apply the standard of "just compensation." Likewise, in the 1926 *Chorzow Factory* judgment, the PCIJ ruled that reparation must, so far as possible, wipe out all the consequences of the illegal act. The famous *Lena Goldfields* arbitration in 1930 also upheld the classical view,[15] and in the 1933 *de Sabla* claim,[16] concerning the expropriation of a country estate in Panama, the U.S.-Panama General Claims Commission ruled that nonpayment of compensation or inadequate compensation was a violation of international law. Then there took place the famous exchange of correspondence in the late 1930s between the Mexican minister of foreign affairs and Secretary of State Hull, which led to the restatement by Secretary Hull of the classical formula that expropriations required for their lawfulness the payment of "prompt, effective and adequate" compensation.[17]

Besides an attempt at multilateral treaty regulation in the Havana Charter,[18] the debate subsequently came to a head during a period that roughly coincided with the UN decolonization era, when the USSR and third-world states had attempted to use the UN General Assembly to rewrite global economic rules.[19] Initially, there was compromise. General Assembly Resolution 1803, in 1962, had earlier rejected a Soviet proposal that would have precluded international law's application. The United States's view that international contracts "shall" be honored was also upheld. However, paragraph four of the resolution went on to provide for the exhaustion of local remedies as a precondition to making an international claim, while at the same time allowing access to international tribunals if the sovereign and "other parties concerned" agreed to international arbitration. Resolution 1803 also provided for "appropriate" compensation to be paid for the taking of alien property. As we know, Resolution 1803's delicate diplomatic balance did not hold. In 1974, with the passing of General Assembly Resolution 3281, third-world states had attempted to further revise the standard for the international protection of property rights by seeking the application of national standards of protection by

14 PCA, 13 October 1922.

15 *The Times* (Sept. 3, 1930) 13. *See* Arthur Nussbaum, "The Arbitration Between the Lena Goldfields, Ltd. and the Soviet Government," 36 CORNELL L.Q. 31 (1950).

16 *Marguerite de Joly de Sabla (United States) v. Panama*, VI R.I.A.A. 358 (1933).

17 ANDREAS LOWENFELD, *INTERNATIONAL ECONOMIC LAW* 470–478 (2d ed. 2008); C. L. Lim & Bryan Mercurio, "The Fragmented Disciplines of International Economic Law after the Global Financial and Economic Crisis: An Introduction, in *INTERNATIONAL ECONOMIC LAW AFTER THE GLOBAL CRISIS: A TALE OF FRAGMENTED DISCIPLINES* 17 et seq. (C. L. Lim & Bryan Mercurio eds., 2015).

18 Art. 11(1)(b).

19 *See, e.g.*, C. L. Lim, "Neither Sheep nor Peacocks: T.O. Elias and Postcolonial International Law," 21 *LEIDEN J. INT'L L.* 295, 302–10 (2008).

their own national courts.[20] Whether Resolution 3281 reflected a change in the customary international law standard of property protection became the central question in the *Texaco* arbitration in 1977.[21] The reality, however, as the late Professor Lowenfeld had explained, was that, between 1945 and 1970, assertions of right under international law were avoided in practice.[22] By 1964, in *Banco Nacional de Cuba v. Sabbatino*,[23] the U.S. Supreme Court, by 8-1, ruled that there was substantial disagreement over the customary international law rule. Instead, bilateral treaty rules came to replace debate over the precise content of customary rules.

The debate over the continued relevance of customary international law took a new turn, beginning with the 1990s, when the NAFTA – in Article 1105 – linked the FET and FPS treaty standards with the international, customary MST norm. Custom had re-emerged to assert itself in a new treaty-dominated era by intertwining itself with treaty-making. But before all that, there was the emergence of another, intermediate device for the protection of property rights – the garden-variety investment contract, foreshadowing a circuitous route for the maintenance of the classical standard of protection.

In the midst of a search for an alternative to vague custom, there had been a contemporaneous debate concerning the "internationalization" of concession agreements in the extractive industries. Such contracts would ordinarily have been considered mere domestic law-based contracts. First-echelon writers such as Lord McNair and Alfred Verdross mounted an attack against this view, Verdross being particularly remembered for his defense of "quasi-international" contracts in the Athens meeting of the Institut de Droit in 1979.[24] But it was McNair who – in the 1957 *British Yearbook of International Law* – had popularized the idea that concession contracts were governed by the general principles of law recognized by civilized nations (i.e., by international law).[25] Arthur von Mehren pitched in with an article in the *Harvard Law Review* along similar lines.[26] These were theoretical attempts to adapt the device of contract to the protection of investors' property rights under the rubric of international or transnational law. But these were merely invocations of the "low order sources" of international law.[27] Eventually, there were efforts to

[20] LOWENFELD, *supra* note 17, at 496–93.

[21] *Supra* note 11.

[22] LOWENFELD, *supra* note 17, at 483–85.

[23] 376 U.S. 398 (1964).

[24] Ignaz Seidl-Hohenveldern, "Recollections of Alfred Verdross," 5 *EUR. J. INT'L L.* 98, 102 (1994). Verdross had published his theory in 1964 in the *Yearbook of World Affairs*.

[25] A. McNair, "The General Principles of Law Recognised by Civilised Nations," 33 *BRIT. Y.B. INT'L L.* 1 (1957).

[26] *Cf.* Arthur von Mehren, "Special Substantive Rules for Multistate Problems," 88 *HARV. L. REV.* 347 (1974) (discussing derivation of special substantive rules from competing domestic law rules).

[27] M. Sornarajah, "Power and Justice: Third World Resistance and International Law," 10 *SINGAPORE Y.B. INT'L L.* 19, 30 (2006).

ground the legal validity of such contracts in custom instead – that is, a "higher" source of international law – through, as we have seen, paragraph eight of General Assembly Resolution 1803. Paragraph eight encapsulates, as noted, the U.S. proposal that foreign investment agreements "shall be observed in good faith" but was, at the same time, intended to reflect the general consent of states.

Subsequently, however, the Abs-Shawcross Draft went on to conceive of treaty protection of contracts in the form of so-called umbrella clauses.[28] Starting with Germany's 1959 bilateral investment treaty (BITs) with Pakistan, the preferred solution was to replace customary international law regulation altogether, and it appeared that BITs would and have done so in relation to the protection of both investors' contractual and property rights. In short, custom was widely perceived to have been an ineffectual lawmaking tool and as a source of behavioral rules. Contracts were "better," and eventually these contracts were themselves sought to be "internationalized" either through general principles of international law or custom and, eventually, by way of treaty umbrella clauses.

Thus, in the case of both the international law protection of property, and of contracts, treaty rules came to be the preferred source of protection either on their own or in combination with international investment contracts. Hermann Abs had originally envisaged a multilateral treaty – a global *magna carta* – for the international law protection of foreign investment. Eventually, we saw a proliferation of BITs and, in more recent times, investment chapters in FTAs. Nonetheless, customary international law technique and reasoning found a renewed role amidst the new treaty landscape.

There are, broadly, two scenarios. First, the challenge of designing the FET norm and the MST norm in international investment law caused states such as the parties to the NAFTA to fall back upon customary norms to supply the content of the intended treaty rule. Today, there are BITs and FTA investment treaty chapters such as NAFTA chapter 11, the 2004 Canadian and 2012 U.S. Model BITs, and the France-Argentina BIT, which state the relevant FET or Minimum Standard of Treatment (MST) rule in such a way, by referring to the customary rule. In effect, the treaty rule results in a *renvoi*.[29] Second, even where the treaty clause is unlike the

[28] The origin of the idea may be traced even earlier to Elihu Lauterpacht's advice to the Anglo-Iraninan Oil Company in connection with the dispute over Iranian nationalization, and to his subsequent, similar advice predating the Abs-Shawcross Draft, as well as to Art. IV(4) of the Abs-Doelle Draft, which was a precursor to the Abs-Shawcross Draft; *see, e.g.*, Thomas Wälde, "The 'Umbrella' (or Sanctity of Contract/Pacta Sunt Servanda) Clause in Investment Arbitration: A Comment on Original Intentions and Recent Cases," 1 *Transnat'l Disp. Management* 1, 32 (2004).

[29] The opposite approach toward custom, in the Rome Statute of the International Criminal Court, is discussed in Larissa van den Herik, "The Decline of Customary International Law as a Source of International Criminal Law" (in this volume).

first situation, and is silent on custom, tribunals have read the treaty rules in light of custom's requirements.

An illustration of the first situation mentioned is NAFTA Article 1105(1): "Each party grants to investors of another Party, treatment in accordance with international law, including fair and equitable treatment and full protection and security." In the NAFTA Free Trade Commission's (FTC's) Interpretation in 2001, which resulted from the NAFTA parties' alarm at the expansive interpretation given by tribunals to Article 1105(1), the FTC sought to clarify the scope of the FET and FPS standards in that clause by stating that Article 1105(1) prescribes the MST rule as it had evolved as a customary rule, and that the references to FET and FPS therefore do not provide protection "in addition to or beyond that" of the customary MST rule.

There followed allegations that the NAFTA parties had amended NAFTA in light of recent tribunal awards.[30] In any case, several tribunals subsequently challenged or circumvented the FTC's ruling. *Mondev v. USA* is one such case,[31] in which the tribunal ruled that the relevant customary international law standard was that in existence no earlier than the date NAFTA came into force. The apparent intention underlying that view was that the NAFTA FTC was therefore referring to a *contemporary* and highly developed international law standard that had been, and continued to be, shaped by more than 2,000 BITs – in other words, the customary, and therefore also the treaty standard that referred to it, had been evolving upward.[32] Likewise, in *ADF v. USA*,[33] the tribunal ruled that the customary standard is constantly in the process of development.[34]

Jason Yackee has argued that international investment law is therefore "embedded" in international custom. He argues, however, that because investment law "privileges its articulation" by nonstate actors (tribunals and international law scholars whose awards and writings constitute evidence of custom), scholars who appear as counsel and serve as arbitrators "are among the most important articulators of international custom," and that the awards themselves have a *de facto* precedential value. Through such evidence of custom, custom itself as a lawmaking device is not

[30] Charles N. Brower, "Why the FTC Notes of Interpretation Constitute a Partial Amendment of NAFTA Article 1105," 46 VA. J. INT'L L. 347 (2006).

[31] ICSID Case No. ARB(AF)/99/2, Award, 11 Oct. 2002.

[32] *Ibid.*, para. 125

[33] ICSID Case No. ARB(AF)/00/1, Final Award, 9 Jan. 2003.

[34] *Ibid.*, para. 179. *See, further*, the more recent Award in *Merrill & Ring Forestry LP v. Canada*, 48 I.L.M. 1038 (2010). For the seminal tribunal ruling that the customary MST rule had evolved to a higher standard over time, see *Pope & Talbot Inc v. Government of Canada*, (2000–2002), Award on Damages of 31 May 2002, 7 ICSID Reports 43, paras. 58–59. *Contra Glamis Gold, Ltd v. United States of America* (Award), 48 I.L.M. 1039, para. 615 (2009).

only alive but has become the principal formal source used for shaping the content of international investment law.[35]

Turning to the second situation, what if, unlike the "qualified" NAFTA treaty clause in Article 1105(1), there is no reference at all in the treaty to "international law" – that is, to customary international law? Some commentators balk at the prospect of applying the *Mondev* tribunal's view in such cases.[36] But in *Siemens v. Argentina*, the tribunal did precisely that. It held that there was "no reference to international law or to a minimum standard," but that "in applying the Treaty, the Tribunal is bound to find the meaning of these terms under international law bearing in mind their ordinary meaning, the evolution of international law and the specific context in which they are used."[37] This passage recalls Ian Brownlie's point that the plain meaning of some words we use even in everyday language find their meaning in the law.[38] Some treaty words find their plain meaning in customary international law – such treaty rules, albeit not "embedded" in, are inseparable from custom.

Recourse to custom in cases such as *Siemens* may be analyzed under Article 31(3)(c) of the Vienna Convention on the Law of Treaties (VCLT). That clause states that:

> 3. There shall be taken into account, together with the context:
>
> ...
>
> (c) any relevant rules of international law applicable in the relations between the parties.

Thus, in *Saluka Investments BV v. Czech Republic*,[39] the tribunal applied Article 31(3)(c) and referred to the customary rules of state responsibility in interpreting the meaning of the word "deprivation" in the treaty.

Furthermore, Article 31(3)(c) may be the only available tool to reconcile problems concerning the systemic interrelationship between treaty and customary norms,[40] a

35 Jason Webb Yackee, "Controlling the International Investment Law Agency," 53 *HARV. INT'L L.J.* 391, 428–29 (2012).

36 Patrick Dumberry, "The Emergence of a Consistent Case Law: How NAFTA Tribunals Have Interpreted the Fair and Equitable Treatment Standard," KLUWER ARBITRATION BLOG (Oct. 30, 2013), *retrieved from* http://kluwerarbitrationblog.com/blog/2013/10/30/the-emergence-of-a-consistent-case-law-how-nafta-tribunals-have-interpreted-the-fair-and-equitable-treatment-standard/.

37 ICSID Case No ARB/02/08, Award, 6 Feb 2007, para 291. *Contra Tecmed v. Mexico*, ICSID Case No. ARB(AF)/00/2, Award, 29 May 2003, para. 155 (concerning a claim brought under the Spain-Mexico BIT for the nonrenewal of a landfill license, where the tribunal had ruled that the treaty clause in the Spain-Mexico BIT was autonomous).

38 Ian Brownlie, "The Reality and Efficacy of International Law," 52 *BRIT. Y.B. INT'L L.* 1 (1981).

39 UNCITRAL Partial Award, 17 March 2006, paras. 254–94.

40 Philippe Sands, "Treaties, Custom and the Cross-fertilisation of International Law," 1 *YALE HUM. RTS. & DEV. L.J.* 85, 87 (1999).

view that has been adopted by the International Law Commission (ILC).[41] One recent argument by Martins Paparinskis is that,[42] ideally, custom's role under Article 31(3)(c) should be akin to that played by the context of the treaty – that is, that where no treaty reference is made to a customary rule, relevant customary rules should only serve to "contextualize" the treaty rule.[43] On this view, custom only plays a subordinate, confirmatory role.

But in *Saluka*,[44] the tribunal concluded that the treaty rule governing expropriation incorporated or "imported" the customary law definition of "deprivation." According to the ILC, treaty rules do employ terms the meaning of which are also defined by customary international law, even though "special regimes" might reflect *lex specialis* rules.[45] But, according to Paparinskis,[46] the customary rule can only serve to clarify the meaning of the treaty rule by comparing and contrasting the treaty rule with the customary rule. It cannot be used to limit the treaty rule, or to support a conflicting meaning.

The two views are nonetheless consistent in cases of "silence" on the part of the treaty. The true question – as Paparinskis views it – is the customary rule's proper "weight" once it is admitted. In some cases custom may simply be inadmissible, but in others, where it is admissible, its weight would vary from case to case.[47] There appears to be some support for this flexible view in tribunal pronouncements, dealing with a variety of treaty formulations that fall short of express importation of a customary rule.[48]

In a series of arbitrations involving the scope of Argentina's non-precluded measures clauses in its BITs, a number of tribunals concluded that the treaty

41 Report of the Study Group on Fragmentation of international law: Difficulties arising from the diversification and expansion of international law 2006, A/CN,4/L.702, 18 July 2006, 13 et seq.

42 Martins Paparinskis, "Investment Treaty Interpretation and Customary Investment Law: Preliminary Remarks," in EVOLUTION IN INVESTMENT TREATY LAW AND ARBITRATION 73, 89 (Chester Brown & Kate Miles eds., 2011).

43 *See also* Article 31(1) ("A treaty shall be interpreted in good faith in accordance with the ordinary meaning to be given to the terms of the treaty in their context and in the light of its object and purpose.").

44 *Supra* note 39, at para. 254.

45 *Supra* note 41, at 12, 15.

46 Paparinskis, *supra* note 42, at 89.

47 Paparinskis, *supra* note 42, at 89.

48 *Sempra v. Argentina*, ICSID Case No. ARB/02/16, Award, 28 September 2007, para 302 (stating that much may still depend upon which rule is clearer or more specific); *Azurix v. Argentina*, ICSID Case No. ARB/01/12, Award, 14 July 2006, para. 361 (which, in fact, involved an express treaty reference to "international law," stating that the treaty's reference to international law makes no difference because the meaning of FET and FPS is substantially similar to the treaty standard, whereas the intent of such treaty language is to ensure that the treaty standard does not fall below – i.e., conflict with – the treaty standard); *Compañiá de Aguas del Aconquija S.A. and Vivendi Universal S.A. v. Argentine Republic*, ICSID Case No. ARB/27/3, Award, 20 August 2006, para. 7.4.7 (where the tribunal construed a treaty reference to "principles of international law" did not simply refer to the customary MST norm).

clauses were to be limited in their scope by the narrower customary international law defense of necessity, because they were inseparable from it.[49] However, such rulings, which imposed more stringent customary international law requirements, have not survived annulment and criticism under ICSID's annulment procedure. In *Sempra v. Argentina*, for example, the earlier ruling of the tribunal was annulled.[50] The annulment committee held that the treaty clause did not require that the Argentinian policy response need have presented the only available option in order to be justified. The annulment committee therefore allowed the treaty clause to operate autonomously.[51]

In all these examples, the success of third-party dispute settlement – which features no more strongly than in the organizationally based, institutionalized domains of international economic law – appears to correlate with the evermore imaginative application of customary international law in the new treaty contexts. A simple explanation for this may be that our treaty regimes are not as self-contained as we would think, and there is no reason for a tribunal or court to limit its resort to all the sources of international law, notwithstanding current views about the functional inutility of custom as a lawmaking tool.[52]

PROTECTING COMMERCIAL EXPECTATIONS: THE RESTRICTIVE IMMUNITY DOCTRINE

States themselves have also engaged in detailed, highly intricate customary international economic law creation, through their legislatures and domestic courts, in the legal enforcement of contractual and property rights. The restrictive doctrine of immunity – according to which a sovereign that behaves like a merchant is treated as such – is customary international law. It is fashioned not only by treaty lawmaking

[49] Under the customary law view, in order to invoke necessity, the state's actions that form the basis for claiming necessity as a defense must have been the *only* available course of action which that state could have taken. The disputes concerned Argentina's pesoization or pesofication policy as a response to the Argentine crisis. *See CMS Gas Transmission Co v. Argentina*, ICSID Case No. ARB/01/8, Award, 12 May 2008, paras. 354–55; *Enron Corp. v. Argentina*, ICSID Case No. ARB/01/3, Award, 22 May 2007, paras. 395, 306, 308, and particularly 333; *Sempra v. Argentina*, 2007, ICSID Case No. ARB/02/16, Award, 28 September 200, paras. 376–77.

[50] *Sempra v. Argentina*, 2007, ICSID Case No. ARB/02/16, Annulment Proceeding, 29 June 2010, para. 208.

[51] The Annulment Committee in *Enron Corp. v. Argentina* (ICSID Case No. ARB/01/3, Annulment Proceeding, 30 July 2010) reached the same result on the basis that the customary standard had not been correctly applied in the first place, on which the tribunal's conclusion that Art. XI of the treaty did not apply had rested; paras. 378, 495; *see, further*, Daniel R. Kalderimis, "Investment Treaties and Public Goods," in MULTILATERALISM AND REGIONALISM IN GLOBAL ECONOMIC GOVERNANCE: TRADE, INVESTMENT AND FINANCE 149 (J. Nakagawa ed., 2012), and more generally, 147–49.

[52] *See* Campbell McLachlan, "The Principle of Systemic Integration and Article 31(3)(c) of the Vienna Convention," 54 INT'L & COMP. L.Q. 279 (2005).

(e.g., the European Convention on State Immunity), but also by domestic, often democratically enacted, legislation as well as domestic judicial decisions.[53] Various national laws are framed with such a high level of precision and detail that the differences between national laws are differences of very fine detail. Professor Damrosch, for example, has argued that changes in national law inject a progressive element into *customary* law-creation.[54]

Here lies an example of how custom – as evidence of a general practice accepted as law – copes with the need for individual tailoring or close customization, a characteristic that – as we shall later see – is sometimes denied by custom's critics. Indeed, custom is so highly susceptible to detailed customization that, in the example of state immunity, not only is there a secondary level of detailed differences between national legislation and regional treaties that subscribe to the restrictive doctrine, there is also a primary level of differentiation between many Western, industrialized states that adopt that doctrine and others, such as China, which do not. The exact details of such differences are often fleshed out in the close reasoning of national courts.[55]

This example shows that custom's continued salience lies not just in the adjudicative contexts of international economic law, but also in some areas of international economic law where custom continues to be the predominant source of international law. It also responds to the sweeping charge that custom's vagueness makes it obsolete, particularly in fields such as the international economic field. The state immunity example also demonstrates that detailed rules may result not only from direct cooperation, but also functional need.[56]

INTEGRATING PREEXISTING CUSTOM WITH INTERNATIONAL ECONOMIC TREATY REGIMES

A third and final illustration of custom's role in international economic law, which is distinct from the adjudicative and legislative illustrations described, lies in a "systemic" need to integrate existing customary rules with treaty regimes that have encroached upon the space previously occupied by custom.

53 Lori Fisler Damrosch, "Changing the International Law of Sovereign Immunity Through National Decisions," 44 VAND. J. TRANSNT'L L. 1185 (2011).

54 Damrosch, *supra* note 53, at 1192–96.

55 *See, e.g.*, *Republic of Argentina v. NML Capital, Ltd.*, 134 S. Ct. 2250 (2014) in the United States; and the Hong Kong Court of Appeal's ruling (later reversed on appeal) in *FG Hemisphere Associates LLC v. Democratic Republic of Congo* [2010] 2 HKLRD 66.

56 David J. Bederman, "Acquiescence, Objection and the Death of Customary International Law," 21 DUKE J. COMP. & INT'L L. 31, 40 (2010).

This is a systemic issue. Where significant portions of international economic law have been treaty based, they have sometimes conveyed the impression of being perfectly self-contained. This may be the result of the very high level of specificity of the treaty rules in question. But true self-containment has only been a polite fiction where tribunals have the mandate to apply the full range of international law sources. Because specific customary rules that preexist the creation of treaty regimes may have simply been replaced, or impinged upon, by new treaty rules in a less than systemically integrated fashion, the relationship between treaty and custom may not have been fully worked out. The problem has come to the fore especially with increased rulemaking and adjudication, and the growth of global administrative regulation.[57]

Recall Article 31(3)(c) of the VCLT, which, according to Philippe Sands, "appears to be the only tool available"[58] with "a potentially generic application" in resolving such issues of imperfect systemic integration.[59] This has been the experience of commentators on the General Agreement on Tariffs and Trade-World Trade Organisation (GATT-WTO) system who question whether the requirement of the Sanitary and Phyto-Sanitary Measures (SPS) Agreement that SPS measures must be taken on the basis of a risk assessment allows for the operation of a customary precautionary principle.[60] Such a role for custom has never been accepted by WTO panels and the appellate body. However, in the recent *Rare Earths* case as well as in the earlier *Raw Materials* case, concerning China's WTO obligation not to impose duties and other restrictions on various exports,[61] China argued that notwithstanding its GATT-WTO obligations, it possesses a customary right to permanent sovereignty over natural resources. The existence of such a customary right was not denied by the relevant panels in those cases. Instead, the panels considered that China had curtailed its sovereign rights by its subsequent treaty commitments regarding the imposition of export restrictions.[62] These examples recall similar attempts to resort to a

[57] Sands, *supra* note 40, at 85.

[58] *Contra US-Gasoline*, in which the appellate body hung the need to interpret the treaty rules of the GATT against the broader framework of international law on the preamble to the WTO agreement.

[59] Sands, *supra* note 40, at 87.

[60] WT/DS26/AB/R, 16 January 1998.

[61] C. L. Lim and J. H. Senduk, "You Don't Miss Your Water 'Til Your River Runs Dry: Regulating Industrial Supply Shortages After 'China Raw Materials,'" 18 STAN. J.L. BUS. & FIN. 72 (2012); C. L. Lim and J. H. Senduk, "The Wages of Belonging: Rare Earths from China and the Return of GATT à la carte," 9 GLOBAL TRADE & CUSTOMS JO 380 (2014).

[62] The applicability of that customary doctrine alongside WTO treaty obligations was acknowledged by the panels in the earlier *Raw Materials* case, although China's argument was ultimately rejected under a proper interpretation of China's treaty commitments; see WT/DS394/R, WT/DS395/R, WT/DS/398/R, 5 July 2011, paras. 6.35, 7.157, 7.265, 7.356, 7.381–87.383, 7.407. The appellate body does not appear to have responded to these arguments explicitly. *See, also*, Sonia E. Rolland, "China-Raw Materials: WTO Rules on Chinese Natural resources Export Dispute," 16:21 *ASIL INSIGHTS* (Jun. 19,

preexisting customary MST norm in the interpretation and application of FET treaty clauses in the investment field, and – as with the *Beef Hormones* case in trade law – where commentators on international investment law have also asked if investment treaties should take a customary precautionary principle into account.[63]

The WTO appellate body has thus far resisted the application of the precautionary principle, and in the case of investment law, there remains only stern debate about how investment norms might accommodate non-investment (e.g., environmental) norms without the need for treaty revision. Such issues exist nonetheless, and will continue to arise, regarding the precise interrelation between trade and investment treaties and preexisting customary rules. They include questions about the relationship of treaty trade rules with customary rules of both a trade and "non-trade" nature, and about the relationship between investment treaty rules and both investment and "non-investment"-related customary rules. Such questions reflect the dynamic needs of global international governance where treaty regimes are imperfect because they are and will always be incomprehensive. As for why WTO panels and the appellate body have been slow to employ custom,[64] part of that explanation lies in the way the WTO Dispute Settlement Understanding frames the relevant legal sources for the settlement of trade disputes.[65] Another part of the explanation lies in the fact that what investment tribunals and trade panels do has been disjointed because of the fragmentation of the subfields of international economic law.[66] A further part of that explanation may have to do with how trade dispute settlement grew out of a diplomatic process, where trade diplomats traditionally stuck to the express rules of the GATT text.[67] The restrictive

2012). *See, further*, the report of the panels in the more recent Rare Earths dispute; WT/DS431/R, WT/DS432/R, WT/DS433/R, 26 March 2014, para. 7.457 et seq.

63 C. E. Foster, "Adjudication, Arbitration and the Turn to Public Law 'Standards of Review': Putting the Precautionary Principle in the Crucible," 3 *J. Int'l Dispute Settlement* 525, 529, 532–34, 538–44 (2012); cited in Gus Van Harten, *Sovereign Choices and Sovereign Constraints: Judicial Restraint in Investment Treaty Arbitration* 158 (2013).

64 *See* Tomer Broude, *Nothing to Declare: The Silence of Custom in International Trade Law* (manuscript, on file).

65 WTO Dispute Settlement Understanding (DSU), art. 3(2), which refers only to the WTO's "covered agreements" and the "customary rules of interpretation of public international law." *See further* Joost Pauwelyn, *Conflict of Norms in Public International Law: How WTO Law Relates to Other Rules of International Law* 460 et seq. (2003); Henry Gao & C. L. Lim, "Saving the WTO from the Risk of Irrelevance," 11 *Jo Int'l Econ L* 899, 911 *et seq.* (2008).

66 *See, e.g.*, R. P. Alford, "The Convergence of International Trade and Investment Arbitration," 12 *Santa Clara JIL* 35, 44 (2013); R. Howse & E. Chalamish, "The Use and Abuse of WTO Law in Investor-State Arbitration: A Reply to Jürgen Kurtz," 20 *Eur. J. Int'l L.* 1087, 1094 (2009); Jürgen Kurtz, "The Use and Abuse of WTO Law in Investor-State Arbitration: Competition and its Discontents," 20 *Eur. J. Int'l L.* 749 (2009).

67 Michael K. Young, "Dispute Resolution in the Uruguay Round: Lawyers Triumph over Diplomats," 29 *International Lawyer* 389 (1995).

approach taken by trade dispute settlement panels toward the sources of international law has long been remarked upon, and may yet change.

It is hard to imagine custom simply falling away as a source of international law. Some of the criticisms in this volume have relied only on inquiries into fact and function. Perhaps the question is framed too narrowly as a simple factual question by asking if tribunals have actually acknowledged custom's relevance; for example, in the WTO dispute settlement context.[68] That approach risks saying that if custom's function has not been acknowledged in a particular tribunal's practice by now, it therefore has no function at all. At the same time, the "obsolescence of custom" theory is a product of legal pragmatism. It lies at the opposite end of the jurisprudential spectrum and tends to ignore the increasing significance of doctrinal debates over the role of custom in international adjudication and third-party settlement.[69]

I now turn to Joel Trachtman's recent restatement of the "obsolescence" theory.[70]

THE OBSOLESCENCE THEORY AND "INTERTWINATION"

According to Trachtman:[71]

(1) Custom operates *ex post*, and cannot be made in a tightly coordinated manner in advance of events.[72]
(2) Custom is vague and it cannot be made with sufficient detail. That would require, as Wolfgang Friedmann had once put it, "more articulate and specific instruments of lawmaking."[73]

[68] *See* Tomer Broude, *Nothing to Declare: The Silence of Custom in International Trade Law* (manuscript, on file).

[69] There is, admittedly, also a prescriptive element in advocating greater resort to custom in adjudicating disputes under treaty-based regimes. For example, Sands has argued that the investment tribunal in *Compania del Desarrollo de Santa Elena SA v. Republic of Costa Rica* ought to have taken environmental principles into account when determining full and fair market value. *See* Sands, "Litigating Environmental Disputes: Courts, Tribunals and the Progressive Development of International Environmental Law", OECD Global Forum on International Investment, 27–28 March 2008, at 9.

[70] Trachtman, "The Growing Obsolescence of Customary International Law" (in this volume).

[71] *See* Trachtman, *Reports of the Death of Treaty Are Premature*, *supra* note 4; Trachtman, The Growing Obsolescence of Customary International Law" (in this volume).

[72] *See also* John K. Setear, "Treaties, Custom, Iteration, and Public Choice," 5 CHI. J. INT'L L. 715, 718–22 (2004); Jack L. Goldsmith & Eric A. Posner, "Theory of Customary International Law," 66 *UNIV OF CHI L REV* 1113, 1129–31, 1170–72 (1999); JACK L. GOLDSMITH & ERIC A. POSNER, *THE LIMITS OF INTERNATIONAL LAW* 35–38 (2005).

[73] *See also* WOLFGANG FRIEDMANN, *THE CHANGING STRUCTURE OF INTERNATIONAL LAW at* 122–23 (1964); ERIC A. POSNER & ALAN O. SYKES, *ECONOMIC FOUNDATIONS OF INTERNATIONAL LAW at* 63–64 (2013).

(3) Custom results from "in-kind reciprocity" and cannot be made with sufficiently heterogeneous tradeoffs between states.[74]
(4) Custom lacks bespoke organizational support.[75]

In the present volume, Trachtman advances a number of additional arguments about how custom lacks both democratic legitimacy (i.e., legislative oversight) and norm legitimacy – which he defines as the absence of support from consenting states and the old problem of "auto interpretation."

What the obsolescence theory understates is the fact that customary lawmaking and treaty lawmaking are so intertwined today that the rise of modern treaty lawmaking (which is said to signal the death of custom) is itself helping to overcome custom's traditional weaknesses. Where state practice today – particularly in the international economic field – occurs against a backdrop of treaties, treaty making, and treaty application by adjudicative bodies, it has become more difficult to distinguish purely customary features that might support claims of custom's efficiency in accomplishing what states want. The features of customary and treaty lawmaking are now common where the making and application of customary and treaty rules have become intertwined and practically – though not doctrinally – identical processes, as Professor Alvarez has explained.[76] Oversight by national parliaments of treaty negotiations therefore also becomes oversight of a parallel, intertwined customary lawmaking process. At the same time, the interpretation of resultant treaty rules has required institutionalized bodies and tribunals in the international economic law field to address the content of parallel customary rules.[77]

Let us first address advanced coordination. Trachtman's scenario largely assumes non-intertwined treaty and customary lawmaking. His is a study primarily of custom as such. He poses the example of achieving a carbon emissions reduction rule. According to Trachtman, in such a scenario, the treaty negotiators sit eyeball to eyeball, negotiating simultaneously: "Each state's best outcome is to abstain from agreement, while others form a stable coalition that will generate the relevant public good. The second-best outcome is to adhere while others adhere. The worst outcome is if no state adheres."[78]

There are, Tracthman explains, two Nash equilibria: "Object when others adhere, and adhere when others object." According to Trachtman, treaties can get around this chicken-game-type coordination problem by having treaty-signing conferences,

[74] *See further* J. Patrick Kelly, "The Twilight of Customary International Law," 40 *Virginia Jo of Int'l L* 449, 532–33 (2000); John O. McGinnis, "The Appropriate Hierarchy of Global Multilateralism and Customary International Law: The Example of the WTO," 44 *Va. J. Int'l L.* 229, 240 (2003).
[75] *See further* McGinnis, *supra* note 74, at 249, 265.
[76] Jose Alvarez, *International Organizations as Law-Makers* 387, 390 (2005).
[77] And thus auto-interpretation of customary rules has also been kept in check.
[78] Trachtman, "The Growing Obsolescence of Customary International Law" (in this volume).

for example, where each state signs simultaneously with the others. In contrast, custom does not form in that manner. Although the sight of others veering away from an emergent rule may be a call to action, and mirrors the chicken game in the treaty context, it is harder to imagine why the first state would have borne the cost of initiating the practice in the first place.[79]

Trachtman's example is largely about providing global public goods, and their associated costs, although his general point is a broader one. Professors Goldsmith and Posner have explained the broader point in the following way, which is that custom requires "the information and monitoring structures needed to overcome a multilateral prisoners' dilemma." In the treaty context, this is achieved "through formal negotiation and specification of what counts as cooperation."[80] States then solve collective action problems by being able to monitor and punish errant conduct, relying on information usually unavailable in the contexts of customary law formation.[81]

Nonetheless, parallel customary lawmaking could mirror the moves made in the multilateral treaty-making context, parasitically overcoming the first-mover and chicken-game problems. It may be objected that there is no perfect temporal coincidence since the point at which custom forms is unclear. While that may be true, Goldsmith and Posner acknowledge that the multilateral treaty-making context or some similar "high profile international event" – one thinks of a General Assembly debate – "might supply a focal point that would be the basis for multilateral coordination."[82] Moreover, one might add that the moment at which states might consciously seek to influence the content of parallel custom is at least clear to them and to their legal advisors.[83]

In short, the very fact that a multilateral lawmaking conference can generate both treaty rules and evidence of customary rules, or may crystallize customary rules, makes it difficult to draw easy distinctions between primitive customary law formation and sophisticated treaty making. This was Friedrich Kratochwil's insight, which predated the latest behavioral line of attack on custom,[84] where

79 Trachtman, "The Growing Obsolescence of Customary International Law" (in this volume).

80 Goldsmith & Posner, *Limits of International Law*, *supra* note 72, at 35–38.

81 Goldsmith & Posner, *Limits of International Law*, *supra* note 72, at 86.

82 Goldsmith & Posner, *Limits of International Law*, *supra* note 72, at 38.

83 Take NAFTA chapter 11 as an example, in which the FET and FPS rules were equated with the MST standard. Its confirmation that the treaty rule granting FET and FPS adheres to the MST standard was a fudge, but it was also a conscious accomplishment. In the NAFTA example, at least one tribunal has ruled expressly that the specific parallel customary rule that the NAFTA parties claimed to exist was that formed no earlier than when NAFTA came into force. *Mondev v. USA*, ICSID Case No. ARB(AF)/99/2, Award, 11 Oct. 2002, para. 125.

84 Friedrich V. Kratochwil, *Rules, Norms and Decisions: On the Conditions of Practical and Legal Reasoning in International relations and Domestic Affairs* 90, 250–251 (1991).

he sought to defend custom against similar claims about its obsolescence at the time.[85]

Another assumption the obsolescence theory makes is that states engaged in making new customary rules do not "converse," and therefore do not engage in any prior coordinated action that could lay the groundwork for sophisticated lawmaking. This assumption was long ago questioned by the English scholar Bin Cheng, who instead saw customary lawmaking as, potentially, a "conscious, open and deliberate" process.[86] According to Cheng, states can "discuss" customary international law formation and, if that is so, customary international lawmaking need not be the result of the mechanical and unintelligent operation of a hidden, unconscious machine. It is interesting in this regard to observe, as will be discussed, that Professors Helfer and Wuerth, as with Professor Kelly, similarly attribute custom's shortcomings to this alleged "uncommunicative" or "inarticulate" aspect of customary law formation. But as Alvarez has also observed, in a similar vein to Cheng, intertwined treaty and customary lawmaking in "universalistic fora" means that customary lawmaking, too, "is as much an outcome of a conscious, result-oriented negotiation process."[87] Finally, in contrast to Trachtman's example about providing public goods, we can think of other examples of customary lawmaking such as the Truman Proclamation, where the United States claimed a new right (to the seabed and subsoil resources of its continental shelf), and other states duly followed suit over time because it was straightforwardly in their interests to do so.

Trachtman's second criticism is that custom lacks specificity.[88] He frames its practical consequence largely as a rule predictability issue – treaties are simply a better guide to conduct *ex ante*, and provide greater predictability when contemplating litigation.[89] Customary norms reflect more general and therefore vague standards. This general juxtaposition of the perceived nature of treaty and customary rules is problematic. Much ought to depend on the content of the rule itself, and its circumstances. Admittedly, the treaty text may be more readily available, and the exact formulation of the treaty rule is at least stated in an authoritative way by virtue of its location in the treaty text. But neither of these factors means that a treaty rule is always more detailed or clearer. In the case where the treaty rule in a treaty text incorporates a customary rule in order to supply the content of the

85 Other than Friedman's work, discussed further, *see, e.g.*, Bruno Simma, "Consent: Strains in the Treaty System," in *The Structure and Process of International Law* 485 (R. St. J. Macdonald & Douglas M. Johnston eds., 1983).

86 Bin Cheng, "Custom: The Future of General State practice in a Divided World," in *Structure and Process*, *supra* note 85, at 535–36.

87 Alvarez, *International Organizations*, *supra* note 76, at 388.

88 The most well-known statement of the problem is probably Friedmann, *Changing Structure*, *supra* note 73, at 122–23.

89 Trachtman, "The Growing Obsolescence of Customary International Law" (in this volume).

treaty rule, custom's lack of clarity would be precisely the same as that in the treaty. This is shown in the many NAFTA cases in which the content of the customary minimum standard of treatment has been controversial in the various attempts to interpret NAFTA Article 1105. But in case it might be thought that this rather proves Trachtman's point – that recourse to custom has had a negative effect on treaty clarity – cases in which the FET clause in BITS do *not* incorporate custom have arguably been *even less* clear.[90] In yet other – "systemic" – cases, it is because it is unclear in the first place that a treaty rule actually applies autonomously of preexisting customary rules that account must be taken of preexisting customary rules in the interpretation of the scope and application of the treaty rule itself (under Article 31[3][c] of the VCLT).

Helfer and Wuerth have argued that custom's vagueness and generality are because custom "is produced through a decentralized, largely non-negotiated process." Compared to treaties and some soft law norms, custom – being largely non-negotiated – "does not provide rules that can be tailored such as those dealing with exit, conditional acceptance, monitoring or compliance."[91] We have seen this criticism before. It ignores the detailed and highly convergent manner in which states have approached the doctrine of restrictive foreign state immunity. It also underestimates the extent to which custom is no longer capable of being addressed in isolation from its complex interaction with treaty rules and organizational settings. And it fails to take account of the fact, as Bin Cheng had observed, that custom may still be created through conscious, deliberate discussion between states.

In his third criticism, Joel Trachtman argues that customary lawmaking depends upon "in kind" forms of barter or reciprocity, presumably because customary law formation is seen to consist of repeated state practice in comparison with "bargained for" treaties,[92] whereas treaty lawmaking allows for more complex and diverse trade-offs – for example, carbon emissions reductions can be traded for a lowering of computer tariffs. This criticism, however, should not be overgeneralized.

Returning to the example of the Truman Proclamation,[93] there, a unilateral act eventually led the great preponderance of states to adopt the United States's position. While it took two multilateral conferences to finally establish the rule, the Third United Nations Conference on the Law of the Sea (UNCLOS III) is a famous

90 See, further, on the uncertain benefits of "treatification" in the international investment law field, that international tribunals cannot dispense with custom when interpreting treaties and that, as Andreas Lowenfeld had once cautioned, it is more important to have recourse to adjudication, arbitration and other legitimate dispute resolution processes than the illusion of treaty certainty; Alvarez, "A BIT on Custom," *supra* note 2, at 72–73.

91 Laurence R. Helfer & Ingrid Wuerth, "Custom in the Age of Soft Law" (working paper), at 4.

92 Helfer & Wuerth, *supra* note 91, at 6, 8–12.

93 To be clear and fair, Trachtman is *not* talking about this situation – his point is about how lawmaking can provide public goods.

example of the use of the "package deal" approach and codified *new* custom. The continental shelf regime was, for all appearances, a product of simple, in-kind barter. But there was "diverse performance reciprocity," as Trachtman calls it, in other areas of ocean regulation. The diverse interests of coastal states and global powers were accommodated by trading a twelve-mile territorial sea for rights of transit passage – that is, a case of "simple diverse performance reciprocity" at play. Complex diverse performance reciprocity is shown instead in the international economic law field; specifically, in the creation of the Exclusive Economic Zone (EEZ) regime. In the case of the EEZ, humanity's interest in conservation, the need of coastal states for an abundant, affordable source of protein, the interests of nations with distant fishing fleets, and the interests of landlocked and geographically disadvantaged states were *all* reconciled. It is true that a treaty conference was needed, but it is also true that a new customary regime was created and the old order of the oceans perished into history.[94]

Trachtman acknowledges that it does not even take a treaty conference to do all of that. Similar trade-offs occurred in the drafting of resolutions of General Assembly Resolution 1803, to use our previous example, in which the doctrine of permanent sovereignty was traded for the binding nature of international contracts, while the exhaustion of local remedies was traded off acceptance of the binding nature of agreements to arbitrate or adjudicate disputes.[95]

True, it took a treaty conference to do what UNCLOS did, and General Assembly resolutions are neither treaty nor custom of themselves. The contention here is, simply, that to the extent that the absence of diverse performance reciprocity is also attributed to the uncommunicative formation of custom,[96] here are examples of instances in which customary rule formation was part of a highly communicative process even though bound up with treaty making and conference diplomacy.[97] There is no apparent reason to think that such continued use of multilateral treaty-making conferences and other multilateral, institutional fora such as the General Assembly will cease. The apparent reluctance on the part of states such

94 Professor Trachtman has questioned this oceanic international economic law example, in our correspondence, as being an example of the last gasp of customary international law. But the practical importance of existing parallel custom could ebb and flow. A recent example involves states exiting from the ICSID Convention. Assuming that past participation in ICSID has contributed toward the crystallization of a customary rule – as the arbitrator in *Texaco v. Libya* ruled – and that investment contract disputes may be submitted to international arbitration by agreement, how, now, would an exiting state propose to contract out of that customary rule?

95 See UNGA res 1803, paras. 4 and 8.

96 Kelly, *supra* note 74, at 532.

97 Alvarez sees negotiations in such "universalistic fora" to be less "vague" and more "structured" than the making of non-treaty-intertwined custom in the past; ALVAREZ, *INTERNATIONAL ORGANIZATIONS*, *supra* note 76, at 387.

as China to accept custom that is *not* formed by such deliberate discussion is noteworthy in this respect.[98]

We can imagine an even more radical argument than Trachtman's – that, where treaty parties have traded off different things in order to reach agreement, a customary rule is thereby precluded from being formed.[99] But it is difficult to see how this can be said with certainty. Much will depend upon the facts of each case. Nothing turns on the motivations of states, as opposed to finding the usual elements of custom – state practice and *opinio juris*. At most, it might be possible to say that where a state binds itself to a treaty rule *only* because of a diverse reciprocal trade-off, it does not therefore evince a sense of legal obligation for the purposes of forming a parallel customary international law rule.

Trachtman's final functional criticism is that treaties will be required to create this kind of highly customized international organizational support in the first place. But while this may be true of the UN Conferences on the Law of the Sea and the UN General Assembly, the San Francisco Conference, which created the United Nations Charter and Organization, was more akin to the Philadelphia Convention that drafted the U.S. Constitution. In the international economic law field, one thinks of such informal devices as the Basel Committee (albeit housed for convenience in the Bank for International Settlements), the various "Gs," including the G20, and Asia-Pacific Economic Cooperation forum.

The more difficult question is why these non-treaty-based fora, which exhibit some of the organizational characteristics of treaty-based organizational support, have generated soft law but not, apparently, custom.[100] Various answers have been suggested – ranging from the growing irrelevance of hard law, which is now being displaced by soft law,[101] to Tim Meyer's argument that it is precisely those similarities between custom and soft norms – for example, low contracting costs and the avoidance of politically charged domestic debates over ratification – that have enabled soft law to replace custom to the extent that soft law has become "the new custom."[102]

This is an untidy comparison. First, nonlegal organizations do not ordinarily set out to make law, even if some forms of soft law can be made as easily as the rules of a club. In some cases, their processes and procedures may even be unique, such as,

[98] Pan Junwu, "Chinese Philosophy and International Law," 1 *Asian Jo Int'l L* 233, 241–42 (2011), citing Wang Tieya, "International Law in China: Historical and Contemporary Perspectives," 1 *Chinese Yrbk Int'l L* 1 (1991).

[99] McGinnis at one juncture seems to come close to saying this; McGinnis, *supra* note 74, at 240.

[100] Helfer & Wuerth, *supra* note 91.

[101] This is where Trachtman's defence of the law of treaties is situated.

[102] Timothy Meyer, "Codifying Custom," 160 *U. Pa. L. Rev.* 995 (2012); Curtis A. Bradley & Mitu Gulati, "Customary International Law and Withdrawal Rights in an Age of Treaties," 21 *Duke J. Comp. & Int'l L.* 1 (2010).

for example, the interaction between the Basel Committee's processes and national administrative processes. Attention naturally turns to the possibility of new legal forms, such as soft laws, but without, I think, the need to reject more traditional conceptions of law.[103] In contrast, the organs of legally based or legally constituted organizations tend, as was the case with the UN General Assembly, to become consciously used or adapted toward shaping general, customary law as well as various UN-sponsored treaties in the international economic field (such as the New York Convention on the Enforcement of International Arbitral Awards, and the UN Convention on the International Sale of Goods).

Secondly, saying custom has been displaced by soft law probably goes too far – one view is that "though the Basel Committee does not operate as a legislative body with formalized law-making power, it nonetheless creates customary international law because most States act in concert with its rules and feel a perceived legal obligation or requirement to do so."[104] It may be questioned whether that perception of a legal obligation actually exists, but the more basic, general point here is that you do not have to set out to make custom to do so. As with treaty bodies that "co-generate" parallel custom, soft law bodies, too, can generate custom insofar as the normal requirements for doing so have been met.

Turning, finally, to Trachtman's criticism in this volume of custom's "democratic" shortcomings, I have already mentioned that where custom is made in parallel with treaty lawmaking, legislative oversight of the latter amounts to oversight of the former. But Trachtman is also concerned about legitimate legal rule formation on the international plane. He argues that we cannot proclaim the existence of new customary rules without grounding them in states' actual or presumed consent. If so, either the consent of all states is required, or if not that would simply amount to majoritarianism. According to Trachtman, treaty lawmaking avoids this "holdout" problem.[105] Put simply, treaties are easier to make because they consist of contractual rules, not general rules. As contracts that apply only to the parties, their legitimacy is not called into question, not, at least, by those to whom they would apply.

Whatever its general merits, Trachtman's claim of illegitimacy does not apply so easily in the international economic field where famous controversy had ensued over the very legitimacy of the world economic order. Third-world legal scholars urged their nations to shape custom in the UN General Assembly during the 1960s and 1970s while developed, capital-exporting nations had turned toward bilateral,

[103] The debate, in the case of Basel-generated norms governing international cooperation in banking supervision is ongoing. *See* Michael S. Barr & Geoffrey P. Miller, "Global Administrative Law: The View from Basel," 17 *Eur. J. Int'l L.* 15 (2006).

[104] David A. Curtiss, "The Interaction Between the Basel Committee in International Banking Regulation and the Opening of the Chinese Banking Market," (2009) *Bus. L. Rev.* 6.

[105] *See also* McGinnis, *supra* note 74, at 250 (drawing a linkage with the problem of diverse performance reciprocity, addressed earlier).

"contractual" treaty making in safeguarding the contractual and property rights of foreign investors. It is important to notice that for such third-world scholars, making new customary rules was a democratic tool, one which was to be used to counter undemocratic attempts to internationalize state contracts, reactionary arbitral awards, and the writings of publicists who were too closely allied to the demands of international capital.[106] For them, a neoconservative global agenda had succeeded instead on the back of treaties and treaty regimes.[107] The attempt to fashion new customary rules, indeed a customary "New International Economic Order," had failed, but would have been more "democratic" precisely because forming custom through organs such as the UN General Assembly required the participation, deliberation, and consent of third-world countries where their strength in numbers – their democratic vote – would count for something more, even more than the "contractual" arrangements that by then were emerging in the form of BITs between industrialized and developing nations.[108] The developing nations had intended to outvote power itself as wielded by the industrialized nations.[109] I have already mentioned that these views linger in contemporary descriptions of China's approach toward the formation of custom.[110]

Such views were not shaped by the kind of democratic discourse about international lawmaking that we would find in the writings today of scholars in the United States. There was no great concern with domestic legislative oversight of international lawmaking, although there has been, amongst some scholars and members of the activist community, a sense that they are targeting special minority interests in the capital-exporting states. To the extent that customary international law doctrines – ranging from the precautionary principle to assertions of an emergent rule against cultural genocide, and the established doctrine of permanent sovereignty over natural resources – are invoked in today's international economic and commercial disputes, that way of thinking may be traced to the older attempts, described, to invoke custom as a progressive tool.

In any event, the modalities of customary law formation also allow for "contractual" customary law formation. Special or regional customary rules, and the persistent objector rule, are evidence of these,[111] putting aside more esoteric, contractualist

[106] Sornarajah, *supra* note 27, at 30.

[107] Sornarajah, *supra* note 27, at 31.

[108] Alvarez makes a similar point about the legitimacy of certain "universalistic fora" accruing to custom made through proceedings in organizations such as the UN General Assembly; ALVAREZ, *INTERNATIONAL ORGANIZATIONS AS LAW-MAKERS*, *supra* note 76, at 388.

[109] Lim, "Neither Sheep nor Peacocks," *supra* note 19, at 303–10.

[110] Pan, *supra* note 98, at 241–42.

[111] For acknowledgement of the similarity between contractual lawmaking and the persistent objector rule, see Timothy Meyer, "From Contract to Legislation: The Logic of Modern International Lawmaking," 14 *CHI. J. INT'L L.* 559, fn 185 at 607 (2014).

accounts of the formation of custom generally.[112] Moreover, progressive or *avant garde* customary international law rules that are insufficiently supported by state practice are perhaps not so likely to emanate from international tribunals and domestic courts after all, which is where pressing problems of legitimacy might tend to lie.[113]

We now come full circle. Trachtman's argument is a response to events at the turn of the millennium that have challenged the legitimacy of international treaty regimes, in the face of sustained deadlock in various multilateral treaty negotiations. The 2001 NAFTA FTC Interpretation was intended to reorientate NAFTA's FET and FPS standards toward a presumed customary MST norm against the backdrop of a democratic backlash against investor-state dispute settlement.[114] But the most visible backlash against treaty regimes have of course been Seattle and Cancun in the late 1990s and early 2000s, and the collapse of the effort to establish a multilateral agreement on investment in 1998.[115]

CONCLUSION

The truth in what Trachtman says – that treaties and treaty making are generally more convenient and efficient tools of lawmaking – cannot be fully denied. What this chapter tries to show and explain, however, is that custom is nonetheless alive in the international economic law field and that we should expect custom to have a future role even in the lawmaking sphere. Part of that explanation has to do with some of the reasons that are cited for custom's demise – increased treaty lawmaking and the greater institutionalization of international law, not least in the establishment of third-party settlement mechanisms. These have led to both "strongly" intertwined treaty and customary lawmaking and application, and the "weakly" intertwined application of customary international law by international investment tribunals in the course of applying treaties.

[112] *See* Curtis A. Bradley & Mitu Gulati, *supra* note 102; C. L. Lim & Olufemi Elias, "Withdrawing from Custom," 20 *Duke J. Comp. & Int'l L.* 143 (2010), restating our earlier view that "special" or "particular" custom and the requirement of non-dissent in the persistent objector rule are ultimately more revealing about the consensual underpinnings of customary international lawmaking; O. A. Elias & C. L. Lim, *The Paradox of Consensualism in International Law* 130–34 (1998).

[113] Some would argue that one area of exception lies in the practice of international investment tribunals, where the acceptance of broad claims of protection for foreign investors has led to a backlash against investor-state dispute settlement.

[114] Nigel Blackaby & Caroline Richard, "Amicus Curiae: A Panacea for Legitimacy in Investment Arbitration?" in *The Backlash against Investment Arbitration: Perceptions and Reality* 261 (Michael Waibel et al. eds., 2010).

[115] Sornarajah, *supra* note 27, at 34.

Intertwined lawmaking, in particular, addresses customary law formation's alleged deficiencies, ranging from difficulties with prospective lawmaking, to custom's lack of detailed predictability, its perceived inability to accommodate complex trade-offs, and the presumed absence of bespoke organizational support for customary lawmaking. Trachtman's thesis understates this intertwination that helps to answer some of custom's weaknesses. Lest it be thought that intertwination is simply a doctrinal construct, Choi and Gulati's latest research on the practice of courts and tribunals shows that, however puzzling this may be, treaties are somehow seen to provide the best, or at least the most popular, evidence of custom.[116] Intertwination is the principal lens through which custom is now viewed, and here we enter into one of the central themes of the present book – tribunals may tend to identify the *contents* of custom by looking to what states prefer, or what they are most willing to accept,[117] but if so, the best way to do so may be to look at the treaties to which states have already bound themselves.[118]

As for the debate in the economic field about the demise of international hard law, spurred by the current widespread failure of multilateral treaty initiatives, we should not ignore the growth in arbitral and other forms of third-party settlement, particularly since the 1990s. In the foreseeable future, custom's presence, and also that of treaties, is likely to be felt more rather than less in these "adjudicative aspects" of international economic law, and in the application of parallel as well as prior customary rules that will have to be systemically integrated – and thus reformulated in tandem – with the continued operation of existing, and new, treaty rules. In so many ways, custom's fate is bound to that of treaty law.[119]

[116] *See* Stephen Choi & Mitu Gulati, "Customary International Law: How Do Courts Do It?" (in this volume).

[117] *See* Curtis Bradley, "Customary International Law Adjudication as Common Law Adjudication" (in this volume).

[118] In contrast, Bradley and Gulati consider the use of treaties as evidence of custom to be a mystery; Bradley, "Customary International Law Adjudication as Common Law Adjudication" (in this volume); Choi & Gulati, "Customary International Law: How Do Courts Do It?" (in this volume).

[119] With the usual caveat, I am indebted to Joel Trachtman for our correspondence, to Curt Bradley, Tomer Broude, Mitu Gulati, Brian Lepard, and Ingrid Wuerth for their helpful comments, and to my former student Mr. Zachary Korman for his valuable research assistance.

9

The Decline of Customary International Law as a Source of International Criminal Law

Larissa van den Herik

This chapter revisits customary international law (hereafter CIL) as a source of international criminal law (hereafter ICL) in the era of the International Criminal Court (hereafter ICC). Most contemporary reflections on the role of CIL in ICL and most observations regarding creative judicial identification processes still tend to zero in on the International Criminal Tribunal for the former Yugoslavia (hereafter ICTY).[1] True, the 1990s were a decade-long experiment, or perhaps, indeed, an extended "Grotian Moment."[2] In any event, it was a period of accelerated, and at times artificial, CIL creation on a variety of ICL issues by the *ad hoc* tribunals. The ICTY's innovative methodologies nourished theories about the fragmentation of sources and ensuing claims regarding the existence of autonomized regimes.[3] Being predominantly based on ICTY experiences, these claims may well have become outdated. The coming into operation of the ICC Statute has gradually engendered a

[1] For instance, in a recent 2014 publication on the methods of interpreting custom used by international criminal tribunals, the vast majority of case law analyzed was ICTY case law. *See* NOORA ARAJÄRVI, *THE CHANGING NATURE OF CUSTOMARY INTERNATIONAL LAW: METHODS OF INTERPRETING THE CONCEPT OF CUSTOM IN INTERNATIONAL CRIMINAL TRIBUNALS* (2014).

[2] The idea of connecting the concept of a "Grotian Moment" to international criminal law comes from Michael Scharf, who defines the Grotian Moment as "an instance in which there is such a fundamental change to the international system that a new principle of customary international law can arise with exceptional velocity." Michael P. Scharf, "Seizing the 'Grotian Moment': Accelerated Formation of Customary International Law in Times of Fundamental Change," 43 *CORNELL INT'L L.J.* 442 (2010). *See also* MICHAEL P. SCHARF, *CUSTOMARY INTERNATIONAL LAW IN TIMES OF FUNDAMENTAL CHANGE: RECOGNIZING GROTIAN MOMENTS* (2013).

[3] Jean d'Aspremont, "An Autonomous Regime of Identification of Customary International Humanitarian Law: Do Not Say What You Do or Do Not Do What You Say?" in *DROIT INTERNATIONAL HUMANITAIRE: UN RÉGIME SPÉCIAL DE DROIT INTERNATIONAL?* (Raphael van Steenberghe ed., 2013); and Igo Venzke, "Understanding the Authority of International Courts and Tribunals: On Delegation and Discursive Construction," 14 *THEORETICAL INQUIRIES IN LAW* 381, 408 (2013).

new momentum that is best described as the "Codified Moment."[4] In this new ICC era, CIL plays a much less prominent role.

This chapter explores the decline of CIL in ICL. It begins by delimiting the concept of "international criminal law" and highlights some consequences of this delimitation for a discussion on the role of CIL. It argues that the essence of this area of law is its judicial application as part of a process to determine guilt of individuals and to impose a sentence. Hence, only judicial agents truly apply ICL, and they do so vis-a-vis individuals. By implication, ICL does not operate in a traditional interstate setting, a feature that also has consequences for CIL formation and identification.

The chapter then presents the causal claim that this decline is a direct consequence of the introduction of the ICC Statute and the move toward a universal and more consent-based regime. As will be demonstrated, the ICC Statute's design reflects the drafters' explicit resolve to marginalize CIL. Subsequently, this chapter argues that the move away from CIL in ICL can itself be explained by the difficulties inherent in grounding rules of ICL in CIL. This normative claim is developed through an analysis of different types of rules, that is, procedural rules, substantive rules regarding the definition of crimes, and rules regarding modes of liability. In doing so, this chapter sheds light on the distinct methodological challenges that may arise in attempts to construe a customary basis for such rules.

THE CONCEPT OF "INTERNATIONAL CRIMINAL LAW" AND CONSEQUENCES FOR THE ROLE OF CIL

ICL is the law on the core international crimes. Yet beyond this substantive demarcation, there is no solid and monolithic concept of ICL. Rather, there are mutually reinforcing claims that ICL is a complex, pluralist,[5] or internally fragmented area of law,[6] or even an area that is subject to an identity crisis.[7] In this

4 This term was coined by Joe Powderly. He has argued that the Rome Statute effectively represents the end point of attempts to codify international criminal law undertaken since Nuremberg. However, he also submits that this end point is normatively unsatisfying. *See* Joseph Powderly, "The Rome Statute and Attempted Corseting of the Interpretative Judicial Function: Reflections on Sources of Law and Interpretative Techniques," in *The Law and Practice of the ICC* (Carsten Stahn ed., 2d ed. forthcoming 2015).

5 *See, e.g.*, *Pluralism in International Criminal Law* (Elies van Sliedregt & Sergey Vasiliev eds., 2014); Alexander K. A. Greenawalt, "The Pluralism of International Criminal Law," 86 *Ind. L.J.* 1063 (2011).

6 Carsten Stahn and Larissa van den Herik, "'Fragmentation,' Diversification and '3D' Legal Pluralism: International Criminal Law as the Jack-in-the-Box," in *The Diversification and Fragmentation of International Criminal Law* (Larissa van den Herik & Carsten Stahn eds., 2012).

7 Darryl Robinson, "The Identity Crisis of International Criminal Law," 21 *Leiden J. Int'l L.* 925 (2008).

context, some introductory remarks on the ambit of the area of ICL are warranted, as these can inform the further observations and reflections in this chapter. It is the premise of this chapter that ICL is both a jurisdictional design for international judicial intervention as well as a body of substantive norms that can be applied and implemented at various levels, both by international and domestic criminal courts.[8]

Within the substantive law on the prohibition and definition of the core crimes, the concept of ICL is delineated by the applying authority, and relatedly, the nature and consequences of application of this field of law. The essence of criminal law is that the commission of certain pre-criminalized behavior is condemned by a criminal court following a meticulously regulated procedure geared toward guaranteeing the rights of the defense. The institution and procedure are thus determining factors of any type of criminal law. The adjective "international" implies the creation of international jurisdiction for the crimes that international criminal law is concerned with. Being a subspecies of criminal law, ICL can only be effectively applied by criminal courts. Those courts can either be international in character or national courts acting as international agents on the basis of extraterritorial jurisdictional grounds.

Because the judiciary has an exclusive role in the application of ICL, international criminal adjudicators are also the prime identifiers of CIL. It has been argued elsewhere in this volume[9] that the exercise of finding or identifying CIL is not much different from CIL making, and that, in particular in the area of international humanitarian and criminal law, courts have participated in the actual formation of CIL in often more prominent ways than states.The *Tadić* case and its finding on the existence of war crimes in non-international armed conflict[10] is being cited in support of the argument that the ICTY heavily engaged in lawmaking while purporting to merely find CIL. Disregarding the question whether *Tadić* is still (or has ever been) truly representative for CIL processes in ICL, an overall perspective on this case may put the ICTY's lawmaking capabilities into perspective. *Tadić* offered several innovative CIL creations, most prominently on war crimes in internal armed conflict and joint criminal enterprise (JCE).[11] It is widely agreed that both findings

[8] This chapter thus principally engages with the so-called modern or vertical ICL, which is the law on the core crimes of genocide, crimes against humanity, and war crimes (which is more distinctively labeled by the French and German terms of *droit international pénal* or *Völkerstrafrecht*). Traditional or horizontal ICL, also called transnational criminal law or the law on extradition and mutual legal assistance, is left out of the equation in light of its inherently different character. *See also* Claus Kress, "International Criminal Law," in MAX PLANCK ENCYCLOPEDIA OF PUBLIC INTERNATIONAL LAW (Oxford University Press).

[9] Monica Hakimi, "Custom's Method and Process: Lessons from Humanitarian Law" (in this volume).

[10] *Prosecutor v. Tadić*, Case No. IT-94-1-1, ¶ 96–136 (ICTY Decision on Defence Motion for Interlocutory Appeal on Jurisdiction, Oct. 2, 1995).

[11] *Prosecutor v. Tadić*, Case No. IT-94-1-1, ¶ 185–237 (ICTY Appeals Chamber, Jul. 15, 1999), as confirmed in *Prosecutor v. Ojdanić*, Case No. IT-99-37-AR72, (ICTY Appeals Chamber Decision on Dragoljub Ojdanić' Motion Challenging Jurisdiction, May 21, 2003).

went beyond existing CIL.[12] Yet, when the reception of the *Tadić* finding on war crimes is contrasted with that of the creation of the concept of JCE, it can be noted that the first pronouncement has indeed been codified in the Rome Statute, while the latter was eventually replaced by different perpetration and participation models in Article 25 of the ICC Statute.[13]

While it is thus undoubtedly true on a general level that international criminal courts, and in particular the ICTY, have been pioneers in the development of ICL, we also find that, ultimately, states retain an important role in determining whether given CIL identifications radiate beyond the institutional setting in which they were pronounced. States thus continue to be critical actors in the lawmaking exercise, also in the area of ICL. Inventive law creations by *ad hoc* tribunals can only gain universal relevance, and thus, true CIL status, if accepted by states through some form of initial, ongoing, or *ex post facto* consent.[14] How such consent should be construed or identified is another matter. In any event, lack of express opposition cannot always be understood as implicit consent or acquiescence.

In relation to the early expansive ICTY findings, absence of state protest or criticism may well be explained through the limited jurisdiction of the ICTY. States simply did not immediately realize the potential applicability of such findings on a grander scale going beyond the ICTY jurisdiction. The institutional architecture thus did not invite direct engagement by the broader community of states to certain ICTY CIL findings. However, as ICL became more established in the international legal society, several scholars have recognized a role of state acceptance for truly innovative CIL discoveries by criminal courts and tribunals. For instance, when the Special Tribunal for Lebanon promulgated a new CIL definition of terrorism in peacetime, it was argued that the further grounding of this CIL articulation ultimately depended on the position of states.[15]

12 For war crimes in internal armed conflict, *see, for example*, Christoper Greenwood, "International Humanitarian Law and the Tadić Case," 7 EUR. J. INT'L L. 265, 278 (1996); and for JCE, *see, for example*, Steven Powles, "Joint Criminal Enterprise: Criminal Liability by Prosecutorial Ingenuity and Judicial Creativity?" 2 J. INT'L CRIM. JUST. 606 (2004).

13 Subsequently, the CIL status of the most far-reaching form of JCE, JCE III, was also challenged by the Pre-Trial Chamber of the Extraordinary Chambers in the Courts of Cambodia, in *Prosecutor v. Jeng Sary et al.*, (Case Nos. 002/19-09-2007-ECCC/OCIJ, ECCC, May 20, 2010), as confirmed by the Trial Chamber (Case File No. 002/19-09-2007/EEEC/TC, ¶ 29 [EEEC/TC, Decision on the Applicability of Joint Criminal Enterprise, Sep. 12, 2011]). The ICC Pre-Trial Chamber refused to adopt the ICTY jurisprudence on modes of liability and particularly the concept of JCE. *See Lubanga, Situation in the Democratic Republic of the Congo*, ICC-01/04-01/06-803-tEN, ¶ 322–341(ICC PTC I, Decision on the Confirmation of the Charges, Jan. 29, 2007).

14 *Cf.* Anthea Roberts & Sandesh Sivakumaram, "Lawmaking by Non-State Actors: Engaging Armed Groups in the Creation of International Humanitarian Law," 37 YALE J. INT'L L. 108, 117–18 (2012).

15 Claus Kress & Nikolaos Gazeas, "Terrorismus," in EUROPÄISCHES STRAFRECHT 336 (Ulrich Sieber et al. eds., 2011).

Hence, even if one agrees with the overall argument that international criminal courts and tribunals have greatly developed the law, the role of states in the validating process should not be underestimated. Indeed, it seems that criminal courts and tribunals have also come to acknowledge this foundational role of states.[16] This realization has gradually inspired an overall trend toward greater justification of potential controversial or innovative findings and the presentation of more evidence to support such findings. Such realization is reinforced by dynamics between coexisting international courts challenging each other's findings. In this sense, it is noteworthy that in the heated jurisprudential battle on the question whether specific direction was an element of the *actus reus* of aiding and abetting, the SCSL compelled the ICTY to prove and substantiate its CIL findings more convincingly.[17] There are thus some indications that international courts and tribunals can act in a disciplined manner in actually proving CIL identifications, and they may especially be inclined to do so with a view to enticing states and the broader international legal community to accept and certify their findings.[18]

A second feature or delimitation of ICL is that its rules are directed to individuals, and not to states. This may have ramifications for the concept of "practice," and at the most extreme and theoretical level, it could even invite the question whether the practice of individuals could also be relevant to identify primary norms.[19] More importantly, the delimitation requires that a differentiation be made between the primary norms that underlie state responsibility versus those related to individual responsibility for the core crimes.[20] Given the definitive dismissal of the notion of state crime from the Articles of State Responsibility,[21] only the latter form of responsibility, that is, individual

[16] This greater role for states corresponds to some extent with the common law account informed and shaped by state preferences as presented in Curtis A. Bradley, "Customary International Law Adjudication as Common Law Adjudication" (in this volume).

[17] *See, e.g.*, SCSL, *Prosecutor v. Taylor*, Case No. SCSL-03-01-A ¶ 466–81 (SCSL, Appeal Judgment, Sep. 23, 2013); ICTY, *Prosecutor v. Šainović et al.*, Case No.IT-05-87-A, ¶ 1617–51 (ICTY, Appeal Judgment, Jan. 23, 2014).

[18] For a different view arguing that tribunals are methodologically undisciplined, see Hakimi, *supra* note 9.

[19] This question is mostly posed in relation to the question regarding the relevance of practice of armed opposition groups to the formation of rules of customary international humanitarian law. *See* Roberts & Sivakumaram, *supra* note 15. For further references as regards the practice of individuals specifically, *see Second Report on the Identification of Customary International Law*, Special Rapporteur Michael Wood, UN Doc. A/CN.4/672, 22 May 2014, fn. 139.

[20] This argument has been made by, for example, Paola Gaeta, "On What Conditions Can a State Be Held Responsible for Genocide," 18 EUR. J. INT'L L. 631 (2007).

[21] Commentary to the Articles on State Responsibility, Commentary to Chapter III, in particular para. (7). *See also* Case Concerning Application of the Convention on the Prevention and Punishment of the Crime of Genocide (*Bosnia and Herzegovina v. Serbia and Montenegro*), ¶ 167 (ICJ Feb. 26, 2007) (Judgment). *But see* Alain Pellet, "Can a State Commit a Crime? Definitely Yes!," 10 EUR. J. INT'L L. 425 (1999).

responsibility, is a matter of international criminal law.[22] In contrast, questions of state responsibility for the commission, prevention, or prosecution of core crimes come within the realm of general international law or they are matters of international humanitarian law and human rights law.[23] It may thus well be that a certain norm has customary status to the extent that it is addressed to a state, but that the corollary international criminalization addressed to individuals lacks such status. This distinction has for instance been made in relation to apartheid and enforced disappearances.[24] It may also be that different types of evidence are required to make a claim on a customary prohibition directed to states versus a claim on the existence of an international crime directed at individuals. Following this delimitation, the discussion of CIL and ICL is intrinsically limited to questions where criminal responsibility is involved, leaving other international law issues related to international crimes, and most notably questions of state responsibility, outside the scope of this chapter.

THE RISE AND FALL OF CIL IN ICL: FROM THE ICTY TO THE ICC

The accepted role for CIL in ICL thus varies per temporal and institutional episode. This section juxtaposes the ICTY and the ICC engagement with CIL and examines how, in the ICC era of codification, CIL operates in parallel with the ICC statute. It first explores the reasons and extent of the ICTY's use of CIL and, relatedly, the ICTY's innovative methods of CIL identification. As already observed, this creativity was facilitated by the ICTY's institutional framework and its limited jurisdiction. The ICC, by contrast, is treaty based and thus by nature more driven by state consent. Moreover, its universal aspirations in a geographical sense have curbed the toleration of creative CIL. After all, states may be less amenable to accept progressive CIL findings that can potentially be applied against their own interests. In addition,

[22] The separation between the two forms of responsibility is underscored by Article 25(4) of the ICC Statute, which reads: "No provision in this Statute relating to individual criminal responsibility shall affect the responsibility of States under international law." For a vision that supports the duality of responsibility, which implies different systems of responsibility and the related autonomy of the regime of individual responsibility, *see* Pierre Marie Dupuy, "International Criminal Responsibility of the Individual and International Responsibility of the State," in *The Rome Statute of the International Criminal Court: A Commentary* volume II 1085–99 (Antonio Cassese et al. eds., 2002).

[23] For instance, the obligation to prosecute the grave breaches of the Geneva Conventions under Articles 49, 50, 129 and 146 of the Geneva Conventions or the duty to prosecute as developed under human rights law.

[24] John Dugard & John Reynolds, "Apartheid, International Law, and the Occupied Palestinian Territory," 24 *Eur. J. Int'l L.* 883 (2013); C. Hall & Larissa van den Herik, "Commentary on Article 7(1)(i) and (j) and (2)(h) and (i), "Enforced Disappearance of Persons," and " Apartheid" in *Commentary on the Rome Statute of the International Criminal Court; Observers' Notes, Article by Article*, margin 94 and 150 (Otto Triffterer & Kao Ambos eds., 3d ed., 2015).

the ICC is built on the substantive foundations of the ICTY's pioneering work. The Rome Statute codified some, but certainly not all, of the ICTY's CIL case law. The ICC thus operates in a more sophisticated normative environment, which reduces the need to resort to CIL for other than interpretive purposes.

Heyday for CIL in the Age of the Ad Hoc *Tribunals*

In the 1990s, ICL revived as an area of international law through the establishment of the non-consent-based and geographically limited *ad hoc* tribunals for the former Yugoslavia and Rwanda.[25] Encouraged by the UN secretary general in his report of the establishment of the ICTY,[26] these tribunals resorted to CIL with great frequency. The official function of CIL was to counter claims regarding infringements of the legality principle and to fill legal gaps. In practice, CIL was used for truly expansionist purposes and it operated as the principal device of intense law development.[27] The paucity of refined norms required such activity to realize the idea of international criminal justice. The creation of new law through CIL occurred for all facets of ICL, both of substantive as well as procedural nature. Hence, CIL was used to erase the distinction between international and non-international armed conflict and to transplant treaty-based crime definitions and modes of liability that were exclusively applicable in international armed conflicts to the setting of non-international armed conflict.[28] Beyond war crimes, CIL was used to develop new criminal acts, particularly for crimes against humanity, or to give substance to existing crime definitions.[29] In addition, the articulation of

[25] Subsequently, other temporally and geographically limited international and hybrid courts were established, such as the Special Court for Sierra Leone, the Extraordinary Criminal Chambers for Cambodia, and the Special Tribunal for Lebanon. In the next section, all these entities are subsumed under the general heading of *ad hoc* tribunals.

[26] UN Doc. S/25704, 3 May 1993, para. 34.

[27] JEAN D'ASPREMONT & JEROME DE HEMPTINNE, *DROIT INTERNATIONAL HUMANITAIRE* 27 (2012).

[28] For example, in the *Hadžihasanović and Kubura* case, the ICTY created the crimes of wanton destruction of towns or villages, the plunder of public or private property, and the destruction of institutions dedicated to religion in non-international armed conflict, and it also determined that command responsibility was an integral part of custom and could also be applied in settings of non-international armed conflict. *See Prosecutor v. Hadžihasanović and Kubura*, Case No. IT-01-47-T, ¶ 38 and 65 (ICTY Judgment, Mar.15, 2006). Notably, the ICTY excluded acts committed by subordinates prior to the commander's assumption of command from the scope of his responsibility arguing that there was insufficient state practice and *opinio juris* for this. *See Prosecutor v. Hadžihasanović and Kubura*, Case No. IT-01-47-AR72, ¶ 45 (ICTY, Decision on Interlocutory Appeal Challenging Jurisdiction in Relation to Command Responsibility, Jul. 16, 2003). *Cf.* Birgit Schlütter, "Constitutionalisation at Its Best or at Its Worst? Lessons from the Development of Customary International Criminal Law," 2006 ESIL Conference paper 7, *retrieved from* http://www.esil-sedi.eu/sites/default/files/Schleutter.PDF (last access Mar. 8, 2015).

[29] The criminal acts of persecution and other inhumane acts were most prone for further development and inclusion of new prohibited acts, such as hate speech as proposed by the ICTR Trial Chamber in

novel modes of liability was grounded on,[30] and in later stages also rejected for absence of,[31] a basis in CIL. As indicated previously, CIL was also used as an argument in the acute precedential dispute that arose at the end of the ICTY's lifespan on the question whether specific direction was an element of the *actus reus* of aiding and abetting.[32] CIL has also been invoked in relation to questions on defenses. In one of the first ICTY cases, the *Erdemović* case, the existence in CIL of a complete defense of duress for soldiers charged with a crime against humanity and/or a war crime involving the killing of innocent human beings was tested.[33] And, finally, on several occasions matters of procedure have been settled by reference to CIL.[34]

The academic writing that notes, acclaims, or criticizes the ICTY's at times original engagement with CIL cites the *Kupreškić* judgment with particular frequency.[35] Indeed, this judgment stands out. In *Kupreškić*, the ICTY departed from the traditional two-element approach and proposed a conception of CIL that would hinge almost exclusively on the identification of *opinio juris*, which in turn could be directly derived from a general principle in the form of the Martens clause.[36] In

Prosecutor v. Ferdinand Nahimana, Jean-Bosco Barayagwiza and Hassan Ngeze, Case No. ICTR-99-52-T, ¶ 1076 (ICTR, Judgment, Dec. 3, 2003). But this was opposed by Judge Meron in a dissent on appeal, Case No. ICTR-99-52-A, ¶ 5 (ICTR, Partial Dissenting Opinion of Judge Meron), with reference to the *Kordić* Trial Judgment, *Prosecutor v. Kordić & Cerkez*, Case No.IT-95-14/2-T, ¶ 209 (ICTY, Judgment, Feb. 26, 2001).

30 Most famously of course, the development of the concept of joint criminal enterprise in *Tadić*. *See supra* note 11.

31 For example, Shahabuddeen argued against the introduction of co-perpetratorship in addition to the theory of joint criminal enterprise since the two were mutually exclusive, and "since several states adhere to one theory while several other states adhere to the other theory, it is possible that the required state practice and opinio juris do not exist so as to make either theory part of customary international law." *Prosecutor v. Gacumbitsi*, Case No.ICTR-2001-64-A, ¶ 51 (ICTR, Appeal Judgment, Separate Opinion of Judge Shahabuddeen, Jul. 7, 2006).

32 *See supra* text accompanying note 17.

33 *Prosecutor v. Erdemović*, Case No.IT-96-22-A, ¶ 19 (ICTY, Appeal Judgment, with reference to the Joint Separate Opinion of Judge McDonald and Judge Vohrah and the Separate and Dissenting Opinion of Judge Li, Oct. 7, 1997).

34 For example, regarding the customary status of ICRC privileges as a bar to disclosure, see *Prosecutor v. Simic* (ICTY, Decision on the Prosecutor's Motion under Rule 73 for a Ruling concerning the Testimony of a Witness, Case No. IT-95-9-PT, July 27, 1999).

35 *See, e.g.*, Schlütter, *supra* note 29; d'Aspremont & de Hemptinne, *supra* note 28; Venzke, *supra* note 3, at 408.

36 In *Prosecutor v. Kupreškic*, Case No.IT-95-16-T, ¶ 527 (ICTY, Trial Chamber, Jan. 14, 2000), the court stated:

> Admittedly, there does not seem to have emerged recently a body of State practice consistently supporting the proposition that one of the elements of custom, namely *usus* or *diuturnitas* has taken shape. This is however an area where *opinio juris sive necessitates* may play a much greater role than *usus*, as a result of the aforementioned Martens Clause. In the light of the way States and courts have implemented it, this Clause clearly shows that principles of international humanitarian law may emerge through a customary process under the pressure

scholarly literature, this judgment has been projected as a harbinger of a new natural law attitude to CIL that is unique to ICL and even further reaching than theories on modern or instant CIL.[37] As such, the *Kupreškić* judgment overhauls accepted theories of CIL and functions as a springboard for ruminations on the fragmentation of sources[38] and constitutionalist tendencies in international lawmaking.[39] Other decisions of the ICTY and fellow *ad hoc* international criminal courts and tribunals are also based on quite wild and non-corroborated proclamations on the existence of CIL, generating instances of *coutume bien sauvage*.[40]

Such decisions of the *ad hoc* tribunals have inspired and nourished speculations on the need for an autonomized regime. Yet, the validity of such ICTY-based theses on "criminal custom" and related claims regarding the ensuing fragmentation of sources may well have expired with the coming into force of the ICC statute and the more limited role for CIL in that setting. Moreover, the *Kupreškić*-case could also be regarded as an incidental exception, while the *ad hoc* tribunals adhered to the theory of the two-pronged approach, at least nominally, in other instances.[41] The ICTY's engagement with CIL is thus more varied than sometimes proclaimed in scholarly writing. Overall, the following main methods of CIL finding can be discerned in the practice of the ICTY and other international criminal courts and tribunals: (i) (nominal) adherence to traditional approach,[42] (ii) ultramodern

of the demands of humanity or the dictates of public conscience, even where State practice is scant or inconsistent. The other element, in the form of *opinio necessitates*, crystalizing as a result of the imperatives of humanity or public conscience, may turn out to be the decisive element heralding the emergence of a general rule or principle of humanitarian law.

37 *Kupreškic* not only proposed to disregard the *usus* element and to focus on *opinio juris*. It also suggested a method of deducing *opinio juris* from a preexisting general principle of law. It thus fully bypassed any need for state practice or state consent. This goes well beyond the modern deductive approach, which only takes *opinio juris* as a starting point. Anthea Roberts, *See* "Traditional and Modern Approaches to Customary International Law: A Reconciliation," 95 AM. *J.* INT'L *L.* 757 (2001). It even goes beyond Bin Cheng's instant custom theory, which still required input from states for the existence of *opinio juris*, for instance in the form of their votes for UNGA Resolutions. *See* Bin Cheng, "United Nations Resolutions on Outer Space: 'Instant' International Customary Law?" 5 *INDIAN J. INT'L L.* 23 (1965).

38 *See, e.g.*, Van den Herik & Stahn (eds.), *supra* note 6 (arguing that risks of fragmentation should not be exaggerated and that pluralism or diversification are often better terms to capture differentiation dynamics). It is also submitted in this work that diversification generally presents an opportunity to improve quality of norms and to enhance universal application and enforcement.

39 Schlütter, *supra* note 28.

40 As concluded in an earlier treatise, the Rwanda Tribunal often made findings on custom without any reasoning or corroborating evidence to show the existence of *opinion juris* and state practice. *See* LARISSA VAN DEN HERIK, *THE CONTRIBUTION OF THE RWANDA TRIBUNAL TO THE DEVELOPMENT OF INTERNATIONAL LAW* 275 (2005).

41 Theodor Meron, "Revival of Customary Humanitarian Law," 99 AM. *J.* INT'L *L.* 817, 821 (2005).

42 *Id.*

approach with exclusive reliance on *opinio juris*,[43] (iii) international precedent (i.e., relying on previous CIL findings),[44] (iv) comparative criminal law approach under the guise of CIL.[45]

The greatest controversy about the tribunals' resort to CIL generally arose from their tendency to make bold claims of existing CIL without offering any or scant evidence, or even worse in some instances, the evidence provided did not actually substantiate the claim made.[46] Those lax evidentiary practices were heavily criticized, and, along with other factors, they eventually were a reason for the negotiators of the Rome statute to revisit the role of CIL in ICC context. Rome was also a watershed moment for other reasons. It was the place and moment where states came back into the lawmaking arena. In Rome, states validated and codified certain progressive pronouncements of the ICTY and the ICTR, for example, regarding war crimes in internal armed conflict and sexual crimes, while rejecting others, such as those on joint criminal enterprise and the absence of a policy element for crimes against humanity.[47] In addition, the ICC negotiations in Rome were the place and moment where ICL became an area of law rather than a peace and security mechanism. In this sense, it was a moment of emancipation and transformation.[48] Initially, the ICTY's limited jurisdiction and its nonconsensual institutional architecture had rendered it a perfect laboratory for exploration in the absence of states. In Rome, ICL as such became a universal rather than an *ad hoc* exercise, which also had repercussions for CIL making in the context of the ICTY, given the higher potentiality of application of any form of ICL beyond the ICTY setting.

43 *See Kupreškić* Judgment as cited at *supra* note 36. *See also Prosecutor v. KAING Guek Eav alias Duch*, Case No. 001/18-07-2007-ECCC/SC, ¶ 93 (Feb. 3, 2012).

44 *See, e.g., Prosecutor v. Perišić*, Case No.IT-04-81-A, ¶ 25–36 (ICTY, Appeal Judgment, Feb. 28, 2013). On the question whether decisions of courts can constitute practice relevant for the formation of custom, *see also Second Report on the Identification of Customary International Law*, Special Rapporteur Michael Wood, UN Doc. A/CN.4/672, 22 May 2014, para. 46.

45 This approach is further explored by Mireille Delmas-Marty, "The Contribution of Comparative Law to a Pluralist Conception of International Criminal Law," 1 *J. Int'l Crim. Just.* 13 (2003).

46 The most notorious example is footnote 118 in the *Kunarac* judgment, claiming that the policy requirement did not exist under CIL, *Prosecutor v. Dragoljub Kunarać*, Case No. ICTY-96-23& IT-96-23/1-A, A. ¶ 98 and fn. 114 (ICTY, Judgment, Jun. 12, 2002). It has been noted that the references given in support of the finding on custom were selective and included judgments that had been overturned as well judgments coming from administrative proceedings. *See* William A. Schabas, "State Policy as an Element of International Crimes" 98 *J. Crim. L. & Criminology* 953 (2008); and Larissa van den Herik, "Using Custom to Reconceptualize Crimes Against Humanity," *in Judicial Creativity at the International Criminal Tribunals* (Shane Darcy & Joseph Powderly eds., 2010).

47 For additional discussion, *see* Darryl Robinson, "Defining 'Crimes Against Humanity' at the Rome Conference," 93 *Am. J. Int'l L.* 43 (1999).

48 On the idea of emancipation of ICL from the Security Council, *see also* Frederic Megret, "A Special Tribunal for Lebanon: The UN Security Council and the Emancipation of International Criminal Justice," 21 *Leiden J. Int'l L.* 485 (2008).

A More Marginal Existence for CIL in the ICC Constellation

In the ICC constellation, CIL has retained relevance, albeit more marginally. In this respect, it is noteworthy that the ICC statute does not mention the term "customary international law" anywhere as such. It rather refers more generically to "rules of international law."

The two pivotal provisions for an inquiry on the ICC's attitude to CIL are Article 10 and Article 21. These provisions govern the ICC's relationship with CIL externally and internally. Respectively, Article 10 arranges the external relationship of the ICC statute as a treaty with CIL as an alternative source of international law, whereas Article 21 instructs the ICC judiciary on the possible application of CIL in the internal setting of the ICC.

Different visions exist on the question whether and to what extent the ICC statute codifies or is constitutive of CIL. Generally, the relationship between treaty and CIL as coexisting sources of international law,[49] and in particular, the circumstances under which treaties or specific treaty provisions may be constitutive or declaratory of CIL or when treaty negotiations may have crystalizing effects have been articulated in scholarly writing and by the International Court of Justice.[50] In deviation from this doctrine and case law, the ICC statute manages its own relationship with CIL through the untitled Article 10.

Article 10 provides: "Nothing in this Part shall be interpreted as limiting or prejudicing in any way existing or developing rules of international law for purposes other than this Statute." This is a rather unique provision. It was included in the ICC statute upon the proposal of Bassiouni and with strong support of the ICRC. The primary reason for including Article 10 was to alleviate concerns about the exclusion of certain war crimes in Article 8, in particular those concerning the use of chemical, biological, and nuclear weapons. As such, Article 10 preempts possible ramifications that the omissions of these war crimes could have for their status in general international law.[51] Article 10 thus serves a function that is similar to the Martens clause, namely to preclude that adverse consequences on legality

[49] Case Concerning the Military and Paramilitary Activities in and Against Nicaragua (Nicaragua v. USA), ¶ 172–182 (ICJ Jun. 27, 1986) (Judgment).

[50] North Sea Continental Shelf Cases (Federal Republic of Germany/Denmark) ¶ 61 (ICJ Feb. 20, 1969) (Judgment); Arthur M. Weisburd, "Customary International Law: The Problem of Treaties," 21 *VAND. J. TRANSNAT'L L.* 1 (1988); Gary L. Scott & Craig L. Carr, "Multilateral Treaties and the Formation of Customary International Law," 25 *DENV. J. INV'L. L. & POL'Y* 71 (1996); R. R. Baxter, "Multilateral Treaties as Evidence of Customary International Law," 41 *BRIT. Y.B. INT'L. L.* 275 (1965–1966).

[51] Leila Nadya Sadat, "Custom, Codification and Some Thoughts About the Relationship Between the Two: Article 10 of the ICC Statute," 49 *DEPAUL L. REV.* 909, 916 (2000); WILLIAM A. SCHABAS, *THE INTERNATIONAL CRIMINAL COURT; A COMMENTARY ON THE ROME STATUTE* 270 (2010).

are being drawn from noninclusion. Through its formulation, the provision has an inherently progressive character, as it is geared toward preventing that regressive ICC codifications and interpretations have external effects.[52] Article 10 does not impede that progressive ICC codifications, such as the criminalization of war crimes in internal armed conflict or the inclusion of new criminal acts in the definition of crimes against humanity, radiate beyond the ICC setting.

In principle, this provision only governs part II of the statute, which is the part on jurisdiction, admissibility, and applicable law. Through the letter of its text, Article 10 does not concern the other parts, which include rules on modes of liability, defenses, procedure, penalties, and state cooperation. These other parts are therefore subjected to more general theories of interrelationship, and consequently it could be argued that the ICC provisions and interpretations on all rules not belonging to part II can more freely affect the development of CIL.[53] However, the argument has also been made that Article 10 applies analogously to other provisions of the ICC statute such as Article 25.[54]

Despite being an ICC provision, Article 10 is not addressed to the ICC. After all, the court, its actors, and state parties are directly bound by the statute and therefore the statute's customary nature beyond ICC confines is not of immediate importance to them.[55] Instead, Article 10 offers guidance to external actors that look at the ICC statute and ICC interpretations as potential evidence of custom.[56] These actors are *ad hoc* international or domestic criminal courts, and also human rights jurisdictions and other human rights bodies including commissions of inquiry established by the Human Rights Council.[57] Given synergies in applicable law and mandate, coexistent international criminal courts and tribunals are the primary destination of Article 10's message. Indeed, the ICTY has made explicit reference to this provision so as to uphold its own more progressive case law in deviation from

52 As observed by Sadat, *supra* note 51, at 918.

53 As also suggested by the ICTY in *Prosecutor v. Tadić, supra* note 11, fn. 282.

54 *Prosecutor v. Laurent Gbagdo*, case No. ICC-02/11-01/11-656-Anx (ICC, Pre-Trial Chamber I, Dissenting Opinion of Judge Christine Van den Wyngaert, June 12, 2014).

55 The Pre-Trial Chamber's reference to Article 10 in the Bashir arrest warrant decision to reject arguments that in cases of discrepancy between the statute and the elements of crimes, it has to interpret the statute in light of customary international law was thus inadequate. This article does not govern applicable or rules of interpretation, as these are regulated by other provisions in the statute, *Prosecutor v. Bashir*, Case No. ICC-02/05-01/09, ¶ 127(ICC, Decision on the Prosecutor's Application for a Warrant of Arrest against Omar Hassan *Ahmad Al Bashir, Mar. 4, 2009)*.

56 *Cf.* SCHABAS, *supra* note 51, at 269; *Prosecutor v. Charles Ghankay Taylor*, Case No. SCSL-03-01-T-1294, ¶ 435, 451 (SCSL, Trial Chamber II, Jun. 19, 2012); *Prosecutor v. Germain Katanga*, Case No. ICC-01/04-01/07, ¶ 508 (ICC, Trial Chamber II, Sep. 30, 2008).

57 *See* Larissa van den Herik & Catherine Harwood, "Commissions of Inquiry and the Charm of International Criminal Law. Between Transactional and Authoritative Approaches," in *HUMAN RIGHTS FACT FINDING IN THE 21ST CENTURY* (Philip Alston & Sarah Knuckey eds., 2016).

the ICC statute.[58] On other occasions, it has indicated more generally that the ICC statute does not necessarily represent CIL in its entirety.[59]

Despite these instances of critical engagement with the ICC statute, the ICTY's overall attitude is marked by respect and it has attributed significant legal value to the statute "being adopted by an overwhelming majority of the States attending the Rome Diplomatic Conference and was substantially endorsed by the Sixth Committee of the United Nations General Assembly."[60] As such, the ICC statute was said to reflect the *opinio juris* of states. Other international agents that engage with ICL concepts are generally even more deferential to the ICC statute as an authoritative statement of customary international law and they tend to ignore Article 10.[61] It must be observed, though, that such other international bodies, including human rights courts and commissions of inquiry, mainly resort to the statute to confirm the CIL nature of given overall prohibitions and definitions of crimes, and very occasionally superior responsibility,[62] and not so much for other matters regulated in the statute.

As for the potential engagement with CIL by the ICC itself, Article 21 is instructive. It reads:

> 1. The Court shall apply:
>
> (a) In the first place, this Statute, Elements of Crimes and its Rules of Procedure and Evidence;
>
> (b) In the second place, where appropriate, applicable treaties and the principles and rules of international law, including the established principles of the international law of armed conflict;
>
> (c) Failing that, general principles of law derived by the Court from national laws of legal systems of the world including, as appropriate, the national laws of States that would normally exercise jurisdiction over the crime, provided that those principles are not inconsistent with this Statute and with international law and internationally recognized norms and standards.
>
> 2. The Court may apply principles and rules of law as interpreted in its previous decisions.
>
> 3. The application and interpretation of law pursuant to this article must be consistent with internationally recognized human rights, and be without any adverse distinction founded on grounds such as gender as defined in article 7,paragraph 3, age, race, colour, language, religion or belief, political or other opinion, national, ethnic or social origin, wealth, birth or other status.

58 *Prosecutor v. Kupreškić*, *supra* note 37, para. 580.

59 *Prosecutor v. Šainović* et al, Case No.IT-05-87-A, ¶ 1648 (ICTY, Appeal Judgment, Jan. 23, 2014).

60 *Prosecutor v. Furundžija*, Case No.: IT-95-17/1-T, ¶ 227 (ICTY, Trial Chamber II, Dec. 10, 1998).

61 Schabas, *supra* note 51, at 271.

62 *See, e.g.*, Independent International Commission of Inquiry on the Syrian Arab Republic (COI Syria), *UN Doc.* A/HRC/21/50, 16 August 2012, Annex II.

In contrast with Article 10, Article 21 has internal effects. It stipulates the applicable law for the ICC organs. The inclusion of such a provision on applicable law deviates from practices regarding the *ad hoc* tribunals and even more generally the statute differs in this respect from many other constitutive documents mandating international courts and tribunals that are silent on applicable law.[63] As observed by Verhoeven, the insertion of a provision on applicable law is not self-evident. In principle, applicable law logically and naturally flows from the delimitation of jurisdiction and judges do not need to be expressly instructed hereon.[64] In the context of the ICC, the question regarding applicable law was perhaps slightly more convoluted in light of arguments supporting direct application of domestic laws on certain matters or in given instances. Article 21 settled this part of the discussion in favor of the exclusive application of international law, with only indirect references to the national laws of states.[65]

The true utility of Article 21, however, or perhaps even its *raison d'être*, lies in the guidance that it offers to the ICC judiciary. Because of its instructive nature, it has been set forth by several scholars and also former judges that Article 21 actually aims to curtail the ICC judiciary.[66] Pursuant to such a line of thought, it could be argued that the ultimate aim of Article 21 is to preempt judicial lawmaking through overreliance on CIL in a similar zealous spirit as the ICC's *ad hoc* predecessors.[67] This seems confirmed by the failure to mention CIL expressly, which can be seen as an indication of the negotiators' unease with this source of law. Moreover, Article 21(1)(b),which only implicitly refers to CIL under the heading of "rules of international law," emphasizes

[63] As also noted by Gilbert Bitti, "Article 21 and the Hierarchy of Sources of Law before the ICC," in *The Law and Practice of the ICC* (Carsten Stahn ed., 2d ed. 2015).

[64] Joe Verhoeven, "Article 21 of the ICC Statute and the Ambiguities of Applicable Law," 33 *Neth. Y.B. Int'l L.* 2, 15 (2002).

[65] Margaret McAuliffe de Guzman, "Commentary on Article 21 'Applicable Law,'" *in Commentary on the Rome Statute of the International Criminal Court; Observers' Notes, Article by Article* (Otto Triffterer ed., 2d ed. 2008).

[66] The view that the drafting of the ICC statute presents an "overwhelming exercise in legal positivism" has been expressed, amongst others, by Judge David Hunt, "The International Criminal Court: High Hopes, 'Creative Ambiguity' and an Unfortunate Mistrust in International Judges," 2 *J. Int'l Crim. J.* 56 (2004). Pellet has also observed that Article 21 aims to deprive judges of their inherent powers. *See* Alain Pellet, "Applicable Law," *in The Rome Statute of the International Criminal Court: A Commentary*. Volume II 1053, 1056 (Antonio Cassese et al. eds., 2002). For a related argument specifically regarding Article 21(3), *see* Mahnoush Arsanjani, "The Rome Statute of the International Criminal Court," 93 *Am. J. Int'l L.* 22, 29 (1999).

[67] "In a different context, Judge Blattman also indicated that the ICC system was more rigid than the ICTY and meant to limit judicial discretion. *Prosecutor v. Lubanga*, Case No. ICC-01/04-01/06-2707, ¶ 7 (ICC, Trial Chamber I, Situation in the Democratic Republic of Congo, Separate Opinion of Judge René Blattman to the Decision on the Defence Request to Reconsider the "Order on Numbering of Evidence" of 12 March 2010, Mar. 30, 2011).

the secondary and complementary nature of CIL vis-à-vis the statute in several ways.[68] Indeed, the ICC Chambers have acknowledged the superiority of the statute over external sources of law.[69] They have confirmed that the external sources including CIL can only be applied when there is a lacuna in the statute.[70] Hence, in the *Katanga and Ruto* Decisions on the Conformation of Charges, the Pre-Trial Chambers refused to resort to CIL for questions on modes of liability as these were properly captured by Article 25 and there was thus no gap.[71] Only where statutory provisions refer more directly to the applicable law of Article 21 can there be an opening for direct engagement with CIL. An example hereof is Article 31 on defenses, which explicitly allows for the consideration of grounds excluding criminal responsibility that are not named in the statute but that exist under the other applicable law of Article 21.[72]

Beyond the setting of crime definitions and modes of liability, some chambers and dissenting judges have explored the concept of customary criminal procedural law.[73]

[68] This is done through the words "in the second place," "where appropriate," and by mentioning rules only after principles, whereas generally principles are regarded as more indeterminate than rules and therefore inherently of complementary nature. The order of Article 21(1)(b) seems to reverse this relationship. However, the Appeals Chamber has indicated that there is a strict hierarchy whereby general principles as mentioned in para. 1(c) are ranked after the sources mentioned in 1(b), *Prosecutor v. Lubanga*, Case No. ICC-01/04-01/06-772, ¶ 34 (ICC, Appeals Chamber, Judgment on the Appeal of Mr. Thomas Lubanga Dyilo against the Decision on the Defence Challenge to the Jurisdiction of the Court Pursuant to Article 19 [2] [a] of the Statute of 3 October 2006, AC, ICC, Dec. 14, 2006).

[69] *Lubanga, Situation in the Democratic Republic of the Congo*, ICC-01/04-01/06-772, ¶ 34 (ICC, Appeals Chamber, Judgment on the Appeal of Mr. Thomas Lubanga Dyilo against the Decision on the Defence Challenge to the Jurisdiction of the Court Pursuant to Article 19 [2][a] of the Statute of 3 October 2006, AC, ICC, Dec. 14, 2006).

[70] *See also* Verhoeven, *supra* note 64, at 17.

[71] *Prosecutor v. Germain Katanga*, Case No. ICC-01/04-01/07, ¶ 508 (ICC, Trial Chamber II, Sep. 30, 2008). *Prosecutor v. Ruto et al.*, Case No. No. ICC-01/09-01/11, ¶ 289 (ICC, Pre-Trial Chamber II, Decision on the confirmation of charges pursuant to Article 61[7][a] and [b] of the Rome Statute, Jan. 23, 2012).

[72] This also means that if CIL was to extend the defenses mentioned in Article 31 of the statute, the court would be entitled to apply those broader defenses going beyond its statute, as also observed by Dapo Akande, "Sources of International Criminal Law," in THE OXFORD COMPANION OF INTERNATIONAL CRIMINAL JUSTICE 50 (Antonio Cassese ed., 2009). It has also been argued that CIL could be given a greater role for *mens rea* questions through the phrase "unless otherwise provided" in Article 30(1) of the ICC Statute. *See, e.g.*, Gerhard Werle & Florian Jessberger, "'Unless Otherwise Provided': Article 30 of the ICC Statute and the Mental Elements of Crimes Under International Criminal Law," 3 J. INT'L CRIM. J. 35 (2005); Steffen Wirth, "Co-Perpetration in the Lubanga Trial," 10 J. OF INT'L CRIM. J. 971 (2012). This argument was explicitly rejected as unconvincing by Judge Van den Wyngaert in her Opinion to the Chui Judgment, *Prosecutor v. Chui*, Case No. ICC-01/04-02/12, fn. 16 (ICC, Trial Chamber II, Concurring Opinion Judge Van den Wyngaert, Dec. 18, 2012).

[73] *See, e.g.*, *Prosecutor v. Ruto and Sang*, Case No.ICC-01/09-01/11-1274-Corr2, ¶ 88–92 (ICC, Trail Chamber V[a], Decision on Prosecutor's Application for Witness Summonses and Resulting Request for State Party Cooperation, Apr. 17, 2014), referring to practices of other international criminal courts and tribunals; *Prosecutor v. Muthaura and Kenyatta*, Case No.ICC-01/09-02/11, (ICC, Trial Chamber V, Decision on the content of the updated document containing the charges, Dec. 28, 2012); and *Prosecutor v. Ruto and Sang*, Case No. ICC-01/09-01/11–522, ¶ 17–25 (ICC, Trial Chamber

For instance, in cases within the Kenya situation, the question came up whether the ICC has the power to subpoena unwilling witnesses to testify. This includes the question whether states can be obligated to compel and ensure the appearance of witnesses. To answer this question, the ICC Trial Chamber surveyed what it called, "customary international criminal procedure." In its view, such CIL was comprised of the practice of other international criminal courts. This reliance on "customary international criminal procedure" goes well beyond any doctrinal understanding of CIL.[74] Moreover, it did not bear fruit, given the different origins of the procedural framework of the ICTY and the ICC, as further explained.

The reluctance of the negotiators to include CIL in more direct terms, and consequently the precise formulation and construction of Article 21, have been severely criticized by commentators. The provision is said to be unduly dismissive of CIL. It forestalls CIL's potential to counter freezing effects that treaty making tends to have. In this respect, commentators have emphasized that CIL is often a useful device to prevent that the law becomes rapidly outmoded.[75] They have submitted that the drafting of Article 21 was inspired by an overly demanding reading of the legality principle as if this principle requires fastidious and punctilious definitions.[76] In contrast, others have welcomed the gradual disappearance of CIL as a source of ICL. They see this as the natural consequence of a proper application of the legality principle.[77] In any event, given the more sophisticated normative framework, and in

V, Decision on the content of the updated document containing the charges, concurring separate opinion Judge Eboe-Osuji to both decisions, Dec. 28, 2012), invoking "customary international law in the area of criminal procedure," but perhaps rather referring to common practice and usage at other international criminal courts and tribunals, and thus conceptualizing custom in a rather creative way.

74 The Special Tribunal for Lebanon also relied on this notion, in a perhaps even more creative manner granting itself inherent jurisdiction on a given procedural matter, stating:

> The extensive practice of international courts and tribunals to make use of their inherent powers and the lack of any objection by States, non-state actors or other interested parties evince the existence of a general rule of international law granting such inherent jurisdiction. The combination of a string of decisions in this field, coupled with the implicit acceptance or acquiescence of all the international subjects concerned, clearly indicates the existence of the practice and opinio juris necessary for holding that a customary rule of international law has evolved." Case No. CH/AC/2010/02, 10 ¶ 47(STL Appeals Chamber, Decision on Appeal of Pre-Trial Judge's Order Regarding Jurisdiction and Standing, Nov. 10, 2010). The paragraph has been qualified as "extravagant," *see for comments* http://dovjacobs.com/2010/11/12/its-alive-judicial-activity-activism-at-the-special-tribunal-for-lebanon/ (last accessed Mar. 08, 2015). *See also* Bitti, *supra* note 64, who argues that the reference to customary international criminal procedural law "is not only misplaced but also in complete contradiction with what Article 21 dictates to the ICC judges."

75 *See supra* text accompanying note 67.

76 Luigi Condorelli, "Présentation de la IIème Partie," in *Droit international pénal* 246 (Herv Ascencio et al. eds., 2000), as cited by Alain Pellet, *supra* note 66, fn. 39.

77 Dov Jacobs, "International Criminal Law," in *International Legal Positivism in a Post-Modern World* 458–66 (Jorg Kammerhöfer & Jean d'Aspremont eds , 2015).

particular the detailed Elements of Crimes, there is decisively less need for resort to CIL in the ICC setting as compared to the ICTY.

CIL AND ICL – SOME DILEMMAS

Despite all the criticisms, the omission of Article 21 to overtly refer to CIL corresponds with dilemmas that are inherent in the notion of customary ICL. Legality concerns have been at the forefront of critiques against CIL. Essentially, it is argued that allowing for too much judicial latitude is at odds with the requirement of fair notice and the prohibition of arbitrariness that underlie the legality principle. Hence, legality critiques are principally concerned with the inherent vagueness of CIL as well as with loose identification methods. With a view to reshaping or preserving any potential future role of CIL in ICL, this section considers some distinct methodological challenges that have arisen in the practice of ICL. It argues, first, that the capacity of CIL to be a ground for rules on procedure and modes of liability is limited because of the domestic origins of these rules. Furthermore, these rules are jurisdiction specific, and particularly procedural rules have no separate existence beyond their jurisdiction. This jurisdiction specificity has implications for the formation of *opinio juris*.[78] On another level, the common law–civil law divide also complicates CIL formation. The coexistence of two different criminal law systems grounded in fundamentally distinct theories very often inhibits the emergence of consistent and general state practice on a certain matter and therefore the common-civil law divide also practically impedes CIL formation.

Substantive ICL, that is, the law on the core crimes, is not confronted with these divides as these rules have a primary international origin. Dilemmas in this context rather relate to the need to ground new CIL rules in sufficient practice and the limited relevance of certain types of practice. Hence, this section presents as a second argument that the strict delimitation of the area of ICL as put forward in the preliminary section has implications for the types of practice that can prove the existence of CIL on substantive ICL, that is, the crime definitions.

Procedure, Modes of Liability, and CIL

More than other areas of international law, ICL is premised on a domestic pedigree and it utilizes concepts that have a domestic origin. Whereas the crime definitions, such as

[78] *See*, for a more general and not-ICL related argument that technical, institutional, and jurisdiction-related rules cannot be customary in nature, the arguments of Eritrea before the Ethiopia/Eritrea Claims Commission, as discussed in Pierre d'Argent and Jean d'Aspremont, "La Commission des Réclamations Érythrée – Éthiopie: un Premier Bilan," 53 *ANNUAIRE FRANÇAIS DE DROIT INTERNATIONAL* 347, 361 (2007).

those for genocide and crimes against humanity, have an international origin, rules on procedure and modes of liability are derived from domestic concepts.[79] The interplay between the domestic and the international and the dual roots of ICL has special bearing on the formation of CIL, in particular for rules on procedure and modes of liability.[80] The setting is further complicated by the coexistence of distinct domestic systems of criminal law in the form of the German-Romanic or civil law system and the Anglo-Saxon or common law system.

ICL's domestic–international law divide thus refers to the domestic origin of some ICL concepts, mainly rules of procedure and modes of liability. This divide has been exacerbated by the general implementing practice of state parties to the ICC statute. Most state parties insert only the ICC crime catalogue and they do not incorporate the ICC rules of procedure or modes of liability into their domestic codes, with a possible exception of superior responsibility. In so doing, states recognize the international character of the crime definition while preserving their own general part on *mens rea*, modes of liability, justifications, and excuses.[81] In the same spirit, domestic international crime prosecutions are generally governed by domestic codes of criminal procedure, as these procedures equally apply to domestic crime prosecution. This setting complicates the formation of CIL on procedure or modes of liability. Whereas the crime definitions may have a double life, as they are international in origin and domesticized by implementation,[82] modes of liability and procedure have remained more jurisdiction specific. They have thus a separate existence at the domestic and the international level with limited interaction. These rules thus do not easily transcend their own jurisdiction. Domestic courts attaching domestic modes of liability to international crimes do not, in principle, contribute to CIL formation on modes of liability. Similarly, when these courts apply domestic rules of procedure as they equally govern other domestic crime prosecutions, there is no international element enabling

[79] An exception to the above are the so-called international modes of liability, such as command responsibility, which have international roots. These international modes of liability have specifically been developed to be attached to international crimes. Command responsibility is enshrined in the 1949 Geneva Conventions and Additional Protocol I and was developed without reference to a concrete jurisdiction. The Genocide Convention also articulates modes of liability in its article 3, but these are so indeterminate that they should be regarded as instructions to states rather than as full-fledged international modes of liability. *Cf.* Elies van Sliedregt, *Individual Criminal Responsibility in International Law* 11 (2012).

[80] *See, generally*, Andre Nollkaemper, *National Courts and the International Rule of Law* (2011); and Larissa van den Herik, "International Criminal Law and Domestic Courts," *retrieved from* http://legal.un.org/avl/ls/Herik_CP.html (last accessed on Mar. 08, 2015).

[81] In favor of this development is Greenawalt, who has indicated that striving for uniformity at the international level would cause fragmentation at the domestic level. *See* Alexander K. A. Greenawalt, "The Pluralism of International Criminal Law," 86 *Ind. L.J.* 1063 (2011).

[82] *See* Nollkaemper, *supra* note 80, 218–20, and Karen Knop, "Here and There: International Law in Domestic Courts, 32 *N.Y.U. J. Int'l L. & Pol.* 501 (2000).

CIL formation. Because domestic courts thus apply domestic rules of procedure and domestic modes of liability, it will be difficult to prove any *opinio juris*, that is, belief that it is international law that requires a certain application of the law. After all, in such instances these courts are merely applying their own domestic law. In this sense, the mixed implementation methods of ICC state parties and adherence to domestic categories has not been conducive to the development of shared rules of procedure or the idea of international modes of liability.

Over time, this could possibly change if a practice emerged of implementing states copy-pasting the ICC modes of liability or a practice of domestic courts that, while adjudicating international crimes, reinterpret their own modes of liability in an ICC spirit.[83] If courts adjudicating international crimes indeed model their modes of liability on the ICC statute or other international law, this could be regarded as evidence of *opinio juris* – a belief that there is a CIL rule giving shape to that specific mode of liability and thus contributing to an international general part of ICL.[84] Similarly, if a coherent approach would emerge of courts using distinct procedural rules that are tailored to international crime adjudication, either through special domestic legislation or judicial reinterpretation inspired by international law, this could indeed, in theory, lead to the formation of customary international procedural law.

However, another great stumbling block to the formation of CIL in ICL relates to the identification of state practice. As indicated, the civil–common law divide often inhibits the formation of CIL. Since foundational theories on modes of liability are split over the same two schools of thought of civil versus common law, no consistent state practice can easily emerge. Judge Shahabuddeen observed this in the ICTR *Gacumbitsi* case, when he contrasted co-perpetration with joint criminal enterprise. These two theories cannot coexist, and so within each legal system one theory must prevail. In the words of Judge Shahabuddeen, "Since several states adhere to one theory while several other states adhere to the other theory, it is possible that the required state practice and *opinio juris* do not exist so as to make either theory part of customary international law."[85] This observation was corroborated by Judge Van

83 The Dutch *travaux préparatoires* of the ICC Implementation Act specifically instruct the Dutch judge to apply domestic modes of liability. However, in situations where this preference for domestic modes of liability over ICC modes of liability leads to diametrically opposed results, as was the case in the Van Anraat case, the domestic mode of liability may be given an attuned interpretation, *Kamerstukken II* 2001/02, 28 337, nr. 3, p. 26–27. For additional discussion of this case, *see* Harmen G. van der Wilt, "Genocide, Complicity in Genocide and International v. Domestic Jurisdiction; Reflections on the van Anraat Case," 4 *J. Int'l Crim. Just.* 239 (2006).

84 *See further* on the need for reflection on the general part of ICL, Kao Ambos, "Remarks on the General Part of International Criminal Law," 4 *J Int'l Crim. Just.* 660 (2006).

85 *Prosecutor v. Gacumbitsi*, Case No.ICTR-2001-64-A, ¶ 51(ITCR, Appeal Judgment, Separate Opinion of Judge Shahabuddeen, July 7, 2006).

den Wyngaert, who highlighted the radical fragmentation of national legal systems regarding the control over the crime theory.[86]

The same divide has had mitigating effects on questions of procedural law. Moreover, given that the drafters of the ICC statute have designed a procedural framework that is more civil law oriented, whereas the ICTY's procedural regime is overall common law based,[87] this divide has also limited the precedential value of ICTY findings for the ICC on questions of procedure and modes of liability. In the *Lubanga* case, when discussing whether witness proofing should be permitted, the ICC chamber reasoned:

> The ICC Statute has, through important advances, created a procedural framework which differs markedly from the ad hoc tribunals, such as, for example, in the requirement in the Statute that the prosecution should investigate exculpatory as well as incriminatory evidence, for which the Statute and Rules of the ad hoc tribunals do not provide. Also, the Statute seemingly permits greater intervention by the Bench, as well as introducing the unique element of victim participation. Therefore, the Statute moves away from the procedural regime of the ad hoc tribunals, introducing additional and novel elements to aid the process of establishing the truth. Thus, the procedure of preparation of witnesses before trial is not easily transferable into the system of law created by the ICC Statute and Rules. Therefore, while acknowledging the importance of considering the practice and jurisprudence at the ad hoc tribunals, the Chamber is not persuaded that the application of ad hoc procedures, in the context of preparation of witnesses for trial, is appropriate.[88]

In this decision, the trial chamber also noted the absence of evidence of any relevant practice from German-Romanic systems on the question of witness preparation.[89]

The civil–common law divide has impeded the existence of consistent state practice on other procedural questions and modes of liability as well. As state practice is very often scattered over the two great families of criminal procedural law, instances of harmonized or consistent practice are rare.

The Core Crimes and CIL

In contrast to rules of procedure and modes of liability, international crimes are inherently constructs of international law. They have been conceived at the

86 *Prosecutor v. Chui*, Case No. ICC-01/04-02/12, ¶ 17 (ICC, Trial Chamber II, Concurring Opinion Judge Van den Wyngaert, Dec. 18, 2012). She made this observation in relation to an exploration of general principles of law, as she had already questioned whether the ICC could draw on custom for modes of liability.

87 *Prosecutor v. Lubanga*, Case No. No.: ICC-01/04-01/06, ¶ 44–45 (ICC, Decision Regarding the Practices Used to Prepare and Familiarize Witnesses for Giving Testimony at Trial, Nov. 30, 2007).

88 *Prosecutor v. Lubanga*, *supra* note 87, paras. 44–45.

89 *Id.* para. 41.

international plane and international intervention is their *raison d'être*. Therefore, they are more susceptible to having CIL status. Nevertheless, given the delimitations of ICL as set out in the preliminary section and in particular ICL's special character as not being an area of law that regulates interstate relations, some observations are in order regarding the manifestations of state practice and *opinio juris*, and particularly the question of a hierarchy of various forms of practice.

It was already proposed by the ICTY in early case law to adopt a differentiated approach to manifestations of practice in ICL matters given the special nature of this field.[90] If it is indeed accepted to move beyond actual or physical state behavior and to accept other forms of practice, it would seem that in the case of ICL national legislation and domestic court decisions constitute the most significant types of practice. Those forms of practice can be important drivers for the further development of customary substantive ICL beyond the ICC statute. Concrete claims in this regard have been that the policy element of crimes against humanity does not exist under CIL and that the customary crime of genocide also protects political groups.[91] When evaluating such claims and the evidence presented to support them, it is important to keep the delimitation of ICL in mind. Not all national court decisions carry the same weight. In the context of ICL, it is particularly important to distinguish between types of courts, that is, criminal courts, civil courts, and administrative courts. Not only criminal courts may resort to ICL concepts. At the domestic level, civil courts may also deal with international crimes for reparatory purposes,[92] and administrative asylum courts may invoke them in the context of 1F procedures.[93] Decisions from these courts should not be given the same weight as decisions from domestic criminal courts. Such an ICL-specific instruction could thus complement the more generic observation of the ILC Special Rapporteur Wood, who underlined the importance of differentiating between courts on the basis of hierarchy. In this sense, he pointed out that judgments of the highest courts carry most weight, whereas cases that have been overturned should not be considered as contributing to practice.[94]

90 *Prosecutor v. Tadić*, *supra* note 11, para. 99. *See also id.* para. 194.

91 Beth van Schaack, "The Crime of Political Genocide: Repairing the Genocide Convention's Blind Spot," 106 YALE L.J. 2259 (1997).

92 *See, e.g.*, Larissa van den Herik, "Addressing "Colonial Crimes" Through Reparations?" 10 *J. INT'L CRIM. JUST.* 693 (2012)

93 Elies van Sliedregt, "International Crimes Before Dutch Courts: Recent Development," 4 *LEIDEN J. INT'L L.* 895 (2007).

94 *Second Report on the Identification of Customary International Law*, Special Rapporteur Michael Wood, UN Doc. A/CN.4/672, 22 May 2014, para. 41.

Going beyond the domestic level, human rights courts and commissions of inquiry have also been engaged with international crimes in their operations.[95] Furthermore, states and other nonjudicial organs may utilize international crime labels for political or other purposes. For instance, in Resolution 1973 the Security Council concluded that the widespread and systematic attacks in Libya could amount to crimes against humanity.[96] The General Assembly used the same term to condemn the prison camps in North Korea and to urge the Security Council to refer the matter to the ICC.[97] However, on a sliding scale from judicial to nonjudicial, such engagement with ICL by actors other than criminal courts can only have peripheral effects on the formation of CIL. It is thus argued here that the pluriformity of actors that engage with international criminal law notions must have a bearing on evidentiary questions, and that, in fact, for ICL there should be a certain hierarchy in the manifestations of practice that are accepted as relevant for the formation of CIL.

CONCLUSION

Clearly, the role of CIL in ICL is not static. After a preponderance of CIL findings in the first period of ICL's revival in the 1990s and the start of the twenty-first century, the ICC statute prompted a codified era. The codification moment and the gradual maturation of ICL made resort to CIL less necessary in a practical sense. The pragmatism thus coincided with more foundational concerns regarding the use of CIL in ICL.

Today, the future of ICL is said to be domestic.[98] If so, it shifts the focus to domestic courts as prime CIL identifiers. As such, domestic courts will act with a double identity as both lawmakers and law identifiers, which renders this distinction even more complicated. Or in the words of the special rapporteur, "Decisions of national courts may play a dual role in relation to customary international law: as State practice, but also as a means for the determination of rules of customary international law."[99] The Special rapporteur also observed that in the latter capacity, domestic judicial decisions should be appraised with caution.[100] To some extent

95 *Jorgic v. Germany*, Application No. 74613/01 (ECtHR, Oct. 12, 2007) (Judgment); and *Korberly v. Hungary*, Application No.9174/02 (ECtHR, Sep. 19, 2008). Independent International Commission of Inquiry on the Syrian Arab Republic (COI Syria).

96 UN Doc. S/RES/1973 (2011).

97 *Situation of Human Rights in the Democratic People's Republic of Korea*.UN Doc. A/RES/69/188.

98 Anne-Marie Slaughter & William Burke-White, "The Future of International Criminal Law *is* Domestic (or, The European Way of Law), 47 HARV. INT'L L.J. 327 (2006).

99 *See* ILC, *Third Report on Formation and Evidence of Customary International Law*, ILC Special Rapporteur Michael Wood, UN Doc. A/CN.4/682, 27 May 2015, para. 58.

100 *Id.*

this ties into suggestions made in this chapter in favor of differentiating between different types of domestic judicial decisions.

As for the domestic level, the actual role of CIL in those settings ultimately depends on jurisdictional schemes and openness to CIL. Domestic legality requirements may be more stringent, thus reducing the space for CIL to play a meaningful role. Nonetheless, domestic courts may refer to CIL for interpretative purposes and thus contribute to the further gradual development of ICL. In these exercises, domestic courts may well advance identification methods that are tailored to the specificity of ICL.[101] Given the nature of ICL, CIL identification on ICL rules is likely to become more and more a comparative law exercise. Hence, there is a future for CIL in ICL, albeit reduced and modified.

[101] This discussion has been taken up by the ILC, and its reports are meant to offer guidance to domestic courts. *See* ILC, *First Report on Formation and Evidence of Customary International Law*, ILC Special Rapporteur Michael Wood, UN Doc. A/CN.4/663, 17 May 2013, para. 22.

10

Customary International Law and Public Goods

Niels Petersen

The protection of global public goods is one of the most pressing challenges we are facing today.[1] The mitigation of climate change is the most prominent example.[2] In the next decades, global warming is feared to cause rising sea levels, melting ice and snow, increasing precipitation, extreme weather events, and widespread coastal flooding. By now, it has been almost universally recognized that these climate changes are at least influenced by human action.[3] Since the early 1990s, the international community has, in principle, recognized this challenge. In 1992, the United Nations Framework Convention on Climate Change (UNFCCC) was concluded at the end of the Rio summit. It has been ratified by 196 parties.

While there are only two international treaties, the Vienna Convention for the Protection of the Ozone Layer and the corresponding Montreal Protocol, which have more ratifications than the UNFCCC, the effectiveness of the Framework Convention has been limited. In the Kyoto Protocol, which was adopted in 1997 and entered into force in 2002, the states agreed for the first time on concrete targets for the reduction of greenhouse gas emissions. Nevertheless, the Kyoto obligations have often been criticized as too modest.[4] But even these modest targets have failed

[1] *See* Curtis A. Bradley & Mitu Gulati, Withdrawing from International Custom, 120 *Yale L. J.* 202, 264 (2010).

[2] *See* Nico Krisch, "The Decay of Consent: International Law in an Age of Global Public Goods," 108 *Am. J. Int'l L.* 1, 15 (2014); Alexander Thompson, "Rational Design in Motion: Uncertainty and Flexibility in the Global Climate Regime," 16 *Eur. J. Int'l Rel.* 269, 269–70 (2010); Joel P. Trachtman, "Persistent Objection, Cooperation, and the Utility of Customary International Law," 21 *Duke J. Comp. & Int'l L.* 221, 223 (2010).

[3] *See* *Climate Change 2013: The Physical Science Basis. Working Group I: Contribution to the Fifth Assessment Report of the Intergovernmental Panel on Climate Change* (Thomas F. Stocker *et al.* eds., 2013). *See also* Jody Freeman & Andrew T. Guzman, "Climate Change and U.S. Interests," 109 *Colum. J. Transnat'l L.* 1531, 1544–45 (2009) (with further references).

[4] *See* Jutta Brunnée & Stephen J. Toope, *Legitimacy and Legality in International Law* 137 (2010); Thompson, *supra* note 2, at 277. *But see* Charlotte Streck, "Innovativeness and Paralysis in

to prevent the significant increase in greenhouse gas emissions during this century.[5] Major actors, such as the United States, have not ratified the Kyoto Protocol, while other states have failed to comply with their reduction goals.

Therefore, progress in establishing a cooperative framework for tackling climate change can, at best, be described as very slow.[6] Some authors have suggested that treaty mechanisms might not be the optimal instrument for climate regulation, as they involve high decision-making costs.[7] Instead, they propose customary international law as an alternative, which might, due to its flexibility, be better suited to the achieve the regulatory goals.[8]

This chapter aims to examine this hypothesis. It explores the potential of customary international law to contribute to the protection of global public goods. The analysis proceeds in the three steps. First, it will have a closer look at the concept of public goods and common pool resources. Economic theory predicts that public goods are usually underprovided and common pool resources overexploited.[9] Experimental evidence confirms the general trend, even though the level of contributions is higher than economic theory predicts. Most individuals are not purely self-interested free riders, but conditional cooperators who are willing to contribute if everybody else contributes.

On this basis, the second section examines the formation of customary international law. The analysis shows that sustaining cooperation in multilateral settings through customary law is difficult. With regard to the mitigation of climate change, it is unlikely that states will coordinate on an equilibrium that will lead to a reduction of greenhouse gas emissions. The section then examines two further ways of identifying customary international law through moral interpretation and judicial

International Climate Policy," 1 *Transnat'l Envt'l L.* 137, 142–51 (2012) (pointing out innovations of the Kyoto Protocol).

5 David G. Victor, *Global Warming Gridlock: Creating More Effective Strategies for Protecting the Planet* 3 (2011). *See also* Sujata Gupta & Dennis A. Tirpak, Policies, Instruments and Co-Operative Arrangements," in *Climate Change 2007: Mitigation. Contribution of Working Group III to the Fourth Assessment Report of the Intergovernmental Panel on Climate Change* 745 (Bert Metz, Ogunlade Davidson, Peter Bosch, Rutu Dave, & Leo Meyer eds., 2007) (with further references).

6 Streck, *supra* note 4, at 139.

7 Scott Sullivan, "Networking Customary Law," 61 *Kan. L. Rev.* 659, 662–65 (2013).

8 Pierre-Marie Dupuy, "Formation of Customary International Law and General Principles," in *The Oxford Handbook on International Environmental Law* 449, 463–65 (Daniel Bodansky, Jutta Brunnée, & Ellen Hey eds., 2008); Sullivan, *supra* note 7, at 661–68 (making special reference to international negotiations on climate change). *See also* Francesco Parisi, "Toward a Theory of Spontaneous Law," 6 *Const. Pol. Econ.* 211, 214 (1995) (arguing that "processes of individual and social adaptation" may still work when explicit bargaining is unlikely to occur or is too costly).

9 Katharina Holzinger, "Common Goods, Matrix Games and Institutional Responses," 9 *Eur. J. Int'l Rel.* 173, 176 (2003).

lawmaking. However, the potential of these two avenues to protect global public goods effectively is rather limited.

The final section analyzes the protection of global public goods through the initially unilateral extension of authority. One problem of global public goods is that states have shared authority over them. A solution might be to divide authority by extending the jurisdiction of the nation states. I will draw from an example of the protection of common pool resources: the protection of fish stocks through the extension of the exclusive economic zone to 200 miles. In a second step, I will analyze whether this example contains any lessons for the mitigation of climate change.

THE CONCEPT OF PUBLIC GOODS AND OF COMMON POOL RESOURCES

Public Goods as an Economic Concept

Traditionally, public goods are characterized by two properties: non-excludability and non-rivalry in consumption.[10] This means that nobody can be excluded from using the good, and the quality and the availability of the good does not depend on the number of people using it. The classic example is national defense. The existence and the quality of an army do not depend on the number of people who are living in the territory defended by the army. Furthermore, no person living in the defended territory can be excluded from the protective effect of the army. Other examples are lighthouses, street lighting, fresh air, or the ozone layer. Public goods have positive externalities and thus give incentives to free ride. For each individual, it is rational not to contribute to the provision or maintenance of a public good and to hope that other members of the community will finance it. If four roommates in a shared apartment decide on whether to buy a shared TV, each roommate may want to understate the utility of the TV. Once the other three have bought the TV for the common space, it would be difficult to exclude the fourth member from using it even though she has not contributed. However, if every roommate understates her utility, it is less likely that the TV will be bought in the end. Therefore, market mechanisms provide fewer public goods than would be economically efficient.[11]

Some public goods have the opposite problem: Their use does not create positive, but negative external effects. These goods are often called "public bads." The

[10] Richard Cornes & Todd Sandler, *The Theory of Externalities, Public Goods, and Club Goods* 144–45 (2d ed. 1996); Holzinger, *supra* note 9, at 176. *See also* the seminal contribution of Paul A. Samuelson, "The Pure Theory of Public Expenditure," 36 *Rev. Econ. & Stat.* 387 (1954) (focusing on the non-rivalry of consumption).

[11] *Seminally*, Mancur Olson, *The Logic of Collective Action: Public Goods and the Theory of Groups* (1965).

primary example of a public bad is the pollution of the atmosphere. Nobody can be excluded from polluting the atmosphere and one's ability to pollute does not depend on whether other people are polluting, as well. In effect, public bads are the mirror image of public goods and can thus be modeled in the same way as the latter. We formulate the previous example in a positive way and talk of the protection of clean air instead of the pollution of the atmosphere, which would transform the problem from one of preventing a public bad to one of maintaining a public good.

Finally, there are goods that are usually called common goods or common pool resources. With public goods, they share the characteristic that nobody can be excluded from using them; but there is rivalry in consumption. The use of a common pool resource diminishes its quality and decreases its availability. Garret Hardin has shown in his classic contribution on the tragedy of the commons that such shared resources are usually overexploited.[12] Every individual who uses a common pool resource obtains the full benefit of the use, while the costs of the use are socialized in that they are borne by the entirety of the potential users. Examples for common pool resources are global fish stocks or other natural resources. The border between public goods and common pool resources is more salient in theory than in practice. Usually, the incentive structure is the same: Public goods are underprovided and common pool resources are overexploited.[13]

In the context of the nation-state, this problem is often met through public regulation. National defense, lighthouses, and street lighting are financed through taxes. Every taxpayer has to contribute to the provision of the public good, regardless of the extent to which they value it. In other cases, the external effects of public goods are internalized through regulatory measures. One example is knowledge, which is protected through copyrights and patents. These measures do not limit the access to knowledge, but their commercial use. Negative external effects that occur through the use of common pool resources are reduced through regulation and taxation of certain conduct. A measure that requires factories to limit their emissions of polluting substances aims at reducing the social costs of private production processes. In turn, imposing taxes on gasoline is a way to internalize the social costs of driving.

[12] Garrett Hardin, "The Tragedy of the Commons," 162 SCIENCE 1243 (1968).

[13] External effects do not necessarily lead to market inefficiencies. Ronald Coase explained in a seminal article that trade can resolve the matter effectively if there are no transaction costs. *See* Ronald H. Coase, "The Problem of Social Cost," 3 *J. L. & ECON.* 1 (1960). However, contracting about public goods and common pool resources often entails significant transaction costs because of the amount of affected people, so that a Coaseian bargaining solution does not work. *See* Alexander Morell, "Demand, Supply, and Markets," in *ECONOMIC METHODS FOR LAWYERS* (Emanuel Towfigh & Niels Petersen eds., forthcoming 2015).

However, regulatory measures that aim at the provision of public goods and the protection of common goods against overexploitation are not easily available on the international plane. This is problematic if the scale of a specific public good transgresses the border of a nation state. The incentive structure for the mitigation of climate change is, basically, the same as for the provision of national defense. The decisive difference is the scale of the problem. While national defense can effectively be provided through public regulation on the nation-state level, climate change is a global problem. Greenhouse gas emissions not only affect the climate on a specific territory, but they affect the global climate. Consequently, national legislatures often do not have incentives to internalize costs through public regulation.

On the international level, there is no central legislature that could impose taxes to finance the provision of a global public good. Certainly, states can coordinate to provide jointly for the public good. However, in such a situation, we reencounter the free-rider problem. As contributions to the public good are costly, every state has incentives to understate its own utility.[14] The same applies to the reverse situation of protecting common pool resources. If the individual benefits obtained by a national economy from using a shared resource are higher than the costs that are imposed on the economy, states have incentives not to impose regulation against overexploiting the resource. The use of common pool resources is thus often not regulated, even if their use has socially detrimental effects on the global level.

Experimental Evidence

Since the 1970s, there has been extensive experimental research on the predictions of economic theory regarding public goods and the conditions under which public goods are provided through cooperation.[15] Public goods experiments are usually played in groups of at least four participants. These participants get a certain initial endowment and have the choice between either keeping their endowment or contributing all or part of it to a specific public good. The money that is contributed to the public good is multiplied by a factor *x* that is greater than 1, but smaller than the number of players, and then equally distributed among all players.

[14] However, public goods may be provided if one state has such a high interest in the provision of the public good that its benefits to that state outweigh the costs; *see* Christopher Bliss & Barry Nalebuff, "Dragon-Slaying and Ballroom Dancing: The Private Supply of a Public Good," 25 *J. PUBL. ECON.* 1 (1984).

[15] For a review of the experimental literature, *see* John. O. Ledyard, "Public Goods: A Survey of Experimental Research," in *THE HANDBOOK OF EXPERIMENTAL ECONOMICS* 111 (John H. Kagel & Alvin E. Roth eds., 1995); Ananish Chaudhuri, "Sustaining Cooperation in Laboratory Public Goods Experiments: A Selective Survey of the Literature," 14 *EXP. ECON.* 47 (2011).

If all players contribute all of their endowment to the establishment of the public good, the overall payoff is higher than if everybody keeps their endowment and contributes nothing. Contributing to the public good is thus the socially optimal solution. Individually, however, it is rational to keep the endowment because everybody receives a payoff of less than 1 for each unit contributed to the public good.[16]

The main finding of the early public goods experiments was that contributions to the public good are higher than would be expected by economic theory.[17] On average, players contribute between 40 to 60 percent of the socially optimal level.[18] However, there is a steady decline in cooperation over time.[19] Early attempts to explain this observation have found that participants are not only guided by pure self-interest, but that most of them base their contributions on fairness considerations even though the individual fairness norms may vary.[20] A closer look roughly distinguishes several types of participants.[21] Only a minority of participants are free riders who are solely interested in personal gains and never contribute significant amounts to the public good. In contrast, the majority of participants are conditional cooperators.[22] These conditional cooperators contribute to the public good, as long as others contribute, as well. They are happy to advance the common cause, but they do not want to be exploited by the other participants.

For conditional cooperators, cooperation thus depends on their beliefs of whether others will cooperate, as well.[23] These expectations are usually formed by observing

[16] *See* Christoph Engel & Michael Kurschilgen, "The Coevolution of Behavior and Normative Expectations: An Experiment," 15 *Am. L. & Econ. Rev.* 578, 579 (2013).

[17] *See*, e.g., Gerald Marwell & Ruth E. Ames, "Experiments and the Provision of Public Goods: I. Resources, Interest, Group Size, and the Free-Rider Problem," 84 *Am. J. Soc.* 1335 (1979); Elinor Ostrom, James Walker, & Roy Gardner, "Covenants With and Without a Sword: Self-Governance is Possible," 86 *Am. Pol. Sc. Rev.* 404 (1992); Vernon L. Smith, "Experiments with a Decentralized Mechanism for Public Good Decisions," 70 *Am. Econ. Rev.* 584 (1980).

[18] Matthew Rabin, "Incorporating Fairness into Game Theory and Economics," 83 *Am. Econ. Rev.* 1281, 1283 (1993); Chaudhuri, *supra* note 15, at 48.

[19] Chaudhuri, *supra* note 15, at 48. *See also* Christoph Engel, "The Emergence of a New Rule of Customary Law: An Experimental Contribution," 7 *Rev. L. & Econ.* 767, 773 (2011).

[20] Marwell & Ames, *supra* note 17, at 1357. For an attempt to model such fairness preferences, *see* Rabin, *supra* note 18.

[21] *See* Urs Fischbacher, Simon Gächter, & Ernst Fehr, "Are People Conditionally Cooperative? Evidence from a Public Goods Experiment," 71 *Econ. Letters* 397 (2001).

[22] Fischbacher, Gächter, & Fehr, *supra* note 21; Urs Fischbacher & Simon Gächter, "Social Preferences, Beliefs, and the Dynamics of Free-Riding in Public Goods Experiments," 100 *Am. Econ. Rev.* 541 (2010); Claudia Keser & Frans van Winden, "Conditional Cooperation and Voluntary Contributions to Public Goods," 102 *Scand. J. Econ.* 23 (2000).

[23] Martin Dufwenberg & Georg Kirchsteiger, "A Theory of Sequential Reciprocity," 47 *Games & Econ. Behav.* 268 (2004); Armin Falk & Urs Fischbacher, "A Theory of Reciprocity," 54 *Games & Econ. Behav.* 293 (2006).

the contributions of other participants.[24] The mean contribution in the first round sets the stage for later rounds: If the group makes high average contributions in the first round, it will continue to make rather high contributions in the following rounds.[25] However, even a small percentage of free riders or imperfect conditional cooperators may lead to a decline in average contributions and to the breakdown of cooperation.[26]

One mechanism that is usually proposed to prevent free riding and imperfect conditional cooperation is decentralized punishment. Several studies show that cooperation is reinforced if individual participants have the ability to punish those participants whose contributions are less than average.[27] Nevertheless, decentralized punishment mechanisms have two problems. On the one hand, punishment is costly. A punishment mechanism is itself a second-order public good.[28] Participants will be reluctant to punish free riders and rather hope that others will implement the costly punishment. Even if this does not mean that there is a total absence of punishment,[29] the punishment of free riders will be the harder to maintain the bigger the group size is. With increasing participants, free riding is more difficult to detect, and the public-good character of the punishment mechanism will become more pronounced. On the other hand, participants may be deterred from punishing free riders if the latter have means for counter-punishment.[30] Just the fear that punished participants might retaliate significantly reduces the willingness to sanction free riders.

Climate Change as an Example: Additional Challenges

In the context of climate change, the incentive structure is not the only obstacle to effective mitigation. Instead, there are several additional challenges. First, there is a

[24] Fischbacher & Gächter, *supra* note 22, at 548–49. *See also* ELINOR OSTROM, *GOVERNING THE COMMONS: THE EVOLUTION OF INSTITUTIONS FOR COLLECTIVE ACTION* 183–84 (1990) (discussing the importance of trust and communication for the success of decentralized cooperation).

[25] Christoph Engel *et al.*, "First Impressions Are More Important Than Early Intervention: Qualifying Broken Windows Theory in the Lab," 37 *INT'L REV. L. & ECON.* 126, 129–30 (2014); Engel, *supra* note 19, at 780–81; Keser & Winden, *supra* note 22, at 32.

[26] *See* Fischbacher, Gächter & Fehr, *supra* note 21, at 403; Fischbacher & Gächter, *supra* note 22, at 550–54.

[27] *See* Ernst Fehr & Simon Gächter, "Cooperation and Punishment in Public Goods Experiments," 90 *AM. ECON. REV.* 980 (2000); Ostrom, Walker, & Gardner, *supra* note 17; Toshio Yamagishi, "The Provision of a Sanctioning System as a Public Good," 51 *J. PERSONALITY & SOC. PSYCH.* 110 (1986).

[28] Yamagishi, *supra* note 27, at 111.

[29] *See* Fehr & Gächter, *supra* note 27 (arguing that participants have fairness preferences that lead them to punish even if it were individually rational to refrain from punishing).

[30] Nikos Nikiforakis, "Punishment and Counter-Punishment in Public Good Games: Can We Really Govern Ourselves?" 92 *J. PUBL. ECON.* 91 (2008).

discounting effect. Human beings usually value future utility less than their present utility.[31] From an individual perspective, it is rational to focus more on the present utility than on the utility in five or ten years, because there is a slight (or, depending on the age, not so slight) probability that one might not be alive anymore. From a social perspective, however, discounting should play a less prominent role. Even if certain individuals have died, it is unlikely that this will affect the society as a whole, as the deceased are replaced by the newly born. In the discussion on climate change, discounting is a significant factor. The costs caused by climate change will increase over the coming years and decades. However, the costs of measures to combat climate change have to be borne now. For this reason, societies may be reluctant to undertake measures that will only reap benefits for future generations.

Second, the costs of climate change are distributed asymmetrically.[32] Coastal regions and regions with a low altitude above sea level will suffer first. China, Europe, and the United States, the principal originators of climate change, may be reluctant to adopt measures for climate protection if the main effects of climate change are first experienced in developing countries, such as Bangladesh or the Comoros.[33] This differential effect shows that combating climate change involves difficult distributive issues. Who should pay what for saving the climate? Large distributive effects change the incentives for certain individual actors to contribute to a public good.[34] In our case, it makes the voluntary protection of the climate even more unlikely.[35]

Finally, there is another distributive issue. The discussion on the reduction of greenhouse gases usually includes proposals that countries have to reduce their emissions by a certain degree. The starting point for the calculation of these reduction targets is the current level of emissions. However, richer countries generally have a higher level of emissions and thus benefit from such a system.[36] Therefore, developing countries fear that their development could be hampered

[31] *See* Marco Percoco & Peter Nijkamp, "Estimating Individual Rates of Discount: A Meta-Analysis," 16 *Appl. Econ. Letters* 1235 (2009).

[32] *See* Niels Petersen, "The Role of Consent and Uncertainty in the Formation of Customary International Law," in *Reexamining Customary International Law* (Brian Lepard ed., forthcoming 2015). *See also* Bradley & Gulati, *supra* note 1, at 265 (generally discussing the challenge of asymmetry and heterogeneity); George Norman & Joel P. Trachtman, "The Customary International Law Game," 99 *Am. J. Int'l L.* 541, 554 (2005) (same).

[33] *See* Freeman & Guzman, *supra* note 3, at 1594.

[34] *See* Theodore Bergstrom, Lawrence Blume, & Hal Varian, "On the Private Provision of Public Goods," 29 *J. Publ. Econ.* 25 (1986).

[35] *See* Todd Sandler, *Global Collective Action* 221–25 (2004) (arguing that the unequal distribution of consequences is the main obstacle to global collective action in mitigating climate change).

[36] Brunnée & Toope, *supra* note 4, at 127–28.

by emission-reduction targets. This is, today, the most hotly contested issue in the negotiations under the UNFCCC.[37] The result of this dispute was the invention of the principle of Common But Differentiated Responsibility.[38] However, until there is agreement on how to operationalize this principle, it is mere lip service that does not contribute to finding a solution to the distributional problem of who bears the costs for mitigating climate change.

PUBLIC GOODS AND THE FORMATION OF CUSTOMARY INTERNATIONAL LAW

This section takes a closer look at the likelihood that norms of customary international law will emerge to protect global public goods. Traditionally, scholars dealing with the formation of customary law examine patterns of state practice: Under which conditions is it likely that states coordinate on an equilibrium that leads to the emergence of a new customary norm?[39] However, the analysis of state practice may not be the only way of identifying a customary norm. There is an important strand of international legal scholarship that observes and advocates a "modern" approach to identifying customary law.[40] This section will look at two different versions of

37 *See* Daniel Bodansky, "The Copenhagen Climate Change Conference: A Postmortem," 104 AM. J. INT'L L. 230, 232 (2010) ("Although the U.S.-EU negotiations were always difficult ... the split between the United States and the European Union pales in comparison to the gulf between developed and developing countries."); Lavanya Rajamani, "The Cancun Climate Agreements: Reading the Text, Subtext and Tea Leaves," 60 *INT'L & COMP. L. Q.* 499 (2011).

38 On this principle, see Duncan French, "Developing States and International Environmental Law: The Importance of Differentiated Responsibilities," 49 *INT'L & COMP. L. Q.* 35 (2000); Paul G. Harris, "Common But Differentiated Responsibility: The Kyoto Protocol and United States Policy," 7 *N.Y.U. ENVT'L L.J.* 27 (1999); Graham Mayeda, "Where Should Johannesburg Take Us? – Ethical and Legal Approaches to Sustainable Development in the Context of International Environmental Law," 15 *COLO. J. INT'L ENVT'L L. & POL'Y* 29 (2004); Christopher D. Stone, "Common But Differentiated Responsibilities in International Law," 98 *AM. J. INT'L L.* 276 (2004).

39 The classic contribution in this sense is Jack L. Goldsmith & Eric A. Posner, A Theory of Customary International Law, 66 *U. CHI. L. REV.* 1113 (1999). *See also* Mark A. Chinen, "Game Theory and Customary International Law: A Response to Professors Goldsmith and Posner," 23 *MICH. J. INT'L L.* 143 (2001); Edward T. Swaine, "Rational Custom," 52 *DUKE L. J.* 559 (2002).

40 *See, e.g.*, Roozbeh B. Baker, "Customary International Law in the 21st Century: Old Challenges and New Debates," 21 *EUR. J. INT'L L.* 173 (2010); Rudolf Geiger, "Customary International Law in the Jurisprudence of the International Court of Justice: A Critical Appraisal," in *FROM BILATERALISM TO COMMUNITY INTEREST: ESSAYS IN HONOUR OF BRUNO SIMMA* 673 (Ulrich Fastenrath *et al.* eds., 2011); Anthea Elizabeth Roberts, "Traditional and Modern Approaches to Customary International Law: A Reconciliation," 95 *AM. J. INT'L L.* 757 (2001); William Schabas, "Customary Law or 'Judge-Made' Law: Judicial Creativity at the UN Criminal Tribunals," in *THE LEGAL REGIME OF THE INTERNATIONAL CRIMINAL COURT. ESSAYS IN HONOUR OF PROFESSOR IGOR BLISHCHENKO* 77 (José Doria, Hans-Peter Gasser, & M. Cherif Bassiouni eds., 2009); John Tasioulas, "*Customary International Law and the Quest for Global Justice*," in *THE NATURE OF CUSTOMARY LAW* 307 (Amanda Perreau-Saussine & James Bernard Murphy eds., 2007).

alternative methods and their potential to contribute to the protection of global public goods. First, it will examine an approach that tries to include some kind of moral element into the identification process. Second, it will analyze elements of lawmaking in the jurisprudence of the International Court of Justice (ICJ).

Customary Law and Patterns of State Practice

It is beyond doubt that cooperation between different individuals is possible even if each participant has short-term incentives to defect. The tension between socially optimal and individually rational behavior is best expressed in the prisoner's dilemma.[41] In a one-shot prisoner's dilemma, each individual is better off if they defect, even though cooperation is the socially optimal strategy. If the game is played repeatedly for an indefinite number of times, participants in a bilateral prisoner's dilemma situation have incentives to cooperate in order to induce cooperation from the counterpart.[42] Nevertheless, it is unclear whether cooperation can also be maintained in multilateral prisoner's dilemma situations.

According to Jack Goldsmith and Eric Posner, cooperation is virtually impossible in multilateral settings.[43] While reciprocity is the driving force for cooperation in the bilateral situation, it is, they argue, much less effective in a multilateral setting: If one state overexploits a common pool resource, the reciprocal overuse by all other states is no effective response to protect the resource.[44] Furthermore, the costs of monitoring increase with the number of states, which could lead to more undetected free riding.[45]

41 A linear public good is characterized as a prisoner's dilemma of more than two players and a continuous space of action. However, the provision of public goods does not necessarily imply the incentive structure of a multiplayer prisoner's dilemma. *See* Holzinger, *supra* note 9; Stefan Magen, "Game Theory and Collective Goods," in ECONOMIC METHODS FOR LAWYERS (Emanuel Towfigh & Niels Petersen eds., forthcoming 2015).

42 *See* Goldsmith & Posner, *supra* note 39, at 1125–27; Norman & Trachtman, *supra* note 32, at 565–66. Technically, a repeated prisoner's dilemma has multiple Nash equilibria. According to the so-called folk theorem, we can expect cooperation in such situations if the players have a sufficiently low discount-rate. *See* Magen, *supra* note 41.

43 Goldsmith & Posner, *supra* note 39, at 1129–30. *But see* Robert Pahre, "Multilateral Cooperation in an Iterated Prisoner's Dilemma," 38 *J. CONFL. RESOLUTION* 326 (1994) (arguing that multilateral cooperation can, in some cases of the iterated prisoner's dilemma, be even more effective than bilateral cooperation).

44 Goldsmith & Posner, *supra* note 39, at 1129–30.

45 *Id.* at 1130. This argumentation was partly disputed by George Norman and Joel Trachtman; *see* Norman & Trachtman, *supra* note 32. They point out that reciprocity must not be impossible in multilateral settings (*id.* at 556–59). This is particularly the case if multilateral obligations can be divided into multiple bilateral obligations. But even in situations involving the protection of a public good, cooperation is – even if less likely – not impossible. Norman and Trachtman highlight reputation effects and possibilities of punishing defecting states by suspending cooperation in other fields (*id.* at 567).

This analysis is based on the assumption that states in international relations are only motivated by self-interest. If all states were potential free riders, multilateral cooperation would indeed be difficult to sustain. However, the experimental evidence suggests that most individuals are conditional cooperators.[46] We certainly have to be cautious to extrapolate from individual behavior to the conduct of state officials. There might be a selection effect that uncooperative individuals are more likely to assume public office. Furthermore, in democratic states, incentives to cater to the national electorate might also lead to more uncooperative conduct. However, there are at least serious doubts whether the assumption that states are free riders is plausible, *per se*. If we assume that at least some states are conditional cooperators, cooperation is more easily achievable.[47]

Furthermore, it is important to keep in mind the asymmetric importance of different states for the maintenance of a public good or a common pool resource. There are usually certain core states whose cooperation is more important for the protection of a global public good than the conduct of other states.[48] If China, the United States, and the European Union are responsible for the great majority of greenhouse gas emissions, any cooperative arrangements to reduce emissions have to include these actors. If states are conditional cooperators, their willingness to reduce greenhouse gas emissions will crucially depend on the cooperation of the principal originators.[49] Their efforts to reduce emissions have little effect on the global climate if the mainly responsible states do not contribute.

By contrast, this also means that not every defection severely undermines overall cooperation. The defection of non-lead actors will have less of an impact than the defection of leading states.[50] The good news is therefore that only the cooperation of a small number of core states is vital to sustain a high level of protection.[51] An

46 *See supra* text accompanying notes 15–30.

47 *See* Engel & Kurschilgen, *supra* note 16, at 585–88 (showing that the mere normativity of cooperation has a positive effect on contributions to public goods).

48 *See* Pierre-Hugues Verdier & Erik Voeten, "Precedent, Custom and Change in Customary International Law: An Explanatory Theory," 108 AM. J. INT'L L. 389, 404 (2014).

49 For experimental evidence showing that the example of stronger participants in a public goods game significantly affects the behavior of other participants, *see* Andreas Glöckner *et al.*, "Leading With(out) Sacrifice? A Public Goods Experiment with a Privileged Player," 49 ECON. INQUIRY 591 (2011).

50 However, this might lead some states to free ride if they know that their deviant conduct does not affect overall cooperation: They would reap the benefits of the cooperation of the lead actors while not bearing the costs; *see* Joel P. Trachtman, "The Growing Obsolescence of Customary International Law" (in this volume).

51 *See* Krisch, *supra* note 2, at 16. *See also* Victor, *supra* note 5, at 242–43 (advocating the formation of small clubs as a starting point for international commitments); Robin Eckersley, "Moving Forward in the Climate Negotiations: Multilateralism or Minilateralism," 12 GLOB. ENVT'L POL. 24 (2012) (advocating an inclusive "minilateralism" for the climate change negotiations); Sandler, *supra* note 35, at 217 (noting that the concentration of key emitters was one of the success factors for international efforts to combat CFC emissions and to protect the ozone layer).

example of how such cooperation could work is the 2009 Copenhagen Accord, which was reached by the heads of states of twenty-eight member states of the UNFCCC, including all major emitters.[52] Even though the Copenhagen Accord failed to achieve a formal status under the Framework Convention, it is often considered to be a promising step in the evolution of the climate regime.[53] Still, the accord focuses more on self-selected targets, actions, and procedural obligations. It is therefore far from being a breakthrough in establishing cooperation for mitigating climate change.

Moral Interpretations of Customary Law

Focusing on patterns of state practice is not the only way of identifying rules of customary law. States often have incentives to pursue short-term interests even though such conduct entails significant social costs and may run counter to the long-term interest of the state.[54] This divergence between short-term and long-term interests often becomes manifest if declarations and resolutions are adopted by states in multilateral settings and express desirable social goals that are not backed by a corresponding state practice. This has led many scholars to propose approaches to customary law that concentrate on *opinio juris* or moral principles instead of state practice.[55]

Some authors have proposed to renounce state practice as a constitutive element of customary international law and to regard it merely as an indication for the existence of an *opinio juris*.[56] Others want to dispense with practice at least for those principles that supposedly reflect the consensus of the international community[57]

52 UNFCCC, Decision 2/CP.15, 'Copenhagen Accord' (March 30, 2010), UN Doc FCCC/CP/2009/11/Add.1 (Copenhagen Accord).

53 Duncan French & Lavanya Rajamani, "Climate Change and International Environmental Law: Musings on a Journey to Somewhere," 25 *J. Envt'l L.* 437, 447 (2013).

54 *See* Vincy Fon & Francesco Parisi, "International Customary Law and Articulation Theories: An Economic Analysis," 2 *Int'l L. & Mgmt. Rev.* 201, 214 (2006) (proposing a stronger emphasis on declared state intentions instead of actual practice in such circumstances).

55 These approaches have often been coined with the term "modern approaches" to customary law. *See* Michael Wood, *First Report on Formation and Evidence of Customary International Law*, UN Doc. A/CN.4/663 (May 17, 2003), para. 98 (with further references).

56 *See* Brian D. Lepard, *Customary International Law. A New Theory with Practical Applications* 98 (2010); Curtis A. Bradley, "Customary International Law Adjudication as Common Law Adjudication" (in this volume); Bin Cheng, "United Nations Resolutions on Outer Space: 'Instant' International Customary Law?" 5 *Indian J. Int'l L.* 23, 35–40 (1965); Bin Cheng, "On the Nature and Sources of International Law," in *International Law. Teaching and Practice* 203 (Bin Cheng ed., 1982); Andrew T. Guzman, "Saving Customary International Law," 27 *Mich. J. Int'l L.* 115 (2005); Brian D. Lepard, "Customary International Law as a Dynamic Process" (in this volume). *See also* John Tasioulas, "Custom, *Jus Cogens*, and Human Rights" (in this volume) (arguing that *opinio juris* can establish a customary norm in the absence of state practice).

57 Louis B. Sohn, "The Human Rights Law of the Charter," 12 *Tex. Int'l L. J.* 129, 133 (1977).

or fundamental moral principles.[58] Bruno Simma and Philip Alston have proposed to concede the status of general principles to fundamental human rights norms.[59] Finally, there are approaches that do not totally dispense with state practice, but aim to diminish the importance of state practice in cases where we can identify a strong and consistent *opinio juris*,[60] or which relate to norms of a considerable moral importance.[61] While these approaches primarily try to justify the normative character of human rights norms because of their moral importance, they could equally be applied to the protection of public goods.[62] In cases where there is a divergence between short-term incentives of states and long-term social benefits, we could identify customary rules based on *opinio juris* rather than on actual state practice.

There are two objections against such an approach. Two decades ago, Martti Koskenniemi warned that an overly idealist interpretation of international law easily gets caught in a utopian trap, which deprives the legal norm of any practical significance.[63] According to this position, customary rules not reflected in state practice are nothing more than ineffective pipe dreams of international lawyers. Second, even if we all agree that the protection of the atmosphere is an important goal, it is unclear how this goal should be implemented, as implementation strategies usually involve complex compromises.[64] Strategies to combat climate change or to protect fish stocks include the reduction of a certain activity without renouncing it completely. States agree to reduce their carbon emissions or their fishing quotas, but they do not set them to zero.

While these objections have some merits, they underestimate the dynamic nature of the process of legal interpretation. It is often difficult to distinguish between the processes of lawmaking and law interpretation. Every dispute about the content of

[58] Fernando R. Tesón, "Two Mistakes about Democracy," 92 *ASIL Proceedings* 126, 127 (1998).

[59] Bruno Simma & Philip Alston, "The Sources of Human Rights Law: Custom, *Jus Cogens*, and General Principles," 12 *Austl. Yb. Int'l L.* 82 (1992). *Similarly*, Albert Bleckmann, "Zur originären Entstehung gewohnheitsrechtlicher Menschenrerchtsnormen," in *Menschenrechtsschutz durch Gewohnheitsrecht* 29, 42–43 (Eckart Klein ed., 2002); Oscar Schachter, "New Custom: Power, *Opinio Juris* and Contrary Practice," in *Theory of International Law at the Threshold of the 21st Century: Essays in Honour of Krzysztof Skubiszewski* 531, 539 (Jerzy Makarczyk ed., 1996). *See also* Niels Petersen, "Customary Law Without Custom? – Rules, Principles, and the Role of State Practice in International Norm Creation," 23 *Am. U. Int'l L. Rev.* 275, 303–05 (2008), who proposes to recur to general principles in the case of norms relating to public goods and to human rights.

[60] *See* Frederic L. Kirgis, "Custom on a Sliding Scale," 81 *Am. J. Int'l L.* 146 (1987).

[61] *See* John Tasioulas, "In Defence of Relative Normativity: Communitarian Values and the Nicaragua Case," 16 *Oxford J. Legal Stud.* 85 (1996).

[62] *See* Petersen, *supra* note 59, at 303–05.

[63] *See* Martti Koskenniemi, "The Politics of International Law," 1 *Eur. J. Int'l L.* 4 (1990).

[64] *See also* Trachtman, *supra* note 50 (discussing the lack of sufficient precision of customary law).

a rule is, at the same time, also a process of developing this legal rule.[65] Monica Hakimi has shown in her contribution to this volume how claims about customary norms can change the expectations of other actors regarding the content of these rules.[66] An interpretation of customary norms that relies on an *opinio juris* not supported by state practice may thus be a factor of changing this rule and thus state conduct in the long term.[67]

Furthermore, customary norms gain in precision if they originate from soft law instruments. It has long been recognized that certain soft law instruments, such as resolutions of the UN General Assembly, may be an expression of *opinio juris*.[68] In the context of climate change, the already-mentioned Copenhagen Accord might be an example of a soft law instrument expressing an *opinio juris* of the participating states. The targets agreed upon in the accord can lead to the formation of expectations by other states with regard to climate change mitigation.

Nevertheless, even under the conditions developed in the last two paragraphs, the role of customary law is limited. On the one hand, it is dependent on the existence of soft law instruments. Up to date, the scope of soft law instruments in the field of climate change is limited. On the other hand, one may ask whether labeling a soft law instrument as customary law adds anything to the normative force of soft law. In situations without a dispute resolution mechanism, the difference is probably marginal. The soft law instrument itself already creates normative expectations. It is hard to see how calling these norms customary norms makes any difference. However, there may be a difference in situations in which we have a dispute settlement system. This consideration will be addressed in more detail in the following section.

Judicial Lawmaking

It is a commonplace today that judicial decisions are not exclusively derived from legal norms. Instead, the social, cultural, and political background of judges influences their decision making, as do strategic considerations.[69] Courts not only

65 *See* Wayne Sandholtz & Alec Stone Sweet, "Law, Politics, and International Governance," in *The Politics of International Law* 238 (Christian Reus-Smit ed., 2004); Wayne Sandholtz, "Dynamics of International Norm Change: Rules Against Wartime Plunder," 14 *Eur. J. Int'l Rel.* 101 (2008).

66 Monica Hakimi, "Custom's Method and Process: Lessons from Humanitarian Law" (in this volume).

67 Niels Petersen, "Rational Choice or Deliberation? – Customary International Law between Coordination and Constitutionalization," 165 *JITE* 71, 81–82 (2009).

68 Alan Boyle & Christine Chinkin, *The Making of International Law* 212 (2007); Laurence R. Helfer & Ingrid Wuerth, "Customary International Law in the Age of Soft Law" (working paper, 2014); Jan Wouters & Linda Hamid, "Custom and Informal International Lawmaking" (in this volume).

69 There has been significant academic debate over which factors influence judicial decision making on the U.S. Supreme Court. However, there seems to be consensus that legal methods are not

find law, they also make law.[70] The ICJ is no exception in this respect.[71] Because of its vagueness, customary international law seems to be particularly prone to judicial lawmaking. If we analyze the approach of the ICJ to the identification of customary international law, it seems indeed that the jurisprudence is, at least in part, driven by strategic considerations.[72] Consequently, this section will analyze whether the ICJ might help the cause of mitigating global climate change by using its strategic discretion in order to identify customary norms favorable to the protection of the global climate.

Customary law often has the function of filling in gaps left by treaty law. According to conventional wisdom, situations that are not governed by legal rules are governed by the *Lotus* principle.[73] In *Lotus*, the Permanent Court of International Justice basically established a freedom of action for states.[74] According to the *Lotus* principle, states are only bound by legal rules to which they have consented. In some situations, however, the *Lotus* principle will not help to resolve a dispute between states because the dispute is precisely about competing claims of sovereignty.[75]

In such cases, the ICJ often recurs to pragmatic strategies of identifying customary norms. In some judgments, the court refers to principles of equity.[76] The most famous example is probably the *North Sea Continental Shelf* case.[77] In *North Sea Continental Shelf*, the court had to deal with the delimitation of continental shelf areas between Germany and Denmark as well as the Netherlands. The court found that there was no customary rule supported by sufficient state practice governing the case. The predominant equidistance principle had not been sufficiently supported

the only factor determining the results of judgments. On the different positions, *see* MICHAEL A. BAILEY & FORREST MALTZMAN, *THE CONSTRAINED COURT: LAW, POLITICS, AND THE DECISIONS JUSTICES MAKE* (2011); Thomas Brennan, Lee Epstein, & Nancy Staudt, "The Political Economy of Judging," 93 *MINN. L. REV.* 1503 (2009); Tracey E. George & Lee Epstein, "On the Nature of Supreme Court Decision Making," 86 *AM. POL. SC. REV.* 323 (1992); Mark J. Richards & Herbert M. Kritzer, "Jurisprudential Regimes in Supreme Court Decision Making," 96 *AM. POL. SC. REV.* 305 (2002); Jeffrey A. Segal & Albert D. Cover, "Ideological Values and the Votes of U.S. Supreme Court Justices," 83 *AM. POL. SC. REV.* 557 (1989).

70 Armin von Bogdandy & Ingo Venzke, "Beyond Dispute: International Judicial Institutions as Lawmakers," in *INTERNATIONAL JUDICIAL LAWMAKING* 3 (Armin von Bogdandy & Ingo Venzke eds., 2012).

71 Niels Petersen, "Lawmaking by the International Court of Justice," 12 *GERMAN L. J.* 1295 (2011).

72 *Similarly*, Bradley, *supra* note 56.

73 Lotus (*Fr. v. Turk.*), 1927 P.C.I.J. (ser. A) No. 10, at 18 (Sept. 7).

74 *See* Albert Bleckmann, "Das Souveränitätsprinzip im Völkerrecht," 23 *ARCHIV DES VÖLKERRECHTS* 450, 464–68 (1985).

75 MARTTI KOSKENNIEMI, *FROM APOLOGY TO UTOPIA: THE STRUCTURE OF THE INTERNATIONAL LEGAL ARGUMENT* 257 (2005).

76 *See* Emily Kadens & Ernest A. Young, "How Customary is Customary International Law?" 54 *WM. & MARY L. REV.* 885, 911 (2013).

77 North Sea Continental Shelf (*F.R.G. v. Denmark, F.R.G. v. the Netherlands*), 1969 I.C.J. 3 (Feb. 20).

by state practice and *opinio juris*.[78] In order to resolve the case, the court referred to equitable considerations.[79] It imposed a procedural obligation to negotiate on the parties and gave some directions for these negotiations.

In other cases, the court relies on treaties or resolutions of international organizations as indicators of customary international law.[80] This reliance predominantly concerns the secondary rules of international law, such as the rules on treaty interpretation, where the ICJ frequently refers to the Vienna Convention on the Law of Treaties,[81] or the norms regarding state responsibility, which are usually identified through reference to the ILC Articles on State Responsibility.[82] In some cases, however, the court also refers to treaties in order to establish substantive norms of international law. In *Nicaragua*, the ICJ relied on the UN Charter to establish a customary prohibition of the use of force in international relations.[83] In its advisory opinion on the use of nuclear weapons, the court stated that the rules of international humanitarian law established in the Hague and Geneva Conventions form part of customary international law.[84] Finally, in the dispute between Belgium and Senegal on the extradition of the former Chadian president Hissène Habré, the court mainly referred to treaty law to identify a customary prohibition of torture.[85]

These examples show that the ICJ sometimes utilizes its discretion to "identify" customary norms without proving a consistent state practice. However, the value of this judicial practice for the protection of global public goods is rather limited. Appeals to equity as in the *North Sea Continental Shelf* judgments rarely have a precedential value that transgresses the individual case because they are usually fitted to the circumstances of the case in question.[86]

78 *Id.* at ¶ 82.

79 *Id.* at ¶¶ 83–99.

80 Alberto Alvarez-Jiménez, "Methods for the Identification of Customary International Law in the International Court of Justice's Jurisprudence: 2000–2009," 60 *INT'L & COMP. L. Q.* 681, 690 (2011). *See also* the empirical analysis of Stephen Choi and Mitu Gulati, "Customary International Law: How Do Courts Do It?" (in this volume), according to whom treaties are the main evidence for custom in international adjudication.

81 *See, e.g.*, Application of the Convention on the Prevention and Punishment of the Crime of Genocide (*Bosnia v. Serbia*), 2007 I.C.J. 43, at ¶ 160 (Feb. 26); Navigational and Related Rights (*Costa Rica v. Nicaragua*), 2009 I.C.J. 213, at ¶ 47 (July 13); Pulp Mills on the River Uruguay (*Argentina v. Uruguay*), 2010 I.C.J. 14, at ¶¶ 64–65, 121, 145 (Apr. 20); Questions relating to the Obligation to Prosecute or Extradite (*Belgium v. Senegal*), 2012 I.C.J. 422, at ¶ 113 (July 20).

82 *See, e.g.*, *Bosnia v. Serbia*, *supra* note 81, at ¶¶ 385, 388, 398, 420; Jurisdictional Immunities of the State (*Germany v. Italy*), 2012 I.C.J. 99, at ¶ 58 (Feb. 3).

83 Military and Paramilitary Activities in and Against Nicaragua (*Nicaragua v. U.S.*), 1986 I.C.J. 14, at ¶ 181 (June 27).

84 Legality of the Threat or Use of Nuclear Weapons, 1996 I.C.J. 226, at ¶ 79 (July 8).

85 *Belgium v. Senegal*, *supra* note 81, at ¶ 99.

86 Kadens & Young, *supra* note 76, at 912.

The main source of identifying customary law seems to be the reference to treaties and to UN General Assembly resolutions.[87] Basing the identification of custom on treaties or soft law may have three practical consequences. First, it allows the judges to apply a norm contained in a treaty even in cases in which it does not have jurisdiction over the treaty. Second, the customary norm may be used in order to extend treaty obligations to states that have not ratified a specific treaty. Third, it may give normative force to soft law instruments. The first two options presuppose the existence of near-universal treaties. At least, with regard to the protection of the atmosphere, such a prospect is still rather distant. The ICJ might play a useful role in stabilizing expectations with regard to international soft law. However, the problem here is that the court will often not have jurisdiction over the corresponding issues.

PUBLIC GOODS AND THE UNILATERAL EXTENSION OF AUTHORITY

The analysis in the previous section has shown that the prospect for protecting global public goods through customary international law is not especially positive.[88] In this section, I will argue that the most promising avenue is the extension of authority through initially unilateral actions.[89] Joel Trachtman has argued that the primary function of jurisdictional rules is the internalization of externalities.[90] The problem of global public goods is that all states have shared authority about their maintenance and protection so that externalities cannot easily be internalized.

One example of how the initially unilateral extension of authority might protect public goods and common pool resources is the development of the protection of fish stocks.[91] While it is socially optimal not to overexploit the existing fish stocks, every state has incentives to free ride.[92] States tried to combat these exploitation incentives by progressively extending their jurisdiction of fishing rights. During the negotiations of the 1958 Geneva Conference on the Law of the Sea, participating states discussed whether to establish a twelve-mile exclusive fishing zone that extended six miles beyond the six-mile territorial sea. The proposal failed by one vote. However, soon after the conference, several states started to issue declarations claiming a twelve-mile exclusive fishing zone.[93] The development did not stop there. Famous is the case of Iceland, which unilaterally extended its exclusive fishing zone

[87] *See* Choi & Gulati, *supra* note 80.

[88] *Similarly*, Bradley & Gulati, *supra* note 1, at 264–66.

[89] *See also* Daniel Bodansky, "What's So Bad *about* Unilateral Action to Protect the Environment?" 11 EUR. J. INT'L L. 339, 343–47 (2000) (pointing out the importance of unilateralism in lawmaking concerning international environmental law).

[90] JOEL P. TRACHTMAN, *THE ECONOMIC STRUCTURE OF INTERNATIONAL LAW* 28–29 (2008).

[91] *See also* Andrew T. Guzman, "Against Consent," 52 VA. J. INT'L L. 747, 769–72 (2012).

[92] Guzman, *supra* note 92, at 768.

[93] LOTHAR GÜNDLING, *DIE 200 SEEMEILEN-WIRTSCHAFTSZONE* 22 (1983).

to fifty nautical miles – an act that was challenged by Germany and the United Kingdom before the International Court of Justice.[94]

However, Iceland's conduct was no exception. A survey of the UN Food and Agricultural Organization in 1969 showed that, while most states claimed an exclusive fishing zone of 12 nautical miles, some countries had extended their exclusive fishing zone up to 200 nautical miles.[95] This development picked up speed in the early 1970s. In 1970 and 1972, several Latin American countries adopted declarations in which they claimed exclusive rights of jurisdiction within 200 nautical miles from their coast.[96] In 1973, the Organisation of African Unity, the precursor to the African Union, declared that African states recognized the right of each coastal state to establish an exclusive economic zone not exceeding 200 nautical miles.[97]

This development was also taken up in Europe. Even though the ICJ had confirmed the position of the UK and Germany in the dispute with Iceland on the extent of exclusive fishing zone,[98] Iceland did not comply with the judgment.[99] On the contrary, it even extended its exclusive fishing zone to 200 nautical miles in July 1975. In 1976, the European Community followed suit. In a council resolution, it asked all member states to extend their exclusive fishing zones to 200 miles.[100] This position was finally confirmed in the United Nations Convention on the Law of the Sea from 1982, which recognized the right of all coastal states to establish a 200-mile exclusive economic zone.[101]

This extension of the fisheries jurisdiction contributed to the protection of fish stocks. In order to protect its own national fisheries, every state had incentives to establish and implement quota systems within its own jurisdiction – a mode of protection that could not have been implemented in a system of shared authority. What helped the cause was the distributional impact of the extension of the fisheries

94 *See* Fisheries Jurisdiction (*U.K. v. Iceland*), 1974 I.C.J. 3 (July 25); Fisheries Jurisdiction (*F.R.G. v. Iceland*), 1974 I.C.J. 175 (July 25).

95 F.A.O., "Limits and Status of the Territorial Sea, Exclusive Fishing Zones, Fishery Conservation Zones and the Continental Shelf," 8 *Int'l Leg. Materials* 516 (1969).

96 *See* "Montevideo Declaration on the Law of the Sea" (May 8, 1970), 64 *Am. J. Int'l L.* 1021 (1970); "Declaration of Latin American States on the Law of the Sea" (Aug. 8, 1970), *reprinted in* 1 New Directions in the Law of the Sea 237 (S. Houston Lay, Robin R. Churchill, & Myron H. Nordquist eds., 1973); "Declaration of Santo Domingo (June 9, 1972), *reprinted in* 1 New Directions in the Law of the Sea 247 (S. Houston Lay, Robin R. Churchill, & Myron H. Nordquist eds., 1973).

97 Declaration of the Organization of African Unity on the Issues of the Law of the Sea (July 19, 1974), UN Doc. A/CONF.62/33.

98 Fisheries Jurisdiction, *supra* note 95.

99 Constanze Schulte, *Compliance with Decisions of the International Court of Justice* 151 (2004).

100 European Communities, "Council Resolution on Certain External Aspects of the Creation of a 200-Mile-Fishing Zone in the Community with Effect from 1 January 1977" (Nov. 3, 1976), 15 *Int'l Leg. Materials* 1425 (1976).

101 United Nations Convention on the Law of the Sea, Art. 57 (Dec. 10, 1982), 1833 U.N.T.S. 3.

zone.[102] States dependent on their coastal fisheries had particularly strong incentives to extend their fisheries jurisdiction.[103] The technological development after the Second World War allowed technologically advanced states to exploit fish stocks at greater distance from the coast. However, as fish stocks migrate, this significantly impacted the coastal fisheries. It is thus not surprising that the development of extending the fisheries jurisdiction first originated in developing countries with a significant coastal subsistence fishing industry.[104] At some point the technologically advanced states had to follow suit. Opposing the extension of fishing zones was too costly.[105] Consequently, they wanted to reap at least the benefits of extending their own fishing zones.

The protection of fish stocks through the extension of the fisheries jurisdiction is certainly not perfect.[106] Notably, fish stocks on the high seas beyond the 200-mile exclusive economic zone are still unprotected. Furthermore, the protection depends on countries adopting effective schemes within their own jurisdiction. Nevertheless, the example shows how at least some level of protection for global public goods can be achieved through the initially unilateral extension of authority.

Does this also entail lessons for the protection of the global climate? *Prima facie*, the situation is more complex. Fishing still has a partly local component. Even though fish stocks migrate, the fishing industry on the east coast of the United States is largely unaffected by what happens in the Indian Ocean. Therefore, if the United States establishes an effective scheme to protect fish stocks, this scheme has certain positive external effects, but the United States will reap the greatest benefits. By contrast, the locality of the emission is irrelevant for the climate.[107] Sea levels in Bangladesh will continue to rise regardless of whether the emissions stem from South Asia or from Western Europe. For this reason, if the European Union (EU) adopts measures to mitigate climate change, the positive effect will not necessarily be the strongest in Europe.

Nevertheless, a unilateral extension of authority that could at some point lead to the emergence of a customary rule does not seem impossible.[108] The primary vehicles of internalizing external effects that are generated through the pollution of the atmosphere are the taxation of polluting conduct or trading schemes that require emission permits for polluting activities. If a country wants to establish such a system

[102] On the distributional impact, see Petersen, *supra* note 71, at 1315.

[103] Verdier & Voeten, *supra* note 48, at 406.

[104] Petersen, *supra* note 71, at 1315.

[105] Verdier & Voeten, *supra* note 48, at 407.

[106] Guzman, *supra* note 92, at 771.

[107] Victor, *supra* note 5, at 4.

[108] *See also*, Elinor Ostrom, "Polycentric Systems for Coping with Collective Action and Global Environmental Change," 20 GLOB. ENVT'L CHANGE 550 (2010) (advocating a polycentric approach for mitigating climate change).

unilaterally, it has to fear detrimental effects for its economy.[109] Emission taxes or the requirement of emission permits impose additional costs on the domestic industries and thus reduce their competitiveness on global markets.

One solution could be to include foreign producers or service providers into the taxing or trading scheme. It is sometimes argued that such an extraterritorial extension of taxing or trading schemes could violate the principle of the territoriality of jurisdiction. When the EU extended the EU Emissions Trading System (ETS) to aircraft operators, it also required non–EU airlines to purchase emission permits for flights to and from the EU.[110] These permits were required for the whole distance of the flight even if only part of it took place over European territory. This regulation was challenged by the Air Transport Association of America and some major U.S. airlines before the European Court of Justice (ECJ).[111] The applicants claimed that the regulation violated the customary principle of territoriality. The ECJ disputed the argument. In its view, the decisive connection for the effects of the regulation was the arrival at or departure from a European airport.[112]

Even if one accepts the view of the ECJ, such an inclusion of transboundary transport services is in need of coordination in order to avoid the service carrier being charged several times. The EU attributes the responsibility for charging the air carrier to the departure state. However, if the departure state does not apprehend its responsibility to include air carriers in a scheme for the mitigation of climate change, the EU reserves the right to step in.[113] While foreign service providers can thus usually be directly included in an emission trading system, such an inclusion is not possible for foreign producers. But countries adopting a tax- or trade-based emission-reduction scheme could levy an environmental tax on imports from countries that do not have a comparable emission reduction scheme.[114]

However, the judgment of the ECJ did not end the controversy. There are several states that oppose the attempt of the EU to include international routes of foreign airlines in the ETS. Such unlikely allies as Russia, China, India, and the United States threatened retaliatory measures.[115] Arguably, the extent of the principle of territorial jurisdiction is thus unclear, to say the least. The regulation of the EU

109 Freeman & Guzman, *supra* note 3, at 1594.

110 *See* Directive 2008/101/EC of the European Parliament and the European Council (Nov. 19, 2008), 2009 O.J. L 8/3.

111 *See* Case C-366/10, Air Transport Association of America, 2011 E.C.R. I-13755.

112 *Id.* at paras. 121–30.

113 *See* Joanne Scott & Lavanya Rajamani, "EU Climate Change Unilateralism," 23 *Eur. J. Int'l L.* 469, 475 (2012).

114 For details of how to design such a border tax, *see* Stéphanie Monjon & Philippe Quirion, "Addressing Leakage in the EU ETS: Border Adjustment or Output-Based Allocation?" 70 *Ecol. Econ.* 1957 (2011).

115 *See* Scott & Rajamani, *supra* note 114, at 473.

can be seen as an attempt to push the development into a specific direction, that is, to change an existing norm of customary international law by extending the territoriality principle.

The result of this controversy is still open. The resistance against the inclusion of international air traffic in the ETS has induced the EU to postpone the inclusion of flights to and from non-EU member countries in the EU ETS for four years after the International Civil Aviation Organization agreed to propose a global, market-based emission-reduction mechanism by 2016.

But it is important to note that there was also retaliation when countries initially started to extend their exclusive fishing zones. The more countries join the effort to establish effective emission reduction systems, the more costly retaliation will become. As with the fisheries jurisdiction case, there will be a tipping point,[116] at which resistance has more costs than benefits so that countries have incentives to adopt their own emission reduction schemes because of pure self-interest. If their companies have to pay environmental taxes for the exports into the most important markets, so that they do not reap the benefits of low production costs, it may be more rational for them to establish their own taxation system in order to generate at least some income for the national budget.

The main problem will be to find an actor to initiate the process. In the fish stock example, it was the countries that depend most on their coastal fisheries that had an interest in extending their fisheries jurisdiction. The countries that are most affected by climate change and that would have a strong interest in initiating a process of greenhouse gas reduction probably do not have enough weight – in terms of their share in the global greenhouse gas emissions and their economic strength. The EU has already established an emission trading scheme. However, the EU ETS has so far been largely ineffective, as economic concerns seem to trump ecological considerations. This will probably not change until environmental concerns rise in the agenda of domestic interest groups.

CONCLUSION

The potential of customary international law to protect global public goods is limited. The advantage of customary law over treaty mechanisms is the increased flexibility. Especially in the field of mitigating climate change, some authors have put hope into custom as a remedy when formal treaty negotiations fail. However, the protection of public goods through customary law has its own problems. Economic theory and experimental evidence show that the establishment of

[116] On this tipping point theory regarding the emergence and change of customary international law, *see* Verdier & Voeten, *supra* note 48.

public goods through decentralized cooperation is difficult. The most promising avenue for protecting public goods on the global plane is the extension of authority through initially unilateral action. The protection of fish stocks through the initially unilateral extension of the exclusive fishing zone to 200 nautical miles provides a good example for how the protection of public goods through the initially unilateral extension of authority can work. Whether this example entails lessons for the mitigation of climate change remains to be seen.[117]

[117] *See* James Bushnell, Carla Peterman, & Catherine Wolfram, "Local Solutions to Global Problems: Climate Change Policies and Regulatory Jurisdiction," 2 *Rev. Envt'l Econ. Pol'y* 195 (2008); Krisch, *supra* note 2, at 20 (expressing skepticism with regard to unilateral action to cope with the challenges of climate change).

11

Reinvigorating Customary International Law

Andrew T. Guzman and Jerome Hsiang

For decades now, some legal commentators have suggested that customary international law (CIL) is dying.[1] Looking out at the world, one might be excused for agreeing with them. As a general trend, states have been codifying the contents of CIL through bilateral and multilateral consent-based agreements.[2] This trend toward codification could be interpreted as diminishing the general impact of CIL by making most of its content redundant if not irrelevant. Furthermore, the content and the process by which CIL comes about is often considered to be largely outside the direct control of states, making it difficult for states to interact with and shape CIL in purposive ways. In a world where urgent global problems require timely and coordinated solutions, CIL seems inadequate as a useful legal form. Finally, it is often difficult to determine the legal content of CIL, leading to an impression that modern CIL is more speculative than functional. It is little wonder, then, that some have argued that CIL will soon be largely irrelevant.[3]

But reports of CIL's demise are greatly exaggerated. At the base of it, CIL is an integral and necessary part of the international legal system.[4] Perhaps more importantly, it is also a tool that can promote cooperation in situations where

[1] *See, e.g.*, J. Patrick Kelly, "The Twilight of Customary International Law," 40 *Va. J. Int'l L.* 449 (2000).

[2] *See* R. R. Baxter, "Multilateral Treaties as Evidence of Customary International Law," 41 *Brit. Y.B. Int'l L.* 275 (1965); Tim Meyer, "Codifying Custom," 160 *U. Pa. L. Rev.* 995 (2012).

[3] *See* Joel P. Trachtman, "The Growing Obsolescence of Customary International Law" (in this volume); Joel P. Trachtman, "Reports of the Death of Treaty Are Premature, but Customary International Law May Have Outlived Its Usefulness," AJIL Unbound, *retrieved from* http://www.asil.org/blogs/reports-death-treaty-are-premature-customary-international-law-may-have-outlived (last accessed Apr. 29, 2014).

[4] *See* Brigitte Stern, "Custom at the Heart of International Law," 11 *Duke J. Comp. & Int'l L.* 89, 89 (2001) ("It is not contested that custom enjoys privileged status in the international order: 'custom is even more central than the treaty'" [quoting and translating Paul Reuter, *Introduction au Droit des Traités* 38 (1972)]); Michael Byers, *Custom, Power and the Power of Rules* 107 (1999)

consent-based rulemaking proves impractical. Indeed, while there are many well-known advantages to the current international legal system predicated on consent, there are also substantial drawbacks. Efforts at consent-based rulemaking can often be bogged down by recalcitrant states. States can avoid new legal obligations either by refusing to consent or simply dragging their feet. For global problems that require at least the acquiescence – if not the direct participation – of specific states this kind of strictly consent-based system can create delays or even outright breakdowns in cooperation. These, in turn, can reduce global well-being.

In this chapter, we argue that one potential solution to the challenges raised by consent-based rulemaking is to lean more heavily on nonconsensual international law such as CIL as a way to promote global cooperation. We believe that, while CIL in its current form is not entirely coherent conceptually, it can nevertheless be reinvigorated for the cooperative benefit of modern states.

CHARACTERISTICS OF CIL

CIL is a much-maligned area of international law. Commentators, including one of us, have attacked its foundations, calling it incoherent, irrelevant, or an anomalous form of legal rulemaking.[5] We do not dispute that CIL is problematic at a theoretical level. In this chapter, however, we are interested in a more pragmatic perspective. CIL is just one form of nonconsensual rulemaking in the international context. Others include *jus cogens* rules, resolutions made by the UN Security Council under its Chapter VII powers, and the nonbinding soft law guidance that emanates from international organizations (IOs) and international tribunals (ITs).[6] Each of

("Treaty rules . . . are based on the general customary rule of *pacta sunt servanda*, which requires that treaty obligations be upheld in good faith.").

5 *See* Andrew Guzman, "Saving Customary International Law," 27 *Mich. J. Int'l L.* 115 (2005); *see also* Anthony A. D'Amato, *The Concept of Custom in International Law* 130 (1971); Jack L. Goldsmith & Eric A. Posner, *The Limits of International Law*, 23–26 (2005); Mark A. Chinen, "Game Theory and Customary International Law: A Response to Professors Goldsmith and Posner," 23 *Mich. J. Int'l L.* 143 (2001); Kelly, *supra* note 1; Michael Akehurst, "Custom as a Source of International Law," 47 *Brit. Y.B. Int'l L.* 1, 1 (1977); Edward T. Swaine, "Rational Custom," 52 *Duke L.J.* 559, 563 (2002); David P. Fidler, "Challenging the Classical Concept of Custom: Perspectives on the Future of Customary International Law," 39 *German Y.B. Int'l L.* 198 (1996); Anthea Elizabeth Roberts, "Traditional and Modern Approaches to Customary International Law: A Reconciliation," 95 *Am. J. Int'l L.* 757, 757 (2001).

6 Although in this chapter we discuss ITs in their soft lawmaking capacity, IT decisions are clearly not only soft law. In fact, the same tribunal decision can be considered both hard and soft law, depending on the identity of the observing party. For example, The International Court of Justice's (ICJ) decisions are binding upon the parties that voluntarily submit a dispute before it. States can also accept compulsory ICJ jurisdiction over matters of international law and treaties. Finally, states can also agree through treaty to submit certain disputes arising under those treaties to the ICJ. But for states not party to an ICJ case, the tribunal's decisions are not facially binding, as per Art. 59 of the ICJ

TABLE 11.1. *Legality/direct state involvement matrix*

	High direct state involvement	Low direct state involvement
Binding law	• Security Council resolutions • CIL	• *Jus cogens* • CIL
Nonbinding soft guidance	• IO & IT speech	• IO & IT speech

these forms of rulemaking can be located along two dimensions: legality and direct state involvement. Table 11.1 situates the various forms of nonconsensual rules a legality/direct state involvement typology. Where a form of nonconsensual rule falls in the typology can helps shape the ways in which those rules affect state behavior.

Legality is a binary trait.[7] A nonconsensual rule can provide either binding law or nonbinding guidance.[8] For our purposes, Article 38(1) of the Statute of the International Court of Justice (ICJ), Chapter VII of the UN Charter, and the Vienna Convention on the Law of Treaties defines which nonconsensual rules are binding: *jus cogens* rules are fully binding on all states, Security Council Chapter VII resolutions are binding on all UN members, and CIL is theoretically binding on all states but allows for persistent objectors and is sometimes unclear as to its content.[9] Soft law emanating from IOs and ITs are not binding but nevertheless implicates states' reputations for compliance, thus making it functionally similar to binding international law in some ways.[10] The main difference between binding law and nonbinding guidance is that nonbinding guidance in theory implicates less reputational consequences for states than comparable binding law.[11]

Statute. *See* Charter of the United Nations and Statute of the International Court of Justice art. 59, June 26, 1945. Nonetheless, states and private parties not before the tribunal often make great efforts to behave in ways to comport with the contents of tribunal decisions. The effect of this is that tribunal decisions begin to look a lot like some kind of international "common law." However, the lack of a formal *stare decisis* doctrine in international law means that, in effect, much of the influence that tribunals wield is technically soft. On a more general and related point, soft law, while not creating binding legal obligations, nevertheless serves an important role in the regulation and management of the international legal system.

7 *See* Kal Raustiala, "Form and Substance in International Agreements," 99 *Am. J. Int'l L* 581 (2005).

8 It is important to note here that whether a rule is binding is not the sole determinant of its compliance pull on states and other international actors.

9 Charter of the United Nations and Statute of the International Court of Justice, 7, 38(1), June 26, 1945; Vienna Convention on the Law of Treaties art. 53, May 23, 1969.

10 Again, we note that there is no controversy that tribunals can issue hard law on parties before the court on that particular matter.

11 For more thorough treatments of the reputational consequences of soft and hard law in the international context, see Andrew T. Guzman, *How International Law Works: A Rational Choice Theory* (2008); Rachel Brewster, "Unpacking the State's Reputation," 50 *Harv. Int'l L.J.*

Direct state involvement should be thought of as a continuum and refers to the amount of purposive influence individual states have over the creation and content of a nonconsensual rule. Different states may have varying levels of power over each of the four forms of nonconsensual international rules. To start, we know that *jus cogens* rules largely defy state manipulation. That is to say, it is extraordinarily difficult for a state to intentionally change the content of a *jus cogens* rule once it is created. It is also likely that individual states have relatively little power to shape a *jus cogens* rule as it is developing.[12] By contrast, states with membership on the Security Council are directly involved in producing Security Council resolutions; the very nature of the Security Council maximizes the role of politicking in this form of international rulemaking. However, this intimate role that direct state involvement plays also means that the Security Council is often ineffective. It is sometimes simply too difficult to overcome the vetoes held by the permanent members. Soft law emanating from IOs and ITs falls somewhere between the two extremes of the continuum, depending on the procedural rules and organizational history of the specific IO or IT.[13] On the whole, IOs and ITs tend to have more freedom to act on their own accord than the Security Council, especially since they are tasked with filling in the gaps left by incomplete underlying hard law that states have consented to. Nevertheless, states limit the authority delegated to these institutions, often retaining a large amount of oversight power.[14] Casual observation suggests that it is very likely that the more politically powerful a state, the more directly involved it will be in the affairs of IOs and ITs it is party to.

231 (2009); George W. Downs & Michael A. Jones, "Reputation, Compliance, and International Law," 31 *J. Legal Stud.* S95 (2002).

[12] Consider what might differentiate a "mere" global custom from a *jus cogens* rule. *Jus cogens* rules often arise to reflect the most deeply and widely held convictions shared across human societies. As an evidentiary matter, we only know that such convictions are shared if we can observe the substantial and material participation of non-state actors in the process, e.g., civil society and social commentators advocating for a rule. All this is only to suggest that states, while remaining a player in determining the existence and content of *jus cogens* rules, have less purposive influence over it than many other forms of international rule making.

[13] For example, the International Labour Organization's tripartite structure is sufficiently different from the World Trade Organization as to suggest that the level of state involvement in each may be radically different.

[14] *See* Oona A. Hathaway, "International Delegation and State Sovereignty," 70 *Law & Contemp. Probs.* 115, 122 (2008) ("States remain free from external control in any meaningful sense, for they are controlled by the decisions of the international body only so long as they agree to be. Once their agreement ceases, the control over them ceases as well."); Andrew T. Guzman & Jennifer Landsidle, "The Myth of International Delegation," 96 *Calif. L. Rev.* 1693, 1694 (2008) ("States have had no shortage of challenges and reasons to cooperate; yet, to date, they have delegated legislative or decision-making authority to supra-national entities in extraordinarily few instances. Moreover, in virtually every such instance, the delegation is narrowly cabined in terms of its scope, its importance, and its ability to influence state conduct.").

On the continuum of direct state involvement, CIL sits closer to *jus cogens* rules than to Security Council resolutions. But it would be wrong to say that states categorically do not have any direct influence on the contours of CIL. In fact, the amount of direct state involvement in shaping CIL can vary greatly depending on the context.[15] It is, of course, true that the creation of CIL involves exogenous elements beyond the immediate control of individual states. Nevertheless, there are many ways through which states can proactively shape CIL, either by providing acute shocks to the system or by engaging in a longer term effort to introduce a rule to the international system (e.g., the Truman Declaration). Furthermore, since the relationship between treaty law and CIL can be recursive, states can effectively show support for a rule of CIL by signing treaties that will ultimately be seen as evidence of *opinio juris*.[16] By contrast, states can also work against what they see as a crystalizing custom through persistent objections. And on the most basic conceptual level, if enough states simply choose to not follow an alleged rule of customary law, it is unclear whether it remains CIL.[17] While it is admittedly difficult to code the exact level of direct state involvement these states actions represent, it is enough for our purposes to show that CIL occupies a unique middle-ground niche on the continuum of direct state involvement. And this sort of middle ground makes CIL "just right" for certain cooperative purposes.

Now that we have introduced the dual dimensions of legality and direct state involvement, we can describe how CIL is different from other forms of nonconsensual international rulemaking. First, once a rule of CIL exists, it is recognized as international law and is, therefore, binding. This gives it a form of compliance pull that is absent in the case of "mere" norms and recommendations. Second, while we just noted that states have some level of influence over the content of CIL, that level of influence waxes and wanes, sometimes unpredictably. This is in stark contrast to the level of control that states have over the form and content of consent-based

[15] An additional wrinkle: CIL relies on *opinio juris*, which in our minds refers to the beliefs of observing states, rather than the acting state. This means that states not only have the direct involvement to make an affirmative argument for a specific customary rule, they also have the power to directly refuse to acknowledge someone else's proposed custom. We will discuss this in further detail.

[16] There is some debate on whether treaties can be evidence of CIL. *See, e.g.*, Jonathan I. Charney, "International Agreements and the Development of Customary International Law," 61 Wash. L. Rev. 971 (1986); Arthur M. Weisburd, "Customary International Law: The Problem of Treaties," 21 Vand. J. Transnat'l L. 1 (1988). Beyond the theoretical debate, Stephen J. Choi and Mitu Gulati conducted an empirical analysis that found that international tribunals rely heavily on treaties as evidence for CIL. *See* Stephen J. Choi & Mitu Gulati, "Customary International Law: How Do Courts Do It? (in this volume).

[17] We leave aside for a moment the question of whether customary law can change after it has crystalized and whether states can withdraw from customary law by choice (mandatory vs. default view). *See* Curtis A. Bradley & Mitu Gulati, "Withdrawing from International Custom," 120 Yale L.J. 200 (2010).

international cooperation. In some circumstances, states' relative lack of direct control over the content and enforcement of CIL gives it a special valence over the other forms of nonconsensual international rulemaking. States may be less wary of CIL because it does not have the imprimatur of specific powerful states.[18] Politically, this may make CIL more palatable to domestic audiences and legislatures. It also assuages some concerns that an international rule is simply a way to accrue benefits to specific influential states.[19] Relatedly, CIL also affords states greater leeway when it comes to specificity of content and applicability. For example, in theory CIL allows states to limit the geographical application of a rule, through what some commentators and the ICJ have called regional or special custom.[20]

CIL, then, is a unique tool in the states' toolbox for securing cooperation. It is important to note here that our defense of CIL relies only on the claim that it is different from other legal tools, not that it is "better." Our point is that this particular legal tool should not be abandoned and, indeed, may deserve to be revitalized in the face of global challenges that have stretched the limits of consent-based lawmaking. It will not be appropriate or effective for all forms of cooperation; but in some situations, it can be useful.[21]

Thus our task in this chapter is to explain why CIL remains a good thing in an anarchic world increasingly governed by consent-based rules, why CIL can help states realize cooperative gains, and why an overly formal contractarian understanding of international law can prevent much-needed cooperation on global challenges. In short, we argue that states should continue to take CIL seriously and even consider using it proactively to solve global problems.

The remainder of the chapter proceeds as follows: we first discuss the benefits and drawbacks of requiring unanimous consent on international rules. We then lay out the traditional and rational-choice approach to CIL, focusing on what we consider common conceptual confusions that plague the understanding of *opinio juris* and state practice. In addition, we argue for the importance of regional custom to the general theory on CIL. Next, we analyze the ways in which the legality and direct

18 Note, however, that some commentators have argued that modern CIL lacks procedural and democratic legitimacy, *See, e.g.*, Kelly *supra* note 1, at 517–30.

19 We suspect that the alleged incoherence and decentralized nature of CIL formation also makes it harder for powerful states to entirely commandeer the process.

20 *See* Right of Passage (*Port. v. India*), 1960 I.C.J. 6, 39 (Apr. 12) ("The Court sees no reason why long continued practice between two States accepted by them as regulating their relations should not form the basis of mutual rights and obligations between the two States."); Anthony A. D'Amato, "The Concept of Special Custom in International Law," 63 AM. J. INT'L L. 211 (1969).

21 There is an important debate over the effectiveness of CIL as a tool to promote cooperation. For a small sampling of this debate, see Niels Petersen's chapter in this volume, arguing that CIL, as understood traditionally, is ill-suited as a tool to address many problems involving global public goods, as it is difficult to sustain cooperation with conditional cooperators over time. *See* Niels Petersen, "Customary International Law and Public Goods" (in this volume).

state involvement dimensions impact the rational-choice understanding of CIL. Finally, we take a first step toward a functional theory of how customs originally held by a subset of states can eventually affect global behavior as CIL.

THE BLESSING AND CURSE OF CONSENT

The current international legal system's strong commitment to consent is both a blessing and a curse. Consent as a baseline requirement for legal rules to apply to states both protects and hamstrings the international community's ability to solve some global problems. On one hand, the consent requirement protects the sovereignty of individual states, thus giving states some level of assurance that they will not be subjected to detrimental rules. On the other hand, a small subset of states can delay effective global cooperation by refusing to participate.

Therefore, the heart of the matter is whether the international system operates with an appropriate amount of consent given the problems that the world faces. We suggest in this chapter that a rigid adherence to consent creates some intractable problems. Perhaps the greatest of these problems is that generating consent on an international level is a slow, difficult, and cumbersome process. The labyrinthine process of international lawmaking creates endless bottlenecks, provides many veto points, and invites strategic holdout behavior, all of which make agreement among states more difficult. Even when states can approach consensus, consent is often reached only by weakening the content of an agreement.[22] As one commentator noted, "The prevailing practice of seeking consensus or near-unanimity to adopt a convention has not only led to drawn-out negotiations, but also to highly ambiguous or empty provisions, undermining what is needed to ensure the establishment of an effective international legal regime."[23] It is, in fact, unclear whether such a rigid consent requirement promotes optimal outcomes for states.

Another way of stating the issue is that the consent requirement will often block Kaldor-Hicks improvements that increase the overall welfare of the international system.[24] Requiring unanimous consent means that states that will suffer negative

22 *See* Guzman, *supra* note 11, at 172–73.

23 Eric Rosand, "The Security Council as 'Global Legislator': *Ultra Vires* or *Ultra Innovative*?" 28 *Fordham Int'l L.J.* 542, 575–76 (2005). While in the nineteenth and early twentieth centuries, conference decisions on issues of substance were taken by unanimity, now, except where states can agree on a majority-voting rule, all that is required for the adoption of a rule at an international convention is consensus. As Cassese points out, "Consensus is different from unanimity, for in the latter case there exists full agreement on a given text and in addition the general consent is underscored by a vote." Antonio Cassese, *International Law in a Divided World* 196 (1987).

24 Kaldor-Hicks welfare improvements are those that increase the overall aggregate welfare of the community, but do not necessarily increase the individual welfare of any given community member.

welfare consequences will often effectively be able to prevent other states from realizing greater welfare gains. This may indeed be the purpose of the consent requirement as a means to protect the interests of states in the abstract. But the result is an unwieldy system that puts undue emphasis on achieving Pareto improvements, an increasingly difficult proposition in a world with many powerful states of diverging interests.[25]

Most importantly, these problems with consent both reduce the likelihood that states will enter into agreements that solve difficult global problems and increase the costs involved when such agreements are completed. Making matters worse, the growing complexity of the international system (in the form of more states, more non-state actors, deeper integration, and more global problems) means that achieving agreement is increasingly difficult. Simply put, effective solutions for many of the world's most serious challenges may be inconsistent with our current commitment to consent.

Of course, none of this is to suggest that the international system should wholly abandon consent as a foundational rule. Rather, in this chapter we argue that the international system should move toward a more nuanced view of states and the way they interact with international rules, with effective cooperation being the ultimate goal, rather than a commitment to doctrine. Ultimately, having the tools to effectively solve global problems that plague states represents a greater expansion of states' sovereignty than adherence to a formal doctrine that does not grant states the expansive capability to solve problems beyond their myopic means.

We believe that it is possible to expand the role of CIL without losing the substantial benefits that come with having a consent-based system of international law. Since states ultimately control the international legal system, it seems unlikely that they will tolerate nonconsensual rules running roughshod over the protections established by a consent-based system. It is far more likely that states will use CIL as necessary to cooperate on appropriate problems. That way, states will continue to enjoy the three main advantages of a system that emphasizes consent: compliance, legitimacy, and welfare.

Compliance and Consent

The signature feature of international law is its lack of coercive enforcement.[26] This reality makes compliance with international law a matter of constant

While individual states that gain welfare can transfer some to those states that lose welfare, side payments are often difficult and costly to negotiate and make.

25 Pareto welfare improvements are those that increase the overall aggregate welfare of the community without imposing any losses on any member of the community.

26 There are some rare exceptions to this statement. Most importantly, the Security Council has the authority under the UN Charter to authorize the use of force for the purpose of maintaining international peace and security. *See* UN Charter art. 42. Even these authorizations, however, rely

concern.[27] Imposing rules on states without their consent creates the risk that the rules will be ignored. The consent requirement, then, promises to reduce the frequency of noncompliance by limiting international law rules to those that states have agreed to accept. Unanimity ensures, at a minimum, that every affected state prefers the new arrangement to the available alternatives.

But should we care if some rules are largely ignored? Consent may increase the overall level of compliance, but it does so by changing the content of international law. There would be perfect compliance with international law if the only rule were that people must breathe in oxygen and breathe out carbon dioxide, but surely a high rate of compliance is not what we are trying to achieve.

A rule that fails to achieve perfect compliance may nevertheless generate "compliance pull" and change the behavior of states.[28] For this reason, incomplete compliance with a desirable rule is better than not having the rule at all. By way of example, consider the international law prohibition of war crimes. War crimes are a common feature of war. Indeed, one would be hard pressed to find a significant conflict that did not feature war crimes. Despite this highly imperfect level of compliance, the prohibition is both desirable and effective if it causes a reduction in the number of war crimes committed. Less than perfect compliance is not a good reason to eliminate the rule.

Furthermore, the consent of a state to a rule hardly ensures compliance. Giving consent does not eliminate a state's incentives to violate the law or necessarily increase the compliance pull of the rule. When developing countries agreed to the Agreement on Trade Related Aspects of Intellectual Property Rights (TRIPs), for example, they did so because it was a necessary condition for membership in the WTO.[29] Most of these countries would have preferred to live without the TRIPs obligations, and, predictably, compliance with the TRIPs Agreement has been a major issue.

Overall, consent is a highly imperfect proxy for state willingness to comply with legal rules. Examples of noncompliance despite state consent abound, including with respect to the prohibition on the use of force, many multilateral environmental agreements, human rights commitments, and investment agreements.[30]

on individual states to take action. The Security Council itself has no enforcement arm, and it can only "authorize" the use of force. It cannot order it. The rarity of authorizations of use of force by the Security Council illustrates the very limited sense in which this represents an exception to the statement that no coercive force exists to compel compliance with international law.

27 *See* Guzman, *supra* note 11.

28 *See* Thomas M. Franck, "Legitimacy in the International System," 82 *Am. J. Int'l L.* 705 (1988).

29 *See* Anu Bradford, "When the WTO Works, and How it Fails," 51 *Va. J. Int'l L.* 1, 21 (2010).

30 Some international commitments are supported by reciprocity – each state complies only as long as the other state does so, and this reality motivates both to honor their commitments. Reciprocity can create a strong and stable arrangement, and is visible in many international trade agreements, some

Legitimacy and Consent

The prevailing view of international law is that a rule made without the consent of affected states faces legitimacy problems that speak to both the desirability and practicability of nonconsensual rules.[31] Stated simply, a rule that is judged illegitimate may also be undesirable as a normative matter. In other words, legitimacy may serve as a stand-in for good rulemaking.[32]

We wish to focus on the possibility that legitimacy can serve as a sort of filter to block harmful rules. Take, for example, Franck's familiar indicators of legitimacy: determinacy,[33] symbolic validation,[34] coherence,[35] and adherence.[36] At least the first two indicators increase the likelihood that a rule will serve the interests of the governed. Determinacy allows those affected to know the meaning of a rule and so gives them greater opportunity to voice any concerns or objections.[37] Symbolic validation includes evidence that authority is being exercised "in accordance with right process."[38] This procedural feature gives potential opponents of a proposed rule an opportunity to object, making adoption of a welfare-reducing rule less likely.

arms control agreements, and so on. This sort of reciprocity is more likely to be present when states have consented to a rule. *See* GUZMAN, *supra* note 11, at 42–45.

31 To be sure, there is something slightly circular in this argument. To a significant degree, the reluctance of states to surrender their commitment to consent does appear to undermine the legitimacy of nonconsensual rulemaking. While the strong form of the argument for consent presented here is weighty, we believe there is some room at the margins for nonconsensual rulemaking to make a positive contribution to solving global problems.

32 To the extent that states' willingness to comply is related to a rule's perceived legitimacy, unanimity may increase the legitimacy of a rule and so improve the likelihood of compliance. *See, generally*, THOMAS FRANCK, *FAIRNESS IN INTERNATIONAL LAW AND INSTITUTIONS* (1995). As long as states can be provided with sufficient incentive to comply, they will do so regardless of whether they have consented to the rule or, for that matter, view the rule as legitimate. It should be added that in some circumstances, consent and legitimacy may actually work against one another. Where an existing rule has ceased to serve the interests of most states, our commitment to unanimity makes changes to that rule difficult. If this dissatisfied majority is unable to change the rule because the minority refuses to consent, the existing rule may be perceived as less legitimate than the would-be alternative. Greater levels of legitimacy may also suggest greater amounts of collective wisdom, higher level of efficaciousness, and state participation associated with a customary rule. *See* John Tasioulas, "Custom, *Jus Cogens*, and Human Rights" (in this volume).

33 Determinacy refers to the extent to which a meaning is transparent through its reference text, for example, the meaning of a rule through the language of law. *See* FRANCK, *supra* note 32, at 30–46.

34 Symbolic validation refers to the rule's ability to communicate or signal authority. *Id.*

35 Coherence refers to the degree to which the rule is applied coherently and consistently. *Id.*

36 Adherence is the relationship between a primary rule of obligation and the secondary rules governing its creation, interpretation, and application. *Id.*

37 *Id.* at 30–46.

38 *Id.* at 25–46; *see also* THOMAS M. FRANCK, *THE POWER OF LEGITIMACY AMONG NATIONS* 27–40 (1990); Thomas M. Franck, "Legitimacy in the International System," 82 AM. J. INT'L L. 705, 705–11 (1988).

Any move toward nonconsensual rulemaking must at least pause to consider these issues. But one should not take this point too far. There is no reason to think that unanimity is a necessary condition to achieve legitimacy. Indeed, most observers would agree that a rule requiring unanimity is illegitimate in many – probably most – governance contexts. Requiring unanimity for domestic legislation, for example, would be rightly thought to give too much power to small groups: it is undemocratic. Demanding unanimity would also create an enormous bias in favor of the status quo even when superior alternatives are available. This, too, might be described as illegitimate.

Concern about legitimacy, then, is certainly not enough to justify an unyielding commitment to consent. Not only is it possible to develop legitimate nonconsensual governance systems, but it is also hard to think of many such systems (other than international law) that rely on consent and are considered legitimate. Legitimacy and consensus may in fact be inconsistent.

Welfare and Consent

The third justification for requiring the consent of states for the creation of international rules is more functional than either a desire to encourage compliance or a perceived connection between consent and legitimacy, although it does have a connection to the legitimacy discussion.

If states are not required to agree to the rules that affect them, they cannot be confident that those rules will serve their interests. A requirement of state consent protects against international legal rules that do more harm than good. Unanimity signals, as well as anything can, that all states believe a proposed rule to be in their interest.[39] This is the best evidence we can hope for that a rule will increase global well-being.[40]

39 A requirement of consent cannot, of course, protect a state from agreements to which that state is not a party. For example, the member states of the Organization of the Petroleum Exporting Countries (OPEC) have consented to the practices of that organization and benefit from them. Most other states, however, are harmed by the existence of this cartel. Because OPEC operates without seeking the consent of this second group of states, the latter are not protected by the consent requirement.

40 If domestic governments fail to represent the interests of their population, the consent of a state need not signal that a rule is good for the people of the country. This is a serious problem that plagues international law. Absent a willingness to inquire into the functioning of domestic governments, however, the international system has no choice but to accept the authority of domestic governments to represent the interests of their citizens. The study of diplomacy has long been understood to involve the pursuit of some objective ascertainable political values encompassed by the principle of *raison d'etat*. This assumption was as questionable in mid-seventeenth-century France as it is in today's complex interdependent state system. A key assumption to conventional rational analysis of international affairs is that states are rational actors interested in maximizing the welfare of their citizens. In some circumstances, a decision may be the consequence of a bargain conducted by groups

Consent, then, generally protects states and the international system from changes that will reduce the joint welfare of states. But providing such a robust protection against welfare-reducing international law comes at a large cost. A requirement of consent defeats many welfare-enhancing forms of international law. So while a relaxation of the consensus requirement would risk the adoption of rules that reduce global welfare, it would also open the door to many welfare-increasing rules that are currently frustrated by the requirement of consent.[41] Another way to describe the same reality is to observe that the unanimity requirement creates an extremely conservative regime with a powerful preference for the status quo. There is no reason to think that the current focus on consent strikes the right balance between avoiding the development of harmful rules and allowing the development of beneficial ones.

A Curious Side Effect of Consent: Aspirational Law

One unusual feature of international law is that some of its content is understood as aspirational. That is to say, states will at times make international law that expresses some kind of moral support for a position. States are likely to have little expectation that aspirational law will be complied with (or at least little chance of imposing sanction if states do not comply). One example of aspirational law is the International Covenant on Economic, Social and Cultural Rights (ICESCR). Although ICESCR is a multilateral treaty, the rights enshrined in it are sufficiently controversial that it prompted the drafters to put in the principle of progressive realization.[42] This principle gives states wide latitude in determining if and how they will implement the rights recognized in ICESCR. Indeed, the principle makes state obligations under ICESCR indeterminate enough that some have questioned the

with very different interests rather than as some single unitary actor. *See* MICHAEL NICHOLSON, *RATIONALITY AND THE ANALYSIS OF INTERNATIONAL CONFLICT* 56 (1992). Distortions in rational decision making based on a unitary actor model are highlighted, for example, by the literature on public choice. *See, e.g.*, GOLDSMITH & POSNER, *supra* note 5; Kenneth W. Abbott, "Enriching Rational Choice Institutionalism for the Study of International Law," 2008 *U. ILL. L. REV.* 5 (2008) (discussing two-level, domestic-international games).

> From a normative or prescriptive point of view the argument is more powerful. Decision making should be rational in so far as this is possible. The problems are analyzed with a view to overcoming them, not submitting to them. If groups do not develop procedures which lead to rational decisions, then it is desirable to design them so that they do. NICHOLSON, *supra*, at 57.

41 Such is the strength of the sovereignty norm in international affairs, according to Stephen Krasner, that the attributes of international legal sovereignty, Westphalian (political) sovereignty and domestic sovereignty, are often confounded with consequences for the effectiveness of states to resolve the biggest international problems. *See* Stephen D. Krasner, "The Hole in the Whole: Sovereignty, Shared Sovereignty, and International Law," 25 *MICH. J. INT'L L.* 1075, 1077–78 (2004).

42 International Covenant on Economic, Social and Cultural Rights, art. 2(1) (1976).

extent to which ICESCR sets forth any obligations at all.[43] This type of lawmaking is not easily explained by most rational actor approaches, but nonetheless deserves some mention here.

Although it may seem incongruous for rational states to make rules with little expectation of compliance, this behavior is far more understandable if viewed in the light of CIL. States may make aspirational rules in order to influence *opinio juris* beliefs of other states. That is to say, a subset of states may intend an international agreement to serve as a basis or precursor for more robust rules later on. This process is possible because *opinio juris* can be shaped by any number of factors. For example, states can certainly hold *a priori* assumptions about the rules that govern state behavior. But they can also be influenced by specific declarations. This is where aspirational lawmaking has the greatest influence on CIL.

Interestingly enough, this approach to aspirational lawmaking also begins to explain some of the dissonance between what states say they will do and what they actually do. States may aspire to a world where a rule is taken seriously, but would prefer to not be in a world where they are the only adherents to the rule. To take one prominent example, we know that almost all states publicly agree that torture is against CIL and, indeed, many multilateral treaties. Nonetheless, many states engage in activity that could plausibly be determined to be torture, even among signatories of the International Covenant on Civil and Political Rights and the Convention against Torture.

What this suggests to us is that states may indeed prefer to live in a world where governments never torture. But they are also unwilling to live in a world where they themselves refrain from torture while other states engage in it. A corollary to this is that states may wish to retain the prerogative to torture even if all other states refrain from it in order to capture some kind of potential benefit. Of course, both of these outcomes are only plausible if torture yields some kind of payoff to the state (n.b.: not obvious this is actually true). If torture does yield payoffs in intelligence, then it is conceivable that states do not want to be in the situation of losing a potential tool to use against their enemies. However, since torture is almost always conducted in secrecy, states cannot easily monitor their counterpart states' compliance. This means that even if states manage to negotiate an international agreement denouncing or banning torture, states have no reliable way of applying reputational sanctions for violations. It goes without saying that in this kind of situation, the incentives to cheat are enormous, particularly in times of conflict. Furthermore, even states that genuinely find torture distasteful may wish to hedge

[43] *See generally* Ellen Wiles, "Aspirational Principles or Enforceable Rights? The Future for Socio-Economic Rights in National Law," 22 *Am. U. Int'l L. Rev.* 35 (2006).

their bets against future developments. States may find themselves in conflicts of unanticipated brutality. The chances of a state breaching an agreement to not torture may in fact be quite high when considered over time. As such, even states that have a genuine desire to live in a world without torture can only reasonably hope that a moral condemnation of torture, along with some minimal level of international legal framework supporting such an aspirational set of rules, would eventually change the *opinio juris* of states. It would not be unreasonable to conclude that aspirational laws are often the result of a recognition that the current environment would not support a truly effective rule.

A FUNDAMENTAL QUESTION: WHAT IS CIL?

One of the main difficulties with any analysis of CIL is that legal commentators have long argued that CIL is conceptually problematic. It is best, then, to lay out the traditional conception of CIL and then briefly summarize the main critiques toward it. After that we present a rational-choice conception of CIL that, we believe, helps move the discussion forward.

Traditional Approach to CIL

The traditional conceptual springboard of CIL comes from Article 38 of the Statute for the ICJ, which recognizes "international custom, as evidence of a general practice accepted as law," as one of the sources of international law.[44] Article 38 itself, however, neither defines international custom nor specifies how to approach evidentiary challenges of determining the existence and contents of international custom. The Restatement (Third) of Foreign Relations Law of the United States gives us more to work with by defining CIL as "result[ing] from a general and consistent practice of States followed by them from a sense of legal obligation."[45]

From these two sources comes the dual-element definition of CIL that is widely accepted today: CIL requires both state practice as well as *opino juris*. State practice is an objective requirement that deals with observable state behavior. *Opinio juris*, by contrast, is a subjective requirement; one that focuses on the states' "belief" that they are legally bound by the rule in question. As it might be clear from the very start, both state practice and *opinio juris* suffer from severe evidentiary problems.

[44] Here we ignore conceptions of custom that predate the ICJ statute for sake of brevity. For an in-depth discussion of the different conceptions of custom before the passage of the ICJ statute, see Curtis A. Bradley, "Customary International Law Adjudication as Common Law Adjudication" (in this volume).

[45] RESTATEMENT (THIRD) OF THE FOREIGN RELATIONS LAW OF THE UNITED STATES §102(2) (1987).

The dual-element definition of CIL has come under increasing attack from various commentators.[46] Here we can classify these critiques under two primary categories: conceptual critiques and practical critiques. Conceptual critiques have been weighty and consistent. Some scholars have argued that the rules of CIL are both circular and indeterminate.[47] For example, for a rule of CIL to exist, states must have the requisite *opinio juris* and follow the rule from a sense of legal obligation.[48] But a tremendous logical conundrum arises here: how can a rule of CIL arise if states must already be acting out of a sense of legal obligation? The traditional conception of CIL offers little guidance here.

As for state practice, it is entirely unclear how much state practice is necessary for CIL to become law. While it is evident that universal practice is not the requisite threshold, there is little consensus over the lower end of the threshold. Moreover, it is also controversial whether practice should be considered from a comprehensive global, regional, or power perspective. Can states within a self-defined region look to each other for evidence of state practice? Or must they look to the world as a whole?[49] Should more powerful states' practices be accorded more weight when determining state practice? These questions are not satisfactorily answered by the traditional approach to CIL.

There is also a conceptual critique of CIL that is rooted in rational-choice theory, arguing that CIL has, at best, *de minimis* impact on state behavior.[50] This critique is founded on the argument that CIL is not an exogenous factor that shapes state behavior. Rather, it is endogenously created from states acting rationally and in pursuit of their own interests. This, coupled with power disparities and historical accidents, ultimately determine the contours of CIL.[51]

While this rational-choice critique may succeed in shedding some light on how CIL comes about, it does not therefore follow that CIL has no effect on subsequent state behavior. States are interested in cooperative behavior only when it will bring

46 *See* Bradley, *supra* note 44, at 1–8.

47 *See* D'AMATO, supra note 5, at 53, 66; BYERS, supra note 4, at 136–37; JEAN D'ASPREMONT, *FORMALISM AND THE SOURCES OF INTERNATIONAL LAW* 164 (2011).

48 *See, e.g.*, Andrew T. Guzman & Tim L. Meyer, "Customary International Law in the 21st Century," in *PROGRESS IN INTERNATIONAL LAW*, 197, 199 (Russell A. Miller & Rebecca M. Bratspies eds., 2008).

49 As a matter of legal history, Western states have generally looked to their own counterparts for guidance on CIL. *See, e.g.*, The Paquete Habana, 175 U.S. 677 (1900) (focusing entirely on European and American conduct except for a single mention of Japan, which the court at the time deemed "the last state admitted into the rank of civilized nations"). There are many reasons for the Western-centric approach. But the inequities of this approach have not been lost on some critics of CIL.

50 *See* GOLDSMITH & POSNER, *supra* note 5, at 42–43; Jack L. Goldsmith & Eric A. Posner, "Understanding the Resemblance Between Modern and Traditional Customary International Law," 40 *VA. J. INT'L L.* 639, 640 (2000) ("The faulty premise is that CIL . . . influences national behavior.").

51 *See* GOLDSMITH & POSNER, *supra* note 5. *See also* Jack L. Goldsmith & Eric A. Posner, "A Theory of Customary International Law," 66 *U. CHI. L. REV.* 1113 (1999).

benefits. But cooperation over repeated rounds of interactions will require states to nurture some sort of reputation for compliance. This is true whether or not the rules in question are treaty based or CIL. Failure to comply with what is seen as a rule will negatively affect a state's future ability to enter cooperative agreements. Whether a state changes its behavior due to CIL (or any other form of international law) then depends on how much that said state values future cooperation. States that devalue future cooperation will be far more likely to ignore CIL, as well. Of course, this kind of devaluation can have severe consequences for the state's welfare in the future. A good reputation for compliance will allow the state to extract future concessions and lower negotiating costs. A state with a bad reputation will predictably be asked to provide more assurances that it will not renege on a future agreement. If the reputation is bad enough, the additional assurance may outweigh the potential benefits, thus leading to a cooperative breakdown. The key takeaway here is that CIL does not fundamentally differ from treaty law when it comes to how CIL interacts with states' reputation for complying with international agreements. Thus, even taking the argument that CIL arises from rational states acting in their own self-interest as true, it is unclear how this creates different incentives for compliance than with treaty law. At most one could argue that states put less reputation on the line when it comes to CIL than treaty law. But that is hardly an argument for *de minimis* impact.

Turning to practical critiques, the flaws inherent in the traditional conception of CIL are quickly exposed. There are truly vexing evidentiary problems with determining the existence of *opinio juris* and state practice. *Opinio juris* is, by definition, a narrowly focused attempt to look into a highly abstract notion of "intent" of states. Leaving aside the obvious question of whether it makes sense to ascribe such anthropomorphic concepts as intent to an entity such as a state, there is the problem that it is quite difficult to determine the precise reasons that a state takes an action. It is even harder to then determine whether a state took an action because it believed there was a legal obligation. Any determination is, at best, inferred from disparate evidence. Even more vexing is whether a lack of evidence allows an observer to conclude anything at all.[52] Since we are primarily interested in states prospectively using CIL as a tool to solve global problems, these evidentiary problems are weighty.

State practice, by contrast, aims to capture as broad a swathe of observable phenomenon as possible. That is to say, state practice can entail everything from treaty law, domestic laws, diplomatic correspondence, and statements from high-level officials.[53] The fact that state practice as a concept could potentially

[52] *See* S.S. "Lotus" (*Fr. v. Turk.*), 1927 P.C.I.J. (ser. A) No. 9 (Sept. 7).

[53] *See* Akehurst, *supra* note 5.

capture the universe of activities engaged by states suggests that determining what counts often amounts to an exercise in arbitrarily drawing lines. Indeed, the amount of evidence that could potentially be marshalled for any given custom is potentially enormous. Decision makers, such as international tribunals, are often under severe resource constraints; they simply cannot possibly review the records in a truly comprehensive way. One particularly problematic implication of this is that evidence of state practice stemming from more powerful or traditionally influential states are more likely to be taken up as persuasive evidence of custom. While this might have indeed been the original intention of the CIL system, it is increasingly unclear whether this is an equitable way to make CIL.

RATIONAL-CHOICE APPROACH TO CIL

As the previous section suggests, the state of commentary on CIL is fragmented and sometimes contradictory. Our task in this section, then, is to clarify and understand the basic building blocks of CIL. Here we argue that the rational-choice approach to CIL does not comport with the traditional dual-element understanding of CIL. Instead, *opinio juris* stands for the subjective beliefs of observing states regarding the legality of a rule. Furthermore, state practice is entirely unnecessary as an element.

Opinio Juris

At the heart of any rational-choice approach to international law must be the interplay between legal rules and state action. This means that any international rule, including CIL, must affect state behavior. If it does not, then we can safely say that it is ineffectual. We have previously suggested that CIL is not fundamentally different from other forms of hard international law when it comes to reputational consequences and payoffs; violating CIL should result in negative reputational sanctions from observing states. From this basic premise we can extrapolate a more theoretically coherent conception of *opinio juris*.

Consider that the traditional understanding of *opinio juris* is that it must be "accepted as law."[54] This means that *opinio juris* is a subjective element – an attempt to ascertain the "psychological" state of states when it comes to the rule in question. A major tension arises here because the traditional understanding, which is largely spelled out by the ICJ, does not specify whether the "states concerned" are the group of affected states or just the violating state.[55] This means that CIL begins to

[54] Statute of the International Court of Justice, art. 38, para. 1(b).

[55] See the analysis of North Sea Continental Shelf (*F.R.G. v. Den., F.R.G. v. Neth.*), 1969 I.C.J. 3, 44 (Feb. 20), which suggests that the "states concerned" are affected states rather than the violating state.

encroach on the consent requirement that most of international law is built upon. As a conceptual matter this can rankle some commentators, some of whom have spent much effort to "rescue" CIL by advancing arguments about inferred consent and secondary consent.[56]

However, in practice, it is clear that the consent of the violating state is not necessary for there to be a rule of CIL.[57] Instead, CIL can bind states as long as there is a general sense of legal obligation among the observing states. When a state violates a rule of CIL, it will suffer negative reputational or direct consequences at the hands of those observing states. It is likely that these observing states will only impose these consequences if they subjectively believe that there was a violation of CIL.[58] Once this reputational pathway is understood by all states, then it is clear that the subjective belief of the violating state is not relevant at all. Indeed, even if the violating state does not believe there to be a rule of CIL, it is enough for it to know the observing states' belief toward a rule for the rule to change the violating state's incentives for behavior.

Thus, *opinio juris* must refer to the belief of states that observe and interact with a potential violator. To the extent that these states believe there exists a legal obligation, the potential violator faces a rule of CIL. In this regard, we believe that our approach to *opinio juris* and CIL is compatible with the arguments advanced by Bradley and Lepard separately in this volume. Both Bradley and Lepard argue that *opinio juris* can be reconceptualized to encapsulate desirable rules in the immediate present or the future, rather than only concerning rules that allegedly already exist.[59]

One particularly important corollary of this understanding of *opinio juris* is that an observing state should update its understanding of another state's reputation for complying with CIL even if the observing state does not believe itself subject to that rule of CIL. This should not be a particularly controversial conclusion. Observing states frequently update their understanding of another state's reputation for compliance based on the obligations that the other state takes upon. In the instance of treaty law, State A updates its own understanding of State B's reputation for compliance with treaty law even if State A is not a party to the treaty between B and C. Strictly speaking, it does not even matter if B and C never concluded a valid treaty, only that A believes there to be some sort of legal commitment between B

56 *See* BYERS, *supra* note 4, at 142–46; MARK VILLIGER, *CUSTOMARY INTERNATIONAL LAW AND TREATIES* 18–22 (1985). *See also* Vaughan Lowe, "Do General Rules of International Law Exist?" 9 *REV. OF INT'L STUD.* 207, 208–10 (1983).

57 *See* BYERS, *supra* note 4, at 142–46; Akehurst, *supra* note 5, at 23 ("A state can be bound by a rule of customary international law even if it has never consented to that rule.").

58 Although it is also possible that even a state that does not possess the requisite *opinio juris* updates its view that the infracting state has a general willingness to break what others believe is CIL.

59 *See* Bradley, *supra* note 44, at 20–21; Brian D. Lepard, "Customary International Law as a Dynamic Process" (in this volume).

and C. There is no particular reason why states should treat compliance with CIL differently. This corollary will prove essential when we sketch out the mechanisms by which nonconsensual rules such as CIL can actually help overcome intractable problems that are difficult to solve through a consent-based system.

State Practice

There has been a large amount of controversy surrounding the state-practice element of CIL. Commentators have disagreed on just about every aspect of state practice. Such disagreements have involved issues such as the number of states required, the consistency of state practice, which behaviors count as state practice, and the timeframe of state practice.[60] Two particularly important implications of these debates are whether there can be instant practice and whether there can be regional custom.[61] While we are not directly interested in instant custom in this chapter, we will discuss the role of regional custom in promoting a coherent theoretical landscape for CIL as a whole.

For now we return to our basic premise that states are incentivized to act in certain ways by reputational consequences attached to compliance with international law. As the previous section on *opinio juris* makes clear, the key element that determines whether CIL changes state behavior vis-à-vis reputational consequences is the subjective belief of observing states. We can thus reason that state practice is irrelevant. Reputational sanctions attach as long as observing states believe there to be legal obligation. The only place for state practice is as evidence that states hold some kind of belief about a rule. Although to be precise, even this is an imperfect proxy for understanding the subjective belief of observing states. After all, it is conceivable that states hold a view about a rule but do not engage in practice that would confirm or deny their viewpoint.[62]

Under the rational-choice conception of CIL, then, state practice can be evidence for the existence of *opinio juris*. This approach begins to alleviate some of the evidentiary problems that plagued traditional conceptions of CIL by eliminating the urgency surrounding determinations of state practice. On a practical level, whether

[60] *See, e.g.*, IAN BROWNLIE, *PRINCIPLES OF PUBLIC INTERNATIONAL LAW* 3 (4th ed. 1990); D'Amato, *supra* note 5, at 88; KAROL WOLFKE, *CUSTOM IN PRESENT INTERNATIONAL LAW*, 84 (2d ed. 1993); Akehurst, *supra* note 5, at 9.

[61] For more on instant custom, see Bin Cheng, "Custom: The Future of General State Practice in a Divided World," in *THE STRUCTURE AND PROCESS OF INTERNATIONAL LAW: ESSAYS IN LEGAL PHILOSOPHY DOCTRINE AND THEORY* 513, 532 (R. St. J. MacDonald & Douglas M. Johnston eds., 1983).

[62] Note that lack of action may not actually reveal significant information about a state's beliefs. A state may not engage in waterboarding because it does not see the present need to; this does not necessarily mean the state believes it is illegal under CIL.

a decision maker will find a particular practice to be evidence of *opinio juris* would be an interpretive question. For example, international tribunals may decide that a UN resolution that reflects the belief of parties may be excellent evidence of *opinio juris*. By contrast, statements by high-level government officials may not be good evidence, particularly when the officials' beliefs may not reflect the practice of government apparatuses. Regardless, the evidentiary issues are not the focus of this chapter and we only wish to make clear that state practice cannot be considered an element of CIL under a rational-choice approach.

Regional and Special Custom

Conceptually, regional and special customs pose a challenge to the general framework of CIL. Under the classical approach, CIL as a legal form binds all states. But if the same process that produces CIL also produces regional or special custom, then it is conceivable that this will lead to fragmentation and a dilution of CIL's power to govern state behavior. Less pessimistically, some commentators have suggested that regional custom is a natural consequence of a heterogeneous world.[63] While we are more sympathetic to the latter view, we must still take seriously the suggestion that widespread acceptance of regional or special custom would fragment CIL to the point that it is no longer an effective form of global governance; since our primary motivation is to show that CIL can be part of states' toolbox for solving global problems, this is a serious challenge.

First, it is worth addressing the most basic way that regional and special custom fit into the rational-choice framework we set forth. As long as there is *opinio juris* among the relevant states, there is no reason why custom cannot be limited to a certain set of states. So long as the observing states in this set of states believe there is a rule of custom and possess the necessary *opinio juris*, there is custom between those states. This formulation will naturally lead to a question of whether it is possible for there to be a special custom between just two states; since there is only one observing state, it seems awkward if not illegitimate to say that only one of the states needs to have the necessary *opinio juris*. The answer to this question is evidentiary. In a two-state situation, it is possible to look at state practice to see if both states have followed and expected a pattern of behavior. The classic ICJ case on the question of bilateral custom is the Right of Passage case between Portugal and India, where the court found little reason to disallow customary rules that involve the two parties.[64] Another example is the Anglo-Norwegian Fisheries case, where Norway had used a system of

[63] *See, e.g.*, George Norman & Joel P. Trachtman, "The Customary International Law Game," 99 AM. J. INT'L L. 541, 568–69 (2005).

[64] [1960] I.C.J. Rep. 4, at 39.

delineating maritime borders since 1869. Importantly, the UK did not object to this system for sixty years. Thus, the ICJ found that the UK had no grounds to claim that there was not a special custom between Norway and the UK.[65]

There is a second, more consequential issue for us to address, however. This is whether the widespread recognition of regional and special customs will ultimately fragment the global CIL system and whether this dilution would ultimately help or hurt the global CIL project.

Under the rational-choice conception of CIL, there is little reason to believe that CIL as a form would be in danger because of regional and special custom. Indeed, we believe that regional and special custom can be a potential building block for strengthening the CIL system; it is the mechanism by which more globalized CIL can emerge. Recall that a rule is only considered law among sets of states if they have the requisite *opinio juris*. What is implicit here is that *opinio juris* must be created through some definable process. That is to say, generally speaking, states do not instantly form similar beliefs about any issue (although instant custom is possible under the rational-choice conception of CIL). Instead, states are influenced by myriad factors, including history, public opinion, political debate, international expectations, and more. This definable process should reflect the competing influence of these many factors.

One particularly important factor is that states are influenced by competing customary norms. Assuming that competing customary norms on the same subject matter are fundamentally incompatible, then states that are interested in a particular form of customary norm will be heavily incentivized to advocate for that norm. Advocacy in this case could take the form of explicit diplomatic negotiations with partner states to agree to act in similar ways. It could also take the form of moral opprobrium toward states that do not hold the same norms.

The result of such a process is a world of competing customary norms – a veritable marketplace of norms. This, of course, is the crux of the criticism that some legal scholars have toward fragmentation. But consider for a moment which issues would likely see customary norms fragment. If states are driven by rational self-interest, then individual states should only oppose and promulgate a differing norm if it is in their interest to do so. For issues on which there is minimal difference in interests, there should be little fragmentation. We suspect that the more cabined the customary norm is when it comes to issue applicability, the more likely there will be fragmentation. After all, there is little reason to think that a special custom between two countries over specific land usage rights should be more generally applicable to other states. However, it may in fact be in every country's interests to have an ascertainable CIL regarding territorial waters. Of course, in the case of territorial

[65] *Fisheries* (*U.K. v. Nor.*), 1951 I.C.J. 116, 131 (Dec. 18, 1951).

waters, it would indeed be easy to see why there might be a competition over which rule ultimately becomes a global rule. But it would be hard to see a mechanism by which the international community could determine the best global rule without the fundamental ability for customs to fragment and compete with each other.

In our view, the viability of regional custom is directly tied to the viability of CIL in general for two main reasons. The first is that recognizing regional custom is a way for CIL to respond to criticism. CIL has increasingly come under attack for being a reflection of traditionally powerful states' interests. This has been particularly true for proposed CIL on human rights. This is a serious critique that undermines the legitimacy and ability of CIL to actually solve problems. There is no reason to believe that the preferences of powerful states would necessarily lead to the greatest increases in global welfare. A recognition that marketplace of norms exists will ultimately strengthen compliance with CIL as well as reinforcing the theoretical coherence of CIL.

The second reason is that it puts CIL on a sounder conceptual foundation. What we now call CIL is often simply the historical regional custom of European nations that then became globalized through myriad forces. Indeed, it is impossible to understand CIL as a legal form and as a process without recognizing its pedigree. What this pedigree says to us, however, is that regional custom can spring up from any groupings of states. And through various legal, political, cultural, and normative processes, these customs can ultimately become global CIL.[66]

HOW THE LEGALITY AND DIRECT INVOLVEMENT AFFECT THE ABILITY OF CIL TO CHANGE STATE BEHAVIOR

In previous work, one of us argued that relaxing the international system's rigid adherence to the consent requirement for making international law would likely increase states' ability to solve certain cooperation problems.[67] Here we explain how CIL's unique place on the continuum of legality and direct state involvement helps give it traction as potentially an important source of nonconsensual international law. Although traditionally CIL is seen as arising organically from underlying state practice, we argue here that the rational-choice approach we articulated allows for more direct state involvement. Indeed, states may even engage in efforts to influence the *opino juris* of other states. Giving states this much control of custom means that a group of interested states can autonomously create a custom that applies to

[66] Undoubtedly, power and influence will largely determine which region's customs become accepted globally. But since relative power and influence change, there is no reason to believe global CIL will not also change.

[67] *See* Andrew T. Guzman, "Against Consent," 52 *Va. J. Int'l L.* 747 (2011).

themselves relatively easily. This is all the more likely since there is no theoretical bar to instant custom under our rational-choice conception of CIL. In short, CIL retains enough elements of direct state involvement for it to be used as a purposive tool. But it also brings in the community consensus form of rulemaking that allows any community to overcome bottlenecks that arise from requiring explicit consent.

The Legality Dimension

Legality is a binary trait. A rule is either binding or it is not. States specifically agree that some rules are legally binding and others are "soft." One question that often arises is why this might make a difference to compliance. The traditional answer would be that states believe that there is something special about a rule being legally binding. Perhaps states simply prefer to comply with legal rules.[68] Alternatively, states may believe that there is a doctrine that calls upon compliance as a matter of course.[69]

A rational-choice theory of legality must reject these assumptions and argue that states have no preference for either compliance or noncompliance. Instead, states comply with rules that are in the states' self-interests. For the purposes of the present discussion on legality, this means that states can modulate the amount of reputational capital to commit to a legal rule. The more reputational capital a state commits, the higher the sanction costs if the state breaches the rule in the future. Sometimes the additional compliance pull of a binding legal rule will be worth the additional sanction costs imposed for breach. But if the potential costs of sanctions are so high as to wipe out the anticipated gains that additional compliance due to bindingness would create, then it is likely that states would prefer the rule remain soft.[70] A binding law, *ceteris paribus*, creates additional compliance pull when compared to a nonbinding rule.

Thus, legality is best seen as a first-order differentiation between two tiers of reputational commitments. Perhaps the most obvious reason for the existence of this mechanism is that some rules are simply more important than other rules. When making a particularly important rule, states want to know that their counterpart states will potentially suffer larger sanctions for future breaches. While subsequent development may tie a state to its international commitments in various ways, including through domestic politics, public opinion, or even simple bureaucratic inertia, *ex ante* states can only count on the setting up of reputational sanctions

[68] *See* Abram Chayes & Antonia Handler Chayes, "On Compliance," 47 INT'L ORG. 175, 178 (1993).
[69] That is, taking *pacta sunt servanda* seriously.
[70] *See* Andrew T. Guzman, "The Cost of Credibility: Explaining Resistance to Interstate Dispute Resolution Mechanisms," 31 *J.* LEGAL STUD. 303 (2002).

through commitments. In other words, states pledge that breaching certain rules will be more costly than other rules. The result is that, when negotiating a rule, states must then weigh the benefits of additional compliance pull due to the binding nature of a rule with the additional costs that sanctions would inflict on the parties. For nonconsensual lawmaking, bindingness and legality creates greater compliance pull.

CIL is in fact binding law. This means that using CIL as a form of cooperation involves a higher level of commitment than a "mere" international norm. Whether CIL is the precise level of commitment that is appropriate for any given international cooperation problem is an empirical question. Here we simply wish to address some predictions as to how the compliance pull generated by CIL might affect outcomes.

Compliance Pull

As a preliminary point it is important to make clear that reputation and the compliance pull it generates operates on the margins of state decision making toward nonconsensual international lawmaking. A plethora of interests can override the concerns of reputation. For example, states are almost certainly willing to break CIL if it means that they will survive an acute emergency. Put slightly differently, then, reputation is one of many factors that states consider when weighing the costs and benefits of complying with nonconsensual rules.

That being said, we can predict that compliance pull is variable depending on observing states' views of the circumstances. In the aforementioned situation where a state violates CIL because its supreme interests are under threat, it is unlikely that observing states will then believe the state will prospectively violate CIL for other reasons. This is particularly true if the state already has a reputation for complying generally with CIL. Essentially, under the rational-choice conception of nonconsensual lawmaking, compliance pull is not a simple black and white determination.

That a nonconsensual rule can be deemed legal should mean that it has more baseline compliance pull. Legality is, after all, a costly signal that states send to ensure that its counterpart states understand the seriousness of the commitment. But the normal conception of reputation is predicated on states willingly committing to a legal rule. Does being subject to a nonconsensual but legal rule such as CIL continue to signal a greater commitment than if the rule were nonbinding? Would it make sense for states to rationally update their beliefs about the likelihood of future compliance if the original rule was "imposed" from outside? We answer these questions by disaggregating states into three camps: those that advocated for a new nonconsensual rule, those that remained silent, and those that actively opposed the rule. Each of these camps should have different reputational consequences attached to compliance with future nonconsensual rules such as CIL.

First, it seems likely that if a state or set of states actively advocated for the establishment of a new nonconsensual rule, those states will be judged by roughly the same reputational standard as if they had created a regular hard law rule. This is because observing states know that these advocate states believed that a new nonconsensual rule is in their self-interest and are willing to commit reputational and resource capital into making it a reality. Thus, for advocate states, breaching a nonconsensual rule that they advocated for implicates both their reputation for general compliance with nonconsensual rules and their reputation for complying with rules they agreed to generally. Also note that for advocate states, it is not obvious the rule should even be considered nonconsensual to begin with.

Second, for states that did not speak out on a new rule because they were not aware that the new rule was in the process of being created or did not find it worth the time and effort, there should be smaller reputational penalties for noncompliance. At a minimum, breaching would still subject these states to a sanction to their reputation for general compliance with nonconsensual rules. But it is unlikely that they will be judged as harshly for breaking a rule they explicitly agreed to. Of course, this does not mean that the consequences of breach would necessarily be trivial. There are other sanctions, such as those stemming from bureaucracy (e.g., being omitted from IO functions), domestic politics, and public opinion. These are all powerful incentives for even states that did not speak out to comply with nonconsensual rules. Furthermore, as time passes after the nonconsensual rule becomes binding, it is possible that these types of states are judged to have implicitly accepted the new rule. At that point, the reputational sanctions for a deliberate breach would largely look like the same kind imposed on the advocate states.

Third, for states that actively opposed the creation of a new nonconsensual rule, the reputational sanctions for breach should be further limited to just their reputation for complying with general compliance with nonconsensual rules. Indeed, current CIL doctrine means that persistent objectors should suffer no reputational sanctions because they would simply be exercising their right of objection. In this sense, the current CIL doctrine recognizes the cabining effect of a rule being nonconsensual when it comes to reputational consequences for breach.

That CIL should create additional compliance pull over a mere norm should not be surprising. Whether this additional amount of compliance pull is sufficient to make it more attractive to states as a legal form than IO/IS soft law is up for debate. A brief examination of the state of the world suggests that soft law emanating from IOs and ITs and CIL tend to deal with different cooperative problems.

The Direct Involvement Dimension

The direct state involvement measures the amount of control individual states may have over the content of a nonconsensual rule. As we mentioned previously, this dimension is not binary and ranges widely depending not just on the specific form of nonconsensual rule, but also the specific forum chosen to make those rules. On one end of the spectrum are *jus cogens* norms, which can be described as being completely immune to direct state involvement. On the other end of the spectrum are Security Council resolutions, which are purposefully designed to only come about with the consensus of all five permanent members of the Security Council. CIL is situated between these two extremes.

Why would the level of direct state action be important to states when it comes to making nonconsensual laws such as CIL? The answer is that, like the legality dimension, states want to be able to modulate how much reputational capital to commit to each rule. Some rules are more important than other rules to states. Some rules are also more likely to be breached in the future than others. States want to be able to commit the right amount of reputational capital. Direct state involvement, then, is like legality, a mechanism by which states can signal various gradations of commitments. These gradations of commitments stem from the costs that a state must accrue in the process of making the nonconsensual rule. The intuition here is that the more involved a state is in the process of constructing a nonconsensual rule (thereby expending more resource and political capital), the more reputational capital a state pledges to the commitment to comply. This should not be surprising to anyone who observes state behavior when it comes to the treaty-making process. States that take a particularly large role in setting the negotiation agenda and content of the final treaty are also judged harshly if they do not comply or even ratify the treaty once it is signed.[71]

As we have mentioned before, CIL and soft laws emanating from IOs and ITs are the two main mechanisms by which states can modulate their commitments. In both instances there is a sort of delegation to a non-state party to do the hard work of producing rules, thus giving these rules a veneer of neutrality. For IOs and ITs, states assign the task of coming up with new soft laws to organizations specifically tasked for it. Depending on the subject matter, states may retain a very direct control over the outcome. The International Monetary Fund comes to mind as an IO that ultimately is directly controlled by the (most powerful) states. For CIL, the non-state

[71] Consider, for example, the United States and its relationship with the Convention on the Elimination of All Forms of Discrimination Against Women, the International Covenant on Economic, Social and Cultural Rights, the Convention on the Rights of Child, or the Rome Statute of the International Criminal Court.

party is more diffuse. It encompasses the entirety of legal commentators and state organs that can speak on custom. States can, of course, also directly speak toward CIL through declarations, unilateral action, or perhaps even treaty making. But on the whole CIL anticipates the input of many third parties. The key here is that the CIL form gives states many levers by which they can commit more or less resources to a rule.

This leads us to perhaps the most unique advantage of nonconsensual rules that have little direct state involvement: that of legitimacy. Rules that are considered to arise from the community rather than any overt bargaining between state actors have a special imprimatur of legitimacy. Let us take the most obvious example. A *jus cogens* rule such as the prohibition against genocide cannot and should not be changed. The human community has decided that it is simply illegal no matter what. While states no doubt played a role in implementing this rule, one could make an argument that ultimately no amount of state action could have prevented this rule from crystalizing. Indeed, one could plausibly argue that many states have historically worked against the development of such a rule. The take-away here is that the prohibition against genocide derives much of its power from the fact that it is seen as arising from the human community, rather than states.

What this means is that some rules may be most effective if they are seen as community rules rather than explicitly state-made rules. The choice between the different forms of nonconsensual law, then, is an attempt to choose a form that best suits the cooperation problem at hand. Since community norms and expectations change over time, it is impossible to predict which form of nonconsensual lawmaking is appropriate without targeted study. This is undoubtedly a somewhat unsatisfying position because it implies a constantly moving target. But ultimately, we believe that this perhaps represents the outer limits of the rational-choice approach to international nonconsensual lawmaking and CIL.

One important question remains: when might states prefer to use CIL as opposed to other forms of nonconsensual law to promote international cooperation?

The kind of circumstances under which states might prefer CIL as the "just right" level of nonconsensual rulemaking is one where the gains that accrue to individual states might be small. The only way for the benefit to accrue is if there is widespread understanding that such a rule exists. Furthermore, the cost of trying to produce the rule on a global basis is high. There are many examples of such situations that have already been addressed by CIL. For instance, while the Vienna Convention on Diplomatic Relations of 1961 codified the treatment of diplomats, much of the content of this convention would have been considered CIL prior to 1961. In a hypothetical world where there were no rules governing diplomatic behavior, the cost of realizing a new custom would be extremely high. With no legal basis, a state wishing to establish a new norm would likely have to rely on politics and power. Certainly, in this situation the costs are likely so high as to overshadow the short-term gains.[72] The point we want to make here is that while every state benefits from the existence of CIL standardizing diplomatic behavior, the potential cost explicitly setting out to establish a new norm by a state, or set of states, could be extremely high. It may not make sense for a state, or set of states, to explicitly try to enact a new norm on a global basis. But it may make sense to slowly nudge the process along through, first, bilateral customs, and then more wide-ranging declarations. Similarly, the situation may be that only a small number of similar states are willing to act in accordance to a custom. If these states can establish a long-term commitment to a certain rule, perhaps as a regional custom, then other states may begin to see it in their own self-interest to commit to the rule, as well. This is particularly true if states that originally reject the custom based on fears that counterpart states would not reliably reciprocate, thereby making the adoption costly and with little benefit.

DOES IT WORK? TOWARD A FUNCTIONAL THEORY

In previous sections we discussed how a rational-choice approach to CIL helps us make more sense of the chaotic theory and practice of CIL. We have established that CIL may be useful as a "just right" form of nonconsensual international rulemaking because it is both binding and not overly determined by direct state involvement. We also suggested that all CIL originates as some form of regional custom that ultimately gains traction among states not in the originating region. In this section, we will briefly sketch out the mechanism by which a regional custom becomes a global custom. This will move us toward understanding how the CIL form can ultimately help states overcome the problems associated with a system overly preoccupied with consent.

CIL is only useful as a tool if it changes state behavior. But how can a custom held by a discrete region change the payoffs for countries outside of that region? To answer this question we must remember that the rational-choice approach to *opinio juris* does not restrict states to only believing that a custom applies to states that have also accepted the custom. To the extent that observing states believe there exists a legal obligation, the potential violator faces the reputational consequences of violating CIL. The crux of the matter here is that states that are within the subset

[72] This suggests that a sufficiently powerful hegemon can "unilaterally" establish new norms as long as it has the political and power capital to spare. This seems to largely reflect the experiences of empires such as Rome, Britain, or China during their more powerful eras. Context, of course, is important. Britain had enough naval power in the nineteenth century to dominate the world's oceans and thus the ability to unilaterally set maritime norms. However, Britain did not have the same level of influence over terrestrial affairs.

of states that accept a specific regional custom can then extend that belief to states that are not part of that original subset of states. We henceforth refer to states within the original subset as "insider states"; other states are "outsider states."

Because an insider state's *opinio juris* regarding a custom is largely up to itself, there is nothing keeping such a state from expanding its subjective belief so that it believes that the custom applies to outsider states, as well. It does not seem like such a far-fetched idea that states may routinely expand their beliefs about the applicability of custom. After all, the payoffs of cooperation often increase with the amount of cooperative partners; successfully expanding the reach of a regional custom may bring benefits. However, whether insider states purposively begin to apply their own beliefs about the applicability of a customary rule to outsider states is an empirical question that deserves much more study.

The direct consequence of this is that those outsider states are taking a reputational loss vis-à-vis the insider states for violating a regional custom. This, in turn, leads outsider states to evaluate whether it is worth taking the reputational sanction vis-à-vis the insider states in order to hold onto a different rule set. As might be expected, if the insider states are powerful or indispensable members of international society, then it may be much harder for outsider states to resist changing their behavior. Nonetheless, if the outsider state is powerful or a part of another regional custom bloc, then the reputational sanctions may not rise to an actionable level. Generally speaking, then, whether a regional custom spreads into global CIL is a function of power and influence. We believe this largely mirrors the actual historical development of CIL. Since international law as we know it was defined and spread across the world by European powers, it is expected that the contours of CIL largely reflects that historical reality.

CONCLUSION

Perhaps the greatest barrier to CIL being taken up by states as a serious and purposive part of the international law toolbox is that the current international legal environment is chaotic. In the current system, CIL is plagued by problems of indeterminacy. First, it is often unclear what CIL entails. States and non-state actors alike often disagree as to the specific content of CIL. Moreover, these actors can also disagree as to the very existence of CIL. For example, while some human rights norms have been slowly introduced as CIL, they are hardly universally accepted. Second, it is often unclear how CIL comes about. Although scholars have tried to theorize about the process by which mere norms become CIL, it is hard to observe any predictable pattern. Some CIL comes about via long-held historical behavior. Such is the case for CIL involving diplomatic behavior. But other CIL comes about via acute shocks to the system. Still stranger is the case when treaties and custom form a feedback loop. Treaties could be evidence for an emerging custom, although it is unclear what makes some treaties evidence of custom and others evidence of nothing more than a treaty.

The crux of the matter, then, is that states cannot easily ascertain the content and existence of CIL. Coming back to our definition of CIL, we see that a natural corollary is that states can only ascertain CIL if they can observe the *opinio juris* of other, observing states. It does not matter whether the state itself feels any obligation; if observing states believe an action violates CIL, the acting state will suffer reputational consequences for breach. The belief of a state, of course, is extremely difficult to ascertain. States have any number of reasons to misrepresent their public statements. States will also likely make statements that support a customary rule but have no intent on observing or sanctioning other states for violations.

However, states do have one method of ascertaining the *opinio juris* of other states: by publicly staking out a legal position and inviting dissent. In situations where the position that a state takes does not infringe upon any other states' belief about the state of CIL, the staking state should provoke no response. But if the staking state violates some held belief about CIL, then dissenters should publicly object. States that object to a staking state's legal position have nothing to gain from remaining silent on the matter. Indeed, prolonged silence can mean that the new position gradually becomes accepted as CIL. States with contrary views of CIL could (indeed, should) stake out their own positions of CIL. Although this may seem to increase the chaos and indeterminacy of CIL in the short term, in the long term it should decrease confusion as less reasonable and accepted legal positions are abandoned. A marketplace of norms, so to speak, should tend to produce rules that increase global welfare. This means that, over a period of time, states can construct the true contours of CIL, as well as its contents. It also means that a certain amount of room for experimentation is necessary in order for this dialectical model of CIL to properly function, perhaps to the chagrin of some commentators.

12

The Evolution of Codification

A Principal-Agent Theory of the International Law Commission's Influence

Laurence R. Helfer and Timothy Meyer

INTRODUCTION

The International Law Commission (ILC or the Commission) has a mandate from the United Nations (UN) General Assembly (the UNGA or the General Assembly) to codify and progressively develop international law. During most of the ILC's history, the lion's share of its work product took the form of draft articles adopted by the UNGA as the basis for multilateral conventions. The ILC's activities received their principal legal effect during this period through the UN treaty-making process, rather than directly on the basis of the ILC's analysis of what customary international law (CIL or custom) does or should require.

In recent decades, however, the ILC has self-consciously limited its efforts to codify or progressively develop international law in the form of multilateral conventions. Instead, it has turned to other outputs – such as principles, conclusions, and draft articles, that it does not recommend be turned into treaties. Significantly, the Commission often claims that these outputs reflect CIL. For example, despite recommending that the General Assembly not base a treaty on the Draft Articles on State Responsibility, the ILC as well as many states and commentators assert that the draft articles largely reflect CIL.

This change in behavior presents a puzzle. If the ILC is still engaged in codification and progressive development, why has it changed the form of the work it produces? In this chapter, we argue that increasing political gridlock in the General Assembly – by which we mean a division of views over the substance of international norms and lack of enthusiasm for convening multilateral diplomatic conferences – has led the Commission to modify the form of its work to preserve its influence in shaping the evolution of international law. More specifically, we argue that the reduced likelihood of the General Assembly adopting draft articles as treaties closes off the primary mechanism of ILC influence. In addition, if the UNGA or member states reject an ILC recommendation that its draft articles become treaties, that rejection may suggest that the work product does not reflect existing custom – an alternative mechanism of ILC influence. To avoid these negative outcomes, we expect the ILC to turn to other outputs that allow it to continue to influence CIL without the General Assembly's approval.

This chapter makes three contributions to the study of the ILC and custom.[1] First, we use principal-agent theory to model the relationship between the General Assembly and the Commission and how that relationship has evolved over time. Second, drawing upon a new data set that codes all ILC outputs since 1947, we show that the Commission began to favor non-treaty outputs beginning in the early 1990s. This followed a decade when ILC treaty recommendations were not adopted by the UNGA or, if adopted, did not garner sufficient ratifications for the treaties to enter into force. Third, we argue that the shift away from draft treaties increases the salience of the methodology that the ILC uses to enhance its influence when the traditional constraint of UNGA review is unavailable due to gridlock.

The remainder of this chapter proceeds as follows. We begin by briefly reviewing the ILC's mandate to codify and progressively develop international law. We then use principal-agent theory to analyze the relationship between the General Assembly and the Commission. We argue that the ILC chooses the work product that maximizes its influence among states and other actors, subject to its expectations of how the UNGA will respond to its work. Our core claim is that, as gridlock in the General Assembly has limited its ability either to adopt treaties or decisively reject non-treaty outputs, the Commission has had both the incentive and the discretion to choose other outputs that do not require General Assembly approval. We proceed to provide empirical support for this claim, documenting the ILC's shift away from recommending to the UNGA that draft articles be adopted as treaties.

After establishing this empirical claim, we explain why methodology acts as a constraint on the ILC even if it does not similarly limit other actors' claims about the content of custom. More specifically, we argue that the methodology that the ILC uses to prepare non-treaty outputs functions as a *de facto* substitute for the political

[1] We are not the first to examine changes in the ILC's work product over time. Previous studies that offer a range of normative approaches and illustrations include: Frank Berman, "The ILC Within the UN's Legal Framework: Its Relationship with the Sixth Committee," 49 GERMAN Y.B. INT'L L. 107 (2006); David D. Caron, "The ILC Articles on State Responsibility: The Paradoxical Relationship Between Form and Authority," 96 AM. J. INT'L L. 857 (2002); Jacob Katz Cogan, "The Changing Form of the International Law Commission's Work" in *EVOLUTIONS IN THE LAW OF INTERNATIONAL ORGANIZATIONS* 275, 278–79 (Robert Virzo & Ivan Ingravallo, eds., 2015); Sean D. Murphy, "Codification, Progressive Development, or Scholarly Analysis? The Art of Packaging the ILC's Work Product" in *THE RESPONSIBILITY OF INTERNATIONAL ORGANIZATIONS: ESSAYS IN MEMORY OF SIR IAN BROWNLIE* 29 (Maurizio Ragazzi ed. 2013).

blessing associated with the UNGA's adoption of draft treaty articles. Adherence to methodology increases the likelihood that a wider audience – government officials, international judges, national courts, and non-state actors – will accept the ILC's non-treaty work products as valid statements of CIL. We thus expect the Commission to select a methodological approach that it expects will be supported by the audience(s) it hopes to persuade. We find suggestive evidence of this behavior in the ILC's recent project on foreign official immunity. We conclude by identifying what we should expect to observe in future ILC work if our claims are correct.

GRIDLOCK AND ITS WAGES

Article 13 of the UN Charter tasks the General Assembly with, among other things, "encouraging the progressive development of international law and its codification."[2] The General Assembly, in turn, has delegated this responsibility to the ILC in the Statute of the International Law Commission (hereinafter the ILC Statute or the Statute).[3] The ILC is therefore an agent of the General Assembly, operating within both the legal limits and political constraints imposed upon it by that UN body.[4] Principal-agent theory can help explain the changing role of the ILC in the codification and progressive development of international law. In this section, we analyze the relationship between the General Assembly and the ILC in light of principal-agent theory. Our central claim is that the increasing gridlock in the General Assembly over time has constrained the ILC's ability to carry out what has historically been its primary function of drafting treaties. These constraints have led the ILC to look for new ways to influence the formation and codification of CIL. We demonstrate this claim empirically with a comprehensive review of how the ILC's output has changed since its inception in 1947, as well as concrete anecdotes drawn from current efforts to prepare draft articles on the immunity of foreign officials from criminal jurisdiction.

The ILC's Mandate

Mirroring the language in Article 13 of the UN Charter, the ILC Statute charges the Commission with "the promotion of the progressive development of international law

[2] UN Charter, art. 13.

[3] Statute of the International Law Commission, adopted by the General Assembly in resolution 174 (II) of 21 November 1947, as amended by resolutions 485 (V) of 12 December 1950, 984 (X) of 3 December 1955, 985 (X) of 3 December 1955, and 36/39 of 18 November 1981.

[4] *Cf.* Cogan, *supra* note 1, at 278–79 ("The ILC's role, therefore, was to assist States in their lawmaking function by thoroughly researching and debating topics in need of codification and drafting texts that could then be used as starting points by States in their treaty negotiations.").

and its codification."[5] As a formal matter, the Statute treats progressive development and codification differently. The Statute defines progressive development as the "preparation of draft conventions on subjects which have not yet been regulated by international law or in regard to which the law has not yet been sufficiently developed in the practice of states."[6] Codification, on the other hand, means "the more precise formulation and systematization of rules of international law in fields where there already has been extensive State practice, precedent and doctrine."[7] The Statute underscores this distinction by establishing somewhat different procedures for the ILC to follow depending on its task.[8]

These definitions are connected to the output the General Assembly expects. Where the ILC engages in progressive development, the Statute contemplates draft conventions; where it engages in codification, the ILC is given wider latitude with regard to the form of its product. More specifically, when engaged in codification the Statute directs the ILC to submit a report to the General Assembly with a recommendation as to whether to: (1) take no action; (2) adopt the report; (3) adopt draft articles submitted by the ILC; or (4) convene a diplomatic conference to negotiate a convention.[9]

No matter the task it is engaged in or its final product, however, the Statute envisions a supervisory role for the General Assembly. Proposals for progressive development come from the General Assembly, UN members, or other international organizations.[10] Where codification is concerned, the ILC has leeway to select its own topics, but it is required both to submit its recommendations for topics to the General Assembly, and to give priority to requests from the General Assembly.[11] The Statute mandates an opportunity for UN member states to comment on the ILC's drafts or provide information about their practices.[12]

Whichever track the ILC chooses, its final product is reported annually to the General Assembly – in particular the Sixth Committee of the General Assembly, which deals with legal affairs – for review and action. The Sixth Committee considers these reports each fall. This review of these reports gives states an additional opportunity to opine on the ILC's progress and provide input and direction on an

[5] ILC Statute, art. 1(1).
[6] *Id.*, art. 15.
[7] *Id.*
[8] *See id.*, arts. 16–24. The principal difference is that the procedures for codification focus on collecting information from governments about state practice and reaching conclusions on the basis of that practice, while progressive development involves consulting with expert and technical bodies and other international organizations with relevant expertise.
[9] *Id.*, art. 23.
[10] *Id.*, art 16–17.
[11] *Id.*, art. 18.
[12] *Id.*, arts. 16(h), 17(b), 19.2, & 21.2.

ongoing basis. Even where the Sixth Committee does not formally endorse or reject the ILC's work product, member state comments provided on the Commission's annual reports and after initial readings of projects implicitly shape the ILC's activities by indicating the kind of reception that its final products will receive in the UNGA (and, in the case of proposed conventions, the likelihood of ratification and entry into force). Ultimately, however, the General Assembly "has the final say on the disposition of the Commission's drafts. It can approve them, send them back for redrafting, or reject them."[13]

Principal-Agent Theory and the UNGA-ILC Relationship

Principal-agent theory helps to understand the relationship between the ILC and the General Assembly, and in particular how that relationship has changed over time. The theory, which has been widely applied to the study of legal institutions,[14] including international organizations,[15] builds upon rational-choice theories of domestic and international politics. It posits that self-interested actors involved in governance and policymaking – such as voters or legislators at the domestic level, and states at the international level – delegate power to other actors – such as domestic administrative agencies or international organizations – to provide benefits that the principals could not achieve on their own.

Delegating may be desirable for a number of reasons – because the agent can perform a task more efficiently, has superior knowledge or experience, or because the principal has limited time or resources. For example, the General Assembly delegated to the ILC its responsibilities under Article 13 of the UN Charter because of the ILC's distinctive expertise in international law. The ILC can draw on its members' specialized knowledge and experience to prepare draft texts that more accurately assess existing state practice and *opinio juris*. The ILC might also be able to prepare these texts more efficiently than the General Assembly due to its significantly smaller size and the fact that its members serve in their personal capacities rather than as representatives of states.

[13] H. W. Briggs, "The Work of the International Law Commission," 17 *JAG J.* 56, 80 (1963).

[14] *See, e.g.*, John Ferejohn & Barry Weingast, "Limitation of Statutes: Strategic Statutory Interpretation," 80 *Geo. L.J.* 565 (1992); Randall, Calvert, Mathew McCubbins, & Barry Weingast, "A Theory of Political Control and Agency Discretion," 33 *Am. J. Pol. Sci.* 588 (1989).

[15] *See, e.g.*, Joel P. Trachtman, "The Economic Structure of the Law of International Organizations," 15 *Chi. J. Int'l L.* 162 (2014); Manfred Elsig, "Principal-Agent Theory and the World Trade Organization: Complex Agency and 'Missing Delegation,'" 17 *Eur. J. Int'l Rel.* 495 (2011); Andrew T. Guzman & Jennifer Landsidle, "The Myth of International Delegation," 96 *Cal. L. Rev.* 1693 (2008); Daniel L. Nielson & Michael J. Tierney, "Delegation to International Organizations: Agency Theory and World Bank Environmental Reform," 57 *Int'l Org.* 241 (2003).

The advantages of delegation are substantial, but they can also entail considerable costs.[16] In particular, in carrying out their assigned tasks agents may seek to further their own interests rather than those of the principals. The mechanisms used to deter such behavior are a central focus of principal-agent theory. These tools can be applied *ex ante* (for example, by controlling the appointment of agents and precisely defining their mandate), contemporaneously (by reviewing agents' activities and decisions), and *ex post* (via reporting requirements, revisions to the terms of the delegation, and, in extreme cases, sanctions). All of these mechanisms are costly, however. The principal is thus unlikely to devote the resources required to ensure that an agent is perfectly faithful to the principal's charge.[17]

Numerous provisions of the ILC Statute reveal the mechanisms by which the General Assembly can influence the Commission and its work. The UNGA is responsible for electing ILC members (Article 3), proposing topics for its consideration (Article 16), approving topics submitted by member states and IOs (Article 17), evaluating draft articles (Article 16, para. [j]), reviewing topics proposed by the Commission (Article 18), commenting on draft Articles (Article 20), reviewing final drafts (Article 22), and returning such drafts to the Commission for further consideration (Article 23).[18] In addition to or instead of these formal oversight mechanisms, the General Assembly can use informal mechanisms, such as providing feedback on topics the Commission has proposed, to exercise *de facto* oversight over the ILC.

Characterizing the UNGA-ILC relationship as one of principal and agent does not, however, imply that the Commission does or should merely carry out the General Assembly's wishes. To the contrary, principal-agent theory, which assumes that agents have their own preferences, is fully compatible with international delegations that give agents broad independence and discretion. "Far from presupposing that agents slavishly follow the preferences of their principals, PA [principal-agent] analyses give us a theoretical language for problematizing and generating testable hypotheses about the sources and the extent of agents' autonomy and influence."[19]

[16] Curtis A. Bradley & Judith G. Kelley, "The Concept of Delegation," 71 *L. & Contemp. Probs.* 1 (2008).

[17] *See* Pablo T. Spiller & Emerson H. Tiller, "Decision Costs and the Strategic Design of Administrative Process and Judicial Review," 26 *J. Leg. Stud.* 347 (1997).

[18] Jeffrey S. Morton, *The International Law Commission of the United Nations* 3 (2000); *see also* B. G. Ramcharan, *The International Law Commission: Its Approach to the Codification and Progressive Development of International Law* 30 (1977) ("The [ILC] is in a subordinate relationship to the General Assembly" and "subject to [its] supervision and control.").

[19] Mark Pollack, "Principal-Agent Analysis and International Delegation: Red Herrings, Theoretical Clarifications, and Empirical Disputes," *Bruges Political Research Papers No. 2*, at 5–6 (Feb. 2007); *see also* Carmen Pavel, *Divided Sovereignty: International Institutions and the Limits of*

Nearly seven decades of codification and progressive development work, which we will describe in detail, reveal how the Commission's independence, expertise, and discretion operate in practice. The Sixth Committee and the General Assembly have frequently approved the draft articles and other legal texts that the Commission has prepared with relatively little modification. But not always. In 1953, for example, when the ILC urged the adoption of draft articles on the Continental Shelf and High Seas Fisheries, "the General Assembly declined to comply with the Commission's recommendation" and directed it to prepare a comprehensive text on the law of the sea.[20] Five years later, the ILC produced a draft Convention on Model Rules on Arbitration Procedures. "The U.N. General Assembly rejected the draft and merely took note of these 'Model Rules,' which were never implemented."[21] Similarly, ILC's work in preparing the UN Watercourses Convention "was not linear: different approaches were tested and some were rejected, in part as a result of information received through the annual interaction between the Commission and the U.N. General Assembly."[22] These examples reveal that the UNGA was not merely a rubber stamp for the ILC's work.

That the instances of outright UNGA rejection are few suggests that the Commission is highly adept at performing the expert tasks entrusted to it by the ILC Statute. But such a pattern is also consistent with an agent operating in the shadow of what it anticipates the principal will accept. As Jeffrey Morton has explained:

> The commission generally seeks to ascertain rules which are likely to be useful to states in the conduct of their relations, bears in mind what rules and formulations states are likely to agree to and, on the basis of its assessment of these two questions, proceeds to examine and deal with each topic. The pragmatic approach of the [ILC] limits, to a large degree, both the sources and subjects of international law that are considered legitimate.[23]

As the conditions under which the principals' review and control of agents change, however, so too do the activities and decisions undertaken by the agents, especially those whose mandate gives them a high degree of autonomy.

STATE AUTHORITY 54 (2015) ("nothing in the principal-agent model entails a dominating relationship"). Agents with extensive independence, such as international courts, are often referred to as "trustees." *E.g.*, Manfred Elsig & Mark Pollack, "Agents, Trustees, and International Courts: The Politics of Judicial Appointment at the World Trade Organization," 20 *EUR. J. INT'L REL.* 391 (2014).

[20] Briggs, *supra* note 13, at 58.

[21] Jonathan I. Charney, "Third Party Dispute Settlement and International Law," 36 *COLUM. J. TRANSNAT'L L.* 65, 70 (1997).

[22] Stephen C. McCaffrey, "The 1997 U.N. Watercourses Convention: Retrospect and Prospect," 21 *PAC. MCGEORGE GLOBAL BUS. & DEV. L.J.* 165, 166 (1997).

[23] MORTON, *supra* note 18, at 2–3 (discussing RAMCHARAN, *supra* note 18).

Expected ILC Behavior Under Conditions of Increasing UNGA Gridlock

As we have explained, principal-agent relationships give agents a range of discretion to act independently. The degree of discretion is a function of the terms of the delegation and the principal's supervision costs. As supervising the agent becomes more costly, the principal is less likely to do so. As such costs decline, an agent becomes more constrained. It follows that changes in the principal's supervision costs can result in shifts in the relationship between the principal and the agent.

The principal's decision-making rules are a key determinant of its supervision costs.[24] In collective bodies, such as the General Assembly, rejecting action by the ILC requires the affirmative vote of no more than a majority of voting members. Where a body operates by consensus, as the UNGA Sixth Committee does in practice,[25] as few as one dissenting vote can result in rejection. Moreover, the ILC may self-discipline, even without a formal vote, where a minority of states strongly opposes a particular proposal.[26] These decision rules raise the cost of supervision, but they tend to remain constant over time.[27] Supervision costs do increase, however, in response to a rise in the number of members of a collective body and in the heterogeneity of their preferences. We thus expect that as the UNGA becomes larger and more diverse, and as its members' views on particular international law topics diverge, its costs of supervising the ILC will also rise.

These insights help to explain changes in how the ILC has pursued its delegated task of codifying and progressively developing international law. The ILC's influence depends on the extent to which states and other actors accept and follow the legal instruments it promulgates. As explained, the ILC Statute presents the Commission with a basic choice that shapes its ability to maximize this influence. The ILC can seek to have its work adopted by the General Assembly and, ultimately, states as a treaty. Alternatively, it can submit other work products – such as "principles" or "conclusions" – or it can prepare draft articles without recommending that the UNGA adopt them as a treaty. These latter outputs resemble restatements of U.S. law. They synthesize, memorialize, and sometimes shift the meaning of existing international rules, but the instrument itself does not take on legally binding status. Critically, turning the ILC's final product into a treaty requires affirmative

24 Other supervision costs include the cost of informing oneself about the agent's actions.

25 *See, e.g.*, Press Release, UN GAOR, *Approving 16 Draft Resolutions, Legal Committee Continues Tradition of Consensus in Session's Final Debate*, GA/L/3473 (Nov. 15, 2013).

26 We are grateful to Sean Murphy for this point.

27 Similarly, the time and effort devoted to analyzing the ILC's annual reports – which can run to 300 pages, cover a wide range of legal issues, and must be done in the two-month period between the ILC's submission of the report and the debate in the Sixth Committee – represents a supervision cost. These costs may have risen over time to the extent that the length of reports has increased, an issue outside the scope of this chapter.

approval by the General Assembly[28] – something that becomes less likely as gridlock increases. The ILC's "restatements," however, can influence CIL without UNGA action. For example, states and international tribunals may cite the ILC's work as authoritative, as has often occurred for the Draft Articles on State Responsibility. To be sure, the General Assembly retains the power to disapprove of the Commission's non-treaty outputs. But under conditions of gridlock, such disapprovals will be rare.

We thus expect the ILC to choose the form of work product that maximizes its influence among states, subject to its expectations of how the UNGA will respond to its work. We are not alone in making this claim. As David Caron has argued, "It is entirely proper for the ILC to consider the endgame of its work product, and to take account of possible dysfunctions in the state system generally or relating to a particular topic."[29] To our knowledge, however, we are the first to put forward a theory of ILC influence using principal-agent theory.

Specifically, we expect the ILC to recommend that the General Assembly adopt its draft articles as a treaty if it believes the treaty is likely to be adopted and enter into force. The political endorsement of the ILC's work by the UNGA that comes with the treaty-making process makes subsequent adherence by states more likely, all else equal. Similarly, the markers of state acceptance that accompany treaty ratifications remove the uncertainty as to the existence and scope of a legal obligation that often accompany CIL.

By contrast, as gridlock increases the likelihood of General Assembly inaction, we expect the ILC to turn to other legal outputs. This enhanced risk of inaction could flow from political polarization within the General Assembly. Alternatively, it could be the result of the ILC considering issues on which states have more heterogeneous preferences. Arguably, the ILC's current work program includes topics that are more difficult than its earlier work codifying, for example, the law of treaties or diplomatic relations.

The causes of polarization do not affect our prediction, however. Rather, our focus is only on the effects of UNGA polarization – whatever its causes – on ILC behavior. Specifically, we predict that polarization increases the probability of inaction by the UNGA. Inaction, in turn, induces the ILC to increase the influence of its outputs through means other than General Assembly approval. Restatements, principles,

[28] Although creating international obligations via a treaty ultimately requires states individually consent to the agreement, the adoption of a treaty by an international organization or diplomatic conference is a critical prior step. Adoption represents a collective decision by the member states to proceed with the instrument. Adoption gives all member states – including those that have no intention of signing or ratifying the agreement – the chance to decide whether the treaty will even be opened for signature, as well as to influence its content. *See* Timothy Meyer, "From Contract to Legislation: The Logic of Modern International Lawmaking," 14 *Chi. J. Int'l L.* 559 (2014).

[29] Caron, *supra* note 1, at 866.

and other nonbinding codifications of particular international law topics allow the ILC to achieve this result.

In sum, principal-agent theory leads us to predict that the ILC – an institution that seeks to influence the content of international law – will have greater control over the form and content of its work product as the likelihood of gridlock and inaction in the General Assembly rises. If the Commission expects the UNGA to approve its work, we expect the ILC to prefer the treaty-making route. In contrast, if it expects that gridlock will prevent a treaty's adoption, the ILC will seek other pathways to enhance its influence. As we explain, the historical record provides support for our claim, revealing that the ILC has shifted over time from treaty making to "restatements" and other non-treaty outputs.

The Shift from Treaties to Nonbinding Codifications

For most of its history, the ILC has submitted draft articles to the General Assembly with a recommendation that the articles form the basis for a convention.[30] This preference for proposing treaties reflects, in part, the fact that the distinction between codification and progressive development has been elusive in practice.[31] As Hersch Lauterpacht has written:

> There is very little to codify if by that term is meant no more than giving [in the language of ILC statute Article 15] precision and systematic order to rules of international law.... Once we approach at close quarters practically any branch of international law, we are driven, amidst some feeling of incredulity, to the conclusion that although there is as a rule a consensus of opinion on broad principle ... there is no semblance of agreement in relation to specific rules and problems.[32]

Most topics that the ILC undertakes involve a mix of identifying abstract rules on which states agree (codification) and attempting to propose solutions about rules on which they do not agree (progressive development).[33] Because the distinction has been difficult to maintain in practice, the ILC in its first half-century usually opted for treaties as the preferred outcome, generally eschewing the "reports" and other outputs that the ILC Statute contemplates when the Commission is engaged in codification. The political blessing that came with recommending treaties papered

30 *See* Cogan, *supra* note 1, at 278.

31 Timothy Meyer, "Codifying Custom," 160 *U. Pa. L. Rev.* 995, 1008 (2012).

32 Hersch Lauterpacht, "Codification and Development of International Law," 49 *Am. J. Int'l L.* 16, 17 (1955).

33 Meyer, *supra* note 31, at 1008 (arguing that in practice "codification has been an exercise in identifying areas of custom and attempting to fill in the gaps").

TABLE 12.1.

	1947–1999	1947–1974	1975–1999	2000–2014
Projects completed	30	21	9	12
Convention recommended	20	14	6	2
Convention recommended and in force	14	12	2	0
Success rate	70%	86%	33%	0%

over the question of whether the draft articles were in fact CIL supported by state practice and *opinio juris* (as required for codification).

Table 12.1[34] illustrates the dominance of treaty recommendations prior to the turn of the century. It displays three sets of data: (1) the number of projects completed in roughly each third of the ILC's existence, classifying projects into time periods based on the date the ILC completed the project; (2) the ILC's recommendation as to whether to pursue a treaty; and (3) whether a treaty has in fact entered into force. Between 1947 and 1999, the Commission's consistent practice was to "recommend to the General Assembly the elaboration of a convention on the basis of the draft articles."[35] During that period, the ILC completed thirty projects. The ILC concluded twenty of those projects by recommending a convention. By and large, the General Assembly collectively, and states individually, agreed. Fourteen of the twenty recommended conventions were adopted and have entered into force – a success rate of 70 percent. This success rate is even higher in the first twenty-five years of the Commission's work, from 1949 to 1974. During that period the Commission recommended fourteen conventions out of twenty-one completed projects. Twelve of those fourteen entered into force, for a success rate of 86 percent (see Table 12.1).

The twenty-first century, by contrast, has seen the ILC shift away from work products that are intended to become treaties. From 2000 to 2014, the Commission completed twelve projects. Of those, it recommended a treaty in only two instances – the Prevention of Transboundary Harm from Hazardous Activities, and Diplomatic

34 The tables and figures in this section are based on authors' coding of all ILC outputs since 1947. The coding is based on data obtained from a four-volume treatise that comprehensively analyzes the Commission's work. Arthur Watts, *The International Law Commission, 1949–1998* (1999) (3 vols.); Michael Wood & Arnold Pronto, *The International Law Commission, 1999–2009* (2011). More recent information was retrieved from the ILC Web site, http://www.un.org/law/ilc/.

35 ILC Statute, art. 13.

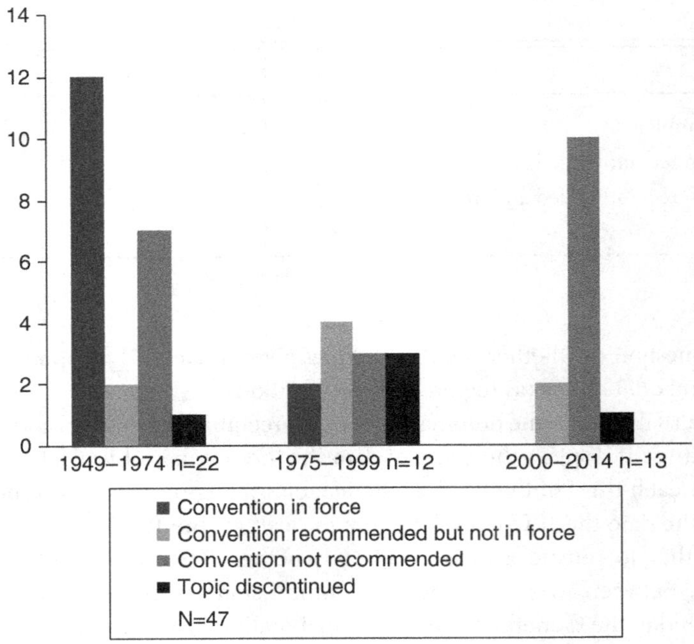

FIGURE 12.1. The ILC's output by era.

Protection. As of March 2015, the UNGA had not adopted either recommendation – a success rate of 0 percent.[36]

Figure 12.1 reproduces the data in a graphical format with two additions – the number of projects for which the Commission did not recommend a treaty, and discontinued projects. Figure 12.1 illustrates the stark decline in recommended treaties and the concomitant rise in outputs that the Commission does not contemplate becoming treaties. Since 1975 the Commission has proposed a total of eight treaties, of which only two are in force – a 25 percent success rate (see Figure 12.1). In contrast, more than two-thirds of ILC projects ending in the same period, including discontinued topics, concluded without the ILC recommending a treaty (seventeen of twenty-five, or 68 percent). This percentage rises to a striking 85 percent (eleven of thirteen) for projects ending between 2000 and 2014.

A more fine-grained look at our dataset of ILC outputs sheds light on the interaction between the General Assembly's behavior and the Commission's recommendations. In the 1980s, the Commission completed only three projects (see Figure 12.2). In

[36] The UNGA adopted the Convention on Jurisdictional Immunities of the State in 2004. However, the ILC completed its work on jurisdictional immunities in 1991, and thus we include this project in the 1974–1999 period.

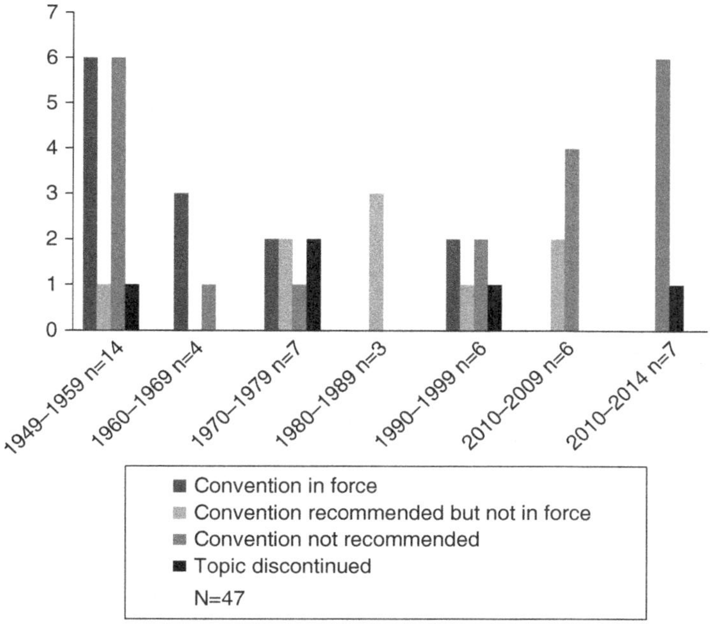

FIGURE 12.2. The ILC's output by decade.

each instance, the ILC recommended a convention, and in each instance the proposed convention failed to enter into force.[37] While the number of observations is too small to draw firm conclusions, the small number of projects completed – and their lukewarm reception from states – suggests that the ILC may have concluded that it could no longer expect to influence the evolution of international law via treaty proposals.

Figure 12.2 reveals that the slow pace of the Commission's work continued through 1990s, a decade during which the ILC completed five projects, recommending conventions in only three instances. Of those three, only two have entered into force. But the way in which they have done so is suggestive of dissensus among states even when the General Assembly adopts a draft treaty. The Convention on the Non-Navigational Uses of International Watercourses, for example, was adopted in 1997 but only obtained the minimum necessary thirty-five ratifications in 2014.

Given this declining success rate, it is hardly surprising that the ILC has turned almost exclusively to non-treaty outputs. Perhaps the most prominent example

[37] The two conventions adopted by the UNGA that did not enter into force are State Succession in Respect of Property, Archives, and Debt, and Treaties with International Organizations. The General Assembly did not adopt the third proposed treaty on Diplomatic Couriers.

of this shift is the ILC's Draft Articles on State Responsibility. The ILC adopted the draft articles in 2001 and submitted them to the General Assembly with a recommendation only that the UNGA "take note" of the draft articles. This recommendation followed guidance from a number of countries (including China, Japan, and the United Kingdom, and the United States) that, in the words of Special Rapporteur James Crawford, worried that states would not:

> see it in their interests to ratify an eventual treaty, rather than relying on particular aspects of it as the occasion arises. They [the countries opposing a draft treaty based on the ILC's draft articles] note the destabilizing and even "decodifying" effect that an unsuccessful convention may have. In their view it is more realistic, and is likely to be more effective, to rely on international courts and tribunals, on State practice and doctrine to adopt and apply the rules in the text. These will have more influence on international law in the form of a declaration or other approved statement than they would if included in an unratified and possibly controversial treaty.[38]

As Crawford makes clear, divisions among states made the adoption and entry into force of a convention on state responsibility unlikely. Given this fact, the ILC had to consider how to enhance the influence of its work. Bypassing the treaty-making process was deemed the best approach, both because it denied states the opportunity to formally express disagreement with aspects of the draft articles and because it opened other avenues – the disaggregated and relatively uncoordinated actions of states and international tribunals and organizations – for the ILC's work to gain traction.[39]

More recently, other conflicts have arisen in the General Assembly regarding ILC outputs. In its 2013 end-of-year review of the Commission's work, a number of states in the Sixth Committee challenged the form of two sets of draft articles – on Transboundary Harm from Hazardous Activities and on the Law of Transboundary Aquifers. The United States, for example, argued that both texts "went beyond the present state of international law and practice ... [and] were designed as resources to encourage national and international action in specific contexts, rather than to form the basis of a global treaty. Therefore, [the United States] expressed support for retaining them in their current form."[40] Similarly, in discussing the Draft Articles on Transboundary Aquifers, Bahrain (on behalf of the Arab Group) called for more

38 James Crawford, *Fourth Report on State Responsibility*, UN Doc. A/CN. 4/517, para. 23 (2001).

39 *See* Caron, *supra* note 1, at 864 ("Weaving through this debate [in the ILC] on the likely outcome of traditional lawmaking processes were estimations that the ILC's articles would have as much, if not more, influence if the attempt were not made to put them in a treaty form.").

40 Press Release, UN GAOR, "Praising Draft Texts on Transboundary Harm, Aquifers, Allocation of Loss, Delegates Disagree over Final Forms, Seek Further Examination," GA/L/3464 (Oct. 22, 2013).

scientific research on the topic and further study of state practice before the draft articles could be turned into a treaty.[41]

The Commission's ongoing projects continue this trend toward nonbinding codifications. Because the ILC has not yet made a recommendation to the General Assembly regarding these projects, we omit them from the data reported in Figures 12.1 and 12.2. Nevertheless, the Commission already appears committed to pursuing nonbinding codifications in its current projects on the Identification of Customary International Law[42] and the Subsequent Agreement and Subsequent Practice in relation to the Interpretation of Treaties,[43] both of which will result in "conclusions." Moreover, the project on the Protection of the Atmosphere is likely to end in nonbinding "Guidelines,"[44] while the ILC's second study of the most-favored nation clause will conclude with just a report.[45]

One ongoing project for which the Commission is preparing draft articles – the immunity of foreign officials from criminal jurisdiction (foreign official immunity) – provides an especially apt example of how disagreement in the General Assembly can stall treaty making in the ILC.[46] The Commission added foreign official immunity to its work program in 2006. Its work began in earnest with the appointment of Roman Kolodkin of Russia as the special rapporteur. During his tenure in that position, Kolodkin wrote three reports on the immunity of state officials: a *Preliminary Report* in 2008,[47] a *Second Report* in 2010,[48] and a *Third Report* in 2011.[49] In 2012, with Kolodkin's departure from the ILC, the Commission appointed a new special rapporteur, Concepcion Escobar Hernandez. To date, Escobar Hernandez has produced a *Preliminary Report*,[50] a *Second Report*,[51] and a *Third Report*.[52]

41 *Id.*
42 *See* Report of the Int'l Law Comm'n ¶¶ 66–76, UN Doc. A/69/10 (2014).
43 *See* Report of the Int'l Law Comm'n ¶¶ 133–85, UN Doc. A/69/10 (2014).
44 *See* Report of the Int'l Law Comm'n ¶¶ 218–28, UN Doc. A/69/10 (2014).
45 *See* Report of the Int'l Law Comm'n ¶¶ 262–64, UN Doc. A/69/10 (2014).
46 It is worth recalling that although it is preparing draft articles, the Commission does not have to choose whether to recommend a treaty until it completes its work.
47 *Preliminary Report on Immunity of State Officials from Foreign Criminal Jurisdiction*, Int'l Law Comm'n, UN Doc. A/CN.4/601 (May 29, 2008) (by Roman Anatolevich Kolodkin).
48 *Second Report on Immunity of State Officials from Foreign Criminal Jurisdiction*, Int'l Law Comm'n, UN Doc. A/CN.4/631 (June 10, 2010) (by Roman Anatolevich Kolodkin).
49 *Third Report on Immunity of State Officials from Foreign Criminal Jurisdiction*, Int'l Law Comm'n, UN Doc. A/CN.4/646 (May 24, 2011) (by Roman Anatolevich Kolodkin).
50 *Preliminary Report on the Immunity of State Officials from Foreign Criminal Jurisdiction*, Int'l Law Comm'n, UN Doc. A/CN.4/654 (31 May 2012) (by Concepcion Escobar Hernandez) [hereinafter *Hernandez Preliminary Report*].
51 *Second Report on the Immunity of State Officials from Foreign Criminal Jurisdiction*, Int'l Law Comm'n, UN Doc. A/CN.4/661(4 April 2013) (by Concepcion Escobar Hernandez).
52 *Third Report on the Immunity of State Officials from Foreign Criminal Jurisdiction*, Int'l Law Comm'n, UN Doc. A/CN.4/673 (2 June, 2014) (by Concepcion Escobar Hernandez) [hereinafter *Escobar Hernandez Third Report*].

Additionally, Escobar Hernandez proposed several draft articles in connection with her second and third reports. The Commission referred these articles to the drafting committee. After some modifications by the drafting committee, the ILC Plenary adopted the draft articles.[53]

Two related issues have animated debates within the ILC on foreign official immunity. The first is the scope of, and any exceptions to, the immunity of foreign officials. The second is the extent to which the Commission should engage in codification versus progressive development. Kolodkin, the first Special Rapporteur, generally favored codifying existing state practice – an approach that many members of the Commission indicated would result in broad immunity for foreign officials. Escobar Hernandez's approach was different. Following the suggestion of many members who were critical of Kolodkin's reports, Escobar Hernandez indicated that she did not intend to undertake a codification exercise limited to extant custom, but would include elements of progressive development, as well.[54] She also stressed the need to base the Commission's work on "the current values of the international community."[55]

The sharp divisions and strongly held beliefs on these questions are also apparent in the General Assembly's review of the ILC's work. State representatives have taken conflicting positions on virtually every key issue with which the Commission is grappling, making it difficult to prepare draft articles that are likely to win favor in the General Assembly. For example, on the question of codification versus progressive development, several delegates urged the Commission to focus on "reviewing and summarizing relevant practices and rules" of immunity instead of establishing new immunity rules.[56] By contrast, others urged the Commission to follow Escobar Hernandez's recommendation that the Commission consider the issue from both the perspective of *lex lata* and *lex ferenda*.[57] States also divided on the scope of

[53] Text of Draft articles 1, 3, and 4 provisionally adopted by the Drafting Committee at the sixty-fifth session of the International Law Commission, UN Doc. A/CN.4/L.814 (4 June 2013); Text of Draft articles 2(e) and 5 provisionally adopted at the sixty-sixth session of the International Law Commission, UN Doc. A/CN.4/L.850 (15 July 2014).

[54] *Preliminary Summary Record of the 3143rd Meeting* [10 July 2012], UN Doc. A/CN.4/SR.3143, 4.

[55] *Hernandez Preliminary Report*, *supra* note 50, at para. 67.

[56] Press Release, UN GAOR, "Delegations Urge Expanded 'Foreign Criminal Jurisdiction' Immunity for Broader Range of State Officials," GA/L/3448 (Nov. 5, 2012); *see also* Press Release, UN GAOR, "Natural Disaster Relief Draft Articles Need Clearer Parameters, Argue Delegates as Legal Committee Continues Review of International Law Commission Report," GA/L/3447 (Nov. 2, 2012).

[57] Press Release, UN GAOR, "Natural Disaster Relief Draft Articles Need Clearer Parameters, Argue Delegates as Legal Committee Continues Review of International Law Commission Report," GA/L/3447 (Nov. 2, 2012) (for example, the Austrian representative urged that "the starting point must be the identification of existing norms of international law, followed by possible progressive development in accordance with the present needs of the international community").

official immunity. Some delegations, such as Greece and South Korea, urged the Commission to include exceptions to immunity for the most serious international crimes.[58] African countries such as Kenya and Algeria expressed greater skepticism that state practice supported such exceptions.[59] Delegations similarly took opposing views on what the outcome of the ILC's process should be, with some suggesting a binding treaty and others urging a nonbinding instrument.[60]

The most telling example of the gridlock within the General Assembly on foreign immunity questions, however, is not what state representatives have said about the ILC's work on foreign official immunity. Rather, it is the fact that the General Assembly has initiated its own parallel investigation into the topic of universal jurisdiction – an issue that is inextricably linked to the immunity of foreign officials from domestic criminal prosecution. The General Assembly took up this investigation at the urging of Tanzania, which in July 2009 submitted a request to include "the subject and scope of universal jurisdiction" on the General Assembly's agenda.[61] In so doing, Tanzania acted on behalf of the African Group, which had previously adopted a resolution calling for "exhaustive" discussions at the UN in response to the indictments "issued in some European States against African leaders."[62] The Tanzanian request noted similar concerns, including the "ad hoc and arbitrary application [of the principle], particularly towards African leaders."[63]

The General Assembly assigned the matter to the Sixth Committee – the same body that reviews the ILC's work.[64] In meetings in 2009 and 2010, state representatives expressed widely divergent views, with a number expressly linking universal jurisdiction to the scope of foreign official immunity. Delegates from African and Asian countries stressed that the application of universal jurisdiction must be consistent with the principles of sovereign and foreign official immunity.[65] In contrast, delegates from Europe and Latin America generally praised universal

58 Press Release, UN GAOR, "As Review of International Law Commission Report Continues, Legal Committee Tackles Definitions, Interpretations of Draft Texts," GA/L/3466 (Oct. 29, 2013).

59 Press Release, UN GAOR, "Immunity of State Officials Remains at Issue as Committee Ends Review of International Law Commission Report," GA/L/3428 (Nov. 4, 2011).

60 *Id.*

61 Letter from the Permanent Representative of the United Republic of Tanzania to the United Nations Secretary-General, Request for the Inclusion of an Additional Item in the Agenda of the Sixty-Third Session, A/63/237/Rev.1 (July 23, 2009).

62 Decision on the Abuse of the Principle of Universal Jurisdiction, Doc. Assembly/AU/11 (XIII) at ¶¶ 3, 4 (Jul. 3, 2009).

63 Press Release, UN GAOR, Sixth Committee, "Legal Committee Delegates See Principle of Universal Law as Safeguard Against Impunity for Major Crimes; Some Caution on Risk of Abuse," GA/L/3371 (Oct. 20, 2009) [2009 Sixth Committee Meeting].

64 Press Release, UN GAOR, "Sixth Committee Delegates Discuss Ways to Further Review Complex, Controversial Issues of Applying Universal Legal Jurisdiction: Some Say Principle Subject to Abuse, Violating Existing Law," GA/L/3392 (Oct. 13, 2010).

65 *Id.*

jurisdiction as a backstop to other national prosecutions and a tool to prevent impunity for international crimes.

In subsequent years, delegates have debated the appropriate division of labor between the Sixth Committee and the ILC. These debates reveal a wide dissensus among member states and suggest the likelihood of gridlock should the ILC recommend a treaty based on draft articles. Several delegates called for referring the universal jurisdiction agenda item to the Commission for further study, stressing "the fundamentally juridical and technical nature of the subject."[66] Others opined that "because the Committee was a political body and the [ILC] an expert one, it was more appropriate for the Commission to study the topic further."[67] In further discussions in 2013, Uganda's representative echoed this sentiment, asking "If not in this forum [the UNGA's Sixth Committee], then where?"[68]

The disagreements in the General Assembly over foreign official immunity and universal jurisdiction have sent conflicting signals to the ILC. On the one hand, state views on the scope of immunity are polarized. Perhaps for this reason, the Commission has to date deferred difficult questions about whether exceptions to immunity exist. At the same time, the General Assembly has opened another front on a closely related issue. Not only have similar divisions among states emerged in the Sixth Committee's discussion of universal jurisdiction; the practice of forum shopping within the General Assembly as a means of fracturing opinion, rather than building consensus in favor of a single approach, has called into question the reception the ILC's work will receive. In sum, this example illustrates the dangers of trying to win consensus on a treaty, and the advantages for the Commission of looking beyond the General Assembly to find an audience for its work.

THE RENEWED IMPORTANCE OF CUSTOMARY INTERNATIONAL LAW METHODOLOGY

Our primary claim in this chapter is that gridlock in the UNGA affects the ILC's choice of work product. In this section, we make a second claim: that the shift away from preparing draft treaties increases the salience of the methodology that the ILC uses to enhance its influence. As the ILC's ability to promulgate texts that will be

66 General Assembly of the UN, Sixth Comm., Sixty-Sixth Session: *The Scope and Application of the Principle of Universal Jurisdiction; see also* UN GAOR, 67th Sess., 13h mtg., at ¶¶ 15, 22, UN Doc. A/C.6/67/SR.13 (Oct. 18, 2012).

67 *Id.*

68 Press Release, UN GAOR, "At Conclusion of Sixth Committee Debate, Delegates Cite Abuse of Universal Jurisdiction, 'Lip Service' to Fight Against Impunity," GA/L/3462 (Oct. 18, 2013); *see also* UN GAOR, 67th Sess., 13th mtg., at ¶9, UN Doc. A/C.6/67/SR.13 (Oct. 18, 2012); Press Release, UN GAOR, "Delegations Urge Clear Rules to Avoid Abuse of Universal Jurisdiction Principle, Seek Further Guidance from International Law Commission," GA/L/3341 (Oct. 17, 2012).

adopted as treaties narrows, the Commission has an increased interest in having states and other actors accept its nonbinding texts as CIL. Yet, these actors may have doubts about the persuasive value of ILC texts that lack the General Assembly's blessing. Adherence to a methodology can alleviate these worries by providing an alternative mechanism to limit the ILC's discretion when the traditional constraint of UNGA review is unavailable due to gridlock. The Commission can use the constraints of methodology, in other words, to persuade other actors that its non-treaty outputs possess the properties of rule legitimacy – determinacy, symbolic validation, coherence, and adherence – that Thomas Franck famously identified.[69] But what methodology has this effect? The traditional sources of CIL are one plausible choice. At the same time, the ILC may wish to adopt approaches that give it relatively more discretion.

In what follows, we first discuss the rising importance of methodology as a constraint on the ILC. We then explore the new light that methodology sheds on the sources the Commission has used in its ongoing codification of the law of foreign official immunity.

Can Methodology Constrain the ILC?

At the outset, we confront the question of whether CIL methodology matters at all. As much scholarship has emphasized, the traditional test for CIL is plagued by indeterminacy. This indeterminacy, in turn, has led scholars to question whether methodology does any work in shaping the content of the CIL. In her contribution to this volume, Monica Hakimi writes that "because the CIL process is so unstructured, it lacks the formal controls that might inhibit the participants from pushing hard for particular norms – making the opportunistic claims that methods for finding CIL are supposed to weed out."[70] And as Curtis Bradley suggests in his analysis of the adjudication of CIL, the content of custom is determined by a process analogous to the common law, in which judges make choice that are shaped by "assessments of state preferences as well as social and moral considerations."[71] If the preferences of states, judges, and other actors about substantive norms are all that matter, methodology would appear to exert little if any force in constraining competing claims about the content of CIL.

We disagree with these approaches, at least as applied to the ILC's articulation of custom. To be clear, we are not saying that the methodology the ILC adopts

69 Thomas Franck, *The Power of Legitimacy Among Nations* (1990).

70 Monica Hakimi, "Custom's Method and Process: Lessons from Humanitarian Law" (in this volume) [manuscript p. 24].

71 Curtis A. Bradley, "Customary International Law Adjudication as Common Law Adjudication" (in this volume) [manuscript p. 2].

will produce determinate answers in all, or even in many, circumstances. But a methodology does not have to produce determinate answers in order to narrow the set of claims that actors may make about the content of rules.[72] Our claim is thus a more modest one: the methodology that the ILC uses to identify custom limits the set of rules that the Commission can plausibly advance to influence the content of CIL through non-treaty outputs. To be sure, the ILC retains discretion in interpreting what custom requires, but that discretion is limited by how faithfully the Commission adheres to the methodology it espouses. Indeed, as we explain, expert bodies such as the ILC are more constrained by methodology than are other international actors.

The following thought experiment helps to illustrate these points. If method exerted no constraining influence whatsoever, the Commission could simply submit a list of proposed rules to the UNGA, perhaps backed by policy justifications. The rules' success would not need to be justified in terms of practice (state or otherwise), *opinio juris*, or other claims to legality, and we would not expect the ILC or other actors to spend any time canvassing these sources or couching their arguments in those terms. In reality, however, we do see actors, and especially the ILC, discussing CIL in terms of doctrine.

Consistency in adhering to a methodology is also in the Commission's self-interest. When the ILC engages in progressive development, it is not limited to examining the two traditional requisites of custom – widespread and extensive state practice and *opinio juris*. It is instead free to advance proposals for new international rules or extensions of existing rules that have yet to garner general support from states. For codification, in contrast, the Statute contemplates reliance on "extensive State practice, precedent and doctrine," an arguably stricter evidentiary standard that imposes somewhat greater constraints on the ILC's norm-generating activities.

As we explained in the previous section, the ILC in its early years demonstrated a preference for draft articles that it anticipated would be approved by the UNGA as multilateral treaties. This choice of work product gave the Commission relatively wide discretion to support draft texts with whatever evidence was available – whether or not that evidence was sufficient, or even relevant, to prove an existing custom – and to elide the often abstruse distinction between codification and progressive development.[73] Yet the ILC exercised this discretion in the shadow of UNGA review. If its members strayed too far from what the political winds would bear, they risked having the draft articles rejected or returned for further study and revision.

[72] *Cf.* Richard A. Primus, "When Should Original Meanings Matter?" 107 *Mich. L. Rev.* 165, 213 (2008) (noting, in a discussion of the relative constraining effect of different theories of constitutional interpretation, that "no sensible legal system eliminates discretion entirely").

[73] *See, e.g.*, Lauterpacht, *supra* note 32, at 17.

In response to increasing gridlock in the UNGA and pressure to take on more unsettled or contentious topics, the ILC has shifted to work products that do not contemplate the adoption of new conventions. Crucially, these more malleable outputs – unlike the Commission's progressive development work – do not require formal approval by the UNGA to influence the content of international law (although such approval is of course desirable). Yet because these work products are generated pursuant to the ILC's delegated authority to codify extant custom rather than extend the law in new directions, their influence with other international and domestic actors depends upon the extent to which the Commission supports its proposals with persuasive evidence that a putative custom in fact exists.[74]

Seen from this perspective, the methodology that the ILC employs to create its work product may function as a *de facto* substitute for the political blessing traditionally associated with the UNGA's adoption of draft articles and preparation of multilateral conventions. That is, under conditions of UNGA gridlock, the Commission's choice of the codification track gives it greater discretion to select its preferred output (principles, conclusions, freestanding draft texts, and so forth), minimizing the possibility of rebuke or sanction from a hamstrung UNGA. Yet if those outputs are in fact to influence international law through uptake by actors other than the UNGA, the ILC must convince those actors that the methodology it employs to produce those outputs is itself valid. The need to persuade these actors serves as a check on the ILC that cabins its discretion to shape the content of international rules. In Franck's terms, the ILC gives its non-treaty outputs legitimacy through adherence to methodology, because the legitimacy available through the political approval process is foreclosed.

Significantly, the Commission's use of methodology as a mechanism of persuasion need not apply to other actors who advance claims about CIL. Some bodies may enjoy such widespread legitimacy and political support that they need not adhere to a particular methodology in order to have their pronouncements followed. The International Committee of the Red Cross (ICRC), which makes claims about customary international humanitarian law (IHL), may provide an example of this kind of institution. As Hakimi says, the ICRC's studies have had broad influence on customary IHL even though the extent to which they accurately reflect custom is contested.[75] Well-established and widely respected tribunals, such as the German

74 *Cf.* Murphy, *supra* note 1 (noting that the "persuasive value" of nonbinding codifications "would turn on their intrinsic merits").

75 Hakimi, *supra* note 70, at 35–38. It bears noting, however, that the ICRC study gave careful and extensive attention to methodology in the introductory chapter of its monumental study of customary IHL. *See* International Committee of the Red Cross, 1 Customary International Humanitarian Law xxxvii–xlviii (Jean-Marie Henckaerts & Louise Doswald-Beck eds., 2005).

Constitutional Court or the Inter-American Court of Human Rights, provide another example.[76]

By contrast, institutions in newly emerging or fragile democracies – such as the U.S. Supreme Court in the early nineteenth century, Eastern European courts in the 1990s, intellectual property agencies in the Andean Community in the late 1980s – must develop legitimacy over time.[77] One strategy for doing so is to apply methodology evenhandedly to a range of problems. Developing a reputation for applying methodology consistently may insulate a judicial, administrative, or other expert body from claims that it is simply resolving distributive conflicts in accordance with its (or some other actor's) preferences.[78] Adherence to methodology, in other words, can enhance the influence of an institution seeking to garner widespread legitimacy.

This is true as well for the ILC under conditions of UNGA gridlock. Of course, the Commission has been codifying and progressively developing CIL for decades and, as a result, has built a reputation as a highly respected expert body. But that expertise – and the discretion that comes with it – was developed in light of certain expectations about the review of its work by UN political bodies. With the membership of those bodies increasingly divided, the opportunity for meaningful review has diminished. It should not be surprising, therefore, that the Commission would turn to other mechanisms – in particular, methodology – as a way to maintain and even expand its influence under very different circumstances.

The Role of Methodology in the ILC's Recent Work

The ILC thus has an interest in adhering to a methodology in how it pursues its work. That methodology, however, need not be the traditional inquiry into state practice and *opinio juris*. Rather, the Commission may augment its influence by consistently employing an approach that deviates from the traditional mode of identifying CIL. We consider that possibility here.

[76] *See, e.g.*, Bradley, *supra* note 71, at [manuscript pp. 46–59] (discussing factors relating to judicial authority to develop custom).

[77] *See, e.g.*, Lee Epstein, Jack Knight, & Olga Shvetsova, "The Role of Constitutional Courts in the Establishment and Maintenance of Democratic Systems of Government," 35 *L. & Soc'y Rev.* 117 (2001) (discussing constitutional courts); Laurence R. Helfer, Karen J. Alter, & M. Florencia Guerzovich, "Islands of Effective International Adjudication: Constructing an Intellectual Property Rule of Law in the Andean Community," 103 *Am. J Int'l L.* 1(2009) (discussing administrative agencies).

[78] *Cf.* Martin Shapiro, *Courts: A Comparative and Political Analysis* (1981) (analyzing how courts maintain legitimacy in a triadic structure that requires them to take sides between two opposed parties).

At first blush, the idea of tinkering with methodology to expand discretion and broaden influence may seem inconsistent with the idea of using methodology as a constraint. Although there is undeniably a tension, the two approaches are not incompatible. A judge, for example, may select a preferred method of constitutional interpretation because she believes it reaches results that accord most of the time with her preferred interpretation of the law. Yet to maintain credibility with litigants and other actors, the judge may still feel compelled to adhere to that methodology even in cases in which it produces a result the judge does not like.

More specifically, we expect the ILC to select and, relatively consistently, adhere to a methodological approach aimed at attracting the support of the audience it hopes to persuade. Actors who are convinced by a strict approach to codification outside of treaty making will be most convinced by ILC work products that closely adhere to the traditional requisites of custom. If this view predominates, we would expect the Commission to carefully investigate all potentially relevant sources of custom – including affirmative state practice by specially affected countries, acquiescence by other nations, and evidence and inferences of *opinio juris*. The ILC may even solicit the views of governments to bolster its findings based on these traditional sources.[79]

On the other hand, executive branch officials are acutely aware that both their actions and their utterances in international law-related venues will be scrutinized for any hints about the existence or emergence of custom. They may, as a result, be reluctant to express a position that could later be used against them and instead couch their statements in deliberately vague or ambiguous language. To the extent that the Commission sticks to the canonical test for CIL, this reticence creates a challenge for the ILC's ability to make claims about extant custom. Similarly, the rise of countries such as India and China, and their growing willingness and capacity to contest the emergence of new customary norms contrary to their interests,[80] makes establishing the veneer of a "consistent state practice" all the more difficult, precisely because more states may seek to prevent a consensus about a putative legal rule from emerging.

If traditional indications of state practice and *opinio juris* are scarce or conflicting, the Commission may seek to expand the sources and materials it considers beyond those that states can manipulate directly. By giving weight to sources such as

79 Indeed, when the ILC engages in progressive development it is required to solicit governments' views. *See* ILC Statute, art. 16(c).

80 *See, e.g.*, Zhu Lijiang, "Chinese Practice in Public International Law: 2013," 13 CHINESE J. INT'L L. 395 (2014); Sunil T. Rao, "India," in SOURCES OF STATE PRACTICE IN INTERNATIONAL LAW 262–67 (Ralph Gaebler & Alison Shea eds., 2014). The dispute between India and Italy over the shooting of two Indian fishermen by Italian military guards aboard the ship *Enrica Lexie* provides an apt illustration. *See* Hari Sankar, "Jurisdictional and Immunity Issues in the Story of *Enrica Lexie*," *EJILTalk!* (Mar. 25, 2013).

international and national court decisions, expert studies, resolutions of international organizations, and the Commission's own past work, the ILC can expand its ability to prepare codification-related work products not supported by canonical evidence of custom. Moreover, since the actors who can informally bless these outputs are not limited to states, concerns about deviating from a strict approach to custom may be reduced. This methodological move may thus liberate the ILC from some of the constraints imposed by conflicting state practice or political factions.

The ILC's study of foreign official immunity provides more support for the latter position – that the ILC relies today on evidence beyond canonical state practice and *opinio juris* – than the former. The Kolodkin and Escobar Hernandez reports – and the heated debates they engendered – do not contain detailed, country-by-country analyses of state practice as expressed, for example, in official government actions or pronouncements regarding immunity rules or their exceptions. The documents and discussions evince even less treatment of *opinio juris*. To the extent that traditional indicia of custom are referenced at all, they appear in the form of national legislation and executive branch positions taken in domestic litigation.

By far the most common evidence cited are ICJ judgments and national court decisions, as well as the views expressed by state representatives during debates in the Sixth Committee of the UNGA and other international forums. The *Escobar Hernandez Third Report* provides an illustrative example. In considering evidence of how the term "official" is defined in international law, the report considers, in order, national judicial practice, international judicial practice, treaty practice, and other works of the Commission.[81] Of these, only treaty practice – listed third – is classic state practice geared toward foreign relations, although it may reveal little about whether the treaty has passed into customary international law.[82]

With respect to the views expressed by state representatives, these statements might be considered a type of state practice, but their declarative content gives them greater weight as *opinio juris*, although it is an *opinio juris* inflected through the distinctive political environment of the international venues in which they are made and the recognition by states of the legal value (or lack thereof) of such statements. In addition, the extensive focus on methodological questions about codifying versus

[81] *See Escobar Hernandez Third Report*, *supra* note 52.

[82] The status of national court judicial decisions as state practice was long viewed with skepticism. Today, most, but not all, commentators accept such decisions as state practice at least where they represent the authoritative statements of a nation's position about custom in a particular area. *See* Ingrid Wuerth, "International Law in Domestic Courts and the Jurisdictional Immunities of the State Case," 13 *Melb. J. Int'l L.* 819, 821–22 (2012); *see also* Anthea Roberts, "Comparative International Law? The Role of National Courts in Creating and Enforcing International Law," 60 *Int'l & Comp. L.Q.* 57, 62 (2011) (noting that such treatment is complicated by the fact that national court decisions "play a distinctive dual role in the doctrine of sources: as evidence of State practice … and as a subsidiary means of determining the existence and content international law").

progressively developing international law suggests a desire by ILC members to be transparent about the Commission's role as it prepares draft articles and other work products for consideration by a wider community of actors.

In this way, the ILC's recent work product reveals that a strong form of the critique of customary international law is misguided – at least when applied to the Commission. While existing CIL methodology may be imprecise relative to the clear procedures that accompany treaty making, institutions such as the ILC have an incentive to clarify custom's methodology as a way to increase their own influence. Indeed, the ILC has undertaken this task with its ongoing project on the "Identification of Customary International Law."[83] The outcome of that process promises to limit, and by so limiting also expand, the ILC's importance in a post-treaty world.

The ILC faces a twin challenge. On the one hand, its influence likely requires working with materials beyond the canonical approach to CIL that privileges the actions and words of states. In a multipolar world, such inquiries will increasingly reveal sources that are conflicting or unclear. On the other hand, the ILC must devise methodological limits on its discretion to win acceptance for its more diverse work products. We therefore expect the ILC to surrender voluntarily some amount of discretion as the price of achieving legitimacy in the eyes of critical stakeholders, including the international bar, international judges, NGOs, and IOs. A renewed dedication to methodology by the ILC, in other words, functions as a costly commitment device. The ILC surrenders some discretion in customary lawmaking in exchange for furthering the influence of its non-treaty outputs in shaping international law in the twenty-first century. In short, the benefits of methodological constraints provide a reason internal to the operation of law – and consistent with the Commission's self-interest – for it to adhere to methodology.

Of course, these arguments do not allow us to show empirically that the ILC is in fact constrained by methodology. Our hope, however, is to test our hypothesis in future work. The simultaneous consideration of foreign official immunity in the ILC and the closely related topic of universal jurisdiction in the political Sixth Committee creates a kind of natural experiment. If our hypothesis regarding methodology is correct, we would expect the ILC, constrained to some degree by methodology, to reach different results than the Sixth Committee.

CONCLUSION

The International Law Commission is an institution at a crossroads. Since the 1980s its tried-and-true method of lawmaking – preparing draft treaty articles that skirt

[83] For ILC documents relating to this project, *see* http://legal.un.org/ilc/guide/1_13.htm.

the line between codification and progressive development – has been stalled by gridlock in the General Assembly. To reassert its influence, the Commission has increasingly turned to nonbinding codifications of customary international law. Yet since the legal force of these nonbinding codifications turns on their status and acceptance as CIL, the methodology by which the ILC justifies its work has become increasingly important. Methodology, in effect, confers a legitimacy on the Commission's work product that the political treaty-making process no longer can.

These shifts in the ILC's practice may reflect larger trends in international lawmaking. We therefore conclude this chapter with some preliminary reflections on the generalizability of two aspects of our study: (1) the decline of broad-based multilateral treaty making, and (2) the distinction between lawmaking by political and expert bodies.

The General Assembly's slow pace of treaty making is in line with broader trends in international lawmaking. On difficult issues ranging from climate change to closing the Doha Round at the World Trade Organization, governments have been unable to sustain the pace of multilateral lawmaking that characterized the second half of the twentieth century. Indeed, in many issue areas lawmaking seems to be shifting to regional fora.[84] This broader trend suggests that gridlock within the UNGA may not simply be a reflection of "easy" topics having already been codified. Rather, it may indicate a more fundamental breakdown of political consensus about critical legal issues, as well as a destabilization of power relationships that have long shaped the international lawmaking agenda.

The combination of these factors may have increased the transaction costs of multilateral treaty making to such a degree that states are forced to look for alternative and less costly avenues of lawmaking, including soft law, customary international law, international tribunals, and regional organizations. But if the General Assembly's gridlock is a symptom of these broader trends, further shifts in interstate power relationships or the emergence of a new consensus on particular topics may reanimate the UNGA. Our findings in this chapter suggest that, should these changes occur, the ILC will pivot back toward treaty making, preferring the political endorsement of its work to more amorphous and diverse pathways to legitimacy and acceptance.

The relationship between the UNGA and the ILC also highlights an increasingly important distinction between expert and political bodies. Expert bodies such as the ILC have come to play an increasingly important role in lawmaking in recent decades. The judicialization of some areas of international law is one manifestation

84 See, for example, the plethora of regional trade agreements as well as the increased activity of regional and subregional organizations such as the African Union, ASEAN, and ECOWAS.

of this trend.[85] In other areas, however, intergovernmental institutions comprised of experts who generally lack formal lawmaking authority have assumed a greater role in coordinating state behavior and in laying the groundwork for interstate negotiations. Examples include the International Renewable Energy Agency, the Intergovernmental Panel on Climate Change (IPCC), or the International Commission on Non-Ionizing Radiation Protection.[86] Expert bodies differ from political bodies such as the UNGA or diplomatic conferences in that they generally operate by producing information – such as reports and recommendations – rather than directly generating new legal rules. This information often serves as the basis of subsequent international regulation.

The rise of expert bodies is often attributed to the increasing complexity of international problems and their potential solutions. In responding to this complexity, technocratic expertise may produce better governance outcomes than political bargaining. Yet the relationships between political and expert bodies remain poorly understood. In particular, we lack a well-developed account of whether delegating international lawmaking – or governance functions more generally – to expert bodies such as the ILC produces different substantive outcomes or alters distributional issues. Questions to be explored include the following: When and why do states delegate responsibility for international governance to expert, as opposed to political, bodies? Do expert bodies that operate under intergovernmental supervision produce systematically different results from intergovernmental political bodies themselves? Are certain states systematically advantaged by the use of expert bodies? What factors influence the decision to delegate to one type of body or another? And what factors influence their effectiveness? Our study of the ILC suggests that widespread political support for expert bodies may be a sufficient but not a necessary condition for their effectiveness. Expert bodies will continue to seek opportunities to influence the evolution of international law even in the absence of a political consensus on the content of their work.

85 *See, e.g.*, Cesare P. Romano, "The Proliferation of International Judicial Bodies: The Pieces of the Puzzle," 31 *N.Y.U. J. Int'l L. & Pol.* 709 (1999); Eyal Benvenisti & George W. Downs, "Prospects for the Increased Independence of International Tribunals," 7 *German L.J.* 1057 (2011); Laurence R. Helfer & Karen J. Alter, "Legitimacy and Lawmaking: A Tale of Three International Courts," 14 *Theoretical Inq. in L.* 479 (2013).

86 *See* Timothy Meyer, "Epistemic Institutions and Epistemic Governance in International Environmental Law, 2 *Transnat'l Environ. L.* 15 (2013) (analyzing the decision to divorce the International Renewable Energy Agency from multilateral lawmaking institutions); Oren Perez, "The Hybrid Legal-Scientific Dynamic of Transnational Scientific Institutions," 26 *Eur. J. Int'l L.* 391 (2015).

13

Custom and Informal International Lawmaking

Jan Wouters and Linda Hamid

Informal international lawmaking, a phenomenon involving new actors, processes, and outputs, is an important and increasingly common phenomenon of contemporary international life. It has not, however, entirely replaced traditional international lawmaking. In fact, informal international law exists alongside treaties and custom. But while the interaction between treaties and informal law has been consistently addressed in the academic literature, the relationship between customary international law and informal law remains largely neglected, despite the complex questions it raises. First, what explains the rise of informal international law to the detriment of custom? What place is there still for custom in a world of increasing informality? Second, how do custom and informal law interact in practice? Do they merely serve as alternatives in states' cooperation endeavours, or do they also interact in a complementary manner?

This chapter will attempt to answer these questions. We first examine and explain the rise of informal international lawmaking and the functions it serves. Next, we turn to the place of custom in an age of informality. We then focus on the complementary interaction between custom and informal international law. Finally, we examine the possibility of stretching the boundaries of the traditional international system in order to incorporate informal international lawmaking.

THE RISE OF INFORMAL INTERNATIONAL LAWMAKING

Many authors have recently indicated that international law finds itself in impasse.[1] Indeed, there is a rather broad acknowledgement among international lawyers that traditional forms of multilateralism are facing a deep crisis, and that

[1] *See, e.g.*, Joel P. Trachtman, "The Crisis in International Law," 44 CASE W. RES. J. INT'L L. 407 (2011).

new multilateral processes have emerged.[2] "What crisis?" one might curiously ask. Contemplate, for instance, all the great challenges to modern international relations: globalization, economic development, global environmental protection, international public health, technology, cybersecurity, and, why not, financial cataclysms. Would you imagine that customary international law (CIL) could effectively address these challenges? Arguably not. Treaty-based hard law, on the other hand, might fit the bill much better.[3] But even the development of treaties is in decline.[4]

The limited scope of Article 38 (1) of the Statute of the International Court of Justice (ICJ Statute) has become ill-suited to address the complexities of global governance. Therefore, in order to circumvent the rigidities and inadequacies of the traditional framework of international law, states have been progressively turning toward informal cooperative fora, radically transforming the international legal order.[5] The ever-changing demands of international society, it seems, have brought into the limelight new, nontraditional types of international lawmaking.

Recent academic literature has shown that international law is not only stagnating in terms of both quantity and quality, but is also increasingly superseded by informal international lawmaking (IN-LAW), a phenomenon involving new outputs, processes, and actors.[6] A fairly rich tapestry of novel forms of international cooperation, ostensibly outside the confines of the traditional framework of international law, is booming. The nomenclature used is increasingly diverse and departs from the formal labels of custom, treaty, or international organization (IO).[7] We have witnessed the creation of the International *Conference* on Harmonization (in respect of pharmaceuticals), the Kimberly *Scheme* on conflict diamonds, the Proliferation Security *Initiative*, the Copenhagen *Accord* on Climate Change, the International Organization for Standardization's 26000 *Standard* on corporate social responsibility, the Ruggie *Guiding Principles* on Business and Human Rights, the Financial Stability *Board*, the Internet Engineering *Task Force*, or the World Health Organization's Global *Strategy* on Diet.[8] In this respect, Jan Klabbers goes so far as

2 Joost Pauwelyn, Ramses A. Wessel, & Jan Wouters, "When Structures Become Shackles: Stagnation and Dynamics in International Lawmaking," 25 EUR. J. INT'L L. 733, 737 (2014).

3 Joel P. Trachtman, "The Obsolescence of Customary International Law", *retrieved from* http://ssrn.com/abstract =2512757, 1–2 (2014).

4 *See* Pauwelyn, Wessel, & Wouters, *supra* note 2, at 734–38.

5 *Id.* at 737.

6 *Id.* at 733.

7 Joost Pauwelyn, Ramses A. Wessel, & Jan Wouters, "Informal International Lawmaking: An Assessment and Template to Keep it Both Effective and Accountable," in *INFORMAL INTERNATIONAL LAWMAKING* 500, 501 (Joost Pauwelyn et al. eds., 2013).

8 *Id. See also* the many cases discussed in AYELET BERMAN ET AL. (EDS.), *INFORMAL INTERNATIONAL LAWMAKING: CASE STUDIES* (2012).

to state that "globalization seems to have bypassed the discipline of international law completely."[9] Indeed, traditional international law is stagnating, while informal international lawmaking is on the rise.

The Concept of Informal International Lawmaking

The concept of informal international lawmaking was introduced by Joost Pauwelyn, Ramses A. Wessel, and one of us in the context of a research project conducted by the Graduate Institute in Geneva, the Leuven Centre for Global Governance Studies, the University of Twente, and the Hague Institute for the Internationalisation of Law. The main objective of this research project was to draw attention to a phenomenon that is omnipresent in global governance, yet largely neglected by international lawyers. They defined IN-LAW as follows:

> Cross-border cooperation between public authorities, with or without the participation of private actors and/or international organizations, in a forum other than a traditional international organization (process informality), and/or as between actors other than traditional diplomatic actors (such as regulators or agencies) (actor informality), and/or which does not result in a formal treaty or traditional source of international law (output informality).[10]

Obviously, the term "informal international lawmaking" is used in stark opposition to "traditional international lawmaking," as it dispenses with certain formalities traditionally linked to international law. This circumvention of formalities generally occurs in relation to three dimensions: output, process, and actors involved.

First, in terms of output, international cooperation is informal in the sense that it does not normally lead to a formal treaty or any other source of traditional international law, but rather to a guideline, standard, declaration, or even informal policy coordination or exchange. Second, in terms of process, international cooperation is considered informal when it takes place in a loosely organized network or forum rather than traditional, treaty-based IOs.[11] Forum informality does not, however, prevent the existence of detailed procedural rules, permanent staff, or physical headquarters. Even more importantly, process informality does not exclude IN-LAW in the context or under the broader auspices of a more formal organization.[12] Third, in terms of actors involved,

[9] Jan Klabbers, "The Idea(s) of International Law," in *The Law of the Future and the Future of Law* 71 (Sam Muller et al. eds., 2011).

[10] For a comprehensive introduction to the concept, *see* Joost Pauwelyn, "Informal International Lawmaking: Framing the Concept and Research Questions," in *Informal International Lawmaking*, *supra* note 7, at 13.

[11] Examples in the area of financial regulation, for instance, would include the Basel Committee on Banking Supervision, or the Financial Action Task Force.

[12] *See* Pauwelyn, *supra* note 10, at 17 (noting that "a lot of IN-LAW occurs, for example, under the auspices of the OECD").

international cooperation is seen as informal because it does not engage traditional diplomatic actors (heads of states, foreign ministers, or ambassadors), but rather other public authorities, such as ministries, domestic regulators, agencies, sub-federal entities, the legislative and judicial branch, or even private actors alongside public actors or international organizations. The actors involved in informal international lawmaking are not traditional subjects of international law. Still, purely private cooperation with no public authority involvement is not covered by the concept of IN-LAW.[13] In essence, the process of informal international lawmaking remains state based.[14]

Many of these new developments, considered in isolation, are not exactly new, and have accrued, with ups and downs, over time. The emergence of new actors on the international plane has been discussed for years, and some are fairly old. To give but one example, a *sui generis* entity such as the International Committee for the Red Cross was created more than a century ago, in 1863. In the same vein, alternative forms of output – other than custom and treaty – such as standards, guidelines, and nonbinding recommendations of IOs, are not new. Indeed, debates on international "soft law" have simmered since the 1970s, when newly independent states sought to instumentalise, in particular, United Nations General Assembly (UNGA) resolutions.[15] IOs as we know them today, such as the United Nations (UN) or the World Trade Organization (WTO), are relatively new as compared to the nineteenth-century transnational administrative unions, congresses, or commissions (such as the Universal Postal Union created in 1874) – entities that are similar to some of the "new" cooperation forms we witness today.[16]

Ever since Anthony Aust, nearly three decades ago, defined an informal international instrument as "an instrument which is not a treaty because the parties to it do not intend it to be legally binding,"[17] one cannot help but wonder to what extent IN-LAW differs from the concept of "soft law," which, from a lawmaking perspective, is similarly defined as "a convenient description for a variety of non-legally binding instruments used in contemporary international relations."[18] Soft law, much

[13] Jean D'Aspremont, "From a Pluralization of International Norm-Making Processes to a Pluralization of the Concept of International Law," in *Informal International Lawmaking*, *supra* note 7, at 185, 188.

[14] Joost Pauwelyn, Ramses A. Wessel, & Jan Wouters, "Informal International Lawmaking: An Assessment and Template to Keep it Both Effective and Accountable," in *Informal International Lawmaking*, *supra* note 7, at 500, 504.

[15] Alan Boyle & Christine Chinkin, *The Making of International Law* 211 (2007).

[16] Christian Tietje, "History of Transnational Administrative Networks," in *Transnational Administrative Rule-Making: Performance, Legal Effects and Legitimacy* 23 (Olaf Dilling et al. eds., 2011).

[17] Anthony Aust, "The Theory and Practice of Informal International Instruments," 35 *Int'l & Comp. L.Q.* 787 (1986).

[18] *See* Boyle & Chinkin, *supra* note 15, at 212.

in the same vein as informal international lawmaking, is seen by its proponents as an almost ubiquitous phenomenon that comes in an endless variety of forms, such as: treaties with imprecise, subjective, or indeterminate language; interstate conference declarations such as the 1972 Stockholm Declaration on the Human Environment or the 1992 Rio Declaration on Environment and Development; various UNGA resolutions dealing with issues as diverse as disarmament, outer space, or decolonization; codes of conduct, guidelines, or recommendations of IOs; and finally, some possibly controversial claimants to soft law status, such as common international standards adopted by transnational networks of national regulatory bodies, nongovernmental organizations (NGOs), expert groups, professional and industry associations, or other instruments that do not emanate directly from states.[19] In this last case, it is argued, the use of a non-legally binding form is dictated not by the intent not to be bound, but by their makers' lack of formal lawmaking capacity.[20] These types of agreements, which some are leery to include in the soft law category,[21] roughly correspond with the phenomenon of IN-LAW.

The rise of informal international lawmaking goes well beyond soft law. To begin with, as indicated by Joost Pauwelyn, the definition of IN-LAW does not necessarily equate output informality with the lack of legal "bindingness."[22] Indeed, while IN-LAW may be informal, that is, take the form of a press communiqué or agreed statement rather than a formal treaty, it can still be construed as legally binding under traditional international law.[23] The IN-LAW project also points out that there is nothing "soft" – that is, vague, aspirational, or deeply contested (as with the effect of UNGA resolutions) – about most informal lawmaking, such as the standards relating to the Internet, medical devices, or financial norms developed in the past few years.[24] What characterizes informal international lawmaking is not so much that it is nonbinding under international law (the hallmark of soft law), but that it generally seems to fall outside traditional international law altogether. The reason, arguably, lies in the fact that IN-LAW actors, be they agencies, corporations, or NGOs, lack the capacity to make international law. Jean D'Aspremont goes as far as stating that "whatever the influence of these non-state actors may be, states

[19] Christine Chinkin, "Normative Development in the International Legal System," in COMMITMENT AND COMPLIANCE: *THE ROLE OF NON-BINDING NORMS IN THE INTERNATIONAL LEGAL SYSTEM* 21, 29 (Dina Shelton ed., 2000).

[20] *Id.*

[21] *See* Boyle & Chinkin, *supra* note 15, at 213.

[22] Pauwelyn, *supra* note 10, at 15.

[23] Aegean Sea Continental Shelf Case (*Greece vs. Turkey*) ¶ 96 (ICJ, Dec. 19, 1978) (Judgment), where the ICJ interpreted an unsigned press communiqué as if it were a binding document.

[24] Pauwelyn, Wessel, & Wouters, *supra* note 7, at 506.

and international organizations remain the exclusive international law-makers."[25] In any event, as Pauwelyn points out, the fact that informal international lawmaking includes non-state actors does not preclude IN-LAW from being international law if its makers at least include some recognized lawmakers such as states or IOs.[26] This, again, indicates a partial overlap with the concept of soft law.

A Slowdown in Traditional International Lawmaking

The crisis in international law is measured not only by the increase in informal international lawmaking, but also by the decline in the making of traditional international law. Indeed, evidence shows that there has been a decline in the generation of new treaties since the early 2000s. Since the 1950s, for each decade, the number of new multilateral treaties deposited with the UN Secretary General was around thirty-five. In the ten years between 2000 and 2010, this number declined to twenty, while in the preceding five decades, it had never been below thirty-four. Between 2005 and 2010, only nine new multilateral treaties were deposited, whereas in 2011, 2012, and 2013, not a single one was.[27] The broader UN Treaty Series database confirms this downward trend as of the 2000s, both for bilateral, as well as multilateral, treaties.[28] In this regard, another indicator that states are increasingly shifting away from traditional international lawmaking is the follow-up given to draft texts prepared by the International Law Commission (ILC). Indeed, whereas the ILC's work previously resulted in new multilateral treaties, the most recent time this has happened was more than a decade ago, in 2004.[29]

Some may object that the picture described here fails to take into consideration various other dynamics of international law, such as the continuing evolution of custom, or the activity of IOs. As for customary law, presently, the great majority of customary international rules are codified, either in treaties or in ILC instruments. In this regard, Joel Trachtman argues that "due to a growth of treaty law incorporating

[25] Jean D'Aspremont, "Non-State Actors from the Perspective of Legal Positivism," in *Participants in the International Legal System: Multiple Perspectives on Non-State Actors in International Law* 23, 25 (Jean D'Aspremont ed., 2011).

[26] Joost Pauwelyn, "Is it International Law or Not, and Does it Even Matter?" in *Informal International Lawmaking*, *supra* note 7, at 125, 144.

[27] For a more comprehensive review of the evidence concerning the slowdown in formal international lawmaking, see Pauwelyn, Wessel, & Wouters, *supra* note 2, at 734–37.

[28] *See* UN Treaty Series, *retrieved from* http://treaties.un.org/pages/UNTSOnline.aspx?id=1. Multilateral treaties in this database include closed multilateral treaties, multilateral treaties deposited with the UN Secretary General, multilateral not deposited with the UN Secretary General, and open multilateral treaties.

[29] This was the United Nations Convention on Jurisdictional Immunities of States and their Property, following the Commission's 1991 draft articles on this matter.

the same norms, if CIL were abolished today, most of the international legal system would remain intact."[30] Indeed, treaty law has increasingly pushed custom to the point of redundancy.[31] Treaty law, however, is also partly in decline. Informal international law, therefore, has a role in replacing both treaties and custom.

Although we agree that intensified international cooperation may lead to a more rapid formation of customary rules in specific instances,[32] today's preference of states for informal cooperation undoubtedly affects CIL as well, since the essence of custom's *opinio juris* component relates precisely to the legally binding character of an obligation. Additionally, with fewer multilateral conventions generated, it is becoming increasingly hard to find evidence of *opinio juris* confirmed by state practice.[33] As to the output of IOs, while they continue to function and produce plenty of resolutions, statements, and decisions, many, if not the majority, of normative instruments produced by IOs are legally nonbinding.[34] Not only this, but a great number of IOs currently face serious challenges, ranging from profound divisions at the UN Security Council to an almost paralyzed WTO to severe budget cuts and membership dropout, as well as extraordinarily difficult reform processes.[35]

Whether measured by the decline in treaties or the rise in informal cooperation in many issue areas, there is a general consensus that informal international lawmaking is rapidly supplanting traditional, "hard" forms of international law – that is, custom and treaties. Multiple factors, many issue specific or country specific, may explain this trend. It is, therefore, methodologically difficult to prove that one or the other is more pervasive. The shift from traditional to informal international lawmaking can partly be explained by the saturation with the existing treaties and changed policy preferences of states. However, at a more fundamental level, the multiple case studies conducted within the IN-LAW framework[36] converge around deep societal changes that are not unique to international law, but affect both international and national legal systems, in particular: the transition toward an increasingly diverse network society and an increasingly complex knowledge society.

[30] Trachtman, *supra* note 3, at 1.

[31] *Id.*, appendix A, where the author shows that in only very few circumstances are there customary rules that have not been reflected in treaty or other codification.

[32] Michael P. Scharf, "Seizing the 'Grotian Moment': Accelerated Formation of Customary International Law in Times of Fundamental Change," 43 CORNELL INT'L L.J. 439 (2011).

[33] Pauwelyn, Wessel, & Wouters, *supra* note 2, at 736.

[34] A compelling example is the Framework Convention on Tobacco Control produced by the World Health Organization in 2003, which was put into practice largely through legally nonbinding "Guidelines for Implementation."

[35] Pauwelyn, Wessel, & Wouters, *supra* note 2, at 736–37.

[36] Berman, *supra* note 8.

Saturation and Changed Policy Preferences of States

Some of the slowdown, especially in treaty making, can be explained by the fact that multilateral treaties now exist on major policy issues. Not only this, but most of these treaties have also codified preexisting custom.[37] Where an area is covered by existing agreements, new treaty making appears to be very difficult (for example, negotiating a successor to the Kyoto Protocol), either because of the diversity of interests involved or the burdensome procedures that apply. A certain saturation, or treaty fatigue, has descended upon states. Policy preferences expressed by a number of states also confirm the stagnation hypothesis. The United States and other countries have even formally signalled a growing preference for informal rather than traditional international commitments.[38] In this regard, the 2010 U.S. National Security Strategy refers to the "shortcomings of international institutions that were developed to deal with the challenges of an earlier time" and calls on U.S. authorities to "spur and harness a new diversity of instruments, alliances, and institutions."[39]

The saturation thesis can further be supported by the following findings. To begin with, many multilateral treaties have created their own governance regimes in which nonbinding rulemaking takes place through Conferences of the Parties (CoPs), Meetings of the Parties (MoPs), or other committees or working groups.[40] Within the area of multilateral environmental agreements, particularly, such bodies are very active in developing nonbinding rules. To give but a few telling examples, the CoPs of the Biodiversity Convention and the UN Framework Convention on Climate Change, have literally produced hundreds of decisions and recommendations, and the MoPs of the 1987 Montreal Protocol on Substances that Deplete the Ozone Layer have taken hundreds of decisions on policy, legal, noncompliance, science, technology, and technical issues.[41] The tremendous level of detail and technicality of the norms concerned, as well as the constant need for adjusting, expanding, and updating them, make traditional/formal international lawmaking a *de facto* less preferred option.[42]

37 Trachtman, *supra* note 3; Trachtman, *supra* note 31.

38 *See* Pauwelyn, Wessel, & Wouters, *supra* note 2, at 739 (describing similar evidence from Germany and Canada).

39 U.S. National Security Strategy (May 27, 2010), *retrieved from* www.whitehouse.gov/sites/default/files/rss_viewer/national_security_strategy.pdf.

40 Pauwelyn, Wessel, & Wouters, *supra* note 2, at 740.

41 The decisions and recommendations of these CoPs and MoPs are available at http://www.cbd.int/decisions/, http://unfccc.int/ documentation/documents/items/3595.php, and http://ozone.unep.org/new_site/en/index.php.

42 *See, e.g.*, Jonathan B. Wiener, "Global Environmental Regulation: Instrument Choice in Legal Context," 108 YALE L.J. 677 (1999).

Emergence of a Transnational Network/Knowledge Society

The shift from traditional international law to informal international lawmaking can also be explained by deep societal changes that are not unique to international law, but affect both international and national legal systems.[43] These trends, referred to in the literature as "the emergence of a transnational network/knowledge society,"[44] go far beyond the functional explanations for informal cooperation traditionally discussed in the literature.[45] However, while they are helpful in providing better context to the rise of informal international lawmaking, they still remain very much grounded in the functionalist approach.

First, we are witnessing a shift from societies of individuals (national level) and a society of territorial states (international level), to a "transnational society of networks."[46] These networks both disaggregate and transcend the state, thereby multiplying the types of actors and processes involved in cross-border cooperation. The state remains a pivotal entity of interest, aggregation, legitimation, and control. However, it is supplemented, assisted, corrected, and continuously challenged by a variety of new actors, such as regulators, national and international agencies, transnational corporations, or NGOs.[47] Seemingly outside the traditional framework of international law, these new actors are unable to legally conclude a treaty or join IOs. Therefore, they are continuously pushing international lawmaking from the formal to the informal arena.[48]

Second, besides the new actors and processes driving these interactions, the output or type of cooperation emerging has completely changed and diversified. It used to be carefully negotiated treaties consented to by states, or resolutions issued by IOs set up by those same states, on the assumption that state representatives most legitimately represent the people. However, in an increasingly complex society, at levels ranging from the political to the technological (scientific or regulatory), authority flows from other sources, too, both public and private – in particular expertise, knowledge, or acceptance by affected stakeholders.[49] In addition, complexity and the resulting uncertainty and rapid change that come with it require more flexible norms or guidelines (IN-LAW), grounded in practical experience and

43 Pauwelyn, Wessel, & Wouters, *supra* note 7, at 502.

44 *Id.* at 503. *See also* Pauwelyn, Wessel, & Wouters, *supra* note 2, at 741–43.

45 *See infra* text accompanying note 55.

46 *See, e.g.*, Kenneth W. Abbott & Duncan Snidal, "Strengthening International Regulation Through Transnational New Governance," 42 VAND. J. TRANSNAT'L L. 1 (2009).

47 Pauwelyn, Wessel, & Wouters, *supra* note 7, at 504.

48 Pauwelyn, Wessel, & Wouters, *supra* note 2, at 742.

49 HELMUT WILKE, *SMART GOVERNANCE: GOVERNING THE GLOBAL KNOWLEDGE SOCIETY* 8 (2011).

expertise, which may be continuously adapted and corrected to take account of new developments and learning.[50]

In sum, the emergence of an increasingly diverse and complex network/knowledge society is transforming the actors, processes, and outputs at work or required to deliver international cooperation. The actors (central state authorities), processes (formal lawmaking in IOs), and outputs (rigid treaties or IOs decisions) recognized within the traditional international law framework are not adapted.[51] This is what explains the rise in informal international lawmaking. In many cases, these new forms of transnational cooperation develop out of technical necessity, because the actors involved are unable, legally or technically, to conclude a treaty, or because traditional international lawmaking would simply not be appropriate to the circumstances of a rapidly changing field with a lot of uncertainty, where adaptable, practice-based norms are needed.[52]

A Functional Approach to the Role of Informal International Lawmaking

Undoubtedly, informal international lawmaking is an important and increasingly common phenomenon of contemporary international life. Nonetheless, a simple empirical observation stands out: IN-LAW has not replaced traditional international law, but exists alongside it. Indeed, across a huge variety of issues, ranging from climate change to cultural diversity, nuclear weapons, or humanitarian interventions, states have opted to address modern challenges through multiple fora, ranging from traditional/formal international lawmaking to IN-LAW procedures featuring nontraditional outputs, processes, and actors.[53] This empirical observation raises, in turn, a simple question: what explains the choice of traditional/formal lawmaking versus informal international agreements? Lipson, for example, argues that "informality is best understood as a device for minimizing the impediments to cooperation, at both the domestic and international levels."[54] But what exactly informs this statement? Within the functionalist tradition of international relations theory,[55] a growing corpus of academic literature, also known as the legalization

[50] Pauwelyn, Wessel, & Wouters, *supra* note 2, at 742–43.

[51] *Id.* at 743.

[52] Pauwelyn, Wessel, & Wouters, *supra* note 7, at 505.

[53] Climate change, for example, is an issue characterized by the existence of a traditional/formal lawmaking forum – the United Nations Framework Convention on Climate Change – as well as by a wide range of other formal and informal fora. *See* Liliana Andonova & Manfred Elsig, "Informal International Lawmaking: A Conceptual View from International Relations," in INFORMAL INTERNATIONAL LAWMAKING, *supra* note 7, at 63.

[54] Charles Lipson, "Why Are Some Agreements Informal?" 45 INT'L ORG. 495, 500 (1991).

[55] This tradition dates back to Robert Keohane's seminal work, which takes a functional approach, explaining the nature and the specific features of international institutions and international law in

literature,[56] has been examining the choice between traditional/formal versus informal international agreements.[57] More recently, this literature has been focused on the distinction between hard and soft lawmaking processes,[58] identifying the benefits and downsides of each. Since informal international law roughly overlaps with soft law,[59] we can therefore draw on this literature to provide further explanations on the rise of IN-LAW.[60]

To begin with, these scholars argue that traditional/formal lawmaking features have many advantages. First, it is generally contended that formal, hard law instruments allow states to commit themselves to international agreements in a more credible manner, since they increase the cost of reneging, either on account of legal sanctions or reputational costs.[61] Second, traditional/formal international lawmaking instruments are more credible because they normally have direct legal effect in national jurisdictions, or require domestic implementation, thereby increasing the credibility of their commitments.[62] Third, while it remains true that informal settings can provide a useful forum for exchanging information and best practices in the implementation of commonly agreed upon rules, formal, treaty-based cooperation can provide more extensive opportunities to monitor compliance and settle disputes, including through the use of dispute settlements bodies such as courts.[63] Consequently, it could be concluded that states tend to use traditional/formal lawmaking procedures when the benefits of cooperation are great, and the potential for opportunism and its costs are quite high.[64]

terms of the functions that these institutions perform and the types of problems they are designed to solve. *See* ROBERT O. KEOHANE, *AFTER HEGEMONY: COOPERATION AND DISCORD IN THE WORLD POLITICAL ECONOMY* (1984).

56 *See, e.g.*, Laurence R. Helfer & Ingrid Wuerth, "Custom in the Age of Soft Law" (draft), at 6, *retrieved from* http://iilj.org/ courses/documents/ CustomintheAgeofSoftLaw.pdf.

57 Lipson, *supra* note 54.

58 Kenneth W. Abbott & Duncan Snidal, "Hard and Soft Law in International Governance," 54 *INT'L ORG.* 421 (2000); *COMMITMENT AND COMPLIANCE: THE ROLE OF NON-BINDING NORMS IN THE INTERNATIONAL LEGAL SYSTEM* (Dinah Shelton ed., 2000); Mark A. Pollack & Gregory C. Shaffer, "Hard vs. Soft Law: Alternatives, Complements, and Antagonists in International Governance," 94 *MINN. L. REV.* 706 (2010).

59 *See supra* text accompanying notes 18, 19, 26.

60 In this regard, see also Mark A. Pollack & Gregory C. Shaffer, "The Interaction of Formal and Informal International Lawmaking," in *INFORMAL INTERNATIONAL LAWMAKING*, *supra* note 7, at 241, where the authors build on their previous literature on hard vs. soft law in order to explain how IN-LAW interacts with traditional/formal international lawmaking.

61 *See generally* Abbott & Snidal, *supra* note 58; Lipson, *supra* note 54, at 508.

62 This difference, however, should not be overstated, since informally produced norms can also be internalized in the form of domestic laws and regulations. *See, e.g.*, Pierre-Hugues Verdier, "U.S. Implementation of Basel II: Lessons for Informal International Lawmaking," in *INFORMAL INTERNATIONAL LAWMAKING*, *supra* note 7, at 437.

63 Abbott & Snidal, *supra* note 58, at 430.

64 *Id.* at 429.

However, it is also argued that traditional/formal lawmaking, with its legally binding rules and strenuous diplomatic negotiation procedures, also entails significant costs, and may restrict the behaviour of states, infringing on their national sovereignty in potentially sensitive areas. This, in turn, encourages states to bargain fiercely over legally binding commitments. In addition, formal, hard law instruments can be more difficult to adapt to changing circumstances.[65] In this context, IN-LAW procedures can offer significant advantages.

First, the legalization literature contends that informal agreements are not only easier, but also far less costly to negotiate, and impose lower sovereignty costs on states. Second, it is suggested that it might be more convenient for some governments to adhere to informal instruments, since these instruments allow states to circumvent lengthy and often difficult domestic ratification procedures. Third, informal international law instruments provide greater flexibility and are generally easier to supplement, amend, or replace. Fourth, it is also argued that reaching an agreement at the international level when the instrument in question is informal and nonbinding is more likely, since it encourages states to agree to more ambitious and detailed provisions than they would if they had to deal with the consequences of enforcement and noncompliance.[66] All in all, Lipson, as well as Abbott and Snidal, argue that states are likely to turn to informal international lawmaking when contracting costs increase, whether because of the number of parties involved, factual uncertainty, the requirements of domestic ratification and its related challenges, politically charged issue areas, or, in certain situations, distributional asymmetries among states.[67]

The functional approach described here accurately explains some of the reasons behind the rise in IN-LAW, as well as the considerations that inform a state's choice between a formal, hard law instrument, and an informal, soft law one. However, almost without exception, the literature frames this choice in terms of treaties and informal agreements, with little or no mention of custom, one of the two forms of hard international law. Consider, for example, Abbott and Snidal's seminal work, which only looks at one form of hard law: treaties. In their work, custom is relegated to a mere footnote.[68] In the same vein, the literature on CIL tends to focus on custom's internal definitional problems, and says little or nothing about informal,

65 *Id.* at 433.

66 *Id.* at 38–39. *See also* Shelton, *supra* note 58, at 12; Lipson, *supra* note 54, at 514–27; Pollack & Shaffer, *supra* note 60, at 246.

67 Lipson, *supra* note 54, and Abbott & Snidal, *supra* note 58.

68 Suggesting that their analysis also applies to custom. *See* Abbott & Snidal, *supra* note 58, at note 19 (mentioning that "customary law is also an important element of the international legal system, but we do not address it systematically here").

soft law instruments and their relationship to custom.[69] But custom, one of the two forms of hard international law, actually shares many of the advantages of informal commitments. Indeed, as compared to treaties, custom costs little to generate, it need not go through a domestic ratification process, and, similar to informal agreements, it is flexible in its content.[70] The same attributes that have triggered the rise in IN-LAW should, supposedly, also determine a growth in CIL. But custom, like treaties, is in decline. The question thus becomes: why has informal international law proliferated, while custom now languishes at the fringes of international lawmaking?

CUSTOM IN AN AGE OF INFORMALITY

If states want an alternative to treaties that provides a cheaper and faster way of generating international cooperation, avoids domestic ratification procedures, and is less costly to violate, they do not need custom. Informal international law seems to fit the bill perfectly. In this regard, some scholars have suggested that, because of their similarities, informal law may have rendered custom irrelevant.[71] However, the rise of IN-LAW is not the only reason behind the decline in custom. CIL is marred by both doctrinal and practical limitations that make it ill-equipped to address the modern challenges to international cooperation.

Custom in Crisis?

Almost a decade ago, Andrew Guzman wrote that custom, one of the two main sources of international law, is "in trouble."[72] This is as valid today as it was then, if not even more so, as CIL is under attack from all sides. While some scholars complain that custom is "wrapped in mystery and illogic,"[73] others assert that it has become "obsolete,"[74] and most, if not all, agree that the traditional doctrine of customary international law is, for lack of a better word, "a mess."[75] The ICJ Statute defines custom as "evidence of a general practice accepted as law."[76] Therefore, CIL is normally said to have two elements: state practice and *opinio juris*. State

69 *See, e.g.*, Andrew T. Guzman, "Saving Customary International Law," 27 *Mich. J. Int'l L.* 115, 116 (2005).
70 Helfer & Wuerth, *supra* note 56, at 2.
71 Guzman, *supra* note 69, at 119.
72 *Id.* at 116.
73 Anthony D'Amato, *The Concept of Custom in International Law* 4 (1971).
74 *See generally* Trachtman, *supra* note 3.
75 *See, e.g.*, J. Patrick Kelly, "The Twilight of Customary International Law," 40 *Va. J. Int'l L.* 449, 450 (2000).
76 ICJ Statute, Article 38 (1) (b).

practice refers to general and consistent practice by states, while *opinio juris* requires that the practice be followed out of a sense of legal obligation.[77] But custom, albeit seemingly simple to define, comes with a laundry list of problems. Many authors have considered these problems, so we will not dwell on them. However, a brief overview of these doctrinal disputes is necessary if one wants to understand why custom finds itself in crisis.

According to Anthony D'Amato, one of the most vexing problems with CIL is the inherent circularity in the requirement that states must act out of a legal sense of obligation in order to create that obligation.[78] But apart from the *opinio juris* prerequisite, D'Amato also points out the uncertainties associated with the state practice norm. Indeed, it still remains unclear what exactly counts as state practice, what is the amount or consistency of state practice required for the formation of custom,[79] or, for that matter, what forms of evidence may be used to demonstrate state practice.[80] For these reasons, custom has also been characterized as a highly imprecise and vague source of international law.[81] As Professor Henkin famously mentioned, the state practice that produces CIL is "informal, haphazard, not deliberate, even partly unintentional and fortuitous ... unstructured and slow."[82] In other words, custom is generated through a decentralized, informal, largely non-negotiated, and relatively flexible process.[83] While this is often portrayed as a flaw, some scholars have suggested that, on the contrary, custom's flexibility is precisely one of the reasons why CIL remains, despite its ingrained problems, a somewhat valuable tool of international lawmaking.[84]

Another central problem with custom, some contend, is that it does not satisfy conventional notions of state consent.[85] Indeed, as Guzman has rightfully indicated, custom is probably one of the oldest forms of nonconsensual international law.[86] More specifically, custom, as is well known, also binds states that did not consent to it, but nonetheless failed to object to its formation, as well as states that have come

77 On the definition of custom and its elements, *see* First Report on Formation and Evidence of Customary International Law by Michael Wood, Special Rapporteur, UN Doc. A/CN. 4.663 (17 May 2013).

78 D'Amato A., *supra* note 73, at 53.

79 *Id.* at 58.

80 Guzman, *supra* note 69, at 125.

81 Karol Wolfke, *Custom in Present International Law* xiii (1993).

82 Louis Henkin, *How Nations Behave: Law and Foreign Policy* 34 (1979).

83 Helfer & Wuerth, *supra* note 56, at 4. *See also* Trachtman, *supra* note 3, at 4.

84 Jeremy Pearce, "Customary International Law: Not Merely Fiction or Myth," *Aust. Int'l L.J.* 125 (2003). Similarly, Anthea Roberts has also suggested that, since it is fluid in nature, custom can easily develop and change in light of new circumstances. In this regard, *see* Anthea Roberts, "Traditional and Modern Approaches to Customary International Law," 95 *Am. J. Int'l L.* 757, 784 (2011).

85 *Id.* at 4.

86 Anthony T. Guzman, "Against Consent," 52 *Va. J. Int'l L.* 747, 775 (2012).

into being after the establishment of a customary rule.[87] Since many commentators believe that international law is built on the consent of states,[88] custom's nonconsensual nature is generally viewed as a failing. However, there are others who contend that some form of nonconsensual lawmaking is actually essential to the international system.[89]

Finally, in addition to the issues related to the process of custom formation, Trachtman adds to the plethora of criticism by arguing that custom lacks legitimacy because it is normally not dependent on parliamentary approval or, for that matter, domestic ratification of any sorts.[90] Furthermore, as Steven Weathley indicates, since CIL is, by definition, unwritten, the task of identifying, interpreting, and applying customary rules is more than often left to non-state actors, judges, or even academics.[91] This, in turn, as Trachtman indicates, provides excessive discretion for interpretation by courts and, oftentimes, even auto-interpretation by states.[92]

Notwithstanding the vigorous criticism exposed here, custom has also been widely recognized in the literature as a fundamental source of international law, and defended as having a prominent and indisputable role in international lawmaking.[93] In this sense, Michael Wood, in his first report to the ILC, made the following statement:

> It is sometimes suggested that treaties are now a more important source of international law than customary international law. Such generalizations are neither particularly illuminating nor necessarily accurate. Even in fields where there are widely accepted "codification" conventions, the rules of customary international law continue to govern questions not regulated by the conventions and continue to apply between non-parties. Rules of customary international law may also fill possible lacunae in treaties, and assist in their interpretation. An international court may also decide that it may apply customary international law where a particular treaty cannot be applied because of limits on its jurisdiction.[94]

Indeed, it is indisputable that CIL retains importance in a few specific areas. First, custom remains relevant in areas where codification is absent. For instance, in international humanitarian law, the provisions of CIL have a continuous relevance in the regulation of non-international armed conflicts.[95] Second, CIL, as the only

[87] Guzman, *supra* note 69, at 172.
[88] *See, e.g.*, ANTHONY AUST, *THE HANDBOOK OF INTERNATIONAL LAW* 4 (2005).
[89] Guzman, *supra* note 86, at 787–90.
[90] Trachtman, *supra* note 3, at 18.
[91] STEVEN WEATHLEY, *THE DEMOCRATIC LEGITIMACY OF INTERNATIONAL LAW* 150 (2010).
[92] Trachtman, *supra* note 3, at 4, 20–21.
[93] *Id.* at 116.
[94] First Report on Formation and Evidence of Customary International Law, *supra* note 77, ¶ 35.
[95] Trachtman, *supra* note 3, at 33.

source of international law that is binding on all states, remains important even in areas where one or more treaties exist, since some of the rules contained in these treaties have customary status and apply to non-state parties as well. Third, custom is also relevant in areas where it is codified in an international instrument, but the instrument is not legally binding. This is the case, for example, with the ILC Articles on State Responsibility.[96] Fourth, given that certain international conventions sometimes refer to customary rules, CIL retains its important role as a tool in the interpretation of treaties.[97] But while the foregoing speaks of the continuing relevance of custom in a few limited domains, it does not, unfortunately, refute the claim that custom is, in fact, in decline.

Contrasting Custom and Informational International Lawmaking

The doctrinal disputes surrounding CIL bring to the fore its "glitches."[98] In certain circumstances, however, CIL has some features that may be understood as attractive. First, it is flexible in content, which may allow it to be more responsive to change. Second, it is formed in a decentralized manner, and therefore unconstrained by procedural requirements and bureaucracy. Third, lack of national parliamentary control might make it attractive to states that, for various reasons, prefer not to engage in lengthy domestic ratification processes. Fourth, since it is formed in ways not entirely consistent with state consent, it may further be useful in solving certain types of international cooperation problems.[99]

CIL, it seems, is not that different from IN-LAW. However, as opposed to informal law, the attributes of custom also come with a host of deficiencies. First, even though described as flexible, its vagueness and lack of precision reduce custom's ability to shape expectations of future behavior.[100] Second, albeit formed in a decentralized manner that would arguably allow it to respond quickly to new challenges, CIL, in fact, changes too slowly, since it needs consistent, general practice, that must further be supplemented by evidence of *opinio juris*.[101] Therefore, CIL cannot be formed in a coordinated manner in advance of certain events. This, Trachtman argues, "will prevent it from being made at all in an important set of circumstances."[102] Third, custom does not allow for withdrawal. Therefore, when a state's interests change,

[96] *Id.* at 33–34.
[97] Guzman, *supra* note 69, at 119–20.
[98] Trachtman, *supra* note 3, at 4.
[99] *Id.* at 21–22.
[100] Helfer & Wuerth, *supra* note 56, at 18.
[101] Harlan G. Cohen, "Finding International Law: Rethinking the Doctrine of Sources," 93 IOWA L. REV. 65, 77 (2008).
[102] Trachtman, *supra* note 3, at 5.

it cannot opt out of custom. Fourth, since custom is binding, hard law, the costs of violation rise significantly.[103] Consequently, despite their similarities, informal international lawmaking appears as the more attractive alternative in addressing modern challenges.

Although similar in certain respects, CIL and IN-LAW are also fundamentally different in others. These differences, Helfer and Wuerth argue, are precisely the features that might open the way forward for custom and preserve its relevance in a world of increasing informality.[104] First, custom is universal, in the sense that once formed, it is universally binding upon all states (except, of course, those objecting to its formation, and in case of a regional or local custom). On the other hand, IN-LAW, in the same vein as treaties, applies only between the parties that participated to its negotiation.[105] Second, CIL is normally produced through a non-negotiated process, whereas the manner in which treaties or informal instruments are developed is highly negotiated.[106] Third, custom, as opposed to treaties and IN-LAW, is, by its very nature, an unwritten source of international law.[107] Fourth, custom also differs from treaties and informal law in terms of entrenchment. More precisely, custom does not permit exit, but instead states must violate the customary rule and pursue the development of new rules. The foregoing attributes distinguish CIL from both treaties and IN-LAW. Therefore, they bolster custom's uniqueness and help explain why custom, at least in certain issue areas, remains a valuable tool for generating international law.[108] What, then, is custom's domain?

Custom's (Limited) Domain

Helfer and Wuerth contend that, viewing custom as a universal, largely non-negotiated, unwritten, and entrenched form of law suggests that states might choose this form of cooperation in two situations: (i) where the distributional impacts of the norms are low among states; and (ii) when states choose to entrench structural and normative features of the international system as a whole.[109]

First, custom may be an attractive means of cooperation when distributional or other substantive differences between states are low, or, for that matter, nonexistent.[110] In this sense, much of today's customary rules are formed around issue

[103] Helfer & Wuerth, *supra* note 56, at 17–18.
[104] *Id.* at 4.
[105] *See, e.g.*, Cohen, *supra* note 101, at 76.
[106] Helfer & Wuerth, *supra* note 56, at 15–16. *See also* Pauwelyn, Wessel, & Wouters, *supra* note 7, at 506.
[107] First Report on Formation and Evidence of Customary International Law, *supra* note 77, ¶ 21.
[108] *See generally* Helfer & Wuerth, *supra* note 56.
[109] *Id.* at 26.
[110] On the distributive approach on hard/informal law, *see* Pollack & Shaffer, *supra* note 60, at 247–50.

areas that govern state-to-state interaction, such as *jus in bello*, the interpretation and application of treaties, state and diplomatic immunities, the rules on state responsibility, and even access to the high seas.[111] Arguably, at least from an *ex ante* perspective, all states benefit from these rules. In this regard, Anthony D'Amato also indicates that the validity of these rules "is reinforced by the consensus that the rules are in the subject's best interest."[112] Accordingly, acceptance or acquiescence is possible without negotiation.[113] Indeed, these rules may have some distributional consequences, but as long as they are low, Helfer and Wuerth contend that the lack of ability to negotiate could expedite agreement "because it prevents states from using negotiation tactics to steer the content of custom in ways that benefit them."[114]

Second, the entrenchment features of custom – it does not permit exit, it often binds states without their consent, some rules have attained the status of *jus cogens* norms – may be useful to states in the same way that constitutions are useful domestically. In this respect, Anthea Roberts argues that CIL may be understood as framing "the ground rules for the international system."[115] Indeed, in the same vein as constitutions, custom binds the entire community of states, even those that join the community after the customary rule has been formed. Furthermore, similarly to constitutions, many CIL rules create structural features of the international legal system, or reflect basic normative or aspirational values. However, custom also has some features that might seem slightly inconsistent with its status as constitution, namely its vague, unwritten, and non-negotiated character. In this regard, it may be argued that constitutional provisions are often vague, too, and much like CIL, they defer interpretative questions to future decision makers. Moreover, constitutionalism need not necessarily take a written form.[116] Even the non-negotiated character of custom may, in certain situations, be seen as consistent with a broader view on constitutionalism. Indeed, non-negotiation and universality may help persuade states that CIL is "the product of basic normative values and structural features that benefit all states, rather than an effort of some states to benefit at the expense of others."[117] Therefore, CIL will develop, in part, around norms that states view as particularly significant and in need of entrenchment.[118]

[111] For an overview of the major international law areas governed by custom, *see* Trachtman, *supra* note 3, appendix A.

[112] D'Amato, *supra* note 73, at 31.

[113] Helfer & Wuerth, *supra* note 56, at 26.

[114] *Id.* at 27. Custom, however, becomes more difficult to create as distributional costs increase (for example, the shift from absolute to restrictive immunity).

[115] Anthea Roberts, "Who Killed Article 38 (1)(b)? A Reply to Bradley & Gulati," 21 *Duke J. Int'l & Comp. L.* 173 (2010).

[116] Richard Primus, "Unbundling Constitutionality," 80 *U. Chi. L. Rev.* 1079, 1082 (2013).

[117] Helfer & Wuerth, *supra* note 56, at 28.

[118] *Id.*

The limited view of custom advanced here indicates that there is a domain for both informal law as well as custom. However, it also remains true that, in an age of increasing informality, custom's domain is quite limited. Nonetheless, it should not be underestimated. Providing basic rules for state-to-state interaction not only improves stability and predictability, but also opens the way for more extensive cooperation. Additionally, the entrenchment of normative values and aspirational principles allows custom to fulfil a constitutional role in the international legal system. In a fragmented, multipolar world, these well-established values of custom may continue to play a limited, albeit important, role.[119]

INFORMAL INTERNATIONAL LAWMAKING AND CUSTOM AS COMPLEMENTS

Informal international lawmaking can be seen as both an alternative as well as a part in the process of traditional international lawmaking. Kirton and Trebilcock argue, in no uncertain terms, that soft law is not a replacement for hard law, but "at best, it is a complement".[120] Building once again upon the functionalist approach and the legalization literature on hard and soft law, we argue that traditional/formal international lawmaking and IN-LAW can also interact in a complementary, even harmonious manner. Hard and soft law, it is contended, can build upon each other in two ways. First, informal international lawmaking can lead the way to more formal law and lawmaking. To give but one example, Abbott and Snidal have identified a range of "pathways to cooperation," some of which explicitly involve the hardening of soft law.[121] They observe, first, how states may sometimes start with a framework convention that over time deepens in the precision of its coverage. They also note that the use of an informal or soft law instrument can subsequently lead to binding legal commitments.[122] Christine Chinkin also contends that soft law can be both elaborative of hard law – by providing guidance in its interpretation – and be subsequently accepted as "emergent hard law," facilitating the building of custom.[123] In the same vein, David Trubek and others argue that soft law instruments "can eventually harden into binding rules once uncertainties are reduced and a higher

[119] *Id.* at 30.

[120] John J. Kirton & Michael J. Trebilcock, "Introduction: Hard Choices and Soft Law in Sustainable Global Governance," in *HARD CHOICES. SOFT LAW: VOLUNTARY STANDARDS IN GLOBAL TRADE, ENVIRONMENT, AND SOCIAL GOVERNANCE* 3 (John J. Kirton & Michael J. Trebilcock eds., 2004).

[121] Kenneth W. Abbott & Duncan Snidal, "Pathways to Cooperation," in *THE IMPACT OF INTERNATIONAL LAW ON INTERNATIONAL COOPERATION: THEORETICAL PERSPECTIVES* 50, 54 (Eyal Benvenisti & Moshe Hirsch eds., 2004).

[122] *Id.*

[123] Chinkin, *supra* note 19, at 30–31.

degree of consensus ensues."[124] Second, where a body of hard law already exists, informal lawmaking processes may provide an avenue to flexibly elaborate and fill in any gaps that form when preexisting hard law encounters unexpected circumstances. In this respect, traditional/formal and informal international lawmaking processes serve as complements in dynamic mechanisms of legalization, arguably leading to the progressive development of international law.[125]

As far as the specific interaction between informal international lawmaking and custom is concerned, it has been rightly observed that, given the appropriate circumstances, a soft law instrument may not only provide evidence of existing CIL, but may also become a catalyst for the development of hard law, prompting the making of new treaties, as well as the formation of customary international law.[126] In fact, to many scholars, this is the *raison d'être* of soft or informal law, and perhaps even its opening wedge into the traditional sources of international law. To them, informal international lawmaking not only furnishes a conceptual frame of reference for future international agreements, but also facilitates the crystallization process that gives rise to customary international law.[127] The fact that the repetition of the same norm in successive soft or informal law instruments can give rise to state practice and then express the *opinio juris* of the international community is also reinforced by certain developments at the international level – for example, the fact that international rules concerning transboundary environmental relations, although initially formulated in informal instruments, are currently seen as constituting custom. Indeed, in this respect, the ICJ recognized a norm stemming from Principle 21 of the Stockholm Declaration and Principle 2 of the Rio Declaration as being "part of the corpus of international law relating to the environment."[128]

What precisely is, therefore, the role that informal international lawmaking plays in the formation of custom? While practice has shown that informal international lawmaking may, in time, lead to the formation of treaties or be incorporated in subsequent treaties,[129] the question of whether informal international lawmaking

[124] David M. Trubek, M. Patrick Cottrell, & Mark Nance, "Soft Law, Hard Law, and European Integration: Toward a Theory of Hybridity," *in* NEW GOVERNANCE AND CONSTITUTIONALISM IN EUROPE AND THE US 65, 89 (Gráinne de Búrca & Joanne Scott eds., 2006).

[125] *See generally* Christine Chinkin, "The Challenge of Soft Law: Development and Change in International Law," 38 INT'L & COMP. L.Q. 850 (1989).

[126] Boyle & Chinkin, *supra* note 15, at 212.

[127] *See, e.g.*, Christophe Joyner, "The Legal Status and Effects of Antarctica Recommended Measures," in COMMITMENT AND COMPLIANCE: THE ROLE OF NON-BINDING NORMS IN THE INTERNATIONAL LEGAL SYSTEM 163, 176 (Dinah Shelton ed., 2000).

[128] Legality of the Threat or Use of Nuclear Weapons ¶ 29 (ICJ, July 8, 1996) (Advisory Opinion).

[129] To provide but one example, Principle 21 of the Stockholm Declaration on the Human Environment, repeated almost verbatim in the Rio Declaration on Environment and Development, is included in the preambles of many multilateral treaties, as well as Article 3 of the Convention on Biological Diversity.

could somehow elaborate on custom or contribute to its formation is much more sensitive. Indeed, whether informal agreements could somehow codify or reflect previous custom, generate state practice, or provide evidence of the *opinio juris* necessary for the formation of custom will depend on various factors that must be assessed in each case.

Informal International Lawmaking as a Reflection of Custom

Some scholars contend that informal instruments may be elaborative of hard law and provide guidance to its interpretation, elaboration, and application.[130] This assertion is most often associated with a soft law instrument that refers back to treaty obligations, with the inference that the former derives authority from and extends the meaning of the latter.[131] However, according to Dinah Shelton, such interdependence does not only exist between soft law and treaties, but also between soft law and custom. More precisely, she argues that a nonbinding normative instrument may codify preexisting CIL and even help provide greater precision through the written text.[132] In this respect, she gives the example of the commentary to the UN Basic Principles and Guidelines on the Right to Remedy and Reparation for Victims of Gross Violations of International Human Rights Law and Serious Violations of Humanitarian Law, approved by the Commission on Human Rights and endorsed by the UNGA, which claims that the principles and guidelines in question contain no new norms, but actually reflect existing law scattered among a large number of treaties and widespread state practice. It seems, therefore, that the soft law instrument in question purports to do nothing more than set down in written form preexisting rights and duties.[133] Yet another relevant example that reinforces this argument is the 1992 Rio Declaration on Environment and Development, which both codifies existing international law and develops new law.[134]

Authors such as Boyle and Chinkin further suggest that, in certain situations, soft or informal law instruments appear to be, alongside treaties, a useful means of codifying custom.[135] The ILC's Draft Articles on the Law of State Responsibility

130 *See, e.g.*, Dinah Shelton, "Soft Law," in ROUTLEDGE HANDBOOK OF INTERNATIONAL LAW 63 (David Armstrong ed., 2008).

131 In this respect, General Assembly Declaration on Principles of International Law Concerning Friendly Relations and Co-Operation Among States in Accordance with the Charter of the United Nations, UN Doc. A/Res/25/2625 (Oct. 24, 1970) is, as its title correctly points out, elaborative of the UN Charter.

132 Shelton, *supra* note 130, at 72.

133 *Id.*

134 Boyle & Chinkin *supra* note 15, at 215.

135 *Id.* at 182.

are certainly an example in this sense. When discussing the eventual form that the draft articles on state responsibility would take, both the ILC and the Sixth Committee recommended that the UNGA initially "take note" of the articles, in the expectation that in such a form they would "exert an influence on the crystallization of the law of state responsibility through application by international courts and tribunals and state practice," and deferred for future consideration a decision on whether to proceed with a codification treaty or not.[136] Likewise, the Articles on Prevention of Transboundary Harm and the Principles on the Allocation of Loss in the Case of Transboundary Harm Arising out of Hazardous Activities were both drafted in treaty form by the ILC and similarly taken note of by the UNGA.[137] These examples reinforce the idea that soft or informal law instruments may be evocative of preexisting custom, as well as catalysts of new developments in international lawmaking.

Informal International Lawmaking as State Practice

Grnchalla-Wesierski has argued that, as a result of its abstract nature, soft or informal law cannot satisfy the material requirement of custom.[138] More precisely, the author suggested that informal agreements may not be held to provide the state practice that the language of such agreements describes, and that the mere adoption of such instruments cannot be interpreted as state practice leading to the formation of custom.[139] To develop a bit more on this point, one could look at the argument made by some scholars that soft law instruments, such as UNGA resolutions and declarations, could create "instant" customary international law.[140] These authors generally argue that the mere adoption of such resolutions and declarations may, in certain situations, afford evidence of a new rule of customary international law, without further state practice on behalf of the states voting in favor of such instruments.[141] However, this must be approached with caution, as one must also take into account the number of states that cast their vote or whether there was a strong consensus or even unanimity in favour of the adoption of such resolutions and declarations, as well as the wording of the text, which should, to a certain

[136] James Crawford & Simon Olleson, "The Continuing Debate on a UN Convention on State Responsibility," 54 *INT'L & COMP. L.Q.* 959, 960 (2005).

[137] Boyle & Chinkin, *supra* note 15, at 182–83.

[138] Tadensz Grnchalla-Wesierski, "A Framework for Understanding 'Soft Law,'" 30 *MCGILL L. J.* 37, 53 (1984).

[139] *Id.*

[140] Bing Cheng, "United Nations Resolutions on Outer Space: 'Instant' Customary International Law?" 5 *INDIAN J. INT'L L.* 23 (1965).

[141] *Id.*

degree, articulate normative rules.[142] Indeed, a resolution or declaration adopted by consensus or even unanimous vote will arguably carry more weight than one supported by, let us say, a wide majority.[143] As far as wording is concerned, one must notice that even a soft or informal law instrument that is appropriately worded, such as, for instance, the Universal Declaration of Human Rights, although undeniably having contributed to "harder" lawmaking, did not have an instantaneous impact. To add to this, international jurisprudence also disapproves of the notion of "instant" customary international law and actually stresses the need for subsequent confirmatory state practice.[144]

While it is clear that the single act of adopting a soft law instrument may not amount to state practice as a constituent element of custom, the assertion that acts subsequent to the adoption of an informal agreement might have a role in the creation of custom as manifestations of state practice is supported by academic literature. In this regard, Christine Chinkin readily accepts the proposition that a soft or informal law instrument may harden over time if subsequent state practice changes the status of that instrument.[145] In the same vein, Dinah Shelton contends that a nonbinding normative instrument may "crystallize a trend toward a particular norm, overriding the views of dissenters and persuading those who have little or no relevant state practice to acquiesce in the development of the norm."[146] She further indicates that, for a soft or informal law instrument to cross into the realm of hard law, subsequent "conforming state practice is needed among states representing different regions and the major legal, economic and political systems."[147]

Contemporary realities come to reinforce the foregoing assertions. To give but one example, the World Charter for Nature, adopted by the UNGA in 1982, although an informal instrument in its incipient form, has hardened over time, at least with respect to some of its provisions (in particular, the environmental impact assessment process and participatory rights), which are considered to have become part of CIL.[148] While it is self-evident that the World Charter for Nature in and of

142 Boyle & Chinkin, *supra* note 15, at 227.

143 See, for example, the case of the Charter for Economic Rights and Duties of States, UN Doc. A/Res/29/3281 (Dec. 12, 1974), where 120 states voted for the resolution, 6 voted against and 10 abstained, illustrating the inability of a majority of states to legislate for minorities in this manner.

144 Case Concerning Military and Paramilitary Activities in and Against Nicaragua (*Nicaragua vs. United States of America*) ¶ 184 (ICJ, June 27, 1986) (Judgment).

145 Chinkin, *supra* note 125, at 856–57.

146 Shelton, *supra* note 130, at 72.

147 *Id.*

148 Sumudu Atapattu, "International Environmental Law and Soft Law: A New Direction or a Contradiction?" in *Non-State Actors, Soft Law and Protective Regimes: From the Margins* 200, 212 (Cecilia M. Bailliet ed., 2012). *See also* Pulp Mills on the River Uruguay (*Argentina vs. Uruguay*) ¶ 204 (ICJ, Apr. 20, 2010) (Judgment).

itself did not create binding law, it nevertheless influenced subsequent state practice in a manner that finally lead to the creation of custom. Accordingly, it might be argued that, even though soft or informal law does not *per se* create binding rules, it still operates in an indirect way, by means of persuasion rather than by means of coercion. Undoubtedly, therefore, such instruments have a real influence on the practice of states by encouraging them to implement the common standards proposed and, in the long term, as practice has shown, by creating binding norms, either in the form of a treaty or by being recognized as custom.

Informal International Lawmaking as Evidence of Opinio Juris

Qualifying subsequent acts as acts that may be seen as state practice leading to the creation of international law should not be confused with the assertion that the adoption of a soft or informal law instrument is an expression of *opinio juris*. As *opinio juris*, the instrument in question would normally need to qualify the practice as legally binding, which, given the nonbinding nature of informal international lawmaking in many cases, might seem quite puzzling. In this regard, Anthony D'Amato contends that the idea "that soft law is nothing other than the alleged component of custom known as *opinio juris* has to be one of the wildest ideas in recent scholarship."[149] Nevertheless, some scholars share the view that soft or informal law may, in certain circumstances, provide the requisite and perennially elusive psychological element of custom.[150] Even the ICJ, in the *Nicaragua* case, held that the *opinio juris* of states with respect to the prohibition against the use of force could be inferred from their attitude toward the relevant UNGA resolutions.[151] According to Christine Chinkin, the judgment in question "appears to represent a willingness by the Court to accept the transformation of soft law principles into hard law."[152]

However, the foregoing claims are not beyond any dispute. For example, one could argue that, since informal international lawmaking is not legally binding *per se*, the intent of the parties adopting such informal agreements is, therefore, not to be bound by a conforming practice. More exactly, if *opinio juris* is said to imply evidence that states consider a norm legally binding (i.e., "accepted as law"), such informal instruments could not be evocative of *opinio juris*. Another troublesome issue with the idea that informal agreements can be expressive of a state's *opinio juris* is, undoubtedly, the nature of the actors involved. As discussed, informal

149 Anthony D'Amato, "Softness in International Law: A Self-Serving Quest for New Legal Materials: A Reply to Jean D'Aspremont," 20 Eur. J. Int'l L. 896, 906 (2009).

150 *See, e.g.*, Boyle & Chinkin, *supra* note 15, at 212; Grnchalla-Wesierski, *supra* note 138, at 53.

151 *Nicaragua Case*, *supra* note 144, ¶ 188.

152 Chinkin, *supra* note 125, at 858.

international lawmaking draws into its processes other actors than the traditional ones, and therefore, apart from the situations where the makers of informal law are states or IOs, IN-LAW could not be said to embody a state's *opinio juris*.[153] But in an increasingly changing international landscape, are such constraints still valid? Is the label attached to the instrument more important than its substance? Should the nonbinding status of an instrument be a sufficient reason to reject it as establishing or reflecting evidence of custom? Does the classical definition of custom fully exclude new actors? Some may argue to the contrary.

First, it may be argued that IN-LAW is, by definition, composed of norms that, expressly or impliedly, are not only enacted with the purpose of being followed, but also raise certain expectations. In this regard, even Prosper Weil, one of the staunchest critics of soft law, has recognized that informal instruments "do create expectations and exert on the conduct of states an influence that in certain cases may be greater than that of rules of treaty or customary law."[154] Indeed, many of these new types of outputs, while not legally binding *per se*, are nevertheless supported and accepted by a large spectrum of stakeholders well beyond nation states, and are often adopted following inclusive, transparent, and predictable procedures that go much further than traditional international lawmaking.[155] Furthermore, one should also note that these new outputs benefit, in most cases, from higher compliance rates as they are based on thick stakeholder consensus, rather than the thin state consent of traditional international lawmaking.[156] Actors in various fields, such as Internet regulation, finance, and health, take informal norms, such as those emanating from the Internet Engineering Task Force, the Basel Committee on Banking Supervision, and the International Conference on Harmonization of Technical Requirements for Registration of Pharmaceuticals for Human Use, extremely seriously. In fact, Joost Pauwelyn even contends that "soft law seems to have equal and sometimes higher compliance rates than hard law."[157] Arguably, therefore, "accepted as law" does not necessarily mean the same as recognizing that something is legally binding. For these reasons, we believe that these informal instruments, when linked to such high compliance rates and subsequent state practice, should be given proper weight in the creation of custom and may, in the appropriate circumstances, provide the psychological element of CIL. As Boyle and Chinkin also mention, "The non-binding form of an instrument is of relatively limited relevance in the context of customary international lawmaking."[158]

[153] Grnchalla-Wesierski, *supra* note 138, at 53–55.
[154] Prosper Weil, "Towards Relative Normativity in International Law?" 77 AM. J. INT'L L. 413, 415 (1983).
[155] Pauwelyn, Wessel, & Wouters, *supra* note 7, at 512.
[156] *Id.* at 530.
[157] Pauwelyn, *supra* note 26, at 160.
[158] Boyle & Chinkin, *supra* note 15, at 215.

Second, with respect to the involvement of nontraditional actors, it may be contended that custom seems to offer the way forward for informal international lawmaking. Note, indeed, that pursuant to Article 38(1)(a), conventions must be recognized by "states," apparently excluding any type of new actors. In contrast, Article 38(1)(b) does not contain a specific reference to states in relation to custom. The ICJ Statute defines custom as "evidence of a general practice accepted as law."[159] Accordingly, the ICJ Statute does not specify who must have accepted this practice as law. Recent academic literature has argued that this might indicate that states are not necessarily the only actors at play in the creation of CIL.[160] Given its less outspoken reliance on states, this provision might well offer a way forward for IN-LAW. It seems that, at least conceptually, custom leaves the door open to new actors. Therefore, the fact that these informal instruments are not always the product of official state action in the traditional sense should not necessarily be seen as an impediment to acknowledging that they might just as well be evocative of *opinio juris*.

Informal International Lawmaking as a Form of "New" Custom

Even those arguing that informal international lawmaking may not constitute or establish custom in the traditional sense would not disagree that these informal instruments have had and will most likely continue to have an important impact on present-day international lawmaking. Even if not "law," or "legally binding," soft or informal instruments have legal effects.[161] While the legal effects of instruments such as guidelines, resolutions, standards, declarations, or even more informal policy coordination is not the same in all cases,[162] it nevertheless is characteristic of most, if not all, to be carefully negotiated, attentively drafted, and extremely precise instruments. In such circumstances, it seems hard to dismiss the idea that these informal instruments are expected to have some sort of normative significance.[163] All of these instruments, informal as they may be, contain an element of good-faith commitment, convey the desire to influence state practice, and generally express at least some measure of lawmaking intent. To the extent that such norms are consistently followed by stakeholders around the globe, could they therefore emerge as the "new" custom of the twenty-first-century global governance, gradually replacing state-centered notions of custom?

159 ICJ Statute, Article 38 (1)(b).

160 Pauwelyn, Wessel, & Wouters, *supra* note 7, at 530–31.

161 Pauwelyn, *supra* note 26, at 155–57.

162 *Id.* at 128.

163 *See, e.g.*, Jan Klabbers, "International Courts and Informal International Law", in INFORMAL INTERNATIONAL LAWMAKING, *supra* note 7, at 219, 220.

Reasonably speaking, two options present themselves. First, international law can entrench itself and adhere to its traditional typologies, acknowledging that it is increasingly just one form of cooperation within a broader legal universe.[164] Second, and ideally, international law can adapt and somehow incorporate informal international lawmaking, thereby increasing its societal relevance.[165] This option would require a radical transformation of international law, both procedurally and substantively.[166] In this sense, it is generally argued that sudden or deliberate change is unlikely.[167] However, history teaches us that international law is dynamic and has gone through equally dynamic changes. Over the ages, the subjects and lawmaking processes and outputs of the legal order have dramatically evolved. As the ICJ indicated in the *Reparations* case, "Throughout its history, the development of international law has been influenced by the requirements of international life." This had led to new "instances of action upon the international plane by certain entities which are not states" and may, in turn, lead to "new subjects of international law."[168] In terms of actors, although states are currently still the principal subjects and creators of international law, there is no fixed list of subjects of international law. In terms of output and processes, there is general agreement that Article 38 of the ICJ Statute does not offer an exhaustive list of the sources of international law, nor does international law require that a specific process be followed to create international norms or that international law, for that matter, can only emerge out of particular fora or IOs.[169]

The constituent elements of custom are notoriously vague.[170] Even the definition of conventions or treaties is deeply contested and open to interpretation.[171] Therefore, even if it is hard to imagine that the states parties to the ICJ Statute would amend Article 38 to include new sources of international law, or that the UN Charter would be rewritten to explicitly allow for new actors, no such formal decisions are required for international law to evolve.[172] After all, whether or not states (or IOs) like it,

[164] Pauwelyn, Wessel, & Wouters, *supra* note 2, at 744.

[165] *See, e.g.*, Paul Schiff Berman, "A Pluralist Approach to International Law," 32 YALE J. INT'L L. 301, 302 (2007) (contending that ignoring the value of soft or informal law does "is not a useful strategy").

[166] *See, e.g.*, Chinkin, *supra* note 125, at 857–58 (explaining why it is problematic to claim that soft law principles have come to represent custom and that a "revolutionary change in the sources of international law" should take place first in order for these norms to be considered as part of CIL). *See also* ASIL Proceedings, "New Voices: Rethinking the Sources of International Law," 103 AM. J. INT'L L. 71 (2009).

[167] Chinkin, *supra* note 125, at 857.

[168] Reparations for Injuries Suffered in the Service of the United Nations ¶ 174 (ICJ, Apr. 11, 1949) (Advisory Opinion).

[169] *See* Pauwelyn, Wessel, & Wouters, *supra* note 2, at 745.

[170] *See supra* text accompanying notes 78, 79, 81.

[171] JAN KLABBERS, *THE CONCEPT OF TREATY IN INTERNATIONAL LAW* (1996).

[172] Pauwelyn, Wessel, & Wouters, *supra* note 7, at 508.

or explicitly allow for it, new actors, processes, and outputs will continue to play an increasingly important role. The conceptual boundaries of how international law will look in the future are, it seems, wide open. The international legal system must become more inclusive and complex, in order to take on new responsibilities and adequately respond to the great modern challenges confronting the twenty-first century.

14

Custom's Bright Future

The Continuing Importance of Customary International Law

Omri Sender and Michael Wood

The conference leading to the present volume was entitled "Custom in Crisis: International Law in a Changing World." That was also the original working title for the volume; it was suggested that customary international law was reaching a crisis point given contemporary challenges exacerbated by continuing uncertainty surrounding its nature and function, and that in these circumstances there was much controversy with respect to both the content of customary rules and the processes by which they evolve and are ascertained. While the title eventually chosen for the book is more neutral about the future of customary international law and does not strike such an alarming note, at least some of the contributors continue to question custom's role and relevance, treating it as a source of law plagued with difficulties and having only limited utility. It is argued, for example, that customary international law is by now mostly obsolete or at the very least in need of "reinvigoration";[1] that its nature "remains stubbornly opaque or conceptually problematic";[2] and that its vitality (in the international protection of contracts, property, and commerce) is "strange," and its role (as a source of international criminal law) "declining."[3]

Such a bleak outlook, we would argue, does not correspond to reality: in fact, customary international law has probably never been in better shape. Not only is it not in crisis in this (ever-)changing world in which it has to operate, but its role in the international legal system continues to be fundamental. Furthermore, recognition of its significance and confidence in its operation are evident in the recent work of the International Law Commission on the topic "Identification of

1 *See* Joel P. Trachtman, "The Growing Obsolescence of Customary International Law" (in this volume); Andrew T. Guzman & Jerome Hsiang, "Reinvigorating Customary International Law" (in this volume).

2 *See* John Tasioulas, "Custom, *Jus Cogens*, and Human Rights" (in this volume).

3 *See* C. L. Lim, "The Strange Vitality of Custom in the International Protection of Contracts, Property, and Commerce" (in this volume); Larissa Van den Herik, "The Decline of Customary International Law as a Source of International Criminal Law" (in this volume).

customary international law." While academic disputes surrounding the nature and function of custom may indeed endure, in practice some long-standing questions have by now been settled, and others do not prevent custom from continuing to play an important role as positive international law in force.

QUESTIONING CUSTOM

Custom's prominent position in the international legal system has not gone unchallenged in the modern era. The oldest source of international law has been subjected to several acute challenges, each of which was explicitly aimed at reducing its influence. For a start, the codification movement dating from the nineteenth century attached much importance to replacing customary international law by *lex scripta*.[4] Custom was considered too uncertain and incomplete, in need of clarification and systemization: "The temple of international justice on near inspection appear[ed] to be rather a storehouse filled with lumber of the ages – a medley of things new and old showing as yet little evidence of order or purpose."[5] More precise law, it was hoped, would promote international stability and cooperation. In the post–WWI era, in particular, treaties were perceived as a principal instrument for such purpose, and an international project was launched to reduce much of custom to written codes. Such codification, as Verdross argued, was to push customary international law somewhat into the background.[6]

Custom again found itself in the line of fire in the 1920s, when stark opposition to it was voiced by Soviet legal theorists who sought to challenge the international order and were "disposed to regard customary international law as a body of ideas and practices imposed on weak nations by capitalist imperialism."[7] They were joined by some lawyers from newly independent states emerging in the postcolonial era, who attacked customary international law as comprising rules that "express relationships of domination, inequality or privilege" in the service of Western states and regarded it as law "which has done nothing to help the poor countries."[8] For both those in the

4 *See also* Hugh W. A. Thirlway, *International Customary Law and Codification* 1–15 (1972).

5 Roland S. Morris, "Codification of International Law," 74 *U. Pa. L. Rev.* 452 (1925–1926).

6 Alfred Verdross, *Völkerrecht* 137 (1964).

7 Charles De Visscher, *Theory and Reality in Public International Law* 161 (P. E. Corbett trans., 1968). Vladimir M. Koretsky, the Soviet member of the International Law Commission, argued in 1949 that customary international law "was backward and belonged to the period of the "white man's burden", the period of the domination by a few powerful States who had disregarded the national sovereignty of weaker States ... conventional law, not customary law, was the basic source of international law": *Yearbook of the International Law Commission* 1949, 232.

8 Mohammed Bedjaoui, *Towards a New International Economic Order* 63, 132 (1979) (also referring to customary international law, at pages 102, 134, and 137–38, as a "kind of international law [that] only perpetuates the law of the jungle and the Faustrecht, or law of the first," whose "historical formation has occurred in a manner totally independent of [the developing States], and for the

Socialist camp and in the developing world, in Cassese's words, "Custom had come to represent the quintessence of the outlook they oppose,"[9] and thus they sought to repudiate its content and have it declared deficient.

Authors, too, suggested that in the modern world the role of customary international law ought to be diminished. Referring to it as "plainly inadequate as a major source of modern international law,"[10] several argued that it "has been of too slow growth to keep pace with the changing relations of the states which it endeavors to regulate" and that it is "unable to reorganize a system which is defective as a whole, and cannot so much as map out progressive lines of reform for the future."[11] Some writers criticized it, moreover, as inherently inefficient[12] or illegitimate,[13] and others described "in loving detail"[14] the fatal conundrums they observed in its conceptual foundation.[15] While some of the authors sought to reframe custom as a source of international law,[16] others called bluntly to do away with it.[17] Academic

most part against [them].... Backward looking, conservative because static, iniquitous in its content, ponderous in its formation"). *See also* S. Prakash Sinha, "Perspective of the Newly Independent States on the Binding Quality of International Law," 14 *INT'L AND COMP. LAW QUARTERLY* 121, 122 (1965).

9 ANTONIO CASSESE, *INTERNATIONAL LAW IN A DIVIDED WORLD* 181 (1987).

10 WOLFGANG FRIEDMANN, *THE CHANGING STRUCTURE OF INTERNATIONAL LAW* 121–23 (1964).

11 Charles G. Fenwick, "Notes on International Affairs," 12 *AM. POL. SCI. REV.* 301, 302 (1918). *See also* De Visscher, *supra* note 7, at 161–62; G. J. H. VAN HOOF, *RETHINKING THE SOURCES OF INTERNATIONAL LAW* 114 (1983).

12 *See, e.g.*, W. Michael Reisman, "The Cult of Custom in the Late 20th Century," 17 *CAL. W. INT'L L.J.* 133, 145 (1987); Samuel Estreicher, "Rethinking the Binding Effect of Customary International Law," 44 *VA. J. INT'L L.* 5, 9, 11, 14 (2003); Eugene Kontorovich, "Inefficient Customs in International Law," 48 *WM. & MARY L. REV.* 859, 921 (2006); Vincy Fon & Francesco Parisi, "International Customary Law and Articulation Theories: An Economic Analysis," 2 *INT'L L. & MGMT. REV.* 201, 202 (2006); Geoffrey Palmer, "New Ways to Make International Environmental Law," 86 *AM. J. INT'L L.* 259, 266 (1992); Jack L. Goldsmith & Eric A. Posner, "Understanding the Resemblance Between Modern and Traditional Customary International Law," 40 *VA. J. INT'L L.* 639, 672 (2000).

13 For the suggestion that customary international law might be incommensurable with basic human rights *see, e.g.*, Jan Klabbers, "The Curious Condition of Custom," 8 *INT'L LEGAL THEORY* 29, 34 (2002); Jan Wouters & Cedric Ryngaert, "Impact on the Process of the Formation of Customary International Law," *in THE IMPACT OF HUMAN RIGHTS LAW ON GENERAL INTERNATIONAL LAW* 111, 129, 130 (Menno T. Kamminga & Martin Scheinin eds., 2009). For the argument that customary international law suffers from a significant democratic deficit, *see, e.g.*, STEVEN WHEATLEY, *THE DEMOCRATIC LEGITIMACY OF INTERNATIONAL LAW* 150 (2010); William S. Dodge, "Customary International Law and the Question of Legitimacy," 120 *HARV. L. REV. FORUM* 19, 26 (2007); John O. McGinnis, "The Comparative Disadvantage of Customary International Law," 30 *HARV. J.L. & PUB. POL'Y* 7, 8 (2006).

14 *See* Edward T. Swaine, "Rational Custom," 52 *DUKE L.J.* 559, 569 (2002).

15 *See, e.g.*, Hiroshi Taki, "*Opinio Juris* and the Formation of Customary International Law: A Theoretical Analysis," 51 *GERMAN Y.B. INT'L L.* 447 (2008).

16 *See, e.g.*, Roozbeh B. Baker, "Customary International Law in the 21st Century: Old Challenges and New Debates", 21 *EUR. J. INT'L LAW* 173 (2010).

17 *See, e.g.*, N. C. H. Dunbar, "The Myth of Customary International Law," 8 *AUST. Y.B. INT'L L.* 1 (1980); J. Patrick Kelly, "The Twilight of Customary International Law," 40 *VA. J. INT'L L.* 449 (2000).

criticism has indeed often been fierce, so much so that a number of writers thought it necessary to attempt "saving customary international law"[18] or "restoring it, at the least, to its conventional position among the walking wounded."[19]

"CUSTOM LIVES"

But the "twilight of customary international law"[20] has not fallen, and "rumours of [custom's] death"[21] were certainly exaggerated. Far from ceasing to be influential, customary international law has all along retained a central place in international legal discourse, as it continues to do at present. As one author has put it, "custom lives."[22]

To begin with, the spectacular rise of treaties in the twentieth century was not accompanied by a significant decline in the importance of custom: as Lim argues in this volume, increased treaty lawmaking may indeed afford a greater role for customary international law.[23] After decades of codification there are only a few conventions that have achieved universal or virtually universal participation, and custom still operates alongside treaties, often by virtue of express or implied provisions in the treaties themselves. Even in highly codified fields such as the law of the sea, the rules of customary international law continue to govern questions not regulated by the conventions,[24] and apply in relations with and between non-parties;[25] customary international law may also fill possible lacunae in treaties, and assist in

[18] Andrew T. Guzman, "Saving Customary International Law," 27 *MICH. J. INT'L L.* 115 (2005). *See also* Jeremy Pearce, "Customary International Law Not Merely Fiction or Myth," 2003 *AUSTL. INT'L LJ.* 125 (2003).

[19] Swaine, *supra* note 14, at 564.

[20] Kelly, *supra* note 17.

[21] Lauri Hannikainen, "The Collective Factor as a Promoter of Customary International Law," 6 *BALTIC Y.B. INT'L L.* 125 (2006).

[22] Detlev F. Vagts, "International Relations Looks at Customary International Law: A Traditionalist's Defence," 15 *EUR. J. INT'L LAW* 1031, 1040 (2004). *See also* Tullio Treves, "Customary International Law," in *MAX PLANCK ENCYCLOPEDIA OF PUBLIC INTERNATIONAL LAW* paras. 90–91 (Rudiger Wolfrum ed., 2012).

[23] *See* Lim, *supra* note 3.

[24] *See, e.g.*, Vienna Convention on the Law of Treaties, final preambular paragraph and Article 4 (non-retroactivity). The Martens clause in the preamble to the Hague Conventions on the Laws and Customs of War on Land was an early example of the continuing importance of custom, notwithstanding a treaty. *See* Jochen von Bernstorff, "Martens Clause," in *MAX PLANCK ENCYCLOPEDIA OF PUBLIC INTERNATIONAL LAW* (Rudiger Wolfrum ed., 2009). In article 1 of Additional Protocol I to the Geneva Conventions the expression "the usages established between civilised nations" was replaced by "established custom."

[25] For example, the Vienna Convention on the Law of Treaties only applies in relations between the States Parties thereto; the rules of customary international law on the law of treaties apply in relations between States not Party to the Convention, and between a State Party and a non-Party. *See* E. W.

their interpretation.[26] A court may moreover seek to apply customary international law where treaty law cannot be applied because of limits on its jurisdiction or applicable law.[27] It still holds true that, as Parry wrote in the early 1960s, "One can have a very fair idea of international law without having read a single treaty; and one cannot have a very coherent idea of the essence of international law by reading treaties alone."[28]

Customary international law continues to play a significant role in uncodified fields, as well, where it has proven itself able to adapt to the ways of modern international life. In fields such as international humanitarian law or international space law, for example, "The reality is that customary law can develop rapidly when the need arises."[29] A significant advantage of custom remains that "if the international community is unhappy with a particular law it can be changed relatively quickly without the necessity of convening and successfully completing a world conference."[30] This might explain, at least in part, why those in the Socialist camp and in the developing world who contested customary international law eventually abandoned that campaign; and why such claims have not been echoed

Vierdag, "The Law Governing Treaty Relations Between Parties to the Vienna Convention on the Law of Treaties and States Not Party to the Convention," 76 *Am. J. Int'l L.* 779 (1982).

26 Vienna Convention on the Law of Treaties, Article 31.3(c).

27 As, for example, in *Military and Paramilitary Activities in and against Nicaragua (Nicaragua v. United States of America), Merits, Judgment, I.C.J. Reports* 1986, p. 14, at pp. 92–96. A more recent example may be found in the Supreme Court of Ghana's judgment of 20 June 2013 (arising out of the ARA *Libertad* case) applying the customary international law on warships that (unlike the United Nations Convention on the Law of the Sea) formed part of Ghanian law. The judgment is available on the Web site of the Permanent Court of Arbitration at http://www.pca-cpa.org/showpage.asp?pag_id=1526.

28 Clive Parry, *The Sources and Evidence of International Law* 34–35 (1965). *See also* James Crawford, "General Course on Public International Law," 365 *Recueil des cours* 49 (2013) ("International law is a customary law system, despite all the treaties; even the principle of *pacta sunt servanda*, the obligation to comply with treaties, is a customary law obligation."); Christopher Gregory Weeramantry, Remarks, Proceedings of the ASIL/NVIR Third Joint Conference 34 (1996):

> It is well to remind ourselves that, in a sense, treaty law is the daughter of customary international law. *Pacta sunt servanda* is, after all, a rule of customary international law, the interpretation of treaties takes place on the basis of customary international law, the validity of treaties is determined on the basis of customary international law. So, it is still a product of customary international law, still heavily dependent upon it. And I do not subscribe to the view that treaties will strangle or diminish the influence of customary international law.

Peter Tomka, "Custom and the International Court of Justice," 12 *Law & Practice of International Courts and Tribunals* 195, 215 (2013) ("In many ways … Lord Phillimore's observation in 1920 before the Committee of Jurists, that custom 'constitutes in the main international law' continues to be relevant."); Thirlway, *supra* note 4, at 35 (suggesting that "it would be unwise, in the light of history, to assume or claim that we are witnessing, or are likely to witness in the near future, any sort of total eclipse of custom as a source of law, as a result of the encroachment of the law-making treaty").

29 Ian Brownlie, *The Rule of Law in International Affairs: International Law at the Fiftieth Anniversary of the United Nations* 19 (1998).

30 Malcolm N. Shaw, *International Law* 52 (7th ed. 2014).

thereafter, including in the early 1990s by the new states of Central and Eastern Europe.[31] Many domestic constitutional orders, both old and new, recognize customary international law as an applicable source of law.

From legal opinions by government legal advisers to diplomatic correspondence and official statements by states, custom features regularly in the everyday practice of international law. It is frequently invoked before international courts and tribunals, particularly in interstate disputes, and it is of course enshrined in Article 38(1) of the Statute of the International Court of Justice, a provision that is widely regarded as setting out the sources of international law and is binding on 193 states. The International Court still "relies on customary international law constantly and as a matter of course":[32] among the recent cases that have turned, at least in part, on custom, one may find the *Diallo* case,[33] *Jurisdictional Immunities of the State*,[34] and *Territorial and Maritime Dispute (Nicaragua v. Colombia)*.[35] New cases on the Court's docket, such as *Obligations concerning Negotiations relating to Cessation of the Nuclear Arms Race and to Nuclear Disarmament*, likewise raise questions of customary international law. Other international courts and tribunals, too – even those that principally apply a particular treaty, such as the International Tribunal for the Law of the Sea – resort to customary rules not infrequently.[36] As Meron wrote, customary international law "now comes up in almost every international court and tribunal, in almost every case, and frequently has an impact on the outcome."[37] It is increasingly raised in national courts, as well.

Despite the academic controversies surrounding customary international law and the difficulty of some authors to accept a source of law that by its very nature defies exact definition, the theoretical torment that accompanies custom in the books simply does not impede it in action. The question of proof, in particular, "does not present as much

31 *See also* Weeramantry, *supra* note 28, at 32.

32 Hersch Lauterpacht, *The Development of International Law by the International Court* 392 (1958). *See also* Rudolf H. Geiger, "Customary International Law in the Jurisprudence of the International Court of Justice: A Critical Appraisal," in *From Bilateralism to Community Interest: Essays in Honour of Judge Bruno Simma* 673–94 (Ulrich Fastenrath et al. eds., 2011). For the suggestion that the International Court has played an important part in the "contemporary resurrection" of customary international law *see also* E. Jiménez de Aréchaga, "Custom," in *Change and Stability in International Law-Making* 1, 2–4 (Antonio Cassese & Joseph H.H. Weiler eds., 1988).

33 *Ahmadou Sadio Diallo (Republic of Guinea v. Democratic Republic of the Congo), Preliminary Objections, Judgment, I.C.J. Reports* 2007, p. 582, at pp. 614–16, paras. 87–93.

34 *Jurisdictional Immunities of the State (Germany* v. *Italy: Greece intervening), Judgment, I.C.J. Reports* 2012, p. 99, at pp. 122–45, paras. 54–107.

35 *Territorial and Maritime Dispute (Nicaragua v. Colombia), Judgment, I.C.J. Reports* 2012, p. 624, at pp. 666, 673, paras. 114, 137.

36 *See also* Theodor Meron, "Customary Humanitarian Law: From the Academy to the Courtroom," in *The Making of International Criminal Justice: A View from the Bench* 29–30 (2011).

37 *Id.* at 29. *See also* Theodor Meron, "The Continuing Role of Custom in the Formation of International Humanitarian Law," 90 Am. J. Int'l L. 238 (1996).

difficulty as the writers have anticipated,"[38] and a settled methodology for ascertaining the existence of a rule of customary international law as "a general practice accepted as law" is indeed observable in practice.[39] While many writings have contributed greatly to better understanding customary international law, others exemplified the concern that "especially wherever scholarly discussion starts to feed on itself, it loses touch with reality."[40] In any case, it ought to be clear that "given persistent recognition of custom as a legal institution, one might be equally skeptical of claims that custom *cannot* exist."[41]

A RENEWED VOTE OF CONFIDENCE: THE INTERNATIONAL LAW COMMISSION AND THE TOPIC "IDENTIFICATION OF CUSTOMARY INTERNATIONAL LAW"

None of this is to say that customary international law is without its problems – some real, some imagined, and mostly nothing new. It was partly these problems, but also a reflection of the importance of customary international law in many walks of life, that led the UN International Law Commission (ILC) to take up in 2012 its present topic on "Identification of customary international law."[42] There was felt to be a need to secure a common understanding of the process of identifying customary international law among all those who are called upon to apply it – not least given the considerable differences of approach amongst writers. Given its role under the UN Charter as well as its composition and working methods, the ILC is well placed to make a contribution to such an understanding that, it is hoped, would advance further the unity and stability of international law.

In deciding to take up the topic, the Commission was well aware of the difficulties inherent in an attempt to "codify the relatively flexible process by which rules of customary international law are formed."[43] But it was also aware of the need for authoritative guidance on how to identify rules of customary international law in concrete cases, especially at a time when "questions of customary international

[38] Brownlie, *supra* note 29, at 21 (referring in particular to the constituent element of *opinio juris*).

[39] *See also* Omri Sender & Michael Wood, "The Emergence of Customary International Law: Between Theory and Practice," in *Research Handbook on the Theory and Practice of International Law-Making* (C. Brölmann & Y. Radi eds., forthcoming).

[40] Hans W. Baade, "Codes of Conduct for Multinational Enterprises: An Introductory Survey," in *Legal Problems of Codes of Conduct for Multinational Enterprise* 407, 413 (Norbert Horn ed., 1980). *See also* Jörg Kammerhofer, "Orthodox Generalists and Political Activists in International Legal Scholarship," in *International Law in a Multipolar World* 138–157 (Matthew Happold ed., 2012).

[41] Swaine, *supra* note 14, at 627 (discussing rational choice scepticism with respect to customary international law claims).

[42] The topic was originally entitled "Formation and evidence of customary international law."

[43] "Report of the Study Group on the Future Work of the International Law Commission," in *The International Law Commission and the Future of International Law* 42 (M. R. Anderson et al. eds., 1998).

law increasingly fall to be dealt with by those who may not be international law specialists, such as those working in the domestic courts of many countries, those in government ministries other than Ministries for Foreign Affairs, and those working for non-governmental organizations."[44] Members of the Commission agreed that the outcome should be of an essentially practical nature, not seeking to resolve largely theoretical controversies. In the words of the Chinese member, "The importance of the topic lies with providing unified and clear guiding principles to international law practitioners, for them to identify and apply customary international law in their practice."[45]

The Commission took up the topic on the basis of a syllabus published in its 2011 Report to the UN General Assembly.[46] This was, of course, far from being the first occasion on which it had looked into customary international law; an earlier methodological foray into this field was mandated by Article 24 of its Statute ("Ways and means of making the evidence of customary international law more readily available"), leading the Commission to call on states already in 1950 to make evidence of their practice more accessible.[47] So far (2015), the Commission has made good progress with the topic, which was debated in 2012 and in each subsequent year, both in the Commission (on the basis of a note and three reports by the Special Rapporteur[48]) and within the Sixth (Legal) Committee of the UN General Assembly (on the basis of the relevant chapters in the Commission's annual reports[49]). One commentator has described the Commission's work as being "pursued with vigour, designed as a broad and inclusive process involving States, international organisations, and academia alongside the Commission, and of course it benefits from the ILA's prior work. All this suggests that it may become a successful exercise in clarification and consolidation."[50]

Towards the end of the Commission's 2015 session, its Drafting Committee provisionally adopted a set of sixteen draft conclusions concerning "the way in which the existence and content of rules of customary international law are to be determined".[51] Draft conclusion 2 ("Two constituent elements") makes clear

[44] Annex A to the Commission's 2011 report, UN document A/66/10, para. 3.

[45] Provisional summary record of the 3185th meeting of the Commission (A/CN.4/SR.3185).

[46] *See supra* note 44.

[47] *Yearbook of the International Law Commission* 1950, Vol. II, 367–74 (Report of the ILC for 1950, UN document A/1316, paras. 24–94, especially paras. 90–94); *see also* the 2013 memorandum by the UN Secretariat (A/CN.4/659).

[48] UN documents A/CN.4/653; A/CN.4/663; A/CN.4/672; A/CN.4/682.

[49] U.N. documents A/68/10 (2013), ch. VII; U.N. Doc. A/69/10 (2014), ch. X; U.N. Doc. A/70/10 (2015), ch. VI.

[50] Christian J. Tams, "Meta-Custom and the Court: A Study in Judicial Law-Making," 14 LAW AND PRACTICE OF INTERNATIONAL COURTS AND TRIBUNALS 51, 54 (2015).

[51] Draft conclusion 1 ("Scope"), as provisionally adopted by the Drafting Committee at the 2015 session. *See also* the 2015 report of the Chairman of the Drafting Committee, available at

that a single approach – one that remains loyal to the two elements found in the formula, now almost a century old, of "a general practice accepted as law" – ought to be applied in identifying any customary rule of international law. The draft conclusions further address, *inter alia*, the nature of each element and its possible manifestations; the process of assessing evidence for the two elements; the significance of particular evidence commonly resorted to in this process (treaties, resolutions, judicial decisions, writings); the "persistent objector rule"; and particular customary international law (customary rules applying only among certain states). It is expected that at its session in 2016 the Commission will adopt, on first reading, a full set of draft conclusions with commentaries, the latter forming an essential part of the overall product. A second reading could then take place in 2018, in light of the reactions from states and others during the intervening two years.

The ILC's debates on the topic to date have demonstrated a common appreciation of custom as a paramount source of international law. The Commission, whose members come from all principal legal systems of the world and "have very varied backgrounds, often as holders of eminent positions of academic authority, or as former diplomats or international officials with practical experience of international law,"[52] has so far also found no great difficulty in reaching broad consensus regarding the processes by which customary rules emerge and are to be ascertained. This, too, attests to customary international law's solid position in today's international legal system, and will hopefully provide it with renewed and greater coherence. As the Jordanian member remarked, "Even if the Commission merely describes the current state of the law, through adopting a set of conclusions, such conclusions will definitely advance the rule of law and a clear understanding of what is part of customary international law and what is not."[53] Most importantly, states reacting to the Commission's work have thus far expressed their general approval of the approach adopted and the output coming into being.

Work on the topic has shown, moreover, that several long-standing theoretical controversies related to customary international law have by now been put to rest. It is no longer contested, for example, that verbal acts, and not just physical conduct, may count as "practice." Nor is it seriously questioned that no particular duration of

http://legal.un.org/docs/index.asppath=../ilc/documentation/english/statements/2015_dc_chairman_statement_cil.pdf&lang=EF&referer=http://legal.un.org/ilc/, and in U.N.document A/CN.4/L.869.

52 Arthur Watts, "Codification and Progressive Development of International Law," in MAX PLANCK ENCYCLOPEDIA OF PUBLIC INTERNATIONAL LAW para. 12 (Rudiger Wolfrum ed., 2006).

53 Provisional summary record of the 3183rd meeting of the Commission (A/CN.4/SR.3183). Mr. Hmoud went on to say that "the complexities associated with the vagueness in determining the law undermine legal stability and certainty," and emphasized that the Commission's output could lead to "the avoidance of dispute and assist in reaching legal certainty that otherwise may only be reached through judicial pronouncements."

a certain practice is required for the emergence of a rule of custom (if the practice is general and accepted as law). The "persistent objector" rule has likewise met little opposition. While some theoretical questions may very well remain, that is not to be regretted: those engaged in the practice of law may benefit much from such debates, as did members of the ILC. The present volume, too, contains very valuable contributions to the doctrinal discussions.

CUSTOM'S BRIGHT FUTURE

Whatever sceptics may think, customary international law is very much still with us. It has withstood the test of time, as well as significant political and doctrinal challenges, and remains both useful and credible. Abundant international practice as well as the ILC's current work on the "Identification of customary international law" suggest that the once-fashionable notion that custom is in crisis, in decline, or even "dead" was never much more than an academic (or political) conceit: it had no basis in reality, no support among states, and no following among practitioners. It seems mainly to have been held by those who had little or no experience of foreign relations in practice, or who viewed anything "international" with knee-jerk suspicion. Even if theoretical difficulties endure, "it would be a mistake to under-rate the part that customary law plays, and will continue to play … merely because many problems of law – particularly of 'lawyers' law' – do not find in general international custom any clear and definitive answer."[54] This is particularly so given that to a large extent, it is precisely custom's flexibility that has afforded it its durability.

This is just as well, for if custom were truly in crisis, the stability of the international legal system as a whole would be endangered. Even if, admittedly, "it may not be the ultimate source of law, [custom] is still the most basic source of rules to govern the activities of States"[55] and "the principal construction material for general international law" (in the sense of its capability to generally bind all States).[56] Being a central means of bringing under one legal regime all members of a large and heterogeneous international community, and underpinning as it does the other sources of international law, not least treaties, customary international law is now more necessary and important than ever.

[54] Thirlway, *supra* note 4, at 145. *See also* M.H. Mendelson, "The Formation of Customary International Law," 272 *Recueil des Cours* 172–176 (1998).

[55] Thirlway, *supra* note 4, at 2. *See also supra* note 28.

[56] V. I. Kuznetsov & B. R. Tuzmukhamedov, *International Law: A Russian Introduction* 77 (William E. Butler ed., trans., 2009).

Bibliography

Akehurst, Michael, "Custom as a Source of International Law," 47 *Brit. Y.B. Int'l L.* 1 (1974–75).

Aust, A., *Modern Treaty Law and Practice* (3rd ed.) (Cambridge: Cambridge University Press 2013).

Baxter, R. R., "Multilateral Treaties as Evidence of Customary International Law," 41 *Brit. Y.B. Int'l L.* 275 (1965–66).

Bederman, David J., *Custom as a Source of Law* (2010).

Bradley, Curtis A. & Goldsmith, Jack L., "Customary International Law as Federal Common Law: A Critique of the Modern Position," 110 *Harv. L. Rev.* 815 (1997).

Bradley, Curtis A., & Gulati, Mitu, "Withdrawing from International Custom," 120 *Yale L.J.* 202, 213 (2010).

Byers, Michael, *Custom, Power and the Power of Rules* (1999).

Charney, Jonathan I., "The Persistent Objector Rule and the Development of Customary International Law," 56 *Brit. Y.B. Int'l L.* 1 (1985).

Corten, O., and Klein, P. (eds.), *The Vienna Conventions on the Law of Treaties: A Commentary*, Vol. I (Oxford: Oxford University Press 2011).

The Vienna Conventions on the Law of Treaties: A Commentary, vol. II (Oxford: Oxford University Press 2011).

Crawford, J., *Brownlie's Principles of Public International Law* (8th ed.) (Cambridge: Cambridge University Press 2012).

Crawford, J., Pellet, A., and Olleson (eds.), *The Law of International Responsibility* (Oxford: Oxford University Press 2010).

D'Amato, Anthony A., *The Concept of Custom in International Law* (1971).

Fitzmaurice, M., Ong, D. M., and Merkouris, P., (eds.) *Research Handbook on International Environmental Law* (Cheltenham: Edward Elgar 2010).

Freestone, D., Barnes, R, and Ong, D. (eds.), *The Law of the Sea, Progress and Prospects* (Oxford: Oxford University Press 2006).

Gaja, G., "*Jus Cogens* Beyond the Vienna Convention, " *Recueil des Cours* 1981.

Goldsmith, Jack L. & Posner, Eric A., *The Limits of International Law* (2005).

Guzman, Andrew T., "Saving Customary International Law," 27 *Mich. J. Int'l L.* 115 (2005).

Henckaerts, J.-M., "Study on Customary International Humanitarian Law: A Contribution to the understanding and Respect for the Rule of Law in Armed Conflict," *IRRC* vol. 87:857 (2005).

ICRC, Study on Customary International Law, retrieved from: http://www.icrc.org/customary-ihl/eng/docs/home.
Int'l Law Ass'n, Final Report of the Committee, *Statement of Principles Applicable to the Formation of General Customary International Law* (2000).
Jennings, R. and Watts, A. (eds.) *Oppenheim's International Law* (9th ed.).
Kammerhofer, Jörg, "Uncertainty in the Formal Sources of International Law: Customary International Law and Some of its Problems," 15 *Eur. J. Int'l L.* 523 (2004).
Kelly, J. Patrick, "The Twilight of Customary International Law," 40 *Va. J. Int'l L.* 449 (2000).
Kirgis, Frederic L., Jr., "Custom on a Sliding Scale," 81 *Am. J. Int'l L.* 146 (1987).
Kontorovich, Eugene, "Inefficient Customs in International Law," 48 *Wm. & Mary L. Rev.* 859 (2006).
Koskenniemi, Martii, *From Apology to Utopia: The Structure of International Legal Argument* (2005) (orig. 1989).
Kunz, Josef L., "The Nature of Customary International Law," 47 *Am. J. Int'l L.* 662 (1953).
Lepard, Brian D., *Customary International Law: A New Theory with Practical Applications* (2010).
Max Planck Encyclopedia on Public International Law, edited by Rüdiger Wolfrum, Oxford University Press, retrieved from: http://opil.ouplaw.com/home/epil.
Meijers, Herman, "How Is International Law Made? The Stages of Growth of International Law and the Use of its Customary Rules," 9 *Neth. Y.B. Int'l L.* 1 (1978).
Mendelson, Maurice H., "The Formation of Customary International Law," 272 *Recueil des Cours* 155 (1998).
Norman, George, & Trachtman, Joel P., "The Customary International Law Game," 99 *Am. J. Int'l L.* 541 (2005).
The Nature of Customary Law: Legal, Historical and Philosophical Perspectives (Amanda Perreau-Saussine & James Bernard Murphy eds., 2007).
Parry, Clive, *The Sources and Evidences of International Law* (1965).
Reinisch, A., *Standards of Investment Protection* (Oxford: Oxford University Press 2008).
Restatement (Third) of the Foreign Relations Law of the United States (1987).
Roberts, Anthea E., "Traditional and Modern Approaches to Customary International Law: A Reconciliation," 95 *Am. J. Int'l L.* 757 (2001).
Sands, P., and Peel, J., *Principles of International Environmental Law*, (3rd ed.) (Cambridge: Cambridge University Press 2012).
Scharf, Michael P., *Customary International Law in Times of Fundamental Change: Recognizing Grotian Moments* (2013)
Schutter, O. De, *International Human Rights Law* (Cambridge: Cambridge University Press 2010).
Shaw, M., *International Law* (6th ed.) (Cambridge: Cambridge University Press 2008).
Shawkat, A., Bhuiyan, J. H., Chowdury, T., and Techera, E. (eds.) *Routledge Handbook of International Environmental Law* (New York: Routledge 2013).
Stern, B., La Succession d'États, *Recueil des Cours* 1996.
Thirlway, H. W. A., *International Customary Law and Codification* (1972).
Thirlway, Hugh, *The Sources of International Law* (2014).
van Hoof, J. H., *Rethinking the Sources of International Law* (1983).
Verdier, Pierre-Hughes, & Voeten, Erik, "Precedent, Compliance, and Change in Customary International Law: An Explanatory Theory," 108 *Am. J. Int'l L.* 389 (2014).

Villiger, M., "The 1969 Vienna Convention on the Law of Treaties – 40 Years Later," *Recueil des Cours* 2009.
Villiger, M. E., *Commentary on the 1969 Vienna Convention on the Law of Treaties* (Boston: Leiden 2009).
Walden, Raphael M., "The Subjective Element in the Formation of Customary International Law," 12 *Isr. L. Rev.* 344 (1977).
Wolfke, Karol, *Custom in Present International Law* (2d ed. 1993).

Index

Abbott, Kenneth and Duncan Snidal, 343
Alvarez, Jose, 220, 222
Anglo-Norwegian Fisheries case, 294
Arrest Warrant case, 54, 71
aspirational norms, 2, 80, 87, 91, 124, 128, 129–34, 145, 286–88, 349–50
Aust, Anthony, 335

Balduinus, Jacobus, 23–24
Banco Nacional de Cuba v. Sabbatino, 210
Barrett, Scott, 175–78
Bartolus, 13
Basel Committee, 226
Beaumanoir, Philippe de, 18
Bederman, David, 11, 44
bilateral investment treaties (BITs), 207, 211–12, 214, 223
Biodiversity Convention, 339
Blackstone, 44, 65
Boyle, Alan and Christine Chinkin, 352, 356
Bradley, Curtis, 1, 5, 34, 83, 85, 92, 123, 124, 134, 292, 323
Brierly, James, 48

Caron, David, 313
Cassese, Antonio, 362
centralized adjudication, tension with customary law, 59
changes to CIL; *see* evolution of CIL
Charter, United Nations, 3, 71, 76, 80, 82, 91, 225, 268, 277, 307, 309, 366
Cheng, Bin, 57, 75, 222
China, 118, 119, 216, 217, 225, 227, 260, 263, 272
Chinkin, Christine, 350, 354, 355
Choi, Stephen and Mitu Gulati, 6, 86, 117, 229
Chorzow Factory case, 209
chronological paradox, 2, 15, 40–41, 56–57, 67, 69–70, 90, 100–1, 122
climate change; *see* global warming
common law, nature of, 22, 31, 49
common law account of CIL, 34–61, 147
common pool resources, 254
compliance with CIL, 53, 106, 155, 176–79, 184, 191–92, 287, 290, 293, 296–99
compliance pull, 298–99
consent to be bound, 24–26, 103, 186–87, 276, 281–83, 291–92
consequentialist considerations in identifying CIL, 54–55
Convention on Climate Change, United Nations Framework, 253, 339
Convention Against Torture, 287
cooperation, CIL's ability to promote, 174, 178–79, 185–86, 189–90, 254, 258–59, 262–64, 275–76, 280–82
coordination problems, 67, 70, 73, 83, 84, 106, 175
Copenhagen Accord, 264
Crawford, James, 199, 318
crisis, whether CIL is in, 4, 5, 10, 63, 94, 333, 344, 360
customization, amenability of CIL to, 198, 206

D'Amato, Anthony, 345, 355
D'Aspremont, Jean, 336
decentralized nature of CIL, 152, 173, 189, 345
decline of custom, 30, 172–204, 230–52, 275
democracy and CIL, 11, 38, 69, 72, 187, 226
Descamps, Baron, 44, 85
Digest of Justinian, 13, 14
diplomatic immunity, 51, 66–67
discretion in determining CIL, 188–89
double counting, 53, 97, 98 n.4, 151

dual-element definition of CIL; *see* two-element definition of CIL
Draft Articles on State Responsibility, 165, 200–2, 234, 268, 305, 313, 318, 347, 352
Duke Law School
  Duke-Geneva Institute on Transnational Law, 4
  project on "Custom and Law", 3
Dumoulin, Charles, 25–26
dynamic process, CIL as, 81–82

Emissions Trading System (ETS), 272–73
empirical assessment of CIL application, 2, 36, 86, 117–47
enquête par turbe procedure, 28–29
Erdemovic case, 237
ethical principles and CIL, 81
European Court of Human Rights, 72 n.21, 162, 200
European Court of Justice, 189, 272
European Union, 186, 263
evidentiary uncertainties relating to CIL, 1, 35, 53, 121, 123, 288, 290
evolution of CIL, 38, 56, 70–71
Exclusive Economic Zone, 224
expropriation of property, 208–9

"fake custom", 91, 152
fiat theory of CIL development, 42
fisheries jurisdiction, 270–71
fragmentation of international law, 59, 350
Franck, Thomas, 284
free riding, 178, 259, 262
Friedmann, Wolfgang, 172

game theory, 175–80, 192, 220–21, 262
General Assembly resolutions, United Nations, 74, 75, 80, 180, 266
general principles category of ICJ jurisdiction, 60–61
general principles versus specific rules, 50
Geneva Conventions, 153, 268
Gény, François, 12–13, 14, 43–44
Germany v. Italy (ICJ), 3, 60, 365
global public goods, 174–75, 177–78, 190, 193, 221–22, 253–74
global warming, 175, 180, 253–61
Goldsmith, Jack and Eric Posner, 221, 262
Great Britain, 3, 42, 119, 201, 270, 302 n.72, 318
Grnchalla-Wesierski, Tadensz, 353
"Grotian Moments", 58, 73, 180, 230
Grotius, Hugo, 32, 172
Guzman, Andrew, 39, 86, 344, 345
Guzman, Andrew and Jerome Hsiang, 8, 275

Hakimi, Monica, 6, 76, 77, 81, 148, 266, 323, 325
Havana Charter, 209
Helfer, Laurence and Ingrid Wuerth, 222, 223, 348, 349
Helfer, Laurence and Timothy Meyer, 9, 305
Henkin, Louis, 345
Herik, Larissa van den, 8, 69, 76, 90, 230
Hernandez, Concepcion Escobar, 319–20, 328
historical development of custom as source of law, 12
historical school of jurisprudence, 45–47
Hudson, Manley, effort to define CIL, 46–48
Hull, Cordell, 209
Human Rights Council, United Nations, 91, 241
human rights norms, 68, 71, 107–15, 139
humanitarian intervention, 3, 79–80, 90–91

ICSID, 215–16, 224 n.94
immunity of officials, 319–21, 328
immunity of states, 51, 56, 215
India, 119, 191, 272, 294, 327
Indonesian law, 19
informal international lawmaking,
  comparison with CIL, 347–48
  complement to CIL, 350–59
  concept of, 334–37
  effect on CIL, 344–47
  generally, 332–59
  rise of, 332–44
"instant custom", 22, 57, 62
International Committee of the Red Cross
  creation of, 335
  study on customary international humanitarian law, 3, 159–62, 169, 199, 325
International Court of Justice; *see also* Statute of the International Court of Justice
  applications of CIL, 3, 71, 117–47, 164, 267–69
  endorsement of two-element definition of CIL, 35
  failure to apply two-element definition in practice, 36, 75, 168
International Covenant on Civil and Political Rights (ICCPR), 287
International Covenant on Economic, Social and Cultural Rights (ICESCR), 286, 287
International Criminal Court (ICC), 63, 230–52
international criminal law and CIL, 230–52

International Criminal Tribunal for Former Yugoslavia (ICTY), 69, 76, 156, 163–66, 168, 169, 170, 200, 230–52
International Criminal Tribunal for Rwanda (ICTR), 239, 248
international criminal tribunals generally, 36, 63, 76, 230–52
international humanitarian law, 76–77, 148–71
international investment law, 196, 205
International Law Association, statement of principles on CIL, 2–3, 74, 151
International Law Commission
  as agent of General Assembly, 310–14
  changing nature of work, 305–31
  decline of treaty recommendations, 315–17
  development of draft conventions, 305
  generally, 165, 197, 214, 305–31
  gridlock in General Assembly, effect on, 307, 313
  influence of General Assembly on, 310
  mandate of, 307–8
  methodology as constraint on, 306–7, 323–26
  project on CIL, 3, 40–41, 66, 126, 153, 194, 329, 346, 360–69
  shift to nonbinding codifications, 314–22
International Monetary Fund, 300

Jennings, Robert, 60
Johnston, Jason, 183
Jurisdictional Immunities of the State; *see Germany v. Italy*
jus cogens norms, 59, 60, 68, 103, 107–15, 277–79, 301

Kadens, Emily, 5, 11, 65
Kant, Immanuel, 95
Kaplow, Louis, 183
Kelly, J. Patrick, 188
Kelsen, Hans, 42, 64
Keohane, Robert, 185, 341 n.55
Kirgis, Frederick, 54–55, 74
Kirton, John and Michael Trebilcock, 350
Klabbers, Jan, 333
Kolodkin, Roman, 319, 320, 328
Kontorovich, Eugene, 55
Koskenniemi, Martii, 265
Kratochwil, Friedrich, 221
Kupreskic case, 76, 237, 238 n.37
Kyoto Protocol, 253–54

Lauterpacht, Hersch, 314
Law of the Sea Convention, 224
Law of the Sea Tribunal, 186, 365
lawmaking, CIL identification as form of, 19, 50, 58, 155–56, 265–66
legality concerns about CIL, 246
legitimacy of CIL, 38, 41, 98–101, 104–7, 152, 186–89, 226, 284
Lepard, Brian, 5, 62, 105 n.20, 157, 292
lex lata and *lex ferenda*, 58, 77, 80, 83, 84, 87
Lim, Chin, 7, 200, 205
limitations of CIL, 173–89
Lipson, Charles, 341
Lotus case, 51–52, 267
Lubanga case, 249

Maggi, Giovanni, and Massiono Morelli, 190
Maine, Henry, 203
Malynes, Gerard, 31
Mann, F.A., 207–8
maritime custom, 17–18
Martens clause, 237, 240, 263 n.24
McNair, Lord, 207
medieval jurists, 11, 21
memory of custom, 16–17, 22
Mendelson, Maurice, 157
Meron, Theodor, 365
Meyer, Timothy, 225
minimum standard of treatment, 207
mistake theory of CIL development, 41
Montreal Protocol, 253, 339
moral account of CIL, 95–116, 265
most favored nation clauses, 207

natural law, 31, 45, 109
New International Economic Order, 227
Nicaragua case, 75, 101, 167–68, 194, 268, 355
Nicaragua v. Colombia, 365
nonconsensual rules, 277
nongovernmental organizations (NGOs), 39, 81, 87, 102, 164, 204, 336, 340
nonstate actors, 1, 76–77, 81, 102, 149, 163
North American Free Trade Agreement (NAFTA), 210, 212
North Sea Continental Shelf case, 75, 86, 267
Norwegian Shipowners' Arbitration, 209
nuclear proliferation, 37, 124, 129, 152
nuclear weapons, use of, 73
Nuclear Weapons case, 101, 365

objective values, 105–6
objectivity of finding CIL, 148
odious debts, 117–21

old-style versus new-style CIL; *see* traditional versus modern approaches to CIL
opinio juris
  nature of, 1, 3, 12, 37–38, 63, 66, 96, 291–92
  reputation and, 292
  sense of legal obligation, 6, 35, 42, 45, 53, 56, 99, 291
  state practice as evidence of, 97–98
  theories of CIL focused on, 79
  variants of, 98–100

Parry, Clive, 364
Paust, Jordan, 76
Pauwelyn, Joost, 334, 336–37
Permanent Court of International Justice, 126
persistent objector doctrine, 59, 103, 107
Petersen, Niels, 8, 89, 253
Pistoria, Cynus de, 27
pollution, 55, 70, 175, 256, 271
practical difficulties in ascertaining CIL, 35–36
pragmatism, 95
precedential effect of international decisions, 59
pre-legal societies, 16
principal-agent theory, 309–10
prisoners' dilemma, 67, 70, 73, 84, 89, 176
procedural evidence of CIL, 128, 130–31, 145
Pulp Mills on the River Uruguay case, 181

rational-choice approaches to CIL, 39, 280, 289, 291–96
Raz, Joseph, 105
Rebuffus, Petrus, 25
reciprocity, 185
regional custom, 280, 293, 294–96, 302–3, 348
Restatement (Third) of Foreign Relations Law, 40, 114, 124, 288
Revigny, Jacques de, 22
Right of Passage case, 294
Rio Declaration, 336, 351, 352
Rivier, Alphonse, 46
Roberts, Anthea, 157, 349
Roman law, 13–14, 20–21, 172
Rome Statute of the International Criminal Court
  Article 10, 240–42
  Article 21, 242–45
  generally 71, 159, 230, 233, 236, 239
rules versus standards, 181–83
Russia, 118, 121, 208, 272, 319

Saluka Investments BV v. Czech Republic, 213
Sands, Philippe, 217
Savigny, Carl von, 45–46
Schooner Exchange, 51
Scharf, Michael, 73, 180
S.D. Myers v. Government of Canada, 207
Security Council, United Nations, 80, 104, 121, 251, 278–79, 282 n.26, 300, 338
Sender, Omri and Michael Wood, 10, 360
Shelton, Dinah, 352, 354
Siemens v. Argentina, 213
Simma, Bruno and Philip Alston, 265
sliding scale approach to identifying CIL, 54, 74, 80, 92
"soft law", 1–2, 37, 68–69, 202, 205, 226, 266, 276–77, 297, 335–36
Soviet legal theorists, 361
space law, 77
Special Tribunal for Lebanon, 233
state involvement in international lawmaking, levels of, 278–79
state practice; *see also* chronological paradox
  duration, 66, 75
  evidence of *opinio juris*, 293
  generality of description of, 49
  national court decisions as, 328 n.82
  requirement of, 64–65, 96
  uncertainties surrounding, 1, 35–36, 293
state preferences and CIL, 49, 54, 57–58, 92, 98–99
state responsibility, 200, 234–35, 318, 352–53
state succession, 201
Statute of the International Court of Justice
  Article 38 of 43, 44, 47, 56, 84, 96, 288, 333, 334, 357, 358, 365
Statute of the Permanent Court of International Justice, 44, 65
Stockholm Declaration, 336, 352
Suárez, Francisco, 31, 172

Tadic case, 76, 156–57, 158–59, 165, 232
Tasioulas, John, 6, 41, 75, 76, 91, 92, 95, 123 n.16
taxonomy of CIL, 198
Tesón, Fernando, 77, 152
Thirlway, Hugh, 37
toolbox of international law, 280
torture, 68, 78, 85, 114, 124, 268, 287–88
Trachtman, Joel, 7, 37, 63, 172, 208, 219–29, 269, 337, 346, 347
Trail Smelter arbitration, 55

traditional versus modern approaches to CIL, 39, 54, 62, 124, 261
transition theories of CIL development, 42
transparency, value of, 100
treaties
  advantages over CIL, 1, 37, 173–74, 206
  displacing CIL, 194–96
  as evidence of CIL, 1, 53–54, 78, 124, 132–35, 151, 196–97, 268
  proliferation of, 1, 37, 63, 363
  slowdown in creation of, 337
Trubeck, David, 350
Truman Proclamation, 52, 58, 73, 223, 279
two-element definition of CIL, 1–2, 34–35, 44–46, 50, 62, 66, 96–97, 118, 148, 288–89, 344

United Kingdom; *see* Great Britain
United States, 49 n.70, 52, 55, 58 n.114, 118, 119, 120, 154, 191, 207, 209, 222, 223, 227, 254, 260, 263, 271, 272, 318, 339
Universal Declaration of Human Rights, 71, 78, 80, 82, 354
universal jurisdiction, 321–22, 329
usefulness of CIL 37, 189, 275, 346–47, 360

Verdross, Alfred, 361
Vienna Convention for the Protection of the Ozone Layer, 253
Vienna Convention on Diplomatic Relations, 301
Vienna Convention on the Law of Treaties, 68, 213, 268, 277
Vienna Conventions on the Succession of States, 117, 201
volksgeist, concept of, 45
voluntarist approach, 103–4
von Mehren, Arthur, 210

Weil, Prosper, 356
welfare of states, 285
Wessel, Ramses, 334
Wittgenstein, Ludwig, 96
Wolfke, Karol, 66
Wood, Sir Michael, 66, 194, 250, 346
World Charter for Nature, 354
World Trade Organization (WTO), 186, 191, 217–18, 335, 338
Wouters, Jan and Linda Hamid, 9, 332
Wuerth, Ingrid, 4

Yackee, Jason, 212

AF848626

A FAMILY GUIDE TO LIVING WELL WITH DEMENTIA

By Liz Leach Murphy and Jayna Patel

Routledge
Taylor & Francis Group
LONDON AND NEW YORK

First published in 2023 by Critical Publishing Ltd.

Published 2025 by Routledge
4 Park Square, Milton Park, Abingdon, Oxon OX14 4RN
605 Third Avenue, New York, NY 10017

Routledge is an imprint of the Taylor & Francis Group, an informa business

Copyright © 2023 Liz Leach Murphy and Jayna Patel

All rights reserved. No part of this book may be reprinted or reproduced or utilised in any form or by any electronic, mechanical, or other means, now known or hereafter invented, including photocopying and recording, or in any information storage or retrieval system, without permission in writing from the publishers.

Trademark notice: Product or corporate names may be trademarks or registered trademarks, and are used only for identification and explanation without intent to infringe.

British Library Cataloguing in Publication Data
A CIP record for this book is available from the British Library

ISBN: 9781041053903 (hbk)
ISBN: 9781915713063 (pbk)
ISBN: 9781041053910 (ebk)

The rights of Liz Leach Murphy and Jayna Patel to be identified as the Authors of this work have been asserted by their in accordance with the Copyright, Design and Patents Act 1988.

Cover design by Out of House
Text design by Out of House
Typeset by Deanta Global Publishing Services, Chennai, India

DOI: 10.4324/9781041053910

Contents

Meet the authors *iv*
Glossary *v*
Introduction *vi*

1 Innovation within dementia care 1
2 Personalisation: What does it mean for people with dementia? 27
3 The dementia journey 46
4 Great support in the later stages of dementia 65
5 The shape of care and support 83
6 Ways to plan 106
7 Understanding behaviour and communication 146
8 Developing and embedding person-centred cultures 174
Conclusion 182
Appendix: Useful organisations 183
Resources 187

Bibliography *188*
Index *195*

Meet the authors

Liz Leach Murphy has worked within health and social care for the past 25 years, and within this time she has been committed to improving the experiences of those accessing the care and support they need, which has involved working alongside people living with dementia and their families. She has been trained in and delivered person-centred planning and support brokerage, both of which are approaches to enable people to direct their own care and support in a manner that is rooted in the community rather than services. Liz's work and passion led her to set up Imagineer Development UK CIC in 2010, a social enterprise based in the north of England with a national reach, which specialises in self-direction.

With a degree in journalism, **Jayna Patel** enjoyed using her investigation skills to help Liz create a practical and valuable resource for informal carers supporting someone living with dementia. As a Content Writer for health and wellbeing app, Evergreen Life, she understood the importance of balancing the need for the book to be thoroughly researched, with the requirement for it to written accessibly. Jayna had more of a marketing background than a health and social care one, having previously worked as an account manager at a web design and digital marketing agency for six years, meaning she was ideally placed to ensure the content was conveyed in an easy-to follow, jargon-free way – as well as tidy up all the references!

Glossary

DCT Dementia Care Triad: the relationship between the person living with dementia, their family carers, and paid professionals.

DH The Department of Health leads, shapes and funds health and care in England. Their aim is to make sure people have the support, care and treatment they need, with the compassion, respect and dignity they deserve.

DP Direct Payment is a method a person can choose to manage their personal budget or personal health budget. The budget is deposited into a bank account that is managed by the individual and/or their representatives.

GP A general practitioner is a doctor based in the community who treats patients with minor or chronic illnesses and refers those with serious conditions to a hospital.

HP A health professional or health care provider is an individual who provides preventive, curative, promotional or rehabilitative health care services in a systematic way to people, families or communities.

IB Indicative budget is presented to an individual as an amount to work with when developing the support plan. As the name suggests, this is only intended as an indication and is open for discussion and negotiation. The actual budget is agreed on completion of the person's care and support plan.

ICB Integrated Care Board, a statutory organisation bringing the NHS together locally to improve population health and establish shared strategic priorities within the NHS. When ICBs were legally established, clinical commissioning groups (CCGs) were abolished.

ISF Individual service fund is a mechanism for controlling a person's personal budget or personal health budget within an organisation. It is managed specifically on behalf of the person, and it can be spent flexibly to achieve the best possible outcomes for a person.

PB Personal budget is allocated to the person by a social work team in a local authority. It can be utilised to meet a person's assessed social care need.

PHB Personal health budget is allocated to the person by a clinical team within an ICB to enable a person to meet their assessed health care needs.

Introduction

Over the past 20 years I have worked within health and social care, with a broad range of people in a variety of circumstances: living in their own home, supported living, residential care and hospital settings, and attending daytime support and respite provision. I have worked with people diagnosed with dementia and, with Down's syndrome and dementia, in applying the person centred and Self-Directed Support principles.

In this time I have provided direct care and support, managed teams working in care and support settings, delivered district-wide projects to implement person-centred approaches and personalisation and been involved in the development of national strategy. I have worked alongside individuals who required care and support in their lives, their families and those who work in the provision of health and social care, including direct support workers, care workers, team leaders, managers, social workers, directors of adult social care, clinicians, leaders in the delivery of personal budgets and personal health budgets, commissioners and members of the community.

It is this accumulation of learning and experience that has brought me to writing this book, based on a true belief that the provision of care and support can be positive and can be done in a manner that upholds the person's identified choices and wishes. It is possible as I have been part of making this possible and I have seen it work well.

I hope you enjoy reading the book. It is not meant to be read from cover to cover although, if this is how you would like to digest the information, please go ahead. It begins with an understanding of how developments have made a difference since the 1990s, moves on to where we are now in the delivery of care and support and, finally, gives a strong sense of direction for the future – all supported with some really useful tools and approaches.

This book is absolutely directed at dementia care and support although I am sure you will appreciate that many of the ideas and approaches are relevant to anyone wanting to achieve a good life.

Liz Leach-Murphy

1 Innovation within dementia care

The dementia picture

There are around 900,000 people living with dementia in the UK. The condition affects one in six people over 80 and one in 14 people over the age of 65 live with dementia in the UK (NHS, 2021).

People are living longer which is increasing the number of people who are diagnosed with dementia. It is estimated that by 2025, the number of people with dementia in the UK will exceed one million (NHS, 2020).

It is likely that dementia will touch the lives of every citizen in the UK either directly or via the people they are related or connected to. At this moment in time, there are 700,000 carers of a person living with dementia in the UK (Dementia Carers Count, n.d.).

The content of this book is intended to be a source of support and guidance for those caring for a person with dementia.

This chapter covers:

- **A definition of personalisation, providing an initial introduction to the approach in practice**
- **An introduction to Self-Directed Support as a important element of personalisation**
- **The Dementia Care Triad – recognising the importance of the triadic relationship between the person living with dementia, you and paid professionals**

- **Person-centred care and Tom Kitwood, the first major influence in shifting dementia care practice**
- **Dementia Care Mapping – an additional approach developed to map the level of person-centred care provided within a setting**
- **The Values of Inclusion as the foundation for working in a person centred way**
- **Person-centred planning – an introduction to a number of the tools and approaches to planning that places the person in the driving seat of their life and their plan**
- **The VIPs framework – a further and more recent development that supports staff within care settings to recognise the needs of the person being supported and the needs of the setting as a whole. This framework can also be applied within the home.**
- **The Senses Framework which sees quality care arising out of mutual and reciprocal relationships between those that form the Dementia Care Triad**
- **Enriched Care Planning – an assessment and planning package which is based on person-centred care.**
- **Dementia Friendly Communities – an overview of the national and international developments of communities that are inclusive and welcoming**
- **Age Friendly Communities – broadening out the topic of communities to recognise that to be inclusive means thinking about everyone.**

Getting the support right for the person

You may have heard the term 'personalisation' used in the context of the provision of health and social care. This term is also used in lots of other industries and generally refers to providing a product or a service that the customer is specifically looking for; an individualised and bespoke approach. This is also what it refers to in the provision of health and social care.

The term 'personalisation' was first acknowledged in 2007 within a ministerial concordant called 'Putting People First', which was the first document to suggest that the whole approach to the provision of health and social needed to change in order to have a greater focus on the person receiving the care and support. The Department of Health described this as: *'every person who receives support, whether provided by statutory services or funded by themselves, will have choice and control over the shape of that support in all care settings'* (Department of Health, 2010).

Personalisation focuses on providing support and care in a way that's unique to each person. It makes sure that they, and those close to them, can exercise full choice and control over their life choices and experience.

Self-Directed Support

Self-Directed Support is a term that has been adopted in how personalisation is achieved. It means that the person with dementia, their family and carers have the opportunity to direct the care and support; this can include who provides the care and support, when it is provided and how. It can also look at other ways of being cared for or supported which don't rely on paid people, exploring alternative options, such as changing and adapting the home environment, and considering the equipment and technology and also the community-based options available with the overall aim of making sure the arrangements put in place work well for both the person living with dementia and their carers.

Self-Directed Support is a complementary term to personalisation and describes a situation when the person is central to making the decisions about how to use the resources available to them to meet their care and support needs. This is regardless of whether the person is self-funding or accessing public monies to meet those care and support needs.

In summary, Self-Directed Support provides the opportunity for the person with dementia, their family and carers to think about what *'good care and support'* would look like and where this can be found.

Personalised Services

Formal care and support services include home care, agency support, day services, supporting living, residential care, and nursing homes. These formal services need to be able to demonstrate how they are responding to the wants and needs of each person they support, being flexible in their approach to make sure that the people they provide care and support to can maintain their identity, individuality, routines and relationships, ensuring they are exercising choice and control.

A Personalised service would structure the way that care and support is provided around the preferences of the person. A service that's not Personalised would follow routines that are designed to suit the smooth running of the service rather than the preferences of each individual cared for by the service. For example, in a care home, each resident would be supported to get up at a time that fits into a morning routine that makes sure everyone is ready to have breakfast at the same time.

Dementia Care Triad

The Dementia Care Triad (Adams and Gardiner, 2005) is commonly referenced in dementia care practice and it highlights the importance of the relationship between the person with dementia, informal carers, and those from the health and social care profession, with the focus on building trust within these relationships. Health and social care professionals are trained to recognise the importance of the Dementia Care Triad within their professional practice, and how a relationship between all three parties that's based on trust and respect benefits the person being supported. When in conversation about how to achieve good care for the person, the individual providing the care and support on an unpaid basis is just as entitled to have their perspective listened to as professional and paid carers are. Each member of the Dementia Care Triad is just as important and their opinions, identity and experiences just as valid, so both have a place in the conversation. It is only through co-operative and open conversations between each of these three parties that personalisation can be achieved for the person being cared for and supported. With that in mind, anyone in the Dementia Care Triad can instigate a conversation about the person's care and support.

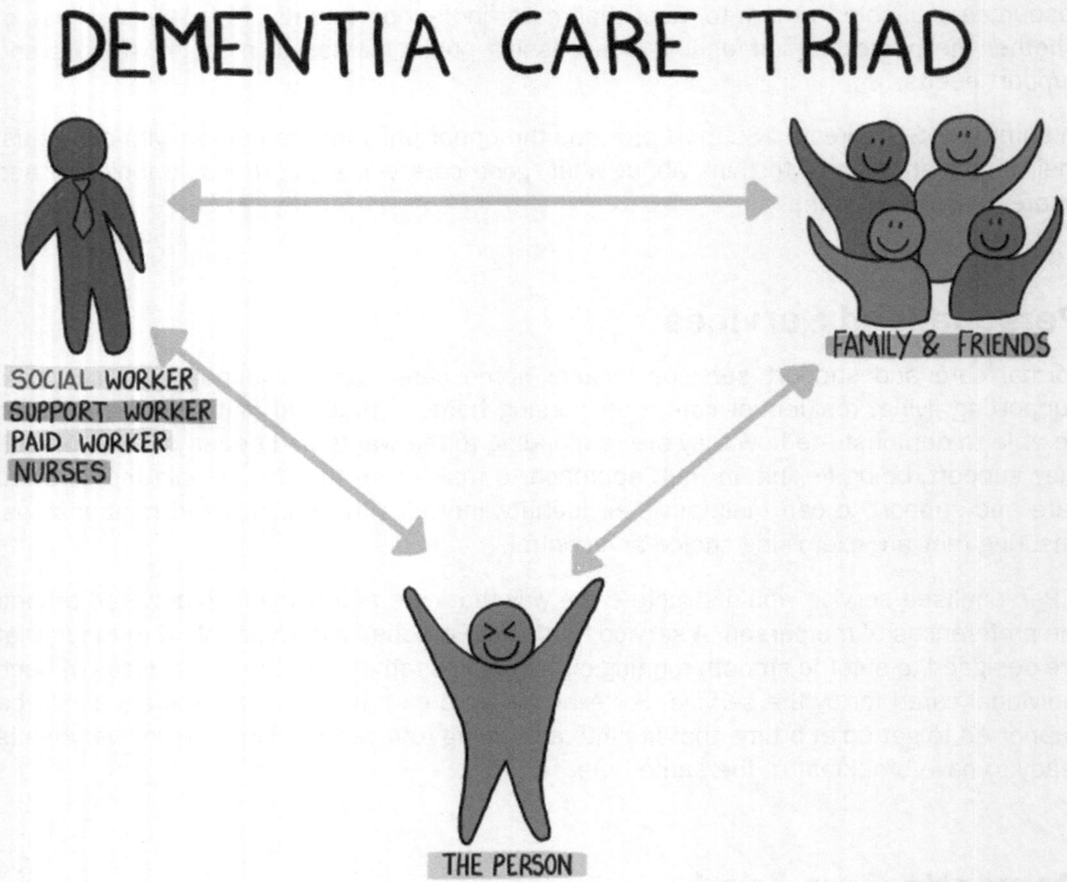

Complementary approaches to personalisation and person-centred care

In this section, we'll explore person-centred care in greater detail with an introduction to the approaches within dementia that have helped to steer the provision of care and support to have a greater focus and person centred response to the person being cared for and their carers / family members.

Approach 1: Person-centred care

Person-centred care was introduced by Tom Kitwood in the 1980s and recognises all aspects of the person and enhances people's sense of self. This includes respecting the person's culture, faith, religious practices, values and lifestyle choices. Person-centred care has a focus on the provision of care within a care setting for example, residential care, nursing homes and hospitals.

As a result of the introduction of person-centred care within these settings, practices have changed. According to Health Innovation Network South London (2016) this has given rise to an increased quality of care and a decreased level of agitation and distress in the people cared for in that setting (Plos Medicine, 2018).

The central ideas within person-centred care state that: the person living with dementia should come first, people with dementia have personhood, and that this should be maintained, despite the continuing presence of dementia. Kitwood defines personhood as: *'a standing or status that is bestowed upon one human being, by others, in the context of relationship and social being'.* (Kitwood, 1997, p.8) Kitwood states that a person having 'personhood' *'implies recognition, respect and trust'* (Kitwood, 1997, p.8).

Kitwood outlines numerous ways staff within service settings can interact with people who have dementia that allows for better interpersonal engagement to promote personhood. These ways of interacting with people living with dementia can be applied to the person being cared for at home/privately too.

Approach 2: Person-centred planning

Being Person Centred

Despite having a similar name, 'person-centred planning' has a different focus from 'person-centred care':

The focus of **person-centred care** is enhancing the personhood and subjective well-being of the person living with dementia within a dementia care setting.

The focus within **person-centred planning** is to recognise what a good life looks like from the perspective of the person being cared for and to learn how they wish to be supported or cared for and what they want their day-to-day life to look and feel like.

Being person centred asks those involved in the planning process, which could include people from the Dementia Care Triad, to relate to each other in the following ways:

- as individuals;
- as equal partners;
- respectfully;
- without assumptions.

Person-centred planning enables those involved to understand:

- the person's story; their history;
- what's important to the person being cared for; this can include the people they'd like to see, the places they like to spend time in and the routines they wish to follow;
- the person's identity, how they like to see themselves and the roles they wish to fulfil;
- how the person communicates and makes decisions;
- how to provide support and care for the person in a way that's meaningful;
- the person's aspirations and fears for the future;
- what's working well in their life;
- areas of the person's life that are not working well and that are a source of upset and frustration for them.

Person-centred approaches also support the person being cared for and supported to:

- have control over their life;
- have their voice heard and responded to;
- identify ways to remain connected to the community;
- draw together a network of support;
- maintain relationships as they wish to;
- create a picture of what good support and care would look like;
- create a sense of direction for their future.

If the person being cared for is being supported in or living in a service setting, such as a day centre or a residential home, a person-centred approach within this setting would support the delivery of a Personalised service by:

- creating a focus on the person being cared for and an opportunity for the their voice to be heard;
- encouraging the conversations that support people to identify and share what they'd like to achieve and how they would like to be supported to work towards these goals;
- talking about what the person would like to change about the way they are supported or cared for;
- opportunities for the person, their carers and the other people around them to consider how these changes can be achieved.

Person-centred planning

Person-centred planning is a range of methods and approaches that place the person in the driving seat of their lives and at the forefront of any planning about their life. Duffy (2004, pp.7–13) identified that families would like to have more control in the planning and delivery of care, support and the services provided.

This approach to planning involves the people who are most important to the person living with dementia. This includes carers, family members, friends, neighbours or other members of the community alongside the professionals or paid people who are involved in the provision of the person's care and support, especially those who the person has a good relationship with or who can bring useful knowledge and insight to support the planning process.

The plan could focus on the immediate future so that the day-to-day experience of the person and their quality of life improves. It can also focus on the long term plans regarding where they want to live and what they'd like to achieve.

Person-centred planning is built on the principles and Values of Inclusion (Inclusion, 2019).

Values of Inclusion

The Values of Inclusion are listed below with a description of what each value means. These Values have been adopted by families who are seeking inclusion and continued connections with communities and friends. A carer may find these useful when considering how to continue to support the person they care for, and these values may be helpful in day-to-day caring life.

Everyone belongs within their community

Everyone is born into and is part of a community, it is only later that they are removed and placed within a segregated service setting. It shouldn't be assumed that, due to a person's diagnosis, they'd be better off somewhere other than the community that they are familiar with. Everyone has a need to be a part of society, to be present with other members of the community and no one is too difficult, too old, too poor or too disabled.

Everyone can communicate

Communication is not reliant on the use of words; people communicate regularly by many other means, such as facial gestures and body language. When there's a challenge to understand one another this is a shared challenge, and they are important to make a commitment to learn about how people can get better at communicating, recognising that communication is the foundation stone of any relationship and conversations about how the person wishes to be supported, and equally, about the support needed as a carer.

Everyone has the right to determine their life experience

Irrelevant of the person's stage in life or their circumstances, they should be able to determine their day-to-day life experience as well as their future wishes.

A community is stronger together

A community functions better if its citizens are all able to be actively involved and contribute. No one present in the community has to meet a certain level or pass a test before they are able to belong. Rather, communities are stronger if people are accepted for who they are; with their unique gifts and qualities they have available to share.

Everyone needs support

Everyone needs support, from help with understanding the detailed letter that has just been received, and support to wash the car, to assistance to set up the internet connection. Sometimes some people need more support than others and the level of support people need may change over time.

Everyone can learn

As human beings all people grow, change and make mistakes and are all able to learn. This learning doesn't refer to formal based learning found within education systems. Instead, this learning happens when people spend time together and talk about what is happening in their local area, learning things about one another, their culture, experiences or areas of knowledge.

Everyone can contribute

Contributions to the community don't have to be grand for them to have meaning and value; it could be the person that picks the litter up on the way home or knocks on the neighbour's door to check in on them. Each person's contributions can be recognised and valued.

Everyone can have aspirations and dreams

It is important to recognise that people have aspirations and dreams. Listening to these can help to deepen understanding of what's important to the person, what motivates them and makes them enthusiastic. This then leads to identifying opportunities to connect and form meaningful relationships around the person and in their community.

People are the most qualified experts in their own lives

The person is not just the person who is seen today; their story of their lived experience provides an insight into what matters to them. Bringing a story together that can be shared with other people can help build understanding in those who'll be alongside the person and also for those who'll be paid to provide care and support.

Anything is possible

The way a person is supported and cared for is often determined by what's available in the local care market, which includes agencies and support providers. This, however, doesn't need to be the only option available to people. Listening to the person's ideas, including their dreams and aspirations, can lead to exploring how to make this possible beyond what's currently available within the local service provision, looking into what's available from within the community or bringing people together; friends, family and paid professionals to find a way to make it happen.

Never say never

No one has the right to say 'no' to a person's ideas, dreams and aspirations. An open dialogue, explorative conversations, involving the right people and having a commitment to the person to find a way to move towards their ideas, dreams and aspirations on the basis of being positive and finding what is possible.

The Values of Inclusion have informed the development of person-centred approaches that place the person at the centre of planning and decision-making. The Values were first written in the early 1980s, and, although they have evolved since then, they've largely remained the same. This demonstrates how relevant and important these statements are in building understanding of inclusion.

Person-centred planning can be applied in many different scenarios, for instance: when the person is wanting to plan for their future, addressing day-to-day routines, determine what good support looks like for them, and supporting the person to maintain or develop connections within the community. Person-centred planning achieves real and positive outcomes for people and their families.

In summary, these ideas help people to recognise the person's relationships, and connections within their community, and enable them to learn about their history, skills, interests, passions, and their ideas of what a good life would look like. Any support provided to the person, either unpaid or paid, would be enriched by understanding the person better.

The information introduced so far also empowers those from within the Dementia Care Triad (the person, their family, carers and paid professionals) to learn and understand which relationships and resources are significant to the person and helps to initiate discussions about how to maintain these or develop them further.

Person-centred planning tools

Person-centred planning comprises a range of planning tools which are described below. These include:

- PATH;
- MAP;
- Essential Lifestyle Planning;
- Community Mapping;
- Relationship Mapping.

Developing a plan supports the person to be in the driving seat of their own plan where they may consider how they want to live their life, what quality of life means to them and what they'd like to achieve. The plan is developed at a pace that makes sense to them. When the plan is complete, the content can be shared with other people to build understanding and relationships. It can be shared in many ways and this can be determined by how the person would prefer to share their plan with others. People have used a range of ways of achieving this, such as short video recordings where the person or the people who care for them talk to camera about the content of the plan and what has been learnt, drawings using clear images and computer presentations that can be shared online.

Here's an explanation of the planning approaches:

PATH stands for Planning Alternative Tomorrows with Hope. This planning approach develops a clear appreciation of a good future from the person's perspective and helps to identify the steps that need to be taken to get there, including who else needs to be involved.

PATH

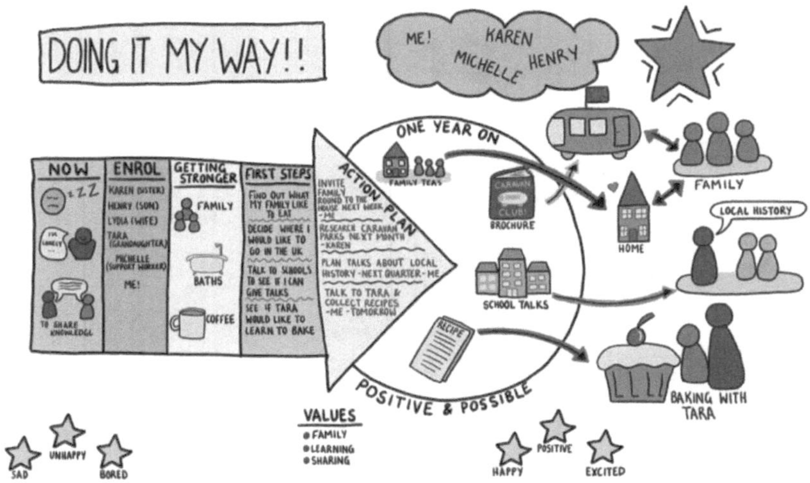

MAP stands for Making Action Plans. This planning approach develops an appreciation of the person's life story, the people who are important in their lives, what the person would like to see happening in their life, and what they'd like to avoid, with a clear action plan of how to move forward.

MAP

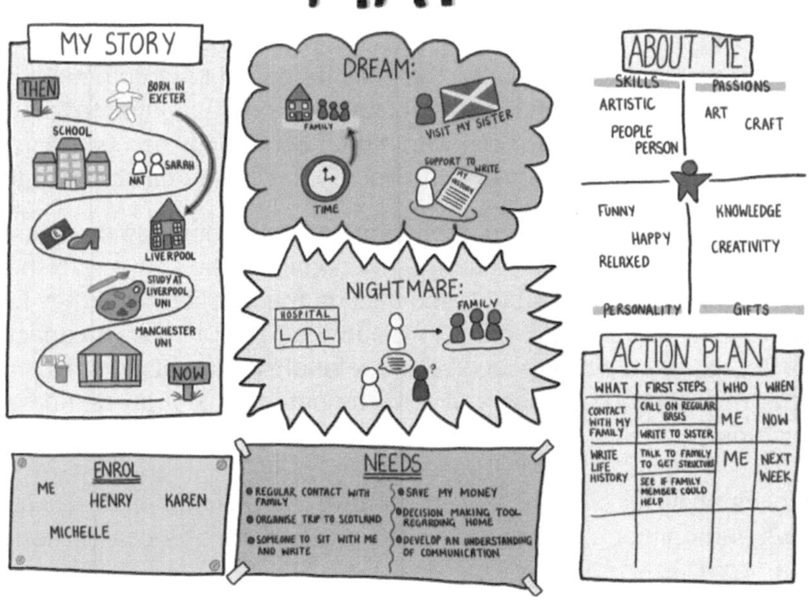

Essential Lifestyle Planning is a planning approach that develops an understanding about what's important to the person in their day-to-day life, and outlines their important routines and preferences.

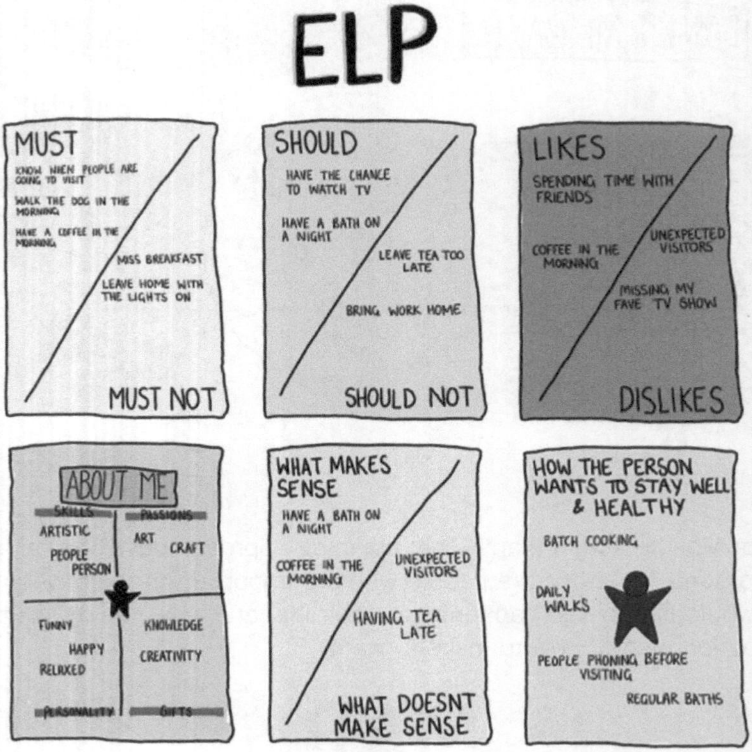

Community Mapping is a method to recognise and record the resources and connections within the community; this may be based on a community of interest or an area where the person lives. Through capturing this information on a map, the connections they already have within their community can be appreciated and also the potential for new connections and opportunities can be recognised.

Relationship Mapping recognises and documents the relationships that the person has in their life; this would include the relationships they have with their carers, close family and friends. It would also include the relationships they have within their community and with people who are in their lives in a paid capacity. A Relationship Map helps to recognise the relationships that are important to them that may need protecting and also the potential to grow and develop new relationships.

These approaches allow for creative and imaginative, as well as practical, ways to plan, in which the views and opinions of the person living with dementia can be heard, and their choices known.

Person-centred planning encourages people to come together to listen and develop and share in their understanding, experiences and learning. This gathering of people creates a network around the person where the collective resources, skills, knowledge, extended connections, and time can be pooled together to support the implementation of the person-centred plan.

Approach 3: The VIPS framework by Dawn Brooker

Dawn Brooker, an early student of Tom Kitwood who developed person-centred care, continues to contribute towards the development and progress of the approach and is responsible for developing the VIPS framework, published in 2004.

The VIPs framework helps those paid to support people living with dementia to consider their needs as an individual which can then influence practice at an organisational level. Following this consideration, this approach can be applied to settings, such as home care, supported living, residential care, and nursing care, plus family and unpaid carers can apply them in their caring role too.

The initials 'VIPS' refers to four key elements within person-centred care:

Valuing people with dementia

Individualised care that treats the person with dementia as a complex person

Personal perspective of the person with dementia

Social environment that supports the person's psychological needs

For each of these key elements there are six key indicators.

V = Valuing People

Every person has value regardless of age or cognitive status.

Time to share knowledge and understanding

Identify skills

Recognise passions

Explore personal connections

I = Individualised Care

Using an approach that promotes the uniqueness of the individual.

Care planning	Individual preferences
Regular reviews	Life history
Activity and occupation	

P = Personal Perspectives

Acknowledge and understand life from the perspective of the person with dementia.

Communication with the person	Physical health
Empathy and acceptable risk	Perspectives
Behaviours of concern as communication	Advocacy

S = Social Environment

Providing a social environment designed to meet psychological needs of the individual.

Inclusion validation	Everyday places
Respect	Warmth
Enabling	Community

(Brooker, 2004).

At the time of writing, the VIPS framework continues to be used in the measurement of the quality of care provision for people living with dementia (Skills for Health and Skills for Care, 2018).

If the person is receiving care or support from a care provider or a service setting, the VIPS framework is a useful reference for discussions about how well they are receiving that care and areas that could be better. It is a good idea to have the VIPS framework to hand ahead of conversations with the care provider, to be able to consider what to contribute to the conversations with professionals and the team providing the support.

VIPS framework and personalisation

The VIPS framework provides further aspects of achieving a quality of life and personalisation in the provision of care and support and broadens what Persona Centred Care considers as important to the individual.

The heading 'Social Environment' within the VIPS framework encourages carers to consider the importance of community connections and inclusion when measuring the quality of care and support a person is receiving. The rights of the person are addressed with the recognition of Advocacy as part of the 'Personal Perspectives' section.

Approach 4: Relationship Centred Care by Mike Nolan

Person-centred care emphasised the importance of creating positive engagement and interaction between the person working within a dementia care setting and the residents or patients within that setting, which can be equally applied when receiving care and support in the home environment.

Relationship Centred Care broadens this thinking beyond the immediate relationship between the caregiver and the care receiver and considers the relationships that exist around the person living with dementia.

This Relationship Centred approach to care states that relationships exist at many levels, including among patients, their families, staff from all disciplines, and the wider community (Nolan et al., 2006).

The quality of the care and support offered to people with dementia depends on the quality of relationships between each of these levels. This is similar to the point raised regarding the Dementia Care Triad, where the quality of support shared between people with dementia, informal carers and health and social care professionals depends on good mutual and reciprocal communication, as well as trusted and respectful relationships.

Nolan et al. (2002) suggests person-centred care portrays communication as a process that occurs from the caregiver to the person living with dementia, which could prevent or overlook the need for the person to have the opportunity to express their preferences and views, negatively impacting on their level of choice and control. In contrast, Nolan et al. (2002) believes Relationship Centred Care highlights the importance of open two-way or multi-way communication.

Through the development and implementation of Relationship Centred Care, it is encouraging that people with dementia are being seen as having agency. 'Agency', is the ability of people to undertake *'actions, activities, decisions and behaviours that represent some measure of meaningful choice'* (Deacon and Mann, 1999, pp.413-435).

Supporting people with dementia to express their views and display agency should be a key consideration in all their daily interactions, including those with the people involved in their care or support. That way, people living with dementia can experience a better quality of life, where they have the opportunity to have control and choice.

People living with dementia, over time, will experience a diminished cognitive ability. It is, however, possible to capture a person's wishes and preferences to help inform the provision of care and support through all stages of dementia. In Chapter 6 a range of approaches will be introduced to show how it is possible to support people with dementia, to share their views and opinions.

In Nolan's work on Relationships Centred Care, he worked with a team at Sheffield Hallam University to develop 'The Senses Framework', which sees quality care arising out of mutual and reciprocal relationships between those that form the Dementia Care Triad. These relationships lead to each party developing subjective 'senses' comprising a sense of security, continuity, belonging, purpose, achievement and significance.

Sense of security

To feel safe and free from threat, to receive care and support for all areas where this is needed in the person's life and to be treated as a competent person who's understood and that their opinions are heard.

For older people: attention to essential psychological needs, to feel safe and free from threat, harm, pain and discomfort. To receive complete and sensitive care. To have the opportunity to share what's important to them and to be reassured when needed.

For staff: to feel free from physical threat, rebuke or censure. To have secure conditions of employment. To have the emotional demands of work recognised. To work within a supportive culture where they have the opportunity to learn about the support people require.

For family carers: to feel confident about their knowledge and ability to provide good care without detriment to their own personal well-being. To have adequate support networks and timely help, when required. To be able to relinquish care when appropriate. To be fully involved in all conversations and to be treated respectfully.

Sense of continuity

To understand the past and how that's reflected in day-to-day life. To have a sense of long term relationships. To experience positivity in all aspects of the person's life and for everyone to know what's expected of them.

For older people: recognising the importance of their own personal history. To have the opportunity to share what has happened in the past and the meaningful aspects of their history. Skilled use of knowledge of the past to help understand the present and future. To recognise what's significant for life now and into the future. Seamless, consistent care delivered by people they know well.

For staff: positive experience of work with older people from an early stage of their career. Exposure to good role models and environments of care. Expectations and standards of care communicated clearly and consistently, so people know what's expected of them and how they can do well in their role. To have the opportunity to hear the person's story and identify how this is significant for the person's life now. To ensure that what's learnt from this is reflected in how the support is provided.

For family carers: to maintain shared pleasures/pursuits with the person they care for. To be able to provide competent standards of care, whether delivered by self or others. To ensure that personal standards of care are maintained by others. To have the opportunity to share in the life story and the aspects of the story that are significant to them. To have the opportunity to bring the important aspects of what makes a good life into the person's day-to-day life now.

Sense of belonging

To feel part of the community and to have meaningful relationships within it with valued roles. To not feel alone.

For older people: to have the opportunity to have conversations about the places and the people that are important to them and how they'd like to stay involved. Opportunities to maintain and/or create meaningful relationships where they have the chance to continue to contribute.

For staff: to feel part of a team where they are recognised and valued for their contribution. To belong to a community of practitioners.

For family carers: to be able to maintain/improve valued relationships. To be able to confide in people they have trusted relationships with, to feel that they are 'not in this alone'. To have the opportunity to talk about the people and places that have meaning to them and how the person wishes to remain involved.

Sense of purpose

To be able to have choice and to take part in purposeful activities; this reflects goals and aspirations. For care to be reciprocal and maintain dignity, integrity, personhood and well-being.

For older people: opportunities to engage in purposeful activity. To be able to identify and pursue goals and challenges. To exercise choice.

For staff: to have a sense of therapeutic direction, a clear set of goals to aspire to that makes sense to the person they are caring for. To maintain the dignity and integrity, well-being and personhood of the person they are supporting.

For family carers: to experience constructive and reciprocal care. Family members share with each other what goals and aspirations each person has for themselves and the family member being cared for. This is an opportunity to notice where people share similarities and also to discuss differences.

Sense of achievement

To have a sense of progress towards achieving goals and to use abilities and skills to the full whilst developing new ones. To have a sense that '*we've done our best*'.

For older people: opportunities to meet meaningful and valued goals. To feel satisfied with one's efforts and to have the opportunity to celebrate. To make a recognised and valued contribution. To make progress towards personal and therapeutic goals as appropriate.

For staff: to be able to provide good care. To feel satisfied with one's efforts. To make progress towards therapeutic and personal goals. To use their skills and abilities to the full. To recognise how each person has played a part in ensuring that achievements and goals are reached and having the chance to share in the celebrations.

For family carers: to feel that they've provided the best possible care. To know they've done their best. To meet challenges successfully and with the right support. To have the opportunity and to be able to develop new skills and abilities. Creating the time to talk about the positive things that have happened for themselves, for the person they care for, and for others involved in the person's life. Recognising the difference these achievements make.

Sense of significance

To feel valued and appreciated and that efforts made have a significance.

For older people: to feel recognised and valued as a person of worth, that one's actions and existence are of importance, that they 'matter'.

For staff: to feel that working alongside older people is valued and important, that their work and efforts 'matter'. Recognising that their function in a person's life has importance in defining their day-to-day experience. Keeping motivated ensures the quality of care and support provided and that enjoyment of the role can continue to be experienced.

For family carers: to feel that one's caring efforts are valued and appreciated, to experience an enhanced sense of self.

Thinking about the Dementia Care Triad, introduced earlier in this chapter, the Senses Framework shows how each person involved can experience interactions with one another. It highlights the different groups of people in a person's life, and recognises the fact that the needs of each – professionals, the person living with dementia and family carers – need to be respected and considered in order to nurture healthy relationships.

Approach 5: Enriched Care Plan

This approach has been developed by Dawn Brooker, Hazel May and Paul Edwards (2009) and it is based on person-centred care. An Enriched Care Plan is there to support people living with dementia and their carers. The Plan looks at the social and psychological aspects of life for the person at the time it is completed. A completed Enriched Care Plan aims to identify how to meet the person's care and support needs.

The five components of an Enriched Care Plan are explained below.

Life history

Life Story: Knowing something about the person's life can help others to know why they are behaving or reacting to current events in a particular way. The Enriched Care Plan package includes a profile relating to 'Life Story', which gains important information about their previous years and the present time. It also includes information about their lifestyle and future wishes.

Lifestyle: This section of the Enriched Care Plan gathers information about what's important for the person regarding their lifestyle choices. This includes their food and drink, clothes, daily living activities, work-like activities, relaxing, social activities, personal attachment figures or objects and spirituality preferences.

Future wishes: This part captures details about what the person wants in relation to the treatments, health interventions and support they receive which includes end-of-life and funeral planning. End-of-life support and planning are covered in more detail in Chapters 3 and 4.

Personality

The person living with dementia will experience their condition in a unique way. Their life experience will be influenced by a number of factors. This includes their personality, their environment and their interactions with other people around them. When completing an Enriched Care Plan, assessing the personality of the person is based on whether the person is:

Sensitive/Nervous

A more sensitive or nervous person is less likely to be emotionally stable, whereas a person who's confident and secure is going to have better control over their emotions.

Introvert/Extrovert

An extroverted person is more likely to be outgoing, active, and assertive whereas an introverted person tends to be less sociable and shy.

Inventive/Curious

Inventive people are open to new experiences and tend to have a more liberal outlook. In contrast, people who are 'preservers' tend to be more conservative, conventional and prefer familiarity.

Friendly/Compassionate

This section is linked to the traits of friendliness, which include compassion, caring, warmth and kindness in contrast to competitiveness, jealousy and spitefulness.

Efficient/Organised

The more conscientious a person, the more competent, dutiful, orderly, responsible and thorough they are. People with this trait are more likely to control their impulses whereas people who are easy going are more likely to be impulsive.

Health

People living with dementia may not be able to give a good account of their physical and mental health issues, which may lead to preventable or treatable conditions being missed. The health profile looks at different areas of the person's health, including weight, height, medication use, diet and fluid intake. Any issues regarding safety, emotional and mental health are also addressed. Memory issues are discussed and recorded.

Neurological impairment

Capacity for Doing – While people living with dementia are cognitively impaired, they can, to varying degrees, engage with the world, whether engagement comprises higher level activities, such as driving a car, or lower level activities, like watching the news. People with dementia are at particular risk of disengaging from the world as a person's dependency on other people to initiate engagement increases, or they may begin to lose the skills, coordination and language needed to fulfil activities they once did, which can impair their well-being. The 'Capacity for Doing' profile allows others to know what the person with dementia wants to do to remain fully engaged with the world.

Life at the moment

This profile assesses the quality of life that the person living with dementia currently experiences. Here, the focus is on the person's psychological needs for comfort, occupation, attachment, identity and inclusion. Quality of life is profoundly affected by interactions with other people and the 'feel' of the person's social environment. The profiling process provides a structure that helps reveal and quantify how satisfied the person currently is with their life and the extent to which their psychological needs are being met. The Enriched Care Plan then identifies any clear actions that should be taken to improve their quality of life.

Enriched Care Planning is ongoing and takes place throughout the duration of care. It comprises five-stages: profiling, identifying needs, documenting needs, implementing the plan and reviewing, and takes a structured approach to considering a broad range of issues that influence the person's quality of life and day-to-day experience. The Enriched Care Plan helps to identify areas of a person's support and care that need further consideration for improvement and creates an action plan to implement the ideas.

Evolution in thinking

The development of these five approaches in dementia care demonstrates a progression in what's deemed as important in the provision of care and support for people living with dementia, and places a much greater emphasis on quality of life and the person's voice being heard.

It was once considered that the development of information and planning to enable person-centred care was the responsibility of those paid to work within dementia care. Personalisation has begun to significantly shift this emphasis, recognising that often people, alongside their families, friends and those paid to support them, are best working in collaboration to plan, gather and record information. This shifts the emphasis of expertise to the person with lived experience and those in their network of support

Life in the community

To achieve a good quality of life for people with dementia, consideration must be given to how people can be supported to continue to live in their community, if this is what they choose. Understanding the opportunities that exist within the person's community and their existing connections within it is important.

It may help to ask: who and what makes up the person's community now and what else can be explored to identify other potential opportunities to get involved and ways to be supported that have not been considered previously?

This works well when looking for community-based associations or groups that align with the person's interests, hobbies and passions.

The conversations to be held and the connections to be developed can become clearer when seeking the answers to the following question:

'*What does the person living with dementia and their family need and want to ensure that they can keep living a quality life and a life of their choosing?*'

The assets that exist in communities include the:

- availability of skills from those within the community;
- the time the local people have to share with others;
- people's passion and commitment to making a difference in their local area;
- knowledge of the local area;
- relationships and connections that exist within the community;
- local services and businesses;
- small community groups, associations and initiatives;
- charities and voluntary organisations;
- statutory provision, such as the police force, education, and local MPs.

As a carer, it can be helpful to try and develop an understanding of what the local community can offer. This can lead to a greater scope of ideas and solutions to help achieve the outcomes that are right for the person, their carers and their family.

A community-focused approach supports the person living with dementia and members of their family to maintain aspects of their life that mean the most to them and encourages everyone in the Dementia Care Triad to have a focus on what's possible, not just what's currently available through service-based solutions.

Why not take this opportunity to think about what is in the local community that is available for the person?

Participating in social networks and recognising citizenship

Continuing to be a part of the local community and surrounded by people who are known or connected to the person can help to maintain their personhood and their sense of belonging, security, safety, purpose, achievement and significance, as covered by The Senses Framework above.

There are two prevailing models for communities that have been specifically developed in response to Dementia and ageing: Dementia Friendly Communities and Age Friendly Communities.

Dementia Friendly Communities

Dementia Friendly Communities are communities whose residents understand dementia. They are places where those living with dementia continue to be recognised as members of the community, are welcome, safe and continue to fulfil meaningful roles.

Reflecting the diverse range of communities in society: a person's; local community, local faith group or community of interest. Communities as a geographical area can include cities, towns and villages. Since the launch of Dementia Friendly Communities, a range of areas have received formal recognition as being a Dementia Friendly Community.

The idea of Dementia Friendly Communities was first introduced in the 2010 to 2015 Government Policy: Dementia (Department of Health and Social Care, 2015). The Prime Minister of the time, David Cameron, issued a national challenge at the Dementia 2012 conference committing to:

- invest further in dementia research;
- improve the quality of dementia care;
- increase professional and public understanding of dementia;
- implement initiatives to achieve Dementia Friendly Communities.

> *'A dementia friendly community can be defined as: a place or culture in which people with dementia and their carers are empowered, supported and included in society, understand their rights and recognise their full potential'*
>
> (Alzheimer's Disease International, 2016, p.10)

Dementia Friendly Communities is an international initiative with a commitment in many countries to engage in the development of a society that understands dementia and supports people living with the condition to continue living an active and full life.

A key component to achieving Dementia Friendly Communities is initiatives like the Dementia Friends Programme, which aims to increase awareness and understanding of dementia across all levels of communities.

The shape and form Dementia Friendly Communities take can vary greatly. For example, banks in the UK have appointed dementia champions who are trained to spot the signs of people living with dementia who may be victims of financial abuse. These banks implemented recommendations from people living with dementia to make banking easier and more secure, including creating a photo ID bank card and a chip to store information to discreetly indicate to the cashier that the individual may need extra support or a side room for a private conversation.

Another example is the Northumberland Dementia Forum which prompted a dementia awareness training package for local bus drivers, as part of their Certificate of Professional Competency (Local Government Association, 2015). This means the many older adults living with dementia who use public transport in rural and urban areas across the UK benefit from having a well-trained and sensitive point of contact in their bus drivers.

Over in Ireland, Dementia Aware Donegal's Dementia Friendly Community programme has 5 project components, each of which cover several key aspects of Dementia Friendly Communities (Alzheimer's Disease International, 2017). These are:

- 'Be a Sport for Dementia,' which encourages people living with dementia to continue to participate in local sports clubs;
- 'Ordinary Man, Extraordinary Memories,' a play intended to raise awareness and understanding of dementia;
- 'You've Got a Friend in Me,' a bid to establish Dementia Champions to recruit and train new Dementia Friends;
- 'Hearth and Home,' an initiative designed to provide support for those living with dementia in the community;
- 'Faith, Hope and Charity,' a development of best practices for clergy to better support members of the congregation living with dementia and their families.

What all these forward-thinking initiatives have in common is that they aim to make every day life as respectful, dignified and inclusive as possible for those living with dementia, regardless of the setting. This demonstrates how Dementia Friendly Communities transcend one physical place and instead become a values-based approach practised by and applied in different communities (Alzheimer's Disease International, n.d.).

For a list of towns, cities and other areas in the UK that have been recognised as Dementia Friendly, visit the website of Dementia Friends (Dementia Friends, n.d.) and for information regarding Dementia Friendly Communities in Scotland please visit Alzheimer's Scotland (Alzheimer's Scotland, n.d.).

The increase in the number of local areas, towns and organisations receiving the Dementia Friendly status demonstrates that there is a strong commitment to improving communities for people living with dementia and that those people and organisations that make up a community are prepared to invest resources to engage with the process.

While the term 'Dementia Friendly Community' is used to describe a village, town or city, it is appropriate to also think of the support and care a person receives as being Dementia Friendly. Hospitals and care homes can be committed to being Dementia Friendly as can other service settings.

When assessing if a service the person plans to use is Dementia Friendly, here are the things to expect:

- The physical structures and environment is accessible and welcoming. There's equipment and resources available to create accessible and adaptable accommodation.
- Visits from family and friends are encouraged and welcomed.
- Time is allocated to ensure 'people time' between the people providing support and those they are supporting.
- There is active engagement between the organisation/service, the people supported, their families, friends and the local community.
- The community around the person living with dementia and those providing care and support have an insight into their skills, abilities, interests and passions.
- It is a place where people are recognised for the life that they have led and the people they are today.
- There's a good range of activities and opportunities available and people living with dementia are involved in planning and deciding which activities take place.
- There's a culture that recognises relationships and associations outside the immediate care/support environment and things in place that facilitate the continuation of these.

The concept of Dementia Friendly Communities improves the experience of living with dementia and identifies how to build a stronger community for everyone within it. A community that can work to protect and support the people within it is a better place for everyone.

Age Friendly Communities

An Age Friendly Community is a place where people of all ages can lead healthy and active lives.

Dementia Friendly Communities have, in effect, been a progression of work previously committed to developing Age Friendly Communities, as supported by the Centre for Ageing Better (Centre for Ageing Better, n.d.). An Age Friendly Community is one that's respectful and enables everyone to actively participate, regardless of age or background.

The communities that are age friendly support people to continue to live in their homes, take part in the activities that are important to them and contribute to their community for as long as possible.

The Centre for Ageing Better is the organisation that supports communities to achieve Age Friendly status. There's a growing UK network of Age Friendly Communities covering Scotland, Northern Ireland, Wales and England. A list of the recognised communities can be found on the Centre for Ageing Better website. (Centre for Ageing Better, n.d).

The World Health Organisation worked with older people to develop the framework for Age Friendly Communities and it is built on the evidence of what supports active and healthy ageing in a community. This led to eight domains of an Age Friendly Community which can be found on the Centre for Ageing Better website, referenced at the end of this book.

Dementia Friendly/Age Friendly Communities and personalisation

The values that underpin Dementia Friendly and Age Friendly Communities, such as inclusiveness, respect and accessibility, are complementary to personalisation, especially in the sense that:

- progress should be shaped around the needs and opinions of older people and people living with dementia, together with input from carers;
- older people and people living with dementia can engage in existing community activities, rather than only specialised provision in a service setting;
- older people and people living with dementia can remain in their homes and within their communities.

Living a good life

The life of a person living with dementia doesn't have to be determined by the services available to them or by the opinions of others.

The approaches introduced in this chapter place the focus on hearing the person's voice about what would work best for them. The ideas covered encourage everyone involved to consider and actively listen to the person and to recognise that some of the ways of achieving what they and their carers want, can be found within the local community.

In this context, the choice of the care and support available moves away from the narrow view of existing service provision to thinking about what's possible in the community. Embracing this approach involves some creative thinking and working together with everyone involved to collectively find new ideas and initiatives where the person is in the driving seat of their lives, and support is based on the person's preferences, interests, identity and connections.

Key lessons learnt

- The development in the provision of care and support called personalisation and Self-Directed Support places a greater emphasis on listening to the person making sure the care and support provided is based on their preferences.
- The Dementia Care Triad encourages people living with dementia, families, carers and paid professionals to work alongside each other as equals.
- Person-centred care, approaches and planning are some of the practical ways to record the plan of the person.
- Relationships are important in achieving quality of life with a focus on building and maintaining trusted and respected relationships where people are equal and supportive.
- The Values of Inclusion can be applied when considering person centred approaches and planning with a focus on remaining connected to the community.
- Good and meaningful connections can be found in the community. Some communities are recognised as Dementia Friendly and Age Friendly Communities.

Questions to aid reflection

1) What do people need to know to make sure that you and the person you care for feel in control and understood?
2) From what has been introduced in this chapter, what's the most important and useful aspect of planning that stands out to you?
3) What purpose would this planning serve for you or the person you care for?
4) How can you move forward with putting these plans in place?
5) Who would be helpful in supporting the development of these plans?

2 Personalisation

What does it mean for people with dementia?

This chapter covers:

- **An introduction to Direct Payments and personal budgets**
- **Support Planning**
- **The options available to manage a budget**
- **What barriers people living with dementia face, as outlined by the World Health Organisation**
- **Overcoming barriers**
- **The Mental Capacity Act and people's right to have control and choice in decision making**
- **Co-Production and its impacts**
- **The difference between 'Outcomes' and 'Outputs'.**

The development of Direct Payments

Direct Payments are a payment made available from the Local Authority or the Integrated Care Board which can then be used to purchase the support, care and services the person wants to meet their care and support needs.

These payments are made directly to the person either into a bank account in their name or in the name of someone who is helping them to manage their Direct Payment.

People also have the option to use the services of an organisation or accountancy firm to manage their Direct Payment; this is called a managed account.

Personal budgets

In 2005, the 'Improving the life chances of disabled people Final Report' talked about Personalised support and Direct Payments (Prime Minister's Strategy Unit, 2005) and the 2007 document titled 'Putting People First' outlined the Government's commitment to offer personal budgets to those who are eligible for social care. (HM Government, 2007)

A personal budget is different to a Direct Payment as it is calculated at the point when the person has been assessed as being eligible for health or social care support. There are a range of ways that a personal budget can be accessed which will be explained later in the chapter, with a Direct Payment as mentioned above, being only one of these options.

Direct Payments under personal budgets

Older people have been accessing Direct Payments across the UK since 2000 and personal budgets since 2005; in this time stories of positive experiences of accessing Direct Payments and personal budgets have emerged.

The Personal Budgets Outcomes Evaluation Tool (POET) was developed by In Control and has been utilised across the social care sector. Here are some key findings:

- The majority of older people said personal budgets had a positive impact on their lives
- 70 per cent or more of older people reported personal budgets have resulted in a good level of dignity within the support they receive, experiencing quality of life and feeling safe and secure
- 60 per cent to 69 per cent of older people reported having access to information to help with making decisions about the support they receive and being happy where they live
- 50 per cent to 59 per cent of older people reported a positive impact of personal budgets on their relationships and everyday activity such as work, leisure and volunteering
- The majority of older people reported a positive impact of personal budgets in regard to their health

(In Control, 2017)

How to access a Direct Payment or a personal budget

Firstly, a referral would need to be made to the local social work team who will then complete an assessment to determine if the person is eligible for social care support. The referral can be made by the person themselves, a family member or their GP.

If the person who has been assessed is deemed as eligible, they may be told the level of budget available to them as an indication of the budget they could receive for their ongoing support as outlined in their Support Plan (see below). As this is only an indication at this stage, this is often called the 'indicative budget'. This budget will be made available to the person once their Support Plan is completed and it has been agreed by the social work team. This is then called the personal budget. Personal budgets for carers are explored further in Chapter 4, whilst personal budget for people living with dementia are covered in more depth in Chapter 5.

With the knowledge of the indicative budget, the person can then develop their Support Plan, which outlines what a good life looks like for them.

Support Plans

The Support Plan demonstrates how the person wants to be supported. It can include what could happen within their family, friend and community networks, and how they can make use of the budget available to them.

The Support Plan is developed with the person's full involvement and with them taking the lead in the content, alongside people they choose to help them. This could include family members, friends, support workers, carers, advocates, and professionals involved in their care and support, such as their social worker.

A Support Plan describes:

- who they are and what is important to them;
- what they wish to change or achieve;
- how they want to be supported;
- how their support will be managed;
- how they will stay in control;
- how the Support Plan will be put into practice;
- how the person would want to manage the budget available to them – the options available are explained below.

The budget section of the Support Plan clearly shows the level of budget available to the person and how this will be used. Creative use of the budget is possible as long as the intended use will reach the best outcomes for them and meet their needs as identified in their social care assessment. The budget could be used to buy equipment and adaptations to improve their ability to get around their home, to purchase items to be able to pursue interests and hobbies or to be able to stay in communication with friends and family, to employ a support/care worker, or to attend activities or events that interest them.

A Support Plan provides the person with the opportunity to consider all the options available and what would work best for them, placing them in the driving seat of their plan.

The personal budget is to be used flexibly to support the person to achieve the outcomes they have set in their Support Plan. The only requirements of use are that the budget:

- is being used to meet the needs and outcomes as identified in their social care assessment;
- isn't intended to be used to purchase or do anything illegal;
- will not put them or anyone close to them at risk;
- is not intended to purchase a service that is usually free at the point of delivery, especially in the case of health care provision.

Once the Support Plan is complete, it is sent to the person's social worker for discussion and then approved by a member/s of their social work team. After the budget is agreed, it will be made available to the person to help make their plan happen.

How can a personal budget be managed?

Within the Support Plan, the person would also decide how they'd like to manage their personal budget. The person has a number of options including:

- **Individual Service Fund** (ISF) – (available to those receiving funding through Social Care). The budget is managed by an organisation on the person's behalf in an account which can be only used for the purposes of meeting the needs of the person. This can be spent flexibly to achieve the best possible outcomes for them. There would be clear communication with the organisation to ensure everything is working well for the person and the budget is being spent appropriately. Changes can be made to how the budget is spent and these changes can be discussed and agreed between the person and the organisation holding the ISF. The authority to make these decisions is passed from the Local Authority to the person and the organisation with a clear agreement in place on how this will be managed. A percentage of the budget or an overall charge from the budget is utilised by the organisation to cover the cost of administering the budget.
- **Direct Payment** – The budget is paid directly into a bank account managed by the person or their representatives / nominees. This has to be a specific account for the sole purpose of the Direct Payment. The money is then used to purchase the care and support as set out in the person's Support Plan.
  Changes to the use of the budget can still be organised in agreement with the person's social worker or with the local Direct Payments team. The bank statements, invoices and receipts would need to be kept for the purposes of the Local Authority or Integrated Care Board's financial audit.

- **Third Party Managed** – (available to people receiving funding through the local ICB). This is similar to an ISF but the fund comes from the local ICB.
- **Managed Account** – An organisation would manage the budget on behalf of the person; the organisation would provide payroll for those who are employed to support them and organise the payments of invoices. This organisation would manage the record keeping for the purposes of the ICB's financial audit. This organisation may be specifically set up to provide this service to Direct Payment recipients and some accountancy firms have added this to the range of services they offer.
- **Local Authority/ICB Managed Budget** – The Local Authority or ICB manages the budget. With this arrangement, the person would choose their services and support from the approved providers list managed by the Local Authority or ICB which will include a range of home care services, respite provision and daytime opportunities.
- **Trust Fund** – The budget is held and managed by a legal Trust in the person's name. The trustees of this fund can be people known to the person or it may be a legal representative, such as a solicitor.
- **Mixed Package** – It is possible to choose from a mix of the above options. For instance, if the person wants to employ their own personal assistant, a Direct Payment could be set up to cover this cost and if they also want to attend the local day centre, a Local Authority or ICB managed budget could cover the cost of this aspect of their support.

Barriers faced by people with living with dementia in the community

The World Health Organization (WHO) (2011) describes barriers as being more than just physical obstacles like steps, curbs, and narrow doors. WHO recognises that other factors create barriers to people's participation and people can experience multiple barriers which can make it impossible for them to go about their daily lives. The most common barriers include:

- communication – not taking into consideration factors that affect communication such as hearing, speaking, reading, writing and / or understanding;
- transportation – a lack of accessible and convenient transport;
- people's attitudes – stereotypes, stigma, assumptions, discrimination and prejudice;
- policy – barriers to accessing the appropriate support and services;

- system shortcomings – inconvenient scheduling of appointments, lack of access to appropriate equipment, short appointment times, and professional language and jargon;
- financial restraints – the level of income available to a person and the additional cost of living with a disability.

Tommy's story

Tom is a young man with dementia who talks about his experience of living with the condition and the reactions he has had from other people in his life.

> Living with dementia means you can still follow – and enjoy – a conversation but often no one gives us a chance to answer, and the questions they ask are often just things like '*Are you okay?*' or statements like '*You're looking well*'.
>
> I find that people will also often talk over me, around me and about me, but never to me – even in a room of people I know. Sometimes the conversation will just stop when I walk into a room.
>
> If I'm with people that don't know me, the conversation will flow until I mention I've got dementia, and then it is like I've put on an invisibility cloak. I think people just don't know how to talk to people with dementia, in a similar way that people talk to the wheelchair rather than the person with disabilities or the guide dog rather than the sight-impaired person.
>
> *(Tommy Dunne, Elder, n.d.)*

Reflection

- Have you or the person you care for experienced similar situations?
- If so, what was this like for you?
- How could these social situations be improved? What would be a better experience for Tommy, one that would improve his quality of life?
- After reading Chapter 1, is there anything you could bring into these conversations to support you and the person you care for to be more involved and understood?

Overcoming barriers

People with lived experiences are arguably best placed to identify effective ways of overcoming the barriers they personally face to improve their quality of life. Here are some of the ideas that have been put forward by people who live with dementia:

- to have more control over the way support and care are provided, with a greater range of choice regarding where to find care and support;
- to be able to utilise the health or social care funding to find solutions within their local community;
- to have access to information that is accurate, easy to understand, accessible and appropriate, with time to discuss this with a knowledgeable person;
- to create greater awareness and understanding amongst paid professionals, benefit application assessors, employers, health and social care professionals, as well as the wider community;
- to provide training and raise awareness, delivered by those who experience the barriers;
- to promote positive attitudes where people are treated with greater sensitivity and understanding.

(Copestake, et al., 2014)

Hidden barriers

The barriers experienced by people living with dementia are often hidden. The Alzheimer's Disease International report, titled 'World Alzheimer Report 2012 Overcoming the stigma of dementia', stated that one in four people with dementia hide their diagnosis, with stigma cited as the main reason. The global survey involved 2,500 participants from 50 countries and highlighted that 40 per cent of people with dementia have been avoided, excluded from everyday life or treated differently. The report produced the following recommendations to tackle stigma surrounding dementia:

- educate the public;
- reduce isolation of people with dementia;
- give people with dementia a voice;
- recognise the rights of people with dementia and their carers;
- involve people with dementia in their local communities;
- support and educate informal and paid carers;
- improve the quality of care at home and in care homes;

- improve dementia training of primary healthcare physicians;
- call on governments to create national Alzheimer's disease plans;
- increase research into how to address stigma.

(Batsch, Mittelman and Alzheimer's Disease International, 2012, p.75)

In 2019, BBC News shared an article about the barriers faced by dementia patients from ethnic minority groups which revealed how a lack of awareness of language, faith and culture are additional barriers faced (BBC, 2019).

- Consider the barriers of day-to-day life; which of the barriers identified are familiar for you?
- What are the barriers that are unique to you?
- Barriers are not always physical; they may relate to the person's quality of life in other areas, for instance, being able to make fundamental decisions or the way people's attitudes have changed towards them. Have all the barriers been considered - what else could be a barrier?
- What would make a real difference in your life and what can be done to help make this happen?

Personal budgets to overcome barriers

Think Local Act Personal, an organisation that is dedicated to achieving Personalised approaches, wrote a report in 2017 called 'The National Personal Budget Survey'. This was the third survey organised by Think Local Act Personal and the results of this survey showed that the majority (75 per cent) of older people reported that they were more positive about the impact on their support in relation to their quality of life (2017, p.7), stating that the freedom and control that is made possible through a personal budget enables them to maximise the resources available, often achieving the best outcomes for themselves and their family.

The survey provides an insight into the difference personal budgets have made, but often to hear a person's own story about the difference it has made for them brings this to life. In 2010, the Department of Health shared stories in 'Putting People First: Personal budgets for older people – making it happen', which highlighted how personal budgets have made a difference for people living with dementia and their families (Department of Health Putting People First Programme Personalisation, 2010). One story is about a friend (Sheila) who supported a woman (Joan), living with dementia, to return home after an unhappy stay at a care home where Joan had become agitated. Joan no longer had the capacity to manage her personal budget and Sheila held enduring Power of Attorney on

Joan's behalf. The local authority appointed Sheila to act in the capacity to receive Joan's personal budget. Although initially Sheila was daunted by the additional responsibilities, she eventually found that the task of managing Joan's personal budget hadn't been a problem in practice and the outcome has been a much-improved quality of life for her friend. (Department of Health Putting People First Programme Personalisation, 2010, pp. 26-27).

Mental Capacity Act

The person's right to be heard and to make their own decisions is protected by the Mental Capacity Act. This Act also provides clear guidance on what to do when a person has reached a stage where they are no longer in a position to or able to make their voice heard and to make the relevant decisions regarding their health and their care and support.

The Mental Capacity Act is an act of parliament of the United Kingdom (*Mental Capacity Act 2005*). Adopted in 2005, it is especially relevant to those involved in the care, treatment and support of people aged 16 and over living in England and Wales.

The Scottish equivalent is the Adults with Incapacity Act 2000 (*Adults with Incapacity (Scotland) Act 2000*), and the equivalent legislation for Northern Ireland is the Mental Capacity Act (Northern Ireland) 2016 (*Mental Capacity Act (Northern Ireland) 2016*).

All these Acts are designed to protect and restore power to people who have been assessed, using a Capacity Assessment, as lacking capacity to make a decision at a particular time in their life.

The Mental Capacity Act in particular sets out five core principles:

1. A person is assumed to have mental capacity until it is proven otherwise.
2. All practical steps must be taken to help the person understand the decision they are making and the options available to them.
3. People with mental capacity have the right to make 'unwise' decisions.
4. When it is assessed, using a Capacity Assessment, that the person does not have capacity to make a decision, these decisions will then be made in the person's 'best interests'.
5. Any decision or action must be done in a way that is the 'least restrictive' of the person's rights and freedoms.

(*Mental Capacity Act 2005*. s.1.(1)(1-6))

If there's any concern that a person lacks capacity to make a decision then a Capacity Assessment would be completed; this is usually led by the social worker. Anyone caring for or supporting the person, either in a paid or unpaid capacity, can be involved in this assessment of capacity.

When completing a Capacity Assessment, the assessor must use the person's preferred way of communicating and any communication aids they use to support them to make their own decisions and maintain choice and control. The person's preferred style of communication as well as any communication aids they use can be recorded in their Support Plan, and the people who know the person well should be asked to share how the person communicates with the social worker completing the Capacity Assessment.

The Mental Capacity Act states someone who is deemed as lacking capacity cannot do one or more of the following four things:

- understand information given to them;
- retain that information long enough to be able to make a decision;
- weigh up the information available to make a decision;
- communicate their decision.

(*Mental Capacity Act 2005*. s.1(3)(1))

When a Capacity Assessment is completed it is only:

- for the decision to be made at that time and
- relevant at the time of completion

Therefore, if a completed Capacity Assessment concludes that the person does not have capacity, this is only relevant to the decision to be made at that time. It cannot be assumed that they do not have capacity for any other decision they need to make in their life at the time of the assessment. It also cannot be assumed that they will lack capacity from that point onwards as *a person's capacity can fluctuate over time.*

If the Capacity Assessment concludes the person you care for does not have capacity the following will be considered.

Best interest

A best interest meeting would involve other people to make the decision the person is not deemed as having capacity to make themselves.

A best interest meeting may be needed to make decisions about:

- where a person wants to live;
- what care services and support they are to receive at home;
- what care services and or support should they receive away from their home, i.e., in a service setting.

A best interest meeting could also be held to make a decision about serious medical treatment, although some decisions like these can only be granted by the Court of Protection (West and Glover, 2014, p.7).

During a best interest meeting, the person would remain central to the decision-making process and the significant people involved in their life, such as their family, would take part in the meeting.

The best interest meeting would be run with a clear focus on the purpose of the meeting, which would be the decision needing to be made and everyone present would have the opportunity to be involved in the decision-making. A written record of the meeting would be made and shared with those who attended.

Independent Mental Capacity Advocate

The *Mental Capacity Act (2005)* established the legal right to advocacy for those who lack mental capacity. This means people living with dementia can be supported by an Independent Mental Capacity Advocate during their Capacity Assessment and best interest meeting, if they don't have family or friends to support or represent them or where there are concerns regarding safeguarding (*Mental Health Capacity Act 2005*. s.1(35)(1–6).

Power of Attorney

Anybody over the age of 18 and who is able to make decisions for themselves can select a person to have Power of Attorney over their affairs should they lose capacity. The person who holds Power of Attorney for an individual would then be able to make decisions on their behalf, following the best interest principles.

There are two types of lasting Powers of Attorney:

1. Personal Welfare – this applies when decisions need to be made regarding medical interventions.
2. Property and Affairs – this applies when decisions are to be made regarding property, assets and financial affairs.

When establishing a Power of Attorney, a person can be specific regarding the areas of decision-making they would like their Power of Attorney to hold, i.e. dietary considerations, clothes they wish to wear, medical treatment they either do or do not want to receive.

To support understanding around the Mental Capacity Act in relation to dementia, Alzheimer's Society (n.d.a; n.d.b) has created a hub of online resources and information.

Mental Capacity Act and personalisation

Personalisation is concerned with people having rights, choice and independence. People living with dementia have the same human and legal rights as people who do not have dementia; this includes the right to have choice, control and freedom.

A diagnosis of dementia does not mean that the person cannot make a particular decision at a particular time.

This section on the Mental Capacity Act, on the assumption of capacity, has highlighted the absolute right a person has to be at the centre of any decision-making that is about them. When one thinks about how care and support can be provided in a way that's in line with personalisation, one soon realises that the person living with dementia, and those close to them, have a good sense of the kind of care and support that will work best for them. This has led to a recognition that those who are in receipt of health and social care support are well positioned to give input and steer regarding the services and support they need within their local area. The idea of co-production puts this thinking into practice.

What is co-production?

'Co-production means delivering public services in an equal and reciprocal relationship between professionals, people using services, their families and their neighbours. Where activities are co-produced in this way, both services and neighbourhoods become far more effective agents of change' (Boyle and Harris, 2009, p. 11).

Co-production is an approach to the development and design of support services available in communities and for people who access health and social care support. The model's introduction to public services first took place just after 2005. A discussion paper published by Nesta in 2009, 'The Challenge of Co-production', stated a key finding: *'The opportunities are huge, but there are plenty of remaining barriers to creating the conditions for co-production'* (Boyle and Harris, 2009, p. 23).

The barriers include public service reform which has been based on years of austerity which began in late 2008. This has resulted in a heavy focus on cost-efficiencies and delivery decisions often being led by selecting the lowest cost option. A consequence of this is a contradictory position to personalisation and a person's opportunity to self-direct their support.

In the context of dementia care and support, co-production could either be focused on an individual approach:

- a support solution that would suit an individual, where the person, family, friends and professionals have worked together to learn what would work best and have collectively found a way to make this possible.

or a larger, community-wide approach:

- a support solution that would work for groups of people in the community which would include people living with dementia, for example, setting up a new social group to attend, resolving a transport issue, or employing a team of people to work as support workers.

Co-production, the Values of Inclusion and Collective Agency

The Values of Inclusion and Agency, as covered in Chapter 1, have principles in common with Co-production. The principles of all these approaches could help guide and inform how to provide good care and support:

- it is important for the person being cared for and supported, their family and the people providing the paid for care and support, to get to know one another;
- it is good to aim to have equality within these relationships. No one group, or person is more important than any other group or person;
- a problem is more likely to find a solution if there is a group of actively interested people involved in the conversation;
- bringing people together from a variety of backgrounds to support each other to exercise their ability to gain voice, choice and control creates a stronger picture of the strengths and assets of a local area;
- collectively, people can build on the picture of what is currently available in the community, by bringing their own knowledge, skills and connections to build a stronger place for everyone.

Outputs and Outcomes

When helping the person to make plans for the future or deciding what needs to change to improve their quality of life, it is useful to understand the different between Outputs and Outcomes:

Outputs relate to: what is delivered. **Outcomes** refer to: the difference made.

Outputs

Outputs have a focus on describing and counting activities:

- the number of times the person attended an activity;
- how long they were there.

For instance, the person may have attended a day centre 20 times and they took part in the cooking workshop which lasted two hours 15 out of these 20 times.

Outcomes

Outcomes focus on what has changed in the person's life as a result of the activity or the care and support they have received, focusing on the subjective experience of the person and the difference it has made for them. For example, as a result of attending the cooking classes the person now knows how to prepare a fresh and nutritious meal that they enjoy safely.

Developing an understanding of the Outcomes of an approach can provide insight into what is needed to improve the person's experience of accessing care and support.

People living with dementia and quality outcomes

In 2009, the 'Living well with dementia: A National Dementia Strategy' was launched with the intention of making sure that considerable enhancements were made to dementia services across three key areas: improved awareness, earlier diagnosis and intervention, and a higher quality of care (Department of Health, 2009). The Department of Health (2010) then developed quality measures in reference to the provision of care and support for people with dementia in England: 'Quality outcomes for people with dementia: building on the work of the National Dementia Strategy'. Within this paper a set of **'I' statements** were presented which were co-produced by members of the Department of Health, people with living dementia and their family carers. These statements captured what people wanted to see in their own lives.

National Dementia Declaration

In the same year as the 'Quality outcomes for people with dementia: building on the work of the National Dementia Strategy', the National Dementia Declaration for England was published, which presented a set of actions to improve dementia care and support.

The declaration was signed by over 100 organisations to show their commitment to these actions. The declaration was successful in providing momentum to address the issues surrounding dementia and has ensured that it remains a public policy priority (Dementia Action Alliance, 2010).

The National Dementia Declaration for England put forward a set of seven outcomes for healthcare professionals and service providers to work towards in the drive to improve the experience for people living with dementia. These seven outcomes were written by those living with dementia and their families and they affirm the wish of people with dementia to have choice and control about the services they receive.

The outcomes are written as seven 'I' statements, which built on and furthered the nine 'I' statements created by the National Dementia Strategy in 2010.

The All-Party Parliamentary Group on Dementia

The All-Party Parliamentary Group on Dementia (APPG on Dementia) is a cross-party group of MPs and peers with an interest in dementia. Its aim is to raise awareness of dementia within parliament and to influence legislation and policymaking with the intention of improving the lives of people with dementia and their carers. In 2014, the APPG on Dementia wrote a report that commended the continued focus on the provision of quality dementia care and support (Alzheimer's Society, 2014).

The report noted areas of practice that had improved including the support people received at the time of diagnosis, which had become more focused on the needs of the person. It had been noted that this made a positive difference to people's quality of life, although this continued to vary throughout the country.

The report, however, did call for improvements in crisis prevention and funding for services that support people with dementia and their carers to live independently in their own homes for as long as they are able and wish to.

A lot had changed in the provision and dementia care and support between 2010 and 2016. To reflect this, the Alzheimer's Society led a review of the 'I' statements in 2016 on behalf of the Dementia Action Alliance, a collective of organisations and people from across England that take action on dementia and work together to build Dementia Friendly Communities.

The review was conducted alongside people living with dementia and their carers to make sure it shared people's opinions around what needed to be updated to reflect their current life experience.

This led to a change of focus to 'We' statements rather than 'I' statements, to signify that people with dementia and carers were equally represented by them.

The 'We' Statements below were endorsed by the Dementia Programme Board that monitored and supported the implementation of the Prime Minister's Challenge on Dementia 2020 across the NHS, social care, the research sector and wider society.

- We have the right to be recognised as who we are, to make choices about our lives including taking risks, and to contribute to society. Our diagnosis should not define us, nor should we be ashamed of it.
- We have the right to continue with day-to-day and family life, without discrimination or unfair cost, to be accepted and included in our communities and not live in isolation or loneliness.
- We have the right to an early and accurate diagnosis, and to receive evidence-based, appropriate, compassionate and properly funded care and treatment, from trained people who understand us and how dementia affects us. This must meet our needs, wherever we live.
- We have the right to be respected, and recognised as partners in care, provided with education, support, services, and training which enables us to plan and make decisions about the future.

- We have the right to know about and decide if we want to be involved in research that looks at the cause, cure and care for dementia and be supported to take part.

(Alzheimer's Society, n.d.c)

> *'The new Dementia Statements reflect the things people with dementia say are essential to their quality of life. Grounded in human rights law, they are a rallying call to improve the lives of people with dementia and to recognise that they shouldn't be treated differently because of their diagnosis.'*
>
> *(Dementia Action Alliance, 2017)*

If you wish to know more about the full Dementia Action Alliance report, the link is available in the Appendix: Useful organisations section at the end of the book.

The work in improving the provision of care and support for people with dementia and their carers continues with a government announcement made in May 2022 of a new 10-year plan. The 10-year plan focuses on new medicines, emerging science and technology to improve outcomes for those living with dementia. It will also focus on supporting people with their specific health and care needs. This is also complemented by the new Health and Care Act 2022, which *'will put the person at the centre of care, with local systems designed to deliver seamless care and support people in retaining their independence, health and wellbeing'* (Department of Health and Social Care and The Rt Hon Sajid Javid MP, 2022).

- Do you and the person you care for relate to the 'We' statements and are they a reality in your lives?
- If yes, what would you share with others about how this has been achieved?
- If not, which ones don't hold true for you and the person you care for?
- What could be done to help say all 'We' statements with confidence?

A summary of personalisation

Table 2.1 provides a summary of what has been learnt about personalisation and what it means for people with dementia.

Table 2.1 *Summary: Personalisation*

Personalisation means...	**What personalisation means to people with living dementia:**
Tailoring support to people's individual needs and aspirations.	The carer and the person will receive support that they have both played an active role in determining and in a way that works best for both of them.
Ensuring people have access to information, advice and advocacy to support them to be able to make informed decisions about their care and support.	The carer and the person will have the information and support they both require to understand what options are available to them and to make effective choices.
Finding new collaborative ways of working (which includes co-production) that support people to actively engage in the design, delivery and evaluation of services.	The carer, the person and everyone else providing support will imaginatively and creatively work together to build an effective package of support and care that corresponds to the wishes of the person and the carer.
Developing local partnerships to co-produce a range of services and options for people to choose from which promote opportunities for social inclusion and community development.	Services from statutory, private, community and voluntary sectors work together, alongside the carer and the person being cared for, so that they are both fully included with a local community that enables the full participation of people living with dementia and their carers.
Developing the right leadership and organisational systems to enable staff to work in creative, person-centred ways.	Investment is made into the culture of organisations and leadership to realise person-centred approaches. People with lived experiences, such as the person and their carer help to shape and inspire leadership.
Embedding early intervention, re-enablement and prevention so that people are supported as soon as it is needed and in a way that's right for them.	People with dementia receive a diagnosis as early as possible, which is clearly explained to them. At the time of diagnosis the person is provided with information and advice which supports them to make choices and plans about the life they want to live.
Recognising and supporting carers in their role, while enabling them to maintain a life beyond their caring responsibilities.	The contribution of informal carers is fully understood and valued, its effect on their physical and emotional health is recognised and, when appropriate, suitable support is offered. They are seen as having a life outside their caring role, which they should have every opportunity to enjoy.
Ensuring all citizens have access to universal community services and resources – a total system response.	People with dementia continue to be and are included in all aspects of community life without experiencing any barriers to participation and contribution.

Key lessons learnt

- Direct Payments and personal budgets are different from one another, where personal budgets is the level of funding available to a person from the local authority or ICB and the Direct Payment is one of the options available for how to access and manage that funding.
- Direct Payments are one option of how to manage a personal budgets, which sits alongside other options available, such as a Managed Account, Individual Service Fund, Local Authority or ICB managed budgets.
- People living with dementia will face barriers to leading a full and inclusive life; these are based on how accessible a place is and also the attitudes of other people, but there are ways to overcome these barriers when people work together to find a solution.
- Self-Directed Support, as mentioned in Chapter 1, and personalisation can help people with living dementia to overcome barriers and achieve a better quality of life.
- The Mental Capacity Act protects people's right to make their own decisions about their life including the care and support they receive and how they receive it.
- When someone is not able to make a decision there is clear guidance, as set out in the Mental Capacity Act, which states that the people who know the person the best, including family and friends, are to be fully involved in the decision making process.
- Co-production brings people together to discuss, explore and identify how communities can become stronger for everyone and the types of investments that would serve the best purpose for local people.
- 'Outputs' relate to 'what we do' whereas 'Outcomes' refer to 'what difference can we make?'. The person in receipt of the care should decide what a good outcome looks like.
- Personalisation has opened up the opportunity to recognise a bigger picture of what it takes to achieve an outcome of 'wellbeing' and 'quality of life'.

Questions to aid reflection

1) What would you say if you were to have a conversation with someone about the areas of your own life that contribute to your wellbeing and quality of life?
2) What really matters to you in your life? What can you find within your:
   network of friends and family?
   community?
3) What really matters to the person you care for in their life? What can be found within their:
   network of friends and family?
   community?

3 The dementia journey

This chapter covers:

- **The different types of dementia**
- **The importance of engaging with the person with dementia**
- **The dementia journey, focusing on the early, mid and late stages in the context of the individual's experience**
- **Personalisation at each stage of dementia**
- **Before a person receives a diagnosis of dementia**
- **The importance of an early and timely diagnosis**
- **Future decision-making using advanced care planning.**

What is dementia?

The word *dementia* is defined by the Alzheimer's Society (2017a, pp 1–3) as '*a set of symptoms that may include memory loss and difficulties with thinking, problem-solving or language. ... Dementia is caused when the brain is damaged by diseases, such as Alzheimer's disease or a series of strokes. ... Dementia is progressive, which means the symptoms gradually get worse over time*'.

The definition provided by the Alzheimer's Society explains the reality of dementia; it is a chronic and progressive condition with, as of yet, no cure. People who have a diagnosis of dementia will experience the condition in different ways, with progression and the changes in the person's life experience varying from one person to the next.

While dementia is seen as a memory disorder, people living with dementia will experience changes in their ability to manage daily tasks and their emotions and maintain their relationships. The decline in ability will vary and fluctuate; some people will retain their skills and abilities, including making decisions for themselves.

Relating to the person on the basis that they are as capable now as they were before their dementia diagnosis is paramount to their sense of well-being and health. Ensuring that the person's ability to make decisions about their life and how they wish to be supported and cared for is not underestimated is key to supporting them to maintain and achieve quality of life, as well as keeping with the principles of the Mental Capacity Act 2005. It is essential that people only change their way of relating to the person when it becomes clear that their ability to make decisions has changed and they need more support.

Retaining control

Providing care and support in a manner that takes away the person's opportunity to do things for themselves or make their own decisions also takes away their control and choice, however well-meaning the intention, it does not consider the negative consequences. This approach has the potential to lead to negative outcomes for the person living with dementia where they may become passive, dependent, frustrated or disconnected from others.

Ken is a person living with dementia who shares his insight into what good support looks like and his need to retain control over his own life.

> *Decay sets in at the moment of diagnosis – medically nothing can be done – but you can do a lot for yourself. Well-meaning people can destroy your confidence. My wife is technically my carer but you need to continue to do things for yourself. There is a fine line between doing too much for someone and giving them the help they need.*
>
> (Ken Howard, Young Dementia UK, n.d.)

Reflection

- What would help the person living with dementia to retain control and self-determination?

People diagnosed with dementia will, most likely, experience a gradual diminished cognition and an impaired ability to make choices. Despite this, the person must not be treated as though they have lost their ability to fully participate in conversations and to make decisions. Remembering that dementia is a fluctuating condition and that it may be necessary to consider when to hold a conversation, choosing a time that works best for the person is important. If, during a conversation, the person is finding it difficult to be involved, try again a day or two later, or at a different time of day, finding a time when the person's more engaged and more involved in making the decision.

The risk of not engaging

The dementia journey happens in stages, from pre-diagnosis and receiving the diagnosis to the middle stages and later stages of the condition, which is referred to as the dementia journey. If a person is not engaged with positively at every stage of their dementia journey, their sense of self-worth is at risk, and this will have a marked impact on their sense of well-being and self-identity, as Seligman (1975) states: '*When this* [not appropriately engaging with people who have dementia] *happens, people with dementia are discouraged and after repeatedly trying to get their voice heard and their wishes granted they will give up and learned helplessness occurs together, perhaps with associated depression*'.

How dementia develops

Dementia is a progressive disease with patterns and stages. The progression will differ for each individual. This progression can be seen in the context of three stages.

- Early or mild – minor changes in a person's ability, behaviour and memory, for instance:
  - mislaying items around the house;
  - forgetting recent conversations or events;
  - struggling to find the right word in conversation or losing the thread of what is being said.
- Middle or moderate – changes in a person's ability, behaviour and memory become more marked, for example:
  - becoming confused about where they are and becoming lost;
  - becoming confused about the time and getting up at night because they believe it is the day time.
- Late or severe – pronounced changes in a person's ability, behaviour and memory and the person becoming increasingly weak, for instance:
  - having difficulty eating and sometimes swallowing;
  - experiencing considerable weight loss, or eating too much and putting on weight.

Health professionals utilise a variety of scales to measure the stage of dementia a person has reached. These scales address a person's:

- mental ability;
- daily living skills;
- behaviours;
- overall functioning and ability to get on with day-to-day skills.

The Mini-Mental State Examination is the most commonly used test for complaints of memory loss and other cognitive functioning. This test would be conducted with consideration of the person's history, symptoms and alongside a physical examination.

The forms of dementia

The main forms of dementia include:

- Alzheimer's disease;
- frontotemporal dementia (including Pick's disease);
- vascular dementia;
- dementia with Lewy bodies (DLB);
- mixed dementia.

Alzheimer's disease

This is the most common form, accounting for between 60 per cent and 80 per cent of cases (Alzheimer's Association, 2020), and it has therefore been studied in the greatest detail.

Various factors may contribute to the onset and development of Alzheimer's disease. The older a person is, the more likely they are to develop it, affecting 1 in 14 people over the age of 65 and 1 in 6 over the age of 80 (NHS, 2021). Genetics also plays a part, and, in a few families, there is an increased chance of developing the condition.

Alzheimer's disease is a physical disease caused by the development of *plaques* and *tangles* in the brain. These plaques and tangles lead to the death of brain cells and give rise to various symptoms. In addition, people with Alzheimer's disease experience a reduction in the level of chemicals involved in transmitting messages within the brain essential for body functioning and movement.

Alzheimer's disease usually develops slowly in people. People in the early stages may experience lapses of memory and have problems finding the right words as well as:

- becoming slower at grasping new ideas and unwilling to try out new things;
- becoming confused or losing track of the day or date;
- showing poor judgement or finding it harder to plan or make decisions;
- having problems judging distance or seeing objects in three dimensions (when navigating stairs or parking the car, for instance);
- losing interest in other people or activities.

(Alzheimer's Society, 2015, p 5)

Seeking a diagnosis at an early stage can help determine if these behaviours are a result of early-stage Alzheimer's or if they can be explained by other life factors. As the condition progresses, the person may start to get up at night as they confuse night and day, go outside in their night clothes, become very agitated or unknowingly behave in socially inappropriate ways.

In the late or severe stage, the person will become heavily dependent on the people around them to maintain their quality of life and to meet their basic needs. They are likely to be physically weak, their memory will have declined, and may find it difficult to recognise everyday objects, surroundings and the people in their lives.

During the final stages, pronounced and fluctuating changes in a person's behaviour are likely, and they may become, even for short periods of time, subdued or frustrated. Any change is due to the heightened confusion and disorientation the person will be experiencing. It can be common for people to lash out aggressively while being supported with close personal care. It is important to remember the person will not be behaving in this way intentionally and to recognise the behaviour as a response to their condition and, at times, that this may also be the only way the person has to express their confusion and frustration.

Frontotemporal dementia (including Pick's disease)

This less-common form of dementia is caused by damage to the frontal lobes and/or the temporal lobes. These areas of the brain are responsible for emotions, feelings and language. For example, someone may become less sensitive to other people's emotions, perhaps making them seem cold and unfeeling. They may also lose some of their inhibitions, leading to behaviour that is out of character, such as making tactless or inappropriate comments. There are three types of frontotemporal dementia: one that affects behaviour first and two that affect language first.

When someone is diagnosed with frontotemporal dementia marked changes can be seen in their behaviour, personality or their command of language. These changes would be recognised by those who know the person well but they themselves may not be aware of these changes. The person may develop repetitive habits, gestures and phrases. This can include developing obsessions or hoarding. The person may also appear uncharacteristically:

- selfish and unattached;
- unfeeling;
- rude;
- easily distracted;
- less interested;
- low in motivation.

Other symptoms of this form of dementia include:

- acquiring a sweet tooth;
- uninhibited behaviour;
- compulsions.

As dementia progresses, people whose behaviour and personality changed initially will begin to have trouble with language and, likewise, those who initially had trouble with language will see changes in their behaviour and personality. In the later stages, the symptoms become similar to the later stages of Alzheimer's disease.

Vascular dementia

Vascular dementia is a deficiency in blood flow to the brain due to diseased blood vessels. The brain requires a constant supply of blood to deliver oxygen and nutrients which are delivered through a network of blood vessels. If these vascular systems within the brain become damaged, the blood vessels leak or become blocked, preventing the blood cells from reaching the brain. The symptoms of vascular dementia can sometimes develop suddenly and quickly get worse, although they can also develop gradually over many months or years. People with vascular dementia may also experience stroke-like symptoms, including muscle weakness or paralysis on one side of their body. If the brain does not receive the right level of oxygen and nutrients, it can affect three elements known as cognition:

- Thinking
- Memory
- Reasoning

Vascular dementia is the loss of ability to manage these three elements, and this begins to affect daily life. There are different types:

- Stroke-related dementia

  A stroke happens when the blood supply to the brain is suddenly cut off. In most strokes, a blood vessel to the brain is blocked or narrowed. The severity depends on whether the blockage is temporary or permanent and where the blockage has occurred.

- Post-stroke dementia

  Twenty per cent of people who have had a stroke are at risk of developing post-stroke dementia within the ensuing 6 months (Alzheimer's Society, 2014, p 3). A person who has had a stroke is at higher risk of having further strokes, which then means the person is at greater risk of developing vascular dementia.

- Single-infarct and multi-infarct dementia

  This is caused by one or many smaller strokes as a result of a blood vessel that becomes blocked by a blood clot which quickly releases itself. The stroke is so small that the person does not notice the symptoms and can often believe they have just been a little bit unwell. The blood clot, even triggering a temporary blockage of a few minutes, can cause permanent damage to a small area of brain tissue known as an *infarct*. It is common in these circumstances for many small strokes of this type to occur over a period of weeks or months, which then leads to infarcts spreading around the brain; this then results in multi-infarct dementia.

- Subcortical dementia

  This is the most common form of vascular dementia, and is caused by a disease of the smaller blood vessels that reside deeper within the brain tissue. The disease causes these smaller blood vessels to have hardened walls and to become twisted, reducing blood flow within this area of the brain. This disease can also affect the nerve fibres that carry signals around the brain, also known as *white matter*. It can cause infarcts at the base of the brain.

Dementia with Lewy Bodies (DLB)

This form of dementia gets its name from microscopic deposits that are in the brain, which can cause damage and eventually the death of the brain's nerve cells. People with DLB will usually have some features of Alzheimer's disease and Parkinson's disease and may experience the following:

- Impaired attention and alertness
- Spatial disorientation causing difficulty in certain activities, such as planning and coordinating mental activities
- Memory is often affected, though less so than in people with Alzheimer's disease

- Slowness, muscle stiffness, trembling of the limbs, a tendency to shuffle when walking, loss of facial expression and changes in the strength and tone of their voice (features of Parkinson's disease)
- Detailed and convincing visual hallucinations (seeing things that are not there), often of people or animals
- Fluctuating abilities day by day
- Falling asleep very easily by day and having restless nights with confusion, nightmares and hallucinations
- Fainting, falling or having funny turns

Mixed dementia

People can have a diagnosis of several forms of dementia, and at least 10 per cent of those diagnosed have a combination of Alzheimer's disease and vascular dementia (Dementia Matters, 2018). In these circumstances, people will experience a mixture of the symptoms inherent to each form of dementia.

Before the diagnosis of dementia

The onset of dementia can be different for each person. Multi-infarct dementia may start suddenly and then progress in a stepwise manner as a repeated series of mini-strokes occur. The person often has periods of lucidity when they are more engaged with others and can take part in everyday activities. Whereas when someone has Alzheimer's disease, the onset and progression of dementia are usually more gradual, though in many ways it is often indistinguishable from multi-infarct dementia.

The period between the onset of dementia and its diagnosis is a difficult time for people living with the condition and their family. It can take a long time to reach a diagnosis with many factors influencing this, for instance:

- the recognition and acceptance, on behalf of the person and their family, that an appointment with a general practitioner (GP) is required;
- the ability and confidence to communicate the person's experiences to a GP;
- the length of time available to speak to a GP during an appointment period;
- the GP being alert to the signs of early-onset dementia.

Early and timely diagnosis

Early diagnosis involves methods to determine the nature of a disease or disorder at its early stage of progression. Receiving a diagnosis of dementia needs to be carefully managed and delivered with the person's well-being, rights and circumstances as a primary focus.

Receiving a diagnosis at any stage of the dementia journey means that the person can begin to receive good and timely advice, guidance, support and treatment.

The National Dementia Strategy (Department of Health, 2009) placed greater emphasis on early dementia diagnosis, prevention and risk reduction. Dementia is not currently screened within the whole population. Health and social care in England has identified the following groups who are at higher risk of developing dementia, including:

- people over 75 years of age, as age is the strongest risk factor for dementia;
- those with high vascular risk;
- people with Parkinson's disease;
- people with learning disabilities.

(Bamford et al, 2007)

Providing an early diagnosis offers an opportunity to:

- allow for the delivery of therapy and the prescription of medication;
- signpost people to sources of information, local and national support groups;
- provide advice about driving;
- provide advice regarding financial matters and making appropriate plans for the future;
- provide guidance regarding the benefits available, when a person would be eligible to apply and the application process;
- start the process of gathering information and planning to ensure that the person continues to experience a good life and maintains optimum control;
- provide guidance on advanced care planning (ACP) (see below);
- provide help with setting up a lasting power of attorney.

Referral to other professional input can also be provided once a diagnosis is confirmed; this could include referrals to community mental health teams, speech and language therapists, dietitians and occupational therapists.

Diagnostic process

Receiving a diagnosis can take time. Depending on the person's presenting stage of dementia, a diagnosis can take between 4 weeks to 12 months, especially if the person is in the early stages.

A wife's story

Here is a story that has been shared by a woman whose husband was diagnosed with early-onset dementia.

My name is Janet. I am a retired Finance Director and I am a world expert on dementia ... more correctly, my husband's dementia. This complicated disease affects everybody in different ways and everybody's journey is different. This is the story of our journey.

First of all, my husband was, and is, a lovely man and that makes it both easier and also so much harder.

How our journey started

Very slowly at first as Martin instinctively compensated for his failing ability to cope for a while before I noticed

- *Avoidance of handling change*
- *Difficulty telling the time*
- *Forgetting how to use various computer programmes*
- *Blaming the tax man for increasing complexity*
- *Not remembering how to play crib*
- *Becoming less able to manage monetary affairs*
- *Forgetting our anniversary (my emotional low!)*
- *Stopping smoking after 40 years because the taste changed*
- *Becoming quieter and more withdrawn.*

In retrospect I think it was more stressful before we got an official diagnosis and 'came out'! I started helping Martin to mask his increasing confusion and spent a lot of time planning how to remove the risk of failure from his life, without upsetting him further.

But Martin wouldn't accept that there was anything wrong and we avoided the subject, until a couple of crisis points after which Martin agreed to speak to the doctor (but for my sake, not his!).

(Janet, Young Dementia UK, n.d.)

An accurate diagnosis during the early stages of dementia can be difficult as a person's *dementia-like* symptoms can be caused by other health conditions. Appropriate assessment and investigations can detect the root cause of the symptoms. The appropriate treatment for the health condition can then be provided to the individual, and the symptoms become fully or partially reversed, depending on the root problem. For instance, the symptoms experienced by the person could be due to the side effects of the medication they have been prescribed, a combination of medications they are taking, an infection or a brain tumour. When seeking a diagnosis, the person and their family may find that they have to be persistent with tests and assessments to eliminate any other factors and to confidently reach a fully informed and accurate diagnosis, followed by appropriate support.

A thorough assessment can also help to accurately diagnose not only dementia but the form of dementia the person has. There are a number of ways that a person may begin to be assessed for and receive their diagnosis of dementia. During a visit to hospital, most people aged over 75, who have been admitted urgently, will have an initial assessment on levels of confusion and memory loss. The outcome of this initial assessment can lead to a full assessment if there are any concerns and if the person has given their consent for this to take place. This is deemed good practice within hospital settings, and it has been put in place to make sure that an early diagnosis is possible. Caution is needed while undertaking an assessment within a hospital setting, especially if the person has been admitted urgently. They may be experiencing confusion and memory loss as a result of an illness or accident. The confusion may also be due to the sudden change in their environment rather than as a symptom of dementia. During a routine visit, a GP may now ask about memory problems, especially if the person is over 60 or had a stroke, heart condition or diabetes.

Screening and assessment

After the initial screening assessment and discussion at the hospital or GP, further assessments and tests will be carried out.

Stage 1: Visit the general practitioner

Generally the first step for anyone who is concerned about memory loss, confusion or changes in behaviour or mood would be a visit to their GP. This visit could take place in the surgery, or the GP may prefer to make a home visit as it is easier to observe changes and assess someone's behaviour in their own home. The person will be in familiar surroundings, and any problems with managing or navigating these surroundings will become clear. The GP will use the following procedures for the assessment:

- **Mental ability tests –** These are designed to test thinking, memory and orientation. This could be in the form of an interview, a written test or a series of questions that test cognitive ability and often comprises a Mini-Mental State Examination.

- **Physical examinations and tests –** These are likely to happen if another medical condition is suspected, that is, stroke or Parkinson's disease. The GP will also take blood and possibly urine samples for testing. Through these examinations, they may identify other conditions that are the cause of the symptoms.
- **Looking at history –** The GP will spend time with the person to talk about when the symptoms began and how they are influencing the person's life now. The GP may also talk with someone who knows them well. They will read the person's medical history and those of close relatives and will do a full review of any medication being taken.

Once the assessment is completed the GP would explain their findings. If they are confident and experienced in the diagnosis of dementia, they would provide the diagnosis at this stage and, although this is not common. It is more likely that the GP, if they suspect dementia, will make a referral to a specialist service, such as a memory clinic, or a consultant.

The GP is more likely to be able to provide the diagnosis if the person has reached the advanced stages of dementia or if they are displaying symptoms of a common form of dementia, such as Alzheimer's.

Stage 2: Referral to a specialist

It is usually the GP who would make the referral to the consultant. If the person or their carer feels that this referral is needed, but it is not offered by the GP, they can ask for a referral to be made, to either one of the following:

- The Memory Clinic
- The Community Mental Health Team

Consultants have specific training knowledge and experience to be able to diagnose dementia. Due to their areas of specialism, they have access to specific investigations based on the person's age and symptoms.

- Neurologists specialise in diseases of the brain and nervous system and some who have specialisms in diagnosing dementia. It is typical for neurologists to see younger people, below 65, who are displaying symptoms of less-common forms of dementia.
- Geriatricians specialise in the care of older people in relation to physical illnesses and disabilities. If the person is in poor general health, the GP may refer them to a geriatrician to eliminate the possibility of the symptoms being connected to another illness or to establish if the person has a physical illness as well as dementia.

- Old-age psychiatrists specialise in the mental health of older people, including dementia. They can also offer support to younger people with dementia if a suitable specialist is not available for the younger age group in the person's area.
- General adult psychiatrists diagnose and treat a range of mental health issues and generally work with younger people who are under the age of 65, they would also be involved in supporting the diagnosis of young dementia, diagnosed in people under the age of 65.

Consultants work with doctors. It is possible that the person may not always have appointments with their consultant, but the consultant is responsible for the care of the person and they will talk to the doctor about the person in their care.

The consultant and their team will work in a multi-disciplinary team with other health care professionals to ensure the best care for the person who is being assessed. This team would include psychologists, occupational therapists, social workers, speech and language therapists and dementia advisers, and it would involve any health care professional particular to the person's circumstances and symptoms they are presenting.

Stage 3: Assessment

The assessment for dementia can be confusing and daunting for the person and the person caring for them. It is good for people to have the opportunity to talk about the assessment process with a professional; memory clinics offer this support. These are often referred to as pre-diagnostic counselling sessions. The sessions are available to the person and their family and carers, and are a good opportunity to talk about what people already know about dementia and to chat about concerns. These sessions can broaden people's understanding of:

- why the person has been referred for an assessment;
- what the assessment process involves;
- what support is available if a diagnosis of dementia is given.

An assessment can happen in a hospital while the person is there as an inpatient. This assessment would be similar to one conducted by a GP and includes looking at the person's history, physical examinations, mental ability tests and scans. However, the test of mental ability will be more detailed, looking at aspects of memory orientation, language and visuospatial skills (for example, copying shapes). The results of these tests can be utilised as a way to measure future changes in the person's memory, which can further aid the diagnosis in the early stages of dementia. These tests are usually conducted with a mental health nurse or occupational therapist. If the experiences of the person are quite complicated, a neuropsychologist would be involved.

A diagnosis of mild cognitive impairment may be given if the symptoms are mild, for instance, trouble thinking and remembering, and they could indicate depression if symptoms are not severe enough to be diagnosed as dementia.

Stage 4: Receiving the diagnosis

The final diagnosis will be given by the consultant or other professionals within the team, though occasionally the consultant will write a letter to the GP and the GP will make an appointment with the person to give the diagnosis.

The diagnosis will be based on the information gathered during the assessment, and should be communicated in a manner that is sensitive to the person's situation and needs. It is important that the person and those supporting them understand the diagnosis and are provided with the opportunity to talk through this and ask any questions they have.

The person has a right to know the outcome of the assessments for dementia, although they should be asked if they wish to know the outcome before being provided this information. Ideally this conversation should be held during pre-diagnostic counselling sessions that usually take place with experienced counsellors at a Memory Clinic.

If the consultant believes the person would find their diagnosis too distressing, or they may not understand the meaning of the diagnosis, these concerns should be discussed with their close family and friends. The purpose of this discussion would be to reach a conclusion, collectively, about what would be in the person's best interests regarding how they should be informed and when. This does not commonly happen but may be the case when:

- the person may also have another serious mental health or physical health condition;
- they have found the assessment particularly distressing;
- they have recently been bereaved.

During the diagnosis process, people often feel considerable uncertainty. On receipt of a confirmed diagnosis, people can be concerned about what will happen to them as dementia progresses. They, alongside their family and carers, can also feel unsure about the support they will need and what they will receive and who will provide this support.

Those providing the diagnosis support will answer any questions asked sensitively and at a pace that works for the person, explaining any terminology and answering questions to the best of their knowledge, whilst signposting to other sources of help when this is more appropriate.

Most memory services offer sessions, running for several weeks after diagnosis, at which the person, and those close to them, can talk through the next steps in more detail and receive further written information. These sessions usually cover medication, living well, driving, benefits, local support services, planning ahead, how the person is feeling and their emotions.

Stage 5: Ongoing assessment

The GP will remain responsible for the ongoing care of the person, and in the initial stages of the diagnosis, people will have access to a local memory service. These are either private clinics or run by the NHS and can include the Community Mental Health Teams for Older People. The memory service is usually short term, and once the person is discharged, their main contact will continue to be their GP. A referral can be made back to the memory service if the person's symptoms change, and for further advice and support.

After receiving a diagnosis of dementia, it is important that the person continues to receive support from knowledgeable and trained professionals. Continued medical and psychiatric consultation is an important source of help to identify the changes that are occurring and recognise them as part of the stages of dementia.

Although professional input is important, it is possible for this to be delivered in a way where the person is listened to and at the centre of all decisions and actions.

Medication

During the early stages, medication that has proven to slow the development of the condition may be prescribed. Good standards of practice determine that the person living with dementia must be involved in any discussions and decisions regarding accessing and receiving treatment, including the prescription of medication. As part of the decision-making process, accurate, clear and detailed information and advice regarding the treatment, including possible side effects, need to be shared with the person in a way they can understand.

Initially, the specialists involved during the assessments will prescribe the medication with the ongoing and routine prescriptions coming from the GP. Follow-up reviews should then take place every six months.

Support during assessment and diagnosis

The person who is undergoing examinations and receiving the diagnosis may find them difficult, and, if the person is diagnosed with dementia, they may find hearing the news and its implications stressful. '*Different people will have different experiences of receiving a diagnosis and therefore they will have different ways of coping afterwards*' (Ken, Social Care Institute for Excellence, 2020). The person may experience a mix of emotions at the point of diagnosis.

The news may be devastating: '*The first time they said, "You've got Alzheimer's disease", it was a hell of a shock. When they said, "There's no cure", I defy anyone to say that's not a shock*' (David, Social Care Institute for Excellence (SCIE), 2020).

The experiences in the early stages of dementia are not fully determined by the biomedical changes that happen; they are also influenced by what people do and say. A person who has recently received a dementia diagnosis may:

- fall into a *sick role* and disengage from all the activities they were capable of before the diagnosis;
- have concerns about how they will be cared for in the future;
- be aware of how the relationships they have in their life may change;
- be concerned about becoming a burden on the people close to them;
- be in denial regarding the diagnosis;
- become anxious and fearful about what the future holds.

It is difficult to accept the diagnosis and to come to terms with what this means for life now and in the future. '*No one has asked me about my feelings ... 'Are you feeling frightened?' We need emotional counsellors, there needs to be tests for emotional wellbeing alongside blood pressure etc.*' (Pete, Social Care Institute for Excellence (SCIE), 2020).

Denial of the diagnosis is not uncommon, including questioning whether the diagnosis is accurate and whether the tests were conducted correctly. The person could be of the opinion that what they are experiencing is simply the natural signs of ageing. Equally a diagnosis of dementia can cause frustration at how life is changing and upset at the loss of skills and abilities.

There can be a sense of shock, anger, isolation, loneliness, helplessness, sadness, despair, bitterness, shame, guilt and grief. Emotions may also fluctuate, going from one extreme to another. Not all the emotional responses are negative, as this example illustrates: '*Well you know for me it was a big relief – knowing that I could give it a name*' (Peter, Social Care Institute for Excellence (SCIE), 2020). For some, being diagnosed places a reason and explanation for the changes they have experienced. Research demonstrates that, although there are negative reactions and feelings at the stage of receiving the diagnosis, people can go on to experience positive emotions.

Realising it is not the end

I was diagnosed with Alzheimer's disease about five years ago in my mid-50s. Being diagnosed with dementia is brutal. It is horrendous. Only someone who has been through it can understand. You are given a mental death sentence.

I felt the doctor left me in that frame of mind and left me to pick up the pieces of my life. Someone needs to give you hope, and tell you that you won't be a burden. You have to realise that this is not the end. You walk away and nothing's changed but everything is different.

(Ken Howard, Young Dementia UK, n.d.)

Reflection

- What can be done to support the person receiving a diagnosis of dementia to remain positive, in control and well connected to their community?
- What do you think is important for you and the person you care for to receive from health and social care professionals and the other people in your life to ensure you both feel positive about managing dementia and the person's future?

Responding to the diagnosis of dementia

It may be wrongly assumed that the person who has received the diagnosis is no longer able to make decisions for themselves, do things for themselves or participate in their usual everyday activities, including attending the clubs and organisations that have been a big part of their life. Those close to the person can start making decisions on their behalf and doing everything for them. This may be due to concerns about the person's ability to carry on doing things for themselves and their safety. However, this response can result in a negative for the person. They would lose their involvement in their regular activities and the opportunity to maintain their skills and abilities. This poses the risk of them becoming detached from those around them and losing their confidence to do things for themselves, potentially leading to dependency on others.

It is important for the person with dementia to:

- continue to do all the things that they are capable of;
- participate in the activities they are interested in;
- maintain the relationships that are important to them;
- have the opportunity to express their concerns and talk these through;
- make plans for the future that are positive.

In the early stages of the dementia journey, it is essential to have a focus on what the person can do and what they want to do. Making time to discuss what good care and support would look like during the early stages of the dementia journey is vital to ensure that any support and care planned and provided going into the future is in keeping with the person's wishes and preferences. These discussions are better held at a time when the person can engage and those involved in the conversation can focus on the discussion to hear the person's voice, ideas and preferences. The person living with dementia should be, wherever possible, taking the lead in shaping their future and making decisions about the care and support they wish to receive. Others could be involved in the discussions at their discretion.

Advanced care planning (ACP)

The Prime Minister's Challenge on Dementia 2020 stated that every person living with dementia in England should be able to agree with the following statement: '*I am confident my end life wishes will be respected. I can expect a good death*' (Department of Health, 2015, p 6). This aim for dementia end-of-life care is currently adopted globally, with initiatives in Europe, the USA and Australia.

One of the quality standards set by the UK based National Institute for Health and Care Excellence highlights the importance of personalised care plans, which play a vital role in making the statement set out earlier a reality: '*People with dementia are given the opportunity to discuss advance care planning at diagnosis and at each health and social care review*' (National Institute of Health and Care Excellence (NICE), 2019).

ACP is an approach to end-of-life care planning that supports the achievement of this aim and standard.

An ACP is a record of the person's future preferences and decisions about their care and treatment, including the circumstances in which they want to refuse treatment. The completed plan can then be shared with those who are responsible for the person's care or treatment. It is best to complete the ACP at a time when the person is able to be fully involved and in control of the information added to their ACP and if possible soon after receiving a diagnosis.

Chapter 6 looks at how the techniques and tools from person-centred planning can complement ACP.

Key lessons learnt

- The dementia journey is covered by three stages: early/mild, middle/moderate and late/severe, with symptoms associated with each stage.
- Each person living with dementia will experience the dementia journey differently.
- Receiving a diagnosis can take a long time due to a number of different reasons; for example, access to specialist teams, or the delay that can sometimes occur

between the person recognising the signs of dementia and then seeking a diagnosis.

- There are standard assessment tools and approaches for diagnosis.
- Good, practical and emotional support at the time of diagnosis is extremely important.
- Advanced Care Planning is a method to capture a person's wishes for their future care and support.

Questions to aid reflection

1) If you are to be actively involved in the support or care of a person with dementia, what would you want to know, or need to know, about them to make sure that the support you offer is based on what the person wants?

2) If you know the person well; how they communicate and their preferences for the future, how would you share this with other people to make sure this knowledge and insight was not lost along the dementia journey?

3) When supporting a person to plan for the immediate or distant future, how would you ensure they were fully informed about all the options available to them and aware of the level of choice and control they have?

4 Great support in the later stages of dementia

This chapter covers:

- **The carers' perspective and rights**
- **Carers' eligibility for support**
- **Finding the right support**
- **Advocacy**
- **Beyond service-based respite, what taking a break looks like**
- **The dementia journey, the mid and late stages in the context of the individual's experience.**

Continued support at home

It may be the wish of the person with dementia and their close family for them to remain living in the family home. With planning and consideration for everyone involved, the right support can be put in place to make this possible. This could come from immediate family, friends, members of the community, community groups, paid-for support and professionals. While living within a family setting, the person with dementia may already be receiving support from family members. This support could cover a range of practical, physical and emotional support, which may include:

- checking that the person has everything they need with them before they leave the house;
- making notices so the person can remember where they have put their items;

- physically supporting the person to use the stairs;
- helping the person to get dressed;
- assisting with preparing and eating meals;
- ensuring everything has been left safely, for instance locking the door or switching the gas off;
- encouraging and supporting the person to take part in everyday activities.

It does not have to be the family members living in the same home who provide the support; other members of the family, friends and neighbours may provide worthwhile support too, such as:

- calling around for a cup of tea;
- ringing for a chat over the phone;
- helping with shopping;
- arranging and attending appointments.

A carer's circumstances will be unique, as it is for everyone, for example, family arrangements, friends, and how frequently there is contact and how well the person and their carer is connected to the local area. Circumstances and the availability of support will be dependent on a range of factors, which include:

- the relationships that exist amongst family members;
- the physical environment, home and surrounding area;
- existing caring arrangements, for example, children, grandchildren, and other elderly relatives;
- work commitments;
- the carers own health conditions.

When a family come together to provide support, each family member is likely to have their own unique perspective on the circumstances and the support and care they are able to provide.

It is important to appreciate that, although the terminology *personalisation* could lead to the assumption that it requires a focus on the person with dementia and getting it right for the individual, in practice it is important to have a much wider perspective that considers the person and the people around them.

Family and friends

It can be challenging for those close to the person to accept the person's diagnosis of dementia. It is helpful to help the person's network of support to build an understanding of what the diagnosis means, how the condition will affect the person now and the support they will need going forward. This can be achieved through sharing the information provided by the professionals involved when it is appropriate to do so, as well as making time for the conversations necessary to build this understanding.

The network of people around the person can build a better understanding of the help they can provide, if the person living with dementia, with friends and family, begin to document details about their life. This can also help inform future decisions about care, support and treatments. Chapter 6 provides guidance on how to record information and what type of information is useful.

As a family or support network, when in discussion, remember to keep communication positive. Focus the conversation on what is possible. Think about what arrangements and approaches would make sense for everyone and talk about what people would like to do. Positive communication can be challenging as people can find the change of circumstances difficult to come to terms with, so make time for people to share how they are feeling. It may be difficult for them to express themselves, and their emotions may prevent them from thinking or acting rationally. Recognise this and give the time necessary for people to voice their thoughts and feelings.

If conversations are focusing on negative aspects or if the content of the conversations are very emotional it might be useful to:

- talk about the feelings being experienced;
- offer support whilst also talking about the other sources of support available;
- give the time and space needed for the conversation;
- avoid saying things like 'do not worry' or 'there is no need to be upset';
- keep the conversation open so everyone involved feels able to come back to the conversation.

Recognising carers

Carers' organisations are recognising the increase in the number of people who are caring for a family member – with an estimated 10.6 million unpaid carers in the UK at the time of writing (Carers UK, 2022, p.1), and 10 per cent of carers supporting someone living with dementia (Carers UK, 2019, p.8). It is important to recognise the support carers need too: to have someone to listen to them and provide advice whilst also having some time to relax, as well as being able to continue to pursue careers, relationships and interests, and meet other responsibilities. There are a number of carers' organisations across the UK already providing invaluable support, which you can find in the Appendix: Useful organisations section at the end of this book.

The Care Act 2014 has improved carers' rights and the likelihood of carers receiving support from their local authority. More information about carers' support from the local authority is available later in this chapter.

The stresses of caring

Family members and carers can find looking after a family member with dementia worthwhile and fulfilling yet stressful. There are difficulties that are commonly experienced in the early stages of dementia, which include their relative with dementia:

- leaving the house and becoming disoriented;
- becoming lost when they are accompanied shopping or visiting public places;
- regularly having restless and sleepless nights;
- having repetitive behaviour, for instance continually asking the same question.

The boundaries and dynamics of the relationship can also change significantly; for instance providing personal care to a relative, or changing the financial responsibilities in the household.

These difficulties and increased responsibilities as a result of caring lead family carers to dedicate much more time to their relative. Over time the caring role can become consuming, and it can place restrictions on the caregiver's life, preventing them from:

- having some time for themselves;
- maintaining other relationships;
- being able to take care of other responsibilities;
- maintaining work commitment;
- following interests and hobbies;
- being able to exercise and take care of their own health.

The family member providing the care may reach a stage where they need to assess the demands in their life and prioritise them. This may result in reaching the conclusion that they will need to give up other commitments, such as work, hobbies, friendships or spending time with other family members. If they feel they are no longer able to maintain work commitments, either having to reduce their hours or give up their working role completely, this will impact their income and therefore the finances of the family home.

It needs to be recognised that providing care to a loved one is often something people find themselves doing, because of the circumstances they have found themselves in. When supporting a relative with dementia to plan for the future, the wants and needs of those providing care need to be taken into consideration too.

When professionals, such as social workers or GPs, become involved and begin to assess the person, it cannot and should not be assumed that a family member is prepared or able to provide the care and support their relative needs on an ongoing basis. Within the assessments it needs to be recognised that providing long-term care can lead to carers struggling with their own health, identity and sense of wellbeing and this must be prevented.

Recent reports by Carers UK, highlighted that:

- 30 per cent of carers had seen a drop of £20,000 or more a year in their household income due to caring. This illustrates that the pressure of providing care to a loved one is not only physical and emotional but also financial.
- Two in five (42 per cent) carers felt they had missed out on financial support as a result of not getting the right information and advice early enough.
- 75 per cent of carers found it hard to maintain relationships and social networks because other people do not understand the impact that caring has. Therefore, many carers may gradually begin to feel isolated and lonely, which, in turn, has a negative impact on their wellbeing.

Capturing memories

Understandably, witnessing a loved one experience the symptoms of dementia, such as memory loss, can be painful for a carer and family member. There are a number of ways to support the person to hold onto their memories and to have time reminiscing. Having photos and objects that the person can have to look at can help them to hold onto and recollect memories. Capturing memories, important moments, events and occasions can be valuable for people with dementia and their carers. If someone enjoys reminiscing and having the opportunity to do this, it is worth spending time bringing memories together in the form of photographs, stories, keepsakes and videos.

If the person living with dementia likes using technology, for instance iPads and tablets, it can be a good way to develop a memory/reminiscence aid. There are a number of Apps and resources available on the internet for this purpose and sometimes the standard features can be helpful, such as the photo gallery.

Some people find it useful to put together a memory box, where they can store all the items to look at when they want to, or when they are in conversation with a friend or family member.

Within the collection of memories, include information about important dates/special occasions to encourage the person to remember and celebrate these.

The middle stages of dementia

As dementia progresses, the person with dementia will become more physically and emotionally dependent. They need more support to ensure good physical care, such as healthy nutrition and hydration, and may need help dressing and maintaining good personal hygiene.

Some families will be willing and able to continue to provide this increased level of care, but this cannot be assumed or expected. For the person or people providing care, it can become tiring and stressful and will impact their quality of life and wellbeing to a greater degree.

A balance needs to be made between the level of care the family members want to and can provide and the provision of support from other sources, which could include other family members, people from the community or paid support.

There are many reasons why a family member may not be able to continue to provide the care and support required, some of which include:

- being physically unable to provide the care required;
- a decline in mental health, which could be linked to the impact of providing care/support;
- being at risk of losing a job/work or income;
- being unable to maintain relationships and at risk of becoming isolated.

It is during the middle stages of dementia that further support from the Local Authorities Older People team or Community Mental Health Team may be required. A LocalGov report found '*two-thirds of people experiencing dementia live in ordinary houses and 85% of them say they should like to stay there for as long as possible*' (Phillipa, 2019). There is a focus on ensuring that people can remain within the family home and the first form of support a person receives should start from this perspective if this is what they and their carer/family want.

If it does reach a stage where it becomes impossible for the person to remain living in the family home and, after all avenues have been explored to achieve this, for example, making adaptations within the home or paid support coming into the home, the local authority's team for older people will guide people through the options available to them, for instance sheltered accommodation, extra care housing, care and nursing homes.

Carers' rights

The number of carers within the country far supersedes the number of people who are employed in the health and social care systems. The 2011 census figures for the UK showed an 11 per cent rise in the number of carers since the 2001 census – increasing by over 620,000 to 6.5 million in just ten years and this number continues to rise (Carers UK, 2012, p 2).

Every year over 2.1 million adults become carers, and almost as many people find that their caring responsibilities come to an end (Hirst, 2014, p 5). This means that caring will touch the lives of the majority of the population as people all need or provide care or support to a family member at some stage.

The figures gathered and presented by the 2011 census bring to the forefront how heavily dependent the health and social care systems are on those who provide care on an unpaid basis within people's own homes. If all carers were to state that they could no longer meet their caring responsibilities, there would be a huge issue around capacity and sustainability in the provision of health and social care. Carers deserve the support and recognition for the care they provide to loved ones, and professional interactions should be respectful and considerate.

Carers have rights that are enshrined in law, covering England, Scotland and Wales. In England, the rights of family carers are identified in the 'Care Act 2014', which received Royal Assent on 14 May 2014. In Scotland, carers' rights are identified in the 'Carers (Scotland) Act 2016', which commenced in 2017–2018. In Wales, the 'Social Services and Well-Being (Wales) Act 2014' represents carers and their rights. As a result of these acts, local authorities have a duty to provide:

- support to carers based on the identified needs of the carer;
- relevant and timely information, advice and guidance;
- appropriate access to support for adult carers and young carers;
- access to advocacy.

In England, the Care Act 2014 grants carers the right to request support; upon this request the local authority also has a duty to complete a Carer's Assessment with the carer.

In England, anybody identified as a carer should be offered a Carer's Assessment. These are designed to capture the circumstances of the carer and areas of their life where support is needed, followed by a discussion regarding the support that would work well for them. The approach taken to complete it should follow the principles of personalisation; the carer should be central to the process, the assessment should capture the carer's perspective and the carer should be fully aware that an assessment of them is being undertaken with an explanation about what will happen after the assessment is completed.

The carer can invite other people to be involved in the assessment, such as other family members and friends. Upon completion, the information gathered determines whether the carer is eligible for social care support and they have a right to a copy of their final assessment. Those who are assessed as eligible have the right to access support that is available or to request a personal budget. The intention of the support being provided to the carer is to meet their own support and care needs and to reduce the stress of providing care.

Eligibility for carers in England

The Care Act 2014 removes the requirement for the carer to be providing substantial care on a regular basis. The only requirement is that the carer has a need for support, either now or in the future. The carer's eligibility threshold is based on identifying:

- whether a carer's needs are a consequence of providing necessary care for an adult;
- to what extent the carer's needs affect their ability to achieve specific outcomes, as covered later;
- to what extent their caring responsibilities are putting their health at risk;
- whether, and to what extent, the caring role they fulfil impacts on their wellbeing.

The eligibility decision is based on the carer's needs and should not reflect the needs of the person they care for, even if the person they care for is not eligible for social care support.

The eligibility of the carer is determined by the local authority, using a set of specific outcomes to establish entitlement. This set of specific outcomes is outlined in the Care Act 2014 Carers' Eligibility Social Care Institute of Excellence (SCIE) factsheet (Social Care Institute for Excellence (SCIE) factsheet (2020). If these outcomes, such as the carer being unable to engage in work, training, education or volunteering, apply to the carer due to their caring role, they then become eligible for support.

If the person providing care is deemed as eligible for social care support, they would be able to make decisions about how they would like to be supported, through the services available, via a personal budget or a mixture of the two.

A carer can utilise the personal budget available to them to shape up the support in a manner that would work for them and they could use part of their personal budget alongside service provision. As an example, a carer may wish to engage with a local agency to provide support in their home while utilising a personal budget to cover the cost of attending a weekly activity. Before the support is put into practice, a care and support plan would be completed which should cover the following areas:

- what the carer wishes to change and achieve;
- what support the carer wants to receive;
- how the budget available to the carer will be used;
- how the budget available to the carer will be managed.

Here are some examples of how a budget could be utilised:

- to employ someone to come and spend time with their family member so they can have a rest;
- to pay for someone to do household tasks;
- to pay for an item of equipment to reduce worry or concern, such as a non-intrusive monitor;
- to pay for something that helps them to relax, for example, a pond in the garden or yoga sessions.

Throughout the process of completing a Carer's Assessment and a care and support plan, a carer has the right to access and be supported by an advocate.

Advocacy

Mind (2018) states that advocacy '*means getting support from another person to help you express your views and wishes, and help you stand up for your rights*'. Advocacy Focus (2020) adds to this by furthering the definition: advocacy is a process that supports people to:

- *Have their voice heard on issues that are important to them*
- *Protect and promote their rights*
- *Have their views and wishes genuinely considered when decisions are being made about their lives*
- *Express their views, thoughts and concerns*
- *Access information, advice and guidance*
- *Explore choices and options for services and care*

Meanwhile, Wells (2006) notes that there are many definitions and approaches towards advocacy. The Care Forum (n.d.) sees advocacy as:

'taking action to help people say what they want, secure their rights, represent their interests and obtain services they need. Advocates and advocacy schemes work in partnership with the people they support and take their side. Advocacy promotes social inclusion, equality and social justice.'

Killeen (1996) alternatively highlights advocacy as a process of empowerment in which it:

'means making the case for someone or a group of people or helping them to represent their own views, usually to defend their rights or to promote their interests. The concept has a special relevance to people who are disadvantaged and as a consequence are less able to speak for themselves.'

These definitions demonstrate that advocacy is underpinned by human rights, social justice and equality. Advocacy is a vehicle to achieve outcomes that align with the values of inclusion, self-determination and personalisation.

An advocate can provide valuable support when people are overwhelmed about their circumstances. Accessing health and social care support can be a daunting and complex experience for people and it can help to be supported by a knowledgeable advocate, especially one who has a focus on reaching the best outcome for the person and can act as a guide through the health and social care processes. The input from an advocate can provide reassurance, especially for people who feel that they have not been heard.

Involving an advocate can be valuable for people with dementia, whether they are living at home, in residential care, in a hospital or in a nursing home. The involvement of an advocate to support people with dementia and their carers can make a considerable difference and provide the much-needed support at certain crucial times of the person's life, for instance:

- when someone requires health care intervention;
- during Assessments and care and support planning;
- during a Capacity Assessment or Best Interest meeting;
- when a person is moving, for example, from home to a supported living scheme or care home.

Advocacy is available in most local areas, and information about accessing advocacy should be available from GPs, social workers or through searching on the internet.

Older People's Advocacy Alliance is the only UK-based national organisation supporting independent advocacy services for older people. Their website provides up-to-date information regarding advocacy and rights. International organisations also exist for advocacy, including Dementia Alliance International. The contact details for both these organisations can be found in the Appendix: Useful organisations section at the end of this book.

Living in the family home

A number of factors will determine someone's ability to continue to live in their own home, including:

- if they live on their own or with other people;
- the ability, capacity and willingness of the people they live with to offer practical and emotional care and support;
- if the right level of paid or unpaid support can be offered from outside the family home to make it possible to remain living there;
- if the environment is suitable or can be adapted to be made more appropriate;
- if there are any significant risks identified regarding the person continuing to live at home and no practical way can be found to manage these risks.

To remain living at home, the person will require care and support to varying degrees, dependent on their circumstances. The introduction of personalisation and personal budgets means that there is now the freedom to create a support arrangement that would work best for the carer and the person they care for. This could include utilising the budget available to employ a support worker directly or to cover the cost of equipment to aid independence in the home (if the equipment is not already available from the occupational therapist team) or to purchase aids for communication which are not available through the local speech and language therapy team.

Respite/Short breaks

The term '*respite*' is not always used in practice now, and over recent years it has been replaced by terms such as '*short breaks*'. What respite and short breaks refer to is a source of funding or service that provides the opportunity for the carer to have a rest or a break from caring. Respite applies when people continue to live at home supported by a family member(s). In these circumstances, the family member may need a break to take care of their own wellbeing, to rest and recuperate. Respite is designed to meet this need, where the person with dementia can continue to receive support and care while their carer has a break. The length of respite support time should be determined by what the carer needs, and the period of time for respite could range from a few hours a week to a number of nights/weeks away. Traditionally, receiving respite has meant that the person with dementia has left the family home for a period to stay somewhere, which was usually a local residential care home offering respite accommodation.

Families of people with dementia often reported that respite, following this traditional model, causes stress and anxiety, rather than alleviating it, as the carer continues to be concerned about the welfare of the person they care for. They may '*feel torn between the necessity to have a break and their anxiety about the impact of institutional respite care on the person with dementia*' (Maayan, Soares-Weiser and Lee, 2014, p 6).

For the person receiving the care and support, the change in environment when staying at a respite service, getting accustomed to where everything is and receiving care from strangers can be a frustrating and confusing experience. In large and institutional respite settings, there is usually a high turnover of carers and different carers working each day. This level of constant change can be frightening and frustrating for a person with dementia. They may then find it challenging to adapt back into their familiar routine when they return home, especially if things have changed in their home environment while they have been away. This situation poses insurmountable difficulties for their family member(s) who are caring for them.

Carers and people with dementia often report that the anticipation of going into a traditional respite setting, such as a care home, can lead to feelings of anxiety, confusion, guilt, disorientation, concern and frustration (Leocadie et al, 2018). Respite, when provided in this way, can be more exhausting for the carer than having no break at all. When a carer has reached a stage where they need a break, it may seem that the only option is for the person they care for to attend a respite service, even if this is not what the person or their family wants.

Approaches to respite

Day care

Day Care centres usually open each day of the working week, depending on the area, and tend to be open during the day time. They offer care, support and activities, and people find them a good option for a number of reasons:

- They offer a link between the person with dementia, their family and health and social care employees.
- There is sometimes specialist support available in the setting, such as Occupational Therapists and Speech and Language Therapists.
- A varied range of recreational activities, some of which will be in keeping with the person's interests and hobbies, are often available.
- They present a chance to meet new people and form new relationships for the person with dementia and their carer.

Day care is a planned provision that has become the main source of community-based support for people with dementia. *Day care* is a term that embraces a range of activities delivered outside the home in a number of different settings, including day centres, community centres, residential care and hospitals.

Home care services

A care or support worker will come to the family home and spend time with the person with dementia while their carer has a break. The person with dementia does not have to leave the home to spend an extended period of time in a care setting, therefore reducing their potential level of anxiety and distress. Home care can be provided flexibly, whether this is for a few hours or a few days and the support can focus on achieving the best outcome for the person and their carer; this could include providing support with personal care, preparing meals, following hobbies and interests in the home or going out of the home to spend time in the community. This option also works well for people who are uncertain of service provision and feel more confident if they can be supported in their own home. In this situation, the carer or support worker is most likely to be employed by a voluntary organisation, through a local home care agency or directly by the family (Levin et al, 1994).

Sitting service

An alternative to paid-for respite is a service usually called a *sitting service*. This is provided through a network of local volunteers who will come to the family home and spend a few hours with the person, while their main carer has the chance to have a break. Due to the nature of this support, it can be assumed that the benefits of this approach will be similar to Home Care services, however as this is provided by volunteers, the range of what they can provide support with may be more limited.

Shared lives schemes or host family respite

In a shared lives or host family scheme, the person with dementia would stay with a specially trained carer in the carer's home. The person with dementia and the short break carer have the opportunity to build a positive relationship, and the environment remains consistent with the aim of providing a greater sense of stability. This provides peace of mind as the person with dementia is receiving the care and support they need in an environment they can become familiar with and people they can develop a trusted relationship with, as well as one-to-one interaction. '*Care recipients preferred a break in a homely environment to a stay in a residential home*' (Arksey et al, 2004, p 5).

Video respite

This approach was developed in the 1980s as a method to provide a short break of a few hours. It introduced the idea of the video tape as a form of engagement for the person with dementia which then provided a break for the carer. This approach can still be practiced using a DVD, a streaming service or an online download. '*The tape was well received by carers and care recipients, and was used regularly to create respite time*' (Arksey et al, 2004, p 6).

The idea of utilising technology as a resource for people with dementia has progressed a great deal. Technology is now used as a way of providing support, entertainment and information. This approach is especially useful when the person with dementia lives a greater distance away from their main carer. '*Used sensitively and thoughtfully, technology enhances rather than replaces human relationships and interactions*' (Social Care Institute for Excellence, 2017).

Multi-dimensional carer

The term *multi-dimensional care* means providing carers with a range of support services in addition to respite care, for example, education about dementia-related issues, training in lifting and handling techniques, support groups for carers or accessing individual counselling. This is designed to address the multi-dimensional problems and needs that carers are likely to experience and provide practical community-based services. This approach gives carers a choice regarding what support they wish to access. When asked, carers believed they had benefitted from this approach, and, whilst results of a research report conducted in 2004 suggested no gains in terms of phycological health of the person with dementia, the report revealed multi-dimensional care had positive effects on the person's behaviour (Arksey et al, 2004, p 6).

Vouchers

Carers may be offered vouchers rather than a service. Vouchers can then be utilised to choose from a range of options available that have been set up for the scheme; this could include gardeners, cleaners and home care. Often the list of services that can be accessed are listed on the vouchers. The voucher scheme for respite is not available in all places. The principle with the vouchers is to offer more choice to the carer, although this is unlikely to offer the same level of choice as a personal budget for respite (Think Local Act Personal, n.d.).

A personal budget and respite

People who access a personal budget can make decisions about the care and support that they receive and this applies to respite too.

People can choose to use their personal budget to cover the cost of having a break in a way that would work best for them. The budget could be used to go to the coast for the day or for going out for the evening with a neighbour or a friend or to pay for the carer to go for a break or holiday away from the family home while someone stays with the person they care for. Direct Payments as a method to receive a personal budget were welcomed as a way of increasing the flexibility of respite provision, and there was support from carers, respite providers and support workers for greater use of these options (Arksey et al, 2004, p 7).

Having a break does not always mean the carer and the person with dementia having to spend time apart. The opportunity to spend time together, away from the house, or visiting a place of interest or importance could make the best use of the respite funding available. Personal budgets provide the flexibility to make this possible.

Coping mechanisms for carers

It is important to recognise approaches to coping with the caring role and what helps to reduce the level of stress experienced. Research has shown that a reduction in expectation appears to help carers manage their situation, while having time to watch TV and read helped the proportion of carers involved in the study to relax and manage any distress (Saad et al, 1995, p 496).

Carers employ different coping styles; for example, adopting a problem-solving approach to address difficult situations that arise or adopting mindfulness or mediation to feel less overwhelmed by situations. Making time to follow interests or do the things that the carer finds relaxing, such as reading, meditation, watching TV or a hobby like gardening or exercise can be extremely beneficial to their wellbeing.

The provision of guidance to develop these mechanisms needs to be considered as part of the support carers are offered. Groups that may be able to offer some assistance in these areas include carers, support groups and dementia support groups, as mentioned in the Appendix: Useful organisations section at the end of the book.

Later stages of dementia – moving on from home

In the last chapter, the indications of the later stages of dementia were introduced, here are some of the other symptoms a person is likely to experience:

- memory loss;
- loss of communication skills;
- loss of movement;
- eating and weight loss;
- problems with continence.

Memory loss

Memory loss is often very severe in the later stages of dementia. The person may still appreciate and respond to stimuli such as music, scent and touch, and this stimulus may trigger related memories. The person with dementia may:

- find it difficult to remember the people closest to them;
- struggle to navigate around environments that were previously familiar to them;
- have flashes of clarity for short periods of time;
- be reliving previous life experiences and be looking for people or things that have not been a part of their life for a long time.

Loss of communication skills

The person with dementia is likely to find it difficult to communicate with others. It is important that they continue to experience a warm, respectful and reassuring connection with those around them. The communication will move from responding to words and clear signs of understanding to conveying thoughts in different ways, such as visible cues, body language, expressions and gestures.

Movement

The ability to walk will gradually decline, and performing everyday tasks unaided will become challenging. The person may begin to shuffle, walk unsteadily and seem slow, which are the first indications that their mobility is declining. Bumping into things, dropping objects or falling are common and may be more severe in people who have had a stroke or arthritis. Input from a physiotherapist and occupational therapist is important at this stage to be able to identify the adaptations that are needed in the home environment, and the pieces of equipment and exercises that would offer most assistance to the person and those caring for them.

Eating and weight loss

Towards the later stages of dementia, people are likely to lose weight. This can lead to a reduced ability for the immune system to function effectively, as well as an increased likelihood of falls. It is important to support the person's nutrition and seek advice on choosing and preparing nutritious food, maintaining hydration and how to offer physical assistance with eating. The person may find it difficult to swallow due to the deterioration of muscles and reflexes. This can lead to choking, which, in turn, can result in chest infections. At this stage, the GP can make a referral for a speech and language therapist or nutritional specialist, who can assess the person's diet and identify what is going to work well for them, signposting them to specialist diets or equipment.

Problems with continence

During the later stages, many people lose control over their continence. This is not inevitable, but there are many reasons why it may be experienced:

- severe constipation;
- not remembering to go to the toilet;
- side effects of medication;
- urinary tract infections;
- not recognising the physical indicators that they need to go to the toilet.

The GP can make a referral to a community nurse or a continence specialist, who can provide advice on personal care and the equipment and products available.

It is in the later stages of dementia that it becomes important to draw on what has been learnt in the earlier stages of the dementia journey about the person's wishes and preferences.

End of life

Problems with capacity and communication in the later stages of dementia mean that it would be difficult for the person to make their wishes and preferences known.

It is at this stage where the information gathered in any future planning documents, completed in the earlier stages of the dementia journey, would help guide those involved in the person's life. This includes information gathered on person-centred plans, advanced decisions and advanced care plans.

These plans can capture the following information:

- Where would the person want to be?
- Who would they want to have with them?
- What treatment would they receive?
- What kind of environment would they want to be in?

Understanding and respecting the individual's wishes and preferences is as important for them as it is for those around them. Loved ones and carers can experience a sense of ease during an emotionally challenging time if they know what to provide and how to provide their care and support.

This is a difficult time, emotionally and physically, for everyone involved. To be able to approach and experience the end of life in a manner that upholds respect for one another and the dignity of the person, while creating an environment that brings people together and provides comfort, is of the greatest importance.

Key lessons learnt

- Carers have rights and may also be eligible for support.
- The importance of recognising the contribution made by carers has been addressed, as has the need to provide appropriate and timely support, including emotional support and access to information and advice.
- It cannot be assumed that carers are prepared or able to continue to provide the care required.
- Respite provides support to carers, enabling them to have a rest and a break from caring. There are now a range of options to access respite including creative and personalised approaches.
- The Care Act 2014 defines the eligibility for carers' support and access to a personal budget.
- People have a right to access independent advocacy, and this is available in most local areas.

Questions to aid reflection

1) People with dementia are equal citizens with rights. Advocacy can support people to have a voice and to have their rights protected. In what areas would advocacy be able to support the person you care for?
2) How would respite support you as a carer and how would you make use of the respite budget available to you?

5 The shape of care and support

This chapter covers:

- **The settings available for the provision of care at the later stages of dementia**
- **Legislation and eligibility for health and social care support**
- **Personal budgets and personal health budgets**
- **Personalisation in care settings and services for people with dementia.**

Caring at the later stages of dementia and end of life

There are a range of settings where people living with dementia could be receiving care and support during the later stages of dementia, such as:

- their own home;
- their own house/flat or bungalow that is part of a sheltered living complex;
- a specially designed residential setting;
- a supported living scheme;
- a residential care home;
- a nursing home;
- temporarily on hospital wards, offering acute provision of care, which is neither intended for nor appropriate for people with dementia.

Care and support in a hospital setting

During emergency or crisis situations, people living with dementia can find themselves admitted to units on hospital wards when it has not been possible to access appropriate alternative accommodation. A report by Public Health England published in March 2015 (p 1) shared the key message that there had been a '*48% increase in emergency admissions involving people identified as having dementia since 2008/09*'. In February 2020, the proportion of patients aged 75 and over who were initially identified as potentially having dementia and who were admitted as an emergency for more than 72 hours was still worryingly high at 81.7 per cent (NHS England, 2020, p 2).

For a person with dementia, being in hospital or preparing for a hospital visit can be frightening and disorientating. The hospital environment can be difficult to navigate, noisy loud and busy. To make the stay at hospital less stressful, it is important that the hospital staff get to know the person and how they need to be supported or cared for whilst in that environment. Sharing information with the hospital staff can help them get to know the person, for instance, providing information about:

- how the person communicates;
- the decisions the person will make, and how they will make these decisions;
- the important people in a person's life with their contact details;
- the person's important day-to-day routines.

Approaches to recording this information are available in Chapter 6.

The Alzheimer's Society (2017) produced a leaflet called 'This Is Me' which can be used to record useful information about the person with dementia that can be shared with hospital staff. This includes:

- the name by which they wish to be known;
- brief details of their normal routines, including whether they need reminders or support with washing, dressing, going to the toilet, eating and drinking or taking medication;
- details about food they particularly like or dislike, or any difficulties they have eating;
- facts about illness or pain they experience;
- details about any cultural or religious needs;
- knowledge about sleeping patterns;
- levels of activity, past times and interests;

- what helps to create calm during periods of agitation;
- whether they have an advanced decision or a personal welfare lasting power of attorney.

Deciding where to live

Personalisation promotes the idea that people with dementia should have control and choice and that they should be supported to make those choices a reality.

This includes the choice of where to live, or whether to move to another setting or to remain within familiar surroundings in the family home and within their local community. Other factors play a part in influencing this, for instance, the onset of other health issues, changes in family dynamics, a reduction in the support and care available from family and friends or the physical environment posing difficulties. In these circumstances, the person with dementia and their family may have to reconsider the choices available to them.

Involving people such as family, friends, advocates, attorneys, social workers and occupational therapists in the conversation about the best option available can help in the decision making process. Those involved in such conversations should refer to any past discussions and information that has been recorded about the person's preferences of a place to live while the person had capacity.

Being prepared for the eventuality of moving to another setting earlier on in the dementia journey and having discussions about a person's preferences can help to inform any decisions that need to be made at a later stage.

These discussions are better if the person is aware of the range of options available to them and that their decisions are recorded for future reference.

The public perception of residential care

High-profile media coverage of abuse within care home settings, such as the 'Nursing Homes Undercover' (BBC, 2016) scandals, drew attention to the neglect and abuse of older people. This is likely to have contributed to a change in public perception of care homes; a poll of 2050 people by think tank Demos (2014, p 26) revealed that:

- more people associated a care home with a place of abuse rather than a place of safety;
- only one in four people said they would consider moving into a care home if they became frail in older age.

In addition to the Demos think tank findings above, a YouGov survey (2013) commissioned by the Alzheimer's Society demonstrated that 70 per cent of people fear going into a care home. The serious and devastating insights into abuse within residential settings have prompted closer inspection of the quality of provision and the following areas of weakness have been identified:

- Hospitals and care homes are organised through regimes and routines that often take precedence over the wishes of individual residents.
- There is a lack of understanding regarding dementia among people within residential settings.
- The arrangements regarding sharing information about residents are not adequate.
- In the worst cases, the practice of the provision of care is not adequate or person-centred. People with dementia experience being treated like objects and are moved about the care home, for meals for instance, without any discussion or agreement.
- The safeguarding and whistleblowing measures lack influence in practice. In extreme cases, people find themselves emotionally and physically abused.
- There is very little time, if any, for the staff working within the care home to have quality time with the people living there.
- Basic needs are the main consideration and, even if these are met, there is very little consideration for how the people want these needs to be met or for any other need a person may have.

Residential care settings vary in their quality of provision. The national drive for improvements within care home settings should have now made a difference in how people experience the care provided. If the care home has good leadership and is managed well, the opportunity for residents to get involved in activities they enjoy and spend meaningful time with others is more likely to occur. If the home has a person-centred approach, the skills, interests and passions of the residents will be taken into consideration.

So, what are the positives of care homes that are worthy of recognition?

The following examples shared by professionals, people with dementia and their families demonstrate examples of good person-centred support within a care home setting:

> *'We saw that the trust had made efforts to increase the profile of people with dementia and improve their care by encouraging involvement of non-medical staff, relatives and volunteers from the local community. They visited patients on wards during mealtimes to help people to eat their meals and have a 'chat'. Open visiting times had also been introduced and this enabled more flexible and enhanced contact for people with dementia.'*
>
> (CQC, 2014, p 25)

> *'Mum now has an appointed carer who liaises with all the carers who were on duty, her care plan is clearly updated [and] we have a 'go to' person. There are lots of*

> *activities and the home is run as a 'family', the carers are her family and mum's wellbeing and dignity is maintained. She is nurtured.'*
>
> *(CQC, 2014, p 33)*

Person-centred support demonstrates that good quality care and support can be achieved within a care home. The examples given show how important good communication is along with an environment that is welcoming to families, friends and members of the community.

Here are some further considerations regarding residential care settings:

- The provision within a care home setting is externally inspected and regulated by the CQC who are committed to training all inspectors in dementia.
- According to regulation 18 of the Health and Social Care Act 2008 (Regulated Activities) Regulations 2014, the staff within a residential setting must receive the support, training, professional development, supervision and appraisals that are necessary for them to carry out their role and responsibilities CQC (2015, p 70).
- Medication is prescribed by a GP and dispensed by a pharmacist who is involved in the residential care practice and should be administered and managed by trained and competent staff.
- Specialist clinical input can be provided from physiotherapists, speech and language therapists, occupational therapists and dental hygienists.
- Residential care settings should provide a choice of nutritious, appetising meals that meet the needs and choices of individuals and support with eating where needed (CQC, 2017).
- The person should be able to make their room into an environment that they are happy with.
- There are no concerns regarding utility bills and upkeep as this is all managed by the residential setting; these costs are incorporated into the overall residential fee, which can range significantly – some care homes do charge the residents for additional services and a report published by the Citizens Advice Bureau titled 'Hidden Charges in Care Homes' found that 40 per cent of the 464 homes contacted for the report charge for the installation of a telephone and the phone calls made (Greenhalgh and Whitham, 2016, p19).
- The person living in the care setting will have access to 24-hour care and support.

It is important for anyone considering a care home option to have the opportunity to visit different places and to decide which home best suits them.

The negative findings surrounding residential care homes, nursing homes and hospitals are not unique to these settings; issues can also occur in day care settings, shared lives schemes and when someone employs their own team.

There is the potential for a positive or negative outcome from all the options available to a person with dementia and their family. The outcome is dependent on building trusted relationships with good communication around expectations and what is needed as well as the provision of training and support.

The following points are recommended to achieve a positive outcome:

- The person with dementia is actively involved and taking the lead in deciding where and how they want to be cared for and supported.
- The person and their carers/family know that if the care and support they receive is not working for them it can be discussed and changed.
- Everything about the person is understood; how to communicate with them meaningfully, their interests, background, culture, important relationships in their life, how they follow, and would like to be supported to follow, day-to-day routines and their preferences on lifestyle choices.
- Everyone involved continues to talk about how things are and can address any areas of concern or areas that can be improved.
- The person is treated with dignity and respect always, no matter where they are and who is providing the care and support.
- The people providing the care and support experience a good environment to work in with a culture that is supportive and encouraging; they are also given the opportunity to be trained in the areas that would make a difference to the people they provide care and support to, and they are given the support needed to fulfil their role well and confidently via team meetings and supervisions.

Personal budgets

In the previous chapter, the concept of personal budgets for carers was explored. In this section, personal budgets for people living with dementia will be introduced.

People accessing health and social care support should experience choice and control over how they are supported with access to information and advice about the options available to them. A personal budget or personal health budget means that the person and their family have access to the funding available to them to be able to determine the best use of this resource.

With the budget available, the person can purchase services or utilise the budget to cover other costs to support any continued connections within their community.

A personal budget comes from a local authority social work team and a personal health budget comes from the clinical team in the local Integrated Care Boards (these have replaced the clinical commissioning groups).

People with dementia who are eligible for support from their local authority or Integrated Care Boards have had the opportunity to access a personal budget or a personal health budget for a number of years. In this time people have shared stories about how they have utilised their budget and the difference this approach has made to their quality of life; here is one such story:

> *We didn't want to consider Mum going into a home and agreed as a family that she had the right to stay in her own home. We visited Residential Care Homes and Nursing Homes on the advice of the Social Worker but found none we felt would be good enough. Equally I had past experience of managing homes and knew Mum would not get the 1:1 attention she needed in those environments. Use of local respite and homecare services also reconfirmed to us that 1:1 support would give Mum the quality of life she deserved.*
>
> (Leslie, 2016)

Overall, people have shared how accessing a personal budget or a personal health budget has enabled them to:

- maintain the relationships that are important to them;
- make choices about how they are supported;
- have a break in a way that works for them;
- live in their own home;
- continue to fulfil meaningful roles in the community;
- plan for a future that matches their aspirations and goals;
- pursue their interests.

The difference between being person-led and service-led

Person-led	Service-led
Planning *with* the person	Planning *for* the person
Talking *with* the person	Talking *about* the person
Doing things *with* the person	*Following* a programme
Having a life and the person is living somewhere and in a way that they have chosen	The person is living where they 'fit' or where this is a 'space'
Starting with *what matters* to the person	Starting with *what's wrong* with the person

Addressing health and safety to enable the person to achieve what they wish to accomplish in life	Health and safety determining what happens and restricting a person's choices or options
The team 'supporting', 'assisting', 'helping'	The team 'letting, 'allowing', 'placing'
The team suggesting and the person deciding	The team setting the person's goals
Living plans – changing with the person as their circumstances and needs change	Determining the person's life by an outdated plan – 'updated annually'

Eligibility for health and social care

Understanding the eligibility criteria for health and social care is essential in assisting someone to access the support available to them.

Eligibility for social care

The Care Act 2014 states the eligibility threshold for people living in England. Under this Act local authorities in England now have a duty to make sure that people living in their areas receive: preventative services, the information and advice they need and the option of a range of provision of high-quality, appropriate services (Department of Health and Social Care, 2016).

The aim of these duties is to ensure that people receive the support they require at the time they require it, moving away from lengthy processes that cause people to wait too long to receive the support they need. Prevention is the key message, working in a way that avoids the need for people to access care home settings or hospitals by offering responsive and person-centred support.

In Scotland, the national eligibility criteria for social care were agreed upon by the Scottish government and the Convention of Scottish Local Authorities (COSLA) in 2009, which were designed to be applied across the whole of adult social care in Scotland. Each local authority has their own eligibility criteria for managers and practitioners to refer to. The Scottish government's National Eligibility Framework employs a four-criterion approach, categorising risk as being critical, substantial, moderate or low:

- **Critical risk** – Indicates that there are major risks to a person's health, well-being and independent living, which calls for immediate or imminent provision of social care services.
- **Substantial risk** – Indicates that there are significant risks to a person's health, well-being and independence, which calls for immediate or imminent provision of social care services.

- **Moderate risk** – Indicates some risks to a person's health, well-being and independence, which calls for some provision of social care services.
- **Low risk** – Suggests that there may be issues surrounding quality of life, but these are low risk to the person's health, well-being and independence, which may require limited provision from social care services, possibly alongside alternative support or advice.

In Wales, the eligibility to access social care is determined by the statutory document 'The Care and Support (Eligibility) (Wales) Regulations 2015'. This document states that an adult would meet the eligibility criteria if the need arose from the adult's physical or mental ill health, age, disability, dependence on alcohol or drugs or if the need related to one or more of the following:

- ability to carry out self-care or domestic routines;
- ability to communicate;
- protection from abuse or neglect;
- involvement in work, education, learning or in leisure activities;
- maintenance or development of family or other significant personal relationships;
- development and maintenance of social relationships and involvement in the community;
- fulfilment of caring responsibilities for a child.

The eligibility for social care support is determined in the instance where the need is such that the adult is not able to meet that requirement, either alone, with the care and support of others who are willing to provide that care and support, or with the assistance of services in the community to which the adult has access.

In Northern Ireland, the eligibility for social care is defined by the Department of Health, Social Services and Public Safety. The system in Northern Ireland is similar to that in Scotland, where the eligibility criteria are determined in each local area for managers and practitioners from that area to refer to. This is based on a national eligibility model centred on the four thresholds of low, moderate, substantial and critical.

Once eligibility has been determined, the exploration around the person's preferences regarding how they utilise the support available can begin; including the option to access a personal budget to set up their care and support in their own unique way.

Well-being principle

The Care Act 2014 establishes the 'well-being principle' – an overarching approach that local authorities should take when exercising their responsibilities under the Act.

Well-being covers a range of outcomes such as:

- physical well-being;
- mental well-being;
- emotional well-being;
- participation in work, education and training;
- social and economic well-being.

(Care Act 2014. s.1(2) (b)(e)(f))

Local authorities must ensure that all aspects of well-being are taken into account when supporting a person through assessment and care and support planning. To establish whether a person is eligible for social care in England, firstly a referral would need to be made to the social work team under the appropriate local authority, which could be the:

- older people's team;
- mental health team;
- generic/neighbourhood team.

Most local authority areas now have a single point of access with one telephone number to call for all social care support; this is the number that would be called to request an assessment. The team would refer this to the appropriate internal team to complete an assessment. The referral could be received directly from the person with dementia, somebody on their behalf or from another professional, such as a general practitioner.

The Care Act 2014 gives local authorities a duty to carry out a Care Act Assessment for all people presenting a need for care and support. The assessment process would determine whether an adult is eligible for care and support. The assessment should be carried out regardless of the person's financial position and any initial opinion the local authority has regarding the person's level of need (Department of Health and Social Care, 2016). The assessment process must address the outcome areas, as identified below, and the completion of the assessment must fully involve the person, while utilising appropriate communication skills and involving their family.

The Care Act calls for strengths and assets-based approaches to assessment, which aims to capture the skills and capabilities of the person, as well as the people and places that are important to them.

Assets – In this context, assets are the relationships the person has and their connections within the local community, which could lead to finding ways they wish to be supported.

Strengths – Here, strengths refer to the skills and knowledge the person has and how they would like to make use of these within their life and ongoing support.

The Care Act 2014 stipulates that people can complete their assessment, either on their own or with their family and friends, with support as they require from a local authority-based, appropriately trained assessor:

> *'If the person agrees and has capacity, they may also carry out a self-assessment, where the person takes the lead in identifying their needs and outcomes. The local authority will still be involved to help support the process, and to be satisfied that the person has identified all of their needs, but the person can take more control.'*
>
> (Department of Health and Social Care, 2016b)

This provides the opportunity for the person to decide what needs to be recorded on their assessment and the flexibility to complete this at a time and pace that works best for them, involving the people of their choice to complete the information as accurately as possible, providing the level of detail necessary to capture their circumstances.

People are not always aware that they have the option available to them to fill in their own assessment. Consequently, the assessment process can still be experienced as a professional-led exercise. Those working within local authority settings can often be pushed to capacity and therefore the time needed to complete an assessment thoroughly and at a pace that works best for the person is not available. This can result in the information collected for the assessment being gathered at times and at a pace that are not good for the person, resulting in the information being sparse and inaccurate. On the other hand, those accessing social care for the first time can find it reassuring to have support and guidance from a professional through the assessment process; therefore, in this situation, a self assessment may not be the option to choose.

Assessments can be challenging to complete as a person's abilities can change and fluctuate each day. One day a person with dementia may be completely confident in their home and go about day-to-day activities and the very next day they may feel disoriented and confused. It is important that an assessment captures this, the person's recent history and how their day-to-day experiences fluctuate.

After completion of the assessment, the local authority will determine whether the individual has *eligible needs*. Under the Care Act 2014, a person is eligible for social care support if the following are true for them:

- They have needs due to a physical or mental impairment or illness.
- Those needs mean that they are unable to achieve two or more specified outcomes (which are detailed below).
- As a consequence, there is, or is likely to be, a significant impact on their well-being.

The Care and Support (Eligibility Criteria) Regulations 2015. s.2(1)(a-c).

The *specified outcomes* are:

- managing and maintaining nutrition

  (The assessment should consider whether the adult has access to food and drink to maintain nutrition, and that they are able to purchase, store, prepare and consume the food and drink safely.)

- maintaining personal hygiene

  (The assessment should take into account the adult's ability to manage their own hygiene, maintain hygienic standards and routines and other activities relating to hygiene, such as washing their clothes.)

- managing toilet needs

  (The assessment should consider the adult's ability to access and use a toilet and manage their toilet needs consistently, hygienically and safely.)

- being appropriately clothed

  (The assessment should review the adult's ability to dress themselves and to be appropriately dressed, for instance, in relation to the weather and to maintain their dignity.)

- being able to make use of the home safely

  (The assessment should examine the adult's ability to move around the home safely, which could, for example, include using the stairs, kitchen facilities or accessing the bathroom. This also includes the immediate environment around the home, such as access to the property, for instance, steps leading up to the home.)

- maintaining a habitable home environment

  (The assessment should consider whether the adult can maintain a clean and safe home. A habitable home also has the essential amenities, such as water, electricity and gas.)

- developing and maintaining family or other personal relationships

  (The assessment should observe whether the adult is lonely or isolated, either because their needs prevent them from maintaining the personal relationships they have or because their circumstances stop them from developing new relationships.)

- accessing and engaging in work, training, education or volunteering

  (The assessment should consider whether the adult has an opportunity to involve themselves and contribute to society through work, training, education or volunteering on a basis of their interests and the opportunity being meaningful to the person. This includes considering the physical access to any facility and support the person needs in order to participate in the work, training, education or volunteering of their choice.)

- making use of necessary facilities or services in the local community, including public transport and recreational facilities or services

  (The assessment should take into account the adult's ability to get around in the community safely and consider their ability to use such facilities as public transport, shops or recreational facilities.)

- carrying out any caring responsibilities the adult has for a child

  (The assessment should consider any parenting or other caring responsibilities the person has. The adult may, for example, be a step-parent with caring responsibilities for their spouse's children.)

An adult is to be regarded as being *unable to achieve* an outcome if they:

- is *unable* to achieve it without assistance or

  (This includes situations where an adult would be unable to do so even when assistance is provided. It also includes scenarios where the adult may need prompting, for example, they may be physically able to wash but they need reminding to do so.)

- can achieve it without assistance *but doing so*:

  – causes them significant pain, distress or anxiety

  (For example, an older person with dementia may be able to prepare a meal, but may become confused with where items are kept in the kitchen which leads to anxiety and distress.)

  – endangers or is likely to endanger health or safety or

  (If the health or safety of the adult or another member of the family, including any child, could be endangered when an adult attempts to complete a task or an activity without relevant support. For instance, if the person is likely to leave the gas on after making some food.)

  – takes significantly longer than would normally be expected

  (For example, an adult with a physical disability can dress themselves in the morning, but it takes them a long time to do this, leaves them exhausted and prevents them from achieving other outcomes.)

If the management of the above outcomes are a concern for the individual and they cannot achieve two or more of these, then they would be deemed as being eligible for social care support.

The Care Act also provides local authority discretion to meet needs that are not covered under the eligibility definition stated above. This would be in circumstances where the person is at risk or support would be preventative to a person's circumstances or condition worsening. The decision to provide support in these circumstances is at the discretion of the local authority, if enough evidence and details are provided to support the decision.

Once a person's eligibility for social care support is established, a social worker is assigned to work with them and their family. The social worker will guide the person and their family in the development of a care and support plan, which is written once the eligibility has been determined. During the development of this plan, the person with dementia, with support from any other person they would like to have involved, will be actively deciding how they would like to be cared for and supported. Information published by the Department of Health and Social Care in 2016 stated that the local authority must do everything it reasonably can to reach an agreement with the person as to how their needs should be met.

As part of the care and support planning process, the local authority will inform the person with dementia how much budget is available to them to meet their social care needs. This will be calculated based on the information provided in the person's assessment and set at an amount that the local authority believes will meet the identified needs of the person. At this stage, the level of budget communicated to the person is known as the indicative budget, an indication of the budget the person will receive. The indicative budget provides a guide for the person and their family when developing the care and support plan; knowing the indicative budget helps:

- make it clear what options are available to the person;
- create the opportunity to make the best use of the resources available to them;
- consider what is possible rather than just what is available.

The plan can then be written demonstrating how the person wants to be cared for and supported and how the resources available to them will help achieve this. The resources would include support from family and friends, for example a friend may be happy to take the person's shopping or a neighbour might assist them with their garden. The person may also include the connections they have in the community, such as groups, activities and grants that they are accessing.

The person is entitled to a copy of their assessment and their care and support plan. Having access to a copy of the completed assessment will prove useful when completing the care and support plan as reference can be made to how the person's outcomes and needs, as identified in the assessment, are going to be met.

Financial assessment

Access to social care support is means tested; once a person's eligibility has been determined this would be followed up with a financial assessment. This is usually done by the financial team within the local authority and they would determine the person's position to make financial contributions towards the cost of their care and support.

The financial assessment should take into consideration all aspects of the person's finances. A person's assets, which would include their income, savings, property and pension, would be calculated as part of this process. Their expenses would also be calculated and this would include any disability-related expenditure (an expense the person has as a direct result of their disability). This could include the cost for: equipment and maintenance, additional travel expenses, home support, supplements, extra training, and seeing specialists. When recording the expenses on a financial assessment, this would encompass all the expenses a person has – utility bills, debt payments, hire purchase payments and any other payments that leave their account. The local authority may ease the administration of a financial assessment by placing figures on these amounts that is deemed as the average household cost for laundry and utilities, the amounts used in the financial assessment has to reflect the true cost to the person for these items. Local authority financial assessment teams will ask for evidence of these expenses and this can be provided through receipts, bills or bank statements.

Local authorities have to apply the rule of affordability when completing financial assessments and calculating the contributions a person needs to make to their own care; these contributions should be affordable and not place people in a position of not being able to afford a standard of living that is acceptable.

What is the difference between a personal budget and a personal health budget

A personal budget comes from social care and is available to meet assessed social care needs and outcomes.

A personal health budget is provided by the NHS and is available to meet identified health care needs.

Health funding and personal health budgets

The 2010 whitepaper 'Equity and Excellence – Liberating the NHS' outlined new initiatives to improve outcomes and to transform NHS culture by increasing choice and control through the offer of personal health budgets to those eligible for Continuing Healthcare, often those with complex, chronic and long-term health conditions (Department of Health, 2010). The guidance on the legal rights to have a personal health budget highlighted that, from October 2014, people have had the right to ask for a personal health budget as an alternative to receiving service provision (NHS England and NHS Improvement, 2019, p 4).

Due to successful outcomes achieved via personal health budgets, the NHS is now rolling out this approach and offering personal health budgets to people accessing wheelchair services, people with learning disabilities or people with mental health conditions. This is relevant for people with dementia as those with long-term health conditions that require a consistent level of health intervention and care may be eligible for Continuing Healthcare funding. To access this, a referral to the local clinical team is required for an assessment of eligibility. The referral needs to be sent to the clinical team by the person's GP or social worker.

Checklist tool

For most people, the first step to establishing whether a person would be eligible is to have an assessment with a health or social care professional using a screening tool called the checklist tool. This helps health and social care professionals determine whether the person's care needs may be of a level or type that indicates that they may be eligible for NHS Continuing Healthcare. If this screening suggests that they may be eligible, the person's local ICB will write to them to confirm that they will be referred for a full assessment for NHS Continuing Healthcare, which is called a decision support tool.

Decisions support tool

In this assessment, information will be gathered regarding the person's circumstances and diagnosed conditions. The person should be fully involved in the assessment and decision-making process. Their views about their needs for care and support should be taken into account. If they need help to explain their views, they may want to involve an advocate or ask a friend or relative for support.

The full assessment will be done by a multidisciplinary team made up of a minimum of two different health or care professionals who are already involved in the person's care. The involvement from professionals helps build an overall picture of their needs. In some cases, more detailed specialist assessments may be required from these professionals.

The decision support tool is used to assess whether the person's main or primary care needs relate to their health, by looking at the following areas of care need, referred to as the 12 domains:

- behaviour;
- cognition (understanding);
- communication;
- psychological/emotional needs;

- mobility (ability to move around);
- nutrition (food and drink);
- continence;
- skin (including wounds and ulcers);
- breathing;
- symptom control through drug therapies and medication;
- altered states of consciousness;
- other significant needs.

Once the assessment is completed, the multidisciplinary team will make a recommendation about the person's eligibility for NHS Continuing Healthcare.

The person taking the lead

The Care Act and the Right to Ask scheme state that the person must be at least involved in developing their plan, and they must take the lead where appropriate and possible for them to do so. Local authorities must also provide an independent advocate to help the person participate in the planning process, if that person would otherwise have substantial difficulty in doing so. Where it has been determined that the person lacks the capacity, through the appropriate capacity assessment process, any person who is interested in the adult's welfare should be involved in the development of the care and support plan, while involving the person directly in the most appropriate way.

A care and support plan should provide information about the following:

- Who the person is and what is important to them – It should be clear within the wording of the plan that it has been developed with the person as the main focus. The plan should be written in a way that represents the individual, their personality and character. The reader of the plan should be able to get a real sense of who the person is.
- What the person would like to change and achieve – This provides an exploration of the aspirations the person has in regard to what they would like to achieve; this could include maintaining aspects of their life as it is now, or achieving a goal or making changes in their life where the current arrangement is not working well for them.
- How the person wants to be supported and cared for – All the ideas about what good care and support would look like should be presented in the plan. Ideas should be written down about what would work well for the person and where this support would come from. When developing the plan, the range of options

regarding good care and support can be explored, with the details of the exploration and the conclusion of the best option presented in the plan. Care and support could be found from a range of options including technology, equipment, care and support agencies, employing a personal assistant (PA), activities and community association, within the person's informal community and the support from their family and friends.

- How the budget will be utilised – This section should clearly outline how the person is going to utilise the budget available to them, this can include all the resources available to the person, the personal budget/personal health budget, pensions, benefit income, grants etc. The costs should be represented as yearly amounts and weekly amounts for ongoing costs, such as support or activity fees. Any costs that are a one-off, such as purchasing equipment or the cost of training, should also be recognised under a section titled 'one-off costs'. If the person has identified a way of being supported that has no cost attached to it, this can also be represented here with a nil amount attached to it.
- How their support and care will be managed – If the person has chosen an agency to provide all the support they require then one would expect the agency to provide some of the ongoing management of the team they employ. If the person is going to utilise the budget they have available to them to commission the service, then how is the person, or their family, going to have some involvement in managing this? Would it be possible to have regular meetings with the agency to discuss progress, and to be involved in the recruitment process for the care or support workers? If the person has decided to employ their own PAs, what needs to be put in place to manage this arrangement, for instance, training, supervisions and meetings?
- How their budget will be managed – There are a range of options available which are described in Chapter 2. If the management option is stated on the care and support plan it means that steps needed to transfer the budget can begin as soon as the care and support plan is agreed, reducing any potential delay in setting up the support.
- Contingency plans to cover when things could go wrong – A contingency plan provides the opportunity for the person, their family and others involved in the provision of care and support to consider how to manage a situation that doesn't go to plan. For instance, if support workers are going to be employed would it be an idea to employ a few people so that back up support is available when needed or would a family member be happy to be available at these times?
- A risk plan – This would consider when the person feels, or others around them feel, that they could be at risk and what measures to put in place to reduce the likelihood of this risk occurring or the severity of this risk if it does occur.

- An action plan for how the care and support plan is going to be put into practice. It is also helpful for everyone involved to know what they need to do once the plan is agreed on. This may include:
  - recruiting PAs or support workers;
  - liaising with the chosen agency;
  - organising the activities;
  - setting up the community-based arrangements.

The people involved in supporting the person to implement their plan could be varied and it does not have to be dependent on one person to oversee everything. For instance, the person may take the lead on certain parts of their plan, while the social worker, friends and family may help with other aspects. This would be agreed with the people involved at the time of writing the plan.

When developing the action plan, it is useful to consider people's strengths, skills and knowledge to help to decide who would be best to take on each task.

Care and support plan templates

The local social work and health teams have template care and support plan documents and these would be the final versions that are submitted to the panel for a decision about the release of the personal budget or personal health budget. However, it is important to remember that this is the person's care and support plan and they should have the opportunity to develop it in whatever format suits them; this could be on big sheets of paper using colourful pens or it could be typed up on a computer. It is important to find a way of developing the plan so the person can be fully involved and has a sense of ownership over their own plan. The information that has been put together using this approach can be transferred onto the template document at a later date, once the person and their family feel that they have captured everything they have wanted to within that plan.

In 2014, National Voices, a coalition of health and social care charities, worked with a wide range of people who use services to develop a common understanding of what is meant by care and support planning. The guide they produced highlights a four-step approach:

Care and support planning guide

Prepare – getting ready for the care and support planning discussion.

This stage is about:

- thinking about what matters most to the person to whom the care and support plan belongs;
- getting any useful and important information before the discussion;
- taking time to think and talk to other people about the different choices the person has.

Discuss – talking with those who care for and support the person.

This stage is about:

- the person and those who care for and support them working together;
- talking about staying well and living well;
- talking about what they are going to do;
- talking about what care and support the person might need from other people.

Document – writing down the main points from the discussion.

This stage is about:

- writing down all the main points that the person talked about;
- making sure that the plan belongs to the person and is easy for them to understand and use.

Review – checking on how things are going.

This stage is about:

- seeing how the person is getting on;
- reviewing the plan with those providing care and support and making any changes.

Reflection

- How can the plan be developed and recorded in a way that makes sense to and is accessible to the person you care for?
- What would 'good' look like for the person you support?
- Consider who would be the best people to involve in the care and support planning for the person you care for?

Reviewing and monitoring

The local authority or health team will arrange a first review to take place within the first three months and would consider:

1. if the person's assessed needs and outcomes are being met, whether they are receiving good quality support and if the arrangement is working well for them;
2. if the budget is being utilised well and records are being kept.

At the first review, everyone involved will get the chance to share how they feel things are progressing. The focus will be on the person, their family and any other people involved in providing the care and support. If this review demonstrates that things are going well, the ongoing reviews will then take place on a regular basis, usually bi-annually or yearly depending on the circumstances. If issues and concerns are discussed at the review, the review meetings will continue to take place on a more frequent basis until these are resolved.

The person and their family may also find it useful to find their own way of checking how things are going, arranging to get together with the specific focus on discussing what is working well for the person and if any areas need to be addressed. Sometimes these gatherings are given a name, such as a circle of support or a circle of friends. More information about circles of support can be found in Chapter 6.

How a budget can be used

A personal budget or personal health budget can bring a great deal of flexibility regarding how someone is supported and the person can use the budget as creatively as they wish. For instance, they may choose to use it on the following:

- the services of a care agency;
- to employ a PA/Support Worker/Carer directly;
- to visit a community or social activity, such as going to a club, concert or sport event;
- to participate in an educational activity;
- to purchase an item of equipment to meet an outcome shown in the care and support plan;
- to access complementary therapies, such as aromatherapy or music therapy, to improve well-being;
- to pay for transport to visit family and friends or attend a leisure or educational activity;
- to be supported to get to a café to meet friends, or to go to a place of worship;
- to take a break with a carer.

A budget cannot be used for expenses that would be deemed as everyday living costs, for example:

- food;
- rent;
- utility bills;
- toiletries.

A personal budget or personal health budget cannot be used:

- for any illegal or unlawful purpose;
- to buy services from the NHS which are delivered on the founding principles of free at the point of delivery;
- to pay for anything that is not directly related to meeting an eligible social care need as set out in the assessment.

Key lessons learnt

- There are eligibility criteria for Adult Social Care and Continuing Healthcare, which outlines when a person would become eligible.
- An assessment would be completed to determine if someone is eligible for health or social care support. When accessing social care support, there is the option to complete a self assessment.
- When people become eligible for health and social care support, they do have the option of receiving a personal budget or a personal health budget.
- Residential care is one of the options for people with dementia and the strengths and weaknesses of this option have been explored in this chapter.
- To achieve a personalised approach to care and support, it is also important to have insight about the following:
  - the services and support available;
  - the rights of the person and what they are entitled to;
  - the range of people and professionals who will be involved;
  - the options available to a person;
  - supporting the person to make decisions;
  - what to include in a care and support plan.

Questions to aid reflection

1. If you are to be actively involved in the support or care of a person with dementia, what would you want to know, or need to know, about them to make sure that the approach offered to them was thoughtful and personalised?
2. If you know a person, how they communicate and their preferences for the future, how would you share this with other people to make sure this knowledge and insight is included in the person's assessment and care and support plan?
3. When supporting a person to plan for the immediate or distant future, how would you ensure they were fully informed about all the options available to them and aware of the level of choice and control they have?

6 Ways to plan

This chapter covers:

- **A toolkit to achieve personalised support**
- **Enhancing positive relationships**
- **The promotion of respect and dignity**
- **Hearing the voices of the person**
- **Person-centred approaches to gathering information**
- **Promoting creative thinking and self-determination**
- **Person-centred planning.**

People first

So far, this book has recognised people with dementia as *people first*, people who have their own unique expectations and wishes.

Although not used previously, the term *people first* shifts the focus to recognise the individual, the person. This position requires an understanding of how dehumanising the experience of being diagnosed and then receiving health care intervention and prescribed services can be. It encourages everyone to continue to relate to the person as an individual first with their own identity and preferences.

Planning

Good planning:

- has the person at the centre of the conversation, which also includes people who care for and love them and is written and recorded in a way that is meaningful to them;
- recognises and builds on the person's community and meaningful connections, this is sometimes referred to as the circle of support or circle of friends (which are explained later in this chapter); and
- goes beyond making decisions about the services and support available to the person to a much wider conversation about achieving a good life.

A plan that is developed on these principles is a useful source of information for those who will provide care and support in the future, whether paid or unpaid. It helps those providing care and support to do so in a person-centred, respectful and dignified way. Having the opportunity to plan ahead means that the person's voice is central to the delivery of care and support going into the future, helping to prevent errors, misunderstandings and even avert crisis.

Thoughtful, sensitive and sometimes intuitive approaches lead to a well-connected and informed network of support, consisting of the person, their family and friends, paid professionals and the community.

These planning approaches have been separated into three *sections* below.

Positive interactions

This section contains a range of basic skills initially outlined by Tom Kitwood (1997) in '12 Positive Interactions'. The aim is to prevent an approach to supporting the person that leads to them feeling like an object. The intended outcome of adopting the 12 positive interactions is for people to feel valued and respected.

1. Celebration

Taking time to celebrate and recognise the things that are happening in a person's life that bring happiness, comfort and a sense of purpose. This requires mindfulness when spending time with the person and embracing a cheerful and conversational approach, which helps them recognise what is happening in their life and the good things that are taking place. It is about being observant of what creates a state of happiness, comfort and a sense of purpose within the person and about mindfully creating opportunities for this to feature in their life.

2. Collaboration

A recognition that the purpose of the collaboration is to work together to achieve the best outcomes and life experience for the person with dementia and their family. The person is treated as an equal with a right to be heard, and their opinions are understood by those around them. Such an approach creates a balance in which family, paid-for carers and the person are equal.

3. Creativity

Opportunities to get creative and to be engaged in activities that feed the creative spirit – finding creative opportunities that suit the person: singing, dancing, painting, baking, etc. Involvement in what the person finds enjoyable is supportive of their well-being and expression, and it can unlock creative thinking for other areas of their life.

4. Facilitation

To really know a person is to understand their subtle gestures and simple body language and to have the level of insight and knowledge to respond positively. Good facilitation to lead a good life day to day is to provide just enough support for the person to spend time with other people and to complete the activities of their day-to-day life.

5. Giving/Accepting

Giving and accepting gifts of goodwill or help from the person. The size of the gift or the level of goodwill given and received is of little importance.

It is the sharing of these moments that hold significance and create meaning and memories within relationships.

6. Creating closeness

Spending dedicated time with the person and, while with them, creating a safe space which is soothing and comforting for them. This could include considering the immediate environment, such as music, lighting and temperature.

7. Negotiation

The person will have decisions to make about day-to-day life as well as longer term decisions. Negotiation in this sense means focusing on the decision of the person then discussing, with others, how feasible this is and how to make it happen. This requires meaningful interaction and conversation based on listening and acting on the person's responses.

8. Playfulness

Based on the principle that everyone likes to have fun, people with dementia can play games, tell stories and interact in ways that they find enjoyable. This may be planned or spontaneous and is a great opportunity for self-expression.

9. Recognition

This means having an open and unprejudiced attitude and recognising the person in an open and positive way. It is important to make sure the way they are supported and cared for is done from their point of view and perspective of their life and the world around them. Anything that is an assumption about what the person would want or any judgement on what they do actually want would go against this positive interaction.

10. Relaxation

People may want to relax by themselves or alongside other people. How relaxation time is spent will be determined by the person.

Relaxation is a physical sensation a person feels when they feel like they are *floating away*. This may be achieved by doing something soothing like having a warm soapy bath, having hands massaged or being in a multi-sensory room that stimulates the senses.

11. Timalation

This interaction emphasises finding what is pleasurable for the person. It could include engaging the person on their different sensory levels, such as providing different taste sensations during meals, smell sensations via aromatherapy or, where appropriate, kinaesthetic sensations via massage therapy. These types of interactions provide contact, reassurance and pleasure without placing any demands on the person.

12. Validation

This involves being sensitive to when the person is experiencing heightened emotional states and demonstrating empathy at these times. This is observed by making each experience for the person with dementia a robust one, accepting and acknowledging their reality and the power of that experience to them on an emotional level.

> *It also involves developing a high degree of empathy for them and trying to understand the person's frame of reference, no matter how chaotic, paranoid or hallucinatory. For example, if the person with dementia is distressed about a person who is absent, acknowledge their distress and ask them about the person or what about this person is upsetting them.*
>
> (De Bellis et al, 2009, p 28)

Achieving the outcomes aimed for through the positive interactions set out by Kitwood (1997) requires those providing care and support to do so sensitively.

To support someone to celebrate would require an insight into what has been happening in that person's life, an understanding of how the person would like to celebrate and an ability to share this information with others to ensure that they can also join in the celebrations.

Collaboration would require people to connect and communicate with others in a way that supports the relationships between the person with dementia, their family and paid carers and professionals involved.

Through his work on person-centred care, Kitwood encouraged those supporting and caring for people with dementia to develop an understanding of the effects of negative and positive interaction on the person. Studies have demonstrated that the person's quality of life and well-being increase significantly when the person is supported in a person-centred way (Matthews, Farrell and Blackmore, 1996; Burgener and Dickerson-Putman, 1999).

Supporting people to make choices

This section covers how to understand a person's preferences, support them to make choices and help them to maintain control.

All About Me profile

An All About Me profile is a way to record and share essential information that anyone providing care and support would need to know. The profile provides the foundation of information to those caring for and supporting the person, and helps inform people of the most important aspects of providing good care and support.

When developing an All About Me profile, it is important to maintain focus on creating a profile that records and shares the information that the person with dementia would like to share about themselves and what they see as the essential information that somebody new in their life would need to know.

The profile should be short and concise, no more than two pages long. It could contain all or some of the following information:

- what name the person prefers;
- important parts of their daily routine;
- what has to happen in their day-to-day life;
- what needs to happen for the person to be and feel safe;
- what makes the person happy;
- what can cause upset for the person;
- what others like and admire about them;
- their skills, interests and passions;
- the important people in their life.

When supporting someone to create a profile, remain mindful of the purpose, which is a short and concise introduction to the person, providing the essential information to support a good life for them.

Focusing on the positives can significantly change the way people spend time with the person and how they experience the care and support they receive. If their All About Me profile includes information about what to avoid, or what upsets them, this information can be used in a positive way to avoid these circumstances. For instance, if loud noises upsets the person being cared for, those supporting them know to avoid noisy places where possible.

Doughnut Sort

The Doughnut Sort was designed by Charles Handy (2002) to support companies to organise roles and responsibilities.

It can be useful to identify and define the roles and core responsibilities of all the people involved in the care and support a person receives. This would include the person themselves, carers, family members, paid people and professionals.

As the Doughnut Sort creates clarity regarding roles and responsibilities, it can, in turn, help to improve the achievement of the desired outcomes and improve the effectiveness of the contributions people make towards these outcomes. This approach can increase the sense of satisfaction and purposefulness people feel within the role they have for the person with dementia.

The doughnut provides a method to clarify what a person has to do within their role, the areas where they have scope to indulge in thinking and acting creatively and the jobs and tasks that are someone else's responsibility.

How to complete a Doughnut Sort:

Imagine an American-style doughnut with jam in the middle:

1. The jam in the centre represents the tasks and responsibilities a person has to do within their job or role to fulfil it successfully. The tasks mentioned in the 'jam' of the doughnut would be a person's core responsibilities and what they are expected to do well. These are the areas where people may need to follow guidance and instruction to achieve a good outcome. The guidance and instruction may be provided by the person with dementia, their immediate family or professionals involved in their care.
2. The jam is not the whole of the doughnut, and the surrounding doughy areas represent the opportunities people have to be creative and go beyond the bounds of duty to live up to their full potential. This would include opportunities for people to use judgement and creativity and to try out new things to see what could work. These are areas where people can use their imagination when working alongside the person with dementia to help find new ways of doing things that work well for the person.
3. Areas that are not people's responsibility are signified by the area outside the doughnut. This determines the responsibilities or tasks people should not be involved in as it is not their area of knowledge, skill or passion. Someone else who is better placed would have already agreed to take on these responsibilities.

Using the Doughnut Sort helps to build a better understanding of the roles people have and when they will be involved. It helps to build a stronger and connected network of people who are making sure, together, that all aspects of the support and care the person with dementia wants and requires will be provided for and that this is shaped by them as much as possible. It can also provide clarity when there are a number of individuals and/or organisations involved.

Relationship mapping

At the root of person-centred approaches, there is an appreciation of the importance of relationships and connection. Spending time together to map out the relationships in a person's life helps everyone to recognise the important connection they have, and the people with whom they have had long-standing and meaningful relationships. Finding ways to support those relationships can lead to a greater sense of well-being, now and into the future. Keeping relationships alive and maintaining and enriching these connections results in a support network that is enduring and purposeful.

The people who form this relationship network also offer a wealth of knowledge and information about the person's life, which can be found through holding meaningful conversations.

Conversations can lead to identifying ways to meet the outcomes the person, and the people around them, would like to achieve, including ways to set up and organise care and support in a way that makes sense to the person.

On the reverse side, not supporting the important and meaningful relationships in the person's life could cause a greater sense of loneliness and isolation for the person, leading to negative effects on their health and well-being.

Conversations help to identify the important people and the key relationships in a person's life, and when these are focused on the person's life rather their 'needs', a lot can be learned about their interests, past achievements, celebrations, recent events and who shared these moments with them.

The relationships will be current as well as from their past. The relationship map may include:

- those who have an intimate relationship with the person, whom they have known over a long period of time or with whom they feel the most comfortable and the most understood;
- good friends; this includes those the person sees or contacts on a regular basis or has known for a long time – perhaps from their school days. These friends will be people they have built memories with or have a shared mutual interest;
- those they see and spend time with through association, for example, through sharing an interest, hobby, sport or faith group;
- those paid to be in the person's life, for instance their GP, nurse, optician, carer, support worker, personal assistant or cleaner, all of whom serve a role in maintaining their quality of life.

A completed relationship map can help to identify:

- the number of meaningful relationships the person has and the close network of support they have around them;
- who the person wants to have involved in planning and future decision-making;
- who would be good to have involved in the context of representing the person's best interests, especially if the person has been assessed as not having the capacity for a particular decision (see more detail regarding capacity in Chapter 7);

- any issues that exist within the relationships a person has and what support is needed or wanted to resolve these;
- the friendships and relationships the person wishes to maintain, for which they may need support to continue in the future;
- the people who have been identified on the relationship map whom the person would like to develop a closer relationship with; this may be someone who they see, or used to see, within a community setting whom they would like to build a friendship with based on their mutual interest;
- the people who would like to invite onto their circle of support;
- those who would be good to talk to and involve in gathering information and planning for the future.

Developing a relationship map can be a really empowering experience as people can see the number of people they have in their life. It is, however, important to be considerate as, for some people, this can emphasise the lack of connections they have. This is especially true for someone who has already been isolated or segregated from their community and relationships as a result of the way care and support has been provided to them. In these circumstances, a person may find themselves in a position where they have lost contact or have no close family, friends or connections with people in the community. For instance, if a person has moved away from their community to a care home setting, they may not have been able to maintain the connections that were important to them, especially if they have not been supported to do so.

If developing a relationship map would highlight a person's lack of connections or connections that used to be much closer, these questions may be worth exploring:

- If the person wishes to, how can they regain the relationships they have lost and how can they be supported to do so?
- How can meaningful relationships be built from the people who are in the person's life right now?
- What opportunities are available to a person to develop friendships and connections?

Circles of support

Circles of support, sometimes called circles of friends, consist of a group of people who form a connection between themselves as an intentional support network around the person. Together, the circle of support holds the purpose of supporting the person to achieve a good quality of life, reach their full potential and overcome the challenges they face along the way.

In an article for Indiana University, Kim Davies (2005) stated that, from birth, people create an intricate network of relationships in their lives consisting of people who are there in times of trouble, sorrow and celebration. Davies suggests that these friendships are born from shared common threads that bind people to one another. It is these people who become part of a circle of support or a circle of friends as these are the people who are valued in a person's life and who offer support to achieve dreams or to provide help when needed. The relationships that develop into this level of meaning can be formed in many ways and from meeting people in different places and environments, which often occurs naturally and throughout life. In her article Davies wrote about:

- the **circle of intimacy** – the name of the person whom the circle belongs to would be written, along with a photograph of them, if wanted. Within this area of the inner circle, the names of the people who are most important to them are written down;
- the **circle of friendship** – where the person's friends would be recorded;
- the **circle of participation** – record all the people the individual knows through a shared interest, hobby or place where they spend time in this third circle. For instance, if the person goes to the gym, it may be the people they see there. These are the people where a close friendship has not been forged, but perhaps the potential is there for this to happen;
- the **circle of exchange** – all the people who are involved in the person's life in a paid capacity would be recorded in this fourth circle. This could include paid support workers, opticians and gym instructors.

Negotiables and non-negotiables

The negotiable and non-negotiable conversations helps us to identify what is:

Negotiable – what would be good to have present in a person's day-to-day experience but is not essential.

Non-Negotiable – the essential elements of a person's day-to-day life.

Developing an insight into what is negotiable or non-negotiable for the person provides the details required to inform those around them, including those providing support or care, about what works well for the person and how to strive towards achieving a good quality of life.

This approach recognises that the small things really matter and make a difference in achieving a quality of life and happiness. Paying attention to the finer details of a person's daily experience and respecting the preferences they have can make the difference between a good day and great support and a bad day and support that does not fully recognise or respect the person.

Gathering this information requires engagement in a conversation that enables the person with dementia to share what really matters to them. Listen to what the person and people around them share about what really matters to the person. When the information has been gathered, it can be sorted under the headings of 'non-negotiable' and 'negotiable'.

If the person with dementia is not able to fully engage with a conversation, it is possible to achieve this insight through gaining information from the people who know the person well; this would include the people who would sit within the person's circle of intimacy and circle of friendship (as covered above in the 'Circle of Support' section). Involve the person with dementia in the conversation wherever possible, engage them in a meaningful way and recognise their contribution to the conversation through listening to and observing their reactions.

The information gathered as a result of the conversations should provide the details needed by those caring or supporting a person to be able to fulfil their roles well. For instance, if this approach was utilised when building an understanding of the person's preferred morning routine, it could include asking and exploring the following questions:

- What makes a good start to the day?
- How does the person prefer to get ready for their day?
- What is missing at the moment that is important to the person?
- How could the person's experience be better than it is now?
- Is there anything about now that needs to be maintained into the future?

The attention to the smaller details can often be overlooked. Honouring simple everyday preferences makes a difference to how a person feels: their well-being, self-esteem and sense of security. The following story illustrates this point:

> *While on clinical placement on a general medical ward in my first year of training, I helped to care for an older patient, whom I will call William.*
>
> *The nursing staff had described William as confused and 'wondersome'. He seemed anxious and was touching his face constantly. Frequently, he could be found walking towards the bathroom in his dressing gown, so the nursing staff would help him back to his bed.*
>
> *My dementia training at university, along with my role as a 'dementia friends champion', led me to believe that William was not wandering without cause, and that there was something of interest to him in the bathroom.*
>
> *After talking to him, it became apparent that all he wanted was a shave. William was accustomed to shaving every Sunday but had not shaved in weeks. He had been wearing his dressing gown to go to the bathroom, which had a shaving brush and a safety razor in the pocket but had not yet managed to have a shave.*
>
> *William was not aware that he was able to shave, or that the staff could help him, so I found some shaving foam and walked with him to the bathroom. I helped him*

to foam his face, and then left him to have a shave, checking in on him regularly to make sure he was okay.

Although William was with dementia, he was not at risk of falling or a danger to himself with the razor. He was clear on what his needs were, and I helped him to meet them. For the rest of the day William was calm and chatty, and there was no wandering or anxiety.

This experience reaffirmed the importance of treating patients as individuals and supporting their choices so that they can be cared for with compassion, dignity and respect. I also have a better understanding of how my actions affect others and am committed to developing my nursing practice with the patient at the centre of all care decisions.

(Perez-Merino, 2014)

From this story, it is clear that a 'non-negotiable' for William is to be able to have a shave on a Sunday morning. It would also be a 'non-negotiable' that William receives support in the bathroom to get set up and ready to have a shave. The 'negotiable' would be discovered through having further conversations with William or observing his responses to situations or options given to him. The 'negotiables' may include the brand of shaving foam he uses or the time when he has a shave on a Sunday morning.

Working/not working

The working/not working conversation helps build an understanding of the aspects of a person's life that are:

- 'working' for them – where they are happy and satisfied. The areas that need to be maintained;
- 'not working' for them – the areas where they are frustrated and dissatisfied. The areas of the person's life that need to improve and a commitment is needed to address these areas.

When holding the conversations to gather the information to learn about what is working or not working, it may start with focusing on an area of the person's life, for instance, where the person is living or how they are being supported. The conversation can also have a wider holistic view of the person's life to achieve a big picture understanding.

Such conversations can also build an understanding of the perspective of other people around the person, such a family, friends, paid people and professionals. The unique perspective of each contribution to the conversation is valuable in creating a deeper understanding of what is happening in the person's life and where changes need to be made.

It is then for those who have contributed to share ideas and thoughts about:

- making sure the information has been sorted accurately;
- how to maintain aspects of a person's life that are 'working' for them;
- how to address the elements of a person's life that are currently 'not working'.

This may involve resolving differing opinions; for example, the person may really enjoy the morning walk they have, the fresh air and the chance to see people, so this is working for them, but, likewise, the person's carer becomes anxious when the person is away from home for a long time so for them it is not working. This would then open a discussion to find a solution to the question 'how can the person be supported to continue their morning walks whilst reducing the anxiety of the person who is providing care?'

Addressing the areas that are not working in a person's life provides an opportunity to generate ideas and solutions. Often it is the creative and individualised ideas that suit the person and the people around them. So it is important to explore all the options available, including those that are based within communities or are yet to be created, and identifying which ones would provide the best solution.

Here is an example: A person who is living at home and being supported by a support worker can no longer get out of bed independently, and they are offered additional home or domiciliary care so there are two people with them each morning. This provision is often the standard solution offered in these circumstances, yet there are other options that could be considered. Exploring this issue further and discussing the details of what is not working in the person's morning routine could lead to a different solution; for instance, changing the height of the bed, investing in a rise and fall bed, setting the warmth of the room for when the person rises, having a chair positioned closer to the bed, the person making changes to their night time routine or their sleeping environment.

Once some good solutions and ideas have been found, an action plan can be developed to put them into practice.

Good days and bad days

Everyone experiences good days where they have enjoyed the company of others and what they have done. Equally, people can have bad days.

Developing an understanding of what a good day and bad day looks like from the person's perspective can be achieved through purposeful conversation, observing how they interact and respond to what is happening around them. Exploring and discovering the answers to the following questions helps to build a picture of a good day:

- What was enjoyable about the day?
- What were the interactions like with other people?

- Who was present?
- What was the source of fun?
- What provided meaning and purpose?

In this time, recognising the details of what makes a good day can help everyone to learn more about the person, what has purpose and meaning for them and what they enjoy. When this understanding has been developed, everyone can work together to take steps to incorporate the enjoyable aspects into more days.

Likewise, these conversations and time spent together can also recognise and observe the elements that make a bad day:

- What caused frustration and was this communicated to others?
- Who was around at the time?
- How were people interacting with each other?
- What was happening, and how were people involved?
- Which routines were followed?

Utilising what has been learnt enables everyone involved in supporting the person to recognise what changes need to be made to achieve better experiences. Creating an action plan can help identify how these changes are going to be made and by whom, with the intention to prevent bad days and move towards more good days. This understanding can help those providing care and support to know what needs to be present and what needs to be avoided in someone's life, while offering ideas that ensure a good or a better experience for the person with dementia on a daily basis.

Life stories

Personal history makes a person who they are today, and an awareness of a person's life and an understanding of them in their own biographical context is important to maintain individuality in the provision of person-centred care (Stokes, 2000). If a person loses their connection with their history, they also lose something of themselves. This is especially the case for people living with dementia as their memory declines as their dementia journey progresses. People may find themselves trying to make sense of who they are, and it is within this context that a documented life story can prove to be valuable and a guide for those in the person's life.

Life stories document a person's history and provide valuable insight into their identity: their experiences, relationships, how they have contributed to the community and society, their joys, sorrows and the current reality. This depth and detail of information can provide clear direction and clues when providing person-centred care and support. A person's life story provides other people with ideas and valuable information and can spark meaningful and worthwhile conversations. This includes talking about the person's achievements, their experiences and relationships. Life stories can also be a source of inspiration when exploring the activities available to the person with dementia, helping them to find things that they would enjoy and, in turn, enhance their quality of life and well-being.

Elements of the concept are demonstrated well in the documentary 'Alive inside', which explores how people with dementia respond to music, while capturing astounding reactions in people who have otherwise found it difficult to recollect past events, remember current events, interact with other people, move, speak or recognise loved ones. Learning about the music that has relevance through listening to their life story – which would include music from their childhood or from special occasions – then creating opportunities to listen to this music can ignite positive and enlivened reactions.

People involved in the documentary, especially family members, noticed that their loved ones responded better by listening to music that had meaning for them over any other therapy or intervention, including pharmaceuticals.

Supporting someone to write their life story can provide the detail and insight to share with others to build a better understanding about what matters in their life. The way the story is recorded and shared can also be based on the person's preferences, for instance, they may want a written story or to use other media such as pictures and videos.

When supporting a person to capture their story, it is worth asking for different perspectives. Naturally, the person themselves is best placed to share their own story as they remember it, including the parts of the story that are most important to them. Other people with an insight into the person's history can also make valuable contributions and add detail.

This work can help to:

- maintain the relationships that mean a great deal to them;
- make it possible to participate in activities that are enjoyable;
- make sure that what means a great deal to the person is available to them. This could include music, sounds, photos, gifts, smells, jewellery or anything else that holds a memory;
- spend time together to reminisce and talk about the past.

Appreciations

Developing appreciations identifies the qualities and characteristics that other people value and admire about the person. This helps to recognise the person's uniqueness and acknowledges their:

- gifts – the aspects of a person's personality and character that draw others to them, the things they remember about a person and what inspires people to spend more time with them;
- qualities – what others admire about a person, their achievements and uniqueness;
- talents – the things that a person shows an ability and where they are naturally talented.
- skills – that have been learnt, the knowledge and abilities a person has that they have received training and education to attain.

Appreciations can be enlightening for the person, and those around them, to hear how they are regarded by others. They can be gathered individually from people in the person's life and then collated and shared.

A round of appreciations at the start of a gathering can form a positive opening. Each person present is asked to share something positive about the person whom the meeting is for, something they like and admire about them. To record people's contributions, a piece of paper is placed on a wall or on a flip chart stand, with the person's name at the centre and the headings: 'gifts', 'qualities', 'talents' and 'skills' written around the person's name and people's contributions are added to this.

This exercise does not take long and helps to set the tone for the rest of the conversation, bringing a positive perspective for everyone involved. This proves useful when the purpose of the gathering is to find ideas and solutions. Appreciation can challenge people who have previously tended to focus on the negative perspective of the person or their situation. It encourages them to develop a positive outlook, regarding what the person can do and what they do have to offer others.

Matching staff profile

Producing a matching staff profile identifies the skills, characteristics, shared interests and practicalities that would be good for any supporter or carer, paid or unpaid, to have in order to develop a good relationship with the person they are supporting or caring for.

Completing one encourages the person with dementia, and those around them, to think about what qualities they would like people who provide care and support to have. It is a good idea to complete this prior to recruiting care and support, whether this is via an agency, employing directly or via natural support in the community. The information gathered can help create the wording for an advert so the right people can be found.

Investing this time in creating such a profile can help to ensure that:

- the person has a good quality of life;
- they enjoy spending time with those who are providing care or support;

- long-lasting relationships between the person and those providing support or care are created as there is a good match in the relationship they have that goes beyond care and support and into shared interests, passions and
- the care or support worker and the person with dementia enjoy spending time together.

The details needed to complete a matching staff profile can be gathered from conversations about what good care or support would look like. This would include a conversation about:

- what skills do the people providing care or support need to have? Do they need to be good at cooking particular meals and involving the person in the preparation; do they need to be confident in using health-related equipment in a way that maintains respect and dignity; do they need to have skills in communication and a creative approach to communication to support the development of a relationship with the person with dementia?
- what interests does the person with dementia have, and which of these interests would be good to have in common with the person who is providing care or support? Pursuing a shared interest can help build relationships; and this is the same for the person with dementia and the people who are supporting or caring for them. Interests can be pursued together helping to develop the relationship further;
- what type of character and personality traits are needed in the people who will be providing care and support to increase the chances of the person getting on with them? Discussing and thinking about the people the person has around them now and from their past relationships helps to identify what makes a positive relationship, for example, does the person tend to get on better with someone who is softly spoken and calm, or do they get on better with someone who is bubbly and loud, or does this change depend on how the person feels?
- what practical requirements does the person have of their carers and supporters? It is important to consider if there are certain needs that must be met. This would include times of the day or days of the week that the person needs care or support; does the person need someone who can drive and who has access to a vehicle; do they need someone who can use computers or someone trained in specific areas such as communication?

My Care and Support Profile

When people want or need care and support, it is good to be clear on what areas of their life they want care and support with and how they would like this to be provided. Developing an understanding of this can help the carer or support worker to be informed and confident in their approach. The information that is gathered can be shared in the My Care and Support Profile, which records this detail, as well as the areas of a person's life that they wish to maintain themselves.

Treasure chest

Developed by the company New Possibilities, a treasure chest brings people together to discover what opportunities and resources exist for the person within their network of friends and family. This approach opens up the idea that a person's quality of life can be improved through developing closer connections with people around them and the wider community. It captures information about skills, interests, passions and what people would love to learn and is usually gathered at a meeting with all the people present contributing to it.

When developing a treasure chest, it helps to have a big sheet of paper divided into four sections with the following headings for each section:

- useful people to know;
- passions, interests and hobbies;
- skills and knowledge;
- something you have always wanted to learn.

Someone can then take the lead in asking people in the group questions to help complete each section of the paper. It is good for the person leading the meeting to have some useful examples to share to get the conversation started and to encourage others to contribute, for instance:

- useful people to know – 'I know a man who delivers free papers', 'I know someone who is great at woodwork' or 'I know a solicitor';
- passions, interests and hobbies – 'I am passionate about baking/dancing/live theatre/shopping';
- skills and knowledge – 'I am a good driver', 'I can knit' or 'I'm a good cook';
- something you have always wanted to learn – 'I would like to learn a new language/play the guitar/dance the Tango';

The person leading will record all the contributions or ask people to come and write their contributions onto the sheet of paper. The connections between people will begin to emerge, which enables the development of stronger relationships and a sense of community among those involved.

The section of the treasure chest that recognises good people to know helps build the potential to extend a person's network. When people begin to identify the people in their own networks who are 'good people to know' and consider how to link people together, it broadens the possible connections available to the person.

When people share something they have always wanted to learn, there may be connections with other people in the room who wish to learn something similar. This creates an opening for these people to come together to self-organise and pursue this together; for example,

'I love dancing and if Mary would like to have the opportunity to dance on a regular basis, we could go together'.

A treasure chest is ever-evolving as new connections emerge and can be explored, providing the opportunity to gather and record rich information and to find ideas and opportunities to create stronger connections. Once completed, it supports people to recognise how they can play a meaningful role in supporting the person to lead a life they are happy with, which, in turn, strengthens their relationship.

Solution circles

A solution circle is a facilitated process that helps those involved identify new solutions to issues and challenges they are facing. This is intended to be short and focused, usually lasting less than half an hour. It can prove effective in addressing long-standing problems, especially areas that are preventing a person from achieving a good quality of life.

The focus person of the solution circle is the person who presents the problem or issue that they are experiencing. At the end of the process, they can select the idea or ideas that have been generated through the conversation which they would like to take forward. The underpinning principle is that those who participate have the potential to help regardless of their background, experience or the role they have in the person's life.

When planning to hold a solution circle, it is best to involve between five to nine people. Once you have the people, find a space where it would be possible to hold the gathering. Choose somewhere with enough space to be comfortable, with no distractions and that has somewhere to record the conversation in a clear visual way, either on a flip chart, whiteboard or a room with blank wall space where big sheets of paper can be stuck to the wall (from experience, use an adhesive that does not leave marks – masking tape is a good option).

There are four main roles that need to be fulfilled during the development of the solution circle; these are:

- problem presenter – also known as the focus person;
- process facilitator – who will manage the people involved and be a timekeeper;
- note taker or graphic recorder who will clearly record all the information shared for everyone involved to see;
- and everyone else – the people who will be supporting the focus person, actively engaging, listening and contributing thoughts and ideas.

Here are the steps, with a recommended six minutes being given to each one:

Step One: The problem presenter outlines the problem they are experiencing. At this stage, the process facilitator makes sure the problem presenter does not run over time and is not interrupted. The note taker takes notes, and everyone else listens. If the problem presenter finishes early, the group stays silent until the six minutes elapses. It is really important to honour this uninterrupted time.

Step Two: The process facilitator opens up the session to the rest of the group and asks them to share their ideas and thoughts. It is important that everything is recorded and the focus remains on sharing the ideas; this is not the time to ask questions, give advice or analyse any of the ideas. The process facilitator should keep everyone focused on sharing ideas and make sure that everyone gets a chance to contribute, ensuring that no one dominates the conversation. The note taker records all the ideas, making sure they use the actual words given and not placing their interpretation onto them. The role of the problem presenter is to listen to the contributions and not to interrupt.

Step Three: The group enters into a dialogue led by the problem presenter, which creates the time to explore the ideas and further clarify the problem. The process facilitator needs to listen to make sure that the content is covering the only positive points and to guide the group away from negativity and anything that relates to what cannot be done.

Step Four: The problem presenter identifies the ideas that stand out as having potential and the ones they would like to put into action. Together with the group, they decide what actions are doable within the next three days and at least one action that can be initiated within 24 hours. It is critical to set actions that can be achieved within a short timeframe, such as researching information on the internet or making a phone call. Unless something is achieved soon after a planning meeting of this kind, people lose momentum and get 'stuck'. Someone from the group volunteers to contact the problem presenter within three days to check if they have completed their first action. To bring the solution circle to a close, the group does a round of words to describe the experience and the recorder hands the finished record to the problem presenter.

A solution circle does not guarantee the perfect solution, but it usually gets people 'unstuck' from the problem and helps the problem presenter think about how to address it, with the momentum to keep going until the solution is found (Forest and Pearpoint, 1996).

Consideration when utilising these tools

Creating the space and opportunity for the person with dementia to take the lead in conversations and planning meetings can be challenging for other people involved. This can be because during a planning session or a conversation, the person may need the time to consider how they want to respond to questions or what they want to contribute to the conversation. This time to think and consider may lead to a silence or pause in the flow of the conversation. During these quiet moments of conversations, people can feel a sense of discomfort and awkwardness and their way of addressing this feeling can be to break the silence, filling the space with their own thoughts and ideas. Ideally, everyone involved should be encouraged to recognise that silence is good and that it provides the much-needed time and space for people to think about their responses and to find the right words. It is essential to support everyone to appreciate the need to go at the pace of the person and to recognise that the periods of silence in the conversation may be where the best answers and insights can be found.

The importance of hearing the voice of the person can be set at the beginning of the meeting, or the conversation and people present can be reminded of this as the conversation develops.

The person is in the driving seat

In this context, the 'driving seat' refers to a person taking the lead, steering the course of direction of their own life and determining the destination.

Self-Directed Support supports the idea of the person directing and making decisions about the care and support they receive and how they receive it. This approach places emphasis on those receiving care and support firmly taking the position of the 'driver', while personalisation focuses on how services and health and social care personnel can respond in a person-centred and bespoke way to those in the 'driving seat'.

Before the person with dementia can take the 'driving seat', it may need to be recognised that other people from their support network may have assumed a lead role in their life. An open discussion with all the significant people is needed to emphasise the importance of the person being able to express themselves, share what is important to them and make decisions for themselves, discussing how they could take a lead role in any conversations and planning meetings about their life and the care and support they receive.

Here are some ideas to help maintain the focus on the person during conversations and meetings or to shift the focus, so the person is taking the lead role:

Use a talking stick – This provides a simple approach to creating focus in a conversation, preventing people from talking over one another. With the understood principle that to contribute to a conversation you need to be holding the talking stick while everyone else present listens. It will also be explained that the talking stick is mainly held by the person who the conversation is focused on. They will pass this onto people who want to contribute; after which it will be passed back to the person who then hands this to the next person who wishes to speak.

A talking stick can be something from the immediate environment that is easy to hold and pass on to other people, a spoon, a pen. It is best to choose something that stands out visually so that people taking part in the conversation can easily identify and be reminded of who is holding it at any one time.

An item serving as a talking stick can be prepared in advance of the meeting if it is expected that there may be a problem in maintaining the focus on the person with dementia and with them taking a lead in the conversation. This advanced preparation provides an opportunity to personalise the talking stick to make it significant to the individual; for example, for someone who loves art the talking stick could be a paint brush, or for someone who loves to sing it could be a microphone. It can be designed and made by the person to help to build a connection with the talking stick and the purpose of it. It serves to remind everyone that the conversation is first and foremost about the person and their life, aspirations and wants.

When a talking stick is used, it is done so with some rules attached:

- The talking stick starts with the person.
- People who wish to contribute to the meeting must signal that they wish to do so and wait until they are given the talking stick before they make their contribution.
- The person holding the talking stick is the only person who can talk.
- Once the person holding the talking stick has finished speaking, it should be handed back to make sure that the person maintains some control over the proceedings at the meeting.
- It is up to the person whom the meeting is for who takes the talking stick next, and the person facilitating the conversation will support in making sure that everyone present is involved and has an opportunity to be heard.

The use of a talking stick brings some order to a conversation; it avoids people talking over others and makes it clear as to whose turn it is to talk. This approach encourages people to listen to one another, and can create a greater level of understanding between those present, encouraging a respectful approach to the proceedings (Seeds for Change, 2020).

Have a facilitator – It is important to have someone who will facilitate the conversation and make sure that everyone present has had the opportunity to contribute when they have requested to do so. The facilitator would make sure that no one person is taking too much of the time available for the meeting to make their contribution. The person taking on the role of the facilitator should be someone who understands the principles of facilitation and should not have an active interest in the conversation taking place, making sure that the person the meeting is for remains involved in and leading the conversation. As the conversation progresses, the facilitator should acknowledge everyone's contributions and redirect ideas and thoughts shared by the group back to the person for them to consider and make the final decision. The facilitator would aim for the conversation to be smooth, energetic and positive.

Here are some principles regarding facilitation that originate from person-centred planning:

- Support the group to maintain positivity in their conversation, while recognising that circumstances may lead to addressing negative aspects of a situation. In this instance, support those present to reframe the negative into a positive wherever possible.
- Arrange the environment in such a way that people can equally contribute and participate, where people can feel on the same level as everyone else.
- Plan the seating so everyone can see each other, remove any barriers that mean people cannot see or hear each other and remove the formality of meeting/ conference tables by arranging the seating in a circle.
- If the meeting is being recorded using visuals on a flipchart or a big sheet of paper, make sure everyone can see the recording.
- A facilitator should position themselves to be able to see everyone to check if people are engaged or if they want to contribute.
- Make sure that all contributions are heard and that no one remains excluded from the conversation.
- Before starting the conversation, set the scene; remind people about the purpose of the gathering, that the meeting is a commitment to listening and developing a greater understanding of the person.
- Encourage people to create a positive experience and explain that the meeting is not the place for using jargon or fixing, judging, digging or obsessing; these are explained later in the chapter.

Preparing for a meeting

The success of a meeting, or a conversation, is largely dependent on having the time to plan ahead with the person with dementia. Preparation can help to build understanding regarding the purpose of the meeting and then for them to exercise control over who they would like to have involved, where they would like the meeting to be held and when would be a good time to bring people together. Here are some areas the person might like to consider:

- the purpose of the meeting, what they would like to achieve and what they would like to discuss;
- what approaches to communication they would like to use;
- who should talk on their behalf if they struggle at any stage;
- specific points they want to make clear to the group; for instance, the person may wish to talk about aspects of their care that aren't working well for them and how they would like this to change.

Time given for preparation can help the person be confident about having their views and ideas heard. The person is likely to gain greater value from this preparation time if they are supported by someone they trust.

Good questions

The type of questions asked and how they are phrased make a huge impact on the quality of the conversations held. Being conscious of using questions that assist a person to take the lead, such as those that are open, positive and supportive, is key.

Assessing the quality of questions being asked

When considering the impact some questions, such as the ones in the reflection task later, can have on the person, also consider how they would feel about themselves as a result of being asked these questions. It comes as no surprise that the outcome of a conversation can be negative for the person if the questions they are asked and how they are asked are negative in content and approach.

Reflection

When talking with someone about what a good start to the day is for them one may incorporate a number of questions.

An open question would be:

'What does a good start to the day look like for you?'

A positive question would be:

'To start the day well, what needs to happen for you?'

A supportive question would be:

'To help you to have a good start to the day, what support would you like to have in place?'

A closed question would be:

'Do you need support to get out of bed in the morning?'

A directive question would be:

'It seems you have reached a stage where you need someone to come and support you in the mornings, is that right?'

A leading question would be:

'*Would you like me to arrange for someone to come and support you in the mornings?*'

- Consider the following questions: are they open, positive and supportive, or are they closed, directive and leading?
  - If you could do anything, what would it be?
  - Would you like to join the day care group?
  - The respite provision available in the local town has a great reputation; would you like to visit?
  - Often people find it difficult to control their behaviours; is this something you are also experiencing?
  - If you were to go away for a break, what would you like to do?
  - If you could live anywhere, where would you like to live?
- How could some of the closed, directing and leading questions be rephrased to talk to the person in an open, positive and supportive way?

Person-centred planning

Person-centred planning provides a range of approaches that help a person to record detailed and meaningful information about themselves. These approaches provide the opportunity to think about the past, present and future and share preferences about how life is now and how they would like it to be in the future (Inclusive Solutions, 2015).

Developing a person-centred plan provides an opportunity to invite people to come and get to know the person better and to learn about what matters to the person.

During a person-centred planning meeting, those present are invited to make a commitment to act upon what they have learnt and to support the person in addressing what they want to change or achieve in their life.

Person-centred planning is built on the values of inclusion, which were covered in Chapter 1, and puts these statements into practice so that they begin to have meaning and purpose in a person's life.

The facilitation of a person-centred plan

Good facilitation of a person-centred plan is crucial in achieving a positive experience where the person whose plan it is has a sense of control and is honoured and respected as an individual. Within person-centred planning, there are two distinct facilitation roles:

1. Process facilitator – guides everyone through the planning process, maintaining momentum and enthusiasm. They aim to acknowledge the finer elements of communication within the group, for instance, the nonverbal communication such as nods of approval or signs that someone is uncertain or unsure and addresses these as they arise. Everyone is guided through the planning process by the process facilitator to reach an inspiring image and direction for a future designed by and with the person whose plan it is. This is accompanied by clear commitments to put the plan into action. The process facilitator also makes sure that the pace and the atmosphere is positive and buoyant.
2. Graphic facilitator – focuses on recording the content of the meeting using big sheets of paper and colourful, bold pens and pastels, just like the role of the graphic recorder in a solution circle exercise. The role involves recording the conversation using images that represent the words but also capturing the emotion of those present. The graphic facilitator utilises the opportunity presented by the natural pauses of the conversation to recap on what has been recorded to ensure accuracy. They also invite people to add to the visual record and they photograph the completed person-centred plan at the end of the meeting, so this can be shared with those present. The original recording of the plan is carefully taken down, rolled up and handed to the person in a gesture that clarifies the ownership of the plan.

Facilitators of person-centred planning can consider the following to make sure the planning process is positive:

- Listening – This skill can be demonstrated by remaining focused on what people are sharing, making sure people have the opportunity to finish what they are saying without being interrupted, reaffirming what people have said, creating opportunities to recap what has been recorded on the person-centred planning graphic record and making changes where the original meaning has been lost.
- Recording – Making sure the information is recorded in such a way that people can clearly see that their contributions have been recognised, capturing exactly what has been shared using either the person's or the contributor's own words. If the contribution has been lengthy and therefore needs summarising, make sure the person or the contributor is involved in deciding on a suitable summary.
- A commitment to finding a solution – Process facilitators should make sure that time is given to consider solutions to some of the issues, especially long-standing problems, and issues the person may face in the future. The aim of planning is to

make a positive difference and to encourage people to think about the range of solutions available to them, considering:

- voluntary groups within the local area;
- people who are within the neighbourhood who may have specific, useful skills and knowledge;
- the skills and knowledge available in the person's immediate network of relationships;
- funding available via the wide range of funding streams, including local authority, Integrated Care Boards, grants and the Department of Welfare and Pensions;
- technology and equipment;
- adaptations to the person's home to enable them to continue to live there;
- setting something up, this could include recruiting support workers or setting up an interest group in the person's local area.

Example: How thinking about what is possible works in practice

Brenda is a keen arts and crafts person; she would like to continue to enjoy these pastimes and would like to find other people to share this interest with. She spent a bit of time with her daughter having a walk around the local area to look at the community notice boards but there was nothing about arts and crafts groups. They decided to pop into the local library and have a look on the internet. They found a group that met on a Wednesday in the community centre in the next village.

Brenda wants to go to the group but does not want to travel on her own. For the first few weeks, Brenda went with her daughter until she got to know someone in the group, Enid. As they got to know each other they learned that Enid drives past Brenda's house on her way to the community centre. After having a chat, Enid starts to give Brenda a lift to the group. Brenda then attends with Enid and gradually gets to know the rest of the people who attend. She is really enjoying her time there and the group have started to sell some of the crafts to raise money for a local charity. As Brenda has settled into the group, she now feels comfortable getting a taxi there when Enid is on holiday; Brenda uses some of her personal budget to cover the cost of a taxi.

The aim of a personalised approach is for the person to still have choice and control in their lives and that they have the liberty to decide how and where they are supported and cared for.

For those involved in facilitating and developing person-centred plans, the following guidelines need to be followed:

- No jargon – Use everyday language that everyone will understand and words to fully explain the meaning rather than acronyms or initials;
- No fixing – Work from the assumption that people are not broken; they have the capacity to know how to best resolve the issues they are facing, they just need some support, space and time to think it through; even if a person is in a position in their life where they are no longer able to express what is important to them, someone close to them will know what is best.
- No obsessing – Give each topic the time and space it deserves but also know when it is time to move on, making sure that the person feels in control of their own plan and that they are setting the pace of what needs to be discussed, as well as how much time to spend discussing it.
- No digging – Create an environment where the person and the people close to them feel able to talk about their circumstances openly and honestly. It is important to recognise when someone is being pushed out of their comfort zone by the questions they are being asked or when people have started digging for more detail than they feel comfortable to share, and to then move the group onto another topic.
- No judging – Everyone has preferences and makes lifestyle choices; person-centred planning encourages everyone to respect each other's decisions irrespective of personal preference and asks people to see life from the person's perspective and to respect whatever the person decides is right for them.

Person-centred planning is a creative and open process that can be adapted to work specifically for the person; this can be achieved through:

- altering the layout of the planning tool; for example, for a person who loves steam engines the planning sections can be related to a steam engine, such as the fuel depicting what the person needs to stay strong and keep going, the wheels representing the direction the person is going in, and the passengers signifying the people they would like to include on their journey;
- holding the meetings in a different way to create an atmosphere where everyone feels welcome and clear on their purpose within the planning meeting. This can be achieved through making the meeting relaxed and fun, holding the meeting in a place that has meaning to the person, somewhere they enjoy spending time and, where people will be comfortable, making sure everyone coming to the meeting is clear on the purpose of the gathering and their role within it before it takes place.

Who to invite to the planning meeting

The main priority is for the person to feel comfortable and at ease to talk openly. Achieving this will largely be determined by how supported the person is by the people present; therefore, it is important to invite the 'right' people. This could include anyone whom the person wishes to have there, for example, friends, neighbours, work colleagues and members of the community. The person may write an invite list or make reference to photographs of the people they would like to invite.

In instances where they may not be able to make this decision, those involved in helping to organise the meeting can consider who the 'right' people to invite would be and who the person would want to have there. Here are a few things to consider:

- Does the person enjoy spending time with them; do they enjoy each other's company?
- Do they have the person's best interests at heart?
- Will they listen to the person or the ideas contributed without being judgemental?
- Will they concentrate on the person and the intention to achieve the best outcome for them?

If, after thinking through these questions, the general feeling is that it would be positive to invite a person, consider:

- What skills or knowledge do they have that they could contribute to the person's plan?
- Do they have the time and energy to contribute to putting the plan into action?
- Are they well connected within their local community?
- Have they got a positive outlook and would they be keen to make a difference?
- Do they recognise the skills, positive characteristics and interests of the person and what they have to offer other people and the community?

The range of people to invite could be very different for people who have already moved into a service setting which may have created a distance in some of the important relationships they had while they remained within their community. In these circumstances, developing a relationship map can identify the relationships that have been lost and generate the discussions needed to find ways to re-establish these.

Professionally led meetings that focused on addressing a person's assessed care or support needs would include a range of professional people:

- social workers;
- nurses;

- physiotherapists;
- occupational therapists;
- speech and language therapists.

It can be difficult and daunting for anyone to feel comfortable talking openly about themselves during a meeting when the attendees are people they do not know well. In these circumstances, people can become so overwhelmed that they do not feel able to talk confidently about themselves, especially when the people present may make decisions about their future care and support.

In recognising this it is then possible to understand that information gathered at meetings that are led heavily from a professional perspective may not lead to developing a good level of insight and understanding of the person and their circumstances. If these meetings form the foundation for the services and support to be put in place for the individual it is fair to suggest that there is a real potential for support and care packages to not work well for the person.

The methods introduced in this chapter regarding person-centred planning could be the answer to ensuring that future care and support work well for the person.

Person-centred planning approaches

Every approach to person-centred planning encourages people to be:

- positive;
- creative in generating ideas and finding solutions;
- rooted in the values of inclusion;
- confident about what they have to bring to the planning conversations and the potential opportunities that could be found through their involvement in the conversation;
- clear on what they have to offer in terms of skills, knowledge and connections.

There are a number of different methods and processes that sit within the family of person-centred plans.

PATH – planning alternative tomorrows with hope

This planning approach was developed by Pearpoint et al (2001). A PATH is a future-focused planning process that has the intention of recognising and honouring a person's desired future and collectively finding a way to move towards achieving this. It must be facilitated by a process facilitator and a graphic facilitator. The graphic facilitator focuses on recording the content of the meeting, while the process facilitator guides the person and the people present through the planning process. People are invited to gather together to support the development of the person's PATH. As part of being present, people take an active role in the conversations and discussions within the PATH meeting.

The meeting begins by welcoming everyone and inviting people to contribute to an appreciation, what they like and admire about the person. This helps everyone present to adopt a positive mindset. Once the group is settled, the process facilitator encourages the person to explore what a good future would look like for them. The person is encouraged to share what their thoughts and ideas are based on the questions 'if nothing was going to get in your way, what would you like to do?' 'If nothing could hold you back, what would you like to do?'

Once a clear picture of someone's dream is established, the PATH process supports them to bring their dream into a time frame of what can be achieved, '*it is now a year on from today – what have you managed to achieve that has taken you towards your dream?*' This enables people's aspirations to be placed in a time frame that feels positive and possible – realistic, manageable and achievable. The process facilitator encourages the person and the rest of the group to talk as though they had already achieved these goals and as though it had already happened, by opening up the conversation using phrases along the lines of '*imagine that this is now a year on from today, tell me what have you achieved towards reaching your dream*'. The importance of this stage is that the person begins to recognise what their achievement would feel like; this can, in turn, build momentum, motivation and commitment. Once a clear picture has emerged, the questions would then be repeated in the context of a six-month time frame. At this stage, it is good to start to create some links between what has been achieved in six months and what has been achieved in a year so that people can start to see the connections and the progression. Producing a graphic record, using arrows or colour to group topics can help emphasise connections.

Once the dream is captured and people have developed an appreciation of what this means for the individual, the person is asked what the dream feels like for them and this is recorded on the PATH. The person and those around them can begin to feel a sense of enthusiasm and energy.

At this stage, a small amount of time is spent on the 'now' section of the PATH. This is intended to capture what life looks and feels like for the person now. It helps to give some measure between the 'dream' and 'now' and begins to highlight any disparity, tension and difference between the two. Only a short length of time is dedicated to this section, five or ten minutes at the most, as the present can be difficult to discuss if there is a big difference between the 'now' and the 'dream'. Listening to the person's current situation can build a sense of urgency towards making the person's dream a reality. Dwelling on this stage of the PATH beyond what the person wishes to share can begin to negatively affect the positive energy that was created at the beginning of the PATH planning process. Time spent discussing the 'now' is not going to create the change that may be desperately needed so it is best to move on.

The fifth step is 'enrolling'; this recognises the people with the commitment, skills, knowledge, understanding and connections to help realise the dream. These are the people who are vital to ensure the success of the plan.

First, and most importantly, the person signs up to their own plan, whenever possible. It is important that the person makes a commitment to work towards achieving their own dreams and aspirations in whichever way they can. They then invite other people to 'enrol' in their PATH to support them in the journey towards achieving their dreams and their desired future. When looking at who to 'enrol' in a plan, it is also possible to think about the people who are not present at the planning meeting who would be valuable in helping to achieve the plan. The planning group could then discuss how to invite people to get involved.

The sixth step is 'building strengths'. This stage recognises that it can take a lot of strength and energy to achieve the goals and aspirations within the plan. Recognising what lifts the feelings and mood of the person and the people involved in their PATH, and what gives them the energy to move forward is going to be an important feature of keeping people's energies high. This could include:

- having lunch once a week with a close friend;
- going to church every week;
- making time to read a favourite book one evening a week;
- going for a walk in the countryside on a Sunday.

When working towards a goal, it is possible to overlook making time for the areas that give the person, and the other people involved in their PATH, the enthusiasm and energy to continue, and it is important to make sure that these areas of people's lives are recognised and not forgotten as part of the journey.

The final stage recognises the 'first steps' that can be taken straight after the planning meeting with the emphasis on identifying something that can be done that day or within the next 48 hours. The intention is to make the most of the momentum generated during the planning meeting and to make sure that this is harnessed and maintained to enable the person and the planning group to work towards making a difference and achieving the steps towards the person's 'dream'.

At the end of the PATH planning process, there is usually a lot of energy and commitment to moving towards the vision of a good future that has been created. People who attended the planning session leave with a clearer insight into what would make a good future for the person and the steps they have committed to take to enable them to work towards this.

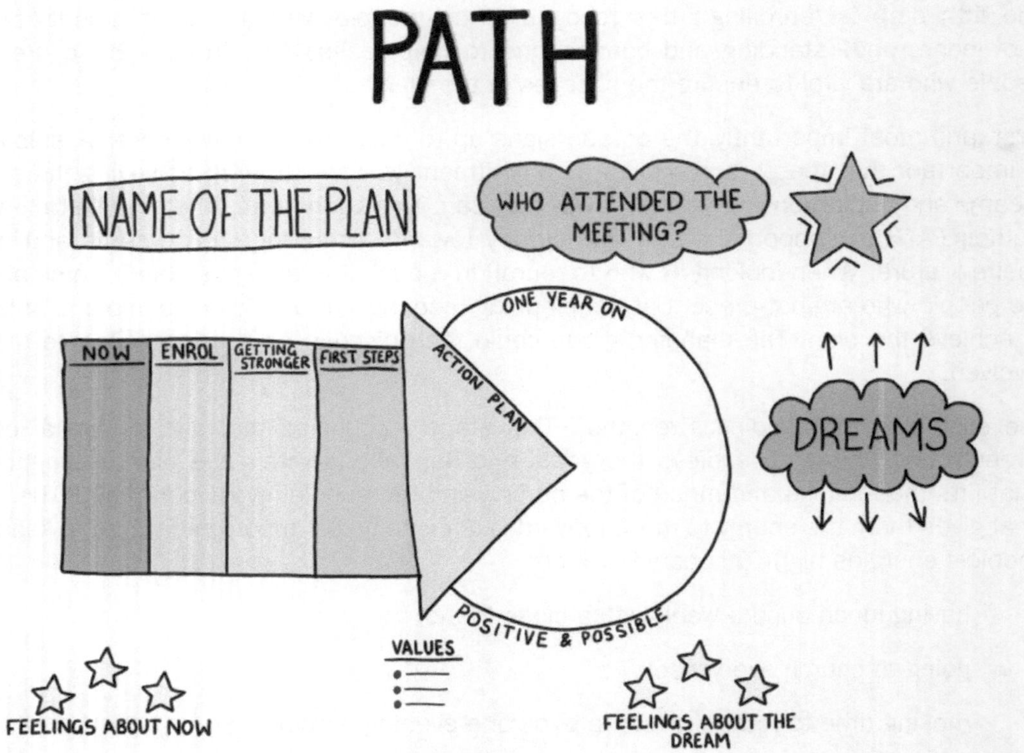
PATH
NAME OF THE PLAN
WHO ATTENDED THE MEETING?
ONE YEAR ON
NOW
ENROL
GETTING STRONGER
FIRST STEPS
ACTION PLAN
DREAMS
POSITIVE & POSSIBLE
VALUES
FEELINGS ABOUT NOW
FEELINGS ABOUT THE DREAM

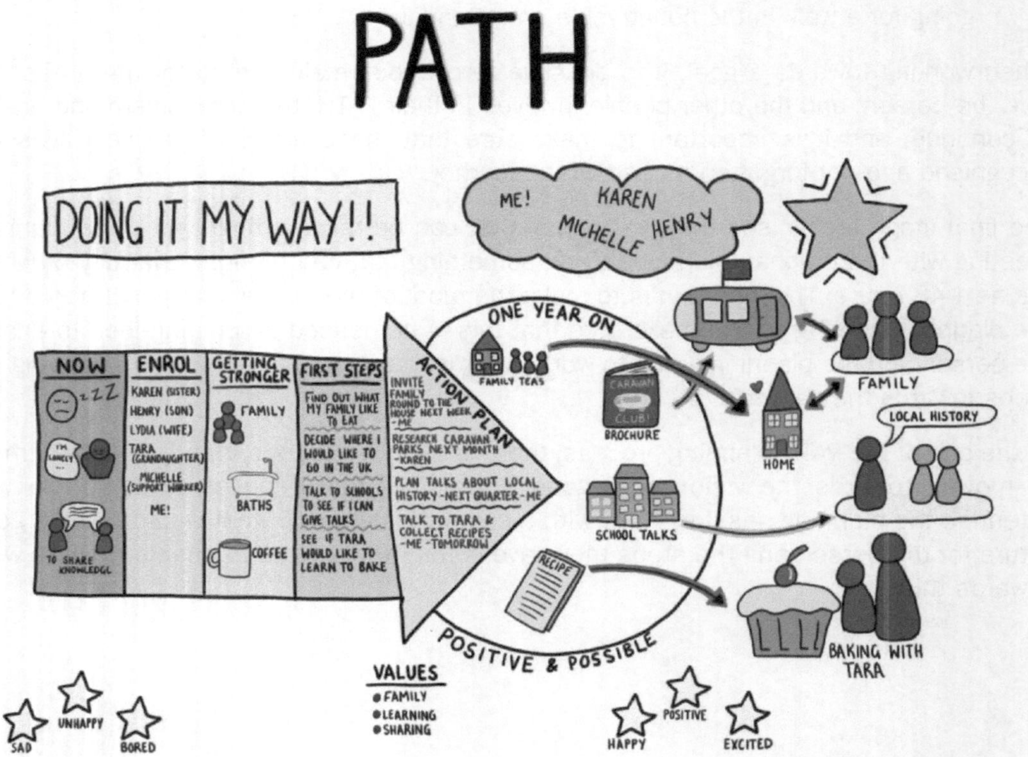
PATH
DOING IT MY WAY!!
ME!
KAREN
MICHELLE
HENRY
ONE YEAR ON
NOW
I'M LONELY ...
TO SHARE KNOWLEDGL
ENROL
KAREN (SISTER)
HENRY (SON)
LYDIA (WIFE)
TARA (GRANDAUGHTER)
MICHELLE (SUPPORT WORKER)
ME!
GETTING STRONGER
FAMILY
BATHS
COFFEE
FIRST STEPS
FIND OUT WHAT MY FAMILY LIKE TO EAT
DECIDE WHERE I WOULD LIKE TO GO IN THE UK
TALK TO SCHOOLS TO SEE IF I CAN GIVE TALKS
SEE IF TARA WOULD LIKE TO LEARN TO BAKE
ACTION PLAN
INVITE FAMILY ROUND TO THE HOUSE NEXT WEEK -ME
RESEARCH CARAVAN PARKS NEXT MONTH -KAREN
PLAN TALKS ABOUT LOCAL HISTORY -NEXT QUARTER -ME
TALK TO TARA & COLLECT RECIPES -ME -TOMORROW
FAMILY TEAS
CARAVAN CLUB!
BROCHURE
HOME
FAMILY
LOCAL HISTORY
SCHOOL TALKS
RECIPE
BAKING WITH TARA
POSITIVE & POSSIBLE
VALUES
FAMILY
LEARNING
SHARING
SAD
UNHAPPY
BORED
HAPPY
POSITIVE
EXCITED

MAP – making action plans

This planning process was developed by O'Brien et al (2010). The MAP process works well when a group of people are invited to come together with a commitment: to listen to the person, learn about what they wish to change and achieve and be actively involved in creating a better future. Similar to a PATH, a MAP is facilitated by both a process facilitator and a graphic facilitator.

At the start of a MAP planning process, everyone attending is welcomed and invited to contribute something they appreciate about the person. To achieve this, the process facilitator asks each person to share something they like and admire about the person.

This first step usually takes about five to ten minutes, depending on the size of the group, and sets the positive tone for the rest of the planning meeting. Opening a meeting this way can fundamentally change how the person views themselves, supporting them to be confident and engaged in their own plan.

The second step of a MAP is where the person shares their story:

- proud moments and achievements;
- important and sentimental memories;
- places where they have lived and spent time;
- influential and significant people and loved ones;
- the roles they have fulfilled;
- difficult times, loss and grief.

Through listening to the person's story, those present can learn a great deal about what was, and still is, important to the person. The story builds a bigger picture about the person and can open a discussion regarding their past and whether they wish to re-establish connections with friendships, family, interests, places or a vocation.

For people who have been supported within a service setting for a number of years their recent personal story and history can become a chronology of the services and medical input they have received, including diagnosis and specialist treatment. When a MAP is developed, this aspect of a person's life may be recognised but should not dominate or overshadow their own personal story. Instead their journey that has got them to where they are today should stand out more than any service-based chronology.

The third and fourth steps of the MAP planning process consider the person's 'dreams' and 'nightmares':

'Dreams' – Understanding a person's 'dreams' gives everyone an insight into what is important to them; this includes the things they would like to strive to achieve and the things they would like to have in place on a day-to-day basis.

Listening to a person's 'dreams' builds an awareness of their interests, passions and what has meaning for them.

'Nightmares' – Understanding a person's 'nightmares' provides an insight into what they would not want to happen; situations that would cause them fear, anxiety, stress or upset. A person's 'nightmares' may refer to negative experiences they have had in the past that they would not want to re-live. Recording the nightmares as part of the MAP planning process helps to inform people about what needs to be avoided and to plan accordingly.

Step five focuses on the question: 'what does the person need to and want to happen on the basis of what we have learnt today?' The intention is to identify what needs to happen to make a positive difference in the person's life. Once areas of need and preferences have been identified, an action plan, involving the person and the other people present at the planning meeting, is created to progress the plan. One of the actions would be establishing a time that the planning group can get back together with the person to check in on progress towards their plan.

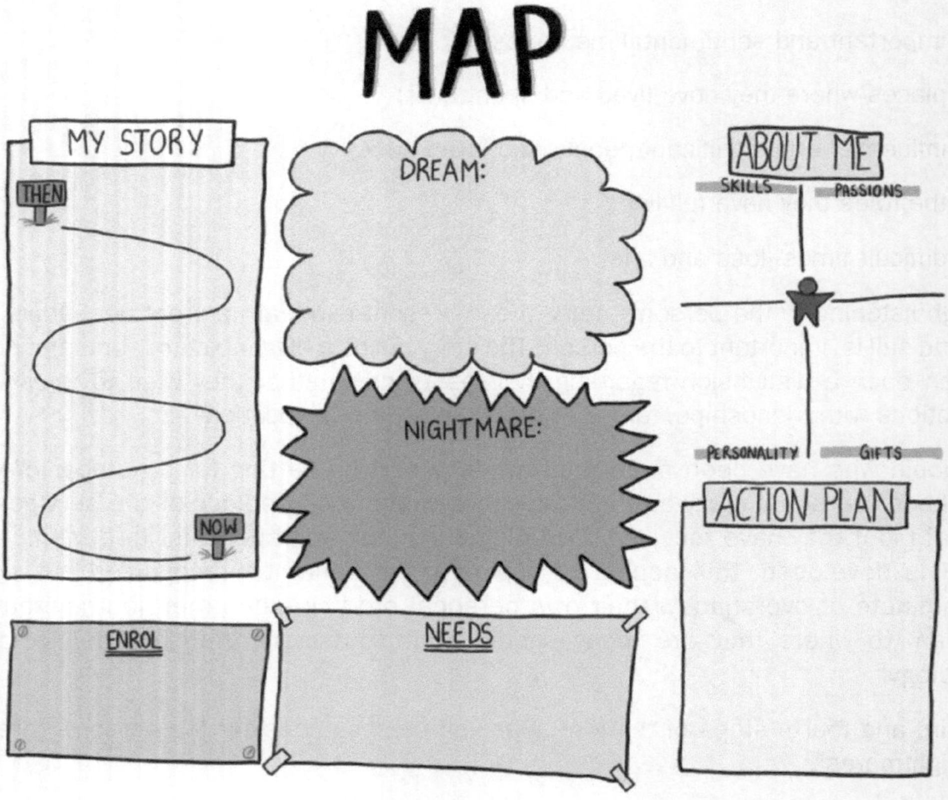

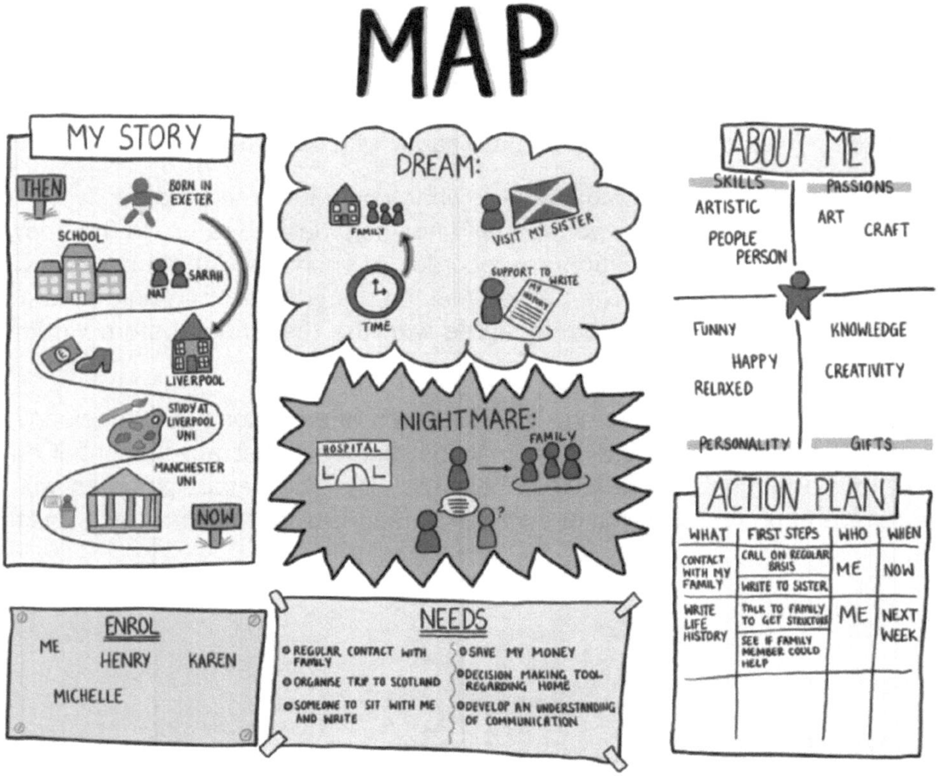

Essential lifestyles planning

Essential lifestyle planning was developed by Michael Smull and Susan Burke Harrison (n.d.). This approach gathers valuable information about how a person wishes to be supported on a day-to-day basis and the finer details of what matters to them in everyday life.

An essential lifestyle plan gathers details about the person's important daily routines and rituals. Once gathered, this detail can then be shared to inform those who are providing support or care in the person's day-to-day life, enabling them to provide it in a way that honours the person's preferences.

The first step of an essential lifestyle plan is to identify the important people who know what works well for the person and to hold conversations with them to gather the details of what is important to the person. However, the first conversation should always be held with the person whose plan it is.

The second step is to do an initial sort of the information gathered under these headings:

- 'Essential and non-negotiable';
- 'Negotiable and not essential';
- 'Would be good if it did happen but the person could live without it'.

Once the initial sort has been completed, a planning meeting would take place. This would be with the person and the people they have chosen to invite, including those who have contributed to the initial information gathering. At the meeting, the information would be checked to determine if it has been prioritised correctly. Information may need to be moved around until the person and those present are happy with how it is presented.

The second stage addresses the question: 'what would need to be done to make sure the person's preferences, both negotiable and non-negotiable, feature in their day-to-day life?' The responses to this question are recorded in such a way that everyone can see them using graphic facilitation. The person making the written or graphic record needs to first check with the person that they agree with the responses before writing them down.

The final stage of the planning meeting is the creation of an action plan that clearly states what is going to be done to make sure that the elements that are 'essential and non-negotiable' to the person feature in their day-to-day life. The details gathered will inform everyone involved about how to make sure the person is supported well and positively in the future.

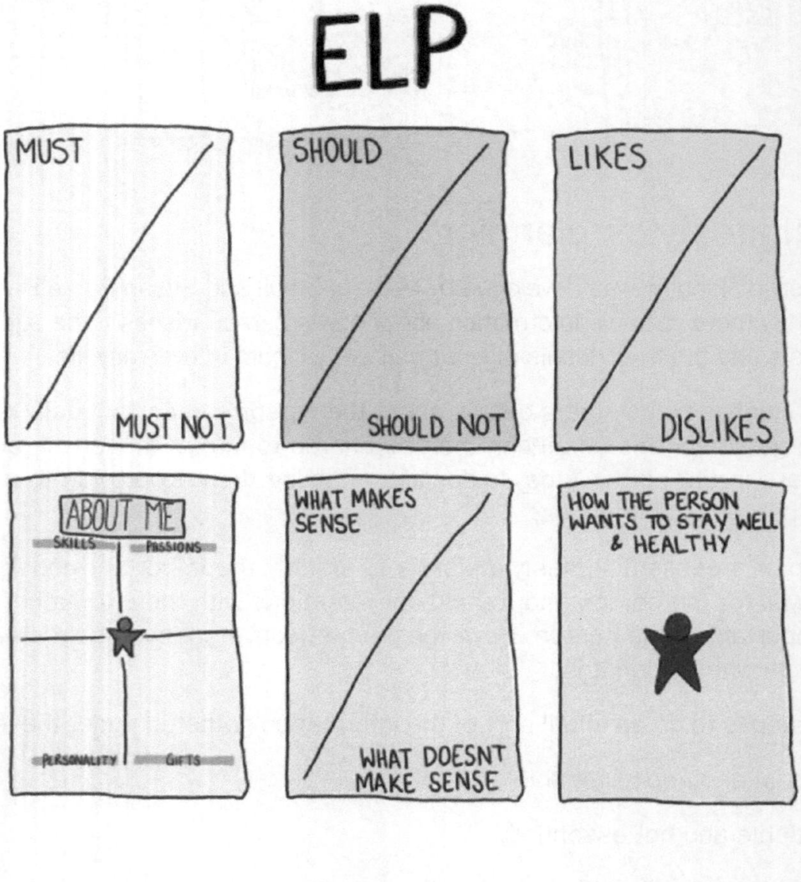

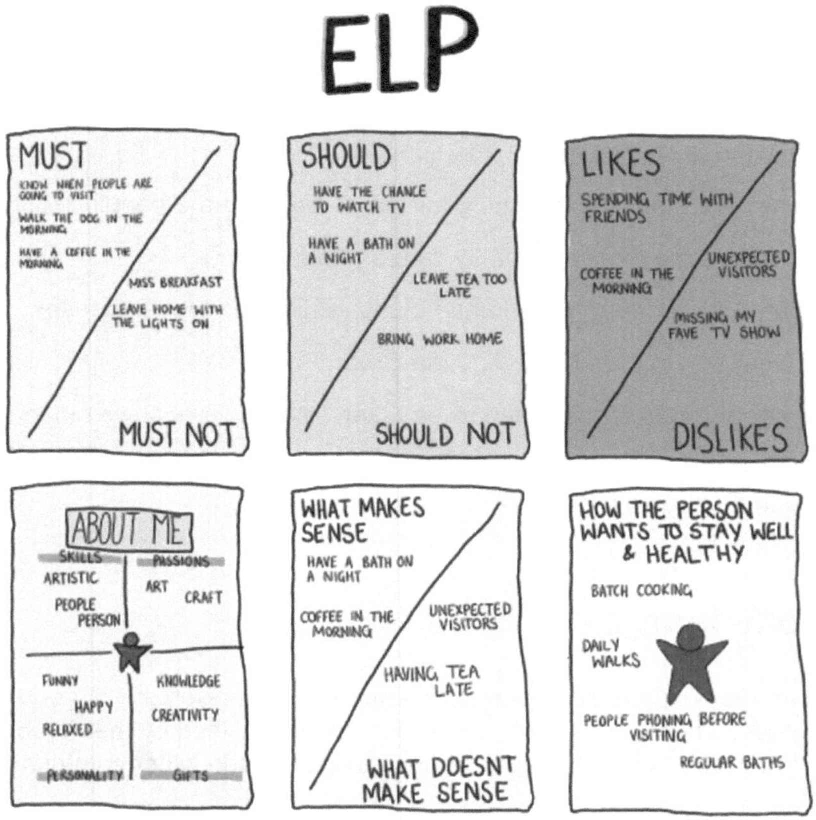

Other approaches for planning meetings

The following approaches can also be utilised during the planning meetings:

- A question board – this records questions that the planning process has generated that cannot be answered in the meeting and would also record ideas to address these questions after the planning session.
- An ideas poster – this records ideas that are shared during the meeting, which the planning group can then refer to when the action plan is developed.
- Deciding 'when will we meet again?' – to keep the plan alive, it is important that the planning group think about when they will reconvene to discuss progress and consider how to overcome barriers.
- Action planning - agree actions using the following points; what needs to be done, how the action is going to be achieved, who is doing what and when things need to be done.

The tools and approaches covered in this chapter serve a clear purpose to:

- raise awareness and understanding;
- develop good relationships between the person and everyone involved in their life, especially those providing care and support;
- notice changes that are happening for the person and the people around them;
- achieve consistent and high-quality care and support;
- be prepared for future changes and how to respond to these changes;
- have a clear sense of purpose and direction;
- clearly identify actions that need to be taken to achieve positive outcomes;
- identify where some of the solutions can be found and the people who could be involved in making things happen.

Key lessons learnt

- There are many approaches that can support the achievement of good care and support. These skills and approaches can be utilised by the person with dementia, the network of support around them and the wider community.
- The approaches can be utilised in a creative and dynamic way to achieve the best outcome for the person with dementia and the people around them.
- Practised for over 50 years, the approaches introduced in this chapter have been proven to work in a whole range of settings. They make a real difference in enabling the voices of people to be heard; promoting positive interactions, respect and dignity; and protecting people's right to make decisions about their own lives.
- Having a detailed understanding of the person will help to ensure that their preferences and wishes are respected at each stage of the dementia journey. This will lead to care and support or professional input being appropriate and personalised in approach.
- Putting the approaches introduced in this chapter into practice will help everyone involved develop a detailed understanding of the person with dementia.
- The approaches presented also represent a method of gathering, sorting and presenting information which can create a reference point for anyone involved in caring for or supporting the person.

Questions to aid reflection

1) Consider the time it would take to utilise the approaches presented in this chapter. When would be the best time in the dementia journey to introduce these?
2) What difference could these approaches make to someone's quality of life?
3) What difference could utilising these approaches make for the people who are paid to provide care and support or for any professional involved?
4) What difficulties could be faced in developing an understanding of the person with dementia during the later stages of dementia?
5) What would need to happen to develop the confidence to utilise the approaches presented in this chapter?

7 Understanding behaviour and communication

This chapter covers:

- **Behaviour as a form of communication**
- **What can influence behaviour**
- **A problem-solving and solution-focused approach**
- **The approaches and skills to aid good approaches to communication.**

Behaviour and dementia

People with dementia may begin to behave differently in various stages of their dementia journey and how a person responds and behaves may depend on the form of dementia they have been diagnosed with. Chapter 3 provides information about the different forms of dementia and the dementia journey.

The person may behave in ways that are out of character, for instance, they might display restlessness, repetitive behaviour, shouting, aggression, screaming, walking about a lot more, disturbed sleeping and waking during the night.

People can become confused and what otherwise would be deemed as familiar surroundings become frightening and unfamiliar. They can also often experience a sense of grief, loss, frustration, anger and hopelessness which can all be emotional states that directly influence how the person responds or behaves in given circumstances. Behaviour can be extreme and interpreted negatively.

Behaviour that can be deemed negatively or pose a challenge includes any '*actions that detract from the well-being of the individual due to the physical or psychological distress they cause*' (James, 2011).

This behaviour may be:

- a direct result of the symptom of dementia, especially in the case of Lewy bodies;
- a result of the difficulties related to dementia, for instance, language, disorientation or memory loss;
- a response to how the person is feeling due their experiences and interactions with others;
- a demonstration of boredom or frustration, especially if the person's social interaction and engagement in enjoyable activities has declined;
- related to physical pain the person is experiencing but is unable to express;
- a reaction to a threat even if this is a perceived threat.

If there has been a sudden change in behaviour, particularly where there is confusion and distress, a visit to the GP would rule out or identify any potential physical problems, side effects of medication or changes to mental health.

Explaining all aspects of behaviour as an outcome of dementia may not be accurate as it is possible that the behaviour is a way of communicating an unmet need.

> *The times when I have felt aggressive it was because of the frustration of living with Alzheimer's – it was about how I felt, not a symptom. Let me give you an example. I have limited mobility – bad hip joint and can't walk easily – and that makes me angry and annoyed sometimes. But no one would say that was a symptom of a broken hip would they? They do that with dementia all the time too though don't they, and it can be incredibly frustrating.*
>
> (Roy, Social Care Institute for Excellence (SCIE), 2020)

Sundowning

The term 'sundowning' is associated with dementia and refers to a behaviour that occurs at a particular time of day, usually late afternoon and early evening. At these times, a person with dementia may become more agitated, confused and aggressive. There maybe a number of reasons for this, such as a lack of sleep or disturbed sleep, a loss of a familiar routine, lots of noise, a response to medication or medication wearing off.

Sundowning usually occurs in the later stages of dementia. It is useful to keep a record of how the person is responding and the time of day this takes place. This can help to identify any potential causes and to find solutions in reducing the person's distress at this time of day.

Communication

Behaviour is often a way for the person to communicate. A commitment to discover what the person is communicating and acting upon is a positive way to respond to behaviours that are otherwise regarded negatively.

The first step in this journey is to see the world and the situation from the person's perspective. Adopting this perspective can support the identification of what is happening around or for the person, what is causing the upset and what solutions would address this. Through a commitment to this approach, people involved in the person's life can begin to learn what experiences, interactions and environmental factors influence a good or a negative emotional state for the person.

Once an appreciation of the person's perspective has been developed, there may be some long-standing issues identified. Finding a solution to these issues can be supported by a solution circle, as covered in Chapter 6.

It can take time to fully understand what a person is communicating and sharing through their behaviour. It may require adopting ways to respond to the person on the basis of what is believed to be what they need or want at the time. It will take time to try ways to support the person and learn from them, adapting or changing what is done with the intention of finding what works best for the person.

Challenging and difficult times

It is recognised that everyone experiences challenging and difficult times:

> *'The other day my sister lost her temper with me when I couldn't remember something – she's usually so good – and it was really really horrible – I cried and cried after she had gone, it was horrendous. It must be so hard for her to put up with me with a smile all the time.'*
>
> (Sandy, Social Care Institute for Excellence (SCIE), 2020)

People with dementia and their families may experience difficult times as they come to terms with how circumstances are changing along with the experience of day-to-day life and long term plans.

Challenging behaviour?

The language used to describe a person's behaviour must also be carefully considered:

- 'difficult behaviours';
- 'challenging behaviour';
- 'behaviours that challenge'.

These phrases promote the impression that the person is the root cause of the difficulty as it is their behaviour that is posing the challenge. However, in reality, it is most likely that the person's behaviour is a response to an unmet need or want, or a representation of a disconnection between the person and those around them.

The use of this type of terminology when describing a person's behaviour is negative and may lead to a may a negative reputation.

Challenging circumstances

It is good to develop an appreciation that it may not be the person who has 'challenging behaviour', but the circumstances the person finds themselves in are challenging for them.

> *'People need to understand that you're frightened and alone. You start out on the pathway and often you're on your own. You can't understand why things are happening. Understanding the details of the problem removes the fear.'*
>
> (Peter H, Social Care Institute for Excellence (SCIE), 2020)

Research has shown when a person-centred approach is developed and delivered within dementia care and support there is a substantial reduction in negative experiences and therefore fewer instances of 'challenging behaviour'. At the same time, there are fewer drug interventions needed (McMaster University, 2019; Fazio et al, 2018).

When a person-centred approach is adopted and time is given to watch, listen and learn, it is possible to begin to appreciate what the person is communicating:

- Unwanted changes to their day-to-day routine that may be causing upset? For instance, a change in the people they are spending time with or to how they are spending their time?
- Are the person's day-to-day routines dictated by the practices and routines of the care setting or of another person rather than built around their preferences and what matters to them?
- Recent experiences that have caused upset that have not been resolved?

Learning what the person is communicating through their behaviour can help to resolve some of the issues they are experiencing while also supporting people to understand the person's feelings and the emotional support they need.

If the person's day-to-day life is working well for them and they have a good network of care and support around them, this is often reflected in how they are feeling, which, in turn, will reflect in the way they are with themselves and the people around them.

Discovering what the person is communicating

Working through some explorative questions can help to gain an insight into what is happening to the person and pinpoint the root cause of any upset they are experiencing. Below is a list of questions that can aid the exploration:

- What is it that is happening around the person or to the person that is causing upset or discomfort?
- Was there anything that happened recently that could have caused them upset?
- Is there something that the person is wanting to know that they have not yet been told?
- Are they feeling insecure about anything, do they need reassurance?
- Is the person comfortable or are they experiencing discomfort?
- Is the person in any pain, is there anything that could indicate a secondary health issue? Weight loss or gain? Change of skin tone?
- Has there been a change to their daily routine?
- Are their preferences, those captured in the person-centred planning work, being honoured?
- Is something that matters to the person being missed?
- Has there been any significant changes in their relationships?

What can be learnt from working through exploratory questions can lead to identifying what can be changed or put in place to better support the person.

It is then possible for people providing any care or support to respond to what they have learnt, leading to the person with dementia being better understood and supported to continue to have choice and control in their life.

This could be demonstrated through involving the person in changing the environment to better suit them, creating the opportunity to spend quality time with someone who the person wants to see or changing day-to-day routines so someone can get up and ready for the day in a way that suits them or to include the activities that are important to the person.

A problem-solving approach

Taking a problem-solving approach can help identify the cause or what the person is communicating to find a positive solution.

A problem-solving approach covers three steps:

1. Define the problem.

   What is happening at the moment that is causing concern?

   What element of the behaviour is concerning?

   Is the impact of the behaviour significantly affecting the person or others around them?

   Is the reaction from other people also causing a problem?

2. Look at the situation and circumstances.

   When does the person behave in this way?

   Is there a pattern around when the behaviour occurs?

   Where does the behaviour happen?

   Is there a pattern relating to the place when the person behaves in this way?

   Who is present when the person behaves in this way?

   Are any patterns in the person's behaviour noticeable?

   What else is noticeable about the person when they behave this way?

3. Think about the solution.

   Can the person express how they felt at the time?

   Is there an approach that has not yet been tried that could be supportive?

   Is there something that could be tried to aid communication?

   Is there a change needed in the approach to how care and support is provided?

   How can the ideas of what to try be communicated with the network of support around the person?

Forward planning

Thinking ahead and planning for what good care and support would look like for the person in the current and later stages of dementia can help to reduce the likelihood of negative experiences.

Planning ahead in this way can avoid upset and frustration and this then being communicated through behaviour.

Other considerations when supporting someone who is behaving in a manner that is having a negative impact on themselves and others include:

- ensuring daily routine is based on their preferences and is clearly communicated;
- introducing an activity that soothes the person when they are agitated, for instance, hand massage, aromatherapy, music or light therapy;
- planning for quality time with people, sharing a task together or doing something together that both parties equally enjoy;
- supporting and encouraging the person to keep active doing something that they enjoy, such as going for a walk, swimming or dancing;
- creating opportunities to have a change of scenery and to spend time somewhere they enjoy.

Taking care of the carers

When a person's behaviour significantly changes, families, carers and people providing support in a paid capacity may find it useful to have some strategies to hand.

Here are some ideas that might help in these situations:

- Try to distract the person, see what is in the immediate environment and what would capture their attention. If there is something there, use this as an opportunity to gently distract them.
- Have something on you that is of interest to re-focus their attention – a picture or a book, for instance.
- Become familiar with what could cause a negative reaction; noisy or busy places, certain smells or sounds, and avoid these places whenever possible.
- Consider the potential cause of the behaviour, tiredness, hunger, pain, frustration and boredom. If this could be the root cause of the behaviour, consider what can be done to alleviate this.
- If the person is being rude or objectionable, do not attempt to correct them. Remember that the person is living their reality and perspective of the world. To correct them or try to change their current reality could lead to more confusion or agitation. Again, think about how the person could be distracted and employ this strategy.

- If the person has become aggressive, provide them some space to help them calm down and anything that is known to help them reduce their heightened state. Remain safe to avoid being harmed at this time and do not respond with aggression.
- Remain as calm as possible, try to not get angry and upset. Remember that the person is not behaving this way on purpose.
- Consider in each situation, is the person's behaviour really a problem? Try to remain rational in the situation and in response.
- Carers should make sure they take some time out and have some rest to be able to recuperate from a situation that they have found difficult to manage. They should have an opportunity to talk to people from their network of support as they may also have some useful suggestions.
- Remove everything from your life which causes you stress (early morning appointments, hospital visits etc).
- Carry on with all the things you enjoy for as long as you can, especially the things you enjoy doing together.
- Caring is an all-consuming onerous task. It remains an overriding responsibility and cannot be denied, delegated or ignored but the emotional and physical support from family is of immeasurable value.
- Friends can be a source of relief, laughter and provide perspective.

Questions to aid reflection

1) Based on the information provided so far in this chapter, how would you approach behaviour differently in the future?
2) Upon reflection, are there any experiences you have had of managing a person's unexpected behaviour that you would now respond to differently?
3) How could you share the information provided in this chapter to others supporting people with dementia?
4) What impact would sharing this information have on people who are confronted with unexpected behaviour shown by the people they care for?

Effective communication: tools and approaches to use to support people with dementia

There are a range of tools and approaches to support better communication which can be used as a means of building a better understanding in relationships. Using what is presented here can help when aiming to build an insight into how a person is feeling and what they are experiencing. The approaches that are relevant for aiding communication include:

- removing unwanted distractions so that those involved in the conversations or interaction can better focus on the conversation;
- introducing items that aid communication, for instance, pictures or objects that hold meaning for the person;
- giving communication the time it needs, as what is learnt during conversation achieves a good understanding on how to provide good support and care.

Information needs to be shared with the person with dementia in a manner that enables them to be involved in the discussion and to exercise control over the decisions that are being made about their future, including their care and support.

People providing care and support can explore approaches to communication to identify what works best and helps to maintain day-to-day decision making and future-based decisions.

There are a number of established approaches to communication which can be explored, including objects of reference, symbols and Talking Mats that are all introduced later in this chapter.

Example: Philip found it difficult to respond to direct questions and would easily become distracted by the activities going on around him. When asked his opinion about something, he found it difficult to express what he liked and what he didn't like, and this could lead to him feeling frustrated. We wanted to find a way to communicate with Philip in a manner that worked for him and didn't create frustration. We had tried using pictures to explore what he would like, we had tried symbols and we had attempted looking on the internet to find images to look at on the screen.

When we thought we had exhausted all the options, one of the person-centred planning facilitators suggested we meet in the local town where Philip had lived most of his adult life.

The meeting was arranged in a local café that Philip used to go to on a regular basis when he lived in the area. On the way to the café, the person with Philip noticed how much he was looking around, looking into shop windows and how much he seemed to enjoy being back in the town.

When Philip met with the small group in the cafe, they talked about this and in the small group meeting he decided that, rather than spend a long time in the café, it would be good to make the most of the opportunity to revisit some places that he used to go to on a regular basis, such as the post office, greengrocers, charity shops and the local pub.

So much was learnt on this visit about the friendships Philip had in the area, the shops he really liked, the things he looked for in these shops and the items that captured his attention.

Sometimes building an understanding of the person and what matters to them does not require talking; instead, spending time together, watching and learning can work just as well.

Good tips for good communication

Communication is critical to ensuring that the person with dementia can maintain a quality of life and that their wishes and needs are understood fully and responded to. A person's identity is maintained through good communication as it enables them to share their preferences and wishes for all aspects of their life.

Communication happens during: small conversations throughout the day or during meetings and get togethers when detailed and focused conversations take place. Here are some tips that help to facilitate good communication:

- make sure that the person with dementia can see other people clearly;
- make eye contact, as this helps the person with dementia focus on the person they are communicating with;
- minimise background noises such as the radio, TV, mobile phones or other conversations.

When communicating;

- speak clearly and calmly in short sentences using everyday language and avoiding jargon;
- maintain a steady pace, allowing time between sentences for the person to process the information and to respond;
- talk to the person in a positive, respectful and patient way;
- do not alter the tone of your voice or adopt a tone that would appear patronising;
- at all times during group conversations, do not talk about the person as if they are not there.

Making choices

When supporting the person to make a choice, be mindful of communicating with them in a way that they can engage with, for instance, offer a manageable amount of options so as to avoid overload which can lead to confusion.

Where there are a number of questions to address, ask the person these one at a time over a comfortable timeframe or over a number of conversations, phrasing them in such a way that allows for a full answer or, where needed, a 'yes' or 'no' answer.

When communicating for the purpose of learning and developing a greater understanding of the person and what is important to them, actively listen to their responses, their spoken responses and those given through facial expressions, gestures, body language and the use of communication aids. To gain clarity and avoid misunderstanding, repeat back to the person what you believe you have understood and check with them to see if you are right.

Be mindful of body language, facial expressions and gestures when communicating with the person with dementia. The person could read these negatively which may cause upset or distress.

Those supporting someone with dementia should never stand too close or over them to communicate as this will feel intimidating. Instead, they should respect the person's personal space and position themselves so they are at the same eye level or below. This will help the person to feel more comfortable during the conversation.

At times, communicating can be frustrating, if there are signs those frustrations are beginning to occur, take a break and come back to the conversation later when it is a better time.

Preparing for a conversation

Before entering any conversation, allow enough time to reflect on past discussions and what has been learnt from these regarding what creates a positive communication exchange. Use what has been discovered to prepare for the next opportunity to have a conversation and learn more about the person.

If a conversation with a person with dementia is approached or conducted with no preparation or consideration for creating a positive experience, the communication exchange can leave the person feeling frustrated and confused. As such, a conscious commitment should be made to ensure that all communication leads to a positive exchange that does not have a negative impact on the person.

Physical barriers to communication

The success of a communication exchange is not solely dependent on the words chosen, the tone of voice or the approach used. Physical factors can either aid or hinder communication for people with dementia. Think about the environment:

- Are there a lot of distractions?
- Are there other people using the same space?
- Is it a busy space?
- Are there other conversations taking place at the same time?
- Is there music or a television playing in the background?
- What is the lighting like?
- Is it bright enough to be able to see well?

Think about how the person is feeling:

- Have they slept well?
- Are they hungry?
- Are they feeling well at the moment?
- Are they in any pain or discomfort?
- Was it a positive experience the last time you spent time together?
- Is there anything that needs to be addressed to rebuild the rapport with them?

Consider the timing of the conversation:

- Is this a good time for the person?
- Does the conversation or meeting change an important part of the person's routine?
- Is there something else that should be happening right now that the person is expecting?
- Will it cause confusion, or frustration, if this does not happen?
- Would they prefer to be doing something else?

Consideration of these factors and planning accordingly can make a significant difference leading to a positive experience and good communication.

Hearing loss

Most people experience hearing loss over the age of seventy. This may be a factor that is affecting a person's ability to engage fully in verbal communication. The Alzheimer's Society has produced useful guidance about communicating with a person with dementia who may also be experiencing hearing loss (Alzheimer's Society, n.d.):

Communicating with a person who has hearing loss

- Speak slightly slower than usual but try to keep the natural rhythms of your speech.
- Do not shout or over-exaggerate words as this will distort your speech.
- Do not cover your mouth as this will interfere with lip-reading. The person should be able to see your face clearly – get onto the same level as them so that they are looking at your face straight on.
- Consider the physical environment – make sure the area is quiet and well-lit.
- Use visual clues – write your message down if the person is able to read and use objects or pictures to help them understand what you're trying to convey. For example, show the person the meals they can choose from.

Reflection

- If you were struggling to understand what a person was communicating to you, how would you want them to respond and what would be useful for them to do?
- What type of environment might make it difficult for a person with dementia to participate in a conversation in a meaningful way? Can you think of places that you would avoid?
- How do you engage in a conversation with someone with dementia that is respectful and not patronising?

Involving the right people in communication

Family carers who hold a longstanding relationship with the person with dementia have a particularly important role to play in supporting communication. People who have a close relationship with the person are more likely to be able to represent their wishes. The role of representing a person requires a great deal of trust as the interpretation of a person's communication needs to be without any bias.

Good and regular communication between the people who form the person's support network or their circle of support is an important part of achieving a quality of life and what matters to the person.

To achieve the principles of good communication, as developed by the Care Service Improvement Partnership (CSIP), the following should be considered and practiced:

- always believe that communication is possible;
- focus on the non-verbal signs as well as the verbal;
- avoid making assumptions; check things out with the person;
- make communication a two-way process that engages the person;
- avoid the use of jargon or complicated explanations;
- keep the style of conversation 'down to earth';
- do not become patronising or sound childish;
- avoid questions that contain the word 'why?' as these are 'justification' questions which can be unreasonable to expect a person to answer;
- be a good listener; focus fully on the person and resist the temptation to finish their sentence or talk for them;
- talk at a slower pace so that the person has the opportunity to understand what is being said;
- maintain a calm and unhurried approach;
- above all, don't be afraid to say: 'I don't understand'.

(Care Services Improvement Partnership (CSIP) National Older People's Mental Health Programme, 2007, p 19)

Recognising that the ability to communicate changes over time

In the early stages of dementia, if an early diagnosis has been given, the person may find it possible to share their views and opinions, should they be given the right opportunity to do so. The person's communication may vary day to day and, as people reach the later stages of dementia, this may rely heavily on body language, gestures and facial expressions.

Example: Mr M was diagnosed with dementia eight years ago and he is now in the later stages of the condition. Mr M has lost all his strength in his body, is often in bed and is cared for 24 hours a day. He remains living in the family home and he has support workers/care workers with him each day. When Mr M is happy he lays still, his eyes look soft and his face is relaxed. People caring for Mr M know how to read the signs that he is happy. At other times, he can be agitated, and his family and support workers have learnt that when he is restless in bed he is too warm, when his face is tense he wants to move position and when his eyes are alert he is in pain.

During the middle to later stages of dementia, it may be increasingly difficult for the person to use words to communicate. At this time, other ways of communicating can be found that do not rely on words, this is known as 'total communication'.

Total communication

Total communication refers to an approach to communicating that recognises all methods of communication:

- speech;
- lip-reading;
- spelling;
- gestures;
- facial expressions;
- formal signs;
- behaviour;
- body language.

Total communication supports the belief that everyone can communicate and all behaviour is a form of communication. Developing an awareness of each one of these methods of communication, then exploring and finding the one, or combination of methods, that works best will support the maintenance and development of strong and positive relationships.

Resources to support communication

There are numerous resources available to support communication:

- **Symbols** – These can represent every day objects, feelings and people, to aid decision-making and conversation. These symbols can be available during conversation so that they can be shared to represent what is being discussed. Symbols can also be visible in the environment, on wall charts or in clear plastic wallets so the person can make reference to them to express what they need or want.

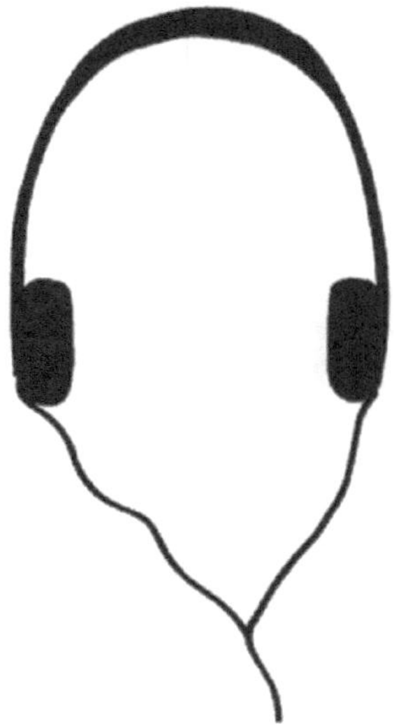

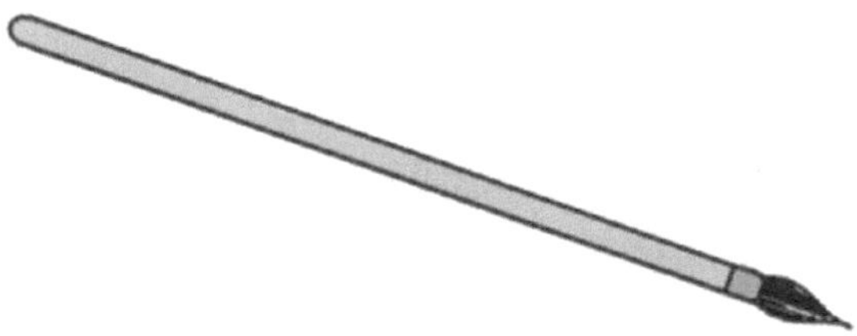

- **Talking Mats** – This tool utilises a combination of symbols to create a story or a question. The symbols are positioned on a mat in an order that represents the conversation or question, for instance, if someone has been swimming the story about going swimming will be organised on the mat to aid a conversation about the experience to establish important information such as: was it enjoyable? Would the person like to go again? Does anything about the experience need to change for it to work better for the person?

The symbols are available from the organisation called Talking Mats. They are based in Scotland and provide practical advice and guidance on communication. Talking Mats have trained families of people with dementia and the participants found that Talking Mats helped them appreciate that their family member was satisfied with many aspects of their life. The participants found this reassuring as many had made the assumption that the person with dementia was frustrated or discontent (Mackay, 2016).

Talking Mats shared the positive feedback from care workers, based at two residential care homes in Denmark, regarding using the Talking Mats communication tool:

- *Talking Mats helps people to understand their options and thereby motivates* [them] *to social activities, eating, physical exercise, etc.*
- *Talking Mats increases the possibility of active participation in caregiving plans*
- *The Talking Mats conversations support a closer relation between* [the] *elderly and care giver*

(Matthews, 2016)

- **Pictures** – Images that are relevant to the person with dementia can aid communication. Choosing images carefully is important to ensure that the person responds to them positively. Having an understanding of the person's interests, passions and the people and times that have been/are important to them can help in selecting the right pictures.

  Pictures to Share (n.d.) is a community interest company working with people with dementia who have, through their work, revealed the power of an image and storytelling to engage a person with the condition. They advocate that this lies not only in the visual and neurological accessibility of the image, but also when the image offers the following:

  - beautiful and vivid colours or subjects;
  - strong photographic or painting style;
  - human or animal subjects looking directly out at the viewer;
  - a strong sense of narrative that encourages storytelling;
  - a strong line or sense of texture that will encourage tactile exploration;
  - a depiction of positive emotions, such as happiness;
  - an element of humour that does not demand sophisticated interpretation.

  Images to be avoided include anything that could be disturbing for someone with dementia, such as people in an aggressive or distressed state, or showing vulnerable people or animals at risk. These can lead to feelings of anxiety or distress that people with dementia may be unable to rationalise.

- **Objects of reference** – Objects that are relevant to the person or to the subject being communicated are useful to aid conversation. It is good to use objects consistently each time the same topic is discussed, and to think creatively about how to use smaller items to represent bigger objects or objects to represent an activity or emotion. For example, referring to a keyring to represent leaving the house or a small towel to represent the choice to have a shower. Objects of reference over time become a form of two-way communication where the person uses the object to express to someone what they want when they are not able to use the words to express this. The objects that have been used to aid communication need to be kept in an accessible place for the person, as well as people providing care and support.

- **Timelines** – These are a useful way to support someone to remember, or become less anxious about, their daily routines. They are also useful in preparing for a future event.

  The person may become confused about the following:

  - the order for daily tasks and routines, for example, when getting ready in the morning;
  - when a holiday or short break will be happening, when they are leaving, how they are getting there, who they are going with where they are going and why;
  - when their loved ones or home care staff are intending to arrive for an expected visit, who is coming and why.

A timeline is a reference point to remind a person of what will be taking place, when it will be happening and some of the finer details to support their understanding.

Visual timelines form a representation of a routine or a visual countdown to when an event is happening, whether it is to be a small event, such as going shopping or a bigger event, such as a holiday or house move.

They provide a greater sense of control in a person's daily life, while reducing confusion and anxiety.

Here are some examples of timelines which demonstrate how they can be implemented:

- A countdown in days to when a person is going to visit a family member, go somewhere for a break or a holiday or when an event is happening. Using a calendar which clearly marks the dates when they have passed helps to show when the person can expect the event to happen.

– A daily task broken down into manageable steps, such as the example shown below, that presents the steps of a person's morning routine. These routines will be particular to the individual. The visual timeline will need to be changed if the person's routines or preferences change, so the timeline is relevant to the person.

- **iPad and touchscreen technology** – Embracing technology can be a supportive means of communicating with people with dementia as well as opening up new avenues to record people's lives, achievements and to support decision-making.

  A report study by the University of Worcester in 2011 (Upton et al, 2011) showcased the many benefits of the use of touchscreen technology within a care service setting. People with dementia can benefit from touchscreen technology within their home too. These benefits included:

- supporting reminiscence;
- aiding recall;
- increasing interpersonal interaction;
- intergenerational communication;
- staff–resident relationships;
- improving quality of life;
- ease of use;
- positive impact on the wider care environment.

'There is now a wide array of apps available for iPads and touchscreen devices that are intended to aid communication. Using the terminology 'communication tool' in the App Store on the device will lead to a list of potential apps to be utilised, for instance, "Know Me – Communicating Personhood" by Know Me Inc.' (Know Me Inc., 2016)

Many speech and language therapists are now trained in touchscreen technology and the applications available to aid communication and can therefore provide valuable advice. Some speech and language therapy departments can also loan the equipment to trial before deciding whether the investment of purchasing a device will be worthwhile.

- **Communication maps** – These record two-way communication:

1. how a person is communicating;
2. how to communicate with the person.

They provide an insight into how a person is expressing themselves and communicating with others, as well as strategies to successfully communicate with them in a way they will understand.

The information added to a communication map is gathered from the person, the people who are important to them and the people who have a good understanding of how they communicate. This tool is especially useful when a person has stopped using words and is relying heavily on the people around them to be able to interpret their body language, behaviour, gestures, sounds and other non-verbal forms of communication.

Communication maps are intended to be used when:

- there are a number of people caring for or supporting the person and a consistent approach would be beneficial across the team;
- when there is an effort to identify what the person is communicating.

The completion of a communication map can provide a focus to those involved and assist the process of learning about how the person is communicating. While developing it, those involved may realise that they had previously made assumptions about what the person was communicating and, through adopting this approach, these assumptions can be tested and reviewed. Therefore, a completed communication map helps to fine tune the understanding held by those supporting and caring for the person with dementia.

Communication maps need to be reviewed regularly and amended to make sure the content reflects the current understanding of how the person is communicating. Regular reviews also help to ensure that anyone following the information provided in the communication map will achieve the best possible results when communicating with the person.

HOW I COMMUNICATE

WHEN JOE BLOGGS DOES THIS	WE THINK IT MEANS/IT DOES MEAN	WE SHOULD RESPOND BY?	IF WE DO NOT LISTEN OR RESPOND THIS WILL HAPPEN	WHAT COULD MAKE THE SITUATION WORSE?
SMILES	HAPPY, LIKES WHAT IS GOING ON AROUND HIM	CARRYING ON THE MOOD IN A POSITIVE WAY		IF THE MOOD IN THE ROOM CHANGED
CRIES	UPSET, ANGRY	TRYING TO CHEER JOE UP WITH MUSIC.	JOE WILL GET MORE UPSET AND DISTRESSED	CARERS/PEOPLE IGNORING JOE.

HOW TO COMMUNICATE WITH ME

WHEN WE NEED TO TELL THE FOCUS PERSON THIS, OR LET THEM KNOW THIS...	WE SHOULD SAY OR DO THIS...	WE KNOW THE PERSON UNDERSTANDS WHEN THEY DO OR SAY...	AND WE SHOULD RESPOND BY...
THEIR LUNCH IS READY	"YOUR LUNCH IS READY, ARE YOU READY TO EAT?"	SMILE THEY ARE READY TO START THE MEALTIME ROUTINE. WHEN THEY SHOUT, THEY DO NOT WANT TO START THEIR MEAL.	GETTING THE FOCUS PERSON READY FOR THEIR MEAL. PUT THEIR MEAL SOMEWHERE APPROPRIATE TO REHEAT LATER

- **Decision-making tables** – While it is important to recognise that people with dementia may have an impaired ability to make decisions, it is also essential to realise that their decision-making ability will vary. This will largely depend on them as an individual, their circumstances, how they are feeling at the time of deciding and the stage of the dementia journey that they have reached. At all times, it must be assumed that the person has the capacity to make decisions, as stated in the principles of the Mental Capacity Act 2005 (Mental Capacity Act 2005). It is the responsibility of those supporting the person to utilise a range of methods of communication to engage with the person to enable them to be involved in the decision-making process. More information about mental capacity can be found in Chapter 2.

A decision-making table helps to record the important decisions in a person's life and how these decisions are made.

The level of 'importance' of the decision is judged by the impact this decision has on the person's quality of life, no matter how big or small the choice is. This could include decisions about where the person chooses to live, what specialist health treatment they wish to receive and it can also capture day-to-day decisions, such as what the person wants to wear and what they choose to eat.

Decision-making tables help to record:

- the decisions the person wants to be in control of;
- they want to make these decisions;
- what support they need to make the decisions;
- the information to be provided to support the person to make the decision;
- the communication approaches / aids to adopt in the decision making process;
- whom they would like to make the decision on their behalf if, and when, they are unable to do so.

DECISION MAKING AGREEMENT

IMPORTANT DECISIONS IN MY LIFE	HOW I MUST BE INVOLVED	WHO WILL MAKE THE FINAL DECISION
WHAT I HAVE TO EAT	I MUST BE SHOWN OPTIONS TO HELP ME MAKE A DECISION	I WILL
WHAT MEDICATION I TAKE	I MUST ATTEND ALL MEDICAL APPOINTMENTS. WHEN I AM NOT HAVING A GOOD TIME I WILL NOT ENGAGE IN THESE DECISIONS	MY BROTHER

A completed decision-making table can ensure that:

- the person with dementia maintains as much control as possible;
- the person's preferences are known;
- everyone involved is clear on the important decisions and how these need to be made;
- everyone is clear about their role in supporting the person to make a decision.

A decision-making table can continue to support a person through their dementia journey and can include some of the decisions the person has made earlier to inform the care and support they wish to receive during the later stages of the condition. The content of a decision-making table should be checked at regular intervals to ensure the information is still accurate and representative of the person and their preferences.

- **Timelines of change** – This is a useful summary of significant changes, achievements and events. The information recorded on a timeline of change can be helpful in aiding conversation with the person, and those who are in their network of support, to be able to talk about the changes that are happening in a person's life and how to be respond to these. This keeps a focus on the person being able to determine what their preferences would be and to make decisions about the care and support they want and need to address these changes. The timeline of change is also a useful summary to take to meetings with professionals to help inform them of recent changes. A record of the person's recent events can also help to observe changes in the person's memory and ability to recollect.

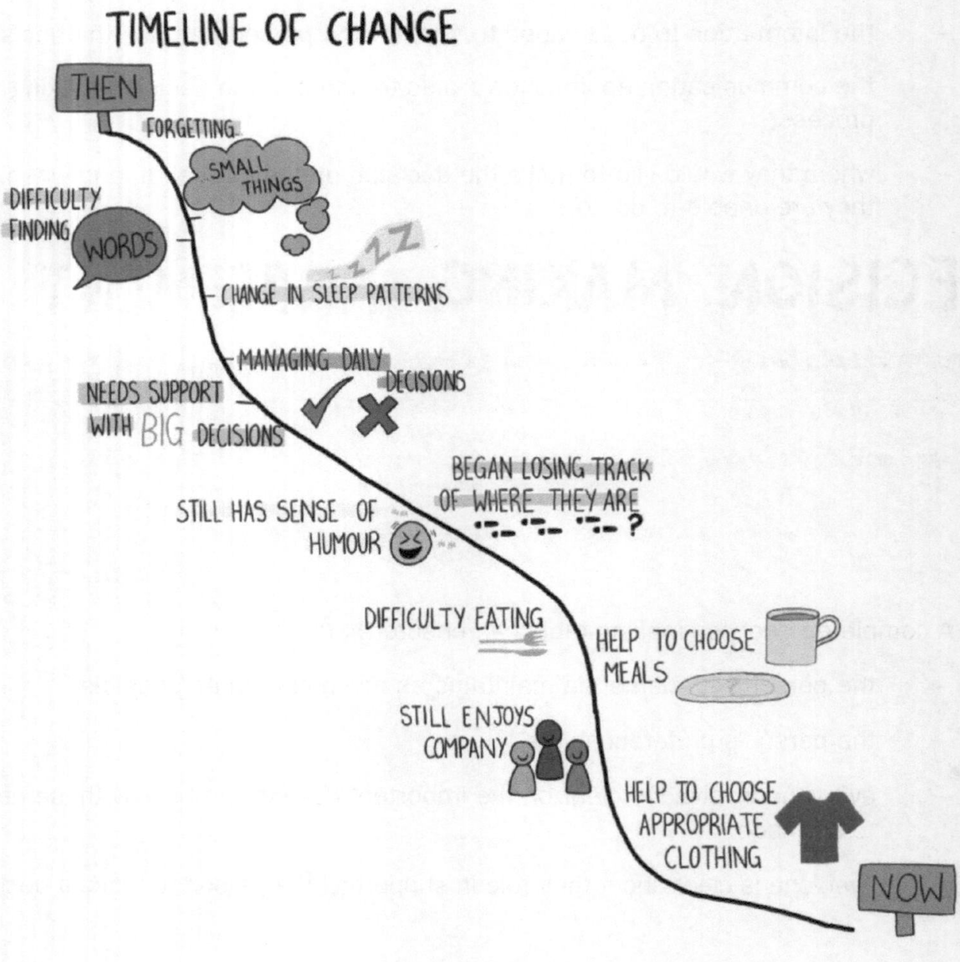

Timeline of achievement – This is a useful visual summary of what the person has recently achieved. The information shared on the timeline can be valuable as a reference point to aid conversation.

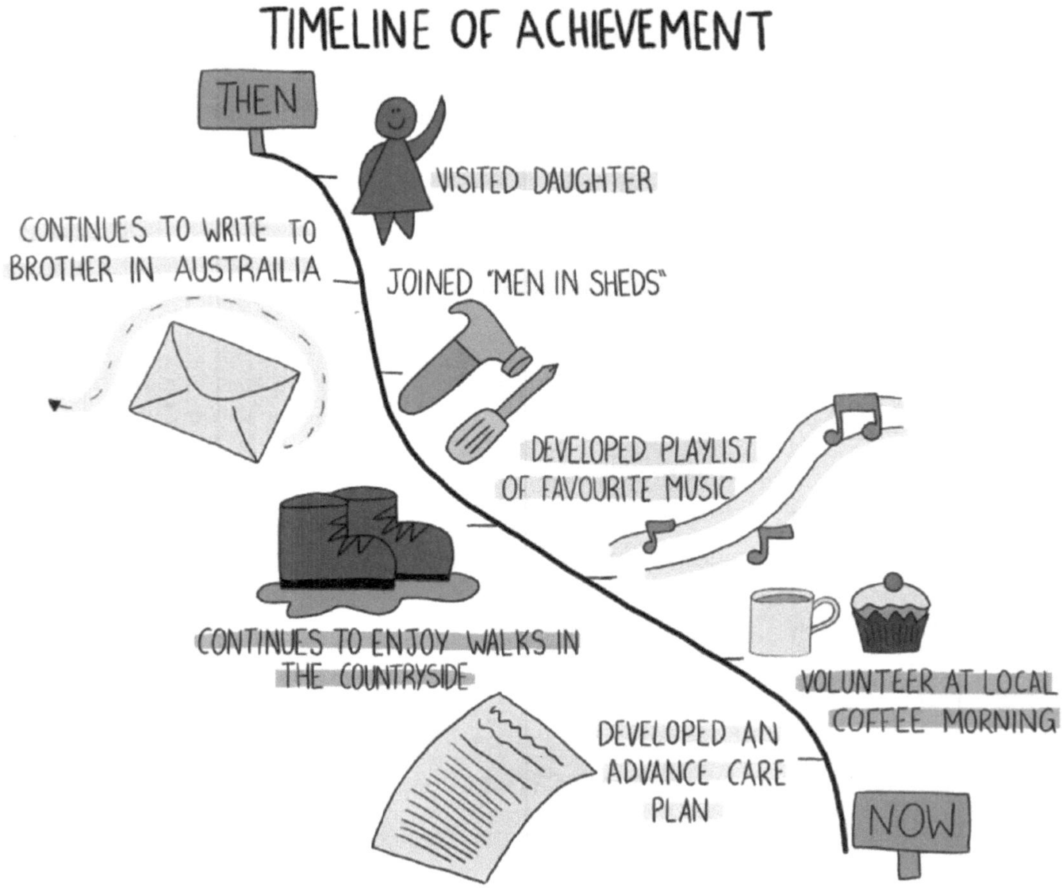

Having a voice

The importance of hearing the views and opinions of people with dementia was highlighted when, in 1996, Malcom Goldsmith wrote *Hearing the Voice of People with Dementia: Opportunities and Obstacles*. Malcolm Goldsmith talked about people with dementia having a voice and that, in listening to this voice and responding to what had been learnt, their care and support could be provided in a manner that improved their quality of life. Later writers, such as Christine Bryden (2015) and Lucy Whitman (2015), have explored this further, and they also found that a positive difference is made to a person's day-to-day experience if those around them truly listened to them.

True and real involvement

People with dementia should be actively involved in any decisions that are to be made about their life, their care and their support. This can be challenging in the later stages of dementia as the person may no longer be using words to communicate and they may find it difficult to comprehend what is happening around them and to be involved in conversations. Even in these circumstances, the principle still stands that the person should be involved. To achieve this, think about what would support the person's involvement.

Reflection

- What could you do to prepare for a discussion to ensure the person is involved and part of the decision-making?
- During the conversation, what would support the person to be involved in the discussion and what would assist them in making their own decisions?
- Considering what has already been covered in this book, what could you utilise to support a positive outcome?
- Is there any time when the person with dementia should not be involved on ethical and moral grounds?

Hearing the voice of the person

Here are some considerations for improving the chances of the person's voice being heard.

Removing unwanted stimuli

This form of enabling communication occurs when the number of stimuli that are impeding the person's ability to communicate are reduced.

Example: Mr E, who has dementia, was not involved in the conversation because the sight of his daughter gardening through the patio doors was distracting him. However, Mrs E and the Community Mental Health Nurse knew it was important to hear what he had to say. The nurse suggested changing positions with Mr E so that he was not distracted and Mr E agreed. He then became more actively involved in the conversation.

Aiding communication

If there is something that encourages a person to communicate, try to make sure that this is available.

Example: Mr W had a fun-filled and lively lifestyle; he loved spending time with his children and was always singing, dancing and going to see shows. His daughter remembers how he used to dance and sing as they walked along the street together. These days Mr W is often in his 'own world' and does not recognise his daughter. The nurse in the home where he lived helped to organise some music on an MP3 player with a set of headphones for Mr W to listen to music that he enjoys. When Mr W listens to this music his posture changes, he recognises people and engages fluently in conversation, expressing himself and his emotions.

This approach has been documented in the film *Alive Inside*.

Getting in the right position

The position in which people place themselves affects the ability of the person with dementia to communicate. It is important to know what positioning will support communication and to organise the space to enable this to happen.

Example: Mrs D invited the nurse to sit next to Mr D, who has dementia. However, Mrs D took her seat on the opposite side of the room from them and was sat at a distance. Throughout the conversation Mr D continually looked to Mrs D for support before responding to the conversation or to questions. The nurse noticed this and realised that Mr D was not feeling comfortable and they then suggested a change in seating arrangement. The nurse exchanged seats with Mrs D and moved a chair closer so that the three of them sat at an equal distance from each other. Mr D appeared more relaxed with this and contributed more fluently to the conversation, he could hear clearly what was being said and therefore was more involved.

Giving communication the time it needs

When a person is having trouble with communicating it is important not to rush them. Giving people the time to express themselves is important in making sure that they are fully understood.

Example: Mrs C can take a long time to think about what she wants to say and to find the right words when she is replying to people. Her support worker had noticed that people in her life were answering the questions for her when there were long gaps between a question being asked of her and her response. The support worker felt that people were answering for her as a way to help her, that some people felt uncomfortable with the length of silence and that other people were answering as they had reached the conclusion that

Mrs C was no longer able to answer the questions. The support worker was concerned that Mrs C was losing her opportunity to express herself and that the other people in her life were losing their chance to fully understand her and what she wanted. Mrs C's support worker talked to the group about this and encouraged everyone to give Mrs C the time to answer and, by doing this, the group began to appreciate that some of the answers that had been given on Mrs C's behalf by others were not quite what she wanted.

Providing opportunities to talk

Life can get incredibly busy, with appointments, household tasks and family commitments. People living in a service setting, for example, in a supported living scheme or a care home, can find themselves in an environment with other people and staff members focused on their daily routines and tasks. In whichever setting a person with dementia finds themselves, making time to talk is vital, whether the conversation happens via having a chat over a cup of tea or through spending time with the person during one of their favourite activities.

Structured or timetabled time for conversations, such as review meetings, should not become the only times for meaningful conversation. Instead, opportunities for a conversation should be captured as they occur.

Example: Miss D lives in a care home with 39 other residents and, to manage everything that needs to be done within the home, everyone living there has to follow set routines. Everyone gets up at a set time, has breakfast at the same time and continues to follow the routines applied in the care home setting until people go to bed in the evening.

There are eight care assistants/nurses on duty at any one time and they are often busy providing personal care to people, record keeping, tidying and making appointments for people. There is very little chance to sit and have a chat. Miss D is often sat watching television and gets very little chance to do anything else.

Apart from knowing that Miss D has a favourite set of nightwear, what she likes to eat and her favourite flower, the carnation, the people working at the care home know very little else about her and especially how she is feeling each day. This is a very lonely existence for Miss D even though she is surrounded by so many people.

The home where Miss D is a resident recognised that they needed to know much more about the people they supported and cared for to be able to offer an increased quality of life for the residents. Time was made to get to know the residents and to listen to what mattered to them, their stories and their everyday preferences. The family and friends of the residents were also invited into these conversations based on whom each individual resident wished to invite.

Underpinned by what was learnt through these conversations, the routines in the home were changed to be able to meet each person's preferences, the options for food and mealtimes became more flexible, as well as opportunities to get involved in activities inside and outside the home that were set up based on people's interests, skills and hobbies.

Key lessons learnt

Behaviour is a form of communication; to understand a person's behaviour is to get to the root of what the person is communicating.

It is important to remember that the reasons underpinning a person's behaviour could be explained by several factors:

- an expression of how they are feeling;
- a response to the medication they are taking;
- an expression of confusion or frustration if their surroundings or environment has changed;
- a representation of physical ill health or pain.

The key to providing a positive experience for the person with dementia is investing time and energy in understanding the meaning behind their behaviour and in responding to this rather than the behaviour itself.

The Alzheimer's Society has produced some useful information and advice regarding good communication on their 'Communicating and Dementia' and 'How to communicate with a person with dementia' webpages (Alzheimer's Society, n.d.; Alzheimer's Society, n.d.). These approaches can support the person with dementia and the people around them to engage in positive two-way communication.

Questions to aid reflection

1) Could any of the above resources to aid communication help the voice of the person you support to be heard?

2) How can you ensure the the person you support has the opportunity and time to talk with you and others involved in their care?

8 Developing and embedding person-centred cultures

This chapter covers:

- **The implementation of personal budget and Direct Payments**
- **The barriers experienced by those wishing to access a personal budget or a Direct Payment**
- **Recognising the person's perspective and working with the person to achieve the best outcome**
- **Recognising the need to continually focus on the person.**

How many people have been reached with Direct Payments?

The ASCOF (Adult Social Care Outcomes Framework) for 2018–19 reported that:

> - 17.6 per cent of adults aged 65 and over, in England, accessing long-term support at the year-end 31st March 2019 were receiving Direct Payments
> - 90.3 per cent of adults aged 65 and over in England accessing long-term community support at year-end 31st March 2019 were accessing Self-Directed Support.
>
> (NHS Digital, 2019)

Since their implementation, Direct Payments have provided people with greater choice and control. For example, in her book *Where Memories Go: Why Dementia Changes Everything* by Sally Magnusson (2014) explains the experience she and her two sisters had as a family caring for their mum with dementia. Within this story they explained how having access to a Direct Payment supported them as a family and enabled them to employ people to care for their mum within the home.

To achieve a personalised approach and Self-Directed Support, it is of the utmost importance that people understand the full picture of what is available to them. This includes an awareness that the option for a personal budget, if funded by the local authority, or a personal health budget, if funded by the local health authority, exists and that these can be managed as a Direct Payment.

The Alzheimer's Society (2010, cited in Feltoe and Orellana, 2013, p 9) reported that a key barrier to older people accessing a personal budget was: '*a personal budgets system that has not yet adapted to the needs of people with dementia and their carers, and is overly complex and burdensome*'.

An example of this is the assessment process. Often people experience long waits for their assessment by the local authority which is required to determine a person's eligibility for health and social care support. For example,

> *'In Cambridgeshire, people told their local Healthwatch that waiting times vary hugely across the country with older people in particular waiting increasingly long times for their assessments, in some cases waiting up to 47 weeks.'*
>
> (healthwatch, 2016)

Often how quickly a person receives the support they require post-assessment depends on their circumstances and that of their family and service providers in their area. Yates et al (2019) state that: '*Receipt of support soon after diagnosis can facilitate better adjustment and ongoing management of dementia*'.

If someone who has been referred for health and social care support waits a long time for the input from a social worker for an assessment, their circumstances may have changed significantly within that time frame. It may have reached the stage where the person needs support putting in place quickly which wouldn't leave enough time to explore what would work best for the person, the options available to them and their preferences around self-directed support. If support is implemented quickly the social worker will be referring to the services already available within the person's area and identifying the services which have capacity to work with that person. This means the social worker moves away from the potential of a broad range of options to a very narrow picture of what is available.

It is, however, possible to respond to the immediate support needs of the person with an intermediary arrangement with a local care or support provider in order to overcome the immediate crisis situation and then continue the discussions with the person about how they would like to live, what good support and care would look like for them and the options available to them with the intention to move towards this over a period of time.

Further information about the implementation of personalisation for people with dementia can be found in the 2013 Think Local Act Personal report, 'Improving Personal Budgets for

Older People' (Routledge and Carr, 2013). This report highlights what helps and hinders those living with dementia in accessing a Direct Payment. It emphasises the need for people to have access to good information, advice and guidance throughout the personal health budget/personal budget process to support them to make decisions at each stage.

Age UK reported that £160 million had been cut from the social care system by the year 2015 and the cuts have continued to be made to the present day. In real terms in the year the report was written:

> *'Age UK's analysis shows there are now nearly 1.2 million people (1,183,900) aged 65+ who don't receive the help they need with essential daily living activities. This represents a 17.9 per cent increase on last year and a 48 percent increase since 2010.'*
>
> (Age UK, 2017, p 4)

This trend suggests the social care system no longer meets the needs of people as it did in 2010 due to the sharp change in funding and resources available as a result of the austerity measures imposed throughout the country.

People with dementia and their families can, however, still expect to receive support, if they are assessed as being eligible. When a person is determined as not being eligible, the local authority has a duty to provide relevant and accurate information, advice and guidance as well as signposting to community-based organisations.

The support made available via the social care system must be personalised with the person deciding how they want it to be provided. This approach to supporting the person can be accompanied by, and work well alongside, the support available to them from within their networks and community.

Key things to remember to do

It is important for people to be able to develop a clear idea of what would work well. This can be achieved through the following ways:

- Look at a wide range of options available to the person. This includes people and places in the community, groups and associations in the local area, equipment or technology and the current services available.
- Think about places where the person would like to spend time; for instance, the café where other friends meet, the garden centre where they can indulge in an interest, a walk around the lake to get out of the house and have a chance to exercise. Consider places that capture the person's interest and imagination. There is growing evidence that keeping active and getting regular exercise can slow down the progression of dementia (Alty et al, 2019). Eating well and keeping socially active also positively slow down the symptoms.

- When looking to employ people, consider the recruitment process, based on the 'matching staff' exercise introduced in Chapter 6. Look for people on the basis of shared interests, a personality or character that would suit the person. Also consider the practical elements to make sure the arrangement will work well, such as the times the person would need to be available, or if they need to be able to drive.
- Ensure that time is made to have important discussions with the people the person is close to, talk about the future and what the person would like to happen.
- Consider developing a person-centred plan, an advanced care plan or an advanced directive. These conversations and discussions can help the people close to the person manage situations as they arise and it will also provide them with some emotional support and reassurance that they are supporting the person in a way that respects their wishes.
- Use some of the ways of recording discussions and decisions that have been introduced in Chapter 6.
- Access the information and advice that is available, and, when needed, access advice and support that is impartial and independent of the social care and health system.
- Be positive and in control; the person has the right to make decisions about how they are supported, make use of this right to ensure that their life is how they want it to be.
- Ask questions and make sure the person and the people supporting them understand as much as possible; the people involved from a professional capacity should want the person to understand their rights and what is available and possible for them.
- Share how the person wants to live their life with others, such as family, paid care and support and professionals, as the person is the expert in their own life.

Key questions to keep asking

One of the factors that prevent a person or a family from making a referral for support from the health and social care system is a concern about what will happen. Will they be listened to? Will they have any say over the support they receive? Will they have any control over what happens? Will they receive the support they need in a way that makes a positive difference?

The emphasis is now for the person to have a sense of control over how they live their life and the support and care that is provided.

Here are some questions that help to consider how much this has been achieved:

- Is the person taking the lead in discussions and decision-making?
- Does the person have support from other people which helps them to take the lead, without people taking control from them?
- Is the person's voice heard?
- Is the person answering the questions that are about them?
- Has there been the opportunity to fully understand the person's preferences, the routines that are important to them and their future wishes?
- Is the support network around the person functioning from a place of full understanding about what the person wants and needs, or are assumptions being made about what would work best for them?
- Is the person receiving support from those that are from their own networks while they are going through their diagnosis, assessments and the development of their plans about ongoing care and support?
- Is the person being informed about the care and support available through social care and health commissioned provision?
- Have they also had the opportunity to think about the support available within their community?
- Has everyone involved in the person's support network had the opportunity to learn about dementia and do they understand the condition?
- Is the information learnt from discussions being recorded in a meaningful way such that members of the person's support network can refer to this afterwards?

Community

Each person is part of a community; how they interact with this community is dependent on where they live, the relationships they have with the people around them, the interests they have and the places they go to pursue these interests.

If a community is to open and inclusive, members of that community can play a part in creating a welcoming place that takes care of one another.

Here are some suggestions of how this can be achieved:

- Build awareness of dementia, make sure there is access to information so people can read about it and create opportunities to share this knowledge and stories of personal experiences.

- Groups, associations and organisations providing a community service, for example, the local shop, library, doctor's practice, should provide a welcoming environment with people running the space who are dementia aware.
- Encourage the use of the community spaces, community centres, sports grounds and schools, as a means of bringing people together and building relationships.
- Consider creating a 'Dementia Friendly Community' where the standards set by the BSI are recognised in how the community functions (BSI Group, n.d.). More information on this is available in Chapter 1.
- Support the development of more Safe Places, a national scheme that recognises places within the community that are safe for people to go to if they are feeling scared, vulnerable, confused or lost. More details about Safe Places National Network can be found at the eend of the book in Appendix: Useful organisations.

For those providing the care and support

- Listen to the person who is being cared for and supported and respond to what has been learnt.
- Read the information that has been gathered and documented about the person's preferences and future wishes.
- Review routines to ensure that the care and support provided meets the needs, preferences and choices of the person receiving the service.
- Look at what is recorded and the paperwork and assess whether it is all needed. Can it be reduced, merged or removed altogether so more time can be spent with the person with dementia rather than on completing paper-based tasks?
- If support workers are employed, create a person-centred culture within the team. Creating opportunities to develop this culture is crucial in achieving a personalised approach. Consider how this is reflected in the recruitment process, in how team members are supported through supervisions, in team meetings and in day-to-day conversations.
- Team meetings create opportunities for team members to get to know each other better and this is where the culture can really begin to be established. Consider utilising some of the person-centred approaches introduced in Chapter 4 within these team gatherings to achieve this.

Helpful questions for conversations and relationships

Here are some questions that could be seen as great questions to bring into conversations with people with dementia and carers:

- What really matters to you?
- What is significant about your story, your achievements and good memories?
- What makes a good day for you?
- What is important in your life and what does a good life look like for you?
- What would make the biggest difference to you right now?
- What would you like to happen now?

Here are some useful questions regarding maintaining and enhancing meaningful relationships:

- Tell me about the people that matter to you?
- What is it about the connections with the people in your life that mean the most to you?
- Who inspires you?
- Who do you look forward to seeing?

Key lessons learnt

- The uptake of Direct Payments have been low in the past, yet they have provided people with dementia with greater choice and control.
- People need to be made fully aware of what support is available to them and to fully understand their options to enjoy a personalised approach and Self-Directed Support.
- Support for people living with dementia can come from their networks and community (providing the community is open and inclusive) as well as from the health social care system.
- Efforts should always be made to keep the focus on the person with dementia, listen to them and to help them retain control over how they live their life.

Questions to aid reflection

1) How can you help the person you care for understand all the options and support available to them?
2) Are there any ways you could support the person you care for to find solutions to any challenges they face from within their community or network?
3) What are some of the ways you could ensure the person you care for remains in control of how they live their life, with their preferences honoured?

Conclusion

This book has introduced a number of ideas and approaches that can help in living a life that makes sense when living with dementia or caring for someone with the condition. These approaches have included Self-Directed Support, where the person being cared for, and their carer remain in the driving seat of their care and support.

The content has explored how to plan for the future, explaining that this is better done earlier in the dementia journey, and creating a better quality of life through developing person centred plans and making use of person centred approaches. The flexibility of personal budgets and personal health budgets have been demonstrated as being successful in taking a personalised approach to dementia support, which people with dementia respond positively to. Community and support networks have been introduced as an often-overlooked source of creative solutions and support where perhaps traditional service-based health and social care solutions alone may struggle to help the person with dementia in a way that makes the most sense to them.

The detail around communication and understanding that behaviour is a form of communication, intended to help all involved build a better understanding of each other and encourage the development of valuable and meaningful relationships to ensure that there are positive experiences along the dementia journey.

The hope is that this book has outlined practical tools and inspired a fresh perspective to make living with dementia a little less challenging, where life is still a source of joy.

Appendix: Useful organisations

Here is a list of useful organisations. Some provide helpful and inspirational insight into approaches to supporting people living with dementia, and others will be available to help people through the complex social care and health systems.

ACAS – Advisory, Conciliation and Arbitration Service
ACAS provides information, advice, training, conciliation and other services for employers and employees to help prevent or resolve workplace problems.
Website: www.acas.org.uk
Tel: 0300 123 1100

Alive Inside
Transforming the lives of people living with dementia through music.
Website: www.aliveinside.org
Tel: (212) 533-8116
Email: info@aliveinside.org

Alzheimer's Research UK
Alzheimer's Research UK states: *'Our vision is a world where people are free from the fear, harm and heartbreak of dementia"*
Website: www.alzheimersresearchuk.org
Telephone: 0300 111 5333
Email: reception@alzheimersresearchuk.org

Alzheimer's Society
A leading charity representing people with dementia and their families.
Website: www.alzheimers.org.uk
Dementia Contact support line: 0333 150 3456
Email: enquiries@alzheimers.org.uk

Care Quality Commission
The CQC monitors, inspects and regulates health and social care services. They publish what they find, including ratings to help people choose care.
Website: www.cqc.org.uk
Telephone: 03000 616161

Carers Trust
Providing support to carers across the UK.
Website: www.carers.org
Tel: 0300 772 9600
Email: info@carers.org

Carers UK
Making life better for carers.
Website: www.carersuk.org
Carers UK Tel: 020 7378 4999
Carers Wales Tel: 029 2081 1370
Carers Scotland Tel: 0141 445 3070
Carers Northern Ireland Tel: 02890 439 843
Email: info@carersuk.org

Chartered Society of Physiotherapy
The professional, educational and trade union body for chartered physiotherapists, physiotherapy students and assistants. Provides contact details of private physiotherapists in your area.
Website: www.csp.org.uk
Tel: 020 7306 6666
Email: enquiries@csp.org.uk

Citizens Advice Bureau (CAB)
Your local CAB can provide information and advice in confidence or point you in the right direction. To find your nearest CAB, look in the phone book, ask at your local library or look on the website.
Website: www.citizensadvice.org.uk
www.adviceguide.org.uk (online information portal)
Advice Link (England) Tel: 0800 144 8848

College of Occupational Therapists
Provides details of independent occupational therapists in your local area.
Website: www.cot.co.uk
Tel: 020 3141 4600
Email: hello@rcot.co.uk

Dementia Action Alliance
Dementia Action Alliance represents organisations across the UK who are committed to transforming the lives of people living with dementia and their carers.
Website: www.dementiaaction.org.uk
Email: ndaa@alzheimers.org.uk

Dementia Care International
An independent Australian based organisation supporting leaders around the world to provide excellence in dementia care.
Website: https://dementiacareinternational.com/
Tel (international): +61 (0) 3 9727 2744

Dementia UK
Providing specialists one-to-one support and expert advice and providing admiral nurses who work hand-in-hand with families, helping them cope with the fear, uncertainty and the difficult everyday reality of dementia.
Website: www.dementiauk.org
Tel: 0300 365 5500
Email: info@dementiauk.org

Forget Me Not Book
Forget Me Not Dementia is a resource hub of publications aiming to provide updates and training for support workers and staff. They are fully committed to improving the lives of people living with dementia.
Website: https://www.forgetmenotdementia.co.uk/
Email: chris@forgetmenotdementia.co.uk
Tel: 01223 207855

Imagineer Development UK CIC
A passionate and creative locally rooted community interest company in the north of England, with a strong national reach and respected reputation for helping organisations, individuals and families to live fulfilling lives, contributing to a fairer society. Offers support brokerage, brokerage management, support planning and graphic facilitation services, along with consultancy and training.
Website: https://www.imagineer.org.uk/
Tel: 07572 322200
Email: info@imagineer.org.uk

In Control
A national charity working for an inclusive society where everyone has the support they need to live a good life and make a valued contribution.
Website: www.in-control.org.uk
Tel: 0121 474 5900
Email: admin@in-control.org.uk

Inclusion Network
Working towards an inclusive society, inspiring and providing practical ideas to encourage inclusion.
Website: https://inclusion.com/
Tel: (416) 658-5363
Email: inclusionpress@inclusion.com

NHS Choices
NHS website that allows you to search for health services near you.
Website: www.nhs.uk

NICE – National Institution for Health and Clinical Excellence
NICE provides national guidance and advice to improve health and social care.
Website: https://www.nice.org.uk/
Tel: 0300 323 0140
Email: nice@nice.org.uk

Older People's Advocacy Alliance
A charity that promotes and develops independent advocacy services for older people, and those who support them – in whatever way they engage with older people.
Website: https://opaal.org.uk/
Email: colette.isaaks@opaal.org.uk

Royal College of Speech and Language Therapists
The professional body for speech and language therapists and support workers. Promotes excellence in practice and influences health, education and social care policies.
Website: www.rcslt.org
Tel: 020 7378 1200
Email: info@rcslt.org

Safe Places National Network
A safe place helps vulnerable people if they feel scared or at risk while they are out in the community and need support right away.
Website: https://www.safeplaces.org.uk/
Email: info@safeplaces.org.uk

Skills for Care
The employer-led workforce development body for adult social care in England. It has a dedicated section of its website to provide advice and guidance to people who are employing their own support.
Website: www.skillsforcare.org.uk
Tel: 0113 241 1275
Email: information.team@skillsforcare.org.uk

Society of Chiropodists and Podiatrists
The professional body and trade union for registered podiatrists. Provides patient information on common foot problems and details of private practice podiatrists.
Website: https://cop.org.uk/
Tel: 020 7234 8620

Talking Mats
A social enterprise whose vision is to improve the lives of people with communication difficulties.
Website: www.talkingmats.com
Tel: 01786 479511
Email: info@talkingmats.com

Resources

All the resources mentioned in the book are available to download free of charge at https://www.imagineer.org.uk/templates

Bibliography

Adams, T and Gardiner, P (2005) Communication and Interaction within Dementia Care Triads: Developing a Theory for Relationship-Centred Care. *Dementia*, [Article] 4 (2). Abstract only. Available through: Gale Academic OneFile. [online] Available at: https://go.gale.com/ps/anonymous?id=GALE%7CA140962504&sid=googleScholar&v=2.1&it=r&linkaccess=abs&issn=14713012&p=AONE&sw=w (accessed 9 May 2020).

Adults with Incapacity (Scotland) Act 2000. (c.4) Edinburgh: HMSO.

Advocacy Focus (2020) What Is Advocacy? [online] Available at: https://www.advocacyfocus.org.uk/understanding-advocacy (accessed 24 June 2020).

All-Party Parliamentary Group on Dementia (2014) *Building on the National Dementia Strategy: Change, Progress and Priorities*. [pdf] London: All-Party Parliamentary Group. Available at: www.alzheimers.org.uk/download/downloads/id/2249/building_on_the_national_dementia_strategy_change_progress_and_priorities.pdf (accessed 18 January 2018).

Alty, J, Farrow, M and Lawler, K (2019) How to Understand it Exercise and Dementia Prevention. *Practical Neurology,* [e-journal] 20 (3). Abstract only. Available through: British Medical Journal website. [online] Available at: https://pn.bmj.com/content/20/3/234 (accessed 29 January 2021).

Alzheimer's Association (2020) What Is Alzheimer's Disease? [online] Available at: https://www.alz.org/alzheimers-dementia/what-is-alzheimers (accessed 26 June 2020).

Alzheimer's Disease International (2016) *Dementia Friendly Communities Key Principles.* [pdf] London: Alzheimer's Disease International. Available at: www.alzint.org/u/dfc-principles.pdf (accessed 11 January 2021).

Alzheimer's Scotland (n.d.) Dementia Friendly Communities. [online] Available at: https://www.alzscot.org/our-work/dementia-friendly-communities (accessed 13 January 2021).

Alzheimer's Society (2014) *What is Vascular Dementia?* [pdf] Alzheimer's Society. Available at: https://www.alzheimers.org.uk/sites/default/files/migrate/downloads/factsheet_what_is_vascular_dementia.pdf (accessed 26 June 2020).

Alzheimer's Society (2017) *This Is Me*. [pdf] Alzheimer's Society. Available at: https://www.alzheimers.org.uk/download/downloads/id/3423/this_is_me.pdf(accessed 3 July 2020).

Bamford, C, Eccles, M, Steen, N and Robinson, L (2007) Can Primary Care Record Review Facilitate Earlier Diagnosis of Dementia? *Family Practice*. [online] Available at: https://academic.oup.com/fampra/article/24/2/108/468369 (accessed 10 June 2020).

Batsch, N L, Mittelman, M S, and Alzheimer's Disease International (2012) *World Alzheimer Report 2012 Overcoming the Stigma of Dementia.* [pdf] London: Alzheimer's Disease International (ADI). Available at: https://www.alz.co.uk/research/WorldAlzheimerReport2012.pdf (accessed 23 May 2020).

Boyle, D and Harris, M (2009) *The Challenge of Co-Production. How Equal Partnerships between Professionals and the Public Are Crucial to Improving Public Services*. London: Nesta. Available from https://media.nesta.org.uk/documents/the_challenge_of_co-production.pdf (accessed 25 May 2020).

Brooker, D (2004) What Is Person-Centred Care in Dementia? *Reviews in Clinical Gerontology*, [e-journal] 13 (3): 215–222. Available through: My Seniors Weebly website. [online] Available at: http://myseniors.weebly.com/uploads/2/1/8/5/21859032/what_is_person-centered_care.pdf (accessed 13 May 2020).

Brooker, D, May, H and Edwards, P (2009) *Enriched Care Planning for People with Dementia: A Good Practice for Delivering Person-Centred Care*. London: Jessica Kingsley Publishers.

Bryden, C (2015) *Nothing about Us, Without Us! 20 Years of Dementia Advocacy*. London: Jessica Kingsley Publishers.

BSI Group (n.d.) *Living with Dementia*. [pdf] BSI Group. Available at: https://www.bsigroup.com/LocalFiles/en-GB/consumer-guides/resources/BSI-Living-With-Dementia-Brochure-UK-EN.pdf (accessed 11 January 2021).

Burgener, S C and Dickerson-Putman, J (1999) Assessing Patients in the Early Stages of Irreversible Dementia. The Relevance of Patient Perspectives. *Journal of Gerontological Nursing,* [e-journal] 25 (2): 33–41. Abstract only. Available through: National Library of Medicine website. [online] Available at: https://pubmed.ncbi.nlm.nih.gov/10347435/ (accessed 18 July 2020).

Care Act 2014. (c.23). London: HMSO.

Care Services Improvement Partnership (CSIP) National Older People's Mental Health Programme (2007) *Strengthening the Involvement of People with Dementia. A Resource for Implementation.* [pdf] York: Care Services Improvement Partnership (CSIP) National Older People's Mental Health Programme. Available at: https://dementiapartnerships.com/wp-content/uploads/sites/2/strengthening-involvement.pdf (accessed 15 July 2020).

Carers (Scotland) Act 2016. (asp.9). Edinburgh: Scottish Government.

Carers UK (2014) *Caring and Family Finances Inquiry - UK Report - (Full Report).* [pdf] London: Carers UK. Available at: https://www.carersuk.org/for-professionals/policy/policy-library/caring-family-finances-inquiry (accessed 19 June 2020).

Carers UK (2019) *Facts about Carers.* [pdf] Carers UK. Available at: https://www.carersuk.org/images/Facts_about_Carers_2019.pdf (accessed 20 September 2020).

Centre for Ageing Better (n.d.) 'Soul-Destroying' Ageism in Recruitment Could Be Final Straw for Over 50s. [online] Available at: https://www.ageing-better.org.uk/ (accessed 9 March 2021).

Copestake, P, Sheikh, S, Johnston, S and Bollen, A (2014) Removing Barriers, Raising Disabled People's Living Standards. Available at: https://docplayer.net/16691383-Removing-barriers-raising-disabled-people-s-living-standards.html (accessed 23 May 2020).

Davies, K (2005) *Creating a Circle of Support*. [online] Available at: https://www.iidc.indiana.edu/irca/articles/creating-a-circle-of-support (accessed 22 July 2020).

De Bellis, A, Bradley, S L, Wotherspoon, A, Walter, B, Guerin, P, Cecchin, M and Patterson, J (2009) *Come into My World.* [pdf] Adelaide: Flinders University of South Australia. Available at: http://nursing.flinders.edu.au/comeintomyworld/docs/Come%20Into%20My%20World%20-%20How%20to%20Interact%20with%20a%20Person%20who%20has%20Dementia.pdf (accessed 17 July 2020).

Deacon, A and Mann, K (1999) Agency, Modernity and Social Policy. *Journal of Social Policy*, 28 (3): 413–35.

Dementia Action Alliance (2010) *National Dementia Declaration for England.* [pdf] Dementia Action Alliance. Available at: https://www.dementiaaction.org.uk/assets/0001/1915/National_Dementia_Declaration_for_England.pdf (accessed 25 May 2020).

Dementia Alliance International (2017) Dementia Alliance International. [online] Available at: https://www.dementiaallianceinternational.org/ (accessed 29 June 2020).

Dementia Friends (n.d.) List of Dementia Friendly Communities. [online] Available at: https://www.dementiafriends.org.uk/WebArticle?page=dfc-public-listing#.X_y5xOj7TIU (accessed 11 January 2021).

Dementia Matters (2018) Mixed Dementia. [online] Available at: https://www.dementiamatters.net/mixed-dementia (accessed 8 June 2020).

Demos (2014) *"A Vision for Care Fit for the Twenty-First Century..." The Commission on Residential Care*. [pdf] London: Demos. Available at: https://www.demos.co.uk/files/Demos_CORC_report.pdf?1409673172 (accessed 3 July 2020).

Department for Constitutional Affairs (2007) *Mental Capacity Act 2005 Code of Practice.* [pdf] London: The Stationery Office. Available at: https://assets.publishing.service.gov.uk/government/uploads/system/uploads/attachment_data/file/497253/Mental-capacity-act-code-of-practice.pdf (accessed 30 September 2020).

Department of Health (2010) *Equity and Excellence. Liberating the NHS.* (Cmnd. 7881) London: TSO (The Stationery Office).

Department of Health (2010) Personalisation. [online] Available at: https://webarchive.nationalarchives.gov.uk/+/http://www.dh.gov.uk/en/SocialCare/Socialcarereform/Personalisation/index.htm (accessed 16 May 2020).

Department of Health (2010) *Quality Outcomes for People with Dementia: Building on the Work of the National Dementia Strategy*. [pdf] Department of Health. Available at: https://assets.publishing.service.gov.uk/government/uploads/system/uploads/attachment_data/file/213811/dh_119828.pdf (accessed 25 May 2020).

Department of Health and Social Care (2018) NHS Continuing Healthcare Decision Support Tool [online] Available at: https://www.gov.uk/government/publications/nhs-continuing-healthcare-decision-support-tool (accessed 11 November 2020).

Department of Health and Social Care (2019) *Dementia 2020 Challenge 2018: Review Phase 1.* [pdf] Department of Health and Social Care. Available at: https://assets.publishing.service.gov.uk/government/uploads/system/uploads/attachment_data/file/780777/dementia-2020-challenge-2018-review.pdf (accessed 4 May 2020).

Department of Health Putting People First Programme Personalisation (2010) *Putting People First: Personal Budgets for Older People – Making it Happen.* [pdf] Department of Health. Available at: https://www.choiceforum.org/docs/pb.pdf (accessed 25 May 2020).

Duffy, S J (2004) In Control. *Journal of Integrated Care*, 12 (6).

Fazio, S, Pace, D, Flinner, J and Kallmyer, B (2018) The Fundamentals of Person-Centered Care for Individuals with Dementia. *The Gerontologist,* [e-journal] 58 (1): 10–19. [online] Available at: https://academic.oup.com/gerontologist/article/58/suppl_1/S10/4816735 (accessed 22 November 2020).

Forest, M and Pearpoint, J (1996) SOLUTION CIRCLE Getting Unstuck a Creative Problem Solving Tool. [online] Available at: https://inclusion.com/inclusion-resources/training-tools/solution-circle/ (accessed 3 August 2020).

Goldsmith, M (1996) *Hearing the Voice of People with Dementia: Opportunities and Obstacles*. London: Jessica Kingsley Publishers.

Greenhalgh, L and Whitham, B (2016) *Hidden Charges in Care Homes Exploring Consumer Protections within the Care Home Market*. [pdf] London: Citizens Advice Bureau. Available at: https://www.citizensadvice.org.uk/Global/CitizensAdvice/Public%20services%20publications/Hidden%20charges%20in%20care%20homes.pdf (accessed 5 November 2016).

Handy, C (2002) *The Empty Raincoat: Making Sense of the Future*. London: Arrow.

Health Innovation Network South London (2016) *What Is Person-Centred Care and Why Is it Important?* [pdf] Health Innovation Network South London. Available at: https://healthinnovationnetwork.com/system/ckeditor_assets/attachments/41/what_is_person-centred_care_and_why_is_it_important.pdf (accessed 5 May 2020).

Hirst, M (2014) *Transitions Into and Out of Unpaid Care*. [pdf] York: University of York. Available at: https://www.york.ac.uk/inst/spru/pubs/pdf/CUK2644.pdf (accessed 17 June 2020).

HM Government (2007) *Putting People First a Shared Vision and Commitment to the Transformation of Adult Social Care*. [pdf] London: HM Government. Available at: https://www.basw.co.uk/system/files/resources/basw_102631-4_0.pdf (accessed 9 March 2021).

In Control (2017) *POET Personal Outcomes Evaluation Tool for Adults in Receipt of Social Care Support 2017 Report*. [pdf] Think Local Act Personal. Available at: http://s557941885.websitehome.co.uk/wp-content/uploads/2017/07/poetpb2017-emb-final.pdf (accessed 17 January 2021).

Inclusive Solutions (2015) Inclusive Solutions. [online] Available at: https://inclusive-solutions.com/person-centred-planning/ (accessed 16 October 2020).

James, I A (2011) *Understanding Behaviour in Dementia that Challenges: A Guide to Assessment and Treatment*. London: Jessica Kingsley Publishers.

Killeen, J (1996) *Advocacy and Dementia*. Edinburgh: Alzheimer Scotland Action on Dementia.

Kitwood, T (1997) *Dementia Reconsidered: The Person Comes First*. Buckingham and Philadelphia: Open University Press.

Kitwood, T (1997) The Experience of Dementia. *Aging and Mental Health*, 1 (1): 13–22.

Leocadie, M-C, Roy, M-H and Rothan-Tondeur, M (2018) Barriers and Enablers in the Use of Respite Interventions by Caregivers of People with Dementia: An Integrative Review. *Archives of Public Health*. [online] Available at: https://www.ncbi.nlm.nih.gov/pmc/articles/PMC6249779/ (accessed 1 July 2020).

Leslie, M (2016) *Discussing Real-Life Examples I Could Use for This Book, Which Included Leslie's Mum's Care Situation*. [letter] (Personal, communication, n.d. 2016).

Levin, E, Moriarty, J and Gorbach, P (1994) *Better for the Break*. London: HMSO.

Mackay, M (2016) Talking Mats Training for Families Living with Dementia. [online] Available at: https://www.talkingmats.com/talking-mats-training-for-families-living-with-dementia/ (accessed 3 February 2021).

Matthews, E A, Farrell, G A and Blackmore, A M (1996) Effects of an Environmental Manipulation Emphasizing Client-Centred Care on Agitation and Sleep in Dementia Sufferers in a Nursing Home. *JAN*, [e-journal] 24 (3): 439–37. Abstract only. Available through: Wiley Online Library. [online] Available at: https://onlinelibrary.wiley.com/doi/abs/10.1046/j.1365-2648.1996.02102.x (accessed 18 July 2020).

Matthews, R (2016) Using Talking Mats to Include People with Dementia in Their Care Plan. [online] Available at: https://www.talkingmats.com/using-talking-mats-to-include-people-with-dementia-in-their-care-plans/ (accessed 23 January 2021).

McMaster University (2019) 3 Simple Ways to Manage Challenging Behaviours Associated with Dementia. [online] (Last updated 20 December 2019) Available at: https://www.mcmasteroptimalaging.org/blog/detail/blog/2016/11/15/3-simple-ways-to-manage-challenging-behaviours-associated-with-dementia (accessed 22 November 2020).

Mental Capacity Act (Northern Ireland) 2016. (c.18). Belfast: HMSO.

Mental Health Capacity Act 2005. (c.9). London: HMSO.

Mind (2018) Advocacy in Mental Health. [online] Available at: https://www.mind.org.uk/information-support/guides-to-support-and-services/advocacy/what-is-advocacy/ (accessed 24 June 2020).

National Institute for Health and Care Excellence (NICE) (2019) *Dementia.* [pdf] National Institute for Health and Care Excellence (NICE). Available at: https://www.nice.org.uk/guidance/ng97/chapter/Person-centred-care (accessed 17 June 2020).

National Voices (2014) *Care and Support Planning Guide.* [pdf] National Voices. Available at: https://www.nationalvoices.org.uk/sites/default/files/public/publications/guide_to_care_and_support_planning_0.pdf (accessed 11 November 2020).

New Possibilities (n.d.) New Possibilities Approach. [online] Available at: https://www.newpossibilities.co.uk/ (accessed 29 July 2020).

NHS Digital (2019) 1C - The Proportions of Users and Carers Receiving Self-Directed Support, and Self-Directed Support via Direct Payments. [online] Available at: https://digital.nhs.uk/data-and-information/publications/statistical/adult-social-care-outcomes-framework-ascof/current/indicator-files/1c--the-proportions-of-users-and-carers-receiving-self-directed-support-and-self-directed-support-via-direct-payments (accessed 17 January 2021).

NHS England (2020) *Dementia Assessment and Referral Data Collection – February 2020.* [pdf] NHS England. Available at: https://www.england.nhs.uk/statistics/wp-content/uploads/sites/2/2020/05/Dementia-Statistical-Summary-February-2020-PDF-126KB.pdf (accessed 12 July 2020).

NHS England and NHS Improvement (2019) *Personalised Care. Guidance on the Legal Rights to Have Personal Health Budgets and Personal Wheelchair Budgets.* [pdf] NHS England and NHS Improvement. Available at: https://www.england.nhs.uk/wp-content/uploads/2014/09/guidance-on-the-legal-rights-to-personal-health-budgets.pdf (accessed 12 July 2020).

Nolan, M R, Brown, J, Davies, S, Nolan, J and Keady, J (2006) *The Senses Framework: Improving Care for Older People through a Relationship-Centred Approach. Getting Research into Practice (GRiP)*

Report No 2. [pdf] Sheffield Hallam University. Available at: https://core.ac.uk/download/pdf/99946.pdf (accessed 10 May 2020).

Nolan, M R, Davies, S, Brown, J, Keady, J and Nolan, J (2002) *Longitudinal Study of the Effectiveness of Educational Preparation to Meet the Needs of Older People and Carers: The AGEIN (Advancing Gerontological Education in Nursing) Project*. London: English National Board for Nursing, Midwifery and Health Visiting.

Nursing Homes Undercover (2016) Panorama, Series 64. [TV Programme]. BBC, BBC1, 25 November 2016, 20.30.

O'Brien, J, Pearpoint, J and Kahn, L (2010) *The PATH and MAPs Handbook. Person-Centred Ways to Build Community*. Toronto: Inclusion Press International.

Pearpoint, J, O'Brien, J and Forest, M (2001) *PATH A Workbook for Planning Positive and Possible Futures*. Toronto: Inclusion Press.

Perez-Merino, R (2014) Starting Out - By Listening to Patient, I Understood His Needs and Enhanced His Care. *Nursing Standard,* [e-journal] 29 (9). Available through: Royal College of Nursing Institute website. [online] Available at: http://journals.rcni.com/doi/full/10.7748/ns.29.9.29.s33 (accessed 22 July 2020).

Pictures to Share (n.d.) Our Books. [online] Available at: https://www.picturestoshare.co.uk/ (accessed 1 February 2021).

Plos Medicine (2018) Impact of Person-Centred Care Training and Person-Centred Activities on Quality of Life, Agitation, and Antipsychotic Use in People with Dementia Living in Nursing Homes: A Cluster-Randomised Controlled Trial. [online] Available at: https://journals.plos.org/plosmedicine/article?id=10.1371/journal.pmed.1002500 (accessed 5 May 2020).

Prime Minister's Strategy Unit (2005) *Improving the Life Chances of Disabled People Final Report.* [pdf] Department for Work and Pensions, Department of Health, Department for Education and Skills and Office of the Deputy Prime Minister. Available at: https://webarchive.nationalarchives.gov.uk/+/http:/www.cabinetoffice.gov.uk/media/cabinetoffice/strategy/assets/disability.pdf (accessed 15 March 2021).

Public Health England (2015) *Reasons Why People with Dementia Are Admitted to a General Hospital in an Emergency*. [pdf] Public Health England. Available at: https://webarchive.nationalarchives.gov.uk/20170302124526/http://www.yhpho.org.uk/default.aspx?RID=207311 (accessed 4 November 2016).

Saad, K, Hartman, J, Ballard, C, Kurian, M, Graham, C and Wilcock, G (1995) *Coping by the Carers of Dementia Sufferers.* [pdf] Oxford University Press. Available at: https://www.researchgate.net/profile/Candida_Graham/publication/14611088_Coping_by_the_Carers_of_Dementia_Sufferers/links/563a510e08ae45b5d284a953.pdf (accessed 3 July 2020).

Seeds for Change (2020) *Facilitation Tools for Meetings Workshops.* [pdf] Seeds for Change. Available at: https://seedsforchange.org.uk/tools.pdf (accessed 7 August 2020).

Seligman, M E P (1975) *Helplessness: On Depression, Development, and Death*. San Francisco: W. H. Freeman.

Skills for Health and Skills for Care (2018) *Managing Success in Dementia Care.* [pdf] Leeds: Leeds Beckett University. Available at: https://www.hee.nhs.uk/sites/default/files/documents/Managing%20Success%20v3.pdf (accessed 17 May 2020).

Smull, M and Harrison, S B (n.d.) *An Overview of Essential Lifestyle Planning*. [pdf] Annapolis: Support Development Associates. Available at: http://www.allenshea.com/brochure.pdf (accessed 7 August 2020).

Social Care Institute for Excellence (SCIE) (2020) Dementia. My Life Is Changing because of Dementia. [online] (Last updated October 2020) Available at: https://www.scie.org.uk/dementia/after-diagnosis/knowing-the-person/life-is-changing.asp (accessed 22 November 2020).

Social Services and Well-Being (Wales) Act 2014. (anaw.4.). Cardiff: National Assembly for Wales.

Stokes, G (2000) *Challenging Behaviour in Dementia : A Person-Centred Approach*. Milton Keynes: Routledge.

The Care and Support (Eligibility Criteria) Regulations 2015 (2015) *SI 2015/313*. London: HMSO.

The Care and Support (Eligibility) (Wales) Regulations 2015. 2015. *SI 2015/1578*. Cardiff: HMSO.

The Health and Social Care Act 2008 (Regulated Activities) Regulations 2014. (2014) *SI 2014/2936*. London: HMSO.

Think Local Act Personal (2014) *Personalisation Action Plan: Key Activity Underway in Advance of the Care Bill.* [pdf] Think Local Act Personal. Available at: http://www.thinklocalactpersonal.org.uk/_assets/PersonalisationActionPlanFINAL.pdf (accessed 13 May 2020).

Think Local Act Personal (n.d.) Paying for Respite Care - Solutions. [online] Available at: https://www.thinklocalactpersonal.org.uk/Personal-Budgets-Minimum-Process-Framework/Respite-care-provided-through-vouchers/#prc6-solutions (accessed 12 July 2020).

Upton, D, Upton, P, Jones, T, Jutlla, K and Brooker, D (2011) *Evaluation of the Impact of Touch Screen Technology on People with Dementia and Their Carers within Care Home Settings*. [online] (Last updated 17 June 2020) Available at: https://eprints.worc.ac.uk/2162/ (accessed 24 January 2021).

Wells, S (2006) *Developments in Dementia Advocacy: Exploring the Role of Advocates in Supporting People with Dementia*. London: WASSR.

West, S and Glover, H (2014) *Best Interest Meeting Guidance*. [pdf] Lincolnshire County Council. Available at: https://www.scie.org.uk/files/mca/directory/best-interests-meetings-guidance.pdf?res=true (accessed 25 September 2020).

Whitman, L (2015) *People with Dementia Speak Out: Creative Ways to Achieve Focus and Attention by Building on AD/HD Traits*. London: Jessica Kingsley Publishers.

World Health Organization (WHO) (2011) *World Report on Disability*. [pdf] Geneva: World Health Organization. Available at: https://www.who.int/disabilities/world_report/2011/report.pdf (accessed 23 May 2020).

YouGov (2013) Care Home Concerns. [online] Available at: https://yougov.co.uk/topics/politics/articles-reports/2013/03/06/care-home-concerns (accessed 12 July 2020).

Index

ACP, *see* advanced care planning
action planning, 143
Adult Social Care Outcomes Framework (ASCOF), 174
advanced care planning (ACP), 54, 63
Age Friendly Communities, 4–25
  personalisation and, 25
agency, 3, 15, 39, 73, 77, 100, 121
Alzheimer's disease, 22, 23, 33, 34, 37, 41, 46, 49–53, 61, 62
Alzheimer's Disease International, 22, 23, 33, 34
Alzheimer's Society, 37, 41, 42, 46, 50, 52, 84, 85, 158, 173, 175
appreciations, 120–121
  consideration, 125–126
  driving seat, 126–128
  matching staff profile, 121–122
  My Care and Support Profile, 122
  preparing for meeting, 128–129
  quality of questions being asked, 131–132
  solution circles, 124–125
  treasure chest, 123–124
ASCOF, *see* Adult Social Care Outcomes Framework

Bamford, C, 54
behaviour, 146–147
  challenging and difficult times, 148
  challenging behaviour, 149
  challenging circumstances, 149–150
  communicating through behaviour, 148
  finding time to talk, 150
  forward planning, 151–152
  problem-solving approach, 151
  solution-focused approach, 151
  solutions circle, 148
  sundowning, 147–148
  taking care of carers, 152–153
being person-led *vs.* service-led, 89–90

Care Act 2014, 68, 71, 72, 90
care and support plan, 101–102, 136
  guide, 104
  in hospital setting, 84–85
  person taking lead, 99–101
  residential care, 85–88
  reviewing and monitoring, 103
  templates, 101
Care and Support (Eligibility Criteria) Regulations 2015. s.2(1)(a-c), 93
Care Quality Commission (CQC), 86, 87
Care Service Improvement Partnership (CSIP), 159
CCGs, *see* clinical commissioning groups
Centre for Ageing Better, 24–25
checklist tool, 98
circle of exchange, 115
circle of friendship, 115
circle of intimacy, 115, 116
circle of participation, 115
circles of support, 114–115
clinical commissioning groups (CCGs), 88
  managed budget, 31
CMHN, *see* Community Mental Health Nurse
communication, 8, 31, 80, 148
  ability to communicate changes over time, 159–160
  communicating through behaviour, 150
  discovering what person is communicating, 150
  hearing loss, 158
  involving right people in, 158–159
  learning logs, 150
  making choices, 156
  and personalisation, 27–44
  physical barriers to, 157
  preparing for conversation, 156
  support resources
    communication maps, 165–166
    iPad and touchscreen technology, 164–165
    pictures, 162
    reference objects, 163
    symbols, 160
    talking mats, 162
    timelines, 163–165
  tips, 155
  tools and approaches, 154–155
  total communication, 160
communication maps, 165–166
community, 178
community mapping, 12
Community Mental Health Nurse (CMHN), 170
contingency plan, 100
Convention of Scottish Local Authorities (COSLA), 90
co-production, 38–39, 44
  approaches, 38

collective agency, 39
values of inclusion, 39
COSLA, *see* Convention of Scottish Local Authorities
Court of Protection, 36
CQC, *see* Care Quality Commission
critical risk, 90
CSIP, *see* Care Service Improvement Partnership

Davies, K, 115
decision-making table, 166–167
decision support tool (DST), 98–99
dementia, 46–47, *see also individual entries*
advocacy, 73–74
capturing memories, 69–70
carers' rights, 71–72
continued support at home, 65–66
coping mechanisms for carers, 79
development, 50–51
diagnostic process, 56–58
early and timely diagnosis, 55–56
eligibility for carers, England, 72–73
end of life, 81
family and friends, 67
forms of, 49
good communication, 155, *see also* communication
having voice, 169
hidden barriers, 33–34
later stages
care settings, 86, 87
eating and weight loss, 80
loss of communication skills, 80
loss of movement, 80
memory loss, 80
problems with continence, 81
medication, 59–60
middle stages of, 70–71
onset of dementia and diagnosis, 53
overcoming barriers, 33
quality outcomes for people, 40
recognising carers, 68
reflection, 170
responding to diagnosis, 62–63
risk of not engaging, 48
seeking diagnosis, 63–64
stresses of caring, 68–69
support during assessment and diagnosis, 60–62
true and real involvement, 170
dementia care triad, 3–4
collective agency, 39
recognising citizenship, 21–22
social networks participation, 21–22
Dementia Friendly Communities, 22–24
principles, 22
dementia with Lewy bodies (DLB), 52–53
Department of Health, 34, 40
Department of Health, Social Services and Public Safety, 91
Department of Health and Social Care, 96
Direct Payments, 174–176
development of
under personal budgets 2005 onwards, 28–29

DLB, *see* dementia with Lewy bodies
Doughnut Sort, 111–112
DST, *see* decision support tool
Duffy, S J, 7

enriched care plan, 18
health, 19
life at the moment, 20
life history, 18
neurological impairment, 19
personality, 18–19
essential lifestyle planning, 12, 141–143

forward planning, 151–152
frontotemporal dementia, 50–51

general adult psychiatrists, 58
geriatricians, 57
Glover, H, 36
Goldsmith, M, 169
graphic facilitator, 131, 135–136, 139

Handy, C, 111
Harrison, S B, 141
Health Innovation Network South London, 5
hearing loss, 158
Hearing the Voice of People with Dementia: Opportunities and Obstacles, 169
helplessness, 48, 61
hopelessness, 146

impairment, 19, 59, 93
Incapacity Act 2000, 35
Individual Service Fund (ISF), 30
infarct, 52
ISF, see Individual Service Fund

Killeen, J, 74
Kitwood, T, 5, 13, 107, 110

local authorities, 31, 35, 71, 90, 92, 93, 96, 97, 99, 176
low risk, 91

Magnusson, S, 174
making action plans (MAP), 11, 139–141
MCA, *see* Mental Capacity Act
Mental Capacity Act (MCA), 35–38, 44, 47, 166
best interest meeting, 36–37
independent mental capacity advocate, 37
personalisation in, 37–38
power of attorney, 37
mild cognitive impairment, 59
Mini-Mental State Examination, 49, 56
mixed dementia, 53
moderate risk, 91
multi-infarct dementia, 52, 53

National Dementia Declaration, 40
National Dementia Strategy, 40, 54

neurologists, 57
Nolan, M, 14–18
Northumberland Dementia Forum, 23

O'Brien, J, 139
old-age psychiatrists, 58
outcomes, 39–40
  of personalisation, 40–41
outputs, 39

PATH, *see* planning alternative tomorrows with hope
Pearpoint, J, 135
people first, 106–107
  planning, 107
personal budgets, 28–31, 34–35, 72, 75, 78–79, 88–89, 97, 103, 104, 175
  vs. personal health budget, 97
  of support plan
    Direct Payment, 28–30
    Individual Service Fund (ISF), 30
    mixed package, 31
    trust fund, 31
Personal Budgets Outcomes Evaluation Tool (POET), 28
personal health budgets, 97
  health funding and, 97–98
  complementary approaches
    enriched care plan, 18, 20
    evolution in thinking, 20
    person-centred care, 119
    person-centred planning and practice, 130–135
    values of inclusion, 7–13
    VIPS framework, 13–14
  dementia care triad, 3–4
  in Mental Capacity Act, 37–38
  national drive, 86
  personalised services, 3
  person-centred care, 119
  Self-Directed Support and, 3, 126
personalised services, 3
person-centred approach, 86, 112, 149, 179
person-centred cultures development
  being person-led *vs.* service-led, 89–90
  clinical commissioning groups, 88
  community, 178–179
  Direct Payments, 174–176
  possible works in practice, 132
person-centred planning, 63, 127, 135–144
  approaches, 135–144
    for planning meetings, 143–144
  facilitation of, 131–133
  invite to planning meeting, 134–135
personhood, 5, 21
planning alternative tomorrows with hope (PATH), 135–138
POET, *see* Personal Budgets Outcomes Evaluation Tool
post-stroke dementia, 52
problem-solving approach, 151
process facilitator, 124–125, 131, 136, 139

quality of life, 7, 10, 14, 20, 26, 28, 33–35, 39, 41, 42, 44, 47, 50, 70, 89, 91, 110, 113, 115, 120, 121, 123, 124, 159, 167, 169, 172, 182

residential care, 3, 5, 13, 74, 76, 83, 85, 89, 104, 162
  public perception of, 85–88
respite, 75–76
  approaches
    day care, 76
    home care services, 77
    host family respite, 77
    multi-dimensional carer, 78
    sitting service, 77
    video respite, 77–78
  personal budget, 78–79
risk plan, 100

SCIE, *see* Social Care Institute for Excellence
screening and assessment, GP
  assessment, 58–59
  ongoing assessment, 60
  receiving diagnosis, 59–60
  referral to specialist, 57–58
  visiting general practitioner, 56–57
    providing diagnosis, 57
Self-Directed Support, 3
  personalisation and, 3
Seligman, M E P, 48
The Senses Framework, 15, 18, 21
  sense of achievement, 17
  sense of belonging, 16–17
  sense of continuity, 16
  sense of purpose, 17
  sense of security, 16
  sense of significance, 17
skills
  enabling people voice, 169
    getting in right position, 171
    giving communication, 171–172
    providing opportunities to talk, 172–173
    removing unwanted stimuli, 170
    stimuli that aids communication, 171
  helping people to make choices
    All About Me profile, 110–111
    circles of support, 114–115
    Doughnut Sort, 111–112
    good days and bad days, 118–119
    life stories, 119–120
    negotiables and non-negotiables, 115–117
    relationship mapping, 112–114
    working/not working conversation, 117–118
  promoting respect and dignity
    celebration, 107
    collaboration, 108
    creating closeness, 108
    creativity, 108
    facilitation, 108
    giving/accepting, 108
    negotiation, 107–108
    playfulness, 109
    recognition, 109
    relaxation, 109
    timalation, 109
    validation, 109–110
Smull, M, 141

AF848627

Evaluation of Ambient Air Quality by Personnel Monitoring

SECOND EDITION

Volume I
Gases and Vapors

Author

A. L. Linch
Consultant
Environmental Health
Everett, Pennsylvania

CRC Press
Taylor & Francis Group
Boca Raton London New York

CRC Press is an imprint of the
Taylor & Francis Group, an informa business

First published 1981 by CRC Press
Taylor & Francis Group
6000 Broken Sound Parkway NW, Suite 300
Boca Raton, FL 33487-2742

Reissued 2018 by CRC Press

© 1981 by CRC Press, Inc.
CRC Press is an imprint of Taylor & Francis Group, an Informa business

No claim to original U.S. Government works

This book contains information obtained from authentic and highly regarded sources. Reasonable efforts have been made to publish reliable data and information, but the author and publisher cannot assume responsibility for the validity of all materials or the consequences of their use. The authors and publishers have attempted to trace the copyright holders of all material reproduced in this publication and apologize to copyright holders if permission to publish in this form has not been obtained. If any copyright material has not been acknowledged please write and let us know so we may rectify in any future reprint.

Except as permitted under U.S. Copyright Law, no part of this book may be reprinted, reproduced, transmitted, or utilized in any form by any electronic, mechanical, or other means, now known or hereafter invented, including photocopying, microfilming, and recording, or in any information storage or retrieval system, without written permission from the publishers.

For permission to photocopy or use material electronically from this work, please access www.copyright.com (http://www.copyright.com/) or contact the Copyright Clearance Center, Inc. (CCC), 222 Rosewood Drive, Danvers, MA 01923, 978-750-8400. CCC is a not-for-profit organization that provides licenses and registration for a variety of users. For organizations that have been granted a photocopy license by the CCC, a separate system of payment has been arranged.

Trademark Notice: Product or corporate names may be trademarks or registered trademarks, and are used only for identification and explanation without intent to infringe.

Library of Congress Cataloging in Publication Data

Linch, A L
Evaluation of ambient air quality by personnel monitoring.
Includes bibliographies and index.
CONTENTS: v. 1. Gases and vapors.—v. 2. Aerosols, monitor pumps, calibration, and quality control.
1. Air sampling apparatus. I. Title.
TD890.L56 1979 628.5'3'028 79-12153
ISBN 0-8493-5293-2 (v. 1)
ISBN 0-8493-5294-0 (v. 2)

A Library of Congress record exists under LC control number: 79012153

Publisher's Note
The publisher has gone to great lengths to ensure the quality of this reprint but points out that some imperfections in the original copies may be apparent.

Disclaimer
The publisher has made every effort to trace copyright holders and welcomes correspondence from those they have been unable to contact.

ISBN 13: 978-1-315-89284-9 (hbk)
ISBN 13: 978-1-351-07194-9 (ebk)

Visit the Taylor & Francis Web site at http://www.taylorandfrancis.com and the
CRC Press Web site at http://www.crcpress.com

PREFACE

Definition of Personnel Monitoring

Personnel monitoring is a term designating the determination of the inhaled dose of an airborne toxic material or of an air-mediated hazardous physical force by the continuous collection of samples in the breathing or auditory zone, or other appropriate exposed body area, over a finite period of exposure time. A personnel monitor is a self-powered device worn by the monitored individual to collect a representative sample for laboratory analysis, or to provide accumulated dose or instantaneous warning of immediately hazardous conditions by visible or auditory means while being worn.

Why Personnel Monitoring?

The case for control of exposure to toxic chemicals or injurious physical forces by personnel monitoring rests on the answers to six fundamental questions: (1) why personnel monitoring? (2) for whom? (3) when? (4) where? (5) for what? and (6) how? On the basis that with few exceptions even the most toxic substances have a no-effect level below which exposure can be tolerated by most workers for a lifetime without incurring any physiological impairment, and further that most hazardous agents are mediated by the air that surrounds us, then analysis of that surrounding air should provide sufficient warning to avoid dangerous concentrations of hazardous substances. Since the reliability of any analytical result is no better than the degree to which the sample analyzed truly represents the environment encountered by the worker, sampling technique is the controlling factor in exposure control. Direct continuous removal of a portion of the inhaled air as it enters the respiratory tract would satisfy the criteria for representative sampling. However, in most cases, physical hindrances to the performance of useful work render this approach impractical.

The concentration of most industrial air pollutants is a highly variable factor with respect to both time and space. Therefore, the workman in the course of his work assignments frequently encounters concentrations that may fluctuate in magnitude by orders of 10 or more over short time intervals. The dilemma faced by the industrial hygienist in attempts to assess exposure experience based on fixed station monitoring, or grab sampling, was illustrated by the results collected from a survey of the potential carbon monoxide (CO) hazard in a large warehouse in which a number of gasoline-powered forklift trucks operated continuously.

Even a cursory examination of the data clearly illustrated the futility of attempting to evaluate this typical industrial environment on the basis of one or two or even ten grab samples (3-min maximum-sampling period) taken with the gas detector tubes now in common use.

In other cases the contribution of skin absorption to the total dose absorbed must be established. In a study of the physiological effects on man, exposure to methylene-*bis*-orthochloroaniline was clearly established by the urinary excretion of the amine and its metabolites. However, airborne amine concentrations disclosed by fallout plate, mobile fixed-station, and personnel monitor sampling were below levels that could account for the observed urinary concentrations. Therefore, assignment of a TLV (threshold limit value) for this compound in air would not be appropriate for health control. In this case biological monitoring was the program of choice.

Personnel monitoring for those substances that exert their effects on a cumulative basis (CT value) should be relegated to a supporting role for biological monitoring after the dose-response relationship is established. Among applications that may be cited are lead exposure control by blood and/or urine analysis, mercury by urine analysis, trichloroethylene by breath analysis, and many others well established in industry. On the other hand, those agents that have assigned "ceiling" TLV's, such as benzene, formaldehyde, the isocyanates, and nitrogen dioxide, by their very nature require indicator response time within a range of a few seconds, as in the NDIR or colorimetric detector for CO.

For Whom Is Personnel Monitoring Designed?

First and foremost, consideration must be given to conservation of the employee's health by establishing tolerable, respirable air levels that correlate directly with absorbed dose and medical history. For some hazardous agents that produce injury that becomes manifest only after a prolonged induction period in man, or that produce chronic effects in test animals, personnel monitoring must be continued for as long as the chemical is present in the work place or until its benign character has been established. The importance of the determination of total absorbed dose cannot be overemphasized.

Personnel monitoring is not necessarily limited to the industrial environment. Monitoring for CO exposure has been extended to traffic police in vehicular tunnels, taxicab drivers, and garage attendants. At least one governmental agency issued detector tabs to their employees to carry in their automobiles and homes to detect exhaust and furnace leaks.

Protection for industrial management also enters the overall exposure control program. To meet the increasing pressure from governmental enforcement agencies, management must demonstrate conformance with TLV's promulgated by the Occupational Safety and Health Administration (OSHA) and the Environmental Protection Agency (EPA). Definitive information relative to the 8-hr day, 40-hr week, time-weighted average TLV's can be obtained only by personnel monitoring. As an added incentive, data collected consistently over a period of months can be used with adequate medical histories to initiate the revision of unrealistic TLV's. The revision of the TLV's for tetraethyl and tetramethyl lead provides an example.*

The production and engineering cost penalties to comply with a TLV set at an unreasonably low value can be sufficient to eliminate some products from the marketplace; to comply with the newer concepts of hazardous agents control, the development of TLV's, both airborne and biological, for new compounds is mandatory. Only with inhaled dose information can the contribution of skin absorption to the body burden be estimated. Adequate monitoring records also establish a favorable legal liability status on which successful defenses against unreasonable claims for alleged injury can be mounted.

When Should Personnel Monitoring Be Used?

General guidelines are suggested:

1. For preliminary surveys to define the problem, to delineate the limits of an airborne exposure problem, and to determine compliance with established TLV's. Short-term sampling, i.e., a minimum of 30 min, preferably 4-hr, or better, an 8-hr shift, would suffice.
2. For follow-up surveys to confirm the effectiveness of equipment or process revisions designed to correct deficient exposure control disclosed by the preliminary survey.
3. To confirm the relevancy of established TLV's by correlation with monitoring and medical records.
4. To establish new TLV's.
5. For permanent routine monitoring programs such as the semiannual survey for asbestos fiber exposure control.

* Linch, A. L., Wiest, E. G., and Carter, M. D., Evaluation of tetraalkyl lead exposure by personnel monitoring surveys, *Am. Ind. Hyg. Assoc. J.*, 31, 170, 1970.

Where Should Personnel Monitoring Be Used?

The answer is simple and straightforward; wherever the potential for exposure to toxic chemicals might be encountered. Workmen assigned to manufacturing areas should be monitored; transportation facilities must not be forgotten; and the consumer is obligated to monitor his employees until the offending material has been converted to a benign form or relegated to an adequate disposal system. On delivery of drum quantities of toxic chemicals in boxcars or tractor trailers, prudence dictates analyzing the atmosphere within the enclosure for the presence of the material, both upon opening the vehicle and during removal of the cargo, to detect leakage before exposure of the cargo handlers has occurred. In addition, facilities should be immediately available for monitoring during decontamination and cleanup after spills and releases of toxic agents. Mechanics who must break into pipelines, enter process vessels, or weld on equipment used in toxic chemical service should be monitored unless they are totally isolated from their environment by protective clothing or unless respiratory protection suffices for exposure control.

What Should be Monitored?

Again, the answer is clear and unequivocal: all substances known to be hazardous for man, those producing toxic effects in test animals, and those of unknown injury potential. However, from a practical consideration, attention would be focused on the most dangerous first. A cue can be taken from the National Institute of Safety and Occupational Health priorities published from time to time in the Federal Register. Not only the products themselves, but their combustion and pyrolysis derivatives generated by fire or overheating may generate serious exposure problems (e.g., phosgene from the pyrolysis of chlorine containing aliphatic hydrocarbons) and must be considered in a health surveillance program.

How Is Personnel Monitoring Carried Out?

Both the hardware and the technique employed will be determined by the purpose of the survey, the physical state of the offending material, the physiological effects of exposure, available equipment, and the analytical procedure to be applied after sample collection. The gravimetric determination of the total atmospheric dust loading by filtration provides the most direct air analysis.

For asbestos fibers a special membrane filter that "clears" (dissolves) when a mixture of dimethyl phthalate and dimethyl oxalate is applied to permit counting by phase contrast microscopy has been adopted by OSHA. A rechargeable battery-operated pump samples continuously for 8 hr in the 1 to 2 Lpm range.

Not all dust entering the nose reaches the lungs, due to aerodynamic size limitations imposed by the nasal passages. The "respirable" fraction is defined as a percentage according to the size of the particles:

Aerodynamic diameter (μm)	% Passing selector
2	90
2.5	75
3.5	50
5.0	25
10	0

Therefore, if the monitor is intended to be used as a respirable fraction sampler, then all particles larger than 10μ and 75% of those larger than 5μ must be rejected before the sample reaches the filter.

A readily portable miniature cyclone separator that approximates the above criteria adopted by the Department of Labor for coal dust, silica-containing dusts, and inert or nuisance dusts can be obtained commercially. Miniature cascade impactors are also commercially available for further classification of the particulates into discrete ranges (5 stages). The tangential flow filter provides uniform distribution of the dust collected over the entire filter surface, and the filter membrane need not be removed from the capsule for weighing.

Gases and vapors can be collected from the industrial atmosphere either by absorption on a solid medium such as silica gel or activated charcoal for laboratory analysis (usually by gas chromatographic procedures) or in a liquid solvent or reagent system for either laboratory or field analysis. The personnel monitoring system developed to derive the data used successfully in a petition to revise upward the TLV's for tetraethyl and tetramethyl leads can be cited as a model.* In this case collection of particulate inorganic lead separately from organic lead vapors was necessary.

Sampling was started when the employee left the change house and continued until he returned for lunch break. After lunch, the monitor was replaced and sampling continued until shift end. Six and seven monitors were operated simultaneously on each day shift in five work areas selected to include the widest possible range of activities. Each operator selected was monitored for 5 days on his day shift rotation. This rotation presented a maximum of individual differences in lead metabolism.

Linear and multiple regression analysis of daily individual results disclosed no correlation between fixed-station and personnel monitors, or between individual paired urine and monitor results on a daily basis. With few exceptions, personnel monitor results were significantly higher than corresponding, fixed-station results. Displacement of the urine results by 24- or 48-hr intervals to account for the lag in urinary excretion after exposure likewise produced no correlation.

An approximate linear relationship was found by converting the weekly averaged personnel monitor tetramethyl lead (TML) and inorganic lead results to TLV coefficients.* The sum of the coefficients (TLVC's) then was related to the averaged urine analysis for the operator who wore the monitor.

A different approach was required for the solution of the potential carbon monoxide problem within a consolidated warehouse where a number of gasoline-powered forklift trucks are employed for drum storage within bays and for loading boxcars and truck trailers. Fixed point alarms, grab sampling, and infrared monitoring did not provide the data necessary to determine compliance with the TLV for CO. None of the available colorimetric or instrumental methods could be adapted to personnel monitoring. Therefore, the length of stain CO detector tube was recalibrated for 4-hr continuous sampling.** Although the best accuracy that could be attained lay in the $\pm 25\%$ range of the true value, the method sufficed to determine compliance with the TLV (0.5, 1, and 2×50 ppm).

For the first time the capability for at least semiquantitative correlation of CO body burden with industrial exposure in the TLV range became available. In such surveys CO analysis of breath samples collected at the end of the 4-hr sampling period would provide correlation with the time-weighted average dose inhaled by the employee during the time spent in his work place. The most significant feature of this type of monitoring is the capability for instantaneous readout. At any time the accumulated dose

* Linch, A. L., Wiest, E. G., and Carter, M. D., Evaluation of tetraalkyl lead exposure by personnel monitor surveys, *Am. Ind. Hyg. Assoc. J.*, 31, 170, 1970.

** Linch, A. L., and Pfaff, H., Carbon monoxide — evaluation of exposure potential by personel monitor surveys, *Am. Ind. Hyg. Assoc. J.*, 32, 745, 1971.

can be quickly estimated from the 1- , 2- , 3- , and 4-hr calibration charts. Contact with a pocket of high CO concentration immediately turns the entire tube from amber to black to give the wearer ample warning to leave the area.

Surveys carried out during a 4-week period under restricted ventilation conditions imposed by cold weather confirmed CO levels that conformed with the 50-ppm CO TLV in this warehouse. Another survey completed in late 1972 again confirmed conformance with governmental regulations and the advantages derived from this type of monitoring with a minimum expenditure of time and cost. The recalibration of colorimetric length of stain detector tubes for other gases and vapors, such as SO_2, NO_2, H_2S, Cl_2, the halogenated hydrocarbons, aromatic hydrocarbons, solvents, etc., for 4-hr sampling has, to a significant degree, revolutionized the concept of personnel monitoring by providing both accumulated dose and ceiling value results on the spot in the field.

Since release of the first edition of this book in 1974, the National Institute for Occupational Safety and Health has published the *"NIOSH Manual of Sampling Data Sheets"* (Stock No. 017-033-00226-6, 1977). This manual describes 256 sampling methods for 245 substances. Most of these methods have been recommended to the Occupational Safety and Health Administration (OSHA) for use in their compliance program. NIOSH Sampling Data Sheets (SDS's) were originally written as guidelines for OSHA compliance Safety and Health offices. Each SDS recommends a sampling procedure for evaluating occupational exposure to toxic air contaminants. With the exception of asbestos and vinyl chloride, occupational expsoure to these contaminants is regulated by 29CFR1910.1000 (Federal Register Reference). Occupational exposure to asbestos is regulated by 29CFR1910.1001, and occupational exposure to vinyl chloride is regulated by 29CFR1910.1017.

NIOSH first published a manual for the analysis of airborne toxic substances in 1974 to provide guidance with the determination of compliance with exposure control standards for the nearly 400 chemical substances included in the standards adopted by the U.S. Department of Labor in 1972 (29CFR1910.1000). The first edition contained 39 analytical methods for measuring 130 toxic substances. The first edition was expanded to 110 methods with the addition of a second part. During 1977, a second edition (017-033-00267-3) in three volumes was released. The 1977 edition contains 337 sampling and analytical methods for use in industrial hygiene environmental monitoring. Some of the procedures are useful for measuring toxic constituents in blood, urine or body tissues, but the majority are intended for use in personnel air exposure monitoring.

Some confusion has been introduced into the Threshold Limit Concept for exposure limits for airborne toxic substances by Federal regulatory agencies. NIOSH in the preparation of *Criteria for Recommended Standards,* and *Recommended Standards for Occupational Exposure* refer to Threshold Limit Values (TVL's) as Federal Standards (FS's). In many cases the recommended FS is lower than the values originally adopted in the 1971 OSHA Safety and Health Standards (29CFR1910.1000), and the TLV's adopted annually by the American Conference of Governmental Industrial Hygienists (ACGIH). In order to minimize this confusion the original ACGIH definitions will be retained in this second edition of *Evaluation of Ambient Air Quality by Personnel Monitoring,* and the exposure limits will be referred to as TLV's rather than FS's. The term Time Weighted Average (TWA) has the same meaning in both ACGIH and Federal Standards.

The "gray" area of instrumentation, which lies between the personnel monitors and fixed station (area) nonportable instruments, was not explored, with a few exceptions, in the first edition of *Evaluation of Ambient Air Quality by Personnel Monitoring.* In this second edition, portable (hand held) equipment that in many cases can be carried on the person in the industrial environment are included under the appropriate sections. The NIOSH physical characteristics specifications were followed in the determination of portability in each case: the instrument should weigh less than 4.55 kg (10 lb), volume less than 0.028 m^3 (1 ft^3), be battery powered and be direct reading. Explosive vapor meters are included for those instruments calibrated in the parts per million (vol/vol) range. However, these detectors respond to any combustible (carbon tetrachloride) gases, or vapors, and are not specific for any particular component in a mixture.

Metric system of measurements are used throughout the second Edition of *Evaluation of Ambient Air Quality by Personnel Monitoring.* Conversions from English units have been rounded off to the nearest significant figure.

The design and evaluation of Industrial Hygiene personnel monitoring equipment and portable instruments has expanded tremendously since the release of the first edition of *Evaluation of Ambient Air Quality by Personnel Monitoring* in 1974. Developments in the following areas are especially significant:

1. Production of sophisticated metallic oxide semiconductor "buttons" that can be calibrated for specific gases such as carbon monoxide, hydrogen sulfide and a great number of organic vapors has provided an electronic detector system suitable for personnel monitoring.
2. A memory cell which accumulates the electronic response generated by exposure of the sensor element to polluted atmospheres provides TWA monitoring as well as ceiling value (instantaneous) read-out capability.
3. The detector badges briefly mentioned in the first edition have undergone extensive improvement and application. The small size, simplicity and reliability of this device has revived passive sampling as first choice for personnel monitoring. No moving parts, no pumps, no batteries, and no flow rate calibration are required for operation. With the introduction of the charcoal badge, we now can predict this sampling mode will replace most of the current TWA procedures for gas and vapor sampling.
4. The continuous tape samplers have been reduced in size to dimensions that are compatible with personnel monitors. Coupled with an integrating photometer the self-indicating colorimetric tape yields both ceiling concentrations and TWA total exposure dose in the form of a permanent record.
5. The most widely used of all survey instruments, the colorimetric length of stain gas and vapor detector tube has been recalibrated for continuous long-term TWA sampling. A number of tubes (eleven in 1977) that cover the common air pollutants have appeared on the market, and submitted for NIOSH certification. Personnel monitoring as well as grab sampling in this mode now is available.
6. The introduction of the paired membrane filter cassette has advanced significantly the art of dust sampling. The bottom (down-stream) filter membrane provides an exact tare for the top exposed membrane. Now the somewhat equivocal before and after weighing procedure has been completely eliminated.
7. Two noteworthy improvements in the design of battery powered personnel monitor pumps has been long overdue. Total volume read-out and self-compensation for changes in pressure drop across the sampler have been incorporated in commercially available models.
8. A comprehensive manual of Industrial Hygiene sampling and analytical methods manual was released during 1977 by NIOSH.

To fill in gaps left in the first edition of *Ambient Air Quality by Personnel Monitoring* several new sections have been added, and several other sections expanded:

1. Portable instruments, as defined by NIOSH, are added to the Passive Sampling, Dynamic Sampling, and Aerosol sections.
2. A chapter on calibration is included to provide information relative to this most important phase of personnel monitoring.
3. Evacuated container, and bag sampling techniques are now included to round-out the available methods.
4. The application of charcoal tubes has been expanded to accomodate the current status-of-the-art.

A comprehensive review of technical information provided by manufacturers and vendors of personnel monitoring equipment, and portable instruments (as defined by NIOSH) has been attempted. Failure to include any particular source or product would be an unintentional oversight which could not be avoided from a consideration of the volume of material, and multitude of sources which had to be consulted. Written requests for any information not already included in the trade bulletins were sent to as many manufacturers and vendors as could be found in the American Industrial Hygiene Association Journal, and contacted at the American Industrial Hygiene Conferences during the period 1973 to 1978. In a few cases, no response was received, and in a few other cases the manufacturer or vendor was reluctant to supply material suitable for inclusion in the second edition of *Evaluation of Ambient Air Quality by Personnel Monitoring.* Those sources which were liberal in their response can be identified by the completeness with which the material is presented in the text of the second edition.

The major, if not the most challenging, problem faced today by the Industrial Hygienist and certainly by the manufacturer is provided not by insufficient technical facilities, but rather by the pressure of rapidly shifting governmental rules, regulations and standards. For example, the recent drastic reduction in the Threshold Limit Value (TLV) for vinyl chloride from 200 ppm to 1 ppm has eliminated completely several types of electronic detectors, notably the combustible gas and vapor detector that employ combustion on a hot catalytic filament to generate a signal and have been calibrated in terms of hundreds of parts per million. With the introduction of the action level at one-half of the TLV-0.5 ppm vinyl chloride, additional strain has been placed on the performance of sample collecting and pollutant detection systems. As a general principle the system adopted must have a sensitivity of at least one-half of the adopted standard; i.e. 0.25 ppm of vinyl chloride. An increasing tendency for regulatory bodies to reduce the TLV's to the sensitivity limits of the sampling and analytical methods has been noted. This policy not only challenges sampling strategy and analytical capability, but also creates a dilemma for the designers and suppliers of equipment. The reluctance to invest limited financial resources in the development and manufacture of sampling and detection devices based on such uncertainty in the marketplace can be appreciated. Delivery of industrial hygiene hardware, likewise, has become a major problem as the vendors also are quite reluctant to maintain an inventory which tomorrow will be obsolete with no chance to liquidate their loses.

This environment of regulatory uncertainty also creates additional problems for the Industrial Hygienist who must keep himself informed, and prepared to quickly revise or develop new sampling, evaluation, and control strategies that are compatible with, in some cases such as vinyl chloride, drastically revised compliance conditions.

THE AUTHOR

Adrian L. Linch retired from E. I. du Pont de Nemours and Company as laboratory supervisor for the industrial hygiene and clinical laboratories of the medical division at Chambers Works in Deepwater, New Jersey on August 1, 1973.

Mr. Linch holds a bachelor's degree in chemical engineering (1933) and a master's degree in biochemistry (1934) from the University of Denver.

During his 39-year career with du Pont, Mr. Linch conducted research and development in such areas as the manufacture of dyes, dye intermediates, rubber chemicals, organic mercurials, tetraethyl lead, fluorocarbons, stabilization of aromatic amines, and the design of laboratory glassware. In addition to the medical laboratories, his career has included supervision of a Manhattan project laboratory for a plant producing fluorocarbons and uranium derivatives.

In 1952, he entered full-time practice in industrial hygiene. Biological monitoring procedures for the control of exposure to the cyanogenic aromatic nitro and amino compounds were developed under his supervision: personnel monitoring programs for alkyl lead and mercury derivatives, asbestos, radiation, silica-bearing dusts, and carbon monoxide also were established. Additional specialties involved development of air sampling and analysis techniques, design of personnel monitoring equipment, and protective clothing.

Mr. Linch is a member of the American Chemical Society, the American Industrial Hygiene Society, and the American Academy of Industrial Hygiene, and is a Fellow of the American Association for the Advancement of Science and the Franklin Institute.

Mr. Linch's bibliography includes over 70 publications, 40 of which are in the field of industrial hygiene. In addition to *Evaluation of Ambient Air Quality by Personnel Monitoring,* he has also written *Biological Monitoring for Industrial Chemical Exposure Control,* and chapters in the 2nd edition of the *Handbook of Safety* for CRC Press.

Professional qualifications include certification in the comprehensive practice of industrial hygiene (certified Industrial Hygienist) and certified Safety Professional. The annual award for the best industrial hygiene publication during 1971 was received from the Michigan section of the American Industrial Hygiene Association (*AIHA J.*, 32, 803).

TABLE OF CONTENTS

Volume I
Gases and Vapors

I. Passive Sampling 1
- A. Self-Indicating Devices — Semiquantitative 1
  - 1. Impregnated Paper 1
    - a. Convection Activated Detectors 1
    - b. Tape Samplers 7
    - c. Reagent Ampule Activated Samplers 8
    - d. Applications to Continuous Monitoring 9
  - 2. Inorganic Support Systems 9
    - a. Chalk and Crayons 9
      - i. Cyanogen Chloride 10
      - ii. Hydrogen Cyanide 10
      - iii. Phosgene 10
      - iv. Lewisite [Dichloro (2-chlorovinyl) arsine] 11
      - v. Gases 11
    - b. Badge Monitors 12
  - 3. Portable Instruments — Electronic 17
    - a. Memory Cell 17
    - b. Electrical Conductance Types (Resistance Circuits) 23
      - i. Hot Platinum Filament — Catalytic Combustion 23
      - ii. Metallic Oxide Semi-Conductor 24
    - c. Galvanic Cell (Redox type) — "Fuel Cell" 30
      - i. Oxygen 30
      - ii. Carbon Monoxide (CO) — Polarographic 37
    - d. Combination Instruments 40
    - e. Ultraviolet Photoionization Instrument 40
      - i. Principle of Operation 40
        - a) Methane 40
    - f. Evaluation of Portable Instruments 42
- B. Grab Samples for Laboratory Analysis 47
  - 1. Evacuated Continers 47
  - 2. Syringe Collection 49
- C. Personnel Monitoring — Collection for Laboratory Analysis 49
  - 1. Wet (Liquid Phase) Collection 49
    - a. Liquid Reagent System 49
    - b. Liquid on a Solid Support 50
      - i. Principle of Operation 51
      - ii. Sampler Analysis 52
      - iii. Experimental Results 52
      - iv. Effects of Temperature 52
      - v. Effects of Pressure 52
      - vi. Effects of Wind Velocity 53
  - 2. Solid Adsorption — Collection by Binary Diffusion 55
    - a. Introduction 55
    - b. Theory and Conceptual Illustration 55
    - c. Accuracy and Precision 58
    - d. Applications and Performance 58

i. Mercury Monitor 58
ii. Organic Vapors (Solvents) 59
iii. Acidic Gases 67
3. Evacuated Container Collection — Limiting Orifice Control 73

II. Dynamic Sampling 75
A. Grab Sampling — Short Term 75
1. Direct Reading 75
a. Paper Detectors 75
i. Phosgene 75
ii. Hydrogen Cyanide 76
iii. Arsine 76
iv. Sulfur Dioxide 76
v. Hydrogen Sulfide 77
vi. Mercury 78
vii. Lead 78
viii. Fluorides 79
ix. Dimethylhydrazine, Unsymmetrical 80
x. Boron Hydrides 80
xi. Chromic Acid 80
xii. Aniline 80
xiii. Quantitative Photometric Instruments 80
b. Granular Inorganic Solid Support Systems — Direct Reading Length of Stain Gas Detecting Tubes 81
i. Historical 81
ii. General Operating Procedure 81
iii. Availability and Limitations 82
iv. Applications 83
a) Testing Atmospheres Within Vessels and Manholes 85
b) Work Area Monitoring 85
c) Emergency Air Sampling 85
d) Personnel Monitoring 86
e) Product Control Testing 132
f) Troubleshooting and Technical Development
g) Fire Fighting 132
h) Detection of Explosive Hazards 132
i) Breath Analysis — Personnel Monitoring and Medical Diagnosis 132
j) Blood Analysis 133
k) Ambient Atmosphere Analysis for Air Quality Criteria Enforcement 133
l) Remote Sampling 133
m) Source Emission Sampling 133
n) Oxygen Deficiency Determination 133
v. Certification Criteria 134
vi. Accuracy-Controlling Factors 141
a) Uniformity of Granular Support 141
b) Tube Construction 143
c) Packing Variations — Pressure Drop Effects
d) Sampling Rate 146

e) Volume of Air Samples 149
f) Temperature Effects 151
g) Stability of the Chemical Reagent System — Storage Life 151
h) Specificity and Interferences 152
i) Observational Errors 168
j) Repetitive Sampling 169
k) Interchangeability of Indicator Tubes and Pumps 172
vii. Concentration of Reagents 173
viii. Constant Rate Sampling 173
ix. Reagent Systems — Identification 177
x. Testing Gas and Vapor Detector Pumps 180
B. Extended Period Sampling — More Than 15 Minutes 187
1. Direct Reading Length of Stain Tubes 187
2. Continuous Tape Samplers 208
a. Continuous Personnel Monitor 208
b. Plastic Bag Sampling 213
c. Impinger Systems — Liquid Reagents 219
i. The Midget Impinger 219
ii. The Microimpinger 222
a) Calibration 222
b) Efficiency Evaluation 225
c) Reagents 225
d) Summary 229
iii. The Wilson Impinger 234
iv. Porous Glass Bubblers 235
v. Traps 236
C. Solid Adsorbents 236
1. Silica Gel 236
a. Factors Affecting Adsorption 237
i. Previous History of the Gel 237
ii. Size of Gel Particles 237
iii. Size Range of Gel Particles 237
iv. Mass of Gel 238
v. Depth of Gel Bed 238
vi. Superficial Velocity of Airstream Through Gel Bed 238
vii. Temperature of Gel Bed 238
viii. The Molecular Species Being Adsorbed 238
ix. Concentration of Air Contaminant 270
x. Humidity of Air Contacting Gel Band 270
xi. Duration of Sampling 270
xii. Sampling Applications 270
a) Nitrogen Compounds 270
b) Ammonia 271
c) Phosphine 271
b. Practical Air-Sampling Details 271
i. U-Tubes 271
ii. Sintered Fiberglass Adsorbers 272
iii. Other Methods 272
c. Desorption of Vapors and Gases 272

i. Thermal Desorption Methods 272
ii. Solvent Leaching Methods 273
2. Alumina Gel 278
3. Carbon 278
a. Introduction 278
b. General Properties of Charcoal 278
c. Limitations 279
i. Capacity 279
ii. Humidity 288
iii. Sample Stability With Time 288
iv. Stability With Heat 290
v. Stability With Light 290
vi. Stability Under Vacuum 290
vii. Variations Among Lots of Charcoal 290
viii. Variation With Contaminant Loadings 290
d. Uses Relative to Sampling 291
i. Specific Application 295
e. Analysis 295
i. Elutian Techniques 295
a) Steam Displacement 295
b) Vacuum Desorption 296
c) Thermal Desorption 296
d) Solvent Extraction (Desorption) 296
ii. Analysis Procedure 299
f. Evaluation of Charcoal Tubes 302
g. Evaluation — Precision and Accuracy 302
4. Other Adsorbents 303
a. Sample Collection System 323
5. Liquid Reagents Carried on Solid Adsorbents 325
6. Sorbent Specifications 326
a. Sampling Procedure 326
i. Collection 326
ii. Desorption — Recovery 327
b. Criteria for Successful Validation 328
7. Equipment 329
D. Portable Dynamic Sampling Instruments Containing Battery Powered Pumps or Manual Squeeze Bulbs 329
1. Electronic 329
2. Gas Chromatographs (GC's) 332
E. The Portable Field Kit Laboratory 340
1. Aromatic Isocyanates in Air — Toluene Diisocyante (TDI) and Methylene-*bis*-(4-phenyliscocyante) (MDI) 340
2. Tetraalkyl Lead Compounds in Air — Tetraethyl Lead (TEL) and Tetramethyl Lead (TML) 341
3. The Unsaturated Perfluoroalkenes — Perfluoroisobutylene (PFIB) and Hexafluoropropene (HFP) 343
4. Inorganic Lead 343
5. Water Analysis Kits 343
F. Daco Toxic Substances Detector Kits for Use With Impingers 345

References 345

Index 355

EVALUATION OF AMBIENT AIR QUALITY BY PERSONNEL MONITORING

CONTENTS

Volume I

Gases and Vapors

Volume II

Aerosols (Dusts, Mists, and Fumes)

Battery Powered Air Samplers

Calibration

Quality Control For Sampling and Laboratory Analysis

Determination of Compliance and Classification of Violations for Air Contaminants

Skin Absorption Monitoring Wipe Sample Policies and Procedures

Control of Workspaces

I. PASSIVE SAMPLING

A. Self-Indicating Devices — Semiquantitative

1. Impregnated Paper

The use of a paper support for reagents that change color on contact with traces of reactive ions or compounds in air or water was undoubtedly among the first methods to detect and monitor hazardous air pollutants. Minimal cost, ready portability, color development without application of liquid reagents, rapidity of response, and simplicity continue to provide commercial incentive for exploitation not only for "go-no go" application in locations where acute exposure to highly hazardous gases or vapors is a threat to health, but also for ambient air monitoring outside of the industrial environment (Figure 1). However, without some provision for measuring the air sample volume and calibration with known concentration standards, only semiquantitative results can be obtained. An uncalibrated detector furnishes little more information than the human nose, which is a notoriously ineffective quantitative analytical instrument unless the contaminant concentration falls below the odor threshold. For those highly toxic gases such as phosgene, for which the threshold limit value (TLV)[1] is well below the odor threshold, a qualitative response which discloses the presence of the hazard can serve as a warning device even though no positive information relative to concentration is available.

a. Convection Activated Detectors

A reasonably acceptable estimate for concentration can be obtained by application of the CT concept: the product of concentration multiplied by time, also referred to as "dosage" or "time-weighted average". Since the first appearance of color, rate of color development, and color intensity (also referred to as depth of color, optical density, or color saturation) are related to concentration, it follows that an approximate calibration can be attained by noting the time required to reach an arbitrarily selected color hue or intensity when the test paper is freely suspended in an atmosphere generated from a known concentration of the contaminant. A good example of this approach is illustrated by the commercially available hydrogen sulfide detectors. The test paper may be exposed either for a preselected time, such as 3 to 5 min, and the spot matched with a segment of the color chart supplied with the paper (Figure 1, Parts 1 and 3), or the stain may be compared to a graded saturation series that is related to a conversion table based on the CT or dosage value. In this latter case, if the time is known, the concentration in ppm can be calculated. For example, a lead acetate impregnated circular pad mounted in the center of a surrounding color density chart imprinted on stiff paper is exposed for a finite time interval (Figure 1, Part 1). The color intensity (grade) read from the chart is then referred to a conversion table (Table 1) to obtain the dosage (CT value), i.e.:

Color grade	=	2.5
Exposure time	=	24 hr
Dosage (from Table 1)	=	0.56 ppm × hr
0.56 ppm × hr/24 hr	=	0.023 ppm H_2S
or for an 8-hr day	=	0.07

The lower limit is 0.03 ppm, and complete saturation is reached at 4 ppm for a 1-hr exposure. On this basis, the TLV of 10 ppm would be reached in 24 min. Although intended primarily for stationary air pollution analysis, the detector could be used directly for personnel monitoring by attaching the tab to the workman's shirt under his chin. A transparent plastic mounting would be necessary to protect the rather fragile impregnated pad and surrounding color chart.

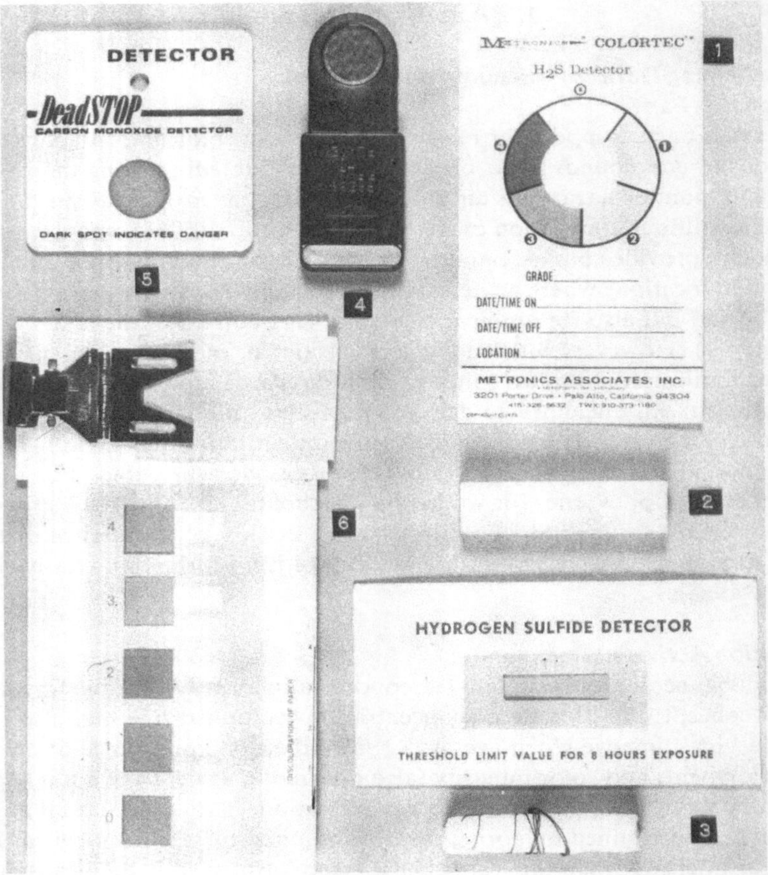

FIGURE 1. Paper-supported, self-indicating hazardous gas detectors for personnel monitoring application. (1) Metronics Colortec® hydrogen sulfide detector. (Courtesy of Metronics Associates, Inc., 3201 Porter Dr., Palo Alto, Calif.) (2) Metronics Ceramtec® tile hydrogen sulfide detector. (Courtesy of Metronics Associates, Inc., 3201 Porter Dr., Palo Alto, Calif.) (3) Davis hydrogen sulfide detector. (4) Portable carbon monoxide alarm, Unico® Model 555. (Courtesy of National Environmental Instruments, Inc., Pilgrim Station, Warwick, R.I.) (5) Deadstop carbon monoxide detector. (Courtesy of Devco Engineering, Inc., 36 Pier Ln. W., Fairfield, N.J. and Deco Systems, Inc., P.O. Box 55562 Houston, Tex., 1968.) (6) Kitigawa® sulfur dioxide test paper. (Courtesy of Bendix Corp., Environmental and Process Instruments Division, Lewisburg, W. Va.)

This principle also has been applied to the more durable ceramic (bisque) tiles as a substrate for the lead reagent. A chart of dosage (D) plotted vs. tile grade (color intensity) on semilogarithmic graph paper indicated a 90% probability that the true D will fall between 0.67 D and 1.5 D (Figure 2). For example, D estimated for grade 2 is equivalent to 0.33 ppm × hr with the probability that the true dosage lies between 0.22 and 0.50 ppm × hr. In this case the precision has been further improved by quantitating the relative color intensity by reflectance measurement at 425 nm. If the accuracy can be controlled consistently within these limits of ±40% relative deviation (0.36 ± 0.14 ppm × hr), then the reliability of this system would be about equivalent to the performance of most length of stain detector tubes (interim performance level ± 50%, except

TABLE 1

Conversion Table — Relationship of Color Density to SO_2 Exposure Dosage (Concentration × Time)

Color grade	0	0.5	1.0	1.5	2.0	2.5	3.0	3.5	4.0	4.5
Dosage (ppm × hr)	0[a]	0.07	0.12	0.20	0.33	0.56	0.93	1.56	2.59	4.33

[a] Zero grade indicates dosage less than 0.03 ppm × hr.

Courtesy of Metronics Associates, Inc., Palo Alto, Calif.

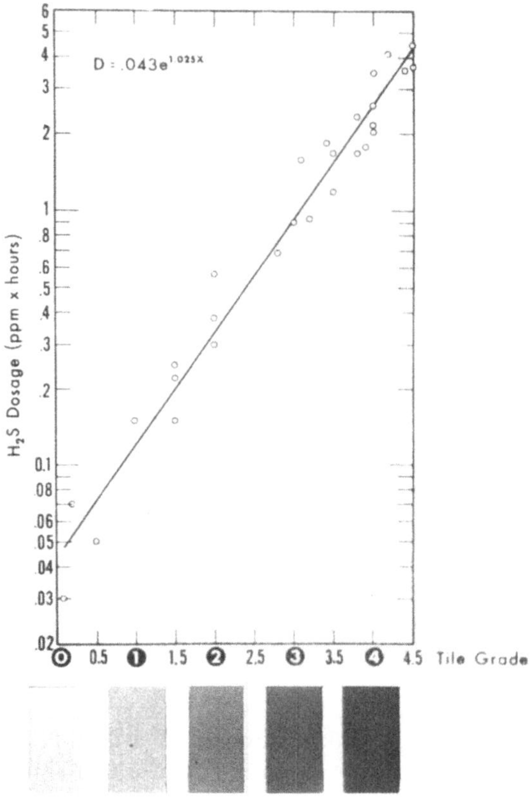

FIGURE 2. Ceramtec® tile calibration chart for hydrogen sulfide. (Courtesy of Metronics Associates, Inc., 3201 Porter Dr., Palo Alto, Calif.)

for those which must meet American Conference of Governmental Industrial Hygienists-American Industrial Hygienists Association (ACGIH-AIHA) criteria of ±25% for certification).[2,3]

A somewhat different approach has been developed for the estimation of sulfur dioxide (SO_2) in air, based on the fading of a blue-red, pH-sensitive dye to a yellow-orange on filter paper in four graded steps (Figure 1, Part 6). In this case, a gas phase

TABLE 2

Relationship between Exposure Time and Concentration of SO_2 Required to Produce Color Grade 3

Exposure time (min)	SO_2 concentration (ppm)
3	0.6
5	0.4
7	0.3
9	0.2
12	0.15
17	0.1
25	0.07
35	0.05
40	0.04
55	0.03

TABLE 3

Relationship between Exposure Time and Concentration of SO_2 Required to Produce Color Grade 4

Exposure time (min)	SO_2 concentration (ppm)
2.0	1.0
3.0	0.8
4.0	0.7
5.0	0.6
6.0	0.5
8.0	0.4
11.0	0.3
15.0	0.2
20.0	0.15
30.0	0.1

titration can be employed, i.e., the time required to attain a definite end point color (grade 3 or 4) is related to the concentration in ppm of SO_2 (Tables 2 and 3). A humectant must be present to retain water for the acid reaction. A chart that relates other color grades and exposure times to SO_2 concentration is furnished also (Figure 3). The colors can be reproduced by referring to Munsel's code (Table 4).

The potential utility of this approach to personnel monitoring was suggested by the report of a study conducted in Yokohama, Japan, during the summer of 1968. Although the mode of sampling was fixed station, the survey could have been carried out just as well by selected individuals wearing detectors while engaged in their normal daily activities.[4]

Using pieces of the SO_2 paper developed by Prof. Dr. Tetsuzo Kitagawa and his collaborators in the Department of Engineering at Yokohama National University, the concentration of atmospheric SO_2 was measured eight times at each of 56 schoolgrounds in Yokohama City between 9 and 10 a.m., and between 12 and 1, 3 and 4, and at 5 p.m. on July 25 and August 3, 1968. On both days a clear sky prevailed, and the wind blew from the southeast, the south, or southwest at speeds up to 7.3 mi/sec.

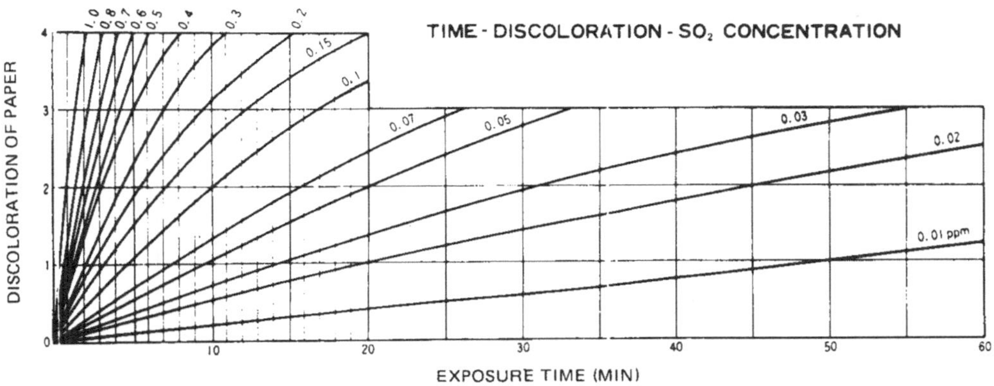

FIGURE 3. Sulfur dioxide detector paper calibration charts showing the relationship between time, discoloration, and SO_2 concentration.[6]

TABLE 4

Color Change of Acidic Gas Detector Paper

Degree of color change	Munsel color code[a]	
0 (Original color)	7.5 RP	5/11
1	2.5 R	5.3/9
2	7.5 R	5.8/8
3	2.5 YR	6.2/8
4	7.5 YR	6.8/8.5

[a] R = red; P = purple; Y = yellow.

The usually employed methods of analysis for sulfur oxides in the atmosphere included the West-Gaeke method, the barium molybdate method, and electroconductivity measurement. These methods required heavy equipment and involved extensive manipulation, which was considered to be impractical for survey analysis. Therefore, the lightweight, self-indicating indicator paper was chosen to obtain rapidly results relative to (1) geographical distribution of SO_2 pollution (Figure 4) by simultaneous measurements at the various locations, (2) variations of pollution concentration with time, and (3) the efficiencyof abatement equipment by comprison of before and after analyses. Comparisons between replicate determinations obtained from the classical wet chemical procedures confirmed the validity of the detector paper results (Figures 5 and 6 and Table 5).

Three measurement techniques were used.

1. The paper strips were exposed continuously, observed at the end of consecutive, discrete time intervals (10 to 15 min), and the color graded 0, 1, 2, 3, or 4 by comparison with the reference color chart supplied with the kit. The color grades then were plotted vs. time on the calibration chart (Figure 3) and the concentration located on one of the calibration curves or extrapolated between adjacent curves. If the plotted results disclosed scattered points, the last reading was used as a time-weighted average to compensate for the changes in SO_2 concentration that occurred during the observation period. This technique gave the best accuracy.

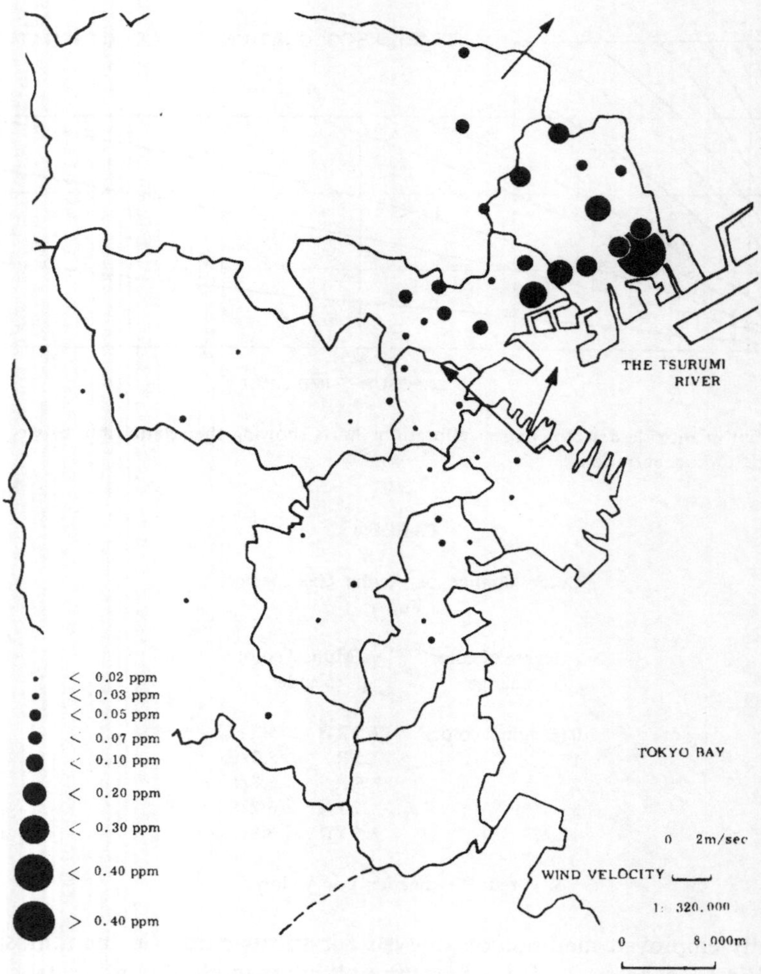

FIGURE 4. SO_2 concentration distribution map from Yokohama, Japan, on July 25, 1968.[6]

2. The time required to produce a given stain density (grade 3 or 4, Tables 2 and 3) by gas titration was calculated. This approach was considered suitable for concentrations above 0.2 ppm.
3. The degree of color change after a fixed exposure period was determined. The range of measurement varied with time. This was considered the best technique for SO_2 concentrations less than 0.03 ppm.

The variables that influenced the results included wind speed, temperature, and humidity. If the wind speed was less than 1.5 mi/sec, low results were noted, but when the velocity exceeded 2 mi/sec, the effect disappeared. Temperature was without influence in the range 0 to 40°C. Although a humectant was added to the paper, relative humidity below 30% RH contributed to low results. The magnitude of this error increased as the temperature rose.

Since the SO_2 test paper is not affected by the ionizable salt content of the ambient atmosphere, the results checked more closely with the West-Gaeke method than with electroconductivity (Figures 5 and 6, and Table 5). When SO_2 and SO_2-containing aer-

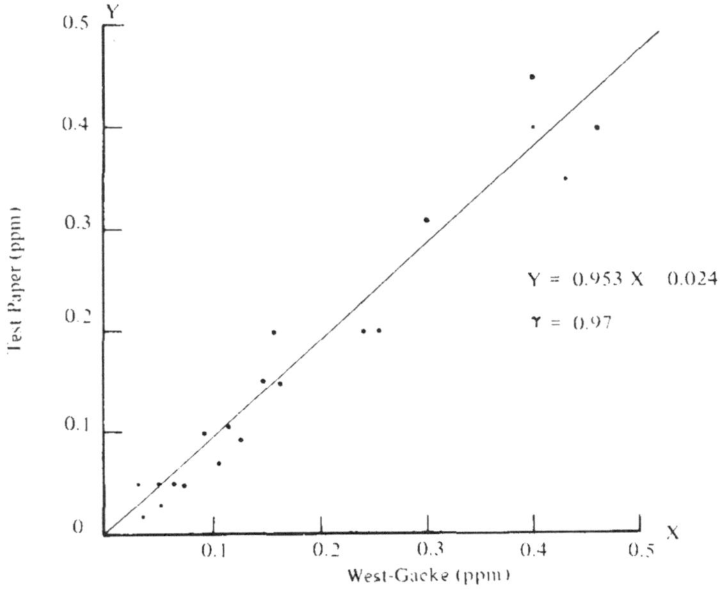

FIGURE 5. Correlation of test paper discoloration with West-Gacke analysis results at the Yokohama Marine Tower.[6]

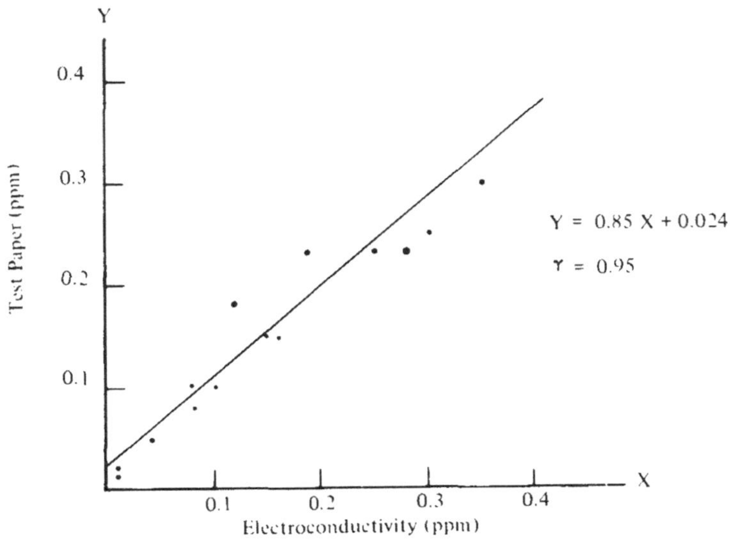

FIGURE 6. Correlation of test paper discoloration with electroconductivity at the Yokohama Marine Tower.[6]

osols were present, the color change on the test paper was not uniform, but rather presented a yellow speckled appearance. This SO_2-sensitive paper is not available from a domestic source.

b. Tape Samplers

A miniature portable tape sampler (20 × 58 × 125) that provides over 1,000 spots 8 mm in diameter on a single roll of filter paper has been developed and exploited com-

TABLE 5

Comparison of the Results by West-Gaeke Method, Electroconductivity Method, and Test Paper Method

Date (June)	West-Gaeke (ppm)	Electroconductivity (ppm)	Test paper (ppm)
11	0.139	0.25	0.23
	0.148	0.35	0.30
14	0.058	0.15	0.15
16	0.020	0.01	0.02
	0.010	0.01	0.015
17	0.140	0.30	0.25
	0.095	0.28	0.23
	0.042	0.10	0.10
18	0.090	0.187	0.23
	0.072	0.160	0.15
	0.100	0.120	0.18
20	—	0.08	0.10
	0.150	0.13	0.15
	0.040	0.08	0.15
23	0.032	0.08	0.08
	0.030	0.04	0.05

FIGURE 7. Hydrogen sulfide-sensitive continuous paper tape detector. (Courtesy of Del Mar Products, P.O. Box 486, Addison, Tex.)

mercially (Figure 7). The tape, instructions, and color charts are housed together in a transparent plastic case, and the sampler weighs 110 g complete. Only lead acetate impregnated tape for H_2S detection has been supplied. The unit is suitable for personnel monitoring if a mounting bracket is attached and the tape recalibrated in terms of dosage (CT values).

c. Reagent Ampule Activated Samplers

Although not strictly a paper-supported reagent system, mention should be made of the wet wick method. An ampule that can easily be crushed is contained in a cotton covering (Figure 1, Part 3). Exactly 1 min after saturation of the cotton by the lead acetate solution, the color is compared with the standard by placing the ampule directly behind a color-bordered slot in a card. A lighter shade of brown indicates less than 10 ppm H_2S, a darker color more than 10 ppm. The system could be adapted to personnel monitoring as an alarm device where intermittent or casual exposure would be expected, since the reagent-soaked wick remains active for approximately 6 days after the ampule is crushed.

d. Applications to Continuous Monitoring

Although the performance of the foregoing gaseous pollutant detectors is at best, as stated in the introductory paragraph, only semiquantitative (deviation range greater than ± 25%), application to personnel monitor surveys designed to determine order of magnitude concentrations (0.5, 1, 2, and 4 times the TLV) will in many cases supply sufficient information to define the problem. With minimum cost and technical skill, large areas can be surveyed for long periods of time to locate pollution focal points and to indicate trends. The possibility that a major source of pollution could be corrected without establishing a costly, comprehensive air surveillance program adds further incentive to this approach.

If the results fall consistently below one half of the TLV, then a more sophisticated analytical system is not necessary. In the 0.5 to 1.5 × TLV range, an extensive, refined analytical survey must be undertaken to evaluate the compliance factor fully. When results collected from any given location in the area under investigation consistently exceed twice the TLV, then corrective action should be initiated and completed before additional funds are expended on improving the reliability of the air quality monitoring system.

Although, after a short briefing, the individual wearing the badge could read off the concentration values from the color comparator furnished with the detector, more consistent results would be expected if the color matching were limited to a few trained technicians to minimize variations in visual acuity. Of course, if the stains are not sufficiently stable to permit forwarding to a central collection point, then reliance must be placed on the veracity of the individual observer.

This type of survey could be applied to ambient air monitoring by the populace at large to delineate pollution areas, as well as for in-plant exposure control. Also, for localized plant conditions the badges could provide instantaneous warning (within reagent response time limits) of TLV ceilings, or excursions into immediately hazardous concentration ranges, in time to permit evacuation of the area before irreparable physiological damage can occur.

The approach of an accumulated dose limit (carbon monoxide, for example) can be anticipated by periodic observation of the stain intensity as it develops during the wearer's work assignment in time to take appropriate corrective action before excessive exposure can occur (by eliminating the pollutant at the source, or transferring the workman to an area of less exposure for the remainder of the shift).

2. Inorganic Support Systems

a. Chalk and Crayons

Detector chalk and crayons have been formulated from active chromogenic agents, inert binders, and fillers. When these chalks or crayons are rubbed on a suitable surface, a mark that will change color in the presence of the air contaminant for which the mixture was sensitized will be deposited. Detector paper can be produced by rubbing with chalk, or a powder can be prepared by crushing the chalk. Since the crayons present a relatively small surface area, the stability is superior to that of impregnated paper, they are easily carried, and they can be used to prepare on-the-spot personnel monitoring tabs.

Development work in this field has been confined to detector crayons for the war gases — phosgene, hydrogen cyanide, cyanogen chloride, Lewisite, and the "G gases".[5,6] The following description of the crayon preparation is a summary of the work reported by Witten and Prostak.[6]

i. Cyanogen Chloride

Ingredients

4-Benzylpyridine — 10%
Barbituric acid — 4%
Blanc fixe (precipitated barium sulfate)
neutral, amorphous, and dry — 86%

Procedure

After thorough mixing, 15-g aliquots of this dry mixture were compacted to 422 kg/cm^2 compression in a 16-mm diameter cylinder held in a Carver laboratory hydraulic press. The piston and cylinder were lubricated with a thin film of silicone grease. White markes from this crayon turned first red and then blue on contact with traces of cyanogen chloride or bromide. In 1 min a concentration of 0.4 ppm was detectable. Storage stability for 3 years were indicated.

ii. Hydrogen Cyanide

Hydrogen cyanide was first converted to cyanogen chloride with a chlorinating agent such as Chloramine-T, and then detected by crossmarking with the cyanogen chloride crayon.

Ingredients

Chloramine-T(*N*-Chloropara-toluene-sulfonamide sodium salt) — 14%
Blanc fixe — neutral, amorphous, and dry — 86%

Procedure

Preparation followed the cyanogen chloride crayon technique. The mixed mark formed by superimposing this crayon's mark on the cyanogen chloride mark turned red-blue in the presence of HCN. A 1-min exposure detected 5 ppm. The Chloramine-T crayon was stable for at least 3 years at room temperature, but decomposed at 65°C. By substituting dibromantin (1,3-dibromo-5,5-dimethylhydantoin), stability at 65°C was attained.

iii. Phosgene

The highly sensitive reagent system 4-(para-nitro-benzyl)pyridine + *N*-phenylbenzylamine, previously described for detector papers in Section I.A.1, "Impregnated Paper," was employed for this system.

Ingredients

4-(para-nitro-benzyl)pyridine — 2%
N-phenylbenzylamine — 5%
Sodium carbonate — 5%
Blanc fixe — neutral, amorphous, and dry — 88%

Procedure

A benzene solution of the first two ingredients was mixed with the blanc fixe. Sufficient benzene to completely wet the inert substrate was suggested. After drying overnight, an aqueous solution of the sodium carbonate was thoroughly mixed in. Again, sufficient solvent (water) to completely wet the blanc fixe mixture was specified, and cooling was applied to avoid melting the *N*-phenylbenzylamine. The dried mixture was regrourd and pressed into crayons as described above.

The light yellow marks from this crayon turned bright red in the presence of phosgene. High concentrations of acid-producing gases (HCl) interfered by consuming the alkaline reserve (Na_2CO_3) required for color formation. At room temperature, storage stability for at least 1 year was indicated. A 1-min exposure detected 5 ppb of phosgene, but the color tended to fade at high concentrations. However, the color returned in fresh air.

iv. Lewisite [Dichloro(2-chlorovinyl)arsine]

Thio-Michler's ketone reacts with Lewisite or high concentrations of phosgene to produce a green color.

Ingredients

4,4-Bis(dimethylamino)thiobenzophenone — 5%
Blanc fixe — neutral, amorphous, and dry — 95%

Procedure

A benzene-chloroform solution of the organic component was mixed thoroughly with blanc fixe, and most of the solvent was allowed to evaporate spontaneously overnight. Drying was completed under vacuum for 12 hr. The dried powder was reground and pressed into crayons as previously described. Light tan marks from this crayon exhibited color changes for several irritant gases as follows:

Lewisite: intense green-blue
Ethyldichloroarsine: green-blue
Phosgene: purple (high concentrations)
Chlorine: gray
Cyanogen bromide: gray

v. Gases

In U.S. Patent 2,929,791, a crayon for detecting the G agents, chemical warfare gases — GB or sarin (isopropyl-methyl-phosphorofluoridate), GA or tabun (ethylmethylphosphoramide cyanidate), GD (pinacolylmethylphosphonofluoridate), and GF (cyclohexylmethylphosphonofluoridate) — was described.[5] The chromogenic detector system included a mixture of disodium diisonitroacetone and tolidine that turned to a magenta or purple color on contact with traces of one or the other of the G agents. The recommended composition:

Disodium diisonitrosoacetone	20%
o-Tolidine	20%
Urea	15%
Lithium stearate	27%
Lithium chloride	3%
Calcium oxide	15%

was compressed into crayons by the usual procedure. The CT value was given as CT = 6 μg min/L.

The concept of CT values (concentration × time) has been applied to crayon marks as well as to paper detectors and ceramic tiles, as discussed in Section I.A.1, "Impregnated Paper." If the crayon marks are exposed to equivalent air turbulence, temperature, humidity, and incident light, the time lapse to the first detectable color change was inversely proportional to the concentration of the contaminant within approximate limits. In other words, the CT value equals a constant that is a measure of sensitivity. In most cases the chromogenic reaction rate is both temperature and humidity dependent, and corrections are required when working outside of normal room temperature and RH limits. The CT values developed by Witten and Prostak for the chemical warfare gases are summarized in Table 6. The crayons are commercially available.[7]

TABLE 6

Detector Crayon Sensitivity at 32°C[57]

Gas detected	Gas concentration (mg/m³)	CT	Gas concentration (mg/m³)	CT	% RH	Air movement
Cyanogen chloride	0.369		1.045			
		1.2		1.3	20	Turbulent
		1.0		0.7	69	Turbulent
		2.2		2.1	20	Static
Hydrogen cyanide	2.5		120			
		6		8	20	Turbulent
		3		4	70	Turbulent
		8		15	20	Static
Phosgene	0.064		0.611			
		0.02		0.02	20	Turbulent
		—		0.03	70	Turbulent
		—		0.04	20	Static

b. Badge Monitors

Paper as a support for chromogenic reagents has very definite limitations. Strong acids, alkalines, or oxidizing agents are not compatible. Parasitic side reactions and rapid deterioration of the reagents often render the system impractical. Packaging in sealed containers to prevent premature contact with the ambient atmosphere presents further problems often difficult to resolve. Fortunately, several inert, highly adsorptive solid support materials are commercially available in granular forms that are readily adaptable to a wide variety of color-generating techniques. Silica and alumina gels probably are the most widely used, on the basis of high adsorption capacity, mechanical strength, and an essentially white background, which provides maximum contrast for visual evaluation.

However, with the exception of the impregnated bisque tiles developed for monitoring H_2S, as described in Section I.A.1, "Impregnated Paper," application to diffusion (spontaneous) monitoring has been restricted to reagents available for the detection of carbon monoxide (CO), H_2S, and phosgene. The potassium pallado sulfite on silica gel reagent developed by Silverman for CO detection[8] has been adapted to a wafer prepared by cementing a layer of this impregnated gel on a 15-mm diameter recess in a 55-mm square by a 2-mm thick plastic tab (Figure 1, Part 5). The tab may be worn either on the clothing within the person's line of vision or may be suspended in fixed locations such as on the instrument panel of a truck or crane cab. As supplied, the detector is a "dead stop" warning device that indicates dangerous concentrations of CO when the orange-brown "spot" turns black.

These carbon monoxide detector tabs were used quite successfully by Argonne National Laboratory personnel to monitor nonoccupational exposures. Of the 454 persons who received these tabs and reported results, 124 (27%) reported positive indications and 21 (nearly 5%) detected sources that were emitting sufficient quantities of carbon monoxide to have some physiological effect.[9] Sixteen of this group followed through with corrective actions which included automobile exhaust system maintenance; automobile manifold heater repairs; automobile replacement; home furnace replacement; cleaning, maintenance, or replacement of furnace heat exchangers; and improved exhaust ventilation of home gas heaters.

When exposed over a period of "a few hours", the nominal ranges of response in known atmospheric concentrations of CO were found to be

Slight darkening	30—70 ppm
Gray	80—120 ppm
Black	130+ ppm

Unexposed tabs were furnished as a "reagent blank" reference for visual detection of the first stage, slight darkening. Although a 50-ppm TLV has been assigned to CO for an 8-hr time-weighted average, in a 40-hr work week the nonoccupational exposure for the remaining 128 hr should not exceed 30 ppm if adverse health effects, as manifested by impaired time interval discrimination, certain other psychomotor functions, and physiological distress for patients with heart disease, are to be avoided.[10] Therefore, even the first stage (30 to 70 ppm) must be considered hazardous for individuals who also encounter occupational CO exposure. "Normal traffic" in many cases produced stage one conditions for passengers in automobiles. In those cases for which a follow-up confirmatory analysis was conducted with an indicator tube, the tab results were found to indicate somewhat higher CO concentrations, i.e., slight darkening indicated 30 to 70 ppm, whereas the detector tube showed 15 to 50 ppm. No instance in which a tab showed less than the true concentration or in which the tab failed to respond to an environment where the presence of detectable CO was known or highly suspected was reported. On this basis the detector tab would be classified as a fail-safe device. Although considerable variance between samples taken at separate times in different ways, especially if one sample was integrated over a period of time and the other was a grab sample of short duration, was expected, the data collected from the two methods showed considerable agreement.

Soon after the release of the first edition of *Evaluation of Ambient Air Quality by Personnel Monitoring* the design was revised from the square form shown in Figure 1, "Dead Stop", to the triangle shape shown in Figure 8, "Lifesign". Note the name also was changed for the new model that is more sensitive to low concentrations of carbon monoxide in the TLV range (50 ppm). Two types, thresholds of 50 ppm and 200 ppm, are available. Confirmatory evaluation by an independent laboratory in September 1974 gave the following response times for ten indicators of the two types:

		Carbon monoxide concentration (ppm)		
Model	Threshold value	50	100	200
"Red Rim"	50 ppm	20 min	7.5 min	5.0 min
Standard	200 ppm	27 min	8.0 min	5.5 min

Note: Every sample from each type responded to the CO atmosphere in about the same time. Standard deviation figures were not supplied.

These results are somewhat different from those supplied with the information sheets forwarded by the vendor for the "ultrasensitive" ("Red Rim") indicator:

CO concentration	Weak reaction time	Distinct reaction time
100 ppm	ca. 2 min	ca. 4 min
50 ppm	ca. 4.5 min	ca. 9 min

These differences probably arise from end-point observational errors. Reaction times are given as approximations because reaction rates will decrease at low temperatures and increase in altitude (reduced pressure).

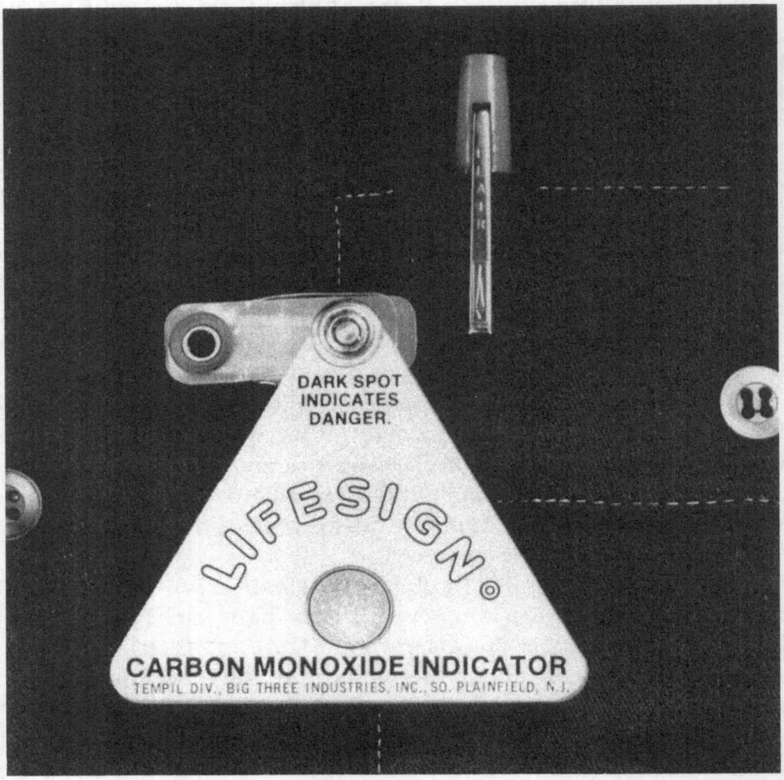

FIGURE 8. New Lifesign® indicator by Tempil® warns of carbon monoxide in shop, home or vehicle. (From Tempil Division, Big Three Industries, Inc., Hamilton, Blvd., South Plainfield, N.J. With permission.)

Shelf (storage) life for each type was claimed to be:

Model	Sealed — as received	Exposed — after breaking seal
"Red Rim"	12 months	10—15 days
Standard	6—12 months	30 days

Note: Expiration date printed on the package.

Within the useful life of the exposed reagent disk the unit "recovers" after exposure to fresh air for a minimum of one hour and may be re-used. On prolonged exposure to CO concentrations above 800 ppm a bleaching effect takes place. The indicators are "poisoned" by ammonia, carbon bisulfide, and hydrogen sulfide to produce a nonreversible discoloration. Nitrogen oxides bleaches the color.

Since the gradual change in color from the initial yellow or straw-color to a pale-to-dark gray to brown-black requires good visual acuity and judgment relative to an end-point, the user should expose several units to known CO in air concentrations to "calibrate" his vision in order to obtain even approximate results.

Automation has been applied to the detection of the color change. By retaining the impregnated gel between two closely spaced fine mesh metal screens, a beam of light can be passed through the CO sensitive "spot" to a phototube, which monitors the

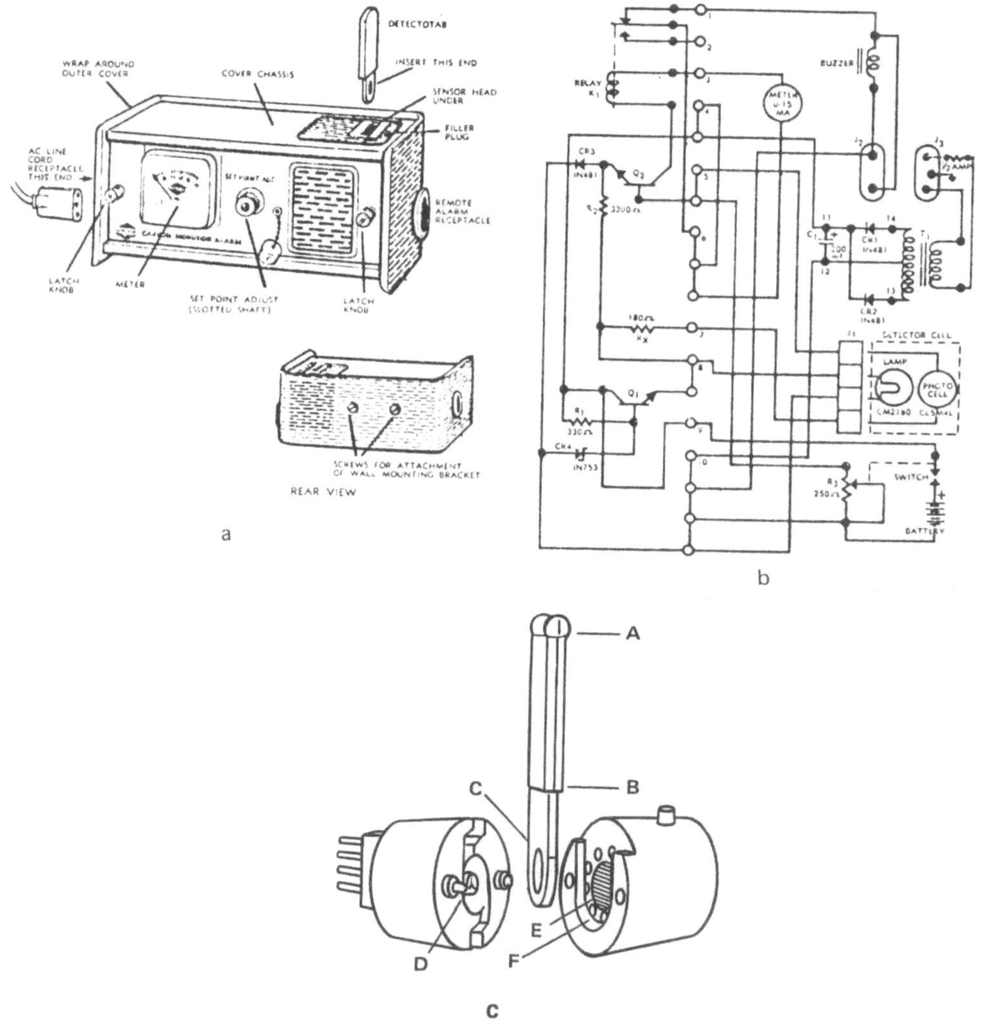

FIGURE 9. Use of a carbon monoxide detector tab in an alarm system. (a) Assembly. (b) Schematic. (c) The detector head. The Unico® Detectotab® (U.S. Patent 3,245,917) is the activating sensor in all Unico carbon monoxide detector alarms. Molded of high impact polystyrene, it provides a chamber and reservoir (A) for the sensitized silica gel reagent. This chamber (C) is defined by two parallel screens of corrosion resistant mesh which allows light to pass through. Constructed to assure proper seating, an indentation in the Detectotab (B) together with a mechanical detent (F) in the module sensor cell retains the Detectotab securely in the proper position relative to the optical path between lamp (D) and photocell (E). When exposed to carbon monoxide, the chemical reagent changes from its initial translucent straw color to a brownish black stain. The color change in the detecting silica gel is measured photoelectrically in a sensor head containing a zinc cadmium photocell and an exciter lamp (ordinary tungsten filament light bulb). The rugged, molded Zytel® sensor head is a plug-in module and can be remotely located up to 1000 ft from the instrument. The molded chamber in the sensor head which holds the Detectotab is so constructed that heat from the exciter lamp creates a "chimney" effect, drawing air over the reactive chamber surface of the Detectotab. An airflow of 200 cc/min is obtained assuring proper sampling with no maintenance or possibility of pump failure or wear. (Courtesy of Bendix Corp., National Environmental Instruments Division, Lewisburg, W. Va.)

beam intensity (Figure 1, Part 4). When the light intensity is reduced to a preset value by darkening of the translucent silica gel reagent, a relay circuit is tripped, and a visual as well as an audible alarm is activated (Figure 9). The alarm point may be adjusted by means of a potentiometer in the detector circuit and is usually set at the TLV (50 ppm for CO). Under low humidity conditions, the reaction rate of CO with the palladium complex decreases to a level which prevents effective functioning of the alarm. Therefore, a humidifying accessory (water reservoir and wick) is furnished as an integral part of the phototube assembly.

Some idea of the CO concentration can be obtained by calibrating the monitor as a gas titration end-point detector. That is, the time required to reach the alarm point is directly related to the CO concentration. Therefore, calibration can be carried out in atmospheres of known concentration (glove box) by recording the time to reach the alarm point. Then, in field use, as soon as the alarm is activated, a fresh tab is immediately inserted, time until it alarms again is noted, and the approximate CO concentration read off from a calibration chart. Field experience has indicated that a good level of reliability can be attained in this manner (see Section II.A.1.b, "Direct Reading Length of Stain Gas Detecting Tubes").[11]

This monitor is small enough (16.2 × 9.5 × 5.1 cm) to be carried as a personnel monitor and is powered by a 10-V rechargeable nickel-cadmium battery (Figure 9A). The detectortabs regenerate themselves (the black color "bleaches" back to the original orange-brown color) on standing in a CO-free atmosphere (usually overnight) unless overexposed to high CO concentrations. However, after repeated cycles, the regenerative capacity deteriorates to a point of no return when balancing of the optical-electronic system is no longer possible.

Tabs are available also for H_2S detection in the TLV range (down to 5 ppm with Model 40). Tabs could be made up from any one of the granular reagents used in filling the length of stain gas and vapor detector tubes currently available on the open market (see Section II.A.1.b, "Direct Reading Length of Stain Gas Detecting Tubes"). Each of these granular reagents would require individual calibration within the TLV range for the particular contaminant for which the detector tube was designed. The nonspecificity of several of the detector systems imposes limitations on their use. Chromic acid, for example, would respond to most reducing agents, and therefore, would be highly nonspecific. This approach offers an attractive solution involving a minimum expenditure of time and money for development to the alarm problem associated with the nearly instantaneous detection of those hazardous gases and vapors for which ceiling TLVs have been assigned.

A novel design for a colorimetric personnel dosimeter for hydrazine fuel handlers incorporated a filter to protect a light-sensitive reagent adsorbed on a thin layer chromatogram sheet.[12] A No. 8 Wratten® filter was mounted as a window flush with the upturned edges (6 mm) of the badge at a distance sufficiently above the hydrazine-sensitive plate to permit free circulation of air through the annular space (approximately 4 mm). The filter absorbed the destructive actinic light while allowing the color development to be observed. The window could be raised for direct comparison of the developed color with the three color standards that surrounded the reagent area. The sensitive plate (14 × 41 mm) was cut from an Eastman® chromatogram sheet type K301R2 (a 100-μ layer of silica gel bonded to a film base with polyvinyl alcohol binder) that had been impregnated with bindone and dried at 65°C. The strip of sensitive material was held in place with springs so that an 8 × 14-mm rectangle was exposed, and an equivalent area was held tightly against a clear plastic (polyacrylate) retainer plate to serve as an unexposed blank section for visual detection of very faint stains. The color change from off-white to a purple provided a very distinct and sensitive response to hydrazine and methyl hydrazine exposure in very low concentration ranges. However, the detector will respond to most alkaline-reacting, volatile bases, including cigarette smoke.

Corrosion and darkening of metallic silver by H_2S have been suggested for ambient air quality measurements.[13] Air contaminated with H_2S or mercaptans on contact forms silver sulfide on the surface of the silver in the presence of oxygen and moisture. The decrease in reflectance of the silver surface as determined in a reflectance meter (Photovolt, Model 610)[14] is proportional to the H_2S-mercaptan dose. When air is drawn through the filter, a cellulose membrane filter is inserted upstream to prevent darkening and light scattering from dust, which would otherwise collect on the filter. This method is specific, is not sensitive to the usual interferences (O_3, SO_2, NO_x, etc.), and avoids the inaccuracy due to the fading of PbS stains on exposure to light in the lead acetate procedures.[15]

A hydrogen sulfide dosimeter, which incorporates an infrared optoelectric system that continuously scans a paper card impregnated with lead acetate for staining by H_2S, has been offered commercially. The manufacturer claims the degree of staining is time-weighted in the same way as H_2S affects the human body (Table 7, No. 17).

3. Portable Instruments — Electronic

a. *Memory Cell*

Probably the most significant breakthrough in the design of portable instruments of a size compatible with personnel monitoring was the addition of a gas dosimeter cell which enables the instrument to record the accumulated gas concentrations (Table 7, No. 10). At the end of the monitoring (sampling) period the totalized exposure is read-out in a digital computer. This innovation provides time weighted average (TWA) as well as instantaneous (within the response time of the detector system) read-out as total accumulated dose — TWA. The electrolytic cell probably contains two silver electrodes sealed in an electrolyte. When a current derived from the sensor passes through the memory cell, silver is dissolved from one electrode and deposited on the other electrode. In the read-out circuit, the electrolytic plating is reversed, and the current consumed computed in terms of the quantity of pollutant encountered.

A miniature computer-based personal alarm dosimeter monitoring system which provides a unitized sensor, microprocessor chip to store electrical impulses, and a bi-modal read-out system was shown for the first time in the U.S. at the American Industrial Hygiene Conference in Chicago, Ill. on May 29, 1979. The microprocessor-controlled data logging dosimeter is worn on the employee's belt and is connected through a six-wire cable to the sensor unit. The schematic arrangement of the unit is shown in Figure 9A. For those monitors such as the continuous tape sampler ("Chromatox") the miniature battery operated pump also is enclosed in the belt pack. The information stored in the belt pack (microprocessor "memory" unit) is read out in a base console, which provides not only the total exposure dose (TWA), but also a profile of the concentration of pollutant gas vs. time on semi-log chart paper. Battery voltage is also included to confirm the electrical integrity of the system. A chlorine "Datogram" sample sheet is shown in Figure 9B. Sensor units for this system include the "Monitox"

TABLE 7

Instruments for the Detection of Oxygen, Combustible Gases and Toxic Gases by Diffusion (Passive) Sampling

Manufacturer or vendor	Model number	Dimensions L × W × D (cm)	Weight kg	Gas or vapor detected	Range (ppm)	Principal of operation
Bacharach[a]	51—7026[b-f]	11.4 × 8.4 × 4.1	0.54	Combustible	0—100% LEL	Hot Pt filament
	Sentox 2[b-d,f]	5.1 × 6.4 × 12.7	0.41	Oxygen	0—40%	Galvanic cell
	Sentox[d,f]	26.6 × 17.1 × 16.5	3.2	H_2S	3—50	MOS-conductor
	51-702[d-f]	11.4 × 8.4 × 4.1	0.54	H_2S	—	MOS-conductor
Biomarine[g]	911[d-f]	8.3 × 14.6 × 5	0.54	Combustible 0—4% CH_4	0—100% LEL	Pt-ceramic bead
	OA222R[d,e,h]	5.1 × 6.4 × 12.7	0.41	Oxygen	0—40%	Galvanic cell
	OM322	5.1 × 6.4 × 12.7	0.41	Oxygen	0—40%	Galvanic cell
	900, 922[b-f]	8.3 × 17.8 × 5	0.68	Oxygen		Galvanic cell
				Combustible		Pt ceramic bead
	500	8.2 × 17.8 × 5	0.68	CO	0—500	
	700	8.3 × 17.8 × 5	0.68	H_2S		?
Chemtrix[i]	30	—	1.37	Oxygen	0—15 ppm	Galvanic cell
Draeger[j]	25D[d,e]	—	—	Oxygen	0—25%	Galvanic cell
Dynamation Combo[k]	75-300[b-f,l]	20.4 × 10.2 × 7	0.85	Oxygen	Alarm—19.5%	Galvanic cell
				Combustible	Alarm—20% LEL	MOS-conductor
				H_2S	Alarm—10	MOS-conductor
	H_2S—8[d-f]	15.1 × 8.2 × 5.1	0.68	H_2S	0—50	MOS-conductor
	CO—8[d-f,l]	15.1 × 8.2 × 5.1	0.68	CO	0—300	MOS-conductor
Dictaphone life-guard[m]	H_2S Detector[d]	7.6 × 15.2 × 3.1	0.40	H_2S	1—100	Electrochemical
Enmet[n]	CGS-8[b-f]	12 × 9.5 × 6.3	0.62	Combustible	0—100% LEL	MOS-conductor
				CO	50—500/TLV	
Enmet[n]	CGS-10[d-f]	12 × 9.5 × 6.3	0.77	Oxygen	18—19.5%	Galvanic cell
				Combustible	0—100% LEL	MOS-conductor
Enmet-Cont[n]	OMA-10	12 × 9.5 × 6.3	0.56	Oxygen	0—25%	Galvanic cell
				CO	50—100	
Tritector[n]	CGS-10[b-f,l]	12 × 9.5 × 6.3	0.77	H_2S	10—100	MOS-conductor

Edmont[o]	60-620 to 60-626	8 × 13 × 13	0.45	xygen	0—25%	Galvanic cell (polarographic)
MDA[p]	60-400[d-f]	30.5 (diameter) × 7.6	3.6	Combustible	0—100% LEL	Galvanic
				Oxygen	0—25%	Hot Pt filament
Gas-Pro	LP-BGA-g[d,r,e]	76 × 51 × 151	0.46	Combustible	0—100% LEL	MOS-conductor
Lumidor[q]				CO	100	
Gas Tech[s]	1177	30 × 9.5 × 14	2.27	Combustible	0—100% LEL	Hot Pt filament
	Protector[b-d,f]	9.7 × 6.4 × 20	0.97	Combustible	0—100% LEL	Hot Pt filament
	Gastechtor[b-d,f]	30 × 9.5 × 14	2.27	Oxygen	0—25%	Galvanic cell
				Combustible	0—100% LEL	Hot Pt filament
Gen Elect.[t]	Area dosimeter	12.7 × 10.2 × 5.1	0.64	CO	0—250	Meter readout TWA coulometer
IST[u]	2500[v,c,d]	10.3, 13.5, 18.9	2.5	Chlorine	0—10, 20, 50	MOS-conductor
	2900-NH[c,d,v]	10.3 × 13.5 × 18.9	2.5	Ammonia	0—200	MOS-conductor
	2400[c,d,v]	10.3 × 13.5 × 18.9	2.5	SO_2	0—5, 0—25	MOS-conductor
	2100[c,d,v]	10.3 × 13.5 × 18.9	2.5	H_2S	0—500, 0—100	MOS-conductor
	2800[c,d]	10.3 × 13.5 × 18.9	2.5	Acrylonitrile	0—200	MOS-conductor
	AG5100	7.6 × 5.1 × 5.2	0.8	Hydrogen	0—100	MOS-conductor
	AG6000[d,e]	7.6 × 5.1 × 5.2	0.8	Combustible	0—100 % LEL	MOS-conductor + semi-permeable membrane
				H_2S	0—200	
Kitagawa[w]		11 × 6 × 3	0.34	Hexane	500—1000	Hot Pt filament
Sipin[w]	FPA40-D			Isobutane	800	
MDA[p] Dosimeter	777[d,e,x]	12.1 × 7.0 × 3.5	0.28	CO	100, 200	Electrochem. oxid.
MSA[y]	100[b,d,f]	25.4 × 17.9 × 9.5	3.2	Combustible	0—100% LEL	Hot Pt filament
	250[b,f]	19 × 13.9 × 8.1	—	Combustible	0—100% LEL	Hot Pt filament
Biomarine[g]				Oxygen	0—25%	Galvanic cell
Biomarine[g]	244, 244R	5.1 × 6.4 × 12.7	0.31	Oxygen	0—25%	Galvanic cell
NEI-Bendix[z]	Micro-Methano Meter-800	800	0.26	Methane	0—5%	?
		3.6 × 6 × 4.7				
	Guardian[d,f] 2450 series	14.6 × 5.7 × 10.8	—	Combustible	1/3 LEL	Hot Pt filament
National Mine Safety[aa]	COMonitor[d-f]	5.1 × 2.5 × 12.7	0.34	CO	100, 200, 300	MOS-semiselect cat. combustion

TABLE 7 (continued)

Instruments for the Detection of Oxygen, Combustible Gases and Toxic Gases by Diffusion (Passive) Sampling

Manufacturer or vendor	Model number	Dimensions L × W × D (cm)	Weight kg	Gas or vapor detected	Range (ppm)	Principle of operation
Neutronics[bb]	Otox-90-91[c,d]	17.8 × 21.6 × 5.1	1.36	Oxygen	0.1—35%	Galvanic cell
	H_2S dosimeter	7.6 × 20.3 × 5.1	1.36	H_2S	10—150	Colorimetrictape
	Otox-80[d,e]					IR sensor
	AGM Series[d,e]	7.6 × 20.3 × 5.1	1.36	Combustible gases	below LEL	Catalytic
Photo-ionization[cc]	PHO1	24 × 21 × 13	4.1	See table in the reference/calib-benzene	0.1—600	UV ionization
	PI-101[d,f,h,i]	24 × 21 × 13	4.1	See table in the reference	0.1—600	UV ionization
Texas Analytical Controls[dd]	701[d,e]	12 × 8.9 × 5.7	0.68	H_2S	0—100, 0—50, 0—25	MOS-conductor
IST[u]	AG 6000[c,r,3,x,l]	7.6 × 5.1 × 5.2	0.8	Combustible	0—100% LEL	MOS-conductor + semipermeable membrane
				H_2S	0—200	
	AG 5100	7.6 × 5.1 × 5.2	0.8	Hydrogen	0—100	
				CO		
				Hydrocarbons	0—1000	
				Ammonia	0—400	
				Chlorine	0—50	

Note: Dimensions = L × W × D = Length × Width × Depth (Wide); kg = kilograms; ppm = parts per million as volume per volume (V/V); LEL = lower explosive limit; Pt = Platinum; MOS = metallic oxide semi-conductor; UV = ultraviolet light

[a] Bacharach Instrument Co., Division of AMBAC Industries, Inc., 625 Alpha Drive, Pittsburgh, Pa.
[b] Also available in the dynamic mode-squeeze bulb.
[c] Also available in the dynamic mode-battery operated pump.
[d] Built-in alarm.
[e] Personnel monitor type.
[f] Calibration gas kit available.

[g] Biomarine Industries, Inc., 45 Great Valley Center, Malvern, Pa.
[h] Chart recorder available.
[i] Chemtrix, Inc., Giangarlo Scientific Co., 3237 Dawson Street, Pittsburgh, Pa.
[j] Schinkman, M., Some aspects of measuring the oxygen concentration in gas mixtures — using fuel cell sensors, *Draeger Rev.*, 39, 34, 1977; see also National Draeger, Inc., Parkway View Dr., Pittsburgh, Pa.
[k] Dynamation Gas Detection Instruments, P.O. Box 225, Ann Arbor, Mich.
[l] Also available for other toxic gases on the TLV list.
[m] Gas Detection Products, Dictaphone, Audio/Electronics Division, 475 Ellis Street, Mountain View, Calif.
[n] Enmet, 2308 Industrial Highway, Ann Arbor, Mich.
[o] Edmont-Wilson, Division of Becton, Dickinson and Co., Cochocton, Ohio, U.S. Patent No. 3,666,650.
[p] MDA Scientific, Inc., 808 Busse Highway, Park Ridge, Ill.
[q] Lumidor Products Corp., Safety Division, 5364 N. W. 167th St., Miami, Fla.
[r] Calibrated with methane (CH_4).
[s] Gas Tech, Inc., Johnston Instrument Division, 331 Fairchild Dr., Mountain View, Calif.
[t] Carbon Monoxide Detectors, General Electric, Aircraft Equipment Division, 50 Fordham Rd., Wilmington, Mass.
[u] International Sensor Technology, 3201 South Halladay St., Santa Ana, Calif.
[v] Formerly E. D. Bullard Company (1977).
[w] "Sipin", Anatole J. Sipin Co., Inc., 425 Park Ave. S., New York, N.Y.
[x] Gas dosimeter cell records and accumulated Time Weighted Average (TWA).
[y] Mine Safety Appliances Co., Inc., 400 Penn Center Blvd., Pittsburgh, Pa.
[z] Micro-Methanometer, Model 800, Bendix Environmental and Process Instruments Division, Lewisberg, W. Va.
[aa] National Mine Service Company, 3000 Koppers Bldg., Pittsburgh, Pa.
[bb] Oxygen, Hydrogen Sulfide and Flammable Vapor Analyzers, Neutronics, Inc., 450 Drew Ct., King of Prussia, Pa.
[cc] HNU Systems, Inc., 383 Elliot St., Newton Upper Falls, Ma; Representative: Giangarlo Scientific Co., 3237 Dawson Street, Pittsburgh, Pa.
[dd] Texas Analytical Controls, 7707 Bissonnet, P.O. Box 42183, Houston, Tx. 77042-Phone (713) 777-4062

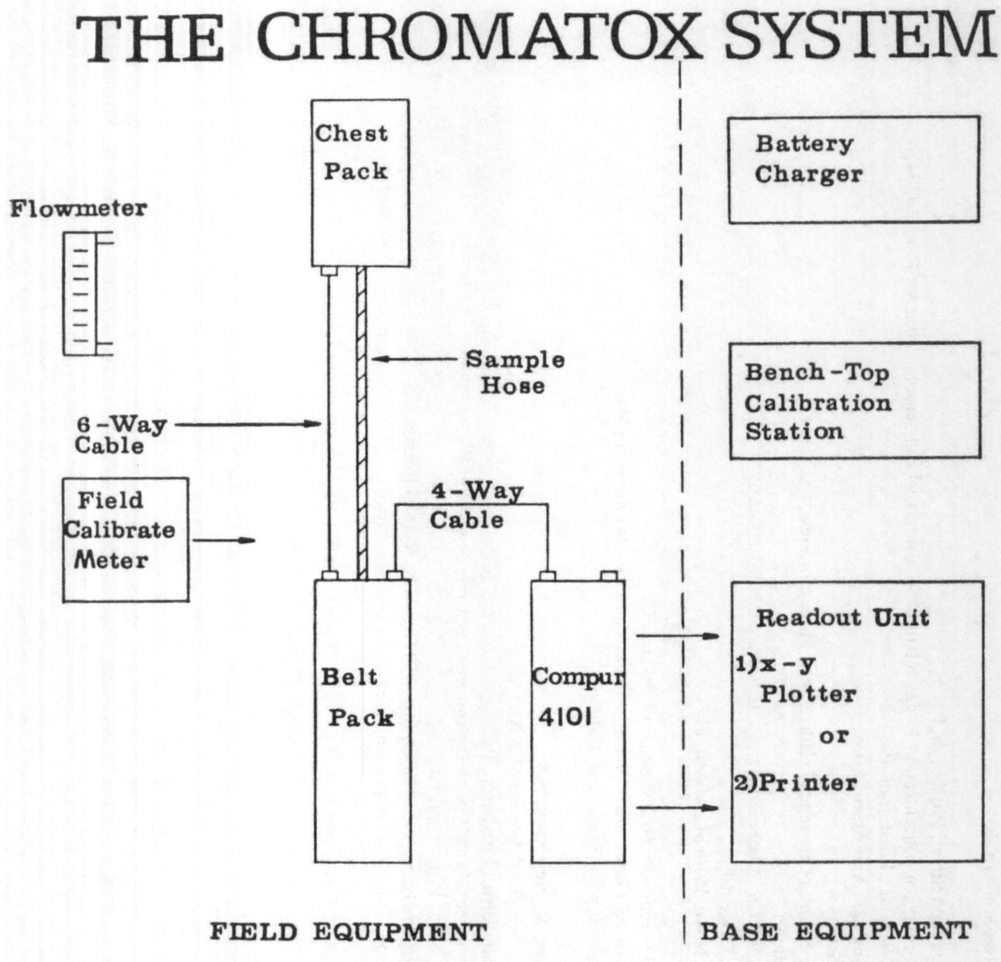

FIGURE 9A. The MDA "chromatox" system with the "chromatox" sensor in place. (Courtesy of MDA scientific, Ina., 808 Busse Hwy., Park Ridge, Ill.)

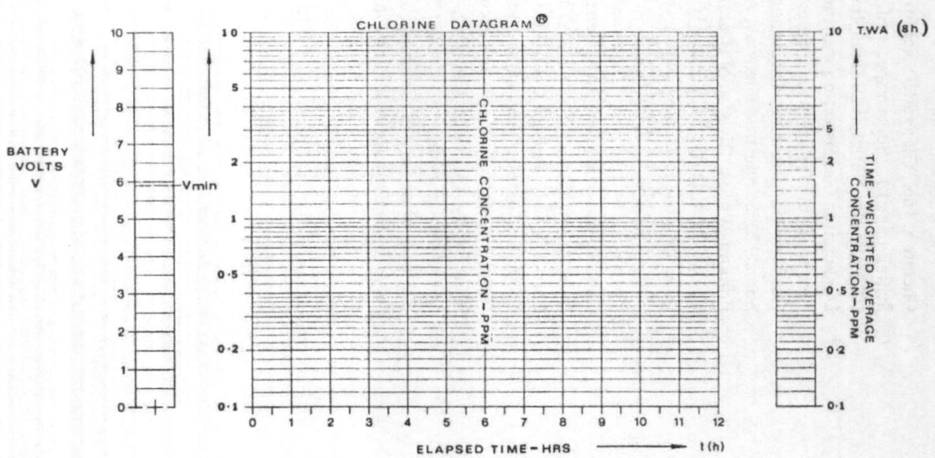

FIGURE 9B. The MDA "chromatox" system "Datagram" read-out chart form. (Courtesy of MDA Scientific, Inc., 808 Bussee Hwy., Park Ridge, Ill.)

fuel-cell monitor, carbon monoxide detector, and the "Chromatox" tape personal monitor.

The sophistication of the microprocessor-controlled data acquisition system is such that a complete profile of exposure together with total dose can be extracted from the output of any sensor. Interchargeable PROMS (programmable read only memory) provide the flexibility of programming to handle a wide variety of input signals and produce data in a common format. The sophisticated software back up available includes linearization, data reduction, and communication with the sensor. Hence, the system can be updated as new sensors or monitoring techniques become available so the system will not become obsolete in the future. In addition to providing accumulated data required for governmental compliance, the Chromatox systems also provide concentration and total dose alarms which are visible and audible at the point of measurement to ensure the worker's safety.

b. Electrical Conductance Types (Resistance Circuits)

i. Hot Platinum Filament — Catalytic Combustion

Although the combustible gas meter is not strictly an industrial hygiene personnel monitor, these instruments are so widely used that some mention should be made in passing. In some cases these meters have been calibrated in parts-per-million (ppm) range by increasing the sensitivity of the electrical circuits that respond to very low voltage changes (see Table 7). However, the user should be aware that, regardless of the calibration, these meters will respond to any gas or vapor which will burn in the presence of atmospheric oxygen.

The classical hot filament principle has been retained in many currently available instruments. All operate on the balanced Wheatstone electrical bridge circuit in which two opposing arms are platinum filaments. One of the filaments heated to a controlled temperature is exposed to the air sample. Any combustible gas present will be catalytically oxidized (burned), which changes the temperature and, consequently, the electrical resistance of the filament. This change in resistance unbalances the bridge and is measured by an indicating meter whose scale provides a read-out in terms of the percentage of lower explosive limit of the flammable gas in air, or in terms of parts per million of the gas for which the instrument was calibrated. Certain limitations must be observed. For methane, the temperature of the filament must be increased; the platinum filament can be poisoned by lead compounds or destroyed by exposure to chlorinated hydrocarbons (carbon tetrachloride, chloroform, etc.). The second platinum coil, which is sealed from contact with any combustible gas, serves as the adjacent leg of the Wheatstone bridge and compensates for variations in ambient temperature and voltage fluctuations.

Recent improvements include imbedding the platinum coil in a ceramic bead to protect the filament from poisoning or damage from contact with corrosive materials (Figure 10). One bead is coated with a catalyst which initiates combustion on the surface of the bead. The sensor filament is connected in series with the second, uncoated, reference bead which operates at the same voltage while no combustible gas is present in the sampled air. As oxidation of combustible gas on the catalyst-coated filament bead raises the temperature and thereby increases its resistance, the voltage applied to the two filaments in series divides so that a greater voltage drop that occurs across the sensor unbalances the Wheatstone bridge circuit. The bridge circuit output voltage taken across connections between the two filaments in series and between two parallel connected equivalent resistors in series with each other then will be proportional to

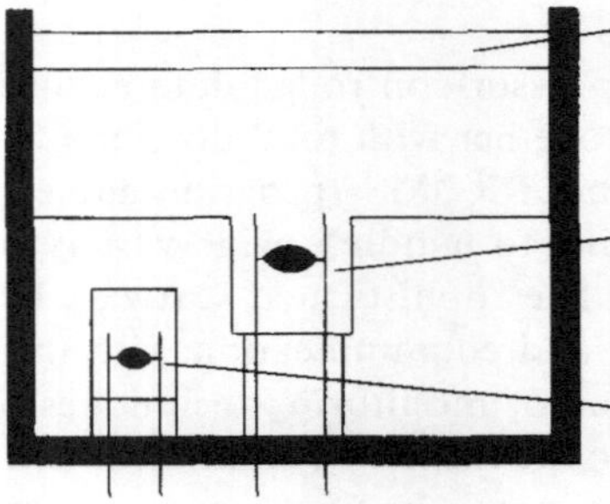

FIGURE 10. Stable combustible gas sensor. (Courtesy of Biomarine Industries, Inc., 45 Great Valley Center Malvern, Pa.)

the concentration of combustible gas in contact with the coated bead. The uncoated bead compensates for changes in ambient temperature and humidity.

Since the heat of combustion varies from one compound to another, the combustible gas meter should be calibrated in terms of the compound for which survey information is required. Figure 11 illustrates the magnitude of differences involved.

The resulting signal (voltage drop) can be amplified and compared to a zero-level reference. When the compared voltages from signal and reference circuits reach a predetermined potential difference, a signal output switches on an oscillator and speaker circuit to produce an audible or visible warning. The greater the potential difference, the higher the frequency of the warning signal; thus the pitch of the speaker response is relative to the percent LEL of gas at the sensor. Typical circuit diagrams are presented in Figure 12.

Representative models available from commercial sources are shown in Figures 13 and 14.[16]

A novel innovation is in the circuitry of the catalytic combustible vapor meter. In this model, the read-out, visible alarm and audio-alarm are energized for 10 sec during each four minutes of detector operation. The supplier claims that this "pulsed mode" insures long term stability of the sensor and completely maintenance free life without recalibration for up to 5 years under "normal" environmental conditions. Battery-powered requirements also are "drastically" reduced. Remote sample heads are available as an accessory (Table 7, Reference bb).

ii. Metallic Oxide Semi-conductor
(See Table 7)

A technological breakthrough in gas detection announced in the first edition of *Evaluation of Ambient Air Quality by Personnel Monitoring* has been promoted and developed into a number of commercially available instruments (Table 7).

The gas-detecting device uses a mixed metal oxide semi-conductor element chosen from the n-type metal and non-metal oxides of the transition Groups III and IV (iron, zinc, and tin families) elements of the periodic table. The electrical resistance of the sintered bead (see Figure 15) varies when exposed to gases such as hydrogen, hydrocarbons (methane, gasoline vapor, etc.), carbon monoxide, alcohols, ammonia, chlorine, sulfur dioxide and hydrogen sulfide (see Tables 7 to 9). The gas molecules are dissociated into charged ions or complexes of ions which are collected as an electrical signal. A unique capability of this system is the detection of very low concentrations of toxic

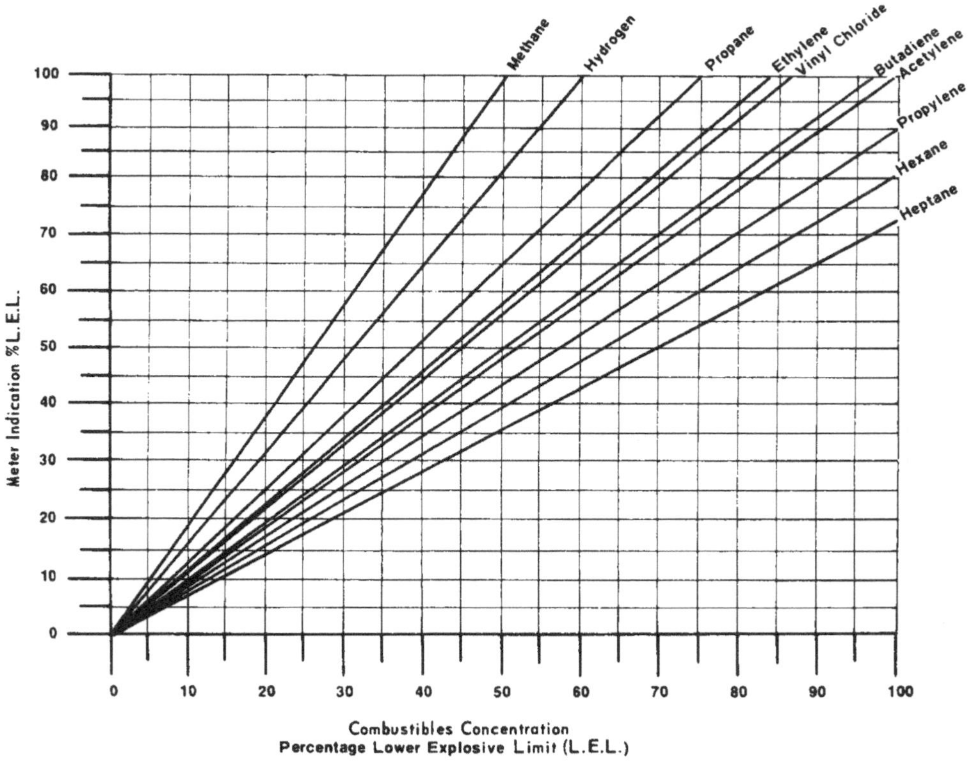

FIGURE 11. Chart for determining percent-of-lower-explosive-limit concentrations for specific gases. (Courtesy of Bacharach Instrument Co., Division of AMBAC Industries, Inc., 625 Alpha Dr., Pittsburgh, Pa.)

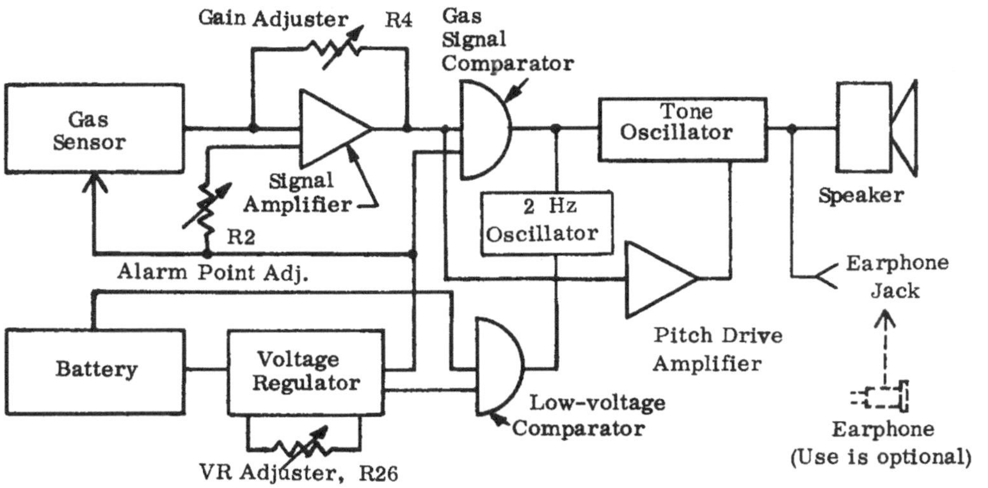

FIGURE 12. Simplified diagram of combustible gas detector showing functional relationships of major components. (Courtesy of Bacharach Instrument Co., Division of AMBAC Industries, Inc., 625 Alpha Dr., Pittsburgh, Pa.)

FIGURE 13. Bacharach combustible gas personal monitor (51-7026). (From Bacharach Instrument Co., Division of AMBAC Industries, Inc., 625 Alpha Dr., Pittsburgh, Pa., With permission.)

or flammable gases in the 0 to 10 ppm range as well as the higher concentrations in the explosive vapor ranges up to 200% of the lower explosive range (% LEL) (Table 7, Reference n and u).

Two electrodes are embedded in the solid-state sensor cell. A constant voltage is applied and then the current flow is measured in terms of gas concentration presented to the detector cell. The system is stable within a wide range if temperature and humidity, is unaffected by vibration or position, and operates on low current and voltage. Most instruments include an alarm circuit which responds to a predetermined reduction in resistivity of the semiconductor unit. Since the unit does not operate on the combustion principle, the sensor will not "burn out" in high gas concentrations. In addition, the instrument will detect gases in the absence of oxygen (Table 7, No. 10).

Both the manufacturer and the user must recognize the limitations of this system, which is not completely selective for any one particular gas (Table 10). Consider the application to carbon monoxide (CO). The physical and chemical properties are employed to detect its presence. However, other gases also have the same or similar properties and will interfere with the instrument response. The interference normally follows two courses. In proportionial interference, for every X ppm of interference gas, the instrument will show Y ppm in terms of gas for which the meter was calibrated. For example, 100 ppm of an interference Gas X may read 10 ppm as Gas Y. In tolerance interference, the instrument will not respond to the interference Gas X until a critical concentration is reached; then a large response in terms of Gas Y will occur. In the example of 100 ppm of Gas X, the instrument will not respond until this concentration is exceeded; then a high reading in terms of Gas Y will be obtained. In some cases, a "filter" can be inserted to reduce the interference effects.

A heater winding imbedded in the oxide semiconductor bead is powered by direct current at 0.75 to 1.75 V (Figure 15) to raise the MOS surface to operating temperature

FIGURE 14. Sipin-Kitagawa Model FPA40-D personal combustible gas alarm. (Courtesy of Anatole J. Sipin Co., Inc., 425 Park Ave. S., New York, N.Y.)

in the 150 to 300°C range. At the higher surface temperature (heater voltage equals 1.25 to 1.75 V), the conductivity measured between the metallic heater coils increases remarkably on exposure to dangerous levels of hydrocarbon gases and vapors. A 10 to 20-fold decrease in electrical resistance occurs from contact with several thousand ppm of explosive gas molecules. Furthermore, for certain sensor metallic oxide mixtures at this higher temperature, CO response is greatly reduced (Figure 16). When the heater power is reduced to the 0.75 to 0.80 V range, the semiconductor becomes highly specific to CO with greatly reduced response to hydrocarbons and virtually no response from methane, even in the several thousand ppm range (Table 7, Reference n).

Because of the elevated signal levels generated directly from this sensor, bridge circuits are not required for simple portable signaling devices. A single electronic stage of isolation and amplification can be used to trigger a variety of intrinsically safe audio and visual alarms. The influence of interferences — CO vs. methane — is further illustrated in Figure 17. Note that the device is failsafe to CO in the combustible mode. That is, dangerously high concentrations of CO will produce a "false" response. However, cool sensor operation in the "toxic" mode at 50 ppm CO will not produce a

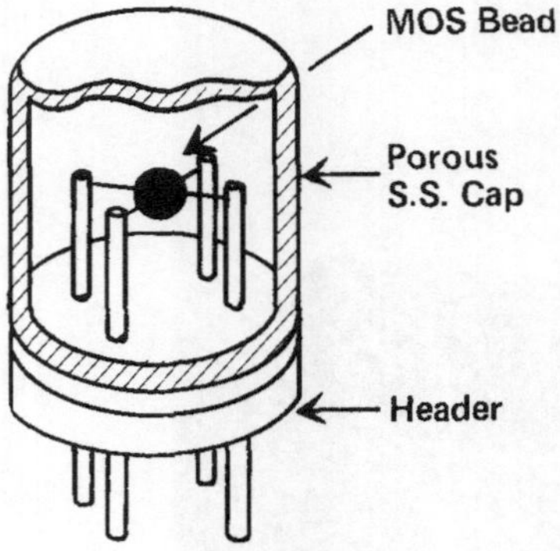

FIGURE 15. Metallic oxide semiconductor (MOS) gas sensor. Specifications: Mechanical—size, 1.8 cm diameter × 2.5 cm high; weight approximately 35 g; electrical: heater resistance, 2 ohms — nominal; heater voltage range, 0.75—1.75; bulk resistance, 2—50k ohms operating range. (Courtesy of Enmet, 2308 S. Industrial Hwy., Ann Arbor, Mich.)

TABLE 8

Metallic Oxide Semiconductor Calibration Gases Operating Ranges

Gases	Range
Acetone	0—200, 500, 1000 ppm and %LEL
Acetylene	%LEL and ppm
Allyl alcohol	0—50, 200 ppm and %LEL
Allyl chloride	0—100 ppm and %LEL
Ammonia	0—75, 100, 150, 200, 400, 1000 ppm and %LEL
Vinyl cyanide or acrylonitrile	0—60, 200 ppm and %LEL
Benzene	0—75 ppm and %LEL
Butadiene	%LEL and ppm
Butane	0—1000 ppm and %LEL
Butanol	0—100 ppm and %LEL
Carbon disulfide	0—60 ppm
Carbon monoxide	0—100, 150, 100, 1000, 3000 ppm and %LEL
Chlorine	0—10, 20, 50 ppm and %LEL
Dichloroethane	%LEL and ppm
Epichlorohydrin	50, 100, 200, 1000 ppm and %LEL
Ethyl acetate	%LEL and ppm
Ethyl benzene	200 ppm and %LEL
Ethylene	%LEL and ppm
Ethylene oxide	0—50, 150, 200, 300, 1500 ppm and %LEL
Freon-22	100, 1000 ppm and %LEL
Hexane	200 ppm and %LEL
Hydrogen	100, 1000 ppm and %LEL
Hydrogen sulfide	10, 20, 30, 50, 100 ppm and %LEL (specify type A or B sensor)
Hydrogen cyanide	50, 100, 200 ppm and %LEL
JP-4	%LEL and ppm
Methane	100 ppm and %LEL
Methanol	200 ppm and %LEL

Methyl bromide	20, 100 ppm and %LEL
Methyl chloride	200, 300 ppm and %LEL
Naphtha	%LEL and ppm
Propane	1000 ppm and %LEL
Propylene	100 ppm and %LEL
Tetrachloroethylene	200 ppm and %LEL
Styrene	200 ppm and %LEL
Tetrahydrofuran	%LEL and ppm
Toluene	200, 500 ppm and %LEL
Vinyl chloride	20, 100, 400 ppm and %LEL
Xylene	ppm range to 1%

Note: Gases listed above are subject to interference by other gases.

TABLE 9

Gases and Vapors Detected and Calibrated with ENMET Products[a-b]

Hydrocarbons and their derivatives	Methane, ethane, propane, butane, pentane, hexane, heptane, octane, decane, petroleum ether, petroleum benzene, gasoline, kerosene, petroleum naphtha, acetylene, ethylene, propylene, butadiene, butylene, benzene, toluene, *o*-xylene, *m*-xylene, ethylene oxide
Halogenized hydrocarbons	Methyl chloride, methylene chloride, ethyl chloride, ethylene chloride, ethylidene chloride, trichloro ethane, vinylidene chloride, trichloroethylene, methyl bromide, vinyl chloride, Freons
Alcohols	Methanol, ethanol, *n*-propanol, iso-propanol, *n*-butanol, iso-butanol
Ethers	Methyl ether, ethyl ether
Ketones	Acetone, methyl ethyl ketone, methyl isobutyly ketone
Esters	Acetate, ethyl acetate, *n*-propyl acetate, iso-propyl acetate, *n*-butyl acetate, iso-butyl acetate
Nitrogen compounds	Nitro methane, mono methyl amine, dimethylamine, trimethyl amine, mono ethyl amine, diethyl amine
Inorganic gases	Ammonia, carbon monoxide, hydrogen, hydrogen cyanide, hydrogen sulfide
Miscellaneous	Jet fuel ("A", JP4), pesticides, styrene, chlorine (contact ENMET)

Gases and Vapors NOT Detected by ENMET Products

Inert gases	Nitrogen, helium, neon, argon
Miscellaneous	Carbon dioxide, nitrogen dioxide, sulfur dioxide

[a] ENMET, 2308 Industrial Highway, Ann Arbor, Mich.

[b] Typical calibrations can range from 10—40% LEL for combustible and 20—100 ppm for toxic gases.

TABLE 10

Typical Interference Data — Hydrogen Detector (IST)[a]

Gas	Without filter	With filter
Methane	500:1[b]	20,000:1
Carbon monoxide	20:1	25:1
Butane	—	50,000:1
Methanol	30:1	10,000+:1
Benzene	30:1	10,000+:1
Acetylene	2:1	200:1
Ammonia	30:1	10,000+:1

[a] International Sensor Technology, 3201 S. Halladay St., Santa Ana, Calif.

[b] 0—100 ppm H_2.

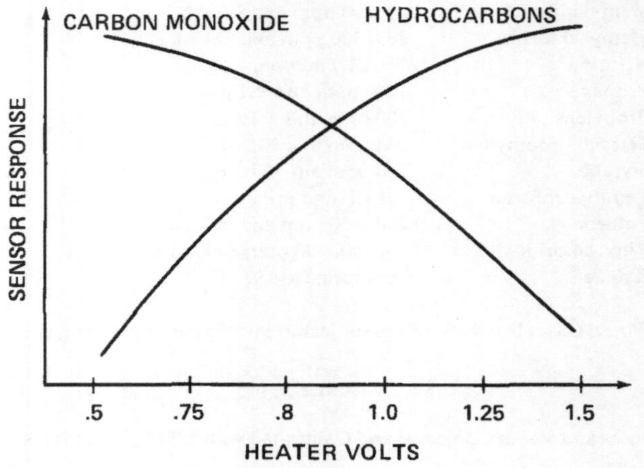

FIGURE 16. Metallic oxide semiconductor (MOS) gas response vs. heater power. (Courtesy of Enmet, 2308 S. Industrial Hwy., Ann Arbor, Mich.)

methane "false" alarm even when CH_4 concentration rises well into the explosion hazard range (Table 7, Reference n).

The electronics associated with the MOS sensor are well compensated over the ambient operating range 0 to 65°C. Increasing temperature produces a slight increase in the MOS electronics response to both CO and CH_4. In the absence of more refined thermal conpensation circuits, the instrument has a ± 10% of full scale alarm capability (Figure 18).

c. Galvanic Cell (Redox type) — "Fuel Cell"

i Oxygen

Although not strictly classified as hazardous gas or vapor detectors, some mention should be made of the pocket-size oxygen deficiency meters. The threat to life for workmen who must enter locations such as sewers, mine tunnels and shafts, silos, wells, fermentation vessels, storage tanks, chemical process equipment areas, oil refineries, etc., where oxygen may be reduced below respirable levels by bacterial action or spontaneous oxidation (coal mines), can be more serious than contact with toxic gases. Without the assistance of a reliable instrument, man's senses cannot provide a warning in time to avoid the fatal consequences incurred by entering an oxygen-deficient atmosphere.

The model developed by Bio-Marine Industries incorporates most of the desirable features that a personnel monitor should furnish: it displays a continuous percent O_2 readout; provides audible as well as visual alarm when O_2 concentration falls below 20%; is of small size (5.1 × 6.4 × 12.7 cm) and light weight (294 g); has rapid response (90% in less than 20 sec), high accuracy (± 1% of full scale), and negligible temperature compensation error (maximum ± 5% in the range 0 to 40°C); is not affected by humidity or CO_2; is rugged, remote sensing as well as self-contained; requires no warm-up; requires no battery or other electrical supply (two small batteries necessary for audible alarm model); and provides safe operation in explosion hazard areas (intrinsically safe per ISA-S.P.12.2 — Factory Mutual approval). Although the sensing element is expendable, replacement after 6,000 hr at 20% O_2 exposure (warranty) poses no problems (Figure 19).

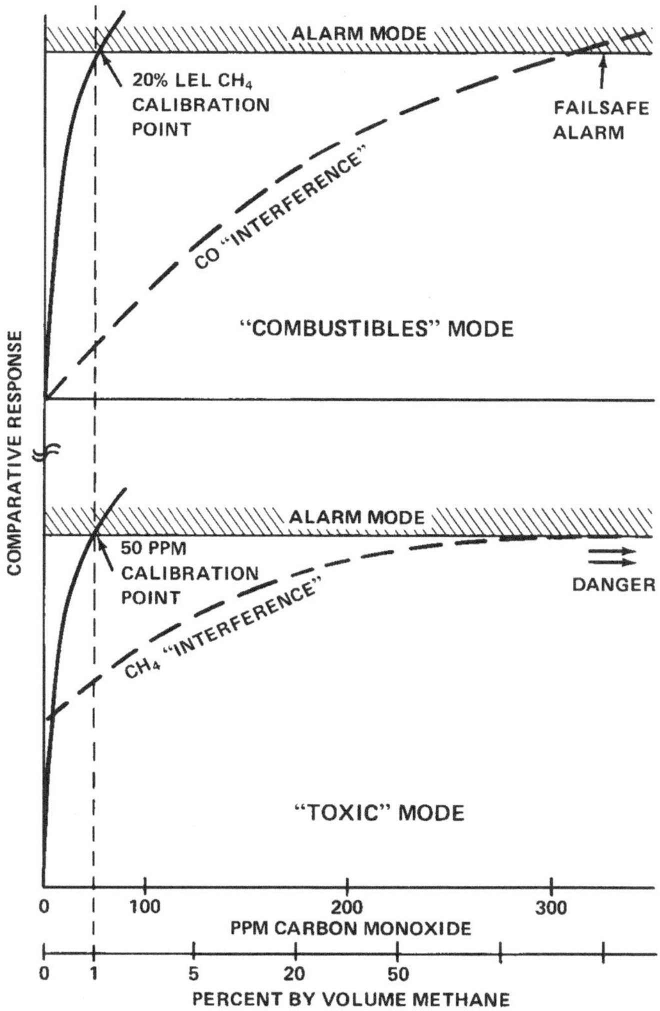

FIGURE 17. Enmet metallic oxide semiconductor (MOS) gas interference data. (Courtesy of Enmet, 2308 S. Industrial Hwy., Ann Arbor, Mich.)

The principle of operation is derived from the galvanic action of oxygen in a cell composed of a gold anode and a lead cathode in an alkaline electrolyte. The entire cell is encapsulated in an inert plastic with a fluorocarbon polymer face that permits O_2 to diffuse into the electrolyte. The redox reaction at the lead cathode generates a minute current proportional to the O_2 partial pressure. A temperature-compensated circuit converts the current to a proportional voltage that is displayed directly on a meter as oxygen concentration. In another model, the signal is amplified through a battery-powered circuit (Figure 20). From the standpoint of compactness, light weight, ruggedness, continuous readout, calibration simplicity, and reliable operation with a minimum of attention, this instrument would be classified as an ideal personnel monitor.

The electrochemical cell contents are enclosed between two polytelrafluorothylene (PTFE) membranes (Fig. 20 and Table 7, No. 2). The external membrane is fabricated from 30-μm thick PTFE sheet, which not only retains the electrolyte, but also selec-

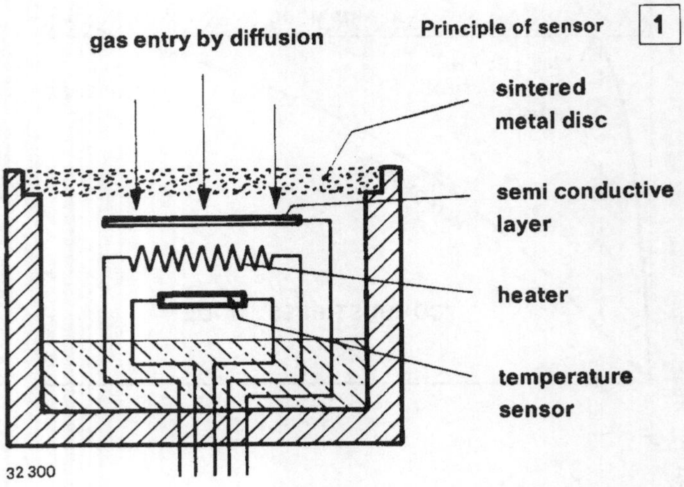

FIGURE 18. The Dräger-General Motors hydrogen sulfide monitor. Measuring principle: the measuring element of the DRAGER-General motors hydrogen sulfide monitor Model 2100 is located in a measuring head, which is closed by a sintered metal disc. The air or gas mixture to be monitored diffuses in through this disc, so that the measuring head can be fitted directly at the point to be monitored and no suction line and pump is necessary. The actual measuring element consists of a ceramic basic body, in which a layer of carbon is embedded as a heating element. The thin semiconductor layer, which is specific to hydrogen sulfide, is applied on this carbon layer. For satisfactory performance, the temperature of the semiconductor should be held constant. Therefore a temperature sensor is fitted to the measuring element to regulate the heating of the control unit. In this way the measuring element will operate satisfactorily over a wide range of ambient temperature. (The manufacturer is General Motors Inc., Costa Mesa, Calif.)[265]

tively permits diffusion of oxygen molecules into the electrolyte and rejects other molecules such as CO_2. The selectivity of this membrane is an essential feature of high-quality sensors.

The electric voltage of 800 mV between the electrodes can be determined by application of Nernst's law. The cathode is gold, the anode is lead, and the electrolyte is aqueous potassium hydroxide (KOH). The electrode reactions are characterized by the following electrode reactions:

$$\text{Cathode:}\quad O_2 + 2H_2O + 4\bar{e} \rightarrow 4OH^-$$

$$\text{Anode:}\quad 4OH^- + 2Pb \rightarrow 2PbO + 2H_2O + 4\bar{e}$$

The O_2 in the gas sample diffuses through the PTFE membrane into the electrolyte. The O_2 reacts on the gold cathode to produce a positive charge. The formation of PbO and water produces a negative charge on the anode. This reaction produces an electrical current which is proportional to the O_2 concentration between the electrodes. The average response time is 10 sec with a ± 1% accuracy. In this reaction system the lead anode is consumed, thus the term "fuel cell". The lead supply thus is the factor which

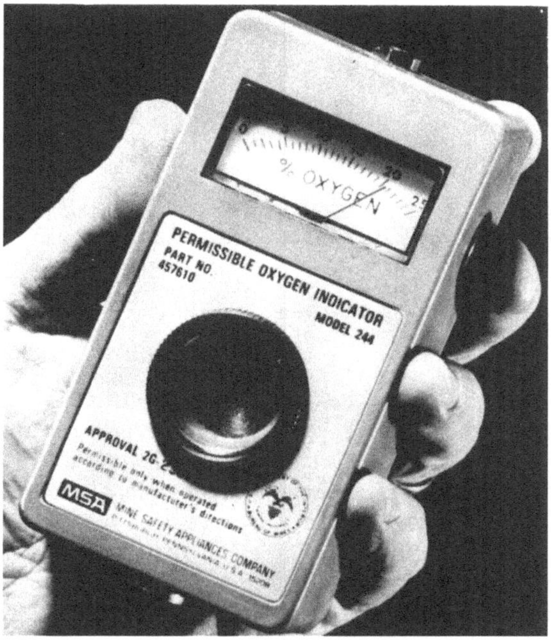

FIGURE 19. Galvanic cell continuous oxygen monitor for personnel monitoring. (Courtesy of Biomarine Industries, Inc., 45 Great Valley Center, Malvern, Pa., and Mine Safety Appliances Co., Inc., 400 Penn Center Blvd., Pittsburgh, Pa.)

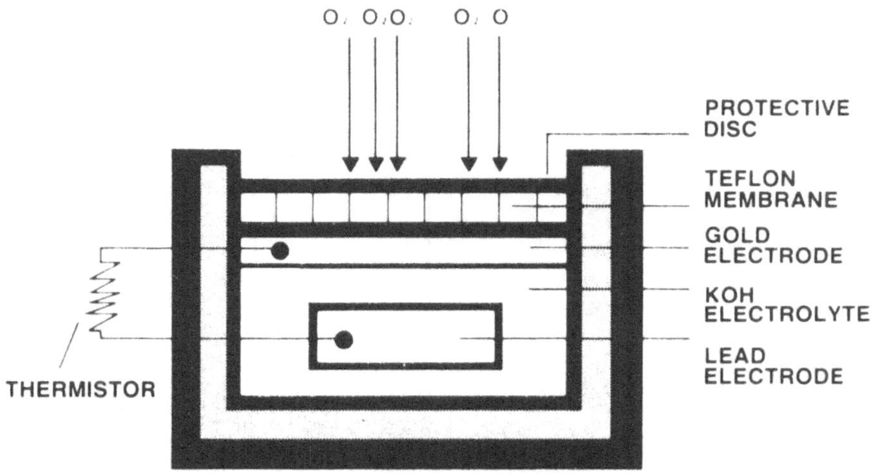

FIGURE 20. Long life oxygen sensor. (Courtesy of Biomarine Industries, Inc., 45 Great Valley Center, Pa.)

controls the useful life of the cell, about 12 months in air. Further, the system is temperature dependent in three areas,

(1) the permeability of the membrane increases as temperature rises;
(2) the mobility of the ions in the electrolyte;
(3) the rate of chemical reaction increases as the temperature rises.[17]

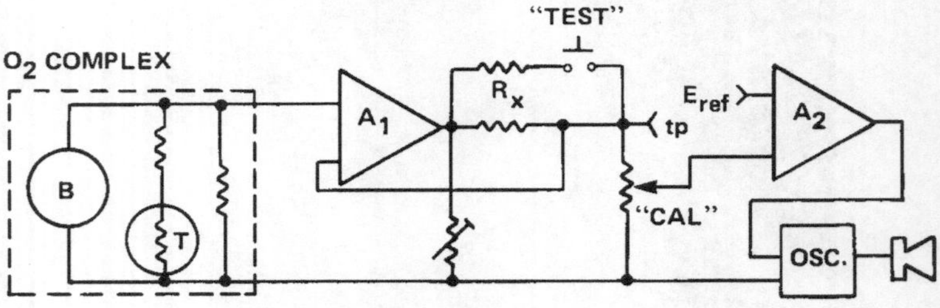

FIGURE 21. Oxygen detector cell electronics. (Courtesy of Enmet, 2308 S. Industrial Hwy., Ann Arbor, Mich.)

These temperature coefficients can be compensated by introducing an NTC-resistor in the circuitry. In the temperature range 0 to 40°C a temperature-induced error of less than ± 5% of the full-scale reading is usually achieved.

Diffusion-controlled sensors' performance is independent of relative humidity (RH) in the range below 100%. However, at an RH of 100% condensation sets in with the formation of a film of water, which has the effect of increasing the thickness of the membrane and reducing the rate of O_2 diffusion into the electrolyte.

Condensation can be avoided by either heating the sensor to a temperature above the dew point or passing the air sample through a cooled condenser to remove the excess water vapor.[17] Disposable high humidity adapters are commercially available (Table 7, No. 2). Since all galvanic O_2 cells are thermally sensitive and, to some degree, altitude dependent, the instrument should be referenced with each use to ambient fresh air (21% O_2). The temperature coefficient is 0.017 V/°C.[18] A typical schematic wiring diagram is presented in Figures 15 and 21. The rate of O_2 cell voltage that decreases with temperature very closely matches the rate of decrease in the reference voltage and alarm circuit. The functioning of the alarm circuit, which is usually set at 19% O_2, can be confirmed by deep breath exhalation (16% O_2) across the O_2 sensor element (Table 7, No. 3). One model employs potassium chloride for the electrolyte in place of the more corrosive potassium hydroxide (Table 7, No. 1). Representative samples are shown in Figures 19, 22, and 23.

A series of electrolytic conductance type detectors for H_2S, phosgene, HCN, and NO_2 was announced by MDA at the American Industrial Health Conference in Chicago, Ill. on May 29, 1979. This "Monitox" detector is quite small (133 × 65 × 40 mm) light weight (150g), has a service life of 2000 hr, and is equipped with a built-in alarm at a preset concentration level. Gas molecules diffuse through a membrane and react on the measuring electrode in the sensor cell (nonaqueous electrolyte). Reaction with the electrolyte produces an electric current which either can trigger the alarm or the electrical output stored in a microcomputer chip in the "Chromatox" personal alarm dosimeter unit for read out at the end of the work shift. Another novel feature is the replaceable sensor unit which can be removed from the basic unit and exchanged for another unit for a different gas (Figure 24). A calibration unit which generates a small quantity of calibration gas when the sensor is inserted in the cavity provided is supplied for in the field use. This function permits the user a quick and convenient way to functionally test the detector immediately before or during use. $AgNO_3 + H_2S \rightarrow AgS + H^+ + -NO_3$.

A very recent release incorporates a diffusion barrier in the form of a simple capillary in place of the usual thin plastic membrane. Improved temperature and pressure coefficients with increased robustness at appreciable cost reduction are claimed. Further, the sensor is described as "substantially independent of changes in atmospheric pressure, temperature and relative humidity." A sensing head on an extension cord is

FIGURE 22. Biomarine Industries oxygen detector remote sensor. (From Biomarine Industries, Inc., 45 Great Valley Center, Malvern, Pa. With permission.)

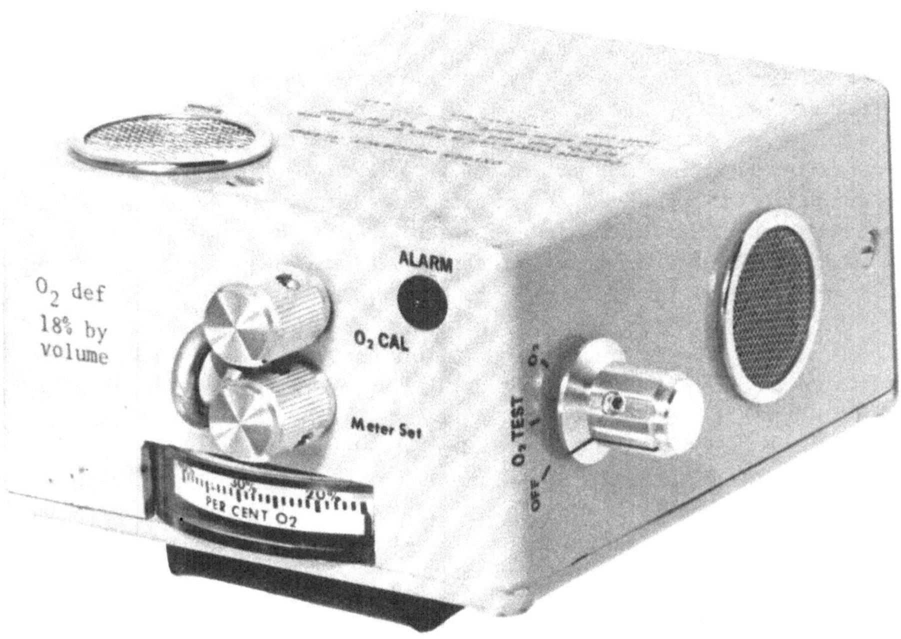

FIGURE 23. Enmet Corporation OM-10 oxygen detector. (From Enmet, 2308 S. Industrial Hwy., Ann Arbor, Mich. With permission.)

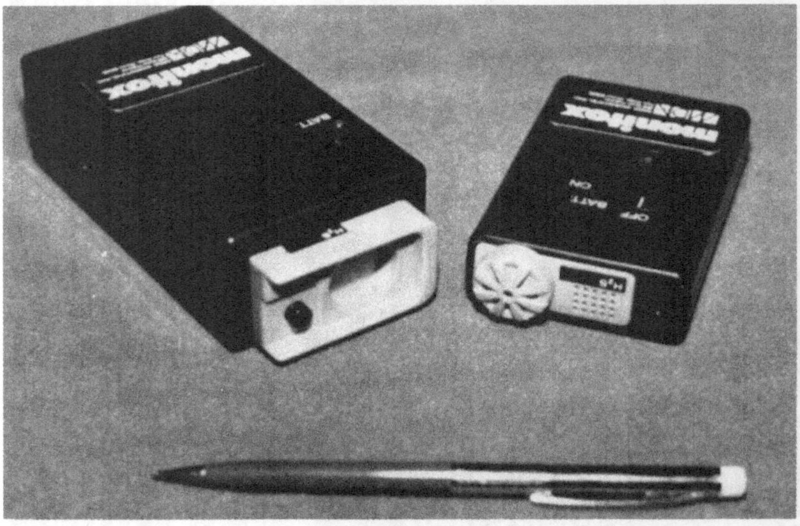

FIGURE 24. The MDA "Monitox" personal gas detection alarm system. Present alarm levels: H_2S — 10 ppm; HCN — 10 ppm; NO_2 — 5 ppm; $COCl_2$ — 0.1 ppm. Temperature range — 0—45°C. (Courtesy of MDA Scientific, Inc., 808 Busse Hwy., Park Ridge, Ill.)

available. Temperature coefficient in the range is −5 to +50°C equals 0.17%/°C, Pressure range equals 1 atmosphere +20% and 6 months cell life.

An entirely different principle for oxygen monitoring has been described recently.[19] Thin films of zinc oxide are used to measure oxygen pressure as a function of conductance. The assembly, which is simple, requires low power (approximately 7 W) and covers a wide dynamic range (1 mm to over 760 mm Hg); consists of a heater, electrical readout, and temperature controller; weighs about 30 g; and occupies about 30 ml of enclosure space. No failures were observed over several months of laboratory tests. Although this instrument has not gone into commercial production, the capabilities indicate that its exploitation as a personnel monitor would seem to offer advantages at least as favorable as the galvanic cell type.

The principle of thin metal film dosimeters has been extended to the detection of fluorine, nitrogen dioxide, and unsymmetrical dimethylhydrazine (UDMH).

The detection system is based upon the change of electrical resistivity of thin metal films when exposed to these gases. Silver metal films coated with appropriate salts proved to be applicable to the detection of all three gases; however, the following sensitized metal films were found to be optimum: for N_2O_4, silver; for F_2, copper; and for UDMH, gold. Using the best film and salt combinations found to date, N_2O_4 can be monitored over the range of 0.1 to 50 ppm, F_2 over the range 1.0 to 50 ppm, and UDMH over the range 10 to 100 ppm, with a standard deviation of about 20%. The effects of temperature over the range 50 to 90°C and of humidity from 10 to 90%, on the response characteristics of the thin film sensors were found to be significant but within the tolerance limits. Means for reducing these effects were suggested which, if successful, would, in effect, make this detection system practically independent of changes in the environment. A portable breadboard readout instrument was designed and fabricated for use with the sensors to form an integrated detection system for personnel protection.[20]

A gas-detecting device was invented that uses a metal oxide semiconductor element, the resistance of which varies when exposed to a reducing gas such as hydrogen, carbon

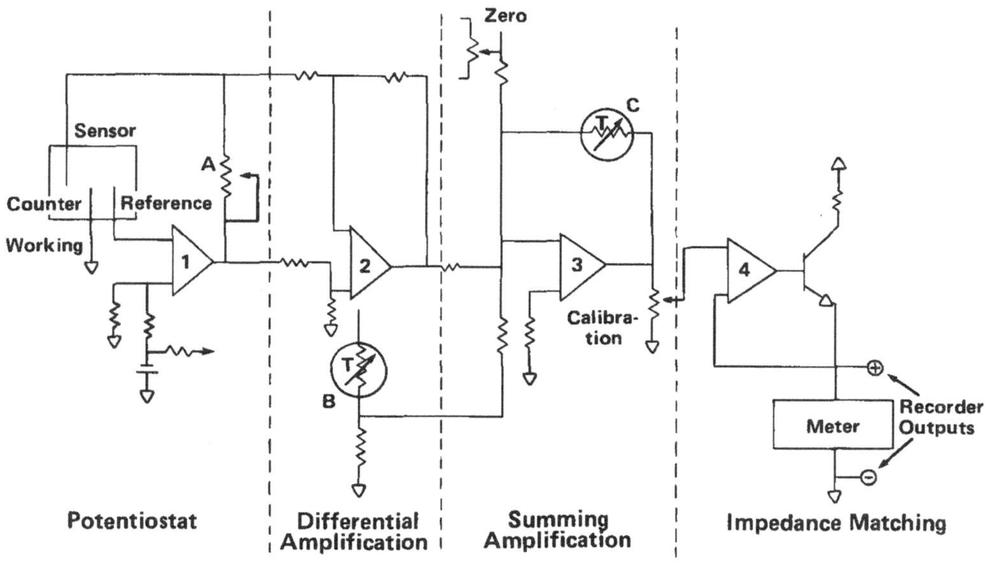

FIGURE 25. Polarographic carbon monoxide detector electronic control circuit. (From Blurton, K. R. and Bay, H. W., *Am. Lab.*, July 1974. With permission.)

monoxide, alcohol vapor, gasoline vapor, or smoke. It also contains an alarm circuit for producing an alarm in response to a predetermined reduction of resistivity of the semiconductor element. Another component is an impedance circuit coupled to a pair of electrodes through a power supply. A thermal switch operates in response to a predetermined temperature rise for the semiconductor. The switch consists of a bimetal-operated, single-pole, double-throw switch with two fixed contacts: the first resistor and the second resistor. The first resistor has a positive temperature coefficient of resistivity and is connected to the impedance circuit. The second resistor has a negative temperature coefficient of resistivity and is connected to the alarm circuit.[21]

ii. Carbon Monoxide (CO) — Polarographic

An instrument based on the controlled-potential electrochemical analysis of trace concentrations of CO has been developed and exploited commercially (see Table 7, References o and t).[22,23] The opearting principle depends on the electrochemical oxidation of CO to CO_2 on a potential controlled electrode which is maintained at a voltage below the critical potential for O_2 evolution. Platinum catalyzed electrodes are immersed in aqueous sulfuric acid retained between PTFE membranes. The overall electrochemical reaction:

$$CO + H_2O \rightarrow CO_2 + 2H^+ + 2e$$

$$O_2 + 4H^+ + 4e \rightarrow 2H_2O \text{ – counterelectrode}$$

The basic components include an electrochemical sensor, a sampling system (pump), a sample preconditioning unit (water bottle and filter), an electronic control circuit (potentiostat and amplifying circuit), and a measuring circuit. The schematic electronic circuit is presented in Figure 25. The operation of this system requires the use of a three-electrode sensor which consists of a sensing electrode held at controlled voltage, a counter electrode, and a reference electrode. The sample flow-rate is maintained constant at 700 cm^3/min to (1) minimize change in signal when flow-rate varies, (2) minimize volume of calibration gases, and (3) maximize the lifetime of the interference filter. For continuous operation the inlet air is humidified by passing over water contained in a bottle of 11.2 cm^2 cross-sectional area. The RH of the air leaving the bottle is independent of the inlet air RH and approximates the vapor pressure of the H_2SO_4 electrolyte.

Some common atmospheric pollutants are oxidized at the sensor-operating potential and would normally produce a signal when introduced into the sensor. The pre-filter eliminates or greatly reduces this interference effect. Relatively high concentrations of O_2 (20%), CH_4 (5%), H_2S (50 ppm), SO_2 (100 ppm), and nitrogen oxides (25 to 103 ppm) were not electrochemically active at the electrode potential selected for CO (0.15 V). Since both the zero (blank) current and the span current due to CO are temperature dependent, compensation is achieved by inserting appropriate thermistor circuitry which provides a "bucking" potential (polarity opposite to the cell output). The operational characteristics of CO monitors are presented in Table 11.

By changes in voltage across the voltametric sensor cell, other compounds, or elements, can be either oxidized (CO), or reduced (NO_2). The effects obtained by increasing the potential across the electrode system is explained in Figure 26. Electrochemical current is plotted as ordinate, and the sensing electrode potential as abscissa, relative to the reversible hydrogen electrode (RHE) at zero potential. Increasing oxidation potential, or more positive, is towards the right. Oxidation current is arbitrarily set as positive current. The working SE potential range is the region between O_2 reduction at more negative potentials and O_2 evolution (oxidation of water) at more positive potentials. Oxidation of the noble metal electrocatalyst (the sensing electrode) begins just prior to O_2 evolution.

Voltametric curves for two gases are shown. Gas A is electro-oxidized at a lower (more negative) potential than Gas B. The sensing electrode in the Gas A sensor is held at a constant potential SEP, within the limiting diffusion current plateau a to b, but below the oxidation potential at which Gas B would be oxidized. Gas B, therefore, would not produce an interference.

Similarly, the sensing electrode in the Gas B sensor is held at a constant potential somewhere within the diffusion plateau c to d. Since Gas A has a lower oxidation potential than Gas B, an interference could occur in the Gas B sensor. The extent of interference is a function of the sensing electrode electrocatalyst, the reaction kinetics of the interfering gas, and the solubility of the gas in the thin liquid film at the electrode. A chemical pre-scrubber can reduce this effect when interferences could be significant.

The only requirement for the electrochemical voltametric method is fixed potential on the sensing electrode. Sensor sensitivity, response time, drift, and interferences are functions of what occurs at the sensing electrode only. A potential two-electrode configuration has several features not found in three-electrode sensors using air-depolarized counter-electrodes. A voltage bias maintains the sensing electrode at a constant potential relative to a second electrode in the sensor. Being non-polarizable, this electrode serves as reference and counter-electrode for the corresponding opposite electrochemical reaction: electro-reduction when the sensing reaction is electro-oxidation, and vice versa. The sensor electrolyte is immobilized as in the dry battery construction. Cell rupture, acid spillage, and damage to the instrument is eliminated, and the analyzer can be operated in any position. In this case the reservoir is sealed to provide longer life and resistance to excessive pressure. Furthermore, the sealed reservoir eliminates the possibility of reference electrode contamination, and does not require air to operate. This system can be operated under anaerobic condition (O_2 free sampling).

TABLE 11

Operational Characteristics of CO Monitors

	Personnel monitor	Restricted area monitor	Continuous monitor
Set-up time, hr	←	24	→
Warm-up time	←	0	→
Lag time, sec	5	5	10
Response and recovery time, sec	←	95% in 20 sec	→
Zero drift		<±½% of full scale/24 hr	
Span drift		<−0.9% of full scale/24 hr	
Response		Linear response over entire range	
Range	←	Various; from 0—50 ppm to 0—4000 ppm	→
Output	←	Meter, 0—1 V for recorder	→
Detection limit		½% of full scale	
Flow rate, cm^3/min	←	700	→
Dimensions, in.	6 × 5 × 5	7 × 11 × 7	15 × 12 × 6
Weight, lb	4	9	32
Power supply	4 "C" Ni-Cd batteries	105—125 V 60 Hz or 4 "C" and 1 "C" Ni-Cd batteries	105—125 V 60 Hz and 4 "C" and 4 "C" Ni-Cd batteries

From Blurton, K. R. and Bay, H. W., *Am. Lab.*, July 1974. With permission.

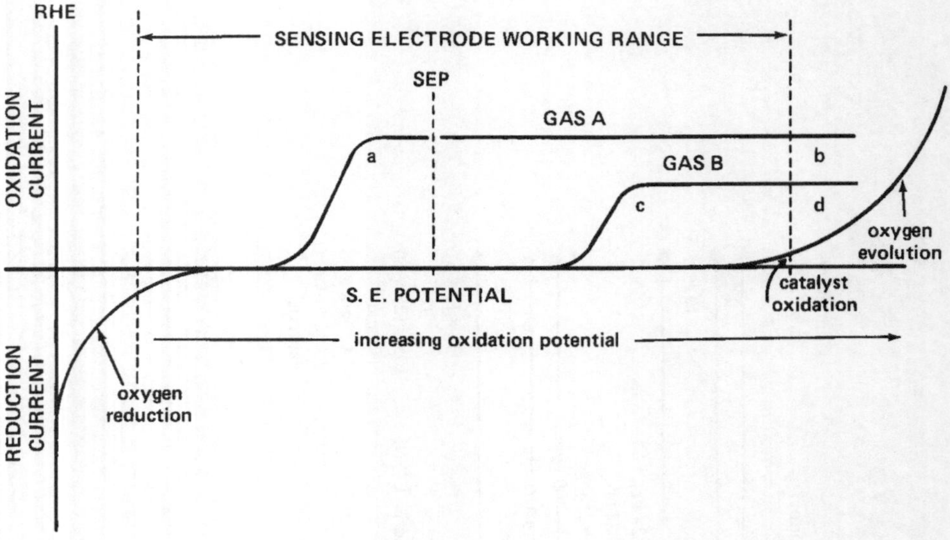

FIGURE 26. Sensor voltametric curve. (From Interscan Corp., 9614 Cozycroft Ave., Chatsworth, Calif. With permission.)

d. Combination Instruments

Recent trends are moving toward incorporating both combustible gas and oxygen detectors in the same meter case. A few even include CO and/or H_2S capability (Figures 23 and 27 to 30).

e. Ultraviolet Photoionization Instrument

(Figure 31)

i. Principle of Operation

The photo-ionizer is a trace gas analyzer used to measure the concentration of a wide variety of species in industrial atmospheres. The analyzer employs the principle of photoionization for detection. The process is termed photoionization since the absorption of ultraviolet light by a molecule leads to ionization via $R + h\nu \rightarrow R^+ + e^-$ where R^+ is the ionized species and $h\nu$ represents a photon which has an energy $\geqslant$ the ionization potential of the species.

The sensor consists of a sealed ultraviolet light source that emits photons which are energetic enough to ionize many trace species (particularly organics) but do not ionize the major components of air such as O_2, N_2, CO, CO_2, or H_2O. A chamber adjacent to the ultraviolet source contains a pair of electrodes. When a positive potential is applied to one electrode, the field created drives any ions formed by the absorption of UV light to the collector electrode, where the current (proportional to the concentration) is measured. Typical calibration curves showing the relative response of toluene and methyl ethyl ketone (at the same gain setting) are shown below (Figure 32).

Photoionization response of a selected list of common industrial air pollutants is summarized in Table 12.

a) Methane

The development of a pocket-size, handheld, battery-operated methane detector should be mentioned as a personnel monitor to detect the presence of explosive gases in the ambient work environment. The Model 800 Micro Methanometer® measures

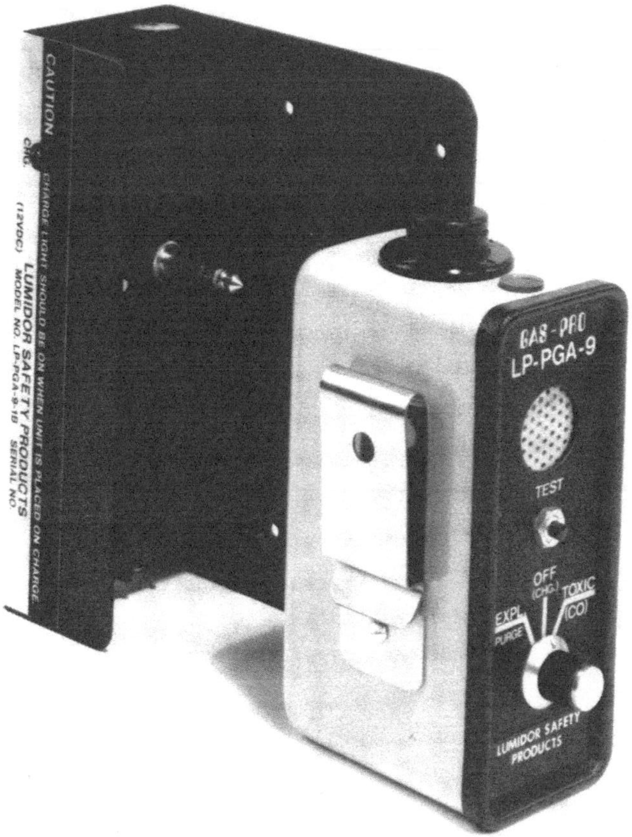

FIGURE 27. Lumidor "gas-pro" detector/alarm for explosive vapor and carbon monoxide detector. (Courtesy of Lumidor Products Corp., Safety Division, 5364 N.W. 167th St., Miami, Fla.)

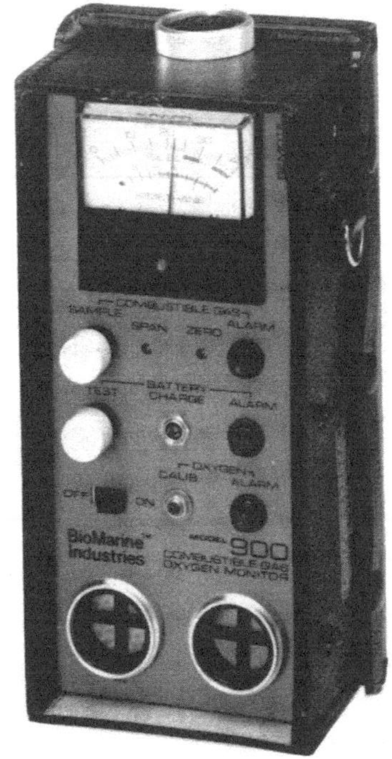

FIGURE 28. Biomarine Industries Model 900 combination combustible gas/oxygen detector with alarms. (Courtesy of Biomarine Industries, Inc., 45 Great Valley Center, Malvern, Pa.)

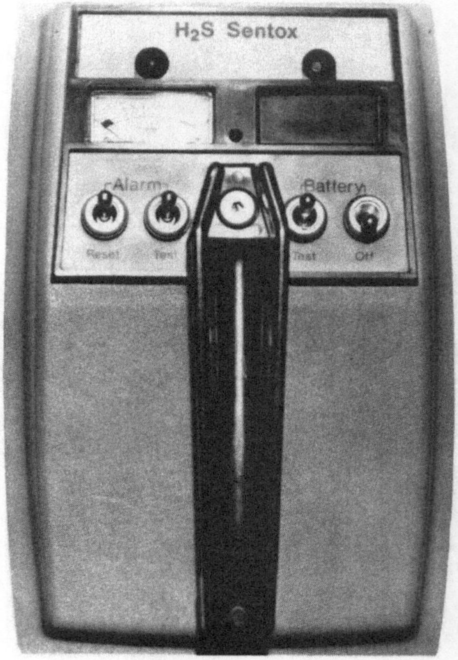

FIGURE 29. Bacharach (J-W) Sentox personal H_2S monitor. (Courtesy of Bacharach Instrument Co., Division of AMBAC Industries, Inc., 625 Alpha Dr., Pittsburgh, Pa.)

36 × 60 × 117 mm, weighs 255 g, and operates on rechargeable batteries (Figure 33). The all solid-state circuitry provides a direct readout as percent (0 to 5%) methane on an illuminated dial.

f. Evaluation of Portable Instruments

It is not the intent of the author to attempt an objective evaluation of currently available instruments. Even if the necessary expertise and laboratory data were present, the conclusions would be obsolete within a year or two at the rate of changes in design currently in progress. However, for those who would pursue this subject the following list of NIOSH evaluations will provide an available source of reliable and objective evaluations of portable instruments:

1. A summary of the NIOSH Evaluation Program of Portable Direct-Reading Meters, *American Industrial Hygiene Association Journal,* 37, 289, 1976. Criteria and specifications for CO, SO_2, combustible gas, and NO_2 instruments discussed. The range of concentration to be covered shall measure 1/5 to 5 times the 8-hr TWA concentration required by Federal or ACGIH standard for the gas analyzed. Further, the instrument should be free from interference from commonly occurring substances. Accuracy should fall in the range ± 5% at the 8-hr TWA concentration. Dimension and weight limits are given in the Preface as follows: weight less than 4.55 k (10 lbs) and occupy a volume less than 0.028 m^3 (1 ft^3).
2. Combustible gas and vapor meters — Evaluation of Portable Direct Reading Combustible Gas Meters, C. S. McCommon, NIOSH Technical Information, HEW Publication No. (NIOSH) 74-107 (1974).

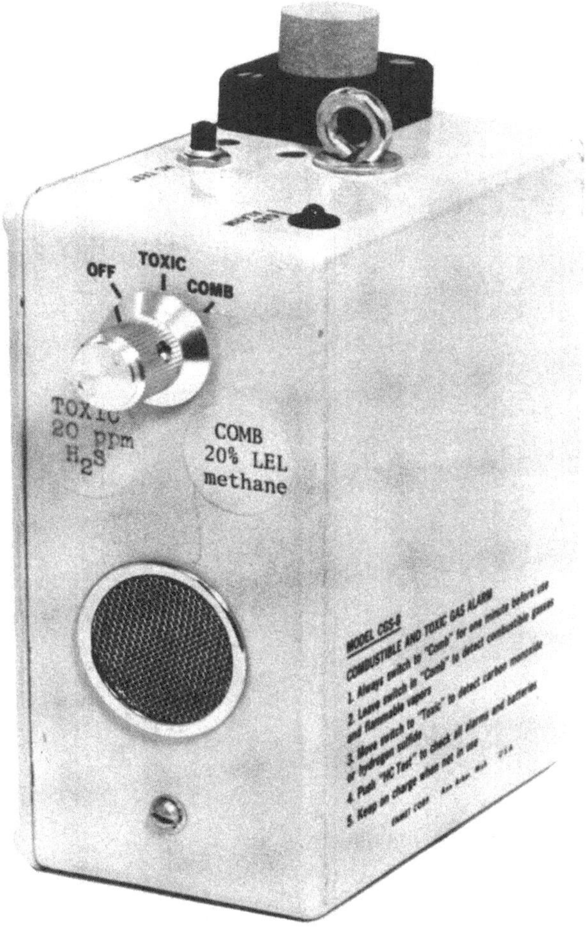

FIGURE 30. Enmet model CGS-8 combination combustible and toxic gas H_2S alarm. (Courtesy of Enmet, 2308 S. Industrial Hwy., Ann Arbor, Mich.)

3. Carbon monoxide — Evaluation of Portable, Direct Reading Carbon Monoxide Meters, C. D. Parker and R. B. Strong, Technical Information, HEW Publication (NIOSH) 75-106 (1974).
4. Hydrogen sulfide — An Evaluation of Portable Direct Reading Hydrogen Sulfide Meters, NIOSH Research Report, F. C. Thompkins and J. H. Becker, HEW Publication No. (NIOSH) 77-137 (1976).
5. Hydrocarbons — Evaluation of Portable Direct Reading Hydro-Carbon Meters, NIOSH Research Report, M. A. Willey and C. S. McCommon, HEW Publication No. (NIOSH) 76-166 (1976), flame ionization, photoionization, and infrared detectors.
6. Mercury — Evaluation of Portable Direct Reading Mercury Vapor Meters, NIOSH Research Report, C. S. McCommon and W. J. Woodfin, HEW Publication No. (NIOSH) 76-167 (1976), one instrument studied.
7. Nitrogen Dioxide — Evaluation of Portable, Direct Reading Nitrogen Dioxide Meters, NIOSH Technical Information, B. A. Johnson, HEW Publication No. (NIOSH) 74-108 (1974).

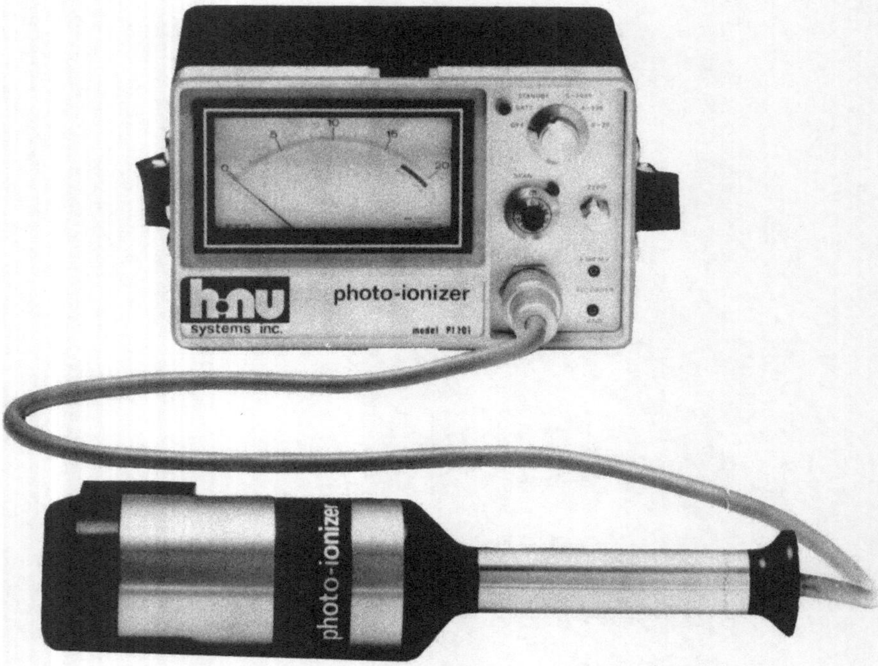

FIGURE 31. HNU systems photoionizer. (Courtesy of HNU Systems, Inc., 30 Ossippee Road, Newport, Upper Falls, Mass. With permission.

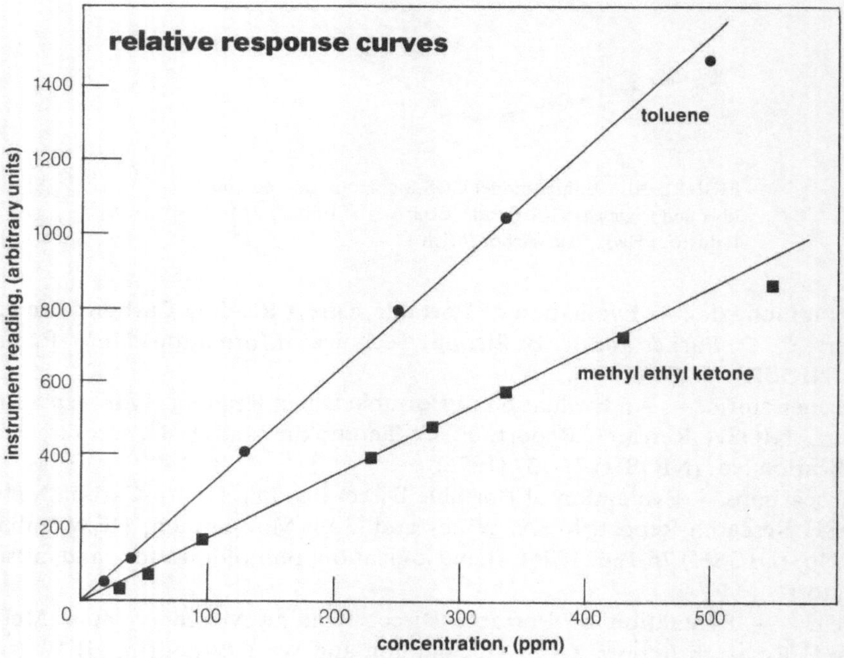

FIGURE 32. HNU systems photoionizer calibration chart. Operating range 0.1—600 ppm, maximum sensitivity 0—2 ppm full scale.

TABLE 12

Photoionization Response of a Selected List of Common Industrial Air Pollutants

Class species	Photoionization response 9.5 eV lamp	10.2 eV lamp	11.7 eV lamp
Paraffins and unsaturated hydrocarbons			
Methane	NR[a]	NR	NR
Ethylene	NR	L[b]	H[c]
Acetylene	NR	NR	H
1-Butene	H	H	H
Hexane	NR	L	H
Chlorinated hydrocarbons			
Methyl chloride	NR	NR	H
Carbon tetrachloride	NR	NR	H
Chloroform	NR	NR	H
Dichloroethane	NR	NR	H
Vinylidene chloride	L	H	H
Vinyl chloride	L	H	H
Trichloroethylene	H	H	H
Heterocyclics and aromatics			
Phenol	H	H	H
Pyridine	H	H	H
Benzene	H	H	H
Toluene	H	H	H
Xylene	H	H	H
Styrene	H	H	H
Aniline	H	H	H
Chlorobenzene	H	H	H
Nitrobenzene	NR	L	H
Nitrogen compounds			
Formamide	NR	H	H
Ammonia	NR	L	H
Hydrazine	H	H	H
Methyl amine	H	H	H
Acetonitrile	NR	NR	NR
Acrylonitrile	NR	NR	H
Sulfur compounds			
Sulfur dioxide	NR	NR	NR
Hydrogen sulfide	NR	H	H
Carbonyl sulfide	NR	NR	H
Carbon disulfide	H	H	H
Methyl mercaptan	H	H	H
Dimethyl sulfide	H	H	H
Dimethyl disulfide	H	H	H
Aldehydes, ketones, alcohols, acids, esters			
Formaldehyde	NR	NR	H
Acetaldehyde	NR	H	H
Propionaldehyde	L	H	H
Acrolein	L	H	H
Crotonaldehyde	L	H	H
Acetone	L	H	H
Methanol	NR	NR	H
Ethanol	NR	L	H
Formic acid	NR	NR	H
Acetic acid	NR	L	H
Methyl methacrylate	L	H	H

TABLE 12 (continued)

Photoionization Response of a Selected List of Common Industrial Air Pollutants

	Photoionization response		
Class species	9.5 eV lamp	10.2 eV lamp	11.7 eV lamp
Others			
Ethylene dibromide	NR	H	H
Ethylene oxide	NR	L	H
Tetraethyl lead	H	H	H
Phosphine	NR	H	H
Arsine	NR	H	H
Iodine	H	H	H

[a] NR = no response.
[b] H = high response.
[c] L = low response.

From HNU Systems, Inc., 30 Ossipee Road, Newton Upper Falls, Mass. With permission.

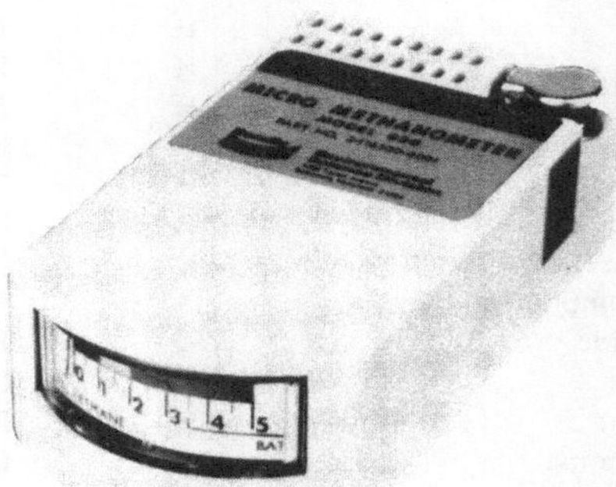

FIGURE 33. The Micro Methanometer®, Model 800. (Courtesy of Bendix Environmental Science Division, 1400 Taylor Ave., Baltimore, Md.) See Reference note **

8. Nitrogen Oxides — Evaluation of Portable Direct Reading No-No_x Meters, C. D. Parker, NIOSH Research Report, HEW Publication No. (NIOSH) 76-161 (1975). Calibration: NO and NO_2 in nitrogen pressurized in cylinders. Enolyzer was the only truly portable instrument available. Principle: electrochemical reaction between NO and NO_2 in aqueous H_2SO_4 electrolyte.
9. Ozone (Oz) — Evaluation of Portable Direct Reading Ozone Meters, NIOSH Technical Information, C. D. Parker, HEW Publication No. (NIOSH) 76-132 (1975). One instrument (chemilumenescent type) conforms to NIOSH specifications (AID, Model 560).
10. Sulfur Dioxide (SO_2) — Evaluation of Portable Direct Reading Sulfur Dioxide Meters, NIOSH Technical Information, C. D. Parker and R. B. Strong, HEW Publication No. (NIOSH) 75-105 (1974).

TABLE 13

Electronic Gas and Vapor Detection Meters — Capability Evaluation

Manufacturer	Combustible	Oxygen def.	Toxic	Approx. cost ($)
ENMET "Tritector"	Yes (MOS sensor)	Yes (Galvanic)	Yes (CO, H_2S, etc.)	385.00
Edmont Wilson 60-400	Yes (Hot wire)	Yes (polarographic)	No	1100.00
Biomarine Model 900 Series	Yes (Hot wire)	Yes (Galvanic)	No	485.00—625.00
Johnson William Bacharach "Sentox"	Yes (Hot wire)	Yes (Polarographic)	No, except H_2S Sentox	1000.00
MSA Explosimeter Model 250	Yes (Hot wire)	Yes (Galvanic)	No	750.00—900.00

Mailing address for HEW publications:
Department of Health, Education and Welfare
Public Health Service, Center for Disease Control
National Institute for Occupational Safety and Health
Robert A. Taft Laboratories
4676 Columbia Parkway, Cincinnati, Ohio 45226

One supplier of portable instruments has published a comparison chart (Table 13) to assist the industrial hygienist with the selection of electronic equipment suitable for personnel monitoring (Table 7, Reference u).

B. Grab Samples for Laboratory Analysis

1. Evacuated Containers

Collection of ambient air samples in evacuated glass flasks and gas sample tubes reaches back into the antiquity of air sampling history. The air sample may be carried to the laboratory without alteration, or an absorbing solution may be present during sampling.[24] This technique has been commercialized in the form of an evacuated metal can that was adapted from the one-pound disposable container for "Freon"-12® refrigerant (Figure 34).[25] A rubber septum has been added adjacent to the vacuum release valve to permit withdrawal of a sample with a gas-tight syringe (Figure 35). Since there is a residual nitrogen gas in the "Vacu-Sampler," the apparent contamination concentration must be corrected.

$$\text{Factor} = 3\frac{T_1}{T_2}$$

where

$$T_1 = \text{sampling temperature} \quad (K)$$

and

$$T_2 = \text{analysis temperature} \quad (K)$$

FIGURE 34. Evacuated container grab sampler.

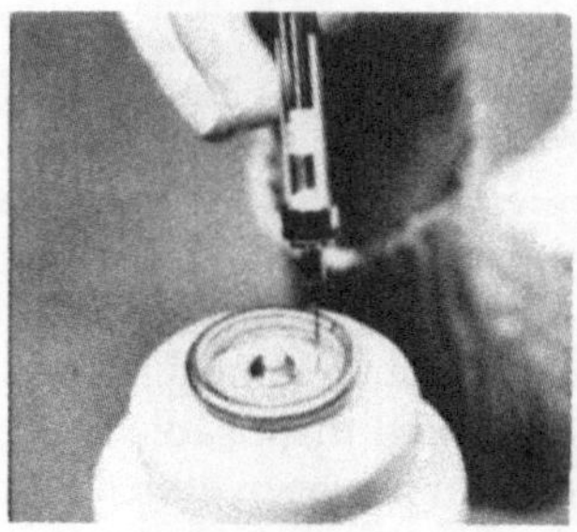

FIGURE 35. Removal of a sample for laboratory analysis. Use a gas-tight syringe to withdraw a sample for chromatographic or other types of gas analysis. (From MDA Scientific, Inc., 808 Busse Hwy., Park Ridge, Ill. With permission)

The "Vacu-Sampler" is a partially evacuated to total volume of 370 mL. The container is partially back-filled with 246.7 mL nitrogen. At standard temperature and pressure, a 123.3-mL sample will be inhaled. The nitrogen back-fill (535 mm mercury pressure) eliminates problems with high dew point samples (Table 7, Reference p).

When considering this technique, either glass or metal containers, the limitations imposed by loss of the required sample on the container walls by adsorption or chemical reactions will prevail.[24] The losses of phosgene or toluene diisocyante from samples

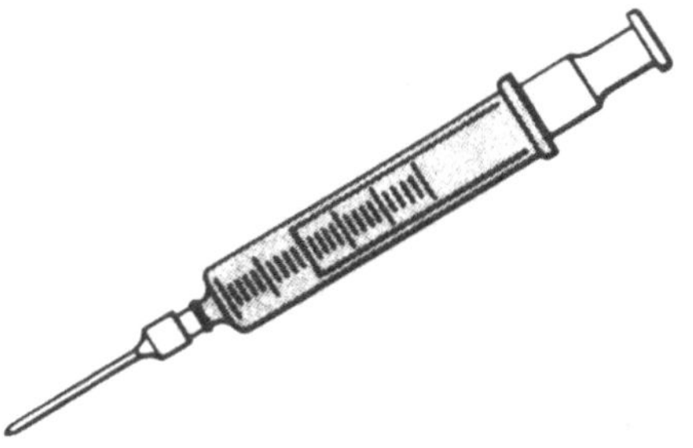

FIGURE 36. Gas chromatograph sampling syringe.

taken by this technique, for example, would completely vitiate the results. Collection of corrosive gases such as Cl, HCl, SO_2, etc., in metallic containers would not be recommended. In general, collection of samples in evacuated containers has a rather limited application in industrial hygiene surveys.

2. Syringe Collection

This technique actually is a modified collection of samples in an evacuated container. The modification lies in the location of evacuation of the sampling container.

The procedure described in the preceding section is based on evacuation of the container in the laboratory before transport to the field, whereas the syringe technique depends upon "on-site" vacuum application. A small syringe for gas chromatograph (GC) analysis is shown in Figure 36. Syringes to collect volumes in the range of microliters to one liter are available.[24] Under some circumstances this technique serves a useful purpose, especially where GC analysis supports the industrial hygiene sampling program. Again, the limitations imposed on pre-evacuated containers must be considered before this technique for grab-sampling is recommended.

Colorimetric reagents can be added to the syringe before sampling is carried out. For example, 5 mL of Saltzman's reagent can be added to a 100-mL glass syringe to obtain nitrogen oxide in air samples.[24] The developed red color can be estimated visually in the field or quantitated in the lab.

C. Personnel Monitoring — Collection for Laboratory Analysis

1. Wet (Liquid Phase) Collection

a. Liquid Reagent System

An unusual gaseous contaminant sampling system was described by Palmes and Gunnison during the May, 1972, AIHA conference.[26] The design incorporated the principle of molecular gas diffusion into a miniature personnel monitoring device which requires no power source for operation. The diffusion chamber was a flat-bottomed cylinder, 10 mm diameter × 60 mm length (5 mL volume), which was fitted with a capillary sealed through the flat top. Two mL of an appropriate collection reagent or solvent was used. Passive sampling with SO_2 was selected for measuring time-weighted average exposure over a concentration range of 0.1 to 10 ppm. Calculated and observed values agreed very well over a considerable range of orifice dimensions (length and diameter are controlling variables). Water vapor collection also demonstrated good precision and accuracy.

The quantity of gas transferred by diffusion from the environment through an orifice of known dimensions into a chamber maintained at zero concentration by a suitable collecting medium can be used as the basis for calculating average concentrations during the time the sampler remains in the environment.

$$J - Ddx/dc = \text{diffusion flux in mol/cm}^2$$

Although solid adsorbents such as silica gel were not specifically mentioned, this approach to sample collection would appear to be an ideal solution to the power supply problem for positive displacement samples.

As would be expected, other investigators took up the development of this idea and extended the application to the collection of oxides of nitrogen (NO_x) from the ambient atmosphere. Reizner and West developed a badge-type monitor which contained the West-Gaeke reagent for the collection of SO_2 which diffused through a semi-permeable membrane in place of the capillary tube. Compared with direct diffusion, use of such a membrane has an advantage in selective passage of gases at a rate proportional to their concentration at the face of the membrane.[47]

For construction of the NO_x dosimeter, a membrane that had a permeation rate for NO_2 about 12 times the rate of NO was selected. If total NO_x determination is required, an oxidizing layer is placed over the permeable membrane to oxidize the NO to NO_2. NO can be determined by exposing two dosimeters side by side, one with the oxidizing wafer, the other without. The absorbing solution (4mL) contains 5×10^{-4} molar H_2SO_4 and 4×10^{-3} molar H_2O_2. The NO_2 concentration in the absorber solution was determined by the selective ion electrode procedure. The body and retaining ring of the 106-g dosimeter were fabricated from PTFE. The permeable membrane is a disc (56 mm O.D. × 0.025 mm thick) of unbacked dimethyl silicone polymer. This disc was cemented to an O-ring of the same O.D. and 3 mm thick. Tightening this assembly produces a gas- and liquid-tight seal. The liquid reagent completely fills the 4-mL cavity to ensure full contact with the membrane, regardless of orientation. The oxidation accessory is composed of a 7-cm glass filter impregnated with acidic sodium bichromate.

The liquid permeation rate through the membrane is quite low-5.3% over 6 hr at 16°C, and temperature corrections are not necessary. Hydrogen and chloride ions interfere strongly with the selective ion analysis. These interferences could be avoided by using the colorimetric (Saltzman) analysis procedure.

The precision for NO_2 at 3.7 to 8.0 ppm was ± 0.57 ppm, and for NO in the range 2.8 to 10.3 ppm it was + 0.25 ppm.[27]

b. Liquid on a Solid Support

To avoid the problems associated with the leakage or spillage of corrosive fluid from the personnel monitor, the liquid phase reagent can be coated onto a solid support. Based on this innovation, a dosimeter for the collection nitrogen oxides by triethanolamine (TA) supported on a stainless steel screen was developed by Ed Palmes and his group at the New York University Medical Center during the period 1973 to 1976.[28] The sampler depends on the transfer for NO_2 by diffusion to a TA coated collector at the sealed end of an open tube constructed from a 7.1-cm length of 1.27 cm O.D. × 0.95 cm I.D. commercial acrylic tubing. One end was fitted with a polyethylene cap which also retained three 40 × 40 stainless steel grids punched from 0.025-cm diameter wire cloth. This fixed cap was secured with one turn of plastic electrician's tape. In use, the dosimeter was worn on the belt or hard hat open end down to avoid picking up dust and other foreign material. The open end was sealed with a second polyethylene cap until used (Figure 37). The devices are accurate, light, simple to use, and have very good shelf life before (39 days) and after (27 days) sampling.

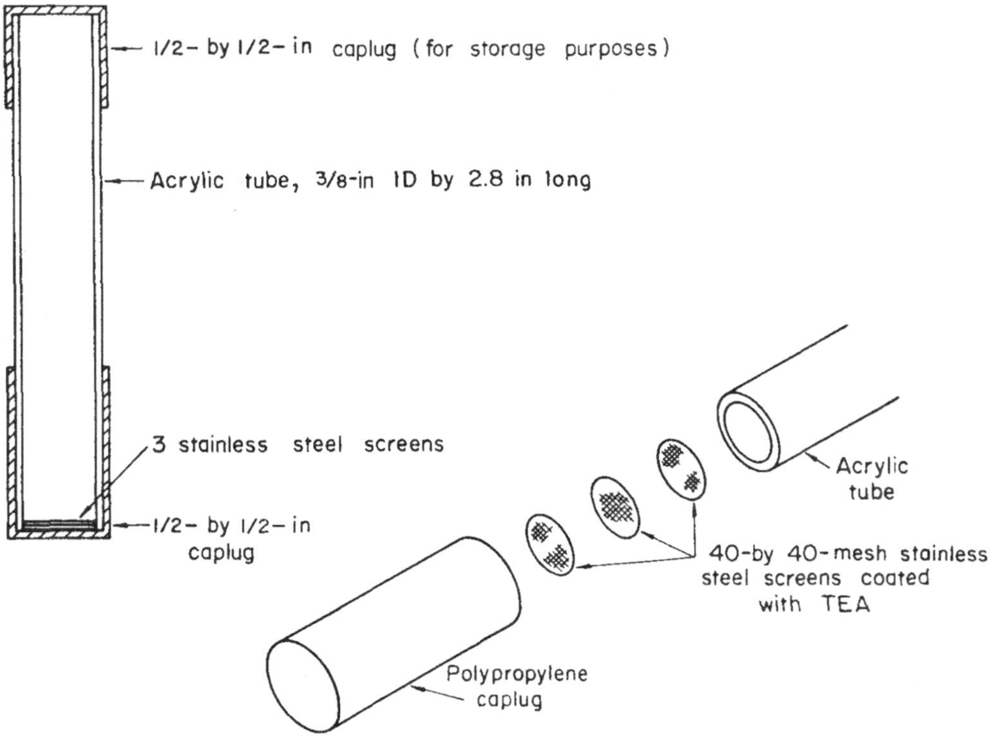

FIGURE 37. Exploded view of a diffusion sampler. (From Palmes, E. D., Gunnison, A. F., Di Mattio, J., and Tomczyk, C., *Am. Ind. Hyg. Assoc. J.*, 37, 570, 1976. With permission.)

The sampler has been successfully demonstrated in the field, as well as in the laboratory.[29] The overall accuracy is approximately ± 5% of reading to a precision of about ±1%. The effects of wind velocity and relative humidity are negligible. The temperature coefficient is close to 0.055%/per °C. Duration of exposure and approximate temperature are the only data required from the field operation. A two-step analysis by Saltzman's colorimetric procedure in the field directly, or in a service laboratory, is applied to determine the total NO_2 exposure dose which can be converted to the TWA concentration of NO_2 in ppm by dividing by the time duration of exposure.

i. Principle of Operation

The basis for this personnel sampler is that the transfer of gas through an orifice or tube can be predicted if the dimensions of the orifice, the concentration gradient across the orifice, and the coefficient of diffusion of the gas in air are known. The collecting medium must maintain the sampler concentration of the test gas at essentially zero so that the gradient becomes numerically equal to the ambient concentration (C_a); that is, $C_a - O = C_a$.

The equation

$$J_a = D\frac{A}{L}C_a$$

where J_a = diffusion flux (mol/sec), D = coefficient of diffusion (cm²/sec), A =

cross-sectional area of the orifice (cm^2), L = the length of the orifice or tube (cm), and C_a = ambient concentration (mol/cm^3), was used to calculate diffusion flux.

Coefficients of diffusion were calculated from Reid and Sherwood's equation 11-11. (0.154 cm^2/sec at 21 °C).[30]

The collecting medium is of such a nature that it can be analyzed readily. One method of analysis involves the diazotization of sulfanilamide with a subsequent coupling of an aromatic amine to give a pink color. This is the basis of a modification of the Saltzman method published by Blacker.[31]

ii. Sampler Analysis

When the samplers are ready for analysis, 2.1 mL of fresh reagent is added directly to the sampler tube containing the stainless steel screens. The sampler is shaken, and after 10 min the absorptance of the resulting pink solution is read at 540 nm wavelength in a spectrophotometer. For rapid analysis in the field, a pocket-sized Bausch and Lomb Spectronic Mini-20 could be used. In a typical calibration, approximately 40 nmol of NO_2 would represent about 17.5 ppm NO_2 × hrs, equivalent to an average exposure over 8 hr of about one-half the TLV (and ceiling level) of 5 ppm NO_2. If the color is too dark, it is possible to dilute with more of the reagent (with the appropriate change in calibration factor) without incurring any significant error. The combined reagent is prepared as follows:

Sulfanilamide solution: 2 g sulfanilamide + 5 mL concentrated H_3PO_4 diluted to 100 mL with water.

NEDA solution: 70 mg NEDA (*N*-1-naphthyl-ethylenediamine dihydrochloride) diluted to 50 mL with water.

Reagent solution: Mix in the proportion of 10 mL sulfanilamide solution + 10 mL water + 1 mL NEDA solution.

iii. Experimental Results

A series of determinations was carried out in which groups of samplers were exposed in the laboratory to atmospheres generated by passing dry nitrogen over NO_2 permeation tubes. The permeation tube was maintained at constant temperature by having the container for the tube itself, as well as a long coil of copper tubing (to carry the nitrogen to the permeation tube), immersed in a thermostated water bath. Weight loss of the permeation tube was used as the primary standard for comparison with results obtained via the sampler. In addition, the NO_2 sampler was compared with a modification of Blacker's triethanolamine (TEA) method, utilizing grab samplers in TEA-coated vessels. Figure 38 demonstrates the results of the comparison. There is good agreement between results obtained with the NO_2 sampler and the reference methods; the dashed line indicates ideal agreement between the various methods of analysis. A coefficient of diffusion, D_{NO_2}, of 0.1541 cm^2/sec at 70°F is considered to be the most accurate estimate possible, considering the current state of the art, for estimating D_{NO_2} in air.

iv. Effects of Temperature

The effect of temperature on sampler performance was calculated according to diffusion theory, and it was concluded that the rate of transfer by diffusion increases approximately as the square root of the absolute temperature in the range of chief concern around 21 °C. The net effect of a temperature increase is to increase the rate of transfer by approximately 1%/5.5 °C; in most cases this correction is negligible.

v. Effects of Pressure

According to diffusion theory, the rate of transfer by diffusion should be insensitive to pressure changes when the pressure remains close to 1 at. The reason for this is that

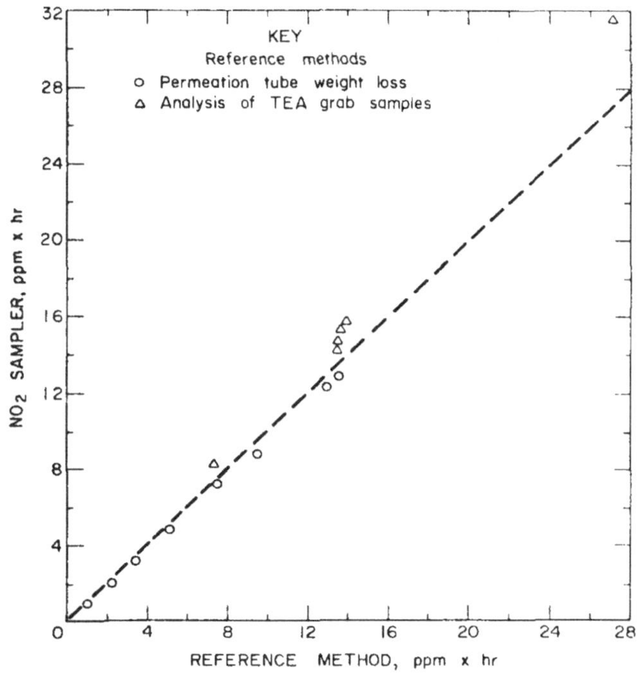

FIGURE 38. Comparison of NO_2 sampler with reference methods.[29]

the coefficient of diffusion increases inversely while concentration increases directly (both as the first power of the pressure); thus, the two terms cancel and make the sampler independent of barometric pressure.

vi. Effects of Wind Velocity

Work was conducted to measure the effect of wind velocity on sampler performance. This was accomplished by rotating groups of samplers at various distances along an arm from the axis of rotation and measuring the uptake of water vapor using a molecular sieve as the dessicant. The results are shown in Table 14. There is an increase in sampling rate as wind velocity increases to several hundreds of linear feet per minute. This effect is direction sensitive, and the highest recovery seen is 135% of the still air value at an angle of 45° and 515 linear ft/min. It should be noted, however, that even at this high airflow, a randomly oriented sampler (the condition we would expect when the sampler is worn by the miner) is only about 14% too high.

Agreement between results from the samples and stain tube measurements is excellent. Simultaneous exposure of two samplers on the same person show identical results (Table 15).

This passive sampler based on the design proposed by Dr. Palmes and modified to collect gaseous phase pollutants on stainless steel screens which support the liquid reagent are now commercially available. An additional innovation places the liquid reagent in a cup retained in the upper part of the diffusion chamber. A wick which is inserted in the bottom of this cup provides a solid carrier for the liquid reagent. However, no data relative to results or calibration of this system were available at press time.*

DuPont entered the "wet" collection passive monitoring field with the introduction of the Pro-Tek® colorimetric air sampling badges to the American Industrial Hygiene

* Daco Products, Inc., Montclair, N.J.

TABLE 14

Effect of Wind Velocity and Direction[a] on Water Vapor Uptake by Personal Samplers

Wind velocity (ft/min)	Water vapor uptake (% of still air control)					
	0°	45°	90°	135°	180°	Av.
103	104	104	106	100	98	102
206	101	109	107	106	100	105
309	103	117	110	105	101	107
412	104	127	121	107	103	112
515	109	135	118	108	102	114

[a] Angle of exposure (degrees from direction directly into wind).

Palmes, E. D., Gunnison, A. F., DiMattio, J., and Tomczyk, C., *Am. Ind. Hyg. Assoc. J.*, 37, 570, 1976. With permission.

TABLE 15

Results on Samplers Exposed at ERDA's Bartlesville Laboratory

Sampler	Exposure (hr)	Weighted average NO_2 concentration (ppm) ERDA	New York University
172	7.9	0.60	0.59
173	7.8	.26	.28
174	7.7	.26	.28
175	7.9	.60	.59
178	7.9	.60	.59
179	7.8	.26	.28
180	7.7	.26	.28
181	Not exposed	Blank	.01[a]
182	7.9	.60	.63
183	Not exposed	Blank	.00[a]
Average 1[b]		.26	.28
Average 2[c]		.60	.60

[a] Based on 7.8-hr exposure.
[b] Bartlesville trial 1: average of 173, 174, 179, and 180.
[c] Bartlesville trial 2: average of 172, 175, 178, and 182.

Palmes, E. D., Gunnison, A. F., DiMattio, J., and Tomczyk, C., *Am. Ind. Hyg. Assoc. J.*, 37, 570, 1976. With permission.

Conference in Chicago, Illinois on May 29, 1979. The collecting reagent and the liquid reagents required for the development of a color specific for the air pollutant collected are retained in compartments in a clear plastic envelope (Figure 38A). When the blister

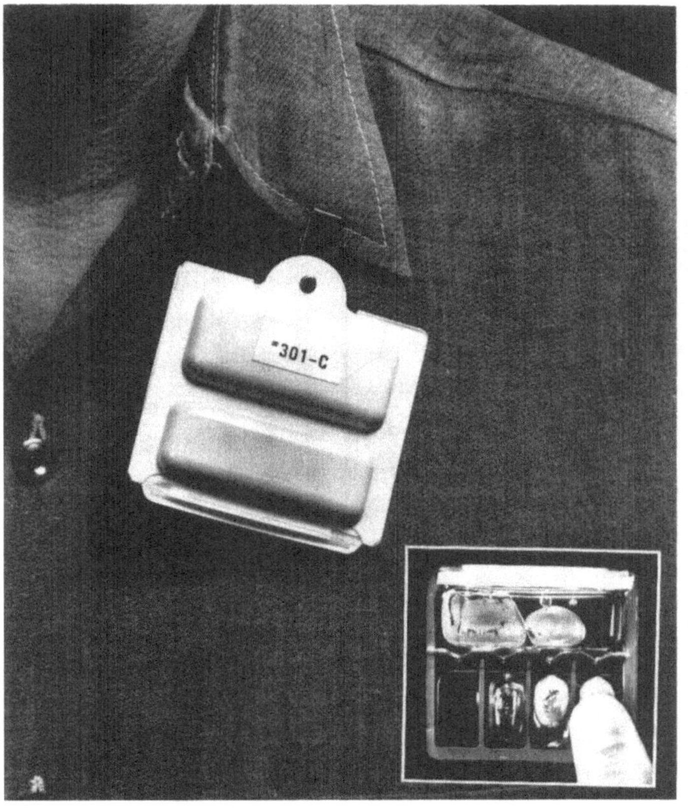

FIGURE 38A. The DuPont Pro-Tek® colorimetric air monitoring badge in use. (From Du Pont, Wilmington, Del. With permission.)

packs are ruptured by finger pressure, the chemicals are mixed within the badge, and the color intensity of the resulting solution is measured in the portable Pro-Tek® colorimetric read-out meter (Figure 38B) by inserting the plastic badge in the slot provided. The digital read-out displays the concentration in ppm/hr. Diffusion samplers for ammonia, sulfur dioxide, and nitrogen dioxide have been developed to date. Contaminant gas uptake is controlled by a unique proprietary diffuser element rather than an organic semipermeable membrane or gas attentuating sheet.

*2. Solid Adsorption — Collection by Binary Diffusion**

a. Introduction

In recent years, significant progress has been made in technology for diffusional monitoring of gases and vapors. The monitors utilize Brownian motion to control sampling into a collection medium. This technology is particularly well suited for personal monitoring devices, resulting in lightweight, unobtrusive, and economical monitors which required no external power source. Palmes and Gunnison[26] described an SO_2 monitor based on molecular diffusion during the May, 1972, AIHA Conference. At the same meeting, Braun described a mercury vapor monitor based on atomic diffusion.[32,37]

b. Theory and Conceptual Illustration

Diffusional monitors rely on a concentration gradient across a static or placid layer of air to induce a mass transfer. The equation below, based on Fick's Law, gives the steady-state relationship for the rate of mass transfer.

* Prepared by David L. Braun, Monitoring Supervisor, 3M Co., St. Paul, Minn., 1978.

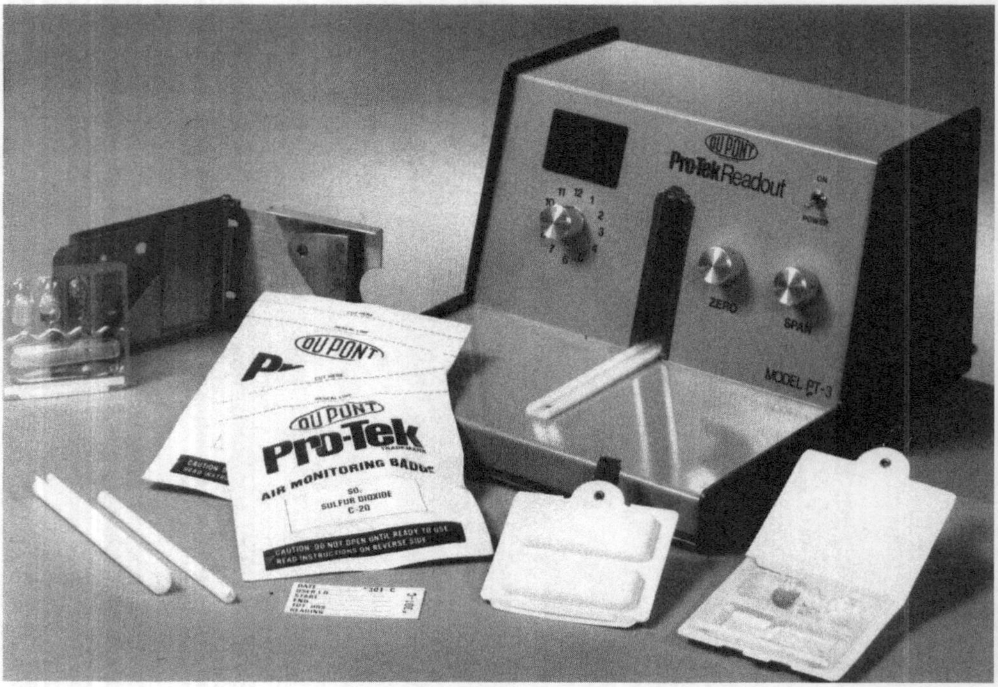

FIGURE 38B.

$$W = D\frac{A}{L}(C_1 - C_o) \quad (1)$$

Where W = mass transfer rate, ng/sec, D = diffusion coefficient, cm^2/sec, A = frontal area of static layer, cm^2, L = length or depth of static layer, cm, C_1 = ambient concentration, ng/cm^3, and C_o = concentration at collection surface.

It can be seen that by choosing an effective collection surface, such that C_o is essentially zero, the mass transfer or collection rate is proportional to the ambient vapor concentration, C_1. It may also be noted that the units of D A/L are cubic centimeters per second, the same as for the volumetric flow in a pump monitoring system. The rate of sampling of the contaminant is then the product of the D A/L term and the ambient concentration. Figure 39 illustrates the construction of a monitor for mercury vapor in which the collection surface is a gold layer.

Mercury vapor in the atmosphere being monitored passes through the barrier film and continues through the static air column according to equation (1). The average time required to reach the collection surface is found by equation (2) below.

$$T = \frac{L^2}{2D} \text{ sec} \quad (2)$$

For example, the average time for a mercury atom to progress to the collection surface when the depth L is 0.65 cm can be shown, using D = 0.12 cm^2/sec, to be 1.75 sec. Thus, the sampling occurs rapidly even though static air columns are used. The mercury vapor sampling rate, D(AL), for a frontal area A of 6 cm^2 and a depth L of 0.65 cm is about 30 cm^3/min. This flow rate is in the range of small personal sampling pumps.

The diffusion coefficient is a vapor property that is temperature and pressure dependent. Studies by Gilliland[34] on the Maxwell derivation of diffusion coefficients resulted in

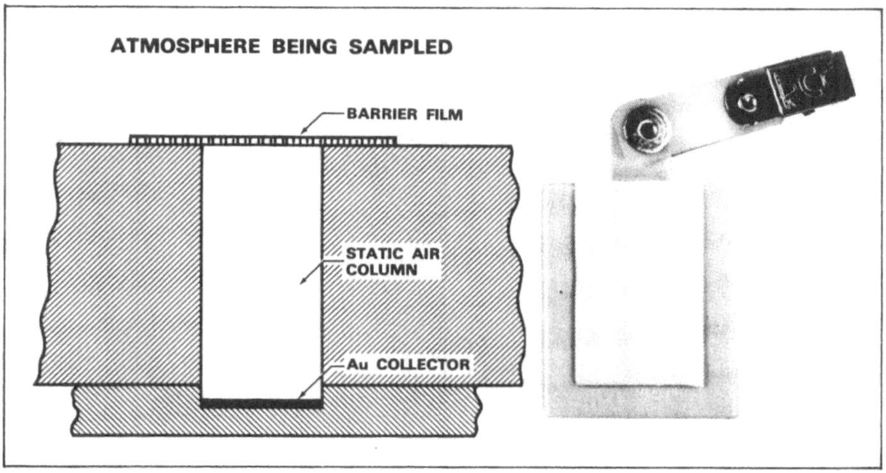

FIGURE 39. Cross-sectional construction and illustration of a diffusion monitor. (Courtesy of 3M Company, St. Paul, Minn.)

$$D = \frac{0.0043\, T^{3/2} \sqrt{1/M_a + 1/M_b}}{P\,(V_a^{1/3} + V_b^{1/3})^2} \qquad (3)$$

Where D = diffusion coefficient, cm^2/sec, T = abs temp, P = total pressure in atmospheres, M_a, M_b = mol wt of the two gases, and V_a, V_b = molal volumes at normal boiling point of the two gases.

Equation (3) shows that the diffusion coefficient is inversely proportional to the atmospheric pressure and directly proportional to the absolute temperature raised to the 3/2 power.* Palmes notes that as pressure increases the decrease in the diffusion coefficient is balanced by an equal increase in concentration so that sampling rate is independent of pressure.[28] Similarly, the increase in the diffusion coefficient with temperature is partly offset by a decrease in concentration so that the effect of the temperature on net sampling rate, W, is small. Coefficients of diffusion for many compounds of industrial hygiene interest are given by Nelson.[33]

Gases and vapors to be collected by transport through the static air column pass through an attached boundary layer adjacent to the barrier films. If the thickness of the boundary layer is small in relation to the thickness L of the static air column, the effects of the attached boundary layer are correspondingly small. Air velocity in the vicinity of the monitor controls the boundary layer thickness. Table 16 illustrates velocity independence for a static layer thickness of 0.65 cm when air velocity is within usual personal environmental limits.

The time-weighted worker exposure level is calculated using the following equations:

$$mg/m^3 = \frac{\text{corrected weight on badge (nanograms)}}{\text{sampling rate } (cm^3/sec) \times \text{sampling time (sec)}}$$

$$ppm = \frac{mg}{m^3} \times \frac{22.4 - L/mole}{mw - g/mole} \times \frac{T^oK}{273K} \times \frac{760\ mmHg}{Pmm\ Hg}$$

* $D \propto \frac{T^{3/2}}{P}$ The increase in temperature from 5°C to 35°C would be: $\left\{\left(\frac{308}{275}\right)^{3/2} - 1\right\} \times 100 = 16\%$

TABLE 16

Velocity Independence for Monitors with Normal Angle of Incidence, L = 0.65 cm

Velocity at face of monitor	Test concentration (mg Hg/m³ air)	Measured concentration (mg Hg/m³ air)	2σ of measurement (mg Hg/m³ air)	Velocity effect (%)
35 ft/min at 90°	0.098	0.097	0.008	‡1.03
100 ft/min at 90°	0.098	0.098	0.014	0.00
250 ft/min at 90°	0.097	0.099	0.004	+ 2.06

Courtesy of 3M Company, St. Paul, Minn.

The sampling rate for each vapor is obtained from its badge calibration bulletin.

For the dosimeter to accurately sample the pollutant of interest, calculations indicate a minimum air velocity is required across the face of the dosimeter.[35] When this critical value is exceeded, the performance will be insensitive to velocity over a wide range. Experimental evaluation indicated no significant effect on dosimeter response at face velocities as low as 15 ft/min. Face velocities encountered in most industrial environments are above a minimum of 50 to 75 ft/min. Therefore, for most applications, the influence of external boundary layer resistance upon accuracy will be negligible. However, we must emphasize that the dosimeter will not yield reliable results in an essentially stagnant atmosphere.

c. Accuracy and Precision

Maintaining acceptable accuracy and precision for diffusional monitors requires attention to three distinct categories where errors can potentially be introduced. These encompassing areas are (1) the calibration procedure, (2) the sampling process, and (3) the analysis.

Calibration of diffusional monitors is accomplished by passing a known concentration of a gas or vapor over the monitor. The concentration and time of exposure are known, but the flow rate need not be accurately controlled so long as it results in a velocity within specified limits.

In the sampling process, accuracy depends upon the exposure time, velocity effects if present, and the aforementioned temperature effect. Precision in sampling is dependent on maintaining close dimensional control on sampler parts, particularly those controlling the static layer geometry.

In the analysis, accuracy is a function of the usual analytical standards and the degree of control over the procedure used. Precision is aided by minimizing the number of operations performed and the amount of handling of the sample itself.

For a diffusional monitor, measurement of total exposure time replaces the measurement of the pumped volume in a conventional flow monitor. Other factors controlling precision and accuracy are potentially common between the two methods and one must evaluate each case individually.

d. Applications and Performance

References by McCammon and Woodfin[36] and Tompkins and Goldsmith[35] describe applications of diffusional exposure monitors under laboratory and field conditions. Braun[37] describes use of a mercury vapor monitor in industrial and dental applications. Diffusional monitors are normally used for determination of exposures over an entire work shift. They can also be used as shorter term sampling devices if the sampling rate is taken into account. Diffusional exposure monitors may be termed "dosimeters" since they measure exposure dose. The data can be used as an estimate of absorbed dose, provided metabolic and other physiological factors can be measured by biological monitoring.[38]

i. Mercury Monitor

Figure 40 gives results from NIOSH accuracy testing of the 3M brand Mercury Vapor Monitor No. 3600. Table 17 shows results of further NIOSH testing and compares the 3M monitor to several pump methods. The precision of the 3M monitor is noted to be higher than that of the pump methods.

An evaluation of this mercury vapor dosimeter in the NIOSH laboratories in Cincinnati, Ohio concluded that excellent accuracy and precision was attained. The coefficient of variation throughout the tests was consistently between 0.003 and 0.005. Face velocities in the range of 50 to 250 ft/min produced no significant

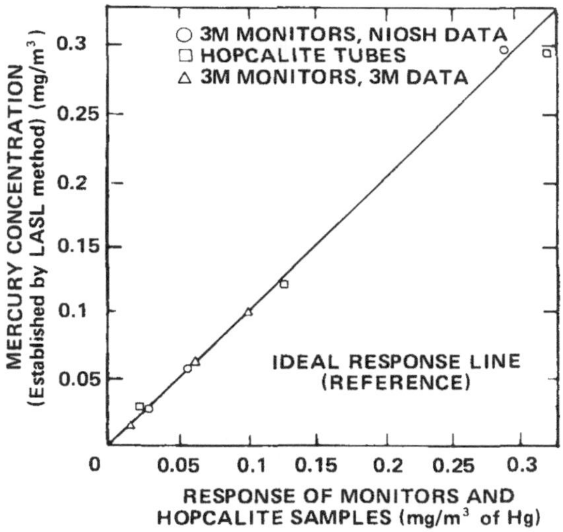

FIGURE 40. Correlation of mercury vapor monitoring methods. (Courtesy of 3M Company, St. Paul, Minn.)

TABLE 17

Results Showing 95% Confidence Limits (±2σ) for Mercury Vapor Monitors

Test concentration	3M Monitor No. 3600	Hopcalite sampling tubes	Charcoal sampling tubes	LASL[a] sampling tubes
1	0.060 ± 0.002[b]	0.059 ± 0.006	0.072 ± 0.044	0.049 ± 0.006
2	0.122 ± 0.002	0.119 ± 0.012	0.115 ± 0.038	0.109 ± 0.020
3	0.218 ± 0.012	0.221 ± 0.012	0.199 ± 0.012	0.219 ± 0.010

From 3M Company, St. Paul, Minn. With permission.

[a] Silver on Chromasorb P with Thermal Desorption.
[b] Values shown are milligrams Hg/m³ air ± 2 S.D.

variations in collection efficiency. High concentrations of chlorine did affect the results, but the magnitude of the effect in the Cl_2 TLV region was reduced. High concentrations of SO_2 and H_2S slightly altered the results. The monitors are temperature dependent, but a simple correction can be applied to correct the data for changes caused by temperature.[36]

ii. Organic Vapors (Solvents)

Figure 41 shows a diffusional organic vapor monitor where the sample elution occurs in the body of the monitor through use of a special ported cap. Eluant is added directly to the monitor and, after a standard elution period, the eluant is decanted directly into an auto-analysis vial. Results for this monitor are given in Tables 18 and 19.

In Table 18, the organic vapor monitor exhibits a coefficient of variation (σm) of between 1.3 and 3.3 % and the apparent accuracy of the measurements lies between 0.25% for measurements near 4 ppm of 7 to 8% for measurements near 1 ppm. The slight negative bias that increases as the concentration drops is the result of a diminishing recovery coefficient as the absolute mass of the analyte is reduced.

Table 19 shows excellent sub TLV agreement for benzene between the 3M diffusional organic vapor monitor and charcoal tube measurements. In work of this kind it is necessary to exercise strict control over all phases of sampling and analysis and data reduction.*

Another group has been active in the development of a similar dosimeter for organic vapors.[35] Although the design is somewhat different in minor details, the principles involved are identical (Figures 42 to 44).

* Prepared by David L. Braun, Monitoring Supervisor, 3M Co., St. Paul, Minn., 1978.

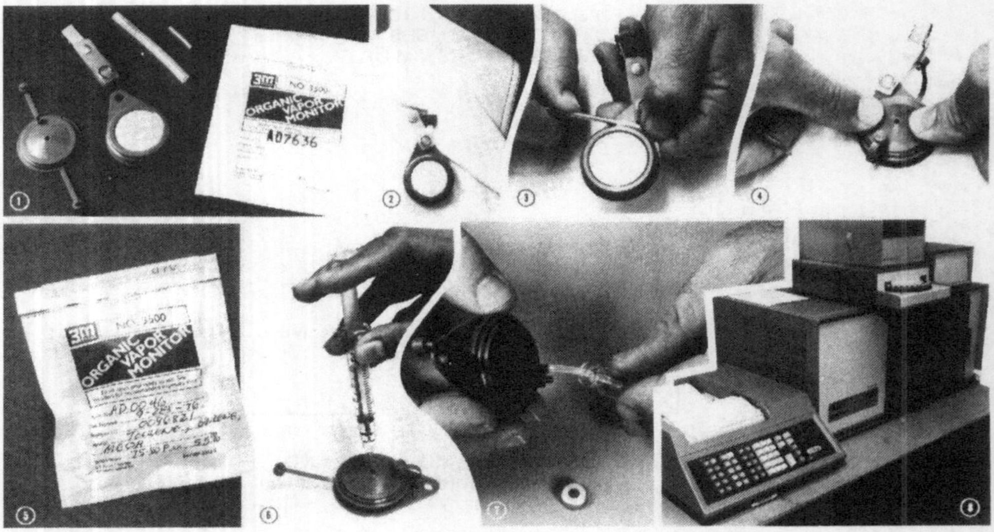

FIGURE 41. Sample collection and analysis with the 3M organic vapor. (1) The 3M brand organic vapor monitor NO. 3500; (2) monitoring near breathing zone; (3) front member is removed; (4) exposure is terminated; (5) monitor repacked for analysis; (6) eluent is added; (7) decanting into autoanalysis vial; (8) chromatographic analysis. (Courtesy of Environmental Monitoring Systems, 3M Company, Bldg. 218-3, 3M Center, St. Paul, Minn.)

TABLE 18

Laboratory Test Results Comparing the 3M Brand Organic Vapor Monitor No. 3500 to Test Concentrations

Test concentration (ppm)	3M Monitor No. 3500 measurement (ppm ± 2σ)
7.80	7.75 ± 0.43
4.13	4.12 ± 0.23
2.27	2.22 ± 0.06
1.32	1.23 ± 0.04
0.99	0.91 ± 0.06

Courtesy of 3M Company, St. Paul, Minn.

Figure 44 presents a schematic diagram of the Walden GASBADGE® dosimeter. Its component parts consist of 1) a removable protective cover designed to prevent contamination or exposure during shipment and/or storage, 2) a badge front with an opening to allow diffusion of nitrogen dioxide into the device, 3) a draft shield made from nonreactive porous material, 4) an open grid to define diffusion geometry and to minimize internal mixing, 5) a collection element formed from specially prepared carbon-impregnated collection substrate (replaceable), and 6) a back to which is attached a spring clip allowing it to be conveniently affixed to the individual to be monitored.

TABLE 19

Field Test Results Comparing the 3M Brand Organic Vapor Monitor No. 3500 to Charcoal Tube Measurements

Test description	Exposure (time-hrs)	3M No. 3500 (ppm)	Charcoal tubes (ppm)
Personal sample	5.90	0.424	0.400
Area sample No. 1	5.63	0.408 ± 0.020[a]	0.401 ± 0.06
Area sample No. 2	5.85	0.454 ± 0.010	0.450 ± 0.02
Grab sample	0.15	0.36	0.50

[a] ±2σ presented when applicable.

Courtesy of 3M Company, St. Paul, Minn.

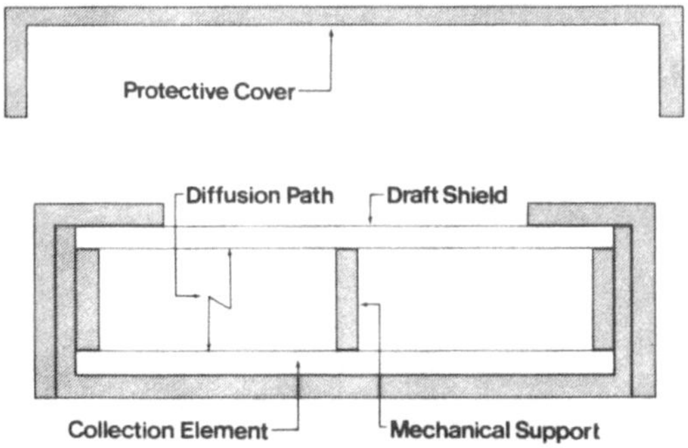

FIGURE 42. Walden Gasbadge® dosimeter schematic diagram. (Courtesy of Abcor Development Corp., 850 Main St., Wilmington, Mass.)

Specifications

Dimensions:	5.1 cm × 6.5 cm × 1.6 cm
Weight:	28 gms
Analysis Method:	Modification of NIOSH method POCM No. 127 (gas chromatograph)
Sampling Time:	Normally an 8-hr work shift
Sample Range:	¼ to 2 times TLV or Federal Standard
Shelf Life:	180 days
Interferences:	Refer to NIOSH method POCM No. 127. Classes of organics sampled: aromatic hydrocarbons, halogenated hydrocarbons and ketones. Mixtures in the 0.5-150 ppm range.

The Walden Gasbadge® for organic vapors was first described in a paper presented to the 1976 American Industrial Hygiene Conference in Atlanta, Ga.[35] Preliminary results included methyl ethyl ketone, ethyl acetate, styrene and benzene in the 13 to 140 ppm range (Table 20).

Later (June 1977) a detailed report in the form of a product bulletin that described a detailed study of benzene in the concentration range 1 to 5 ppm was released.[39]

A. Method: Three sets of 20 dosimeters containing activated carbon collection elements were exposed to nominal concentrations of 1, 2, and 5 ppm of benzene for 8 hrs in a dynamic test chamber. Benzene vapor at a concentration of 95.2 ppm was diluted with air to generate the desired exposure levels.

After exposure the collection elements were removed from the dosimeters, placed in capped vials, and desorbed with carbon disulfide. An aliquot of each desorbed sample was injected into a gas chromatograph. The area of the resulting peak was determined and compared with areas obtained from the injection of standards to determine the amount of benzene collected.

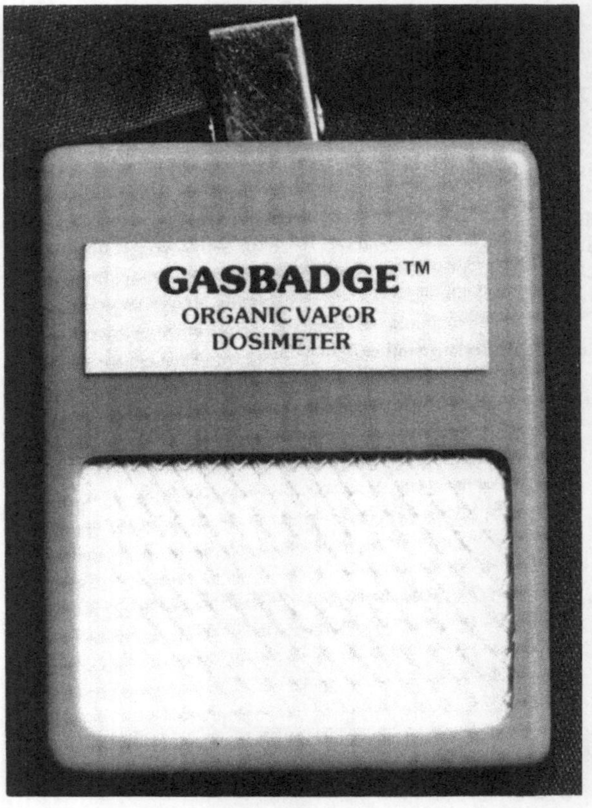

FIGURE 43. The Gasbadge® dosimeter in sampling position. (Courtesy of Abcor Development Corp., 850 Main St., Wilmington, Mass.)

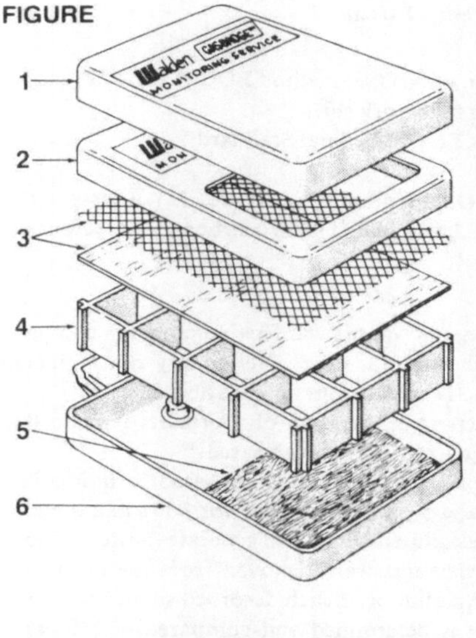

FIGURE 44. Schematic diagram of the Walden Gasbadge®. (Courtesy of Walden Research Division of Abcor Development Corp., 850 Main St., Wilmington, Mass.)

TABLE 20

Preliminary Gasbadge® Organic Vapor Results: Summary

Organic	$\overline{C}_\infty$ delivered (ppm)	Dosimeter value (ppm)
Methyl ethyl ketone	90.3	85.5
		92.0
		83.5
		90.0
Ethyl acetate	138.7	141.2
		131.9
		143.6
Styrene	21.4	21.7
		21.1
Benzene	13.4	10.9
		10.8
		12.1
		16.2
Benzene	13.0	13.5
		12.0
		16.4
		14.4

From Tompkins, F. C. and Goldsmith, R. L., *Am. Ind. Hyg. Assoc. J.*, 38, 371, 1977. With permission.

Results

The data indicated that the Gasbadge® organic vapor dosimeter meets the accuracy requirements of the OSHA Emergency Temporary Standard for benzene and the proposed permanent standard (Sections e-5 and e-6).[268] Ninety-eight % of the results fell within ± 25% of the true concentration for concentration greater than or equal to 1 ppm. Reference to Table 21 shows that the dosimeter yields substantially better results than obtained with charcoal tubes using an air sampling pump. The regression plot in Figure 45 for 55 data points with a correlation coefficient of 0.994 confirms the good performance obtainable with this type of personnel monitor.

By decreasing the volume of desorbent (CS_2) from 4.0 mL to 2.0 mL and increasing the aliquot size injected into the GC from 2.0 μL to 5.0 μL, the average recovery for 8-hr exposure samples was increased from 76 to 91% ± 3%.[40]

The U.S. Army Environmental Hygiene Agency evaluated the Gasbadge® Organic Vapor Dosimeter (Special Study No. S2-0002 77) for a variety of organic solvents including benzene.[41] Exposures were conducted in a 7-m^3 environmental chamber in which the predetermined liquid phase solvent was evaporated and the air circulated continuously by a small blower during evaporation and dosimeter exposure. After exposure the charcoal-impregnated glass wool pads were removed and sealed in glass vials to prevent loss of sample from the charcoal. Two different sizes of dosimeters were studied. The small model is approximately one-third the size of the larger one. Desorption with CS_2 was complete in less than 1.5 min.

Vapor retention for methyl chloroform (MC) was not complete. Dosimeters first exposed to 700 ppm MC for 1 hr and exposed for 5 hrs in MC-free air lost 17% of the adsorbed MC. These results indicate that if a worker is exposed to relatively high concentrations (peak exposure) early in his shift and sampling is continued for the balance of the shift, some adsorbed vapor may be lost before the charcoal pad can be transferred to the sealed storage vial.

The experimental data disclosed both minimum and maximum loading on the charcoal for acceptable recovery. For example, with benzene as the mg, absorbed (loading) dropped from the 0.37 to 0.26; the % recovery declined from 94 to 82 in an ambient 10-ppm concentration. Average recovery for methyl chloroform dropped from 90% at a 19.5 mg loading to 58% at a 25.5 mg loading. The critical point in the mass loading is a limiting factor in the use of these dosimeters as it is in the charcoal tube version. The relationship for five solvents is demonstrated by reference to Figure 46.[270]

It is obvious from these results that a back-up section is required in the dosimeter to conform to the construction of the charcoal sampling tube (Figure 47). This revision is under active study by the 3M investigators. In the Walden Gasbadge® the addition of a second carbon pad, separated by a suitable diffusion membrane, would suffice. After termination of the sampling, the two pads would be sealed in separate storage vials, and analyzed separately as is currently required for charcoal tubes.

TABLE 21

Performance of the Walden Gasbadge® Organic Vapor Dosimeter vs. Charcoal Tube Sampling for 8 Hours

Sampling procedure	Concentration (ppm) Known	Found mean ($\bar{x}$)	No. of samples (n)	Deviation — precision Standard (s)	Relative (%)	Accuracy systematic error[a] (%)
Gasbadge®	0.97	0.98	20	0.04	4.1	1.0
Gasbadge®	1.90	2.24	15	0.10	4.5	17.9
Gasbadge® (outside lab)	1.90	2.09	5	0.11	5.2	10.0
Gasbadge®	4.69	5.25	20	0.30	5.7	11.9
Charcoal tube	25	22.8	10	2.7	11.6	8.8

Gasbadge® Dosimeter for Organic Vapors, Technical Bulletin No. 0377, U.S. Patent 3,985,01f, Abcor Development Corp., 850 Main St., Wilmington, Mass. With permission.

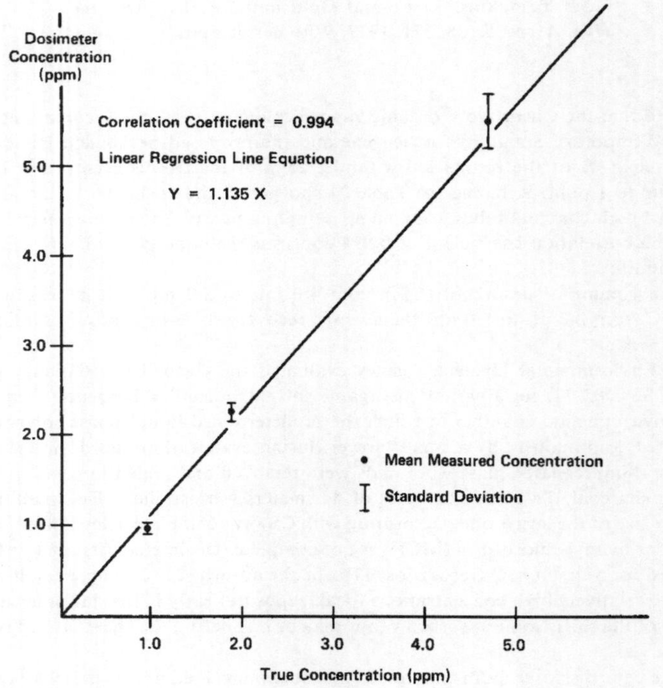

FIGURE 45. Linearity of Gasbadge® organic vapor dosimeter for monitoring benzene exposures. (Courtesy of Abcor Development Corp., 850 Main St. Wilmington, Mass.)

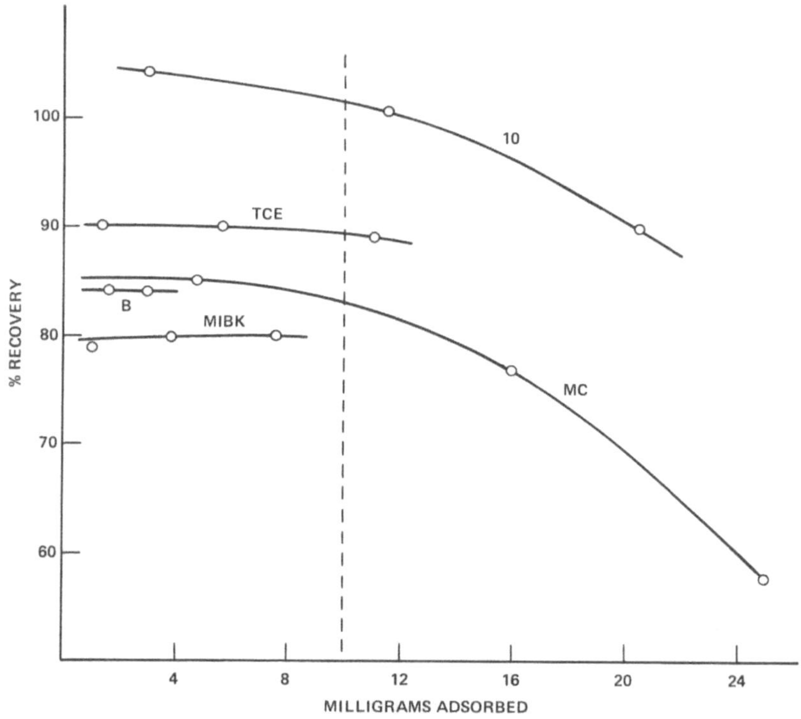

FIGURE 46. Walden Gasbadge® recovery vs. milligrams adsorbed of solvent vapor. IO = isooctone; TCE = trichlorethylene; MC = methyl chloroform; B = benzene; MIBK = methyl isobutyl ketone. (From Mazur, J. R., Bamberger, R. I., Esposito, G. G., Podolak, G., and Jacobs, B., *Am. Ind. Hyg. Assoc. J.*, 39, 701, 1978. With permission.)

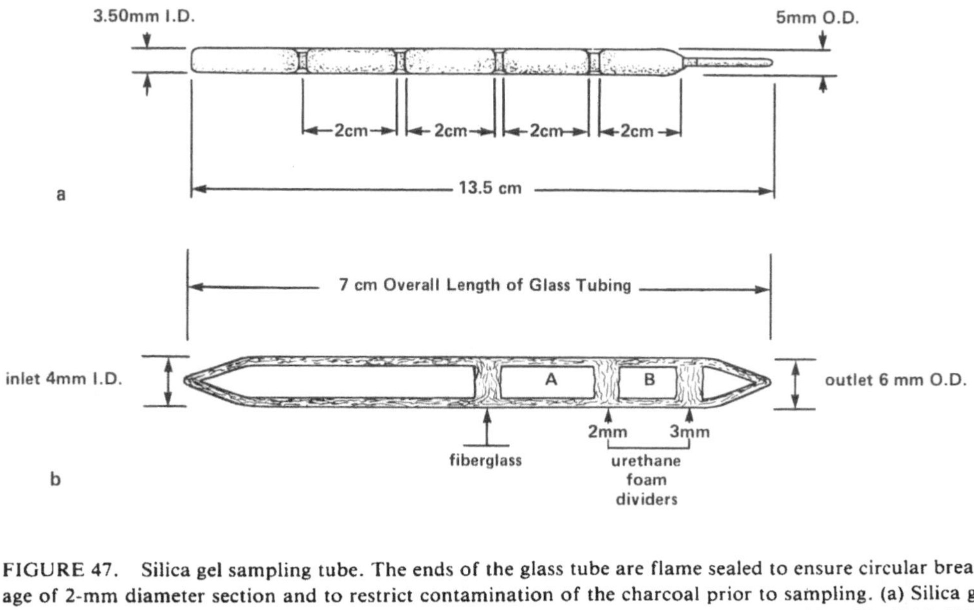

FIGURE 47. Silica gel sampling tube. The ends of the glass tube are flame sealed to ensure circular breakage of 2-mm diameter section and to restrict contamination of the charcoal prior to sampling. (a) Silica gel sampling tube. (From Campbell, E. E. and Ide, H. M., *Am. Ind. Hyg. Assoc. J.*, 27, 323, 1966. With permission.) (b) The charcoal sampling tube. (From Kupel, R. E. and White, L. D., *Am. Ind. Hyg. Assoc. J.*, 32, 456, 1971. With permission.)

TABLE 22

Walden Gasbadge® — Percentage Recovery vs. Relative Concentration of Solvent Vapor for 5-hour Exposure Time Interval

Compound	Ambiant conc. (ppm)	Total mg adsorbed	Average % recovery	Charcoal tube
Benzene	2.5	0.076[a]	76	
	10.0	0.37	94 ± 4	97
	20.0	0.71	89 ± 4	
Isooctane	75	3.0	104 ± 3	
	300	11.5	101 ± 3	NA
	600	20.5[b]	90	
n-Butanol	12.5	0.28[a]	64	
	50.0	1.5	84 ± 3	88 ± 5[c]
	100.0	2.9	84 ± 5	
MIBK	25	1.0	79 ± 4	
	100	3.8	80 ± 6	93
	200	7.5	80 ± 2	
n-Butyl acetate	37.5	1.5	93 ± 3	
	150.0	5.8	88 ± 4	95[c]
	300.0	10.5	84 ± 4	
Methyl chloroform	87.5	4.7	85 ± 2	
	350.0	15.8[b]	77 ± 3	100
	700.0	25.5[b]	58	
Trichloroethylene	25	1.4	90 ± 2	
	100	5.7	90 ± 3	95
	200	11.1	89 ± 3	

[a] Too low for accurate quantitation;
[b] Overload conditions;
[c] From the literature.

From Mazur, J. R., Bamberger, R. I., Esposito, G. G., Podolak, G., and Jacobs, B., *Am. Ind. Hyg. Assoc. J.*, 39, 706, 1978. With permission.

Storage of "sealed" dosimeters in a methyl chloroform contaminated (700 ppm) atmosphere disclosed inward leakage, i.e., recoveries exceeded 100%. However, the dosimeter cover did exclude benzene vapor during storage at 10 ppm concentration.

Recovery with the Gasbadge® was compared with comparable charcoal tubes for seven solvents (Table 22) in the one-quarter, one, and two times the TLV concentration ranges. Although percent recovery was somewhat less for the dosimeter, the performance is sufficiently reproducible to justify application of a correction factor.

A mixture of four solvents was sampled for 4 hrs and the recoveries compared in Table 23. Recoveries for each component compare favorably with individual component recoveries obtained previously.

An investigation of the temperature, dosimeter size, and charcoal lot variation parameters is summarized in Table 24. In a 700-ppm methyl chloroform atmosphere the recovery efficiency appeared to be independent of temperature in the 24 to 35°C (75 to 95°F) range and independent of variations in charcoal lots. Small badge efficiency was found to be better than the large badge performance (14%), but the authors discounted the significance as due to determinant errors.

Storage of the charcoal pads in the closed dosimeter case for two weeks disclosed significant loss of solvent vapor collected during monitoring (21% of methyl chloroform). Greater losses occurred under similar conditions with the smaller Gasbadge® (62%). Unless this leakage can be corrected, the exposed charcoal pads must be transferred to sealed glass vials as soon as sampling is terminated.

A recent newcomer, DuPont, into the passive monitoring for organic vapors by adsorption on charcoal was announced during the American Industrial Hygiene Conference in Chicago, Illinois on May 29, 1979. The unique design of the Pro-Tek® sampling badge permits vapor collection at two different rates. The elongated capsule is fitted with two covers (Figure 47A). With one cover removed, the sampling rate is

TABLE 23

Walden Gasbadge® — Percentage Recoveries for a Four Component Mixture Solvent Vapor Mixture — 4-hour Sampling Period

Components	Ambient conc. (ppm)	% Recovery	Reference[a] % recoveries
Isooctane	46.1	102	101—104
Benzene	80.0	86	89—94
MIBK	56.6	75	79—80
Butyl acetate	54.0	90	84—93

[a] % recovery in the absence of other components.

From Mazur, J. R., Bamberger, R. I., Esposito, G. G., Podolak, G., and Jacobs, B., *Am. Ind. Hyg. Assoc. J.*, 39, 701, 1978. With permission.

TABLE 24

Walden Gasbadge® Temperature, Dosimeter Size, and Charcoal Lot vs Percentage Recovery of Methyl Chloroform

Exper.	Temp. (°F)	Dosimeter size	Charcoal lot	Total mg adsorbed	% recovery
A	75	Large	A	15.7	98
	95	Large	A	16.4	101
B	75	Small	A	5.4	103
	95	Small	A	5.6	104
C	75	Large	A	15.0	90
	75	Small	A	5.2	96
D	75	Large	A	14.3	90
	75	Large	B	13.4	85

From Mazur, J. R., Bamberger, R. I., Esposito, G. G., Podolak, G., and Jacobs, B., *Am. Ind. Hyg. Assoc. J.*, 39, 701, 1978. With permission.

approximately 50 mL/min and with both covers removed 100 mL/min. Organic vapors are collected through a multi-cavity diffuser element and adsorbed onto the activated charcoal collection element. Diffusion coefficient data are available from DuPont. Draft shields or attenuating sheets which are normally needed for protection against windage effects are unnecessary with this badge. The collected solvent vapor is desorbed for laboratory GC analysis by conventional procedures. The small lightweight unit is delivered in a sealed pouch fully assembled with covers. To activate the badge for use, remove the protective covers and clip the badge to the employee's shirt collar. After exposure the badge is deactivated by replacing the protective covers and resealing in the original pouch with the enclosure provided.

A four-fold sensitivity is claimed for the Pro-Tek® badge. This improved sensitivity allows for sampling times as low as 15 min, with improved accuracy.[270a] The active charcoal strip contains approximately 300 mg of activated charcoal that provides sampling time up to 16 hr. The charcoal strip, because of its tensile strength and elasticity, can be transferred readily to a 2.0 mL automatic sampler vial for desorption and GC analysis without any loss of charcoal.

A recent (June 4, 1979) Abcor product bulletin (79-201a) was issued to correct previously published dosimeter constants as shown in Table 24A. The most recent model of the Gasbadge® is fitted with a sliding protective cover to prevent contamination or exposure for short periods of time (Figure 47B). Specifications also have been included (Table 24B).

iii. Acidic Gases[269]

An alkaline substrate (component 5 in Figure 44) collects the acidic gases — NO_2, SO_2, and H_2S — that diffuse into the internal cavity of the Gasbadge®.[35] The exposed dosimeter then is returned to Walden for

FIGURE 47A. The Du Pont Pro-Tek® organic vapor monitoring badge. (From Du Pont, Applied Technology Division Wilmington, Del. With permission.)

TABLE 24A

Dosimeter Constants from the Abcor Product Bulletin No. 79-201a

Solvent	Dosimeter sampling constant	To update Gasbadge® results calculated using published diffusion coefficients, multiply results by
Benzene	0.0816	1.142
Heptane	0.0626	1.086
Hexane	0.0692	1.058
Pentane	0.0773	1.089
Styrene	0.0691	1.014
Toluene	0.0715	1.187
Xylene	0.0642	1.075

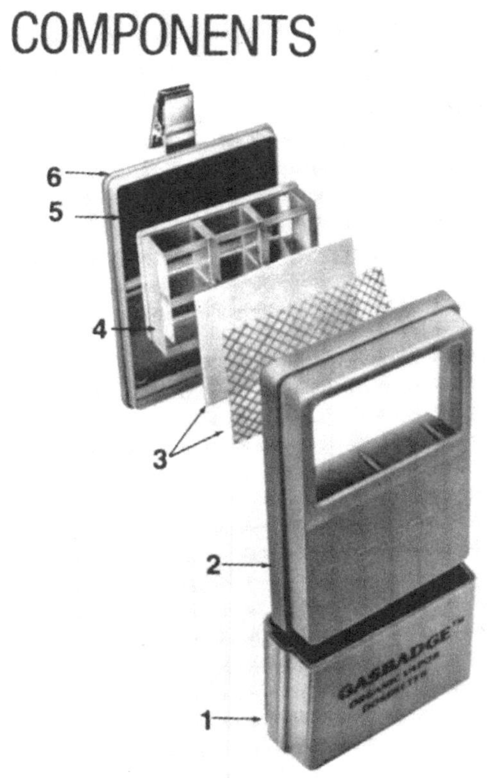

FIGURE 47B. The improved Abcor Gasbadge® fitted with a sliding cover.

TABLE 24B

Gasbadge® Specifications

Dimensions:	2 × 2 9/16 × 5/8 in. (5.1 × 6.5 × 1.6 cm)
Weight:	1.5 oz (43 g)
Power:	None
Collection element:	Activated carbon
Analysis method:	NIOSH P & CAM 127 (modified)
Sampling time:	8 hr nominal
Sampling range:	0.2—160 ppm/8-hr TWA (benzene)
Accuracy:	25%
Shelf life:	2 years
Air Velocity:	15 fpm (0.08 m/s) minimum, normal industrial plant air velocity exceeds 20 fpm (0.1 m/s)
Patent:	U.S. patent 3,985,017

TABLE 25

Walden Gasbadge® Dosimeter Performance Characteristics—Statistical Summary

Application	Average accuracy %[a]	Average precision % (S)	Percent within ±25% of true value
SO_2	13.4	13.8	100
NO_2	7.4	21.7	85
H_2S	11.1	15.3	80

[a] Systematic error (see Table 21).

From Tompkins, F. C. and Goldsmith, R. L., *Am. Ind. Hyg. Assoc. J.*, 38, 371, 1977.

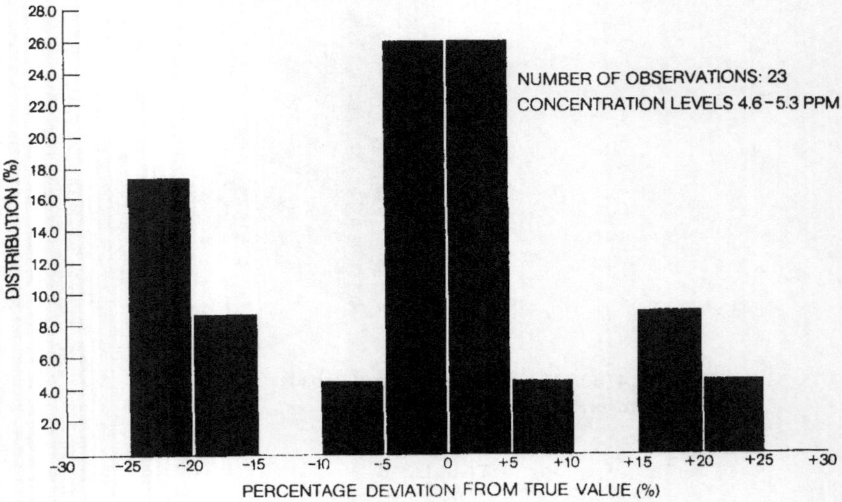

FIGURE 48. Walden Gasbadge® dosimeter: SO_2 distribution function. (From Tompkins, F. C. and Goldsmith, R. L., *Am. Ind. Hyg. Assoc. J.*, 38, 371, 1977. With permission.)

analysis and a report which satisfies OSHA record-keeping regulations. (See Technical Data Sheets supplied for these three acidic gases) Analytical procedures recommended by NIOSH are followed:[42] date of measurement, operation being monitored, sampling and snalytical method used, name, social security number, and exposure of the employee monitored. Except for the collection medium, the badges are identical to the Organic Vapor Gasbadge® (Figures 42 to 44).

The inherent performance capabilities of the dosimeter are illustrated in the Statistical Summary of Table 25. The distribution of results is shown in Figures 48 to 50. The distribution of results for the SO_2 dosimeter (Figure 48) is far from normal and indicates determinate errors. Likewise, the distribution found for NO_2 monitoring is not normal and indicates determinate errors (Figure 49). The distribution of H_2S results shows a log-normal, nonlinear, relationship (Figure 50). Although regression analyses for SO_2 and NO_2 dosimeter results are linear, the slopes confirm this observed bias in the collection data (Figures 51 and 52).*

Application of the Organic Vapor Gasbadge® for acrylonitrile (AN) monitoring has been validated.[43] Both charcoal tubes and dosimeters were confirmed to be acceptably accurate for monitoring exposures to AN within the limits of ambient temperature (18 to 36°C), relative humidity (30 to 87%), and known interferences. Statistical analysis of results are summarized in Table 26. The evidence certainly supports a conclu-

* Furnished by John C. Gillespie of Abcor, 1977. With permission.

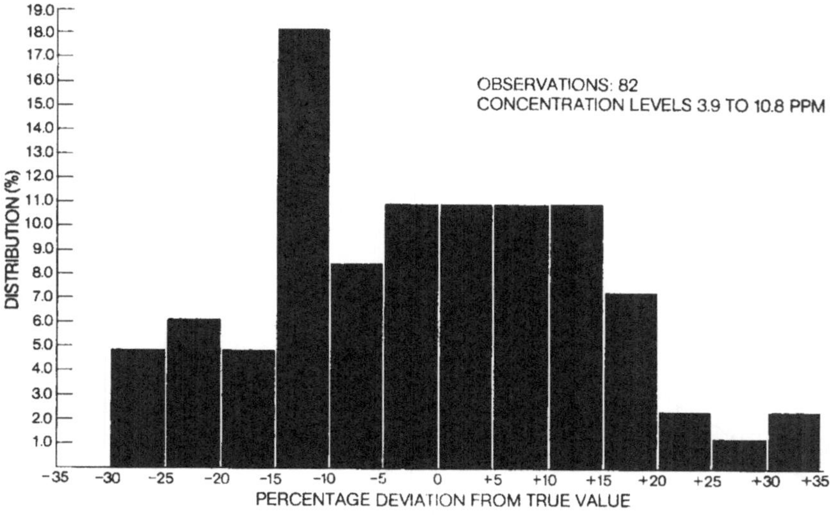

FIGURE 49. Walden Gasbadge® dosimeter: NO_2 distribution function. (From Tompkins, F. C. and Goldsmith, R. L., *Am. Ind. Hyg. Assoc. J.*, 38, 371, 1977. With permission.)

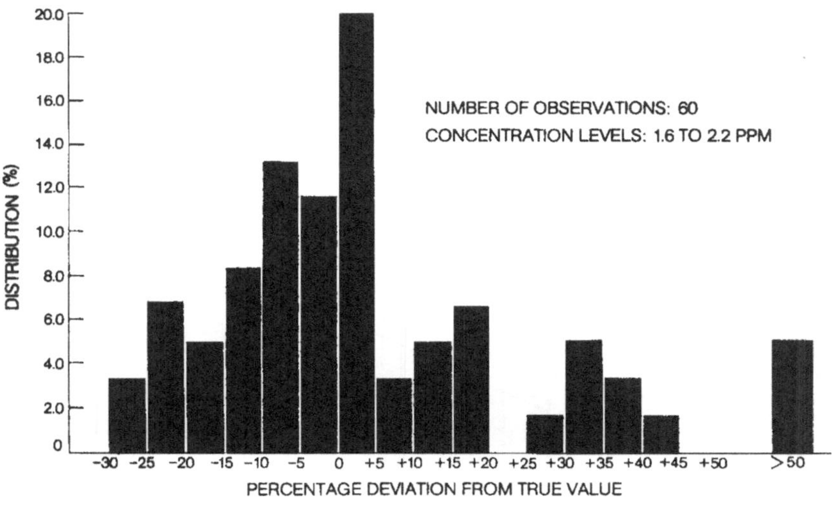

FIGURE 50. Walden Gasbadge® dosimeter: H_2S distribution function. (From Tompkins, F. C. and Goldsmith, R. L., *Am. Ind. Hyg. Assoc. J.*, 38, 371, 1977. With permission.)

sion that this activated carbon technique is a highly reliable procedure for the collection of acrylonitrite vapor from industrial atmospheres in concentrations down to less than 1 ppm. To increase desorption efficiency, 2% acetone was added to the CS_2 (increased from 58 to 94%).

Application of the Gasbadge principle has been investigated also in England. Studies conducted by Bailey and Hollingdale-Smith with polypropylene and silicon rubber membranes for collection of carbon tetrachloride, styrene and dichlorodifluoromethane on charcoal indicated these devices are ideally suited for occupational hygiene

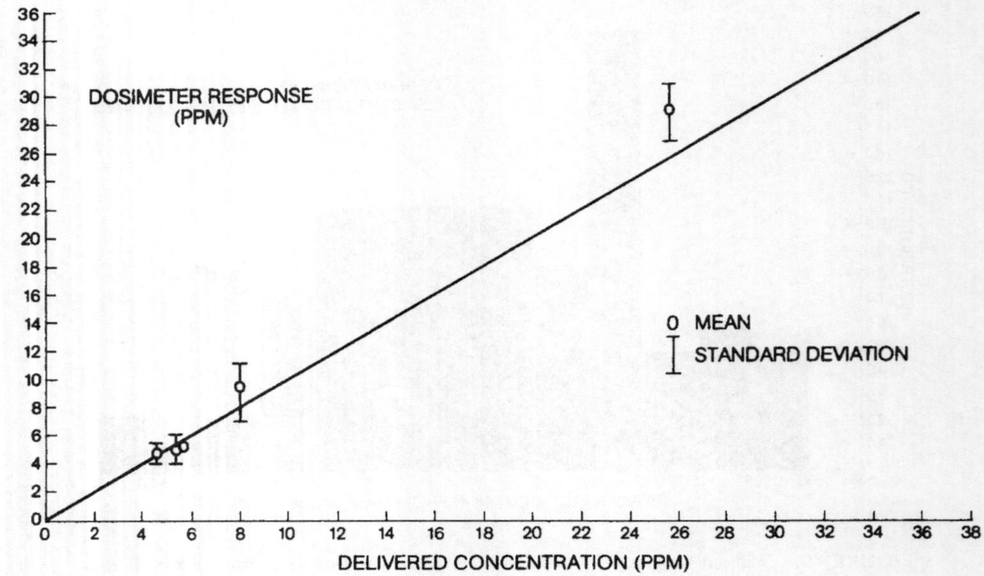

FIGURE 51. Regression analysis. Linearity plot: SO_2 Gasbadge® dosimeter. (From Tompkins, F. C. and Goldsmith, R. L., *Am. Ind. Hyg. Assoc. J.*, 38, 371, 1977. With permission.)

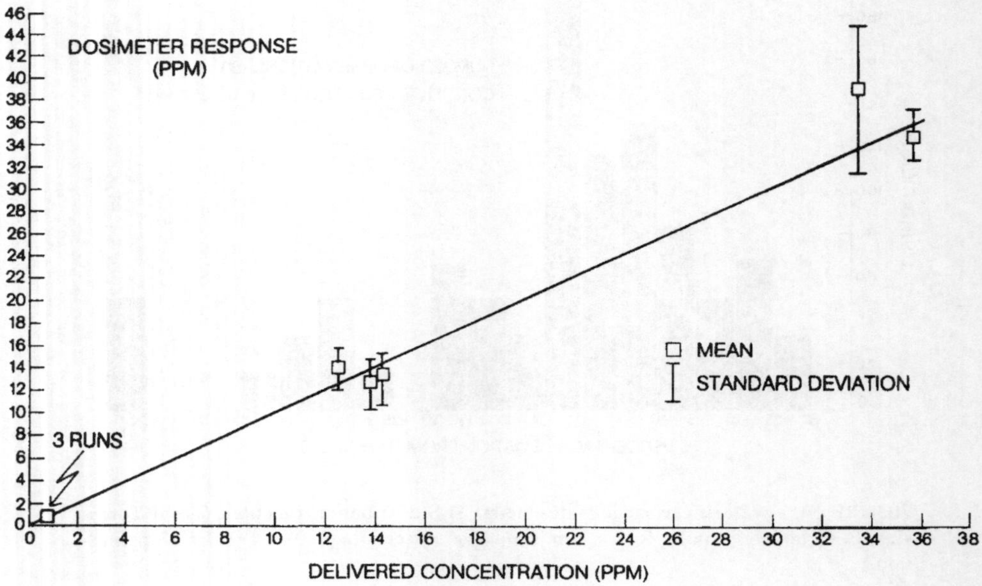

FIGURE 52. Regression Analysis. Linearity plot: NO_2 Gasbadge® dosimeter. (From Tompkins, F. C. and Goldsmith, R. L., *Am. Ind. Hyg. Assoc. J.*, 38, 371, 1977. With permission.)

and air pollution monitoring. Although experimental work was not presented, the authors mention the use of glass filter paper soaked in sodium tetrachloromecurate and dried for quantitative collection of SO_2 and lead acetate paper for H_2S.

Carbon tetrachloride in the concentration range of 2 to 300 ppm was collected on charcoal cloth discs after diffusion through silicone rubber membranes (Silescol sili-

TABLE 26

Validation of the Gasbadge® for Acrylonitrite Monitoring — Summary of Data

	Concentration (ppm)			Deviation — precision		
Sampling procedure	Known	Found mean (x)	No. of samples	Standard (S)	Relative (%)	Accuracy systematic error (%)
Gasbadge®	8.90	8.67	16	6.4—9.7	—	2.5
Gasbadge®	19.0	19.1	7	15—20	—	0.71
Gasbadge® group	8.9—19.0	—	23	1.28	10.1	3.3
Gasbadge®	0.74	0.75	10	0.03	4	1.4
Charcoal tubes	—	0.8—3.8	18	—	—	Reference
Gasbadge®	—	1.6—4.1 2.73	18	—	—	+ 28

cone rubber reinforced with polyester net). Permeation rates for concentrations over the range of 2 to 420 ppm were rather constant (0.059 to 0.069 μg/min/ppm) and practically no loss occurred in storage over five days (94% recovery). Plots of exposure time vs. weight of sample collected were linear.

Similar results were obtained with styrene diffused through a porous polypropylene membrane in the range of 2 to 60 ppm (Celgard 4500 from Amcel Ltd.). The diffusion rate in this case was 0.073 μg/min/ppm with good storage stability. Surprisingly good results were obtained with the collection of difluorodichloromethane vapor through porous polypropylene membranes on charcoal in the concentration range of 120 to 480 ppm. The diffusion rate fell in the range of 0.38 to 43 μg/min/ppm, and storage stability was acceptable (94% recovery after six days). At a dose of about 10,000 ppm (CT value), a change in the rate of adsorption occurs. This critical value probably is close to the saturation capacity of the charcoal.

The linearity of the results and the constancy of the normalized uptake rates confirm that the rate of sampling is diffusion controlled over the time/concentration conditions likely to be encountered in the industrial atmosphere. The authors also obtained excellent results for toluene, trichloroethylene, methylethylketone, dichloromethane, halothane, and tri-*n*-butylphosphate.[44]

Data on diffusion coefficients, solubility, and permeability for a wide range of substances have been made available by General Electric, Schenectady, N. Y.[45]

A dosimeter which combines membrane permeation and adsorption of vinyl chloride vapor on charcoal for laboratory analysis was developed by Nelms and his co-workers and described in 1977.[46] The system is insensitive to temperature and humidity variations and is relatively free from ordinary interferences (SO_2, NO_2, ozone and Cl_2). Single-backed dimethylsilicone rubber membrane (General Electric Co.) provided satisfactory selective diffusion. The system was dynamically calibrated and evaluated in a continuous flow system in which a gravimetrically calibrated permeation tube supplied the vinylchloride. The adsorbed vinyl chloride was analyzed by gas chromatography. The only interference reported for ethylene dichloride was eliminated by switching from thermal desorption to carbon bisulfide extraction. The design of the dosimeter was the same as previously described except charcoal was substituted for the original liquid absorption medium.[47]

3. Evacuated Container Collection — Limiting Orifice Control

Development work carried out at the Naval Research Laboratory culminated in a

TABLE 27

Maximum Sampling Intervals for Various Orifice Diameters and Sample Bottle Volumes Times for Attaining 1/2 Atmosphere Within the Sampler in Air[199a]

Volume of the sampler at STP Cm^3	Sampling duration for various orifice diameters—orifice diameter in micrometers (μm)					
	1	2	3	5	7	10
50	44.2	11.0	4.9	1.8	0.9	0.4
100	88.3	22.1	9.8	3.5	1.8	0.9
150	132.5	33.1	14.7	5.3	2.7	1.3
200	176.6	44.2	19.6	7.1	3.6	1.8
500	441.5	110.4	49.1	17.1	9.0	4.4

[a] The sampling unit is commercially available from MDA Scientific, Inc., 808 Busse Highway, Park Ridge, Ill.

personnel monitor based on the flow of the air sample through a rate controlling critical orifice into an evacuated stainless steel container. The sample flow remains constant through a small orifice as long as the internal pressure of the evacuated container remains less than 50% of the external atmospheric pressure during the sampling period. Maximum sampling intervals for selected sampler volumes and orifice diameters are presented in Table 27. The dimensions of the recommended sampler were 6.35 cm wide × 7.32 cm high × 2.54 cm deep and approximately 100 mL capacity, less two valve stems. The evacuation and sampling valve stood 5.41 cm high, and the tube which housed the critical orifice and the glass-fiber guard filter stood 6.35 cm high. The "box" was constructed of 0.1 cm stainless steel sheet with square corners and center internal reinforcing to provide dimensional stability. Details of construction are provided in the first reference.[48] The 175-L environmental test chamber employed for evaluation of the sampler is also described in detail.

A typical analysis on a time-weighted-average (TWA) basis of a variable, but known, carbon monoxide concentration in air indicated very close agreement with direct analysis of the environmental test chamber atmosphere.

Direct analysis	Sampler (7.5 mm orifice)
138 ppm	140 ppm
438 ppm	446 ppm

No low molecular weight bias (diffusion across the orifice) was found from the collection of a mixture of hydrogen and hexane in air. However, high adsorption losses on the walls of the test chamber and the sampler were encountered with high molecular weight and high boiling compounds (n-decane), 55 ppm from the heated container vs. a 70-ppm calculated concentration. This finding is another example of the major deficiency encountered when air samples are collected in rigid containers — high adsorption losses. For those relatively inert low molecular weight gases and vapors such as methane, carbon monoxide, vinyl chloride, and hexane, excellent agreement with direct chamber analysis was obtained with sampling through a 3-mm orifice over a 9.8 hr period. Accuracy was satisfactory for a 100-mL sampler equipped with a 12.5 mm orifice for carbon monoxide in the range 4 to 168 ppm (± 4.7% deviation in the TLV range). Mean deviation at 18 to 19 ppm was 0.2 (precision).

Major revisions in the gas chromatograph analytical system were required to accommodate analysis under reduced pressure conditions.[49] Procedures for ethane, dichlorodifluoromethane, vinyl chloride, trichloromonofluoromethane, vinylidene chloride, hexane, benzene, total hydrocarbons, methane, and carbon monoxide were developed for use with the reduced pressure contents of the personnel monitoring "can".

II. DYNAMIC SAMPLING

A. Grab Sampling — Short Term

1. Direct Reading

a. Paper Detectors

A closer approach to quantification is attainable if the sample volume is known precisely. Earlier literature described manually operated piston type exhausting pumps that draw a known volume of air through test paper clamped in a circular metal holder. The most sophisticated design was developed by the Department of Scientific and Industrial Research in England during the period 1938 to 1943 for the semi-quantitative detection of hydrogen sulfide,[50] hydrogen cyanide,[51] sulfur dioxide,[52] phosgene,[53] and arsine.[54] The pump had a 32-mm bore and a 126-mL volume and could be fitted with a stroke counter. A 50-mm square of the test paper was clamped in the threaded aluminum holder under a rubber sealing gasket. The circular retaining ring encompassed a 38-mm diameter field for collection of the contaminant as the air sample was drawn into the pump through the detector paper. Equidiameter color standards printed on heavy paper were supplied with each copy of the method. The concentration range was extended by multiple strokes of the pump, i.e., in the case of SO_2, as many as 10 strokes to detect less than 4 ppm.[52] The sampling rate approximated 2.5L/min (3 sec/126-mL stroke).[52]

i. Phosgene

Phosgene derived considerable attention as a poison gas during World War I, and its detection in the ambient atmosphere has been rather intensively studied. A comprehensive literature review disclosed at least 48 references to procedures developed for its detection in the trace ranges. To obtain sensitivity in the 1-ppm range for phosgene with the British hand pump, 85 piston strokes were required. On the basis of 4 sec/stroke (3 to retract, 1 to return) the sampling time would have been approximately 6 min and the volume 10L. In 1966, the TLV was lowered from 1 to 0.1 ppm; therefore, this manual system became highly impractical. However, if the color generating reagent were sufficiently stable, this system could serve for personnel monitoring with a miniature battery-driven pump to draw the air sample throughout the detector paper (see Chapter III). Then the reduced TLV would be within practical sampling limits, that is, 1 hr and a 107-L sample; or, if the sampling rate were reduced to 320 mL/min, 8-hr shift monitoring could be established. However, Harrison's reagent (diphenylamine + 4-dimethylaminobenzaldehyde) is not stable for more than 1 hr, and calibration of alternative reagents must be considered.[55]

Greater sensitivity was derived by substituting either *N,N*-dimethylaniline[56] or *N*-ethyl-*N*-2-hydroxyethylaniline (blue at 0.25 ppm) in place of diphenylamine as the color-forming component with the 4-dimethylaminobenzaldehyde, but again, these detector papers were not sufficiently stable for extended sampling and were sensitive to acidic gases and chlorine.[55] Probably the most promising reagent with respect to sensitivity and specificity for phosgene was described by Lamouroux in 1956.[58] The system, 4-(4′-nitrobenzyl)pyridine (NBP) + *N*-benzylaniline, produced a sharp, distinct, and easily detected color change from yellow to pink to brilliant red on contact with phosgene concentrations as low as 5 ppb.[55,59]

Whatman® No. 1 filter paper was saturated with 5% aqueous sodium carbonate, dried, then saturated with a solution of 4% *N*-benzylaniline and 2% NBP in benzene, and dried. The sodium carbonate was added to provide greater tolerance for volatile acids which would otherwise interfere at acid concentrations greater than 500 ppm. This paper, prepared, dried, and stored in an atmosphere free of acid vapors and phosgene, had a shelf life in excess of 5 months.[55] An airflow of 190 mL/min for each cm^2 of exposed filter paper area for 5 min ±10 sec provided color response proportional to the amount of phosgene in the range 0.002 to 2 ppm. A somewhat higher range, 0.25 to 10 ppm, was obtained a higher rate for a shorter period, 230 $mL/cm^2/min$ for 40 sec. Permanent glass color standards for this procedure were available from abroad.[59] These results indicate that application to personnel monitoring would require only adjustment of rates and volumes and recalibration. The color formation related to sampling rate as well as to concentration; therefore, if the rate were reduced to 19 to 20 $mL/cm^2/min$, the time required for detection of the 0.1.ppm TLV within the range 0.05 to 0.2 ppm probably would fall in the range of 7 to 8 hr.

The NBP test paper may be applied also to gas titration by determining the volume of air required to produce the first perceptible pink coloration. This coloration, when monitored by a miniaturized photoelectric sensing circuit, provides an integrated alarm system in the personnel monitor (see Unico® Portable Personnel Carbon Monoxide Alarm). For example, 5 ppb can be detected by passing a 10-L sample through an indicator disk. By equating the volume of air required to duplicate this color, estimates of unknown concentrations of phosgene can be made.[55]

ii. Hydrogen Cyanide

Other detectors for use in the hand pump include hydrogen cyanide sensitive paper impregnated with either benzidine and copper sulfate (blue) or Congo red and silver nitrate (blue).[51,60,61] These papers must be moist when used, are not stable in storage, and are subject to interference from H_2S, SO_2, and acidic and alkaline reacting gases, however. The Prussian blue test (ferrous sulfate + sodium hydroxide) is specific for HCN but requires immersion in 30% H_2SO_4 or dilute (1:4) HCl for color development (blue).[61,60-62] Although not sufficiently sensitive for grab sampling, on an 8-hr basis this detector could provide satisfactory cumulative results for personnel monitoring, since a 360-mL air sample is sufficient to detect 2.5 ppm HCN, and 10 ppm yields a bright blue color.[60] The major weakness of this method is the requirement for a wet reagent treatment to develop color.

iii. Arsine

Arsine may be detected on filter paper impregnated with H_gCl_2 (yellow to orange).[54,61] A hand pump-filter assembly which was designed to draw a reproducible volume of air through an arsine sensitive detector filter has been developed and offered for sale on the domestic market.[63]

iv. Sulfur Dioxide

Sulfur dioxide on contact with a mixture of starch-potassium iodate-potassium, and iodide-glycerine supported on filter paper will produce a brown stain in the 4- to 10-ppm range. However, close attention to details of preparation are necessary to ensure papers which are sensitive, yet stable in storage. Inasmuch as this reagent appears promising for personnel monitoring, but reference to its preparation is not readily available, details are included here.

To 60 mL distilled water, add 12 mL of approximately *N*/10 aqueous barium hydroxide. Heat to boiling and add 1 g potassium iodate and 2 g potassium iodide, both of AR grade.[64] When completely dissolved, add a thin paste of 1 g of soluble starch in 10 mL distilled water and continue boiling for a few minutes. Cool to room temperature, add 30 mL AR glycerine, dilute to 100 mL, and mix thoroughly. The pH should be approximately 9.5; if it is not, add more barium hydroxide solution dropwise. Cut extra-thick white filter paper (18 × 24 to 60 lb) into strips 50 mm wide, immerse 30 sec in the reagent, drain excess liquid, pre-dry by pressing between sheets of dry filter paper two or three times, and dry 10 min in a well ventilated oven at 50°C. The papers should not be limp, but should have dried sufficiently to regain some of their original stiffness; otherwise the stain will be spotty and not evenly distributed on the exposed paper. Discard 1 in. length from top and bottom of the strip and cut into 75-mm lengths. Preferably the test papers should be used within a few hours, but may be stored in a well-stoppered bottle in the dark for 1 to 2 days. The reagent is stable for about 7 days.[52]

A more recent procedure[72] recommended by the British Ministry of Labor[65] employs ammoniacal zinc prusside reagent, which turns red on contact with SO_2. Between 1 and 20 ppm SO_2 can be detected in a 360-mL air sample (three pump strokes) at a rate of 90 to 180 mL/min. This method is highly specific for SO_2, and only very slight interference from H_2S is encountered. Color standards are available,[65,66] preparation is simple, and the papers may be stored for at least a month. After mixing 50 mL of 6% aqueous zinc sulfate heptahydrate solution with 50 mL freshly prepared 10% aqueous sodium nitroprusside solution, ammonium acetate is added under agitation until the zinc nitroprusside is dissolved. Then 20 mL glycerine are added and mixed thoroughly, and the filter strips are dipped immediately — before zinc prusside again precipitates. Excess reagent is drained off, and the strips are air-dried at room temperature.[60]

If obtainable in sufficient size (area) from Kitigawa, the paper impregnated with the bluish-red dye, which bleaches to a brownish-yellow color (see Section I.A.1.a, "Convection Activated Detectors"), would provide an ideal personnel monitor with a battery-powered pump to draw through a known volume of air.[4] However, storage stability may be a problem that remains to be solved.

Alkali-treated filter paper has been used to collect SO_2 for quantitative analysis in the laboratory by the standard West-Gaeke procedure (reacton of SO_2 + pararosaniline + formaldehyde in the presence of sodium tetrachloromercurate).[60,62] The impregnating solution contains 20% potassium hydroxide and 10% triethanolamine or glycerine. The uptake, after removing excess reagent by dabbing with dry filter paper and drying at 110°C, is approximately 13 mg of solution per cm^2 of filter area. The air sample is drawn through at a rate not exceeding $1L/cm^2/min$. The air sample should contain not less than 30% relative humidity and not more than 5 ppm SO_2 (not more than 10% of the available alkali should be consumed) at this flow rate, which should be reduced proportionately for higher concentrations. After washing into dilute aqueous sodium tetrachloromercurate, the colorimetric analysis procedure is applied.[60,62] Recoveries that exceed 95% have been reported, and the method can be used for short-term or routine daily and weekly analyses.

v. Hydrogen Sulfide

Lead acetate impregnated papers also have been used for hydrogen sulfide (H_2S) estimation with the hand pump and could be applied directly to personnel monitoring. In this application glycerine must be added as a humectant.

vi. Mercury

Test paper for the semiquantitative detection of mercury vapor has been developed commercially for stationary monitoring (General Electric Co.).[61] The active ingredient is selenium sulfide powder applied by rubbing into the pores of the filter paper. The stability and sensitivity are excellent (0.12 ppm in 8 min). The color change from yellow to black is considered satisfactory; however, application is limited to an optimum temperature of 70°C.[61]

Possible application of the dithizone colorimetric technique to the estimation of mercury was suggested by the development of quantitative (90 + %) collection on filter paper impregnated with potassium triiodide [from aqueous KI + I_2]). The excess iodine is reduced with sodium sulfite. In the reference procedure the mercury was quantitated by UV spectroscopy after reduction with $SnCl_2$ and collection on a copper coil. However, spraying with the dithizone reagent after destruction of the excess iodine would develop the red mercury dithizonate which could be compared visually with known standard stains.[67]

vii. Lead

Although wet chemical color development methods are not entirely desirable for field application where laboratory facilities are not provided, their utility when available in readily portable kit form should not be ignored. In some cases, no other alternative is practical. For example, a colorimetric semiquantitative field method for the determination of inorganic lead fumes and dust in air was derived from the pink- to-red complex that forms when tetrahydroxyquinone comes in contact with lead.[68] The insoluble compounds such as lead sulfide, arsenate, carbonate, chromate, and oxides, as well as the water soluble salts, react quantitatively. The intensity of the red stain on the filter paper is proportional to the amount of lead collected and is estimated by visual comparison with color standards.

The lead-containing dust is collected on No. 42 Whatman filter paper by drawing the air sample through a 2.0-cm diameter exposed area at a 4 to 5L/min rate for 10 min (1.25 to 1.60L/cm²/min). To the center of the filter field then are added two drops of freshly prepared orange colored (0.04 mL) tetrahydroxyquinone reagent (0.3 g of the disodium salt dissolved in 0.5 mL distilled water + 1 mL acetone). After approximately 20 sec, two drops of pH 2.8 buffer, prepared by dissolving 15 g tartaric acid and 18.9 g sodium bitartrate in 100 mL distilled water, are added. Since the air sample volume is known, the concentration can be estimated by comparing the color density of the developed stain with color standards prepared by analyzing a series of exposed filters by the dithizone method. This calibration technique compensates for the typical speckled effect produced by insoluble lead dust, but not by soluble lead compounds, when applied by the usual methods for standard preparation.

The original reference[68] supplies the following description:

> A series of atmospheric samples was collected close to a source of lead fumes at different intervals of time. Stains were developed on each filter as described above, and grouped according to the color intensity. Six filters showing the same stain intensity were grouped together. Five of these were analyzed for lead content by the dithizone procedure and the sixth was set aside for a visual color standard. The lead in air concentration was calculated from the volume of air sampled, and from these groups four standards corresponding to 0.100, 0.200, 0.300 and 0.400 mg Pb/m³ were selected. For convenience in estimating lead concentrations, the four color standards were arranged in a radial pattern on a circular disk which contained a circular center aperture. Unknown stains were positioned in this aperture for color comparison. These standards were found to be stable for at least six months. However, in the commercially available field kits, the color standards are made up from stable printing inks.*

* Bendix lead-in-air detector, Bendix Corp., Environmental and Process Instruments Division, Lewisburg, W. Va.

Experience in the field indicated that the error of visual estimation of lead concentrations was not greater than 0.05 (±0.025) mg/m³. Beyond 0.4 mg/m³ the stain intensity becomes too dark for visual comparisons; therefore, the lead content must be determined by laboratory procedures or a smaller air sample taken. The Whatman No. 42 filter dissolved easily by heating in a mixture of 0.2 ml H_2SO_4 and 1 ml HNO_3 in a silica evaporating dish for analysis by either atomic absorption,[69] polarographic analysis,[70,71] dithizone colorimetry,[73] or the non-destructive X-ray emission spectrography[74] methods. In this fashion the colorimetric estimation could be confirmed or precisely quantitated at a later date if desired.

Blanks should be carried out along with the samples to confirm satisfactory level of contamination control in handling the filter discs and for background lead in the paper itself. Concentrations of lead intermediate between the 0.10 color standard intervals were estimated by interpolation in 0.025 increments. No interference was detected from Zn, Sb, As, Cr, Ca, Bi, Cu, or Fe in concentrations equal to the lead content. When concentrations of Cd were twice the lead content, a slight interference was observed.*

The need for adjustment of the sample volume for personnel monitoring is obvious, as the color intensity otherwise would be saturated in 10 min (0.40 mg/m³ limit). Therefore, the sampling rate must be reduced to 52 mL/min/cm² (equivalent to 164 mL/min for a 2.0-cm diamter filter) for a 4-hr sample or 25 mL/min/cm² (82 mL/min for a 2.0-cm diameter filter) for an entire 8-hr day in order to bracket the current TLV for inorganic lead dust in air (0.15 mg/m³).[1]

The red color of lead dithizonate also has been adapted to colorimetric paper analysis. Lead-containing dust is collected on No. 54 Whatman filter paper, solubilized by treatment with a mixture of *n*-butanol saturated with 3 *N* aqueous hydrochloric acid, and then sprayed with 0.1% dithizone in chloroform reagent and 0.1 *N* aqueous sodium carbonate solution in turn. The change in background color from green to pale orange provides a built-in indicator for full development of the red lead dithizonate complex. The stains are stable for more than 6 months after development. Other heavy metal ions that also produce colored complexes include Bi, Cd, Co, Mn, Hg, Ag, and Sn. If lead or any one of these other metals occurs singly in the air sample, this procedure can be applied directly. For mixtures, a paper chromatographic separation directly on the paper with the butanol-HCl developer has been described.[68] Furthermore, when the sample is concentrated onto a narrow paper strip, the stain length becomes proportional to the amount of lead (or any other color-forming metal) present. Then concentration can be estimated by reference to calibration charts in the same fashion as employed for length of stain detector tubes described in Section II.B.1, "Impinger Systems — Liquid Reagents,"[75] and confirmed by laboratory analysis as described previously for the tetrahydroxyquinone procedure.[68]

viii. Fluorides

In at least one case, a double-impregnated paper has been recommended for collecting separately the particulate and gaseous pollutant phases. The upper filter is treated with sodium citrate to retain particulate fluorides and the lower (downstream) paper with sodium hydroxide to absorb gaseous fluorides (HF).[76,77,77a] Although no specific reference to a direct color development system for either layer was described, references to support direct reading reagents based on the fading of the violet zirconium alizarin lake to the yellow color of the dissociated dye could be applied directly. Another color development system is based on the bleaching by fluoride ion of the intense red color of alizarin sulfonate absorbed on thorium hydroxide, which can be equated to concentration.[80]

A squeeze bulb type hand pump fitted with a filter holder has been offered for detection of fluorides in grab samples. The air sample is drawn across a replaceable impregnated paper that changes color on contact with HF. Concentration is determined by comparison with the reference chart delivered with the kit. However, the reagent was not identified, nor was a reference to extended sampling mentioned.

* From Quino, E. A., *Am. Ind. Hyg. Assoc. J.*, 20, 134, 1959. With permission.

ix. Dimethylhydrazine, Unsymmetrical

This squeeze bulb filter assembly has been adapted also to the detection of unsymmetrical dimethylhydrazine, which was developed for use as a fuel in rocket engines.[81] Again, application to personnel monitoring was not suggested. The color-forming reagent (purple on exposure) was not identified, but probably was bindone — ($\Delta^{1,2}$-biindon)-1,3,3′-trione — which was described in a later reference by authors affiliated with the manufacturer of these kits.[12] This reagent was applied in the development of a dosimeter designed to be worn as a detector badge.

x. Boron Hydrides

An adaptation of the constant-volume piston hand pump developed in England during the 1930s has been offered on the domestic market for chromic acid mist and boron hydride detection. In this model, the filter paper retaining assembly became an integral part of the pump cylinder. The boron hydride kit was developed to meet the needs for monitoring in locations where this highly toxic rocket fuel creates a serious health problem for industries manufacturing and handling volatile boranes.[82] The procedure would be classified as a gas titration, since 600-mL aliquots of air were drawn through the freshly impregnated paper (wet reagent system) until the color change matched a reference standard and the concentration determined by the number of piston strokes. A sensitivity of 1 ppm (pentaborane) to 3 ppm (diborane) was claimed. The reagent composition was not identified, but at least three have been described.[83] Adaptation to personnel monitoring in this case probably would require only calibration for 4- or 8-hr sampling.

xi. Chromic Acid

This same equipment has been offered in a kit form for quick check for chromic acid mist over the range 0.1 to 0.2 mg/m^3 in electrolytic chromeplating rooms. The colorimetric reagent has been identified as the usual phthalic anhydride + *S*-diphenylcarbazide couple, which turns blue to violet (lilac color) on contact with chromates. The shelf life of the impregnated paper appeared to be approximately 6 months from date of manufacture.[84]

xii. Aniline

Aniline vapor in concentrations of about 5 ppm (by volume) in air is known to be toxic. Current applications of aniline make it imperative that a fast and simple field method be developed. The concentration of aniline vapor in the air can be determined easily and rapidly by use of paper strips impregnated with the liquid or vapor of a furfural-acetic acid mixture (4%, by volume, in glacial acetic acid). The test paper turns from white to pink or red in the presence of aniline vapor, having a range of determination from 5 to approximately 150 ppm. The speed and intensity of color development increases with the concentration of aniline vapors. The test is specific for aniline in the presence of many other organic vapors. The reagent mixture has a life of at least 1 month at room temperature.[85]

xiii. Quantitative Photometric Instruments

Photometric instruments that have been designed to determine the color density of stains, dyes, and light-absorbing dusts, such as smoke, on paper and other filter media can be applied to personnel monitoring paper impregnated with color-generating reagents. Nearly quantitative results are obtained when the paper is sufficiently uniform in texture and thickness to deliver consistently precise blank values required to correct the total absorbance for the absorbance of the unexposed paper. Usually absorbance measurements are made through either the maximum or the average density area of the color zone and the unexposed background directly adjacent to the boundary of the color zone. The blank is then subtracted either mechanically, by adjusting the photometer (densitometer) to 0% absorbance, or arithmetically in the final calculation. In some cases optical density ratios are more convenient to apply.

Several easily portable precision densitometers that provide a high degree of sensitivity have been adapted to the scanning of paper chromatograms.[14,86,87] Tape samplers that automatically draw successive air samples through 25- or 50-mm wide paper tape at predetermined intervals through a sampling head incorporate an integral photometer and provide a zero reference for each sample.[76,77] For applications where transmittance measurements are impractical, reflectance measurement is available.[77] Since this is a modular instrument, the detector unit could be removed or purchased separately for reading personnel monitor spots in the field.

b. Granular Inorganic Solid Support Systems — Direct Reading Length of Stain Gas Detecting Tubes

i. Historical

Glass tubes filled with a solid, granular material, such as silica or alumina gel, which has been impregnated with a color-forming reagent, are by far the most universally used portable toxic gas or vapor detectors.

Hoover and Lamb designed the first detector tube at Harvard University during 1917 for the rapid analysis of CO with Hoolamite mixture (IO_5 + SO_3). This detector tube probably was the first of the color-indicating devices, and it survived the test of time for over 25 years. During the interval from World War I to 1935, few detectors were developed for field use. In 1935 the U.S. Bureau of Mines designed the first length of stain type tube to appear on the open market for the determination of H_2S in coal mines. The reliability and sensitivity were improved in 1946 by reducing the diameter of the glass tube. At the National Bureau of Standards, Martin Shepherd and his staff produced the first high-sensitivity detector tube for determining CO concentrations in military vehicles and on aircraft carrier decks during World War II. This CO detector is still used, but it is being superseded rather rapidly by the length of stain reagent system, developed by the Royal Air Force Research Establishment during World War II and later improved by Leslie Silverman in 1965.[8]

The first detector tube for organic solvents (benzene, toluene, xylene, and other aromatic solvents) derived from research conducted at Harvard by Silverman's groups.[62a] Since then, other tubes have appeared in Japan, Sweden, Germany, and the United States. The American Industrial Hygiene Association in 1976 issued a manual for Detector Tube use.[88]

ii. General Operating Procedure

The use of detector tubes is quite simple. After the two sealed ends are snapped off, the tube is inserted in the holder, which is fitted with a calibrated squeeze bulb, bellows, or piston pump. The specified volume of the air sample then is drawn through the tube by the operator; and after adequate time is allowed for each stroke to draw completely its full volume of air, the contaminant concentration is determined by examining the exposed tube. The concentration can be determined by one of three procedures.

1. By comparing the shade of color observed with a set of color standards. The accuracy of this visual judgment is limited by the color acuity of the observer and by the lighting conditions.
2. By measuring the length of stain produced and relating the length to a calibration graph or chart. The scale may be printed directly on the tube or on a chart which is provided.

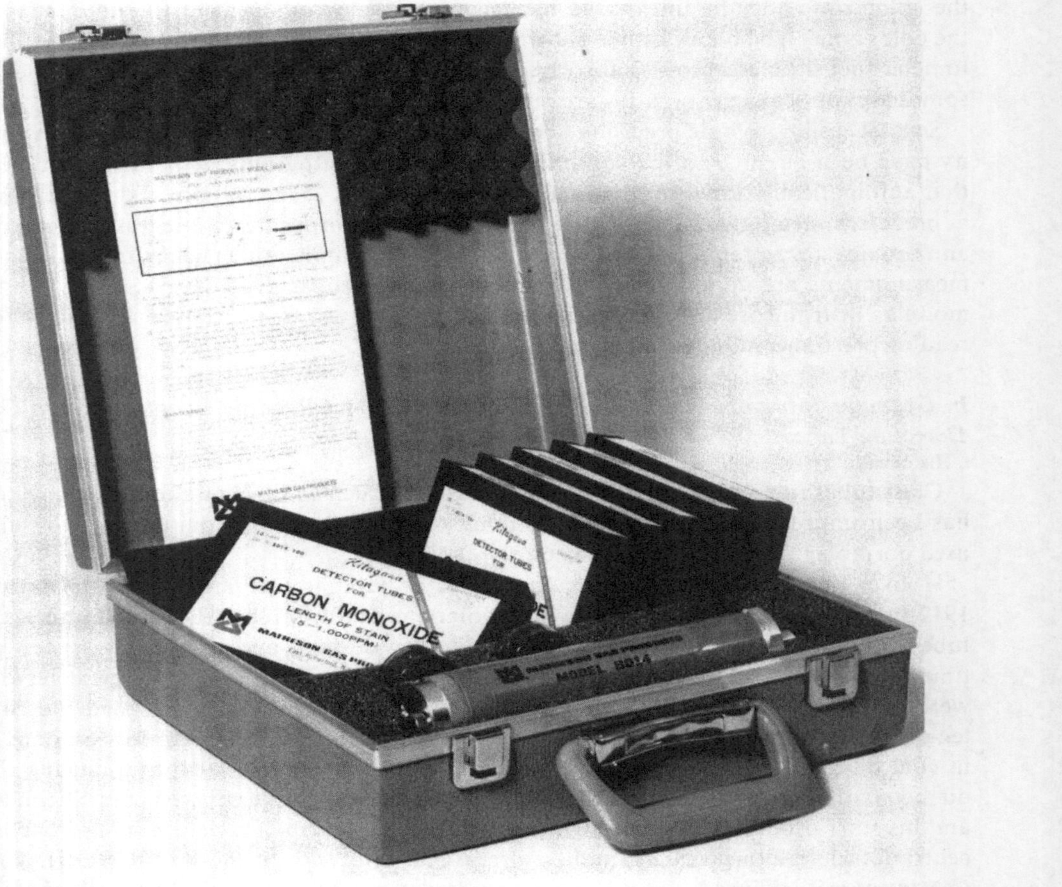

FIGURE 53. The Kitagawa® gas detector tube kit. (Courtesy of Matheson, 932 Paterson Plank Rd., East Rutherford, N. J.)

3. By measuring the volume of air required to produce a visible color change and relating the sample volume to a calibration chart (gas phase titration).

The length of stain technique (2) is used most commonly and has with a few exceptions displaced the other two methods. However, under the right conditions, the gas phase titration (3) is fundamentally capable of producing the best precision and accuracy. All three types have at one time or another over the past 25 years been commercially available.

iii. Availability and Limitations

Examples of commercially available kits are shown in Figures 53 to 60. These direct-reading indicator tubes provide a very rapid, convenient, and inexpensive analytical tool with which a semiskilled operator can perform air analysis. However, this very simplicity creates problems when inexperienced, untrained operators misuse the technique. The design of commercially available, field-type equipment has been simplified to a degree that almost everyone considers himself qualified to carry out analyses and interpret the data. The results may be dangerously misleading unless the sampling procedure is supervised and the findings interpreted by an adequately trained industrial hygienist.[89,89a] Available tubes and their operational parameters are summarized by manufacturer in Tables 28 to 32.

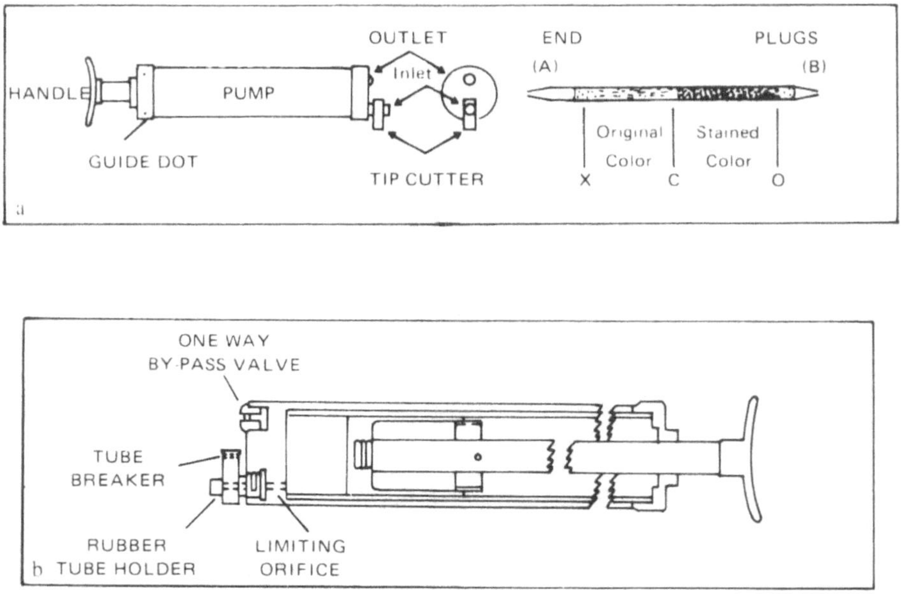

FIGURE 54. Unico® Model 400 precision gas detector pump — cutaway section. (Courtesy of Matheson, 932 Paterson Plank Rd., East Rutherford, N.J.)

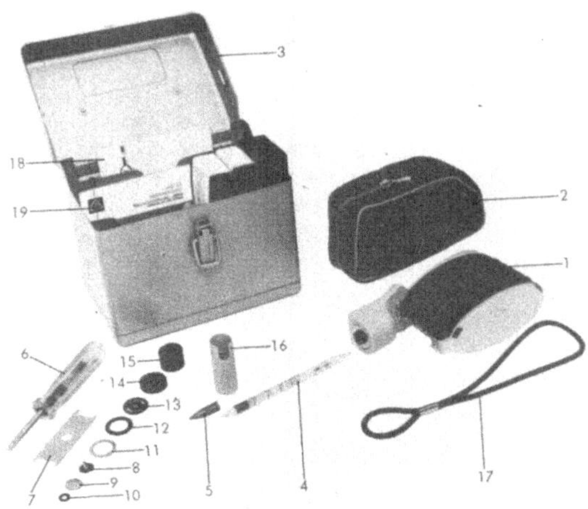

FIGURE 55. The Dräger gas detector tube kit. (Courtesy of Drägerwerk, U.S. Division, National Dräger, Inc., Parkway View Dr., Pittsburgh, Pa.)

iv. Applications

A review of the common applications of grab sampling with detector tubes will convey some idea of their utility and broad problem-solving potential.[90] Even greater application potential can be expected with recalibration for continuous personnel monitoring.

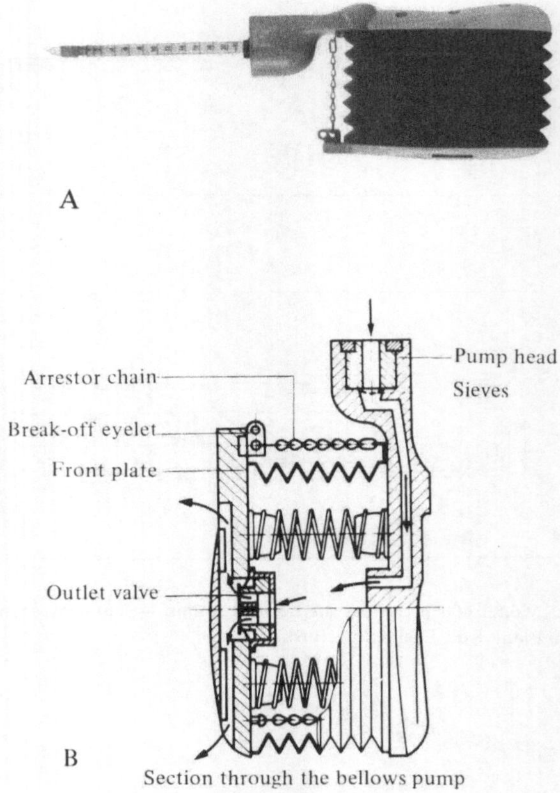

FIGURE 56. The Dräger gas detector pump. (A) the Dräger gas detector consisting of the gas detector pump and a Dräger tube; (B) cross-section through the gas detector pump. (Courtesy of Drägerwerk, U.S. Division, National Dräger, Inc., Parkway View Dr., Pittsburgh, Pa.)

FIGURE 57. The Bacharach gas hazard indicator. (Courtesy of Bacharach Instrument Co., Division of AMBAC Industries, Inc., 625 Alpha Dr., Pittsburgh, Pa.)

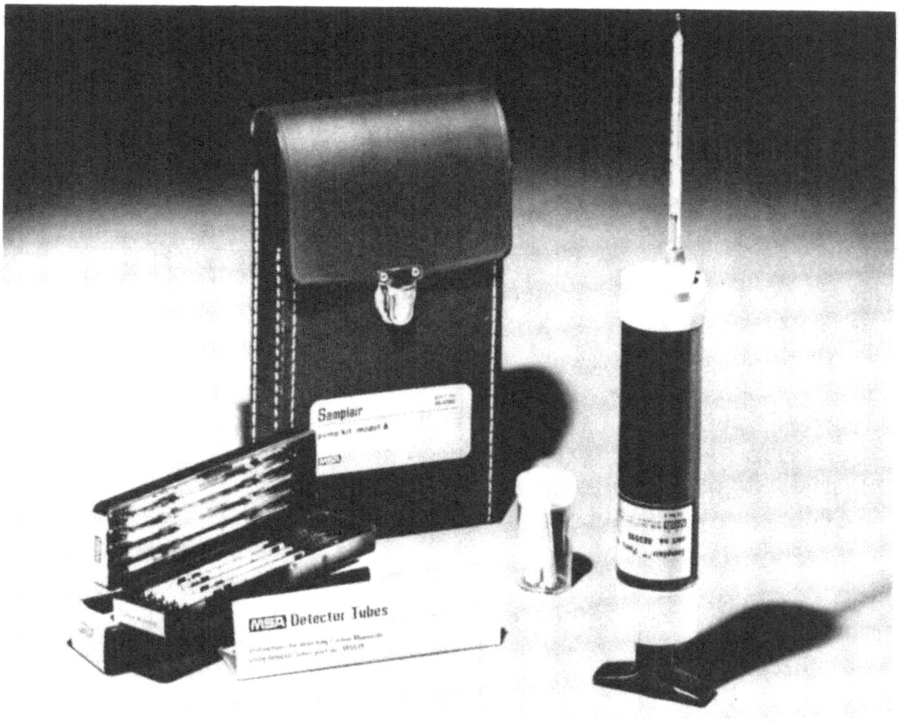

FIGURE 58. Samplair® pump, Model A, with carrying case and accessories. (Courtesy of Mine Safety Appliances Co., 201 N. Braddock Ave. Pittsburgh, Pa.)

a) Testing Atmospheres Within Vessels and Manholes

Clearance before workmen enter enclosures that may contain toxic gases or explosive vapors or may be deficient in oxygen is required by law in some states (New Jersey, for one). Handheld, clothing or harness supported, self-indicating field-type instruments that provide quick, on-the-spot results offer the most practical means for testing atmospheres inside vessels, pipes, manholes, or other confined spaces into which entrance is necessary.

b) Work Area Monitoring

Problems that arise in this traditional area for industrial hygiene surveys and exposure control programs can best be solved by application of readily portable, field test equipment, which is described in other sections.

c) Emergency Air Sampling

Almost without exception, the rapidly reacting, self-indicator type field instrument must be available to delineate the high-risk hazard area created by a chemical spill, accidental release of toxic gas or vapor, or abnormal operating conditions. Serious psychological problems can be encountered if the sensitivity range of the instrument chosen does not include the odor threshold, or if the instrument sensitivity is greater than the odor threshold. Hydrogen sulfide furnishes an example of the first condition: the TLV is 10 ppm[1] and the odor threshold less than 1 ppm. As a result, many unjustified complaints must be investigated only to confirm the nuisance value of H_2S in concentrations below 10 ppm. Phosgene may be cited as an example of the alternative condition. In this case the TLV is 0.1 ppm,[1] whereas the odor threshold is above 0.5

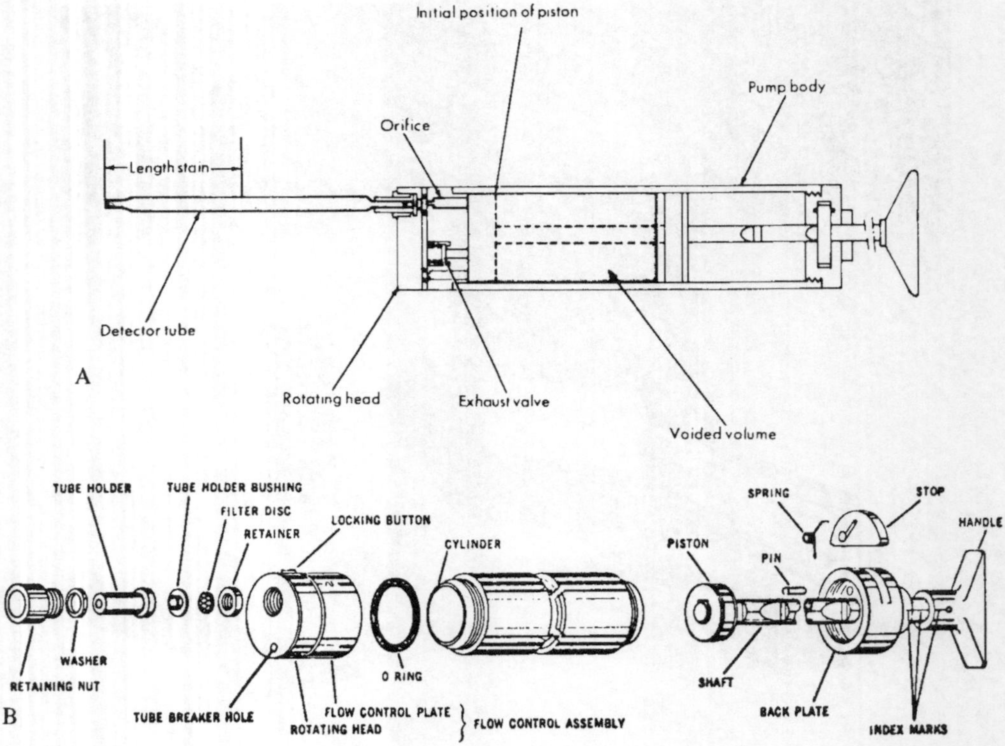

FIGURE 59. The M-S-A® sampling pump. (A) Cutaway view. Sampling begins with adjustment of orifice index marks to assure proper flow rate. Tips then are broken off glass detector tube, and it is fitted into rubber tube holder. Handle is drawn back until proper volume number appears on pump shaft. At this point spring-loaded stop will drop into place, preventing handle from being drawn forward into evacuated pump body. After minimum time has elapsed, handle is rotated 90° and is pushed forward to exhaust sample air. Concentration of contaminant is indicated by length (or, in some cases, color) of stain. (B) Exploded view. Sampling pump components are of stainless steel. Pump body is plastic covered to eliminate possible damage from knocks encountered in ordinary service. Interesting design feature is use of Teflon® cap over standard O-ring in piston seal. O-ring provides required resiliency while Teflon provides minimum friction and good wear characteristics. Piston stop is of nylon. Bottom hole in rotating head is for breaking tips of sealed glass detector tubes. (Courtesy of Mine Safety Appliance Co., 201 N. Braddock, Ave., Pittsburgh, Pa.)

ppm. Persistent odor of phosgene, then, indicates a hazardous condition that requires correction.

d) Personnel Monitoring

Although the current application of detector tubes to short interval or grab sampling (3 min or less per sample) by drawing a discrete volume of air through the tube by means of a manually operated piston or bellows pump cannot be classified strictly as personnel monitoring, the technique falls in a gray zone between fixed station and mobile industrial atmosphere sampling. The equipment is readily portable, in some cases pocket size; results are obtained quickly on the spot; and these tubes can be used for practically instantaneous warning of hazardous gas concentrations before ceiling limits are reached. If a sufficient number of samples is taken in the breathing zone during residence in an exposure hazard area, the procedure does in fact approach continuous personnel monitoring.

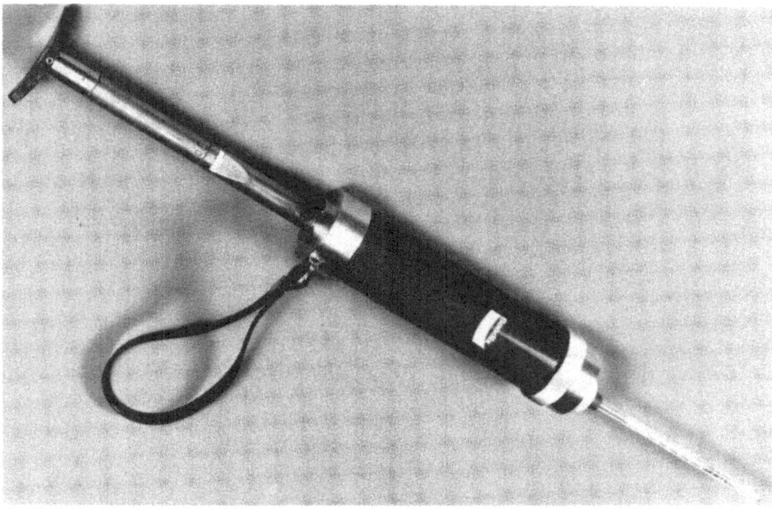

FIGURE 60. The Gastec® gas and vapor detector — pump and tube. (Courtesy of the Bendix Corp., Environmental and Process Instruments Division, Lewisburg, W. Va.)

However, grab sampling will reduce efficiency because the operator or mechanic must interrupt his assigned duties to draw the sample through the sampling device and then interpret the results. However, for those gases and vapors that exhibit an odor threshold below the TLV, this factor alone does not detract from the advantages that can be derived from quick assessment of the unusual condition. One might consider this a "tool box" instrument to be carried on the job for confirmation of safe or hazardous conditions when abnormal odors are encountered. However, olfactory fatigue becomes a limiting factor whenever the sense of smell is relied upon for protection.

The following procedure is used to determine compliance with the 8-hr, time-weighted average standard based on a small number of instantaneous (grab) samples collected at random intervals during the workday.

The National Institute for Occupational Safety and Health has included in an Industrial Hygiene Measurements training course (No. 550) a plan for the collection of random short-term sampling based on four to seven individual measurements. A maximum of 10 samples was proposed. The choice of time intervals is made on the basis of a random numerical series as illustrated in the following example drawn from a phosgene detector tube survey.

A total of ten 5-min samples are to be taken during an 8-hr work shift. To obtain the number of possible 5-min intervals, 480 min (8 × 60) is divided by 5 to provide 96 possible choices. The intervals are numbered from 1 to 96 for identification:

Measuring time #1-	from 8:00 to 8:05 a.m.
#2-	from 8:05 to 8:10 a.m.
#3-	from 8:10 to 8:15 a.m.

and so on through the 96th interval. Then 10 numbers are chosen from a random number table. In this example the numbers were 70, 22, 34, 75, 67, 25, 74, 73, 31, and 96. These numbers then correspond to the time interval at which the work room air is to be sampled. The mean value (average) is calculated and evaluated statistically to

TABLE 28

Detector Tubes Used for Gases and Vapors

Tube No.	Detector tube	Measurable concentration (ppm)	TLV[a] (ppm)	No. of strokes	Minimum[b] detectable concentration (ppm)	Interferences[c]	Shelf[d] life (yr)
133	Acetaldehyde	0.004 ~ 1.6%	100	1	5	Acetone (20), acrolein (25), methyl ethyl ketone (200), methyl isobutyl ketone (600), sulfur dioxide (10)	3[e] months
102a	Acetone	0.05 ~ 5.0%		1	50	Other organic vapors except halogenated hydrocarbons (50)	3[e]
102Sa[f]	Acetone	0.05 ~ 5.0%	1000	1	50		3[e]
102b	Acetone	200 ~ 8,000		5	10		1[e]
102Sb[f]	Acetone	200 ~ 8,000		5	10		1[e]
102c	Acetone — in presence of other hydrocarbons	0.01 ~ 4.0%	1000	1	10	Acetaldehyde (15), acrolein (20), methyl ethyl ketone (150), methyl isobutyl ketone (350), sulfur dioxide (10)	3 months
102Sc[f]	Acetone — in presence of other hydrocarbons	0.01 ~ 4.0%		1	10		3 months
101	Acetylene	50 ~ 1,000		1	10	Butadiene, butane, butylene (10), carbon monoxide (10), ethylene (10), hydrogen sulfide (10), methylacetylene (10), pentadiene, pentane, propylene (10)	3
101S[f]	Acetylene	50 ~ 1,000	1000	1	10		3
136	Acrolein	0.005 – 2.5%	0.1	1	5	Acetaldehyde (25), acetone (80), methyl ethyl ketone (300), methyl isobutyl ketone (600), sulfur dioxide (10)	3 months
128	Acrylonitrile — high range	0.1 ~ 3.5%		1	10	Other organic vapors except halogenated hydrocarbons (10)	3
128Sa[f]	Acrylonitrile — high range	0.1 ~ 3.5%	20	1	10		3
128b	Acrylonitrile — low range	10 ~ 500		1	1		3
105a	Ammonia	1 ~ 25.0%		1	500		3[g]
105Sa[f]	Ammonia	0.5 ~ 20.0%		1	500		3[g]
105b	Ammonia — high range	25 ~ 900	25	1	5	Hydrogen sulfide (1000)	1[g]
105Sb[f]	Ammonia — high range	25 ~ 900		1	5		1[g]

105c	Ammonia NIOSH Cert. No. TC-84-033	10 ~ 260		1	1	Diethylamine (50), dimethylamine (50)	1[g]
		5 ~ 130		2			
105Sc[f]	Ammonia NIOSH Cert. No. 84-032	10 ~ 260	25	1	1		1[g]
		5 ~ 130		2			
140	Arsine	5 ~ 160	0.05	1	1	Hydrogen selenide, hydrogen sulfide, phosphine	2
118a	Benzene	10 ~ 500		1	5	Toluene, xylene, hexane, CO	1
118Sa[f]	Benzene	25 ~ 500		1	5		1
118c	Benzene	2 ~ 70	10	5	0.5		1
118Sc[f]	Benzene	2 ~ 70		5	0.5		1
118b	Benzene — in presence of gasoline and/or other aromatic hydro-carbons	20 ~ 400	10	1	5	Toluene (500)	1[g]
114	Bromine	1 ~ 50	0.1	1	0.1	Chlorine (1), chlorine dioxide (1), iodine (1), nitric oxide, nitrogen dioxide (1), ozone (5)	1
168a	Butadiene — high range	0.03 ~ 2.6%	1000	1	10	Acrylonitrile, butane, butylene, ethylene, pentane, propylene, styrene	3
168Sa[f]	Butadiene — high range	0.03 ~ 2.6%	1000	1	10		3
168b	Butadiene — low range	30 ~ 600		1	5	Benzene, butylene, carbon monoxide, ethylene, hydrogen sulfide, pentane, propylene	3
138	Butyl acetate	0.01 ~ 1.0%	150	1	10	Other organic vapors except halogenated hydrocarbons (50)	3
138S[f]	Butyl acetate	0.01 ~ 1.0%		2	10		3
126a	Carbon dioxide NIOSH Cert. No. TC-84-026	1000 ~ 26,000	5000	1	100	Chlorine, hydrogen cyanide, hydrogen sulfide, nitrogen dioxide	2
126Sa[f]	Carbon dioxide NIOSH Cert. No. TC-84-026	1000 ~ 26,000		1	100		2
126b	Carbon dioxide — low range	0.03 ~ 0.7%	5000	1	50	Chlorine (50), hydrogen cyanide (200), hydrogen sulfide (20), nitrogen dioxide (10), sulfur dioxide (100)	2
		100 ~ 1,500		3	20		
126Sb[f]	Carbon dioxide — low range	0.05 ~ 1.0%		1	50		2
141	Carbon disulfide	10 ~ 200	20	1	5	Hydrogen sulfide (150), sulfur dioxide (50)	1[g]
100	Carbon monoxide — length of stain NIOSH Cert. No. TC-84-045	25 ~ 1,000	20	1	1	Acetylene (10), hydrogen sulfide (100), nitrogen dioxide (50), sulfur dioxide (50)	1
		10 ~ 300		3	1		
106S[f]	Carbon monoxide — NIOSH Cert. No. TC-84-019	10 ~ 250	50	3	1		1

TABLE 28 (continued)

Detector Tubes Used for Gases and Vapors

Tube No.	Detector tube	Measurable concentration (ppm)	TLV[a] (ppm)	No. of strokes	Minimum[b] detectable concentration (ppm)	Interferences[c]	Shelf[d] life (yr)
106a	Carbon monoxide — color intensity	Measurement for 30 sec 100—1000, measurement for 300 sec 10—100	50	1	10	Acetylene (10), ammonia (1000), butadiene, butane (1000), butylene (0.1), ethylene (0.1), hydrogen chloride, hydrogen cyanide (100), hydrogen sulfide (1000), methylacetylene (10), nitrogen dioxide (1), pentadiene, pentane (1000), propylene (0.1)	3
106b	Carbon monoxide — in presence of ethylene color intensity	Measurement for 30 sec 100—1000, measurement for 300 sec 10—100	50	1	10	Ethylene (5), hydrogen sulfide (1000), nitrogen dioxide (1)	3
106c	Carbon monoxide — in presence of ethylene and nitrogen oxides, color intensity	Measurement for 30 sec 100—1000, measurement for 300 sec 10—100	50	1	1	Ethylene (5), hydrogen sulfide (1000)	1
147	Carbon tetrachloride	5 ~ 300	10	1	3	Chloroform, hydrogen chloride, phosgene	3[g] months
109	Chlorine	1 ~ 40		1	0.1		1[g]
109Sa[f]	Chlorine	1 ~ 40		1	0.1		1[e]
109b	Chlorine	0.5 ~ 10	1	1	0.1	Bromine (1), chlorine dioxide (1), iodine (1), nitric oxide, nitrogen dioxide (1), ozone (5)	1[e]
		0.3 ~ 4.5		2			1[e]
109Sb[f]	Chlorine	0.3 ~ 6.0		2	0.1		
116	Chlorine dioxide	1 ~ 40	0.1	1	0.1	Bromine (1), chlorine (1), iodine (1), nitrogen dioxide (1), ozone (3)	1[e]
152	Chloroform	75 ~ 600	25	1	50	Bromine (1), chlorine (1), hydrogen chloride, methylene chloride (500)	3[g] months

172	Chloropicrin	0.1 ~ 15.0	0.1	1	0.01	Chloropicrin, carbon tetrachloride, phosgene, chlorine	3[g] months
169	Chloroprene	5 ~ 800	25	1	1	Bromine, chlorine, 1, 2-dichloroethylene, hydrogen chloride, methyl bromide, nitrogen dioxide, perchloroethylene, trichloroethylene	10[g] months
115	Cyclohexane	0.01 ~ 0.6%	300	1	10	Other organic vapors except halogenated hydrocarbons (50)	3
145	1,2-dichloroethylene	10 ~ 400	200	1	1	Bromine (1), chlorine (1), hydrogen chloride, nitrogen dioxide (1), perchloroethylene (5), trichloroethylene (5)	10[g] months
123	Dimethyl ether	0.01 ~ 1.2%	400	1	10	Other organic vapors except halogenated hydrocarbons (50)	3
154	Dioxane	0.01 ~ 2.5%	50	1	10	Other organic vapors except halogenated hydrocarbons (50)	3
154S[f]	Dioxane	0.01 ~ 2.5%		2	10		3
107	Ether (diethyl ether)	0.04 ~ 1.4%	400	1	10	Other organic vapors except halogenated hydrocarbons (50)	3
107S[f]	Ether	0.04 ~ 1.4%		1	10		3
111	Ethyl acetate	0.01 ~ 5.0%	400	1	10	Other organic vapors except halogenated hydrocarbons (50)	3
111S[f]	Ethyl acetate	0.01 ~ 5.0%		1	10		3
104a	Ethyl alcohol	0.04 ~ 5.0%	1000	1	20	Other organic vapors except halogenated hydrocarbons (50)	3
104Sa[f]	Ethyl alcohol	0.01 ~ 5.0%		1	20		3
108b	Ethylene — low range	0.5 ~ 100 0.1 ~ 20	50	1	0.01	Benzene, butane, butylene, carbon monoxide (10), hydrogen sulfide (10), nitrogen dioxide (1), pentane, propylene	3
166	Ethylenedibromide (di-brome ethane)	1 ~ 100	20	1	0.1	Bromine (1), chlorine (1), methyl bromide (5), methylchloroform (5)	6[g] months
122	Ethylene oxide	0.01 ~ 3.5%	50	1	10	Other organic vapors except halogenated hydrocarbons (50)	3
122S[f]	Ethylene oxide	0.01 ~ 3.5%		1	10		3

TABLE 28 (continued)

Detector Tubes Used for Gases and Vapors

Tube No.	Detector tube	Measurable concentration (ppm)	TLV[a] (ppm)	No. of strokes	Minimum[b] detectable concentration (ppm)	Interferences[c]	Shelf[d] life (yr)
165	Ethyl mercaptan	5 ~ 160 1 ~ 26	0.5	1 5	1 0.2	Acetylene (20), carbon monoxide (150), chlorine (0.2), ethylene (100), hydrogen sulfide (40), methyl mercaptan (1), methyl sulfide (1), nitrogen dioxide (1)	8[e] months
171	Formaldehyde	1 ~ 35	2	1	1	Styrene, trichloroethylene (500), ethyl acetate (1000), ethyl ether	3
161	Furan (furfuran)	0.05 ~ 1.6%	2	1	100	Other organic vapors except halogenated hydrocarbons (50)	3
110	Gasoline	0.01 ~ 0.6%	—	1	10	Gasoline, 3% acetylene, 0.2% propane, other organic vapors except halogenated hydrocarbons (50)	3
113	*n*-Hexane	0.01 ~ 0.6%	100	1	10	Other organic vapors except halogenated hydrocarbons (50)	3
113S[f]	*n*-Hexane	0.01 ~ 0.6%		1	10		3
173	Hydrogen chloride	2 ~ 20 1 ~ 10	5	1 2	0.5	Hydrogen chloride, 20 ppm NO_2, 150 ppm H_2S, 500 ppm SO_2, 1 ppm chlorine	1
112a	Hydrogen cyanide — high range	0.01 ~ 3.0%	10	1	10	Acetone, dicyan, hydrogen sulfide, sulfur dioxide	1.5
112Sa[f]	Hydrogen cyanide	0.01 ~ 3.0%		1	10		1.5
112b	Hydrogen cyanide — low range	0.25 ~ 100	10	1—4	0.2	Ammonia, chlorine, hydrogen sulfide (2), sulfur dioxide (1)	1

112Sb[f]	Hydrogen cyanide — low range	0.5 ~ 100		1—4	0.2		1
167	Hydrogen selenide	10 ~ 600 1 ~ 100	0.05	1 5	1 0.5	Arsine, carbon monoxide (1000), hydrogen sulfide, iron carbonyl, mercury vapor, nickel carbonyl, sulfur dioxide	8 months
120a	Hydrogen sulfide — high range	0.01 ~ 0.17%	10	1	10	Sulfur dioxide (50)	3
120Sa[f]	Hydrogen sulfide	100 ~ 1800		1	10		3
120b	Hydrogen sulfide — low range, NIOSH Cert. No. TC-84-037	5 ~ 160 2 ~ 40	10	1 4	0.5	Nitrogen dioxide (5), sulfur dioxide (20)	1.5
120Sb[f]	Hydrogen sulfide	1.5 ~ 165		4	0.5		1.5
120c	Hydrogen sulfide in presence of SO_2	0.005 ~ 0.16%	10	1	2	Acetylene (1), butylene (10), carbon monoxide (10),	3
120Sc[f]	Hydrogen sulfide in presence of SO_2	0.005 ~ 0.16%		1	2	ethylene (10), hydrogen cyanide, methyl acetylene (10), sulfur dioxide (6%)	3
120h	Hydrogen sulfide — extra high range	0.02 ~ 0.3%	10	1	10	Sulfur dioxide (50)	3
153	Isobutyl acetate	0.01 ~ 1.4%	150	1	10	Other organic vapors except	3
153S[f]	Isobutyl acetate	0.01 ~ 1.4%		2	10	halogenated hydrocarbons (50)	3
149	Isopropyl acetate	0.01 ~ 1.2%	250	1	10	Other organic vapors except	3
149S[f]	Isopropyl acetate	0.01 ~ 1.2%		2	10	halogenated hydrocarbons (50)	3
150	Isopropyl alcohol	0.01 ~ 2.0%	400	1	10	Other organic vapors except	3
150S[f]	Isopropyl alcohol	0.01 ~ 2.0%		2	10	halogenated hydrocarbons (50)	3
142a	Mercury	0.5 ~ 12.0 mg/m^3, 0.1 ~ 2.0 mg/m^3	0.05 mg/m^3	1 5	0.1 mg/m^3, 0.02 mg/m^3 0.1 mg/m^3, 0.02 mg/m^3	Ammonia (100), chlorine (.01), hydrogen sulfide (0.5), nitrogen dioxide (0.1)	1
142b	Mercury	0.5 ~ 12.0 mg/m^3, 0.1 ~ 2.0 mg/m^3		1 5			1

TABLE 28 (continued)

Detector Tubes Used for Gases and Vapors

Tube No.	Detector tube	Measurable concentration (ppm)	TLV[a] (ppm)	No. of strokes	Minimum[b] detectable concentration (ppm)	Interferences[c]	Shelf[d] life (yr)
148	Methyl acetate	0.01 ~ 3.0%	200	1	10	Other organic vapors except halogenated hydrocarbons (50)	3
119	Methyl alcohol	0.01 ~ 6.0%	200	1	20	Other organic vapors except halogenated hydrocarbons (50)	3
119S[f]	Methyl alcohol	0.01 ~ 6.0%		1	20		3
157	Methyl bromide	10 ~ 500	15	1	1	Bromine (1), chlorine (1), chloroform (50), ethyl bromide (5), ethylene dibromide, hydrogen chloride, methylene chloride (500), nitrogen dioxide (1), perchloroethylene, trichloroethylene	8[g] months
157S[f]	Methyl bromide	10 ~ 500		1	1		8[g] months
160	Methyl chloroform (1,1,1-trichloroethane)	50 ~ 800	350	1	10	Bromine (1), chlorine (1), chloroform (50), ethyl bromide (5), hydrogen chloride, methyl bromide (1), methylene chloride (500)	6[g] months
139a	Methyl ethyl ketone — high range	0.2 ~ 5.0%	200	1	100	Other organic vapors except halogenated hydrocarbons (50)	3
139Sa[f]	Methyl ethyl ketone	0.2 ~ 5.0%		1	100		3
139b	Methyl ethyl ketone — low range	0.01 ~ 1.4%		1	10		3
139Sb[f]	Methyl ethyl ketone — low range	0.01 ~ 1.4%		2	10		3
155	Methyl isobutyl ketone	0.005 ~ 1.0%	100	1	10	Other organic vapors except	3

155S^f	Methyl isobutyl ketone	0.005 ~ 1.0%		1	10	halogenated hydrocarbons (50)	3
164	Methyl mercaptan	5 ~ 120	0.5	1	1	Acetylene (20), carbon	8^c months
		1 ~ 20		5	0.2	monoxide (150), chlorine	8^c months
154S^f	Methyl mercaptan	5 ~ 140		1	1	(0.2), ethylene (200), ethyl mercaptan (1), hydrogen sulfide (40), methyl sulfide (1), nitrogen dioxide (1)	
129	Nickel carbonyl	20 ~ 700	0.001	1	10	Arsine, carbon monoxide (1000), hydrogen sulfide, iron carbonyl, mercury vapor, sulfur dioxide	6^c months
174	Nitrogen oxide and dioxide — separately measurable	NO 10 ~ 300	25 (NO),	1	1	Nitric oxide compound, 10	1
		NO_2 1 ~ 40	5 (NO_2)		0.5	ppm chlorine	
117	Nitrogen dioxide	1 ~ 1000	5	1	0.1	Bromine (1), chlorine (1), chlorine dioxide (1),	1^c
117b	Nitrogen dioxide	1 ~ 100		1	0.5	iodine (3), nitric oxide, ozone (5)	1
117Sb^f	Nitrogen dioxide	1 ~ 100		1	0.5		1
159b	Oxygen	2.0 ~ 30.0%	—	1	1000	Carbon dioxide (1%), chlorine (1%), hydrogen sulfide (1%), sulfur dioxide (1%)	1.5
135	Perchloroethylene	10 ~ 400	100	1	1	Bromine, chlori ne, 1,2-dichloroethylene (5),	10^2 months
135S^f	Perchloroethylene	10 ~ 600		1	1	hydrogen chloride, methyl bromide (5), nitrogen dioxide, trichloroethylene	10^c months
146	Phosgene	0.1 ~ 50	0.05	1	0.05	Chlorine (10), hydrogen chloride (50),	8^g months
		0.05 ~ 10		5	0.02	nitrogen dioxide (20)	
121a	Phosphine in acetylene — high range	20 ~ 800	0.3	1	5	Arsine, hydrogen sulfide (10)	3
121Sa^f	Phosphine in acetylene	20 ~ 800		1	5		3
121b	Phosphine in acetylene — low range	5 ~ 90	0.3	1	1	Arsine (10), hydrogen sulfide (1)	3
121Sb^f	Phosphine in acetylene — low range	5 ~ 90		1	1		3
121c	Phosphine — high range	50 ~ 700	0.3	1	10	Arsine (30), hydrogen selenide (50), hydrogen	2
121Sc^f	Phosphine	50 ~ 700		1	10	sulfide (40)	2
121d	Phosphine — low range	10 ~ 160	0.3	1	0.3	Arsine, carbon monoxide (7000), hydrogen	6^c months
		2 ~ 30		5	0.1	sulfide (5), mercury vapor, nickel carbonyl,	6^c months
121Sd^f	Phosphine — low range	10 ~ 160		1	0.3	sulfur dioxide (50)	

TABLE 28 (continued)

Detector Tubes Used for Gases and Vapors

Tube No.	Detector tube	Measurable concentration (ppm)	TLV[a] (ppm)	No. of strokes	Minimum[b] detectable concentration (ppm)	Interferences[c]	Shelf[d] life (yr)
151	Propyl acetate	0.01 ~ 1.4%	200	1	10	Other organic vapors except halogenated	3
151S[f]	Propyl acetate	0.01 ~ 1.4%		2	10	hydrocarbons (50)	3
163	Propylene oxide	0.1 ~ 4.0%	100	1	100	Other organic vapors except halogenated hydrocarbons (50)	3
158	Styrene	1 ~ 300	100	1	0.1	Acrylonitrile (20), butadiene (20), butyl acetate (20), ethyl acetate (20), ethyl alcohol (20), propyl acetate (20)	3
103a	Sulfur dioxide — high range	0.1 ~ 4.0%	5	1	100	Hydrogen sulfide (400)	3
103Sa[f]	Sulfur dioxide	1.0 ~ 3.0%		1	100		3
103b	Sulfur dioxide — middle range	0.02 ~ 0.3%	5	1	10	Hydrogen sulfide (1000)	3
103Sb[f]	Sulfur dioxide	0.02 ~ 0.3%		1	10		3
103bf	Sulfur dioxide — in flue gas	0.02 ~ 0.3%	5	1	10	Hydrogen sulfide (1700)	3
103c	Sulfur dioxide — low range	5 ~ 300	5	1	1	Hydrogen sulfide, nitrogen dioxide	1[h] 1[h]
103Sc[f]	Sulfur dioxide	10 ~ 300		1	1		
103d	Sulfur dioxide NIOSH	2 ~ 80	5	1	0.5	Hydrogen sulfide (1000), nitrogen dioxide (5)	1[c]
	Cert. No. TC-84-028	1 ~ 20		3			
103Sd[f]	Sulfur dioxide NIOSH	2 ~ 80		1	0.5		1[c]
	Cert. No. TC-84-028	1 ~ 20		3			
162	Tetrahydrofuran	0.1 ~ 5.0%	200	1	500	Other organic vapors except halogenated hydrocarbons (50)	3
124	Toluene	10 ~ 700	100	1	5	Benzene, hexane, xylene, CO	1
124S[f]	Toluene	25 ~ 700		1	5		1
124b	Toluene	5 ~ 520		2	2.5		1
124Sb[f]	Toluene	5 ~ 550		2	2.5		1

134	Trichloroethylene	10 ~ 400	100	1	1	Bromine, chlorine, 1,2-dichloroethylene (5),	1[g]
134S[f]	Trichloroethylene	10 ~ 600		1	1	hydrogen chloride, methyl bromide (5), nitrogen dioxide, perchloroethylene	1[g]
132b	Vinyl chloride	1 ~ 500	1	1	0.5	Other organic vapors except halogenated	10[g] months
132c	Vinyl chloride	1 ~ 20		1	0.5	hydrocarbons (50), trichloroethylene,	1[g]
		0.2 ~ 3		5	0.1	perchloroethylene (50)	
143	Xylene	25 ~ 1000	100	1	5	Benzene, toluene, hexane, CO	1
143S[f]	Xylene	25 ~ 1000		1	5		1

Detector Tubes Used for Dissolved Substances in Solution

Tube No.	Detector tube	Chemical formula	Measurable concentration[b] (ppm)	Sampling: Volume (mℓ)	Sampling: Time (sec)	Sampling method	Interferences[c]	Shelf life[d] (yr)	Use of detector kit
200	Sulfide ion	S^{2-}	2 ~ 1000	0.2		Immersion	Carbonate ion (1500), chloride ion (1500), chromate ion (500), ethyl mercaptan, ferric ion (2000), methyl mercaptan	1	
201	Chloride ion	Cl^-	50 ~ 200	0.2	300	Immersion	Bromide ion, cyanide ion, iodide ion	1	Injector (2 mℓ)
			5 ~ 40	2.0	300	Injection			
202	Iron ion	Fe^{3+}	50 ~ 400	1.0	250	Injection	Iron ion, 10 ppm copper ion, 10 ppm aluminum ion, 10 ppm cobalt ion, 100 ppm mercury ion	1	Injector (1 mℓ)
203	Copper ion	Cu^{2+}	1 ~ 150	1.0	250	Injection	Copper ion, 10 ppm zinc ion, 10 ppm cadmium ion, cobalt ion, nickel ion, lead ion, manganese ion	1	Injector (1 mℓ)
204	Cyanide ion	CN^-	1 ~ 20	2.0	40	Diffusion	Chlorine (), sulfide (1)	1	Cyanide ion detector

[a] Threshold Limit Values (TLV) for 1970: the American Conference of Governmental Industrial Hygienists;

[b] minimum detectable limit is defined as the minimum concentration at which the stain in the length-of-stain detector tubes, or the discoloration in the colorimetric tubes, can be visually detected. A length of stain or a color change less than the lowest graduated concentration on the concentration chart is only qualitative not quantitative;

[c] since the listing of all possible interferences is difficult, only compounds which commonly occur and affect accurate readings are listed in this column. An interference will affect the accuracy of a test in proportion to the ratio of the interfering compound to the compound to be measured. If the

TABLE 28 (continued)

concentration to be measured is high relative to the interfering compound, the error is minimal. If the reverse is true, the reading can be seriously affected. The figure shown after the name of the interfering compound is ppm unless otherwise indicated. This value causes the following conditions (1) the interfering compound produces a minimum detectable stain by itself (see note, [d]); (2) the interfering compound causes discoloration over the entire tube which can be detected visually; (3) the minimum concentration of an interfering compound causes greater than a 10% error in the reading when it exists with the compound to be measured at a concentration of half value on the calibration chart;

[d] the expiration date stamped on each box of tubes is determined from the date of manufacture. Tube accuracy prior to the expiration date is ±10%. The reading of the detector tubes gradually becomes inaccurate after the expiration date. Even expired tubes are still usable for qualitative purposes;

[e] reagents in tubes so marked will slowly deteriorate after the expiration date, giving a slight discoloration which can make the stain demarcation indistinct. These tubes are still usable with excellent accuracy after the expiration date as long as the demarcation line is distinct. The aging of the reagents does not affect the length of the stain;

[f] all "S" tubes have direct reading scales.

[g] tubes so marked must be stored in a refrigerator (20—40°F); otherwise the shelf life will be shortened.

[h] tubes so marked must be stored in a dark area; otherwise the shelf life of these tubes will be shortened.

From Eckstein, W., *Dräger Rev.*, 38, 18, 1976. With permission.

TABLE 29

Bendix-Gastec Precision Gas Detector Tubes (1978)

Gas or vapor to be measured	TLV (ppm)	Tube no.[a]	Detector tube to be used[b]	Measurable concentration	Pump strokes[c] for measurable concentration
Acetic acid	10	81	Acetic acid	1—80 ppm	2 — 1/2
Acetic anhydride	5	81	Acetic acid	0.5—40 ppm	2 — 1/2
Acetone	1000	151	Acetone	0.01—2%	2 — 1
Acetylene		171	Acetylene	0.1—4%	2 — 1/2
Acetylene dichloride	See 1,2-Dichloroethylene				
Acrylonitrile	20	191	Acrylonitrile	10—500 ppm	5
Ammonia	25	3H	Ammonia	0.2—32%	5 — 1/2
Ammonia	25	3M	Ammonia	10—1000 ppm	5 — 1/2
Ammonia	25	3L	Ammonia	1—60 ppm	2 — 1/2
Aniline	5	181	Aniline	1.25—60 ppm	5 — 2
Benzene	10	121	Benzene	5—120 ppm	4 — 1
Benzyl bromide		136	Methyl bromide	25—850ppm	1
Benzyl chloride	1	132L	Trichloroethylene	0.5—25 ppm	2
Bromine	0.1	9L	Nitrogen dioxide	2—23 ppm	2
Bromoform	0.5	136	Methyl bromide	0.5—50 ppm	2 — 1
Butadiene	1000	174	Butadiene	50—800 ppm	1
Butane	600	104	Butane	50—1400 ppm	1
n-Butanol	See *n*-Butyl alcohol				
Butyl acetate	150	142	Butyl acetate	0.01—0.8%	2
n-Butyl alcohol	50	112	Ethyl alcohol	100—1500 ppm	4
Carbon dioxide	5000	2H	Carbon dioxide	1.0—20%	1 — 1/2
Carbon dioxide	5000	2L	Carbon dioxide	0.13—6%	2 — 1/2
Carbon dioxide	5000	2LL	Carbon dioxide	300—5000 ppm	1
Carbon disulfide	20	13	Carbon disulfide	5—50 ppm	1
Carbon monoxide	50	1H	Carbon monoxide	0.2—10%	1 — 1/2
Carbon monoxide	50	1M	Carbon monoxide	0.1—4%	1 — 1/2
Carbon monoxide	50	1L	Carbon monoxide	50—2000 ppm	1 — 1/2
Carbon monoxide	50	1La	Carbon monoxide	8—1000 ppm[e]	3 — 1/2
Carbon monoxide	50	1LL	Carbon monoxide	5—50 ppm	2
Carbon tetrachloride	10	134	Carbon tetrachloride	1—60 ppm	5 — 1

Table 29 (continued)

Bendix-Gastec Precision Gas Detector Tubes (1978)

Gas or vapor to be measured	TLV (ppm)	Tube no.[a]	Detector tube to be used[b]	Measurable concentration	Pump strokes[c] for measurable concentration
Chlorine	1	8La	Chlorine	0.3—16 ppm	3 — 1/2
Chlorine dioxide	0.1	8L	Chlorine	0.3—16 ppm	3 — 1/2
Chlorobenzene	75	121	Benzene	5—350 ppm	2 — 1
Chlorobromomethane	200	136	Methyl bromide	8—80 ppm	1
Chlorobromomethane	200	135	Methyl chloroform	40—350 ppm	1
Chloropicrin	0.1	134	Carbon tetrachloride	1—60 ppm	5 — 1
Cumene	50	122	Aromatic hydrocarbon	5—2400 ppm	2 — 1/2
Cyclohexane	300	102H	Hexane	0.015—1.2%	2 — 1/2
1,1-Dibromoethane		136	Methyl bromide	7—70 ppm	1
1,2-Dibromoethane	20	136	Methyl bromide	6—80 ppm	1
o-Dichlorobenzene	50	121	Benzene	10—700 ppm	2
1,2-Dichloroethylene	200	132H	Trichloroethylene	10—450 ppm	1
Diethyl benzene		122	Aromatic hydrocarbon	25—2400 ppm	2
Diethylamine	25	3L	Ammonia	1.6—48 ppm	2 — 1
Dimethyl aniline	5	181	Aniline	2.5—45 ppm	3
Dimethylamine	10	3L	Ammonia	2—30 ppm	1
Dimethyl ether		161	Ethyl ether	0.03—0.9%	1
Dioxane	50	163	Ethylene oxide	0.6—5.6%	1
Ethanol	See Ethyl alcohol				
Ether	See Ethyl ether				
Ethyl acetate	400	141	Ethyl acetate	0.04—1.5%	1
Ethyl alcohol (ethanol)	1000	112	Ethyl alcohol	0.05—5%	2 — 1/2
Ethyl benzene	100	122	Aromatic hydrocarbon	7—700 ppm	2 — 1/2
Ethylene (ethene)		172	Ethylene	50—800 ppm	1
Ethylene		172L	Ethylene	0.2—50 ppm	4
Ethylenediamine	10	3L	Ammonia	4—120 ppm	1 — 1/2
Ethylene dibromide	See 1,2-Dibromoethane				
Ethylene oxide	50	163	Ethylene oxide	0.1—3.0%	1
Ethyl ether	400	161	Ethyl ether	0.04—1.0%	1

Ethylidine dibromide	See 1,2-Dibromoethane				
Ethyl mercaptan	0.5	72	Ethyl mercaptan	5—120 ppm	2 — 1
Formaldehyde	2	91	Formaldehyde	2—20 ppm	2
Formic acid	5	81	Acetic acid	1—80 ppm	2 —1/2
Gasoline		101	Gasoline	0.015—1.2%	2 — 1/2
n-Heptane	400[d]	101	Gasoline	0.015—1.2%	2 — 1/2
n-Hexane	100[d]	102H	Hexane	0.015—1.2%	2 — 1/2
n-Hexane	100[d]	102L	*n*-Hexane	50—1200 ppm	1
Hydrazine	1	3L	Ammonia	0.5—35 ppm	2 — 1
Hydrochloric acid	See Hydrogen chloride				
Hydrocyanic acid	See Hydrogen cyanide				
Hydrogen chloride	5	14L	Hydrogen chloride	0.2—40 ppm	10 — 1/2
Hydrogen cyanide	10	12H	Hydrogen cyanide	0.05—2%	1
Hydrogen cyanide	10	12L	Hydrogen cyanide	2.5—60 ppm	6 — 3
Hydrogen fluoride	3	17	Hydrogen fluoride	0.5—20 ppm	7 — 4
Hydrogen sulfide	10	4HH	Hydrogen sulfide	0.1—4%	1 — 1/2
Hydrogen sulfide	10	4H	Hydrogen sulfide	100—3200 ppm	1 — 1/2
Hydrogen sulfide	10	4L	Hydrogen sulfide	10—240 ppm	1 — 1/2
Hydrogen sulfide	10	4LL	Hydrogen sulfide	0.5—60 ppm	10 — 1
Isobutyl acetate	150	142	Butyl acetate	.005—66%	2
Isobutyl alcohol	50[d]	112	Ethyl alcohol	100—3000 ppm	4 — 2
Iso-octane		101	Gasoline	0.015—1.2%	2 — 1/2
Isopropanol	See Isopropyl alcohol				
Isopropyl acetate	250	151	Acetone	0.05—0.75%	2
Isopropyl alcohol	400	113	Isopropyl alcohol	0.02—5%	2 — 1/2
LP gas	1000	100A	LP gas (propylene)	0.02—0.8%	1
Methanol	See methyl alcohol				
Methyl alcohol (methanol)	200	111	Methyl alcohol	0.01—3%	2 — 1/2
Methyl bromide	15	136	Methyl bromide	10—200 ppm	1 — 1/2
Methyl chloroform	350	135	Methyl chloroform	100—500 ppm	1
Methyl ethyl ketone	200	152	Methyl ethyl ketone	0.02—0.6%	2
Methyl ether	See Dimethyl ether				
Methyl isobutyl ketone	100	153	Methyl isobutyl ketone	0.01—0.6%	2
Methyl mercaptan	0.5	71	Methyl mercaptan	0.5—70 ppm	10 — 1
Monochlorobenzene	See chlorobenzene				
Monoethylamine	10	3L	Ammonia	2—30 ppm	1

Table 29 (continued)

Bendix-Gastec Precision Gas Detector Tubes (1978)

Gas or vapor to be measured	TLV (ppm)	Tube no.[a]	Detector tube to be used[b]	Measurable concentration	Pump strokes[c] for measurable concentration
Monomethylamine	10	3L	Ammonia	2.4—36 ppm	1
Monomethyl aniline	2	181	Aniline	2.5—45 ppm	3
Monostyrene	See Styrene				
Nitric acid	2	15L	Nitric acid	0.2—40 ppm	10 — 1/2
Nitric oxide	25	10	Nitric oxide	NO:2—200 ppm, NO_2: 5—200 ppm	5 — 1
Nitrogen dioxide	5	9L	Nitrogen dioxide	0.2—100 ppm	8 — 1
Nitrogen dioxide	5	10	Nitric oxide	NO:2—200 ppm, NO_2:5—200 ppm	5 — 1
Nitrogen oxides	NO-25, NO_2-5	11	Nitrogen oxides (NO + NO_2)	25—600 ppm	1
Nitrogen oxides	NO-25, NO_2-5	10	Nitric oxide	NO:2—200 ppm, NO_2:5—200 ppm	5 — 1
Octane	300[d]	101	Gasoline	0.015—1.2%	2 — 1/2
Ozone	0.1	18L	Ozone	0.05—3 ppm	10 — 1
Perchloroethylene	100	133	Perchloroethylene	5—200 ppm	10 — 1
Phosgene	0.1	16	Phosgene	0.1—90 ppm	5 — 1/2
Phosphine	0.3	7	Phosphine	5—200 ppm	2 — 1/2
n-Propyl acetate	200	151	Acetone	0.06—0.9%	2
n-Propyl alcohol	200	112	Ethyl alcohol	0.02—0.8%	2
Propylene		100A	LP gas (propylene)	0.02—0.8%	1
Propylene oxide	100	163	Ethylene oxide	0.3—3.6%	1
Styrene	100	124	Styrene	100—1000 ppm	2 — 1/2
Sulfur dioxide	5	5H	Sulfur dioxide	0.25—8%	2 — 1/2
Sulfur dioxide	5	5M	Sulfur dioxide	100—3600 ppm	1 — 1/2
Sulfur dioxide	5	5L	Sulfur dioxide	5—200 ppm	1 — 1/2

Sulfur dioxide	5	5La	Sulfur dioxide	1—90 ppm	4 — 1
Tetrachloroethylene	See Perchloroethylene				
Tetrahydrofuran	200	161	Ethyl ether	100—2500 ppm	5
Toluene	100	122	Aromatic hydrocarbon tube calibrated for toluene	10—600	2 — 1/2
o-Toluidine	5	181	Aniline	2.5—35 ppm	3
Tribromomethane	See Bromoform				
1,1,1-Trichloroethane	See Methyl chloroform				
Trichloroethylene	100	132H	Trichoroethylene	20—560 ppm	2 — 1/2
Trichloroethylene	100	132L	Trichloroethylene	2—50 ppm	1 — 1/2
Trichloronitromethane	See Chloropicrin				
Triethylamine	25	3M	Ammonia	3.5—140 ppm	4 — 2
Trimethylamine		3M	Ammonia	2.5—100 ppm	4 — 2
Vinyl acetate	10	141	Ethyl acetate	0.01—1.0%	1
Vinyl chloride low range	1	131L	Vinyl chloride	0.5—28 ppm	7 — 2
Vinylidene chloride		131L	Vinyl chloride	0.3—17 ppm	7 — 2
Water vapor		6	Water vapor	0.5—32 mg/ℓ	2 — 1/2
Xylene	100	123	Xylene	10—500 ppm	2 — 1/2

[a] All tube boxes contain complete information including quality control number, expiration date, interferences, temperature and humidity corrections, and storage instructions as specified in *Fed. Reg.*, 38, 88, 1973.

[b] All tubes should be refrigerated below 10°C (50°F).

[c] One half (½) pump stroke (50 cc) is used when sampling for higher concentrations. One (100 cc) or more pump strokes may be used to sample for lower concentrations. Refer to instruction sheet found in each box of tubes.

[d] Proposed TLV, 1974 ACGIH.

[e] When using carbon monoxide Tube ILa to measure 8 ppm, you take 3 strokes. To measure 1000 ppm, you take one half (½) stroke.

TABLE 30

Dräger Detector Tubes (1976)

Gases and vapors to be measured	Dräger tube to be used	Order code for 1 pack[a]	Range of measurement 20°C, 760 mmHg (1013 mbar)	No. of pump strokes	Threshold limit value[b] (USA 1975)
Acetaldehyde	Acetaldehyde 100/a	67 26665	100—1000 ppm	20	100 ppm
Acetic acid	Acetic acid 5/a	67 22101	5—80 ppm	3	10 ppm
Acetone	Acetone 100/b	CH 22901	100—12,000 ppm	10	1000 ppm
Acetylene	Hydrocarbons 0.1%	CH 26101	500—3000 ppm	12—1	—
Acrylonitrile	Acrylonitrile 5/a	CH 26901	5—30 ppm	5	20 ppm (S)[b]
Ammonia	Ammonia 5/a	CH 20501	5—70 ppm,	10	25 ppr
			50—700	1	—
	Ammonia 25/a	CH 25501	25—700 ppm	10	—
	Ammonia 0.5%/a	CH 31901	0.5—10 vol.%,	1	—
			0.05—1 vol.%	10	—
Aniline	Aniline 5/a	CH 20401	1—20 ppm	25—5	5 ppm (S)[b]
Arsine	Arsine 0.05/a	CH 25001	0.05—3 ppm,	20	0.05 ppm
			1—60 ppm	1	—
Benzene	Benzene 5/a	67 18801	5—40	15—2	10 ppm (S)[b]
	Benzene 5/b	67 28071	50 ppm	20	—
	Benzene 0.05	CH 24801	15—420 ppm,	20—2	10 ppm (S)[b]
			(0.05—1.4 mg/L)		
Bromine	Chlorine 0.2/a	CH 24301	0.2—3 ppm,	10	0.1 ppm
			2—30 ppm	1	—
Butadiene-(1,3)	Olefines 0.05%/a	CH 31201	0.01—0.12 vol.%	13—2	1000 ppm
n-Butane	Hydrocarbons 0.1%	CH 26101	0.1—1 vol.%	20—2	600 ppm
n-Butyl alcohol	Alcohol 100/a	CH 29701	100—3000 ppm	10	50 ppm (S)[b]
sec-Butyl alcohol	Alcohol 100/a	CH 29701	100—3000 ppm	10	150 ppm
Butylene-(1)	Olefines 0.05%/a	CH 31201	1—50 mg/L	20—1	—
Carbon dioxide	Carbon dioxide 0.01%/a	CH 30801	0.01—0.3 vol.%	10	—
	Carbon dioxide 0.1%/a	CH 23501	0.1—1.2 vol.%,	5	5000 ppm
			0.5—6 vol.%	1	—
	Carbon dioxide 0.5%/a	CH 31401	0.5—10 vol.%	1	—
	Carbon dioxide 1%	CH 25101	1—20 vol.%	1	—

	Carbon dioxide 5%/A	CH 20301	5—60 vol.%	1	—
	Simultan tube CO 200/a + CO_2 2%/a	67 18301	200—2500 ppm CO,	2	—
			(2—12 vol.% CO_2)	—	
Carbon disulfide	Carbon disulfide 0.04	CH 26001	13—288 ppm, (0.04—0.9 mg/L)	23—1	20 ppm (S)[b]
	Carbon disulfide 30/a	CH 23201	0.1—10 mg/L	6	—
Carbon monoxide	Carbon monoxide 5/c	CH 25601	5—150 ppm,	10	50 ppm
			100—700 ppm	2	—
	Carbon monoxide 8/a (only for CO in H_2)	CH 19701	8—150 ppm	10	—
	Carbon monoxide 10/a	CH 28900	10—300 ppm,	10	—
			100 —3000 ppm	1	—
	Carbon monoxide 10/b	CH 20601	10—300 ppm,	10	—
			100—3000 ppm	1	—
	Carbon monoxide 0.1%/a	CH 29601	0.1—1.2 vol.%	1	—
	Carbon monoxide 0.3%/a	CH 24901	0.3—4 vol.%	1	—
	Carbon monoxide 0.5%/a	CH 29901	0.5—7 vol.%	1	—
	Simultan tube CO 200/a + CO_2 2%/a	67 18301	200—2500 ppm CO	2	—
	Carbon pretube	CH 24101	—	—	—
Carbon tetrachloride 5/c	Carbon tetrachloride 5/c	CH 27401	5—50 ppm	5	10 ppm (S)[b]
Chlorine	Chlorine 0.2/a	CH 24301	0.2—3 ppm,	10	1 ppm
			2—30 ppm	1	—
	Chlorine 50/a	CH 20701	50—500 ppm	1	—
Chlorine dioxide	Chlorine 0.2/a	CH 24301	0.1—1.5 ppm	10	0.1 ppm
Chloroform	Methyl bromide 5/b	CH 27301	—	—	25 ppm
Chloroformates	Chloroformates 0.2/b	67 18601	0.2—10 ppm	20	—
Chloroprene	Chloroprene 5/a	67 18901	5—60 ppm	3	25 ppm (S)[b]
			7.5—90 ppm	2	—
Coolant R 11 (trichlorofluoromethane)	Carbon tetrachloride 10/b	CH 27401	—	—	1000 ppm
Cyanogen chloride	Cyanogen chloride 0.25/a	CH 19801	0.25—5 ppm	20—1	—
Cyclohexane	Cyclohexane 100/a	67 25201	100—1500 ppm	10	300 ppm
Diborane	Diborane 0.05/a	67 18101	0.05—3 ppm	20	0.1 ppm
1,2-Dichloroethane	Methyl bromide 5/b	CH 27301	—	—	50 ppm
1,1-Dichloroethylene (vinylidene chloride)	Vinyl chloride 1/a	67 28031	2.5—25 ppm	5	—

TABLE 30 (continued)

Dräger Detector Tubes (1976)

Gases and vapors to be measured	Dräger tube to be used	Order code for 1 pack[a]	Range of measurement 20°C, 760 mmHg (1013 mbar)	No. of pump strokes	Threshold limit value[b] (USA 1975)
1,2-Dichloropropane	Methyl bromide 5/b	CH 27301	—	—	75 ppm
1,2-Diethyl benzene	Toluene 25/a	CH 27801	—	—	—
Diethyl ether (ethyl ether)	Diethyl ether 100/a	67 30501	100—4000 ppm	10	400 ppm
Dimethyl acetamide	Dimethyl acetamide 10/a	67 28011	10—40 ppm	20	10 ppm (S)[b]
Dimethyl formamide	Dimethyl formamide 10/b	67 18501	10—40 ppm	10	10 ppm (S)[b]
1,1-Dimethyl hydrazine	Hydrazine 0.25/a	CH 31801	0.25—3 ppm	10	0.5 ppm (S)[b]
Dimethyl sulfate	Dimethyl sulfate 0.2/b	67 18701	0.2—5 ppm	22—2	—(S)
Ethyl acetate	Ethyl acetate 200/a	CH 20201	200—3000 ppm	20	400 ppm
Ethyl alcohol (ethanol)	Alcohol 100/a	CH 29701	100—3000 ppm	10	1000 ppm
Ethyl glycol acetate	Ethyl glycol acetate 50/a	67 26801	50—700 ppm	10	—
Ethyl mercaptan	see Mercaptan	—	—	—	—
Ethylene (ethene)	Hydrocarbons 0.1%	CH 26101	0.5—2 vol.%	28—7	—
	Ethylene 50/a	67 28051	50—2500 ppm	3	—
Ethylene imine	Hydrazine 0.25	CH 31801	0.25—3 ppm	10	0.5 ppm (S)[b]
Ethylene oxide	Hydrocarbons 0.1%	CH 26101	1.5—5 vol.%	24—7	50 ppm
Extraction solvents	Hydrocarbons 2	CH 25401	—	—	—
Formaldehyde	Formaldehyde 0.002	CH 26401	2—40 ppm	5	2 ppm
	Formaldehyde 0.5/a	67 26760	0.5—10 ppm	16—1	2 ppm
Formic acid	Formic acid 1/a	67 22701	1—15 ppm	20	5 ppm
Hydrazine	Hydrazine 0.25/a	CH 31801	0.25—3 ppm	10	1 ppm (S)[b]
Hydrocarbons	Hydrocarbons 0.1%	CH 26101	0.1—1 vol.%	15—3	—
	Hydrocarbons 2	CH 25401	2—25 mg/L	24—3	—
Hydrochloric acid (hydrogen chloride)	Hydrochloric acid 1/a	CH 29501	1—10 ppm	10	5 ppm
			2—20 ppm	5	—
Hydrocyanic acid (hydrogen cyanide)	Hydrocyanic acid 2/a	CH 25701	2—30 ppm	5	10 ppm (S)[b]
			10—150 ppm	1	
Hydrogen	Special tube, hydrogen 0.5%/a	CH 30901	0.5—3 vol.%	5	—
Hydrogen chloride	see Hydrochloric acid	—	—	—	—

Hydrogen cyanide	see Hydrocyanic acid	—	—	—	—
Hydrogen fluoride	Hydrogen fluoride 1.5/b	CH 30301	1.5—15 ppm	20	3 ppm
Hydrogen sulfide	Hydrogen sulfide 1/c	67 19001	1—20 ppm	10	10 ppm
			10—200 ppm	1	—
	Hydrogen sulfide 5/b	CH 29801	5—60 ppm	10	—
			50—600 ppm	1	—
	Hydrogen sulfide 0.5/a	67 28041	0.5—15 ppm	10	—
			25—75 ppm	2	—
	Hydrogen sulfide 100/a	CH 29101	100—2000 ppm	1	—
	Hydrogen sulfide 0.2%/A	CH 28101	0.2—7 vol.%	1	—
	Hydrogen sulfide + sulfur dioxide 0.2%/A	CH 28201	0.2—7 vol.%	1	—
			0.02—0.7 vol.%	10	—
Kerosene	Hydrocarbons 2	CH 25401	—	—	—
Liquid gas	See Butane and Propane				
Mercaptan	Mercaptan 2/a	CH 20801	2—100 ppm	10	0.5 ppm
Mercury vapor	Mercury 0.1/b	CH 23101	0.1—2 mg/m^3	20—1	0.05 mg/m^3
Metasystox	Systox 1/a	CH 27501	—	—	—
Methacrylonitrile	Methacrylonitrile 1/a	67 30101	1—10 ppm	20	1 ppm (S)
Methane	See Natural gas	—	—	—	—
Methyl alcohol (methanol)	Alcohol 100/a	CH 29701	100—3000 ppm	10	200 ppm (S)[b]
Methyl bromide	Methyl bromide 5/b	CH 27301	5—50 ppm	5	15 ppm (S)[b]
Methyl chloroform	See Trichloroethane	—	—	—	—
Methyl ethyl ketone	Acetone 100/b	CH 22901	—	—	200 ppm
Methyl mercaptan	See Mercaptan	—	—	—	—
Methylene chloride (dichloromethane)	Methylene chloride 100/a	67 24601	100—2000 ppm	10	200 ppm
Monostyrene (styrene)	Monostyrene 10/a	67 23301	10—200 ppm	15—2	100 ppm
	Monostyrene 50/a	CH 27601	50—400 ppm	11—2	100 ppm
Natural gas (methane)	Drager natural gas test	CH 20001	Qualitative	—	—
Nickel tetracarbonyl	Nickel tetracarbonyl 0.1/a	CH 19501	0.1—1 ppm	20	0.1 ppm (S)[b]
Nitrogen dioxide	Nitrogen dioxide 0.5/c	CH 30001	0.5—10 ppm,	5	5 ppm
			5—25 ppm	2	—
	Nitrogen dioxide 2/c	67 19101	2—50 ppm,	10	—
			5—100 ppm	5	—
Nitrogen monoxide	See Nitrous fumes	—	—	—	—

TABLE 30 (continued)

Dräger Detector Tubes (1976)

Gases and vapors to be measured	Dräger tube to be used	Order code for 1 pack[a]	Range of measurement 20°C, 760 mmHg (1013 mbar)	No. of pump strokes	Threshold limit value[b] (USA 1975)
Nitroglycol	Nitroglycol 0.25/a	67 18201	0.25 ppm	20	—
Nitrous fumes (NO + NO_2)	Nitrous fumes 0.5/a	CH 29401	0.5—10 ppm	5	5 ppm (NO)
	Nitrous fumes 2/a	CH 31001	2—50 ppm,	10	—
			5—100 ppm	5	—
	Nitrous fumes 20/a	67 24001	20—500 ppm	2	—
	Nitrous fumes 100/c	CH 27701	100—1000 ppm,	5	—
			500—5000 ppm	1	—
Olefines (butylene, propylene)	Olefines 0.05%/a	CH 31201	1—50 mg/L	20—1	—
Oxygen	Oxygen 5%/B	67 28081	5—23 vol.%	1	—
Ozone	Ozone 0.05/a	CH 31301	0.05—1.4 ppm,	10	0.1 ppm
			0.5—14 ppm	1	—
	Ozone 10/a	CH 21001	10—300 ppm	1	—
n-Pentane	*n*-Pentane 100/a	67 24701	100—1500 ppm	5	600 ppm
Perchloroethylene (Tetrachloroethylene)	Perchloroethylene 10/a	CH 30701	10—500 ppm	3	100 ppm
	Perchloroethylene 5/a	67 26699	5—50 ppm	10	—
	Perchloroethylene 0.1%/a	67 28021	0.1—1.4 vol.%	5	—
Petroleum hydrocarbons	Petroleum hydrocarbons 100/a	67 30201	100—2500 ppm, (*n*-Octane)	2	300 ppm (Octane)
	Hydrocarbons 2	CH 25401	—	—	—
Phenol	Phenol 5/a	CH 31501	5 ppm	10	5 ppm (S)[b]
Phosgene	Phosgene 0.05/a	CH 19401	0.05—1.2 ppm	26—1	0.05 ppm
	Phosgene 0.25/b	CH 28301	0.25—15 ppm,	5	—
			1.25 —75	1	—
Phosphine	Phosphine 0.1/a	CH 31101	0.1—4 ppm,	10	0.3 ppm
			1—40 ppm	1	—
	Phosphine 50/a	CH 21201	50—1000 ppm,	3	—
			15—300 ppm,	10	—
			150—3000 ppm	1	—

(Polytest)	Drager polytest	CH 28401	—	—	—
Propane	Hydrocarbons 0.1%	CH 26101	0.2—1 vol.%	25—6	—
n-Propyl alcohol (propanol)	Alcohol 100/a	CH 29701	100—3000 ppm	10	200 ppm (S)[b]
iso-Propyl alcohol (isopropanol)	Alcohol 100/a	CH 29701	100—3000 ppm	10	400 ppm (S)[b]
Propylene	Olefines 0.05%/a	CH 31201	1—50 mg/L	20—1	—
Styrene	See monostyrene	—	—	—	—
Sulfur dioxide	Sulfur dioxide 0.1/a	67 27101	0.1—3 ppm	100	—
	Sulfur dioxide 1/a	CH 31701	1—25 ppm	10	5 ppm
	Sulfur dioxide 20/a	CH 24201	20—200 ppm,	10	—
			200—2000 ppm	1	—
	Sulfur dioxide 50/a	67 26289	50—500 ppm,	10	—
			500—5000 ppm	1	—
Systox	Systox 1/a	CH 27501	0.5 mg/m³	20	0.1 mg/m³
Tetrahydrofuran	Alcohol 100/a	CH 29701	30—400 ppm	20	200 ppm
Tetrahydrothiophene	Systox 1/a	CH 27501	—	—	—
Toluene	Toluene 5/a	CH 23001	5—400 ppm	5	100 ppm (S)[b]
	Toluene 25/a	CH 27801	25—1860 ppm, 0.1—7 mg/L	10	—
Toluylene diisocyanate (TDI)	TDI 0.02/A	67 24501	0.02—0.2 ppm	25	0.02 ppm
Town gas	Drager Polytest	CH 28401	—	—	—
	Carbon monoxide 10/a	CH 28900	—	—	—
	Carbon monoxide 10/b	CH 20601	—	—	—
1,1,1-Trichloroethane (methyl chloroform)	Trichloroethane 50/b	CH 21101	50—350 ppm,	10	350 ppm
			100—700 ppm	5	—
Trichloroethylene	Trichloroethylene 10/a	CH 24401	10—500 ppm	5	100 ppm
Triethylamine	Triethylamine 5/a	67 18401	5—60 ppm	5	25 ppm
Vinyl chloride (chloroethylene)	Vinyl chloride 0.5/a	67 28061	0.5—3 ppm,	10	1 ppm
			0.25—1.5 ppm	20	—
	Vinyl chloride 1/a	67 28031	1—10 ppm,	20	—
			5—50 ppm	5	—
	Vinyl chloride 100/a	CH 19601	100—3000 ppm	18—1	—
Vinylidene chloride	See 1,1-Dichloroethylene	—	—	—	—

TABLE 30 (continued)

Dräger Detector Tubes (1976)

Gases and vapors to be measured	Dräger tube to be used	Order code for 1 pack[a]	Range of measurement 20°C, 760 mmHg (1013 mbar)	No. of pump strokes	Threshold limit value[b] (USA 1975)
Water vapor	Water vapor 0.1	CH 23401	0.1—40 mg/L	10	—
o-Xylene	Toluene 25/a	CH 27801	0.1—7 mg/L	10	100 ppm (S)[b]
Ethylene oxide	Ethylene oxide	67 28241	25—500 ppm	30	50 ppm
Hydrocarbon propane	Hydrocarbon	CH 26101	0.5—1.3%	1	—
Hydrocarbon butane			0.1—0.8%		
Methyl acrylate	Methyl acrylate	67 28161	5—200 ppm	20	
Methyl bromide	Methyl bromide	67 28211	10—100 ppm,	2	
			3—35 ppm	5	
Methyl methacrylate	Methyl methacrylate	67 28171	50—500 ppm	10	100
Tetrhydrothiophene	THT	67 28251	1—10 ppm,	30	
			1—16 ppm	20	

[a] With a few exceptions, 1 pack contains 10 Drager tubes (the shelf life of Drager tubes is 2 years).

[b] Substances which pass into the body not by inhalation, but by skin resorption in the liquid or vapor are marked with an (S) in parentheses in the TLV column.

From Drägerwerk, A. G., Lubeck-D-24, Lubeck 1, Moislinger Allee 53/55, P.O. Box 1339, Federal Republic of Germany, U.S. Representative, National Mine Service Co., 3000 Koppers Building, Pittsburgh, Pa. With permission.

TABLE 31

Detector, Reagents, Accessories for Universal Tester® and Samplair® Pumps and Testing Kits

Detectable hazard		Measurable range (ppm)	1978 ACGIH TLVs[a] (ppm)	Pkg of detector tubes[b]	Reagent kit	Box of filter papers	Required pyrolyzer	Interferences[c]
Acetaldehyde		25—1000	100	93963				Turpentine, other aldehydes
Acetic acid		0—50	10	460022				SO_2
Acetone	NIOSH (TC-84-54)	125—6500	1000	460423				Ethanol
Acetonitrile		10—200	40	91624			460325	H_2S, halogen, nitrogen and hydrocarbon compounds
Acetylene		3—600		82802				Ammonia, H_2S, CO, hydrocarbons
Acrylonitrile		1—150	(20)	91624			460325	H_2S, halogen, nitrogen and hydrocarbon compounds
Ammonia	NIOSH (TC-84-034	10—1500	25	92115				Acid gases and other amines
		5—1600	25	460103				Acid gases and other amines
n-Amyl alcohol (pentanol 1)		25—2000		95097				Other alcohols
sec-Amyl alcohol (pentanol 2)		25—2000		95097				Other alcohols
tert-Amyl alcohol (2 methyl-butanol 2)		25—2000		95097				Other alcohols
n-Amyl mercaptan (pentanethiol-1)		0.5—100		454206[d]				H_2S
Arsine[e,f]		0.025—1.0	0.05		81101[g]	81220[h]		Stibine, phosphine
Benzene		5—200	10	460754				Aromatic, aliphatic hydrocarbons
		0—150	10	467050				Aromatic, aliphatic hydrocarbons
Bromine		5—175	0.1	82399				H_2S, NH_3, NO_2, ethylene, halides
Bromobenzene (Mono)		10—800		93074				Aromatic, aliphatic hydrocarbons
Butadiene (1,3-butadiene)		100—4000	1000	93962				—
2-Butoxy ethanol (butyl Cellosolve®)		30—900	50	95097				Other alcohols
n-Butyl alcohol (butanol 1)		50—4000	50	95097				Other alcohols
sec-Butyl alcohol (butanol 2)		50—4000	150	95097				Other alcohols
tert-Butyl alcohol (2 methyl propanol 2)		100—1000	100	95097				Other alcohols

TABLE 31 (continued)

Detector, Reagents, Accessories for Universal Tester® and Samplair® Pumps and Testing Kits

Detectable hazard		Measurable range (ppm)	1978 ACGIH TLVs[a] (ppm)	Pkg of detector tubes[b]	Reagent kit	Box of filter papers	Required pyrolyzer	Interferences[c]
n-Butylamine		2—100	5	92115				Acid gases
Butyl mercaptan (butanethiol-1)		0.5—100	0.5	454206[d]				H_2S
tert-Butyl mercaptan (2-methyl propanethiol-2)		0.5—100		454206[d]				H_2S
Carbon dioxide	NIOSH (TC-84-025)	200—120,000	5000	85976				—
Carbon disulfide		0—100	(20)	463527				—
Carbon monoxide	NBS color change[i]	10—1000	50	47134				NO_2, ethylene, acetylene
	NIOSH (TC-84-067)	0—3000	50	465519				Hydrogen, H_2S, acetylene
Carbon tetrachloride		0—70	10	460548			460325	Other halogen compounds and halogens
Chlorine	NIOSH (TC-84-042)	0.5—100	1	82399				H_2S, NH_3, NO_2, ethylene, halides
		0.1—25	1	460225				H_2S, NH_3, NO_2, ethylene, halides
Chlorine dioxide		0.05—16	0.1	82399				H_2S, NH_3, NO_2, ethylene, halides
Chlorobenzene (monochlorobenzene)		10—200	75	85834[j]				Other halogen compounds and halogens
		10—800	75	93074				Aromatic, aliphatic hydrocarbons
Chlorobromomethane		50—400	200	85834[j]			460325	Other halogen compounds and halogens
		25—500	200	88536				Other halogen compounds and halogens
1-Chloro-1,1-difluoroethane (Genetron* 142B)		100—4000		88536			460325	Other halogen compounds and halogens
		100—2500		92030			460325	HNO_3 hydrocarbons
Chlorodifluoromethane (Freon® 22)		200—2000	1000	88536			460325	Other halogen compounds and halogens

Chloroform (trichloromethane)	25—400	10	88536			460325	Other halogen compounds and halogens
1-Chloro-1-nitropropane	2—75	20	91624			460325	H_2S, halogen, nitrogen and hydrocarbon compounds
Chloropentafluoroethane (Freon® 115)	200—2000		88536			460325	Other halogen compounds and halogens
Chlorotrifluoromethane (Freon® 13)	250—2500		88536			460325	Other halogen compounds and halogens
	200—3500		92030			460325	HNO_3, hydrocarbons
Chromic acid mist[e,f]	0.05—1.4 mg/m³	0.1	79014[g]		79010		—
Cumene	0—1000	50	460422				Butyl mercaptan and unsaturated hydrocarbons
Cyclohexanol	25—1000	50	95097				Other alcohols
Cyclohexylamine	2—100	10	92115				Acid gases
Decarborane[e,f]	0.01—1.0	0.05		82099[g]	82388		Hydrazine, UDMH, H_2S
Diborane[e,f]	0.1—3.0	0.1		82099[g]	82388		Hydrazine, UDMH, H_2S
1,2-Dibromoethane (ethylene dibromide)	10—200	20	88536			460325	Other halogen compounds and halogens
	5—100	20	85834[j]				Other halogen compounds and halogens
o-Dichlorobenzene	10—200	50	85834[j]				Other halogen compounds and halogens
Dichlorodifluoromethane (Freon® 12)	25—3000	1000	88536			460325	Other halogen compounds and halogens
1,1-Dichloroethane	25—400	200	85834[j]				Other halogen compounds and halogens
1,2-Dichloroethylene	10—300	200	85834[j]				Other halogen compounds and halogens
	100—500	200	88536			460325	Other halogen compounds and halogens
Dichloroethylene (*trans*-1,2)	25—1000	200	92030			460325	HNO_3, hydrocarbons
Dichloroethyl ether	5—75	5	85834[j]				Other halogen compounds and halogens
1,1-Dichloro-1-nitroethane	2—75	10	91624			460325	H_2S, halogen, nitrogen and hydrocarbon compounds

TABLE 31 (continued)

Detector, Reagents, Accessories for Universal Tester® and Samplair® Pumps and Testing Kits

Detectable hazard	Measurable range (ppm)	1978 ACGIH TLVs[a] (ppm)	Pkg of detector tubes[b]	Reagent kit	Box of filter papers	Required pyrolyzer	Interferences[c]
Dichlorotetrafluoroethane (Freon® 114)	200—2000	1000	88536			460325	Other halogen compounds and halogens
Diethylamine	5—200	25	92115				Acid gases
Diethyl sulfate	5—50		95300				—
Diisopropylamine	1—100	5	92115				Acid gases
Dimethyl acetamide	5—80	10	91624			460325	H_2S, halogen, nitrogen and hydrocarbon compounds
Dimethylamine	5—200	10	92115				Acid gases
Dimethylformamide	5—150	10	91624			460325	H_2S, halogen, nitrogen and hydrocarbon compounds
Dimethyl sulfate	1—50	0.1	95300				—
Dimethyl sulfoxide	10—250		95739			460325	SO_2, organic sulfur compounds
Di-*n*-propylamine	2—70		92115				Acid gases
2-Ethoxyethanol	50—1000	100	95097				Other alcohols
Ethyl alcohol (ethanol)	200—10,000	1000	95097				Other alcohols
Ethylamine	10—500	10	92115				Acid gases
Ethyl benzene	10—500	100	463202				Aromatic, aliphatic hydrocarbons
Ethyl bromide	25—400	200	85834[j]				Other halogen compounds and halogens
Ethyl chloride	100—2000	1000	85834[j]			460325	Other halogen compounds and halogens
	100—2000	1000	92030				HNO_3, hydrocarbons
Ethylene	0.5—100		82802				NH_3, H_2S, CO, hydrocarbons
Ethylene dichloride 1,2-Dichloroethane NIOSH (TC-84-058)	0—600	(50)	461863[j]				Other halogen compounds and halogens

Ethylenimine	2—100	0.5	92115				Acid gases
Ethyl mercaptan	0.5—100	0.5	454206[d]				H_2S
N-Ethylmorpholine	2—100	20	92115				Acid gases
Fluorotrichloromethane (Freon® 11)	250—2000	1000	88536			460325	Other halogen compounds and halogens
	250—2000	1000	92030			460325	HNO_3, hydrocarbons
Formaldehyde	1—100	2	93963				Turpentine, other aldehydes
Furfuryl alcohol	25—500	5	95097				Other alcohols
Hexane	25—5000	100	463838				Aliphatic and some aromatic hydrocarbons
Hydrazine	0.5—40	0.1	460425				Acid gases and other amines
Hydrogen chloride	0—500	5	91636				HNO_3
	2—170	5	466612				HNO_3
Hydrogen cyanide	0—80	(10)	93262				Ammonia, H_2S
Hydrogen fluoride	0—25	3	462019				—
Hydrogen sulfide NIOSH (TC-84-022)	1—1000	10	460058				SO_2
	10—2000	10	463875				SO_2
Isoamyl alcohol (2 methyl-butanol 4)	50—1000	100	95097				Other alcohols
Isobutyl alcohol (2 methyl propanol)	50—4000	50	95097				Other alcohols
Isopropylamine	2—100	5	92115				Acid gases
Lead dust/fume[e, f]	0.05—6.3 mg/m³	0.15 mg/m³		78399[g]	78693		—
Mercury, except alkyl	0.05—2.0 mg/m³	0.05 mg/m³	83089				—
Methacrylonitrile	10—80		91624			460325	H_2S, halogen, nitrogen and hydrocarbon compounds
Methyl alcohol (methanol)	100—10,000	200	95097				Other alcohols
Methylamine	10—500	10	92115				Acid gases
Methyl bromide	2.5—90	15	462135				Other halogen compounds and halogens
Methyl chloride	25—1000	100	92030			460325	HNO_3 hydrocarbons
Methyl chloroform	100—700	350	88536			460325	Other halogen compounds and halogens

TABLE 31 (continued)

Detector, Reagents, Accessories for Universal Tester® and Samplair® Pumps and Testing Kits

Detectable hazard	Measurable range (ppm)	1978 ACGIH TLVs[a] (ppm)	Pkg of detector tubes[b]	Reagent kit	Box of filter papers	Required pyrolyzer	Interferences[c]
Methylcyclohexanol	25—800	50	95097				Other alcohols
Methylene chloride (dichloromethane)	0—3000	(200)	462516				Other halogen compounds and halogens
Methyl isobutyl carbinol	25—1000	25	95097				Other alcohols
Methyl mercaptan	0.5—100	0.5	454206[d]				H_2S
Monomethyl hydrazine	0.5—50	0.2	460425				Acid gases and other amines
Nitric oxide NIOSH (TC-84-059)	1—150	25	460424				H_2S and halogens
Nitrogen dioxide NIOSH (TC-84-040)	0—30	5	83099				H_2S, halides
Nitromethane	5—300	100	91624			460325	H_2S, halogen, nitrogen and hydrocarbon compounds
1-Nitropropane	2—100	(25)	91624			460325	H_2S, halogen, nitrogen and hydrocarbon compounds
2-Nitropropane	5—150	25	91624			460325	H_2S, halogen, nitrogen and hydrocarbon compounds
Ozone	0.05—5.0	0.1	93865				NO_2SO_2
Pentaborane[e,f]	0.01—1.0	0.005		82099[g]	82388		Hydrazine, UDMH, H_2S
Perchloroethylene NIOSH (TC-84-065)	0—800	100	460467				Other halogen compounds and halogens
Phosgene (carbonyl chloride)	0.1—10	0.1	89890				—
Phosphine[e,f]	0.025—1.0	0.3		81101[g]	81220[h]		Arsine, stibine
Propyl alcohol (propanol-1)	100—10,000	200	95097				Other alcohols
iso-Propyl alcohol (propanol-2)	100—10,000	400	95097				Other alcohols
n-Propyl mercaptan (propanethiol-1)	0.5—100		454206[d]				H_2S
n-Propyl nitrate	10—100	25	91624			460325	H_2S, halogen, nitrogen and hydrocarbon compounds

Propylene	1—400		82802				NH_3, H_2S, CO, hydrocarbons
Propylene dichloride (1,2-dichloropropane)	25—250	75	85834[j]			460325	Other halogen compounds and halogens
	25—500	75	92030				HNO_3, hydrocarbons
Propylene imine	2—300	2	92115				Acid gases
Pyridine	2—50	5	91624			460325	H_2S, halogen, nitrogen and hydrocarbon compounds
Stibine[e,f]	0.025—1.0	0.1		81101[g]	81220[h]		Arsine, phosphine
Styrene, monomer (phenylethylene)	0—500	100	461781				—
Sulfur dioxide NIOSH (TC-84-046)	1—400	(5)	92623				Acetic acid
1,1,2,2-Tetrabromoethane	2—50		85834[j]				Other halogen compounds and halogens
1,1,2,2-Tetrachloroethane	2—50	5	85834[j]				Other halogen compounds and halogens
1,1,3,3-Tetrachloropropene	5—100		85834[j]				Other halogen compounds and halogens
Toluene (toluol) NIOSH (TC-84-057)	12.5—700	100	461371				Aromatic, aliphatic hydrocarbons
1,1,2-Trichloroethane (beta)	25—700	10	85834[j]			460325	Other halogen compounds and halogens
	50—500	10	88536				Other halogen compounds and halogens
Trichloroethylene NIOSH (TC-84-055)	25—600	100	460328				Other halogen compounds and halogens
1,2,3-Trichloropropane	10—500	50	85834[j]				Other halogen compounds and halogens
1,1,2-Trichloro-1,2,2-trifluoroethane (Freon® 113)	100—2000	1000	88536			460325	Other halogen compounds and halogens
	100—4000	1000	92030			460325	HNO_3, hydrocarbons
Triethylamine	2—100	25	92115				Acid gases
Trifluoromonobromomethane (Freon® 13B1)	100—4000	1000	88536			460325	Other halogen compounds and halogens
Trimethylamine	5—250		92115				Acid gases
Unsymmetrical dimethylhydrazine	0.5—50	0.5	460425				Acid gases and other amines

TABLE 31 (continued)

Vinyl chloride (chloroethylene)	0—100	1	462534	HNO_3, hydrocarbons
Vinylidene chloride	10—300	10	85834[j]	Other halogen compounds and halogens
Xylene (*o*-, *m*-, *p*-isomers)	10—500	100	463201	Aromatic, aliphatic hydrocarbons

[a] Refers to Threshold Limit Values as established by the American Conference of Governmental Industrial Hygienists, 1978. (TLVs refer to time-weighted-average concentrations for a 7- or 8-hr workday and 40-hr work week). Threshold Limit Values are reviewed annually by the Conference. For current TLV data, contact: Secretary-Treasurer, American Conference of Governmental Industrial Hygienists, P.O. Box 1937, Cincinnati, Ohio 45201.

[b] If remote sampling is desired, also order a sampling line of the desired length: #73067 Sampling Line, 10 ft long; #73068 Sampling Line, 25 ft long; #73069 Sampling Line, 50 ft long. *Note:* when sampling point is remote from pump, detector tube should be placed in remote end of sampling line.

[c] Interferences and temperature correction data shown on individual instruction sheets.

[d] If desired, a gas-sampling collecting kit is available for measuring the mercaptan content of a sample taken from a natural gas line. Order #454500 Gas-Sample Collecting Kit.

[e] Accessory Kit #87030 required for detecting Chromic acid mist, decaborane, diborane, lead dust/fume, and pentaborane. Accessory kit #87031 required for arsine, phosphine, and stibine.

[f] Requires adaptor (#467063) when using Samplair® Pump.

[g] Also order #82389 vial of charcoal (required for filling filter retainer of accessory kit to protect sampling pump from reagent vapors and/or droplets).

[h] Also order #81316 package of pre-filter discs to remove interfering H_2S and SO_2 (if present). A pre-filter effectively absorbs these gases in concentrations up to 100 ppm.

[i] Order sample conditioning kit #87028. When measuring CO content of flue gases or in presence of NO_2 in concentrations from 7 to 100 ppm, also order #73139 Gasorbent Refill (sufficient for 10 refills of sample conditioning kit "prescrubber").

[j] Also order #58238 pinch clamp (necessary to restrict contents of reactor tube).

Note: This data sheet contains a general description of certain detectors, reagents and accessories for use with the Universal Tester® and Samplair® Pumps from MSA and includes a list of representative hazards that can be detected; however, under no circumstances should the products be used except by qualified, trained personnel and not until the instructions, labels, or other literature accompanying them have been carefully read and uderstood and the precautions therein set forth followed. Only they contain the complete and detailed information concerning these products.

Courtesy of Mine Safety Appliances Co., 600 Penn Center Blvd., Pittsburgh, Pa., 15235.

TABLE 32

Detector Tube Procurement Sources — Summary[87]

Compound	Bendix-Gastec	Dräger	Matheson-Kitagawa	MSA
	Detector tube identification, tube number, and detection range			
Acetaldehyde	—	6726665 (100—1000 ppm)	8014-133 (0.004—1.6 %)	93963 (25—1000 ppm)
Acetic acid	81 (1—80 ppm)	6722101 (5—80 ppm)	—	92623 (5—400 ppm)
Acetic anhydride	81 (0.5—40 ppm)	—	—	—
Acetone	151 (0.01—2%)	CH 22901 (100—1200 ppm)	8104-102A (0.05—5.0%) 8104-102C (0.01—4.0%)	—
Acetonitrilie	—	—	—	91624 (10—200 ppm)
Acetylene	171 (0.1—4%)	CH 26101 (500—3000 ppm)	8104-101 (50—1000 ppm)	82802 (3—600 ppm)
Acritet	—	—	—	91624 (8—70 ppm)
Acrolein	—	—	8014-136 (0.005—2.5%)	—
Acrylonitrile	191 (10—500 ppm)	CH 26901 (5—30 ppm)	8014-128A (0.1—3.5%) 8014-128B (10—500 ppm)	91624 (5—150 ppm)
Alcohol	—	CH 29701 (100—3000 ppm)	—	—
Ammonia	3L (1—60 ppm) 3H (0.2—32%) 3M (10—1000 ppm)	CH 25501 (25—700 ppm) CH 31901 (0.5—10%) (0.05—1%) CH 20501 (5—70 ppm) (50—700 ppm)	8104-105B (20—700 ppm) 8014-105A (1—25%)	92115 (10—1500 ppm)
n-Amyl alcohol	—	—	—	95097 (25—2000 ppm)
Iso-Amyl alcohol	—	—	—	95097 (50—1000 ppm)
sec-Amyl alcohol	—	—	—	95097 (25—2000 ppm)
tert-Amyl alcohol	—	—	—	95097 (25—2000 ppm)
n-Amyl mercaptan (1-pentanethiol)	—	—	—	454206 (0.5—100 ppm)
Aniline	181 (1.25—60 ppm)	CH 20401 (1—20 ppm)	—	—
Arsine	—	CH 25001 (0.05—3 ppm) (1—60 ppm)	804-140 (5—160 ppm)	—
Benzene	121 (5—120 ppm)	67 18801 (5—40 ppm) CH 24801 (15—420 ppm) (0.05—1.4 mg/L)	8014-1186 (2—70 ppm) 8014-118A (10—310 ppm) 8014-118B (25—345 ppm)	93074 (5—200 ppm)

TABLE 32 (continued)

Detector Tube Procurement Sources — Summary[87]

Compound	Detector tube identification, tube number, and detection range			
	Bendix-Gastec	Dräger	Matheson-Kitagawa	MSA
Benzyl bromide	136 (25—850 ppm)	CH 24401 (qual)	—	—
Benzyl chloride	132L (0.5—25 ppm)	67 28031 (1—10 ppm) (5—50 ppm)	—	—
Bromine	9L (2—23 ppm)	CH 24301 (0.2—3 ppm) (2—30 ppm)	8014—114 (1—50 ppm)	82399 (5—75 ppm) 87042 (5—200 ppm) 93074 (10—800 ppm)
Bromobenzene (mono)	—	—		
Bromoform	136 (0.5—50 ppm)	CH 24401 (qual)	—	—
Butadiene	174 (50—800 ppm)	CH 31201 (0.01—0.12%)	8014-168A (0.03—2.6%) 8014-168B (30—600 ppm)	93962 (100—4000 ppm)
Butane	104 (50—1400 ppm)	CH 26101 (0.1—1%)	—	—
Butyl acetate	142 (0.01—0.8%)	—	8014-138 (0.01—1%)	—
2-Butoxy ethanol (butyl cellosolve)	—	—	—	95097 (30—900 ppm)
n-Butyl alcohol	112 (100—1500 ppm)	CH 29701 (100—3000 ppm)	—	95097 (50—4000 ppm)
Iso-Butyl alcohol	—	—	—	95097 (50—4000 ppm)
sec-Butyl alcohol	—	CH 29701 (100—3000 ppm)	—	95097 (50—4000 ppm)
tert-Butyl alcohol	—	—	—	95097 (100—1000 ppm)
		—	—	92115 (2—100 ppm)
n-Butyl mercaptan (1-butane thiol)	—	— —	454206	(0.5—100 ppm)
tert-Butyl mercaptan (2-methyl-2-propanethiol)	—	—	—	454206 (0.5—100 ppm)
Butylene-(1)	—	CH 31201 (1—50 mg/ℓ)	—	—
Carbon dioxide	2LL (0.03—0.5%) 2L (0.13—8%) 2H (1—20%	CH 30801 (0.01—0.3%) CH 23501 (0.1—1.2%) (0.5—6.5%) CH 25101 (1—20%)	8014-126B (100—7,000 ppm)	85976 (0.1—10%)

		CH 31401 (0.5—10%) CH 20301 (5—60%)		
Carbon disulfide	13 (5—50)	CH 26001 (13—228 ppm) (0.04—0.9 mg/L) CH 23201 (0.1—10 mg/L) (32—3200 ppm)	8014-141 (10—200 ppm)	95297 (5—500 ppm)
Carbon monoxide	1LL (5—50 ppm) 1La (8—1000 ppm) 1L (50—2000 ppm) 1M (0.1—4%) 1H (0.2—10%)	CH 28900 (10—300 ppm) (100—3000 ppm) CH 19701 (8—150 ppm) CH 25601 (5—150 ppm) (100—700 ppm) CH 20601 (10—300 ppm) (100—3000 ppm) CH 29601 (0.1—1.2%) CH 24901 (0.3—4%) CH 29901 (0.5—7%)	8014-100 (20—1000 ppm) 8014-106A (10—6000 ppm) 8014-106B (10—6000 ppm) 8014-106C (10—6000 ppm)	91229 (10—3000 ppm) 47134 (0.01—0.1 %)
CO200/a + CO_2 (simultaneous tube)	—	67 18301 (200—2500 ppm CO) (2—12% CO_2)	—	—
CO (supplementary part for the respiratory test)	—	CH 270	—	—
Carbon tetrachloride	134 (1—60 ppm)	CH 27401 (10—100 ppm)	8014-147 (5—300 ppm)	88536 (10—200 ppm)
Chlorine	8LA (0.3—16 ppm)	CH 20701 (50—500 ppm) CH 24301 (0.2—3 ppm) (2—30 ppm)	8014-109 (1—40 ppm)	87042 (3—100 ppm) 82399 (0.5—20 ppm)
Chlorine dioxide	8LA (0.3—16 ppm)	CH 24301 (0.1—1.5 ppm)	8014-116 (1—40 ppm)	82399 (0.05—16 ppm)
Chlorobenzene	121 (5—350 ppm)	—	—	85834 (10—200 ppm) 93074 (10—800 ppm)
Chlorobromomethane	136 (8—80 ppm) 135 (40—350 ppm)	—	—	85833 (50—500 ppm) 88536 (25—500ppm)
Chloroform (trichloromethane)	—	CH 27301 (5—50 ppm)	8014-152 (75—600 ppm)	88536 (25—400 ppm)

TABLE 32 (continued)

Detector Tube Procurement Sources — Summary[87]

Compound	Bendix-Gastec	Dräger	Matheson-Kitagawa	MSA
	Detector tube identification, tube number, and detection range			
Chloropicrin	134 (1—60 ppm)	—	—	—
1-Chloro-1,1-difluoroethane (Genetron 142B)	—	—	—	88536 (100—2500 ppm)
				92030 (100—400 ppm)
Chlorodifluoromethane (Freon® 22)	—	—	—	88536 (200—2000 ppm)
1-Chloro-1-nitropropane	—	—	—	91624 (2—75 ppm)
Chloropentafluoroethane (Freon® 115)	—	—	—	88536 (200—2000 ppm)
Chloroprene	—	67 18901 (5—60 ppm)	8014-169 (5—800 ppm)	—
		(7.5—90 ppm)		
Chlorotrifluoromethane (Freon® 13)	—	—	—	88536 (250—2500 ppm)
				92030 (200—3500 ppm)
Cumene	122 (5—2400 ppm)	CH 27801 (0.1—7 mg/L)	—	—
		(25—1860 ppm)		
Cyanogen	—	—	—	91624 (2—100 ppm)
Cyanogen chloride	—	CH 19801 (0.25—5 ppm)	—	—
Cyclohexane	102H (0.015—1.2%)	67 25201 (100—1500 ppm)	8014-115 (0.01—0.6%)	—
Cyclohexanol	—	—	—	95097 (25—1000 ppm)
Cyclohexylamine	—	—	—	92115 (2—100 ppm)
Diborane	—	67 18101 (0.05—3 ppm)	—	—
1,1-Dibromoethane	136 (7—70 ppm)	—	—	—
1,2-Dibromoethane	136 (6—80 ppm)	CH 24401 (qual)	—	88536 (10—200 ppm)
				85834 (5—100 ppm)
o-Dichlorobenzene	121 (10—700 ppm)	—	—	85834 (10—200 ppm)

Dichlorodifluoromethane	—	—	—	88536 (25—3000 ppm)
(Freon® 12)				
1,1-Dichloroethane	—	—	—	85833 (25—500 ppm)
1,2-Dichloroethylene	132H (10—450 ppm)	—	8014-145 (10—400 ppm)	85833 (cis)
				(10—300 ppm)
				88536 (cis)
				(100—500 ppm)
				85833 (trans)
				(25—400 ppm)
				92030 (trans)
				(25—1000 ppm)
1,2-Dichloroethane	—	CH 27301 (5—50 ppm)	—	85834 (25—400 ppm)
(ethylene dichloride)				92030 (25—450 ppm)
Dichloroethyl ether	—	—	—	85834 (5—75 ppm)
1,1 Dichloro-1-nitroethane	—	—	—	91624 (2—75 ppm)
Dichlorotetrafluoroethane	—	—	—	88536 (200—2000 ppm)
(Freon® 114)				
Diethyl benzene	122 (25—2400 ppm)	CH 27801 (0.1—7 mg/L)	—	—
		(25—1860 ppm)		
Diethylamine	3L (1.6—48 ppm)	67 18401 (5—60 ppm)	—	92115 (5—200 ppm)
Diethyl ether	161 (0.04—1.0%)	67 30501 (100—4000 ppm)	8014-107 (0.04—1.4%)	—
Dimethyl aniline	181 (2.5—45 ppm)	—	—	—
Dimethyl sulfate	—	—	—	95300 (5—50 ppm)
Dimethyl acetamide	—	67 28011 (10—40 ppm)	—	91624 (5—80 ppm)
Dimethylamine	3L (2—30 ppm)	67 18401 (5—60 ppm)	—	92115 (5—200 ppm)
Dimethyl ether	161 (0.03—0.9%)	—	8014-123 (0.01—1.2%)	—
Dimethylformamide		67 18501 (10—40 ppm)	—	91624 (5—150 ppm)
1,1-Dimethyl hydrazine	—	CH 31801 (0.25—3 ppm)	—	—
Dioxane	163 (0.5—5.6%)	—	8014-154 (0.01—2.5%)	—
Dimethyl sulfate	—	67 18701 (0.2—5 ppm)	—	95300 (1—50 ppm)
(ester)		67 18901 (5—60 ppm)		
Dimethyl sulfoxide	—	—	—	95739 (10—250 ppm)
Di-N-propylamine	—	—	—	92115 (2—70 ppm)
Ethyl acetate	141 (0.04—1.5%)	CH 20201 (200—3000 ppm)	8014-111 (0.01—5.0%)	—
2-Ethoxyethanol (Cellosolve)	—	—	—	95097 (50—1000 ppm)

TABLE 32 (continued)

Detector Tube Procurement Sources — Summary[87]

Compound	Bendix-Gastec	Dräger	Matheson-Kitagawa	MSA
	Detector tube identification, tube number, and detection range			
Ethyl alcohol (ethanol)	112 (0.05—5%)	CH 29701 (100—3000 ppm)	8014-104A (0.04—5.0%)	95097 (200—10,000 ppm)
Ethylamine	—	67 18401 (5—60 ppm)	—	92115 (10—500 ppm)
Ethyl benzene	122 (7—700 ppm)	CH 27801 (0.1—7 mg/L) (25—1860 ppm)	—	—
Ethyl bromide	—	—	—	85833 (25—400 ppm)
Ethyl chloride	—	—	—	85833 (100—2000 ppm)
				92030 (100—2000 ppm)
Ethylene (ethene)	172 (50—800 ppm)	CH 26101 (0.5—2%)	—	—
Ethylene	172L (0.2—50 ppm)	67 28051 (50—2500 ppm)	8014-108B (0.01—100 ppm)	82802 (0.5—100 ppm)
Ethylenediamine	3L (4—120 ppm)	—	—	—
Ethyl mercaptan	72 (5—120 ppm)	CH 20801 (2—100 ppm)	8014-165 (1—160 ppm)	454206 (0.5—100 ppm)
Ethylene dibromide	136 (6—80 ppm)	—	8014-166 (1—100 ppm)	—
Ethylene imine	—	CH 31801 (0.25—3 ppm)	—	92115 (2—100 ppm)
Ethylene oxide	163 (0.1—3.0%)	CH 26101 (1.5—5%)	8014-122 (0.01—3.5%)	—
n-Ethyl morpholine	—	—	—	92115 (2—100 ppm)
Fluorotrichloromethane (Freon® 11)	—	CH 27401 (5—50 ppm)	—	88536 (250—2000 ppm)
				92030 (250—2000 ppm)
Formaldehyde	91 (2—20 ppm)	CH 26401 (2—40 ppm) (0.002—0.05 mg/L)	8014-171 (1—35 ppm)	93963 (1—100 ppm)
Formic acid	81 (1—80 ppm)	67 22701 (1—15 ppm)	—	—
Fumi-1-Gate	—	—	—	91624 (10—200 ppm)
Furan (furfuran)	—	—	8014-161 (0.05—1.6%)	—
Furfuryl alcohol	—	—	—	95097 (25—500 ppm)

Gasoline	101 (0.015—1.2%)	—	—	—
n-Heptane	101 (0.015—1.2%)	—	—	—
n-Hexane	102H (0.015—1.2%)	67 25201 (100—1500 ppm)	8014-113 (0.01—0.6%)	—
	102L (50—1200 ppm)			
Hydrazine	3L (0.5—35 ppm)	CH 31801 (0.25—3 ppm)	—	—
Hydrocarbon	—	CH 26101 (0.1—1 vol.%)	—	—
	—	CH 25401 (2—25 mg/L)	—	—
Hydrochloric acid	14L (0.2—40 ppm)	CH 29501 (1—10 ppm)	—	91636 (2—500 ppm)
		(2—20 ppm)		
Hydrocyanic acid	12H (0.05—2%)	CH 25701 (2—30 ppm)	8014-112A (0.01—3.0%)	93262 (1—65 ppm)
	12L (2.5—60 ppm)	(10—150 ppm)	8014-112B (1—100 ppm)	
Hydrogen (special tube)	—	CH 30901 (0.5—3 vol.%)	—	—
Hydrogen fluoride	17 (0.5—20 ppm)	CH 30301 (0.5—15 ppm)	—	
Hydrogen selenide	—	—	8014-167 (1—600 ppm)	—
Hydrogen sulfide	4HH (0.1—4%)	CH 28101 (0.2—7 vol.%)	8014-120H (0.02—0.3%)	—
	4H (100—3200 ppm)	CH 29101 (100—2000 ppm)	8014-120A (0.01—0.17%)	
	4LL (0.5—60 ppm)	67 19001 (1—20 ppm)	87414 (1—800 ppm)	
	4L (10—240 ppm)	(10—200 ppm)		
		CH29801 (5—60 ppm)		
		(50—600 ppm)		
Hydrogen sulfide + sulfur dioxide	—	CH 28201 (0.2—7 vol.%) (0.02—0.7 vol.%)	8014-120C (0.005—0.16%)	—
Insect-O-Fume	—	—	—	91624 (8—70 ppm)
Isobutyl acetate	142 (0.005—66%)	—	8014-153 (0.01—1.4%)	—
Isobutyl alcohol	112 (100—3000 ppm)	—	—	—
Iso-octane	101 (0.015—1.2%)	—	—	—
Isopropyl acetate	151 (0.05—0.75%)	—	8014-149 (0.01—1.2%)	—
Isopropyl alcohol	113 (0.02—5%)	CH 29701 (100—3000 ppm)	8014-150 (0.01—2%)	—
Isopropylamine	—	—	—	92115 (2—100 ppm)
Kerosene	—	CH 25401 (2—35 mg/L)	—	—
LP gas	100A (0.02—0.8%)	—	—	—
Mercaptan	—	CH 20801 (2—100 ppm)	—	—
Mercury vapor	—	CH 23101 (0.1—2 mg/m^3)	8014-142 (0.1—2.0 mg/m^3)	83089 (0.05—2.0 mg/m^3)

TABLE 32 (continued)

Detector Tube Procurement Sources — Summary[87]

Compound	Detector tube identification, tube number, and detection range			
	Bendix-Gastec	Dräger	Matheson-Kitagawa	MSA
Methyl acetate	—	—	8014-148 (0.01—3.0%)	—
Methacrylonitrile	—	67 30101 (1—10 ppm)	—	91624 (10—80 ppm)
Methyl alcohol (methanol)	111 (0.01—3%)	CH 29701 (100—3000 ppm)	8014-119 (0.01—6.0%)	95097 (100—1000 ppm)
Methyl amine	—	67 18401 (5—60 ppm)	—	92115 (10—500 ppm)
Methyl bromide	136 (10—200 ppm)	CH 27301 (5—50 ppm)	8014-157 (10—500 ppm)	85834 (10—100 ppm) 88536 (10—100 ppm)
Methyl chloride	—	—	—	92030 (25—1000 ppm)
Methyl chloroform 1,1,1-trichloroethane	135 (*100—500 ppm*)	CH 24401 (10—500 ppm)	8014-160 (50—800 ppm)	88536 (100—700 ppm)
Methyl chloroformate	—	67 18601 (0.2—100 ppm)	—	—
Methylene chloride	—	67 24601 (100—2000 ppm)	—	88536 (100—1000 ppm) 85834 (100—2000 ppm) 92030 (50—2000 ppm)
2-Methyl isobutyl carbinol (methyl amyl alcohol)	—	—	—	95097 (25—1000 ppm)
Methyl diethylamine	—	67 18401 (5—60 ppm)	—	—
Methyl ethyl ketone	152 (0.02—0.6%)	CH 2290 (100—2000 ppm)	8014-139B (0.01—1.4%)	—
Methyl isobutyl ketone	153 (0.01—0.6%)	CH 22901 (100—12,000 ppm)	8014-155 (0.005—1.0%)	—
Methyl mercaptan	71 (0.5 —70 ppm)	—	8014-164 (1—20 ppm)	454206 (0.5—100 ppm)
Monoethylamine	3L (2—30 ppm)	—	—	92115 (10—500 ppm)
Monomethyl aniline	181 (2.5—45 ppm)	—	—	—
Monostyrene	124 (10—1000 ppm)	67 23301 (10—200 ppm) CH 27601 (50—400 ppm)	8014-158 (1—300 ppm) —	— 93962 (10—900 ppm)
Natural gas (Drager test)	—	CH 20001	—	—
Nickel tetracarbonyl	—	CH 19501 (0.1—1 ppm)	8014-129 (20—700 ppm)	—
Nitric acid	15L (0.2—40 ppm)	—	—	—

Nitric oxide	10 (NO:2—200 ppm) (NO_2:5—200 ppm)	—	—	—
Nitroethane	—	—	—	91624 (10—300 ppm)
Nitrogen dioxide	9L (0.2—100 ppm)	CH 30001 (0.5—10 ppm)	8014-117 (1—1000 ppm)	83099 (0.1—50 ppm)
	10 (5—200 ppm)	67 19101 (2—50 ppm) (5—100 ppm)		
Nitrous fumes	—	CH 29401 (0.5—10 ppm)	—	—
		CH 31001 (2—50 ppm) (5—100 ppm)		
		67 24001 (20—500 ppm)		
		CH 27701 (100-1000 ppm) (500—5000 ppm)		
Nitroglycerine	—	67 18201 (5—100 ppm)	—	—
Nitroglycol	—	67 18201 (5—100 ppm)	—	—
Nitromethane	—	—	—	91624 (5—300 ppm)
1-Nitropropane	—	—	—	91624 (2—100 ppm)
2-Nitropropane	—	—	—	91624 (5—150 ppm)
Octane	101 (0.015—1.2%)	—	—	—
Olefine 0.05	—	CH 31201 (1—50 mg/L)	—	—
Oxygen	—	CH 31601 (5—21 vol.%)	8014-159 (2—30%)	—
Ozone	18L (0.05—3 ppm)	CH 31301 (0.05—1.4 ppm) (0.5—14 ppm)	—	93865 (0.05—5.0 ppm)
	—	CH 21001 (10—300 ppm)	—	—
n-Pentane	—	67 24701 (100—1500 ppm)	8014-170 (0.01—0.5%)	—
Perchoroethylene (tetrachloroethylene)	133 (5—200 ppm)	CH 30701 (10—400 ppm)	8014-135 (10—400 ppm)	85833 (10—400 ppm) 88536 (25—400 ppm)
Phenol	—	CH 31501 (5 ppm)	—	—
Phosgene	—	CH 19401 (0.05—1.2 ppm)	—	—
Phosgene (carbonyl chloride)	16 (0.1—90 ppm)	CH 28301 (0.25—15 ppm) (1.2—75 ppm)	8014-146 (0.05—50 ppm)	89890 (0.1—10 ppm)
Phosphine	7 (5—200 ppm)	CH 31101 (0.1—4 ppm) (1—40 ppm)	8014-121B (5—90 ppm)	—
		CH 21201 (50—1000 ppm) (15—300 ppm) (150—3000 ppm)	8014-121A (20—800 ppm) 8014-121C (50—700 ppm) 8014-121D (2—160 ppm)	

TABLE 32 (continued)

Detector Tube Procurement Sources — Summary[87]

Compound	Detector tube identification, tube number, and detection range			
	Bendix-Gastec	Dräger	Matheson-Kitagawa	MSA
Propane	—	CH 26101 (0.1—1.3 vol.%)	—	—
-n-Propyl acetate	151 (0.06—0.9%)	CH 20201 (200—3000 ppm)	8014-151 (0.01—1.4%)	—
n-Propyl alcohol	112 (0.02—0.8%)	CH 29701 (100—3000 ppm)	—	95097 (10—10,000 ppm)
iso-Propyl alcohol	—	CH 29701 (100—3000 ppm)	—	95097 (100—10,000 ppm)
n-Propyl mercaptan (1-oropa-nethiol)	—	CH 20801 (2—100 ppm)	—	454206 (0.5—100 ppm)
n-Propyl nitrate	—	—	—	91624 (10—100 ppm)
Propylene	100A (0.02—0.8%)	CH 31201 (1—50 mg/L)	—	82802 (1—400 ppm)
Propylene dichloride (1,2-dichloro-propane)	—	CH 27301 (5—50 ppm)	—	85834 (25—250 ppm) 92030 (25—500 ppm)
Propylene imine	—	—	—	92115 (2—300 ppm)
Propylene oxide	163 (0.3—3.6%)	—	8014-163 (0.1—4.0%)	—
Pyridine	—	—	—	91624 (2—50 ppm)
Sulfur dioxide	5L (5—200 ppm) 5M (100—3600 ppm) 5LA (1—90 ppm) 5H (0.25—8%)	CH 24201 (20—200 ppm) (200—2000 ppm) 67 27101 (0.1—3 ppm) CH 31701 (1—20 ppm)	8014-103C (5—300 ppm) 8014-103A (0.1—4.0%) 8014-103B (0.02—0.3%) 8014-103BF (0.02—0.3%)	92623 (1—400 ppm)
Xystox (tetrahydrotheop hene)	—	CH 27501 (0.5 mg/m³)	—	—
Termi-Gas	—	—	—	91624 (10—200 ppm)
Termi-Nate	—	—	—	91624 (8—70 ppm)
1,1,2,2-Tetra-bromoethane	—	—	—	85834 (2—50 ppm)

1,1,2,2-Tetra-chloroethane	—	—	—	85834 (2—50 ppm)
1,1,3,3-Tetra-chloropropane	—	—	—	85834 (5—100 ppm)
Tetrahydrofuran	161 (100—2500 ppm)	CH 29701 (30—400 ppm)	8014-162 (0.1—5.0%)	—
Toluene	122 (10—600 ppm)	CH 23001 (5—400ppm) CH 27801 (0.1—7 mg/L) (25—1860 ppm)	8014-124 and 124B (1—1000 ppm)	93074 (10—800 ppm)
Toluylene diisocyanate	—	67 24501 (0.02—0.2 ppm)	—	—
o-Toluidine	181 (2.5—35 ppm)	—	—	—
Trichlorethane	—	CH 21101 (50—350 ppm) (100—700 ppm)	—	85834 (25—700 ppm) 88536 (50—500 ppm)
Trichlorethylene	132L (2—50 ppm) 132H (20—560 ppm)	CH 24401 (10—400 ppm)	8014-134 (10—400 ppm)	85833 (25—600 ppm) 88536 (25—400 ppm)
1,2,3-Trichloro-propane	—	—	—	85833 (10—500 ppm)
1,1,2-Trichloro-1,2,2-Trifluor-oethane (Freon® 113)	—	—	—	88538 (*100—2000 ppm*) 92030 (100—4000 ppm)
Triethylamine	3M (3.5—140 ppm)	67 18401 (5—60 ppm)	—	92115 (2—100 ppm)
Trifluoromonobro momethane (Freon® 1381)	—	—	—	88536 (100—4000 ppm)
Trimethylamine	3M (2.5—100 ppm)	—	—	92115 (5—250 ppm)
Vinyl acetate	141 (0.01—1.0%)	—	—	—
Vinyl chloride (chloroethylene)	131L (0.5—28 ppm)	CH 19601 (100—3000 ppm)	8014-132 (0.05—1.0%) 8014-132B (1—500 ppm)	85833 (100—1250 ppm) 92030 (100—2000 ppm)
Vinylidene chloride	131L (0.3—17 ppm)	—	—	—
Vinyl pyridine	—	—	—	91624 (1—100 ppm)
Water vapor	6 (0.5—32 mg/L)	CH 23401 (0.1—40 mg/L)	—	—

TABLE 32 (continued)

Detector Tube Procurement Sources — Summary[87]

Compound	Detector tube identification, tube number, and detection range			
	Bendix-Gastec	Dräger	Matheson-Kitagawa	MSA
Xylene	123 (10—500 ppm)	CH 27801 (0.7—7 mg/L)	8014-143 and 143B (50—5000 ppm)	93074 (10—800 ppm)

Courtesy of Bendix Corporation, Environmental and Process Instruments Division, Lewisburg, W. Va. National Drager, Inc., Parkway View Dr., Pittsburgh, Pa.; Matheson, P.O. Box 85, 932 Paterson Plank Rd., E. Rutherford, N.J.; Kittagawa Precision Laboratories 1-8-24, Chu-Cho, Meguro, Tokyo, 152, Japan; Toxic Gas Detectors, Mine Safety Appliances Co. (MSA), 400 Penn Center Building, Pittsburgh, Pa.

determine whether or not the area is within Federal compliance limits for the contaminant collected.[91]

The range of concentrations may be quite large. Generally this variability is due to actual fluctuations of the contaminant concentration with respect to both time and space in the environment and is not due to imprecision or inaccuracy of the detector tubes (within ± 25%). This large variability in concentration readings produces relatively wide confidence limits in the estimate of true air concentrations. The contaminant concentrations usually are best described by a log normal distribution. Leidel and Busch have presented a procedure for determining compliance or noncompliance with a particular TLV or Federal Standard.[91-92]

The procedure is conservative from the enforcement standpoint in that noncompliance is decided only if the lower confidence limit on the average exposure exceeds the standard. The procedure also can be used to determine compliance when a "no action" decision is concluded only if the upper confidence limit of the average exposure does not exceed the standard. More samples yield "narrower" confidence limits that permit a better estimate of the true air concentration, but from a practical application standpoint a point of diminishing returns suggests a limit to the number of samples. Leidel and Busch have developed curves showing the effect of sample size (number of samples) on the relative widths of these confidence intervals.[91-92] Beyond 10 determinations the reduction of error in the mean (average) concentration is hardly worthwhile.

The employee's time-weighted average (TWA) exposure can be calculated as follows:

$$E = \frac{C_a T_a + C_b T_b + \dots \; C_n T_n}{8}$$

where E = equivilant exposure for the 8-hr work shift, C = concentration during and time period T, and T = duration in hours of exposure at C.

In the case of the 5-min "grab sample" program, T_a, T_b and T_n = 5/60 = 0.083 hr.

If the detector-tube samples are taken for comparison with an occupational exposure ceiling standard which allows a maximum employee exposure for a short time interval, then a somewhat different strategy is involved. Leidel and Busch have developed four types of ceiling standards in Federal Regulations 29CFR1910.93 (Occupational Safety and Health Standards), and statistical procedures for handling the data. In most cases the appropriate sampling period is 15 min (total), which would require three 5-min "grab" samples (detector tubes) for the best estimate of the average exposure during the 15-min period. In this case, the sampling period is chosen in a nonrandom mode. All avialable knowledge relating to the area, employee, and process being sampled should be applied to the selection of the sampling site during periods of maximum expected concentrations of contamination sought.[88]

A more detailed example of this procedure is given by the following reprint of OSHA Sampling Data Sheet No. 1.[93]

Substance:
Carbon monoxide.
Standard: 29 CFR 1910. 93(b) Table G-1, of August 13, 1971
8-hr time-weighted average: 50 ppm (55 mg/m^3).

Sampling Equipment:

1. CO meter (Portable, Hopcalite — reagent type) — Follow directions. (Primary Method.)
2. Certified or NIOSH approved detector tubes and a hand operated pump (not squeeze bulb) may be used for compliance purposes. (Secondary Method.)

Analytical Procedure:

1. Direct reading instrument based on measurement of heat or reaction produced by oxidation of CO to CO_2. Follow instructions.
2. Detector tubes are read for length of stain, strokes of pump are noted and appropriate corrections made for pressure (or altitude) above 5000 feet (1500 m). The tubes are not to be used when below 0°F (−18°C) or above 125°F (52°C).

Sampling Period:

1. Hopcalite: maximum feasible number of readings should be averaged to determine average exposure. Time required dependent on variability and cyclical nature of exposures.
2. Detector tubes: direct reading, length-of-stain detector tubes (certified or NIOSH approved) may be used for compliance purposes, if at least one sample per hour is taken per exposed worker or close group of workers; in an 8-hr day more frequent samplings per hour are preferred, especially for exposures over 50 ppm. Eight different samplings (one per hour is considered the minimum acceptable number).

Recent studies have indicated that at least some of these detector tubes can be recalibrated and adapted directly to personnel monitoring for periods of at least 4 hr.[11] Tubes for NH_3, CO_2, CO, Cl_2, HCl, H_2S, Hg, NO_2, and SO_2 have been offered for this purpose, but information relative to calibration details and operational limits has not been made available in technical publications (see Section II.B.1).

e) Product Control Testing[90]

In some cases a contaminant in an otherwise inert gas, e.g., phosphine in acetylene, is the controlling hazard. Self-indicating devices offer a satisfactory solution to this problem, provided the "carrier" gas or vapor does not interfere with the detector's reagent system.

f) Troubleshooting and Technical Development[90]

Detector tubes are quite suitable for leak testing. However, if the tube selected does not include the TLV range, potential problems for health protection can be created by their misuse. A nickel carbonyl tube, for example, has been available for the concentration range 20 to 700 ppm, which is adequate for source emission (leak) surveying, but must not under any circumstances be used for health conservation[94] (TLV = 0.0001 ppm).[1]

g) Fire Fighting[90]

Detector tubes have demonstrated usefulness for establishing the presence or absence of certain toxic gases derived from the thermal decomposition or burning of combustible materials, especially plastics (HCN, $COCl_2$, Cl_2, CO, SO_2, etc.). Fire fighters, especially, could benefit from routine use of these tubes to detect near-lethal concentrations of those hazardous combustion products that offer little or no odor warning or that exhibit a strong olfactory fatigue potential.

h) Detection of Explosive Hazards

Detector tubes for detection of flammable gases and vapors in the explosive mixture range have been designed.[89,95] This particular application should be exploited further as it shows considerable promise, since no electrical circuits which might provide an ignition source are involved. Such tubes also have been used to detect gasoline in soil and to clarify cases of suspected incendiarism.[89]

i) Breath Analysis — Personnel Monitoring and Medical Diagnosis

Carbon monoxide poisoning may be confirmed by determining CO in the exhaled breath.[96-98] Also, indicator tubes have been employed for law enforcement purposes to determine the amount of alcohol in the breath, the presence of which is then related to blood concentration by means of well-established charts and time decay curves.[97,99-101]

j) Blood Analysis

Volatile organic compounds such as CO or alcohol, after release by acidification, oxidation, or other appropriate chemical treatment, have been analyzed in the blood by means of detector tubes.[89,94]

k) Ambient Atmosphere Analysis for Air Quality Criteria Enforcement

To date, a few tubes have been developed that are sufficiently sensitive to determine the pollutants for which criteria have been established (SO_2, CO, photochemical oxidants, hydrocarbons, lead, and nitrogen oxides)[102] (see Section II.B.1).

l) Remote Sampling

When sampling an inaccesible location, the detector may be placed directly in the location to be examined, and the pump operated at some conveniently remote spot. A connecting flexible (rubber) tube of the same internal diameter as the inner diameter of the glass tube connects the pump to the exhaust end of the detector tube. Under no circumstances should the sample be drawn through the flexible hose, as serious errors can ensue from losses of the contaminant on the walls of the sampling line. Lengths as great as 60 ft can be used without appreciable error, provided more time is allowed between strokes of the pump to compensate for the reservoir effect and obtain the full sample volume. This method has a major disadvantage; the detector tube cannot be observed during the sampling.

m) Source Emission Sampling

Cooling the sample is essential when analyzing hot gases from stacks or engine exhaust. Otherwise, the calibration will be invalid and the sample volume uncertain. A glass or metal probe may be attached to the inlet end of the detector tube with a short length of flexible tubing (silicone rubber or polyperfluoroethylene). If this probe is cold initially, a length as short as 4 in. outside of the hot duct will cool the gas sample from 250C to about 30°C.[89] This equipment assembly must be used with caution since under some conditions serious absorption losses occur on the tube walls or in the condensed moisture. The dead volume of the probe should be negligible with respect to the sample volume. Condensible vapors should not be sampled by this method, as serious condensation interferences and losses will be encountered when the dew point (saturated vapor temperature) is above the temperature of the sample as it is drawn through the detector tube.[95]

n) Oxygen Deficiency Determination[103]

A detector tube for oxygen analysis is available commercially. The principle is unique and shows great promise as a "vest pocket" instrument that requires no electrical circuiting for operation (Table 32).[95] Hydrochloric acid is generated in the detection tube. The acid is absorbed in a "guard" tube and the pump. The oxygen content is determined over the range 0 to 23%.

This novel application suggests tnat the principle perhaps could be applied to major constituents in other gas mixtures, and that a volumetric readout could be substituted for measurement of length of stain. Absorbing columns to eliminate interferences have been coupled with certain detector tubes,[89] especially for removal of water to protect moisture-sensitive reagents (see "Specificity and Interferences," Section II.A.1.b.vi.h).

v. Certification Criteria

These tubes are in a continuing state of development, and highly variable results have been obtained. Close attention to seemingly trivial details is required to obtain even semiquantitative results. The failure to understand the limitations of the technique, to follow instructions meticulously, to be aware of interferences, and to apply corrections when necessary have accounted for most of the unsatisfactory performances turned in by semi- or unskilled users. To attain the ultimate precision and accuracy, frequent recalibration must be carried out by the user.

Accuracy ultimately depends on the attainment of a high degree of skill on the part of the manufacturer who, even more than the user, must be aware of the controlling variables and limitations, as no one else is in a better position to eliminate the sources of variability and error that limit the application of currently available tubes to semiquantitative range finding. The best accuracy that can be expected lies in the range of ± 20%,[89] and many tubes do not deliver a performance better than ± 50%. The reliability of these tubes has so often been questioned that the U.S. Public Health Service, Bureau of Occupational Safety and Health, was prompted to undertake a performance study of the four most commonly used tubes: CO, benzene, SO_2, and CCl_4[104-108] Of these, only the SO_2 tubes consistently delivered results that conformed to the criteria established by the ACGIH-AIHA Joint Committee on August 9, 1965: "Each calibration point listed by the manufacturer shall be correct to within ± 25% of the stated concentration over the working range of the tube."[107,109,110] The calibrations were checked at 0.5, 1.0, and 5 times the TLV for the gas under investigation. Some of the tubes could not meet an alternate performance level of ± 50% at the 95% confidence limit (CCl_4),[108,109] and others (CO and benzene) did conform[104-106] to the alternate ±50% criteria, but not to the recommended ± 25% limits. Later testing indicated that the Unico® Kitagawa® tubes (catalog No. 100) did indeed perform within the ± 25% criteria (unpublished reports). Similar reports relative to the inaccuracies encountered with styrene vapor,[111] ozone,[112] and other detector tubes[113] have appeared in the technical literature.

Further testing by the National Institute for Occupational Safety and Health (NIOSH) confirmed the accuracy of the Unico-Kitagawa and Dräger tubes for perchloroethylene to be within the ± 25% criteria limits.[109,110,114] As a further consequence of these findings, the mounting dissatisfaction, uncertainty voiced by professional industrial hygienists,[90,115] and recommendations presented by the Joint ACGIH-AIHA Committee,[110] the creation of a Federal Certification Agency was initiated during June 1971.[114] The preliminary draft of the Detector Tube Certification Schedule included purpose, gas detector unit, construction and performance, conduct of tests (demonstrations), applications, certification of approval, approval tables, material required for record, changes after certification, withdrawal of certification, and quality control.

In September 1972 an announcement in the *Federal Register* proposed certification of gas detector tube units for gas and vapor measurements.[117] In response to this directive, a certification laboratory was established by NIOSH in Morgantown, W.Va.[118] OSHA inspectors are required to use NIOSH-certified instruments when available. In May 1973 the specifications for certification were published in the *Federal Register*.[119] The program was started in June 1973. Manufacturers may apply for certification by doing the following things:

1. Assuring that the gas detector tube and aspirating pump form a single unit and belong together
2. Paying a fee for the testing of the complete gas detector tube unit and for each further detector tube type

3. Taking the gas detector tubes and aspirating pump, which are submitted to the Institute, from normal batch production
4. Submitting a complete set of drawings of the gas detector tube unit.
5. Submitting a detailed description of all the measures taken by the manufacturer for quality assurance (control plan).

The following paragraphs and illustrations cover quality standards for gas detector tubes as set out in the *Federal Register.*

The measuring range of the gas detector tube in question is to cover the following concentrations:

0.5 test standard (TLV)
1 test standard (TLV)
2 test standard (TLV)
5 test standard (TLV)

Example for carbon monoxide: test standard (TLV): 50 ppm

Requirement:

0.5 TLV ≙ 25 ppm
1 TLV ≙ 50 ppm
2 TLV ≙ 100 ppm
5 TLV ≙ 250 ppm

Measuring range: 25 to 250 ppm; carbon monoxide detector tubes, in which the measuring range begins below 25 ppm and goes beyond 250 ppm, also satisfy the requirements.

The deviation from the actual value at 0.5 times the test standard (TLV) must not be greater than ± 35%. At 1, 2, and 5 times the test standard (TLV), ± 25% is admitted as the maximum deviation. These data relate to the 95% confidence interval.

At the test standard (TLV), the length of the stain is to be at least 15 mm (Figure A); if this requirement cannot be met and the stain is shorter, the end point of the stained indicating reagent must be so sharp that the following condition is satisfied:

$$\frac{\sigma}{\bar{x}} \leqq 0.10$$

where σ = the standard deviation of the tube readings obtained from three or more independent tube readers when reading an individual stained tube, and $\bar{x}$ = Mean value of the tube readings

$$\sigma = \sqrt{\frac{\Sigma \Delta^2}{n-1}}$$

where Δ^2 = the square of the deviation of the individual value from the mean value $\bar{x}$ and n = number of measurements. Example 1st tube reader 210 ppm 2nd tube reader 205 ppm 3rd tube reader 215 ppm

$$\frac{\sigma}{\bar{x}} = \frac{5}{210} = 0.024$$

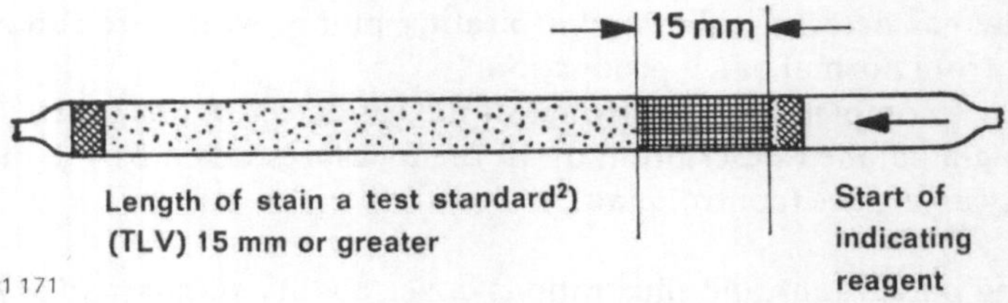

FIGURE A. At the test standard (TLV), the length of staining of the indicating reagent is to be at least 15mm. Shorter stains are also permissible, if the transition between the stained and unstained indicating reagent is sharply defined.[120]

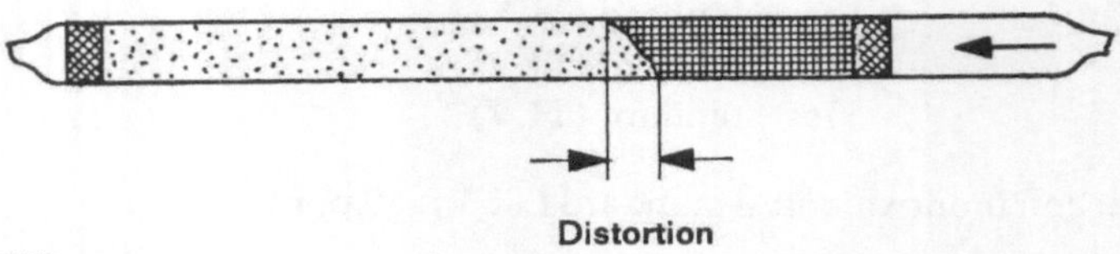

FIGURE B. Distorted indication in detector tube.[119]

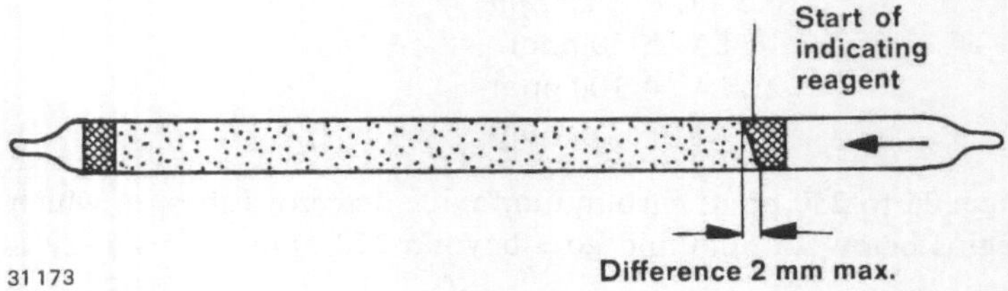

FIGURE C. The start of the indicating reagent does not form a right angle to the axis of the detector tube.[119]

The result of 0.024 is less than 0.10. With that the detector tubes satisfy the requirements.

The term distortion (channeling) is applied to the non-uniform course of the stain length (Figure B), when the length of the stain is not equal in the whole range. The maximum difference between the longest and shortest indication, related to the whole circumference of the detector tube, is not to be greater than is stated by the following relationship:

$$\frac{\Delta l}{M} \leqq 0.20$$

where $\Delta l = l_2 - l_1$, l_2 = longest indication, l_1 = shortest indication, and M = median value between l_2 and l_1 This formula means, for example, that with an indication of 20 mm, the distortion must not be more than 4 mm.

In order to insure uniformity, the surface at the start of the indicating reagent is to form, as far as possible, a right angle to the longitudinal axis of the detector tube. 2 mm is permitted as the maximum difference between the highest and the lowest point (Figure C). (See Ref. 120 for all illustrations.)

The detector tube manufacture has to carry out calibration tests at the following concentrations:

0.5 test standard (TLV)
1 test standard (TLV)

2 test standard (TLV)
5 test standard (TLV)

To produce the specified concentrations, one of the following methods is to be used: 1) dynamic method (the concentrated gas is mixed continuously with an air stream), 2) test gas of known composition from a pressurized cylinder, 3) static method (the substance to be measured is mixed uniformly with air in a sealed container), 4) the test concentration (4.1, 4.2, or 4.3) is to be verified by a chemical or physical laboratory method.

After the pump, closed with a plug, has been fully compressed, the leakage rate per minute must not be greater than 3% of the stroke capacity. For example, the DRÄGER Gas Detector Pump, with a stroke capacity of 100 cm^3, has a maximum permissible leak-rate of 3 cm^3/min. The pumps must be so calibrated by the manufacturer that the flow rate does not deviate by more than ± 10% from the nominal value. The actual stroke capacity must not deviate from the nominal value by more than ± 5%. For example, the DRÄGER Gas Detector Pump, with a nominal value of stroke capacity of 100 cm^3, has a maximum permissible deviation of ± 5 cm^3.

The gas detector tube manufacturer must summarize all measures for quality control in a control plan. Among other things this control plan must include:

1. Quality standards for production.
2. Checking of drawings and supervision of drawings changes
3. Adjustment, calibration, and control of measuring and test equipment
4. Control of purchased material to include incoming inspection
5. Batch identification, control of process, manufacturing, fabrication, and assembly work conducted in the applicant's plant
6. Audit of final inspection of the completed product
7. The organizational structure necessary to carry out the provisions and an organizational diagram, on which the quality control procedure is based.

Sampling is to be carried out in accordance with Military Standard MIL-STD-105-D[121,122] or MIL-STD-414. (These standards have been compiled by the U.S. Defense Department and are widely used in American industry.) The sampling plan is to contain a list of the product characteristics which are verified by the manufacturer. The product characteristics are to be classified according to the defects which may possibly occur. A distinction is to be made between the following types of defect:

1. Special defects, which relate to the tube reading accuracy
2. Critical defects, which make the detector tube completely inoperative
3. Major defects, which can result in failure of the detector tube or defects by which its usability will be reduced
4. Minor defects, which will hardly reduce the usability of the detector tube, but because of deviations from the nominal characteristics its effectiveness may be slightly affected.

The permissible proportion of defective items will be laid down by AQL (Acceptable Quality Level). The AQL is the specified limit of defects in the production rate, up to which release for the intended purpose can take place with high probability. According to the *Federal Register*, the following AQLs apply for detector tubes: special defects 6.5%; critical defects 0% (where tests are non-destructive); critical defects 1.0% (where tests are destructive); major defects 2.5%; minor defects 4.0%.

TABLE 33

NIOSH Certified Dräger Detector Tubes (May 1977)

NIOSH test number	Dräger tube	Order No.
TC-84-012	CO 5/c	CH 25601
TC-84-013	CO 10/b	CH 20601
TC-84-016	NO_2 0.5/c	CH 30001
TC-84-024	H_2S 1/c	67 19001
TC-84-029	CO_2 0.1 %/a	CH 23501
TC-84-030	SO_2 1/a	CH 31701
TC-84-031	NH_3 5/a	CH 20501
TC-84-039	Trichlorethylene 10/a	CH 24401
TC-84-044	Benzene 5/b	67 28071
TC-84-048	NO + NO_2 2/a	CH 31001
TC-84-050	Toluol 5/a	CH 23001
TC-84-051	HCN 2/a	CH 25701
TC-84-056	Methyl bromide 3/a	67 28211
TC-84-061	Methylen chloride 100/b	67 28331
TC-84-062	H_2S 5/b	CH 29801[a]
TC-84-063	*n*-Hexane 100/a	67 28391
TC-84-064	Athyl benzine 30/a	67 28381

[a] Certification of the Dräger tube H_2S 5/b relates to the measuring range 50 to 500 ppm, feasible for technical gas analysis.

From Drägerwerk, A. G., Lübeck-D-24, Lübeck 1, Moislinger Allee 53/55, P.O. Box 1339, Federal Republic of Germany, U.S. Representative, National Mine Service Co., 3000 Koppers Building, Pittsburgh, Pa. With permission.

The gas detector tube unit is to include detailed operating instructions with the following data: instruction that the pump and the gas detector tube form a single unit, instructions for the user on how he can carry out a leak test on the aspirating pump, and instructions for carrying out simple repairs. Every box of detector tubes is to be labelled. The label must state the manufacturer's batch number and the date until which the detector tube manufacturer will maintain a sample of the corresponding manufacturing batch. Up to this time, the detector tubes — provided they are stored in accordance with the manufacturer's instructions — are to conform, in their characteristics, with the standards set out in the *Federal Register*. The date printed on the label, therefore, corresponds with the expiration date for use of the tubes. The label must also give instructions on storage conditions. Each box of detector tubes is to be accompanied by operating instructions giving the temperature, pressure, and humidity at which the tubes were calibrated; the effect of temperature, pressure, and humidity on the use of the tubes; instructions for use of the pump and the tubes; interference effects from other gases and vapors (cross-sensitivity); and basic chemical reaction and reagent used, provided the manufacturer is able to do so without revealing trade secrets. The detector tube manufacturer must furnish an assurance that he will make known to the user, on request, the methods used for the calibration of the tubes. He must also undertake to inform the user of the effects of particular environmental conditions, where these are known.

Commercially available certified gas and vapor detector tubes are listed in Tables 33 to 37.

TABLE 34

NIOSH Certified Dräger Tubes

Gas or vapor	National Drager tube to be used	Dräger Ref. No.	Range	Color change
Ammonia	Ammonia 5/a (10)[a]	CH 20501	5—70 ppm, 50—700 ppm	Orange to dark blue
Benzene	Benzene 5/b (20)[a]	6728071	5—50 ppm	White to brown, green
Carbon dioxide	Carbon dioxide 0.1% o/a (1, 5)[a]	CH 23501	0.1—1.2 vol %	White to violet
Carbon monoxide	Carbon monoxide 5/c (2, 10)[a]	CH 25601	5—150 ppm, (100—700 ppm)	White to brown, green
	Carbon monoxide 10/b (1, 10)[a]	CH 25601	10—300 ppm, (100—3000)	White to brown, green
Ethyl benzene	Ethyl benzene 30/a	6728381	30—400 ppm	—
U-hexane	U-hexane 100/a	6728391	500—3000 ppm	—
Hydroganic acid	Hydroganic acid 2/a (5)[a]	CH 25701	2—30 ppm, 10—150 ppm	Yellow to red
Hydrogen sulfide	Hydrogen sulfide 1/c (1, 10)[a]	6719001	1—20 ppm, 10—200 ppm	White to brown
	Hydrogen sulfide 5/b (10)[a]	CH 29801	5—60 ppm, 50—600 ppm	White to brown
Methyl bromide	Methyl bromide 5/b (5)[a]	CH 27301	5—50 ppm	White to brown
Methylene chloride	Methylene chloride 100/b (10)[a]	6728331	100—3000 ppm	White to brown, green
Nitrogen dioxide	Nitrogen dioxide 0.5/c (2, 5)[a]	CH 30001	0.5—10 ppm	White to blue, green
NO + NO_2	Nitrous fumes 2/a (2)[a]	CH 31001	2—50 ppm	Green to red, brown
Sulfur dioxide	Sulfur dioxide 1/a (10)[a]	CH 31701	1—25 ppm	Blue to white
Toluene	Toluene 5/a (5)[a]	CH 23001	5—400 ppm	White to brown
Trichloroethylene	Trichloroethylene 10/a (5)[a]	CH 24401	10—400 ppm	White to orange

Note: Number in parenthesis = pump strokes.

From Drägerwerk, A. G., Lübeck-D-24, Lübeck 1, Moislinger Allee 53/55, P.O. Box 1339, Federal Republic of Germany, U.S. Representative, National Mine Service Co., 3000 Koppers Building, Pittsburgh, Pa. With permission.

TABLE 35

NIOSH Certified Mine Safety Appliances (MSA) Tubes

Part No.	Detector tube for	Certification No.
83099	Nitrogen dioxide	TC-84-040
460103	Ammonia	TC-84-034
85976	Carbon dioxide	TC-84-025
460058	Hydrogen sulfide	TC-84-022
91229	Carbon monoxide	TC-84-015
460225	Chlorine	RC-84-042
92623	Sulfur dioxide	TC-84-046
460423	Acetone	TC-84-054
460328	Trichloroethylene	TC-84-055
461371	Toluene	TC-84-057
461863	Ethylene dichloride	TC-84-058
460424	Nitric oxide	TC-84-059

From Toxic Gas Detectors, Mine Safety Appliances Co., (MSA), 400 Penn Center Building, Pittsburgh, Pa. With permission.

TABLE 36

NIOSH Certified Kitagawa Detector Tubes

Gas or vapor	Tube No.	NIOSH certification number	No. pump strokes	Time (min)	Range (ppm)	Color change	Reagent
CO	100	TC-84-045	1	3	25 — 1000	Yellow to dark brown	Palladium sulfite
			3	9	10 — 300	Yellow to dark brown	
CO	106S[a]	TC-84-019	3	9	10 — 250	Yellow to dark brown	Palladium sulfite
SO_2	103D	TC-84-035	1	3	2 — 80	Purple to white	Dye bleaching
			3	9	1 — 20	Purple to white	Dye bleaching
SO_2	103SD[a]	TC-84-028	3	9	1 — 30	Purple to white	Dye bleaching
H_2S	120B	TC-84-037	1	3	5 — 160	White to brown	Lead salt
			4	12	2 — 40	White to brown	Lead salt
CO_2	126A	TC-84-027	1	5	1000 — 26,000	Purple to pink	pH indicator
CO_2	126SA[a]	TC-84-026	1	5	1000 — 26,000	Purple to pink	pH indicator
NH_3	105C	TC-84-033	1	3	10 — 260	Pink to pale yellow	pH indicator
			2	6	5 — 130	Pink to pale yellow	pH indicator
NH_3	105SC[a]	TC-84-032	1	3	10 — 260	Pink to pale yellow	pH indicator
			2	6	5 — 130	Pink to pale yellow	pH indicator
HCN	1125B[a]	TC-84-052	1	3	0.5 — 100	Yellow to pink	pH indicator
			4	12	0.5 — 100	Yellow to pink	pH indicator

Note: Tubes with S designation are supplied with correction chart, scale applied directly onto the glass tube.

[a] Data derived from Kubota, Ryvichi, Kitagawa Precision Gas Detectors, Komyo Research Laboratories, 1-8-24, Chuo-Cho, Meguro, Tokyo, Japan; Eckstein, W., *Dräger Rev.*, 138, 18, 1976.

TABLE 37

NIOSH Certified Gastec® Detector Tubes

Gas or vapor	Tube No.	NIOSH certification number	Concentration range	Pump strokes
Carbon monoxide	1La	TC-84-014	8 — 1000	3 — 1/2
Sulfur dioxide	5La	TC-84-017	1 — 90	4 — 1
Nitrogen dioxide	9L	TC-84-018	0.2 — 100	8 — 1
Hydrogen sulfide	4LL	TC-84-020	0.5 — 60	10 — 1
Carbon dioxide	2L	TC-84-021	0.13 — 6%	2 — 1/2
Ammonia	3M	TC-84-023	10 — 1000	5 — 1/2
Trichloroethylene	132H	TC-84-038	20 — 560	2 — 1/2
Chlorine	8La	TC-84-041	0.3 — 16	3 — 1/2
Benzene	121	TC-84-043	5 — 20	4 — 1
Nitrogen oxides	10	TC-84-049	NO: 2 — 200, NO_2: 5 — 200	5 — 1 5 — 1
Toluene	122	TC-84-053	10 — 600	2 — 1/2
Vinyl chloride	131L	TC-84-060	0.5 — 28	7 — 2

Data derived from Bendix Corp., Environmental and Process Instruments Division, Lewisburg, W. Va.

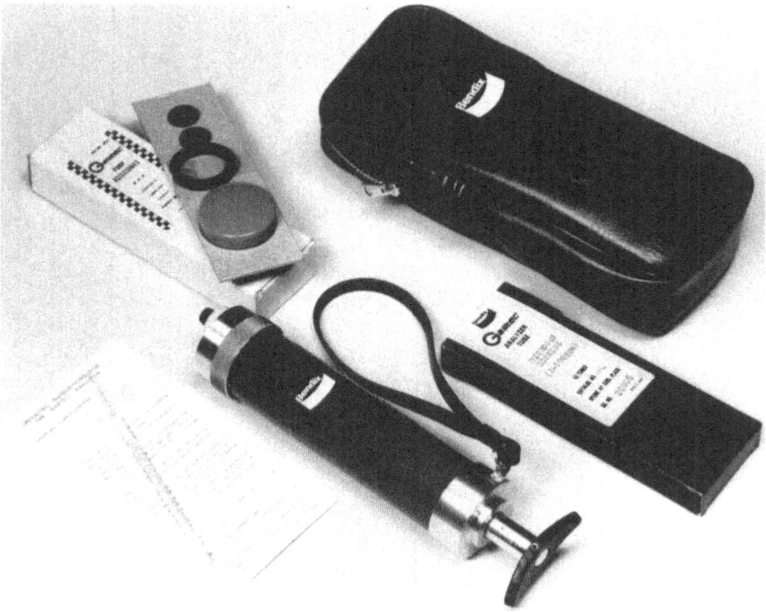

FIGURE 61. The Gastec® gas and vapor detector kit — complete. (Courtesy of the Bendix Corp., Environmental and Process Instruments Division, Lewisburg, W. Va.)

vi. Accuracy-controlling Factors

a) Uniformity of Granular Support

In general one of three particle-size ranges (Tyler mesh 20-40, 40-60, or 60-80) of silica or alumina gel is selected for the reagent support. Pumice, ground glass, and organic resins have been restricted to specific applications in which some special property is required. The stain length increases as the particle size increases, but the demarcation of the stain front becomes obscure. Conversely, demarcation of the stain front becomes more definite as the mesh size decreases, but the sensitivity also decreases due to shortening of the reaction zone (stain length), and the pressure drop (ΔP) increases to a point beyond which airflow practically ceases. The distribution of particle sizes in the chosen range should be sufficiently uniform to provide, on a unit-length basis, uniform ΔP, homogeneous reagent distribution, and stability in transport. When the size analysis includes an appreciable range, the fines may segregate to one side of the tube with the result that the cross-sectional airflow resistance will not be uniform and airflow will vary from side to side within the granular column (channeling). This non-uniform sample velocity frequently produces oval, spiked, or trailing (sometimes referred to as "tailing") stain fronts that are not perpendicular to the tube bore and are particularly difficult to delineate visually. If the gel support is friable, attrition may produce troublesome fines.

For those detectors which depend on ΔP rather than on a limiting orifice to control the flow rate, the uniformity of particle size distribution is especially critical. The diameter of the column has been increased as partial compensation for this factor, but this alternative approach involves a greater risk of channeling. By variations in the "charging preparation" used, a range of 3 to 40 sec in sampling time can be obtained with a constant tension spring bellows hand pump[97] (Figure 61).

The ideal solid support system should contribute no color of its own, that is, it should be a "pure white" background. Refinements in gel production have for all

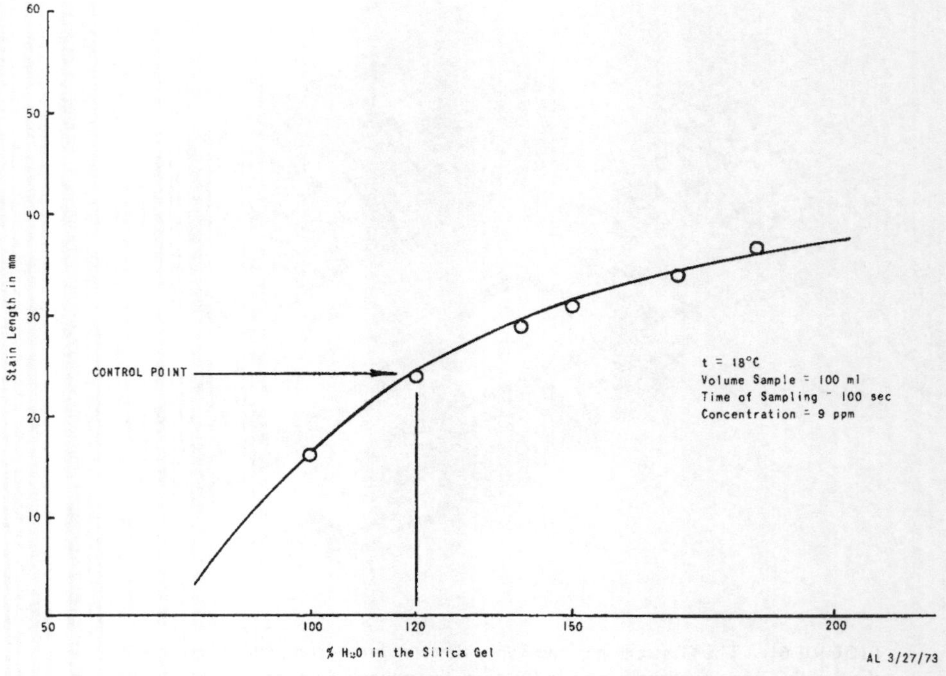

FIGURE 62. Effect of moisture in the silica gel on stain length in the arsine detector tube.[116]

practical purposes permitted attainment of this goal. However, very wide and unpredictable variations in the physical and chemical properties of different batches of indicating gels is the rule rather than the exception. Since the major portion of the chemical reaction occurs on the surface, the number of active centers that are highly sensitive to traces of impurities affects the reaction rate. These problems are well recognized in the manufacture of catalysts.

Here the manufacturer must exercise very close quality control on the purity and activity of materials going into the processes, the method of preparation, the cleanliness of the atmosphere in the production and assembly areas, and the particle size distribution of the finished product. Each batch of indicating gel must be individually calibrated, and adequate quality control procedures must be applied (see Chapter IV, Volume II). The preparation of "active" gels that satisfy specific requirements is largely empirical, and details of their manufacture are closely guarded trade secrets. In some cases the properties of the gel itself contribute as much as the color-forming reagent to the success of the detector.

Moisture may have a significant effect on both the formation of the colored derivative and the physical properties of the gel itself. As an example, consider the lengthening of the colored zone produced by arsine in contact with mercuric chloride adsorbed on silica gel as the water content of the gel increases[116] (Figure 62). When the moisture content decreases, not only does the stain length diminish, but the color reaction changes from dark brown to yellow. Furthermore, above about 120% water content (by solid absorbant weight), the gel tends to shrink in storage, thereby loosening the column and creating voids. A compromise between maximum stain response and gel shrinkage was reached by maintaining water content close to 120%.[116]

The detector tube method of measurement should meet the required confidence level of 95% for overall accuracy when analyzing industrial atmospheres for standards compliance. This means that simple sample results will lie within the stated required accuracy (± 25%) at least 95% of the time. If normally distributed errors for this method and unbiased procedures are summed, the coefficient of variation (CV, or relative standard deviation) can be used as a parameter to determine if the method is accurate enough to meet the standard. The CV in percentage units is defined as 100 times the ratio of the standard deviation of the method, divided by the true concentration of the analyte. The required CV of the method is obtained by dividing the required accuracy by 1.96 (statistical standard normal deviation for 95% two-sided confidence limits [Z value].)

Typical required CV's:

Concentration	Required accuracy (±)	Required %CV less than
Above permissible exposure	25%	12.8
At or below permissible exposure and above action level	35%	17.9
At or below action level (½ the TLV or Fed. Std.[123])	50%	25.5

b) Tube Construction

In general, the narrower the column, the better the definition of the stain front and the smaller the risk of channeling. However, here again, as in decreasing the particle size to sharpen the leading edge, this advantage is attained at the expense of increasing ΔP. Therefore, a compromise between grain size, column diameter, channeling, and ΔP must be reached. One manufacturer has successfully reduced the diameter to 2 to 4 mm (I.D. of the glass tube),[94,124] while others have continued to use the larger diameter (8 to 10 mm).[97,125]

In the manufacture of the narrow indicator tubes, variations of internal diameter produce significant variations in cross-sectioned (transverse) area that result in calibration errors as great as 50%. This error is possible because the volume of sample per unit cross-section differs from that established under standard test conditions. In addition, constant flow rate through the variable cross-sectioned areas cannot be maintained within critical limits. If an exactly equal quantity — by weight or by volume — of indicating gel is added to each tube, then variations in the transverse area will be disclosed by corresponding deviations in total column lengths. One manufacturer furnishes correction charts, which reduce this error to within ± 5%, on which the tube is positioned according to column length before reading the stain length from the compensated scale[94] (see Figure 63). Although the theoretical corrections are rather complex, practically linear approximations can provide corrections that reduce the error to within 10%. In the larger diameter tubes, errors from tube diameter variations are ignored. A generalized correction factor can be derived from the geometry of the system.

$$l_0 = l\,(L + a)/(L_0 + a^{\circ})$$

in which l = stain length in mm, l_0 = corrected stain length in mm, L = total length of detecting reagent column in mm, L_0 = standard total reagent column length in mm, and a° = a constant determined experimentally for each detecting reagent and each kind of gas to be measured.[94] And

$$L = w/\pi r^2 d$$

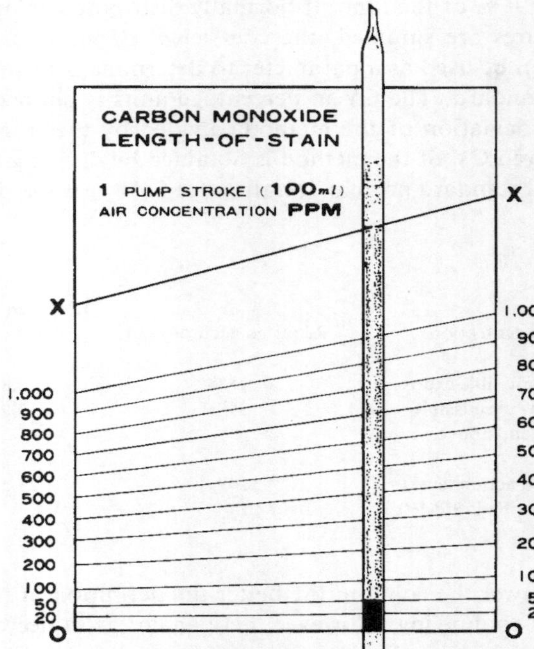

Temperature Correction					
Chart Readings (ppm)	Correct Concentrat[ion] (ppm)				
	0°C (32°F)	10°C (50°F)	20°C (68°F)	30°C (86°F)	40°C (104°F)
1.000	800	900	1.000	1.060	1.140
900	720	810	900	950	1.030
800	640	720	800	840	910
700	570	640	700	740	790
600	490	550	600	630	680
500	410	470	500	520	560
400	340	380	400	420	440
300	260	290	300	310	320
200	180	200	200	200	210
100	100	100	100	100	100

FIGURE 63. Carbon monoxide length-of-stain detector tube calibration chart. (From Eckstein, W., *Dräger Rev.*, 38, 18, 1976. With permission.)

in which w = total weight of detecting reagent in g, r = radius of the tube in mm, d = density of the detecting reagent in g/mL, and π = 3.1416. And

$$l = w/\pi r^2 d$$

in which w = weight of stained detecting reagent in mg.

If the sample volume is fixed (100 mL) and the sampling rate held constant, then

$$l/L = w/W = Kc/w$$

in which K = a constant and c = gas concentration when w is directly proportional to c. Correction charts can be constructed also from calibration data.

Glass tubing with internal diameters controlled within tolerances that would eliminate the need for a correction factor (± 0.01 mm) is available (Ace Glass, Vineland, N.J.), but the cost is prohibitive for use in detector tubes.

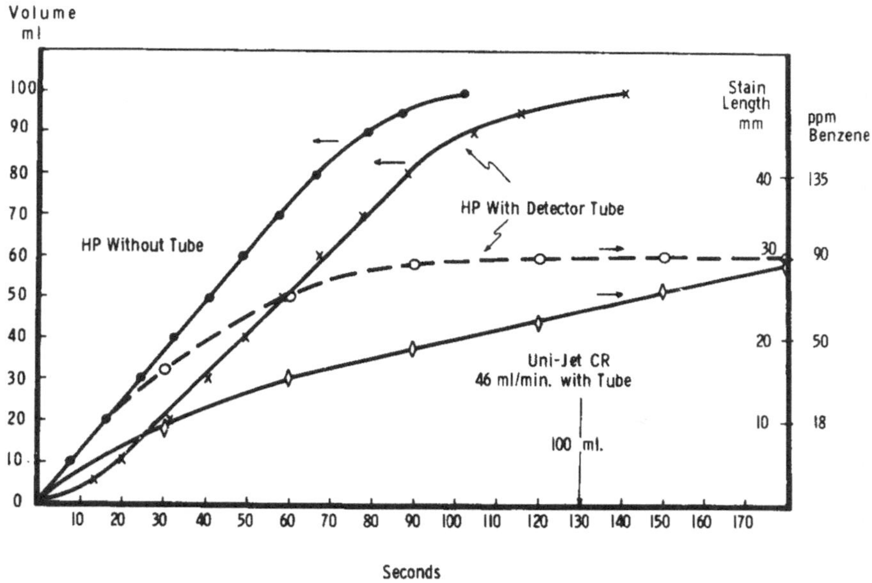

FIGURE 64. Hand pump (HP) vs. constant rate (CR) air sampling.

c) Packing Variations — Pressure Drop Effects

Not only is the relationship of particle size distribution to pressure drop (Section II.A.1.b.vi.a, "Uniformity of Granular Support") a critical factor in controlling variations in flow rate, but the skill exercised in filling and packing the reagent granules in the tubes, and in inserting the reagent granule retainers (usually cotton or glass wool plugs), becomes a major factor when improvements in precision and accuracy are considered. On this basis, the use of reagent systems that, for reasons of stability of the chemical reagent system, must be mixed and loaded into the glass tubes by the user just before use in the field cannot be condoned for even "approximation analysis." Also, the addition of a liquid reagent to an already packed tube before sampling constitutes a highly questionable practice.

Although the major source of sampling rate variability is located in the tube itself, the overall characteristics of the system must be considered whenever grab sampling is carried out. In most of these devices, the flow of air created by the vacuum is controlled by a limiting orifice. The notable exception is the Dräger system, which relies on the tube packing controls to regulate the flow rate.[97] The dimensions of these fine orifices are critical both in the manufacturing precision, to ensure reproducible rates from one instrument to another, and in the reduction of effective cross-section by particulate matter, grease, or corrosion. Fine-pore porous polyethylene or fluorocarbon prefilters can prevent mechanical plugging. Only recalibration or microscopic examination will reveal minor changes due to corrosion, erosion, or mechanical damage. Extreme care must be exercised when cleaning these orifices to avoid causing enlarging by reaming with a fine wire or by etching in cleaning solutions.

Contrary to widely held opinion, these single-stroke manual pumps do not deliver a uniform, constant sample flow through the detector tubes. The sampling characteristics of a 100-mL metal syringe-type hand pump,[94] as determined by the soap bubble technique,[126-129] is illustrated in Figure 64 (HP curves). With no coupled resistance, the rate remained linear (upper curve) for approximately 70 mL, but then became exponential as the total 100-mL volume was approached. With a benzene detector tube[94] in place, (lower curve) linearity occurred only over a limited span near the mid-

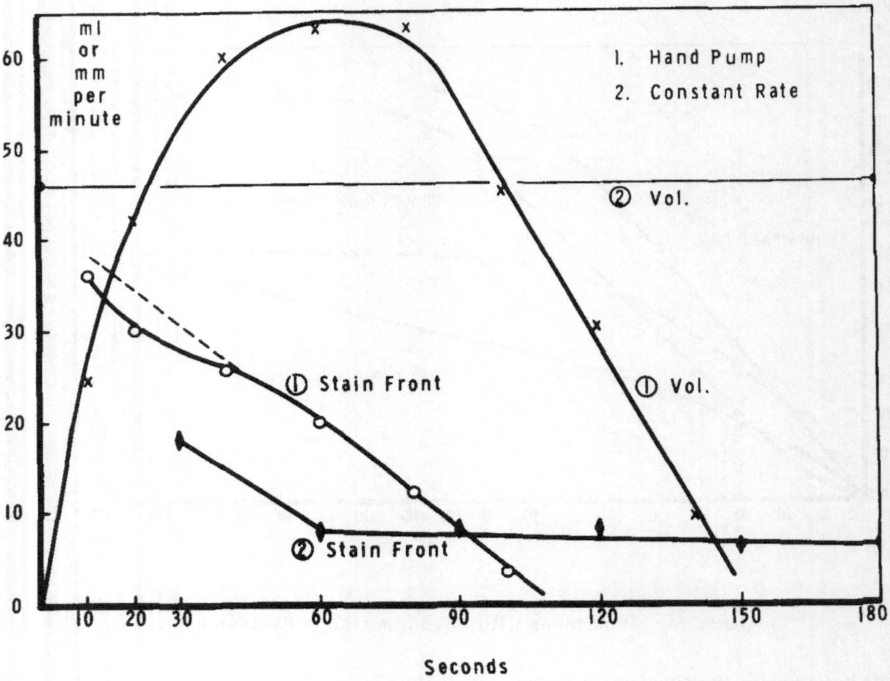

FIGURE 65. Sampling and stain production rates vs. time for benzene analysis.

point. A plot of sampling rate vs. time more graphically illustrates this variable flow pattern (Figure 65, curve 1). The rate reached a maximum within 60 sec, then remained relatively constant only for 20 sec. At the point of maximum rate, only 40% of the sample had passed through the tube, and the pressure drop had reached a maximum equivalent to 27 in. of water.

The resistance measured as pressure drop across the indicator tube must be held constant if reproducible airflow precision is to be attained. However, this is not a realistic goal in routine manufacturing operations. As an example, the variations in pressure drop across benzene detector tubes selected at random from the same lot of one commercially available type were within ± 10% (95% C.L.) of the averaged results (Figure 66). This range of flow resistance would be expected on the basis of known variations in tube diameter and packing density.[94] However, variables other than pressure drop must influence the response of this system, as recalibration by an independent agency indicated an accuracy of only ± 50% (95% C.L.).[105] The chemical nature of the reagent system (formaldehyde + concentrated H_2SO_4), rather than the geometry of the system, probably limits the performance.

d) Sampling Rate

Sampling rate is the most critical controlling variable encountered in the calibration and use of these detectors and should not exceed ± 5% relative deviation from tube to tube.[114]

Again referring to the Kitagawa benzene detector tube as an example, the critical effects of sampling rate can be demonstrated by reference to changes in stain front velocity (Figure 65). As a result of the changing flow rate, the stain front progressed at a nonlinear rate to a maximum at midpoint, then decelerated rapidly as a linear function of time (Figure 64). During this deceleration phase, movement of the stain front ceased when the sample flow dropped to 37 mL/min. The "effective" sample

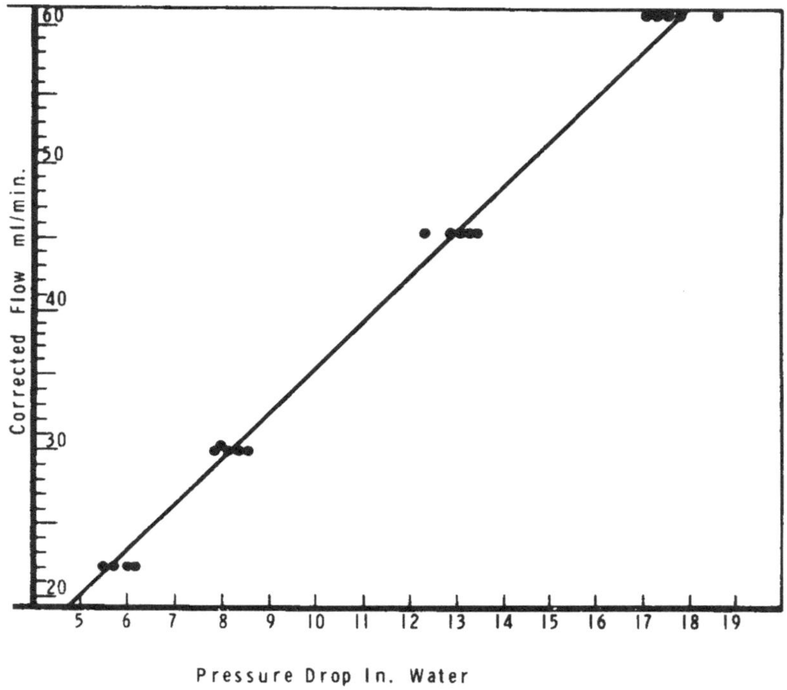

FIGURE 66. Kitagawa® benzene detector tube — airflow vs. pressure drop.

volume turned out to be 90 mL, and the "effective" time to draw this "effective" sample fell in the range 100 to 110 sec (Figure 65, HP — dashed curve). A plot of stain front velocity vs. time (first derivative) accentuated this nonlinear response (Figure 65, curve 1 — stain front).

The effect of air sample velocity was further amplified if the rate was held constant (Figure 64, lower curve — Uni-jet® CR). A 138-mL sample was required to match the stain length of the 90-mL aliquot drawn by hand pump, and the time was extended to 180 sec. However, after a 60-sec "conditioning" period, the stain front advanced at a near linear rate (Figure 65, curve 2 — stain front). The effect of a change in sampling rate expressed as a deviation from the supplied calibration provides the most significant incentive to operate within the calibration parameter limits (Figure 67).

The same consideration must be paid to precision control of these variables when constant rate sampling is considered. When the data are presented in terms of the effect of sampling rate on stain length development, a family of curves can be derived either to determine the precision of flow rate control required to reduce the analytical deviation range within acceptable limits or to select conditions for monitoring for extended periods (Figure 68 and Table 38). For example, at a 20-mL/min rate, the sampling period can be extended to 9 min without deviating more than ± 10% from the calibration chart supplied with the hand pump.

Included in Figure 68 are data published by Kinosian and Hubbard for a nitrogen dioxide detector tube.[130] These samples were drawn by squeeze bulb under controlled conditions. The critical effect of concentration is worthy of note. At 5 ppm NO_2, the stain front ceases to move after about 4 min, whereas the 25-ppm curve continues to rise steeply.

The variable sample flow rate produced by hand pump (Figures 64 and 65) has been claimed to be an advantage since the initial high rate produces a long stain and the

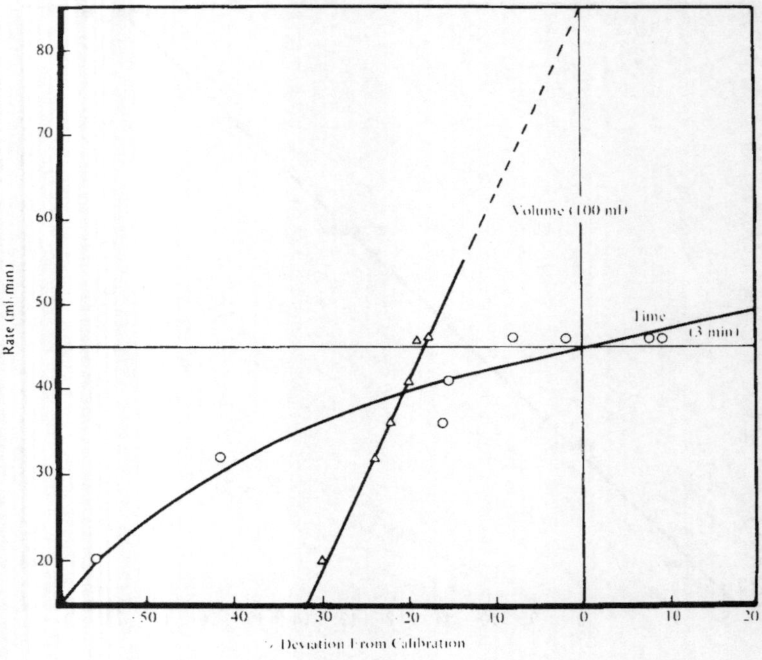

FIGURE 67. Sampling rate vs. deviation from hand pump calibration (benzene).

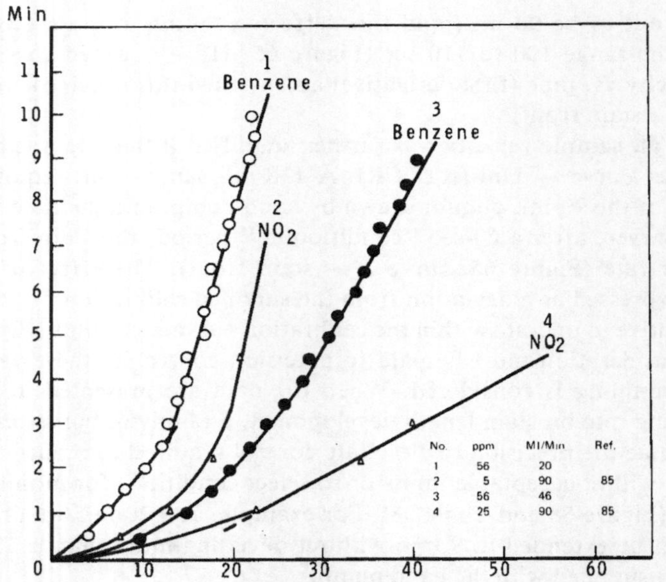

FIGURE 68. Sampling rate vs. stain development rate (benzene and nitrogen dioxide).

TABLE 38

Calculated Slopes for Log-log Curves for the Plots of Stain Length vs. ppm Benzene

	b_0	b_1	95% C.L. for b_1
First sample (CR and HP)	−0.117	1.396	±0.171
Second and third (CR)	−0.165	1.437	±0.270 Same curve
(Composite)	−0.196	1.453	±0.138
Second sample (CR and HP)	0.435	1.165	±0.165
Third sample (CR and HP)	0.664	1.008	±0.304 Same curve
(Composite of second and third)	0.665	0.982	±0.159
Nitrogen dioxide[6]	−2.010	2.231	±0.730

Key:
$(\log Y) = b_0 + b_1 (\log X)$

Y = Benzene concentration
X = Stain length
b_1 = Slope
CR = Constant rate
CCR = Continuous CR
HP = Hand pump
b_0 = Intercept

final tapering off sharpens the stain front. However, even if this effect is valid, accuracy requires precise reproduction of the flow rate pattern to maintain valid calibrations.[89]

Serious pressure drop defects can be detected easily in the field by testing for residual vacuum in the hand pump, or incomplete filling of a squeeze bulb or bellows pump after the specified sampling time interval has elapsed. For example, after sampling a 49-ppm-CO-in-air calibration mixture, the piston of the hand pump, when rotated from the locked position at the end of the designated 3 min was pulled back into the cylinder by the residual vacuum a distance equivalent to 10 mL. The 90-mL of sample which had been drawn through the tube at a nonstandard rate indicated a 30- to 35-ppm CO concentration instead of the 45- to 50-ppm range registered by the other tubes of the same lot.

e) Volume of Air Sampled

The volumetric error of repetitive aliquots delivered by the sampling pump should be less than ± 5%, and with less than ± 5% deviation from the volume used to calibrate the detector tubes.[114] Saltzman established the mathematical basis that equates volume and flow rate with the other variables required for calibration. The following excerpt is quoted from his article prepared for the ACGIH air sampling instruments handbook:

The variables which can affect the length of stain are: concentration of test gas, volume of air sample, sampling flow rate, temperature, and pressure, as well as a number of factors related to tube construction. There is a striking similarity in the fact that most of the calibration scales are logarithmic with respect to concentration in spite of the widely differing chemicals employed in different tube types. Although very little data are available for these relationships, a basic mathematical analysis was made by Saltzman.[131] These theoretical formulae discussed below will, of course, have to be modified as more data become available.

In the usual case, although the test gas is completely sorbed, equilibrium is not reached between the gas and the absorbing indicator gel, because the sampling period is relatively short and the flow rate relatively high. The length of stain is determined by the kinetic rate at which the gas either reacts with the indicating chemical or is adsorbed on the silica gel. The theoretical analysis shows that the stain length is proportional to the logarithm of the product of gas concentration and sample volume as shown in the following simplified equation:

$$L/H = \ln (CV) + \ln (K/H) \quad (4)$$

where L is the stain length in cm, C is the gas concentration in parts per million (ppm), V is the air sample volume in mL, K is a constant for a given type of indicator tube and test gas, and H is a mass transfer proportionality factor having the dimension of centimeters, and known as the height of a mass transfer unit.

The factor H varies with the sampling flow rate raised to an exponent of between 0.5 and 1, depending on the nature of the process that limits the kinetic rate of sorption. This process may be diffusion of the test gas through a stagnant gas film surrounding the gel particles, the rate of surface chemical reaction, or diffusion in the solid gel particles. If the indicator tube follows this mathematical model, a plot of stain length, L, on a linear scale, versus the logarithm of product CV (for a fixed constant flow rate) will be a straight line of slope H. The equation indicates the importance of controlling flow rate, as it may affect stain lengths more than gas concentration.

If larger samples are taken at low concentrations and the value of L/H exceeds 4, the gel approaches equilibrium saturation at the inlet end, and calibration relationships are modified. The solution to the equations for this case has been presented by Saltzman[131] graphically in a generalized chart. However, there is little advantage to be gained in greatly increasing the sample size, since the stain front is greatly broadened and various errors are greatly increased.

For some types of tubes such as hydrogen sulfide and ammonia the reaction rate is fast enough so that equilibrium can be attained between the indicating gel and the test gas. Under these conditions there is a stoichiometric relationship between the volume of discolored indicating gel and the quantity of test gas absorbed. In the simplest case the stain length is proportional to the product of concentration and volume sampled:

$$L = K'CV \quad (5)$$

If adsorption is important, the exponent of concentration may differ from unity:

$$L = K''C^{(1-n)}V \quad (6)$$

The value of n is the same as that in the Freundlich isotherm equation for equilibrium adsorption, which states that the mass of gas adsorbed per unit mass of gel is proportional to the gas concentration raised to the power n. If the value of n is unity, which is not unusual, equation 6 indicates that stain length is proportional to sample volume, but is independent of concentration. The physical meaning of this is that all concentrations of gas are completely adsorbed by a fixed depth of gel. Such a tube is obviously of no practical value.

Equilibrium conditions may be assumed for a given type of indicator tube if stain lengths are directly proportional to the volume of air sampled (at a fixed concentration), and are not affected by air sampling flow rate. A log-log plot then may be made of stain length versus concentration for a fixed sample volume. A straight line with a slope of unity indicates equation 5 applies; if another value of slope is obtained, equation 6 applies.*

The sample volume must be corrected for temperature and pressure when ambient conditions differ significantly from the calibration conditions — 20°C and 760 mm (1 at pressure). Detector tubes are usually used in a temperature range between 0 and 40°C. At lower temperatures systematic low indications can be expected. At higher temperatures excessively high results may be obtained. By utilizing simple warm-up methods (body heat — hand-held) or cooling devices (metal tubes for stack or exhaust sampling) the temperature of the sample can be adjusted to the optimum range for the detector tube reaction. Volumetric corrections can be estimated on the basis of 3.5% for a 10°C temperature change.

* Reprinted from Saltzman, B. E., Direct Reading Colorimetric Indicators, in *Air Sampling Instruments for Evaluation of Atmospheric Contaminants*, 4th ed., ACGIH, Cincinnati, Ohio, 1972. With permission.

Pressure corrections should be applied by gas law calculation when significantly above or below standard conditions — 760 mm. This correction becomes necessary when the sampling is carried out at altitudes either above or below sea level. Consider the hyperbaric effect on a 50-ppm CO reading (underwater vessel). At 2 at the reading would increase to 100 ppm, at 5 at to 250 ppm; if the detector is carried to a 2000-m altitude, the ambient pressure drops from 1 at pressure to 0.78 at resulting in a detector reading of 39 ppm CO.[132]

f) Temperature Effects

In addition to the volumetric correction required by the gas laws (PV = NRT) to compensate for temperatures significantly different from the calibration standard, 25°C and 760 mm Hg pressure, correction for changes in reaction kinetics must be applied in some cases. If the manufacturer furnishes temperature correction tables or charts, the recommended practice should be closely followed. For example, the reaction system employed for CO determinaton is temperature dependent, and the correction factors should be applied whenever samples are taken below 15°C or above 40°C (Figure 63). However, at low concentrations the magnitude of error is small enough to be ignored. The effect on the response of mercuric chloride to arsine is similar.[116] Failure on the part of the supplier to include temperature corrections does not necessarily mean that a significant temperature corefficient is not involved. The magnitude of the coefficient should be determined by the user before assuming that this variable can be ignored.[103, 132] Correction charts are available.[132]

When the reaction velocity between the reagent and the contaminant sought is relatively low, colorimetric rather than stain length measurement must be applied. The first CO tube was in this category (color change from buff yellow to blue through a graded series of greens). A relatively large temperature coefficient was involved. The rate of color change was doubled for each 10°C rise in temperature. Therefore, even small deviations from the calibration temperature produced significant errors if correction was ignored. The colorimetric reagents for methyl and ethyl alcohol behave in a similar fashion.

g) Stability of the Chemical Reagent System — Storage Life

The significance of this factor can best be appreciated by the importance assigned to prominently displaying the expiration date on each box of tubes.[114] Very few of the reagents currently employed have an indefinite shelf life. The decomposition of the reagents during storage has had a profound effect on the production of gas detectors and has severely limited the variety of reagents that can be considered for air analysis. In general, the most sensitive colorimetric systems fail in this respect. Attempts to avoid this limitation have entailed either mixing the two components of an inpregnated gel system just before use, or adding a liquid reagent either before or after sampling. As previously stated, this expedient is not without its pitfalls and cannot be considered good practice.

Storage temperature has a profound effect on shelf life, as would be expected. For example, the temperature coefficients for the Dräger carbon monoxide tubes are shown here:

Temperature (°C)	25	50	80	100	125	150
Shelf life	2 years	6 months	6 weeks	1 week	3 days	1 day

Note: The time indicates the period for which calibration accuracy remains within ± 30%.

This data, which reveals a rate, typical of a simple chemical reaction, describes an approximately straight line when the logarithm of time is plotted against a linear scale of the reciprocal of absolute temperature. Of course, more complex relationships will be encountered with other systems.[89] From this relationship, it is obvious that tube life can be greatly extended by storage under refrigeration. Indeed, the life expenctancy of most of the otherwise unstable detector tubes can be at least doubled by storage at temperatures below 4°C. In the absence of stability data from the manufacturer, the user of these tubes should determine for himself the useful life by setting aside controls of each lot for observation and recalibration. Most of the limited shelf life reagents indicate decomposition by definite, characteristic color changes. "Off-color" tubes should be discarded. In any case, tubes should not be used after the expiration date without confirmation of the calibration.

A shelf life of at least 2 years is highly desirable from the standpoints of both the distributor and the customer. It is not reasonable to expect a distributor to absorb the financial losses from the storage of short life detectors. In most cases the supplier can estimate the service life of new reagent systems by accelerated aging at elevated temperatures and extrapolation to room temperature or refrigeration based on the relationship of time to temperature for simple chemical reactions (see Dräger CO tube data above). Small variations in gel composition, such as moisture content, may exert a significant effect on stability of a given reagent system from batch to batch. The responsiblity for this quality control rests with the manufacturer, but the customer should beware.

When multilayer tubes are involved, shorter shelf life can be expected, as diffusion of the "guard" reagent may accelerate deterioration. This condition usually is quite apparent from a "pseudo" stain front or liquid (tar) droplets that develop in the sealed tube. Again, the importance of calibration and recalibration cannot be overemphasized. The user of colorimetric analysis systems must satisfy himself that what is seen actually is the result produced by the material that is sought and that the color nuances are not due to extraneous causes.

h) Specificity and Interferences

Probably no other factor in the application and interpretation of results from detector tube surveys taxes the expertise of the investigator more than the recognition and allowance for the contribution of coexistent components that are present in addition to the one under study in the contaminated environment. Few industrial pollution problems involve a single contaminant, and a majority of the detector tube types are nonspecific. For example, not less than 35% of the detector tubes offered by one manufacturer are based on the reduction of either a chromate or chromic acid to a colored chromium salt (Tables 39 to 42). It is rather obvious that most reducing agents will produce this effect. Therefore, the investigator must determine what components are to be expected from an inventory of chemicals entering the area under investigation.

Paint and Varnish Industry Survey — A preliminary survey carried out in a small paint and varnish manufacturing facility to detect the presence of health hazards and establish conformity with TLVs illustrates the difficulties that can be encountered when using a chromate reagent system.

Procedure — An inventory of the volatile solvents incorporated into the products disclosed more than twenty organic compounds that would reduce hexa-valent chromium (Table 43). Detector tubes calibrated for 12 of the solvents were available, but except those for toluene and xylene, all were derived from the same chromic acid reagent. However, both toluene and xylene depend upon the reaction with formaldehyde in the presence of sulfuric acid to produce a brown stain, and therefore these tubes are not specific either.

Under the circumstances, a decision was reached to analyze only for the vapors of the major components, methylethylketone (MEK or 2-butanone) and toluene, in use

TABLE 39

Classification of Detecting Reactions in Kitagawa® Gas Detector Tubes

Production of colored complex compounds	
105A	Ammonia, high range, cobaltous salt, complex of NH_3
142	Mercury vapor
Production of metallic compounds (salts)	
103A	Sulfur dioxide, high range, bichromate, chromic compound
120A	Hydrogen sulfide, high range, lead salt
120B	Hydrogen sulfide, low range, lead salt
120HH	Hydrogen sulfide, extra-high range
121A	Phosphine, high range, $CuSO_4$ + mercuric compound
121B	Phosphine, low range
164	Methyl mercaptan
165	Ethyl mercaptan
200	Sulfide ion in water, lead salt
201	Chloride ion in water
Ionic color change by oxidation-reduction of metallic salts	
103B	Sulfur dioxide, middle range, KIO_3, iodine
103BF	Sulfur dioxide in flue gas
129	Nickel carbonyl, auric compound
167	Hydrogen selenide
140	Arsine, auric compound
Production of molybdenum blue from ammonium molybdate and palladium sulfate	
101	Acetylene (palladium catalyzed)
106A	Carbon monoxide (palladium catalyzed)
106B	Carbon monoxide and ethylene coexisting (palladium catalyzed)
106C	Carbon monoxide, olefin hydrocarbon, and nitrogen oxides coexisting — chromic acid guard layer
108B	Ethylene (palladium catalyzed)
120C	Hydrogen sulfide and sulfur dioxide coexisting (palladium catalyzed)
168B	Butadiene, low range
Ionic color change with reduction of chromic acid	
102A	Acetone
104A	Ethyl alcohol
107	Ether
111	Ethyl acetate
113	*n*-Hexane
115	Cyclohexane
119	Methyl alcohol
122	Ethylene oxide, acidified bichromate
123	Dimethyl ether
128A	Acrylonitrile, high range
128B	Acrylonitrile, low range
132	Vinyl chloride
138	Butyl acetate
139B	Methyl ethyl ketone
148	Methyl acetate
149	Isopropyl acetate
150	Isopropyl alcohol
151	Propyl acetate
153	Isobutyl acetate
154	Dioxane
155	Methyl isobutyl ketone
161	Furan
162	Tetrahydrofuran
163	Propylene oxide
168A	Butadiene, high range
Color change of pH indicators	
105B	Ammonia, low range, thymol blue
112B	Hydrogen cyanide, low range

126A	Carbon dioxide, high range, NaOH + indicator
126B	Carbon dioxide, low range, NaOH + indicator
133	Acetaldehyde, hydroxylamine hydrochloride, thymol blue
Bleaching of dyestuff	
103C	Sulfur dioxide, low range
103D	Sulfur dioxide, D-type
Formation of dyestuffs	
109	Chlorine, o-tolidine
112A	Hydrogen cyanide, high range, alkaline picrate, isopurpurate
114	Bromine, o-tolidine
116	Chlorine dioxide, o-tolidine
117	Nitrogen dioxide, nitroso-o-tolidine
141	Carbon disulfide
145	1,2-Dichloroethylene
134	Trichloroethylene and perchloroethylene
146	Phosgene
157	Methyl bromide
160	Methyl chloroform
166	Ethylene dibromide
Formation of colored condensed polymer from formaldehyde + H_2SO_4 reagent	
118A	Benzene
118B	Benzene and other aromatic hydrocarbons coexisting
124	Toluene
143	Xylene
158	Styrene
Oxidation of basic pyrogallol	
159	Oxygen

TABLE 40

Classification of Detecting Reactions in Dräger Analyzer Tubes

Detector tube	Reaction principle	Measuring range	Number of strokes
Acetone 100/b	Dinitrophenylhydrazine	100—12,000 ppm	10
Acrylonitrile 5/a	Via hydrogen cyanide	5—30 ppm	5
Alcohol 100/a	Chromate and sulfuric acid	100—3000 ppm	10
Ammonia 5/a	Bromphenol blue and acid	5—70 ppm, 50—700 ppm	10, 1
Ammonia 25/a	Mercuric nitrate	25—700 ppm	10
Ammonia 0.5%/a	Bromphenol blue and acid	0.5—10 vol.%, 0.05—1 vol.%	1, 10
Aniline 5/a	Furfural + acid	1—20 ppm	25—5
Arsine 0.05/a	Gold salt	0.05—3 ppm, 1—60 ppm	20, 1
Benzene 5/a	Formaldehyde + sulfuric acid	5—40 ppm	15—2
Benzene 0.05	Formaldehyde + sulfuric acid	15—420 ppm	20—2
Carbon dioxide 0.01%/a	Hydrazine + crystal violet	0.01—0.3 vol.%, 0.5—6 vol.%	10, 1
Carbon dioxide 0.1%/a	Hydrazine + crystal violet	0.1—1.2 vol.%	5
Carbon dioxide 0.5%/a	Hydrazine + crystal violet	0.5—10 vol.%	1
Carbon dioxide 1%	Hydrazine + crystal violet	1—20 vol.%	1
Carbon dioxide 5%/A	Hydrazine + crystal violet	5—60 vol.%	1
Carbon disulfide 0.04	Copper salt + amine	13—228 ppm (0.04—0.9 mg/L)	23—1

Carbon disulfide 30/a	Copper salt + amine	32—3.200 ppm (0.1—10 mg/L)	6
Carbon monoxide 5/c	Iodine pentoxide + fuming sulfuric acid	5—150 ppm 100—700 ppm	10 2
Carbon monoxide 8/a (only for CO in H_2)	Iodine pentoxide + fuming sulfuric acid	8—150 ppm	10
Carbon monoxide 10/a	Iodine pentoxide + fuming sulfuric acid	10—300 ppm 100—3000 ppm	10 1
Carbon monoxide 10/b	Iodine pentoxide + fuming sulfuric acid	10—300 ppm, 100—3000 ppm	10, 1
Carbon monoxide 0.1%/a	Iodine pentoxide + fuming sulfuric acid	0.1—1.2 vol.%	1
Carbon monoxide 0.3%/a	Iodine pentoxide + fuming sulfuric acid	0.3—4 vol.%	1
Carbon monoxide 0.5%/a	Iodine pentoxide + fuming sulfuric acid	0.5—7 vol.%	1
Carbon tetrachloride 10/b	Via phosgene	10—100 ppm	3
Chlorine 0.2/a	*o*-Tolidine	0.2—3 ppm, 2—30 ppm	10, 1
Chlorine 50/a	*o*-Tolidine	50—500 ppm	1
Chloroprene 5/a	Permanganate	5—60 ppm, 7.5—90 ppm	3,2
Cyanogen chloride 0.25/a	Pyridine + barbituric acid	0.25—5 ppm	20—1
Cyclohexane 100/a	Chromic acid	100—1500 ppm	10
Diborane 0.05/a	Gold complex	0.05—3 ppm	20
Diethyl ether 100/a	Chromate + sulfuric acid	100—4000 ppm	10
Dimethylformamide 10/b	Via amine	10—40 ppm	10
Ethyl acetate 200/a	Chromate + sulfuric acid	200—3000 ppm	20
Formaldehyde 0.002	Xylene + sulfuric acid	2—40 ppm (0.002—0.05 mg/L)	5
Hydrazine 0.25/a	Bromphenol blue + acid	0.25—3 ppm	10
Hydrocarbon 0.1%	Iodine pentoxide + fuming sulfuric acid	0.1—1 vol.%	—
Hydrocarbon 2	Selenium dioxide + fuming sulfuric acid	2—25 mg/L	—
Hydrochloric acid 1/a	Bromphenol blue	1—10 ppm, 2—20 ppm	10, 5
Hydrogen 0.5%/a	Via H_2O	0.5—3 vol.%	5
Hydrogen cyanide 2/a	$HgCl_2$ + methyl red	2—30 ppm, 10—150 ppm	5, 1
Hydrogen fluoride 1.5/b	Zirconium quinalizarinate	1.5—15 ppm	20
Hydrogen sulfide 1/c	Lead salt	1—20 ppm, 10—200 ppm	10, 1
Hydrogen sulfide 5/b	Lead salt	5—60 ppm, 50—600 ppm	10, 1
Hydrogen sulfide 100/a	Lead salt	100—2000 ppm	1
Hydrogen sulfide 0.2%/A	Copper salt	0.2—7 vol.%	1
Hydrogen sulfide + sulfur dioxide 0.2%/A	Iodine	0.2—7 vol.%, 0.02—0.7 vol.%	1, 10
Mercaptan 2/a	Copper salt + sulfur	2—100 ppm	10
Mercury vapor 0.1/b	Copper(I)-iodide	0.1—2 mg/m^3	20—1
Methyl bromide 5/b	Via bromine	5—50 ppm	5
Methylene chlorde 100/a	Iodine pentoxide + selenium dioxide + fuming sulfuric acid	100—2000 ppm	10
Monostyrene 10/a	Sulfuric acid	10—200 ppm	15—2
Monostyrene 50/a	Sulfuric acid	50—400 ppm	11—2
Natural gas test	Via CO	Qualitative	2
Nickel carbonyl 0.1/a	Iodine + dioxine	0.1—1 ppm	20
Nitrogen dioxide 0.5/c	Diphenylbenzidine	0.5—10 ppm	5
Nitrogen dioxide 2/c	Diphenylbenzidine	2—50 ppm, 5—100 ppm	10, 5

TABLE 40 (continued)

Classification of Detecting Reactions in Dräger Analyzer Tubes

Detector Tube	Reaction Principle	Measuring Range	Number of Strokes
Nitrous fumes 0.5/a	Diphenylbenzidine	0.5—10 ppm	5
Nitrous fumes 2/a	Diphenylbenzidine	2—50 ppm, 5—100 ppm	10, 5
Nitrous fumes 20/a	*o*-Dianisidine	20—500 ppm	2
Nitrous fumes 100/c	*o*-Dianisidine	100—1000 ppm, 500—5000 ppm	5, 1
Olefin 0.05%/a	Permanganate	1—50 mg/L	20—1
Ozone 0.05/a	Indigo	0.05—1.4 ppm, 0.5—14 ppm	10, 1
Ozone 10/a	Indigo	10—300 ppm	1
Oxygen 5%/A	Via CO	5—21 vol.%	1
n-Pentane 100/a	Chromic acid	100—1500 ppm	5
Perchloroethylene 10/a	Via chlorine	10—400 ppm	3
Phenol 5/a	Dibromoquinonimide	5 ppm	10
Phosgene 0.05/a	Dimethylaminobenzaldehyde + diethylaniline	0.05—1.2 ppm	26—1
Phosgene 0.25/b	Dimethylaminobenzaldehyde + dimethylaniline	0.25—15 ppm, 1.25—75 ppm	5, 1
Phosphine 0.1/a	Gold salt	0.1—4 ppm, 1—40 ppm	10, 1
Phosphine 50/a	Gold salt	50—1000 ppm, 15—300 ppm	3, 10
Polytest tube	Iodine pentoxide + fuming sulfuric acid	Qualitative	5
Simultaneous tube	See carbon dioxide and carbon monoxide tubes	200—2500 ppm CO, 2—12 vol.% CO_2	2
Sulfur dioxide 0.1/a	Disodium tetrachloromercurate	0.1—3 ppm	100
Sulfur dioxide 1/a	Iodine + starch	1—20 ppm	10
Sulfur dioxide 20/a	Iodine	20—200 ppm, 200-2000	10, 1
Systox 1/a	Chloramide + gold salt	0.5 mg/m³	20
Toluene 5/a	Iodine pentoxide + sulfuric acid	5—400 ppm	5
Toluene 25/a	Iodine pentoxide + sulfuric acid	25—1860 ppm (0.1—7 mg/L)	10
Toluylenediisocyanate 0.02/A	Pyridylpyridiniumchloride	0.02—0.2 ppm	25
Triethylamine 5/a	Bromphenol blue + acid	5—60 ppm	5
Trichloroethane	Via chlorine	50—350 ppm, 100—700 ppm	10, 5
Trichloroethylene 10/a	Via chlorine	10—400 ppm	5
Vinyl chloride 100/a	Permanganate	100—3000 ppm	18—1
Water vapor 0.1	Selenium dioxide + sulfuric acid	0.1—40 mg/L	10

Courtesy of Drägerwerk AG, Lübeck, West Germany.

TABLE 41

List of Gastec® Detector Tubes

Catalog tube no.	Detector tube description	Calibration scale	Measuring range
81	Acetic acid	2—40 ppm	0.2—80 ppm
151	Acetone	0.01—0.8 vol %	0.01—2 vol %
171	Acetylene	0.2—2.0 vol %	0.1—4 vol %
191[a]	Acrylonitrile	10—500 ppm	10—500 ppm
3H	Ammonia, high range	1—16 vol %	0.2 —32 vol %
3M	Ammonia, middle range	50—500 ppm	5—1000 ppm
3L	Ammonia, low range	2—30 ppm	1—60 ppm
121[b]	Benzene	10—200 ppm	10—200 ppm
142	Butyl acetate	0.01—0.8 vol %	0.01—0.8 vol %
2H	Carbon dioxide, high range	1.0—10 vol %	0.1—10 vol %
2L	Carbon dioxide, low range	0.2—3.0 vol %	0.02—6 vol %
2LL	Carbon dioxide, extra low range	300—5000 ppm	300—5000 ppm
13[a]	Carbon disulfide	5—50 ppm	5—50 ppm
1H	Carbon monoxide, high range	0.2—5.0 vol %	0.1—10 vol %
1L	Carbon monoxide, low range	50—1000 ppm	5—2000 ppm
1La	Carbon monoxide, low range	50—600 ppm	5—1200 ppm
1LL	Carbon monoxide, extra low range	5—50 ppm	5—50 ppm
8L	Chlorine, low range	1—30 ppm	0.5—60 ppm
103	Cyclohexane	0.03—0.6 vol %	0.015—1.2 vol %
141	Ethyl acetate	0.04—1.5 vol %	0.04—1.5 vol %
112	Ethyl alcohol (ethanol)	0.1—2.5 vol %	0.05—5 vol %
125	Ethylbenzene	20—250 ppm	10—500 ppm
161	Ethyl ether (ether)	0.04 —1.0 vol	0.04—1.5 vol %
163	Ethylene oxide	0.1—3.0 vol %	0.1—3.0 vol %
91[a]	Formaldehyde	2—20 ppm	2—20 ppm
101	Gasoline	0.03—0.6 vol %	0.015—1.2 vol %
102H	*n*-Hexane, high range	0.03—0.06 vol %	0.015—1.2 vol %
102L	*n*-Hexane, low range	50—1200 ppm	50—1200 ppm
14L[c]	Hydrogen chloride, low range	2—20 ppm	0.2—40 ppm
12H	Hydrogen cyanide, high range	0.05—2 vol %	0.05—2 vol %
12L[b]	Hydrogen cyanide, low range	5—15 ppm	2.5—30 ppm
4H	Hydrogen sulfide, high range	100—1600 ppn	10—3200 ppm
4L	Hydrogen sulfide, low range	10—120 ppm	1—240 ppm
113	Isopropyl alcohol (isopropanol)	0.04—2.5 vol %	0.02—5 vol %
111	Methyl alcohol (methanol)	0.02—1.5 vol %	0.01—3 vol %
136[a]	Methyl bromide	10—100 ppm	10—200 ppm
135[a]	Methyl chloroform	100—500 ppm	100—500 ppm
152	Methyl ethyl ketone	0.02—0.6 vol %	0.02—0.6 vol %
153	Methyl isobutyl ketone	0.01—0.6 vol %	0.01—0.6 vol %
15L[c]	Nitric acid, low range	2—20 ppm	0.2—40 ppm
9L[b]	Nitrogen dioxide, low range	2—100 ppm	2—100 ppm
11L[b]	Nitrogen oxides, low range	2—100 ppm	2—100 ppm
133	Perchloroethylene	50—200 ppm	5—200 ppm
16	Phosgene	0.1—10 ppm	0.1—10 ppm
7[b]	Phosphine	50—500 ppm	5—1000 ppm
124[b]	Styrene	20—500 ppm	10—1000 ppm
5H	Sulfur dioxide, high range	0.5—4.0 vol %	0.05—8 vol %
5M	Sulfur dioxide, middle range	100-1800 ppm	10—3600 ppm
5L[b]	Sulfur dioxide, low range	5—100 ppm	0.5—200 ppm
122[b]	Toluene	20—250 ppm	10—500 ppm
132H	Trichloroethylene, high range	50—200 ppm	5—200 ppm
132L	Trichloroethylene, low range	2—25 ppm	1—50 ppm
131	Vinyl chloride	0.05—1.0 vol %	0.025—2 vol %
6	Water vapor	2—18 mg/L	1—36 mg/L

[a] Consisting of an analyzer tube for indicating gas concentration and a primer tube for pretreating sample air. A pair of the tubes must be connected using a rubber tubing (contained in each box);[b] Tubes must be kept below 10°C (50°F) for stability of reagent contained;[c] Containing five analyzer tubes and five water vapor tubes (No. 6). Since the indication of the tubes is based on water content in the test air, measurement of water content is required before test for humidity correction.

TABLE 42 (continued)

Classification of Detecting Reactions in Gastec® Detector Tubes

1. No. 1H carbon monoxide, high range
   CO reduces iodine pentoxide to liberate iodine, which produces a blackish green stain. Coexisting hydrocarbons are eliminated in the brown precleaning layer consisting of chromic acid and sulfuric acid.
   $CO + I_2O_5$ — $I_2 + CO_2$
2. No. 1L carbon monoxide, low range
   CO reduces potassium palladosulfite to liberate metallic palldium, which produces blackish brown stain.
   $CO + K_2Pd(SO_3)_2$ — $K_2 + Pd + SO_2 + CO_2$
3. No. 1La carbon monoxide, low range
   See No. 1L carbon monoxide, low range
4. No. 1LL carbon monoxide, extra-low range
   See No. 1L carbon monoxide, low range
5. No. 2H carbon dioxide, high range
   CO_2 neutralizes potassium hydroxide and changes the color of pH indicator (alizarin yellow).
   $CO_2 + KOH$ — $K_2CO_3 + H_2O$
6. No. 2L carbon dioxide, low range
   See No. 2H carbon dioxide, high range
7. No. 2LL carbon dioxide, extra-low range
   CO_2 reacts with hydrazine to form carbonic acid monohydrazide, which changes the color of redox indicator (crystal violet).
   $CO_2 + N_2H_4$ — $NH_2 \cdot NH \cdot COOH$
8. No. 3H ammonia, high range
   NH_3 reacts with cobalt chloride to form a complex or a coordinate compound.
   $NH_3 + CoCl_2$ — $aCoCl_2 \cdot bNH_3$
9. No. 3M ammonia, middle range
   NH_3 neutralizes phosphoric acid to change the color o pH indicator (thymol blue).
   $NH_3 + H_3PO_3$ — $(NH_3)_2\ PO_3$
10. No. 3L ammonia, low range
    NH_3 neutralizes sulfuric acid to change the color of pH indicator (thymol blue).
    $NH_3 + H_2SO_4$ — $(NH_4)_2SO_4$
11. No. 4H hydrogen sulfide, high range
    H_2S reacts with lead acetate to form lead sulfide.
    $H_2S + Pb(CH_3COO)_2 \cdot 3H_2O$ — $PbS + CH_3COOH$
12. No. 4L hydrogen sulfide, low range
    See No. 4H hydrogen sulfide, high range
13. No. 4LL hydrogen sulfide, extra-low range
    See No. 4H hydrogen sulfide, high range
14. No. 5H sulfur dioxide, high range
    SO_2 reduces chromic ion of orange yellow to trivalent chromous ion (green).
    $SO_2 + CrO_3$ — $Cr_2(SO_4)_3 \cdot 18H_2O + K_2O$
15. No. 5M sulfur dioxide, middle range
    SO_2 neutralizes sodium hydroxide to change the color of pH indicator (phenol red).
    $SO_2 + NaOH$ — $Na_2SO_3 + H_2O$
16. No. 5L sulfur dioxide, low range
    See No. 5M sulfur dioxide, middle range
18. No. 6 water vapor
    H_2O is absorbed by magnesium perchlorate to produce an alkaline, which changes the color of Hammett indicator (crystal violet).
    $H_2O + Mg(ClO_4)_2$ — $Mg(ClO_4)_2 \cdot aH_2O$
19. No. 7 phosphine
    PH_3 reacts with mercuric chloride to produce mercuric phosphochloride of yellow color.
    $PH_3 + HgCl_2 + H_2O$ — $Hg_3P_2 \cdot 3HgCl_2 \cdot 3H_2O$
20. No. 8L chlorine, low range
    Cl_2 reduces *o*-tolidine to form nitroso-*o*-tolidine of orange color.
    $Cl_2 + H_2O$ + *o*-tolidine — Nitroso-*o*-tolidine + HCl
21. No. 9L nitrogen dioxide, low range
    NO_2 reduces *o*-tolidine to form nitroso-*o*-tolidine of bright yellow color.
    $NO_2 + H_2O$ + *o*-Tolidine — Nitroso-*o*-tolidine
22. No. 10 nitrogen oxides (NO and NO_2)
    NO_2 is absorbed in the first tube (NO_2 tube) and reduces *o*-tolidine to form nitroso-*o*-tolidine of bright yellow color. NO passes through the first tube and is absorbed in the second tube (NO tube), then oxidized

TABLE 42 (continued)

Classification of Detecting Reactions in Gastec® Detector Tubes

by orange prelayer consisting of chromic and sulfuric acids to NO_2, which reduces o-tolidine to form nitroso-o-tolidine of bright yellow color.

NO_2 tube – (first tube): NO_2 + o-Tolidine + H_2O — Nitroso-o-tolidine

NO tube – (second tube): NO + CrO_3 + H_2SO_4 — NO_2

NO_2 + o-Tolidine — Nitroso-o-tolidine

23. No. 11L nitrogen oxides (NO + NO_2), low range

Coexisting NO is oxidized by chromic and sulfuric acids in orange prelayer to form NO_2, which combining with NO_2 in sample, reduces o-tolidine in white indicating layer to form nitroso-o-tolidine of bright yellow color.

NO + CrO_3 — NO_2

NO_2 + H_2O + o-Tolidine — Nitroso-o-tolidine

24. No. 12H hydrogen cyanide, high range

HCN reacts with potassium palladosulfite to form a white complex or coordinate compound, the chemical formula of which is still unknown.

HCN + $K_2Pd(SO_3)_2$ — White compound

25. No. 12L hydrogen cyanide, low range

HCN reacts with mercuric chloride to liberate HCl, which changes the color of pH indicator (methyl orange) to pink. Other acidic gases are eliminated by benzeneazodiphenylamine in precleaning layer.

HCN + $HgCl_2$ — HCl + $Hg(CN)_2$

26. No. 13 carbon disulfide

CS_2 is oxidized by nascent oxygen, which is yielded by the reaction of chromic and sulfuric acids in brown prelayer in the primer tube, to form SO_2. This SO_2 neutralizes sodium hydroxide to change the color of pH indicator (phenol red) to dull yellow. Coexisting H_2S is removed by cupric sulfate in white cleaning layer in the primer tube.

Primer tube:

CrO_3 + H_2SO_4 — (O)

(O) + CS_2 — SO_2 + CO_2

Analyzer tube:

SO_2 + NaOH — $Na_2(SO_3)$ + H_2O

27. No. 14L hydrogen chloride, low range

HCl is absorbed in silica sand to change the color of Hammett indicator (4-phenylazodiphenylamine).

28. No. 15L nitric acid, low range

HNO_3 is absorbed in silica sand to change the color of Hammett indicator (4-phenylazodiphenylamine).

29. No. 16 phosgene

$COCl_2$ reacts with nitrobenzylpyridine to form urea derivative, which reacts with benzylaniline to produce a red dye. Chemical formulas of the produced compounds are still unknown.

$COCl_2$ + nitrobenzylpyridine — Urea derivative

Urea derivative + benzylaniline — Red dye

30. No. 81 acetic acid

Acetic acid neutralizes sodium hydroxide to change the color of pH indicator (phenol red).

CH_3COOH + NaOH — $Na_2(CH_3COO)_2$ + H_2O

31. No. 91 formaldehyde

HCHO reacts with o-xylene in the primer tube and when dehydrated by sulfuric acid forms a condensation polymer in the analyzer tube.

$(HCHO)_n + [2\cdot C_6H_4(CH_3)_2]_n$ — $[C_6H_4(CH_3)_2 - CH_2 - C_6H_4(CH_3)_2]_n + H_2O$

32. No. 100A LP gas (propylene)

Propylene or butane reduces bichromic ion of potassium bichromate to trivalent chromous ion.

C_3H_6 or C_4H_{10} + $K_2Cr_2O_7$ + H_2SO_4 — $Cr_2(SO_4)_3$

33. No. 101 gasoline

Gasoline reduces chromic ion of chromic acid to trivalent chromous ion.

C_nH_m + CrO_3 + H_2SO_4 + H_2O — $Cr_2(SO_4)_3$

34. No. 102H n-Hexane, high range

n-Hexane reduces chromic ion of chromic acid to trivalent chromous ion.

C_6H_{14} + CrO_3 + H_2SO_4 + H_2O — $Cr_2(SO_4)_3$

35. No. 102L n-Hexane, low range

n-Hexane reduces bichromic ion of bichromic acid to trivalent chromous ion.

C_6H_{14} + $K_2Cr_2O_7$ + H_2SO_4 — $Cr_2(SO_4)_2$

36. No. 103 Cyclohexane

Cyclohexane reduces chromic ion of chromic acid to trivalent chromous ion.

C_6H_{12} + CrO_3 + H_2SO_4 + H_2O — $Cr_2(SO_4)_3$

TABLE 42 (continued)

Classification of Detecting Reactions in Gastec® Detector Tubes

37. No. 104 butane

Butane reduces bichromic ion of bichromic acid to trivalent chromous ion.

$C_4H_{10} + K_2Cr_2O_7 + H_2SO_4 \qquad Cr_2(SO_4)_3$

38. No. 111 methyl alcohol (methanol)

Methyl alcohol reduces bichromic ion of bichromic acid to trivalent chromous ion after water content in sample is removed by the white precleaning layer consisting of potassium carbonate. The indicating reagent is not reduced by other organic vapors except alcohol.

$CH_3OH + K_2Cr_2O_7 \qquad Cr_2(SO_4)_3$

39. No. 112 ethyl alcohol (ethanol)

Ethyl alcohol reduces bichromic ion of bichromic acid to trivalent chromic ion. Water content in sample is eliminated by the white precleaning layer consisting of potassium carbonate. The indicating reagent is not reduced by other organic vapors except alcohol.

$C_2H_5OH + K_2Cr_2O_7 \qquad Cr_2(SO_4)_3$

40. No. 113 isopropyl alcohol (isopropanol)

Isopropyl alcohol reduces bichromic ion of bichromic acid to trivalent chromous ion after water content in sample is eliminated in the white precleaning layer consisting of potassium carbonate. The indicating reagent is not reduced by other organic vapors except alcohol.

$(CH_3)_2CHOH + K_2Cr_2O_7 \qquad Cr_2(SO_4)_3$

41. No. 121 benzene

Benzene reduces iodine pentoxide to liberate iodine, which produces a green stain.

$C_6H_6 + I_2O_5 + H_2SO_4 \qquad I_2 + CO_2$

42. No. 122 toluene

Toluene reduces iodine pentoxide to liberate iodine, which produces a blackish brown stain. Water content in sample is eliminated in the rose precleaning layer of molecular sieves.

$C_6H_5CH_3 + I_2O_5 + H_2SO_4 \qquad I_2 + CO_2$

43. No. 123 xylene

Xylene reduces iodine pentoxide to liberate iodine, which produces a black stain. Water content in sample is removed in the rose precleaning layer of molecular sieves.

$C_6H_5(CH_3)_2 + I_2O_5 + H_2SO_4 \qquad I_2 + CO_2$

44. No. 124 styrene

Styrene is dehydrated by sulfuric acid to form a condensation polymer.

$C_6H_5CH{:}CH_2 + H_2SO_4 \qquad C_6H_5CH{:}CH{-}CH{:}CHC_6H_5$

$C_6H_5CH{=}CH_2$

45. No. 125 ethylbenzene

Ethylbenzene reduces iodine pentoxide to liberate iodine, which produces a blackish brown stain. Water content in sample is removed in the rose precleaning layer of molecular sieves.

$C_6H_5O_2H_5 + I_2O_5 + H_2SO_4 \qquad I_2 + CO_2$

46. No. 131 vinyl chloride

Vinyl chloride reduces chromic ion of chromic acid to trivalent chromous ion.

$CH_2{=}CHCl + CrO_3 + H_2SO_4 + H_2O \qquad Cr_2(SO_4)_2$

47. No. 132H trichloroethylene, high range

Trichloroethylene is oxidized by nascent oxygen, which is generated by the reaction of lead dioxide and sulfuric acid in the black oxidizing agent, to liberate HCl. This generated HCl changes the color of Hammett indicator (benzeneazodiphenylamine) to pink.

$PbO_2 + Pb(NO_3)_2 + H_2SO_4 \qquad (O)$

$CHCl{:}CCl_2 + (O) \qquad HCl$

48. No. 132L trichloroethylene, low range

See No. 132H trichloroethylene, high range

50. No. 133 perchloroethylene

Perchloroethylene is oxidized by nascent oxygen, which is generated by the reaction of lead dioxide and sulfuric acid in the black oxidizing agent, to liberate HCl. This HCl changes the color of Hammett indicator (benzeneazodiphenylamine).

$PbO_2 + Pb(NO_3)_2 + H_2SO_4 \qquad (O)$

$CCl_2{:}CCl_2 + (O) \qquad HCl$

51. No. 135 methylchloroform (1,1,1-trichloroethane)

Methylchloroform is oxidized by nascent oxygen, which is liberated by the reaction of iodine pentoxide and sulfuric acid in the white oxidizing agent, to generate chlorine in the primer tube. This generated chlorine reduces *o*-tolidine to form nitroso-*o*-tolidine of yellow color.

TABLE 42 (continued)

Classification of Detecting Reactions in Gastec® Detector Tubes

Primer tube:

$CH_2{=}CHCN + CrO_3 + H_2SO_4$ $\qquad$ $HCN + Cr_2(SO_4)_3$

Analyzer tube:

$HCN + HCl + Hg(CN)_2$ $\qquad$ $HCl + Hg(CN)_2$

Cl_2 + *o*-Tolidine $\qquad$ Nitroso-*o*-tolidine + HCl

52. No. 136 methyl bromide

Methyl bromide is oxidized by nascent oxygen, which is liberated by the reaction of iodine pentoxide and sulfuric acid in white oxidizing agent, to generate bromine in the primer tube. This bromine reduces *o*-tolidine to form nitroso-*o*-tolidine in the analyzer tube.

Primer tube:

$I_2O_5 + H_2SO_4$ $\qquad$ (O)

CH_3Br + (O) $\qquad$ $Br_2 + CO_2 + H_2O$

Analyzer tube:

Br_2 + *o*-Tolidine $\qquad$ Nitroso-*o*-tolidine + HBr

53. No. 138 chloropicrine

Chloropicrine is decomposed by nascent oxygen, which is generated by the reaction of iodine pentoxide and sulfuric acid in white pre-oxidizing agent, to liberate phosgene in the primer tube. This liberated phosgene reacts with nitrobenzylpyridine to form urea derivative, which reacts with benzylaniline to produce a red dye.

Primer tube:

$CCl_3NO_2 + I_2O_5 + H_2SO_4$ $\qquad$ $COCl_2$

Analyzer tube:

$COCl_2$ + nitrobenzylpyridine $\qquad$ Urea derivative

Urea derivative + benzylaniline $\qquad$ Red dye

54. No. 141 ethyl acetate

Ethyl acetate reduces bichromic ion of bichromic acid to trivalent chromous ion. Water content is eliminated in white precleaning layer of potassium carbonate.

$CH_3COOC_2H_5 + K_2Cr_2O_7 + H_2SO_4$ $\qquad$ $Cr_2(SO_4)_3$

55. No. 142 butyl acetate

Butyl acetate reduces bichromic ion of bichromic acid to trivalent chromous ion after water content is removed in white precleaning layer of potassium carbonate.

$CH_3COOC_4Hg + K_2Cr_2O_7 + H_2SO_4$ $\qquad$ $Cr_2(SO_4)_3$

56. No. 151 acetone

Acetone reduces bichromic ion of bichromic acid to trivalent chromous ion. Water content is removed in white precleaning layer of potassium carbonate.

$(CH_3)_2CO + K_2Cr_2O_7 + H_2SO_4$ $\qquad$ $Cr_2(SO_4)_3$

57. No. 152 methyl ethyl ketone (MEK)

Methyl ethyl ketone reduces bichromic ion to trivalent chromous ion. Water content in sample is eliminated in white precleaning layer of potassium carbonate.

$CH_3COC_2H_5 + K_2Cr_2O_7 + H_2SO_4$ $\qquad$ $Cr_2(SO_4)_3$

58. No. 153 methyl isobutyl ketone (NIBK)

Methyl isobutyl ketone reduces bichromic ion of bichromic acid to trivalent chromous ion. Water cntent in sample is removed in white precleaning layer of potassium carbonate.

$CH_3COCH_2CH(CH_3)_2 + K_2Cr_2O_7 + H_2SO_4$ $\qquad$ $Cr_2(SO_4)_3$

59. No. 161 ethyl ether (ether)

Ethyl ether reduces bichromic ion of bichromic acid to trivalent chromous ion. Water content in sample is removed in white precleaning layer.

$(C_2H_5)_2O + K_2Cr_2O_7 + H_2SO_4$ $\qquad$ $Cr_2(SO_4)_3$

60. No. 171 acetylene

Acetylene reduces iodine pentoxide to liberate iodine, which produces a greenish brown stain.

$C_2H_2 + I_2O_5 + H_2SO_4$ $\qquad$ $I_2 + CO_2$

61. No. 172 ethylene

Ethylene reacts with palladium sulfate to form an additive compound, which reacts with ammonium molybdate to yield molybdenum blue. Chemical formula of this reaction is still unknown.

$C_2H_2 + (NH_4)_2MoO_4 + PdSO_4$ $\qquad$ Mo_3O_8

62. No. 191 acrylonitrile

Acrylonitrile is decomposed by brown oxidizing agent consisting of chromic acid and sulfuric acid in the primer tube to liberate HCN. This HCN reacts with mercuric chloride to generate HCl, which changes the color of Hammett indicator (benzeneazodiphenylamine). Coexisting acidic gases are removed in white precleaning layer consisting of NaOII in the primer tube.

TABLE 43

Paint and Varnish Manufacturing Solvent Inventory

Chemical compound	Usage: Major	Usage: Minor	TLV (ppm)	Catalog tube no.[a]	Remarks
Alphatic hydrocarbons (as *n*-hexane)		X	500	113	Mineral spirits, V.M. and P. naphtha
Methanol		X	200	119	
Ethanol		X	1000	104A	Synasol® solvent
Isopropanol		X	400	150	
Butanols		X	100	—	C.S. solvent
2-Methoxyethanol		X	25	—	Methyl Cellosolve®
2-Ethoxyethanol		X	100	—	Cellosolve
2-Butoxyethanol		X	50	—	Butyl Cellosolve
Acetone		X	1000	102A	
Methyl ethyl ketone (MEK)	X		200	139B	
Methyl *n*-butyl ketone		X	100	—	
Methyl isobutyl ketone		X	100	155	
Methyl isoamyl ketone		X	100	—	
n-Butyl acetate		X	150	138	
Isobutyl acetate		X	200	138	
sec-Butyl acetate		X	200	138	
Methyl amyl acetate		X	—	—	
Toluene	X		100	124	Limited exposure
Xylene		X	100	143	

[a] Unico®-Kitagawa® Model 400 and Unico gas detector tubes, 1956.

on the day of the survey, and to ignore the contribution of the very minor amounts of the other solvents. Furthermore, since the concentrations of toluene vapor were relatively minor compared to the "apparent" MEK results, which were well above the TLV limit (Table 44), no attempt was made to correct the contribution of toluene to the reduction of the chromate reagent in the MEK tube.

Air was drawn through the detector tubes as specified in the instructions enclosed with each box of tubes. Approximately 3 min was required to draw the 100-mL air sample through each tube. Leak tests were made on the pump, and leakage was found to be within acceptable limits (less than 5 mL/3 min). Stain lengths were determined in subdued daylight near outside windows and related to concentration on the charts supplied with the tubes.

Samples were taken in the breathing zone as close as possible to the operator's nose without interfering with his work assignments or in the general room area at nose level. Doors and windows were closed, and ventilation was set to simulate cold-weather operating conditions from about 10 a.m. to shift end.

Results — The results are summarized in Tables 44 and 45.

Basis for conclusions —

1. Any given operator would, on the average, spend the amount of time required to complete the operation at each of the stations sampled.
2. The operations observed represented the average daily activity in the building.
3. The ventilation was minimal, as would be expected during cold-weather operation.
4. Interferences in the determination of MEK were minimal.
5. The operator spent his entire 8-hr work shift in the manufacturing building.

TABLE 44

Air Analysis Results

Test number	Time	Temperature (°F)	Location	Occupancy	Material	Results (ppm)	Comments
	a.m.						
1	10:30	65	Paint dept., 10 in. from rim of Hockmeyer mixer	5 min/day	MEK	100	
1A	10:35	62	As above	5 min/day	Toluene	50	
2	10:42	66	Stain dept., pigment stain mixer	5 min/day	Naphtha	30	Hexane tube
3	10:50	63	Lacquer dept., fill operator	30 min, 3 times/day, 2 times/month	Toluene	10	
4	10:55	64	Lacquer dept., center	30 min, 3 times/day, 2 times/month	MEK	750	
5	11:01	65	Lacquer dept., top of tank No. 8 on platform	10—15 min, 4 times/day	Toluene	15	
6	11:06	65	Paint dept., No. 2 mill, paste		Mineral spirits	—	No odors
7	11:11	65	Paint dept., 10 in. from rim of old Cowles® mixer	5 min/day	Toluene	120	
7A	11:15	65	As above, dumping N.C. into tank No. 8	5 min/day	MEK	1500	
8	11:20	66	Lacquer dept., 40 in. from floor at opening	5 min, 12 times/day	Toluene	60	
9	11:26	63	1st floor, main aisle, south end, 20 ft. from door	Varied	Toluene	40	
10	11:30	63	As above	Varied	MEK	50—100	
11	11:33	62	1st floor, main aisle, center	Varied	Toluene	2	
12	11:35	61	As above	Varied	MEK	50	
13	11:40	64	Men filtering batch of 30 gal of solvent and putting into 5-gal cans, adjacent to Cowles®	10 min 10—15 times/day	Toluene	160	
14	11:43	64	As above	10 min, 10—15 times/day	MEK	100	
	p.m.						

TABLE 44 (continued)

Air Analysis Results

Test number	Time	Temperature (° F)	Location	Occupancy	Material	Results (ppm)	Comments
15	2:40	65	Drum cleaning bldg., north room	240 min/day	MEK	1500	
16	2:53	65	1st floor, main aisle, north end, 20 ft from door	Varied	Toluene	8	
17	2:55	64	Paint dept., center, at 55-gal drum mixer	15 min/day	Toluene	20	
18	2:58	64	As above	15 min/day	MEK	200	
19	3:08	64	Paint dept., new Cowles mixer	10 min, 3 times/week	MEK	1200	
20	3:13	64	Lacquer dept., on platform, 3 mixers running	15 min, 12 times/day	Toluene	5	
21	3:30	63	Paint dept., old Cowles mixer changing mixer from one tank to another	5 min/day	Toluene	40	
22	3:37	63	1st floor, main aisle, center and area around mixers, paint dept. to east	Varied	MEK	1000	

TABLE 45

Exposure Summary for Methyl Ethyl Ketone (MEK) (2-Butanone)[a]

Test number	Results (ppm)	Length of exposure (min)	Exposures/day	CT value
1	100	5	1	10
4	750	30	3	1130
7	1500	5	1[b]	150
10	75		1	520
Av	65		1	520
12	50		1	520
14	100	10	12	240
18	200	15	1	50
19	1200	10	0.6	140
22	1000	1[b]	1[b]	1000
Total				3240

Notes: Major sources: test number 4 (Lacquer dept., center at fill operation); test number 22 (1st floor, center, main aisle, around paint mixer perimeter). Background, overall levele: test numbers 10—12.

[a] TLV/1971 = 200 ppm; CT/8 hr = 8 × 200 = 1600 ppm hr.

[b] Estimated.

This preliminary survey indicated that an integrated 8-hr average exposure could exceed the 1971 TLV by at least a factor of two, based on the CT value concept (ppm × hr). However, since considerable uncertainty is inherent in a preliminary survey of this kind, a more definitive study is required to determine whether or not the exposure potential for any given operator actually exceeds the 200-ppm TLV for MEK during an 8-hr work shift.

A personnel monitor survey would provide the most reliable in-depth evaluation of exposure severity and would furnish the data necessary to establish conformance with the federal TLV.[133]

Briefly, this would require an operator to carry on his belt a small battery-operated pump that draws a determinable volume of air through a charcoal bed in a glass tube to adsorb the solvent vapor.[220] The tube is then capped and forwarded to a laboratory where, by gas chromatography, the weight of the individual solvents collected is determined. From the flow rate and sampling time, the volume, and then the concentration as ppm, are calculated. In order to average out daily variations in work activity and production schedules, this personnel monitoring should be carried on for at least 1 month (20 shifts). After a short training period, local supervisers would be able to conduct the sampling. Gas chromatography capability is available on location. To demonstrate continuing conformity with governmental regulations, periodic (for example, one 8-hr shift per month) monitoring may be required.

Conclusions —

1. This preliminary survey disclosed no potential health problems under minimum ventilation conditions employed to conserve heat during severe cold-weather conditions.

2. Two categories of airborne contamination are involved, (a) hazardous vapors (toluene, xylene, cellosolve,® and methanol) and (b) nuisance vapors (ketones and acetic acid esters).
3. Based on the CT (concentration × time = ppm/hr) concept for the evaluation of TLVs derived from an integrated average for an 8-hr workday, exposure to toluene vapor was not significant. However, MEK vapor may present a nuisance problem. The other solvents used in comparatively minor quantities were not analyzed.
4. Personnel monitoring will be necessary to establish conformance with the TLV for MEK. The grab sampling survey indicated violations, but a more definitive sampling and analysis program is needed to establish the magnitude of the problem before corrective actions are considered.
5. If the gas chromatograph results confirm the assumption that the major fraction (75% or more) of the solvent vapor was indeed MEK, then recalibration of the MEK tube for 4-hr monitoring, as described for the CO detector tube,[11] should be initiated to minimize cost and provide quick results directly in the field.

From the foregoing example, one crystal clear conclusion can be drawn: the investigator must be able to identity the reagents in the detector tubes and understand the chemical reaction principles before he can evaluate the significance of his findings, or, indeed, decide whether available detector tubes could solve his problem. The manufacturer must supply at least a list of the components of the reagent system, if not the relative concentrations, to guide the user in choosing an appropriate analytical method and to avoid interferences. Without this information, the investigator should not attempt a survey, as his conclusions can be entirely vitiated by coexistent components that may either mask or enhance the true result. Some manufacturers have been willing to provide this information, which is summarized in Tables 39 to 42. If the reagent cannot be identified, the user is advised either to recalibrate with the interference present, if known, or better, to reject the system as a valid analytical tool. In any event, the preliminary survey carried out with detector tubes should be followed by a definitive, specific method such as gas chromatography if the apparent results fall in the 0.5- to 1.5-TLV region, as recommended in the paint and varnish plant report. This follow-up is critically important if a large expenditure for the correction of a nonexistent hazard is to be avoided,[11] and if the validity of preliminary inspections based on detector tube surveys is to be successfully challenged. (See Chapter IV, Volume II for a statistical approach to interference evaluation). In Tables 39 to 42 are listed a number of reported interferences, but the list is by no means complete and is offered only as a warning gleaned from the experience of others.

Some interference problems have been solved by incorporating a granular "guard" column in front of the reagent zone to trap the "impurities" quantitatively without removing a significant amount of the target component. In some cases the scavenger layer is packed in a separate tube and connected to the detector tube by a flexible collar[97] (Table 46), and in other cases the guard is an integral part of the detector tube and separated from the reagent granules by a porous plug. Generally this first granular layer is impregnated with a chemical such as sulfuric acid on silica gel, which reacts preferentially with the extraneous substance but retains only to a negligible degree the substance sought. Several examples illustrate this principle:

Passed to detector layer	Retained by guard layer
Carbon monoxide	Olefines (ethylene) by H_2SO_4, and nitrogen oxides

TABLE 46

Moisture Removal from Saturated (100% RH) Air Samples Before Analysis with Detector Tubes[a]

Kitagawa detector tubes suitable for stack sampling	
No. 103a	Sulfur dioxide, high range
No. 103b	Sulfur dioxide, middle range
No. 103c	Sulfur dioxide, low range
No. 106a	Carbon monoxide
No. 106b	Carbon monoxide
No. 106c	Carbon monoxide
No. 106g	Carbon monoxide length-of-stain
No. 112a	Hydrogen cyanide, high range
No. 112b	Hydrogen cyanide, low range
No. 120a[b]	Hydrogen sulfide, high range
No. 120c	Hydrogen sulfide in coexistence of SO_2
No. 120hh[b]	Hydrogen sulfide, extra high range
No. 126a	Carbon dioxide, high range
No. 126b	Carbon dioxide, low range
No. 142	Mercury vapor
No. 159	Oxygen (not very sensitive to moisture)
Moisture-sensitive tubes that cannot be used with the desiccating guard tube	
No. 105a and 105b	Ammonia
No. 103d	Sulfur dioxide, D type
No. 109	Chlorine
No. 114	Bromine
No. 116	Chlorine dioxide
No. 117	Nitrogen dioxide
No. 120b	Hydrogen sulfide, low range

[a] Temperature range = 0—40°C; guard layer (Kitagawa® tube No. 103BF) — H_2SO_4 on 40—60 mesh diatomaceous earth.

[b] Can be used without guard tube.

Carbon bisulfide	H_2S by lead acetate
Hydrogen cyanide	H_2S and HCl by lead compound, and NO_2, SO_2
Benzene	Toulene and xylene by acid and aldehyde
Phosphine and arsine	H_2S, mercaptans, NH_3, and HCl by copper compound
Sulfur dioxide	H_2S by copper compound

Many detector reagents are sensitive to water vapor and must be protected by passing the air sample through a desiccant layer before the colorimetric reaction place. Some examples are listed in Table 46. Limitations imparted by the use of sulfuric acid supported on silica gel should be carefully considered when applying this desiccant to other mixtures, especially those that contain reactive components of analytical interest. Ammonia, aliphatic amines, ethylene, etc. would be completely removed. Silica or alumina gel and anhydrous calcium sulfate (Drierite®) are effective desiccating agents in themselves, but are not specific for water and usually absorb a significant fraction of the substance under analysis. Calibration with the selected guards in place offers the only final answer relative to retention of the component for which an analysis is desired.[132]

A rather limited number of indicator tubes requires a preliminary reaction to convert the compound to a detectable derivative. The chlorinated hydrocarbons fall into this class. The first layer, which contains a strong oxidizing agent such as potassium permanganate or chromic acid, converts the chlorine atoms to elemental chlorine, which then reacts with the color-forming compound (dianisidine or tolidine). Carbon tetrachloride is converted to phosgene by fuming H_2SO_4, nitric oxide to nitrogen dioxide by chromic acid, and acrylonitrile to HCN by a chromate before passage into the detection zone in other examples.[97] While such multiple-layer tubes serve a useful purpose when properly constructed, they frequently have a shorter shelf life than single-layer tubes because the chemicals diffuse between layers and produce an accelerated deterioration.

Detector tube systems have been developed to eliminate the diffusion between layers in storage and to improve the stability of the color-generating system by incorporating one of the reactive components as a liquid or an impregnated gel in an integral fragile glass vial that is broken just prior to use. The Drager methyl bromide detector, for example, retains the sulfur trioxide required for preoxidation sealed in the foresection of the glass detector tube. A plastic tube is shrunk on over this section to serve as a seal and to protect the operator after the tube is broken at this point.[97]

Many other complications can be expected when deviations from calibration conditions are introduced. Thus, for nitrogen dioxide the proportion of side reactions changes with flow rate. Altering sample volumes from calibration conditions cannot be assumed to be without error unless the detector reagent is known to be thoroughly free from interfering gases and humidity in the air.

The lack of specificity may be used to an advantage for determining substances other than those intended by the manufacturer. Thus, the trichloroethylene tube[97] can be applied also to the detection of chloroform, *o*-dichlorobenzene, dichloroethylene, ethylene chloride, methylene chloride, and perchloroethylene when present as sole chlorine-containing constitutents in the atmosphere. The methyl bromide tube may be used for chlorobromomethane and methylchloroform. The chlorine tube has been used for bromine and chloride dioxide. Such conversion requires specific knowledge of the feasibility and proper corrections to calibration scales (recalibration with known mixtures of the desired compound)[89] (see also section II.A.1.b.ix, "Reagent Systems — Identification").

i) Observational Errors

When all of the foregoing precautions have been observed and effects of the variables compensated, the investigator arrives at the final stage, examination of the exposed tube, with a reasonable assurance that the best possible result is available. This final judgment on his part is to a great degree dependent upon his visual acuity. If he has carried out his own calibration, this variable can be minimized by practice. Some detectors are provided with color charts to be matched with the stain on the granules in the indicating portion of the tube. In others a prestained section is included in the detector tube.[97] Judgment depends not only upon the observer's color discrimination, but also on lighting conditions. This is the case when the reaction with test gas is relatively slow and color is produced throughout the length of the tube, since the gas is incompletely absorbed and the concentration at the exiting end is an appreciable fraction of that at the inlet. This type includes the Drager tubes for benzene, nickel carbonyl, carbon bisulfide, cyanogen chloride, styrene, phosgene, and vinyl chloride.[97] Subtle color changes are difficult to detect, particularly under artificial light; and very often the standard color chart does not have the same tint or hue as the stain obtained. Subdued daylight through a north window provides the best comparison conditions. Exposure of the stain to direct sunlight should be avoided.

Other tubes, such as arsine, stibine,[89] aniline, mercury, phenol,[97] ethylene oxide, and propane[94] require a variable volume of sample be drawn through until the first visible color is noted (gas titration). This technique requires the most acute judgment and should be made retrospectively by drawing additional sample through the system to deepen the stain and thereby confirm the validity of the first observation as being from the gas sought and not from an artifact.

One manufacturer based his entire product line on this gas titration principle.[89] For reasons other than technical considerations, the producer of these gas detectors recently went out of business. However, a short description of the product is presented here to acquaint the reader with the possibilities for application to "instantaneous" alarm meters of the type described in Section I.A.2., "Inorganic Support Systems,"[2] as well as for hand-held grab sampling. The air sample was drawn through the short (approximately 40 mm) detector tube at a manually regulated rate of 100 mL in 20 sec by a calibrated precision piston pump. The pump itself was quite similar in construction to those currently available models, but was not fitted with a flow-limiting orifice or a check valve in either the inlet or exhaust ports, which has to be manually regulated.[94,125] The volume of sample required to produce the first color change in the sensitive layer of the tube was related to the concentration of the test gas. Each detector tube was packed in a lead foil cylinder which was opened by bending the crimped portion sharply backward over the end of the detector tube.

The detector tube for phosgene serves as a good example of the application to field monitoring. The change from yellow to green indicated the titration end-point, which was related to concentration by reference to the standardization chart:

mL for first color change	350	100	50	25
Gas concentration (ppm)	0.1	1	5	10

The presence of 38 different gases and vapors could be determined with this detector system.[89]

To reduce the errors inherent in the variations between observers (see the study reported by Kusnetz, Saltzman, and Lanier for a statistical analysis of observer differences in estimating CO by the NBS detector tube[113]), most detector tubes have been converted from color shade matching to produce a variable length of stain on the indicator gel, although in a few tubes a variable volume of sample is collected until a standard length of stain is obtained (alcohol in breath samples[94]). This technique may be considered to be a variation of gas titration, as volume of "titrant" is related to quantity sought.[97] In most cases a fixed volume of sample is passed through the tube and the stain length measured against a calibration scale which may be printed directly on the tube[97] or on a chart.[94]

If the stain fronts are not well defined or are not perpendicular to the axis of the tube, again a broad range in the interpretation of the results by different observers will be encountered. Experience in sampling known concentrations is indispensible in training the observer to decide whether to measure the length up to the beginning or end of the stain front and what portion of an irregularly shaped stain to use as the limit. Stains may change with time, and therefore the reading should not be delayed. Precision can be improved by taking several readings (four or five) and averaging the results, or better, by averaging the readings taken by two or more observers.

j) Repetitive Sampling

Repeated use of detector tubes as an economy measure may be commendable, but in the hands of the uninformed must be condenmed as a habitual practice. Certain detector tubes may be used again after a negative test, but the investigator must bear in mind that some detector reagent systems are adversely affected by humidity, acid

TABLE 47

Recalibration of the Kitagawa® No. 100 Carbon Monoxide Length of Stain Detector Tube[a]

Standard CO (ppm)		Tube no.	Results (ppm) 1 stroke	Results (ppm) 2 strokes	Reversed ends (ppm) 1	Reversed ends (ppm) 2
25	In air	1	25	25	30	20
49	In air	1	50	45	—	—
		2	50	55	60	—
		3	60	50	—	—
99	In air	1	85	95	90	—
		2	90	—	—	—
		3	90	—	85	—
		4	90	85	90	75
50±5	In N_2	1	50	50	50	—
		2	50	—	50	—
100±10	In N_2	1	125	113	110	—
		2	110	100	90	—

[a] Kitagawa tube No. 100, 3-min stroke, leak test: 0 mL/min.

gases, alkaline gases, and other extraneous interferences likely to be present in the atmosphere sampled. For example, traces of ammonia will completely inhibit the color response of the mercury vapor detector tube even when used in a saturated atmosphere. Water vapor will have a depressing effect on stain formation in the benzene reagent mixture by diluting the sulfuric acid required in the condensation with formaldehyde (Table 39). Other false negative reactions can be expected under high humidity conditions also (see Table 46). Tubes that require a guard layer to scavenge interferences are especially vulnerable to repetitive sampling. The palladium reduction for carbon monoxide detection will produce false positive results in the presence of ethylene if the guard layer is exhausted. Other examples should be consulted before decisions relative to reuse of any particular tube are made (Tables 39 to 42).[94]

Information relative to the presence of interferences can be obtained by reversing a tube supplied with a guard layer after a negative result has been obtained from a sample drawn through in the designed direction. Color development in the unguarded end on reversal indicates the presence of a reactive component in the mixture other than the one sought. Such a color development with the carbon monoxide detector tube would indicate the presence of ethylene, acetylene, benzene, or H_2S.[97]

Symmetrical tubes, those filled with a homogeneous reagent composition from end to end, can be reversed and a second determination made. Results obtained from recalibration of the length-of-stain carbon monoxide detector tube are typical (Table 47). At least one supplier has recommended this technique for those tubes that require guard layers. The thrifty investigator will quickly note the cost savings in this recommendation.

However, the temptation either to reduce cost or to increase sensitivity by drawing additional aliquots of air through the same detector tube after a short, positive response has been noted can lead to significant errors. Calibration results indicate that at least two samples can be taken with the carbon monoxide (Table 47) or the benzene detector tube if the second aliquot is drawn immediately after the 3-min period required for the first sample (final result divided by 2). With the benzene tube, a time lapse of more than 10 or 15 min between aliquots produced errors that exceed 15%,

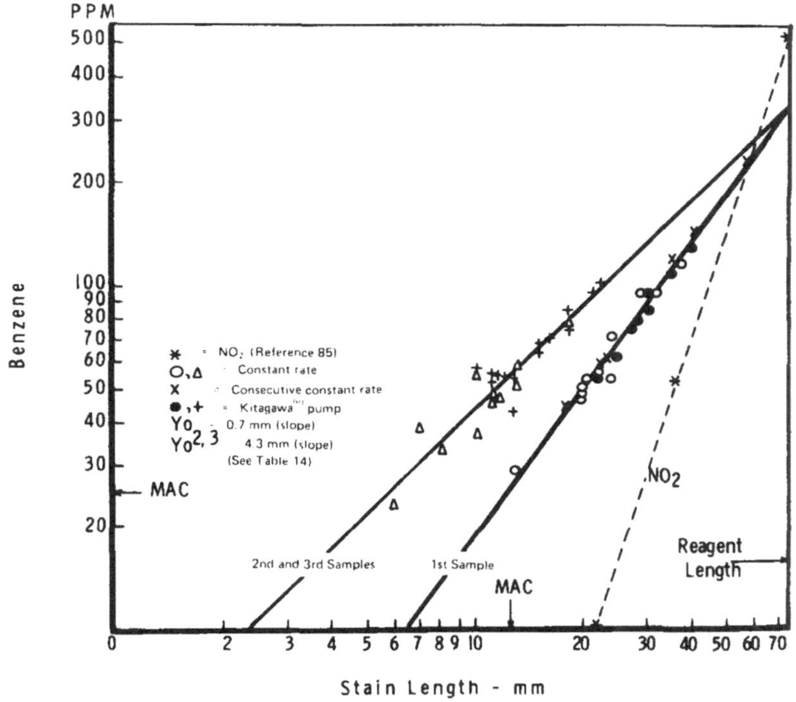

FIGURE 69. Stain length vs. benzene concentration.

and the wide deviation in the results (± 40% at 95% C.L.) renders a correction factor questionable at best. Even when taken in quick succession, a third sample through the same tube may be in error by as much as 20%. If recalibration for repetitive sampling is considered, deviations greater than those experienced with a single sample must be expected. Both graphic plotting and calculation of the slopes, (log Y) = $b_o + b_1$ (log X), confirm a significant difference between the slopes of the second and third delayed aliquot curves and the slope of the initial sample curve (Figure 69). Detailed results are summarized in Table 48.

Some detector tubes have been designed for repetitive aliquot sampling to obtain the desired sensitivity. Reference to available detector tubes offered by the four major manufacturers (Tables 28 to 31) will reveal tubes that require as many as 30 pump strokes (Dräger-ethylene oxide) to produce results in the TLV range. In other words, the sample volume must be orders of magnitude (11 × 30 = 3000 mL) greater than the 100-mL volume obtained with a single pump stroke to produce a readable stain length.

The Dräger Company has designed and promoted an electrically operated pump which duplicates the hand pump action to relieve operator fatigue and attention time involved in this multiple aliquot sampling.[134] The "Quantimeter" must be considered a portable instrument rather than a personnel monitor. The number of pump strokes can be preset so the sampling proceeds unattended. Ambient air as well as industrial atmospheres is recommended for NH_3, Cl_2, CO, mercaptans, nitrogen oxides, O_3, SO_2, and H_2S. An accessory stroke counter for mounting on the hand pump also is available to reduce counting errors in field operations (Figure 85).

Suggestions have been offered to use two different tubes in series, such as passing the air sample first through a hydrogen sulfide tube and then through a phosgene tube to obtain two simultaneous determinations, or to remove interferences.[89] In selected

TABLE 48

Multiple Sampling with the Kitagawa® Benzene Detector Tubes

Tube no.	Pressure drop (maximum) (cm H_2O)	1st 100 mL (ppm)	2nd 100 mL ppm	2nd 100 mL % decrease	3rd 100 mL ppm	3rd 100 mL % decrease	Error[a] (% decrease) 2nd	Error[a] (% decrease) 3rd
21[b]	—	107	105	2	106	1	1	0
15	—	84	66	21	57	32	11	18
9	—	16	14	12.5	16	0[c]	6	0[c]
7	72	67	56	16	52	22	7	13
12	—	79	69	13	57	28	6	14
17	—	78	64	18	58	26	9	14
19[b]	67	87	83	5	70	20	2	8
23[b]	71	128	117	9	—	—	4	—
25[b]	66	56	53	5	49	13	2	5
Average	69	—	—	16	—	27	8	15
Rapid succession[d]				5		11	2	4
S.D.[e]	1.2	—	—	3	—	4	4	2
Rapid succession[d]				3[b]				
S.D. at 95% C.L.	8.5	—	—	37	—	29	—	26

[a] Calculated error based on averaged results; total benzene found was divided by 2nd or 3rd sample taken through the same tube.
[b] Tubes 19 to 25 used in rapid succession.
[c] Excluded from averages.
[d] These averages are from only a fraction of the total results.
[e] Standard deviation.

cases this technique will resolve mixtures, but recalibration to establish the compatibility of the reagent systems and to compensate for the added resistance (Δ P) that will reduce sample flow rate must be carried out before even semiquantitative results can be obtained. In the foregoing example, serious losses of phosgene may occur in the hydrogen sulfide zone. Again, the identity of the reagent mixtures must be known in order to decide whether a given pair of different tubes can be expected to deliver the desired separation (Tables 39 to 42). If a lead compound is incorporated in the gel to remove the hydrogen sulfide, then significant losses of phosgene in this zone would not be anticipated.[97] Also, the granular support may absorb the gas under investigation, i.e., the absorption of arsine on silica gel.[116] Hydrogen sulfide is a common interference in the reaction of mercuric chloride with arsine and must be removed by a "cleansing" layer. The traditional scavenger — lead acetate on silica gel — also removes the arsine. Copper sulfate crystals (40-60 mesh) solve this problem by retaining the hydrogen sulfide without retaining the arsine fraction. Arsine produces no color in the lead acetate tube; therefore, simultaneous determinations can be carried out.[116]

k) Interchangeability of Indicator Tubes and Pumps

The interchange of various manufacturers' gas detector tubes with various manufacturers' sampling pumps can introduce serious systematic (determinate) errors.[118] Each manufacturer's gas detector tube is calibrated for a fixed sampling volume and air flow rate compatible with its sampling pump. Connection of one gas detector tube to some other manufacturer's sampling pump will usually yield only qualitative results. These systematic errors added to the already inherent errors in any given manufacturer's tube and sampling pump will increase the total error beyond the range permitted by the NIOSH certification specification. The manufacturer's gas detector tube and

TABLE 49

Results of CO Determinations Using Commercially Available Indicator Tubes with Various Pumps,[a,b] Summary of Reference Data[82]

	Tube identification — source									
	Bacharach		Drager		Gastec		Kitagawa		MSA	
Pump manufacturer	A.C.[c]	% dev.[d]	A.C.[c]	% dev.[d]	A.C.[c]	% dev.[d]	A.C.[c]	% dev.[d]	A.C.[c]	% dev.[d]
Bacharach — orifice A	81.5	—	65.4	−31.9	62.5	−42.8	62.4	−37.7	112.4	−26.9
S.D.[e]	5.4		4.0		14.0		16.2		11.8	
Dräger	261.7	+221	96.0	—	97.7	−10.5	88.8	−11.3	244.0	+58.8
S.D.[e]	19.4		3.3		4.7		9.4		36.6	
MSA — orifice No. 1	235.3	+189	64.5	−32.9	101.4	−7.1	132.0	+31.9		
S.D.[e]	6.7		11.3		5.2		28.8		24.8	
Orifice No.2	299.7	+268	78.5	−19.3	103.5	−5.2	123.7	+23.6	213.2	+38.7
S.D.[e]	0.9		4.9		4.3		43.0		27.1	
Orifice No. 3	300	+268	87.5	−8.9	112.3	+2.9	154.2	+53.9	339.6	+120.9
S.D.[e]	0.0		4.9		11.7		31.1		22.2	
Orifice No. 4	300	+268	115.8	+20.6	114.3	+4.7	190.2	+90	523.6	+241
S.D.[e]	0.0		3.7		7.2		9.5		72.8	
Kitagawa	156.9	+92.5	63.8	−33.6	92.8	−15.0	100.1	—	135.0	−12.2
S.D.[e]	21.2		3.3		10.6		26.8		7.2	
Gastec	285.7	+251	115.9	+20.7	109.1	—	184.0	+83.8	168.8	+9.8
S.D.[e]	22.9		7.6		8.5		30.8		71.2	

[a] Nominal carbon monoxide concentration = 100 ppm.
[b] Ten determinations (tubes) for each tube/pump combination.
[c] A.C. = Average concentration in ppm.
[d] % dev. = % deviation from specified use value.
[e] S.D. = standard deviation of 10 test results.

pump must be regarded as a single unit and components should not be interchanged with any other unit. To do so invalidates the detector tube calibration, unless the user recalibrates the "hybrid" assembly for use otherwise. An example of the errors to be expected are illustrated in Table 49.

vii. Concentration of Reagents

In general, as the concentration of color-forming reagent adsorbed on the substrate gel particles decreases, the length of the stained zone increases. However, the color becomes lighter and the stain front more diffuse. Therefore, a compromise between improved sensitivity gained by reduced reagent concentration and inaccuracies involved in determination of the zone boundary must be reached. The relationship for an arsine detecting system (mercuric chloride) is presented in Figure 70.[116]

viii. Constant Rate Sampling

Introduction of the Kitagawa gas detector in 1958 provided a wide range of significantly improved, granular reagent-filled detector tubes for quantitative analysis of the most frequently encountered airborne toxic gases and vapors by stain length measurement. Reproducible sample volumes and flow rates supplied by the 100-mL, single-stroke, manual piston pump fitted with a rate-limiting orifice assured volumetric precision. The increased utility attainable by extending the sampling period beyond the 3-min grab sample limit and by improving the sensitivity by increasing the sample volume soon became apparent. Consideration of the critical relationship between sample flow and stain development rates clearly indicated that recalibration or delineation of time-rate factors to produce results equivalent to existing calibration charts would be necessary.

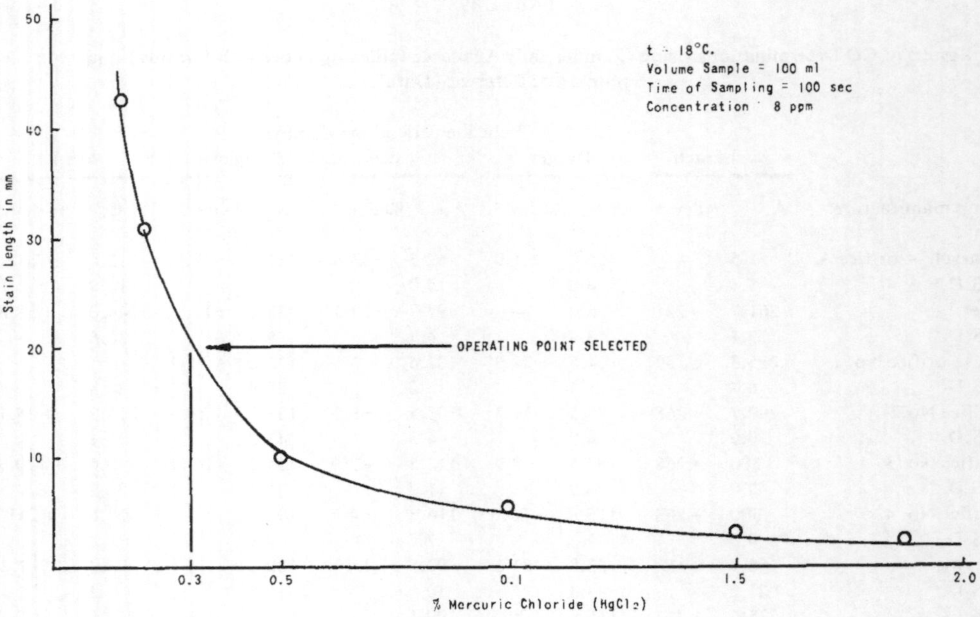

FIGURE 70. Arsine detector tube — relationship of reagent concentration to stain length.[116]

Adaptation to a readily portable, self-powered pump unit from a manual operation offered the most practical approach to extended interval sampling. A sampler that derived its power by expansion of dichlorodifluoromethane or chlorotrifluoromethane through an aspirator without generating static spark potential was chosen on the basis of surge-free constant airflow at low sampling rates and suitability for use in explosion hazard areas.*[135-139]

Benzene, which had presented industrial hygiene exposure control problems, was selected on the basis of what appeared to be easily prepared standard air mixtures and because it is an example of a reagent system (formaldehyde and sulfuric acid) especially sensitive to sampling rate variations. The following procedure provides the reader with one approach to recalibration for extended periods at constant flow rates. The airflow characteristics of the Kitagawa hand pump (HP) have been described in Section II.A.1.b.vi.c, "Packing Variations — Pressure Drop Effects," and the effect of variable flow rate on stain front velocity in Section II.A.1.b.vi.d, "Sampling Rate." A low range rotameter was inserted in the vacuum line between the pump and the detector tube, and fine control was provided by a needle valve built into the rotameter body. The rotameter was calibrated by the soap bubble technique.[126] In passing, a "trick of the trade" is worthy of note: the soap film can be stabilized by adding 10% glycerine and 3% gelatin to the soap solution. Quite durable bubbles can be produced from this mixture.

The mixing chamber and the sampling trains are shown in Figure 71. Detector tubes were held in place by rubber tubing sleeves through which the detector tubes were inserted to a depth that provided an essentially glass-to-glass connection. The rubber stopper was covered completely with aluminum foil. Two midget impingers, fitted with glass ball joints and packed with glass beads, were connected in series. The impingers

* Since completion of the initial study of the adaptation of detector tubes to constant rate sampling, the manufacture of the Uni-Jet® constant rate sampler was discontinued in 1964. In its place a miniature battery-operated piston pump was developed and is currently available (see Chapter 2, Volume II).

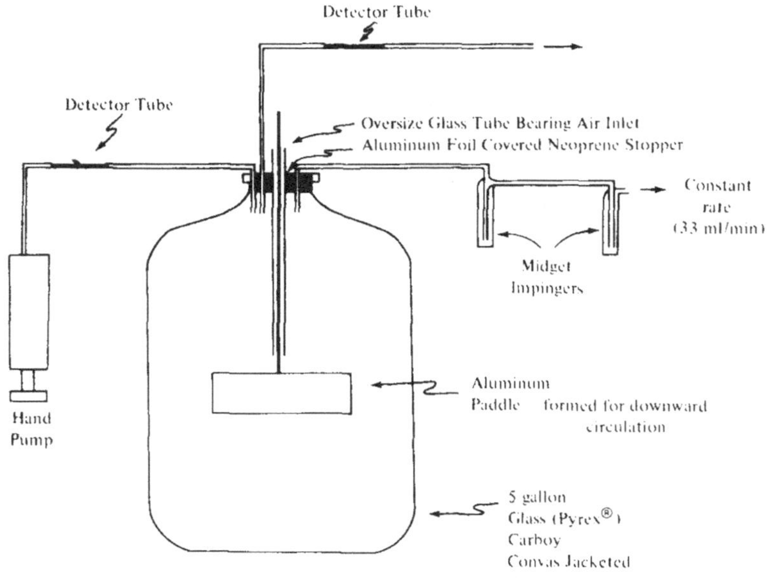

FIGURE 71. Apparatus for calibration of detector tubes on the Uni-Jet® constant rate air sampler.

were immersed in a low temperature bath composed of a freezing mixture of liquid nitrogen and ethanol (− 125°C).[136] In each freeze-out collection, 10 mL of isoocatane was added to the cold impinger as soon as the run was finished to avoid bezene losses, and the optical density determined in 10-mm silica cells at 254 nm. The detector tubes were selected in pairs for identical reagent packing lengths.

A borosilicate glass carboy was conditioned before use by overnight equilibration with several hundred ppm of benzene. The excess benzene was removed by sweeping with fresh air for 20 min. Sampling ports and the agitator bearing were then sealed off and the carboy partially evacuated. The calculated amount of benzene in isooctane (302 mg benzene per mL isooctane) was introduced by a 0.02-mL pipette through a hypodermic needle into a sampling port. After equalizing to atmospheric pressure, the benzene-air mixture was agitated vigorously for 5 to 10 min.

Three samples were withdrawn simultaneously: one through a detector tube for the required 3 min by HP; a second through another tube for the time required for 100 mL, or a 3-min interval at constant rate; and the third 100-mL aliquot through the midget impinger train at a rate of 33 mL/min. Isooctane solutions were analyzed spectrophotometrically at 254 nm to detect benzene. The stain lengths were measured at 30-sec intervals.

A stain length equivalent to one stroke of the HP was obtained in 3 min at a constant sampling rate of 46 mL/min (Figure 64 and Table 50, column 8). Close control of aspirating rate, within ± 6% (± 2.5 mL/min), was required to maintain an overall precision of ± 10% (Figure 67). Operation within these limits required recalibration of the rotameter to compensate for the pressure drop across the detector tube. The sample volume required at this constant rate increased to 138 mL. On an equivalent 100-mL HP volume basis, extrapolation of the calibration deviation curve (Table 50, column 11) indicated that an equivalent stain length would be obtained in 1.2 min at a sampling rate of 85 mL/min (Figure 67). This linear adjustment curve had a much more favorable slope than the 3-min time curve; therefore, considerably greater latitude in flow rate deviation could be tolerated. The estimated pressure drop (69 cm

TABLE 50

Hand Pump (HP) vs. Constant Rate (CR) Sampling

	Hand pump (HP)			Constant rate (CR)			% Deviation on time basis (3 min/sample) [a]			% Deviation on volume basis (100 mL) [b]					
Tube no.	Stain (mm)	ppm	Error ±% [c]	Tube no. [d]	Flow rate (mL/min)	Pressure drop (cm H_2O)	1st	2nd	3rd	1st	2nd	3rd	Mass rate (μg benzene/min) [e]	Time (min) [f]	Vol (mL) [g]
21	35	107	5	22	20	14	−56	−65	−78	−28	−55 [h]	—	6.8	8.3	166
15	30	84	5	16	32	—	−42	—	—	−24	—	—	8.6	4.3	138
7	25	67	6	8	36	20	−16	—	—	−22	−31	−45	7.7	3.4	122
12	28	79	5	11	41	31	−15	—	—	−20	−40	−52	10.3	3.6	148
25	22	56	7	26	46	34	+8	−21	−32	−18	−33	−55	8.2	2.5	115
17	27	78	5	18	46	—	+9	−17	−30	−19	−31	−45	11.6	2.5	115
19	30	87	5	20	46	31	−2	−18	−28	−18	−33	−46	12.8	3.2	147
23	40	128	3	24	46	33	−8	−47 [h]	—	−21	−55	—	18.8	3.5	161
Average	—	—	—	—	46	—	+2	−19	−30	−19	−32	−49	—		
	—	—	—	—											
S.D. [i]	—	—	—	—	46	—	9	2 [j]	2 [j]	2.0	1	—	—		
	—	—	—	—	All	—	—	—	—	3.4	4	4	—		
S.D. at 95% C.L.	—	—	—	—	46	—	—	1	7	11	3	—	—		
	—	—	—	—	All	—	—	—	—	16	12	8	—		

[a] % difference in stain lengths at end of 3 min.
[b] % difference in CR vol vs. 100 mL (HP vol).
[c] Error — % difference vs. calculated ppm benzene in calibration mixture.
[d] CR tubes 20 to 26 — continuous sampling.
[e] Mass rate $= \frac{\text{ppm} \times 0.00319 \times 100 \times \text{mL/min}}{1000} = \text{ppm} \times 0.00319 \times \text{mL/min}$.
[f] To produce stain length equivalent to HP result.
[g] Vol required to produce stain length equivalent to 100 mL by HP.
[h] Not used in averages.
[i] S.D. = standard deviation at C.L.
[j] 46 mL/min rate.

H_2O) would pose no problems. However, this alternative was not confirmed by laboratory trials because the experimental phases of this project were curtailed.

When the constant sampling rate was reduced to 20 mL/min, benzene concentrations, after a 3-min sampling period, could be obtained by doubling the observed HP chart reading. However, no correction was required when the sampling time was extended to 9 min (Figure 68). The ppm benzene readings on the right-hand ordinate axis are equivalent to the stain lengths taken from the HP calibration chart furnished with the detector tubes (Figure 64). Some sensitivity (mm stain length per ppm benzene) was lost, but for most purposes the results were satisfactory within a ± 10% range for time intervals of less than 5 min. For extended periods, 9 min or more, sensitivity actually improved. The stain front development rate passed through an inflection point at between 3 and 4 min to a significantly lower rate. Another tube exposed to the same benzene in air mixture at 46 mL/min did not exhibit such an inflection (Figure 68). Results that required only 15% correction were obtained in 3 min at the flow rate of 46 mL/min (Figure 68 and Table 50).

Increased sensitivity was obtained by operating at a 46 mL/min constant rate for periods up to 10 min. Although the 9-min results averaged 15% less than three times the 3-min value, the much greater stain length compensated for the observational error in locating the color front; and at low benzene concentrations, the averaged value would be more reliable than a single 3-min or 100-mL analysis. However, additional data is needed to determine whether or not the correction factor developed for the 50- to 150-ppm range is directly applicable to the 10- to 40-ppm region. Better accuracy in the preparation of the standard benzene in air mixtures and more extensive calibration undoubtedly would reduce the uncertainty factor to within the desired ± 5% range. This technique certainly offers the most promise for increasing the reliability, sensitivity, and accuracy of granular reagent tubes for application to personnel monitoring.

Plots of the data as a log-log function of stain length vs. benzene concentration (Figure 69) disclosed a significant difference between the slope of the original Kitagawa calibration and the slope of the second and third aliquot results from the same detector tube after a lapse between analyses of 5 min or more. Since this slope was unity within reasonable limits, this reversion suggested complete sorption with a stoichiometric relation between quantity of sorbed gas and volume of the colored indicator that was independent of gas concentration.[131] Results reported by Kinosian and Hubbard for nitrogen dioxide have been included for comparison with an entirely different system (Figures 68 and 69).[130]

Adsorption of benzene from the standard mixtures in air by the Pyrex® glass container walls occurred at a rate that reduced the benzene concentrations in the 100- to 200-ppm range by as much as 50% in less than an hour (Table 51). Only by equilibration overnight with mixtures that approximated the concentration required could reasonably stable conditions be attained. Simultaneous determinations with hand pump, constant rate, and control analysis were carried out to provide comparative results to minimize this variable.

ix. Reagent Systems — Identification

In the foregoing sections only a few of the better known colorimetric reagents that have been developed to detect trace materials that could be present in the industrial atmosphere have been presented. Selection was based on availability, familiarity, extent of field application, published performance evaluations, and consensus standards. However, many potentially valuable detectors have been described in the trace chemistry literature. Campbell and Miller published a comprehensive review of this literature in the form of annotated abstracts.[140] Although this publication has long been

TABLE 51

Benzene Loss by Adsorption on Container Walls

Tube no.	Benzene added (ppm)	Benzene found (ppm) Hours elapsed	Kitagawa®	Cold trap	% Loss	Remarks Temperature = (75 ± 5°F)
6	100	0.5	56	—	44	
7	100	0.5	67	—	33	
9	100	16	16	—	84	
10	—	192	0	—	100	
11	200	16	100	—	50	
12	100	0.5	79	—	21	Flask conditioned overnight
15	100	0.5	84	80	16	Flask conditioned overnight, 100 ppm
17	—	1	78	80, 88	25	No additional benzene added
19	100	0.5	87	—	13	Flask conditioned overnight, 200 ppm
21	100	0.5	107	—	−7	Flask swept; 2nd 100 ppm added
23	200	0.5	128	—	36	100 ppm added after tube 21
25	50	0.5	56	—	−12	Flask swept; then 50 ppm benzene added
—	100	0.1	75	—	25	No conditioning for 30 days
		0.25	72	—	28	
		0.5	55	—	45	
		1	55	—	45	
		1.5	60	—	40	
		4.5	48	—	52	
		7.5	63	—	37	
		24	42	—	58	

out of print, the contents have been made available by the commercial press.[83]

A description of the preparation, calibration, and use of detector tubes for phenol, formaldehyde, SO_2, and NH_3 that might not ordinarily come to the attention of the field inspector would add two additional (phenol and formaldehyde) capabilities to his surveys.

The tubes are suitable for measurements up to 40°C, so that waste gases above this temperature must be cooled before being tested. Phenol, formaldehyde, sulfur dioxide, and ammonia are the substances of principal interest in connection with oven drying. Test tubes for phenol show a blue discoloration and have a measuring range up to 5 ppm. For formaldehyde, the discoloration is reddish, and the measuring range of the tubes is 2 to 40 ppm. The results obtained with this method in a plant for sheet metal packing materials coated with varnish, where oven drying and catalytic afterburning are included in the operation, are reported by Merz.[141]

The industrial hygiene investigator must identify the identity of the colorimetric reagent incorporated in the detector tube chosen for an industrial survey. Failure to recognize this critical factor can lead to very serious errors produced by either tube deficiences or co-existing interferences. Few tubes are specific for the trace gas or vapor for which use is intended. For example, more than 30% of all detector rubes are based on the reduction of either a chromate salt or chromic acid to a differently colored (green) chromium salt.

This deficiency has provided the basis for a qualitative test colorimetric tube by the Dräger Company for warning of the presence of a number (16) of hazardous gases in confined spaces such as manholes and sewers (Table 52).

The following procedure has been recommended.

1. Determine the carbon dioxide concentration with the Dräger detector tube (Table 40).

TABLE 52

Color Reactions Obtained with Drager Polytest Detector Tube

Vapor or gas	Concentration (ppm)	Color reaction
Acetone	2000	Brown to green
Acetylene	10	Brown to green
Arsine	1	Green
Benzene	50	Brown
Carbon disulfide	1	Green
Carbon monoxide	5	Green to brown
Ethylene	50	Green
Hydrogen sulfide	2	Green
Liquid petroleum vapors (propane, butane)	100	Green
Styrene monomer	10	Brown to green
Nitric oxide (NO)	20	Green
Perchoroethylene	20	Green
Petroleum distillate (gasoline)	50 $\mu g/m^3$	Brown to green
Toluene, xylene	10	Violet
Producer gas (2% + CO)	100	Brown to green
Trichloroethylene	10	Green

From National Dräger, Inc., Parkway View Dr., Pittsburgh, Pa. With permission.

2. Determine the oxygen concentration with the Dräger colorimetric length of stain tube or an electronic detector (Table 7).
3. With the "Polytest" detector, determine the presence or absence of one or more of the gases or vapors listed in Table 41. This test reuires five strokes of the bellows pump. A positive result (color formation) provides adequate warning of the presence of a hazardous environment. The sensitivity of the detector for the presence of these contaminants provides confidence in the safety of negative (no color change) results (absence of any of the 16 compounds listed).
4. If a positive indication is obtained, then a process of elimination will provide a clue to the identity of the offending agent by additional detector tube analysis: a) test for carbon monoxide, b) test for hydrogen sulfide, c) test for nitric oxide (NO) and nitrogen dioxide, d) test for combustible gases with any reliable combustible gas meter (see Table 7).

The positive results should fall within one of these four groups. If only combustible gases in class d are found, further analysis will be necessary for identification by laboratory analysis (charcoal or silica gel collection and GC analysis). The "Polytest" tube does not detect methane, ethane, carbon dioxide, sulfur dioxide, or hydrogen. Most reducing agents will produce a positive reaction. The same sulfuric acid-formaldehyde reagent system is employed for the detection of benzene, toluene, and xylene. A true reading will not be obtained if two or all three are present together, or if homologues such as trimethyl benzene, cumene, styrene, or other alkylated benzenes are also present (see Tables 39 to 42).

Negative interferences also must be anticipated. For example, in the presence of ammonia, mercury vapor produces no color change in the mercury vapor detector tube. Other interferences are well enough recognized to justify warnings in the instructions for the proper use of the detector tubes. In some cases, guard layers are incorporated directly in the detector, in the detector tube, or furnished as an accessory to be attached to the inlet end of the detector tube. Examples include ethylene in the presence of CO, nitrogen oxides present with CO, SO_2 in the presence of H_2S, benzene in mixtures with other hydrocarbons (see Tables 28 to 35), and removal of water from systems which are humidity sensitive (Table 46). A limited number of tubes incorporates a preliminary reaction zone to convert the pollutant to a detectable derivative. The chlorinated hydrocarbons are representative. The preliminary reactive usually involves a strong oxidizing agent to convert combined chlorine atoms to elemental chlorine which produces a color in the following color-forming reagent (dianisidine or tolidine). Multilayer tubes frequently possess a short shelf life because of diffusion between the two layers.

The investigator must determine before sampling what components would be expected to be present from an inventory of chemicals entering the area under investigation (see "Paint and Varnish Industry Survey"). However, in preliminary surveys, a tube containing a reagent which reacts with a family of organic vapors such as the sulfuric acid-formaldehyde reagent for aromatic hydrocarbons can be used if the stain length is read in terms of the most hazardous member (benzene) of the mixture anticipated. If the results are above the action level, then a follow-up (evaluation) survey based on a specific, quantitative method (charcoal tube collection and GC analysis) must be carried out.

Some detector tubes can be recalibrated in terms of some other contaminant. For illustration, the Dräger methyl bromide detector tube (5/b) has been calibrated for chloroform, 1,1-dichloroethylene, 1,2-dichloroethane, and 1,2-dichloropropane.[142] The revised conditions (pump strokes) and calibrations are summarized in Table 53. In this case, the incoming vapor is oxidized to produce Cl_2 which is detected in the following detector tube (*o*-tolidine). Following are some other examples.

Desired component	Recalibrated tube	No. pump strokes	Range (ppm)
Dioxane (32)	Ethyl glycol acetate, 50/A	10	12—180
Acetic anhydride (37)	Acetic acid, 5/a	3	20—320
Hexane (37)	Cyclohexane, 100/a	10	100—1500
Cumene (36)	Toluene, 25/a	10	25—1860
Toluene (36)	Toluene, 25/a	50	5—372
Benzyl chloride (37)	Vinyl chloride, 1/a	20	1—10
		5	5—50
Acrolein (37)	Formaldehyde,0.5/a	5	1—20

Notes: The number in parenthesis is Dräger Review volume number; color shade comparison chart for acrolein.[120]

A detector tube for the detection of methyl isocyanate has been developed by private investigators.[143] The color forming reagent paraaminoazobenzene was supported on silica gel and changes color from orange to yellow on contact with methyl isocyanate vapor in concentrations above 1 ppm. Heating is required to accelerate the reaction. Alkyl isocyanates can be differentiated from aromatic isocyanates by passing HCl vapor through the tube after exposure. The yellow zone produced by methyl isocyanate changes from the yellow color to red-orange, whereas an aromatic isocyanate (phenyl isocyanate) yellow zone will turn violet. The unreacted zone changes to red-orange. Ammonia and methylamine interfere (orange to yellow color change). At 115 to 125°C (electric heater coil) 60 mL of sample must be drawn through the tube to attain sensitivity in the 1- to 30-ppm range.

x. Testing Gas and Vapor Detector Pumps

The critical factors governing the response of chemical detector tubes are (1) the volume of air sampled (contaminant mass, or concentration) and (2) the airflow rate

TABLE 53

Recalibration of the Drager 5/b Methyl Bromide Detector Tube for Chlorinated Alkyl Hydrocarbon Vapors

Compound	Formula	TLV (ppm)	No. of strokes	Concentration (ppm)								
				20	25	50	75	100	150	200	250	300
Chloroform (trichloromethane)	$CHCl_3$	25	10		ca.8	ca.10		ca.20	ca.30	ca.40	ca.50	
1,1-Dichloroethylene (vinylidene chloride)	$CH_2:CCl_2$	200	3					ca.7	ca.14	ca.21	ca.28	ca.35
1- 2-Dichloroethane (ethylene dichloride)	$CiCH_2 \cdot CH_2Cl$	50	20	ca.5				ca.10		ca.30		ca.50
1-2-Dichloropropane (propylene dichloride)	$CH_3 \cdot CHCl \cdot CH_2 Cl$	75	10			ca.10	ca.20	ca.40				

(contact time with the detecting reagent). Therefore, the detector pump must be calibrated when procured and periodically thereafter. Three common types of pumps are available: (1) squeeze bulbs with or without limiting orifices,[89] (2) bellows-type pumps with limiting orifices (Figures 55 and 56), and (3) precision piston pumps with limiting orifices (Figures 53, 54, 58, and 59). Only types (2) and (3) are recognized as permissible by OSHA.

Two procedures are presented: (1) testing for leaks and (2) determining airflow characteristics by measuring the volume of air sampled and the flow rate. In leak testing, the manufacturer usually describes the recommended procedure in his instructions. In general, to test for leakage, a sealed detector tube is placed in the pump inlet port and the time to reinflate the bulb or bellows, or to reach atmospheric pressure with a piston pump, is determined.

The tentative "Detector Tube Certification Schedule" proposed by the National Institute for Occupational Safety and Health delineates pump performance as follows:

(1) Aspirating pumps for drawing the air to be sampled with detector tubes shall be free from leakage which would result in erroneously low tube readings, and shall be calibrated by the manufacturer to sample at whatever flow rate is deemed appropriate by the manufacturer in order to assure accurate measurements using the respective manufacturer's detector tubes.

After the pump has been evacuated, leakage shall not be more per minute than 2% of the pump's volume capacity in a single stroke, if a sealed detector tube is used to plug the pump inlet.

The flow rates which are set by the tube manufacturer are regulated by limiting orifices or other restrictions of flow at the pump inlet. These flow control devices shall regulate the flow rate to within ± 5% of the rate stated by the manufacturer for each flow rate control device. The choice of the flow rate into the pump is the manufacturer's prerogative.

Leakage test and flow rate tests shall be made at room conditions of temperature, pressure, and humidity.

Pumps shall be calibrated by the manufacturer to insure that they are capable of sampling accurately the stated volume at the stated flow rate. Subsequent to a check of proper flow rate and volume, the pump shall be capable of drawing 100 full-capacity strokes of air without deviating more than ± 5% from the calibration flow rate.

(2) The pump shall be capable of drawing the volume stated by the manufacturer to within ± 5% of said volume. The pump shall be able to continue to draw this volume throughout its normal working lifetime, with minor repairs only.

(3) The pump shall be capable of drawing ten consecutive strokes using one detector tube with no more than ± 5% error from the theoretical sampling volume.

(4) The manufacturer shall provide printed instructions with each detector tube pump so that users may verify by a simple field test the absence of leakage and capability for accurate flow rate sampling. The manufacturer shall also provide instructions for making simple adjustments to the pump and replacing minor components.*

In the field the test may be carried out rapidly after inserting an unopened detector tube into the inlet port of the piston pump by pulling the handle back completely, locking in position by rotation of the handle, and then 3 min later releasing the piston cautiously. The piston should return to within 5% of its original totally advanced position, i.e., out no more than 5% of original withdrawn position. In a 100-mL pump this would be equivalent to 5 mL leakage in 3 min. If the residual volume exceeds 5 mL, replacement of check valves, tightening connections (especially the inlet port-limiting orifice assembly), replacement of tube connections, greasing piston, or advancing the piston expansion nut (Kitagawa pump) is indicated. This test should be performed prior to each use of the device.

In normal use, the pump should be taken apart, cleaned and re-lubricated after about 100 to 200 samples have been drawn. To disassemble, unscrew the cylinder end cover and slowly pull the pump handle unitl

* From Roper, P., Detector Tube Certification Schedule, first draft, NIOSH, Health Services and Mental Health Admin., Public Health Service, Department of Health, Education, and Welfare, Cincinnati, Ohio, June 1971; *Fed. Register,* 37 (No. 84), 19643, 1972; 42CFR, part 84.

the piston leaves the cylinder. Remove the inlet and pressure-relief valve from the front end of the pump, and clean all the bearing surfaces carefully using non-abrasive wipers and a good degreasing solvent such as trichloroethylene or perchloroethylene. Lubricate all surfaces which make metal-to-metal contact with the special silicone grease supplied with the pump. Reassamble the pump carefully, making sure that abrasive particles which could score the walls do not get inside the pump.

If excessive leakage is found, it usually takes place either at the pump inlet, or between the piston and cylinder walls. The latter source of leakage is much less likely than the former, and can usually be eliminated simply by cleaning and relubrication. Leakage at the inlet may result from a poor seal between the detector tube and the rubber inlet tip, or between the flange of the rubber inlet and the pump body. To check for leakage at the tip, reposition the sealed detector tube or replace it with another one, and repeat.... If the leakage persists, it is probably at the second location, and simply tightening down the inlet clamping nut may suffice to eliminate it. If not, remove the rubber inlet and examine it for cracks and/or foreign matter on its bearing surface. If it is clean and undamaged, replace it in the pump inlet, and repeat the leak test. If necessary, replace the worn rubber inlet with a fresh one.*

In some instances, the manufacturer's instructions include measurement of flow rate and the volume of air sampled by their equipment. These procedures may furnish only a rough estimate of performance. A more accurate method of measuring these parameters, a simple procedure employing a soap bubble flowmeter, follows. This calibration should be performed when the equipment is procured and at frequent intervals during use thereafter. When experience has been gained with a particular detector kit, the user will recognize changes in performance that indicate a need for rechecking sampling characteristics.

Basic equipment required consists of a 100-mL laboratory burette, several short pieces of rubber and glass tubing, a solution of synthetic liquid detergent in water, and a stopwatch. Since this apparatus will be used, in some cases, to measure air volumes slightly greater than 100 mL, it is necessary to increase the scale on the burette by 6 to 8 mL in increments of 1 mL. This is done by filing the burette with water to the 100-mL mark adding successively 1 mL of water from a 1-mL volumetric pipette, and marking the burette after the addition of each 1-mL portion.

A solution of synthetic liquid detergent and water is prepared, and the foam is allowed to subside. Soap films may be stabilized by adding 10% glycerine and 3% gelatin to the soap (1%) or synthetic detergent. The solution is poured into the burette carefully to prevent additional foaming, and then poured out in a manner that will coat the inner surface of the burette with a thin film of the solution. The open end of the burette is placed momentarily in contact with the surface of the detergent solution. When the burette is withdrawn, a soap film will adhere to the mouth of the burette. Upon the application of suction to the tip of the burette, the soap bubble can be drawn through the length of the burette. In the calibration procedure, start the bubble at a reference point, draw air through the burette by means of the detector pump being tested, and note the total volume after the bubble has come to rest. The filling time of the pump can be determined also by measuring with a stopwatch the time it takes the bubble to move from the reference point to the rest position. Further details have been described in the technical literature,[126] and several commercial sources supply both the glassware and complete kits in size ranging from 2 mL to 2,000 mL (Figures 72 and 73).[127-129, 144]

It can be determined from the manufacturer's instructions whether or not the detector pump should be calibrated with an open detector tube in place. In general, pumps equipped with limiting orifices may be calibrated by the soap bubble technique without a detector tube. Where the rate of airflow is governed only by the resistance to flow of the detector tube, the device should be calibrated with the tube in place. With some pumps, it will be necessary to use short pieces of glass and/or rubber tubing to connect

* From Kohn, R. J., Unico® Carbon Monoxide Detector Kit (Model 400 Precision Gas Detector Pump with Model 100 Carbon Monoxide Detector Tubes), Bendix Corp., Environmental and Process Instruments Division, Lewisburg, W.Va.

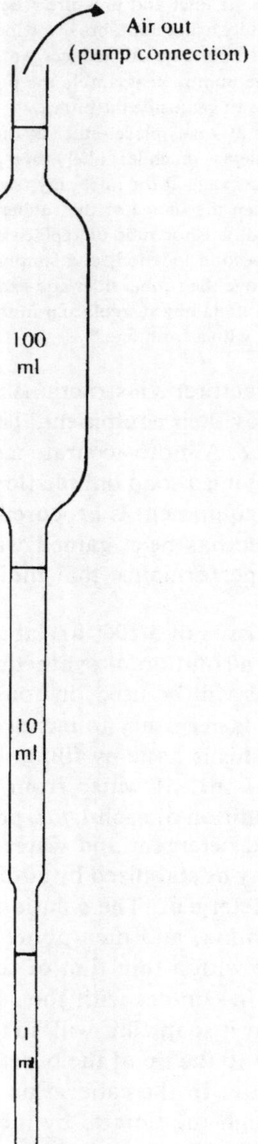

FIGURE 72. Soap film flowmeter — representative design. (Courtesy of Bubble-O-Meter,® Soap Film Flowmeter, Box 338, Temple City, Calif.)

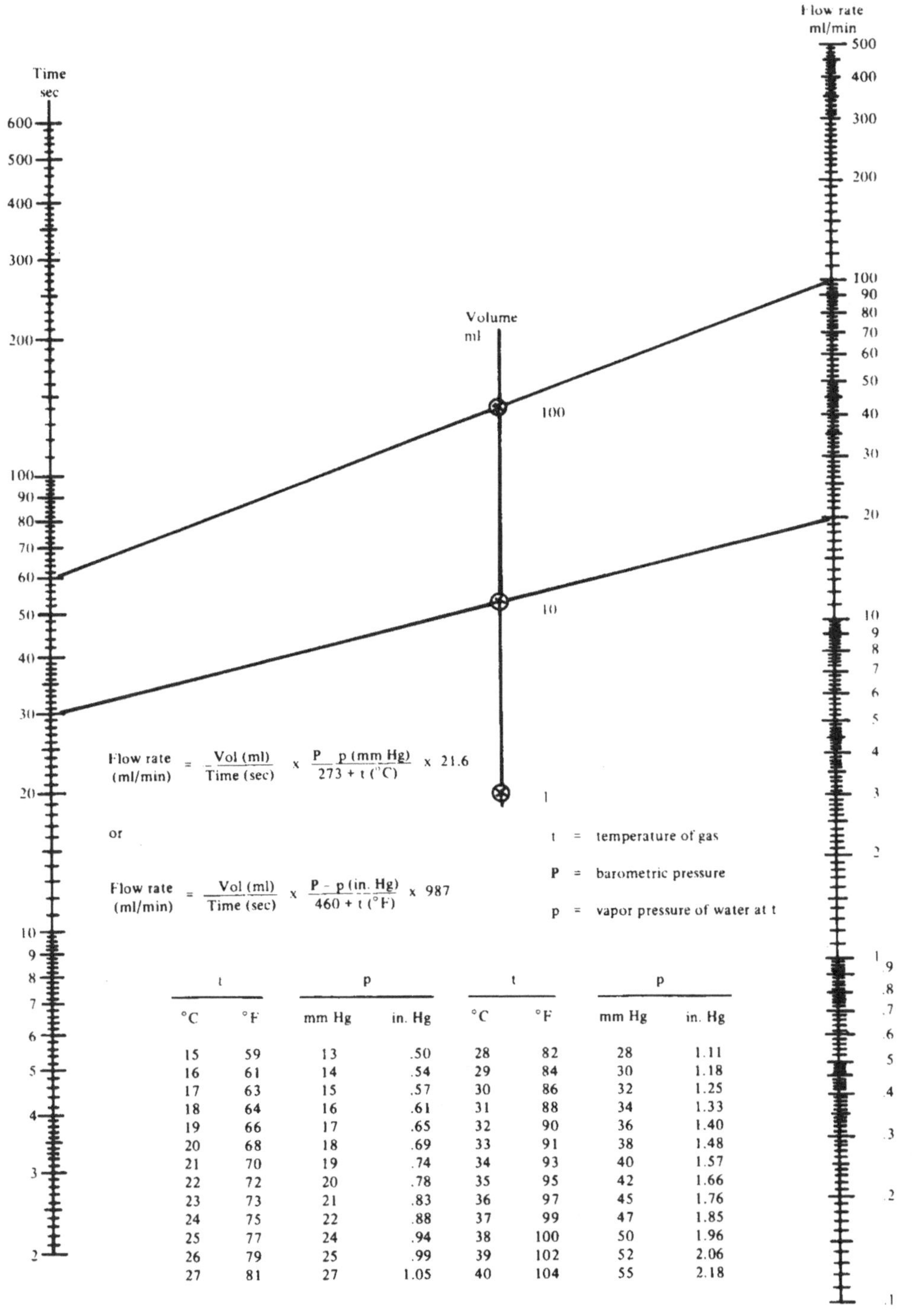

$$\text{Flow rate (ml/min)} = \frac{\text{Vol (ml)}}{\text{Time (sec)}} \times \frac{P - p\ (\text{mm Hg})}{273 + t\ (^\circ C)} \times 21.6$$

or

$$\text{Flow rate (ml/min)} = \frac{\text{Vol (ml)}}{\text{Time (sec)}} \times \frac{P - p\ (\text{in. Hg})}{460 + t\ (^\circ F)} \times 987$$

t = temperature of gas

P = barometric pressure

p = vapor pressure of water at t

t		p		t		p	
°C	°F	mm Hg	in. Hg	°C	°F	mm Hg	in. Hg
15	59	13	.50	28	82	28	1.11
16	61	14	.54	29	84	30	1.18
17	63	15	.57	30	86	32	1.25
18	64	16	.61	31	88	34	1.33
19	66	17	.65	32	90	36	1.40
20	68	18	.69	33	91	38	1.48
21	70	19	.74	34	93	40	1.57
22	72	20	.78	35	95	42	1.66
23	73	21	.83	36	97	45	1.76
24	75	22	.88	37	99	47	1.85
25	77	24	.94	38	100	50	1.96
26	79	25	.99	39	102	52	2.06
27	81	27	1.05	40	104	55	2.18

FIGURE 73. Nomograph for computing flow was using soap film flowmeter. Nomograph gives flow rates at ambient temperature and pressure, uncorrected for the vapor pressure of water. To correct observed values to standard conditions of temperature (0°C, 32°F) and pressure (760 mm Hg, 29.92 in. Hg), use either of the equations and the table of vapor pressures.

the air intake of the device to the burette. In some devices the major resistance to the airflow is in the chemical packing of the tube (Figures 55 and 56). Therefore, each batch should be checked before use. Error in the flow rate indicates a partially clogged strainer or orifice which should be cleaned or replaced.

The flow rate for bellows and squeeze bulb type pumps can be checked expediently in the field simply by timing the period for complete expansion. The rate is quickly calculated by dividing the known volume by the time observed. Alternatively, a used detector tube may be mounted in the holder and connected by a short piece of rubber tubing to a low flow, precision rotameter, and the air sample then drawn through the assembly. The rate than is read from the calibration chart furnished with the flowmeter. One particular model (Visitrol®) is provided with 6 interchangeable tubes to cover the range 5 to 140,000 mL/min with stainless steel, or borosilicate glass floats. The unit may be rotated 180° to permit reversal of flow through the system. Precision has been rated at ± 5%.[145]

With the piston-type pump, a field check for flow rate can be made as follows:

a) With no detector tube in the pump inlet, pull the handle all the way out and lock it.

b) Wait exactly 30 seconds. Release the pump handle. When the handle is released, it should spring back part way toward its initial position. It will come to rest at an intermediate position at which the pressure on both sides of the piston will be equal. If the pump is operating properly, and the orifice at the inlet is free of obstructions, approximately 35-40 cc of air should enter the pump in 30 seconds. The volume that entered can be read from the markings on the pump handle. If less than 35 cc was drawn in, it will be most likely due to the orifice being blocked by a stray particle or fiber. It will usually be possible to dislodge this foreign matter simply by flushing the pump several times with air. If the obstruction remains, the orifice must be taken out and be cleaned or replaced with a clean one.

The leakage and flow-rate tests just described should be performed whenever the pump is used after an extended idle period, and in routine use should be repeated after every 10-12 samples to insure accuracy and reliability.*

Restriction in either the orifice or the detector should be checked. The first can be determined by retracting the piston and locking without a detector tube in place. When the handle is rotated off of locking position, no residual vacuum should remain, i.e., the piston will not voluntarily move back into the pump (less than 100 mL sample indicated on the shaft) after 3 min. A defective indicator tube which has high internal pressure drop will exhibit the same residual vacuum after 3 min. For example, a CO tube under test indicated that only 90 mL of a 49-ppm CO in-air standard mixture had been drawn through the tube. The indicated concentration was 30 and 35 ppm instead of the 45 to 50 ppm found with other tubes of the same lot.

Pumps should be considered defective if either of the following conditions is found: leakage that cannot be corrected by the usêr following the simple procedures prescribed by the manufacturer, or deviation of the airflow characteristicvs characteristics by more than ± 5% from the manufacturer's specifications. New pumps that are found defective should be returned to the manufacturer in accordance with the terms of the warranty.

The use of factors to adjust observed air volumes or rates of airflow to the values stated by the manufacturer should not be considered. This manipulation assumes a linear relationship between air volume and/or rate of airflow and the response of the detector tube. This assumption is either not valid at all or is applicable only over a narrow range of operating conditions that must be determined by recalibration.

* From Kohn, R. J., Unico® Carbon Monoxide Detector Kit (Model 400 Precision Gas Detector Pump with Model 100 Carbon Monoxide Detector Tubes), Bendix Corp., Environmental and Process Instruments Division, Lewisburg, W.Va.

B. Extended Period Sampling — More than 15 Minutes

1. Direct Reading Length of Stain Tubes

The linear relationship between stain length and sampling time at constant sampling rate for benzene (Figure 68) (10 min) indicated that length of stain detector tubes could be recalibrated for long term sampling in the industrial atmosphere. A 1973 report of work conducted in Germany with Dräger tubes confirmed these results.[146] The sensitivity of the H_2S detector tube was increased to 3 ppm for ambient air analysis by drawing 10 L of contaminated air through the detector tube at a constant continuous rate with an aspirator. Numerous discussions relative to the design and production of detector tubes for long-term sampling were held with the major manufacturers of this equipment, Kitagawa, Bendix (Gastex) and Dräger, during the period 1969 to 1974 to promote commercial development of a personnel monitor based on length of stain colorimetric detectors.

During 1968 a study was initiated to recalibrate the carbon monoxide tube (Kitagawa) for use as a personnel monitor to evaluate a potential exposure problem in a large warehouse where gasoline powered fork-lift trucks were in constant use.

The report of this investigation, "Carbon Monoxide Evaluation of Exposure Potential by Personnel Monitor Surveys", by A. L. Linch and H. V. Pfaff is reproduced here by permission of the editor of the *American Industrial Hygeine Association Journal.*

INTRODUCTION

Downward revision of the Threshold Limit Value for carbon monoxide (CO) from 100 ppm to 50 ppm in 1967 by the American Conference of Governmental Industrial Hygienists (ACGIH)[95] and incorporation of the 1967 TLV's into Safety Regulation No. 3, issued December 1, 1967, by the State of New Jersey Commissioner of Labor and Industry,[97] focused attention on an area of potential exposure within the Consolidated Warehouse where a number of gasoline powered fork lift trucks are employed for drum storage within bays, and for loading box cars and truck trailers. Although no health problems had been encountered since completion of this very large building in 1955, reduction in ventilation capacity during cold weather when ventilators and the large loading ramp doors were closed to conserve heat could allow CO to accumulate beyond permissible levels. During January, 1968, fixed station CO alarms (see Figure 74) set to trip at 50 ppm were installed at the two centrally located "write-up" desks located in the general traffic area and in the fluorocarbon shipping areas where maximum exposure to the exhaust emissions from two to four continually circulating fork lift trucks would be expected and where at least one employee was in constant attendance while preparing invoices. Although these alarms did sound occaisonlly, the problem was not considered sufficiently acute to justify more extensive investigation while the more urgent concurrent study of tetraalkyl lead exposure in another area was in progress.[148]

The project was reopened in December, 1968. Frequent sounding of the alarms indicated concentrations of CO above the TLV at these locations for significant periods of time. Grab samples taken with length of stain indicator tubes[149] whenever the alarm sounded disclosed concentrations in the range of 60 to 80 ppm for 26 out of 50 determinations taken during the period 8 a.m. to midnight over a ten consecutive day period. Only once did the indicated concentration reach 100 ppm. In one closely spaced series of five grab samples taken over a two hour period, the alarm tripped once in the 25-30 ppm concentration range, but not below 25 ppm. Since the alarm operates by electronic monitoring of the optical density of the stain which develops from CO exposure in a thin circular wafer of the same reagent used in the detector tube, this activation below the 50 ppm preset limit probably was derived from either an accumulation of stain as the concentration of CO slowly approached 50 ppm or from short term excursions of relatively high CO concentrations just preceding the grab sample. No correlation between traffic patterns, open or closed doors, nor operation of the ventilating fans could be established. However, on an averaged 8 hour work day basis, exposure to CO concentrations above the TLV was not indicated.

To further delineate the probable average concentration range and excursion limits, a recording nondispersive infrared (NDIR) analyzer was assembled in the Instrument Shop and installed at the general shipping area write-up desk. The chart tracing disclosed an extremely variable ambient atmosphere with respect to CO concentration. Not only did 30 minute averages differ by a factor of ten, but the short period excursions fluctuated as much as twenty-five ppm within the response time of the recorder (12 seconds). Three typical nine minute tracings taken from the eight hour night shift recording of May 20, 1969, illustrate this extreme variability in Figure 74.

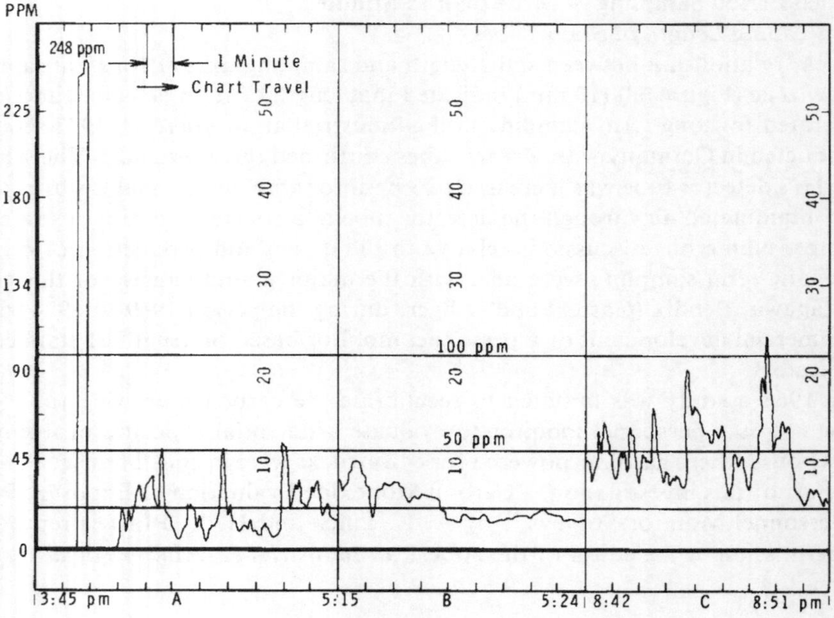

FIGURE 74. Carbon monoxide in air by nondispersive infrared (NDIR) analysis (From Linch, A. L. and Pfaff, H., *Am. Ind. Hyg. Assoc. J.*, 32, 745, 1971. With permission.)

The recorder disclosed by "spike" count CO concentrations which exceeded 50 ppm during 16% of the time and 75 ppm for 2.5% of the same period. The range was 5 ppm to 158 ppm (single 12 second "spike").

Blood analyses by the pyrotanic acid colorimetric method[150] during this period disclosed less than 5% carboxyhemoglobin was present in the blood of personnel selected at random from the crew assigned to this building. The reliability of the colorimetric procedure had been confirmed earlier in our laboratory by comparison with the microgasometric procedure developed by Scholander and Roughton.[151] These results indicated the 8 hour average CO level must have been less than 25 ppm.[152, 153]

By this time the conclusion that only by personnel monitoring could a true exposure level be determined was obvious. This conclusion was reinforced by the results obtained from the concurrent tetraalkyl lead in air study which clearly demonstrated a complete absence of correlation between personnel and fixed station air sampling.[148] Further emphasis on the urgency relative to establishing the true exposure potential was provided by an engineering design estimate for $100,000 to provide additional ventilation required to maintain CO levels below 50 ppm.

Attempts to adopt either the clorimetric aqueous palladium chloride[154] or the silver para-sulfaminobenzoate[155] procedure to the microimpinger ampling technique met with complete failure due to the slow reaction kinetics. Other colorimetric methods were ruled out on one or more counts,[156] and the Hopcalite gravimetric Procedure[157] required an impractically large sample. The highly accurate iodine pentoxide procedure required a reagent temperature of 150°C[156] which obviously cannot be considered for personnel sampling. Likewise, instrumental methods for measurement of physical parameters such as infrared absorption and combustion calorimetry have not been sufficiently miniaturized to permit use as personnel monitors.

A telephone call to Paul Toth at the Ford Motor Company plant in Dearborn, Michigan in early December, 1969 confirmed the soundness of our proposal to attempt recalibration of a length of stain CO detector tube for four-hour sampling based on our earlier experience with a benzene-in-air detector tube. The Dearborn application employed an MSA Catalog No. BY-47134 detector tube recalibrated for a one-hour sampling period at approximately one cubic foot per hour (470 ml/min) rate. Later in the month a reply from an earlier inquiry submitted to the manufacturer relative to the use of the Kitagawa No. 100 length of stain detector tube for personnel monitoring included a calibration chart for four-hour sampling at a 9.6 ml per minute rate. A laboratory investigation was initiated immediately to confirm the calibration furnished by the manufacturer.

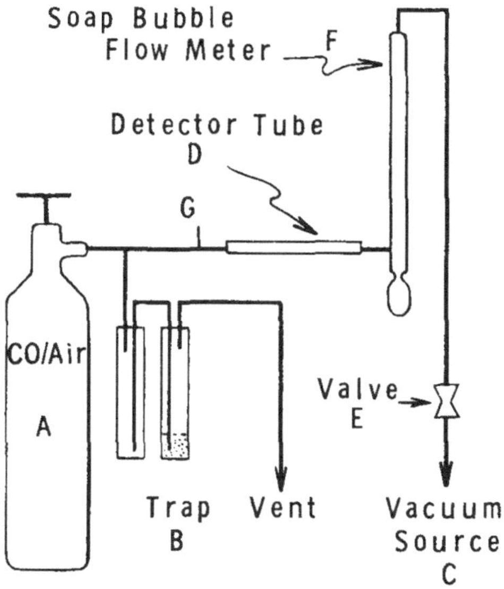

FIGURE 75. Carbon monoxide in air calibration system. (From Linch, A. L. and Pfaff, H., *Am. Ind. Hyg. Assoc. J.*, 32, 745, 1971. With permission.)

PROCEDURE

Calibration — A 10-L (approximate) carbon steel cylinder which had been steamed and dried in accordance with specifications established for packaging refrigerant grade fluorocarbons was evacuated. The calculated quantity of carbon monoxide was injected from a glass syringe through a rubber septum connected to the inlet of the cylinder. The cylinder was then pressurized to 200 psi with purified air. The concentration of CO was confirmed by vapor phase chromatography on a column with the following specifications:

Column Packing	—	13 × molecular sieve
Length	—	4 m
Temperature	—	40°C
Carrier Gas	—	Helium
Gas Flow	—	110 ml/min
Sample Size	—	5.175 ml

The schematic drawing shown in Figure 75 is largely self-explanatory. The valve on the calibration gas supply cylinder A is opened sufficiently to provide a definite gas overflow through trap B and to prevent introduction of outside ambient air into the system by the vacuum source C. After insertion of the detector tube D into the gum rubber tubing connections, the gas flow rate is adjusted to the desired rate by needle valve E as determined by the soap bubble flow meter F. (see References 126 and 127 for operational details). If secondary dilution is required, purified air is metered by a second soap bubble flow meter through a tee at G. The length of stain was read hourly and recorded in millimeters. The ambient temperature was 25 ± 2°C. Since the manufacturer disclaimed interference from humidity effects unless saturation conditions were approached, no humidity control was attempted.

A dynamic system assembled as shown in Figure 75, except cylinder A contained purified air and pure CO was introduced at G from a syringe mounted in a constant speed drive which incorporated a wide selection of rates and volumes[158] was abandoned before the criticality of concentration and flow rate relative to stain production rate was recognized. In retrospect, however, we have considered dynamic calibration as the better approach and will go back to this system when future calibration work becomes necessary.

The stain lengths were corrected to 60 mm and 80 mm reagent column lengths: corrected mm = mm found × 80/mm reagent length. Calibration charts similar to those supplied for grab samples were constructed on mm cross section paper. A 100 mm (vertical) × 75 mm rectangle was laid out. Then a line was ruled from a point 60 mm above the base on the left hand vertical edge to a point 80 mm above the base on the right hand side. The slope of this line then represented the total reagent column length variation and

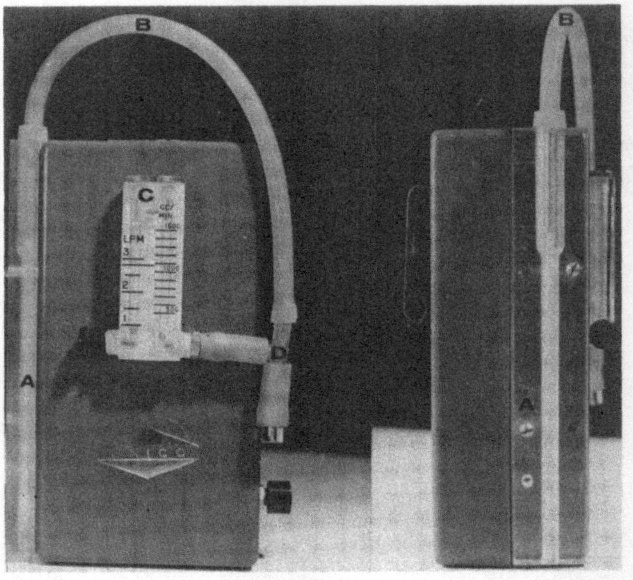

FIGURE 76. The carbon monoxide personnel monitor. (From Linch, A. L. and Pfaff, H., *Am. Ind. Hyg. Assoc. J.*, 32, 745, 1971. With permission.)

determined the position of an exposed tube for CO concentration readout. Calibration lines were constructed by plotting appropriate calibration stain lengths on the 60 mm and 80 mm axes.

Collection of the Sample — The personnel monitor employed in this study is shown in Figure 76. The detector tube holder A was constructed from a section of polyacrylate plastic plate 135 mm long by 20 mm wide by 9 mm thick by milling a 5 mm semi-circular channel on the face. When fastened to the side of the miniature, portable, air sampler[159] by two brass machine screws with the channeled surface against the case, a secure, transparent holder was provided for the detector tube. If the connecting tube B was kept short as shown, the detector tube remained securely in place during field use.

In order to provide the constant, slow, air flow through the detector tube and yet allow the pump to operate at the 700 to 1200 ml/min optimum rate (rotameter C), a bypass capillary (0.737 mm inside diameter) was fitted to a tee inserted at D. The total air flow through the sampler was maintained at 1100 ± 25 ml/min to maintain the sampling rate within tolerable limits (8.0 ± 1.2 ml/min). Volumetric calibration was carried out with the soap bubble flow meter previously described on a weekly basis. Performance is summarized in Table 54.

After insertion of a freshly charged battery in the sampler, and connection of the detector tube located in the holder A to the rotameter C through tube B, the flow rate was adjusted to the calibration point. The assembly then was fastened to the employee's belt in a position which would least interfere with his work activities. The monitor was checked hourly, the pumping rate adjusted if necessary, and the detector tube examined for stain length progress. After four hours, a fresh detector tube was installed and the sampling continued for the balance of the 8-hour shift. A completely charged battery provided ample power for 8-hour operation without running to complete discharge. The CO concentration in ppm was read off the calibration chart recorded, and the tubes capped for future reference.

Results

Calibration — Attempts to calibrate the detector tubes with 50 ppm CO in air at 2.5 ml/min flow rate or with 13 ppm CO at 9.6 ml/min in the dynamic system met with uniform failure. A stain 2 to 5 mm in length would develop in the first 20-30 minutes and then remain stationary for the balance of the 4-hour period. This failure was attributed to some undetected malfunction of the assembly and the system abandoned before operational range limitations had been recognized.

Carbon monoxide calibration mixtures in pressurized nitrogen were immediately available; therefore, the static system shown in Figure 75 was employed. A plot of the calibration stain lengths gave a linear curve entirely different from the one indicated by the data submitted by the manufacturer. Not only was the slope different, but the rate had to be reduced from 9.6 to 2.5 ml/min to remain "on chart".

TABLE 54

Precision Limits for Calibration of Carbon Monoxide Length of Stain Detector Tubes

Variable	Number of observations	Average	Standard deviation at 95% C.L.
Reagent length (mm)	33	70	±5 mm
Flow rate through same tube for ±25mL/min total flow rate (ratio = 0.017 mL/ml total)	10	8.0	±0.4 mL/min ±2.5 mm stain ±7 ppm
Total flow 1100 ± 25 mL/min (0.737-mm orifice) through different tubes	24	8.0	±1.2 mL/min ±7 mm stain ±18 ppm
Stain length at 8.0 mL/min	5	28 ppm	±4 mm stain ±7 ppm

From Linch, A. L. and Pfaff, H., *Am. Ind. Hyg. Assoc. J.*, 32, 745, 1971. With permission.

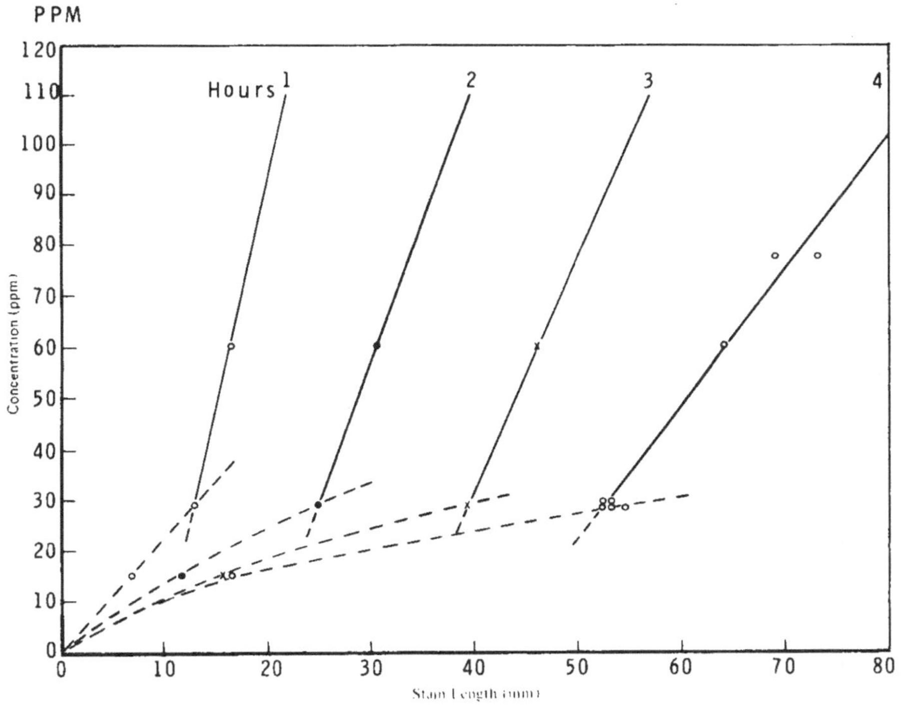

FIGURE 77. The carbon monoxide in air calibration chart for personnel monitoring. (From Linch, A. L. and Pfaff, H., *Am. Ind. Hyg. Assoc. J.*, 32, 745, 1971. With permission.)

Obviously, oxygen had a profound effect on the stain production rate; therefore, calibration mixtures were reconstituted in air. In order to include the 25 to 100 ppm CO range on a 4-hour sampling chart, the flow was reduced from 9.6 ml/min to 8.0 ml/min (Figure 77). The effect of flow rate on the stain development rate was found to be critical in the 6.5 ml/min region (Figure 78). Only the range 7 to 9 ml/min could be used for 4-hour sampling.

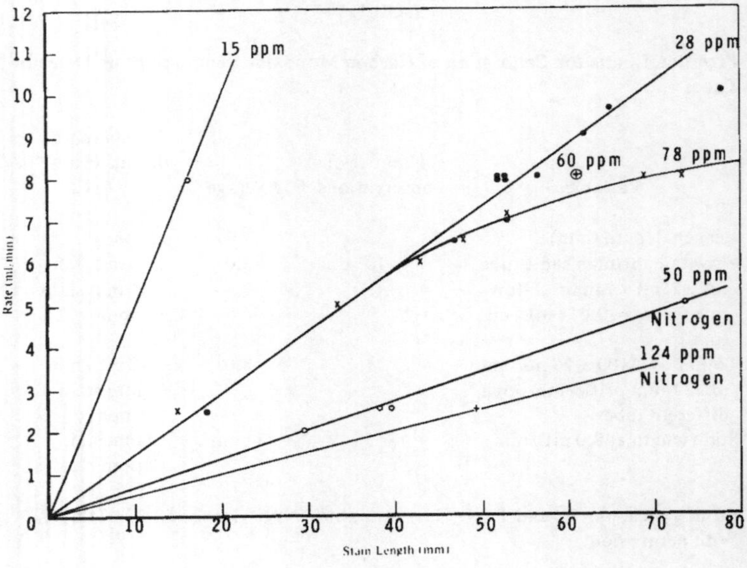

FIGURE 78. Carbon monoxide in air — flow rate vs. stain length. (From Linch, A. L. and Pfaff, H., *Am. Ind. Hyg. Assoc. J.*, 32, 745, 1971. With permission.)

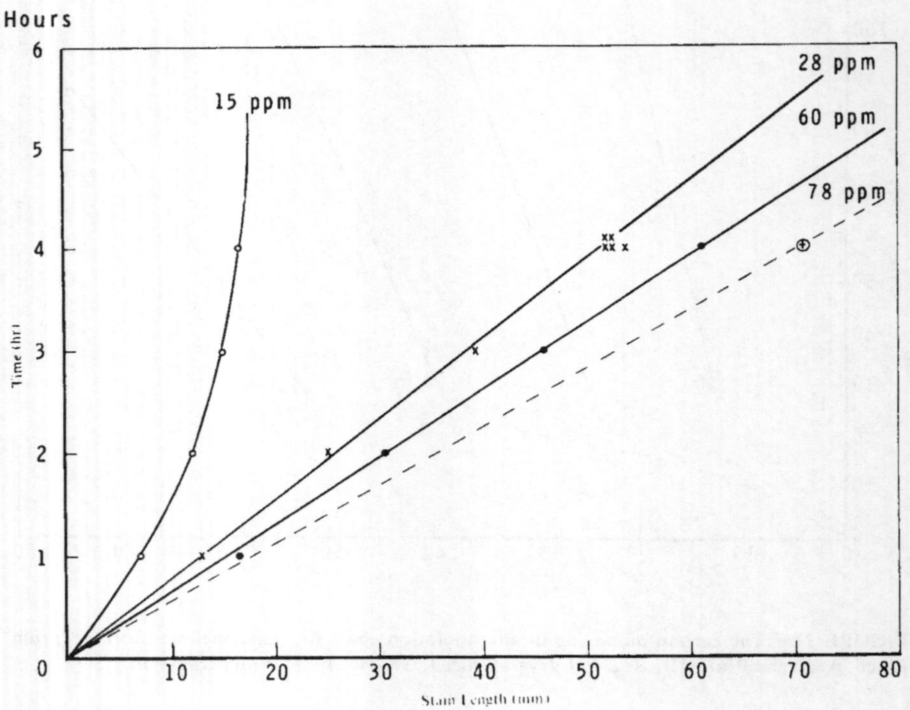

FIGURE 79. Carbon monoxide in air — stain length vs. time. (From Linch, A. L. and Pfaff, H., *Am. Ind. Hyg. Assoc. J.*, 32, 745, 1971. With permission.)

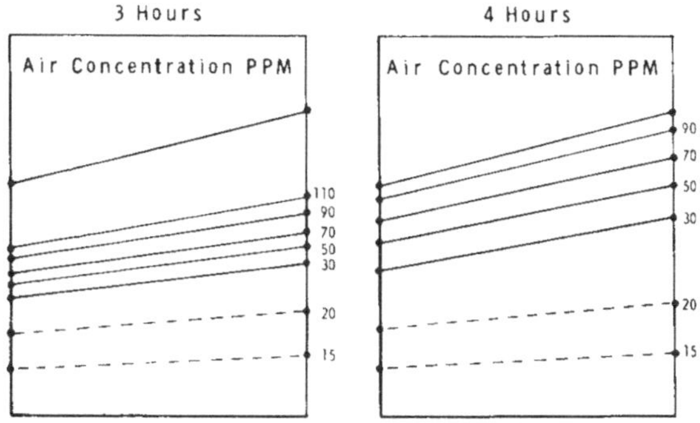

FIGURE 80. Carbon monoxide in air calibration charts — Kitagawa® length of stain tube No. 100 (flow rate = 8.9 mL/min). (From Linch, A. L., and Pfaff, H., *Am. Ind. Hyg. Assoc. J.*, 32, 745, 1971. With permission.)

The criticality of concentration is clearly demonstrated by plotting stain length against concentration (Figure 77), sampling rate (Figure 78), or time (Figure 79). The 30 ppm CO concentration region was found to be the lower operational limit in the construction of calibration charts in the range 0 to 100 ppm for 1-, 2-, 3- or 4-hour monitoring (Figure 77). Below 30 ppm the stain front progressed at a very slow rate and practically stopped after 2 hours (Figure 79). Above 30 ppm the stain development velocity became linear and useable to the 100 ppm limit of the chart (Figures 79 and 80). The incremental increase in stain length vs. concentration (ppm/min = 2.8) was satisfactory for the short remaining range (25 mm) equivalent to ½ to 2 times the current TLV. Calibration charts prepared from the 1- and 2-hour calibration data indicated CO concentrations above 100 ppm could be accommodated, but this possibility was not explored further. Sharper delineation of the critical area between 20 and 30 ppm by additional calibration study might provide an extension of the useful region down to 20 ppm (dashed calibration line, Figure 80).

Field Survey — A personnel sampling survey carried out in the Consolidated Warehouse on the 4-12 p.m. shift from September 11 to 22, 1970 when all ventilation sources were wide open disclosed no 4-hour periods during which the time-weighted average CO concentration exceeded 30 ppm. The "apparent" average was 7 ppm in a 5 to 11 ppm range. However, reliability in this region is equivocal. A similar survey over a 13-day period in early December with one exception also indicated less than 30 ppm CO in the breathing zone. One sample taken from 4 to 8 p.m. on December 8, 1970, by a fork lift operator's monitor while loading drums in a tractor trailer reached 50 ppm. The other 12 samples averaged 21 ppm in a range of 10 to 50 ppm apparent concentration. However, again the equivocal quality of all estimations below 30 ppm must be recognized.

DISCUSSION

Transitions in curves relating stain length to gas or vapor concentration had been reported previously.[130] Two entirely unrelated reagent systems for benzene and nitrogen dioxide are summarized in Figure 68. Although the total time intervals are much shorter than four hours, the significance of these transitions for limiting the useful range of detector tubes for continuous monitoring is clearly implied. In each case these transition zones occurred beyond the usual limit for grab sampling — 5 minutes or less — imposed by the squeeze bulb or hand pump employed to draw the sample through the detector tube. Extension of the mathematical models developed by Saltzman[131] for short term sampling to periods beyond one hour undoubtedly will shed more light on the implications of the observations when more data become available.

Our failure to find conditions which would permit long-term monitoring for CO in concentrations below 25 ppm confirmed Ingram's earlier conclusion that currently available detector tubes for CO, SO_2, NO_2, and hexane do not possess sufficient sensitivity for use as personnel monitors in the range of 3 to 10 ppm under ambient air conditions.[102] However, Saltzman's calculations indicate possible success at higher flow rates since mass transfer rate appears to control these systems. Ingram's work indicated this rate must be greater than 12 ml/min.[102] Later studies at a much higher sampling rate (40 ml/min) for "a few hours" confirmed Ingram's conclusion that the state of the detector tube art does not support their use for the detection of air contaminants in the ambient air quality (non-industrial) concentration ranges, at least not for carbon monoxide and nitrogen dioxide.[160]

The black stain produced by CO reduction of the palladium sulfite complex to palladium metal on the silica gel support[8] can be bleached by atmospheric oxygen. This reversible property which provides the basis for recovery of the CO alarm detector tabs may not be well recognized. Thus, yet another variable which is

TABLE 55

Carbon Monoxide Detector Tube Flow Resistance Calibration — Reagent Length vs. Flow Rate[a]

	Column length (mm)			Flow rate through pump (rotameter mL/min)			Flowmeter rate for 8 mL/min through tube	
Tube no.	Reagent	Retainer	Total	(510)[b] 700	(900)[b] 1100	(1160)[b] 1400	Nominal	Actual
1	68	9	77	4.0	9.6	13.8	1000	800
2	70	10	80	3.0	7.0	9.6	1200	980
3	74	10	84	—	6.1	8.6	1325	1100
4	72	14	86	—	6.3	8.6	1325	1100
5	68	10	78	3.7	9.6	13.8	1000	800
6	65.5	9.5	75	3.2	8.1	10.9	1125	920
7	76.5	9.5	85	3.6	8.6	11.9	1075	870
8	66	16	82	3.5	8.6	11.9	1075	870
9	69	12	81	3.5	8.6	11.3	1100	900
10	75	12	87	3.0	7.8	10.4	1175	960
Average	70.5	11	81.5	3.4	8.0	11.1	1140	930
Variance	±3.8				±1.2			

[a] By-pass capillary size (I.D.) = 0.737 mm.
[b] Values ( ) = Calibrated flow rate in mm/min vs. nominal calibration scale.

unique for CO is introduced into the length of stain detector tube system. If the two reaction rates converge, the overall effect would be catalytic oxidation to CO_2 without stain formation.

CONCLUSIONS

The length of stain carbon monoxide detector can be recalibrated for semi-quantitative aplication within restricted sampling limits as a personnel monitor for time intervals up to four hours. The flow rate must be maintained at 8.0 ± 1.2 ml/min to contain the standard deviation within a ± 18 ppm range. The useful range is limited to 30 to 100 ppm. However, this range does include the TLV within the range of one-half to twice the current 50 ppm limit.[148] With sampling rates in the 6.5 to 10 ml/min region, response of the reagent stain is not proportional to CO concentrations below 30 ppm.

Improved flow rate control would significantly reduce the range of analytical uncertainty. Part of this improvement must come from either reduction of pressure drop variability within the individual detector tubes or introduction of a compensator within the air sampler system. Future studies should include simultaneous sampling and analysis in parallel with a standard method such as gravimetric Hopcalite[157] or the iodine pentoxide[156] procedure for "in place" calibration at a fixed station site in the highly variable warehouse atmosphere. For the first time the capability for at least semi-quantitative correlation of CO body burden with industrial environmental exposure in the TLV range has become available. Therefore, in future surveys, CO analysis of breath samples collected at the end of the 4-hour sampling period for correlation with the personnel monitor results should be included.

Surveys in a large Consolidated Warehouse served by gasoline-powered fork lift trucks under restricted ventilation conditions imposed by cold weather confirmed CO levels which were in conformance with the current 50 ppm TLV. A recording NDIR instrument centrally located in the general traffic area indicated transient concentrations equivalent to one and one-half hours continuous CO exposure above the TLV per 8-hour shift. Again, the necessity for personnel air sampling or biological monitoring for evaluation of personnel exposure in highly variable industrial environments has been vindicated.*

After the foregoing paper had been published, the study was reopened in an effort to improve the reliability of these carbon monoxide detector tubes for field use. First to be considered was the relationship between reagent length and pumping rate (total rate indicated by the pump rotameter with by-pass capillary in place — Figure 76) to determine whether a chart could be developed for control of the detector tube flow rate at 8.0 ± 0.4 mL/min (± 5% volumetric error) based on reagent length. The results were disappointing. Reference to Table 55 and Figure 81 indicated an intolerable scat-

* Reproduced from Linch, A. L. and Pfaff, H., *Am. Ind. Hyg. Assoc. J.*, 32, 745, 1971. With permission.

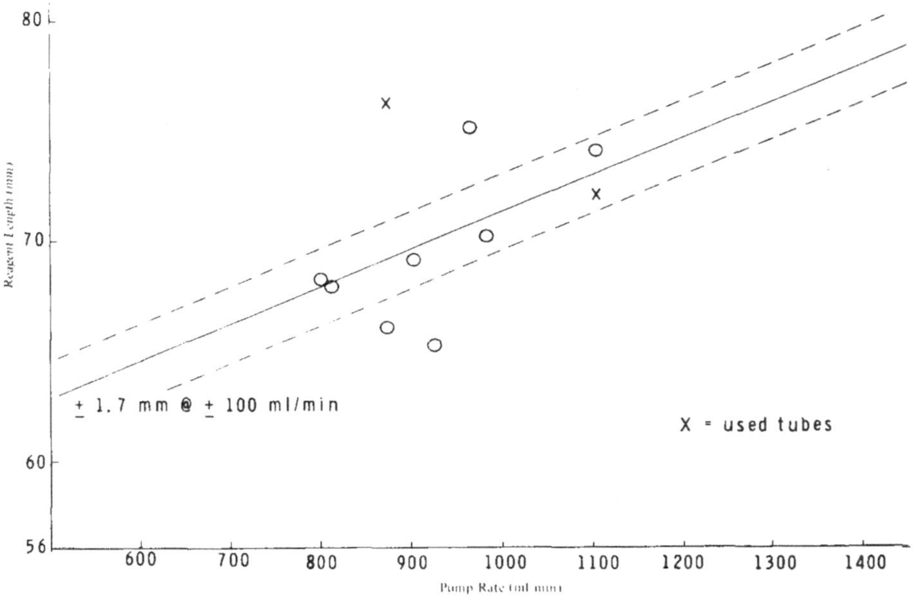

FIGURE 81. Pump rotameter rate (calibrated) vs. reagent length. (By-pass capillary size [I.D.] = 0.73 mm;rate = 8 mL/min.)

TABLE 56

Pump Rotameter Calibration[a]

Nominal	Rate (mL/min) actual	Pressure drop (cm H_2O)
1400	1160	34
1100	900	36
700	510	39

[a] Bypass capillary size = 0.737 mm.

ter of data. Obviously other factors, such as packing density and uniformity of particle size distribution, were overriding the granular column length as the prime source of variations in flow resistance. Pressure drop (differential) data are presented in Table 56. Estimates of the magnitude of permissible variable tolerances are summarized in Table 57.

The final solution to the sampling rate control problem was suggested by the development of a miniature precision glass rotameter designed for the range 3 to 30 mL/min airflow rate.[145] The original model was modified to fit into the carbon monoxide monitor previously described (Figure 83). This innovation provided sample flow rate control within the ± 5% to 8.0 ± 0.4 mL/min goal specification.

The previous descriptions of constant rate sampling through detector tubes over prolonged periods of time have been limited to systems using battery power driven pumps to draw the air sample through the tubes. In at least one instance an evacuated rubber squeeze bulb was employed to create the vacuum needed to draw the air sample. The slow sampling apparatus, which used a 120-mL deflated thick-walled rubber bulb, was designed primarily to draw air through a trichloroethylene indicator tube at a constant rate over a period of 8 hr. The method depends upon the "reasonably" con-

TABLE 57

Flow Rate Tolerances for Carbon Monoxide Detector Tube No.100[a,b]

		Result		
Variable	Magnitude	Measured	Magnitude	Reference
Pump rate	±100 mL/min	Tube flow	±0.7 mL/min	Table 17 (tubes 3 and 4)
			±1.5 mL/min	Table 17 (tubes 1 and 5)
	±25 mL/min	Tube flow	±1.2 mL/min	53
Reagent length	±1.7 mm	Pump rate	±100 mL/min	Figure 36
Stain length	±2 mm	Concentration	±6 ppm	Figure 32 (at 50 ppm)
Stain length	±2 mm	Tube flow	±0.6 mL/min	Figure 33
Tube flow	±1.2 mL/min	Concentration	±12 ppm at 50 ppm	

[a] Bypass capillary size (I.D.) = 0.737 mm.
[b] tube flow rate = 8.0 mL/min.

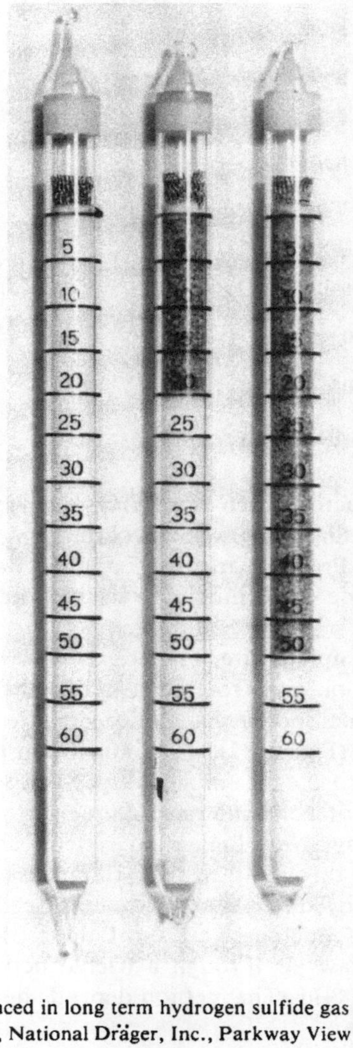

FIGURE 82. Stain length produced in long term hydrogen sulfide gas detector tubes — Dräger. (Courtesy of Drägerwerk, C., U.S. Division, National Dräger, Inc., Parkway View Dr., Pittsburgh, Pa.)

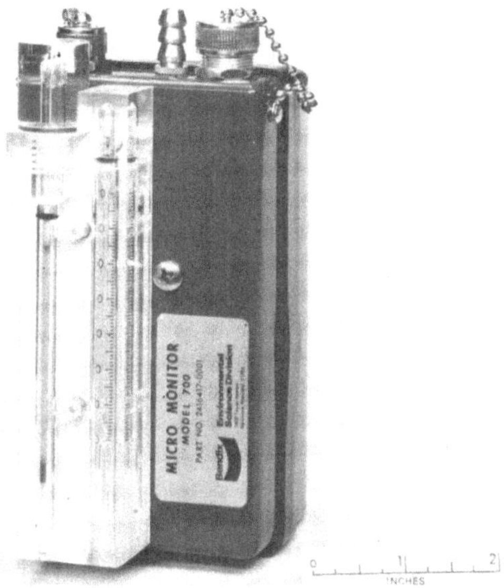

FIGURE 83. Micro Monitor. Model 700. This model utilizes a Bendix-Gastec® reagent type gas and vapor detector tube which is inserted in a transparent holder mounted on a Micronair® sampling pump. Total exposure to a contaminant is clearly and instantly indicated by the length of color stain in the the detector tube. More than 80 different gas detector tubes are readily available to monitor the TLV of most common industrial contaminants. This monitor also functions as a universal collector by sampling with a silica gel or charcoal-filled tube to collect a sampler for subsequent analysis by gas chromatography or other suitable means. The micro monitor indicates accumulated exposure and features a built-in flow-meter and flow-adjustment valve. (Courtesy of Bendix Environmental Science Division, 1400 Taylor Ave., Baltimore, Md.)

stant pressure exerted by the rubber wall irrespective of the degrees of deflation. Against a suction head of "about" 120 mm of mercury, a sample flow of 15 mL/hr was obtained through a 20-mm long × 0.05-mm bore (I.D.) glass tube (polarography capillary). The capillaries were protected from dust by porous glass disks fitted to either end. The authors also found that longitudinal air "leakage" through a 100-mm length of 13-mm diameter electrode carbon rod sealed in a glass tube furnished a suitable sampling rate.[161]

Active development work for commercial production was first announced in 1973 by Drägerwerk.[146, 146a] Further study disclosed erratic behavior of the CO detector tube based on the reduction of palladium chloride when used for 4-hr continuous sampling. Secondary reactions which produce a bleaching effect (probably oxidation of the catalytically active palladium metal by atmospheric oxygen) resulted in a negative interference, i.e., indicated CO concentrations less than known to be present. These results came as no surprise. The "grab" sample detector tubes were designed for relatively high flow rates in the range of 20 to 80 L/hr, but a low sample volume (0.1 to 2). Reduction in flow rate to levels compatible with long term sampling in the range 6 to 15 mL/min would provide an extended contact time between the gas and the reagent which could cause secondary reactions compounded by absorption effects which are difficult to control and quantify. Therefore, further attempts to develop a stable palladium chloride reagent was abandoned in favor of the iodine pentoxide, selenium dioxide-fuming sulfuric acid reaction:

$$5CO + I_2O_5 \xrightarrow[H_2S_2O_7]{SeO_2} I_2 + 5CO_2$$

Extensive evaluation of this CO detector tube confirmed its reliability under most conditions to be expected in either the industrial or ambient atmospheres.[120] Large fluctuations in CO concentration did not affect the response of the reagent to collection of the total weight of CO presented to the system. A CO standard in air was drawn through each of five detector tubes at 15 mL/min for 67 min (except No. 3) as follows:

Tube No.	Time of sampling	CO concentration	Calculated μL CO
1	67 min	200 ppm	200
2	6.7 min	2000 ppm	200
	60.3 min	0 ppm	200
3	268 min	50 ppm	200
4	60.3 min	0 ppm	—
	6.7 min	2000 ppm	200
5	6.7 min	400 ppm	40
	6.7 min	0 ppm	—
	6.7 min	400 ppm	40
	6.7 min	0 ppm	—
Continued alternations until 67 minutes sampling time was accumulated			200
Range			190-215
% Relative standard deviation			5-10

Seven other common gases and vapors in air have been calibrated for two 8-hr sampling periods (Table 58).[162] The prediction that the precision and accuracy would be improved with constant rate sampling was confirmed, i.e., % relative standard deviation of 5 to 15% vs. grab sampling specification of 25%. The results obtained with the hydrogen sulfide monitor are illustrated in Figure 82.

Preparations by NIOSH for certification of these long-term sampling detector tubes is in progress currently in their Morgantown, W.Va. laboratories. Since Table 58a was published, Drager has added the following tubes to the growing list of gas and vapor detectors:[95]

Pollutant	Reference Number	Measuring Range μL	Measuring Range ppm
Chlorine—1a/L	6728421	1—20	1—20
Chloroprene—Sa/L	6728431	5—100	1—100
Hydrocyanic acid — 10a/L	6728441	10—120	1—120
Nitrogen dioxide — 10a/L	6728281	10—100	1—100
Toluene — 200a/L	6728271	200—4000	20—4000

TABLE 58

Gases and Vapors Measurable by Long-term Tubes

Substance	Measuring range (20°C at 760 mm Hg)[a] In absolute units (μL)	Measuring range (20°C at 760 mm Hg)[a] Related to concentration (ppm)	Size of sample (L)	Maximum sampling time (hrs)	Reagent	Color change	Indicating error (relative SD) (%)
Ammonia	10—100	5—50	2	2	Bromophenol blue	Yellow to blue	10—5
		10—100	1				
Benzene	20—200	10—100	2	2	Jodic acid	White to brown-green	15—10
		20—200	1				
Carbon monoxide	50—500	10—100	5	8	Jodine pentoxide	White to brown	10—5
		50—500	1				
Hydrogen sulfide	5—60	1—12	5	8	Lead acetate	White to brown	10—5
		5—60	1				
Nitrogen oxides ($NO + NO_2$)	50—350	100—700	5	4	Diphenyl benzidine	Yellow to blue-gray	15—10
		50—50	1				
Sulfur dioxide	5—50	1—10	5	4	Mercury chloride + indicator	Yellow to red	15—10
		5—50	1				
Trichloroethylene	50—500	25—250	2	2	(Via Cl) *o*-tolidine	White to orange	15—10
		50—500	1				
Vinyl chloride	10—50	2—10	5	8	(Via Cl) *o*-tolidine	White to pale yellow	15—10
		10—50	1				

Note: Air flow 10 cm/20 min.

From Leichnitz, K., *Nat. Saf. News*, April, p. 59, 1977. With permission.

TABLE 58A

Dräger Long Term Tubes

Long term detector tube	Catalogue no.	Maximum period of use (hr)	Measuring range absolute unit (20°C, 1013 m bar)	Flushing air after the measurement (desorption)	Measuring range in ppm for maximum period of use	Measuring range in ppm for shorter period of use
Ammonia 10/a-L	6728231	4	10—100μL	No	2.5—25 ppm (volume 4L in about 4 hr)	20— 200 ppm (volume 0.5 L in about ½ hr)
Benzene 20/a-L	6728221	4	20—200μL	No	5—50 ppm (volume 4L in about 4 hr)	20—200 ppm (volume 1L in about 1 hr)
Carbon dioxide 1000/a-L	6728611	4	1000—6000μL	No	250—1500 ppm (volume 4L in about 4 hr)	1000—6000 ppm (volume 1L in about 1 hr)
Carbon monoxide 50/a-L	6728121	8	50—500μL	No	6.3—63 ppm (volume 8L in about 8 hr)	100—1000 ppm (volume 0.5L in about ½ hr)
Chlorine 1/a-L	6728421	8	1—20μL	No	0.13—2.5 ppm (volume 8L in about 8 hr)	0.5—10 ppm (volume 2L in about 2 hr)
Chloroprene 5/a-L	6728431	4	5—100μL	500 cm^3	1.3—25 ppm (volume 4L in about 4 hr)	5—100 ppm (volume 1L in about 1 hr)
Hydrocarbons 100/a-L (calibrated for *n*-octane)	6728571	4	100—3000μL	1000 cm^3	25—750 ppm (volume 4L in about 4 hr)	100—3000 ppm (volume 1L in about 1 hr)
Hydrochloric acid 10/a-L	6728581	8	10—50μL	500 cm^3	1.3—6.3 ppm (volume 8L in about 8 hr)	10—50 ppm (volume 1L in about 1 hr)
Hydrocyanic acid 10/a-L	6728441	8	10—120μL	No	1.3—15 ppm (volume 8L in about 8 hr)	10—120 ppm (volume 1L in about 1 hr)

Hydrogen sulfide 5/a-L	6728141	8	50—60μL	No	0.63—7.5 ppm (volume 8L in about 8 hr)	10—120 ppm (volume 0.5L in about ½ hr)
Nitrogen dioxide 10/a-L	6728281	8	10—100μL	No	1.3—13 ppm (volume 8L in about 8 hr)	10—100 ppm (volume 1L in about 1 hr)
Nitrogen monoxides 50/a-L ($NO + NO_2$)	6728191	(4)	50—350μL	No	13—88 ppm (volume 4L in about 4 hr)	50—350 ppm (volume 1L in about 1 hr)
Sulfur dioxide 5/a-L	6728151	4	5—50μL	No	1.3—13 ppm (volume 4L in about 4 hr)	5—50 ppm (volume 1L in about 1 hr)
Toluene 200/a-L	6728271	8	200—4000μL	No	25—500 ppm (volume 8L in about 8 hr)	100—2000 ppm (volume 2L in about 2 hr)
Trichloroethylene 10/a-L	6728291	4	10—200μL	500 cm^3	2.5—50 ppm (volume 4L in about 4 hr)	10—200 ppm (volume 1L in about 1 hr)
Vinyl chloride 10/a-L	6728131	8	10—50μL	No	1.3—6.3 ppm (volume 8L in about 8 hr)	10—50 ppm (volume 1L in about 1 hr)

Dräger announced formation of a U.S.-based production facility in mid 1977.*

The basic assumption that the concentration of atmospheric pollutants will be subject to wide fluctuations with respect to time and location during the sampling period determines the specifications for long term detector tubes. Therefore, the stain length must be a direct measure of the absolute mass of the gas introduced into the reaction zone. In other words the stain length must increase in proportion to the volume of gas drawn into the detector tube. At the termination of the sampling period, the average concentration of the gas or vapor collected can be calculated from the total stain length and total volume of air pumped through the system. However, since the total exposure dose is the industrial health controling factor, calculation of the average concentration (TWA) is not necessary. Therefore, the total sample volume is not necessary so long as the sampling rate (15 ± 5 mL/min) is maintained within established limits.

The calibration curves for long-term detector tubes can be expressed by the following equations[162]

$$\ell = \frac{m}{c} + \frac{v}{A}\left(1 - e^{-1\frac{A}{v}}\right)$$

where ℓ = stain length, m = gas mass reacted, c = absorption capacity of one layer element, v = flow rate, and A = a constant derived from several different constants in terms of an inverse time constant which indicates the rate reaction between the tube reagent and the gas.

$$\ell = 2.718$$

A typical calibration curve is presented in Figure 84.

Samples of both unexposed (1) and exposed hydrogen sulfide tubes (2 and 3) are shown in Figure 82. The indicated stain lengths correspond to 20 and 50 μL, respectively.

A peristaltic pump was developed by the Dräger engineers to draw a continuous sample at a uniform rate of 15 ± 5 mL/min through these long-term sampling detection tubes. The pump-detector tube assembly in the portable instrument mode is shown in Figure 85 and mounted for personnel monitoring in Figure 86. Charging the battery with the case open is illustrated in Figure 87. The pump rotation counter which can be calibrated in terms of volume pumped appears in Figures 85 and 87. Diaphragm pumps that will operate reliably within this rate range also are available.[166,167,172,250] One model will maintain a constant rate which compensates for changes in pressure drop (Δ P) across the sampling assembly.[167] However, if the exposure is related to total exposure dose, then constant rate sampling is not required as long as the rate remains with the range 15 ± 5 mL/min (33% tolerance).

See Table 59 for long-term detector tubes supplied by the Mine Safety Appliances Company.

At the time of the preparation of this second edition, one other competitor entered the long-term detector-tube field. Kitagawa, represented by Anatole J. Sipin, announced in late 1976 five tubes — carbon monoxide, ammonia, hydrogen sulfide, sul-

* Since preparation of this material, distribution of Dräger products has been taken over by National Dräger, Inc., Parkway View Drive, Pittsburgh, Pa.

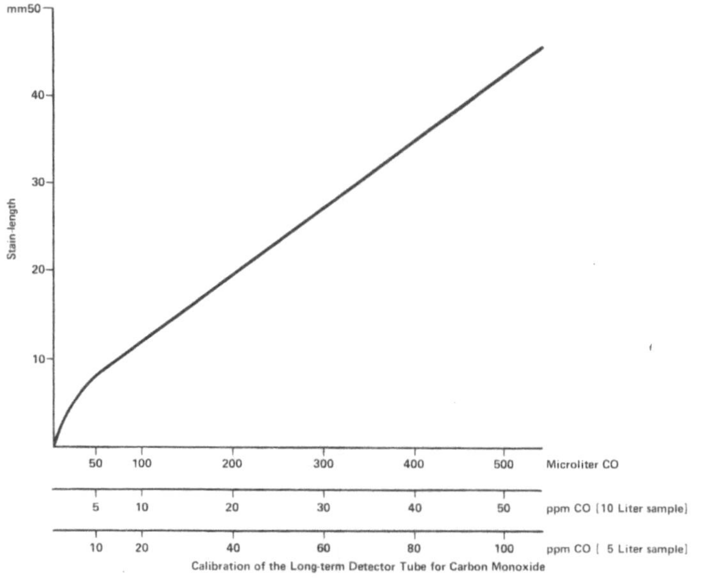

FIGURE 84. Calibration of the long-term detector tube for carbon monoxide. (From Leichnitz, K., *Nat. Saf. News*, April, 59, 1977. With permission.)

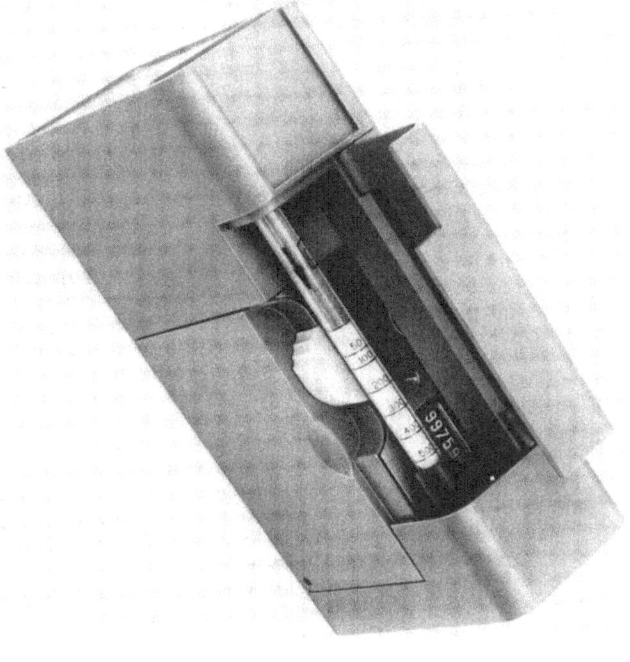

FIGURE 85. The Dräger "polymeter" persistaltic rechargeable battery operated sampling assembly in the portable instrument mode with a long-term detector tube in place. (Courtesy of Drägerwerk, U.S. Division, National Dräger, Inc., Parkway View Dr., Pittsburgh, Pa.)

FIGURE 86. The Dräger "polymeter" in position for personnel monitoring with a long-term length-of-stain detector tube. (Courtesy of Drägerwerk, U.S. Division, National Dräger, Inc., Parkway View Dr., Pittsburgh, Pa.)

fur dioxide, and toluene.[166,168] Additional tubes for vinyl chloride, chlorine, benzene and trichloroethylene were under development for release in the near future.

The battery operated diaphragm pump developed for charcoal tube sampling was adapted to the 6 mL/min flow rate required for the Kitagawa TWA indicating tubes (Figure 88). From outward physical appearances these detector tubes closely resemble their "grab sampling" counterpart (Figure 53 and 63). However, in the case of the CO tube, a brown ring progresses through the reagent column rather than producing a uniform stain throughout the reagent zone proportional to the mass of CO absorbed (Figure 89). A chart which relates TWA in terms of ppm versus sample volume and ring position is furnished (Figure 90). A guard layer (air cleaning reagent, Figure 89) is incorporated into the tube to reduce interferences.

The calibration is based on a total sample volume of 2880 mL (8 hours at 6 mL/min). For other sample volumes, the actual volume is related to

$$2880 \text{ mL: ppm} = \text{indicated ppm} \times \frac{\text{sample}}{2880}$$

Or the time of sampling can serve as a correction factor: for 4-hr sampling, multiply the observed reading by two. Accuracy within 10% has been claimed. No correction

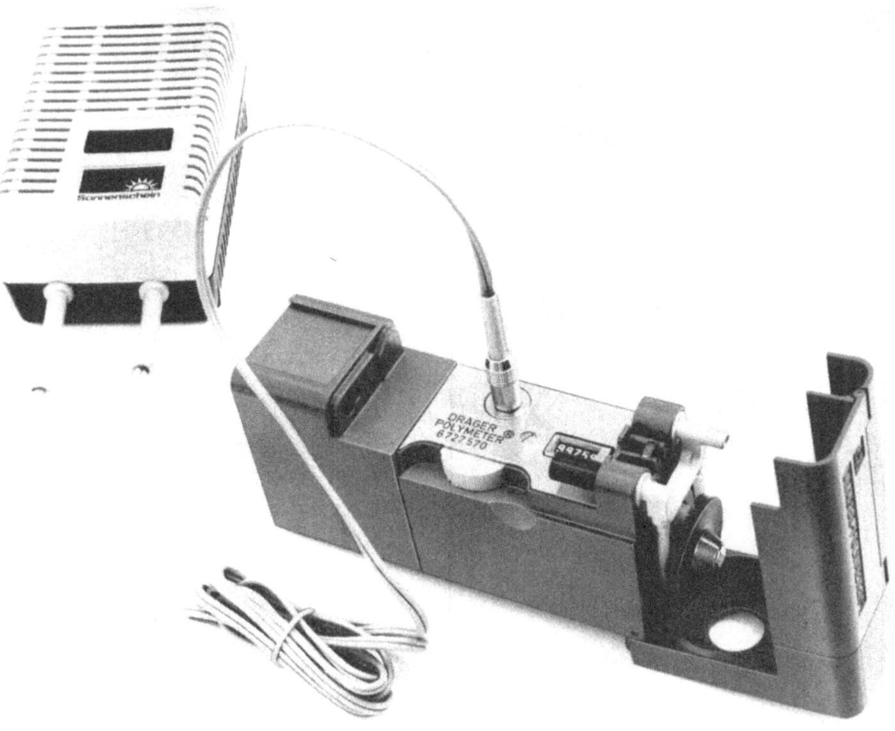

FIGURE 87. The Dräger "polymeter" long-term detector tube-pump assembly in the battery recharge mode. (Courtesy of Drägerwerk, U.S. Division, National Dräger, Inc., Parkway View Dr., Pittsburgh, Pa.)

TABLE 59

Long-term Detector Tubes Offered by Mine Safety Appliances Company (MSA) Sampling Time = 0.4 Hours with 0 — 2 Monitor TD

Part no.	Measured component	TLV for 1975 (ppm)	Calibration range (ppm)
460103	Ammonia	25	5—200
85976	Carbon dioxide	5000	0.03—1.0%
91229	Carbon monoxide	50	5—200
460225	Chlorine	1	0.5—5
91636	Hydrogen chloride	5	2—20
93262	Hydrogen cyanide	10	1—35
14063	Hydrogen sulfide	10	1—100
83089	Mercury vapor	0.01 mg/m^3	0.05—1.0μg/μm
83099	Nitrogen dioxide	5	0.05—5
93865	Ozone	0.1	0.03—2
92623	Sulfur dioxide	5	0.5—40

Note: The following information is the sum total available to the author.

From Toxic Gas Detectors, Mine Safety Appliances Co. (MSA), 400 Penn Center Building, Pittsburgh, Pa. With permission.

FIGURE 88. Sipin-Kitagawa TWA indicating continuous gas detection system. (Courtesy of Anatole J. Sipin Co., Inc., 425 Park Ave. S., New York, N.Y.)

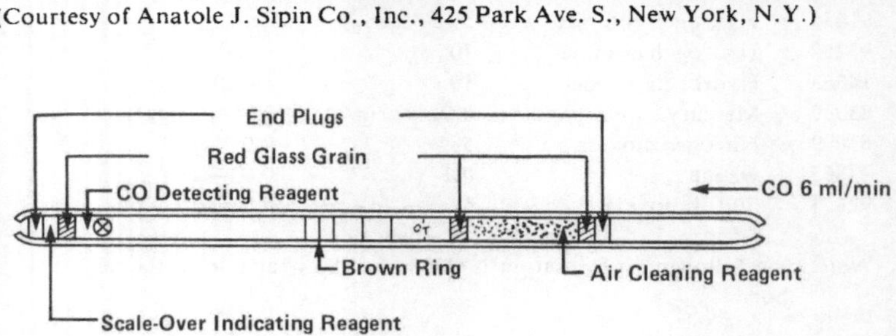

FIGURE 89. Sipin-Kitagawa time-weighted-average (TWA) carbon monoxide detector tube. (Courtesy of Anatole J. Sipin Co., Inc., 425 Park Ave S., New York, N.Y.)

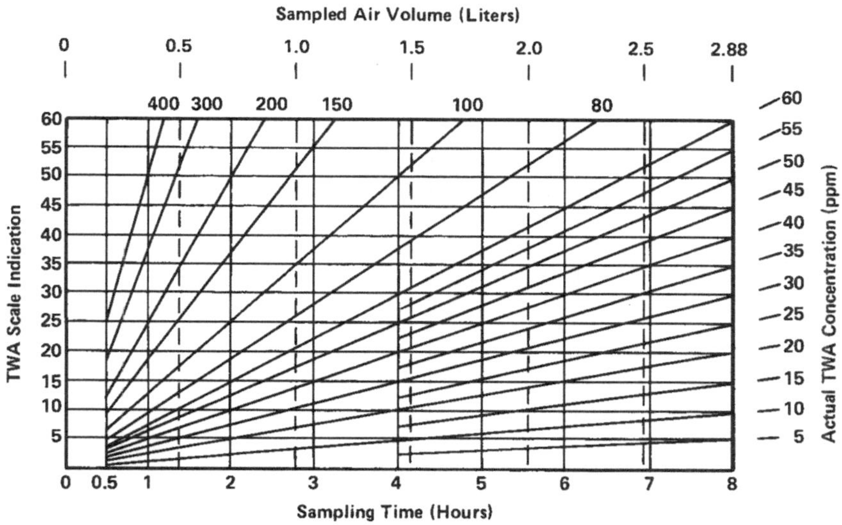

FIGURE 90. Kitgawa time-weighted-average (TWA) CO gas detector tube scale conversion chart.

for temperature is required in the range 10 to 40°C. Interruptions in sampling or excursions in concentration do not influence the results.

By means of a news release at the American Industrial Hygiene Conference in Chicago, Ill., on May 29, 1979, Anatole J. Sipin Co. presented a photoelectric scanner to detect the stain front as it develops during extended period sampling (Figure 90A). The initial design is intended to provide an alarm by locating the optical sensor at the TLV (TWA) required for the particular pollutant gas or vapor being measured. When the stain progresses to this fixed setting, a visual and/or audible alarm responds. By manually moving the optical sensor up or down on the detector tube the "Dosalert" can be used as a stain front detector to read any accumulated exposure dose. Early indications are that visual acuity errors can be significantly reduced by this technique. Also announced in this news release was the availability (off the shelf) of long term tubes for H_2S, SO_2, NH_3, toluene, and trichloroethylene as well as for CO previously provided. The "Dosalert" measures 15.2 cm × 5.1 cm × 1.6 cm and is battery powered. The pump is shut off at the alarm point, and the time or air volume registered on the pump counter.

Another news release from Sipin described a multi-gas dosimeter which is a manifold to retain two or more Sipin-Kitagawa gas dosimeter tubes for simultaneous measurement of dosage by the same detector tubes to determine the precision of the measurement, or by different detector tubes to measure two or more atmospheric components. The accumulated dosage for each tube is indicted by stain front which is expressed as the 8-hr time weighted average concentration in ppm. A new Sipin model SP-34 personal sampler pump (3 mL/min to 45 mL/min) provides optimal operation for any tube, singly as in the new "Dosalert", or any combination of the two tubes. Model SP-15 pump can also be used.

The tubes are retained in a special holder which measures 19.7 cm × 2.8 cm and includes an adapter fitting to normalize the flow rate through the two tubes. The adapter contains a restriction in series with one tube to provide the appropriate flow rate through each tube at the precalibrated pump rate. Then both tubes will be direct reading at its calibrated flow rate. Different adapters are available to permit use of any combination of two simultaneously.

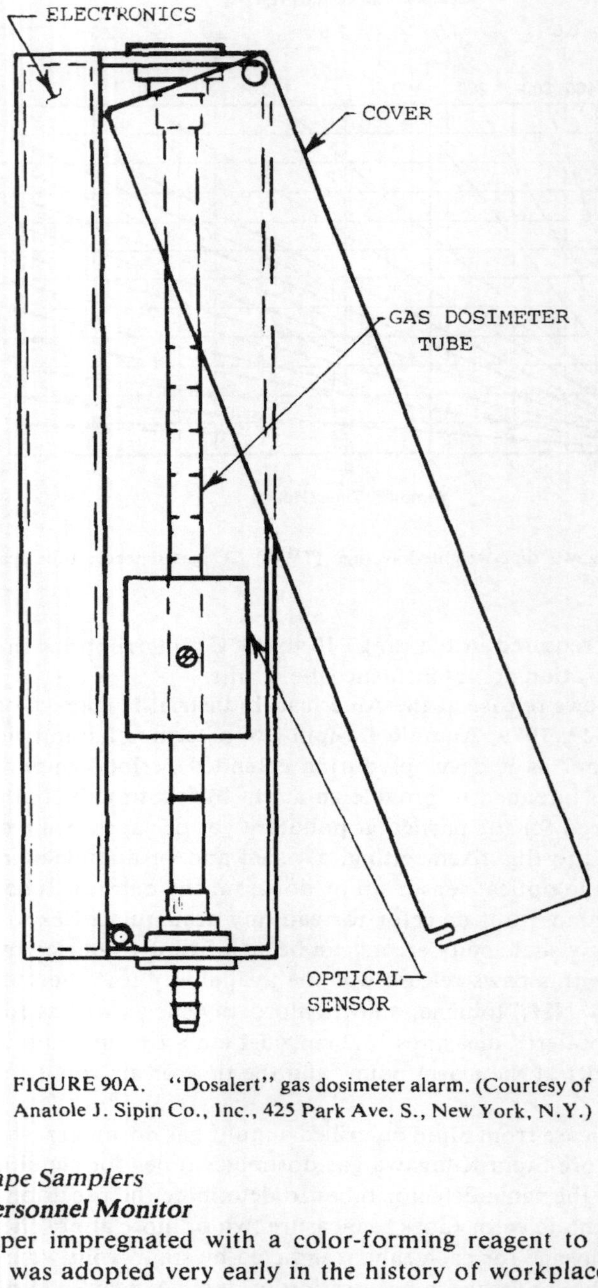

FIGURE 90A. "Dosalert" gas dosimeter alarm. (Courtesy of Anatole J. Sipin Co., Inc., 425 Park Ave. S., New York, N.Y.)

2. Continuous Tape Samplers

a. Continuous Personnel Monitor

The use of paper impregnated with a color-forming reagent to detect hazardous gases and vapors was adopted very early in the history of workplace monitoring (see Section I.A.1, "Impregnated Paper"). Development of continuous tape samplers on a commercial basis began in the early 1950s.[76,77] Photometric instruments that had been designed to determine the color density of stains, dyes, and light-absorbing dusts and smoke on paper were included in later designs to provide semiquantitative readout as the tape progressed continuously through the sampling head. Such tape samplers were calibrated for gaseous and particulate fluorides and hydrogen sulfide.[76,77] The weight and bulk of these assemblies precluded their use as personnel monitors or even as readily portable field instruments (see Section I.A.3, "Portable Instruments", for a definition of portability). Soon after publication of the works carried out by D. A.

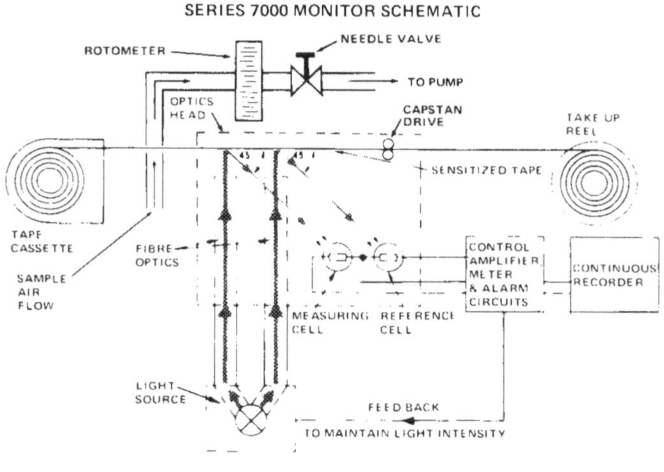

FIGURE 91. MDA series 7000 tape sampler schematic. (Courtesy of MDA Scientific, Inc., 808 Busse Hwy., Park Ridge, Ill.)

Reilly in England on the development of a test-paper procedure for the detection of toluene-di-isocyanate vapor in air, active development of a breathing zone continuous tape sampler was initiated.[164] Utilizing Reilly's test paper, a group at Dunlop Rubber in Wales designed a continuous monitor based on a continuous reel of paper tape rather than paper strips. In the early 1970s the technique was licensed to Universal Environmental Instruments, Ltd., where the current model 7000 series was put in production. Rights to distribution in the United States were obtained by MDA Scientific, Inc. (Figure 91). The instruments were portable by NIOSH definition and after correction of several mechanical deficiencies proved to be highly reliable. The personnel monitor mode (MCM series 4000) was announced September 14, 1975.[165]

The continuous tape samplers currently available appear to satisfy the ideal performance characteristics established for personnel monitors more closely than any other type so far developed. These characteristics follow.

1. They should be specific. Only the compound or element of interest should be measured even in a background of other substances normally present in the industrial environment.
2. They should be sufficiently sensitive to permit adequate detection of the required concentration and provide an adequate signal-to-noise ratio.
3. They should provide a continuous readout with adequate response time and resolution.
4. They should provide long-term unattended operation.
5. They should be low in both initial and operating costs.
6. They should be simple to use and operable by nontechnical personnel.

In the case of TDI, the test paper is prepared by treating Whatman® No. filter paper with a solution of sodium nitrite, 2-hydroxy-11H-benzo(a)carbazole-3-carboxy-*p*-anisidide (Brenthol GB), ammonium acetate, and diethylphthalate in methanol. The treated paper is stable for at least three months in the dark. Reaction with TDI or MDI produces a red-gray color. In the Series 4000 miniature continuous monitors (MCM), ambient air is drawn at 100 cc/min through the continuously moving tape by means of an integral, double diaphragm battery powered pump (Figure 91 and 92).

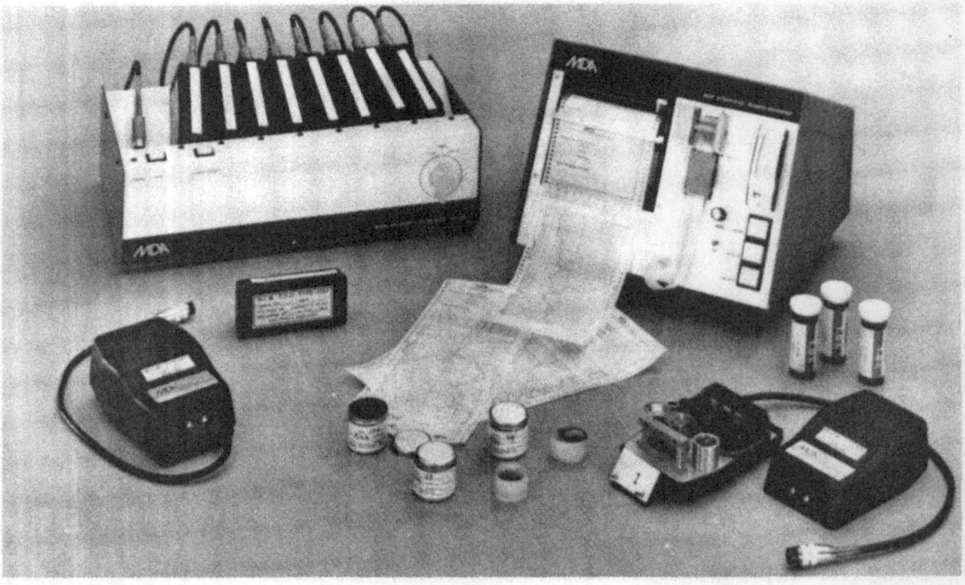

MCM SPECIFICATIONS

Ranges:	0-0.08 ppm TDI
	0-0.4 ppm Phosgene
Tape Speed:	2 cm/hour
Sampling Rate:	100 cc/minute
Running Time:	10 hours per battery charge
Dimensions:	6" x 2¼" x 4"
Weight:	1½ pounds including pump, excluding battery pack

MCM INTEGRATING READER/RECORDER SPECIFICATIONS

Range:	0-0.08 ppm TDI, 0-0.4 ppm phosgene
Tape Speed:	1 cm/second
Power Supply:	115 volt/60 cycle or 220 volt/50 cycle
Dimensions:	10" x 11" x 12"
Weight:	20 pounds

FIGURE 92. MDA personal monitoring system — tape sampler. (Courtesy of MDA Scientific, Inc., 808 Busse Hwy., Park Ridge, Ill.)

The tape is transported past the exposure orifice at a rate of 2 cm/hr while the contaminant in the sampled air reacts with the color-forming reagent. The tape is then wound onto the take-up reel while retained between two pressure plates, magnetically coupled, to prevent longitudinal leakage. Tape movement is provided by a coupled gear system through a governed DC motor. Power is supplied by a rechargeable battery pack worn on a belt or chest harness and provides 8-hour continuous operation. Because of size and weight limitations, the direct readout function is not incorporated in the monitor unit. Instead after the work-shift exposure (up to 8 hr), evaluation of the stained tape is conducted in a specially designed (Series 4100) integrating reader/recorder (Figure 92). The exposed tape is positioned at the optical drive and transported by a drive mechanism directly coupled to the chart drive. By directing light beams of equal intensity from a common source through matched fiber optics to both the top and bottom portions of the tape (exposed and unexposed regions), and mounting matched photoelectric detectors at a 45-° angle, the difference between the two reflected beams is measured. The photoelectric cell outputs produce a differential signal to the strip recorder which is logarithmic in response. A sample readout is shown by Figure 93. During the readout cycle of approximately 20 sec, the real time vs. concentration profile is recorded while a linearizing, integrating circuit computes the exposure integral, i.e., the area under the time/concentration curve. The accumulated total dose is transcribed in the form of a bar graph (total exposure dose).

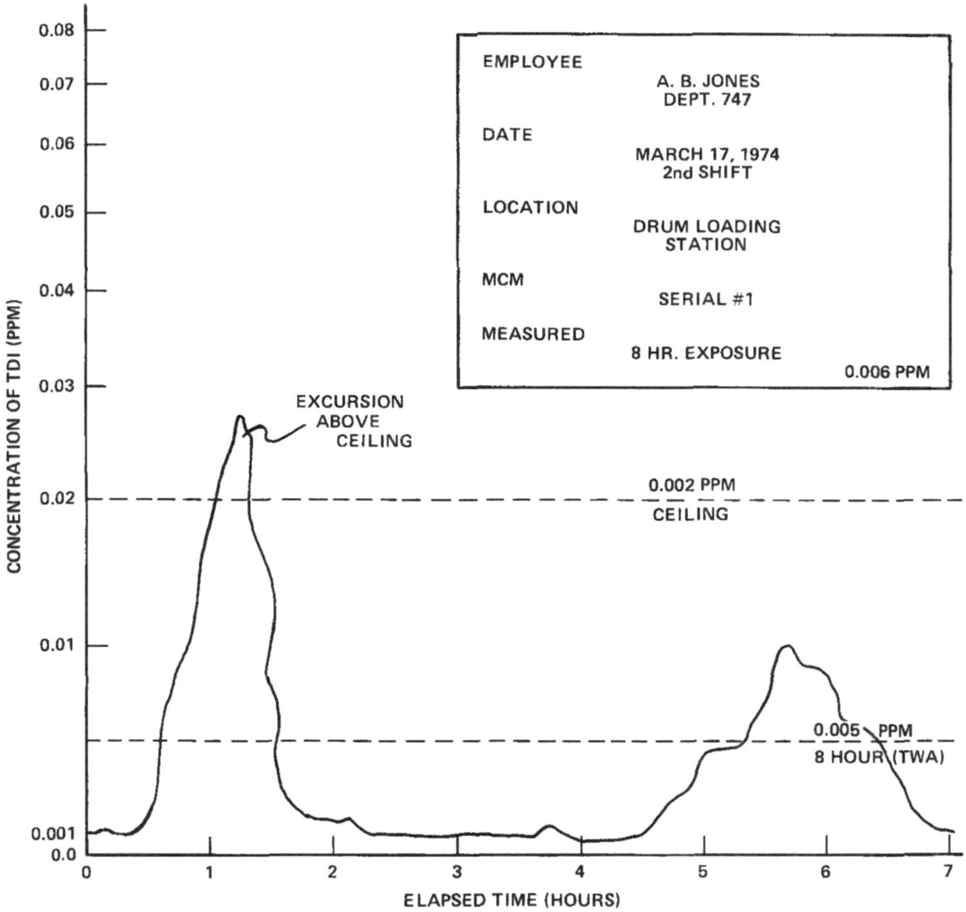

FIGURE 93. Typical readout from MCM (personal profile monitor). (Courtesy of MDA Scientific, Inc., 808 Busse Hwy., Park Ridge, Ill.)

The utility of this principle is clearly illustrated by the results from a 20-min sampling of a TDI vapor location simultaneously by the tape recorder and the Marcali method which uses a "wet" reagent in a midget impinger.[169] The Marcali method indicated 0.01 ppm for 20 min. The tape monitor disclosed 0.001 ppm during the first 5 min, the next 5 min at 0.078 ppm, and the last 10 min at 0.001 ppm. Integration of the area under the curves indicated an 0.02 ppm average TDI concentration, same as the Marcali procedure. An employee exposed under these conditions would have inhaled 100 to 140 times during the excursion at 0.078 ppm, or four times the TLV. A further advantage of the tape monitor is the provision of a permanent record which can be filed for future reference. Portable monitors can be supplied with an alarm which is activated as soon as the TWA limit, or the excursion limit, has been exceeded. Tapes are available for other common contaminants in addition to TDI vapor.

A study conducted by NIOSH confirmed good agreement between the Marcali method and the tape sampler for TDI.[170] Miller and Mueller also found good correlation.[171] New instruments developed jointly with Dow Chemical Co., Union Carbide, Moboy Chemical Co., and Olin Corporation for toluene diamine (TDA), ammonia, aromatic amines, aliphatic amines, and HCl were released in late 1977.[172]

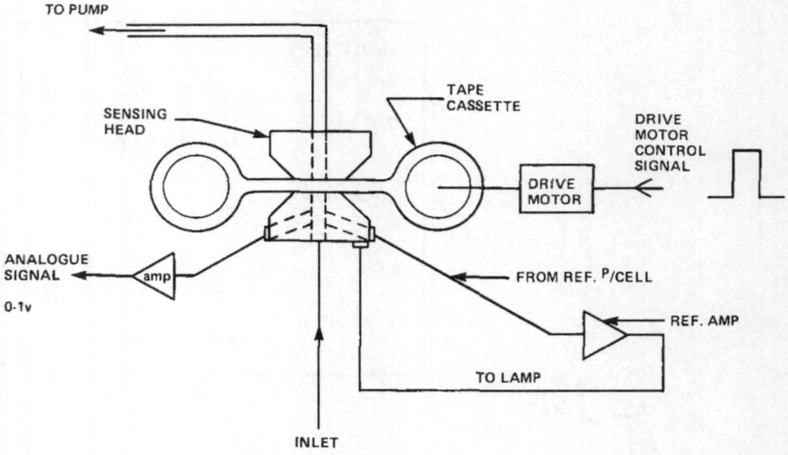

FIGURE 93A. "Chromatox" chest pack functional diagram. (Courtesy of MDA Scientific, Inc., 808 Bussee Hwy., Park Ridge, Ill.)

The Compur engineers in Germany have further miniaturized the continuous tape sampler that was presented to the American Industrial Hygiene Conference in Chicago, Ill. on May 29, 1979 by MDA. Tapes are available for the same gases as previously provided for the MDA MCM series of personal monitors, and all are interchangeable in the form of sealed cassettes, which are similar to photographic film cassettes. The Chromatox system consists of two basic units, the chest pack, which is worn in the breathing zone and the belt pack, which contains the battery operated pump and electronic control units.

Data storage and read out is provided by the Chromatox 4101 microprocessor-controlled dosimeter and the bimodal chart console which delivers the accumulated total exposure dose (TWA), a profile of concentration time, and a battery voltage record to confirm electrical system integrity (Figures 9A and 9B). The chest pack functional diagram is shown in Figure 93A and belt pack schematic in Figure 93B. An alarm system is built into the system to provide a warning relative to accumulated dose and concentration excursions above the permissible ceiling limit.

The air sample is drawn continuously through a defined area of the tape, which is advanced in time intervals preset to advance the tape every 2 min to deliver 240 readings during an 8 hr shift. Other intervals can be programmed also. The tape actually is stationary during the 2 min-sampling interval. In operation, the Chromatex provides a sampling data control pulse to the dosimeter which commands the photocell unit to read the reflectance immediately after the tape has been advanced (background or blank reading) and stores the signal in a temporary memory. After the sampling period the stain on the tape is read and transmitted to the dosimeter. The background memory is electronically subtracted from the sample reading to deliver a true stain intensity value to the microprocessor storage unit.

Sizes and Weights

Chest pads — dimensions: 11.1 cm × 51.1 cm × 31.8 cm
weight: 170 g
Belt pack — dimensions: 17.5 cm × 9.2 cm × 3.8 cm
weight: not given

Calibration accessories are available and include analog meter, battery voltage check, air flow meter, and precalibrated standard stain to be inserted in place of the tape.

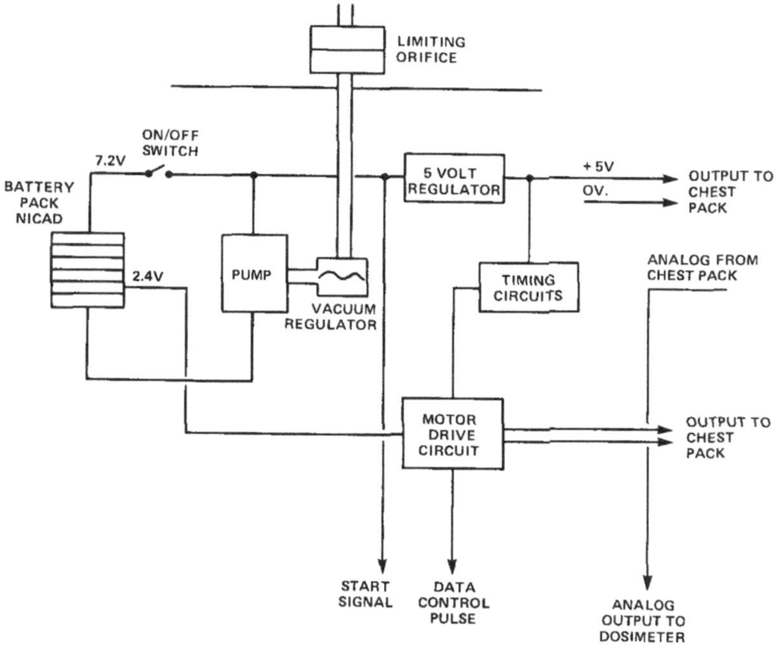

FIGURE 93B. "Chromatox" belt pack functional diagram. (Courtesy of MDA Scientific, Inc., 808 Busse Hwy., Park Ridge, Ill.)

b. Plastic Bag Sampling

Although plastic bags, which are flexible but not elastic, have been employed for the collection of air samples for laboratory analysis and to prepare calibration mixtures of pollutants in air,[24,267] application to personnel monitoring was not in general use until 1974 when the OSHA emergency standard for vinyl chloride monomer (VCM) was lowered to a 50-ppm ceiling, and a further final standard reduction to 1 ppm with an action level of 0.5 ppm.[173,174] Plastic bags offer several advantages: (a) when properly sealed, they behave as rigid containers in that the volume is constant at constant (atmospheric pressure), and therefore, can be calibrated with respect to retained volume; (b) they are easily transported in the field when empty; (c) they offer a practical facility for preparation of calibration mixtures of pollutants in air in the field; (d) they are not fragile as gas containers would be; and (e) sample aliquots can be removed without introducing diluent air, as in rigid containers.

However, certain disadvantages are involved: (a) permeability of the thin plastic membrane permitting loss of sample or accumulation of impurities by diffusion after sample collection, (b) adsorption of the air pollutant on the inside wall of the plastic bag, (c) reaction of the pollutant on the inside surface, (d) intereaction among constituents present in the air sample after collection, and (e) bulk after being filled. In view of these limitations, the use of plastic bags for either sample collection or standards preparation requires the determination of the half-life of the pollutants in the bag if the sample is stored for more than an hour before analysis. That is, sequential analyses should be determined in a geometric progression (1, 2, 4, 8, 16, 32, etc. hr intervals) until enough data is collected to extrapolate or estimate the half-life (time elapsed until 50% of the pollutant has disappeared.)

Although considerable data relative to diffusion rates of gases and vapors through plastic films has been developed, the information is dispersed widely through the jour-

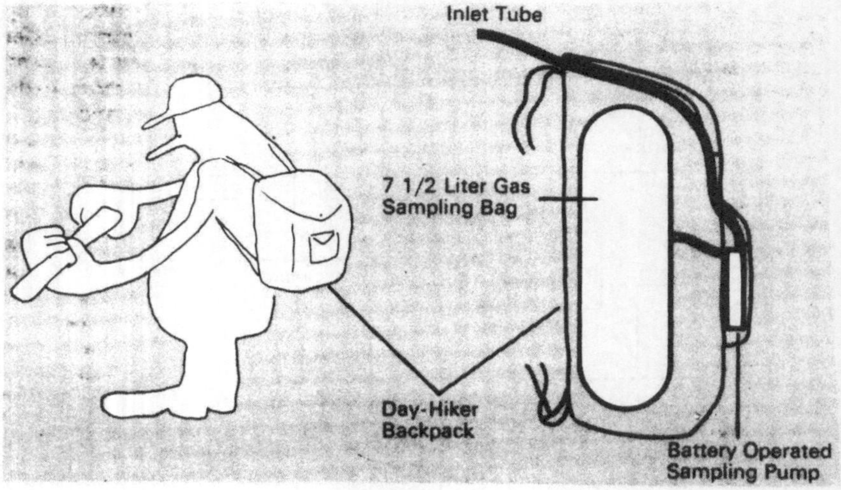

FIGURE 94. Vinyl chloride personnel sampling unit — back pack. (From Lucchesi, C. A., *Anal. Chem.*, 47, 1075A, 1975. With permission.)

nal literature and trade bulletins. Some quite useful data will be found in Nelson.[33] However, technical developments in the plastics field move along so rapidly as to render technical data obsolete within a few years, and therefore, the investigator must either carry out his own evaluation, or depend upon manufacturer's information. In one case, polytetrafluoroethylene (PTFE), rather complete permeative data is available.[24]

In response to the very restrictive VCM standard adopted by OSHA, Calibrated Instruments, Inc., staff, in collaboration with Stauffer Chemical Company, undertook development of an air-sampling bag which would solve the difficulties previously encountered.[174] PTFE was abandoned when a VCM loss of 20%/day was encountered. Development work was concentrated on aluminized three-layer Scotchpak plastic film when no loss of VCM in the range of 0.1 to 1.1 ppm VCM was established for at least 8 weeks.[174] Later a five-layer bag material was developed and is currently offered to customers (USP 3,907,164).

The NIOSH charcoal tube sampling procedure was used as a basis for comparison with the bag sampling technique for personnel monitoring.[42] Two improvements in the analytical procedure were found, desorption with CS_2 at "Dry Ice" bath temperatures and GC column packing Durapak (Carbowax 400 on Parasil) instead of SE-30.

The basic bag sampling procedure involves collection and storage of the sample gases without concentration by pumping the air sample into the bag with a battery operated pump from the employee's breathing zone. The contents of the bag then are analyzed directly by gas chromatography (GC) or any other suitable technique such as infrared (IR) spectrophotometry. The pump and bag are assembled in small backpacks (Figures 94 and 95a) or a side-pack (Figure 95).

Gas sampling bags which are commercially available[272] usually are fitted with a metal-twist lock (Figure 96) valve or with a filling snout (Figure 97). Some also are equipped with a permanent or replaceable septum for insertion of a syringe needle (Figure 97). For GC analysis, there is little need for a septum since maximum precision can be achieved more easily by using gas sampling valve injection rather than injection by gas syringe. The possibility of leaks in gas sampling bags requires leak testing with clean compressed air for a period of several hours before use or reuse. In actual field use, a 10% "mortality" rate for aluminized Scotchpak bags after repeated use was found.

FIGURE 95. Air sampling — personnel bag collection — side pack. (Courtesy of Calibrated Instruments, Inc., 731 Saw Mill River Rd., Ardsley, N.Y.)

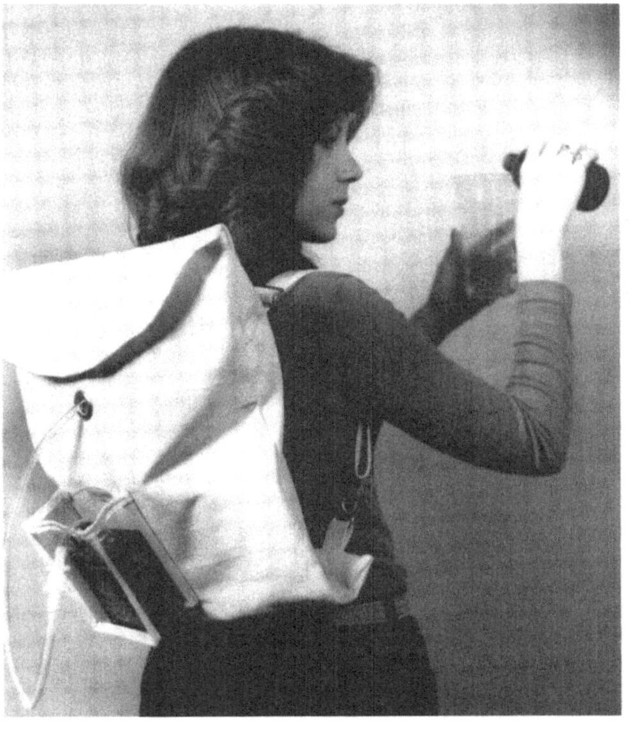

FIGURE 95a. The "no hands" personel pollution pack. (Courtesy of Calibrated Instruments, 731 Saw Mill River Rd., Ardsley, N.Y.)

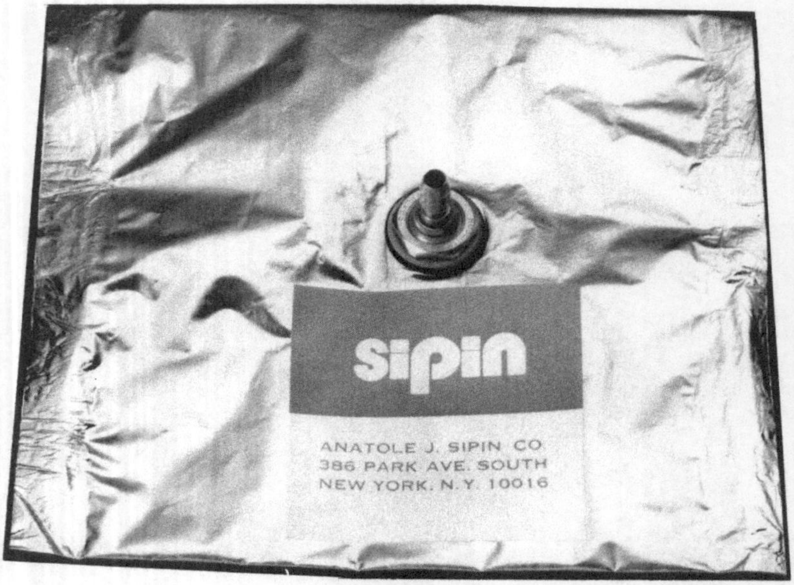

FIGURE 96. Gas collection bag. (Courtesy of Anato J. Sipin, Co., Inc., 425 Park Ave. S., New York, N.Y.)

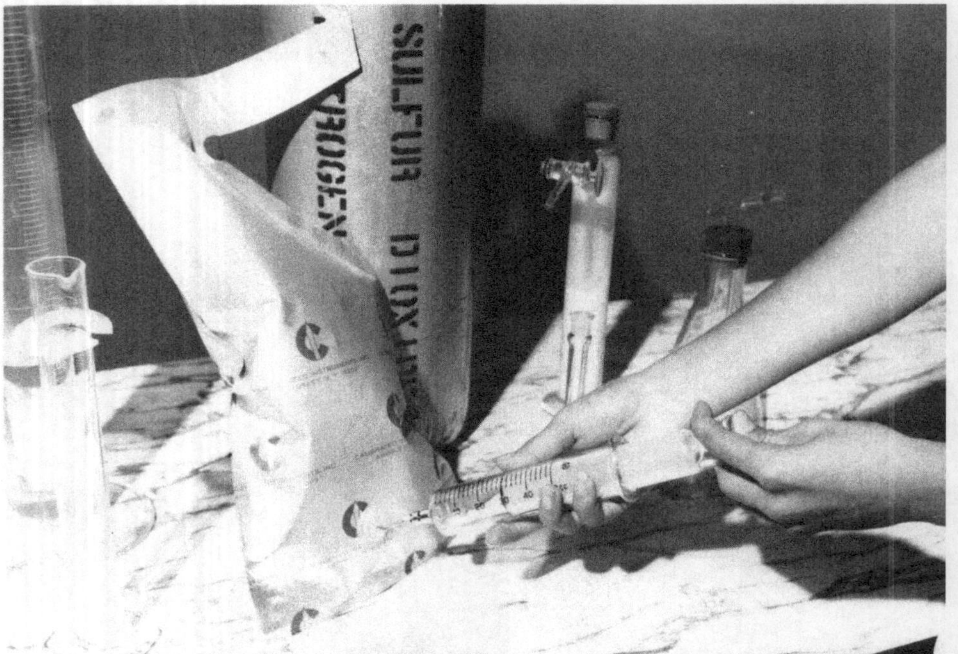

FIGURE 97. Bag sampler — removing sample aliquot for analysis or injecting contaminant for preparation of calibration mixture. (Courtesy of Calibrated Instruments, Inc., 731 Saw Mill River Rd., Ardsley, N.Y.)

FIGURE 98. Ambient air bag sampling. (Courtesy of Calibrated Instruments, Inc., 731 Saw Mill River Rd., Ardsley, N.Y.)

Bags filled with 1.0 to 10 ppm VCM in air presented no detectable amount (less than 0.03 ppm) of retained VCM after two repeated vacuum pumpings of the bags and refilling with compressed air. Therefore, three pump-and-fill cleaning cycles are recommended before reuse of the bag.

The performance of the sampling analytical system was evaluated by pumping 1.0 ppm VCM in air from a full gas sampling bag through PTFE-lined neoprene tubing into an empty bag to simulate field conditions. A recovery of 96 to 97% of the initial VCM was obtained. Application to collection of other gases and vapors should be evaluated in this manner to ensure minimal losses in the pumping system.

Entry into the bag can be made through either the on-off valve or the silicone rubber septum, or by snipping open the snout of the bag (Figure 97). The opened snout of the bag can be connected to "bubble" (Tygon) tubing for flushing the bag with clean air prior to reuse. The snipped snout can be reclosed easily by folding over and holding in place with a paper clip, or it can be resealed by heating. Pressure tests indicated an appropriate bursting strength at 0.46 at.

Application to portable ambient air sampling is shown in Figure 98. Usually a 5-L bag is employed to collect an air sample of an hour (pump rated at 5 L/hr) duration in either the portable or personnel monitor mode. A cycle timer is available to deliver discrete samples over a 24-hr period. The batteries provide approximately 20 hrs operation. Two 22- and 44-L bags can be supplied.

Dilution losses of trace hydrocarbons in 5-L bags over a 10-wk storage at room temperature were less than experimental analytical errors when compared to storage in a steel cylinder for CH_4, C_2H_6, C_2H_2, and C_2H_4 in a concentration range of 9 to 10 ppm each. Annual (365 days) diffusion volumes for four gases in helium were:

	Argon	CO_2	CH_4	H_2
Concentration after storage	14—18	120—127	21—24	170
Diffusion Volume* — mL	1.5	10.8	2	14.7

* From the air retained in a second bag in which the test bag was enclosed.

When placed side by side, carbon monoxide results from bag samples and continuous electrochemical analyzers failed to correlate well. However, if the samples were drawn from a common manifold, concentrations agreed within ±1 ppm for at least 90% of the data points. To determine sample loss in storage, calibrated gas standards contained in bags fabricated from various materials were analyzed after 0, 24, 48, and 100 hr. Polyvinyl chloride bags were satisfactory for sample storage for about 15 hr, whereas Tedlar® bags were in general unsatisfactory for sample storage. Bags fabricated from layers of polyester, polyvinyl chloride, aluminum foil, polyamide, and polyethylene were rated excellent for long term storage of CO, but completely unacceptable for nonmethane hydrocarbons. Aluminized polyester bags made up of a layer of polyester on both sides of the aluminum sheet rated excellent for long term storage of both CO and nonmethane hydrocarbons. However, these bags are no longer available since the laminated sheet was discontinued by the manufacturer (Minnesota Mining — 3M). Tests confirmed the losses of components in the sample in rubber diaphragm pump and gum rubber tubing (0.6 in 10 ppm).[174a]

Direct reading visual comparators such as indicator papers, chemical dosimeters, and detector tubes provide an unquestionably valuable service in the hands of the qualified industrial hygienist for obtaining immediate semiquantitative measurements from screening surveys. However, as previously emphasized (see Section II.A.1.b.vi.h, "Specificity and Interferences"), the preliminary survey should be followed by a quantitative evaluation, especially when results fall within the one-half to one and one-half TLV range, or when the results will provide the basis for assigning a TLV to a hazardous substance. If a mixture of similar compounds is known to be present, or interferences are anticipated, separation procedures must be carried out in the laboratory to minimize errors inherent in nonspecific methods. Furthermore, by reserving a portion of each field sample for later analysis, constituents other than those initially investigated can be determined without returning to the location to secure additional samples. Also, the validity of the study can be improved by replicate analysis by the same or different methods (see Chapter IV). Laboratory analysis of collected samples minimizes the loss of valuable data missed when the direct reading instrument or spot detector responds erroneously or not at all. Most analytical laboratory methods quantitate color densities in a spectrophotometer, thereby eliminating subjective interpretation, a major source of analytical error. Sensitivity can be greatly increased through concentration of the target component, or by application of sophisticated instruments that are not amenable to field application (gas chromatography, thin layer chromatography, ultraviolet or infrared spectroscopy, atomic absorption spectroscopy, etc.).

The most obvious solution to the complete analysis problem would be to transport a sample of the atmosphere itself back to the laboratory. However, as applied to personnel monitoring, this technique is at best impractical due to the large volumes required for most analyses. In some few cases where sufficient sensitivity is available to permit the analysis of small sample volumes, conceivably a sample could be collected over a period of hours by a fine capillary "bleed" into a small evacuated rigid vessel (glass tube) or by slowly evacuating a rigid container in which a plastic bag (bladder) is sealed by means of a bulkhead capillary fitting. The sample would be drawn as the vacuum created within the rigid container expanded the bag.

With the exception of a rubber squeeze bulb restricted by a limiting orifice to draw an air sample through a trichloroethylene detector tube at a constant 15 ml/min rate over an 8-hr period,[110] equipment of this type is not commercially available, nor has it been described in the literature. Furthermore, the relatively large surface-to-volume ratio greatly increases the errors due to adsorption and chemical interactions of the air contaminants on the container walls and losses from leaks.

c. Impinger Systems — Liquid Reagents

1. The Midget Impinger

The alternative to transporting the entire air sample is to collect the contaminants from a known volume of the atmosphere in a small volume of liquid absorbant or on a solid adsorbant. One of the most widely used and thoroughly studied collection devices suitable for application to personnel monitoring is the midget impinger. The essential specifications and design features of a currently available model are shown in Figures 99 and 100. The original model (1938) employed a rubber stopper to retain the impinger tube,[175] and an all-glass counterpart in which a ring seal replaced the rubber stopper, and was introduced commercially shortly thereafter (1944).[176] A Teflon® (Registered U.S. patent for Du Pont's fluorocarbon resins and film) sleeve was fitted to the inner member of the joint as shown in Figure 100.[138] The elimination of contamination derived from reagent contact with either the rubber stopper or the joint lubricant is especially critical if the analytical procedure is based on ultraviolet or infrared spectrometry.

Although the midget impinger was designed primarily for aerosol (dust and liquid) collection, its application to gases and vapors has been well documented.[177, 178] Collection efficiency in the range of 85 to 95% is usually attained, and may reach 95 to 99% if the gas or vapor to be collected is highly soluble in the collecting liquid or if the gas reacts rapidly with a component of the liquid reagent, for example, ammonia collected in dilute sulfuric acid. The major factors that influence this system are[179]

1. Solubility of the contaminant in the collecting liquid.
2. Rate of diffusion of contaminant into the liquid.
3. Vapor pressure of contaminant at the sampling temperature.
4. Chemical reactivity of the contaminant with the reagent (including hydrogen bonding as in collecting acetone in chloroform).
5. Contact area (size of air bubble) and time of contact (flow rate) of the contaminated air through the liquid reagent.
6. Volatility of the collecting liquid.
7. Analytical techniques to be applied after the sample has been collected.

Without a reagent to react with the gas in the liquid medium, the rate of absorption may be slower, and a portion of the gas that does dissolve may be swept out by subsequent aeration (escape factor). The ratio of the air sample volume to the volume of collecting liquid is a controlling variable in collection efficiency, and beyond the critical limit of this ratio the efficiency declines significantly. For acetone in water, the ratio would be approximately 5-L air per 20 mL water distributed between two "bubblers" connected in series. Therefore, if an attempt were made to collect a 20-L air sample in this system at room temperature, the collection efficiency would be very poor, in fact, unacceptable. Methylethylketone (MEK) would be even more poorly absorbed than

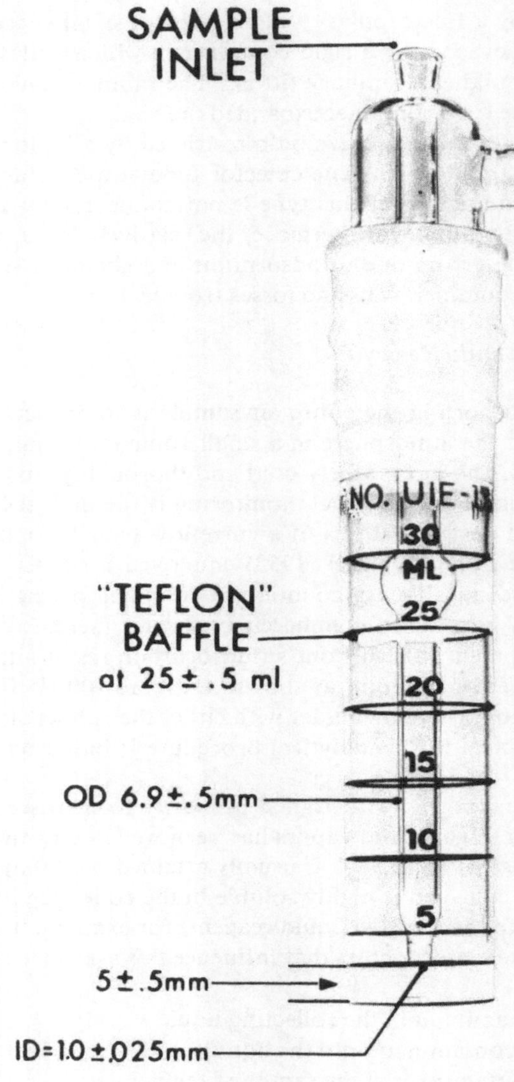

FIGURE 99. The standard midget impinger specifications. (From Linch, A. L., Davis, R. B., Stalzer, R. F., and Anzilloti, W. F., *Am. Ind. Hyg. Assoc. J.*, 25, 81, 1964. With permission. Supplied by Ace Glass, Inc., Vineland, N.J.)

absorbed than acetone in water.[180] In some cases this escape factor can be calculated.[182]

The close-tolerance specifications (Figure 99) required to maintain a reproducible flow rate of 2.8 L/min at 305 mm pressure drop (Δ P) across the impinger nozzle and an impinging velocity of 60 m/sec intended for dust collection are not critical for gas and vapor collection. In fact, excessive liquid entrainment at the 2.8 L/min rate may reduce the collection efficiency.[181] Considerable latitude in flow rate (0.56 to 5.6 L/min) without sacrificing more than 5% efficiency is permissible[178] unless the limiting orifice effect of the nozzle is required to maintain a reproducible, constant flow rate on which to base sample size. In systems provided with a calibrated rotameter and a

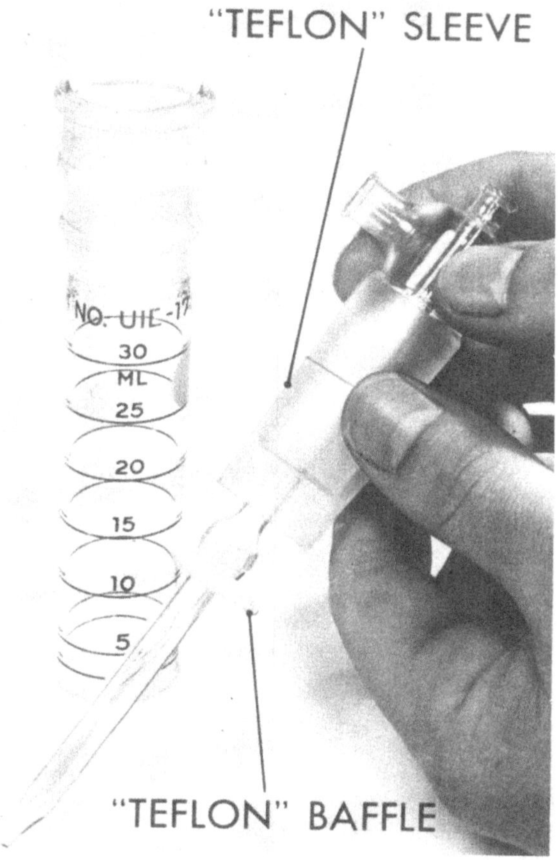

FIGURE 100. The standard midget impinger — exploded view. (From Linch, A. L., Davis, R. B., Stalzer, R. F., and Anzilloti, W. F., *Am. Ind. Hyg. Assoc. J.*, 25, 81, 1964. With permission.)

timing device to calculate the total volume collected, the impinger nozzle must be matched with the pump assembly or the geometric tolerances held within limits consistent with the established limits of sampling error.[138,183] In any event, the distance from the orifice to the cylinder base is not at all critical for gas or vapor collection.

Two impingers connected in series will increase the overall collection efficiency when the rate of absorption is slow or higher sampling rates are desired. In developing a collection method, series collection always should be employed to establish the average and range of collection efficiency. The amount collected in the first midget impinger divided by the total collected multiplied by 100 equals percentage of efficiency. If the amount collected in the second impinger is greater than 10%, then the performance cannot be considered satisfactory. The added Δ P introduced by the second impinger nozzle must be taken into consideration when calculating the sample volume. The Δ P effects are not directly additive; therefore, the most prudent approach is to calibrate volumetrically with all components of the system in place.[184] The following section is included here by permission of the copyright holder.

Replacement of the rubber stopper in the original midget impinger by an all-glass ring seal to support the impinger tube did eliminate undesirable contact of reagent with rubber, reduced spray carryover and provided more convenience in assembling, but at the cost of flexibility in choice of orifice size or replacement of damaged nozzles. The introduction of elastic "O" ring seals for support of glass mercurial thermometers in standard-taper holders offered an alternative design which would retain the advantages of an all-glass system with the flexibility of the original model.

The fragile glass ring seal in the midget impinger head was replaced with a pair of "O" ring seals in an indented tube to provide a flexible support for the impinger tube (Figure 101). The distance between the orifice and the cylinder base is adjustable by sliding action, but lateral movement of the orifice is prevented by spacing the "O" rings, which limits motion to rotation and longitudinal movement. Tubing to be connected to these seals is slightly tapered and conservatively lubricated — usually water is sufficient — and then inserted with a rotating motion to prevent displacement of the "O" ring from its channel. The optional porous glass bubbler (Figure 101) illustrates one of many possible alternative inlet tube forms which can be substituted for the easily removed impinger tube. Because the distance between orifice and cylinder base was adjustable, and impinger orifices were ground within tolerances that permit interchange without preliminary calibration, the matching of midget impinger components was not necessary for reproducible performance. The rapid replacement of damaged inlet tubes or changeover to a different orifice size not only reduced cost but also facilitated field sampling. Tolerances required for proper fit in the inlet tube "O" ring seal (tube outside diameter, 6.4 to 6.5 mm; "O" ring retainer inside diameter, 7.0 to 1 mm) did not pose a problem for competent glass shops.

The addition of a third "O" ring in the outlet tube provided essentially glass-to-glass connections for series assembly or for direct coupling to other collection components (Figure 101), without sacrificing the freedom associated with flexible tubing connections. However, addition of a right-angle bend in the inlet tube for series connection limited the tip to a diameter that would pass through the "O" ring.

The relatively inexpensive impinger tubes were interchanged or replaced easily and permitted selection of orifices within Δ P-flow rate tolerances which made the matching of assemblies and individual calibrations unnecessary. Tubes fabricated with a variety of orifice sizes, porous glass bubblers, spiral construction or other modifications could be used with the same head and cylinder.

ii. The Microimpinger

The recent development of a self-powered aspirator[138, 183] and battery-powered pumps[148, 159] small enough to be worn as personnel monitors by workmen during their routine job assignments created an immediate need for further miniaturization of the midget impinger. Consideration of several proposed designs[185, 186] indicated that consolidation of desirable features and incorporation of "O" ring seals could provide a versatile microimpinger one third the size of the standard midget impingers, without sacrificing the performance characteristics of the original impinger design.

Although every effort was made to scale down the midget impinger on the basis of critical design parameters — that is, air velocity and pressure drop — this was not considered sufficient basis for assumption of equivalent performance. Therefore, the collection efficiencies for aniline vapor, SO_2 gas, and silica dust were selected as measures of probable field utility.

Miniaturization of the standard midget impinger to one-third of the original size resulted in a microimpinger which operated at one-fifth of the sampling rate and consumed one-fifth of the reagent required with the larger model (Figure 102). Since all physical parameters were scaled down by one-fifth, it is important to note that it is not necessary to recalibrate or to revise analytical procedures established for the standard midget impinger.

In our design, "O" ring seals provided the same flexibility and interchangeability features described for the revised midget impinger.[184] Replacement of the conical ground orifice with a short length of precision-bore capillary tubing eliminated the need for adjustment of the orifice to provide matching impinger assemblies.[145]

a) Calibration

The diameters of the orifices were measured with a microscope equipped with a low-power objective (8 ×) and a 10 × eyepiece with a calibrated graticule. The body of the microscope was tilted to the horizontal position, and the substage condenser was removed to permit insertion of a one-hole rubber stopper to permit the operator to focus on the orifice plane. The measured capillary diameters agreed with the manufacturer's values to within the error of measurement. From these results, air velocity in the nozzle was calculated for flow rates of 0.01, 0.02, and 0.03 cfm where

$$V = 0.0283\, q/60A$$

where V = air velocity, meters per second (m/sec); q = air flow rate, cubic feet per minute (cfm); and A = surface area, square meters (m^2).

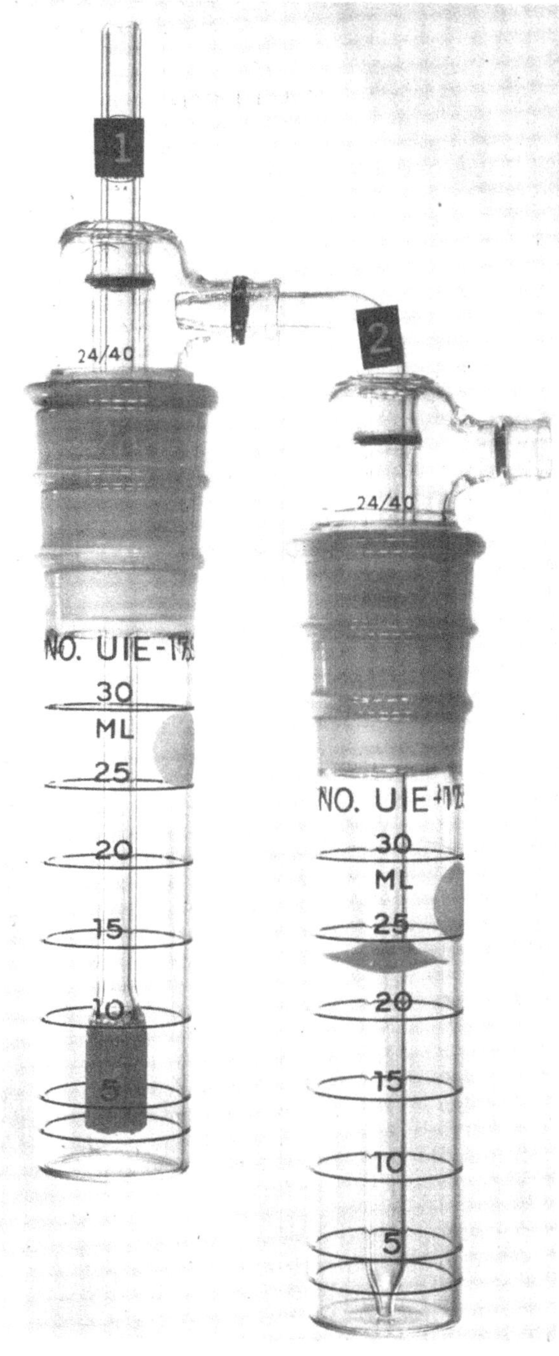

FIGURE 101. The improved midget impinger and bubbler in series. (From Linch, A. L. and Corn, M., *Am. Ind. Hyg. Assoc. J.*, 26, 601, 1965. With permission. Supplied by Ace Glass, Inc., Vineland, N.J.)

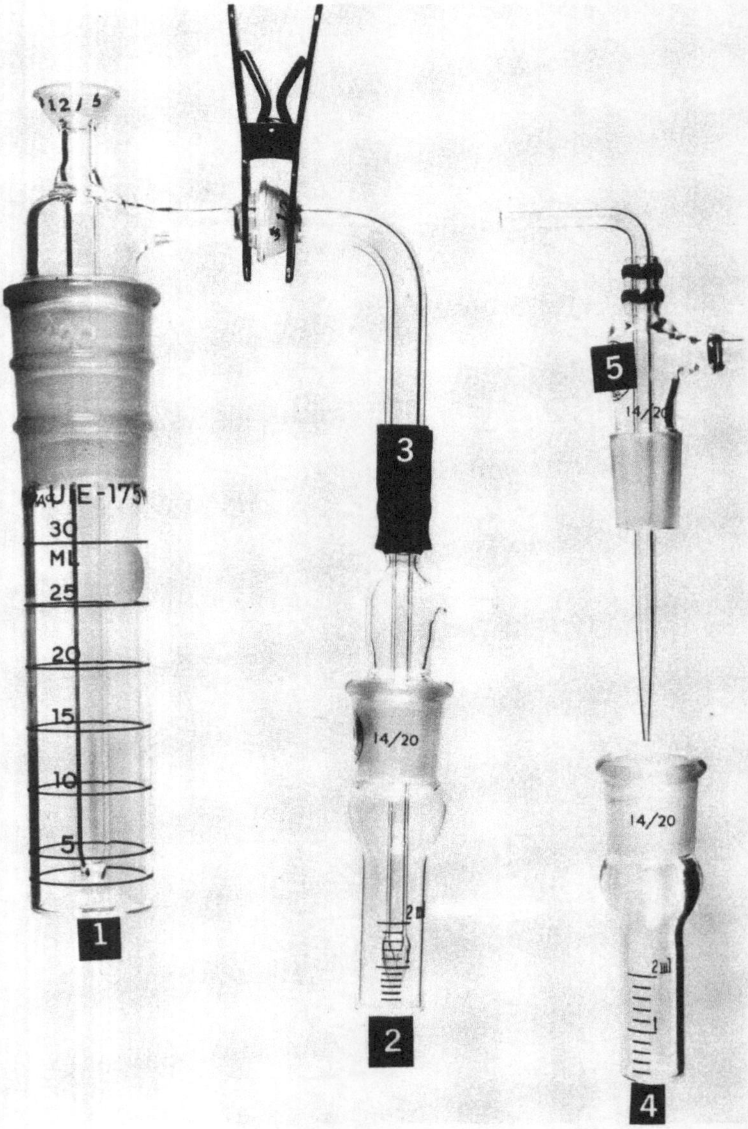

FIGURE 102. The standard midget impinger vaporizer coupled to the microimpinger gas absorber. (From Linch, A. L. and Corn, M. *Am. Ind. Hyg. Assoc. J.*, 26, 601, 1965. With permission. Supplied by Ace Glass, Inc., Vineland, N.J.)

Relationships of pressure drop (Δ P) to flow rate for orifices of different diameters were determined with an open-end water manometer inserted between the vacuum source control valve and the vent of the impinger. A dry test meter was connected to the impinger tube (American Meter Company, Model 2-40) to measure air volume. All runs were made with 10 ml of water in the midget impinger or 2 ml of water in the microimpinger.

Figure 103 illustrates the nozzle velocities obtained with some commercially available capillary tips. Figure 104 summarizes the measured pressure drops across the microimpinger with three different capillary tips. The reproducibility of flow characteristics with different impingers made it possible to operate simultaneously two or more units in parallel without the need for individual flow meters, control valves, or volume correction. As many as five microimpingers have been operated in parallel without increasing the vacuum capacity required for a single midget impinger.

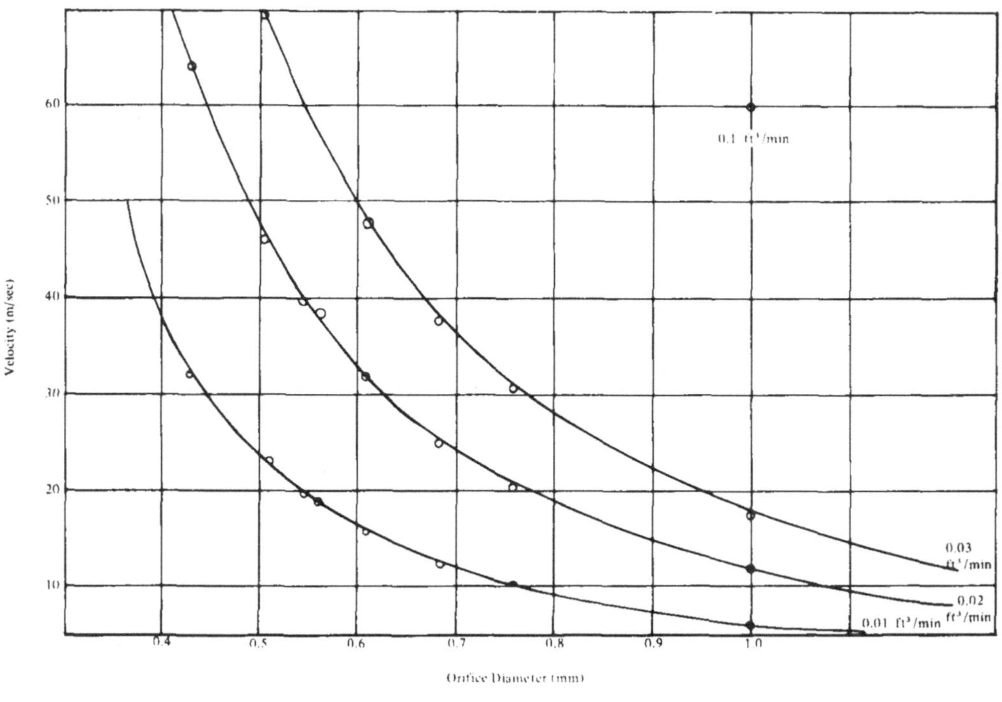

FIGURE 103. Calculated air velocities for commercially available capillary nozzles. (From Linch, A. L. and Corn, M., *Am. Ind. Hyg. Assoc. J.*, 26, 601, 1965. With permission.)

At rates as high as 0.04 cfm, physical entrainment of the absorber liquid did not create a problem if care was taken to exclude detergents, greases, and other foam generating materials. Consistently good results were obtained by soaking units in 1% aqueous trisodium phosphate in water for several hours and following this with 1:1 nitric acid and distilled water rinse.

b) Efficiency Evaluation

Aniline Vapor — Sampling — Two to three drops of reagent-grade aniline were dropped into a 5-gallon "Pyrex" glass bottle fitted with a rubber stopper covered with aluminum foil. Two sampling lines and a glass sleeve for the shaft of a "Teflon" bladed agitator were inserted through the stopper. The air in the bottles was agitated until the aniline was completely vaporized and then for an additional hour to ensure thorough mixing. Two impingers connected in series, each containing 10 ml of spectrograde ethanol and chilled in an ice pack, were used for sampling.

Recovery of aniline in alcohol was more consistent both in the laboratory and in the field than was recovery in dilute (1% or 10%) aqueous hydrochloric acid. Therefore, alcohol was adopted as the collecting reagent of choice for both aromatic amino and nitro-compound collection from air samples. Air flow through the midget impinger was held at the 0.1-cfm rate which Roberts and McKee found to be the mid-point of maximum collection efficiency.[178] The contents of the impingers were analyzed separately to determine efficiency of collection by the first impinger. Aniline was determined as follows.

c) Reagents

Hydrochloric Acid, 10% — Dilute 200 ml of analytical reagent grade 37% hydrochloric acid with 340 ml of distilled water.

Chicago Acid — Dissolve 0.5 gm of 8-amino-1-naphthol-5, 7-disulfonic acid in 1 ml of 37% hydrochloric acid and 45 of distilled water, and dilute to 50 ml with distilled water. Store in refrigerator and avoid exposure to light. Discard after the third day.

Sulfamic Acid — Dissolve 10 gm of 99% sulfamic acid in distilled water and dilute to 100 ml.

Sodium Nitrite, 3% Aqueous Solution — Discard when color appears any darker than a faint lemon-yellow. This reagent is stable for about 2 weeks.

Sodium Acetate Buffer — Dissolve 320 gm of analytical reagent grade sodium acetate trihydrate in 1000 ml of distilled water.

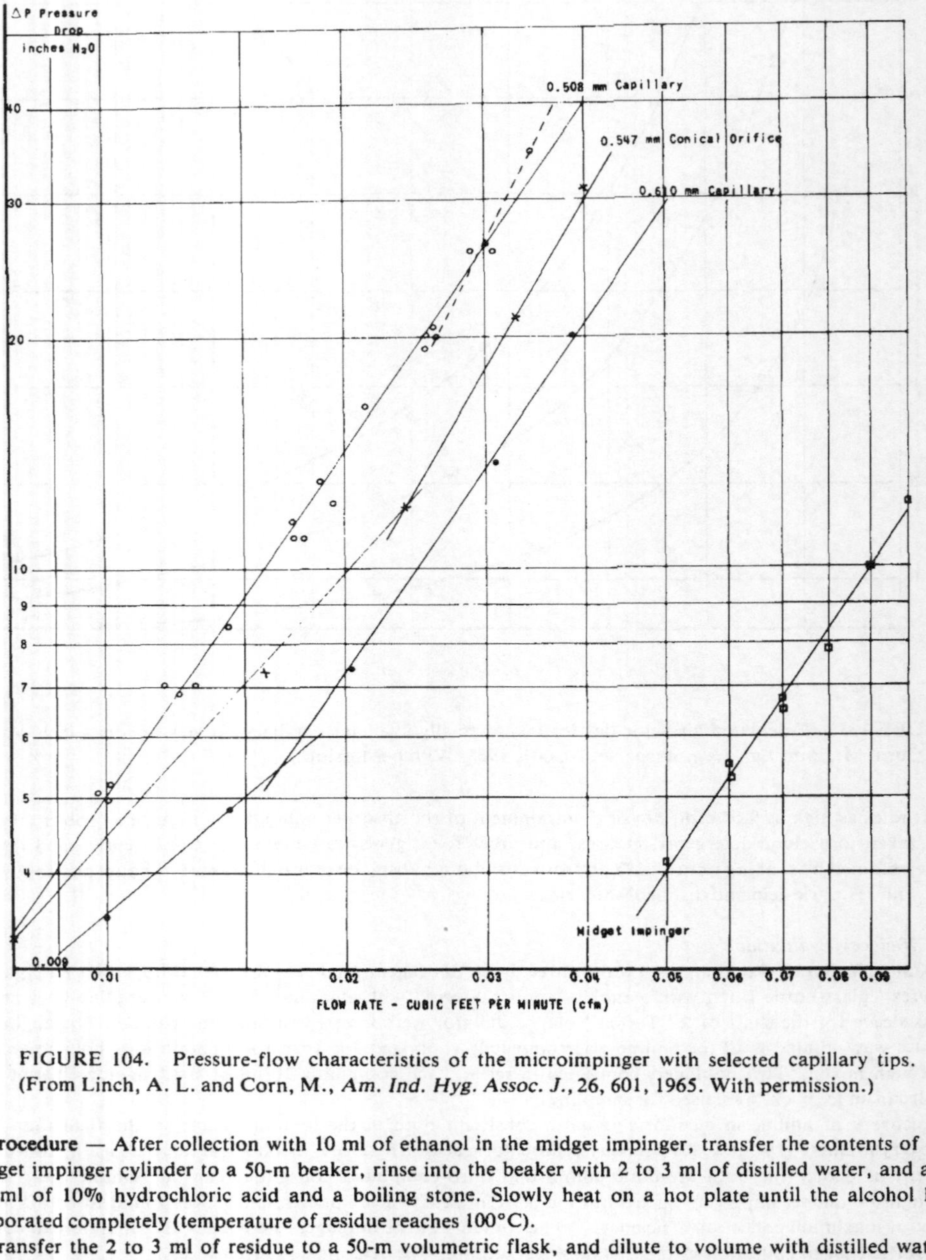

FIGURE 104. Pressure-flow characteristics of the microimpinger with selected capillary tips. (From Linch, A. L. and Corn, M., *Am. Ind. Hyg. Assoc. J.*, 26, 601, 1965. With permission.)

Procedure — After collection with 10 ml of ethanol in the midget impinger, transfer the contents of the midget impinger cylinder to a 50-m beaker, rinse into the beaker with 2 to 3 ml of distilled water, and add 1.5 ml of 10% hydrochloric acid and a boiling stone. Slowly heat on a hot plate until the alcohol has evaporated completely (temperature of residue reaches 100°C).

Transfer the 2 to 3 ml of residue to a 50-m volumetric flask, and dilute to volume with distilled water. Mix thoroughly and pipet 5 ml of this solution into each of two 25-ml volumetric flasks. Chill to 0 to 5°C in an ice bath, add 10 ml of distilled water, and diazotize by adding 1 ml of sodium nitrite reagent to each flask. Mix by swirling, and after 1 minute destroy excess nitrous acid by adding 1 ml of sulfamic acid reagent. Mix thoroughly, let stand at 0°C for 10 to 15 minutes, and test for nitrous acid by spotting on cadmium iodide-starch test paper. If a positive test appears, add another 1 ml aliquot of sulfamic acid, and let stand for another 20 minutes.

Add 0.5 ml of Chicago Acid reagent to one flask only; and 5 ml of sodium acetate reagent to each flask. Mix by swirling, heat to 50 ± 5°C for 5 to 10 minutes to ensure complete color development, and dilute to volume with distilled water. Within 4 hours after coupling, determine light transmission at 520 mμ. The micrograms of aniline are read from a calibration chart (semilogarithmic graph paper) based on known quantities of aniline in aliquots taken from a standard solution (100 μg/ml) in 1% aqueous hydrochloric acid.

$$\text{ppm Aniline} = \frac{10 \times \mu g \times 263}{1000 \times 28.3 \times \text{vol. (ft}^3)}$$

$$= \frac{\mu g \times 0.0093}{\text{vol.}}$$

TABLE 60

Aniline Losses in Gum Rubber Tubing Connected to the Midget Impinger Vaporizer[a,b]

Exposed tubing length (in.)	Age of tubing (days)	Aniline recovery (%)
0	—	100
2.5	—	91
9.5	0	83
9.5	1	76
9.5	7	65
9.5	8	57
19.5	0	80
19.5	7	58
19.5	8	50
29.0	10	41

[a] Collected in 95% ethanol in midget impinger.
[b] Vaporization finished at 60°C for 5 min.

From Linch, A. L. and Corn, M., *Am. Ind. Hyg. Assoc. J.*, 26, 601, 1965. With permission.

The sensitivity for trace concentrations can be increased by repeating the analysis with a 20-ml aliquot from the initial 50 ml of dilution mixture, and revising the volumetric correction in the final calculation (factor 2.5 instead of 10).

Analysis of the 2 ml of alcohol used to collect aniline in the microimpinger is identical except that all volumes are divided by 5, and final color dilution is made in 5-ml graduated cylinders calibrated to 0.1-ml intervals.

When significant losses of aniline by absorption in gum rubber tubing connections were confirmed (see Table 60), the essentially all-glass system shown in Figures 101 and 102 was adopted for efficiency evaluation. In this case, "clean" air was drawn through the system at 0.1 cfm for the midget impinger or 0.02 cfm for the microimpinger. A known amount of aniline in the range of 10 to 100 μg was then pipetted into the impinger tube (part 1, Figure 102) and allowed to vaporize into the collecting impinger. The last traces of aniline were vaporized, probably from absorption on the glass, by heating the vaporizing impinger (part 1, Figure 102) for 5 minutes in a 60°C water bath. A solution of 100 μg/ml in alcohol was used.

Results — Serious aniline losses were encountered when the contaminated air sample was drawn through gum rubber tubing. Furthermore, this loss increased as the age and number of uses increased (Table 60). Even 3-inch connections introduced significant losses, but the loss was not entirely proportional to hose length. With an all-glass system (ball joints or "O" ring seals), 100% recoveries could be attained. However, at room temperature some adsorption on the glass surfaces of the vaporizer was encountered which could be cleared by heating to 60°C.

Based both on sampling from the prepared air mixture and use of the midget impinger as a vaporizer the efficiency of the microimpinger operated at 0.02 cgm was equivalent or superior to the performance of the midget impinger at 0.1 cfm (Tables 61 and 62). Near quantitative recoveries (98 to 100%) were attained with either impinger when all-glass systems were employed. Increasing the microimpinger flow rate from 0.02 to 0.04 cfm did not significantly reduce collection efficiency, but increasing the midget impinger rate from 0.1 to 0.5 cfm lowered the efficiency by more than 15%.[178]

Sulfur Dioxide Procedure — Standard atmospheres containing sulfur dioxide in the concentration range 0.5 to 5 ppm were prepared by means of a dynamic dilution system modeled after the one described by Kusnetz et al.[113] Compressed gas in a stainless-steel tank at a nominal SO_2 concentration of 300 ppm in

TABLE 61

Recovery of Aniline Vapor from Mixtures in Air by Two Impingers in Series[a,b]

Pressure drop (in. H_2O)	Sampling rate (ft^3/m)	Efficiency of first impinger (%)	Aniline (ppm)
Microimpinger			
12.3	0.02	90.0	11.9
22.3	0.02	93.7	11.8
16.4	0.02	96.0	9.6
19.0	0.02	95.8	5.5
19.6	0.02	88.0	3.9
23.7	0.03	89.5	1.8
25.0	0.04	91.7	1.1
Average		92.6	
19.5	0.02	83.3[c]	1.4
Midget Impinger			
12	0.10	89.7	11.8
12	0.10	93.8	3.2
12	0.08	95.5	3.2
—	0.15	85.2	1.4
—	0.16	81.0	1.2
Average		89.0	
12	0.10	94.5[c]	1.0

[a] Sampling time = 10 min.
[b] Absorbed in 95% ethanol.
[c] Absorbed in 10% aqueous HCl.

From Linch, A. L. and Corn, M., *Am. Ind. Hyg. Assoc. J.*, 26, 601, 1965. With permission.

TABLE 62

Aniline Recovery in the Midget Impinger Vaporizer[a]

Aniline added (μg)	Vaporization time (min) 20°C	60°C	Aniline recovered (%) Micro-impinger	Midget impinger
Rubber tubing connection (3 in.) from vaporization to absorber				
110	10	—	42.2[b]	65.0[b]
92	10	—	61.0	85.6
103	20	—	80.8	85.3
103	10	5	95.0	91.0
Ball joint or O-ring seals				
100	5	5	—	101.0
105	5	5	98.1	96.0
11	5	5	109.0	100.0
13	5	5	96.2	108.0
1150	5	5	102.6	102.2

[a] Absorbed in 95% ethanol.
[b] Absorbed in 10% aqueous HCl.

From Linch, A. L. and Corn, M., *Am. Ind. Hyg. Assoc. J.*, 26, 601, 1965. With permission.

TABLE 63

Comparison of Midget and Microimpinger Efficiencies for SO_2 in Air[a]

SO_2 concentration by midget impinger collection (ppm)	Microimpinger performance	
	Sampling time (min)	Collection efficiency (%)
0.069	30[b]	99.5
0.10	10	100.0
0.20	10	99.1
0.65	10	99.8
1.0	10	100.0

[a] Midget impinger collection efficiency = 100%, Midget impinger reagent volume = 10 mL, Microimpinger reagent volume = 2 mL, Midget impinger sampling rate = 0.099 ± 0.001 ft^3/m (25°C and 760 mm Hg), Microimpinger sampling rate = 0.0194 ± 0.004 ft^3/m (25°C and 760 mm Hg), two samples at each rate.

[b] Final volume = 1.8 mL.

From Linch, A. L. and Corn, M., *Am. Ind. Hyg. Assoc. J.*, 26, 601, 1965. With permission.

nitrogen was mixed with purified dilution air to achieve the above concentrations of SO_2. The midget impinger and microimpinger, each containing 0.1 M disodium tetrachloromercurate (II) scrubbing solution, were run simultaneously and used to sample standard concentrations from a flask at the final stage of the dilution system. The scrubbing solution and method of analysis were those of West and Gaeke.[187] As reported by Roberts and McKee,[178] collection of 1 ppm of SO_2 in air in disodium tetrachloromercurate is essentially complete with the midget impinger at a flow rate of 0.1 cfm. Therefore, in this evaluation the microimpinger was compared to the midget impinger; that is, collection efficiency for the midget impinger was considered to be 100%.

Results — Table 63 summarizes results of tests with the micro and midget impingers at 735 to 760 mm Hg and 20° to 25°C with different concentrations of SO_2 and at a sampling rate of approximately 0.02 cfm. Collection of SO_2 at concentrations equal to or less than 1 ppm was essentially complete, with deviations from 100% efficiency being attributed to experimental errors.

Summary

A threefold reduction in the dimensions of this "O" ring model midget impinger gave a microimpinger which performed efficiently at one-fifth the flow rate and with only one-fifth of the sample and reagent volumes required for the larger counterpart. Results for aniline vapor, sulfur dioxide, and dust collection confirmed performance equivalent or superior to that of the standard midget impinger. Replacement of the conical ground orifice with a short section of precision-bore capillary permitted fabrication precision consistently within tolerances which permitted interchangeability without calibration to ensure matched performance. As many as 5 units could be operated in parallel with no increase in aspirating capacity over that required for a single midget impinger, or need for individual flow control meters. Combined with a recently available battery-operated miniature pump, the microimpinger provides a compact personnel monitoring instrument which delivers standard midget impinger results.

An all-glass vaporizer assembly was used to determine impinger performance after significant losses of aniline by adsorption on gum rubber connecting hoses were confirmed. Collection efficiency in alcohol from a static mixture through impingers in series varied from 81 to 96% for the midget model in a concentration range of 1 to 12 ppm. Under similar conditions the microimpinger gave 88 to 96% recoveries. Both impingers gave essentially quantitative recoveries (96 to 100%) when assembled with the all-glass vaporizer system. Microimpinger performance was not significantly altered by increasing the sampling rate from 0.02 to 0.04 cfm, but a 50% rate increase (0.10 to 0.15 cfm) through the midget impinger decreased the efficiency by more than 15%.

Collection of sulfur dioxide from air in a concentration range of 0.07 to 1.0 ppm was essentially complete for both impingers when operated at the recommended flow rates.*

* From Linch, A. L. and Corn, M., *Am. Ind. Hyg. Assoc. J.*, 26, 601, 1965. With permission.

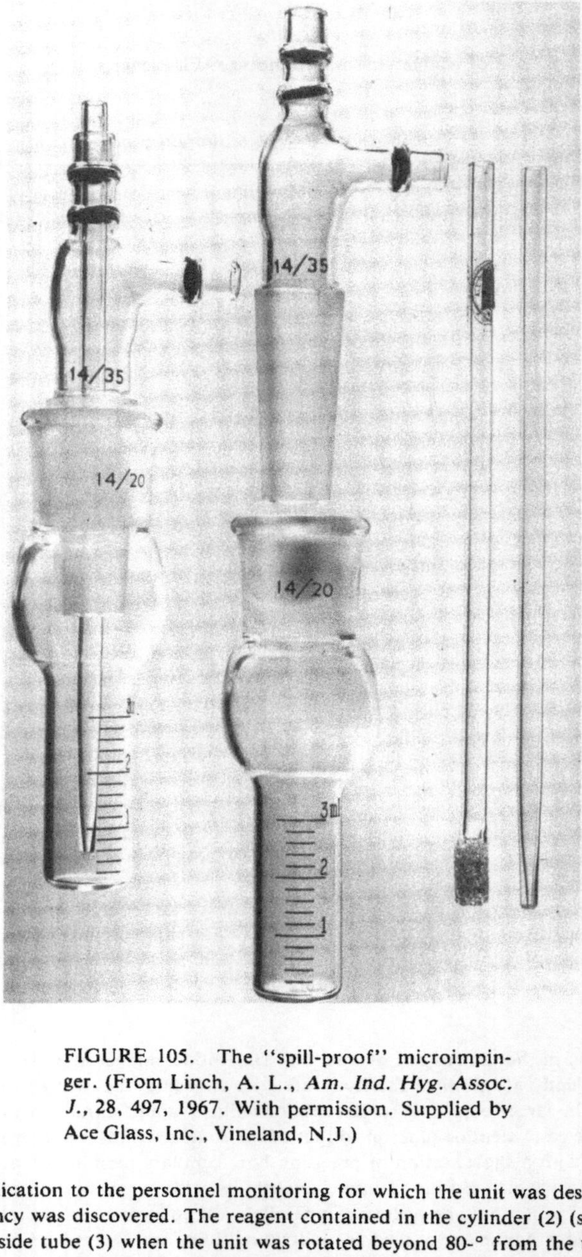

FIGURE 105. The "spill-proof" microimpinger. (From Linch, A. L., *Am. Ind. Hyg. Assoc. J.*, 28, 497, 1967. With permission. Supplied by Ace Glass, Inc., Vineland, N.J.)

However, when application to the personnel monitoring for which the unit was designed was attempted, a major design deficiency was discovered. The reagent contained in the cylinder (2) (see Figure 102) would drain out through the side tube (3) when the unit was rotated beyond 80-° from the upright position. Not only was the absorbing solution lost, but, unless an effective trap was installed, the sampling rate meter and the pump or aspirator would be badly damaged. Since the employee is expected to perform his normal work activities while wearing the monitor, rotation beyond the critical angle cannot be avoided.

To correct this design imperfection, the cylinder (4) length was increased slightly, and the upper part was expanded sufficiently to accommodate 3 ml of liquid when the cylinder was placed in a horizontal position. Then, to form a trap for 180-° rotation, the length of the tapered section of the head (5) was extended by using a standard-taper 14/35 joint inner member in the original standard-taper 14/20 joint outer member on the cylinder (Figure 105).

A second trap is formed by extending a glass tube through the side arm until the open end nearly touches the impinger tube. As long as air flows through the impinger, no liquid can be lost through the inlet. For storage or shipping, a rubber cap can be used to seal this opening. In the field, to avoid accidental expulsion of a drop of corrosive reagent through the impinger the tube may be recessed by shortening down to the first O ring in the head, and a cylinder of porous Teflon or polyethylene pressed in place on top. This filter

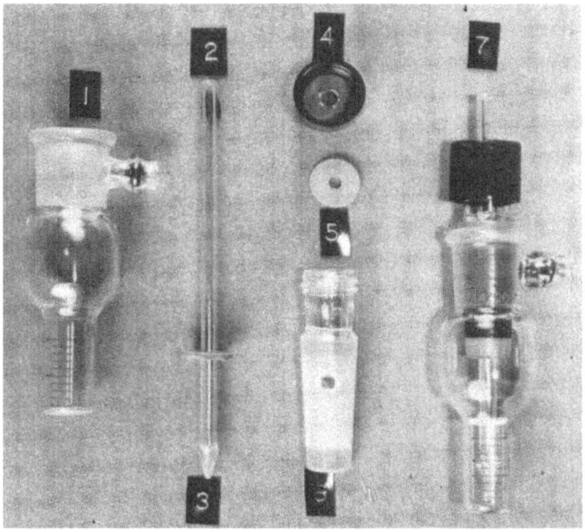

FIGURE 106. The improved "spill-proof" microimpinger. (Supplied by Ace Glass, Inc., Vineland, N.J.)

also provides removal of dust which might otherwise plug the impinger orifice. Replacement of the impinger tube with a porous glass diffusion tube will also minimize this possibility.* [188]

This is the model that has been used extensively for personnel monitoring for organic and inorganic lead[148] and methylene-*o*-chloroaniline[189] exposure control. Late in 1971 the design was further miniaturized (Figure 106) to permit reduction in the overall dimensions of the personnel monitor assembly (Figure 107). The head height of the microimpinger (Figure 106) was decreased by locating the exhaust port on the side of the ground joint and drilling a hole through the side tube seal to engage a matching hole in the head (6). This has the further advantage in providing a closure by rotating the head which is fitted with a Teflon® sleeve as a sealing "lubricant". Part 2 is the impinger tube fitted with a Teflon® baffle and precision bore nozzle (3). Part 5 is a rigid Teflon® gasket which retains an O-ring against the cap (4) to produce a compression seal when the cap is screwed down tightly (7).

No liquid entrainment or carry-over was found when the impinger was operated at flow rates up to 600 ml/min with 2 mL of either water, ethanol, or 0.01 *N*-sodium hydroxide in the cylinder (1). The spill-proof feature was confirmed by rotating the assembly through 180-° without any loss of the liquid contents. Insertion of the impinger tube through the head (6) also has been simplified as compared to the sometimes difficult manipulation through the two O-rings of its predecessor (Figure 105).

A plastic midget impinger also has been developed and promoted commercially. Again, operating characteristics were not available. However, the one-piece construction would introduce operational difficulties, and the instrument is not spill-proof. Therefore, application to personnel monitoring would be minimal.**

During 1976 another microimpinger of somewhat different design was announced by Daco Products (see Figure 108), and in late 1978 a second generation Mini Spill-Proof Impinger (see Figure 109). The size of the unit is about equivalent to a 2-oz bottle and can be rotated 360-° without loss of liquid contents. A plastic-coated bottle can be furnished to reduce breakage hazard and retain the liquid reagent in case breakage should occur. The impinger orifice is supplied in the 1-mm diameter mode. After

* From Linch, A. L., *Am. Ind. Hyg. Assoc. J.*, 28, 497, 1967. With permission.

** MDA Scientific, Inc., 808 Busse Highway, Park Ridge, Ill.

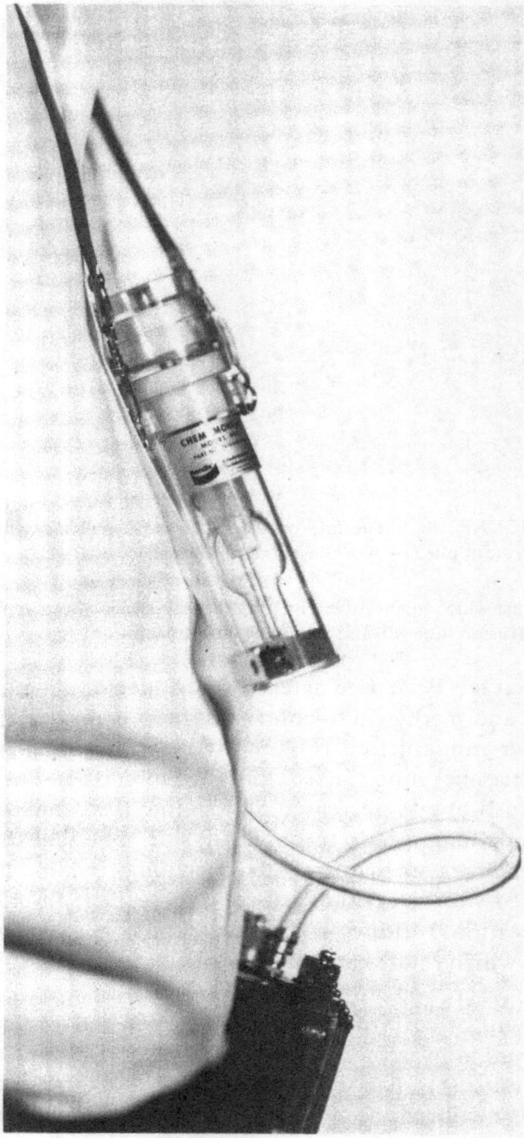

FIGURE 107. The Chem Monitor, Model 900. This personal monitor has a 37-mm membrane filter for the collection of aerosols and a spill-proof microimpinger for the collection of gases and vapors. (Size = 40 × 146 mm; weight = 156 g.) (Courtesy of Bendix Environmental Science Division, Baltimore, Md.)

collection of the sample, the impinger head is removed and the bottle closed with a screw cap. In the second generation model, the stabilizing and impinging well breaks the air stream into small bubbles as they impinge on the well bottom even at 80-° incline. The height of the impinger tube from the bottom is easily adjustable.

The new Daco personnel pump includes a compartment to retain the Mini Spill-Proof Impinger (Figure 110) (see Chapter 2, Volume II for pump specifications). Figure 111 shows the unit in place for personnel monitoring.

FIGURE 108. Daco spill-proof impinger Model SI-10. (Courtesy of Daco Products, Inc., 12 S. Mountain Ave., Montclair, N.J.)

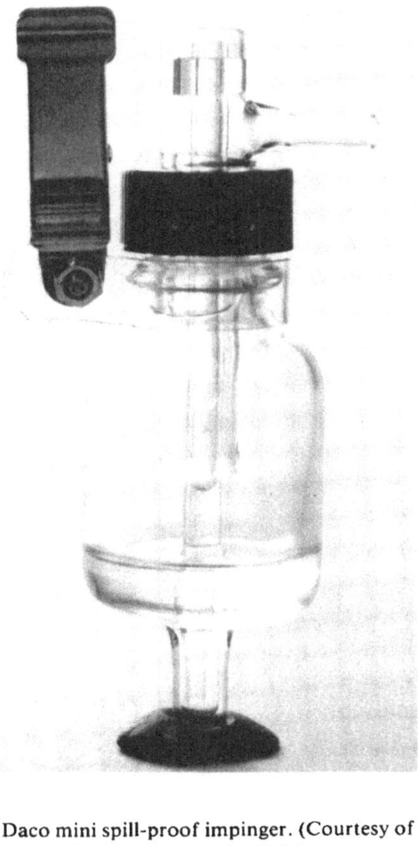

FIGURE 109. Daco mini spill-proof impinger. (Courtesy of Daco Products, Inc., 12 S. Mountain Ave., Montclair, N.J.)

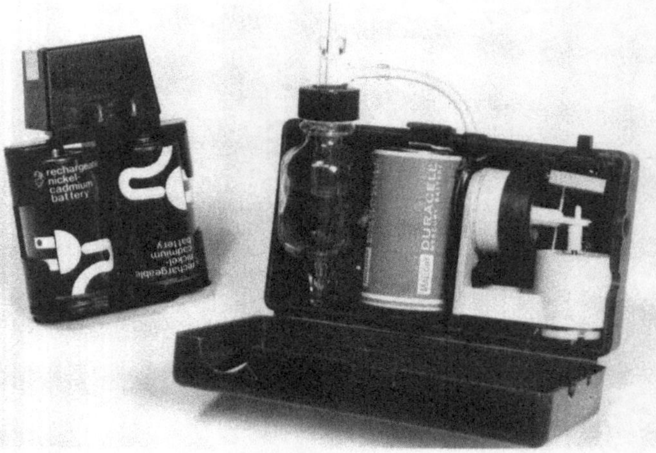

FIGURE 110. Daco Model PO-302 impinger-pump personnel monitor. (Courtesy of Daco Products, Inc., 12 S. Mountain Ave., Montclair, N.J.)

FIGURE 111. Daco personnel impinger monitor in use. (Courtesy of Daco Products, Inc., 12 S. Mountain Ave., Montclair, N.J.)

iii. The Wilson Impinger

This modification collects at a 5 L/min flow rate under a 107-cm Δ P. The nozzle velocity is approximately 100 m/sec, which is equivalent to the velocity attained by the larger (Greenburg-Smith) counterpart. This higher nozzle velocity provides a greater collection efficiency for fine particulate matter. The greater height, which would be undesirable for personnel monitoring, lessens the possibility of loss through liquid entrainment. Collecting liquid volume is the same as for the midget impinger — 10 mL. This design is commercially available.[145]

iv. Porous Glass Bubblers

The following description is included by permission of the copyright holder.

Fritted (sintered) bubblers designed for sampling gases and vapors provide fine, dispersed streams of bubbles capable of increasing the diffusion so that contaminants, sparingly soluble in the collecting medium, are retained with acceptable efficiency. In general, the fritted glass bubbler improves efficiency by increasing the contact area between the air stream and the absorbing medium.[190]

Frits come in various grades, usually designated as fine, medium, coarse and extra coarse. Although fine bubbles are desirable, the fine frits have no advantage because a high pressure drop is required. The effective gas bubble size is practically independent of pore size when the bubble population is high. If a useful flow rate is maintained through a fine frit, numerous bubbles will coalesce near the surface of the frit, making the effective gas bubbles essentially the same size as those from a coarse frit.[182]

The contact time of the bubbles can be increased either by increasing the height of the liquid or increasing the diameter of the sampling flask holding the liquid. An increase in the diameter of the flask lowers the velocity of the air passing through the liquid because of the greater cross-sectional area. For a 22 mm diameter flask (midget impinger flask) a flow rate of about 2 L/min is satisfactory; the contact time drops off rapidly at flow rates above this.

A coarse frit is usually best if the gas or vapor is appreciably soluble or reactive. A medium porosity frit may be used for gases and vapors difficult to collect, but the sampling rate must be adjusted to maintain a flow of discrete bubbles (fairly low bubble population). This necessitates an increase in sampling time.

If fritted bubblers are used, the gas or vapor sample must not form a precipitate when it reacts with the collecting medium, and the air sample must contain very little dust. If the air being sampled contains particulate matter which might clog the pores or interfere with the analysis, a non-reactive and non-absorbing pre-filter, such as dry fibrous glass, porous plastic, cellulosic paper or asbestos[179] should be used.

Calibration of fritted bubblers is neither so convenient or reliable as that of single orifice systems. There is considerable variation among bubblers caused mainly by the multiple orifices in the frit which are not of a consistent size. Each bubbler, containing the same volume of the chosen medium required for the field sampling procedure, must be calibrated individually. Air flow rates can be measured directly with a spirometer or wet test meter or, indirectly, with any of the following devices, previously calibrated with a reliable primary standard:

1. A manometer or vacuum gauge to measure pressure drop across the fritted bubbler(s)[191]
2. A limiting orifice between the bubbler(s) and the pump (this requires a high pressure drop)
3. A rotameter between the bubbler(s) and the pump. It may be necessary to dry the air entering the rotameter to prevent condensation of moisture therein.
4. A rotameter at the inlet of the bubbler(s) is not recommended.

Strongly alkaline media will attack glass frits and alter the calibration, but the effect is not serious unless the alkali remains in contact with the frit for extended periods. Alkaline solutions should be poured into the bubbler just prior to sampling and removed promptly after collection. Glass fritted bubblers, used regularly in alkaline media, should be recalibrated frequently.*

In most cases the bubbler has little to offer as the collection efficiency of the impinger exceeds 90%. A typical example is illustrated by the results from the collection of tetraethyl lead vapor. Three midget impingers, each of which contained 10 mL of 1 *N* aqueous iodine and two of which were packed with glass wool, were operated side by side at the standard 2.8 L/min rate for 40 min in an atmosphere known to contain organic lead vapors. Results were as follows:

1.	Without glass wool	— 14 μg Pb
2.	With glass wool	— 13 μg Pb
3.	With glass wool	— 15 μg Pb
	Average	— 14 μg Pb

The glass wool was used primarily to reduce frothing, but the results indicate no increased lead absorption from increasing the glass surface area.

* From Monkman, J. L., *Analytical Guides*, AIHA, Akron, Ohio, 1965. With permission.

The exception to this generalization would be collection of contaminants that are dependent upon surface reaction with the reagent for collection. In this case an extended glass surface area is desirable. The collection of nitrogen dioxide in Saltzman's reagent serves as an example of this condition.[190] As applied to the microimpinger, even the coarse frits cannot be sealed to the impinger tube without incorporating an excessive degree of airflow resistance which requires an excessive Δ P for operation even at low flow rates.

Packing the impinger with glass beads or spheres offers a better solution to the extended surface area problem.[136] Three-mm glass beads packed to the 25-mL graduation in a standard midget impinger, will increase the collection efficiency significantly. The beads contribute an extended surface area for absorption, act as baffles to extend the air bubble liquid contact time, and increase the liquid turbulence to promote mixing. A smaller volume of liquid absorbent is needed when beads are used. Micro-glass spheres are available for packing the microimpinger.[192] This type of packing also greatly reduces liquid entrainment at high flow rates or at maximum liquid fill level.

v. Traps

Even for water, a protective trap should be inserted between the pump and the liquid absorbant. Even though the absorbant may be inert insofar as the aspirating train is concerned, a liquid film can foul the rate meter (rotameter floats especially), reduce the pump's efficiency, or interfere with valve action. Activated charcoal is probably the best all-purpose scavenger for solvent and corrosive vapors, and silica gel serves well for water vapor. A mixture of the two would be suitable for certain aqueous systems. A polyethylene tube 110 mm (75-mm packed section) × 75 mm diameter packed with 8/10 mesh activated charcoal provided protection for 40 hr of application at 420 mL/min when used with the iodine monochloride reagent to collect tetraalkyl lead vapors.[145]

C. Solid Adsorbents

1. Silica Gel

The following is taken from *Analytical Guides* by permission of the copyright owners.

Advantages of Silica Gel — Silica gel is a valuable medium for the collection of gases and vapors from air. While silica gel does not have as great an affinity for organic materials as does activated charcol, it has a decided advantage as an analytical tool in that adsorbed materials can be removed easily, consistently and quantitatively. It is a clean, white, granular solid and it is easy to work with. Silica gel generally surpasses liquid substrates on solid supports in its ability to hold materials and thus has gained wide acceptance as the preferred adsorbent or carrier for the direct indicating gels.

For the industrial hygienist interested in quantitative air analysis adsorption on silica gel offers a variety of advantages. This technique is, primarily, one of concentrating the air contaminant of interest. Secondly, only the concentration step needs to be carried out in the field, leaving the delicate manipulations of analysis for the laboratory. With the use of a granular adsorbent such as silica gel quantitative transfers from the sampling vessel to a sample bottle in the field are simple and rapid. Finally, because the phenomenon of adsorption is quite nonspecific, a single air sample taken on silica gel may be used to determine quantitatively several different materials coexisting in the sampled air.

Mode of Adsorption on Silica Gel — When a gas or a vaporized liquid is brought into contact with a solid, there is a tendency for the formation of a solution of the gas or vapor in the solid, as well as a tendency for the vapors to collect on the surface of the solid. Although the first tendency is usually slight because of the low solubility of vapors and gases in most solids, the second tendency of adsorbing vapors and gases at the surface is exceedingly high for a number of solids.*[193,194]

Many theories have been proposed to account for the adsorption of vapors on materials such as silica gel. At present, none of these theories appear to explain the phe-

* From Monkman, J. L., Ed., *Analytical Guides,* AIHA, Akron, Ohio, 1965. With permission.

nomenon of adsorption in such a manner that accurate predictions can be made without trial runs and experimental data. However, despite the lack of a good theoretical base, it is known that silica gel does adsorb many materials quantitatively under the proper conditions and will release these adsorbed materials quantitatively under the proper desorption conditions.

a. *Factors Affecting Adsorption*

Factors that affect the dynamic adsorption of materials from air on silica gel include the following: previous history of the gel, size of the gel particles, size range of the gel particles, mass of gel used, depth of gel used, superficial velocity of the airstream through the bed of gel, temperature of the gel bed, the molecular species being adsorbed, concentration of the air contaminant, humidity of air contacting the gel bed, and duration of sampling. These factors vary in importance, and some appear to interact with others. Until more information is obtained, use of silica gel for air sampling must be based on empirical data.

i. Previous History of the Gel

Experience has shown that silica gel must be used in a standardized manner if quantitative results are to be obtained. The first consideration is the particular gel to use. Silica gel is not difficult to make, but making it in exactly the same manner each time over a period of years is a problem. Therefore, most users purchase their gel to rather exacting specification. The Davison Chemical Company (Baltimore) manufactures gels well-suited for analytical methods. Good results have been obtained from "part No. 05-08-05-215, 6-16 mesh, type 3," "PA100, 14-28 mesh," and "PA400, 14-28 mesh."[193,195]

As purchased, these gels are almost completely free of contaminants but may contain small amounts of ammonia. In addition, they may contain more moisture than desirable for use of the gel as a quantitative adsorbent. Therefore, all purchased gel should undergo a standardized pretreatment before use.[196,197]

One method for removing all ammonia and water is baking the gel in an oven overnight at 350 to 400°C in a dry nitrogen atmosphere. Enough nitrogen is allowed to flow through the oven to maintain a slight positive pressure and to sweep out any contaminants driven off the gel. After the baking has been completed, the gel is allowed to cool in uncontaminated room air until warm to the touch. It is then stored in sealed containers. This pretreatment is necessary only once. If the gel is used for studies involving thermal desorption, each successive desorption reactivates the gel; if the gel is used for studies involving leaching, it is discarded.

ii. Size of Gel Particles

Investigations on the use of silica gel as a quantitative adsorbent have been carried out with gel ranging in size from 200 mesh to about 6 mesh. In general, the larger mesh sizes have proved easier to handle, but they have a somewhat reduced adsorption capacity. The reduction in capacity is not serious, however, and gel in the size range of 6 to 28 mesh performs quite satisfactory.

iii. Size Range of Gel Particles

For best results the range of particle sizes of gel used for quantitative adsorption should be kept small. Investigators have used 6 to 16 mesh and 14 to 28 mesh gel with success,[193,195] but wider ranges than these should be avoided. Gel that has been used repeatedly will gradually produce fines. As soon as fines accumulate, the gel should be discarded.

iv. Mass of Gel

With other variables remaining the same, doubling the mass of gel used will increase the breakthrough time by a factor of about four. The breakthrough time is the time required before the first traces of the air contaminant can be detected downstream from the gel bed. It is the duration of time that adsorption is quantitative.

v. Depth of Gel Bed

Preliminary data have shown that the depth of the silica gel bed has an effect on the breakthrough time that is somewhat independent of the mass of gel used. With all other conditions kept constant, increasing the bed depth increases the breakthrough time and the total amount of contaminant adsorbed on the gel. The ideal container for air sampling is a relatively long tube with a relatively small diameter.

To avoid channeling, the minimum diameter of the adsorbing column probably should be several times the diameter of the largest gel particles.

Limited bed depth studies have indicated that channeling, wall effects, and nonuniform adsorption become important as the bed depth for 6 to 16 mesh gel is decreased below 2 cm. Less bed depth is satisfactory for smaller gel sizes.

vi. Superficial Velocity of Airstream Through Gel Bed

The superficial velocity is the velocity that would be observed if the gel were not present in the adsorber. It is determined by dividing the airflow rate (cm^3/min) by the cross-sectional area of the adsorber (cm^2). Doubling the superficial velocity will result in a decrease of the breakthrough time by at least a factor of four. This relationship, however, is not linear and is dependent also upon the concentration of the contaminent in the air. Superficial velocities normally encountered in sampling operations range from 50 to 1000 cm/min.

vii. Temperature of Gel Bed

Few, if any, experimental data are available relating bed temperature to retention or breakthrough time. Although most of the work with silica gel has been done at room temperature, 20 to 35°C, some investigators have used an ice bath when sampling for extremely volatile materials.[198] However, halogenated hydrocarbons, with boiling points as low as −28°C, have been quantitatively adsorbed on silica gel at room temperature.[193]

During the sampling period the temperature of the gel bed is elevated from the heat of adsorption contributed both by the contaminant and by water vapor in the airstream. A "front" can easily be detected passing through the bed of gel simply by feeling the adsorber. This lack of uniformity of the bed temperature has undoubtedly complicated the investigation of that variable.

viii. The Molecular Species Being Adsorbed

The affinity of silica gel for water and for water vapor has long been known; normally, it is used as a desiccant. Silica gel apparently will adsorb water in preference to any other material. Some investigations have shown, however, that its affinity for ammonia and a few other materials may be almost as great as for water. Both because of structure and known affinity for water, silica gel should best adsorb materials possessing a hydroxyl group.

Most of the commonly encountered halogenated hydrocarbons, benzene, many other hydrocarbons, and other organic polar and nonpolar materials have been determined quantitatively by collection of silica gel.

Materials with high boiling points appear to adsorb somewhat better on silica gel than do those with low boiling points. Attempts to determine methyl chloride in the

Table 64
NIOSH Sampling and Analytical Methods

Analyte chemical name	Federal standard: TWA, ppm (mg/m³)[b]	Federal standard: Ceiling limit, ppm (mg/m³)[b]	Method: Number	Method: Class[c]	Method: Matrix	Sampling: Collection medium[d]	Sampling: Volume (L)	Sampling: Rate[e] (mℓ/min)	Analysis: Range mg/m³	Analysis: Method procedure[f]	Analysis: % CV_T[g]
Acetic anhydride	5 (20)		S170	—	Air	MI-alk, NH_2OH	100	1000	9.35 — 37.4	Colorimetric—Fe Cl_3	6
Acetone	1000 (2380)		S101	B	Air	Charcoal	2	200	1200 — 4500		
			S1							GC/ CS_2—FID—1	8.2
Acetonitrile	40 (70)		S165	B	Air	Charcoal	10	200		GC/ benzene—FID —1	7
Acetylenetetra-bromide	1 (14)		S117	B	Air	Silica gel	100	1000	7.04 — 28.1	GC/ THF—FID—2	9.6
Acrylamide[h]	(0.3)	[h]		[h]	Air	MI-H_2O	[h]	1000	0.005 — 0.20	DPP	—
Acrylonitrile	20		S156	B	Air	Charcoal	20	200	40 — 1100	G/C methanol—FID —1	7.3
	(45)		202	E		Carbo sieve				G/C methanol—FID	10
Allyl alcohol	2 (4.75)		S52	B	Air	Charcoal	10	200	1.8 — 8.4	GC/CS_2 + 5% ipa—FID—3	11
Allyl chloride[h]	1 (3)	3[h] (9.4)	S116	B	Air	Charcoal	100	1000	1.8 — 7.19	GC/ benzene—FID —1	7.1
Alpha-methyl styrene		100 (483)	S26	B	Air	Charcoal	3	200	236 — 943	GC/CS_2—FID—3	5
n-Amyl acetate	100 (525)		S51	B	Air	Charcoal	10	200	208 — 871	GC/CS_2—FID—4	5.1
s-Amyl acetate	125 (650)		S31	B	Air	Charcoal	10	200	349 — 1460	GC/CS_2—FID—4	7.1
Antimony and compounds	(0.5)		S2	—	Air	CEF	360	1500	0.258 — 1.08 (0.05 — 3.0)	AA/21716 nm	5.9
	(0.5)		189	E	Air	Filter	100	—	0.005	ASV	1.0
			107	A	Urine	—	—	—	1 — 10 μg	Colorimetric	5.0
			197	E	Urine	—	—	—	2 — 500 μg	ASV	8
Antimony	(0.5)		193	E	Blood	—	—	—	0.003 — 0.5 μg/mL	ASV	10

Table 64 (continued)
NIOSH Sampling and Analytical Methods

Analyte chemical name	Federal standard: TWA, ppm (mg/m³)[b]	Federal standard: Ceiling limit, ppm (mg/m³)[b]	Method: Number	Method: Class[c]	Method: Matrix	Sampling: Collection medium[d]	Sampling: Volume (L)	Sampling: Rate[e] (mℓ/min)	Analysis: Range mg/m³	Analysis: Method procedure[f]	Analysis: % CV,[g]
Acrolein	0.1 (0.25)		118	C	Air	MI—hexyl-resorcinol	1	2000	1—30 μg in 10 mL	Colorimetric	—
			211	C	Air	MI—NaH SO₃	2	2000	Cf. 118	Colorimetric	5
Aldrin	(0.25)		S275	V	Air	GF + MI	180	1000	0.15—0.50	i-octane,GC-ECC-5	
Arsenic[d]			139	D	Air	CEF	30	1700	0.05—2.0 μg/sample	AsH₃/AA	10.7
	—	—			Urine	—	0.075	—			
	(0.004)[h]	(0.01)[h]	140	A	Air	CEF	30	1700	1.0—25 μg/sample	AsH₃—Colorimetric	4—9
					Urine	—	0.075	—			
	(0.05)		173	D	Air	CEF	3	1000—2000	420—2100	AA/193.7—0	3
			180	E	Air	CEF	25	1500—2000	0.1—1	Catalytic-I-I₂	2
			188	E	Air	CEF	100	1500—2000	0.0005—0.005	ASV—Au	8
			192	E	Blood	—	0.001	—	0.016—0.6 μg	ASV—Au	8
			196	E	Urine	—	0.001	—	0.016—0.6 μg	ASV—Au	8
			S309	V	Air	CEF	85	1500	0.198—0.903	FAA—UV	5.8
Arsine	0.05 (0.2)		S229	V	Air	Charcoal	10	200	0.094—0.404	FAA—UV	5.6
Benzene solubles	—	—	217	D	Air	GF	300	1500	0.02/sample	Benzene/gravimetric	2—23
Benzidine and sulfate	Cancer[i]		243	E	Air	GF + Silica gel	50	200	0.003—0.13	TEA/methanol—LCG/UV	7
Benzo (b) fluoranthene	—	—	183	D	Air	GF or SMF	500	2000	0.001—0.5	GC/benzene/UV—12—FID	—
Benzo (j) fluoranthene	—	—	183	D	Air	GF or SMF	500	2000	0.001—0.5	GC/benzene/UV—12—FID	—
Benzo (k) fluoranthene	—	—	183	D	Air	GF or SMF	500	2000	0.001—0.5	GC/benzene/UV—12—FID	—
			184	D	Air	GF or SMF	500	2000	0—3	LCG benzene—9—FID	—

Benzo (g,h,i) perylene	—	—	183	D	Air	GF or SMF	500	2000	0.001 — 0.5	GC/benzene/UV—12	—
Benzo (a) pyrene	—	—	183	D	Air	GF or SMF	500	2000	0.001 — 0.5	GC/benzene/UV—12	—
			184	D	Air	GF or SMF	500	2000	0 — 3	LCG/benzene—9	—
			186	D	Air	GF or SMF	500	2000	0.02 — 0.8 μg	TLC/SF	7
			206	E	Air	GF	1.5×10^5	2000	0.005 — 0.2	LCG/cyclohexane/UV	1.3
			251	E	Air	GF	500	2000	6 — 240 ng	LCG/acetonitrile	6
Benzo (e) pyrene	—	—	183	D	Air	GF or SMF	500	2000	0.001 — 0.5	GC/benzene/UV—12	—
			184	D	Air	GF or SMF	500	2000	0 — 3	LCG/benzene—9	—
Benzyl chloride	1 (5)		S115	V	Air	Charcoal	10	200	2 — 8	GC/CS_2—3	9.6
Beryllium[a]		(25 μg)[h]	121	D	Air—B	CEF	HV[i]	1.5 m^3	0.03 — 8.0 μg	AA/234.86/r	
			173	D	Air	CEF	3	1000 —2000	4.2 — 84	TM—AA	3
	(2 μg)	(5 μg)	S339	V	Air	CEF	40	1500	2.7 — 11.8 μg	AA/graphite	6.4
BiBenzyl	—	—	183	D	Air	GF or SMF	500	2000	0.001 — 0.5	GC/benzene/UV—12	—
bis(chloromethyl) ether	Cancer[i]		213	E	Air	C-101	5	80	0.5 — 10 ppb	GC/thermal/MS	10
			220	E	Air	MI	60	500	0.5 — 7.5 ppb	GC/hexane—13—EC	10
Bismuth	—	—	173	D	Air	CEF	3	1000 —2000	0.042 — 1.25	TM—AA	—
Boron oxide	(15)		S349	V	Air	PVCF	100	1000	7.8 — 27.7	Gravimetric	5.6
Boron trifluoride[a]		(3)	[a]	[a]	Air	MI—*NaOH	—	—	0.27 — 4.98	Ion spec[a] electrode	—
Bromoform	0.5 (5)		S114	V	Air	Charcoal	10	200	3 — 10	GC/CS_2—3—FID	7.1
Butadiene	100 (2200)		S91	V	Air	Charcoal	1	50	1065 — 4590	GC/CS_2—3—FID	5.8
2-Butanone (MEK)	200		127	B	Air	Charcoal	5—13	1000	0.01 — 3000	GC/CS_2—3—FID	0.5
	(590)		S3	V	Air	Charcoal	10	200	380 — 1240	GC/CS_2—3—FID	7.2
2-Butoxy ethanol (butyl cellosolve)	50 (240)		S76	V	Air	Charcoal	10	200	124 — 490	GC/CH_2CL_2—FID —3	6.0
s-Butyl acetate	200 (950)		S46	V	Air	Charcoal	10	200	478 — 2005	GC/CS_2—4—FID	5.4
t-Butyl acetate	200 (950)		S32	V	Air	Charcoal	10	200	424 — 1780	GC/CS_2—4—FID	9.1
n-Butyl acetate	150 (710)		S47	V	Air	Charcoal	10	200	352 — 1475	GC/CS_2—4—FID	6.9
s-Butyl alcohol	150 (455)		S53	V	Air	Charcoal	10	200	270 — 850	GC/CS_2 + 1% ipa—3—FID	6.6

Table 64 (continued)
NIOSH Sampling and Analytical Methods

	Federal standard		Method			Sampling			Analysis		
Analyte chemical name	TWA, ppm $(mg/m^3)^b$	Ceiling limit, ppm $(mg/m^3)^b$	*Number*	*Class*[c]	*Matrix*	Collection *medium*[d]	*Volume (L)*	Rate[e] *(mℓ/min)*	Range mg/m^3	Method *procedure*[f]	% CV_T[g]
t-Butyl alcohol	100 (305)		S63	V	Air	Charcoal	10	200	165—600	GC/CS_2 + 1% 2 butyl alcohol FID—3	7.5
n-Butyl alcohol	100 (305)		S66	V	Air	Charcoal	10	200	170—610	GC/CS_2 + 1% ipa-3-FID	6.5
n-Butyl glycidylether	50 (270)		S81	V	Air	Charcoal	10	200	133—542	GC/CS_2—3—FID	7.4
Butyl amine	—	5 (15)	221	E	Air	Silica gel	10	200	1—2400	GC/H_2SO_4/ NaOH —FID—6	3
p-*t*-Butyl toluene	10 (60)		S22	V	Char-coal Air	Charcoal	10	200	29—119	GC/CS_2—3	6.7
Cadmium[h]	(0.05)		173	D	Air	CEF	3	1000—2000	4.2—84	TM—AA	—
	(0.04)[h]	(0.20)[h]	191	E	Air	CEF	100	1500—2000	0.00008—0.01	ASV	1.3
			223	E	Blood	—	—	—	0.002—1 mμ/mL	ASV	5
	(0.2)	(0.6)	S312	V	Air	CEF	25	1000—2000	0.12—0.98	AA/228.8—0	6.4
	(0.1)	(0.3)	S313	V	Air	CEF	25	2000	0.12—0.57	AA/228.8—0	6.2
Calcium (oxide)	(5)		173	D	Air	CEF	1.5	1500	4.2—210 μg	TM—AA	—
Calcium arsenate	(1)		173	D	Air	CEF	1.5	1500	0.42—2.1	TM—AA	—
Calcium oxide	(5)		173	D	Air	CEF	1.5	1500	0.0042—0.21	TM—AA	—
	(5)		S205	V	Air	CEF	85	1500	2.6—10.16	AA/239.9	6.3
Camphor	2 (12.5)		S10	V	Air	Charcoal	20	200	6—27	GC/CS_2—FID—3	7.4
Carbaryl[h] (Sevin)	(5)[h]		S273	V	Air	CEF	90	1500	1.96—13.43	Colorimetric-diazo	5.7
Carbon black	(3.5)		S262	V	Air	PVCF	200	1700	1.86—7.7	Gravimetric	5.6
Carbon dioxide[h]	(10,000) (18,000)	(30,000) (54,000)									
Carbon dioxide	5000 (9000)		S249	V	Air	Bag	3.5	50	4100—18,000	GC/TC	1.4
Carbon disulfide[h]	20 (60)	30	179	E	Air	Charcoal	3	50	5—100	AA/i—Amyl ace-	4

										tate + Cw/324.7	
	20	30	S248	V	Air	Charcoal	6	200	44.1 — 176.4	GC/benzene—14 —FP	5.9
	1[a] (3)	10[a] (30)									
Carbon monoxide[a]	35[a]	200[a]	112	C	Air	Bag	5	—	10 — 500	IR/10 meter	10
	(38.5)	(220)	113	D	Blood	—	—	—	1 — 50%	GC/MS	
Carbon tetrachloride	10	25	S314	V	Air	Charcoal	17	200	65 — 299	GC/CS_2—FID—4	9.2
	(65)	(160)	127	A	Air	Charcoal	10—60	1000	0.01 — 325	GC/CS_2—3—FID	10.5
Chloride	—	—	115	D	Air	MI	200	2500	0.1 — 1750	Spec. ion electrode	2
Chlorinated camphene (toxaphene)	(0.5)	—	S67	V	Air	CEF	15	1000	0.225 — 1.16	GC/pet ether—14—EC	7.6
Chlorinated diphenyl oxide	(0.5)		S119	V	Air	CEF	90	1500	0.10 — 0.965	GC/i—Octane—ECC	7
Chlorine	1 (3)		209	C	Air	MI—bubbler	30	1000—2000	0.15 — 3.0	Colorimetric—methyl orange	5
Monochlorbenzene	75 (345)		S133	V	Air	Charcoal	10	200	183 — 736	GC/CS_2—3—FID	5.6
Chlorobromomethane	200 (1050)		S113	V	Air	Charcoal	5	200	640 — 2655	GC/CS_2—3—FID	6.1
Chlorodiphenyl	(0.5)		S121	V	Air	CEF	100	1500	0.322 — 1.255	GC/pet ether—15—EC	6.3
Chloroform[a]		50	127	A	Air	Charcoal	0.5—13	1000	0.1 — 600	GC/CS_2—FID—3	10.5
		(240)	S351	V	Air	Charcoal	15	1000	99.8 — 416	GC/CS_2—FID—4	5.7
	10[a] (48.9)	50[a] (244)									
Chloromethyl methyl ether	Cancer[c]		220	E	Air	MI	60	500	0.5 — 7.5 ppb	GC/hexane—EC—13	10
Chloroprene (2-chloro 1-3 butadiene)	25 (90)		S112	V	Air	Charcoal	3	10—50	44 — 174	GC/CS_2—FID—1	7.1
Chromic acid[a]	(0.1)		152	D	Air	CEF	100	200	0.01 — 0.4	AA/357.9/—r	5
			169	D	Air	PVCF	100	2000	0.004 — 0.2	Colorimetric/540	4
	(0.05)[a]	(0.10)[a]	S317	V	Air	PVCF	22.5	1500	0.052 — 0.2	Colorimetric/540	8.4
Chromium			152	D	Air	CEF	100	200	0.01 — 0.4	AA/357.9—r	5
Carcinogenic[a] (VI)		(0.001)[a]	173	D	Air	CEF	1.5	1500	21 — 210	TM—AA	—
Noncarcinogenic[a] (VI)	(0.025)[a]	(0.05)[a]	182	E	Air	PVCF	100	1.5—2	0.01 — 0.25	Catalytic/H_2O_2	3
Soluble	(0.5)		S323	V	Air	CEF	90	1500	0.28 — 0.95	AA/357.9/r	8.5
Metal-insoluble	(1)		S352	V	Air	CEF	90	1500	0.49 — 1.83	AA/357.9/r	7.6

Table 64 (continued)
NIOSH Sampling and Analytical Methods

Analyte chemical name	Federal standard: TWA, ppm (mg/m^3)[b]	Federal standard: Ceiling limit, ppm (mg/m^3)[b]	Method: Number	Method: Class[c]	Method: Matrix	Sampling: Collection medium[d]	Sampling: Volume (L)	Sampling: Rate[e] (mℓ/min)	Analysis: Range mg/m^3	Analysis: Method procedure[f]	Analysis: % CV_T[g]
Chrysene	—	—	183	D	Air	GF or SMF	500	2000	0.001 — 0.5	GC/benzene/ UV—12	—
			184	D	Air	GF or SMF	500	2000	0 — 3	LCG/benzene—9	—
(T_pAH) (total particulate aromatic hydrocarbons)	—	—	206	E	Air	GF	1.5×10^5	2000	0.005 — 0.2	LCG/ cyclohexane— UV	1.3
Chrysotile	See asbestos		245	E	Bulk	—	—	—	0.1 — 10 mg/sample	Thermal analysis	12
Coal tar pitch volatiles	(0.2)										
			183	D	Air	GF or SMF	500	2000	0.001 — 0.5	GC/benzene/ UV—12	—
			184	D	Air	GF or SMF	500	2000	0 — 3	LCG/benzene—9	—
			206	E	Air	GF	1.5×10^5	2000	0.005 — 0.2	LCG/ cyclohexane— UV	1.3
Cobalt	(0.1)		173	D	Air	CEF	1.5	1500	21 — 210	TM—AA	—
Cotton dust[h] (Elutriator) respirable fraction	(1)	—	[h]	[h]	Air	PVCF	2664	7400	—	Gravimetric	—
Copper — dusts	(1)		173	D	Air	CEF	1.5	1500	21 — 210	TM—AA	—
	(1)		S186	V	Air	CEF	90	1500	0.47 — 1.83	AA/324.7—0	5.1
Cresol — all isomers	5 (22)		S167	V	Air	Silica gel	20	200	10.5 — 42.2	GC/ acetone—FID—3	6.8
Cristobalite — respirable fraction			109	D	Air	PVCF	—	—	0.005 — 0.2	X-Ray diffraction	5 mg
			259	D	Air	PVCF	—	1700	0.02 — 2.0	X-Ray diffraction	10
Respirable fraction	Calc.[k]		S315	V	Air, Bulk	SMF	200	1700	0.167 — 0.846	X-Ray diffraction	6.5
Di-chloro fluoro methane	1000 (4200)		S109	V	Air	Charcoal	3	10—50	1730 — 7600	GC/CS_2—FID—4	6.1
Di-chlorotetra-fluoro ethane	1000 (7000)		S108	V	Air	Charcoal	3	10—50	3500 — 14,100	GC/CH_2Cl_2—FID —18	6.4

Dieldrin	(0.25)		S283	V	Air	GF	180	1500	0.12—0.59	GC/i—Octane—ECC—5	8.6
Diethyl amine	(75)		221	E	Air	Silica gel	10	200	1—2400	GC/H_2SO_4/NaOH—FID—6 (19)	3
	25		S139	V	Air	Silica gel	50	1000	37.6—165		7.0
Difluorodi-bromo methane	100 (860)		S107	V	Air	Charcoal	10	200	470—1875	GC/ipa—FID—3	9.0
Difluorodi-chloro methane	1000 (4950)		S111	V	Air	Charcoal	3	10—50	2940—10,5000	GC/CH_2CL_2—FID—18	6.4
Di-i-butyl ketone	50 (290)		S358	V	Air	Charcoal	12	200	145.5—582	GC/CS_2—FID—3	7.0
Di-i-propyl amine	—	—	221	E	Air	Silica gel	10	200	1—2400	GC/H_2SO_4/NaOH—FID—6	3
Dimethylacetamide	10 (35)		S254	V	Air	Silica gel	50	1000	18.2—106	GC/methanol—FID—20	6.7
Dimethylformamide	10 (30)		S255	V	Air	Silica gel	50	1000	10.6—61.1	GC/methanol—FID—20	5.6
Dimethyl amine			221	E	Air	Silica gel	10	200	1—2400	GC/H_2SO_4/NaOH—FID—6	3
	(18) 10		S142	V	Air	Silica gel	50	1000	7.0—29.5	GC/H_2SO_4—methanol—FID—19	6.2
2,4 dimethyl amino benzene (xylidene)	5		168	D	Air	Silica gel	5—3600	See 168	0.01—14		9
	(25)		S162	V	Air	Silica gel	20	200	12.5—50	GC/ethanol—FID—6	5.7
*N,N*Dimethylanaline	5		168	D	Air	Silica gel	5—3600	See 168	0.01—14	GC/ethanol—FID—7	9
	(25)		S164	V	Air	Charcoal	20	200	10.5—46.7	GC/CS_2—FID—3	5.2
1,1 Dimethyl hydrazine	(0.5)		248	E	Air	Silica gel	96	200	0.04—120	GC/H_0—FID—21	4
	(1)		S143	V	Air	MI—HCl	100	1000	0.50—2.22	Colorimeter—phosphomolydic	6.2
Dioxane	100 (360)		127	A	Air	Charcoal	1—18	1000	0.05—900	GC/CS_2—FID—3	10.5
			S360	V	Air	Charcoal	10	200	155—651	GC/CS_2—FID—3	5.4

Table 64 (continued)
NIOSH Sampling and Analytical Methods

	Federal standard		Method			Sampling			Analysis		
Analyte chemical name	TWA, ppm (mg/m^3)[b]	Ceiling limit, ppm (mg/m^3)[b]	Number	Class[c]	Matrix	Collection medium[d]	Volume (L)	Rate[e] (mℓ/min)	Range mg/m^3	Method procedure[f]	% CV_T[g]
di-Phenyl methane, di-Isocyanate	0.02 (0.2)		142	D	Air	MI—acid	20	1000	0.07—0.73 ppm	Colorimetric—hydrolysis—azo	6.4
di-Propylene glycol methyl ether	100 (610)		S69	V	Air	Charcoal	10	200	314—1197	GC/CS_2—FID	6.4
2,6 di-*t*-Butyl-*p*-cresol	—	—	226	D	Air	Silica gel	10	200	1—500	GC/CS_2—methanol—FID—5	5
Epichlorohydrin[h]	0.5[h] (2)[h]	5[h] (19)[h]	S118	V	Air	Charcoal	20	200	11.7—43.1	GC/CS_2—FID—3	5.7
EPN (*o*-Ethyl-*o*-*p* nitrophenyl phosphorothioate	(0.5)		S285	V	Air	GF	120	1500	0.3—1.16	GC/*i*—*Octane*—FP—15	5.9
Ether, ethyl	400 (1210)		S80	V	Air	Charcoal	3	200	606—2400	GC/ethyl acetate—*FID*—*1*	5.3
2-Ethoxy ethyl acetate	100 (540)		S41	V	Air	Charcoal	10	200	262—1100	GC/CS_2—FID—4	6.2
Ethyl acetate	400 (1400)		S49	V	Air	Charcoal	6	200	704—2950	GC/CS_2—FID—4	5.8
Ethyl acrylate	25 (100)		S35	V	Air	Charcoal	10	200	50—210	GC/CS_2—FID—4	5.4
Ethyl alcohol (ethanol)	1000 (1885)		S56	V	Air	Charcoal	1	50	900—3300	GC/CS_2 + 1% butyl alcohol FID—3	6.5
Ethyl benzene	100 (435)		S29	V	Air	Charcoal	10	200	222—884	GC/CS_2—FID—3	4.1
Ethyl bromide	200 (890)		S106	V	Air	Charcoal	4	200	550—2200	GC/ipa—FID—3	5.4
Ethyl butyl ketone	50 (230)		S16	V	Air	Charcoal	10	200	100—460	GC/CS_2 + 1% methanol—FID—3	8.6
Ethyl formate	100 (300)		S36	V	Air	Charcoal	10	200	171—666	GC/CS_2—FID—3	7.8
Ethyl silicate	100 (850)		S264	V	Air	XAD—2 resin	9	50	377—1620	GC/CS_2—FID—17	5.6
Ethylene dichloride[h]	50	100	127	D	Air	Charcoal	1—18	1000	0.05—1000	GC/CS_2—FID—3	10.5
	(200)	(405)	S122	V	Air	Charcoal			195—819	GC/CS_2—*FID*—*17*	7.9
	5[h] (20)	15[h] (60)									

Ethylene glycol dinitrate + nitroglycerine	0.2		203	E	Air	Tenax —GC	10		0.01 — 100	GC/ethanol —EC—22	10
	(1)	(0.2)	S216	V	Air	Tenax —GC	15	1000	0.51 — 1.84	GC/ethanol —EC—22	8.9
	0.2 (1)	(0.2)		V	Air	Tenax —GC			0.56 — 3.21		10.4
Ethyl amine	10		221	E	Air	Silica gel	10	200	1 — 2400	GC/H_2SO_4/ NaOH— FID—6	3
	(18)		S144	V	Air	Silica gel			7.41 — 31.8	GC/H_2SO_4/ NaOH— FID—19	11.3
Ethylene chlorohydrin	5 (16)		S103	V	Air	Charcoal	20	200	7 — 30	GC/CS_2 + 5% ipa—FID—	7.6
Ethylene oxide	50 (90)		S286	V	Air	Charcoal	5	50	41 — 176	GC/CS_2—FID—1	10.3
N-Ethylmorpholine	20 (94)		S146	V	Air	Silica gel	10	200	57 — 206	GC/H_2SO_4/ NaOH— FID—19	10.1
Fluoranthene	—	—	183	D	Air	GF or SMF	500	2000	0.001 — 0.5	GC/benzene— UV—12	—
			184	D	Air	GF or SMF	500	2000	0 — 3	LCG/benzene—9	—
Fibrous glass[h]	3 fibers/mL		[h]	[h]	Air	Filter				Visual count	
Fluoride	(2.5)[h]	[h]	114	D	Urine	—	—	—	0.19 + mg/L	Ion spec. elect.	—
			117	D	Air	MI— NaOH	40—200	2500	0.05 — 475	Ion spec. elect.	6.5
			212	E	Air	CEF	250	2500	0.005 +	Ion spec. elect.	—
Fluorotrichloromethane	1000 (5600)		S102	V	Air	Charcoal	4	10—50	2390 — 10,500	GC/CS_2—FID—4	7.2
Formaldehyde[h]	1[h]		125	C	Air	MI	25	1000	0.15 — 3.0	Colorimetric-chromotropic acid	5
	(1.2)		235	E	Air	Alumina gel	6	200	0.4 — 52	Colorimetric-chromotropic acid	6
Formic acid	5 (9)		232	D	Air	MI— NaOH	10—100	2500	3.8 — 75	As ethyl formate GC/FID—23	11
Gallium	—	—	177	E	Air	CEF	100	1500 —2000	0.0002 — 0.01	ASV—Hg/carbon	15
	—	—	198	E	Urine	—	—	0.002	10 — 500 mg/L	ASV	12

Table 64 (continued)
NIOSH Sampling and Analytical Methods

	Federal standard		Method			Sampling			Analysis		
Analyte chemical name	TWA, ppm (mg/m³)[b]	Ceiling limit, ppm (mg/m³)[b]	Number	Class[c]	Matrix	Collection medium[d]	Volume (L)	Rate[e] (mℓ/min)	Range mg/m³	Method procedure[f]	% CV_T[g]
Glycidol (2,3 epoxy, propanol)	50 (150)		S70	V	Air	Charcoal	50	1000	73 — 310	GC/ THF—FID—3	8.0
Heptane	500 (2000)		S89	V	Air	Charcoal	4	200	968 — 4060	GC/ CS_2—FID—17	5.6
Hexachloroethane	1 (10)		S101	V	Air	Charcoal	10	200	5 — 25	GC/ CS_2—FID—22	12.1
Hexane	500 (1800)		S90	V	Air	Charcoal	4	200	877 — 3679	GC/CS_2—FID—3	6.2
Hexachloronaphthalene	(0.2)		S100	V	Air	CEF	30	1000	0.104 — 459	GC/hexane—EC—17	6.3
2-Hexanone (MBK)	100		S178	V	Air	Charcoal	10	200	188 — 190	GC/CS_2—FID—3	5.3
Hexone (MIBK)	100 (409)		S18	V	Air	Charcoal	10	200	208 — 836	GC/CS_2—FID—3	6.4
Ethylene chlorohydrin	5 (16)		S103	V	Air	Charcoal	20	200	7 — 30	GC/CS_2 + 5% ipa—FID—	7.6
Ethylene oxide	50 (90)		S286	V	Air	Charcoal	5	50	41 — 176	GC/ CS_2—FID—1	10.3
N-Ethylmorpholine	20 (94)		S146	V	Air	Silica gel	10	200	57 — 206	GC/H_2SO_4/ Na-OH—FID—19	10.1
Fluoranthene	—	—	183	D	Air	GF or SMF	500	2000	0.001 — 0.5	GC/ benzene—UV—12	—
			184	D	Air	GF or SMF	500	2000	0 — 3	LCG/ benzene—9	—
Fibrous glass[h]	3 fibers/mL		[h]	[h]	Air	Filter				Visual count	
Fluoride	(2.5)[h]	[h]	114	D	Urine	—	—	—	0.19 + mg/L	Ion spec. elect.	—
			117	D	Air	MI—NaOH	40—200	2500	0.05 — 475	Ion spec. elect.	6.5
			212	E	Air	CEF	250	2500	0.005 +	Ion spec. elect.	—
Fluorotrichloromethane	1000 (5600)		S102	V	Air	Charcoal	4	10—50	2390 — 10,500	GC/ CS_2—FID—4	7.2

Formaldehyde[h]	1[h]		125	C	Air	MI	25	1000	0.15 — 3.0	Colorimetric-chromotropic acid	5
	(1.2)		235	E	Air	Alumina gel	6	200	0.4 — 52	Colorimetric-chromotropic acid	6
Formic acid	5 (9)		232	D	Air	MI—NaOH	10—100	2500	3.8 — 75	As ethyl formate GC/FID—23	11
Gallium	—	—	177	E	Air	CEF	100	1500—2000	0.0002 — 0.01	ASV—Hg/carbon	15
	—	—	198	E	Urine	—	—	0.002	10 — 500 mg/L	ASV	12
Glycidol (2,3 epoxy, propanol)	50 (150)		S70	V	Air	Charcoal	50	1000	73 — 310	GC/THF—FID—3	8.0
Heptane	500 (2000)		S89	V	Air	Charcoal	4	200	968 — 4060	GC/CS_2—FID—17	5.6
Hexachloroethane	1 (10)		S101	V	Air	Charcoal	10	200	5 — 25	GC/CS_2—FID—22	12.1
Hexane	500 (1800)		S90	V	Air	Charcoal	4	200	877 — 3679	GC/CS_2—FID—3	6.2
Hexachloro-naphthalene	(0.2)		S100	V	Air	CEF	30	1000	0.104 — 459	GC/hexane—EC—17	6.3
2-Hexanone (MBK)	100		S178	V	Air	Charcoal	10	200	188 — 190	GC/CS_2—FID—3	5.3
Hexone (MIBK)	100 (409)		S18	V	Air	Charcoal	10	200	208 — 836	GC/CS_2—FID—3	6.4
S-Hexylacetate (1,3 dimethylbutylacetate)	50 (300)		S37	V	Air	Charcoal	10	200	143 — 601	GC/CS_2—FID—4	5.8
Hydrazine	1		248	E	Air	Silica gel—H_2SO_4	96	200	0.002 — 6.0	Derivative GC—FID—21	—
	(1.3)		S237								
Hydrogen bromide (HBr)	3 (10)		S175	V	Air	Bubbler — NaOH	100	1000	6.14 — 20.4	Specific ion electrode	6.7
Hydrogen chloride (HCl)		5 (7)	S246	V	Air	Bubbler — NaAC	15	1000	3.5 — 14	Specific ion electrode	6.4

Table 64 (continued)
NIOSH Sampling and Analytical Methods

	Federal standard		Method			Sampling			Analysis		
Analyte chemical name	TWA, ppm (mg/m³)[b]	Ceiling limit, ppm (mg/m³)[b]	Number	Class[c]	Matrix	Collection medium[d]	Volume (L)	Rate[e] (mℓ/min)	Range mg/m³	Method procedure[f]	% CV_T[g]
Hydrogen cyanide[h]		4.7[h] (5)[h]	116	D	Air	MI—NaOH	10—200	2500	0.013—13	Specific ion electrode	—
Hydrogen fluoride (HF)			117	D	Air	MI—NaOH	40—200	2500	0.05—475	Ion spec. electrode	6.5
	3		212	E	Air	CEF—alkaline	250	2500	0.005 +	Ion spec. electrode	—
	(2.45)		S176	V	Air	Bubbler — NaOH	45	1500	1.33—4.50	Ion spec. electrode	5.7
Hydrogen sulfide[h] (H_2S)		10[h] (15)	126	C	Air	MI—Cd $(OH)_2$	18	1500	0.012—75	Colorimetric—methylene blue	(3.5)
	50 (70)	20 (30)	S4	V	Air	MI—Cd $(OH)_2$	2	200	8.5—63	Methylene blue	12.1
Hydroquinone	(2)		S57	V	Air	CEF	90	1500	0.84—4.05	HAc/LCG/UV	6.1
Indium	(0.1)		173	D	Air	CEF	1.5	1500	.021—0.21 (21—210 μg)	TM—AA	—
			190	E	Air	CEF	100	1500—2000	0.002—0.03	ASV	10
			194	E	Blood	—	0.2 mL	—	0.01—5 μg	ASV	10
			199	E	Urine	—	2 mL	—	1—500 μg	ASV	10
Iron	(5)		173	D	Air	CEF	1.5	1500	(21—210) 0.021—0.21	TM—AA	—
Isoamyl acetate	100 (525)		S45	V	Air	Charcoal	10	200	208—874	GC/CS_2—FID—4	5.6
Isoamyl alcohol	100 (360)		S58	V	Air	Charcoal	10	200	195—680	GC/5 % ipa—FID—3	7.7
Isobutyl acetate	150 (700)		S44	V	Air	Charcoal	10	200	306—1280	GC/CS_2—FID—4	6.5
Isobutyl alcohol	100 (305)		S64	V	Air	Charcoal	10	200	180—620	GC/CS_2 + 1% ipa—FID—3	7.3
Isophorone	25 (140)		S367	V	Air	Charcoal	12	200	67.5—283	GC/CS_2—FID—3	5.8
S-Hexylacetate (1,3 dimethylbutylacetate)	50 (300)	S37	V	Air	Charcoal	10	200	143	—601GC/ CS_2—FID—4	5.8	

Hydrazine	1		248	E	Air	Silica gel — H_2SO_4	96	200	0.002 — 6.0	Derivative GC—FID—2 1	—
	(1.3)	S237									
Hydrogen bromide (HBr)	3 (10)		S175	V	Air	Bubbler — NaOH	100	1000	6.14 — 20.4	Specific ion electrode	6.7
Hydrogen chloride (HCl)		5 (7)	S246	V	Air	Bubbler — NaAC	15	1000	3.5 — 14	Specific ion electrode	6.4
Hydrogen cyanide[h]		4.7[h] (5)[h]	116	D	Air	MI—NaOH	10—200	2500	0.013 — 13	Specific ion electrode	—
Hydrogen fluoride (HF)			117	D	Air	MI—NaOH	40—200	2500	0.05 — 475	Ion spec. electrode	6.5
	3		212	E	Air	CEF—alkaline	250	2500	0.005 +	Ion spec. electrode	—
	(2.45)		S176	V	Air	Bubbler — NaOH	45	1500	1.33 — 4.50	Ion spec. electrode	5.7
Hydrogen sulfide[h] (H_2S)		10[h] (15)	126	C	Air	MI—Cd($OH)_2$	18	1500	0.012 — 75	Colorimetric— methylene blue	(3.5)
	50 (70)	20 (30)	S4	V	Air	MI—Cd($OH)_2$	2	200	8.5 — 63	methylene blue	12.1
Hydroquinone	(2)		S57	V	Air	CEF	90	1500	0.85 — 4.05	HAc/LCG/UV	6.1
Indium	(0.1)		173	D	Air	CEF	1.5	1500	.021 — 0.21 (21 — 210 μg)	TM—AA	—
			190	E	Air	CEF	100	1500 —2000	0.002 — 0.03	ASV	10
			194	E	Blood	—	0.2 mL	—	0.01 — 5 μg	ASV	10
			199	E	Urine	—	2 mL	—	1 — 500 μg	ASV	10
Iron	(5)		173	D	Air	CEF	1.5	1500	(21 — 210) 0.021 — 0.21	TM—AA	—
Iso-amyl acetate	100 (525)		S45	V	Air	Charcoal	10	200	208 — 874	GC/ CS_2—FID—4	5.6
Iso-amyl alcohol	100 (360)		S58	V	Air	Charcoal	10	200	195 — 680	GC/5 % ipa— FID—3	7.7

Table 64 (continued)
NIOSH Sampling and Analytical Methods

Federal standard			Method		Sampling				Analysis		
Analyte chemical name	TWA, ppm (mg/m³)[b]	Ceiling limit, ppm (mg/m³)[b]	Number	Class[c]	Matrix	Collection medium[d]	Volume (L)	Rate[e] (mℓ/min)	Range mg/m³	Method procedure[f]	% CV_T[g]
Isobutyl acetate	150 (700)		S44	V	Air	Charcoal	10	200	306 — 1280	GC/ CS_2—FID—4	6.5
Isobutyl alcohol	100 (305)		S64	V	Air	Charcoal	10	200	180 — 620	GC/CS_2 + 1% ipa—FID—3	7.3
Isophorone	25 (140)		S367	V	Air	Charcoal	12	200	67.5 — 283	GC/ CS_2—FID—3	5.8
Lead[h]			102	A	Blood	—	0.10	—	3 — 250 mg	Colorimetric —dithizonate	6
					Urine				12 — 1000 mg		
	(0.15)[h]		173	D	Air	CEF	1.5	1500	0.042 — 840	TM—AA	—
			191	E	Air	CEF	100	1500 —2000	0.0016 — 0.01	ASV	9.0
			195	E	Blood	—	0.10 mL	—	0.04 — 5 µg	ASV	5
			200	E	Urine	—	0.001	—	4 — 1000 µg/L	ASV	6
			208	E	Blood	—	0.001	—	0.05 — 1.5 µg/mL	Chelation—ext —AA	5
					Urine						
			214	E	Air	CEF	5	15	1500	AA—graphite furnace	5.7
					Blood						
			262	E	Blood	—	0.10	—	5 — 250 µg	Chelation—ext —AA	6.4
					Urine		1		10 — 500 µg		
	(0.2)		S341	V	Air	CEF	180	1500	0.128 — 0.399	AA	7.2
Lindane	0.5		S290	V	Air	GF + bubbler	90	1500	0.28 — 1.66	GC/i—Octane—ECC —5	8.6
Lithium hydride	(0.025)		173	D	Air	CEF	1.5	1500	0.28 — 84 µg	TM—AA	—
LPG (liquid petroleum gas)	1000 (1800)		S93	—	Air	—	—	—	478 — 2006 ppm	Combustible gas meter	—
Magnesium			173	D	Air	CEF	1.5	1500	2.1 — 21 µg	TM—AA	—
Magnesium oxide fume	(15)		S369	V	Air	CEF	150	1500	7.48 — 286	AA	6.2

Isopropyl acetate	250 (950)		S50	V	Air	Charcoal	9	200	446 — 1870	GC/CS_2—FID—4	6.7
Isopropyl alcohol[a]	400[a] (985)	800[a] (1968)	S65	V	Air	Charcoal	3	200	505 — 1890	GC/CS_2 + 2 butyl alcobol—FID—3	6.4
Isopropyl glycidyl ether	50 (240)		S77	V	Air	Charcoal	10	200	121 — 484	GC/CS_2—FID—3	6.7
Isopropyl amine	5		221	E	Air	Silica gel	10	200	1 — 2400	GC/H_2SO_4/ NaOH— FID—6	3
	(12)		S147	V	Air	Bubbler — H_2SO_4	1000	1000	8.9 — 21.5	GC/NaOH— FID—19	6.7
Isopropyl ether	500 (2100)		S368	V	Air	Charcoal	3	50	992 — 4260	GC/CS_2—FID—3	5.6
Kepone[a]		(0.001)[a]	225	E	Air	CEF	120	1000	0.002 — 1.0	GC/EC—24	2.3
Ketene	(0.9)		S92	V	Air	Bubbler — NH_2OH	50	1000	0.40 — 1.74	Colorimetric— $FeCl_3$	6.4
Malathion	15		S370	V	Air	GF			8.0 — 35	GC/i— Octane—FP—25	6.3
Manganese		(5)	173	D	Air	CEF	1.5	1500	0.021 — 0.125	TM—AA	—
			S5	V	Air	CEF	0.38	1500	2.5 — 10	AA	6.5
Mercury[a]			145	D	Urine	—	0.10	—	0.04 — 0.15 mg/L	Colorimeter— dithizonate?	
			165	B	Urine	—			0.003 — 0.3	Flameless AA	3
			167	D	Blood	—	0.02	—	0.005 — 4.5	Flameless AA	—
Organic	(0.01)	(0.04)	175	E	Air	CEF—car-bo—B	60	1000	0.001 — 1.0 mg	Flameless AA	7.3
Inorganic	(0.05)										
Mesityl oxide	25 (100)		S12	V	Air	Charcoal	10	200	45 — 210	GC/FID—3—CS_2 + 1% ethanol	7.1
Methyl (*n*-amyl) Ketone (2-heptan-one)	100 (465)		S15	V	Air	Charcoal	10	200	200 — 925	GC/CS_2 + 1% ethanol—FID—3	6.6
5-Methyl 3-heptan-one	25 (130)		S13	V	Air	Charcoal	10	200	60 — 270	GC/CS_2 + 1% methanol—FID —3	10.4
Methyl acetate	200 (610)		S42	V	Air	Charcoal	7	200	343 — 1330	GC/CS_2—FID—3	5.5
Methyl acrylate	10 (35)		S38	V	Air	Charcoal	5	200	13.9 — 58.4	GC/CS_2—FID—4	6.6
Methyl alcohol[a] (methanol)	200[a]	800[a]	247	D	Air	Silica gel	5	200	135 — 534	GC/ H_2O—FID—3	6.5
	(260)[a]	(1048)[a]	S59	V	Air	Silica gel	5	200	140 — 540	GC/ H_2O—FID—3	7.0
Methyl bromide		20 (80)	S372	V	Air	Charcoal	11	0.75	35 — 150	GC/CS_2—FID—4	10.3

Table 64 (continued)
NIOSH Sampling and Analytical Methods

	Federal standard		Method		Sampling				Analysis		
Analyte chemical name	TWA, ppm (mg/m³)[b]	Ceiling limit, ppm (mg/m³)[b]	Number	Class[c]	Matrix	Collection medium[d]	Volume (L)	Rate[e] (mℓ/min)	Range mg/m³	Method procedure[f]	% CV_T[g]
Methyl cellosolve (2-methoxy ethanol)	25 (80)		S79	V	Air	Charcoal	50	1000	44—160	GC/CH_2Cl_2 + 5% MeOH—FID—3	6.8
Methyl cellosolve acetate	25 (120)		S39	V	Air	Charcoal	20	200	51—214	GC/CS_2—FID—4	6.8
Methyl chloride	—	—	201	E	Air	Charcoal	10	1000	2—1600	GC/ methanol—FID —1	13
Methyl chloroform (1,1,1 trichloro ethane)	350		127	B	Air	Charcoal	0.5—13	1000	0.05—9500	GC/CS_2—FID—3	10.5
	(1900)		S328	V	Air	Charcoal	6	200	904—3790	GC/ CS_2—FID—17	5.4
Methyl iodide	5 (28)		S98	V	Air	Charcoal	50	1000	17—52	GC/toluene—FID —26	7.0
Methyl-i-amyl acetate	50 (300)		S37	V	Air	Charcoal	10	200	143—601	GC/CS_2—FID—4	5.8
Methyl-i-butyl carbinol	25 (105)		S60	V	Air	Charcoal	10	200	45—175	GC/CS_2 + 5% ipa—FID—3	8.0
Methylal (di-methoxy methane)	1000 (3110)		S71	V	Air	Charcoal	2	200	1684—6670	GC/ hexane—FID—1	6.0
Methylamine	10 (12)		221	E	Air	Silica gel	10	200	1—2400	GC/H_2SO_4/ NaOH— FID—6	3
Methylcyclohexane	500 (2000)		S94	V	Air	Charcoal	4	200	940—3941	GC/CS_2—FID—3	5.2
4,4′ Methylene-*bis*-(2 chloro aniline)	Cancer[i]		236	E	Air	GF + silica gel	3—480	200 —1000	0.003—0.2	LCG/ methanol—UV	10
Methylene-*bis*-phenyl-di-isocyanate (MDI)		0.02 (0.2)	142	D	Air	MI—acid reagent	20	1000	0.07—0.73	Colorimetric—diazo reaction—blue dye	—
Methylene chloride[k]	500 (1740) 75[k] (261)	1000 (3475) 500[k] (1740)	127	D	Air	Charcoal	0.5—3.8	1000	0.05—3600	GC/CS_2—FID—3	10.5
			S329	V	Air	Charcoal	1—2.2	50—200	1700—7100	GC/CS_2—FID—4	7.3

Methyl hydrazine	0.2 (0.35)		248	E	Air	Silica gel—H_2SO_4	96	200	0.04 — 120	GC/H_2O—derivative—FID—21	4
	0.2		S149	V	Air	Bubbler — HCl	22.5	1500	0.169 — 0.78	Colorimetric — phosphomolybdate	10.6
Molybdenum	(5)		173	D	Air	CEF	1.5	1500	0.625 — 1.65	TM—AA	—
Insoluble	(15)		S376	V	Air	CEF	90	1500	6.3 — 30.6	AA/313.3—r	5.6
Soluble	(5)		S193	V	Air	CEF	90	1500	1.96 — 9.75	AA/H_2O—313.3—r	8.9
Monomethylaniline	2 (9)		S153	V	Air	Bubbler — H_2SO_4	100	1000	4.38 — 17.5	GC/NaOH—FID—6	8.9
Monomethylhydrazine		0.2 (0.35)	248	E	Air	Silica gel — H_2SO_4	96	200	0.09 — 90	GC/H_2O—derivative—FID—21	
			S149	V	Air	Bubbler—HCl	22.5	1500	0.169 — 0.78	Colorimetric—phosphomolybdate	10.6
Methylparathion[a]	(0.2)		[a]	[a]	Air	*Ethylene—MI—glycol	50[a]	2800[a]	0.025 — 1.25	GC/hexane—FP—29	—
Morpholine	20 (70)		S150	V	Air	Silica gel	20	200	28.5 — 108.4	GC/H_2SO_4—FID—6	5.7
Naphtha (coal tar)	100 (400)		S86	V	Air	Charcoal	10	200	193 — 809	GC/CS_2—FID—27	5.1
Naphthalene	10 (50)		S292	V	Air	Charcoal	200	1000	19.3 — 83	GC/CS_2—FID—17	5.5
Nickel[a]	(0.1)		173	D	Air	CEF	1.5	1500	0.021 — 0.21	TM—AA	—
	(1)		S206	V	Air	CEF	90	1500	0.45 — 2.1	AA/acid—232.0—0	5.7
	(0.015)[a]										
Nicotine	(0.5)		S293	V	Air	XAD—2 resin	100	1000	0.31 — 1.20	GC/Et—ac/alkaline FID—17	
Nitric oxide	25 (30)		231	D	Air	13 × mol. sieve—TEA	1.0	50	0.96 — 36	TEA ext—SF/540 nm	6—7
P-Nitroaniline	1 (6)		168	D	Air	Silica gel	80—12,60	See 168	0.01 — 14	GC/ethanol—FID—7	9
Nitrobenzene	1 (5)		S217	V	Air	Silica gel	50	1000	3.11 — 12.45	GC/methanol—FID—3	5.8

Table 64 (continued)
NIOSH Sampling and Analytical Methods

	Federal standard		Method		Sampling				Analysis		
Analyte chemical name	TWA, ppm (mg/m³)[b]	Ceiling limit, ppm (mg/m³)[b]	Number	Class[c]	Matrix	Collection medium[d]	Volume (L)	Rate[e] (mℓ/min)	Range mg/m³	Method procedure[f]	% CV_T[g]
P-nitrochlorobenzene	(1)		S218	V	Air	Silica gel	50	1000	0.43 — 2.75	GC/methanol—FID—3	10.3
Nitrogen dioxide	5 (9)		108	C	Air	Bubbler—Sallzman	4—12	400	0.01 — 10	Colorimetric—azo Sallzman's reagent	5
NO_2 NO			231	D	Air	TEA/mol sieve oxidizer	1	50	0.96 — 36	TEA ext—colorimetric—Sallzman's reagent	6—7
Nitric acid[h]	2[h] (5)	[h]	[h]	[h]	Air	MI—H_2O	200	2800	—	UV/220 nm (−NO_3)	5
Nitroglycerine		0.2	203	E	Air	Tenax GC	0.1—1000	1000	0.1 — 100	GC/ethanol—EC—22	10
		(1.0)	S216	V	Air	Tenax GC	15	1000	0.56 — 3.21	GC/ethanol—EC—22	10.4
Nitroglycol	0.2 (2)		203	E	Air	Tenax GC	10—15	1000	0.01 — 100	GC/ethanol—ED—22	10
N-Nitrosodimethyl amine	Cancer[i]		252	E	Air	Tenax GC	79—385	10—30,000	0.5 ppt — 10 ppb	GC/thermal—28—MS	10
Nitrotoluene	5 (30)		S223	V	Air	Silica gel	20	200	11.4 — 58.2	GC/methanol—FID—2	6.2
Octachloro-naphthalene	(0.1)		S97	V	Air	CEF	30	1000	0.055 — 0.215	GC/hexane—EC—17	6.7
Octane	500 (2350)		S378	V	Air	Charcoal	4	200	1050 — 4403	GC/CS_2—FID—3	6.0
Di-s-octylphthalate	(5)		S40	V	Air	CEF	30	1000	2.03 — 10.9	GC/CS_2—FID—17	5.7
Oil mist	(5)		159	D	Air	CEF	100	2000	0.005 — 0.50	Fluorescence—emission	
	(5)		S272	V	Air	CEF	100	1500	2.5 — 11.7		6.5

Organo (alkyl) mercury	(0.01)	(0.04)									
			175	E	Air	CEF—carbo—B	60	1000	0.001 — 1.0 μg	Flameless AA	7.3
Ozone	0.1		153	B	Air	MI—KI	60	1000 —2000	0.02 — 20	Colorimetric	5
	0.1		154	C	Air	MI—KI	45	1000	0.1 — 0.4	Alkaline—colorimetric	8
	(0.2)		S8	V	Air	MI—KI	45	1000	0.1 — 0.4	Alkaline—color-	8.1
Palladium	NL		173	D	Air	CEF	1.5	1500	0.17 — 0.63	TM—AA	—
Parathion[a]	(0.11)		158	C	Air	MI—EG	25—30	1000 —2000	0.005 — 0.25	GC/Hexane—FP—29	—
	(0.05)[a]		S295	V	Air	GF	120	1500	0.073 — 0.264	GD/i—Octane—FP—15	7.8
Pentachloronaphthalene	(0.5)		S96	V	Air	GF + Bubbler	250	1300	0.293 — 1.54	GC/i—Octane—ECC—5	6.7
PCB (Polychlorinated biphenyls)			244	E	Air	Florisil	50	200	0.01 — 10	GC/Hexane—EC—15	4.4
			253	D	Air	Florisil	50	200	0.01 — 10	GC/Hexane—EC—30	2.8
	(0.5)		S121	V	Air	CEF	100	1500	0.32 — 1.26	GC/Pet. ether—EC—15	6.3
Pentachlorophenol air	(0.5)		230	E	Urine	—	0.002	—	0.02 — 0.18	Benzene ext—GC—ED	15
Pentane	1000 (2590)		S379	V	Air	Charcoal	2	50	1476 — 6190	GC/CS_2—FID—3	5.5
2-Pentanone	200 (705)		S20	V	Air	Charcoal	10	2000	395 — 1570	GC/CS_2—FID—3	6.3
Perylene	—	—	184	D	Air	GF or SMF	500	2000	0 — 3	GC/Benzene—UV—12	
(TpAH)			206	E	Air	GF	1.5×10^5	2000	0.005	LC/Cyclohexane—UV	1.3
Petroleumn naphtha	500										
	(2000)		S380	V	Air	Charcoal	4	200	937 — 3930	GC/CS_2—FID—17	5.2
Phenathene	—	—	206	E	Air	GF	1.5×10^5	2000	.005 — 0.2	LCG/cychohexane—UV	1.3

Table 64 (continued)
NIOSH Sampling and Analytical Methods

	Federal standard		Method			Sampling				Analysis	
Analyte chemical name	TWA, ppm (mg/m^3)[b]	Ceiling limit, ppm (mg/m^3)[b]	Number	Class[c]	Matrix	Collection medium[d]	Volume (L)	Rate[e] (mℓ/min)	Range mg/m^3	Method procedure[f]	% CV_T[g]
Phenol[h]	5[h] (19)	15[h] (60)	S330	V	Air	Bubbler—-NaOH	100	1000	9.46—37.8	GC/H_2SO_4—FID	6.8
Phenyl ether	1 (7)		S72	V	Air	Charcoal	10	200	3—13	GC/CS_2—FID—3	7.0
Phenylether:biphenyl mixture	1 (7)		S73	V	Air	Silica gel	10	200	3.86—15.7	GC/benzene—FID—31	8.9
Phenylglycidyl ether	10 (60)		S74	V	Air	Charcoal	50	1000	31—121	GC/CS_2—FID—3	5.7
Phenylhydrazine	5		248	E	Air	Silica gel—H_2SO_4	96	200	0.005—220	GC/H_2O—FID—21	—
	(22)		S160	V	Air	Bubbler—-HCL	100	1000	10.4—44.8	Colorimetric—phosphomolybdate	6.0
Phosgene[h]	0.1[h] (0.4)	0.2[h] (0.8)	219	E	Air	MI—color reagent	50	1000	0.02—20	Colorimetric/nitrobenzylpyridine	—
Phosphate	—	—	215	D	Urine	—	0.025	—	0.6—6 mg/L	Colorimetric + sbphosphomolybdate	7.0
Phosphoric acid			216	D	Air	CEF	50	1500—	0.2—4.0	Molybdenum blue	4.3
	(1)		S333	V	Air	CEF	90	2000	0.1—2.0	Colorimetric	5.5
Palladium	NL		173	D	Air	CEF	1.5	1500	0.17 — 0.63	TM—AA	—
Parathion[h]	(0.11)		158	C	Air	MI—EG	25—30	1000—2000	0.005 — 0.25	GC/Hexane—-FP—29	—
	(0.05)[h]		S295	V	Air	GF	120	1500	0.073 — 0.264	GD/i—Octane—-FP—15	7.8
Pentachloronaphthalene	(0.5)		S96	V	Air	GF + Bubbler	250	1300	0.293 — 1.54	GC/i—Octane—-ECC—5	6.7
PCB (Polychlorinated biphenyls)			244	E	Air	Florisil	50	200	0.01 — 10	GC/Hexane—-EC—15	4.4

			253	D	Air	Florisil	50	200	0.01 — 10	GC/Hexane—-EC—30	2.8
	(0.5)		S121	V	Air	CEF	100	1500	0.32 — 1.26	GC/Pet. ether—EC—15	6.3
Pentachlorophenol air	(0.5)		230	E	Urine	—	0.002	—	0.02 — 8.18	Benzen ext—-GC—ED	15
Pentane	1000 (2590)		S379	V	Air	Charcoal	2	50	1476 — 6190	GC/ CS_2—FID—3	5.5
2-Pentanone	200 (705)		S20	V	Air	Charcoal	10	2000	295 — 1570	GC/ CS_2—FID—3	6.3
Perylene	—	—	184	D	Air	GF or SMF	500	2000	0 — 3	GC/ Benzene—-UV—12	
(TpAH)			206	E	Air	GF	1.5×10^5	2000	0.005	LC/ Cyclohexane —UV	1.3
Petroleumn naphtha	500										
	(2000)		S380	V	Air	Char coal	4	200	937 — 3930	GC/ CS_2—FID—17	5.2
Phenathene	—	—	206	E	Air	GF	1.5×10^5	2000	.005 — 0.2	LCG/ cychohexane —UV	1.3
Phenol[a]	5[a] (19)	15[a] (60)	S330	V	Air	Bubbler—NaOH	100	1000	9.46 — 37.8	GC/ H_2SO_4—FID	6.8
Phenyl ether	1 (7)		S72	V	Air	Charcoal	10	200	3 — 13	GC/ CS_2—FID—3	7.0
Phenylether:biphenyl mixture	1 (7)		S73	V	Air	Silica gel	10	200	3.86 — 15.7	GC/ benzene—FID —31	8.9
Phenylglycidyl ether	10 (60)		S74	V	Air	Charcoal	50	1000	31 — 121	GC/ CS_2—FID—3	5.7
Phenylhydrazine	5		248	E	Air	Silica gel—H_2SO_4	96	200	0.005 — 220	GC/ H_2O—FID—21	—

Table 64 (continued)
NIOSH Sampling and Analytical Methods

	Federal standard		Method			Sampling			Analysis		
Analyte chemical name	TWA, ppm (mg/m^3)[b]	Ceiling limit, ppm (mg/m^3)[b]	Number	Class[c]	Matrix	Collection medium[d]	Volume (L)	Rate[e] (mℓ/min)	Range mg/m^3	Method procedure[f]	% CV_T[g]
	(22)		S160	V	Air	Bubbler—HCL	100	1000	10.4 — 44.8	Colorimetric—phosphomolybdate	6.0
Phosgene[h]	0.1[h] (0.4)	0.2[h] (0.8)	219	E	Air	MI—color reagent	50	1000	0.02 — 20	Colorimetric/nitrobenzylpyridine	—
Phosphate	—	—	215	D	Urine	—	0.025	—	0.6 — 6 mg/L	Colorimetric + sbphosphomolybdate	7.0
Phosphoric acid			216	D	Air	CEF	50	1500—	0.2 — 4.0	Molybdenum	4.3
	(1)		S333	V	Air	CEF	90	2000	0.1 — 2.0	Colorimetric	5.5
Phosphorus (white and yellow)	(0.1)		242	E	Air	GF + MI xylene	24—96	500—1000	0.005 — 5.0	GC—FP—15	3.3
			257	E	Air	Tenax—GC	20	50—200	0.01 — 1.0	GC/xylene—FP—15	6
Phthalic anhydride	2 (12)		S179	V	Air	CEF	100	1500	5.5 — 28	Hydrolyze NH_3—LCG ethyl acetate	8.9
Platinum soluble salts		(0.002)	S191	V	Air	CEF	720	1500	0.0079 — 0.031	AA/graphite atm.	6.2
PNAs (polynuclear aromatics)	—	—	183	D	Air	GF or SMF	500	2000	0.001 — 0.05	GC/benzene—UV—12	—
			184	D	Air	GF or SMF	500	2000	0 — 3	LCG/benzene—9—	—
Gen-methods			186	D	Air	GF or SMF	500	2000	0.02 — 0.8	TLC/SF	7
Hydrocarbons			206	E	Air	GF	1.5×10^5	2000	0.005	LCG/cyclohexane—UV	1.3
Polymethylsiloxane	—	—	227	E	Air	CEF	400	1700	0.01 — 0.1	Flameless AA/$CHCl_3$ extract	15
Potassium as KOH	(2)		173	D	Air	CEF	1.5	1500	0.0042 — 0.084	TM—AA	—

Propane	1000 (1800)		S87	—	Air	Meter	—	—	481 — 2016	Combustible gas meter	—
n-Propyl acetate	200 (840)		S48	V	Air	Charcoal	10	200	384 — 1610	GC/CS_2—FID—4	5.6
Propyl alcohol (propanol)	200 (490)		S62	V	Air	Charcoal	10	200	225 — 835	GC/FID—3—CS_2 + 1% ipa	7.5
n-Propyl nitrate	25 (110)		S227	V	Air	Charcoal	70	1000	50 — 209	GC/CS_2—FID—3	5.4
Propylene dichloride	75 (350)		S95	V	Air	Charcoal	10	200	135 — 665	GC/CS_2—FID—3	5.6
Propylene oxide (1,2 epoxy propane)	100 (240)		S75	V	Air	Charcoal	5	200	121 — 482	GC/CS_2—FID—1	8.5
Pyrene	—	—	183	D	Air	GF or SMF	500	2000	0.001 — 0.05	GC/benzene—UV—12	—
			184	D	Air	GF or SMF	500	2000	0 — 3	LCG/benzene—9	—
Pyridine	5 (15)		S161	V	Air	Charcoal	100	1000	7.59 — 30.4	GC/CH_2Cl_2—FID—23	5.9
Rhodium (fume and dust)	(0.1)		S188	V	Air	CEF	720	1500	0.057 — 0.21	Acid/AA/343.5	7.9
Soluble	(0.001)		S189	V	Air	CEF	370	1500	0.004 — 0.0018	Acid/AA—graphite	6.9
Rubidium	Not listed		173	D	Air	CEF	1.5	1500	0.021 — 0.21	TM—AA	—
Selenium			124	D	Urine	—	0.3	—	0.017 — 0.50/L	Se colloid	4
			181	E	Air	CEF	90	1500	0.0006 — 0.01	Catalytic red. MB	3
	(0.2)		S190	V	Air	CEF	360	1500	0.1 — 0.5	Acid/AA—196.0	9.0
Silicon	(10)		173	D	Air	CEF	1.5	1500	2.1 — 6.3	TM—AA	—
Silica crystalline	Calc.[a]		106	B	Dust Air	CEF	100	2000	0.01 — 2.5	H_3PO_4 digestion—colorimetric—mo blue	9.3
Quartz, cristobalite, tridymite	Calc.[a]		109	D	Dust Air	PVCF	Respirable fraction	See 109	0.005 — 0.20	X-ray diffraction	5 μg
Quartz	Calc.[a]		110	B	Coal Dust	CEF	Respirable fraction	See 110	0.01 — 0.100	Infrared—spectrography	15
Quartz, cristobalite, tridymite	Calc.[a]		259	D	Air	CEF	Respirable fraction	1700	0.02 — 2.0	X-ray diffraction	10
Quartz, cristobalite, tridymite	Calc.[a] (0.05)[a]	[a]	S315	V	Air	SMF	Respirable 200	1700	0.167 — 0.846	X-ray diffraction	6.5
Silver	(0.01)		173	D	Air	CEF	1.5	1500	0.021 — 0.170	TM—AA	—
Sodium (NaOH)[a]	(2.0)[a]		173	D	Air	CEF	1.5	1500	0.002 — 0.04	TM—AA	—
Sodium hydroxide	(2)		241	E	Air	Bubbler—HCl	350	1000	1 — 4	Electro. titration	6

Table 64 (continued)
NIOSH Sampling and Analytical Methods

	Federal standard		Method			Sampling			Analysis		
Analyte chemical name	TWA, ppm (mg/m³)[b]	Ceiling limit, ppm (mg/m³)[b]	Number	Class[c]	Matrix	Collection medium[d]	Volume (L)	Rate[e] (ml/min)	Range mg/m³	Method procedure[f]	% CV$_T$[g]
Stoddard solvent	500 (2950)		S382	V	Air	Charcoal	3	200	1417—5940	GC/ CS_2—FID—27	5.2
Strontium	—	—	173	D	Air	CEF	1.5	1500	0.021—0.210	TM—AA	—
Styrene (vinyl benzene)	100	200	127	E	Air	Charcoal	1.5—34	1000	0.1—2100	GC/CS_2—FID—3	10.5
	(425)	(850)	S30	V	Air	Charcoal	5	1000	426—1710	GC/CS_2—FID—3	5.7
Sulfur dioxide[h]	2[h]	[h]	146	C	Air	MI—H_2O	10—100	1000—2000	0.026—26	Titration—NaOH	10
	(5)[h]		160	C	Air	MI	4—5	200—1000	0.008—13	Colorimetric—TCM	5
			163	D	Air	MI—H_2O_2	10—100	1000—2000	0.01—5	Titration—Ba^{++}	4
			204	E	Air	Mol sieve 5A	See 204	200	2—625	Thermaldes/MS	0.2
Sulfuric acid[h]			187	E	Air	MI—H_2O	100	1500	0.1—4.0	Turbidimetry	9
	(1)[h]		S174	V	Air	CEF	180	1500	0.561—2.58	Titrate—Ba^{++}	8.2
Tellurium	(0.1)		S204	V	Air	CEF	670	1500	0.05—0.240	Acid digestion—AA	5.5
	(0.1)		173	D	Air	CEF	1.5	1500	0.021—1.05	TM—AA	—
Terphenyl		1	183	D	Air	GF or SMF	500	2000	0.0007—1.0	GC/benzene—UV—12	—
		(10)	S27	V	Air	CEF	15	1000	4.5—20	GC/CS_2—FID—5	9.7
Tellurium hexafluoride	0.02 (0.2)	Not listed	S187	V	Air	Charcoal	400	1000	0.094—0.37	AA/NaOH	5.4
1,1,1,2 Tetrachloro, 2,2 difluoro ethane	500 (4170)		S131	V	Air	Charcoal	2	35	2160—9020	GC/CS_2—FID—3	6.9
1,1,2,2, Tetrachloro 1,2 difluoro ethane	500 (4170)		S132	V	Air	Charcoal	2	50	1883—8060	GC/CS_2—FID—3	5.4
1,1,2,2 Tetrachloro ethane	5 (35)		S124	V	Air	Charcoal	10	200	16—70	GC/CS_2—FID—3	5.7

Tetrachloro ethylene (perchloro ethylene)	100	200	127	B	Air	Charcoal	1—25	1000	0.06 — 2100	GC/CS_2—FID—3	10.5
	(680)	(1355)	S335	V	Air	Charcoal	3	200	655 — 2749	GC/ CS_2—FID—17	5.2
	50* (339)	100* (678)									
Tetrachloro naphthalene	(2)		S130	V	Air	GF bubbler	100	1300	0.912 — 4.04	GC/i—Octane ECC—5	9.7
Tetrahydrofuran	200 (590)		S78	V	Air	Charcoal	9	200	323 — 1240	GC/CS_2—FID—1	5.5
Tetramethyl succinonitrile	0.5 (3)		S155	V	Air	Charcoal	50	1000	1.8 — 8.20	GC/CS_2—FID—5	7.5
Tetranitro methane	1 (8)		S224	V	Air	MI—Ethyl acetate	250	1000	2.7 — 11.5	GC/alkaline—FID —17	7.6
Tetryl (2,4,6 trinitro phenyl methyl nitramine)	(1.5)		S225	V	Air	CEF	100	1500	0.88 — 4.51	Colorimetric/420 nm—*N,N*—diethyl ethanol amine	6.5
Thalium	(0.1)		173	D	Air	CEF	1.5	1500	0.21 — 0.84	TM—AA	—
			S306	V	Air	CEF	540	1500	0.034 — 0.15	Acid digestion—AA	6.0
Thiophene	—	—	255	E	Air	Charcoal	70	1000	0.05 — 7	GC/toluene—FP—32	3.2
Thiram (tetramethyl thiuram disulfide)	(5)		228	D	Air	CEF	50	1700	0.1 — 5	$CHCl_3$ + CuI —Colorimetric/ 440 nm	2.3
Tin			173	D	Air	CEF	1.5	1500	—	TM—AA	—
			176	D	Air	CEF	5	1700	0.1 — 2	Colorimetric—pyrocatechol violet	8.7
	(2)		S183	V	Air	CEF	250	1500	0.94 — 4.32	Acid digestion—AA	
Organic*	(0.1)		176	D	Air	CEF	—	—	0.1 — 2	Color—pyrocatechol violet	3.0 —5.7
Titanium dioxide	(15)	200*	S385	V	Air	CEF	100	1500	8.1 — 29.5	Acid digestion—AA	11.2
Toluene	100* (375)	200* (750)	127	B	Air	Charcoal	1—25	1000	0.01 — 3750	GC/CS_2—FID—3	10.5
	200 (750)	300 (1125)	S343	V	Air	Charcoal	2—12	200	145.5 — 582	GC/CS_2—FID—1	5.2
TDI (Toluene 2,4 diisocyanate)	0.005* (0.036)	0.02* (0.14)	141	D	Air	MI—acid reagent	20	1000	0.05 — 1.0	Colorimetric—diazo—dye	—

Table 64 (continued)
NIOSH Sampling and Analytical Methods

Analyte chemical name	Federal standard: TWA, ppm (mg/m^3)[b]	Federal standard: Ceiling limit, ppm (mg/m^3)[b]	Method: Number	Method: Class[c]	Method: Matrix	Sampling: Collection medium[d]	Sampling: Volume (L)	Sampling: Rate[e] (mℓ/min)	Analysis: Range mg/m^3	Analysis: Method procedure[f]	Analysis: % CV_T[g]
O-Toluidine	5 (22)		168 S168	D	Air	Silica gel	5—300	See 168	0.01 — 14	GC/ ethanol—FID—7	9
Tributyl phosphate	(5)		S208	V	Air	CEF	100	1500	2.7 — 12.6	GC/ ether—FP—17	7.6
1,1,2 Trichloro 1,2,2 trifluoro ethane	1000 (7660)		S129	V	Air	Charcoal	1.5	50	3300 — 14,200	GC/CS_2—FID—1	7.0
1,1,2 Trichloro ethane	10		127	B	Air	Charcoal	10—97	1000	0.05 +	GC/CS_2—FID—3	10.5
	(55)		S134	V	Air	Charcoal	10	200	26 — 111	GC/ CS_2—FID—17	5.7
Trichloro ethylene[h]	100[h]	150[h]	127	A	Air	Charcoal	1—17	1000	0.05 +	GC/CS_2—FID—3	10.5
	(535)	200 (1075)	S336	V	Air	Charcoal	3—10	200	519 — 2176	GC/ CS_2—FID—17	8.2
1,1,1 Trichloro ethane[h]	350 (1900)	330[h] (1910)	[h]	[h]	Air	Charcoal				GC/CS_2	
Trichloro naphthal-ene	(5)		S128	V	Air	GF bubbler	100	1300	1.65 — 8.8	GC/i—Octane—ECC —5	11.7
1,2,3 Trichloro pro-pane	50 (300)		S126	V	Air	Charcoal	10	200	163 — 629	GC/CS_2—FID—3	6.8
Tridymite—R			109	D	Dust	PVCF	—	—	5 — 200 $\mu g/cm^2$	X-ray diffraction	±5 μg
			259	D	Dust	PVCF	Respirable	1700	20 — 2000 μg	X-ray diffraction	10
	Calc.[h]		S315	V	Air	SMF	200	1700	0.167 — 0.846	X-ray diffraction	6.5
Triethylamine	25		221	E	Air	Silica gel	10	200	1 — 2400	GC/H_2SO_4/ NaOH—FID —6	3
	(100)		S152	V	Air	Bubbler—H_2SO_4	1000	1000	50 — 193	GC/alkaline—FID —19	6.4
Trifluorobromo methane	1000 (6090)		S125	V	Air	Charcoal	1	50	2890 — 11,560	GC/ CH_2Cl_2—FID—1	6.5

Tri-o-cresyl phosphate	(0.1)		S209	V	Air	CEF	100	1500	0.029—0.170	GC/ ether—FP—17	8.6
Tri-phenyl phosphate	(3.0)		S210	V	Air	CEF	100	1500	1.25—6.99	GC/ ether—FP—17	6.6
Turpentine	100 (560)		S88	V	Air	Charcoal	10	200	264—1107	GC/ CS_2—FID—27	5.5
Vanadium fume	—	(0.5)	173	D	Air	CEF	1.5	1500	0.42—6.3	TM—AA	—
		(0.1)	S391	V	Air	CEF	25	1700	0.24—0.90	NaOH—AA (flameless)	6.5
Vinyl chloride[a]	Not listed[a]	Not listed[a]	178	B	Air	Charcoal	5	50	0.008—5.2	GC/ CS_2—FID—33	8
	Detectable[a]		[a]	[a]	Air	Glass tube[a]	0.25[a]	1000[a]	—	GC/	—
Vinyl toluene	100 (480)		S25	V	Air	Charcoal	10	200	256—970	GC/CS_2—FID—3	5.8
Xylene (xylol[a], dimethyl benzene)	100[a]	200[a]	127	A	Air	Charcoal	0.5—31	1000	0.02 +	GC/CS_2—FID—3	10.5
	(435)[a]	(868)	S318	V	Air	Charcoal	12	200	218—870	GC/CS_2—FID—1	6.0
Yttrium	(1)		S200	V	Air	CEF	500	1500	0.53—2.21	Digestion—AA	5.4
Zinc	(5)		173	D	Air	CEF	1.5	1500	0.0042—0.042	TM—AA	—
Zinc oxide[a]	(5)[a]	(15)[a]	222	E	Air	PVCF	25	1500	1—80	X-ray diffraction	5—15
Zirconium compounds	(5)		S185	V	Air	CEF	720	1500	2.1—9.0	Acid digestion—AA/ 360.1	5.3
Toluene	100[a] (375)	200[a] (750)	127	B	Air	Charcoal	1—25	1000	0.01 — 3750	GC/ CS_2—FID—3	10.5
	200 (750)	300 (1125)	S343	V	Air	Charcoal	2—12	200	145.5 — 582	GC/ CS_2—FID—1	5.2
TDI (Toluene 2,4 diisocyanate)	0.005[a] (0.036)	0.02[a] (0.14)	141	D	Air	MI—acid reagent	20	1000	0.05 — 1.0	Colorimetric—diazo—dye	—
O-Toluidine	5 (22)		168 S168	D	Air	Silica gel	5—300	See 168	0.01 — 14	GC/ethanol—FID—7	9
Tributyl phosphate	(5)		S208	V	Air	CEF	100	1500	2.7 — 12.6	GC/ether—FP—17	7.6
1,1,2 Trichloro 1,2,2 trifluoro ethane	1000 (7660)		S129	V	Air	Charcoal	1.5	50	3300 — 14,200	GC/ CS_2—FID—1	7.0
1,1,2 Trichloro ethane	10		127	B	Air	Charcoal	10—97	1000	0.05 +	GC/ CS_2—FID—3	10.5

Table 64 (continued)
NIOSH Sampling and Analytical Methods

	Federal standard		Method			Sampling			Analysis		
Analyte chemical name	TWA, ppm (mg/m^3)[b]	Ceiling limit, ppm (mg/m^3)[b]	Number	Class[c]	Matrix	Collection medium[d]	Volume (L)	Rate[e] (mℓ/min)	Range mg/m^3	Method procedure[f]	% CV_T[g]
	(55)		S134	V	Air	Charcoal	10	200	26 — 111	GC/ CS_2—FID—17	5.7
Trichloro ethylene[h]	100[h]	150[h]	127	A	Air	Charcoal	1—17	1000	0.05 +	GC/ CS_2—FID—3	10.5
	(535)	200 (1075)	S336	V	Air	Carcoal	3—10	200	519 — 2176	GC/ CS_2—FID—17	8.2
1,1,1 Trichloro ethane[h]	350 (1900)	330[h] (1910)[h]	[h]	Air	Charcoal				GC/CS_2		
Trichloro naphthalene	(5)		S128	V	Air	GF bubbler	100	1300	1.65 — 8.8	GC/ i—Octane—-ECC—5	11.7
1,2,3 Trichloro propane	50 (300)		S126	V	Air	Charcoal	10	200	163 — 629	GC/ CS_2—FID—3	6.8
Tridymite—R			109	D	Dust	PVCF	—	—	5 — 200 μg/cm^2	X-ray diffraction	±5 μg
			259	D	Dust	PVCF	Respirable	1700	20 — 200 μg	X-ray diffraction	10
	Calc.[k]		S315	V	Air	SMF	200	1700	0.167 — 0.846	X-ray diffraction	6.5
Triethylamine	25		221	E	Air	Silica gel	10	200	1 — 2400	GC/H_2SO_4/ NaOH—FID—6	3
	(100)		S152	V	Air	Bubbler—H_2SO_4	1000	1000	50 — 193	GC/alkaline—-FID—19	6.4
Trifluorobromo methane	1000 (6090)		S125	V	Air	Charcoal	1	50	2890 — 11,560	GC/ CH_2Cl_2—FID—1	6.5

Tri-o-cresyl phosphate	(0.1)		S209	V	Air	CEF	100	1500	0.029 — 0.170	GC/ether—-FP—17	8.6
Tri-phenyl phosphate	(3.0)		S210	V	Air	CEF	100	1500	1.25 — 6.99	GC/ether—-FP—17	6.6
urpentine	100 (560)		S88	V	Air	Charcoal	10	200	264 — 1107	GC/ CS_2—FID—27	5.5
Vanadium fume	—	(0.5)	173	D	Air	CEF	1.5	1500	0.42 — 6.3	TM—AA	—
		(0.1)	S391	V	Air	CEF	25	1700	0.24 — 0.90	NaOH—AA (flameless)	6.5
Vinyl chloride[a]	Not listed[a]	Not listed[a]	178	B	Air	Charcoal	5	50	0.008 — 5.2	GC/ CS_2—FID—33	8
		Detectable[a]		[h]	[h]	Air	Glass tuve[h]	0.25[h]	1000[h] — GC/	—	
Vinyl toluene	100 (480)		S25	V	Air	Charcoal	10	200	256 — 970	GC/ CS_2—FID—3	5.8
Xylene (xylol[a], dimethyl benzene)	100[a]	200[a]	127	A	Air	Charcoal	0.5—31	1000	0.02 +	GC/ CS_2—FID—3	10.5
	(435)[h]	(868)	V	Air	Charcoal	12	200	218	—870GC/ CS_2—FID—6.0		
Yttrium	(1)		S200	V	Air	CEF	500	1500	0.53 — 2.21	Digestion—AA	5.4
Zinc	(5)		173	D	Air	CEF	1.5	1500	0.0042 — 0.042	TM—AA	—
Zinc oxide[a]	(5)[h]	(15)[h]	222	E	Air	PVCF	25	1500	1 — 80	X-ray diffraction	5—15
Zirconium compounds	(5)		S185	V	Air	CEF	720	1500	2.1 — 9.0	Acid digestion—AA/360.1	5.3

[a] Federal standards, unless noted by [h], are those given in the Code of Federal Regulations, Title 29, Chapter XVII, Part 1910, 37FR22139 (October 19, 1972), Subpart G-1910.93;

[b] mg/m^3 = figures in parentheses.

[c] Class: A = recommended: fully evaluated and collaboratively tested, B = accepted: evaluated in the NIOSH laboratory, C = tentative: recommended by one of several professional societies, D = operational: in general use, E = proposed: new, improved or suggested method, V = B: equivalent to the B classification established by NIOSH for Volume 1.

Table 64 (continued)

NIOSH Sampling and Analytical Methods

[d] Collection medium: TEA = triethanolamine, MI = midget impinger, CEF = cellulose ester filter, GF = glass fiber filter, PVCF = polyvinyl chloride filter, SMF = silver membrane filter, KI = potassium iodide;

[e] Sampling rates are recommended maximum rates.

[f] Method procedure:GC = gas chromatography,

/	= desorption solvent — example: GC/CS_2 — desorb, material from adsorbent with CS_2 for GC,
FID	= flame ionization detector,
THF	= tetrahydrofurane,
ipa	= isopropyl alcohol,
AA	= atomic absorption spectroscopy,
ASV	= anodic stripping voltammetry,
ECC	= electrolytic conductivity cell,
LCG	= liquid column chromatography,
UV	= ultraviolet detection,
VIS	= visible range spectroscopy,
TM	= trace metals — general method,
TLC	= thin layer chromatography/alumina gel,
SF	= spectrophotometric determination,
MS	= mass spectrograph,
EC	= electron capture,
TC	= thermal conductivity
IR	= infrared,
MS	= molecular sieve — 5A,
DPP	= differential pulse polarograph,
FP	= flame photometric,
bu.alc.	= butyl alcohol.

Numbers following AA and colorimetric = wave length in nm,

o	= oxidizing flame,
r	= reducing flame,
ext	= extract,
ac	= acid,
MB	= methylene blue,
des	= desorption.

Numbers following GC detector identification = column packing:

1) Poropac Type Q,
2) 5% SE-30/Gas Chrom W,
3) 10% FFAP/Chromosorb W-AW,
4) 5% FFAP/Supelcoport,
5) 5% SE-30/DMCS Chromosorb W,
6) Chromosorb 103 (ascarite pre-column),
7) 10% Silicone OV-25 on Supelcoport,
8) 3% Dexsil 300 GC on Chromosorb G,
9) Alumina column/pentane + diethyl ether,
10) Corrasil II — High speed — cyclohexane elution,
11) SP2100 (6%) on Supelcoport — derivatized — trimethyl silyl ester,
12) Dexsil (3%) on Chromosorb G,
13) Coated glass beads (0.1% QF-1 and 0.1% OV-17),
14) Gaschrom Q,
15) 3% OV-1 on Gas Chrom Q,
16) 5% SP-1000 on Supelcoport,
17) 5% OV-101 on Supelcoport (also 10% OV-101),
18) Chromosorb 102,
19) 4% Carbowax 20M + 0.8% KOH on Carbosieve B,
20) 20% Ucon-50 LB-550X on Chromosorb P (20% KOH treated),
21) 10% Silicon OV-7 on Supelcoport,

air by silica gel adsorption at room temperature have been unsuccessful; methyl bromide, on the other hand, adsorbs quite readily. Variables affecting the adsorption of these materials have not been examined thoroughly.

ix. Concentration of Air Contaminant

Limited investigation has shown that the concentration of the contaminant in air may have very little effect on its adsorption on silica gel below a certain critical concentration. For carbon tetrachloride the critical concentration appears to be about 400 ppm. That this level varies appreciably for different materials has been established, but a thorough study has not been conducted. With other conditions held constant, an increase in concentration above the critical level causes a decrease in the breakthrough time.

x. Humidity of Air Contacting Gel Bed

Because of its affinity for water, the usefulness of silica gel for air-sampling could be quite limited when the absolute humidity of the air is high. Most materials other than water adsorb very poorly, or not at all, on drying agents such as magnesium perchlorate or Drierite. Silica gel in a sampling train may be preceded by a drying agent to remove the water.[61,199] This allows the silica gel to be used to full capacity for the air contaminant of interest. This technique should be used with caution. Under normal circumstances, silica gel has been used without a drying agent to sample air for most of the halogenated hydrocarbons.[193,200]

xi. Duration of Sampling

The duration of sampling should, in all cases, be somewhat less than the breakthrough time under the conditions encountered. The most practical solution to a particular problem is to sample known concentrations under conditions similar to those anticipated in field sampling. If the time necessary to obtain sufficient contaminant for analysis is greater than the breakthrough limit, the other variables will have to be altered. Breakthrough times ranging up to several hours are attained by proper manipulation of the sampling variables.

xii. Sampling Applications

a) Nitrogen Compounds

The major application for silica gel currently is directed to the collection of nitrogen-containing compounds — aliphatic amines, aromatic amines, acid amides, hydrazine, alkylhydrazines, aromatic hydrazines, heterocyclic nitrogen compounds (morpholine), aromatic nitro-compounds, cresols, phenyl ether and biphenyl, methanol, and acetylene tetrabromide. Silica gel also has been used as a support for sulfuric acid in the collection of methyl hydrazines from the industrial atmosphere. A summary of sampling and analytical procedures is presented Table 64, NIOSH Sampling and Analytical Methods. The data in this table were derived from the second edition of the NIOSH Manual of Analytical Methods, in three volumes, published in 1977.[201] A total of 337 methods provide 426 procedures (a number of general methods which include a number of substances are described for 322 different specific elements and compounds.) Of the 426 procedures, 41, or 9.6%, are based on sampling with silica gel. Of the 200 procedures which specify collection on a solid adsorbent, 41, or 20.5%, employ silica gel as the collecting medium of choice.[24]

Silica gel tubes are available commercially in several sizes and with minor modifications in construction relative to the original design shown in Figure 47 (Table 88). The construction details are detailed in the appropriate NIOSH Analytical Method included in Table 64. Further details will be found in Katz, page 865 (Analytical Method for Aromatic Amines in Air).

TABLE 65

Sorption Capacity of Silica Gel for Aniline Derivatives

Compound	Weight (mg) on/50 mg silica gel at 2% B.T.[a]	Vapor pressure at 25°C in MmHg	Recommended range of Sample Volumes	
			Min no. L	Max no. L
Aniline	1.8 ± 0.2	0.671	5	150
N,N-dimethyl aniline	3.0 ± 0.1	0.557	5	197
O-toluidine	4.1 ± 0.1	0.317	5	300
2,4-Xylidine	6.8 ± 0.1	0.154	5	442
P-anisidine	10.3	0.050	250	33,200
O-anisidine	11.2	0.037	250	35,900
P-Nitroaniline	42.3	6.4×10^{-6}	80	12,100

[a] B.T. = breakthrough

Data derived from Wood, G. O. and Anderson, R. G., *Am. Ind. Hyg. Assoc. J.*, 36, 538, 1975.

Results from a study of the collection efficiency for aniline compounds (*N, N* diethylaniline, *O*-toluidine, 2-4 xylidene, *p*-anisidine, *O*-anisidine and *p*-nitroaniline) on silica gel confirmed identical collection with ethanol in a bubbler. The rates were 200 mL/min. for 8 hr, or 1000 mL/min for short-term sampling. The collection tube measured 8 mm diameter by 125 mm length, was compartmented into three sections which contained consecutively 150, 150, and 700 mg of silica gel. The sections were retained by 100-mesh stainless steel screens which were spaced by 12-mm glass tube spaces. Results are summarized from the study in Table 65.[202] The amines were eluted into 95% ethanol and analyzed by gas chromatography. Desorption efficiencies ranged from 95.6 to 101.8 ± 7.2%.

b Ammonia

Silica gel has a high affinity for ammonia. However, aliphatic derivatives of ammonia also are strongly sorbed and interfere in the analysis by the Nessler procedure. These interferences can be avoided by passing the air sample through charcoal coated with sodium hydroxide which absorbs ethylenediamine and monoethanolamine (guard layer) before the air-stream contacts the silica gel. The NH_3 is then stripped from the silica gel in dilute H_2SO_4 and quantitated by the Nessler colorimetric method.[203]

c) Phosphine

A review of the literature indicated the most reliable method for sampling air for the presence of phosphine would be collection on silica gel coated with mercuric cyanide followed by analysis by chromatography with a flame photometric detector. Concentrations in the range 0.02 to 1.64 mg/m^3 have been studied.[204,205]

b. Practical Air-sampling Details

i. U-Tubes

Several different types of adsorbing vessels may be used with silica gel. A common type is the U-tube with silica gel in both arms of the tube. A train of two or more U-tubes has been used,[61] but experience has shown that one such tube is usually sufficient.[195] A single U-tube with silica gel in one arm and a desiccant in the other may also be employed.[199] U-tubes should hold at least 10 g of gel with a bed depth of more than 2 cm. Add the silica gel to the U-tube slowly with a slight tapping to obtain uniform distribution, which limits airflow channeling through the gel bed. Vessels constructed of 16-mm tubing, 150 cm long, have worked satisfactorily when airflow rates have not exceeded 2 L/min.[200]

ii. Sintered Fiberglass Adsorbers

Straight tubes with a sintered glass support for the gel bed have also been used. Usually, the sintered glass is of a medium or coarse porosity to ensure efficient distribution of air throughout the bed without causing an excessive pressure drop. Adsorber diameters are commonly 30 or 40 mm. With 20 g or more of silica gel in such an adsorber, flow rates up to 3 L/min may be used.[193]

iii. Other Methods

A technique for using the adsorption properties of silica gel, but avoiding the difficulties of controlling the variables encountered in a dynamic flow system, has been developed. It might be called a "static adsorption system." In this method, 10 g of silica gel is added to a 3- or 5-L saran bag that is subsequently filled with air and sampled by means of a glass hypodermic syringe pump equipped with two glass check valves. The bag is shaken occasionally, and a few hours later the gel will have absorbed the air contaminant quantitatively.

c. Desorption of Vapors and Gases

Very little adsorbed vapor is lost if the gel is stored at room temperature in a closed container (provided that colder temperatures were not found necessary for sampling). Air samples collected on silica gel can be held for at least 3 to 10 days without loss of contaminant. This retentive property of silica gel is very useful when considerable time must elapse between sampling and analysis.

Adsorbed materials can be determined on the gel or after removal. Determination on the gel is the basis for the detector tube technology. Usually the detecting reagents are adsorbed onto the gel before exposure to the contaminant, but the reverse technique also has been used. Many factors render these tubes acceptable only for semiquantitative analysis. For quantitative analysis the adsorbed material is removed from the gel.

Removal of adsorbed vapors or gases from silica gel is not difficult when the proper conditions are selected. Desorption is accomplished thermally or by solvent leaching. The method of choice is determined by the analytical technique selected and by the efficiency of removal.

i. Thermal Desorption Methods

All thermal desorption methods require elevated temperatures and a carrier gas to collect the desorbed material. Usually the gel is transferred to a tube which, in turn, is placed in a tube furnace. As the gel temperature rises, desorbed material is swept along with the carrier gas into a collecting device. The carrier gas instead of the gel may be heated, for example, superheated steam. Heating the gel alone, without a flowing carrier gas, produces a hysteresis effect that prevents good recoveries.[194]

Desorption temperatures and the carrier gas composition have been varied widely with good results. Using helium, Altshuller[198] found that acetylene could be desorbed at 50°C; low molecular weight sulfur compounds (H_2S, SO_2, CH_3SH) at 100°C; hydrocarbons such as *n*-butane, isobutane, isopentane, propane, and propylene at 150°C. Jacobs reports benzene can be freed from silica gel at 100°C with air.[206] Peterson, Hoyle, and Schneider, in experiments with trichloroethylene, found that the time required for complete desorption varied with temperature, but that, at 350°C with a flow rate of 2 L/min of dry air, complete desorption was assured.[193] Halogenated hydrocarbons with boiling points ranging from −28 to 190°C were recovered quantitatively under these conditions. Erley used a temperature of 400°C and about 7 L dry nitrogen at 0.5 L/min for the desorption of several materials.[196]

After the material has been desorbed from the gel as a vapor into the carrier gas,

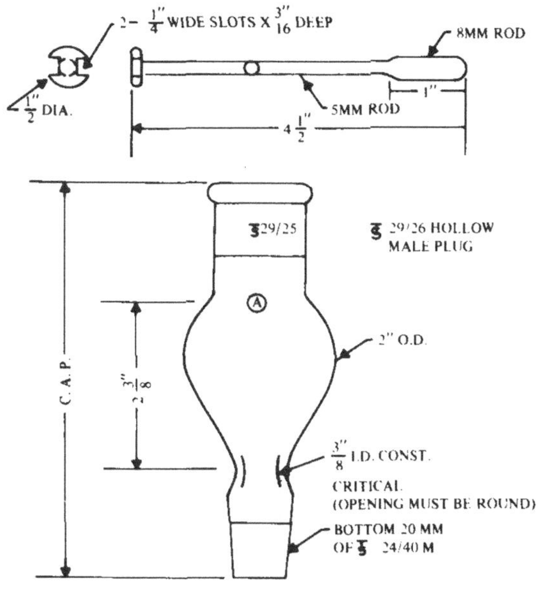

FIGURE 112. Silica gel dispenser. (From Monkman, J. L., Analytical Guides, AIHA, 1965. With permission.)

the concentration is greater than in the originally sampled air. If the carrier gas is superheated steam, the analysis proceeds after condensation as if the material had been trapped in water. When the carrier gas is noncondensable at ordinary temperatures, the mixture may be fed into a second furnace at a higher temperature for direct combustion, or may be collected in a saran or polyethylene bag for analysis by various instrumental techniques. Both gas chromatography and long-path infrared spectrometry have been employed with excellent results, especially when more than one molecular species has been adsorbed on the gel. Methods of analyzing the concentrated vapor are limited only to the ingenuity of the analyst.

ii. Solvent Leaching Methods

Some analytical techniques require the contaminant to be in solution, such as UV absorption methods, gas chromotography techniques, or wet chemical determinations. The collected contaminant may be removed quantitatively by immersing the gel in a solvent. If interferences are introduced by the gel, or if removal is not quantitative, the contaminant should be collected in a solvent and not on silica gel.

When removing the contaminant, the silica gel should be added to the solvent slowly, almost grain by grain, to avoid a build-up of heat in the solution. If the solvent is added to the gel, the heat of reaction may be sufficient to drive off some contaminants; moreover, since the gel is also an active catalyst, a polymerization or a breakdown of some substances could occur. Decrepitation of the gel may produce fine fragments that will cause interferences in UV absorption analysis.

A convenient dispenser, shown in Figure 112, is attached to a 50-mL round-bottom flask containing the absorbing solution. The flask with dispenser is placed in an ice bath on a laboratory shaker, and as the flask and dispenser are shaken, the glass rod with the slots swirls around in the dispenser allowing a few grains of silica gel at a time to drop into the absorbing solution.

Much of the information on the desorption of micro amounts of contaminants from silica gel by leaching with solvents is unpublished. Members of the AIHA Analytical

Chemistry Committee have reported that the following materials can be leached quantitatively by the agent indicated:

Materials	Leaching agent	Materials	Leaching agent
Benzene Divinyl benzene Alpha methyl styrene Isopropyl benzene Toluene	Isooctane plus water	Aniline Benzene Isophorone Nitrobenzene Styrene Toluene	Ethanol
Allyl alcohol Allyl chloride Propylene dichloride	*n*-Hexanol	Dinitrophenol Ethyl benzene	Isopropanol
3,4-Dichloronitrobenzene Nitrobenzene	Methanol		

When the materials listed above are sampled separately and leached with the appropriate solvent, satisfactory results can be expected. Difficulties have been encountered, however, when sampling mixtures with poor recovery of one or all of the components are present. Recoveries of solvents in mixtures must be checked in the laboratory before attempting to use this technique for atmospheric sampling.

Likewise, recoveries of some individual materials were found to be low and erratic; for example, recovery of acetone leached from silica gel with ethanol was about 50%; the same solvent used for methylethylketone resulted in recoveries varying from 60 to 80%. In any case, when definite information is not available, recoveries must be verified with known concentrations before field sampling is undertaken.

In UV absorption methods, particulate matter released from the gel can interfere; these particulates are not removed by centrifuging. Marked interference has occurred with grade 05 silica gel. It is advisable to use either the PA100 or PA400 grade gels mentioned earlier; PA400 gives the least difficulty. Interference may be serious at wavelengths below 250 nm. Silica gel is not recommended when the adsorption peak is below 250 nm.

Available information on solvents indicates that the greatest interference occurs with ethanol. Relative interference, in descending order, was found to be ethanol, methanol, water, isopropanol, isooctane. Generally, isooctane is a very satisfactory solvent, even at low wavelengths. However, a small amount of water to obtain satisfactory release of the adsorbed contaminants is necessary.

Additional factors that contribute to interference are leaching time and temperature. Interferences increase as time and temperature increase.

In 1966 Campbell and Ide contributed one of the most significant developments for the use of silica gel in collecting aromatic hydrocarbons from the atmosphere.[207]

A column of silica gel, divided into four sections, was held in a glass tube 3 mm I.D. by 10 cm long. The gel was in four 2 cm lengths separated by glass wool; each section contained approximately 125 mg of silica gel. Air contaminated with an aromatic hydrocarbon was drawn through the tube at 60 ml per minute with a commercial, hand syringe-type pump.

After the air sample was taken, the tube was cut into portions representing the first through fourth sections, and the silica gel from each section was transferred to a separate 15 ml centrifuge tube. The silica gel then was eluted with 5 ml ethanol for the 30 min standing time (isopropanol may be substituted), the mixture centrifuged at 18,000 rpm for 5 min and the alcohol extract analyzed by UV spectrophotometry. The alcohol interferences noted previously were not encountered by these investigators.

The resulting O.D.-wavelength plot was converted to amount of aromatic hydrocarbon, and the efficiency of the tube determined. The series of silica gel sections readily compensated for factors which affect retention of the contaminant, such as ambient temperature at the time of sampling, relative humidity, and interfering substances.

Polar substances cause aromatic hydrocarbons to migrate through the sampling tubes by displacement.

If c is the concentration of the substance in the sample, and c_1, c_2, c_3, and c_4 are the amount of substance in each section, experiment shows that Equation 7, which is the sum of a standard geometrical progression, gives the true concentration of the substances in the air sample.

$$c = c_1 + [c_2 (k^n - 1)/k - 1)] \qquad (7)$$

The amount of the substance in section one, c_1, was found always higher than was expected from the amounts in the remaining sections, probably because of impingement, condensation, and other factors influencing equilibrium. The constant k is defined as

$$k = c_{n+}/c_n \text{ and } c = c_1 + c_2 (K^{n-1})/(k^{-1})$$

where n realistically should be less than 10 (a theoretical number of sections).

If the concentration in section four, c_4, is zero, the amount of substance in the air sample is the arithmetical sum of the sections. If section four, c_4, is equal to or nearly approaches section three, c_3, no calculation can be made. An estimation is possible, however, if the difference is less than 30% and if similar sampling conditions can be duplicated in the laboratory for comparison.*

The background optical absorbance was corrected by base line calculations. The UV absorbance curve of the ethanol solution between 210 and 340 nm was plotted and a line drawn tangent to the minimum points. The absorbance on this base line at the wavelength of maximum total absorbance was then subtracted from the total absorbance and the difference related to a calibration curve constructed in a similar fashion. Each section of the recommended tube design contained 115 to 120 mg of silica gel. Therefore, the background optical absorption due to the fine silica gel fragments that arise from the shattering of the gel during desorption can be so controlled that the simple centrifugation and base line construction minimizes this problem. Only *o*-dichlorobenzene, nitrobenzene, methylaniline, and aniline failed to desorb completely in ethanol during the 30-min extraction period.

Pretreatments of the silica gel did not affect its retention characteristics. The four-section tube eliminated most extrinsic (relative humidity, temperature, sampling rate variations, etc.) variability as total recovery was the same under all conditions studied. The first section did not completely remove nonpolar materials such as benzene, toluene, cyclohexane, or most aliphatic hydrocarbons. For example, 88% of the benzene in a 100-mL air sample was retained on the first section and the balance (12%) on the second section. From these values, a silica gel saturation of 0.045 mg benzene per mg silica gel was calculated. Polar compounds — aniline, pyridine, nitrobenzene — were retained completely on the first section.

The silica gel tubes were prepared as follows:

The silica gel is poured into the tube through a small funnel or from a filling jar through a glass tube in a one-hole rubber stopper, then packed uniformly by tapping the tube on a hard surface. Glass cloth or glass wool pledgets secure the ends of the column and separate the sections. The first section is filled to a level 0.25 cm above the 2-cm mark, tapped to pack the gel just to the 2-cm level, and secured with a glass wool pledget. The second 2-cm mark is measured from the surface of the pledget, and the second section is gel filled to 0.25 cm above the mark and compacted, as before. The third and fourth sections are prepared similarly. Routinely, the tubes are left open and stored over silica gel, but for field use they may be capped with rubber (tip of rubber policeman) or with Parafilm® or other suitable covering.*

The pressure drop (ΔP) across these tubes packed with 42 to 80-mesh silica that had been activated by heating to 150°C and had an acid equivalent titer of 9.1×10^{-6} in Eq of H^+ per mg at various flow rates were

30 mL/min – 22 cm H_2O or 8.6 in. H_2O

60 mL/min – 46 cm H_2O or 18.2 in. H_2O

* From Campbell, E. E. and Ide, H. M., *Am. Ind. Hyg. Assoc. J.*, 27, 323, 1966. With permission.

90 mL/min – 70 cm H_2O or 27.5 in. H_2O

120 mL/min – 94 cm H_2O or 37 in. H_2O

$$J = D\, dx/dc = \text{diffusion flux in mol/cm}^2$$

The tubes may be used to collect air samples for vapor chromatography analysis. The silica gel may be placed in special sample holders and the vapor removed by thermal displacement, or the tube itself may be placed in line with the carrier gas and the contaminant removed by polar-solvent vapor displacement. Gas chromatographic analysis is more sensitive than the method described here, but is more time-consuming. The silica gel tubes may well serve as reactant beds for field or laboratory procedures similar to those described by the Chemical Warfare Service.[208] The silica gel tube approximates a critical orifice; therefore, the tube may be carried onto the field on routine surveys and used with aspirators, house vacuum systems, or any other convenient vacuum source capable of overcoming a pressure drop of 34 mm of mercury, to sample atmospheres for qualitative or semiquantitative evaluation in the laboratory. This convenience permits the industrial hygienist to analyze any air contaminant without having to carry a large number of specific sampling media.*

Loss of nonpolar contaminants from the silica gel was found to be very nearly constant. For example, an average 13.8% of the benzene retained on the first section desorbed into each 100-mL aliquot of clean, dry air passed through the tube after the initial 100 mL of nearly saturated benzene vapor sample collection. However, under "normal" sampling conditions, all test compounds, including benzene, were quantitatively retained at concentrations up to 500 ppm even when 1 L of clean air was drawn through the tube after sample collection. The microcolumns may be shipped without sample loss and may be stored for some time with only slight migration of the aromatic hydrocarbons to other sections of the column.

The sensitivity limits that determine the parameters for sample size and optimum concentrations for 11 aromatic derivatives in the eluting solvent are presented in Table 66. Although the system was designed for grab sampling with a 100-mL hand-operated syringe pump, direct adaptation to personnel monitoring would require only the addition of a small electrically driven pump and accessories to determine flow rate and sample size.

Silica gel columns also have been used to determine the accuracy of length of stain gas detector tubes.[209] Grab samples were collected simultaneously in the field on silica gel and detector tubes for benzene, toluene, and xylene. The aromatic hydrocarbons were desorbed from the silica gel in solvent and analyzed by UV spectroscopy as in the Campbell reference.[207] The collection efficiency was 93%. The three sources of detector tubes read ± 60% of the true value in the range 15 to 40 ppm. Isooctane in a midget impinger did not collect quantitatively. Sherwood used silica gel cartridges to collect benzene vapor from the breathing zone with a battery-operated pump (personnel monitor). Although the sampling period extended only over a 15-min interval, the application to 8-hr time-weighted sampling was obvious. Gas chromatographic analysis of an ethanol elute was elected for analysis.[210]

Peterson and co-workers developed procedures for the analysis of airborne halogenated hydrocarbon contaminants by collection on silica gel and combustion of the collected material to yield a mixture of halogen and halogen acid, which is then titrated by the micro Volhard procedure.[193] Desorption is accomplished by passing air at a 1 to 3 L/min rate through the collection tube in a furnace at 350 to 450°C for 10 min. The gel may be reused without loss in absorption capacity. The technique was applied to 14 halogenated hydrocarbons including CCl_4, chlorotrifluoroethylene, dichloroben-

* From Campbell, E. E. and Ide, H. M., *Am. Ind. Hyg. Assoc. J.*, 27, 323, 1966. With permission.

TABLE 66

Sensitivity of Silica Gel Sorption and Subsequent Spectral Determination

Substance	TLV (760 mm Hg)		Spectral sensitivity	Optimum concentration	Wavelength	Minimum air volume needed to determine mL	
	ppm in air	mg/m^3 in air	(μg in 5 mL EtOH)	(μg/mL in 5 mL EtOH)	(maximum)	TLV (ppm in 5 mL EtOH)	Optimum concentration (ppm in 5 mL EtOH)
Aniline	5	19	0.4	3.5	233	100	920
Benzene	25	80	10	175	253.9	625	11,000
Chlorobenzene	75	350	15	175	263.3	214	2500
m-*p*-Cresol	5	22	2	35	273.5	454	8000
o-Dichlorobenzene	50	300	17	175	168.8	283	2920
Ethylbenzene	100	870	13	177	260.7	150	1018
Methylaniline	∿2	∿9	0.3	3.5	244	167	1950
Nitrobenzene	1	5	0.5	8.75	258	500	8750
Styrene	100	420	0.2	3.5	247	2.5	0.42
Toluene	100	750	10	175	261.2	134	1170
Xylene	200	870	1	175	264.5	6	1000

From Campbell, E. E. and Ide, H. M., *Am. Ind. Hyg. Assoc. J.*, 27, 323, 1966. With permission.

zene, methyl bromide, and vinyl chloride with 100 ± 10% recovery. The anomalous results encountered with methyl chloride suggested that silica gel at room temperature should not be used for halogenated hydrocarbons boiling below 0°C. The pyrolysis gases were absorbed in aqueous mixture of 1% Na_2CO_3 and 1% sodium formate to reduce the evolved halogen to halide ion. Other useful source material can be found in the references.[196,211-214]

To remove water vapor from the air sample before sample adsorption takes place, 5A molecular sieves have been installed in front of the silica gel bed.[214] However, the possibility that the target contaminant may be partially if not completely absorbed in the surface of the molecular sieve, which usually is organic in nature, must be considered when attempts are made to avoid the desorbing effect of water vapor. Whitman and Johnston reported on an investigation in which contaminated air was drawn at 1 L/min through an 8-mm I.D. × 245 mm tube containing 5 g 5A molecular sieve and 5 g 12 to 28-mesh silica divided by a pledget of glass wool. The contaminants were displaced by adding first 1 mL isopropyl benzene to the silica gel followed by 50 mL cold water. The lighter solvent layer was decanted and analyzed by gas chromatography (GC).

Recoveries of benzene, toluene, and xylene were in the range 95 to 100%. Concentrations were confirmed by drawing a simultaneous sample through an isopropanol scrubber at 0.5 L/min and analysis by UV spectroscopy.

2. Alumina Gel

Although alumina gel is an excellent adsorbent for water vapor, very little information relative to collection of air contaminants is available. Application to the collection of alkaline components such as methylamines, pyridine, morpholine, piperidine, etc. should be worth investigating, especially in those applications where the acidic nature of silica gel interferes with desorption. Tests of the efficiency of 80 to 200 mesh adsorption alumina indicated that columns of alumina not only absorbed a higher percentage of parathion and gamma benzene hexachloride insecticides than glass wool, Florisil,® or Dry Ice®-acetone, but also desorbed in acetone or 95% ethanol more readily.[215]

A procedure has been developed and proposed by NIOSH for the collection of formaldehyde on alumina gel. However, due to significant losses after one hour, the sampling period is limited to 30 min, and immediate elution into 1% aqueous methanol is required. The eluted sample is stable for at least 30 days. There is very little interference from other aldehydes, but ethanol and high molecular weight alcohols and olefines will produce a negative interference (Table 64).[201]

3. Carbon

a. Introduction

The use of activated carbon or charcoal to adsorb undesirable components from the atmosphere was established prior to World War I and employed during and after that period in the canisters of gas masks to remove toxic chemical vapors from inhaled air to protect the respiratory tract. Application to air analysis as well as to air purification has been widely employed, and no attempt will be made to review the subject here. The most recent developments will serve to illustrate the utility of activated charcoal for the collection of air contaminants for subsequent analysis. The following summary quoted from the methods for air sampling prepared by the Intersociety Committee serves as an excellent introduction to the subject.

b. General Properties of Charcoal

Activated carbon is electrically nonpolar and is consequently capable of adsorbing organic gases and vapors in preference to atmospheric moisture. In fact, even activated carbon with previously adsorbed moisture will lose such moisture by displacement with organic gases and vapors from contaminated atmospheres. Such displacement of moisture is not exhibited to any significant extent by siliceous adsorbents.

The following comparison between activated carbon and silica gel will further illustrate the differences between the two types of adsorbents for purposes of sampling atmospheric gases and vapors:

1. Activated carbon has a greater overall capacity for adsorbing and retaining atmospheric gaseous and vaporous contaminants, but silica gel exhibits a greater selectivity among them.
2. Activated carbon is more suitable for collecting a conglomerate mixture of gases and vapors, especially when these are present in very small concentrations and must, therefore, be collected over relatively long periods of time.
3. Desorption of gases and vapors from silica gel is easier than from activated carbon; and it is in fact feasible to obtain some separation of gases and vapors by selective desorption from silica gel.
4. Each adsorbent can be suitably impregnated to render it more suitable for chemisorption of specific gases or vapors.

To determine which gases and vapors may be sampled by physical adsorption using activated carbon, the following approximate criteria, based on available information, may be used. True gases, usually having critical temperatures below −50°C and boiling points below −150°C are virtually nonadsorbable at ordinary temperatures by physical means. These include, for example, hydrogen, nitrogen, oxygen, carbon monoxide, and methane. Low boiling vapors having critical temperatures between approximately 0 and 150°C and boiling points between −100 and 0°C are moderately adsorbable. The adsorption efficiency at ordinary temperatures and concentrations, however, is not quantitative. Such vapors can be concentrated from the atmosphere by using a "thick carbon bed", preferably refrigerated. These vapors include ammonia, ethylene, formaldehyde, hydrogen chloride, and hydrogen sulfide. Chemisorption methods may also be used for such low boiling vapors. The heavier vapors (boiling above 0°C) are readily adsorbed and retained by activated carbon at ordinary temperatures; these include most of the odorous organic and inorganic substances.*

C. Limitations

Performance testing of the NIOSH charcoal technique for the determination of air concentrations of organic vapors disclosed major limitations.[216]

i. Capacity

Charcoal will absorb most organic vapors quantitatively for a time with a steady flow of vapor through the bed. Eventually, a small amount of the contaminant will start to pass through. This point is defined as "breakthrough". If sampling is continued the charcoal will absorb some vapor, but the amount (concentration) "breaking through" will increase until it equals the amount (concentration) entering the bed. At this point the charcoal is saturated and will absorb no more vapor. The recommended charcoal sampling tube (Figures 47 and 113), therefore, is divided into two sections to detect breakthrough. The fore section through which the air sample enters contains 100 mg of charcoal and the backup section contains 50 mg of the same charcoal. Breakthrough is usually accepted as the point at which 1% of the sample is found in the backup section.

The length of sampling time during which quantitative retention is attained can be measured as time, volume, or weight of the contaminant collected. The results from the determination of breakthrough for six typical solvent vapors are shown in Table 67.

A plot of the logarithm of the weight in the backup section of the volume sampled exhibits a linear relationship between these two variables. The equation for the least squares regression line may be used to solve for the 1% "breakthrough" point. Up to this point, the cumulative sorption efficiency is 99%. In Table 68, the OSHA Federal Standard and Test Concentration are used since the breakthrough volume will vary with concentration. The significantly lower volumes for 2-butanone, trichloroethane, and trichloroethylene are primarily due to the higher vapor pressures. This effect would be expected as adsorption appears to be related to condensation.

* From Katz, M., Ed., Tentative method for preparation of carbon monoxide standard mixtures (42101-01-69T), *Lab. Health Sci.*, Suppl. 7, 72, 1970. With permission.

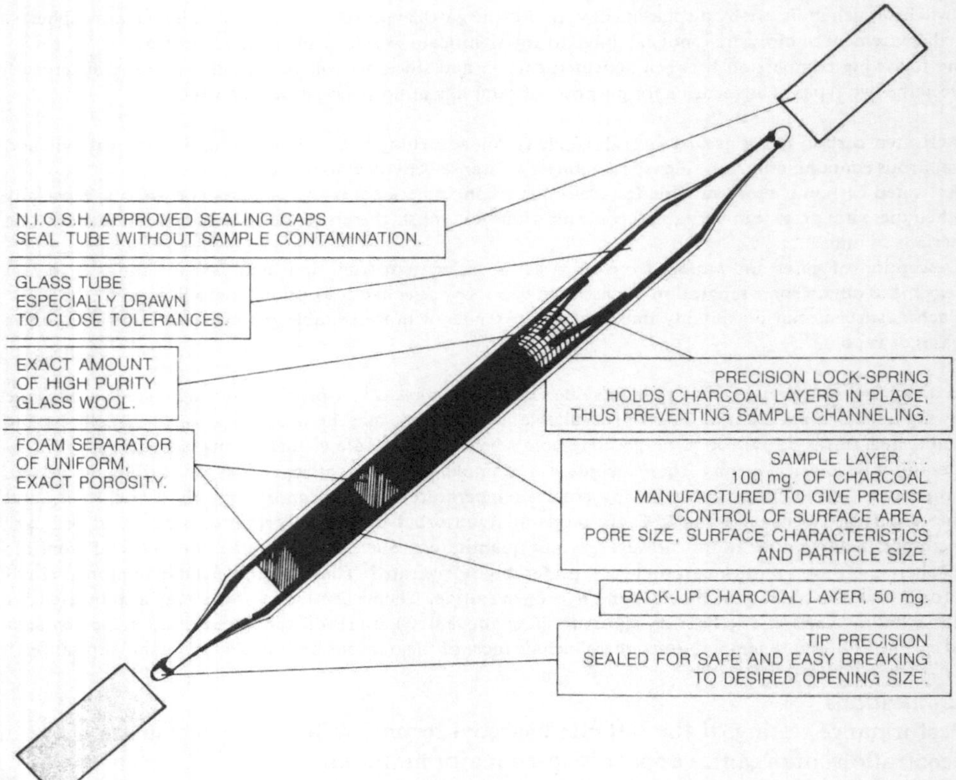

FIGURE 113. NIOSH recommended charcoal sampling tube for organic vapors.

For reliable sampling, not more than 10% of the contaminant should be collected in the backup section. Results less than this amount can be added to the fore section for total collection calculation. However, the major value of the backup section resides in the detection of breakthrough and should not be used as a sampling device.

Early breakthrough can be avoided by taking a shorter sample (less volume) or by sampling at a lower rate. Lower flow rates may increase the sample volume since the vapors have a longer time to equilibrate over the entire surface of the charcoal, and the longer sampling time is more economical since fewer samples are required to determine 8-hr TWAs.

Breakthrough and other data for 121 solvent vapors and gases, aromatics, alcohols, acetate esters, alkanes, ketones, amines, and chlorinated compounds, will be found in Nelson and Harder.[217] The data gathered is summarized in Table 68. Although this investigation was carried out to evaluate respirator cartridge efficiency, the results can be used to estimate sampling parameters which can be easily confirmed by a minimum number of samplings. Furthermore, the conditions were considerably more rigorous than would be encountered in compliance monitoring, i.e., 1000-ppm concentrations at 50% relative humidity (RM) and 53.5 L/min flowrate through 55 to 70 g of charcoal. Initial breakthrough can be calculated:

$$t_b = \frac{10^6 \, V_m \, wW_c}{MQC}$$

where: t_b = initial breakthrough time in minutes; v_m = molar volume at the system temperature and pressure (24.1 L/mol at 20°C and 760 torr); w = weight of solvent

TABLE 67

Organic Vapor Breakthrough Parameters at the 1% Level on BC-580-26 Charcoal (150 mg Tube) (Figure 113)

Compound	OSHA standard (mg/L)	Test concentration (mg/L)	Total weight collected (mg)	% On the backup section	Breakthrough (vol. L)
2-Butanone	0.59	0.88	7.2	0.97	8.2
Styrene	0.43	0.62	18.0	1.00	29.1
Tetrachloroethylene	0.68	1.04	34.5	0.98	33.2
Toluene	0.75	0.68	15.7	1.02	23.1
1,1,1 Trichloroethane	1.90	1.96	11.4	0.96	5.8
Trichloroethylene	0.54	1.46	17.4	0.98	11.9

Summarized from Saalwalchter, A. T., McCammon, C. S., Roper, C. P., Carlberg, K. S., *Am. Ind. Hyg. Assoc. J.*, 38, 476, 1977.

TABLE 68

Solvent	BP (C°)	Vapor pressure at 20°C (Torr)	Diffusion coefficient at 25°C (cm²/sec)	Volume charcoal (mL)[a]	Weight charcoal (g)[b]	Experimental breakthrough times			Wt. solvent adsorbed/wt. of carbon			Weight water adsorbed at t_{100}%/wt. carbon[f] (g/g)
						t_1% (min)	t_{10}% (min)	t_{99}% (min)	t_1%[c] (g/g)	t_{10}%[d] (g/g)	t_{100}%[e] (g/g)	
Aromatics[g]												
Benzene	80.1	72.4	0.0932	146	59.1	73.3	88.6	170	0.214	0.249	0.327	0.035
Toluene	110.6	21.8	0.0849	144	57.5	94.3	114	196	0.334	0.397	0.473	—
Ethyl benzene	136.2	7.08	0.0755	148	55.9	83.7	105	223	0.352	0.432	0.573	0.021
m-Xylene	138.4	6.16	0.0670	148	59.7	98.7	116	193	0.388	0.450	0.536	—
Cumene	152.4	3.34	0.0677	146	60.0	81.2	103	153	0.359	0.447	0.538	0.022
Mesitylene	164.7	1.73	0.0663	147	59.5	85.5	105	189	0.382	0.460	0.565	—
p-Cymene	176.7	1.28	0.0630	148	56.9	75.6	92.9	253	0.394	0.476	0.594	—
Alcohols[g]												
Methanol	64.7	96.8	0.1520	148	55.4	0.2	3.2	47.0	0.0003	0.004	0.018	0.091
Ethanol	78.4	43.6	0.1181	146	55.5	28.0	45.3	207	0.051	0.079	0.168	0.135
Isopropanol	82.3	32.4	0.1013	145	55.7	54.3	81.8	247	0.129	0.188	0.314	
Allyl alcohol	97.0	21.4	0.1021	145	55.4	65.5	105	280	0.152	0.234	0.362	0.117
Propanol	97.1	14.4	0.0993	144	55.7	70.4	111	250	0.168	0.255	0.384	0.065
sec-Butanol	99.5	12.6	0.0891	144	55.9	96.0	121	196	0.281	0.347	0.438	0.050
Butanol	117.7	5.65	0.0861	143	55.5	115	141	235	0.340	0.409	0.512	0.016
2-Pentanol	119.9	4.36	0.0728	147	56.7	86.8	111	277	0.298	0.373	0.517	0.020
3-Methyl-1-butanol	131.2	2.4	0.0709[h]	147	55.0	97.0	121	195	0.344	0.421	0.525	0.047
4-Methyl-2-pentanol	131.8	3.9	0.0653[h]	148	55.7	75.4	96.1	243	0.306	0.382	0.545	0.041
Pentanol	137.9	1.82	0.0716	148	55.0	102	130	208	0.362	0.451	0.552	0.019
2-Ethyl-1-butanol	146.8	1.56	0.0653[h]	150	55.7	76.5	101	257	0.310	0.400	0.582	0.030
Monochlorides[g]												
Methyl chloride	−24.2	3460	0.1140[h]	148	55.7	0.05	0.7	14.6	0.0001	0.001	0.008	0.074
Vinyl chloride	−13.9	2310	0.0995[h]	147	55.7	3.8	6.6	46.3	0.009	0.016	0.038	0.070

Ethyl chloride	12.3	1110	0.0950[h]	148	56.1	5.6	10.7	38.7	0.014	0.026	0.048	0.096
2-Chloropropane	35.2	430	0.0819[h]	148	56.8	26.1	35.9	109	0.080	0.107	0.0167	0.018
Allyl chloride	44.5	300	0.0975	141	55.5	30.5	44.6	108	0.093	0.132	0.194	0.040
1-Chloropropane	46.7	277	0.0829[h]	146	55.8	24.5	34.8	141	0.076	0.105	0.196	0.080
2-Chloro-2-methyl-propane	50.8	240	0.0737[h]	147	57.0	37.4	52.3	168	0.134	0.183	0.279	0.010
1-Chlorobutane	77.5	80.0	0.0745[h]	145	55.8	72.3	88.1	145	0.265	0.317	0.391	0.008
2-Chloro-2-methyl-butane	85.7	60.3	0.0675[h]	148	57.0	58.8	79.3	194	0.243	0.320	0.439	—
1-Chloropentane	108.4	23.9	0.0681[h]	147	56.8	74.7	96.6	197	0.310	0.392	0.505	—
Chlorocyclopentane	113	—	0.0718	146	55.6	77.5	106	211	0.322	0.429	0.562	0.050
Chlorobenzene	132	9.11	0.0747	146	55.8	107	131	205	0.478	0.574	0.688	—
1-Chlorohexane	134.5	7.15	0.0631[h]	150	55.8	77.3	95.8	189	0.370	0.449	0.591	0.013
o-Chlorotoluene	159.2	2.66	0.0688	146	55.5	102	122	192	0.514	0.605	0.741	0.005
1-Chloroheptane	159.2	2.2	0.0591[h]	150	55.5	81.5	101	143	0.437	0.531	0.625	0.013
3-(Chloromethyl) heptane	172	—	0.0557[h]	148	55.7	63.4	80.5	183	0.374	0.465	0.614	0.032
Dichlorides[g]												
Dichloromethane	40.2	346	0.1037	144	55.4	10.1	15.8	63.7	0.034	0.052	0.101	0.132
trans-1,2-Dichloroe-thylene	49	263	0.0828[h]	145	55.7	33.0	50.3	124	0.127	0.187	0.268	0.065
1,1-Dichloroethane	56.5	158	0.0919	143	55.5	23.3	40.1	225	0.092	0.152	0.327	0.034
cis-1,2-Dichloroethy-lene	59.8	142	0.0828[h]	144	55.3	29.8	42.6	165	0.116	0.160	0.298	0.063
1,2-Dichloroethane	83.5	60.2	0.0907	144	55.3	54.0	79.7	186	0.214	0.305	0.456	0.036
1,2-Dichloropropane	96.4	42.3	0.0794	145	55.8	65.0	90.3	200	0.291	0.393	0.554	0.034
cis, trans-1,3-Dichlo-ropropene	108	—	0.0763[h]	147	56.7	85.5	110	208	0.319	0.402	0.606	0.036
1,4-Dichlorobutane	162	3.10	0.0666[h]	146	55.6	108	129	215	0.505	0.637	0.755	—
o-Dichlorobenzene	180.4	0.86	0.0646[h]	146	57.3	109	132	239	0.618	0.736	0.911	—
Trichlorides[g]												
Chloroform	61.2	159.3	0.0888	144	55.6	33.2	52.4	474	0.158	0.240	0.415	0.050
Methyl chloroform	74	105	0.0794	144	56.2	40.4	58.9	197	0.212	0.299	0.550	0.036

TABLE 68 (continued)

Solvent	BP (C°)	Vapor pressure at 20°C (Torr)	Diffusion coefficient at 25°C (cm^2/sec)	Volume charcoal (mL)[a]	Weight charcoal (g)[b]	Experimental breakthrough times t_1% (min)	t_{10}% (min)	t_{99}% (min)	Wt. solvent adsorbed/wt. of carbon t_1%[c] (g/g)	t_{10}%[d] (g/g)	t_{100}%[e] (g/g)	Weight water adsorbed at t_{100}%/wt. carbon[f] (g/g)
Trichlorides[g] (continued)												
Trichloroethylene	86.5	58.6	0.0875	145	54.8	55.3	83.0	195	0.293	0.428	0.625	0.038
1,1,2-Trichloroethane	113.6	17.5	0.0792	145	56.1	71.8	112	206	0.377	0.568	0.773	0.045
1,2,3-Trichloropropane	156	2.1	0.0672[h]	147	56.2	111	132	223	0.643	0.754	0.915	—
Tetrachlorides[g]												
Carbon tetrachloride	76.8	92.1	0.0828	144	55.4	77.0	90.0	147	0.473	0.545	0.677	
Perchloroethylene	121.2	14.0	0.0797	146	55.7	107	129	209	0.704	0.835	1.01	
1,1,2,2-Tetrachloroethane	146	4.73	0.0722	146	55.3	104	131	216	0.809	0.861	1.07	
Tentachlorides[g]												
Pentachloroethane	161	—	0.0673	145	56.1	93.0	117	187	0.742	0.914	1.13	
Acetates[g]												
Methyl acetate	57.3	170.2	0.0978	146	55.6	32.8	46.5	143.8	0.097	0.133	0.214	0.073
Vinyl acetate	72.5	91.2	0.0793[h]	148	55.6	55.0	81.1	235	0.188	0.269	0.391	0.006
Ethyl acetate	72.2	74.4	0.0861	145	60.6	66.8	84.7	172	0.215	0.267	0.350	0.008
Isopropyl acetate	87.5	47.3	0.0770	146	55.6	64.5	85.6	166	0.262	0.339	0.453	0.036
Isopropenyl acetate	96	30.0	0.0718[h]	145	57.1	80.6	106	193	0.323	0.409	0.507	
Propyl acetate	101.3	24.9	0.0768	145	55.9	78.8	99.0	164	0.318	0.392	0.507	0.020
Allyl acetate	103.5	—	0.0718[h]	148	56.1	75.8	95.6	246	0.299	0.369	0.517	0.043
sec-Butyl acetate	111.5	—	0.0648	145	55.9	82.8	101	164	0.381	0.456	0.537	—

Acetates (continued)												
Butyl acetate	126.1	8.0	0.0672	148	55.0	77.3	96.9	228	0.361	0.443	0.595	0.023
Isopentyl acetate	141.5	3.8	0.0603[h]	148	55.8	70.9	88.3	177	0.366	0.447	0.584	0.020
1,3-Dimethylbutyl acetate	146.2	4	0.0567[h]	149	55.8	60.6	76.0	188	0.346	0.426	0.588	0.025
Pentyl acetate	148.4	2.95	0.0610	148	55.7	72.6	87.3	225	0.375	0.444	0.616	0.014
Hexyl acetate	169	—	0.0567[h]	148	55.5	67.0	85.3	202	0.385	0.480	0.658	—
Ketones[i]												
Acetone	56.2	307	0.1049	164	60.8	37.1	46.0	119	0.078	0.095	0.135	0.109
2-Butanone	79.6	70.6	0.0903	160	66.0	81.9	94.4	239	0.198	0.225	0.295	0.056
2-Pentanone	102.3	30.5	0.0793	159	58.0	104	121	231	0.341	0.392	0.483	0.016
3-Pentanone	102.7	26.8	0.0740[h]	162	58.4	93.5	114	175	0.305	0.365	0.449	0.055
4-Methyl-2-pentanone	115.5	16	0.0677[h]	160	58.1	96.1	111	211	0.367	0.418	0.511	0.007
Mesityl oxide	129.7	7.9	0.0760	169	60.2	122	139	224	0.440	0.495	0.578	—
Cyclopentanone	130.7	—	0.0796[h]	167	64.3	141	161	308	0.408	0.460	0.585	—
2,4-Pentanedione	140.4	6.5	0.0727[h]	176	70.6	130	144	214	0.408	0.447	0.527	0.009
3-Heptanone	147.3	—	0.0628[h]	160	58.1	91.0	105	186	0.396	0.451	0.548	0.014
2-Heptanone	151.2	2.8	0. 0628[h]	187	61.9	101	114	237	0.412	0.460	0.560	0.005
Cyclohexanone	155.6	5.4	0.0729[h]	166	64.7	126	144	249	0.423	0.477	0.585	0.003
5-Methyl-5-heptanone	159.5	—	0.0588[h]	165	62.9	86.4	99.3	183	0.389	0.442	0.550	0.024
3-Methylclohexanone	168	—	0.0669[h]	162	63.5	101	123	216	0.395	0.472	0.595	—
Diisobutyl ketone	169.4	—	0.0554[h]	162	60.4	70.8	82.3	171	0.369	0.427	0.556	0.005
4-Methylcyclohexanone	171.3	—	0.0669[h]	172	67.9	111	126	245	0.463	0.520	0.580	0.009
Alkanes[i]												
Pentane	36.1	424	0.0842	166	62.6	60.7	71.3	147	0.155	0.179	0.228	0.020
2,3-Dimethylbutane	58.0	191	0.0689[h]	167	63.5	72.0	81.1	192	0.216	0.241	0.300	0.021
Hexane	68.7	121	0.0732[h]	160	55.0	52.3	64.6	178	0.181	0.220	0.334	0.036
Methylcyclopentane	71.8	110	0.0734[h]	164	66.7	62.2	76.1	174	0.174	0.209	0.278	0.036
Cyclohexane	80.7	77.5	0.0743[h]	161	59.3	68.7	82.3	179	0.216	0.254	0.337	0.024
Cyclohexane	83.3	70.4	0.0765[h]	159	57.3	85.8	100	193	0.272	0.313	0.401	—

TABLE 68 (continued)

Solvent	BP (C°)	Vapor pressure at 20°C (Torr)	Diffusion coefficient at 25°C (cm^2/sec)	Volume charcoal (mL)[a]	Weight charcoal (g)[b]	Experimental breakthrough times: t_1% (min)	t_{10}% (min)	t_{99}% (min)	Wt. solvent adsorbed/wt. of carbon: t_1%[c] (g/g)	t_{10}%[d] (g/g)	t_{100}%[e] (g/g)	Weight water adsorbed at t_{100}%/wt. carbon[f] (g/g)
Alkanes (continued)												
2,2,4-Trimethylpentane	96.5	38.6	0.0594[h]	161	57.5	68.3	80.4	166	0.300	0.348	0.440	0.019
Heptane	98.5	35.4	0.0636[h]	160	57.9	78.2	89.8	188	0.299	0.339	0.432	0.019
Methylcyclohexane	100.9	36.2	0.0679[h]	160	57.5	68.5	80.5	191	0.259	0.300	0.440	0.045
1,3,5-Cycloheptatriene	115.5	—	0.0742[h]	168	63.4	121	137	246	0.389	0.435	0.536	0.003
2,2,5-Trimethylhexane	124.5	12.5	0.0559[h]	166	64.2	67.6	80.0	168	0.266	0.310	0.444	0.001
5-Ethylidene-2-norbornene	147.5	4.2	0.0644[h]	165	67.4	86.5	101	221	0.341	0.393	0.526	0.007
Cyclooctane	150	—	0.0639[h]	172	70.3	96.8	113	220	0.345	0.397	0.497	0.027
Nonane	150	3.15	0.0559[h]	168	63.5	76.2	89.3	198	0.340	0.393	0.490	
Decane	174	0.06	0.0530[h]	164	57.7	70.8	81.5	158	0.386	0.439	0.553	0.017
Amines[i]												
Methylamine	−6.7	2160	0.1300[h]	161	58.5	12.4	17.9	91.5	0.015	0.020	0.041	0.150
Dimethylamine	6.7	1285	0.1023[h]	164	70.0	17.1	21.7	94	0.024	0.030	0.050	0.291
Ethylamine	16.6	872	0.1032[h]	161	59.8	40.5	49.7	270	0.068	0.081	0.179	0.268
Isopropylamine	31.5	478	0.0879[h]	160	60.4	65.6	75.8	194	0.142	0.162	0.275	0.136
Propylamine	47.8	247	0.0879[h]	172	65.5	90.0	111	330	0.180	0.217	0.335	0.252
Diethylamine	55.5	188	0.0993	160	62.1	88.0	105	213	0.229	0.269	0.351	0.100
Butylamine	77.5	75	0.0872	162	63.9	110	125	278	0.278	0.312	0.429	0.031
Triethylamine	89.4	46.2	0.0754	160	59.1	81.1	91.0	174	0.307	0.341	0.426	0.032
Dipropylamine	110	18.2	0.0647[h]	170	68.2	93.0	105	226	0.305	0.340	0.439	0.060
Diisopropylamine	110.5	—	0.0647[h]	171	67.0	77.0	87.1	185	0.257	0.288	0.370	0.069
Cyclohexylamine	134	—	0.0664[h]	169	64.3	112	128	279	0.382	0.431	0.564	0.141
Dibutylamine	159	1.61	0.0567[h]	164	64.6	75.5	84.8	186	0.335	0.372	0.480	0.009

Miscellaneous

Methyl iodide[i]	42.4	336	0.0900[h]	160	58.2	11.6	17.7	94.5	0.063	0.092	0.187	0.223
Acrylonitrile[i]	77.3	67.8	0.1059	160	58.0	48.5	61.1	168	0.098	0.121	0.174	0.016
Dibromomethane[i]	98.6	33	0.0826	160	62.8	82	121	279	0.502	0.717	1.06	0.006
Pyridine[i]	115.2	15.4	0.0824[h]	160	64.3	119	134	292	0.324	0.360	0.460	0.076
Epichlorohydrin[g]	116.9	12.5	0.0821[h]	146	56.9	85.5	110	288	0.308	0.387	0.566	0.014
2-Methoxyethanol[g]	124.4	9.0	0.0884	146	55.5	116	145	249	0.352	0.431	0.540	0.096
1,2-Dibromoethane[i]	131.5	8.6	0.0840[h]	161	61.9	141	165	303	0.946	1.09	1.336	—
1-Nitropropane[i]	131.6	7.2	0.0781[h]	160	63.2	143	164	322	0.443	0.504	0.638	—
2-Ethoxyethanol[g]	135.5	—	0.0788	148	55.4	77.0	123	283	0.277	0.426	0.593	0.011
Acetic anhydride	139.6	3.7	0.0755[h]	161	61.0	124	138	200	0.459	0.506	0.592	0.070
2-Methoxyethylacetate[g]	144.5	3.6	0.0666[h]	150	55.7	93.3	113.2	303	0.438	0.521	0.707	—
Bromobenzene[i]	156	3.0	0.0661[h]	160	58.1	142	159	297	0.849	0.940	1.126	—
2-Ethoxyethyl acetate[g]	156.3	2.0	0.0619[h]	148	55.8	79.5	96.5	248	0.417	0.497	0.689	0.010

[a] Average volumes for a cartridge pair are 146.3 ± 1.9 mL for the Type 1 and 163.8 ± 4.3 mL for the Type 2

[b] Average weights for a cartridge pair are 56.1 ± 1.2 g for the Type 1 and 62.2 ± 3.9 g for the Type 2;

[c] Calculated from $W_1\% = t_1\%\ MQC/24.1\ 10^6\ W_c$;

[d] Estimated from $W_{10}\% = [t_1\% + 0.9\,(t_{10}\% - t_1\%)]\ [MQC]/24.1\ 10^6\ W_c$;

[e] See Ref. 4, "Data Processing," calculation of W_s;

[f] Calculated from $W_{H2O} = W_{total} - W_s$;

[g] Used Type 1 cartridges;

[h] Calculated from Gilliland's equation $D_{25} = [0.0043T^{1.5}\,(1/M_{air} + 1/M_{vap})^{0.5}]/(v_{air}^{0.33} + v_{vap}^{0.33})^2P$, where v_{air} is 29.9 mL/mol and v_{vap} is calculated from the LeBas approximation;[7]

[i] Used Type 2 cartridges.

From Nelson, G. O. and Harder, C. A., *Am. Ind. Hyg. Assoc. J.*, 35, 391, 1974.

TABLE 69

Adsorption of Water Vapor on Charcoal

Relative humidity at 25°C	MgH_2O adsorbed	% of H_2O/air adsorbed
95	12.4	25
75	6.8	20
50	2.6	11
25	0.8	3.4

Adapted from Burnett, R. D., *Am. Ind. Hyg. Assoc. J.*, 37, 37, 1976.

sorbed per g of charcoal (g/g); W_c = weight of charcoal (g); M = molecular weight of the contaminant (g/mole); Q = airflow rate (L/min); C = upstream gas concentration (ppm.)

All terms can be measured or calculated except w, which can be determined by measuring the saturation weight (downstream = upstream concentration). Estimations also can be calculated by the more complicated Mecklenburg equation.[217] However, calculated adsorption capacities (W_s = g/g) were as much as 169% greater and 20% lower (decane) than the observed results (ethyl chloride). The deviations appeared to be related to volatility (boiling point). The breakthrough at 1% times were considerably lower than predicted by the Mechlenberg equation.

ii. Humidity

Although early reports indicated that because charcoal is a nonpolar material the adsorption efficiency is not affected by water vapor, later investigations disclosed a significant reduction in collection efficiency for organic vapors at high RHs. With toluene, increasing the RH from 7 to 52% did not reduce collection efficiency (breakthrough less than 1%). However, in a series of five determinations, significant breakthrough (4.2%) did occur in every case. A substantial amount (36%) of toluene was not collected in the presence of condensed moisture. A comparison of the toluene data from Table 67 with the 94% RH data discloses the severity of the reduction in the breakthrough volume — 23 to 10 L at 0.68 to 0.75 m/L concentration. The sample volume must be reduced by at least 50% to obtain reliable samples under high humidity conditions. The amount of water vapor sorbed at 50% RH while collecting various organic vapor is shown in Table 68. The RH decreases the activated carbon's affinity for volatile or water-soluble solvents.[217]

Humidity effects also were evaluated to determine whether gravimetric determination of single solvent systems would provide reliable results. The answer was negative. Above 25% relative humidity (25% RH) significant quantities of water were adsorbed (Table 69).

iii. Sample Stability with Time

Time lapse between collection and analysis frequently becomes the controlling variable in the evaluation of the industrial atmosphere. Therefore, an evaluation of storage stability must be made to delineate permissible limits. The principle of half-life widely recognized in atomic radiation exposure control can be applied, that is, the time required for 50% of the sample to be lost from the system. If this factor is known and is reproducible, then allowances and corrections can be applied to the final result calculations. However, calculations must be based on the total weight collected (sum of the fore and backup sections) as migration of the more volatile compounds can become a large factor which obscures breakthrough indications. Storage stability test results for five typical solvent vapors are summarized in Table 70.

Refrigeration at 4°C slows down the migration of vapor from the fore to the backup section. On the 28th day of storage, 27 vs. 33% of the dichloromethane was found in

TABLE 70

Stability of Solvent Vapors Sorbed on Charcoal Related to Time after Collection

Compound	Time (days)		No. of samples (μ)	Concentration (mg/L)	Volume sampled (L)	Recovered %	
	Total	To equilibrium				Total	Back-up section
Dichloromethane	1		31	1	5	96	4
	57	22				98	33
2-Butanone	1		46	0.8	10	100[a]	2[a]
	78	40				76	23[a]
Trichloroethylene	1		48	9	10	100	1–
	92	92 +				100 +	12.7
Toluene	35 +	—	—	—	—	100	—
Styrene	35 +	—	—	—	—	100	—

[a] Estimated from figures.

Data compiled from Saalwachter, A. T., McCammon, C. S., Roper, C. P., and Carlberg, K. S., *Am. Ind. Hyg. Assoc. J.*, 38, 476, 1977.

the backup section. Sample decomposition also should be retarded. NIOSH now routinely stores charcoal tube samples at 4°C.

iv. Stability with Heat

Exposure to heat cycles might produce apparent sample loss due to vaporization and redeposition of the vapor into deeper pores. To determine the magnitude of this effect for toluene and styrene, 20 replicate charcoal tube samples of each were stored at 48°C for varying periods of time. The differences between the control and exposed groups were insignificant, and no discernible trends were noted over seven 8-hr periods at 48°C followed by 16-hr periods at 25°C. Other more reactive compounds may be heat labile, however.

v. Stability with Light

Sample loss from actinic light catalysis presents another storage variable. Styrene, which is known to be polymerized by light, was chosen for study. Heat had been ruled out as a source of loss previously. No styrene loss was observed when groups of four replicate tubes were exposed to sunlight for 19 days, to heat only, to sunlamp for 24 hrs, and the control group held in the dark. The concurrent presence of catalyst during field collection may alter the stability and must be taken into account when conducting surveys.

vi. Stability Under Vacuum

Shipment of tube samples by air in unpressurized air cargo space from the sampling site to the laboratory could contribute to sample losses. Actual air shipment and simulated shipment at 450 mm Hg pressure for 4 hr produced no discernible loss of toluene in groups of 10 sample tubes.

vii. Variations Among Lots of Charcoal

As previously indicated in Section II. C. 3. e. i, "Elution Techniques," different batches of activated charcoal activity may vary to a degree that will introduce serious errors into the final results. These variations involve both retention and desorption properties. Data for the nonpolar vapors (trichloroethylene, styrene, and toluene) revealed no significant variation from one batch to another of seven lots of charcoal. However, variations did appear among the polar solvent vapors (*n*-butanol, 2-ethoxyethyl acetate, and 2-butanone). As much as 20% difference was found between the extremes in the seven lots. These findings again emphasize the importance attached to the determination of desorption efficiency in the analysis protocol.

viii. Variation with Contaminant Loadings

In at least one case, the amount in μL determined the desorption efficiency. The efficiency for 2-butanol increased from 70 to 84% as the amount sorbed increased from 1.8 to 7 μL. The loading effect also was considered in the preparation of known charcoal tube standards. Data collected are presented in Table 71.

The evidence indicates that whether the solvent sample is added directly as the liquid phase to the charcoal or by adsorption from the vapor phase has no effect on retention or desorption. However, equilibration from a solution in the eluent onto the charcoal added to the solution has profound effect on recovery. Considerably less hexanone is sorbed from the solvent-eluent system than is retained from the spiked charcoal at low loading values (28% difference at 0.2 mg/100 mg charcoal).

These results introduce a note of caution into the use of internal standards in analysis by gas chromatography. Either the internal standard must not be removed from the elution solvent, or the internal standard must be added *after* the contaminant has been

TABLE 71

Effect of Technique Relative to Desorption Efficiency for Cyclohexanone of Three Loading Levels

Technique	Desorption efficiency (%) .02 mg	1 mg	10 mg
Direct injection of the liquid into the charcoal	42	64	91
Injection of the liquid on glass wool below the charcoal	42	64	92
Addition of charcoal to the spiked elution solvent	70	76	96

Data derived from Saalwachter, A. T., McCammon, C. S., Roper, C. P., and Carlberg, K. S., *Am. Ind. Hyg. Assoc. J.*, 38, 476, 1977.

eluted and separated in solution from the charcoal matrix. Otherwise, errors of unknown and serious magnitude will be introduced.

Evaluation of seven solvents by collaborative testing in 15 laboratories under NIOSH support indicated the NIOSH charcoal tube gas chromatographic method is versatile and precise. The overall method precision is 10.5% relative standard deviation (RSD).[218]

d. Uses Relative to Sampling

The collection of organic solvent vapors on activated carbon for analysis by gas chromatograph and the delineation of instrumental operating parameters were rather completely reported by Franst and Hermann in 1966.[219] Variations in recovery efficiency of methyl ethyl ketone, toluene, trichloroethylene, butyl acetate, 2-methylcyclohexanol, and styrene, and desorption into carbon bisulfide were evaluated in the TLV to five times TLV concentration range. The experimental collection tubes were 6 mm I.D. × 83 mm long and were filled with ordinary water purification charcoal that had been size graded by sieving. A 50-mm fill required approximately 0.7 g charcoal. Desorption was carried out in 10 mL of CS_2 and allowed to stand for approximately 1 hr. The following conclusions were drawn from the experimental results:

1. The optimum carbon particle size was found to be that of those passing through 10 mesh and retained on a 30-mesh sieve.
2. In general, the optimum sample rate was 100 mL/min for the sampling tube fabricated to the above dimensions. Although most reproducible at this rate, the efficiency of recovery was not always at a maximum.
3. When sampling at 100 mL/min the recovery efficiency was relatively independent of the volume of vapor collected.
4. Higher sampling rates did not always provide the same results as higher concentrations, which indicated that the amount of vapor collected might not be the controlling factor.

An investigation reported later in 1968 by Reid and Halpin confirmed the reliability and flexibility of charcoal as a collecting medium.[220] The conditions employed were quite similar to those reported by Franst and Hermann end later recommended by NIOSH (Table 72). Excellent recoveries were obtained with relatively small sample volumes in the one-half to two times TLV range (Table 73). Trichloroethylene recov-

TABLE 72

Operational Parameters for Charcoal Absorption Tubes

Parameter	Franst and Herman[146]	Reid and Halpin[146]	NIOSH recommendations[148,151]
Tube diameter (I.D.) (mm)	6	6	4
Tube length (mm)	83	127	70
Charcoal sieve size	10/30	12/20	20/40
Charcoal type	Water purified	Darco®	—
Charcoal weight (gm)	0.7	0.6	0.15 (2 sections)
Pressure drop adjustments (mm Hg)	—	35—40/2 L/min	25/1 L/min
Optimum sampling rate (L/min)	0.100	1—2	1
Desorption liquid	CS_2	CS_2	CS_2
Concentration range (TLVs)	1—5	0.5—2	0.5—2
Recoveries (%)	30—100	88—100	100
GC column packing	3% SF96/Celite®	20% Carbowax® 1500/60—80 mesh GC-22	10% FFAP/80-100 mesh acid-washed DMCS Chemosorb® W
Sample size (ℓ)	—	—	10

TABLE 73

Parameters for Determination of Several Halogenated and Aromatic Hydrocarbons[a]

Compound	Chromatographic settings								
	Oven temperature (°C)	Detector temperature (°C)	Injection block temperature (°C)	Helium flow rate (mL/min)	Attenuation	Minimum air samples (L)	Efficiency[b] (%)	Retention time (min)	Conversion factor (to ppm)
Benzene	50	50	100	175	× 2	10	89	5.8	313
Carbon tetrachloride	50	50	100	175	× 1	10	95	4.1	158.8
Ethylene dichloride	75	75	150	156	× 1	4	96	6.3	247
Methyl chloroform	50	50	100	175	× 2	4	100	4.2	183.3
Methylene chloride	50	50	100	175	× 2	4	97	4.8	288
Perchloroethylene	75	75	150	156	× 1	4	95	4.6	147.4
Toluene	75	75	150	156	× 2	4	93	5.7	266
Trichlorotrifluoroethane	50	50	100	175	× 4 and × 16	2	92	0.8	130.5
Trichloroethylene	75	75	150	156	× 1	4	95	4.2	186
Xylene	100	100	200	140	× 1	4	95	5.4	230.7

[a] In all cases the column inlet pressure is 24 psi, and the detector is set at 8 V.
[b] Efficiences are based on one half to twice TLV concentrations and sampling rates of 1 to 2L/min.

From Reid, F. H. and Halpin, W. R., *Am. Ind. Hyg. Assoc. J.*, 29, 390, 1968. With permission.

TABLE 74

Results Obtained with Other Compounds[a]

Compound	Chamber concentration (ppm)	Air volume (L)	Range of recoveries (ppm)	Average recoveries (ppm)	Average recoveries (%)
Perchloroethylene	50	4	45—47	46	92
	100	4	95—96	95	95
	200	4	192—193	193	97
Methyl chloroform	175	4	168	168	96
	350	4	355—363	358	102
	700	4	682—721	697	100
Ethylene dichloride	25	4	24	24	96
	50	4	45—49	47	94
	100	4	96—100	98	98
Carbon tetrachloride	5	10	4	80	80
	10	10	12	12	120
	25	10	20—25	23	92
Methylene chloride	250	3.8	234—235	234	94
	500	3.8	484	484	97
	1000	3.8	950—1010	988	99
Benzene	12.5	10	12	12	96
	25	10	21	21	84
	50	10	46—48	47	94
Toluene	100	3.8	90	90	90
	200	3.8	180—189	186	93
	400	3.8	382—391	386	96
Xylene	50	3.8	48—49	48	98
	100	3.8	90—92	91	91
	200	3.8	194	194	97
Trichlorotrifluoroethane	25	0.5	26—29	27	108
	50	2	49—51	50	100
	500	2	476—509	491	98
	1000	2	900—959	939	94
	2000	2	1756—1835	1797	90

[a] In each case three samples were collected at a sampling rate of 2 L/min.

From Reid, F. H. and Halpin, W. R., *Am. Ind. Hyg. Assoc. J.*, 29, 390, 1968. With permission.

eries ranged from 88 to 100% in the concentration range 50 to 200 ppm, sampling rate range 1.5 to 2.0 L/min and sample size 3 to 10 L. The GC parameters are summarized in Table 73. This technique is ideally suited for direct application to personnel monitoring, especially for chlorinated hydrocarbons, aromatic hydrocarbons, ketones, alcohols, and esters, with a minimum of time and effort to calibrate a GC system selected from those already established (Table 74).

The collection of mercury vapor on charcoal has been shown to be both rapid and complete.[221] Glass tubes 4 mm I.D. × 152 mm long were packed with two 25-mm sections (180 mg each) of 20 to 40-mesh impregnated charcoal, which were separated and retained by fiberglass plugs, activated by heating at 600 to 800°C for 1 hr. The tubes gave quantitative recoveries of mercury from air or vaporized from urine, tissue, and water. Analysis of the sections separately indicated complete absorption in the first section in every case. The mercury was quantitated directly from the absorbant in a tantalum sample boat that was introduced into an atomic absorption spectrometer oxidizing air-acetylene flame. The minimum detectable quantity of mercury was 0.02 μg.

TABLE 75

Passive vs. Dynamic Charcoal Sampling — Accuracy and Precision Data

	Passive	Dynamic
Known concentration (ppm)	1	25
Mean ($\bar{x}$ ppm) (no. determinations)	0.98	22.8
	20	10
Standard deviation (S) (ppm at 95% confidence limits)	0.04	2.7
Accuracy — systematic error	1.0%	8.8%
Precision (relative S)	4.1%	11.6%

i. Specific Application

Benzene — A rather exhaustive study of the collection of benzene on charcoal by passive vs. dynamic adsorption (Table 75) indicated that at low concentrations the passive sampling technique was superior to the dynamic tube method.[222]

Phenol[223] — A 10-L sample was collected at 0.2 to 1.0 L/min on activated charcoal. The percentage of recovery in the range 25 to 1000 mg was 96 to 100%. The phenol was desorbed with diethyl ether and analyzed by GC as the trimethylsilyl derivative. Regression analysis showed close agreement with collection in 5% aqueous NaOH followed by colorimetric analysis.

Carbon Bisulfide[224] — Carbon bisulfide adsorbed on charcoal was eluted with benzene and quantitated by gas chromatographic analysis. A sensitivity of 1 mg was demonstrated.

Furfural[225] — An atmosphere containing coexisting furfuryl alcohol was sampled on charcoal. Methylene chloride was used to desorb the collected sample. Oxidation during storage was solved by desorption immediately after collection.

Ethylene Oxide[226] — After collection on charcoal at 25 to 500 mL/min, the ethylene oxide is desorbed with CS_2 and analyzed by GC. With a 10-ℓ sample, the low detection limit was 0.15 ppm. Desorption efficiency averaged 91% with an overall recovery of 97 ± 3.8%. An evaluation of 10 common interferences (aldehydes, water vapor, etc.) indicated minimal interferences.

Vinyl Chloride[227,228] — Thermal desorption was used to recover vinyl chloride (VC) from charcoal. A detection limit of 20 ppb with a standard deviation at the 1-ppm level of ±0.04 ppm was attained. Replicate field samples were taken with an MSA pump in a four-phase manifold. Flow rate through each tube was determined before sampling was started. Standard concentrations in air were made up in a 100-mL glass syringe.

e. Analysis

i. Elution Techniques

Adsorbed gases and vapors may be desorbed from activated carbon by (a) displacement of the adsorbed material by superheated steam, (b) heating the carbon under vacuum and distilling the desorbed material into cold traps, (c) thermal desorption, and (d) solvent extraction.

a) Steam Displacement

For steam displacement, the saturated carbon is flushed with superheated steam at 300°C or higher. The effluent steam and displaced vapors and condensate are held for analysis. Flushing is continued until the condensate is substantially odorless. Upon standing the condensate may separate into an oily and an aqueous layer. Both oily and aqueous fractions may then be subjected to qualitative chemical or spectrometric analysis for detection of inorganic ions and organic functional groups. The steam displacement method, although very effective as a means of desorption, suffers from the disadvantage that steam will hydrolyze some organic compounds, such as esters, thereby causing qualitative changes. This effect can be circumvented by using the vacuum method described in the following paragraph.

b) Vacuum Desorption

For vacuum desorption, the saturated carbon is connected to a train of three traps in series, which are immersed in ice-salt, Dry Ice-Cellosolve and liquid nitrogen, respectively. The entire system is evacuated and pumping is continued at a pressure of about 1 μ until distillation practically ceases. During the pumping, the carbon sample is held at a temperature of 200 to 250°C.

At the end of the pumping period, the system is slowly returned to atmospheric pressure by bleeding in air or nitrogen. The traps are then removed and their contents examined. Generally, most of the water and some nonaqueous matter will be found in the ice-salt trap, most other atmospheric vapors and some water in the Dry Ice trap, and some of the lighter gases in the liquid nitrogen trap. These materials, with or without further fractionation, may be subjected to chemical or spectrometric qualitative analysis.*

c) Thermal Desorption

Recent technology favors thermal desorption directly into a gas chromatograph (GC) or desorption (extraction) with a solvent and injection of an aliquot of the extract into a gas chromatograph column. Thermal desorption performs well with the loosely sorbed and low boiling compounds, but thermal decomposition at the higher temperature required to displace tightly sorbed and high boiling constituents limits this recovery technique. Also, resolution of some mixtures is not as good as obtained by chromatographing a solvent extract. For example, vinyl chloride and ethylene dichloride were not resolved when field samples were thermally desorbed directly into the GC, but did separate when a CS_2 extract was injected.[46] When thermal desorption is considered, then desorption efficiency must be determined, i.e.:

$$\% \text{ efficiency} = \frac{\text{amount recovered} \times 100}{\text{amount on charcoal}^{**}}$$

Of the 200 GC methods included in the NIOSH Analytical Manual (Table 64), only two employ thermal desorption. However, reusable metal tubes for direct coupling to the chromatograph furnace are commercially available (Figure 114)[167,229]

d) Solvent Extraction (Desorption)

By far the most widely used technique for recovery of sorbed compounds from charcoal is solvent to desorption; and carbon bisulfide (CS_2) is the solvent of choice for most analysts (125 of the 142 charcoal-GC procedures listed in the NIOSH Analytical Manual). Other solvents included methanol, benzene, methylene chloride, isopropyl alcohol, ethyl acetate, tetrahydrofurane, toluene, hexane, and mixtures of these solvents (Table 64). In a number of extraction procedures, the sorbed material may not

* From Katz, M., Ed., *Methods of Air Sampling and Analysis* 2nd ed., Intersociety Committee on Methods for Air Sampling and Analysis, Am. Public Health Assoc. Publication Service, Washington, D.C., 1977.

** Amount collected from an air stream of known concentration

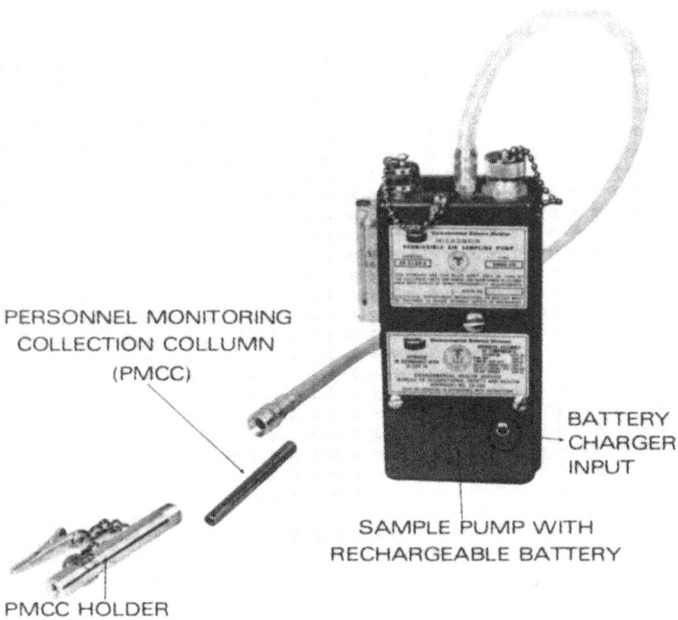

FIGURE 114. The Bendix "Flasher" H/S 10 sample collection system. (Courtesy of Bendix Environmental Science Division, 1400 Taylor Ave., Baltimore, Md.)

TABLE 76

Recovery of Solvent Vapor Adsorbed on Activated Charcoal[a]

Solvent	Concentration range (ppm)	No. determinations	Recovery average % overall	CV (% coefficient variation)	Range (% recovery)
Trichloroethylene — EHL	85—245	8	95	±5	80—100
MSA	85—245	7	95	±8	80—106
Methylene chloride — EHL	100—500	8	91	±9	77—170
MSA	100—500	8	87	±6	75—94
Methyl ethyl ketone	100—300	9	82	±1	80—85
	100—400	9	81	±5	74—88
Ethylacetate	200—600	10	107	±7	99—120
Toluene	50—200	8	105	±3	101—110
Isopropyl alcohol	200—400	9	44	±7	32—57
Acetone	200—600	9	66	±10	50—76
Cyclohexane	150—450	9	84	±3	79—90
Ethyl cellosolve acetate	33—132	13	58	±7	43—69
Methyl i-butyl ketone	49—188	9	94	±7	86—104
Styrene monomer	50—200	9	85	±9	76—102

[a] MSA tubes unless otherwise noted.

be completely desorbed by CS_2 (Table 76). Quantitative recovery was attained by a small quantity — 1 to 5% — of a more polar solvent such as methanol or isopropanol. The magnitude of the improved desorption efficiency produced by adding 5% methanol to the CS_2 is shown in Table 77.[271] These compounds were present as a mixture on the charcoal. The importance for quality control of determining the desorption effi-

TABLE 77

Effect Produced by Adding 5% Methanol to Carbon Bisulfide on Charcoal Desorption Efficiency — Summary

	CS_2 desorption		Mixture desorption		
Compound	%[a] Recovery	CV[b] (%)[a]	%[a] Recovery	CV[b] (%)[a]	%[a] Improvement
Toluene	94	2.3	99	0.0	6.4
N-butanol	49	12.2	70	2.0	43.0
Cellosolve acetate	73	8.6	97	1.4	35.7
N-butyl cellosolve	24	8.8	85	1.7	254

[a] % = Average results from 5 tubes.
[b] CV = coefficient of variation × 100.

From Fracchia, M., Pierce, L., Graul, R., and Stanley, R., *Am. Ind. Hyg. Assoc. J.*, 38, 144, 1977. With permission.

TABLE 78

Effect of Component Interaction on Desorption Efficiency

		Alone			In the mixture		
Compound	% In the mixture	Average (5 tubes)	Range	CV (%)	Average (5 tubes)	Range	CV (%)
Toluene	42.3	102	100—104	1.6	94	92—97	2.3
N-butanol	26.7	34	28—44	20.6	49	43—59	12.2
Cellosolve acetate	17.9	47	39—54	13.4	73	63—79	8.6
N-butylcellosolve	13.1	7	3—11	45.8	24	21—26	8.8

Data derived from Fracchia, M., Pierce, L., Graul, R., and Stanley, R., *Am. Ind. Hyg. Assoc. J.*, 38, 144, 1977.

ciency is illustrated by the interaction of the components in the mixture on the individual desorption efficiencies is illustrated in Table 78.

It is obvious from these results that desorption efficiencies must be determined not only for individual compounds by themselves, but also when present in mixtures of compounds which exhibit varying polarities. Also, selection of a suitable solvent, or solvent mixture, to obtain quantitative desorption is required. In the quoted example, perhaps increasing the methanol content to 10% would further improve the desorption of *n*-butanol.

Further studies along these lines provide an insight into methods by which the solvent system can be evaluated and applied. The phase equilibrium principle has been applied to the determination of desorption efficiencies for organic compounds sorbed on charcoal. The method is based on the true equilibrium distribution between the adsorbed phase and the solution phase of the organic compounds present in the system. Improved and more consistent desorption efficiencies are obtained by determining the effects of variations in the adsorbent/solvent rates and equilibrium temperature at the time of analysis. The technique also provides early detection of interaction, or reaction, of the sorbate on the adsorbent surface.

The technique consist in comparing concentration in the liquid phase from spiked charcoal and spiked solvent phase. At equilibrium the two should be equivalent. However, the ratio of solvent volume to charcoal weight must be kept constant. Temperature control also must be observed. For example, the effect of temperature for methyl ethyl ketone was found to be:

At-78°C, desorption efficiency = 15.8%
0°C, desorption efficiency = 66.8%
24°C, desorption efficiency = 89.5%[230]

If the spiked tube results are less than the spiked solvent, decomposition or interaction on the charcoal matrix is indicated.

Other investigations have shown that desorption efficiency varies from lot to lot of charcoal and with the loading (amount of material sorbed). For example, with *n*-butanol, one lot of charcoal yielded 63% recovery while another gave only 47%, and declined from 91 to 47% as the loading increased from 0.9 to 3.8 mL/100 mg charcoal. The source of charcoal — coconut vs. petroleum — made little difference in recovery of solvent vapor.[216]

ii. Analysis Procedure

Atmospheric solvent vapors collected on activated charcoal are desorbed with carbon disulfide prior to separation and analysis using a gas chromatograph with a flame ionization detector.

1. A straight 4-mm I.D. glass tube is packed with two sections of 20- to 60-mesh activated charcoal.[231] Charcoal source: Barneby-Chaney, Columbus, Ohio.[232]
2. Each section of activated charcoal is placed in a small glass stoppered microcentrifuge tube. Carbon disulfide (0.5 mL) is added to the charcoal and, after desorption is complete (30 min with agitation), a known aliquot is withdrawn with a microliter syringe and introduced into the gas chromatograph.
3. Calibration curves are established by injecting known amounts of each identified material into the gas chromatograph and measuring the areas under the peaks. A curve is established using concentration vs. peak area.

Sampling precautions and instructions follows.

1. The ends of the tubes should be broken so as to provide an opening at least 2 mm in diameter (i.e., ½ the I.D. of the tube). A smaller opening causes the pressure drop to exceed 1 in. of mercury.
2. The smaller section of charcoal is used as a backup and should therefore be positioned nearest the sampling pump. Sampling should be done with the charcoal tube in a vertical position.
3. These charcoal tubes have been tested at flow rates of up to 2.5 L/min and have demonstrated 100% trapping efficiency. The recommended sampling rate is 1 L/min.
4. In most cases, a 10-L sample of ambient air is adequate.
5. Treat one charcoal tube in the same manner as the sample tubes (break, seal, and transport) except for the taking of an air sample. This tube will be used as a blank.
6. After breaking and sampling, the polyethylene caps provided should be used for capping these charcoal tubes. Masking tape is the only acceptable substitute; rubber caps should never be used.
7. Return the tubes with a sample of the solvent(s) considered to be the source of contamination. If this is impossible (e.g., the contaminant is from an unknown

TABLE 79

Retention Data and Boiling Points

Compound	Retention time (sec)	Boiling point (°C)	Retention volume (mL)
Ethyl ether	89	35	125
Isooctane	97	99	136
Carbon disulfide	120	46	168
Carbon tetrachloride	162	77	227
2-Butanone	181	80	253
Ethanol	190	79	266
Benzene	213	80	298
Trichloroethylene	248	87	347
Chloroform	266	61	372
Perchloroethylene	293	121	410
Toluene	321	111	450
n-Butyl acetate	344	127	482
p-Dioxane	362	101	507
p-Xylene	445	138	620
m-Xylene	466	139	648
o-Xylene	522	144	725
Pyridine	560	115	775

From White, L. D., Taylor, D. G., Maner, P. A., and Kupel, R. E., *Am. Ind. Hyg. Assoc. J.*, 31, 225, 1970. With permission.

source or is a gas at room temperature), saturate a charcoal tube by drawing a large volume of contaminant-laden air through it.

8. Liquid samples and charcoal tube samples (with broken ends) should never be transported or mailed in the same container.

The foregoing procedure was developed by Kupel, White, and co-workers at the Bureau of Occupational Safety Health in Cincinnati, Ohio.[233] A good bibliography of charcoal sampling prior to 1970 was included in the publication of their findings. Since a solvent vapor is seldom found by itself in the industrial environment, the analytical method should be capable of simultaneously separating and quantitatively measuring all compounds in the solvent mixture with a minimum of time and effort. Optimum GC operating conditions for separation and analysis of 14 selected solvent vapors of occupational health significance were developed. The 20 ft × 1/8 in. stainless steel column was packed with 10% FFAP stationary phase on 80 to 100 mesh acid-washed DMCS Chromosorb® W solid support, programmed from 100 to 140°C (10 min elapsed time per sample) and flushed with helium as carrier gas. Retention data are summarized in Table 79.

The 14 selected solvent vapors were collected with 100% efficiency on activated charcoal, either alone or in any combination, including all 14 in a mixture, in the concentration range one half to two times TLV for each solvent. The overall efficiency was equivalent to the efficiency of the desorption phase for each solvent vapor and was over 90% for each of the solvents except ethanol and pyridine. Polar compounds tended to be displaced from the first into the second charcoal section of the collecting tube by high concentrations of nonpolar organic solvents. For example, approximately 25% of acetonitrile collected was displaced when an equal concentration (100 ppm) of petroleum ether was present, regardless of sampling rate (0.5 to 2 L/min). Chloroform, 2-butanone, and ethanol were similarly affected. Sampling rate and volume did not alter the adsorption (collection) efficiency.

TABLE 80

Limits of Detection

Compound	Parts per million	Micrograms
Isoamyl acetate	10	2.4
Benzene	20	2.9
2-Butanone	20	2.6
n-Butyl acetate	20	4.2
Isobutyraldehyde	10	1.5
n-Butyraldehyde	25	3.3
Carbon tetrachloride	20	5.6
Chloroform	20	4.4
Crotonaldehyde	10	1.3
1,2-Dichloroethane	10	1.8
p-Dioxane	20	3.2
Ethanol	50	4.2
Ethyl acetate	25	4.0
Ethyl benzene	10	1.9
Ethyl ether	10	1.3
Heptaldehyde	25	5.2
Heptane	10	1.8
Hexane	20	3.1
Methylene chloride	20	3.1
Perchloroethylene	20	6.1
Propionaldehyde	25	2.7
Isopropyl alcohol	20	2.2
Toluene	25	4.2
Trichloroethylene	20	4.9
Isovaleraldehyde	10	1.6
n-Valeraldehyde	25	3.9

From Cooper, C. V., White, L. D., and Kupel, R. E., *Am. Ind. Hyg. Assoc. J.*, 32, 383, 1971. With permission.

Further refinements in the charcoal sampling technique contributed by this group of investigators included trapping the GC peak in a charcoal capillary tube and transfer to a time-of- flight mass spectrometer via the hot filament tube for qualitative identification.[234] The limits of detection were also further improved (Table 80). As described, the system is a completely integrated unit for collection, identification, and quantitation of solvent vapors from industrial atmospheres.

The relative effectiveness of activated charcoal for the removal of gases and vapors for clean room filter installations as classified by N. R. Rowe of the Barneby-Chaney Company should be of interest to air pollution investigators.[235]

1. High capacity group — acetic acid, asphalt fumes, some Freons®, acetone, some amines, engine exhausts, chlorinated hydrocarbons, some mercaptans, phosphine, CS_2, HCl, methyl ethyl ketone, HNO_3, smog, $COCl_2$, tobacco pyrolysis vapors, and oil vapor.
2. Low capacity group — some Freons,® formic acid, SO_2, some amines, some mercaptans, arsine, stibine, CO, CO_2, NO, NH_3, and CH_2O.

Monitoring pumps that have been adpated to sampling periods up to 8 hr with the standard NIOSH charcoal tube are commercially available (see Chapter 2, Volume II).

TABLE 81

Orifice Flow Rate Calibration — ΔP = 584 mmHg[225]

Orifice number	Flow rate (L/min)
1	0.786 ± 0.002
2	0.806 ± 0.003
3	0.745 ± 0.004
4	0.753 ± 0.005
Range	0.745 ± 0.806
Difference	0.061
Average	772.5
Difference (% of average)	7.8

f. Evaluation of Charcoal Tubes

Evaluation of seven solvents by collaborative testing in 15 laboratories under NIOSH support confirmed that the NIOSH charcoal tube-gas chromatograph procedure is versatile and precise. The overall method precision was 10.5% relative standard deviation (RSD).[218]

The concentration range included 60% of the OSHA standard and excursion limits. The total relative error was 12.72%, which included 9.49% analytical error and 8.01 sampling deviation. At high concentrations, the backup sections varied widely in solvent content. Desorption efficiencies also exhibited wide variations in the range 0.87 to 1.00, which contributed a large portion of the analytical error. Large errors at the 5% OSHA standard level were attributed to instrument and laboratory operator sources. Percentage of relative errors for single components was 8.01, for mixtures was 10.62, and for pump calibration was 9.70 (major source in sampling). A significant number of pumps exhibited volumetric errors greater than 1.0 ± 0.1 L/min.[236]

An extensive evaluation of sampling variables confirmed no significant difference in collection efficiency between two different lots of charcoal and two different charcoal tube designs, MSA tubes containing 150 mg charcoal (Table 88) and USAF Environmental Health Laboratory tubes containing 600 mg charcoal (EHL).[237] However, the MSA tubes consistently exceeded the pressure drop specification suggested by NIOSH (average found was 35.6 ± 4.4 mmHg vs. 25.4 mmHg at 1 L/min flow rate). The solvent vapors were eluted with CS_2 and analyzed by gas chromatography, which employed a flame ionization detector and a 10% FFAP on a DCMS Chromosorb column.

Known calibration standards in air were prepared either by injecting the calculated volume of liquid phase from a micro syringe into a -100 mL glass syringe or in a dynamic system using a vapor saturator to provide the source of solvent vapor. The coefficient of variation for the syringe method was found to be 3 ± 2%.

Four samples were collected simultaneously in a four-place manifold. Flows were controlled with limiting orifices designed to provide a flow rate of 1 L/min at 584 mm Hg pressure drop (ΔP). Actual calibrated flow rates were as shown in Table 81.

A study conducted to determine the filtration efficiency of organic vapor charcoal tubes for diocetyl phthalate and latex aerosols disclosed collection in the range of 70 to 90%. However, only 20% of the particulate was deposited on the charcoal granules. Penetration through the tubes also was highly variable. The authors recommended the addition of a cassette pre-filter to retain the aerosols.[238]

g. Evaluation — Precision and Accuracy

In order to minimize the variability of results obtained with charcoal sampling,

NIOSH has recommended tentative standard dimensions for the charcoal tube, which are reproduced as follows:

The charcoal sampling tube used by the National Institute of Occupational Safety and Health, Cincinnati, Ohio, for the collection and concentration of solvent vapors has been modified to meet the requirements for a personal sampler. The packing size and configuration of the tube assure that the pressure drop across the tube will not exceed one inch of mercury when sampling is accomplished at a flow rate of 1 L/min. This feature allows battery-operated pumps to be used thus providing the advantage of complete portability, particularly necessary in personal-type sampling.

The tubes are fabricated of glass tubing 7 cm long with a 6-mm O.D. and a 4-mm I.D. The absorbent consists of two sections of 20/40 mesh-sized activated charcoal. In the tube a 100 mg adsorbent section is separated by a 2 mm portion of urethane foam from a 50 mg backup, blank adsorbent section. A 3 mm portion of urethane foam is placed between the 50 mg section and the outlet end of the tube. A plug of glass wool is placed in front of the 100 mg section at the inlet end of the tube. Both ends of the tube are immediately flame-sealed after being packed to restrict contamination of the charcoal prior to sampling (Figures 47 and 113).

This modified tube has also been used for atmospheric sampling of mercury vapor exposures and has potential use in sampling for contaminants other than solvent vapors and mercury vapor. The tube was developed in a continuing effort to assist industrial hygienists and other interested persons in carrying out inspection responsibilities and will be of particular interest to those involved in inspection activities under the recent Occupational Safety and Health Act of 1970 (PL 91-596).*

An additional tentative specification has been added to denote charcoal quality based on retention of the target material at a concentration of two times the TLV from a 10-L total sample volume. Under these conditions there should be less than 0.1 mg of the target material retained in the backup charcoal column.

Most currently available organic vapor charcoal tubes do meet the suggested standard. However, the user now is responsible for confirming the quality of the tubes before application to field monitoring. Assistance with the selection of the proper sampling tube will be found in Tables 64 and 82.

In addition to the small NIOSH personnel monitoring charcoal, two larger sizes, "large" and "jumbo" have been "standardized" by the trade (Table 83) for the collection of larger samples. These (organic, and mercury vapor) tubes are commercially available (see Table 88).[166,167,232,239]

4. Other Absorbents

The AIHA-ACGIH Joint Committee on Respirators in 1971 compiled a list of solid absorbents for the substances listed in the 1970 TLV table.[240] Although primarily intended as guidelines for use in the selection of respiratory system protective devices, many of the recommendations could be applied directly to air sampling. Therefore, the entire list is reproduced here with the permission of the copyright holder (Table 84).

A prototype contaminant sensor based on three sorbents — Poropak® Q, charcoal, and molecular sieve 5A — was developed by the Perking-Elmer Corporation for detection of contaminants during space flight.[241] The sorbents are contained in an accumulator cell for extraction of contaminants from air at room temperature. By heating the cell, the contaminants are desorbed into a mass spectrometer for identification and determination. A palladium-coated charcoal operated successfully as a CO detector. The lower range of detectable concentrations varied from 0.5 to 5 ppm. This system has sufficient advantages to justify adaptation to personnel monitoring.

The use of gas chromatograph (GC) column packing has been exploited to an increasingly greater extent since publication of the first edition of *Evaluation of Ambient Air Quality by Personnel Monitoring* in 1974. Reference to the NIOSH Manual of

* From Kupel, R. E. and White, L. D., *Am. Ind. Hyg. Assoc. J.*, 32, 456, 1971. With permission.

TABLE 82

Guide to Sampling, Analytical Procedure and Sample Tubes

Chemical with established procedure	Reference[a]	OSHA standard: TWA (ppm)[b]	OSHA standard: Ceiling (ppm)[c]	OSHA standard: Action level (ppm)[d]	Sample volume (L)[e]: Ceiling	Sample volume (L)[e]: TWA	Sample rate (mL/min): Ceiling	Sample rate (mL/min): TWA	Sample time: Ceiling (min)	Sample time: TWA (hr)	Desorption solvent (eluent)[f]	Analytical method[g]	GC column[h]	Sample tube and cat. no.
Acetic anhydride	K													
Acetone	A	1000		500	2	2	100	20	20	1.5	a	GC	V	Charcoal 226-01
Acetonitrile	K													
Acetylene tetrabromide	I	1		0.5	100	100	1000	200	100	8	k	GC	XI	Silica Gel 226-10
Allyl alcohol	E	2		1	2	10	200	20 (50)	10	8 (3.3)	e	GC	II	Charcoal 226-01
Allyl chloride	I	1		0.5	100	100	1000	200	100	8	m	GC	V	Charcoal 226-01
Alpha-methyl styrene	C		100		2	3	200	20	10	2.5	a	GC	II	Charcoal 226-01
Aniline	2	5		2.5	6	48	200	100	30	8	c	GC	III	Silica Gel 226-14
Arsine	J	0.05		0.025	2	10	200	20 (50)	10	8 (3.3)	q	AA	—	Charcoal 226-01
Benzene	6	10	25	5	2	10	200	20 (50)	10	8 (3.3)	a	GC	I	Charcoal 226-01
Benzyl chloride	H	1		0.5	2	10	200	20 (50)	10	8 (3.3)	a	GC	II	Charcoal 226-01
Bromoform	H	0.5		0.25	2	10	200	20 (50)	10	8 (3.3)	a	GC	II	Charcoal 226-01
Butadiene	G	1000		500	1	1	50	20	20	0.8	a	GC	I	Charcoal 226-01
2-Butanone	A	200		100	2	10	200	20 (50)	10	8 (3.3)	a	GC	II	Charcoal 226-01

2-Butoxy ethanol	F	50		25	2	10	200	20	(50)	10 8	(3.3)	h	GC	II	Charcoal 226-01
Butyl acetate	D	150		75	2	10	200	20	(50)	10 8	(3.3)	a	GC	VII	Charcoal 226-01
Butyl alcohol	E	100		50	2	10	200	20	(50)	10 8	(3.3)	s	GC	II	Charcoal 226-01
Butylamine	3	5		2.5	6	48	200	100		30 8		d	GC	IV	Silica Gel 226-14
Camphor	B	2		1.0	2	10	200	20	(50)	10 8	(3.3)	b	GC	II	Charcoal 226-01
Carbon tetra-chloride	J	10	25	5	5	15	1000	50		5 5		a	GC	I	Charcoal 226-01
Chlorodi-phenyl (54% chlorine)	I	0.5 mg/m³		0.25 mg/m³		100		1500		1.1		i	GC	IX	Filter 225-5-1
Chlorobenzene	I	75		37.5	10	10	200	20	(50)	50 8	(3.3)	a	GC	II	Charcoal 226-01
Chlorobromo-methane	H	200		100	2	5	200	20	(50)	10 4	(1.5)	a	GC	II	Charcoal 226-01
Chloroform	J		50		2	10	200	20	(50)	10 8	(3.3)	a	GC	I	Charcoal 226-01
Chlorinated camphene (toxaphene)	F	0.5 mg/m³				15		1000		0.25		i	GC	IX	Filter 225-5-1
Cresol (all iso-mers)	K														
Cumene	C	50		25	2	10	200	20	(50)	10 8	(3.3)	a	GC	II	Charcoal 226-01
Cyclohexane	C	300		150	2	2.5	200	20		10 2		a	GC	V	Charcoal 226-01
Cyclohexanol	E	50		25	2	10	200	20	(50)	10 8	(3.3)	e	GC	II	Charcoal 226-01
Cyclohexanone	A	50		25	2	10	200	20	(50)	10 8	(3.3)	a	GC	II	Charcoal 226-01
Cyclohexene	F	300		150	2	5	200	20	(50)	10 4	(15)	a	GC	V	Charcoal 226-01
Diacetone alco-hol	E	50		25	2	10	200	20	(50)	10 8	(3.3)	e	GC	II	Charcoal 226-01

TABLE 82 (continued)

Guide to Sampling, Analytical Procedure and Sample Tubes

Chemical with established procedure	Reference[a]	OSHA standard: TWA (ppm)[b]	OSHA standard: Ceiling (ppm)[c]	OSHA standard: Action level (ppm)[d]	Sample volume (L)[e]: Ceiling	Sample volume (L)[e]: TWA	Sample rate (mL/min): Ceiling	Sample rate (mL/min): TWA	Sample time: Ceiling (min)	Sample time: TWA (hr)	Desorption solvent (eluent)[f]	Analytical method[g]	GC column[h]	Sample tube and cat. no.
Diazomethane	J	0.2		0.1		10		200		0.8	a	GC	XII	Coated Resin 226-23
Dibutyl phthalate	D	5 mg/m³		2.5 mg/m³		30		1000		0.5	a	GC	VIII	Filter 225-5-1
1,1-Dichloroethane	I	100		50	2	10	200	20 (50)	10	8 (3.3)	a	GC	II	Charcoal 226-01
1,2-Dichloroethylene	H	200		100	2	3	200	20 (50)	10	2.5 (1.0)	a	GC	II	Charcoal 226-01
Diethylamine	J	25		12.5	50	50	1000	100	50	8	r	GC	XIII	Silica Gel 226-10
Di-2-ethylhexyl phthalate	D	5 mg/m³		2.5 mg/m³		30		1000		0.5	a	GC	VIII	Filter 225-5-1
Difluorodibromomethane	H	100		50	2	10	200	20 (50)	10	8 (3.3)	n	GC	II	Charcoal 226-01
Diisopropylamine	3	5		2.5	6	48	200	100	30	8	d	GC	IV	Silica Gel 226-14
Dimethylamine	J	10		5	50	50	1000	100	50	8	r	GC	XIII	Silica Gel 226-14
Dioxane	G	100		50	2	10	200	20 (50)	10	8 (3.3)	a	GC	I	Charcoal 226-01
Dimethyl analine	K													
Dipropylene glycol methyl ether	F	100		50	2	10	200	20 (50)	10	8 (3.3)	a	GC	II	Charcoal 226-01

Epichlorohydrin	I	5		2.5	20	20	200	50		100	8		a	GC	II	Charcoal 226-01
2-Ethoxyethyl acetate	D	100		50	2	10	200	20	(50)	10	8	(3.3)	a	GC	VII	Charcoal 226-01
Ethyl acetate	E	400		200	2	6	200	20	(50)	10	5	(2)	a	GC	VII	Charcoal 226-01
Ethyl acrylate	D	25		12.5	2	10	200	20	(50)	10	8	(3.3)	a	GC	VII	Charcoal 226-01
Ethyl alcohol	E	1000		500				50		20	0.33		f	GC	II	Charcoal 226-01
Ethylamine	3	10		5	6	48	200	100		30	8		d	GC	IV	Silica Gel 226-14
Ethyl benzene	C	100		50	2	10	200	20	(50)	10	8	(3.3)	a	GC	II	Charcoal 226-01
Ethyl bromide	H	200		100	2	4	200	20	(50)	10	3.3	(1.3)	n	GC	II	Charcoal 226-01
Ethyl butyl ketone	B	50		25	2	10	200	20	(50)	10	8	(3.3)	b	GC	II	Charcoal 226-01
Ethylene chlorohydrin	H	5		2.5	2	10	200	20	(50)	10	8	(3.3)	e	GC	II	Charcoal 226-01
Ethylene dibromide	H	20	30	10	2	10	200	200		10	0.8		a	GC	II	Charcoal 226-01
Ethylene dichloride	J	50	100	25	3	10	200	20	(50)	15	8	(3.3)	a	GC	XV	Charcoal 226-01
Ethyl ether	F	400		200	2	3	200	20	(50)	10	2.5	(1)	j	GC	V	Charcoal 226-01
Ethyl formate	D	100		50	2	10	200	20	(50)	10	8	(3.3)	a	GC	I	Charcoal 226-01
Glycidol	F	50		25	2	10	200	20	(50)	10	8	(3.3)	k	GC	II	Charcoal 226-01
Heptane	G	500		250	2	4	200	20	(50)	10	3.3	(1.3)	a	GC	XV	Charcoal 226-01
Hexachloronaphthalene	H	0.2 mg/m³				30		1000			0.5		l	GC	VIII	Filter 225-5-1
Hexane	G	500		250	2	4	200	20	(50)	10	3.3	(1.3)	a	GC	I	Charcoal 226-01

TABLE 82 (continued)

Guide to Sampling, Analytical Procedure and Sample Tubes

Chemical with established procedure	Reference[a]	OSHA standard			Sample volume (L)[e]		Sample rate (mL/min)		Sample time		Desorption solvent (eluent)[f]	Analytical method[g]	GC column[h]	Sample tube and cat. no.
		TWA (ppm)[b]	Ceiling (ppm)[c]	Action level (ppm)[d]	Ceiling	TWA	Ceiling	TWA	Ceiling (min)	TWA (hr)				
Hoxone	A	100		50	2	10	200	20 (50)	10	8 (3.3)	a	GC	II	Charcoal 226-01
Isoamyl acetate	D	100		50	2	10	200	20 (50)	10	8 (3.3)	a	GC	VII	Charcoal 226-01
Isoamyl alcohol	E	100		50	2	10	200	20 (50)	10	8 (3.3)	e	GC	II	Charcoal 226-01
Isobutyl acetate	D	150		75	2	10	200	20 (50)	10	8 (3.3)	a	GC	VII	Charcoal 226-01
Isobutyl alcohol	E	100		50	2	10	200	20 (50)	10	8 (3.3)	s	GC	II	Charcoal 226-01
Isopropyl acetate	E	250		125	2	9	200	20 (50)	10	7 (3)	a	GC	VII	Charcoal 226-01
Isopropyl alcohol	E	400		200	2	3	200	20 (50)	10	2.5 (1)	f	GC	II	Charcoal 226-01
Isopropylamine	3	5		2.5	6	48	200	100	30	8	d	GC	IV	Silica Gel 226-14
Isopropyl gylcidyl ether	F	50		25	2	10	200	20 (50)	10	8 (3.3)	a	GC	II	Charcoal 226-01
Ketone	G	0.9 mg/m^3				50		1000		0.8	—	Color	—	Impinger 225-9
Mercury	4	0.1 mg/m^3			10		1000		10			AA		Charcoal 226-17
Mesityl oxide	B	25		12.5	2	10	200	20 (50)		8 (3.3)	b	GC	II	Charcoal 226-01
Methyl acetate	D	200		100	2	7	200	20 (50)	10	5.5 (2.3)	a	GC	I	Charcoal 226-01
Methyl acrylate	D	10		5	2	5	200	20 (50)	10	4 (1.5)	a	GC	VII	Charcoal 226-01

Methyl alcohol	E	200		100	2	5	200	20	(50)	10	4	(1.5)	g	GC	II	Silica Gel 226-15
Methylal (di-methoxyme-thane)	F	1000		500	2	2	200	20	(50)	10	1.5	(0.6)	l	GC	V	Charcoal 226-01
Methylamine	3	10		5	6	48	200	100		30	8		d	GC	IV	Silica Gel 226-14
Methyl cello-solve	F	25		12.5	2	10	200	20	(50)	10	8	(3.3)	h	GC	II	Charcoal 226-01
Methyl cello-solve acetate	D	25		12.5	2	10	200	20	(50)	10	8	(3.3)	a	GC	VII	Charcoal 226-01
Methyl chloro-form	J	350		175	2	6	200	20	(50)	10	5	(2)	a	GC	XV	Charcoal 226-01
Methyl cyclo-hexane	G	500		250	2	4	200	20	(50)	10	3.3	(1.3)	a	GC	I	Charcoal 226-01
Methylene chloride	J	500	1000	250	1	2.2	200	20	(50)	5	1.8	(0.7)	a	GC	I	Charcoal 226-01
Methyl ethyl ketone (see 2-butanone)																
5-Methyl-3-heptanone	B	25		12.5	2	10	200	20	(50)	10	8	(3.3)	b	GC	II	Charcoal 226-01
Methyl iodide	H	5		2.5	2	10	200	20	(50)	10	8	(3.3)	o	GC	XV	Charcoal 226-01
methyl-iso-amyl-acetate	D	50		25	2	10	200	20	(50)	10	8	(3.3)	a	GC	VII	Charcoal 226-01
Methyl isobutyl carbinol	E	25		12.5	2	10	200	20	(50)	10	8	(3.3)	e	GC	II	Charcoal 226-01
Methyl *n*-amyl-ketone	B	100		50	2	10	200	20	(50)	10	8	(3.3)	b	GC	II	Charcoal 226-01
N-amyl acetate	E	100		50	2	10	200	20	(50)	10	8	(3.3)	a	GC	VII	Charcoal 226-01
Naphtha, coal tar	G	100		50	2	10	200	20	(50)	10	8	(3.3)	a	GC	XVI	Charcoal 226-01
N-butyl glyci-dyl ether	F	50		25	2	10	200	20	(50)	10	8	(3.3)	a	GC	II	Charcoal 226-01

TABLE 82 (continued)

Guide to Sampling, Analytical Procedure and Sample Tubes

Chemical with established procedure	Ref-er-ence[a]	OSHA standard			Sample volume (L)[e]		Sample rate (mL/min)		Sample time		Desorption solvent (eluent)[f]	Analytical method[g]	GC column[h]	Sample tube and cat. no.
		TWA (ppm)[b]	Ceiling (ppm)[c]	Action level (ppm)[d]	Ceiling	TWA	Ceiling	TWA	Ceiling (min)	TWA (hr)				
N,n-dimethylaniline	K													
N-propyl acetate	D	200		100	2	10	200	20 (50)	10	8 (3.3)	a	GC	VII	Charcoal 226-01
O-anisidine	2	0.1		0.05	240	480	1000	1000	4 hrs	8	c	GC	III	Silica Gel 226-14
Octachloronaphthalene	G	0.1 mg/m³				30		1000		0.5	1	GC	VIII	Filter 225-5-1
Octane	G	500		250	2	4	200	20 (50)	10	3.3 (1.3)	a	GC	I	Charcoal 226-01
O-dichlorobenzene	J			50	3	3	200	20	15	2.5	a	GC	XV	Charcoal 226-01
O-toluidine	K													
P-anisidine	2	0.1		0.05	240	480	1000	1000	4 hrs	8	c	GC	III	Silica Gel 226-14
Pentane	G	1000		500	2	2	50	20	40	1.5	a	GC	I	Charcoal 226-01
2-Pentanone	A	200		100	2	10	200	20 (50)	10	8 (3.3)	a	GC	II	Charcoal 226-01
Petroleum distillate	G	500		250	2	4	200	20 (50)	10	3.3 (1.3)	a	GC	XV	Charcoal 226-01
Phenol	K													
Phenyl ether (vapor)	F	1		0.5	2	10	200	20 (50)	10	8 (3.3)	a	GC	II	Charcoal 226-01
Phenyl ether-biphenyl	F	1		0.5	2	10	200	20 (50)	10	8 (3.3)	m	GC	X	Silica Gel 226-10
Phenyl glycidyl ether	F	10		5	2	10	200	20 (50)	10	8 (3.3)	a	GC	II	Charcoal 226-01

Phenyl hydrazine	K															
P-nitroaniline	2	1		0.5	90	96	1000	200		90	8		c	GC	III	Silica Gel 226-14
Propyl alcohol	E	200		100	2	10	200	20	(50)	10	8	(3.3)	s	GC	II	Charcoal 226-01
Propylene dichloride	G	75		37.5	2	10	200	20	(50)	10	8	(3.3)	a	GC	II	Charcoal 226-01
Propylene oxide	F	100		50	2	5	200	20	(50)	10	4	(1.5)	a	GC	V	Charcoal 226-01
P-tert-butyltoluene	C	10		5	2	10	200	20	(50)	10	8	(3.3)	a	GC	II	Charcoal 226-01
Pyridine	K															
Sec-amyl acetate	D	125		62.5	2	10	200	20	(50)	10	8	(3.3)	a	GC	VII	Charcoal 226-01
Sec-butyl acetate	D	200		100	2	10	200	20	(50)	10	8	(3.3)	a	GC	VII	Charcoal 226-01
Sec-butyl alcohol	E	150		75	2	10	200	20	(50)	10	8	(3.3)	s	GC	II	Charcoal 226-01
Stoddard solvent	G	500		250	2	3	200	20	(50)	10	2.5	(1)	a	GC	XVI	Charcoal 226-01
Styrene	C	100	200	50	2	10	200	20	(50)	10	8	(3.3)	a	GC	II	Charcoal 226-01
Terphenyls (ortho-terphenyl)	C		1		15		1000			15			a	GC	VI	Filter 225-5-1
Tert-butyl acetate	D	200		100	2	10	200	20	(50)	10	8	(3.3)	a	GC	VII	Charcoal 226-01
Tert-butyl alcohol	E	100		50	2	10	200	20	(50)	10	8	(3.3)	f	GC	II	Charcoal 226-01
1,1,2,2-Tetrachloro-1,2,-difluoroethane	I	500		250	2	2	50	20		40	1.6		a	GC	II	Charcoal 226-01
1,1,1,2-Tetrachloro-2,2-difluoroethane	I	500		250	2	2	35	20		55	1.6		a	GC	II	Charcoal 226-01

TABLE 82 (continued)

Guide to Sampling, Analytical Procedure and Sample Tubes

Chemical with established procedure	Reference[a]	OSHA standard			Sample volume (L)[e]		Sample rate (mL/min)		Sample time		Desorption solvent (eluent)[f]	Analytical method[g]	GC column[h]	Sample tube and cat. no.
		TWA (ppm)[b]	Ceiling (ppm)[c]	Action level (ppm)[d]	Ceiling	TWA	Ceiling	TWA	Ceiling (min)	TWA (hr)				
1,1,2,2-Tetrachloroethane	I	5		2.5	10	10	200	20 (50)	50	8 (3.3)	a	GC	II	Charcoal 226-01
Tetrachloroethylene	J	100	200	50	3	10	200	20 (50)	15	8 (3.3)	a	GC	XV	Charcoal 226-01
Tetrahydrofuran	F	200		100	2	9	200	20 (50)	10	7 (3)	a	GC	V	Charcoal 226-01
Toluene	6	200	300	100	2	10	200	20 (50)	10	8 (3.3)	a	GC	I	Charcoal 226-01
1,1,2-Trichloro-1,2,2-trifluoroethane	I	1000		500	1.5	1.5	50	20	30	1.25	a	GC	XIV	Charcoal 226-01
1,1,2-Trichloroethane	J	10		5	2	10	200	20 (50)	10	8 (3.3)	a	GC	XV	Charcoal 226-01
Trichloroethylene	J	100	200	50	3	10	200	20 (50)	15	8 (3.3)	a	GC	XV	Charcoal 226-01
1,2,3-Trichloropropane	I	50		25	2	10	200	20 (50)	10	8 (3.3)	a	GC	II	Charcoal 226-01
Triethylamine	3	25		12.5	6	48	200	100	30	8	d	GC	IV	Silica Gel 226-14
Trifluoromonobromomethane	I	100		500	1	1	50	20	20	0.8	p	GC	V	Charcoal 226-01 and 226-09
Turpentine	G	100		50	2	10	200	20 (50)	10	8 (3.3)	a	GC	XVI	Charcoal 226-01
Vinyl chloride	5										a	GC	I	Charcoal 226-01

Vinyl toluene	C	100	50	2	10	200	20 (50)	10 8 (3.3)	a	GC	II	Charcoal 226-01
Xylene	6	100	150 50	2	10	200	20 (50)	10 8 (3.3)	a	GC	I	Charcoal 22-01
2-4-Xylidine	K											
Acrolein		0.1		6		200		0.5	f			Alumina 226-11
Allyl glycidyl ether					200	20	10	8 (3.3)	a	GC	II	Charcoal 226-01
Alpha pinene						50						PP Tube 226-26
Carbon disulfide		6		10	200	20(50)	10	8 (3.3)	m	GC	XVII	Charcoal 226-01
Catechol		1										PP Tube 226-26
Chloroprene		25	1	1	50	20	20	0.8	a	GC	I	Charcoal 226-01
Chlorotoluene		50	10	10	200	20(50)	50	8 (3.3)	a	GC	II	Charcoal 226-01
Cyclohexanone		50	2	10	200	20(50)	10	8 (3.3)	a	GC	II	Charcoal 226-01
Decane		400	2	4	200	20(50)	10	3.3 (1.3)	a	GC	I	Charcoal 226-01
Dichlorobenzene		50	10	10	200	20(50)	50	8 (3.3)	a	GC	II	Charcoal 226-01
Diisobutyl ketone		50	2	10	200	20(50)	10	8 (3.3)	a	GC	II	Charcoal 226-01
Ethyl acrylate		25	2	5	200	20(50)	10	4 (1.5)	a	GC	VII	Charcoal 226-01
Ethyl benzene		100	2	10	200	20(50)	10	8 (3.3)	a	GC	I	Charcoal 226-01
Ethyl chloride		1000	2	10	200	20(50)	10	8 (3.3)	a	GC	II	Charcoal 226-01
Ethylene glycol		100										PP Tube 226-26
Ethylene oxide		50	2	5	200	20(50)	10	4 (1.5)	a	GC	V	Charcoal 226-01
Ethyl mercaptan		0.5										
Formaldehyde		2		6		200		0.5	f	Color	—	Alumina 226-11
Fufural												Alumina 226-11
Furfuryl alcohol		50	2	10	200	20(50)	10	8 (3.3)	s	GC	II	Charcoal 226-01
Isooctane		400	2	4	200	20(50)	10	3.3 (1.3)	a	GC	I	Charcoal 226-01
Isophorone			2	10	200	20(50)	10	8 (3.3)	b	GC	II	Charcoal 226-01
2,6-Lutidine												
Methyl bromide		15	2	10	200	20(50)	10	8 (3.3)	a	GC	II	Charcoal 226-01
Methyl isobutyl ketone			2	10	200	20(50)	10	8 (3.3)	b	GC	II	Charcoal 226-01
Methyl methacrylate		100	2	10	200	20(50)	10	8 (3.3)		GC		Charcoal 226-01
Methyl *n*-butyl ketone			2	10	200	20(50)	10	8 (3.3)	b	GC	II	Charcoal 226-01
Naphthalene		10										NP Tube 226-34
Nonane		400	2	4	200	20(50)	10	3.3 (1.3)	a	GC	I	Charcoal 226-01
3-Octane			2	10	200	20(50)	10	8 (3.3)	b	GC	II	Charcoal 226-01
P-dichlorobenzene		50	3	3	200	20	15	2.5	a	GC	XV	Charcoal 226-01
Propyl acetate		200	2	10	200	20(50)	10	8 (3.3)	a	GC	VII	Charcoal 226-01
Resorcinol												PP Tube 226-26

TABLE 82 (continued)

Guide to Sampling, Analytical Procedure and Sample Tubes

[a] References: ANIOSH Analytical Methods for Set A Standards Completion Program. NTI Service No. PB-245-850,
BNIOSH Analytical Methods for Set B Standards Completion Program. NTI Service No. PB-245-851,
CNIOSH Analytical Methods for Set C Standards Completion Program. NTI Service No. PB-245-852,
DNIOSH Analytical Methods for Set D Standards Completion Program. NTI Service No. PB-245-935,
ENIOSH Analytical Methods for Set E Standards Completion Program. NTI Service No. PB-246-148,
FNIOSH Analytical Methods for Set F Standards Completion Program. NTI Service No. PB-246-149,
GNIOSH Analytical Methods for Set G Standards Completion Program. NTI Service No. PB-246-150,
HNIOSH Analytical Methods for Set H Standards Completion Program. NTI Service No. PB-246-151,
INIOSH Analytical Methods for Set I Standards Completion Program. NTI Service No. PB-246-152,
JNIOSH Analytical Methods for Set J Standards Completion Program. NTI Service No. PB-246-153,
LNIOSH Analytical Methods for Set L Standards Completion Program. NTI Service No. PB-246-155,
2Wood, G. O. and Anderson, R. G., Personal air sampling for vapors of aniline compounds, *Am. Ind. Hyg. Assoc. J.*, 36, 538, 1975,
3Wood, G. O. and Anderson, R. G., Sampling and Analysis of Aliphatic Amine Vapors. Presented at the American Industrial Hygiene Conference, June, 1975,
4Moffitt, Jr., A. E. and Kupel, R. E., A rapid method employing impregnated charcoal and atomic absorption spectrophotometry for the determination of mercury in atmospheric, biological, and aquatic samples, *At. Absorpt. Newsl.*, Vol. 9, No. 6, November—December 1970,
5Wood, G. O. and Anderson, R. G., Air Sampling of Formaldehyde With a Solid Sorbent Tube, Presented at the American Industrial Hygiene Conference, June, 1975,
6McCammon, Jr., C. S., Quinn, P. M., and Kupel, R. E., *Am. Ind. Hyg. Assoc. J.*, 36, 618, 1975,

[b] TWA (ppm): time weighted average is the average contamination level over an 8-hr period in one million parts of air to which a worker can be exposed continuously for 8 hrs without damage to his health. Government regulations explicitly require that the TWA be kept below this level and monitored;

[c] ceiling (ppm): the ceiling level is the maximum concentration in one million parts of air to which a worker can be exposed for short periods. Note that this is above the TWA. The TWA must still not be exceeded so that there must be a corresponding period below TWA to compensate for high exposure periods. Where the chart is blank, no ceiling has been established;

[d] action level: this is a level that is currently being considered. It is the level at which a monitoring program must be established. An action level of approximately ½ TWA is being considered for most compounds;

[e] sample volume is a suitable sample volume that will give accurate analysis. Normally this volume can be taken for concentrations up to twice the TWA without breakthrough. Other volumes and sampling rates can sometimes be used. Check literature reference for limitations on other volumes and rates;

[f] desorption solvents (eluent):

a carbon disulfide — cat. no. 226-05-1a,
b carbon disulfide containing 1% methanol —226-05-1b,
c ethyl alcohol,
d 1 N sulfuric acid and regenerated for gas chromatography analysis by making aliquote of each eluent basic with equal volumes of 1.1 N sodium hydroxide — 226-05-1d,
e carbon disulfide containing 5% 2-propanol — 226-05-1e,
f carbon disulfide containing 1% 2-butanol — 226-05-1f,
g distilled water,
h 5% methanol in methylene chloride — 226-05-1h,
i petroleum ether — 226-05-1i,
j ethyl acetate — 226-05-1j,
k tetrahydrofuran — 226-05-1k,
l hexane — 226-05-1l,
m benzene — 226-05-1m,
n isopropyl alcohol — 226-05-1n,
o toluene — 226-05-1o,
p methylene chloride — 226-05-1p,
q nitric acid, 0.01 m — 226-05-1q,
r 0.2 N H_2SO in 10% methanol — 226-05-1r,
s carbon disulfide containing 1% 2-propanol — 226-05-1s,
t water containing 1% methanol 226-05-1t;

[g] analytical method: in most cases, GC (gas chromatography) is used, except where noted. Where GC is used, the desorption solvent (eluent) and column are listed in [h];

[h] gas chromatography columns:

I 20-ft. × 1/8 in. stainless steel packed with 10% FFAP on Chromosorb — cat. no. 226-08-1,
II 10-ft. × 1/8 in. stainless steel packed with 10% FFAP on Chromosorb — 226-08-2,
III (1.22 mm × 3 mm O.D.) stainless steel packed with 10% Silicone OV-25 on Supelcoport — 226-08-3,
IV 6-ft. × ¼ in. glass packed with Chromosorb column and Ascarite precolumn — 226-08-4,
V 4-ft. × ¼ in. stainless steel packed with Porapak Q — 226-08-5,
VI 10-ft. × 1/8 in. stainless steel packed with 5% SE-30 on Gas Chrom — 226-08-6,
VII 10-ft. × 1/8 in. stainless steel packed with 5% FFAP on Supelcoport — 226-08-7,
VIII 6-ft. × 1/8 in. stainless steel packed with 5% OV-101 on Supelcoport — 226-08-8,
IX 6-ft. × ¼ in. glass packed with 3% OV-1 on Gas Chrom — 226-08-9,
X 10-ft. × 1/8 in. stainless steel packed with 5% OV-17 on Chromosorb — 226-08-10,

TABLE 82 (continued)

Guide to Sampling, Analytical Procedure and Sample Tubes

XI 3-ft. × ¼ in. glass packed 5% SE-30 on Gas Chrom — 226-08-11,
XII 10-ft. × 1/8 in. stainless steel packed with 5% SP — 1000 stationary phase on Supelcoport, preceded by a short 6 × 1/8 in. stainless steel precolumn of Carbowax — 226-08-12,
XIII 6-ft. × ¼ in. glass packed with 4% Carbowax — 20 m + 0.8% KOH on Carbosieve B — 226-08-13,
XIV 6-ft. × ¼ in. stainless steel packed with Porapak — 226-08-14,
XV 10-ft. × 1/8 in. stainless steel packed with 10% OV-101 on Supelcoport — 226-08-15,
XVI 6-ft. × 1/8 in. stainless steel packed with 1.5% OV-101 on Chromosorb — 226-08-16,
XVII 6-ft. × ¼ in. O.D. glass column packed with 5% OV-17 on Gas Chrom — 226-08-17;

[f] TLV or threshold limit value is the concentration in parts per million of air that has been established by the American Conference of Governmental Industrial Hygienists' Association as a safe working level continuously for eight hours a day.

Courtesy of SKC, P.O. Box 8538, Pittsburgh, Pa.

TABLE 83

Available Adsorbent Tube Sizes

Adsorbent packing material	Standard size (mg)		Large size (mg)		Jumbo size (mg)		Special
	1st Section	2nd Section	1st Section	2nd Section	1st Section	2nd Section	
Alumina	100	50	400	200	800	200	Aldehydes
Coated resin	100	50	—	—	—	—	
Charcoal	100	50	400	200	800	200	
Mercury collection	100	50	—	—	—	—	
Molecular sieve	100	50	400	200	800	200	3, 4, 10, and 1 A pores
Polar partition	100	50	—	—	—	—	Silicone coating
Porous aromatic polymer	100	50	—	—	—	—	Polar molecules
Non-polar partition	100	50	—	—	—	—	Non-polar molecules
Silica gel	—	—	—	—	—	—	Three sections
	130	65	520	260	1000	260	
Custom made	—	—	—	—	—	—	Customer's choice of packing

Courtesy of SKC, P.O. Box 8538, Pittsburgh, Pa.

TABLE 84

Sorbents for Contaminants Listed in ACGIH 1970 Threshold Limit Values

List of Contaminants with Sorbent Code[a]

Substance	Sorbent	Substance	Sorbent
Abate	9	*tert*-Butyl alcohol	1
Acetaldehyde	1	Butylamine — skin	7,6,4
Acetic acid	1,2	*tert*-Butyl chromate (as CrO_3) — skin	8
Acetic anhydride	3,2	*n*-Butyl glycidyl ether (BGE)	1
Acetone	1	Butyl mercaptan	1
Acetonitrile	3,1	*p-tert*-Butyltoluene	1
Acetylene dichloride, see 1,2-Dichloroethylene		Cadmium (metal dust and soluble salts)	9,6
Acetylene tetrabromide	1	Cadmium oxide fume	8
Acrolein	1	Calcium arsenate	9
Acrylamide — skin	1,6	Calcium oxide	9
Acrylonitrile — skin	3,1	Camphor	9 + 1
Aldrin — skin	8 + 1	Carbaryl (Sevin®)	9 + 1
Allyl alcohol — skin	1	Carbon black	9
Allyl chloride	1,3	Carbon dioxide	2
Allyl glycidyl ether (AGE)	1	Carbon disulfide — skin	1
Allyl propyl disulfide	3	Carbon monoxide	5
2-Aminoethanol, see Ethanolamine		Carbon tetrachloride — skin	1
2-Aminopyridine	1,7	Chlordan — skin	9 + 1
Ammonia	7,6,4	Chlorinated camphene — skin	1
Ammonium sulfamate (ammate)	9	Chlorinated diphenyl oxide	1
n-Amyl acetate	1	Chlorine	3,6
sec-Amyl acetate	1	Chlorine dioxide	2,7
Aniline — skin	1,7	Chlorine trifluoride	2,7
Anisidine (*o,p*-isomers) — skin	9 + 1	Chloroacetaldehyde	1
Antimony and compounds (as Sb)	9	Chloroacetophenone (phenacyl chloride)	1
ANTU® (alpha naphthyl thioura)	9 + 1	Chlorobenzene (monochlorobenzene)	1
Arsenic and compounds (as As)	6 + 9	*O*-chlorobenzylidene malononitrile (OCBM)	3,1
Arsine	6,7	Chlorobromomethane	1
Azinphosmethyl — skin	9 + 1	2-Chloro-1,3-butadiene, see Chloroprene	
Barium (soluble compounds)	9 + 6	Chlorodiphenyl (42% chlorine) — skin	1,3
Benzene (benzol) — skin	1	Chlorodiphenyl (54% chlorine) — skin	1,3
Benzidine — skin	8 + 1	1-Chloro-2,3-epoxypropane, see Epichlorohydrin	
p-Benzoquinone, see Quinone		2-Chloroethanol, see Ethylene chlorohydrin	
Benzoyl peroxide	9 + 1	Chloroethylene, see Vinyl chloride	
Benzyl chloride	1,3	Chloroform (trichloromethane)	1
Beryllium	8	1-Chloro-1-nitropropane	1
Biphenyl, see Diphenyl		Chloropicrin	1
Boron oxide	9	Chloroprene (2-chloro-1,3-butadiene) — skin	1
Boron tribromide	5 + 3	Chromic acid and chromates (as CrO_3)	8 + 2
Boron trifluoride	5 + 3	Chromium, soluble chromic or chromous salts (as Cr)	9
Bromine	3	Metal and insoluble salts	9
Bromine pentafluoride	2,7	Coal tar pitch volatiles (benzene soluble fraction — anthracene,	9 + 1
Bromoform — skin	1		
Butadiene (1,3-butadiene)	1		
Butanethiol, see Butyl mercaptan			
2-Butanone	1		
2-Butoxy ethanol (Butyl Cellosolve®) — skin	1		
Butyl acetate (*n*-Butyl acetate)	1		
n-Butyl acetate	1		
sec-Butyl acetate	1		
tert-Butyl acetate	1		
Butyl alcohol	1		
sec-Butyl alcohol	1		

TABLE 84 (continued)

Sorbents for Contaminants Listed in ACGIH 1970 Threshold Limit Values
List of Contaminants with Sorbent Code[a]

Substance	Sorbent
BaP, phenanthrene, acridine, chrysene, pyridine)	
Cobalt, metal fume and dust	8
Copper fume	8
Dusts and mists	9
Cotton dust (raw)	9
Crag herbicide	9 + 1
Cresol (all isomers) — skin	1
Crotonaldehyde	1
Cumene — skin	1
Cyanide (as CN) — skin	9 + 2
Cyanogen	3
Cyclohexane	1
Cyclohexanol	1
Cyclohexanone	1
Cyclohexene	1
Cyclopentadiene	1
2,4-D	9 + 1
DDT — skin	9 + 1
DDVP — skin	9 + 1
Decarborane — skin	9 + 7
Demeton — skin	8 + 1
Diacetone alcohol (4-methyl-2-pentanone)	1
1,2-Diaminoethane (diazomethane), see Ethylenediamine	
Diborane	7
1,2-Dibromoethane (ethylene dibromide) — skin	1
Dibutyl phosphate	8 + 1
Dibutyl phthalate	1
Dichloroacetylene	3,7
o-Dichlorobenzene	1
p-Dichlorobenzene	1
Dichlorodifluoromethane	1
1,3-Dichloro-5-dimethylhydantoin	3
1,1-Dichloroethane	1
1,2-Dichloroethane	1
1,2-Dichloroethylene	1
Dichloroethyl ether — skin	1
Dichloromethane, see Methylene chloride	
Dichloromonofluoromethane	1
1,1-Dichloro-1-nitroethane	3
1,2-Dichloropropane, see Propylenedichloride	
Dichlorotetrafluoroethane	1
Dieldrin — skin	9 + 1
Diethylamine	7,6,4
Diethylaminoethanol — skin	7
Diethylenetriamine — skin	7,6,4
Diethylether, see Ethyl ether	
Difluorodibromomethane	1
Diglycidyl ether (DGE)	1
Dihydroxybenzene, see Hydroquinone	
Diisobutyl ketone	1
Diisopropylamine — skin	7,6,4
Dimethoxymethane, see Methylal	
Dimethyl acetamide — skin	1
Dimethylamine	7,6,4
Dimethylaminobenzene, see Xylidene	
Dimethylaniline (*N*-dimethylaniline) — skin	7
Dimethylbenzene, see Xylene	1
Dimethyl-1,2-dibromo-2,2-dichloroethyl phosphate (Dibrom)	9 + 1
Dimethylformamide — skin	1
2,6-Dimethylheptanone, see Diisobutyl ketone	
1,1-Dimethylhydrazine — skin	7,6,4
Dimethyl phthalate	1
Dimethyl sulfate — skin	1,3
Dinitrobenzene (all isomers) — skin	9 + 3
Dinitro-*o*-cresol — skin	9 + 3
Dinitrotoluene — skin	9 + 3
Di-*sec*-octyl phthalate [Di(2-ethylhexyl) phthalate]	1 + 8,9
Dioxane (Diethylene dioxide) — skin	1
Diphenyl	1
Diphenylamine	9 + 1.7
Diphenylmethane diisocyanate, see Methylenbisphenyl isocyanate (MDI)	
Dipropylene glycol methylether — skin	1
2,6-Dimethylheptanone, see Diisobutyl ketone	
1,1-Dimethylhydrazine — skin	7,6,4
Dimethyl phthalate	1
Dimethyl sulfate — skin	1,3
Dinitrobenzene (all isomers) — skin	9 + 3
Dinitro-*o*-cresol — skin	9 + 3
Dinitrotoluene — skin	9 + 3
Di-*sec*-octyl phthalate [Di(2-ethylhexyl) phthalate]	1 + 8,9
Dioxane (Diethylene dioxide) — skin	1
Diphenyl	1
Diphenylamine	9 + 1,7
Diphenylmethane diisocyanate, see Methylenebisphenyl isocyanate (MDI)	
Dipropylene glycol methylether — skin	1
Endosulfan (Thiodan®)	8 + 1
Endrin — skin	8 + 1
Epichlorohydrin — skin	1
EPN — skin	9 + 1
1,2-Epoxypropane, see Propylene oxide	
2,3-Epoxy-1-propanol, see Glycidol	

TABLE 84 (continued)
Sorbents for Contaminants Listed in ACGIH 1970 Threshold Limit Values
List of Contaminants with Sorbent Code[a]

Substance	Sorbent
Ethanethiol, see Ethyl mercaptan	
Ethanolamine	7,6,4
2-Ethoxyethanol — skin	1
2-Ethoxyethyl acetate (Cellosolve® acetate) — skin	1
Ethyl acetate	1
Ethyl acrylate — skin	1
Ethyl alcohol (Ethanol)	1
Ethylamine	7,6,4
Ethyl *sec*-amyl ketone (5-methyl-3-heptanone)	1
Ethyl benzene	1
Ethyl bromide	1
Ethyl butyl ketone (3-heptanone)	1
Ethyl chloride	1
Ethyl ether	1
Ethyl formate	1
Ethyl mercaptan	1
Ethyl silicate	1
Ethylene chlorohydrin — skin	1
Ethylenediamine	7,6,4
Ethylene dibromide, see 1,2-Dibromoethane	
Ethylene dichloride, see 1,2-Dichloroethane	
Ethylene glycol dinitrate and/or nitroglycerin — skin	1
Ethylene glycol monomethyl ether acetate, see Methyl Cellosolve® acetate	
Ethyleneimine — skin	7,6,4
Ethylene oxide	1
Ethylidene chloride, see 1,1-Dichloroethane	
N-ethylmorpholine — skin	1
Ferbam	9 + 3
Ferrovanadium dust	9
Fluoride (as F)	9 + 3
Fluorine	3,2
Fluorotrichloromethane	1
Formaldehyde	1
Formic acid	1
Furfural — skin	1
Furfuryl alcohol	1
Gasoline	1
Glycidol (2,3-epoxy-1-propanol)	1
Glycol monoethylether, see 2-Ethoxyethanol	
Guthion®, see Azinphos methyl	
Hafnium	9
Heptachlor — skin	9
Heptane (*n*-Heptane)	1
Hexachloroethane — skin	1
Hexachloronaphthalene — skin	9 + 1
Hexane (*n*-Hexane)	1
2-Hexanone	1
Hexone	1
sec-Hexyl acetate	1
Hydrazine — skin	7,6,4
Hydrogen bromide	2,3 + 9
Hydrogen chloride	2,3 + 9
Hydrogen cyanide — skin	2,7
Hydrogen fluoride	2,3,7
Hydrogen peroxide, 90%	2,7
Hydrogen selenide	6
Hydrogen sulfide	3,6
Hydroquinone	9 + 1
Indene	1
Indium and compounds (as In)	8
Iodine	9 + 1
Iron oxide fume	9
Iron salts, soluble (as Fe)	9
Isoamyl acetate	1
Isoamyl alcohol	1
Isobutyl acetate	1
Isobutyl alcohol	1
Isophorone	1
Isopropyl acetate	1
Isopropyl alcohol	1
Isopropylamine	7,6,4
Isopropyl ether	1
Isopropyl glycidyl ether (IGE)	1
Ketene	1
Lead	9
Lead arsenate	8
Lindane — skin	9
Lithium hydride	8
L.P.G. (liquefied petroleum gas)	1
Magnesium oxide fume	9
Malathion — skin	9 + 1
Maleic anhydride	9 + 3
Manganese	9
Mercury — skin	6,7
Mercury (organic compounds) — skin	8 + 6,7
Mesityl oxide	1
Methanethiol, see Methyl mercaptan	
Methoxychlor	9 + 1
2-Methoxyethanol, see Methyl Cellosolve®	
Methyl acetate	1
Methyl acetylene (propyne)	1
Methyl acetylene-propadiene mixture (MAPP)	1
Methyl acrylate — skin	1
Methylal (dimethoxymethane)	1
Methyl alcohol (methanol)	1
Methylamine	7,6,4
Methyl amyl alcohol, see Methyl isobutyl carbinol	
Methyl *n*-amyl ketone (2-Heptanone)	1
Methyl bromide — skin	1
Methyl butyl ketone, see 2-Hexanone	

TABLE 84 (continued)
Sorbents for Contaminants Listed in ACGIH 1970 Threshold Limit Values
List of Contaminants with Sorbent Code[a]

Substance	Sorbent
Methyl Cellosolve® — skin	1
Methyl Cellosolve acetate — skin	1
Methyl chloride	1
Methyl chloroform	1
Methylcyclohexane	1
Methylcyclohexanol	1
o-Methylcyclohexanone — skin	1
Methyl ethyl ketone (MEK), see 2-Butanone	
Methyl formate	1
Methyl iodide — skin	1
Methyl isoamyl ketone	1
Methyl isobutyl carbinol — skin	1
Methyl isobutyl ketone, see Hexone	
Methyl isocyanate — skin	1
Methyl mercaptan	1
Methyl methacrylate	1
Methyl propyl ketone, see 2-Pentanone	
Methyl silicate	1
Methyl styrene	1
Methylenebisphenyl isocyanate (MDI)	8 + 1
Methylene chloride (dichloromethane)	1
Molybdenum, soluble compounds,	9
insoluble compounds	9
Monomethylaniline — skin	1,7
Monomethylhydrazine — skin	7,6,4
Morpholine — skin	1
Naphtha (coal tar)	1
Naphthalene	1
β-Naphthylamine	7,6,4
Nickel carbonyl	3 + 5
Nickel, metal and soluble compounds	9
Nicotine — skin	1
Nitric acid	2,3
Nitric oxide	3
p-Nitroaniline — skin	1
Nitrobenzene — skin	1
p-Nitrochlorobenzene — skin	1
Nitroethane	1
Nitrogen dioxide	2,7
Nitrogen trifluoride	2,7
Nitroglycerin — skin	1
Nitromethane	1
1-Nitropropane	1
2-Nitropropane	1
N-Nitrosodimethylamine (Dimethylnitrosoamine) — skin	7,6,4
Nitrotoluene — skin	1
Nitrotrichloromethane, see Chloropicrin	
Octachloronaphthalene — skin	1
Octane	1
Oil mist (mineral)	9 + 1
Osmium tetroxide	8
Oxalic acid	9 + 3
Oxygen difluoride	3,6
Ozone	7
Paraquat — skin	9 + 7
Parathion — skin	8 + 1
Pentaborane	7,1
Pentachloranaphthalene — skin	9 + 1
Pentachlorophenol — skin	9 + 1
Pentane	1
2-Pentanone	1
Perchloroethylene	1
Perchloromethyl mercaptan	1
Perchloryl fluoride	2,7
Petroleum distillates (naphtha)	1
Phenol — skin	1
p-Phenylenediamine — skin	7
Phenyl ether (vapor)	1
Phenyl ether-biphenyl mixture (vapor)	1
Phenylethylene, see Styrene	
Phenylglycidyl ether (PGE)	1
Phenylhydrazine — skin	7,6,4
Phosdrin® (Mevinphos) — skin	8 + 1
Phosgene (carbonyl chloride)	3,6
Phosphine	6
Phosphoric acid	9 + 1
Phosphorus (yellow)	8 + 2
Phosphorus pentachloride	9 + 3
Phosphorus pentasulfide	9 + 3
Phosphorus trichloride	9 + 3
Phthalic anhydride	3
Picric acid — skin	8 + 3
Pival® (2-Pivalyl-1,3-indandione)	8
Platinum (soluble salts)	8
Polytetrafluoroethylene decomposition products	8 + 3
Propane	1
β-Propiolactone	1
Propargyl alcohol — skin	1
n-Propyl acetate	1
Propyl alcohol	1
n-Propyl nitrate	1,7
Propylene dichloride	1
Propylene oxide	1
Propylenimine — skin	1
Propyne, see Methylacetylene	
Pyrethrum	9
Pyridine	1
Quinone	1
RDX® — skin	8
Rhodium, metal fume and dusts,	8
soluble salts	8
Ronnel	9 + 1
Rotenone (commercial)	9 + 1
Selenium compounds (as Se)	8

TABLE 84 (continued)
Sorbents for Contaminants Listed in ACGIH 1970 Threshold Limit Values
List of Contaminants with Sorbent Code[a]

Substance	Sorbent
Selenium hexafluoride	3
Silver, metal and soluble compounds	8
Sodium fluoroacetate (1080) — skin	8
Sodium hydroxide	9
Stibine	6
Stoddard solvent	1
Strychnine	8
Styrene monomer (phenylethylene)	1
Sulfur dioxide	2
Sulfur hexafluoride	3
Sulfuric acid	2
Sulfur monochloride	9 + 3
Sulfur pentafluoride	3
Sulfuryl fluoride	7,3
Systox®, see Demeton	
2,4,5-T	9
Tantalum	9
TEDP — skin	8 + 1
Teflon® decomposition products	8 + 3
Tellurium	8
Tellurium hexafluoride	3
TEPP — skin	8
Terphenyls	1
1,1,1,2-Tetrachloro-2,2-difluoroethane	1
1,1,2,2-Tetrachloro-1,2-difluoroethane	1
1,1,2,2-Tetrachloroethane — skin	1
Tetrachloroethylene, see Perchloroethylene	
Tetrachloromethane, see Carbon tetrachloride	
Tetrachloronaphthalene — skin	9 + 1
Tetraethyllead (as Pb) — skin	8 + 1
Tetrahydrofuran	1
Tetramethyllead (TML) (as Pb) — skin	8 + 1
Tetramethyl succinonitrile — skin	1
Tetranitromethane	3
Tetryl (2,4,6-Trinitrophenylmethylnitramine) — skin	9 + 3
Thallium (soluble compounds) — skin	8
Thiram	9 + 1
Tin (inorganic compounds, except SnH_4 and SnO_2)	9
Tin (organic compounds)	8 + 1
Titanium dioxide	9
Toluene (toluol)	1
Toluene-2,4-diisocyanate (TDI)	9 + 1
o-Toluidine — skin	1
Toxaphene, see Chlorinated camphene	
Tributyl phosphate	9 + 3
1,1,1-Trichloroethane, see Methyl chloroform	
1,1,2-Trichloroethene — skin	1
Trichloroethylene	1
Trichloromethane, see Chloroform	
Trichloronaphthalene — skin	9 + 1
1,2,3-Trichloropropane	1
1,1,2-Trichloro-1,2,2-trifluoroethane	1
Tri-*o*-cresyl phosphate	8 + 3
Triethylamine	7,6,4
Trifluoromonobromomethane	1
Trimethyl benzene	1
2,4,6-Trinitrophenol, see Picric acid	
2,4,6-Trinitrophenylmethylnitramine, see Tetryl	
Trinitrotoluene — skin	9 + 1
Triphenyl phosphate	9 + 3
Tungsten and compounds, water soluble	9
Insoluble	9
Turpentine	1
Uranium (natural) soluble and insoluble compounds (as U)	8
Vanadium, V_2O_5 dust	8
V_2O_5 fume	8
Vinylbenzene, see Styrene	
Vinyl chloride	1
Vinyl cyanide, see Acrylonitrile	
Vinyl toluene	1
Warfarin	8
Xylene (xylol)	1
Xylidine — skin	1
Yttrium	9
Zinc chloride fume	9
Zinc oxide fume	9
Zirconium compounds (as Zr)	9

[a] Sorbent code: 1. activated carbon, 2. soda lime, 3. combination of activated carbon and soda lime, 4. silica gel, 5. hopcalite (with drier), 6. impregnated carbon, e.g., carbon treated with metallic salt to absorb ammonia, 7. special sorbent (manufacturer's proprietary mix), 8. filter, for highly toxic dusts, 9. filter, for toxic dusts; where two or more sorbent code numbers are listed after a substance, the sorbents are in decreasing order of preference. Where a plus (+) sign is used, both are required. Very strong oxidizing substances in low concentrations may be sorbed by activated or impregnated carbon;

however, carbon should not be used because of a possible explosion hazard with these combinations. For respiratory protection against very strong oxidizers, a self-contained breathing apparatus should be used.

From AIHA-ACGIH Joint Committee on Respirators, *Am. Ind. Hgy. Assoc. J.*, 32, 404, 1971. With permission.

Analytical Methods (Table 64) will show 17 sampling procedures which are based on GC packings of various types. Parts-per-billion (ppb) concentration range of 12 representative solvent vapors has been evaluated extensively.[242] Contaminated air was drawn at 100 mL/min through a 7.6-cm long × 0.64-cm diameter stainless steel tubing packed with either purchased charcoal (tubes packed by supplier)[243] or 200 mg of 80/100 Chromosorb® 101. Prior to use for sampling, the tubes were preconditioned for 30 min at 25°C above the temperature used for thermal desorption of the collected sample directly into the GC column in the "Bendix Flasher"® unit supplied with the metal tubes (Figure 114). Standard in-air calibration atmospheres were prepared by injecting the liquid solvent from a 1-μL syringe into a 30-L (61 × 61 cm) Tedlar® bag fitted with GC septum and a Roberts® valve.[244] Charcoal was used to collect aromatic hydro-carbons and vinyl chloride, whereas the aliphatic ketones were adsorbed on the Chromosorb® 101. Bischloromethyl ether collected on Chromosorb® was detected at the 1-ppb level with flame ionization detector. Separation of a mixture of ten components was attained by temperature programing of the GC column. For collection of vinyl chloride, methylene chloride and chloroform charcoal was superior to Chromosorb®. Charcoal did not deliver acceptable results with bischlormethyl ether, but adequate recoveries were obtained from the chromosorb® packing. The desorption efficiencies exceed 85% for all compounds studied. Results are summarized in Table 85. The "flasher" sample collection system is shown in Figure 114.

Further studies have been reported for a collection on Tenax® GC and Porapak®N.[245] Sampling tubes were fabricated from 6.3-mm O.D. × 3.2-mm I.D. × 152 mm-long Pyrex® glass tubing packed with 180-mg 60/80 mesh Tenas® GC or 310 mg 50/80 mesh Porapak® N which was held in place with silanized glass wool. Prior to use the tubes were thermally conditioned. The samples were thermally desorbed directly into a temperature-programed GC column. Standards in air were prepared by injecting μL quantities of the liquid phase into a 125-mL gas sampling bulb fitted with PTFE stopcocks and a septum port. Then the required volume of dilution air was passed through the collection tube connected to the sampling bulb outlet. For example, for a 10mg/L concentration in 5L of air, 50 mg of the liquid is injected into the bulb and 5L of air drawn through the system by a personnel monitoring pump at 10 to 150 mL/min. Recoveries by this technique were equivalent to those obtained from 50-L PTFE bags, permeation tubes, or three large glass containers in series. Results from 19 organic chemical recoveries are summarized in Table 86.

Other investigations include the collection of mercury vapor of Hopcalite® in a small glass tube. The mercury is then recovered by dissolving in aqueous $HCl\text{-}HNO_3$ mixture and quantitated by flameless atomic absorption analysis.[246,247]

a. *Sample Collection System*

The Sample Collection System consists of the Micronair Permissible Air Sampling Pump, the Personnel Monitoring Collection Column (PMCC), and the PMCC Holder Assembly. The interconnecting tubing can be tailored to accommodate any wearing requirements.

The Sample Pump has an on-off switch, a flow regulating control, and flow rate indicator. The on-off switch and flow control have screw-on cap covers to prevent accidental manipulation. As an accessory, the sample pump has a battery charger which will allow charging directly from any 110-vac source.

TABLE 85

Collection Efficiency and Approximate Sensitivity Limits for Airborne Organic Compounds on Activated Charcoal and Chromosorb 101[a]

Compound	Collective tube packing	% Collected: Fore section	Backup	ppb[b] Detection limits: Thermal desorp.	CS_2 desorp.	CV[c] — % GC peak height
Benzene	Activated charcoal	99	1	0.2	40	1.5
Toluene	Activated charcoal	99	1	0.1	20	3.0
Xylene	Activated charcoal	99	1	0.2	40	2.6
Styrene	Activated charcoal	99	1	0.2	40	4.3
Vinyl chloride	Activated charcoal	99	1	0.3	60	3.4
Methylene chloride	Activated charcoal	99	1	0.7	140	2.2
Chloroform	Activated charcoal	99	1	0.5	100	2.7
Ethylene dichloride	Activated charcoal	99	1	0.3	60	1.1
Tetrachloroethylene	Activated charcoal	99	1	0.5	100	3.0
Acetone	Chromosorb-101	96	4	0.5	100	2.7
Methylethyl ketone	Chromosorb-101	99	1	0.4	80	2.0
3-Pentanonl	Chromosorb-101	99	1	0.4	80	2.4
Methyli-butyl ketone	Chromosorb-101	99	1	0.3	60	2.7
3-Hexanol	Chromosorb-101	99	1	0.3	60	1.7
2-Heptanol	Chromosorb-101	97	3	0.3	60	0.9
Cyclohexanol	Chromosorb-101	96	4	0.3	60	2.3
Bischloromethyl ether	Chromosorb-101	99	1	0.5	NG	3.2

[a] 10L of air containing 5 mg/L sampled at 200 mL/min.
[b] ppb = parts per billion v/v.
[c] CV = coefficient of variation — CS_2 desorption by NIOSH methods.

TABLE 86

Recovery of Organic Vapors on Porous Polymer Beads

Organic compound	Collection tube packing	TLV or federal standard	% Recovery — levels: TLV	0.5 TLV	0.2 TLV
Acetic acid	Porapak N	10	95	—	101
Benzene	Porapak N	10	105	99	—
Benzyl chloride	Porapak N	1	93	83	—
bis-B-chloroethyl ether	Porapak N	5	98	106	—
h-Butanol	Porapak N	50	100	—	102
Butylbenzyl phthalate	Tenax GC	5	96	91	—
Dichloropropanel	Porapak N	75	88	—	89
Dimethyl acetamide	Tenax GC	10	102	104	—
Methylene chloride	Porapak N	200	98	—	91
Mono chlorobenzene	Porapak N	75	102	—	104
P-nitroanline	Tenax GC	1	93	106	—
Phenol	Tenax GC	5	93	94	—
Polychlorinated bi-phenyl	Tenax GC	5	91	—	114
Styrene monomer	Tenax GC	100	92	—	101
Toluene	Porapak N	100	100	—	101
Trichloroethylene	Porapak N	100	93	—	93
Triethylamine	Porapak N	25	101	—	98
Xylene	Porapak N	100	102	—	101
Precision	Porapak N				97 ± 12
Phenol precision	Tenax GC	5	98 ± 15	106 ± 11	—

Data derived from Dietrich, M. W., Chapman, L. P., and Miene, J., *Am. Ind. Hyg. Assoc. J.*, 37, 311, 1976.

The Personnel Monitoring Collection Column (PMCC) is easily installed in the PMCC holder by simply unscrewing the holder at the pump-end. The PMCC is held firmly in line with the pump flow by being seated in a grommet. This arrangement provides mechanical support for the PMCC and also serves to maintain a closed flow line to the sample pump, thereby ensuring that the total sample flow is through the PMCC.

A standard length of interconnecting tubing is supplied and is sufficient for almost any wearing configuration. The tubing is of a construction that will not collapse when bent and is a diameter that fits snugly over the connectors, thereby precluding the necessity for clamps to ensure flow integrity. Also, the tubing is downstream of the PMCC, which eliminates the possibility of the sample tube material contaminating or interfering with the collected sample.

5. Liquid Reagents Carried on Solid Adsorbents

Nitrogen dioxide can be collected quantitatively on molecular sieve (MCB No. MX 1583-1) impregnated with triethanolamine (TEA). Results compared well with values obtained with the liquid reagent in a fritted glass bobbler (Table 87). Samples were collected in two absorbers, either liquid or solid, in series to detect breakthrough. Known concentration standards were prepared dynamically from a permeation tube. Field results compared closely with length-of-stain detector tubes from one source, but those from another manufacturer were consistently 50% lower than the collection of TEA-solid absorbent results indicated.[248]

The TEA sorbent collection system has been further extended to the recovery of both nitrogen dioxide and nitric oxide simultaneously[249] in the range of 0.5 to 5 ppm NO_2 and 9 to 50 ppm NO. The collection train consists of a 400-mg section of TEA on molecular sieve section to collect the NO_2, then a 1200-mg section of solid supported chromium trioxide oxidizer section to convert the NO to NO_2, followed by a second 400-mg section of the TEA on molecular sieve to absorb the NO_2 produced in the oxidizer section. The glass tubes containing these three sections separated by fiberglass wool plugs (7 mm O.D. × 15 cm overall length) are commercially available.[250] The oxidizer section will permit sampling of a 50-ppm NO-in-air concentration at 25 mL/min for two hours. The NO_2 in the two TEA sections is analyzed separately by the classical Griese-Saltzman colorimetric procedure. This method was developed to sample NO_2 in the range 0.5 to 5 ppm and NO over the range 9 to 50 ppm in a 15-min sampling period. The collection efficiency (average) for NO_2 was 96%. The collection efficiency for NO was concentration dependent:

at 9 ppm =	97%
11 ppm =	106%
24 ppm =	84%
50 ppm =	67%

A complete description of the permeation tube procedure for generating NO_2 in-air standards, and stepdown dilution from a pressurized cylinder for NO is included in the reference. The collected samples were stable in storage for at least 2 weeks. The flow rate was approximately 50 mL/min for 20 min.

An acid-coated solid support system has been developed for the collection of diethylamine from contaminated air. Chromosorb® W (30/60 mesh) coated with 17% prosphoric acid contained in a 15.2-cm glass drying tube served as a personnel monitor. Sampling through a 15 mL volume of the acid coated pellets was carried out a 2L/min for two hours. The diethylamine was extracted in an aqueous solution of *n*-buty-

TABLE 87

Nitrogen Dioxide Absorption in Triethanolamine Reagents[a]

Parameter	Collection matrix	
	Liquid	Solid
Sensitivity, vol. of sample to 1 ppm	1.7 L	4.2
Collection efficiency, average, 1st unit	96.8%	97.3%
Stoichiometry, moles NO_2 ion formed	0.506—0.800	0.513—0.725
(49 trials), average	0.63	0.63
Adsorbing capacity, μg NO_2/mg	—	1.12
Sampling rate, L/min	0.20	0.22
Amount of collection matrix	20 mL	770 mg

[a] Range = 3—11 ppm.

Data derived from Blacker, J. H., *Am. Ind. Hyg. Assoc. J.*, 34, 390, 1973.

lamine, made alkaline (pH above 10) with NaOH, and then analyzed by gas chromatography. Recoveries were 97 to 102% in the range of 9 to 18 mg.

6. Sorbent Specifications

[251] A group working at the Dow Chemical Laboratory in Midland, Mich. has developed a set of specifications which should be most useful for guidance in evaluating solid sorbent applications and selection of a suitable field collection substrate.

a. Sampling Procedure

i. Collection

Sorption Process — At the start of collection the pollutant is distributed through the front section of the sorbent column. As collection continues, this frontal zone becomes saturated with respect to the pollutant, that is, an equilibrium has been established between the incoming concentration and the sorbed gas. At this stage sorbtion and desorption rates are equal. As the saturation front moves through the column of sorbent, a point is reached when breakthrough into the back-up section occurs. Since saturation is not attained in the advancing sorbtion front, this breakthrough is gradual. Therefore, an arbitrary cut-off limit has been recommended. Breakthrough has been defined in the terms of the detection of the pollutant in the effluent as a percentage of the inlet concentration. The usually accepted limit is 5%.[252] The breakthrough volume obtained as the product of the sampling rate and the sampling time is more indicative of the collection efficiency. In general, if less than 10% of the total amount collected is found in the backup section, the sampling may be considered valid. If more than 25% is detected, then significant loss probably has occurred and the results should be reported as "minimum amount present".

Properties of the Compound — Usually direct experimental determination of breakthrough is much more efficient than attempting to calculate this parameter by equations which are at best quite involved and require data not readily available. By relating boiling point of an untested compound some predictions can be made (see Table 68).

Nature of Sorbent — The collection efficiency must be high, but equally important is high desorption efficiency — both should be above 90%. Decreasing the sorbent particle size increases collection efficiency due to increased surface area, but pressure drop (ΔP) also increases beyond the limitations of the sampling pump.

Size of Collection Tube — In general, doubling the amount of sorbent will double the sampling capacity (breakthrough volume). Usually the capacity is somewhat more than doubled as the sorbent is not at full saturating capacity at the break through point.

Flow Rate — The effect of flow rate varies with the sorbent. Frequently, an optimum flow rate is reached, below which no increase in breakthrough volume can be attained by reduction in flow rate — about 100 mL/min for coconut charcoal. Linear velocity should be considered when extrapolating data to different size tubes.

Concentration — Breakthrough will occur sooner at higher concentrations. For charcoal, the empirical Freundlich isotherm can be applied:

$$\log T_B + \log a + b \log c = \text{break through time}$$
$$c = \text{concentration.}$$

A straight line plot is obtained when log T_B is plotted vs. log c with a slope b and intercept a. Equations developed for the collection of amines on silica gel indicate that collection is not dependent on initial concentration or flow rate.

Humidity — Usually breakthrough is increased as the humidity increases, but the magnitude of the effect depends upon the properties of the sorbate and sorbent. Although charcoal absorbs only small amounts of water, the high ratio of H_2O molecules to sorbate molecules may increase breakthrough as much as 50%, especially at low concentrations of sorbate.

Polar sorbates such as silica and alumina gels sorb water more strongly than most organic compounds. This property not only reduces the effective capacity, but also contributes to the displacement of organic compounds already collected.

Co-Adsorption — When two or more sorbents are adsorbed, the one most strongly adsorbed will displace other compounds down the length of the sorbent tube. Polar sorbents adsorb compounds with the largest dielectric constant and di-pole moment most strongly. The nonpolar sorbents, on the other hand, select compounds with the highest boiling point and molecular volume.

Temperature — Breakthrough increases as the temperature rises. For charcoal an approximation of 1 to 10% increase in breakthrough for each 10°C temperature rise can be used.

Migration — False breakthrough may be indicated for highly volatile compounds (vinyl chloride) or when collected samples are stored for long periods (33.3% in the backup section at equilibrium). This effect can be reduced by refrigeration or eliminated by physical separation of the two sections.

ii. Desorption — Recovery

Efficiency as a fraction of total compound collected — This factor probably is the most significant consideration in defining sorption-desorption systems. Experimental evaluation must be carried out before field collection is attempted. Organic compounds are partitioned between the solid sorbent and desorption solvent:

$$\frac{1}{D} = K \frac{\text{Mg sorbent}}{\text{Mg solvent}} + 1$$

where 1 D = desorption efficiency (× 100 = % efficiency) and K = equilibrium constant.

This equation clearly shows that the solvent: sorbent ratio cannot be altered without changing the efficiency. As an example, if 0.5 mL of solvent desorbs 50% of the organic sorbate, then an increase to 5 mL of solvent should increase the efficiency to

91%. Even more important is the effect on the backup section if extracted with the same volume of solvent as the fore section with the result that 5 to 15% positive error could be produced.

Temperature — The magnitude of the effect depends on the compound sorbed. While vinyl chloride exhibits no effect in the temperature range 0 to 25°, methyl ethyl ketone efficiency increases from 66.8 to 89.5%. This parameter must be carefully evaluated before field collection is started. If high collection temperatures are encountered, sample loss through decomposition, polymerization, or other chemical reactions may take place on the surface of the adsorbent phase.

Humidity — Low recoveries of hydrolysable compounds collected under high humidity conditions may be encountered. Large amounts of water present may interfere in contact between the nonpolar (hydrophobic) solvent and sorbent interface or shift the desorption equilibrium by dissolving in the extraction solvent.

Co-adsorption — The presence of other compounds adsorbed on the sorbent can, and often do, influence the recovery efficiency and must be evaluated for each system mixture.

Time — Thirty to sixty minutes with good agitation usually provides maximum desorption, equilibrium conditions between sorbate and sorbent and solvents. In some cases (styrene), extension of the desorption time beyond the optimum actually may produce a loss of the desorbed compound. This effect usually is greater at low concentrations. This parameter can be anticipated from the chemical reactivity of the adsorbed compound and, in any case, must be evaluated.

The catalytic effect of the charcoal surface may introduce problems also. For example, if 1% methanol added to CS_2 to improve desorption efficiency is allowed to remain in contact with the activated charcoal for 4 hrs, several reaction products from the reaction of the two solvents on each other are detected in chromatograms derived from the mixture.

Storage — In general, samples are not analyzed the same day as collected and, in some cases, remain in storage, sometimes at elevated temperatures, for one or two weeks. Serious losses of collected material can be encountered. For example, formaldehyde collected on alumina gel must be desorbed within a few hours after collection. When instability is a problem, desorption directly in the field after collection must be considered.

Concentration — For most collection of organic compounds on charcoal, the desorption efficiency does not vary more than ±5% for concentrations of interest to industrial hygienists. Some compounds (styrene monomer) tend to produce lower recoveries as the concentration decreases.

Predictions — Great care must be exercised in extrapolations from existing data before predictions can be made from known physical and chemical properties to untested systems. For example, the desorption efficiency in CS_2 from a charcoal substrate was found to be 55% for biphenyl, 84% for paramethyl biphenyl, 98% for benzene, but only 43% for naphthalene.

b. Criteria for Successful Validation

1. Accuracy and precision of combined analytical and collection procedures for concentration range of 0.1 to 2 TLV = ±16% relative at the 95% confidence level
2. Total recovery efficiency greater than 74%
3. Bias between total recovery and desorption efficiency less than ±10% relative
4. Capable of taking 10 to 15-min samples at ceiling and excursion levels
5. Minimum sampling time 1 hr, preferred 4 to 8 hr for TWA

6. Storage samples within ±10% relative to initial samples after 14-day storage period
7. The flow rate of the sampling pump controlled within an accuracy limit of ±5%.

7. Equipment

Battery operated personnel monitoring pumps are described in Chapter 2, Volume II, — Battery Powered Air Samplers. This chapter includes the models designed specifically for low flow rates, i.e., 2 to 200 mL/min as well as the higher flow rate units. Available equipment needed for air sampling with solid sorbents is summarized in Table 88. The following paragraphs discuss the most notable among the useful accessories.

Multiple Tube Manifolds — Holders for two, three, and four sampling tubes mounted in parallel are available. The air sampling rate is controlled by limiting orifices, which are capable of delivering satisfactory flowrates through each tube if the pressure drops across the individual tubes is uniform. Otherwise, each branch must be individually calibrated with collector tubes in place (see Table 81). One of the commercially available replicate sampling heads is shown in Figure 115. This model also is available in the 3- and 4-place mode. (Figure 115A).

Holders for Personnel Monitoring Tubes — Representative solid sorbent tube holders are shown in Figure 116. In item 1 the holder is arranged for single tube collection, while item 2 is arranged for series collection. Item 3 contains the large-size charcoal tube, and item 4 is the holder for long-term detector tube sampling. The personnel monitor assembly in place is shown in Figure 117. Charcoal tubes in the shipping carton appear in Figure 118.

Calibration — Frequent volumetric calibration of the sampling assembly with all units in place is mandatory. The recommended procedure, which is a simple, expedient primary standard method based on the soap film (or soap bubble) flowmeter, can be carried out directly in the field before and after sampling if necessary. Both air flowrate and total volume calibration are obtained simultaneously by this technique (see Section II.A.1.b.x., "Testing Gas and Vapor Detector Pumps,") for further details. Other soap film flowmeter designs appear in Figure 119. Commercial units also are available (consult Table 88) in kit form as shown in Figures 120 to 122 (in action for calibrating a charcoal tube).

D. Portable Dynamic Sampling Instruments Containing Battery Powered Pumps or Manual Squeeze Bulbs

With the exception of the portable gas chromatographs, these instruments fall into the same classes as those already presented in Section I.A.3, "Portable Instruments," passive mode. The only significant difference is the addition of a positive air-displacement pump, manual or battery operated, to draw the air sample into and across the detector (see Table 89). The classes are electronic and gas chromatographs.

1. Electronic

1. Oxidation of the combustible gas or vapor on contact with a hot platinum filament — the flammable and explosive gas and vapor meters
2. Catalytic ceramic bead activated by platinum — similar to 1., but better durability and ruggedness
3. Electrochemical oxidation in a galvanic cell to produce an electric current for CO, H_2S and SO_2
4. Electrochemical reduction in a galvanic cell to produce an electrical detector current for O_2
5. Ultraviolet light (UV) absorption for mercury

TABLE 88

Suppliers of Solid Sorbent Tubes, Pumps, Accessories, Midget Impingers, and Dust Sampling Equipment

Vendor — supplier	Pumps	Pump conversion	Charcoal tubes	Silica gel tubes	Alumina gel tubes	GC packing tubes	Custom packing	Multiple manifold	Soap film flowmeter	Tube holders	Midget impinger	Detector tubes	Dust filters	Respirable dust filters	Dust sampling bags
MDA Scientific, Inc.[a]	X	X	X	X				X	X	X	X	X	X	X	
National Environmental Instruments — Bendix[b]	X	X	X							X		X	X	X	X
National Drager, Inc.[c]	X											X			
Research Appliance Co.[d]	X		X	X						X	X		X	X	
SKC, Inc.[e]	X		X	X	X	X	X		X	X	X		X		
Du Pont[f]	X	X	X						X	X					
Anatole J. Sipin[g]	X	X	X	X	X	X			X	X		X			X
Mine Safety Appliances[h] (MSA)	X	X	X	X				X	X	X	X	X	X	X	
Calibrated Instruments[i]	X														X
Environmental Compliance[j]			X						X						
Century Systems, Inc.[k]	X		X			X	X			X					
National Mine Service Co.[l]			X	X	X	X	X			X					
Spectrex[m]	X														

[a] MDA Scientific, Inc., 808 Busse Highway, Park Ridge, Ill.
[b] The Bendix Corp., Environmental and Process Instruments Division, Lewisburg, W. Va.
[c] National Drager, Inc., Parkway View Dr., Pittsburgh, Pa.
[d] "RAC", Research Appliance Co., Route 8, Gibsonia, Pa.
[e] SKC, Inc., P.O. Box 8538, Pittsburgh, Pa.
[f] "Du Pont", E. I. du Pont de Nmours, Inc., Applied Technology Division, Wilmington, Del.
[g] "Sipin", Anatole J. Sipin Co., Inc., 425 Park Ave. S., New York, N.Y.
[h] Toxic Gas Detectors, Mine Safety Appliances Co. (MSA), 400 Penn Center Building, Pittsburgh, Pa.
[i] Calibrated Instruments, Inc., 731 Saw Mill River Rd., Ardsley, N.Y.
[j] "EC", Environmental Appliance Corp., P.O. Box 55, Venetia, Pa.
[k] Century Systems, Inc., P. O. Box 133, Arkansas City, Kan.
[l] National Mine Service Co., 3000 Koppers Building, Pittsburgh, Pa.
[m] Spectrex, 3594 Haven Ave., Redwood City, Calif.

6. Thermal conductivity for high concentrations of natural gas above the lower explosive range (LEL).

Several of these instruments may be used for personnel monitoring, notably the Energetics Science "Hipster." All instruments included in Table 89 meet NIOSH specifications for size and weight.

The research group at Harvard University developed a personnel monitor for carbon monoxide in the range 20 to 100 ppm based on the catalytic oxidation of CO to CO_2. Hopcalite was used as the catalyst and the CO_2 was absorbed on Ascarite beads. The

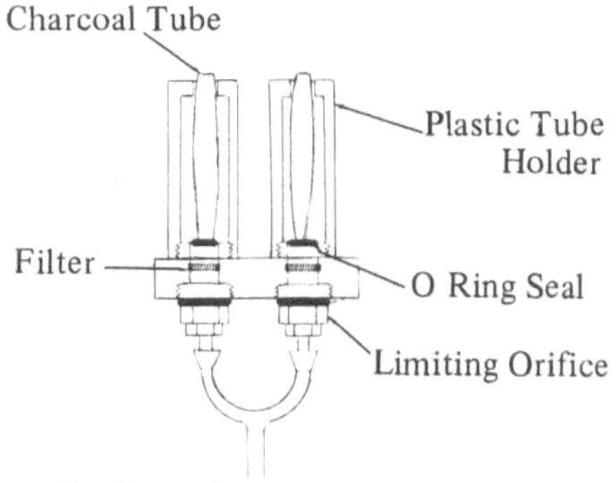

FIGURE 115A. MDA low-flow multiple sampling kit. (Courtesy of MDA Scientific, Inc., 808 Busse Hwy., Park Ridge, Ill.)

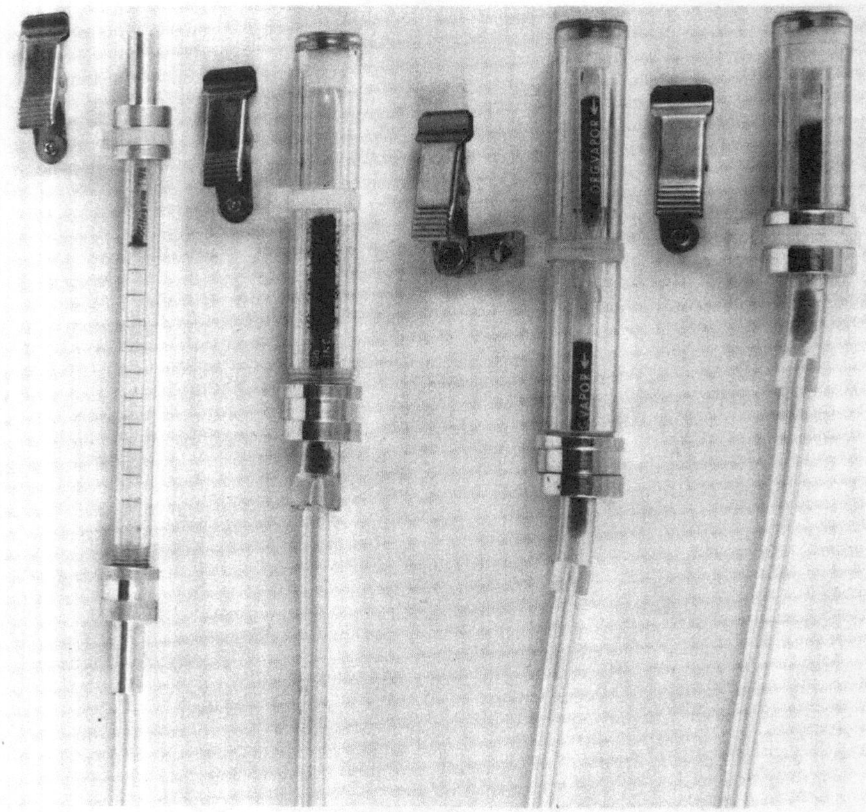

FIGURE 116. Solid sorbant organic vapor monitor tubes assembled for sampling. (Courtesy of Anatole J. Sipin, Inc., 425 Park Ave. S., New York, N.Y.)

increase in weight was then a direct measure of the CO in the atmospheric sample. Since the analysis was gravimetric, this method could be considered a primary standard. Air was sampled for 4 to 8 hr at 0.5 L/min. Recovery of CO as CO_2 was above 90%. Catalyst activity was checked after each 8-hr use by injecting 20 mL pure CO into an airstream moving through the assembly. If the weight gain fell below 36 mg (90%) recovery the drying in a stream of dry nitrogen at 280°C for 4 hr is continued. Precision was demonstrated to be ±2.4 ppm CO with duplicate samples taken in the field. Since water interferes NaOH on asbestos drying facilities must be added. Therefore, the air sample first passes through a polyethylene tube containing activated molecular sieve, then Ascarite, then granular anhydrous magnesium perchlorate before entering the sampling U-shaped tube which contains, in order, Hopcalite, anhydrous magnesium perchlorate, Ascarite and anhydrous magnesium perchlorate (reverse diffusion guard layer). The entire assembly weighs approximately 1.5 kg and can easilly be worn on a man's belt.[253]

2. Gas Chromatographs (GCs)

One commercial unit that is borderline with respect to weight is available (see Table 89).[229] However, no "off-the-shelf" models suitable for personnel monitoring has been offered. Work carried out under an NIOSH contract and a study in progress in the

TABLE 89

Instruments for the Detection of Gases and Vapors by Dynamic (Forced Air Sample Flow) Sampling

Manufacturer or vendor	Model number	Dimensions L × W × D cm	Weight Kg	Gases of vapors detected	Range ppm	Principle of operation
Bacharach-J.W.[a]	MV-2	28.9 × 12.1 × 11.1	2.7	Mercury	0—0.2, 0—1.0 mgm^3	UV absorption at 253.7 nm
	H[b]	14.6 × 7.6 × 11.1	2.5	Combustible	0—100% gas	Cat. combustion
	G[b]	14.6 × 7.6 × 11.1	1.8	Combustible	—	Cat. combustion
	L[b]	14.6 × 7.6 × 11.1	1.8	Combustible	0—100% LEL	Cat. combustion
	TLV[b-d]	2.2 × 9.5 × 16.8	2.6	Hydrocarbons	0—10,000	Cat. combustion
	K	14.6 × 7.6 × 11.1	1.1	Oxygen	0—25%	Electrolytic
	GPK	10.3 × 14 × 19	2.6	Combustible + O_2	0—25% O_2	Combination electrolytic and cat. combustion
	Sentox 2[c]	26.7 × 17.2 × 16.5	3.4	Combustible + O_2	0—100% LEL	Combination electrolytic and cat. combustion
Biomarine[e]	500[b]	21.9 × 11.8 × 12.4	1.6	CO	0—100, 0—500	Electrochemical oxid.
	501[b]	21.9 × 11.8 × 12.4	1.6	CO	0—100, 0—500	Pulsed operation
	701	21.9 × 11.8 × 12.4	1.5	H_2S	0—20, 0—100	Electrochemical oxid. diffusion membrane
	911	21.9 × 11.8 × 12.4	1.5	Combustible	0—100% LEL	Pt ceramic bead
	901	21.9 × 11.8 × 12.4	1.6	Combustible	0—100% LEL	Pt ceramic bead
	900	21.9 × 11.8 × 12.4	1.6	Combustible + O_2	0—100% LEL 0—40% O_2	Electrochemical
Bendix[f]	GA-1C	14.6 × 11.4 × 13.4	—	Combustible	% LEL	Cat oxid. Pt filament
Century[g]	OVA	21.9 × 29.5 × 10.8	5.0	Organic	1—100,000	Gas chromatography
	88 and 98	21.9 × 29.5 × 10.8	5.0	Vapors	1—10,000	Flame ionization
Energetics Science[h]	Ecolyzer 2100—2800	18.7 × 33.0 × 18.7	4.1	CO	0—50, 0—4000	Electrochemical
	Hipster 6000 series[c,d,j,i]	—	1.6	CO	0—100, 0—500	3 Electrode electrochemical
	Hipster 6000 series[c,d,j,i]	—	1.6	H_2S	0—320	Oxid.

TABLE 89 (continued)

Instruments for the Detection of Gases and Vapors by Dynamic (Forced Air Sample Flow) Sampling

Manufacturer or vendor	Model number	Dimensions L × W × D cm	Weight Kg	Gases or vapors detected	Range ppm	Principle of opration
Enmet[k]	[l]	—	—	Combustible + O_2	1—100% LEL 0—25% O_2	Hand operated squeeze bulb
Gas Tech[m]	1177—1314[c]	—	—	Combustible + O_2	0—100% LEL 0—25% O_2	Cat. oxid. Pt filament
	GV-3[b,c,j]	—	—	Combustible + O_2	0—100% LEL 0—25% O_2	Electrochemical O_2
	GP-204[b]	—	—	Combustible	0—100% LEL	Cat. oxid. Pt filament
	XP-204[b]	—	—	O_2	0—25%	Electrochemical
	1238[b]	30.5 × 9.5 × 14.0	2.7	Hydrocarbons	0—500 as toluene	Cat. combustion 0—100% LEL
Interscan Corp.[n]	1140—1148	15.2 × 17.8 × 25.4	3.2	CO	0—600	Electrochemical oxid.
	1240—1249	15.2 × 17.8 × 25.4	3.2	SO_2	0—20	Electrochemical oxid.
	1170—1175[c,d]	15.2 × 17.8 × 25.4	3.2	H_2S	0—50	Electrochemical oxid.
IST[o]	[l]	—	—	Combustible oxygen	0—100% LEL, 0—25%	Hand operated squeeze bulb, cat. oxid.
MSA[p]	2A[j,q]	9.5 × 13.6 × 14.0	1.8	Explosimeter	0—100% LEL	Pt filament, cat. oxid.
	3[j]	9.5 × 13.6 × 14.0	1.8	H_2 + O_2 mixture	0—100% LEL	Pt filament, cat. oxid.
	4[j]	9.5 × 13.6 × 14.0	1.8	Acetylene	0—100% LEL	Pt filament, cat. oxid.
	5[j]	9.5 × 13.6 × 14.0	1.8	Leaded gasoline	0—100% LEL	Pt filament, cat. oxid.
	20—40[r]	9.5 × 13.6 × 14.0	1.8	5 Gases	0—100% LEL	Pt filament, cat. oxid.
	Gascope[j] natural gas	15.2 × 10.2 × 14.6	2.9	%/Volume combustible	0—100%	Thermal conductivity
	100[c,j]	25.4 × 17.8 × 9.5	3.2	Combustible	0—100% LEL	Pt filament, cat. oxid.
	250	19.1 × 13.7 × 8.6	—	Combustible + O_2	0—100% LEL	Pt filament, cat. oxid.
	77600[q]	17.2 × 10.5 × 14.6	2.6	O_2	0—25%	Electrolytic
	87762[q]	17.2 × 10.5 × 14.6	2.6	O_2	5—40%	Electrolytic
Neutronics[s]	BZ/3007-AA[t]	—	—	Combustible	—	Conversion kit battery operated pump

Notes: Dimensions = LxWxD = Length × Width × Depth (wide); Kg = kilograms; ppm = parts per million as volume per volume (V/V); LEL = lower explosive limit; Pt = platinum; MOS = metallic oxide semiconductor; UV = ultraviolet light; HC = hydrocarbons (aliphatic and aromatic); CO = carbon monoxide; Cat. = catalytic; Oxid. = oxidation.

[a] Bacharach Instrument Co., Division of AMBC Industries, Inc., 625 Alpha Dr., Pittsburgh, Pa.
[b] Also available in the passive mode.
[c] Built-in alarm.
[d] Chart recorder available.
[e] Biomarine Industries, Inc., 45 Great Valley Center, Malverne, Pa.
[f] The Bendix Corp., Environmental and Process Instruments Division, Lewisburg, W. Va.
[g] Century Systems, Inc., P.O. Box 133, Arkansas City, Kan.
[h] "Ecolyzer", Energetics Science, Inc., 85 Executive Blvd., Elmsford, N.Y.
[i] Personnel monitor type.
[j] Calibration kits (gas) available.
[k] Enmet, 2308 Industrial Hwy., Ann Arbor, Mich.
[l] Conversion kit for passive instruments.
[m] Gas Tech, Inc., Johnston Instrument Division, 331 Fairchild Dr., Mountain View, Calif.
[n] Interscan Corp., 9614 Cozycroft Ave., Chatsworth, Calif., U.S. Patent No. 4,017,373.
[o] International Sensor Technology, 3201 S. Halladay St., Santa Ana, Calif.
[p] "MSA", Mine Safety Appliances Co., 400 Penn Center Bldg., Pittsburgh, Pa.
[q] All MSA models hand squeeze bulb operated except models 100 and 2.
[r] Four models available for five specific gases: pentane, octane, xylene, and combustible vapors.
[s] Neutronics, Inc., 450 Drew Ct., King of Prussia, Pa.
[t] Battery operated pump for conversion to dynamic sampling mode for passive instruments.

7690	Nitrogen dioxide
7814	Nitrogen dioxide refill
7969	Phenol
7970	Phenol refill
7067	Silica
6627	Silica refill
7671	Sulfide (H_2S)
7815	Sulfide refill
7691	Sulfur dioxide
7816	Sulfur dioxide refill
7702	TDI
7818	TDI refill

[u] Kit consists of absorbing solution, reagents, instructions, and color comparator. Approximately 25 tests per kit.

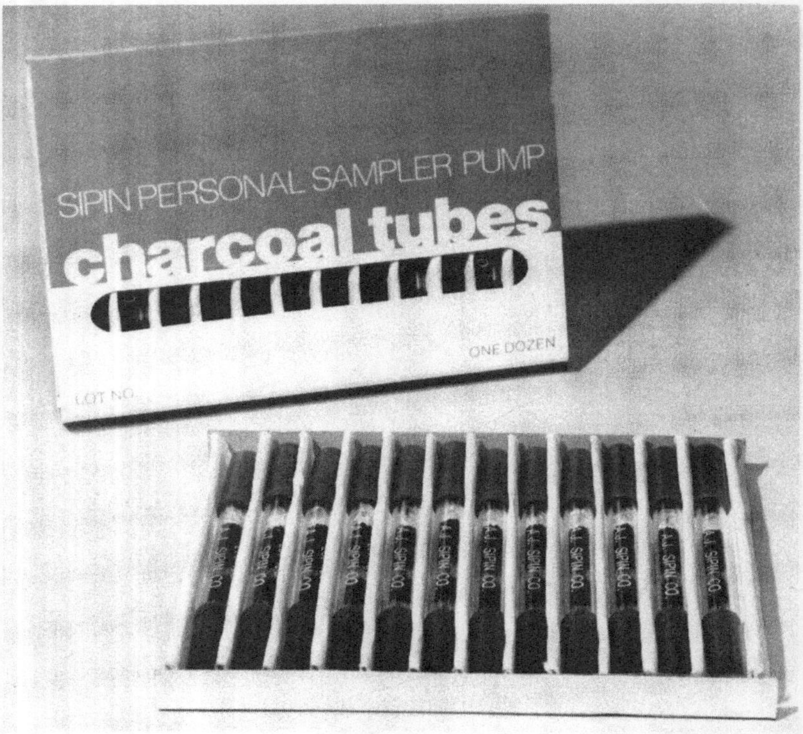

FIGURE 118. Charcoal tubes in shipping carton. (Courtesy of Anatole J. Sipin, Inc., 425 Park Ave. S., New York, N.Y.)

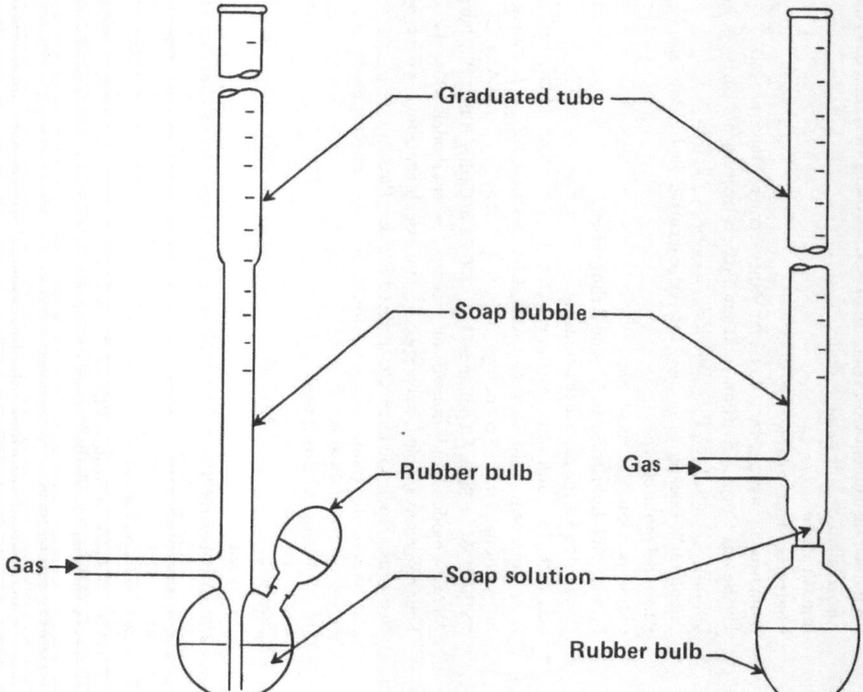

FIGURE 119. Soap film flowmeters.

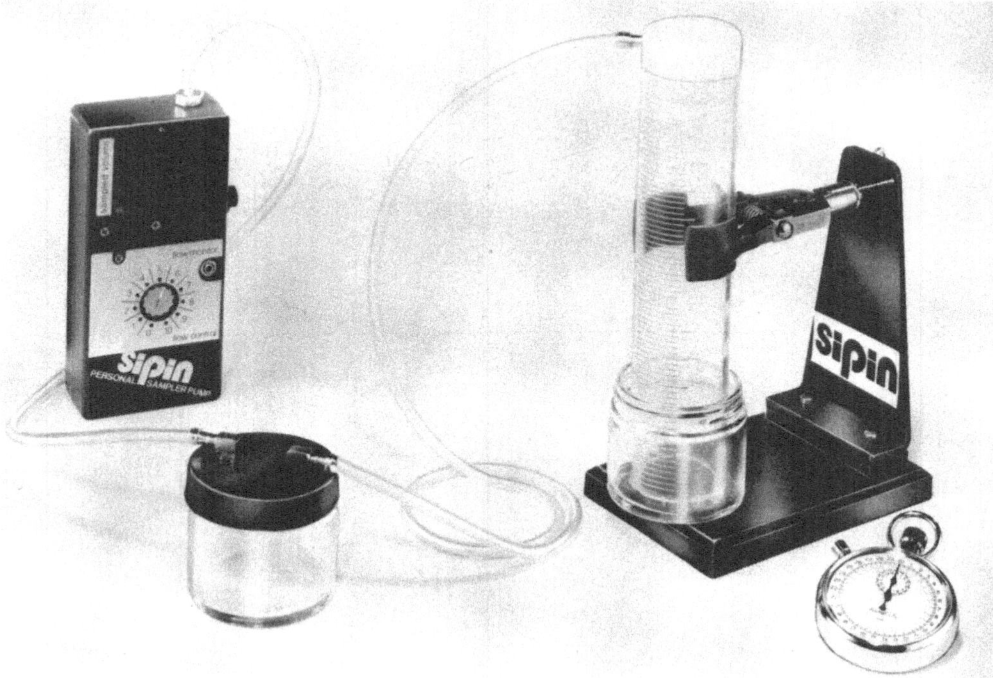

FIGURE 120. Calibration kit assembled for field calibration. (Courtesy of Anatol J. Sipin, Inc., 425 Park Ave. S., New York, N.Y.)

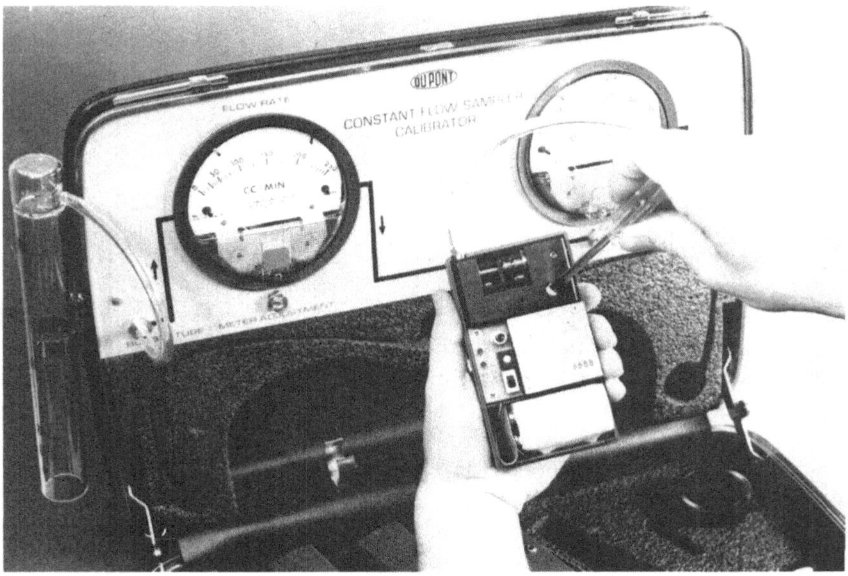

FIGURE 121. The calibrator pack in use with a low-flow sampling pump. (Courtesy of E. I. du Pont de Nemours, Inc., Applied Technology Division, Wilmington, Del.)

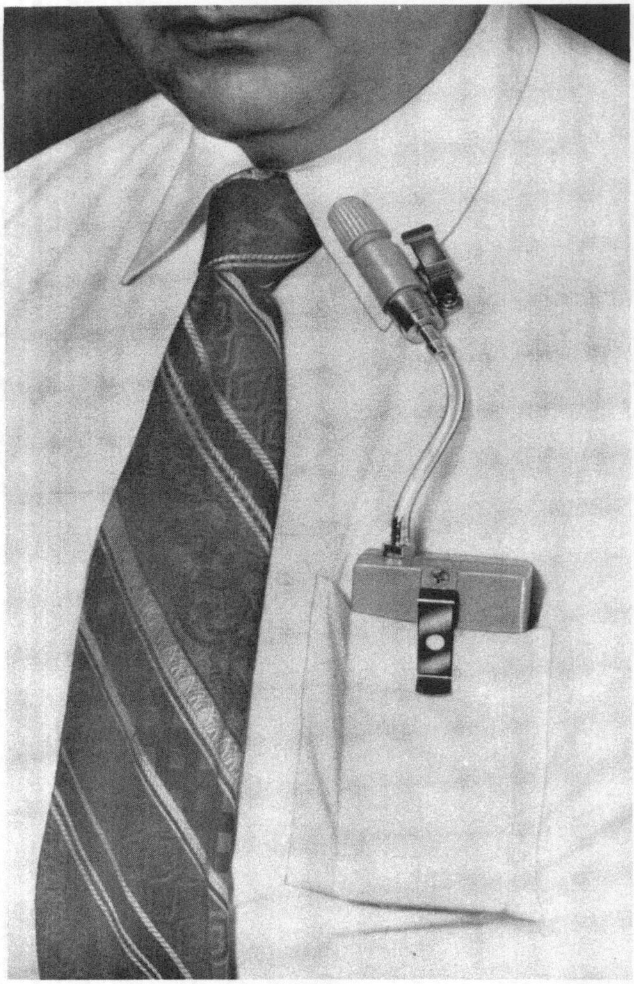

FIGURE 117. Sipin organic vapor sampler in place for personnel monitoring. (Courtesy of Anatole J. Sipin, Inc., 425 Park Ave. S., New York, N.Y.)

NIOSH Cincinnati, Ohio laboratory should in the near future provide a micro-chromatograph designed for personnel monitoring.[254] The pocket-sized instrument is fully self-contained in an 8 × 4 × 15 cm case weighting approximately 0.6 kg. The operational units are built around a miniature integrated GC consisting of a capillary column 3 meters long × 200 μm wide × 20 μm deep, a sample injection valve, and a detector. A state-of-the-art integrated circuit microcomputer handles control and data processing functions. The proposed instrument will be capable of sampling automatically once a minute for 8 hr measuring the concentrations of as many as 10 different vapors simultaneously to within 10% accuracy, calculating and storing the time weighted-average (TWA) and peak concentrations for each of the gases, measuring concentrations upon demand, and sounding an alarm when any concentration exceeds a predetermined value.

The integrated capillary column is formed by etching a long spiral groove into the surface of a 5-cm diameter silicon wafer and then hermetically sealing the wafer to a

FIGURE 122. Calibration in progress with kit components in place. (Courtesy of E. I. du Pont de Nemours, Inc., Applied Technology Division, Wilmington, Del.)

Pyrex® glass cover plate. The cross-section of the enclosed glass channel is roughly rectangular. Holes etched through the wafer at the ends of the gas channel serve as feed and exit for the carrier gas. Fabrication proceeds in a sequence of photo-lithography and etch steps very similar to photo-lithography and diffusion procedure for standard IC device processing. A layer of SiO_2 approximately a μm thick is thermally grown on the wafer to serve as the photoresist mask. After cleaning and stripping off the SiO_2 the wafer is anodically bonded to the Pyrex® cover glass. The wafers are held together and heated to 400°C and a 600-V potential applied between the glass and silicon. Under these conditions complete bonding occurs almost instantaneously.

Stationary phase linings were then added by first conditioning the groove with concentrated hydrochloric acid and then coating with an organosilane compound of the form $RSiX_3$ or R_3SiX, where R is an alkyl group and X is either chlorine or an alkoxide group, to form an Si-O-Si linkage between the silicon and the silane. This surface is now wettable by coatings such as the very polar Carbowax® 400 as the stationary phase.

Because of size limitations, this study was conducted almost exclusively with the thermal conductivity detector (thermistor), which could be assembled directly in the exit port of the column. The miniature solenoid-actuated sampling system employs a simple microdiaphragm valve mounted directly on the intake of the GC column. The injected volume of air sample was in the range 1 to 5 μL. The helium carrier gas was contained in a 10-mL cylinder at approximately 35 $kg/cm.^2$

Space does not permit detailed presentation of the construction and performance of the prototype instrument, but the report from which this material was extracted is complete and in great detail. Although more development and investigation is required to improve the separation efficiency and sensitivity of the detector, the prospect of a personnel monitor of this capability is exciting indeed.

E. The Portable Field Kit Laboratory

The "laboratory in a suitcase" has been a concept that has intrigued the investigator in the field who finds himself beyond the reach of even the basic analytical necessities such as distilled water, electrical power, balances, spectrophotometers, etc. A number of innovations have been proposed, but few have stood the test of time and even fewer have been applied successfully to personnel or biological monitoring. Those models that have survived have done so on their ability to deliver results intermediate in accuracy and precision between the relatively imprecise length of stain gas detector tubes and highly precise laboratory analysis, and on their convenience with on-the-spot results. Length-of-stain gas detector tube kits is not included in this category as, almost without exception, the colorimetric indicator is self-contained and requires no chemical manipulations (see Section II.A.b, "Direct Reading Length of Stain Gas Detecting Tubes"). The field kits available currently are described below.

1. Aromatic Isocyanates in Air — Toluene Diisocyanate (TDI) and Methylene-bis-(4-phenylisocyanate) (MDI)

A kit based on the method developed by Marcali as modified by Grim and Linch for the detection of airborne TDI and MDI in the concentration range 0.01 to 0.08 ppm has produced consistently reliable results in the field and is commercially available.[255] The diisocyanate absorbed in the acid reagent (aqueous HCl and acetic acid) as the air sample is drawn through the sampler is hydrolyzed quantitatively to toluene diamine. The addition of sodium nitrite reagent diazotizes the aromatic amine which is then coupled to "diamine" (1-*a*-naphthylethylenediamine) to produce a blue-red azo dye. The color density is translated to ppm of either TDI or MDI by visual comparison with hermetically sealed, nonfading liquid color standards retained in a comparator block. Although intended primarily for use with the midget impinger, by volumetric adjustments the assembly can be used directly with the microimpinger recommended for personnel monitoring. Personnel monitoring kits complete with battery-operated samplers, microimpingers, reagents, and color standards also are available (Figure 123).[255]

A measured sample of air, 2 ft^3, is drawn at a constant rate of 0.1 ft^3/min through a scrubber containing an iodine-in-methanol solution. As the air sample is scrubbed, organic lead compounds will combine with the iodine to form lead iodide when the reaction is complete. The amount of lead in the scrubber solution is determined by a dithizone colorimetric analysis.

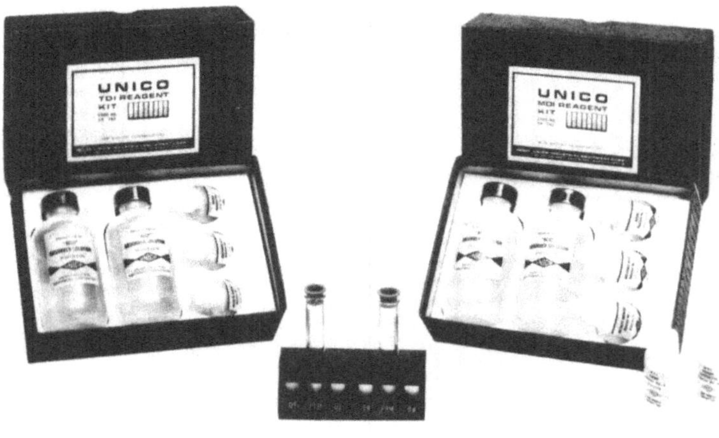

FIGURE 123. TDI/mbl reagents and color standards. Developed by the E. I. du Pont de Nemours Company, the Unico® TDI/mbl reagents and color standards kit provides the sensitivity required for the rapid colorimetric determination of atmospheric concentrations of toluene diisocyanate in the threshold limit range. Results within 0.01 to 0.08-ppm limits can be obtained in less than 20 min. (Courtesy of Bendix Corp., Environmental and Process Instruments Division, Lewisburg, W. Va.)

A similar TDI in air kit based on the orange-brown color formed by the reaction of the toluene diamine that is produced by quantitative hydrolysis of the diisocyanate in the acidic absorbing reagent with glutaconic aldehyde generated *in situ* has been reported.[256] The dye is concentrated by absorption on ion-exchange beads and compared visually with colorimetric standards. The instantaneous readout of the color, which forms as rapidly as TDI is absorbed and permits detection of short duration excursions of TDI concentrations above the TLV, is a distinct advantage. Application to MDI was not mentioned. However, the investigator loses the option of laboratory confirmation of the color density by spectrophotometric analysis. A commercial kit has been offered to the trade.[257] Application to personnel monitoring would be quite simple and would provide immediate warning of excessive concentrations to the water and his associates. Sensitivity as related to sampling time was reported to be 0.005 ppm in 10 min at 2.8 L/min sampling rate and 0.16 ppm in 1¼ min at 2.8 L/min sampling rate.

A test paper method also has been described, but little is known relative to its performance in the field.[258] A 5 sample was drawn at 1 L/min through a 1-cm diameter paper treated with sodium nitrite and 2-hydroxy-11H-benzo(*a*) carbazole-3-carboxy-*p*-anisidide (Brenthol® GB) to produce a red-gray stain at 0.1 ppm (see Section II.B.2.a).

2. Tetraalkyl Lead Compounds in Air — Tetraethyl Lead (TEL) and Tetramethyl Lead (TML)

The dithizone procedure on which the commercially available Telmatic 150® TEL/TML lead-in-air analyzer was based involves an immiscible solvent (chloroform) extraction to isolate the lead as the red lead dithizone complex, which requires a somewhat more elaborate suitcase laboratory[259] (Figures 124 and 125). The Telmatic sampling unit was designed for "walk-around" surveys to evaluate the efficiency of lead gasoline storage tank cleaning operations. The "laboratory" is complete from deionized water to permanent glass color standards. The legends for Figures 124 through 125 tell the story rather completely.

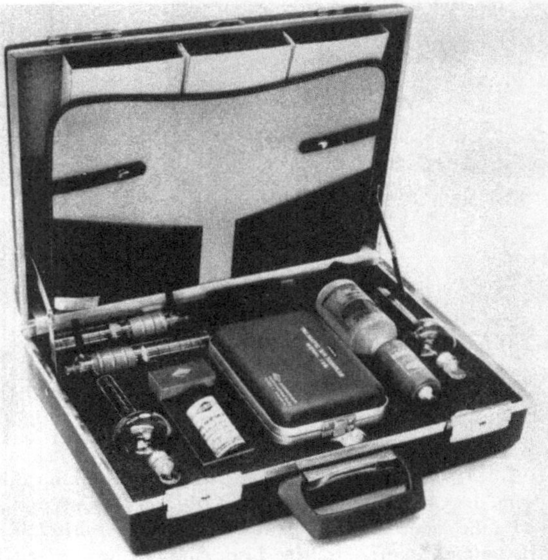

FIGURE 124. The Telmatic 150® lead in air analyzer. The outer case is open to show compact storage and accessibility of equipment. (Courtesy of Bendix Corp., Environmental and Process Instruments Division, Lewisburg, W. Va.)

FIGURE 125. The Telmatic 150® complete kit components. (Courtesy of Bendix Corp., Environmental and Process Instruments, Lewisburg, W. Va.)

To insure a complete reaction to lead iodide, a 2.0 *N* iodine solution in aqueous potassium iodide is added to the methanol solution following collection of the sample. This mixture must then be warmed to at least 26.67°C and held at that temperature for 5 min before proceeding with the analysis. Unless these steps are followed, low readings may be obtained.

The excess iodine is reduced to a colorless compound in the comparator tube by the addition of sulfite solution "A." The resulting aqueous solution is then shaken with a chloroform solution of dithizone which reacts with the lead compounds to form colored complexes. The concentration of lead is determined by comparing the color density with permanent glass color standards in a Unico comparator which is calibrated to read directly in $\mu g/ft^3$ when a 2-ft^3 sample is collected. (The color discs are arranged so that determinations can be made quickly in the field within ± 1 $\mu g/ft^3$ in the range between 0 and 10.)

By substituting the personnel monitor assembly employed for the evaluation of tetraalkyl lead exposure in the manufacturing environment and iodine monochloride reagent for the alcoholic iodine, direct conversion from "walk-around" to personnel monitoring could be attained with a minimum of volumetric changes.[148] No equipment modifications would be necessary.

Conversion to personnel monitoring for mercury and its compounds, including alkyl mercury, also would require only a minimum of adjustments to the reagents and volumetric requirements.[260] The only limitation to the flexibility of this basic field kit is the ingenuity of the investigator, as many other "wet" colorimetric analysis procedures can be adapted with but few alterations in the glassware and calibration of additional color standards. For example, the Reilly procedure for MDI in air would fit into this system with only recalibration of the color standards needed.[261,262]

3. The Unsaturated Perfluoroalkenes — Perfluoroisobutylene (PFIB) and Hexafluoropropene (HFP)

The yellow color produced when PFIB or HFP reacts with a mixture of pyridine and piperidine in methanol provided the basis for a laboratory kit, but this development was not commercially exploited.[263] A 28 L sample collected at a 2.8 L/min rate was sufficient to detect 0.1 ppm PFIB with certainty. Permanent liquid visual color standards were made up from aqueous $K_2Cr_2O_7$. The original specifications were based on the midget impinger for collection of the sample, but only minor volumetric adjustments would be necessary for conversion to personnel monitoring with the microimpinger.[184]

4. Inorganic Lead

A technique for the detection of lead-containing particulate retained on filter paper by reaction with tetrahydroxyquinone to produce a pink to red color[68] has been incorporated into a self-contained air sampler and is available commercially.[264]

5. Water Analysis Kit

This area of application has been exploited to a much greater extent than air analysis. One manufacturer in particular has developed a number of ingenious kits, a suitable electronic colorimeter, and a continuous color scale visual comparator. Complete units for acidity, alkalinity (pH), halogens, halides, CO_2, Ca, Ba, Al, Cr, Cu, cyanide,

TABLE 90

Toxic Substance Detection Kits[a] for Use with Impingers

7616	Alkalinity
7735	Ammonia
7807	Ammonia refill
7971	Bromine
7972	Bromine refill
7793	Cadmium
7796	Cadmium refill
7782	Carbon monoxide
7708	Carbon monoxide refill
7753	Chloride
7809	Chloride refill
7973	Chlorine
7974	Chlorine refill
7700	Chromic acid
7810	Chromic acid refill
7975	Chromium
7976	Chromium refill
7977	Copper
7978	Copper refill
7691	Cyanide
7962	Cyanide refill
7752	Fluoride
7812	Fluoride refill
C-101	Formaldehyde
C-102	Formaldehyde refill
7963	Hydrazine
7964	Hydrazine refill
7753	Hydrogen chloride
7809	Hydrogen chloride refill
7994	Hydrogen peroxide
7997	Hydrogen peroxide refill
7995	Iodine
7998	Iodine refill
7965	Iron oxide (fume)
7966	Iron oxide refill
7667	Lead
7813	Lead refill
7967	Manganese
7968	Manganese refill
7690	Nitrogen dioxide
7814	Nitrogen dioxide refill
7969	Phenol
7970	Phenol refill
7067	Silica
6627	Silica refill
7671	Sulfide (H_2S)
7815	Sulfide refill
7691	Sulfur dioxide
7816	Sulfur dioxide refill
7702	TDI
7818	TDI refill

[a] Kit consists of absorbing solution, reagents, instructions, and color comparator. Approximately 25 tests per kit.

detergents, hydrazine, H_2S, Fe, Mn, Ni, NH_3, $-NO_3$, $-NO_2$, O_2, phenol, phosphate, Se, Si, Ag, $-SO_4$, $-SO_3$, tannin, turbidity, and Zn are available.[273] Collection of any one of these common contaminants in an appropriate reagent in a microimpinger combined with visual color comparison readout would provide a very convenient "package" for in-the-field monitoring.

F. Daco Toxic Substances Detector Kits for Use with Impingers

A liquid phase detector reagent system for over 26 toxic vapors and aerosols designed for use with liquid filled impinger systems was introduced during 1978. These kits include absorbing solution, reagents, instructions, and a colorimeter (110-V AC power source) that reads both absorbance and transmittance with eight filters on a rotating wheel (350, 420, 490, 530, and 610 mm). Tests are basically designed to provide results at 0.5, the TVV and 2 × TLV. Dense color solutions may be diluted to provide on-range readings which are then multiplied by the dilution factor to obtain results above the 2 ×-TLV range. The available kits are tabulated in Table 90.

REFERENCES*

1. **Kelly, W. D.,** Threshold Limit Values of Airborne Contaminants and Physical Agents with Intended Changes Adopted by ACGIH for 1977, Committee on Threshold Limits, Cincinnati, O., 1978.
2. **Morgenstern, A. S., Ash, R. M., and Lynch, J. R.,** The evaluation of detector tube systems. I. Carbon monoxide, *Am. Ind. Hyg. Assoc. J.,* 31, 630, 1970.
3. Joint ACGIH-AIHA Committee on Direct Reading Gas Detecting Systems, Direct reading gas detecting tube systems, *Am. Ind. Hyg. Assoc. J.,* 32, 488, 1971.
4. The Results of Survey of the SO_2 Concentration Distribution in Yokohama City by Kitagawa Test Paper, Yokohama Municipal School Pharmacists Assoc., 1968.
5. **Pfeil, R. W.,** Crayon for Detection of G Agents (Chemical Warfare Gases), U.S. Patent 2,929,791, 1960.
6. **Witten, B. and Prostak, A.,** Sensitive detector crayons for phosgene, hydrogen cyanide, cyanogen chloride and lewisite, *Anal. Chem.,* 29, 885, 1957.
7. Aromil Detector Crayons, Aromil Chemical Company, 5646 Belle Ave., Baltimore, Md., 1957.
8. **Silverman, L. and Gardner, G. R.,** Potassium pallado sulfide method for carbon monoxide detection, *Am. Ind. Hyg. Assoc. J.,* 26, 97, 1965.
9. **McFee, D. R., Lavine, R. E., and Sullivan, R. J.,** Carbon monoxide, a prevalent hazard indicated by detector tabs, *Am. Ind. Hyg. Assoc. J.,* 31, 749, 1970.
10. Air Quality Criteria for Carbon Monoxide, Publication No. AP-62, Natl. Air Pollut. Control Admin., Environmental Health Service, Public Health Service, Department of Health, Education and Welfare, Washington, D.C., March 1970.
11. **Linch, A. L. and Pfaff, H.,** Carbon monoxide — Evaluation of exposure potential by personnel monitor surveys, *Am. Ind. Hyg. Assoc. J.,* 32, 745, 1971.
12. **Plantz, C. A., McConnaughey, P. W., and Jenca, C. J.,** Colorimetric personal dosimeter for hydrazine fuel handlers, *Am. Ind. Hyg, Assoc. J.,* 29, 162, 1968.
13. **Falgout, D. A. and Harding, C. I.,** Determination of H_2S exposure by dynamic sampling with metallic silver filters, *J. Air Pollut. Control Assoc.,* 18, 15, 1968.
14. Photovolt Photometers, Models 200-M, 502-M, 520-M, and 610. Bulletin 810, Photovolt Corp., 1115 Broadway, New York, 1968.
15. **Sanderson, H. P., Thomas, R., and Katz, M.,** Limitations of the lead acetate impregnated paper tape method for hydrogen sulfide, *J. Air Pollut. Control Assoc.,* 16, 329, 1966.
16. **Wolff, G.,** New measuring principle for the determination of hydrogen sulfide, *J. of Drägerwerk A. G.,* 36, 14, 1975.
17. **Chinkman, M.,** Some aspects of measuring the oxygen concentration in gas mixtures using fuel-cell sensors, *Draeger Rev.,* 39, 34, 1977.
18. **Brown, V. R.,** New Solid State Electronic Sensor Developments for Portable Hazardous Gas Detecting Instruments, Enmet, Ann Arbor, Mich.

* Since this reference list was prepared, National Environmental Instruments was acquired by the Bendix Corporation, Environmental and Process Instruments Division, Lewisburg, W. Va.; distribution of Dräger products has been taken over by National Dräger, Inc., Parkway View Drive, Pittsburgh, Pa.

19. Thin film devices as oxygen sensors, *Am. Ind. Hyg. Assoc. J.*, 32, 672, 1971: NASA-CR-1-182 (N68-35149) Clearinghouse for Federal Scientific and Technical Information, Springfield, Va.
20. **Townsend, C. R., Giarrusso, G. A., and Silverman, H. P.,** Thin Film Personal Dosimeters For Detecting Toxic Propellants, Research and Development Division, Magna Corp., Redondo Beach, Cal., Contract AF-33(615)-1751, Project 6302, Task 630203, AMRL-TR-66-231, February 1967, CFSTI, DDC: Ad 652849.
21. **Taguchi, N.,** Gas Detecting Device, British Patent 1,259,566, January 5, 1972.
22. **Blurton, K. R. and Bay, H. W.,** Controlled-potential electrochemical analysis of carbon monoxide, *Am. Lab.*, July, 1974; *Anal. Chem.*, 46, 1837, 1974.
23. Ecolyzer, Energetics Science, Inc., 85 Executive Blvd., Elmsford, N.Y.
24. **Katz, M., Ed.,** *Methods of Air Sampling and Analysis,* 2nd ed., Intersociety Committee on Methods for Air Sampling and Analysis, Am. Public Health Assoc. Publication Service, Washington, D.C., 1977.
25. **Miller, R. W., Stroz, D., and Principe, A. H.,** Vacu-samples, a simple method for taking accurate quantitative gas and vapor samples using a partially evacuated steel containers, *AIHA J.*, 37, 315, 1976.
26. **Palmes, E. D. and Gunnison, A. F.,** Personal monitoring device for gaseous contaminants, *Am. Ind. Hyg. Assoc. J.*, 34, 78, 1973.
27. **Ferber, B. I., Sharp, F. A., and Freedman, R. W.,** Dosimeter for oxides of nitrogen, *Am. Ind. Hyg. Assoc. J.*, 37, 32, 1976.
28. **Palmes, E. D., Gunnison, A. F., DiMattio, J., and Tomczyk, C.,** Personal sampler for nitrogen dioxide, *Am. Ind. Hyg. Assoc. J.*, 37, 570, 1976.
29. **Schnakenberg, G. H.,** A Passive Sampler for Nitrogen Dioxide, Bur. of Mines Tech. Progress Report 95, April 1976, U.S. Dept. Interior, Pittsburgh, Pa.
30. **Reid, R. C. and Sherwood, T. K.,** *The Properties of Gases and Liquids,* 2nd ed., McGraw-Hill, New York, 1966.
31. **Blacker, J. H.,** Triethanolamine for collecting nitrogen dioxide in the TLV Range, *Am. Ind. Hyg. Assoc. J.*, 34, 390, 1973.
32. **Braun, D. L.,** Detection of mercury vapors as an adjunct to respiratory protective devices, presented at the American Industrial Hygiene Conference, San Francisco, May, 1972.
33. **Nelson, G. O.,** *Controlled Test Atmospheres — Principles and Techniques,* Ann Arbor Science Publishers, Ann Arbor, Mich., 1971.
34. **Gilliland, E. R.,** Diffusion coefficients in gaseous systems, *Ind. Eng. Chem.*, June 1934.
35. **Tompkins, F. C. and Goldsmith, R. L.,** A new personal dosimeter for the monitoring of industrial pollutants, *Am. Ind. Hyg. Assoc. J.*, 38, 371, 1977.
36. **McCammon, C. S. and Woodfin, J. W.,** An evaluation of a passive monitor for mercury vapor, *Am. Ind. Hyg. Assoc. J.*, 38, 378, 1977.
37. **Braun, D. L.,** Personal monitoring of mercury vapor exposures, *Proc. Int. Conf. Env. Sens. Assess.*, IEEE Catalogue No. 75-CH 1004-1, also available upon request from 3M Company, St. Paul, Minn.
38. **Linch, A. L.,** *Biological Monitoring for Industrial Chemical Exposure Control,* CRC Press, West Palm Beach, Fla., 1974.
39. Abcor Gasbadge® dosimeter, Product Bulletin No. 1, Abcor Development Corp., Wilmington, Mass.
40. Gasbadge® dosimeter for organic vapors, Product Bulletin No. 2, Abcor Development Corp., Wilmington, Mass.
41. **Mazur, J. R., Bamberger, R. I., Esposito, G. G., Podolak, G., and Jacobs, B.,** A new personal sampler for organic vapors, *Am. Ind. Hyg. Assoc. J.*, 39, 701, 1978.
42. **Taylor, D. G.,** NIOSH Manual of Analytical Methods, Vol. 1-3, Publication No. 77-157-A, National Institute for Occupational Safety and Health, U.S. Department of Health, Education and Welfare, Cincinnati, Ohio, 1977.
43. **Silverstein, L. G.,** Validation of Abcor Gasbadge® for acrylonitrile and improved desorption efficiency, *Am. Ind. Hyg. Assoc. J.*, 38, 412, 1977.
44. **Bailey, A. and Hollingdale-Smith, P. A.,** *A Passive Sampler for Gases and Vapors,* Chemical Defence Establishment, Poston Down, Salisburg, Wiltshire, England, preprint.
45. **Stannet, V. et. al.,** *TAPPI Monogr. Ser.*, 23, 1962.
46. **Nelms, L. H., Reiszner, K. D., and West, P. W.,** Personal vinyl chloride monitoring device with permeation technique for sampling, *Anal. Chem.*, 49, 994, 1977.
47. **Reiszner, K. D. and West, P. W.,** Collection and determination of sulfur dioxide incorporating permation and West-Gaeke procedure, *Environ. Sci and Technol.*, 7, 526, 1973.
48. **Eaton, H. G., Stone, J. P., and Williams, F. W.,** Personal Atmospheric Gas Sampler with a Critical Orifice, Part I, Development and Evaluation, NRL Report 7963, Naval Research Lab., Washington, D.C., 1976.

49. **Eaton, H. G., Stone, J. P., and Williams, F. W.,** Personal Atmospheric Gas Sampler With a Critical Orifice, Part II, System for Handling and Analyzing the Gas from the Sampler, NRL Report 7980, Naval Research Lab., Washington, D.C.
50. Hydrogen Sulfide — Methods for the Detection of Toxic Substances in Air, Booklet No. 1, Her Majesty's Stationery Office, London, 1958.
51. Hydrogen Cyanide Vapour — Methods for the Detection of Toxic Gases in Industry, Leaflet No. 2, Her Majesty's Stationery Office, London, 1938.
52. Sulfur Dioxide — Methods for the Detection of Toxic Gases in Industry, Leaflet No. 3, Her Majesty's Stationery Office, London, 1939.
53. Phosgene — Methods for the Detection of Toxic Substances in Air, Booklet No. 8, Her Majesty's Stationery Office, London, 1939.
54. Arsine — Methods for the Detection of Toxic Gases in Industry, Leaflet No. 9, Her Majesty's Stationery Office, London, 1939.
55. **Linch, A. L., Lord, S. S., Kubitz, K. A., and DeBrunner, M. R.,** Phosgene in air — development of improved detection procedures, *Am. Ind. Hyg. Assoc. J.*, 26, 465, 1965.
56. **Moureu, H., Chovin, P., and Triffert, L.,** Sur la transformation photochemique de la chloropicrin en phosgene, *C.R.*, 228, 1954, 1949.
57. **Liddel, H. F.,** A detector paper for phosgene, *Analyst (London)*, 82, 375, 1957.
58. **Lamouroux, A.,** A new specific color reaction of phosgene, *Mem. Poudres*, 38, 383, 1956.
59. **Dixon, B. E. and Hands, G. C.,** A field method for the determination of phosgene, *Analyst (London)*, 84, 463, 1959.
60. **Leithe, W.,** *The Analysis of Air Pollutants*, Ann Arbor Science, Ann Arbor, Mich., 1970.
61. **Jacobs, M. B.,** *The Analytical Toxicology of Industrial Inorganic Poisons*, Interscience, New York, 1967.
62. Intersociety Committee for a Manual of Methods for Air Sampling and Analysis, Tentative method of analysis for sulfur dioxide content of the atmosphere — Colorimetric 42401-01-69T, *Health Lab. Science*, 7, 4, 1970 (See also Reference 24).
63. Arsine Detector Kit, Catalog No. DA-81100, Mine Safety Appliances Co., 201 N. Braddock Ave., Pittsburgh, Pa., 1962.
64. *Reagent Chemicals — ACS Specifications*, 4th ed., American Chemical Society, Washington, D.C., 1968.
65. British Ministry of Labor, Sulfur Dioxide, Booklet No. 5, Her Majesty's Stationery Office, London.
66. Tintometer Ltd., Salisbury, Wiltshire, England.
67. **Hemeon, W. C. and Haines, G. F.,** Automatic sampling and determination of micro-quantities of mercury vapor, *Am. Ind. Hyg. Assoc. J.*, 22, 75, 1961.
68. **Quino, E. A.,** Field method for the determination of inorganic lead fumes in air, *Am. Ind. Hyg. Assoc. J.*, 20, 134, 1959.
69. **Yeager, D. W., Cholak, J., and Henderson, E. W.,** Determination of lead in biological and related material by atomic absorption spectrophotometry, *Environ. Sci. Technol.*, 5, 1020, 1971.
70. **Matson, W. R., Allen, H. E. and Mancy, K. H.,** Trace metal characterization in aquatic environments by anodic stripping voltammetry, *J. Water Pollut. Control Fed.*, 42, 573, 1970.
71. The Multiple Anodic Stripping Analyzer, MASA 2014, Environmental Sciences Associates, 49 Hampshire St., Cambridge, Mass., 1971.
72. **Hands, G. C. and Bartlet, A. F. F.,** A field method for the determination of sulfur dioxide in air, *Analyst (London)*, 85, 147, 1960.
73. **Keenan, R. G.,** Determination of Lead in Air and in Biological Materials: The "USPHS" Double Extraction, Mixed Color Dithizone Method, ACGIH, Cincinnati, Ohio, 1963; also *Am. Ind. Hyg. Assoc. J.*, 24, 481, 1963.
74. **Leroux, J. and Mahmud, M.,** Flexibility of X-ray emission spectrography as adapted to microanalysis of air pollutants, *J. Air Pollut. Control Assoc.*, 20, 402, 1970.
75. **Judd, S. H. and Tebbens, B. S.,** Air sampling and analysis of lead on paper, *Am. Ind. Hyg. Assoc. J.*, 22, 86, 1961.
76. AISI Samplers, Bulletins 2322A and 2331, Research Appliance Co., Craighead Rd., Allison Park, Pa., 1971.
77. Tape Samplers, 2800 Series, Models 80TS-1, H_2S, and HF, National Environmental Instruments, Inc., P.O. Box 590, Pilgrim Station, Warwick, R.I., 1971.
77a. **Katz, M., Ed.,** Tentative Method of Analysis for Particulate Fluorides in the Atmosphere (Separation and Collection with a Double Paper Tape Sampler), *Intersociety Committee on Methods for Air Sampling and Analysis*, Am. Public Health Assoc., Washington, D.C., 1972 (See also Reference 24).
78. **Stone, I.,** Detection of fluorides using the zirconium lake of alizarin, *J. Chem. Educ.*, 8, 347, 1931; see also Mavrodineanu, R., U.S. Patent 2,823,984.

79. **Leithe, W.**, *The Analysis of Air Pollutants*, Ann Arbor Science, Ann Arbor, Mich., 1970, 210.
80. **Bennet, E. L., Gould, C. W., and Swift, E. H.**, Systematic qualitative tests for certain acid elements in organic compounds, *Anal. Chem.*, 19, 1035, 1947.
81. Unsymmetrical Hydrazine Detector Kit, Catalog No. DH-82050, Mine Safety Appliances Co., 201 N. Braddock Ave., Pittsburgh, Pa., 1962.
82. Boranes-In-Air Detector, Catalog No. DA-82100, Mine Safety Appliances Co., 201 N. Braddock Ave., Pittsburgh, Pa., 1962.
83. **Ruch, W. E.**, *Chemical Detection of Gaseous Pollutants*, Ann Arbor Science, Ann Arbor, Mich., 1966.
84. Chromic Acid Mist Detector, Catalog No. DA-79011, Mine Safety Appliances Co., 201 N. Braddock Ave., Pittsburgh, Pa., 1962.
85. **Riehl, W. A. and Hager, K. F.**, Rapid detection of aniline vapors in air, *Anal. Chem.*, 27, 1768, 1955.
86. Paper Tape Densitometer, Catalog No. 1401, National Environmental Instruments, Inc., P.O. Box 590, Pilgrim Station, Warwick, R.I., 1971.
87. RAC Spot Evaluators, Bulletins 2296, 2322A, and 2348 (models numbered same as bulletins), Research Appliance Co., Craighead Rd., Allison Park, Pa., 1971.
88. **Klinsky, J. W., Ed.**, Direct Reading Colorimetric Indicator Tubes Manual, American Industrial Hygiene Association, Akron, Ohio, 1976.
89. **Saltzman, B. E.**, Direct Reading Colorimetric Indicators, in *Air Sampling Instruments for Evaluation of Atmospheric Contaminants* 4th ed., ACGIH, Cincinnati, Ohio, 1972.
89a. **Silverman, L.**, Panel discussion of field indicators in industrial hygiene, *Am. Ind. Hyg. Assoc. J.*, 23, 108, 1962.
90. **Ketcham, N. H.**, Practical experience with routine use of field indicators, *Am. Ind. Hyg. Assoc. J.*, 23, 127, 1962.
91. **Leidel, N. A., Busch, K. A., and Lynch, J. R.**, Occupational Exposure Sampling Strategy Manual, DHEW (NIOSH) Publication No. 77-173, National Institute of Occupational Safety and Health, U.S. Department of Health, Education and Welfare, Cincinnati, Ohio, 1977.
92. **Leidel, N. A. and Busch, K. A.**, Statistical Methods for the Determination of Noncompliance with Occupational Health Standards, NIOSH Technical Publication No. 75-159, National Institute of Occupational Safety and Health, U.S. Department of Health, Education and Welfare, Cincinnati, Ohio, 1975.
93. Criteria for a Recommended Standard Occupational Exposure to Carbon Monoxide, HSM 73-11000, National Institute of Occupational Safety and Health, Department of Health, Education and Welfare, Rockville, Md., 1972.
94. **Kitagawa, T.**, The rapid measurement of toxic gases and vapors, *Trans. 13th Int. Congr. Occup. Health*, New York, 1960.
95. The National Draeger System of Detecting and Measuring Hazardous Gases and Vapors, BGI Inc., Waltham, Mass.
96. **Ringold, A., Goldsmith, J. R., Helwig, H. L., Finn, R., and Scheute, F.**, Estimating recent carbon monoxide exposures, a rapid method, *Arch. Environ. Health*, 5, 38, 1962.
97. Concentration Determination With Dräger Tubes — Gases and Vapours — Physical and Toxicological Data, Table 2340e, 4th ed., Drägerwerk AG, Lübeck, Germany, May 1971. (U.S. representative: National Mine Service Co., 300 Koppers Bldg., Pittsburgh, Pa.)
98. **Beatty, R. L.**, Methods for Detecting and Determining Carbon Monoxide, Bulletin 557, Bureau of Mines, Department of the Interior, 1955.
99. Drunko-Tester, National Environmental Instruments, Inc., P.O. Box 590, Pilgrim Station, Warwick, R.I., 1959.
100. **Coldwell, B. B. and Smith, H. W.**, Alcohol levels in body fluids after ingestion of distilled spirits, *Can. J. Biochem.*, 37, 43, 1959.
101. **Turner, R. F., Bennett, W. B., Cestaric, E. S., and Meuhlberger, C. W.**, Evaluating Chemical Tests for Intoxication, report of the Committee on Tests for Intoxication, Natl. Safety Council, 425 N. Michigan Ave., Chicago, 1952.
102. **Ingram, W. T.**, Personal air-pollution monitoring devices, *Am. Ind. Hyg. Assoc. J.*, 25, 298, 1964.
103. **Linch, A. L.**, Oxygen In air analysis — evaluation of a length of stain detector, *Am. Ind. Hyg. Assoc. J.*, 26, 645, 1965.
104. **Morgenstern, A. S., Ash, R. M., and Lynch, J. R.**, The evaluation of gas detector tube systems: Carbon monoxide, *Am. Ind. Hyg. Assoc. J.*, 31, 630, 1970.
105. **Ash, R. M. and Lynch, J. R.**, The evaluation of gas detector tube systems: Benzene, *Am. Ind. Hyg. Assoc. J.*, 32, 410, 1971.
106. **Hay, E. B.**, Exposure to aromatic hydrocarbons in a coke oven by-product plant, *Am. Ind. Hyg. Assoc. J.*, 25, 386, 1964.

107. **Ash, R. M. and Lynch, J. R.,** The evaluation of gas detector tube systems: Sulfur dioxide, *Am. Ind. Hyg. Assoc. J.,* 32, 490, 1971.
108. **Ash, R. M. and Lynch, J. R.,** The evaluation of gas detector tube systems: Carbon tetrachloride, *Am. Ind. Hyg. Assoc. J.,* 32, 552, 1971.
109. **Roper, C. P.,** An evaluation of perchloroethylene tubes, *Am. Ind. Hyg. Assoc. J.,* 32, 847, 1971.
110. Direct reading gas detecting tube systems, report of the Joint Am. Conf. Governmental Industrial Hygienists, Am. Ind. Hyg. Assoc. Committee on Direct Reading Gas Detecting Systems, *Am. Ind. Hyg. Assoc. J.,* 32, 488, 1971.
111. **McKarns, J. S., Hill, F. N., and Bolton, P. R.,** Monostyrene vapor concentrations from plastic concrete, *Am. Ind. Hyg. Assoc. J.,* 28, 414, 1967.
112. **Larsen, L. B. and Hendricks, H.,** An evaluation of certain direct reading devices for the determination of ozone, *Am. Ind. Hyg. Assoc. J.,* 30, 620, 1969.
113. **Kusnetz, H. L., Saltzman, B. E., and Lanier, M. E.,** Calibration and evaluation of gas detecting tubes, *Am. Ind. Hyg. Assoc. J.,* 21, 361, 1960.
114. **Roper, P.,** Detector Tube Certification Schedule, first draft, NIOSH, Health Services and Mental Health Admin., Public Health Service, Department of Health, Education and Welfare, Cincinnati, O., June 1971; *Fed. Register,* 37 (No. 84), 19643, September 21, 1972; 42CFR, part 84.
115. **Kusnetz, H. L.,** Evaluation of Chemical Detector Tubes, in *Trans. 1965 Natl. Safety Congr.,* Vol. 5, Chicago, 1965, 28.
116. **Kobayashi, Y.,** Rapid quantitative determination of a minute quantity of arsine by means of a detector tube, *J. Chem. Soc. Jpn., Ind. Chem. Sec.,* 55, 544, 1952.
117. Department of Health, Education and Welfare, Gas detector tube units, proposed approved, 42CFR Part 84, *Fed. Regist.,* Vol. 37, No. 184, 1972.
118. **Colen, F. H.,** A Study of the interchangeability of gas detector tubes and pumps, *Am. Ind. Hyg. Assoc. J.,* 35, 696, 1974.
119. Certification of gas detector tube units, procedure for evaluation and certification, Title 42, Public Health, Part 84, *Fed. Regist.,* Vol. 38, No. 88, 1973.
120. **Leichnitz, K.,** *Dräger Rev.,* 33, 1, 1974.
121. Military Standard, Sampling Procedures and Tables for Inspection by Attributes, MIL-STD-105D, U.S. Government Printing Office, Washington, D.C., 1963.
122. Military Standard, Sampling Procedures and Tables for Inspection by Variables for Percent Defective, MIL-STD-414, U.S. Government Printing Office, Washington, D.C., 1957.
123. **Leidel, N. A., Busch, K. A., and Crouse, W. E.,** Exposure Measurement Action Level and Occupational Environmental Variability, NIOSH Technical Information, Division of Laboratories and Criteria Development, U.S. Department of Health, Education and Welfare, Cincinnati, Ohio, 1975.
124. **Kitagawa, T.,** Rapid method of quantitative gas analysis by means of detector tubes, *J. Chem. Soc. Jpn.,* 6, 386, 1952.
125. MSA Monitair Colorimetric Detector Tube Assembly, Catalog Nos. 92943 and 93481, Mine Safety Appliances Co., 201 N. Braddock Ave., Pittsburgh, Pa., 1964.
126. **Kusnetz, H. L.,** Air flow calibration of direct reading colorimetric gas detecting devices, *Am. Ind. Hyg. Assoc. J.,* 21, 340, 1960.
127. Field Calibration Flowmeter, Inspector Model No. 302, and Soap Film Flowmeters, Catalog Nos. 307-02 to 313-100 (2-ml to 100-ml capacity), SKC, Inc., 402 Greentree Rds., Pittsburgh, Pa., 1972; National Mine Service Co., 3000 Koppers Bldg., Pittsburgh.
128. Hastings Mini-Flo Calibrator (0.1 to 2000 cc/min), Hastings-Raydist, Teledyne Company, 1500 Newcombe Ave., Hampton, Va., 1971.
129. Soap Film Flowmeter, John P. Mickel Associates, Inc., 158 S. Rochester Rd., Clawson, Mich., 1970.
130. **Kinosian, J. R. and Hubbard, B. R.,** Nitrogen oxide indicator, *Am. Ind. Hyg. Assoc. J.,* 19, 453, 1958.
131. **Saltzman, B. E.,** Basic theory of gas indicator tube calibrations, *Am. Ind. Hyg. Assoc. J.,* 23, 112, 1962.
132. **Leichnitz, K.,** Use of detector tubes under extreme conditions (humidity, pressure, temperature), *Am. Ind. Hyg. Assoc. J.,* 38, 702, 1977.
133. Air contaminants (gases, vapors, fumes, dusts, and mists), Part II, National Consensus Standards and Established Federal Standards, Occup. Safety and Health Standards, Occup. Safety and Health Admin., Department of Labor, *Fed. Regist.,* 36(No. 105), 1910.93, May 29, 1971.
134. **Eckstein, W.,** Dräger quantimeter, *Dräger Rev.,* 38, 18, 1976.
135. **Charsha, R. C. and Linch, A. L.,** "Freon" powered portable air sampling kit, *Am. Ind. Hyg. Assoc. J.,* 18, 135, 1957.
136. **Linch, A. L. and Charsha, R. C.,** Development of a freeze-out technique and constant sampling rate for the portable Unit-Jet Air Sample®, *Am. Ind. Hyg. Assoc. J.,* 21, 325, 1960.

137. **Linch, A. L., Charsha, R. C., and Wetherhold, J. M.,** The "Uni-Jet Sampler"® operating characteristics and applications, *Proc. 13th Int. Congr. Occup. Health,* 580, 1961.
138. **Linch, A. L., Davis, R. B., Stalzer, R. F., and Anzilotti, W. F.,** Method for lead-in-air determinations and use with an improved self-powered portable sampler, *Am. Ind. Hyg. Assoc. J.,* 25, 81, 1964.
139. **Kohn, R. J.,** The "Uni-Jet"® Constant Rate Air Sampler, National Environmental Instruments, Inc. (formerly Union Industrial Equipment Corp.), P.O. Box 590, Pilgrim Station, Warwick, R.I., 1960.
140. **Campbell, E. E. and Miller, H. M.,** *Chemical Detectors — A Bibliography for the Industrial Hygienists with Abstracts and Annotations,* Part 1, May 1961, Part 2, December 1963, LAMS 2378, Office of Technical Services, Department of Commerce, U.S. Government Printing Office, Washington, D.C.
141. **Merz, O.,** Orientierende Messungen mit Gaspruefroehrchen bei Luft- und Ofentrocknung (Informative measuring with gas test tubes in air and oven drying), *Staub Reinhalt. Luft,* 31(10), 399, 1971.
142. **Leichnitz, K.,** *Dräger Rev.,* 33, 38, 1974.
143. **Neff, J. E. and Ketcham, N. H.,** A detector tube for the analysis of methyl isocyanate in air or nitrogen purge gas, *Am. Ind. Hyg. Assoc. J.,* 35, 468, 1974.
144. Bubble Flowmeters, Catalog Nos. K-627100 and K-627150, Kontes Glass Co., Vineland, N.J., 1972.
145. **Klein, A. H.,** Ace Glass, Inc., Vineland, N.J.
146. **Leichnitz, K.** Das Prüfrörchenverfahen und Seine Entrvicklungstendenzen, *Chem. Ztg.,* 97, 638, 1973.
146a. **Leichnitz, K.,** Luftunterschung mit Langzeit, Prüfröhrchen, Sonderdruck GT 38 aus dem Dragerheft 304, April-June 1976.
147. Safety Regulation No. 3, Establishing Threshold Limit Values for Dust, Vapors, Fumes, Gases and Mists, Bureau of Engineering and Safety, New Jersey Department of Labor and Industry, Trenton, N.J., 1967.
148. **Linch, A. L., Wiest, E. G., and Carter M. D.,** Evaluation of tetraalkyl lead exposure by personnel monitor surveys, *Am. Ind. Hyg. Assoc. J.,* 31, 170, 1970.
149. **Kohn, R. J.,** Unico® Carbon Monoxide Detector Kit (Model 400 Precision Gas Detector Pump with Model 100 Carbon Monoxide Detector Tubes), National Environmental Instruments Corp. (formerly Union Industrial Equipment Corp.), Box 590, Pilgrim Station, Warwick, R.I., 1971.
150. **Sayers, R. R., Yant, W. P., and Jones, G. W.,** The pyro-tannic acid method for the quantitative determination of carbon monoxide in blood and air, Reprint No. 872, U.S. Public Health Service, Treasury Department, *Public Health Rep.,* 38(No. 40), 2311, 1923; Mine Safety Appliances, Catalog No. CB-4598.
151. **Scholander, P. F. and Roughton, F. J. W.,** Micro gasometric estimation of blood gases, II, Carbon monoxide, *J. Biol. Chem.,* 148, 551, 1943.
152. **Stokinger, H. E., Ed.,** *Documentation of Threshold Limit Values,* 3rd ed., ACGIH, Cincinnati, Ohio, 1971, 41.
153. **Patty, F. A.,** *Industrial Hygiene and Toxicology,* Vol. II, 2nd ed., Interscience, New York, 1962, 930.
154. **Allen, T. H. and Root, W. S.,** Colorimetric determination of carbon monoxide in air by an improved palladium chloride method, *J. Biol. Chem.,* 216, 309, 1955.
155. **Levaggi, D. A. and Feldstein, M. F.,** The colorimetric determination of low concentrations of carbon monoxide, *Am. Ind. Hyg. Assoc. J.,* 25, 64, 1964.
156. **Beatty, R. L.,** Methods for Detecting and Determining Carbon Monoxide, Bulletin 557, Bureau of Mines, Department of the Interior, 1955.
157. **Katz, M., Ed.,** Tentative method for preparation of carbon monoxide standard mixtures (42101-01-69T), *Lab. Health Sci.,* Suppl. 7, 72, 1970 (See also Reference 24).
158. Compact Infusion Pump, Model No. 975, Harvard Apparatus Co., Millis, Mass.
159. **Koh, R. J.,** "Mighty-Mite" Air Sampler, Model 440-B, National Environmental Instruments Corp. (formerly Union Industrial Equipment Corp.), Box 590, Pilgrim Station, Warwick, R.I.
160. **McFee, D. R. and Fish, H. R.,** Exhaust entry monitored automatically with flame ionization and indicator tubes, *Am. Ind. Hyg. Assoc. J.,* 32, 826, 1971.
161. **Little J. and Penketh, F. G.,** The use of a constricted rubber bulb for obtaining an average sample of an atmosphere during a working period, *Analyst (London),* 88, 973, 1963.
162. **Leichnitz, K.,** Detector tubes and prolonged air sampling, *Natl. Saf. News,* p. 59, 1977; *Drager Rev.,* 40, 9, 1977.
163. **Demenberg, B. A. and Kriesel, R.,** Chemically impregnated paper tapes as a detection method in continuous personal and area monitors, *Am. Ind. Hyg. Assoc. J.,* 37, 246, 1976, MDA Scientific, Inc., Park Ridge, Ill.

164. **Reilly, D.A.,** A test paper method for the determination of tolylene di-isocyanati in air, *Analyst (London)*, 93, 178, 1968.
165. **Denenberg, B. A., Kriesel, R. S., and Miller, R. W.,** A Miniature Continuous Monitoring System for Determination of Breathing Zone and Ambient Air Concentrations of Toxic Substances, paper presented to the Intl. Conf. on Environ. Sensing and Assessment, Sept. 14-19, 1975, Las Vegas, Nev; (copies available from MDA Scientifice).
166. Sipin, Cenatole J. Sipin Company, Inc., 425 Park Ave. S., New York, N.Y.
167. Du Pont, E.I. du Pont de Nemours, Inc., Applied Technology Division, Wilmington, Del.
168. **Kubota, R.,** Kitagawa Precision Gas Detectors, Komyo Research Laboratories, 1-8-24 Chu-Cho, Meguro, Tokyo.
169. **Grim, K. E. and Linch, A. L.,** Recent isocyanate-in-air analysis studies, *Am. Ind. Hyg. Assoc. J.*, 25, 285, 1964.
170. **Momberg, H. E. and McCannon, C. S.,** A Comparison of Two Methods for the Determination of Toluene-di-isocyanate in Air, NIOSH Technical Information, HEW Publ. No. (NIOSH) 760168, June 1976.
171. **Miller, J. A. and Mueller, F. X.,** Evaluation of a continuous instrument method for the determination of isocyanates, *Am. Ind. Hyg. Assoc. J.*, 36, 477, 1975.
172. MDA Scientific Inc., New instruments, *Am. Ind. Hyg. Assoc. J.*, 38, A-22, 1977.
173. *Fed. Regist.*, 39, 194, 1974.
174. **Lucchesi, C. A.,** The analytical approach — industrial analytical chemists and OSHA regulations for vinyl chloride, *Anal. Chem.*, 47, 1075A, 1975.
174a. **Polasek, J. C. and Bullin, J. A.,** Evaluation of bag sequential sampling technique for ambient air analysis, *Environ. Sci. Technol.*, 12, 708, 1978.
175. **Littlefield, J. B. and Schrenk, H. H.,** Bureau of Mines Midget Impinger for Dust Sampling, R.I. 3387, Bureau of Mines, Department of the Interior, 1938.
176. **McAfee, J.,** All-glass midget impinger unit, *Ind. Eng. Chem. Anal. Ed.*, 16, 346, 1944.
177. **Lippman, M., Ed.,** *Air Sampling Instruments for Evaluation of Atmospheric Contaminants*, 4th ed., ACGIH, Cincinnati, Ohio, 1972.
178. **Roberts, L. R. and McKee, H. C.,** Evaluation of absorption sampling devices, *J. Air Pollut. Control Assoc.*, 9, 51, 1959.
179. Recommended practices for sampling atmospheres for analysis of gases and vapors, ASTM designation D, *Am. Soc. Test. Mater.*, 1605, 1960.
180. **Elkins, H. B.,** *The Chemistry of Industrial Toxicology*, 2nd ed., John Wiley and Sons, New York, 1959.
181. **Monkman, J. E., Ed.,** *Analytical Guides*, American Industrial Hygiene Assoc., Akron, Ohio, 1965.
182. **Calvert, S. and Workman, W.,** The efficiency of small gas absorbers, *Am. Ind. Hyg. Assoc. J.*, 22, 318, 1961.
183. **Dowling, T., Davis, R. B., Charsha, R. C., and Linch, A. L.,** Lead-in-air analyzer — a specialized air sampling and analysis kit, *Am. Ind. Hyg. Assoc. J.*, 19, 330, 1958.
184. **Linch, A. L. and Corn, M.,** The standard midget impinger — design improvement and miniaturization, *Am. Ind. Hyg. Assoc. J.*, 26, 601, 1965.
185. **Neale, E. and Perry, B. J.,** The sampling and analysis of toxic vapors in the field, *Analyst (London)*, 84, 226, 1959.
186. **Yaffee, C. D., Beyers, D. H., and Horsey, A. D.,** *Encyclopedia of Instrumentation for Industrial Hygiene*, University of Michigan, Ann Arbor, 1956, 211.
187. **West, P. W. and Gaeke, G. C.,** Fixation of sulfur dioxide as disulfitomercurate (II) and subsequent colorimetric estimation, *Anal. Chem.*, 21, 361, 1960.
188. **Linch, A. L.,** The spill-proof microimpinger, *Am. Ind. Hyg. Assoc. J.*, 28, 497, 1967.
189. **Linch, A. L., O'Connor, G. B., Barnes, J. R., Killian, A. S., and Neeld, W. E.,** Methylene-bis-ortho-chloro-aniline (MOCA®): Evaluation of hazards and exposure control, *Am. Ind. Hyg. Assoc. J.*, 32, 802, 1971.
190. **Katz, M., Ed.,** Tentative method of analysis for nitrogen dioxide content of the atmosphere (Griess-Saltzman reaction) (42602-01-68T), *Health Lab. Sci.*, 6, 106, 1972: **Katz, M.,** *Methods of Air Sampling and Analysis*, Intersociety Committee, American Public Health Assoc., Washington, D.C., 1972.
191. **McCaldin, R. D. and Hendrickson, E. R.,** Use of a gas chamber for testing air samplers, *Am. Ind. Hyg. Assoc. J.*, 20, 509, 1959.
192. Micro-glass-spheres ("Uni-Spheres"), Class 4, No. 203, Microbeads Division, Cataphote Corp., Jackson, Miss., 1969.
193. **Peterson, J. E., Hoyle, H. B., and Schneider, E. J.,** The analysis of air for halogenated hydrocarbon contaminants by means of absorption on silica gel, *Am. Ind. Hyg. Assoc. Q.*, 17, 429, 1956.
194. **Schetlan, L. and Jacobs, M. B.,** *The Handbook of Solvents*, Van Nostrand Reinhold, New York, 1953.

195. **Moffet, D. A., Doherty, T. F., and Monkman, J. L.,** Collection and determination of micro amounts of benzene and toluene in air, *Am. Ind. Hyg. Assoc. Q.*, 17, 186, 1956.
196. **Erley, D. S.,** Infrared analysis of air contaminants trapped on silica gel, *Am. Ind. Hyg. Assoc. J.*, 23, 388, 1962.
197. **Urone, P. and Smith, J. E.,** Analysis of chlorinated hydrocarbons with the gas chromatograph, *Am. Ind. Hyg. Assoc. J.*, 22, 36, 1961.
198. **Altshuller, A. P., Bellar, T. A., and Clemons, C. A.,** Concentration of hydrocarbons on silica gel prior to gas chromatographic analysis, *Am. Ind. Hyg. Assoc. J.*, 23, 164, 1962.
199. **Pernell, C.,** The collection and analysis of halogenated hydrocarbon vapors employing silica gel as an absorbing agent, *J. Ind. Hyg. Toxicol.*, 26, 331, 1944.
200. **Fahy, J. P.,** Determination of chlorinated hydrocarbons in air, *J. Ind. Hyg. Toxicol.*, 30, 207, 1948.
201. **Taylor, D. G.,** Manual Coordinator NIOSH Manual of Analytical Methods, U.S. Department of Health, Education and Welfare, National Institute for Occupational Safety and Health, Center for Disease Control, Public Health Service, Cincinnati, O., 1977.
202. **Wood, G. O. and Anderson, R. G.,** Personal air sampling for vapors of aniline compounds, *Am. Ind. Hyg. Assoc. J.*, 36, 538, 1975.
203. **Vincent, W. J., Hohn, K. J., and Ketcham, N. H.,** Determining ammonia in the presence of airborne concentrations of ethylenedramine and monoethanolamine, *Am. Ind. Hyg. Assoc. J.*, 39, 502, 1978.
204. **Verstuyft, A. W.,** Sampling and analytical methods for phosphine, a review, *Am. Ind. Hyg. Assoc. J.*, 39, 431, 1978.
205. **Dillon, J. K. and Barrett, W. J.,** Determination of Elemental Phosphorous and Phosphine in Air, NIOSH Contract No. 210, 75-0047, final report 50R1-EAS-77-060.
206. **Jacobs, M. B.,** *The Analytical Chemistry of Industrial Poisons, Hazards and Solvents*, 2nd ed., Interscience, New York, 1949.
207. **Campbell, E. E. and Ide, H. M.,** Air sampling and analysis with micro columns of silica gel, *Am. Ind. Hyg. Assoc. J.*, 27, 323, 1966.
207a. **Kring, E. V., Douglas, J. J., Hoffman, R. A., and Lautenberger, R. A.,** A New Passive Personal Sampler for Inorganic Vapors — Ammonia, Sulfur Dioxide and Nitrogen Dioxide, Paper No. 115, Am. Ind. Hyg. Conf., Chicago, Ill., May 29, 1979.
208. Individual Protective and Detection Equipment, Technical Manual TM3-290, Department of the Army, 56, 1953.
209. **Hay, E. B.,** Exposure to aromatic hydrocarbons in a coke oven by-product plant, *Am. Ind. Hyg. Assoc. J.*, 25, 386, 1964.
210. **Sherwood, R. J.,** The monitoring of benzene exposure by air sampling, *Am. Ind. Hyg. Assoc. J.*, 32, 840, 1971.
211. **Feldstein, M., Bolestrieri, S., and Leoaggi, D. A.,** The use of silica gel in source testing, *Am. Ind. Hyg. Assoc. J.*, 28, 381, 1967.
212. **Van Mourik, J. H. C.,** Experiences with silica gel as adsorbent, *Am. Ind. Hyg. Assoc. J.*, 26, 498, 1965.
213. Charcoal tubes for OSHA compliance of certain chemicals in air, Environmental Compliance Corp., Venetia, Pa., 1972.
214. **Whitman, N. E. and Johnston, A. E.,** Sampling and analysis of aromatic hydrocarbon vapors in air: A gas-liquid chromatographic method, *Am. Ind. Hyg. Assoc. J.*, 25, 464, 1964.
215. **Durham, W. F. and Wolfe, H. R.,** Measurement of the exposure of workers to pesticides, *Bull. W.H.O.*, 26, 75, 1962.
216. **Saalwalchter, A. T., McCammon, C. S., Roper, C. P., and Carlberg, K. S.,** Performance testing of the NIOSH charcoal tube technique for the determination of air concentrations of organic vapors, *Am. Ind. Hyg. Assoc. J.*, 38, 476, 1977.
217. **Nelson, G. O. and Harder, C. A.,** Respirator cartridge efficiency studies: V. Effect of solvent vapor, *Am. Ind. Hyg. Assoc. J.*, 35, 391, 1974.
218. Collaborative Testing of Activated Charcoal Sampling Tubes for Seven Solvents, HEW Publication No. (NIOSH) 75-184, National Institute for Occupational Safety and Health, U.S. Government Printing Office, Washington, D.C., 1975.
219. **Franst, C. L. and Hermann, E. R.,** Charcoal sampling tubes for organic vapor analysis by gas chromatography, *Am. Ind. Hyg. Assoc. J.*, 27, 68, 1966.
220. **Reid, F. H. and Halpin, W. R.,** Determination of halogenated and aromatic hydrocarbons in air by charcoal tube and gas chromatography, *Am. Ind. Hyg. Assoc. J.*, 29, 390, 1968.
221. **Moffitt, A. E. and Kupel, R. E.,** A rapid method employing impregnated charcoal and atomic absorption spectrophotometry for the determination of mercury, *Am. Ind. Hyg. Assoc. J.*, 32, 614, 1971.
222. **D'Agostino, R. B. and Gillespie, J. C.,** Comments on the OSHA accuracy of measurement requirements for monitoring employee exposure to benzene, *Am. Ind. Hyg. Assoc. J.*, 39, 590, 1978.

223. **Yrjanheikli, E. A.,** A new method for personal sampling and analyzing of phenol, *Am. Ind. Hyg. Assoc. J.*, 39, 326, 1978.
224. **McCammon, C. S., Quinn, P. M., and Kupel, R. E.,** A charcoal sampling method and a GC analytical procedure for carbon bisulfide, *Am. Ind. Hyg. Assoc. J.*, 36, 618, 1975.
225. **Greinke, R. A.,** Determination of atmospheric furfural, a non-ideal compound, after adsorption on charcoal tubes, *Am. Ind. Hyg. Assoc. J.*, 35, 809, 1974.
226. **Qazi, A. H. and Ketcham, N. H.,** A new method for monitoring personal exposure to ethylene oxide in the occupational environment, *Am. Ind. Hyg. Assoc. J.*, 38, 635, 1977.
227. **Myers, S. A., Quinn, H. J., and Zook, W. C.,** Determination of vinyl chloride monomer at the sub-parts per million level using a personal monitor, *Am. Ind. Hyg. Assoc. J.*, 36, 332, 1975.
228. **Amey, G. C.,** Vinyl chloride — monitoring exposures, *Nat. Saf. News.*, August, p. 81, 1975.
229. Century Systems, Inc., Arkansas City, Kan.
230. **Dommer, R. A. and Melcher, R. G.,** Phase equilibrium method for determination of desorption efficiencies, *Am. Ind. Hyg. Assoc. J.*, 39, 240, 1978.
231. **Kupel, R. E. and White, L. D.,** Report on a modified charcoal tube, *Am. Ind. Hyg. Assoc. J.*, 32, 456, 1971.
232. Charcoal Sampling Tubes, Barneby-Chaney, Columbus, Ohio.
233. **White, L. D., Taylor, D. G., Maner, P. A., and Kupel, R. E.,** A convenient optimized method for the analysis of selected solvent vapors in the industrial atmosphere, *Am. Ind. Hyg. Assoc. J.*, 31, 225, 1970.
234. **Cooper, C. V., White, L. D., amd Kupel, R. E.,** Qualitative detection limits for specific compounds utilizing gas chromatographic fractions, activated charcoal and a mass spectrometer, *Am. Ind. Hyg. Assoc. J.*, 32, 383, 1971.
235. **Rowe, N. R.,** Removal of contaminant gases by adsorption, *Ind. Hyg. News Rep.*, No. 6, 3, June 1963; Second Annu. Meeting Am. Assoc. Contamination Control, May 1-3, 1963.
236. **Larkin, R. L., Crable, J. V., Catlett, L. R., and Seymore, M. J.,** Collaborative testing of a gas chromatographic charcoal tube method for seven organic solvents, *Am. Ind. Hyg. Assoc. J.*, 38, 543, 1977.
237. **Burnett, R. D.,** Evaluation of charcoal sampling tubes, *Am. Ind. Hyg. Assoc. J.*, 37, 37, 1976.
238. **Fairchild, C. I. and Tillery, M. I.,** The filtration efficiency of organic vapor sampling tubes against particulate, *Am. Ind. Hyg. Assoc. J.*, 38, 277, 1977.
239. National Mine Service Company, 300 Koppers Bldg., Pittsburgh, Pa.
240. AIHA-ACGIH Joint Committee on Respirators, Sorbents for contaminants listed in ACGIH 1970 threshold limit values, *Am. Ind. Hyg. Assoc. J.*, 32, 404, 1971.
241. Perkin-Elmer Corp., Laboratory Contaminant Sensor, Final Report, Contract NAS-1-7266, NASA CR-66606, CFSTI: N68-23461, March 1968.
242. **Parker, D. G., Ganz, C. R., Polinsky, A., and Schulze, J.,** A simple gas chromatographic method for the analysis of trace organics in ambient air, *Am. Ind. Hyg. Assoc. J.*, 37, 165, 1976.
243. **McMorran, R. E.,** The Bendix Corporation, Environmental and Process Instruments Div., Baltimore, Md.
244. Tedlar Bags, Pollution Measurement Corp., Chicago, Ill.
245. **Dietrich, M. W., Chapman, L. P. and Miene, J.,** Sampling for organic chemicals in workplace atmospheres with porores polymer beads, *Am. Ind. Hyg. Assoc. J.*, 39, 385, 1978.
246. **Rathje, A. D. and Marcero, D. M.,** Improved hopcalite procedure for the determination of mercury vapor in air by flameless atomic absorption, *Am. Ind. Hyg. Assoc. J.*, 37, 311, 1976.
247. **Rathje, A. O., Marcero, D. H., and Datillo, D.,** Personal monitoring technique for mercury vapor in air and determination by flameless atomic absorption, *Am. Ind. Hyg. Assoc. J.*, 35, 57, 1974.
248. **Blacker, J. H.,** Triethanolamine for collecting nitrogen dioxide in the TLV range, *Am. Ind. Hyg. Assoc. J.*, 34, 390, 1973.
249. **Willey, M. A., McCammon, C. S., and Doemeny, L. J.,** A solid sorbent personal sampling method for the simultaneous collection of nitrogen dioxide and nitric oxide in air, *Am. Ind. Hyg. Assoc. J.*, 38, 358, 1977.
250. SKC, Pittsburgh, Pa.
251. **Melcher, R. G., Langner, R. R., and Kagel, R. O.,** Criteria for the evaluation of methods for the collection of organic pollutants in air using solid sorbents, *Am. Ind. Hyg. Assoc. J.*, 39, 349, 1978.
252. Documentation of NIOSH Validation Tests, NIOSH Contract No. CDC-99-74-45.
253. **Lynch, J. J. and Burgess, W. A.,** A personal exposure sampler for carbon monoxide, *Am. Ind. Hyg. Assoc. J.*, 35, 354, 1974.
254. **Terry, S. C. and Jerman, J. H.,** A Feasibility Study of a Pocket Sized Gas Chromatographic Air Analyzer, Final Report, (SEL77-027), July 1977, NIOSH Contract No. 210-76-0140, Integrated Circuits Lab., Dept. of Electrical Engineering, Stanford Univ., Stanford, Calif.

255. TDI/MDI Reagents and Color Standards, Unico, Catalog No. 5965, National Environmental Instruments, Inc., P.O. Box 590, Pilgrim Station, Warwick, R.I.,
256. **Belisle, J.**, A portable field kit for the sampling and analysis of toluene diisocyanate in air, *Am. Ind. Hyg. Assoc. J.*, 30, 41, 1969.
257. The TDI Analyzer, John A. Pendergrass Co., 2643 S. Riviera Dr., White Bear Lake, Minn.
258. **Reilley, D. A.**, Test paper method for the determination of toluene diisocyanate vapour in air, *Analyst (London)*, 93, 178, 1968.
259. "Telmatic 150" TEL/TML Lead-in-Air Analyzer, National Environmental Instruments, Inc., P.O. Box 590, Pilgrim Station, Warwick, R.I.
260. **Linch, A. L., Stalzer, R. F., and Lefferts, D. T.**, Methyl and ethyl mercury compounds — recovery from air and analysis, *Am. Ind. Hyg. Assoc. J.*, 29, 79, 1968.
261. **Reilly, D. A.**, A field method for determining 4,4′ di-isocyanatodiphenylamine (MDI) in air, *Analyst (London)*, 92, 513, 1967.
262. **Meddle, D. W. and Wood, R.**, A method for the determination of aromatic isocyanates in air in the presence of primary aromatic amines, *Analyst (London)*, 95, 402, 1970.
263. **Marcali, K. and Linch, A. L.**, Perfluoroisobutylene and hexafluoropropene determination in air, *Am. Ind. Hyg. Assoc. J.*, 27, 360, 1966.
264. National Environmental Instruments, Inc. (formerly Union Industrial Equipment Corp.), P.O. Box 590, Pilgrim Station, Warwick, R.I., 1960.
265. General Electric Co., Aircraft Equipment Div., 50 Fordham Rd., Wilmington, Mass.
266. **Roper, P.**, The NIOSH detector tube certification program, *Am. Ind. Hyg. Assoc. J.*, 35, 438, 1974.
267. Plastic bags for collection of gas samples, in *Atmospheric Environment*, Pergamon Press, Elmsford, N.Y., 1, 967, 1.
268. Occupational Exposure to Benzene, Table IV-4, NIOSH Criteria Document, HEW Publication No. (NIOSH) 74-137, 67, 1974.
269. **Ferber, B. I., Sharp, F. A., and Freedman, R. W.**, Dosimeter for oxides of nitrogen, *Am. Ind. Hyg. Assoc. J.*, 37, 32, 1976.
270. Gasbadge® Dosimeter for Organic Vapors, Technical Bulletin No. 0377, U.S. Patent 3,985,017, Abcor Development Corp., 850 Main St., Wilmington, Mass.
270a. **Lautenberger, W. J. and Kring, E. V.**, A New Personal Badge Monitor for Organic Vapors, Paper No. 114, Am. Ind. Hyg. Conf., Chicago, Ill., May 29, 1979.
271. **Fracchia, M., Pierce, L., Graul, R., and Stanley, R.**, Desorption of polar vapors greatly increased by adding 50% CH_3OH to the CS_2, *Am. Ind. Hyg. Assoc. J.*, 38, 144, 1977.
272. Saran Gas Sampling Bags, The Anspec Company, P.O. Box 704, 4126 Packard Rd., Ann Arbor, Mich.
273. Hach Chemical Company, P. O. Box 907, Ames, Iowa; also P.O. Box 389, Loveland, Colorado, 80837.

PREFERRED LITERATURE FOR INDUSTRIAL HYGIENE SURVEYS

1. *American Industrial Hygiene Association Journal.*
2. *Archives of Environmental Health* (AMA).
3. *Journal of Occupational Medicine.*
4. *The Analyst* (British).
5. *Analytical Chemistry* (ACS).
6. **Stokinger, H. E., Ed.**, *Documentation of Threshold Limit Values*, ACGIH, Cincinnati. Ohio, 1977.
7. Hygienic Guides, AIHA, Akron, Ohio.
8. Analytical Guides, AIHA, Akron, Ohio.
9. *Air Pollution Manual*, Parts I and II, AIHA, Akron, Ohio.
10. *Journal of the Air Pollution Control Association.*
11. **Linch, A. L.**, *Evaluation of Ambient Air Quality by Personnel Monitoring*, CRC Press, Cleveland, Ohio, 1973.
12. **Steere, N. V., Ed.**, *The Handbook of Laboratory Safety*, 2nd ed., The Chemical Rubber Co., Cleveland, Ohio, 1971.
13. **Katz, M. Ed.**, *Methods of Air Sampling and Analysis*, 2nd ed., Intersociety Committe, American Public Health Assoc., Washington, D.C., 1977.
14. **Stern, A. C., Ed.**, *Air Pollution*, Vol. 2; *Analysis, Monitoring and Surveying*, 2nd ed., Academic Press, New York, 1978.
15. *Industrial Hygiene Digest — Environmental and Occupational Health News* (monthly publication), Industrial Health Foundation, Pittsburgh, Pa.
16. *Air Sampling Instruments for Evaluation of Atmospheric Contaminants*, 5th ed., American Conference of Governmental Industrial Hygienists, Cincinnati, Ohio, 1978.
17. **Linch, A. L.**, *Biological Monitoring for Industrial Chemical Exposure Control*, CRC Press, Boca Raton, Fla., 1974.

INDEX

A

Accuhaler personal sampling pump, II: 178
Acetate ester membrane as filter, II: 17
Acetylene
  collection by silica gel, I: 270
  desorption, I: 272
Acrylic copolymer filter, II: 65
Acrylonitrite, detection instrument, I: 19
Activated carbon, see Charcoal
Adsorbents, see Charcoal; Silica gel; Solid adsorbents
Aerosol
  collection by midget impinger, I: 219
  generators for study, II: 132
  physics, tables for use in, II: 140
Air Chek - SKC air sampler, II: 156
Air quality standards
  complianc determination of, II: 273
  use of detector tubes in enforcement, I: 133
  violations, classification of, II: 284
Air sampling systems, see Sampling systems
Alcohol
  breath analysis by detector tube, I: 132
  collection on charcoal, I: 294
  detection devices, I: 24, 37
Alumina gel
  desorption of formaldehyde collected on, I: 328
  potential use in collection, I: 278
Amines
  collection by silica gel, I: 270
  tape samplers for, I: 211
Ammonia
  adsorption by charcoal, I: 279
  detection instruments, I: 19, 20, 24
  detector tube
    calibration, I: 150
    description, I: 178
  diffusion samplers for, I: 55
  in silica gel, I: 237, 238, 271
  tape sampler for, I: 211
Amiline
  compounds, silica gel collection, I: 271
  detector tube use, I: 168
  microimpinger collection, I: 222, 225
  Vapor detection method, I: 80
Antimony particles, collection and analysis, II: 116
Arsenic particles, collection and analysis, II: 117
Arsine detector, I: 76, 168
Asbestos
  air sampling methods, II: 78
  collection by filter, II: 17
  evaluation of samples, II: 80
  industrial dust control proposal, II: 85
  laboratory proficiency testing program, II: 82
  medical examination of employees, II: 78
  monitoring program, II: 78
  OSHA procedure, II: 81

B

Battery powered air samplers, II: 149
Bendix C115 personnel air sampler, II: 158
Benzene
  collection on charcoal, I: 295
  collection on silica gel, I: 238, 276
  constant rate sampling, I: 174
  desorption, I: 272, 278
  detector tube for, I: 81
    performance study, I: 134, 168, 170
  reagent system for detection, I: 179
Beryllium
  collection and analysis, II: 119
  exposure limits, II: 273
Biphenyl tetrabramide, collection by silica gel, I: 270
Bischloromethyl ether, detection, I: 323
Blood
  analysis
    by detector tube, I: 133
    during CO study, I: 188
  lead content, II: 239, 255
Boron hydrides, detection method, I: 80
Breath analysis, CO poisoning or alcohol detected by, I: 132
Butane, desorption, I: 272
Butanol, variation in charcoal tube samples, I: 290
Butanene, variation in charcoal tube samples, I:290
Butylacetate, collection on charcoal, I: 291

C

Cadmium particles, collection and analysis, II: 120
Calibration
  aerosol generators for, II: 132
  aerosol personnel monitor, II: 206
  aspirator bottle, II: 195
  capillary diffusion manifold, II: 214
  charcoal sampling tubes, II: 227
  chemical protocol, II: 220
  detector tube pumps, II: 202
  diffusion-permeation chamber, II: 215
  dry test meter, II: 204
  dynamite systems, II: 212
  evaporation technique, II: 226
  exponential dilution flask, II: 216
  flowmeter, II: 195
  flowrate, II: 204
  gas laws, II: 194
  interference verification, II: 225
  mixing manifold, II: 216, 225
  motor-driven syringes, II: 227
  organic-free air system, II: 215
  permeation tubes, II: 212, 221

pressurized cylinders, II: 227
pump-stroke counter, II: 203
rotameter, II: 204, 230
sample bags, II: 200, 208
sampling manifold, II: 216
soap film flowmeter, II: 195, 218, 230
spirometer, II: 195
static systems, II: 206
TWA sampling, II: 226
volumetric, II: 194
wet test meter, II: 204
CAP-130 air sampling pump, II: 181
Carbon, activated, see Charcoal
Carbon bisulfide
detector tube observation, I: 168
adsorption on charcoal, I: 295
Carbon monoxide
detection instruments, I: 18, 19, 20, 24, 36, 37, 43
detection reagents, I: 12
detector tubes, I: 81
performance study, I: 134, 135, 149
pump, I: 171
false positive results in detection, I: 170
nondispersive infrared analyzer, use in study, I: 187
OSHA sampling procedure, I: 131
plastic bag sampling, I: 218
poisoning, determination by breath analysis, I: 132
recalibration of detector tube, I: 187
study of detector tubes, I: 187—194
Carbon tetrachloride
adsorption on silica gel, I: 270, 276
collection device, I: 71
conversion in detection process, I: 168
detector tube performance study, I: 134
Cascade impactor analysis, II: 103
Casella personal sampler, II: 154
Cellulose acetate filters, II: 73
Cellulose-ester membranes as filters, II: 17
Cellulose nitrate filters, II: 66
Century Systems Corporation portable air sampling pump, II: 181
Chalk, use in inorganic support systems, I: 9
Charcoal
actimic light catalysis of samples, I: 290
analysis of samples collected, I: 299
breakthrough point of adsorption, I: 279, 327
calibration of sampling tubes, II: 227
capacity for adsorption, I: 279
catalytic effect, effect on desorption, I: 328
desorption from, I: 279, 295, 327
elution techniques, I: 295
evaluation of tubes, I: 302
humidity affecting adsorption efficiency, I: 288, 327
palladium coated, as CO detector, I: 303
properties, I: 278
sample stability, I: 288, 290
sampling use report, I: 291
solvent vapor collection, II: 193, 230
storage stability of samples, I: 288, 290
use as adsorbent, I: 278, 323
use in personnel monitor, II: 188
variation with contaminant loadings, I: 290
variations among lots, I: 290
Chloranilic acid method of zirconium collection and analysis, II: 123
Chlorine detection instruments, I: 19, 20, 24, 171
Chloroform, collection on charcoal, I: 323
Chlorotrifluoroethylene, desorption, I: 276
Chromatox system, I: 17, 22
Chromic acid detection method, I: 80
Coal dust
collection by impingers, II: 7
counting errors, II: 88
grarimetric personnel monitoring, II: 108, 109
Coal mines, detector tubes for use in, I: 81
Coal for pitch volatiles, filter collection, II: 66
Collection of gases and vapors, see Sampling systems
Collision nebulizer, II: 136
Combustible gases
detection instruments, I: 18, 19, 20, 42
identification, I: 179
Combustion, gases derived from, detection of, I: 132
Compliance limits, calculation of, I: 87, 131
Convection activated detectors, I: 1
Copper, use in gas detection, I: 36
Cotton dust
exposure limits, II: 273
OSHA regulations for control, II: 108
Crayons, use in inorganic support systems, I: 9
Cresols, collection by silica gel, I: 270
Cristobalite
inhalation hazard, II: 112
respirable dust standards, II: 93
Cyanogen chloride
detector crayons for, I: 10, 12
detector tube observation, I: 168
Cyclone, use in dust sampling, II: 94—103, 110

D

Daco Products, Inc., personal sample pump PO-100, II: 185
Dermotitis risk, II: 287
Desorption of gases and vapors, I: 272, 276, 279, 295, 327
Detector tubes
accuracy-controlling factors, I: 141—173
adsorption of benzene, I: 177, 179
air quality criteria enforcement, use of tube for, I: 133
applications, I: 83
blood analysis by, I: 133
breath analysis by, I: 132
calculation for federal compliance limits, I: 87, 131, 132
calibration, I: 149, 174
certification criteria, I: 134—140

charcoal tube, diaphragm pump for, I: 204
combustion products detected by, I: 132
concentration of reagents, I: 173
constant rate sampling, I: 173
construction standards, I: 143
conversion of compounds to detectible derivatives, I: 167
defects, classification of, I: 137
design, I: 81
dessicant layer, I: 167
determination of other substances
diaphragm pump, I: 204
direct reading length of steim tubes, I: 187
emergency air sampling, I: 85
explosive hazards detected by, I: 132
extended period sampling, I: 187
flow rate
control, I: 141, 145
measurement, I: 183, 186
granular support uniformity, I: 141
guard column for impurities, I: 166
guard layers, I: 179
identification of colorimetric reagent, I: 178
interchangeability of tubes and pumps, I: 172
length of stain technique, I: 81, 82
mathematical basis for calibration, I: 149
moisture removal, I: 167
multi-gas dosimeter, I: 207
observational errors, I: 168
operating procedure, I: 81
OSHA procedure, I: 131
oxygen deficiency determination, I: 133
paint and varnish industry survey, I: 152, 162
performance study, I: 134
peristaltic pump, I: 202
personnel monitoring, I: 86
photoelectric scanner, I: 207
Polytest tube, I: 179
pressure drop effects, I: 145
pumps
electrical, operation, I: 171
peristaltic, I: 202
testing of, I: 180
quality control, I: 137, 142
Quantimeter instrument, I: 171
reagent systems
selection, I: 177
stability of, I: 151
recalibration, I: 180, 187
remote sampling, I: 133
repetitive sampling I: 169
sampling period, I: 131, 132
sampling plan, I: 137
sampling rate, I: 146
variability, I: 145
scavenger layer, I: 166
simultaneous determination by two tubes, I: 171
soap bubble calibration techniques, I: 174
soap bubble flowmeter, use of, I: 183
source emission sampling, I: 133
specificity, I: 152
stability of chemical reagent system, I: 151
storage life, I: 151
study of carbon monoxide exposure, I: 187—194
temperature effects, I: 151
variables affecting length of stain, I: 149
volumetric error, I: 149
water vapor, protection against, I: 167, 170
Dichlorobenzene desorption, I: 276
Dichlorodifluoromethane, collection device, I: 71
Dichloromethane, storage of sample, I: 288
Diisocyanates, analysis by Telmatic analyzer kit, II: 153
Dimethylhydrazine detection system, I: 36, 80
Diocetyl phthalate, collection on charcoal, I: 302
Dosimeters, I: 58, 60
DuPont personal monitor pumps, II: 171
Dust, sampling of, see Filter collection; Impinger systems, Sampling systems
Dynamic sampling, I: 75—345
Dyspnea, silica inhalation causing, II: 112

E

Electronic portable instruments, see Portable instruments
Electron microscopy of samples, II: 138
Emergency air sampling, I: 85
Emphysema, silica inhalation causing, II: 112
Esters
collection on charcoal, I: 294
use as filter membranes, II: 17
Ethoxyethyl acetate, variation in charcoal tube samples, I: 290
Ethylene, adsorption by charcoal, I: 279
Ethylene oxide
collection on charcoal, I: 295
detector tube use, I: 168
Explosive hazards, detection by detector tubes, I: 132
Exposure limits, II: 273
Extended period sampling
continuous tape samplers, I: 208—236
length of stain tubes, I: 187—207

F

Federal compliance limits, determination of, I: 87, 131
Fiber filters, II: 62
Field kit laboratory, I: 340
Filter collection
acrylic copolymer filter, II: 65
analysis monitors, II: 73
asbestos fiber monitoring, II: 78
calibration of airflow rate, II: 75
cellulose-ester membranes, II: 17
cellulose nitrate filters, II: 66
"Celotate" type, II: 73

chemical analysis, II: 111
counting, II: 76
dichotomous membrane sampler, II: 56
dissolution of fiber, II: 111
dust counting, II: 76
electron microscope, evaluation by, II: 77
evaluation of samples, II: 80, 88
exhaust gas measurement, II: 62
fiber filters, II: 62
flowrate meter, II: 75, 81
flue gas measurement, II: 62
"Fluropore" type, II: 69
glass fiber filters, II: 62
gravimetric analysis, II: 106, 131
industrial hygiene sampling, II: 68
leaching for analysis, II: 111
lead collection and analysis, II: 115
manual counting of particulate, II: 76
matched weight filters, II: 75
media characteristics, II: 16, 76, 111
membrane filter characteristics, II: 69, 76
mercury intrusion method of determining pore size, II: 66
"MF-Millipore" type, II: 69
micro quartz fiber filters, II: 62
"Mitex" type, II: 73
moisture effects II: 109
mounting of samples, II: 80
"Nucleopore" process, II: 50, 66
oil mist, II: 106
open face and close face sampling, II: 100
open face filter holder, II: 79
particle size analysis, II: 77
"Pasadena" particle distribution, II: 56, 61
Poisson distribution of fiber count, I: 80
polycarbonate film Nucleopore process, II: 50
polytetrafluorethelyne, use in filter, II: 61, 73
"Polyric" type, II: 73
polyvinyl chloride membranes, II: 17
pore sizes, II: 76
pressure drop aeross filter, II: 75, 109
preweighed filter control, II: 109
radiometric techniques, II: 111
respirable fraction sampling II: 89
sampling method, II: 79
selection, II: 17, 76, 109
silica dust collection, II: 112
silver membrane filters, II: 65
two-stage dust samplers, II: 94
types, II: 16, 17, 69
X-ray diffraction analysis, II: 50
Fire fighting, use of detector tubes in, I: 132
Fluorides, sampling method, I: 79, 208; II: 121
Fluorine detection system, I: 36
Formaldehyde
adsorption by charcoal, I: 279
desorption from alumina gel, I; 328
detector tube description, I: 178
Fritted bubbler sampling systems, I: 235
Furfural, collection on charcoal, I: 295

G

Gas chromatograph
Column packing sampling procedure, I: 303
portable dynamic sampling instruments, I: 332
Gas detecting tubes, see
Gas laws, II: 194
Gasoline vapor detection device, I: 37
Generators for aerosols, II: 132
G gases, detector crayons for, I: 11
Glass bubbler sampling system, I: 235
Glass fiber filters, II: 62
Gold, use in gas detection, I: 36
Grab sampling
detector tubes, see Detector tubes
evacuated containers technique, I: 47
for compliance determination, II: 279
inorganic solid support systems, I: 81—186
length of stain gas detecting tubes, see Detector tubes
paper detectors, I: 75—81
photometric instruments to determine color density, I: 80
syringe collection, I: 49
Greenburg-Smith impinger, II: 1

H

Hazardous substances, skin effects, II: 287
Heyane detection instrument, I: 19
Hopcalite
mercury vapor collection, I: 323
OSHA rules for, I: 131
Hydrazines, collection by silica gel, I: 270
Hydrocarbons
collection on charcoal, I: 294, 323
desorption, I: 272
detection instruments, I: 20, 24, 43, 167, 180
halogenated, adsorption on silica gel, I: 238, 276
plastic bag sampling, I: 218
Hydrogen chloride
adsorption by charcoal, I: 279
fuel cell detectors for, I: 34
tape sampler for, I: 211
Hydrogen cyanide
detectors, I: 10, 12, 76
imminent danger, II: 278
Hydrogen detection instruments, I: 19, 20, 24, 36
Hydrogen sulfide
adsorption by charcoal, I: 279
collection for analysis, I: 67, 72
detection instruments, I: 18, 19, 20, 24, 34, 43, 77
detector tabs, I: 16
detector tube, I: 81
calibration, I: 150
pump, I: 171
detectors, CT concept, I: 1
effect on silver as detection method, I: 17
imminent danger condition, II: 278
long-term sampling, I: 187, 196, 198
odor threshold, I: 85

tape samplers for, I: 208

I

Impinger systems
  airflow measurement, II: 4
  collection medium, II: 10
  Coulter counter techniques to determine efficiency, II: 10
  counting cells, standard, II: 12
  counting technique, II: 10
  dust counters, training requirement, II: 15
  efficiency as collectors, II: 4
  glass bubblers, I: 235
  Greenburg-Smith impinger, II: 1
  kits for use with, I: 344, 345
  microimpingers
    alkyl lead compound collection, II: 193
    design, I: 222; II: 2
    evaluation, II: 6
    pump used with, II: 152
  midget impingers
    battery pack used with, II: 163
    calibration, II: 230
    design, I: 219; II: 2
    use in asbestosis study, II: 78
  series collection, I: 221
  silica dust collection, II: 112
  tolerances, II: 2
  traps, I: 236
  types, II: 2
  Wilson impinger, I: 234; II: 2
Impregnated paper
  continuous monitoring, I: 9, 208
  convection activated detectors, I: 1
  CT concept, application, I: 1
  gas phase titration, use in sulfur dioxide estimation, I: 3
  indication of TLV, I: 1
  reagent ampute activated samplers, I: 8
  sulfur dioxide estimation, I: 3
  tape samplers, I: 7
  use as self-indicating device, I: 1
  use in continuous personnel monitor, I: 208
Infrared spectrophotometry sample analysis, II: 88
Ingestion hazard, wipe sampling to determine, II: 283
Inorganic support systems
  badge monitors, I: 12
  carbon monoxide detection, I: 12
  chalk and crayons, I: 9
  CT values of gases, I: 11
  detector tubes, see Detector tubes
  gases detected by crayons, I: 10, 11, 12
  hydrogen sulfide detector tabs, I: 16
  paper as support for reagents, I: 12
Isobutane
  desorption, I: 272
  detection instrument, I: 19
Isocyanates
  detection of, I: 180
  field kit for detection, I: 340
Isopentane, desorption, I: 272

K

Ketones, collection on charcoal, I: 294

L

Lead
  alkyl, microimpinger collection, II: 193
  analysis by Telmetic analyzer kit, II: 153
  analysis method, II: 115
  concentrations in blood, II: 239, 255
  concentrations in urine, II: 240, 242, 245, 255
  detection methods, I: 78
  field kit laboratory for detection, I: 341, 343
  impinger efficiency for collection, II: 4
  impinger monitoring system, I: 231, 235
  polyvinyl chloride membrane for collection, II: 17, 73
  tetraalkyl, evaluation of exposure to, II: 188
Length of stain gas detecting tubes, see Detector tubes
Lewisite, detector crayons for, I: 11
Limestone dust, counting errors, II: 88
Lung fibrosis, silica inhalation causing, II: 112

M

Manganese particles, collection and analysis, II: 119
Mass monitor, portable, II: 127
MDA Scientific Company Accuhaler pump, II: 178
Mercaptan, detector tube pump for detection, I: 171
Mercury vapor
  collection, I: 323
  detection, I: 43, 73, 163, 170, 179
Methane detection instruments, I: 19, 23, 24, 40
Methanol tetrabromide, collection by silica gel, I: 270
Methyl bromide
  desorption, I: 278
  detector tube recalibration, I: 180
Methylcyclohexanol, collection on charcoal, I: 291
Methylene, impinger monitoring system, I: 231
Methylene chloride, collection on charcoal, I: 323
Methylene-o-dichloroaniline, personnel monitor to determine source, II: 188
Methyl ethyl ketone
  collection on charcoal, I: 291
  desorption, effect of temperature on, I: 328
  detector tube surrey, I: 152, 162

response to photoionization, I: 40
Methyl hydrazines, collection by silica gel, I: 270
Methyl isocyanate, detector tube for, I: 180
Mica, collection by filter, II: 94
Microimpinger collection system see Impinger systems
Micronair piston pump, II: 152
Midget impinger, collection system, see Impinger systems
Mineral dusts, TLV, II: 1
Misco lapel sampler, II: 155
Molybdenum particles, collection and analysis, II: 119
Monitair-MSA air sampler, II: 156
Morpholime, collection by silica gel, I: 270

N

Neutron activation technique of sample analysis, II: 89
Nickel carbonyl, detector tube observation, I: 168
Nitrate ester membrane as filter, II: 17
Nitric oxide
conversion in detection process, I: 168
sample collection, I: 325
Nitrogen compounds, collection by silica gel, I: 270
Nitrogen dioxide
collection for analysis, I: 67
detection systems, I: 34, 36, 43, 46
detector tube sampling rate, I: 147
diffusion samplers for, I: 55
dosimeter specifications, I: 50
impinger collection, I: 236
sample collection, I: 325
Nitrogen oxides
collection for analysis, I: 50
detector tube pump for detection, I: 171
evaluation of detection device, I: 46
Nucleopore collection process, II: 50, 66
Nuisance particulates, TLV, II: 1

O

Oil mist collection, II: 106
Organic vapors, monitoring of, I: 59
OSHA sampling procedure, I: 131
Oxygen
deficiency determination, I: 133
deficiency meters, I: 30
detection instruments, I: 18, 19, 20
pressure measurement, I: 36
Ozone
detector tube inaccuracies, I: 134
meters, evaluation, I: 46

P

Paint solvents
collection on charcoal, II: 193
survey of, I: 152, 162
Paper, see also Impregnated paper
as support for reagents, I: 12
detectors for gases, I: 75—81
treatment with detector chalk or crayons, I: 9
Particle counter, portable, II: 127
PAS-2000 personal air sampler, II: 183
Perfluoroalkenes, laboratory kit for detection, I: 343
Permissible exposure limits, II: 273
Personnel monitoring
acidic gases, I: 67
battery powered air samplers, II: 149
binary diffusion, collection by, I: 55
calibration, see Calibration
continuous tape samplers, I: 208
dosimeters, I: 58, 60
evacuated container collection, I: 73
gravimetric, for coal and cotton dust, II: 108
in use of detector tubes, I: 86
liquid phase collection, I: 49
mercury vapor monitor, I: 55, 58
organic vapors, I: 59
orifice controlling rate of flow, I: 74
plastic bag sampling, I: 213
sampling criteria, II: 194
solvents, I: 59
Petrographic techniques of silica dust determination, II: 112
Phenol
collection on charcoal, I: 295
detector tube description, I: 178
detector tube use, I: 168
Phenyl ether, collection by silica gel, I: 270
Phosgene
detectors, I: 10, 12, 34, 75, 168, 169
odor threshold, I: 85
Phosphine, silica gel collection, I: 271
Photoionization instrument, I: 40
Piezoelectric microbalance collector, II: 128
Plastic bag sampling, I: 213
Pneumoconiosis
counting procedure for dust related to, II: 11
"respirable dust" definition, II: 91, 92
Polystyrene latex, impinger sampling, II: 15
Polytetrafluorethylene, use in filters, II: 61, 73
Polyvinyl chloride filters, II: 17, 73
Portable field kit laboratory, I: 340
Portable instruments see also Personnel monitoring
battery powered air samplers, II: 149
capillary diffusion barrier, I: 34
catalytic combustion, I: 23
combination gas detectors, I: 40
dose alarms, I: 23
dynamic sampling
electronic, I: 329
gas chromatographs, 332
electrical conductance types, I: 23
electro micro-balances, II: 131
evaluation, I: 42, 47

fuel cell type detectors, I: 34
galvanic cell, I: 30
impaction collection method, II: 128
mass monitor, II: 127
memory cell, I: 17
metallic oxide semi-conductor, I: 24, 28, 31
oxygen deficiency meters, I: 30
oxygen pressure measurement, I: 36
particle counter, II: 127
piezoelectric microbalance collector, II: 128
polarographic device for CO detection, I: 37
reducing gas-detection device, I: 36
thin metal film dosimeters, I: 36
tyndalometer, II: 127
ultraviolet photoionization device, I: 40, 45
zinc oxide use in oxygen monitoring, I: 36
Potassium pallado sulfite, use in CO detection, I: 12
Propane
desorption, I: 272
detector tube use, I: 168
Propylene, desorption, I: 272
Public, air monitoring by, I: 9
Pulsation-free sampling, II: 96
Pumps for battery powered air samplers, II: 152

Q

Quality control
accuracy concept, II: 235
accreditation requirements, II: 258
air quality standards, II: 273
control charts, II: 239, 255, 260, 263
determinate error, II: 235
graphic analysis for correlations, II: 252
indeterminate error, II: 248
interference, II: 244
interlaboratory reference systems, II: 258
internal standards technique, II: 244
methodology, II: 240
precision concept, II: 235
proficiency testing, II: 258
programs, II: 258
routine analysis control, II: 255
sample size, effect of, II: 240
sampling criteria, II: 266
"spiked" sample procedure, II: 236
statistical evaluation of error, II: 248
Quantimeter instrument, I: 171
Quartz
fiber filters, II: 62
inhalation hazard, II: 112
respirable dust standards, II: 92
sampling system for, II: 103

R

Radiation, imminent danger condition, II: 278
Raney nickel, use in sulfur analysis, II: 124
Remote sampling I: 133
Respirator standards, wipe sampling to determine, II: 289
Respirable fraction sampling, II: 89

S

Sampling systems, see also Charcoal; Silica gel; Solid adsorbents
battery powered air samplers, II: 149
calibration, II: 193
calibration mixtures, II: 243
criteria, II: 193, 266
desorption, I: 272, 279, 295, 327
dynamic sampling, I: 75—345
error
correction, II: 240
determinate, II: 235
indeterminate, II: 248
sampling and analytical errors, II: 278
statistical evaluation, II: 248
filter collection, see Filter collection
"flasher" system, I: 297, 323
forced circulation effects, II: 243
full-shift sampling, II: 273
impinger systems, see Impinger systems
interferences, II: 244
internal standards technique, II: 249
liquid reagents, I: 219
location of sampling sites, II: 243
methods for calculating confidence limits, II: 280
optical density measurements, II: 240
passive sampling, I: 1—75
plastic bag collection, I: 213
quality control, II: 235
removal of collected contaminants, I: 272, 273
silica gel, see Silica gel
specifications for solid adsorbent applications, I: 326
stratification of pollutant, II: 243
tape samplers, I: 208
temperature coefficients, II: 240
Selenium particles, collection and analysis, II: 118
Self-indicating devices, I: 1—46
Silica
collection and analysis, II: 112
conditions caused by inhalation, II: 112
crystalline, dust-count technique, II: 15
filter collection, II: 66
microimpinger collection, I: 222
respirable dust standards, II: 92, 94
Silica gel
adsorption methods I: 271, 272
as collection medium, I: 236
comparison with charcoal as adsorbent, I: 279
desorption, I: 272
factors affecting adsorption, I: 237
Silicosis, development, II: 112

Silver
membrane filters, II: 65
use in gas detection, I: 36
Sipin Personal Sampler Pump, II: 164
Skim absorption monitoring, II: 287
Smoke detection device, I: 37
Solid adsorbents, see also Charcoal; Silica gel
alumina gel, I: 278
co-adsorption, I: 327, 328
contaminents, adsorbents for, I: 318
desorption, I: 327
equipment for use, I: 329
"flasher" collection system, I: 297, 323
gas chromatograph column packing sampling procedure, I: 303
liquid reagents carried on, I: 325
specifications, I: 326
validation criteria, I: 328
Solonoid type pump for air sampler, II: 170
Solvents
detector tubes for, I: 81
monitoring of, I: 59
used in paint and varnish, detection procedure, I: 152, 162
vapor collection on charcoal, II: 193, 230
Source emission sampling, I: 133
Spectrex Corporation personal air sampler, II: 183
Stibine, detector tube use, I: 168
Styrene
collection device, I: 71
desorption time, I: 328
detector tube inaccuracies, I: 134, 168
stability of charcoal tube samples, I: 290
Sulfur, analysis of sample, II: 124
Sulfur dioxide
collection for analysis, I: 67, 72
detection instruments, I: 19, 24, 46, 76
detector tube for
description, I: 178
performance study, I: 134
pump, I: 171
diffusion samplers for, I: 55
estimation by impregnated paper devices, I: 3
microimpinger collection, I: 222
Sulfuric acid
aerosol generation, II: 137
impinger efficiency for collection, II: 4

T

Talc dust, collection by impingers, II: 7
Tape samplers, see also Impregnated paper
dosimeter operation, I: 212
evaluation of exposed tape, I: 210
operation, I: 209, 210
portable, development of, I: 7, 208, 211
types, I: 211, 212
Telmatic piston pump, II: 153
Tetrabromides, collection by silica gel, I: 270
Threshold limit value
for dusts and particulates, II: 1
indication by convection activated detectors, I: 1
revision for carbon monoxide, I: 187
TLV, see Threshold limit value
Toluene
calibration calculation, II: 211
detector tube for, I: 81, 152
humidity affecting collection efficiency of charcoal, I: 288
reagent system for detection, I: 179
recovery from silica gel, I: 278
response to photoionization, I: 40
stability of charcoal tube samples, I: 290
tape sampler for, I: 211
test-paper detection procedure, I: 209
Toxic substances detector kits, I: 344, 345
Trichloroethylene
desorption, I: 272
variation of charcoal tube samples, I: 290
Tridymite
inhalation hazard, II: 112
respirable dust standards, II: 93
Triethanolamine, use in nitrogen dioxide collection, I: 325
Trioxice, detector tube pump for, I: 171
Tyndalometer, use of, II: 127

U

Unit density spheres, table, II: 140
Uranium samples, collection and analysis, II: 121
Urine, lead concentrations in, II: 240, 242, 245, 255

V

Varnish solvents
collection on charcoal, II: 193
survey of, I: 152, 162
Vinyl chloride
collection on charcoal, I: 323
desorption, I: 278, 295
detector tube observation, I: 168
plastic bag sampling, I: 213

W

Water analysis, field kit for, I; 343
Wilson impinger collection system, I: 234
Wipe sampling policies and procedures, II: 287
Workspaces, control of, II: 291

X

X-ray fluorescence spectrometry sample analysis, II: 89

Xylene
detector tube for, I: 81, 152
reagent system for detection, I: 179
recovery from silica gel, I: 278

Z

Zinc oxide, use in oxygen monitoring, I: 36
Zirconium samples, collection and analysis, II: 123

AF848628

Databases for Data-Centric Geotechnics

Databases for Data-Centric Geotechnics forms a definitive reference and guide to databases in geotechnical and rock engineering, to enhance decision making in geotechnical practice using data-driven methods. This first volume pertains to site characterization. The opening chapter presents an in-depth analysis of site data attributes, including the establishment of a new taxonomy of site data under "4S" (site generalizations, spatial features, sampling characteristics, and smart data) to provide a novel agenda for data-driven site characterization. Type 3 machine learning methods (disruptive value) are possible as sensors become more pervasive and more intelligent. A comprehensive overview of site characterization information is also presented with a focus on its availability, coverage, value to decision making, and challenges. The remaining 13 chapters cover databases of soil and rock properties and the application of these databases to rock socket behavior, rock classification, settlement on soft marine clays, permeability of fine-grained soils, and liquefaction among others. The databases were compiled from studies undertaken in many countries including Austria, Australia, Brazil, Canada, China, France, Finland, Germany, India, Iran, Japan, Korea, Malaysia, Mexico, New Zealand, Norway, Singapore, Sweden, Thailand, the United Kingdom, and the United States.

This volume on site characterization is a companion to the volume on geotechnical structures. *Databases for Data-Centric Geotechnics* represents the most diverse and comprehensive assembly of database research in a single publication (consisting of two volumes) to date. It follows from *Model Uncertainties for Foundation Design*, also published by CRC Press, and suits specialist geotechnical engineers, researchers, and graduate students.

Kok-Kwang Phoon is Cheng Tsang Man Chair Professor and President of the Singapore University of Technology and Design. He was awarded the ASCE Norman Medal twice in 2005 and 2020, the Humboldt Research Award in 2017, and the ASCE Alfredo Ang Award in 2024. He is the Founding Editor of *Georisk* and *Geodata and AI*. He is a Fellow of the Academy of Engineering Singapore and Singapore National Academy of Science.

Chong Tang is Professor at Dalian University of Technology in China and a former senior research fellow at the National University of Singapore. He was awarded the ASCE Norman Medal in 2020.

Challenges in Geotechnical and Rock Engineering

Chief Editor Kok-Kwang Phoon, *Singapore University of Technology and Design*
Assistant Editor Dongming Zhang, *Tongji University*

Challenges in Geotechnical and Rock Engineering presents advanced level books which focus on state of the art methods for handling problems across geotechnical engineering. Typical books are in include measuring and understanding material or structural behaviours, handling special environmental conditions, performing complex numerical simulation, design and construction for novel structures, life cycle and risk/reliability informed management, data-driven algorithms and AI-based decision making, resilience engineering, addressing climate change and sustainability, and digital twin and smart infrastructure.

Evolutionary Process of a Steep Rocky Reservoir Bank in a Dynamic Mechanical Environment
Luqi Wang and Wengang Zhang

Uncertainty, Modelling, and Decision Making in Geotechnics
Edited by Kok-Kwang Phoon, Takayuki Shuku and Jianye Ching

Bayesian Machine Learning in Geotechnical Site Characterization
Jianye Ching

Databases for Data-Centric Geotechnics: *Site Characterization (Volume 1)*
Edited by Kok-Kwang Phoon and Chong Tang

Databases for Data-Centric Geotechnics: *Geotechnical Structures (Volume 2)*
Edited by Chong Tang and Kok-Kwang Phoon

For more information about this series, please visit: www.routledge.com/Challenges-in-Geotechnical-and-Rock-Engineering/book-series/CGRE

Databases for Data-Centric Geotechnics

Site Characterization

Edited by

Kok-Kwang Phoon and Chong Tang

CRC Press
Taylor & Francis Group
Boca Raton London New York

CRC Press is an imprint of the
Taylor & Francis Group, an **informa** business

Cover image: BEST-BACKGROUNDS/Shutterstock

MATLAB® and Simulink® are trademarks of The MathWorks, Inc. and are used with permission. The MathWorks does not warrant the accuracy of the text or exercises in this book. This book's use or discussion of MATLAB® or Simulink® software or related products does not constitute endorsement or sponsorship by The MathWorks of a particular pedagogical approach or particular use of the MATLAB® and Simulink® software.

First edition published 2025
by CRC Press
2385 NW Executive Center Drive, Suite 320, Boca Raton FL 33431

and by CRC Press
4 Park Square, Milton Park, Abingdon, Oxon, OX14 4RN

CRC Press is an imprint of Taylor & Francis Group, LLC

© 2025 selection and editorial matter, Kok-Kwang Phoon and Chong Tang; individual chapters, the contributors

Reasonable efforts have been made to publish reliable data and information, but the author and publisher cannot assume responsibility for the validity of all materials or the consequences of their use. The authors and publishers have attempted to trace the copyright holders of all material reproduced in this publication and apologize to copyright holders if permission to publish in this form has not been obtained. If any copyright material has not been acknowledged please write and let us know so we may rectify in any future reprint.

Except as permitted under U.S. Copyright Law, no part of this book may be reprinted, reproduced, transmitted, or utilized in any form by any electronic, mechanical, or other means, now known or hereafter invented, including photocopying, microfilming, and recording, or in any information storage or retrieval system, without written permission from the publishers.

For permission to photocopy or use material electronically from this work, access www.copyright.com or contact the Copyright Clearance Center, Inc. (CCC), 222 Rosewood Drive, Danvers, MA 01923, 978-750-8400. For works that are not available on CCC please contact mpkbookspermissions@tandf.co.uk

Trademark notice: Product or corporate names may be trademarks or registered trademarks and are used only for identification and explanation without intent to infringe.

ISBN: 978-1-032-57895-8 (hbk)
ISBN: 978-1-032-57988-7 (pbk)
ISBN: 978-1-003-44194-6 (ebk)

DOI: 10.1201/9781003441946

Typeset in Sabon
by Deanta Global Publishing Services, Chennai, India

Access the Support Material: www.routledge.com/9781032578958

Contents

Challenges in geotechnical and rock engineering *vii*
Foreword *viii*
Preface *xi*
List of contributors *xx*

1 Role of site characterization information in data-centric geotechnics **1**
KOK-KWANG PHOON, JIANYE CHING, AND CHONG TANG

2 Rock hydromechanical parameters for rock foundation design: A database approach **50**
POUYAN ASEM AND PAOLO GARDONI

3 Evaluation of soil/rock properties using databases **102**
JIANYE CHING

4 Undrained shear strength of Finnish soft clays: A database perspective **135**
MARCO D'IGNAZIO AND TIM LÄNSIVAARA

5 Role of databases in the evaluation of soil properties **152**
ABOLFAZL ESLAMI AND PIN-QIANG MO

6 New laboratory database of hydraulic conductivity measurements on fine-grained soils **210**
SHUYIN FENG AND PAUL J. VARDANEGA

7 Normalised active undrained shear strengths of soft Scandinavian clays: A data-centric and geomechanical approach **229**
SØLVE HOV, JEAN-SÉBASTIEN L'HEUREUX, KATHARINA KAHRS, KARIN LUNDSTRÖM, AND DAVID GAHARIA

8 Prediction for the mechanical response of gravels **244**
JINGMAO LIU, DEGAO ZOU, DUO LI, FANWEI NING, CHENGUANG ZHOU, AND KAIYUAN XU

9 Evaluation of compressibility properties for soft marine clays 280
MONICA S. LÖFMAN AND LEENA KORKIALA-TANTTU

10 An engineering geological parameter database of tunnel surrounding rock and its application 304
JUNJIE MA AND TIANBIN LI

11 Exploring challenges via analysis of multivariate geotechnical properties: Insights from large-scale local sampling of Japanese marine clay 332
YU OTAKE, TAIGA SAITO, WU STEPHEN, IKUMASA YOSHIDA, AND DAIKI TAKANO

12 In situ test-based evaluation of soil effective stress strength properties and stress history 344
ZHONGKUN OUYANG, PAUL W. MAYNE, AND SHEHAB AGAIBY

13 Mechanical–statistical evaluation of soil properties 369
CHANGHONG WANG, KUN WANG, DAOFEI TANG, AND TIAN FANG

14 Data-centric seismic soil liquefaction assessment: Approaches, data, and tools 427
CHAOFENG WANG, QIUSHI CHEN AND PEDRO ARDUINO

15 Prediction for soil design properties based on a multivariate database for Shanghai soft clay 452
YELU ZHOU, DONGMING ZHANG, AND HONGWEI HUANG

Index *473*

Challenges in geotechnical and rock engineering

Geotechnical and rock engineering studies have made significant strides in response to different challenges and opportunities that include measuring and understanding material/structural behaviors, handling special environmental conditions, performing complex numerical simulations, design, construction methods, and circular roles for low-carbon materials, novel structures, green construction and operation, life cycle and risk/reliability informed management, data-driven algorithms and AI-based decision making, autonomous systems, digital twin and smart infrastructure, resilience engineering, climate change and sustainability, among others. These challenges are interrelated. Although some of the challenges are common to all industries, it is not meaningful to engage them in an abstract manner outside of the practice context. One example is data-centric geotechnics that takes a "data first" approach to decision making, but the data are actual "ugly" field observations (multisource, sparse, incomplete, spatially variable, corrupted, etc.) rather than ideal abstract numbers. Data-centric geotechnics should deploy digital technologies in cognizance of the context of geotechnical and rock engineering that includes physics, empirical knowledge, experience, and engineering judgment. How decision making in geotechnical and rock engineering can be revolutionized through human–machine collaboration is one grand challenge that epitomizes the motivation for launching this book series. This book series presents exciting emerging solutions in geotechnical and rock engineering that are expected to transform practice and meet fast-evolving environmental/societal trends in the twenty-first century. It is a timely response to the changing technological, environmental, and societal landscape presented in the Institution of Civil Engineers (ICE) State of the Nation Report on "Digital Transformation" and the American Society of Civil Engineers (ASCE) "Future World Vision: Infrastructure Reimagined" paper.

Book series editors
Kok-Kwang Phoon
Dongming Zhang

Foreword

Many nations have identified that the development of digital capabilities is a key strategic priority, such as the development of intelligent systems to assess the performance of infrastructure assets and inform maintenance requirements (the National Academy of Sciences 2019, Communist Party of China Central Committee (CPCCC) 2019, the Institution of Civil Engineers (ICE) 2019, and Australian Department of Infrastructure and Regional Development (DIRD) 2016). This includes big data analytics and machine learning concepts. At the same time, there is uncertainty about whether artificial intelligence will just automate repetitive tasks or make work more precarious, and how climate action and AI could create more and better jobs (World Economic Forum; weforum.org).

I do not share a fear of technological disruption. The personal computer revolutionalized engineering in the 1990s. Instead of calculating factors of safety for slope stability problems by manually drawing a circle and doing the calculations over several hours, an engineer could do thousands of calculations in 5 min. That did not mean that engineers lost their jobs, it just meant they could do many more calculations in a shorter period of time. Similarly, drafters used to create drawings by hand. They now create digital models. The role of the drafter has changed, but the number of jobs currently available in Australia is much larger than the number of people available to fill them.

Productivity in the construction industry has stagnated for 30 years (Infrastructure Australia 2022). Increasing populations demand more infrastructure but there are competing needs for investment, such as energy transition, which means less money may be available for other uses. Increasing productivity means that we can provide our communities with more of the infrastructure they desire. Automation in various forms including machine learning will increase in scope and capability to help address productivity requirements.

It is my view that in the (near) future people with a deep understanding of their subjects will survive and thrive and everything else will become automated. Hence, it is of vital importance that geotechnical professionals develop a deep understanding of machine learning as it applies to geotechnical engineering. Machine learning has two components that all engineers should understand to some extent in order to be able to effectively use emerging techniques. These are (1) the algorithms and mathematical optimization that are used to interrogate data and (2) the data. I have dabbled for some time with the use of Bayesian updating to predict the behavior of embankments constructed on soft clay. This is a physics-based application of machine learning where measurements of settlement at the surface, at various points in the ground, and pore pressures are compared to predictions from conventional consolidation theory and used to update the predictions. Typically, we have been overpredicting measured pore pressures at the end of filling and underpredicting pore pressures after a period of time. The reason for this is that real soil typically does not generate much pore pressure until the yield pressure is reached, whereas consolidation theory states that the entire applied stress increment is immediately transferred to the pore water. The likelihood function within the Bayesian algorithm tries to minimize the error between prediction and measurement, and because the consolidation model overpredicts initial pore pressure generation, the function compensates by underpredicting pore pressures later in time so that the net difference between measurement and theory is almost zero. This is both an example of the need to know how the algorithms work and an example of the need to understand what the data means relative to the consolidation model. The latter issue can potentially be eliminated using purely data-driven methods advocated by the editors of *Databases for Data-Centric Geotechnics: Site Characterization* and *Databases for Data-Centric Geotechnics: Geotechnical Structures*.

While knowledge of algorithms and data are both important, the geotechnical professional will find that they need to spend most of their time understanding, cleaning, and storing data in a format that can be interrogated by algorithms, which will mostly be written by others. Anecdotally, up to 80% of the effort is applied to data and the rest to training the algorithms and using the data. Consequently, the focus of these two books on databases for site characterization and predictive performance is of fundamental importance. All geotechnical engineers will know the challenge of collecting data, let alone managing the data, as well as the challenges posed by the MUSIC-X concept (Phoon et al, 2019). MUSIC-X means Multivariate, Uncertain and Unique, Sparse, Incomplete, and potentially Corrupted, with "X" denoting the spatial/temporal dimension. Commercial and insurance issues relating to whether an engineer can rely on data collected by other people further complicate matters. The types of data that are collected will also change in the future as methods of digital surface and subsurface measurement are developed. These changes will enhance the use of machine learning techniques and should bring improvements in productivity.

There is a view that the ability of our profession to characterize sites has regressed in recent years (DeGroot and DeJong, 2020). While it is possible that data-driven machine learning approaches can compensate for the "potentially Corrupted" element of MUSIC-X, it is also possible that the widespread use of machine learning will drive greater efforts to collect high-quality data. This has the potential to transform the practice of site characterization into the future.

Arguably, the key skill of the geotechnical professional is to make decisions, or help other people make decisions, in the face of uncertainty. I am very pleased to see the editors focus on the ability of machine learning to help make decisions in their books. In my view, decisions will be made by people who have confidence in the advice provided by geotechnical professionals who have a deep understanding of their subjects. Ideally, decisions are made as early as possible to take advantage of opportunities or efficiently manage risks, and machine learning will become a powerful tool to aid geotechnical professional in providing appropriate advice.

In focusing on site investigation and field performance, these two books provide many examples of how machine learning is being used to manage and interpret data. They provide our profession with insight into the current state-of-the-art and what the future of geotechnical practice might become. They will become invaluable reference material as the field matures. I congratulate the editors and the authors for their work and commend these books as required reading for all geotechnical professionals.

Dr Richard Kelly
General Manager of Technical Excellence & Chief Technical Principal for Geotechnical Engineering, SMEC Australia.
Conjoint Professor of Practice, University of Newcastle, Australia

REFERENCES

Australian Government Department of Infrastructure and Regional Development. 2016. Trends Transport and Australia's Development to 2040 and beyond

DeGroot, D.J. and DeJong, J.T. 2020. Best Practices for Geotechnical Site Characterization: Have we regressed from decades past? Geostrata magazine, 24(2), 50-57.

Infrastructure Australia (2023). Infrastructure market capacity 2022 report, https://www.infrastructureaustralia.gov.au/publications/2022-market-capacity-report

Institution of Civil Engineers UK 2019 What should be in the National Infrastructure Strategy? https://www.ice.org.uk/getattachment/news-and-insight/policy/what-should-national-infrastructure-strategy/What-should-be-in-the-National-Infrastructure-Strategy-ICE-July-2019.pdf.aspx

National Academies of Sciences, Engineering, and Medicine 2019. Critical Issues in Transportation 2019. Washington, DC: The National Academies Press. https://doi.org/10.17226/25314

Phoon, K. K., Ching, J., and Wang Y. 2019. Managing risk in geotechnical engineering – from data to digitalization. Proceedings, 7th International Symposium on Geotechnical Safety and Risk (ISGSR 2019), Taipei, Taiwan, 13-34. https://www.issmge.org/uploads/publications/96/97/SL.pdf

The Communist Party of China Central Committee and the State Council 2019. Program of Building National Strength in Transportation. http://www.gov.cn/xinwen/2019-09/19/content_5431432.htm (in Chinese)

Preface

An initiative to promote public sharing of databases, called 304dB, was initiated by Professor Kok-Kwang Phoon and Jianye Ching in 2017: http://140.112.12.21/issmge/tc304.htm. Despite its success (304dB has contributed to the development of data-driven methods by situating these methods in the context of real data over the past seven years), it is timely to take stock and discuss how this effort can be expanded to cover more types of databases such as multi-modal databases, more life cycle stages beyond design, more countries, and more projects. The purpose of this book which is published in two volumes covering "site characterization" and "geotechnical structures" is to expand the frontiers of 304dB, to share new data-driven methods (e.g. machine learning) and applications for data-centric geotechnics, and to identify opportunities and challenges in research and practice.

The publication of *Databases for Data-Centric Geotechnics* in 2025 is timely. The breakout of powerful generative AI such as ChatGPT in 2023 should give new impetus to collecting data, understanding data, protecting data as a critical asset, developing learning algorithms to transform data into intelligence, and reimagining infrastructure as a service to end-users through deployment of smart technologies. The vision for taking a data-centric perspective to geotechnical engineering (data-centric geotechnics) to march toward these top-level goals is more compelling than before. An initial study on large language models was conducted recently (Wu et al. 2024). Geo-everything-dB can be a hashtag for collecting, understanding, and protecting geotechnical data across the entire life cycle of the structure.

Data-centric geotechnics is distinctive from other fields in geotechnical engineering and machine learning (ML) (Phoon and Ching 2021; Phoon et al. 2022a). It is a new interdisciplinary field involving the integration of information, data, techniques, tools, perspectives, concepts, theories, and/or experiences from geotechnical engineering and machine learning (Phoon and Zhang 2023). A broader development in "data-centric engineering" (Girolami 2020) and data-driven discovery in geosciences (Chen et al. 2023) are taking place elsewhere in parallel. It is incorrect to say that geotechnical engineering has always been data-centric in the strict sense. It has been experience-centric or physics-centric, but to the knowledge of the authors, data has never supported decision making by itself. At present, data only supports physics or experience in decision making. Data centricity does not imply that experience and physics are not relevant. If data can support experience and physics, the reverse argument is also true. In machine learning, experience can be regarded as one type of "thick data" to distinguish it from the more well-known quantitative "big data." Physics-informed machine learning (PIML) methods offer the advantages of "explainability" and "interpretability" (Karniadakis et al. 2021). Large language models can be trained on the entire body of geotechnical engineering literature. One should not equate data to numerical data alone.

Many research papers on machine learning in geotechnics (MLIG) are "method first" rather than "data and practice first." Numerous foundational challenges beyond methodological ones can be identified when research is pursued under a more balanced data-centric geotechnics agenda covering data centricity, fit for (and transform) practice, and geotechnical context (Phoon and Zhang 2023; Phoon et al. 2023b). Research in the context of data-centric geotechnics must articulate its value proposition clearly and should be directed to compete within a benchmarking framework (Phoon et al. 2022b; Phoon et al. 2023a) to hasten the rate of digital transformation in the industry. Some of the databases in *Site Characterization* and *Geotechnical Structures* can provide the training and validation datasets for new benchmark examples to improve specific machine learning tasks. The value of MLIG research can be classified as follows: (1) Type 1 (incremental value) involving available data and existing conventional applications, (2) Type 2a (potentially high value) involving available data and new applications, (3) Type 2b (high value) involving new data and existing applications, and (4) Type 3 (disruptive value) involving new data and completely novel applications (e.g., precision construction, autonomous construction, human-machine teaming) (Phoon and Zhang 2023). Type 1 MLIG provides an alternate approach to an existing problem. It is only "good to have." Type 2 MLIG outperforms existing approaches (if they exist) to such an extent that it will replace them. Type 3 MLIG will make practice unrecognizable by present-day engineers. These MLIG methods are distinctive from reliability methods, although they are both data-driven. MLIG methods can improve by learning from data and may eventually even modify or propagate themselves (an AI programming a better AI). Reliability methods address uncertainty explicitly, but the classical ones do not contain a learning step. Hence, it is possible to classify reliability methods as Type 0.

Notwithstanding rapid progress in the last two or more years, there are major gaps in the current data-centric geotechnics agenda. For example, the agenda should not focus purely on maximizing the value to practice but also on minimizing the leakage of sensitive data. A proactive approach to address potential data privacy and security risks enhances the trust and confidence of users and stakeholders. Hence, Chapter 1 of "site characterization" suggested that the "data first practice central" agenda in data-centric geotechnics can be expanded to "data first, practice central, and privacy enhanced." Phoon (2025) advocated updating "data-centric geotechnics" to "trustworthy data-centric geotechnics" to manage the risk of ML/AI more systematically in response to the EU AI Act. Many exciting Type 2 MLIG research projects can be conducted. One example is data-driven site characterization (DDSC) (Phoon et al. 2022c; 2025). The construction of a quasi-site-specific ground model that includes learning cross-correlations and spatial correlations from generic databases under MUSIC-3X (Multivariate, Uncertain and Unique, Sparse, Incomplete, and potentially Corrupted with "3X" denoting 3D spatial variability) is within reach in a few years. Interest in combining site characterization information with monitoring information collected during construction to guide decision making in real time is emerging rapidly. This agenda is called the machine learning-guided observational method (MLOM). Its value is arguably higher than DDSC, and it can lead to Type 3 applications in practice. It is uncertain if a new discipline called geotechnical data science is needed to provide a stronger theoretical underpinning for data-centric geotechnics, given the large interconnected challenges.

A commonly encountered reaction to data-centric geotechnics is "data is not enough." Although this reaction is understandable, it is not helpful for two reasons: (a) there are no insights on what, where, and why data is not enough and (b) a data-poor environment is commonly seen as a showstopper rather than as a driver for data-driven innovations. It is recognized that MUSIC-3X is a basic challenge to DDSC, MLOM, and other broad application agendas in data-centric geotechnics. In fact, sparsity, uncertainty, and spatial variability are only three out of seven attributes. Current MLIG research is already handling data with much greater realism than prevailing deterministic and probabilistic methods. In fact, current MLIG research handles data "as is" with almost no idealizations. Chapter 1 of "site characterization" proposed a new taxonomy of site data under "4S" to provide a novel agenda for future MLIG research. The "4S" are site generalizations, spatial features, sampling characteristics, and smart data. The topics in the "4S" agenda are important to practice, but no work has been done testifying to the nascency of data-centric geotechnics. In addition to an in-depth analysis of geotechnical data attributes, this chapter also presents an overview of site characterization information with a focus on its availability, coverage, value to decision making, and challenges. Chapter 1 of "geotechnical structures" presents a survey of performance databases and the effectiveness of our prediction models in matching the field measurements in these databases based on three significant sources of information in the literature: (1) full-scale field tests, mainly foundation load tests, (2) prediction exercises conducted as part of international conferences, and (3) comparison between numerical analyses and in situ or field measurements.

The current book constitutes a modest step forward in data-centric geotechnics, although the database compilation efforts undertaken by the invited authors are significant. This book also represents the most diverse and most comprehensive assembly of database research in a single publication (consisting of two volumes) to date. The difference between a database assessment and a case history analysis is rarely emphasized. The current literature is replete with case history analyses, but they are primarily conducted to elucidate or improve our qualitative understanding of the governing physical mechanisms. They are not intended to quantify prediction errors in a representative way. For those studies that do compute several prediction errors, they do not attempt to characterize the prediction errors statistically with sufficient robustness for reliability-based design (sample size is too small). Reporting a prediction error for a single or a handful of case histories does not assist an engineer in decision making for a new problem. The prediction error for a new problem can be larger or smaller. It can be conservative or unconservative. In short, the engineer is none the wiser. The need to understand prediction errors statistically for a population of problems (not a single problem) is not well appreciated in a practice that remains deterministic in nature. The prediction error for a population of problems is a random variable (or it can be transformed into a random variable after removing dependencies on input parameters). Statistical characterization of the prediction error *probability distribution* requires a systematic compilation of data in the form of a database that remained a relative rarity in the past. Contrary to common perception that occasionally surfaces even in a design code, it is not possible to ensure that a calculation model is conservative with complete certainty. In the presence of a random prediction error, it is only meaningful to control the *probability* of a model being unconservative below a fractile, say 5%. The contributions in this book are already useful for design in the presence of the above model bias and imprecision. One example is the statistical characterization of the model factor to support uncertainty-informed design practice in parts of the Eurocode 7, in AASHTO's load and resistance factor design (LRFD), and in the first-order reliability-based design method permitted as an alternate design check in new design codes.

In the first volume *Site Characterization*, the properties of a wide variety of soils/rocks are covered:

- Overview – Chapter 1 (Phoon et al. 2024) presents the role of site characterization information in data-centric geotechnics regarding data availability, coverage, data attributes, value to decision making, and challenges.
- Chapter 2 (Asem and Gardoni 2024) presents the use of large databases in the development and analysis of rock socket behavior. The first database is related to full-scale field load tests on rock-socketed piles (e.g., shaft shearing, end bearing) and the other is associated with rock permeability from laboratory and in situ tests.
- Chapter 3 (Ching 2024) presents an extensive review of soil/rock property databases and their application in determining site-specific statistics for the inherent variability of soil/rock properties (e.g., mean, coefficient of variation, and scale of fluctuation) using the hierarchical Bayesian model.
- Chapter 4 (D'Ignazio and Länsivaara 2024) presents the collection and use of a high-quality database ANI-F-CLAY/7/87 to statistically establish the transformation models for estimating the anisotropic undrained shear strength of Finnish clays. They are validated against existing databases (e.g., F-CLAY/7/216 and S-CLAY/7/168) that cover a broader range of basic properties, particularly pre-consolidation stress and index properties.
- Chapter 5 (Eslami and Mo 2024) underscores the importance of soil characterization and classification in geotechnical engineering, discussing methods, uncertainties, and advancements. It presents the AUT:CPTu-Geo-Marine database that consists of 398 CPTu records from 58 different locations in 18 countries. The database is used to develop a novel triangular chart for soil behavior evaluation.
- Chapter 6 (Feng and Vardanega 2024) presents the use of a new database FG/KSAT-1358 to evaluate the variability of fine-grained soil hydraulic conductivity and study the accuracy of some established transformation models for prediction of hydraulic conductivity.
- Chapter 7 (Hov et al. 2024) presents the use of a bivariate database (Norwegian and Swedish clays) to study the correlation between normalized undrained shear strength and liquid limit for normally to slightly over-consolidated clay (OCR $\leq$ 1.5), with particular attention to uncertainties and sample quality.
- Chapter 8 (Liu and Zou 2024) presents 1,102 triaxial compression test records for 332 types of gravels with 15 parameters affecting the gravel behavior. The database is termed Geo-Gravel/15/1102, which is used to predict the shear strength and stress–strain–volume response of gravels with empirical and deep learning models.
- Chapter 9 (Löfman and Korkiala-Tanttu 2024) presents the use of an incomplete multivariate database FI-CLAY/14/856 to evaluate the compressibility of soft marine clays and its effect on settlement response.
- Chapter 10 (Ma and Li 2024) presents the use of the rock mass database SurroundingRock/8/286 in the development of an intelligent and accurate rock classification model using machine learning (e.g., convolutional neural network, random forest, Gaussian process, and support vector machine).
- Chapter 11 (Otake et al. 2024a) presents a large clay database from a single site in Japan – Tokyo-CLAY/14/67760 (2,972 boreholes covering a site area of 10 km^2) that can be used as a benchmark example for data-driven site characterization.

- Chapter 12 (Ouyang et al. 2024) presents statistical analyses of laboratory and in situ testing data to evaluate the pre-consolidation stress, over-consolidation ratio, and effective strength parameters of soil.
- Chapter 13 (Wang et al. 2024a) presents Bayesian-random field modeling of soil parameters (e.g., cohesion, friction angle, and Young's modulus) required for geotechnical analysis and design.
- Chapter 14 (Wang et al. 2024b) presents the assessment of seismic soil liquefaction with 155 CPT profiles from the New Zealand Geotechnical Database (NZGD).
- Chapter 15 (Zhou et al. 2024) presents the development and use of a large database SH-CLAY/12/4191 to evaluate the stress history (K_0), stiffness (E_s), cohesion (c), and friction angle (ϕ) of Shanghai soft clay from index parameters such as the water content, Atterberg limits, and void ratio.

In the second volume *Geotechnical Structures*, the performances of a wide variety of geotechnical structures are covered:

- Overview – Chapter 1 (Phoon and Tang 2024) presents a survey of performance databases and the effectiveness of our prediction models in matching the field measurements in these databases based on (1) full-scale field tests, (2) 39 prediction exercises organized as part of international conferences, and (3) comparison between numerical analyses and in situ or field measurements conducted by the French LCPC. The geotechnical structures include embankments, shallow and deep foundations, tunnels, and retaining walls. The focus is on the evaluation of the *statistical* degree of confidence in predicting various quantities of interest such as capacity and deformation.
- This volume covers foundations that distribute the loads from the superstructure (e.g., high-rise building, highway bridge, offshore working platform) to the bearing layer, including:
  - Spudcan foundations – Chapter 2 (Tang et al. 2024) presents an extensive review of the study (experimental, analytical, and numerical) on the punch-through behavior of spudcan in stiff-over-soft clay and clay with sand, along with the development and use of a centrifuge test database – DUT/Foundation/Punch-Through/242.
  - Deep foundations – Chapters 3–11 present the development and use of load test databases worldwide. These databases can be categorized into two types. The first is national where almost all of load tests were performed in a single country or region, e.g., France in Chapter 3 (Burlon and Frank 2024), South Africa in Chapter 5 (Dithinde and Moriasi 2024), the United Kingdom in Chapter 7 (Flynn and McCabe 2024), Malaysia in Chapter 8 (Ong et al. 2024), Japan in Chapter 9 (Otake et al. 2024b), United Kingdom in Chapter 10 (Othman et al. 2024), and Brazil in Chapter 11 (Tsuha et al. 2024). The others in Chapter 4 (Chen et al. 2024) and Chapter 6 (Eslami and Heidarie Golafzani 2024b) are generic in the sense that the databases contain load tests from many countries or regions.

These databases are widely used by researchers for the evaluation and revision of existing analysis methods that are recommended in design codes or manuals (e.g., Belgium, France, Germany, Japan, and the United States), development of more accurate and economical design methods (e.g., implementation of load and resistance factor design especially in the United States), and by engineers for the improvement of foundation design in their project (e.g., Florida and Louisiana Departments of Transportation).

- Anchors and pipelines
  - Chapter 12 (Fu et al. 2024) presents a comprehensive collection of 464 datasets obtained from centrifuge and field model tests of dynamically installed anchors in development of data-driven design methods.
  - Chapter 13 (Najjar et al. 2024) presents the collection and use of 143 uplift laboratory tests on pipes in sand and 70 field anchor pullout tests in the data-driven quantification of uncertainty and design of offshore pipelines and anchored shoring systems.
- Retaining systems and excavations
  - Chapter 14 (Bathurst and Miyata 2024) presents an extensive review of database assessment of model uncertainty in the internal stability limit states of mechanically stabilized earth walls constructed with extensible geogrids and polyester straps, and inextensible steel strip and grid soil reinforcement materials.
  - Chapter 15 (Lin 2024) presents the use of five databases to characterize the variability of predictions regarding global failure, nail bonding strength, nail load, facing tensile force, and nail displacement of soil nail walls.
  - Chapter 16 (Tan and Fan 2024) presents the use of a large database of 606 case records during 1995–2018 to address what are the main features of deep excavations in Shanghai soft clays and what are the key parameters affecting this behavior.
- Landslides – Chapter 17 (Zhang et al. 2024) presents the development of probabilistic methods to predict time to slope failure with a database of 55 landslides.

Additional supplementary resources are made available at: www.routledge.com/9781032578958 (*Site Characterization*) and www.routledge.com/9781032579108 (*Geotechnical Structures*).

The contributions of the invited authors are deeply appreciated. The editors would also like to thank Mr Tony Moore and Ms Aimee Wragg from CRC Press & Routledge for their tireless support and patient guidance.

Editors
Kok-Kwang Phoon
Chong Tang

REFERENCES

Asem, P. and Gardoni, P. 2024. Selection of rock hydromechanical parameters for rock foundation design: a database approach. Chapter 2 in *Databases for Data-Centric Geotechnics: Site Characterization*, edited by K. K. Phoon and C. Tang. Boca Raton, FL: CRC Press.

Bathurst, R. and Miyata, Y. 2024. Mechanically stabilized earth (MSE) walls. Chapter 14 in *Databases for Data-Centric Geotechnics: Geotechnical Structures*, edited by C. Tang and K. K. Phoon. Boca Raton, FL: CRC Press.

Burlon, S. and Frank, R. 2024. Development and use of LCPC pile database. Chapter 3 in *Databases for Data-Centric Geotechnics: Geotechnical Structures*, edited by C. Tang and K. K. Phoon. Boca Raton, FL: CRC Press.

Chen, G., Cheng, Q., and Puetz, S. 2023. Special issue: data-driven discovery in geosciences: opportunities and challenges. Mathematical Geosciences, 55, 287–293.

Chen, Y. J., et al. 2024. Development and use of CYCU pile load test database. Chapter 4 in *Databases for Data-Centric Geotechnics: Geotechnical Structures*, edited by C. Tang and K. K. Phoon. Boca Raton, FL: CRC Press.

Ching, J. 2024. Evaluation of soil/rock properties using databases. Chapter 3 in *Databases for Data-Centric Geotechnics: Site Characterization*, edited by K. K. Phoon and C. Tang. Boca Raton, FL: CRC Press.

D'Ignazio, M. and Länsivaara, T. 2024. Undrained shear strength of Finnish soft clays: a database perspective. Chapter 4 in *Databases for Data-Centric Geotechnics: Site Characterization*, edited by K. K. Phoon and C. Tang. Boca Raton, FL: CRC Press.

Dithinde, M. and Moriasi, H. Pile load test database for Southern Africa and evaluation of direct SPT-based pile design methods. Chapter 5 in *Databases for Data-Centric Geotechnics: Geotechnical Structures*, edited by C. Tang and K. K. Phoon. Boca Raton, FL: CRC Press.

Eslami, A. and Mo, P. Q. 2024a. Role of databases in the evaluation of soil properties. Chapter 5 in *Databases for Data-Centric Geotechnics: Site Characterization*, edited by K. K. Phoon and C. Tang. Boca Raton, FL: CRC Press.

Eslami, A., and Heidarie Golafzani, S. 2024b. Relevant data base pile design approach. Chapter 6 in *Databases for Data-Centric Geotechnics: Geotechnical Structures*, edited by C. Tang and K. K. Phoon. Boca Raton, FL: CRC Press.

Feng, S. and Vardanega, P. 2024. New laboratory database of hydraulic conductivity measurements on fine-grained soil. Chapter 6 in *Databases for Data-Centric Geotechnics: Site Characterization*, edited by K. K. Phoon and C. Tang. Boca Raton, FL: CRC Press.

Flynn, K. N. and McCabe, B. A. 2024. Piling insights from a data-centric approach. Chapter 7 in *Databases for Data-Centric Geotechnics: Geotechnical Structures*, edited by C. Tang and K. K. Phoon. Boca Raton, FL: CRC Press.

Fu, Y., Ding, K. and Han, C. C. 2024. Experimental database and data-driven design method of dynamically installed anchors. Chapter 12 in *Databases for Data-Centric Geotechnics: Geotechnical Structures*, edited by C. Tang and K. K. Phoon. Boca Raton, FL: CRC Press.

Girolami, M. 2020. Introducing data-centric engineering: an open access journal dedicated to the transformation of engineering design and practice. Data-Centric Engineering, 1:e1.

Hov, S., L'Heureux, J. S., Kahrs, K., Lundström, K. and Gaharia, D. 2024. Normalised active undrained shear strengths of soft Scandinavian clays – a data-centric and a geomechanical approach. Chapter 7 in *Databases for Data-Centric Geotechnics: Site Characterization*, edited by K. K. Phoon and C. Tang. Boca Raton, FL: CRC Press.

Karniadakis, G. E., Kevrekidis, I. G., Lu, L., Perdikaris, P., Wang, S. and Yang, L. 2021. Physics-informed machine learning. Nature Reviews Physics, 3, 422–440, DOI: 10.1038/s42254-021-00314-5.

Lin, P. Y. 2024. Statistical evaluation and calibration of model uncertainty for reliability-based design of soil nail walls. Chapter 15 in *Databases for Data-Centric Geotechnics: Geotechnical Structures*, edited by C. Tang and K. K. Phoon. Boca Raton, FL: CRC Press.

Liu, J., Zou, D. 2024. Prediction for the mechanical response of gravels. Chapter 15 in *Databases for Data-Centric Geotechnics: Site Characterization*, edited by K. K. Phoon and C. Tang. Boca Raton, FL: CRC Press.

Löfman, M. S. and Korkiala-Tanttu, L. 2024. Evaluation of compressibility properties for soft marine clays. Chapter 9 in *Databases for Data-Centric Geotechnics: Site Characterization*, edited by K. K. Phoon and C. Tang. Boca Raton, FL: CRC Press.

Ma, J. and Li, T. 2024. An engineering geological parameter database of tunnel surrounding rock and its application. Chapter 10 in *Databases for Data-Centric Geotechnics: Site Characterization*, edited by K. K. Phoon and C. Tang. Boca Raton, FL: CRC Press.

Najjar, S., Sadek, S., Ismail, S. and Chahbaz, R. 2024. Pipes and anchors. Chapter 13 in *Databases for Data-Centric Geotechnics: Geotechnical Structures*, edited by C. Tang and K. K. Phoon. Boca Raton, FL: CRC Press.

Ong, Y. H., Toh, C. T. and Chee, S. K. 2024. Bored piles in tropical soils and rocks – a database approach to design. Chapter 8 in *Databases for Data-Centric Geotechnics: Geotechnical Structures*, edited by C. Tang and K. K. Phoon. Boca Raton, FL: CRC Press.

Otake, Y., Saito, T., Wu, S., Yoshida, I. and Takano, D. 2024a. Exploring challenges via analysis of multivariate geotechnical properties: insights from large-scale sampling of Japanese marine clay. Chapter 11 in *Databases for Data-Centric Geotechnics: Site Characterization*, edited by K. K. Phoon and C. Tang. Boca Raton, FL: CRC Press.

Otake, Y., Nakamura, T., Fujita, T. and Nishida, H. 2024b. Development and Use of Databases of Pile Foundation Load Tests in Japan. Chapter 9 in *Databases for Data-Centric Geotechnics: Geotechnical Structures*, edited by C. Tang and K. K. Phoon. Boca Raton, FL: CRC Press.

Othman, M., Voyagaki, E., Crispin, J. J., Ntassiou, K., Gilder, C. E. L., De Luca, F., Mylonakis, G. E. and Vardanega, P. J. 2024. The DINGO database of axial pile load tests for the UK: determination of ultimate load. Chapter 10 in *Databases for Data-Centric Geotechnics: Geotechnical Structures*, edited by C. Tang and K. K. Phoon. Boca Raton, FL: CRC Press.

Ouyang, Z. and Agaiby, S. 2024. In situ test-based evaluation of soil effective stress strength properties and stress history. Chapter 12 in *Databases for Data-Centric Geotechnics: Site Characterization*, edited by K. K. Phoon and C. Tang. Boca Raton, FL: CRC Press.

Phoon, K. K. 2025. Trustworthy data-centric geotechnics. Geodata and AI.

Phoon, K. K. and Ching, J. 2021. Project DeepGeo – data-driven 3D subsurface mapping. Journal of GeoEngineering, 16(2), 61-74.

Phoon, K. K., and Zhang, W. G. 2023. Future of Machine Learning in Geotechnics. Georisk: Assessment and Management of Risk for Engineered Systems and Geohazards, 17(1), 7-22.

Phoon, K. K., Ching, J. and Cao, Z. J. 2022a. Unpacking data-centric geotechnics. Underground Space, 7(6), 967–989.

Phoon, K. K., Shuku, T., Ching, J. and Yoshida, I. 2022b. Benchmark examples for data-driven site characterization. Georisk: Assessment and Management of Risk for Engineered Systems and Geohazards, 16(4), 599-621.

Phoon, K. K., Ching, J. and Shuku, T. 2022c. Challenges in data-driven site characterization. Georisk: Assessment and Management of Risk for Engineered Systems and Geohazards, 16(1), 114–126.

Phoon, K. K., Shuku, T., Ching, J., and Yoshida, I. 2023a. Editorial for Special Collection on "Benchmarking data-driven site characterization methods", ASCE-ASME Journal of Risk and Uncertainty in Engineering Systems: Part A, 9(2), 02023001.

Phoon, K. K., Zhang, L. M., Cao, Z. J. 2023b. Editorial for Special Issue on "Machine learning and AI in geotechnics", Georisk: Assessment and Management of Risk for Engineered Systems and Geohazards, 17(1), 1-6.

Phoon, K. K. and Tang, C. 2024. Role of performance information in data-centric geotechnics. Chapter 1 in *Databases for Data-Centric Geotechnics: Geotechnical Structures*, edited by C. Tang and K. K. Phoon. Boca Raton, FL: CRC Press.

Phoon, K. K., Ching, J. and Tang, C. 2024. Role of site characterization information in data-centric geotechnics. Chapter 1 in *Databases for Data-Centric Geotechnics: Site Characterization*, edited by K. K. Phoon and C. Tang. Boca Raton, FL: CRC Press.

Phoon, K. K., Cai, Y. and Tang, C. 2025. Geotechnical "facial recognition" challenge. Intelligent Geoengineering.

Tan, Y. and Fan, X. Z. 2024. Statistical analyses on a database of deep excavations in Shanghai soft clays. Chapter 16 in *Databases for Data-Centric Geotechnics: Geotechnical Structures*, edited by C. Tang and K. K. Phoon. Boca Raton, FL: CRC Press.

Tang, C., Feng, X. and Phoon, K. K. 2024. Variability of predictions for punch-through of foundation in layered soils. Chapter 2 in *Databases for Data-Centric Geotechnics: Geotechnical Structures*, edited by C. Tang and K. K. Phoon. Boca Raton, FL: CRC Press.

Tsuha, C. H. C., Sousa Oliveira, C. C., Silva, B. O., Santos Filho, J. M. S. M., Schiavon, J. A. and Tang, C. 2024. Development and use of tensile loading test databases for analysis and design of helical piles. Chapter 11 in *Databases for Data-Centric Geotechnics: Geotechnical Structures*, edited by C. Tang and K. K. Phoon. Boca Raton, FL: CRC Press.

Wang, C., et al. 2024a. Mechanical-statistical evaluation of soil properties. Chapter 13 in *Databases for Data-Centric Geotechnics: Site Characterization*, edited by K. K. Phoon and C. Tang. Boca Raton, FL: CRC Press.

Wang, C., et al. 2024b. Seismic soil liquefaction assessment: approaches, data, and tools. Chapter 14 in *Databases for Data-Centric Geotechnics: Site Characterization*, edited by K. K. Phoon and C. Tang. Boca Raton, FL: CRC Press.

Wu, S., Otake, Y., Mizutani, D., Liu, C., Asano, K., Sato, N., Saito, T., Baba, H., Fukunaga, Y., Higo, Y., Kamura, A., Kodama, S., Metoki, M., Nakamura, T., Nakazato, Y., Shioi, A., Takenobu, M., Tsukioka, K. and Yoshikawa, R. 2024. Future-proofing geotechnics workflows: accelerating problem-solving with large language models. Georisk: Assessment and Management of Risk for Engineered Systems and Geohazards. https://doi.org/10.1080/17499518.2024.2381026.

Zhang, J., Yao, H., Wang, Z. and Dai, M. 2024. Probabilistic methods for slope failure time prediction. Chapter 17 in *Databases for Data-Centric Geotechnics: Geotechnical Structures*, edited by C. Tang and K. K. Phoon. Boca Raton, FL: CRC Press.

Zhou, Y., Zhang, D. and Huang, H. 2024. Prediction for soil design properties based on a multivariate database for Shanghai soft clay. Chapter 15 in *Databases for Data-Centric Geotechnics: Site Characterization*, edited by K. K. Phoon and C. Tang. Boca Raton, FL: CRC Press.

List of contributors

Shehab Agaiby
Cairo University
Giza, Egypt

Pedro Arduino
University of Washington
Seattle, WA, United States

Pouyan Asem
University of Minnesota
Minneapolis, MN, United States

Qiushi Chen
Clemson University
Clemson, SC, United States

Jianye Ching
National Taiwan University
Taipei, Taiwan

Marco D'Ignazio
Tampere University
Tampere, Finland

Abolfazl Eslami
Amirkabir University of Technology
Tehran, Iran

Shuyin Feng
Birmingham City University
Birmingham, England

David Gaharia
LabMind
Nacka, Sweden

Paolo Gardoni
University of Illinois Urbana-Champaign
Champaign, IL, United States

Sølve Hov
Norwegian Geotechnical Institute
Trondheim, Norway

Hongwei Huang
Tongji University
Shanghai, China

Katharina Kahrs
Norwegian Geotechnical Institute
Trondheim, Norway

Leena Korkiala-Tanttu
Aalto University
Espoo, Finland

Tim Länsivaara
Tampere University
Tampere, Finland

Jean-Sébastien L'Heureux
Norwegian Geotechnical Institute
Trondheim, Norway

Duo Li
Dalian University of Technology
Dalian, Liaoning, China

Tianbin Li
Chengdu University of Technology
Chengdu, Sichuan, China

Jingmao Liu
Dalian University of Technology
Dalian, Liaoning, China

Monica S. Löfman
Aalto University
Espoo, Finland

Karin Lundström
Swedish Geotechnical Institute
Linköping, Sweden

Junjie Ma
Chengdu University of Technology
Chengdu, Sichuan, China

Paul W. Mayne
Georgia Institute of Technology
Atlanta, GA, United States

Pin-Qiang Mo
China University of Mining and Technology
Xuzhou, Jiangsu, China

Fanwei Ning
Dalian University of Technology
Dalian, Liaoning, China

Yu Otake
Tohoku University
Sendai, Japan

Zhongkun Ouyang
Tsinghua Shenzhen International Graduate School
Nanshan, Shenzhen, China

Taiga Saito
Tohoku University
Sendai, Japan

Daiki Takano
Kumamoto University
Kumamoto, Japan

Daofei Tang
Shanghai University
Shanghai, China

Paul Vardanega
University of Bristol
Bristol, United Kingdom

Changhong Wang
Shanghai University
Shanghai, China

Chaofeng Wang
University of Florida
Gainesville, FL, United States

Kun Wang
Shanghai University
Shanghai, China

Stephen Wu
The Institute of Statistical Mathematics
Tokyo, Japan

Kaiyuan Xu
Xidian University
Xi'an, Shaanxi, China

Ikumasa Yoshida
Tokyo City University
Tokyo, Japan

Dongming Zhang
Tongji University
Shanghai, China

Chenguang Zhou
Dalian University of Technology
Dalian, Liaoning, China

Yelu Zhou
Tongji University
Shanghai, China

Degao Zou
Dalian University of Technology
Dalian, Liaoning, China

Chapter 1

Role of site characterization information in data-centric geotechnics

Kok-Kwang Phoon, Jianye Ching, and Chong Tang

This chapter presents an overview of site characterization information with a focus on its availability, coverage, data attributes, value to decision making, and challenges. In terms of availability, it is accurate to say that site characterization databases mainly reside in industry. However, the majority is not shared for confidentiality reasons ("dark data"). The data-centric geotechnics agenda cannot focus purely on maximizing the value of practice but also on minimizing the leakage of sensitive data. It is suggested that the "data first practice central" agenda in data-centric geotechnics can be expanded to "data first, practice central, and privacy enhanced." Data-centric geotechnics now encompasses four elements: (1) data centricity, (2) fit for (and transform) practice, (3) geotechnical context, and (4) privacy enhancement. In terms of coverage, publicly available site characterization databases are mainly geotechnical in nature containing borehole, laboratory test, and field test data. Detailed geological data such as soil stratification, rock mass features (weathering, discontinuities, etc.), and real-time data such as geoenvironmental processes (subsurface flow, contaminant transport, etc.) are lacking. This "paucity of ground truth" is a rate-limiting step in the development of machine learning algorithms for geo-related disciplines, such as soil science, geotechnical engineering, rock engineering, tunnel engineering, earthquake engineering, geoenvironmental engineering, engineering geology, mining, geo-hazard, geophysics, remote sensing, and others in geosciences. In terms of data attributes, MUSIC-3X (Multivariate, Uncertain and Unique, Sparse, Incomplete, and potentially Corrupted with "3X" denoting the spatial/temporal dimension) is a basic challenge to data-driven site characterization (DDSC) research. This chapter proposes a new taxonomy of site data under "4S" to expand the agenda for future machine learning research. The "4S" are site generalizations, spatial features, sampling characteristics, and smart data. In terms of value to design, the majority of the research is focused on the prediction of soil properties (spatial variability) and soil types (stratification) at a target site. One particularly interesting and significant challenge called "site recognition" has attracted attention in recent years. The site recognition challenge is fundamental to geotechnical engineering because it attempts to quantify the "uniqueness" (or site specificity) attribute of a target site. Recent research has demonstrated that quasi-site-specific transformation models can be developed notwithstanding the well-known limitations – site-specific data is sparse, and a generic database containing data from other sites is not directly applicable to a target site. There are three notable achievements: (1) the site recognition problem is tractable even under MUSIC, (2) inference uncertainty produced by a quasi-site-specific transformation model is smaller than that produced by a conventional generic or site-specific model, and (3) increasingly effective quasi-site-specific transformation models have been constructed, particularly those that exploit clustering. The construction of a quasi-site-specific ground model that includes learning cross-correlations and spatial auto-correlations from generic databases under MUSIC-3X is an

DOI: 10.1201/9781003441946-1

ongoing research problem. A more difficult problem is to address MUSIC-3X-G, where the attribute "G" refers to geologic uncertainty such as stratification. Research on other site characterization databases such as geophysical and remote sensing data and data fusion are limited. Interest in combining site characterization information with monitoring information collected during construction to guide decision making in real time is emerging rapidly. This new agenda is called the machine learning-guided observational method (MLOM). Its value is arguably higher than DDSC, possibly leading to a Type 3 (disruptive) outcome that can transform geotechnical practice. Other challenges such as explainability and interpretability, transferability, geo-compatibility, frugal HBM, thick data, and data protection are briefly discussed. It is uncertain whether a new discipline called geotechnical data science is needed to create novel theoretical concepts and methods for data-centric geotechnics. But trustworthiness needs to be addressed to comply with the EU AI Act and other regulatory frameworks that should be introduced soon to govern AI systems.

1.1 DATA-CENTRIC GEOTECHNICS

Foundations and underground structures have been built since antiquity. In all likelihood, the builders learnt from successful case histories. Recurring observations that are useful to practice are eventually codified as rules of thumb or passed down from masters to apprentices as traditions. Data does support decision making even when geotechnical engineering is practised as an art, except it mainly appears as a qualitative experience base. The modern form may appear as table 54.1 of Terzaghi and Peck (1967) (citing the *Kidder-Parker Architects' and Builders' Handbook* dated 1931 as the source) or table 1 of BS8004:1986 (BS 1986). These tables provide presumptive allowable bearing values that are based on local experience. The introduction of science to geotechnical engineering may be traced to Coulomb's failure wedge analysis behind a retaining wall using the laws of friction and cohesion. Data does support decision making when geotechnical engineering is practised as a science, but data is restricted to a set of inputs required by a theoretical model (e.g., friction, cohesion). It is incorrect to say that geotechnical engineering has always been data-centric in the strict sense. It has been experience-centric or physics-centric, but to the knowledge of the authors, data has never supported decision making *by itself*. Data-driven decision making is only gaining some interest because machine learning (ML) and artificial intelligence (AI) have transformed many industries.

Data-centric geotechnics is distinctive from other fields in geotechnical engineering and machine learning (Phoon and Ching 2021; Phoon et al. 2022a). It is a new interdisciplinary field involving the integration of information, data, techniques, tools, perspectives, concepts, theories, and/or experiences from geotechnical engineering and machine learning (Phoon and Zhang 2023). Broader developments in "data-centric engineering" (Girolami 2020) and data-driven discovery in geosciences (Chen et al. 2023) are taking place elsewhere in parallel. Bilal et al. (2016) and Munawar et al. (2022) surveyed the application of "big data engineering" in construction. Research in machine learning in geotechnics (MLIG) in the context of data-centric geotechnics must articulate its value proposition clearly to hasten the rate of digital transformation in the industry. For example, "ML supremacy" projects are those involving large datasets (perhaps real-time data) from multiple sources covering large spatial domains that cannot be handled effectively and efficiently using conventional methods, and there is a strong demand for

Table 1.1 Value of machine learning in geotechnics as a function of data and method

		Method	
		Available generic ML (e.g., artificial neural network, support vector machine, deep learning, transfer learning, meta learning)	*Novel geotechnical ML*
Data	Available data (e.g., site investigation data, proof test data, monitoring data)	**Type 1** Incremental value – applications of available ML to conventional geotechnical structures, e.g., foundations, slopes ("ML advantage")	**Type 2a** Potentially high value – development of new ML to address distinctive challenges in geotechnical engineering resulting in new applications, e.g., DDSC ("ML advantage/supremacy")
	Emerging data (e.g., wireless sensor network, laser scanning, drone photogrammetry, satellite imagery, etc.)	**Type 2b** High value – applications of available ML to current problems that cannot be addressed effectively using conventional methods, e.g., hazard/risk mapping ("ML advantage/supremacy")	**Type 3** Disruptive value – development of new "fit for practice" geotechnical ML to transform research and practice resulting in new products, systems, and services and may even redefine the role of geotechnical engineering

Source: table 1, Phoon and Zhang (2023).

implementation of ML methods in these projects (Shahri et al. 2019; Abubakar et al. 2021; Christensen et al. 2021; Wang et al. 2021a; Wang et al. 2021b; Lin et al. 2022; Phoon et al. 2023a). Physics cannot support decision making in these projects because it is absent or grossly incomplete. Classical statistics may entail insupportable assumptions on the attributes of the data. The value of MLIG research can be classified as follows: (1) Type 1 (incremental value) involving available data and existing conventional applications, (2) Type 2a (potentially high value) involving available data and new applications, (3) Type 2b (high value) involving new data and existing applications, and (4) Type 3 (disruptive value) involving new data and completely novel applications (e.g., precision construction, autonomous construction control) (Phoon and Zhang 2023). Details are given in Table 1.1. Type 1 MLIG provides an alternate approach to an existing problem. It is only "good to have." Type 2 MLIG outperforms existing approaches (if they exist) to such an extent that it will replace them. Type 3 MLIG will make practice unrecognizable by present-day engineers. These MLIG methods are distinctive from reliability methods (Ji and Cao 2024), although they are both data-driven. MLIG methods can be improved by learning from data and may eventually even modify or propagate themselves (e.g., an AI programming a better version of itself). Reliability methods address uncertainty explicitly, but the classical ones do not contain a learning step. Hence, it is possible to classify reliability methods as Type 0 in the context of Table 1.1.

Interest in MLIG has been gathering momentum in recent years (Faramarzi, 2020; Jaksa and Liu 2021; Morgenroth et al. 2022; Wang et al. 2022; Yuen et al. 2021; Zhang and Phoon 2022; Du et al. 2023; Phoon et al. 2023b; Phoon et al. 2023c; Zhang et al. 2023). However, the emphasis remains "method first" rather than "data and practice first." Numerous foundational challenges beyond methodological ones can be identified when research is pursued under a more balanced data-centric geotechnics agenda covering three elements (Phoon and Zhang 2023; Phoon et al. 2023b):

(1) Data centricity – The goal is to develop methods that make sense of all real-world data sets. The prevailing strategy is the opposite. The engineer screens/interprets data to fit a particular theory/paradigm. For example, only some specific constitutive model parameters are relevant to a numerical model. Another example is to reduce a dataset to a single number to fit the deterministic paradigm [e.g., characteristic value in Eurocode 7 (CEN 2004), Länsivaara et al. 2022]. The third example is to accept/reject a dataset based on its compliance with a set of conditions imposed by classical frequentist analysis (e.g., minimum sample size). These examples exhibit a "method first, data second" philosophy. The central tenet in data-centric geotechnics is that data has value if it is not fake, completely corrupted (bad data), or completely irrelevant. Hence, it may not be economical to use data selectively. Data centricity does not imply that experience and physics are not relevant. Data has supported prevailing experience-centric or physics-centric practice. In the same way, experience and/or physics can support data-centric geotechnics. In machine learning, experience can be regarded as one type of "thick data" to distinguish it from the more well-known quantitative "big data." Physics-informed machine learning (PIML) methods offer the advantages of "explainability" and "interpretability" (Karniadakis et al. 2021; Li et al. 2023; Shioi et al. 2023; Tan et al. 2023; Yan et al. 2023). Large language models such as GPT-4o are already capable of using multi-modal data.

(2) Fit for (and transform) practice – The goal is to bring significant value to critical real-world decisions (not decisions for an ideal world or decisions of minor concern to geotechnical engineers just to showcase the latest method developed in computer science). One classification of value is shown in Table 1.1. Data-centric geotechnics should, therefore, work on real-world datasets from real-world problems only and should enhance decision making where it matters most to the practitioner. Autonomous construction is one example where data can lead to intelligence that takes over routine decision making at a construction site from a human supervisor.

(3) Geotechnical context – Datasets are not abstract numbers divorced from their physical basis and their physical origin. As such, direct application of machine learning developed in other fields with cursory reference to the geotechnical context is not appropriate. The geotechnical context that gives rise to the data is important. The context can be related to statistics, physics, or experience. Statistics refer to the attributes of geotechnical data that depart significantly from the assumptions in classical statistics (large sample size, spatial/temporal/parametric independence, homogeneity, normality, etc.). Physics refers to a body of rational knowledge that associates a "number" with "meaning and casuality." An engineer distinguishes between material and state parameters, between effective and total stress parameters, and between input and output parameters from a numerical model. These distinctions exist when one approaches data from the lens of physics. Experience refers to a body of empirical knowledge accrued from deliberate practice. In contrast to statistics and physics, it is mainly subjective and qualitative. Nonetheless, many engineers regard experience as critical. Burland (2012) termed well-winnowed experience as "experience that results from a rigorous sifting of all the facts that relate to a particular empirical procedure or case history" in the geotechnical triangle. The key limitation of experience is that it cannot be transferred from one engineer to the next efficiently, certainly nothing close to the speed that one can share data and algorithms in the cloud. In addition, it is rare for an engineer to accumulate global experience. This means an engineer can only make sense of regional data that he/she is familiar with, but would struggle to do the same for the entire global database.

Given the nascent stage of development in data-centric geotechnics, it may be more realistic for this chapter to examine whether there is a plausible pathway to Table 1.1 for site investigation information. The value of performance information is discussed in the companion volume (Phoon and Tang 2025). In all likelihood, Type 3 methods will appear in the context of Industry 4.0. At the end of this chapter, it is proposed that "privacy enhancement" be added to the agenda for data-centric geotechnics.

1.2 SITE CHARACTERIZATION INFORMATION

The first step in data-centric geotechnics is not to deep dive into one source of information because it is the most convenient or most tractable. Since all data are regarded as potentially useful, the first step is to take stock of all available sources (including emerging ones related to the Internet of Things or Everything) and to lay out a strategy on how to bring the entire *geodata estate* to bear on decision support or to transform practice in novel ways. Data gaps may be identified that can motivate the development and deployment of new sensors in a more purposeful way. This requires an understanding of the traits or attributes of geotechnical information that are likely to change in tandem with the growth and convergence of digital technologies. Below are examples of classifying the attributes of site investigation information in relation to data-driven site characterization (DDSC) and the larger Industry 4.0 paradigm to push for Type 2 or 3 machine learning research. Huang et al. (2022) explored the progress, challenges, and opportunities in big earth data analytics in the same spirit as this chapter. "Big earth data" refers to data about the planet: the ocean, the land, the atmosphere, and climate.

1.2.1 Big indirect data (BID)

Phoon et al. (2019) coined the term "big indirect data" (BID) to refer to any data that are potentially useful but not directly applicable to the decision at hand. *Generic* databases containing soil/rock properties (Phoon et al. 2024), load test data (load-movement curves, load transfer curves) (Tang et al. 2024), or monitoring data from multiple sites involving different structures would belong to BID. A representative list of soil/rock property databases is shown in Table 1.2. There are several national databases (large) for public access via the Internet. The New Zealand geotechnical database (NZGD), curated by the Earthquake Commission, holds approximately 23,000 cone penetration test (CPTs), 10,000 borehole log records, 1000 piezocone (CPTu) records, and 6000 laboratory test records. The United States Community shear wave velocity profile database (VSPDB) consists of geophysical (3845 V_s/V_p profiles and 1993 spectral ratio records), in situ (1212 boreholes and 1342 CPTs), and laboratory tests (26 sieve analyses and 845 Atterberg limits) conducted at 4327 sites, where V_s and V_p are shear and compression wave velocities (Kwak et al. 2021). Another similar compilation work is the Next-Generation Liquefaction (NGL) database, which provides researchers with a substantially larger, more consistent, and more reliable source of liquefaction data than what has existed previously (Ulmer et al. 2023). The NGL database also consists of geophysical (199 V_s/V_p profiles), in situ (941 boreholes and 747 CPTs), and laboratory tests (29,023 sieve analyses and 1665 Atterberg limits). The US Geological Survey (USGS) and other governmental agencies funded the studies to compile V_{30} values for 4389 sites in the United States, where V_{30} is the time-averaged V_s in the upper 30 m (McPhillips et al. 2020). The National Geotechnical Properties Database (NGPD) maintained by the British Geological Survey (BGS) may be the most

Table 1.2 Soil/rock property databases

	Database	*Reference*	*Soil/rock parameters*	*# Data points*	*# Sites/ studies*	*% complete*
Univariate	CLAY/16	Phoon and Kulhawy (1999a)	γ, γ_d, w_n, PL, LL, PI, LI, ϕ', s_u, s_u^{FV}, q_c, q_t, SPT-N, DMT (A, B), PMT p_L		a	
	SAND/11	Phoon and Kulhawy (1999a)	ϕ', D_r, q_c, SPT-N, DMT (A, B, I_D, K_D, E_D), PMT (p_L, E_{PMT})		b	
	ROCK/8	Prakoso (2002)	γ (or γ_d), n, R, S_h, σ_{bt}, I_s, σ_c, E		c	
	ROCK/13	Aladejare and Wang (2017)	ρ, G_s, I_{d2}, n, w_c, γ, R_L, S_h, σ_{bt}, I_{s50}, σ_c, E, ν		d	
	USGS	McPhillips et al. (2020)	V_{s30}		4389 sites in the USA	Web portal
Multivariate	CLAY/5/345	Ching and Phoon (2012)	LI, s_u, s_u^{re}, σ'_p, σ'_v	345	37 sites	100%
	CLAY/7/6310	Ching and Phoon (2013)	s_u from seven different test procedures	6310	164 studies	17.7%
	CLAY/6/535	Ching et al. (2014)	s_u/σ'_v, OCR, q_{tc}, q_{tu}, $(u_2–u_0)/\sigma'_v$, B_q	535	40 sites	100%
	CLAY/10/7490	Ching and Phoon (2014a)	LL, PI, LI, σ'_v/P_a, σ'_p/P_a, s_u/σ'_v, S_t, q_{tc}, q_{tu}, B_q	7490	251 studies	34.1%
	FI-CLAY/7/216	D'Ignazio et al. (2016)	s_u^{FV}, σ'_v, σ'_p, w_n, LL, PL, S_t	216	24 sites	100%
	FI-CLAY/14/856	Löfman and Korkiala-Tanttu (2021)	w_n, e_0, LL, PL, PI, γ, C_l, s_u, S_t, σ_p', OCR, C_c, C_R, C_s	856	33 sites	
	JS-CLAY/5/124	Liu et al. (2016)	M_r, q_c, f_s, w_n, γ_d	124	16 sites	100%
	JS-CLAY/7/372	Zou et al. (2017)	σ_v, σ'_v, q_{tc}, f_s/σ'_v, B_q, V_{s1}, s_u/σ'_v	372	25 sites	100%
	SAND/7/2794	Ching et al. (2017)	D_{50}, C_u, D_r, σ'_v/P_a, ϕ', q_{t1}, $(N_1)_{60}$	2794	176 studies	60.0%
	EMI-ROCK/8/26000+	Kim and Hunt (2017)	σ_c, σ_{bt}, ρ, CAI, PPI, cohesion, direction shear, triaxial confining	26,000+	–	–
	FG/5/1000	Kootahi and Moradi (2017)	e, w_n, LL, PI, C_c	1000	170 sites	100%
	ROCK/9/4069	Ching et al. (2018)	γ, n, R_L, S_h, σ_{bt}, I_{s50}, V_p, σ_{ci}, E_i	4069	184 studies	34.2%
	FG-KSAT/6/1358	Feng and Vardanega (2019)	e, k, LL, PL, PI, G_s	1358	33 studies	91.4%
	CG/KSAT/7/1278	Feng et al. (2023)	e, k, D_{10}, D_{50}, G_s, C_u, C_z	1278	> 50 studies	
	RFG/TXCU-278	Beesley and Vardanega (2020)	s_u/σ_{v0}', $\gamma_{50\ CIU}$, $\gamma_{50\ CKU}$, b_{CIU}, b_{CKU}	278	278 CU tests	100%
	SH-CLAY/11/4051	Zhang et al. (2020)	LL, PI, LI, e, K_0, σ'_v/P_a, s_u/σ'_v, S_t, q_c/σ'_v	4051	50 sites	39.5%
	CLAY/8/12225	Ching (2020)	LL, PI, w, e, σ'_v/P_a, C_c, C_{ur}, c_v	12,225	427 studies	–

(*Continued*)

Table 1.2 (Continued) Soil/rock property databases

	Database	*Reference*	*Soil/rock parameters*	*# Data points*	*# Sites/ studies*	*% complete*
	CLAY/12/3997	Ching (2020)	LL, PI, LI, σ'_v/P_a, σ'_p/P_a, s_u/σ'_v, K_0, E_u/σ'_v, B_q, q_{t1}, $N_{60}/(\sigma'_v/P_a)$	3997	237 studies	–
	SAND/13/4113	Ching (2020)	e, D_r, σ'_v/P_a, σ'_p/P_a, K_0, E_{dn}, q_{c1n}, B_q, $(N_1)_{60}$, K_{DMT}, E_{DMTn}, E_{PMTn}, M_{dn}	4113	172 studies	–
	ROCKMass/9/5876	Ching et al. (2021a)	RQD, RMR, Q, GSI, E_m, E_{em}, E_{dm}, E_i, σ_{ci}	5876	225 studies	29.3%
	CLAY-C_c/6/6203	Ching et al. (2022a)	LL, PI, w_n, e, C_c, C_{ur}	6203	429 studies	61%
	SOIL-DMT/8/7186	Ching and Kuo (2023)	LI, D_{50}, s_u^{FV}/P_a, N_{60}, $(q_t-\sigma_v)/P_a$, B_q, E_{PMT}/P_a, $(1-\nu^2) \times E_{DMT}/P_a$	7286	237 studies	28%
	Global-CPT/3/1196	Ching et al. (2023a)	q_t, f_s, u_2	1196	59 sites	100%
	Geo-Marine-CPT/3/398	Eslami et al. (2023)	q_t, f_s, u_2	398	58 sites	100%
	ROCK/10/4025	Muzamhindo and Ferentinou (2023)	γ, n, R_L, I_{s50}, V_p, σ_{bt}, BPI, m_i, E_i, σ_c	4025	96 studies	50%
	SAND-Small/9/939	Lo et al. (2021)	D_{50}, C_u, e_{min}, e_{max}, σ'_3, e_c, G_{max}, σ'_{1p}, ϕ'	939	15 studies	79%
National	NGL	Ulmer et al. (2023)	Geophysics – velocity profiles (199), in situ tests (941 boreholes and 747 CPTs), and laboratory tests (29,023 sieve analyses and 1665 Atterberg limits)			Web portal
	NGPD	Self et al. (2012)	Boreholes (178,436), in situ tests (3,617,186), laboratory tests (5,180,330)		7370 sites (UK)	Unavailable
	NZGD		In situ tests (23,000 CPTs, 10,000 borehole logs, 1000 CPTu records) and laboratory tests (6000)		6000 sites (New Zealand)	Web portal
	VSPDB	Kwak et al. (2021)	Geophysics – velocity profiles (3845) and spectral ratio records (1993), in situ tests (1212 boreholes and 1342 CPTs), and laboratory tests (26 sieve analyses and 845 Atterberg limits)		4327 sites	Web portal

(Continued)

Table 1.2 (Continued) Soil/rock property databases

	Database	*Reference*	*Soil/rock parameters*	*# Data points*	*# Sites/ studies*	*% complete*
SWCC	MEPDG	Zapata (2010)	20,526 soil units with enough information to compute the SWCC parameters		US	Web portal

Source: updated from Phoon et al. (2024).

Notes:

Basic – C_u = coefficient of uniformity; D_{50} = median grain size; ρ = density; G_s = specific gravity; γ = unit weight; γ_d = dry unit weight; D_r = relative density; e = void ratio; e_{min} = minimum void ratio; e_{max} = maximum void ratio; e_c = void ratio under isotropic consolidation; n = porosity; w_n (or w_c) = water content; PL = plastic limit; LL = liquid limit; PI = plasticity index; LI = liquidity index; GSI = geological strength index.

Stress – σ_v = total vertical stress; σ'_v = effective vertical stress; σ'_p = pre-consolidation stress; σ'_3 = effective confining stress under isotropic consolidation; σ'_{1p} = effective axial stress at peak state under drained triaxial compression; OCR = over-consolidation ratio; P_a = atmospheric pressure = 101.3 kPa; K_0 = at-rest lateral earth pressure coefficient.

Strength – ϕ' = effective friction angle; s_u = undrained shear strength; s_u^{FV} = field vane s_u; s_u^{re} = remoulded s_u; S_t = sensitivity; σ_{bt} = Brazilian tensile strength; σ_{ci} (or σ_c) = uniaxial compressive strength of intact rock; BPI = block punch index; m_i = Hoek–Brown constant.

Deformation – C_c = compression index; C_{ur} = unloading–reloading index; modulus; E_u = undrained modulus of clay; E_d = drained modulus of sand; E_{dn} = $(E_d/P_a)/(\sigma'_v/P_a)^{0.5}$; E_{dm} = dynamic modulus of rock mass; E_{em} = elasticity modulus of rock mass; E_i (or E) = Young's modulus of intact rock; E_m = deformation modulus of rock mass; ν = Poisson ratio; M_r = subgrade resilience modulus; M_d = effective constrained modulus determined by oedometer; M_{dn} = normalized M_d = $(M_d/P_a)/(\sigma'_v/P_a)^{0.5}$; G_{max} = small-strain shear modulus; $\gamma_{50\ CIU}$ = shear strain to mobilize $0.5s_u$ under isotropically consolidated undrained condition; $\gamma_{50\ CKU}$ = shear strain to mobilize $0.5(s_u - \tau_0)$; τ_0 = initial shear stress; b_{CIU} and b_{CKU} = coefficients to describe nonlinearity.

Permeability – k = hydraulic conductivity; c_v = coefficient of consolidation.

Dynamic – V_p = P-wave velocity; V_s = S-wave velocity; V_{s30} = time-averaged V_s to a depth of 30 m; V_{s1} = $V_s(P_a/\sigma'_v)^{0.25}$.

Field test – SPT-N = standard penetration test blow count; N_{60} = corrected SPT-N; $(N_1)_{60}$ = $N_{60}/(\sigma'_v/P_a)^{0.5}$; q_c = cone tip resistance; q_t = corrected cone tip resistance; f_s = sleeve frictional resistance; q_{tc} = $(q_t/P_a)/(\sigma'_v/P_a)^{0.5}$; q_{t1} = $(q_t-\sigma_v)/\sigma'_v$ = normalized cone tip resistance; q_{tu} = $(q_t-u_2)/\sigma'_v$ = effective cone tip resistance; q_{c1n} = $(q_c/P_a)/(\sigma'_v/P_a)^{0.5}$; B_q = pore pressure ratio = $(u_2-u_0)/(q_t-\sigma_v)$; $(u_2-u_0)/\sigma'_v$ = normalized excess pore pressure; u_2 = pore pressure behind cone tip; u_0 = hydrostatic pore pressure; PMT (p_L, E_{PMT}) = pressuremeter limit stress, modulus; E_{PMTn} = normalized E_{PMT} = $(E_{PMT}/P_a)/(\sigma'_v/P_a)^{0.5}$; DMT ($A$, B, I_{DMT}, K_{DMT}, E_{DMT}) = dilatometer A and B readings, material index, horizontal stress index, modulus; E_{DMTn} = normalized E_{DMT} = $(E_{DMT}/P_a)/(\sigma'_v/P_a)^{0.5}$; CAI = Cerchar abrasivity index; PPI = punch penetration index; Q = Q-system; RMR = rock mass rating; RQD = rock quality designation; R = Schmidt hammer hardness (R_L = L-type Schmidt hammer hardness); S_h = Shore scleroscope hardness; I_{d2} = slake durability index; I_s = point load strength index (I_{s50} = I_s for diameter 50 mm).

a – The no. of data groups varies between 2 and 42 depending on the clay parameter. Statistics are calculated at the data group level. The average no. of data points/data group varies between 16 and 564. Details given in tables 1–3, Phoon and Kulhawy (1999b).

b – The no. of data groups varies between 5 and 57 depending on the sand parameter. Statistics are calculated at the data group level. The average no. of data points/data group varies between 15 and 123. Details are given in tables 1–3, Phoon and Kulhawy (1999b).

c – The no. of data groups varies between 30 and 174 depending on the rock parameter with no differentiation of rock type [igneous (intrusive, extrusive, pyroclastic), sedimentary (clastic, chemical), metamorphic (foliated, non-foliated)]. Statistics are calculated at the data group level. The average no. of data points/data group varies between 3 and 161 for σ_c (Prakoso, 2017). Details are given in table 4.4, Prakoso (2002).

d – The no. of data groups varies between 2 and 47 depending on the rock parameter and rock type (igneous, sedimentary, or metamorphic). Statistics are calculated at the data group level. The average no. of data points/data group varies between 7 and 92. Details are given in tables 2–4, Aladejare and Wang (2017).

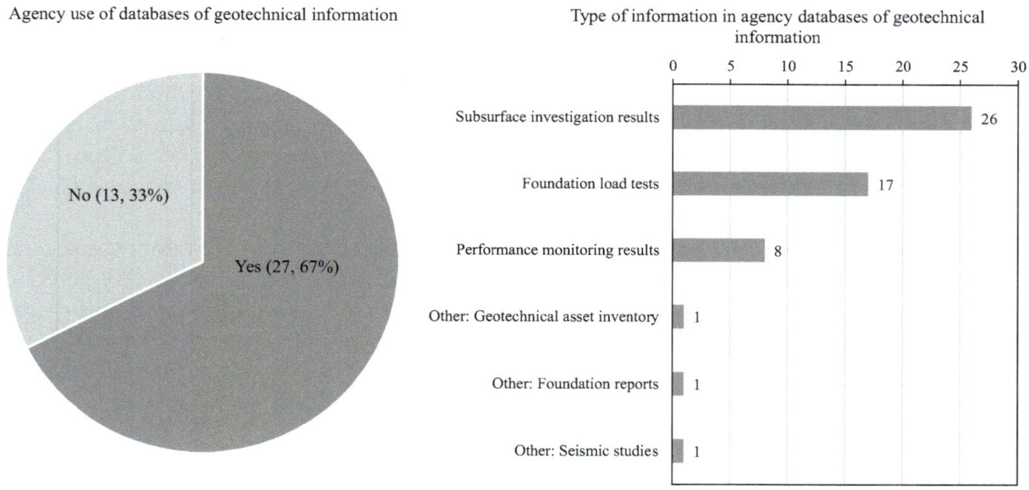

Figure 1.1 Types of information included in state DOTs' databases of geotechnical information (Source: data from Boeckmann and Loehr 2023).

comprehensive database, holding the data and information from commercial site investigations (Self et al., 2012). By 2019, the NGPD contains records for 7370 projects. Linked to the site investigation reports are 178,436 boreholes. There are a total of 3,617,186 in situ testing records and 879,293 samples linked to the boreholes, with a total of 5,180,330 laboratory testing records. Most recently, Otake et al. (2024) presented a large clay database from a single site in Japan – Tokyo-CLAY/14/67,760. It contains 14 soil parameters determined from 2972 boreholes covering a site area of approximately 10 km^2. The NCHRP Synthesis 601 indicated that 27 state Departments of Transportation (DOTs) across the United States maintain a database of geotechnical information (Boeckmann and Loehr 2023). The types of information included in the database are presented in Figure 1.1. The most common is subsurface investigation results (96%) and the next is foundation load tests (63%). Compilation, interpretation, and application of subsurface investigation results for geotechnical analysis and design constitute the main content of this volume. Development and use of performance databases for geotechnical structures (e.g., shallow/deep foundations, offshore spudcans, anchors/pipelines, slopes/retaining walls, and excavations) are the primary objectives of the companion volume (Tang and Phoon 2025), where foundation load tests are the most prevalent.

There is a widespread perception that geotechnical data is scarce and no data-driven methodologies are applicable. This is only true for site-specific data based on current site investigation technology. To insist that only site-specific data is relevant to the project at hand is not aligned to current practice. Generic transformation models that relate a field test parameter (e.g., cone tip resistance) to a design parameter (e.g., undrained shear strength) are prevalent (Kulhawy and Mayne 1990; National Academies of Sciences, Engineering, and Medicine 2019). Clause 2.4.5.2(10) of Eurocode 7 (CEN 2004) "Characteristic values of geotechnical parameters" takes a more cautious position on pooling data from different sites: "If statistical methods are employed in the selection of characteristic values for ground properties, such methods should differentiate between local and regional sampling and should allow the use of a priori knowledge of comparable ground properties." However, no data-driven method exists to differentiate between "local and regional sampling" until recently (Ching and Phoon 2020a; Ching et al. 2021a, 2021b, 2022a; Sharma et al. 2022, 2023; Cai et al. 2024a). A cursory examination of the word "local" would already reveal some complications. For example, would "local" be restricted to one site or can it encompass an adjacent site, a block, a precinct, or a city? It is also not well appreciated that drawing upon "comparable ground properties" is using some relevant information from *past* regional sampling. Vardanega and Bolton (2016) essentially raised the same concern in the selection of the coefficient of variation (COV):

> Although reliable estimates of the mean and standard deviation are easier to ascertain than the shapes of the pdfs, there remains an unjustified tendency to rely solely on published COV values from other soil deposits. Because variability in a soil deposit is a function of the processes of geological deposition and geomorphological change that have influenced the site (e.g., Hutchinson 2001), intensive efforts would be necessary to draw parallels between a new site that lacks such information and previously explored sites for which COV values have been established.

The majority of the engineers would take a more practical position that inferences drawn from BID can be useful when they are moderated by engineering judgment ("reality check"). As mentioned above, the drawback of judgment is that it cannot be applied beyond the experience base of an engineer. A global database containing sites that are unfamiliar to an engineer is not accessible to judgment. The purpose of classifying geotechnical big data as BID is to sharpen the research statement: between site-specific small data that cannot be used for machine learning training and BID that cannot be used directly without accounting for site differences, is there a quasi-site-specific machine learning solution better than both extremes? Current research shows that the concern regarding the relevance of BID to decision making at a specific site is a moot point, because it has since been addressed directly in the site recognition challenge (Phoon and Ching 2022; Phoon et al. 2025). Research motivated by BID shows that the conceptual demarcation of local and regional sampling and the assumption that only local sampling is useful for the decision at hand are overly simplistic and do not lead to fruitful research directions.

1.2.2 MUSIC-3X

Phoon et al. (2019) presented a useful mnemonic, MUSIC-X (Multivariate, Uncertain and Unique, Sparse, Incomplete, and potentially Corrupted with "X" denoting the spatial/temporal dimension) to highlight seven common ugly attributes in *site-specific* data. One example is illustrated in Table 1.3. It is common to conduct multiple tests at one site. Different test data may be cross-correlated when they are measured in close proximity (*M*). Site uniqueness is recognized as a distinctive feature of geotechnical engineering (*U*). Site-specific data are frequently sparse and incomplete in the sense that tests are conducted at limited depths and few locations (*S*), and any one test is conducted at selective depths and/or locations only (*I*). This incompleteness attribute covers both missing values in a test parameter depth profile (incomplete record) and missing tests at a particular location (entire records are missing for some test parameters) as shown in Figure 1.2. A sufficiently large database must contain potentially corrupted data (*C*). A manual process of screening out corrupted data will be defeated by the sheer volume of data ultimately. A yottabyte (YB) is 2^{80} bytes, or approximately a million trillion megabytes (MB). It is likely that this inconceivably large number will be breached in less than 5 years. The

Table 1.3 Data-driven methods for site characterization in Project DeepGeo

Method	*No. of parameters*	*Spatial variability*		*Use generic database?*	*Other limitations*	*Reference*
		Trend	*Autocorrelation*			
Sparse Bayesian learning (SBL)	Single	Yes	1X	No	Stationary autocorrelation	Ching and Phoon (2017)
Sparse Bayesian learning (3D SBL)	Single	Yes	3X	No	Stationary separable autocorrelation; vertically dense; lattice	Ching et al. (2020a)
Sparse Bayesian learning (3D SBL)	Single	Yes	3X	No	Stationary separable autocorrelation; vertically dense	Ching et al. (2021c)
Gaussian process regression (GPR-MUSIC)	Multiple	–	–	No	No vertical autocorrelation; perfect horizontal autocorrelation	Ching and Phoon (2019)
Gaussian process regression (GPR-MUSIC-X)	Multiple	No	1X	No	Prescribed stationary vertical autocorrelation; perfect horizontal autocorrelation	Ching and Phoon (2020b)
Gaussian process regression (GPR-MUSIC-3X)	Multiple	No	3X	No	Prescribed stationary separable autocorrelation	Ching et al. (2021d)
Gaussian process regression + hierarchical Bayesian model (HBM-MUSIC)	Multiple	–	–	Yes	No vertical autocorrelation; perfect horizontal autocorrelation	Ching et al. (2021a, 2021b)
Gaussian process regression + hierarchical Bayesian model (HBM-MUSIC-3X)	Multiple	No	3X	Yes	Prescribed stationary separable autocorrelation	Ching et al. (2021d)
Full Gaussian process regression (FGPR-MUSIC-3X)	Multiple	Yes	3X	No	Stationary separable autocorrelation	In progress

Source: table 2, Phoon and Ching (2021).

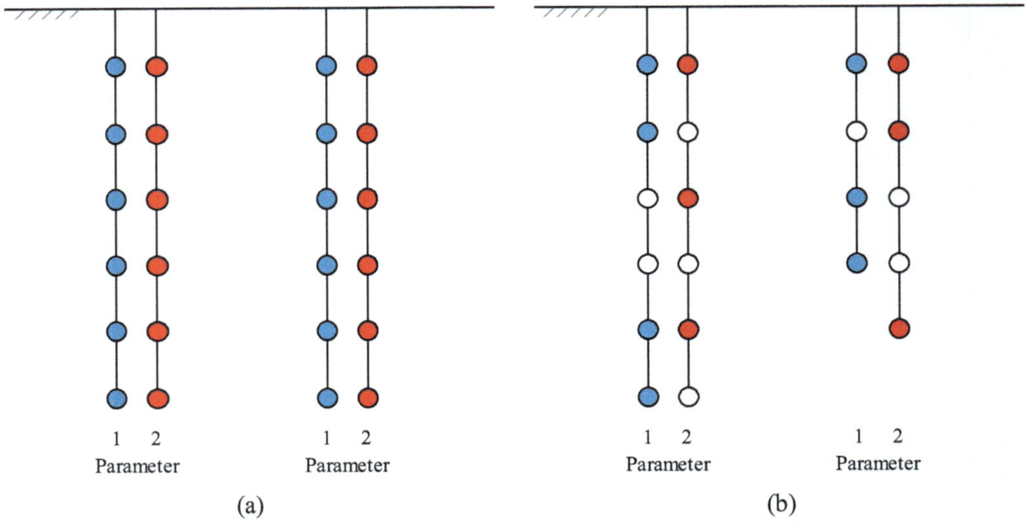

Figure 1.2 Bivariate site data: (a) complete (ideal case) and (b) incomplete (realistic case) (Source: figure 2.27, Phoon et al. 2024). Note: Open circle denotes unmeasured location.

only feasible strategy is to develop data-driven methods that can identify corrupted data automatically. Corrupted data can be bad data, which is defined as data that mislead the decision maker. In short, no data is better than bad data. Recent research on outlier analysis in data-driven site characterization takes a different perspective that inference can be improved by screening out "irrelevant" data (Ching et al. 2024; Cai et al. 2024b). Irrelevant data is not necessarily bad or corrupted data. It can be good data that so happens not to contribute to the inference of a specific property at a target site. In this chapter, the labels "outlier", "irrelevant", "bad", and "corrupted" are used interchangeably. The "outlier" label is neutral. The "irrelevant" label focuses on the impact to decision making. The "bad" and "corrupted" labels are possible explanations for outlier data. Finally, site data are spatially variable or heterogeneous (X). They can vary with time, possibly within a project duration for improved ground. Phoon (2025) has raised the ugly data challenge to MUSIC-3X-G to cover geologic uncertainty ("G"). Geologic uncertainty refers to the distribution of soil types, rock level, anomalies or other geologic features.

It is an open question whether DDSC can solve real-world 3D subsurface mapping problems based on real-world MUSIC-3X data (3X refers to 3D spatial variability as defined by Phoon et al. 2022b) with minimum ad hoc assumptions. The successes shown in the literature are based on data that addresses some attributes of MUSIC-3X only. Research in site recognition attacks the "U" attribute (Phoon and Ching 2022) (see Sections 1.3.1 and 1.4). The first attempt to address "C" in the full context of MUSIC-3X (not in the typical context of ideal data) was made only recently (Ching et al. 2024; Cai et al. 2024b) (see Section 1.4.3). The computational challenges are also very significant given the size of the 3D ground volume, but some reasonable partial solutions have been obtained recently in Project DeepGeo, which constitutes one major research effort in the emerging field of data-centric geotechnics (Phoon and Ching 2021). Phoon et al. (2022c) suggested an initial "small" ground volume of a 20 m long × 20 m wide × 10 m deep cuboid for benchmark examples. The authors suggested that "medium" and "large" ground volumes can be defined as a 40 m long × 20 m wide × 20 m deep cuboid and an 80 m long × 40 m wide × 40 m deep cuboid, respectively. A summary of the data-driven methods in Project DeepGeo is given in Table 1.3. There are other research initiatives attacking this 3D subsurface mapping challenge (Phoon et al. 2023c). Although it is easy to imagine a wider set of attributes for site investigation data such as those presented in Section 1.2.3, it is worth pointing out that MUSIC-3X has already served to drive DDSC research in a fruitful way to meet the second element of data-centric geotechnics "fit for (and transform) practice."

1.2.3 New taxonomy of site data

This section proposes a new taxonomy of site data under "4S" to expand the agenda for future MLIG research. The "4S" are site generalizations, spatial features, sampling characteristics, and smart data. This expanded agenda covers more than site investigation data and attempts to go beyond conventional applications in current practice.

1.2.3.1 Site generalizations

The concept of a "site" is fundamental in geotechnical engineering, but the full extent of its complexity is still unfolding. The current understanding of a "site" is mainly related to geology, which looks at how the earth is formed, its structure and composition, and the types of processes acting on it. The argument that every site is unique is founded on an intrinsic feature of geology, namely its natural variability. One can draw three conclusions from this line of argument: (1) differences between sites are measured by structural and/or compositional aspects of the ground, (2) the degree of difference (or similarity) between two sites is absolute in the sense that the structural and compositional aspects do not change within the lifetime of a structure for a natural ground, and (3) for a local geology that varies somewhat smoothly, an adjacent site is likely to be more similar than a distant site.

From a data-centric geotechnics perspective, one can imagine generalizing the concept of a "site" in three dimensions: (1) broaden the understanding from geology to the performance of a geotechnical structure, (2) dynamic data-driven resizing to achieve homogeneity, and (3) precision characterization and construction. For the first dimension, data on soil behaviours and soil–structure interactions can inform the similarity between two sites. The soil behaviour type index derived from the cone penetration test or the dilatometer material index is not based on compositional data, although they are used for soil classification. Monitoring, pile load tests, or other performance data have never been used for site comparison. This is surprising because an engineer is ultimately more interested in performance rather than geology. This general definition of a "site" is no longer absolute because it may depend on the soil behaviour and the limit state of interest. For example, two sites may be more similar in terms of the compression index but less similar in the small strain stiffness. The load-settlement behaviour of a shallow foundation at both sites may thus be similar. However, the deflection behaviour of a sheet pile wall at these sites may be different.

The second dimension relates to an assumption that an arbitrarily bounded "site" containing a geotechnical structure is statistically homogeneous. The volume of a "site" is currently defined by the project boundaries and the depth of its deepest foundation elements. This assumption is made when an engineer uses a single local transformation model in design to cover the whole site. However, it is possible that a "site" is governed by two or more distinct transformation models. An engineer may feel that the ground conditions are the "same" within a site because its dimensions are typically much smaller than the scale of geological structures. For linear infrastructures such as a tunnel or a levee, an engineer is certainly aware that ground conditions will change. For example, Lacasse (2015) modelled the New Orleans Levees as a series system of 560 reaches. A levee reach is defined as a portion of a levee system (usually a length of a levee) that may be considered for analysis purposes to have approximately uniform representative properties. The current definition of a reach is based on judgment rather than data-driven. A reach is a "site" in the above sense. Data-driven methods may be able to adjust site boundaries automatically to satisfy a given homogeneity requirement (e.g., foundation length for one part of a foundation layout, soil nail spacing/length for a slope section, rock bolt spacing/length for a tunnel section). Given the current data-poor environment, it goes without saying that this dynamic resizing will be probabilistic. As an insightful illustration, Otake et al. (2024) treated each borehole in the database Tokyo-CLAY/14/67,760 (from a 10 km^2 site in Japan) as a single "site" and showed that the cross-correlation scatterplots obtained are similar to those in CLAY/10/7490 (a generic database from 251 studies carried worldwide). It indicates that the prevailing physical definition of a "site" is only nominal. For DDSC, it is more meaningful to divide a physical site into statistically homogeneous sites using this dynamic resizing concept. This chapter proposes replacing the physical definition of a site boundary with a data-driven definition and terms this a "data-discovered site."

The third dimension is "precision characterization," which is implied by the "precision construction" challenge first proposed by Phoon (2018). This is unquestionably a "holy grail" pursuit in site characterization. For a critical structure such as a bridge foundation or an offshore wind turbine foundation, the current practice is to mandate sampling and testing at its location. Clearly, precision characterization is a challenge only if there is no observation at the location of interest. It is important to emphasize that precision characterization does not imply that the ground condition is known accurately at every location. "Precision" refers to spatial resolution, not an understanding of the ground. Hence, precision characterization and construction refer to taking different actions or making different decisions due to variations in ground conditions at different data-discovered "sites" that can be as small as a single pile or pile group location. These ideas are inspired by precision medicine that customizes treatment to suit the unique DNA of each individual. Precision understanding refers to a significant reduction in the range of plausible ground conditions at each data-discovered site, possibly in real time. This precision understanding may be achievable through advances in ground sensing technologies and machine learning-based data fusion (Meng et al. 2020). With this distinction in mind, precision characterization can be defined as a DDSC methodology that can produce a plausible range of ground conditions at a given location or one part of a site. The range can be narrow or broad depending on the layout of the site investigation plan, i.e., it is consistent with the available data. Kriging is the standard approach in geostatistics (Matheron 1963; Journel and Huijbregts 1978), but it does not seek to optimize its inference using data from other sites. HBM-MUSIC-3X is the first method to combine site-specific data and BID (Ching et al. 2021d). Kriging can combine data from multiple sources (Wackernagel 2003), but it does not consider performance data that may shed light on the ground condition through inverse analysis. There are other well-known limitations in cokriging (Guan et al. 2024a). The authors believe that novel machine learning or artificial intelligence solutions are available when one approaches precision characterization from a "fit for (and transform) practice" perspective rather than a method-centric perspective such as making incremental improvements to kriging.

1.2.3.2 Spatial features

The most analysed feature in geotechnical engineering is spatial variability. It was popularized by Vanmarcke (1977)'s classic paper on "Probabilistic modelling of soil profiles" and subsequently his book *Random Fields: Analysis and Synthesis* (Vanmarcke 1983). However, it is well known among engineers that uncertainty in the geological model is possibly larger (Einstein and Baecher 1982; Fookes 1997; Hoek 1999; Bárdossy and Fodor 2001; Bock 2006; Parry et al. 2014; Juang et al. 2018; Juang et al. 2019). This is denoted by "G" in MUSIC-3X-G. The spatial variability of soil/rock properties (Lumb 1966; Vanmarcke 1977, 1983) is termed "geotechnical uncertainty" to distinguish it from "geological uncertainty" (Phoon et al. 2022d). This is denoted by "3X" in MUSIC-3X-G.

Bárdossy and Fodor (2004) categorized geological uncertainty as "structured" and "unstructured." The former refers to relatively regular spatial changes that can be described by a trend-surface analysis. Examples include gradual compositional transitions of one strata/rock into another and cyclic repetitions of sedimentary features in a sequence of layers. The latter refers to features that occur unexpectedly in a stratum. The detection of anomalies would require grappling with unstructured uncertainty (Tang 1986; Tang and Halim 1988; Halim and Tang 1993). Soil stratigraphy can be structured or unstructured depending on its complexity. The characterization of soil stratigraphy has attracted some attention (Liao and Mayne 2007; Cao and Wang 2013; Li et al. 2013; Wang et al. 2014; Ching et al. 2015; Qi et al. 2016; Cao et al. 2019; Wang et al. 2019; Shuku et al. 2020; Shi and Wang 2021a, 2021b; Shuku and Phoon 2023a).

This section proposed a more general term "spatial features" to encompass geotechnical properties, geological features, ground improvement structures, and geoenvironmental processes (hydro, thermal, transport, etc.). They are basically 3D functions of space (or 4D when time is included) that influence the behaviour of a geotechnical structure constructed on or in a spatial domain. They are dominantly natural and permanent apart from ground improvement or geoenvironmental processes that can be transient. It is safe to say that data-driven site characterization focuses mainly on CPT-based soil stratigraphy, in part because richer datasets capturing more spatial features are currently unavailable and/or physics is incomplete and poorly understood.

To promote the development of DDSC methods that can infer increasingly complex spatial features (one or more), one important step is to create reference geotechnical datasets and standard performance metrics to train and measure different DDSC methods on a uniform basis. This benchmark testing or benchmarking is widely used in machine learning (ML) to support unbiased and competitive evaluation of emerging ML methods (Thiyagalingam et al. 2022). In one of the earlier attempts at establishing a benchmarking framework for DDSC, Phoon et al. (2022c) recommended the following considerations: (1) benchmark examples must be 3D, realistic in scale, stratigraphy, and properties; (2) cover a sufficient range of ground conditions; (3) restrict training datasets to only data that an engineer has at his/her disposal (MUSIC-3X or other data attributes); and (4) selection of validation datasets and performance metrics to compare methods fairly and showcase the value of DDSC to decision making in practice.

Phoon et al. (2022c) proposed the first set of eight benchmark examples for DDSC as summarized in Table 1.4. One component of a benchmark example is the virtual ground that represents the complete state of "reality." Everything is known within this virtual ground because it is generated by assuming stratigraphy and populating the properties in each layer using a random field generator (Figure 1.3). However, the rule of the game in

Table 1.4 Benchmark examples for data-driven site characterization

"Small" virtual ground	Description	Training dataset	
		T1 (3 CPT soundings)	T2 (6 CPT soundings)
S-VG1	Horizontal layers with constant property trend in each layer	S-VG1-T1	S-VG1-T2
S-VG2	Inclined layers with constant property trend in each layer	S-VG2-T1	S-VG2-T2
S-VG3	Inclined layers with linear property trend in each layer	S-VG3-T1	S-VG3-T2
S-VG4	Mixture of continuous and discontinuous layers with constant trend in each layer	S-VG4-T1	S-VG4-T2

Source: table 2, Phoon et al. (2022c).

DDSC is to assume that only the training dataset (containing CPT data from various sounding layouts) is available. Hence, the training dataset represents measured "reality" or what is known. Figure 1.3 shows two training datasets, T1 and T2. The purpose of a DDSC method, after it has been suitably trained, is to predict the soil profile and the properties at a prescribed set of validation soundings that are distinct from the set of training soundings (Figure 1.4). Note that the CPT data in the validation soundings are strictly restricted to cross-validation. They are assumed to be "unobserved" and cannot be used for training. Different DDSC methods can be compared uniformly using a set of performance metrics.

Jessell et al. (2022) generated 1 million 3D geological models for ML training. Each model is the outcome of superimposing user-defined kinematic events that represent idealized geological events such as folds, faults, unconformities, dykes, plugs, shear zones, and tilts on a base stratigraphy. An example is shown in Figure 1.5. It is effectively a "virtual ground" similar to Figure 1.3. The purpose of generating these synthetic "geologically plausible" models is to overcome challenges in applying ML to geoscience problems. Karpatne et al. (2018) presented nine commonly encountered attributes of geoscience data that should be accounted for in the design of ML algorithms: (1) spatial-temporal structure, (2) heterogeneity in space and time, (3) high dimensionality, (4) lack of concise object definitions, (5) rare classes, (6) multi-source multi-resolution data, (7) poor quality of data, (8) small sample size, and (9) paucity of ground truth. The training of a facial recognition algorithm is far simpler, because the "face" is a well-defined object, the number of facial features is small, and the training sets (photographs) are extremely large. Spatial-temporal structure and heterogeneity in space and time are two aspects of spatial features that arise from naturally occurring processes. It refers to "3X" in MUSIC-3X. Hence, some of the attributes are comparable to those highlighted in DDSC (Phoon et al. 2022b).

1.2.3.3 Sampling characteristics

MUSIC-3X is an initial attempt to describe the attributes of geotechnical data found in a typical site investigation report. However, it does not consider geotechnical data produced by a broader range of investigation methods (e.g., geophysical methods, laser scanning, drone photogrammetry, satellite imagery) and instrumentation methods (e.g., fibre optic sensors, wireless sensor networks, Internet of Things). It is useful to look at the contrasting characteristics of these data sources because they can guide the development of more general data-driven fusion methods that can tap into a wider spectrum of multi-source data multi-modal beyond those encountered in a routine site investigation.

(a) VG1 (soil type)

(b) VG2 (soil type)

(c) VG3 (soil type)

(d) VG4 (soil type)

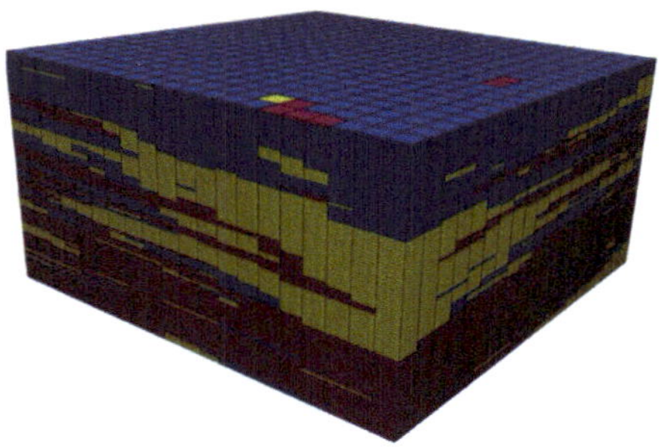

(e) VG5 (soil type)

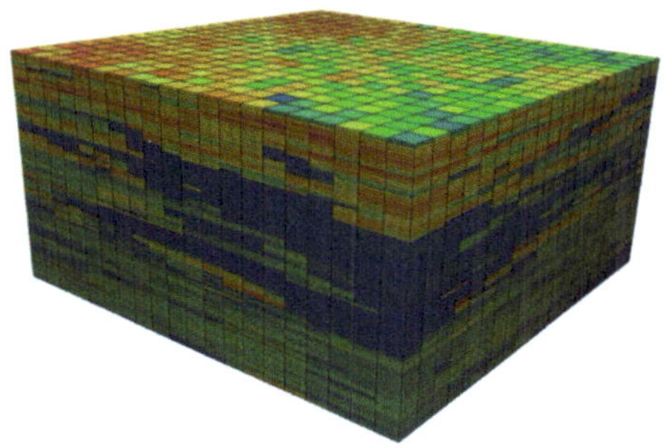

(f) VG6 (soil type)

Figure 1.3 Virtual grounds (VGs) with increasing stratigraphic complexity: (a) VG1: horizontal layers with constant property trend, (b) VG2: inclined layers with constant property trend, (c) VG3: inclined layers with linear property trend, (d) VG4: mixture of continuous and discontinuous layers with constant property trend, (e) VG5: random soil type distribution based on coupled Markov chain with constant property trend, and (f) spatially varying soil property in VG5 (Source: VG1, VG2, VG3, and VG4 from figure 2, Phoon et al. 2022c; VG5 from figures 6 and 7, Shuku and Phoon 2023b).

(1) Univariate versus multivariate data – A conventional site investigation typically involves conducting different tests in a small neighbourhood such as the bivariate example shown in Figure 1.2. However, the more general multivariate data contains a spatio-temporal structure as noted by Karpatne et al. (2018). When viewed in the context of the World Wide Web, this characteristic may be a special case of networked data (Zhou and Song 2016).

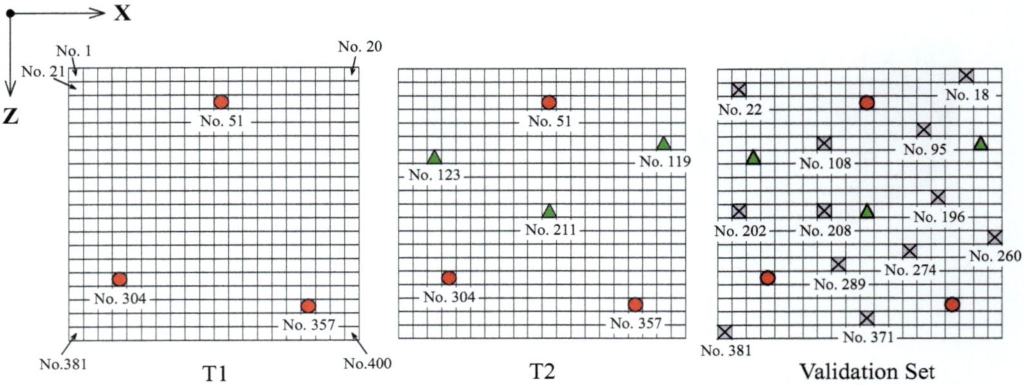

Figure 1.4 CPT soundings that produce the training and validation datasets (Source: figure 4, Phoon et al. 2022c).

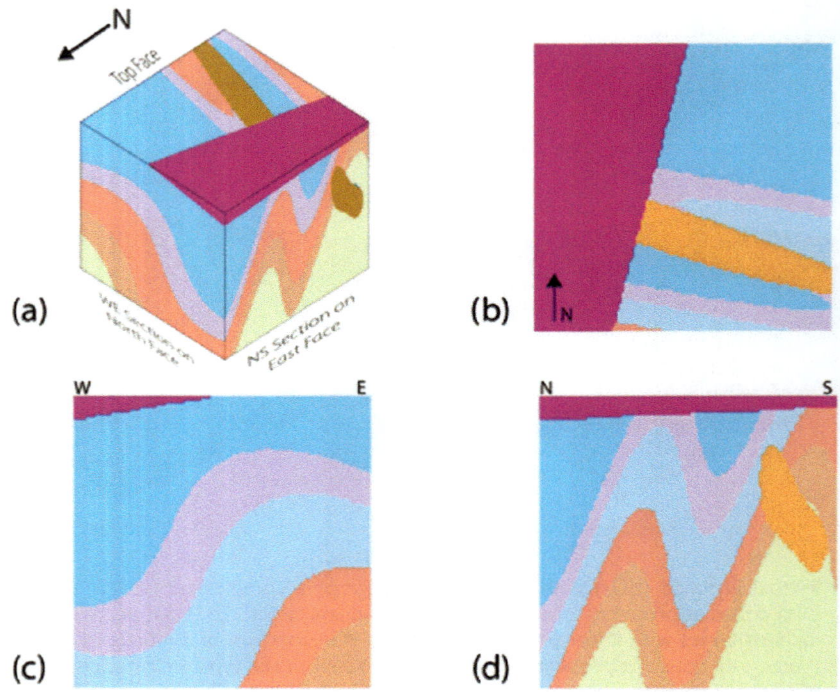

Figure 1.5 Example of simulated geological model: (a) 3D visualization, (b) top surface, (c) E–W section, and (d) N–S section (Source: figure 1, Jessell et al. 2022).

(2) Single site versus multiple sites – In the absence of the time dimension, this characteristic is addressed by the site recognition challenge (Phoon and Ching 2022). If one defines a "data-discovered site" as a spatially bounded volume of ground that is statistically homogeneous in some well-defined sense related to the limit state (e.g., governing properties, responses to a loading condition), it is possible for a

site boundary to change with time if the limit state changes with time to maintain homogeneity within a "data-discovered site" as discussed in Section 1.2.3.1. This characteristic is related to the heterogeneity in space and time attributes noted by Karpatne et al. (2018).

(3) Low-density versus high-density data – A video is high-density data in space and time, but there is no method to image a 3D subsurface continuously at low cost. It may not be necessary for the purpose at hand. A single image is high-density data in 2D space, but it does not provide any information on the time evolution of the quantity of interest. An image is also a slice of 3D space. Many field tests such as the standard penetration test (SPT) produce low-density data in 1D space, and it is usually conducted once in the lifetime of the project unless there is value in repeating these tests. For example, tests can be repeated before and after ground improvement for quality control or after a major failure in forensic engineering. It is not straightforward to combine data with different sampling densities in space and/or time (e.g., construction of a shear wave velocity profile from geophysical and borehole data). If a database is tabulated based on the source with the highest sampling density, there will be many empty cells for sources with lower sampling density. In other words, the database will be highly incomplete. If the opposite approach is taken, sources with higher sampling density will be down-sampled to match the source with the lowest sampling density (Shuku et al. 2023). This is one challenge in data fusion that has found new solutions in machine learning and artificial intelligence.

(4) Low-fidelity versus high-fidelity data – In geotechnical engineering, data produced by a field/laboratory test that can replicate in situ boundary conditions, and the tested volume is undisturbed and sufficiently representative, are regarded as high fidelity. Data produced by simple index tests on disturbed samples, such as the Atterberg limit tests, are regarded as low fidelity, even if the tests are highly repeatable (small measurement errors). Performance data produced by physical modelling could be of medium fidelity. Basically, the criterion is how closely the measurement/observation conditions match the operational context of interest.

(5) "Input" versus "output" parameters – All site investigation data are "inputs" while instrumentation data are "outputs" when viewed through the lens of physics. ML can be successful without invoking physics. In the first place, physics may not be available. In geotechnical engineering, there are problems that are poorly understood and the physics is grossly incomplete. The coefficient of variation of the model factor is very high (greater than 90%) for such problems (Tang and Phoon 2021). Hence, the distinction between inputs and outputs may not be possible. However, for problems with well-understood physics, synthetic data can be generated to supplement limited real-world data for ML training. One approach is the application of transfer learning to bridge the gap between simulation and the real world.

(6) Categorical versus numerical data – Geotechnical properties are mainly numerical. This is not the case for rock mass properties. Harrison (2019) pointed out that statistical analysis is not meaningful for some data types in rock mass properties. For example, it is not possible to determine the mean and coefficient of variation of nominal and ordinal data. Both data types are categorical, except the latter is amenable to ranking. The genetic group – sedimentary, igneous, and metamorphic – is an example of nominal data. He cited the joint set number (J_n) in the Q-system as an example of ordinal data. The reader can refer to ISO 14689-1:2003 Geotechnical investigation and testing – Identification and classification of rock – Part 1:

Identification and description for a more complete coverage of categorical data. Phoon et al. (2022b) opined that handling a mixture of categorical and numerical data of varying proportions in high random dimensional space could be a challenge. ChatGPT-4o can handle multi-modal data, which is arguably the most varied in terms of data types.

The above is a partial list. There are possible relationships between the characteristics. For example, high-fidelity data are likely to have a low sampling density due to cost.

1.2.3.4 Smart data

"Smart data" is defined as actionable data at the point of collection. In contrast, conventional data are compiled and analysed in batches. Clearly, timely decisions cannot be made in the batch processing mode. The strategy to maximize the value of data in the presence of the Internet of Things (IoT) is not likely the same as conventional monitoring instruments. The collection of all possible data and storing them in the cloud with the intent to organize and analyse big data in one batch at a later point in time is becoming less effective as the volume and complexity of data from networked sensors increase. One alternate strategy is to implement algorithms (intelligence) at every step of the data processing chain, from smart sensors with limited low-power computing resources through gateways and edge computers to large cloud computers. There are advantages to implementing algorithms as close as possible to the point of data collection (sensors). Besides enabling time-critical interventions, the refinement of data at an early stage will reduce power, storage, transmission, and subsequent analysis costs.

Smart data can be regarded as processed big data that leads to a meaningful action (or decision). The intelligence of the algorithm must be a function of the level in the data processing chain. For a single sensor collecting data in a small neighbourhood, intelligence is low. For an algorithm with access to all sensors (and connected to lower-level intelligence), trained on past data produced by similar projects, and possibly interacting continuously with human-in-the-loop, intelligence is much higher. A simple example is the action of a hand jerking away from a heat source. The intelligence resulting in this action is much lower than that of the affected person who realizes the heat source is a fire outbreak and thus takes a higher-order decision to run away or put out the fire. The value of information that defines smart data depends in part on the complexity of the data (e.g., single versus networked sensor data), an appreciation of the context (from understanding or experience), and the level of decision making (e.g., temperature exceeds threshold versus risk management considering complex trade-offs). In other words, the value of information is relative to the level of intelligence in the data processing chain. It is reasonable to speculate that there must be a balance of autonomy and interaction in intelligence at different levels because decision making is taking place at different scales (local versus global, fast versus slow) and different levels of intelligence. One example of an intelligence hierarchy is the DIKW (Data, Information, Knowledge, Wisdom) pyramid (Rowley 2007). The current direction of travel is embodied AI that combines large language models with robotics or other devices that interact with the real world in real time. Finally, communication itself is becoming smart as well. Luo et al. (2022) explained the concept of semantic communication: "The core of semantic communication is to extract the 'meanings' of sent information at a transmitter, and with the help of a matched knowledge base (KB) between a transmitter and a receiver, the semantic information can be 'interpreted' successfully at a receiver. Therefore, semantic communication essentially is a communication scheme based largely on AI."

In civil engineering, the value of information (VoI) is typically founded on Bayesian decision analysis and utility theory (Raiffa and Schlaifer 1961; Zhang et al. 2021). The purpose of evaluating VoI is to decide how to collect data to maximize its value for a given purpose before the data is measured. Hence, it is also called pre-posterior analysis by Bayesians. The underlying motivation for VoI is that additional information does not always produce sufficient benefits for the purpose at hand to compensate for the attendant costs. While the concept is useful, Straub (2014) observed that there are practical difficulties in relation to the design of monitoring systems: (a) proper probabilistic modelling of the monitored process and the monitoring itself, (b) the modelling of the action alternatives following the monitoring results, and (c) the computational efforts in evaluating the VoI. The design of monitoring systems plays a central role in the observational approach in geotechnical engineering. Hu et al. (2021) also pointed out that the application of VoI to a slope problem can be computationally prohibitive. The application of VoI in geotechnical engineering is limited at present (Yoshida et al. 2018; Hu et al. 2021). The relation between VoI and smart data is unclear. However, it is safe to say that the VoI framework needs to be revisited to guide data collection for data-driven decision making in the modern context of Industry 4.0, which consists of new hardware (IoT, 5G connectivity, cloud, edge computing, autonomous systems) and new software (AI, digital twin, blockchain, web3). The contours of a possible Value of Smart Information (VoSI) framework are not defined at this point, but a good start is to imagine how to extract more value from conventional monitoring instruments that can talk to each other by building intelligence into every stage of the data processing chain. VoSI is expected to be an integral part of machine learning-guided observational method (Section 1.5).

1.3 DATA-DRIVEN SITE CHARACTERIZATION

Site characterization is a cornerstone of geotechnical and rock engineering. It is hard to imagine a safe and economical practice in its complete absence. The classic example is the table of allowable bearing values in Terzaghi and Peck (1967) (table 54.1). No site investigation and no site-specific data are needed other than a broad qualitative appreciation of soil type and the name of the city. However, a practice based on qualitative knowledge of the site is very conservative. Note that it is possible to consider site characterization being conducted in tandem with construction. This is typical in tunnelling, say using a computer jumbo (Shuku et al. 2023). This "characterize, construct, and learn as we go" adaptive strategy is the basis of the observational method. It is arguably a more robust risk management strategy for a practice that operates in a spatially variable but data-poor environment (with occasional undetected "geologic surprises").

"Data-driven site characterization" (DDSC) refers to any site characterization methodology that relies solely on measured data, both site-specific data collected for the current project and existing data of any type collected from past stages of the same project or past projects at the same site, neighbouring sites, or beyond (Phoon et al. 2022b). This is a "data first" definition in line with data-centric geotechnics. It is silent on the role of physics and experience. As noted above, this does not mean insights from other sources are not valuable and cannot be included in DDSC. Karniadakis et al. (2021) noted that physics-informed machine learning (PIML) has two advantages: (a) it can integrate data and physics, even in partially understood, uncertain, and high-dimensional contexts, and (b) it is effective and efficient for ill-posed and inverse problems. One can speculate that PIML can operationalize the observational method in a more practical way. To the knowledge of the authors, no research has been conducted in machine learning-guided observational method (MLOM) within the agenda of data-centric geotechnics. However, some limited aspects have been studied. For example, Tan et al. (2023) mapped the spatial stress distribution over a coalmine roof by training a transfer convolutional neural network (TCNN) model using physics (finite difference method) and stress sensor data collected in the roof of the Dongtan coalmine. Tian and Wang (2023) and Tian et al. (2024) developed sparse dictionary learning methods using physics (finite element analysis) with sparse site characterization data and settlement monitoring data for improving settlement prediction in real time and continuously, particularly at locations without monitoring data and/or subsequent time steps. Section 1.5 briefly discusses the challenges of MLOM.

Thick data is qualitative and able to associate numbers with meaning and context. Hong et al. (2022) described thick data in the context of urban research:

> Thick data is descriptively rich and intensively detailed, collected from or with research participants on issues relating to their values, visions, knowledge, life experiences, and opinions, typically obtained by observing or interacting with participants in their daily lives and through in-depth interview techniques. Thick data may be collected by using and analysing a variety of formats including photographs, videos, sound recordings, sketches, and written ethnographic field notes.

Experience and other qualitative knowledge accumulated in more than half a decade of modern practice can be regarded as "thick data" in geotechnical engineering. Phoon et al. (2022b) suggested that engineers can be asked to tag a data-driven prediction with "agree," "disagree," or "not sure." These tags are valuable because an AI can learn and improve its predictions through continuous "reality checks" by human experience, but this process is expensive. With the emergence of powerful generative AI such as ChatGPT, it is reasonable to hypothesize that thick data found in the issues of the *ISSMGE International Journal of Geoengineering Case Histories* and in the proceedings of the "International Conference on Case Histories in Geotechnical Engineering" may be assimilated. In addition, the ISSMGE Online Library is open source and can serve as a training dataset for ChatGPT (https://www.issmge.org/publications/online-library). To the knowledge of the authors, there is no research on thick data or ChatGPT in geotechnical engineering although an initial study on large language models was conducted recently (Wu et al. 2023).

1.3.1 Site recognition challenge

At present, there are no physics-based models that can predict ground conditions or properties at a particular location. It is also unlikely that the geologic history at one site (which has a significant bearing on the current ground conditions and properties) can be obtained with sufficient quantitative accuracy to specify the boundary conditions of any physics-based model. There are attempts to explain transformation models more rationally, but field measurements are needed (Lim et al. 2018; 2019). Hence, DDSC is a promising direction of research (e.g., Huang et al. 2018; Cao et al. 2019; Ching and Phoon 2019; Wang et al. 2019; Ching and Phoon 2020a; Ching and Phoon 2020b; Shuku et al. 2020; Ching et al. 2021a; Ching et al. 2021b; Shi and Wang 2021a, 2021b; Wang et al. 2021; Yoshida et al. 2021; Xiao S. et al. 2021a, 2021b; Xu et al. 2021; Ching et al. 2022b; Sharma et al. 2022, 2023; Wu et al. 2022; Ching et al. 2023b, 2024; Shuku and Phoon 2023a; Guan et al. 2024a, 2024b; Hu et al. 2024). Phoon et al. (2022b) outlined three challenges in DDSC: (1) ugly data, (2) site recognition, and (3) stratification. An example of ugly data called MUSIC-3X is presented in Section 1.2.2. No data-driven research in geotechnical engineering is meaningful if it does not address real-world data. The methods proposed in many papers are data driven in taxonomy, but not in practice. The ugly data challenge is foundational and it is an integral part of all other challenges in DDSC and more broadly in data-centric geotechnics. Section 1.2.3 can be regarded as an expansion of the ugly data challenge.

The site recognition challenge is related to a fundamental feature in geotechnical practice, namely all sites are different to some extent (site specificity) (Phoon et al. 2025). For example, Clause 2.4.5.2(10) of Eurocode 7 (CEN 2004) "Characteristic values of geotechnical parameters" advises against pooling data from different sites into a single population for statistical analysis. The stratification challenge requires identification of soil type and a mapping of the spatial distribution of each soil type based on field tests typically conducted at limited discrete locations/depths in a soil mass. The stratification challenge has not been expanded to address more general spatial features discussed in Section 1.2.3.2. An example of a site investigation programme is shown in Figure 1.6. Section 1.4.3 presents an HBM-MUSIC-3X method that attempts to address the site recognition and stratification challenges together. The authors argue that the ugly data and site recognition challenges underlie all DDSC applications in geotechnical and rock engineering.

1.3.2 Data-informed decision support index (DIDI)

The site recognition and stratification challenges are not new. An engineer encounters them routinely in practice, but they are not regarded as challenges because judgment-based solutions are widely accepted as satisfactory. For instance, an engineer applies experience gained from other sites to interpret limited data at the current site of interest. However, he/she will not regard all sites as equally relevant – again, experience (occasionally complemented by some broad knowledge of local geology) is needed to recognize that some sites are more relevant or "similar." For the construction of a 2D geologic section (or fence diagram), it is not uncommon to connect the same soil type (interpreted by an engineering geologist from disturbed samples) between two adjacent boreholes using a straight line or to insert a question mark with a speculative dashed line to indicate "not sure" if the sequence of soil types does not match between two adjacent boreholes. Geotechnical failures are rare despite the adoption of a practice that is arguably more art than science.

There is a growing contrast between judgment-based site characterization that remains relatively unchanged for decades and increasingly sophisticated physics-based modelling of soil–structure interaction such as the material point method. Phoon et al. (2022d) opined that:

> Currently, numerical analysis is conducted using deterministic inputs and simple soil profiles that are inconsistent with sparse site data. This deterministic mapping (one

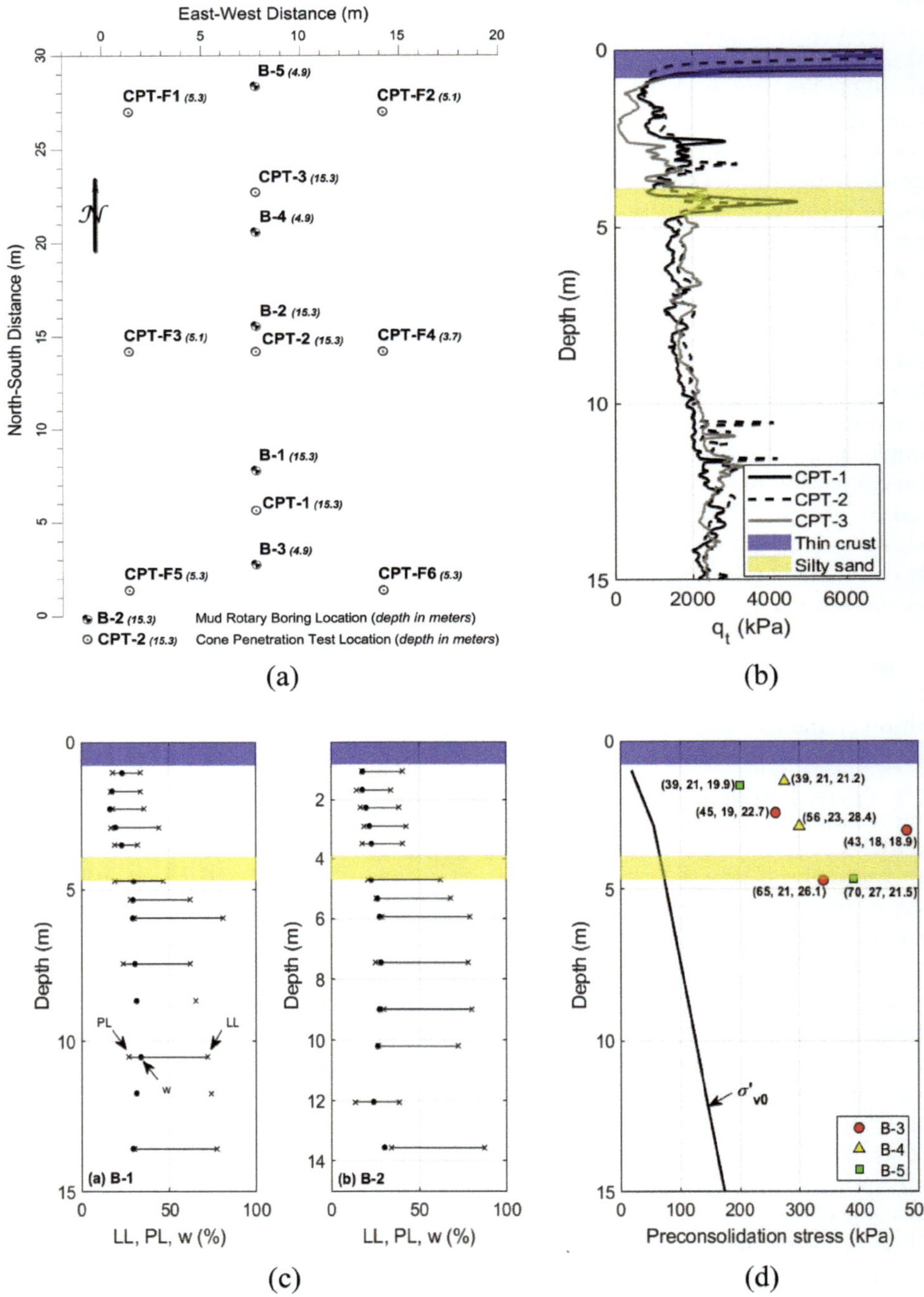

Figure 1.6 Typical site investigation programme at Baytown, Texas: (a) exploration (borings and CPT soundings), (b) some CPT test results, (c) some index test results (LL = liquid limit, PL = plastic limit, and *w* = natural water content), and (d) pre-consolidation stress results (numbers in parenthesis refer to LL, PL, *w*) (Stuedlein et al. 2012).

set of inputs to one set of outputs) does provide valuable physical insights, but cannot support risk-informed decision making on its own without appealing to engineering judgment.

In the opinion of the authors, the most important question is whether a pure judgement-based site characterization can fit Industry 4.0. Phoon (2023b) posed this question in the context of broader emerging trends: (1) rapid surge in computing power and digital technologies in general, (2) availability of more efficient and more powerful probabilistic and other data-driven methods (Phoon and Ching 2021; Phoon et al. 2022b), (3) sharing of large databases [termed big indirect data (BID) by Phoon et al. 2019], (4) incorporation of probabilistic analysis in geotechnical software (refer to table 10, Phoon 2023b), (5) increasing emphasis on risk-informed decision making in design codes (Phoon and Retief 2016; Phoon et al. 2016; Retief et al. 2016), (6) innovations should be the norm in practice to address urgent challenges such as sustainability, but current technology acceptance criteria lean towards risk-averse rather than risk-informed (ASCE 2019), and (7) the deterministic paradigm is at odds with digital transformation because it pre-dates the personal computer era (Phoon et al. 2022a). Phoon (2023b) opined that the "shift to data-informed decision making is a near certainty given the increasing power, ubiquity, and convergence of digital technologies." Phoon et al. (2022a) presented a data-informed decision support index (DIDI) (Table 1.5) to explain the expanding role of data in geotechnical engineering decisions that spans from:

(1) Almost no data (DIDI = 1) in pre-Terzaghi years.
(2) Empirical or simple closed-form solutions (DIDI = 2) found in Terzaghi (1943).
(3) Physics-centric solutions with deterministic site-specific inputs (DIDI = 3) such as those reviewed by Salgado et al. (2008) for foundation engineering.
(4) Physics-centric solutions with more attention on data representation using a frequentist interpretation of uncertainty such as those reviewed by Chwała et al. (2023) (DIDI = 3 to 4).
(5) Uncertainty-informed decision making that is more comprehensive and with increasing emphasis on the Bayesian interpretation of uncertainty (DIDI = 4).
(6) Data-centric decision making such as those reviewed by Phoon and Ching (2021), Phoon et al. (2022a), Phoon and Zhang (2023), Phoon et al. (2025) (DIDI = 5).

The arc of history in decision making can be viewed as an evolution from qualitative methods, to quantitative methods using closed-form theoretical solutions, to numerical methods, to probabilistic methods, and possibly bending to data-driven methods that include machine learning and artificial intelligence. The advent of powerful generative AI, such as GPT-4o in 2024, may lead to an elevation of DIDI beyond the current state of 3 to 4 in the near future. Human–machine teaming is now a distinct possibility. The convergence of AI, robotics, and connectivity alone is likely to disrupt many industries, including geotechnical engineering.

1.4 APPLICATIONS

Phoon and Ching (2022) and Phoon et al. (2025) presented some state-of-the-art methods in the site recognition challenge. A summary is given in Table 1.6. It is possible to

Table 1.5 Data-informed decision support index (DIDI)

Example	*Data to decision*	*Data-informed decision support index (DIDI)*
Allowable bearing values	Conservative range of design values based on qualitative soil type and name of city	1
Generic empirical correlations	Non-site-specific empirical relationship between a design value (e.g., end bearing) and an indirect field test (e.g., SPT)	2
Deterministic engineering calculations with observational approach	Design value produced by a physical model with site-specific input parameters from generic transformation models (e.g., undrained shear strength versus cone tip resistance), possibly updated by site-specific observations (e.g., load tests, monitoring)	3
Reliability analysis with Bayesian observational approach	Design value represented by a probability distribution that is related to uncertainties in input parameters, model idealizations, and possibly updated by field observations utilizing site-specific and generic data	4
Precision design and construction	Design value updated by explainable AI that: (1) processes all big and small data from generic and site-specific sources, respectively, (2) recognizes site specificity, (3) "explains" its outputs, and (4) improves from continuous interaction with engineers	–
Partially smart infrastructure lifecycle management	Data-informed management of one or more lifecycle phases using BIM, digital twin, AI, etc.	
Completely smart infrastructure that fits circular economy and delivers service to end-users and community from project conceptualization to decommission/reuse with full integration to smart city and smart society	Value engineering at system level city scale to fulfil sustainability, resilience engineering, and other societal goals through ethical human-centric deployment of digital technologies	–

Source: table 8, Phoon et al. (2022a).

organize the methods in Table 1.6 under six broad strategies with increasing emphasis on the quantification of "similarity" between sites:

I. Use the entire global database with every record assigned an equal weight independent of its similarity (at record or site level) to the target site.
II. Use the entire global database with every record assigned a weight based on its similarity (at record or site level) to the target site.
III. Use a subset of the global database ("quasi-regional" database) with every record assigned an equal weight.
IV. Use a subset of the global database ("quasi-regional" database) with every record assigned an unequal weight.

Table 1.6 Methods to construct a quasi-site-specific transformation model from MUSIC-3X data and BID

Method	*Big indirect data*	*Site-specific data*	*Quasi-site-specific model* ***M***	*Site uniqueness*	*Explainable*
Probabilistic multiple regression (Ching and Phoon 2012)	G	G	$M = G$	No	
Hybrid (Ching and Phoon 2019)	G	S	$M = G \times S$	Yes, but empirical	No
Record similarity (Ching and Phoon 2020a)	$G = (G_1, G_2)$	S	Select G_2 (list of records) $\approx S$ $M = S + G_2$	Yes, at record level	Yes
Two-step Gibbs (Ching et al. 2020b)	G_2	S	G_2 (regional) $\approx S$ $M = G_2$ updated by S	Maybe	No
Hierarchical Bayesian model (Ching et al. 2021b)	$G = (G_1, G_2, \ldots, G_n)$	G_{n+1}	$M = G$ updated by G_{n+1}	Yes, at site level	No
Site similarity (Sharma et al. 2022)	$G = (G_1, G_2)$	S	Select G_2 (list of sites) $\approx S$ $M = S + G_2$	Yes, at site level	Yes
Site clustering (Sharma et al. 2023)	$G = (G_1, G_2, \ldots, G_n)$	S	Select $S \in G_i$ (cluster of sites) $M = S + G_i$	Yes at cluster level	Yes
Site clustering - tailored (Cai et al. 2024a)	$G = (G_1, G_2)$	S	Select G_2 (tailored cluster) $\approx S$ $M = S + G_2$	Yes at cluster level	Yes

Source: updated from table 1, Phoon and Ching (2022).

V. Use a subset of the global database ("quasi-regional" database) and a subset of the soil/rock parameters that is relevant to one or more governing limit states.
VI. Do not use the global database, in part or in full, to inform the target site because they are "dissimilar."

Past research studies addressing the site recognition challenge can be broadly classified under Strategies I to VI. The sub-sections below briefly review five studies to showcase the range of applications and value of DDSC for real-world decision making. Real-time decision making requires a combination of site characterization and monitoring information. Research in this machine learning-guided observational method (MLOM) is largely absent at present, although its value is arguably higher than DDSC. It is uncertain whether a new discipline called geotechnical data science is needed to provide a stronger theoretical underpinning for data-centric geotechnics, given the large interconnected challenges highlighted in the previous sections. It is accurate to say that current data-driven research in geotechnical engineering is fragmented in terms of its theoretical foundation and largely follows concepts and methods developed in the wider machine learning community.

1.4.1 Strategy I – Generic model for clay parameters

Ching and Phoon (2014b) compiled a generic database called CLAY/10/7490. Each record in the database constitutes a set of multivariate clay properties (e.g., liquid limit, plasticity index, water content, undrained shear strength, cone tip resistance, etc.). In Ching and Phoon (2014b), it is assumed that there is only a "generic population" for all records in the database. Furthermore, the paper assumed that the target site belongs to the same generic population and the (transformed) multivariate soil properties of each record follow a multivariate normal distribution with mean vector = μ_g and covariance matrix = $\mathbf{C}_g$. The training of the model parameters (μ_g, $\mathbf{C}_g$) is straightforward: all records in the database as well as all known records for the target site are considered to be equally relevant. Hence, they are adopted to update (μ_g, $\mathbf{C}_g$) with equal weights. The updated parameters (μ_g, $\mathbf{C}_g$) (namely, the quasi-site-specific model) are then used to predict the unknown records at the target site. This strategy of constructing a quasi-site-specific model is referred to as the probabilistic multiple regression (PMR). One limitation is already clear. If there are 7490 records in the database and only 10 records at the target site, the effect of the target-site records on (μ_g, $\mathbf{C}_g$) is marginal. The quasi-site-specific is almost the same as the generic model (site-specific data not included).

For illustration, consider the $(q_t-\sigma_v)/\sigma_v'$ versus s_u/σ_v' data in CLAY/10/7490 (grey dots in Figure 1.7a) (q_{t1} denotes $(q_t-\sigma_v)/\sigma_v'$). Also, consider the q_{t1} versus s_u/σ_v' data for an Onsøy site (Norway) (Lacasse and Lunne 1982). Some target-site records have complete q_{t1} and s_u/σ_v' data (red dots), yet some records only have q_{t1} data available (yellow dots) with their s_u/σ_v' data deliberately regarded as unknown for validating the model prediction. The solid and dashed lines in the figure show the mean trend and 95% confidence interval (CI) of the quasi-site-specific model constructed by the records in the database (grey dots) and the known records of the target site (red dots). This mean trend and 95% CI can then be used to predict the s_u/σ_v' values of the known records of the target site (yellow dots) based on their q_{t1} information, as shown in Figure 1.7b.

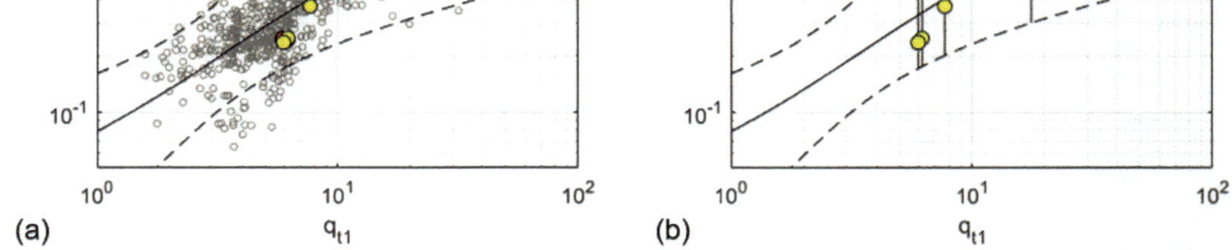

Figure 1.7 (a) The q_{t1} versus s_u/σ_v' data (grey dots are from CLAY/10/7490; red dots are from the Onsøy site; yellow dots are also from the Onsøy site, yet the s_u/σ_v' values are deliberately regarded as unknown); the dark and dashed lines are the mean trend and 95% CI of the quasi-site-specific model; (b) the prediction results for the s_u/σ_v' values of the known records of the Onsøy site.

1.4.2 Strategy II (without spatial variability) – Quasi-site-specific model for MUSIC clay parameters

Ching et al. (2021b) proposed a different strategy for modelling a generic soil database, called the hierarchical Bayesian model (HBM). In this model, the records of each site in the database are assumed to follow a distinct local population rather than a common generic population. To be more specific, the (transformed) multivariate soil properties of the records of the *i*th site are assumed to follow the multivariate normal distribution with mean vector = μ_i and covariance matrix = $\mathbf{C}_i$. Moreover, (μ_i, $\mathbf{C}_i$) and (μ_j, $\mathbf{C}_j$) are different if $i \neq j$ to account for site uniqueness. More specifically, (μ_1, μ_2, ..., μ_{ns}) (n_s is the number of sites in the database) are assumed to follow the multivariate normal distribution with (hyper) mean vector = μ_0 and (hyper) covariance matrix = $\mathbf{C}_0$, and ($\mathbf{C}_1$, $\mathbf{C}_2$, ..., $\mathbf{C}_{ns}$) are assumed to follow the inverse-Wishart distribution with (hyper) scale matrix = $\mathbf{\Sigma}_0$ and (hyper) degree of freedom = ν_0. The parameters for the Onsøy site, denoted by (μ_s, $\mathbf{C}_s$), are also assumed to follow the same (hyper) distributions. The HBM can learn the inter-site and intra-site variabilities in CLAY/10/7490. Note that the HBM is one plausible method to model site uniqueness.

Figure 1.8a shows the database data (records from the same site are in the same colour), whereas Figure 1.8b shows the behaviour of the HBM trained by CLAY/10/7490. The trained HBM can simulate (μ_s, $\mathbf{C}_s$) of a hypothetical target site (each skewed ellipse represents such a hypothetical target site) whose variabilities are similar to those of the database sites. The trained HBM can be further updated by the Onsøy-site data. Many ellipses in Figure 1.9a are incompatible with the Onsøy-site data (shown as grey ellipses). A few ellipses are compatible (shown as coloured ellipses). The coloured (compatible) ellipses illustrate the quasi-site-specific model of (μ_s, $\mathbf{C}_s$) after the Bayesian updating. The solid and dashed lines in Figure 1.9b show the mean trend and 95% confidence interval (CI) of the quasi-site-specific model. This mean trend and 95% CI can then be used to predict the s_u/σ_v' values of the known records of the Onsøy site (yellow dots) based on their q_{t1} information, as shown in Figure 1.9b. Figure 1.9b can be compared with Figure 1.7b. The comparison shows that the HBM (Strategy 2) is more effective than the PMR (Strategy 1)

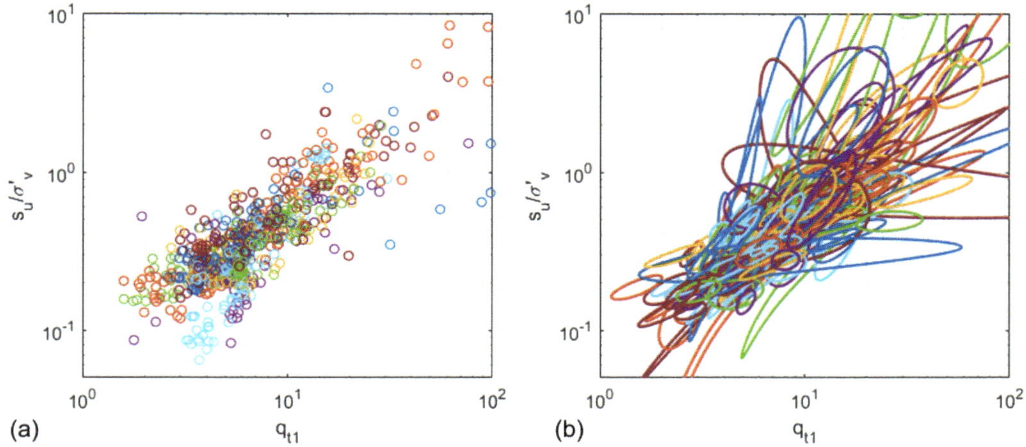

Figure 1.8 (a) The database data (records from the same site are with the same colour); (b) the behaviour of the HBM trained by CLAY/10/7490 (each skewed ellipse represents such a hypothetical target site).

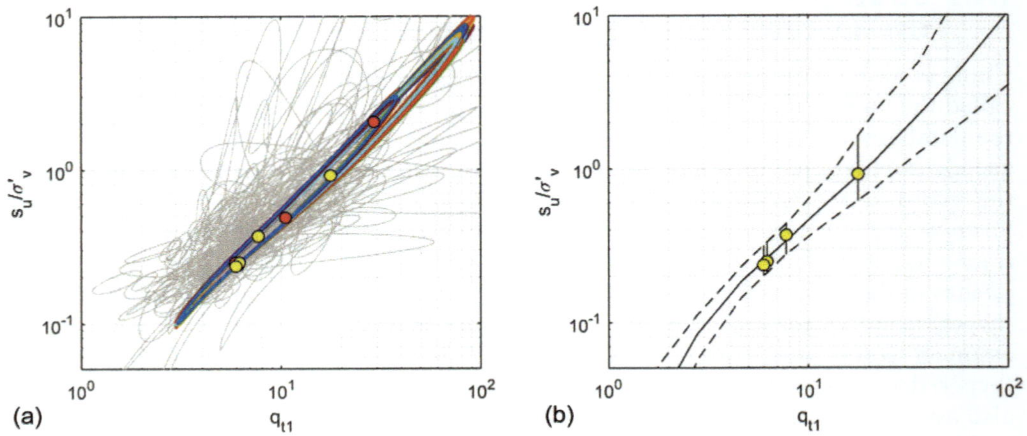

Figure 1.9 (a) Illustration of the quasi-site-specific model (grey ellipses are incompatible to the Onsøy-site data; coloured ellipses are compatible); (b) the mean trend and 95% CI of the quasi-site-specific model as well as the prediction results for the s_u/σ_v' values of the known records of the Onsøy site.

with narrower 95% CIs. Note that in the real application of HBM, the simulation of the skewed ellipses is not necessary: the quasi-site-specific model is automatically constructed by Bayes' rule. These skewed ellipses are shown herein to explain the basic idea of the HBM. The HBM has been adopted by Ching et al. (2021a) to make quasi-site-specific predictions for the deformation modulus of rock mass and by Ching et al. (2022a) to make quasi-site-specific predictions for the compressibility of clay. The key limitation of the HBM is that it is not an "explainable" method. Most engineers will feel more comfortable if the sites contributing to the quasi-site-specific inferences are real and judged by them as comparable or similar to the target site.

1.4.3 Strategy II (with spatial variability) – Quasi-site-specific model for MUSIC-3X clay parameters

In Section 1.4.2, the HBM trained by the soil database was updated by the Onsøy-site data into a quasi-site-specific model. The Onsøy-site data was assumed to be spatially independent because the locations (depths) of q_{t1}–s_u/σ'_v data are far apart. Ching et al. (2022b) showed that the HBM trained by the soil database can be updated by MUSIC-3X target-site data as well. Here, "3X" denotes spatially variable site data. The target site considered by Ching et al. (2022b) was the Baytown site (Texas, USA) (see Figure 1.6). The spatial correlation of the target site is characterized by the autocorrelation matrix **R**. By assuming that all soil properties follow the same autocorrelation structure, the total covariance matrix can be expressed as the Kronecker product between the autocorrelation matrix **R** and the cross-correlation matrix $\mathbf{C}_s$. This assumption allows the HBM trained by the soil database to be updated by MUSIC-3X site data into a quasi-site-specific model with affordable computational cost. This quasi-site-specific model, denoted by the HBM-MUSIC-3X model, can further simulate the conditional random fields for all properties because it addresses both spatial correlation and cross-correlation. Figure 1.10

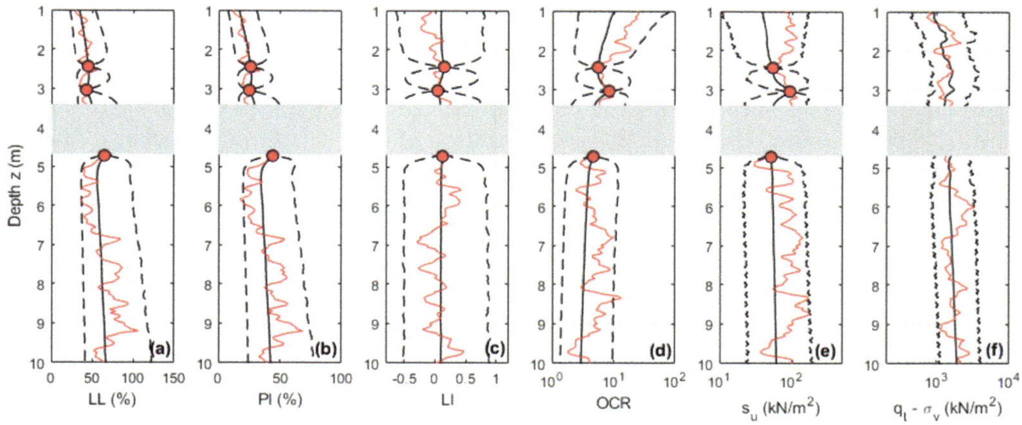

Figure 1.10 The conditional random fields simulated at the B-3 location at the Baytown site for several soil properties.

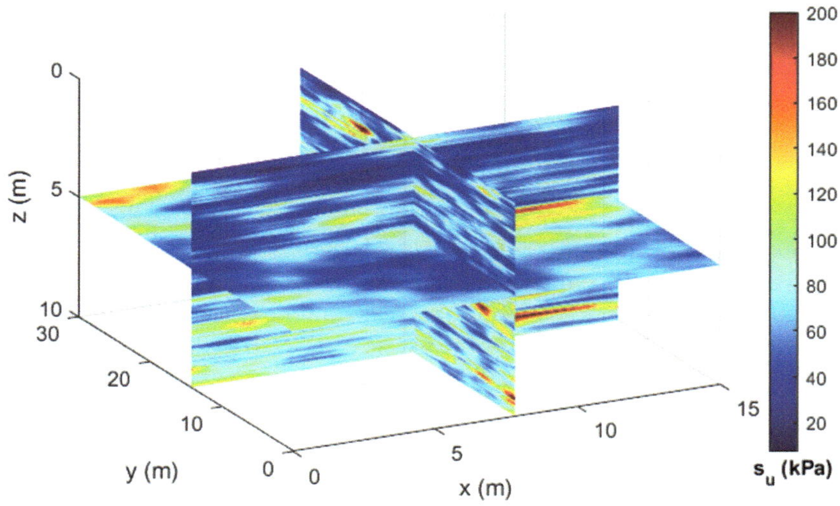

Figure 1.11 One conditional random field realization of s_u for the 3D underground space of the Baytown site.

shows the conditional random fields simulated at the B-3 location at the Baytown site (see Figure 1.6a) for several soil properties. The red lines are realizations of the conditional random fields, whereas the dark and dashed lines are the medians and 95% CIs. The red dots are the observed data at the B-3 location. Figure 1.11 shows one conditional random field realization of s_u for the 3D underground space of the Baytown site. The limitation is that only cross-correlation information in the database is updated in HBM-MUSIC-3X. Guan et al. (2024b) proposed a dictionary learning method to infer spatial variability at a target site from CPT profiles at neighbouring sites.

1.4.4 Strategy IV – Quasi-regional cluster model for MUSIC clay parameters

Ching et al. (2020b) showed that a quasi-site-specific model constructed by a Shanghai municipal database outperforms the one constructed by a generic database if the target site is a Shanghai site. Therefore, there is a question of whether the entire generic database should be adopted to train the HBM because some database sites may not be relevant to the target site. It is possible that a subset of the generic database with sites similar to the target site can train a more suitable HBM. This is known as the "regional advantage" (Phoon and Ching 2022). Sharma et al. (2022) proposed a hierarchical Bayesian similarity index that measures the similarity between (μ_i, C_i) and (μ_j, C_j) of two sites. This similarity measure can be used to construct the similarity matrix (size = $n_s \times n_s$) among the n_s sites in the generic database. Based on this similarity matrix, Sharma et al. (2023) further proposed a spectral clustering algorithm to divide the generic database into clusters. Sharma et al. (2023) showed that the HBM trained by a cluster more relevant to the target site [i.e., (μ_i, C_i) of the sites in the cluster are "closer" to (μ_s, C_s)] tends to outperform the HBM trained by the entire generic database.

1.4.5 Strategy IV – Tailored cluster model for MUSIC clay parameters

The limitation in classical clustering (CC) is that the number of clusters is small while the number of target sites can be very large. It is possible that some target sites do not fit any of the clusters well. Cai et al. (2024a) suggested a tailored clustering (TC) approach that addresses this limitation. In comparison to CC, TC is more flexible because the clustering results are dependent on the target site information. In the TC algorithm proposed by Cai et al. (2024a), a modified inter-dataset similarity measure (MIDSM) is used to calculate the similarity (*Sim*) between the target site and each database site. For each database site, MIDSM quantifies the value of *Sim* by considering the difference in geometries between the 95% confidence regions of datasets collected from the target site and the database site. The database sites that have high *Sim* values are considered to qualify as members of the desired cluster. Consequently, each target site is assigned to a cluster tailored to it.

Cai et al. (2024b) conducted a real example based on the generic database CLAY/10/7490 to show the advantage of TC in inferring soil parameters at a target site. The target site is selected from CLAY/10/7490 (#211 site) and this target site is removed from CLAY/10/7490. The two soil properties, limit liquid (LL, in %) and plasticity index (PI, in %), are considered in this example. The TC proposed by Cai et al. (2024a) and the CC proposed by Sharma et al. (2023) were adopted to identify two clusters (termed as CC cluster and TC cluster, respectively) from CLAY/10/7490 that most closely resemble the target site. Then, three inference methods [e.g., probabilistic multiple regression (PMR) (e.g., Kulhawy and Mayne 1990; Ching and Phoon 2012), classical Bayesian model (CBM) (e.g., Ching and Phoon 2019), and hierarchical Bayesian model (HBM) (e.g., Ching et al. 2021b)] were applied to the TC and CC clusters to infer clay parameters at the target site. Five PI data at the target site were assumed to be unknown and removed for checking the inference error arising from each combination of clustering method and inference method. The inference error was quantified by the root mean square error (RMSE). The value of RMSE for each combination is plotted against the corresponding computational time in Figure 1.12. Results show that the TC cluster tends to outperform the CC cluster in terms of reducing the inference error and computational

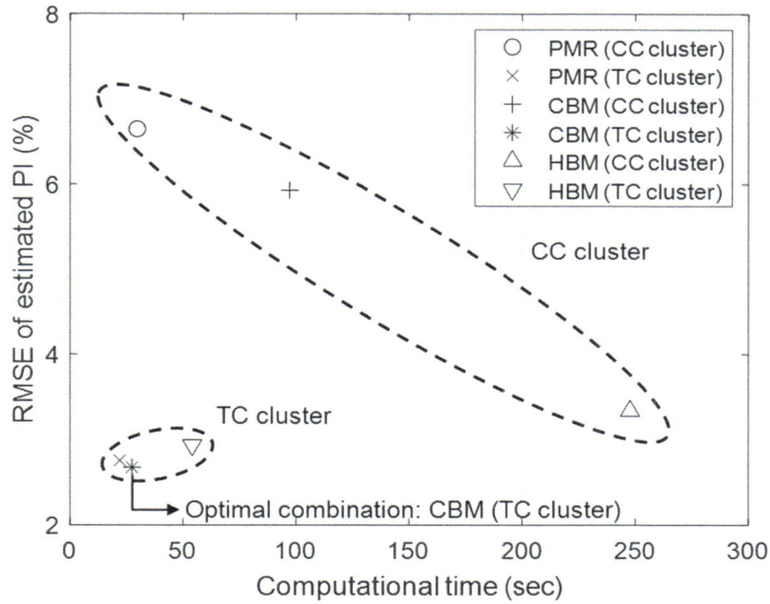

Figure 1.12 Computational time and RMSE for different combinations of cluster and inference method.

time for each of the investigated combinations. In this example, TC with CBM is considered to be the optimal approach because it can achieve the smallest inference error with the least computational cost. However, recent research showed that further improvement is possible (Cai et al. 2024c; Phoon et al. 2025).

1.5 CHALLENGES

The development of DDSC methods that are tractable and computationally efficient for large 3D spatial domains in the presence of MUSIC-3X and site uniqueness is far from mature. Research on more complex data attributes such as MUSIC-3X-G has not received the attention it deserved. For site characterization or other geotechnical applications where physics and experience are encountering difficulties in extracting useful insights from an increasing volume and complexity of data, it is clear that the role of machine learning and artificial intelligence in decision making will grow in importance. Although data science is a rapidly evolving field, research specific to geotechnical engineering is limited. The same can be said of machine learning and artificial intelligence, which are important "learning" algorithms in data science. Data-centric geotechnics is an agenda to transform geotechnical engineering practice to Geo 4.0 (geotechnical realization of Industry 4.0). DDSC is one broad application domain in data-centric geotechnics. One can easily imagine autonomous construction as another application domain, although there are complex ethical dimensions (Assaad and Boshuijzen-Van Burken 2023). This application is outside the scope of this chapter. What is missing is a geotechnical data science foundation that can guide the development of novel machine learning or artificial intelligence algorithms to support applications in data-centric geotechnics. The research questions raised in this chapter are use-inspired to meet the "fit for (and transform) practice" element in data-centric geotechnics. They may inform the agenda for geotechnical data science in the future. In the meantime, for the specific instance of DDSC, the following research questions are open:

(1) **Explainability and interpretability** – The finite element method is the most prevalent embodiment of physics in geotechnical engineering. Using finite element analysis, an engineer understands cause and effect (interpretability) and knows which input parameters affect the outputs (explainability). Decisions supported by physics-informed machine learning (PIML) are arguably more "explainable" and "interpretable." Xu et al. (2023) provide a review of the state of the art of physics-informed machine learning methods in enhancing the reliability and safety of complex systems from both technical and application perspectives. Explainability and interpretability are important in the trustworthy data-centric geotechnics agenda.
(2) **Transferability** – One can expect a generic database to expand with the addition of new sites on a regular basis. Learning is deemed "transferable" if past training outcomes can be updated without requiring brute force training from scratch. This is already implemented in ChatGPT, where responses are more relevant in the presence of few-shot learning or fine-tuning.
(3) **Geo-compatibility** – The generic databases in Table 1.2 are currently restricted to a single geomaterial type such as clay and sand. However, the target site data can contain records belonging to different geomaterial types because of the presence of different layers at different depths. All the methods listed in Table 1.7 require the target site data and the generic databases to be "geo-compatible," meaning that only layers belonging to the same geomaterial type as the records in the generic database can be analysed.
(4) **Frugal HBM** – A typical building regulation mandates a minimum three boreholes on a project site. Another typical guideline is to conduct one boring every 300 m^2. The "medium" and "large" ground volumes suggested by Phoon et al. (2022c) are a 40 m long × 20 m wide × 20 m deep cuboid and a 80 m long × 40 m wide × 40 m deep cuboid, respectively. The number of boreholes to cover these ground volumes

Table 1.7 Literature survey of site recognition research organized according to six broad strategies

Strategy	*Description*	*References*
I	Use entire global database with every record assigned an equal weight independent of its similarity to the target site.	Ching and Phoon 2012, 2013, 2014b; Ching et al. 2014, 2017, 2018, 2020b
II	Use entire global database with every record assigned a weight based on its similarity to the target site.	Ching and Phoon 2020a; Ching et al. 2021a, 2021b, 2022a; Sharma et al. 2022
III	Use a subset of the global database ("quasi-regional" database) with every record/site assigned an equal weight.	
IV	Use a subset of the global database ("quasi-regional" database) with every record/site assigned an unequal weight.	Sharma et al. 2023; Cai et al. 2025
V	Use a subset of the global database ("quasi-regional" database) and a subset of the soil/rock parameters that is relevant to one or more governing limit states.	Cai et al. 2024c
VI	Do not use the global database, in part or in full, to inform the target site because they are "dissimilar."	Ching et al. 2023b; Cai et al. 2024b

is approximately 3 and 10, respectively. In addition to borings, field tests are almost always conducted. In comparison, a full-scale performance evaluation may be conducted only once on a project site. Typical building regulations for foundation load tests are: (1) "1 number or 0.5% of the total piles, whichever is greater" for an ultimate load test on a preliminary pile (preferably instrumented) and (2) "2 numbers or 1% of working piles installed or 1 for every 50 m length of proposed building, whichever is greater" for a working load test. While site-specific soil property data are sparse, site-specific performance data are extremely sparse because they are expensive. There is no hierarchical Bayesian model (HBM) that can extract inter-site variability from a database where the site-specific data are extremely sparse. The proposed concept of a frugal HBM is to prune a conventional HBM such that it can deliver meaningful inter-site variability insights in the presence of a performance database. Even if the site-specific data are sufficient to train a conventional HBM, it is also useful to ask if it is the most frugal.

(5) **Thick data** – It is not possible to ignore the rapid pace of development in generative AIs. The ability to generate insights from multi-modal data such as unstructured text, images, and videos appears to have a direct bearing on the question of how human experience can add value to quantitative data analytics. The authors hypothesize that generative AIs such as ChatGPT offer interesting opportunities to further advance data-centric geotechnics, especially because of their strong ability to handle multi-modal data.

(6) **Machine learning-guided observational method (MLOM)** – The combination of site characterization and monitoring information can support real-time decision making during construction. This MLOM agenda is in its infancy, although its value is arguably higher than DDSC. After all, for geotechnical engineering, the purpose of understanding a site is to understand the performance of a structure. Some studies have emerged, and a representative sample is briefly presented below to illustrate the current state-of-the-art:

   a. Ma et al. (2023) proposed a real-time intelligent classification method for blasting and drilling tunnels installed in rock using a machine learning algorithm. A new database is compiled from 286 case studies in China. The data include rock hardness, weathering degree, rock mass structure, structural plane integrity, rock mass integrity, in situ stress condition, groundwater condition, and surrounding rock grades. An intelligent classification method for the surrounding rock is first established using machine learning. A tunnel information management system for drilling and blasting tunnels is next built by combining the intelligent classification method and cloud technology, and it is applied to the Lexi Expressway project. The tunnel information management system can provide a timely and accurate reference for the dynamic design and construction of tunnels.

   b. Yan et al. (2023) noted that developing a generalized model for axial displacement prediction for immersed tunnel joints is challenging because of the complex geo-environment and limited monitoring data. This is a typical small data problem in geotechnics since only a few features are available for training machine learning models. The one-year monitoring data in this study, including axial displacement, structure temperature, and water level, show seasonal trends and their correlations vary spatially along the tunnel. The study proposed a hybrid physical data (HPD) informed deep neural network (DNN) to improve spatial generalization for axial displacement prediction. The HPD is created based on physical analysis and contributes to the DNN as a substituting feature rather than an additional feature. The proposed model is shown to outperform conventional regression methods in terms of spatial generalization with improved physical interpretation.

c. Zhou et al. (2023) proposed a tunnel lining crack detection algorithm named improved You Only Look Once version X (YOLOX), which can detect tunnel cracks in a complex environment efficiently and accurately. The images taken during the damage detection phase of the study conducted in several tunnels are compiled and expanded to obtain a tunnel crack damage image dataset that can represent the complex physical state of the tunnel. These crack images can be used for model training and testing, and after experiments, the improved YOLOX model can identify these images and can achieve high speed, high accuracy, and real-time dynamic detection of tunnel cracks.
d. Morgenroth et al. (2023) observed that stress model updating for mine-scale finite difference models is generally time-consuming and tedious, requiring a large volume of engineering hours and computational power. Their study presents a novel machine learning approach, using a microseismic database, to update a stress model at Garson Mine in Ontario, Canada. A long-short term memory (LSTM) network is trained on the microseismic events and the geomechanical parameters from a previously manually calibrated FLAC3D model. Two LSTM networks are trained for comparison: one to predict the principal stresses and the other to predict the six-component stress tensor. The predicted values are compared to the values computed by the FLAC3D model ("ground truth").
e. Tan et al. (2023) mapped the spatial stress distribution over a coalmine roof by using a transfer convolutional neural network (TCNN) model. A convolutional neural network (CNN) model is first pre-trained based on the simulated data from the finite difference method. This CNN can be refined for a specific coalmine roof using limited structural health monitoring (SHM) data through transfer learning. The SHM data are recorded by a stress sensor installed in the roof of the Dongtan coalmine over a range of 120 m ahead of the working face. This study demonstrated that practical site-specific predictions can be made even in the presence of limited field measurements by adopting a physics-informed rather than pure data-driven approach.
f. Tian and Wang (2023) proposed a data-driven and physics-informed Bayesian learning framework that generates a project-specific training dataset by the random finite element method (reflecting the embedded physics) using sparse site investigation data. The generated training dataset is then used in Bayesian sparse dictionary learning of settlement monitoring data for improving model prediction in real time and continuously, particularly at locations without monitoring data and/or subsequent time steps. Uncertainty in settlement prediction may also be properly quantified. The proposed approach can also be applied to real-time fusion of numerical modelling results with multi-source monitoring data (Tian et al. 2024).

(7) **Data protection** – If geotechnical data is viewed as a critical asset and a repository of significant value, the data-centric geotechnics agenda cannot focus purely on maximizing the value to practice but also on minimizing the leakage of sensitive

data. Data protection is widely recognized in the machine learning community, but to the authors' knowledge, no research has been conducted in the geotechnical engineering community. This chapter proposes adding "privacy enhancement" to the agenda for data-centric geotechnics. Data protection may cover the following aspects:

a. Data minimization – The risk of data breaches, unauthorized access, or misuse of data is reduced if the minimal training set is used. This minimization principle will also encourage the use of relevant data sets only, which can improve accuracy and computational efficiency.
b. Data encryption – Data encryption is a technique that transforms data into an unreadable form using a secret key or algorithm. Homomorphic encryption is regarded as the "holy grail" in privacy-enhancing technologies (PETs). It allows operations to be performed on encrypted data such that the final results exactly match the output of the operations on unencrypted data.
c. Data protection by design – Data protection by design means implementing privacy-enhancing technologies (PETs) into the ML solutions from the earliest stages of design and throughout the entire life cycle of the system or application. PETs include homomorphic encryption differential privacy, generative adversarial networks (GANs), synthetic data generation, and federated learning. A proactive approach to address potential data privacy and security risks enhances the trust and confidence of users and stakeholders.
d. Data governance – The purpose of data governance is to define the roles, responsibilities, policies, and standards for managing and using data in an organization. This covers the rules of the game for data collection, processing, storage, sharing, and disposal. In the presence of a data governance framework, data ethics, compliance, and accountability could be enforced.

1.6 CONCLUSIONS

One of the key findings of the ISSMGE 2017 SOA/SOP Survey is that geotechnical databases, which include information on inherent spatial variability, soil and rock properties, and risk databases, should be made more readily available to the profession. Thus, compiling geotechnical databases represents one of the key missions of the ISSMGE TC304 (Engineering Practice of Risk Assessment & Management). An initiative to promote public sharing of databases, called 304dB, was initiated by Professors Kok-Kwang Phoon and Jianye Ching in 2017: http://140.112.12.21/issmge/tc304.htm. Despite its success (304dB has contributed to the development of data-driven methods by situating these methods in the context of real data over the past seven years), it is timely to take stock and discuss how this effort can be expanded to cover more types of databases, more multi-modal and real-time data, more countries, and more projects. In short, what do we need to do in research and practice to go from 304dB to Geo-everything-dB? The breakout of powerful generative AI such as ChatGPT in 2023 should give new impetus to collecting data, understanding data, protecting data as a critical asset, developing learning algorithms to transform data and devices into embodied intelligence, and reimagining infrastructure as a service to end-users through the deployment of smart technologies. The intent of taking a data-centric perspective to geotechnical engineering (data-centric geotechnics) is to march towards these top-level goals. Geo-everything-dB can be a hashtag for collecting, understanding, and protecting data across the entire life cycle of the structure. The current book and its companion volume (Tang and Phoon 2025) are edited in the spirit of expanding the boundaries of Geo-everything-dB.

This chapter presents an overview of site characterization information with a focus on its availability, coverage, data attributes, value to decision making, and challenges. In terms of availability, it is accurate to say that site characterization databases mainly reside in industry. They were compiled to support the design and construction of a particular project and as part of compliance. Their value is almost zero after the project is completed as they are shelved in some storage, physical or virtual. Phoon (2020) called this "dark data." There are several reasons for geotechnical databases remaining "dark," but one can argue that data belong to the clients and there are non-disclosure or confidentiality agreements in many projects. Nonetheless, it is worthwhile to remind ourselves that this issue is not unique to geotechnical engineering. There are different strategies in secure data sharing being developed to extract insights from databases without exposing confidential details or breaching legal agreements. They include homomorphic encryption, differential privacy, and federated learning. To the authors' knowledge, there is no systematic research on the protection of geotechnical data. The "data first practice central" agenda in data-centric geotechnics can be expanded to "data first, practice central, and privacy enhanced." Data-centric geotechnics now encompasses four elements: (1) data centricity, (2) fit for (and transform) practice, (3) geotechnical context, and (4) privacy enhancement. Phoon (2025) advocated updating "data-centric geotechnics" to "trustworthy data-centric geotechnics" to manage the risk of ML/AI more systematically in response to the EU AI Act. There is also "dirty data" (Phoon and Zhang 2023). It is very time-consuming and labour-intensive to clean, label, and organize dirty data from legacy systems. Current databases are compiled manually (Tang and Phoon 2021; Phoon and Tang 2025; Tang and Phoon 2025). There is a recent trend of using more flexible and efficient database query technology based on a large language model and vector database. Users can extract relevant and useful data from the database using natural language, which allow more interactive and meaning-based queries to the database.

In terms of coverage, publicly available site characterization databases are mainly geotechnical in nature containing borehole, laboratory test, and field test data. Detailed geological data such as soil stratification, rock mass features (weathering, discontinuities, etc.), and real-time data such as geoenvironmental processes (subsurface flow, contaminant transport, etc.) are lacking. Karpatne et al. (2018) identified the "paucity of ground truth" as one attribute of geoscience data as well. In terms of data attributes, Phoon et al. (2019) posed MUSIC-X (Multivariate, Uncertain and Unique, Sparse, Incomplete, and potentially Corrupted with "X" denoting the spatial/temporal dimension) as a basic challenge to data-driven site characterization (DDSC) research. It has since been expanded to MUSIC-3X-G. It is clear that sparsity, uncertainty, and spatial variability are only three out of seven attributes. This chapter proposed a new taxonomy of site data under "4S" to expand the agenda for future machine learning research. The "4S" are site generalizations, spatial features, sampling characteristics, and smart data. Sampling characteristics include time. No real-time picture of the subsurface is available and no one knows how to achieve this real-time understanding with available data.

In terms of value to design, the majority of the research is focused on the prediction of soil properties (spatial variability) and soil types (stratification) at a target site. Phoon et al. (2022b) argued that these predictions are intrinsically probabilistic because statistical uncertainties are significant at the current sampling density (e.g., volume fraction for logging at a Brent Field site in the North Sea is 1×10^{-6}, Chilès and Delfiner 1999). The prediction of soil types and properties is related, and the volume to be characterized is 3D. Phoon et al. (2022c) proposed the first set of eight benchmark examples to accelerate the pace of data-driven site characterization (DDSC) research. One component of a benchmark example is the virtual ground (VG) that represents the complete state of "reality." Everything is known within this virtual ground because it is generated by assuming stratigraphy and populating the properties in each layer using a random field generator. Phoon et al. (2022c) suggested an initial "small" ground volume of a 20 m long × 20 m wide × 10 m deep cuboid for benchmark examples. The authors suggested that "medium" and "large" ground volumes can be defined as a 40 m long × 20 m wide × 20 m deep cuboid and a 80 m long × 40 m wide × 40 m deep cuboid, respectively. The key point here is that there are computational challenges in terms of time and cost. In addition, there are at least two limitations to this initial set of benchmark examples. First, the soil layers are too idealized. Shuku and Phoon (2023b) proposed a more realistic VG5 model (Figure 1.3e and f) with spatially varying and discontinuous soil layers simulated using a coupled Markov chain (Qi et al; 2016). Second, real examples are needed. The database Tokyo-CLAY/14/67,760 (Otake et al. 2024) contained developed 2792 boreholes covering a site area of more than 10 km^2 from a real project in Japan. Each borehole has 29 data records on average. This database can be adopted as the basis for establishing a benchmark example for a real site.

One DDSC challenge called "site recognition" has attracted attention in recent years. The site recognition challenge is fundamental to geotechnical engineering because it attempts to quantify the "uniqueness" (or site specificity) attribute of a target site. The argument that geotechnical engineering practice is more art than science and engineering judgment is indispensable arises in large part from the recognition that site specificity is not sufficiently accounted for in our calculation models. One classical example is the use of generic transformation models to estimate design parameters from field test parameters. Recent research has demonstrated that quasi-site-specific transformation models can be developed notwithstanding the well-known limitations that site-specific data is sparse and a generic database containing data from other sites is not directly applicable to a target site (Phoon and Ching 2022). There are three notable achievements: (1) the site recognition problem is tractable even under MUSIC, (2) inference uncertainty produced by a quasi-site-specific transformation model is smaller than the conventional generic or site-specific model, and (3) increasingly effective quasi-site-specific transformation models have been constructed, particularly those that exploit clustering (Cai et al. 2024a; Phoon et al. 2025). The construction of a quasi-site-specific ground model that includes learning cross-correlations, spatial auto-correlations, and geologic uncertainties from generic databases under MUSIC-3X-G is an ongoing research problem.

Research on other site characterization databases, such as geophysical and remote sensing data and data fusion, is limited (Xie et al. 2022; Xu et al. 2022; Guan et al. 2024a). Interest in combining site characterization information with monitoring information collected during construction to guide decision making in real time is emerging rapidly (Shuku et al. 2023; Tian and Wang 2023; Tian et al. 2024). This agenda is called the machine learning-guided observational method (MLOM). Its value is arguably higher than DDSC. It may evolve to human–machine teaming, which is outside the scope of this chapter. Other challenges such as explainability and interpretability, transferability, geo-compatibility, frugal HBM, thick data, and data protection are briefly discussed. It is uncertain whether a new discipline called geotechnical data science is needed to provide a stronger theoretical underpinning for data-centric geotechnics, given the large interconnected challenges highlighted in this chapter. But trustworthiness needs to be addressed to comply with the EU AI Act and other regulatory frameworks of AI systems.

ACKNOWLEDGEMENTS

The authors gratefully acknowledge the assistance of Dr Yongmin Cai and insightful comments from Professors Zjiun Cao, Yu Otake, Yu Wang, Stephen Wu, and Tengyuan Zhao. The authors also would like to thank the members of the TC304 Committee on Engineering Practice of Risk Assessment & Management of the International Society of Soil Mechanics and Geotechnical Engineering for developing the database 304dB used in this study and making it available for scientific inquiry.

REFERENCES

Abubakar, A., Brædstrup, M. J., Di, H., Diaz, A. T., Freeman, S., Hviid, S., Karkov, K. H., Kriplani, S., Manikani, S., Salun, G., and Zhao, T. 2021. Deep learning applications for wind farms site characterization and monitoring. SEG Technical Program Expanded Abstracts, 3009–3013.

Aladejare, A. E. and Wang, Y. 2017. Evaluation of rock property variability. *Georisk: Assessment and Management of Risk for Engineered Systems and Geohazards*, 11(1), 22–41.

ASCE 2019. *Future World Vision: Infrastructure Reimagined*. American Society of Civil Engineers, Reston, VA.

Assaad, Z. and Boshuijzen-Van Burken, C. 2023. Ethics and safety of human-machine teaming. *Proceedings, First International Symposium on Trustworthy Autonomous Systems (TAS '23)*, July 11–12, 2023, Edinburgh, United Kingdom. ACM, New York.

Bárdossy, G. and Fodor, J. 2001. Traditional and new ways to handle uncertainty in geology. *Natural Resources Research*, 10(3), 179–187.

Bárdossy, G. and Fodor, J. 2004. *Evaluation of Uncertainties and Risks in Geology*. Springer, Berlin.

Beesley, M. and Vardanega, B. J. 2020. Parameter variability of undrained shear strength and strain using a database of reconstituted soil tests. *Canadian Geotechnical Journal*, 57(8), 1247–1255.

Bilal, M., Oyedele, L. O., Qadir, J., Munir, K., Ajayi, S. O., Akinade, O. O., Owolabi, H. A., Alaka, H. A., and Pasha, M. 2016. Big Data in the construction industry: A review of present status, opportunities, and future trends. *Advanced Engineering Informatics*, 30(3), 500–521.

Bock, H. 2006. Common ground in engineering geology, soil mechanics and rock mechanics: Past, present and future. *Bulletin of Engineering Geology and the Environment*, 65(2), 209–216.

Boeckmann, A. and Loehr, J. E. 2023. *Practices for local calibration of LRFD geotechnical resistance factors. NCHRP Synthesis 601*. Transportation Research Board, Washington, DC.

BS 1986. *BS8004: Code of Practice for Foundations*. British Standard, London.

Burland, J. B. 2012. The geotechnical triangle. *ICE Manual of Geotechnical Engineering: Geotechnical Engineering Principles, Problematic Soils and Site Investigation*, Eds. John B Burland, T. J. P. Chapman, H. Skinner, M. J. Brown, Vol. 1, pp. 17–26. Emerald Publishing Limited, Leeds, England.

Cai, Y., Ching, J., and Phoon, K. K. 2024a. Tailored clustering method to identify quasi-regional sites. *Engineering Geology*, 333, 107490.

Cai, Y., Pan, Q., Phoon, K. K., and Luo, W. 2024b. Modifying tailored clustering enabled regionalization (TCER) framework for outlier detection and inference efficiency. *Engineering Geology*, 335, 107537.

Cai, Y., Phoon, K. K., Ching, J., and Wang, Y. 2024c. Dimension reduction of MUSIC database for tailored clustering using Lasso. *Canadian Geotechnical Journal*, Under review.

Cai, Y., Phoon, K. K., Otake, Y., and Wang, Y. 2025. Efficient dictionary learning for constructing quasi-local transformation models. *Computers and Geotechnics*. Under review.

Cao, Z. and Wang, Y. 2013. Bayesian approach for probabilistic site characterization using cone penetration tests. *Journal of Geotechnical and Geoenvironmental Engineering*, 139(2), 267–276.

Cao, Z. J., Zheng, S., Li., D. Q., and Phoon, K. K. 2019. Bayesian identification of soil stratigraphy based on soil behavior type index. *Canadian Geotechnical Journal*, 56(4), 570–586.

CEN 2004. EN 1997-1: *Geotechnical design – Part 1: General rules*. Comité Européen de Normalisation.

Chen, G., Cheng, Q., and Puetz, S. 2023. Special issue: data-driven discovery in geosciences: opportunities and challenges. *Mathematical Geosciences*, 55, 287–293.

Chilès, J-P. and Delfiner, P. 1999. *Geostatistics: Modeling Spatial Uncertainty*. John Wiley & Sons, New York.

Ching, J. 2020. Unpublished databases.

Ching, J. and Kuo, M.C. 2023. Data-centric quasi-site-specific prediction for soil modulus. *Journal of GeoEngineering*, 18(4), 215–223.

Ching, J. and Phoon, K. K. 2012. Modeling parameters of structured clays as a multivariate normal distribution. *Canadian Geotechnical Journal*, 49(5), 522–545.

Ching, J. and Phoon, K. K. 2013. Multivariate distribution for undrained shear strengths under various test procedures. *Canadian Geotechnical Journal*, 50(9), 907–923.

Ching, J. and Phoon, K. K. 2014a. Transformations and correlations among some clay parameters – The global database. *Canadian Geotechnical Journal*, 51(6), 663–685.

Ching, J. and Phoon, K. K. 2014b. Correlations among some clay parameters – The multivariate distribution. *Canadian Geotechnical Journal*, 51(6), 686–704.

Ching, J. and Phoon, K. K. 2017. Characterizing uncertain site-specific trend function by Sparse Bayesian learning. *Journal of Engineering Mechanics, ASCE*, 143(7), 04017028.

Ching, J. and Phoon, K. K. 2019. Constructing site-specific multivariate probabilistic distribution model by Bayesian machine learning, *Journal of Engineering Mechanics, ASCE*, 145(1), 04018126.

Ching, J. and Phoon, K. K. 2020a. Measuring similarity between site-specific data and records from other sites. *ASCE-ASME Journal of Risk and Uncertainty in Engineering Systems, Part A: Civil Engineering*, 6(2), 04020011.

Ching, J. and Phoon, K. K. 2020b. Constructing a site-specific multivariate probability distribution using sparse, incomplete, and spatially variable (MUSIC-X) data. *Journal of Engineering Mechanics, ASCE*, 146(7), 04020061.

Ching, J., Phoon, K. K., and Chen, C. H. 2014. Modeling piezocone cone penetration (CPTU) parameters of clays as a multivariate normal distribution. *Canadian Geotechnical Journal*, 51(1), 77–91.

Ching, J., Wang, J. S., Juang, C. H., and Ku, C. S. 2015. Cone penetration test (CPT)-based stratigraphic profiling using the wavelet transform modulus maxima method. *Canadian Geotechnical Journal*, 52(12), 1993–2007.

Ching, J., Lin, G. H., Chen, J. R., and Phoon, K. K. 2017. Transformation models for effective friction angle and relative density calibrated based on a multivariate database of coarse-grained soils. *Canadian Geotechnical Journal*, 54(4), 481–501.

Ching, J., Li, K. H., Phoon, K. K., and Weng, M. C. 2018. Generic transformation models for some intact rock properties. *Canadian Geotechnical Journal*, 55(12), 1702–1741.

Ching, J., Huang, W. H., and Phoon, K. K. 2020a. 3D probabilistic site characterization by Sparse Bayesian Learning. *Journal of Engineering Mechanics, ASCE*, 146(12), 04020134.

Ching, J., Phoon, K. K., Khan, Z., Zhang, D. M., and Huang, H. W. 2020b. Role of municipal database in constructing site-specific multivariate probability distribution. *Computers and Geotechnics*, 124, 103623.

Ching, J. Y., Phoon, K. K., Ho, Y. H., and Weng, M. C. 2021a. Quasi-site-specific prediction for deformation modulus of rock mass. *Canadian Geotechnical Journal*, 58(7), 936–951.
Ching, J., Wu, S., and Phoon, K. K. 2021b. Constructing quasi-site-specific multivariate probability distribution using hierarchical Bayesian model. *Journal of Engineering Mechanics, ASCE*, 147(10), 04021069.
Ching, J., Yang, Z. Y., and Phoon, K. K. 2021c. Dealing with non-lattice data in three-dimensional probabilistic site characterization. *Journal of Engineering Mechanics, ASCE*, 147(5), 06021003.
Ching, J., Phoon, K. K., Yang, Z., and Stuedlein, A. W. 2021d. Constructing a quasi-site-specific multivariate probability distribution model for soil properties using sparse, incomplete, and three-dimensional (MUSIC-3X) data. *Georisk*, 16(1), 53–76.
Ching, J., Phoon, K. K., and Wu, C. T. 2022a. Data-centric quasi-site-specific prediction for compressibility of clays. *Canadian Geotechnical Journal*, 59(12), 2033–2049.
Ching, J., Phoon, K. K., Yang, Z. Y., and Stuedlein, A. W. 2022b. Quasi-site-specific multivariate probability distribution model for sparse, incomplete, and three-dimensional spatially varying soil data. *Georisk: Assessment and Management of Risk for Engineered Systems and Geohazards*, 16(1), 53–76.
Ching, J., Uzielli, M., Phoon, K. K., and Xu, X. 2023a. Characterization of autocovariance parameters of detrended cone tip resistance from a global CPT database. *Journal of Geotechnical and Geoenvironmental Engineering, ASCE*, 149(10), 04023090.
Ching, J., Yoshida, I., and Phoon, K. K. 2023b. Comparison of trend models for geotechnical spatial variability: sparse Bayesian learning vs. Gaussian process regression. *Gondwana Research*, 123, 174–183.
Ching, J., Phoon, K. K., and Huang, P. 2024. Detection of outliers with respect to a MUSIC geotechnical database. *Canadian Geotechnical Journal*, 61(7), 1275–1293.
Christensen, C. W., Harrison, E. J., Pfaffhuber, A. A., and Lund, A. K. 2021. A machine learning–based approach to regional-scale mapping of sensitive glaciomarine clay combining airborne electromagnetics and geotechnical data. *Near Surface Geophysics*, 19(5), 523–539.
Chwała, M., Phoon, K. K., Uzielli, M., Zhang, J., Zhang, L., and Ching, J. 2023. Time capsule for geotechnical risk and reliability. *Georisk: Assessment and Management of Risk for Engineered Systems and Geohazards*, 17(3), 439–466.
D'Ignazio, M., Phoon, K. K., Tan, S. A., and Lansivaara, T. 2016. Correlations for undrained shear strength of Finnish soft clays. *Canadian Geotechnical Journal*, 53(10), 1628–1645.
Du, B., Ye, J., Zhu, H., Sun, L., and Du, Y. 2023. Intelligent monitoring system based on spatio-temporal data for underground space infrastructure. *Engineering*, 25, 194–203.
Einstein, H. and Baecher, G. 1982. *Probabilistic and statistical methods in engineering geology I. Problem statement and introduction to solution. Ingenieurgeologie und Geomechanik als Grundlagen des Felsbaues/Engineering Geology and Geomechanics as Fundamentals of Rock Engineering*, Springer, pp. 47–61.
Eslami, A., Heidarie Golafzani, S., and Naghibi, M. H. 2023. Developed triangular charts; deltaic CPTu-based soil behavior classification using AUT:CPTu-Geo-Marine Database. *Probabilistic Engineering Mechanics*, 71, 103380.
Faramarzi, A. 2020. Editorial for themed issue on "Application of machine learning". *Proceedings of the Institution of Civil Engineers-Smart Infrastructure and Construction*, 173(4), 73.
Feng, S. and Vardanega, P. J. 2019. A database of saturated hydraulic conductivity of fine-grained soils: probability density functions. *Georisk: Assessment and Management of Risk for Engineered Systems and Geohazards*, 13(4), 255–261.
Feng, S., Barreto, D., Imre, E., Ibraim, E., and Vardanega, P.J. 2023. Use of hydraulic radius to estimate the permeability of coarse-grained materials using a new geodatabase. *Transportation Geotechnics*, 41, 101026.
Fookes, P. 1997. Geology for engineers: The geological model, prediction and performance. *Quarterly Journal of Engineering Geology and Hydrogeology*, 30(4), 293–424.

Girolami, M. 2020. Introducing data-centric engineering: an open access journal dedicated to the transformation of engineering design and practice. *Data-Centric Engineering*, 1, e1.

Guan, Z., Wang, Y., and Phoon, K. K. 2024a. Fusion of sparse non-co-located measurements from multiple sources for geotechnical site investigation. *Canadian Geotechnical Journal*, 61(8), 1574–1592.

Guan, Z., Wang, Y., and Phoon, K. K. 2024b. Dictionary learning of spatial variability at a specific site using data from other sites. *Journal of Geotechnical and Geoenvironmental Engineering, ASCE*, 150(9), 04024072.

Halim, I. S. and Tang, W.H. 1993. Site exploration strategy for geologic anomaly characterization. *Journal of Geotechnical Engineering*, 119(2), 195–213.

Harrison, J. 2019. Challenges in determining rock mass properties for reliability-based design. *Proceedings, 7th International Symposium on Geotechnical Safety and Risk (ISGSR 2019)*, Taipei, Taiwan, pp. 35–44.

Hoek, E. 1999. Putting numbers to geology- an engineer's viewpoint. *Quarterly Journal of Engineering Geology and Hydrogeology*, 32(1), 1–19.

Hong, A., Baker, L., Prieto Curiel, R., Duminy, J., Buswala, B., Guan, C., and Ravindranath, D. 2022. Reconciling big data and thick data to advance the new urban science and smart city governance. *Journal of Urban Affairs*, 45(10), 1737–1761.

Hu, J. Z., Zhang, J., Huang, H. W., and Zheng, J. G. 2021. Value of information analysis of site investigation program for slope design. *Computers and Geotechnics*, 131, 103938.

Hu, Y., Wang, Y., Phoon, K. K., and Beer, M. 2024. Similarity quantification of soil spatial variability between two cross-sections using auto-correlation functions. *Engineering Geology*, 331, 107445.

Huang, J., Zheng, D., Li, D., Kelly, R., and Sloan, S. 2018. Probabilistic characterization of two-dimensional soil profile by integrating CPT with MASW data. *Canadian Geotechnical Journal*, 55(8), 1168–1181.

Huang, T., Vance, T. C., and Lynnes, C. 2022. *Big data analytics in earth, atmospheric, and ocean sciences*. AGU Special Publications, American Geophysical Union.

Hutchinson, J. N. 2001. Reading the ground: Morphology and geology in site appraisal. *Quarterly Journal of Engineering Geology and Hydrogeology*, 34(1), 7–50.

Jaksa, M. B and Liu, Z. Q. 2021. Editorial for Special Issue on "Applications of artificial intelligence and machine learning in geotechnical engineering". *Geosciences*, 11(10), 399.

Jessell, M., Guo, J., Li, Y., Lindsay, M., Scalzo, R., Giraud, J., Pirot, G., Cripps, E., and Ogarko, V. 2022. Into the Noddyverse: a massive data store of 3D geological models for machine learning and inversion applications. *Earth System Science Data*, 14(1), 381–392.

Ji, J. and Cao, Z.J. 2024. Geotechnical reliability analysis for practice. Chapter 6 in *Uncertainty, Modeling, and Decision Making in Geotechnics*, edited by Kok-Kwang Phoon, Takayuki Shuku, and Jianye Ching. CRC Press, Boca Raton, FL.

Journel, A. G. and Huijbregts, C. J. 1978. *Mining Geostatistics*. Academic Press, London.

Juang, C. H., Zhang, J., Shen, M., and Hu, J. 2018. Probabilistic methods for unified treatment of geotechnical and geological uncertainties in a geotechnical analysis. *Engineering Geology*, 249, 148–161.

Juang, C. H., Ge, Y., and Zhang, J. 2019. Geological uncertainty: a missing element in geotechnical reliability analysis. *Proceedings, 7th International Symposium on Geotechnical Safety and Risk (ISGSR 2019)*, Taipei, Taiwan, pp. 1–12. http://rpsonline.com.sg/proceedings/isgsr2019/pdf/WT.pdf

Karniadakis, G. E., Kevrekidis, I. G., Lu, L., Perdikaris, P., Wang, S., and Yang, L. 2021. Physics-informed machine learning. *Nature Reviews Physics*, 3, 422–440.

Karpatne, A., Ebert-Uphoff, I., Ravela, S., Ali Babaie, H., and Kumar, V. 2018. Machine learning for the geosciences: challenges and opportunities. *IEEE Transactions on Knowledge and Data Engineering*, 31(8), 1544–1554.

Kwak, D.Y., Ahdi, S.K., Wang, P., Zimmaro, P., Brandenberg, S.J., and Stewart, J.P. 2021. Web portal for shear wave velocity and HVSR databases in support of site response research and applications. UCLAgeo, Los Angeles.

Kim, E. and Hunt, R. 2017. A public website of rock mechanics database from Earth Mechanics Institute (EMI) at Colorado School of Mines (CSM). *Rock Mechanics and Rock Engineering*, 50(12), 3245–3252.

Kootahi, K. and Moradi, G. 2017. Evaluation of compression index of marine fine-grained soils by the use of index tests. *Marine Georesources and Geotechnology*, 35(4), 548–570.

Kulhawy, F. H. and Mayne, P. W. 1990. *Manual on Estimating Soil Properties for Foundation Design.* Report EL-6800, Electric Power Research Institute, Palo Alto, CA.

Lacasse 2015. *55th Rankine Lecture: Hazard, Risk and Reliability in Geotechnical Practice.* Presentation at the Institution of Civil Engineers, London.

Lacasse, S. and Lunne, T. 1982. Penetration tests in two Norwegian clays. *Proceedings, 2nd European Symposium on Penetration Testing*, Amsterdam, pp. 661–670.

Länsivaara, T., Phoon, K. K., and Ching, J. 2022. What is a characteristic value for soils? *Georisk: Assessment & Management of Risk for Engineered Systems & Geohazards*, 16(2), 199–224.

Li, X., Li, P., and Zhu, H. 2013. Coal seam surface modeling and updating with multi-source data integration using Bayesian Geostatistics. *Engineering Geology*, 164, 208–221

Li, X., Li, H., Du, S., Jing, J., and Li, P. 2023. Cross-engineering utilization of Tunnel Boring Machines (TBM) construction data: a case study using big data from Yin-Song Diversion Project in China. *Georisk: Assessment and Management of Risk for Engineered Systems and Geohazards*, 17(1), 127–147.

Liao, T. and Mayne, P. 2007. Stratigraphic delineation by three-dimensional clustering of piezocone data. *Georisk*, 1(2), 102–119.

Lim, Y. X., Tan, S. A., and Phoon, K. K. 2018. Application of press-replace method to simulate undrained cone penetration. *International Journal of Geomechanics*, ASCE, 18(7), 04018066.

Lim, Y. X., Tan, S. A., and Phoon, K. K. 2019. Interpretation of horizontal permeability from piezocone dissipation tests in soft clays. *Computers and Geotechnics*, 107, 189–200.

Lin, Q., Ci, T., Wang, L., Mondal, S. K., Yin, H., and Wang, Y. 2022. Transfer learning for improving seismic building damage assessment. *Remote Sensing*, 14, 201.

Liu, S., Zou, H., Cai, G., Bheemasetti, T. V., Puppala, A. J., and Lin, J. 2016. Multivariate correlation among resilient modulus and cone penetration test parameters of cohesive subgrade soils. *Engineering Geology*, 209, 128–142.

Lo, M. K., Wei, X., Chian, S. C., and Ku, T. 2021. Bayesian network prediction of stiffness and shear strength of sand. *Journal of Geotechnical and Geoenvironmental Engineering*, 147(5), 04021020.

Löfman, M.S. and Korkiala-Tanttu, L.K. 2021. Transformation models for the compressibility properties of Finnish clays using a multivariate database. *Georisk: Assessment and Management of Risk for Engineered Systems and Geohazards*, 16(2), 330–346.

Lumb, P. 1966. The variability of natural soils. *Canadian Geotechnical Journal*, 3(2), 74–97.

Luo, X., Chen, H.-H., and Guo, Q. 2022. Semantic communications: overview, open Issues, and future research directions. *IEEE Wireless Communications*, 29(1), 210–219.

Ma, J., Li, T., Yang, G., Dai, K., Ma, C., Tang, H., Wang, G., Wang, J., Xiao, B., and Meng, L. 2023. A real-time intelligent classification model using machine learning for tunnel surrounding rock and its application. *Georisk: Assessment and Management of Risk for Engineered Systems and Geohazards*, 17(1), 148–168.

Morgenroth, J., Kalenchuk, K., Moreau-Verlaan, L., Perras, M., and Khan, U. T. 2023. A novel long-short term memory network approach for stress model updating for excavations in high stress environments. *Georisk: Assessment and Management of Risk for Engineered Systems and Geohazards*, 17(1), 196–216.

McPhillips, D.F., Herrick, J.A., Ahdi, S., Yong, A.K., and Haefner, S. 2020. Updated compilation of Vs30 data for the United States. U.S. Geological Survey data release.

Matheron, G. 1963. Principles of geostatistics. *Economic Geology* 58, 1246–1266.
Meng, T., Jing, X., Yan, Z., and Pedrycz, W. 2020. A survey on machine learning for data fusion. *Information Fusion*, 57, 115–129.
Morgenroth, J., Unterlaß, P. J., Sapronova, A., Khan, U. T., Perras, M. A., Erharter, G. H., and Marcher, T. 2022. Practical recommendations for machine learning in underground rock engineering – On algorithm development, data balancing, and input variable selection. *Geomechanics and Tunnelling*, 15(5), 650–657.
Munawar, H.S., Ullah, F., Qayyum, S., and Shahzad, D. 2022. Big data in construction: current applications and future opportunities. *Big Data and Cognitive Computing*, 6, 18.
Muzamhindo H. and Ferentinou, M. 2023. Generic compressive strength prediction model applicable to multiple lithologies based on a broad global database. *Probabilistic Engineering Mechanics*, 71, 103400.
National Academies of Sciences, Engineering, and Medicine. 2019. *Manual on Subsurface Investigations.* The National Academies Press, Washington, DC. https://doi.org/10.17226/25379.
Otake, Y., Saito, T., Wu, S., Yoshida, I., and Takano, D. 2025. Exploring challenges via analysis of multivariate geotechnical properties: insights from large-scale local sampling of Japanese marine clay. Chapter 11 in *Databases for Data-Centric Geotechnics: Site Characterization*, edited by K. K. Phoon and C. Tang. CRC Press, Boca Raton, FL.
Parry, S., Baynes, F.J., Culshaw, M.G., Eggers, M., Keaton, J.F., Lentfer, K., Novotny, J., and Paul, D. 2014. Engineering geological models: an introduction: IAEG commission 25. *Bulletin of Engineering Geology and the Environment*, 73(3), 689–706.
Phoon, K. K. 2018. Editorial for special collection on probabilistic site characterization. *ASCE-ASME Journal of Risk and Uncertainty in Engineering Systems, Part A: Civil Engineering*, 4(4), 02018002.
Phoon, K. K. 2020. The story of statistics in geotechnical engineering. *Georisk: Assessment and Management of Risk for Engineered Systems and Geohazards*, 14(1), 3–25.
Phoon, K. K. 2023a. Uncertainty-informed decision making in Burland Triangle. Chapter 1, *Uncertainty, Modelling, and Decision Making in Geotechnics*, CRC Press, Boca Raton, FL.
Phoon, K. K. 2023b. What geotechnical engineers want to know about reliability. *ASCE-ASME Journal of Risk and Uncertainty in Engineering Systems, Part A: Civil Engineering*, 9(2), 03123001.
Phoon, K. K. 2025. *Trustworthy data-centric geotechnics.* Geodata and AI.
Phoon, K. K. and Kulhawy, F. H. 1999a. Characterization of geotechnical variability. *Canadian Geotechnical Journal*, 36(4), 612–624.
Phoon, K. K. and Kulhawy, F. H. 1999b. Evaluation of geotechnical property variability. *Canadian Geotechnical Journal*, 36(4), 625–639.
Phoon, K. K. and Retief, J. V. 2016. *Reliability of Geotechnical Structures in ISO2394*, CRC Press/Balkema, London.
Phoon, K. K. and Ching, J. 2021. Project DeepGeo –Data-driven 3D subsurface mapping. *Journal of GeoEngineering*, 16(2), 61–74.
Phoon, K. K. and Ching, J. 2022. Additional observations on the site recognition challenge. *Journal of GeoEngineering*, 17(4), 231–247.
Phoon, K. K. and Tang, C. 2025. Role of performance information in data-centric geotechnics. Chapter 1 in *Databases for Data-Centric Geotechnics: Geotechnical Structures*, edited by Tang, C. and K. K. Phoon, CRC Press, Boca Raton, FL.
Phoon, K. K. and Zhang, W. G. 2023. Future of machine learning in geotechnics. *Georisk: Assessment and Management of Risk for Engineered Systems and Geohazards*, 17(1), 7–22.
Phoon, K. K., Retief, J. V., Ching, J., Dithinde, M., Schweckendiek, T., Wang, Y., and Zhang, L. M. 2016. Some observations on ISO2394:2015 Annex D (Reliability of Geotechnical Structures). *Structural Safety*, 62, 24–33.

Phoon, K. K., Ching, J., and Wang Y. 2019. Managing risk in geotechnical engineering – From data to digitalization. *Proceedings, 7th International Symposium on Geotechnical Safety and Risk (ISGSR 2019)*, Taipei, Taiwan, pp. 13–34. https://www.issmge.org/uploads/publications/96/97/SL.pdf

Phoon, K. K., Ching, J., and Cao, Z. J. 2022a. Unpacking data-centric geotechnics. *Underground Space*, 7(6), 967–989.

Phoon, K. K., Ching, J., and Shuku, T. 2022b. Challenges in data-driven site characterization. *Georisk: Assessment and Management of Risk for Engineered Systems and Geohazards*, 16(1), 114–126.

Phoon, K. K., Shuku, T., Ching, J., and Yoshida, I. 2022c. Benchmark examples for data-driven site characterization. *Georisk: Assessment and Management of Risk for Engineered Systems and Geohazards*, 16(4), 599–621.

Phoon, K. K., Cao, Z., Ji, J., Leung, Y. F., Najjar, S., Shuku, T., Tang, C., Yin, Z. Y., Yoshida, I., and Ching, J. 2022d. Geotechnical uncertainty, modeling, and decision making. *Soils and Foundations*, 62(5), 101189.

Phoon, K. K., Cao, Z., Liu, Z., and Ching, J. 2023a. Report for ISSMGE TC309/TC304/TC222 Third ML dialogue on "Data-Driven Site Characterization (DDSC)". *Georisk: Assessment & Management of Risk for Engineered Systems & Geohazards*, 17(1), 227–238.

Phoon, K. K., Zhang, L. M., Cao, Z. J. 2023b. Editorial for Special Issue on "Machine learning and AI in geotechnics", *Georisk: Assessment and Management of Risk for Engineered Systems and Geohazards*, 17(1), 1–6.

Phoon, K. K., Shuku, T., Ching, J., and Yoshida, I. 2023c. Editorial for special collection on "Benchmarking data-driven site characterization methods", *ASCE-ASME Journal of Risk and Uncertainty in Engineering Systems: Part A*, 9(2), 02023001.

Phoon, K. K., Ching, J., and Tao, Y. 2024. Soil and rock parametric uncertainties. Chapter 2 in *Uncertainty, Modelling, and Decision Making in Geotechnics*, CRC Press, Boca Raton.

Phoon, K. K., Cai, Y., and Tang, C. 2025. *Geotechnical "facial recognition" challenge. ASCE-ASME Journal of Risk and Uncertainty in Engineering Systems: Part A.*

Prakoso, W. A. 2017. Personal communication.

Prakoso, W. A. 2002. Reliability-based design of foundations on rock for transmission line and similar structure. PhD Thesis, Cornell University, New York.

Qi, X. H., Li, D. Q., Phoon, K. K., Cao, Z., and Tang, X. S. 2016. Simulation of geologic uncertainty using coupled Markov chain. *Engineering Geology*, 207, 129–140.

Raiffa, H. and R. Schlaifer. 1961. *Applied Statistical Decision Theory.* MIT Press, Cambridge, MA.

Retief, J. V., Dithinde, M., and Phoon, K. K. 2016. General principles on reliability according to ISO2394. Chapter 2 in *Reliability of Geotechnical Structures in ISO2394*, CRC Press, Balkema, pp. 33–48.

Rowley, J. 2007. The wisdom hierarchy: representations of the DIKW hierarchy. *Journal of Information and Communication Science*, 33(2), 163–180.

Salgado, R., Houlsby, G. T., and Cathie, D. N. 2008. Contributions to Géotechnique 1948–2008: foundation engineering. *Géotechnique*, 58(5), 369–375.

Self, S.J., Entwisle, D.C., and Northmore, K.J. 2012. The structure and operation of the BGS National Geotechnical Properties Database. Version 2. British Geological Survey, Nottingham, England.

Sharma, A., Ching, J., and Phoon, K. K. 2022. A hierarchical Bayesian similarity measure for geotechnical site retrieval. *Journal of Engineering Mechanics, ASCE*, 148(10), 04022062.

Sharma, A., Ching, J., and Phoon, K. K. 2023. A spectral algorithm for quasi-regional geotechnical site clustering. *Computers and Geotechnics*, 161, 105624.

Shahri, A. A., Spross, J., Johansson, F., and Larsson, S. 2019. Landslide susceptibility hazard map in southwest Sweden using artificial neural network. *Catena*, 183, 104225.

Shi, C. and Wang, Y. 2021a. Nonparametric and data-driven interpolation of subsurface soil stratigraphy from limited data using multiple point statistics. *Canadian Geotechnical Journal*, 58(2), 261–280.

Shi, C. and Wang, Y. 2021b. Development of subsurface geological cross-section from limited site-specific boreholes and prior geological knowledge using iterative convolution XGBoost. *Journal of Geotechnical and Geoenvironmental Engineering*, 149(9), 04021082.

Shioi, A., Otake, Y., Yoshida, I., Muramatsu, S., and Ohno S. 2023. Data-driven approximation of geotechnical dynamics to an equivalent single-degree-of-freedom vibration system based on dynamic mode decomposition. *Georisk: Assessment and Management of Risk for Engineered Systems and Geohazards*, 17(1), 77–97.

Shuku, T. and Phoon, K. K. 2023a. Data-driven subsurface modeling using a Markov random field model. *Georisk: Assessment and Management of Risk for Engineered Systems and Geohazards*, 17(1), 41–63.

Shuku, T. and Phoon, K.K. 2023b. Comparison of data-driven site characterization methods through benchmarking: methodological and application aspects. *ASCE-ASME Journal of Risk and Uncertainty in Engineering Systems, Part A: Civil Engineering*, 9(2), 04023006.

Shuku, T., Phoon, K. K., and Yoshida, I. 2020. Trend estimation and layer boundary detection in depth-dependent soil data using sparse Bayesian lasso. *Computers and Geotechnics*, 128, 103845.

Shuku, T., Phoon, K. K., Ishii, M., Kumagai, T., Yokota, Y., and Date, K. 2023. A probabilistic transformation model between two rock mass properties: specific fracture energy and p-wave velocity. *Canadian Geotechnical Journal*, 60(8), 1161–1172.

Straub, D. 2014. Value of information analysis with structural reliability methods. *Structural Safety*, 49, 75–85.

Stuedlein, A. W., Kramer, S. L., Arduino, P., and Holtz, R. D. 2012. Geotechnical characterization and random field modeling of desiccated clay. *Journal of Geotechnical and Geoenvironmental Engineering, ASCE*, 138(11), 1301–1313.

Tan, X., Chen, W., Qin, C., Zhao, W., and Ye, W. 2023. Characterisation for spatial distribution of mining-induced stress through deep learning algorithm on SHM data. *Georisk: Assessment and Management of Risk for Engineered Systems and Geohazards*, 17(1), 217–226.

Tang, C. and Phoon, K. K. 2021. *Model Uncertainties in Foundation Design.* CRC Press, Boca Raton, FL.

Tang, C. and Phoon, K. K. 2025. *Databases for Data-Centric Geotechnics: Geotechnical Structures.* CRC Press, Boca Raton, FL.

Tang, C., Phoon, K. K., and Yuan, J. 2024. Variability of predictions in geotechnics. Chapter 5 in *Uncertainty, Modelling, and Decision Making in Geotechnics*, CRC Press, Boca Raton, FL, pp. 176–216.

Tang, W. H. 1986. Updating anomaly statistics-single anomaly case. *Structural Safety*, 4(2), 151–163.

Tang, W. H. and Halim, I. 1988. Updating anomaly statistics-multiple anomaly pieces. *Journal of Engineering Mechanics*, 114(6), 1091–1096.

Terzaghi, K. 1943. *Theoretical Soil Mechanics.* John Wiley & Sons, New York.

Terzaghi, K. and Peck, R. B. 1967. *Soil Mechanics in Engineering Practice.* John Wiley, New York.

Thiyagalingam, J., Shankar, M., Fox, G., and Hey, T. 2022. Scientific machine learning benchmarks. *Nature Review Physics*, 4, 413–420.

Tian, H. and Wang, Y. 2023. Data-driven and physics-informed Bayesian learning of spatiotemporally varying consolidation settlement from sparse site investigation and settlement monitoring data. *Computers and Geotechnics*, 157, 105328.

Tian, H., Wang, Y., and Phoon, K. K. 2024. Real-time fusion of multi-source monitoring data with geotechnical numerical model results using data-driven and physics-informed sparse dictionary learning. *Canadian Geotechnical Journal*, e-First https://doi.org/10.1139/cgj-2023-0457

Vanmarcke, E. H. 1977. Probabilistic modeling of soil profiles. *Journal of the Geotechnical Engineering Division*, 103(11), 1227–1246.

Vanmarcke, E. H. 1983. *Random Fields: Analysis and Synthesis*, The MIT Press, Cambridge, MA.

Vardanega, P. J. and Bolton, M. D. 2016. Design of geostructural systems. *ASCE-ASME Journal of Risk and Uncertainty in Engineering Systems, Part A: Civil Engineering*, 2(1), 04015017.

Ulmer, K.J., Zimmaro, P., Brandenberg, S.J., Stewart, J.P., Hudson, K.S., Stuedlein, A.W., Jana, A., Dadashiserej, A., Kramer, S.L., Cetin, K.O., Can, G., Ilgac, M., Franke, K.W., Moss, R.E.S., Bartlett, S.F., Hosseinali, M., Dacayanan, H., Kwak, D.Y., Stamatakos, J., Mukherjee, J., Salman, U., Ybarra, S., and Weaver, T. 2023. *Next-generation liquefaction database*, Version 2. Next-Generation Liquefaction Consortium.

Wang, Y., Huang, K., and Cao, Z. 2014. Bayesian identification of soil strata in London clay. *Géotechnique*, 64(3), 239–246.

Wang, X., Wang, H., Liang, R. Y., and Liu, Y. 2019. A semi-supervised clustering-based approach for stratification identification using borehole and cone penetration test data. *Engineering Geology*, 248, 102–116.

Wang, H., Zhang, L. M., Wang, L., He, J., and Luo, H. 2021a. An automated snow mapper powered by machine learning. *Remote Sensing*, 13(23), 4826.

Wang, H., Zhang, L. M., Luo, H., He, J., and Cheung, R. W. M. 2021b. AI-powered landslide susceptibility assessment in Hong Kong. *Engineering Geology*, 288, 106103.

Wang, Y., Hu, Y., and Phoon, K. K. 2021c. Non-parametric modelling and simulation of spatiotemporally varying geo-data. *Georisk: Assessment and Management of Risk for Engineered Systems and Geohazards*, 16(1), 77–97.

Wang, Y., Qi, X., Zhang, W., and Ching, J. 2022. Editorial for Special Issue on "Data analytics in geotechnical and geological engineering", *Georisk: Assessment and Management of Risk for Engineered Systems and Geohazards*, 16(1), 1.

Wackernagel, H. 2003. *Multivariate Geostatistics: An Introduction with Applications*. Springer, Berlin.

Wu, S., Ching, J., and Phoon, K. K. 2022. Quasi-site-specific soil property prediction using a cluster-based hierarchical Bayesian model. *Structural Safety*, 99, 102253.

Wu, S., Otake, Y., Mizutani, D., Liu, C., Asano, K., Sato, N., Saito, T., Baba, H., Fukunaga, Y., Higo, Y., Kamura, A., Kodama, S., Metoki, M., Nakamura, T., Nakazato, Y., Shioi, A., Takenobu, M., Tsukioka, K. and Yoshikawa, R. 2024. Future-proofing geotechnics workflows: accelerating problem-solving with large language models. *Georisk: Assessment and Management of Risk for Engineered Systems and Geohazards*. https://doi.org/10.1080/17499518.2024.2381026

Xiao, S., Zhang, J., Ye, J., and Zheng, J. 2021a. Establishing region-specific N–Vs relationships through hierarchical Bayesian modeling. *Engineering Geology*, 287, 106105.

Xiao, T., Zou, H. F., Yin, K. S., Du, Y., and Zhang, L. M. 2021b. Machine learning-enhanced soil classification by integrating borehole and CPTU data with noise filtering. *Bulletin of Engineering Geology and the Environment*, 80, 9157–9171.

Xie, J., Huang, J., Lu, J., Burton, G. J., Zeng, C., and Wang, Y. 2022. Development of two-dimensional ground models by combining geotechnical and geophysical data. *Engineering Geology*, 300, 106579.

Xu, J., Wang, Y., and Zhang, L. 2021. Interpolation of extremely sparse geo-data by data fusion and collaborative Bayesian compressive sampling. *Computers and Geotechnics*, 134, 104098.

Xu, J., Wang, Y., and Zhang, L. 2022. Fusion of geotechnical and geophysical data for 2D subsurface site characterization using multi-source Bayesian compressive sampling. *Canadian Geotechnical Journal*, 59(10), 1756–1773.

Xu, Y., Kohtz, S., Boakye, J., Gardoni, P., and Wang, P. 2023. Physics-informed machine learning for reliability and systems safety applications: State of the art and challenges. *Reliability Engineering & System Safety*, 230, 108900.

Yan, W., Yan, Y., Shen, P., and Zhou, W. H. 2023. A hybrid physical data informed DNN in axial displacement prediction of immersed tunnel joint. *Georisk: Assessment and Management of Risk for Engineered Systems and Geohazards*, 17(1), 169–180.

Yoshida, I., Tasaki, Y., Otake, Y., and Wu, S. 2018. Optimal sampling placement in a Gaussian random field based on Value of Information. *ASCE-ASME Journal of Risk and Uncertainty in Engineering Systems, Part A: Civil Engineering*, 4(3), 04018018.

Yoshida, I., Tomizawa, Y., and Otake, Y. 2021. Estimation of trend and random components of conditional random field using gaussian process regression. *Computers and Geotechnics*, 136, 104179.

Yuen, K. V., Ching, J., and Phoon, K. K. 2021. Editorial for special collection on "Bayesian learning methods for geotechnical data". *ASCE-ASME Journal of Risk and Uncertainty in Engineering Systems: Part A*, 7(1), 02020002.

Zapata, G.E. 2010. A national database of subgrade soil-water characteristic curves and selected soil properties for use with the MEPDG. *NCHRP Web-Only Document 153*. Transportation Research Board, Washington, DC.

Zhang, W. G. and Phoon, K. K. 2022. Editorial for Special Issue on "Advances and applications of deep learning and soft computing in geotechnical underground engineering". *Journal of Rock Mechanics and Geotechnical Engineering*, 14(3): 671–673.

Zhang, D. M., Zhou, Y., Phoon, K. K., and Huang, H. W. 2020. Multivariate probability distribution of Shanghai clay properties. *Engineering Geology*, 273, 105675.

Zhang, W. H., Lu, D. G., Qin, J. J., Thöns, S., and Faber, M. H. 2021. Value of information analysis in civil and infrastructure engineering: A review. *Journal of Infrastructure Preservation and Resilience*, 2, 16.

Zhang, W., Gu, X., Hong, L., Han, L., and Wang, L. 2023. Comprehensive review of machine learning in geotechnical reliability analysis: Algorithms, applications and further challenges. *Applied Soft Computing*, 136, 110066.

Zhou, Y. and Song, X. K. 2016. Regression analysis of networked data. *Biometrika*, 103(2), 287–301.

Zhou, Z., Yan, L., Zhang, J., and Yang, H. 2023. Real-time tunnel lining crack detection based on an improved You Only Look Once version X algorithm. *Georisk: Assessment and Management of Risk for Engineered Systems and Geohazards*, 17(1), 181–195.

Zou, H., Liu, S., Cai, G., Puppala, A. J., and Bheemasetti, T. V. 2017. Multivariate correlation analysis of seismic piezocone penetration (SCPTU) parameters and design properties of Jiangsu quaternary cohesive soils. *Engineering Geology*, 228, 11–38.

Chapter 2

Rock hydromechanical parameters for rock foundation design

A database approach

Pouyan Asem and Paolo Gardoni

Rock sockets are often constructed in weak rock masses. Understanding the hydromechanical properties of the rock mass is required for the successful design and performance of the rock sockets. This chapter establishes separate databases for the hydromechanical properties of rock masses, with particular attention to weak shales and mudstones. The databases include information on rock socket geometry, load–displacement response, and the rock mass properties and mechanical and transport properties of the rock. The chapter then reviews the available models for the prediction of (i) the axial resistance (compression or uplift), (ii) axial deformation, and (iii) permeability. The databases are used to evaluate the existing predictive models for hydromechanical properties of rock as related to the design of rock sockets. The load test databases are also used to evaluate the effect of rock mass variability on the reliability of the foundations in weak rock. Based on the results of the analysis presented, the applicability of the existing models and approaches to the study of socket behavior in weak rocks is discussed.

2.1 INTRODUCTION

Rock sockets are being increasingly used for the support of structures that carry heavy loads (Horvath and Kenney 1979; O'Neill and Reese 1999; Seidel and Collinwood 2001; Brown *et al.* 2010; Dai *et al.* 2016; AlKhafaji *et al.* 2020), and hence determination of the rock mass properties required for their design (e.g., Asem 2018; Asem *et al.* 2018; Asem *et al.* 2019a,b; Asem and Gardoni 2019a,b,c; Asem 2019a,b; Asem 2020a,b; Asem and Labuz 2020; Guevara-Lopez *et al.* 2020; Tabandeh *et al.* 2020; Asem *et al.* 2021a,b; Asem and Gardoni 2021a,b; Asem and Gardoni 2022) is of utmost interest. Rock-socketed foundations often interact with weak rocks (i.e., a rock mass with unconfined compressive strength, q_u, from ~0.5 MPa to ~30 MPa; Geological Strength Index, GSI ≤ 70 after Hoek and Brown 1997; or equivalently, rock mass rating, RMR ≤ 40). This is because the upper zone of most rock formations hosting foundations is altered by physical and chemical weathering, especially in tropical and humid areas (Goodman 1993). A weak rock mass consists of relatively intact rock blocks that are separated by joint sets (Hoek 1983; Jaeger *et al.* 2007). The behavior of foundations in weak rock, therefore, is complex due to the existence of the rock mass secondary structures and their influence on rock mass mechanical properties.

Some researchers have based their work on the results of rock socket load tests in weak rock (e.g., Teng 1962; Coates 1967; Rosenberg and Journeaux 1976; Horvath and Kenney 1979; Meigh and Wolski 1979; Williams 1980; Gupton and Logan 1984; Rowe and Armitage 1987; Carter and Kulhawy 1988; Toh *et al.* 1989; Kulhawy and Phoon 1993; Zhang and Einstein 1998; Miller 2003; Abu-Hejleh *et al.* 2003; Paikowsky *et al.*

 DOI: 10.1201/9781003441946-2

2010; Stark *et al.* 2013; Stark *et al.* 2017; Asem 2018; Asem *et al.* 2018; Baghdady 2018; Asem *et al.* 2019a,b; Asem 2019a,b; Asem and Gardoni 2019a,b,c; Tabandeh *et al.* 2020; Asem 2020a,b). These studies have led to the development of empirical design methods where the observed side and base resistances and the deformational properties of rock sockets have been related to the rock properties (e.g., unconfined compressive strength, q_u, and rock mass deformation modulus, E_m) and to the rock socket geometry (e.g., rock socket length, L, and rock socket diameter, B). These methods are sometimes site-specific (e.g., Williams 1980; Miller 2003; Abu-Hejleh *et al.* 2003).

Other researchers (e.g., Lam 1983; Haberfield 1987; Johston and Lam 1989; Kodikara 1989; Seidel 1993; Hassan 1994; Collingwood 2000; Seidel and Collingwood 2001; Haberfield and Lochaden 2018; Johnston 2020) have used a combination of theoretical and empirical methods for the analysis of the behavior of rock sockets. These semi-empirical methods are applicable to a broader range of field conditions because they are mechanistic where the axial behavior of drilled shafts is related to the fundamental variables that characterize the weak rock mechanical and engineering properties, the rock socket geometry and properties of mobilized shear surfaces on the rock socket sidewalls, and the boundary conditions at the base of rock sockets.

Another important consideration is the selection of the correct rock parameters, and hence the degree of excess pore pressure dissipation for the rock mass during loading should be determined. Excess pore water pressure is generated in a rock mass during the application of loads (Terzaghi *et al.* 1996). The average degree of consolidation (U) of weak rock, as controlled by excess pore water pressure dissipation, dictates whether a rock mass would behave under drained or undrained conditions. Therefore, hydromechanical properties such as rock permeability are also needed for proper analysis of rock behavior during the application of loads.

This chapter will discuss the use of large databases in the development and analysis of rock socket behavior, particularly those constructed in weak clay-based rock and will present a synthesis of the work of the authors on the subject at hand (e.g., Asem 2018; Asem *et al.* 2018; Asem and Gardoni 2019; Asem *et al.* 2019; Asem and Gardoni 2019a,b; Asem 2019a,b; Asem 2020a,b; Asem and Labuz 2020; Guevara-Lopez *et al.* 2020; Tabandeh *et al.* 2020; Asem *et al.* 2021a,b; Asem and Gardoni 2021a,b,c; Asem and Gardoni 2022). The databases include (i) an in situ side-resistance database, (ii) an in situ base-resistance database, and (iii) a permeability database from laboratory and in situ tests. The databases (i) and (ii) include information on the rock socket geometry, rock socket load–displacement response, and rock mass properties. The available methods for the design of rock sockets are discussed in detail to understand the range and variability in the rock hydromechanical properties involved in design processes. In particular, the databases (i) and (ii) are used to evaluate the existing predictive models for axial resistance and settlement. Once a clear picture of the required hydromechanical rock properties is presented, we use the available data from the literature to propose models for the prediction of rock hydromechanical properties.

2.2 IN SITU LOAD TESTS FOR DETERMINATION OF MECHANICAL ROCK PROPERTIES

2.2.1 Existing databases

The existing load test databases that are relevant to this study are summarized in Table 2.1. The following should be noted:

Table 2.1 Summary of the existing load test databases

Database	*Number of load tests in rocks*	*Location of load tests*	*Rock type*	*Unconfined compressive strength range, q_u (MPa)*	*Rock mass properties reported*	*Load–displacement relationship reported*	*Load test type*
FHWA deep foundation load test database (DFLTD)	Limited	The United States			No	No	
NCHRP report 507, Paikowsky *et al.* (2004)	83					No	
NCHRP report 651, Paikowsky *et al.* (2010)	122	The United States, the United Kingdom, Italy, Russia, Taiwan, Japan, Canada, Australia, Singapore, South Africa	Claystone, mudstone, sandstone, shale, till, chalk, diabase, limestone, mudstone, gypsum, granite, tuff	0.2–103	Yes	No	Tip resistance
Rosenberg and Journeaux (1976)	8		Interbedded shale and sandstone, shale, sandstone, limestone, and andesite	0.5–55	No	No	Side and tip resistances
Horvath and Kenney (1979)	76	Canada, the United States, the United Kingdom, Australia, Czechoslovakia, Germany, South Africa, New Zealand, and Switzerland	Shale, clayshale, mudstone, siltstone, sandstone, chalk, limestone, marl, diabase, basalt, schist, and slate	0.35–110	No	No	Side resistance
Williams (1980)		Australia	Siltstone	0.2–80	No	Yes	Side and tip resistances
Williams and Pells (1981)	71	Australia, Ireland, South Africa, the United States	Mudstone, marl, claystone, clayshale, shale, and sandstone	0.48–62	No	No	Side resistance
Rowe and Armitage (1984)	158	Australia, the United States, Canada, the United Kingdom, South Africa, Czechoslovakia, Ireland	Sandstone, mudstone, shale, claystone, siltstone, chalk, andesite, schist, limestone, and slate	0.412–110	Yes	No	Side resistance

(Continued)

Table 2.1 (Continued) Summary of the existing load test databases

Database	*Number of load tests in rocks*	*Location of load tests*	*Rock type*	*Unconfined compressive strength range, q_u (MPa)*	*Rock mass properties reported*	*Load–displacement relationship reported*	*Load test type*
Hassan (1994)	139	South Africa, the United States, Singapore, Ireland, Canada, Brazil, Australia, the United Kingdom, Hong Kong	Shale, diabase, claystone, marl, till, hard clay, hardpan, basalt, siltstone, mudstone, chalk, marl, and limestone	0.14–37.23	No	No	Side and tip resistances
Zhang and Einstein (1998), Zhang (1999)	39	Australia, the United States, South Africa, Italy, Singapore	Mudstone, shale and clayshale, gypsum, till, diabase, hardpan, sandstone, siltstone, marl, and limestone	0.52–55	No	No	Tip resistance
Collingwood (2000)	162	The United States, Australia, Canada, Ireland, the United Kingdom, South Africa, Singapore	Shale, clayshale, limestone, sandstone, mica schist, mudstone, chalk, hard clay, andesite, diabase, siltstone, marl, and granite	0.375–110	No	No	Side resistance
Stark *et al.* (2013)	54 (side) and 33 (tip)	The United States, South Africa, Australia, Ireland, Canada	Shale, mudstone, claystone, and mudstone	0.12–4.45	No	No	Side and tip resistances
Baghdady (2018)	93 (side) and 62 (tip)	The United States, South Africa, Australia, Ireland, Canada	Shale, mudstone, and claystone	0.12–6.6	No	No	Side and tip resistances

1. *Number of load tests*: The number of load tests in some databases is limited. For example, the database collected by Rosenberg and Journeaux (1976) only contains eight drilled shaft load tests. The accuracy of the estimation of model parameters is dependent on the number of observations (e.g., load tests) (Gardoni *et al.* 2002). The use of small databases for the prediction of model parameters could result in inaccurate predictions and large uncertainties in the estimated parameters. Some databases only focus on side or base resistance (Zhang and Einstein 1998) and not both.
2. *Availability of complete load–displacement information*: In the majority of the current databases (see Table 2.1), the information regarding the load–displacement response (i.e., side shear stress–displacement and base contact pressure–displacement, f_s–δ and q–δ relationships, respectively) is not reported. Therefore, such databases cannot be used for a proper study of the load–displacement mechanism of drilled shafts in weak rock mass, which is a fundamental aspect of drilled shaft behavior.
3. *Lack of basic rock mass mechanical properties*: In the majority of the existing drilled shaft load test databases (e.g., Abu-Hejleh *et al.* 2003; Stark *et al.* 2013; Stark *et al.* 2017; Baghdady 2018), only the unconfined compressive strength (q_u) for weak rock is reported. Moreover, the weathering condition of weak rock is not discussed and the engineering properties of the weak rock "mass" are not reported or estimated. Therefore, the actual problem involving the interaction of a drilled shaft with the weak rock "mass" is simplified to a problem where the rock mass is represented by the properties of intact rock (i.e., unconfined compressive strength, q_u, and intact rock deformation modulus, E_i).
4. *Range of rock compressive strength*: The range of rock unconfined compressive strength (q_u) in some of the existing databases is well beyond the upper bound for the compressive strength of weak rocks (i.e., 30 MPa) (e.g., see database of Zhang and Einstein 1998).
5. *Range of rock socket size*: The socket diameter (B) reported in the existing databases covers a wide range. However, reported diameters are not distributed uniformly. Therefore, the effect of socket diameter (i.e., size) cannot be properly investigated. For example, the data in Williams (1980) consist mostly of small diameter (B) sockets and the rock socket lengths (L) were usually short.

2.2.2 New side-resistance and base-resistance databases

Two new databases are summarized in the Supplemental Data Section and in other publications (Asem 2018; Asem *et al.* 2018; Asem and Gardoni 2019; Asem *et al.* 2019; Asem and Gardoni 2019a,b; Asem 2019a,b; Asem 2020a,b; Asem and Labuz 2020; Guevara-Lopez *et al.* 2019; Tabandeh *et al.* 2020; Asem *et al.* 2021a,b; Asem and Gardoni 2021a,b,c; Asem and Gardoni 2022). One is a side-resistance database, and the other is a base-resistance database. The rock types include weak shale, siltstone, claystone, mudstone, limestone, sandstone, tuff, and granite. The case histories where q_u was larger than 30 MPa (that is, the selected upper bound for weak rock after Deere and Miller 1966; Barton *et al.* 1978; Rowe and Armitage 1987; Kanji 2014) are not used in the development of the database mainly because the aim of development of the databases in this section is to study the performance of the available design models with particular attention to weak rocks. The main contributions of the new databases include reporting: (i) the load deformatiom properties for the rock sockets and (ii) the rock mass properties (e.g., deformation modulus E_m). Additionally, the load test databases only concentrate on the axial behavior of rock sockets in weak rocks. In the following, the side- and base-resistance databases are discussed.

2.2.2.1 Side-resistance database

The side-resistance database includes 317 rock socket load tests. The peak side resistance (f_{sp}) and shear stress–shear displacement (f_s–δ) relationship for each socket sidewall is directly measured using in situ instrumentation. Therefore, it was not necessary to assume a value of base resistance to obtain the f_s–δ relationships. The side-resistance database contains the following information:

1. *Load-transfer function (f_s–δ relationship) for rock socket sidewalls*: The f_s–δ relationships are obtained based on the measured rock socket load–displacement relationship and the load distribution versus depth measurements that are gathered using strain gages or load cells (see Figure 2.1). The displacements reported in f_s–δ relationships exclude the elastic deformation $\left(\frac{PL}{AE_c}\right)$ of the test shaft to give the net rock socket deformation where P is the axial load, L is the rock socket length, A is socket cross-sectional area, and E_c is Young's modulus of concrete. A typical f_s–δ relationship is shown in Figure 2.2b.
2. *Properties of intact rock*: The unconfined compressive strength (q_u) of the intact rock specimens is reported for all cases. The internal friction angle (ϕ_i), shear strength intercept (c_i), and intact rock deformation modulus (E_i) were not available for a majority of case histories.
3. *Properties of the rock mass:* The rock mass modulus is estimated using the method of Pells and Turner (1979) – that is essentially a collection of solutions based on the theory of elasticity for the calculation of socket settlement, based on the numerical integration of Mindlin's equation – as $E_m = 2\pi LK_{si}I$, where L is the rock socket length, I is the embedment influence factor (Pells and Turner 1979), and K_{si} is the initial shear stiffness (Figure 2.2b) that is obtained from load tests. Other properties of the rock mass are estimated using the same approach explained for the base-resistance database.
4. *Side resistance*: Figure 2.2b shows a typical f_s–δ relationship that illustrates the definitions of initial shear stiffness (K_{si}), peak shear stress (f_{sp}), and ultimate shear stress (f_{su}). The initial shear stiffness (K_{si}) is the initial slope of the f_s–δ relationship.
5. *Drilled shaft geometry and material*: The diameter for each rock socket (B) and the depths of embedment from the ground surface (D_{GS}) and from the top of rock formation (D_{TOR}) to the center of the shear profile are obtained based on the idealized site stratigraphy and socket reported dimensions. The length of the rock socket (L) is also provided for each case history. The method of construction is summarized when reported. The shaft concrete compressive strength (f'_c), concrete modulus of elasticity (E_c), and concrete slump are summarized when they are available.

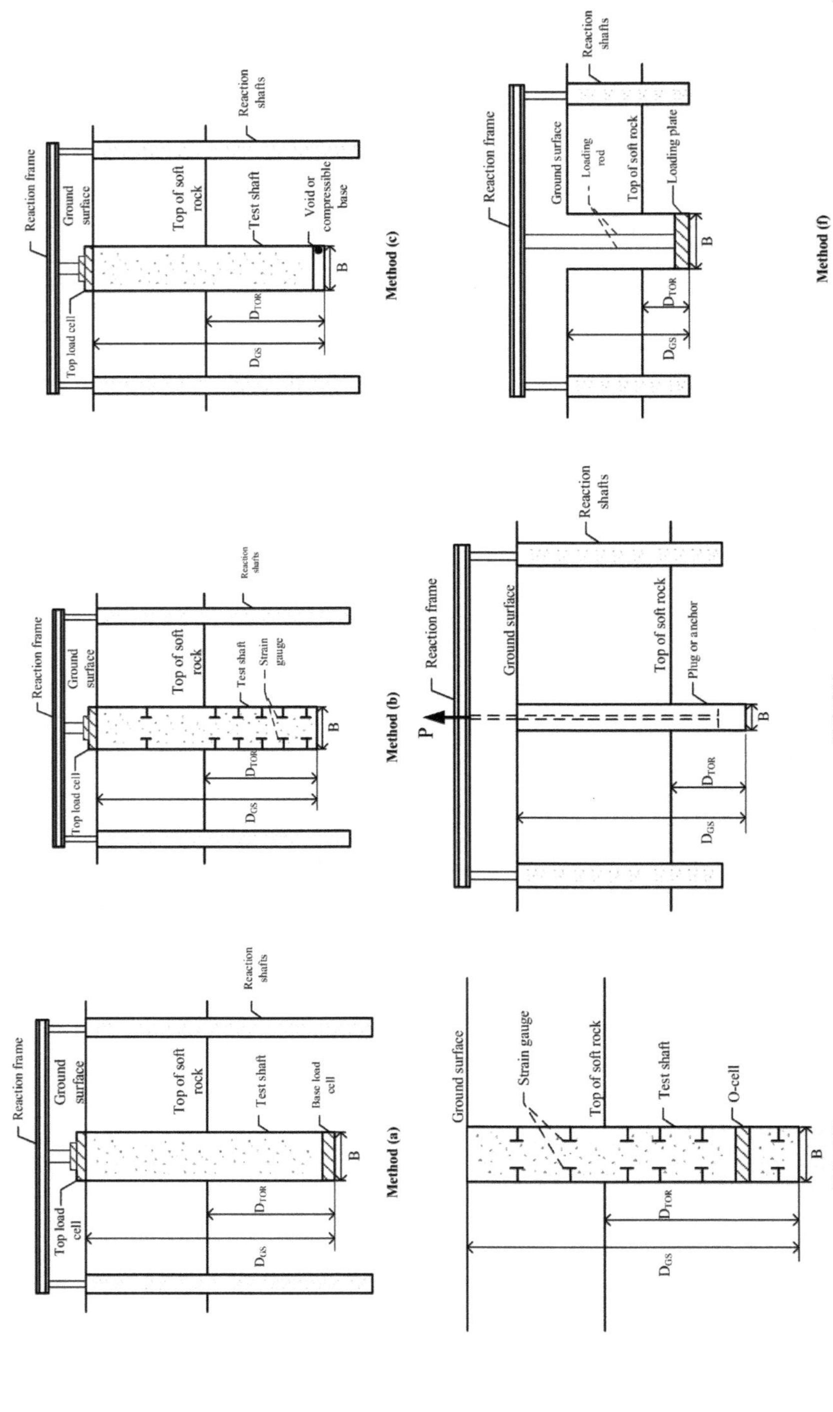

Figure 2.1 Load test methods used in the in situ test database.

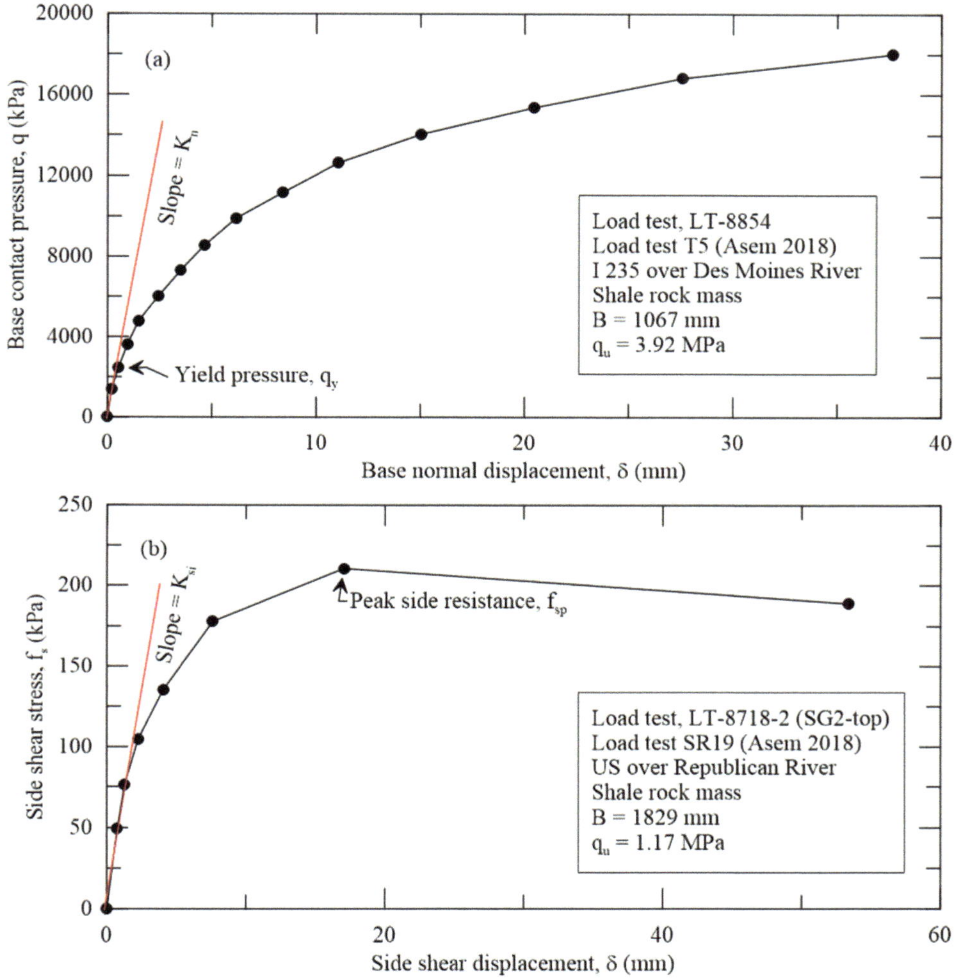

Figure 2.2 Typical side resistance and base displacement relationships for rock sockets in weak rock.

2.2.2.2 Base-resistance database

Figure 2.1 shows the load test methods that are commonly used to measure the base resistance of rock sockets. The base-resistance database includes 190 axial load tests (socket and plate load tests) and contains the following information:

1. *Load-transfer function (q–δ relationship) for socket base*: The q–δ relationships (e.g., Figure 2.2a) are developed based on the top of rock socket load–displacement relationship and load distribution measured by strain gages or tell-tales in conventional top-loaded load tests or the load–displacement relationship obtained from the O-Cell in Osterberg load tests. The elastic compression of the test shaft $\left(\frac{PL}{AE_c}\right)$ is considered and excluded in the determination of deformation of rock mass under the rock socket base in top-loaded load tests.

2. *Properties of the intact rock*: The database contains rock socket case histories that include the measured unconfined compressive strength (q_u) of weak rock specimens obtained from rock cores extracted from each load test site. The unconfined compressive strength (q_u) of the intact rock specimens is reported for all cases. The friction angle (ϕ_i), shear strength intercept (c_i), and modulus of deformation (E_i) for intact rock are not reported in the majority of cases.
3. *Properties of the rock mass*: The modulus of deformation of the rock mass (E_m) is estimated by the authors from the load test results in each case history using the method of Pells and Turner (1979) where the rock mass deformation modulus (E_m) is related to the initial normal stiffness of the rock socket base (K_n) as $E_m = 0.5\pi BK_n$ (B is the rock socket diameter and I is depth influence factor and is based on the results of finite element models Pells and Turner 1979). For design purposes, E_m can be estimated using a model of E_m as a function of q_u (Rowe and Armitage 1984; Asem 2018). This model is developed using linear regression analysis of (i) the results of in situ plate load tests reported by Chern *et al.* (2004) for weak rocks in China and Taiwan, (ii) back-analysis of q–δ relationships from 190 rock socket load tests from the United States, Puerto Rico, Canada, Australia, South Africa, Italy, the United Kingdom, and Singapore, and (iii) back-analysis of 340 f_s–δ relationships from 292 rock socket load tests in weak rock as discussed in subsequent sections. This linear model may be expressed as:

$$E_m = 150q_u^{1.1} \tag{2.1}$$

where E_m and q_u are in units of MPa. The predicted values of E_m and the method of Hoek and Diederichs (2006) – $E_m = \dfrac{100{,}000}{1+e^{[75\text{-}GSI]}}$ – are used to infer the Geological Strength Index (GSI). Rock mass Mohr–Coulomb strength parameters (Hoek and Brown 1997; Labuz and Zang 2012), namely the drained friction angle (ϕ_m) and failure envelope shear strength intercept (c_m) for rock mass are estimated based on (i) the inferred values of GSI as explained above, (ii) approximate material constant (m_i) that relates to the type of rock and frictional characteristics of rock minerals based on values reported in Hoek (1983) and Marinos and Hoek (2001) and (iii) the method of Hoek and Brown (1997) for estimation of ϕ_m and c_m. The predicted values of ϕ_m and c_m are not sensitive to m_i (Marinos and Hoek 2001), and thus estimation of m_i using qualitative description of rock will not significantly affect the predicted values of ϕ_m and c_m, especially for sedimentary and metamorphic rocks in the side-resistance database.
4. *Base resistance*: The q–δ relationships (Figure 2.2a) usually exhibit a strain hardening response and a failure point is not apparent. Therefore, the L1–L2 approach (Hirany 1988) is used to interpret two failure points, i.e., the values of q_y (i.e., contact pressure at an initial yield that is analogous to Hirany's $q1$ stress) and q_f (i.e., the interpreted failure pressure that is analogous to Hirany's $q2$ stress for Terzaghi 1943 failure pressure). When a distinct failure point was reported, it is included in the database.
5. *Initial normal stiffness* (K_n): The initial normal stiffness (K_n) is defined as the slope of the tangent line to the initial portion of q–δ relationship as shown in Figure 2.2.
6. *Rock socket geometry*: As shown in Figure 2.1, the base diameter for each rock socket (B) and the depths of embedment of the rock socket base from the ground

surface (D_{GS}) and from the top of rock formation (D_{TOR}) are obtained based on the reported site stratigraphy and rock socket reported dimensions. The method of construction and the condition of rock socket base are summarized when reported.

2.3 REVIEW OF SIDE RESISTANCE METHODS

Rock sockets carry their loads in side resistance, base resistance, or combination of both (Horvath and Kenney 1979; Rowe and Armitage 1987; Hassan *et al.* 1997; Zhang and Einstein 1998; Seidel and Collingwood 2001; CGS 2006; Turner 2006; Asem 2018; Haberfield and Lochaden 2018). Osterberg and Gill (1973) and Rowe and Armitage (1987) using finite element analysis and Goodman (1980; 1989) and Seo and Prezzi (2008) using solutions based on the theory of elasticity showed that the portion of the load transferred to the rock socket base is only a small percentage of the total applied load when the ratio of socket length to diameter, L/B, is ≥ 4. Therefore, rock sockets with $L/B \geq 4$ and especially those socketed in stiff rock masses (i.e., $E_m/E_c \geq 5$) carry most of their service load using side resistance (Seidel and Collingwood 2001; Seo and Prezzi 2008). Additionally, the design of foundations for base resistance requires inspection procedures that guarantee the cleanness of the rock socket base, which has been proven to be expensive and difficult, especially for the case of deep sockets in weathered rock mass (Seidel and Collingwood 2001; Brown *et al.* 2010). The influence of the construction procedure and the detrimental effects of soft material near the rock socket base are discussed in the literature (e.g., Holden 1984; Turner 2006; Brown *et al.* 2010; Haberfield and Lochaden 2018). Therefore, significant effort has been devoted to the development of predictive models for side resistance. Design methods for side resistance are commonly empirical and have been developed using load test databases that have a limited number of tests in a wide range of rock types with a broad range of rock compressive strengths. Exceptions include those of Rowe and Armitage (1984), Stark *et al.* (2013), and Asem (2018). Some methods, however, are based on laboratory constant normal stress and constant normal stiffness (CNS) direct shear tests on synthetic rock/concrete interfaces where bonding between rock and concrete is prevented (e.g., Seidel 1993; Collingwood 2000; Seidel and Collingwood 2001). These methods commonly relate the measured peak shear stress that is mobilized on the rock/concrete interface to the properties of the rock (e.g., the unconfined compressive strength of intact rock, q_u, and rock mass deformation modulus, E_m) and the properties of the rock/concrete interface such as interface roughness height (h).

2.3.1 Empirical side resistance models

Rosenberg and Journeaux's (1976) study is among one of the first models for peak side resistance. It is based on a total of eight rock socket load tests where intact rock compressive strength for four cases is reported. The load tests were conducted on drilled shafts in sandstone, shale, limestone, and andesite where 0.5 MPa $< q_u <$ 34 MPa. Rosenberg and Journeaux (1976) did not provide a mathematical expression for their proposed method. Kulhawy *et al.* (2005) provided a best-fit equation for the method of Rosenberg and Journeaux (1976) as:

$$\frac{f_{sp}}{\sigma_p} = 1.1\left(\frac{q_u}{\sigma_p}\right)^{0.56} \tag{2.2}$$

where f_{sp} is the peak shear stress and σ_p is the atmospheric pressure (σ_p = 0.101 MPa). In the method of Rosenberg and Journeaux (1976), (i) some of the load tests did not mobilize their peak side shear stress, (ii) the effect of rock type on side resistance is not evaluated, (iii) the effect of construction and shaft geometry on the development of shear stress is not studied, and (iv) the effect of rock mass properties on side resistance is not considered. The method of Rosenberg and Journeaux (1976) is evaluated using the new load test database in Figure 2.3a.

The method of Horvath (1978) is based on the results of six rock socket load tests in Queenston Shale Formation where (i) all test sockets are constructed at one test site near Ontario, Canada, (ii) the shale is horizontally bedded and $4.7 < q_u < 11.1$ MPa, and (iii) B = 710 mm. The effect of socket geometry and embedment depth could not be investigated using the database of Horvath (1978) as all rock sockets had the same diameter and the depth of embedment was 1.37 m in all cases. Horvath (1978; 1982), based on the

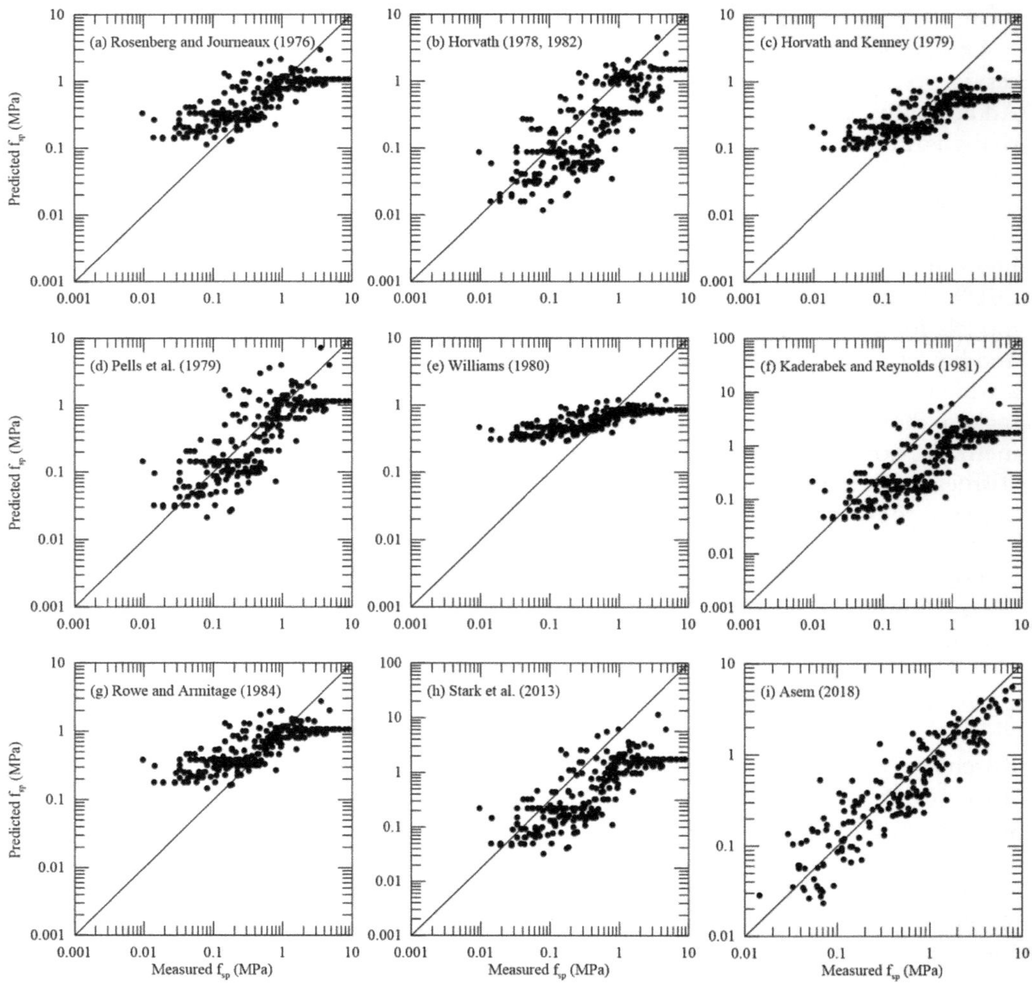

Figure 2.3 Evaluation of the empirical side resistance models using the in situ load test database (Source: data from Asem 2018).

measurement of socket wall roughness, developed a roughness factor (RF) that is used to account for the effect of the socket roughness on the peak side resistance (f_{sp}). The predictive models for RF and f_{sp} are as follows:

$$\mathrm{RF} = \frac{h}{r}\frac{l_t}{L} \tag{2.3}$$

and

$$f_{sp} = 0.8 q_u \mathrm{RF}^{0.45} \tag{2.4}$$

where h represents the average height of asperities that is measured with respect to the surface of the largest imaginary cylinder that can fit inside the rock socket, l_t is the total travel distance along the socket wall that accounts for the irregular texture of the rock socket sidewalls (e.g., see Seidel and Collingwood 2001 for additional discussion on characterization of sidewall roughness), r is the socket radius, and L is the socket length. The roughness of the interface (h) can be evaluated using the following expression (Asem 2018) that was obtained based on rock socket roughness measurements compiled by Collingwood (2000) (see Figure 2.4):

$$h = \frac{50}{q_u + 6.53} \tag{2.5}$$

where h is in mm and q_u is in MPa. The average socket sidewall roughness height (h) is inversely related to the unconfined compressive strength (q_u) because a greater q_u indicates a greater resistance to abrasion, and thus cutting tools are expected to create a smoother sidewall. The method of Horvath is evaluated in Figure 2.3. The results of recent research (e.g., Lam 1983; Haberfield 1987; Johnston and Lam 1989; Kodikara 1989; Seidel 1993; Hassan 1994; Collingwood 2000; Haberfield and Lochaden 2018; Johnston 2020) show that drilled shaft/rock "interface" roughness dominates the behavior of rock sockets as

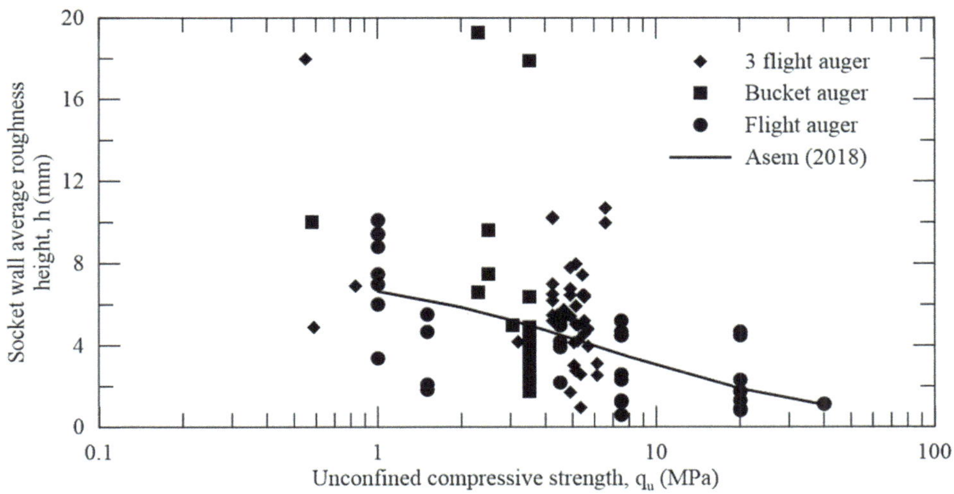

Figure 2.4 Predictive model for the drilled shaft/rock interface roughness (Source: data from Collingwood 2000; Asem 2018).

Table 2.2 Summary of the existing empirical methods for side resistance of rock sockets

Reference	*Equation*	*Comments*
Horvath and Kenney (1979)	$f_{sp} = a\sqrt{q_u(\text{MPa})}$	(i) Large- and small-scale drilled shafts and rock anchors, (ii) rocks include shale, sandstone, limestone, chalk, and igneous and metamorphic rocks, (iii) $0.35 < q_u < 110$ MPa, and (iv) $0.2 < a < 0.25$ (large diameter) and $0.25 < a < 0.33$ (small diameter), where parameter a is an empirical factor to account for the socket sidewall roughness and is obtained by fitting $f_{sp} = a\sqrt{q_u(\text{MPa})}$ to f_{sp} and q_u data.
Pells *et al.* (1979)	$f_{sp} = \alpha q_u$	(i) Load tests in Hawkesbury sandstone in the Sydney, Australia, (ii) $6 < q_u < 40$ MPa, (iii) $75 < B < 710$ mm, (iv) $240 < D_{GS} < 1370$ mm, (v) $1 < h < 10$ mm, and (vi) $0.1 < \alpha$ (= f_{sp}/q_u) < 0.3, where α is the ratio of f_{sp} to q_u.
Williams (1980)	$\frac{f_{sp}}{\sigma_p} = 2.69\left(\frac{q_u}{\sigma_p}\right)^{0.28}$	(i) Weak Melbourne siltstone at Stanley Avenue, Middleborough Road, and Westgate Highway test sites, (ii) $0.4 < q_u < 80$ MPa, (iii) $0.335 < B < 1.35$ m, and (iv) $D_{GS} = D_{TOR} < 2$ m.
Kaderabek and Reynolds (1981)	$f_{sp} = 0.3q_u$	(i) Miami limestone, (ii) $q_u < 8.6$ MPa, and (iii) 50 < rock quality designation (RQD) < 80% in the upper part of the formation and usually is near 0 in the lowest portion of the formation.
Rowe and Armitage (1984)	$f_{sp} = 0.45\sqrt{q_u(\text{MPa})}$ for smooth and $f_{sp} = 0.6\sqrt{q_u(\text{MPa})}$ for rough rock sockets	(i) Sandstone, mudstone, shale, siltstone, chalk, and andesite, (ii) $0.55 < q_u < 32$ MPa, (iii) $0.1 < B < 1.12$ m, and (iv) $0.2 < D_{GS} < 10$ m.
Stark *et al.* (2013), Stark *et al.* (2017), and Baghdady (2018)	$f_{sp} = 0.3q_u$	(i) Stark *et al.* (2013) developed a database of 54 drilled shaft load tests mostly in Intermediate Geomaterial (IGMs) (i.e., shale, mudstone, siltstone, and claystone with $0.13 < q_u < 3.2$ MPa), (ii) $0.3 < B < 2$ m, and (iii) Stark *et al.* (2017) and Baghdady (2018) updated the method of Stark *et al.* (2013) using new load tests (e.g., Vu 2013) where $f_{sp} = 0.31q_u$.

it governs the development and maintenance of the normal stresses on the sidewalls of rock sockets. Field observations (Williams 1980), however, showed that the shear surface most likely forms within the rock mass and not at the shaft/rock interface, especially in weak rocks. Therefore, the shaft/rock interface roughness may not be as important as is implied from the laboratory tests conducted by previous investigators. Other investigators have proposed similar empirical models to that of Rosenberg and Journeaux (1976) for side resistance of rock sockets in weak rocks. These models are summarized in Table 2.2 and evaluated in Figure 2.3.

Asem (2018) proposed two alternative models that account for the rock socket geometry and rock mass properties. The model in Asem (2018) is written as:

$$f_{sp} = \frac{q_u^{0.22} E_m^{0.31}}{12.8L^{0.79}} \tag{2.6}$$

where L is the length of the rock socket. Alternatively, a more refined approach in Asem (2018) is based on the Mohr–Coulomb model and is written as:

$$f_{sp} = \sigma_{np} \tan \phi_m \tag{2.7}$$

where σ_{np} is the mobilized normal stress on the rock socket sidewalls at the instant of mobilization of f_{sp} and may be obtained as:

$$\sigma_{np} = \frac{0.15 E_m^{0.37}}{L^{0.77} B^{0.01}} \tag{2.8}$$

These models (for Asem 2018; Equation 2.7) are evaluated in Figure 2.3.

2.3.2 Semi-empirical side resistance models

Hassan (1994) and Hassan *et al.* (1997) proposed that rock/concrete interface has a sinusoidal roughness pattern when the rock socket is constructed with rock auger in clayshale and distinguished between smooth and rough sockets. Hassan (1994) proposed that for the case of rough sockets, the shear plane is within the weak rock material and thus the unit side resistance (f_{sp}) is equal to the average drained shear strength of the geomaterial. Hassan (1994) and Hassan *et al.* (1997) proposed the following expression for obtaining the peak side resistance (f_{sp}) of rock sockets with rough rock/concrete interfaces:

$$f_{sp} = c_m + \sigma_{no} \tan \phi_m \tag{2.9}$$

where it is assumed that c_m and ϕ_m are the drained shear strength intercept and friction angle of rock mass, and σ_{no} is the initial normal stress on the shear surface. For smooth rock/concrete interfaces, Hassan (1994) and Hassan *et al.* (1997) proposed that the peak unit side resistance (f_{sp}) is mobilized at very small displacements (i.e., local displacements of 5–10 mm). They also assumed that the shear surface for smooth interfaces is at the rock/concrete interface and suggested that f_{sp} in smooth sockets is governed by initial normal stress on the interface and interface friction angle. For smooth rock/concrete interface, Hassan (1994) suggested the following expression:

$$f_{sp} = \alpha q_u \tag{2.10}$$

where α is the adhesion factor written as:

$$\alpha = (5 - 8.8d)\left(\frac{q_u}{\sigma_p}\right)^{d-1} \frac{\tan \phi_{int}}{\tan 30^o} \tag{2.11}$$

in which ϕ_{int} is the interface friction angle of rock/concrete interface, and

$$d = \frac{15 - \dfrac{\sigma_{no}}{\sigma_p}}{27} \tag{2.12}$$

In relation to the effect of stresses on side resistance, Vesic (1963) indicated that side resistance is a "linear function of vertical stress at failure." Vesic (1963) further maintained that "this stress is not necessarily equal nor proportional to the overburden pressure." Contrary to the proposal of Hassan (1994), rock socket load test data show no

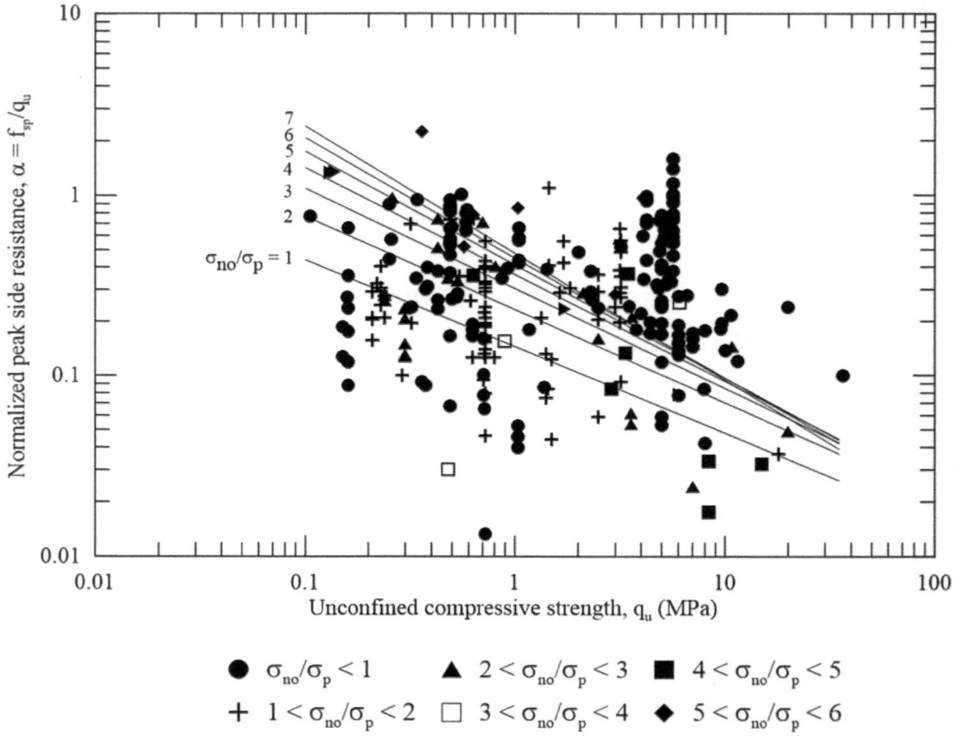

Figure 2.5 Evaluation of the model of Hassan (1994) and Hassan *et al.* (1997) using the in situ drilled shaft load test database (Source: data from Asem 2018).

relationship between peak side resistance and initial normal stress on the mobilized shear surface, especially when the rock socket sidewalls are relatively rough and large dilations are expected to take place as a result. The method of Hassan (1994) is evaluated in Figure 2.5, which shows that data do not provide a clear trend as suggested by the Hassan (1994) method.

Alternatively, the design model of Seidel and Collingwood (2001) is semi-empirical and is based on the analysis using the ROCKET computer program, which is based on the mechanistic models of Seidel (1993) and other researchers (e.g., Lam 1983; Haberfield 1987; Johnston and Lam 1989; Kodikara 1989; Seidel 1993; Hassan 1994; Collingwood 2000; Seidel and Collingwood 2001; Haberfield and Lochaden 2018). These mechanistic models were developed based on the observations of the behavior of natural and synthetic rock/concrete interfaces in drained constant normal stiffness (CNS) direct shear tests. It is noted that the bonding between the synthetic rock and the concrete was sometimes prevented in the preparation of the specimens (e.g., Seidel 1993). The method accounts for the effects of socket geometry (that is modeled using the normal stiffness, K_{ns}, Seidel and Collingwood 2001), q_u, the equivalent elastic properties of rock mass (i.e., modulus of deformation, E_m, and Poisson's ratio, ν), and the socket sidewall average roughness height (h). The method of Seidel and Collingwood (2001) is shown in Figure 2.6. The shaft resistance coefficient (SRC) in Figure 2.6 is computed as:

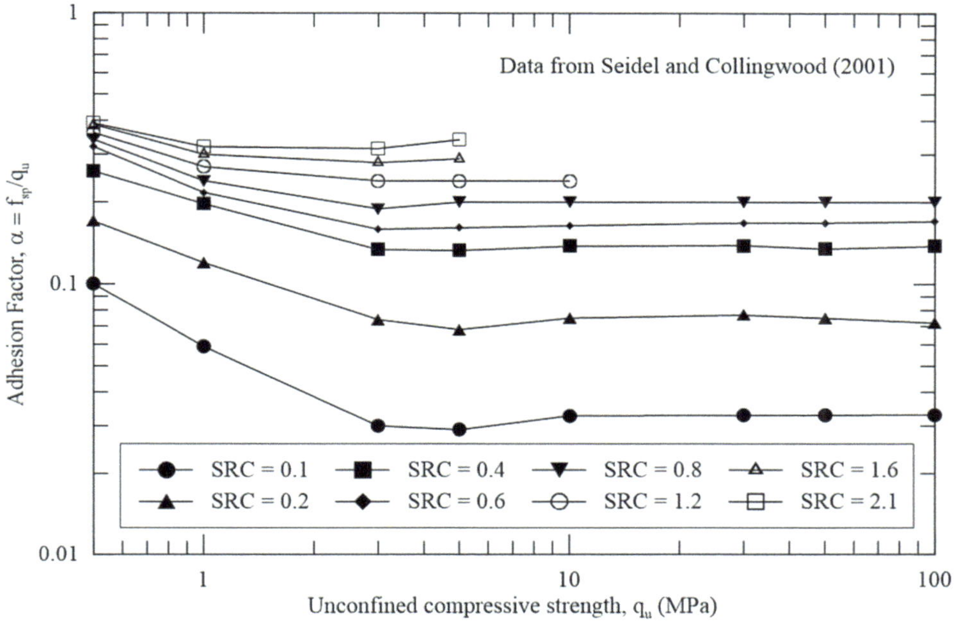

Figure 2.6 Design method of Seidel and Collingwood (2001) (Source: reproduced from Seidel and Collingwood 2001).

$$SRC = \eta \frac{\frac{E_m}{q_u}}{1+\nu} \frac{h}{B} \tag{2.13}$$

where η is the construction method reduction factor. One of the important parameters that is accounted for in this model is the roughness of rock/socket interface. The roughness model is based on idealizing the interface roughness by a series of interconnected cords. The assumption is made that the distribution of the cord angle is Gaussian, and thus the standard deviation of roughness height, cord angle, and roughness angle is related as:

$$\sigma_h = l_c \sin \sigma_i \tag{2.14}$$

where σ_h is the standard deviation of roughness height, l_c is the cord length, and σ_i is the standard deviation of roughness angle. The model of Seidel and Collingwood (2001) is evaluated using the in situ load test data in Figure 2.7, which shows a reasonable degree of agreement between measured and predicted values.

2.3.3 Discussion

The previous sections presented a review of the existing literature on the predictive models for the peak side resistance (f_{sp}) of rock sockets. For each model, we compute the bias (λ) defined as:

$$\lambda = \frac{M'}{P'} \tag{2.15}$$

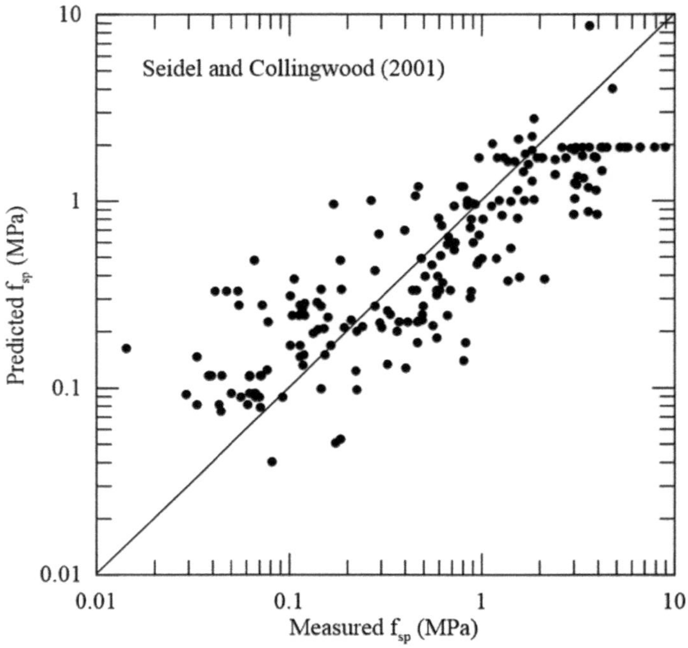

Figure 2.7 Evaluation of design method of Seidel and Collingwood (2001) using the in situ load test data (Source: data from Asem 2018).

where M' is the measured quantity of interest and P' is the predicted quantity of interest. The bias (λ) for the method of Rowe and Armitage (1984) is plotted versus various rock mass and rock socket properties in Figure 2.8 as an example that shows that λ varies with B, q_u, K_{ns}, the shear length (L) and ϕ_m that suggests $\lambda = f(B, q_u, K_{ns}, L)$. Further, because q_u is included in the method of Rowe and Armitage (1984), the variation of λ with q_u indicates that the model form is not appropriate. The existing models are also evaluated using other databases compiled by other researchers (Horvath and Kenney 1979; Rowe and Armitage 1984; Stark *et al.* 2013); the results are presented in Table 2.3. The analyses based on our new database are summarized in Table 2.4. The mean of bias (μ_λ) is:

$$\mu_\lambda = \frac{1}{n}\sum_{i=1}^{n}\left(\frac{M'}{P'}\right)_i \tag{2.16}$$

where n is the number of observations in each database. Similarly, the variance of $\frac{M'}{P'}$ (or other quantities of interest such as initial normal stiffness or yield pressure as defined later in this paper) is estimated using the following expression (Ang and Tang 2007):

$$\sigma_\lambda^2 = \frac{1}{n}\sum_{i=1}^{n}\left[\left(\frac{M'}{P'}\right)_i - \mu_\lambda\right]^2 \tag{2.17}$$

The square root of the variance is the standard deviation (σ_λ) of λ for each predictive model. The coefficient of variation (c.o.v.) for bias (δ_λ) is a measure of "dispersion" of the $\frac{M'}{P'}$ ratio about its mean which is indicative of the model accuracy:

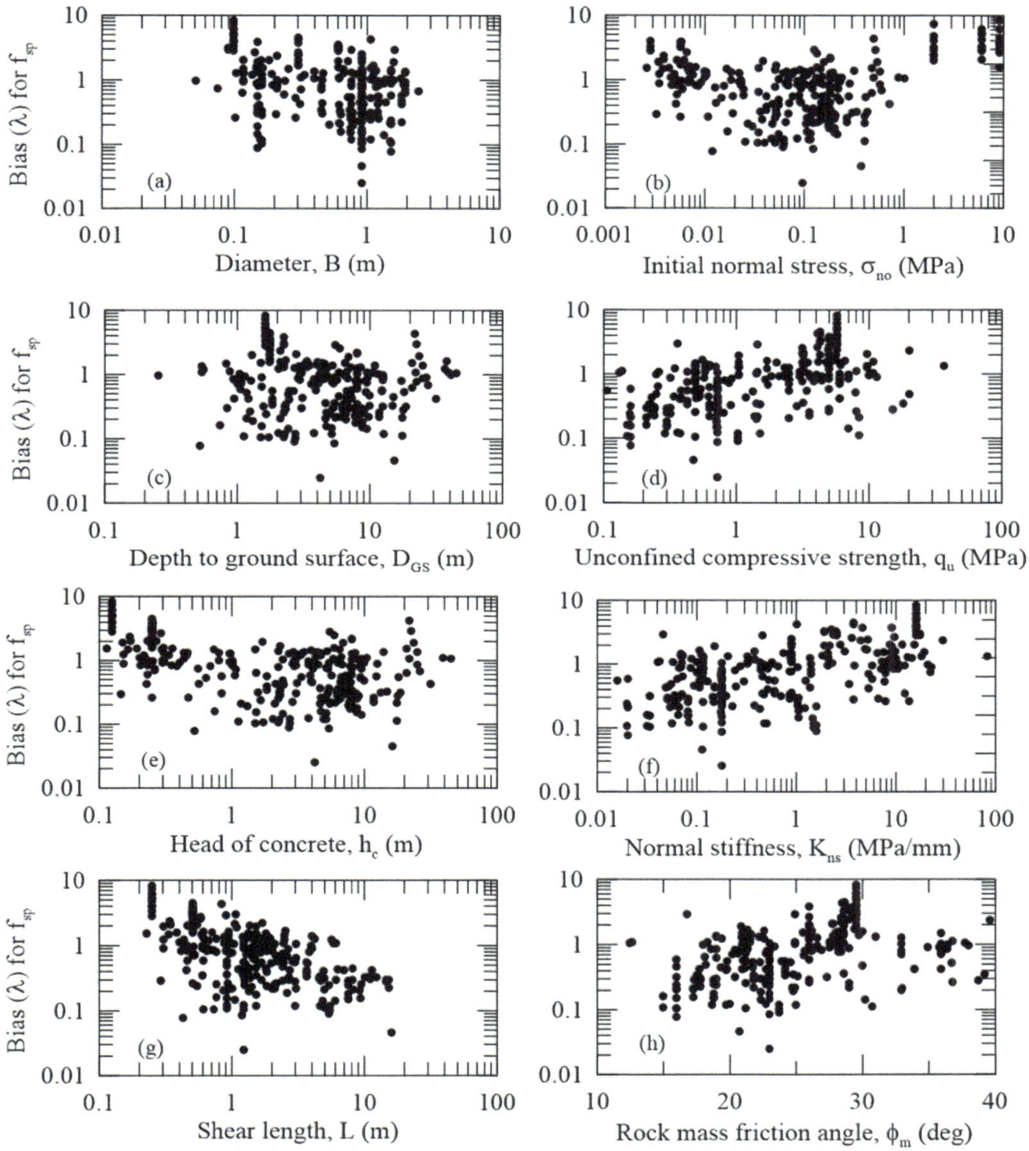

Figure 2.8 Evaluation of the variation of bias (λ) for the design method of Rowe and Armitage (1984) using the in situ load test data (Source: data from Asem 2018).

$$\delta_\lambda = \frac{\sigma_\lambda}{\mu_\lambda} \tag{2.18}$$

The data presented in Tables 2.3 and 2.4 suggest that most of the available models are biased and have high c.o.v.

Table 2.3 Evaluation of the side resistance predictive models using databases of Horvath and Kenney (1979), Rowe and Armitage (1984), and Stark *et al.* (2013)

		Databases		
Design methods		*Horvath and Kenney (1979)*	*Rowe and Armitage (1984)*	*Stark* ***et al.*** *(2013)*
Rosenberg and Journeaux (1976)	μ_λ	0.83	1.15	1.08
	σ_λ	0.45	0.58	0.44
	δ_λ	0.55	0.5	0.41
Horvath and Kenney (1979)	μ_λ	1.24	1.73	1.57
	σ_λ	0.68	0.87	0.65
	δ_λ	0.55	0.5	0.41
Pells *et al.* (1979)	μ_λ	0.89	1.07	1.78
	σ_λ	0.72	0.95	0.88
	δ_λ	0.8	0.89	0.5
Williams (1980)	μ_λ	0.89	1.3	0.96
	σ_λ	0.54	0.71	0.42
	δ_λ	0.6	0.54	0.44
Kaderabek and Reynolds (1981)	μ_λ	0.59	0.71	1.18
	σ_λ	0.48	0.63	0.59
	δ_λ	0.8	0.89	0.5
Rowe and Armitage (1984)	μ_λ	0.69	0.96	0.87
	σ_λ	0.38	0.48	0.36
	δ_λ	0.55	0.5	0.41
Carter and Kulhawy (1988)	μ_λ	1.54	2.16	1.96
	σ_λ	0.85	1.09	0.81
	δ_λ	0.55	0.5	0.41
Hassan (1994) for rough interfaces)	μ_λ	0.36	0.43	0.71
	σ_λ	0.29	0.38	0.35
	δ_λ	0.8	0.89	0.5
Miller (2003)	μ_λ	0.77	1.08	0.98
	σ_λ	0.43	0.54	0.4
	δ_λ	0.55	0.5	0.41
Kulhawy *et al.* (2005)	μ_λ	0.97	1.08	1.24
	σ_λ	0.54	0.54	0.51
	δ_λ	0.55	0.5	0.41
Stark, Long and Assem (2013)	μ_λ	0.59	0.71	1.18
	σ_λ	0.48	0.63	0.59
	δ_λ	0.8	0.89	0.5

2.4 BASE RESISTANCE METHODS

In weak rock masses, a significant portion of the axial load is carried in base resistance in short sockets (e.g., when $E_m/E_c < 0.02$; Seo and Prezzi 2008) after peak side resistance (f_{sp}) is mobilized and full-slip condition has materialized. Mobilization of the failure mechanism in base resistance requires a significant amount of displacement at the base

Table 2.4 Evaluation of the side resistance predictive models using new databases (this study) with 279 measurements at peak side shear stress

		Database (Asem 2018)			
Design methods		*35 > GSI*	*35 < GSI <50*	*50 < GSI <70*	*Entire database*
Rosenberg and Journeaux (1976)	μ_λ	0.77	2.41	1.12	1.40
	σ_λ	0.63	1.94	0.78	1.50
	δ_λ	0.82	0.81	0.69	1.07
Horvath and Kenney (1979)	μ_λ	1.11	3.60	1.71	2.07
	σ_λ	0.91	2.91	1.19	2.25
	δ_λ	0.82	0.81	0.70	1.09
Pells *et al.* (1979)	μ_λ	1.89	2.08	0.62	1.88
	σ_λ	1.55	1.54	0.42	1.54
	δ_λ	1.89	0.74	0.68	0.82
Williams (1980)	μ_λ	0.61	2.57	1.39	1.38
	σ_λ	0.52	2.12	0.99	1.65
	δ_λ	0.85	0.83	0.71	1.20
Kaderabek and Reynolds (1981)	μ_λ	1.26	1.38	0.41	1.25
	σ_λ	1.03	1.03	0.28	1.02
	δ_λ	0.82	0.74	0.68	0.82
Rowe and Armitage (1984)	μ_λ	0.62	2.00	0.95	1.15
	σ_λ	0.51	1.62	0.66	1.25
	δ_λ	0.82	0.81	0.70	1.09
Carter and Kulhawy (1988)	μ_λ	1.39	4.50	2.14	2.58
	σ_λ	1.14	3.64	1.49	2.81
	δ_λ	0.82	0.81	0.70	1.09
Hassan (1994) for rough interfaces)	μ_λ	0.75	0.83	0.25	0.75
	σ_λ	0.62	0.62	0.17	0.61
	δ_λ	0.82	0.74	0.68	0.82
Hassan (1994) for smooth interface)	μ_λ	3.35	9.47	6.47	5.55
	σ_λ	3.59	9.43	6.26	6.87
	δ_λ	1.07	1.00	0.97	1.24
Seidel and Collingwood (2001)	μ_λ	1.25	1.65	0.64	1.36
	σ_λ	1.11	1.14	0.58	1.13
	δ_λ	0.89	0.69	0.90	0.83
Kulhawy *et al.* (2005)	μ_λ	0.87	2.84	1.35	1.63
	σ_λ	0.72	2.30	0.94	1.78
	δ_λ	0.82	0.81	0.70	1.09
Stark *et al.* (2013), Stark *et al.* (2017), Baghdady (2018)	μ_λ	1.26	1.38	0.41	1.25
	σ_λ	1.03	1.03	0.28	1.02
	δ_λ	0.82	0.82	0.68	0.82

of the rock socket that is significantly greater than that required for mobilization of the peak side resistance (f_{sp}). The available design models are of empirical and semi-empirical nature.

Empirical models are based on in situ or laboratory load tests (e.g., Teng 1962; Coates 1967; Rowe and Armitage 1987; Carter and Kulhawy 1988; ARGEMA 1992; Zhang and Einstein 1998; CGS 2006; Stark *et al.* 2013; Stark *et al.* 2017; Baghdady 2018; Asem 2018; Johnston 2020). Due to the large displacement required for the development of the failure mechanism in the base resistance, most of the load tests that were used in the development of empirical models did not reach failure, and thus the maximum mobilized base contact pressure was often used to develop predictive models for base pressure (q) (e.g., Zhang and Einstein 1998). Therefore, the failure criteria vary from one predictive model to another. In many cases, the failure is based on serviceability requirements, where the resulting stresses correspond to different rock socket axial displacements. Thus, the resulting design models cannot be readily compared and the predicted base resistance using these models for one drilled shaft and a given in situ condition can vary by several orders of magnitude (Zhang and Einstein 1998). Other researchers have developed semi-empirical predictive models for the base resistance based on an assumed failure mechanism (e.g., Bishnoi 1968; Turner 2006) or by inconsistently combining an assumed failure mechanism and the results of in situ load tests (e.g., Stark *et al.* 2013; Stark *et al.* 2017; Baghdady 2018). Therefore, a clear understanding of the assumptions and failure criteria (i.e., definitions) in the existing methods is a prerequisite to their evaluation and use. Some of the suggested predictive models are shown in Figures 2.9 and 2.10 and are compared with q_y and q_f. Additionally, the assumed failure mechanism (e.g., general shear failure mode) is often not mobilized due to resistance to the amount of allowable deformation (Terzaghi 1943) or an increase in confining pressure that promotes a local type of failure. Such change in failure mode with change in confining pressure has been observed in the literature.

2.4.1 Empirical base resistance models

Teng (1962) discussed the base resistance of rock sockets in terms of allowable bearing pressure as:

$$q_{all} = \frac{1}{5} to \frac{1}{8} q_u \tag{2.19}$$

where q_{all} is the allowable unit base resistance. Teng (1962) pointed out that the evaluation of the effect of embedment on the base resistance becomes exceedingly difficult in weak rock because of the presence of joints, and bedding planes and fissures. Therefore, the effect of embedment depth is usually conservatively discounted. Additionally, it is unclear what displacement is required for mobilization of the base resistance proposed by Teng (1962). Other investigators have proposed similar models for predicting the bearing capacity of foundations on rocks. These models are reviewed in Table 2.5 and are evaluated in Figure 2.11.

2.4.2 Semi-empirical base resistance models

Coates (1967) developed a model based on Griffith's fracture theory (Griffith 1921; Hoek 1965), suggesting that microcracks in the rock mass migrate and expand when the stresses along their edges grow and become equal to the tensile strength of the material. Coates (1967) used a wedge analysis in combination with Griffith's fracture theory for the determination of the bearing capacity of strip foundations to write:

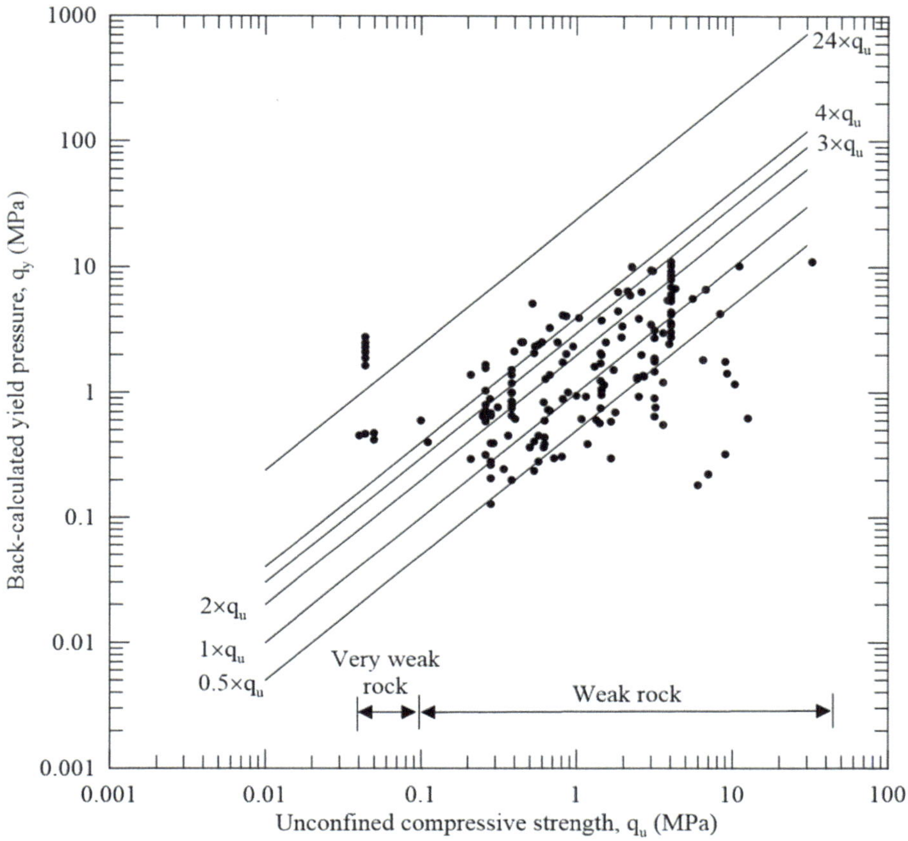

Figure 2.9 Variation of the available theories for base resistance (q) and a comparison with measured yield pressure (q_y) (Source: data from Asem 2018).

$$q = 3q_u \tag{2.20}$$

where q is the base resistance of a long foundation. The model of Coates (1967) is developed for hard rock and is based on the assumption that the strength of rock is mobilized along the entire failure surface at the same time. In reality, however, the failure in brittle materials (e.g., rock) would initiate from regions of stress concentration and subsequently propagate to less severely stressed zones (Hoek 1965; Coates 1967). The method of Coates (1967) assumes that the surcharge loads representing the embedment depth are small compared to the applied loads and thus do not contribute greatly to the bearing capacity and do not account for the effect of foundation depth on the bearing capacity of foundations. Other researchers (e.g., Zhang and Einstein 1998) have made similar conclusions about the tip resistance of sockets in weak rock and in sand (Vesic 1963).

The method of Carter and Kulhawy (1988) is based on the Hoek–Brown failure criterion for jointed rock mass and was developed for foundations placed on the surface of rock mass (Figure 2.12). It is written as:

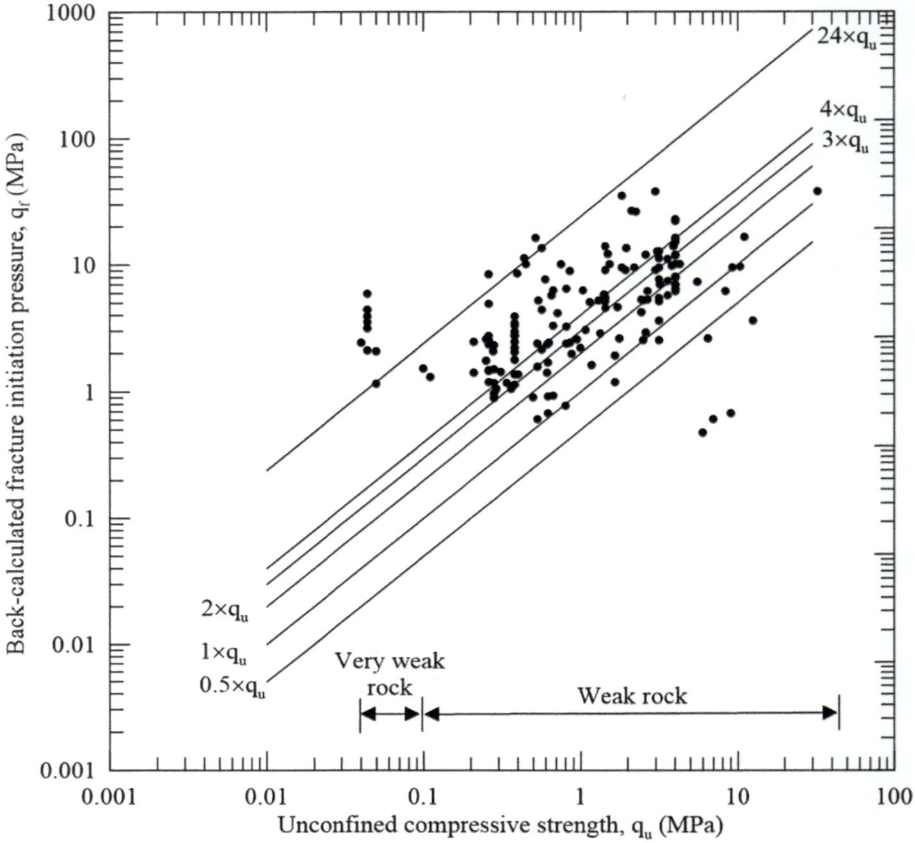

Figure 2.10 Variation of the available theories for base resistance (q) and a comparison with measured fracture initiation pressure (q_f) (Source: data from Asem 2018).

$$q = (m + \sqrt{s})q_u \tag{2.21}$$

where q is the lower bound estimate of base resistance for rock sockets, and "m" and "s" are the material constants from the Hoek–Brown (Hoek and Brown 1980, 2019) criterion. The method in Carter and Kulhawy (1988), however, is not based on rock socket load tests but may be classified as a semi-empirical method only due to material constants "m" and "s," which are determined empirically. As for the previous methods, it is unclear what displacements are required for mobilization of the base resistance proposed by Carter and Kulhawy (1988), and thus compatibility with side resistance cannot be checked. Other investigators have developed semi-empirical models which are summarized in Table 2.6. The base resistance models are evaluated in Table 2.7.

2.4.3 Relationships with rock mass properties

Peck *et al.* (1974) proposed a relationship between the allowable bearing pressure of foundations on unweathered and sound rock masses and the rock quality designation (RQD).

Table 2.5 Summary of empirical base resistance models

Reference	*Equation*	*Comments*
Rowe and Armitage (1987)	$q = 2.5q_u$	(i) Based on in situ load tests by Horvath (1980), Glos and Briggs (1983), and Williams (1980), (ii) $0.3 < B < 1$ m, (iii) $0.65 < q_u < 9.26$ MPa, and (iv) rock types in this database include siltstone, sandstone, and shalely sandstone.
Zhang and Einstein (1998)	$q = 4.8\sqrt{q_u(\text{MPa})}$	(i) Based on a database of 39 drilled shaft load tests in mudstone, shale, gypsum, till, diabase, hardpan, sandstone, siltstone, marl, and limestone, (ii) $0.3 < B < 1.92$ m, (iii) $1 < D_{GS} < 20.7$ m, (iv) $0.52 < q_u < 55$ MPa, (v) 6 load tests reported are model centrifuge tests (Leung and Ko 1993) on synthetic rock, and (vi) 3 load tests from Carrubba (1997) were not instrumented and base resistance had to be back-analyzed using elastic theory.
Stark *et al.* (2013), Stark *et al.* (2017), Baghdady (2018)	$q = 2.5q_ud_c$ depth factor (d_c) after Vesic (1973)	(i) Based on 33 drilled shaft load tests where the base stress–displacement relationships (q–δ relationships) were back-calculated from the results of drilled shaft load tests, (ii) rocks include shale and clayshale, claystone and siltstone, (iii) $0.34 < q_u < 4.2$ MPa, (iv) $0.3 < B < 1.82$ m, and (v) failure mechanism in the base resistance is developed at a displacement of 5% of the drilled shaft tip diameter.

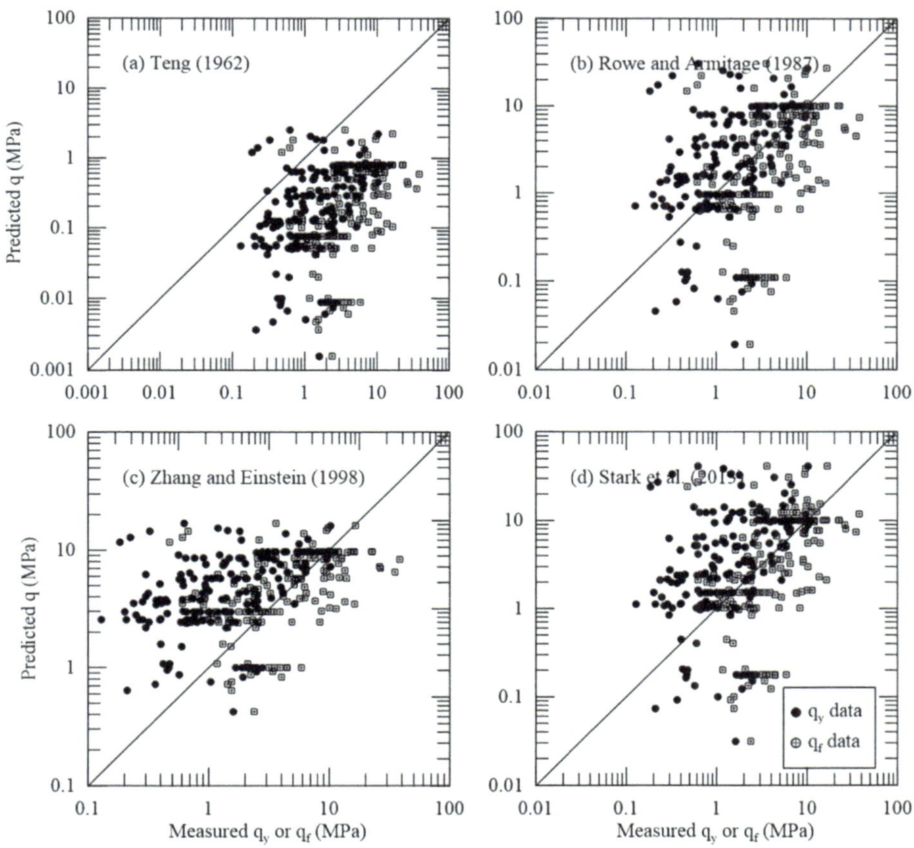

Figure 2.11 Evaluation of empirical models for base resistance with measured base resistance in in situ load tests.

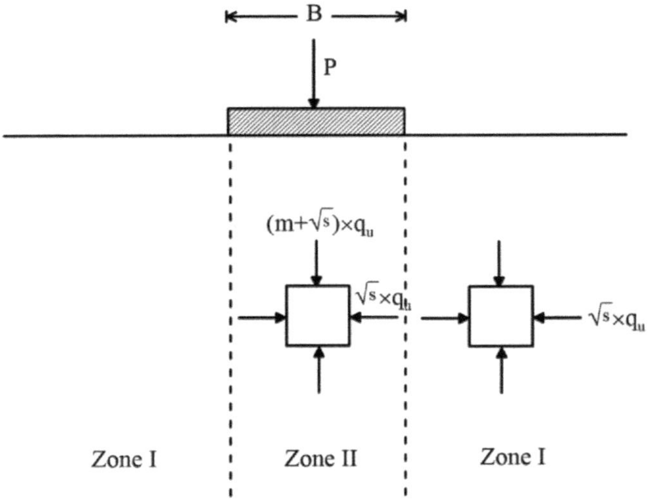

Figure 2.12 Method of Carter and Kulhawy (1988) for the end resistance of foundations on rock mass (Souce: reproduced from Carter and Kulhawy 1988).

Table 2.6 Summary of semi-empirical base resistance models

Reference	*Equation*	*Comments*
Canadian Geotechnical Society (2006)	$q = K_{sp}q_u$ $K_{sp} = \frac{3 + S_J / B}{10\sqrt{1 + 300\omega / S_J}}$	(i) q is the allowable bearing capacity of foundation, (ii) K_{sp} is an empirical coefficient that includes a factor of safety of 3.0 (K_{sp} that ranges from 0.1 to 0.4), and (iii) CGS defines the "favorable conditions" as conditions where "i) the rock surface is perpendicular to the foundation, ii) the foundation load has no tangential component and iii) the rock mass has no open discontinuities." $0.05 < S_J/B < 2$ and $0 < \omega/S_J < 0.02$.
Lee *et al.* (2013)	$q = 79.2\left(\frac{q_u S_J}{\sigma_p B}\right)^{0.315}\sigma_p$	(i) Parametric studies using numerical models and (ii) in situ load test data from South Korea

The use of this correlation (see Table 2.8 or Peck *et al.* 1974; Kulhawy and Goodman 1980) results in foundation settlements that are less than 13 mm (Peck *et al.* 1974; Kulhawy and Goodman 1980). GSI of rock mass for each load test is converted to RQD based on the recommendations of Hoek *et al.* (2013). The relationship proposed by Peck *et al.* (1974) is plotted in Figure 2.13. q_y and q_f from the base-resistance database case histories interpreted using the Terzaghi (1943) and Hirany (1988) methods and the estimated RQD values are also shown. The comparison of the data from 190 load tests (i.e., plate, O-cell, and rock socket load tests) with the proposed method of Peck *et al.* (1974) in Figure 2.13 shows that this method only provides crude estimates of q_f and q_y and that significant scatter exists around the mean correlation proposed by Peck *et al.* (1974). The data scatter in Figure 2.13 indicates that in addition to the RQD, that is mainly a measure of the joint frequency (Peck *et al.* 1974; Pells *et al.*, 2017) in the weak rock mass, other rock mass properties (i.e., rock mass friction angle ϕ_m) as well as joint and fissure alteration, socket geometry (i.e., diameter), and displacement to mobilize q_f may also be

Table 2.7 Evaluation of base resistance models using existing models

Design methods		Databases		
		Paikowsky et al. (2010)	*Stark et al. (2013)*	*Zhang and Einstein (1998)*
Teng (1962)	μ_λ	24.64	20.32	17.79
	σ_λ	24.98	10.97	10.86
	δ_λ	1.01	0.54	0.61
Coates (1967)	μ_λ	1.64	1.35	1.19
	σ_λ	1.67	0.73	0.72
	δ_λ	1.01	0.54	0.61
Rowe and Armitage (1987)	μ_λ	1.97	1.63	1.42
	σ_λ	2	0.88	0.87
	δ_λ	1.01	0.54	0.61
Carter and Kulhawy (1988)	μ_λ	23.46	19.35	16.94
	σ_λ	23.79	10.45	10.34
	δ_λ	1.01	0.54	0.61
ARGEMA (1992)	μ_λ	1.09	0.9	0.79
	σ_λ	1.11	0.49	0.48
	δ_λ	1.01	0.54	0.61
Zhang and Einstein (1998)	μ_λ	1.64	1.01	1.06
	σ_λ	2.82	0.44	0.34
	δ_λ	1.72	0.44	0.32
Canadian Foundation Engineering Manual (2006)	μ_λ	7.51		
	σ_λ	6.92		
	δ_λ	0.92		

Table 2.8 Relationship between allowable foundation pressure and rock quality designation (RQD) for unweathered rock masses

RQD (%)	*Allowable pressure (MPa)*
0	1
25	3
50	6.5
75	12
90	20
100	30

Source: After Peck *et al.*, 1974 or Peck, 1976.

important (Asem 2018; Tang *et al.* 2020). These parameters are missing in the Peck *et al.* (1974) predictive equation. Moreover, it must be noted that the method of Peck *et al.* (1974) was originally developed for "unweathered and sound rock masses" and should be used with caution with weak rocks.

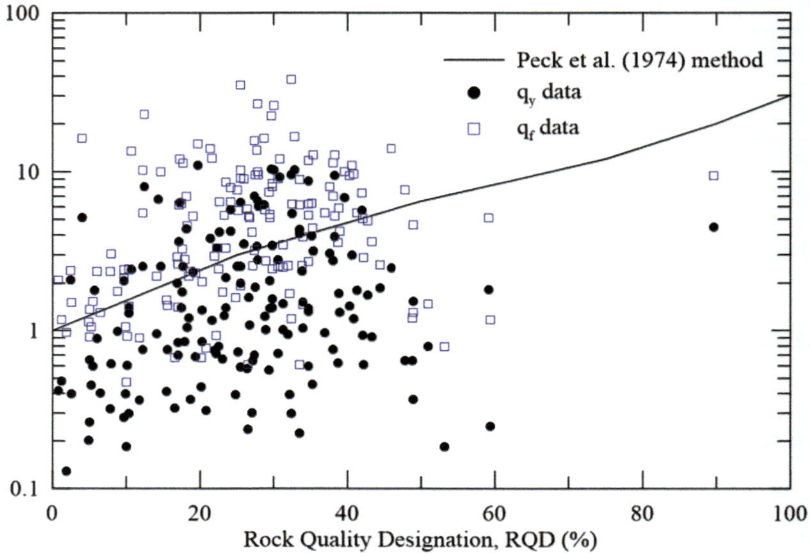

Figure 2.13 Peck *et al.* (1974) method for the prediction of base resistance based on rock quality designation (RQD)

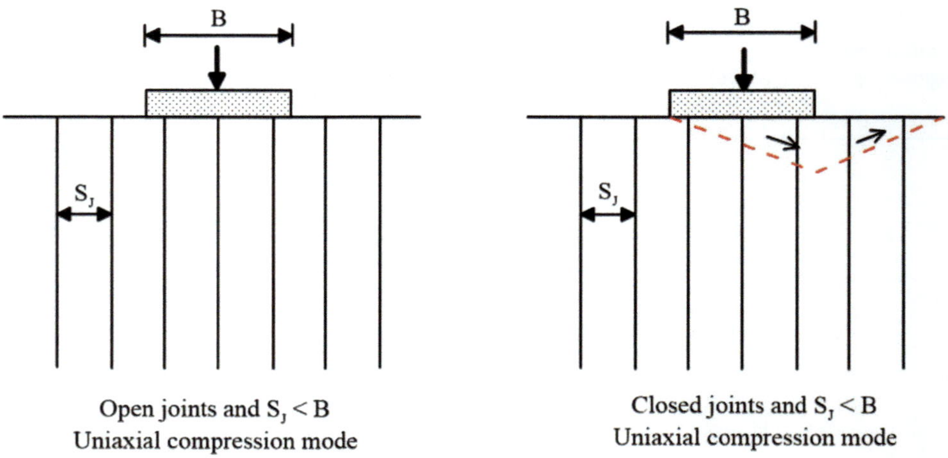

Figure 2.14 Failure modes proposed by Sowers (1979) for shallow foundations on rock masses (Source: reproduced from Kulhawy and Goodman, 1980).

Sowers (1979) solutions for the case of shallow foundations on rock masses for situations where joint spacing (S_J) is less than the foundation size (i.e., rock socket or in general foundation diameter, B) (see Figure 2.14) are discussed below. Accordingly, when rock joints are vertical, open, and $S_J < B$, the bearing capacity is obtained from the following expression:

$$q = q_u = 2c_m \tan\left(45 + \frac{\phi_m}{2}\right) \quad (2.22)$$

and when the joints are vertical, closed, and $S_J < B$, the bearing capacity is evaluated using the following expression:

$$q = c_m N_c s_c + \frac{B}{2}\gamma N_\gamma s_\gamma + \sigma_v N_q s_q \tag{2.23}$$

where all the material parameters in Equations (2.22) and (2.23) are that of the rock mass. N_c, N_γ, and N_q are the bearing capacity factors and s_c, s_γ, and s_q are the shape factors and may be obtained from Sowers (1979). The models proposed by Sowers (1979) are plotted in Figures 2.15 and 2.16 for different values of rock mass cohesion intercept (c_m) and friction angle (ϕ_m). The q_f values (obtained from q–δ relationships collected by Asem 2018) are also superposed on these figures. Examination of Figures 2.15 and 2.16 shows the following:

1. These models cannot properly represent the actual variation of q_f with the rock mass friction angle for sockets in weak rock masses.
2. The significant scatter in the plots of q_f versus friction angle indicates that other parameters in addition to friction angle and cohesion intercept must affect q_f that are missing in the above-proposed models (Asem 2018; Tang *et al.* 2020).

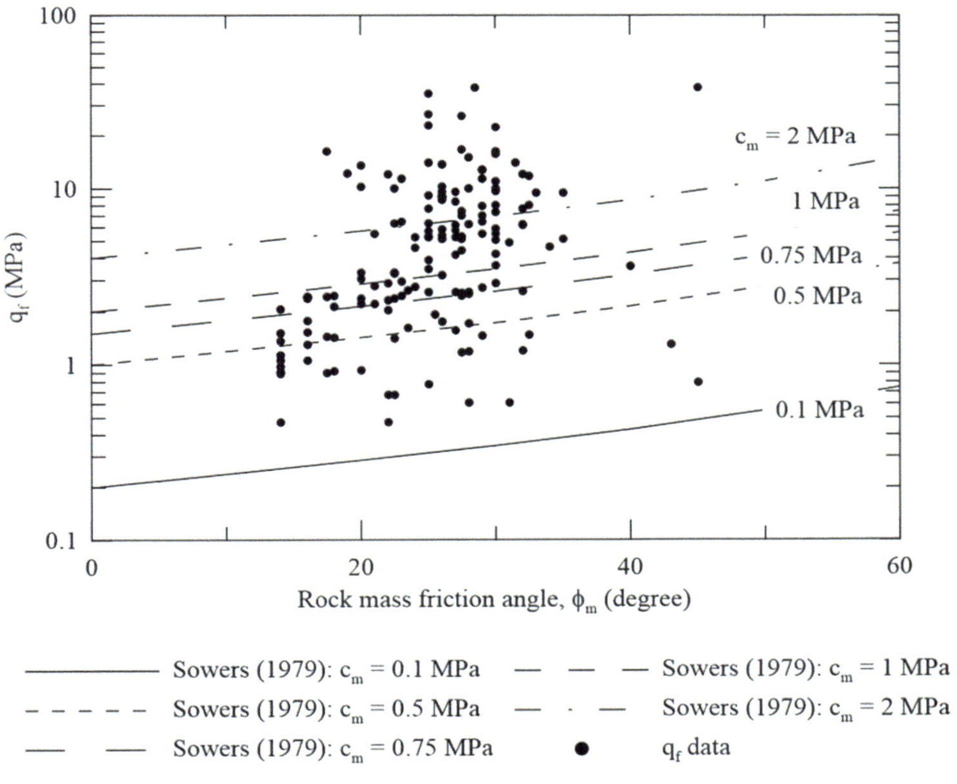

Figure 2.15 Comparison of the model of Sowers (1979) for foundations on rock masses with vertical open joints when $S_J < B$ with the qf data from the drilled shaft tip resistance database.

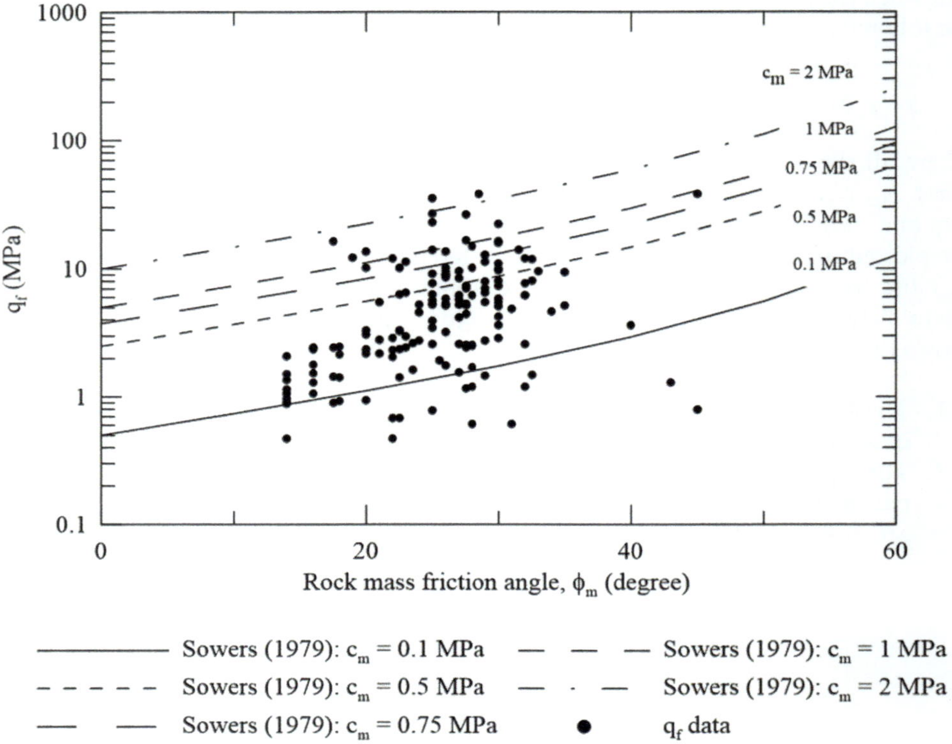

Figure 2.16 Comparison of the model of Sowers (1979) for foundations on rock masses with vertical closed joints when $S_J < B$ with the *qf* data from the drilled shaft tip resistance database.

3. The inability of Sowers (1979) methods to properly predict the base resistance in weak rocks can also be attributed to the assumed failure mechanism that is used to develop the above-proposed design equations, which is likely to be different from that mobilized in the field.

2.5 SETTLEMENT METHODS

2.5.1 Load-transfer methods

The serviceability limit state is important in the design of foundations on rock masses (Goodman 1980) and should be evaluated carefully. Some researchers developed predictive models based on the theory of elasticity and some used the load-transfer approach (Seed and Reese 1957) to develop nonlinear stress–displacement models for the side and the tip of drilled shafts. The models based on the theory of elasticity relate the settlement to rock socket geometry, rock mass properties, and socket concrete properties. In the load-transfer approach, the stiffness of the shear surface mobilized in the socket walls and the stiffness of the tip of rock socket may be modeled using nonlinear springs (i.e., f_s-δ and q–δ relationships). This model is robust because the effect of socket geometry, rock and concrete mechanical properties, local displacement, rock/concrete interface characteristics (e.g., socket wall roughness), and construction methods can be accounted for. The most important advantage of the load-transfer method is that it accounts for the nonlinearity of the rock behavior, which is an important aspect of the behavior of weathered and weak rocks that cannot be properly addressed when the theory of elasticity (e.g., Pells and Turner 1979; Kulhawy and Carter 1992) is used for evaluating the settlement of foundations in soft rocks unless such effects are reflected in rock elastic properties.

Williams (1980) developed a load-transfer function that is based on the results of field load tests in Melbourne siltstone where side and tip resistances were directly measured. The siltstone at the load test sites is commonly weathered to highly weathered, $0.335 < B < 1.35$ m and $D_{TOR} < 2$ m (the ground surface and the top of weak rock coincided in most of the load tests). Williams (1980) proposed a method for normalizing the f_s–δ and q–δ relationships (i.e., load-transfer functions for side and tip resistance, respectively) for sockets in weak rock. The normalization method, with particular attention to tip resistance, is shown in Figure 2.17 and is applied to 190 load tests in weak rock in Figure 2.18. The Williams (1980) method first predicts the load–displacement response of the foundation using an elastic method. Williams (1980) does not propose a method that could be specifically used for drilled shafts in weak rocks. Williams (1980), however, realizes that the behavior of actual drilled shafts in weathered rocks is highly nonlinear. To account for the effect of nonlinearity observed in the field for back-calculated f_s–δ and q–δ relationships, Williams (1980) defined plastic stress, f_p or q_p (for side and tip, respectively), at

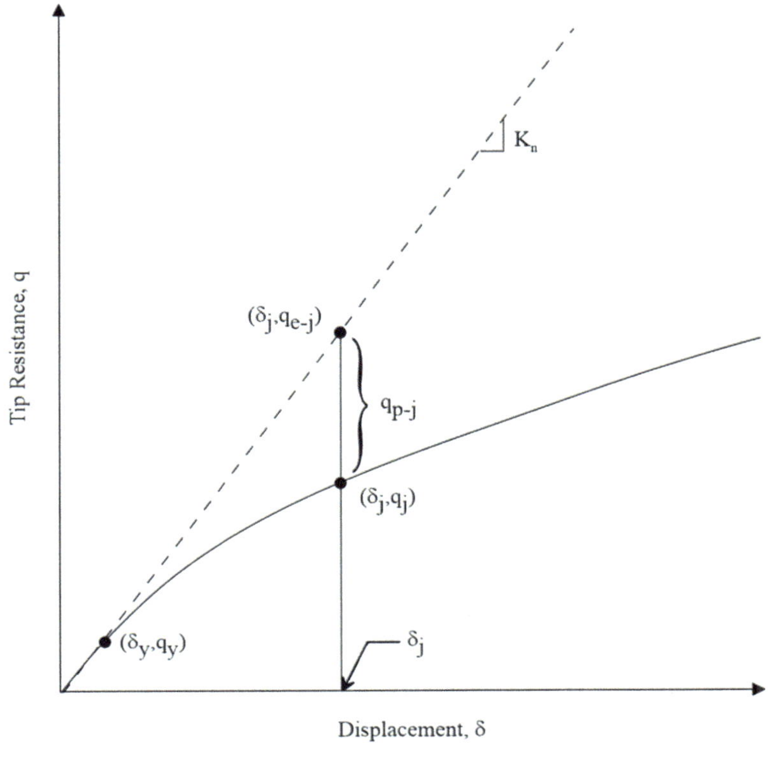

Figure 2.17 Normalization method developed by Williams (1980) (Source: after Williams, 1980, with modifications).

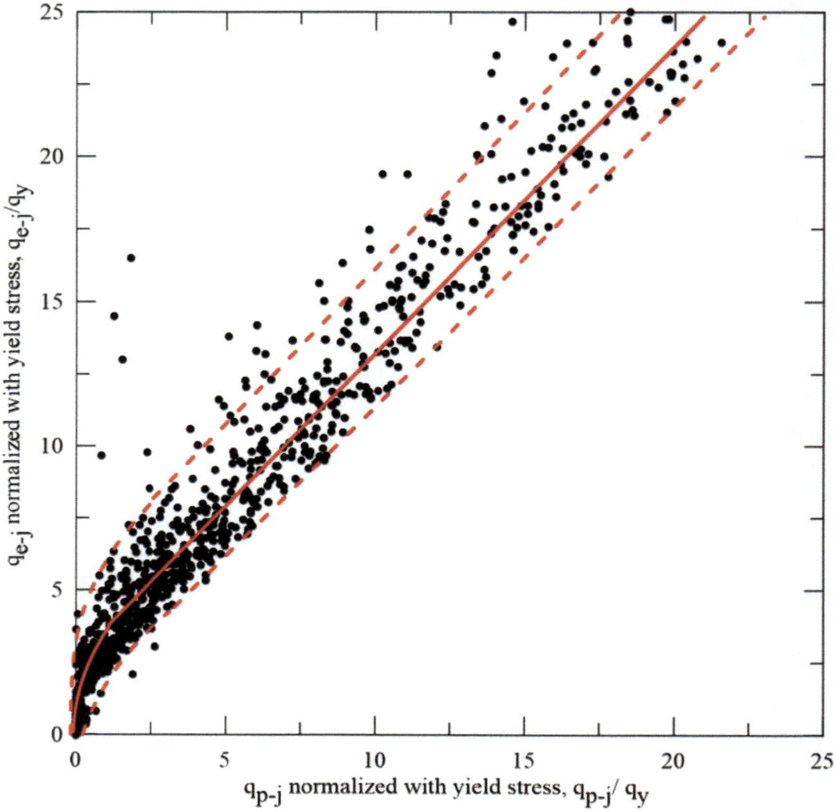

Figure 2.18 Empirical method of Williams (1980) (modified) for prediction of q_p for the purpose of determining nonlinear load-transfer functions for tip resistance (all data from the tip resistance database in Appendix A or Chapter 3 that is compiled from literature, normalization method after Williams, 1980, with modifications and updates).

each increment of axial displacement, which is described in mathematical forms for side and tip resistance in Equations (2.24) and (2.25), respectively as:

$$f_p = f_e - f_s \tag{2.24}$$

and

$$q_p = q_e - q \tag{2.25}$$

where f_e and q_e are the predicted values of side and tip stresses based on the elastic theory at any given value of local displacement along the side and near the tip of the foundation, respectively, and f and q are the corresponding values from the field "nonlinear" load-transfer functions for side and tip resistances, respectively. The q_e values in Figures 2.17 and 2.18 are normalized using the yield pressure (q_y) of the rock mass, which is an alternative to what was proposed by Williams (1980). A similar relationship may be developed for the side resistance f_s–δ relationships. The method of Williams (1980) for normalization of load-transfer functions of sockets in weak rock is a novel approach for the prediction of load-transfer functions. Although this method has been mostly applied to the investigation of load–displacement response of foundations in weak rock, it can be applied to hard rock masses as was done in some of the load tests in relatively hard sedimentary rocks by Williams (1980). Asem (2018) developed a family of equations that could be used to predict the deviation of the tip response from linear elastic behavior written as:

$$\frac{q_{p\text{-}j}}{q_y} = 0.007797\left(\frac{q_{e\text{-}j}}{q_y}\right)^3 + 0.05209\left(\frac{q_{e\text{-}j}}{q_y}\right)^2, \frac{q_{e\text{-}j}}{q_y} \leq 5 \tag{2.26}$$

and

$$\frac{q_{p\text{-}j}}{q_y} = 0.9436\left(\frac{q_{e\text{-}j}}{q_y}\right) - 2.467, \frac{q_{e\text{-}j}}{q_y} > 5 \tag{2.27}$$

The Hirayama (1990) method is based on the modeling of the load-transfer function for unit side resistance and tip resistance using a "two-constant" hyperbolic model written as:

$$f_s = \frac{\delta}{a_f + b_f\delta} \tag{2.28}$$

and

$$q = \frac{\delta}{a_q + b_q\delta} \tag{2.29}$$

where f_s and q are the unit side and base resistances of rock socket and a_f, b_f, a_q, and b_q are model parameters that define the shape of the hyperbolic function. Hirayama relates the model parameters to the "ultimate" side and base resistances of rock sockets. The peak side resistance of rock sockets in weak rock can be estimated using the available predictive models for the unit side resistance of sockets in weak rock. In the case of base resistance, however, the majority of the field load tests have not been carried to failure, and thus reliable methods for the prediction of failure stress in tip resistance of sockets in weak rock are not available. This is an important limitation of Hirayama's method. Hirayama (1990) does not account for the difference between peak side or ultimate base resistance predicted from the hyperbolic model and the actual measured values, and thus does not use a fitting ratio (Mesri *et al.* 1981; Terzaghi *et al.* 1996) (R_f) that has been used to account for this difference between the back-calculated and theoretical values. Other load-transfer models have been developed by other investigators that are summarized in Table 2.9.

2.5.2 Initial shear stiffness and initial normal stiffness models

2.5.2.1 Evaluation of models for initial shear stiffness in side resistance (K_{si})

The initial shear stiffness for rock joints has been studied extensively (e.g., Bandis 1980; Bandis *et al.* 1983; Yoshinaka and Yamabe 1986). The problem of determination of the

Table 2.9 Summary of load-transfer models

Reference	*Equation*	*Comments*
Hassan *et al.* (1997), side resistance	$f_s = \Theta_f f_{sp}$ For pre-peak and $f_s = k_f f_{sp}$ For post-peak	(i) k_f and Θ_f are correction factors that account for mobilization of unit side resistance and are functions of displacement, (ii) f_{sp} is the peak unit side resistance, and (iii) k_f and Θ_f could be obtained from Hassan *et al.* (1997)
Hassan *et al.* (1997), tip resistance	$q = \Lambda \delta^{0.67}$	Based on the results of finite element numerical models (FEM).
Gupta (2012)	$f_s = \dfrac{\frac{\delta}{B}}{\dfrac{\ln\left[\frac{5L(1-\nu)}{B}\right]}{2G_m} + \dfrac{\frac{\delta}{B}R_f}{f_{sp}}}$	Method does not provide any recommendations for the prediction of load-transfer function for tip resistance of drilled shafts.
Stark *et al.* (2013)	$q = \dfrac{3.2\frac{\delta}{B}}{\frac{\delta}{B}+1.3} q_u d_c \le 2.5 q_u d_c$	(i) Based on 33 drilled shaft load tests, and (ii) rock types in the database included shale and clayshale, claystone, and siltstone.
Lee *et al.* (2013)	$q = \dfrac{\delta}{\frac{1}{K_n} + \frac{\delta}{q_f}}$ where $K_n = 0.00037\left(\dfrac{E_m S_J}{\sigma_p B_{ref}}\right)^{0.4058}\left(\dfrac{\sigma_p}{B}\right)$ $q_f = 79.2\left(\dfrac{q_u S_J}{\sigma_p B}\right)^{0.315}\sigma_p$	(i) Based on the parametric studies that were performed using numerical models, (ii) model depends on δ, B, E_m, and the discontinuity spacing (S_j), and (iii) B_{ref} = 1.0 m and σ_p is the atmospheric pressure (0.101 MPa).

initial shear stiffness (K_{si}) in side resistance of sockets in weak rock, however, has not been explored in detail. Therefore, it is necessary that the existing models for the prediction of initial shear stiffness be evaluated against the back-calculated initial shear stiffness (K_{si}) from the results of rock socket load tests. One of the most widely used correlations is that of Randolph and Wroth (1978):

$$K_{si} = \frac{G_m}{r} \tag{2.30}$$

where K_{si} denotes the initial shear stiffness in side resistance of rock sockets, G_m is the shear modulus of geomaterial while accounting for the properties of the rock mass, and r is the radius of the rock socket. The shear modulus of rock mass (G_m) can be related to the modulus of deformation (E_m) of rock mass using the following equation (Goodman 1980):

$$G_m = \frac{E_m}{2(1+\nu)} \tag{2.31}$$

where ν is the Poisson's ratio. It must be noted that Equations (2.30) and (2.31) are based on the theory of elasticity and are only applicable to small axial displacements (i.e., displacements $\leq$0.5% of the socket diameter to ensure shear stresses on the socket sidewalls are significantly less than the peak shear strength of rock socket and hence to ensure socket would remain in the elastic range). Hirayama (1990) proposed an empirical design model for the estimation of initial shear stiffness (K_{si}) in side resistance of drilled shafts that is shown below:

$$K_{si} = \frac{f_{sp}}{0.0025B} \tag{2.32}$$

where f_{sp} is the peak side resistance and B is the diameter of rock socket. Hirayama (1990) recognizes that the value at which peak side resistance is developed can range from 0.1% to 0.5% of the rock socket diameter (B). The peak shear stress must be determined before this model can be used. This introduces additional uncertainty because current predictive methods for peak side resistance are not accurate as suggested by comparison of these methods with the new side resistance database.

These predictive models are evaluated herein in terms of the mean and coefficient of variation (c.o.v.) of the ratios of measured to predicted initial shear stiffness (μ_λ and δ_λ, respectively). It must be noted that the measured (i.e., back-calculated) initial shear stiffness values are obtained from the back-calculated f_s–δ relationships reported in the side-resistance database and following the procedure that is outlined in the preceeding sections of this chapter. The evaluation results are summarized in Table 2.10. The bias (i.e., ratio of the measured to predicted initial shear stiffness) is also plotted against different variables from the side-resistance database and is shown in Figures 2.19 and 2.20. The following observations can be made:

1. Table 2.10 shows the statistics of bias (λ) for initial shear stiffness (K_{si}). The statistics of bias are calculated and grouped based on the condition of soft rock mass in which the rock sockets have been constructed. These are rock masses with 50 < GSI < 70, 35 < GSI < 50, and GSI < 35.

Table 2.10 Evaluation of the predictive models for the initial shear stiffness (K_{si}) of drilled shafts in weak rock mass

		Database (Asem 2018)			
Design methods		*35 > GSI*	*35 < GSI <50*	*50 < GSI <70*	*Entire database*
Randolph and Wroth (1978)	μ_λ	2.70	0.84	0.63	1.62
	σ_λ	5.39	1.23	1.01	3.77
	δ_λ	1.99	1.46	1.61	2.33
Hirayama (1990)	μ_λ	2.14	0.92	2.13	1.6
	σ_λ	7.47	1.32	4.44	5.37
	δ_λ	3.49	1.43	2.08	3.36

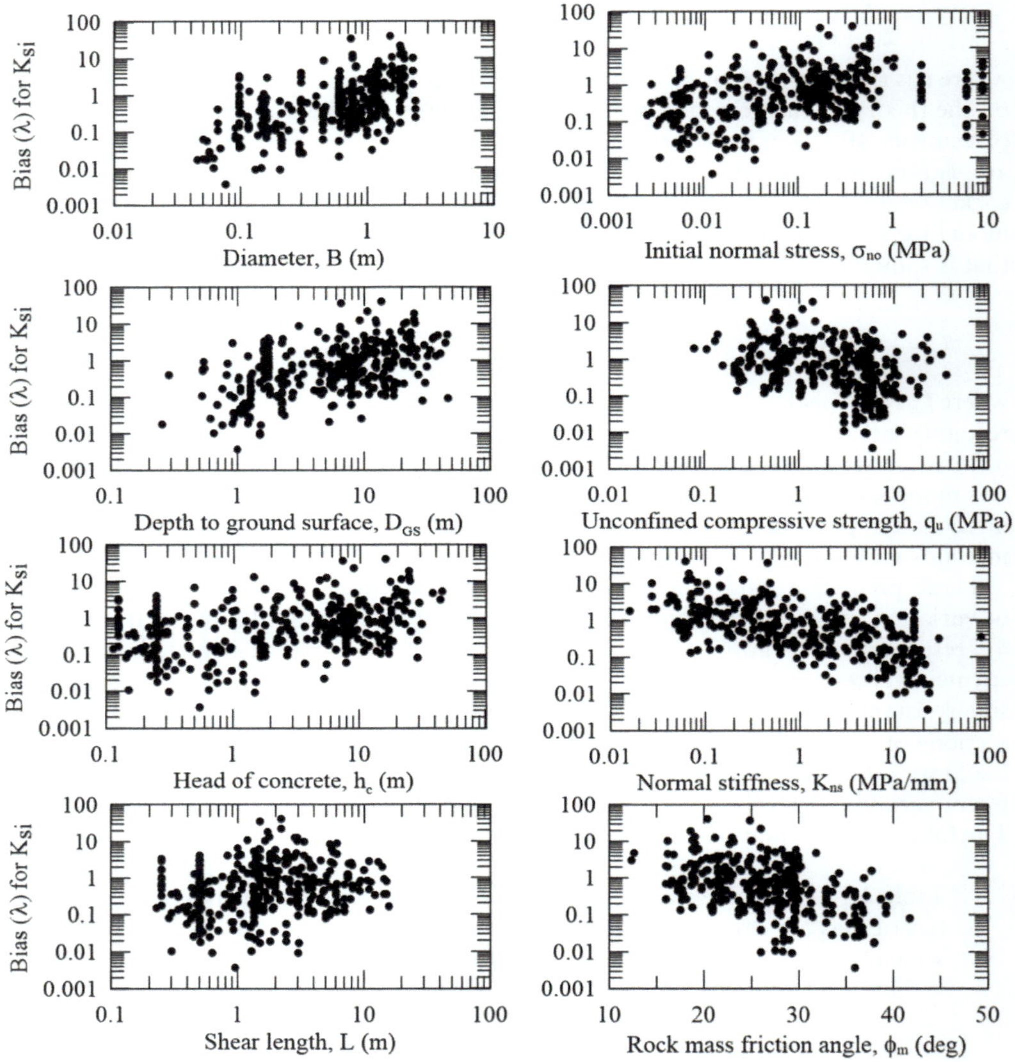

Figure 2.19 Variation of bias (λ) in the predictions of the method of Randolph and Wroth (1978) for initial shear stiffness (K_{si}) with properties of weak rock mass and drilled shaft geometry.

2. Table 2.10 shows that the method of Randolph and Wroth (1978) underestimates the initial shear stiffness (K_{si}) for the sockets in rocks with GSI smaller than 35. The analysis also shows that the variability in the predictions using the method of Randolph and Wroth (1978) is considerably larger for very weathered and jointed weak rock mass (i.e., GSI < 35) as compared to better quality rocks. The predictive method of Hirayama (1990) is unconservative for weak rock masses with GSI between 35 and 50. The variability of the predictions is also considerably larger for more weathered weak rocks as is the case for the method of Randolph and Wroth (1978).

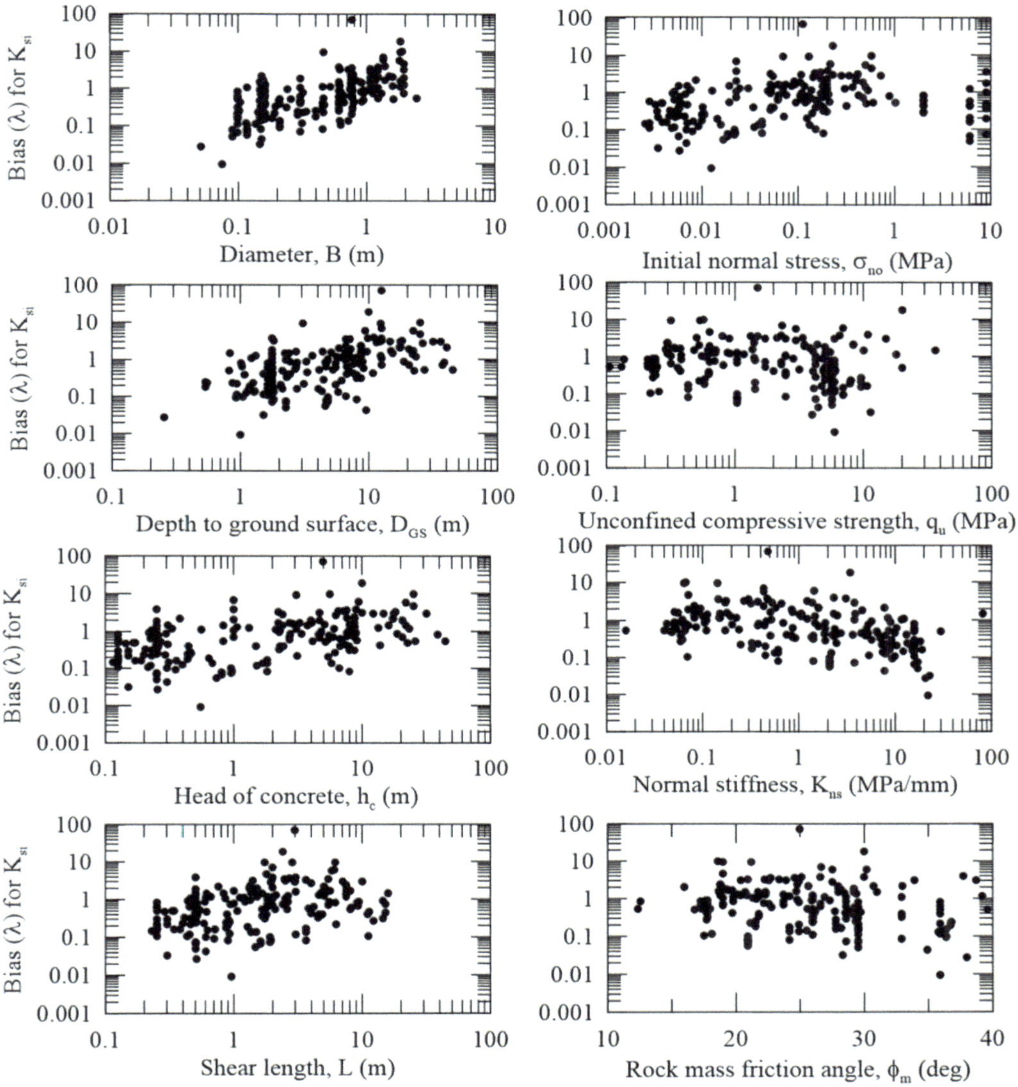

Figure 2.20 Variation of bias (λ) in the predictions of the method of Hirayama (1990) for initial shear stiffness (K_{si}) with properties of soft rock mass and drilled shaft geometry.

3. The variations of bias for the predictions of the methods of Randolph and Wroth (1978) and Hirayama (1990) with different properties of soft rock mass and socket geometry are shown in Figures 2.19 and 2.20. These figures show that the error of prediction of these methods for initial shear stiffness varies significantly with the diameter of the socket (B), depth of embedment (D_{GS}) (or equivalently the degree of confinement of the side of the drilled shaft), the normal stiffness of the surrounding rock mass (K_n), and the unconfined compressive strength of the soft rock (q_u).

4. The correlation between the bias in the predictions of the methods of Randolph and Wroth (1978) and Hirayama (1990) and weak rock mass properties and rock socket characteristics indicates that additional properties of the socket geometry and soft rock mass mechanical properties should be accounted for in the predictive models.
5. In addition to the possibility of the existence of missing parameters in the current predictive models, the strong correlation between the bias and weak rock properties and the drilled shaft geometry may result from model inexactness, meaning that the mathematical form of the current predictive models may not accurately describe the actual variation of initial shear stiffness with rock and drilled shaft properties.

2.5.2.2 Evaluation of models for initial normal stiffness in tip resistance (K_n)

Similar to the initial shear stiffness (K_{si}) of rock sockets, the initial normal stiffness (K_n) for the tip of the socket in soft rock mass has not been studied rigorously. Lee *et al.* (2013) developed a model for the prediction of the initial normal stiffness of socket tip in rock. Pells and Turner (1979), Goodman (1980), Jeong *et al.* (2010), and Lee and Jeong (2016) measured and developed models for the prediction of load–displacement behavior of shallow and socketed foundations in rocks that may be used to obtain correlations for prediction of initial normal stiffness for the tip of rock sockets.

The method of Goodman (1980) may be used to develop the following expression for the prediction of initial normal stiffness (K_n) for the tip of rock sockets:

$$K_n = \frac{E_m}{C\left(1 - \nu^2\right)r} \tag{2.33}$$

where C is a constant that depends on the boundary conditions (Goodman 1980) and is equal to $\pi/2$ if the foundation can be shown to behave as a rigid structure compared to the soft rock or 1.7 if the foundation is a flexible structure compared to the soft rock mass (Goodman 1980), ν is the Poisson's ratio (see Gercek 2007 for typical values for rocks), and r is the radius of the foundation (i.e., $B/2$).

Pells and Turner (1979) also used a similar method to that proposed by Goodman (1980). Pells and Turner (1979), however, add a correction factor to account for the effect of embedment depth on load–displacement behavior of the tip of sockets in rock mass as:

$$K_n = \frac{E_m}{C\left(1 - \nu^2\right)rI} \tag{2.34}$$

where I is the correction factor for embedment. This is a theoretical factor that is a function of Poisson's ratio (ν) of rock and depth of embedment in rock to diameter ratio (D_{TOR}/B). I ranges from 0.4 to 1.0 for D_{TOR}/B ratios of 0 to 10 (Pells and Turner 1979).

Lee *et al.* (2013) proposed a predictive model for the initial normal stiffness (K_n) for the tip of sockets in rock masses. The method is semi-empirical because it was formulated based on parametric studies performed using numerical models and was calibrated using a rock socket load test database that was reported by Lee *et al.* (2013) and is written as:

$$K_n = 0.00037\left(\frac{E_m S_J}{B_{ref}\sigma_p}\right)^{0.4058}\left(\frac{\sigma_p}{B}\right) \quad (2.35)$$

where K_n is in units of MPa/mm, S_J is the discontinuity spacing, B_{ref} = 1.0 m, σ_p is the atmospheric pressure (= 0.101 MPa), and B is the diameter of the rock socket. The mean discontinuity frequency per meter (i.e., $M = 1/S_J$) may be obtained from the following expression proposed by Priest and Hudson (1976):

$$RQD = 100e^{-0.1M}(0.1M + 1) \quad (2.36)$$

where RQD is the rock quality designation and M is the discontinuity frequency. RQD can be converted to GSI using the recommendations of Hoek *et al.* (2013).

These methods are evaluated herein using the mean (μ_λ) and coefficient of variation, c.o.v., (δ_λ) of ratios of back-calculated to predicted initial normal stiffness for the tip of sockets in weak rock masses. It must be emphasized that the back-calculated initial normal stiffness values are obtained from the back-calculated q–δ relationships reported in the tip resistance database. The analysis results are summarized in Table 2.11 and the diagnostic plots are shown in Figures 2.21 and 2.22. The findings and summary of observations are discussed in Tables 2.12–2.14.

1. The review of statistics of bias shows that all three initial normal stiffness models that are evaluated in this section underestimate the initial normal stiffness by a large margin.
2. The c.o.v. of the bias for predictive models for initial normal stiffness does not show a clear trend with the condition of rock mass. However, the study of the statistics of bias for the models of Pells and Turner (1979) and Lee *et al.* (2013) suggests that these models tend to become less accurate as the rock becomes more fractured (i.e., as the GSI values decrease).

Table 2.11 Evaluation of the predictive models for the initial normal stiffness (K_n) of drilled shafts in weak rock mass

		Database (Asem 2018)			
Design methods		*35 > GSI*	*35 < GSI <50*	*50 < GSI <70*	*Entire database*
Goodman (1980)	μ_λ	3.65	5.32	61.39	6.72
	σ_λ	4.43	7.18	78.92	20.54
	δ_λ	1.21	1.35	1.29	3.06
Pells and Turner (1979)	μ_λ	1.91	1.66	1.48	1.82
	σ_λ	0.93	0.44	0.66	0.83
	δ_λ	0.49	0.26	0.44	0.45
Lee *et al.* (2013)	μ_λ	2.27	5.37	30.90	4.68
	σ_λ	1.97	2.96	28.44	9.08
	δ_λ	0.87	0.55	0.92	1.94

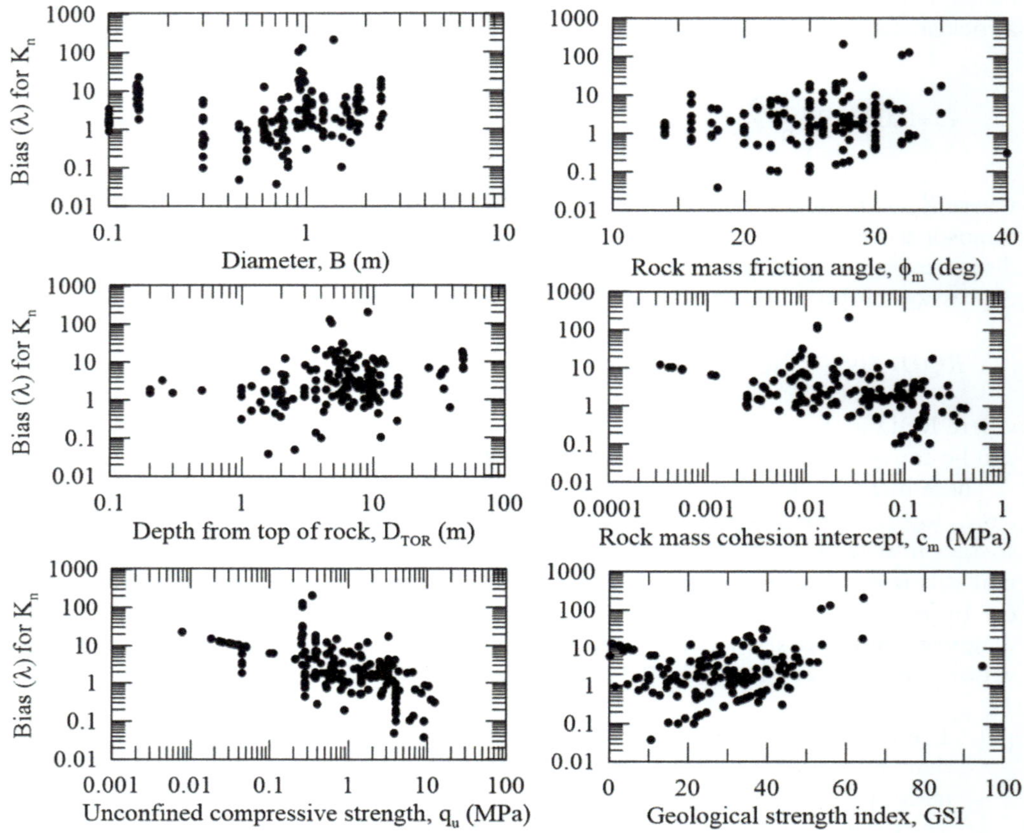

Figure 2.21 Variation of bias (λ) in the predictions of the method of Goodman (1980) for drilled shaft tip normal stiffness (K_n) with properties of weak rock mass and drilled shaft geometry using tip resistance database.

3. The examination of variation of bias (λ) with the properties of rock mass and socket geometry indicates a strong correlation between bias and socket diameter (B), unconfined compressive strength of intact rock (q_u), Geological Strength Index (GSI), and rock mass friction angle (ϕ_m).

2.6 ESTIMATION OF MODEL PARAMETERS

The rock mass deformation modulus (E_m), unconfined compressive strength (q_u), and the degree of consolidation (U) are used when deformations in rock mass are calculated (Bieniawski 1978; Hoek and Diederichs 2006). Previous investigators have used published data to develop predictive models for the strength and elastic properties of rock (e.g., Labuz *et al.* 2018; Guevara-Lopez *et al.* 2019; Asem 2020; Asem *et al.* 2021). Following this approach, we used in situ and laboratory test results to evaluate the existing methods and to recommend predictive models for E_m, q_u, and k.

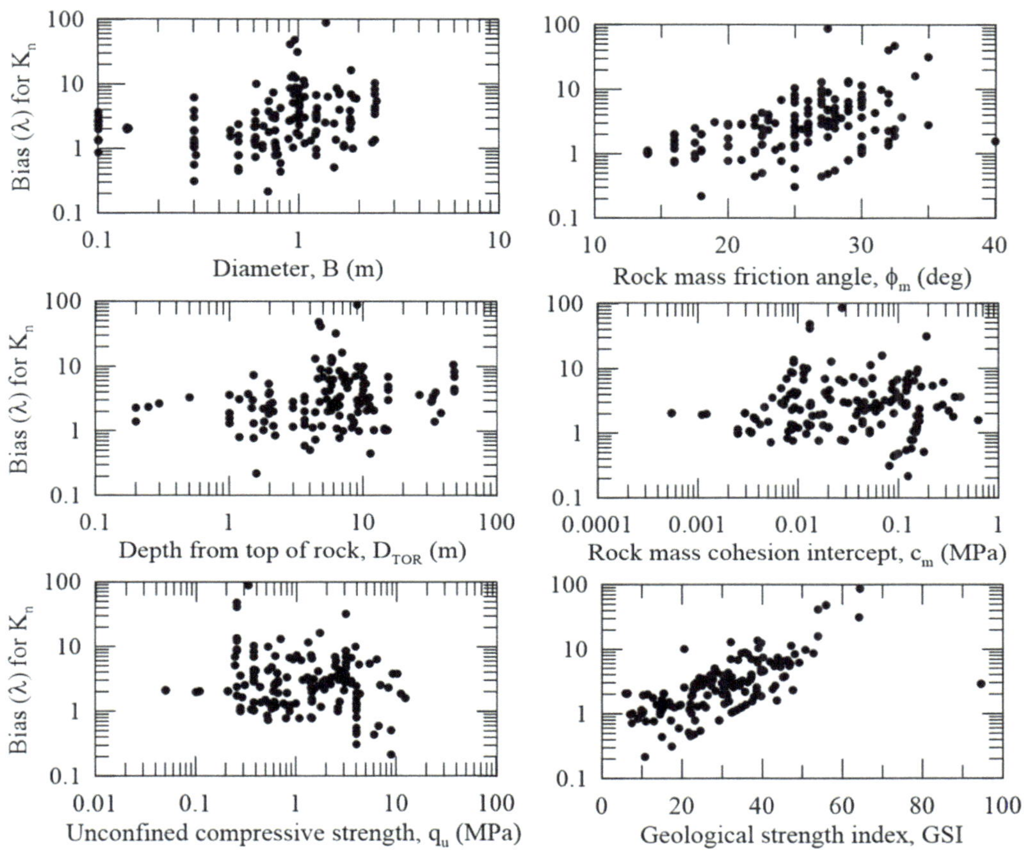

Figure 2.22 Variation of bias (λ) in the predictions of the method of Lee *et al.* (2013) for drilled shaft tip normal stiffness (K_n) with properties of soft rock mass and drilled shaft geometry using the tip resistance database.

Table 2.12 Summary of predictive methods for the modulus of deformation of rock mass

Number	*Empirical methods (E_m in units of GPa)*	*Reference*
1	$E_m = \frac{E_i}{100\left[0.0028RMR^2 + 0.9\exp(\frac{RMR}{22.82})\right]}, E_i = 50GPa$	Nicholson and Bieniawski (1990)
2	$E_m = \frac{E_i\left[1-\cos(\pi.\frac{RMR}{100})\right]}{2}, E_i = 50GPa$	Mitri *et al.* (1994)
3	$E_m = 0.1(RMR/10)^3$	Read *et al.* (1999)
4	$E_m = 10\sqrt[3]{\frac{Qq_u}{100}}, q_u = 100MPa$	Barton (2002)
5	$E_m = E_i\sqrt[4]{\exp\left[(GSI-100)/9\right]}, E_i = 50GPa$	Carvalho (2004)
6	$E_m = 7\sqrt{10\left[RMR-44\right]/21}$	Diederichs and Kaiser (1999)

Table 2.13 Rock mass classification and rock mass deformation modulus database

Rock type	*Joint infilling material*	*Plate diameter (cm)*	*Em (GPa)*	*RQD*	*RMR-76*	*RMR-89*	*GSI*	**Q**	*Reference*
Fault breccia	Sand and gravel with silt	30.2	0.25		15	18	10	0.0078	Keffeler (2014)
Oxidized limestone	Silty sand with gravel	30.2	4.5		41	34	45	0.27	Keffeler (2014)
Argillized rhyolite dike	Clayey sand with gravel	30.2	0.37		16	21	15	0.017	Keffeler (2014)
Argillized rhyolite dike	Clayey sand	25.4	0.031		16	21	10	0.01	Keffeler (2014)
Decalcified limestone	Carbon and iron oxide coatings	25.4	2.5		37	47	40	0.1	Keffeler (2014)
Decalcified limestone	Carbon and iron oxide coatings	30.2	1.7		35	45	45	0.069	Keffeler (2014)
Decalcified limestone	Carbon and non-plastic clay coating	30.2	1.3		34	44	30	0.067	Keffeler (2014)
Decalcified limestone	Carbon coatings	30.2	1.1		40	49	40	1	Keffeler (2014)
Decalcified limestone	Carbon and minor soft white infillings	30.2	1.4		41	49	40	0.83	Keffeler (2014)
Decalcified limestone	Carbon and pyrite coatings	30.2	1.4		40	51	45	1.7	Keffeler (2014)
Decalcified limestone	Carbon coatings and 1-3 mm calcite	30.2	0.71		47	51	50	3.3	Keffeler (2014)
Decalcified and argillized limestone	Clayey sand	30.2	0.9		20	28	15	0.02	Keffeler (2014)
Decalcified limestone	Soft realgar coatings	30.2	4.2		45	54	50	0.67	Keffeler (2014)
Decalcified and argillized limestone	Soft realgar 1–3 mm	30.2	2.3		32	45	45	0.067	Keffeler (2014)
Siltstone and sandstone	Calcite and iron oxide		30	84	59	66			USACE (1988)
Siltstone and sandstone	Calcite and iron oxide		4.8	25	55	57			USACE (1988)
Siltstone and sandstone	Calcite and iron oxide		2.8	63	44	53			USACE (1988)
Siltstone and sandstone	Calcite and iron oxide		18	57	53	54			USACE (1988)
Siltstone and sandstone	Calcite and iron oxide		17	75	68	63			USACE (1988)

(Continued)

Table 2.13 (Continued) Rock mass classification and rock mass deformation modulus database

Rock type	*Joint infilling material*	*Plate diameter (cm)*	*Em (GPa)*	*RQD*	*RMR-76*	*RMR-89*	*GSI*	**Q**	*Reference*
Siltstone and sandstone	Calcite and iron oxide		2.1	18	42	45			USACE (1988)
Siltstone and sandstone	Calcite and iron oxide		34	73	65	66			USACE (1988)
Siltstone and sandstone	Calcite and iron oxide		8.3	26	39	40			USACE (1988)
Sarvak limestone	Calcite and iron oxide	0.971	14.1	5	44	50			Agharazi et *al.* (2012), Agharazi (2013)
Sarvak limestone	Calcite and iron oxide	0.971	16	14	43	52			Agharazi et *al.* (2012), Agharazi (2013)
Sarvak limestone	Calcite and iron oxide	0.971	19.5	3	45	52			Agharazi et *al.* (2012), Agharazi (2013)
Sarvak limestone	Calcite and iron oxide	0.971	15.6	19	44	53			Agharazi et *al.* (2012), Agharazi (2013)
Sarvak limestone	Calcite and iron oxide	0.65	5.5	6	45	52			Agharazi et *al.* (2012), Agharazi (2013)
Sarvak limestone	Calcite and iron oxide	0.65	3.7	16	45	53			Agharazi et *al.* (2012), Agharazi (2013)
Sarvak limestone	Calcite and iron oxide	0.65	3.7	15	45	54			Agharazi et *al.* (2012), Agharazi (2013)
Sarvak limestone	Calcite and iron oxide	0.65	3.2	25	44	54			Agharazi et *al.* (2012), Agharazi (2013)

Table 2.14 Evaluation of empirical models for prediction of rock mass modulus of deformation

Predictive method	*Database used for evaluation*	*Number of cases*	μ_λ	σ_λ	δ_λ
Bieniawski (1978)	Keffeler (2014)	5	1.35	1.04	0.77
Serafim and Pereira (1983)	Keffeler (2014)	30	0.81	0.74	0.91
Nicholson and Bieniawski (1990)	Keffeler (2014)	30	0.83	0.79	0.95
Mitri *et al.* (1994)	Keffeler (2014)	30	0.26	0.26	1.00
Read *et al.* (1999)	Keffeler (2014)	30	0.52	0.45	0.86
Barton (2002)	Keffeler (2014)	14	0.28	0.21	0.75
Diederichs and Kaiser (1999)	Keffeler (2014)	23	0.58	0.49	0.83
Carvalho (2004)	Keffeler (2014)	14	0.16	0.11	0.66
Palmström and Sigh (2001)	Keffeler (2014)	14	5.1	4.58	0.88
Galera *et al.* (2007)	Keffeler (2014)	30	0.65	0.59	0.92
Gokceoglu and Zorlu (2004) RMR	Keffeler (2014)	30	1.86	1.68	0.90
Gokceoglu and Zorlu (2004) GSI	Keffeler (2014)	14	0.92	0.58	0.62
This study, Equation (37)	Asem (2018)	581	0.98	1.1	1.12

Source: Based on data from Keffeler (2014) and Asem (2018).

2.6.1 Model formulation framework

Reliability analysis of structures built in rock masses requires the use of probabilistic models that can properly quantify the design uncertainties (e.g., Gardoni *et al.* 2002; Asem 2018; Asem 2019a,b; Asem *et al.* 2019; Asem and Gardoni 2019a,b,c; Asem 2020; Tan *et al.* 2020; Asem and Gardoni 2021). Following Gardoni *et al.* (2002), a probabilistic model is formulated for the estimation of permeability of rock:

$$T[f(x,\Theta)] = \gamma(x,\theta) + \sigma\varepsilon \tag{2.37}$$

where $T[.]$ is a suitable transformation of f;$\gamma(x,\theta) = \sum_{j=1}^{n} \theta_j h_j(x)$ is a function expressed in terms of observed variable, x, and the unknown parameter $\theta = (\theta_1, \theta_2,\ldots, \theta_n)$; and $\sigma\varepsilon$ is a model error term that captures the model uncertainty resulting from missing model parameters or inappropriate selected model form. In the model error term, σ is the standard deviation of the model error and ε is a random variable with zero mean and unit variance.

A suitable transformation of f should be used to justify the assumptions that $\sigma\varepsilon$ is additive to the model (additivity assumption), σ is constant and independent of the variable x (homoskedasticity assumption), and that ε follows a normal distribution (normality assumption). These assumptions are used in the estimation of the model parameters $\Theta = (\theta, \sigma)$. Here, we use the natural logarithm of f to satisfy these assumptions. The validity of this transformation may be checked using diagnostic plots as described by Rao and Toutenburg (1995).

A Bayesian approach is used to estimate the vector of unknown model parameters $\Theta = (\theta, \sigma)$ using a Markov-chain Monte-Carlo (MCMC) approach. A Bayesian framework allows different types of information to be included in the estimation of $\Theta = (\theta, \sigma)$ such as field data, laboratory test results, and engineering judgment. In a Bayesian approach, the prior probability distribution density function (PDF) for the model parameters Θ is updated following the Bayes' theorem (Box and Tiao 1992):

$$f''(\Theta) = \kappa L(\Theta) f'(\Theta) \tag{2.38}$$

where $f''(\Theta)$ is the posterior PDF of Θ that captures all available information about Θ; the prior distribution PDF, $f'(\Theta)$ is the prior PDF that represents the state of our knowledge about the model parameters Θ before new information become available; $L(\Theta)$ is the likelihood function that captures the information from the data; and κ is a constant used to ensure that $f''(\Theta)$ is a proper PDF (i.e., $\kappa = \frac{1}{\int L(\Theta) f'(\Theta) d(\Theta)}$ such that the area under $f''(\Theta)$ is equal to unity). The effect of $f'(\Theta)$ on $f''(\Theta)$ becomes smaller as the number of new measurements and observations increases. Box and Tiao (1992) suggested that when no previous information on Θ is available, a noninformative prior may be used for $f'(\Theta)$. Gardoni *et al.* (2002) showed that for a univariate model that is linear in parameter θ, the noninformative prior uniform PDF may be represented as:

$$f'(\Theta) \propto \frac{1}{\sigma} \tag{2.39}$$

The information from n sets of measured quantities (k_j, x_j) $j = 1, \ldots, n$ is included in the Bayesian estimation of the model parameters Θ using the likelihood function, $L(\Theta)$ that is the probability of the observation of a given value of Θ and its formulation depends on the type of available data. When the available observations are statistically independent, the likelihood function, $L(\Theta)$, is written as:

$$L(\Theta) \propto \prod_{j=1}^{n} \left\{ \frac{1}{\sigma} \varphi \left[\frac{r_j(\theta)}{\sigma} \right] \right\} \tag{2.40}$$

where φ [.] is the standard normal PDF and $r_j(\theta)$ is the residual term that may be given by $r_j(\theta) = k_j - \gamma(x_j, \theta)$ and is valid for equality data. Equation (2.38) can be used to update the current posterior as new data become available by using the current posterior as the new prior and using the new data to construct the likelihood function.

2.6.2 Rock mass deformation modulus (E_m)

A database of rock mass deformation modulus (E_m) and the unconfined compressive strength (q_u) of weak rock in rock socket load tests was developed by Asem (2018). These load tests were augmented by those collected by Chern *et al.* (2004), which include sedimentary, igneous, and metamorphic rocks from China and Taiwan. These data are plotted in Figure 2.23. Equation (2.41) can be used to predict E_m based on the unconfined compressive strength (q_u):

$$E_m = 150 q_u^{1.1} \tag{2.41}$$

where E_m and q_u are in units of MPa.

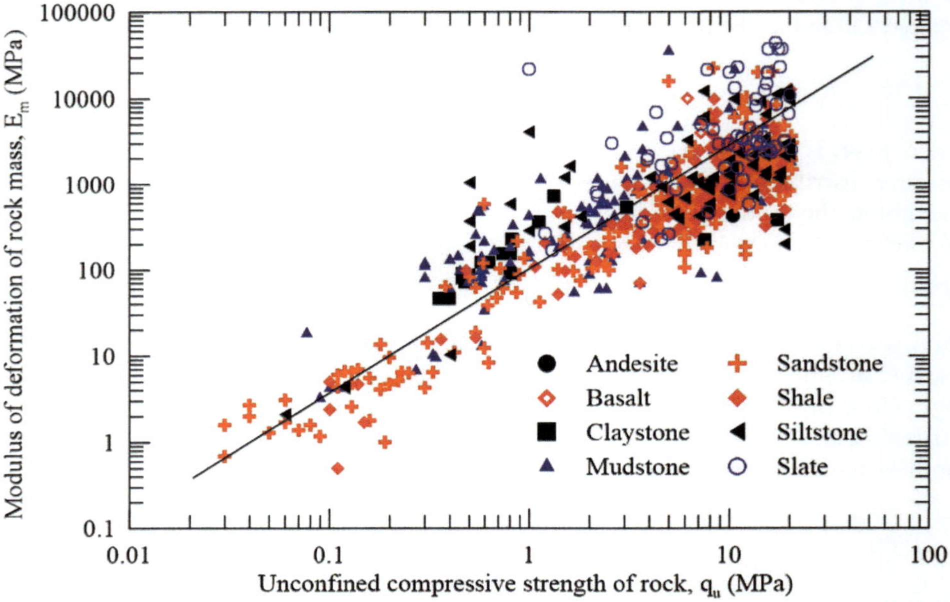

Figure 2.23 Variation of modulus of deformation of rock mass (E_m) with unconfined compressive strength for different rock types and the proposed design trend line. All data from published literature are summarized in Asem (2018).

2.6.3 Unconfined compressive strength (q_u)

The following approach is suggested for the prediction of q_u based on standard penetration test (SPT) results: (i) develop a plot of the blow counts versus penetration at each SPT depth, (ii) use the approach proposed by Stark *et al.* (2013) and Asem (2020) to estimate the penetration rate (Ψ) at each SPT depth, (iii) obtain the corrected Ψ for 85% energy transfer (Ψ_{85}), (iv) correct the corresponding value of Ψ_{85} for the effect of the mean effective stress and obtain $(\Psi_{85})_{\sigma'}$, and (v) use the corresponding values of $(\Psi_{85})_{\sigma'}$ to estimate q_u (Asem 2020):

$$q_u(\text{MPa}) = \frac{10}{4}(\psi_{85})_{\sigma'} \tag{2.42}$$

Equation (2.42) is developed by regression analysis between data for the corrected values of Ψ_{85} for the effect of σ', $(\Psi_{85})_{\sigma'}$, and the measured values of q_u and E at 21 Illinois sites. The corrected value of Ψ_{85} for the effect of the mean effective stress obtained $(\Psi_{85})_{\sigma'}$

$$(\psi_{85})_{\sigma'} = C_\psi \psi_{85} = \frac{0.23}{\sigma'}\psi_{85} \tag{2.43}$$

It must be noted that the correlation in Equation (2.43) is only intended for the range of $0 < \sigma' < 0.8$ MPa (Asem 2020). If this correlation is applied to $\sigma' > 0.8$ MPa, it will lead to unreliable results. Additional data is required to further enhance Equation (2.43) for $\sigma' > 0.8$ MPa. It is also noted that the data scatter for $0 < \sigma' < 0.8$ MPa is significant. This equation should be updated using the approach proposed by Gardoni *et al.* (2002) as more reliable data become available.

The depth correction factor in Equation (2.43), i.e., C_Ψ, was formulated for a reference effective vertical stress of $\sigma'_v = 0.1$ MPa (1 ton/ft^2 after Skempton 1986) or equivalently a mean effective stress of $\sigma' = 7/3\ (0.1) = 0.23$ MPa. C_Ψ may be obtained as (after Skempton 1986).

$$C_\psi = \frac{\frac{a}{b} + 0.23}{\frac{a}{b} + \sigma'} \tag{2.44}$$

For $a/b = 0$ for the case of weak rocks, $C_\Psi = 0.23/\sigma'$. It must be noted that this correlation in only valid for a range of mean effective stress of $0 < \sigma' < 0.8$ MPa that is equivalent to an overall test depth of about 26 m.

2.6.4 Permeability (*k*)

The correlation between some of the parameters affecting the permeability of mudstone was examined by Gardoni and Asem (2022) who showed (i) porosity ϕ and the clay size fraction w_{CL} are not strongly related, and (ii) clay size fraction w_{CL} and mean pore radius r are more or less correlated, as w_{CL} increase, r decreases. This is expected because the size of pore spaces and the size of the particles making the porous medium are inversely related, i.e., as the percentage of small particles increases, the pore size decreases. Based on data from the literature, Asem and Gardoni (2022) presented a simple permeability model that could be formulated:

$$k = f(r, \phi) \tag{2.45}$$

Note that Equation (2.45) excludes the clay size fraction w_{CL} because $r = f(w_{CL})$ and hence the effect of w_{CL} on k is indirectly accounted for by including r in the predictive model. Using the method of Gardoni *et al.* (2002), a probabilistic model for the prediction of permeability of mudstone is proposed

$$k = e^{-41.6}(r\phi)^{0.6} \tag{2.46}$$

For a Terzaghi effective pressure of 10 MPa, with our formulation and results (Asem and Gardoni 2022), the effect of overall model uncertainty $\sigma\varepsilon$ as well as model parameter uncertainty $\Theta = (\theta, \sigma)$ can be readily accounted for that is an improvement over existing probabilistic models.

2.7 CONCLUSIONS

Rock sockets are often constructed in weak rock mass. Understanding the hydromechanical properties of the rock mass is required for the successful design and performance of the rock sockets. This chapter establishes separate databases for the hydromechanical properties of rock masses with particular attention to weak shales and mudstones. The databases include information on rock socket geometry, load–displacement response, and the rock mass properties and mechanical and transport properties of the rock. The chapter then reviews the available models for the prediction of (i) the axial resistance (compression or uplift), (ii) axial deformation, and permeability. The databases are used to evaluate the existing predictive models for hydromechanical properties of rock as related to the design of rock sockets. The load test databases are also used to evaluate the effect of rock mass variability on the reliability of the foundations in weak rock. Based on the results of the analysis presented, the applicability of the existing models and approaches to the study of socket behavior in weak rocks is discussed.

ACKNOWLEDGEMENTS

Professor James H. Long of the University of Illinois at Urbana-Champaign contributed to this manuscript. His suggestions and contributions are acknowledged.

REFERENCES

Abu-Hejleh, N., O'Neill, M.W., Hanneman, D., *et al.* (2003). *Improvement of the geotechnical axial design methodology for Colorado's drilled shafts socketed in weak rocks*. Denver, CO: Colorado Department of Transportation.

Agharazi, A., Tannant, D. D. and Martin, C. D. (2012). Characterizing rock mass deformation mechanisms during plate load tests at the Bakhtiary dam project. *International Journal of Rock Mechanics & Mining Sciences*, 49(2012), 1–11.

Agharazi, A. (2013). *Development of a 3D equipment continuum model for deformation analysis of systematically jointed rock masses*. Ph.D. thesis, University of Alberta.

AlKhafaji, H., Imani, M., and Fahimifar, A. (2020). Ultimate bearing capacity of rock mass foundations subjected to seepage forces using modified Hoek–Brown criterion. *Rock Mechanics and Rock Engineering*, 53, 251–268.

Ang, A. H.-S., and Tang, W. H. (2007). *Probability Concepts in Engineering: Emphasis on Applications in Civil and Environmental Engineering*, 2nd ed. Wiley, New York.

ARGEMA (1992). Design guides for offshore structures: offshore pile design. in P.L. Tirant (ed.), Paris, France: Association de recherché en geotechnique marine.

Asem, P. (2018). *Axial behavior of drilled shafts in soft rock*. Ph.D. Thesis. University of Illinois at Urbana-Champaign, Urbana, Illinois, United States.

Asem, P., Long, J. H., and Gardoni, P. (2018). Probabilistic model and LRFD resistance factors for the tip resistance of drilled shafts in soft sedimentary rock based on axial load tests. *Innovations in Geotechnical Engineering: Honoring Jean-Louis Briaud*. X. Zhang, P. J. Cosentino and M. H. Hussein. Orlando, Florida, American Society of Civil Engineers. *GSP* 299, 1–49.

Asem, P. (2019a). Load-displacement response of drilled shaft tip in soft rocks of sedimentary origin. *Soils and Foundations* 59(12), 1193–1212.

Asem, P. (2019b). Base resistance of drilled shafts in soft rock using in situ load tests: a limit state approach. *Soils and Foundations* 59(6), 1639–1658.

Asem, P. and Gardoni, P. (2019a). Evaluation of the peak side resistance for rock socketed shafts in weak sedimentary rock from an extensive database of published field load tests: a limit state approach. *Canadian Geotechnical Journal* 56(12), 1816–1831.

Asem, P. and Gardoni, P. (2019b). A load-transfer function for the side resistance of drilled shafts in soft rock. *Soils and Foundations* 59(2019), 1241–1259.

Asem, P. and Gardoni, P. (2019c). Bayesian estimation of the normal and shear stiffness for rock sockets in weak sedimentary rocks. *International Journal of Rock Mechanics and Mining Sciences* 124(2019), 104–129.

Asem, P., Jimenez, R. and Gardoni, P. (2019a). Probabilistic prediction of intact rock strength using point load tests using a Bayesian formulation. *7th International Symposium on Geotechnical Safety and Risk (ISGSR 2019)*. Taipei, Taiwan.

Asem, P., Taukoor, V., and Kane, T. (2019b). Departure of soft rock mass from undrained response in drilled shaft and plate load tests. *Soils and Foundations* 59(1), 228–233.

Asem, P. (2020a). Prediction of unconfined compressive strength and deformation modulus of weak argillaceous rocks based on the standard penetration test. *International Journal of Rock Mechanics and Mining Sciences* 133.

Asem, P. (2020b). The effect of expansive concrete on the side resistance of sockets in weak rock. *Soils and Foundations* 60(1), 274–282.

Asem, P. and J. F. Labuz (2020). *A failure mechanism around axially loaded sockets in weak rock. Geo-Congress 2020.* ASCE, Minnesota.

Asem, P., Fuselier, H., and Labuz, J. F. (2021a). On a four-parameter linear failure criterion. *Rock Mechanics and Rock Engineering* 54, 3369–3376. https://doi.org/10.1007/s00603-021-02451-w

Asem, P., X. Wang, C. Hu and J. F. Labuz (2021b). On tensile fracture of a brittle rock. *International Journal of Rock Mechanics & Mining Sciences* 144, 104823.

Asem, P. and P. Gardoni (2021a). A generalized Bayesian approach for prediction of strength and elastic properties of rock. *Engineering Geology* 289, 106187.

Asem, P. and P. Gardoni (2021b). On the use and interpretation of in situ load tests in weak rock masses. *Rock Mechanics and Rock Engineering* 54, 3663–3700.

Asem, P. and P. Gardoni (2022). A probabilistic, empirical model for permeability of mudstone. *Probabilistic Engineering Mechanics* 69, 103262.

Baghdady, A. K. (2018). *Axial behavior of drilled shafts socketed into weak Pennsylvanian shales.* Ph.D. Thesis. University of Illinois at Urbana-Champaign, Urbana, Illinois, United States.

Bandis, S. (1980). *Experimental studies of scale effects on shear strength and deformation of rock joints.* Ph.D. Thesis. University of Leeds, Leeds, United Kingdom.

Bandis, S. C., Lumsden, A. C., and Barton, N. R. (1983). Fundamentals of rock joint deformation. *International Journal of Rock Mechanics and Mining Sciences and Geomechanics Abstracts*, 20(6), 249–268.

Box, G. E. P. and Tiao, G. C. (1992). *Bayesian inference in statistical analysis.* Reading, Massachusetts, Addison–Wesley.

Barton, N., *et al.* (1978). Suggested methods for the quantitative description of discontinuities in rock masses. *International Journal of Rock Mechanics and Mining Sciences and Geomechanics Abstracts*, 15(6), 319–368.

Barton, N. (2002). Some new Q value correlations to assist in site characterisation and tunnel design. *International Journal of Rock Mechanics and Mining Sciences*, 39(2), 185–216.

Bieniawski, Z. T. (1978). Determining rock mass deformability: experience from case histories. *International Journal of Rock Mechanics and Mining Sciences and Geomechanics Abstracts*, 15(5), 237–247.

Bishnoi, B.L. (1968). *Bearing capacity of a closed jointed rock.* Ph.D. Thesis. Georgia Institute of Technology, Atlanta, Georgia, United States.

Brown, D.A., Turner, J.P., and Castelli, R.J. (2010). *Drilled shafts: construction procedures and LRFD design methods.* Washington, DC: National Highway Institute, U.S. Department of Transportation, Federal Highway Administration, 972pp.

Carrubba, P. (1997). Skin friction on large-diameter piles socketed into rock. *Canadian Geotechnical Journal*, 34(2), 230–240.

Carter, J.P. and Kulhawy, F.H. (1988). *Analysis and design of drilled shaft foundations socketed into rock.* Palo Alto: Electric Power Research Institute, p. 190.

Carvalho, J. (2004). Estimation of rock mass modulus. Pers. commun.

CGS (2006). *Canadian foundation engineering manual.*

Chern, J. C., *et al.* (2004). *Correlation study on the deformation modulus and rating of rock mass (R-GT-97-04), from Taipei*, Taiwan.

Coates, D.F. (1967). Rock mechanics principle. *Department of Energy, Mines and Resources, Canada*, 410.

Collingwood, B. (2000). *The effect of construction practices on the performance of rock socketed bored piles*. Ph.D. Thesis, Monash University, Melbourne, Australia.
Dai, G., *et al.* (2016). The effect of sidewall roughness on the shaft resistance of rock-socketed piles. *Acta Geotechnica*, 1–12, 429–440.
Deere, D.U. and Miller, R.P. (1966). *Engineering classification and index properties for intact rock*. Urbana, IL: University of Illinois at Urbana-Champaign/Air Force Weapons Laboratory, p. 327.
Diederichs, M. S. and Kaiser, P. K. (1999). Stability of large excavations in laminated hard rock-masses: the Voussoir analogue revisited. *International Journal of Rock Mechanics and Mining Sciences*, 36(1), 97–117.
Galera, J. M., Alvarez, M., and Bieniawski, Z. T. (2007). Evaluation of the deformation modulus of rock masses using RMR: Comparison with dilatometer tests. In M. Romana, A. Perucho, and C. Olalla (eds.), *Underground works under special conditions.* Taylor & Francis, London, pp. 71–77.
Gardoni, P., Kiureghian, A. D., and Mosalam, K. M. (2002). Probabilistic capacity models and fragility estimates for reinforced concrete columns based on experimental observations. *Journal of Engineering Mechanics*, 128(10), 1024–1038.
Gercek, H. (2007). Poisson's ratio values for rocks. *International Journal of Rock Mechanics and Mining Sciences*, 44(1), 1–13.
Glos, G. H. III and Briggs, O. H. Jr. (1983). Rock Sockets in Soft Rock. *Journal of Geotechnical Engineering*, 109(4), 525–535.
Gokceoglu, C. and Zorlu, K. (2004). A fuzzy model to predict the uniaxial compressive strength and the modulus of elasticity of a problematic rock. *Engineering Applications of Artificial Intelligence*, 17(1), 61–72.
Goodman, R.E. (1980). *Introduction to rock mechanics.* United States: John Wiley & Sons, Inc..
Goodman, R. E. (1989). *Introduction to rock mechanics.* United States: John Wiley & Sons, Inc..
Goodman, R. E. (1993). *Engineering geology: rock in engineering construction* (1st ed.). United States: John Wiley & Sons, Inc.
Guevara-Lopez, F., Jimenez, R., Gardoni, P., and Asem, P. (2019). Probabilistic prediction of intact rock strength using point load tests using a Bayesian formulation. *Georisk: Assessment and Management of Risk for Engineered Systems and Geohazards* 14, 206–215. https://doi.org/10.1080/17499518.2019.1634274
Gupta, R. C. (2012). Hyperbolic model for load tests on instrumented drilled Shafts in intermediate geomaterials and rock. *Journal of Geotechnical and Geoenvironmental Engineering*, 138(11), 1407–1414.
Gupton, C. and Logan, T. J. (1984). Design guidelines for drilled shafts in weak rocks of south Florida. Florida, United States: South Florida Annual ASCE Meeting.
Griffith, A. A. (1921). The phenomena of rupture and flow in solids. *Philosophical Transactions of the Royal Society of London.* Series A, Containing Papers of a Mathematical or Physical Character, 221, 582–593.
Haberfield, C. M. (1987). *The performance of the pressuremeter and socketed piles in weak rock.* Ph.D. Thesis, Monash University, Monash University, Melbourne, Australia.
Haberfield, C. M. and Lochaden, A. L. E. (2018). Analysis and design of axially loaded piles in rock. *Journal of Rock Mechanics and Geotechnical Engineering* 11, 535–548.
Hassan, Khaled M. (1994). *Analysis and design of drilled shafts socketed into soft rock.* Ph.D. Thesis, University of Houston, Houston, TX.
Hassan, K. M., *et al.* (1997). Design method for drilled shafts in soft argillaceous rock. *Journal of Geotechnical and Geoenvironmental Engineering*, 123(3), 272–80.
Hirany, A. (1988). *Test loading of drilled shafts.* Ph.D. Thesis. Cornell University, Ithaca, NY.
Hirayama, H. (1990). Load-settlement analysis for bored piles using hyperbolic transfer functions. *Journal of Soils and Foundations*, 30(1), 55–64.
Horvath, R. G. and Kenney, T. C. (1979). Shaft resistance of rock socketed drilled piers. In F.M. Fuller (ed.), *Symposium on Deep Foundations*, Atlanta, GA, 182–214.

Hoek, E. (1965). *Rock fracture under static stress conditions*. Ph.D. Thesis, University of Cape Town, Cape Town, South Africa.
Hoek, E. and Brown, E. T. (1980). Empirical strength criterion for rock masses. *Journal of the Geotechnical Engineering Division*, 106(9), 1013–1035.
Hoek, E. (1983). Strength of jointed rock masses. *Géotechnique*, 33(3), 187–223.
Hoek, E. and Brown, E. T. (1997). Practical estimates of rock mass strength. *International Journal of Rock Mechanics and Mining Sciences*, 34(8), 1165–1186.
Hoek, E. and Diederichs, M. S. (2006). Empirical estimation of rock mass modulus. *International Journal of Rock Mechanics and Mining Sciences*, 43(2), 203–215.
Hoek, E., Carter, T. G., and Diederichs, M. S. (2013). *Quantification of the geological strength index chart*. San Francisco, CA: American Rock Mechanics Association.
Hoek, E. and Brown, E. T. (2019). The Hoek-Brown failure criterion and GSI — 2018 edition. *Journal of Rock Mechanics and Geotechnical Engineering*, 11(2019), 445–463.
Horvath, R. G. (1982). *Drilled piers socketed into weak shale-methods of improving performance*. Ph.D. Thesis, University of Toronto, Toronto, Ontario, Canada.
Holden, J. C. (1984). *Construction of board piles in weathered rocks*. Victoria, Australia: Road Construction Authority of Victoria, p. 69.
Horvath, R.G. (1978). *Field load test data on concrete-to-concrete rock bond strength drilled pier foundations*. Toronto: University of Toronto, p. 46.
Horvath, R. G. and Kenney, T. C. (1979). Shaft resistance of rock socketed drilled piers. In F.M. Fuller (ed.), *Symposium on Deep Foundations*, Atlanta, GA, pp. 182–214.
Jaeger, J. C., Cook, N. G., and Zimmerman, R. W. (2007). *Fundamentals of rock mechanics*. MA: Blackwell Publishing.
Jeong, S., *et al.* (2010). Point bearing stiffness and strength of socketted drilled shafts in Korean rocks. *International Journal of Rock Mechanics and Mining Sciences*, 47(6), 983–995.
Johnston, I. W. and Lam, T. S. K. (1989). Shear behavior of regular triangular concrete/rock joints-Analysis. *Journal of Geotechnical Engineering*, 115(5), 711–727.
Johnson, I. W. (2020). Revisiting methods for the design of rock socketed piles. *Journal of Geotechnical and Geoenvironmental Engineering*, 146(12), 1–13.
Kaderabek, T. J. and Reynolds, R. T. (1981). Miami limestone foundation design and construction. *Journal of the Geotechnical Engineering Division*, 107(GT7), 859–872.
Kanji, M. A. (2014). Critical issues in soft rocks. *Journal of Rock Mechanics and Geotechnical Engineering*, 6(3), 186–195.
Keffeler, E. R. (2014). *Measurement and prediction of in-situ weak rock mass modulus: case studies from Nevada, Puerto Rico, and Iran*. Ph.D. Thesis, University of Nevada, Reno.
Kodikara, J. K. (1989). *Shear behaviour of rock-concrete joints and side resistance of piles in weak rock*. Ph.D. Thesis, Monash University, Melbourne, Australia.
Kulhawy, F. H. and Carter, J. P. (1992). Socketed foundations in rock masses. In F.G. Bell (ed.), *Engineering in Rock Masses,* Oxford: University of Oxford, pp. 509–529.
Kulhawy, F. H. and Goodman, R. E. (1980). Design of foundations on discontinuous rock. In: *International conference on structural foundations on rock*, Sydney, pp. 209–220.
Kulhawy, F. H. and Phoon, K. K. (1993). Drilled shaft side resistance in clay soil to rock. In P.P. Nelson, T.D. Smith, and E.C. Clukey (eds.), *Design and Performance of Deep Foundations: Piles and Piers in Soil and Soft Rock*, New York: American Society of Civil Engineers, pp. 172–183.
Kulhawy, F. H., Prakoso, W. A., and Akbas, S. O. (2005). Evaluation of capacity of rock foundation sockets. In *40th U.S. Symposium on Rock Mechanics*, American Rock Mechanics Association, Anchorage, AL.
Labuz, J. F., and Zang, A. (2012). Mohr–Coulomb failure criterion. *Rock Mechanics and Rock Engineering*, 45, 975–979.
Labuz, J. F., Zeng, F., Makhnenko, R., and Li, Y. (2018). Brittle failure of rock: a review and general linear criterion. *Journal of Structural Geology* 112(2018), 7–28.

Lam, T. S. K. (1983). *Shear behaviour of concrete-rock joints.* Ph.D. Thesis. Monash University, Melbourne, Australia.

Lee, J., *et al.* (2013. Proposed point bearing load transfer function in jointed rock-socketed drilled shafts. *Soils and Foundations*, 53(4), 596–606.

Lee, J. and Jeong, S. (2016). Experimental study of estimating the subgrade reaction modulus on jointed rock foundations. *Rock Mechanics and Rock Engineering*, 49, 2055–2064.

Leung, C. F. and Ko, H. Y. (1993). Centrifuge model study of piles socketed in soft rock. *Soils and Foundations*, 33(3), 80–91.

Marinos, P. and Hoek, E. (2001). Estimating the geotechnical properties of heterogeneous rock masses such as Flysch. *Bulletin of Engineering Geology and the Environment*, 60(2), 85–92.

Meigh, A.C. and Wolski, W. (1979). Design parameters for weak rock. In *7th European Conference on Soil Mechanics and Foundation Engineering*, Brighton, pp. 59–79.

Mesri, G., Febres-Cordero, E., Shields, D. R., and Castro, A. (1981). Shear stress-strain-time behaviour of clays. *Géotechnique* 31(4), 537–552.

Miller, A.D. (2003). *Prediction of ultimate side shear for drilled shafts in Missouri shales.* M.S. Thesis, University of Missouri-Columbia, Columbia, Missouri, United States.

Mitri, H. S., Edrissi, R., and Henning, J. (1994). Finite element modeling of cablebolted stopes in hard rock ground mines. In *The SME Annual Meeting*, New Mexico, Albuquerque.

Nicholson, G. A., and Bieniawski Z. T. 1990. A nonlinear deformation modulus based on rock mass classification. *International Journal of Minining and Geological Engineering*, 8, 181–202.

O'Neill, M. W. and Reese, L. C. (1999). *Drilled shafts: construction procedures and design methods.* Washington, DC: Federal Highway Administration, 758pp.

Osterberg, J. O. and Gill, S. A. (1973). Load transfer mechanism for piers socketted in hard soils or rock. In *The 9th Canadian Rock Mechanics Symposium*, Montreal, Canada, pp. 235–262.

Paikowsky, S. G., Birgisson, B., McVay, M. C., *et al.* (2004). *Load and resistance factor design (LRFD) for deep foundations (NCHRP 507).* Washington, DC, United States: American Association of State Highway and Transportation Officials.

Paikowsky, S. G., Canniff, M. C., Lesny, K., Kisse, A., Amatya, S., and Muganga, R. (2010). *LRFD design and construction of shallow foundations for highway bridge structures.* Washington DC: American Association of State Highway and Transportation Officials.

Palmström, A. and Singh, R. (2001). The deformation modulus of rock masses - comparisons between in situ tests and indirect estimates. *Tunnelling and Underground Space Technology*, 16(2), 115–131

Peck, R. B., Hanson, W. E., and Thornburn, T. H. (1974). *Foundation engineering.* 2nd ed. New York: John Wiley & Sons.

Peck, R. B. (1976). Rock foundations for strucrures. In: *Rock Engineering for Foundations and Slopes*, Boulder, CO, pp. 1–20.

Pells, P. J. N. and Turner, R. M. (1979). Elastic solutions for the design and analysis of rock-socketed piles. *Canadian Geotechnical Journal*, 16(3), 481–487.

Pells, P. J. N., *et al.* (1979). *Socketed bored piles in Hawkesbury sandstone.* Sydney, Australia: University of Sydney, 138pp.

Pells, P., Bieniawski, Z. T. R., Hencher, S., and Pells, S. (2017). Rock quality designation (RQD): time to rest in peace. *Canadian Geotechnical Journal*, 54, 825–834.

Priest, S. D. and Hudson, J. A. (1976). Discontinuity spacings in rock. *International Journal of Rock Mechanics and Mining Sciences and Geomechanics Abstracts*, 13(5), 135–148.

Randolph, M.F. and Wroth, C. P. (1978). Analysis of deformation of vertically loaded piles. *Journal of the Geotechnical Engineering Division*, 104(GT12), 1465–1488.

Rao, C. R. and Toutenburg, H. (1995). *Linear models: least squares and alternatives.* New York, Springer.

Read, S. A. L., Richards, L. R., and Perrin, N. D. (1999). Applicability of the Hoek-Brown failure criterion to New Zealand greywacke rocks. In *The Ninth International Congress on Rock Mechanics*, Paris, France.

Rosenberg, P. and Journeaux, N. L. (1976). Friction and end bearing tests on bedrock for high capacity socket design. *Canadian Geotechnical Journal*, 13(3), 324–333.

Rowe, R. K. and Armitage, H. H. (1984). *The design of piles socketed into weak rock*. Ottawa, Ontario, Canada: National Research Council Canada, p. 380.

Rowe, R. K. and Armitage, H. H. (1987). A design method for drilled piers in soft rock. *Canadian Geotechnical Journal*, 24(1), 126–142.

Seed, H. B. and Reese, L. C. (1957). The action of soft clay along friction piles. *ASCE Transactions*, 122(2882), 731–754.

Seidel, J. P. (1993). *Analysis and design of pile shafts in weak rock*. Ph.D. Thesis, Monash University, Melbourne, Australia.

Seidel, J. P. and Collingwood, B. (2001). A new socket roughness factor for prediction of rock socket shaft resistance. *Canadian Geotechnical Journal*, 38(1), 138–153.

Seo, H. and Prezzi, M. (2008). *Use of micropiles for foundations of transportation structures (FHWA/IN/JTRP-2008/18)*. West Lafayette, IN.

Serafim, J. L. and Pereira, J. P. (1983). Consideration of the geomechanical classification of Bieniawski. In *Symposium on Engineering Geology and Underground Constructions*.

Skempton, A. W. (1986). Standard penetration test procedures and the effects in sands of overburden pressure, relative density, particle size, ageing and overconsolidation. *Geotechnique*, 36(3), 425–447.

Sowers, G. F. (1979). *Introductory soil mechanics and foundations*. New York, United States: Macmillan Publishing Co., Inc. , pp. 621.

Stark, T. D., Long, J. H., and Asem, P. (2013). *Improvements for determining the axial capacity of drilled shafts in shale in Illinois*. Urbana, IL: Illinois Department of Transportation.

Stark, T. D., Long, J. H., Baghdady, A. K., and Osouli, A. (2017). *Modified standard penetration test-based drilled shaft design method for weak rocks (Phase 2 study)*. Urbana, IL: Illinois Center for Transportation.

Tabandeh, A., Asem, P., and Gardoni, P. (2020). Physics-based probabilistic models: Integrating differential equations and observational data. *Structural Safety* 87(2020), 101981.

Tang, C., Phoon, K. K., Li, D.-Q., and Akbas, S. O. (2020). Expanded database assessment of design methods for spread foundations under axial compression and uplift loading. *Journal of Geotechnical and Geoenvironmental Engineering*, 146(11), 1–21.

Teng, W. C. (1962). *Foundation design*. New Jersey: Prentice Hall, Englewood Cliffs.

Terzaghi, K. (1943). *Theoretical soil mechanics*. New York: John Wiley & Sons.

Terzaghi, K., Peck, R. B., and Mesri, G. (1996). *Soil mechanics in engineering practice*. New York: Wiley-Interscience.

Toh, C. T., Ooi, T. A., Chiu, H. K., Chee, S. K., and Ting, W. H. (1989). Design parameters for bored piles in weathered sedimentary formation. In *12th International Conference on Soil Mechanics and Foundation Engineering*, Rio De Janeiro, pp. 1073–1078.

Turner, J. (2006). *Rock-socketed shafts for highway structure foundations*. Project 20-5, Washington, WA.

USACE. (1988). *Review of consolidation grouting of rock masses and methods for evaluation*. Report No. REMR-GT-8. Washington, DC: US Army Corps of Engineers.

Vesic, A. S. (1963). Bearing capacity of deep foundations in sand. *Highway Research Record*, (39), 112–153.

Williams, A. F. (1980). *The design and performance of piles into weak rock*. Ph.D. Thesis, Monash University, Melbourne, Australia.

Williams, A. F. and Pells, P. J. N. (1981). Side resistance rock sockets in sandstone, mudstone, and shale. *Canadian Geotechnical Journal*, 18(4), 502–513.

Yoshinaka, R. and Yamabe, T. (1986). 3. Joint stiffness and the deformation behaviour of discontinuous rock. *International Journal of Rock Mechanics and Mining Sciences and Geomechanics Abstracts*, 23(1), 19–28.

Zhang, L. and Einstein, H. H. (1998). End bearing capacity of drilled shafts in rock. *Journal of Geotechnical and Geoenvironmental Engineering*, 124(7), 574–584.

Zhang, L. (1999). *Analysis and design of drilled shafts in rock*. Ph.D. Thesis, Massachusetts Institute of Technology, Massachusetts, MA, United States.

Chapter 3

Evaluation of soil/rock properties using databases

Jianye Ching

The proper characterization of the variability of soil/rock properties for a specific site (including inherent variability and transformation uncertainty) is necessary for reliability-based design as well as for the determination of soil/rock characteristic values. This chapter reviews some useful sources that provide site-specific statistics for the inherent variability of soil/rock properties (e.g., site-specific mean, COV, and SOF). The emphasis is placed on the site-specific statistics extracted from some soil/rock property databases compiled by the author of this chapter. The soil/rock property databases can also be used to calibrate generic transformation models, but the transformation uncertainty is usually significant. The hierarchical Bayesian model (HBM) recently proposed by the author of this chapter may reduce the transformation uncertainty significantly. A real example is adopted to illustrate this reduction in the transformation uncertainty.

3.1 INTRODUCTION

The proper characterization of the variability of soil/rock properties for a specific site plays a critical role in the reliability-based design (RBD) of geotechnical structures (e.g., Phoon and Kulhawy 1999a; Baecher and Christian 2003; Fenton and Griffiths 2008). In Eurocode 7 (CEN 2004), the proper characterization of soil/rock property variability also plays an important role in the determination of the characteristic value. Clause 2.4.5.2(4) in Eurocode 7 states a list of factors for the selection of a characteristic value:

1. Geological and other background information, such as data from previous projects.
2. The variability of the measured property values and other relevant information, e.g., from existing knowledge.
3. The extent of the field and laboratory investigation.
4. The type and number of samples.
5. The extent of the zone of ground governing the behaviour of the geotechnical structure at the limit state being considered.
6. The ability of the geotechnical structure to transfer loads from weak to strong zones in the ground.

Factor #2 "the variability of the measured property values" recognizes the role of spatial variability of soil/rock property as well as measurement error. Factor #3 "the extent of the field and laboratory investigation" and Factor #4 "the number of samples" recognize the role of statistical uncertainty. Factor #5 "the extent of the zone of ground governing the behaviour of the geotechnical structure" recognizes the role of spatial average. The role of transformation uncertainty is not explicitly stated in the above list, but the draft version of

 DOI: 10.1201/9781003441946-3

the revised Eurocode 7 (CEN 2020) explicitly recognizes the role of transformation uncertainty (Orr 2017; CEN 2020). The total variability involved during the evaluation of a soil/rock property can be summarized by the following equation (Phoon and Kulhawy 1999a; Schneider and Schneider 2013; Orr 2017):

$$COV_{TOT}^2 = \Gamma_S^2 \cdot COV_{inher}^2 + COV_{meas}^2 + COV_{trans}^2 + COV_{stat}^2 \quad (3.1)$$

where COV_{TOT} is the coefficient of variation (COV) for the total variability; COV_{inher} is for inherent (spatial) ground variability; $\Gamma^2{}_S$ is a factor that considers the variance reduction due to spatial averaging over the zone of ground governing the behaviour of the geotechnical structure; COV_{meas} is for measurement uncertainty; COV_{trans} is for transformation uncertainty; and COV_{stat} is for statistical uncertainty. Schneider and Schneider (2013) adopted the following simplification for the variance reduction factor: $\Gamma_S^2 = \Gamma_x^2\Gamma_y^2\Gamma_z^2$, where Γ_x^2 and Γ_y^2 are the variance reduction factors in the two horizontal directions, x and y, and Γ_z^2 is the variance reduction factor in the vertical direction, z. The approximate value for $\Gamma_i^2 = 1$ when the size of the influence zone in the ith direction (denoted by L_i) is shorter than the scale of fluctuation in the ith direction (denoted by SOF_i). For the case of $L_i > SOF_i$, the approximate value is $\Gamma_i^2 = SOF_i/L_i$ (Vanmarcke 1977). It is customary to assume the soil property of interest follows a normal (Gaussian) population if the total COV is less than 30% (Schneider and Schneider 2013). In this case, the characteristic value can be calculated as

$$X_k = X_{mean}\left(1 - 1.645 \times COV_{TOT}\right) \quad (3.2)$$

where X_{mean} is the (sample) mean value of evaluated soil properties. If the total COV exceeds 30%, it is more reasonable to assume the soil property of interest follows a lognormal population (Schneider and Schneider 2013) to avoid a negative characteristic value:

$$X_k = X_{mean} \cdot \frac{e^{-1.645 \cdot \sqrt{\ln\left(1+COV_{TOT}^2\right)}}}{\sqrt{1+COV_{TOT}^2}} \approx X_{mean} \cdot \frac{0.2^{\sqrt{\ln\left(1+COV_{TOT}^2\right)}}}{\sqrt{1+COV_{TOT}^2}} \quad (3.3)$$

In the above equations, the evaluation of the characteristic value X_k may require site-specific statistics for inherent variability such as site-specific mean, COV_{inher}, and SOFs. However, in geotechnical site investigation, site-specific data are usually sparse and limited, particularly for small- or medium-sized projects. This leads to the difficulty in obtaining meaningful site-specific statistics of soil/rock properties. To deal with these challenges, sparse site-specific data might be integrated with prior knowledge such as typical ranges of site-specific mean, COV_{inher}, and SOFs (e.g., Phoon and Kulhawy 1999a; Wang and Cao 2013). This underlines a need to summarize the typical values of site-specific mean, COV_{inher}, and SOFs for different soil/rock properties from existing databases. The EPRI TR-105000 report (Phoon et al. 1995) and Phoon and Kulhawy (1999a) have complied the statistics for some soil parameters. Kulhawy et al. (2000) summarized the ranges of COV_{inher} for some soil/rock parameters as shown in Table 3.1, whereas Phoon et al. (1995) summarized the ranges of SOFs for some soil parameters as shown in Table 3.2.

Nonetheless, more soil/rock databases have been collected recently, and some databases are shown in Table 3.3, labelled as (soil type)/(number of parameters of interest)/(number of data points). Because these databases cover a broad range, they may be used to update the

Table 3.1 Coefficients of variation for inherent variability for some soil/rock parameters

			COV_{inher} (%)		
Test type	*Property*	*Material type*	*# Groups*	*Range*	*Mean*
Index	γ, γ_d (kN/m^3)	Fine-grained	14	2–20	7.8
	w_n (%)	Fine-grained	40	7–46	18.1
	PL	Fine-grained	23	6–34	15.7
	LL	Fine-grained	38	7–39	18.1
	PI – all data	Fine-grained	33	9–57	29.5
	– ≤20%	Fine-grained	13	16–57	35
	– >20%	Fine-grained	20	9–40	26
	γ, γ_d	Rock	42	0.1–3	0.9
	n (%)	Rock	25	3–71	25.9
Strength	ϕ, tan ϕ	Sand, clay	48	4–50	13.9
		Sand	32	4–15	9
		Clay	16	10–50	23.5
	s_u (kPa)	Clay	100	6–80	31.5
	q_u (kPa)	Rock	184	0.3–61	14.2
	σ_{bt} (kPa)	Rock	74	2–58	16.6
Stiffness	$E_{t\text{-}50}$ (MPa)	Rock	32	7–63	30.7
CPT	q_c (kPa)	Sand, clay	65	10–81	36.6
		Sand	54	10–81	38.2
		Clay	11	16–40	28.4
	q_t (kPa)	Clay	9	2–17	7.9
FV	s_u^{FV} (kPa)	Clay	26	13–36	25.3
SPT	N	Sand, clay	23	19–62	38
DMT	A,B	Sand, clay	56	12–59	27.9
		Sand	30	13–59	34.8
		Clay	26	12–38	19.9
	I_{DMT} (w/o outliers)	Sand	29	8–66	37.7
	K_{DMT} (w/o outliers)	Sand	29	15–67	37.6
	E_D (w/o outliers) (MPa)	Sand	30	7–69	41.1

Source: Table 1, Kulhawy et al. (2000).

Note: γ = total unit weight; n = apparent porosity; γ_d = dry unit weight; w_n = natural water content; PL = plastic limit; LL = liquid limit, PI = plasticity index; ϕ' = effective stress friction angle; s_u = undrained shear strength; q_u = uniaxial compressive strength; σ_{bt} = Brazilian indirect tensile strength; $E_{t\text{-}50}$ = tangent modulus at $0.5q_u$; q_c = cone penetration test (CPT) tip resistance; q_t = corrected tip resistance; s_u^{FV} = undrained shear strength from field vane; N = standard penetration test (SPT) blow count; A, B = dilatometer test (DMT) A and B readings; I_{DMT} = dilatometer material index; K_{DMT} = dilatometer horizontal stress index; E_{DMT} = dilatometer modulus.

statistics reported in the EPRI TR-105000 report. One purpose of this chapter is to present the typical values of site-specific mean and COV_{inher} observed from the seven soil/rock databases, including CLAY/10/7490, SAND/7/2794, ROCK/9/4069, ROCKMass/9/5876, CLAY-C_c/6/6203, SOIL-DMT/8/7186, and SOIL-KSAT/7/2293 in Table 3.3. These databases are compiled by the author of this chapter. The details of these databases can be found in Chapter 1 of ISSMGE-TC304 (2021). Another purpose of this chapter is to present the typical values of site-specific SOFs observed from the literature. The

Table 3.2 Scales of fluctuation for some soil parameters

Direction	*Parameter*	*Soil type*	*# Studies*	*SOF (m)*	
				Range	*Mean*
Vertical	s_u	Clay	5	0.8–6.1	2.5
	q_c	Sand, clay	7	0.1–2.2	0.9
	q_t	Clay	10	0.2–0.5	0.3
	s_u^{FV}	Clay	6	2.0–6.2	3.8
	N	Sand	1	–	2.4
	w_n	Clay, loam	3	1.6–12.7	5.7
	LL	Clay, loam	2	1.6–8.7	5.2
	Γ	Clay, loam	2	2.4–7.9	5.2
Horizontal	q_c	Sand, clay	11	3.0–80.0	47.9
	q_t	Clay	2	23.0–66.0	44.5
	s_u^{FV}	Clay	3	46.0–60.0	50.7
	w_n	Clay	1	–	170.0

Source: Table 4-4, Phoon et al. (1995).

Table 3.3 Soil/rock databases compiled by the author of this chapter

Database	*Reference*	*Parameters of interest*	*# Records*	*# Studies*
CLAY/10/7490	Ching and Phoon (2014a)	LL, PI, LI, σ'_v/P_a, σ'_p/P_a, s_u/σ'_v, S_t, q_{t1}, q_{tu}, B_q	7490	251
SAND/7/2794	Ching et al. (2017)	D_{50}, C_u, D_r, σ'_v/P_a, ϕ', q_{c1n}, $(N_1)_{60}$	2794	176
ROCK/9/4069	Ching et al. (2018)	γ, n, R_L, S_h, σ_{bt}, I_{s50}, V_p, σ_{ci}, E_i	4069	184
ROCKMass/9/5876	Ching et al. (2021a)	RQD, RMR, Q, GSI, E_m, E_{em}, E_{dm}, E_i, σ_{ci}	5784	225
CLAY-C_c/6/6203	Ching et al. (2022)	LL, PI, w_n, e, C_c, C_{ur}	6203	427
SOIL-DMT/8/7186	Ching and Kuo (2023)	LI, D_{50}, s_u^{FV}/P_a, N_{60}, $(q_t-\sigma_v)/P_a$, B_q, E_{PMT}/P_a, $(1-\nu^2) \times E_{DMT}/P_a$	7286	237
SOIL-KSAT/7/2293	Ching et al. (2023)	k_{sat}, e, w, LL, PI, D_{10}, CC	2293	76

Note: ν = Poisson ratio; σ'_p = pre-consolidation stress; σ'_v = vertical effective stress; σ_{ci} = uniaxial compressive strength of intact rock; $(N_1)_{60} = N_{60}/(\sigma'_v/P_a)^{0.5}$; B_q = CPT pore pressure ratio = $(u_2-u_0)/(q_t-\sigma_v)$; CC = clay content (in %); C_c = compression index; C_{ur} = unload/reload index; C_u = coefficient of uniformity; D_{10} = 10% percentile grain size; D_{50} = median grain size; D_r = relative density; e = void ratio; E_{PMT} = soil modulus determined by PMT; E_{dm} = dynamic modulus of rock mass; E_{em} = elasticity modulus of rock mass; E_i = Young's modulus of intact rock; E_m = deformation modulus of rock mass; GSI = geological strength index; I_{s50} = point load strength index for diameter 50 mm; k_{sat} = saturated hydraulic conductivity; LI = liquidity index; N_{60} = corrected SPT N; P_a = atmospheric pressure = 101.3 kPa; Q = Q-system; $q_{c1n} = (q_c/P_a)/(\sigma'_v/P_a)^{0.5}$; $q_{t1} = (q_t-\sigma_v)/\sigma'_v$ = normalized cone tip resistance; $q_{tu} = (q_t-u_2)/\sigma'_v$ = effective cone tip resistance; R_L = L-type Schmidt hammer hardness; RMR = rock mass rating; RQD = rock quality designation; S_h = Shore scleroscope hardness; S_t = sensitivity; s_u^{re} = remoulded s_u; u_0 = hydrostatic pore pressure; u_2 = CPTU pore pressure; V_p = P-wave velocity; w = water content; w_n = natural water content.

SOF typical ranges are extracted from three new sources: Cami et al. (2020), Chapter 3 of ISSMGE-TC304 (2021), and Ching et al. (2023).

The COV_{trans} term in Equation (3.1) is for the transformation uncertainty. Transformation models (Phoon and Kulhway 1999b) are useful to infer soil/rock properties from indirect measurements. Although a site-specific transformation model is desirable, its calibration may require sufficient site-specific data. When site-specific data are sparse (which is usually the case), a generic transformation model calibrated by a soil/rock database may be adopted to infer soil/rock properties. The multivariate soil/rock databases in Table 3.3 can be adopted to calibrate generic transformation models. Another purpose of this chapter is to present the bias (*b*) and COV_{trans} for some useful generic transformation models calibrated by the databases in Table 3.3.

The main disadvantage of a generic transformation model is that it is usually subjected to significant transformation uncertainty (COV_{trans} is large). The final section (Section 3.4) presents a recent theoretical advancement called the hierarchical Bayesian model (HBM), which constructs a "quasi-site-specific" transformation model that combines the sparse target-site data with the experiences learned from a soil/rock database. The transformation uncertainty for a quasi-site-specific transformation model can be significantly less than that for a generic transformation model.

3.2 TYPICAL VALUES OF SITE-SPECIFIC MEAN AND COV_{INHER} OBSERVED IN SOIL/ROCK DATABASES

One purpose of the current chapter is to report the typical range of site-specific mean and COV_{inher} of soil/rock properties observed in the databases in Table 3.3. Another purpose is to present the bias (*b*) and COV_{trans} of some useful generic transformation models calibrated by these databases.

3.2.1 CLAY/10/7490

The CLAY/10/7490 database (Ching and Phoon 2014a) compiles 7490 records for ten dimensionless clay parameters, including liquid limit (LL), plasticity index (PI), liquidity index (LI), effective stress (σ'_v/P_a), pre-consolidation stress (σ'_p/P_a), normalized undrained shear strength (s_u/σ'_v), sensitivity (S_t), normalized CPT cone resistances (q_{t1} and q_{tu}), and CPT pore pressure coefficient (B_q). For s_u/σ'_v, all s_u values in CLAY/10/7490 are converted to the mobilized s_u values defined by Mesri and Huvaj (2007), denoted by s_u(mob). The clay properties cover a wide range of LL (from 18 to 515), PI (from 2 to 363), LI (from −0.75 to 6.5), OCR (from 1 to 60.2), and S_t (from 1 to 1467). Details are given by Ching and Phoon (2014a). Each record in CLAY/10/7490 is stored as one row in the Excel worksheet. There were originally 7490 rows (records) in the worksheet, but the number of records has gradually grown since the CLAY/10/7490 database was first presented by Ching and Phoon (2014a). At the time this chapter is written, there are 12,754 records. The ten parameters are not completely observed for each record, i.e., there are empty entries. If the database is complete, it contains 12,754 × 10 = 127,540 entries. The current CLAY/10/7490 contains 55,976 entries, i.e., it is 44% complete.

The basic marginal statistics of the ten parameters are listed in Table 3.4. The statistics for five derived parameters (s_u/σ'_p, OCR, w_n, q_t, s_u) are also shown. The number of records containing a given parameter (e.g., LL) is shown in the second column of Table 3.4. Note that the statistics in Table 3.4 are not for a specific site but for the entire database (generic), i.e., data from multiple sites are pooled together. In this chapter, the term "site" can be regarded as the footprint below a typical size building or building complex and its peripheral land, occupying a few football fields. Different field and laboratory tests can be conducted at different locations within a site boundary. The records obtained from these tests conducted at

Table 3.4 Summary of generic statistics for (expanded) CLAY/10/7490

Variable	*# Data*	*Mean*	*COV*	*Min*	*Max*
LL (%)	10315	62.9	0.64	18.0	612.7
PI (%)	10720	35.0	0.94	1.7	493.0
LI	8410	0.86	0.91	−2.72	8.20
σ'_v/P_a	8692	1.50	1.46	0.00	35.32
σ'_p/P_a	5608	2.84	1.81	0.01	167.82
s_u/σ'_v	3914	0.50	1.57	0.02	16.00
S_t	3228	51.7	17.08	1.0	26,679.2
B_q	1475	0.50	0.65	−0.10	3.12
q_{t1}	2083	13.29	1.53	0.68	354.50
q_{tu}	1531	7.75	1.76	−7.34	228.24
s_u/σ'_p	2238	0.21	0.46	0.072	0.94
OCR	5362	2.95	1.98	0.045	212.0
w_n (%)	9588	56.5	0.80	5.80	868.7
q_t (kPa)	2209	1474.3	1.06	66.7	15,748.0
s_u (kPa)	3468	58.6	1.48	0.47	1049.8

Table 3.5 Summary of site-specific statistics for (expanded) CLAY/10/7490

	#	*# Records/site*		*Site-specific mean*		*Site-specific COV_{inher}*	
Property	*Sites*	*Range*	*Mean*	*95% CI*	*Mean*	*95% CI*	*Mean*
LL (%)	551	5-61	12.4	27.7–144.2	62.7	0.036–0.41	0.18
PI (%)	575	5-61	12.5	7.7–88.7	35.1	0.048–0.58	0.26
LI	479	5-61	12.6	−0.16–2.29	0.89	0.068–1.63	0.40
σ'_v/P_a	526	5-64	13.2	0.17–6.57	1.40	0.12–1.01	0.44
σ'_p/P_a	326	5-61	12.0	0.27–15.74	2.60	0.058–0.91	0.38
s_u/σ'_v	234	5-63	12.2	0.14–1.80	0.52	0.078–1.62	0.45
S_t	165	5-60	11.7	1.0–175.47	24.2	0.068–1.45	0.50
B_q	98	5-46	13.9	0.028–0.99	0.52	0.065–1.46	0.37
q_{t1}	140	5-64	13.6	2.29–57.6	14.01	0.097–2.11	0.56
q_{tu}	102	5-46	13.8	1.11–27.9	7.72	0.12–2.29	0.64
s_u/σ'_p	148	5-59	10.8	0.088–0.39	0.22	0.052–0.62	0.22
OCR	315	5-60	12.2	0.88–14.67	2.90	0.068–1.54	0.50
w_n (%)	549	5-61	12.9	19.4–141.6	60.9	0.039–0.53	0.19
q_t (kPa)	149	5-64	13.6	274.8–4806.3	1513.4	0.099–1.11	0.42
s_u (kPa)	229	5-63	12.1	8.51–299.3	58.4	0.069–0.88	0.38

different depths and different locations are grouped into a "site." For CLAY/10/7490, the number of records associated with each site varies from 5 to 64. Table 3.5 shows the site-specific statistics. To obtain meaningful second-order site-specific statistics, only sites with no less than five records are considered.

Soil/rock parameters may be correlated to each other. The site-specific bivariate correlation coefficients are shown in Table 3.6. They are computed based on the multivariate

Table 3.6 Summary of site-specific correlations for (expanded) CLAY/10/7490 (Pearson, Spearman, Kendall)

	PI	*LI*	σ'_v/P_a	σ'_p/P_a	s_u/σ'_v	S_t	B_q
LL	0.90,0.85,0.75 (459)	−0.24,−0.22,−0.17 (423)	−0.21,−0.18,−0.13 (358)	−0.16,−0.16,−0.13 (223)	0.04,0.03,0.02 (169)	−0.10,−0.08,−0.06 (104)	0.04,0.02,0.00 (65)
PI		−0.21,−0.18,−0.14 (427)	−0.20,−0.17,−0.13 (369)	−0.13,−0.14,−0.11 (233)	0.03,0.01,0.01 (178)	−0.12,−0.10,−0.09 (107)	0.04,0.03,0.03 (68)
LI			−0.11,−0.13,−0.11 (359)	−0.31,−0.30,−0.24 (217)	−0.10,−0.05,−0.04 (180)	0.29,0.24,0.20 (104)	0..03,0.02,0.01 (68)
σ'_v/P_a				0.47,0.45,0.40 (294)	−0.57,−0.56,−0.47 (228)	0.05,0.04,0.02 (80)	0.41,0.41,0.34 (98)
σ'_p/P_a					−0.15,−0.23,−0.18 (146)	0.04,0.08,0.06 (56)	0.15,0.22,0.16 (54)
s_u/σ'_v						−0.08,0.02,0.03 (57)	−0.49,−0.41,−0.34 (43)
S_t							0.12,0.11,0.09 (22)
B_q							
q_{t1}							
q_{tu}							
s_u/σ'_p							
OCR							
W							
q_t							

Table 3.6 Continued

q_{tl}	q_{tu}	s_u/σ'_p	*OCR*	*w*	q_t	s_u
0.00,0.01,0.01 (86)	−0.01,0.00,0.00 (68)	−0.06,−0.09,−0.08 (103)	0.09,0.11,0.09 (204)	0.61,0.57,0.47 (428)	−0.11,−0.09,−0.07 (94)	−0.22,−0.23,−0.17 (169)
−0.01,0.00,0.00 (90)	−0.02,−0.01,−0.01 (71)	−0.06,−0.09,−0.07 (110)	0.11,0.12,0.10 (211)	0.54,0.51,0.41 (434)	−0.09,−0.09,−0.06 (98)	−0.23,−0.25,−0.19 (175)
−0.11,−0.09,−0.08 (88)	−0.07,−0.05,−0.04 (71)	0.00,0.04,0.04 (110)	−0.12,−0.09,−0.07 (199)	0.40,0.37,0.30 (421)	−0.24,−0.20,−0.17 (96)	−0.28,−0.27,−0.21 (179)
−0.45,−0.45,−0.38 (140)	−0.50,−0.52,−0.44 (102)	0.01,−0.01,0.00 (147)	−0.54,−0.51,−0.43 (296)	−0.29,−0.27,−0.22 (400)	0.63,0.62,0.55 (143)	0.61,0.56,0.49 (228)
−0.14,−0.24,−0.19 (61)	−0.13,−0.25,−0.19 (53)	−0.27,−0.28,−0.22 (146)	0.20,0.14,0.11 (296)	−0.42,−0.41,−0.33 (261)	0.70,0.66,0.57 (62)	0.67,0.63,0.54 (146)
0.69,0.57,0.46 (48)	0.66,0.52,0.42 (43)	0.18,0.28,0.25 (129)	0.73,0.62,0.53 (130)	0.00,0.03,0.01 (182)	−0.35,−0.39,−0.32 (48)	0.02,−0.03,−0.03 (228)
−0.14,−0.21,−0.17 (24)	−0.12,−0.15,−0.12 (22)	0.07,0.01,0.02 (40)	−0.07,−0.03,−0.01 (53)	0.12,0.10,0.07 (109)	−0.03,−0.08,−0.06 (24)	−0.03,−0.02,−0.02 (57)
−0.61,−0.61,−0.52 (98)	−0.76,−0.78,−0.70 (98)	0.05,0.07,0.06 (33)	−0.50,−0.50,−0.42 (54)	0.09,0.05,0.03 (71)	0.02,0.07,0.07 (98)	0.23,0.24,0.20 (43)
	0.92,0.86,0.78 (101)	−0.03,0.01,0.01 (33)	0.61,0.50,0.42 (61)	−0.05,−0.01,−0.01 (94)	0.13,0.02,0.01 (140)	−0.22,−0.27,−0.21 (48)
		−0.10,−0.06,−0.5 (32)	0.68,0.60,0.50 (53)	−0.05,−0.03,−0.02 (74)	0.03,−0.11,−0.10 (102)	−0.22,−0.27,−0.21 (43)
			−0.32,−0.27,−0.23 (130)	−0.02,−0.03,−0.4 (114)	−0.15,−0.17,−0.11 (33)	0.29,0.22,0.19 (143)
				−0.01,0.03,0.02 (235)	−0.27,−0.32,−0.27 (62)	−0.24,−0.22,−0.18 (130)
					−0.30,−0.27,−0.22 (102)	−0.39,−0.37,−0.30 (182)
						0.64,0.60,0.52 (48)

site-specific records in CLAY/10/7490. Three types of correlation coefficients are considered: (a) the Pearson product–moment correlation coefficient; (b) the Spearman rank correlation coefficient (Hotelling and Pabst 1936); and (c) the Kendall's tau rank correlation coefficient (Kendall 1938). The numbers in the parentheses indicate the numbers of sites with no less than five bivariate site-specific records. The correlations shown in the table are the mean values among these sites.

The CLAY/10/7490 database can also be used to calibrate some generic transformation models. A generic transformation model is usually developed by regression based on bivariate/multivariate records in a soil/rock database. The resulting regression equation does not provide an exact fit, and the variability is called the transformation uncertainty. When implemented to a future case that is not within the database, a generic transformation model may exhibit both bias and transformation uncertainty. Ching and Phoon (2014a) defined the bias (b) as the sample mean of the following ratio:

$$\varepsilon = \frac{\text{measured target value}}{\text{prediction made by a transformation model}} \tag{3.4}$$

The COV of the transformation uncertainty (COV_{trans} in Equation 3.1) can be estimated as the sample COV of the ε ratio. The bias (b) and COV_{trans} of a transformation model can be calibrated by a generic soil/rock database. For instance, consider the s_u-CPT transformation model in Table 3.7. The target value is s_u, and the prediction is $(q_t-\sigma_v)/[29.1 \times \exp(-0.513 \times B_q)]$. For all records in CLAY/10/7490 with simultaneous knowledge of s_u, q_t, σ_v, and B_q, their ε ratios can be computed, and b and COV_{trans} are simply the sample mean and sample COV of these ε ratios. In Table 3.7, the values of b and COV_{trans} for some transformation models are calibrated by the records in CLAY/10/7490. The numbers of records used to calibrate b and COV_{trans} are shown in the third column in the table. To implement the calibrated transformation model to a future case with known (q_t, σ_v, B_q), the unbiased prediction for its s_u is simply $b \times (q_t-\sigma_v)/[29.1 \times \exp(-0.513 \times B_q)]$, and the transformation uncertainty has a COV = COV_{trans}.

3.2.2 SAND/7/2794

The SAND/7/2794 database (Ching et al. 2017) compiles 2794 records for seven sand parameters, including the median grain size (D_{50}), coefficient of uniformity (C_u), relative density (D_r), normalized vertical effective stress (σ'_v/P_a), effective stress friction angle (ϕ'), normalized cone tip resistance $q_{c1n} = (q_c/P_a) \times C_N$ ($C_N = (\sigma'_v/P_a)^{-0.5}$ is the correction factor for overburden stress), and normalized N value $(N_1)_{60} = N_{60} \times C_N$ is recorded. The sand properties cover a wide range of D_{50} (0.1 mm to more than 100 mm), C_u (1 to more than 1000), D_r (−0.1% to 117%), and OCR (1 to 15, but mostly 1). Details are given in Ching et al. (2017). The SAND/7/2794 database is 59% complete. The SAND/7/2794 database is dominated by reconstituted sands (79%). Only 21% of the records are in situ sands. In particular, 86% of the q_{c1n} data are based on reconstituted sands (laboratory calibration chamber tests) and 93% of the ϕ' data are for reconstituted sands.

The generic statistics of the seven parameters are listed in Table 3.8. There are only limited in situ records in SAND/7/2794, so there are insufficient sites to obtain meaningful site-specific statistics. The statistics in Table 3.8 are dominated by reconstituted sands. The bivariate correlation coefficients (Pearson, Spearman, and Kendall) are computed

Table 3.7 Some transformation models calibrated by (expanded) CLAY/10/7490

Model	*Literature*	*# Data*	*Transformation model*	*b*	COV_{trans}
$s_u^{re} - LI$	Locat and Demers (1988)	915	$s_u^{re} / P_a \approx 0.0144 \times LI^{-2.44}$	6.68	3.17
$S_t - LI$	Bjerrum (1954)	2329	$S_t \approx 10^{0.8LI}$	1.48	1.27
$\sigma'_p - LI - S_t$	Ching and Phoon (2012a)	794	$\sigma'_p / P_a \approx 0.235 \times LI^{-1.319} \times S_t^{0.536}$	2.24	1.01
$s_u - \sigma'_p$	Mesri (1975, 1989)	2238	$s_u(mob) / \sigma'_p \approx 0.22$	0.97	0.46
$s_u - \text{OCR}$	Jamiolkowski et al. (1985)	2238	$s_u(mob) / \sigma'_v \approx 0.23 \times \text{OCR}^{0.8}$	1.05	0.48
$s_u - \text{CPT}$	Ching and Phoon (2012b)	658	$s_u(mob) \approx \dfrac{q_t - \sigma_v}{29.1 \times \exp(-0.513B_q)}$	1.30	0.45
OCR – CPT	Kulhwawy and Mayne (1990)	856	$\text{OCR} \approx 0.32 \times (q_t - \sigma_v) / \sigma'_v$	1.04	0.67
$\sigma'_p - \text{CPT}$	Kulhwawy and Mayne (1990)	856	$\sigma'_p \approx 0.33 \times (q_t - \sigma_v)$	1.01	0.67

Table 3.8 Summary of generic statistics for SAND/7/2794

Variable	*# Data*	*Mean*	*COV*	*Min*	*Max*
D_{50} (mm)	2356	2.52	3.43	0.080	160.0
C_u	2196	14.7	5.43	1.0	1913.0
D_r (%)	2089	64.6	0.38	−0.071	117.0
σ'_v/P_a	1772	1.80	1.19	0.010	24.7
ϕ' (°)	1135	39.5	0.13	22.8	59.9
q_{c1n}	1541	157.2	0.72	0.75	536.8
$(N_1)_{60}$	611	34.1	0.76	2.15	243.5

Source: Ching et al. (2017).

based on the multivariate generic records in SAND/7/2794. The generic correlations are shown in Table 3.9. The numbers in the parentheses indicate the numbers of bivariate generic records. Table 3.10 shows the values of b and COV_{trans} for some generic transformation models calibrated by the records in SAND/7/2794. It is worth mentioning that the two ϕ'-SPT transformation models are calibrated by undisturbed in situ samples (e.g., extracted by ground freezing).

3.2.3 ROCK/9/4069

The ROCK/9/4069 database (Ching et al. 2018) compiles 4069 records for nine intact rock parameters, including unit weight (γ), porosity (n), L-type Schmidt hammer hardness (R_L), Shore scleroscope hardness (S_h), Brazilian tensile strength (σ_{bt}), point load strength index (I_{s50}), uniaxial compressive strength (σ_{ci}), and Young's modulus (E_i), and P-wave velocity (V_p). Jointed rock masses are not covered by this database. The database is dominated by igneous

Table 3.9 Summary of generic correlations for SAND/7/2794 (Pearson, Spearman, Kendall)

	C_u	D_r	σ'_v/P_a	ϕ'	q_{cln}	$(N_1)_{60}$
D_{50}	0.53,0.37,0.24 (2153)	0.10,0.13,0.08 (1844)	0.53,0.20,0.14 (1483)	0.22,0.40,0.27 (1021)	0.18,0.24,0.16 (1375)	0.29,0.24,0.17 (351)
C_u		0.08,0.09,0.06 (1763)	0.15,0.16,0.10 (1452)	0.25,0.27,0.19 (895)	−0.10,−0.04,−0.03 (1358)	0.14,0.37,0.26 (267)
D_r			0.04,−0.03,−0.02 (1465)	0.53,0.52,0.37 (716)	0.77,0.82,0.62 (1060)	0.60,0.74,0.58 (354)
σ'_v/P_a				−0.07,−0.08,−0.06 (295)	−0.07,0.10,0.00 (1255)	−0.14,−0.19,−0.13 (553)
ϕ'					0.71,0.78,0.56 (376)	0.46,0.44,0.30 (69)
q_{cln}						0.88,0.84,0.66 (188)

Table 3.10 Some generic transformation models calibrated by SAND/7/2794

Model	*Literature*	*# Data*	*Transformation model*	*b*	*COV$_{trans}$*
D_r – SPT	Terzaghi and Peck (1967)	198	$D_r(\%) \approx 100 \times \sqrt{(N_1)_{60} / 60}$	1.05	0.23
	Kulhawy and Mayne (1990)	199	$D_r(\%) \approx 100 \times \sqrt{\dfrac{(N_1)_{60}}{[60 + 25\log_{10}(D_{50})] \times OCR^{0.18}}}$	1.01	0.21
D_r – CPT	Jamiolkowski et al. (1985)	681	$D_r(\%) \approx 68 \times [\log_{10}(q_{c1n}) - 1]$	0.84	0.33
ϕ'– D_r	Bolton (1986)	391	$\phi' \approx_{cv} + 3 \times (D_r[10 - \ln(p'_f)] - 1)$ [b]	1.03	0.052
ϕ'– SPT	Hatanaka and Uchida (1996)	58[a]	$\phi' \approx \begin{cases} \sqrt{15.4 \times (N_1)_{60}} + 20 & (N_1)_{60} \le 26 \\ 40 & (N_1)_{60} > 26 \end{cases}$	1.07	0.090
	Chen (2004)	59[a]	$\phi' \approx 27.5 + 9.2 \times \log_{10}[(N_1)_{60}]$	1.00	0.095
ϕ'– CPT	Robertson and Campanella (1983)	99	$\phi' \approx \tan^{-1}[0.1 + 0.38 \times \log_{10}(q_t / \sigma'_v)]$	0.93	0.056
	Kulhawy and Mayne (1990)	376	$\phi' \approx 17.6 + 11 \times \log_{10}(q_{c1n})$	0.97	0.081

Source: ISSMGE-TC304 (2021).

[a] Based on undisturbed in situ samples (e.g., extracted by ground freezing).

[b] ϕ'_{cv} = critical-state friction angle; p'_f = mean confining stress at failure in kPa.

and sedimentary rocks (27.5% and 59.4%, respectively). The remaining (about 13.1%) data points are metamorphic rocks. The records in ROCK/9/4069 cover a wide range of γ (15 to 35 kN/m^3), n (0.01 to 55%), σ_{ci} (0.7 to 380 MPa), E_i (0.03 to 120 GPa), and V_p (0.4 to 8 km/s). Details are given in Ching et al. (2018). The ROCK/9/4069 database is 34% complete.

The generic statistics of the nine parameters are listed in Table 3.11. Tables 3.12–3.14 show the site-specific mean and COV$_{inher}$ for igneous, sedimentary, and metamorphic sites, respectively. Table 3.15 shows the site-specific bivariate correlation coefficients. The

Table 3.11 Summary of generic statistics for ROCK/9/4069

Variable	*# Data*	*Mean*	*COV*	*Min*	*Max*
γ (kN/m^3)	1288	24.6	0.12	15.0	34.7
n (%)	1371	11.0	1.02	0.01	55.0
R_L	812	41.4	0.29	8.1	81.6
S_h	333	47.6	0.44	8.4	100.0
σ_{bt} (MPa)	854	8.2	0.61	0.07	34.4
I_{s50} (MPa)	1303	4.2	0.69	0.05	17.4
σ_{ci} (MPa)	3226	72.3	0.77	0.68	379.0
E_i (GPa)	1495	26.5	0.99	0.03	116.3
V_P (km/s)	1858	4.0	0.39	0.44	8.0

Source: Ching et al. (2018).

Table 3.12 Summary of site-specific statistics for ROCK/9/4069 (igneous)

		# Records/site		Site-specific mean		Site-specific COV_{inher}	
Property	*# Sites*	*Range*	*Mean*	*95% CI*	*Mean*	*95% CI*	*Mean*
γ (kN/m³)	11	5-20	11.0	21.4–30.1	25.4	0.002–0.18	0.049
n (%)	7	5-47	24.3	0.22–7.78	4.03	0.15–0.73	0.46
R_L	9	5-145	30.9	35.4–62.6	48.4	0.020–0.28	0.13
S_h	1	20	20	–	54.9	–	0.16
σ_{bt} (MPa)	5	5-24	15.0	11.2–19.4	14.8	0.064–0.54	0.19
I_{s50} (MPa)	9	5-145	38.0	1.75–9.02	4.44	0.14–0.81	0.41
σ_{ci} (MPa)	24	5-145	25.4	35.4–244.1	106.8	0.016–0.87	0.35
E_i (GPa)	15	5-62	21.1	4.45–86.0	34.5	0.029–0.60	0.31
V_P (km/s)	17	5-145	27.5	2.54–7.55	4.57	0.002–0.35	0.12

Table 3.13 Summary of site-specific for ROCK/9/4069 (sedimentary)

		# Records/site		Site-specific mean		Site-specific COV_{inher}	
Property	*# Sites*	*Range*	*Mean*	*95% CI*	*Mean*	*95% CI*	*Mean*
γ (kN/m³)	40	5-66	16.0	18.6–28.1	24.6	0.003–0.13	0.037
n (%)	42	5-55	17.5	1.09–37.4	13.1	0.030–0.74	0.33
R_L	14	5-44	13.6	20.4–44.4	31.5	0.026–0.32	0.14
S_h	6	8-31	20.5	13.4–76.1	43.2	0.12–0.35	0.23
σ_{bt} (MPa)	16	5-45	17.8	1.25–14.9	6.81	0.043–0.81	0.26
I_{s50} (MPa)	34	5-44	14.0	0.53–10.7	4.51	0.042–0.74	0.25
σ_{ci} (MPa)	78	5-66	14.3	7.05–134.6	61.5	0.019–0.71	0.26
E_i (GPa)	40	5-66	15.0	0.38–64.7	25.5	0.022–1.15	0.38
V_P (km/s)	58	5-66	13.3	0.89–6.30	3.67	0.004–0.37	0.10

Table 3.14 Summary of site-specific statistics for ROCK/9/4069 (metamorphic)

		# Records/site		Site-specific mean		Site-specific COV_{inher}	
Property	*# Sites*	*Range*	*Mean*	*95% CI*	*Mean*	*95% CI*	*Mean*
γ (kN/m³)	11	5-20	9.5	25.7–30.7	27.1	0.010–0.058	0.025
n (%)	7	5-20	10.3	0.18–6.67	1.47	0.26–1.39	0.56
R_L	2	8	8.0	47.0–50.7	48.9	0.10–0.16	0.13
S_h	–	–	–	–	–	–	–
σ_{bt} (MPa)	7	5-20	8.7	6.67–18.1	11.9	0.13–0.54	0.30
I_{s50} (MPa)	11	5-32	13.3	1.27–6.92	3.96	0.11–0.41	0.30
σ_{ci} (MPa)	17	5-32	11.7	24.1–172.5	81.0	0.068–0.55	0.29
E_i (GPa)	6	5-20	9.0	4.49–102.3	54.6	0.081–0.82	0.25
V_P (km/s)	13	5-32	12.5	2.27–6.27	4.66	0.026–0.31	0.11

numbers in the parentheses indicate the numbers of sites with no less than 5 bivariate site-specific records. Ching et al. (2018) indicated that the correlation behaviours between rock parameters do not seem to depend on the rock class (igneous, sedimentary, and metamorphic), so Table 3.15 does not differentiate the rock class. Table 3.16 shows the values of b and COV_{trans} for some generic transformation models calibrated by the records in ROCK/9/4069.

Table 3.15 Summary of site-specific correlations for ROCK/9/4069 (Pearson, Spearman, Kendall)

	n	R_L	S_h	σ_{bt}	I_{s50}	σ_{ci}	E_i	V_P
γ	−0.42,−0.37,−0.30 (22)	0.57,0.48,0.34 (7)	0.27,0.38,0.30 (3)	0.56,0.53,0.43 (8)	0.65,0.63,0.57 (16)	0.59,0.56,0.48 (43)	0.38,0.37,0.29 (18)	0.47,0.44,0.39 (39)
n		−0.38,−0.38,−0.30 (5)	−0.68,−0.62,−0.47 (3)	−0.55,−0.54,−0.43 (15)	−0.53,−0.56,−0.47 (19)	−0.54,−0.55,−0.44 (37)	−0.56,−0.51,−0.41 (28)	−0.57,−0.53,−0.43 (29)
R_L			0.94,0.93,0.80 (1)	0.90,0.88,0.75 (2)	0.66,0.61,0.49 (8)	0.72,0.71,0.58 (16)	0.64,0.58,0.47 (7)	0.47,0.45,0.38 (15)
S_h				0.69,0.68,0.54 (2)	0.76,0.73,0.58 (3)	0.55,0.51,0.40 (6)	0.55,0.51,0.40 (5)	0.91,0.85,0.72 (2)
σ_{bt}					0.57,0.52,0.44 (15)	0.60,0.57,0.46 (23)	0.42,0.37,0.28 (12)	0.43,0.44,0.35 (14)
I_{s50}						0.72,0.63,0.52 (32)	0.27,0.50,0.39 (13)	0.75,0.62,0.51 (15)
σ_{ci}							0.60,0.57,0.45 (55)	0.69,0.62,0.52 (48)
E_i								0.63,0.67,0.55 (29)

Table 3.16 Some generic transformation models calibrated by ROCK/9/4069

Model	*Literature*	*# Data*	*Transformation model*	*b*	*COV$_{trans}$*
$\sigma_{ci}-n$	Kilic and Teymen (2008)	911	$\sigma_{ci}\approx 147.16\times e^{-0.0835n}$	0.91	0.75
$\sigma_{ci}-R_L$	Karaman and Kesimal (2015)	664	$\sigma_{ci}\approx 0.1383\times R_L^{1.743}$	0.78	0.53
$\sigma_{ci}-S_h$	Altindag and Guney (2010)	297	$\sigma_{ci}\approx 0.1821\times S_h^{1.5833}$	1.15	0.65
$\sigma_{ci}-\sigma_{bt}$	Prakoso and Kulhawy (2011)	525	$\sigma_{ci}\approx 7.8\times\sigma_{bt}$	1.31	0.50
$\sigma_{ci}-I_{s50}$	Mishra and Basu (2013)	1074	$\sigma_{ci}\approx 14.63\times I_{s50}$	1.18	0.45
$\sigma_{ci}-V_P$	Kahraman (2001)	1247	$\sigma_{ci}\approx 9.95\times V_P^{1.21}$	1.26	0.63
E_i-R_L	Katz et al. (2000)	289	$E_i\approx 0.00013\times R_L^{3.09074}$	1.47	0.99
E_i-S_h	Deere and Miller (1966)	197	$E_i\approx 0.739\times S_h+11.51$	0.61	0.71
$E_i-\sigma_c$	Deere and Miller (1966)	1152	$E_i\approx 0.303\times\sigma_{ci}-0.8745$	1.23	0.94
E_i-V_P	Yaşar and Erdogan (2004)	192	$E_i\approx 10.67\times V_P-18.71$	0.90	0.72

Source: ISSMGE-TC304 (2021).

3.2.4 ROCKMass/9/5876

The ROCKMass/9/5876 database (Ching et al. 2021a) compiles 5876 records for nine rock mass parameters, including rock quality designation (RQD) (Deere 1964), rock mass rating (RMR) (Bieniawski 1973), Q-system (Barton et al. 1974), geological strength index (GSI) (Hoek and Brown 1997), deformation modulus of rock mass (E_m), elastic modulus of rock mass (E_{em}), dynamic modulus of rock mass (E_{dm}), Young's modulus of intact rock (E_i), and uniaxial compressive strength of intact rock (σ_{ci}). The database consists of 17% igneous, 37% sedimentary, and 26% metamorphic cases (the remaining 20% of the cases do not contain rock class information). The remaining (about 13.1%) data points are metamorphic rocks. The records in ROCKMass/9/5876 cover a wide range of RMR (0 to 97), *Q* (0.001 to 1000), GSI (8 to 100), and E_m (0.0011 to 104 GPa). Details are given by Ching et al. (2021a). The ROCKMass/9/5876 database is 29% complete.

The generic statistics of the nine parameters are listed in Table 3.17. Tables 3.18–3.20 show the site-specific mean and COV$_{inher}$ for igneous, sedimentary, and metamorphic sites,

Table 3.17 Summary of generic statistics for ROCKMass/9/5876

Variable	*# Data*	*Mean*	*COV*	*Min*	*Max*
RQD	1380	62.5	0.48	0	100
RMR	3399	55.5	0.33	0	97
Q	1933	21.5	2.92	0.001	1000
GSI	871	47.2	0.34	8	100
E_m (GPa)	3212	9.99	1.15	0.0011	104
E_{em} (GPa)	1386	14.50	1.12	0.00014	121
E_{dm} (GPa)	466	31.78	0.60	0.0411	85
E_i (GPa)	592	29.81	0.95	0.24	278
σ_{ci} (MPa)	2071	74.07	0.78	0.15	318

Source: Ching et al. (2021a).

Table 3.18 Summary of site-specific statistics for ROCKMass/9/5876 (igneous)

		# Records/site		Site-specific mean		Site-specific COV_{inher}	
Property	*# Sites*	*Range*	*Mean*	*95% CI*	*Mean*	*95% CI*	*Mean*
RQD	13	5-70	18.2	27.7–91.7	72.5	0.022–1.21	0.31
RMR	13	5-40	20.2	34.9–81.2	61.9	0.064–0.53	0.22
Q	15	7-70	22.3	0.36–74.3	21.1	0.36–3.04	1.13
GSI	6	6-26	12.2	36.3–58.2	50.8	0.016–0.30	0.13
E_m (GPa)	5	6-21	10.4	4.90–20.7	10.6	0.13–1.02	0.62
E_{em} (GPa)	6	5-21	11.7	1.91–25.3	14.1	0.57–1.04	0.71
E_{dm} (GPa)	–	–	–	–	–	–	–
E_i (GPa)	6	5-15	8.0	12.8–77.5	32.5	0.17–1.00	0.69
σ_{ci} (MPa)	11	5-31	14.2	3.00–192.0	76.5	0.13–1.18	0.35

Table 3.19 Summary of site-specific for ROCKMass/9/5876 (sedimentary)

		# Records/site		Site-specific mean		Site-specific COV_{inher}	
Property	*# Sites*	*Range*	*Mean*	*95% CI*	*Mean*	*95% CI*	*Mean*
RQD	22	5-80	14.9	0.37–93.4	47.9	0.061–1.07	0.37
RMR	30	5-80	19.1	27.4–73.2	47.8	0.053–0.53	0.23
Q	19	5-49	16.0	0.30–18.0	5.56	0.22–1.59	0.85
GSI	20	5-93	18.1	22.2–75.6	43.9	0.054–0.57	0.23
E_m (GPa)	20	5-94	20.1	0.12–30.6	9.53	0.14–1.34	0.67
E_{em} (GPa)	8	6-27	10.5	1.02–56.9	12.6	0.36–0.99	0.63
E_{dm} (GPa)	4	5-27	13.3	8.10–54.6	24.0	0.15–0.62	0.33
E_i (GPa)	11	5-53	13.7	2.80–76.0	21.5	0.00–1.20	0.42
σ_{ci} (MPa)	35	5-80	13.0	3.24–184.0	55.4	0.077–1.06	0.40

Table 3.20 Summary of site-specific statistics for ROCKMass/9/5876 (metamorphic)

		# Records/site		Site-specific mean		Site-specific COV_{inher}	
Property	*# Sites*	*Range*	*Mean*	*95% CI*	*Mean*	*95% CI*	*Mean*
RQD	11	5-47	15.8	20.2–81.9	56.4	0.078–1.20	0.40
RMR	12	5-330	43.2	32.4–74.0	49.4	0.12–0.41	0.27
Q	11	6-330	49.1	0.13–55.2	9.75	0.50–5.26	1.54
GSI	3	10-26	16.0	23.1–53.6	39.7	0.054–0.18	0.13
E_m (GPa)	8	5-24	10.1	0.71–17.2	5.86	0.24–0.74	0.50
E_{em} (GPa)	3	5-24	11.3	2.37–9.79	5.95	0.38–0.55	0.46
E_{dm} (GPa)	2	5-5	5.0	2.94–18.7	10.8	0.30–0.45	0.37
E_i (GPa)	5	5-18	8.4	8.64–92.8	47.8	0.15–0.82	0.46
σ_{ci} (MPa)	9	5-30	13.6	30.0–99.9	55.5	0.24–0.72	0.40

respectively. Table 3.21 shows the site-specific bivariate correlation coefficients. The numbers in the parentheses indicate the number of sites with no less than 5 bivariate site-specific records. Ching et al. (2021a) indicated that the correlation behaviours between rock mass parameters do not seem to depend on the rock class (igneous, sedimentary, and metamorphic). Table 3.22 shows the values of b and COV_{trans} for some generic transformation models calibrated by the records in ROCKMass/9/5876.

Table 3.21 Summary of site-specific correlations for ROCKMass/9/5876 (Pearson, Spearman, Kendall)

	RMR	*Q*	*GSI*	E_m	E_{em}	E_{dm}	E_i	σ_{ci}
RQD	0.74,0.74,0.62 (26)	0.45,0.60,0.52 (24)	0.61,0.70,0.55 (5)	0.43,0.36,0.28 (9)	0.21,0.16,0.12 (6)	0.85,0.82,0.72 (3)	−0.03,−0.06,−0.04 (5)	0.25,0.27,0.23 (29)
RMR		0.62,0.69,0.58 (32)	0.76,0.74,0.65 (13)	0.76,0.70,0.58 (11)	0.29,0.20,0.14 (6)	0.60,0.59,0.47 (3)	0.25,0.28,0.22 (9)	0.46,0.45,0.38 (27)
Q			0.64,0.71,0.62 (9)	0.43,0.62,0.51 (4)	0.33,0.65,0.53 (2)	0.67,0.71,0.55 (1)	0.08,0.02,0.00 (7)	0.21,0.24,0.20 (13)
GSI				0.76,0.77,0.63 (4)	0.65,0.63,0.50 (1)	−,-,− (0)	0.66,0.67,0.55 (6)	0.43,0.50,0.41 (13)
E_m					0.89,0.83,0.72 (11)	0.73,0.80,0.71 (4)	0.61,0.58,0.44 (8)	0.46,0.44,0.38 (10)
E_{em}						0.69,0.73,0.65 (5)	0.60,0.62,0.52 (4)	0.34,0.38,0.35 (6)
E_{dm}							0.46,0.30,0.20 (1)	−0.34,−0.35,−0.32 (1)
E_i								0.65,0.68,0.57 (11)

Table 3.22 Some generic transformation models calibrated by ROCKMass/9/5876

Model	*Literature*	*# data*	*Transformation model*	*b*	*COV_{trans}*
$E_m - E_i - \text{RQD}$	Coon and Merritt (1970)	147	$E_m / E_i \approx 0.0231 \times \text{RQD} - 1.32$	1.26	1.09
	Zhang and Einstein (2004)	161	$E_m / E_i \approx 10^{0.0186 \times RQD - 1.91}$	1.54	0.89
$E_m - \text{RMR}$	Bieniawski (1978)	1091	$E_m \approx 2 \times \text{RMR} - 100$	0.57	1.48
	Gokceoglu et al. (2003)	1749	$E_m \approx 0.0736 \times e^{0.0755 \times RMR}$	1.53	1.21
	Serafim and Pereira (1983)	1749	$E_m \approx 10^{\frac{(\text{RMR}-10)}{40}}$	0.51	1.00
$E_m - Q$	Grimstad and Barton (1993)	288	$E_m \approx 25 \times \log_{10}(Q)$	0.58	0.88
$E_m - GSI$	Hoek and Diederichs (2006)	349	$E_m \approx 100 \times \frac{(1 - D/2)}{1 + e^{\frac{75+25D-GSI}{11}}}$	0.83 ($D = 0$)	1.00 ($D = 0$)

Source: ISSMGE-TC304 (2021).

Note: σ_{ci} in MPa; E_m in GPa; and D is the disturbance factor (Hoek and Diederichs 2006).

3.2.5 CLAY-C_c/6/6203

The CLAY-C_c/6/6203 database (Ching et al. 2022) compiles 6203 records for six clay parameters, including liquid limit (LL), plasticity index (PI), natural water content (w_n), void ratio (e), compression index (C_c), and unloading–reloading index (C_{ur}). The unloading–reloading index C_{ur} denotes the average of swelling index (C_s) and reloading index (C_r) (Kulhawy and Mayne 1990). 85% of the records are (natural) undisturbed clays, whereas the rest 15% are (laboratory) reconstituted clays. The records in CLAY-C_c/6/6203 cover a wide range of LL (19 to 613%), PI (5 to 493%), w_n (6 to 560%), and C_c (0.01 to 14.5). Details are given in Ching et al. (2022). The CLAY-C_c/6/6203 database is 61% complete. The generic statistics of the six parameters are listed in Table 3.23. Tables 3.24 and 3.25 show the site-specific mean, COV, and bivariate correlation coefficients. Table 3.26 shows the values of b and COV_{trans} for some generic transformation models calibrated by the records in CLAY-C_c/6/6203.

3.2.6 SOIL-DMT/8/7186

The SOIL-DMT/8/7186 database (Ching and Kuo 2023) compiles 7186 records for eight soil parameters, including liquidity index (LI), median grain size (D_{50}), normalized undrained

Table 3.23 Summary of generic statistics for CLAY-C_c/6/6203

Variable	*# Data*	*Mean*	*COV*	*Min*	*Max*
LL	3702	63.1	0.60	19.0	612.7
PI	3282	34.6	0.91	5.0	493.0
w_n (%)	3752	56.8	0.68	5.8	559.5
e	3610	1.58	0.68	0.13	14.66
C_c	6203	0.79	1.18	0.01	14.50
C_{ur}	2102	0.094	1.14	0.002	2.16

Source: Ching et al. (2022).

Table 3.24 Summary of site-specific statistics for CLAY-C_c/6/6203

		# Records/site		Site-specific mean		Site-specific COV_{inher}	
Property	# Sites	Range	Mean	95% CI	Mean	95% CI	Mean
LL	149	5-136	12.7	33.9–130.1	64.7	0.021–0.56	0.20
PI	149	5-60	11.9	12.5–68.8	36.1	0.035–0.66	0.28
w_n (%)	170	5-93	14.0	21.2–140.8	66.8	0.051–0.60	0.22
e	161	5-72	13.0	0.57–4.82	1.87	0.043–0.51	0.22
C_c	227	5-136	15.7	0.18–3.46	1.03	0.088–0.89	0.40
C_{ur}	118	5-115	14.2	0.024–0.41	0.11	0.083–0.88	0.40

Table 3.25 Summary of site-specific correlations for CLAY-C_c/6/6203 (Pearson, Spearman, Kendall)

	PI	w_n	e	C_c	C_{ur}
LL	0.91,0.86,0.77 (117)	0.59,0.56,0.47 (116)	0.53,0.52,0.42 (103)	0.53,0.50.0.40 (141)	0.46,0.43,0.35 (60)
PI		0.54,0.51,0.41 (118)	0.49,0.48,0.38 (100)	0.50,0.48,0.38 (142)	0.45,0.43,0.34 (60)
w_n			0.88,0.82,0.74 (102)	0.63,0.59,0.48 (161)	0.50,0.46,0.36 (84)
e				0.71,0.67,0.55 (155)	0.58,0.52,0.42 (74)
C_c					0.53,0.49,0.40 (110)

Table 3.26 Some generic transformation models calibrated by CLAY-C_c/6/6203

Model	*Literature*	*# Data*	*Transformation model*	*b*	COV_{trans}
$C_c - LL$	Skempton (1944)	3398	$C_c \approx 0.007 \times (LL - 10)$	1.59	0.90
	Terzaghi and Peck (1967)	3398	$C_c \approx 0.009 \times (LL - 10)$	1.24	0.90
$C_c - PI$	Kulhawy and Mayne (1990)	2964	$C_c \approx PI / 73$	1.31	0.91
$C_c - LL - G_s$	Nagaraj and Murty (1985)	1523	$C_c \approx 0.2343 \times (LL / 100) \times G_s$	1.06	0.81
$C_{ur} - e$	Peck and Reed (1954)	668	$C_{ur} \approx 0.208 \times e + 0.0083$	0.31	0.69
$C_{ur} - w$	Azzouz et al. (1976)	771	$C_{ur} \approx 0.003 \times (w + 7)$	0.47	0.67
$C_{ur} - PI$	Kulhawy and Mayne (1990)	847	$C_{ur} \approx$ PI/385	0.96	0.63
$C_{ur} - LL$	Isik (2009)	846	$C_{ur} \approx 0.0007 \times LL + 0.0062$	1.55	0.64

Source: ISSMGE-TC304 (2021).

shear strength for field vane (s_u^{FV}/P_a), corrected SPT blow count (N_{60}), normalized CPT cone tip resistance [$(q_t-\sigma_v)/P_a$], CPT pore pressure coefficient (B_q), normalized soil modulus obtained by the pressuremeter test (E_{PMT}/P_a), and normalized soil modulus obtained by the dilatometer test [$(1-\nu^2) \times E_{DMT}/P_a$], where ν is the Poisson ratio (ν is assumed to be 0.5 for clays and 0.2 for granular soils). Most soil modulus data are from DMT. E_{DMT} is multiplied by $(1-\nu^2)$ to represent the Young's modulus (E) because according to Kulhawy and Mayne (1990), $E \approx (1-\nu^2) \times E_{DMT}$. In SOIL-DMT/8/7186, 74% of the records are in situ clays, and

Table 3.27 Summary of generic statistics for SOIL-DMT/8/7186

Variable	*# data*	*Mean*	*COV*	*Min*	*Max*
LI	3347	1.03	0.71	−2.17	7.0
D_{50} (mm)	406	0.29	3.77	0.01	11.0
s_u^{FV}/P_a	2984	0.32	0.89	0.002	3.62
N_{60}	1354	18.71	0.89	0.99	124.1
$(q_t-\sigma_v)/P_a$	3198	28.22	1.36	0.38	378.2
B_q	1408	0.50	0.51	−0.10	1.56
E_{PMT}/P_a	514	280.44	1.08	7.73	1888.7
$(1-\nu^2) \times E_{DMT}/P_a$	2966	174.06	1.09	1.12	1733.2

Source: Ching and Kuo 2023).

Table 3.28 Summary of site-specific statistics for SOIL-DMT/8/7186

		# Records/site		*Site-specific mean*		*Site-specific COV_{inher}*	
Property	*# Sites*	*Range*	*Mean*	*95% CI*	*Mean*	*95% CI*	*Mean*
LI	256	5-49	12.0	0.044–2.39	1.03	0.02–6.49	0.37
D_{50} (mm)	36	5-31	10.5	0.016–0.61	0.21	0.07–0.62	0.36
s_u^{FV}/P_a	240	5-50	12.0	0.084–1.10	0.34	0.03–1.82	0.38
N_{60}	102	5-57	12.2	1.98–46.7	17.0	0.08–1.53	0.43
$(q_t-\sigma_v)/P_a$	231	5-50	12.8	2.12–133.6	29.2	0.07–1.61	0.44
B_q	97	5-46	13.5	0.028–0.97	0.49	0.06–2.75	0.36
E_{PMT}/P_a	31	5-57	14.2	15.7–861.4	289.4	0.11–1.07	0.54
$(1-\nu^2) \times E_{DMT}/P_a$	216	5-50	12.9	7.96–529.6	167.2	0.08–1.21	0.46

the rest 26% are in situ granular soils. The records in SOIL-DMT/8/7186 cover a wide range of LI (–2.17 to 7 for clayey records), D_{50} (0.01 to 11 mm for granular soil records), N_{60} (less than 1 to 124), and soil modulus [e.g., $(1-\nu^2) \times E_{DMT}$ ranges from 0.11 to 176 MPa]. Details are given in Ching and Kuo (2023). The SOIL-DMT/8/7186 database is 28% complete. The generic statistics of the eight parameters are listed in Table 3.27. Tables 3.28 and 3.29 show the site-specific mean, COV, and bivariate correlation coefficients. Table 3.30 shows the values of b and COV_{trans} for some generic transformation models calibrated by the records in SOIL-DMT/8/7186.

3.2.7 SOIL-KSAT/7/2293

The SOIL-KSAT/7/2293 database (Ching et al. 2024) compiles 2293 records for seven soil parameters, including saturated hydraulic conductivity (k_{sat}), void ratio (e), water content (w), liquid limit (LL), plasticity index (PI), 10% percentile grain size (D_{10}), and clay content (CC). All k_{sat} data are based on laboratory tests on saturated soils. About 80% of the records are soils with clay content (CC) > 5%, and the remaining 20% are with CC < 5%. For soils with CC > 5%, D_{10} information is usually missing, and for soils with CC < 5%, (LL, PI) information is usually missing. Some records in SOIL-KSAT/7/2293 are extracted from the FG-KSAT/6/1358 and CG-KSAT/7/1278 databases developed by Feng and Vardanega (2019) and Feng et al. (2023). The 2293 records are divided into 272 groups. Each group is

Table 3.29 Summary of site-specific correlations for SOIL-DMT/8/7186 (Pearson, Spearman, Kendall)

	D_{50}	s_u^{FV}/P_a	N_{60}	$(q_t-\sigma_v)/P_a$	B_q	E_{PMT}/P_a	$(1-\nu^2) \times E_{DMT}/P_a$
LI	–,–,– (0)	−0.30,−0.29,−0.23 (199)	−0.10,−0.09,−0.07 (15)	−0.28, −0.27, −0.22 (84)	0.04,0.03,0.02 (64)	0.07,0.22,0.12 (3)	−0.29,−0.17,−0.13 (38)
D_{50}		–,–,– (0)	0.00,0.08,0.08 (8)	0.13,0.35,0.28 (17)	–,–,– (0)	−0.80,−0.63,−0.47 (1)	0.12,0.43,0.36 (19)
s_u^{FV}/P_a			0.09,0.09,0.10 (4)	0.65,0.63,0.53 (50)	0.16,0.20,0.16 (44)	–,–,– (0)	0.58,0.56,0.48 (17)
N_{60}				0.47,0.46,0.38 (59)	−0.10,−0.13,−0.11 (11)	0.52,0.50,0.40 (21)	0.43,0.40,0.32 (59)
$(q_t-\sigma_v)/P_a$					−0.12,−0.04,−0.03 (97)	0.39,0.34,0.26 (11)	0.56,0.53,0.42 (141)
B_q						−0.47,−0.46,−0.43 (1)	0.03,0.04,0.03 (39)
E_{PMT}/P_a							0.40,0.49,0.40 (4)

Table 3.30 Some generic transformation models calibrated by SOIL-DMT/8/7186

Model	*Literature*	*# Data*	*Transformation model*	*b*	COV_{trans}
$E_{PMT} - N_{60}$	Ohya et al. (1982)	94	$E_{PMT} / P_a \approx 19.3 \times N^{0.63}$ (clay)	1.99	0.60
	Ohya et al. (1982)	248	$E_{PMT} / P_a \approx 9.08 \times N^{0.66}$ (sand)	3.73	0.76
$E_{DMT} - N_{60}$	Mayne and Frost (1989)	472	$E_{DMT} / P_a \approx 22 \times N^{0.82}$ (sand)	1.40	0.57
$E_{DMT} - N_{60}$	This chapter	662	$\frac{(1-\nu^2)E_{DMT}}{P_a} \approx 14.2 \times N_{60}^{0.99}$ (all soils)	1.25	0.70
$E_{PMT} - q_t$		126	$\frac{E_{PMT}}{P_a} \approx 7.64 \times \left(\frac{q_t - \sigma_v}{P_a}\right)^{0.76}$ (all soils)	1.30	0.75
$E_{DMT} - q_t$		1796	$\frac{(1-\nu^2)E_{DMT}}{P_a} \approx 4.49 \times \left(\frac{q_t - \sigma_v}{P_a}\right)^{1.00}$ (all soils)	1.18	0.63
$E_{DMT} - s_u^{FV}$		252	$\frac{(1-\nu^2)E_{DMT}}{P_a} \approx 51.7 \times \left(\frac{s_u^{FV}}{P_a}\right)^{1.13}$ (clay)	1.08	0.38

featured with a series k_{sat} versus e (or w) data on the same type of soil (the same minerology, LL, PI, D_{10}, CC, fluid, etc.). Basically, soil specimens of the same type are consolidated to different void ratios (or water contents) in the laboratory, and their k_{sat} values are measured. LL, PI, D_{10}, and CC data for the records in the same group are constant because of the same soil type. The records in SOIL-KSAT/7/2293 cover a wide range of k_{sat} (5.2E–13 to 0.6 cm/s), e (0.19 to 4.81), w (7.1 to 178%), LL (21.7 to 675), PI (11 to 626), D10 (0.0077 to 0.62 mm), and CC (0 to 100%). Details are given in Ching et al. (2023). The SOIL-KSAT/7/2293 database is 71% complete. The generic statistics of the seven parameters are listed in Table 3.31. Tables 3.32 and 3.33 show the group-specific mean, COV, and bivariate correlation coefficients. Note that these are group-specific statistics rather than site-specific ones (there is no "site" in SOIL-KSAT/7/2293). Each group consists of a series of k_{sat} versus e (or w) laboratory data on the same type of soil (the same LL, PI, D_{10}, CC) where e (or w) is controlled by consolidation. Because (LL, PI, D_{10}, CC) within each group are controlled to be constant, their group-specific COVs and correlations are not reported in Tables 3.32 and 3.33. Table 3.34 shows the values of b and δ for some generic transformation models

Table 3.31 Summary of generic statistics for SOIL-KSAT/7/2293

Variable	*# Data*	*Mean*	*COV*	*Min*	*Max*
k_{sat} (cm/s)	2293	1.1E−02	4.62	5.2E−13	0.60
e	2293	1.37	0.59	0.19	4.81
w (%)	1625	54.5	0.60	7.12	178.0
LL	1707	98.7	0.80	21.7	675.0
PI	1256	56.9	1.39	11.0	625.9
D_{10} (mm)	166	0.22	0.78	0.0077	0.62
CC (%)	2091	50.96	0.65	0.0	100.0

Table 3.32 Summary of group-specific statistics for SOIL-KSAT/7/2293

Property	# Groups	# Records/group		Group-specific mean		Group-specific COV	
		Range	Mean	95% CI	Mean	95% CI	Mean
k_{sat} (cm/s)	227	5-49	9.4	1.9E−10–0.20	1.15E−02	0.18–2.07	0.87
e	227	5-49	9.4	0.52–2.82	1.40	0.064–0.56	0.25
w (%)	153	5-49	10.1	19.6–106.0	58.4	0.065–0.57	0.29
LL	164	5-49	9.9	31.4–346.8	108.9	–	–
PI	95	5-49	12.6	12.3–440.9	65.4	–	–
D_{10} (mm)	14	5-17	8.9	0.0077–0.62	0.27	–	–
CC (%)	209	5-49	9.3	0.0–100.0	47.7	–	–

Table 3.33 Summary of group-specific correlations for SOIL-KSAT/7/2293 (Pearson, Spearman, Kendall)

	e	w	LL	PI	D_{10}	CC
k_{sat}	0.90,0.89,0.81 (75)	0.89,0.90,0.81 (58)	–	–	–	–
e		1.00,0.99,0.96 (2)	–	–	–	–
w			–	–	–	–
LL				–	–	–
PI					–	–
D_{10}						–

Table 3.34 Some generic transformation models calibrated by SOIL-KSAT/7/2293

Model	Literature	n_{cal}	Transformation model	b	COV_{trans}
$k_{sat} - D_{10}$	Hazen (1930)	166 (coarse-grained)	$k_{sat}(cm/s) \approx 1.25 \times D_{10}^2$	0.61	1.45
$k_{sat} - e - D_{10}$	Chapuis (2004)	166 (coarse-grained)	$k_{sat}\left(\frac{cm}{s}\right) \approx 2.4622 \times \left[D_{10}^2 \times \frac{e^3}{1+e}\right]^{0.7825}$	0.81	1.41
$k_{sat} - e - PI$	Nishida and Nakagawa (1969)	1256 (fine-grained)	$k_{sat}\left(\frac{cm}{s}\right) \approx 10^{\frac{e}{0.01 \times PI + 0.05} - 10}$	2.14	5.96
$k_{sat} - e - e_L$[a]	Nagaraj et al. (1994)	1468 (fine-grained)	$k_{sat}\left(\frac{cm}{s}\right) \approx 10^{\left(\frac{e}{e_L} - 2.162\right)/0.195}$	9.13	6.51
	Sivapullaiah et al. (2000)		$k_{sat}\left(\frac{m}{s}\right) \approx \left(\frac{e}{1.97 \times e_L}\right)^{4.2}$	1.50 × 10^{-7}[b]	4.68

[a] e_L = void ratio at liquid limit.

[b] The large bias may be due to an error in the unit in Sivapullaiah et al. (2000). According to Sivapullaiah et al. (2000), the unit for k_{sat} is m/s, but it is likely that this unit is incorrect.

Table 3.35 Ranges of site-specific SOF_h and SOF_z for various soil types

	SOF_h (m)				SOF_z (m)			
Soil type	*# Studies*	*Min*	*Max*	*Average*	*# Studies*	*Min*	*Max*	*Average*
Alluvial	9	1.07	49	14.2	13	0.07	1.1	0.36
Ankara clay	–	–	–	–	4	1	6.2	3.63
Chicago clay	–	–	–	–	2	0.79	1.25	0.91
Clay	9	0.14	163.8	31.9	16	0.05	3.62	1.29
Clay, sand, silt mix	13	1.2	1000	201.5	28	0.06	21	1.58
Hangzhou clay	2	40.4	45.4	42.9	4	0.49	0.77	0.63
Marine clay	8	8.37	66	30.9	9	0.11	6.1	1.55
Marine sand	1	15	15	15	5	0.07	7.2	1.43
Offshore soil	1	24.6	66.5	45.6	2	0.48	1.62	1.04
Over consolidated clay	1	0.14	0.14	0.14	2	0.063	0.255	0.15
Sand	9	1.69	80	24.5	14	0.1	4	1.17
Sensitive clay	–	–	–	–	2	1.1	2.0	1.55
Silt	3	12.7	45.5	33.2	5	0.14	7.19	2.08
Silty clay	7	9.65	45.4	29.8	14	0.095	6.47	1.40
Soft clay	3	22.2	80	47.6	8	0.14	6.2	1.70
Undrained engineered soil	–	–	–	–	22	0.3	2.7	1.42
Water content	9	2.8	22.2	12.9	8	0.05	6.2	1.70

Source: Cami et al. (2020).

calibrated by the records in SOIL-KSAT/7/2293. It is clear that the transformation uncertainty (quantified by COV_{trans}) for k_{sat} of fine-grained soils is significant.

3.3 TYPICAL VALUES OF SITE-SPECIFIC SOFS

Cami et al. (2020) summarized the ranges of the site-specific horizontal SOF (SOF_h) and vertical SOF (SOF_z) reported in the literature for various soil types, which are shown in Table 3.35. In practice, a common assumption is that the two horizontal directions have the same SOF: $SOF_h = SOF_x = SOF_y$. There are cases in the literature with simultaneous information on site-specific SOF_h and SOF_z. Chapter 3 of ISSMGE-TC304 (2021) summarized the values and ranges for the site-specific SOFs of these cases as Figure 3.1. Ching et al. (2023) conducted a comprehensive analysis of a large CPT database. From the database, they extracted hundreds of homogeneous soil units from 42 sites to identify the vertical scales of fluctuation of their q_t data. Table 3.36 shows the statistics of the site-specific SOF_z for soil units of various soil behaviour types (SBT) (Robertson 2016).

3.4 QUASI-SITE-SPECIFIC TRANSFORMATION MODEL

A major challenge when evaluating soil/rock properties through generic transformation models is that the transformation uncertainty is usually significant. The significant transformation uncertainty manifests as large COV_{trans} in the tables present above. A large COV_{trans}

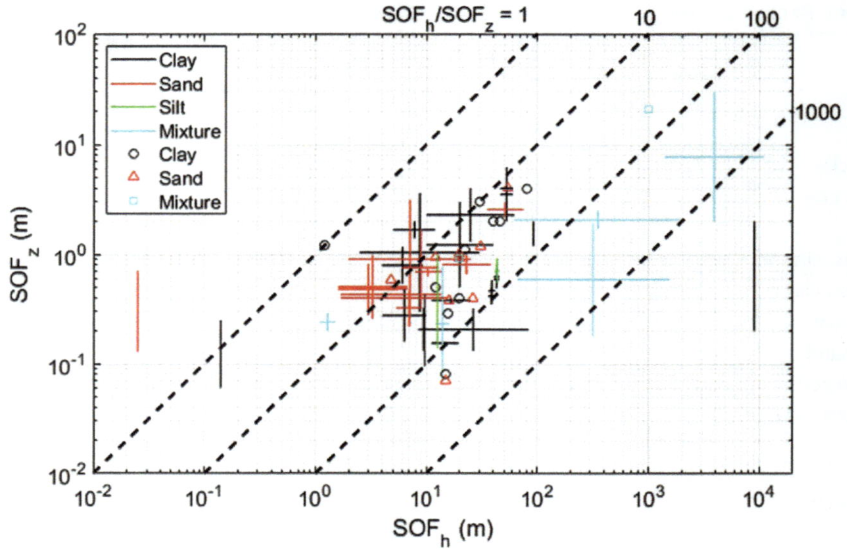

Figure 3.1 Variation of vertical scale of fluctuation (SOF_z) with horizontal scale of fluctuation (SOF_h) for various soil conditions (Source: ISSMGE-TC304 2021).

Table 3.36 Statistics for site-specific SOF_z identified from CPT cone tip resistance (q_t) for soil units of various soil behaviour types

		Site-specific SOF_z (m)	
SBT	*# Sites*	*Range*	*Mean*
2 (organic)	1	0.30	0.30
3 (clay)	7	0.10–0.46	0.26
4 (silt mixtures)	8	0.13–0.42	0.27
5 (sand mixtures)	13	0.084–0.58	0.32
6 (sand)	13	0.31–0.89	0.49
2, 3, 4 (clay-like)	16	0.10–0.46	0.27
5, 6 (sand-like)	26	0.084–0.89	0.41
All	42	0.084–0.89	0.35

Source: Ching et al. (2023b).

leads to a large COV_{TOT} in Equation (3.1), so the resulting characteristic value (X_k) is small (see Equations 3.2 and 3.3), which will further lead to a conservative design. Ching et al. (2021b) proposed the hierarchical Bayesian model (HBM) to construct a quasi-site-specific transformation model. The transformation uncertainty for the quasi-site-specific model can be significantly less than that for the generic transformation model. First, the HBM learns the site-specific statistics in the soil/rock property database. The learning outcome produces a prior model for the target site. Then, this prior model is updated into a posterior model by further conditioning on the target-site data. The posterior model can produce a "quasi-site-specific" transformation model for the target site. The resulting transformation model is quasi-site-specific because it is not entirely based on the sparse site-specific (target-site) data: it also adopts the prior knowledge learned from a database. The transformation uncertainty

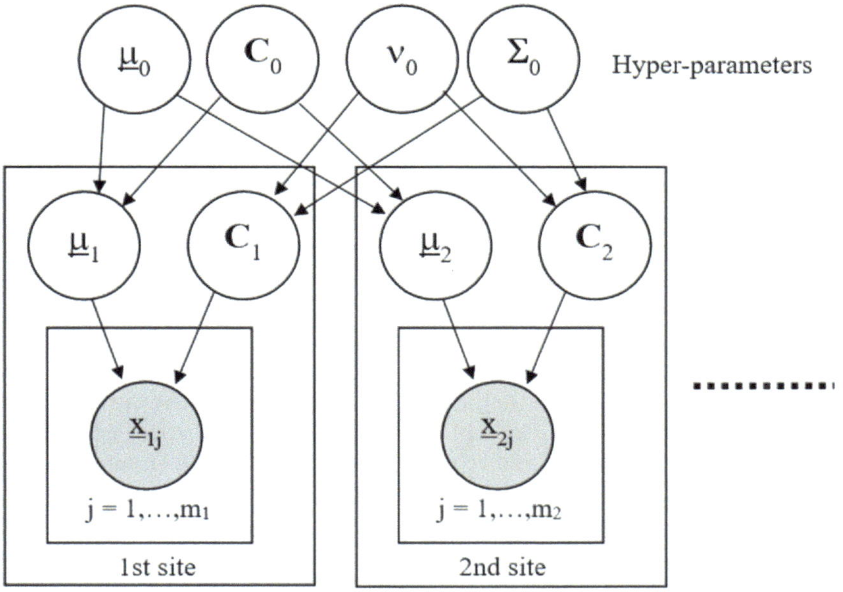

Figure 3.2 Model structure of the HBM (Source: Ching et al. 2021b).

for the quasi-site-specific model can be significantly less than that for the generic transformation model, as illustrated by an example in this section.

Figure 3.2 shows the model structure of the HBM. The (transformed) soil parameters for the jth record at the ith site in a database are denoted by the vector $x_{ij} \in \mathbf{R}^{n\times 1}$ (n is the dimension of the soil/rock parameters of interest, e.g., n = 10 for CLAY/10/7490). The vector x_{ij} is assumed to follow a multivariate normal probability density function (PDF) with site-specific mean vector = $\mu_i \in \mathbf{R}^{n\times 1}$ and site-specific covariance matrix = $\mathbf{C}_i \in \mathbf{R}^{n\times n}$, denoted by $N(\mathrm{x};\mu_i, \mathbf{C}_i)$. The parameters $(\mu_i,\mathbf{C}_i)$ quantify the "intra-site" variability within the ith site. Because of site uniqueness, $\mu_i \neq \mu_j$ and $\mathbf{C}_i \neq \mathbf{C}_j$ if $i \neq j$, i.e., there is also "inter-site" variability. In the HBM, $\{\mu_i$: $i = 1, \ldots, n_s\}$ (n_s is the number of sites) are assumed to follow the same multivariate normal PDF $N(\mu\ ;\mu_0,\mathbf{C}_0)$, where $\mu_0 \in \mathbf{R}^{n\times 1}$ is the (hyper) mean vector and $\mathbf{C}_0 \in \mathbf{R}^{n\times n}$ is the (hyper) covariance matrix. Similarly, $\{\mathbf{C}_i$: $i = 1, \ldots, n_s\}$ are assumed to follow the same inverse-Wishart (IW) distribution (James 1964), denoted by $\mathrm{IW}(\mathbf{C};\Sigma_0,\nu_0)$, where $\Sigma_0 \in \mathbf{R}^{n\times n}$ is the (hyper) scale matrix and $\nu_0 \in \mathbf{R}$ is the (hyper) degree of freedom. The parameters $\mathbf{\Theta} = (\mu_0, \mathbf{C}_0,\Sigma_0,\nu_0)$, called the hyper-parameters of the HBM, quantify both inter-site and intra-site variabilities in the soil/rock property database.

The HBM contains two analysis stages: the learning and inference stages. In the learning stage, the hyper-parameters $\mathbf{\Theta} = (\mu_0,\mathbf{C}_0,\Sigma_0,\nu_0)$ are calibrated to learn the site-specific statistics in the soil/rock property database. The calibrated hyper-parameters can produce a prior model for the target site. In the inference stage, this prior model is updated into a posterior model by further conditioning on the target-site data. The posterior model can produce a "quasi-site-specific" transformation model for the target site. In the following two subsections, the learning and inference stages will be illustrated using a bivariate database $[(q_t - \sigma_v)/\sigma'_v$ versus $s_u/\sigma'_v]$ extracted from CLAY/10/7490 as an illustration example. It will be shown that the HBM learning stage can effectively extract site-specific statistics in the database and that the HBM inference stage can construct a quasi-site-specific transformation model that is more effective than a generic transformation model.

3.3.1 HBM learning stage: extract site-specific statistics in the database

Let us take the CLAY/10/7490 database as an example. For illustration purposes, let us consider two clay parameters $Y_1 = q_{t1} = (q_t - \sigma_v)/\sigma'_v$ and $Y_2 = s_u/\sigma'_v$. In CLAY/10/7490, there are 91 sites with Y_1–Y_2 records. These records (with site-label information) are shown in Figure 3.3a. The (Y_1, Y_2) data are transformed to (roughly) standard normal space (X_1, X_2) by the Johnson transformation (Ching and Phoon 2014b). During the HBM learning stage, the hyper-parameters $\mathbf{\Theta}$ are calibrated by the database (X_1, X_2) data. The calibration is done by performing Bayesian analysis: the posterior samples of $\mathbf{\Theta}$ are drawn by conditioning on the database data. The $\mathbf{\Theta}$ samples can simulate the site-specific mean and covariance matrix of a "hypothetical site" (denoted by μ_h and $\mathbf{C}_h$) whose site-specific statistics "mimic" the 91 database sites. The red cross markers and red histogram in Figure 3.4 show the simulated site-specific statistics (means, standard deviations, and correlation) for numerous hypothetical sites. They are consistent with the actual site-specific statistics for the 91 database sites (dark open circles and dark histograms). The consistency indicates that the HBM learning stage can effectively capture the site-specific statistics in the database. The simulated hypothetical sites can be illustrated in a different way. Each (skewed) ellipse in Figure 3.3b represents the one-standard-deviation confidence region of a hypothetical site. These (skewed) ellipses mimic the site-specific records in Figure 3.3a.

3.3.2 HBM inference stage: construction of quasi-site-specific transformation model

The calibrated hyper-parameters in the HBM learning stage can produce a "prior model" for the site-specific mean and covariance matrix of the target site, denoted by μ_s and $\mathbf{C}_s$. In

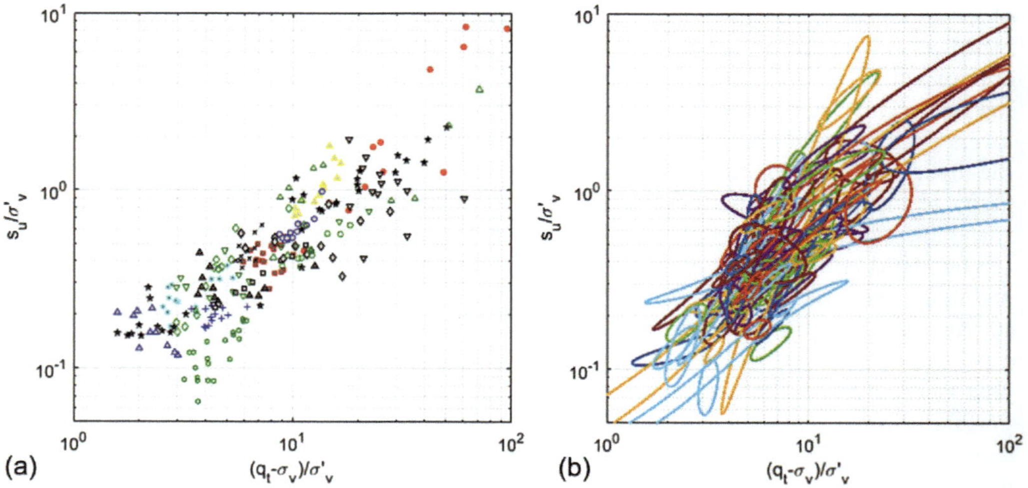

Figure 3.3 Y_1–Y_2 data points in the CLAY/10/7490 database (records from different sites are shown as distinct markers) versus simulated hypothetical sites (Source: Ching 2020).

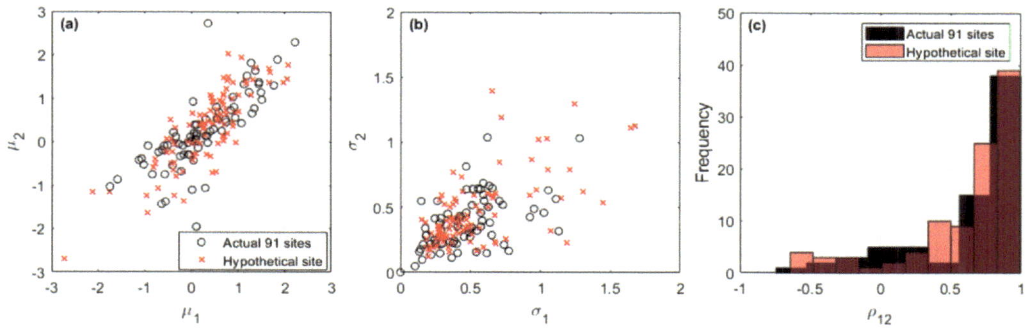

Figure 3.4 Site-specific statistics (μ and σ denote the mean and standard deviation of X, and ρ denotes the correlation coefficient), database versus hypothetical sites: (a) site-specific means; (b) site-specific standard deviations; (c) site-specific correlation (Source: Ching 2020).

the inference stage, this prior model (μ_s,C_s) is updated into a posterior model for (μ_s,C_s) by further conditioning on the target-site data. This posterior model is "quasi-site-specific" because not only the target-site data are used to construct the model but also the database is used to develop its prior.

Now consider the Onsøy (Norway) site (Lacasse and Lunne 1982) as the target site. The Onsøy site is not within the CLAY/10/7490 database. There are many in situ and laboratory tests conducted at this test site, but only those relevant to (Y_1, Y_2) are investigated in this section. The site investigation data are digitalized from Lacasse and Lunne (1982), as shown in Table 3.37. To illustrate the HBM inference stage, the s_u/σ'_v values for some depths are deliberately treated as unknowns (see the numbers in the parentheses in the table). The "unknown" s_u/σ'_v values will later serve as the validation data for the inference stage. The known s_u/σ'_v values are sparse such that a site-specific transformation model may not be reliably constructed. A possible alternative is to adopt a generic transformation model. However, the generic transformation uncertainty is usually large, e.g., $COV_{trans} = 0.45$ for the s_u-CPT transformation model in Table 3.7.

The prior model constructed during the HBM learning stage can be further conditioned by the sparse target-site data in Table 3.37. By conditioning on the target-site data, most ellipses in Figure 3.3b are incompatible with the target-site data. Figure 3.5a shows the incompatible

Table 3.37 Site investigation data for a site in Onsøy (Norway)

Index	*Depth (m)*	q_{t1} (Y_1)	s_u/σ'_v (Y_2)
1	1.0	29.11	2.03
2	1.9	17.69	(0.91)
3	3.5	10.52	0.48
4	5.2	7.70	(0.37)
5	7.6	5.89	0.24
6	9.5	6.19	(0.25)
7	10.8	5.93	0.25
8	13.4	5.95	(0.24)
9	16.3	6.13	0.24

Source: Lacasse and Lunne (1982).

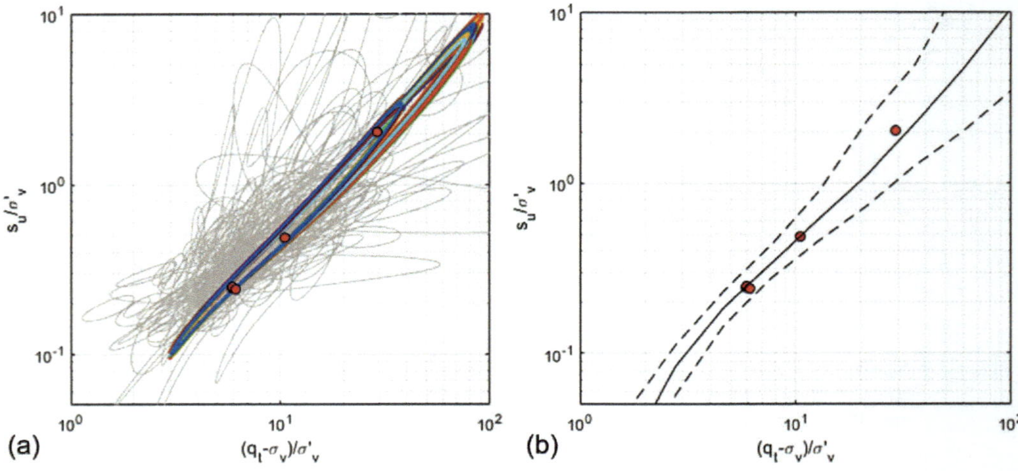

Figure 3.5 Illustration of the quasi-site-specific model.

ellipses in grey and compatible ellipses in colours. The red dots show the target-site data. The coloured (compatible) ellipses illustrate the quasi-site-specific model after the conditioning. Prediction based on the quasi-site-specific model can be carried out by reading the median values as well as the 95% confidence interval (95% CI) for Y_2 at different values of Y_1, as shown in Figure 3.5b.

Figure 3.6a shows the predictive median and 95% CI for the generic transformation model developed based on the database Y_1–Y_2 data (shown as grey dots), whereas Figure 3.6b shows the predictive median and 95% CI for the quasi-site-specific transformation model developed by the HBM inference stage. The validation data (the numbers in the parentheses in Table 3.37) are also shown as yellow dots for comparison. Although both 95% CIs contain the validation data, the transformation uncertainty for the quasi-site-specific model is much less than that for the generic model. Although the superiority of a quasi-site-specific

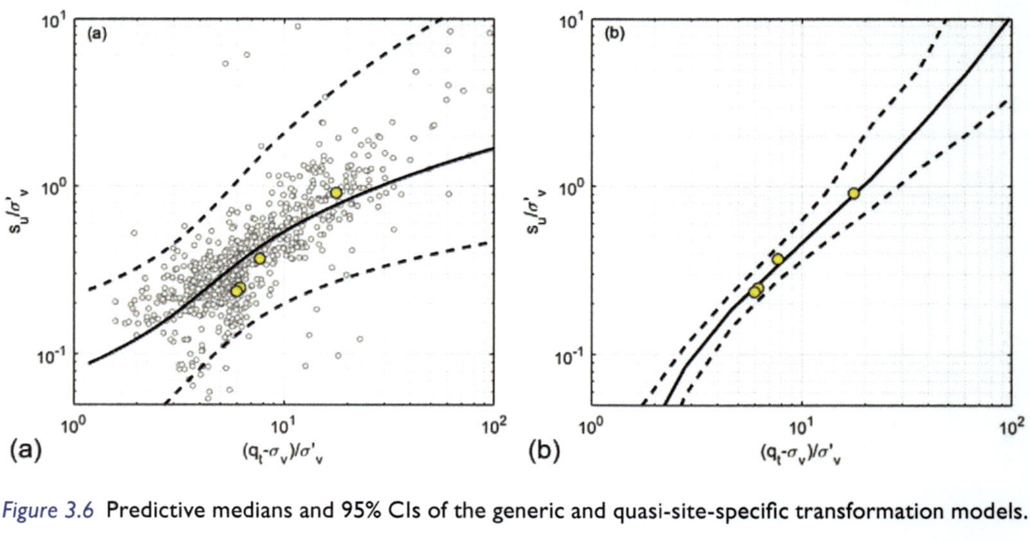

Figure 3.6 Predictive medians and 95% CIs of the generic and quasi-site-specific transformation models.

transformation model constructed by the HBM over a generic transformation model is illustrated above using a bivariate problem [$(q_t - \sigma_v)/\sigma'_v$ versus s_u/σ'_v] for the CLAY/7/7490 database, the similar superiority is found for other databases and for higher dimensions (e.g., Ching et al. 2021a, 2021b, 2022; Meng and Pei 2023).

3.5 CONCLUSION

The proper characterization of the variability of soil/rock properties for a specific site is necessary for reliability-based design as well as for the determination of soil/rock characteristic values. This chapter reviews some useful sources that provide site-specific statistics for the inherent variability of soil/rock properties (e.g., site-specific mean, COV, and SOF). Some sources are relatively old (e.g., Phoon et al. 1995; Phoon and Kulhawy 1999a; Kulhawy et al. 2000), whereas others are relatively new (e.g., Cami et al. 2020; ISSMGE-TC304 2021; Ching et al. 2023b). Some newer results in this chapter are based on the soil/rock property databases compiled by the author of this chapter. The extent of these databases is still limited, so only a limited number of soil/properties are covered. In the opinion of the author of this chapter, there is an urgent need to compile more soil/rock property databases to broaden the coverage. Although the soil/rock property databases can be used to calibrate generic transformation models, one key issue for these generic models is that the transformation uncertainty is usually significant. The large transformation uncertainty may result in a conservative design. The hierarchical Bayesian model (HBM) recently proposed by the author of this chapter may reduce the transformation uncertainty significantly by constructing a quasi-site-specific transformation model.

REFERENCES

Altindag, R. and Guney, A. (2010). Predicting the relationships between brittleness and mechanical properties (UCS, TS and SH) of rocks. *Scientific Research and Essays*, 5(16), 2107–2118.

Baecher, G.B. and Christian, J.T. (2003). *Reliability and Statistics in Geotechnical Engineering*. John Wiley & Sons, Hoboken, NJ.

Barton, N., Lien, R., and Lunde, J. (1974). Engineering classification of rock masses for the design of tunnel support. *Rock Mechanics*, 6(4), 189–236.

Bieniawski, Z.T. (1973). Engineering classification of jointed rock masses. *Transactions of the South African Institution of Civil Engineers*, 15, 335–344.

Bieniawski, Z.T. (1978). Determining rock mass deformability: Experience from case histories. *International Journal of Rock Mechanics and Mining Sciences*, 15, 237–247.

Bjerrum, L. (1954). Geotechnical properties of Norwegian marine clays. *Geotechnique*, 4(2), 49–69.

Bolton, M.D. (1986). The strength and dilatancy of sands. *Geotechnique*, 36(1), 65–78.

Cami, B., Javankhoshdel, S., Phoon, K.K., and Ching, J. (2020). Scale of fluctuation for spatially varying soils: estimation methods and values. *ASCE-ASME Journal of Risk and Uncertainty in Engineering Systems, Part A: Civil Engineering*, 6(4), 03120002.

CEN (2004). EN 1997–1: *Geotechnical Design – Part 1: General Rules*. Comité Européen de Normalisation, Brussels, Belgium.

CEN (2020). *prEN 1997–1 Geotechnical Design – General Rules (PT6) Oct 2020*. Comité Européen de Normalisation, Brussels, Belgium.

Chapuis, R.P. (2004). Predicting the saturated hydraulic conductivity of sand and gravel using effective diameter and void ratio. *Canadian Geotechnical Journal*, 41, 787–795.

Chen, J. R. (2004). *Axial Behavior of Drilled Shafts in Gravelly Soils*. Ph.D. Dissertation, Cornell University, Ithaca, NY

Ching, J. (2020). Value of geotechnical BIG DATA and its application in site-specific soil property estimation. *Journal of GeoEngineering*, 15(4), 173–182.

Ching, J. and Kuo, M.C. (2023). Data-centric quasi-site-specific prediction for soil modulus. *Journal of GeoEngineering*, 18(4), 215–223.

Ching, J. and Phoon, K. K. (2014a). Transformations and correlations among some parameters of clays – the global database. *Canadian Geotechnical Journal*, 51(6), 663–685.

Ching, J. and Phoon, K. K. (2014b). Correlations among some clay parameters – the multivariate distribution. *Canadian Geotechnical Journal*, 51(6), 686–704.

Ching, J. and Phoon, K.K. (2012a). Modeling parameters of structured clays as a multivariate normal distribution. *Canadian Geotechnical Journal*, 49(5), 522–545.

Ching, J. and Phoon, K.K. (2012b). Establishment of generic transformations for geotechnical design parameters. *Structural Safety*, 35, 52–62.

Ching, J., Lin, G.H., Chen, J.R., and Phoon, K.K. (2017). Transformation models for effective friction angle and relative density calibrated based on a multivariate database of coarse-grained soils. *Canadian Geotechnical Journal*, 54(4), 481–501.

Ching, J., Li, K.H., Phoon, K.K., and Weng, M.C. (2018). Generic transformation models for some intact rock properties. *Canadian Geotechnical Journal*, 55(12), 1702–1741.

Ching, J., Phoon, K.K., Ho, Y.H., and Weng, M.C. (2021a). Quasi-site-specific prediction for deformation modulus of rock mass. *Canadian Geotechnical Journal*, 58, 936–951.

Ching, J., Wu, S., and Phoon, K.K. (2021b). Constructing quasi-site-specific multivariate probability distribution using hierarchical Bayesian model, *ASCE Journal of Engineering Mechanics*, 147(10), 04021069.

Ching, J., Phoon, K.K., and Wu, C.T. (2022). Data-centric quasi-site-specific prediction for compressibility of clays. *Canadian Geotechnical Journal*, 59(12), 2033–2049.

Ching, J., Uzielli, M., Phoon, K.K., and Xu, X.J. (2023). Characterization of autocovariance parameters of detrended cone tip resistance from a global CPT database. *ASCE Journal of Geotechnical and Geoenvironmental Engineering*, 149(10), 04023090.

Ching, J., Phoon, K.K., and Huang, P.S. (2024). Detection of outliers with respect to a MUSIC geotechnical database. *Canadian Geotechnical Journal*, 61, 1275–1293.

Coon, R.F. and Merritt, A.H. (1970). Predicting in situ modulus of deformation using rock quality indexes. In *Determination of the In Situ Modulus of Deformation of Rock*, By Committee D-18, ASTM International, West Conshohocken, PA, pp. 154–173.

Deere, D.U. (1964). Technical description of rock cores. *Rock Mechanics Engineering Geology*, 1, 16–22.

Deere, D.U. and Miller, R.P. (1966). *Engineering Classification and Index Properties for Intact Rock*. Technical Report No. AFWL-TR-65-116, Air Force Weapons Lab, Kirtland Air Force Base, Albuquerque, NM.

Feng, S. and Vardanega, P. J. (2019). A database of saturated hydraulic conductivity of fine-grained soils: probability density functions. *Georisk: Assessment and Management of Risk for Engineered Systems and Geohazards*, 13(4): 255–261.

Feng, S., Barreto, D., Imre, E., Ibraim, E., and Vardanega, P.J. (2023). Use of hydraulic radius to estimate the permeability of coarse-grained materials using a new geodatabase. *Transportation Geotechnics*, 41, 101026.

Fenton, G. A. and Griffiths, D. V. (2008). *Risk Assessment in Geotechnical Engineering*. John Wiley & Sons, New York.

Gokceoglu, C., Sonmez, H., and Kayabasi, A. (2003). Predicting the deformation moduli of rock masses. *International Journal of Rock Mechanics and Mining Sciences*, 40(5), 701–710.

Grimstad E. and Barton N. (1993). Updating the Q-System for NMT. In: *Proceedings of the International Symposium on Sprayed Concrete-Modern Use of Wet Mix Sprayed Concrete for Underground Support*, Oslo, Norwegian Concrete Association.

Hatanaka, M. and Uchida, A. (1996), Empirical correlation between penetration resistance and internal friction angle of sandy soils. *Soils and Foundations*, 36(4), 1–9.

Hazen, A. (1930). Water supply. In: *American Civil Engineers Handbook*, eds. Merriman, T. and Wiggin, T., Wiley, New York.

Hoek, E. and Brown, E.T. (1997). Practical estimates of rock mass strength. *International Journal of Rock Mechanics and Mining Science*, 34, 1165–1186.

Hoek, E. and Diederichs, M.S. (2006). Empirical estimation of rock mass modulus. *International Journal of Rock Mechanics and Mining Sciences*, 43(2), 203–215.

Hotelling, H. and Pabst, M.R. (1936). Rank correlation and tests of significance involving no assumption of normality. *American Mathematical Statistics*, 7, 29–43.

ISSMGE-TC304 (2021). State of the art review of inherent variability and uncertainty in geotechnical properties and models. *International Society of Soil Mechanics and Geotechnical Engineering (ISSMGE) Technical Committee TC304 'Engineering Practice of Risk Assessment and Management'*, March 2nd. http://140.112.12.21/issmge/2021/SOA_Review_on_geotechnical_property_variablity_and_model_uncertainty.pdf

Jamiolkowski, M., Ladd, C.C., Germain, J.T., and Lancellotta, R. (1985). New developments in field and laboratory testing of soils. In: *Proceeding of the 11th International Conference on Soil Mechanics and Foundation Engineering*, San Francisco, CA, 1, pp. 57–153.

Kahraman, S. (2001). Evaluation of simple methods for assessing the uniaxial compressive strength of rock. *International Journal of Rock Mechanics and Mining Sciences*, 38(7), 981–994.

Karaman, K. and Kesimal, A. (2015). A comparative study of Schmidt hammer test methods for estimating the uniaxial compressive strength of rocks. *Bulletin of Engineering Geology and the Environment*, 74(2), 507–520.

Katz, O., Reches, Z., and Roegiers, J.C. (2000). Evaluation of mechanical rock properties using a Schmidt hammer. *International Journal of Rock Mechanics and Mining Sciences*, 37(4), 723–728.

Kendall, M. (1938). A new measure of rank correlation. *Biometrika*, 30(1–2), 81–89.

Kılıç, A. and Teymen, A. (2008). Determination of mechanical properties of rocks using simple methods. *Bulletin of Engineering Geology and the Environment*, 67(2), 237–244.

Kulhawy, F. H. and Mayne, P. W. (1990). Manual on estimating soil properties for foundation design. Report EL-6800, Electric Power Research Institute, Palo Alto, CA.

Kulhawy, F. H., Phoon, K. K., and Prakoso, W. A. (2000). Uncertainty in the basic properties of natural geomaterials. In: *Proceedings, First International Conference on Geotechnical Engineering Education and Training*, Sinaia, Romania, 12–14 June 2000, pp. 297–302.

Lacasse, S. and Lunne, T. (1982). Penetration tests in two Norwegian clays. In: *Proceedings, 2nd European Symposium on Penetration Testing*, Amsterdam, pp. 661–670.

Locat, J. and Demers. D. (1988). Viscosity, yield stress, remoulded strength, and liquidity index relationships for sensitive clays. *Canadian Geotechnical Journal*, 25, 799–806.

Mayne, P.W. and Frost, D.D. (1989). Dilatometer experience in Washington, D.C. and vicinity. Research Record 1169, Transportation Research Board, Washington, DC, pp. 16–23.

Meng, F. and Pei, H. (2023). Quasi-site-specific prediction of shear wave velocity from CPTu. *Soil Dynamics and Earthquake Engineering*, 172, 108005.

Mesri, G. (1975). Discussion on "New design procedure for stability of soft clays". *Journal of the Geotechnical Engineering Division, ASCE*, 101(4), 409–412.

Mesri, G. (1989). A re-evaluation of s_u(mob) = 0.22 σ'p using laboratory shear tests. *Canadian Geotechnical Journal*, 26(1), 162–164.

Mesri, G. and Huvaj, N. (2007). Shear strength mobilized in undrained failure of soft clay and silt deposits. ASCE Geotechnical Special Publication 173, Geo-Denver.

Mishra, D.A. and Basu, A. (2013). Estimation of uniaxial compressive strength of rock materials by index tests using regression analysis and fuzzy inference system. *Engineering Geology*, 160, 54–68.

Nagaraj, T.S., Pandian, N.S., and Narasimha Raju, P.S.R. (1994). Stress state permeability relationships for overconsolidated clays. *Geotechnique*, 44(2), 349–352.

Nisihda, Y. and Nakagawa, S. (1969). Water permeability and plastic index of soils. In: *Proceedings of the IASH-UNESCO Symposium*, Tokyo, Japan, pp. 573–578.

Ohya, S. Imai, T., Matsubara, M. (1982). Relation between N value by SPT and LLT pressuremeter results. In: Proceedings 2nd European Symposium on Penetration Testing, Amsterdam, 1, 125–130.

Orr, T.L.L. (2017). Defining and selecting characteristic values of geotechnical parameters for designs to Eurocode 7. *Georisk: Assessment and Management of Risk for Engineered Systems and Geohazards*, 11(1), 103–115.

Phoon, K.K. and Kulhawy, F.H. (1999a). Characterization of geotechnical variability. *Canadian Geotechnical Journal*, 36(4), 612–624.

Phoon, K.K. and Kulhawy, F.H. (1999b). Evaluation of geotechnical property variability. *Canadian Geotechnical Journal*, 36(4), 625–639.

Phoon, K.K., Kulhawy, F.H., and Grigoriu, M.D. (1995). *Reliability-Based Design of Foundations for Transmission Line Structure*, Report TR-105000, Palo Alto, CA, Electric Power Research Institute.

Prakoso, W.A. and Kulhawy, F.H. (2011). Effects of testing conditions on intact rock strength and variability. *Geotechnical and Geological Engineering*, 29(1), 101–111.

Robertson, P.K. (2016). Cone penetration test (CPT)-based soil behaviour type (SBT) classification system – an update. *Canadian Geotechnical Journal*, 53, 1910–1927.

Robertson, P.K. and Campanella, R.G. (1983). Interpretation of cone penetration tests: Sands and clays. *Canadian Geotechnical Journal*, 20, 719–745.

Schneider H.R. and Schneider M.A. (2013). Dealing with uncertainties in EC7 with emphasis on determination of characteristic soil properties. In: *Modern Geotechnical Design Codes of Practice*, eds. P. Arnold, G.A. Fenton, M.A. Hicks, T. Schweckendiek, and B. Simpson, IOS Press, Amsterdam, pp. 87–101.

Serafim, J.L. and Pereira, J.P. (1983). Considerations on the geomechanical classification of Bieniawski. In: *Proceedings of the Symposium on Engineering Geology and Underground Openings*, Lisboa, Portugal, pp. 1133–1144.

Sivapullaiah, P.V., Sridharan, A., and Stalin, V.K. (2000). Hydraulic conductivity of bentonite-sand mixtures. *Canadian Geotechnical Journal*, 37, 406–413.

Terzaghi, K. and Peck, R. B. (1967). *Soil Mechanics in Engineering Practice.* 2nd Ed., John Willey & Sons, New York, 729pp.

Vanmarcke, E.H. (1977). Probabilistic modeling of soil profiles. *Journal of the Geotechnical Engineering Division, ASCE*, 103(11), 1227–1246.

Wang, Y. and Cao, Z. J. (2013). Probabilistic characterization of Young's modulus of soil using equivalent samples. *Engineering Geology*, 159, 106–118.

Yaşar, E. and Erdogan, Y. (2004). Correlating sound velocity with the density, compressive strength and Young's modulus of carbonate rocks. *International Journal of Rock Mechanics and Mining Sciences*, 41(5), 871–875.

Zhang, L. and Einstein, H. H. (2004). Using RQD to estimate the deformation modulus of rock masses. *International Journal of Rock Mechanics and Mining Sciences*, 41(2), 337–341.

Chapter 4

Undrained shear strength of Finnish soft clays

A database perspective

Marco D'Ignazio and Tim Länsivaara

The undrained shear strength (s_u) of clays is a governing parameter in several geotechnical engineering problems. However, s_u is not a unique parameter; it depends on factors like stress paths, loading rates, soil composition, and stress history. This complexity results in significant uncertainty in s_u determination, impacting construction and operational safety. Further, the selection of design s_u from a given test type shall reflect the anticipated field deformation mechanism for a specific structure. To address these uncertainties, global and local databases have emerged as powerful tools. They offer statistical correlations between measured s_u values and basic clay properties, thus reducing uncertainty. In this study, high-quality data from Southern Finland is compared with extensive databases, resulting in a robust statistical correlation for anisotropic undrained shear strength of Finnish soft clays. The proposed correlations are meant for preliminary analyses, where site-specific data is limited or unavailable. Overall, databases provide invaluable resources to assess local data reliability, reduce bias, and enhance confidence in infrastructure design and construction.

4.1 INTRODUCTION

Geotechnical engineering applications such as stability of road and railway embankments, or short-term stability of excavations in soft clays, rely critically on the undrained shear strength (s_u). This parameter serves as the cornerstone for assessing safety during construction and/or under operational loads.

The direct field measurement of s_u can be achieved through various techniques, including the field vane test (FVT), which is popular in Finland (Selänpää et al. 2017), or indirect methods such as cone penetration testing (CPT/CPTU) or dilatometer testing (DMT). In the laboratory, tests like undrained triaxial (TX) and direct simple shear (DSS), as well as the Fall Cone (FC), are used to derive s_u values for unconsolidated or reconsolidated soil specimens.

However, s_u is far from a static, universal value. It is highly dependent on factors such as stress paths induced by a given test type, loading rates, in situ stress history of the soil deposit as well as reconsolidation history in the laboratory, and the specific nature and chemical composition of the soil (e.g., Bjerrum 1973; Länsivaara 1999; D'Ignazio 2016). As a result, accurately determining s_u for a particular location can be a complex endeavour. Given this inherent complexity, s_u for a specific site or soil types might be characterized by a notable degree of uncertainty (e.g., Phoon and Kulhawy 1999; Ching and Phoon 2013).

Moreover, the determination of its design value necessitates a keen consideration of the in-situ deformation mechanism relevant to a specific type of structure. For example, in

DOI: 10.1201/9781003441946-4

embankment design, the pivotal factor is the assessment of the average mobilized shear strength along a potential slip surface (Bjerrum 1973; Larsson et al. 2007; D'Ignazio et al. 2017a). In cases involving uniform soil conditions and a circular or quasi-circular slip surface, the selection of the design undrained shear strength (s_u) aligns with s_{uDSS} or $s_{u(mob)}$ values. The latter can be derived from the measured strength from FVT, s_{uFVT}, which is empirically adjusted for anisotropy and rate effects, according to the soil's plasticity characteristics (e.g., Bjerrum 1973; Liikennevirasto 2018). Notably, the literature often presumes that s_{uDSS} is approximately equal to $s_{u(mob)}$ (Jamiolkowski et al. 1985; D'Ignazio 2016; D'Ignazio et al. 2021). On the other hand, for excavations, the results of triaxial compression tests (anisotropically or isotropically consolidated undrained compression, CAUC/CIUC and extension, CAUE/CIUE) find better applicability.

To address these uncertainties, a database-driven approach emerges as a powerful tool. Such databases, like those established by D'Ignazio et al. (2016) for Finland and Scandinavia (Sweden-Norway), were used to develop statistical correlations between s_{uFVT} and $s_{u(mob)}$ and other soil properties, such as water content, plasticity index, and over-consolidation ratio (OCR), among others. These correlations can be used to estimate, e.g., s_{uFVT} of Finnish clays for local data, considering the site-specific conditions, i.e., the basic soil parameters and stress history. Furthermore, Selänpää (2021) collected data from various sites in Finland and presented regional correlations for s_u from different test types, including triaxial compression (TXC) and extension (TXE), DSS, and FVT. These comprehensive databases allow for cross-comparison and statistical analysis to derive meaningful insights and correlations, reducing the inherent uncertainty associated with local data.

In addition to the aforementioned local databases, comprehensive global databases have been compiled by Ching and Phoon (2012, 2013, 2014a,b); Ching et al (2017, 2018); D'Ignazio et al. (2019) to cover an extensive array of soils and their associated properties, including clays, sands, and rocks. Such studies demonstrated the benefits of employing databases for transformation model assessments. Furthermore, soil databases serve as indispensable resources when dealing with situations where data for a particular site is either incomplete, sparsely available, or not entirely reliable, as highlighted by Ching and Phoon (2020).

In general, databases can be used to improve the determination of a specific soil property, in our case the undrained shear strength (s_u), in a number of ways. For example, databases can be used to:

- **Identify and quantify bias in local data.** This is done by comparing the local data to a larger and more comprehensive set of data including similar soil types, from a variety of sites and conditions. For example, if the local data is consistently lower than the data in the database, this may indicate that the local data is biased low.
- **Assess the variability of s_u at a particular site.** This is done by calculating the standard deviation of the s_u data for that site. This information can be used to estimate the uncertainty associated with s_u predictions for that site, to be compared with the uncertainties from similar test sites where testing has been carried out using similar equipment..
- **Develop site-specific s_u correlations.** This is done by identifying statistical relationships between s_u and other soil properties for the local data. These correlations can then be used to estimate s_u for local data, taking into account the specific conditions at the site.

- **Validate s_u predictions from numerical models.** This is done by comparing the model predictions to the s_u data in the database. This information can be used to assess the accuracy of the numerical models and to identify any areas where the models may need to be improved (e.g. D'Ignazio et al. 2018).

In this study, data from Selänpää (2021) from six sites in Southern Finland is compared with the large F-CLAY/7/216 database by D'Ignazio et al. (2016). The dataset includes triaxial compression (CAUC/CIUC, s_{uC}), extension (CAUE/CIUE, s_{uE}), DSS, and FVT results on soft sensitive clays. The anisotropic s_u data by Selänpää (2021) is collected to derive a new database of Finnish soft clays named ANI-F-CLAY/7/87, following the nomenclature by Ching and Phoon (2014a) for multivariate database as "soil type"/"number of parameters of interest"/"number of data points." For the ANI-F-CLAY/7/87 database, the label "ANI" stands for anisotropic. The database is further compared with an extensive high-quality database of Norwegian clays presented by Karlsrud and Hernandez-Martinez (2013) and Paniagua et al. (2019). This analysis results in the establishment of a robust statistical correlation for anisotropic undrained shear strength of Finnish soft clays. By analysing Finnish and Norwegian soil data from databases exhibiting variations in water content, plasticity, and stress history, this study aims to quantify the uncertainties involved in the determination of s_u for Finnish clays in the absence of site-specific test data.

In essence, these databases serve as valuable repositories of collective knowledge and experience, enabling researchers and engineers to assess the reliability of local data, reduce bias, and gain a more comprehensive understanding of the undrained shear strength of clays in specific regions. The utilization of such databases not only aids in enhancing the precision of calculations but also fosters greater confidence in infrastructure design and construction practices.

4.2 DATABASES OF FINNISH AND SCANDINAVIAN SOFT CLAYS

4.2.1 Large regional databases

In the study conducted by D'Ignazio et al. (2016), two large regional databases were compiled, focusing on clays from Finland and Scandinavia (Sweden-Norway). These databases were denoted as F-CLAY/7/173 and S-CLAY/7/168, each containing 173 and 168 data points, respectively. These comprehensive datasets encompass multivariate information, consisting of seven fundamental parameters measured at equivalent sampling locations and depths. The parameters include in-situ undrained shear strength from FVT (s_{uFVT}), effective vertical stress (σ'_v), vertical pre-consolidation pressure (σ'_p), natural water content (w), liquid limit (LL), plastic limit (PL), and sensitivity index (S_t, calculated as the ratio of s_u to s_{ur}, where s_{ur} represents the remoulded undrained shear strength from the Fall Cone test). Additionally, ten dimensionless parameters were derived from these basic parameters, which encompass the following categories:

1. Index properties, including natural water content (w), liquid limit (LL), plasticity index (PI), and liquidity index (LI).
2. Stress and strengths, including OCR, $s_{u(mob)}/\sigma'_v$, $s_{u(mob)}/\sigma'_p$, s_{uFVT}/σ'_v, s_{uFVT}/σ'_p, $S_t = s_u/s_{ur}$.

The databases based on dimensionless parameters were renamed as F-CLAY/10/173 and S-CLAY/10/168.

The $s_{u(mob)}$ is defined as λ multiplied by s_{uFVT}, where λ is a correction factor based on the liquid limit as $\lambda = 1.5/(1 + LL/100)$ (Liikennevirasto 2018).

The clay properties within the F-CLAY/7/173 dataset exhibit a wide range of S_t values, ranging from 2 (insensitive clays) to 58 (quick clays), a broad spectrum of PI values (ranging from 2% to 95%), w values spanning from 25% to 150%, and OCR values ranging from 1 to 3.7. As for the basic clay properties within the S-CLAY/7/168 dataset, they fall within the following ranges: S_t=3.0–42.5, PI = 4–128, w=17–180, and OCR=1–6.1.

D'Ignazio et al. (2016) presented their findings in terms of normalized s_u as a function of OCR as expressed in Equation (4.1):

$$\frac{s_u}{\sigma'_v} = \left(\frac{s_u}{\sigma'_v}\right)_{NC} \text{OCR}^m = \text{SOCR}^m \tag{4.1}$$

Here, S represents the normalized s_u/σ'_v for the normally consolidated state, and m is an empirical material coefficient. For $s_{u(mob)}$ of Finnish soft clays, D'Ignazio et al. (2016) found $S \approx 0.24$ and $m \approx 0.76$ with coefficient of variation COV≈0.25 (see Figure 4.1). These coefficients align with the literature range of 0.19–0.27 as reported in previous studies by e.g., Jamiolkowski et al. (1985), Ladd and DeGroot (2003), DeGroot et al. (2019), Yang et al. (2019), and Paniagua et al. (2019). The empirical coefficient m is also consistent with a value of $m \approx 0.8$ observed for lightly over-consolidated Scandinavian clays (Larsson et al. 2007). D'Ignazio et al. (2017b) found $m < 1$ for Finnish and Scandinavian clays from statistical analyses. Furthermore, data from Drammen clay in Norway, as presented by Andersen (2004), suggest a value of $m \approx 0.8$ for OCR ranging from 1 to 40. Jamiolkowski et al. (1985) and Andersen et al. (2023) recommended a value of $m \approx 0.8$ for clays.

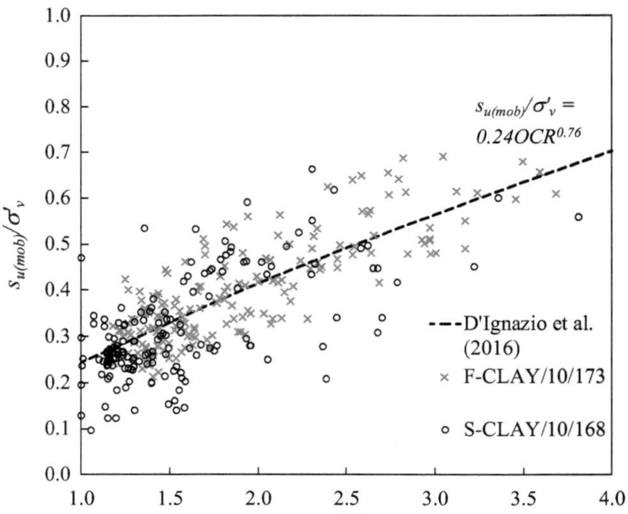

Figure 4.1 $s_{u(mob)}/\sigma'_v$ versus OCR from F-CLAY/10/173 and S-CLAY/10/168 databases and trendline by D'Ignazio et al. (2016) for Finnish clays.

Additionally, the normalized $s_{u(mob)}/\sigma'_v$ and $s_{u(mob)}/\sigma'_p$ were found to be independent of index properties, while the basic clay properties seemed to influence s_{uFVT}/σ'_v and s_{uFVT}/σ'_p. Therefore, in this study, we focus on the more fundamental relationship between s_u and σ'_p or normalized s_u and OCR=σ'_p/σ'_v.

Finally, for the purpose of this study, Equation (4.2) for the OCR-s_{uFVT}/σ'_v transformation model for Finnish clays is derived based on Equation (4.1) as follows:

$$\frac{s_{uFVT}}{\sigma'_v} = 0.27\text{OCR}^{0.78} \qquad COV = 0.34 \tag{4.2}$$

The COV≈0.34 quantifies the uncertainties associated with the validation database S-CLAY/10/168.

4.2.2 Multivariate ANI-F-CLAY/7/87 database from Finland

Selänpää (2021) compiled a high-quality multivariate dataset from six sites in Southern Finland, including Perniö, Paimio, Sipoo, Masku, Lempäälä, and Kotka. This dataset was collected using a modified Laval-type sampler with a diameter of 132 mm (TUT132, Di Buò et al. 2019a), and it includes results of index tests, CAUC/CIUC and CAUE/CIUE (five sites), and DSS (four sites) tests. Specimens were reconsolidated to the in situ stress in order to preserve their original structure. Miniblock (150-mm-diameter) Sherbrook sampler (Emdal et al. 2016) data are available for the Perniö site. Such data is presented for comparison with the TUT132 sampler.

The database also incorporates FVT data points. Notably, the FVT data from Perniö is part of the F-CLAY/7/173 database by D'Ignazio et al. (2016). The clay properties within this dataset exhibit a wide range of PI values (from 17% to 63%), w values ranging from 57% to 117%, and S_t values spanning from 16 to 98. The OCR values range in the 1~2 interval. The OCR values are inferred from constant-rate-of-strain (CRS) oedometer tests carried out in the vicinity of the specimens used for triaxial and DSS tests. The data are collected from depths up to 10 m below ground level.

Sample quality was assessed according to Lunne et al. (1997). Data points fall within the sample quality category "Very good to excellent."

The multivariate TXC (CAUC/CIUC), DSS, TXE (CAUE/CIUE), and FVT data from Selänpää (2021) are presented in Appendix 4.A of this chapter. The database is named ANI-F-CLAY/7/87, as discussed in Section 4.1.

Figure 4.2 illustrates a comparison between s_{uFVT}/σ'_v and OCR from the data points in ANI-F-CLAY/7/87 and D'Ignazio et al. (2016). Figure 4.3 compares s_{uC}/σ'_v versus OCR, s_{uDSS}/σ'_v versus OCR, and s_{uE}/σ'_v versus OCR with $s_{u(mob)}/\sigma'_v$ of Finnish soft clays (D'Ignazio et al. 2016). The trend observed in the FVT data in ANI-F-CLAY/7/87 from four sites appears to align with the general trend and scatter observed in the F-CLAY/7/173 database. Furthermore, the DSS test results are in agreement with the mean trend of $s_{u(mob)}/\sigma'_v$; while TXC and TXE test data are higher and lower than $s_{u(mob)}$, respectively.

The TXC and DSS data present a clear and consistent trend, while the TXE data seem to display a lack of dependence on the OCR. However, it is worth noting that this dataset may contain outliers, as it exhibits greater variability compared to TXC and DSS. For instance, Selänpää (2021) identified n.3 outliers for the TXE dataset. Their removal results in the overall trend of the TXE data points align more closely with the TXC and

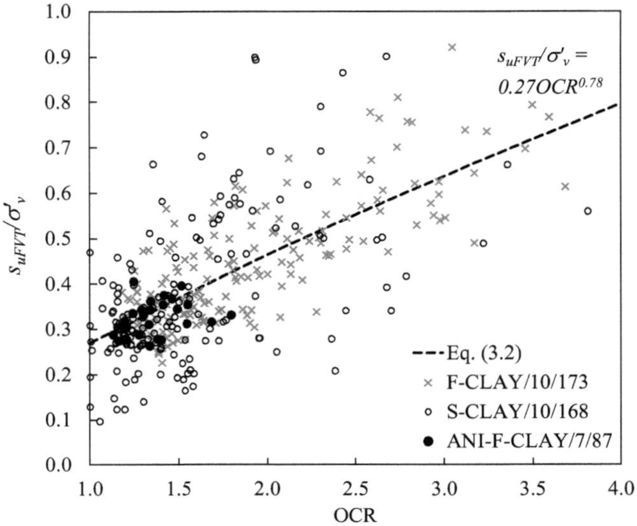

Figure 4.2 s_{uFVT}/σ'_v versus OCR from F-CLAY/10/173, S-CLAY/10/168 and ANI-F-CLAY/7/87 databases and trendline for s_{uFVT}/σ'_v of Finnish clays based on F-CLAY/10/173 database (Equation 4.2).

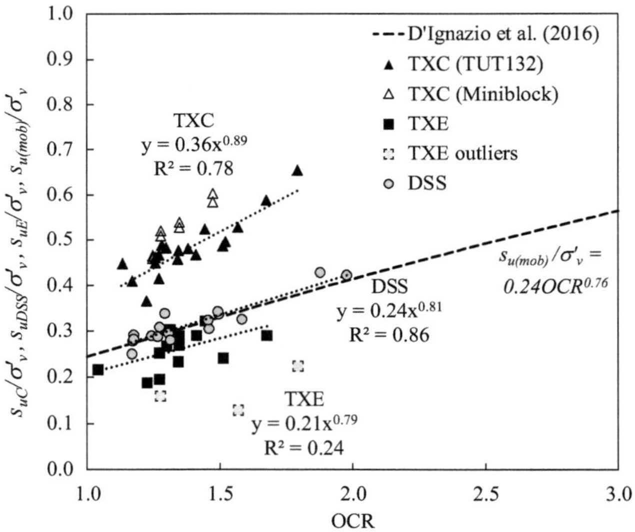

Figure 4.3 s_{uC}/σ'_v, s_{uDSS}/σ'_v, and s_{uE}/σ'_v versus OCR from ANI-F-CLAY/7/87 database and trendline for $s_{u(mob)}/\sigma'_v$ of Finnish clays based on D'Ignazio et al. (2016).

DSS data. Trendlines for TXC, DSS, and TXE of Finnish soft clays with OCR less than 2 are proposed as follows (see Figure 4.3, modified after Selänpää 2021):

$$\frac{s_{uC}}{\sigma'_v} = 0.36\ \text{OCR}^{0.89} \qquad R^2 = 0.78,\ \text{COV} = 0.06 \tag{4.3}$$

$$\frac{s_{uDSS}}{\sigma'_v} = 0.24\text{OCR}^{0.81} \qquad R^2 = 0.86,\ \text{COV} = 0.06 \tag{4.4}$$

$$\frac{s_{uE}}{\sigma'_v} = 0.21\text{OCR}^{0.79} \qquad R^2 = 0.24,\ \text{COV} = 0.23;\ n.3\ \textit{outliers removed} \tag{4.5}$$

The coefficient of variation (COV) is defined later in Section 4.3. The S and m values calculated for the OCR-s_{uDSS}/σ'_v model for Finnish clays closely resemble those obtained by D'Ignazio et al. (2016) for the OCR-$s_{u(mob)}/\sigma'_v$ correlation. Furthermore, based on the information presented above, it is not straightforward to establish anisotropy ratios s_{uDSS}/s_{uC} and s_{uE}/s_{uC} for Finnish clays. The apparent influence of OCR on s_{uDSS}/s_{uC} and s_{uE}/s_{uC} can be attributed to two factors: (i) the limited number of data points available to establish the correlations, and (ii) the potential presence of additional outliers, although this aspect is beyond the scope of this study.

Therefore, given the restricted range of OCR, the relationships between σ'_p and s_u, as depicted in Figure 4.4, are regarded as more suitable for determining anisotropy factors. As shown in Figure 4.4:

$$s_{uC} = 0.36\sigma'_p \qquad R^2 = 0.99,\ \text{COV} = 0.07 \tag{4.6}$$

$$s_{uDSS} = 0.23\sigma'_p \qquad R^2 = 1.00,\ \text{COV} = 0.07 \tag{4.7}$$

$$s_{uE} = 0.18 - 0.19\sigma'_p \qquad R^2 = 0.95 - 0.98,\ \text{COV} = 0.24 \tag{4.8}$$

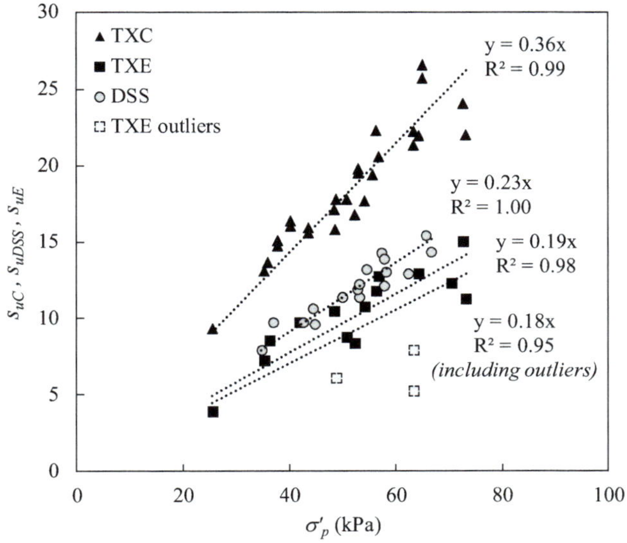

Figure 4.4 Undrained shear strengths s_{uC}, s_{uDSS}, and s_{uE} against pre-consolidation stress σ'_p from ANI-F-CLAY/7/87 database.

Consequently, for Finnish soft clays with OCR values less than 2, the anisotropy factors are estimated as s_{uDSS}/s_{uC} = 0.64 and s_{uE}/s_{uC} = 0.50–0.53. These findings align with those determined for Perniö clay based on CPTU data and finite element back analysis of the Perniö failure test, as presented in the studies by D'Ignazio et al. (2014, 2015, 2017a) and Di Buò et al. (2019b). Selänpää (2021) suggested an average s_{uE}/s_{uC} = 0.62, with a range of calculated values 0.52–0.70. The s_{uE}/s_{uC} calculated by Selänpää (2021) is based on the average of the s_{uE}/s_{uC} calculated for TXC and TXE tests at the same depths and not based on database trendlines. However, the values obtained based on Figure 4.4 are in line with typical values of soft clays from Scandinavia (Larsson et al. 2007; Karlsrud and Hernandez-Martinez 2013).

4.2.3 Comparing regional databases from Finland and Norway

Karlsrud and Hernandez-Martinez (2013) and Paniagua et al. (2019) conducted extensive data collection on block samples (250 mm diameter) of Norwegian clays from seventeen test sites spread across Norway for depths up to 23 m below ground level. Their study involves the evaluation of both TXC (n. 61) and DSS (n. 22) data. The clay properties within this dataset exhibit a wide range of PI values (ranging from 4% to 49%), *w* values spanning from 28% to 72%, and S_t values ranging from 2 to 240. The OCR values within this dataset range from 1 to 6, and the clay content varies between 21% and 65%. The OCR values are inferred from constant-rate-of-strain (CRS) oedometer tests carried out in proximity to the triaxial and DSS specimens.

Sample quality was assessed according to Lunne et al. (1997). Data points fall within sample quality categories "Very good to excellent" and "Good to fair."

In developing the OCR-s_{uC}/σ'_v and OCR-s_{uDSS}/σ'_v transformation models, Karlsrud and Hernandez-Martinez (2013) and Paniagua et al. (2019) observed that the normalized s_u and exponent *m* increase with increasing water content. It is noteworthy that, similar to the Finnish dataset, the tests on Norwegian clay samples involved reconsolidating specimens to their in situ stress levels to preserve the original soil structure. For the purpose of this study, mean trend equations for TXC and DSS, following the format of Equation (4.1), were established through linear regression analyses as follows:

$$\frac{s_{uC}}{\sigma'_v} = 0.34\ OCR^{0.58} \qquad R^2 = 0.71,\ COV = 0.16 \tag{4.9}$$

$$\frac{s_{uDSS}}{\sigma'_v} = 0.25\ OCR^{0.62} \qquad R^2 = 0.73,\ COV = 0.18 \tag{4.10}$$

Figure 4.5 provides a comparison between the TXC data in ANI-F-CLAY/7/87 and the Norwegian block-sample data. While the Finnish TXC data displays a well-defined trend with limited scatter, the Norwegian data appears to exhibit higher uncertainty.

Notably, for OCR values less than 2, the normalized s_{uC}/σ'_v for Finnish clays seems to be higher than that of Norwegian clays, with the Miniblock samples data showing slightly higher values than the TUT132 samples. This observation might be attributed to the lower OCR data points in the Norwegian database being observed at greater depths compared to the Finnish dataset, as discussed by D'Ignazio et al. (2022). This discrepancy could be an indication of the stress-dependent nature of the normalized undrained shear strength of clays. Further, both Miniblock and TUT132 triaxial samples fall into the "Very good to excellent" category by Lunne et al. (1997).

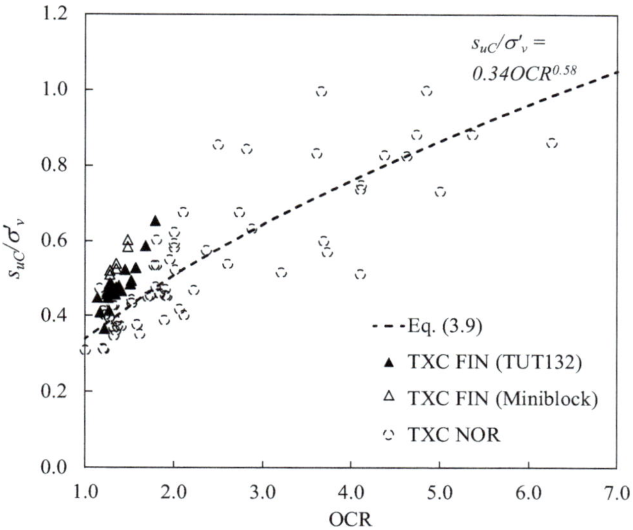

Figure 4.5 s_{uC}/σ'_v versus OCR from ANI-F-CLAY/7/87 database and trendline for s_{uC}/σ'_v of Norwegian clays based on the data in Paniagua et al. (2019) (Equation 4.9).

As discussed earlier, the variability in the Norwegian clay data appears to be linked to the natural water content, while the Finnish data, as reported by Selänpää (2021), seems to exhibit less sensitivity to basic clay properties. Nevertheless, considering that the Finnish data is characterized by higher water content, this might support the findings of Karlsrud and Hernandez-Martinez (2013) and Paniagua et al. (2019).

Furthermore, the increased scatter observed in the Norwegian data could be associated with several factors. Firstly, the Finnish data originates from 6 sites all located in Southern Finland, while the Norwegian database comprises data from seventeen sites across Norway, where regions may have experienced different depositional histories and relative sea level changes. Additionally, the Finnish data was collected using the same sampling equipment and personnel across all sites, whereas the Norwegian data was accumulated over three decades using various types of equipment.

Figure 4.6 presents a comparison between DSS data from Finland and Norway and the F-CLAY/7/173 database. The range of DSS tests on Norwegian clays appears to align with the range of DSS and FVT data from Finland, indicating that s_{uDSS} is approximately equal to $s_{u(mob)}$. As expected, results from triaxial compression tests in both datasets exhibit a generally higher trend compared to DSS and FVT data, hinting at the stress-path-dependent (or anisotropic) behaviour of soft clays.

4.3 STATISTICAL DETERMINATION OF TRANSFORMATION MODELS FOR ANISOTROPIC UNDRAINED SHEAR STRENGTH OF FINNISH CLAYS

Correlations between soil properties are frequently established through linear regression, such as the relationship between undrained shear strength and the OCR – these are known as pairwise correlations. The transformation model, represented by the regression

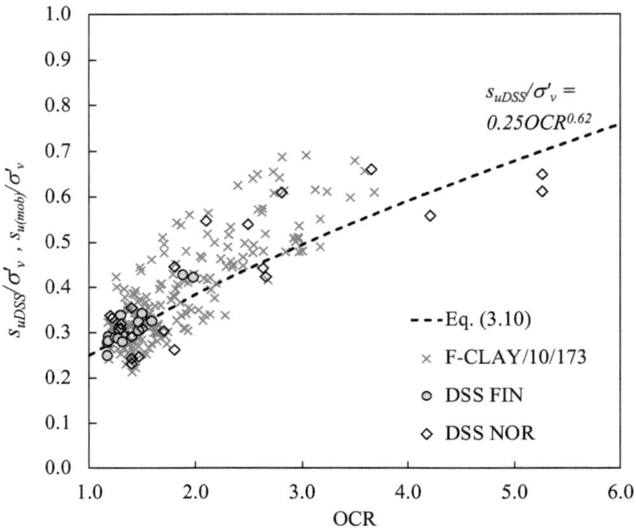

Figure 4.6 s_{uDSS}/σ'_v versus OCR from ANI-F-CLAY/7/87 database, $s_{u(mob)}/\sigma'_v$ versus OCR from F-CLAY/10/173 database, and trendline for s_{uDSS}/σ'_v of Norwegian clays based on the data in Karlsrud and Hernandez-Martinez (2013) (Equation 4.10).

line, is typically considered valid solely for the specific dataset from which it was derived. Such datasets are often tailored to a particular site or region. However, to extrapolate these correlations beyond the local context, it's necessary to validate them against a comprehensive database to quantify bias and uncertainties.

In general, the effectiveness of a transformation model concerning a database can be evaluated in terms of two key parameters: the bias factor (b') and the coefficient of variation (δ), as recommended by Ching and Phoon (2014b). These parameters, b' and δ , express the sample mean and the coefficient of variation, respectively, of the ratio (actual target value/predicted target value) and are defined as follows:

$$b' = \sum_{i=1}^{n} \frac{(\text{actual value})_i / (\text{predicted value})_i}{n} \tag{4.11}$$

$$\delta = \text{COV} = \frac{1}{b'}\sqrt{\frac{1}{n-1}\sum_{i=1}^{n}\left(\frac{(\text{actual value})_i}{(\text{predicted value})_i} - b'\right)^2} \tag{4.12}$$

When b' = 1, the model prediction is considered unbiased. For instance, for the OCR-$s_{u(mob)}/\sigma'_v$ model by D'Ignazio et al. (2016), the actual target value is s_{uDSS}/σ'_v or s_{uC}/σ'_v and the predicted target value is $0.24OCR^{0.76}$. The calibrated bias factors (b') from reference datasets or transformation models for the OCR-s_u/σ'_v models imply $s_u/\sigma'_v = b'S\ OCR^m$ with a coefficient of variation equal to δ .

The ratio of actual to predicted value is widely accepted as a model factor in geotechnical reliability literature to characterize the model uncertainty in calculated responses. For instance, in the context of the hyperbolic limit from a measured load–displacement curve compared to a calculated lateral capacity from limit equilibrium, this ratio is referred to as a model factor (Phoon and Kulhawy 2005). Equations (4.11) and (4.12) can be thought of as representing the mean and coefficient of variation of this model factor. However, in this case, they are applied at the level of soil properties rather than at the level of response. Characterizing the model factor at a soil property level offers the advantage of estimating the model factor of the response, assuming that no other uncertainties emerge from different model idealizations in the adopted calculation model that are more significant than measuring s_u from a specific laboratory test type.

Table 4.1 summarizes the bias factor and COV values for anisotropic undrained shear strength of Finnish soft clays.

Based on the data in Table 4.1, the statistically determined transformation model for s_{uDSS} is unbiased with respect to the mean trend of the $s_{u(mob)}$ of Finnish soft clays. It exhibits a relatively low coefficient of variation (COV) of approximately 0.06, which is noteworthy, considering that it finds its validity over a wider OCR range than the DSS tests alone (specifically, OCR = 1–2 as compared to OCR ≈ 1–4). In contrast, the OCR-s_{uFVT}/σ'_v model expressed in Equation (4.2) tends to overestimate the data points by 9% (with b' = 0.91).

When comparing the OCR-s_{uDSS}/σ'_v model for Finnish clays with the s_{uDSS} of Norwegian clays, the former appears to align well, even though it is based on a dataset with broader OCR range compared to the Finnish one (OCR = 1–2 vs OCR ≈ 1–6). Additionally,

Table 4.1 Bias factors and COV for anisotropic s_u of Finnish soft clays from ANI-F-CLAY/7/87 database.

Actual target value (ANI-F-CLAY/7/87, Selänpää 2021)	*Predicted target value (trendlines from databases)*	*Reference*	*No. of data points*	*b′*	*δ = COV*
s_{uDSS}/σ'_v	$s_{u(mob)}/\sigma'_v = 0.24OCR^{0.76}$	D'Ignazio et al. (2016)	17	1.01	0.06
s_{uDSS}/σ'_v	$s_{uFVT}/\sigma'_v = 0.27OCR^{0.78}$	Equation (4.2), derived from the data in D'Ignazio et al. (2016)	17	0.91	0.06
s_{uDSS}/σ'_v	$s_{uDSS}/\sigma'_v = 0.25OCR^{0.62}$	Equation (4.10), derived from the data in Karlsrud and Hernandez-Martinez et al (2013)	17	1.04	0.06
s_{uC}/σ'_v	$s_{uC}/\sigma'_v = 0.34OCR^{0.58}$	Equation (4.9), derived from the data in Paniagua et al (2019)	19	1.18	0.08
s_{uC}/σ'_v	$s_{u(mob)}/\sigma'_v = 0.24OCR^{0.76}$	D'Ignazio et al. (2016)	19	1.55	0.08
s_{uC}/σ'_v	$s_{uFVT}/\sigma'_v = 0.27OCR^{0.78}$	Equation (4.2), derived from the data in D'Ignazio et al (2016)	19	1.39	0.08
s_{uE}/σ'_v	$s_{uC}/\sigma'_v = 0.34OCR^{0.58}$	Equation (4.9), derived from the data in Paniagua et al (2019)	17	0.60	0.22
s_{uE}/σ'_v	$s_{u(mob)}/\sigma'_v = 0.24OCR^{0.76}$	D'Ignazio et al. (2016)	17	0.80	0.23
s_{uE}/σ'_v	$s_{uFVT}/\sigma'_v = 0.27OCR^{0.78}$	Equation (4.2), derived from the data in D'Ignazio et al (2016)	17	0.72	0.23

the OCR-s_{uC}/σ'_v model for Finnish clays tends to overestimate the mean trend of the Norwegian block-sample database by approximately 18%, as qualitatively discussed in the previous section.

Furthermore, in comparison to $s_{u(mob)}$ and s_{uFVT} of Finnish clays, s_{uC} for OCR values less than 2 is approximately 55% and 39% larger, respectively, and is characterized by a fairly low COV = 0.08.

Finally, the OCR-s_{uE}/σ'_v model underestimates the predicted target values of s_{uC}/σ'_v, s_{uFVT}/σ'_v and $s_{u(mob)}/\sigma'_v$ as anticipated. Additionally, the calculated COV is higher in comparison to the OCR-s_{uC}/σ'_v and the OCR-s_{uDSS}/σ'_v models (COV = 0.22–0.23 as opposed to 0.06–0.08). This is in line with the scatter in the original TXE dataset contained in the ANI-F-CLAY/7/87 database.

4.4 DISCUSSION

The choice of an appropriate testing setup for determining the undrained shear strength of clay depends on the expected deformation behaviour of a geotechnical structure in the field. For instance, when assessing embankment stability, it often relies on the average mobilized shear strength along a potential slip surface, which is typically approximated as $s_{uavg} \approx s_{uDSS}$ or $s_{u(mob)}$. However, it is important to note that the shear strength along a slip surface is influenced by the orientation of the major principal stress, making it stress-path dependent or anisotropic, as described by Bjerrum (1973). Therefore, for accurate modelling of shear strength in embankment stability, results from both triaxial and simple shear tests should be considered.

In the case of excavations or cuts, undrained triaxial compression and extension tests can be conducted to capture the active and passive soil resistances in undrained conditions, yielding s_{uC} and s_{uE}, respectively. For pile design, especially for axial loading, s_{uDSS} or s_{uavg} is typically adopted. However, the deformation behaviour of a pile under lateral loading involves a wide range of stress paths.

In practice, determining anisotropic undrained shear strength typically necessitates a combination of triaxial and direct simple shear tests, which are not commonly conducted in Finland, especially in medium- to low-sized projects. However, field vane tests performed in situ find practical utility in everyday geotechnical applications. To address the need for practical solutions, statistically derived transformation models, as outlined in Table 4.1, can be employed to derive equations suitable for initial assessments when specific site data is unavailable. These equations can also serve as a valuable tool for scrutinizing potentially unreliable data, given their foundation in high-quality datasets.

The following equations are proposed:

$$s_{uC} \approx 1.55 s_{u(\mathrm{mob})} \quad \mathrm{COV} = 0.08,\ OCR < 4 \tag{4.13}$$

$$s_{uDSS} \approx s_{u(\mathrm{mob})} \quad \mathrm{COV} = 0.06,\ \mathrm{OCR} < 4 \tag{4.14}$$

$$s_{uE} \approx 0.80 s_{u(\mathrm{mob})} \quad \mathrm{COV} = 0.23,\ \mathrm{OCR} < 4 \tag{4.15}$$

The equations mentioned earlier are in line with the anisotropy ratios $s_{uDSS}/s_{uC} \approx 0.64$ and $s_{uE}/s_{uC} \approx 0.50$ estimated in Section 4.2.2 for Finnish soft clays. Equations based on $s_{u(mob)}$ are favoured over those based on s_{uFVT}, primarily because of the lower COV reported for $s_{u(mob)}$ by D'Ignazio et al (2016) compared to the COV of Equation (4.2) for s_{uFVT} (COV ≈ 25% versus 34%).

In addition, the stress exponent m obtained from Equations (4.9)–(4.10) is notably lower than the range of m=0.79–0.89 observed in Finnish soft clays, as indicated in Equations (4.3)–(4.5). As a result, although Table 4.1 indicates δ values below 0.1 for s_{uC}/σ'_v and s_{uDSS}/σ'_v associated with the Norwegian clay database, which includes OCR values in the range 1~6, it is important to exercise caution when extrapolating Equations (4.13)–(4.15) to OCR values exceeding 4. Such extrapolation should be undertaken with due consideration for local geology and stress history.

4.5 CONCLUSIONS

This chapter presents statistical transformation models for anisotropic undrained shear strength of Finnish clays based on data from triaxial and direct simple shear tests. The correlations have been validated against comprehensive regional databases that cover a broader spectrum of soil properties.

A high-quality database collected from six distinct sites in Finland, named ANI-F-CLAY/7/87, serves as the foundation for this study. This dataset incorporates test results from laboratory triaxial compression, triaxial extension, direct simple shear tests, and in-situ field vane tests. The study focuses on establishing correlations for anisotropic undrained shear strength, validated against a large database of field vane test results from Finland and a laboratory database of high-quality Norwegian clay block samples. Both validation databases cover a broader range of basic properties, particularly pre-consolidation stress and OCR values, compared to the Finnish laboratory dataset.

The primary focus of this study is the relationship between undrained shear strength and pre-consolidation stress in clays, emphasizing normalized properties. The derived transformation models are suitable for various plasticity, water content, sensitivity, and over-consolidation ratio values in Finnish clays.

Given its relevance in the determination of undrained shear strength of clays in Finnish practice, transformation models for anisotropic shear strength based on field vane strength are proposed. These anisotropic strength formulas find relevance in a broad range of geotechnical applications, including embankment, excavation, and deep foundation design and analysis.

It is important to note that these results are most applicable for preliminary analyses, particularly when site-specific data is unavailable, and the expected in situ over-consolidation ratio is below 4. In more detailed design phases, local correlations should be verified and, if necessary, recalibrated based on specific test data.

As a final remark, for databases characterized by large scatter, using simpler equations often yields results comparable to using more complex ones, as concluded by Länsivaara et al. (2023). They noted that while adding more variables might improve a training set performance, it often worsens testing set predictions, indicating overtraining, especially with incoherent data. Hence, for scattered data, it is advisable to use simpler models with minimal input data to avoid overtraining risks.

REFERENCES

Andersen, K. H. (2004). Cyclic clay data for foundation design of structures subjected to wave loading. In *Intern. Conf. on Cyclic Behaviour of Soils and Liquefaction Phenomena, CBS04,* Bochum, Germany (pp. 371–387). A.A. Balkema, Ed. Th. Triantafyllidis. https://www.researchgate.net/publication/299810172_Cyclic_clay_data_for_foundation_design_of_structures_subjected_to_wave_loading.

Andersen, K. H., Engin, H. K., D'ignazio, M., and Yang, S. (2023). Determination of cyclic soil parameters for offshore foundation design from an existing data base. *Ocean Engineering,* 267, 113180.

Bjerrum, L. 1973. Problems of soil mechanics and construction on soft clays. State-of-the-art report. In *Proceedings, 8th ICSMFE,* Moscow. Vol. 3, pp. 111–159.

Ching J. and Phoon KK. 2012. Modeling parameters of structured clays as a multivariate normal distribution. *Canadian Geotechnical Journal* 49(5): 522–545.

Ching, J. and Phoon, K.K. 2013. Multivariate distribution for undrained shear strengths under various test procedures. *Canadian Geotechnical Journal,* 50(9): 907–923. doi:10.1139/cgj-2013–0002.

Ching, J. and Phoon, K.K. 2014a. Correlations among some clay parameters – the global database. *Canadian Geotechnical Journal,* 51(6): 663–685.

Ching, J. and Phoon, K.K. 2014b. Correlations among some clay parameters – the multivariate distribution. *Canadian Geotechnical Journal,* 51(6): 686–704.

Ching, J. and Phoon, K. K. 2020. Measuring similarity between site-specific data and records from other sites. *ASCE-ASME Journal of Risk and Uncertainty in Engineering Systems, Part A: Civil Engineering,* 6(2), 04020011.

Ching, J., Lin, G. H., Chen, J. R., and Phoon, K. K. 2017. Transformation models for effective friction angle and relative density calibrated based on generic database of coarse-grained soils. *Canadian Geotechnical Journal,* 54(4), 481–501.

Ching, J., Li, K. H., Phoon, K. K., and Weng, M. C. 2018. Generic transformation models for some intact rock properties. *Canadian Geotechnical Journal,* 55(12), 1702–1741.

Di Buò, B., Selänpää, J., Länsivaara, T. T., and D'Ignazio, M. 2019a. Evaluation of sample quality from different sampling methods in Finnish soft sensitive clays. *Canadian Geotechnical Journal,* 56(8), 1154–1168.

Di Buò, B., D'Ignazio, M., Selänpää, J., Haikola, M., Länsivaara, T., and Di Sante, M. 2019b. Investigation and geotechnical characterization of Perniö clay, Finland. *AIMS Geosciences,* 5, 591–616.

D'Ignazio, M. 2016. *Undrained shear strength of Finnish clays for stability analyses of embankments.* PhD Thesis, Tampere University of Technology.

D'Ignazio, M., Mansikkamäki, J., and Länsivaara, T. 2014. Anisotropic total and effective stress stability analysis of the Perniö failure test. In *Proceedings of Conference on Numerical Methods in Geotechnical Engineering (NUMGE),* Delft, Netherlands. Vol. 2, pp. 609–614.

D'Ignazio, M., Di Buò, B., and Länsivaara, T. 2015. A study on the behavior of weathered clay crust in the Perniö failure test. In *Proceedings of XVI ECSMGE,* 13–17 September 2015, Edinburgh, Scotland. 7, pp. 3639–3644.

D'Ignazio, M., Phoon, K.K., Tan, S.A., and Länsivaara, T. 2016. Correlations for undrained shear strength of Finnish soft clays. *Canadian Geotechnical Journal.* DOI: 10.1139/cgj-2016–0037.

D'Ignazio, M., Länsivaara, T. T., and Jostad, H. P. 2017a. Failure in anisotropic sensitive clays: finite element study of Perniö failure test. *Canadian Geotechnical Journal,* 54(7), 1013–1033.

D'Ignazio, M., Phoon, K. K., Tan, S. A., Länsivaara, T., and Lacasse, S. 2017b. Reply to the discussion by Mesri and Wang on "Correlations for undrained shear strength of Finnish soft clays". *Canadian Geotechnical Journal,* 54(5), 749–753.

D'Ignazio, M., Phoon, K. K., Länsivaara, T., and Tan, S. A. 2018. Uncertainties in modeling undrained shear strength of sensitive clays using finite-element method. *ASCE-ASME Journal of Risk and Uncertainty in Engineering Systems, Part A: Civil Engineering,* 4(2), 04018011.

D'Ignazio, M., Lunne, T., Andersen, K. H., Yang, S., Di Buò, B., and Länsivaara, T. 2019. Estimation of preconsolidation stress of clays from piezocone by means of high-quality calibration data. *AIMS Geosciences*, 5(2), 104–116.

D'Ignazio, M., Phoon, K. K., and Länsivaara, T. T. 2021. Uncertainties in modelling undrained shear strength of clays using Critical State Soil Mechanics and SHANSEP. In *IOP Conference Series: Earth and Environmental Science* (Vol. 710, No. 1, p. 012075). IOP Publishing. https://iopscience.iop.org/article/10.1088/1755-1315/710/1/012075/pdf

D'Ignazio, M., Di Buò, B., Länsivaara, T., L'Heureux, J. S., Paniagua, P., and Selänpää, J. 2022. Piezocone testing in Nordic soft clays: Comparison of high-quality databases. In *Cone Penetration Testing* 2022 (pp. 356–362). CRC Press. https://www.taylorfrancis.com/chapters/oa-edit/10.1201/9781003308829-48/piezocone-testing-nordic-soft-clays-comparison-high-quality-databases-ignazio-di-bu%C3%B2-l%C3%A4nsivaara-heureux-paniagua-sel%C3%A4np%C3%A4%C3%A4?context=ubx&refId=eb599372-fb27-4cc0-99fc-837f960be2b8.

Emdal, A., Gylland, A., Amundsen, H.A., Kåsin, K., and Long, M. 2016. Mini-block sampler. *Canadian Geotechnical Journal*, 53(8), 1235–1245.

Jamiolkowski, M., Ladd, C.C., Germain, J.T., and Lancellotta, R. 1985. New developments in field and laboratory testing of soils. In *Proceedings of the 11th International Conference on Soil Mechanics and Foundation Engineering*, San Francisco, CA. Vol. 1, pp. 57–153.

Karlsrud, K. and Hernandez-Martinez, F.G. 2013. Strength and deformation properties of Norwegian clays from laboratory tests on high-quality block samples 1. *Canadian Geotechnical Journal*, 50(12):1273–1293.

Länsivaara, T. 1999. *A study of the mechanical behavior of soft clay.* PhD thesis, Norwegian University of Science and Technology, Trondheim.

Länsivaara, T. T., Farhadi, M. S., and Samui, P. 2023. Performance of traditional and machine learning-based transformation models for undrained shear strength. *Arabian Journal of Geosciences*, 16(3), 183.

Larsson, R., Sällfors, G., Bengtsson, P.E., Alén, C., Bergdahl, U., and Eriksson, L. 2007. *Skjuvhällfasthet: utvärdering I kohesionsjord* (2nd edition), Information 3. Swedish Geotechnical Institute (SGI), Linköping.

Liikennevirasto, 2018. Penkereiden stabiliteetin laskentaohje. *Liikenneviraston ohjeita 14/2018*, Helsinki. (In Finnish, Title in English: Guidelines for embankment stability calculations).

Lunne, T., Berre, T., and Strandvik, S. 1997. Sample disturbance effects in soft low plastic Norwegian clay. In *Proceedings of the Conference on Recent Developments in Soil and Pavement Mechanics*, Rio de Janeiro, Brazil, 25–27 June. Balkema, Rotterdam. pp. 81–102. [Also published in Norwegian Geotechnical Institute, Publication 204.]

Paniagua, P., D'Ignazio, M., L'Heureux, J. S., Lunne, T., and Karlsrud, K. 2019. CPTU correlations for Norwegian clays: an update. *AIMS Geosciences*, 5, 82–103.

Phoon, K.K., and Kulhawy, F.H. 1999. Characterization of geotechnical variability. *Canadian Geotechnical Journal*, 36(4), 612–624. doi:10.1139/t99-038.

Phoon, K. K. and Kulhawy, F. H. 2005. Characterisation of model uncertainties for laterally loaded rigid drilled shafts. *Geotechnique*, 55(1), 45–54.

Selänpää, J. 2021. *Derivation of CPTu Cone Factors for Undrained Shear Strength and OCR in Finnish Clays.* PhD Thesis, Tampere University, Tampere, Finland.

Selänpää, J., Di Buò, B., Länsivaara, T., and D'Ignazio, M. 2017. Problems related to field vane testing in soft soil conditions and improved reliability of measurements using an innovative field vane device. *Landslides in Sensitive Clays: From Research to Implementation*, 109–119.

Yang, S. L., Lunne, T., Andersen, K. H., D'lgnazio, M., and Yetginer, G. 2019. Undrained shear strength of marine clays based on CPTU data and SHANSEP parameters. Proc. XVII ECSMGE-2019, Iceland.

APPENDIX 4.A

Table 4.A1 Basic information of the ANI-F-CLAY/7/87 database (based on Selänpää, 2021)

Site	*Test type*	*Sampler type*	s_u *(kPa)*	σ'_v *(kPa)*	σ'_{pCRS} *(kPa)*	*w*	*LP*	*LL*	S_t
Kotka	DSS	TUT132	9.8	28.6	37.0	106	33	70	49
Kotka	DSS	TUT132	10.6	38.0	44.4	57	19	40	38
Kotka	DSS	TUT132	9.7	38.4	44.8	58	19	39	38
Paimio	DSS	TUT132	12.9	39.5	62.4	66	24	42	98
Paimio	DSS	TUT132	14.3	49.0	57.4	109	31	65	91
Paimio	DSS	TUT132	13.9	49.3	57.8	111	31	65	91
Paimio	DSS	TUT132	15.4	53.0	65.7	102	29	66	73
Perniö	DSS	TUT132	7.9	26.6	34.7	99	28	67	38
Perniö	DSS	TUT132	9.7	33.7	42.6	81	25	46	72
Perniö	DSS	TUT132	13.2	42.9	54.4	98	26	58	49
Sipoo	DSS	TUT132	11.4	26.6	49.9	91	30	56	45
Sipoo	DSS	TUT132	11.4	26.9	53.1	90	30	66	45
Sipoo	DSS	TUT132	11.9	35.3	52.8	109	31	85	23
Sipoo	DSS	TUT132	12.3	35.6	53.2	115	31	89	23
Sipoo	DSS	TUT132	12.1	39.7	57.8	110	31	94	22
Sipoo	DSS	TUT132	13.0	40.0	58.2	107	31	94	22
Sipoo	DSS	TUT132	14.4	51.0	66.7	98	28	79	16
Lempäälä	FV	N/A	6.7	18.8	29.2	74	27	47	25
Lempäälä	FV	N/A	12.2	30.8	46.8	69	28	54	29
Masku	FV	N/A	9.3	29.6	50.0	80	27	66	20
Masku	FV	N/A	14.6	38.9	55.2	117	36	95	18
Masku	FV	N/A	17.8	52.1	69.9	86	28	73	20
Paimio	FV	N/A	12.1	35.3	52.8	67	23	40	82
Paimio	FV	N/A	13.0	41.6	64.5	72	24	42	98
Paimio	FV	N/A	14.4	50.4	57.1	108	30	66	77
Paimio	FV	N/A	15.0	54.7	64.0	94	29	59	67
Paimio	FV	N/A	16.8	61.1	74.0	85	28	53	82
Perniö	FV	N/A	8.7	26.5	34.5	99	28	67	38
Perniö	FV	N/A	9.4	27.4	35.5	99	28	67	38
Perniö	FV	N/A	7.5	28.8	38.6	94	26	54	53
Perniö	FV	N/A	8.9	31.1	39.9	86	25	54	50
Perniö	FV	N/A	10.2	33.5	40.4	71	25	44	72
Perniö	FV	N/A	11.1	34.9	41.9	71	25	44	72
Perniö	FV	N/A	11.2	35.9	42.9	72	24	45	72
Perniö	FV	N/A	16.9	41.8	52.2	107	26	70	55
Perniö	FV	N/A	14.4	43.1	53.7	107	26	70	55
Perniö	FV	N/A	16.6	45.2	66.3	84	26	59	
Perniö	FV	N/A	17.1	47.4	63.7	95	27	64	66
Perniö	FV	N/A	15.4	49.6	66.2	95	27	64	66
Sipoo	FV	N/A	8.9	26.8	48.2	89	30	66	45
Sipoo	FV	N/A	9.8	35.5	49.9	116	31	88	23
Sipoo	FV	N/A	11.0	39.5	54.6	110	31	89	28
Sipoo	FV	N/A	18.1	51.2	72.2	99	28	79	16

Site	Test type	Sampler type	s_u (kPa)	σ'_v (kPa)	σ'_{pCRS} (kPa)	w	LP	LL	S_t
Lempäälä	CIUC	TUT132	9.4	20.1	25.5	74	27	47	25
Lempäälä	CIUC	TUT132	15.8	32.0	48.6	69	28	54	29
Masku	CIUC	TUT132	17.8	30.3	50.7	80	27	66	20
Masku	CIUC	TUT132	20.6	39.4	56.9	117	36	95	18
Paimio	CIUC	TUT132	16.8	34.7	52.3	67	23	40	82
Paimio	CIUC	TUT132	21.3	40.4	63.3	72	24	42	98
Paimio	CIUC	TUT132	22.3	49.8	56.4	108	30	66	77
Paimio	CIUC	TUT132	22.2	54.2	63.3	94	29	59	67
Paimio	CIUC	TUT132	22.0	60.0	73.3	85	28	53	82
Perniö	CAUC	Miniblock	25.7	44.0	65.0	105	30	65	55
Perniö	CAUC	Miniblock	16.4	31.5	40.3	86	25	54	50
Perniö	CAUC	Miniblock	14.7	28.0	37.7	98	26	56	54
Perniö	CAUC	Miniblock	15.1	28.0	37.7	98	26	56	54
Perniö	CAUC	Miniblock	19.8	42.6	53.0	107	26	70	55
Perniö	CAUC	TUT132	16.0	34.7	43.6	71	25	44	72
Perniö	CIUC	Miniblock	26.6	44.0	65.0	105	30	65	55
Perniö	CIUC	Miniblock	16.1	31.5	40.3	86	25	54	50
Perniö	CIUC	Miniblock	19.5	42.6	53.0	107	26	70	55
Perniö	CIUC	TUT132	15.6	34.7	43.6	71	25	44	72
Perniö	CIUC	TUT132	17.7	42.6	54.2	90	26	58	49
Perniö	CIUC	TUT132	13.7	28.1	35.9	94	26	54	53
Perniö	CIUC	TUT132	22.0	48.0	64.4	95	27	64	66
Perniö	CIUC	TUT132	13.1	27.1	35.2	99	28	67	38
Sipoo	CIUC	TUT132	17.8	27.2	48.8	90	30	73	22
Sipoo	CIUC	TUT132	17.1	36.0	48.4	116	31	88	23
Sipoo	CIUC	TUT132	19.4	40.4	55.7	110	31	89	28
Sipoo	CIUC	TUT132	24.1	51.6	72.8	80	27	71	16
Lempäälä	CIUE	TUT132	3.9	20.1	25.5	74	27	47	25
Lempäälä	CIUE	TUT132	9.7	32.0	41.9	69	28	54	29
Masku	CIUE	TUT132	12.7	39.4	56.9	117	36	95	18
Masku	CIUE	TUT132	8.8	30.3	50.7	80	27	66	20
Masku	CIUE	TUT132	12.3	52.7	70.6	86	28	73	20
Paimio	CIUE	TUT132	8.4	34.7	52.3	67	23	40	82
Paimio	CIUE	TUT132	11.8	54.2	56.4	108	30	66	77
Paimio	CIUE	TUT132	11.3	60.0	73.3	85	28	53	82
Paimio	CIUE	TUT132	5.3	40.4	63.3	72	24	42	98
Paimio	CIUE	TUT132	8.0	49.8	63.3	94	29	59	67
Perniö	CAUE	TUT132	10.8	42.6	54.2	90	26	58	49
Perniö	CIUE	TUT132	12.9	48.0	64.4	95	27	64	66
Perniö	CIUE	TUT132	8.5	28.1	36.3	94	26	54	53
Perniö	CIUE	TUT132	7.3	27.1	35.2	99	28	67	38
Sipoo	CIUE	TUT132	10.5	36.0	48.4	116	31	88	23
Sipoo	CIUE	TUT132	15.0	51.6	72.8	80	27	71	16
Sipoo	CIUE	TUT132	6.1	27.2	48.8	90	30	73	22

Chapter 5

Role of databases in the evaluation of soil properties

Abolfazl Eslami and Pin-Qiang Mo

This chapter delves into the pivotal role of databases in assessing soil properties of geotechnical engineering. It underscores the significance of site investigations, methodologies, in-situ testing, and soil variability management in advancing the field of geotechnical engineering. The chapter emphasizes the crucial function of geotechnical databases in furnishing historical data and research experiences, fostering the evolution of research methodologies, and enhancing our comprehension of the underground. The text elaborates on the pivotal importance of in-situ testing methods such as Cone Penetration Testing (CPT), Standard Penetration Testing (SPT), and Field Vane Shear Testing (FVST) in soil identification and classification. Concurrently, it explores the latest research achievements in soil correlation and parameter identification leveraging machine learning and Bayesian algorithms, emphasizing the paramountcy of accounting for uncertainties in geotechnical modeling and decision-making. Furthermore, the chapter introduces the AUT: CPTu-Geo-marine Database developed by Eslami et al. (2022), offering a comprehensive evaluation of the performance of various soil behavior classification methodologies. It presents a novel triangular classification chart aimed at enhancing the accuracy of classifying marine sediments.

5.1 INTRODUCTION

In the realm of geotechnical engineering research, soil characterization and classification stand as foundational elements supporting advancements in design and construction. This chapter emphasizes the role of databases in assessing soil properties within the context of geotechnical engineering research. Soil, as the subject of inquiry, plays a central role, in influencing the success and safety of construction projects across infrastructural development and environmental sustainability. The core of geotechnical research lies in site investigation, where researchers embark on a journey to uncover critical insights into soil characteristics and behaviors, deeply intertwined with their environmental interactions. Various methodologies are explored, including remote sensing, data review, site surveying, and continuous monitoring, thus enriching our understanding of soil mechanics. Acknowledging the inherent variability and uncertainty in soil properties, researchers draw from diverse data sources to construct robust research frameworks, mitigating the unpredictability inherent in decision making. In-situ testing emerges as a pivotal research tool, providing real-time insights into soil behavior, with a focus on tests such as the Standard Penetration Test (SPT), Field Vane Shear Test (FVST), and Cone Penetrometer Test (CPT). Geotechnical databases play a central role, providing access to historical data and research experiences, facilitating the advancement of research methodologies, and deepening our collective understanding of the complex world beneath our feet. This

 DOI: 10.1201/9781003441946-5

chapter highlights the crucial role of databases in evaluating soil properties, emphasizing the significance of site investigation, methodologies, in-situ testing, and the management of soil variability, ultimately paving the way for a comprehensive exploration of the field.

5.2 DATA SOURCES IN GEOTECHNICAL ENGINEERING

Geotechnical site investigation stands as the foremost undertaking in any project, serving as the primary means to acquire critical on-site information. This comprehensive process encompasses activities ranging from drilling boreholes to collect both disturbed and undisturbed samples to conducting laboratory tests and in-situ testing, enabling the observation of soil behavior in authentic stress scenarios and under real boundary conditions.

One of the primary and fundamental phases in project design, known as geotechnical site investigation, involves the evaluation of subsurface soil and rock characteristics, which can be obtained through various approaches:

- Remote sensing by evaluating regular aerial photographs
- Reviewing existing data and published sources, such as geological and geographical maps, geotechnical reports, and geological hazard maps
- Site survey and visual observation
- Local experience and ongoing construction operations
- Geophysical and seismic evaluations
- Drilling, sampling, and groundwater assessment
- In-situ and laboratory testing
- Evaluation and analysis of environmental indices
- Instrumentation, surveillance, and monitoring

Typically, a site investigation program progresses through the following stages:

1. Collecting data and literature review
2. Site visit and conducting non-destructive tests
3. Performing operations in the field including drilling, sampling, and in-situ testing
4. Conducting laboratory testing on collected samples
5. Data synthesis and providing a geotechnical investigation report
6. Instrumentation, surveillance, and monitoring

Geotechnical engineers grapple with the inherent variability and uncertainties in natural geomaterials. To ensure the safe design of foundations and minimize uncertainties, it becomes imperative to rely on a variety of measured and collected data. This primary objective can be realized through the utilization of diverse sources of geotechnical data, which can be acquired using the following approaches:

1. Maps and technical literature review
2. Site visit
3. On-situ testing
4. Geophysical surveying
5. Drilling operations

6. In-situ penetration tests
7. Laboratory element tests
8. Physical modeling
9. Full-size field tests
10. Instrumentation and monitoring

5.3 UNCERTAINTY IN GEOTECHNICAL DESIGNS

Geotechnical design, particularly in the field of foundation engineering, encompasses the application of site investigations, design models, adherence to codes, and computational analysis. This process involves informed decision making regarding site-specific conditions, ultimately leading to an optimal design tailored to the project's significance.

In conventional design practices, engineers typically treat material properties like friction angle, soil undrained shear strength, or the uniaxial compressive resistance of concrete as fixed parameters with known values. However, when delving into borehole data or examining the results of tests like SPT or CPT, it becomes evident that there exists inherent uncertainty in soil properties. Furthermore, uncertainties extend beyond material properties; they also encompass simplified constitutive models and the estimation of material characteristics, introducing significant uncertainties into the design process. Empirical evidence in this field substantiates the notable disparities between measured values and predicted design parameters.

Uncertainty in geotechnical engineering can be broadly categorized into two main groups. The first category arises from the inherent randomness in the nature of materials and is a persistent feature that cannot be entirely eradicated. However, it can be better characterized and managed through increased testing or the use of comprehensive databases. Parameters like undrained shear strength fall into this category, as their values exhibit spatial and temporal variations. Shear strength is typically estimated from a limited number of samples and associated tests, which are then used in the design at locations different from where the samples were obtained. The disparity between the estimated resistance value at one point and its actual resistance constitutes the inherent uncertainty. This variability in soil properties is a consequence of the intricate processes governing geomaterial formation, including factors such as sedimentation, weathering, stress history, and the passage of time.

The second category of uncertainty pertains to our incomplete and constrained understanding of a system or process. In general, this type of uncertainty can be classified into three primary types: model uncertainty, transformation uncertainty, and measurement/random noise error.

Measurement errors arise from various sources, including equipment limitations, operator factors, and random fluctuations during the measurement process. Additionally, empirical methods or other models that integrate laboratory or field measurements into soil design characteristics fall into this category of uncertainty. Consequently, this form of uncertainty can be mitigated by enhancing both the quantity and precision of experiments or adopting more accurate models. Statistical uncertainty, which stems from having limited data, can be addressed by increasing the number of samples tested. However, errors attributed to the sampling or measurement procedures cannot be entirely eradicated merely by expanding the sampling effort.

Transformation error represents an additional source of uncertainty and is often associated with a limited number of case records. Every experimental correlation relies on a database that may have constraints, such as being specific to a particular location or soil type. Moreover, the development of such correlations involves assumptions and specific computational models, which means that inherent errors are implicitly introduced into the estimation of soil properties.

Model error, particularly in the context of geotechnical engineering, encompasses a multitude of uncertainties arising from factors like simplifications in equilibrium and deformation analyses, the neglect of three-dimensional modeling effects, and more. Model uncertainty is quantified as the ratio between the actual quantity and the quantity predicted by a model. The uncertainty associated with the model can be characterized by its mean and coefficient of variation. The disparity between the mean value of the unit quantity reveals a systematic error within the model, while the standard deviation denotes the variability in the model's predicted values.

Moreover, a substantial level of uncertainty surrounds the assessment of live and dynamic loads, including factors like earthquakes, winds, floods, and the like. Unlike uncertainties tied to static loads, uncertainties concerning material parameters may be less prominent. However, it is crucial to recognize that under dynamic loading conditions, material parameters may exhibit substantial variations, which might not be immediately evident. For instance, during cyclic loading, soil shear strength tends to decrease, leading to heightened uncertainty regarding the performance of this geotechnical parameter.

Recognizing the limitations of in-situ testing equipment is essential in geotechnical investigations, especially when considering the inherent variability in soil properties. Among in-situ tests, the Standard Penetration Test (SPT) is notably high in variability, while the dilatometer and electrical cone penetration tests exhibit lower levels of variability. Table 5.1 outlines the variability observed in laboratory tests. Table 5.2 illustrates the variability of in-situ tests. Soil density can be evaluated with high accuracy. The variability of shear strength test results in clays and plasticity index is high.

More recently, Phoon et al. (2022a) discussed the importance of considering uncertainties in geotechnical engineering modeling and decision making. This review paper focuses

Table 5.1 Variability from laboratory tests

Coefficient of variation (COV) (%)			
Mean	*Range*	*Property*	*Soil type*
24	5–51	Plasticity index	Fine-grained
24	7–56	Effective angle of friction, ϕ'	Clay, silt
20	19–20	Shear strength, s_u	Clay, silt
19	8–38	Shear strength, s_u	Clay, silt
14	13–14	Effective angle of friction, ϕ'	Sand
14	6–22	Effective angle of friction, ϕ'	Clay
13	3–29	Effective angle of friction, ϕ'	Clay, silt
10	7–18	Plastic limit	Fine-grained
8	2–22	Effective angle of friction, ϕ'	Clay, silt
7	3–11	Liquid limit	Fine-grained
1	1–2	Density	Fine-grained

Source: Phoon and Kulhawy (1999).

Table 5.2 Variability from in-situ tests

Coefficient of variation (COV) (%)	*Test*
15–45	SPT
15–25	Mechanical CPT
15–25	SBP
10–20	VST
10–20	PBP
5–15	Electric CPT
5–15	DMT

Source: Phoon and Kulhawy (1999).

on uncertainty quantification, uncertainty calculation, and how they enhance the role of modeling in decision making (Ching and Phoon, 2013; Phoon et al., 2016; Ching et al., 2016; Tang and Phoon, 2021). The key output from reliability analysis is the probability of failure, which takes into account both mechanics and statistics and is meaningful for both system and component failures (Phoon et al., 1995; AASHTO, 2020).

The need for system-level analysis in resilience engineering and the importance of considering non-classical failure mechanisms in spatially variable soils are also highlighted (Vessia et al., 2021; Tabarroki et al., 2022). It is emphasized that uncertainty quantification brings decision making closer to reality and supports digital transformation in geotechnical engineering practice (Phoon et al., 2022b).

Phoon et al. (2022a) provided guidance on the statistical characterization of soil properties and transformation uncertainties. Additionally, it discusses the use of machine learning and Bayesian-based algorithms for developing soil correlations and parameter identification (Zhang et al. 2020; Ching et al., 2021; Zhang et al., 2022). Overall, the recent studies can help practitioners understand and use geotechnical risk and reliability methods in conjunction with numerical modeling for improved decision making.

5.4 DESCRIPTIVE ROLE OF IN-SITU TESTS IN SOIL IDENTIFICATION

Visual inspections and a range of laboratory tests serve as standard practices for identifying and classifying geotechnical deposits. Nonetheless, visual identification is primarily suitable for preliminary investigations, while laboratory testing has its limitations, including issues related to sample recovery, size effects, and the modeling of actual stress conditions.

In the early stages of geotechnical exploration, soil type identification was often approximated, and the cone penetrometer was limited in its ability to provide detailed information about the location of soil type boundaries. Soil type confirmation typically relied on traditional borings, with empirical interpretations confined to the geological regions where they had been developed. Measured parameters such as pore pressure, tip resistance (q_c), and sleeve frictional resistance (f_s) were recognized as functions of soil type and behavior. A key application of the Cone Penetration Test (CPT) was profiling for soil type identification. Consequently, various charts were introduced in the literature to classify soil types based on CPT or CPTu data (Eslami and Fellenius, 2004; Ku et al., 2010; Cai et al., 2015).

5.4.1 Soil behavior classification based on in-situ tests

In every foundation design, understanding the pertinent parameters that govern soil behavior is imperative. Soil behavior is intricately complex, influenced by numerous factors including particle size, shape, mineral composition, pore fluid characteristics, particle packing, stress history, and various other considerations.

In recent decades, thanks to significant advancements in in-situ testing methods and the increased certainty and reliability of the obtained results, one of the primary applications of in-situ tests is now the determination of soil profiles and their behavioral classifications based on direct measurements. In cases where direct measurement of essential soil parameters is unfeasible, estimates must be derived from available data, often obtained through in-situ tests. Researchers have proposed numerous correlations between these tests and soil parameters. To effectively utilize these interpreted results, it is vital to establish correlations with data from other sources, both field and laboratory, and to draw from accumulated experience. The selection of properties for any correlation is a critical decision in this process.

5.4.2 Soil classification based on SPT blow counts

The Standard Penetration Test (SPT) is a field test carried out during borehole drilling to estimate the soil's resistance to the penetration of a split-spoon sampler at different depths beneath the ground surface. The SPT offers several advantages, including its long-standing use, which has contributed to a wealth of practical experience. Moreover, it is a relatively straightforward and cost-effective testing method. However, it is crucial to understand that the test results are influenced by a multitude of factors, and there exist various sources of potential errors, despite some correlation between the N value and soil engineering parameters (as shown in Table 5.3).

5.4.3 Soil classification based on FVST records

In fine-grained soils, the accuracy of Standard Penetration Test (SPT) records is compromised by elevated pore water pressure. In contrast, the Field Vane Shear Test (FVST) and small-scale Vane Shear Test (VST) are more reliable and suitable for these soil types. A general correlation between the soil consistency and undrained shear strength is presented in Table 5.4, highlighting the usefulness of these alternative tests in assessing fine-grained soils.

5.4.4 Soil classification based on CPT data

One of the primary objectives of conducting a Cone Penetrometer Test (CPT) in a site investigation project is to delineate the subsoil profile based on the test results. Early attempts at classifying soils using CPT data were made by Begemann (1965), who established a relationship between soil type and the ratio of cone resistance to sleeve friction. Subsequently, researchers like Schmertmann (1978), Robertson and Campanella (1983), and Olsen and Mitchel (1995) have utilized parameters such as cone resistance (q_c) or normalized cone resistance and the friction ratio (Fr) in developing the Soil Classification Chart, as depicted in Figure 5.1a, c, offering valuable tools for soil classification based on CPT results.

Table 5.3 Strength from SPT on different types of soils

Material	*Description*	*SPT (N value)*	*Corrected SPT* $(N_1)_{60}$
Clay	Very soft	≤2	–
	Soft	2–5	–
	Firm	5–10	–
	Stiff	10–20	–
	Very stiff	20–40	–
	Hard	>40	–
Clean sand	Very loose	≤4	≤3
	Loose	4–10	3–8
	Medium dense	10–30	8–25
	Dense	30–50	25–42
	Very dense	>50	>42
Fine sand	Very loose	–	≤3
	Loose	–	3–7
	Medium dense	–	7–23
	Dense	–	23–40
	Very dense	–	>40
Medium sand	Very loose	–	≤3
	Loose	–	3–8
	Medium dense	–	8–25
	Dense	–	25–43
	Very dense	–	>43
Coarse sand	Very loose	–	≤3
	Loose	–	3–8
	Medium dense	–	8–27
	Dense	–	27–47
	Very dense	–	>47

Source: Adapted from Look, (2007).

Table 5.4 General correlation between consistency and undrained shear strength

Consistency	s_u *(kPa)*
Very soft	0–25
Soft	25–50
Medium	50–100
Hard	100–200
Very hard	200–400

Source: Das (2006).

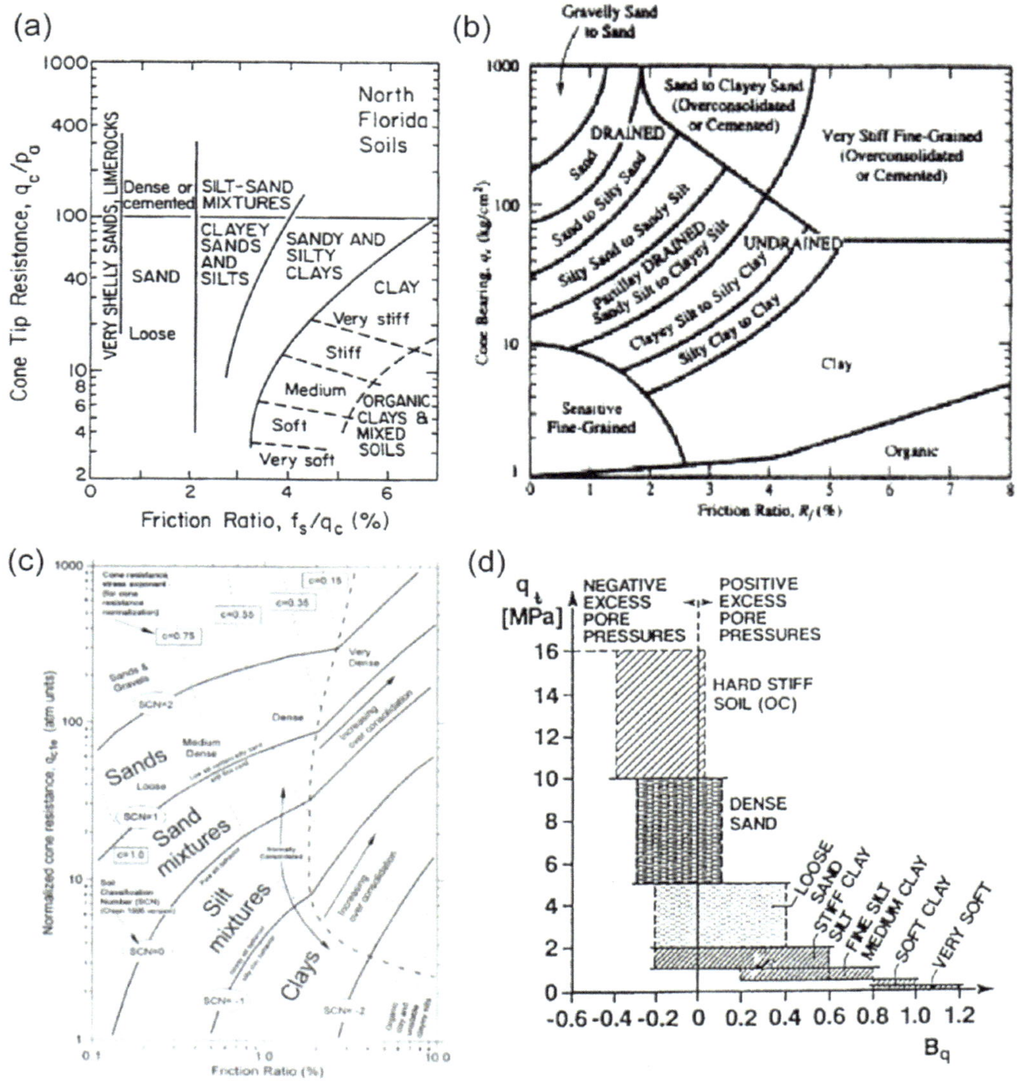

Figure 5.1 Some of the pioneer soil profiling charts: (a) Schmertmann (1978); (b) Campanella (1986); (c) Senneset et al. (1989); and (d) Olsen and Mitchel (1995).

The introduction of the piezocone brought about the capability to measure pore pressure during the test, significantly improving soil classification. The pore pressure ratio B_q, a critical parameter for soil classification, can be calculated using Equation (5.1). Senneset et al. (1989) presented a graph for utilizing this parameter in soil classification, as depicted in Figure 5.1d, enhancing the accuracy and effectiveness of classification based on CPT results.

$$B_q = \frac{u_2 - u_0}{q_t - \sigma'_{v0}} \tag{5.1}$$

where u_0 = pore pressure due to groundwater table, u_2 = pore pressure measured at the shoulder, σ'_0 = total overburden stress, and q_t = corrected tip resistance.

Corrected tip resistance (q_t) can be calculated by using Equation (5.2):

$$q_t = q_c - u_2(1 - a_n) \tag{5.2}$$

where q_c = cone resistance and a_n = net area ratio, approximately equal to the ratio of shaft cross-section and cone cross-section areas (usually measured in a calibration cell, Lunne et al., 1997).

Over the years, numerous researchers have put forth various soil classification methods and charts based on either Cone Penetration Test (CPT) or Cone Penetration Test with pore pressure measurements (CPTu) results. These approaches encompass works by Jones and Rust (1982), Jefferies and Davies (1991), and Ramsey (2002).

Jefferies and Davies (1993) introduced the soil behavior index (I_{SBT}), which delineates the demarcation between distinct soil types and can be computed using Equation (5.3), offering an innovative perspective on soil classification.

$$I_{\text{SBT}} = \left[\left(3.47 - \log\left(q_c/P_a\right)\right)^2 + \left(\log R_f + 1.22\right)^2\right]^{0.5} \tag{5.3}$$

where R_f = friction ratio and q_c = cone resistance.

The values of I_C defined on the boundaries of various soil types on the Robertson (1990) classification chart are depicted in Figure 5.2 (Table 5.5).

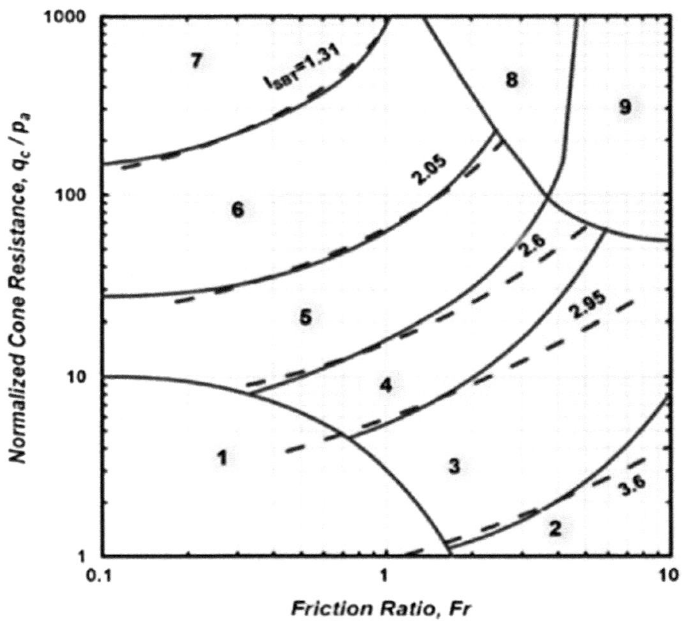

Figure 5.2 Values of I_C on the Robertson (1990) classification chart (Source: Jefferies and Davies, 1991).

Table 5.5 Soil classification based on I_{SBC}

Soil classification	*Zone*	*CPTu index* (I_{SBC})
Gravelly sands	7	$I_{SBC} < 1.25$
Sands – clean sand to silty sand	6	$1.25 < I_{SBC} < 1.90$
Sand mixture – silty sand to sandy silt	5	$1.90 < I_{SBC} < 2.54$
Silt mixture – clayey silt to silty clay	4	$2.54 < I_{SBC} < 2.82$
Clays	3	$2.82 < I_{SBC} < 3.22$
Organic clays	2	$I_{SBC} < 3.22$

Source: Li et al. (2015).

Robertson (1990) introduced an advancement of the profiling chart developed by Robertson et al. (1986). This chart, as depicted in Figure 5.3, features a normalized cone resistance plotted against a normalized friction ratio, providing valuable insights into soil behavior and classification. Additionally, an accompanying pore pressure ratio chart represents the relationship between the normalized cone resistance and pore pressure ratio, further contributing to the understanding of soil properties.

Robertson (2016) presented an update of some popular CPT-based SBT classification systems, since textural-based descriptions, such as sand and clay are used in providing them. Moreover, it includes a method for identifying the presence of microstructures in soils. Figure 5.4 illustrates examples of these updated classification systems, offering a more nuanced and behavior-driven approach to characterizing soil properties.

Eslami and Fellenius (1997) compiled a comprehensive database comprising CPT and CPTu data, which were linked with results from borehole investigations, sampling, laboratory testing, and soil properties. These data points were utilized to create a profiling chart, and distinct envelopes were delineated, encapsulating five soil types, as illustrated in Figure 5.5a. Notably, the plotting of an "effective" cone resistance, defined by Equation (5.4), was observed to offer a more consistent demarcation of these envelopes compared to using solely the cone resistance.

$$q_E = q_t - u_2 \tag{5.4}$$

where q_E = effective cone resistance, q_t = cone resistance corrected for pore water pressure at cone shoulder (Equation 5.2), and u_2 = pore pressure measured at cone shoulder.

A new format of classification chart, i.e., triangular form, is proposed by Eslami et al. (2016) containing cone tip resistance (q_c), sleeve friction (f_s), and pore pressure (u_2). The proposed chart, as shown in Figure 5.5b with more accuracy and less scatter of data than the previous charts, is able to identify soil types, particularly for deltaic soils. Moreover, Eslami et al. (2022) developed a triangular classification chart for deltaic soils by compiling a marine database of the CPTu soundings and their soil profiling in their vicinity. The proposed chart is presented in Figure 5.5c. Overall a qualitative chart is suggested for the general behavior classification of soils, dividing into five generic zones (types) as illustrated in Figure 5.5d.

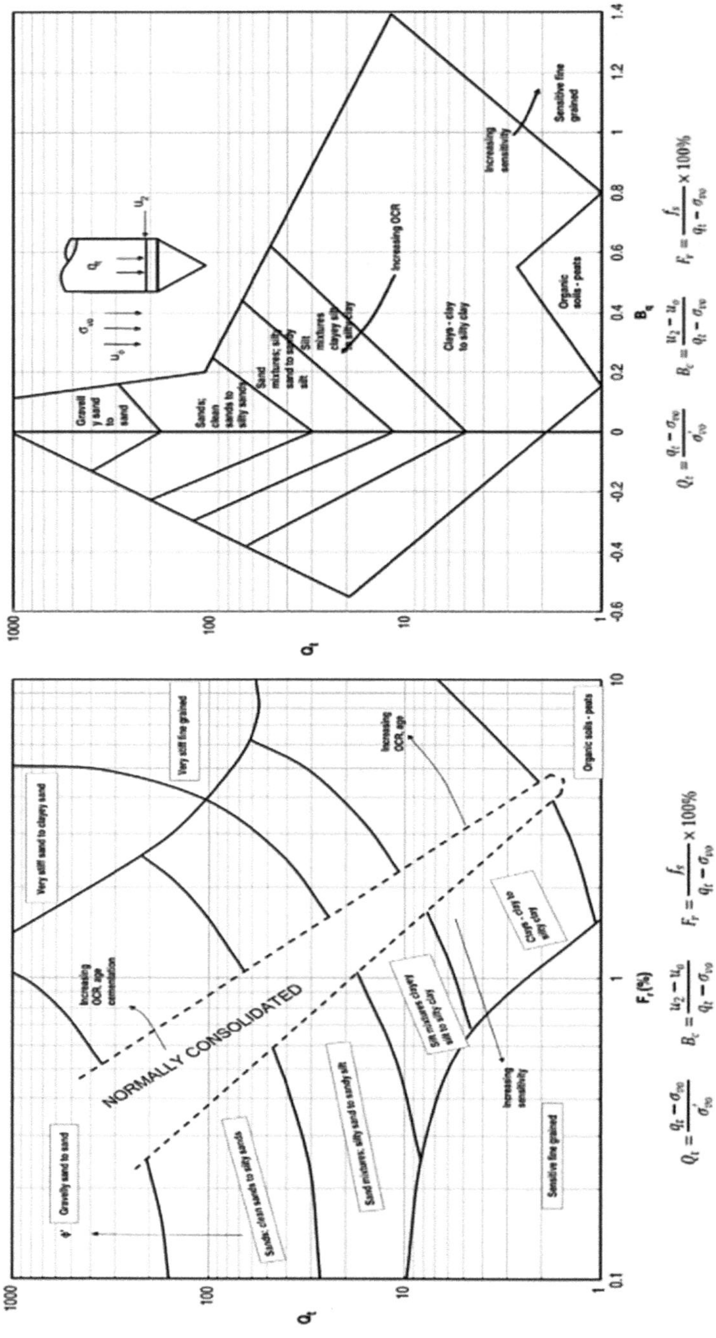

Figure 5.3 Proposed soil behavior type classification system from CPT/CPTu data (Source: adapted from Robertson, 1990).

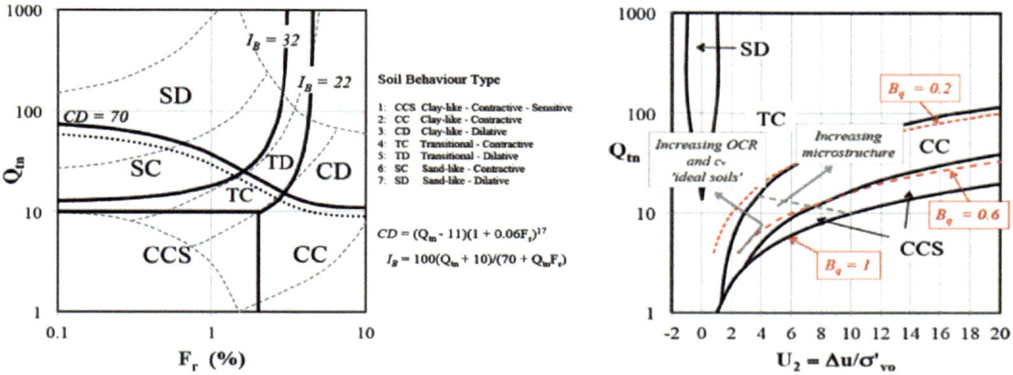

Figure 5.4 Examples of updated classification charts (Source: Robertson, 2016).

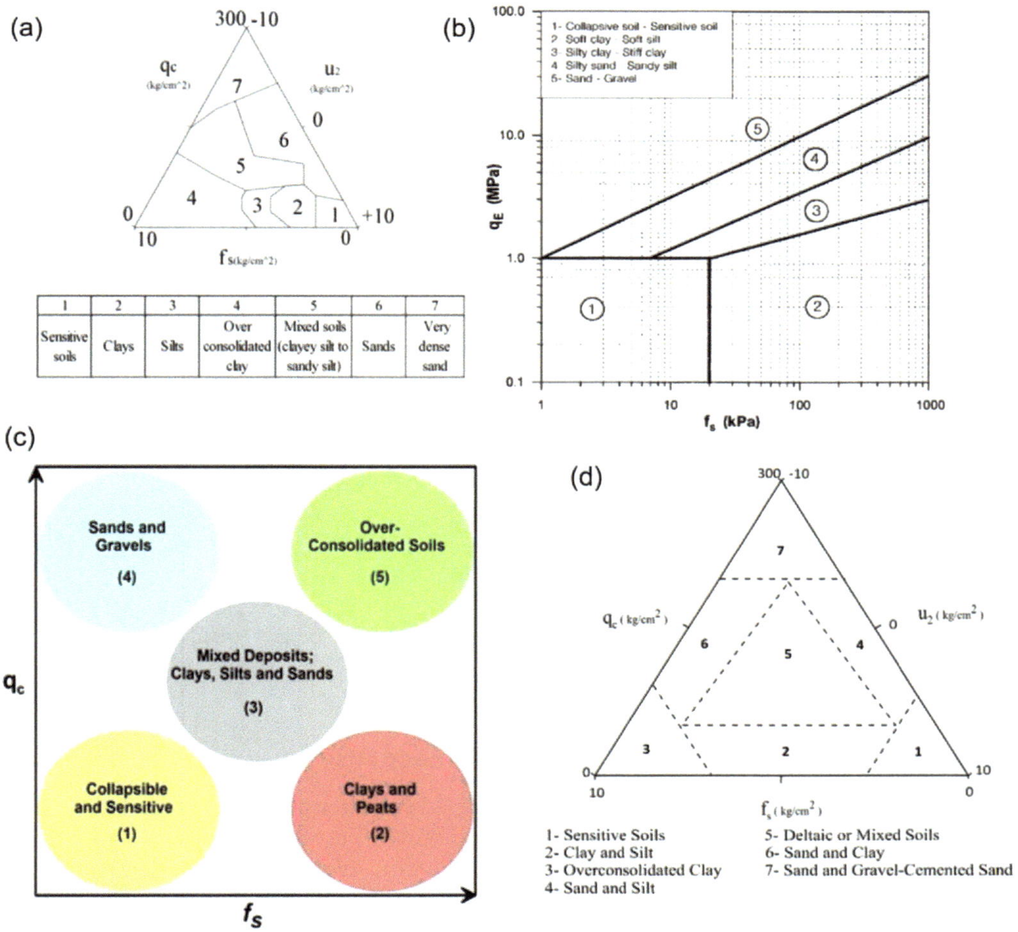

Figure 5.5 (a) Classification chart based on effective cone resistance and shaft resistance (Eslami and Fellenius, 1997), (b) first generation of triangular charts (Eslami et al., 2016), (c) enhanced triangular charts for deltaic soils (Eslami et al., 2022), and (d) general qualitative classification chart.

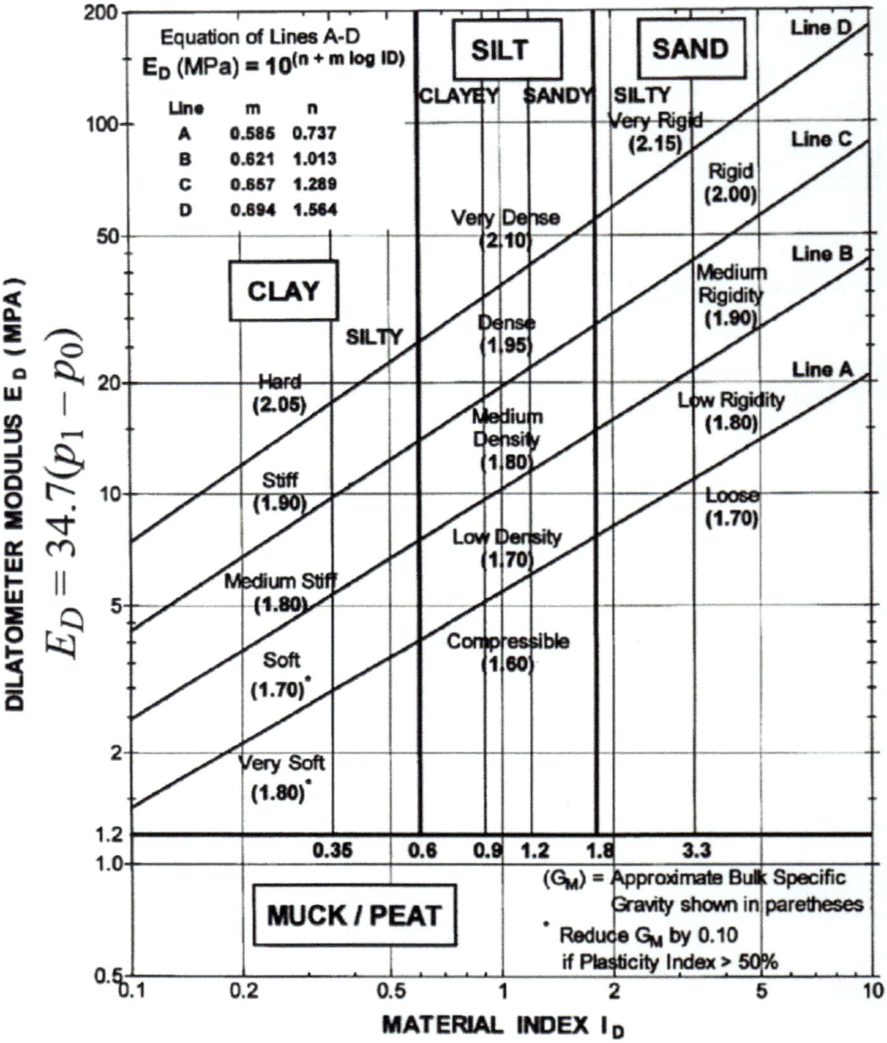

Figure 5.6 Chart for estimating soil type and unit weight for soil based on DMT result (Source: Marchetti, 1980).

5.4.5 Soil classification based on DMT records

In recent years, the static probing method using dilatometer penetration tests (DMT) has become a widely employed in-situ investigation technique, providing essential information for civil engineers involved in design, construction, permitting, and operational control. Diagram charts are commonly used for soil classification based on the dilatometer test (DMT). Marchetti (1980) introduced a diagram chart (Figure 5.6 and Table 5.6) that relies on the dilatometer modulus (E_D) and the material index (ID). This chart serves as a valuable tool for identifying various soil types, including minerals (clay, silt, and sand), as well as organic soils, based on the results obtained from DMT.

Table 5.6 Soil classification based on material index

Soil type		*Material index (I_D)*
Organic soils	Peat/sensitive clays	<0.10
Cohesive soils	Clay	0.10–0.35
	Silty clay	0.35–060
	Clayey silt	0.60–0.90
	Silt	0.90–1.20
	Sandy silt	1.20–1.80
Noncohesive soils	Silty sand	1.80–3.30
	Sand	>3.30

Source: Marchetti (1980).

Table 5.7 Typical Menard pressuremeter values

Type of soil	*Limit pressure (kPa)*	E_m/p_l
Soft clay	50–300	10
Firm clay	300–800	10
Stiff clay	600–2500	15
Loose silty sand	100–500	5
Silt	200–1500	8
Sand and gravel	1200–5000	7
Till	1000–5000	8
Old fill	400–1000	12
Recent fill	50–300	12

Source: Canadian Foundation Engineering Manual (2006).

5.4.6 Soil classification based on PMT measurements

The Borehole Pressuremeter Test (PMT), which was initially developed by Menard in 1956, has gained extensive use in France. While PMT is not commonly employed for soil classification, as indicated in Table 5.7, the typical Menard pressuremeter values can be utilized to help identify soil types, offering valuable insights into the soil's characteristics.

5.5 IN-SITU TESTS AND PREDICTION OF GEO-PARAMETERS

5.5.1 Geotechnical parameters

Every geotechnical design entails the determination of subsoil stratification and the estimation of geotechnical parameter values, often referred to as soil engineering parameters, for each distinct soil layer. Yu (2007) reviews the use of fundamental mechanics in developing rational interpretation methods for deriving soil properties from in-situ test results. The interpretation of in-situ tests is challenging due to complex boundary value problems, poorly controlled drainage conditions, differences in effective stress paths, and the highly nonlinear behavior of soils. Simplifying assumptions need to be made in the interpretation of in-situ tests, including assumptions about geometry and boundary conditions, soil behavior, and water drainage conditions. The cavity expansion methods can be used to analyze the stress–strain relationship and deformation characteristics of the soil and are typically adopted for interpretation of in-situ testing data (Yu, 2000; Mo and Yu, 2017, 2018; Chen and Mo, 2022; Mo et al., 2022). Cavity expansion testing provides valuable information for geotechnical engineers to assess the stability and behavior of soils in different engineering applications (Mo et al., 2017, 2020, 2024).

Soil engineering parameters in the context of foundation engineering can be broadly categorized into two main groups: those directly applied in foundation engineering, such as unit weight, internal friction angle, and relative density, and a few additional parameters, including sensitivity and permeability, which will be elaborated upon further.

5.5.2 Index parameters

5.5.2.1 Unit weight

In geotechnical engineering, a critical property of soil is its unit weight (γ), a parameter essential for various earthwork applications. Unit weight refers to the density of soil, defined as the ratio of the weight to the volume of an undisturbed soil sample, and it is typically determined in laboratory settings. For in-situ measurements of unit weight, methods such as sand cones, rubber balloons, or nuclear densitometer tests are commonly employed.

Table 5.8 shows that there is no disparity between the unit weights of the bulk and saturated weights in cohesive soils. However, in granular soils, the difference is somewhat more pronounced in loose materials compared to dense ones.

There have been several studies linking the unit weight to CPT measurements. The Robertson and Cabal (2010) correlations for calculating γ / γ_w for different soils using CPT records are shown in Figure 5.7.

Table 5.8 Typical values for bulk and saturated unit weights

		Bulk unit weight (kN/m³)		*Saturated unit weight (kN/m³)*	
	Soil type	*Loose*	*Dense*	*Loose*	*Dense*
Granular soils	Gravel	16	18	20	21
	Well-graded sand and gravel	19	21	21.5	23
	Coarse or medium sand	16.5	18.5	20	21.5
	Well-graded sand	18	21	20.5	22.5
	Fine or silty sand	17	19	20	21.5
	Rock fill	15	17.5	19.5	21
	Brick hardcore	13	17.5	16.5	19
	Slag fill	12	15	18	20
	Ash fill	6.5	10	13	15
Cohesive soils	Peat	12		12	
	Organic clay	15		15	
	Soft clay	17		17	
	Firm clay	18		18	
	Stiff clay	19		19	
	Hard clay	20		20	
	Stiff or hard glacial clay	21		21	

Source: Das (2016).

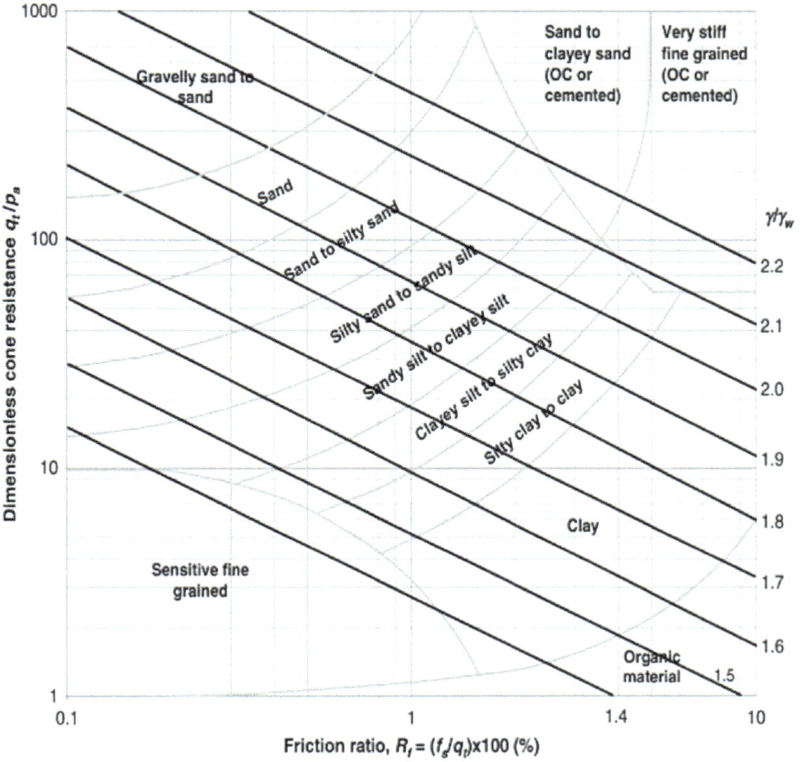

Figure 5.7 Normalized unit weight and R_f (%) (Source: adapted from Robertson and Cabal, 2010).

5.5.2.2 Relative density

As shown in Figure 5.8, Holtz and Gibbs (1979) presented the correlation of N and D_r in a more usable form. Some correlations suggested by researchers are listed in Table 5.9.

The application of in-situ density tests is constrained, as they can only be effectively measured at shallow depths. Therefore, the utilization of Cone Penetration Test (CPT) records to evaluate relative density (D_r) through empirical correlations has gained prominence. Several researchers have observed the influence of parameters like q_c and σ'_v on D_r, as illustrated in Figure 5.9. In this context, a selection of essential correlations has been compiled in Table 5.10, offering valuable tools for estimating relative density using CPT data.

5.5.3 Strength parameters

5.5.3.1 Internal friction angle

The internal friction angle is a fundamental physical property of earth materials, representing the slope of a linear representation of the shear strength of these materials. Typical values for the friction angle of sand and silt can be found in Table 5.11, providing important reference points for these materials.

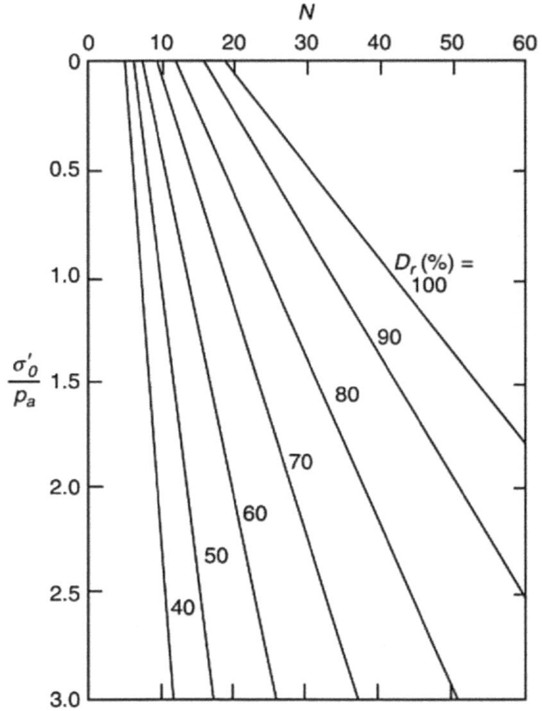

Figure 5.8 Variation of N with σ_0' / p_a and D_r (Source: adapted from Gibbs and Holtz, 1979).

Table 5.9 Proposed correlation for D_r based on SPT number

Reference	*Proposed correlation*	*Remarks*
Meyerhof (1957)	$D_r = 20.4\left(\dfrac{N}{0.7+\left(\sigma_o' / P_a\right)}\right)^{0.5}$	σ_o' = effective overburden pressure P_a = atmospheric pressure
Kulhawy and Mayne (1990)	$D_r = \left(\dfrac{(N_1)_{60}}{(60+25\log D_{50})C_A C_{OCR}}\right)^{0.5}$	C_A = correction factor for aging $C_A = 1.2 + 0.05\log(t/100)$ t = time, in years, since deposition C_{OCR} = correction for OCR $C_{OCR} = OCR^{0.18}$
Yoshida et al. (1988)	$D_r = C_0\left(\sigma_o'\right)^{-C_1} N^{C_2}$	The unit of σ_o' is kN/m^2 $C_0 = (18-25)$ (best fit value = 25) $C_1 = (0.12-0.14)$ (best fit value = 0.12) $C_2 = (0.44-0.57)$ (best fit value = 0.46)

Source: Adapted from Gibbs and Holtz (1957).

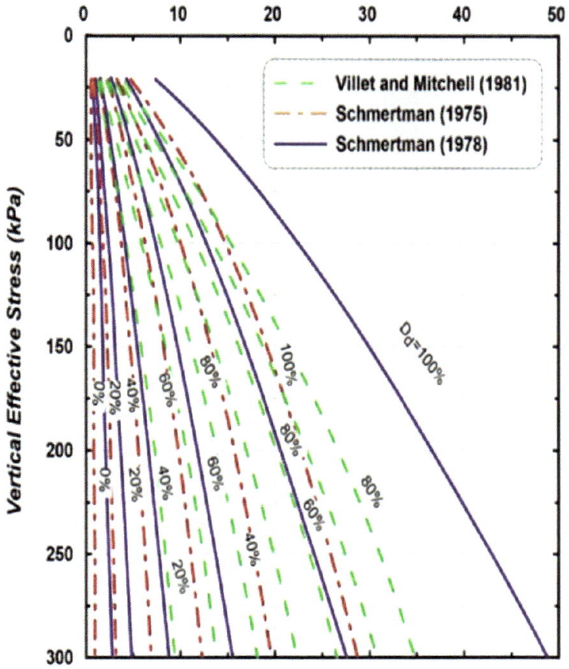

Figure 5.9 Comparison of $D_r - q_c$ relationships.

Table 5.10 Correlations predicting D_r from CPT records

Reference	*Proposed correlation*	*Remarks*
Baldi et al. (1986)	$D_r = \frac{1}{C_2}\ln(\frac{q_c}{C_0(\sigma_v')^{0.55}})$	C_0 and C_2: soil constants, C_0=157 and C_2=2.41 normally consolidated sand. q_c and σ_v' are in the kPa unit
Jamiolkowski et al. (2001)	$D_r = 26.8\ln\frac{q_c / p_a}{(\sigma_v' / p_a)^{0.5}} - b_x$	b_x = 52.5 for high compressibility sands b_x = 67.5 for medium compressibility sands b_x = 82.5 for low compressibility sands
Kulhawy and Mayne (1990)	$D_r = \frac{Q_{cn}}{305Q_cQ_{OCR}}$	$Q_{cn} = q_t / p_a / (\sigma_v' / p_a)^{0.5}$ Q_c = Compressibility factor (0.91 for high, 1.0 for medium, and 1.09 for low). Q_{OCR} = Over-consolidation factor, $OCR^{0.18}$
Mayne (2007)	$D_r = 100\left(0.268\ln(\frac{q_t / p_a}{\sqrt{\sigma_v' / p_a}}\right) - 0.675)$	q_c and σ_v' are in kPa unit

Table 5.11 Common amount of friction angle for sand and silt

Soil type		ϕ (°)
Angular sand	Loose	27–30
	Medium	30–35
	Dense	35–38
Rounded sand	Loose	30–35
	Medium	35–40
	Dense	40–45
Sandy gravel		34–48
Silt		26–35

Source: Das (2006).

Shear strength in coarse-grained soils is typically characterized by the friction angle, denoted as ϕ. Several researchers have proposed methods for determining the friction angle in sands, based on one of the following theories:

- Cavity expansion
- Bearing capacity
- Empirical approaches often derived from calibration chamber tests

The friction angle (ϕ) of granular soils exhibits an increase with factors such as grain angularity, surface roughness, and relative density. It is important to note that poorly graded granular soils generally have lower ϕ values than well-graded ones. Figures 5.10

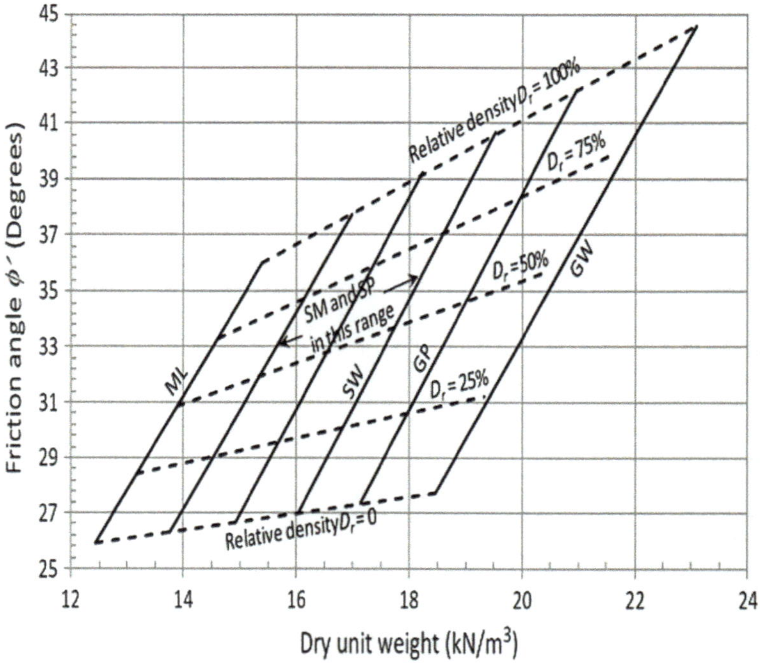

Figure 5.10 Friction angles of granular soils (Source: U.S. Navy, 1982).

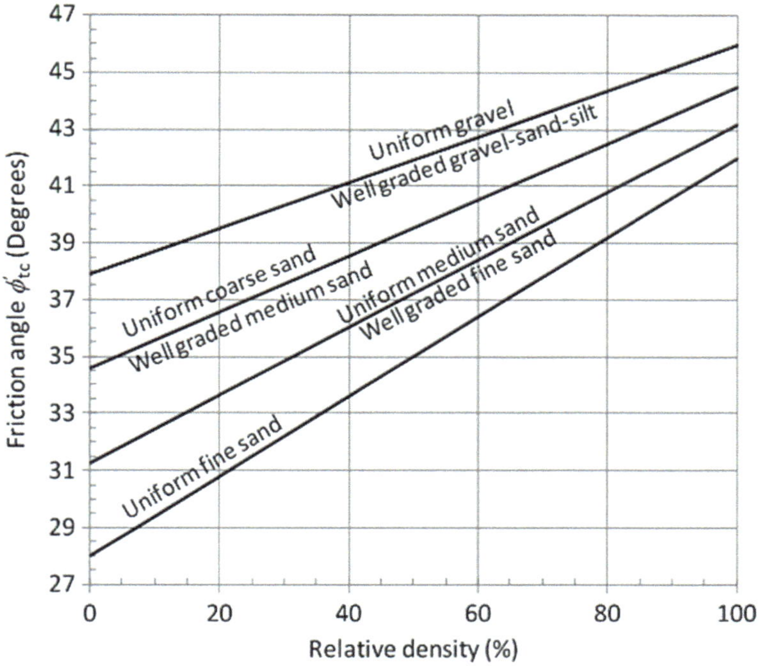

Figure 5.11 $\phi'_{tc} - D_r$ relations (Source: adapted from Schmertmann, 1978).

and 5.11 illustrate the friction angles determined from triaxial compression tests for various granular soils with the absence of plastic fines, providing a visual representation of the relationship between these factors and friction angle values.

Schmertmann (1975) proposed a correlation between effective overburden pressure σ_0, N, and ϕ, and this correlation is shown in Figure 5.12. Some other correlations are listed in Table 5.12.

Bergdahl et al. (1993) proposed a correlation of q_c and relative density with friction angle for cohesionless or mixed soils, as shown in Table 5.13. Mitchell and Durgunoglu (1983) suggested a correlation between cone tip resistance and internal friction angle as depicted in Figure 5.13.

Other researchers (e.g., Meyerhof, 1974; Robertson et al., 1986; Kulhawy and Mayne, 1990; and Uzielli et al., 2013) introduced a couple of equations to predict ϕ, as summarized in Table 5.14. Moreover, Mayne (2014) proposed correlations to predict the friction angle based on empirical correlation and case histories.

By employing two fundamental equations for assessing the bearing capacity of deep foundations, one for the corrected tip resistance q_t (q_E) and the other for the sleeve friction (f_s), and adopting the effective bearing capacity approach instead of the total stress method, it is possible to establish a dual equation system with two unknowns for static loading conditions, as outlined below:

$$CN_C + qN_q + 0.5\gamma BN_\gamma = q_E = q_t - u_2 \tag{5.5}$$

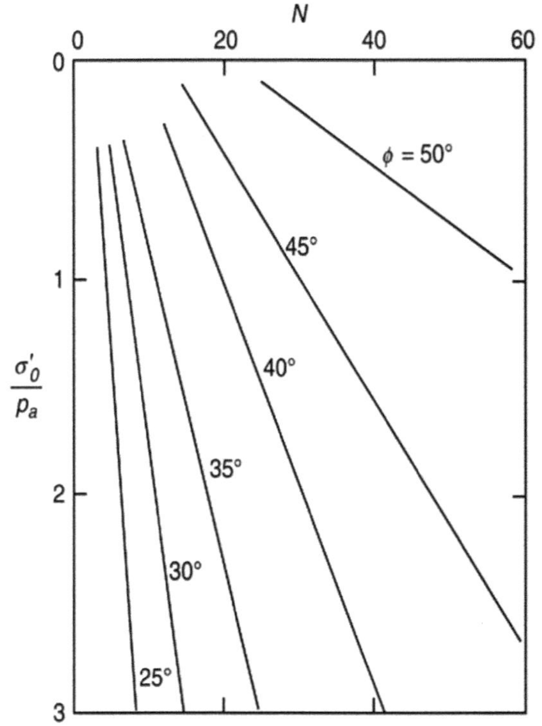

Figure 5.12 Variation of ϕ with SPT number and σ_0' / P_a (Source: adapted from Schmertmann, 1975).

Table 5.12 Variation of ϕ with N

Reference	*Proposed correlation*
Meyerhof (1959)	$\phi = 28 + 0.15\left[20.4\left(\frac{N}{0.7 + (\sigma_o' / Pa)}\right)^{0.5}\right]$
Wolff (1989)	$\phi = 27.1 + 0.3N_1 - 0.00054N_1^2$
Kulhawy and Mayne (1990)	$\phi = tan^{-1}\left[\frac{N}{12.2 + 20.3(\sigma_o' / Pa)}\right]^{0.34}$
Hatanaka and Uchida (1996)	$\phi = \sqrt{20(N_1)_{60}} + 20$

Table 5.13 Correlation of q_c and relative density with friction angle for cohesionless or mixed soils

Relative density	q_c *(MPa)*	ϕ *(°)*
Very weak	0.0–2.5	29–32
Weak	2.5–5.0	32–35
Medium	5.0–10.0	35–37
Large	10.0–20.0	37–40
Very large	>20.0	40–42

Source: Adapted from Bergdahl et al. (1993).

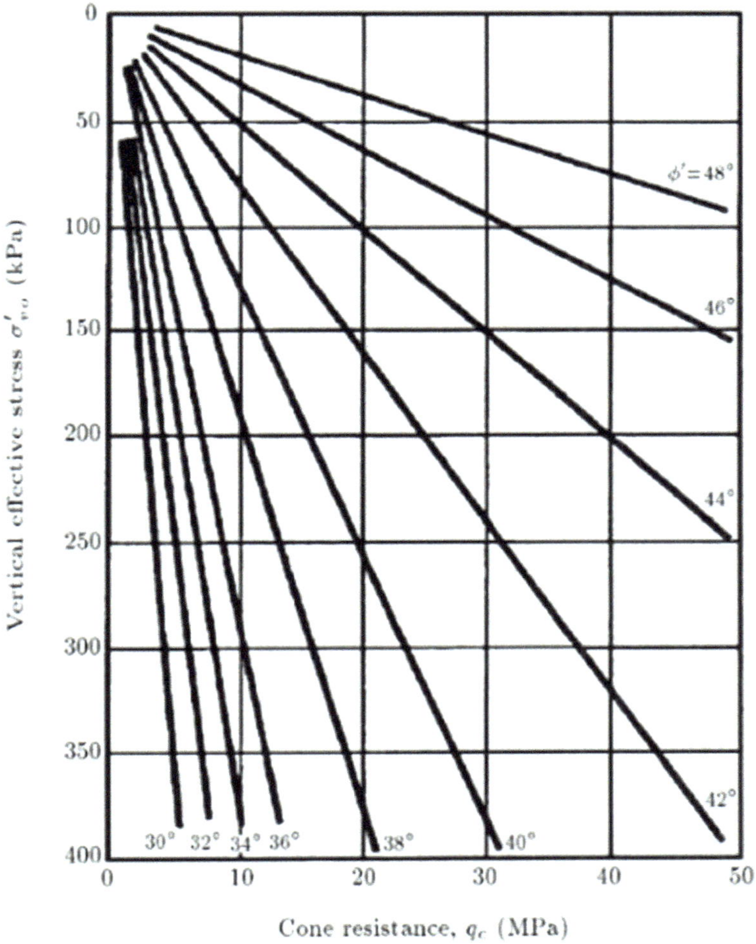

Figure 5.13 Friction angle changes and cone tip resistance (Source: Mitchell and Durgunoglu, 1983).

Table 5.14 Proposed correlations for friction angle based on CPT result

Reference	*Correlations*	*Soil type and Remarks*
Meyerhof (1974)	$\phi = \tan^{-1}(q_c / 0.5N_q)$	Sand, q_c (MPa)
Robertson et al. (1986)	$\phi = \tan^{-1}[0.1 + 0.38\log(q_c / \sigma'_v)]$	Sand
Kulhawy and Mayne (1990)	$\phi = 17.6 + 11\log(q_c / \sqrt{100\sigma'_v})$	Sand, σ_v' and q_c are in kPa unit
Uzielli et al. (2013)	$\phi = 25(q_c / \sqrt{100\sigma'_v})^{0.1}$	Sand, σ_v' and q_c are in kPa unit
Robertson and Cabal (2012)	$\tan\phi' = \frac{1}{2.68}(\log(q_c / \sigma'_v) + 0.29)$	Uncemented, unaged, moderately compressible quartz sands
Mayne (2014)	$\phi = 29.5B_q^{0.121}[0.256 + 0.33B_q + \log Q_t]$	Cohesive soils $Q_t = \frac{q_t - \sigma_v}{\sigma'_v}$ $B_q = \frac{u_2 - u_0}{q_t - \sigma_v}$

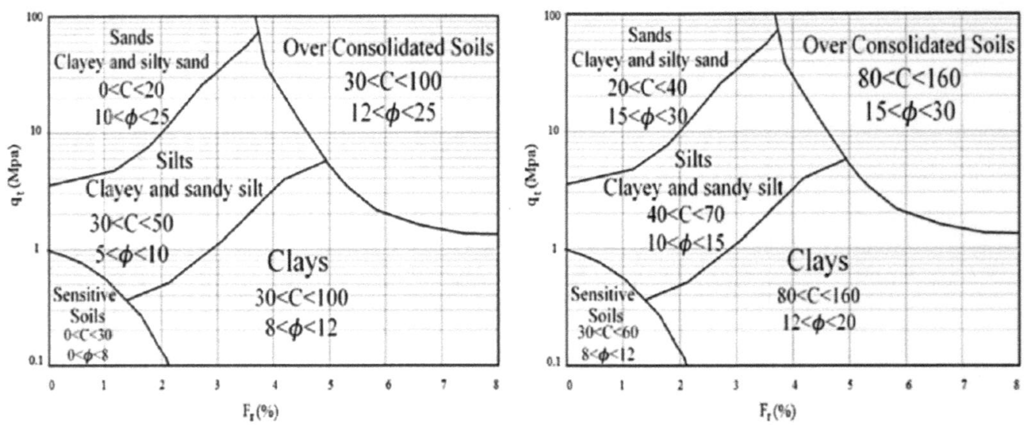

Figure 5.14 Variation range for C (kPa) and ϕ (degree), in soil classification chart: (a) 200 kPa < σ_v' < 300 kPa and (b) 300 kPa < σ_v' < 400 kPa (Source: Eslami and Mohammadi, 2016).

$$C + \sigma' \tan(\delta) = f_s \tag{5.6}$$

Eslami and Mohammadi (2016) have presented approaches to predict ϕ and C of marine deposits from CPTu records based on bearing capacity and cavity expansion theories. They summarized the Soil Classification Chart in five categorized areas based on CPTu test results. For specified q_t and F_r, i.e., friction ratio, the soil type can be determined in every depth interval. However, the variation range of C and ϕ using the proposed method, at 200 kPa < σ_v' < 300 kPa into five categorized soil types, is presented and also for 300 kPa < σ_v' < 400 kPa in Figure 5.14.

Combining bearing capacity theories and direct shear modes of failure considering the pore water pressure, a set of equations is derived. By inputting CPTu at a certain

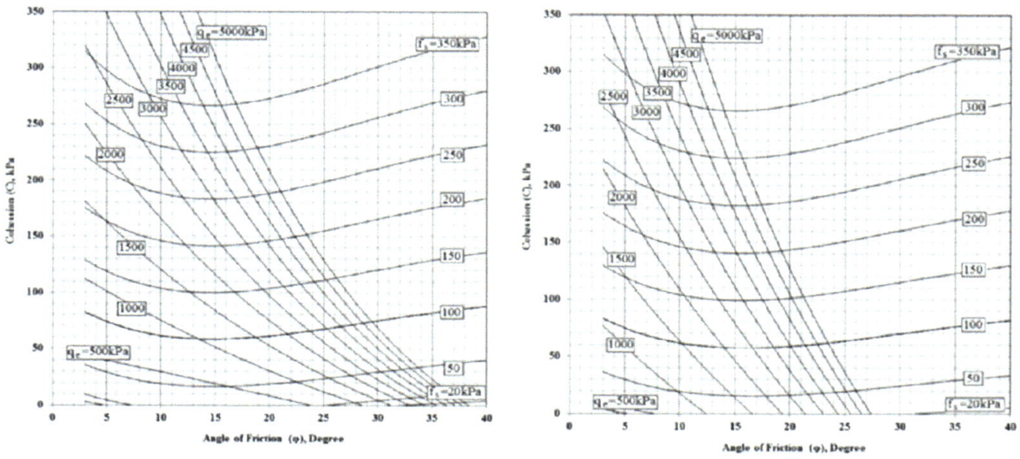

Figure 5.15 Graphic estimation of C and φ : (a) $\sigma_v' = 50$ kPa, (b) $\sigma_v' = 200$ kPa (Source: Eslami and Mohammadi, 2016).

depth, soil shear strength parameters can be calculated simultaneously as presented in Figure 5.15.

5.5.3.2 Undrained shear strength

The shear strength of soil represents the maximum internal resistance to applied shearing forces, and undrained shear strength (s_u) specifically refers to the shear strength of soil when it is subjected to shearing at a constant volume, typically under conditions where the pore water cannot drain or dissipate.

Numerous correlations have been proposed to relate Standard Penetration Test (SPT) results to the shear strength of soil. Terzaghi and Peck (1967) offer approximate relationships between consistency, corresponding N values, and undrained cohesion (s_u) for clay soils, as presented in Table 5.15. However, it is essential to exercise caution when using these values. Furthermore, Figure 5.16 illustrates the relationship between the standard penetration blow count (N) and undrained shear strength (s_u), as suggested by Sowers (1979), providing an alternative perspective on this correlation.

Table 5.15 Approximate variation of consistency, N, and undrained cohesion of clay

Consistency	*SPT (N value)*	s_u *(kPa)*
Very soft	0 – 2	< 15
Soft	2 – 4	12 – 25
Medium	4 – 8	25 – 50
Stiff	8 – 15	50 – 100
Very Stiff	15 – 30	100 – 200
Hard	> 30	> 200

Source: Terzaghi and Peck (1967).

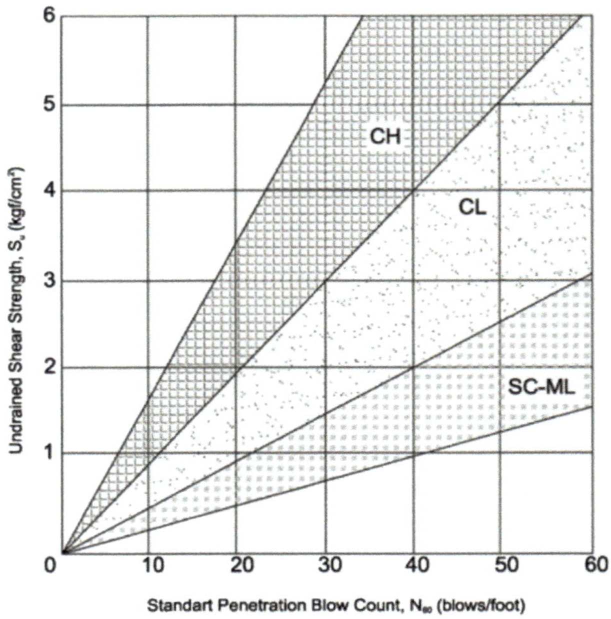

Figure 5.16 Relationship between N, and s_u (Source: adapted from Sowers, 1979).

In addition to soil profile classification, one of the primary engineering parameters derived from Cone Penetration Tests (CPT) for fine-grained soils is the undrained shear strength (s_u) of clayey soil. It is important to note that s_u cannot be directly measured by the CPT; however, empirical correlations can be employed to estimate it, and these estimates should be calibrated using laboratory tests such as triaxial compression or direct shear tests. Numerous researchers have conducted investigations into the relationship between s_u and CPT parameters. Some of the methods used for interpreting s_u from in-situ tests are detailed in Table 5.16, offering various approaches for estimating undrained shear strength using CPT data.

5.5.4 Stiffness and compressibility parameters

5.5.4.1 Stiffness

Soil modulus is a complex parameter in geotechnical engineering due to the various conditions it can be involved in. The Young's modulus (E) is a parameter associated with the deformability of soil, representing its linear elastic behavior. However, it is important to note that soil cannot be treated as an ideal linear elastic material, and as such, the soil modulus is not equivalent to the slope of the stress–strain curve. Instead, it is a stress-dependent parameter, meaning that the soil modulus changes with varying stress levels and must be considered as such in geotechnical analyses and design.

In predicting settlement for a foundation placed on a normally consolidated deposit under its initial loading, the secant modulus (E_s) is a valuable parameter. To calculate E_s , a slope is drawn from the origin to a selected point on the stress–strain curve. In addition, the tangent slope (E_t), associated with the tangent modulus, can be used to estimate

Table 5.16 Methods in s_u interpretation of in-situ tests

References	*Reviews*	*In-Situ Test*
Chandler (1988)	$\mu \cong 2.5(PI)^{-0.3} \leq 1.1$	VST: $s_{uv} = 6T/(7\pi D^3)$, $H/D = 2$
Windle and Wroth (1977) Baguelin et al. (1972)	$N_c = 5.5$	PMT: $s_{uPMT} = dp/d(Ln\varepsilon_v)$ $s_{uPMT} = (P_L - P_O)/N_c$
Stroud (1974) Stroud (1989)	$f_1 = 4.5, PI = 50$ $f_1 = 5.5, PI = 15$	SPT: $s_{u(N60)} = f_1 N_{60} P_a / 100$
Meyerhof (1951) Vesic (1977) Aas et al. (1986) Eslami et al. (2020) and Anagnostopolous et al. (2003)	$N_{kt} = 10(TC)$ $N_{kt} = 15(DSS)$ $N_{kt} = 20(TE)$	CPT & CPTu : $s_{uCPT} = (q_c - \sigma_{vo})/N_c$ $s_{uCPT} = (q_t - \sigma_{vo})/N_{kt}$ $s_{uCPT} = (0.8 - 1.2)f_s$
Marchetti (1980) Schmertmann (1991) Lacasse and Lunne (1988)	TC: $d_s = 0.20$ VST: $d_s = 0.19$ DSS: $d_s = 0.14$	DMT: $s_{uDMT} = 0.22\sigma'_{vo}(0.5K_D)^{1.25}$ $s_{uDMT} = (P_o - u_o)/10$ $s_{uDMT} = d_s\sigma'_{vo}(0.5K_D)^{1.25}$
Meyerhof (1951)	-	PLT: $s_{uPLT} = q_{ult}/6.18$

Source: FHWA (2002).

Table 5.17 Drained elastic modulus for soils using CPT records

Soil		q_c *(MPa)*	*Drained elastic modulus (MPa)*	E_t *from CPT*
Sand	Very loose	<2.5	<10	(2~4)q_c
	Loose	2.5–5	10–20	
	Medium dense	5–10	20–30	(3~7)q_c
	Dense	10–20	30–60	(6~10)q_c
	Very dense	>20	>60	
Clay	Very Soft	<2	<5	(1~2)q_c
	Soft	2–5	5–15	(1~2)q_c
	medium	5–10	15–25	(2~4)q_c
	Hard	10–20	>25	(3~8)q_c

Source: Look (2007).

progressive settlement when incremental loads are applied. Table 5.17 provides proposed relationships between elastic modulus and CPT results, offering a means to estimate these moduli based on CPT data, which can be crucial for settlement predictions in geotechnical engineering.

Table 5.18 Elastic modulus parameter of in-situ tests

SPT	*CPT*	*Soil type*
$E_s = 500(N+15)$ $= 7000\sqrt{N}$ $= 6000N$ - - - - - - - - - - - - - - - $E_s = (15000-22000)\cdot \ln N$	$E_s = (2-4)q_u$ $= 8000q_u$ - - - - - - - - - - - - - - - - $E_s = 1.2(3D_r^2+2)\cdot q_c$ $E_s = (1+D_r^2)\cdot q_c$	Sand
$E_s = 250(N+15)$	$E_s = F\cdot q_c$ e = 1.0 $F = 3.5$ e = 0.6 $F = 7.0$	Saturated sand
$E_s = 40000 + 1050 \times N$ $E_{s(OCR)} \approx E_{s(NC)}\sqrt{OCR}$	$E_s = (6-30)q_c$	OCR sand
$E_s = 1200(N+6)$ $=600(N+6)$ $N \le 15$ $= 600(N+6)+2000$ $N > 15$	-	Sandy gravel
$E_s = 320(N+15)$	$E_s = (3-6)q_c$	Clay sand
$E_s = 300(N+6)$	$E_s = (1-2)q_c$ $q_c < 2500kPa$ $E_s' = 2.5q_c$ $2500 < q_c < 5000kPa$ $E_s' = 4q_c + 5000$	Silty sand
-	$E_s = (3-8)q_c$	Soft clay

Source: Bowles (1996).)

As Baldi et al. (1989) suggested, the stiffness of fine-grained soil is affected by different factors, including mineralogy, grain size, particle shape, grain fabric, drainage conditions, mean effective stress, and stress–strain behavior. In a given soil, cone tip resistance is mainly controlled by the void ratio and the effective stress condition. Some methods in E_s interpretation of in-situ tests are listed in Table 5.18.

5.5.4.2 Constrained modulus

The constrained modulus of clays can be calculated in the laboratory using the consolidation test:

Table 5.19 Constrained modulus coefficient for cohesive soils

Soil type	q_c *(MPa)*	α
Low plasticity clay (CL)	$q_c < 0.7$	$3 < \alpha < 8$
	$0.7 < q_c < 2$	$1 < \alpha < 5$
	$q_c > 2$	$1 < \alpha < 2.5$
Silt, low liquid limit (ML)	$q_c < 2$	$3 < \alpha < 6$
	$q_c > 2$	$1 < \alpha < 2$
High plasticity clay (CH)/ silt, high liquid limit (MH)	$q_c < 2$	$2 < \alpha < 6$
	$q_c > 2$	$1 < \alpha < 2$
Organic silt (OL)	$q_c < 1.2$	$2 < \alpha < 8$
Organic clay (OH)/peat (Pt) with $50 < w < 100$	$q_c < 0.7$	$1.5 < \alpha < 4$
Organic clay (OH)/peat (Pt) with $100 < w < 200$		$1 < \alpha < 1.5$
Organic clay (OH)/peat (Pt) $w > 200$		$\alpha < 4$

Source: Adapted from Sanglerat (1972); adapted from Frank and Magnan (1995).

$$M = \frac{1}{m_v} \tag{5.7}$$

where m_v = the coefficient of volume compressibility.

Lunne et al. (1997) pointed out that most correlations between penetrometer test results and drained constrained modulus refer to the tangent modulus as found from oedometer tests. The correlations are typically represented as:

$$M = \alpha q_c \tag{5.8}$$

where M = the constrained modulus and α = the constrained modulus factor. Relations between α and q_c for cohesive soils are summarized in Table 5.19.

5.5.4.3 Over-consolidation ratio

In geotechnical engineering, in order to describe the stress history condition of a soil deposit, the over-consolidation ratio (OCR) is introduced as in Equation (5.9):

$$\text{OCR} = \frac{\text{maximum past effective vertical stress}}{\text{current effective vertical stress}} = \frac{\sigma'_p}{\sigma'_v} \tag{5.9}$$

The value of over-consolidation ratio (OCR), denoted as σ'_p, can be determined through oedometer tests. However, oedometer tests, especially those involving complex stress path approaches, can be time-consuming. Moreover, various factors, such as sample disturbance and the methods used to identify the maximum past consolidation pressure, can influence the test results. Therefore, it is recommended to conduct

Table 5.20 Proposed correlations for OCR

Reference	*Correlations*	*Remarks*
Mayne and Kemper (1988)	$OCR = 0.37\left(\left(q_c - \sigma_v\right)/\sigma'_v\right)^{1.01}$	–
Mayne and Kemper (1988)	$OCR = 0.193\left(N/\sigma'_o\right)^{0.689}$	σ'_o is in MPa
Trevor and Mayne (2004)	$OCR = 2\left(0.029 + 0.409M\right)\left[\frac{1}{1.95M+1}\left(\frac{q_t - u_2}{\sigma'_v}\right)\right]^{\frac{1}{\theta}}$	$\theta = 0.8 - 0.9$ $M = \frac{6\sin\varphi'}{3-\sin\varphi'}$
Mayne (2007)	$\sigma'_p = k\left(q_t - \sigma_v\right)$	*k*: Pre-consolidation cone factor with an expected range of 0.2–0.5
Robertson (2009)	$OCR = 0.25Q_t^{1.25}$	–
Robertson (2012)	$OCR = \left(2.625 + 1.75\log Fr\right)^{-1.25} Q_t^{1.25}$	–
Chanmee et al. (2017)	$OCR = k\left(\frac{q_t - \sigma_v}{OCR\sigma'_v}\right)$	Under consolidated deposits *k* = 0.14–0.4

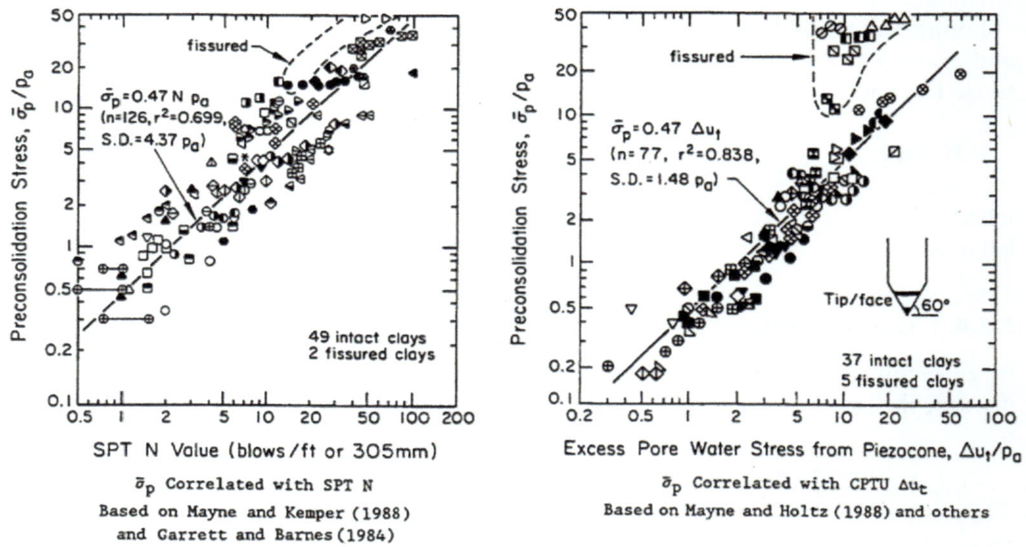

Figure 5.17 SPT and CPT correlation with OCR (Source: Kulhawy and Mayne, 1990).

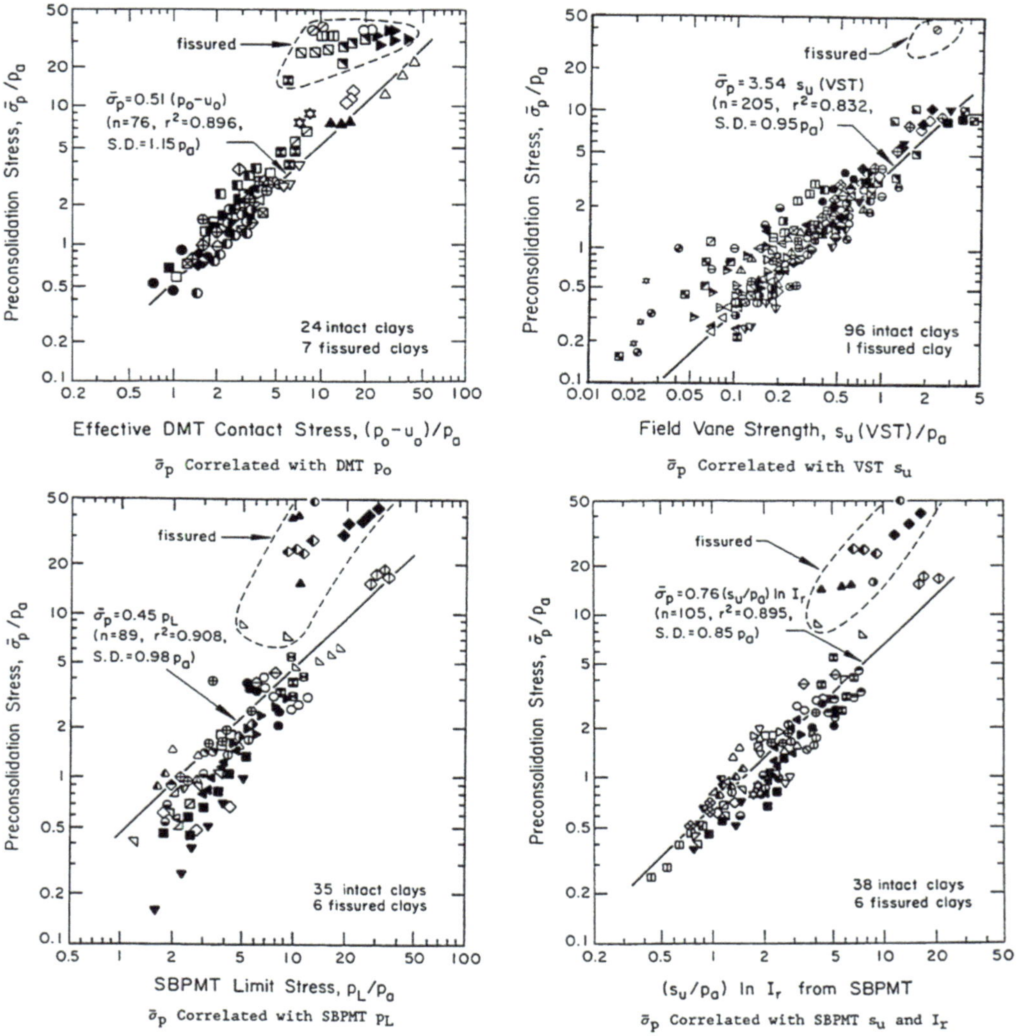

Figure 5.18 DMT, VST, and SBPMT correlation with OCR (Source: Kulhawy and Mayne, 1990).

in-situ tests as a more efficient and cost-effective means to estimate OCR using field data, which can provide valuable insights without the complications associated with oedometer testing.

There are several existing methods based on in-situ tests' data for assessing the OCR. Table 5.20 and Figure 5.17 show some proposed SPT and CPT correlations for OCR. Additionally, Figure 5.18 shows some other correlations for other in-situ tests, such as DMT and VST, based on data collected by Kulhawy and Mayne (1990).

Table 5.21 Empirical correlation for the compression ratio

Soil type	*Correlation for C_r*	*Reference*
Marine clays of Southeast Asia	$0.0043\ \omega_n$	Azzouz et al. (1976)
	$0.0045\ LL$	Balasubramaniam and Brenner (1981)
Bangkok clays	$0.00.463\ LL - 0.013$	Balasubramaniam and Brenner (1981)
	$0.00566\ \omega_n - 0.037$	
French clays	$0.0039\ \omega_n - 0.013$	Balasubramaniam and Brenner (1981)
Indiana clays	$0.0249 + 0.003\ \omega_n$	Lo and Lovell (1982)
	$0.0294 + 0.00238\ LL$	
	$0.0125 + 0.152\ e_0$	
	$0.2037(e_0 - 0.2465)$	Goldberg et al. (1979)
Clays from Greece and parts of the United States	$0.002(LL + 9)$	Azzouz et al. (1976)
	$0.14(e_0 + 0.007)$	
	$0.003(\omega_n =7)$	
	$0.126(e_0 + 0.003\ LL - 0.06)$	
Chicago clays	$0.208\ e_0 + 0.0083$	Azzouz et al. (1976)
Inorganic and organic clays and silty soils	$0.156\ e_0 + 0.0107$	Elnaggar and Krizek (1970)

Source: Ameratunga et al. (2016).

5.5.4.4 Consolidation indexes

From the comprehensive compilation presented by Djoenaidi (1985), several empirical correlations for the compression ratio have been selected. These correlations offer valuable tools for estimating compression ratios in geotechnical engineering analyses and design, as listed in Table 5.21.

The slope of the consolidation line in the $e - \log\sigma_v'$ space is known as the compression index (C_c). When the clay is normally consolidated, the $e - \sigma_v'$ values will be located on this line, irrespective of the stress level. When clay is normally consolidated, C_c is a measure of its stiffness, and it is a key parameter in evaluating the final settlements due to consolidation. It is usually related to the initial void ratio (e_0), in-situ natural water content (w_n), liquid limit (LL), or plasticity index (PI). Some of the suggested empirical correlations for C_c are listed in Table 5.22.

5.5.5 Related parameters

5.5.5.1 Shear wave velocity and shear modulus at small strain

A shear wave velocity (V_s) induced shear modulus is a critical geotechnical parameter corresponding to a small strain, which is essential in earthquake engineering design. Due to difficulties in collecting undisturbed samples, especially in granular soils, in-situ seismic

Table 5.22 Empirical correlations for C_c

Correlation for C_c	*Comment*	*Reference*
$0.009(LL - 10)$	Undisturbed clay of sensitivity <4. Reliability ± 30%	Terzaghi and Peck (1948)
$0.007(LL - 10)$	Remolded clay	Skempton (1944)
$0.0046(LL - 9)$	Sao Paulo, Brazil clays	Cozzolino (1961)
$0.0186(LL - 30)$	Soft silty Brazilian clays	Cozzolino (1961)
$0.01(LL - 13)$	All clays	USACE (1990)
$0.008(LL - 8.2)$	Indiana soils	Lo and Lovell (1982)
$0.21 + 0.008LL$	Weathered and soft Bangkok clays	Balasubramaniam and Brenner (1981)
$0.3(e_0 - 0.27)$	Inorganic silty clay	Hough (1957)
$1.15(e_0 - 0.35)$	All clays	Azzouz et al. (1976)
$0.75(e_0 - 0.50)$	Soils of very low plasticity	Azzouz et al. (1976)
$0.4(e_0 - 0.25)$	Clays from Greece and parts of the United States	Azzouz et al. (1976)
$0.256 + 0.43(e_0 - 0.84)$	Brazilian clays	Cozzolino (1961)
$0.54(e_0 - 0.35)$	All clays	Nishida (1956)
$0.22 + 0.29e_0$	Weathered and soft Bangkok clays	Balasubramaniam and Brenner (1981)
$0.575e_0 - 0.241$	French clays	Balasubramaniam and Brenner (1981)
$0.5363(e_0 - 0.411)$	Indiana soils	Goldberg et al. (1979)
$0.496e_0 - 0.195$	Indiana soils	Lo and Lovell (1982)
$0.40(e_0 - 0.25)$	Clays from Greece and parts of the United States	Azzouz et al. (1976)
$0.01\omega_n$	Chicago clays	Azzouz et al. (1976)
$0.01\omega_n$	Canada clays	Koppula (1981)
$0.0115\omega_n$	Organic soils, peat	USACE (1990) and Azzouz et al. (1976)
$0.012\omega_n$	All clays	USACE (1990)
$0.01(\omega_n - 5)$	Clays from Greece and parts of the United States	Azzouz et al. (1976)
$0.0126\omega_n - 0.162$	Indiana soils	Lo and Lovell (1982)
$0.008\omega_n + 0.20$	Weathered and soft Bangkok clays	Balasubramaniam and Brenner (1981)
$0.0147\omega_n - 0.213$	French clays	Balasubramaniam and Brenner (1981)
$(1 + e_0)[0.1 + 0.006(\omega_n - 25)]$	Varved clays	USACE (1990)

Source: Ameratunga et al. (2016).

Table 5.23 Some correlations between V_s (m/s) and SPT-N value

Source	*Soil type*	V_s *(m/s)*
Imai (1977)	All soils	$91N^{0.337}$
	Sand	$80.6N^{0.331}$
	Clay	$80.2N^{0.292}$
Ohta and Goto (1978)	All soils	$85.35N^{0.348}$
Seed and Idriss (1981)	All soils	$61.4N^{0.5}$
Sykora and Stokoe (1983)	Sand	$100.5N^{0.29}$
Okamoto et al. (1989)	Sand	$125N^{0.3}$
Pitilakis et al. (1999)	Sand	$145N^{0.178}$
	Clay	$132N^{0.271}$
Kiku et al. (2001)	All soils	$68.3N^{0.292}$
Jafari et al. (2002)	Sand	$22N^{0.77}$
	Clay	$27N^{0.73}$
Hasancebi and Ulusay (2007)	All soils	$99N^{0.309}$
	Sand	$90.82N^{0.319}$
	Clay	$97.89N^{0.269}$
Dikmen (2009)	All soils	$58N^{0.39}$
	Sand	$73N^{0.33}$
	Silt	$60N^{0.36}$
	Clay	$44N^{0.48}$

Source: Ameratunga et al. (2016).

tests are preferred to laboratory tests for achieving the V_s. By using down-hole and cross-hole techniques, as well as surface wave velocity determining techniques, a shear wave velocity profile with depth can be established.

Several correlations have been proposed to estimate the shear wave velocity based on SPT and CPT results for different soil types. The basic variables as input parameters can be q_c, f_s, N, and σ_v' in the correlation functions linear, logarithm, and power form. Some of these correlations are listed in Tables 5.23 and 5.24.

In general, it is accepted that the best way of assessing small strain (<0.001%) shear modulus (G_0) is to determine it using shear wave velocity ($G_0 = \rho V_s^2$). However, it is mainly the practice to carry out in-situ penetration tests to assess G_0 using empirical correlations, except that project-specific detailed investigations are performed. Some of the correlations for G_0 prediction based on CPT records are presented in Table 5.25.

Table 5.24 Empirical correlations between V_s and CPT data

Reference	*Proposed correlation (m/s)*	*Soil type*	*Units of parameters*	
			q_c	f_s
Hegazy and Mayne (1995)	$V_s = 12.02(q_c)0.319(f_s) - 0.0466$	Sand	kPa	kPa
	$V_s = 3.18(q_c)0.549(f_s)\ 0.025$	Clay	kPa	kPa
Mayne and Rix (1995)	$V_s = 1.75(q_c)0.627$	Clay	kPa	kPa
Iyisan (1996)	$V_s = 55.3(q_c)0.377$	Clay	kgf/cm²	–
	$V_s = 218 + 0.70\ q_c$	Sand	kgf/cm²	–
Piratheepan (2002)	$V_s = 25.3(q_c)0.163(f_s)\ 0.029D\ 0.115$	Sand	kPa	kPa
	$V_s = 11.9(q_c)0.269(f_s)\ 0.108D\ 0.127$	Clay	kPa	kPa
Tun (2003)	$V_s = 109.29 + 52.674\text{Ln}(q_c)$	All	MPa	MPa
Madiai and Simoni (2004)	(1) $V_s = 211(q_c)0.23$	All	MPa	MPa
	(2) $V_s = 155(q_c)0.29(f_s) - 0.10$			
Mayne (2006)	$V_s = 18.5 + 118.8\log(f_s)$	All	–	kPa
Paoletti et al. (2010)	$V_s = 50[(q_c/P_a)0.43 - 3]$	Sand	kPa	–
MolaAbasi et al. (2015)	$V_s = 100[1.36 - 0.35f_s + 0.15q_c - 0.05f_s^2 - 0.018q_c^2 + 0.39f_sq_c]$	Clay	MPa	MPa
	$V_s = 100[1.73 + 2.74f_s + 0.03q_c - 4.015f_s^2 - 0.00026q_c^2 + 0.007f_sq_c]$	Sand	MPa	MPa
	$V_s = 100[1.47 + 2.07f_s + 0.10q_c + 9.50f_s^2 - 0.0023q_c^2 - 0.034f_sq_c]$	Mixed	MPa	MPa
	$V_s = 100[1.40 + 1.59f_s + 0.09q_c - 1.33f_s^2 - 0.002q_c^2 + 0.05f_sq_c]$	All	MPa	MPa

Source: Ameratunga et al. (2016).

Table 5.25 Empirical correlations for G_0 based on q_c

Reference	*Correlations*	*Remarks*
Mayne and Rix (1995)	$G_{max} = 99.5pa^{0.305}(qc)^{\frac{0.695}{e_0^{1.13}}}$	Cohesive soils, e_0 = initial void ratio
Eslaamizaad and Robertson (1997)	$\left(\frac{G_0}{q_c}\right) = 1634\left(\frac{q_c}{\sqrt{\sigma'_v}}\right)^{-0.75}$	Cohesionless soils
Schnaid (2009)	$G_0 = b(q_c\sigma'_v Pa)^{0.3}$	Cohesionless soils, b = 280 and 110 for an upper and lower bond

Source: Ameratunga et al. (2016).

Table 5.26 Permeability from CPT results

Soil behavior type	*Soil permeability (m/s)*
Sensitive fine-grained	3×10^{-9} to 3×10^{-8}
Organic soils – peats	1×10^{-8} to 1×10^{-6}
Clays – clay to silty clay	1×10^{-10} to 1×10^{-7}
Silt mixtures clayey silt to silty clay	3×10^{-9} to 1×10^{-7}
Sand mixtures; silty sand to sandy silt	1×10^{-7} to 1×10^{-5}
Sands; clean sands to silty sands	1×10^{-5} to 1×10^{-3}
Gravelly sand to sand	1×10^{-3} to 1
Very stiff sand to clayey sand	1×10^{-8} to 1×10^{-6}
Very stiff fine-grained	1×10^{-9} to 1×10^{-7}

Source: Robertson (1990).

5.5.5.2 Sensitivity

Sensitivity (S_t) is defined as the ratio of peak undisturbed shear strength to remolded shear strength (Equation 5.10). Many researchers have reported that the shear strength of remolded samples is affected by skin friction, f_s.

$$S_t = {s_u}/{s_{u(\text{remolded})}} = {(q_t - \sigma_v)}/{N_{kt}} \left({1}/{fs} \right) \quad (5.10)$$

Schmertmann's (1978) suggested formula is based on this and can be simply written as follows:

$$S_t = {N_s}/{R_f} \quad (5.11)$$

where R_f = the friction ratio and N_s = a constant.

5.5.5.3 Permeability

Robertson (1990) proposed a relationship between permeability (k) and the soil behavior type for CPT, as shown in Table 5.26. Robertson and Cabal (2010) suggested Equation (5.12) to evaluate the average permeability (k_{ave}):

$$\begin{cases} k = 10^{(0.952-3.04I_c)} & 1 < I_c \le 3.27 \\ k = 10^{(-4.52-1.37I_c)} & 3.27 < I_c \le 4 \end{cases} \quad (5.12)$$

where $I_c = \sqrt{(3.47 - \log Q_t)^2 + (\log F_r + 1.22)^2}$.

5.6 GEO-MARINE DATABASE AND TRIANGULAR CHARTS

5.6.1 AUT: CPTu-Geo-marine Database

Eslami et al. (2022) proposed a marine database that was meticulously assembled through an extensive examination of approximately 350 geotechnical reports, articles, research

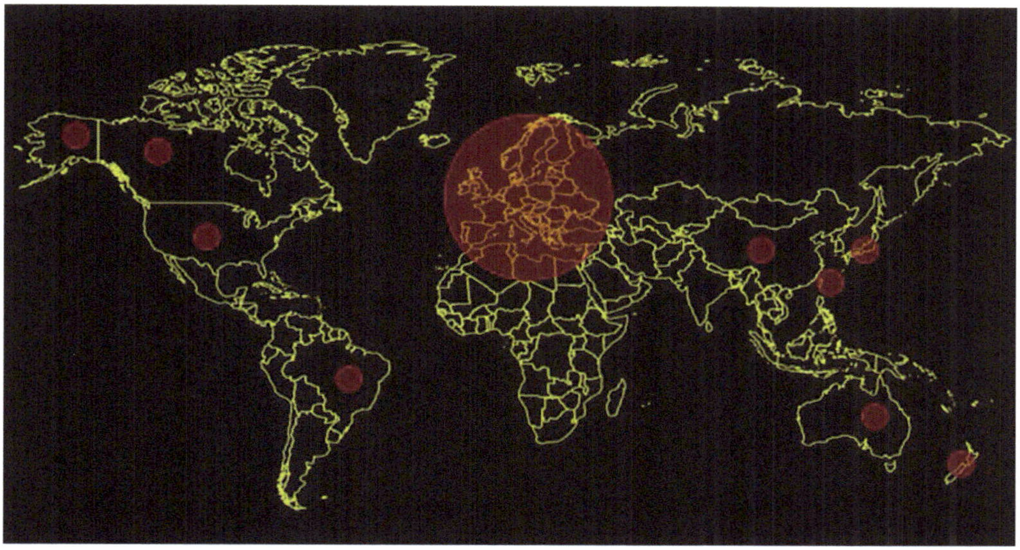

Figure 5.19 The scatter of the AUT: CPTu-GMD case studies.

programs, and doctoral theses. This comprehensive database comprises soil profiles extracted from boreholes and CPTu (Cone Penetration Testing) results obtained in their proximity. It encompasses a total of 398 cases derived from 58 different locations spanning 18 countries, with the majority of data originating from the United States, Europe, China, and New Zealand. The distribution of CPTu-GMD (Geotechnical and Marine Data) cases across these countries is illustrated in Figure 5.19, while Figure 5.20 provides a visual representation of a selection of typical CPTu logs.

The marine database consists of a diverse range of site types, with 21% located onshore along the coastline, 26% in near-shore areas within seas less than 30 m deep, 15% in offshore locations within waters exceeding 30 m, and 38% situated in riverine settings, as reported by the Federal Geographic Data Committee in 2012.

For each penetrometer test conducted in proximity to a borehole, CPT records were digitized at 10 cm intervals, capturing essential parameters such as cone tip resistance (q_c), sleeve friction (f), and excess pore water pressure (u_2). Soil profiles were meticulously characterized based on the available borehole data. Additionally, various other parameters, including soil density, effective stress, and eight additional factors, were deduced from CPT records utilizing relevant correlations as outlined by Robertson and Cabal (2010). This extensive database encompasses approximately 10,000 m of recorded soil profiles, with primary soil types comprising clay, sensitive soils, sand and gravel, overconsolidated clay, as well as mixed or deltaic soils. Figure 5.21 provides detailed insights into the AUT: CPTu-GMD dataset.

5.6.2 Soil Behavioral Classification methods

Soil Behavioral Classification diagrams can be categorized into three distinct groups, each based on the incorporation of CPTu records. The first group, often referred to as "basic diagrams," includes diagrams developed by Douglas and Olsen (1981), Campanella et al.

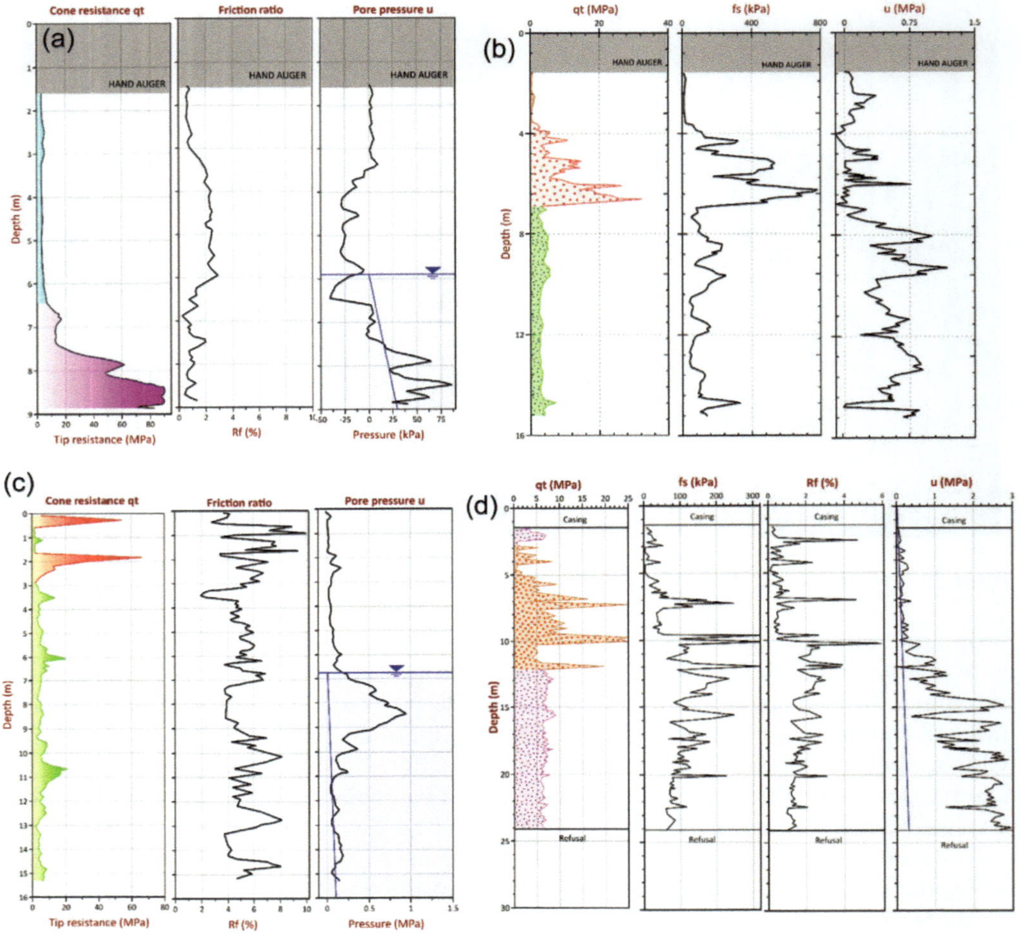

Figure 5.20 Typical CPTu logs: (a) onshore project, California, USA (Geotechnical Report, 2015); (b) nearshore project, New Zealand (Geotechnical Report, 2018); (c) offshore project, Australia (Geotechnical Report, 2019); and (d) riverine project, Canada (Geotechnical Report, 2020).

(1985), and Robertson (2010). These diagrams rely on only two of the three piezocone test records and employ mathematical equations. The most commonly utilized parameters in this group are the cone tip resistance (q_c) and the sleeve friction (f).

The second group, known as "indirect diagrams," encompasses diagrams created by Robertson (1990), Jefferies and Davies (1993), Eslami and Fellenius (1997), and Eslami et al. (2019). These diagrams utilize all three piezocone test parameters, in conjunction with mathematical equations, to provide a comprehensive description of subsurface layers. It is worth noting that the parameters considered in these second-group diagrams may vary, setting them apart from the basic diagrams.

The first set of charts includes those developed by Robertson (1990) and Jefferies and Davies (1993). These charts account for the depth-related variations in piezocone parameters, such as cone tip resistance, sleeve friction, and excess pore water pressure, by specifying the parameters Q_t, B_q, and F_r. In contrast, the second group, which comprises the

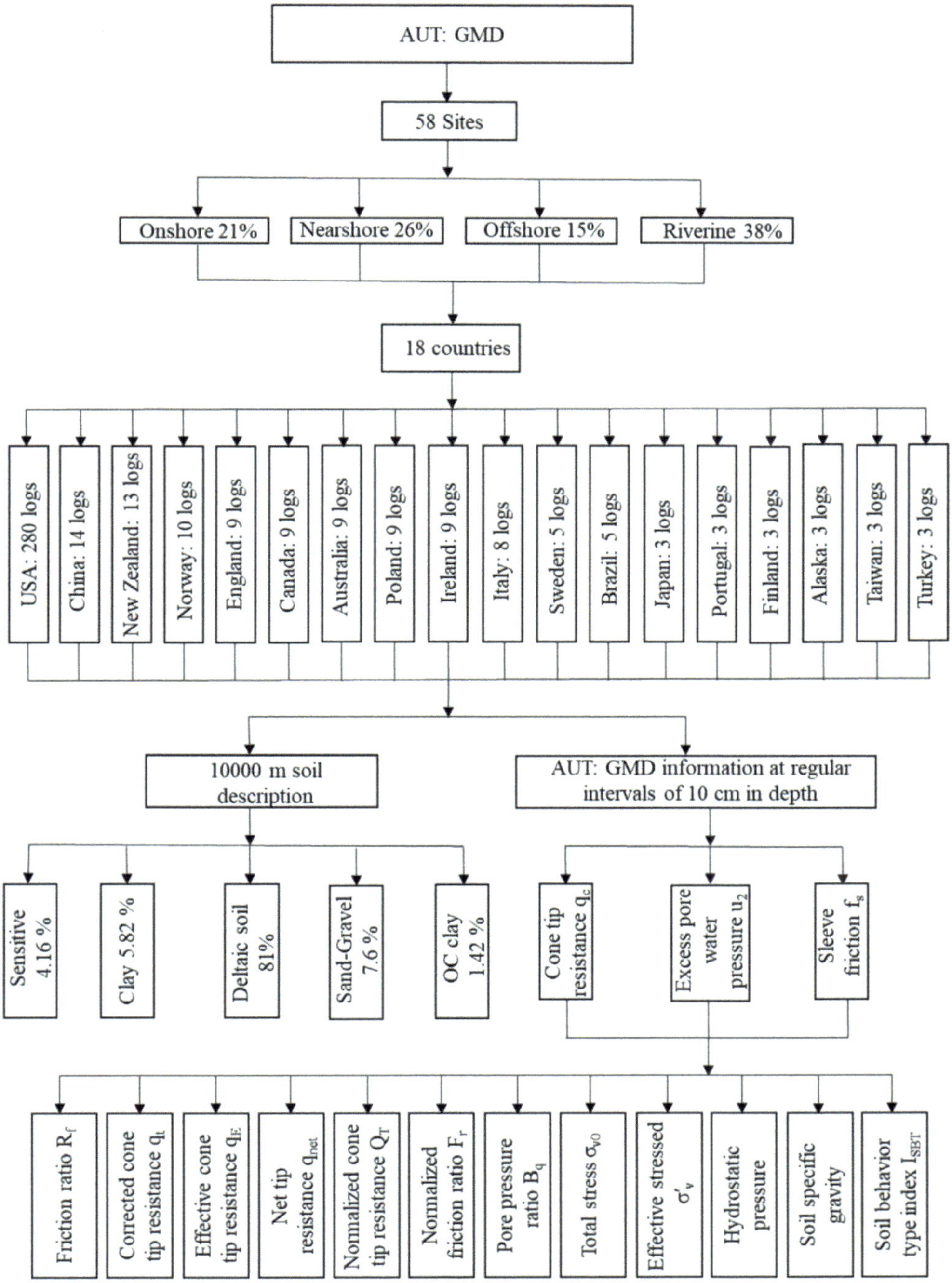

Figure 5.21 The details of AUT: CPTu-GMD.

diagrams created by Eslami and Fellenius (1997) and Eslami et al. (2019), does not address the depth effect separately through mathematical calculations for piezocone parameters.

These diagrams in the second group offer a broader classification of soils, resulting in fewer soil identification zones compared to the first group. As a consequence, each zone covers a larger area and accommodates a wider range of variations. Notably, even at considerable depths, the effect of depth on CPTu records does not prompt the relocation of the subsurface layer to a different zone in the Eslami and Fellenius (1997) and Eslami et al. (2019) diagrams.

Conversely, the diagrams presented by Robertson (1990) and Jefferies and Davies (1993) feature a greater number of zones, resulting in a smaller area allocated to each soil type. These diagrams incorporate normalized parameters, such as Q_t and B_q, to account for depth-related variations. Neglecting these equations leads to inaccuracies in identifying deep-seated soil layers, often resulting in these layers being incorrectly categorized within neighboring zones.

In contrast, the third group, referred to as "direct diagrams," directly employs the three specified parameters (q_c, f_s, and u_2) without the use of mathematical equations to identify subsurface layers. Prominent examples include the work of Eslami et al. (2015). This approach reduces uncertainties and minimizes the reliance on mathematical correlations. The key distinction between the second group of diagrams and the direct diagrams, such as those by Eslami et al. (2015), lies in the application of mathematical relationships.

5.6.3 Results and discussions

The study conducted by Eslami et al. (2022) evaluated the performance of CPT-based Soil Behavioral Classification (SBC) diagrams using 57 case studies sourced from the Marine Database (AUT: CPTu-GMD). These case studies encompassed a selection of widely recognized and efficient methods for soil behavior classification, including Douglas and Olsen (1981), Campanella et al. (1985), Robertson (1990), Robertson (2010), Jefferies and Davies (1993), Eslami and Fellenius (1997), Eslami et al. (2015), and Eslami et al. (2019).

These methods were chosen for their popularity and reliability in classifying soil behavior. Figure 5.21 provides a geographical overview of the analyzed cases, with a distribution as follows: the United States (10 logs), China (5 logs), Canada (3 logs), New Zealand (3 logs), Italy (3 logs), Norway (3 logs), Poland (3 logs), Finland (3 logs), Sweden (3 logs), Ireland (3 logs), Brazil (3 logs), England (3 logs), Japan (3 logs), Portugal (3 logs), Australia (3 logs), and Turkey (3 logs).

Furthermore, the study encompassed an assessment of all four site types found in AUT: CPTu-GMD, which includes 15 onshore locations, 20 near-shore areas, 4 offshore sites, and 18 riverine regions.

The 57 cases under investigation include a total of 1420 m of CPTu testing and soil descriptions obtained from borehole drilling. These data are distributed as follows: 117 m of sensitive soil, 98 m of clay, 1135 m of mixed or deltaic soil, 54 m of sand–gravel, and 16 m of over-consolidated clay.

Figures 5.22–5.25 provide visual representations of the distribution of these four soil types, namely sensitive, clay, mixed, and sand–gravel, on various charts.

To evaluate the accuracy of each diagram, Equation (5.13) is employed. This assessment method involves calculating the ratio of the accurately identified length of each soil

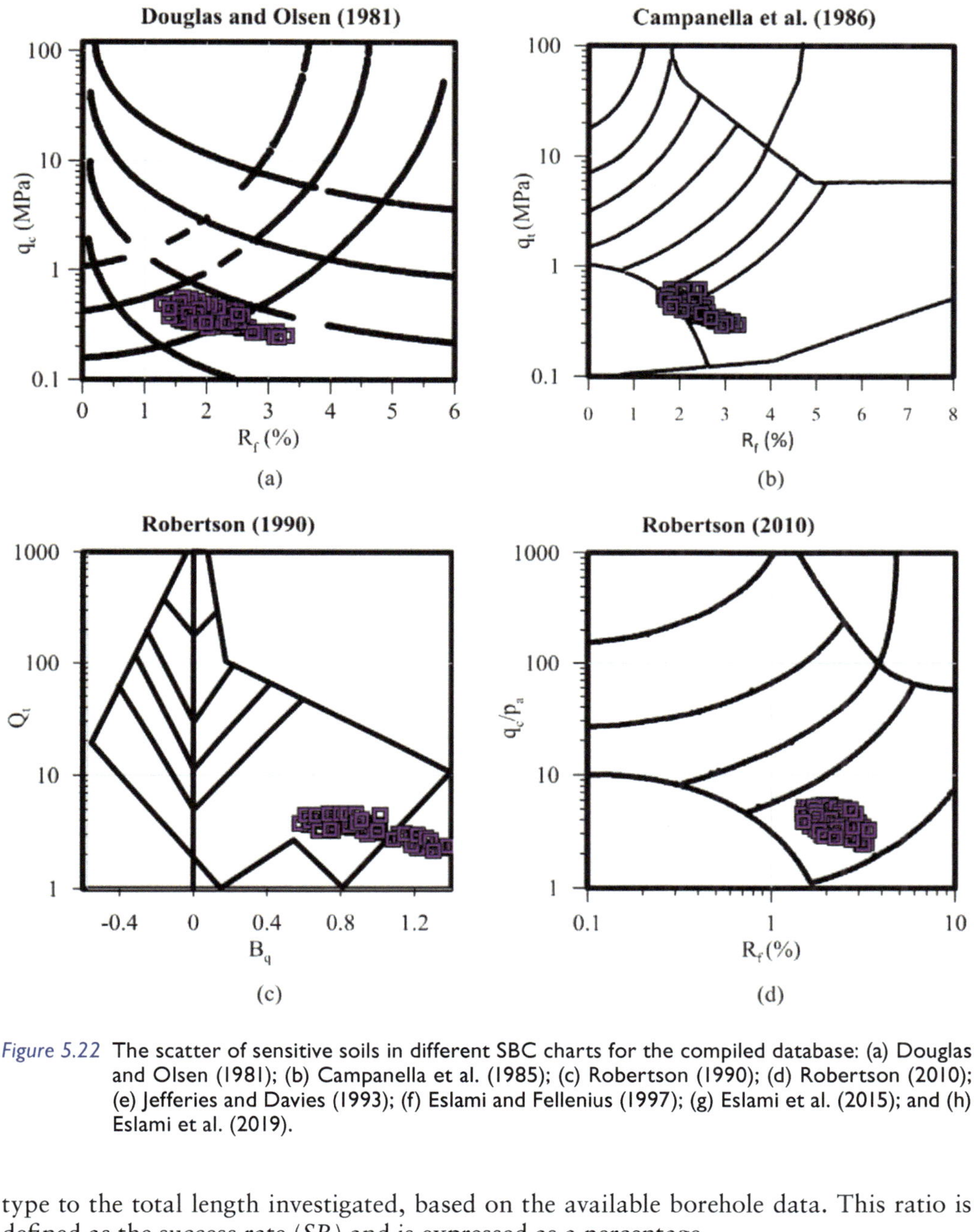

Figure 5.22 The scatter of sensitive soils in different SBC charts for the compiled database: (a) Douglas and Olsen (1981); (b) Campanella et al. (1985); (c) Robertson (1990); (d) Robertson (2010); (e) Jefferies and Davies (1993); (f) Eslami and Fellenius (1997); (g) Eslami et al. (2015); and (h) Eslami et al. (2019).

type to the total length investigated, based on the available borehole data. This ratio is defined as the success rate (*SR*) and is expressed as a percentage.

$$SR = \left(L' / L\right) \times 100 \tag{5.13}$$

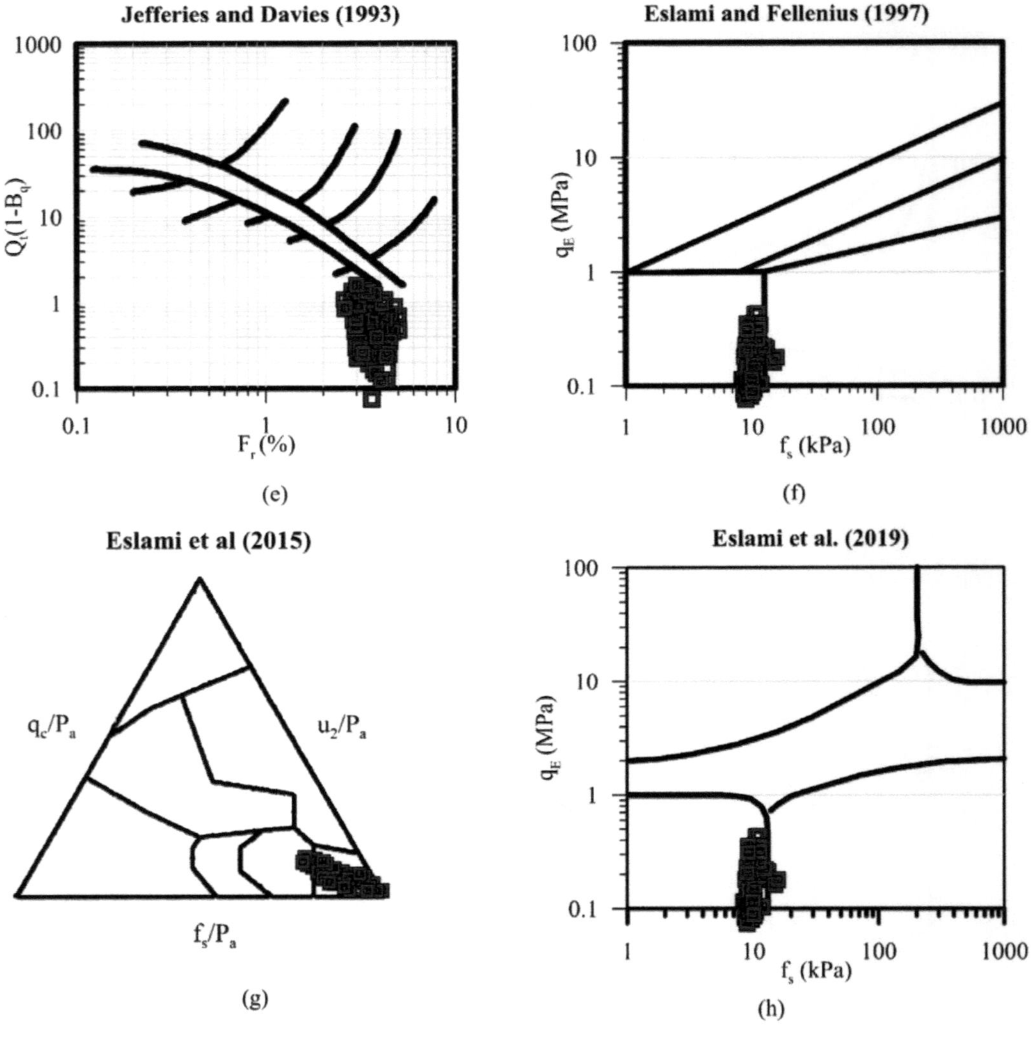

Figure 5.22 (Continued).

where SR = success rate; L' = the correct predicted length of a specific soil type in the investigated borehole; L = the total length of borehole with a particular soil type that exists in the investigated case.

The statistical assessment of each diagram is prominently displayed on a radar chart in Figure 5.26. Each radar chart is equipped with five axes, representing sensitive soils, clay, deltaic soils, sand–gravel, and over-consolidated clay, and these axes are scaled from 0% to 100%. These values effectively quantify the precision with which each diagram identifies soil types associated with their respective axes.

As indicated in Figure 5.26, the capacity to accurately identify sensitive soils has notably improved with the adoption of the second group of diagrams, which consider all

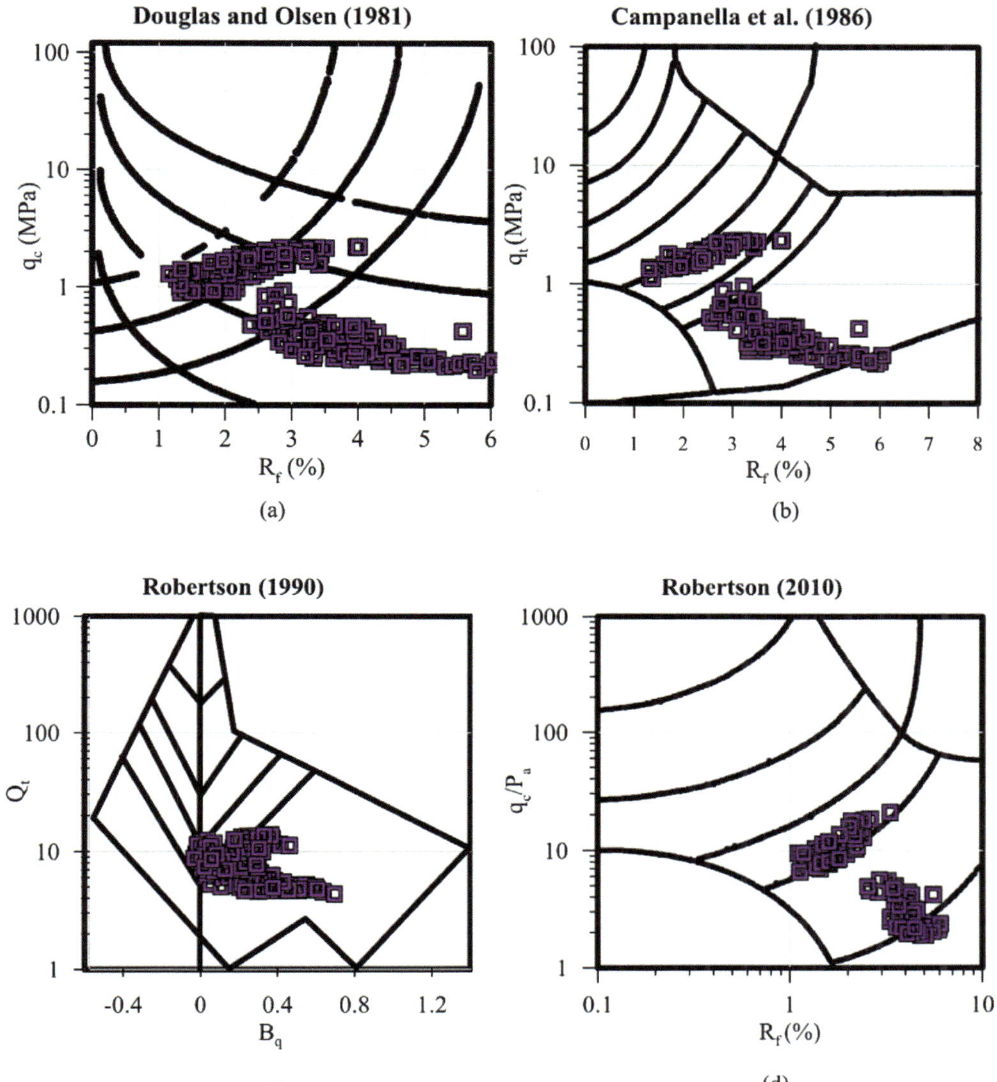

Figure 5.23 The scatter of clayey soils in different SBC charts for the compiled database: (a) Douglas and Olsen (1981); (b) Campanella et al. (1985); (c) Robertson (1990); (d) Robertson (2010); (e) Jefferies and Davies (1993); (f) Eslami and Fellenius (1997); (g) Eslami et al. (2015); and (h) Eslami et al. (2019).

three factors – q_c, f_s, and u_2 – along with mathematical correlations for soil classification. Basic charts, such as those developed by Douglas and Olsen (1981) and Campanella et al. (1985), exhibit less than 40% accuracy in detecting sensitive soils. In contrast, the implementation of indirect charts has led to an enhanced reliability in characterizing sensitive soils. This trend continues until the Eslami and Fellenius (1997) diagram, which achieves a remarkable success rate of over 90% in identifying sensitive soils.

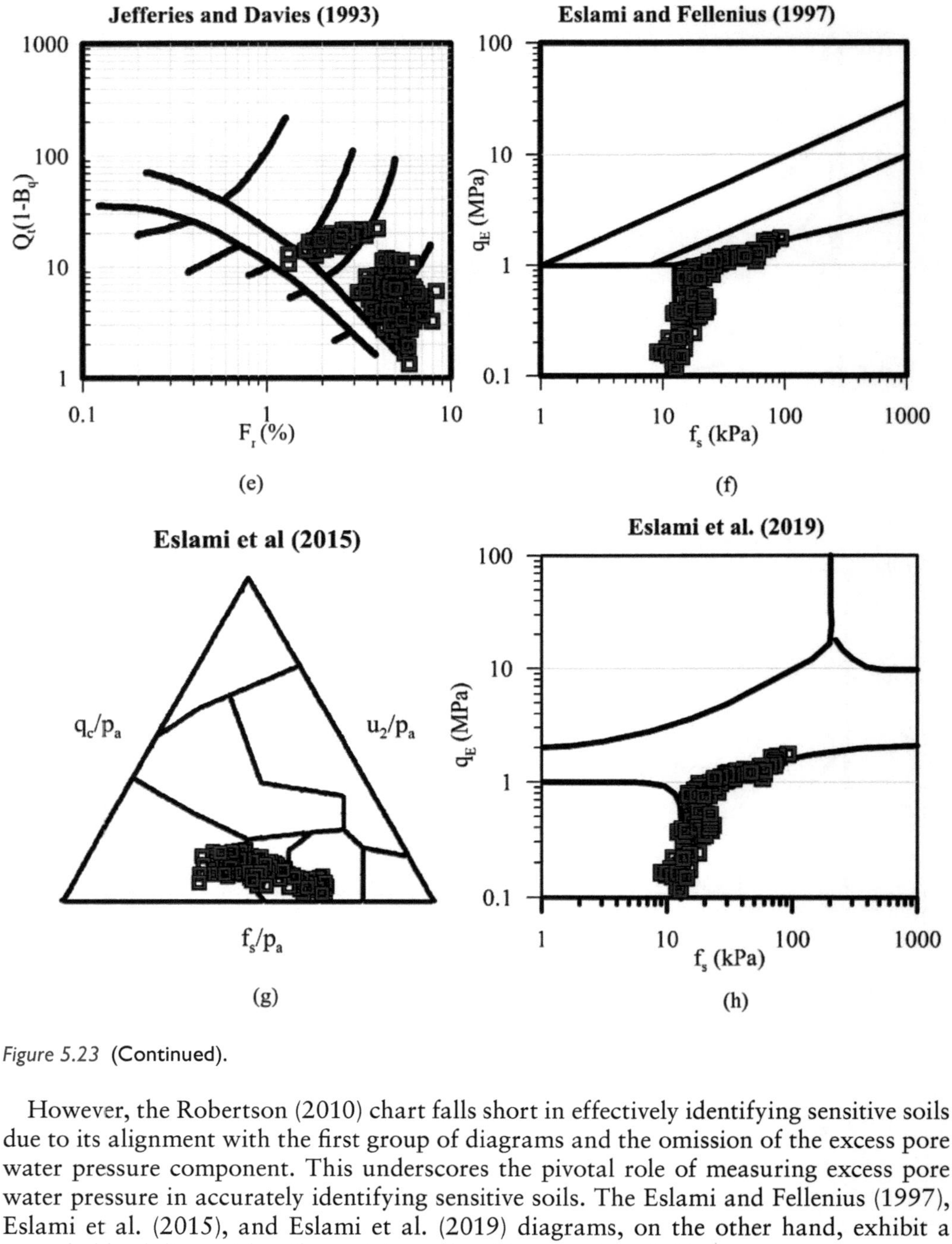

Figure 5.23 (Continued).

However, the Robertson (2010) chart falls short in effectively identifying sensitive soils due to its alignment with the first group of diagrams and the omission of the excess pore water pressure component. This underscores the pivotal role of measuring excess pore water pressure in accurately identifying sensitive soils. The Eslami and Fellenius (1997), Eslami et al. (2015), and Eslami et al. (2019) diagrams, on the other hand, exhibit a remarkable ability to identify sensitive soils with more than 90% accuracy.

Moreover, the first and second groups of diagrams exhibited challenges in accurately identifying normally consolidated clays, achieving less than 60% accuracy. This is attributed to the fact that the cone tip resistance, q_c, tends to be low for silts and

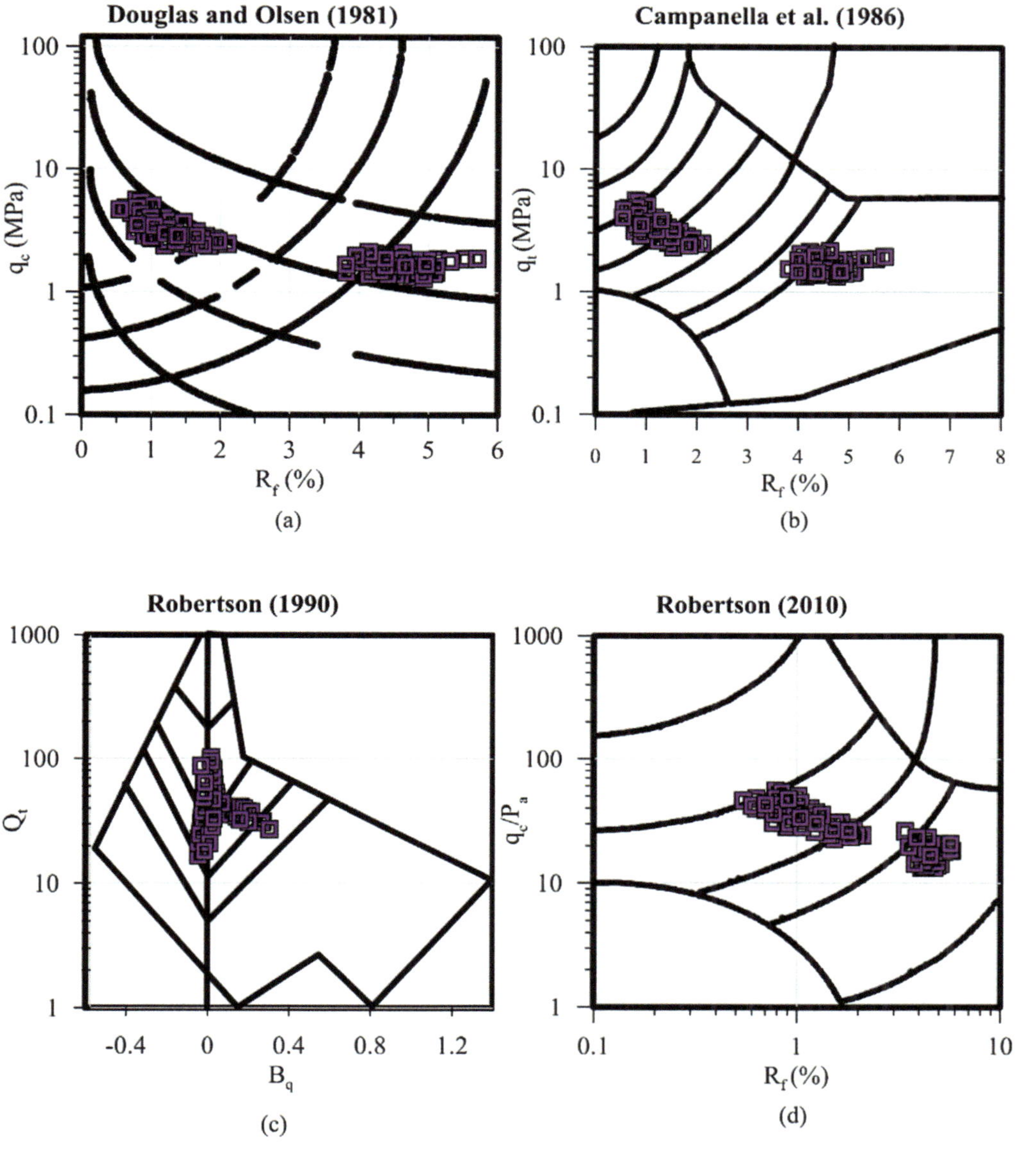

Figure 5.24 The scatter of deltaic soils in different SBC charts for the compiled database: (a) Douglas and Olsen (1981); (b) Campanella et al. (1985); (c) Robertson (1990); (d) Robertson (2010); (e) Jefferies and Davies (1993); (f) Eslami and Fellenius (1997); (g) Eslami et al. (2015); and (h) Eslami et al. (2019).

normally consolidated clays, which limits its effectiveness in distinguishing between these soil types. Consequently, these diagrams now primarily rely on parameters such as sleeve friction and pore pressure for the Soil Behavioral Classification (SBC) of these specific soil types.

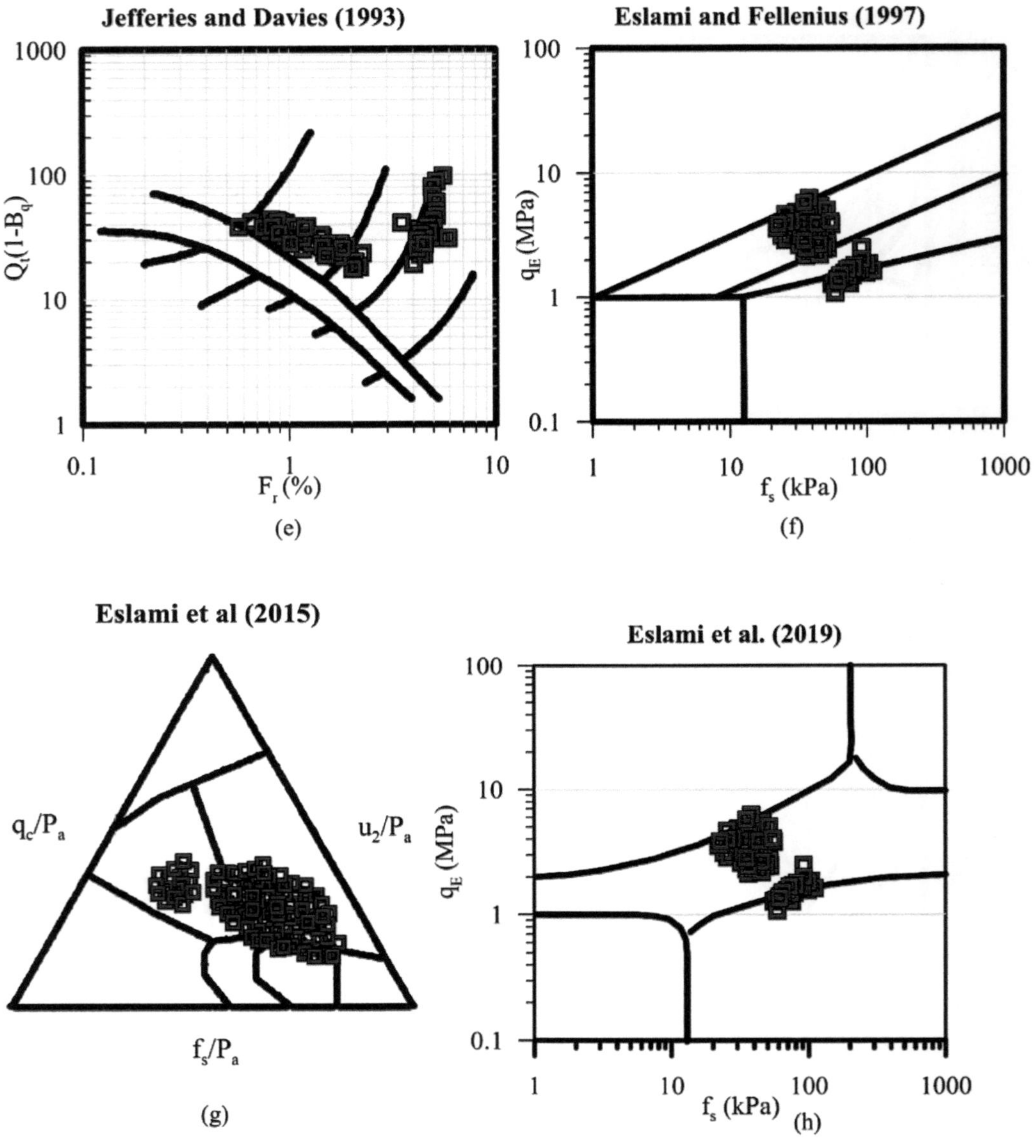

Figure 5.24 (Continued).

With the exception of the Eslami et al. (2019) diagram, other diagrams in the first and second groups displayed an accuracy rate of less than 60% when detecting over-consolidated clays. The Eslami et al. (2019) diagram, however, showed a considerably higher accuracy of approximately 80%. It is worth noting that due to the high cone tip resistance observed in over-consolidated clay, the other diagrams in the first and second groups often misclassify it as sandy soil.

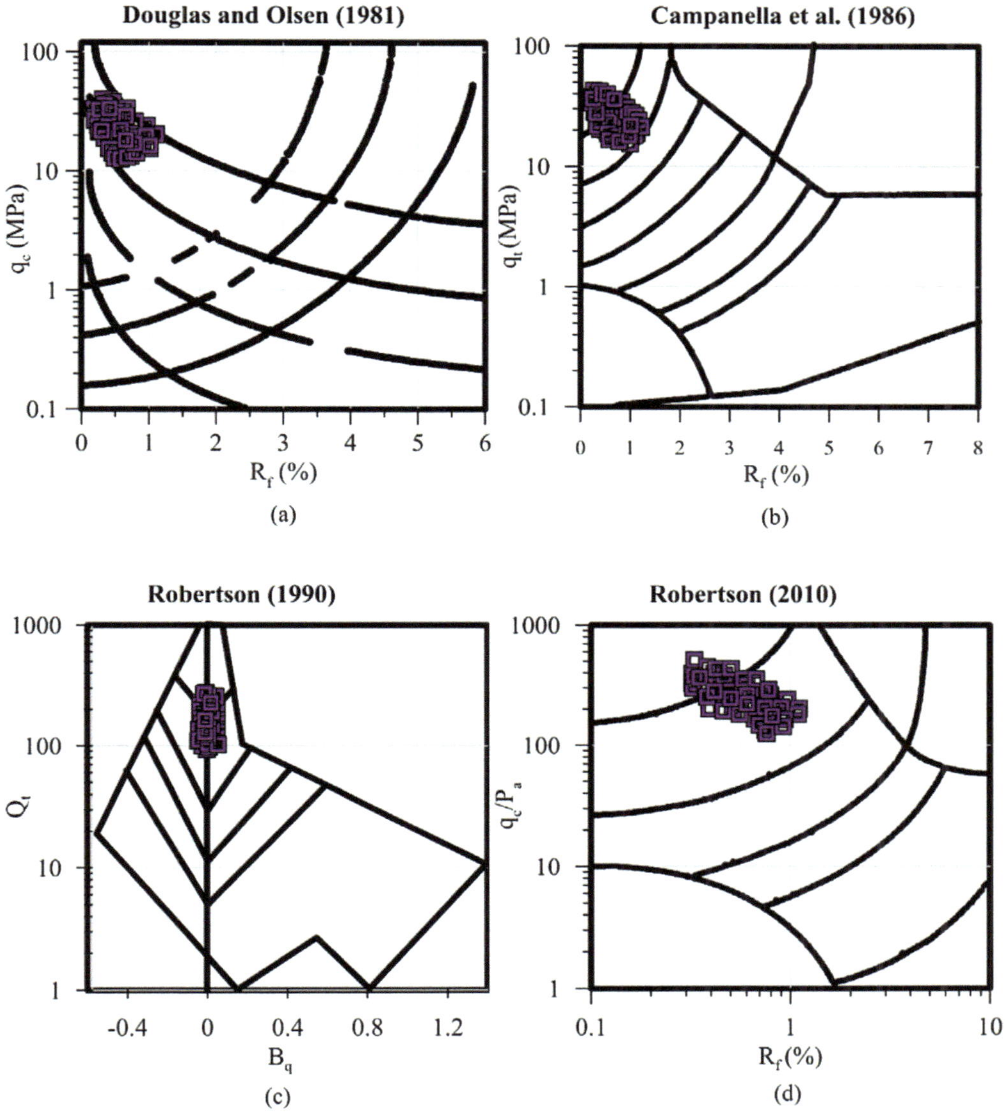

Figure 5.25 The scatter of sand and gravel in different SBC charts for the compiled database: (a) Douglas and Olsen (1981); (b) Campanella et al. (1985); (c) Robertson (1990); (d) Robertson (2010); (e) Jefferies and Davies (1993); (f) Eslami and Fellenius (1997); (g) Eslami et al. (2015); and (h) Eslami et al. (2019).

As depicted in Figure 5.26, the first two types of diagrams demonstrated a good level of accuracy, correctly identifying deltaic soils and sand–gravel mixtures with rates ranging from 80% to 90%. In contrast, the Eslami et al. (2015) triangle diagram, which directly employs all three CPTu records rather than relying on mathematical correlations, exhibited the highest accuracy in predicting sensitive soils, over-consolidated clays, and sand–gravel, with success rates of approximately 96%, 85%, and 94%, respectively.

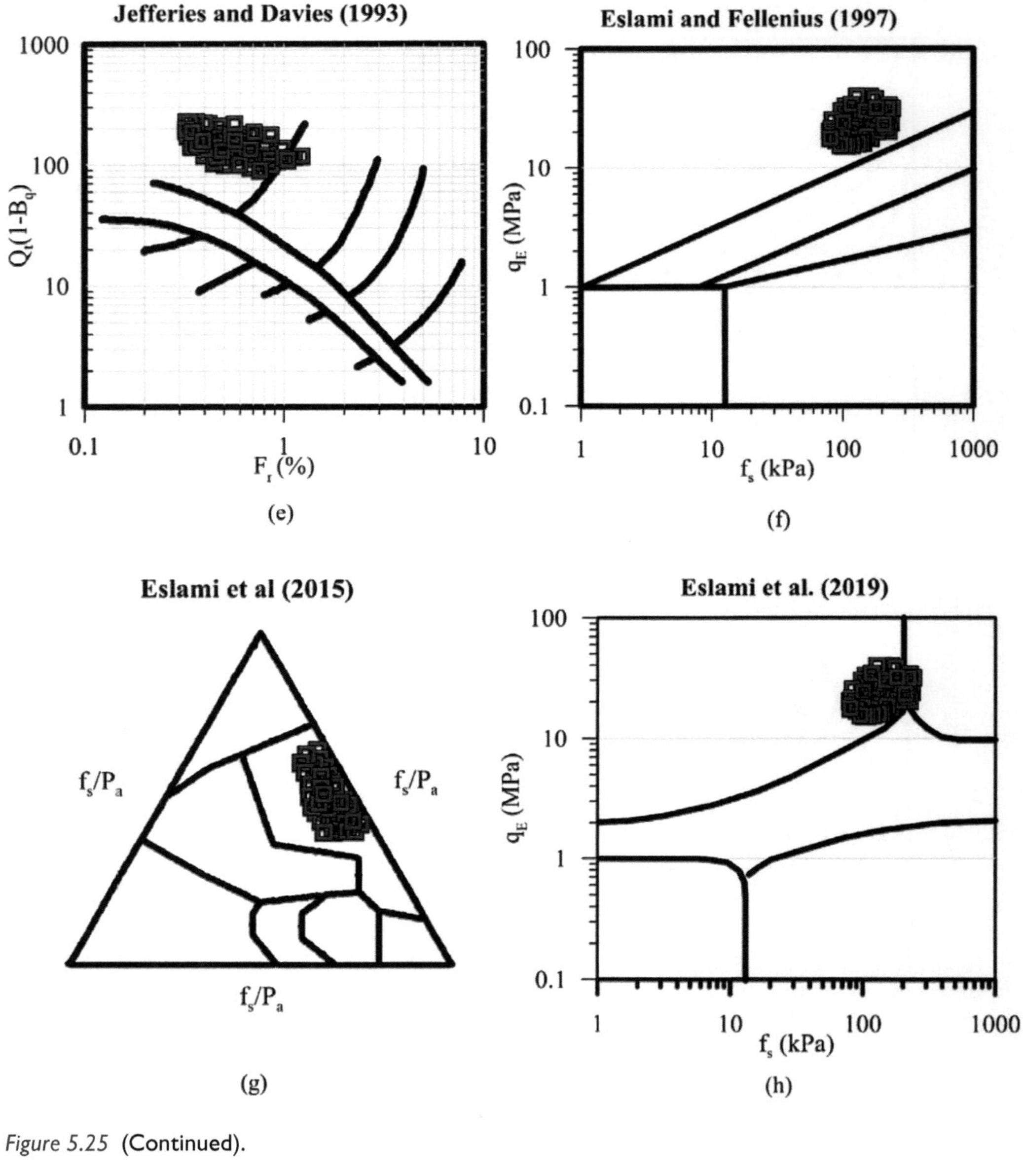

Figure 5.25 (Continued).

Nevertheless, it should be noted that this diagram tended to inaccurately classify deltaic soils and normally consolidated clays. This suggests that by refining the applied model assumptions and minimizing uncertainties introduced by mathematical relationships, the accuracy of SBC for deltaic soils and normally consolidated clay can be significantly improved with minor adjustments.

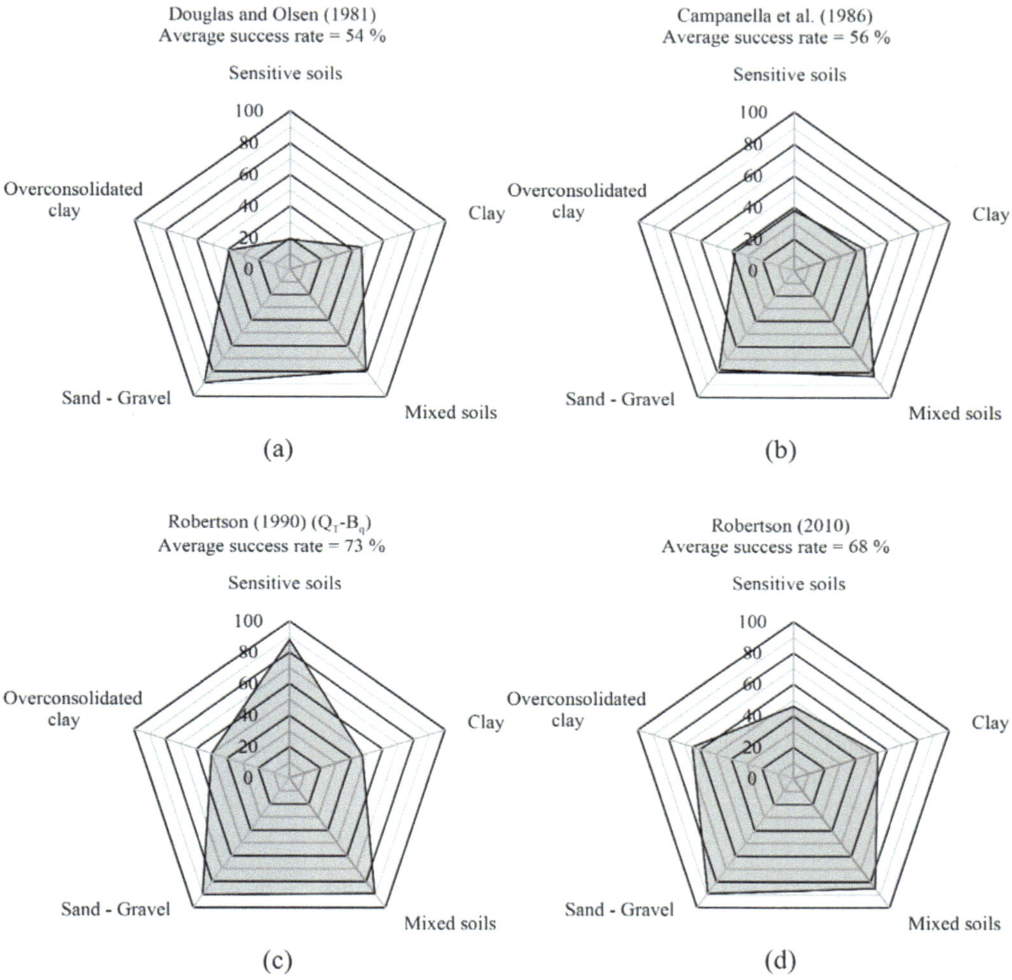

Figure 5.26 The statistical assessment of SBC diagrams via radar charts.

5.6.4 Developed triangular chart

The performance evaluation of the Soil Behavioral Classification (SBC) diagrams in accurately identifying all five deltaic soil types within the examined database highlights the critical role of input parameters, mathematical correlations, and the database source from which these diagrams are derived. This underscores the need for the development of a chart that can effectively classify marine deposits with precision.

The third group of SBC diagrams, exemplified by the Eslami et al. (2015) diagram, demonstrates significant potential in accurately identifying all soil types. In the Eslami et al. (2015) diagram, all three parameters – cone tip resistance, sleeve friction, and pore water pressure — are directly incorporated, eliminating the need for mathematical correlations to classify subsurface layers. This results in a reduction of uncertainties associated with subsurface layer identification, as mathematical formulas, are not relied upon.

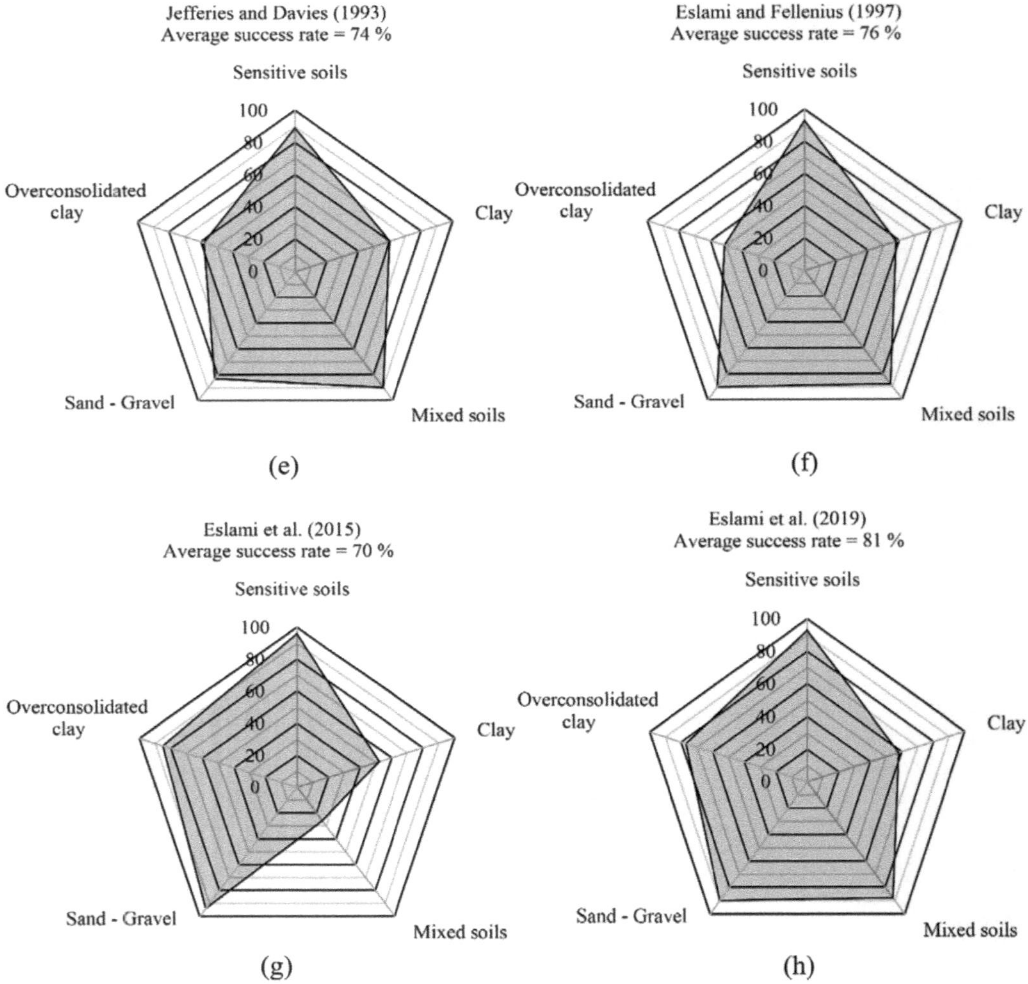

Figure 5.26 (Continued).

In traditional triangular charts, the sum of parameters on the three sides is typically a constant value, usually set at 100. For instance, in the Soil Classification Chart recommended by the United States Department of Agriculture (USDA) for deltaic soil classification, the sum of the three parameters (percentage of silt, clay, and sand) must equal 100. However, the output parameters of Cone Penetration Testing (CPTu) are not interrelated, and their sum cannot be constrained to equal 100. Consequently, the Eslami et al. (2015) diagram introduces a novel approach to triangular diagrams, integrating all three CPTu records within a single chart.

In the case of the Eslami et al. (2015) diagram, the developed triangular chart consists of three axes that correspond to the CPTu parameters, namely cone tip resistance, sleeve friction, and pore water pressure, as depicted in Figure 5.27. The application of this developed triangular chart closely mirrors that of the Eslami et al. (2015) diagram and involves the following steps.

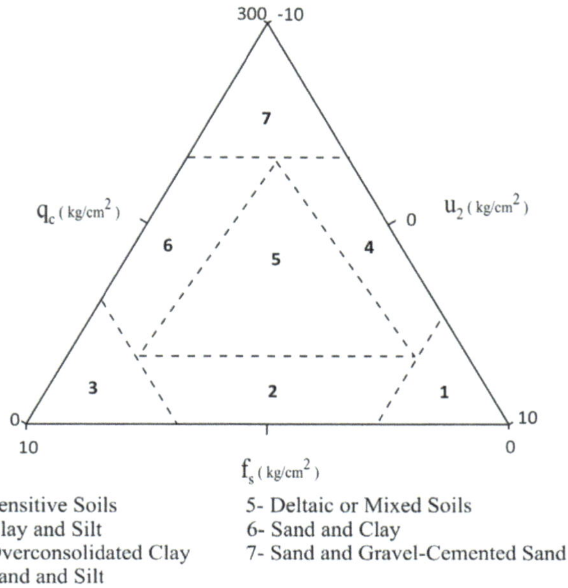

Figure 5.27 The delineated zones for the developed triangular chart using a deltaic deposit.

1. Convert the CPTu records, including q_t, f_s, and u_2, into normalized values by dividing them by a reference stress. In this case, the reference stress is equivalent to atmospheric pressure, which is 1 atm = 1 kg/cm² = 100 kPa = 1 bar.
2. Locate the q_t value on the chart and draw a line parallel to the f_s axis.
3. Locate the u_2 value on the chart and draw a line parallel to the q_t axis.
4. Locate the f_s value on the chart and draw a line parallel to the u_2 axis.
5. The point of intersection of these three lines forms an isosceles triangle.
6. The center of this triangular region on the chart accurately represents the soil condition.

As shown in Figure 5.27, the developed triangular diagram consists of seven zones. The larger middle triangle is the location of mixed or deltaic soils, i.e., clay, silt, and sand, in varying proportions. As the center of the isosceles triangle moves upward, i.e., toward the middle triangle vertex, the clay ratio decreases and the proportion of sand increases. If the ratio of one of the three soil types mentioned is negligible, it will be located in one of the three trapezoidal areas depending on the other two soil types. The three smaller triangles at the edges of the developed triangular chart are used to identify sensitive soil, over-consolidated clay, sand and gravel, and cemented sand. Figure 5.28 shows the mentioned soil types on the developed triangular chart.

5.7 SUMMARY

In the realm of geotechnical engineering, the accurate characterization and classification of soils stand as pivotal factors influencing the success of design and construction

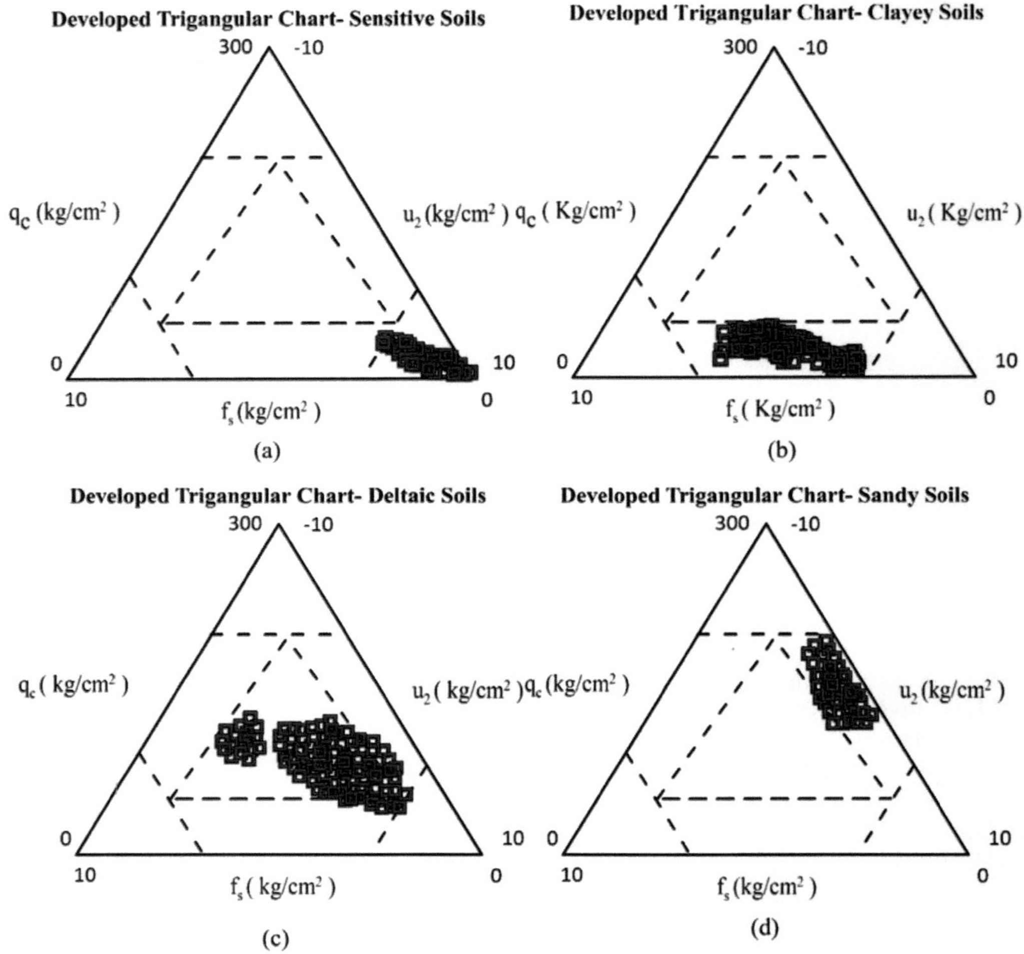

Figure 5.28 Various soil types on developed triangular chart: (a) sensitive soils, (b) clay, (c) deltaic soil, and (d) sand.

projects. This chapter delves into the profound significance of leveraging databases to assess soil properties in geotechnical engineering, highlighting the crucial role of geotechnical site investigation in acquiring vital insights into soil characteristics and behavior. The various methods explored encompass remote sensing, data review, site survey, testing, and monitoring.

The narrative emphasizes the inherent uncertainty and variability intrinsic to soil properties, underscoring the imperative to source geotechnical data from diverse channels. These uncertainties are categorized into natural variations in material properties and an incomplete understanding of the system. Addressing the former requires more testing and robust databases, while the latter involves tackling measurement errors, transformation uncertainties, and model errors through enhanced experiments and accurate models. The narrative underscores recent advancements in uncertainty quantification, machine learning, Bayesian algorithms, and system-level analysis as key elements in enhancing geotechnical decision making, supporting resilience engineering, and advancing digital transformation.

Turning attention to in-situ tests, the chapter elucidates their significant role in soil identification and classification within geotechnical engineering. Notably, the Cone Penetration Test (CPT), Standard Penetration Test (SPT), Field Vane Shear Test (FVST), and Dilatometer Penetration Test (DMT) are underscored as pivotal contributors to understanding soil behavior. The narrative accentuates the development of correlations and charts that aid in classifying soils based on test results, thus elevating the accuracy of soil type identification. Additionally, it sheds light on recent advancements, including triangular classification charts and behavior-driven approaches, providing more precise methodologies for characterizing soil properties and thereby enhancing the reliability of geotechnical investigations and foundation design.

The exploration extends to the prediction of geotechnical parameters through in-situ tests, encompassing soil engineering parameters, index parameters, strength parameters, stiffness, and compressibility parameters. The narrative navigates the intricate interpretation of in-situ test results, elucidating the use of empirical correlations to estimate geotechnical properties. Key parameters such as unit weight, relative density, internal friction angle, undrained shear strength, stiffness, constrained modulus, over-consolidation ratio, and consolidation indexes are meticulously covered. Additionally, the chapter addresses shear wave velocity and its correlation with shear modulus, sensitivity, and permeability. A wealth of empirical correlations and insights are provided to facilitate parameter estimation in geotechnical engineering applications.

Finally, the chapter delves into the AUT: CPTu-Geo-marine Database, meticulously crafted by Eslami et al. (2022). This comprehensive repository comprises 398 cases from 58 locations, offering invaluable insights into geotechnical and marine data. The chapter introduces Soil Behavioral Classification methods, categorizing diagrams into basic, indirect, and direct types. The performance of these diagrams is rigorously evaluated through 57 case studies, with statistical assessments and radar charts providing a comprehensive overview. A notable addition is the introduction of a novel triangular chart, enhancing accuracy in classifying marine deposits based on Cone Penetration Testing parameters. This chapter serves as a thorough exploration of geotechnical data analysis and classification techniques, marking a significant contribution to the field.

REFERENCES

Aas, G., Lacasse, S., Lunne, T., Hoeg, K. (1986). Use of in situ tests for foundation design on clay. In: *Proceedings of In Situ '86: Use of In Situ Tests in Geotechnical Engineering*, Virginia, pp. 1–30.

AASHTO. (2020). *LRFD Bridge Design Specifications*, 9th ed. American Association of State Highway and Transportation Officials, Washington, DC.

Ameratunga, J., Nagaratnam Sivakugan, N., Das, B. M. (2016). *Correlations of Soil and Rock Properties in Geotechnical Engineering*. Springer, India, New Delhi.

Anagnostopoulos, A., Koukis, G., Sabatakakis, N., Tsiambaos, G. (2003). Empirical correlation of soil parameters based on cone penetration tests (CPT) for Greek soils. *Geotechn Geol Eng*, 21(4), 373–387.

Azzous, A. S., Krizek, R. J., Corotis, R. B. (1976). Regression Analysis of Soil Compressibility. *Soils Found*, 16(2), 19–29.

Baguelin, F., Jezequel, J. F., Le Mee, E., Le Mehaute, A. (1972). Expansion of cylindrical probes in cohesive soils. J Soil Mech Found Eng Div ASCE, 98(SM11), 1129–1142.

Balasubramaniam, A. S., Brenner, R. P. (1981). Chapter 7: Consolidation and settlement of soft clay. In: Brand EW, Brenner RP (eds) *Soft Clay Engineering.* Elsevier, Amsterdam, pp. 481–566.

Baldi, G., Bellotti, R., Ghionna, V., Jamiolkowski, M., Pasqualini, E. (1986). Interpretation of CPT's and CPTU's. 2nd Part: drained penetration. In: *Proceedings 4th International Geotechnical Seminar,* Singapore, pp. 143–156.

Baldi, G., Jamiolkowski, M., Lo Presti, D.C.F., Manfredini, G., Rix, G.J. (1989). Italian experience in assessing shear wave velocity from CPT and SPT. In: *Proceedings of Discussion Session on Influence of Local Conditions on Seismic Response*, vol. XII. ICSMFE, Rio De Janerio, pp. 157–168.

Begemann, H. K. S. (1965). The friction jacket cone as an aid in determining the soil profile. In: *Proceedings of the 6th International Conference on Soil Mechanics and Foundation Engineering*, ICSMFE, Montreal, September 8–15, vol. 2, pp. 17–20.

Bergdahl, U., Ottosson, E., Malmborg, B. S. (1993). *Plattgrundlaggning. AB Svensk Byggtjanst,* Stockholm.

Bowles, J. E. (1996). *Foundation analysis and design*, 5th ed. McGraw-Hill, New York.

Cai, G. J., Liu, S. Y., Puppala, A. J., Tong, L. Y. (2015). Identification of soil strata based on general regression neural network model from CPTU data. *Marine Georesources & Geotechnology,* 33(3), 229–238.

Campanella, R. G., Robertson, P. K., Gillespie, D., Grieg, J. (1985). Recent development in in-situ testing of soils. In: *Proceedings of 11th International Conference on Soil Mechanics and Foundation Engineering*, vol. 2, ICSMFE, San Francisco, CA, pp. 849–854.

Canadian Foundation Engineering Manual, CFEM, 4th ed., 2006. Canadian Geotechnical Society, BiTech Publishers, Vancouver, p. 488.

Chandler, R. J. (1988). *The in-situ measurement of the undrained shear strength of clays using the field vane. Vane shear strength testing of soils: field & lab studies*, STP1014, ASTM, West Conshohocken, pp. 13–44.

Chanmee, N., Chai, J., Hino, T., Wang, J. (2017). Methods for evaluating overconsolidation ratio from piezocone sounding results. *Underground Space,* 2(3), 182–194.

Chen, H., Mo, P. Q. (2022). An undrained expansion solution of cylindrical cavity in SANICLAY for K0 consolidated clays. *J Rock Mech Geotechn Eng,* 14, 922–935.

Ching, J. and Phoon, K.K. (2013). Probability distribution for mobilized shear strengths of spatially variable soils under uniform stress states. *Georisk*, 7(3), 209–224.

Ching, J., Li, D.Q., and Phoon, K.K. (2016). *Statistical Characterization of Multivariate Geotechnical Data.* Chapter 4, Reliability of Geotechnical Structures in ISO2394, CRC Press, Balkema, pp. 89–126.

Ching, J., Wu, S., Phoon, K.K. (2021). Constructing quasi-site-specific multivariate probability distribution using hierarchical Bayesian model. *J Eng Mech ASCE* 147(10), 04021069.

Cozzolino, V. M. (1961). Statistical forecasting of compression index. In: *Proceedings of the 5th ICSMFE*, Paris, pp. 51–53.

Das, M. B. (2006). *Principles of Geotechnical Engineering,* 7th ed. Cenghage Learning, United States.

Das, M. B. (2016). *Fundamentals of Geotechnical Engineering,* 5th ed. Cenghage Learning, United States.

Dikmen, U. (2009). Statistical correlations of shear wave velocity and penetration resistance for soils. *J Geophys Eng*, 6(1), 61–72.

Djoenaidi, W. J. (1985). *A compendium of soil properties and correlations.* MEngSc thesis, University of Sydney, Australia.

Douglas, B. J., Olsen, R. S. (1981). Soil classification using electric cone penetrometer. In: American Society of Civil Engineers, ASCE, *Proceedings of Conference on Cone Penetration Testing and Experience*, St. Louis, 1981, October 26–30, pp. 209–227.

Elnaggar, M. A., Krizek, R. J. (1970). Statistical approximation for consolidation settlement. *Highway Research Record*, no. 323, HRB, pp. 87–96.

Eslaamizaad, S., Robertson, P. K. (1997). Evaluation of settlement of footings on sand from seismic in-situ tests. In: *Proceedings of the 50th Canadian Geotechnical Conference,* Ottawa, Ontario, vol. 2, pp. 755–764.

Eslami A., Heidarie Golafzani S., Moshfeghi S. (2020). CPT and pile database for performance-based design of pile axial bearing capacity. In: *Proceedings of the 45th Annual Conference on Deep Foundations, Deep Foundation Institute (DFI),* Oxon Hill, MD, p. 12.

Eslami, A., Alimirzaei, M., Aflaki, E., Molaabasi, H. (2016). Deltaic soil behavior classification using CPTu records-proposed approach and applied to fifty-four case histories. *Marine Georesour Geotechnol* 35(1), 62–79.

Eslami, A., Fellenius, B. H. (1997). Pile capacity by direct CPT and CPTU data. In: *Year 2000 Geotechnics, Geotechnical Engineering Conference,* AIT Bangkok, 18pp.

Eslami, A., Fellenius, B. H. (2004). *CPT and CPTu Data for Soil Profile Interpretation: Review of Methods and a Proposed New Approach.*

Eslami, A., Heidarie Golafzani, S., Naghibi, M. H. (2022). Developed triangular charts; deltaic CPTu-based soil behavior classification using AUT: CPTu-Geo-Marine Database. *Probabilistic Engineering Mechanics*, 71, 103380.

Eslami, A., Mohammadi, A. (2016). Drained soil shear strength parameters from CPTu data for marine deposits by analytical model. *Ships Offshore Struct*, 11(8), 913–925.

Eslami, A., Moshfeghi, S., Molaabasi, H., Eslami, M.M. (2019). *Piezocone and Cone Penetration Test (CPTu and CPT) Applications in Foundation Engineering*, Butterworth-Heinemann.

Federal Geographic Data Committee, *Coastal and Marine Ecological Classification Standard,* Publication# FGDC-STD-018-2012, 2012.

Frank, R., Magnan, J.P. (1995). Cone penetration testing in France: national report. In: *Proceedings CPT'95, Linkoping, Swedish Geotechnical Society*, vol. 3, pp. 147–156.

Gibbs, H. J., Holtz, W. G. (1957). Research on determining the density of sand by spoon penetration testing. In: *Proceedings, 4th International Conference on Soil Mechanics and Foundation Engineering,* London, I, pp. 35–39.

Goldberg, G. D., Lovell, C. W., Miles, R. D. (1979). Use of geotechnical data bank, *Transportation Research Record*, No. 702, TRB, pp. 140–146.

Hasancebi, N., Ulusay, R. (2007). Empirical correlations between shear wave velocity and penetration resistance for ground shaking assessments. *Bull Eng Geol Environ*, 66(2), 203–213.

Hatanaka, M., Uchida, A. (1996). Empirical correlations between penetration resistance and internal friction angle of sandy soils. *Soils Found*, 36(4), 1–10.

Hegazy, Y. A., Mayne, P. W. (1995). Statistical correlations between vs and cone penetration data for different soil types. In: *Proceedings of the International Symposium on Cone Penetration Testing (CPT'95)*, Linkoping, Sweden, 4–5 October 1995, vol. 2. Swedish Geotechnical Society, pp. 173–178.

Holtz, W. G., Gibbs, H. J. (1979). Discussion of "SPT and relative density in coarse sand". *J Geotech Eng Div ASCE*, 105(3), 439–441.

Hough, B. K. (1957). *Basic Soils Engineering.* The Ronald Press Co., New York.

Imai, T. (1977). P- and S-wave velocities of the ground in Japan. In: *Proceedings, 9th International Conference on Soil Mechanics and Foundation Engineering*, Tokyo, Japan, 2, pp. 257–260.

Iyisan, R. (1996). Correlations between shear wave velocity and in-situ penetration test results. Chamber of Civil Engineers of Turkey, *Teknik Dergi*, 7(2), 1187–1199.

Jafari, M. K., Shafiee, A., Razmkhah, A. (2002). Dynamic properties of the fine grained soils in south of Tehran. *J Seismol Earthquake Eng*, 4(1), 25–35.

Jamiolkowski, M., Lo Presti, D. C. F., Manassero, M. (2001). Evaluation of relative density and shear strength of sands from CPT and DMT. In: *CC Ladd Symposium.* MIT, Cambridge, MA. ASCE Geotechnical Special Publication No. 119: 201–238.

Jefferies, M. G., Davies, M. P. (1991). Soil classification using the cone penetration test. *Discussion Canadian Geotechnical Journal*, 28(1), 173–176.
Jefferies, M. G., Davies, M. P. (1993). Use of CPTu to estimate equivalent SPT N60. American Society for Testing and Materials, *Geotechnical Testing Journal* 16(4), 458–468.
Jones, G. A., Rust, E. (1982). Piezometer penetration testing, CPTU. In: *Proceedings of the 2nd European Symposium on Penetration Testing*, ESOPT-2, Amsterdam, May 24-27, vol. 2, pp. 607–614.
Kiku, H., Yoshida, N., Yasuda, S., Irisawa, T., Nakazawa, H., Shimizu, Y., Ansal, A., Erkan, A. (2001). In-situ penetration tests and soil profiling in Adapazari, Turkey. In: *Proceedings, 15th International Conference on Soil Mechanics and Geotechnical Engineering*, TC4 Satellite Conference on Lessons Learned from Recent Strong Earthquakes, Istanbul, Turkey, pp. 259–269.
Koppula, S. D. (1981). Statistical evaluation of compression index. *Geotech Test J ASTM* 4(2), 68–73.
Ku, C. S., Juang, C. H., Ou, C. Y. (2010). Reliability of CPT Ic as an index for mechanical behavior classification of soils. *Géotechnique*, 60(11), 861–875.
Kulhawy, F. H., Mayne, P. W. (1990). *Manual on estimating soil properties for foundation design*, Report EL-6800. Electric Power Research Institute, Palo Alto.
Lacasse, S., Lunne, T. (1988). Calibration of dilatometer correlations. *Proc. ISOPT-1, Orlando*, Vol. I, pp. 539–548.
Li, J., Cassidy, M. J., Huang, J., Zhang, L., Kelly, R. (2015). Probabilistic identification of soil stratification. *Géotechnique*, 66(1), 16–26.
Lo, Y. K. T, Lovell, C. W. (1982). Prediction of soil properties from simple indices, *Transportation Research record*, No. 873, Overconsolidated clays: Shales, TRB, pp. 43–49.
Look, B. G. (2007). *Handbook of Geotechnical Investigation and Design Tables*. 1st ed., Taylor & Francis, London.
Lunne, T., Robertson, P. K., Powell, J. J. M. (1997). *Cone Penetration Testing in Geotechnical Practice*. Blackie Academic & Professional/Chapman-Hall Publishers, London, 312pp.
Madiai, C., Simoni, G. (2004). Shear wave velocity-penetration resistance correlation for Holocene and Pleistocene soils of an area in central Italy. In: Viana da Fonseca, Mayne (Eds.), *Proceedings ISC-2 on Geotechnical and Geophysical Site Characterization*. Mill Press, *Rotterdam*, pp. 1687–1694.
Marchetti, S. (1980). In situ tests by flat dilatometer. *J Geotech Eng Div ASCE*, 106(GT3), 299–321.
Mayne, P. W. (2006). Undisturbed sand strength from seismic cone tests. *Geomech Geoeng* 1(4), 239–257.
Mayne, P. W. (2007). Cone penetration testing State-of-Practice. In: *NCHRP Synthesis. Transportation Research Board Report Project 20-05*, 118pp.
Mayne, P. W. (2014). Interpretation of geotechnical parameters from seismic piezocone tests. In: *3rd International Symposium on Cone Penetration Testing*. Las Vegas, NV.
Mayne, P. W., Kemper, J. B. (1988). Profiling OCR in stiff clays by CPT and SPT. *Geotechn Test J*, 11(2), 139–147.
Mayne, P. W., Rix, G. J. (1995). Correlations between shear wave velocity and cone tip Resistance in natural clays. *Soils Found*, 35(2), 107–110.
Menard, L. (1956). Mesures in situ des proprietes physiques des sols. *Ann Ponts Chaussees Paris*, No. 14, pp. 357–377.
Meyehof, G. G. (1957). Discussion on research on determining the density of sand by spoon penetration testing. In: *Proceedings, 4th International Conference on Soil Mechanics and Foundation Engineering*, London, 3, pp. 110–114.
Meyerhof, G. G. (1951). The Ultimate Bearing Capacity of Foundations. *Géotechnique*, 2, 301.
Meyerhof, G. G. (1959). Compaction of sands and the bearing capacity of piles. *J Soil Mech Found Div ASCE*, 85(6), 1–29.

Meyerhof, G. G. (1974). Ultimate bearing capacity of footings on sand layer overlying clay. *Canad Geotech J*, 11(2), 223–229.

Mitchell, J. K., Durgunoglu, H. T. (1983). Cone resistance as measure of sand strength. *J Geotechn Eng Div ASCE*. 104(GT7), 995–1012.

Mo, P. Q., Yu, H. S. (2017). Undrained cavity expansion analysis in a unified state parameter model for clay and sand. *Géotechnique* 67(6), 503–515.

Mo, P. Q. and Yu, H. S. (2018). Drained cavity expansion analysis with a unified state parameter model for clay and sand. *Canad Geotechn J* 55(7), 1029–1040.

Mo, P. Q., Cai, G. J., Wang, K. J., Eslami, A., Yu, H. S. (2024). Cavity expansion-based interpretation of piezocone penetration test (CPTu) data in clays. *Géotechnique*. Ahead of Print. https://doi.org/10.1680/jgeot.23.00045

Mo, P. Q., Chen, H., Yu, H. S. (2022). Undrained cavity expansion in anisotropic soils with isotropic and frictional destructuration. *Acta Geotechn*, 17, 2325–2346.

Mo, P. Q., Gao, X. W., Yang, W. B. and Yu, H. S. (2020). A cavity expansion–based solution for interpretation of CPTu data in soils under partially drained conditions. *Int J Numer Anal Methods Geomech* 44(7), 1053–1076.

Mo, P. Q., Marshall, A. M. and Yu, H. S. (2017). Interpretation of cone penetration test data in layered soils using cavity expansion analysis. *J Geotechn Geoenviron Eng ASCE* 143(1), 04016084.

Mola-Abasi, H., Dikmen, U., Shooshpasha, I. (2015). Prediction of shear-wave velocity from CPT data at Eskisehir (Turkey), using a polynomial model. *Near Surface Geophys* 13(2), 155–167.

Nishida, Y. (1956). A brief note on compression index of soils. *J Soil Mechan Found Div* ASCE, 82, SM3, 1027.

Ohta, Y., Goto, N. (1978). Empirical shear wave velocity equations in terms of characteristic soil indexes. *Earthquake Eng Struct Dyn*, 6, 167–187.

Okamoto, T., Kokusho, T., Yoshida, Y., Kusuonoki, K. (1989). Comparison of surface versus subsurface wave source for P-S logging in sand layer. In: *Proceedings, 44th Annual Conference*, Japan Society of Civil Engineers, 3, pp. 996–997 (in Japanese).

Olsen, R. S., Mitchell, J. K. (1995). CPT stress normalization and prediction of soil classification. In: *Proceedings of International Symposium on Cone Penetration Testing*, pp. 257–262. CPT95, Link oping, Sweden, SGI Report 3:95, 2.

Paoletti, L., Hegazy, Y., Monaco, S., Piva, R. (2010). Prediction of shear wave velocity for offshore sands using CPT data e Adriatic Sea. In: Second International Symposium on Cone Penetration Testing, Huntington Beach, CA, pp. 1–8.

Phoon, K. K., Kulhawy, F. H. (1999). Characterization of geotechnical variability. *Canad Geotechn J*, 36, 612–624.

Phoon, K. K., Kulhawy, F. H., and Grigoriu, M. D. (1995). *Reliability based design of foundations for transmission line structures*. Report TR-105000, Electric Power Research Institute, Palo Alto, CA.

Phoon, K.K., Cao, Z., Ji, J., Leung, Y.F., Najjar, S., Shuku, T., Tang, C., Yin, Z.Y., Yoshida, I., Ching, J. (2022a). Geotechnical uncertainty, modeling, and decision making. *Soils Found*, 62, 101189.

Phoon, K.K., Prakoso, W.A., Wang, Y., and Ching, J. (2016). *Uncertainty representation of geotechnical design parameters*. Chapter 3, Reliability of Geotechnical Structures in ISO2394, CRC Press: Balkema, pp. 49–87.

Piratheepan, P. (2002). *Estimating shear-wave velocity from SPT and CPT Data*, Master of Science thesis, Clemson University.

Pitilakis, K., Raptakis, D., Lontzetidis, K. T., Vassilikou, T., Jongmans, D. (1999). Geotechnical and geophysical description of euro-seistests using field and laboratory tests, and moderate strong ground motions. *J Earthquake Eng*, 3, 381–409.

Ramsey, N. (2002). A calibrated model for the interpretation of cone penetration tests (CPT's) in North Sea quaternary soils. In: *Proceedings International Conference Offshore Site Investigation and Geotechnics*, SUT 2002, London, pp. 341–356.

Robertson, P. K. (1990). Soil classification using the cone penetration test. *Canadian Geotechnical Journal* 27(1), 151–158.

Robertson, P. K. (2009). Interpretation of cone penetration tests e a unified approach. *Canadian Geotechnical Journal* 46(11), 1337–1355.

Robertson, P. K. (2010). Soil behavior type from CPT. In: *2nd International Symposium on Cone Penetration Testing*, Huntington Beach, CA, pp. 575–583.

Robertson, P. K. (2012). Interpretation of in-situ tests-some insights. Mitchell Lecture-ISC 4, 1–22.

Robertson, P. K. (2016). Cone penetration test (CPT)-based soil behaviour type (SBT) classification system — an update. *Canadian Geotechnical Journal* 53, 1910–1927.

Robertson, P. K., Cabal, K. L. (2012). *Guide to Cone Penetration Testing for Geotechnical Engineering*, 6th edn. Gregg Drilling & Testing, Inc., Signal Hill, CA.

Robertson, P. K., Campanella, R. G. (1983). Interpretation of Cone Penetration Tests. Part I: Sand, *Canadian Geotechnical Journal*, 20(4), 718–733.

Robertson, P. K., Campanella, R. G., Gillespie, D., Grieg, J. (1986). Use of piezometer cone data. In: Clemence, S. (Ed.), *Proceedings of American Society of Civil Engineers, ASCE, In-Situ 86 Specialty* Conference, pp. 1263–1280. Blacksburg, June 23–25, Geotechnical Special Publication GSP No. 6.

Robertson, P.K., Cabal, K. (2010). Estimating soil unit weight from CPT. In: *2nd International Symposium on Cone Penetration Testing*, pp. 2–40.

Robertson, P.K., Campanella, R.G., Gillespie, D., Greig, J. (1986). Use of piezometer cone data. In: *Use of in Situ Tests in Geotechnical Engineering*, ASCE, pp. 1263–1280.

Sanglerat, G. (1972). *The Penetrometer and Soil Exploration*. Elsevier Pub, Amsterdam, 488pp.

Schmertmann, J. H. (1975). Measurement of in-situ shear strength. In: *Proceedings of ASCE Specialty Conference on In-Situ Measurements of Soil Properties*, Raleigh, NC, vol. 2, pp. 57–138.

Schmertmann, J. H. (1978). *Guidelines for Cone Test, Performance, and Design*. Report FHWATS-78209. Federal Highway Administration, Washington, 145pp.

Schmertmann, J. H. (1991). The mechanical aging of soils. *J Geotech Eng ASCE*, 117(9), 1288–1330.

Schnaid, F. (2009). In: *Situ Testing in Geomechanics*. Taylor & Francis, London, 329pp.

Seed, H. B., Idriss, I. M. (1981) Evaluation of liquefaction potential of sand deposits based on observations and performance in previous earthquakes, Preprint No. 81-544, *In situ testing to evaluate liquefaction susceptibility, Session No. 24, ASCE Annual Conference*, St. Louis, MO.

Senneset, K., Sandven, R., Janbu, N. (1989). The evaluation of soil parameters from piezocone tests. In: *Proceedings In Situ Testing of Soil Properties for Transportation Facilities, Research Council*, Trans. Research Board, Washington, DC.

Skempton, A. W. (1944). Notes on the Compressibility of Clays. *Quarterly Journal of Geological Society of London*, 100, 119–135.

Sowers, G. F. (1979). *Introductory Soil Mechanics and Foundations: Geotechnical Engineering*. 4th ed. Macmillan, New York.

Stroud, M. A. (1974). The Standard Penetration test in insensitive clays and soft rocks. *Proc. Eur Semin Penet Test*. Vol. 2:2, pp. 366–375. Stockholm.

Stroud, M. A. (1989). The Standard Penetration Test – its application and interpretation. In: *Proceedings of the ICE Conference on Penetration Testing in the UK*, Thomas Telford, London.

Sykora, D. E., Stokoe, K. H. (1983). Correlations of in-situ measurements in sands of shear wave velocity. *Soil Dyn Earthquake Eng*, 20, 125–136.

Tabarroki, M., Ching, J., Phoon, K.K., Chen, Y.Z. (2022). Mobilisationbased characteristic value of shear strength for ultimate limit states. *Georisk: Assessment and Management of Risk for Engineered Systems and Geohazards* , 16, 413–434. https://doi.org/10.1080/17499518.2020.1859121.
Tang, C., Phoon, K.K. (2021). *Model Uncertainties in Foundation Design*. CRC Press, Boca Raton, Florida, US.
Terzaghi, K., Peck, R. (1948). *Soil Mechanics in Engineering Practice*. Wiley, New York.
Terzaghi, K., Peck, R. (1967). *Soil Mechanics in Engineering Practice*, 2nd edn. Wiley, New York.
Trevor, F. A., Mayne, P. W. (2004). Undrained shear strength and OCR of marine clays from piezocone test results. In: Viana da Fonseca, M. (Eds.), *Proceeding ISC-2 on Geotechnical and Geophysical Site characterization*. Mill Press, Rotterdam, pp. 391–398.
Tun, M. (2003). *Investigation of the Characteristics of Eskisehir Soils Due to Shear Wave Velocity and Determination of Their Fundamental Vibration Periods*. Master Thesis. Anadolu University Institute of Science and Technology Department of Physics (in Turkish).
U.S. Navy. (1982). *Soil mechanics* – design manual 7.1, Department of the Navy, Naval Facilities Engineering Command, U.S. Government Printing Office, Washington, DC.
USACE. (1990). *Engineering and design – settlement analysis*, EM1110-1-1904, Department of the Army, US Army Corps of Engineers.
Uzielli, M., Mayne, P. W., Cassidy, M. J. (2013). Probabilistic assessment of design strengths for sands from in-situ testing data. In: *Modern Geotechnical Design Codes of Practice*, eds. Patrick Arnold, Gordon A. Fenton, Michael A. Hicks, Timo Schweckendiek, Brian Simpson, IOS Press, Amsterdam, Netherlands. pp. 214–227.
Vesic, A. S. (1977). *Design of Pile Foundations, National Cooperative Highway Research Program*, Transportation Research Board, National Research Council, National Academy of Sciences, Washington, Synthesis of Highway Practice No. 42.
Vessia, G., Zhou, Y. G., Leung, Y. F., Pula, W., Di Curzio, D., Tabarroki, M., and Ching, J. (2021). *Numerical evidences for worstcase scale of fluctuation. TC304 State-of-the-art Review of Inherent Variability and Uncertainty*, pp. 204–215.
Windle, D., Wroth, C. P. (1977). The use of a self-boring pressuremeter to determine the undrained properties of clays. *Ground Eng*, 10(6), 37–46.
Wolff, T. F. (1989). Pile capacity prediction using parameter functions. In: *Predicted and Observed Axial Behavior of Piles, Results of a Pile Prediction Symposium*, ASCE Geotechnical Special Publication 23, pp. 96–106.
Yoshida, Y., Ikemi, M., Kokusho, T. (1988). Empirical formulas of SPT blow-counts for gravelly soils. In: *Proceedings*, 1st International Symposium on Penetration Testing, Orlando, Florida, 1, pp. 381–387.
Yu, H. S. (2000). *Cavity Expansion Methods in Geomechanics*, Kluwer Academic, Dordrecht.
Zhang, P., Yin, Z.Y., Jin, Y.F. (2022). Bayesian neural network-based uncertainty modelling: application to soil compressibility and undrained shear strength prediction. *Canad Geotechn J* 59(4), 546–557.
Zhang, P., Yin, Z.Y., Jin, Y.F., Chan, T.H.T. (2020). A novel hybrid surrogate intelligent model for creep index prediction based on particle swarm optimization and random forest. *Eng Geol* 265, 105328.

Chapter 6

New laboratory database of hydraulic conductivity measurements on fine-grained soils

Shuyin Feng and Paul J. Vardanega

Geotechnical databases are vital for engineers wishing to make well-informed, empirical estimations of key soil mechanics parameters. Transformation models linking more difficult-to-measure parameters from readily attainable ones allow engineers to make low-cost a-priori estimations of the more difficult-to-measure parameters. Databases and associated transformation models allow for new test data to be benchmarked against prior knowledge. Other smaller databases (which may represent a specific data subset) can also be benchmarked against larger ones. The use of databases is especially valuable for assessing permeability (or hydraulic conductivity) – an important parameter that exhibits a very large range of variation across different soil types and within datasets obtained from individual laboratory testing campaigns. The issue of parameter uncertainty is often difficult to discount when hydraulic conductivity is a key parameter in a geotechnical model due to said variation. This chapter provides a comprehensive review of the variability and uncertainty of soil hydraulic conductivity with a focus on fine-grained materials, followed by a comparative study on the compilation of fine-grained soil hydraulic conductivity databases and transformation models for hydraulic conductivity making use of the recently established soil hydraulic conductivity database, FG/KSAT-1358. This chapter concludes with recommendations for future development and analysis of soil hydraulic conductivity databases.

6.1 PERMEABILITY OF SOILS

6.1.1 Definitions

Water, as one of the three phases of soil, governs both the physical and mechanical properties of the material. Understanding the distribution and percolation of water (or other permeants) in the soil is of crucial importance in geotechnical engineering design work since it is associated with many geotechnical phenomena (e.g., Taylor 1948, Lambe & Whitman 1969). The percolation of water through soil is often characterised by the hydraulic conductivity k (Length, L. Time, T^{-1}), described in Darcy's empirical law (e.g., Darcy 1856, Taylor 1948, Craig 2004):

$$\frac{Q}{A} = v = ki \tag{6.1}$$

where Q is the flow rate ($L^3.T^{-1}$), A is the area of cross section of the soil (L^2), v is the superficial velocity ($L.T^{-1}$), and i is the hydraulic gradient in the flow direction. The hydraulic conductivity k is governed by both the soil and the permeating fluid (i.e., water), and can be expressed as:

DOI: 10.1201/9781003441946-6

$$k = K\frac{\gamma}{\mu} \tag{6.2}$$

where K is the intrinsic permeability (L^2) which is independent of the permeating pore fluid, γ is the unit weight of the pore fluid (Mass, $M.L^{-2}.T^{-2}$), and μ is the dynamic viscosity ($M.L^{-1}.T^{-1}$). The reader is also directed to the doctoral thesis of Feng (2022) for preliminary work on some of the topics presented in this chapter.

6.1.2 Variation of hydraulic conductivity

Compared to the other commonly used geotechnical parameters, one of the most notable characteristics of k is its extensive range of variation (e.g., Carrier & Beckman 1984, Mbonimpa *et al.* 2002). The k value reported in soil mechanics textbooks generally ranges from 1×10^{-11} m/s to 1×10^{0} m/s, i.e., over 11 orders of magnitude (cf., Terzaghi & Peck 1948, Lambe & Whitman 1969, Craig 2004), and ranges up to 14 orders of magnitude in some compiled hydraulic conductivity datasets (databases) (cf., Ren & Santamarina 2018, Feng 2022, Feng *et al.* 2023).

Table 6.1 summarises the parameter variation range of four global multivariate databases focusing on fine-grained/clay soils in 304dB (the TC304 database of databases that can be found at: http://140.112.12.21/issmge/tc304.htm). The collected soil parameters in the databases encompass basic physical properties (e.g., void ratio (e), specific gravity (G_S), water content (w)), soil classification properties (e.g., liquid limit w_L, plasticity index I_P, liquidity index I_L), mechanical properties (e.g., compression index C_c, unload–reload index C_{ur}, over-consolidation ratio OCR, sensitivity S_t, undrained shear strength s_u), and hydraulic conductivity k. The variation range of k for fine-grained soil (FG/KSAT-1358) spans over 7 orders of magnitude, while the corresponding physical and intrinsic soil parameters within the same database show variations within 2 orders of magnitude. The variation of the soil properties listed in Table 6.1 is mostly within 3 orders of magnitude across four global fine-grained soil/clay databases available from the TC304 website mentioned above (with the exception of k).

6.1.3 Factors influencing hydraulic conductivity

As both γ and μ are temperature-dependent parameters, a specific soil with a specific permeating fluid may exhibit different k levels under different **testing temperatures**, i.e., k measured at 0°C is 1.8 times the value measured at 20°C, and k at 40°C is 0.58 times the value measured at 20°C (data from BSI 1990). The hydraulic conductivity measured at a specific test temperature can be converted to a corresponding permeability at different temperatures based on Equation (6.2), using a correction factor (e.g., BSI 2019). A test temperature of 20°C is often deemed as the standard or reference temperature condition for laboratory measurement (e.g., ASTM 2022, BSI 1990).

The permeant percolation mechanism in most geomaterials exhibits an anisotropic nature. The **flow direction** during testing is another factor that influences the measured level of k. The horizontally measured hydraulic conductivity k_h is generally greater than the vertically measured hydraulic conductivity k_v as reported in various studies (e.g., Al-Tabbaa & Wood 1987, Dewhurst *et al.* 1996, Clennell *et al.* 1999). The reported anisotropy ratio ($r_k = k_h/k_v$) typically falls below 4, and the maximum r_k observed in most engineering soil is around 2.5 (Liu, 2015). The anisotropic behaviour in soil k is often less

Table 6.1 Soil properties variation in some global multivariate fine-grained soil database (see: http://140.112.12.21/issmge/tc304.htm[a] for the downloadable database files, available at the time of writing)

Database	*Reference*	*Soil type*	*Data points*	*No. studies/sites*		*Range of parameters*
FG/KSAT-1358	Feng and Vardanega (2019a, b)	Fine-grained soil	1358	33	k	1.44×10^{-13} to 7.5×10^{-6} m/s
					e	0.19 to 8.57
					w_L	22 to 675%
					I_P	5 to 625.9%
					G_S	2.09 to 2.9
CLAY-C_C/6/6203	Ching *et al.* (2022)	Clay	6203	429	e	0.13 to 14.66
					w_L	19 to 612.69%
					I_P	5 to 493%
					w	5.8 to 559.5%
					C_c	0.01 to 14.50
					C_{ur}	0.002 to 2.16
CLAY/10/7490	Ching and Phoon (2014)	Clay	7490	251	w_L	18.1 to 515 %
					I_P	1.9 to 363%
					I_L	−0.75 to 6.45
					q_{t1}	0.48 to 95.98
					q_{tu}	0.61 to 108.20
					B_q	0.01 to 1.17
					S_t	1 to 1467
					OCR	1.0 to 60.23
CLAY/5/345	Ching and Phoon (2012)	Clay	345	37	I_L	0.14 to 4.85
					s_u^{re}	0.06 to 33.69 kPa
					$s_{u(test)}$	1.97 to 162.95 kPa
					$s_{u(mob)}$	1.19 to 162.52 kPa
					σ_v'	3.70 to 364.09 kPa
					σ_p'	7.94 to 712.71 kPa

Definitions: e = void ratio; w_L = liquid limit; I_P = plasticity index; G_S = Specific gravity; w = water content; C_c = compression index; C_{ur} = unload–reload index; I_L = liquidity index; $q_{t1} = (q_t - s_v)/s'_v$ = normalised cone tip resistance; $q_{tu} = (q_t - u_2)/s'_v$ = effective cone tip resistance; B_q = pore pressure ratio = $(u_2 - u_0)/(q_t - s_v)$; S_t = sensitivity; OCR = over-consolidation ratio; s_u = undrained shear strength; s_u^{re} = remoulded s_u; σ_v' = vertical total stress; and σ_p' = pre-consolidation stress.

[a]This website was last accessed on 10 July 2023.

pronounced at a higher void ratios, e.g., for static compacted sand, $r_k = 1.8$ for $e = 0.4$ and $r_k = 1$ for $e = 0.8$ (Chapuis *et al.* 1989).

The permeability of soil, when tested under different **saturation levels**, is closely dependent on the air–water interface (Baver 1948). The k value measured under partially saturated conditions is generally lower than under saturated conditions (e.g., Corey 1957, Lambe & Whitman 1969, Bear 1972). The unsaturated k may be estimated from the saturated k value in conjunction with the soil water retention curve (correlation between soil suction and its degree of saturation) and hydraulic conductivity function (relationship between k and soil suction) (e.g., Mualem 1976, Van Genuchten 1980, Zhai *et al.* 2021).

In addition to the aforementioned factors, the **chemical constituent**(s) of the permeant can also have a substantial influence on the k measured in soil with high clay content (see also the discussion in Feng 2022). The presence of certain chemical constituents in the permeant may cause the soil particles to either flocculate or disperse, thus leading to changes in hydraulic conductivity (e.g., Lambe & Whitman 1969, Mesri & Olson 1971). The ratio of hydraulic conductivity for the same soil, measured using permeants with different chemical constituents at a constant void ratio, can vary significantly by several orders of magnitude, e.g., measured k of smectite at e = 2 is around 10^{-11} cm/s in water with NaCl, 10^{-9} cm/s in water with $CaCl_2$, 10^{-7}cm/s in ethyl alcohol, and 10^{-5} cm/s (Mesri & Olson 1971).

6.1.4 Uncertainty in permeability test results

Soil permeability can be measured with various laboratory and field tests (e.g., falling head tests, constant head tests, oedometer test, and well pumping tests; see the detailed discussion in Feng 2022). The application of different permeability testing methods has been discussed in various publications (e.g., Tavenas *et al.* 1983, Chapuis 2012, Nagy *et al.* 2013, Daniel *et al.* 1985). While each permeability testing approach has its merits (and limitations), they are all established based on specific assumptions, e.g., Darcian flow, homogeneous soil, and soil saturation. The results obtained from the tests are dependent on the selected testing methodology (Chapuis 2012). The inherent "inaccuracy" of permeability measurement approaches leads to a degree of variation close to an order of magnitude from the same test method (e.g., Pane *et al.* 1983, Nagy *et al.* 2013), and up to three orders of magnitude higher when comparing results from different test methods (cf. Moulton & Seals 1979, Chapuis 2012, Nagy *et al.* 2013). The range of variation of measured permeability data coupled with laborious and time-consuming testing procedures (see Feng 2022 for further discussion of soil testing methodologies) make the determination of soil permeability challenging – this challenge can be partially mitigated by the availability of extensive geotechnical databases.

6.2 TRANSFORMATION MODELS

6.2.1 Predictors of hydraulic conductivity

Transformation models can be used to estimate soil parameters using more easily accessible ones. Similar to all the other geotechnical parameters, the k of soil can also be assessed with more basic soil parameters using empirical correlations established from past database studies. The main soil characteristics that influence the hydraulic conductivity k are summarised as follows:

(a) **Void ratio/structure-related parameters:** Void ratio e, porosity n, water content w, void ratio function $e^3(1+e)$ (e.g., Kozeny 1927, Carman 1937, Carman 1939, Taylor 1948, Terzaghi & Peck 1948, Baver 1948, Lambe & Whitman 1969, Carrier 2003, Chapuis 2004, Liu 2015, Ren & Santamarina 2018, Nagaraj *et al.* 1993, Nagaraj *et al.* 1994, Mbonimpa *et al.* 2002);

(b) **Particle size distribution-related parameters:** Effective particle size D_{10} D_{50} (e.g., Hazen 1895, Taylor 1948, Shahabi *et al.* 1984, Shepherd 1989, Alyamani & Sen 1993, Chapuis 2004), specific surface by volume S_A (Feng 2022, Feng *et al.* 2023), specific surface by mass S_S (Ren *et al.* 2016, Ren & Santamarina 2018, Feng and

Vardanega 2019a), coefficient of uniformity C_U (Shahabi *et al.* 1984), mean constriction size D_c^m (Jaafar & Likos 2014), and normalised grading entropy coordinates A, B (Feng *et al.* 2019, Feng 2022);

(c) **Atterberg Limits:** Liquid limit (w_L) (Feng & Vardanega 2019a, Feng & Vardanega 2019b, Nagaraj *et al.* 1993, Nagaraj *et al.* 1994, Chapuis 2012, Ren & Santamarina 2018), plastic limit (w_p) (Carrier & Beckman 1984, Mbonimpa *et al.* 2002), and shrinkage limit (w_s) (Sridharan & Nagaraj 2005);

(d) **Soil fabric, shape, and composition:** Soil composition (Jones 1955, Shahabi *et al.* 1984, Indrawan *et al.* 2006, Shafiee 2008, Gao *et al.* 2019), soil fabric (Lambe & Whitman 1969, Qiu & Wang 2015), and soil shape (Cabalar & Akbulut 2016, Taiba *et al.* 2019, Katagiri *et al.* 2020, Nguyen & Indraratna 2020).

Burnham and Anderson (2004) suggested that the selection of predictors should always follow the principle of "simplicity and parsimony." The complexity of the prediction model is controlled based on whether the selection of parameters can be assessed from a relatively small number of standard tests (Whittle 1993). There is always a trade-off between model accuracy and the number of constituent parameters – the more the parameters in a model, the more difficult the calibration is.

6.2.2 Model structure

A prediction model can be established using one or multiple predictors, with a generalised form of prediction model allowing further determination of the empirical factors in the model. The generalised form may be structured directly between the predictors and k using simple linear correlation, famous examples include Hazen's formula (Hazen 1985) and the Kozeny–Carman equation (Kozeny 1927, Carman 1937, Carman 1939). More recently developed prediction models often involve the data transformation of k, such as the works of Carrier and Beckman (1984), Nagaraj *et al.* (1994), Mbonimpa *et al.* (2002), and Chapuis (2012).

Feng *et al.* (2023) reviewed some commonly used prediction models for assessing the permeability of coarse-grained soil, with a compiled permeability database (CG/KSAT/7/1278) consisting of 1278 test data sourced from over 50 studies. The model adequacy analysis of the prediction models revealed that all linear-correlating models exhibited a "funnel" pattern in the residuals versus predicted plots, indicating anomalies in prediction (see Figure 6.1, full results are given in the recent paper of Feng *et al.* 2023) when calibrated using CG/KSAT/7/1278. This observation indicates the need for data transformation in the predicted value k (Montgomery *et al.* 2007, p. 303). The need for such data transformation may be attributed to the difference in the range of variation between k and corresponding predictors, as detailed in Table 6.1.

6.2.3 Evaluation of regression strength

Kulhaway and Mayne (1990) suggested that for any correlation for estimating soil properties it is essential to supplement the coefficient of determination (R^2) with other statistical measures, particularly the number of data points used in the regression (n), to properly assess the strength of fitted models (see also e.g. Paradine and Rivett 1953). Phoon and Kulhawy (1999a, 1999b) emphasised the significance of reporting the standard error (SE) to provide a measure of "transformation uncertainty" in correlations. The *p-value* of the

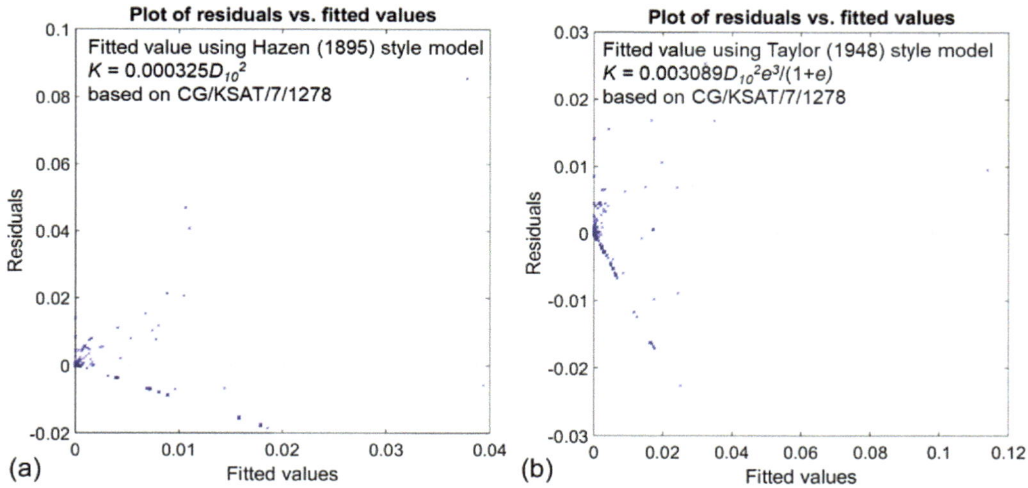

Figure 6.1 Examples of "funnel" pattern residuals versus predicted plot: (a) residual versus predicted plot using Hazen (1895) style model using CG/KSAT/7/1278; (b) residual versus predicted plot using Taylor (1948) style model using CG/KSAT/7/1278 (Source: adapted from Feng *et al.* 2023, used under the terms of the cc-by 4.0 licence).

correlation, which explains the probability of the absence of correlation (i.e., rejecting the null hypothesis, see Montgomery *et al.* 2007), should also be considered. For details on the statistical measures employed in this work, see Feng (2022).

6.2.4 Database subdivision analyses

Further subdivision analysis is often required in model development, especially when dealing with data that varies over an extensive range (e.g., k) to evaluate how generally applicable a developed transformation model is. Correlation may exhibit distinct patterns among data over different value ranges or under various test conditions, as illustrated in Figure 6.2. The subdivision analysis helps to examine the potential influencing

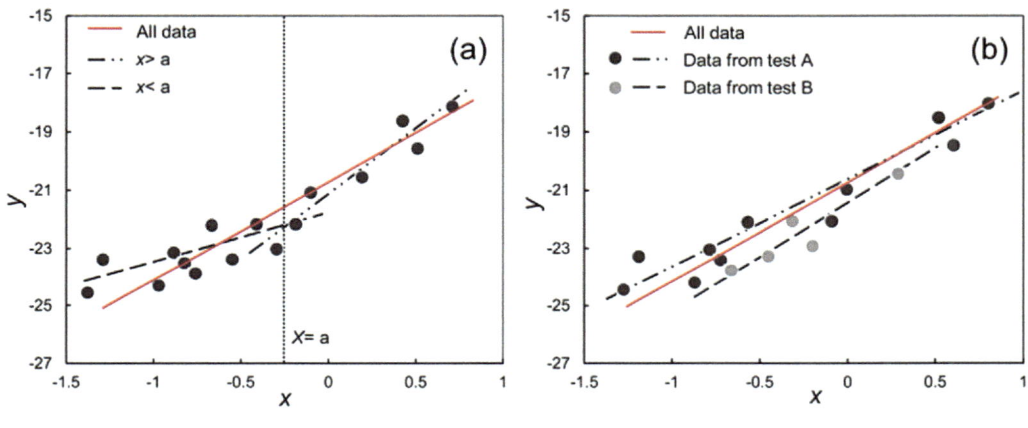

Figure 6.2 Divergency in subdivision analysis: (a) subdivision over different value ranges and (b) subdivision over different test conditions (Source: adapted from Feng 2022).

factors (e.g., permeability testing methods) and uncover underlying trends or correlations that may exist within specific subgroups which may be rather different from the transformation model calibrated with the whole dataset (see the detailed discussion in Feng 2022).

6.3 PERMEABILITY DATABASES FOR FINE-GRAINED SOIL

6.3.1 Database summary

As discussed previously, databases play a crucial role in geotechnical engineering (e.g., Phoon *et al.* 2020). Geodatabases with sufficient and reliable data are often used to improve understanding of geotechnical phenomena and further calibrate proposed models (Phoon & Tang 2019). Geotechnical databases (or datasets) can be populated through direct laboratory/on-site testing (e.g., FI-CLAY/7/216 from D'Ignazio *et al.* 2016) or sourced from geotechnical literature (e.g., RFG/TXCU-278 from Beesley & Vardanega 2020). Some databases are compiled using a combination of literature and industry sources (e.g., the DINGO database on piled foundations, Vardanega *et al.* 2024).

Table 6.2 presents a summary of eight fine-grained soil permeability database/sets used for constructing permeability prediction models, including the recently compiled database FG/KSAT-1358 (Feng & Vardanega 2019a, 2019b). These databases include datasets from individual laboratory campaigns, regional compiled databases, and global compiled databases. Each database has been supplemented with information on the size of database, soil type, property range, number of studies, and details regarding the permeability influencing factors, including the test type, test temperature, flow direction, saturation level, and chemical constituent. The property ranges are given for the soil properties that have been reported or can be digitised from the listed sources.

6.3.2 Database evaluation

All database/set sources reviewed in this chapter have offered detailed information on the soil type, reported/digitisable size, k range, and the number of studies. As discussed earlier in this chapter, different test types can introduce significant uncertainty and lead to variations up to orders of magnitude in the measured k value. However, there is a noticeable information gap concerning the permeability test type, particularly in the compiled databases, i.e., the test type is either missing or partially reported in 3 out of 5 compiled database sources. Similarly, there appear to be gaps in the reported data pertaining to other factors influencing k, including test temperatures, saturation levels, and permeant chemical constituents. Half of the reviewed databases (4 out of 8) provide information on one factor influencing k, and only two databases (Sridharan & Nagaraj 2005, Feng & Vardanega 2019a, 2019b) report relevant information on all four influencing factors.

It was noticed that more recently compiled databases tend to encompass data from earlier databases. For example, Mbonimpa *et al.* (2002) incorporated data from Nagaraj *et al.* (1994), Sivapullaiah *et al.* (2000), while Ren & Santamarina (2018) included data from Sridharan and Nagaraj (2005) and Sivapullaiah *et al.* (2000). The data sources collected to develop FG/KSAT-1258 detailed in Feng and Vardanega (2019a, 2019b) also overlap with the compiled databases in Mbonimpa *et al.* (2002) and Ren and Santamarina (2018).

Table 6.2 Summary of permeability database/set for fine-grained soil

Sources	*Size*	*Soil type*	*Property range*	*No. of studies*	*Test type*	*Test temp.*	*Flow direction*	*Saturation level*	*Chemical constituent*
Dataset from individual laboratory test campaign									
Nagaraj *et al.* (1994)	54	Four local residual soils: red soil, brown soil, black cotton soil, marine soil	k (9.19×10^{-12} to 1.82×10^{-9} m/s); w_L (50% to 106%); G_s (2.68 to 2.9)	1	Falling head test	n/a	Vertical, one dimensional	n/a	n/a
Sivapullaiah *et al.* (2000)	291 (no. digitised, include sand and sand mixture)	Bentonite–sand mixture (includes mixture with over 50% sand)	k (1.21×10^{-12} to 3.76×10^{-8} m/s); w_L (35% to 344%); G_s (2.60 to 2.75)	1	Consolidation test	n/a	One dimensional	n/a	n/a
Sridharan and Nagaraj (2005)	63 (no. digitised)	Red earth, silty soil, kaolinite, Cochin clay, brown soil, illitic soil, B C soil	k (1.20×10^{-11} to 6.89×10^{-8} m/s); e (0.53 to 1.84); w_L (37.0% to 73.5%); w_P (18.0% to 51.9%); w_S (11.9% to 46.4%); G_s (2.58 to 2.70)	1	Falling head test	20°C	One dimensional	Saturated	Distilled water
Regional compiled database									
Chapuis (2012)	76 (no. digitised)	Quebec Champlain Sea clay	k (2×10^{-11} to 5×10^{-9} m/s)	4	More than two types of tests involved (constant head, falling head)	n/a	n/a	Saturated	n/a
Multi-source (global) compiled database									
Carrier and Beckman (1984)	66	22 phosphatic; 13 dredged materials; 26 remoulded natural clays	k (10^{-10} to 10^{-5} m/s)	8	Constant head; consolidation test (stress-controlled and constant rate of deformation)	n/a	One dimensional	n/a	Permeating fluid with same chemical composition as the porewater

(Continued)

Table 6.2 (Continued) Summary of permeability database/set for fine-grained soil

Sources	*Size*	*Soil type*	*Property range*	*No. of studies*	*Test type*	*Test temp.*	*Flow direction*	*Saturation level*	*Chemical constituent*
Mbonimpa *et al.* (2002)	342	Bentonite silts mixture, Kaolin, marine soil, black cotton soil, clay, Louiseville soil, St-Esprit soil, Bächebol soil, Matagami soil, New Liskeard soil, Don Valley soil, and Leda soil	k (1.3×10^{-12} to 6.79×10^{-9} m/s); w_L (33.4% to 436.3%); e (0.51 to 5.96); G_s (2.61 to 2.78)	8	n/a; author's test conducted under constant head or falling head condition.	20°C	n/a	Saturated	n/a
Ren and Santamarina (2018)	1440 (includes sandy soil data)	Natural and remoulded sediments, from coarse sands to fine-grained clays	k (2.55×10^{-14} to 9.33×10^{-4} m/s, range for silty and clayey soil, data digitised from plot); e (0.24 to 9.1, data digitised from plot)	23	n/a	n/a	Most likely to be vertical	n/a	n/a
Feng and Vardanega (2019a)	1358	Natural or laboratory fine-grained soil with less than 50% coarse particles ($d >$ 75 mm) and a measured w_L	k (1.44×10^{-13} to 7.5×10^{-6} m/s); e (0.19 to 8.57); w_L (22% to 675%); I_P (5% to 625.9%); G_S (2.09 to 2.9)	33	Constant head test, falling head test, consolidation test, flow pump test	20 ± 1°C to 26 ± 0.1°C	Vertical	Saturated	Water

There has been an obvious increase in the size of databases for soil *k* due to the aforementioned research efforts. Among the compiled databases, FG/KSAT-1358 by Feng and Vardanega (2019a, 2019b) contains the largest number of *k* test data on fine-grained soil (over 1300), along with comprehensive details on the key factors influencing *k*.

6.3.3 Model calibration

Table 6.3 provides a summary of the calibrated transformation models to assess the permeability of fine-grained soil. The majority of the established models are based on fundamental soil parameters such as e, w, w_L, void ratio at liquid limit e_L, w_L, I_P, and G_S. Sridharan and Nagaraj (2005) used the shrinkage limit (w_s) as a predictor, which is widely used especially in pavement engineering in dry regions. Most of the calibrated prediction models, such as the transformation model developed by Nagaraj *et al.* (1994), Sivapullaiah *et al.* (2000), Chapuis (2012), Ren and Santamarina 2018, and Feng and Vardanega 2019a, 2019b, involve some form of data transformation for *k*.

Statistical measures have been reported in prediction models calibrated using individual laboratory test datasets and regional compiled databases (Nagaraj *et al.* 1994, Sivapullaiah *et al.* 2000, Sridharan & Nagaraj 2005, Chapuis 2012). However, the sample size (n) used in model calibration is missing in some cases (e.g., Sivapullaiah *et al.* 2000, Sridharan & Nagaraj 2005, Chapuis 2012). Statistical measures are largely missing (i.e., unreported) from the model calibration when using compiled databases with relatively larger sample sizes, such as Carrier and Beckman (1984) with n = 66, Mbonimpa *et al.* (2002) with n = 342, and Ren and Santamarina (2018) with n = 1440 (including data on sandy soils).

Transformation models calibrated with larger sample sizes, encompassing a range of soil types and sourced from multiple studies, tend to exhibit a wider range of prediction accuracy up to an order of magnitude, which is to be expected as the model has to account for a wider range of materials and testing conditions. Therefore, this transformation uncertainty may be attributed to the inherent variability among soil types and sources within the database (Phoon & Kulhawy 1999a, 1999b). Subdivision analysis has been conducted in Ren and Santamarina (2018) and Feng and Vardanega (2019a) to further analyse and verify the accuracy and applicability of the established transformation models.

6.3.4 Benchmarking databases

In order to validate and compare the predictive performance of transformation models for *k* calibrated using databases of varying source type, size, and data range, a benchmarking analysis is conducted using the largest fine-grained soil database – FG/KSAT-1358. Figure 6.3 summarises the analysis results for the hydraulic conductivity prediction models of fine-grained soil calibrated using database/sets with a smaller variation range of *k* compared to FG/KSAT-1358. The measured *k* values from FG/KSAT-1358 are plotted against the predicted *k* values based on the examined model with the corresponding calibration dataset range following the method described in Piñeiro *et al.* (2008). It was observed that models established using the global compiled database generally demonstrate better predictive performance for FG/KSAT-1358, with a higher percentage of the data points falling within the one-order-of-magnitude prediction range. Transformation models developed based on regional compiled databases and individual laboratory test campaigns tend to exhibit an overall biased prediction trend for the global compiled database FG/KSAT-1358.

Table 6.3 Summary of calibrated prediction models (data sources as shown in Table 6.2)

Sources	*Database/set*	*Calibrated transformation models*	*Reported statistical measures*	*Prediction accuracy*	*Subdivision analysis*
Dataset from individual laboratory test campaign					
Nagaraj *et al.* (1994)	Regional soil database from laboratory testing dataset on 54 samples of four local residual soils	$\frac{e}{e_L} = 2.162 + 0.195 \log k$	$r = 0.98, n = 54$	n/a	n/a
Sivapullaiah *et al.* (2000)	Laboratory testing dataset on bentonite–sand mixtures, 291 samples (no. of data digitised from plot, includes mixture with over 50% sand)	$\log_{10} k = \frac{e - 0.0535 w_L - 5.286}{0.0063 w_L + 0.2516}$; k in m/s for $w_L > 50\%$	$r = 0.94, SE = 1.486$	n/a	n/a
Sridharan and Nagaraj (2005)	Laboratory testing dataset on 10 soil types, 70 samples (no. of data digitised from plot)	$k_p = Ce^x / (1+e)$; $C = 2.5 \times 10^{-4} (I_S)^{-3.69}$ $x = 4$	$SEE = 24.34 \times 10^{-10} m/s$	Usually within 0.4 to 2.5 times range	n/a
Regional compiled database					
Chapuis (2012)	Regional clay database from 4 studies on 76 Quebec Champlain Sea clays (no. of data digitised from plot); test type not available	$k_{sat}(m/s) = 6.68 \times 10^{-6} \left[\frac{e^3}{(1+e)(w_L^{-1} + z)^2} \right]^{1.339}$; $z = 0.00836$	$R^2 = 0.81; n = 76$ (number digitised)	Within 1/3 to 3 times range	n/a
Global compiled database					
Carrier and Beckman (1984)	Compiled database from 8 studies on 66 clay samples; cover data from 2 test types (constant head, consolidation test)	$k(m/s) = \mu \frac{(e-\delta)^{\nu}}{1+e}$; $\mu(m/s) = \left[\frac{0.389}{PI} \right]^{4.29}$ $\delta = 0.027 \left[w_p - 0.242 I_p \right] \quad \nu = 4.29$	n/a	Within an order of magnitude	n/a

(*Continued*)

Table 6.3 (Continued) Summary of calibrated prediction models (data sources as shown in Table 6.2)

Sources	*Database/set*	*Calibrated transformation models*	*Reported statistical measures*	*Prediction accuracy*	*Subdivision analysis*
Mbonimpa *et al.* (2002)	Compiled database from 6 studies on 342 plastic/cohesive soils sample	$k_{sat}(cm/s) = C_P \frac{\gamma_w}{\mu_w}\frac{e^{3+x}}{1+e}\frac{1}{\rho_s^2 w_L^{2\chi}}$; $C_P = 5.6g^2/m^4$ $x = 7.7w_L^{-0.15} - 3$ $\chi = 1.5$ $\gamma_w \approx 9.8kN/m^3$ $\mu_w \approx 10^{-3} Pa.s$	n/a	Usually within half of one order of magnitude	n/a
Ren and Santamarina (2018)	Compiled database from 23 studies on 1440 natural and remoulded sediment sample, from coarse sand to fine-grained soil	$k(cm/s) = 10^{-5}\left(\frac{S_s}{m^2/g}\right)e^{\beta}$; $S_S = 1.8w_L - 34$ β may be derived from the power relation between k and e for the specific soil types: $\beta = 3 \pm 1$ for coarse-grained soils $\beta = 5 \pm 1$ for fine-grained soils	n/a	Mostly within 0.2 to 5 times, significant scatter below $k < 10^{-5}$ cm/s (silty/clayey soil region).	Prediction model with varying β for each soil type
Feng and Vardanega (2019a)	Compiled database from 33 studies on 1358 fine-grained soil sample	$k_{sat}(m/s) = 1.91\times10^{-9}\left(w/w_L\right)^{4.083}$	$R^2 = 0.62, n = 1352$, SE = 1.58, $p < 0.0001$	Mostly within an order of magnitude	Sub database analysis on Atterberg limits, test types, sample states
Feng *et al.* (2022)	Compiled database from Feng and Vardanega (2019a) with identified outliers removed	$k_{sat}(m/s) = 1.86\times10^{-9}\left(w/w_L\right)^{4.226}$	$R^2 = 0.65, n = 1272$, SE = 1.303, $p < 0.0001$	Mostly within an order of magnitude	

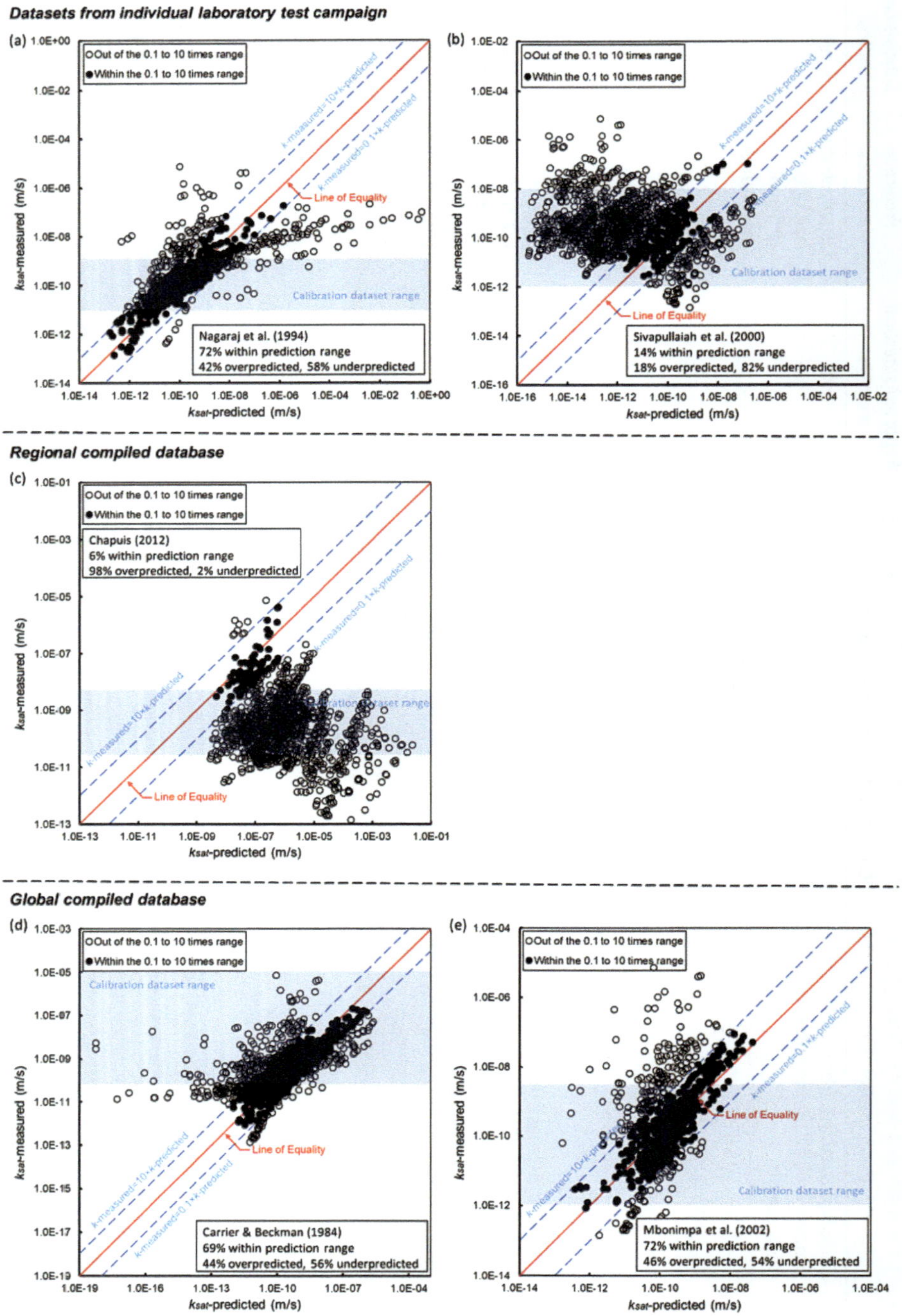

Figure 6.3 Benchmarking analysis result for models with smaller calibration dataset range using FG/KSAT-1358 as the measured values on the plots: (a) predicted values using transformation model from Nagaraj *et al.* (1994); (b) predicted values using transformation model from Sivapullaiah *et al.* (2000); (c) predicted values using transformation model from Chapuis (2012); (d) predicted values using transformation model from Carrier and Beckman (1984); (e) predicted values using transformation model from Mbonimpa *et al.* (2002).

This is to be expected as smaller datasets are unlikely to yield models representing the behaviour of a wide range of soil types/testing approaches, and this observation serves as a reminder that the use of empirical models based only on small datasets should be used with caution (even if they have a high coefficient of determination, for instance) unless calibration of the model form is undertaken with a larger and more extensive dataset.

Among the reviewed fine-grained soil database studies, the database compiled by Ren and Santamarina (2018) stands out due to its extensive variation in *k* value and a database size comparable to FG/KSAT-1358. Figure 6.4 presents the benchmarking analysis results using the model presented in Ren and Santamarina (2018) with $\beta = 4$, 5, and 6 (selection of β based on the range reported by Ren and Santamarina 2018: $\beta = 5 \pm 1$ for fine-grained soils), in comparison with the model developed by Feng and Vardanega using FG/KSAT-1358 (Feng and Vardanega 2019a). The prediction models by Ren and Santamarina (2018) generally underpredict *k* for FG/KSAT-1358, with approximately 50% of the data points within the one order of magnitude prediction range, while the model by Feng and Vardanega (2019a) yields slightly overpredicted results, with 89% of the prediction within the same prediction range. It should also be noted that while Ren and Santamarina (2018) report a β range of 5 ± 1 for fine-grained soil, individual soil types in FG/KSAT-1358 give a variation of β mostly ranging from 2 to 7 (full results in online supplement, Feng & Vardanega 2019a). This discrepancy in the β value range used in the analysis may partially account for the reduced prediction accuracy of *k*.

6.4 SUMMARY AND CONCLUSIONS

This chapter has explored the variability and uncertainty associated with fine-grained soil *k*. This was followed by a comparative analysis of the compilation of fine-grained soil *k* databases and the development of transformation models, using the recently established fine-grained laboratory *k* database, FG/KSAT-1358. Information gaps regarding the factors influencing *k* and transformation model prediction strength have been identified for the reviewed databases and datasets.

6.4.1 Database compilation

It is important to stress the variability and uncertainty in *k* measurements during the database compilation process. The measured *k* value of the soil is highly dependent on factors such as the test type, test condition, and sample state. Therefore, information on the relevant influencing factors, such as *k* testing type, test temperature, sample states, and permeating fluid, should be included in the assembled test database, together with the other corresponding soil properties where possible – this assists the development of new transformation models and also future database expansion and harmonisation efforts.

6.4.2 Transformation models

There remain information gaps in the published literature regarding the prediction strength (i.e., sample size, means of statistical measures) and accuracy (i.e., prediction bandwidth) of transformation models used to predict *k*. Considering the inherent uncertainty and variability of soil *k* assessment, adequate model evaluations that include

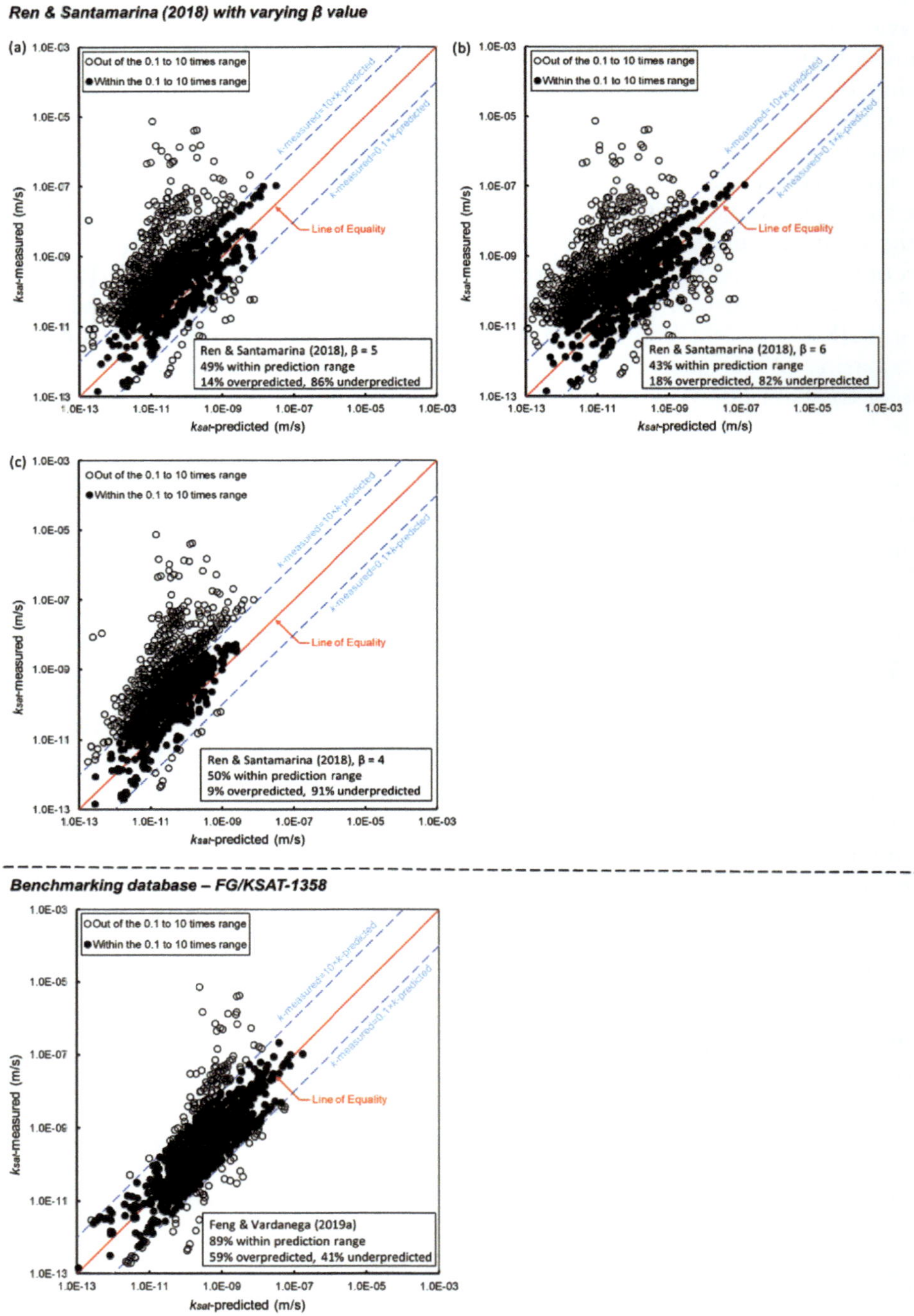

Figure 6.4 Benchmarking analysis results (using data from FG/KSAT-1358 as the measured data on the plots) for Ren and Santamarina (2018) model with varying β: (a) predicted values using transformation model from Ren and Santamarina (2018) with $\beta = 5$; (b) predicted values using transformation model from Ren and Santamarina (2018) with $\beta = 6$; (c) predicted values using transformation model from Ren and Santamarina (2018) with $\beta = 4$; (d) Predicted values using transformation model from Feng and Vardanega (2019a), data from FG/KSAT-1358, plot based on Feng *et al.* (2022).

statistical measures, prediction accuracy, and model adequacy should be provided to measure the validity and reliability of the calibrated model. For compiled databases, further subdivision analysis should also be performed to examine the variations in model prediction performance across different soil types, testing conditions, and sample states to further evaluate the uncertainty in the calibrated transformation model.

With the digitisation of geotechnical resources, significant efforts have been made to compile and publish studies on databases of k measurements. However, even with the availability of such studies, the predictions obtained using existing transformation models still exhibit a considerable range of variation, often spanning up to an order of magnitude, particularly when considering globally compiled databases. High-quality soil sampling and testing programs are still essential to reduce the uncertainty and variation associated with permeability prediction. Complete database compilation and classification are also required to enhance the accuracy of future assessments of soil hydraulic conductivity.

REFRENCES

Al-Tabbaa, A. & Wood, D.M. (1987). Some measurements of the permeability of kaolin. *Géotechnique*, **37**(4): 499–514. https://doi.org/10.1680/geot.1987.37.4.499

Alyamani, M.S. & Sen, Z. (1993). Determination of Hydraulic Conductivity from Complete Grain-Size Distribution Curves. *Groundwater*, **31**(4): 551–555. https://doi.org/10.1111/j.1745-6584.1993.tb00587.x

ASTM International (2022). *Standard test methods for measurement of hydraulic conductivity of coarse-grained soils.* D2434-22, West Conshohocken, PA.

Baver, L. (1948). *Soil Physics.* John Wiley Sons, New York, NY.

Bear, J. (1972). *Dynamics of Fluids in Porous Media.* American Elsevier, New York, NY.

Beesley, M.E.W. & Vardanega, P.J. (2020). Parameter variability of undrained shear strength and strain using a database of reconstituted soil tests. *Canadian Geotechnical Journal*, **57**(8): 1247–1255. https://doi.org/10.1139/cgj-2019-0424

British Standards Institution (BSI) (1990). *Methods of test for soils for civil engineering purposes – Part 5: Compressibility, permeability and durability tests.* BS 1377-5:1990, London, UK.

BSI (2019). *Geotechnical investigation and testing. Laboratory testing of soil. Permeability tests.* BS EN ISO 17892-11:2019, London, UK.

Burnham, K.P. & Anderson, D.R. (2004). Multimodel inference: Understanding AIC and BIC in model selection. *Sociological Methods and Research* **33**(2): 261–304. https://doi.org/10.1177/0049124104268644

Cabalar, A.F. & Akbulut, N. (2016). Evaluation of actual and estimated hydraulic conductivity of sands with different gradation and shape. *SpringerPlus*, **5**(1): 820. https://doi.org/10.1186/s40064-016-2472-2

Carman, P.C. (1937). Fluid flow through granular beds. *Transaction Institution of Chemical Engineers*, **15**: 150–166.

Carman, P.C. (1939). Permeability of saturated sands, soils and clays. *The Journal of Agricultural Science*, **29**(2): 262–273.

Carrier, W.D. & Beckman, J.F. (1984). Correlations between index tests and the properties of remoulded clays. *Géotechnique*, **34**(2): 211–228. https://doi.org/10.1680/geot.1984.34.2.211

Carrier, W.D. (2003). Goodbye, Hazen; Hello, Kozeny-Carman. *Journal of Geotechnical and Geoenvironmental Engineering*, **129**(11): 1054–1056. https://doi.org/10.1061/(ASCE)1090-0241(2003)129:11(1054)

Chapuis, R. P., Gill, D. E. & Baass, K. (1989). Laboratory permeability tests on sand: Influence of the compaction method on anisotropy. *Canadian Geotechnical Journal*, **28**(1): 172–173. https://doi.org/10.1139/t89-074

Chapuis, R.P. (2004). Predicting the saturated hydraulic conductivity of sand and gravel using effective diameter and void ratio. *Canadian Geotechnical Journal*, **41**(5): 787–795. https://doi.org/10.1139/t04-022

Chapuis, R.P. (2012). Predicting the saturated hydraulic conductivity of soils: A review. *Bulletin of Engineering Geology and the Environment*, **71**(3): 401–434. https://doi.org/10.1007/s10064-012-0418-7

Ching, J. & Phoon, K-K. (2012). Modeling parameters of structured clays as a multivariate normal distribution. *Canadian Geotechnical Journal*, **49**(5): 522–545. https://doi.org/10.1139/t2012-015

Ching, J. & Phoon, K-K. (2014). Transformations and correlations among some clay parameters – the global database. *Canadian Geotechnical Journal*, **51**(6): 663–685. https://doi.org/10.1139/cgj-2013-0262

Ching, J., Phoon, K-K. & Wu, C.T. (2022). Data-centric quasi-site-specific prediction for compressibility of clays. *Canadian Geotechnical Journal*, **59**(12): 2033–2049. https://doi.org/10.1139/cgj-2021-0658

Clennell, M.B., Dewhurst, D.N., Brown, K.M. & Westbrook, G.K. (1999). Permeability anisotropy of consolidated clays. In: *Muds and mudstones: Physical and fluid flow properties*, vol. 158, Aplin, A.C., Fleet, A.J. & MacQuaker, J.H.S. (eds.). Geological Society of London, London, UK, pp. 79–96. http://doi.org/10.1144/GSL.SP.1999.158.01.07

Corey, A.T. (1957). Measurement of water and air permeability in unsaturated soil. *Soil Science Society of America Journal*, 21(1): 7–10. https://doi.org/10.2136/sssaj1957.03615995002100010003x

Craig, R.F. (2004). *Craig's soil mechanics*. 7th Edition. Taylor & Francis, Abingdon, UK.

D'Ignazio, M., Phoon, K-K., Tan, S.A. & Länsivaara, T.T. (2016). Correlations for undrained shear strength of finnish soft clays. *Canadian Geotechnical Journal*, **53**(10): 1628–1645. https://doi.org/10.1139/cgj-2016-0037

Daniel, D., Anderson, D., & Boynton, S. (1985). Fixed-wall versus flexible-wall permeameters. In *Hydraulic barriers in soil and rock*. Johnson, A. I., Frobel, R. K., Cavalli, N. J., & Pettersson, C. B., (eds.), American Society for Testing and Materials, Philadelphia, PA, pp. 107–126.

Darcy, H. (1856). *Les fontaines publiques de la ville de Dijon. Victor Dalmont.*

Feng, S. & Vardanega, P.J. (2019a). Correlation of the hydraulic conductivity of fine-grained soils with water content ratio using a database. *Environmental Geotechnics*, **6**(5): 253–268. https://doi.org/10.1680/jenge.18.00166

Feng, S. & Vardanega, P.J. (2019b). A database of saturated hydraulic conductivity of fine-grained soils: probability density functions. *Georisk: Assessment and Management of Risk for Engineered Systems and Geohazards*, **13**(4): 255–261. https://doi.org/10.1080/17499518.2019.1652919

Feng, S., Vardanega, P.J., Ibraim, E., Widyatmoko, I., & Ojum, C. (2019). Permeability assessment of some granular mixtures. *Géotechnique*, **69**(7): 646–654. https://doi.org/10.1680/jgeot.17.T.039

Feng, S. (2022). *Hydraulic conductivity of road construction materials: With a focus on freeze-thaw effects*. Ph.D. thesis, University of Bristol, Bristol, UK.

Feng, S., Ibraim, E. & Vardanega, P.J. (2022). Databases to assess the hydraulic conductivity of road construction materials. In: *Proceedings of twentieth international conference on soil mechanics and geotechnical engineering (ICSMGE 2022), A Geotechnical Discovery Down Under, Sydney, New South Wales, Australia, 1–5 May 2022*, vol. 2, Rahman, M.M. & Jaksa, M. (eds.), Australian Geomechanics Society, Sydney, Australia, pp. 4547–4552. <https://www.issmge.org/uploads/publications/1/120/ICSMGE_2022-777.pdf> [19/10/2024].

Feng, S., Barreto, D., Imre, E., Ibraim, E. & Vardanega, P.J. (2023). Use of hydraulic radius to estimate the permeability of coarse-grained materials using a new geodatabase. *Transportation Geotechnics*, **41**: [101026]. https://doi.org/10.1016/j.trgeo.2023.101026

Gao, Q.F., Dong, H., Huang, R. & Li, Z.F. (2019). Structural characteristics and hydraulic conductivity of an eluvial-colluvial gravelly soil. *Bulletin of Engineering Geology and the Environment*, **78**(7): 5011–5028. https://doi.org/10.1007/s10064-018-01455-1

Hazen, A. (1895). *The filtration of public water supplies*. John Wiley & Sons, New York, NY.

Indrawan, I.G.B., Rahardjo, H. & Leong, E.C. (2006). Effects of coarse-grained materials on properties of residual soil. *Engineering Geology*, **82**(3): 154–164. https://doi.org/10.1016/j.enggeo.2005.10.003

Jaafar, R. & Likos, W.J. (2014). Pore-scale model for estimating saturated and unsaturated hydraulic conductivity from grain size distribution. *Journal of Geotechnical and Geoenvironmental Engineering*, **140**(2): 04013012. https://doi.org/10.1061/(ASCE)GT.1943-5606.0001031

Jones, C.W. (1955). The Permeability and Settlement of Laboratory Specimens of Sand and Sand-Gravel Mixtures. *Symposium on Permeability of Soils*, ASTM, Philadelphia, PA, pp. 68–79.

Kozeny, J. (1927). *Über kapillare leitung des wassers im boden:(aufstieg, versickerung und anwendung auf die bewässerung). Hölder-Pichler-Tempsky* (in German).

Kulhawy, F.H. & Mayne, P.W. (1990). *Manual on estimating soil properties for foundation design*, Electric Power Research Institute, Report No. EL-6800, Palo Alto, CA.

Lambe, T.W. & Whitman, R.V. (1969). *Soil mechanics*. John Wiley Sons, New York, NY.

Liu, Y. (2015). *Permeability and pore structure of granular materials*. Ph.D. thesis. McMaster University, Hamilton. Available from: <https://macsphere.mcmaster.ca/bitstream/11375/18104/2/Dissertation%20final.pdf> [10/07/2023]

Mbonimpa, M., Aubertin, M., Chapuis, R. & Busssière, B. (2002). Practical pedotransfer functions for estimating the saturated hydraulic conductivity. *Geotechnical and Geological Engineering*, **20**(3): 235–259. https://doi.org/10.1023/A:1016046214724

Mesri, G. & Olson, R. E. (1971). Mechanisms controlling the permeability of clays. *Clays and Clay Minerals*, **19**(3): 151–158. https://doi.org/10.1346/CCMN.1971.0190303

Montgomery, D. C., Runger, G. C., & Hubele, N. F. (2007). *Engineering statistics* (4th ed.). John Wiley & Sons Inc, New York, NY.

Moulton, L. & Seals, D. (1979). *Determination of the in situ permeability of base and subbase courses*. Federal Highway Administration, Washington, D.C.

Mualem, Y. (1976). A new model for predicting the hydraulic conductivity of unsaturated porous media. *Water Resources Research*, **12**(3): 513–522. https://doi.org/10.1029/WR012i003p00513

Nagaraj, T.S., Pandian, N.S. & Narashimha Raju, P.S.R. (1993). Stress state–permeability relationships for fine-grained soils. *Géotechnique*, **43**(2): 333–336. https://doi.org/10.1680/geot.1993.43.2.333

Nagaraj, T.S., Pandian, N.S. & Narashimha Raju, P.S.R. (1994). Stress state–permeability relations for overconsolidated clays. *Géotechnique*, **44**(2): 349–352. https://doi.org/10.1680/geot.1994.44.2.349

Nagy, L., Tabácks, A., Huszák, T., Mahler, A. & Varga, G. (2013). Comparison of permeability testing methods. In *18th International Conference on Soil Mechanics and Geotechnical Engineering, Paris, France*, Presses des Ponts, Paris, France, pp. 399–402. Available from: <https://www.issmge.org/uploads/publications/1/2/399-402.pdf> [10/07/2023].

Nguyen, T.T. & Indraratna, B. (2020). The role of particle shape on the hydraulic conductivity of granular soils captured through Kozeny-Carman approach. *Géotechnique Letters*, **10**(3): 398–403. https://doi.org/10.1680/jgele.20.00032

Pane, V., Croce, P., Znidarcic, D., Ko, H.-Y., Olsen, H.W. & Schiffman, R.L. (1983). Effects of consolidation on permeability measurements for soft clay. *Géotechnique*, **33**(1): 67–72. https://doi.org/10.1680/geot.1983.33.1.67

Paradine, C.G. & Rivett, B.H. (1953). Statistics for technologists. English Universities Press, London, UK.

Phoon, K-K. & Kulhawy, F.H. (1999a). Characterization of geotechnical variability. *Canadian Geotechnical Journal*, **36**(4): 612–624. https://doi.org/10.1139/t99-038

Phoon, K-K. & Kulhawy, F.H. (1999b). Evaluation of geotechnical property variability. *Canadian Geotechnical Journal*, **36**(4): 625–639. https://doi.org/10.1139/t99-039

Phoon, K-K. & Tang, C. (2019). Characterisation of geotechnical model uncertainty. *Georisk: Assessment and Management of Risk for Engineered Systems and Geohazards*, **13**(2): 101–130. https://doi.org/10.1080/17499518.2019.1585545

Phoon, K-K., Ching, J. & Wang, Y. (2020). Managing risk in geotechnical engineering – From data to digitalization. In: *Proceedings of the 7th international symposium on geotechnical safety and risk (ISGSR)*, Research Publishing, Taipei, Taiwan, pp. 13–34. Available from: <http://rpsonline.com.sg/proceedings/isgsr2019/index.html> [10/07/2023]

Piñeiro, G., Perelman, S., Guerschman, J. P., & Paruelo, J. M. (2008). How to evaluate models: Observed vs. predicted or predicted vs. observed? *Ecological Modelling*, **216**(3–4), 316–322. https://doi.org/10.1016/j.ecolmodel.2008.05.006

Qiu, Z.F. & Wang, J.J. (2015). Experimental study on the anisotropic hydraulic conductivity of a sandstone-mudstone particle mixture. *Journal of Hydrologic Engineering*, **20**(11): 04015029. https://doi.org/10.1061/(ASCE)HE.1943-5584.0001220

Ren, X., Zhao, Y., Deng, Q., Kang, J., Li, D. & Wang, D. (2016). A relation of hydraulic conductivity — void ratio for soils based on Kozeny-Carman equation. *Engineering Geology*, **213**: 89–97. https://doi.org/10.1016/j.enggeo.2016.08.017

Ren, X.W. & Santamarina, J.C. (2018). The hydraulic conductivity of sediments: A pore size perspective. *Engineering Geology*, **233**: 48–54. https://doi.org/10.1016/j.enggeo.2017.11.022

Shahabi, A., Das, B., and Tarquin, A. (1984). An Emperical Relation for Coefficient of Permeability of Sand. In *Fourth Australia-New Zealand Conference on Geomechanics*, , vol. 1, The Institution of Engineers, Australia, Barton, ACT, pp. 54–57. Available from: <https://www.issmge.org/uploads/publications/89/91/4ANZ_008.pdf> [19/10/2024]

Shepherd, R.G. (1989). Correlations of permeability and grain size. *Groundwater*, **27**(5): 633–638. https://doi.org/10.1111/j.1745-6584.1989.tb00476.x

Sivapullaiah, P.V, Sridharan, A. & Stalin, V.K. (2000). Hydraulic conductivity of bentonite-sand mixtures. *Canadian Geotechnical Journal*, **37**(2): 406–413. https://doi.org/10.1139/t99-120

Sridharan, A. & Nagaraj, H.B. (2005). Hydraulic conductivity of remolded fine-grained soils versus index properties. *Geotechnical and Geological Engineering*, **23**(1): 43–60. https://doi.org/10.1007/s10706-003-5396-x

Taiba, C.A., Mahmoudi, Y., Hazout, L., Belkhatir, M. & Baille, W. (2019). Evaluation of hydraulic conductivity through particle shape and packing density characteristics of sand–silt mixtures. *Marine Georesources and Geotechnology*, **37**(10): 1175–1187. https://doi.org/10.1080/1064119X.2018.1539891

Tavenas, F., Leblond, P., Jean, P. & Leroueil, S. (1983). The permeability of natural soft clays. Part I: Methods of laboratory measurement. *Canadian Geotechnical Journal*, **20**(4): 629–644. https://doi.org/10.1139/t83-072

Taylor, D.W. (1948). *Fundamentals of soil mechanics*. John Wiley & Son Inc. New York, NY.

Terzaghi, K. & Peck, R.B. (1948). *Soil mechanics in engineering practice*. John Wiley & Sons, New York, NY.

Van Genuchten, M.T. (1980). A closed-form equation for predicting the hydraulic conductivity of unsaturated soils. *Soil Science Society of America Journal*, **44**(5): 892–898. https://doi.org/10.2136/sssaj1980.03615995004400050002x

Vardanega, P.J., Othman, M., Voyagaki, E., Crispin, J.J., Gilder, C.E.L. & Ntassiou, K. (2024). *The DINGO database: Summary report: March 2024, v1.2*. University of Bristol, Bristol, UK. https://doi.org/10.5523/bris.1jraem68g7ara21p2oi6hv4z22

Whittle, A. (1993). Evaluation of a constitutive model for overconsolidated clays. *Géotechnique*, **43**(2): 289–313. https://doi.org/10.1680/geot.1993.43.2.289

Zhai, Q., Rahardjo, H., Satyanaga, A., Zhu, Y., Dai, G. & Zhao, X. (2021). Estimation of wetting hydraulic conductivity function for unsaturated sandy soil. *Engineering Geology*, **285**: 106034. https://doi.org/10.1016/j.enggeo.2021.106034

Chapter 7

Normalised active undrained shear strengths of soft Scandinavian clays

A data-centric and geomechanical approach

Sølve Hov, Jean-Sébastien L'Heureux, Katharina Kahrs, Karin Lundström, and David Gaharia

The undrained shear strength is one of the most important engineering parameters of soft clays. This chapter specifically deals with active undrained strengths from triaxial tests and their normalisation with the preconsolidation pressure from oedometer tests (c_{uA}/σ'_p) on normally to slightly overconsolidated clays (OCR ≤ 1.5). A bivariate database is presented containing data from Norwegian and Swedish clays with a large range in plasticity quantified by the liquid limit (LL). Linear regression analyses are performed to show that c_{uA}/σ'_p and LL are poorly correlated; however, there is a clear trend of increasing c_{uA}/σ'_p with increasing LL with respect to the 95% confidence interval. On average, c_{uA}/σ'_p ranges from just below 0.3 at $LL \approx$ 20–30% to around 0.36–0.38 at $LL \approx 80\%$. The results are explained within a geomechanical framework where a simplified yield surface, defined by a drained failure criterion (φ') and the yield stress σ'_p, is used to show that C_{uA}/σ'_p is dependent on φ'. An increase in organic content results in an increase in both φ' and LL, and by pure geometrical constructions of the simplified yield surface it is shown that $\varphi' = 30°$ results in $c_{uA}/\sigma'_p = 0.33$ for low-to-medium plastic clays and $\varphi' = 55°$ results in $c_{uA}/\sigma'_p = 0.45$ for highly plastic organic clays. This agrees with the bivariate database. Specific attention is given to uncertainties and sample quality.

7.1 INTRODUCTION

The undrained shear strength (c_u) of clays is one of the most important parameters in geotechnical engineering and has over the years been researched extensively. The parameter is vital to the assessment of slope stability, embankments, and excavations in clay as well as the bearing capacity of structures founded on clayey soils. The magnitude of c_u depends on several factors, e.g., direction of loading, rate of loading, and temperature; however, it is also found to be inherently dependent on the magnitude and orientation of effective stresses in the clay before and during loading. In particular, it depends on the effective stress at yielding, i.e., the preconsolidation pressure (σ'_p) in oedometer boundary conditions. Normalised undrained shear strengths (c_u / σ'_p), i.e., the ratio between the undrained strength and the drained yield stress, are thus of great value in clay characterisation – both in research and in engineering practice.

This chapter specifically deals with normalised active undrained shear strengths (c_{uA} / σ'_p) of soft Scandinavian clays in a normally to slightly overconsolidated state, i.e., overconsolidation ratio $OCR \leq 1.5$. The c_{uA} is obtained from active undrained triaxial tests. This normalised strength has also been the topic for several previous studies, which have shown that c_{uA} / σ'_p exhibits a fairly constant value of around 0.3–0.35 for

DOI: 10.1201/9781003441946-7

$OCR \leq 1.5$ (Ladd et al. 1977; Larsson 1980; Jamiolkowski et al. 1985; Tavenas and Leroueil 1987; Ladd and DeGroot 2003; Larsson et al. 2007; Karlsrud and Hernandez-Martinez 2013; Sällfors et al. 2017; Lundström and Dehlbom 2019).

Normalised undrained shear strengths obtained through direct simple shear tests (DSS), field vane tests (FV), and fall cone (FC) tests have also previously been dealt with by several researchers. Although c_{uDSS} / σ'_p clearly depends on the OCR, varying results have been found on its dependence of the clay's plasticity. Some studies have for example shown that c_{uFV} / σ'_p is around 0.22 independently of plasticity, typically quantified by the plasticity index (IP) or liquid limit (LL) (Mesri 1975, 1989; D'Ignazio et al. 2016). Other studies have shown the opposite, i.e., values of c_{uDSS} / σ'_p increase with increasing plasticity (Jamiolkowski et al. 1985; Jardine and Hight 1987; Mayne and Mitchell 2011; Hov et al. 2021).

The reason why some data tend to show a dependence of plasticity on c_{uDSS} / σ'_p and others do not is not clear. In fact, the causes may be attributed to various factors such as the intrinsic clay properties or systematic errors in the database due to, e.g., sample disturbance or measurement errors. These are important uncertainties, and with the fast-growing amount of data available within geotechnical engineering, the topic of representative results is more important than ever. Myriads of databases with both simple and advanced statistical analyses are available to the geotechnical community. These works and databases are highly valuable and give new insights into clay behaviour (Ching and Phoon 2013, 2014a, 2014b; D'Ignazio et al. 2016; Beesley and Vardanega 2020; Hov et al. 2021), and it is transforming the understanding of soil mechanics in engineering practice, and provide a huge opportunity for further research (Tang and Phoon 2021; Phoon 2020; Phoon et al. 2022).

However, if care is not taken, a risk of database contamination with non-representative values is imminent. Consequences of such can lead to unrepresentative machine learning models and unreliable geotechnical designs. One could perhaps argue that historically published data likely had a higher degree of quality control and assurance since there were fewer test results to process. As the number of tests and data over time increases, the sheer volume of data makes quality control and assurance more difficult, simply by practical limitations. In addition, data from studies are used exchangeably throughout the world where data is being re-analysed by other researchers, often located on different continents, using different laboratory tools and different interpretation methods. This is most often advantageous, since it avoids any dogmatism that researchers can fall into. On the contrary, there is also an increased likelihood that important preconditions are not addressed since they are unknown.

In the light of all uncertainties linked to the assessment of geotechnical parameters, it could be advantageous to use a geomechanical framework to assess data quality when comparing data from different sources. This is however not always possible. The FC test, as an example, is interpreted by the cone penetration depth which is correlated to c_{uFC}. This is easily modelled with numerical tools; however, the correlation is too dependent on the (unknown) adhesion between the cone and the clay and, subsequently, a correlation cannot be established numerically. As a few more examples, the principal stress rotation in a DSS sample is unknown, and for the FV test, the failure pattern and soil structure disturbance prior to rotation are too uncertain. Similarly, the piezocone (CPTu) test can be challenging to analyse geomechanically since the effective stress paths around the cone are uncertain and depend on many variables. One test that however can be analysed within a geomechanical framework is the triaxial test. Here, the principal total stresses and pore pressure measurements enable the calculation of all effective stresses, and it is known how these change over time and with strain. This is vastly advantageous since its data sets can be analysed not only by statistical analyses but also by assessing credibility from a geomechanical point of view.

The study of normalised active undrained shear strengths (c_{uA} / σ'_p) outlined in this chapter is based on experimental results – the data-centric approach – and by analysing the results by a simplified yield surface in an effective stress plane – the geomechanical approach. A bivariate database is presented which contains the variables c_{uA} / σ'_p and liquid limit, LL. The c_{uA} is obtained from triaxial tests consolidated anisotropically and sheared undrained in an active (compression) loading (CAUC), and the σ'_p is obtained from the constant rate of strain (CRS) oedometer tests. The LL is used as a measure of plasticity, instead ofplasticity index (IP). The database is "only" bivariate since all data points originate from normally to slightly overconsolidated clays ($OCR \leq 1.5$), and hence OCR is not a variable and no other parameter has been shown to affect normalised strengths from CAUC tests. Further, the strength is normalised with σ'_p instead of the in situ effective stress (σ'_{v0}) since σ'_p is interlinked with shear strength, as will be shown.

The database contains test results from both Norwegian and Swedish clays. These clays are similar from a mineralogical and stress history point of view; however, they differ in other aspects. Norwegian clays typically have a lower clay content and a lower organic content compared to Swedish clays, resulting in lower natural water content and lower plasticity. The data thus provides a great opportunity to analyse the impact of clay plasticity on the normalised undrained shear strength of clays.

One however needs to be careful when comparing data from different areas and countries, even from Norway and Sweden which have had extensive collaboration and have had many similarities within geotechnical engineering throughout history. In fact, it turns out that not all parameters are interpreted using the same methods. In this case, although the CRS oedometer tests are performed in a similar way, the interpretation of σ'_p differs, as will be explained. This difference has been in place at least since the beginning of the 1980s when researchers in Sweden introduced the CRS test and implemented an interpretation which still today has a very limited international use. In Norway, the σ'_p interpretation has also varied with time (Janbu 1963, 1989; Karlsrud and Hernandez-Martinez 2013). The difference in interpretation method still exists since there is no ISO or EN standard for CRS oedometer tests. This fact might have resulted in erroneous analyses that have been made when including Norwegian and Swedish data containing σ'_p as a variable.

These differences in interpretation techniques will firstly be highlighted, followed by a presentation of the bivariate database. Afterwards, the results are analysed through a geomechanical framework to attempt a critical validation of the empirical data. The aim is to analyse if and how c_{uA} / σ'_p is affected by the LL. The two approaches result in similar conclusions.

7.2 DATA AND METHODS

7.2.1 Laboratory tests

All laboratory tests were performed according to their respective EN ISO standard if available. The LL was determined by the fall cone test according to ISO 17892-12:2018 and is defined as the water content at 10 mm penetration with a 60 g steel fall cone with an apex angle of 60°. The LL is used since there is a very strong correlation between LL and IP as seen in most databases (Ching and Phoon 2014b; NIFS 2014; D'Ignazio et al. 2016; J. DeGroot et al. 2019), and any of these will thus work to quantify the clay's plasticity. Another reason is that values of *IP* are often missing in Swedish data since the plastic limit (*PL*) is not measured in engineering practice in Sweden. One also needs to remember that the plasticity limits, in total 11 of them originally developed by Atterberg in the early 1910s for agricultural purposes (Atterberg 1911, 1912) and introduced to soil mechanics by Casagrande in the 1930s (Casagrande 1932), are not determined based on any scientific principle. It is instead an arbitrary property of a gradual shift in the consistency of a clay in a remoulded state. Hence, the LL alone is a sufficient measure to describe a clay's plasticity and the word "plasticity" is used hereafter as a common term for LL and IP.

The CAUC tests were performed according to ISO 17892-9:2018. The sample dimensions were typically 50 or 54 mm in diameter with 100 mm in height. The sample is consolidated to its calculated in situ effective stress state, which for these normally consolidated clays is anisotropic with earth pressure at rest coefficients (K_0) values around 0.5–0.7. Samples were sheared actively under undrained conditions with a strain rate of around 0.6–0.7% per hour. The active undrained shear strength (c_{uA}) is interpreted as the peak strength that typically occurs within 0.5–3% axial strain for inorganic clays with relatively low values of LL. For highly plastic organic clays, the peak strength typically occurs at higher strains, but the interpretation of c_{uA} is limited to 15% axial strain.

As previously mentioned, the OCR highly impacts the clay behaviour and, in particular, the effective stress path to failure. To illustrate the effects of OCR on effective stress path type, Figure 7.1 plots a compilation of effective stress paths from the Norwegian Geotechnical Institutes (NGI) database (the database is presented in the following section).

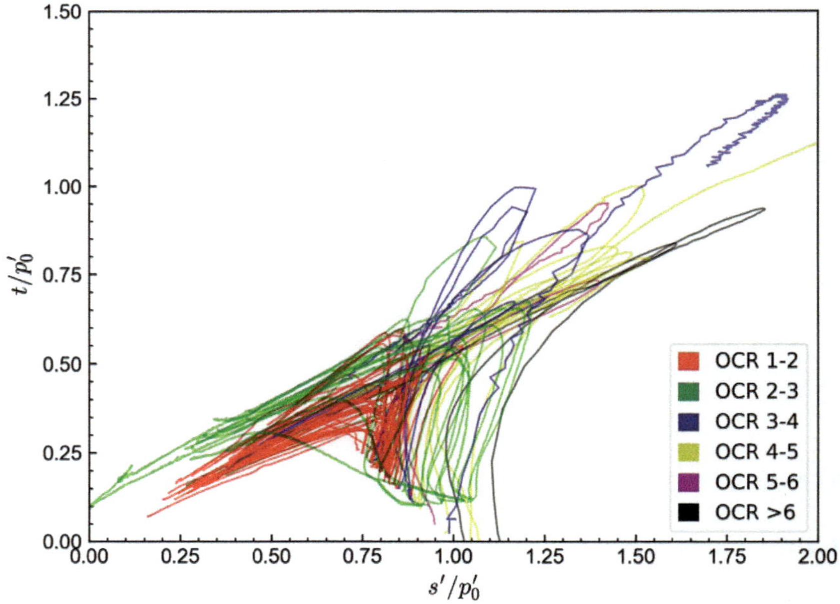

Figure 7.1 Compilation of effective stress paths in the $s' = (\sigma_v' + \sigma_h')/2$ vs. $t = (\sigma_v - \sigma_h)/2$ stress plane, i.e., the MIT plot. The axes in this plot are normalised with respect to in situ effective vertical stress (p_0') for comparison. Internal NGI data (NGI 2017).

It is seen that clays in a normally to slightly overconsolidated state ($OCR \leq 1.5$) tend to share a common type of effective stress path marked in red. From a nearly elastic initial response, stress path inclination 1:3 in this plot, yielding occurs whereupon a contractive behaviour leads to an increased pore pressure development and the effective stress path turning left and towards the origin. Samples with $OCR \gg 1.5$ exhibit a dilatant stress path which turns right and follows a drained failure envelope, e.g., Mohr–Coulomb, before complete failure occurs. The results presented herein only originate from clays in a normally to slightly overconsolidated state ($OCR \leq 1.5$).

CRS oedometer tests are executed similarly in Norway and Sweden. Samples of 50 or 54 mm in diameter with 20 mm height are compressed with a strain rate of around 0.6–0.8% per hour. An undrained boundary at the bottom enables pore pressure measurements, whilst the top boundary is drained. The interpretation of oedometer moduli is also similar in both countries; however, the interpretation of σ'_p differs. In Norway, the most commonly used method today is referred to as the "Karlsrud" method (Karlsrud and Hernandez-Martinez 2013) and is used on all Norwegian test results presented in this chapter. In Sweden, the "Sällfors method" (Sällfors 1975) has been used in Sweden since the early 1980s, and again, all Swedish test results presented in this chapter are based on the Sällfors method.

Figure 7.2 shows two CRS oedometer test results to illustrate the differences between the Norwegian and Swedish methods of interpreting σ'_p . The Karlsrud method, shown in red, interprets the σ'_p value based on the modulus curve, the first derivative of the oedometer curve. A curve-fitting procedure matches linear parts of the modulus curve in the overconsolidated range (M_i), the normally consolidated range (M_L), and after a threshold is reached where the modulus increases linearly after M_L . The σ'_p is interpreted as the average between this threshold and the end of the M_i line. The Sällfors method, shown in blue in Figure 7.2, uses a curve-fitting procedure of the oedometer

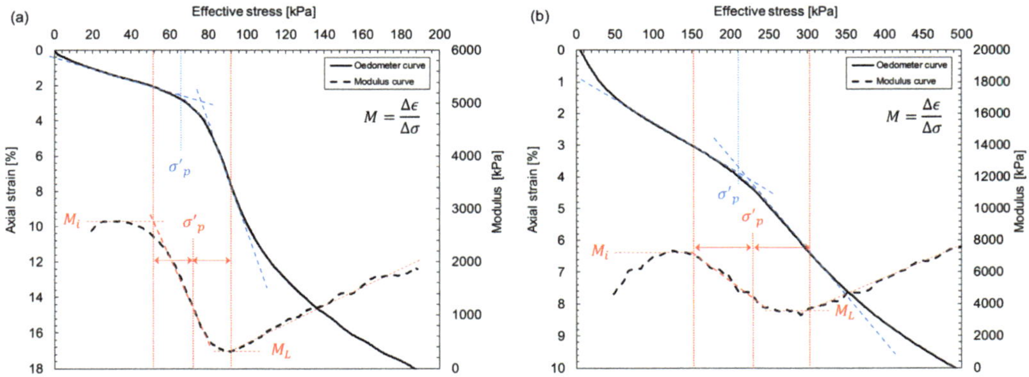

Figure 7.2 Examples of interpretation of preconsolidation pressure (σ'_p) from CRS oedometer tests with the Karlsrud (Norwegian) and Sällfors (Swedish) methods marked in red and blue, respectively: (a) test result on a Swedish clay with $M_i / M_L \approx 8$ giving $\sigma'_{p,NO} \approx 72$ kPa and $\sigma'_{p,SE} \approx 66$ kPa, (b) test result on a Norwegian clay with $M_i / M_L \approx 2$ giving $\sigma'_{p,NO} \approx 230$ kPa and $\sigma'_{p,SE} \approx 210$ kPa.

curve instead of the modulus curve. A similar procedure is performed to find linear parts of the curve representing M_i and M_L. An isosceles triangle is then drawn with its base tangential to the oedometer curve as shown in the figure. The σ'_p is defined as the intersection of the baseline and the M_i line. It has been shown that this approach gives similar values as incremental oedometer tests and field measurements (Sällfors 1975).

Another distinction is that the threshold stress, i.e., the onset of modulus increase after M_i, is directly interpreted by the apparent modulus curve according to the Karlsrud method, whilst a strain rate correction is made in the Sällfors method to take into account the considerably higher strain rate in CRS tests compared to incremental oedometer tests or field behaviour (Sällfors 1975; Tavenas and Leroueil 1987).

The interpretation method is thus different between the two countries, affecting values of c_{uA} / σ'_p. There has been no systematic study of this difference between these methods, although it is known that the Karlsrud method yields roughly similar results to other commonly used methods (Paniagua et al. 2016). For the two examples in Figure 7.2, the Karlsrud method shows σ'_p of 72 and 230 kPa, respectively, whilst the Sällfors method correspondingly shows σ'_p of 66 and 210 kPa. The difference is around 9–10% in both cases, i.e., the normalised strength c_{uA} / σ'_p will be around 9–10% lower if a Norwegian interpretation practice is used compared to Swedish practice. Based on the authors' experience and judgement, this difference of ~10% between the two methods seems reasonable for typical CRS oedometer test results.

7.2.2 Bivariate database

The database, labelled CLAY/2/213 (Ching et al. 2016), is summarised in Table 7.1. In total, it contains 213 bivariate data points of c_{uA} / σ'_p and LL, of which 46 originated from Norway and the remaining from Sweden. The reason that Swedish clays are dominant is because a large proportion of the Norwegian clays are overconsolidated ($OCR > 1.5$) and have thus been removed.

Most of the Norwegian data originates from the NGI database of onshore piston samples (72–75 mm diameter samples) and high-quality block samples (160–250 mm samples). Some of this data has previously been presented but has been supplemented with more recent test data (Karlsrud and Hernandez-Martinez 2013; Paniagua et al. 2019). The data originates from sites distributed in the central and south-eastern parts of Norway. The Swedish data originates from several sources, both from research and commercial projects on organic clays and inorganic clays. The Swedish data includes piston samples (50 mm diameter) and high-quality block samples (160 mm diameter). The data originates from sites distributed around southern Sweden.

The range of LL in the database is large; between 16 and 349, reflecting the range from silty clays, or perhaps clayey silts, to highly organic clays and gyttja (defined as clay with organic content >6%). The range of c_{uA} / σ'_p is between 0.22 and 0.55.

Common for all data is that the test results have been subjected to a thorough quality control by engineers during evaluation and processing of the data. Even though a large proportion of the data originate from commercial projects, e.g., Sällfors and Larsson (2016) and the NGI and LabMind (LM) databases, subsequent analyses have been performed and when needed, data has been removed if found not representative due to, e.g., sample disturbance. The database does not include a sample quality designation since this is not available for all test results, e.g., (Lunne et al. 1997).

Table 7.1 Overview of data ($OCR \leq 1.5$)

			c_{uA} / σ'_p		LL		
Reference	*n*	*Country*	*min*	*max*	*min*	*max*	*Remarks*
NGI database	43	Norway	0.23	0.48	16	107	Data from Karlsrud and Hernandez-Martinez (2013), L'Heureux et al. (2018), and Paniagua et al. (2019) supplemented with new data not previously published. Data from 11 sites in Norway.
LM database	18	Sweden	0.30	0.46	44	83	Data from Hov and Garcia de Herreros (2020) supplemented with new data not previously published. Data from seven sites in South-eastern Sweden.
Lundström and Dehlbom (2019)	16	Sweden	0.28	0.43	36	104	Data from three sites in Sweden
Sällfors and Larsson (2016)	55	Sweden	0.25	0.44	25	107	Data from 15 sites across southern Sweden
Larsson et al. (2007)	51	Sweden	0.26	0.36	25	85	The number of sites uncertain
Westerberg (1999)	19	Sweden	0.26	0.37	25	84	Data from one specific site
Holstad (2016); Holstad and Degago (2021)	3	Norway	0.30	0.55	90	230	Data included to illustrate normalised strengths of organic clays and gyttja
Larsson (1990)	8	Sweden	0.33	0.51	20	349	Data included to illustrate normalised strengths of organic clays and gyttja
All compiled data (CLAY/2/213)	213	–	0.23	0.55	16	349	–

7.3 A DATA-CENTRIC APPROACH

Figure 7.3(a) plots all data of c_{uA} / σ'_p vs. LL available in the database (Table 7.1). The values of c_{uA} / σ'_p tend to increase with increasing LL until a value of $LL \approx 100\%$ is reached. Such a value typically corresponds to a rough threshold used to distinguish inorganic clay from organic clay (gyttjig clay or gyttja). A few test results obtained from organic clay and gyttja with $LL > 100\%$ show that values of c_{uA} / σ'_p are fairly constant at around 0.5.

Looking only at results from inorganic clays, Figure 7.3(b) plots values for $LL < 100\%$. This figure also shows a trend line from a linear regression analysis with a 95% confidence interval of the mean value. The linear trend shows values of c_{uA} / σ'_p ranging from just below 0.3 at $LL \approx 20 - 30\%$ to around 0.36–0.38 at $LL \approx 80 - 100\%$. Despite a large scatter and a very weak correlation ($R^2 = 0.24$), there is clearly a trend of increasing c_{uA} / σ'_p with increasing LL of the 95% confidence interval.

To account for the difference in interpretation method, the c_{uA} / σ'_p for the Norwegian data in Figure 7.3(c) is increased by 10%. It is chosen to correct the Norwegian data

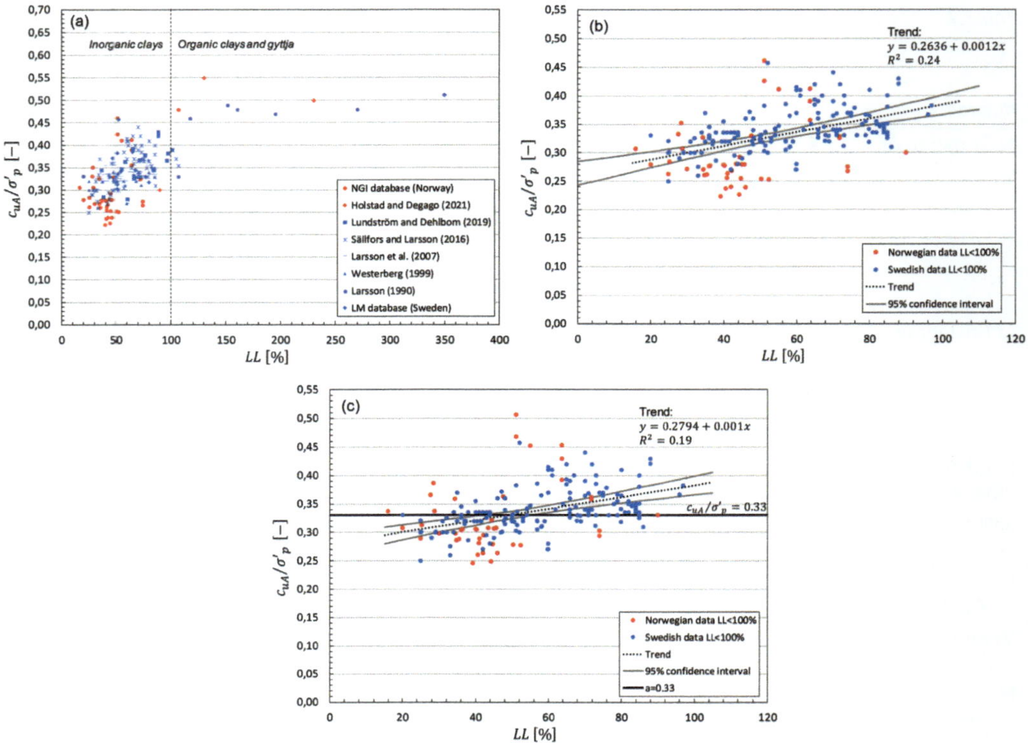

Figure 7.3 Normalised active undrained shear strengths (c_{uA} / σ'_p) vs. liquid limit (LL): (a) all data and all LL , (b) data with $LL < 100\%$, and (c) data $LL < 100\%$ with Norwegian data corrected for σ'_p interpretation (+10%). The line in (a) serves as an approximate transition between inorganic and organic clays. Linear regression analysis and 95% confidence interval are visualised in (b) and (c).

instead of the Swedish because of the reasonable strain rate effects that should be considered in CRS oedometer testing. Again, the linear regression analysis shows a weak correlation with an $R^2 = 0.19$; however, the trend of increasing c_{uA} / σ'_p with increasing LL is still seen.

This trend is contradictory with most other studies of normalised strengths from CAUC (c_{uA} / σ'_p) that show a constant c_{uA} / σ'_p independent of IP or LL (Jamiolkowski et al. 1985; Tavenas and Leroueil 1987; Ladd et al. 1977; Larsson 1980; Larsson et al. 2007; Karlsrud and Hernandez-Martinez 2013; Sällfors et al. 2017; Lundström and Dehlbom 2019). There are however a few published studies, or compilations of data, that show similar trends of increasing c_{uA} / σ'_p with increasing IP or LL, e.g., (Westerberg 1999; Lundström and Dehlbom 2019; Vesterberg and Andersson 2022). Other studies show a very slight increase with plasticity (DeGroot et al. 2019).

Why do we see these differences in data? Is it due to inherent behaviour of different clays, effects of sample disturbance, or simply by chance? From a pure data-centric approach, these effects are difficult to investigate, and hence it is difficult to validate if the data demonstrates real representative soil behaviour or not.

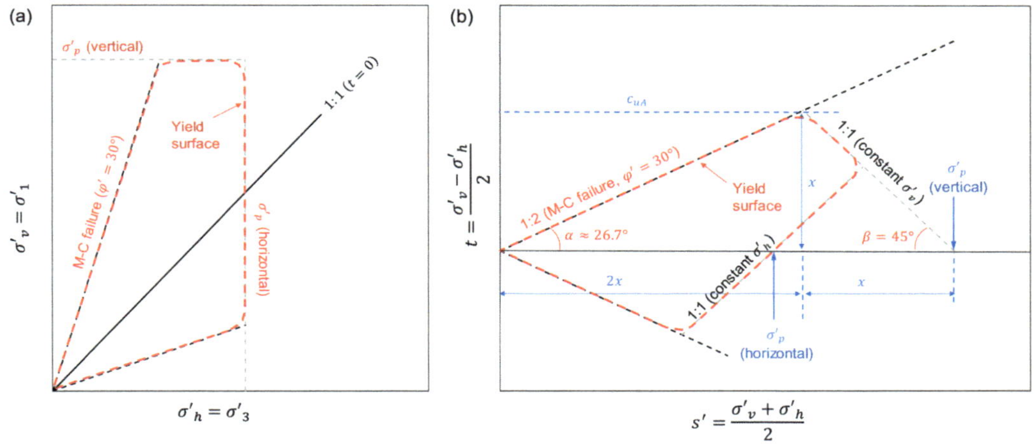

Figure 7.4 Simplified yield surface in (a) σ'_v vs. σ'_h plot and (b) s' vs. t plot (MIT). A friction angle of 30° is used to illustrate typical values for low-to-medium plastic clays.

7.4 A GEOMECHANICAL APPROACH

Active triaxial tests and oedometer tests differ in terms of boundary conditions such as drainage and radial constraints; however, they also share the principal stress orientation and the active (compressive) loading. Thus, yielding in both tests can be assumed to reflect the same vertical soil properties. This is illustrated in Figure 7.4 where (a) shows a σ'_v vs. σ'_h stress plane with a simplified yield surface consisting of the Mohr–Coulomb failure criterion, and the σ'_p in the vertical and horizontal directions. For CRS oedometer tests, the vertical effective stress is continuously increased, resulting in yield around σ'_p. In undrained triaxial shear, however, the effective stress path cannot go beyond this yield surface since the yielding results in an excess pore pressure response and hence a change in direction. This yield surface is illustrated in red in Figure 7.4. It has great value as a basis for explaining and analysing effective stress paths for clays with $OCR \lesssim 1.5$, and has previously been used to analyse both Norwegian and Swedish clays (Larsson and Sällfors 1981; Sällfors et al. 2017).

Figure 7.4(b) illustrates the same yield surface in the $s' = (\sigma'_v + \sigma'_h)/2$ vs. $t = (\sigma_v - \sigma_h)/2$ stress plane, i.e., the MIT plot. Here, constant values of vertical stress (σ'_v) are at a −45° angle, and constant values of horizontal stress (σ'_h) are at a +45° angle, and consequently, the respective σ'_p are also at 45° angles. If the cohesive intercept is negligible, as it often is for these clays, the yield surface is simply defined as a function of the friction angle (φ') and σ'_p in the vertical and horizontal directions. It should be noted that this concept of the simplified yield surface only applies to normally and slightly overconsolidated clays ($OCR \lesssim 1.5$), i.e., where the effective stress paths during triaxial shearing reach the "top" of the yield surface in the MIT plot. Examples of such stress paths were shown in Figure 7.1.

This simplified yield surface allows for a simple geometrical calculation of c_{uA} / σ'_p as a function of φ' (Equation 5.1):

$$\frac{c_{uA}}{\sigma'_p} = \frac{\sin\varphi'}{1 + \sin\varphi'}$$

If the friction angle increases, the "top" of the yield surface will go up, whilst the σ'_p remains constant, and therefore, c_{uA} / σ'_p will increase. An increase in c' instead of φ' also result in an increase in c_{uA} / σ'_p.

Experimental results on low-to-medium plastic clays ($LL \approx 30-50\%$) typically show values of $\varphi' \approx 28-32°$ (Larsson 1977; Tavenas and Leroueil 1987; Karlsrud and Hernandez-Martinez 2013; L'Heureux et al. 2019). If one assumes $\varphi' = 30°$, then $c_{uA} / \sigma'_p \approx 0.33$. In the MIT stress plane, a $\varphi' = 30°$ failure criterion equals a line with $\alpha \approx 26.7°$, and the geometrical construction can then be simplified, Equation (5.2) (see Figure 7.4):

$$\frac{c_{uA}}{\sigma'_p} = \frac{x}{2x+x} = \frac{1}{3}$$

Experimental results on organic clays and gyttja, however, indicates much higher friction angles, up to 55–60° (Larsson 1990; Cheng et al. 2007; Holstad 2016; O'Kelly 2017; Międlarz et al. 2019). For example, $\varphi' = 55°$ results in $c_{uA} / \sigma'_p \approx 0.45$. These higher friction angles are possibly due to embedded fibres in the organic clay (Larsson 1990). Also, due to large strains to failure in organic clays and the undrained (constant volume) shearing, the samples are typically considerably elongated in the radial direct, i.e., in the minor principal stress direction. This can cause cracking even if the minor principal stress is compressive (Janbu 1979). One can thus question if a Mohr–Coulomb failure criterion is meaningful for these types of soils; however, it still reflects real stress paths in triaxial tests, and the measured c_{uA} / σ'_p are as a result higher than those for clays with a lower plasticity and hence a lower (and more meaningful) friction angle.

Figure 7.5 show two examples of effective stress paths in the s' vs. t plot: (a) from a site with low-to-medium plastic clay, and (b) from a site with highly plastic clay, both Swedish clays. All samples are consolidated to around 85% of σ'_p, i.e., $OCR \approx 1.15$. The clay with low-to-medium plasticity, where LL ranges between 40% and 68%, shows a friction angle of around 33°. Simplified yield surfaces can be drawn for each of these stress paths, however, in the figure only that for the sample from 8.0 m depth are shown. This sample seems slightly disturbed resulting a lower σ'_p than expected (giving $OCR = 1.0$), and the effective stress path immediately yields and follows a constant

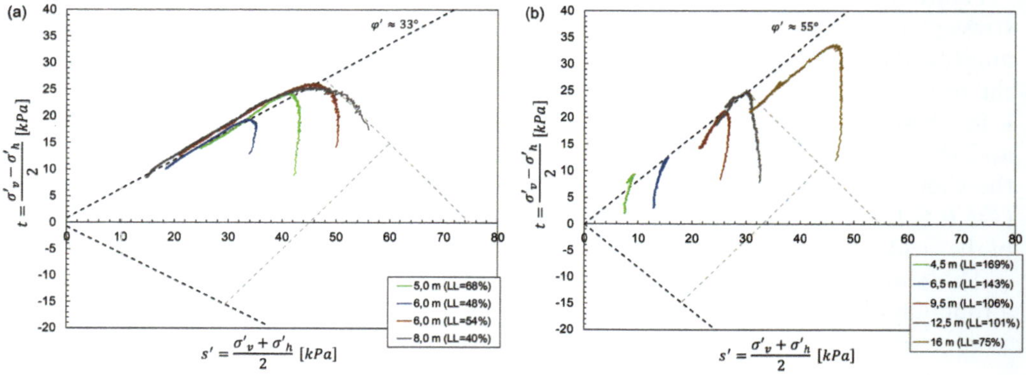

Figure 7.5 Examples of effective stress paths: (a) inorganic clay and (b) organic clay/gyttja.

vertical stress, i.e., along the σ'_p line. Notably, this disturbance does not affect the normalised strength.

Figure 7.5(b) shows the example of high plastic clays with LL ranging from 75% to 169% and where the plasticity decreases with depth. Four of the samples have considerable organic content with LL around or above 100%. For illustration, a simplified yield surface is shown for the 12.5 m sample ($LL = 101\%$). The φ' for the organic clays is interpreted to ~55°. The sample from 16 m ($LL = 75\%$) indicates a lower φ' .

The simplified yield surface and these two examples illustrate an important indirect effect of plasticity: c_{uA} / σ'_p appears to increase with increasing plasticity since the organic content affects both LL and φ' .

Figure 7.6 combines the data-centric and the geomechanical approach. Here, all data points in the database are plotted together with two c_{uA} / σ'_p points calculated based on the simplified yield surface: $c_{uA} / \sigma'_p = 0.33$ for a medium plastic clay ($\varphi' = 30°$) and $c_{uA} / \sigma'_p = 0.45$ for a high plastic organic clay ($\varphi' = 55°$). The magnitudes and trends are similar, serving as illustrative examples of how c_{uA} / σ'_p apparently depends on the apparent φ' , in effect on the plasticity. Further, it seems that there is no significant threshold of organic content with a subsequent shift in values of c_{uA} / σ'_p . The gradually higher φ' values with increasing plasticity thus seem to result in a gradual increase of c_{uA} / σ'_p .

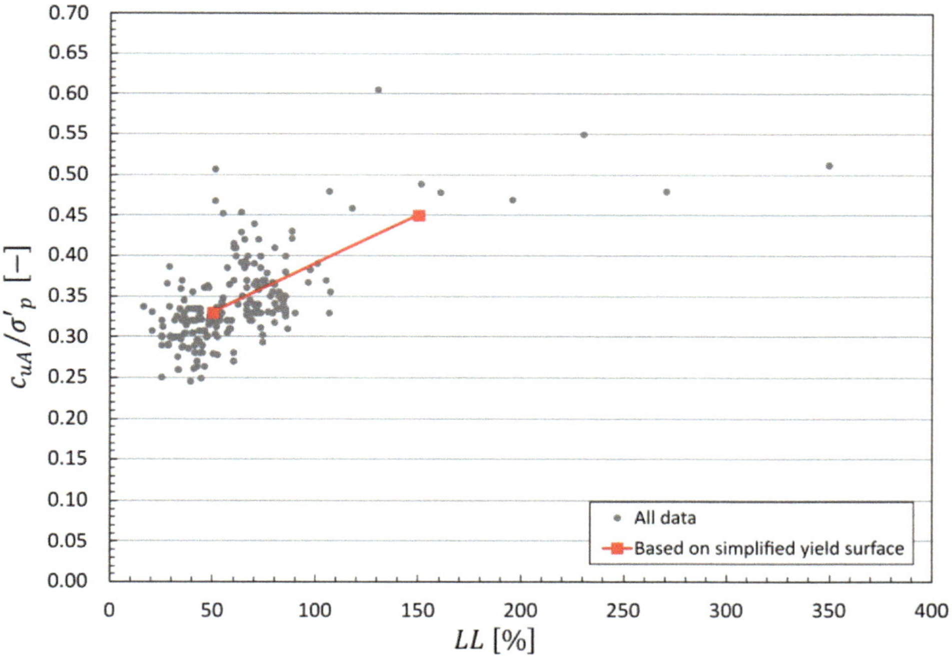

Figure 7.6 Normalised strength c_{uA} / σ'_p vs. plasticity LL . All bivariate data along with c_{uA} / σ'_p calculated from the simplified yield surface (two points $c_{uA} / \sigma'_p = 0.33$ at $LL = 50\%$ and $c_{uA} / \sigma'_p = 0.45$ at $LL = 150\%$ with an interpolation). The figure illustrates the data-centric and geomechanical approach combined.

7.5 UNCERTAINTIES AND SAMPLE QUALITY

The uncertain factors affecting normalised strengths (c_u / σ'_p) are, of course, the same as those affecting the included parameters: shear strength (c_u) and preconsolidation pressure (σ'_p). Examples of these factors are direction of loading, rate of loading, temperature, and sample quality. Naturally, this also applies to normalised strengths from CAUC tests (c_{uA} / σ'_p). The direction of loading can be neglected since it is always the same for oedometer and active triaxial tests. The rate of loading, on the contrary, is shown to affect both c_u (Lacasse 1995; Ladd and DeGroot 2003; Lunne and Andersen 2007) and σ'_p (Larsson and Sällfors 1986; Tavenas and Leroueil 1987). The combined effect is however more difficult to assess, but Ladd and DeGroot (2003) have shown a slight increase in c_{uA} / σ'_p with increased rate of loading, meaning that the effect on c_{uA} is higher than the effect on σ'_p . Furthermore, Lunne and Andersen (2007) concluded that the rate of loading does not impact the relation between strength and plasticity, indicating that the effect of plasticity on rate of loading vs. c_{uA} / σ'_p can be neglected.

Similarly, the effect of temperature on c_{uA} / σ'_p is barely studied; however, recently performed tests on a Swedish clay indicate that c_{uA} / σ'_p could increase with decreasing temperature (Vesterberg and Andersson 2022). If this is reflected in a plasticity effect is unknown, and further, the temperature at the laboratory tests is unknown for many of the test results available in the database. It is therefore not possible to analyse a possible temperature effect on the relationship between c_{uA} / σ'_p and LL .

The effect of sample quality on c_{uA} and σ'_p is extensively researched, and it is shown that low plastic clays are more sensitive to disturbance than high plastic clays (Lacasse et al. 1985; Lunne et al. 2006; Karlsrud and Hernandez-Martinez 2013). But again, the combined effect on c_{uA} / σ'_p is more difficult to assess and has not been studied in detail to the authors' knowledge. The few stress paths in Figure 7.5 indicate that there is no effect of sample quality on c_{uA} / σ'_p , and the simplified yield surface framework can give further indications since sample quality has not been shown to affect φ' . When a clay sample is disturbed, causing a destructuration of the clay fabric, the yield stresses (σ'_p in both vertical and horizontal directions) will typically decrease. In the simplified yield surface, this results in a shift in the 45° σ'_p line affecting also the peak strength in undrained loading (c_{uA}); however, the geometrical construction and the resulting value of c_{uA} / σ'_p remain constant as it is only dependent on φ' . The only effect of sample disturbance would be if the stress path does not reach the "top" of the yield surface, but there are too few studies on this to conclude. Further, there is no apparent effect of different samplers on c_{uA} / σ'_p in the NGI or LM data (Karlsrud and Hernandez-Martinez 2013; Hov and Garcia de Herreros 2020). This together indicates that there is probably no significant effect of sample quality on normalised active strengths c_{uA} / σ'_p .

7.6 CONCLUSIONS

This chapter has presented a data-centric and geomechanical approach to investigate the effect of plasticity on normalised strengths (c_{uA} / σ'_p) of normally to slightly overconsolidated soft clays ($OCR < 1.5$). A few initial remarks were made on the data origins of which one of the most important is knowing any execution or interpretation differences. In this case, the σ'_p interpretation is believed to give around 10% higher values using the

Norwegian practice compared to the Swedish practice. It was also noted that LL alone is a sufficient parameter for describing plasticity.

Linear regression analyses of the bivariate database showed that c_{uA} / σ'_p is weakly correlated to LL with a coefficient of determination (R^2) of 0.19 and 0.24, depending on if the difference in interpretation method of σ'_p was accounted for. Despite the very weak correlation, there was a clear trend of increasing c_{uA} / σ'_p with increasing LL with respect to the 95% confidence interval. On average, $c_{uA} / \sigma'_p \approx 0.30$ at $LL = 20\%$ and $c_{uA} / \sigma'_p \approx 0.38$ at $LL = 100\%$. A few data points show that organic clays, where $LL > 150\%$, have $c_{uA} / \sigma'_p \approx 0.5$. A visual trend over a large LL range is evident.

Furthermore, the data trend was analysed using a geomechanical approach where the drained Mohr–Coulomb failure criterion (c' and φ') and the vertical yield stress, i.e., σ'_p, were combined to form a simplified yield surface. A geometrical construction shows an increase in c_{uA} / σ'_p with increasing apparent φ'. Notably, $\varphi' = 30°$ results in $c_{uA} / \sigma'_p = 0.33$ and $\varphi' = 55°$ results in $c_{uA} / \sigma'_p = 0.45$. As the organic content increases, both LL and φ' increase, linking plasticity and normalised active strengths. The simplified yield surface thus provides a logical explanation of the data.

Having addressed the uncertainties and effects of sample quality, it is worth noting that the geomechanical framework is simplified. However, a useful maxim in science and engineering is "as simple as possible, but no simpler," adhered to in geotechnical engineering by, e.g., (Jardine 2020). Here, although the yield surface is simplified, it nonetheless can explain the behaviour resulting in the normalised strength seen from experimental data. And this is important, since data should be attempted to be explained, not only described, giving more confidence to the data-centric approach which includes real data along with its uncertainties.

REFERENCES

Atterberg, A. 1911. Clay and Its Relation to Water Content and Plasticity Limits [In Swedish]. *Kungliga Lantbruksakademiens Handlingar och Tidskrift* 50(2):132–158.

Atterberg, A. 1912. Soil Consistencies and Stiffness [In Swedish]. *Kungliga Lantbruksakademiens Handlingar och Tidskrift* 51(1):93–123.

Beesley, M.E.W., and Vardanega, P.J. 2020. Parameter Variability of Undrained Shear Strength and Strain Using a Database of Reconstituted Soil Tests. *Canadian Geotechnical Journal* 57(8):1247–1255.

Casagrande, A. 1932. Research on the Atterberg Limits of Soils. *Public Roads* 13(8):121–136.

Cheng, X.H., Ngan-Tillard, D.J.M., and den Haan, E.J. 2007. The Causes of the High Friction Angle of Dutch Organic Soils. *Engineering Geology* 93 (1–2):31–44.

Ching, J., and Phoon, K.K. 2013. Multivariate Distribution for Undrained Shear Strengths under Various Test Procedures. *Canadian Geotechnical Journal* 50(9):907–923.

Ching, J., and Phoon, K.K. 2014a. Correlations among Some Clay Parameters - The Multivariate Distribution. *Canadian Geotechnical Journal* 51(6):686–704.

Ching, J., and Phoon, K.K. 2014b. Transformations and Correlations among Some Clay Parameters - The Global Database. *Canadian Geotechnical Journal* 51(6):663–685.

Ching, J., Li, D.Q., and Phoon, K.K. 2016. Statistical Characterization of Multivariate Geotechnical Data. In: *Reliability of Geotechnical Structures*, CRC Press, Balkema, pp. 89–126.

D'Ignazio, M., Phoon, K.K., Tan, S.A., and Länsivaara, T. 2016. Correlations for Undrained Shear Strength of Finnish Soft Clays. *Canadian Geotechnical Journal* 53(10):1628–1645.

Holstad, Ø.B. 2016. *Shear Strength and Deformation Parameters on Gyttja* [In Norwegian]. Norwegian University of Science and Technology, Trondheim.

Holstad, Ø.B., and Degago, S. 2021. Strength and Deformation Characterization of Norwegian Organic Cohesive Soil (Gyttja). In: *18th Nordic Geotechnical Meeting. IOP Conference Series: Earth and Environmental Science*, Helsinki, 170:012018.

Hov, S., and Garcia de Herreros, C. 2020. Comparison of the Mini Block and Standard Piston Sampler in a Varved East Swedish Clay [In Swedish]. *LabMind*, A2018-A2026.

Hov, S., Prästings, A., Persson, E., and Larsson, S. 2021. On Empirical Correlations for Normalised Shear Strengths from Fall Cone and Direct Simple Shear Tests in Soft Swedish Clays. *Geotechnical and Geological Engineering* 39(7):4843–4854.

Jamiolkowski, M., Ladd, C.C., Germaine, J.T., and Lancellotta, R. 1985. New Developments in Field and Laboratory Testing of Soils. In: *11th International Conference on Soil Mechanics and Foundation Engineering*, 57–153. Balkema.

Janbu, N. 1963. Soil Compressibility as Determined by Oedometer and Triaxial Tests. In: *European Conference on Soil Mechanics and Foundation Engineering*, Wiesbaden. 19–25.

Janbu, N. 1979. Mechanics of Failure in Natural and Artificial Soil Structures. In: *International Symposium on Soil Mechanics*. Oaxaca, Mexico. pp. 95–124.

Janbu, N. 1989. *Fundamentals of Soil Mechanics* [In Norwegian]. Tapir, Trondheim.

Jardine, R.J. 2020. Geotechnics, Energy and Climate Change. *The 56th Rankine Lecture. Géotechnique* 70(1):3–59.

Jardine, R.J., and Hight, D.W. 1987. The Behaviour and Analysis of Embankments on Soft Clay. *Bulletin of the Public Works Research Center*, 159–244, Athens.

DeGroot, D., Lunne, T., Ghanekar, R., Knudsen, S., Jones, C.D., and Yetginer-Tjelta, T.I.. 2019. Engineering Properties of Low to Medium Overconsolidation Ratio Offshore Clays. *AIMS Geosciences* 5(3):535–567.

Karlsrud, K., and Hernandez-Martinez, F.G. 2013. Strength and Deformation Properties of Norwegian Clays from Laboratory Tests on High-Quality Block Samples. *Canadian Geotechnical Journal* 50(12):1273–1293.

Lacasse, S. 1995. Stress-Strain Behaviour: Importance of Mode and Rate of Load Application for Engineering Problems. In: *International Symposium on Pre-Failure Deformation Characteristics of Geo-Materials*. 2:887–907. Sapporo, Japan.

Lacasse, S., Berre, T., and Lefevbre, G. 1985. Block Sampling of Sensitive Clays. In: *11th International Conference on Soil Mechanics and Foundation Engineering*, 887–892. San Francisco.

Ladd, C.C., and DeGroot, D.J. 2003. Recommended Practice for Soft Ground Site Characterization: Arthur Casagrande Lecture. In: *12th Panamerican Conference on Soil Mechanics and Geotechnical Engineering*, 1–60. Cambridge, MA.

Ladd, C.C., Foott, R., Ishihara, K., Schlosser, F., and Poulos, H.G. 1977. Stress-Deformation and Strength Characteristics. In: *9th International Conference on Soil Mechanics and Foundation Engineering*, 421–494. Tokyo.

Larsson, R. 1977. Basic Behaviour of Scandinavian Soft Clays. *Swedish Geotechnical Institute*, Report 4, Linköping.

Larsson, R. 1980. Undrained Shear Strength in Stability Calculation of Embankments and Foundations on Soft Clay. *Canadian Geotechnical Journal* 17: 591–602.

Larsson, R. 1990. Behaviour of Organic Clay and Gyttja. *Swedish Geotechnical Institute*, Report 38, Linköping.

Larsson, R., and Sällfors, G. 1981. Hypothetical Yield Envelope at Stress Rotation. In: *International Conference on Soil Mechanics and Foundation Engineering*, 693–696. Stockholm.

Larsson, R., and Sällfors, G. 1986. Automatic Continuous Consolidation Testing in Sweden. *Consolidation of Soils: Testing and Evaluation*, ASTM STP 892, 299–328.

Larsson, R., Sällfors, G., Bengtsson P.E., Alén, C., Bergdahl, U., and Eriksson, L. 2007. Undrained Shear Strength [In Swedish]. *Swedish Geotechnical Institute, Information* 3: 1–68.

L'Heureux, J.S., Lindgård, A., and Emdal, A. 2019. The Tiller-Flotten Research Site: Geotechnical Characterization of a Very Sensitive Clay Deposit. *AIMS Geosciences* 5(4):831–867.

L'Heureux, J.S., Gundersen, A.S., D'Ignazio, M., Smaavik, T., Kleven, A., Rømoen, M., Karlsrud, K., Paniagua, P., and Hermann, S. 2018. Impact of Sample Quality on CPTU Correlations in Clay—Example from the Rakkestad Clay. In: *Cone Penetration Testing*, 395–400. CRC Press, London.

Lundström, K., and Dehlbom, B. 2019. Change in Properties over Time under Existing Embankments [In Swedish]. *Swedish Transport Authority and Swedish Geotechnical Institute*, Report A2016-10, Linköping.

Lunne, T., and Andersen, K. 2007. Soft Clay Shear Strength Parameters for Deep-Water Geotechnical Design. In: *6th International Site Investigation and Geotechnics Conference, Confronting New Challenges and Sharing Knowledge*, 151–176. London.

Lunne, T., Berre, T., and Strandvik, S. 1997. Sample Disturbance Effects in Soft Low Plastic Norwegian Clay. In: *Recent Developments in Soil and Pavement Mechanics*, 81–102. Rotterdam, Balkema.

Lunne, T., Berre, T., Andersen, K.H., Strandvik, S., and Sjursen, M. 2006. Effects of Sample Disturbance and Consolidation Procedures on Measured Shear Strength of Soft Marine Norwegian Clays. *Canadian Geotechnical Journal* 43(7):726–750.

Mayne, P.W., and Mitchell, J.K. 2011. Profiling of Overconsolidation Ratio in Clays by Field Vane. *Canadian Geotechnical Journal* 25(1):150–157.

Mesri, G. 1975. Discussion on 'New Design Procedure for Stability of Soft Clays.' *Journal of Geotechnical Engineering* 101: 409–412.

Mesri, G. 1989. A Reevaluation of Su(Mob) = 0.22σ 'p Using Laboratory Shear Tests. *Canadian Geotechnical Journal* 26(1):162–164.

Międlarz, K., Konkol, J., and Bałachowski, L. 2019. Effective Friction Angle of Deltaic Soils in the Vistula Marshlands. *Studia Geotechnica et Mechanica* 41(3):143–150.

NGI., 2017. *OCR and Type of Stress Path* [In Norwegian]. Internal Report, Norwegian Geotechnical Institute, Trondheim.

NIFS, 2014. *Choice of Characteristic SuA-Profile Based on Field and Laboratory Investigations* [In Norwegian]. Norwegian Water Resources and Energy Directorate, Oslo.

O'Kelly, B.C. 2017. Measurement, Interpretation and Recommended Use of Laboratory Strength Properties of Fibrous Peat. *Geotechnical Research* 4(3):136–171.

Paniagua, P., L'heureux, J.S., Yang, S.Y., and Lunne, T. 2016. Study on the Practices for Preconsolidation Stress Evaluation from Oedometer Tests. In: *Proceedings 17th Nordic Geotechnical Meeting*, 547–557, Reykjavik.

Paniagua, P., D'Ignazio, M., L'Heureux, J.S., Lunne, T., and Karlsrud, K. 2019. CPTU Correlations for Norwegian Clays: An Update. *AIMS Geosciences* 5(2):82–103.

Phoon, K.K. 2020. The Story of Statistics in Geotechnical Engineering. *Georisk* 14(1):3–25.

Phoon, K.K., Ching, J., and Cao, Z. 2022. Unpacking Data-Centric Geotechnics. *Underground Space* 7(6):967–989.

Sällfors, G. 1975. *Preconsolidation Pressure of Soft, High-Plastic Clays*. Doctoral thesis, Chalmers University of Technology, Gothenburg.

Sällfors, G., and Larsson, R. 2016. *Compilation of Case Records* [In Swedish]. Swedish Transport Authority, Publication 2017:078. Gothenburg.

Sällfors, G., Larsson, R., and Bengtsson, P.E. 2017. *Recommendations for Undrained Shear Strength* [In Swedish]. Swedish Transport Authority, Publication 2017:078. Gothenburg.

Tang, C., and Phoon, K.K. 2021. Model Uncertainties in Foundation Design. Model Uncertainties in Foundation Design. In: *Model Uncertainties in Foundation Design*, CRC Press, Boca Raton, FL.

Tavenas, F., and Leroueil, S. 1987. Laboratory and in Situ Stress-Strain-Time Behavior of Soft Clays I: Laboratory Tests. In: *Simposio Internazionale de Ingegneria Geotecnica*, Mexico City. 7–47.

Vesterberg, B., and Andersson, M. 2022. *Settlement and Strength Properties in Sulphide Soils* [In Swedish]. Swedish Geotechnical Institute, Linköping.

Westerberg, B. 1999. *Behaviour and Modelling of a Natural Soft Clay*. Doctoral thesis, Luleå University of Technology, Luleå.

Chapter 8

Prediction for the mechanical response of gravels

Jingmao Liu, Degao Zou, Duo Li, Fanwei Ning, Chenguang Zhou, and Kaiyuan Xu

This chapter presents an extensive study on predictions for the mechanical response of gravels. Large amounts of laboratory testing data for gravels are not sufficiently utilized to support the prediction of the mechanical properties of gravels. A dataset of gravels from a triaxial test, termed Geo-Gravel, was established to provide the basis for estimating the mechanical response using physical properties and external conditions in the absence of test results. The dataset involved 332 types of gravels with 1102 triaxial compression test records. Fifteen indicators affecting mechanical behaviors were collected, including particle properties (e.g., particle hardness, σr, and roundness ρ), soil mass properties (e.g., initial void ratio e0, typical particle sizes, and gradation parameters), and external condition (e.g., confining pressure σ3). The dataset was employed to predict the shear strength and stress–strain–volume responses. Based on this dataset, empirical formulations and deep learning models were developed to predict shear strength and stress–strain–volume response, respectively. Firstly, the detail of establishing the dataset is presented; subsequently, the methods of predicting shear strength (including empirical formulas and models) are presented; finally, the methods of predicting curves (including empirical formulas and models) are presented. The dataset can supply the possibility for data-driven geotechnical design.

8.1 INTRODUCTION

Gravels are widely used in hydraulic construction, road, and railway projects due to their high strength, low compressibility, and widespread availability (Indraratna et al. 1998, Sevi and Ge 2012, Xiao et al. 2014). It is important to accurately assess the mechanical behavior of gravels for the stability analysis and safety evaluation of engineering. Triaxial test is a commonly used approach to evaluate the mechanical properties of soils, primarily employed to determine the shear strength and deformation characteristics of gravels (Xu et al. 2012; Chu 2014; Huang et al. 2019). However, performing experimental investigations involves high costs and long period, including processes from quarry blasting for extraction, material transportation, to laboratory tests. Moreover, many projects are unable to conduct relevant experiments at the preliminary design phase. Experimental parameters are often obtained through simple analogies with similar projects, resulting in a lack of a reasonable basis for parameter determination. Additionally, the absence of test curves poses a challenge in determining the parameters for advanced constitutive models. To address this, efficiently collecting and leveraging existing experimental data can offer a practical solution. Establishing a reliable dataset is crucial for this purpose, and integrating these data with data-driven approaches supports the evaluation of gravel's mechanical properties and informs engineering decisions.

 DOI: 10.1201/9781003441946-8

Recent years have seen the establishment of valuable datasets focusing on geotechnical materials (Phoon and Retief 2016; Phoon 2020). Notably, Ching and Phoon et al. provided useful multivariate datasets on clay (Ching and Phoon 2012; Ching and Phoon 2014; Ching and Phoon 2019), sand (Ching et al. 2017), and rock (Ching et al. 2018) properties. Tang and Phoon (2018; 2019a; 2019b) further provide a comprehensive survey of datasets on geotechnical structures. These studies provide important support for the assessment of geotechnical variability and uncertainty in site data. Regarding gravel materials, Kaunda (2015) gathered data on 47 types of rockfills with 165 test records, utilizing machine learning for shear strength prediction. Ovalle et al. (2020) collected data on 33 types of gravels with 158 test records.

Further, there are also extensive datasets that have been used to model and estimate the mechanical behaviors of the soils. Datasets regarding the mechanical behavior of soils primarily come from three sources: constitutive models (Javadi et al. 2012; Zhao et al. 2014; Zhang et al. 2020; 2021b; Wu et al. 2023b), discrete element simulations (Li et al. 2017; Wang and Sun 2019; Wu and Wang 2022; Wang et al. 2022; Guan et al. 2023), and laboratory tests (Habibagahi and Bamdad 2003; Nassr et al. 2018; Rashidian and Hassanlourad 2014; Banimahd et al. 2005; Araei 2014). Constitutive models and discrete element simulations can generate up to thousands of data records, through calculations based on specific requirements. These datasets are mainly used to validate and improve the effectiveness of machine learning (ML) methods, as well as to fit and model the mechanical behavior of specific materials, serving as a substitute for complex constitutive models. However, their accuracy is generally limited by the rationality of model assumptions and parameter selections. The data struggle to reliably reflect the true patterns of soil characteristics still depends on test data for validation.

Phoon and Zhang (2023) emphasize placing data at the core and underscore the significance of a data-centric approach in geotechnics. Unlike areas like computer technology and the Internet of Things, where data is readily accessible through automated means, collecting primary gravel test data is challenging due to diverse, non-digital sources. This necessitates extensive manual effort for collection, hindering dataset development, and constraining the advancement of data-driven approaches like machine learning. Thus, establishing a reliable dataset of laboratory tests is crucial.

With the rapid development of artificial intelligence (AI) technology, data-driven methods, particularly machine learning (ML), have demonstrated significant potential in the field of geotechnical engineering due to their powerful capabilities in handling intricate problems (Reich 1997, Vadyala et al. 2022, Aydin et al. 2023). ML techniques efficiently harness available data, directly learning and recognizing patterns without making any assumptions. They discover hidden patterns and capture complex nonlinear relationships (Janiesch et al. 2021). The cornerstone for the success of ML methods in civil engineering lies in the available datasets and efficient modeling algorithms (Schmidhuber 2015).

Machine learning (ML) is an important approach to achieving AI and has been widely applied in civil engineering practice and shown great promise (Lazarevska et al., 2014; Adeli, 2001; Zhu et al., 2021; Wei et al., 2021). Based on these datasets, various ML algorithms were applied to predict the mechanical properties using index properties, such as rocks (Mahmoodzadeh et al., 2021; Liu et al., 2015), soils (Zhang et al., 2021; Hao and Pabst, 2022; Zhang et al., 2023; Zhang et al., 2021), and concretes (Patil et al., 2022; Ly et al., 2019; Ling et al., 2019; Chen et al., 2018). Recently, Kaunda (2015) established a dataset with 165 sets of data for rockfill materials and used index properties to predict the shear strengths of rockfill materials using an artificial neural network (ANN).

Based on this dataset (Kaunda, 2015); Ahmad et al. (2021, 2022) and Zhou et al. (2019) applied support vector machine (SVM), Random forest (RF), AdaBoost, k-nearest neighbor (KNN), Gaussian process regression (GPR), and Cubist models for shear strength prediction.

Additionally, the feasibility of various ML algorithms in capturing the mechanical behavior of soils has been explored and validated (Zhang et al. 2020, 2022b, Qu et al. 2023). Shallow ML algorithms dominate, such as evolutionary polynomial regression (EPR) (Javadi et al. 2012), genetic programming (GP) (Javadi and Rezania 2009), support vector machine (SVM) (Zhao et al. 2014), radial basis function (RBF) (Peng et al. 2008), artificial neural network (ANN) (Wu et al. 2023b), etc., which can effectively utilize simple parameters (such as void ratio) and computation results obtained from constitutive models and discrete element simulation datasets to model the mechanical response of specific soils. However, these algorithms often exhibit limitations when predicting the mechanical behavior of unknown materials through test datasets containing multiple physical parameters and a large amount of data (LeCun et al. 2015). For instance, the ANN model proposed by Penumadu and Zhao (1999) encounters challenges in simultaneously capturing the mechanics behavior of soil under varying confining pressures, prompting the need for distinct models for different pressure stages. This is due to the small capacity of shallow models with limited representational capabilities and inadequate ability to capture complex features and relationships. Furthermore, their performance relies heavily on manual feature engineering, presenting a challenge in handling intricate and extensive test data. There exists a risk of overfitting on large datasets, which may result in a reduction of the model's generalization ability (Schmidhuber 2015; LeCun et al. 2015).

In comparison to shallow ML models, deep learning (DL) models possess the capability to automatically capture essential features and facilitate end-to-end predictions using original data as input. DL, characterized by its flexible structure, robust nonlinear mapping capacity, and powerful generalization, surpasses other shallow ML algorithms in various domains such as computer vision, speech recognition, and bioinformatics (Szegedy et al. 2015; Chauhan and Singh 2018). In the domain of large-scale visual recognition, with the increase of images in the ImageNet dataset and the development from conventional MLs to advanced DLs, the recognition error rate reduced from the initial at least 25% to less than 5% (Russakovsky et al., 2015; Deng et al., 2009), which is lower than that of humans (5.1%). Sun et al. (2017) attribute this significant success to the advanced algorithmic model based on the availability of large-scale data and the benefit of high-performance computing power.

Recurrent neural networks (RNNs), especially a variant Long Short-Term Memory (LSTM) network, demonstrate significant advantages in handling sequential data, such as stress–strain–volume responses. This has proven effective in capturing the mechanics behavior of soils (Zhang et al. 2020; Qu et al. 2021; Benabou 2021; Wu and Wang 2022). However, the challenge lies in fully utilizing complex spatial relationships among multiple physical properties, potentially limiting the model's performance in managing spatiotemporal data. For instance, the study of Zhang et al. (2021b) with LSTM reported unsatisfactory prediction results under low-stress levels. Recognizing the potential of convolutional neural networks (CNNs) in extracting spatial features, hybrid models combining CNNs and RNNs have gained popularity in various applications, including air quality prediction (Zhang et al. 2022a, Wu et al. 2023a), motor fault detection (Mohammad-Alikhani et al. 2023), energy consumption forecasting (Albelwi 2022), and agricultural farming monitoring (Du et al. 2022). Inspired by this approach, the implementation of such a hybrid structure to capture the mechanical behavior of gravels, leveraging both spatial features in physical parameters and sequential information in stress–strain–volume curves, holds promise to significantly enhance predictions of mechanics behavior for gravels. Notably, this approach has yet to be explored in applications for modeling the mechanical behavior of soils.

8.2 ESTABLISHMENT OF GEO-GRAVEL DATASET

8.2.1 Establishment process

This study introduces "Geo-Gravel," a gravel dataset using particle and soil mass properties to estimate peak deviatoric stress. To create a robust dataset, diverse experiments from 63 public papers and 22 practical lab reports, including triaxial compression tests on 332 gravels with 1102 records, were collected. 90.2% of the material is associated with rockfill used in dam construction. The data was sourced from the Web of Science and the China National Knowledge Infrastructure. The triaxial compression tests were performed all under drained conditions. Generally, the peak point of the deviatoric stress is considered as q_{peak} (equal to $\sigma_{1peak} - \sigma_3$, σ_{1peak} is the peak major principal stress, σ_3 is minor principal stress or confining pressure) in triaxial compression tests. If there is no peak point of the deviatoric stress as for strain hardening soils, the corresponding value of deviatoric stress at the strain of 15% is used as q_{peak} (GB/T 50123. 2019; ASTM D7181-11). q_{peak} is the key parameter for assessing soil shear strength. Based on multiple q_{peak}, other strength parameters such as the cohesion (c) and angle of internal friction (φ) in the Mohr–Coulomb criterion (Labuz and Zang, 2012) can be easily converted by fitting. q_{peak} obtained directly from the experiments and avoids any assumptions and simplifications in the fitting process as compared to calculating c and φ . Therefore, the inclusion of q_{peak} contributes to more accurate assessment results.

Previous research (Xiao et al., 2014; Liu and Chen, 2019; Varadarajan et al., 2003) highlights the influence of factors like lithology, density, gradation, particle shape, and stress conditions on gravel mechanics. Based on previous studies (Sadrekarimi and Olson, 2010), the factors affecting q_{peak} are divided into three categories: particle properties, soil mass properties, and external conditions (specifically, confining pressure in our dataset). The 12 indicators for gravels collected in the dataset were as many as possible.

(1) Particle properties

Particle shape. Generally, gravels are primarily obtained by artificially blasting parent rock or collected from natural river alluvium with different shapes. The regularity factor ρ is selected as the indicator of particle shapes, which has been adopted in previous studies (Cho et al., 2007; Krumbein, 1941).

Particle hardness. The uniaxial compressive strength of saturated parent rock σ_r is selected as the indicator of particle hardness.

For the missing data, the value was selected based on descriptions of materials as Table 8.1 (Bareither et al., 2008) and Table 8.2 (GB50021-2001. 2001), and the average value within each category was chosen as the empirical value.

Table 8.1 Roundness categories including the range of roundness and average roundness

Roundness	*Range*	*Average*
Angular	0.10–0.26	0.18
Sub-angular	0.26–0.42	0.34
Sub-rounded	0.42–0.58	0.50
Rounded	0.58–0.74	0.66
Well-rounded	0.74–0.90	0.82

Table 8.2 Hardness categories including the range of hardness, average hardness, and description

Hardness	*Range (MPa)*	*Average (MPa)*	*Description*
Highly hard	>60	75	Fresh granite, diorite, diabase, basalt, andesite, gneiss, quartzite, quartz sandstone, siliceous conglomerate, siliceous limestone, etc.
Sub-hard	30–60	45	Weak-weathered hard rock; fresh marble, slate, limestone, dolomite, calcareous sandstone, etc.
Sub-soft	15–30	22.5	Medium-weathered hard rock or sub-hard rock; fresh tuff, phyllite, marl, sandy mudstone, etc.
Soft	5–15	10	Highly weathered hard rock or sub-hard rock; medium-weathered sub-soft rock; fresh to weak-weathered shale, mudstone, argillaceous sandstone, etc.
Highly soft	≤5	2.5	Completely weathered rock; hypabyssal rock.

(2) Soil mass properties

Initial void ratio. The initial void ratio e_0 is the important control index, especially for rockfill dams. To maintain data integrity, initial void ratio e_0, dry density ρ_d, and specific gravity G_s are collected to facilitate conversion and supplementation.

Particle gradation. The coefficient of uniformity coefficient C_u and curvature coefficient C_c, as well as the typical particle size D_{10}, D_{30}, D_{50}, D_{60}, D_{80}, and D_{max} corresponding to the passing percentage of 10%, 30%, 50%, 60%, 80%, and 100%, respectively, are collected as the comprehensive indicators of particle gradation.

(3) External condition

The confining pressure σ_3 is selected as an indicator of the external conditions for estimating q_{peak}.

Overall, a total of 12 indicators of gravels were collected, including ρ, σ_r, e_0, D_{max}, D_{80}, D_{60}, D_{50}, D_{30}, D_{10}, C_u, C_c, and σ_3.

(4) Stress–strain–volume curves

The equal-length sequences (axial strain values of 10%) of stress–strain–volume curves were intercepted from the test curves that were loaded with different axial strains.

8.2.2 Data statistical analysis

Figure 8.1 visually displays the data's statistical properties using a scatter matrix. The lower triangular section exhibits scatter plots, the diagonal presents data distribution histograms, and the upper triangular section showcases the correlation matrix.

The statistical characteristics of the indicators are as follows:

Regularity factor. The range of ρ is from 0.18 to 0.82, with a mean value of 0.48 and a standard deviation of 0.19. The 25% percentile (Q_1), 50% percentile (Q_2 or median), and 75% percentile (Q_3) values are 0.34, 0.34, and 0.66, respectively. ρ is distributed centrally at both ends with a bimodal distribution, indicating the presence of two main material types (blasting and river alluvium gravels). Sub-angular and angular particles ($\rho < 0.5$) constitute 66.1%, mainly from parent rock blasting. Sub-rounded, rounded, and well-rounded particles ($\rho \geq 0.5$) from river alluvium make up 33.9%.

Particle hardness. σ_r is in a wide range from 2.5 to 131.1MPa. The mean value is 68.1 MPa, with a standard deviation of 16.9 MPa, and Q_1, Q_2, and Q_3 values of 60, 75, and 75 MPa, respectively. The majority (95.3%) has σ_r >30 MPa, indicating highly hard and sub-hard rock gravels. The minority (4.7%) with σr ≤30 MPa represents sub-soft, soft, and highly soft materials. Thus, the dataset is primarily suited to evaluate the materials for highly hard and sub-hard rock gravels.

Initial void ratio. The range of e_0 is from 0.142 to 0.890. The mean value is 0.297, with a standard deviation of 0.136. Q_1, Q_2, and Q_3 values are 0.230, 0.250, and 0.300, respectively. e_0 shows a positive skew distribution and a medium negative correlation with q_{peak}.

Gradation. C_u is in a wide range from 1.0 to 2140.0, with a mean value of 48.67 and a standard deviation of 196.06. Q_1, Q_2, and Q_3 values are 7.0, 14.0, and 34.5, respectively. 13.5% falls under $C_u < 5$ (uniformly graded soil), and 86.5% falls under $C_u \geq 5$. A small portion with clay has $C_u > 500$ values, making up 1.5%. C_c is found in a range from 0.10 to 149.60, with a mean value of 4.60 and a standard deviation of 14.93. Q_1, Q_2, and Q_3 values are 1.1, 1.6, and 2.5, respectively. Continuous graded soil ($1 < C_c < 3$) constitutes 64.4%, while $C_u \leq 1$ and ≥3 make up 35.6%. D_{max} ranges from 10 to 160 mm, with a significant concentration at 60 mm (74.5%), influenced by the commonly employed specimen diameter limit. The typical particle size distributions are concentrated, and share a similar discrete degree with negative skew and leptokurtic distribution (i.e., the shape of the distribution curve is more pointed and steeper). D_{max} and D_{80} have a weak positive correlation with q_{peak}, while D_{30}, D_{10}, and C_u have a weak negative correlation.

External condition. σ_3 spans 0.001–4.0 MPa, encompassing diverse engineering project scenarios. The mean value is 0.985 MPa, respectively. Q_1, Q_2, and Q_3 values are 0.4, 0.8, and 1.4 MPa, respectively. There is a strong positive correlation observed between σ_3 and q_{peak}.

8.3 PREDICTION OF SHEAR STRENGTH

8.3.1 Machine learning model

Three algorithms such as artificial neural network (ANN), support vector machine regression (SVR), and random forest (RF) are used. For a more in-depth understanding

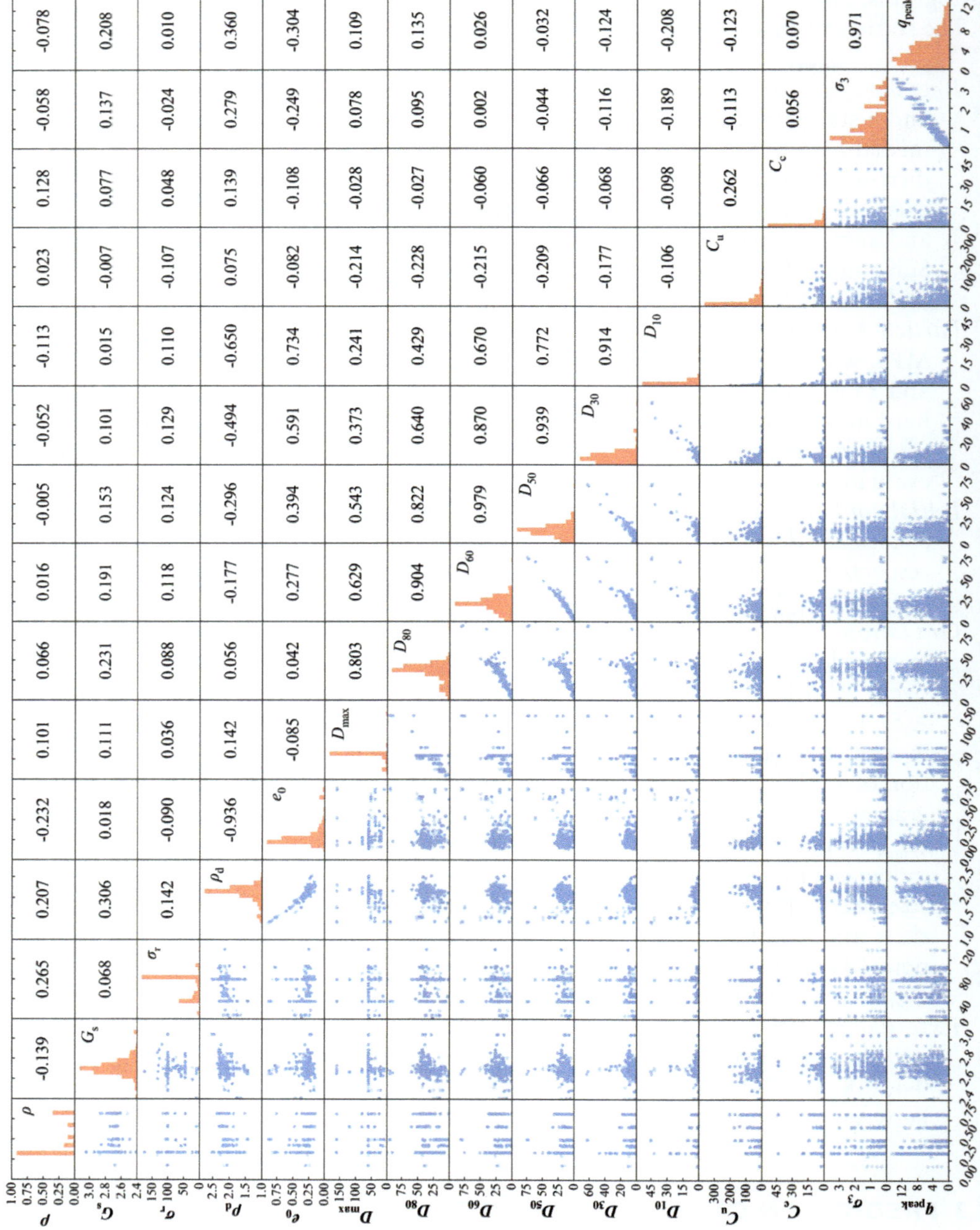

Figure 8.1 Scatter matrix.

of the basic theory behind these models, readers are referred to previous studies, specifically those conducted by Phoon and Zhang (2023), Aydin et al. (2023), and Janiesch et al. (2021). Shear strength prediction was approached as a nonlinear regression optimization task, using 12 input indicators (e_0, C_u, C_c, D_{10}, D_{30}, D_{50}, D_{60}, D_{80}, D_{max}, σ_r, ρ, and σ_3). The single output variable was shear strength (i.e., σ_{1peak}, corresponding to the empirical equations in the later section). The dataset was split into 5:1 training and testing sets. The training set was used to learn patterns, with model parameters fine-tuned through optimization and fivefold cross-validation. This involved dividing the dataset into five subsets, using four for training and one for validation in each iteration. The test set evaluated the model's proficiency on unseen data, gauging its generalization capabilities. Crucial hyperparameters for each model underwent iterative optimization using the trial and error method. The ANN has a hidden layer with 50 neurons and a sigmoid activation function. For the SVR, the regularization parameter (C) is set to 10, the regression loss parameter ε is 0.1, and the kernel type is radial basis function (RBF). The RF consists of 150 trees, and each limits to a depth of five levels. Models for the shear strength of gravels are implemented using the scikit-learn library in Python.

Three performance metrics, i.e., coefficient of determination (R^2), root mean square error (RMSE), and mean estimation error (MAE), were employed to evaluate the predictive results of the models. The calculations for these metrics are as follows:

$$R^2 = 1 - \frac{\Sigma_{i=1}^{n}(A_i - P_i)^2}{\Sigma_{i=1}^{n}(A_i - \bar{A})^2} \tag{8.1}$$

$$MAE = \frac{1}{N}\Sigma_{i=1}^{n}\left|A_i - P_i\right| \tag{8.2}$$

$$RMSE = \sqrt{\frac{1}{N}\Sigma_{i=1}^{n}(A_i - P_i)^2} \tag{8.3}$$

where A_i and P_i are the actual value and predicted value, respectively, $\bar{A}$ is the average of the actual values, and n is the amount of the data.

Table 8.3 presents the evaluation results for the models, highlighting RF's superior performance in both the training set (R^2 = 0.972, MAE = 0.379 MPa, RMSE = 0.516 MPa) and the test set (R^2 = 0.917, MAE = 0.575 MPa, RMSE = 0.698 MPa) compared to ANN and SVR. Based on the optimal prediction performance of RF, the importance of 12 input indicators was assessed by analyzing their contributions to decision trees (Babar et al., 2020). Each indicator's contribution is gauged through the variation in node Gini

Table 8.3 Performance metrics of ANN, SVR, and RF models

Model	*Phase*	R^2	*MAE (MPa)*	*RMSE (MPa)*
ANN	Training	0.948	0.514	0.709
	Testing	0.835	0.633	0.998
SVR	Training	0.956	0.420	0.584
	Testing	0.881	0.625	0.861
RF	Training	0.972	0.379	0.516
	Testing	0.917	0.575	0.698

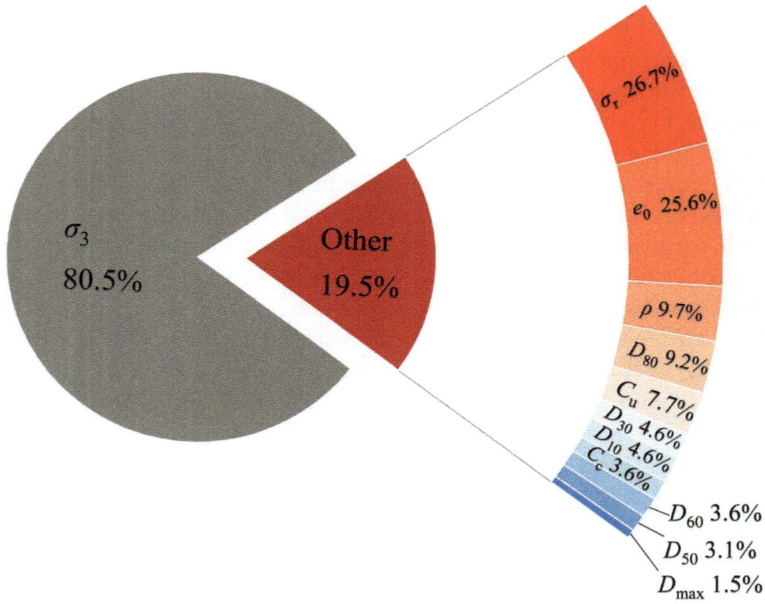

Figure 8.2 **Relative importance for 12 indicators to evaluate q_{peak}.**

impurity before and after branching in decision trees, with total contribution obtained by summing these variations across all trees. Normalization of contributions yields relative importance, allowing for the ranking of indicators (Archer and Kimes, 2008; Gómez-Ramírez et al., 2020). The importance is depicted in Figure 8.2.

The importance of ranking reveals that the external condition σ_3 holds the highest importance (80.5%) in estimating gravel q_{peak}, followed by σ_r, e_0, ρ, D_{80}, C_u, D_{30}, D_{10}, C_c, D_{60}, D_{50}, and D_{max}, totaling 19.5%. Among particle and soil mass properties, σ_r, e_0, ρ, D_{80}, and C_u stand out as the top five indicators, constituting 28.0%, 26.9%, 10.2%, 9.7%, and 8.1%, respectively. These key indicators formed the basis for a statistical model used to create an empirical formula for predicting shear strength.

8.3.2 Deep learning model

8.3.2.1 Seft-Net model architecture

In this chapter, the prediction of q_{peak} using index properties was regarded as a nonlinear regression optimization problem that can be presented as follows:

$$q_{peak} = f^{DL}(\sigma_r, \rho, e_0, D_{max}, D_{80}, D_{60}, D_{50}, D_{30}, D_{10}, C_u, C_c, \sigma_3) \tag{8.4}$$

where q_{peak} is peak deviatoric stress; f^{DL} denotes the relation between the input variables and q_{peak} from the model; a total of 12 input variables of gravels are collected as much as possible, including uniaxial compressive strength of saturated parent rock σ_r, regularity factor ρ, initial void ratio e_0, D_{max}, D_{80}, D_{60}, D_{50}, D_{30}, D_{10}, uniformity coefficient C_u, curvature coefficient C_c, and confining pressure σ_3.

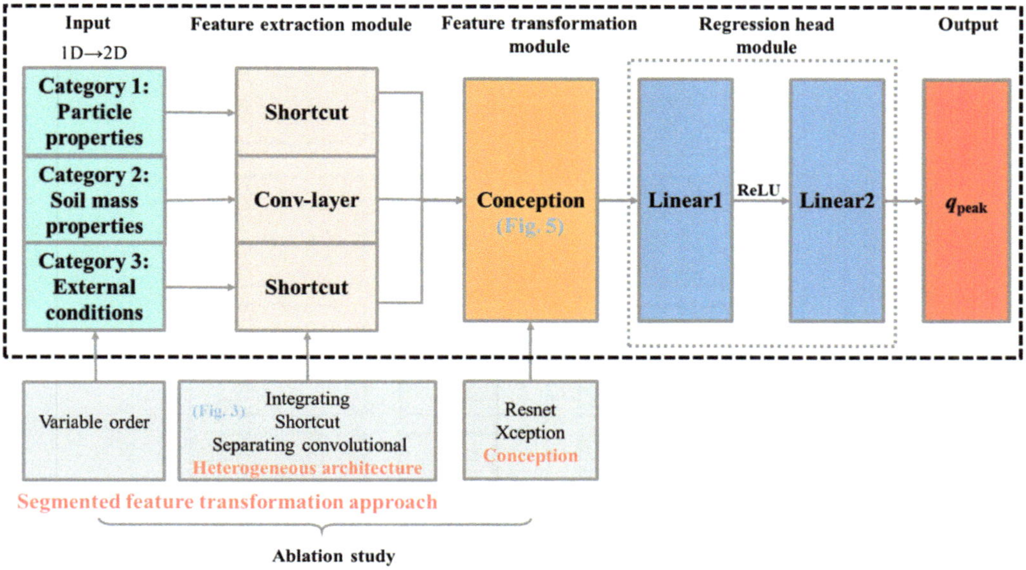

Figure 8.3 Seft-Net (Segmented feature transformation-Network) model architecture.

Thus, a deep learning model to achieve this complex task is proposed, which can abstract deep features and learn the potential mapping relationship of the data with excellent nonlinear transformation capability. The proposed methodology for q_{peak} prediction is based on the Segmented Feature Transform Network model (Seft-Net), which incorporates depthwise separable convolutions and a segmented feature transformation approach adopted for the gravels' physical properties.

The Seft-Net model architecture executes a three-step process: feature extraction module, feature transform module, and regression head module, as shown in Figure 8.3. First, the feature extraction module extracts rough features from the segmented input properties of gravels. Then, the feature transform module takes over and utilizes multi-scale receptive fields and multi-branch architectures to abstract high-level features from the gravel properties. Finally, the regression head module with fully connected layers is responsible for reducing the dimension of learned high-level features and providing the ultimate output as q_{peak} prediction. The utility of the proposed modules in Seft-Net is validated by an ablation study. The implementation process details of the proposed module are as follows.

8.3.2.1.1 Proposed feature extraction module

The Seft-Net model employed 12 input variables in the dataset. In pre-training, it was found that the ordering had a certain impact on the model performance, and the model performance is better when similar attribute variables are ordered together. Inspired by it, a novel segmented feature transformation approach is proposed. Generally, these properties influencing q_{peak} could be segmented into three different categories (Sadrekarimi and Olson, 2010): category 1 is about the particle properties (i.e., parent rock strength σ_{r}, regularity factor ρ), category 2 is about the soil mass properties (i.e., initial void ratio e_0, typical particle sizes $D_{\max}$, D_{80}, D_{60}, D_{50}, D_{30}, D_{10}, and gradation parameters C_{u} and

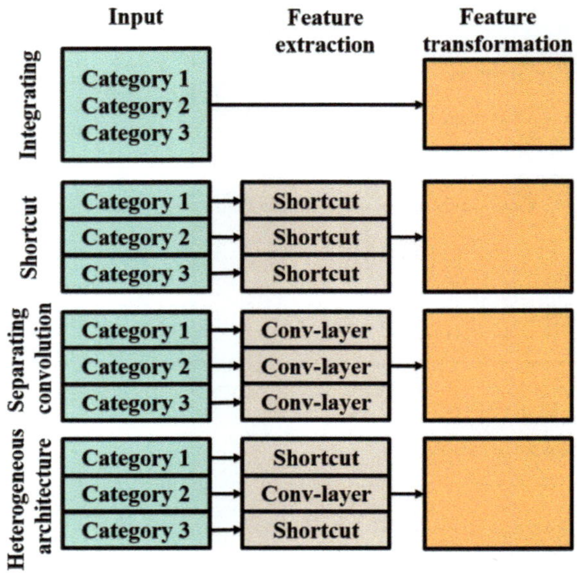

Figure 8.4 Architecture of different feature extraction modules.

C_c), and category 3 is about the external conditions (i.e., confining pressure σ_3). It is the segmenting of different variables based on their intrinsic physical properties, using the experience of long-term human research as a priori knowledge.

The input variables are first arranged in a 1 × 12 1D matrix. This initial arrangement is then subjected to a dimensionality expansion process, resulting in a 3 × 12 2D matrix that effectively expands the input space to optimize the model's efficiency in capturing spatial features associated with gravel properties. Subsequently, the variables within the 2D matrix are segmented into the aforementioned three categories and calculated by the feature extraction module. Each category is processed, respectively, by transmitting them through the heterogeneous architecture of neural networks. The utility of heterogeneous architecture of feature extraction in Seft-Net is validated by an ablation study, which compares the heterogeneous architecture module with three different modules, including the integrating module, shortcut module, and separating convolutional module, whose architectures are shown in Figure 8.4.

8.3.2.1.2 Proposed feature transform module

Depthwise separable convolution

Depthwise separable convolution has been proven to be successful in neural image classification compared to standard convolution. Standard convolution extracts feature both the spatial (width and height) and channel dimensions of the input with a high degree of coupling, where the convolutional kernel needs to describe spatial and cross-channel correlations simultaneously (Shang et al., 2020). In contrast to standard convolution, depthwise separable convolution firstly performs depthwise convolution independently on each channel of the input, followed by a pointwise convolution which processes the output of depthwise convolution with a 1 × 1 kernel to a new channel space, as shown in Figure 8.5.

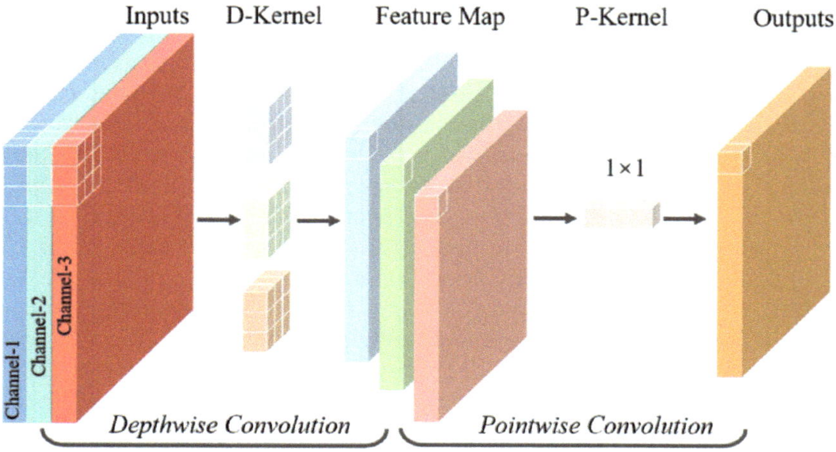

Figure 8.5 Depthwise separable convolution.

In depthwise convolution, a single kernel is applied to each input channel so that each channel can output one feature map, where the number of channels does not change after depthwise convolution. This process (*D-Conv*) can be expressed as Equation (8.2):

$$D\text{-}Conv(W,x)_{(i,j)} = \sum_{m,n}^{M,N} W_{(m,n)} \cdot x_{(i+m,j+m)} \tag{8.2}$$

In pointwise convolution, a 1 × 1 convolution is applied to combine the outputs of the depthwise convolution to extract spatial features. This does not change the spatial size of feature maps but can change the channel number. This process (*P-Conv*) can be expressed as Equation (8.3):

$$P\text{-}Conv(W,x)_{(i,j)} = \sum_{m,n}^{M,N} W_{(k)} \cdot x_{(i,j)} \tag{8.3}$$

where *W* is the weight matrix of convolutional kernels; *x* is the input of the convolutional layer, and (*i*, *j*) is the coordinate point of output feature maps; *m*, *n*, and *k* are the three dimensions of the convolutional kernel.

In further, the overall process (*DP-Conv*) of depthwise separable convolution can be expressed as Equation (8.4):

$$DP\text{-}Conv(W_p, W_d, x)_{(i,j)} = P\text{-}Conv(W_p, x)_{(i,j)}(W_p, D\text{-}Conv(W_d, x)_{(i,j)}) \tag{8.4}$$

Depthwise separable convolution can considerably reduce computational cost, by applying a separate filter to each input channel which reduces redundancy in the parameters. This facilitates better generalization because it reduces the risk of overfitting, and the efficiency of depthwise separable convolution is significantly higher than standard convolution (Chollet, 2017).

Conception module

Convolutional neural network (CNN)-based networks have achieved remarkable success in various image processing domains (Gu et al., 2018). Deep CNNs, in particular, have made noteworthy contributions to tasks like image classification and segmentation (Yamashita et al., 2018). They become widely known standards (Dhillon and Verma, 2020; Ajit et al., 2020). Increasing the depth and width of deep neural networks (DNNs) is a common approach to improving network performance (Khan et al., 2020; Tan and Le, 2019). Residual connections in models like ResNet (He et al., 2016) enable effective training even in deeper networks. Architectures inspired by the Inception module, such as GoogLeNet (Szegedy et al., 2015), have demonstrated the ability to extract multi-scale features. By concatenating different convolutional layers and stitching the resulting matrices together, the depth and width of the network can be efficiently expanded. This approach prevents overfitting and enhances model accuracy (Szegedy et al., 2016; Szegedy et al., 2017). Taking a step further, Xception (Chollet, 2017) is an improved architecture based on InceptionV3 (Szegedy et al., 2016) that separates channel relationships from spatial relationships through depthwise separable convolutions to efficiently extract features and improve performance.

Inspired by Xception (Chollet, 2017) and Densenet (Huang et al., 2017), a modified Xception architecture was used as the backbone of the feature transform module, called Conception. Similar to Xception (Chollet, 2017), the Conception architecture could be divided into three parts: entry flow, middle flow, and exit flow. The architecture of Conception is shown in Figure 8.6. The entry flow, which contains three residual blocks, utilizes two separable convolution layers and one normal convolution layer in parallel and then concretes the results as a bigger matrix. The middle flow simply shortcuts from three repeated separable convolutional layers and finally concatenates them in one matrix. This part is repeated eight times, as a higher count of duplication did not show great improvement in performance while increasing model complexity and calculating cost (Chollet, 2017). The upper part of the exit flow is similar to the residual blocks in the entry flow, then the exit flow uses two separable convolutional layers to adjust the output into vectors. Rectified Linear Unit (ReLU) is used as the activation function to perform nonlinear transforms. The successful part could be modified as a linear layer or convolutional layer to adapt to the demand of downstream tasks.

Unlike the conventional Xception (Chollet, 2017) architecture, Conception incorporates residual blocks (He et al., 2016) by directly adding the shortcut to the residual. The residual and shortcut vectors are concatenated into a two-row matrix, intending to retain more information about the original features from the output of the previous block. This change does show an improvement in final performance while replacing those blocks.

In the Conception module, as the input from the feature extraction module passes through subsequent convolutional layers, the network learns to extract increasingly abstract and high-level features. Lower layers tend to capture low-level features, while higher layers capture more complex features. Fusing hierarchically extracted features through multi-scale receptive fields and multi-branching architectures to abstract high-level features of gravel properties, which refer to abstract representations of the input gravel properties that capture complex and meaningful patterns relevant to the q_{peak} prediction. However, understanding these high-level features is difficult. In deep networks with multiple layers, the relationship between the input data and the high-level features becomes highly nonlinear and intricate. The models learn a huge number of parameters, and understanding the precise meaning and relevance of each parameter or connection becomes extremely difficult, impeding interpretability, as black boxes (Samek et al., 2021). We use an indirect approach to improve the interpretability of the model in a later section.

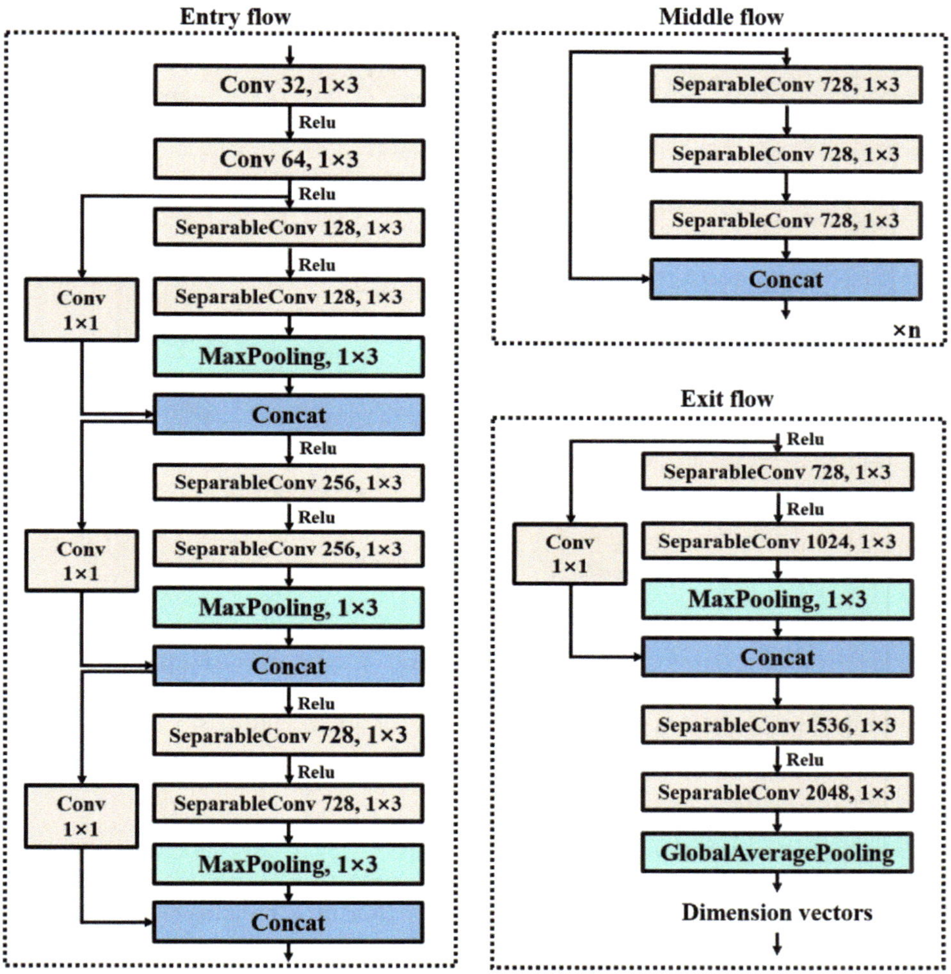

Figure 8.6 Conception module architecture.

Once the Conception module has extracted these high-level features of 12 gravel properties, they are fed into a few fully connected layers as the regression head module to reduce the dimension of features extracted from above and map to learned high-level features of gravel properties, providing the ultimate q_{peak} prediction in regression tasks.

8.3.2.2 Training strategy

In this section, the performance metrics of the evaluation and training strategy of the model are provided. The Seft-Net model can be deployed as below: First, the Seft-Net model is established and initialized. Then, the entire dataset was separated into training data and testing data with a ratio of 5:1. The ratio may vary depending on the specific requirements of the model and the dataset. The training data was used to decide the parameters of the model, and testing data was used to evaluate the model performance.

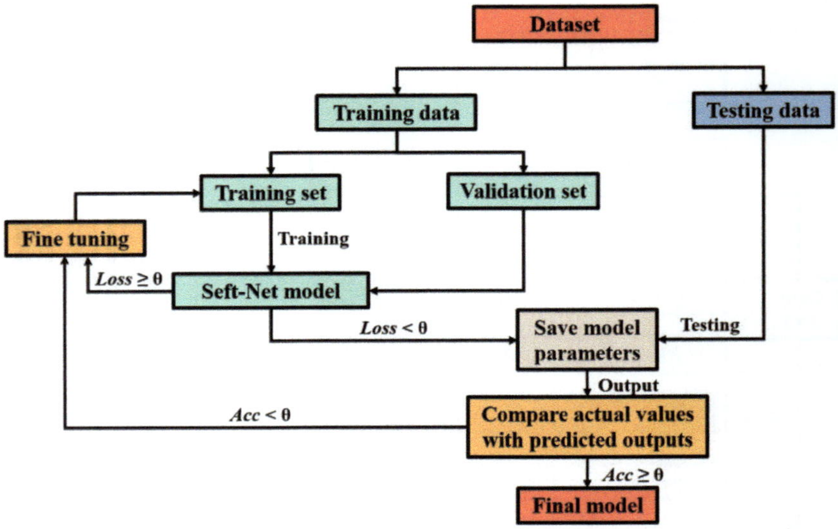

Figure 8.7 Calculation process of model.

During the experiments, Python is utilized as the programming language, and the PyTorch framework is used to perform the neural network. The model has a total of 22,756,959 parameters and was deployed on a workstation equipped with NVIDIA Tesla A100 GPU for training. The complete calculation process is shown in Figure 8.7.

(I) Performance metrics

In this study, the coefficient of determination (R^2), mean absolute error (MAE), root mean square error (RMSE), and mean absolute percentage error (MAPE) between the predicted and actual values were chosen as the criterion for defining the model's accuracy. Additionally, mean square error (MSE) and MAE are used as loss functions. These parameters can be calculated as follows:

$$R^2 = 1 - \frac{\Sigma_{i=1}^{n}(A_i - P_i)^2}{\Sigma_{i=1}^{n}(A_i - \bar{A})^2} \tag{8.5}$$

$$MAE = \frac{1}{N}\Sigma_{i=1}^{n}|A_i - P_i| \tag{8.6}$$

$$MSE = \frac{1}{N}\Sigma_{i=1}^{n}(A_i - P_i)^2 \tag{8.7}$$

$$RMSE = \sqrt{\frac{1}{N}\Sigma_{i=1}^{n}(A_i - P_i)^2} \tag{8.8}$$

$$MAPE = \frac{1}{N}\Sigma_{i=1}^{n}\frac{|A_i - P_i|}{A_i} \quad (8.9)$$

where A_i and P_i are the actual value and predicted value of q_{peak}, respectively, $\bar{A}$ is the average of the actual values, and n is the amount of data.

(2) Model training

During the training phase, the network learns the optimal weights for the convolutional filters and regression mapping layers by minimizing a loss function that measures the discrepancy between the predicted regression outputs and the actual values of q_{peak}. The network's parameters are adjusted through optimization techniques like gradient descent and backpropagation (Lecun et al., 2015) to improve the accuracy of the regression predictions.

During the training phase, the training set is iteratively fed into the Seft-Net model to optimize its parameters. The model parameters are updated and adjusted through the training process until the loss function, which measures the difference between the predicted values and the true values, decreases to a threshold of 1e−6. At this point, the model parameters are considered optimized, and they are saved for further evaluation and testing.

The training data is randomly divided into the training set and validation set. The ratio is set as 9:1, although other rational ratios can also be considered based on the specific requirements of the task. The training set is iteratively fed into the Seft-Net model to optimize its parameters. The model parameters are updated and adjusted through the training process making the loss function decrease. The validation set is used to fine-tune the hyperparameters and helps monitor the model's performance during training and detect signs of overfitting. The initial learning rate η is 0.001 with a rate decay of 0.98 every 10 epochs, and weight decay is 0.99 every 10 epochs. The specific values for these hyperparameters may vary depending on the model, dataset, and training objectives (Bergstra and Bengio, 2012). The settings used in this case are based on our former experiments and experience of the task domain. Two loss functions (MAE and MSE) were chosen for comparison.

(3) Model testing

In the testing phase, the testing data is used to evaluate the model performance with the parameters trained above. Once the accuracy achieved the goal pre-set, the model can be used as the final model; otherwise, the architecture of the model needs to be adjusted and trained again.

8.3.2.3 Experimental evaluation and results

8.3.2.3.1 Ablation study

The ablation study is used to identify the most important features of a model by adding, removing, or altering individual components of the model and observing the effect on the model's performance. It is used to identify which frameworks or features are most useful for the model's accuracy and to identify potential areas of improvement (Wang et al., 2021). According to the ablation study, the usefulness and effectiveness of the proposed modules to improve the accuracy of the model were verified. The evaluation metrics use the performance of the testing phase.

8.3.2.3.2 Input variable order

In the variable order experiments, the effects of different orders of input variables were compared. Due to the numerous ways of arranging and combining the 12 variables, five typical orderings (Table 8.4) were selected for experimentation based on the pre-training: (1) the variables were arranged in the order of the three categories; (2) the order of the variables was changed within the category; (3) the order of the variables was interchanged between category 1 and category 2; (4) the order of the variables was interchanged between category 2 and category 3; (5) the variables were interchanged in the order of all three categories.

As shown in Table 8.5(a), the result indicated that the ordering had a certain impact on the model performance. The model performance is better when similar attribute variables are ordered together as No. (1). No. (2) shows that changes in the order of variables within the category have little effect on model performance. No. (3)–No. (5) show that the interchange of variables between different categories has an effect on model performance, especially the ordering of σ_3 in category 3, which is related to the variable importance discussed later.

Different ordering of the variables may cause changes in the response pattern of the convolution kernel, and a reasonable ordering of the data facilitates the convolution network to capture useful local patterns and features more easily (Zhang et al. 2020, Dosovitskiy et al. 2021). By arranging related variables together, the model is more likely to find correlations and dependencies between these features. As for this task, it makes more sense to incorporate a priori knowledge indicating a certain way of ordering the input variables to help the network converge faster or extract relevant features better.

Table 8.4 Input variable order

No.	Variables											
(1)	σ_r	ρ	e_0	D_{max}	D_{80}	D_{60}	D_{50}	D_{30}	D_{10}	C_u	C_c	σ_3
(2)	ρ	σ_r	e_0	D_{60}	D_{50}	D_{max}	D_{30}	C_c	D_{10}	C_u	D_{80}	σ_3
(3)	D_{50}	ρ	e_0	D_{max}	D_{80}	D_{60}	σ_r	D_{30}	D_{10}	C_u	C_c	σ_3
(4)	σ_r	ρ	e_0	D_{max}	D_{80}	D_{60}	σ_3	D_{30}	D_{10}	C_u	C_c	D_{50}
(5)	D_{80}	C_u	e_0	D_{max}	σ_3	D_{60}	σ_r	D_{30}	D_{10}	D_{50}	C_c	ρ

Table 8.5 Experimental results of input variable order

	(a) Integrating				*(b) Heterogeneous architecture*			
No.	R^2	*MAE (MPa)*	*RMSE (MPa)*	*MAPE (%)*	R^2	*MAE (MPa)*	*RMSE (MPa)*	*MAPE (%)*
(1)	0.909	0.591	0.840	18.551	0.959	0.336	0.445	9.127
(2)	0.905	0.593	0.848	18.922	0.958	0.342	0.450	9.201
(3)	0.902	0.599	0.852	19.556	0.957	0.356	0.464	9.842
(4)	0.888	0.609	0.857	21.438	0.931	0.395	0.576	13.502
(5)	0.878	0.615	0.865	21.929	0.924	0.426	0.591	14.204

Inspired by it, we further propose a segmented feature transformation approach using a heterogeneous architecture in the feature extraction module. We also investigate the effects of variable order in heterogeneous architecture. The results in Table 8.5(b) indicated that the segmentation of variables into the wrong category is detrimental to the network decision making, resulting in some degradation of the model performance, and the degree of influence of the sort is related to the importance of the variable. It is speculated that correct categorization guides the model to enhance the learning of important features and improve the overall performance of the task through a priori knowledge, whereas incorrect categorization can somewhat misguide the model learning. In addition, it is noted that most deep convolutional networks usually automatically learn to adapt to feature representations of different ordering through convolutional operations. Thus, for non-sequentially related tasks, the order of the data usually has a limited impact on performance, and the impacts need to be assessed based on task requirements and data characteristics.

8.3.2.3.3 Feature extraction module

In the feature extraction module experiment, the utility of heterogeneous architecture within the segmented feature transformation approach is validated by comparing the heterogeneous architecture module with three different modules. Figure 8.8 illustrates the performance variance of different feature extraction modules.

The integrating module shows a decaying result due to its integrated architecture. This issue arises from the absence of segmentation in the input, preventing the fully connected (FC) layers from converging efficiently. The shortcut module shows a significant improvement in accuracy, and separating the convolutional module shows a close result, from which our idea comes from. By combining the shortcut module and separating the convolutional module, the heterogeneous architecture achieves optimal performance. The segmentation of variables is the grouping of similar indicator attributes, where a certain correlation or multicollinearity in the same categories may make the model become unstable and the predictive power reduced. By feature extraction from heterogeneous modules, this effect may be alleviated by segmenting to feature extraction, thus facilitating the convergence of the model and providing better feature abstraction for subsequent processing.

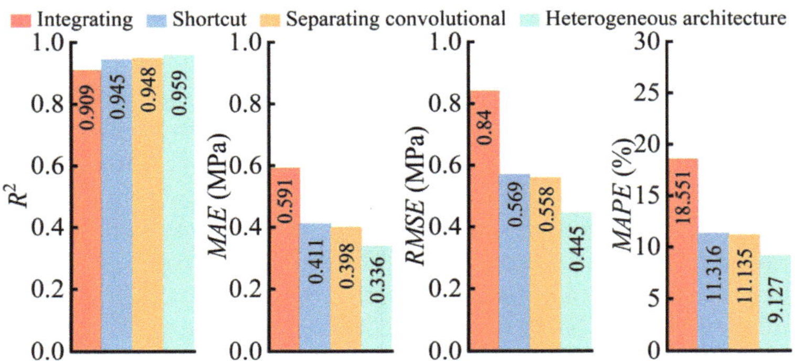

Figure 8.8 Ablation study results of different feature extraction modules.

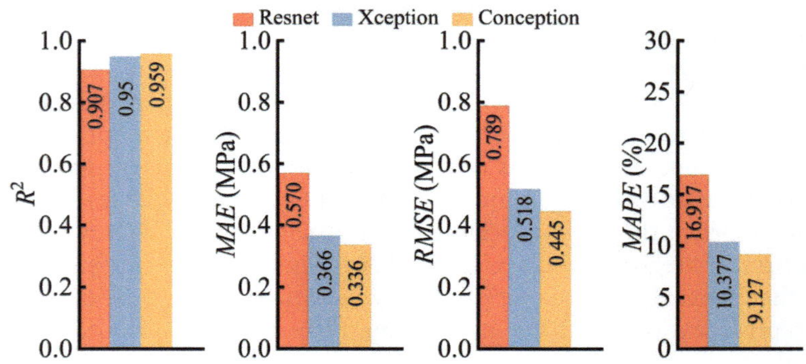

Figure 8.9 Ablation study results of different feature transformation modules.

8.3.2.3.4 Feature transformation module

To verify the utility of the proposed Conception module, compare the Conception with the original Xception module (Chollet, 2017) and ResNet module (He et al., 2016) in the Seft-Net. Figure 8.9 illustrates the performance variance of different feature extraction modules.

ResNet increases the depth of the network, while Xception and Conception expand the depth and width of the network efficiently. Additionally, Xception and Conception are able to extract features at different scales through convolutional operations with depthwise separable convolution and multi-scale fusion, using model parameters more efficiently and allowing the model to learn richer features. Therefore, Xception and Conception modules exhibit significantly superior performance to ResNet. By concatenating the residual and shortcut rather than adding both, the Conception module squeezes more performance by keeping more useful information as compared to Xception.

8.3.2.4 Seft-Net model evaluation

Figure 8.10 shows the progress of Seft-Net model training using the loss function of MAE and MSE for 3000 epochs. The loss value decreases gradually and the loss tends to converge with the increase in the epoch. The model parameters in the minimum loss are saved as the best model. It is indicated that the model using the loss function of MSE presents a faster convergence after about 300 epochs, while the model using MAE tends to converge after more than 1500 epochs.

Additionally, the accuracy of the model using the loss of MSE is higher than MAE, with MAPE of 9.127% and 9.901%, respectively. MSE gives higher weights to larger errors because of the squaring operation, making it more sensitive to outliers or large deviations between the predicted and actual values in this model. On the other hand, MAE treats all errors equally, regardless of their magnitude, but may provide a better understanding of the average prediction error (Goodfellow et al., 2016). Consequently, MSE was chosen as the loss function for the subsequent experiments. It applies to the present model and considers the specific characteristics of the problem.

The accuracy of the Seft-Net model in predicting q_{peak} of gravels is displayed in Figure 8.11, with the scatter plot of the actual and the predicted values for the training phase shown in Figure 8.11(a) and the testing phase in Figure 8.11(b), respectively. The closer

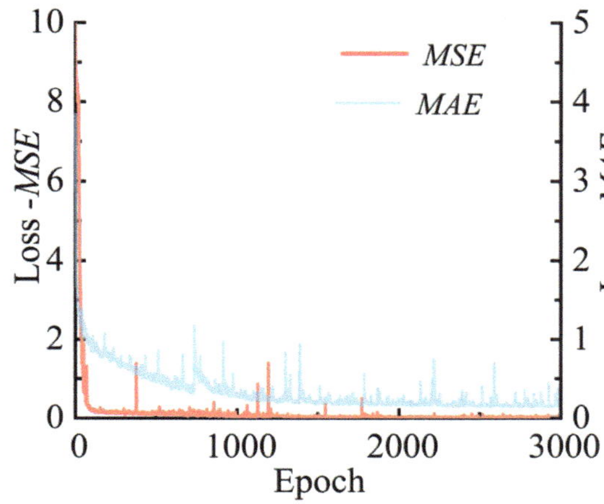

Figure 8.10. Progress of model training for 3000 epochs.

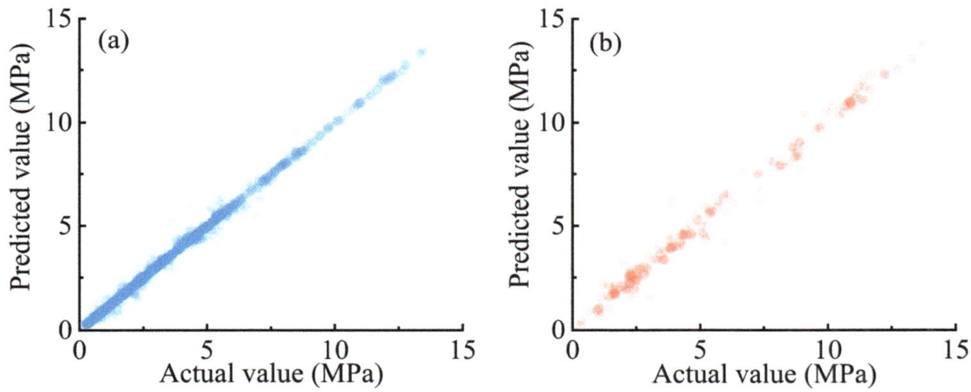

Figure 8.11 Scatter plots of actual vs. predicted q_{peak}: (a) training phase and (b) testing phase.

the points are to the diagonal line, the higher the accuracy. It is indicated that the higher aggregation of the results around the diagonal line both in the training and testing set, except for a few noise points.

The evaluation results in the training and testing phase are summarized in Table 8.6. As expected, the error in the testing phase is higher than that in the training phase, which is in line with the typical behavior of deep learning models (Zhang et al., 2021).

Figure 8.12 illustrates the percentage distribution of MAE and MAPE in both the training and testing phases. Notably, samples with high errors (MAE $\geq$ 1MPa, MAPE $\geq$ 50%) have a low occurrence percentage in both phases. The majority of predictions and fits

Table 8.6 Performance metrics of Seft-Net

Phase	R^2	*MAE (MPa)*	*RMSE (MPa)*	*MAPE (%)*
Training	0.994	0.111	0.232	4.102
Testing	0.959	0.336	0.445	9.127

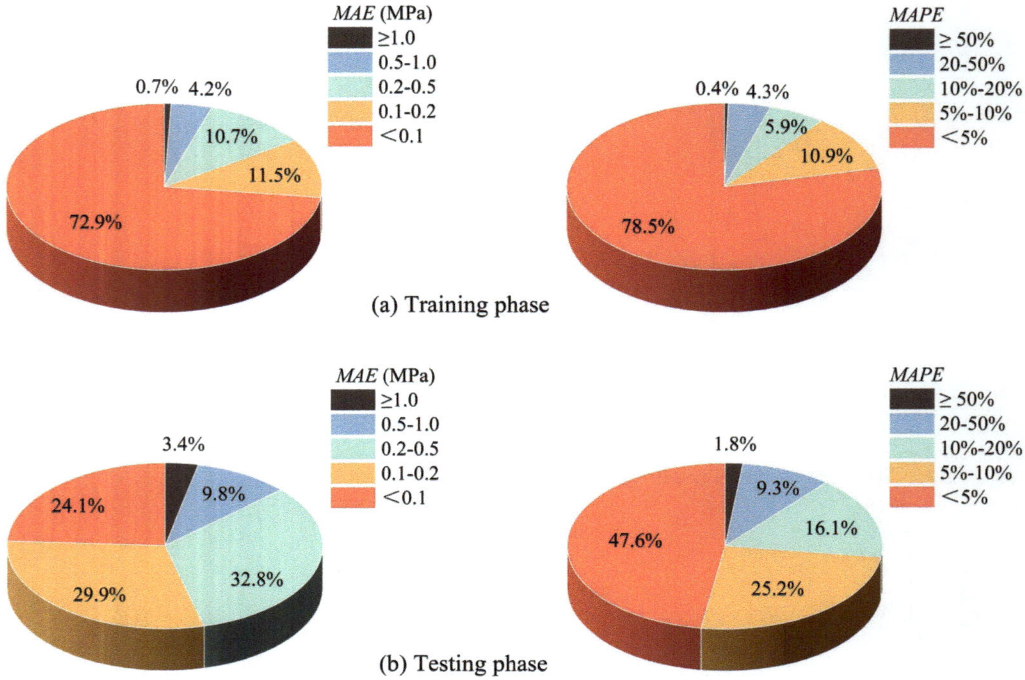

Figure 8.12 Percentage distribution of errors (MAE, MAPE): (a) training phase and (b) testing phase

were satisfactory, indicating that the model was able to accurately capture the intricate nonlinear relationships underlying patterns between gravel properties and q_{peak}. Upon analysis and examination, it was found that the samples with larger errors primarily stemmed from ballast materials and soft rock materials. The dataset included relatively few samples of these materials, resulting in the model being poorly fitted and struggling to accurately predict their properties. To enhance the accuracy of the model, the next step is to collect more data on the ballast materials and soft rock materials and include a more representative sample of them in the dataset.

In addition, three conventional ML models, i.e., naive ANN (Kaunda, 2015), RF (Zhou et al., 2019), and SVM (Ahmad et al., 2021), are used for comparison. The comparison of model performance is presented in Figure 8.13. In the training phase, Seft-Net achieved the best prediction performance. Although the R^2 is closer, other error metrics (i.e., MAE, RMSE, MAPE) of Seft-Net are reduced by more than 50% compared to ANN, RF, and SVM. As well, the Seft-Net achieved the best prediction performance in the testing phase, demonstrating the highest R^2 values, and MAE, RMSE, and MAPE of Seft-Net are reduced by more than 30% compared to ANN, RF, and SVM.

DL models like Seft-Net are superior to traditional ML methods in terms of scalable architecture (Dean et al., 2012) and advanced feature extraction (Schmidhuber, 2015; Deng and Yu, 2014), but are less interpretable (Carvalho et al., 2019; Lecun et al., 2015). In the cases of the established dataset, the developed Seft-Net model provided better performance. Additionally, the division of data in training and testing sets may influence the model performance indicators. Overall, based on the larger dataset and advanced deep learning algorithm structure, the proposed Seft-Net model has better performance.

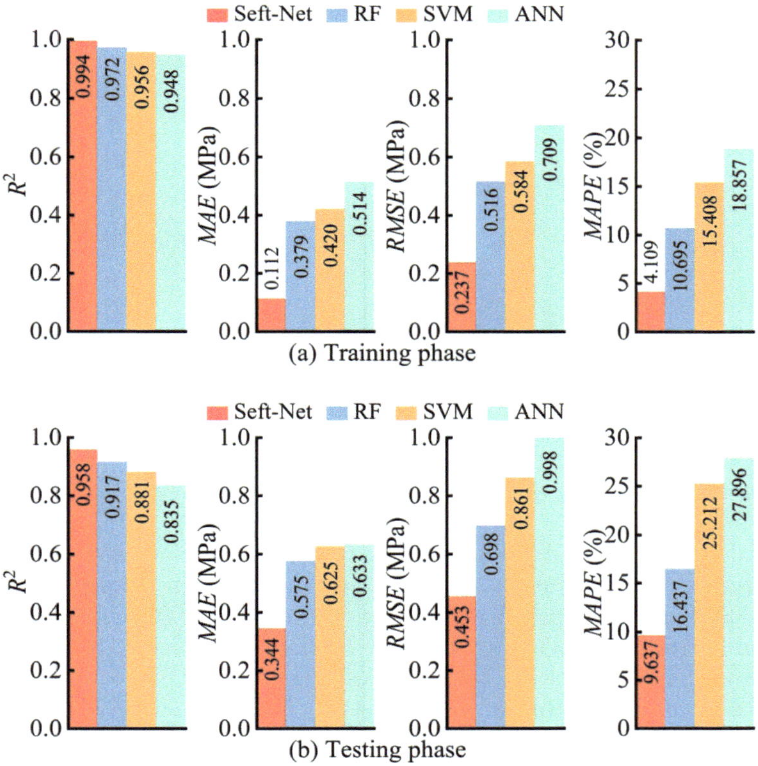

Figure 8.13 Comparison of performance metrics for Seft-Net, RF, SVM, and ANN: (a) training phase and (b) testing phase

8.3.2.5 Variable importance analysis

Deep neural networks are often considered difficult to interrogate as black box (Shrikumar et al., 2017). However, explanations for models play an important role in researching, developing, and using predictive models as information on what features were important for a given output enabling us to better understand, validate, and interpret model decisions (Doshi-Velez and Kim, 2017). To address this apparent dichotomy between performance and interpretability, perturbation-based methods are used for feature importance estimation (Li et al., 2016; Fong and Vedaldi, 2017; Dabkowski and Gal, 2017). It is mainly to mask or replace a feature of all samples and then compare the change in model performance before and after the masking or replacement. For example, a large drop in accuracy indicates that the features are more important.

Specifically, a model is trained in a training set with a specific variable masked, then predicted with the testing set with the same masked variable. The importance of each variable can be individually evaluated by the performance metrics MAE and MAPE of each masking model. The Seft-Net model in the database employed 12 input variables. By comparing and ranking (Figure 8.14), σ_3 is the most influential variable for predicting q_{peak}, where masking σ_3 results in a sharp increase in the mean absolute percentage error (MAPE) up to 69.552%, followed by σ_r, e_0, D_{80}, C_u, ρ , D_{30}, D_{10}, C_c, D_{max}, D_{60}, and D_{50}, respectively.

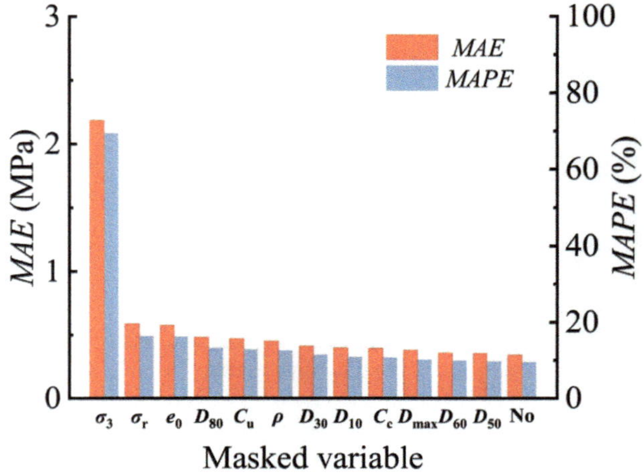

Figure 8.14 Performance metrics of different masking models and ranking.

After analyzing the variable importance, the top six variables that exhibit a MAPE variation greater than 2% have been chosen as inputs to train a simplified model, aiming to enhance convenience in practical application. The simplified Seft-Net model achieved a prediction performance with R^2 = 0.949, MAE = 0.407 MPa, RMSE = 0.562 MPa, and MAPE = 11.03% in testing, which was superior to RF (R^2 = 0.915, MAE = 0.678 MPa, RMSE = 0.816 MPa, and MAPE = 18.41%), SVM (R^2 = 0.877, MAE = 0.737 MPa, RMSE = 0.982 MPa, and MAPE = 27.49%), and ANN (R^2 = 0.831, MAE = 0.753 MPa, RMSE = 1.147 MPa, and MAPE = 29.68%). Although the prediction of the simplified model is also relatively effective as well, its prediction performance is lower than that of the original Seft-Net model due to the discarding of information on some variables. These observations and findings are dependent on the data obtained from the tests, and the amount and quality of data may affect the evaluation results.

8.3.3 Empirical formula for predicting shear strength

The dataset employed ML and DL regression models aimed at predicting shear strength based on particle and soil mass indicators. However, a significant drawback of these methods is their "black box" nature, as they do not provide explicit and straightforward formulas for interpretation. To overcome this limitation, key indicators were identified by ranking their importance. These key indicators were then utilized in a statistical modeling approach to formulate a multivariate nonlinear empirical formula for predicting shear strength.

The relationship between gravel shear strength and various factors like particle sizes (e.g., Dmax), parent rock strength, and particle characteristics has been extensively studied (Abbas, 2003; De Mello, 1977; Indraratna et al., 1993; Honkanadavar and Gupta, 2010; Varadarajan et al., 2003; Varadarajan et al., 2006). However, these studies often lack comprehensive data on particle and soil properties, and certain crucial aspects like particle shape are not widely available or easily measured empirically. Utilizing extensive data from the Geo-Gravel dataset, particularly focusing on key particle and soil mass properties (σ_3, σ_r, e_0, ρ , D_{80}, and C_u), an empirical regression formula is proposed based on Equation (8.4) (Indraratna et al., 1993). In general, confining pressure σ_3 is the controlled experimental condition in performing the triaxial compression tests, and σ_{1peak} and q_{peak} can be derived from each other. The empirical regression formula for σ_{1peak}, derived through multiple nonlinear regression using the Levenberg–Marquardt algorithm (Gavin, 2019), adopts a power exponential form as presented in Equation (8.5).

$$\frac{\sigma_{1peak}}{\sigma_r} = \alpha \left(\frac{\sigma_3}{\sigma_r} \right)^{\beta} \tag{8.4}$$

$$\sigma_{1peak} = 2.372 \sigma_3^{0.826} \sigma_r^{0.089} e_0^{-0.170} \rho^{-0.084} \underbrace{D_{80}^{0.032} C_u^{0.004}}_{\alpha_g} \tag{8.5}$$

where α and β are considered to be characteristic parameters obtained by curve fitting in Equation (8.4); σ_{1peak}, σ_3, and σ_r in units of MPa, D_{80} in units of mm, and the other indicator are dimensionless. In the application of Equation (8.5), σ_3 and e_0 can be taken according to the project design. The use of measured values for other indicators is beneficial for obtaining more accurate predictions. In the absence of measured values, empirical values of ρ and σ_r in Tables 8.1 and 8.2 can be used. For missing data of gradation, the gradation parameter α_g is taken as 1.101.

In Equation (8.5), the training set achieved R^2 = 0.968, MAE = 0.452 MPa, RMSE = 0.721 MPa, and for the test set: R^2 = 0.902, MAE = 0.687 MPa, RMSE = 0.896 MPa. Figure 8.15 illustrates scatter plots of actual versus predicted values. Despite the empirical formula's performance being less than ML and DL models, its simplicity makes it suitable for shear strength estimates.

Subsequently, to further evaluate the prediction performance of the formula, the five top important indicators σ_r, e_0, ρ , D_{80}, and C_u affecting the value of σ_{1peak} under four ranges of σ_3 divided by quartiles (σ_3=0.001–0.4 MPa, 0.4–0.8 MPa, 0.8–1.4 MPa, and 1.4–4.0 MPa) are analyzed.

Clear comparisons are made by categorizing σ_r (>30 MPa: highly hard and sub-hard, ≤30 MPa: sub-soft, soft, highly soft), ρ (≥0.5: sub-rounded, rounded, well-rounded; <0.5: sub-angular, angular), and dividing e_0, D_{80}, and C_u based on median values of 0.250, 37.5 mm, 14.0, respectively.

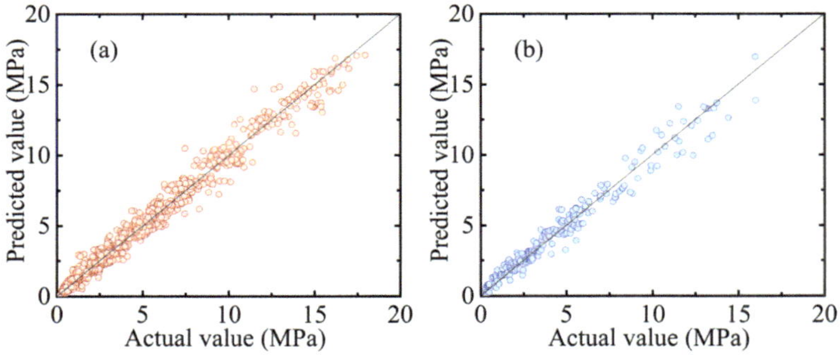

Figure 8.15 Scatter plots of actual value vs. predicted value of Equation (8.5): (a) training set and (b) testing set.

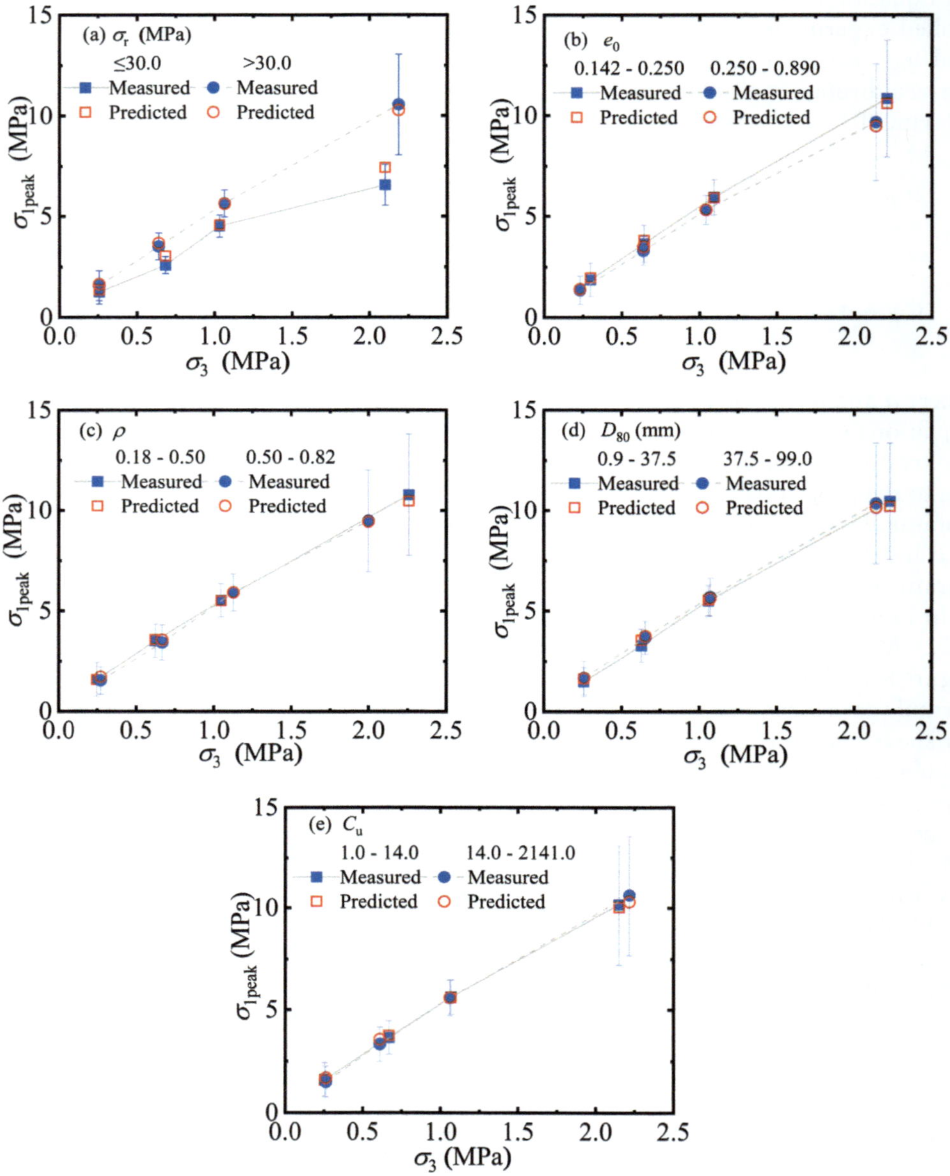

Figure 8.16 **Variation of σ_{1peak} with σ_3 in different indicators ranges: (a) σ_r, (b) e_0, (c) ρ, (d) D_{80}, and (e) C_u.**

In Figure 8.16(a), Equation (8.5) effectively captures the influence of σ_r at various σ_3 levels. Notably, the σ_{1peak} with $\sigma_r > 30$ MPa exceeds that of $\sigma_r \leq 30$ MPa, with the difference amplifying at higher σ_3, as weaker particles are more prone to break under increased confining pressure (Alonso et al., 2016; Dong et al., 2021). Predictions align better with $\sigma_r > 30$ MPa, reflecting the empirical formula's statistical representativeness for this subset, constituting 95.3% of the total data. Additionally, in Equation (8.4), the characteristic parameters α and β are 2.593 and 0.841 with $R^2 = 0.952$, MAE = 0.574

MPa, RMSE = 0.883 MPa in training set, and R^2 = 0.885, MAE = 0.715 MPa, RMSE = 1.094 MPa in test set. Comparing Equations (8.4) and (8.5), it shows that the best-fitting value of the power exponent of σ_r is 0.089, which is different from that of 0.159 (=1–β) in Equation (8.4).

Figure 8.16(b) reveals that the measured and predicted σ_{1peak} values of e_0 in 0.142–0.250 are higher than that of 0.250–0.890, and the difference becomes more pronounced with the increase in σ_3. The closer particle contact and the stronger interlocking forces among particles contribute to the loading transmission in the interaggregate structure at high confining pressure and low e_0 (Xiao et al., 2014).

Figure 8.16(c) shows that the measured and predicted σ_{1peak} of $\rho < 0.5$ (i.e., sub-angular and angular) gravels is slightly higher than that of $\rho \geq 0.5$ (i.e., sub-rounded, rounded, and well-rounded) gravels because the angular particle interlocking could frustrate the rotation, while this difference decreases with the increase in σ_3 (Giang et al., 2017).

Figure 8.16(d) shows that the measured and predicted σ_{1peak} of D_{80} in 37.5–99.0 mm gravels is slightly higher than that of D_{80} in 0.9–37.5 mm gravels, which may be related to the large-size particles having larger contact force and strong interlocking force (Huang et al., 2017).

Figure 8.16(e) indicates that the measured and predicted dots in different C_u ranges are relatively close, owing to the gradations of the gravels being mostly well-graded by design, which may make the differences relatively insignificant (Liu et al., 2021).

8.4 PREDICTION OF STRAIN–STRESS BEHAVIOR

Based on Geo-Gravel, this section develops a simulator for predicting triaxial compression test curves of gravels directly using physical indicators. We developed a DL model, Prior information-based ResNet-LSTM network (PRL-Net), as the core deep learning model in the simulator Figure 8.17.

It employs deep residual networks to extract high-dimensional spatial features of physical indicators, and combined with LSTM (Long Short-Term Memory) networks to extract the time-series information from test curves, as well as introducing prior information constraints in the loss function. Through ablation studies, the optimal architecture of the model was determined, with training and validation predictive accuracy significantly surpassing the baseline. This model effectively captures the intricate nonlinear relationships in the mechanical behavior of gravels. The simulator has been demonstrated to effectively predict the strain–stress and strain–volumetric behaviors of gravels under different physical indicators and confining pressure (Table 8.7), including fundamental characteristics such as softening/hardening and shrinkage/dilatancy (Figure 8.18).

This study provides efficient and accurate support for predicting mechanical behaviors the prediction of stress–strain curves and their mechanical properties of gravels through a data-driven approach.

8.5 SUMMARY AND CONCLUSION

In this chapter, a novel dataset Geo-Gravel for triaxial compression test of gravels was established. Utilizing this dataset, deep learning, machine learning, and empirical formulas were applied to predict the shear strength and strain–stress behavior of gravels. The main contributions and conclusions are summarized as follows:

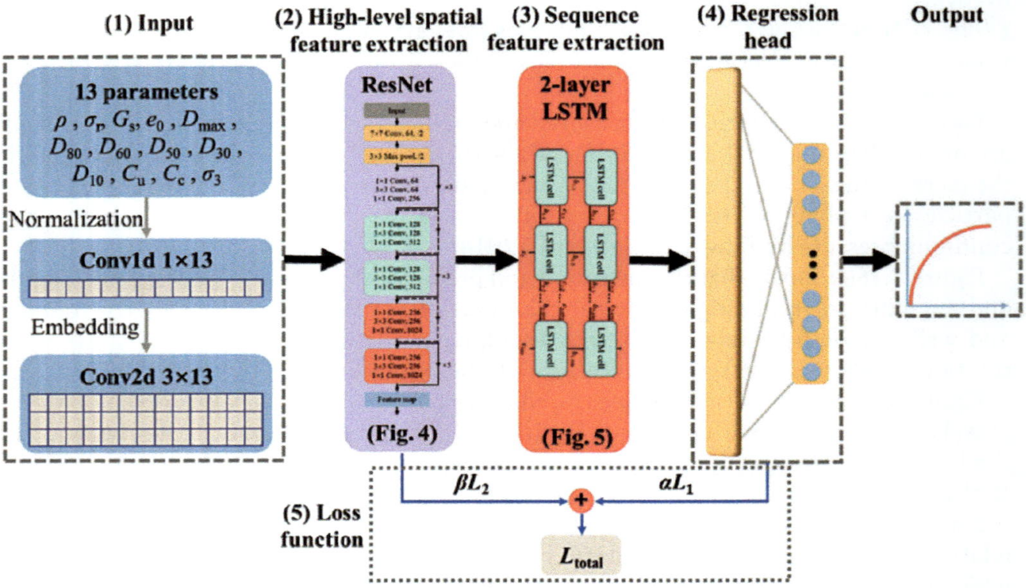

Figure 8.17 PRL-Net architecture.

(a) The Geo-Gravel dataset includes triaxial compression test data for 332 types of gravels with 1102 test records. A total of 12 attributes affecting shear strength and strain–stress behavior were collected, including particle properties (i.e., σ_r, ρ), soil mass properties (i.e., e_0, D_{10}, D_{30}, D_{50}, D_{60}, D_{80}, D_{max}, C_u, and C_c), and external condition (i.e., σ_3). The dataset is useful for stability evaluation and project design involving gravels in the absence of test results.

(b) Three machine learning regression models, i.e., artificial neural network (ANN), support vector regression (SVR), and random forest (RF), were employed to predict shear strength based on particle and soil mass indicators. RF exhibited superior performance among these models. The analysis of the importance of 12 indicators on shear strength revealed that the external condition σ_3 is the most important indicator for the estimation of shear strength and accounts for 80.5% of relative importance, while particle and soil mass indicators make up the remaining 19.5%. For particle and soil mass properties, σ_r, e_0, ρ , D_{80}, and C_u are the top five important indicators, which account for 28.0%, 26.9%, 10.2%, 9.7%, and 8.1%, respectively.

(c) A novel deep learning-based model Seft-Net was proposed to directly predict the peak deviatoric stress (q_{peak}) of gravels using gravel properties. The innovative network architecture incorporates advanced techniques, including a tailored segmented feature transformation approach and depthwise separable convolutions. Ablation studies confirmed the superiority of the proposed modules and approaches, and comparative evaluations with ANN, SVM, and RF further validated the good performance of Seft-Net. It achieved an optimal prediction performance with R^2 = 0.994, MAE = 0.111 MPa, RMSE = 0.232 MPa, and MAPE = 4.102% in the training part, and with R^2 = 0.959, MAE = 0.336 MPa, RMSE = 0.445 MPa, and MAPE = 9.127% in testing part.

Table 8.7 Test materials

No.	Lithology	σ_r (MPa)	ρ	e_0	D_{max} (mm)	D_{50} (mm)	C_u	C_c	Reference
(a)	Micaceous, sandstone, quartzite	77.3	0.82 Well-rounded	0.25	50	10.5	3.52	10.34	Varadarajan et al. (2006)
(b)	Marble, limestone, quartz sandstone	106.5	0.82 Well-rounded	0.189	60	26.4	5.54	1.57	Xiao et al. (2016)
(c)	Weakly weathered fresh tuff	66.1	0.34 Angular	0.290	60	17.8	14.02	1.93	Liu and Chen (2019)
(d)	Weakly weathered fresh tuff	66.1	0.34 Angular	0.205	60	17.8	14.02	1.93	Liu and Chen (2019)

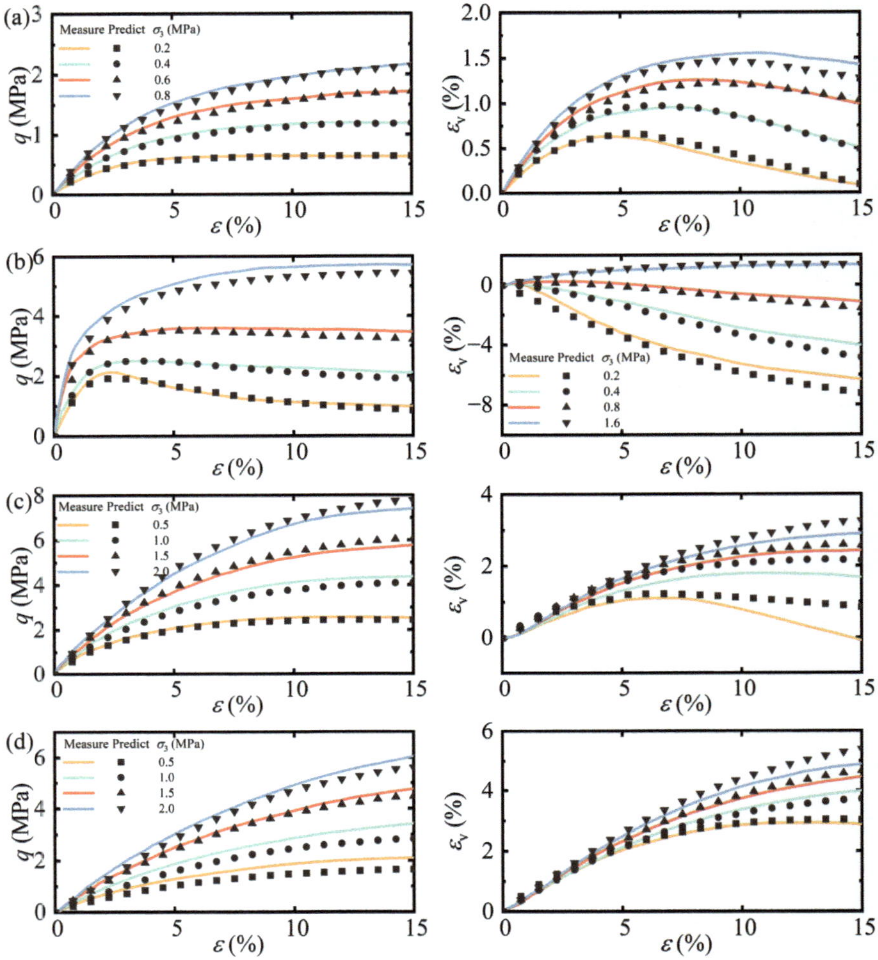

Figure 8.18 Actual curves vs. predicted curves.

(d) The modified empirical formula is proposed for estimating the strength of the gravels by considering the influence of five important particle and soil mass indicators (σ_r, e_0, ρ , D_{80}, and C_u) and external condition σ_3, as well as its simplified form for missing data of particle or soil mass indicators is also given. It also showed that the best-fitting value of the power exponent of σ_r is 0.089, which is different from that of 0.159 in the original empirical formula. The modified empirical formula is more statistically representative for $\sigma_r > 30$ MPa (highly hard and sub-hard) gravels due to its proportion is 95.3% of the total data.

(e) The DL model PRL-Net is proposed by leveraging deep residual networks for extracting spatial features from the physical parameters and LSTM networks for capturing sequential features from curves, as well as introducing prior information constraints in the loss function. The model can efficiently capture the hidden associations and patterns of mechanical behavior of gravels and reasonably predict them. The optimal model architecture is obtained through ablation studies, and its accuracy is significantly superior to the baseline model (LSTM), with the mean absolute percentage error (MAPE) reduced by 46.84%. It exhibits effectiveness in characterizing and predicting the mechanical behaviors of gravels with diverse physical properties, spanning different confining pressures, lithologies, particle shapes, gradations, and void ratios. This capability allows the model to reflect the softening and dilatancy characteristics inherent in various gravel types. Offering a novel approach to data-driven geotechnical engineering design provides a reasonable reference and support for evaluation even in the absence of test data.

REFERENCES

Abbas, S.M., 2003. *Testing and Modeling the Behaviour of Riverbed and Quarried Rockfill Materials*. IIT Delhi.

Adeli, H., 2001. Neural networks in civil engineering: 1989–2000. *Computer-Aided Civil and Infrastructure Engineering* 16 (2), 126–142.

Ahmad, M., Al-Mansob, R.A., Jamil, I., Al-Zubi, M.A., Sabri, M.M.S., Alguno, A.C., 2022. Prediction of Rockfill Materials' Shear Strength Using Various Kernel Function-Based Regression Models—A Comparative Perspective. *Materials* 15 (5), 1739.

Ahmad, M., Kamiński, P., Olczak, P., Alam, M., Iqbal, M.J., Ahmad, F., Sasui, S., Khan, B.J., 2021. Development of prediction models for shear strength of rockfill material using machine learning techniques. *Applied Sciences* 11 (13), 6167.

Ajit, A., Acharya, K., Samanta, A., 2020. A review of convolutional neural networks. In: *2020 International Conference on Emerging Trends in Information Technology and Engineering (ic-ETITE)*. IEEE. pp. 1–5.

Albelwi, S. 2022. A Robust Energy Consumption Forecasting Model using ResNet-LSTM with Huber Loss. International Journal of Computer Science and Network Security, 22(7): 301–307.

Alonso, E.E., Romero, E.E., and Ortega, E. 2016. Yielding of rockfill in relative humidity-controlled triaxial experiments. Acta Geotechnica, 11: 455–477.

Araei, A.A. 2014. Artificial Neural Networks for Modeling Drained Monotonic Behavior of Rockfill Materials. International Journal of Geomechanics, 14(3): 04014005.

Archer, K.J., and Kimes, R.V. 2008. Empirical characterization of random forest variable importance measures. Computational Statistics & Data Analysis, 52(4): 2249–2260.

Aydin, Y., Bekdaş, G., Işıkdağ, Ü., and Nigdeli, S.M. 2023. The State of Art in Machine Learning Applications in Civil Engineering. *In* Hybrid Metaheuristics in Structural Engineering: Including Machine Learning Applications. Edited by G. Bekdaş and S.M. Nigdeli. Springer Nature Switzerland, Cham. pp. 147–177.

Babar, B., Luppino, L.T., Boström, T., and Anfinsen, S.N. 2020. Random forest regression for improved mapping of solar irradiance at high latitudes. Solar Energy, 198: 81–92.

Banimahd, M., Yasrobi, S.S., and P.K.Woodward. 2005. Artificial neural network for stress–strain behavior of sandy soils: Knowledge based verification. Computers and Geotechnics, 32(5): 377–386.

Bareither, C.A., Edil, T.B., Benson, C.H., Mickelson, D.M., 2008. Geological and physical factors affecting the friction angle of compacted sands. *The Journal of Geotechical and Geoenvironmental Engineering* 134 (10), 1476–1489.

Benabou, L. 2021. Development of LSTM Networks for Predicting Viscoplasticity With Effects of Deformation, Strain Rate, and Temperature History. Journal of Applied Mechanics, 88(7): 071008.

Bergstra, J., Bengio, Y., 2012. Random search for hyper-parameter optimization. *Journal of Machine Learning Research* 13 (2), 281–305.
Carvalho, D.V., Pereira, E.M., Cardoso, J.S., 2019. Machine learning interpretability: A survey on methods and metrics. *Electronics* 8 (8), 832.
Chauhan, N.K., Singh, K., 2018. A review on conventional machine learning vs deep learning. In: *2018 International Conference on Computing, Power and Communication Technologies (GUCON)*. Greater Noida, India, IEEE. pp. 347–352.
Chen, H., Qian, C., Liang, C., Kang, W., 2018. An approach for predicting the compressive strength of cement-based materials exposed to sulfate attack. *PLoS One* 13 (1), e0191370.
Ching, J., Li, K.-H., Phoon, K.-K., and Weng, M.-C. 2018. Generic transformation models for some intact rock properties. Canadian Geotechnical Journal, 55(12): 1702–1741.
Ching, J., Lin, G.-H., Chen, J.-R., and Phoon, K.-K. 2017. Transformation models for effective friction angle and relative density calibrated based on generic database of coarse-grained soils. Canadian Geotechnical Journal, 54(4): 481–501.
Ching, J., and Phoon, K.-K. 2012. Modeling parameters of structured clays as a multivariate normal distribution. Canadian Geotechnical Journal, 49(5): 522–545.
Ching, J., and Phoon, K.-K. 2014. Transformations and correlations among some clay parameters — the global database. Canadian Geotechnical Journal, 51(6): 663–685.
Ching, J., and Phoon, K.-K. 2019. Constructing Site-Specific Multivariate Probability Distribution Model Using Bayesian Machine Learning. Journal of Engineering Mechanics, 145(1): 04018126.
Cho, G., Dodds, J., Santamarina, J.C., 2007. Closure to "particle shape effects on packing density, stiffness, and strength: natural and crushed sands" by Gye-Chun Cho, Jake Dodds, and J. Carlos Santamarina. *The Journal of Geotechical and Geoenvironmental Engineering* 133 (11), 1474–1474.
Chollet, F., 2017. Xception: Deep learning with depthwise separable convolutions. In: *Proceedings of the IEEE Conference on Computer Vision and Pattern Recognition.* Honolulu, HI, pp. 1251–1258.
Chollet, F., 2021. *Deep Learning with Python.* Simon and Schuster. Shelter Island, NY.
Chu, F.Y. 2014. Influence Factor on Dilatancy of Coarse-Grained Soil Based on Large-Scale Triaxial Test. Advanced Materials Research, 919–921: 687–692.
Dabkowski, P., Gal, Y., 2017. Real time image saliency for black box classifiers. *Advances in Neural Information Processing Systems* 30, 1–10.
De Mello, V.F., 1977. Reflections on design decisions of practical significance to embankment dams. *Géotechnique* 27 (3), 281–355.
Dean, J., Corrado, G., Monga, R., Chen, K., Devin, M., Mao, M., Ranzato, M., Senior, A., Tucker, P., Yang, K., 2012. Large scale distributed deep networks. *Advances in Neural Information Processing Systems* 25, 1–9.
Deng, J., Dong, W., Socher, R., Li, L., Li, K., Fei-Fei, L., 2009. Imagenet: A large-scale hierarchical image database. In: *2009 IEEE Conference on Computer Vision and Pattern Recognition.* Miami, FL, IEEE. pp. 248–255.
Deng, L., Yu, D., 2014. Deep learning: methods and applications. SpringerBriefs in Computer Science 7 (3–4), 197–387.
Dhillon, A., Verma, G.K., 2020. Convolutional neural network: a review of models, methodologies and applications to object detection. *Progress in Artificial Intelligence* 9 (2), 85–112.
Dong, H., Peng, B., Gao, Q.-F., Hu, Y., and Jiang, X. 2021. Study of hidden factors affecting the mechanical behavior of soil–rock mixtures based on abstraction idea. Acta Geotechnica, 16(2): 595–611.
Doshi-Velez, F., Kim, B., 2017. *Towards a Rigorous Science of Interpretable Machine Learning.* arXiv preprint arXiv:1702.08608.

Dosovitskiy, A., Beyer, L., Kolesnikov, A., Weissenborn, D., Zhai, X., Unterthiner, T., Dehghani, M., Minderer, M., Heigold, G., Gelly, S., Uszkoreit, J., and Houlsby, N. 2021. *An Image is Worth 16x16 Words: Transformers for Image Recognition at Scale.* arXiv.

Du, L., Lu, Z., and Li, D. 2022. Broodstock breeding behaviour recognition based on Resnet50-LSTM with CBAM attention mechanism. Computers and Electronics in Agriculture, 202: 107404.

Fong, R.C., Vedaldi, A., 2017. Interpretable explanations of black boxes by meaningful perturbation. In: *Proceedings of the IEEE International Conference on Computer Vision.* Venice, Italy, pp. 3429–3437.

Gavin, H.P. 2019. The Levenberg-Marquardt algorithm for nonlinear least squares curve-fitting problems. Department of civil and environmental engineering, Duke University, 19.

Giang, P.H.H., Van Impe, P., Van Impe, W., Menge, P., Cnudde, V., and Haegeman, W. 2017. Effects of particle characteristics on the shear strength of calcareous sand. Acta Geotechnica Slovenica, 14(2): 77–89.

Goodfellow, I., Bengio, Y., Courville, A., 2016. *Deep Learning.* MIT press, Cambridge, MA.

Gómez-Ramírez, J., Ávila-Villanueva, M., and Fernández-Blázquez, M.Á. 2020. Selecting the most important self-assessed features for predicting conversion to mild cognitive impairment with random forest and permutation-based methods. Scientific Reports, 10(1): 20630.

Gu, J., Wang, Z., Kuen, J., Ma, L., Shahroudy, A., Shuai, B., Liu, T., Wang, X., Wang, G., Cai, J., 2018. Recent advances in convolutional neural networks. *Pattern Recognition* 77, 354–377.

Guan, S., Qu, T., Feng, Y.T., Ma, G., and Zhou, W. 2023. A machine learning-based multi-scale computational framework for granular materials. Acta Geotechnica, 18(4): 1699–1720.

Habibagahi, G., and Bamdad, A. 2003. A neural network framework for mechanical behavior of unsaturated soils. Canadian Geotechnical Journal, 40(3): 684–693.

Hao, S., Pabst, T., 2022. Prediction of CBR and resilient modulus of crushed waste rocks using machine learning models. *Acta Geotechnica* 17 (4), 1383–1402.

He, K., Zhang, X., Ren, S., Sun, J., 2016. Deep residual learning for image recognition. In: *Proceedings of the IEEE Conference on Computer Vision and Pattern Recognition.* Las Vegas, NV, pp. 770–778.

Honkanadavar, N.P., Gupta, S.L., 2010. Prediction of shear strength parameters for prototype riverbed rockfill material using index properties. In: *Proceedings of the Indian Geotechnical Conference*—2010, Mumbai, India. pp. 16–18.

Huang, G., Liu, Z., Van Der Maaten, L., Weinberger, K.Q., 2017. Densely connected convolutional networks. In: *Proceedings of the IEEE Conference on Computer Vision and Pattern Recognition.* Venice, Italy, pp. 4700–4708.

Huang, Y., Li, J., Ma, D., Gao, H., Guo, Y., and Ouyang, S. 2019. Triaxial compression behaviour of gangue solid wastes under effects of particle size and confining pressure. Science of The Total Environment, 693: 133607.

Indraratna, B., Ionescu, D., & Christie, H. D. 1998. Shear behavior of railway ballast based on large-scale triaxial tests. *Journal of geotechnical and geoenvironmental Engineering*, 124 (5), 439–449.

Indraratna, B., Wijewardena, L., Balasubramaniam, A.S., 1993. Large-scale triaxial testing of grey wacke rockfill. *Géotechnique* 43 (1), 37–51.

Janiesch, C., Zschech, P., and Heinrich, K. 2021. Machine learning and deep learning. Electronic Markets, 31(3): 685–695.

Javadi, A.A., Faramarzi, A., and Ahangar-Asr, A. 2012. Analysis of behaviour of soils under cyclic loading using EPR-based finite element method. Finite Elements in Analysis and Design, 58: 53–65.

Javadi, A.A., and Rezania, M. 2009. Applications of artificial intelligence and data mining techniques in soil modeling. Geomechanics and Engineering, 1(1): 53–74.

Kaunda, R., 2015. Predicting shear strengths of mine waste rock dumps and rock fill dams using artificial neural networks. *International Journal of Mining and Mineral Engineering* 6 (2), 139–171.

Khan, A., Sohail, A., Zahoora, U., Qureshi, A.S., 2020. A survey of the recent architectures of deep convolutional neural networks. *Artificial Intelligence Review* 53, 5455–5516.

Labuz, J.F., and Zang, A. 2012. Mohr–Coulomb Failure Criterion. Rock Mechanics and Rock Engineering, 45(6): 975–979.

Lazarevska, M., Knezevic, M., Cvetkovska, M., Trombeva-Gavriloska, A., 2014. Application of artificial neural networks in civil engineering. *Tehnički vjesnik* 21 (6), 1353–1359.

Lecun, Y., Bengio, Y., Hinton, G., 2015. Deep learning. *Nature* 521 (7553), 436–444.

Li, J., Monroe, W., Jurafsky, D., 2016. *Understanding neural networks through representation erasure.* arXiv preprint arXiv:1612.08220.

Li, Z., Chow, J.K., and Wang, Y.H. 2017. Applying the Artificial Neural Network to Predict the Soil Responses in the DEM Simulation. IOP Conference Series: Materials Science and Engineering, 216: 012040.

Ling, H., Qian, C., Kang, W., Liang, C., Chen, H., 2019. Combination of Support Vector Machine and K-Fold cross validation to predict compressive strength of concrete in marine environment. *Construction and Building Materials* 206, 355–363.

Liu, D., Chen, H., 2019. Relationship between porosity and the constitutive model parameters of rockfill materials. *Journal of Materials in Civil Engineering* 31 (2), 04018384.

Liu, Z., Shao, J., Xu, W., Wu, Q., 2015. Indirect estimation of unconfined compressive strength of carbonate rocks using extreme learning machine. *Acta Geotechnica* 10 (5), 651–663.

Ly, H., Pham, B.T., Dao, D.V., Le, V.M., Le, L.M., Le, T., 2019. Improvement of ANFIS model for prediction of compressive strength of manufactured sand concrete. *Applied Sciences* 9 (18), 3841.

Mahmoodzadeh, A., Mohammadi, M., Ibrahim, H.H., Abdulhamid, S.N., Salim, S.G., Ali, H.F.H., Majeed, M.K., 2021. Artificial intelligence forecasting models of uniaxial compressive strength. *Transportation Geotechnics* 27, 100499.

Mohammad-Alikhani, A., Nahid-Mobarakeh, B., and Hsieh, M.-F. 2023. One-Dimensional LSTM-Regulated Deep Residual Network for Data-Driven Fault Detection in Electric Machines. IEEE Transactions on Industrial Electronics,: 1–10.

Nassr, A., Esmaeili-Falak, M., Katebi, H., and Javadi, A. 2018. A new approach to modeling the behavior of frozen soils. Engineering Geology, 246: 82–90.

Patil, H., Dwivedi, A., Bidkar, K., 2022. Prediction of upgraded properties of the concrete with the wash sand waste. *Journal of Building Pathology and Rehabilitation* 7 (1), 1–12.

Ovalle, C., Linero, S., Dano, C., Bard, E., Hicher, P.-Y., and Osses, R. 2020. Data Compilation from Large Drained Compression Triaxial Tests on Coarse Crushable Rockfill Materials. Journal of Geotechnical and Geoenvironmental Engineering, 146(9): 06020013.

Peng, X., Wang, Z., Luo, T., Yu, M., and Luo, Y. 2008. An elasto-plastic constitutive model of moderate sandy clay based on BC-RBFNN. Journal of Central South University of Technology, 15(S1): 47–50.

Penumadu, D., and Zhao, R. 1999. Triaxial compression behavior of sand and gravel using artificial neural networks (ANN). Computers and Geotechnics, 24(3): 207–230.

Phoon, K.-K. 2020. The story of statistics in geotechnical engineering. Georisk: Assessment and Management of Risk for Engineered Systems and Geohazards, 14(1): 3–25.

Phoon, K.-K., and Retief, J.V. (Editors). 2016. Reliability of geotechnical structures in ISO2394. CRC Press, Taylor & Francis Group, Boca Raton, FL.

Phoon, K.-K., and Zhang, W. 2023. Future of machine learning in geotechnics. Georisk: Assessment and Management of Risk for Engineered Systems and Geohazards, 17(1): 7–22.

Qu, T., Di, S., T. Feng, Y., Wang, M., Zhao, T., and Wang, M. 2021. Deep Learning Predicts Stress–Strain Relations of Granular Materials Based on Triaxial Testing Data. Computer Modeling in Engineering & Sciences, 128(1): 129–144.

Qu, T., Guan, S., Feng, Y.T., Ma, G., Zhou, W., Zhao, J., 2023. Deep active learning for constitutive modelling of granular materials: From representative volume elements to implicit finite element modelling. *International Journal of Plasticity* 164, 103576.

Rashidian, V., and Hassanlourad, M. 2014. Application of an Artificial Neural Network for Modeling the Mechanical Behavior of Carbonate Soils. International Journal of Geomechanics, 14(1): 142–150.

Reich, Y. 1997. Machine Learning Techniques for Civil Engineering Problems. Computer-Aided Civil and Infrastructure Engineering, 12(4): 295–310.

Russakovsky, O., Deng, J., Su, H., Krause, J., Satheesh, S., Ma, S., Huang, Z., Karpathy, A., Khosla, A., Bernstein, M., 2015. Imagenet large scale visual recognition challenge. *International Journal of Computer Vision* 115 (3), 211–252.

Sadrekarimi, A., Olson, S.M., 2010. Particle damage observed in ring shear tests on sands. *Canadian Geotechnical Journal* 47 (5), 497–515.

Samek, W., Montavon, G., Lapuschkin, S., Anders, C.J., Müller, K., 2021. Explaining deep neural networks and beyond: A review of methods and applications. *Proceedings of the IEEE* 109 (3), 247–278.

Schmidhuber, J., 2015. Deep learning in neural networks: An overview. *Neural Network* 61, 85–117.

Sevi, A., Ge, L., 2012. Cyclic behaviors of railroad ballast within the parallel gradation scaling framework. *Journal of Materials in Civil Engineering* 24 (7), 797–804.

Shang, R., He, J., Wang, J., Xu, K., Jiao, L., Stolkin, R., 2020. Dense connection and depthwise separable convolution based CNN for polarimetric SAR image classification. *Knowledge-Based Systems* 194, 105542.

Shrikumar, A., Greenside, P., Kundaje, A., 2017. Learning important features through propagating activation differences. In: *International Conference on Machine Learning*. PMLR. Sydney, Australia, pp. 3145–3153.

Sun, C., Shrivastava, A., Singh, S., Gupta, A., 2017. Revisiting unreasonable effectiveness of data in deep learning era. In: *Proceedings of the IEEE International Conference on Computer Vision*. Venice, Italy, pp. 843–852.

Szegedy, C., Ioffe, S., Vanhoucke, V., Alemi, A.A., 2017. Inception-v4, inception-resnet and the impact of residual connections on learning. In: *Thirty-First AAAI Conference on Artificial Intelligence*, Honolulu, HI, 4278–4284.

Szegedy, C., Liu, W., Jia, Y., Sermanet, P., Reed, S., Anguelov, D., Erhan, D., Vanhoucke, V., Rabinovich, A., 2015. Going deeper with convolutions. In: *Proceedings of the IEEE Conference on Computer Vision and Pattern Recognition*. Boston, MA, pp. 1–9.

Szegedy, C., Vanhoucke, V., Ioffe, S., Shlens, J., Wojna, Z., 2016. Rethinking the inception architecture for computer vision. In: *Proceedings of the IEEE Conference on Computer Vision and Pattern Recognition*. Las Vegas, NV, pp. 2818–2826.

Tan, M., Le, Q., 2019. Efficientnet: Rethinking model scaling for convolutional neural networks. In: *International Conference on Machine Learning*. PMLR. Lugano, Switzerland, pp. 6105–6114.

Tang, C., and Phoon, K.-K. 2018. Statistics of Model Factors and Consideration in Reliability-Based Design of Axially Loaded Helical Piles. Journal of Geotechnical and Geoenvironmental Engineering, 144(8): 04018050.

Tang, C., and Phoon, K.-K. 2019a. Characterization of model uncertainty in predicting axial resistance of piles driven into clay. Canadian Geotechnical Journal, 56(8): 1098–1118.

Tang, C., and Phoon, K.-K. 2019b. Evaluation of Stress-Dependent Methods for the Punch-Through Capacity of Foundations in Clay with Sand. ASCE-ASME Journal of Risk and Uncertainty in Engineering Systems, Part A: Civil Engineering, 5(3): 04019008.

Vadyala, S.R., Betgeri, S.N., Matthews, J.C., and Matthews, E. 2022. A review of physics-based machine learning in civil engineering. Results in Engineering, 13: 100316.

Varadarajan, A., Sharma, K.G., Abbas, S.M., Dhawan, A.K., 2006. Constitutive model for rockfill materials and determination of material constants. *International Journal of Geomechanics* 6 (4), 226–237.

Varadarajan, A., Sharma, K.G., Venkatachalam, K., Gupta, A.K., 2003. Testing and modeling two rockfill materials. *The Journal of Geotechical and Geoenvironmental Engineering* 129 (3), 206–218.

W. C. Krumbein. 1941. Measurement and Geological Significance of Shape and Roundness of Sedimentary Particles. SEPM Journal of Sedimentary Research, 11(2), 64-72.

Wang, K., and Sun, W. 2019. Meta-modeling game for deriving theory-consistent, microstructure-based traction–separation laws via deep reinforcement learning. Computer Methods in Applied Mechanics and Engineering, 346: 216–241.

Wang, M., Qu, T., Guan, S., Zhao, T., Liu, B., and Feng, Y.T. 2022. Data-driven strain–stress modelling of granular materials via temporal convolution neural network. Computers and Geotechnics, 152: 105049.

Wang, W., Xie, E., Li, X., Fan, D., Song, K., Liang, D., Lu, T., Luo, P., Shao, L., 2021. Pyramid vision transformer: A versatile backbone for dense prediction without convolutions. In: *Proceedings of the IEEE/CVF International Conference on Computer Vision*. Montreal, Quebec, Canada, pp. 568–578.

Wei, Y., Xing, Z., Jian, C., Wang, K., Wu, S., Chiam, K., 2021. Use of tree-based machine learning methods for stratigraphic classification in 3D geological modelling. In: *IOP Conference Series: Earth and Environmental Science*. IOP Publishing, 2021. Bristol, UK, p. 072039.

Wu, C., He, H., Song, R., Zhu, X., Peng, Z., Fu, Q., and Pan, J. 2023a. A hybrid deep learning model for regional O3 and NO2 concentrations prediction based on spatiotemporal dependencies in air quality monitoring network. Environmental Pollution, 320: 121075.

Wu, M., and Wang, J. 2022. Constitutive modelling of natural sands using a deep learning approach accounting for particle shape effects. Powder Technology, 404: 117439.

Wu, M., Xia, Z., and Wang, J. 2023b. Constitutive modelling of idealised granular materials using machine learning method. Journal of Rock Mechanics and Geotechnical Engineering, 15(4): 1038–1051.

Xiao, Y., Liu, H., Chen, Y., Jiang, J., 2014. Strength and deformation of rockfill material based on large-scale triaxial compression tests. II: Influence of particle breakage. *The Journal of Geotechical and Geoenvironmental Engineering* 140 (12), 04014071.

Xiao, Y., Liu, H., Chen, Y., Jiang, J., Zhang, W., 2014. Testing and modeling of the state-dependent behaviors of rockfill material. *Computers and Geotechnics* 61, 153–165.

Xiao, Y., Liu, H., Ding, X., Chen, Y., Jiang, J., and Wengang Zhang. 2016. Influence of Particle Breakage on Critical State Line of Rockfill Material. International Journal of Geomechanics, 16(1): 04015031.

Xu, M., Song, E., Chen, J., 2012. A large triaxial investigation of the stress-path-dependent behavior of compacted rockfill. *Acta Geotechnica* 7 (3), 167–175.

Yamashita, R., Nishio, M., Do, R.K.G., Togashi, K., 2018. Convolutional neural networks: an overview and application in radiology. *Insights Imaging* 9 (4), 611–629.

Zhang, B., Zou, G., Qin, D., Ni, Q., Mao, H., and Li, M. 2022. RCL-Learning: ResNet and convolutional long short-term memory-based spatiotemporal air pollutant concentration prediction model. Expert Systems with Applications, 207: 118017.

Zhang, C., Bengio, S., Hardt, M., Recht, B., Vinyals, O., 2021. Understanding deep learning (still) requires rethinking generalization. *Communication of the ACM* 64 (3), 107–115.

Zhang, N., Shen, S.-L., Zhou, A., and Jin, Y.-F. 2021. Application of LSTM approach for modelling stress–strain behaviour of soil. Applied Soft Computing, 100: 106959.

Zhang, P., Yin, Z., Jin, Y., Liu, X., 2021. Modelling the mechanical behaviour of soils using machine learning algorithms with explicit formulations. *Acta Geotechnica*, 17, 1403–1422.

Zhang, P., Yin, Z., Sheil, B., 2023. A physics-informed data-driven approach for consolidation analysis. *Géotechnique*, 74 (7), 620–631.

Zhang, S., Yao, L., Sun, A., Tay, Y. 2020. Deep Learning based Recommender System: A Survey and New Perspectives. *ACM Computing Surveys*, 52(1): 1–38.

Zhang, W., Wu, C., Zhong, H., Li, Y., Wang, L., 2021. Prediction of undrained shear strength using extreme gradient boosting and random forest based on Bayesian optimization. *Geoscience Frontiers* 12 (1), 469–477.

Zhao, H., Huang, Z., and Zou, Z. 2014. Simulating the Stress-Strain Relationship of Geomaterials by Support Vector Machine. Mathematical Problems in Engineering, 2014: 1–7.

Zhou, J., Li, E., Wei, H., Li, C., Qiao, Q., Armaghani, D.J., 2019. Random forests and cubist algorithms for predicting shear strengths of rockfill materials. *Applied Sciences* 9 (8), 1621.

Zhu, X., Chu, J., Wang, K., Wu, S., Yan, W., Chiam, K., 2021. Prediction of rockhead using a hybrid N-XGBoost machine learning framework. *Journal of Rock Mechanics and Geotechnical Engineering* 13 (6), 1231–1245.

Chapter 9

Evaluation of compressibility properties for soft marine clays

Monica S. Löfman and Leena Korkiala-Tanttu

Low-carbon ground improvement methods such as preloading with surcharge require reliable compressibility properties as input for the settlement prediction. This chapter demonstrates how the database approach can be used to evaluate the needed parameters for the reliability-based design of geo-structures dealing with soft marine clays. The special features related to soft and sensitive Nordic clays are investigated and discussed: inherent variability, defined via coefficient of variation or COV, seems to be smaller than that of COV guidelines based on global summaries since such soft soil deposits are often quite homogeneous. Furthermore, most of the general transformation models used to predict compressibility properties from index properties are biased when applied to soft marine clays; Accordingly, smaller transformation uncertainties may be acquired if a regional clay database is used to calibrate existing transformation models and to derive carefully fitted models with outlier detection. Finally, it is shown how the cross-correlation between compressibility parameters may be effectively modeled via database approach and Monte Carlo simulation with copula.

9.1 INTRODUCTION

Given the climate change mitigation goals, it is essential to prefer low-carbon ground improvement over carbon-intensive foundation methods such as piled slabs whenever possible. According to recent studies (Kivi 2022; Perttu 2023), preloading with surcharge embodies lower carbon emissions than that of other ground improvement methods used at soft soil sites: on average, the carbon dioxide equivalent is only 20–40% of the emissions embodied in the lime–cement column stabilization or the use of lightweight fills (Kivi 2022). Optimized design of preloading (with or without vertical drains) requires reliable compressibility properties as an input for the settlement prediction. However, high-quality undisturbed samples are needed for the oedometer tests, and sampling from soft and sensitive clays has been shown to be challenging (e.g., Mataić 2016). Meanwhile, such soft clays are prevalent, especially in the coastal areas of Finland and other Nordic countries, where the pressure to construct on soft soil sites is increasing as the cities are becoming denser. Hence, there is a growing need to properly characterize the compressibility properties – and the related uncertainties – for soft marine clays.

In a conventional serviceability limit state (SLS) verification according to Eurocode 7, the uncertainty in compressibility properties is considered via the determination of characteristic value. Yet, case studies have demonstrated that this method may lead to overdesigning when applied to ground-supported embankments on soft Finnish clays: López Ramirez, Löfman, and Korkiala-Tanttu (2022) used statistically defined characteristic values to compute settlements and observed that the obtained settlement value was in the

 DOI: 10.1201/9781003441946-9

scale of 3–6 standard deviations (SDs) apart from the mean settlement, thus indicating a safety margin that is unnecessarily large for the SLS design of preloading. Moreover, the safety margin varied between study cases, depending on the amount of non-linearity in the stress–strain response of the subsoil. This non-linearity is a result of both the overconsolidation and the stress–strain behavior of soft marine clay (López Ramirez, Löfman, and Korkiala-Tanttu 2022).

Uncertainty quantification and reliability analysis provide a systematic and robust approach to handling various sources of uncertainty in geotechnical design (Lacasse 2016; López Ramirez, Löfman, and Korkiala-Tanttu 2022; Nadim 2017; Phoon 2017). This chapter describes how the database approach can be harnessed to evaluate the compressibility properties of soft marine clays. A database of laboratory tests is used to characterize the inherent variability and transformation uncertainty in the compressibility properties of Finnish clay soils. Moreover, it demonstrates how the database approach provides insight into the behavior of soft sensitive Nordic clays and their differences compared with other marine clays. Further, the chapter discusses the phenomena of cross-correlation between compressibility parameters and demonstrates its significance to the uncertainty quantification of settlement response in soft marine clays.

9.2 COMPRESSION BEHAVIOR OF SOFT MARINE CLAYS

Soft marine clays are characterized by non-linear stress–strain response defined by a clear distinction between the small and elastic strain in the overconsolidated (OC) region and the larger plastic strain in the normally consolidated (NC) region at stresses beyond the preconsolidation pressure σ_p'. The compression index (C_c) method is one of the most widely used methods for estimating the one-dimensional vertical strain ε_v caused by primary consolidation in clay:

$$\varepsilon_v = \begin{cases} \varepsilon_{NC} = \dfrac{C_c}{1+e_0}\log_{10}\left(\dfrac{\sigma_{v0}' + \Delta\sigma_v}{\sigma_p'}\right) & if\,\sigma_p' \approx \sigma_{v0}' \\ \varepsilon_{OC} = \dfrac{C_s}{1+e_0}\log_{10}\left(\dfrac{\sigma_{v0}' + \Delta\sigma_v}{\sigma_{v0}'}\right) & if\,\sigma_{v0}' + \Delta\sigma_v < \sigma_p' \\ \varepsilon_{OC+NC} = \dfrac{C_s}{1+e_0}\log_{10}\left(\dfrac{\sigma_p'}{\sigma_{v0}'}\right) + \dfrac{C_c}{1+e_0}\log_{10}\left(\dfrac{\sigma_{v0}' + \Delta\sigma_v}{\sigma_p'}\right) & if\,\sigma_{v0}' + \Delta\sigma_v \geq \sigma_p' \end{cases} \tag{9.1}$$

where σ_{v0}' is the effective in-situ stress; $\Delta\sigma_v$ is the vertical external load; C_c is the compression index; C_s is the swelling index; and e_0 is the initial void ratio. The compression ratio is defined as CR = $C_c/(1+e_0)$. Compression indices C_c and C_s are analogous to Cam–Clay parameters λ and κ with the following conversions: $\lambda = C_c/\ln(10)$ and $\kappa = C_s/\ln(10)$.

Many Finnish clays, similar to other soft Nordic clays (e.g. Larsson 1977), are characterized by significant compressibility after σ_p'. Indeed, marine clays that have been deposited in seawater are characterized by a flocculated structure, which leads to a high void ratio e_0 (e.g. Rankka et al. 2004). If these marine clays are later subjected to leaching (reduction in pore water salinity), higher sensitivity S_t and compressibility are induced (Bjerrum 1967; Keinonen 1963; Rosenqvist 1953). Figure 9.1 illustrates a typical

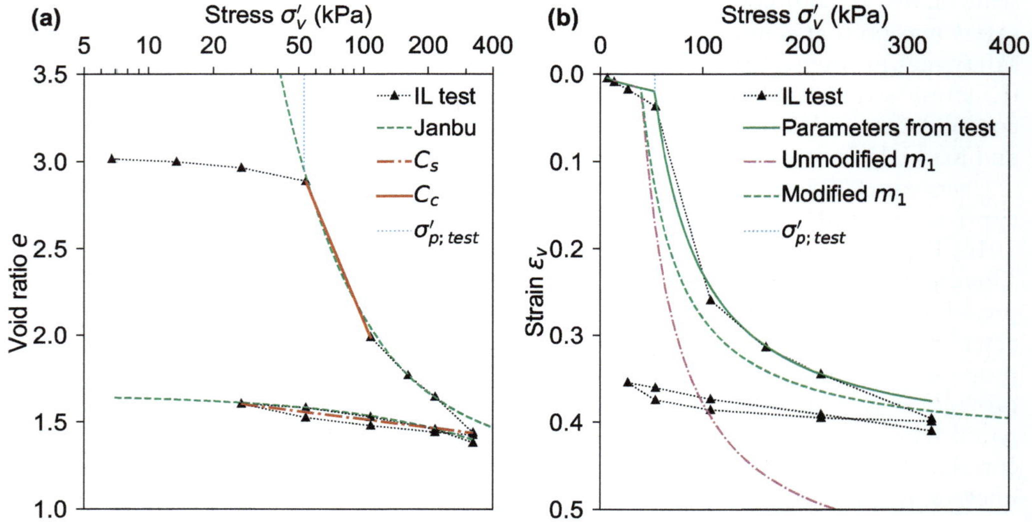

Figure 9.1 Compression behavior of soft Nordic clays: (a) comparison between the C_c method and the Janbu method when stress is in the $\log_{10}$ scale; and (b) modification of Janbu parameter m_1 ($\sigma_{p;calc}{}' = 40$ kPa; $\sigma_{p;test}{}'=53$ kPa; $m_{1;test} = 4.4$; $\beta_1 = -1.1$) (Source: modified from Löfman 2022, with permission).

incrementally loaded (IL) oedometer curve for a Finnish clay specimen: large deformations occur after $\sigma_p{}'$, and the curve is non-linear in semi-logarithmic scale. As a result, the C_c method (which assumes linear stress–strain response in a semi-logarithmic scale) is applicable only for the stress increment, which is used to interpret C_c.

Due to the limitations of the C_c method, Janbu (1963, 1967) proposed a tangent modulus method that would be applicable to all soil types. This method is also known as the "Ohde–Janbu method" as the formulation was first suggested by Ohde (1939). In the Janbu method, the tangential modulus M is defined as:

$$M = m\sigma_{ref}\left(\frac{\sigma'}{\sigma_{ref}}\right)^{1-\beta} \tag{9.2}$$

where m is the modulus number, σ_{ref} is the reference stress (100 kPa), and β is the stress exponent. Strain ε_{NC} is calculated with (Janbu 1963, 1967):

$$\varepsilon_{NC} = \int_{\sigma_p{}'}^{(\sigma_{v0}{}'+\Delta\sigma_v)} \frac{d\sigma'}{M} = \begin{cases} \dfrac{1}{m_1\beta_1}\left[\left(\dfrac{\sigma_{v0}{}'+\Delta\sigma_v)}{\sigma_{ref}}\right)^{\beta_1} - \left(\dfrac{\sigma_p{}'}{\sigma_{ref}}\right)^{\beta_1}\right] & if\ \beta_1 \neq 0 \\ \dfrac{1}{m_1}\ln\left(\dfrac{\sigma_{v0}{}'+\Delta\sigma_v}{\sigma_p{}'}\right) & if\ \beta_1 = 0 \end{cases} \tag{9.3}$$

Janbu parameters m_1 and β_1 are defined from the virgin compression curve using curve-fitting methods: m_1 is the stiffness parameter, while β_1 defines the shape of the curve.

Soft clays often have negative β_1, i.e., concave shape as in Figure 9.1a. Silts often have $\beta_1 > 0$ (i.e., convex oedometer curve, see also Figure 9.3a). If $\beta_1 = 0$, the Janbu method corresponds to the C_c method and the following applies:

$$m_1 = \frac{(1+e_0)\ln(10)}{C_c} = \frac{\ln(10)}{CR} \qquad (\beta_1 = 0) \tag{9.4}$$

Equation (9.3) can also be used to calculate ε_{OC}: OC parameters, referred to as m_2 and β_2, are typically using the recompression (reloading) sequence of the oedometer test (see Figure 9.10a). $\beta_2 = 1$ is often assumed, which corresponds to linear stress–strain dependency.

The Janbu method is flexible but also prone to user errors especially if β_1 is negative: if the value of σ_p' used in the calculation is lower than the one defined while curve-fitting m_1 and β_1, the oedometer curve is extrapolated and hence the settlement is overestimated. The following m_1 modification, suggested by Länsivaara (1995; 2003), may be used to avoid this error:

$$m_{1;calc} = m_{1;test}\left(\frac{\sigma'_{p;test}}{\sigma'_{p;calc}}\right)^{-\beta_1} \tag{9.5}$$

where $m_{1;test}$, $\sigma_{p;test}'$, and β_1 are the parameters defined from the oedometer test, while $\sigma_{p;calc}'$ is the value to be used in settlement calculation. Figure 9.1b illustrates how the m_1 modification preserves the shape of the original oedometer test curve. Meanwhile, much larger strains would be computed if the unmodified m_1 were used with $\sigma_{p;calc}' < \sigma_{p;test}'$ ("unmodified" in Figure 9.1b). Equation (9.5) implies that no modification is needed when $\beta_1 = 0$; that is, the C_c method is not affected by this stress dependency of stiffness parameters. According to the Finnish design guidelines (Finnish Transport Agency 2012), this modulus number modification should be applied, and it has also been incorporated into the Finnish geotechnical software GeoCalc (Civilpoint).

The m_1 modification is typically needed when a slightly overconsolidated clay is assumed to be NC, or when the $\sigma_{p;test}'$ value is from a CRS (constant rate of strain) oedometer test and needs to be rate-corrected to correspond to a lower strain rate (usually IL strain rate). In the case of Finnish clays, the rate correction of σ_p' may be performed using the Sällfors (1975) graphical method or via an empirical method based on Finnish clay data (Finnish Transport Agency 2012):

$$\sigma'_{p;calc} = \frac{\sigma'_{p;test}}{k} \qquad ,where\, k = \left(\frac{\dot{\varepsilon}_{test}}{\dot{\varepsilon}_{calc}}\right)^{B} \tag{9.6}$$

where $\dot{\varepsilon}_{test}$ and $\dot{\varepsilon}_{calc}$ are the CRS strain rate and the target strain rate (10^{-7} 1/s for IL test), respectively. B is the empirical factor, for which the value $B = 0.0728$ is recommended (Finnish Transport Agency 2012).

9.3 DATABASE OF FINNISH CLAYS FI-CLAY/14/856

FI-CLAY/14/856 is a partly multivariate database consisting of 856 odometer tests on Finnish clay soils (undisturbed soil specimens). All the data points include specimen-specific

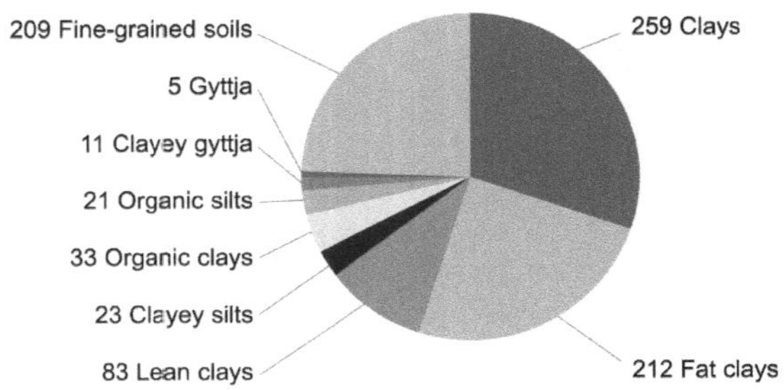

Clay fraction	Soil type
>10...30 %	Clayey silt
>30...50 %	Lean clay
>50 %	Fat clay

Organic content	Soil type
>2...6 %	Organic silt or Organic clay
>6...20 %	Silty gyttja or Clayey gyttja
>20 %	Gyttja

Figure 9.2 **Pie chart of soil types included in the FI-CLAY/14/856 database and the Finnish classification system for fine-grained soils (Source: modified from Löfman and Korkiala-Tanttu 2022).**

water content, void ratio, compression index, and compression ratio. Some are accompanied by additional laboratory test results for the same or nearby soil specimen (up to 14 parameters in total). The included data originated from various sources, although the majority were from research projects. Most of the specimens represent clays, as illustrated in Figure 9.2 (following the Finnish classification system, i.e., Korhonen, Gardemeister, and Tammirinne 1974). Full description of the sources and procedure used to compile the FI-CLAY/14/856 database can be found from Löfman and Korkiala-Tanttu (2022). The database is included in the TC304 compendium of databases known as 304dB (http://140.112.12.21/issmge/tc304.htm).

Since the Janbu method has been the preferred over C_c method in Finland, some of the data in FI-CLAY/14/856 lacked the C_c interpretation. For those oedometer test results, the C_c was calculated from the Janbu parameters (m_1, β_1, and σ_p') by assuming identical strains at stress increase from σ_1' to σ_2' (Löfman and Korkiala-Tanttu 2019):

$$C_c = \frac{(1+e_0)\dfrac{1}{m_1\beta_1}\left[\left(\dfrac{\sigma_2'}{\sigma_{ref}}\right)^{\beta_1} - \left(\dfrac{\sigma_1'}{\sigma_{ref}}\right)^{\beta_1}\right]}{\log_{10}\left(\dfrac{\sigma_2'}{\sigma_1'}\right)} \tag{9.7}$$

The approximate value for C_c was calculated by setting $\sigma_1' = \sigma_p'$ and $\sigma_2' = 2\sigma_p'$ since usually two load increments after σ_p' are used to interpret C_c. Figure 9.3a illustrates this C_c approximation using Equation (9.7). However, if σ_p' was greater than 120 kPa, the final stress was selected to be $\sigma_2' = \sigma_p' + 120$ kPa in order to better replicate the typical infrastructure geometries in Finland (120 kPa corresponds to a 6-m-high embankment). Figure 9.2a presents the verification of the approximation based on those tests where C_c interpretation was available ("actual C_c"): it can be seen that the scatter around the "no bias" line is moderate.

In a study reported by Löfman (2022), the FI-CLAY/14/856 database was extended: some oedometer test results were complemented with Janbu parameters in addition to compression index parameters. This database is referred to as FI-CLAY/14/856-B, and it has been published in an online repository (Löfman and Korkiala-Tanttu 2021).

Table 9.1 summarizes the statistics for the soil properties included in the FI-CLAY/14/856-B database. The table also includes the mean value from a global

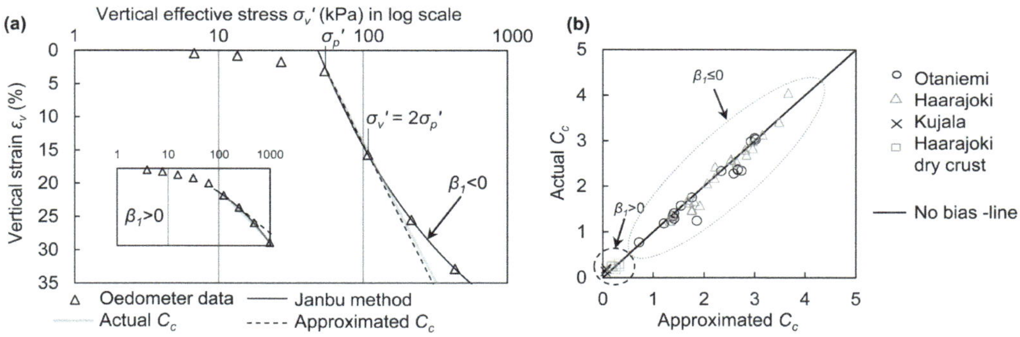

Figure 9.3 Approximation of C_c from Janbu parameters: (a) methodology and (b) verification (Source: modified from Löfman 2022, with permission. Based on Löfman and Korkiala-Tanttu 2019).

Table 9.1 Statistics for the FI-CLAY/14/856-B database and the mean values for the global clay database for comparison

Soil property	*n*	*Min*	*Median*[a]	*Mean*	*Max*	*COV*	*Mean, global database*[b]
Water content, w_n (%)	856	23.0	85.0	84.5	187	0.37	79.1
Void ratio, e_0	856	0.520	2.302	2.309	5.200	0.36	1.992
Liquid limit, w_L (%)	354	19.0	69.0	69.7	203.9	0.39	76.6
Plastic limit, w_P (%)	370	12.0	26.6	27.3	55.6	0.22	N/A
Plasticity index, PI (%)	354	1.0	41.5	42.5	153.7	0.54	41.4
Unit weight, γ (kN/m^3)	799	12.0	15.0	15.3	20.6	0.10	N/A
Organic content, Org (%)	369	0.0	0.8	1.2	7.8	1.26	N/A
Clay content, Cl (%)	263	12.7	54.5	56.6	95.0	0.36	N/A
Undrained shear strength, s_u (kPa)	413	5.2	22.5	30.2	240.0	1.01	N/A
Sensitivity, S_t	281	1.7	17.0	23.2	163.0	0.84	8.0
Preconsolidation pressure, σ_p' (kPa)	843	6.0	49.0	66.7	604.1	0.93	N/A
Overconsolidation ratio, OCR	817	0.3	1.4	2.2	41.3	1.32	2.4
Compression index, C_c	856	0.026	1.287	1.417	9.284	0.70	0.847
Compression ratio, CR	856	0.014	0.378	0.394	2.088	0.56	0.254
Modulus number, m_1	746	0.700	7.60	11.25	339.0	1.43	N/A
Stress exponent, β_1	746	−4.00	−0.31	−0.34	1.23	1.50	N/A
Swelling index, C_s	656	0.002	0.101	0.109	0.611	0.55	N/A
Modulus number, m_2	609	2.25	59.0	77.82	903.3	0.96	N/A
Stress exponent, β_2	609	−1.00	1.00	0.87	3.00	0.48	N/A

Notes: n = number of measurements; COV = coefficient of variation.

[a]50% sample percentile; [b]global database of marine soils compiled by Kootahi and Moradi (2017) for model building.

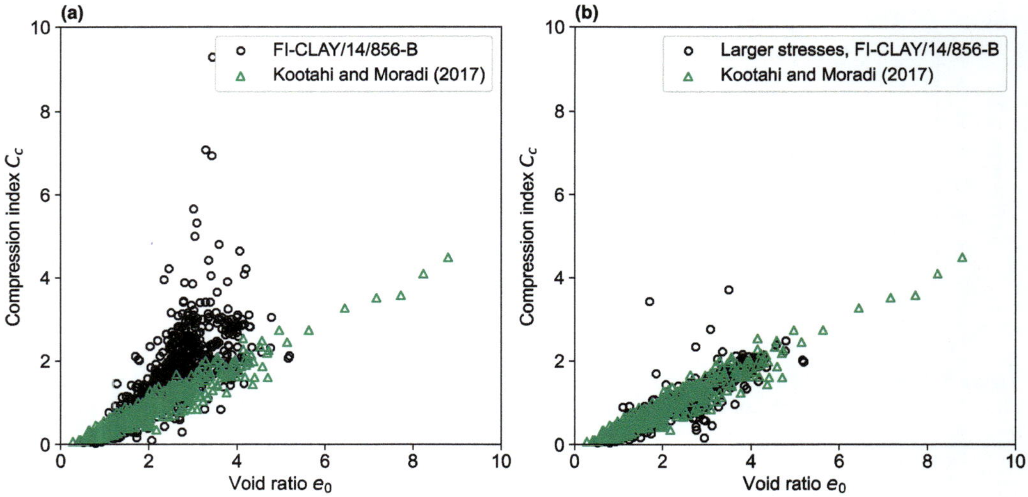

Figure 9.4 Relationship between e_0 and C_c in the global database of marine clays and in (a) Finnish clays (stress range approximately from σ_p' to $2\sigma_p'$) and (b) Finnish clays without structure (stress range from $2\sigma_p'$ to $4\sigma_p'$).

database of marine clays (Kootahi and Moradi 2017) for comparison. On average, the specimens in FI-CLAY/14/856-B have higher values for sensitivity S_t and compressibility (C_c and CR). Indeed, the mean and median values for the stress exponent β_1 are negative, which is typical for soft Nordic clays.

The difference between the Finnish and global clay databases is also visible in the e_0–C_c relationship, as shown in Figure 9.4a. Finnish clays tend to have greater C_c mainly due to their higher sensitivity (Di Buò et al. 2019; Leroueil et al. 1983). However, it is noteworthy that the interpretation of C_c, namely the considered stress range, also contributes to this phenomenon. High C_c values are acquired when the slope is fitted to the stress range right after the preconsolidation pressure σ_p'. On the other hand, if the slope is fitted to the "tail" of the oedometer curve, the acquired C_c values lack the effect of structure on the compressibility (similar to the compression line of reconstituted specimen, see e.g., Mataić 2016; Karstunen et al. 2005). Figure 9.3b shows the C_c values calculated from Janbu parameters using Equation (9.7), when the considered stress range is from $2\sigma_p'$ to $4\sigma_p'$: after removing the effect of structure, the e_0–C_c relationship of soft Nordic clays resembles other marine clays. Further, the data scatter is reduced.

9.4 INHERENT VARIABILITY

Total uncertainty in an estimated soil property is affected by inherent variability, measurement error, statistical uncertainty, and transformation uncertainty (Phoon and Kulhawy 1999). The inherent variability (also known as natural or spatial variability) is a result of geological processes. The magnitude of inherent variability may be estimated from the observed variability within a site or geotechnical layer (while assuming that the measurement error is small by comparison, see Phoon and Kulhawy 1999). The

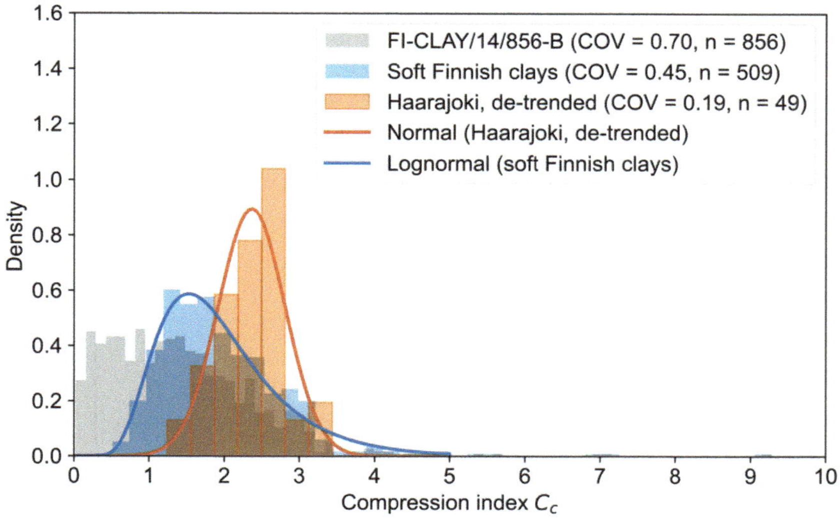

Figure 9.5 **Observed variability in C_c: the effect of scale.**

magnitude of inherent variability (or any other aleatoric uncertainty) may be characterized via coefficient of variation (COV), which is defined as the standard deviation divided by the absolute value of the mean.

Inherent variability is overestimated if data from different geological layers are mixed (Phoon and Kulhawy 1999). Figure 9.5 compares the histograms for C_c: even though all the considered datasets consist of Finnish clay soils, the variability decreases from COV = 0.70 to COV = 0.45 when the regional database FI-CLAY/14/856-B is filtered to contain soft clays only (for the criteria, refer to Section 9.6). Further, within a homogeneous clay layer at a specific site (Haarajoki), the observed COV is only 0.19. It is noteworthy that also the shape of histogram and hence the probability density function (pdf) changes with the considered scale.

If there is a linear trend with depth, de-trending should be applied before assessing COV of inherent variability (Phoon and Kulhawy 1999). De-trended standard deviation, which measures the scatter around the trend line, is defined by Lacasse et al. (2007):

$$SD_{de-trended} = \sqrt{\frac{1}{n-2}\sum_{i=1}^{n}\left(y_i - \left(b_o + b_1 z_i\right)\right)^2} \tag{9.8}$$

where y_i is the measured soil property value at depth z_i and $(b_0+b_1z_i)$ is the mean soil property value predicted by the linear trend (defined by intercept b_0 and slope b_1).

Knuuti and Löfman (2022) have summarized the observed variability (taken as the estimate of inherent variability) of geotechnical properties within homogeneous clay layers in various sites in Finland. Table 9.2 presents those values (relevant to settlement problems) and the corresponding COV values based on the global clay database or studies. It is noteworthy that some Finnish clay sites contain more than one homogeneous soil layer that was analyzed. The studied sites are also included in the FI-CLAY/14/856 database, but

Table 9.2 Inherent variability (observed variability) for Finnish clays and COV based on global study for comparison

Soil property[a]	*Soil types in the dataset*	***n**, sites*	*Ref.*[b]	***n**, sample size*	*Soil property mean, range*	*Soil property COV, range*[c]	*COV, (approx. guideline)*	*COV, mean or guideline (global)*
w_n (%)	Clay, silt, gyttja	8	A, B, C	15–105	40–146	0.03–0.14	0.1	0.15[e]
e_0	Clay, gyttja	5	A, C	10–53	1.4–3.9	0.02–0.13	0.1	0.07–0.30[f]
γ (kN/m^3)	Clay, silt, gyttja	8	A, B, C	11–80	13–18	0.01–0.04	<0.1	<0.1[f]
Fall cone s_u (kPa)	Dry crust clay	1	A	16	103	0.32–0.58	0.4	0.28[e]
	Clay, silt, gyttja	7	A, B	8–46	12–35	0.10–0.27	0.2	
	Clay/silt with sand layers	1	B[d]	12	35	0.61	0.6	
σ_p' (kPa)	Clay, gyttja	4	A	10–32	26–75	0.11–0.27	0.2	0.10–0.35[f]
OCR	Clay, gyttja	5	A, B	9–34	1.2–2.6	0.12–0.31	0.2	0.18[e]
C_c	Clay, gyttja	5	A, B	10–34	0.49–2.70	0.15–0.21	0.2	0.36[e]
CR	Clay, gyttja	5	A, B	10–34	0.19–0.69	0.15–0.20	0.2	N/A
m_1	Clay, gyttja	4	A[d]	10–34	3.5–12.6	0.09–0.32	0.2	N/A
β_1	Clay, gyttja	3	A[d]	9–34	(−0.27)–(−1.3)	0.33–0.55	0.5	N/A
β_1 (≈ 0)	Clay	1	A[d]	15	−0.09	≈1.00	1.0	N/A
C_s	Clay, gyttja	4	A[d], B	8–15	0.08–0.19	0.08–0.43	0.3	0.42[e]
m_2	Clay, gyttja	4	A[d]	8–10	49–97	0.15–0.52	0.3	N/A
β_2	Clay, gyttja	4	A[d]	8–10	0.49–1.12	0.26–0.60	0.3	N/A
c_v (m^2/a)	Clay	3	A	12–19	0.10–0.50	0.28–0.61	0.5	0.33–0.68[f]
k_1 (10^{-9} m/s)	Clay	3	A	10–13	0.45–1.40	0.29–0.56	0.5	0.68–0.90[f]
$C_{\alpha\varepsilon}$ (%)	Clay	3	A	12–25	0.67–2.47	0.26–0.52	0.4	N/A

[a] c_v = coefficient of consolidation, k_1 = permeability at zero strain, $C_{\alpha\varepsilon}$ = coefficient of secondary compression; [b] A = Löfman and Korkiala-Tanttu (2021b), B = Löfman and Korkiala-Tanttu (2019), C = Löfman (2016); [c] range includes COV values with and without de-trending, [d] based on the same dataset, but not reported in the publication; [e] Guan et al. (2021); and [f] Uzielli et al. (2006).

the analyzed data contains more laboratory test results (mainly index test results) than the FI-CLAY/14/856 database, which focuses on oedometer test results. The datasets used in the inherent variability analysis reported by Löfman and Korkiala-Tanttu (2021a) have been published open access in a data repository (Löfman and Korkiala-Tanttu 2021c).

Table 9.2 shows that the COV values for Finnish clays are mostly within the global guideline ranges or close to the global mean value. However, for properties such as s_u for soft soils, compressibility, and consolidation properties, the observed upper bound for COV is systematically lower than the global value. On the other hand, the global summaries and guidelines are typically based on datasets containing various soils and test types. For example, COV for s_u was between 0.3 and 0.6 in the datasets containing stiff Finnish clay soils (i.e., dry crust clay, and silt–clay mixtures with sand layers), while significantly smaller values were observed in soft Finnish clay or gyttja soil layers (COV = 0.10–0.27). Indeed, soft Finnish clays tend to be rather homogeneous (Gardemeister 1975).

Figure 9.5 further illustrates the relationship between the mean and COV values for selected compressibility-related soil properties (defined using the dataset published by Löfman and Korkiala-Tanttu 2021c). Table 9.2 and Figure 9.6 show that the decrease in observed COV via de-trending can be rather significant even though the analysis aimed to identify pseudo-homogeneous clay layers.

The results align with the hypothesis that COV values of the same physical properties are independent of the geological age of the soil (Uzielli et al. 2006). The analysis included various sites in Finland, and the clay sediments have different geological histories (for more detailed discussion on the geological characterization, see Löfman and Korkiala-Tanttu 2019). The inherent variability analyses also showed that the COV values for the compressibility-related properties of Finnish clays were rather consistent with respect to mean values, as noted in earlier summary studies (Guan et al. 2021; Phoon and Kulhawy 1999).

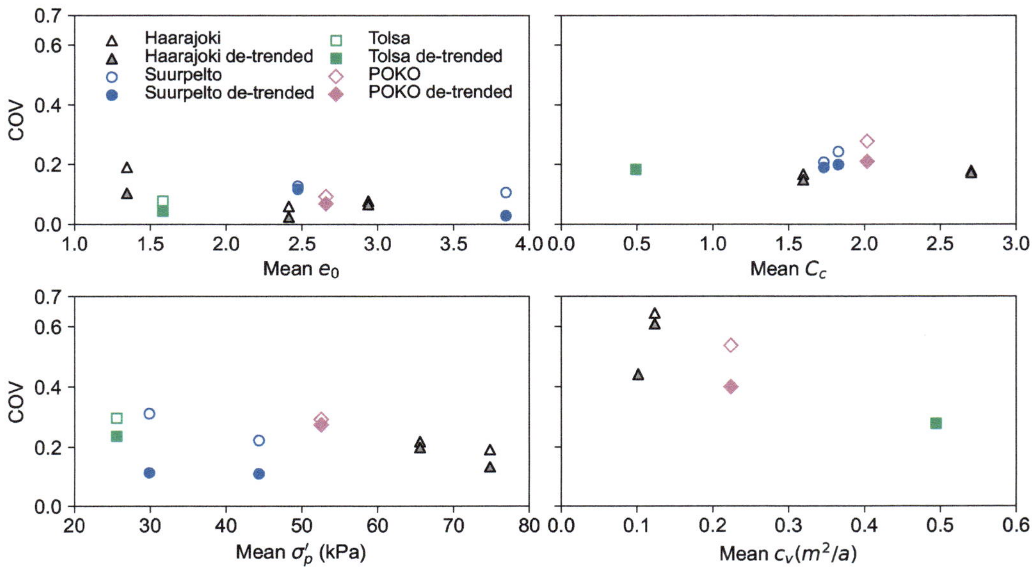

Figure 9.6 Observed variability in compressibility parameters as a function of mean value. The data represent homogeneous clay soil layers (n = 10...42).

9.5 PROBABILITY DENSITY FUNCTIONS

The probability density functions (pdfs) most often used for soil properties have been investigated in a survey by Lacasse and Nadim (1996): normal distribution is often used for index properties such as unit weight or density, while also lognormal distribution is commonly used for strength and overconsolidation. The suitability of different pdfs to represent the observed variability for Haarajoki clay (marine clay deposit from Finland) was investigated in a study by Löfman and Korkiala-Tanttu (2021b). The data was first de-trended (normalized) with depth, so that results from depth z = 2…18 m could be combined to represent one homogeneous clay layer. Figure 9.7 illustrates how the de-trending decreases the COV value for C_c (dataset is from Löfman and Korkiala-Tanttu 2021c). Next, Shapiro–Wilk normality test was used to evaluate the suitability of normal or lognormal distribution. Table 9.3 also shows the results for the homogeneous soft clay layer at depth z = 2…6.5 m, which did not require de-trending with depth (see Figure 9.8).

The results shown in Table 9.3 are consistent with the pdf preference observed in the survey. According to the analysis of Haarajoki clay, pdf suitability seems to be dependent on the considered soil layer and the consideration of trend with depth; For example, normal distribution was suitable for the de-trended C_c, while both normal and lognormal distributions were found appropriate when the homogeneous soft layer was considered (without de-trending).

If the Janbu method is used instead of the C_c method, it is important to apply the necessary corrections prior to the assessment of pdf suitability (or inherent variability). Namely, rate correction is needed when the dataset contains both incremental and CRS oedometer tests. Preconsolidation pressure σ_p' is first rate-corrected with Equation (9.6) to correspond incremental oedometer strain rate, and that lower value is then used to correct m_1 via Equation (9.5) in the case of specimens with negative β_1. It is also

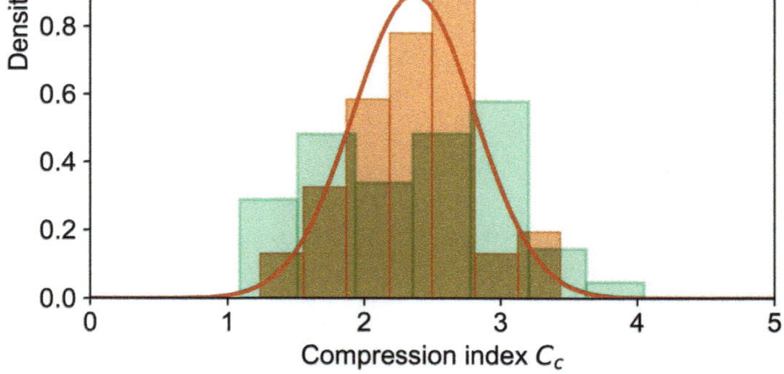

Figure 9.7 Observed variability in C_c: the effect of de-trending (Haarajoki clay layer, depth z = 2…18 m, n = 49).

Table 9.3 Suitable pdfs (N = normal; LN = lognormal) for geotechnical properties relevant to settlement calculations in clay subsoil

Soil property	*Haarajoki soft clay, **z** = 2...6.5 m, pdf*[a]	*Haarajoki clay, **z** = 2...18 m, de-trended pdf*[b]	*Survey, pdf*[c]	*Survey, soil type*[c]
Stress-normalized s_u	N/A	N/LN (n = 121)	N/LN	Clay
γ	N/A	N (n = 121)	N	All soils
e_0	N/LN (n = 39)	N/LN (n = 59)	N	All soils
OCR	N/LN (n = 39)	LN (n = 45)	N/LN	Clay
σ_p' (kPa)	LN (n = 39)	N/LN (n = 53)	N/A	N/A
C_c	N/LN (n = 39)	N (n = 49)	N/A	N/A
CR	N/LN (n = 39)	N/LN (n = 49)	N/A	N/A
m_1	N/LN (n = 39)	N/A	N/A	N/A
β_1	–[d]	N/A	N/A	N/A
Coefficient of creep $C_{\alpha\varepsilon}$	N/A	N/LN (n = 28)	N/A	N/A

[a]Data from homogeneous layer and rate correction of σ_p' and m_1 using Equations (9.5) and (9.6); [b]normality test for laboratory test data de-trended with depth (Löfman and Korkiala-Tanttu 2021b). Rate correction of σ_p' based on Sällfors (1975) method; [c]survey on pdf selection (Lacasse and Nadim 1996); and [d]neither normal or lognormal was suitable (note: β_1 is negative). Johnson's system-bounded (SB) distribution was found to be suitable (see Löfman 2022).

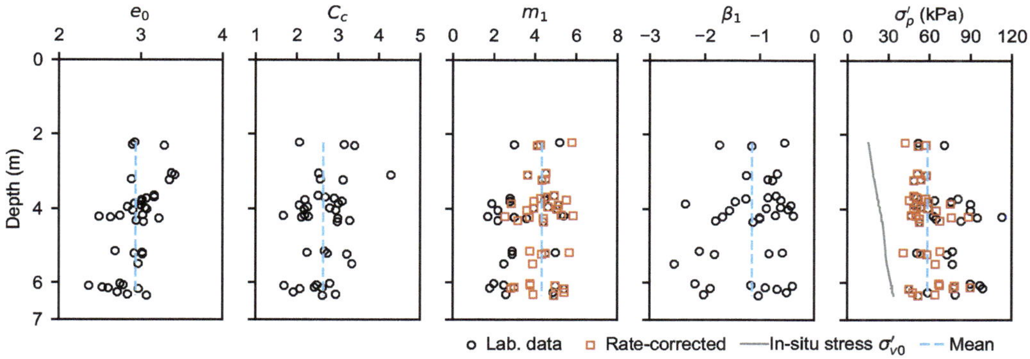

Figure 9.8 Haarajoki clay, homogeneous soft clay layer (z = 2...6.5 m). Preconsolidation pressure σ_p' and modulus number m_1 from CRS test results have been rate-corrected.

noteworthy that since β_1 is usually negative, non-negative pdfs such as lognormal cannot be used.

9.6 TRANSFORMATION UNCERTAINTY IN EXISTING AND NEW TRANSFORMATION MODELS

If there are no site-specific oedometer tests available, index properties (e.g., w_n) can be used to indirectly estimate compressibility properties (e.g., C_c). Such transformation models, which are used to predict a target soil property from one or more predictor variables, are known to be marked by some degree of transformation uncertainty. Transformation models that are defined for given site-specific or regional data may be biased if applied to other sites (e.g., Ching and Phoon 2014), whereas "global" models tend to be less biased on average but also less precise when applied to a specific site (Ching and Phoon 2012). Indeed, regional soil databases may be utilized to calibrate existing transformation models with the approach proposed by Ching and Phoon (2014). This approach incorporates transformation error ε_i of ith data point, which is defined as the actual target value (C_c from the oedometer test) divided by the predicted target value (such as C_c predicted from w_n):

$$\varepsilon_i = \frac{\text{actual target value}}{\text{predicted target value}} \tag{9.9}$$

The arithmetic mean of all errors ε_i is the model bias b while the sample COV of ε_i represents the transformation uncertainty δ. For an unbiased transformation model, which has no data scatter around the trend curve, $b = 1$ and $\delta = 0$.

With the calibration study results, the actual (non-biased) target value is acquired from (Ching and Phoon 2014):

$$\text{actual target value} = \text{predicted target value} \times b \times \varepsilon \tag{9.10}$$

where ε is the variability term for the transformation model with a mean value equal to unity and COV = δ. Lognormal distribution may be assumed for the variability term (e.g., Ching 2017).

Löfman and Korkiala-Tanttu (2022) calibrated existing transformation models for predicting the compressibility of soft Finnish clays using the FI-CLAY/14/856 database. Some selected results for compression index C_c, compression ratio CR, and swelling index C_s are presented in Table 9.4. Transformation models for Janbu parameters m_1 and β_1 were not investigated in the study, because these curve-fitting parameters are dependent on the stress state ($\sigma_{p;test}'$) and should not be treated separately (unless the cross-correlation is modeled, see Section 9.6.)

The bias factors confirmed that almost all the existing models, including those most commonly used in Finland (i.e., Helenelund 1951; Janbu 1998), underestimate the value of C_c or CR of soft Finnish clays ($b > 1$). On average, models for C_s were less biased. Transformation uncertainty δ, which represents the amount of scatter around the trend curve, was $\delta = 0.4–0.8$ (up to 1.47 in the case of the transformation model for C_s). In the classification proposed by Phoon and Tang (2019), this corresponds to medium to high variability.

Due to the high transformation uncertainties, Löfman and Korkiala-Tanttu (2022) used the FI-CLAY/14/856 database to derive new transformation models for the compressibility of Finnish clays. To properly quantify the transformation uncertainties, the other sources of uncertainties (mainly measurement errors) should be minimized (Phoon and Kulhawy 1999). Hence, care was taken to remove disturbed specimens and other outliers from the database: the new transformation models were derived using the filtered database FI-CLAY/14/822. Finally, those data points where the residual error ε_{res} was more than three standard deviations ($\sigma_{\varepsilon;res}$) away from the zero-mean were considered outliers and hence removed before performing the final regression fitting, as illustrated Figure 9.9. Natural logarithm transformation was used to ensure that the residual error around the trend (defined as a polynomial regression curve) is normal distributed (see Löfman and Korkiala-Tanttu 2022 for further details).

Table 9.4 Selected results of the calibration study for some existing models for compressibility against FI-CLAY/14/856

Reference	*Applicability*	*Transformation model*	***n***	***b***	*δ*
Helenelund (1951)	Finnish and Danish clays, silts, and organic soils	$C_c = 0.85\left(\frac{w_n}{100}\right)^{1.5}$	856	1.97	0.44
Di Buò et al. (2019)	Finnish soft sensitive clays (stress range σ_p' + 10…50 kPa)	$C_{c(+10kPa)} = 0.000102(w_n)^{2.284}$	856	0.54	0.51
		$C_{c(+20kPa)} = 0.000289(w_n)^{2.031}$	856	0.57	0.46
		$C_{c(+50kPa)} = 0.0011(w_n)^{1.669}$	856	0.73	0.43
Moran et al. (1958)	Organic soils, peats, organic silts, and clays	$C_c = 0.0115w_n$	856	1.34	0.49
Kootahi and Moradi (2017)	Marine fine-grained soils	$C_c = 0.012(w_n - 7.75)$	856	1.42	0.47
		$C_c = 0.510(e_0 - 0.33)$	856	1.30	0.46
		$C_c = 0.012(w_L - 8)$	354	1.50	0.64
		$C_c = 0.020(PI + 0.65)$	354	1.43	0.80
		$C_c = 0.374(e_0 + 0.01w_L - 0.47)$	354	1.22	0.47
		$C_c = -0.117 + 0.009w_n + 0.004w_L$	354	1.26	0.52
Nakase et al. (1988)	Reconstituted Japanese marine clays	$C_c = 0.0104(PI + 4.42)$	354	2.36	0.67
Nishida (1956)	All clays	$C_c = 0.54(e_0 - 0.35)$	856	1.24	0.45
Ogawa and Matsumoto (1978)	Japanese marine clays	$C_c = 0.015(w_L - 19)$	354	1.63	0.89
Andersen (2012)	Danish clays, silts, and organic soils	$CR(\%) = 28.9\log_{10}\left(\frac{w_n}{11.4}\right)$	856	1.56	0.46
Janbu (1998)	All normally consolidated clays	$CR = \frac{\ln 10}{700} \times w_n \left(m_l = \frac{700}{w_n}\right)$	856	1.40	0.42
Elnaggar and Krizek (1970)	Various clays and silts	$CR = 0.156e_0 + 0.0107$	856	1.04	0.42
Di Buò et al. (2019)	Finnish soft sensitive clays	$C_s = 0.0113(w_n)^{0.67}$	656	0.49	0.44
Kootahi and Moradi (2017)	Marine fine-grained soils	$C_s = 0.0014(w_n - 10)$	656	1.09	0.43
		$C_s = 0.0540(e_0 - 0.3)$	656	1.05	0.41
		$C_s = 0.0017(w_L - 21)$	265	1.68	1.13
		$C_s = 0.0025(PI - 3)$	264	1.64	1.47
		$C_s = 0.0210(e_0 + 0.06w_L - 1.60)$	267	1.16	0.52
		$C_s = 0.0330(e_0 + 0.04PI - 0.55)$	266	1.05	0.44
Kulhawy and Mayne (1990)	All clays	$C_s = 0.0027PI$	267	1.18	0.87

Source: Data from Löfman and Korkiala-Tanttu (2022).

Notes: n = number of data points; b = bias (mean value of errors); and δ = transformation uncertainty (COV of errors).

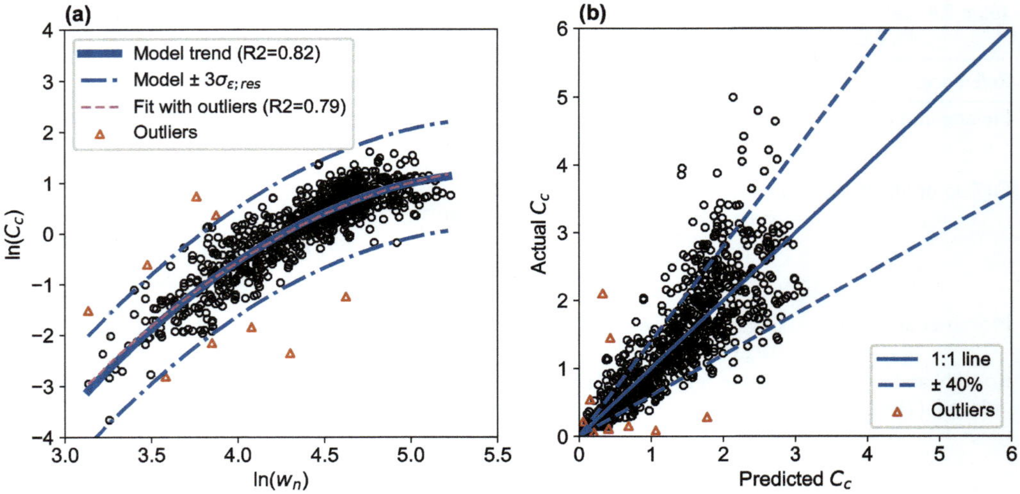

Figure 9.9 Fitting of $C_c(w_n)$-transformation model using the FI-CLAY/14/822 database: (a) polynomial regression curve and outlier detection and (b) predicted versus actual C_c.

Some newly derived transformation models and their statistics are collected in Table 9.5. Standard deviation of residual error $\sigma_{\varepsilon;res}$ quantifies the transformation uncertainty in the model. In addition, the statistics for the transformation error ε_i are provided to allow comparison with existing transformation models. It can be observed that via careful outlier detection and fitting of the transformation model, the transformation uncertainty in predicting C_c and CR for Finnish soft clays can be reduced. Moreover, the transformation uncertainty can be further reduced by incorporating a second predictor; For example, considering plastic limit w_P in addition to e_0 reduced $\sigma_{\varepsilon;res}$ for C_c from 0.346 to 0.317. Further analysis verified that for the same dataset used to derive C_c (e_0, w_P) model, the transformation uncertainty for $C_c(e_0)$ model was indeed greater, and hence not a result of dataset differences (Löfman and Korkiala-Tanttu 2022). In the case of C_s, transformation uncertainty was on average greater, and adding a second predictor did not reduce the performance of the model notably. Löfman and Korkiala-Tanttu (2022) also observed that liquid limit or plasticity index did not perform well as predictors for Finnish clays, contrary to the other studies dealing with marine clays (e.g. Kootahi and Moradi 2017; Kulhawy and Mayne 1990).

Figure 9.10 compares the newly derived and some existing transformation models against the database FI-CLAY/14/856-B. Figure 9.10a illustrates how the newly derived transformation model for C_c properly models the skewed error (shown as ± 2 standard deviations). Similarly, bias statistics in Table 9.5 indicate non-symmetric error distribution (mean ≠ median). This observation aligns with the assumption of lognormal pdf for variability term ε (e.g., Ching 2017). Figure 9.10a also shows how Di Buò et al.'s (2019) transformation models for sensitive Finnish clays form the upper bound for the data, which is due to differences in data and in C_c interpretation: Di Buò et al. (2019) considered a rather small stress range after σ_p' (10–50 kPa), and the data consisted of CRS tests with significant compression after σ_p', thus resulting in very high C_c values.

Figure 9.10b, which compares transformation models for C_s, shows that the general transformation model for marine clays (Kootahi and Moradi 2017) is quite suitable

Table 9.5 Transformation models for compressibility properties derived using a database of Finnish clays FI-CLAY/14/822

Model variables		*Regression model:* $Y = b_0 + b_1X_1 + b_2X_2 + b_3(X_1)^2 + b_4X_1X_2 + b_5(X_2)^2 + \varepsilon_{res}$									*Transformation error statistics*		
X_1	X_2	b_0	b_1	b_2	b_3	b_4	b_5	$\sigma_{\varepsilon;res}$	n	R^2_{adj}	Median	b	δ
Models for $Y = \ln(C_c)$:													
$\ln(w_n)$	N/A	−21.57	8.227	N/A	−0.743	N/A	N/A	0.357	813	0.82	1.00	1.07	0.37
$\ln(e_0)$	N/A	−1.586	2.861	N/A	−0.720	N/A	N/A	0.346	816	0.83	1.01	1.06	0.35
$\ln(e_0)$	$\ln(w_P)$	−7.516	1.911	4.429	−0.365	0.248	−0.807	0.317	357	0.88	1.02	1.05	0.32
Models for $Y = \ln(CR)$:													
$\ln(w_n)$	N/A	−20.75	7.986	N/A	−0.791	N/A	N/A	0.350	814	0.67	1.00	1.06	0.36
$\ln(e_0)$	N/A	−2.279	2.353	N/A	−0.831	N/A	N/A	0.343	816	0.69	1.01	1.06	0.35
$\ln(e_0)$	$\ln(w_P)$	−8.193	1.395	4.420	−0.481	0.252	−0.806	0.317	357	0.78	1.02	1.05	0.32
Models for $Y = \ln(C_s)$:													
$\ln(w_n)$	N/A	−12.339	3.618	N/A	−0.302	N/A	N/A	0.440	619	0.48	1.09	1.09	0.40
$\ln(e_0)$	N/A	−3.253	1.480	N/A	−0.317	N/A	N/A	0.435	619	0.49	1.10	1.09	0.38

Source: Data from Löfman and Korkiala-Tanttu (2022).

Notes: $\sigma_{\varepsilon;res}$ = standard deviation of residual errors; n = number of data points; R^2_{adj} = adjusted *R*-squared (considers the number of predictor variables); b = bias (mean value of errors); and δ = transformation uncertainty.

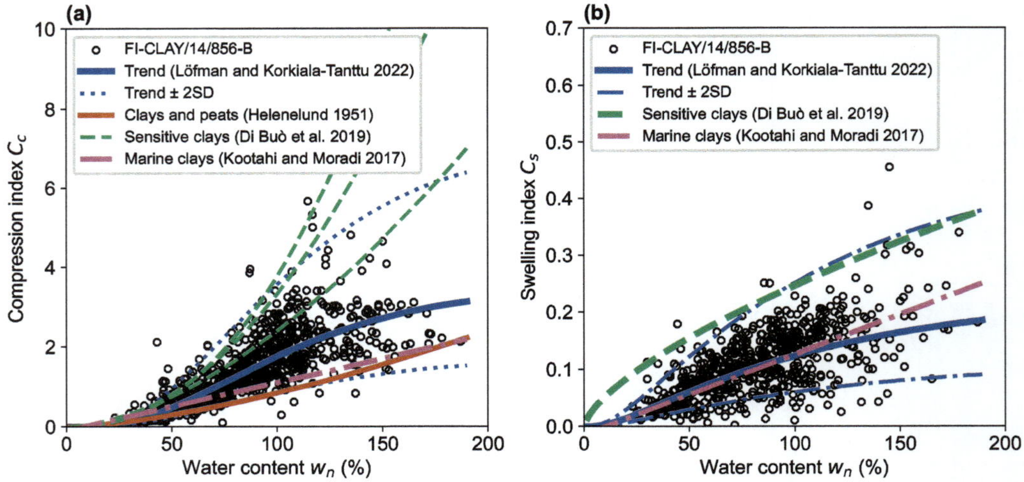

Figure 9.10 Transformation models for the compressibility of Finnish clays: (a) $C_c(w_n)$ models and (b) $C_s(w_n)$ models.

for soft Finnish clays, and the transformation error statistics (Tables 9.4 and 9.5) confirm that there is no notable improvement from the newly derived model. Meanwhile, the model for sensitive clays suggested by Di Buò et al. (2019) overestimates C_s. This bias is due to the differences in the interpretation (Di Buò et al. 2019 used an initial recompression curve, which is affected by sample disturbance) and weak correlation ($R^2 = 0.16$).

9.7 CROSS-CORRELATION AND CORRELATION COEFFICIENTS

Cross-correlation refers to a statistical dependency between two random variables. Ignoring a significant cross-correlation between calculation parameters may result in erroneous estimates of the probability of exceeding the limit state (e.g., Baecher and Christian 2003; Li et al. 2012).

The magnitude of cross-correlation may be measured by means of correlation coefficient ρ . Pearson product–moment correlation coefficient (ρ_{XY}) measures the linear correlation between two random variables, *X* and *Y*. Another common measure is Spearman's rank correlation coefficient. To evaluate the significance of cross-correlation, categories suggested by Evans (1996) may be used (Zhou et al. 2021): $|\rho| \geq 0.8$ means very strong, $0.6 \leq |\rho| < 0.8$ means strong, while $0.4 \leq |\rho| < 0.6$ means moderate cross-correlations. Smaller values indicate weak to very weak cross-correlation.

The measured correlation coefficient ρ varies between considered sub-populations. In a global clay database (containing multiple sites), cross-correlation between certain geotechnical parameters may be weak, while for a specific site or soil layer, strong cross-correlation is sometimes observed. This scale effect is noticeable in Table 9.6, which compares Pearson's ρ defined with site-specific, soil type-specific data, and regional or global data. For example, cross-correlation between m_1 and σ_p' is weak on a regional scale (all Finnish sites and data combined), but within a certain site (soft Haarajoki clay or Suurpelto clay), very strong cross-correlation is observed.

Table 9.6 Pearson's correlation coefficient ρ for certain populations

Considered population	$\rho_{e_0C_c}$	$\rho_{e_0\sigma'_p}$	$\rho_{C_c\sigma'_p}$	$\rho_{C_cC_s}$	$\rho_{m_1\sigma'_p}$	$\rho_{m_1\beta_1}$	$\rho_{\beta_1\sigma'_p}$
Site-specific ρ, Haarajoki clay	0.53	−0.67	0.09	0.50[a]	−0.88	0.86	−0.64
Site-specific ρ, Suurpelto clay	0.54	−0.86	−0.30	0.35[a]	−0.87	0.73	−0.73
Site-specific average ρ, soft Finnish clays	0.72	−0.32	0.05	N/A	−0.73	0.34	−0.28
Regional ρ (based on FI-CLAY/14/856-B database)	0.76	−0.44	−0.27	0.59[a]	0.25	0.42	0.11

[a]To compare, for the global clay database, Zhou et al. (2021) found a median value of 0.46 for site-specific ρ.

Indeed, cross-correlation between m_1 and σ_p' is the most prominent in soft Finnish clays with negative stress exponent β_1. A regional clay database may be utilized to define site-specific values for correlation coefficients, and the average or mean value can then be utilized in a reliability analysis if the amount of site-specific data is too limited. Löfman (2022) reported a study, where the FI-CLAY/14/856-B database was used to define the average correlation coefficients for soft Finnish clays. The database was first filtered to be representative of soft Finnish clays ($\beta_1 < 0$, $m_1 < 10$, $w_n > 75\%$). Only the datasets with $n \geq 10$ variable pairs were considered, as in Zhou et al. (2021). Figure 9.11 illustrates how the most consistent ρ values are related to the variable pairs m_1–σ_p' and e_0–C_c, and others are marked with greater variability between sites.

Monte Carlo simulation (MCS) with Gaussian copula was then used to compare the computed stress–strain behavior for Haarajoki clay and Suurpelto clay in different scenarios (see Löfman (2022) for further details):

1. Case Ind.: cross-correlation ignored ($\rho = 0$).
2. Case Corr.: cross-correlation defined using site-specific oedometer test results.
3. Case Datab.: average site-specific ρ for NC parameters (Haarajoki clay and Suurpelto clay not included in the average).

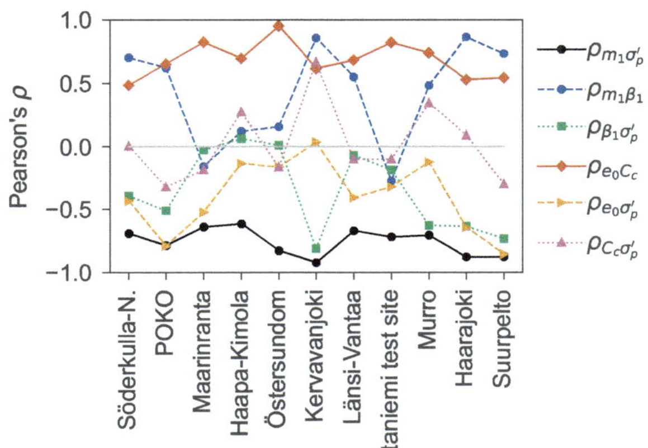

Figure 9.11 Site-specific cross-correlation between compressibility parameters in soft Finnish clays (Source: modified from Löfman 2022, with permission).

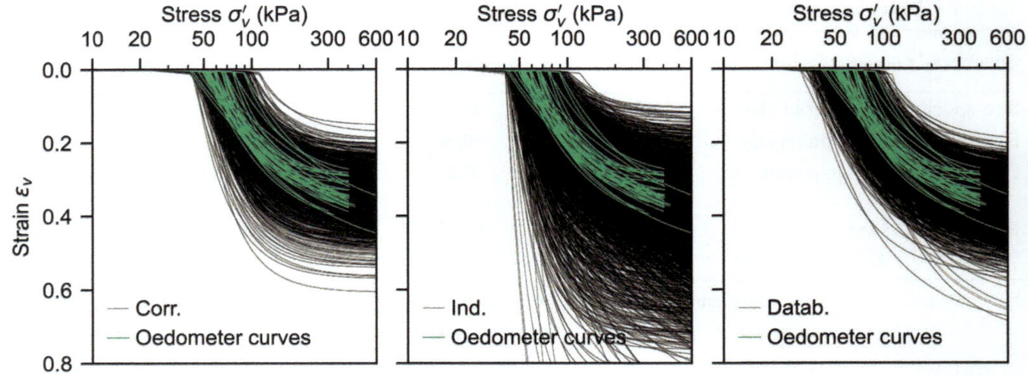

Figure 9.12 Janbu method: simulated oedometer curves for the Haarajoki clay layer in the different calculation scenarios (Source: modified from Löfman 2022, with permission).

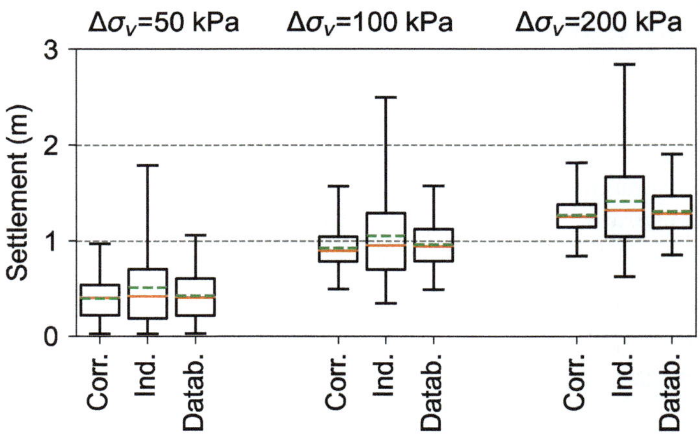

Figure 9.13 Janbu method: settlement boxplots for Haarajoki clay while assuming stress increase of 50, 100, or 200 kPa (Source: modified from Löfman 2022, with permission).

Figure 9.12 illustrates a sub-sample (2000 simulation runs) of stress–strain curves constructed with the simulated Janbu parameter sets. Oedometer curves were reconstructed using the Janbu parameters. Figure 9.13 presents the simulated settlements as boxplots for the Haarajoki clay layer. In the boxplots, the boundaries of the boxes represent the interquartile range, the dashed horizontal line is the mean, and the solid line is the median. The whiskers show the 2nd and 98th percentiles. The number of simulation runs was 10^5 for each calculation depth. It is evident that ignoring the cross-correlation between Janbu parameters (mainly variable pairs m_1–σ_p') will lead to an overestimation of uncertainty in the settlement response in soft clay. On the other hand, the correlation matrix defined via database analysis (Case Datab.) provided a fair replication of the parameter dependence (note that also the COV values were defined using literature values defined for Finnish clays, see Löfman 2022).

Meanwhile, if the C_c method (corresponding to $\beta_1 = 0$) is used to compute the stress–strain response, the influence of cross-correlation is less notable: Figure 9.14 presents the

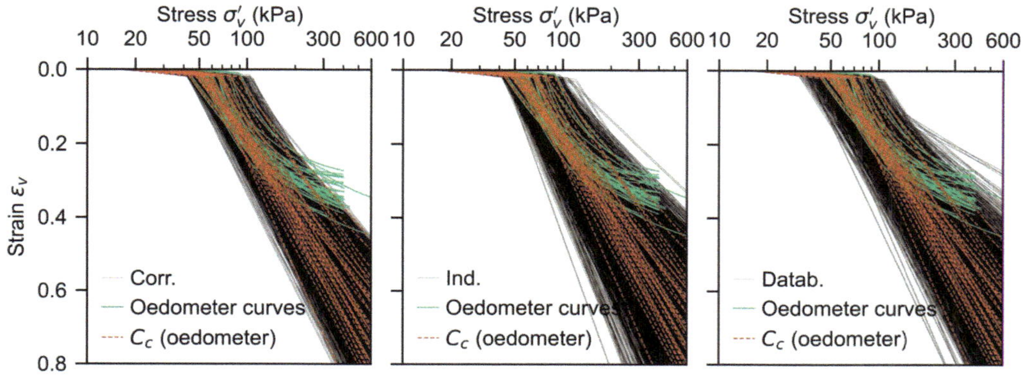

Figure 9.14 C_c method: simulated oedometer curves for the Haarajoki clay layer in the different calculation scenarios (Source: modified from Löfman 2022, with permission).

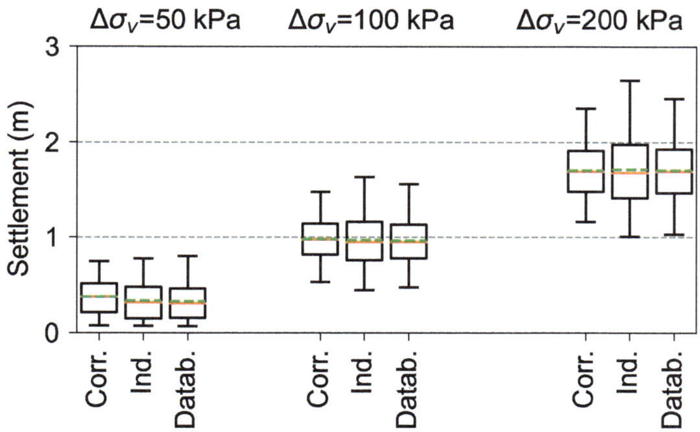

Figure 9.15 C_c method: settlement boxplots for Haarajoki clay while assuming stress increase of 50, 100, or 200 kPa (Source: modified from Löfman 2022, with permission).

oedometer curves constructed using both Janbu and C_c parameters, together with the simulated stress–strain response based on the C_c method. Figure 9.15 shows the simulated settlements as boxplots, and the calculation scenarios are rather similar. Hence, in most cases, the cross-correlation between C_c parameters may be ignored for the sake of simplicity. However, for soft soils with negative β_1 (such as Haarajoki clay), the C_c method will lead to an overestimation of strain (and thus settlement), especially at large load increments as illustrated in Figure 9.14. Hence, if the C_c method is used for soft marine clays, it is crucial that the problem-specific stress range is considered when the value of C_c is interpreted from the oedometer test result.

In a deterministic analysis, Equation (9.5) should be used to calculate the stress-corrected m_1. Using this procedure accounts for the cross-correlation between Janbu parameters. In fact, using Equation (9.5) within MCS allows us to approximate the result of Case Corr., even if the correlation coefficients are set to zero (Löfman 2022; López Ramirez, Löfman, and Korkiala-Tanttu 2022). However, at larger stress increments, such approximation leads to underestimation of the uncertainty in strain response (see Löfman 2022 for further details).

9.8 SUMMARY

This chapter has demonstrated how the database approach can be used to evaluate the needed input for the reliability-based SLS design of geo-structures dealing with soft marine clays. The special features related to soft and sensitive Nordic clays have been investigated and discussed: inherent variability (COV) for soft marine clays was found to be somewhat smaller compared to COV guidelines based on global summaries, since such soft soil deposits are often quite homogeneous. Similarly, smaller transformation uncertainties may be acquired if a regional clay database is used to calibrate existing transformation models and to derive carefully fitted models with outlier detection. Finally, it was shown how the cross-correlation between Janbu parameters may be effectively modeled via a database approach and Monte Carlo simulation with copula.

The COV values describing inherent variability were found to be rather consistent, and the literature values defined for Finnish marine clays may be utilized in the case of too limited data. In a typical situation, the amount of site-specific data is sufficient to determine the mean soil property μ for the considered homogeneous layer. Then, standard deviation (SD) is defined as μ multiplied by COV. Simple pdfs, such as normal or lognormal distribution, may be defined using the first two moments, μ and SD. For example, if C_c from one oedometer test is 2.95 (i.e., $\mu \approx 2.95$), and the specimen represents soft and homogeneous Finnish clay (COV = 0.2, see Table 9.2), SD = 0.2 × 2.95 = 0.59. If needed, the remaining sources of uncertainty are used to assess the total uncertainty COV_{tot} (see Phoon and Kulhawy 1999), and SD is calculated using the acquired COV_{tot} value.

Once the transformation uncertainties have been defined using a regional soil database (see Table 9.5), they can be properly accounted for in the case of indirectly evaluated compressibility properties. For example, if the water content in the studied layer is w_n = 100 %, $C_c = \exp[-21.567 + 8.227\ln(w_n) - 0.743[\ln(w_n)]^2] = 1.75$. In a simplified approach, transformation uncertainty is modeled via Equation (9.10), meaning that the predicted target value (1.75) is multiplied by bias factor b (= 1.07). Hence, the (non-biased) mean value is qual to μ = 1.07 × 1.75 = 1.87. Next, total uncertainty (here statistical uncertainty, measurement error, and variance reduction from averaging are ignored for the sake of simplicity) may be defined via the sum of squared COV values according to Phoon and Kulhawy (1999):

$$COV_{tot} = \sqrt{COV_{wn}^2 + \delta^2} \tag{9.6}$$

where COV_{wn} is the inherent variability of water content (0.10, Table 9.2) and δ is transformation uncertainty (0.37, see Table 9.5). For the indirectly estimated C_c, we acquire COV_{tot} = 0.383. Hence, SD is 0.383 × 1.87 = 0.717.

For C_c within a homogeneous clay layer, lognormal pdf may be assumed (see Table 9.3).

Figure 9.16 presents the pdfs based on one oedometer test result and transformation model. Further, the figure shows the data from soft Haarajoki clay layer (with parameters μ = 2.65 and SD = 0.531), from which the oedometer test result originated. The chosen value for w_n = 100 % is close to the layers' mean value (107 %). In this example, the actual data falls within the pdf defined from the transformation model, but the clay layer in

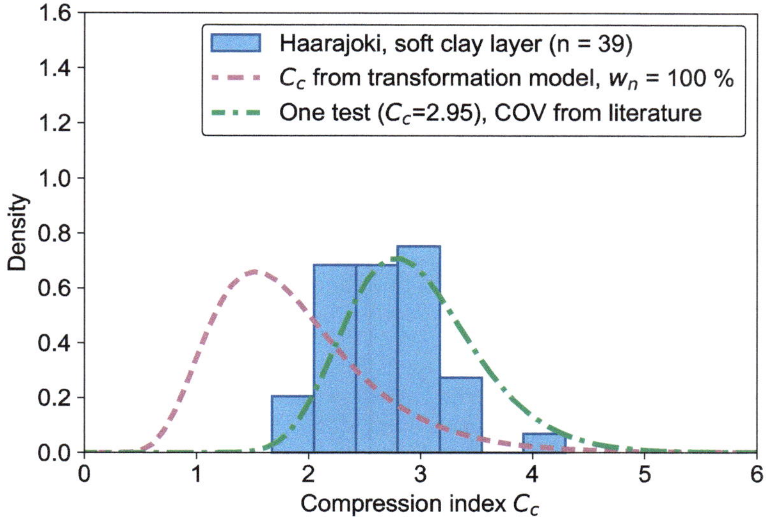

Figure 9.16 Simple example of evaluating the pdf of soft marine clay based on transformation model or an oedometer test and literature COV.

question represents marine clays with higher compressibility; Hence, the settlement prediction will be biased to some extent, even though the pdf of the settlement response will be wider accordingly. The figure also illustrates that even though COV value was lower for the oedometer test case, higher mean C_c will lead to greater SD (and therefore wider pdf).

REFERENCES

Andersen, J. D. 2012. Prediction of compression ratio for clays and organic soils. In *Proceedings of the 16th Nordic Geotechnical Meeting (NGM 2021)*. Copenhagen: Danish Geotechnical Society, 303–310.

Baecher, G. B., and J. T Christian. 2003. *Reliability and Statistics in Geotechnical Engineering*, Chichester: John Wiley & Sons.

Bjerrum, L. 1967. Engineering geology of normally consolidated marine clays as related to settlements of buildings. *Géotechnique* 17(2), 83–119.

Ching, J. 2017. Transformation models and multivariate soil databases. *Joint TC205/TC304 Working Group on "Discussion of statistical/reliability methods for Eurocodes" – Final Report (Sep 2017), ISSMGE.*

Ching, J., and K.-K. Phoon. 2012. Establishment of generic transformations for geotechnical design parameters. *Structural Safety* 35, 52–62.

Ching, J., and K.-K. Phoon. 2014. Transformations and correlations among some clay parameters - The global database. *Can Geotech J* 51(6), 663–685.

Di Buò, B., Länsivaara, T., Franza, A., and P. Mayne. 2019. Compressibility of Finnish sensitive clay. In: *Proceedings of the XVII European Conference on Soil Mechanics and Geotechnical Engineering.* The Icelandic Geotechnical Society.

Elnaggar, H. A., and R. J. Krizek. 1970. Statistical approximation for consolidation settlement. *Highway Res Rec*, 323, 87–96.

Evans, J. D. 1996. *Straightforward Statistics for the Behavioral Sciences.* Pacific Grove, CA: Brooks/Cole Publishing Company.

Finnish Transport Agency. 2012. *Tien geotekninen suunnittelu [Geotechnical design of roads]. Guidelines of Finnish Transport Agency* 10/2012. Helsinki: Finnish Transport Agency.

Gardemeister, R. 1975. *On Engineering-Geological Properties of Fine-Grained Sediments in Finland.* Doctoral thesis. Turku: University of Turku.

Guan, Z., Chang, Y.-C., Wang, Y., Aladejare, A., Zhang, D., and J. Ching. 2021. Site-specific statistics for geotechnical properties. In: *State-of-the-Art Review of Inherent Variability and Uncertainty in Geotechnical Properties and Models*, ed. J. Ching and T. Schweckendiek. TC304, ISSMGE.

Helenelund, K. V. 1951. *Om konsolidering och sättning av belastade marklager. Maa- ja vesiteknillisiä tutkimuksia* 6. Helsinki.

Janbu, N. 1963. Soil compressibility as determined by oedometer and triaxial tests. In: *Proceedings of European Conference on Soil Mechanics and Foundation Engineering,* Wiesbaden, Vol. 1, 19–25.

Janbu, N. 1967. Settlement calculations based on the tangent modulus concept. Three guest lectures at Moscow State University, Bulletin 2, *Soil Mechanics.* Norwegian Institute of Technology.

Janbu, N. 1998. *Sediment Deformations.* Bulletin 35. Trondheim: Department of Geotechnical Engineering, Norwegian University of Science and Technology.

Karstunen, M., Krenn, H., Wheeler, S. J., Koskinen, M., and R. Zentar. 2005. Effect of anisotropy and destructuration on the behavior of Murro test embankment. *Int J Geomech* 5(2), 87–97.

Keinonen, L. 1963. *On the Sensitivity of Water-Laid Sediments in Finland and Fac-Tors inducing sensitivity.* Publication 77. Helsinki: The State Institute for Technical Research.

Kivi, E. 2022. *Pohjanvahvistusmenetelmät Suomessa – käyttömäärät ja hiilijalanjälki [Ground improvement methods in Finland – usage rates and carbon footprint].* Master thesis. Espoo: Aalto University School of Engineering.

Knuuti, M. and M. S. Löfman. 2022. *Epävarmuudet ja luotettavuus geoteknisessä mitoituksessa – Luotettavuus-workshop 2022 [Uncertainty and reliability in geotechnical design - Reliability workshop 2022].* Publications of the FTIA 37/2022. Helsinki: Finnish Transport Infrastructure Agency.

Kootahi, K., and G. Moradi. 2017. Evaluation of compression index of marine fine-grained soils by the use of index tests. *Mar Georesources Geotechnol* 35(4), 548–570.

Korhonen, K.-H., Gardemeister, R., and M. Tammirinne, M. 1974. *Geotekninen maaluokitus [Geotechnical soil classification]. Geotekniikan laboratorio, tiedonanto 14.* Otaniemi: VTT Technical Research Centre of Finland.

Kulhawy, F. H., and P. W. Mayne. 1990. *Manual of Estimating Soil Properties for Foundation Design.* Palo Alto, CA: Electric Power Research Institute.

Lacasse, S. 2016. Hazard, Reliability and Risk Assessment - Research and Practice for Increased Safety. In: *Proceedings of the 17th Nordic Geotechnical Meeting (25–28 May 2016, Reykjavik, Iceland)*, 17–42, Reykjavík: The Icelandic Geotechnical Society,

Lacasse, S., and Nadim, F. 1996. Uncertainties in characterising soil properties." In: *Uncertainty in the Geologic Environment: From Theory to Practice*, Geotechnical Special Publication No. 58, C. D. Shackleford, P. P. Nelson, and M. J. S. Roth, eds., 49–75. New York: ASCE.

Lacasse, S., Nadim, F., Rahim, A., and T. R. Guttormsen. 2007. *Statistical Description of Characteristic Soil Properties.* Offshore Technology Conference.

Larsson, R. 1977. *Basic Behaviour of Scandinavian Soft Clays.* Linköping: Swedish Geotechnical Institute.

Leroueil, S., Tavenas, F., and J. P. L. Bihan. 1983. Propriétés caractéristiques des argiles de l'est du Canada. *Can Geotech J* 20(4), 681–705.

Li, D., Tang, X., Zhou, C., and K.-K. Phoon 2012. Uncertainty analysis of correlated non-normal geotechnical parameters using Gaussian copula. *Sci China Tech Sci* 55(11), 3081–3089.

López Ramírez, A., Löfman, M. S., and L. K. Korkiala-Tanttu. 2022. Safety margin from characteristic values in settlement calculations of embankments on clay. *Proc Inst Civ Eng: Geotech Eng* 175(6): 605–617.

Länsivaara, T. 1995. Stress-strain-strain rate relation in oedometer tests. In: *Proceedings, 70 years of Soil Mechanics*, International symposium, Istanbul.
Länsivaara, T. 2003. Problems related Scandinavian clays to settlement calculations on soft Scandinavian clays." In: *Int. Workshop on Geotechnics of Soft Soils -Theory and Practice*, ed. Vermeer, Schweiger, Karstunen, and Cudny. VGE.
Löfman, M. S. 2016. *Perniön saven parametrien luotettavuuden ja saven eri ominaisuuksien välisten korrelaatioiden arviointi [Estimation of the Reliability of Perniö Clay Parameters and Correlations between Clay Properties]*. Master thesis. Espoo: Aalto University School of Engineering.
Löfman, M. S. 2022. *Uncertainty Quantification for Compressibility and Settlement Response of Clays*. Doctoral thesis. Espoo: Aalto University School of Engineering.
Löfman, M. S. and L. Korkiala-Tanttu. 2019. Variability and typical value distributions of compressibility properties of fine-grained sediments in Finland. In: *Proceedings of the 7th International Symposium on Geotechnical Safety and Risk (ISGSR)*, ed. J. Ching, D.-Q. Li, and J. Zhang, 182–187. Singapore: Research Publishing.
Löfman, M. S., and L. Korkiala-Tanttu. 2021a. *Data and code for 'Influence of cross-correlation on the modelled uncertainty in stress–strain behavior of soft clays.' Zenodo*, https://doi.org/10.5281/zenodo.5040083
Löfman, M. S. and L. Korkiala-Tanttu. 2021b. Inherent variability of geotechnical properties for Finnish clay soils. In: 18th International Probabilistic Workshop. *IPW 2021*, ed. J. C. Matos *et al.*, Lecture Notes in Civil Engineering, vol 153, 431–443. Cham: Springer.
Löfman, M. S. and L. Korkiala-Tanttu. 2021c. Dataset for 'Inherent variability of geotechnical properties for Finnish clay soils'. https://doi.org/10.5281/zenodo.4709003
Löfman, M. S., and L. Korkiala-Tanttu. 2022. Transformation models for the compressibility properties of Finnish clays using a multivariate database. *Georisk* 16(2), 330–346.
Mataić, I. 2016. *On Structure and Rate Dependence of Perniö Clay*. Doctoral thesis. Espoo: Aalto University School of Engineering.
Moran, Proctor, Mueser, and Rutlege. 1958. *Study of Deep Soil Stabilization by Vertical Sand Drains*. Washington, DC: US Department of Commerse OT.S.
Nakase, A., Kamei, T., and O. Kusakabe. 1988. Constitutive parameters estimated by plasticity index. *J Geotech Eng* 114(7), 844–858.
Nishida, Y. 1956. A brief note on compression index of soil. *J Soil Mech Found Div* 82(3), 1–14.
Ogawa, F., and K. Matsumoto. 1978. The correlation of the mechanical and index properties of soils in harbour districts. *Rep Port Harbour Res Instit* 17(3), 3–89.
Ohde, J. 1939. *Zur Theorie der Druckverteilung im Baugrund*. Berlin: Der Bauingenieur.
Perttu, O. 2023. *Pohjanvahvistuksen optimointi osana vähähiilistä esirakentamista [Optimization of Ground Improvement as a Part of Low-Carbon Preconstruction]*. Master thesis. Espoo: Aalto University School of Engineering.
Phoon, K.-K. 2017. Role of reliability calculations in geotechnical design. *Georisk* 11(1), 4–21.
Phoon, K.-K., and F. H. Kulhawy 1999. Characterization of geotechnical variability. *Can Geotech J*, 36 (4), 612–624.
Phoon, K.-K., and C. Tang. 2019. Characterisation of geotechnical model uncertainty. *Georisk* 13(2), 101–130.
Rankka, K., Andersson-Sköld, Y., Hultén, C., Larsson, R., Leroux, V., and T. Dahlin. 2004. *Quick Clay in Sweden*. Report 65. Linköping: Swedish Geotechnial Institute.
Rosenqvist, I. T. 1953. Considerations on the sensitivity of Norwegian quick-clays. *Géotechnique* 3(5), 195–200.
Sällfors, G. 1975. *Preconsolidation Pressure of Soft, High-Plastic Clays*. Doctoral thesis. Göteborg: Chalmers University of Technology.
Uzielli, M., Lacasse, S., Nadim, F., and K.-K. Phoon. 2006. Soil variability analysis for geotechnical practice. In: *Proceedings of the 2nd International Workshop on Characterisation and Engineering Properties of Natural Soils*. Singapore.
Zhou, Y., Zhang, D., and J. Ching. 2021. Site-specific correlations between soil/rock properties. In: *State-of-the-Art Review of Inherent Variability and Uncertainty in Geotechnical Properties and Models*, ed. J. Ching and T. Schweckendiek. TC304, ISSMGE.

Chapter 10

An engineering geological parameter database of tunnel surrounding rock and its application

Junjie Ma and Tianbin Li

A generic rock mass database consisting of eight parameters is compiled from ten tunnels. The eight surrounding rock engineering geological parameters are the rock harness, rock mass integrity, dicontinuity condition, weathering degree, rock mass structure, groundwater condition, in-situ stress condition, and surrounding rock grade. This generic database, labeled as SurroundingRock/8/286, consists of 286 surrounding rock cases. This chapter aims to build an intelligent and accurate surrounding rock classification model for generic drilling and blasting excavation sites by combining SurroundingRock/8/286 with machine learning. Four methods are used to construct the surrounding rock classification models: convolutional neural network, random forest, Gaussian process, and support vector machine. The 5-fold cross-validation, 10-fold cross-validation, precision, recall, F-measure, and 24 actual tunnel validation results show that the support vector machine classification model is the most effective. Finally, a tunnel information management system was developed with cloud technology and the best surrounding rock classification model, which is now well-used in the Leshan–Xichang Expressway.

10.1 INTRODUCTION

Tunnel excavation will reveal new engineering geological conditions ahead. The different engineering geological conditions of the surrounding rock determine the support measures for the current section and the excavation method for the next section. Therefore, the collection and effective use of new engineering geological information revealed by tunnel excavation is essential for determining the tunnel surrounding rock's stability and the tunnel's dynamic excavation and support.

Many scholars have graded the quality of the surrounding rock based on the engineering geological information revealed by tunnel excavation, used this to determine the surrounding rock's stability, and developed corresponding support measures and excavation methods for different grades of surrounding rock quality. For different engineering geological conditions, the quality of the tunnel surrounding rock is graded, which is called surrounding rock classification or rock mass classification (Ma et al. 2023). Bieniawski (1973) collected and analyzed over 300 tunnel cases and proposed a rock mass classification method for jointed rock mass, called the rock mass rating (RMR) system. The engineering geological parameters considered in the RMR method include rock strength, rock quality designation (RQD), joint spacing, joint state, groundwater, and the relationship between joint attitude and tunnel axis. Barton et al. (1974) created a database of 212 tunnel cases and proposed the *Q* system for rock mass classification. The calculated parameters of the *Q* system include RQD, joint set number, joint roughness number, joint alteration number, joint water reduction factor, and stress reduction factor. Barton (2002)

 DOI: 10.1201/9781003441946-10

analyzed the 1050 new tunnel cases and revised the Q system. Hoek (1994) proposed the Geological Strength Index (GSI) method and considered the rock mass structure and weathering conditions. In China, the mainly used rock mass classification method is basic quality (BQ), which is determined by a statistical analysis of 460 cases. The BQ method (GB/T 50218–2014) considers the rock strength, rock mass integrity, groundwater condition, in-situ stress condition, and the relationship between weak structural plane occurrence, and tunnel axis direction. Different rock mass classification methods have been widely used in traffic tunnels, mining tunnels, and engineering slopes. However, with the advent of many deep-buried and ultra-long tunnels, traditional classification methods cannot meet the requirements of real-time and accurate acquisition of surrounding rock grades on site.

In the 21st century, the rapid development of artificial intelligence (AI) technology has brought opportunities for geotechnical engineering (Phoon, Ching, and Cao 2022; Phoon, Ching, and Shuku 2022; Phoon and Zhang 2022). Many scholars build a database based on the engineering geological information of the site and combine AI technology to build an intelligent model to realize intelligent prediction of some characteristics of the site. Santos et al. (2021) collected 3216 samples with RMR input parameters in open pit mines, and four rock mass classifiers were established using naive Bayes (NB), random forest (RN), artificial neural networks (ANNs), and support vector machine (SVM). Wang et al. (2021) built a database based on the 299 samples of the drilling parameters of the intelligent rock drilling trolley and established a classification system for surrounding rock. Bo et al. (2022) built a TBM boring parameter database containing 4464 samples, and a rock mass classification model was established using catboost integrated with sequential model-based optimization. In particular, some scholars use AI technology to extract surrounding rock information and use it for surrounding rock classification. Liu et al. (2018) built an image database containing 6000 tunnel face pictures, which was used to extract the surrounding rock information using deep learning technology. Chen et al. (2021) built a rock tunnel fracture database with 3000 tunnel face images, which was used to extract the rock fracture information using deep learning technology. In summary, building a generic and complete database of engineering geological information is exceptionally challenging. And most of the existing intelligent surrounding rock classification methods are limited to theoretical study and have little practical application.

In summary, most current research focuses on TBM-excavated tunnels and little research on intelligent systems for drilling and blasting excavation. Moreover, there are few studies on constructing a generic surrounding rock engineering geological information database for the traditional drilling and blasting tunnel. The above problems are the focus of this chapter. The novelty of this study is twofold:

1. This chapter compiles a new generic database for surrounding rock parameters named SurroundingRock/8/286. This database consists of 286 surrounding rock cases from ten tunnels for eight surrounding rock engineering geological properties, including rock harness, rock mass integrity, discontinuity condition, weathering degree, rock mass structure, groundwater condition, in-situ stress condition, and surrounding rock grade. Figure 10.1 presented in the following summarizes the generic and site-specific statistics for surrounding rock engineering geological parameters in SurroundingRock/8/286. In particular, each sample in SurroundingRock/8/286 contains eight complete surrounding rock engineering geological parameters.

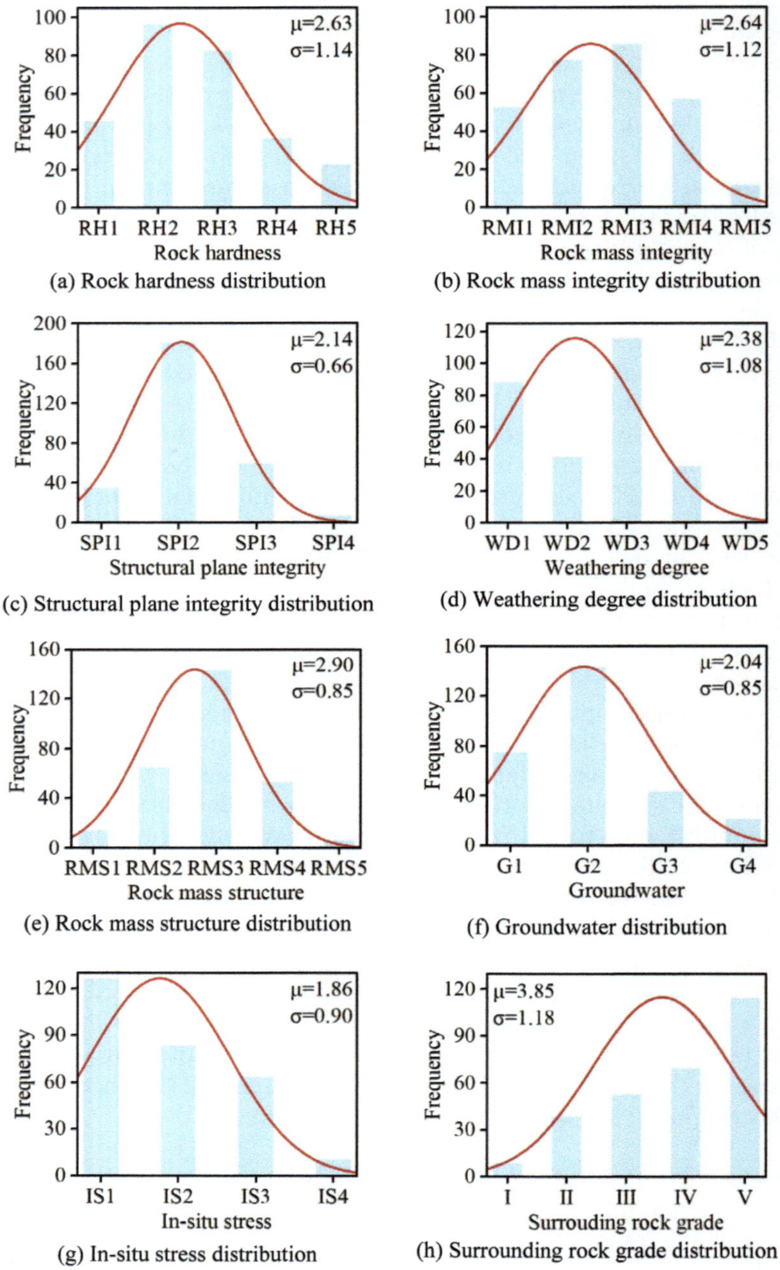

Figure 10.1 Distribution of the engineering geological parameters in SurroundingRock/8/286. Note: The abbreviations for each *x*-axis in Figure 10.1 correspond to the category abbreviations in Table 10.1.

2. This chapter further investigates the four methods (convolutional neural network (CNN), random forest (RF), Gaussian process (GP), and SVM) of constructing the surrounding rock classification model that SurroundingRock/8/286 supports. Further, accuracy, precision, recall, and *F*-measure are used to evaluate model performance comprehensively. Moreover, 24 actual tunnel cases are used further to test the practicability of the surrounding rock classification (SRC) model. Moreover, combined with cloud technology, a tunnel information management system with the best SVM-based classification model as the core is constructed. The tunnel information management system is now well-used in the Leshan–Xichang (Lexi) Expressway.

10.2 DATABASE SURROUNDINGROCK/8/286

This study complies a generic database (SurroundingRock/8/286) from the tunnel construction site for eight surrounding rock engineering geological parameters. In the tunnel construction site, generic databases have been complied for metamorphic, sedimentary, and igneous rocks. The SurroundingRock/8/286 database consists of 286 different surrounding rock cases from ten tunnels, each case with eight complete surrounding rock engineering geological parameters. The ten tunnels from western China include the Niba mountain tunnel, Erlang mountain tunnel, Wangjiangling tunnel, Zaozilin tunnel, Zhegu mountain tunnel, Longxi tunnel, Futang tunnel, Micang mountain tunnel, Miyaluo #3 tunnel, and Shiziping tunnel.

The eight surrounding rock engineering geological parameters in SurroundingRock/8/286 include rock harness, rock mass integrity, discontinuity condition, weathering degree, rock mass structure, groundwater condition, in-situ stress condition, and surrounding rock grade. Since there are quantitative and qualitative parameters in SurroundingRock/8/286, it is necessary to discretize them. In Table 10.1, the parameters in SurroundingRock/8/286 are discretized according to the parameter division range recommended by "Code for Engineering Geological Investigation of Water Resources and Hydropower" and "Standard for Engineering Classification of Rock Mass." The quantitative and qualitative division bases of the eight surrounding rock engineering geological parameters are shown in Tables S10.1–Table S10.7 in the supplementary material.

Figure 10.1 shows the distribution of the engineering geological parameters in SurroundingRock/8/286. The data in our database covers a wide range of values for these eight surrounding rock engineering geological parameters. The SurroundingRock/8/286 database has the following features:

1. Figure 10.1(a) shows that the rock strength in the database is predominantly slightly hard and slightly soft. The database also has 20–50 hard, soft, and extremely soft rock samples.
2. Figure 10.1(b) shows that the distribution of rock mass integrity in the database is relatively balanced, except for the extremely broken surrounding rock samples, which are only 12.
3. Figure 10.1(c) shows that the unbalanced distribution of discontinutiy condition in the database is mainly ordinary.

Table 10.1 Intervals of the engineering geological parameters in SurroundingRock/8/286.

Parameter		*Intervals (category abbreviation)*
Rock hardness (RH)	Intervals (MPa)	(60, ∞), (30, 60], (15, 30], (5, 15], (0, 5]
	States	Hard (RH1), slightly hard (RH2), slightly soft (RH3), soft (RH4), extremely soft (RH5)
Rock mass integrity (RMI)	Intervals (K_v)	(0.75, 1], (0.55, 0.75], (0.35, 0.55], (0.15, 0.35], (0, 0.15]
	States	Complete (RMI1), slightly complete (RMI2), slightly broken (RMI3), broken (RMI4), extremely broken (RMI5)
Dicontinuity condition (Dc)	States	Good (Dc1), ordinary (Dc 2), bad (Dc 3), very bad (Dc 4)
Weathering degree (WD)	Intervals (k_r)	(0.9, 1], (0.8, 0.9], (0.6, 0.8], (0.4, 0.6], [0.2, 0.4]
	States	Fresh (WD1), slight (WD2), medium (WD3), severe, and extreme (WD4)
Rock mass structure (RMS)	States	Integral/extremely thick-layered (RMS1), block/thick-layered (RMS2), fractured block/medium-thick-layered/interlayer/th in-layered/mosaic (RMS3), fragmented (RMS4), granular (RMS5)
Groundwater condition (G)	States	Dry/wet (G1), moist/dripping (G2), rain-like dripping/linear water (G3), tubular water/gushing water (G4)
In situ stress condition (IS)	Intervals [R_c/σ_{max}]	[9, +∞), [7, 9), [4, 7), (0, 4)
	States	Low (IS1), medium (IS2), high (IS3), extremely high (IS4)
[BQ]	Intervals	(550, 710], (450, 550], (350, 450], (250, 350], (0, 250]
	States	I, II, III, IV, V

4. Figure 10.1(d) shows that the weathering degree of rock mass is mainly fresh and medium weathering in the database. There are only three samples of the extreme weathering rock mass.
5. Figure 10.1(e) shows that the database's normal distribution trend of rock mass structure is apparent. There are fewer granular structures in the database.
6. Figure 10.1(f) shows that the groundwater condition in the database is predominantly no or little water, with few samples for tubular water/gushing water.
7. Figure 10.1(g) shows that the in-situ stress condition in the database is predominantly low, medium, and high, with only 11 samples for extremely high in situ stress.
8. Figure 10.1(h) shows that the database is dominated by classes III, IV, and V surrounding rocks, with a small number of classes I and II surrounding rocks. The small and unbalanced surrounding rock database corresponds to the actual situation on the tunnel project site. However, machine learning algorithms such as SVM can perform well in small and unbalanced sample databases.

10.3 INTELLIGENT METHODS FOR SURROUNDING ROCK CLASSIFICATION

10.3.1 Convolutional neural network

Deep learning is a subfield of machine learning (Ketkar and Moolayil 2021). Convolutional neural network (CNN) is one of the representative algorithms of deep learning, which was first reported in 1989 for machine vision (LeCun et al. 1989). In the past 20 years, a series of classification networks with different architectures have been proposed based on the CNN principle, such as LeNet, AlexNet, GoogleNet, VGGNet, etc. (Khan et al. 2020). In particular, VGGNet has significantly improved network depth, computing speed, and complexity.

Simonyan and Zisserman proposed the VGGNet in 2015, which was mainly developed based on the AlexNet. The VGGNet used a small-sized filter to reduce the number of parameters and the computational complexity. In addition, the VGGNet used random inactivation to mitigate the effect of overfitting. Therefore, the VGGNet has a more significant improvement than the AlexNet regarding recognition accuracy (Khan et al. 2020). In particular, the convolutional and fixed pooling layers of the same size allow the VGGNet to be easily applied to various scenarios.

The VGGNet consists of 3 × 3 convolution kernel (stride 1 and padding 1) and pooling layer (stride 2). The distinction between the VGGNet and the AlexNet is that a stack of 3 × 3 convolution kernels is used instead of the large-sized convolution kernels (5 × 5 and 7 × 7). For example, the VGGNet16 stacks two identical 3 × 3 convolution kernels and three identical 3 × 3 convolution kernels to replace one 5 × 5 convolution kernel and one 7 × 7 convolution kernel, respectively. The number of parameters generated by two 3 × 3 convolution kernels is $2(32D2) = 18D2$, and the number of parameters generated by a 5 × 5 convolution kernel is $52D2 = 25\text{D}2$ (D is the depth of the channel of the feature matrix). Therefore, the number of parameters produced by two 3 × 3 convolution kernels is significantly less than that produced by one 5 × 5 convolution kernel. Similarly, the number of parameters generated by three 3 × 3 convolution kernels is much lower than that generated by one 7 × 7 convolution kernel. Thus, the VGGNet with relatively few parameters can achieve a more efficient training process. On the other hand, after the 3

$\times$ 3 convolution kernel replaces the 5 $\times$ 5 and 7 $\times$ 7 convolution kernels, the new network will have a deeper structure, which can further improve the recognition performance of the VGGNet. The distinction between the VGGNet16 and the VGGNet19 is the difference in network depth. The VGGNet16 contains 16 hidden layers (thirteen convolutional layers and three fully connected layers). The VGGNet19 contains 19 hidden layers (sixteen convolutional layers and three fully connected layers). More details about the VGGNet can be found in Simonyan and Zisserman (2015).

10.3.2 Random forest

Random forest (RF) is a statistical learning theory that uses the bootstrap resampling method to take multiple samples from the original sample, model the decision tree for each bootstrap sample, and then combine the predictions of multiple decision trees to obtain the final prediction result by voting (Breiman 2001, Genuer and Poggi 2020).

Random forest classification (RFC) is a combinatorial classification model composed of many decision tree classification models $\{h(X, \Theta_k), k=1, 2, \ldots\}$, and the parameter set $\{\Theta_k\}$ is an independent and homogeneous random vector. Given the independent variable X, each decision tree classification model is selected by one vote to select the optimal classification result.

The basic idea of RFC: first, bootstrap sampling is used to extract k samples from the original training set, and each sample has the same sample size as the initial training set; Secondly, k decision tree models were established for k samples to obtain k classification results. Finally, each record is voted on based on the results of the k classifications to determine its final classification.

RF increases the difference between classification models by constructing different training sets, thereby improving the extrapolation prediction ability of combined classification models. Through k round training, a classification model sequence $\{[h_{11}(X), h_{12}(X), \ldots, h_{1n}], [h_{21}(X), h_{22}(X), \ldots, h_{2n}], \ldots, [h_{k1}(X), h_{k2}(X), \ldots, h_{kn}(X)]\}$ is obtained, and then they are used to form a multi-classification model system. In contrast to the original publication (Breiman 2001), the scikit-learn implementation combines classifiers by averaging their probabilistic prediction instead of letting each classifier vote for a single class.

The final classification decision formula is as Equation (10.1).

$$H(x) = \max\left(\begin{array}{l} \dfrac{h_{11}(X)+h_{21}(X)+\cdots+h_{k1}(X)}{k}, \dfrac{h_{12}(X)+h_{22}(X)+\cdots+h_{k2}(X)}{k}, \\ \ldots, \dfrac{h_{1n}(X)+h_{2n}(X)+\cdots+h_{kn}(X)}{k} \end{array}\right) \tag{10.1}$$

where $H(x)$ represents the combinatorial classification model, k is the number of decision tree classifiers, and n is the number of categories that predict labels. $H_{kn}(X)$ refers to the probability that the kth decision tree classifier predicts the label class number. Equation (10.1) illustrates the use of averaging probability decisions to determine the final classification.

10.3.3 Gaussian processes

A Gaussian process is a generalization of the Gaussian probability distribution. A probability distribution describes random variables that are scalars or vectors (for multivariate distributions), whereas a stochastic process governs the properties of functions (Rasmussen and Williams, 2005).

The mean function and the covariance function uniquely determine a Gaussian process. A classification problem based on a Gaussian process can be described as a given training set (T) as Equation (10.2).

$$T = \{(x_i, y_i), i = 1, 2, \cdots, m\} \tag{10.2}$$

where x_i is the input value and y_i is the category label. When data x^* is the input, the predicted output y^* can be given. For binary classification problems, y is either 1 or −1.

A standard method used in solving binary classification problems in Gaussian processes is to convert the predicted output of a regression model into probability through a response function. Binary classification based on Gaussian processes essentially adds Gaussian processes prior directly to the latent function $f(x)$ and then obtains the prior probability $p(y=1|x) = \sigma\ (f(x))$ through the logistic function, and finally finds the probability distribution of $p(y^*=1|X,\ y,\ x^*)$ (see Equation 10.3). Since the integrable function in Equation (10.3) is not a Gaussian distribution, the Laplace approximation method or the expectation propagation approximation method is usually used to approximate the non-Gaussian joint posterior distribution to obtain an analytical solution. More details about the Laplace approximation can be found in Rasmussen and Williams (2005).

$$p(y^* = 1 | X, y, x^*) = \int p(f^* | X, x^*, f) p(f | X, y) df \tag{10.3}$$

where X is the input vector, y is the output label, f is the latent function, and the posterior probability of the latent function is shown in Equation (10.4).

$$p(f | X, y) = \frac{p(y | f) p(f | X)}{p(y | X)} \tag{10.4}$$

The covariance function in the Gaussian process model is equivalent to the kernel function. Commonly used kernel functions include Constant, Radial basis function (RBF), Matern, RationalQuadratic, Exp-Sine-Squared, and DotProduct kernels. It is also possible to add, multiply, and other operations on the primary kernel function to form a new kernel function of the Gaussian process.

Gaussian process classification (GPC) supports multi-class classification by performing either one-versus-rest or one-versus-one-based training and prediction. In one-versus-rest, one binary Gaussian process classifier is fitted for each class, which is trained to separate this class from the rest. In "one_vs_one," one binary Gaussian process classifier is fitted for each pair of classes, which is trained to separate these two classes. The predictions of these binary predictors are combined into multi-class predictions.

10.3.4 Support vector machine

Support vector machine (SVM) can solve regression and classification tasks with relatively few datasets (Drucker et al. 1996), which was proposed by Cortes and Vapnik in 1995. In particular, SVM can be applied to multi-classification problems. The theory of SVM can be summarized as follows.

Suppose there are training sample sets (x_i, y_i). x_i is the sample features of the i-sample set ($i = 1, 2, \cdots, k$; $i, k \in \mathrm{N}^*$). y_i is the label of the i-sample set ($y_i \in \{+1, -1\}$). The hyperplane and its condition can be defined as Equation (10.5). Obviously, the hyperplane with the largest margin is the optimal hyperplane (Cristianini and Shawe-Taylor 2000).

$$\begin{cases} \boldsymbol{w}^T \boldsymbol{x} + b = 0 \\ y_i\left(\boldsymbol{w}^T \cdot \boldsymbol{x}_i + b\right) - 1 \geq 0 \end{cases} \tag{10.5}$$

where $\boldsymbol{w}^{\mathrm{T}}$ is the weight vector, x is the feature vector, and b represent the bias.

Introducing the Lagrange function, α_i are Lagrange multipliers ($\alpha_i \geq 0$). The linearly separable problem can be rewritten as Equation (10.6).

$$L(\boldsymbol{w}, b, \alpha) = \frac{\|w\|^2}{2} - \sum_{i=1}^{k} \alpha_i \left[y_i\left(\boldsymbol{w}^T \cdot \boldsymbol{x}_i + b\right) - 1 \right] \tag{10.6}$$

Then, the hyperplane of the SVM is determined according to the Karush–Kuhn–Tucker (KKT) conditions (Equation 10.7).

$$\begin{cases} \dfrac{\partial L}{\partial w} = w - \sum_{i=1}^{k} \alpha_i y_i x_i = 0 \\ \dfrac{\partial L}{\partial b} = -\sum_{i=1}^{k} \alpha_i y_i = 0 \end{cases} \tag{10.7}$$

Thus, the primal problem can be transformed into a dual problem. The optimal solution α_i^*, the optimal weight vector $\boldsymbol{w}^*$, and the optimal bias b^* are obtained, where $\boldsymbol{w}^* = \sum_{i=1}^{k} \alpha_i^* y_i \boldsymbol{x}_i$.

Finally, the decision function is shown as in Equation (10.8). The sgn() is a symbolic function.

$$f(x) = \mathrm{sgn}\left[\sum_{i=1}^{k} \alpha_i^* y_i \left(x_i^T \cdot x \right) + b^* \right] \tag{10.8}$$

In particular, the slack variable ξ_i is introduced to seek a balance between the goals of maximum margin and minimum training sample error rate. The condition of Equation (10.5) can be transformed as Equation (10.9).

$$y_i\left(\boldsymbol{w}^T \cdot \boldsymbol{x}_i + b\right) - 1 + \xi_i \geq 0 \tag{10.9}$$

subject to Equation (10.10).

$$\begin{cases} \xi_i \geq 0 \\ \sum_{i=1}^{k} \alpha_i y_i = 0 \\ C \geq \alpha_i \geq 0 \end{cases} \tag{10.10}$$

where C represents the penalty value.

Therefore, the original Lagrange function can be rewritten as Equation (10.11).

$$L(\mathrm{w},b,\alpha,\xi_i)=\frac{\|w\|^2}{2}-\sum_{i=1}^{k}\alpha_i\left[y_i\left(\mathrm{w}^{\mathrm{T}}\cdot \mathrm{x}_i+b\right)-1+\xi_i\right]+\frac{C}{2}\sum_{i=1}^{k}\xi_i^2 \tag{10.11}$$

In order to ensure that the classifier obtained by training has a high generalization ability. The original input space must be mapped into a high-dimensional space when the training samples are non-linearly separable.

In order to avoid complex operations in high-dimensional space, the kernel function $\mathrm{K}(\boldsymbol{x}_i^T, \boldsymbol{x})$ is introduced to calculate the inner product of feature vectors. At this point, the Mercer conditions must be satisfied (Mercer and Forsyth 1909). Equation (10.8) can be rewritten as Equation (10.12).

$$f(\mathrm{x})=\mathrm{sgn}\left[\sum_{i=1}^{k}\alpha_i^* y_i K\left(\mathrm{x}_i^T\cdot \mathrm{x}\right)+b^*\right] \tag{10.12}$$

Commonly used kernel functions include linear kernel, polynomial kernel, RBF kernel, and sigmoid kernel. More details about kernel function can be found in Cervantes et al. (2020).

In practice, the rock mass classification is a typical multi-classification problem. The one-versus-one (OVO) and the one-versus-rest (OVR) are used to transform multi-classification problems into multiple binary classification problems in SVM. Further details about the multi-classification methods in SVM can be found in Cervantes et al. (2020).

10.3.5 Model training and evaluation

This work employs SVM, RF, GPC, and CNN to build surrounding rock classification (SRC) models based on the SurroundingRock/8/286 database. Figure 10.2 depicts the process of building the SRC model. Firstly, the SurroundingRock/8/286 database was established through on-site investigation and literature review. Secondly, the data is discretized by employing OrdinalEncoder or OneHotEncoder methods in scikit-learn. Thirdly, the intelligent SRC model is trained and optimized by adjusting the hyperparameters. Finally, the performance of the SRC model is evaluated by accuracy, precision, recall, and F-measure to select the optimal model.

10.3.5.1 CNN-based SRC model

Four CNN-based SRC models are established by VGGNet16 and VGGNet19, which are the representative algorithm of CNN. In a ratio of 7:3, the SurroundingRock/8/286 database is divided into a training set (200 cases) and a validation set (86 cases).

The four CNN-based SRC models are trained with the Adam optimizer, 16 and 32 batch sizes, categorical_crossentropy loss function, and 200 epochs. And the learning rate is set to 0.0001, which means the model training is terminated when the validation loss value is less than 0.0001, and the model is determined to be optimal. Figure 10.3 depicts the changes in the accuracy of the four established CNN-based SRC models with the number of iterations, respectively.

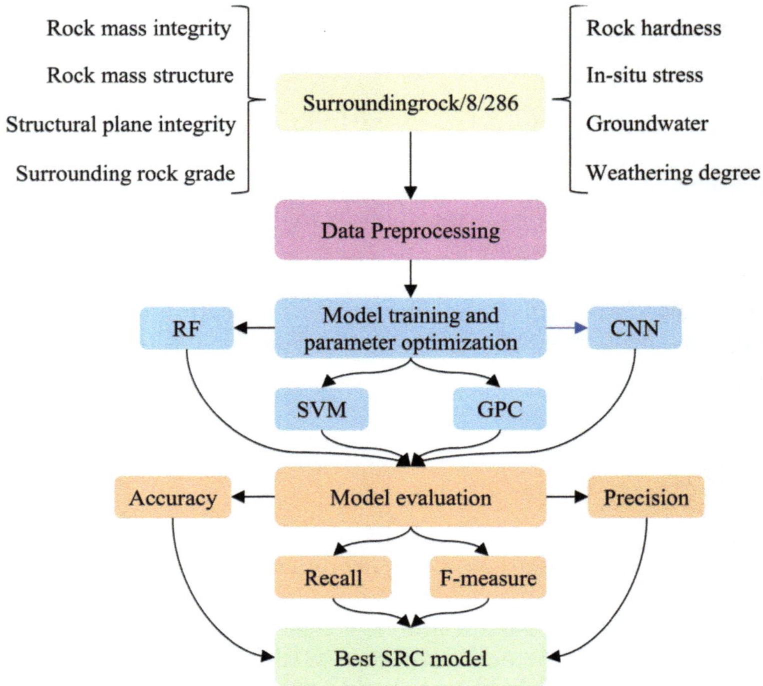

Figure 10.2 The workflow of the intelligent SRC model.

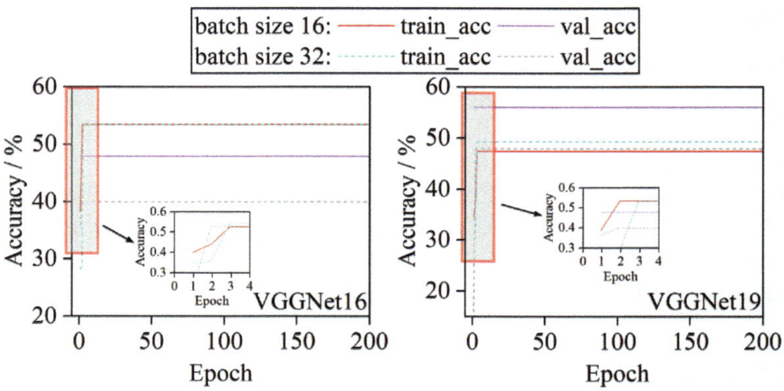

Figure 10.3 Curves of the accuracy of training and validation sets of the VGGNet16-based and VGGNet19-based SRC models.

As is shown in Figure 10.3, the training accuracy and validation accuracy are stabilized after three epochs in VGGNet16-based and VGGNet19-based SRC models. Especially the validation accuracy of the VGGNet19-based SRC model (batch size 16) is stabilized in the first epoch. As a result, the accuracy of the training and validation sets of the four VGGNet16-based SRC models are lower than 60.00% (see Table 10.2), so further model performance evaluation is unnecessary.

Table 10.2 Hyperparameter values and validation set accuracy for intelligent SRC models

Algorithms	*Kernel*	*Gamma*	*Degree*	*Coef0*	C	*Accuracy (%)*	*5-fold (%)*	*10-fold (%)*
SVM	RBF	0.03	–	–	14.09	**93.02**	**93.50**	**94.00**
	Linear	–	–	–	4.30	87.21	90.00	92.50
	Sigmoid	0.01	–	−0.20	18.98	87.21	84.00	84.00
	Poly	10	1	0	0	86.05	88.00	87.00
RF	n_estimators	max_depth	min_samples_ split	class_weight		Accuracy (%)	5-fold (%)	10-fold (%)
	25	None	2	None		83.72	84.50	82.00
				Balanced		80.23	85.00	81.50
				balanced_subsample		83.72	85.00	83.00
	100			None		82.56	85.50	83.50
				Balanced		**83.72**	**87.00**	**83.00**
				balanced_subsample		81.40	84.50	84.00
GPC	Kernel	length_scale	length_scale_bounds	multi_class		Accuracy (%)	5-fold (%)	10-fold (%)
	RBF	2.1	(1e-5, 1e5)	one_vs_rest		82.56	81.00	83.00
		1.5	(1e-5, 1e5)	one_vs_one		**86.05**	**87.00**	**91.50**
		[3,3,2,2,1,1,1]	(1e-5, 1e7)	one_vs_rest		82.56	88.00	86.00
		[3,3,2,2,1,1,1]	(9e-9, 9e21)	one_vs_one		86.05	86.00	88.50
	DotProduct	sigma_0	sigma_0_bounds	multi_class		Accuracy (%)	5-fold (%)	10-fold (%)
		1	(1e-5, 1e5)	one_vs_rest		76.74	75.00	74.00
		1	(1e-5, 1e5)	one_vs_one		86.05	82.50	87.00
	Matern	Length_scale	Nu	multi_class		Accuracy (%)	5-fold (%)	10-fold (%)
		2	1.5	one_vs_rest		83.72	84.50	85.50
		1	1.5	one_vs_one		86.05	87.00	88.00
	RationalQuadratic	Length_scale	Alpha	multi_class		Accuracy (%)	5-fold (%)	10-fold (%)
		1	1	one_vs_rest		84.88	84.00	87.00
		1	1	one_vs_one		86.05	90.50	92.50

(Continued)

Table 10.2 (Continued) Hyperparameter values and validation set accuracy for intelligent SRC models

Algorithms	*Kernel*	*Gamma*	*Degree*	*Coef0*	*C*	*Accuracy (%)*	*5-fold (%)*	*10-fold (%)*
CNN	Model	Batch size		Validation set accuracy (%)			Training set accuracy (%)	
	VGGNet16	16		53.42				48.00
		32		53.42				40.00
	VGGNet19	16		47.49				56.00
		32		49.32				48.00

Note: Accuracy means the accuracy of a random validation set. 5-fold means the average accuracy of five random validation sets. 10-fold means the average accuracy of ten random validation sets. More details about the cross-validation can be found in Rao et al. (2008).

In summary, the four established VGGNet16-based SRC models cannot meet the high accuracy requirements in tunnel construction. However, as more cases are collected, the VGGNet-based models can be further optimized, or the latest CNN architecture is introduced to build the newer SRC model.

10.3.5.2 RF-based SRC model

The RF-based SRC model is trained and validated in a ratio of 7:3. The hyperparameters of the RF-based SRC model include n_estimators, max_depth, min_samples_split, and class_weight. The n_estimators means the number of trees in the forest. The max_depth means the maximum depth of the tree. The default value of the max_depth is None, which means the nodes are expanded until all leaves are pure or until all leaves contain less than min_samples_split samples. The min_samples_split means the minimum number of samples required to split an internal node. The class_weight is used to assign weight to the class. The 5-fold and 10-fold cross-validation are preliminarily used to evaluate the model's performance.

Let the model parameters in condition 1 be n_estimators = 100, max_depth = None, min_samples_split = 2, and class_weight = balanced. Condition 2: n_estimators = 100, max_depth = None, min_samples_split = 2, and class_weight = balanced_subsample. As shown in Table 10.2, in the case of condition 1, the average accuracy of 5-fold cross-validation of the RF-based SRC model is the highest, 87.00%. In the case of condition 2, the average accuracy of 10-fold cross-validation of the RF-based SRC model is the highest, 84.00%. The average accuracy of the 10-fold cross-validation of the model under condition 1 is only 1% lower than in condition 2. The average accuracy of the 5-fold cross-validation of the model under condition 1 is 2.50% higher than in condition 2. The model's random validation accuracy under condition 1 is 2.32% higher than in condition 2. In summary, the RF-based SRC model's performance is best in condition 1.

10.3.5.3 GPC-based SRC model

The GPC-based SRC model is trained and validated in a ratio of 7:3. The mainly hyperparameters of the GPC-based SRC model are kernel and multi_class. Ten GPC-based SRC models were trained using RBF, DotProduct, Matern, and RationalQuadratic kernel functions. The parameter values for multi_class are "one_vs_rest" and "one_vs_one." In "one_vs_rest," one binary Gaussian process classifier is fitted for each class, which is trained to separate this class from the rest. In "one_vs_one," one binary Gaussian process classifier is fitted for each pair of classes, which is trained to separate these two classes. In addition, different kernel functions have their own unique hyperparameters (see Table 10.2).

As is shown in Table 10.2, the accuracy of the random validation set of GPC-based SRC models is generally greater than 80%. Let the model parameters in condition 3 be kernel = RBF, length_scale = 1.5, length_scale_bounds = (1e–5, 1e5), multi_class = "one_vs_one." Condition 4: kernel = RationalQuadratic, length_scale = 1, alpha = 1, multi_class = "one_vs_one." The GPC-based SRC model's accuracy of the random validation set is equal and the highest in conditions 3 and 4, which is 86.05%. The model's average accuracy of the 5-fold and 10-fold cross-validation under condition 4 is higher than that of condition 3. However, the model training has a convergence warning under condition 4. Therefore, the model trained by condition 4 still needs further debugging in practical applications. In addition, under condition 3, the model's average accuracy of the 10-fold cross-verification reached 91.50%. Therefore, the model in the case of condition 3 is the optimal GPC-based SRC model.

10.3.5.4 SVM-based SRC model

The ratio of training set (200 cases) to validation set (86 cases) is 7:3 for SVM-based SRC model. The mainly hyperparameters of the SVM-based SRC model is kernel function, decision_function_shape, and C. The SVM-based SRC model is trained using the default parameter values of decision_function_shape ("ovr"). "ovr" is one_vs_rest, as mentioned in the GPC-based SRC model. *C* is regularization parameter, which must be strictly positive. The strength of the regularization is inversely proportional to *C*. Gamma is the kernel coefficient for RBF, Poly, and sigmoid kernel functions. Degree is the degree of the polynomial kernel function, which must be non-negative and ignored by all other kernels. Coef0 is only significant in "poly" and "sigmoid" kernel functions. Table 10.2 shows the SVM-based SRC model with the highest accuracy under different kernel functions and its corresponding hyperparameters.

As is shown in Table 10.2, the SVM-based SRC model established with the RBF kernel function has the highest accuracy (the average accuracy of 5-fold and 10-fold cross-validation is 93.05% and 94%, respectively). The accuracy of random validation of the SVM-based SRC model established with the linear kernel function did not reach 90%. The average accuracy of cross-validation of the SVM-based SRC model established with the sigmoid and Poly kernel functions reached less than 90%. The SVM-based SRC model established with the RBF kernel function performs better.

10.3.5.5 Performance evaluation of each class of the intelligent SRC model

The accuracy of optimal RF-based, GPC-based, and SVM-based SRC models exceeds 80%. It is necessary to evaluate further the RF-based, GPC-based, and SVM-based SRC models using precision, recall, and *F*-measure. Accuracy measures the model's ability to predict samples correctly. Precision and recall measure the model's ability to correctly predict the majority and minority classes. However, the classification of surrounding rock needs to be as accurate as possible for each type of sample. Therefore, the *F*-measure is a comprehensive indicator considering both precision and recall. See Equations (10.13)–(10.16) for calculating accuracy, precision, recall, and *F*-measure. The meaning of TP, FP, FN, and TN is shown in Table 10.3.

$$\text{Accuray} = \frac{\text{TP} + \text{TN}}{\text{TP} + \text{FN} + \text{FP} + \text{TN}} \tag{10.13}$$

$$\text{Precision} = \frac{\text{TP}}{\text{TP} + \text{FP}} \tag{10.14}$$

$$\text{Recall} = \frac{\text{TP}}{\text{TP} + \text{FN}} \tag{10.15}$$

$$\text{F-measure} = \frac{2\text{Recall} \cdot \text{Precision}}{\text{Recall} + \text{Precision}} \tag{10.16}$$

Prediction of surrounding rock classification is a five-classification problem, and it can be transformed into five binary classification problems. Table 10.3 shows the confusion matrix (taking the class III as an example). Table 10.4 shows the prediction results of the validation set based on the RF-based, GPC-based, and SVM-based SRC models.

As is shown in Table 10.4, the RF-based, GPC-based, and SVM-based SRC models correctly predict class I surrounding rock samples. Both SVM-based and GPC-based SRC models predict class II surrounding rocks correctly. The RF-based and GPC-based SRC model's predictions of class III surrounding rocks have a much higher error than the SVM-based SRC model. The SVM-based SRC model also had the least number of samples to predict class IV surrounding rock errors. The RF-based, GPC-based, and SVM-based SRC models predicted that the class V surrounding rocks were all wrong by only one sample.

According to Table 10.4 and Equations (6.13)–(6.16), the calculation results of precision, recall, and *F*-measure are shown in Figures 10.4–10.6.

Figure 10.4 shows that the precision values of the SVM-based SRC model in class II, class III, class IV, and class V are greater than those of the RF-based and GPC-based SRC models. Figure 10.5 depicts that the recall values of the SVM-based SRC model in class III and class IV are greater than those of the RF-based and GPC-based SRC models. The SVM-based, RF-based, and GPC-based SRC models have equal recall values in class I, class II, and class V. Figure 10.6 shows that the *F*-measure values of the SVM-based SRC model in class II, class III, class IV, and class V are all greater than those of the RF-based and GPC-based SRC models. As a result, the average precision, average recall, and average *F*-measure of the best SVM-based SRC model are 93.87%, 94.41%, and 93.92%, respectively. In summary, the established SVM-based SRC model performs better than RF-based, and GPC-based SRC models.

10.3.6 Validation with new cases

Twenty-four real cases are used to verify the above RF-based, GPC-based, and SVM-based SRC models. The actual verification cases come from nine tunnels of the Leshan–Xichang (Lexi) Expressway, including the Dafengding tunnel, Daliangshan No. 1 tunnel, Daliangshan No. 2 tunnel, Guihua tunnel, Qingheng No. 1 tunnel, Waju No. 2 tunnel, Wuyiwan tunnel, Xiaoliangshan No. 2 tunnel, and Yinchanggou tunnel. In particular, the SurroundingRock/8/286 database does not contain the above 24 actual verification cases. Table 10.5 shows the surrounding rock engineering geological parameters and prediction results of the validation cases.

As is shown in Table 10.5, the SVM-based SRC model predicts correctly in all validation cases, the RF-based SRC model predicts two cases incorrectly, and the GPC-based

Table 10.3 Binary confusion matrix (taking the prediction of class III as an example)

		Predicted	
		III	I/II/IV/V
Actual	III	True positive (TP)	False negative (FN)
	I/II/IV/V	False positive (FP)	True negative (TN)

SRC model predicts one case incorrectly. In summary, the SVM-based SRC model performs better in real verification cases.

10.4 FIELD APPLICATION

Combined with the above-mentioned best SVM-based SRC model and cloud technology, a tunnel information management system is developed. The tunnel information management system can provide data storage, query, and intelligent analysis services for tunnel construction. It is currently applying to a project – the Lexi Expressway project in China.

The Lexi Expressway project is currently using the tunnel information management system (TIMS). The Lexi Expressway starts in Mabian Yi Autonomous County of Leshan City, passes through Leibo County and Meigu County in Liangshan Prefecture, and ends at the south side of Zhaojue County, where it connects with the planned Xichang–Zhaotong Expressway (see Figure 10.7). The total length of the Lexi Expressway is about 152.48 km, and there are 41 tunnels on the whole line. The lithology of the extension line of the project is sandstone, siltstone, mudstone, shale, limestone, basalt, etc. The tunnels are distributed as class III, class IV, and class V surrounding rocks.

The tasks of TIMS can be divided into four categories: management of tunnel design information, real-time intelligent grading of surrounding rock, management of advanced geological forecast information, and dynamic monitoring of surrounding rock deformation. TIMS stores the engineering geological information related to four kinds of tunnel tasks in the server. Users can set filter criteria to query and access the required data. This

Table 10.4 Confusion matrix of SVM-based, RF-based, and GPC-based SRC models

		Predicted (SVM-based)					*Predicted (RF-based)*					*Predicted (GPC-based)*				
		I	II	III	IV	V	I	II	III	IV	V	I	II	III	IV	V
Actual	I	1	0	0	0	0	1	0	0	0	0	1	0	0	0	0
	II	0	15	0	0	0	0	14	1	0	0	0	15	0	0	0
	III	0	3	17	0	0	0	6	13	1	0	0	6	12	2	0
	IV	0	0	1	19	1	0	0	1	16	4	0	0	1	18	2
	V	0	0	0	1	28	0	0	0	1	28	0	0	0	1	28

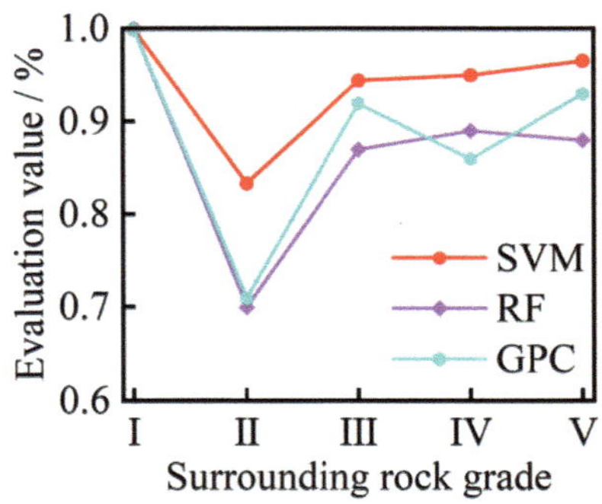

Figure 10.4 The precision of SVM-based, RF-based, and GPC-based SRC models.

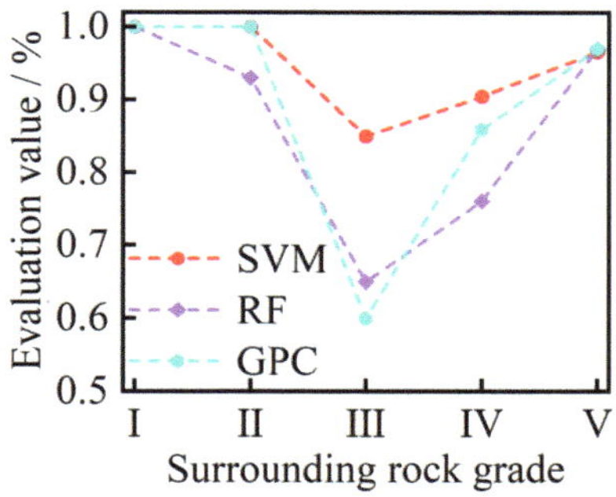

Figure 10.5 The recall of SVM-based, RF-based, and GPC-based SRC models.

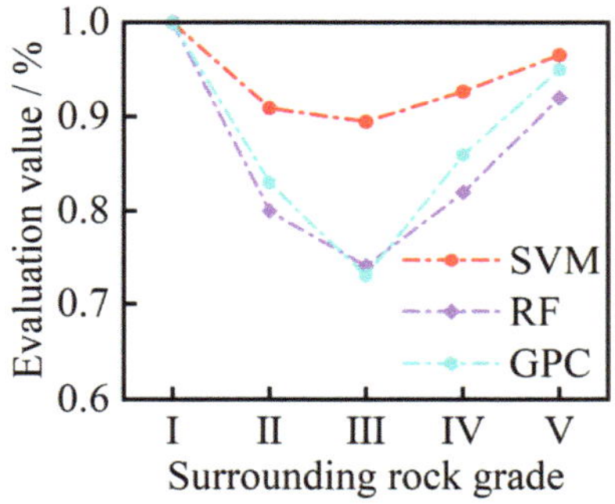

Figure 10.6 The *F*-measure of SVM-based, RF-based, and GPC-based SRC models.

study visualized the existing data of the system and displayed it on the TIMS homepage in the form of tables and graphs (see Figure 10.8). From Figure 10.8, the homepage of TIMS displays the current distribution of surrounding rock quality grades, alterations in design surrounding rock grades, the time distribution of new surrounding rock information, to-do tasks, monitoring and warning of surrounding rock deformation, and system announcements. In particular, user can jump to the details page of the target data by clicking the data on the homepage. In addition, this study have customized standard management procedures for each of the four tunneling tasks in TIMS.

The tunnel information management system was launched in July 2020, and the real-time intelligent surrounding rock classification function is gradually being used in all tunnels of the Lexi project. In particular, more than 5,500 surrounding rock data had been submitted to the system as of December 2022. Figure 10.9 shows that most tunnel excavation sections did not have the design alteration of surrounding rock grade. And 850 excavation sections initiated the online alteration procedure to the designed surrounding rock grade.

More importantly, the online real-time intelligent classification of surrounding rocks dramatically improves the efficiency of on-site office work. (1) The identification of the surrounding rock grade at the construction site is faster. (2) All parties involved in the tunnel construction can view the engineering geological information of the excavation face in real-time in the TIMS system. (3) The online alteration procedure of the surrounding rock grade reduces on-site meetings and increases productivity. Real-time surrounding rock classification is significant for guiding dynamic tunnel design and construction.

Table 10.5 Verification results of the new tunnel cases

Tunnel	*Mileage*	*RH*	*WD*	*RMS*	*Dc*	*RMI*	*IS*	*G*	*Actual*	*SVM*	*RF*	*GPC*
Dafengding	K68 + 586	Slightly hard	Slight	d	Ordinary	Slightly broken	Medium	Rain-like dripping	IV	IV	IV	IV
	K68 + 670	Slightly soft	Medium	e	Good	Slightly complete	Medium	Rain-like dripping	IV	IV	IV	IV
	ZK68 + 536	Slightly hard	Slight	d	Ordinary	Broken	Medium	Dripping	IV	IV	IV	IV
Daliangshan No. 1	ZK82 + 068	Soft	Slight	c	Ordinary	Slightly broken	Low	Dry	V	V	V	IV
	ZK86 + 496	Slightly soft	Medium	e	Ordinary	Broken	Medium	Moist	V	V	V	V
	K75 + 542	Slightly hard	Medium	d	Ordinary	Slightly broken	Low	Rain-like dripping	IV	IV	IV	IV
	PK75 + 417	Soft	Severe	c	Bad	Slightly broken	Low	Dripping	V	V	V	V
Daliangshan No. 2	K107 + 709	Slightly soft	Medium	a	Ordinary	Slightly complete	High	Dripping	IV	IV	IV	IV
	ZK113 + 510	Slightly hard	Fresh	d	Ordinary	Slightly broken	Low	Dripping	IV	IV	IV	IV
	ZK108 + 474	Slightly soft	Medium	a	Ordinary	Slightly complete	High	Moist	IV	IV	IV	IV
	ZK104 + 821	Soft	Medium	c	Very bad	Slightly broken	Medium	Dripping	V	V	V	V
Guihua	K62 + 745	Slightly hard	Medium	e	Bad	Broken	Low	Gushing	V	V	V	V
	K62 + 755	Slightly hard	Medium	e	Bad	Broken	Low	Linear	V	V	IV	V
	K62 + 795	Slightly hard	Medium	f	Bad	Broken	Low	Linear	V	V	IV	V
Qingheng No. 1	K128 + 665	Slightly hard	Slight	b	Good	Slightly complete	Low	Moist	III	III	III	III
	ZK128 + 660	Slightly hard	Slight	b	Good	Slightly complete	Medium	Dry	III	III	III	III

(Continued)

Table 10.5 (Continued) Verification results of the new tunnel cases

Tunnel	*Mileage*	*RH*	*WD*	*RMS*	*Dc*	*RMI*	*IS*	**G**	*Actual*	*SVM*	*RF*	*GPC*
Waju No. 2	K23 + 300	Soft	Slight	e	Bad	Broken	Medium	Dripping	V	V	V	V
Wuyiwan	ZK33 + 131	Soft	Medium	c	Ordinary	Broken	Medium	Dry	V	V	V	V
Xiaoliangshan No. 2	K39 + 240	Slightly hard	Slight	a	Ordinary	Slightly complete	Low	Dripping	III	III	III	III
Yinchanggou	K55 + 452	Hard	Slight	c	Ordinary	Slightly broken	Low	Dry	III	III	III	III
	ZK57 + 600	Slightly soft	Medium	g	Very bad	Extremely broken	Low	Moist	V	V	V	V
	K55 + 509	Hard	Slight	d	Ordinary	Slightly broken	Low	Dripping	III	III	III	III
	ZK55 + 596	Hard	Slight	d	Ordinary	Slightly broken	Low	Moist	III	III	III	III
	ZK57 + 540	Soft	Severe	g	Very bad	Extremely broken	Low	Moist	V	V	V	V

Note: (1) The RH, WD, RMS, SPI, RMI, IS, and GW represent rock hardness, weathering degree, rock mass structure, structural plane integrity, rock mass integrity, in situ stress, and groundwater, respectively. (2) The a, b, c, d, e, f, and g represent block structure, thick-layered structure, medium-thick-layered structure, mosaic structure, fractured block structure, fragmented structure, and granular structure, respectively. (3) The "Actual" column in the table is the surrounding rock grade referenced by the actual support at the tunnel construction site.

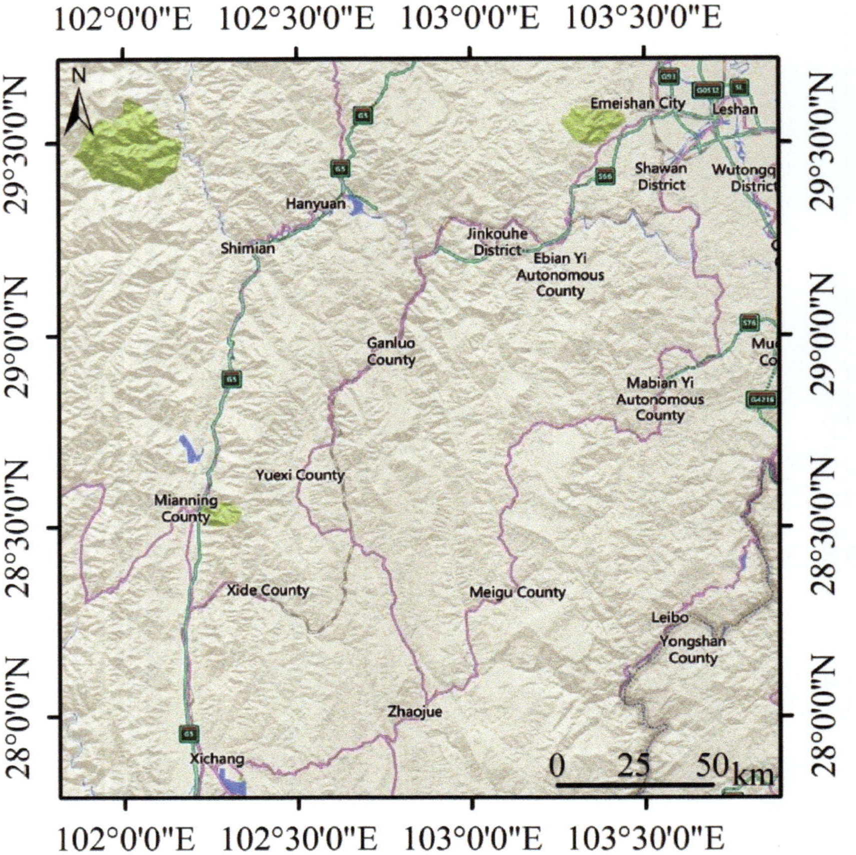

Figure 10.7 Route diagram of the Lexi Expressway.

10.5 CONCLUSIONS

This chapter presents a new generic database for surrounding rock engineering geological parameters named SurroundingRock/8/286. Four types of surrounding rock grade classifiers are established with CNN, RF, GPC, and SVM algorithms. A tunnel information management system is developed with SVM-based SRC model and cloud technology, which is now well-used in the Lexi Expressway project. Based on the results of this study, the following conclusions can be drawn:

1. The SurroundingRock/8/286 database consists of eight parameters (rock harness, rock mass integrity, discontinuity condition, weathering degree, rock mass structure, groundwater condition, in-situ stress condition, and surrounding rock grade), which can well characterize the surrounding rock engineering geology conditions.
2. Based on the SurroundingRock/8/286 database, four intelligent SRC models are built with CNN, RF, GPC, and SVM algorithms, and the SVM-based SRC model has better performance. In particular, the 5-fold cross-validation accuracy, 10-fold cross-validation accuracy, average precision, average recall, and average *F*-measure

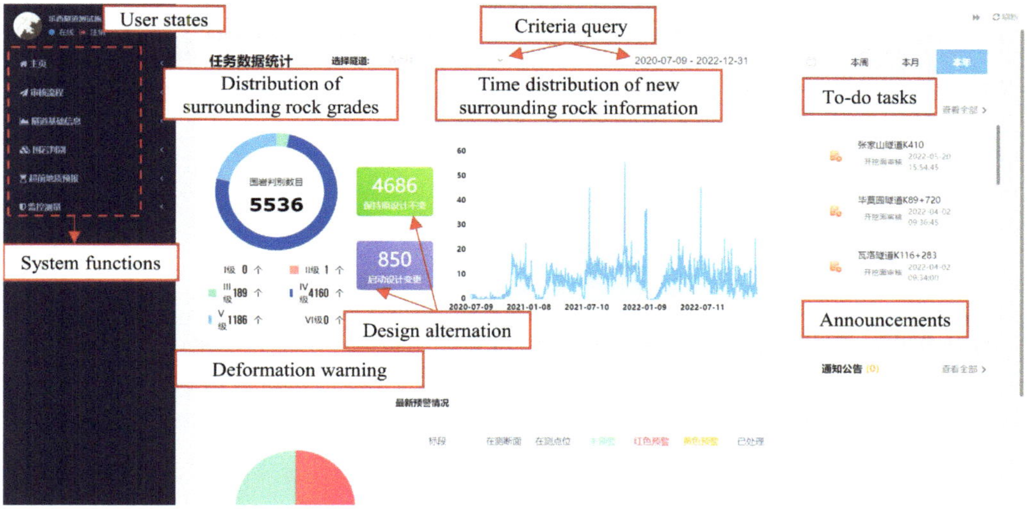

Figure 10.8 Data visualization of the TIMS homepage.

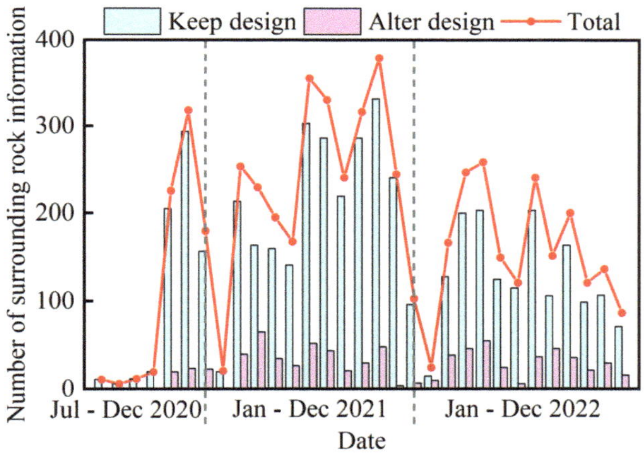

Figure 10.9 Monthly distribution of alteration in the design grade of the surrounding rock.

of the best SVM-based SRC model are 93.50%, 94.00%, 93.87%, 94.41%, and 93.92%, respectively. Further, the SVM-based SRC model predicts all 24 actual tunnel cases correctly.

3. The tunnel information management system can predict the surrounding rock grade accurately and online. And the system used in the Lexi Expressway project has dramatically improved the efficiency of construction and management.

In conclusion, the intelligent SRC models built on the SurroundingRock/8/286 database and machine learning methods have demonstrated excellent performance in other sites. However, the accurate acquisition of surrounding rock engineering geological parameters is an existing problem in the complex environment of the site. Therefore, our study will focus on using deep learning technology to automatically, quickly, and accurately obtain the surrounding rock engineering geological parameters to expand the surrounding rock database further.

DISCLOSURE STATEMENT

No potential conflict of interest was reported by the author(s).

FUNDING

This work was supported by the Science & Technology Department of Sichuan Province, China [Grant No. 2021YFS0317]; and the National Natural Science Foundation of China (NSFC) [Grant No. U19A20111].

REFERENCES

Barton, N. 2002. "Some New Q-Value Correlations to Assist in Site Characterisation and Tunnel Design." *International Journal of Rock Mechanics and Mining Sciences* 39 (2): 185–216. https://doi.org/10.1016/S1365-1609(02)00011-4.

Barton, N., R. Lien, and J. Lunde. 1974. "Engineering Classification of Rock Masses for the Design of Tunnel Support." *Rock Mechanics* 6 (4): 189–236. https://doi.org/10.1007/BF01239496.

Bieniawski, Z. T. 1973. "Engineering Classification of Jointed Rock Masses." *The Civil Engineering in South Africa* 15 (12): 335–343.

Bo, Y., Q. S. Liu, X. Huang, and Y. C. Pan. 2022. "Real-Time Hard-Rock Tunnel Prediction Model for Rock Mass Classification Using CatBoost Integrated with Sequential Model-Based Optimization." *Tunnelling and Underground Space Technology* 124: 104448. https://doi.org/10.1016/j.tust.2022.104448.

Breiman, L. 2001. "Random Forests." *Machine Learning* 45: 5–32. https://doi.org/10.1023/A:1010933404324.

Cervantes, J., F. Garcia-Lamont, L. Rodríguez-Mazahua, and A. Lopez. 2020. "A Comprehensive Survey on Support Vector Machine Classification: Applications, Challenges and Trends." *Neurocomputing* 408: 189–215. https://doi.org/10.1016/j.neucom.2019.10.118.

Chen, J. Y., M. l. Zhou, H. W. Huang, D. M. Zhang, and Z. C. Peng. 2021. "Automated Extraction and Evaluation of Fracture Trace Maps from Rock Tunnel Face Images via Deep Learning." *International Journal of Rock Mechanics & Mining Sciences* 142: 104745. https://doi.org/10.1016/j.ijrmms.2021.104745.

Cortes, C., and V. Vapnik. 1995. "Support-Vector Networks." *Machine Learning* 20 (3): 273–297. https://doi.org/10.1023/A:1022627411411.

Cristianini, N., and J. Shawe-Taylor. 2000. *An Introduction to Support Vector Machines and Other Kernel-Based Learning Methods*. Cambridge: Cambridge University Press. https://doi.org/10.1017/CBO9780511801389.

Drucker, H., C. Burges, L. Kaufman, A. Smola, and V. Vapnik. 1996. "Support Vector Regression Machines." In *Advances in Neural Information Processing Systems,* MIT Press, Cambridge, MA, 779–784.

Genuer, R., and J-M. Poggi. 2020. *Random Forests with R*. Cham, Switzerland: Springer Nature Switzerland AG. https://doi.org/10.1007/978-3-030-56485-8.

Hoek, E. 1994. "Strength of rock and rock masses." *ISRM News* J 2 (2): 4–16.

Ketkar, N., and J. Moolayil. 2021. *Deep Learning with Python: Learn Best Practices of Deep Learning Models with PyTorch.* California: Apress Media. https://doi.org/10.1007/978-1-4842-5364-9.

Khan, A., A. Sohail, U. Zahoora, and A. S. Qureshi. 2020. "A Survey of the Recent Architectures of Deep Convolutional Neural Networks." *Artificial Intelligence Review* 53 (8): 5455–5516. https://doi.org/10.1007/s10462-020-09825-6.

LeCun, Y., B. Boser, J. S. Denker, D. Henderson, R. E. Howard, W. Hubbard, and L. D. Jackel. 1989. "Backpropagation Applied to Handwritten Zip Code Recognition." *Neural Computation* 1 (4): 541–551. https://doi.org/10.1162/neco.1989.1.4.541.

Liu, H. X., W. S. Li, Z. Y. Zha, W. J. Jiang, and T. Xu. 2018. "Method for Surrounding Rock Mass Classification of Highway Tunnels Based on Deep Learning Technology." *Chinese Journal of Geotechnical Engineering* 40 (10): 1809–1817. https://doi.org/10.11779/CJGE201810007 (in Chinese).

Ma, J. J., T. B. Li, G. Yang, K. K. Dai, C. C. Ma, H. Tang, G. W. Wang, J. F. Wang, B. Xiao, and L. B. Meng. 2023. "A real-time intelligent classification model using machine learning for tunnel surrounding rock and its application." *Georisk: Assessment and Management of Risk for Engineered Systems and Geohazards* 17 (1): 148–168. https://doi.org/10.1080/17499518.2023.2182891.

Ministry of Water Resources of the People's Republic of China. 2014. *Standard for Engineering Classification of Rock Mass (GB/T 50218-2014).* Beijing: China Planning Press (in Chinese).

Mercer, J., and A. R. Forsyth. 1909. "Functions of Positive and Negative Type, and Their Connection the Theory of Integral Equations." *Philos Trans R Soc Lond* 209: 415–446. https://doi.org/10.1098/rsta.1909.0016.

Phoon, K. -K., J. Ching, and Z. Cao. 2022. "Unpacking data-centric geotechnics." *Underground Space* 7: 967–989. https://doi.org/10.1016/j.undsp.2022.04.001.

Phoon, K. -K., J. Ching, and T. Shuku. 2022. "Challenges in Data-driven Site Characterization." *Georisk: Assessment and Management of Risk for Engineered Systems and Geohazards* 16 (1): 114–126. https://doi.org/10.1080/17499518.2021.1896005.

Phoon, K. -K, and W. Zhang. 2022. "Future of Machine Learning in Geotechnics." *Georisk: Assessment and Management of Risk for Engineered Systems and Geohazards* 17 (1):7–22. https://doi.org/10.1080/17499518.2022.2087884.

Rao, R. B., G. Fung, and R. Rosales. 2008. "On the Dangers of Cross-Validation. An Experimental Evaluation." *Proceedings of the 2008 SIAM International Conference on Data Mining*, 558–596. https://doi.org/10.1137/1.9781611972788.54.

Rasmussen, C. E., and C. K. I. Williams. 2005. *Gaussian Processes for Machine Learning.* London, England: The MIT Press. https://doi.org/10.7551/mitpress/3206.001.0001.

Santos, A. E. M., M. S. Lana, and T. M. Pereira. 2021. "Evaluation of Machine Learning Methods for Rock Mass Classification." Neural Computing and Applications 34: 4633–4642. https://doi.org/10.1007/s00521-021-06618-y.

Simonyan, K., and A. Zisserman. 2015. "Very Deep Convolutional Networks for Large-Scale Image Recognition." *International Conference on Learning Representations.* https://doi.org/10.48550/arXiv.1409.1556.

Wang, M. N., S. G. Zhao, J. J. Tong, Z. L. Wang, M. Yao, J. W. Li, and W. H. Yi. 2021. "Intelligent Classification Model of Surrounding Rock of Tunnel Using Drilling and Blasting Method." *Underground Space* 6 (5): 539–550. https://doi.org/10.1016/j.undsp.2020.10.001.

SUPPLEMENTARY MATERIAL

Table S10.1 Qualitative and quantitative classification of rock hardness

Hardness	*Qualitative characteristics*	R_c/*MPa*
Hard	The hammering sound is crisp with rebound, shaking hands, harder to break; after soaking in water, usually there is no water absorption reaction.	(60, ∞)
Slightly hard	The hammering sound is slightly crisp with slightly rebound, slightly shaking hands, hard to break; after soaking in water, slight water absorption reaction.	(30, 60]
Slightly soft	The hammering sound is not clear and crisp, no rebound, easy to break; after soaking in water, nails can be imprinted.	(15, 30]
Soft	The hammering sound is dumb, no rebound, dents, easier to break; after soaking in water, the hand can be broken.	(5, 15]
Extremely soft	The hammering sound is dumb, no rebound, deep dents, and the hands can be crushed; after soaking in water, it can be crushed into a ball.	(0, 5]

Source: Modified by the Standard for Engineering Classification of Rock Mass GB/T 50218–2014.

Table S10.2 Qualitative and quantitative classification of rock weathering degree

Degree	*Qualitative characteristics*	k_r	k_f
Fresh	The rock structure does not change, and the rock quality is fresh.	(0.9, 1.0]	(0.9, 1.0]
Slight	The rock structure, mineral composition, and color are basically unchanged, and some fracture surfaces contain iron and manganese or are slightly discolored.	(0.8, 0.9]	(0.8, 0.9]
Medium	The rock structure is partially destroyed; the mineral composition and color are obviously changed, and the fracture surface is severely weathered.	(0.6, 0.8]	(0.4, 0.8]
Severe	Most of the rock structure is destroyed, and the mineral composition and color are obviously changed. The feldspar, mica, and iron–magnesium minerals are weathered and altered.	(0.4, 0.6]	(0, 0.4]
Extreme	The rock structure is completely destroyed, disintegrated, and decomposed into loose soil or sand. All minerals are discolored, and the luster disappears. Most of the minerals except quartz particles are weathered and eroded into secondary minerals.	[0.2, 0.4]	–

Source: After the Standard for Engineering Classification of Rock Mass GB/T 50218–2014.

Table S10.3 Classification of rock mass structure

Type	*Subtype*	*Characteries of rock mass structure*
Block structure	Integral structure	The rock mass is complete with huge block shape and undeveloped structural plane, and the spacing is greater than 1 m.
	Block structure	The rock mass is slightly complete with block shape and slightly developed structural plane, and the spacing is generally 1–0.5 cm.
	Fractured block structure	The rock mass is slightly complete with fractured block shape and moderately developed structural plane, and the spacing is generally 0.5–0.3 m.
Layered structure	Extremely thick-layered structure	The rock mass is complete with extremely thick-layered shape and no developed layers, and the spacing is greater than 1 m.
	Thick-layered structure	The rock mass is slightly complete with thick-layered shape and slightly developed layers, and the spacing is generally 1–0.5 cm.
	Medium-thick-layered structure	The rock mass is slightly complete with medium-thick-layered shape and moderately developed layers, and the spacing is generally 0.5–0.3 cm.
	Interlayer structure	The rock mass is slightly complete or has poor integrity with inter-layered shape and slightly developed or developed layers, and the spacing is generally 0.3–0.1 m.
	Thin-layered structure	The rock mass has poor integrity with thin-layered shape and developed layers, and the spacing is generally less than 0.1 m.
Mosaic structure		The rock mass integrity is poor, the rock blocks are tightly embedded, and the structural plane is slightly developed to developed. The spacing is generally 0.3–0.1 m.
Fragmented structure		The rock mass is broken, the structural plane is well developed, and the spacing is generally less than 0.1 m.
Granular structure		The rock mass is broken. Rocks, cuttings, and mud are mixed together.

Source: After the Code for Engineering Geological Investigation of Water Resources and Hydropower GB 50487–2008.

Table S10.4 Classification and characteristics of the discontinuity condition

Degree	*Characteristics*
Good	1) The opening degree of the structural plane is less than 1 mm, and the filling is siliceous, iron, or calcium cement. Otherwise, the structure plane is rough and there is no filling. 2) The opening degree of the structural plane is 1–3 mm, and the filling is siliceous or iron cement. 3) The structural plane is rough and the opening degree is greater than 3 mm, and the filling is siliceous cement.
Ordinary	1) The structural plane is straight and the opening degree is less than 1 mm, and the filling material is cement with calcareous mud or there is no filling material. 2) The opening degree of the structural plane is 1–3 mm, and the filling is calcium cement. 3) The structural plane is rough and the opening degree is greater than 3 mm, and the filling is iron or calcium cement.
Bad	1) The structural plane is straight and the opening degree is 1–3 mm, and the filling is argillaceous cement or calcium argillaceous cement. 2) The opening degree of the structural plane is greater than 3 mm, and the filling is mostly muddy or rock debris.
Very bad	The filling is muddy or mud with rock cuttings, and the thickness is greater than the undulation difference.

Source: After the Standard for Engineering Classification of Rock Mass GB/T 50218–2014.

Table S10.5 Quantitative classification of rock mass integrity

Rank	*Number of joint sets*	*Joint spacing*	*Degree of major structural plane bonding*	K_v	J_v
Complete	1–2	>1.0	Good or ordinary	(0.75, 1]	[0, 3)
Slightly complete	1–2	>1.0	Bad	(0.55, 0.75]	[3, 10)
	2–3	0.4–1.0	Good or ordinary		
Slightly broken	2–3	0.4–1.0	Bad	(0.35, 0.55]	[10, 20)
	≥3	0.2–0.4	Good or ordinary		
Broken	≥3	0.2–0.4	Bad	(0.15, 0.35]	[20, 35)
		≤0.2	Ordinary or bad		
Extremely broken	Unordered		Very bad	(0, 0.15]	[35, +∞)

Source: After Ma et al. (2023).

Table S10.6 Classification of in-situ stress

Level	*Low*	*Medium*	*High*	*Extremely high*
R_c/σ_{max}	[9, +∞)	[7, 9)	[4, 7)	(0, 4)

Source: After Ma et al. (2023).

Table S10.7 Classification of water outflow from tunnel excavation face

Type	*Qualitative description*	*Quantitative division*
Dry	Excavation face is completely dry.	$p \in [0, 0.01]$ or $q \in [0, 5]$
Damp	Wet marks can be seen in some areas of the excavation face.	
Wet	Water can be seen or leached from the excavation face.	$p \in (0.01, 0.1]$ or $q \in (5, 25]$
Dripping	Water drips from the excavation face.	
Rain-like dripping	There are drops of water resembling rain on the excavation face.	$p \in (0.1, 0.5]$ or $q \in (25, 125]$
Linear-like flowing	The excavation face has similar water pipes, and the water column is smaller (the water volume and water pressure are smaller).	
Tubular-like flowing	The excavation face has similar water pipes, and the water column is relatively thick (the water volume and water pressure are very large).	$p \in (0.5, \infty)$ or $q \in (125, \infty)$
Gushing-like flowing	There is a large water outlet on the excavation face; the flow and velocity are large, which often causes great damage to the project.	

Source: After Ma et al. (2023).

Chapter 11

Exploring challenges via analysis of multivariate geotechnical properties

Insights from large-scale local sampling of Japanese marine clay

Yu Otake, Taiga Saito, Wu Stephen, Ikumasa Yoshida, and Daiki Takano

We are actively engaged in the systematic collection and organization of geotechnical investigation data relevant to civil engineering projects in Japan. This effort is aimed at advancing data-driven methodologies in geotechnical design and construction. Our approach is inspired by the analytical frameworks of Ching and Phoon (2014), Ching et al. (2021), and Wu et al. (2022), with a focus on the detailed analysis of multivariate geotechnical properties. Our study includes a comparative evaluation with key benchmark databases, such as CLAY/10/7490, represented as "Soil Type/Number of Indicator Types/Number of Data Entries." We concentrate on a 10 km² area, marked by uniform geological features in its foundational layers. Notably, the top 30 m of this region contains a soft clay type that is commonly found in Japan. Our dataset, comprising 67,760 samples, aligns partially with the parameters of CLAY/10/7490 but includes some data gaps. In contrast to CLAY/10/7490, a Big Indirect Database (global BID) recognized globally and collected from numerous countries, our dataset, named "Tokyo-CLAY/14/67760," is a localized BID, meticulously assembled through intensive sampling within this specific area. In this chapter, we present a thorough comparison between our high-density local BID and the global BID, aiming to categorize and illuminate the distinctive features of our dataset. We also offer visual representations of statistical dispersion across different sites, operating under the premise that each borehole represents a single site. This methodology uncovers patterns of site-specific diversity that correspond with global trends, yielding insights that are both anticipated and unprecedented. Building on the fundamental statistical analysis of our data, we identify and explore the critical challenges that must be addressed to further the field of data-centric geotechnical engineering.

11.1 INTRODUCTION

This chapter conducts a rigorous investigation of the statistical attributes essential to a geotechnical investigation database, with a primary focus on the soft clays typical of the Tokyo lowlands in Japan (Ishii (1985), and Watabe and Tanaka (2004)). The uniqueness of this database lies in its comprehensive compilation of geotechnical data, encompassing 67,760 samples across a 10 km² area. Our primary objective is to unravel the statistical complexities of multivariate ground parameters derived from this local sampling. Additionally, by comparing this database with global counterparts, particularly CLAY/10/7490 – as detailed by Ching and Phoon (2014) and represented by "Soil Type/Number of Indicator Types/Number of Data Entries" – we aim to provide insights crucial for addressing current challenges in data-centric geotechnical engineering.

 DOI: 10.1201/9781003441946-11

The geotechnical attributes of the Tokyo lowlands, vital for seismic and geological hazard assessments as well as infrastructure development planning, are exhaustively analyzed. The stratigraphic delineations in this region, especially the Nanagō and Yūrakuchō Layers, have evolved due to climatic variations, sea level changes, and the Last Glacial Maximum (LGM), forming key components of the Kanto Plain's geology. The Nanagō Layer, found at depths of 30–50 m below the surface and characterized by its alternating sand and clay sequences, measures 20–30 m in thickness. Its *N*-value, indicative of compaction, varies between 20 and 50, with significant fluctuations due to the layer's stratification, necessitating careful analysis for foundational material applications.

Above the Nanagō Layer is the Yūrakuchō Layer, distinguished by its fine clay composition. This layer, reaching a total depth of around 30 m, is typically divided into upper and lower segments. The upper segment, containing sandy elements and extending 5–10 m, generally has an *N*-value ranging from 2 to 5, which may increase above 10 in some areas. The lower segment, approaching a depth of 20 m, features a significantly lower *N*-value, oscillating between 0 and 2, and is marked by high water retention and substantial porosity. The clays of the Yūrakuchō stratum, known for their high plasticity index, show increased sensitivity to moisture changes.

Globally, sedimentary terrains and their soil properties exhibit remarkable diversity, as evidenced in internationally compiled databases, such as those described by Ching et al. (2021). However, our database is uniquely sourced from a specific geographic area. This discussion aims to spotlight the statistical differences between our local database and global datasets. Through comparative analysis, we seek to shed light on the diverse nature of data in geotechnical assessments .

11.2 OUTLINE OF THE DATABASE

In this chapter, we have conducted a comprehensive examination of various soil parameters, including saturation (S_r), wet density (ρ_t), specific gravity of soil particles (G_s), void ratio (e), liquid limit (LL), plastic limit (PL), water content (W), compression index (C_c), coefficient of consolidation (C_v), deformation modulus (E_{50}), preconsolidation pressure (P_c), uniaxial compressive strength (S_u), and the *N*-value, comprising blow count (N_b) and penetration (N_p). The depths at which these parameters were observed are denoted as "Depth." Our database classifies any collection of data that includes at least one of these indicators as an individual dataset. Following this criterion, the cumulative count of datasets in our study amounts to 67,760. Figure 11.1 graphically represents the database with a scatter plot, where the top-right triangular section displays the pairwise counts between parameters. The frequency of these correlations varies significantly, with some exceeding 10,000 occurrences, while the average count is approximately 5,000.

Tables 11.1 and 11.2 systematically present the basic statistical measures and percentile values for each soil parameter, respectively. Table 11.1 provides the mean, coefficient of variation (COV), minimum, and maximum values of each parameter. We have categorized the parameters into three groups: indicative, mechanical, and sounding – the latter akin to those evaluated in the standard penetration test. Indicative parameters like saturation (S_r), wet density (ρ_t), specific gravity of soil particles (G_s), void ratio (e), liquid limit (LL), plastic limit (PL), and water content (W) essentially describe the inherent properties of soil. Meanwhile, mechanical parameters include compression index (C_c), coefficient of consolidation (C_v), deformation modulus (E_{50}), preconsolidation pressure

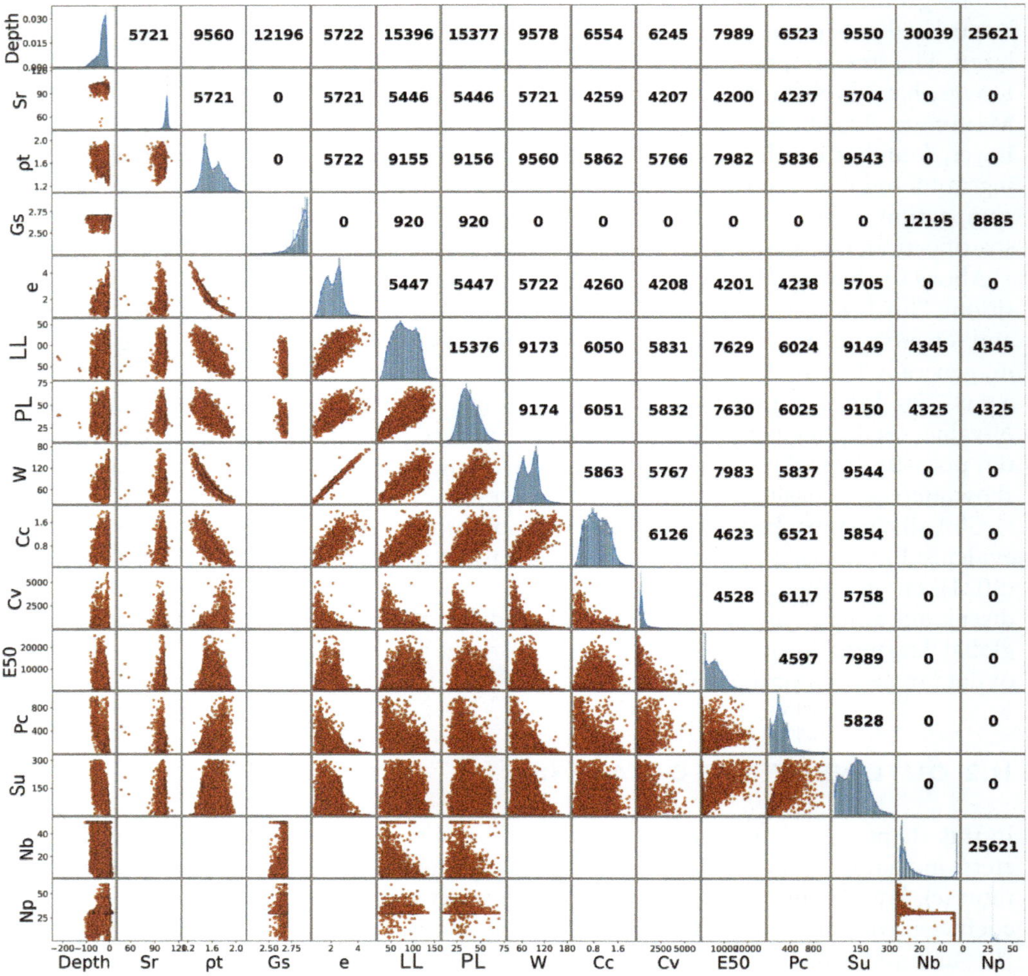

Figure 11.1 Scatterplot matrix of all collected data: pairwise plots between two indicators are shown, as the data includes some missing values.

(P_c), and uniaxial compressive strength (S_u). The N-value, encompassing blow count (N_b) and penetration (N_p), is categorized as a sounding parameter, vital in the Japanese design parameter estimation.

This chapter focuses on analyzing the coefficient of variation across different parameters. Insights from Table 11.1 indicate that certain indicative parameters, particularly saturation (S_r), wet density (ρ_t), and specific gravity of soil particles (G_s), have significantly lower coefficients of variation compared to others. This aligns with established geotechnical views on sub-seafloor clays, supported by research from Phoon (1995), Phoon and Kulhawy (1999a,b), TC304 (2021), and Otake and Honjo (2022). Parameters like LL, PL, and W show coefficients of variation around 0.300, consistent with the higher range observed in the TC304 dataset (LL: 0.034–0.39, PL: 0.029–0.38, W: 0.035–0.46). Mechanical parameters such as C_c, C_v, E_{50}, P_c, and S_u exhibit trends akin to the indicative

Table 11.1 List of basic statistical measures for each parameter

	n	*Mean*	*CV*	*Min*	*Max*
Depth	67,760	−22.69	−0.913	−220.00	24.50
S_r	5721	98.95	0.022	48.60	111.30
ρ_t	9560	1.58	0.075	1.23	1.97
G_s	12,196	2.67	0.010	2.50	2.70
e	5722	1.98	0.285	0.70	4.83
LL	15,396	75.02	0.324	20.30	148.70
PL	15,377	35.34	0.252	10.30	69.90
W	9578	73.15	0.299	22.60	171.50
C_c	6554	0.83	0.397	0.11	1.97
C_v	6245	435.36	1.372	1.68	6000.00
E_{50}	7989	4901.31	0.769	1.02	25,386.96
P_c	6523	205.42	0.740	4.22	983.00
S_u	9550	108.72	0.554	1.32	299.84
N_b	30,039	12.72	1.181	1.00	50.00
N_p	25621	30.15	0.207	1.00	60.00

Table 11.2 List of basic percentile value measures for each parameter

	Percentile value						
	0.025	*0.05*	*0.25*	*0.5*	*0.75*	*0.95*	*0.975*
Depth	−74.30	−65.40	−30.10	−16.50	−8.00	−1.80	−0.90
S_r	94.05	95.67	98.27	99.35	100.00	101.23	101.88
ρ_t	1.40	1.42	1.49	1.56	1.67	1.79	1.83
G_s	2.60	2.62	2.66	2.68	2.69	2.70	2.70
E	1.03	1.12	1.53	2.00	2.41	2.83	3.04
LL	32.80	36.40	55.80	74.40	95.00	113.00	118.00
PL	20.50	22.40	28.70	34.30	41.50	51.00	54.50
W	37.10	40.20	55.66	73.63	89.50	107.51	116.91
C_c	0.29	0.33	0.57	0.81	1.09	1.36	1.45
C_v	66.11	80.02	140.00	220.00	450.00	1550.00	2140.00
E_{50}	5.42	44.42	1971.14	4383.57	7058.34	11,669.72	13,851.89
P_c	13.73	21.57	110.82	174.56	264.78	519.65	646.38
S_u	8.58	14.08	61.34	109.83	150.78	208.39	231.47
N_b	1.00	1.00	3.00	6.00	15.00	50.00	50.00
N_p	11.00	16.00	30.00	30.00	32.00	38.00	42.00

parameters, with their coefficients approaching the upper limits of previous studies (C_c: 0.181–0.473, S_u: 0.159–0.57). The *N*-value stands out, registering at 1.181, significantly exceeding the site-specific variation ranges noted in earlier studies (0.159–0.570).

Figure 11.2 presents a scatter plot highlighting five indices: uniaxial compressive strength (S_u), liquid limit (LL), plastic limit (PL), water content ratio (*W*), and depth of observation. These indices align with those in CLAY/10/7490, a globally recognized database. The figure juxtaposes our database with CLAY/10/7490 using an overlaid

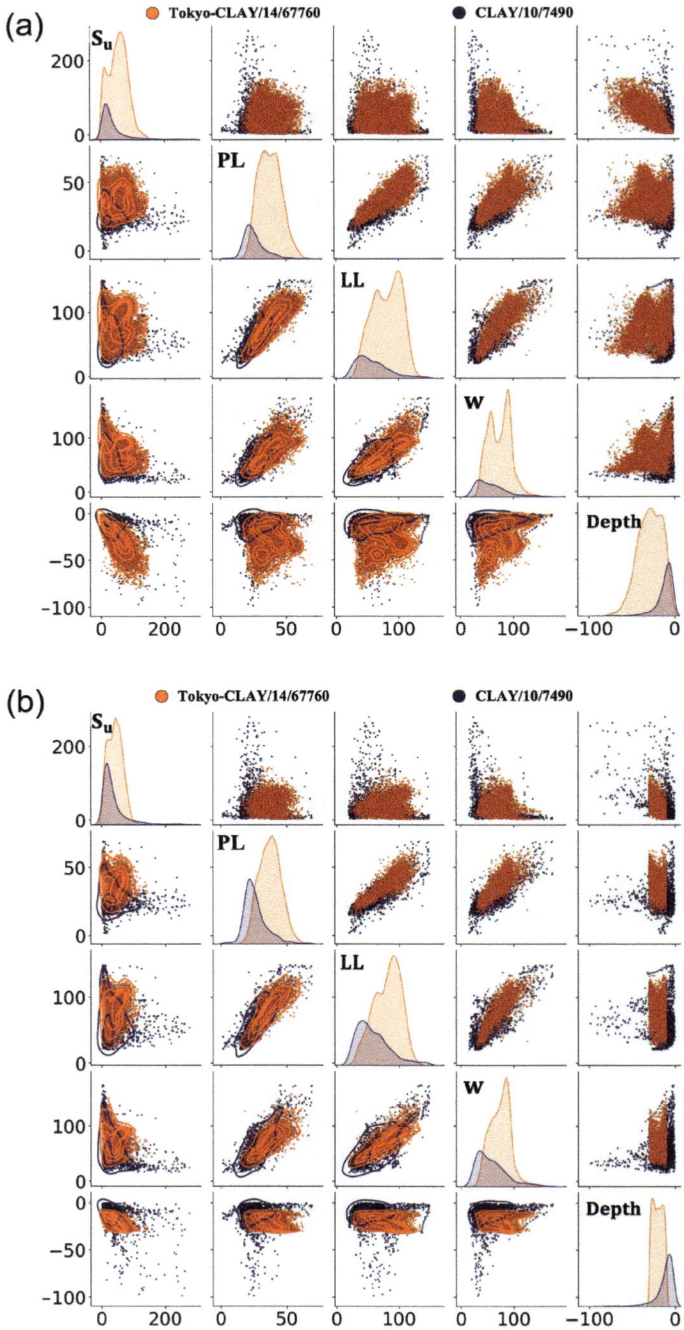

Figure 11.2 Detailed display of scatter plots for indicators competing with CLAY/10/7490.

representation. The diagonal of the figure illustrates the kernel density distribution for each index, with the upper right triangle showing a straightforward scatter plot and the lower left triangle integrating the simultaneous kernel density distribution within the scatter plot. It is important to note that subsets (a) and (b) utilize distinct plotting methods specific to this database: (a) displaying the entire dataset while (b) focusing on depths between 10 and 30 m. The lower section of the Yūrakuchō Layer, located at depths of 10–30 m, is characterized as a pure clay layer with minimal sand content and is distinct from other strata.

While our database is derived from extensive sampling within a specific part of the same soil layer, its variability in ground parameters parallels that in CLAY/10/7490. This similarity is further illustrated in Figure 11.3, which plots data on a plasticity diagram similar to Figure 11.2. However, this representation lacks soil layer distinction in (a) and limits depth to the 10–30 m range in (b). The distribution of our database shows a range that encompasses the intrinsic variation of CLAY/10/7490.

Geotechnical parameter variations fall into two primary categories: intra-subgrade variation (variability within a specific site or soil type) and inter-subgrade variation. The

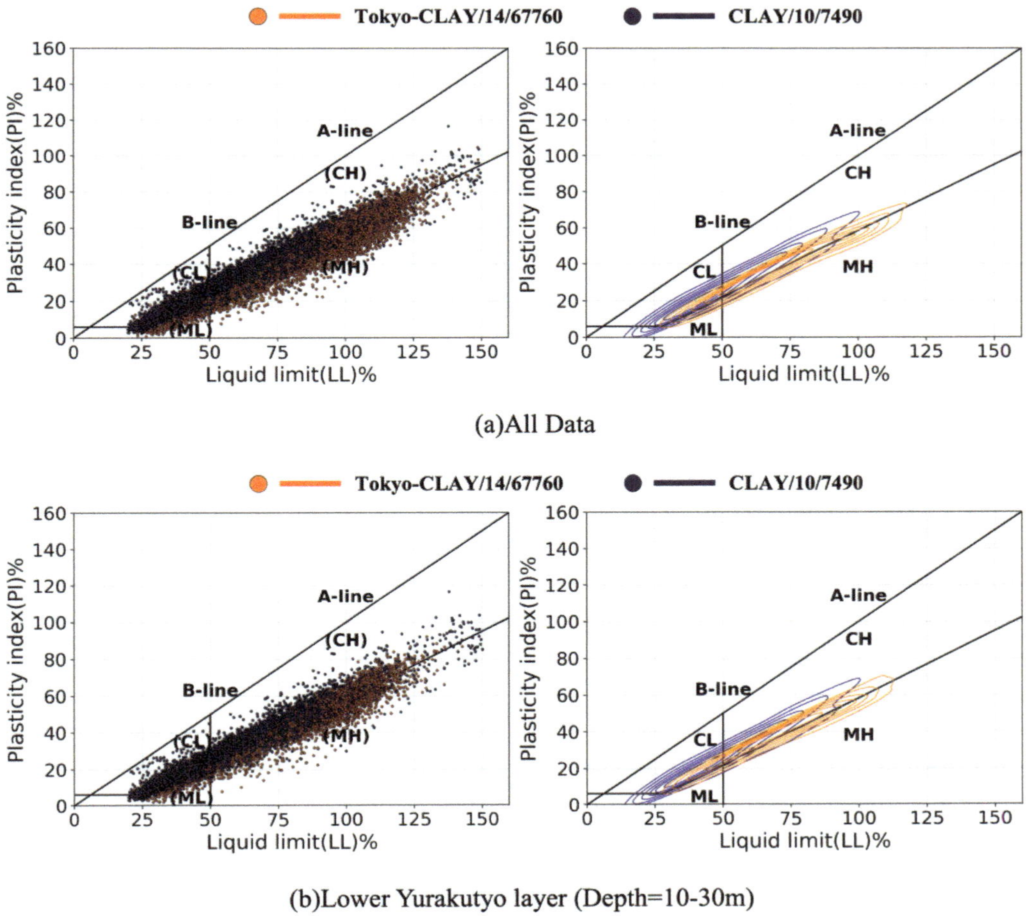

Figure 11.3 Scatter plots on the plasticity diagram.

former encompasses the inherent variability of the soil, representing dispersion specific to a single site. The latter, on the other hand, can be viewed as a bias across different soil types, where unique site characteristics lead to variations in soil parameters once data from multiple locations are aggregated, also known as site-to-site variability.

To explore the effects of inter-subgrade variation, our study focuses on the dispersion of statistics per site, using the hierarchical Bayesian modeling approach proposed by Ching et al. (2021) and Wu et al. (2022). Following a similar trajectory as CLAY/10/7490, we examine the dispersion in mean vectors and covariance matrices computed per site. However, our database, by its nature, limits the discussion on site-to-site variability since it is derived from a single site. Therefore, we adopt a methodology where each borehole is treated as an individual site. Our database includes 2972 boreholes, each containing between 1 and 275 data points, with an average of 29 data points per borehole. For comparison, CLAY/10/7490 (referenced in Ching and Phoon, 2014) encompasses 651 sites, each with 1 to 419 data points and an average of 30 data points per site.

Figure 11.4 displays pairwise scatter plots of the sample means for each site. Following the format of Figure 11.2 and Figure 11.3, these plots distinguish between (a) the comprehensive dataset and (b) data confined to the 10–30 m depth interval. Figures 11.5 and 11.6 further expand on this by presenting scatter plots of the sample means for each site.

A modal difference exists between our database and CLAY/10/7490, suggesting a slight bias. Figures 11.5 and 11.6 show the pairwise scatter plots of the sample standard deviation and sample correlation coefficient, respectively. In these figures, the bias observed in the mean versus scatter plot in Figure 11.4 appears to be reduced. Additionally, the inter-site variability, in terms of standard deviation and correlation coefficient, seems to be consistent between our database and CLAY/10/7490. Notably, the inter-site variability in our database aligns closely with that of the global database, CLAY/10/7490, indicating that the primary differences are likely in the attributes of the sample means (Figure 11.7).

However, these interpretations are fundamentally qualitative. Due to the challenges in visually interpreting correlation coefficients from pairwise scatter plots, we aim to conduct a comprehensive quantitative analysis of these characteristic differences in our future research.

11.3 CONCLUSION

In this chapter, we have examined the statistical properties of multivariate geotechnical parameters derived from extensive sampling within a representative marine clay deposit near Tokyo, Japan. These properties were compared and analyzed in conjunction with CLAY/10/7490, a well-established global database of multivariate geotechnical parameters for clays. The key conclusions from our study are as follows:

1. The overall variability of the geotechnical parameters, gathered from extensive sampling in the localized region, is consistent with the distributions observed in the global database, CLAY/10/7490.
2. We chose to treat each borehole as a single site. This approach revealed primary differences, notably in the fluctuation of mean values between sites. Interestingly, no significant variations were found in the standard deviations or correlation coefficients among the parameters.

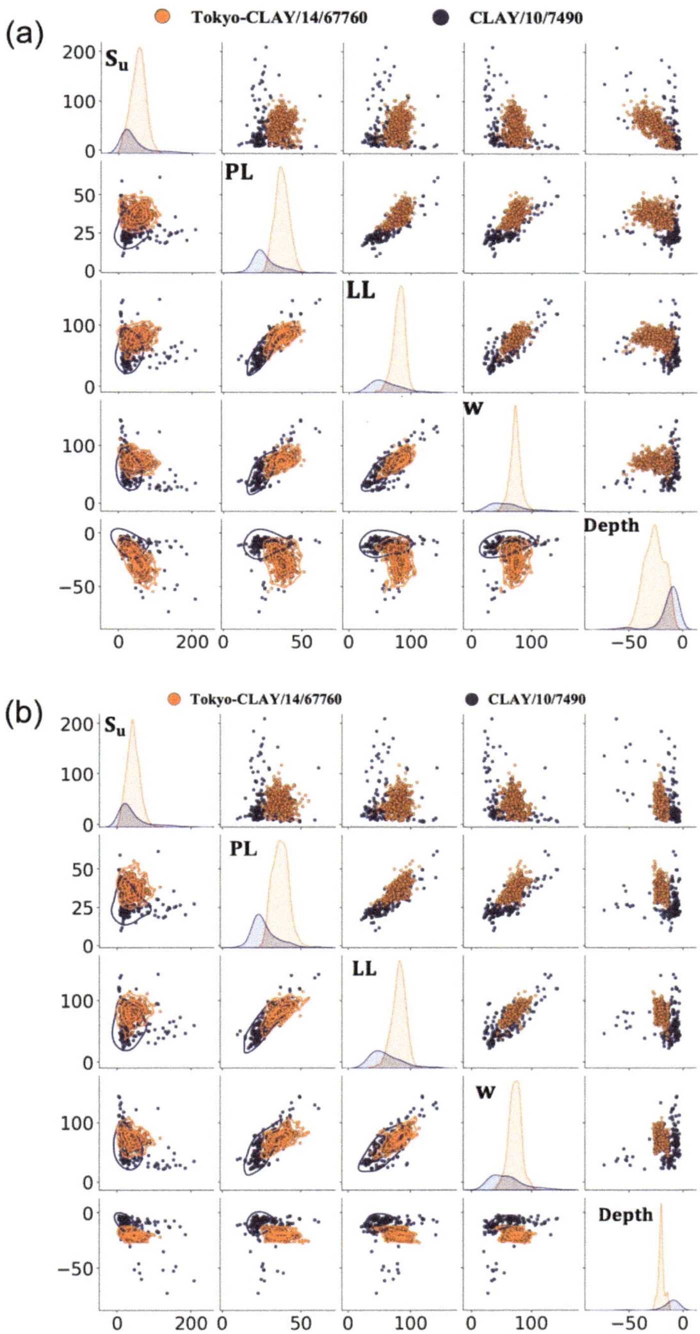

Figure 11.4 Scattering of means per site.

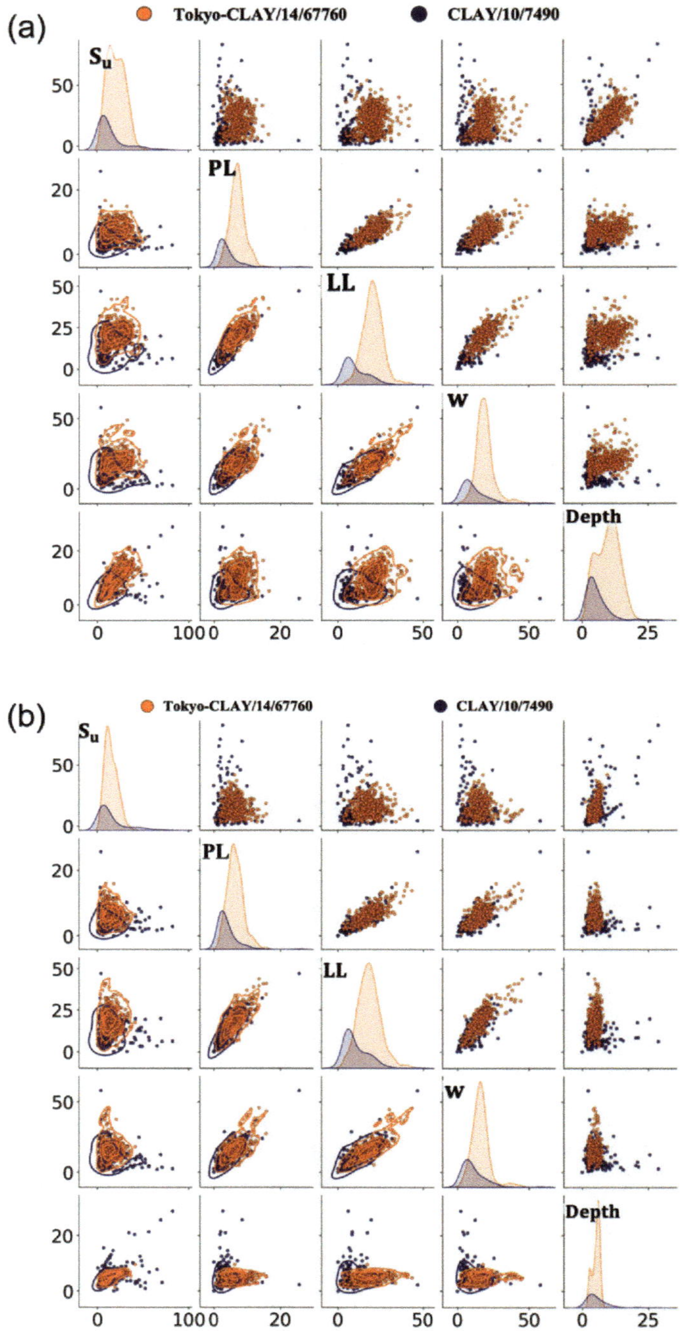

Figure 11.5 Scattering of standard deviations per site.

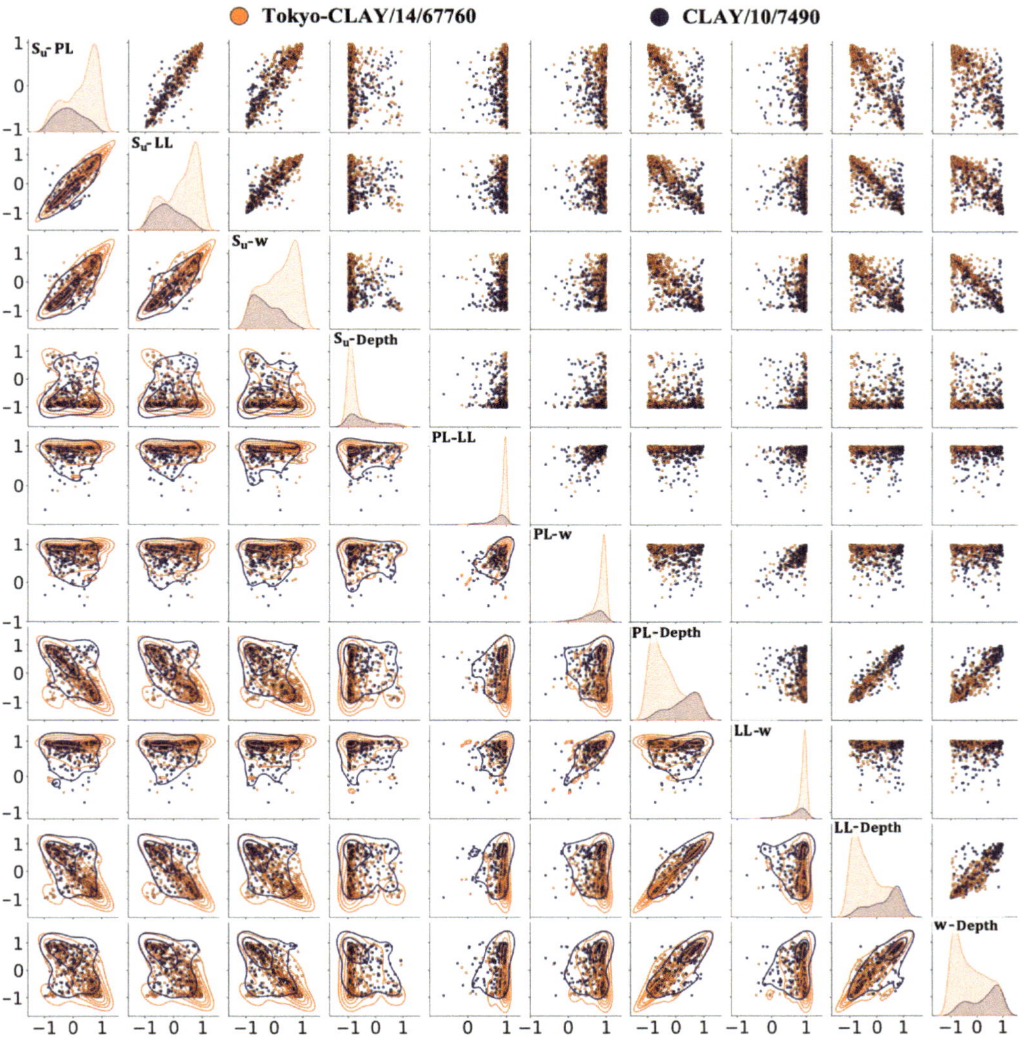

Figure 11.6 Scattering of correlation coefficients per site (all data).

3. Our findings suggest that intensive sampling within a specific area may closely mirror global databases. While the creation of comprehensive databases is vital for advancing data-driven geotechnical engineering, a global scale of data collection may not be essential. Targeted sampling within accessible sites could contribute significantly to building a robust database.
4. Developing methodologies to enhance generic databases is crucial, including techniques like similarity assessment (as referenced in Atma et al., 2022). Identifying hidden patterns in extensive databases is key to effectively filling in missing values.
5. In our analysis, each borehole was treated as an individual site. Standardizing the definition of sites is imperative for constructing a global, generic database. Managing trend components related to soil groupings and depth distributions is also a critical element in developing a universal database.

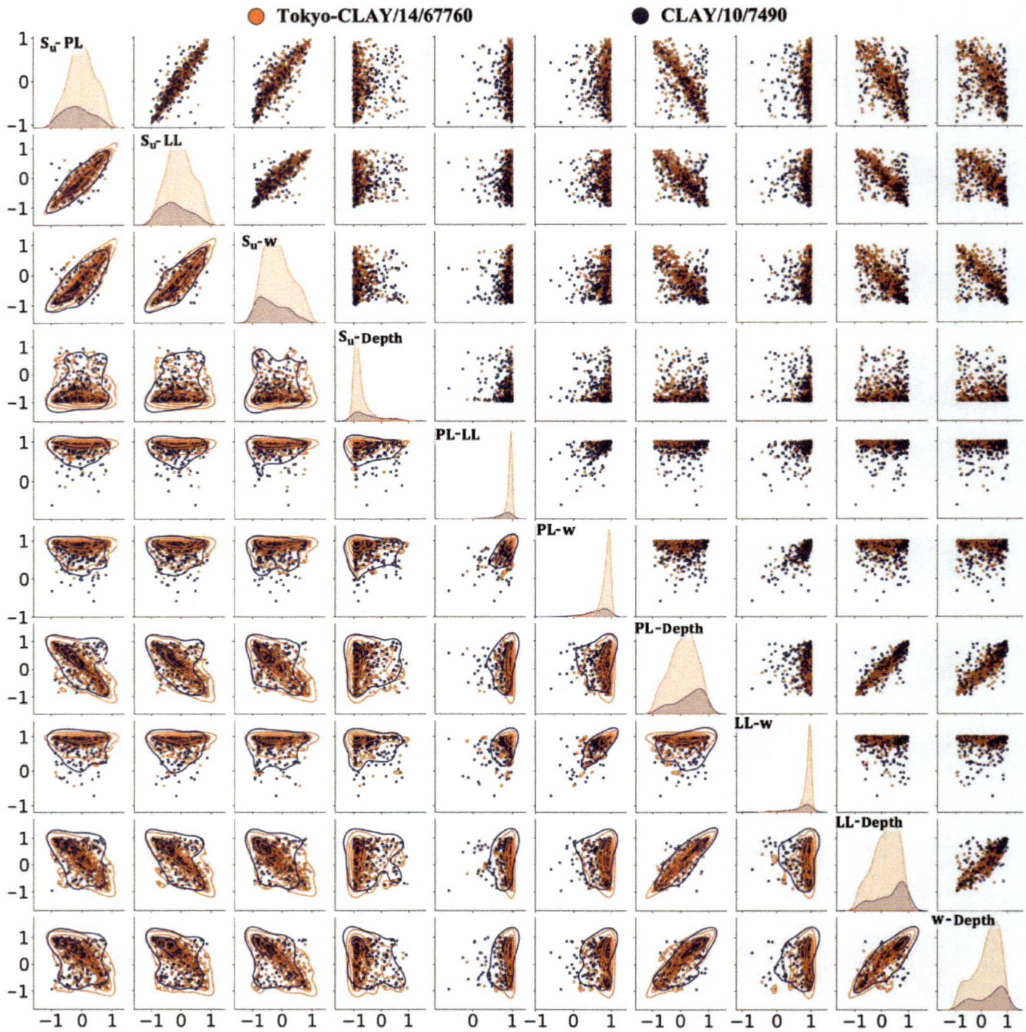

Figure 11.7 Scattering of correlation coefficients per site (Lower Yurakutyo Layer (depth = 10–30 m)).

Through a systematic analysis of results from local sampling, we have arrived at these conclusions and identified directions for future research. Our forthcoming endeavors include meticulous data screening, refined differentiation of soil categories, and reevaluation with consideration of the temporal aspect of surveys. Following comprehensive data management, we aim to develop methods to supplement missing soil parameters. This involves integrating similarity assessments and hierarchical Bayesian modeling to enhance the estimation of multivariate 3D spatial distributions, which will assist in optimally positioning observation points.

REFERENCES

Sharma, A., Ching, J. and Phoon, K. -K. 2022. A hierarchical Bayesian similarity measure for geotechnical site retrieval. *Journal of Engineering Mechanics*, 148(10), 04022062.

Ching, J. and Phoon, K. K. 2014. Transformations and correlations among some clay parameters — the global database. *Canadian Geotechnical Journal*, 51(6): 663–685.

Ching, J., Wu, S. and Phoon, K.-K. 2021. Constructing quasi-site-specific multivariate probability distribution using hierarchical Bayesian model. *Journal of Engineering Mechanics*, 147(10), 04021069.

Ishii, I. 1985. On the development of soil data base system, Technical Note of the Port and Harbour Research Institute, No.515, pp. 1–69.

Otake, Y. and Honjo, Y. 2022. Challenges in geotechnical design revealed by reliability assessment: Review and future perspectives. *Soils and Foundations*, 62(3), 101129.

Phoon, K., Kulhawy, F. and Grigoriu, M. 1995. *Reliability-Based Design of Foundations for Transmisssion Line Structure*. Report TR-105000, Palo Alto, Electric Power Research Institute.

Phoon, K. and Kulhawy, F. 1999a. Characterization of geotechnical variability. *Canadian Geotechnical Journal*, 36, 612–624.

Phoon, K. and Kulhawy, F. 1999b. Evaluation of geotechnical property variability. *Canadian Geotechnical Journal*, 36, 625–639.

Wu, S., Ching, J. and Phoon, K. K. 2022. Quasi-site-specific soil property prediction using a cluster-based hierarchical Bayesian model, *Structural Safety*, 99, 102253.

TC304, S.I., 2021. *State-of-the-Art Review of Inherent Variability and Uncertainty in Geotechnical Properties and Models*. ISSMGE Technical Committee 304.

Watabe, Y. and Tanaka, M. 2004. History of geotechnical database for port and airport construction in Japan. *Proceedings of the International Conference on Engineering Practice and Performance of Soft Deposits*, IS-Osaka 2004, pp. 409–414.

Chapter 12

In situ test-based evaluation of soil effective stress strength properties and stress history

Zhongkun Ouyang, Paul W. Mayne, and Shehab Agaiby

12.1 INTRODUCTION

To investigate the mechanical response of soils due to external loading, one needs to grasp the stress–strain characteristics of the geomaterials, which often could be manifested theoretically via elasticity, plasticity, cavity expansion, and various other models. Such analysis often requires the establishment of a series of parameters to be used in geotechnical problems ranging from stability, foundation-bearing capacity, and deformational investigation (settlement and time rate of consolidation). Additionally, these parameters could be evaluated based on data-driven approaches using regression, statistical, and probabilistic analyses. Comprehensively conducting all available geophysical, in situ, and laboratory testing techniques would be the optimal strategy for obtaining the most accurate and realistic values for these geotechnical parameters. As stated by the first principles of soil mechanics, the effective stress is equal to the total overburden stress minus the porewater pressure, thus soil effective stress strength parameter determination plays a crucial role in geotechnical site investigations involving embankment and slope stability, foundation capacity, and excavations. Also, the stress state analysis of the soils in the ground involves the understanding of soils' unit weight, which is an important soil physical property.

In most geotechnical projects, the degree of preconsolidation of the natural soils is particularly significant as it is a state parameter and represents the current and past stress history. The stress history is commonly represented by the overconsolidation ratio (OCR):

$$OCR = \sigma_p' / \sigma_{v0}' \tag{12.1}$$

where σ_p' = preconsolidation stress and σ_{v0}' = current effective overburden stress. It is well known that OCR governs the undrained shear strength (s_u), lateral stress coefficient (K_0), pore pressure behavior (A_f), elastic moduli (E' and E_u), and small-strain shear modulus (G_{max}) of soils, and serves as a key parameter in critical-state soil mechanics (CSSM) that has been proven to be a valuable and rational framework in which to organize the mechanical response of soils to loading.

This chapter covers evaluating soils' effective stress strength and stress history parameters from two widely used in situ testing techniques, namely the piezocone penetration testing (CPTu) and flat plate dilatometer tests (DMT). The emphasis in this chapter is on parameters commonly used in geotechnical analysis and design concerned with pilings, shallow foundations, slopes, earth-retaining structures, roadways, and embankments. It is not intended to be exhaustive and mainly focuses on geotechnical engineering parameters of usual concern including the following: (1) unit weight (γ_{total}), (2) preconsolidation

DOI: 10.1201/9781003441946-12

stress or effective yield stress (σ'_p = preconsolidation stress = OCR· σ'_{v0}), and (c) effective shear strength parameters (ϕ', c'). Additional details and information may be found in Kulhawy and Mayne (1990), Mayne et al. (2002), Loehr et al. (2016), and Ouyang and Mayne (2017a, 2017b, 2018a, 2018b, 2019, 2023). The emphasis in this chapter is on using the results of in situ tests to estimate soil properties via a data-driven method.

12.2 SOIL UNIT WEIGHT

The behavior of soils depends upon the effective stresses in the ground. The total overburden stress is the accumulation of soil unit weight times soil layer thickness and is caused by gravitational forces acting on the soil particles. As such, the unit weight of soils is needed to calculate the total overburden stresses (also later effective stresses) with depth. For other soil parameters involving stresses in soil in general, soil unit weight often time is a required input.

12.2.1 Unit weight from shear wave velocity

For the following, soil unit weight (γ_{total}) is required to evaluate (1) total (σ_{v0}) and effective overburden (σ'_{v0}) stresses, (2) the small-strain shear modulus (G_{max}) from the shear wave velocity (V_s), and (3) DMT horizontal stress index (K_D), normalized cone penetration test (CPT) parameters, and net cone resistance. According to the following, unit weight and mass density (ρ) are related:

$$\gamma_{total} = \rho_{total} g$$

where g = acceleration due to gravity = 9.8 m/s^2.

For direct-push probes and geophysical measurements, unit weights can be calculated in the lab using weight–volume relationships, or they can be inferred using empirical equations. Mayne et al. suggested a correlation between total unit weight and shear wave velocity (V_s). The following are the regression expressions:

$$\gamma_{total}(pcf) = 49.8\log(V_s) - 26.5 \, with \, V_s in \; ft \, / \, s \tag{12.2}$$

These data come from a range of uncemented soil types, including clays, silts, sands, gravels, and mixed soils, as well as some peats. The measured unit weights came from undisturbed sampling, while the vast majority (92%) of the shear wave velocities were determined using seismic downhole tests. It should be pointed out that the formulas could be applied to soils with a shear wave velocity of less than 3000 m/s.

12.2.2 Unit weight from cone penetration test (CPT)

An empirical association between the total unit weight and the CPT's recorded sleeve friction is proposed by Mayne (2014). The dataset comprises unit weights measured on clays, silts, sands, and mixed soils, and the approach is suitable for uncemented, insensitive, and inorganic soils. This regression analysis's trendline is shown in a dimensionless form as follows:

$$\gamma_{total} = \gamma_w[1.22 + 0.345\log(100 f_s \, / \, \sigma_{atm} + 0.01)] \tag{12.3}$$

12.2.3 Unit weight from flat dilatometer test

Flat dilatometer test (DMT) was developed in Italy by Prof. S Marchetti in 1980. As an important in situ testing method, it gained popular acceptance since its introduction due to its simplicity, repeatability, and quick application in geotechnical explorations. By employing nitrogen gas to inflate a flexible steel membrane with 60 mm diameter at each test depth, the DMT generates two pressure readings (A and B). When the membrane is flush with the flat blade face, the corrected A reading or contact pressure, p_0, occurs. The corrected B reading or expansion pressure, p_1, happens when the membrane extends outward 1.1 mm.

For soft clays with a high groundwater level, the corrected contact pressures (p_0) show a linear trend and increase with depth and the expansion pressures (p_1) in various materials exhibit a similar tendency with depth. Regression analyses can be used to define a new parameter (m_{p0}) and produce the best-fit line (intercept = 0):

$$m_{po} \quad = \quad \Delta p_0 \,/\, \Delta z \tag{12.4}$$

For the majority of the soft to firm inorganic clay soils, it is revealed that the total unit weights track with the slope parameter m_{p0}, as shown in Figure 12.1. A few organic-containing clays displayed varying trends and were offset to the right.

An expression can be developed between the total unit weight and slope parameter m_{p0} for soft to firm clays based on the pattern depicted in Figure 12.1, shown by Equation (12.5):

$$\gamma_t = \gamma_w + 0.22 m_{po} \tag{12.5}$$

This empirical expression can therefore be used to forecast the total unit weight of each of these soft clay soils (Ouyang and Mayne 2016). Figure 12.2 shows a comparison of the laboratory-measured unit weights to the predicted unit weights using the empirical equation for a worldwide DMT clay database, and Figure 12.3 compares the lab unit weight profile to the predicted unit weight using DMT data at the Anacostia soft clay site in Washington, DC. It should be noted that in stiff to hard clays with high OCRs or fissured OC clays, further study is required.

12.3 SOIL EFFECTIVE STRESS STRENGTH PARAMETERS

When penetrometer probes (CPTu and DMT) are advanced into clays and fine-grained soils at a standard push rate of 20 mm/s, the interpretation by geotechnical engineers traditionally centers around a total stress analysis, and consequently, the evaluation is focused on the undrained shear strength (s_u). However, it is well recognized that the fundamental behavior of soils belongs to an effective stress framework, as established by several avenues: (1) laboratory stress path response from triaxial tests (e.g., Lambe and Whitman 1979; Mayne et al. 2009), (2) critical-state soil mechanics (e.g., Schofield and Wroth 1968; Holtz et al. 2010), and (3) numerical modeling of soil behavior in triaxial compression, extension, and simple shear modes (Whittle et al. 1994). Therefore, the fundamental strength of clays is represented by effective stress strength parameters, including ϕ' = effective friction angle and c' = effective cohesion intercept. Most commonly, the effective stress strength of soils is represented by a Mohr–Coulomb criterion:

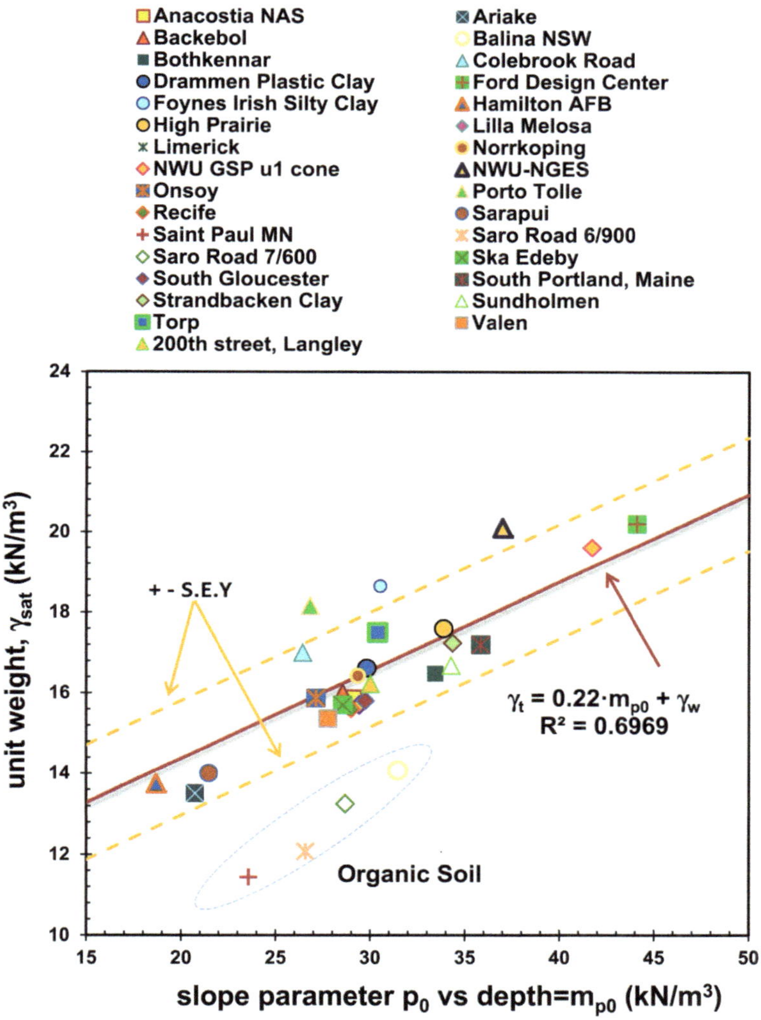

Figure 12.1 Total unit weight versus slope parameter $m_{p0.}$

$$t_{max} = c' + s' \cdot \tan\phi \tag{12.6}$$

where τ_{max} = shear strength (i.e., maximum shear stress), c' = effective cohesion intercept, σ' = effective normal stress, and ϕ' = effective stress friction angle.

A traditional means of evaluating the effective stress strength parameters c' and ϕ' has been via laboratory strength tests, such as the direct shear (DS), isotropically and/or anisotropically consolidated undrained triaxial test with pore pressure measurements (CIU/CAU), and consolidated drained triaxial test (CD). The focus of this section is on evaluating these effective stress strength parameters of clays via in situ tests and examined through comparison with laboratory benchmark values obtained from undrained consolidated anisotropic (CAUC) and undrained compression (CIUC) triaxial tests made on undisturbed samples.

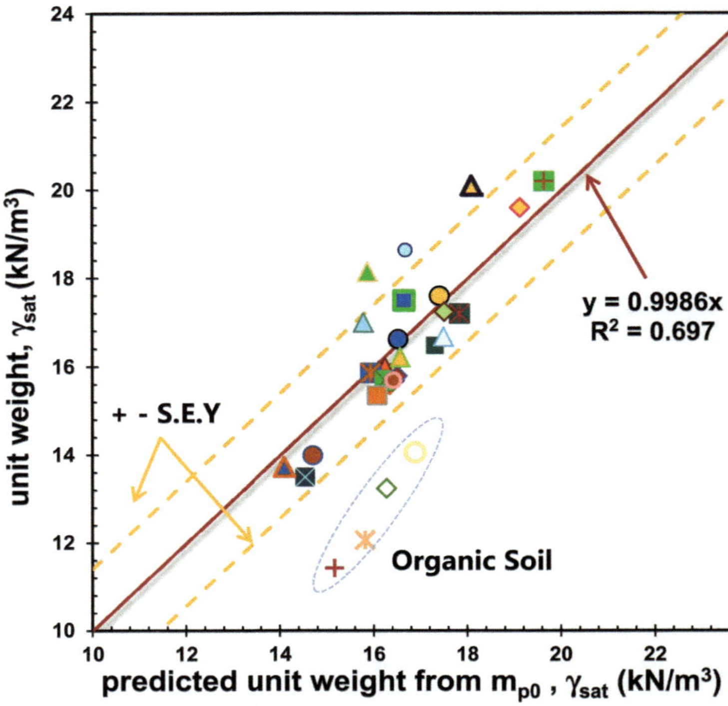

Figure 12.2 Lab soil unit weight versus predicted unit weight (organic clay excluded).

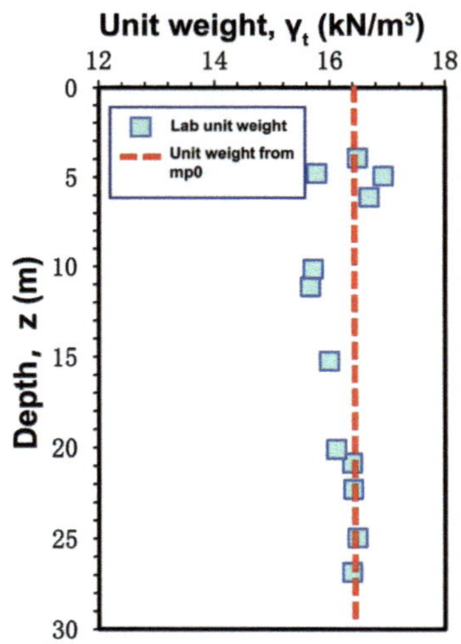

Figure 12.3 Profile of lab-measured unit weight and DMT predicted unit weight for Anacostia soft clay in Washington, DC.

12.3.1 Friction angle criteria

The effective friction angle obtained from triaxial tests can be evaluated on the basis of different criteria, which include: (a) value at maximum deviator stress (q_{max}); (b) value at maximum obliquity $(\sigma_1'/\sigma_3')_{max}$; and (c) value taken at large strains, normally at 15% axial strain, or in some cases, at 20% strain. The latter criterion is sometimes interpreted as the value corresponding to the critical state (ϕ_{cs}'). For CIUC and CAUC tests on soft inorganic clays of low sensitivity and low OCRs, these criteria often occur at similar points on the effective stress path, so only minor differences are notes in their magnitudes (Lade 2016; Ouyang & Mayne 2018a). A study of 19 natural clays generally showed minor differences in values of ϕ' defined at q_{max} and ϕ' at $(\sigma_1'/\sigma_3')_{max}$ (Bjerrum & Simons 1960). For a large series of CK0UC tests on soft Chicago clays, Finno and Chung (1992) showed that $\phi' = 28.3°$ at $(\sigma_1'/\sigma_3')_{max}$; and $\phi' = 27.6°$ at large strains, thus <1° difference. For sensitive and structured clays, however, the values can be of several degrees different (Leroueil & Hight 2003). In most of the database, only a single reported value of ϕ' was provided by the sources of information and the adopted criterion was often not identified. In some cases, an effective stress envelope was provided, yet the criterion used was not noted. In a few cases, a completely effective stress path and/or stress–strain curve was provided, and more than value could be assessed. It is believed that the database primarily contains results from inorganic and insensitive clays where differences in ϕ' from the different criteria are minor.

12.3.1 Effective stress strength parameters for intact clays from piezocone penetration test

Excess pore pressures are observed when the penetrometer is advanced into intact clays and silts at a standard rate of 20 mm/s. In order to assess the effective stress friction angle (ϕ') for a variety of fine-grained soils, ranging from natural lean plastic clays and clayey silts from marine, alluvial, lacustrine, deltaic, and glaciofluvial origins, an effective stress limit plasticity solution for piezocone penetration tests (CPTu) developed at NTH (Norwegian Institute of Technology) by Janbu and Senneset (1974), Senneset et al. (1989), and Sandven (1990) and further modified by Ouyang and Mayne (2018a, 2019, 2023) is illustrated below for the task.

The relationship between soils' effective stress friction angle (ϕ') and the CPTu measurements are given as:

$$N_{mc} = Q' = \frac{\tan^2(45° + \phi'/2) \cdot \exp(\pi \cdot \tan\phi') - 1}{1 + 6 \cdot \tan\phi' \cdot (1 + \tan\phi') \cdot B_q} \tag{12.7}$$

where $N_{mc} = Q'$ = modified resistance number from CPTu= Q/OCR^{Λ}, Q = normalized cone tip resistance = $(q_t - \sigma_{v0})/\sigma'_{vo})$, OCR = overconsolidation ratio of soil, Λ = plastic volumetric strain ratio≈0.7–0.8, and B_q= normalized porewater pressure = $(u_2 - u_0)/(q_t - \sigma_{v0})$.

Typical value of Λ ranges from 0.5 to 0.7 for artificially prepared samples such as kaolin and kaolinitic–sand mixture to around 0.7–0.9 for inorganics and insensitive clays to as high as 1 for structured and sensitive soils.

For intact clays with $18° \leq \phi' \leq 45°$ and $0.05 \leq B_q \leq 1.0$, the above solution for the effective stress friction angle (ϕ') can be numerically approximated by:

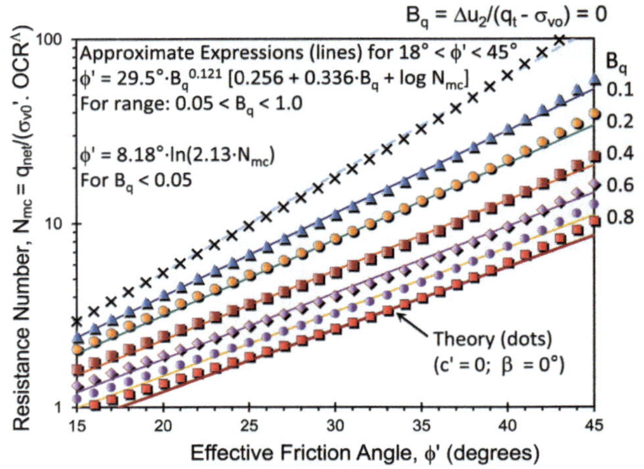

Figure 12.4 NTH method for evaluating ϕ' from CPTu using analytical and approximate solutions.

$$\phi' = 29.5° \cdot B_q^{0.121} \cdot [0.256 + 0.336 \cdot B_q + \log N_{mc}] \tag{12.8}$$

Figure 12.4 displays both the approximation equation and the closed-form solution. The effective cohesion intercept can be taken as $c'= 0$ for uncemented soft to hard clays and silts, especially those that are normally consolidated (NC) to lightly overconsolidated (LOC) with OCRs < 2 (Mayne and Stewart 1988; Mesri and Abdel-Ghaffar 1993). For overconsolidated (OC) clays, a low value of c' may be reasonable for interpreting the effective stress parameters (φ' and c') (Sellountou et al. 2004; Lade 2016). Since c' really refers to the part of the yield surface that extends above the frictional envelope, many, if not most, c' = 0 also applies to OC clays (Leroueil and Hight 2003; Mayne 2016).

12.3.1.1 Case study: CPTu in soft intact NC clays at Sandpoint, Idaho

A case study from Sandpoint, Idaho, is presented to show the process of interpreting effective friction angle using the NTH method. The Idaho Department of Transportation required soils information for design of approach embankments and a new bridge for State Route 95 highway. Figure 12.5 shows a representative CPTu at the site indicating over 80 m of soft silty clays, with occasional sand layers, lenses, and seams.

For the case of these NC clays having OCR = 1, the resistance number Q' is simply the normalized cone tip resistance Q, and Figure 12.6 shows the procedure for determining Q and B_q. The parameter Q is found as the slope from plotting net cone resistance versus the effective overburden stress. In this example, we force the line through the origin (assuming c'=0) to obtain Q =4.2. By the same token, the porewater parameter B_q is determined as the slope of the Δu versus q_{net}, giving the porewater parameter B_q = 0.75 for the Sandpoint site.

Results from 32 laboratory CIUC triaxial compression tests on undisturbed samples taken from 5 to 80 m depths are used as the reference benchmark values. Figure 12.7 shows the laboratory s–t' space of the triaxial tests that provided a lab friction angle of 33.1°. The NTH CPTu outputs an interpreted friction angle of 32.8° that is in excellent agreement.

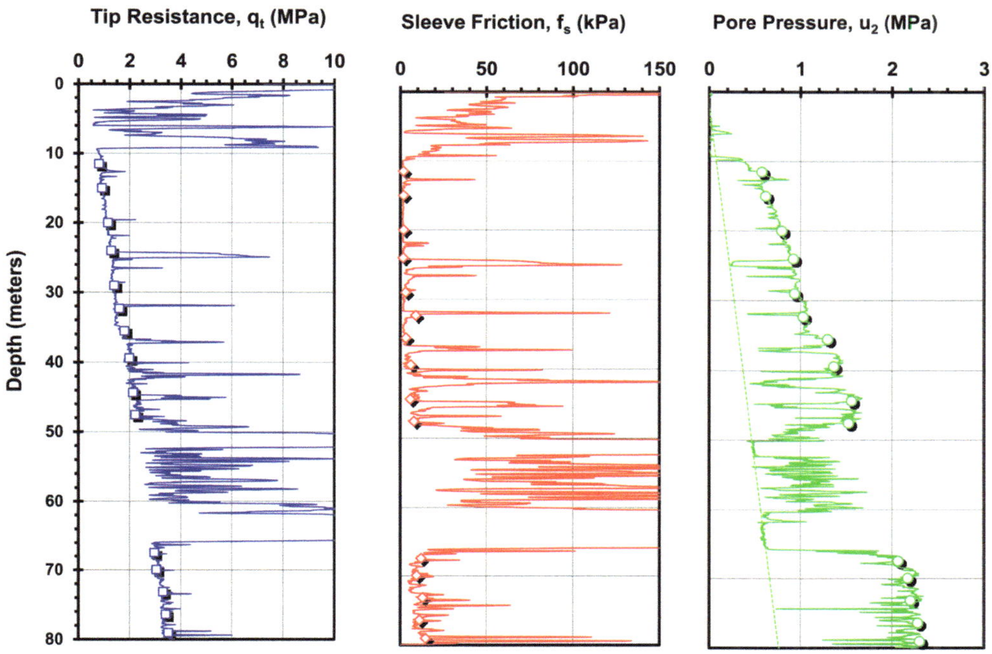

Figure 12.5 Deep CPTu profile at Sandpoint, Idaho with select data points for analysis.

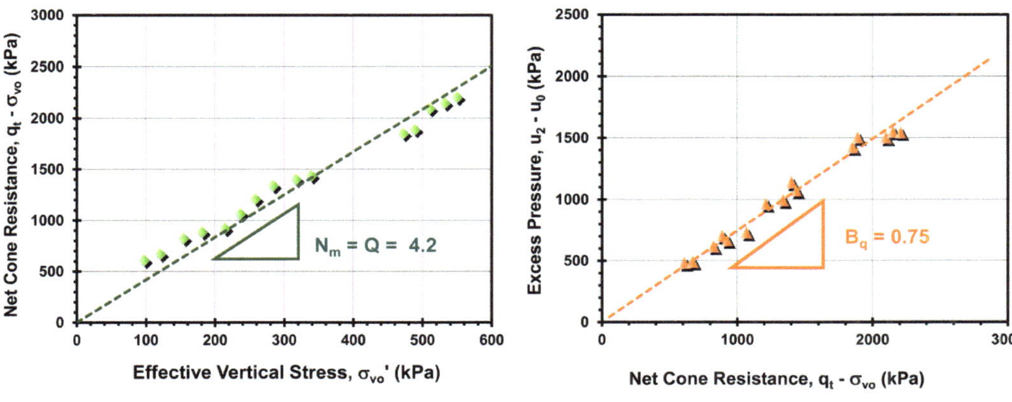

Figure 12.6 NTH postprocessing of CPTu data for *Q* and B_q at Sandpoint

12.3.1.2 Case study: CPTu-ϕ' profile interpretation for soft intact NC Chicago clays

The national geotechnical experimental site at Northwestern University (NWU) in Evanston, Illinois, is underlain by a 10-m-thick sand layer over 12 m of soft Chicago clay, which overlie other deeper soil layers. A variety of in situ and laboratory tests have been conducted here to investigate the properties of the clay soils (Finno et al. 2000). Figure 12.8 shows a representative portion of a CPTu sounding in the soft clay from 10 to 22 m depths with respective q_t, f_s, and u_2 values obtained using the Georgia Tech cone

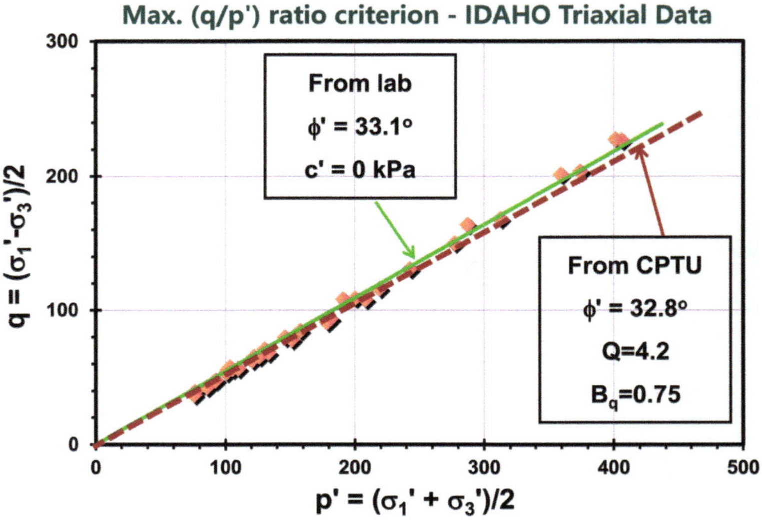

Figure 12.7 Laboratory triaxial effective stress friction angle ϕ' versus interpreted ϕ' from NTH using CPTu data at Sandpoint.

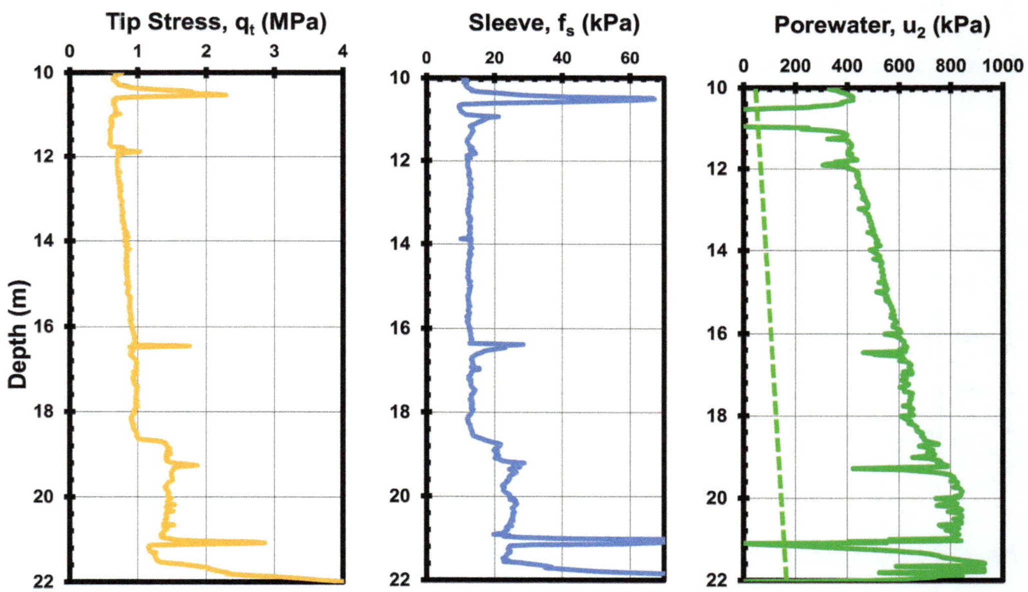

Figure 12.8 CPTu sounding of soft Chicago clays.

truck (Mayne 2007a). The soft clay under investigation has a mean unit weight of 20 kN/m^3, water content of 20%, liquid limit of 38%, and plasticity index (PI) = 12%.

The laboratory value of ϕ' obtained from a series of CAUC triaxial tests conducted on undisturbed samples of these soft clays (Figure 12.9), as reported by Finno and Chung (1992) and Chung and Finno (1992) are adopted to scrutinize the performance of the approximate NTH solution for ϕ' profile determination using Equation (12.8). The

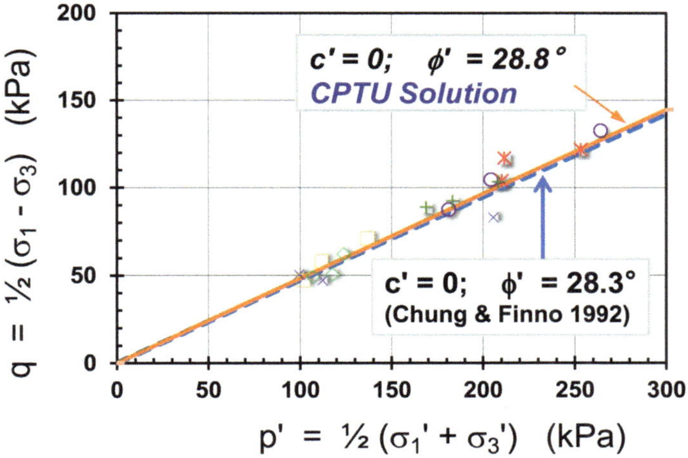

Figure 12.9 Comparison between laboratory triaxial tests showing ϕ′ at q_{max} and NTH friction angle from CPTu at NWU site (Note: triaxial results from Chung and Finno 1992).

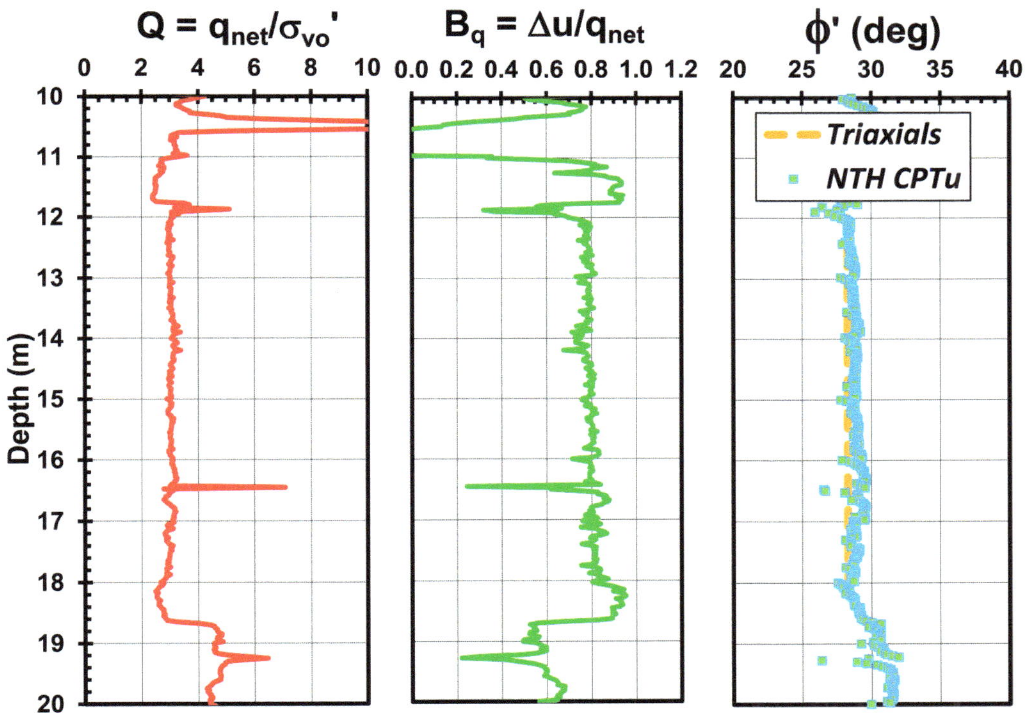

Figure 12.10 Profiles of resistance number (*Q*), normalized porewater pressure (B_q), and evaluated ϕ′ profile for soft Chicago clays.

profiles of *Q* and B_q with depth from the CPTu data are presented in Figure 12.10. Here it can be seen that the approximate NTH solution gives results in agreement with the laboratory triaxial tests.

12.3.1.3 Case study: CPTu in overconsolidated Presumpscot clay, ME

For CPTu in overconsolidated intact clays, the same methodology applies, however, the cone resistance number must be adjusted for stress history effect using Equation (12.7). A field case study with CPTu soundings in natural overconsolidated intact clays is presented using data from Martin's Point Bridge, near Portland, Maine (Hardison and Landon 2015). The general stratification of the site consists of an organic silt layer underlain by firm to stiff OC Presumpscot clay that overlies lower strata comprised of glacial outwash sand and bedrock. Shelby tube samples of Presumpscot clay were collected at the site from different elevations and tested for index properties, consolidation parameters, and strength characteristics. The results of laboratory testing on the OC clay layer gave an average total unit weight γ_t = 16.5 kN/m^3, natural water contents ≈ 30%–40%, liquid limits in the range of 20%–45%, and a plasticity index (PI) between 10% and 20%.

Figure 12.11(a) shows a representative CPTu sounding at the site with profiles of q_t and u_2 with depth. Depths of 2–14 m represent the extent of the OC Presumpscot clay of interest at this location with the groundwater table located at about 2 m. The corresponding profile of overconsolidation ratio (OCR = σ'_p/σ'_{vo}) determined by constant-rate-of-strain (CRS) consolidation tests is shown in Figure 12.11(b) (Hardison and Landon 2015).

The profiles of the cone resistance number (N_m and N_{mc}) using Λ = 0.7) with normalized porewater parameter B_q are shown in Figure 12.12(a). These are utilized to produce depth profiles of ϕ' in direct comparison with the CIUC triaxial data that are shown at their respective sampling depths in Figure 12.12(b). Overall, the values from CIUC results are in good agreement with those from CPTu using the modified NTH solution.

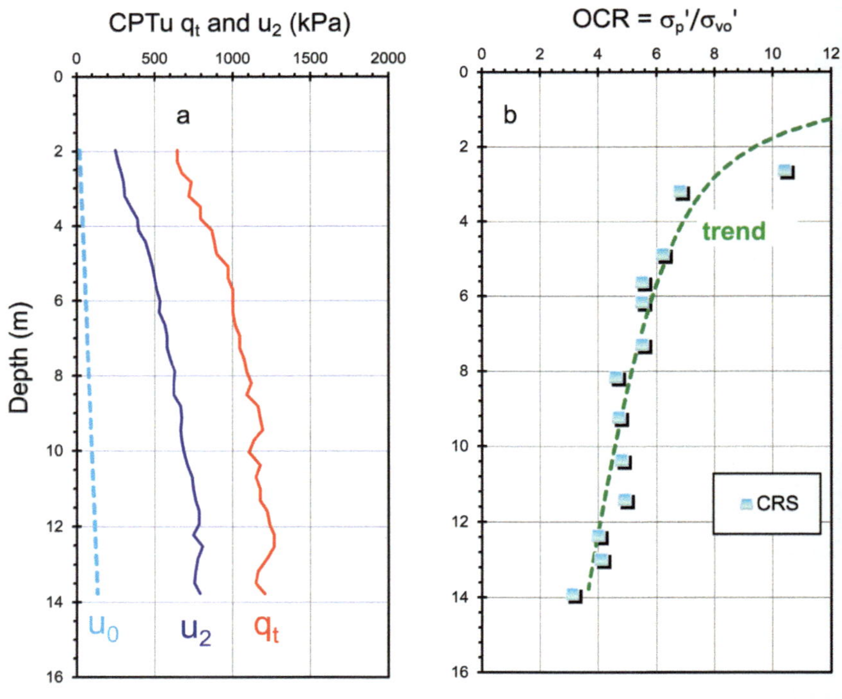

Figure 12.11 Profiles at Martin's Point Bridge, Maine: (a) piezocone readings and (b) OCR from CRS consolidation tests (Source: data from Hardison and Landon 2015).

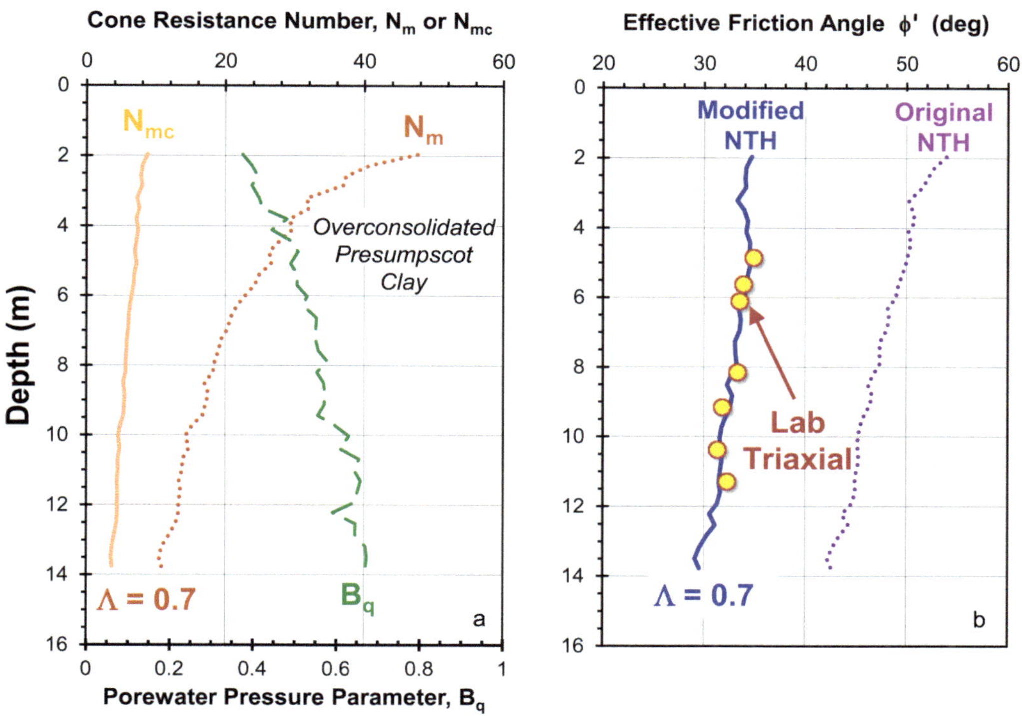

Figure 12.12 Profiles for Presumpscot clay at Martin's Point Bridge: (a) N_m, N_{mc}, and B_q using Λ =0.7; and (b) NTH ϕ′, modified NTH ϕ′, and CIUC triaxial ϕ′.

12.3.2 Effective stress strength parameters for fissured clays from piezocone penetration test

In fissured overconsolidated fine-grained soils, the porewater pressure u_2 registered by CPTu often demonstrates a near zero magnitude, $u_2 \approx 0$ (Ouyang and Mayne 2019, 2020). To interpret the value of effective stress friction angle ϕ′ in OC fissured clays, a practical solution is to take $u_2 \approx 0$, and $B_q < 0.05$ is usually observed, thus ϕ′ in OC fissured clays is given by the following equation:

$$\phi' \approx 8.18° \cdot \ln(2.13 N_{mc}) \quad (12.9)$$

12.3.2.1 Case study: Fissured London Clay at Brent Cross, United Kingdom

Brent Cross serves as an experimental test site involving London Clay (Powell and Quarterman 1988). London Clay is an overconsolidated Eocene marine formation that is notable for the presence of discontinuities in the form of fissures and cracks (Lunne et al. 1986). The characteristic effective strength of the fissured clay can be represented by a lower bound envelope given by c'=0 and ϕ′=19.5° (Hight and Jardine 1993). Laboratory testing on the London Clay at Brent Cross has shown a representative total unit weight

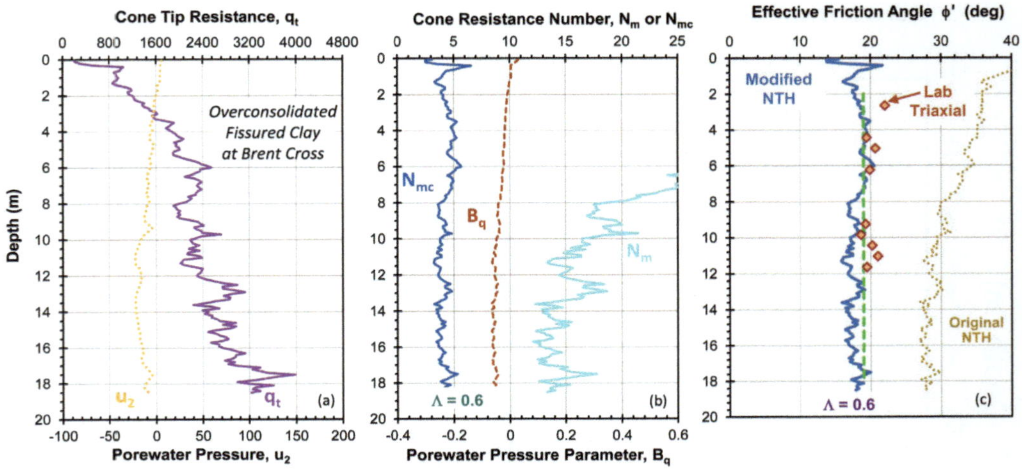

Figure 12.13 Profiles for fissured London Clay at Brent Cross: (a) Piezocone penetration data; (b) N_m, N_{mc}, and B_q using Λ = 0.6; and (c) NTH ϕ' and modified NTH ϕ' and CAUC triaxial ϕ' (Source: data from Lunne et al. 1986; Hight et al. 2003).

γ_t = 20 kN/m^3, natural water contents ≈ 28%–32%, liquid limits in the range of 65%–85%, and plasticity index (PI) between 55% and 65%. A representative exponent Λ = 0.6 has been assigned to the fissured London Clay at Brent Cross. Representative data from piezocone tests at Brent Cross are shown in Figure 12.13(a) with the profiles of q_t and u_2 with depth (Lunne et al. 1986). In Figure 12.13(b), the cone resistance number N_m and modified cone resistance number N_{mc} are presented along with the porewater pressure parameter B_q. It is observed that the value of B_q is essentially zero throughout the entire profile, which again is indicative of a fissured clay. The modified NTH solution using the approximate expression for $B_q = 0$ appears in good agreement with the laboratory triaxial interpretation of the effective friction angle, specifically $\phi' \approx 19.5°$.

12.3.2.2 Triaxial – Piezocone database for NC–OC clays

For the examination of the modified NTH solution for normally consolidated and overconsolidated clays, a master summary plot for the triaxial-measured ϕ' versus the CPTu-determined values via the modified NTH solution is shown in Figure 12.14.

The above statistics generally support that the revised NTH method gives a reasonable evaluation, albeit about 4% conservative, on the effective friction angle when referenced to the benchmark triaxial value.

12.3.2.3 Special considerations for organic soils, sensitive clays, and structured geomaterials

For the data that were considered in these studies, CPTu and triaxial data were obtained from clays and clayey silts that were primarily inorganic soils of low to medium sensitivity. The characteristic porewater pressure response for this group exhibits: 0.5 < Bq < 0.6 (Ouyang and Mayne 2019). However, it is recognized that natural organic clays tend to generate lower normalized porewater pressure (B_q < 0.5) while sensitive and structured

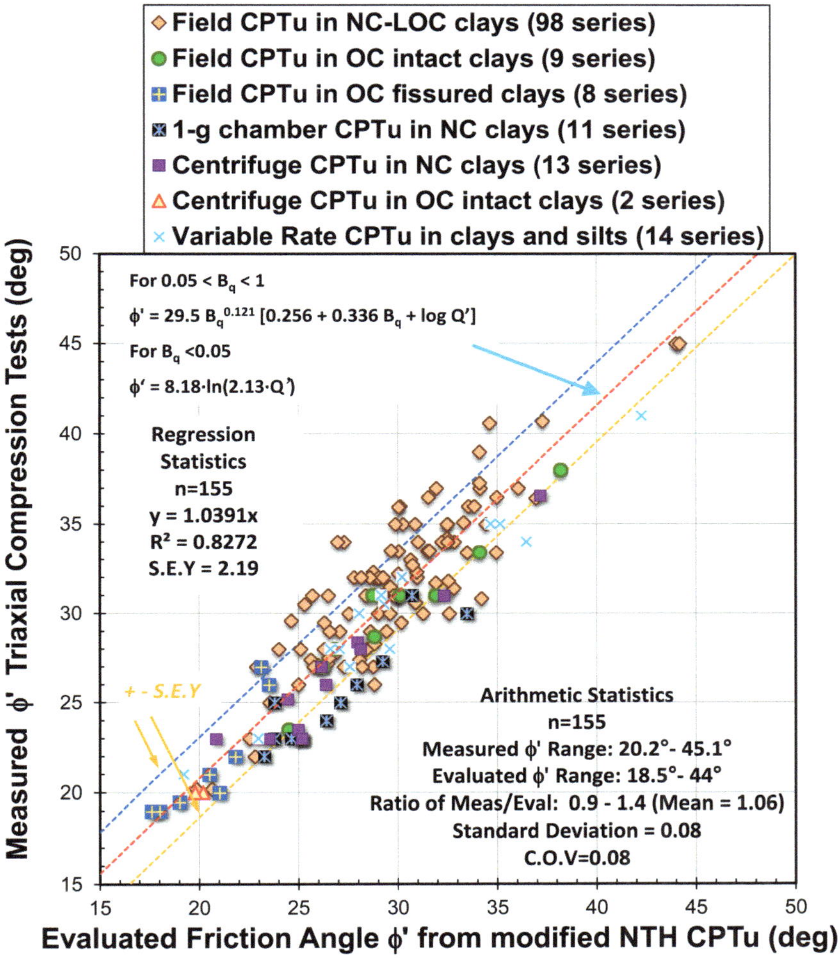

Figure 12.14 Summary plot of laboratory triaxial-measured ϕ' versus CPTu ϕ' from different test series using modified NTH method (Source: modified after Ouyang & Mayne 2023).

clays generally exhibit higher values ($B_q > 0.8$) when compared to normal or well-behaved clays. Therefore, special caution should be taken when evaluating CPTu in organic clays, sensitive or structured soils, or clays of special mineralogy, as the application of the NTH solution should only be undertaken after careful calibration with triaxial tests on high-quality laboratory samples to confirm the appropriate effective stress friction angles are valid in the specific geologic setting of study.

12.3.3 Effective stress strength parameters for intact clays from flat dilatometer test

The flat plate dilatometer test (DMT) was developed in Italy by Silvano Marchetti in the mid-1970s and quickly gained popularity among both practitioners in the geoengineering profession and academicians in geotechnical research who were seeking new and improved methods for site exploration. The DMT is a quick, reliable, and robust in situ test that investigates subsurface characteristics and allows for the interpretation of geoparameters in an efficient, economic, and expedient manner along the soil profile. Details concerning the DMT equipment, field test procedures, and interpretations are given by Marchetti (1980) and ASTM D6635-15, Standard Test Method for Performing the Flat Dilatometer Test (DMT). Updates to the test are discussed by Marchetti et al. (2006) and Marchetti (2015).

For DMT and CPTu soundings in soft to firm clays, a theoretical nexus has been established based on the spherical cavity expansion (SCE) theory by Ouyang and Mayne (2017a, 2017b, 2018). As a result, it is possible to express an equivalent net cone resistance ($q_{\text{net-DMT}}$) and excess porewater pressure (Δu) from DMT soundings as follows:

$$q_{net-DMT} = 2.93p_1 - 1.93p_0 - u_0 \tag{12.10}$$

$$\Delta u_{DMT} = p_0 - u_0 \tag{12.11}$$

where p_0 and p_1 are the contact and expansion pressures from the DMT, respectively.

For soils that are classified as inorganic and insensitive intact clays with OCRs < 2.5, the above relationship is valid. Thus, DMT data can be used to predict the effective friction angle of soft to firm clays using the following equation:

$$\phi' \approx 29.5 \quad (U^*_{DMT}/Q_{DMT})^{0.121}[0.256 + 0.336\,(U^*_{DMT}/Q_{DMT}) + \log Q_{DMT}] \tag{12.12}$$

where

$$U^*_{DMT} = (p_0 - u_0)/\sigma_{v0}' \tag{12.13}$$

$$Q_{DMT} = (2.93p_1 - 1.93p_0 - u_0)/\sigma_{v0}' \tag{12.14}$$

This equation is valid for the following ranges: $20° \le \phi' \le 45°$ and $0 \le U^*_{\text{DMT}} \le 4$.

12.3.3.1 Case studies: Route 17 Bridge, Chesapeake, Virginia

A roadway improvement project was constructed off Route 17 in Chesapeake, Virginia, in the United States between Interstate 64 and Military Highway where the Norfolk Western Railroad crosses the project site. The Chesapeake area is located within the Atlantic Coastal Plain geologic province and underlain by approximately 800 m of marine and estuarine deposits. Standard soil test borings were performed to depths up to 25 m at the site. Recovered undisturbed tube samples were subjected to consolidation testing for soil settlement characteristics. In addition, isotropically consolidated undrained triaxial compression tests (CIUC) were conducted to determine the total and effective stress strength parameters of the primary clay layer. In situ flat dilatometer tests were also carried out at the project site. Field boring performed for this project encountered a subsurface stratigraphy consisting of three basic strata, where upper strata I and bottom strata III show interlayered loose to dense clean to clayey fine sands. The strata II soils consist predominately of soft to firm high to low plasticity clays about 10 m thick (Ouyang and

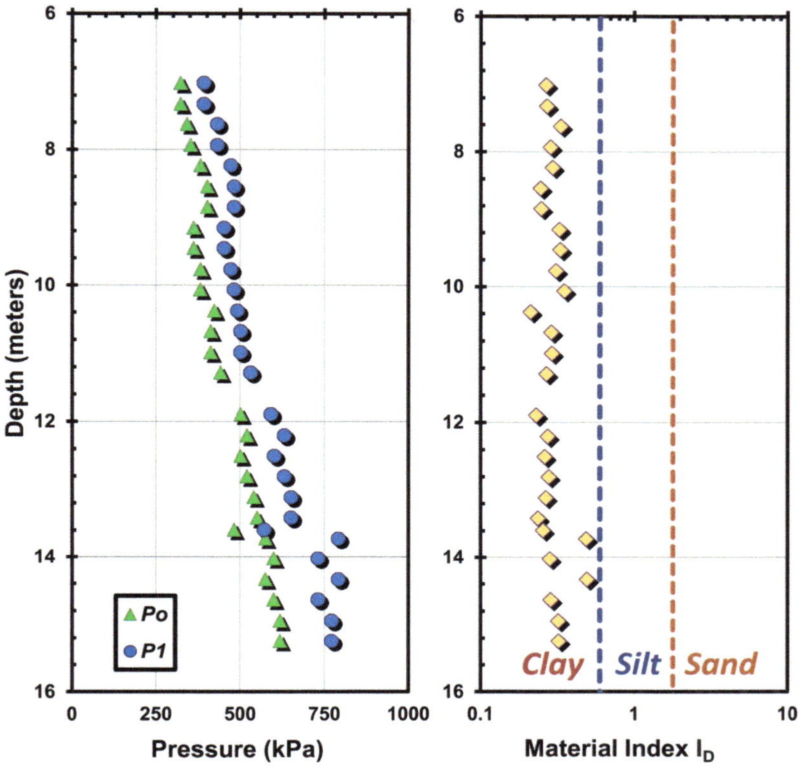

Figure 12.15 Representative DMT sounding at Route 17 bridge site, Chesapeake, VA (Source: after Ouyang and Mayne 2018).

Mayne 2018). This layer created a major problem to approach embankment stability and settlement and characterized by an average unit weight of 15.6 kN/m³. The natural moisture contents ranged from 53% to 92% and the plasticity index (PI) ranged from 31% to 68%. The ground water level was about 1 m below the surface.

Figure 12.15 shows a representative portion of a DMT sounding in the soft clay layer from 6 to 16 m depths with respective contact pressures (p_0) and expansion pressures (p_1). The procedure for determining the equivalent DMT cone resistance number Q_{DMT} is found as the slope from plotting the equivalent DMT net resistance q_{netDMT} versus the effective overburden stress, as illustrated in Figure 12.16a. In this example, we force the line through the origin (assuming $c'=0$) to obtain $Q_{DMT} = 9.73$. By the same token, the equivalent DMT porewater parameter $B_{q\text{-}DMT}$ is determined as the slope of measured $\Delta u_{DMT} = (p_0 - u_0)$ versus q_{netDMT}, giving the value of $B_{q\text{-}DMT} = 0.52$ for the clay layer at this site, as indicated by Figure 12.16b. These values gave an effective friction angle $\phi'=38.6°$ for the soft Chesapeake Clay. The DMT-NTH interpreted ϕ' is then compared with the laboratory value of the effective stress friction angle from a series of CIUC triaxial tests on the soft clay. The results from the NTH method match reasonably with the laboratory data, as shown by Figure 12.17.

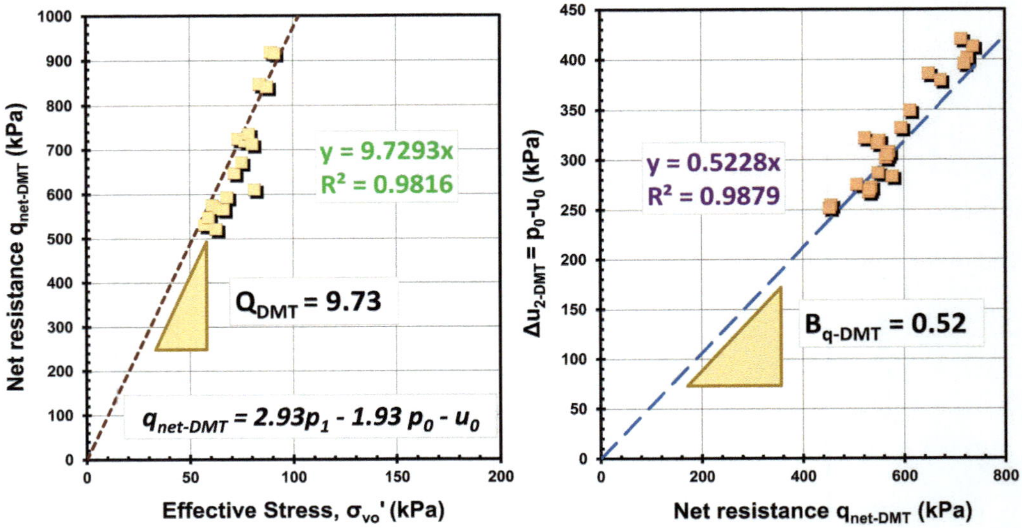

Figure 12.16 Postprocessing of the DMT data in clay layer in Chesapeake, Virginia, for determining (a) DMT-equivalent resistance number, Q_{DMT}; and (b) DMT-equivalent porewater parameter, B_{q-DMT}.

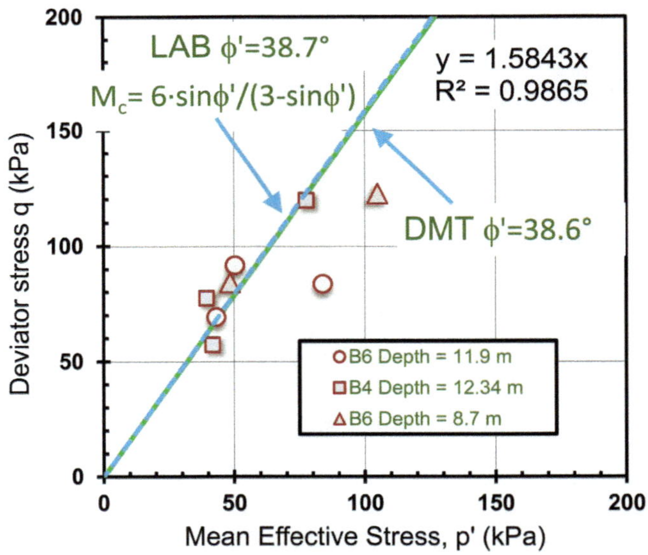

Figure 12.17 Comparison of effective friction angle measured by laboratory triaxial tests and the NTH solution from DMT data in Chesapeake, Virginia.

12.4 DISCUSSION AND APPLICABILITY

12.4.1 Friction angle criteria

For the clays reviewed in the above DMT analysis, the effective stress friction angle ϕ' was obtained from either isotropically and/or anisotropically consolidated undrained laboratory triaxial test (CIUC, CK_0UC, and/or CAUC) on undisturbed samples to establish as the benchmark value. Criteria considering the choice of the laboratory benchmark effective stress friction angle is discussed in Section 12.3. In general, recognized guidelines for the test procedures such as ASTM standards (D4767), Bishop and Henkel (1962), Germaine and Germaine (2009), and Lade (2016) were followed.

12.4.2 Effective cohesion intercept, c′

For this study, the effective cohesion intercept (c') has been taken as zero ($c' = 0$) since the vast majority of the clays considered are NC to LOC with low OCRs < 2.5 (e.g., Finno & Chung 1992; Diaz-Rodriguez et al. 1992). Even for overconsolidated clays, a small value of c' is appropriate (Lade 2016). If a value of c' is necessary, the ratio $c'/\sigma_p' \approx 0.03$ can be used, where σ_p' = effective preconsolidation stress (Larsson et al. 1987; Mayne, 2016).

A recent study by Amundsen and Thakur (2017) investigated the effects of sample disturbance on soft clay parameters. It was found that sample quality, storage time, and testing time between sampling and shearing appreciably affect the magnitude of preconsolidation stress, undrained shear strength, and constrained modulus. However, the magnitude of effective stress friction angle was little affected by these same factors.

12.4.3 Partial drainage effects

The aforementioned approach is valid only for undrained conditions. For a standard penetration rate of 20 mm/s for both CPTu and DMT, it is generally accepted that clay soils with low permeability will experience an undrained penetration process. However, drainage conditions during DMTs in intermediate permeability soils, such as silts, clayey silts, and sandy clays may be different. Schnaid et al. (2016) indicated that the DMT is essentially undrained in soft clay and dominated by penetration pore pressures, whereas the test is drained in clean sands and partially drained in intermediate permeability soils, such as silts.

For the CPTu, special series of tests at variable rates is termed "twitch" testing that allows the demarcation of undrained, partially drained, to fully drained behavior (Randolph 2004). DeJong et al. (2012) give an approximate range of c_v (coefficient of consolidation) that applies to clays, silts, and sands for field operation of CPTu during twitch testing. For clays, the expected range and upper limit for the coefficient of consolidation is approximately $c_v < 0.05$ cm²/s. Thus, for soils with higher values, such as applicable to silts and sands, the SCE nexus in this study would no longer apply.

Also, as noted earlier, CPTu readings q_t and u_2 are taken during penetration at time $t = 0$, whereas the DMT readings p_0 and p_1 are obtained at approximately $t = 15$ s and $t = 45$ s after penetration, respectively. Thus, the method can be invalidated when dissipations occur in soils that do not maintain undrained conditions at constant volume.

Finally, special considerations should also be given when conducting site investigations in more difficult soils, such as highly organic clays, peats, and muskegs, as well as sensitive and structured geomaterials, as the aforementioned relationships may not apply. In highly stratified deposits where clay layers are sandwiched between more permeable silty and sandy layers, then the effects of drainage and partial drainage may also influence the results, especially for thin clay strata. In these cases, caution is warranted.

12.5 SOIL STRESS HISTORY PARAMETERS IN CLAYS

12.5.1 Stress history for intact clays from piezocone penetration test

12.5.1.1 SCE-CSSM solution

A hybrid formulation of spherical cavity expansion (SCE) and critical-state soil mechanics (CSSM) expresses the overconsolidation ratio (OCR) of clays in three separate formulations using net cone tip resistance ($q_{net} = q_t - \sigma_{vo}$), excess pore pressure ($\Delta u = u_2 - u_0$), and effective cone resistance ($q_E = q_t - u_2$). Details of the solution for insensitive inorganic clays are given by Mayne (1991), Chen and Mayne (1994), and Burns and Mayne (1998), whereas the application in sensitive and structured clays is provided by Agaiby and Mayne (2018), Mayne et al. (2018, 2019), DiBuö et al. (2019), Mayne and Benoît (2020), and Agaiby and Mayne (2021).

Three separate algorithms relate the OCR to normalized CPTu parameters: $Q = q_{net}/\sigma_{vo}'$ and $U = \Delta u_2/\sigma_{vo}'$, where $q_{net} = q_t - \sigma_{vo}$ = net cone resistance and $\Delta u_2 = u_2 - u_0$ = excess porewater pressure. These are expressed by the following:

$$OCR = 2 \cdot \left[\frac{Q / M_{c1}}{0.667 \cdot \ln(I_R) + 1.95} \right]^{1/\Lambda} \quad (12.15)$$

$$OCR = 2 \cdot \left[\frac{U^* - 1}{0.667 \cdot M_{c2} \cdot \ln(I_R) - 1} \right]^{1/\Lambda} \quad (12.16)$$

$$OCR = 2 \cdot \left[\frac{Q - \dfrac{M_{c1}}{M_{c2}} \cdot (U^* - 1)}{1.95 \cdot M_{c1} + \dfrac{M_{c1}}{M_{c2}}} \right]^{1/\Lambda} \quad (12.17)$$

where $\Lambda = 1 - C_s/C_c$ = plastic volumetric strain potential, C_s = swelling index, C_c = virgin compression index, $I_R = G/s_u$ = undrained rigidity index, $M = 6 \cdot \sin\phi'/(3 - \sin\phi')$ = frictional parameter in q–p' space. The value of M_{c1} is defined at peak strength (i.e., ϕ' at q_{max}), whereas M_{c2} is the value defined at large strains which occurs at maximum obliquity (i.e., ϕ' when the ratio $[\sigma_1'/\sigma_3']$max). For insensitive clays, the value of ϕ' at q_{max} is equal to ϕ' at $(\sigma_1'/\sigma_3')_{max}$, and thus $M_c = M_{c1} = M_{c2}$. For insensitive clays, the value of $\Lambda = 0.80$, whereas for sensitive clays, a value of $\Lambda \approx 1.0$ is more suitable.

While Equations (12.15) and (12.16) both depend on the I_R of the clay, Equation (12.17) is independent of the I_R and is obtained by combination of the first two formulations. The rigidity index is thus given directly from:

$$I_R = \exp\left[\frac{1.5 + 2.925 \cdot M_{c1} \cdot a_q}{M_{c2} - M_{c1} \cdot a_q}\right] \tag{12.18}$$

where $a_q = (U - 1)/Q = (u_2 - \sigma_{vo})/(q_t - \sigma_{vo})$. The parameter a_q can be determined as a single value for any clay layer or uniform clay deposit by taking the slope of a plot of the parameter $(U - 1)$ versus Q, or alternatively taken as the slope of $(u_2 - \sigma_{vo})$ versus $(q_t - \sigma_{vo})$. Using regression analyses, slightly different slope values for a_q are obtained.

12.5.1.2 Simplified approach for insensitive and inorganic clays

A series of simplifications can be made for insensitive inorganic clays, or "regular" and "normal" clays. For one, Equation (12.16) can be approximated by:

$$OCR \approx 2 \cdot \left[\frac{U}{0.667 \cdot M_{c2} \cdot \ln(I_R)}\right]^{1/\Lambda} \tag{12.19}$$

As noted previously for regular clays, the values $M_c = M_{c1} = M_{c2}$, therefore Equation (12.17) reduces to:

$$OCR = 2 \cdot \left[\frac{1}{1.95M_c + 1}\left(\frac{q_t - u_2}{\sigma_{vo}'}\right)\right]^{1/\Lambda} \tag{12.20}$$

For a first-order estimate of σ_p' in regular clays, further simplifications are achieved by (a) setting the exponent $\Lambda = 1$ to reduce the power law format to linear equations; (b) adopting a characteristic effective friction angle of clay $\phi' = 30°$ ($M_c = 1.2$), and (c) using a default value of $I_R = 100$ (Houlsby & Teh 1988; Mayne 2007). The reduced expressions become:

$$\begin{aligned} s_p' &\approx 0.33 q_{\text{net}} = 0.33(q_t - s_{vo}) \\ s_p' &\approx 0.53 Du_2 = 0.53(u_2 - u_o) \\ s_p' &\approx 0.60 q_E = 0.60(q_t - u_2) \end{aligned} \tag{12.21}$$

$$\sigma_p' \approx 0.53 \Delta u_2 = 0.53\ (u_2 - u_o) \tag{12.22}$$

$$\sigma_p' \approx 0.60\ q_E = 0.60\ (q_t - u_2) \tag{12.23}$$

12.5.2 Stress history for sensitive clays from piezocone penetration test

A review of the CPTu data on different selective sensitive soils as presented in Agaiby and Mayne (2021) show that when the three expressions from Equations (12.21)–(12.23) are used, unmatched OCR profiles among each other and evident disagreement when compared to lab reference values from consolidation tests are present. Moreover, a consistent hierarchy can be observed:

$$0.60\ q_E < 0.33\ q_{net} < 0.53\Delta u_2 \tag{12.24}$$

which gives the signature identification of sensitive clays.

To assess the OCR from CPTu in sensitive clays, Equations (12.15)–(12.17) are utilized with corresponding M_{c1} and M_{c2} that can be obtained from the original NTH expression (Ouyang & Mayne 2018) and modified NTH expression (Ouyang & Mayne 2019), respectively. Results from laboratory triaxial compression tests also confirm and validate these values.

12.5.3 Stress history for organic clays from piezocone penetration test

A review of the CPTu data on different organic soils show that when the three expressions from Equations (12.21)–(12.23) are used, unmatched OCR profiles occur in the following hierarchal order (Mayne & Agaiby 2019):

$$0.53\Delta u_2 < 0.33\ q_{net} < 0.60\ q_E \tag{12.25}$$

Therefore, Equation (12.25) serves as the CPTu signature that is characteristic of organic fine-grained soils.

For evaluating the yield stress of soft organic soils by CPTu, it has been recommended to lower the coefficients of Equations (12.21)–(12.23). This could be justified because organic clays and peats exhibit rather high friction angles ($\phi' > 40°$) when compared to regular clays. By use of a higher friction angle (e.g., $M_c > 1.2$) in Equations (12.15)–(12.17) and 12.20, the resulting set of simplified Equations (12.21)–(12.23) would have smaller coefficients for q_{net}, Δu, and q_E.

For instance, for the soft organic clays of Brazil, modified expressions have been developed (Baroni and Almeida, 2017):

$$\sigma_p' = 0.125\ q_{net} \tag{12.26}$$

$$\sigma_p' = 0.154\ q_E \tag{12.27}$$

In another approach, for CPTu in a variety of soil types, a generalized solution retains the 0.33 coefficient of Equation (12.21) and employs a power law algorithm in the form (Mayne, 2017; Agaiby and Mayne, 2019):

$$\sigma_p' = 0.33 q_{net}^{m'} (units\ of\ kPa) \tag{12.28a}$$

where the exponent m' depends upon the soil behavioral type (= 1.0 intact inorganic clays; 0.90 organic clays; 0.85 silt mixtures; 0.80 silty sands; and 0.72 clean quartz–silica sands). Figure 12.18 shows the yield stresses σ_p' for various soil types plotted versus the net cone resistance. The corresponding expression in dimensionless terms is given by:

$$\sigma_p' = 0.33\ q_{net}^{m'} (s_{atm} / 100)^{1-m'} \tag{12.28b}$$

where σ_{atm} = reference pressure equal to 1 atm ≈ 1 bar = 100 kPa.

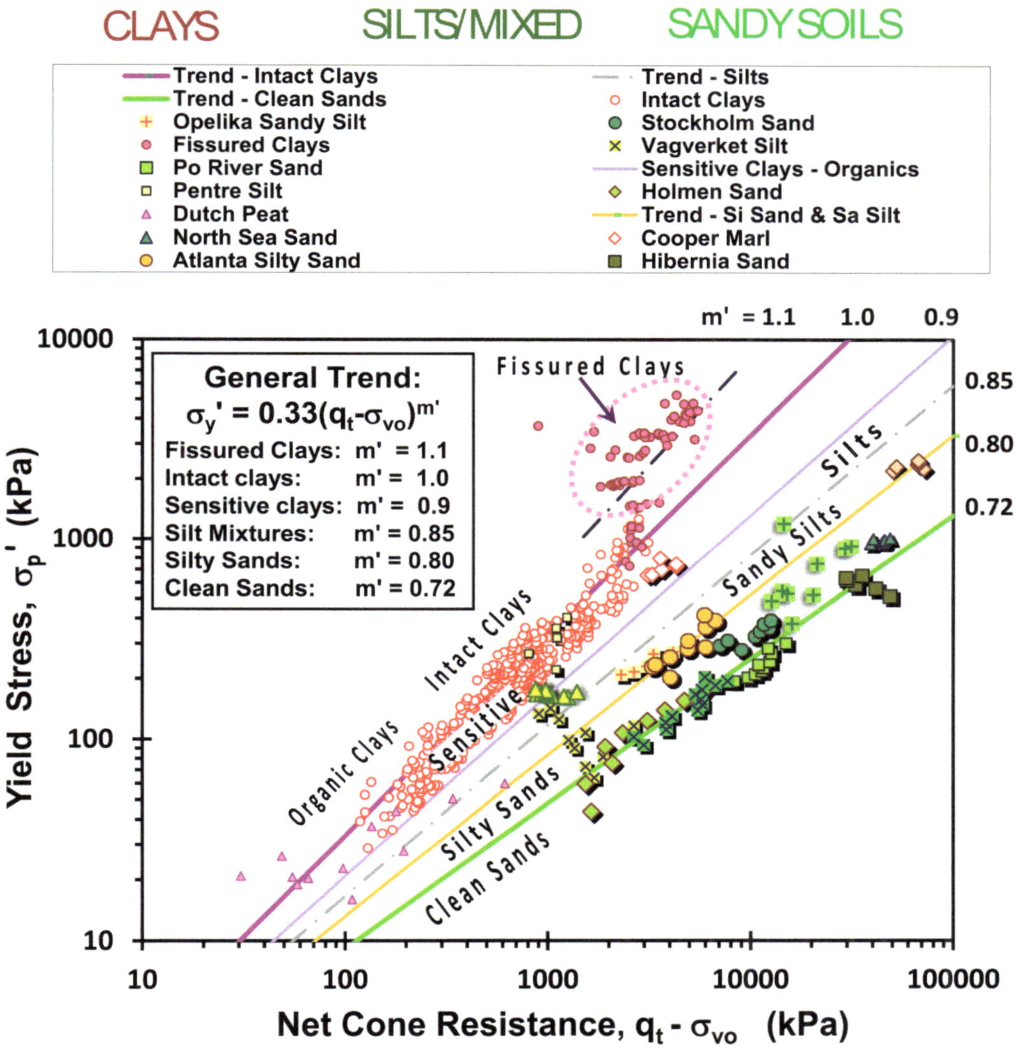

Figure 12.18 General relationship for yield stress of soils and CPT net cone resistance in different geomaterials (Source: modified after Agaiby and Mayne, 2019).

REFERENCES

Agaiby, S. S. and Mayne, P. W. (2018). Evaluating undrained rigidity index of clays from piezocone data. *Proc. 4th Intl. Symposium on Cone Penetration Testing*, TU Delft, Taylor & Francis, CRC Press, 65–72.

Agaiby, S. S. and Mayne, P. W. (2021). CPTU identification of regular, sensitive, and organic clays towards evaluating preconsolidation stress profiles. *AIMS Geosciences*, 7 (4): 553–573.

Amundsen, H. A., Jønland, J., Emdal, A., and Thakur, V., 2017. An attempt to monitor pore pressure changes in a block sample during and after sampling. *Géotechnique Letters,* 7, 2, 119–128. https://doi.org/10.1680/jgele.16.00176

Baroni, M. and Almeida, M. S. S. (2017). Compressibility and stress history of very soft organic clays. *Proceedings of the Institution of Civil Engineers, London—Geotechnical Engineering*, 170 (2): 148–160.

Bishop, A. W. and Henkel, D. J. (1962). *The Measurement of Soil Properties in the Triaxial Test.* 2nd edition, Edward Arnold Publishers Ltd., London. 227 p.

Bjerrum, L. and Simons, N. E. (1960). *Comparison of Shear Strength Characteristics of Normally Consolidated Clays*, presented at the Research Conference on Shear Strength of Cohesive Soils, Boulder, CO. Reston, VA: American Society of Civil Engineers, 711–726.

Burns, S. E. and Mayne, P. W. (1998). Monotonic and dilatory porewater pressures during piezocone dissipation tests in clay. *Canadian Geotechnical Journal*, 35 (6): 1063–1073.

Chen, B. Y. and Mayne, P. W. (1994). *Profiling the Overconsolidation Ratio of Clays by Piezocone Tests*, Report No. GIT-CEE/GEO-94-1 submitted to National Science Foundation by Georgia Institute of Technology, Atlanta, 280pp.

Chung, C. K. and Finno, R. J. (1992). Influence of depositional processes on the geotechnical parameters of Chicago glacial clays. *Engineering Geology*, 32 (4): 225–242.

Di Buò, B., D'Ignazio, M., Selãnpaã, J., Länsivaara, T., and Mayne, P. W. (2019). Yield stress evaluation of Finnish clays based on analytical CPTU models. *Canadian Geotechnical Journal*, 57 (11): 1623–1638.

DeJong, J. T., Jaeger, R. A., Boulanger, R. W., Randolph, M. F., and Wahl, D. A. J. (2012). Variable penetration rate cone testing for characterization of intermediate soils. *Geotechnical and Geophysical Site Characterization 4.* Vol. 1, Taylor and Francis, London: 25–42.

Diaz-Rodriguez, J. A., Leroueil, S., and Aleman, J. D. (1992). Yielding of Mexico City clay and other natural clays. *Journal of Geotechnical Engineering*, 118(7): 981–995. doi:10.1061/(ASCE)0733-9410(1992)118:7(981).

Finno, R. J., Gassman, S. L., and Calvello, M. (2000). *The NGES at Northwestern University. National Geotechnical Experimentation Sites (GSP 93).* Reston, VA: ASCE, 130–159.

Germaine, J. T., and Germaine, A. V. (2009). *Geotechnical Laboratory Measurements for Engineers.* John Wiley & Sons, New York. 359 p.

Hardison, M. A. and Landon, M. (2015). Correlation of engineering parameters of the Presumpscot Formation to the seismic cone penetration test (SCPTu), Technical Report 15-12, Department of Civil Engineering, University of Maine, submitted to Maine Dept. of Transportation: 393 p.

Hight, D. W. and Jardine, R. J. (1993). Small-strain stiffness and strength characteristics of hard London tertiary clays. *Geotechnical Engineering of Hard Soils & Soft Rocks*, Balkema, Rotterdam: 533–552.

Holtz, R. D., Kovacs, W. D., and Sheahan, T. C. (2010). *An Introduction to Geotechnical Engineering*, 2nd ed. Pearson, NJ: Prentice-Hall.

Houlsby, G. T. and Teh, C. I. (1988). Analysis of the piezocone in clay. *Penetration Testing,* Vol. 2 (Proc. ISOPT-1, Orlando). Balkema, Rotterdam, 777–783.

Janbu, N. and Senneset, K. (1974). Effective stress interpretation of in-situ static penetration tests. *Proceedings of the 1st European Symposium on Penetration Testing*, Vol. 2. Stockholm: Swedish Geotechnical Society, 181–193. Download available at: www.usucger.org

Kulhawy, F. H. and Mayne, P. W. (1990). *Manual on Estimating Soil Properties for Foundation Design.* Report EL-6800, Palo Alto, CA: Electric Power Research Institute.

Lade, P. V. (2016). *Triaxial Testing of Soils.* Hoboken, NJ: John Wiley & Sons, Ltd., 402pp.

Lambe, T. W. and Whitman, R. V. (1979). *Soil Mechanics, SI version.* New York: Wiley.

Larsson, R., Bergdahl, U., and Eriksson, L. (1987). Evaluation of Shear Strength in Cohesive Soils with Special Reference to Swedish Practice and Experience. *Geotech. Test. J.*, Vol. 10, No. 3, pp. 105–112, https://doi.org/10.1520/GTJ10942J

Leroueil, S. and Hight, D. W. (2003). Behaviour and properties of natural soils and soft rocks. *Characterisation and Engineering Properties of Natural Soils Workshop*, Vol 1. (Singapore), Balkema, Rotterdam, the Netherlands, 29–254.

Loehr, J. E., Lutenegger, A. J., Rosenblad, B. and Boeckmann, A. (2016). *Geotechnical Site Characterization (GEC 5)*. Report FHWA NHI-16-072, Washington, DC: National Highway Institute, Federal Highway Administration.

Lunne, T., Eidsmoen, T., Powell, J. and Quarterman, R. (1986). Piezocone testing in overconsolidated clays. *Proceedings, 39th Canadian Geotechnical Conference: In-Situ Testing and Field Behavior*. Ottawa, 209–218.

Marchetti, S. (1980). In-situ tests by flat dilatometer. *Journal of the Geotechnical Engineering Division* ASCE 106 (GT3): 299–321.

Marchetti, S. (2015). Some 2015 updates to the TC16 DMT report 2001. *Proc. 3rd Int. Conf. on the Flat Dilatometer DMT'15*. Rome, Italy. www.dmt15.com

Marchetti, S., Monaco, P., Calabrese, M., and Totani, G. (2006). Comparison of moduli determined by DMT and back figured from local strain measurements under a 40 m diameter circular test load in the Venice area. *Proc. 2nd Int. Conf. on the Flat Dilatometer*, Washington DC: pp. 220–230.

Mayne, P. W. (2016). Evaluating effective stress parameters and undrained shear strengths of soft-firm clays from CPT and DMT. *Aust. Geomech. J.* 51 (4): 27–55.

Mayne, P. W. and Stewart, H. E. (1988). Pore pressure response of K0-consolidated clays. *Journal of Geotechnical Engineering*, 114 (11): 1340–1346.

Mayne, P. W. (1991). Determination of OCR in clays by piezocone tests using cavity expansion and critical state concepts. *Soils and Foundations*, 31(2): 65–76.

Mayne, P. W. (2007). *In-Situ Test Calibrations for Evaluating Soil Parameters, Characterization & Engineering Properties of Natural Soils*, Vol. 3. London: Taylor & Francis, 1602–1652.

Mayne, P. W. (2017). Stress history of soils from cone penetration tests. *34th Manual Rocha Lecture, Soils & Rocks*, Vol. 40 (3). Saö Paulo: Brazilian Society for Soil Mechanics, 203–218. http://www.soilsandrocks.com.br/

Mayne, P. W. (2014). Keynote: interpretation of geotechnical parameters from seismic piezocone tests. *Proceedings of the 3rd International Symposium on Cone Penetration Testing*. Las Vegas, NV, 47–73.

Mayne, P. W. and Agaiby, S. (2019). Profiling yield stresses and identification of soft organic clays using piezocone tests. *Proceedings XVI Pan American Conference on Soil Mechanics & Geotechnical Engineering*, Paper 0149, Cancun, Mexican Society of Geotechnical Engineering (SMIG): www.issmge.org

Mayne, P. W. and Benoît, J. (2020). Analytical CPTU Models Applied to Sensitive Clay at Dover, New Hampshire. *Journal of Geotechnical and Geoenvironmental Engineering*, 146 (12): 04020130.

Mayne, P. W., Christopher, B., Berg, R., and DeJong, J. (2002). Subsurface Investigations - Geotechnical Site Characterization. Publication No. FHWA-NHI-01-031, National Highway Institute, Federal Highway Administration (FHWA). Washington, DC: U.S. Department of Transportation.

Mayne, P. W., Greig, J., and Agaiby, S. (2018). Evaluating CPTu in sensitive Haney clay using a modified SCE-CSSM solution. *Proceedings 71st Canadian Geotechnical Conference: GeoEdmonton*, Paper ID No. 279, Canadian Geotechnical Society: www.cgs.ca

Mayne, P. W., Coop, M. R., Springman, S., Huang, A. -B., and Zornberg, J. (2009). State-of-the-art paper (SOA-1): geomaterial behavior and testing. *Proceedings of the 17th International Conference on Soil Mechanics & Geotechnical Engineering*, Vol. 4 (ICSMGE, Alexandria). Rotterdam: Millpress/IOS Press, 2777–2872.

Mayne, P. W., Paniagua, P., L'heureux, J. -S., Lindgård, A., and Emdal, A. (2019). Analytical CPTu model for sensitive clay at Tiller-Flotten site, Norway. *Proc. XVII ECSMGE: Geotechnical Engineering Foundation of the Future*, Paper 0153, Reykjavik, Icelandic Geotechnical Society: www.issmge.org

Mesri, G. and Abdel-Ghaffar, M. E. M. (1993). Cohesion intercept in effective stress-stability analysis. *Journal of Geotechnical Engineering*, 119 (8): 1229–1249.

Ouyang, Z. and Mayne, P. W. (2018a). Calibration of NTH method for friction angle using centrifuge CPTUs in clays. *Proceedings 4th International Symposium on Cone Penetration Testing (CPT'18, Delft)*, London: Taylor & Francis Group, www.cpt18.org

Ouyang, Z. and Mayne, P. W. (2018b). Effective friction angle of clays and silts from piezocone. *Canadian Geotechnical Journal*, 55(9): 1230–1247.

Ouyang, Z. and Mayne, P. W. (2019). Modified NTH method for assessing effective friction angle of normally consolidated and overconsolidated clays from piezocone tests. *ASCE Journal of Geotechnical & Geoenvironmental Engineering*, 145(10): doi.org/10.1061/ (ASCE) GT.1943-5606.0002112

Ouyang, Z. and Mayne, P. W. 2018c. Effective stress strength parameters of clays from DMT. ASTM *Geotechnical Testing Journal*. 41 (5): 851–867.

Ouyang, Z. and Mayne, P. W. (2023). Evaluating friction angles for clays: piezocone tests versus Atterberg limits. *Geotechnical Engineering, Proc. Institution of Civil Engineers* 177(2), 147–157.

Ouyang, Z., Mayne, P. W. (2019). Evaluating rigidity index, OCR, and su from dilatometer data in soft to firm clays. *Proceedings, 16th Panamerican Conference on Soil Mechanics and Geotechnical Engineering*. Cancun: Mexican Geotechnical Society. www.issmge.org

Ouyang, Z. and Mayne, P. W. (2016). New DMT method for evaluating soil unit weight in soft to firm clays. *Geotechnical and Geophysical Site Characterisation*, Lehane, B., Acosta-Martínez, HE, Kelly, R., *Eds*, 785–790.

Ouyang, Z. and Mayne, P. W. (2017a). Spherical cavity expansion nexus between CPTu and DMT in soft-firm clays. *Proceedings of the 19th International Conference on Soil Mechanics and Geotechnical Engineering*, Vol. 1. Seoul: Korean Geotechnical Society, 631–634. www.issmge.org

Ouyang, Z. and Mayne, P. W. (2017b). Effective friction angle of soft to firm clays from flat dilatometer tests. *Geotechnical Engineering*, Vol. 170 (2). London: Proc. Institution of Civil Engineers, 137–147.

Ouyang, Z. and Mayne, P. W. (2020). Modified NTH solution for overconsolidated fissured clays. *Proc. Sixth International Conference on Geotechnical and Geophysical Site Characterization ISC-6*.isc6.org.

Ouyang, Z. and Mayne, P. W. (2021). Variable rate piezocone data evaluated using NTH limit plasticity solution. *ASTM Geotechnical Testing Journal*. 44. (1), 174–190. https://doi.org/10.1520/GTJ20190183.

Powell, J. J. M. and Quarterman, R. S. T. (1988). The interpretation of cone penetration tests in clays with particular reference to rate effects. *Penetration Testing*, Vol. 2 (Proc. ISOPT, Orlando). Rotterdam: Balkema, 903–910.

Randolph, M. F. (2004). Characterization of soft sediments for offshore applications. *Geotechnical & Geophysical Site Characterization*, Vol. 1 (Proc. ISC-2, Porto), Millpress, Rotterdam: 209–232.

Sandven, R. (1990). *Strength and Deformation Properties Obtained from Piezocone Tests*. PhD thesis, Trondheim: Norwegian University of Science & Technology, 342pp.

Schnaid, F., Odebrecht, E., Sosnoski, J., and Robertson, P. K. (2016). Effects of Test Procedure on Flat Dilatometer Test (DMT) Results in Intermediate Soils. *Can. Geotech. J.*, 53, 8, 1270–1280, https://doi.org/10.1139/cgj-2015-0463

Schofield, A. and Wroth, P. (1968). *Critical State Soil Mechanics*. London: McGraw-Hill.

Sellountou, E. A., Vipulanandan, C., and O'Neill, M. W. (2004). Estimation of drained shear strength parameters of Beaumont overconsolidated clay from piezocone data. *Proc., Center for Innovative Grouting Materials and Technology (CIGMAT)*. Houston: Dept. of Civil Engineering, Univ. Houston.

Senneset, K., Sandven, R., and Janbu, N. (1989). Evaluation of soil parameters from piezocone tests. *Transportation Research Record*, Vol. 1235. Washington DC: National Academies Press, 24–37.

Whittle, A. J., DeGroot, D. J., Ladd, C. C., and Seah, T-H. (1994). Model prediction of anisotropic behavior of Boston Blue Clay. *J. Geotech. Eng.* 120 (1): 199–224.

Chapter 13

Mechanical–statistical evaluation of soil properties

Changhong Wang, Kun Wang, Daofei Tang, and Tian Fang

This chapter delves into the mechanical-statistical evaluation of soil properties, emphasizing the utility of in-situ testing techniques like the Piezocone Penetration Test (CPTU) for assessing physical state and mechanical characteristics of soil layers. Despite fruitful applications of the empirical and regression formulas, their universality is constrained by regional soil property variations. The chapter underscores the significance of appropriate constitutive models and accurate geotechnical parameters in soil mechanics analysis, advocating for a more integrated approach leveraging CPTU data. The text outlines methods for evaluating cohesive and cohesionless soil properties using CPTU data, detailing the estimation of state indexes, strength indexes, and deformation indexes. Further, the chapter explores the application of constitutive models like the Modified Cam Clay (MCC) and Clay and Sand Model (CASM) within the cavity expansion theory to interpret CPTU data accurately. The integration of Bayesian-random field theory in geotechnical parameter evaluation is also discussed, highlighting its potential in accounting for natural variability and uncertainty in geotechnical media. The chapter presents stochastic mechanics methods, including Monte Carlo simulation and the Response Surface Method (RSM) to simulate the stochasticity and spatial variability of geotechnical parameters.

13.1 EVALUATION METHOD OF SOIL PROPERTIES BASED ON CPTU

In-situ testing techniques such as the Piezocone Penetration Test (CPTU) provide high-quality, reproducible data, and can directly evaluate the physical state and mechanical properties of soil layers, which is suitable for engineering exploration activities in deep underground space. The interpretation of CPTU data is a long-term important task. At present, CPTU data has achieved fruitful results in soil layer classification and engineering classification, soil physical and mechanical indexes, soil liquefaction discrimination, and other fields. However, these results are mainly based on empirical and regression formulas established from laboratory experiments and field data. Due to the difference in soil properties of the different areas, the universality of the empirical formula is low. Therefore, CPTU test data have not been fully applied in geotechnical engineering.

Proper constitutive models and accurate geotechnical parameters are the two supports of soil mechanics analysis. CPTU data based on the constitutive model can quickly and accurately obtain the geotechnical parameters required for scientific calculation.

DOI: 10.1201/9781003441946-13

13.1.1 Main mechanical characteristics of cohesive soil

CPTU data can be used to evaluate and estimate the major properties of soil, including state indexes, strength indexes, and deformation indexes. The natural weight-specific density and overconsolidation ratio (OCR) are the main state indexes, the undrained shear strength is the main strength index, and the compression modulus and undrained Young's modulus are the main deformation indexes. This section introduces the evaluation method of the major properties of clay based on the CPTU data.

13.1.1.1 Natural weight-specific density of soil

Based on the analysis of experimental data from Switzerland, Norway, and the United Kingdom, Larsson and Mulabdic[1] proposed a rough estimation of soil density by using net cone-tip resistance $(q_t - \sigma_{v0})$ and pore pressure parameter B_q.

Mayne[2] proposed that the natural weight-specific density of soil is related to shear wave velocity and depth, which is described as follows:

$$\gamma_t = 8.63 \lg V_s - 1.18 \lg z - 0.53, \tag{13.1}$$

where γ_t is the natural weight (kN/m³), V_s is the shear wave velocity (m/s), and z is the depth (m). The statistical data of the formula include clay and incohesive soil, and the shear wave velocity measured by the seismic CPTU can be used to estimate the natural weight of the soil and determine the density of the soil.

13.1.1.2 Overconsolidation ratio (OCR)

In general, for the mechanically overconsolidated soil, the OCR is defined as the maximum effective consolidation stress of the soil layer in the history, while for the structural soil layer, OCR may represent the ratio of the yield stress to the current effective stress. The methods for estimating OCR from Cone Penetration Test (CPT) and CPTU data can be roughly divided into three categories: estimating OCR by undrained shear strength S_u, estimating OCR by the range of cone-tip resistance q_t, and estimating OCR directly from CPTU data.

For normally consolidated clays, the normalized cone-tip resistance varies with the plasticity index I_p and generally varies in the following ranges:

$$Q_t = \frac{q_t - \sigma_{v0}}{\sigma'_{v0}} = 2.5 \sim 5.0 (\text{depends on } I_P). \tag{13.2}$$

If the soil layer is beyond the range of the above calculation, the soil layer may be overconsolidated.

13.1.1.3 Undrained shear strength

The undrained shear strength of soil (S_u) depends on the failure mode of the soil, the heterogeneity of the soil, and stress history. Hence, for the same soil, different geotechnical test methods will obtain different S_u values. At present, there are two kinds of methods to obtain undrained shear strength by using CPTU: theoretical method and empirical formula method.

13.1.1.4 Compression modulus

The compression modulus E_s of soil based on CPTU data can be expressed as a function of net cone-tip resistance.

For the overconsolidated clay, Senneset et al.[3][4] suggest a linear model:

$$E_{si} = \alpha_i q_n = \alpha_i (q_t - \sigma_{v0}), \tag{13.3}$$

where α_i is the coefficient, which varies from 5 to 15 for most clays.

For normally consolidated clay, a similar relationship exists:

$$E_s = \alpha_n q_n = \alpha_n (q_t - \sigma_{v0}), \tag{13.4}$$

According to Senneset et al.[4], the value α_n ranges from 4 to 8.

Kulhawy and Mayne[5] proposed a more general relationship:

$$E_s = 8.25(q_t - \sigma_{v0}). \tag{13.5}$$

13.1.1.5 Undrained Young's modulus

When using CPT or CPTU data to estimate the undrained Young's modulus E_u, the empirical formula is generally used to estimate indirectly:

$$E_u = nS_u, \tag{13.6}$$

where n is a constant, depending on factors such as shear stress level, overconsolidation ratio, and soil layer sensitivity.

13.1.2 Main mechanical characteristics of cohesionless soil

CPTU probes are usually inserted in the undrained clay and silt. When the CPTU is performed under the undrained conditions, the pore pressure generated near the probe will affect the measured values of cone-tip resistance q_c and sleeve friction f_s. The CPTU in coarse-grained soil such as sandy soil is penetrated under drainage conditions, and none of the excess pore pressure is generated during penetration, that is, only in-situ static pore pressure can be measured. Therefore, only the q_c and f_s can be used for the interpretation of the test data under the condition of complete drainage. However, due to the high compressive stress near the probe, pore pressure can be measured at the probe tip position u_1 [6]. In very dense sand or silt, negative pore pressure may be recorded at the probe shoulder position u_2 due to the soil dilatancy.

13.1.2.1 Relative compactness

Relative compactness is the main state characteristic of cohesionless soil and is usually used as an intermediate soil parameter. Relative compactness D_r or compactness index I_D is defined as:

$$I_D = D_r = \frac{e_{max} - e}{e_{max} - e_{min}}, \tag{13.7}$$

where $e_{\max}$ and $e_{\min}$ are the maximum and minimum pore ratios, respectively; and e is the in-situ porosity ratio.

The large-scale calibration tank test shows that the cone-tip resistance depends on the density of sand, in-situ vertical stress, horizontal effective stress, and compressibility of sand. According to the sand calibration tank test, different scholars put forward a series of prediction formulas D_r.

13.1.2.2 Effective internal friction angle of sand

There are many methods to evaluate the effective internal friction angle ϕ' of sand based on the cone-tip resistance of CPTU, which can be divided into empirical correlation, bearing capacity theory, and cavity expansion theory. For the pure sand, the effective stress friction angle ϕ' is directly evaluated when the drainage is complete, and the effective cohesion is assumed to be zero (i.e., $c' = 0$). To normalize the stress level of the cone-tip resistance q_t, the following formula[7] is proposed for estimation:

$$\phi' = 17.6^\circ + 11 \log q_{t1}, \tag{13.8}$$

where $q_{t1} = (q_t / \sigma_{\text{atm}}) / (\sigma'_{v0} / \sigma_{\text{atm}})^{0.5}$ is the normalized cone-tip resistance.

13.1.2.3 Young's modulus

The study on the test data of the calibration tank shows that the drained Young's modulus of sand mainly depends on the relative density, overconsolidation ratio, and the current stress level. For the secant Young's modulus (E'_s), when the mean axial strain over the stress history and aging range is 0.1%, the stiffness of the normally consolidated aged sand (i.e., >1000 years) falls between the newly normal consolidated sand and the over-consolidated soil[8].

13.2 CONSTITUTIVE MODEL AND CAVITY EXPANSION THEORY

The dynamic process of CPTU is accompanied by large deformation of the soil around the cone tip, which makes it very difficult to calculate the stress and displacement of the soil. The major approximate theories are bearing capacity theory, moving point dislocation method, pore expansion theory, and strain path method. The pore expansion theory not only considers the elastic deformation, plastic deformation, and initial stress state of soil, but also can describe the mechanical behavior of CPTU more accurately than other theoretical methods.

Cavity expansion theory can be divided into cylindrical cavity expansion and spherical cavity expansion methods. The similarity between cavity expansion theory and CPTU was first proposed by Bishop[9] et al. Gibson and Anderson[10] introduced the theory to geotechnical engineering. The study of cavity expansion theory needs to consider the theoretical knowledge of equilibrium equations, geometric equations, and constitutive models, among which the most important factor is the latter. In recent years, many excellent constitutive models have emerged, from the initial linear elastic constitutive model to the ideal elastic–plastic model, and then to various critical elastic–plastic constitutive models commonly used today, which can more accurately describe the complex stress state of soil. As classical critical state models, the Modified Cam Clay (MCC) model and Clay And Sand Model (CASM) are widely used to describe the mechanical properties of normal consolidated clay and sand, respectively.

13.2.1 Cavity expansion theory of the MCC model

Different from other laboratory tests, CPT does not directly measure the mechanical properties of soil and it should be interpreted. It is required to convert the force on the probe into geotechnical parameters. The penetration process of CPT can be considered as the radial compression of soil under the action of cone indentation, which is highly consistent with the cylindrical cavity expansion theory. Because of the succinct mechanical principle and simple solution form, the cavity expansion theory is widely used in geotechnical engineering, such as in-situ CPTU and pressuremeter tests. According to cylindrical cavity expansion theory, the mathematical relationship between CPTU data and constitutive parameters can be established.

13.2.1.1 Definition of the problem

A cylindrical cavity with an initial radius a_0 in the infinite soil mass is subjected to the radial stress σ_r, the tangential stress σ_θ, and the vertical stress σ_z. In the initial conditions, $\sigma_{r0} = \sigma_{\theta 0} = \sigma_{\mathrm{h}}$, $\sigma_{z0} = \sigma_{\mathrm{v}}$, where σ_{h} and σ_{v} are the in-situ horizontal stress and gravitational stress, respectively. When the inner diameter a_0 of the cylindrical cavity expands to a, the expansion pressure correspondingly increases from σ_{r0} to σ_a. Under the action of expansion pressure, the cavity wall firstly yields and enters the plastic region subsequently. As the inner diameter increases, the radius of the plastic area expands from R_{p0} to R_{p}. At that time, two regions appear around the cavity wall, the internal plastic region and the external elastic region. The schematic diagram of cylindrical cavity expansion is shown in Figure 13.1.

Cylindrical cavity expansion is a simplification of the axisymmetric problem in space, where the radial stress σ_r and tangential stress σ_θ are assumed to be only the functions

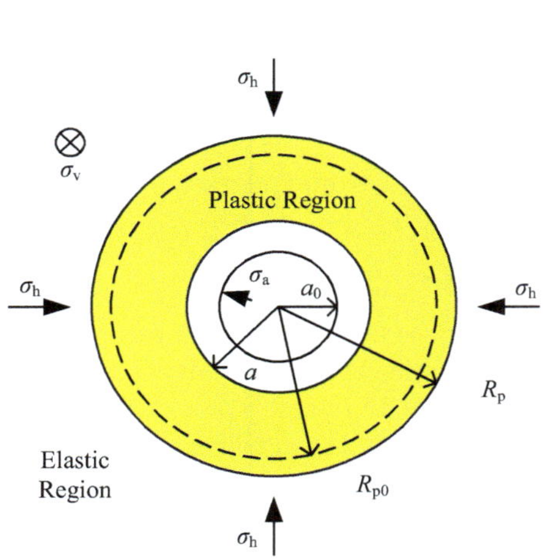

Figure 13.1 Schematic diagram of cylindrical cavity expansion.

of radius r, and the effects of shear stress $\tau_{r\theta}$ are ignored. In both elastic and plastic regions, the equilibrium differential equation scan be written as,

$$\frac{d\sigma_r}{dr} + \frac{\sigma_r - \sigma_\theta}{r} = 0. \tag{13.9}$$

Alternatively,

$$\frac{d\sigma'_r}{dr} + \frac{du_w}{dr} + \frac{\sigma'_r - \sigma'_\theta}{r} = 0, \tag{13.10}$$

where σ'_r and σ'_θ are the radial and tangential effective stress, respectively, and u_w is the pore pressure.

13.2.1.2 Elastic analysis

Cylindrical cavity expansion theory is subjected to the generalized Hooke's law in the elastic region. According to the elastic/plastic boundary conditions $r = R_p$ and $\sigma_r = \sigma_{rp}$, radial stress σ_r, tangential stress σ_θ, and vertical stress σ_z can be easily obtained in the elastic region, which can be written as[11],

$$\begin{aligned} \sigma_r &= \sigma_h + (\sigma_{rp} - \sigma_h)\left(\frac{R_p}{r}\right)^2 \\ \sigma_\theta &= \sigma_h - (\sigma_{rp} - \sigma_h)\left(\frac{R_p}{r}\right)^2, \\ \sigma_z &= \sigma_v \\ u_r &= \frac{\sigma_{rp} - \sigma_h}{2G_0}\frac{R_p^2}{r} \end{aligned} \tag{13.11}$$

where σ_{rp} is the total radial stress at the boundary of the elastic/plastic region. G_0 is the initial shear modulus. Shear modulus G can be calculated by the mean effective stress p', the specific volume υ, Poisson's ratio v, and the slope of the loading–reloading line κ in the $\upsilon - \ln p'$ plane, which is derived from equation[12] $G = \frac{3(1-2v)}{2(1+v)}\frac{\upsilon p'}{\kappa}$.

13.2.1.3 Elastoplastic analysis

The MCC model is established based on the tests of normal and light overconsolidated clay. The MCC model can explain the elastoplastic deformation characteristics of soil from experimental and theoretical analyses. It is widely used in scientific studies and engineering practices. In addition, the MCC model is easy to extend and a series of representative constitutive models are developed based on it, such as the CASM[13] and the Uniform Hardening (UH) model[14]. For sand–clay mixed soil, when the sand content is lower than 70% ~ 80%, it is considered that the soil still shows the mechanical characteristics of clay[15]. In the layer ⑧$_2$ of silty clay with sand, the sand is mainly silty sand and the content is less than 8%. Thus, the MCC model is adopted in Shanghai, China to perform the elastoplastic analysis in this study. The yield function of the MCC model can be written as,

$$f = \frac{\lambda - \kappa}{1 + e_0} \ln \frac{p'}{p'_0} + \frac{\lambda - \kappa}{1 + e_0} \ln \left(1 + \frac{q^2}{M^2 p'^2} \right) - \varepsilon_{\mathrm{v}}^{\mathrm{p}} = 0, \tag{13.12}$$

where λ is the slope of normal compression in the $\upsilon - \ln p'$ plane, M is the slope of the critical state line, and e_0 is the initial void ratio. In the MCC model, the hardening parameter is a function of the plastic volumetric strain $\varepsilon_{\mathrm{v}}^{\mathrm{p}}$.

Deviatoric stress q and mean effective stress p' are defined as[16],

$$\begin{aligned} p' &= \frac{1}{3}\left(\sigma'_r + \sigma'_\theta + \sigma'_z\right) \\ q &= \sqrt{\frac{1}{2}\left[\left(\sigma'_r - \sigma'_\theta\right)^2 + \left(\sigma'_\theta - \sigma'_z\right)^2 + \left(\sigma'_z - \sigma'_r\right)^2\right]}. \end{aligned} \tag{13.13}$$

Drainage can affect the volumetric strain, which in turn affects the excess pore water pressure and the effective stress components around the cylindrical cavity. For the two cases of undrained and drained conditions, Chen and Abousleiman[16][17] adopted the large deformation theory and the associated flow rule to derive the elastoplastic solution for cylindrical cavity expansion in the plastic region.

For the undrained condition, the elastoplastic solution is expressed as[16],

$$\begin{aligned} \frac{\mathrm{d}\sigma'_r}{\mathrm{d}r} &= \frac{a_{11} - a_{12}}{\Delta r} \\ \frac{\mathrm{d}\sigma'_\theta}{\mathrm{d}r} &= \frac{a_{21} - a_{22}}{\Delta r}. \\ \frac{\mathrm{d}\sigma'_z}{\mathrm{d}r} &= \frac{a_{31} - a_{32}}{\Delta r} \end{aligned} \tag{13.14}$$

Moreover, the pore water pressure u_a at the cavity wall is expressed as

$$u_a = \sigma'_{r\mathrm{p}}(r) - \sigma'_a - \int_{R_\mathrm{p}}^{a} \frac{\sigma'_r(r) - \sigma'_\theta(r)}{r} \mathrm{d}r + u_0, \tag{13.15}$$

where σ'_a is the radical effective stress at the cavity wall.

For the drained condition, an auxiliary independent variable ξ is denoted as the ratio of the particle displacement u_r in the radial direction to its present radius r, which is written as $\xi = \dfrac{u_r}{r}$.

The elastoplastic solution is expressed as[17]

$$
\begin{aligned}
&N = 1 - \xi - \frac{\upsilon_0}{\upsilon(1-\xi)} \\
&\frac{d\sigma'_r}{d\xi} = -\frac{\sigma'_r - \sigma'_\theta}{N} \\
&\frac{d\sigma'_\theta}{d\xi} = -\frac{b_{21}}{b_{11}}\left[\frac{\sigma'_r - \sigma'_\theta}{N} + \frac{b_{11} - b_{12}}{\Delta(1-\xi)}\right] - \frac{b_{22} - b_{21}}{\Delta(1-\xi)} \\
&\frac{d\sigma'_z}{d\xi} = -\frac{b_{31}}{b_{11}}\left[\frac{\sigma'_r - \sigma'_\theta}{N} + \frac{b_{11} - b_{12}}{\Delta(1-\xi)}\right] - \frac{b_{32} - b_{31}}{\Delta(1-\xi)} \\
&\frac{d\upsilon}{d\xi} = -\frac{\upsilon\Delta}{b_{11}}\left[\frac{\sigma'_r - \sigma'_\theta}{N} + \frac{b_{11} - b_{12}}{\Delta(1-\xi)}\right]
\end{aligned}
\tag{13.15}
$$

where

$$
\begin{aligned}
&a_r = \frac{p'(M^2 - \eta^2)}{3} + 3(\sigma'_r - p') \\
&a_\theta = \frac{p'(M^2 - \eta^2)}{3} + 3(\sigma'_\theta - p') \\
&a_z = \frac{p'(M^2 - \eta^2)}{3} + 3(\sigma'_z - p') \\
&y = \frac{\lambda - \kappa}{\upsilon p'^3(M^4 - \eta^4)} \\
&b_{11} = \frac{1}{E^2}(1 - v^2 + Ea_\theta^2 y + 2Eva_\theta a_z y + Ea^2{}_z y) \\
&b_{22} = \frac{1}{E^2}[1 - v^2 + Ea_r^2 y + 2Eva_r a_z y + Ea_z^2 y] \\
&b_{33} = \frac{1}{E^2}[1 - v^2 + Ea_r^2 y + 2Eva_\theta a_r y + Ea_\theta^2 y] \\
&b_{12} = b_{21} = \frac{1}{E^2}[-Ea_r(a_\theta + va_z) + v(1 + v - Eva_\theta a_z y + Ea^2{}_z y)] \\
&b_{13} = b_{31} = \frac{1}{E^2}[-Ea_r(va_\theta + a_z)y + v(1 + v + Eva^2{}_\theta y - Ea_\theta a_z y)] \\
&b_{23} = b_{32} = \frac{1}{E^2}[v + v^2 + Eva_r^2 y - Ea_\theta a_z y + Eva_r(a_\theta + a_z)y] \\
&\Delta = -\frac{1+v}{E^3}\begin{bmatrix} \left(-1 + v + 2v^2\right) + E(-1+v)a_r^2 y + E(-1+v)a_\theta^2 y \\ -2Eva_\theta a_z y - Ea_z^2 y + Eva_z^2 y - 2Eva_r\left(a_\theta + a_z\right)y \end{bmatrix}
\end{aligned}
\tag{13.16}
$$

At the elastic/plastic boundary, the effective stress components can be written as

$$
\begin{aligned}
& q(r_p)=Mp'_0\sqrt{OCR\left[1+\left(\frac{q_0}{M+p'_0}\right)^2\right]-1} \\
& \sigma'_r(r_p)=\sigma'_{r0}+\sqrt{\sigma'^2_{r0}-\frac{1}{3}(4\sigma'^2_{r0}+\sigma'^2_{z0}-2\sigma'_{r0}\sigma'_{z0}-q(r_p)^2)} \\
& \sigma'_\theta(r_p)=\sigma'_{r0}-\sqrt{\sigma'^2_{r0}-\frac{1}{3}(4\sigma'^2_{r0}+\sigma'^2_{z0}-2\sigma'_{r0}\sigma'_{z0}-q(r_p)^2)} \\
& \sigma'_z(r_p)=\sigma'_{z0}=\frac{3}{1+2K_0}p'_0
\end{aligned}
\tag{13.17}
$$

where q is the initial deviatoric stress, and p'_0 is the initial mean effective stress.

Δ and b_{ij} are the elastoplastic matrix coefficients, which are all explicit functions of the three effective stress components σ'_r, σ'_θ, σ'_z and model parameters of λ, κ, M, E, v, and υ. q_p is the deviatoric stress on the initial yield locus, which can be calculated by q_0, p'_0, M, and OCR. Please refer the Chen and Abousleiman[16][17] for the detailed derivation process.

For the MCC model, the constitutive parameters mainly include the OCR, the slope of the loading–reloading line κ in the $\upsilon-\ln p'$ plane, the slope of normal compression λ in the $\upsilon-\ln p'$ plane, the slope of the critical state line M, the Poisson ratio v, and the initial void ratio e_0. Taking the constitutive parameters into Equation (13.17), the effective stress components at the elastic/plastic boundary can be obtained. The elastic–plastic boundary displacement and expanded cavity radius a/a_0 are taken as the lower and upper limits of the integral. The fourth-order Runge–Kutta method is used to solve a set of first-order simultaneous differential equations of Equation (13.13) or (13.15) in the plastic zone under undrained and drained conditions. By numerical calculation, the stress distribution and expansion stress in the plastic zone can be obtained.

13.2.1.4 CPTU mechanical model

In recent years, due to the developments of multifunctional sensors, CPT technologies have been further updated. However, single/double-bridged probes and CPTU are the most commonly used CPT specifications in the geotechnical engineering practice. Compared with CPTU, the double-bridged probe can only obtain the tip resistance q_c and side friction f_s, but it cannot obtain the excess pore water pressure u. Double-bridged probes and CPTU are widely used in the shallow soil layers. However, for the deep soil layers, which are located in a high-stress state, the probe faces cost difficulties in penetration and extraction. As well as it will produce a large excess pore water pressure in the clay layers. Excess pore water pressure will exceed the measuring range, making it difficult to obtain the full data. According to the above reasons, double-bridged probes and CPTU are limited in deep geotechnical investigation. The single-bridged probe is a CPT specification unique to China, which can obtain the total stress during penetration. It has been widely used in China[18][19][20], and a large number of empirical formulas have been summarized in engineering practice. In addition, a single-bridged probe is cheaper and simpler, which plays an important role in in-situ tests. Hence, the study of the single-bridged probe test mechanism has a practical significance. Moreover, specific penetration resistance p_s, cone resistance q_c, and side friction resistance f_s share a clear statistical law[21].

A single-bridge probe can only measure the total resistance F during penetration. Assuming that A is the cone cross-sectional area, the specific penetration resistance p_s is defined as

$$p_s = \frac{F}{A}. \tag{13.18}$$

The mechanical behavior of CPT is similar to the cylindrical cavity expansion process. Specific penetration resistance p_s has a particular relation to the stress components at the cavity wall. In the process of penetration, the stresses applied to the probe consist of cone resistance p_c, cylindrical friction f_s, and pore water pressure u_w as shown in Figure 13.2.

According to the principle of effective stress, the total stress of saturated soil consists of effective stress and pore pressure. Different from the effective stress, pore pressure cannot resist shear stress. Research[16][17] indicates that when the expansion radius a reaches 3–5 times of the initial radius a_0, the stress at the cavity wall reaches a limit stress state. The penetration process can be considered as a cylindrical cavity expansion from the initial inner diameter of nearly zero to the probe radius R. In that scenario, the soil around the probe approaches the limit stress state, i.e., $\sigma'_{rc} = \sigma'_{rs} = \sigma'_{rlim}$. Thus, the probe cylindrical friction is defined as

$$f_s = \mu\sigma'_{rlim}, \tag{13.19}$$

where μ is the friction coefficient between soil and steel probe.

The interface between the probe and the soil is a conical surface, which is different from the cylindrical cavity expansion. It assumes that there is a rigid slide block that can only move horizontally and the steel probe indirectly interacts with the surrounding soil as shown in Figure 13.2, Δ_{LMN} is the slide block of soil. As an axisymmetric problem,

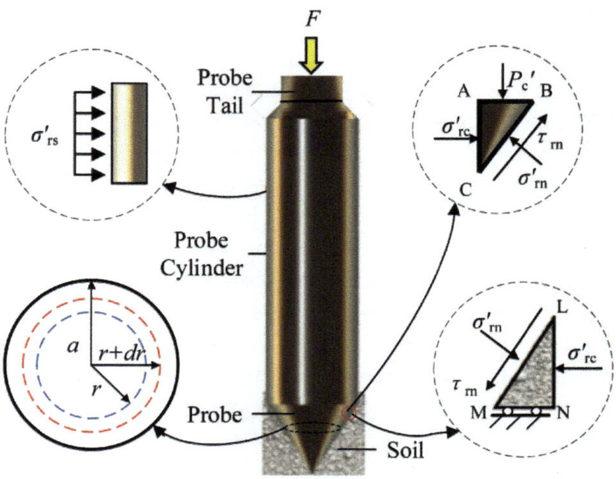

Figure 13.2 Schema of CPT mechanical model.

the forces on the slide block are decomposed vertically and horizontally. Firstly, the stress equilibrium is established on the horizontal plane for slide blocks, which is written as

$$\sigma'_{rc} l_{LN} = (\sigma'_{rn}\cos\alpha - \tau_{rn}\sin\alpha) l_{LM}, \tag{13.20}$$

where τ_{rn} and σ'_{rn} are the shear and the normal stress of the probe–soil interface, respectively. Shear stress τ_{rn} can be calculated by $\tau_{rn} = \mu\sigma'_{rn}$. α is the half of cone angle that constantly equals 30°.

According to Equation (13.20), the relationship between expansion stress σ'_{rc} and normal stress σ'_{rn} of the soil–probe interface can be obtained by

$$\sigma'_{rn} = \frac{l_{LN}}{(\cos\alpha - \mu\sin\alpha) l_{LM}} \sigma'_{rc}. \tag{13.21}$$

Correspondingly, the stress equilibrium is established on the vertical plane Δ_{ABC} for the cone probe, which is written as

$$p_c l_{AB} = (\sigma'_{rn}\sin\alpha + \tau_{rn}\cos\alpha) l_{BC}. \tag{13.22}$$

According to Equation (13.22), the relationship between cone resistance p_c and normal stress σ'_{rn} of the soil–probe interface can be obtained by

$$p_c = \frac{l_{BC}}{l_{AB}} (\sin\alpha + \mu\cos\alpha)\sigma'_{rn}, \tag{13.23}$$

where $l_{BC} = l_{LM}$, substituting Equation (13.21) into Equation (13.23) can get the cone resistance,

$$p_c = \frac{1+\sqrt{3}\mu}{1-\sqrt{3}\mu/3} \sigma'_{rc}. \tag{13.24}$$

The drained condition means that the penetration process does not produce excess pore water pressure and the pore pressure u_a at the cavity wall always equals the initial pore pressure u_0. The undrained condition means that the penetration process produces excess pore pressure u_a without dissipation, which is calculated by Equation (13.16). Under these two drainage conditions, assume that the positions of the probe cone, probe cylinder, and probe tail have equal pore water pressure. The influence of pore water pressure on specific penetration resistance p_s is mainly related to the cross-sectional area A_t of the probe tail. Hence, the static equilibrium formula along the axis of the probe can be written as

$$p_s A = p_c A + f_s A_s + A_t u_a. \tag{13.25}$$

where A_s is the surface area of the probe cylinder.

Taking Equation (13.19) and Equation (13.24) into Equation (13.25), the formula of specific penetration resistance p_s can be obtained by

$$p_s = \left(\frac{1+\sqrt{3}\mu}{1-\sqrt{3}\mu/3} + \mu\frac{A_s}{A} \right) \sigma'_{rlim} + \frac{A_t}{A} u_a. \tag{13.26}$$

In Equation (13.26), the area of A_t, A_s, and A is determined by the CPT probe size. The friction coefficient μ is determined by back-analysis using the parameters of the drilling test near the CPT curve at a specific depth. Assuming that expansion stress at the cavity wall reaches a limit stress state at an expanded cavity radius of $a/a_0 = 10$. According to Equation (13.13)–(13.15), the excess pore pressure u_a and limit expansion stress σ'_{rlim} can be obtained. To sum up, the specific penetration resistance p_s is calculated.

13.2.2 Cavity expansion theory of the CASM model

Cavity expansion theory can simulate the changes in stress, displacement, and pore water pressure caused by the expansion and contraction of a cylindrical or spherical cavity. The expansion principle is a dynamic process under the internal pressure of a small cavity close to radius zero, which is highly consistent with the mechanism of CPTU. Hence, it is particularly suitable to interpret the in-situ test data, such as cone-tip resistance, q_t, sleeve friction, f_s, and pore water pressure, u. Derivations of spherical cavity expansion theory and mechanical transformation equations for CPTU are provided as follows.

13.2.2.1 Constitutive mode

Roscoe et al. proposed the Original Cam Clay (OCC) model based on London Clay, which successfully described the characteristics of normally consolidated clay as elastic–plastic geomaterial[22]. To predict the mechanical properties of clayey soil under specific loads, the OCC model was modified by many scholars, for example, MCC is a well-known improved model for lightly overconsolidated soil.

Yu proposed the CASM that can uniformly consider the mechanical behaviors of clay and sand[23]. Compared with the OCC and MCC models, CASM adopts the non-associated flow law and introduces two new constitutive model parameters, spacing ratio r^* and stress-state coefficient n. As shown in Figure 13.3, r^* is the ratio of p'_c to p'_x, which is used to define the shape of the yield surface, where p'_c is the preconsolidation stress at the isotropic normal consolidation line (NCL) and p'_x is the intersection point of the swell

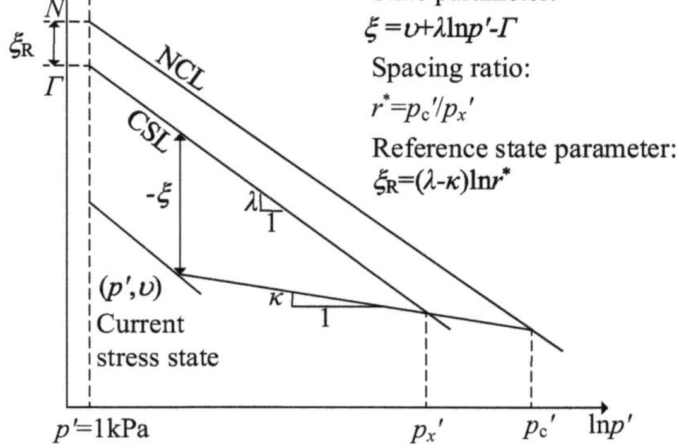

Figure 13.3 Definitions of state parameter, space ratio, and reference state parameter.

line and the critical state line (CSL). n defines the size of the yield surface. r^* is a fixed value for the OCC or MCC models, specifically. It varies with the change of the stress in CASM.

Parameter definitions of CASM are shown in Figure 13.3. υ is the specific volume, $\upsilon = 1+e$, e is the void ratio. κ is the slope of the swell line, and λ is the slope of CSL in $\upsilon - \ln p'$ space. The state parameter ξ is the difference between the specific volume υ and the counterpart at the CSL under the same effective stress p'. The reference state parameter ξ_R is the difference between the specific volumes at the CSL and the NCL, which describes the loosest state that the soil can reach. The concept of state parameter ξ is proposed by Been and Jefferies based on the critical state soil mechanics, it can describe the sandy soils by combining the effects of void ratio and confining pressure[24]. The yield function can be defined by the state parameter as

$$f(p',q,\xi) = \left(\frac{\eta}{M}\right)^n - 1 + \frac{\xi}{\xi_R} = 0, \tag{13.27}$$

where $\eta = q/p'$ is the stress ratio. M is the critical stress ratio, which is related to the internal friction angle. For the OCC model, $n=1$, $r^* = 2.718$. For the MCC model, $n = 1.5 \sim 2$, $r^* = 2$. To compare with the MCC model, the initial values of r^* and n adopted in CASM are $n = 1.667$, $r^* = 2$. The plastic potential function can be defined as

$$g(p',q,\beta) = 3M\ln\frac{p'}{\beta} + (3+2M)\ln\left(\frac{2q}{p'}+3\right) - (3-M)\ln\left(3-\frac{q}{p'}\right) = 0, \tag{13.28}$$

where the parameter β reflects the size of the plastic potential surface, which can be determined easily for any given stress state (p',q) by solving Equation (13.28). Yield surfaces of the CASM, OCC, and MCC models in the normalized p'-q space are shown in Figure 13.4.

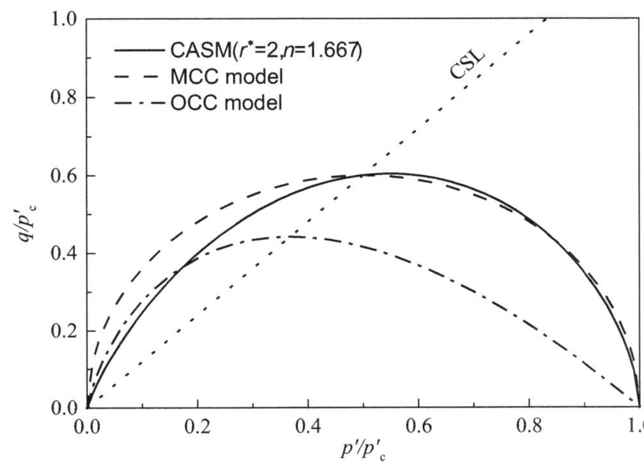

Figure 13.4 Yield surfaces of the CASM, OCC, and MCC models.

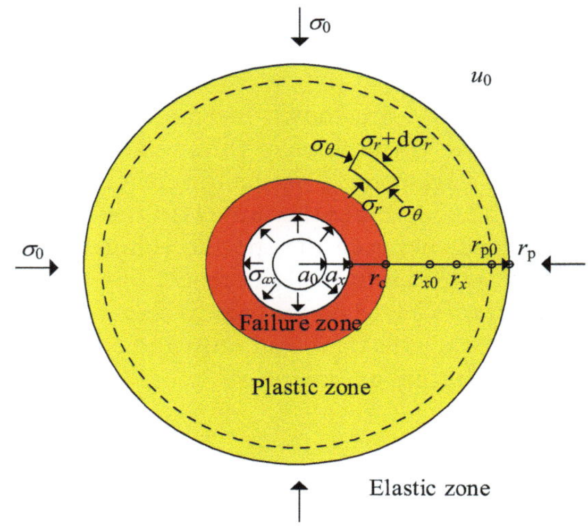

Figure 13.5 Schematic diagram of the spherical cavity expansion.

13.2.2.2 Spherical cavity expansion theory

As shown in Figure 13.5, a schematic process of spherical cavity expansion, which has an initial radius a_0, initial stress σ_0, initial pore water pressure u_0, and initial mean total stress p_0, is illustrated. Positive values should be given for the compressive stress and strain. As the internal cavity pressure increases, the cavity radius will gradually grow. When the pressure of the cavity wall reaches a threshold value, the surrounding soil begins to deform plastically. If the internal pressure continuously increases to a limit value, a failure zone begins to appear. Mean total stress is p, deviatoric stress is q, and the pore water pressure is u at any specific position r_x.

As the internal cavity pressure increases from σ_0 to σ_{ax}, the cavity wall radius gradually expands from a_0 to a_x. While the internal cavity pressure continues to increase, a failure zone is formed in a range of $a_x < r < r_c$, and plastic deformation is occurring in $r_c < r < r_p$, where r_c is the radius of the interface between the failure zone and the plastic zone, r_p is the position of the elastic–plastic boundary, with which the initial position is r_{p0}. Any specific point r_x located in the plastic zone has an initial position of r_{x0}. The margin area outside the plastic zone remains in the elastic state. Due to the spherical symmetry, tangential stress σ_θ is equal to the circumferential stress σ_ϕ. The total mean principal stress and deviatoric stress are represented by p and q, respectively.

$$p = \frac{\sigma_r + 2\sigma_\theta}{3}, \tag{13.29}$$

$$q = \sigma_r - \sigma_\theta, \tag{13.30}$$

where σ_r and σ_θ are the total radial and tangential stresses. The initial radial stress and initial tangential stress $\sigma_{r0} = \sigma_{\theta 0} = \sigma_0 + u_0$, the initial mean stress $p_0 = \sigma_0 + u_0$. In the elastic and plastic zone, the radial and tangential stresses at the position r from the spherical expansion center satisfy the equilibrium differential equation,

$$\frac{\partial\sigma_r}{\partial r}+2\frac{\sigma_r-\sigma_\theta}{r}=0, \tag{13.31}$$

or expressed in the form of effective stress

$$\frac{\partial\sigma_r'}{\partial r}+\frac{\partial u}{\partial r}+2\frac{\sigma_r'-\sigma_\theta'}{r}=0. \tag{13.32}$$

13.2.2.3 Elastic zone analysis

The small deformation theory is adopted in the elastic region, then the radial strain increment $\mathrm{d}\varepsilon_r$ and tangential strain increment $\mathrm{d}\varepsilon_\theta$ at the radius r ($r > r_\mathrm{p}$) can be expressed as follows:

$$\mathrm{d}\varepsilon_r=-\frac{\partial \mathrm{d}u_r}{\partial r},\ \mathrm{d}\varepsilon_\theta=-\frac{\mathrm{d}u_r}{r}, \tag{13.33}$$

where $\mathrm{d}u_r$ is radial displacement increment.

Stress–strain relationship in the elastic zone can be obtained by Hooke's law:

$$\begin{bmatrix}\mathrm{d}\varepsilon_r\\ \mathrm{d}\varepsilon_\theta\end{bmatrix}=\frac{1}{E}\begin{bmatrix}1 & -2v\\ -v & 1-v\end{bmatrix}\begin{bmatrix}\mathrm{d}\sigma_r'\\ \mathrm{d}\sigma_\theta'\end{bmatrix}, \tag{13.34}$$

where v is the Poisson's ratio, E is the Young's modulus, and $\mathrm{d}\sigma_r'$ and $\mathrm{d}\sigma_\theta'$ are the effective radial and tangential stress increments. Combining equations (13.31), (13.33), and (13.34), the total stresses, radial displacement, and the pore water pressure in the elastic region can be expressed as[17]

$$\begin{cases}\sigma_r=\sigma_0+(\sigma_{r\mathrm{p}}-\sigma_0)(\frac{r_\mathrm{p}}{r})^3\\ \sigma_\theta=\sigma_0-\frac{1}{2}(\sigma_{r\mathrm{p}}-\sigma_0)(\frac{r_\mathrm{p}}{r})^3\\ u_r=\frac{\sigma_{r\mathrm{p}}-\sigma_0}{4G_0}(\frac{r_\mathrm{p}}{r})^3 r\\ u=u_0\end{cases}, \tag{13.35}$$

where u_r is the radial displacement at r and $\sigma_{r\mathrm{p}}$ denotes the total radial stress at the elastic–plastic boundary.

13.2.2.4 Plastic zone analysis

Non-associated flow rule is adopted in CASM, and the increment of plastic strain can be expressed as

$$\mathrm{d}\varepsilon_r^\mathrm{p}=\Lambda\frac{\partial g}{\partial\sigma_r'}=\Lambda\left(\frac{\partial g}{\partial p'}\frac{\partial p'}{\partial\sigma_r'}+\frac{\partial g}{\partial q}\frac{\partial q}{\partial\sigma_r'}\right), \tag{13.36}$$

$$d\varepsilon_{\theta}^{p} = \Lambda\frac{\partial g}{\partial \sigma_{\theta}'} = \Lambda\left(\frac{\partial g}{\partial p'}\frac{\partial p'}{\partial \sigma_{\theta}'} + \frac{\partial g}{\partial q}\frac{\partial q}{\partial \sigma_{\theta}'}\right), \tag{13.37}$$

where Λ is a scalar multiplier. Due to the spherical symmetry, $d\varepsilon_{\phi}^{p}$ is known to be equal to $d\varepsilon_{\theta}^{p}$. The plastic volumetric strain $d\varepsilon_{v}^{p}$ can be expressed as

$$d\varepsilon_{v}^{p} = d\varepsilon_{r}^{p} + 2d\varepsilon_{\theta}^{p}. \tag{13.38}$$

It can also be expressed by Yu[25]

$$d\varepsilon_{v}^{p} = \frac{\partial g/\partial p}{H}\times\left[\left(\frac{\partial f}{\partial p'} + \frac{\partial f}{\partial \xi}\frac{\partial \xi}{\partial p'}\right)dp' + \frac{\partial f}{\partial q}dq\right], \tag{13.39}$$

where H denotes the plastic hardening modulus. The elastic–plastic stiffness matrix is obtained by combining equations (13.28), (13.36)–(13.39) and Hooke's law:

$$\begin{bmatrix} d\sigma_r' \\ d\sigma_\theta' \end{bmatrix} = \frac{1}{\Delta}\begin{bmatrix} a_{11} & a_{12} \\ a_{21} & a_{22} \end{bmatrix}\cdot\begin{bmatrix} d\varepsilon_r \\ d\varepsilon_\theta \end{bmatrix}, \tag{13.40}$$

where a_{11}, a_{12}, a_{21}, and a_{22} are the matrix coefficients. Soil deformation is subjected to the large deformation theory, thus logarithmic strain is adopted in the plastic zone. The relationship between strain and displacement is

$$d\varepsilon_r = -\frac{\partial dr}{\partial r}, d\varepsilon_\theta = -\frac{dr}{r}, \tag{13.41}$$

where dr is the infinitesimal deformation at r. the volume change $d\varepsilon_v$ is zero under the undrained conditions, hence the radial strain increment can be given

$$d\varepsilon_r = -2d\varepsilon_\theta = 2\frac{dr}{r}. \tag{13.42}$$

Substitute Equation (13.42) into Equation (13.40), the first-order ordinary differential equations are solved to establish the relationships among the effective radial, tangential stresses, and the radius of spherical cavity expansion in the plastic zone,

$$\frac{d\sigma_r'}{dr} - \frac{2\times a_{11} - a_{12}}{\Delta\cdot r} = 0, \tag{13.43}$$

$$\frac{d\sigma_\theta'}{dr} - \frac{2\times a_{21} - a_{22}}{\Delta\cdot r} = 0. \tag{13.44}$$

Through the integral of Equation (13.32), the pore water pressure at the r_x can be obtained,

$$u(r_x) = -2\int_{r_p}^{r_x}\frac{\sigma_r' - \sigma_\theta'}{r}dr + \sigma_{rp}' - \sigma_r(r_x) + u_0. \tag{13.45}$$

The stress at any specific point r_x can be calculated by Equations (13.43)–(13.45) in the plastic zone, but the precondition is the radius of the plastic zone r_p and the initial stress at the r_x. This question will be discussed below.

13.2.2.5 Elastic–plastic boundary analysis

According to Equations (13.29) and (13.35), the effective radial stress and the effective tangential stress at the elastic–plastic boundary are as follows,

$$\sigma'_{rp} = \sigma'_0 + \frac{2}{3} q_p, \sigma'_{\theta p} = \sigma'_0 - \frac{1}{3} q_p, \tag{13.46}$$

where q_p denotes the deviatoric stress at the elastic–plastic boundary, and $p' = p'_0$. Hence, the deviatoric stress at the elastic–plastic boundary can be defined as

$$q_p = Mp'_0 \sqrt[n]{\frac{\ln OCR}{\ln r^*}}, \tag{13.47}$$

where OCR is the overconsolidation ratio. The plastic zone radius r_p can be obtained by the volume invariable,

$$\frac{r_p}{a} = \left[\left(\left(\frac{a_0}{a}\right)^3 - 1\right) \Big/ \left(\left(1 - \frac{\sigma'_{rp} - \sigma'_0}{4G_0}\right)^3 - 1\right)\right]^{\frac{1}{3}}. \tag{13.48}$$

13.2.2.6 Mechanical transformation model

Pseudo-static equilibrium model between the corrected cone-tip resistance q_t, the limit expansion stress σ_L, and the tangential stress σ_θ is shown in Figure 13.6. While the penetrometer indents into the soil, the radius of the spherical cavity expands from a_0 to a, the soil within the range of r_c is destroyed and reaches to the critical state. Hence, the limit expansion stress σ_L and the shear stress τ_n act on the 1/4 spherical surface BC with the radius r_c, and the tangential stress σ_θ acts on both sides of the cone shoulder. As an axisymmetric problem, the forces on BC are decomposed vertically and horizontally, and the stress equilibrium is established on the vertical plane.

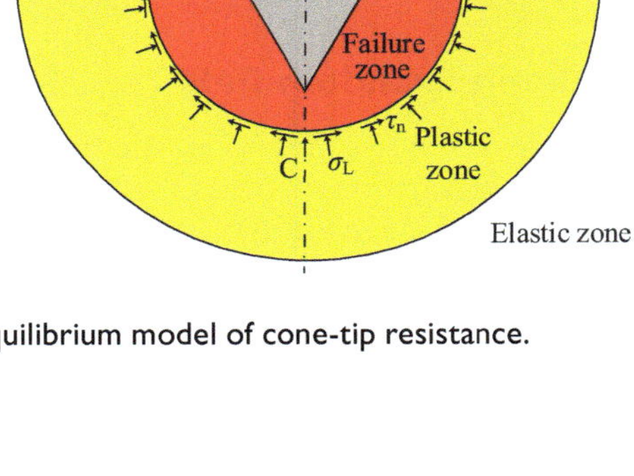

Figure 13.6 Pseudo-static equilibrium model of cone-tip resistance.

$$q_t \cdot a + \sigma_\theta \cdot (r_c - a) = \int_0^{\frac{\pi}{2}} (\sigma_L \sin\theta + \tau_n \cos\theta) r_c d\theta, \tag{13.49}$$

$$q_t = \frac{(\sigma_L + \tau_n) \cdot r_c - \sigma_\theta (r_c - a)}{a}, \tag{13.50}$$

where $\sigma_L = \sigma'_a + u_c$, u_c is the pore water pressure at the radius r_c . $\tau_n = c + \sigma'_L \tan\varphi'$, c is the cohesion which is much smaller than the limit expansion stress, hence it is assumed to be zero. The relationship between the corrected cone-tip resistance q_t and the uncorrected cone-tip resistance q_c is

$$q_t = q_c + (1 - \beta) u \tag{13.51}$$

where β denotes the cone area ratio, which is assumed to be 0.87[26]. Pore water pressure u can be directly determined by the measurement u_a at the radius a,

$$u = u_a \tag{13.52}$$

Cone penetration is regarded as a fast dynamic process in clay; hence, the pore water pressure does not dissipate immediately.

Unknown r_c is the key to derive corrected cone-tip resistance q_t. There are no specific methods to solve r_c. Therefore, the numerical solution of r_c obtained by MATLAB codes may be feasible. As explained above, the effective stress within the radius r_c does not change in the plastic zone. The algorithm can find out the critical point r_c, with which the effective stress alters from constancy to variation.

13.3 BAYESIAN-RANDOM FIELD THEORY

Due to the natural variability of the geotechnical medium, the stochastic nature of geotechnical parameters must be evaluated. Static penetration analysis involving geotechnical parameters, the stochastic description of the geotechnical parameters, and the stochastic simulation of cylindrical cavity expansion theory are anointed stochastic mechanics methods.

Due to the uncertainty of geotechnical parameters, limited site-investigation data and unknown disturbances have a great influence on the acquisition of geotechnical parameters. Therefore, the stochastic mechanics based Bayesian method is an effective tool that helps engineers make a reasonable decision, as it can integrate the prior information with in-situ test data to decrease the uncertainties.

13.3.1 Stochastic mechanics/random field theory

Stochastic mechanics needs to consider the stochasticity of geotechnical parameters, which defines the geotechnical parameters as random variables. Stochastic simulation methods mainly include the Monte Carlo algorithm and Response Surface Method (RSM). In addition, in the study of the statistical characteristics (i.e., mean value and standard deviation) of small samples, Bayesian theory is used to study the stochasticity of geotechnical parameters and the uncertainties of geotechnical parameters are analyzed based on in-situ test data.

13.3.1.1 Monte Carlo simulation

Monte Carlo method is based on probability theory and mathematical statistics, and generally uses the probability density function (PDF) to achieve sampling of random variables. It is necessary to determine the probability distributions of the random variables in advance, such as uniform distribution, normal distribution, and lognormal distribution. The Monte Carlo method randomly takes samples from the prior distributions to simulate the real variables[27][28].

13.3.1.2 Response surface method

For the complicated geotechnical mechanics, it is very hard to directly use the Monte Carlo method for numerical calculation. Therefore, Box and Wilson[29] put forward RSM to determine objective function, which uses Central Composite Design (CCD) to approximate the real state surface. In this research, a quadratic polynomial without cross-terms is established between CPT data and geotechnical parameters instead of numerical analysis. The functional relationship between RSM $g(\psi)$ and the target objective $\mathbf{z}_b$ (i.e., Type B data) is

$$\mathbf{z}_b = g(\psi) + \varepsilon, \tag{13.53}$$

where ψ denote the key geotechnical parameter vector, ε are the model errors.

13.3.1.3 Multivariate spatial random fields

The spatial variability of a geotechnical parameter is simplified as aleatory uncertainty with a random variables assumption. However, the spatial random field theory can not only take the stochasticity into account but also emphasize the spatial correlation. The intrinsic uncertainty of a geotechnical parameter can be depicted by a spatial random field, $\theta(x)$. The letter x defines the Euclidean coordinates. In addition to the random variables (i.e., random variables with only the mean values, μ, and the variance, σ_0^2), spatial random fields use the correlation function[30] to depict the spatial variability,

$$\rho_{i,i}(h) = \frac{E\{[\theta_i(x+h) - \mu_i]\cdot[\theta_i(x) - \mu_i]\}}{\sigma_i^2}, \tag{13.54a}$$

$$\rho_{i,j}(h) = \frac{E\{[\theta_i(x+h) - \mu_i]\cdot[\theta_j(x) - \mu_j]\}}{\sigma_i\sigma_j}, \tag{13.54b}$$

where $\rho_{i,i}(h)$ is the autocorrelation function, $i = 1,\ldots,m$, and m is the number of key geotechnical parameters. The cross-correlation function, $\rho_{i,j}(h)$, $(i,j = 1,\ldots,m, i \neq j)$, depicts the mutually spatial variation of spatial random fields. It may be assumed that $\rho_{i,j}(h) = \rho_{j,i}(h)$. In this chapter, the deterministic drift is a constant. After removing the deterministic drift, the mean value is satisfied $E[\theta(x)] = E[\theta(x+h)] = \mu$.

The exponential model of theoretical function adapts the scenario of a bigger correlation length, while the Gaussian model can depict the reversed curves of a correlation function. The spherical model[31] has the advantages of concision and robustness. Hence, this model is applied widely in the correlation function,

$$\rho(h) = \begin{cases} 1-\left[1.5(h/a_1)-0.5(h/a_1)^3\right] & 0 \le h \le a_1 \\ 0, & h > a_1 \end{cases}, \tag{13.55}$$

where $\rho(h)$ is the correlation coefficient of the lag distance h, and a_1 is the primary correlation length.

The three-dimensional heterogeneity of the spatial random fields can be considered with the weighted lag distance h [32] as,

$$h = \sqrt{\left(\frac{h_1}{\eta_1}\right)^2 + \left(\frac{h_2}{\eta_2}\right)^2 + \left(\frac{h_2}{\eta_3}\right)^2}, \tag{13.56}$$

where η_1 is equal to a constant value of 1.0, and η_2, η_3 are the correlation length ratio of the second correlation length a_2 and the third correlation length a_3 divided by a_1, respectively.

13.3.2 Bayesian inverse modeling method

The spatial variability of key geotechnical parameters cannot be completely described by the initial site investigation. Nevertheless, construction measurements are continuously obtained during excavations. The Bayesian method can dynamically calibrate the statistical characteristics of key geotechnical parameters by assimilating multi-source data.

13.3.2.1 Data classification

Two types of data are defined in the spatial random fields based Bayesian method. The borehole testing data are originally measured for the spatial random fields, $\boldsymbol{\theta}(\boldsymbol{x})$, at spatial positions, $\boldsymbol{x}$, which in turn are defined as the input data, z_a. The tunneling measurements are deductively connected to the spatial random fields, and these measurements are defined as the output data, z_b. The two data types, z_a and z_b, are referred to in the following equations:

$$z_a = f(\boldsymbol{x}_a) + \varepsilon_a, \tag{13.57a}$$

$$z_b = \chi \cdot M(\theta, z_a) + \varepsilon_b, \tag{13.57b}$$

where $f(\cdot)$ is a mathematical operator based on the spatial positions, $\boldsymbol{x}_a$, and the tunnel monitoring data z_b are indirectly derived from the soil–fluid coupling model $M(\theta, z_a)$. ε_a and ε_b are the zero-mean testing errors. The model factors χ are given by the ratios of the real measurements divided by the theoretical predictions, which are defined as independent random variables.

13.3.2.2 Bayesian framework

The unknown objectives, $\Theta = [\theta, \chi]$, include the key geotechnical parameters θ and the model factors χ. The spatial random fields based Bayesian framework is designed to assimilate prior knowledge, $p(\Theta)$, borehole testing data, z_a, and tunnel monitoring data, z_b. Accordingly, the posterior distribution, $p(\Theta \mid z_a, z_b)$, is derived from

$$p(\Theta \mid z_a, z_b) = \frac{L(z_b \mid \Theta, z_a)p(\Theta)}{\int_{\Theta} L(z_b \mid \Theta, z_a)p(\Theta)d\Theta}, \tag{13.58}$$

where the likelihood function $L(z_b \mid \Theta, z_a)$ is a probabilistic distribution of the tunnel monitoring data, z_b, which is also subject to the borehole testing data, z_a, prior distribution, $p(\Theta)$, and soil–fluid coupling model $M(\theta, z_a)$. The posterior distribution $p(\Theta \mid z_a, z_b)$ accounts for the test errors, ε_a and ε_b.

13.3.3 Algorithm implementation

The spatial random fields based Bayesian method is difficult to use to obtain analytical results. Thus, a sampling algorithm is widely used for a sampling solution. The algorithm consists of a numerical analysis, prior distribution, likelihood function, and posterior distribution modules. The analyses are referred to in Figure 13.7.

13.3.3.1 Prior distribution

Before any on-site observations (e.g., z_a and z_b) are used in the Bayesian framework, $p(\Theta)$ summarizes the knowledge that is available on the key geotechnical parameters θ and the model factors χ. Accordingly, the multivariate normal distribution is generalized to depict m geotechnical parameters at n positions, as shown in the following:

$$p(\theta) = |\mathrm{C}|^{-\frac{1}{2}} (2\pi)^{-\frac{m \times n}{2}} \exp\left[-\frac{1}{2}(\theta - \mu)^T \mathrm{C}^{-1}(\theta - \mu)\right], \tag{13.59}$$

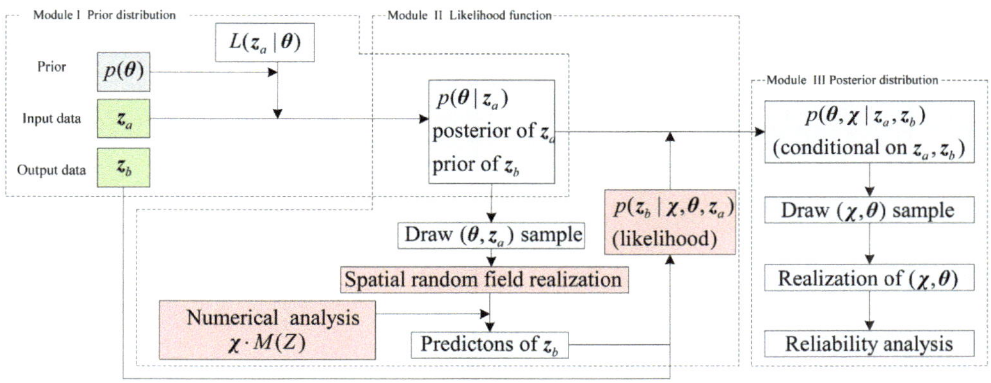

Figure 13.7 Data flow and algorithm schema of spatial random fields-based Bayesian method.

where C is the covariance matrix of the spatial random fields, $C = \rho_n \otimes C_m$, C_m is the covariance matrix of the number m of normal random variables, the matrix ρ_n consists of autocorrelation and cross-correlation functions $\rho_{i,j}(h)$, $(i, j = 1,...,m)$, and the symbol "$\otimes$" denotes the operator of matrix multiplication. In a special scenario, the lag distance $h = 0$, which means that the spatial random fields degenerate into random variables.

The model factors χ are defined as the real measurements divided by the analytical results, which can be assumed to be multiplying independent normal random variables. Examples are given for the case studies. Thus, the prior distribution $p(\Theta)$ is presented as

$$p(\Theta) = p(\theta)\prod_{i=1}^{n_b} p(\chi_i), \tag{13.60}$$

where n_b is the total number of tunneling measurements, and $\prod_{i=1}^{n_b} p(\chi_i)$ denotes the prior joint distribution of the model factors.

13.3.3.2 Likelihood function

The spatial random fields based Bayesian method can assimilate prior knowledge $p(\Theta)$, borehole testing data z_a, and tunnel monitoring data z_b to obtain the posterior distributions of the key geotechnical parameters and model factors, $\Theta = [\theta, \chi]$. The likelihood function calculation includes two key components: the general driver of the soil–fluid coupling model $M(\theta, z_a)$ and the multivariate discretization of spatial random fields.

A general driver is designed to communicate with the Bayesian framework with the software FLAC^{3D} 3.0, as shown in Figure 13.8. This depends on a control file, and several input and output functions are used to configure the stochastic analysis of the soil–fluid coupling mechanics. The command stream of FLAC^{3D} 3.0 can execute numerical analysis and thereby obtain the analytical results for the tunneling measurements.

There are several well-known discretization algorithms of spatial random fields. Examples include the Cholesky decomposition[33], turning band method[34], local average method[35], Karhunen–Loeve decomposition method[36], sequential Gaussian simulation[37], and random harmonic function[38]. These algorithms are effective in accounting for spatial discretization for univariate spatial random fields.

Sequential Gaussian simulation has the advantage of robustness due to the spatial interpolation principle, and it is convenient to apply sequential Gaussian simulation in the conditional multivariate discretization of spatial random fields. As such, the trivariate spatial

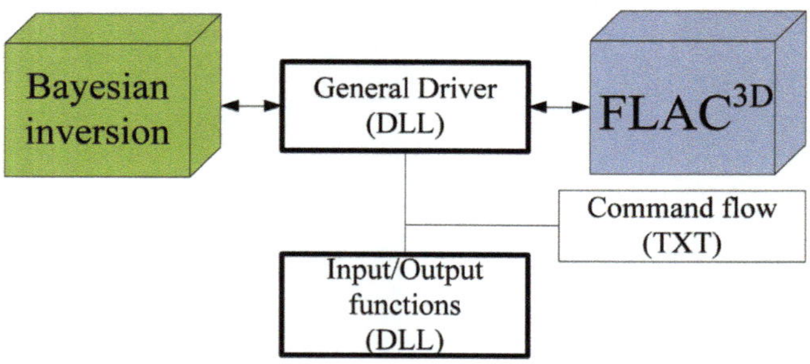

Figure 13.8 General master driver of connecting stochastic discretization and mechanics software.

random fields $\theta_i(x)$, $i=1, 2, 3$, are used for demonstration. In this case, the known quantities include the mean values, μ_i, standard deviation values, σ_i, autocorrelation function, $\rho_{i,i}(h)$, and cross-correlation function, $\rho_{i,j}(h), i=1, 2, 3, i \neq j$. There is n_i conditional data $\theta_i(x_k), k=1,\ldots n_i$ adjacent to point x_0. The mean value $\mu_i(x_0)$ and the standard deviation value $v_i(x_0)$ of the unknown $\theta_i(x_0)$ are calculated with linear co-kriging equations[39].

A sequential Gaussian simulation executes the spatial discretization of the stochastic elements point by point, and individual iteration takes a sample from the joint distribution. The following five key steps involve conditional multivariate discretization.

After defining a random path on the stochastic elements, each midpoint of an element is traversed once in the iteration, and the element edge size is less than the smallest correlation length, which guarantees the depiction of the spatial variation.

The mean values $\mu_1(x_k)$, $\mu_2(x_k)$, $\mu_3(x_k)$ and the standard deviation values $\sigma_1(x_k)$, $\sigma_2(x_k)$, $\sigma_3(x_k)$, at the midpoint, x_k, are predicted. The recognized quantities include the prior mean values, μ_1, μ_2, μ_3, the standard deviation values, σ_1, σ_2, σ_3, autocorrelation functions, $\rho_{1,1}(h)$, $\rho_{2,2}(h)$, $\rho_{3,3}(h)$, and the cross-correlation functions, $\rho_{1,2}(h)$, $\rho_{1,3}(h)$, $\rho_{2,3}(h)$. The conditional data consists of the original borehole testing data z_a (i.e., the errors ε_a are randomly added to the observations), and the recently discretized samples $\theta(x_1),\ldots,\theta(x_{i-1})$.

Three normal distributions with $N[\mu_1(x_k),\sigma_1^2(x_k)]$, $N[\mu_2(x_k),\sigma_2^2(x_k)]$, and $N[\mu_3(x_k),\sigma_3^2(x_k)]$ are built for the midpoint x_k, and one sample is taken. The outcomes, $\theta_1(x_k)$, $\theta_2(x_k)$, and $\theta_3(x_k)$, are saved as the new conditional data for the next midpoint x_{k+1}.

Moving to the next midpoint x_{k+1}, Steps 2 and 3 are repeated until the sampling is done for all the stochastic elements.

One multivariate discretization of the spatial random fields is finished.

The stochastic analysis of the soil–fluid coupling model $M(\theta, z_a)$ provides analytical results for the tunnel monitoring data z_b of the ground surface settlements. Accordingly, the likelihood function $L(z_b \mid \theta, z_a)$ can be constructed for $\varepsilon_b = z_b - \chi M(\psi, z_a)$, which can be expressed by a multivariate normal distribution, as shown in Equation (13.61):

$$L(z_b \mid \psi, z_a) = \frac{\exp[-\frac{1}{2}(z_b - \chi M(\psi, z_a))^T C_{\varepsilon_b}{}^{-1}(z_b - \chi M(\psi, z_a))]}{\sqrt{(2\pi)^{n_b} \mid C_{\varepsilon_b} \mid}}, \tag{13.61}$$

where $C_{\varepsilon b}$ is the covariance matrix of the monitoring errors ε_b and $|C_{\varepsilon b}|$ is the determinant of C_{ε_b}.

Furthermore, the correlation coefficient ρ_{ij} ($i, j = 1,\ldots,k$) among ε_b is simplified into the function of the mutual distance[40], as shown in Equation (13.62),

$$\rho_{ij} = 1 - \frac{D(x_i, x_j)}{\max D}, \tag{13.62}$$

where $D(x_i, x_j)$ is the Cartesian distance between two monitoring points x_i and x_j, and $\max D$ is the maximum distance of all the monitoring points.

13.3.3.3 Posterior distribution

The stochastic analysis of soil–fluid coupling mechanics $\chi M(\theta, z_a)$ can consume a large amount of computational resources. In the current analysis, a Markov chain Monte Carlo simulation is used to obtain a numerical solution quickly. The Markov chain Monte Carlo analysis process consists of the following major six stepwise procedures.

(1) The prior distribution, $p(\Theta = [\theta, \chi])$, is prepared for spatial random fields, which includes the mean value vector μ and the covariance matrix C.
(2) The ith multivariate discretization $\theta^i(x)$ is subjected to the prior distribution $p(\theta)$ and the original condition data z_a are extracted.
(3) The tunneling measurements z_b are predicted using the stochastic analysis of the soil–fluid coupling model, $M(\theta^i, z_a)$, and the threshold, $\gamma_i = \min\left(\frac{L(z_b \mid \theta^i, z_a)p(\theta^i)}{L(z_b \mid \theta^{i-1}, z_a)p(\theta^{i-1})}, 1\right)$, is calculated.
(4) A random number u in the uniform distribution 0~1 is taken and compared with γ_i. If $\gamma_i > u$, the spatial discretization $\theta^i(x)$ is accepted as a posterior sample. Otherwise, Step (2) is repeated.
(5) The prior mean value vector μ and the covariance matrix C from the accepted posterior samples for acceleration purposes are updated regularly, and then Step (2) is repeated.
(6) The predefined maximum samples, $i \leq N$, are reached, and the posterior distribution $p(\Theta \mid z_a, z_b)$ is calculated.

13.4 ENGINEERING APPLICATION

This chapter introduces three engineering application examples of CPTU, two of which are located in Shanghai, China, and another one is located in the Bothkennar site in the United Kingdom.

13.4.1 National Convention and Exhibition Center in Shanghai

The National Convention and Exhibition Center (Shanghai) is the largest single-block building in the world. It is apart 1.5 km from Shanghai Hongqiao Transportation Hub. The primary facilities include exhibition areas *A*, *B*, *C*, and *D* and a transportation center as shown in Figure 13.9A. In this chapter, we focus on the site characterization of exhibition region *A* that is leaf shaped, with the largest dimension equals to 350 m in the Figure 13.9B.

The information available from the geotechnical investigation report includes the interfaces separating the various soil layers, groundwater table, and geotechnical parameters such as compression modulus E_s(MPa), consolidated shear cohesion coefficient c_{cq}(kPa), and internal friction angle $\varphi_{cq}(^0)$ as shown in Figure 13.9C. The silty clay soil (i.e., the third soil layer from the ground surface) is the load-bearing soil layer of the shallow structure foundation, the average thickness is about 2.0 m, and the three-dimensional domain is shown in Figure 13.9D. The 109 CPT data curves are sampling each 0.1 m along the drill bar as shown in Figure 13.10A, total $n_a = 1381$ penetration resistance data p_s(MPa) (i.e., the sum of tip resistance and sleeve friction) covers the whole three-dimensional

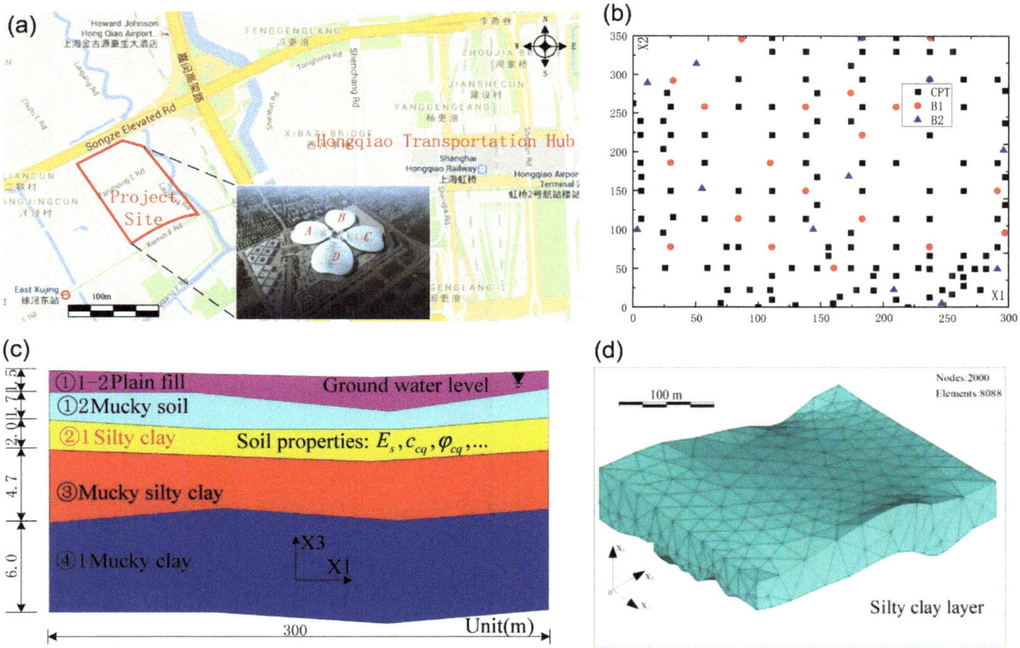

Figure 13.9 National Convention and Exhibition Center (Shanghai), China: (a) the geographic location of the background project, (b) planar geotechnical investigation layout of CPTs and boreholes of the region *A*, B1 provides compression modulus E_s(MPa) samples, B2 provides consolidated shear samples of cohesion coefficient c_{cq}(kPa) and shear internal friction angle $\varphi_{cq}(^\circ)$, (c) typically vertical section of the soil layers of region *A*, and (d) three-dimensional layer of the silty clay soil, the vertical displaying scale has been amplified 30 times.

domain, and histogram plot is shown in Figure 13.10B. To calibrate the transformation model between CPT data and laboratory data, the borehole group "B1" provides $n_{b1} = 21$ laboratory samples for compression modulus E_s(MPa), the histogram plot is shown in Figure 13.10C, and Table 13.1 lists the detailed sampling coordinates; the borehole group "B2" provides $n_{b2} = 12$ laboratory samples for consolidated shear cohesion coefficient c_{cq}(kPa) and internal friction angle $\varphi_{cq}(^0)$, histogram plot is shown in Figure 13.10D, and detailed sampling coordinates are shown in Table 13.2.

CPT and laboratory tests cannot be collocated at the same spatial location since the above geotechnical investigation belongs to destructive testing technologies, and the traditional regression method between CPT data and regular geotechnical parameters ignored the factors of spatial coincidence and uncertainty propagation of geotechnical investigation. The transformation model would be improved within the three-dimensional spatial random field theory. First, the massive CPT data is denoted as a three-dimensional anisotropic spatial random field variable, then the prediction of CPT data will be projected to the borehole sampling location coincidently with the laboratory data; finally, transformation models between CPT data and regular geotechnical parameters would be calibrated by giving spatial variability of CPT data, measurement errors, prior knowledge uncertainty of transformation model, statistical uncertainties of structural parameters, and Bayesian inversely modeling method itself.

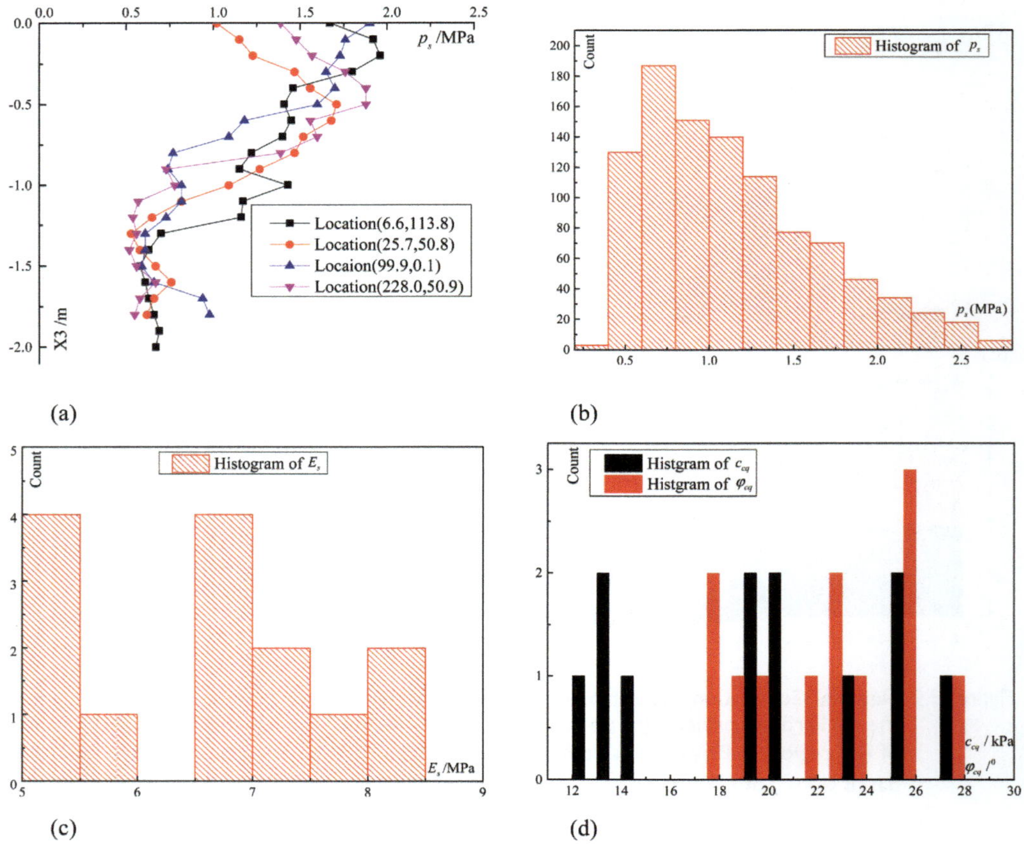

Figure 13.10 CPT data and laboratory data of the silty clay soil in the background project: (a) typical CPT data curves in the in-situ test; (b) histogram plot of whole CPT penetration resistance p_s(MPa); (c) histogram plot of compression modulus E_s(MPa); and (d) histogram plots of consolidated shear cohesion coefficient c_{cq}(kPa) and internal friction angle $\varphi_{cq}(^0)$.

Reversely evaluate transformation models within spatial random field theory and then convert CPT data p_s(MPa) to regular geotechnical parameters E_s(MPa), c_{cq}(kPa), and $\varphi_{cq}(^0)$ of the silty clay soil in the background project. CPT penetration resistance p_s(MPa) is represented by a spatial random variable, and the case studies are divided into two parts: a pseudo-transformation model M between CPT data p_s(MPa) and compression modulus E_s(MPa) is assumed at first, and the impacts of inherent variability and epistemic uncertainty will be taken into account; then, the calibrated transformation models are compared with linear regression models for the geotechnical parameters E_s(MPa), c_{cq}(kPa), and $\varphi_{cq}(^0)$ of the silty clay soil in cross-validation; finally, the calibrated transformation models will be employed in the three-dimensional site characterization.

13.4.1.1 Spatial variability of CPT data

There are 109 CPT data curves with total $n_a = 1381$ penetration resistance p_s(MPa) collected in the silty clay soil, and the classical statistics parameters are the mean value

Table 13.1 Spatial sampling coordinates of compression modulus E_s(MPa) in the laboratory test

Coordinates/m			E_s / *MPa*	*Coordinates/m*			E_s / *MPa*	*Coordinates/m*			E_s /*MPa*
X1	*X2*	*X3*		*X1*	*X2*	*X3*		*X1*	*X2*	*X3*	
30.00	185.8	2.00	6.94	111.10	77.80	2.64	9.43	183.10	113.80	2.42	4.73
30.10	77.80	2.13	8.09	138.00	257.80	1.29	6.93	210.10	257.90	2.12	4.61
32.10	292.10	2.40	6.72	138.10	149.80	2.61	6.75	237.00	293.80	3.29	4.63
57.10	257.80	2.58	7.00	160.60	50.90	1.95	7.62	237.10	77.90	1.03	5.27
84.10	113.90	1.68	8.10	174.00	275.90	1.48	5.39	237.60	347.90	2.72	4.96
86.40	345.80	2.48	4.51	183.00	347.80	1.92	7.15	291.10	149.90	2.27	5.98
109.60	185.80	1.54	4.76	183.10	221.80	1.68	5.05	297.10	95.90	1.91	5.29

Table 13.2 Spatial sampling coordinates of consolidation shear cohesion coefficient c_{cq}(kPa) and internal friction angle φ_{cq}(^{0}) in the laboratory test

Coordinates/m					*Coordinates/m*				
X1	*X2*	*X3*	c_{cq} / kPa	φ_{cq} /°	*X1*	*X2*	*X3*	c_{cq} / kPa	φ_{cq} /°
183.00	347.80	1.92	23.50	18.75	237.00	293.80	3.29	27.00	22.50
172.40	168.80	0.67	19.00	17.50	55.10	152.90	1.71	14.00	22.50
291.30	49.60	1.48	20.00	17.50	144.10	100.30	2.43	20.00	27.50
246.80	4.90	1.25	19.00	19.00	50.50	314.00	2.27	13.00	21.50
296.00	202.40	2.68	25.00	25.00	208.40	22.60	1.96	12.00	25.50
3.60	99.80	2.47	25.00	25.00	11.60	289.10	2.08.	23.75	13.00

1.148MPa without ε_a consideration. According to the research of Tan Phoon and Hight et al.[40], CPT measurement error ε_a is usually depicted by the coefficient of variation (COV), which equals 2~5%, here COV equals to 3% (denoted as ε_a =3%) is adopted in this study, the standard deviation σ_a equals to 3% multiplying 1.148MPa, which will be put into a zero-mean normal distribution $\varepsilon_a \sim N(0, 0.034^2)$. Moreover, CPT penetration resistance p_s(MPa) is denoted as a spatial random field variable Z, its anisotropic structural parameters $\theta = [\mu, n, s, \lambda, \rho_s, \rho_t]$ can be calculated directly by CPT data $z_a \pm \varepsilon_a$ (there are $n_a = 1381$ samples of ε_a, each corresponding to 1000 times simulation), and the structural parameters θ consist of mean value $\bar{\mu} = 1.148\,\text{MPa}$, nugget $\bar{n} = 0.040\,\text{MPa}^2$, partial sill $\bar{s} = 0.211\,\text{MPa}^2$, major range $\bar{\lambda} = 88.616\text{m}$ in the $X1$ coordinate, and other two orthogonal range ratio $\bar{\rho}_s = 0.512$ (i.e., range in the $X2$ coordinate) and $\bar{\rho}_t = 0.031$ (i.e., range in the $X3$ coordinate), the theoretical variogram of the spherical model $\gamma(h)$ is fitted by lag distance h as shown in Figure 13.11. The statistical uncertainty of structural parameters θ is listed in Table 13.3, the maximal uncertainty occurs in the secondary range ratio ρ_s, and its uncertainty is less than 5%, for the reason of relieving the Bayesian

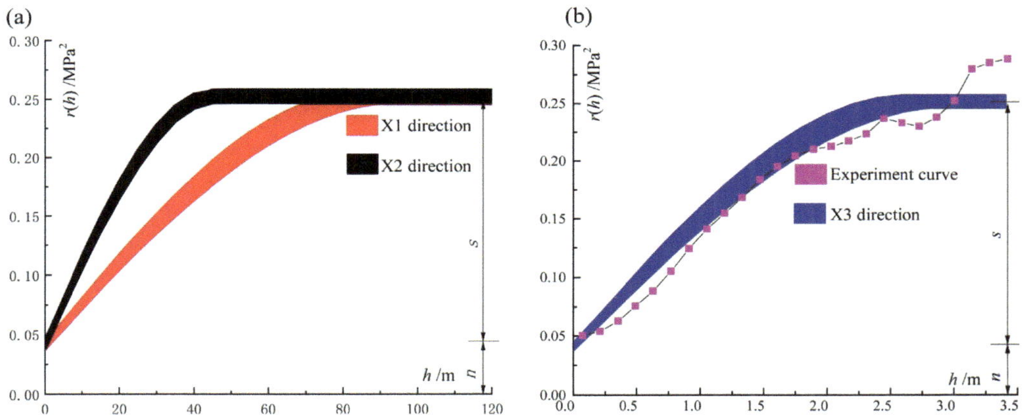

Figure 13.11 Three-dimensional anisotropic variogram of CPT data p_s(MPa), the experimental variogram is simulated with measurement error ε_a consideration. (a) Horizontal variogram of $X1$ $X2$ coordinates and (b) vertical variogram of the $X3$ coordinate.

Table 13.3 Statistical uncertainty of the structural parameters θ of the spatial random field by 1000 times simulation

Structural parameters θ	*Minimum*	*Maximum*	*Mean*	*COV*
μ / MPa	1.152	1.145	1.148	0.001
n / MPa^2	0.043	0.038	0.040	0.022
s / MPa^2	0.216	0.206	0.211	0.008
λ / m	94.310	81.322	88.616	0.030
ρ_s	0.550	0.460	0.512	0.038
ρ_t	0.034	0.029	0.031	0.035

inversely modeling burdens, the structural parameters θ will adopt the mean values to delegate the statistical uncertainty.

Laboratory data z_b consists of $n_{b1} = 21$ compression modulus E_s(MPa) and another $n_{b2} = 12$ consolidated shear cohesion coefficient c_{cq}(kPa), and internal friction angle $\varphi_{cq}(^0)$, respectively. Tan Phoon and Hight et al.[41] suggested that COV equals 10~37% of E_s(MPa), and Thermann Gau and Tiedemann[42] summarized that COV equals 2%–17% of c_{cq}(kPa) and $\varphi_{cq}(^0)$. Therefore, the laboratory measurement error ε_b is uniformly denoted as 10% of E_s(MPa), c_{cq}(kPa), $\varphi_{cq}(^0)$, and then building the zero-mean normal distribution $\varepsilon_b \sim N(0,\sigma_b^2)$, respectively.

13.4.1.2 Transformation model baseline

The spatial random field consists of $n_a = 1381$ CPT data p_s(MPa) and measurement error is $\varepsilon_a \sim N(0,0.034^2)$, the structural parameters of spatial random field adopt the deterministic mean values $\theta = [\bar{\mu},\bar{n},\bar{s},\bar{\lambda},\bar{\rho}_s,\bar{\rho}_t]$, and sequential Gaussian simulation algorithm projects CPT data prediction to collocate with laboratory data. The baseline tries to test the effectiveness of applying the Bayesian approach on transformation model calibration within the spatial random field theory, the linear mathematic formula of converting penetration resistance p_s(MPa) to compression modulus E_s(MPa) is taken into demonstration in the silty clay soil,

$$E_s = k \cdot p_s + q \pm \varepsilon_T\,. \tag{13.63}$$

According to prior knowledge of linear regression between CPT data p_s(MPa) and its spatially closest laboratory data of compression modulus E_s(MPa), the unknown parameters $m = (k,q,\varepsilon_T)$ of the transformation model M is assumed as $k = 4.453$, $q = 0.932$, and $\varepsilon_T \sim N(0,0.2^2)$. Then, taking out one spatial random field realization given z_a, ε_a, θ, and m will generate the pseudo-laboratory data of E_s(MPa), for convenience, the coordinates of $n_{b1} = 21$ laboratory test of the background project are borrowed in Table 13.4. Proposing the COV of ε_b equals 10% (denoted as $\varepsilon_b = 10\%$) of the pseudo-mean value $6.228\,\text{MPa}$, the measurement error is $\varepsilon_b \sim N(0,0.623^2)$. The baseline will be used to evaluate the effectiveness of the Bayesian inversely modeling method.

In the Bayesian framework, the prior knowledge of unknown coefficients and uncertainty $m = (k,q,\varepsilon_T)$ are usually assumed as maximal uncertainty when there is a non-informative knowledge assumption: $k \sim U(0,10)$, $q \sim U(-1,5)$, $\sigma_T \sim U(0,0.7)$, and $\varepsilon_T \sim N(0,\sigma_T^2)$ can cover the potential value intervals of m. The prior joint distribution $p[m = (k,q,\varepsilon_T)]$ is sampled $N_m = 1000$ times from Latin hypercube space (Stein, 1987); furthermore, each sample $m_i, i = 1,\ldots,N_m$ will take $N_a = 200$ spatial random field realizations and $N_b = 100$ calibrations, which is used to calculate the likelihood function value $p(z_b \mid \theta, z_a, m)$. The CPU time for completing the above calculations is 8 h of five threads on a desktop computer that has Intel Xeon (R) dual CPU processors of 2.00 GHz and 4GB RAM.

Figure 13.12 (a–c) gives out the posterior PDF $m = (k,q,\varepsilon_T)$ of the transformation model M baseline, except for the posterior mode of slope k underestimates 4% of the true value, which considers the measurement errors ε_a and ε_b simultaneously as shown in Figure 13.12(d), Bayesian approach captures robustly the true values, the two types of

Table 13.4 Compression modulus E_s(MPa) baseline of spatial random field

Coordinates /m			E_s / MPa	Coordinates /m			E_s /MPa	Coordinates /m			E_s /MPa
X1	X2	X3		X1	X2	X3		X1	X2	X3	
30.00	185.8	2.00	4.75	111.10	77.80	2.64	7.10	183.10	113.80	2.42	4.77
30.10	77.80	2.13	7.72	138.00	257.80	1.29	5.71	210.10	257.90	2.12	4.86
32.10	292.10	2.40	8.09	138.10	149.80	2.61	11.65	237.00	293.80	3.29	6.29
57.10	257.80	2.58	7.53	160.60	50.90	1.95	7.92	237.10	77.90	1.03	3.55
84.10	113.90	1.68	4.24	174.00	275.90	1.48	8.64	237.60	347.90	2.72	3.23
86.40	345.80	2.48	6.97	183.00	347.80	1.92	2.79	291.10	149.90	2.27	6.82
109.60	185.80	1.54	9.27	183.10	221.80	1.68	2.59	297.10	95.90	1.91	6.30

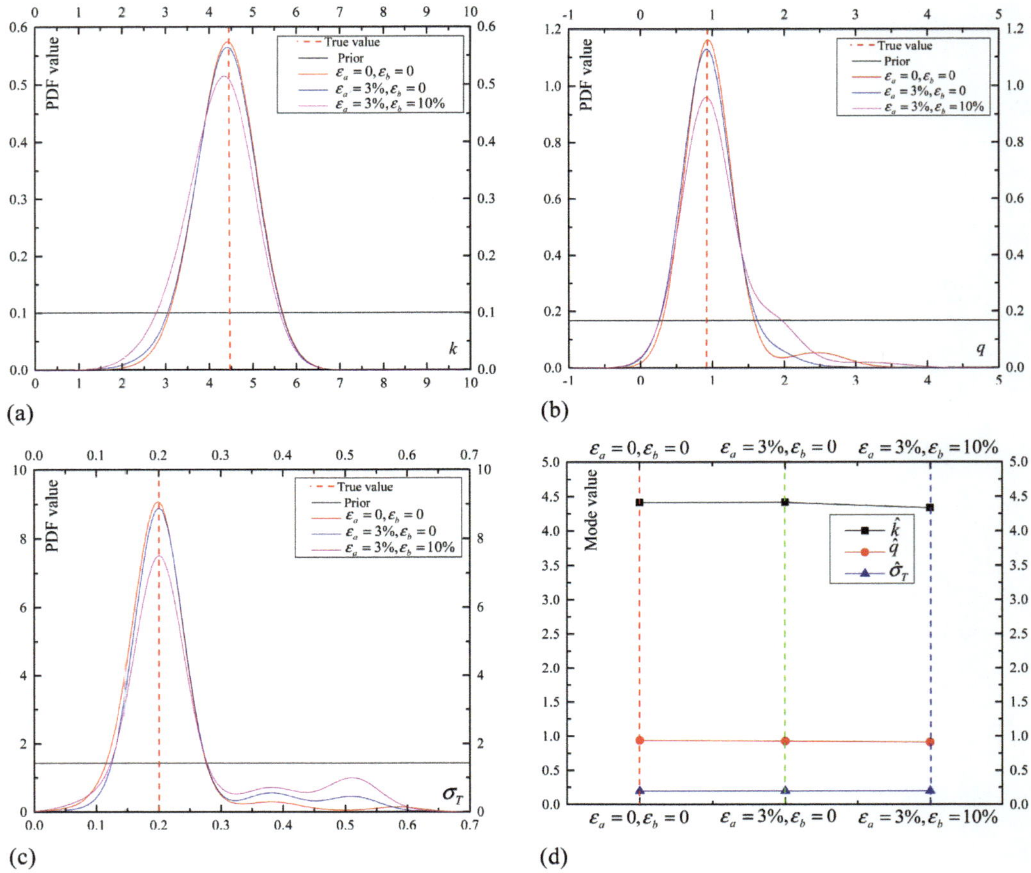

Figure 13.12 Posterior probabilistic characterizations of transformation model baseline within spatial random field theory: (a) posterior PDFs of slope coefficient k, (b) posterior PDFs of intercept coefficient q, (c) posterior PDFs of transformation model uncertainty $\varepsilon_T \sim N(0,\sigma_T^2)$, and (d) posterior mode value comparisons of transformation model baseline.

measurement error impact the posterior variance dramatically, but with subtle influence to the posterior model values, the detailed probabilistic characterizations are listed in Table 13.5.

Sampling-based solutions for the Bayesian framework inevitably bring statistical uncertainty into the system, especially the conditional simulation process of the spatial random field, Latin hypercube sampling algorithm takes $N_m = 1000$ samples, which guarantees to traverse the whole probability space of the joint multivariate distribution $p[m = (k, q, \varepsilon_T)]$. There are $N_a = 200$ spatial random field realizations and $N_b = 100$ calibrations of ε_b corresponding to each sample $m_i, i = 1, \ldots, N_m$. Spatial random field realizations provide the predictions z_b to calculate the joint multivariate likelihood function $p(z_b \mid \theta, z_a, m)$; therefore, the following discussion will focus on the statistical uncertainty of spatial random field realization to impact the posterior results.

Table 13.5 Posterior probabilistic characterizations of transformation model baseline

Transformation model			*Posterior PDF*			
Coefficient	*True value*	*Measurement errors*	*Mean*	*Mode*	*Variance*	*PDF value*
k	4.453	$\varepsilon_a = 0, \varepsilon_b = 0$	4.392	4.415	0.501	0.574
		$\varepsilon_a = 3\%, \varepsilon_b = 0$	4.356	4.415	0.543	0.564
		$\varepsilon_a = 3\%, \varepsilon_b = 10\%$	4.224	4.251	0.652	0.510
q	0.932	$\varepsilon_a = 0, \varepsilon_b = 0$	1.025	0.941	0.230	1.162
		$\varepsilon_a = 3\%, \varepsilon_b = 0$	0.965	0.928	0.165	1.129
		$\varepsilon_a = 3\%, \varepsilon_b = 10\%$	1.101	0.874	0.343	0.961
$\varepsilon_T \sim N(0, \sigma_T^2)$	0.2	$\varepsilon_a = 0, \varepsilon_b = 0$	0.209	0.199	0.006	9.077
		$\varepsilon_a = 3\%, \varepsilon_b = 0$	0.227	0.200	0.008	8.895
		$\varepsilon_a = 3\%, \varepsilon_b = 10\%$	0.248	0.200	0.014	7.514

In the scenario of taking both measurement errors $\varepsilon_a, \varepsilon_b$ into account, calculating the likelihood function $p(z_b \mid \theta, z_a, m)$ given the different number of spatial random field realizations, $N_a = 50,..,200$, increment of 25 realizations, the results of Figure 13.13 show that the posterior mode value $\hat{k}$ varies from 94% ($N_a = 50$) to 99% ($N_a = 175$) comparing with the final result of $N_a = 200$ realizations, the impacts of spatial random field realization keeps in stable value after $N_a = 175$ realizations for the posterior mode value $\hat{k}$; the variation of the posterior mode value $\hat{q}$ is similar as $\hat{k}$, the value approximates

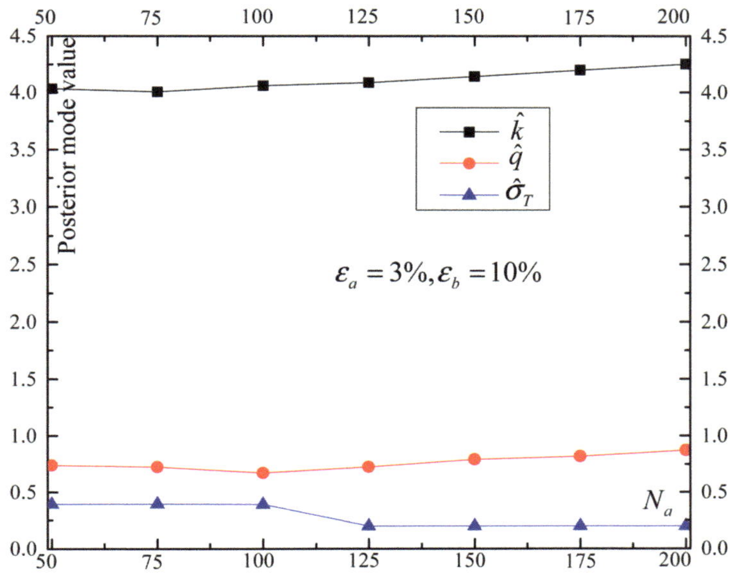

Figure 13.13 Statistical uncertainty of Bayesian reverse modeling method, for example, posterior mode values of transformation model baseline varying with spatial random field realizations.

to 95% ($N_a = 175$) of the final result; but for the transformation model uncertainty, the posterior mode value $\hat{\sigma}_T$ changes significantly from $N_a = 100$ jumping to $N_a = 125$, and then the result keeps in stable value. The statistical uncertainty of the Bayesian inversely modeling method will decrease to 6% when the spatial random field simulations increase $N_a = 175$. Therefore, the realization number $N_a = 200$ would guarantee the Bayesian reversely modeling precision, and the baseline study proves that the Bayesian method is effective for calibrating the transformation model within spatial random field theory.

13.4.1.3 Transformation model calibration

The only difference between the transformation model baseline and engineering application is the latter provides real-life laboratory data, such as geotechnical parameters $E_s(\mathrm{MPa})$, $c_{cq}(\mathrm{kPa})$, $\varphi_{cq}(^{\circ})$ are shown in Tables 13.1 and 13.2, the engineering application is divided into two steps: firstly, the Bayesian inversely modeling method calibrates the transformation model M within spatial random field theory, and comparing with traditional regression models in cross-validation; then applying the calibrated transformation models into three-dimensional site characterization of the background project.

There are $n_a = 1381$ CPT data $p_s(\mathrm{MPa})$, and the COV of measurement error ε_a equals 3%. Laboratory data consists of $n_{b1} = 21$ compression modulus $E_s(\mathrm{MPa})$, $n_{b2} = 12$ consolidation shear parameters $c_{cq}(\mathrm{kPa})$, $\varphi_{cq}(^{\circ})$, and COV of measurement error ε_b equals 10%, respectively. The spatial random field of CPT data still takes the deterministic structural parameters $\theta = [\bar{\mu}, \bar{n}, \bar{s}, \bar{\lambda}, \bar{\rho}_s, \bar{\rho}_t]$, the sampling number of the prior distribution of the transformation model M is $N_m = 1000$, the number of spatial random field realizations is $N_a = 200$, and the number of laboratory data calibrations is $N_b = 100$ corresponding to each sample. The prior knowledge uncertainty of the transformation model M is assumed as non-informative for all the geotechnical parameters in Table 13.6.

All the finally converted geotechnical parameters should be positive values whatever the negative coefficients imported in the transformation models, for example, although the compression modulus $E_s(\mathrm{MPa})$ keeps a significantly negative correlation with CPT data $p_s(\mathrm{MPa})$, the calibrated transformation model should guarantee positive geotechnical parameters; for the internal friction angle, linear-logarithmic transformation model will provide more robust prediction than the linear one, the vertically effective earth pressure σ' of CPT sampling depth is calculated as

$$\sigma' = \sum \gamma_i h_i + \sum \gamma_j' h_j \, , \tag{13.64}$$

Table 13.6 Non-informative prior knowledge of transformation models between CPT data and regular geotechnical parameters

Geotechnical parameters	*Transformation model*	*Slope k*		*Intercept q*		$\varepsilon_T \sim N(0, \sigma_T^2)$	
		Lower	*Upper*	*Lower*	*Upper*	*Lower*	*Upper*
E_s / MPa	$E_s = k_E \cdot p_s + q_E \pm \varepsilon_{T,E}$	0	10.0	−1.0	5.0	0	0.7
c_{cq} / kPa	$c_{cq} = k_c \cdot p_s + q \pm \varepsilon_{T,c}$	−25.0	0	10.0	60.0	0	7.0
φ_{cq} /°	$\varphi_{cq} = k_\varphi \cdot \ln(p_s / \sigma') + q_\varphi \pm \varepsilon_{T,\varphi}$	0	15.0	−6.0	1.0	0	6.0

where γ_i is the ith soil unit weight with the thickness h_i above the groundwater table and γ'_j is the jth soil unit weight with the thickness h_j beneath the groundwater table.

Figure 13.14 (a–c) provides the posterior PDFs of the unknown parameters $m_E = (k_E, q_E, \varepsilon_{T,E})$ converting CPT data p_s(MPa) into compression modulus E_s(MPa), the posterior mean and mode values are close to each other, the posterior PDF curves show symmetric shapes, and the finally adopted mode value is $\hat{k}_E = 4.442$, $\hat{q}_E = 0.901$, $\hat{\sigma}_{T,E} = 0.2$, and $\hat{\varepsilon}_{T,E} \sim N(0, 0.2^2)$. Scatter data plots between predictions and observations are shown in Figure 13.14 (d), Bayesian result $\hat{m}_E = (\hat{k}_E, \hat{q}_E, \hat{\varepsilon}_{T,E})$ proves that the averaged prediction of compression modulus linearly approximates laboratory data, the gradient equals 0.788, and the R-square value equals 0.713 given $N_m = 1000$ samples $(z_a, \varepsilon_a, \theta, \widehat{m_E})$, each sample corresponding to $N_a = 200$ realizations of a spatial random field, although the linear regression method between CPT data and their closest neighbor laboratory data hold the higher gradient, the R-square value displays that there is the relatively lower linear correlation of traditional regression method comparing with Bayesian method. Figure 13.14 (e) proves that Bayesian predictions always cover the laboratory data within one-time standard deviation (SD), but several laboratory data aren't included in the linear regression prediction with one-time SD (see linear regression method in the Appendix).

Figure 13.15 (a–c) provides the posterior PDFs for unknown parameters $m_c = (k_c, q_c, \varepsilon_{T,c})$ to convert CPT data p_s(MPa) to consolidated shear cohesion coefficient c_{cq}(kPa), the posterior PDF curves show non-normal probabilistic characterizations, and the posterior mean and mode values are in the same magnitude, but the uncertainty is larger than other geotechnical parameters, therefore, the mode values are adopted: $\hat{k}_c = -12.176$, $\hat{q}_c = 33.734$, $\hat{\sigma}_{T,c} = 4.199$, and $\hat{\varepsilon}_{T,c} \sim N(0, 4.199^2)$, which guarantees the biggest posterior PDF values. However, according to scattered data plot of $n_{b2} = 12$ laboratory data and prediction in the cross-validation, Figure 13.15 (d) certificates that averaged prediction of consolidated shear cohesion coefficient approximates diagonally to the laboratory data, the gradient equals to 0.978, it has subtle improvements comparing with linear regression method. In Figure 13.15 (e), Bayesian predictions cover the laboratory data within a one-time SD, and four laboratory data aren't included in the linear regression prediction with a one-time SD.

Figure 13.16 (a–c) provides the posterior PDFs for unknown parameters $m_\varphi = (k_\varphi, q_\varphi, \varepsilon_{T,\varphi})$ to convert CPT data p_s(MPa) into consolidated shear internal friction angle $\varphi_{cq}(^\circ)$, according to former research experiences, there are two subtle changes in the transformation model to improve prediction quality, firstly, the vertically effective earth pressure σ'(MPa) is taken into account; then, the linear-logarithmic formula is adopted. The posterior PDF curves show non-normal probabilistic characterizations, but still in the symmetric shapes, and the posterior mode values approximate to the mean values, the ultimately adopted mode values are $\hat{k}_\varphi = 9.048$, $\hat{q}_\varphi = -2.672$, $\hat{\sigma}_{T,\varphi} = 2.391$, and $\hat{\varepsilon}_{T,\varphi} \sim N(0, 2.391^2)$. According to scattered data plot of $n_{b2} = 12$, laboratory data and prediction in the cross-validation, Figure 13.16(d) proves that the averaged prediction of consolidated shear internal friction angle approximates the laboratory data along with 45° degree straight line, the gradient equals 0.989, and the R-square value equals 0.477, it has been substantially improved comparing with traditional regression method. In Figure 13.16 (e), similar to the other two geotechnical parameters, Bayesian predictions

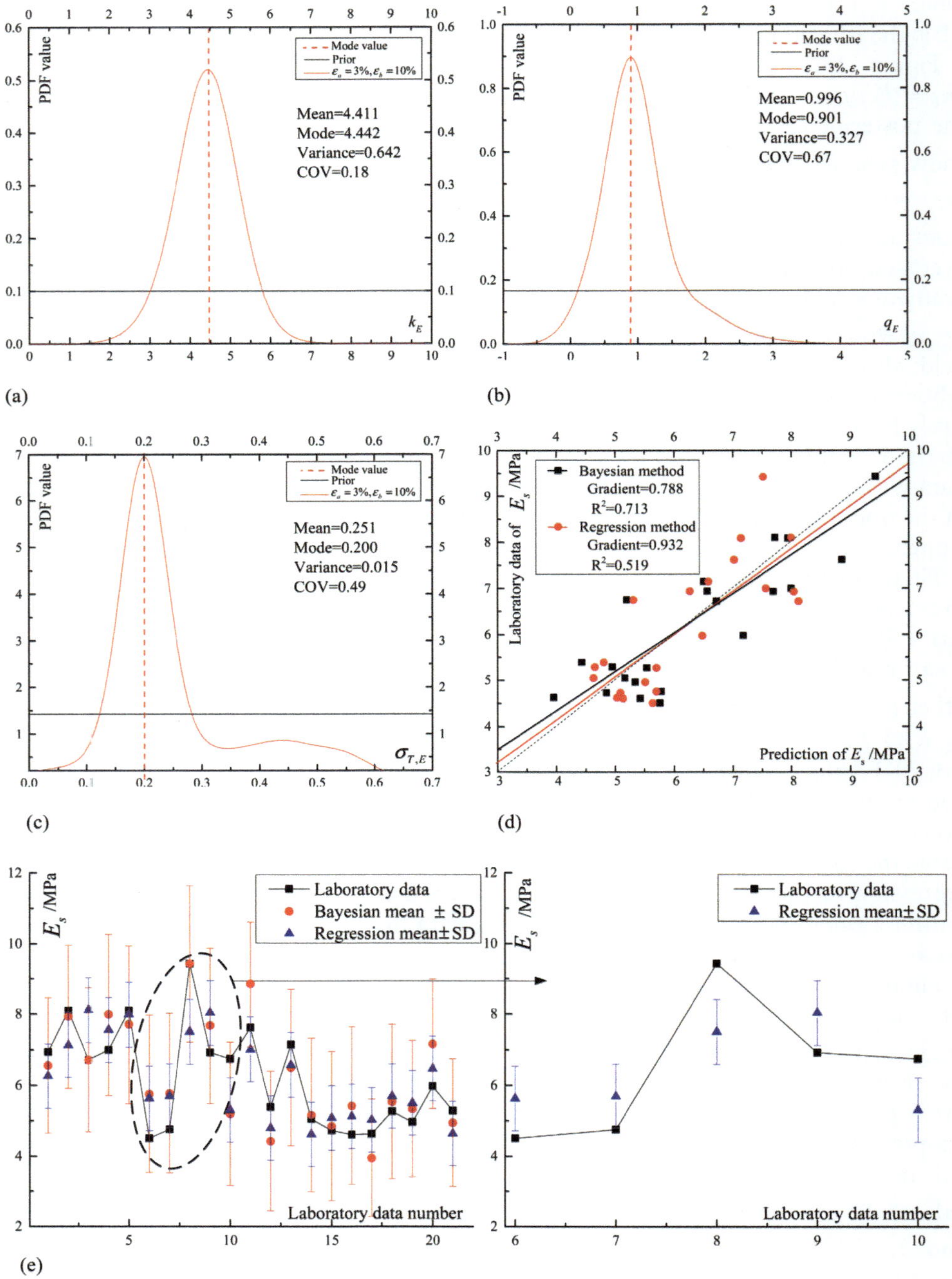

Figure 13.14 Posterior distribution and cross-validation of transformation model M between CPT data p_s(MPa) and compression modulus E_s(MPa). (a) Posterior PDF of coefficient k_E, (b) posterior PDF of coefficient q_E, (c) posterior PDF of transformation model uncertainty $\varepsilon_{T,E} \sim N(0,\sigma^2_{T,E})$, (d) scattered data plot of compression modulus E_s(MPa) in the cross-validation, and (e) predictive intervals comparison of compression modulus E_s(MPa) in the cross-validation.

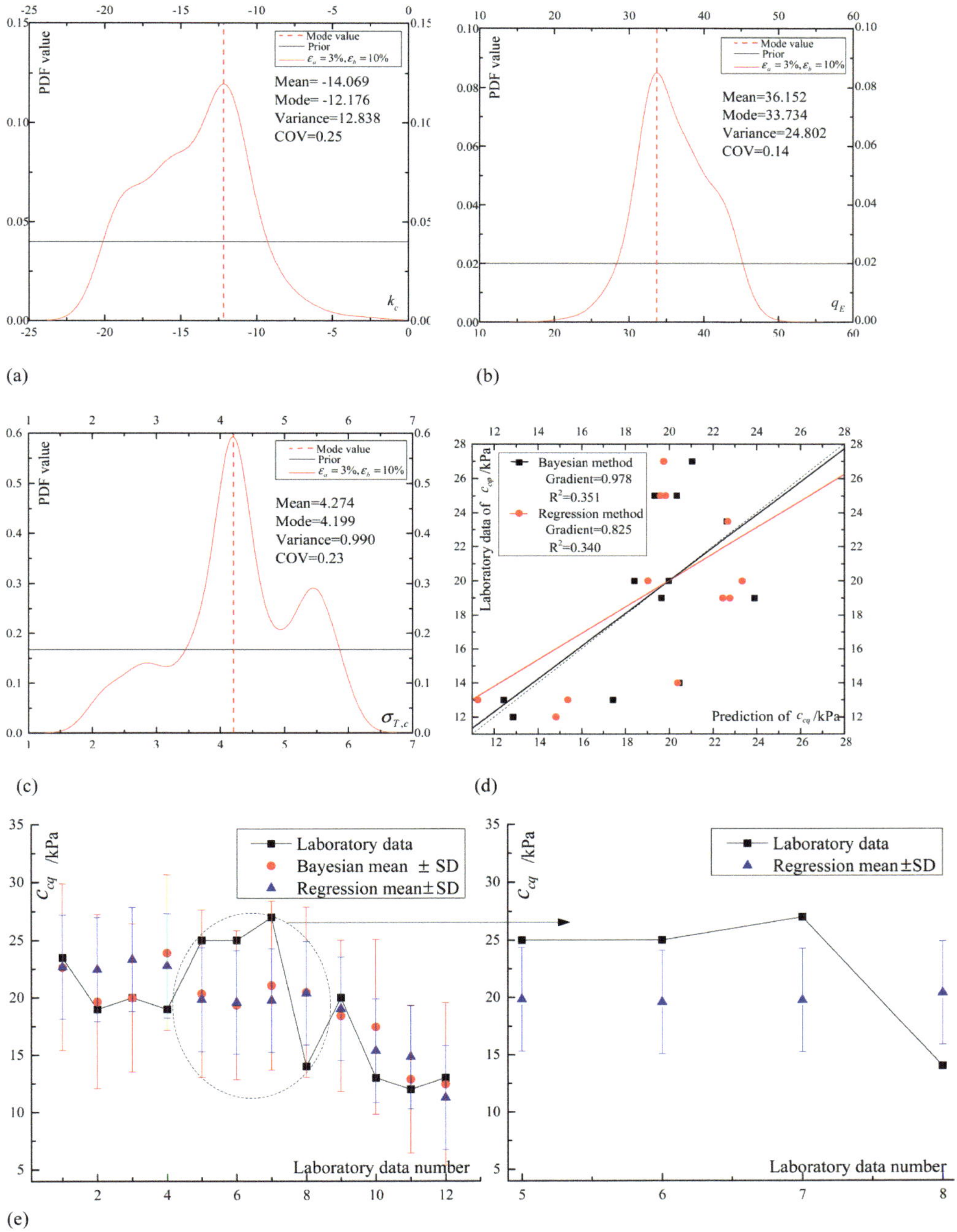

Figure 13.15 Posterior distribution and cross-validation of transformation model M between CPT data p_s(MPa) and consolidated shear cohesion coefficient c_{cq}(kPa). (a) Posterior PDF of coefficient k_c, (b) posterior PDF of coefficient q_c, (c) posterior PDF of transformation model uncertainty $\varepsilon_{T,c} \sim N(0, \sigma_{T,c}^2)$, (d) scattered data plot of consolidated shear cohesion coefficient c_{cq}(kPa)in the cross-validation, and (e) predictive intervals comparison of consolidated shear cohesion coefficient c_{cq}(kPa) in the cross-validation.

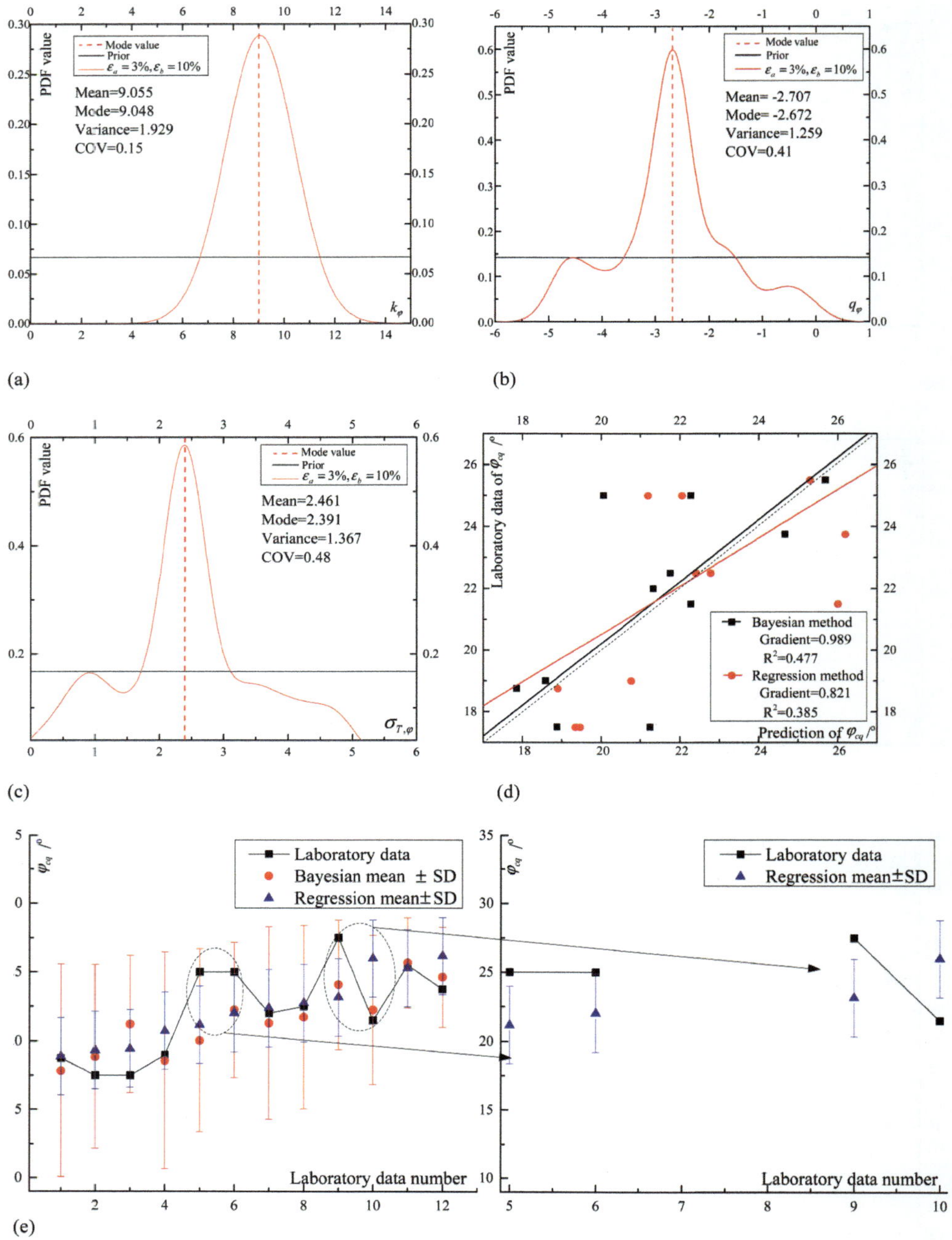

Figure 13.16 Posterior distribution and cross-validation of transformation model M between CPT data p_s(MPa) and consolidated shear internal friction angle φ_{cq}(°). (a) Posterior PDF of coefficient k_φ, (b) posterior PDF of coefficient q_φ, (c) posterior PDF of transformation model uncertainty $\varepsilon_{T,\varphi} \sim N(0, \sigma_{T,\varphi}^2)$, (d) scattered data plot of consolidated shear internal friction angle φ_{cq}(°) in the cross-validation, and (e) predictive intervals comparison of consolidated shear internal friction angle φ_{cq}(°) in the cross-validation.

Table 13.7 Calibrated transformation models between CPT data and regular geotechnical parameters in the silty clay soil

Geotechnical parameters	*Transformation model*	*Slope* $\hat{k}$	*Intercept* $\hat{q}$	$\hat{\varepsilon}_T \sim N(0, \hat{\sigma}_T^2)$
E_s / MPa	$E_s = k_E \cdot p_s + q_E \pm \varepsilon_{T,E}$	4.442	0.901	0.200
c_{cq} / kPa	$c_{cq} = k_c \cdot p_s + q \pm \varepsilon_{T,c}$	-12.176	33.734	4.199
φ_{cq} /°	$\varphi_{cq} = k_\varphi \cdot \ln(p_s / \sigma') + q_\varphi \pm \varepsilon_{T,\varphi}$	9.048	-2.672	2.391

cover the laboratory data within one-time SD, and there are still three laboratory data excluded from the linear regression prediction with one-time SD.

Bayesian reverse modeling method successfully calibrated the transformation models between CPT data p_s(MPa) and geotechnical parameters E_s(MPa), c_{cq}(kPa), and φ_{cq}(°) within spatial random field theory, which takes uncertainty propagation of geotechnical investigation into consideration, the finally adopted parameters of transformation models are listed in Table 13.7, which will be employed into three-dimensional site characterization of the background project.

13.4.1.4 Three-dimensional site characterization

The massive CPT data can support the three-dimensional site characterization of the background project, which provides discrete elements for further soil mechanics analysis. The whole domain of the silty clay layer is meshed into 2000 nodes, and 8088 tetrahedral elements; the maximal element edge size is meshed less than the smallest range of spatial random field, which guarantees the spatial variability of soil would be considered in the mechanical analysis.

The CPT data p_s(MPa), measurement error ε_a, structural parameter θ, and calibrated transformation model M are utilized in the conditional simulation of spatial random field, it needs to be pointed out that the laboratory data is also denoted as conditional observation, Figure 13.17 lists out the three-dimensional discretization (e.g., mean and COV values) given $N_m = 1000$ samples of (z_a, ε_a, θ, m), here, only measurement error ε_a is in variation, other parameters keep in constant values; each sample corresponds to $N_a = 200$ realizations of spatial random field, and then transforming the penetration resistance p_s(MPa) realization of each element's centroid into geotechnical parameters E_s(MPa), c_{cq}(kPa), and φ_{cq}(°).

Three-dimensional site characterizations show that E_s(MPa) c_{cq}(kPa) versus φ_{cq}(°) keeps a globally negative correlation, mean value $\bar{E}_s$ located in the interval [3.24, 10.76], and the COV is confined in [0.15, 0.47] kPa; the variation of c_{cq} is bigger than E_s and φ_{cq}, the mean value $\bar{c}_{cq}$ varies from 6.55 kPa to 27.61 kPa, and the biggest COV value equals to 1.00; linear-logarithmic transformation model is adaptive to consolidated shear internal friction angle, the mean value $\bar{\varphi}_{cq}$ varies from 13.35° to 28.49°, and the COV of φ_{cq} equals to 0.10 ~ 0.66.

In summary, the three-dimensional site characterization depends on the massive CPT data, spatial random field theory, and transformation models, the conditional simulation method would give out the spatial distribution of the regular geotechnical parameters.

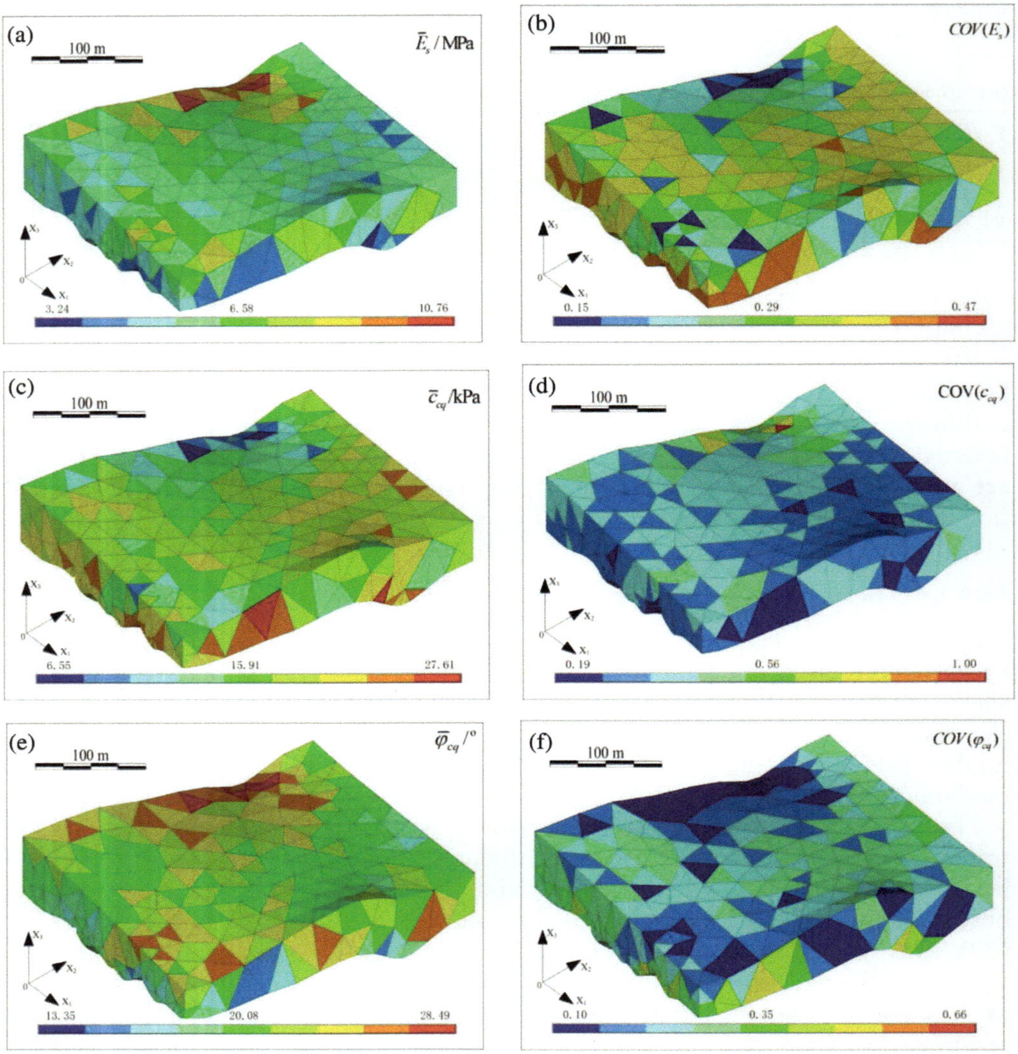

Figure 13.17 Three-dimensional site characterization of geotechnical parameters E_s(MPa), c_{cq}(kPa), and φ_{cq}(°) in the silty clay layer: (a) mean value $\bar{E}_s$(MPa) spatial distribution; (b) COV spatial distribution of E_s(MPa); (c) mean value $\bar{c}_{cq}$(kPa) distribution; (d) COV spatial distribution of c_{cq}(kPa); (e) mean value $\bar{\varphi}_{cq}$(°) distribution; and (f) COV spatial distribution of φ_{cq}(°).

13.4.2 Suzhou river drainage and storage pipeline project

Suzhou River Deep Tunnel Reservoir Project adopts a shield excavating method with a total length of 15 km and a buried depth of 50 m~60 m. This project includes eight extra-deep shafts and complexes. The trial part is about 1.67 km long, which includes Yunling and Miaopu complexes. The depth of the underground diaphragm wall reaches 105 m

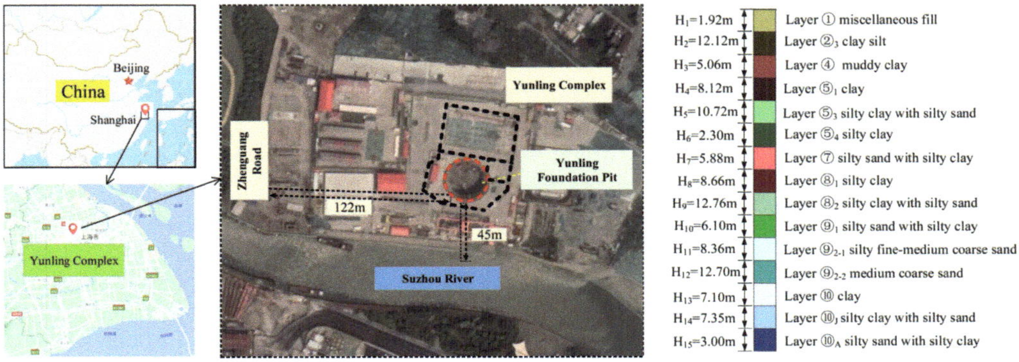

Figure 13.18 Schematic diagram of the surrounding environment of Yunling complex.

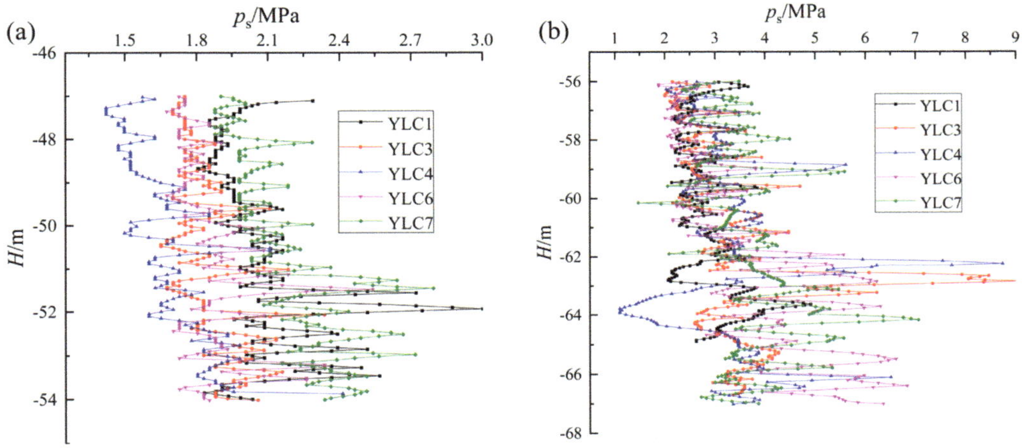

Figure 13.19 The in-situ CPT data in layer ⑧. (a) In-situ specific penetration resistance p_s of layer ⑧$_1$ silty clay; (b) in-situ specific penetration resistance p_s of layer ⑧$_2$ silty clay with sand.

and the excavation depth approaches 57.8 m. The surrounding environment and soil layers are shown in Figure 13.18.

There are five CPT curves in Yunling complexes, as shown in Figure 13.19. In the figure, "YLC" represents the in-situ CPT borehole, which is in Yunling complexes, the numbers "i" (i = 1, 3, 4, 6, 7) represent the borehole number of CPT test.

13.4.2.1 Prior knowledge

Before the stochastic analysis, the key constitutive parameters are necessarily determined by sensitivity analysis. According to the preliminary geotechnical investigation, the constitutive parameters of the soil layer ⑧ are divided into seven parametric levels as shown in Table 13.8. According to the division, the orthogonal table $L_{49}\left(7^8\right)$ is selected for the sensitivity analysis, which has 49 parameter combination. In the calculation, the last two columns of the orthogonal table are regarded as error columns. Substitute each parameter

Table 13.8 Parametric levels of the constitutive parameters

Parametric level	*M*	λ	κ	*OCR*	ν	e₀
1	1.259	0.059	0.007	1.01	0.308	0.602
2	1.276	0.076	0.012	1.26	0.312	0.692
3	1.293	0.093	0.017	1.51	0.316	0.782
4	1.31	0.11	0.022	1.76	0.32	0.872
5	1.327	0.127	0.027	2.01	0.324	0.962
6	1.344	0.144	0.032	2.26	0.328	1.052
7	1.361	0.161	0.037	2.51	0.332	1.142

combination into Equation (13.26) to obtain the corresponding specific penetration resistance p_s. And then, according to Equations (13.30)–(13.32), the critical value F_i is obtained and the key constitutive parameters are determined.

Table 13.9 lists the sensitive orders of constitutive parameters under undrained/drained conditions, which are $OCR > \kappa > \lambda > e_0 > M > \upsilon$ and $OCR > \kappa > e_0 > \lambda > M > \upsilon$, respectively. In general, the overconsolidation ratio OCR and the slope of the loading/reloading line κ have an extreme sensitivity. The sensitivity under drained conditions is greater than that under undrained conditions. Compared with the undrained condition, the slope of normal compression λ and the initial void ratio e_0 have a high sensitivity under drained conditions. In the preliminary geotechnical investigation, a total of 105 initial void ratios e_0 are obtained, which has a strong correlation with the depth. Although the initial void ratio e_0 is an important physical parameter in soil mechanics, sensitivity analysis shows that it has less impact on the specific penetration resistance p_s, as well as it is easy to obtain in engineering practice. Thus, it is not considered the key constitutive parameter. To reflect the difference in the initial void ratio e_0 at a specific depth accurately, the polynomial regression method is used to establish the function between the initial void ratio e_0 and the depth. In addition, as an important constitutive parameter, the slope of the critical state line M should be calibrated. Hence, $\theta = [M, \lambda, \kappa, OCR]$ are denoted as the key constitutive parameters.

Prior distribution is the first step to calibrate constitutive parameters, which is organized based on the former geotechnical investigation. Due to the limitation of cost and time, only a few borehole logs and laboratory test data can be obtained. It is difficult to denote the constitutive parameters completely and accurately. Hence, uniform distribution is regarded as the PDF of prior knowledge. Prior information on the constitutive parameters in soil layer ⑧ is listed in Table 13.10.

13.4.2.2 Posterior distribution

Preliminary data of the geological investigation indicate that silty clay layer $⑧_1$ has a poor permeability. In the process of CPT, pore water pressure could not discharge, which results in a large value. For the silty clay with a sand layer, the amount of sand will affect the permeability. Silty clay with sand layer $⑧_2$ increases the permeability and the excess pore pressure dissipates immediately. At the same time, the CPT data will be affected by the penetration speed. The slower the penetration speed, the greater the excess pore pressure dissipates. According to the permeable condition, soil sub-layers $⑧_1$ and ⑧ adopt the undrained/drained formulas to be calibrated, respectively.

Table 13.9 Variance of sensitivity analysis

Constitutive parameter	The sum of deviation squares S_i^2		Degrees of freedom f_i	Analytic value of F		Critical value of F	Significance	
	Undrained	Drained		Undrained	Drained		Undrained	Drained
M	43.0	216.3	6	1.25	1.51	$F_{0.01}(6,12) = 4.82$	L	L
λ	53.3	441.8	6	1.55	3.09		L	H
κ	767.1	4305.0	6	22.36	30.11	$F_{0.05}(6,12) = 3.00$	E	E
OCR	4488.7	4467.2	6	130.87	31.24	$F_{0.10}(6,12) = 2.33$	E	E
v	34.1	132.5	6	0.99	0.93		L	L
e_0	51.9	470.4	6	1.51	3.29		L	H
ε_e	68.6	286.0	12	-	-		-	-

Table 13.10 Prior information of the constitutive parameters

Soil Sub-layer	***H*/*m***	*M*	λ	κ	*OCR*	*v*	K_0	e_0
⑧$_1$	46–55	1.26–1.33	0.11–0.16	0.015–0.027	1.01–2.51	0.324	0.48	0.904–1.125
⑧$_2$	55–68	1.28–1.36	0.06–0.16	0.007–0.037	1.01–2.51	0.319	0.47	0.619–1.051

Using the MCMC algorithm to analyze every CPT data will require a lot of computational resources. In addition, the extreme value will have a great influence on the data quality. It is reasonable to choose a specific range of CPT curves for calibration. Thus, statistics of all CPT data within 1.0 m are used as the indirect data z_b^i.

The joint prior distribution $p(\theta)$ takes totally $T = 10{,}000$ Markov chain samples. Statistical characteristics of the posterior distribution for the key constitutive parameters $\theta = [M, \lambda, \kappa, OCR]$ are obtained based on the accepted posterior samples. The margin posterior distributions are shown in Figure 13.20 (a) and (b).

Figure 13.20 demonstrates the prior and posterior distributions of the key constitutive parameters for soil layer ⑧. The mean values of the key constitutive parameters $\theta = [M, \lambda, \kappa, OCR]$ in silty clay layer $⑧_1$ are 1.296, 0.134, 0.024, and 1.438, respectively. The posterior mean values of the corresponding parameters in the silty clay with sand layer $⑧_2$ are 1.318, 0.110, 0.019, and 1.699. Except for the critical state stress ratio M, the SDs of the other three parameters in the silty clay with sand layer $⑧_2$ are greater than that of the silty clay layer $⑧_1$, which indicates that the silty clay with sand layer $⑧_2$ has greater variability. The overconsolidation ratio (OCR) is less than 2.0 after the calibration. Hence, both the two soil sub-layers are in a light overconsolidation state, and the MCC model can describe the mechanical characteristics.

13.4.2.3 Prediction of deep shaft pit deformation

To provide accurate constitutive parameters by stochastic mechanics based Bayesian method and verifying whether the posterior statistics characteristics can reflect the mechanical properties of the deep soil layers, the above calibrated constitutive parameters are used to carry out numerical analysis for the Yunling shaft pit of Suzhou River Deep Tunnel Reservoir Project in Shanghai, China. Then, the numerical simulation results are compared with the field measurements.

13.4.2.3.1 Numerical model

The shaft pit of the Shanghai deep tunnel project is a complex engineering practice. If a two-dimensional model is adopted, it needs to set up a middle beam as an internal support. At the same time, the supporting element of the annular beam, framework beam, and lining cannot be simulated. In addition, the anti-seepage wall is installed outside of the shaft pit. To simulate the actual engineering situation, according to the design schema of the Yunling shaft pit, a three-dimensional finite element model consisting of width = 160 m, length = 200 m, and height = 150 m is demonstrated in Figure 13.21. The inner diameter of the shaft is 34 m. The bottom of the shaft is located in silty clay with a sand layer $⑧_2$. The thickness and depth of the diaphragm wall are 1.5 m and 105 m, respectively.

The MCC model can simulate light overconsolidated clay, but it cannot describe the mechanical properties of sand. As a classical plastic model, the Mohr–Coulomb (MC) model can simulate the primary mechanical behavior of clay and sand simultaneously. In addition, the effect of seepage is also considered in the numerical simulation. If the MCC model is used in the clay stratum and the MC model is used in the sand stratum, the mixture of the constitutive models will greatly cause the problem of computational non-convergence. In addition, the preliminary geotechnical investigation only provides limited

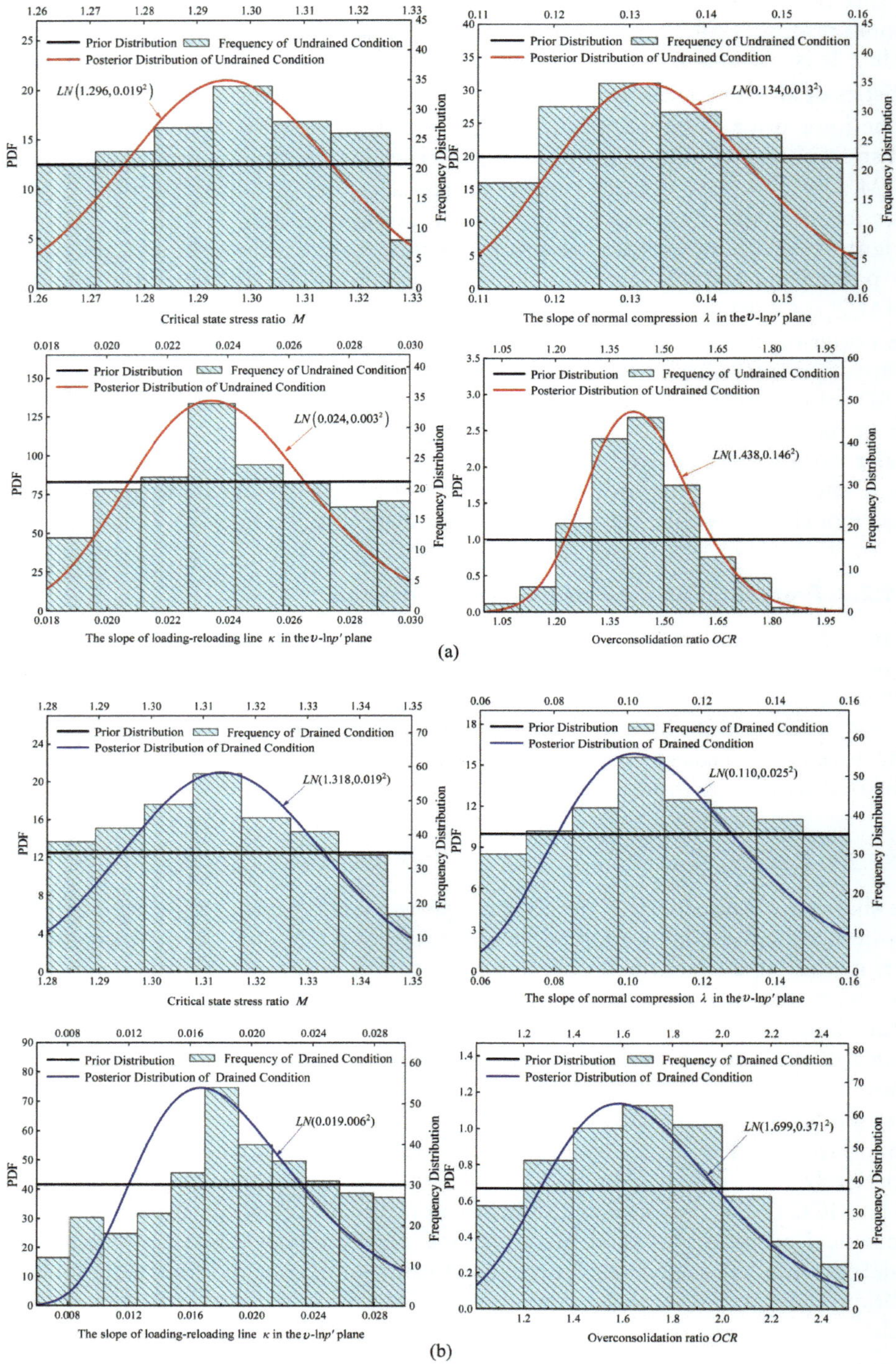

Figure 13.20 Prior and posterior distributions of the key constitutive parameters θ. (a) Silty clay layer ⑧$_1$; and (b) silty clay with sand layer ⑧$_2$.

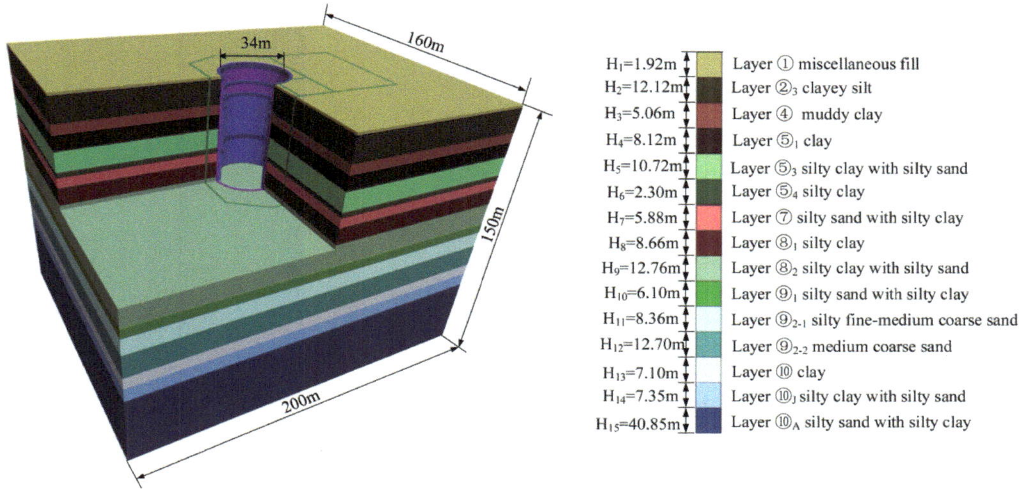

Figure 13.21 Three-dimensional numerical model.

Table 13.11 Basic geotechnical parameters of the MC model

Soil layer	*Density* ρ (kg/m³)	*Compression modulus* E_s (MPa)	*Poisson's ratio* v	*Cohesion c* (kPa)	*Internal frictional angle* φ (°)
①	1900	5.14	0.35	10	10
②$_3$	1860	9.38	0.36	6	27
④	1700	2.68	0.42	3	27.9
⑤$_1$	1760	3.75	0.35	4	30.5
⑤$_3$	1790	5.14	0.36	5	31.6
⑤$_4$	1950	6.38	0.36	16	32.4
⑦	1950	11.58	0.34	2	34
⑨$_1$	1940	14.66	0.31	0	35.5
⑨$_{2\text{-}1}$	2020	15.69	0.3	0	38.4
⑨$_{2\text{-}2}$	1910	17.21	0.31	0	39.8
⑩	1990	11.58	0.34	19	31.6
⑩$_J$	1940	8.56	0.32	16	32
⑩$_A$	1930	12.37	0.32	2	33.5

MCC constitutive parameters. Hence, the MCC model is only used for layer ⑧. For the other soil layers, the MC model is adopted. The basic geotechnical parameters of the MC model are listed in Table 13.11, which are obtained by the preliminary geotechnical investigation. In addition, the dilatancy angle ψ also affects the soil deformation, which adopts the proposed value given by Bolton[43]. Thus, the dilatancy angle ψ of silty clay is set to zero, as well as the dilatancy angle of silty clay with sand is calculated $\psi = \varphi' - 30°$.

For numerical analysis, solid elements are selected for supporting structures such as diaphragm wall, beam, and lining, as well as C30 concrete is used as the elastic material.

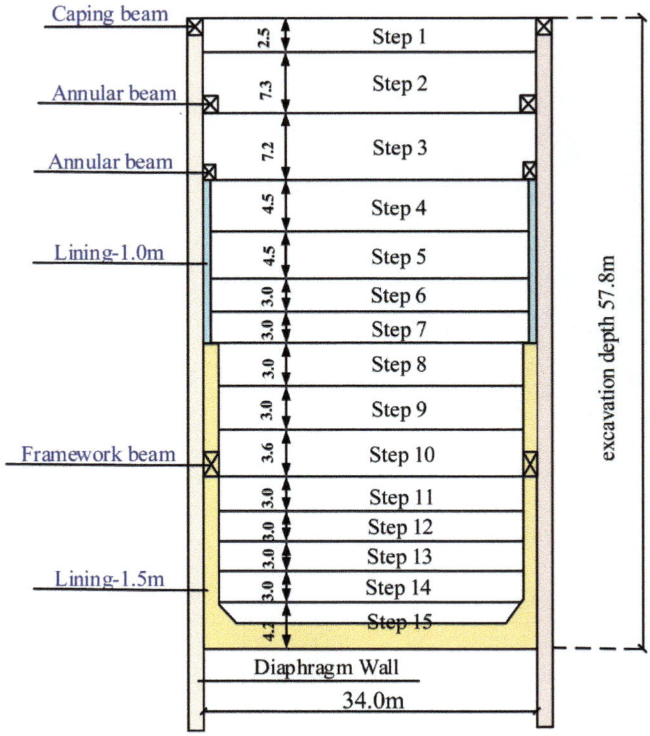

Figure 13.22 Excavation steps of Yunling shaft pit.

The deep shaft is excavated in 15 steps as shown in Figure 13.22. Drainage is carried out according to excavation conditions, and the underground water level is always 3~8 m below the excavation surface.

13.4.2.3.2 Prediction of the pit bottom heave

During foundation pit construction, two monitoring points are set at the depths of 45 m and 50 m, which are arranged at the center of the pit bottom. Figure 13.23 shows the changes in the soil heave at each excavation step, and the "measurement" represents the field measurement. From the perspective of probability, when the normal distribution is followed, the probability of falling within the range of 1.0, 2.0, and 3.0 standard deviations is 68.26%, 95.44%, and 99.74%, respectively. Considering the particularity of geotechnical engineering, 1.0 standard deviation is selected as the value range of geotechnical parameters. In this chapter, the prediction value of the pit bottom heave is calculated by the posterior mean value. Then, according to the range of posterior value, 16 calculated values of the pit bottom heave were obtained after $2^4 = 16$ calculations and the maximum and minimum prediction values were determined. The predicted values are close to the measurements, and the measurements are enclosed in the predicted envelope. It indicates that the stochastic mechanics-based Bayesian method can provide accurate constitutive parameters for the prediction of the pit bottom heave.

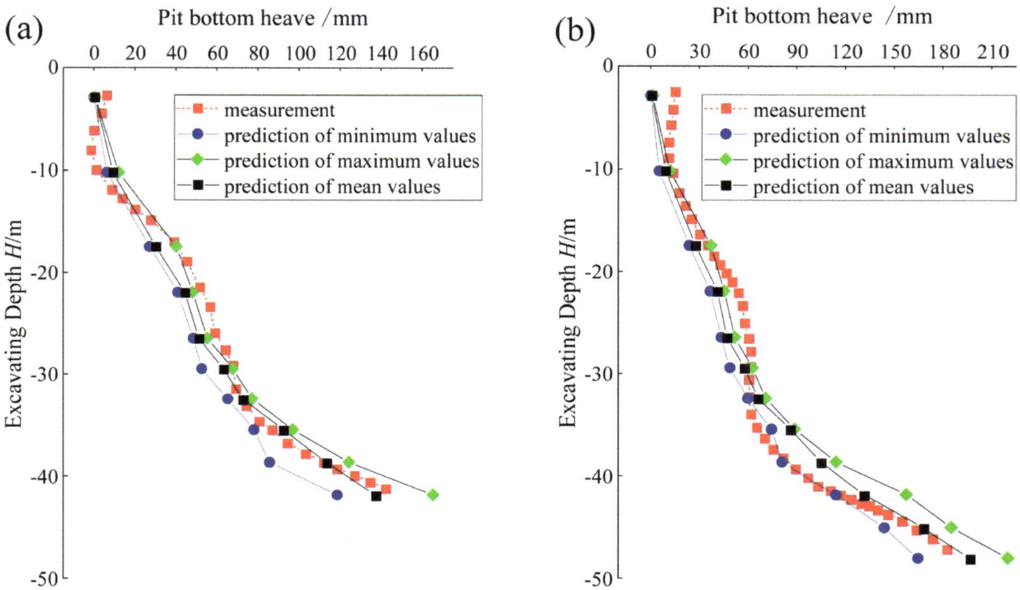

Figure 13.23 Pit bottom heave. (a) Monitoring point at a depth of 45 m; and (b) monitoring point at a depth of 50 m.

13.4.2.3.3 Prediction of the lateral displacement of the diaphragm wall

In addition to soil heave at the center of the pit bottom, the diaphragm wall lateral displacement is another important monitoring item. Figure 13.24 (a, b) shows the lateral displacement of the diaphragm wall when the foundation pit is excavated to 47.6 m and

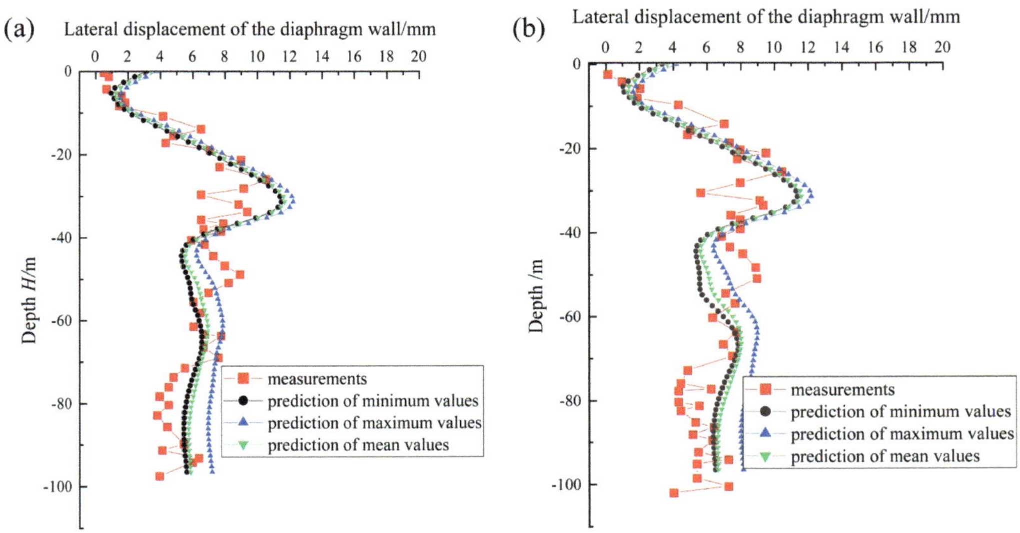

Figure 13.24 Lateral displacements of the diaphragm wall: (a) excavation to 47.6 m; and (b) excavation to 53.6 m.

53.6 m, respectively. The simulated results of the posterior maximum and minimum constitute the upper and lower limits of the lateral displacements for the diaphragm wall. Although the basic geotechnical parameters of other soil layers are not calibrated, the predicted values coincide with the measurements, which further verify the effectiveness of the stochastic mechanics-based Bayesian method.

13.4.3 Bothkennar site in the United Kingdom

Bothkennar Parish is a soft clay testing ground, which is located on the south bank of the Forth River, near the Kincardine Bridge in Scotland, UK. The site is approximately 11 hectares of a low-lying alluvial plain, which is surrounded on three sides by flood dykes. The surface layer is the modern tidal flat sediment with a thickness of 1.5 m. The water table is about 0.75 m below the ground[44]. The geotechnical parameters of the Bothkennar area are listed in Table 13.12. The bottom layer is silty clay with a lower overconsolidation as shown in Figure 13.25.

Many engineers and scholars have carried out geological surveys on the site. Jacobs and Coutts conducted experiments to study the influence of different piezocone types on pore water pressure and found that the filter material has little influence, while the filter location has a greater influence[45]. Two CPTU resistance curves extracted from Hight et al.[44] and Nash et al.[26] are shown in Figure 13.26.

13.4.3.1 Parametric analysis

According to the yield function of CASM, the key constitutive parameters can be selected from M, λ, κ, OCR , r^*, n and e. To determine the influences of constitutive parameters on the CPTU resistances, the single-factor method is utilized to take the parametric sensitivity analysis, and the statistical correlations between the parameters are ignored temporarily. The parametric values at each level are listed in Table 13.13, the mean μ of a single parameter is selected as the baseline of the sensitivity analysis, and the level factors are equally distributed with the baseline, i.e., $\Delta x = (A_{\max} - A_{\min})/6$, where $A_{\max}$ is the maximum, $A_{\min}$ is the minimum of the parameter. Based on the past experimental results and engineering experiences of clayey soil, the value n ranges from 1.0 to 5.0, and the value r^* ranges from 1.5 to 3.0[23]. The values of M, λ, κ, and OCR are listed in Table 13.12. The influences of the constitutive parameters on the corrected cone-tip resistance q_t and pore water pressure u are shown in Figure 13.27 (a, b).

The sensitivity factors of an individual constitutive parameter are listed in Table 13.14. The sensitivity order of a single parameter to q_t is $r^* \approx M > OCR > n > e > \kappa > \lambda$. On the other hand, the sensitivity order to u is $M > OCR > r^* > e \approx \kappa > n > \lambda$. In summary, by comprehensive consideration of parameter acquisition methods and sensitivity results, the key constitutive parameters are selected as r^*, n, M, and OCR

13.4.3.2 Preliminary prediction of CPTU resistances

According to the original geotechnical parameters of the clayey soil as listed in Table 13.13, the CPTU resistances are preliminarily predicted by spherical cavity expansion theory based on CASM. Hence, the theoretical CPTU resistance curve is shown in Figure 13.28 (a, b). The theoretical values of corrected cone-tip resistance q_t and

Table 13.12 Geotechnical parameters of Bothkennar site

Depth **h**(m)	**OCR**	λ	κ	s_u *(kPa)*	G_s *(MPa)*	e	*M*	φ' (°)
2.7~17.3	1.4~1.6	0.2~0.3	0.015~0.025	20.1~50.3	2.6~8.6	0.6~1.2	1.374~1.418	34~35

Notes: Effective unit weight $\gamma = 6.5$ kN/m^3, Poisson ratio ν =0.3.

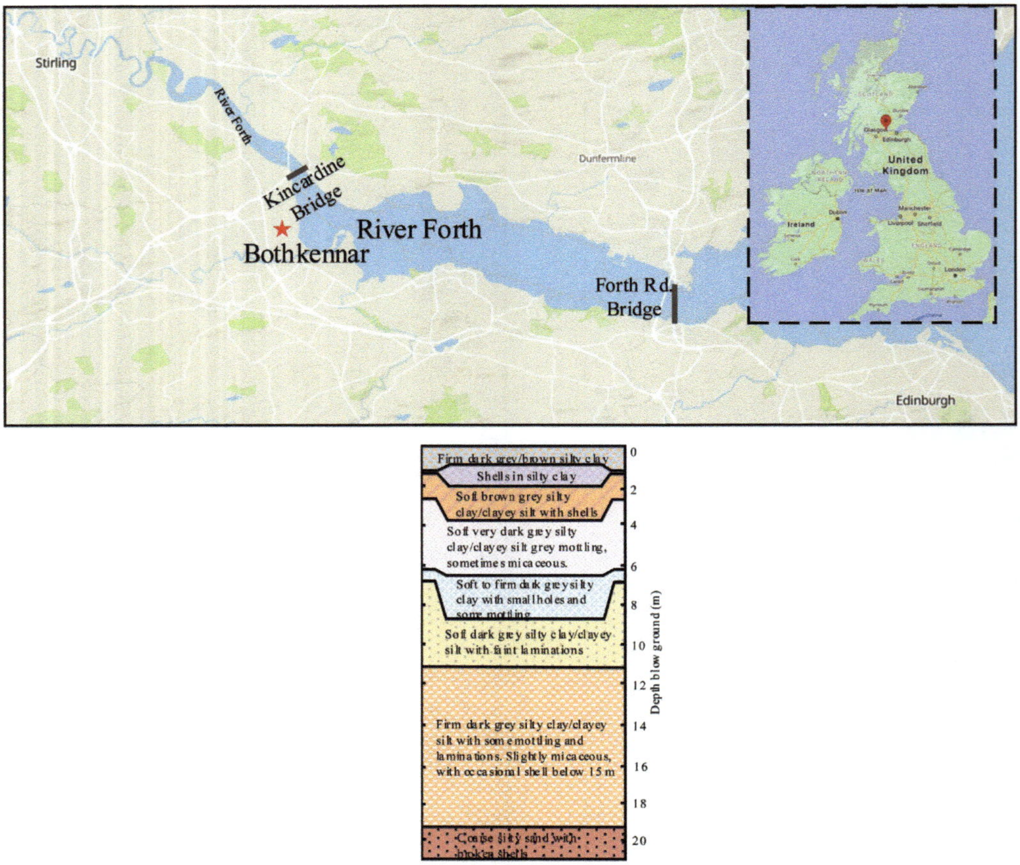

Figure 13.25 Location and soil distribution of the Bothkennar site.

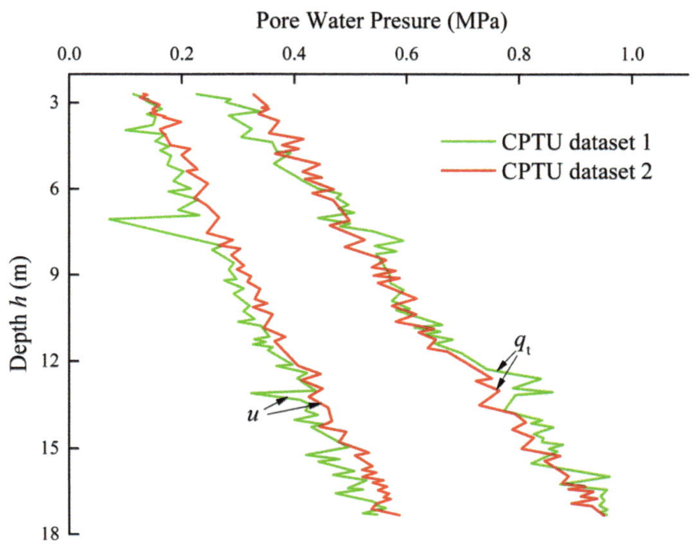

Figure 13.26 In-situ test data of CPTU curves.

Table 13.13 Statistics of parametric sensitivity analysis

Level factor	*M*	λ	κ	r^*	*OCR*	*n*	e
Baseline μ	1.394	0.250	0.020	2.25	1.500	3.00	0.9
$\mu - 3\Delta x$	1.370	0.200	0.014	1.50	1.400	1.00	0.6
$\mu - 2\Delta x$	1.378	0.217	0.016	1.75	1.433	1.67	0.7
$\mu - \Delta x$	1.386	0.233	0.018	2.00	1.467	2.34	0.8
$\mu + \Delta x$	1.402	0.267	0.022	2.50	1.533	3.67	1.0
$\mu + 2\Delta x$	1.410	0.283	0.024	2.75	1.567	4.34	1.1
$\mu + 3\Delta x$	1.418	0.300	0.026	3.00	1.600	5.00	1.2

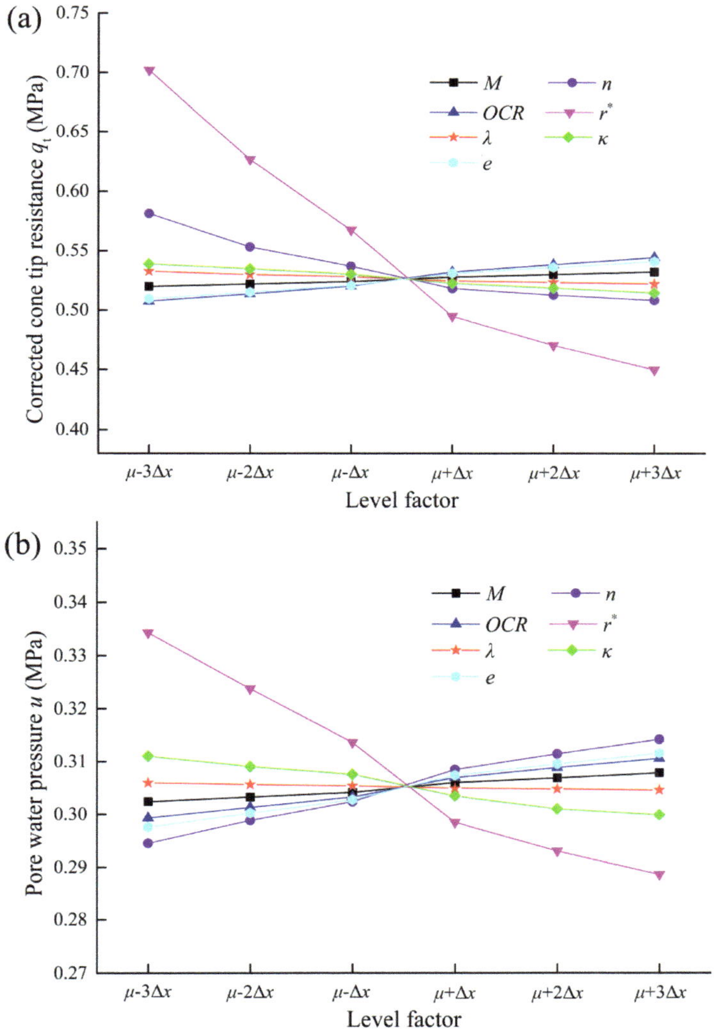

Figure 13.27 Parametric sensitivity of constitutive parameters. (a) Influence on corrected cone-tip resistance q_t, and (b) influence on pore water pressure *u*.

Table 13.14 Sensitivity factors of constitutive parameters

CPTU resistances	Level factor	M	n	OCR	r^*	λ	κ	e
q_t	$\mu \pm 3\Delta x$	0.655	0.103	0.525	0.803	0.049	0.076	0.089
	$\mu \pm 2\Delta x$	0.643	0.086	0.524	0.666	0.045	0.075	0.088
	$\mu \pm \Delta x$	0.602	0.079	0.515	0.621	0.048	0.070	0.086
u	$\mu \pm 3\Delta x$	0.499	0.053	0.277	0.249	0.011	0.060	0.068
	$\mu \pm 2\Delta x$	0.499	0.049	0.276	0.233	0.011	0.065	0.068
	$\mu \pm \Delta x$	0.499	0.047	0.276	0.223	0.011	0070	0.068

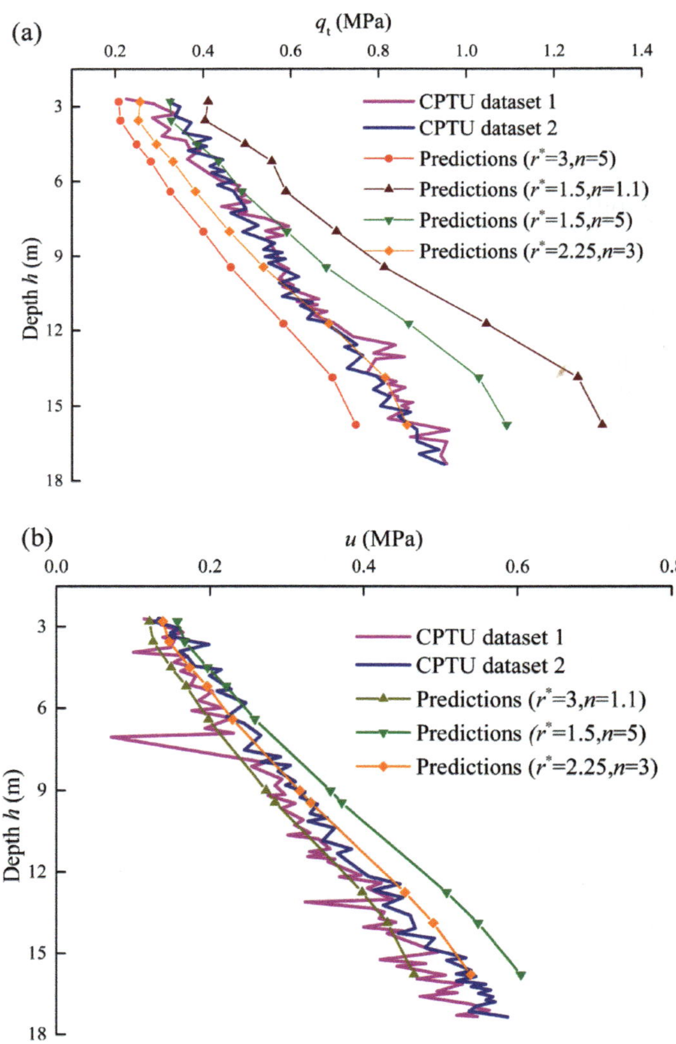

Figure 13.28 Predicted curves of the CPTU resistances by the original geotechnical parameters: (a) cone-tip resistance q_t; and (b) pore water pressure u.

Table 13.15 Prior knowledge of the key constitutive parameters ψ

Prior knowledge	r^*	n	M	OCR
Minimum	1.5	1.0	1.37	1.4
Maximum	3.0	5.0	1.42	1.6

pore water pressure u differ greatly from the in-situ test data because some constitutive parameters are random, fuzzy, and uncertain.

13.4.3.3 Constitutive parameters calibration

To reduce the boundary disturbance of the soil layers, the in-situ CPTU resistances at the depth of 3.5–15.75 m are taken as the z_b . Due to the insufficient test data, the prior knowledge of multiple random variables $\psi=[r^*,n,M,OCR]$ is subjected to uniform distribution at the beginning stage, which is listed in Table 13.15.

The joint prior distribution $p(\psi)$ is randomly sampled 10,000 times. Hence, the posterior results of the key constitutive parameters through the Bayesian calibration by CPTU resistances are shown in Figure 13.29. The posterior distribution of the spacing ratio subjects, $r^* \sim LN(2.447,0.267^2)$, is shown in Figure 13.29 (a), stress-state coefficient subjects, $n \sim LN(2.952,1.092^2)$, is shown in Figure 13.29(b), critical stress ratio subjects,

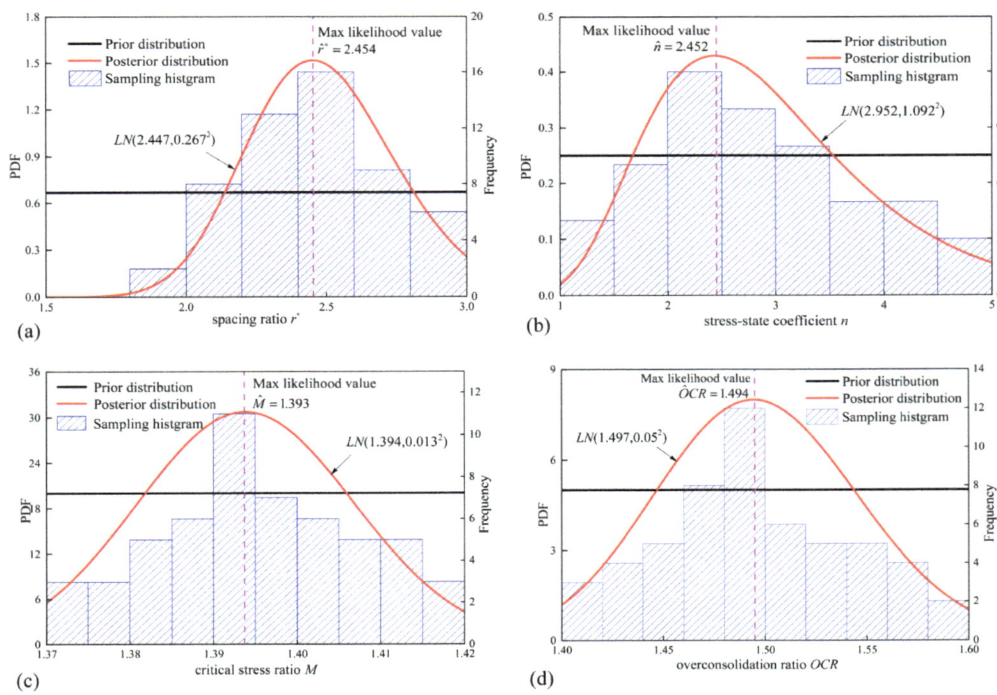

Figure 13.29 Posterior distributions of the key constitutive parameters: (a) spacing ratio r^*; (b) stress-state coefficient n; (c) critical stress ratio M; and (d) overconsolidation ratio OCR .

Table 13.16 Correlation coefficients among the key constitutive parameters

Constitutive parameter	r^*	n	M	OCR
r^*	1.000	−0.162	−0.132	0.266
n		1.000	−0.005	−0.003
M			1.000	0.040
OCR				1.000

$M \sim LN(1.394, 0.013^2)$, is shown in Figure 13.29(c), and overconsolidation ratio subjects, $OCR \sim LN(1.497, 0.05^2)$, is shown in Figure 13.29(d). The coefficient of variation is $\delta_{r^*} = 0.109$, $\delta_n = 0.368$, $\delta_M = 0.009$, and $\delta_{OCR} = 0.033$, respectively. The maximum likelihood value is $\hat{r}^* = 2.454$, $\hat{n} = 2.452$, $\hat{M} = 1.393$, and $\hat{O}CR = 1.494$, individually.

Through the posterior discrete samples of the key constitutive parameters $\psi=[r^*,n,M,OCR]$, the correlation coefficients among them are calculated as listed in Table 13.16.

Table 13.16 indicates that the key constitutive parameters exist cross-correlations between each other. The correlation coefficients between r^* and n, M, OCR are −0.162, −0.132, and 0.266, respectively. There is an unignorable correlation between r^* and OCR. However, there is a subtle correlation between the n and M, OCR, even M and OCR.

ACKNOWLEDGMENT

This work is sponsored by the National Natural Science Foundation of China (51208303), Program for Professor of Special Appointment (Eastern Scholar) at Shanghai Institutions of Higher Learning (TP2018042), and the Shanghai Pujiang Programme (18PJ1403900). Our research is also supported by the Science and Technology Commission of Shanghai Municipality (21DZ1204300). The contributors include Kun Wang, Daofei Tang, Chaoxing Wu, Chengtao Ma, and Tian Fang.

REFERENCES

1. Larsson R, Mulabdic M. Piezocone tests in clay. Report 42. Linköping: Swedish Geotechnical Institute, 1991.
2. Mayne PW. Stress-strain-strength-flow parameters from enhanced in-situ tests. Proceedings International Conference on In-Situ Measurement of Soil Properties and Case Histories, Bali, Indonesia, 2001.
3. Senneset K, Janbu N, Svano G. Strength and deformation parameters from cone penetration tests. Proceedings of the 2nd European Symposium on Penetration Testing, ESOPT-II. Rotterdam: Balkema Publishers, 1982: 863–870.
4. Senneset K, Sandven R, Janbu N. Evaluation of soil parameters from piezocone tests. Journal of Transportation Research, 1989(1): 24–37.
5. Kulhawy FH, Mayne PH. Manual on Estimating Soil Properties for Foundation Design. Electric Power Research Institute, EPRI, Ithaca, 1990.

6. Bruzzi D, Battaglio M. Pore pressure measurements during cone penetration test. ISMES Research Report No. 229. 1987.
7. Kulhawy FH, Mayne PW. Manual on estimating soil properties for foundation design. Electric Power Research Institute, 1990.
8. Baldi G, Bellotti R, Ghionna VN, et al. Modulus of sands from CPTs and DMTs. Proceedings of the 12th International Conference on Soil Mechanics and Foundation Engineering, Rio de Janeiro, 1989.
9. Bishop R, Hill R, Mott NF. Theory of indentation and hardness tests. *Proceedings of the Physical Society* 1945, 57: 147–169.
10. Gibson R, Anderson W. Insitu measurement of soil properties with the pressuremeter. *Civil Engineering and Public Works Reviews*, 1961, 56(658): 615–618.
11. Wood DM. *Soil Behaviour and Critical State Soil Mechanics*. Cambridge University Press, London, 1990.
12. Cao Z, Zheng S, Li D, Phoon KK. Bayesian identification of soil stratigraphy based on soil behaviour type index. *Canadian Geotechnical Journal*, 2019, 56: 570–586.
13. Yu H. CASM: a unified state parameter model for clay and sand. *International Journal for Numerical and Analytical Methods in Geomechanics*, 1998, 22: 621–653.
14. Yao Y. Advanced UH models for soils. *Chinese Journal of Geotechnical Engineering*, 2015, 37: 193–217 (in Chinese).
15. Miller EA, Sowers GF. The strength characteristics of soil-aggregate mixtures & discussion. *Highway Research Board Bulletin*, 1958, 183:16–32.
16. Chen S, Abousleiman YN. Exact undrained elasto-plastic solution for cylindrical cavity expansion in modified Cam Clay Soil. *Géotechnique*, 2012, 62: 447–456.
17. Chen S, Abousleiman YN. Exact drained solution for cylindrical cavity expansion in modified Cam Clay soil. *Géotechnique*, 2013, 63: 510–517.
18. Zhang W, Zou J, Zhang X, Yuan W, Wu W. Interpretation of cone penetration test in clay with smoothed particle finite element method. *Acta Geotechnica*, 2021, 16: 2593–2607.
19. Ma H, Chen Z, Yu S. Correlations of soil shear strength with specific penetration resistance of CPT in Shanghai area. *Rock and Soil Mechanics*, 2014, 35(2): 536–542 (in Chinese).
20. Zheng S, Cao Z, Li D, Au Suikui. Identification of underground stratigraphy based on specific penetration resistance. *Engineering Journal of Wuhan University*, 2018, 51(8): 679–687 (in Chinese).
21. Fan X, Wei G, Ma X. The relationship between single bridge probe and double bridge prob cone penetration test in Shanghai. *Geotechnical Investigation & Surveying*, 2007, 2007(9):10–12.
22. Roscoe KH, Thurairajah A, Schofield AN. Yielding of clays in states wetter than critical. *Géotechnique*, 1963, 13(3): 211–240.
23. Yu H. CASM: a unified state parameter model for clay and sand. *International Journal for Numerical and Analytical Methods in Geomechanics*, 1998, 22(8): 621–653.
24. Been K, Jefferies MG. A state parameter for sands. *Géotechnique*, 1985, 35(2): 99–112.
25. Yu H. *Cavity Expansion Methods in Geomechanics*. Dordrecht: Springer, Netherland, 2000.
26. Nash FT, Powell JJM, Lloyd IM. Initial investigations of the soft clay test site at Bothkennar. *Géotechnique*, 1992, 42(2): 163–181.
27. Gong W, Tien YM, Juang CH, Martin JR, Luo Z. Optimization of site investigation program for improved statistical characterization of geotechnical property based on random field theory. *Bulletin of Engineering Geology and the Environment*, 2017, 76(3): 1021–1035.
28. Qi X, Zhou W. An efficient probabilistic back-analysis method for braced excavations using wall deflection data at multiple points. *Computers and Geotechnics*, 2017, 85: 186–198.
29. Box GEP, Wilson KB. On the experimental attainment of optimum conditions. *Breakthroughs in Statistics*. Springer, New York, 1992: 270–310.
30. Matheron G. Principles of geostatistics. *Economic Geology*, 1963, 58(8):1246–1266.

31. Vanmarcke EH. Probabilistic modelling of soil profiles. *Journal of Geotechnical Engineering Division*, 1977, 103(11): 1227–1246.
32. Wang C, Osorio-Murillo C, Zhu H, Rubin Y. Bayesian approach for calibrating transformation model from spatially varied CPT data to regular geotechnical parameter. *Computers and Geotechnics*, 2017, 85: 262–273.
33. Dereniowski D, Kubale M. Cholesky factorization of matrices in parallel and ranking of graphs. *International Conference on Parallel Processing and Applied Mathematics*, Czestochowa, 2003: 985–992.
34. Mantoglou A, Wilson JL. The turning bands method for simulation of random fields using line generation by a spectral method. *Water Resources Research*, 1982, 18(5): 1379–1394.
35. Zhu H, Griffiths DV, Fenton GA, Zhang L. Undrained failure mechanisms of slopes random soil. *Engineering Geology*, , 2015, 191: 31–35.
36. Huang S, Quek ST, Phoon KK. Convergence study of the truncated Karhunen-Loeve expansion for simulation of stochastic processes. *International Journal for Numerical Methods in Engineering*, 2001, 52(9): 1029–1043.
37. Pebesma EJ. Multivariable geostatistics in S: the gstat package. *Computers and Geosciences*, 2004, 30(7): 683–691.
38. Liang S, Ren X, Li J. A random medium model for simulation of concrete failure. *Science China Technological Sciences*, 2013, 56(5): 1273–1281.
39. Eldeiry AA, Garcia LA. Comparison of ordinary kriging, regression kriging, and cokriging techniques to estimate soil salinity using LANDSAT images. *Journal of Irrigation and Drainage Engineering*, 2010, 136: 355–364.
40. Bhattacharjee A. Distance correlation coefficient: an application with Bayesian approach in clinical data analysis. *Journal of Modern Applied Statistical Methods*, 2014, 13(1):354-366.
41. Tan TS, Phoon KK, Hight DW, Leroueil S. Characterisation and engineering properties of natural soils. *Two Volume Set: Proceedings of the Second International Workshop on Characterisation and Engineering Properties of Natural Soils*, Singapore, 29 November–1 December 2006, Vol. 3. CRC Press, London.
42. Thermann K, Gau C, Tiedemann J. Shear strength parameters from direct shear tests–influencing factors and their significance. *IAEG2006 Paper*, 2006, 484:1–12.
43. Bolton MD. The strength and dilatancy of sands. *Géotechnique*, 1986, 36: 65–78.
44. Hight DW, Bond AJ, Legge JD. Characterization of the Bothkennaar clay: an overview. *Géotechnique*, 1992, 42(2): 303–347.
45. Jacobs PA, Coutts JS. A comparison of electric piezocone tips at the Bothkennar test site. *Géotechnique*, 1992, 42(2): 369–375.

Chapter 14

Data-centric seismic soil liquefaction assessment

Approaches, data, and tools

Chaofeng Wang, Qiushi Chen and Pedro Arduino

This chapter discusses seismic soil liquefaction and the pressing need for enhanced assessment methods to better evaluate this devastating geological phenomenon. The main focus revolves around the shift from traditional, primarily empirical, assessment methods toward more comprehensive, data-centric approaches. While conventional assessment practices have been useful, they exhibit limitations due to the inherent complexity of soil and seismic dynamics. This chapter explores how contemporary approaches harness large-scale, diverse datasets, cutting-edge computational power, and advanced analytical methodologies, such as artificial intelligence, to provide deeper insights into soil liquefaction and enhance prediction accuracy. The multiscale applicability of the data-centric approach is also underscored, demonstrating its potential from microscale studies to macroscale and regional assessments. Finally, the chapter reports on a case study utilizing extensive datasets and advanced AI tools for high-resolution, city-scale liquefaction hazard estimation, demonstrating the effectiveness of this innovative approach. This contributes to our evolving understanding of seismic soil liquefaction and our efforts to predict and mitigate its damaging effects.

14.1 INTRODUCTION

In fluid-saturated, geologically unconsolidated soil, liquefaction is a phenomenon caused by a sudden change in stress due to a rapid earthquake loading, during which the pore-water pressure increases and the effective stress reduces so that the soil loses much of its stiffness and strength and behaves like a liquid. As summarized in the Kavazanjian et al.'s [28] report, the consequences of liquefaction may include vertically or laterally displaced ground, landslides, slumped embankments, foundation failures, and mixtures of soil and water erupting at the ground surface. In turn, these effects may lead to settlement, distortion, and the collapse of buildings; the disruption of roadways; the failure of earth-retaining structures; the cracking, sliding, and over-topping of dams, highway embankments, and other earth structures; the rupture or severing of sewer, water, fuel, and other lifeline infrastructure; the lateral displacement and shear failure of piles supporting bridges and waterfront structures; and the uplift of underground structures.

Liquefaction is a complex phenomenon – a multiscale, multiphysics problem that can be characterized at the microscale (particle-pore), mesoscale (particle cluster), and macroscale (field). Modern liquefaction engineering was initiated in the wake of two devastating earthquakes: the 1964 Niigata (Japan) and the 1964 Great Alaska Earthquakes. Since then, liquefaction research has been a continuous topic. Over the decades that have followed, significant progress has been made.

DOI: 10.1201/9781003441946-14

In the early stage, researchers focus on the assessment of the likelihood of initiation of liquefaction in clean, sandy soils. As observations continue to increase, there is a growing awareness of the problems associated with silty and gravelly soils in the context of earthquakes. Additionally, the issues of post-liquefaction strength and the stress-deformation behavior of soils have also begun to attract more attention. Quantitative assessment of the "triggering"/initiation of liquefaction and the consequencing deformations of soils are two topics of great importance. The dominant approach in common engineering practice is to correlate observed field behavior with various in-situ "index" tests including the standard penetration test (SPT), the cone penetration test (CPT), shear-wave velocity (V_s), etc. These in-situ test methods have now reached a level of sufficient maturity. A plethora of such empirical and semi-empirical methods have been proposed and utilized in the past (e.g., [24, 44]). Despite their effectiveness to a certain degree, these methods tend to oversimplify the intricate dynamics of soil behavior under seismic loading. This is primarily due to their limitations in handling the complexity and diversity of soil and seismic activity data.

To date, state-of-the-art developments in liquefaction predict it at the macroscopic scale. To enable a more complete understanding of this devastating instability, it is imperative that the micro-mechanical origins of liquefaction should be understood. Numerical simulation tools such as finite-element method (FEM) and discrete element method (DEM) have been used to solve geotechnical problems since they were invented. Numerous efforts have been made to use FEM and DEM or other numerical methods for modeling the liquefaction of soils [15, 36, 43, 49, 66]. However, none of these numerical methods could fully address the cross-scale issue independently. The multiscale issue in the numerical simulation of granular media has attracted more and more attention. However, less attention has been paid to the multiscale simulation of the liquefaction phenomenon.

In recent years, there has been a growing interest in the data-centric approach, which leverages advanced data collection technologies, computational power, and analytical methods to improve the understanding of seismic soil liquefaction. At its core is the pattern recognition enabled by large-scale and diverse datasets collected from different sources such as geological surveys, lab testing, and field observations after earthquakes. These growing datasets provide rich and multidimensional information about soil and seismic characteristics. Analyzed by new methods, such as artificial intelligence algorithms, these datasets can potentially shed light on the complexities of seismic soil liquefaction. By identifying patterns and relationships within the data, we can develop more accurate models for predicting soil liquefaction. This is a significant advancement over traditional methods, which often struggle with handling the complexity of soil and seismic data. Moreover, the data-centric approach is not confined to a single scale. It can be employed at various levels – from the particle-pore scale to the regional assessments. This multiscale applicability further amplifies the potential of the data-centric approach in seismic soil liquefaction assessment. With the growth of data, the data-centric approach shows promise in modeling the cross-scale nature of soil liquefaction within a single unified framework.

This chapter first outlines the state-of-the-art liquefaction assessment in terms of the methodology, data, and tools. Then, we present a case study that leverages a large dataset and advanced artificial intelligence tool to estimate the liquefaction hazard at the city scale.

14.2 DATA

Grasping the data that underpins seismic soil liquefaction assessments is crucial for crafting models that effectively predict hazards. The rich and varied data sources feeding into the extensive datasets now increasingly harnessed for predicting seismic soil liquefaction underline the complexity of the task. Essential input data for evaluating the liquefaction risk include earthquake intensity measures (IMs), geological and technical site condition characteristics, and site topography. A comprehensive data would enable a nuanced understanding and prediction of liquefaction hazards.

14.2.1 Ground motion and intensity measures

Ground motion and intensity measures (IMs) are fundamental to both data-driven and mechanics-based approaches in the assessment of liquefaction potential. These approaches necessitate detailed input data regarding the seismic activity expected or experienced at a site.

1. **Data-driven approaches:** These methods leverage simplified empirical models to predict liquefaction susceptibility. They typically require inputs such as the earthquake moment magnitude, which provides a measure of the total energy released by the earthquake, and various intensity measures, including peak ground acceleration (PGA), which indicates the maximum acceleration of ground motion at a site during an earthquake; spectral acceleration ($Sa(T)$), reflecting the maximum acceleration of a structure responding at its own period T; and Arias intensity (A_I), a measure of the earthquake's energy content or severity as it affects the site. The choice of IM is critical and is dictated by the specific analytical model in use, aiming to capture the response of soil layers to seismic waves accurately.
2. **Mechanics-based simulations:** These approaches are more sophisticated, often requiring full ground motion records for analysis. They simulate the soil's physical and mechanical response to seismic waves, offering a detailed perspective on how different layers and conditions of soil contribute to liquefaction risk.

14.2.2 Site characteristics

Site characteristics include geologic and geotechnical site conditions, water table, and topography.

14.2.2.1 Geologic characteristics

A significant portion of the data is derived from geological surveys. These comprehensive studies generate geological maps that show various soil types, their distribution, and the underlying geological formations in different regions. They include extensive data on soil and rock characteristics, stratigraphy, and structural features like faults and folds, essential for assessing soil liquefaction potential. Regional liquefaction assessment depends on such geological information. Youd and Perkins [64] proposed a method for classifying the liquefaction susceptibility of numerous geologic units. This requires the mapping of geological units, which are generally characterized by their depositional environment, physical characteristics, and age. The depth of the groundwater table is also required.

14.2.2.2 In-site geotechnical characteristics

In-situ geotechnical characterization plays an important role in understanding the ground conditions of a site, offering insights critical for the assessment of liquefaction hazards. This characterization involves a comprehensive examination of soil properties directly at the site, providing data that reflects the nature of the soil layers. Key to in-situ testing is the utilization of methods designed to measure soil response and characteristics. The cone penetration test (CPT) is one such technique, known for its ability to provide a continuous profile of soil resistance, from which soil type, shear strength, and stratification can be inferred. Similarly, the standard penetration test (SPT) measures the resistance of soil to penetration by a standard-sized hammer, offering valuable data on soil density and composition. Shear-wave velocity (V_s) measurements, on the other hand, yield information on the elastic properties of the soil, crucial for seismic response analysis. These in-situ tests are indispensable for the accurate modeling of soil behavior under seismic loading, allowing engineers to predict how different soil layers will react to earthquake-induced forces. The choice of tests and the interpretation of their results require expertise in geotechnical engineering, as the data must be analyzed within the context of the site's overall geologic setting and seismic history.

14.2.2.3 Laboratory geotechnical characteristics

Complementing in-situ analyses, laboratory geotechnical testing offers a controlled setting to delve into the fundamental properties of soil samples. These tests are essential for a granular understanding of soil behavior, especially under conditions that might not be directly measurable in the field. Laboratory tests typically involve detailed examinations of undisturbed soil samples, which are carefully extracted from the site to preserve their natural structure and water content. Through these tests, geotechnical engineers can determine a wide range of soil properties, including particle size distribution, which helps in classifying soil types; specific gravity, vital for understanding the density and buoyancy of soil particles; void ratio, indicating the porosity of the soil; permeability, which affects how water flows through soil layers; and shear strength, a key parameter in evaluating the stability of soil and its resistance to sliding or flowing under stress. Despite the invaluable insights provided by laboratory testing, it is important to acknowledge the challenges in extrapolating these controlled results to predict field behavior accurately. The complex and nonlinear nature of soil response, influenced by factors such as heterogeneity, anisotropy, and the dynamic nature of loading conditions, can make it difficult to fully mimic in-situ conditions. Therefore, a combination of laboratory and in-situ tests, along with advanced modeling techniques, is often employed to achieve a comprehensive understanding of soil behavior under both static and dynamic loads. Together, in-situ and laboratory geotechnical characteristics form the backbone of soil mechanics and geotechnical engineering, offering the detailed data necessary for the effective assessment of liquefaction hazards.

14.2.2.4 Water

The water table is a pivotal element in assessing the risk of liquefaction. Sites where the groundwater level is close to the surface are particularly prone to this hazard. In areas with high groundwater levels, this increase in seismic-induced pore water pressure can be more pronounced and rapid, therefore is more effective in turning the soil into a fluid-like state, drastically reducing its ability to support structures. Moreover, the presence of a shallow water table means that even moderate seismic events can trigger significant increases in pore water pressure, leading to widespread soil liquefaction. This susceptibility is exacerbated in soils composed of loose, fine-grained materials, such as silt and sand, which are inherently more prone to rearrangement under shaking. The interaction between the seismic forces and the inherent properties of the soil and groundwater creates conditions ripe for liquefaction, underscoring the importance of water table considerations in geotechnical and seismic risk assessments. Understanding the dynamics of the groundwater table in relation to seismic activity also involves recognizing how changes in the water level, whether seasonal or due to human activities, can affect the liquefaction potential of a site. For example, irrigation, drainage, and changes in land use can alter the depth of the groundwater table, potentially increasing the risk of liquefaction in previously stable areas or vice versa. Given these factors, accurate mapping of the groundwater table is crucial for identifying areas at risk of liquefaction.

14.2.2.5 Topography

Ground topography, referring to the arrangement and features of the ground surface including elevation, slope, orientation, and other aspects of the terrain, has a significant impact on how seismic waves interact with the soil, affecting the risk and extent of liquefaction. Three fundamental topographic parameters serve as crucial inputs in liquefaction assessments: ground slope, free face height, and the distance to a free face. Ground slope, or gradient, is the degree of steepness or incline of a surface. It is typically quantified as the ratio of the vertical change (height) to the horizontal change (distance) over a specific interval. The ground slope influences the behavior of the soil during an earthquake. Generally, the steeper the slope, the higher the risk of slope instability and subsequent soil liquefaction. Therefore, a comprehensive understanding of the ground slope is necessary to estimate the potential for seismic soil liquefaction accurately. A free face refers to an exposed vertical or near-vertical surface of soil or rock, such as cliffs, bluffs, or steep excavation walls, with no material support beyond it. The height of the free face is another important parameter for determining soil stability. Tall free faces are more susceptible to slumping or collapse during a seismic event if the underlying soils undergo liquefaction. Therefore, it is important to consider the free face height in liquefaction assessments, especially when dealing with sloping terrains or engineered structures like embankments and retaining walls. The distance to a free face from the point of interest also plays a significant role in a liquefaction assessment. If a soil mass close to a free face undergoes liquefaction, the loss of strength can trigger a lateral spread or landslide, especially if the soil mass is on a slope. Thus, the proximity to a free face can significantly influence the soil's behavior during an earthquake and the subsequent hazard from soil liquefaction.

The process of modeling soil liquefaction under seismic loading is inherently complex, and the choice of the modeling approach used can greatly affect the types of outputs produced. These outputs offer insights into the nature and magnitude of the potential liquefaction risks at a specific site. The primary outputs that can be generated include liquefaction indices, induced ground displacement, and liquefaction maps.

14.2.3 Liquefaction indices at a site

The liquefaction indices, such as the liquefaction potential index (LPI) and the liquefaction severity number (LSN), are quantitative measures that reflect the likelihood and potential severity of liquefaction at a given site. These indices are calculated based on a range of input data collected from the site, including geotechnical and geological properties, seismic hazard characteristics, and topographic parameters. The LPI, established by Iwasaki et al. [25, 26], offers an initial method to quantify the likelihood and potential severity of liquefaction within a soil profile, based on the depth and characteristics of soil layers and their susceptibility to liquefaction. The LSN [56], a more recent development, refines this approach by incorporating depth-weighted assessments to produce a dimensionless number that better correlates with observed damage from liquefaction. It accounts for the depth to liquefaction layers, emphasizing the impact of liquefaction closer to the surface and considering the initial soil state, which improves the prediction of liquefaction-induced land damage. Both indices serve as critical methods for evaluating liquefaction risks.

14.2.4 Induced ground displacement at a site

Another important output is the induced ground displacement at the site. Liquefaction can cause significant deformations of the ground surface, including vertical settlement and lateral spreading. They can lead to the tilting or collapse of buildings and other structures, and damage to infrastructure like roads, pipelines, and utilities. Liquefaction temporarily increases the pore water pressure in the soil, which in turn reduces the effective stress between soil particles, causing the soil structure to collapse or densify. When the shaking stops and the soil begins to reconsolidate, the soil particles settle into a denser configuration than before, leading to vertical settlement of the ground surface. This settlement can be uneven across a site, depending on the variability of soil conditions and the presence of non-liquefiable layers, leading to differential settlement. Lateral spreading involves the horizontal displacement of the ground, typically observed on gentle slopes or near water bodies like rivers, where the liquefied layer allows the overlying soil to move toward areas of lower elevation or pressure. This movement can be particularly destructive to structures, tearing apart roads, breaking pipelines, and undermining the foundations of buildings and bridges. Lateral spreading is most hazardous in areas where the water table is high, and loosely packed, sandy soils are prevalent, as these conditions are conducive to significant liquefaction-related movements.

14.2.5 Liquefaction maps

Liquefaction maps offer a comprehensive visual overview of areas at risk of soil liquefaction and the potential extent of liquefaction-induced damage. These maps synthesize data from detailed site- and region-specific analyses, including soil composition, groundwater levels, historical seismic activity, and other relevant geotechnical factors, to model the likelihood and potential severity of liquefaction across varying landscapes. The creation of liquefaction maps involves methodologies that take into account the complex interplay between soil properties, seismic forces, and hydrological conditions. By overlaying this data onto geographic maps, engineers and planners can pinpoint zones of varying liquefaction susceptibility, from low to high risk, thus enabling more informed decision-making processes. These maps are invaluable not only for guiding urban and infrastructure development but also for implementing zoning regulations that ensure buildings and critical infrastructure are designed with adequate safeguards against liquefaction effects. In the context of urban planning, liquefaction maps can direct the allocation of resources toward areas requiring ground stabilization efforts or special construction practices, such as deep foundations, soil compaction, or the use of liquefaction mitigation techniques like dynamic compaction or grouting. For existing structures, these maps help prioritize retrofitting efforts to enhance resilience against future seismic events. Beyond planning and development, liquefaction maps are vital for emergency preparedness and response. In the aftermath of an earthquake, these maps offer rapid insight into areas that may have sustained significant ground deformation and infrastructure damage, allowing for more efficient deployment of emergency services, assessment teams, and reconstruction resources. This facilitates quicker and more targeted recovery efforts, ultimately reducing the economic and social impacts of seismic disasters. As research advances and more comprehensive geotechnical data becomes available, the accuracy and utility of liquefaction maps continue to improve. Incorporating technologies such as remote sensing, machine learning, and Geographic Information Systems (GIS) further enhances their precision and usability, making them an indispensable component of seismic risk assessment and mitigation strategies worldwide.

14.3 ASSESSING APPROACHES

Analysis of liquefaction and its consequences is an active area of research and development in geotechnical engineering. Methods for estimating liquefaction triggering and its consequences vary. They fall into two categories: simplified methods and mechanics-based numerical methods.

14.3.1 Simplified methods

In 1998, a consensus was reached within the geotechnical community on the use of an empirical stress-based approach for liquefaction triggering assessment called the "simplified method," first developed by Seed and Idriss [48]. This method is still commonly used in practice today [28, 63].

In a simplified method, a factor of safety (FS) against liquefaction triggering, defined as the ratio between the seismic loading required to trigger liquefaction (i.e., the liquefaction resistance) and the seismic loading expected from the earthquake (i.e., the seismic demand), is computed. Both the seismic demand and the liquefaction resistance are characterized as cyclic stress ratios, defined as the ratio of the equivalent cyclic shear stress to the initial vertical effective stress. The seismic demand is the earthquake-induced cyclic stress ratio (CSR), and the liquefaction resistance is the cyclic resistance ratio (CRR): that is, the cyclic stress ratio required to trigger liquefaction.

Seed and Idriss [48] proposed a simplified equation, based on Newton's second law, to compute a representative CSR for a given earthquake magnitude. This model was later revised by Boulanger and Idriss, Onder Cetin and Seed, and Idriss and Boulanger, and Idriss [4, 10, 22, 23].

The most common approaches used in practice to compute CRR are based on geotechnical field data, e.g., CPT, SPT, and V_s. The most commonly used relationships to estimate CRR from a CPT profile are those developed by Idriss and Boulanger and Wride,

Moss et al., and Robertson, and Robertson [22, 40, 44, 45]. The most commonly used relationships to estimate CRR from SPT blow count are those proposed by Youd and Boulanger and Idriss, and Onder Cetin et al., Idriss [4, 9, 63]. The most commonly used relationships to estimate CRR from a V_s profile were developed by Andrus and Stokoe and Kayen et al. [1, 29].

14.3.2 Mechanics-based numerical simulation

The development and validation of numerical analysis tools and procedures for estimating the effects of liquefaction on the built environment is identified as an overarching research need [7]. Numerical analysis is critical for several reasons, including obtaining insights on field mechanisms that cannot be discerned empirically, providing a rational basis for developing or constraining practice-oriented engineering models, and providing a tool for evaluating complex structures with unique characteristics that are outside the range of empirical observations.

Finite-element (FE) and finite-difference (FD) procedures are the most common procedures used in engineering practice. As pointed out by Bray et al. [7], there are major challenges in developing robust validated numerical analysis procedures for evaluating the effects of liquefaction on civil infrastructure systems due to the variety of multiscale, multiphysics coupled nonlinear interactions that come to the forefront in different scenarios where analytical capabilities for liquefaction effects have not been validated. Currently, research or commercial software platforms have not incorporated the best available solution techniques/options for these challenging problems, such as the coupled, large-deformation analysis of strain-softening, localizations, cracking, and interfaces in two or three dimensions with complex constitutive models. However, significant progresses have been made [3, 35, 37, 39, 41, 50, 60]. An FE method divides the domain into a set of smaller, interconnected elements, where the material behavior is defined at each point within an element, which allows for a more accurate representation of complex behaviors like soil liquefaction. FEM is particularly useful for modeling nonlinear behavior, making it well suited for simulating seismic soil liquefaction. It can handle complex boundary and initial conditions and heterogeneous soil properties. Furthermore, coupled solid–fluid analyses are possible within the FEM framework, enabling the simulation of pore water pressure build-up and dissipation during an earthquake, which is crucial in liquefaction studies. The FD method represents the continuous domain as a set of discrete points and approximates the differential equations using differences at these points. This method is highly efficient and straightforward, making it suitable for large-scale and long-duration simulations like seismic soil liquefaction. By accurately simulating these wave propagations, the method can estimate the pore pressure changes and consequent changes in soil strength, giving valuable insight into the potential for soil liquefaction.

The discrete element method (DEM) and smoothed particle hydrodynamics (SPH) are other powerful approaches for simulating soil behavior subject to liquefaction [16, 51]. DEM provides an excellent framework for studying the micro-mechanics of liquefaction, as it allows for direct simulation of the grain-scale interactions and rearrangements that lead to liquefaction. By incorporating realistic particle shapes and considering the influence of pore water pressure, DEM can effectively simulate the cyclic loading conditions during an earthquake and track the onset and progression of liquefaction. One of the major advantages of DEM is its ability to model large deformations, such as those that occur during the liquefaction process. This makes it a valuable tool for investigating the effects of liquefaction on structures and infrastructures, such as settlement and lateral spreading. SPH allows for modeling of large deformations and material discontinuities, making it well suited for simulating the nonlinear, dynamic behavior of soil during an earthquake. The fluid-like behavior of liquefied soil can be effectively captured using this method.

14.4 ANALYSIS TOOLS AND SOFTWARE

Systems for liquefaction evaluation are divided into two categories according to the method used: simplified empirical methods and mechanics-based numerical methods.

14.4.1 Tools for simplified methods

Simplified methods have been developed for rapid engineering evaluations of site-specific liquefaction. To date, no open-source software is available: but methods mentioned in References [67–70] are all Windows based. These tools provide a user interface that can let the user input the soil profiles and earthquake loading, and then visualize the liquefaction index for each soil layer.

14.4.2 Tools for numerical methods

For mechanics-based numerical methods, creating a constitutive model that captures the soil's behavior under cyclic loads is crucial. Commercial software such as mentioned in References [17, 42] is widely used by the geotechnical community. Both of them are Windows based. OpenSees is the only open-source software identified for dealing with liquefaction. Several well-known liquefaction-capable constitutive models are PM4Sand, PM4Silt, PDMY02, UBCSAND, and DAFALIAS-MANZARI. Except for UBCSAND, they are all available in OpenSees.

The NHERI SimCenter also develops both research and educational tools to facilitate performing site-specific analysis of soil response to earthquakes. The tools currently available focus on the one-dimensional propagation of ground shaking from the bedrock to the free surface.

14.4.3 QS3HARK

The Site-Specific Seismic Hazard Analysis and Research Kit with Uncertainty Quantification (*QS3HARK*), developed by [60], performs site-specific analysis of ground shaking and liquefaction by simulating wave propagation through soil layers. The simulations use the finite-element method to perform the calculations. Several advanced material models are available in QS3HARK (e.g., PM4Sand, PM4Silt, PDMY, PDMY02, PDMY03, ManzariDafalias, Borja-Amies) to support advanced site-response analysis.

14.5 CASE STUDY: MULTISCALE REGIONAL ASSESSMENT

Traditionally, detailed liquefaction hazard assessments have been focused on site-specific analyses, utilizing both empirical methodologies and numerical simulations such as finite-element method (FEM) and discrete element method (DEM). In recent times, the emphasis has expanded toward encompassing high-resolution regional-scale assessments, a shift driven by the need for a broader understanding of seismic risks. Traditional regional hazard maps, which are often qualitative in nature, classify liquefaction susceptibility into categories such as high, medium, or low, relying largely on the geological characteristics of the area. Innovations in this field suggest enhancing these conventional approaches by integrating detailed spatial information regarding soil properties.

While site-specific assessments accurately determine the liquefaction potential at specific points where field tests have been conducted, they fall short in predicting potential in areas without direct testing, due to the inherent spatial variability of soil properties. To bridge this gap, geostatistical methodologies, which model spatial variations as a stochastic process, have been employed effectively. The following sections delve into two primary methodologies that account for the spatial variability of geotechnical properties. The first methodology, termed the local soil property approach, leverages local field data to infer spatial correlations among soil properties. The second methodology, referred to as the averaged index approach, involves calculating a mean soil property or index based on data from tested sites and employing geostatistical techniques to estimate these indices in untested regions.

Acknowledging that spatial variability often occurs across multiple scales, we introduce a multiscale random field model. This innovative approach harmoniously combines multiscale soil property models and averaged index random fields with established empirical liquefaction prediction models. The objective of this study is to explore the impacts of these methodologies on the assessment of liquefaction potential within spatially variable soil environments. Furthermore, the study aims to demonstrate the practical application of these advanced assessment strategies in a region known for its susceptibility to liquefaction, providing valuable insights for future seismic risk management and urban planning efforts.

14.5.1 General framework

In this work, the CPT-based empirical liquefaction model is integrated with geostatistical tools to account for the spatial variability of geotechnical properties for regional liquefaction evaluation. The flow of the general framework is shown in Figure 14.1. Within a liquefaction-prone region, CPT measurements (e.g., the tip resistance and the side friction) and other geotechnical data of interest (e.g., water table, soil unit weight) are first collected and their geostatistical properties are inferred and characterized (e.g., probabilistic distribution, spatial structure). At a CPT sounding, the empirical liquefaction model described in Section 14.5.2 will be used to evaluate the damage potential of liquefaction, quantified here by the liquefaction potential index (LPI). The aforementioned two approaches as detailed in work by [61], i.e., *the averaged index approach* and *the local soil property approach*, are developed to incorporate soil variability into the evaluation of liquefaction over an extended area. Details of multiscale random field model development and implementation will be discussed in Section 14.5.3. Finally, Monte Carlo simulations will be used to generate realizations of the random fields and the results will be used for the probabilistic and spatial assessment of various quantities of interest for liquefaction evaluation over the region.

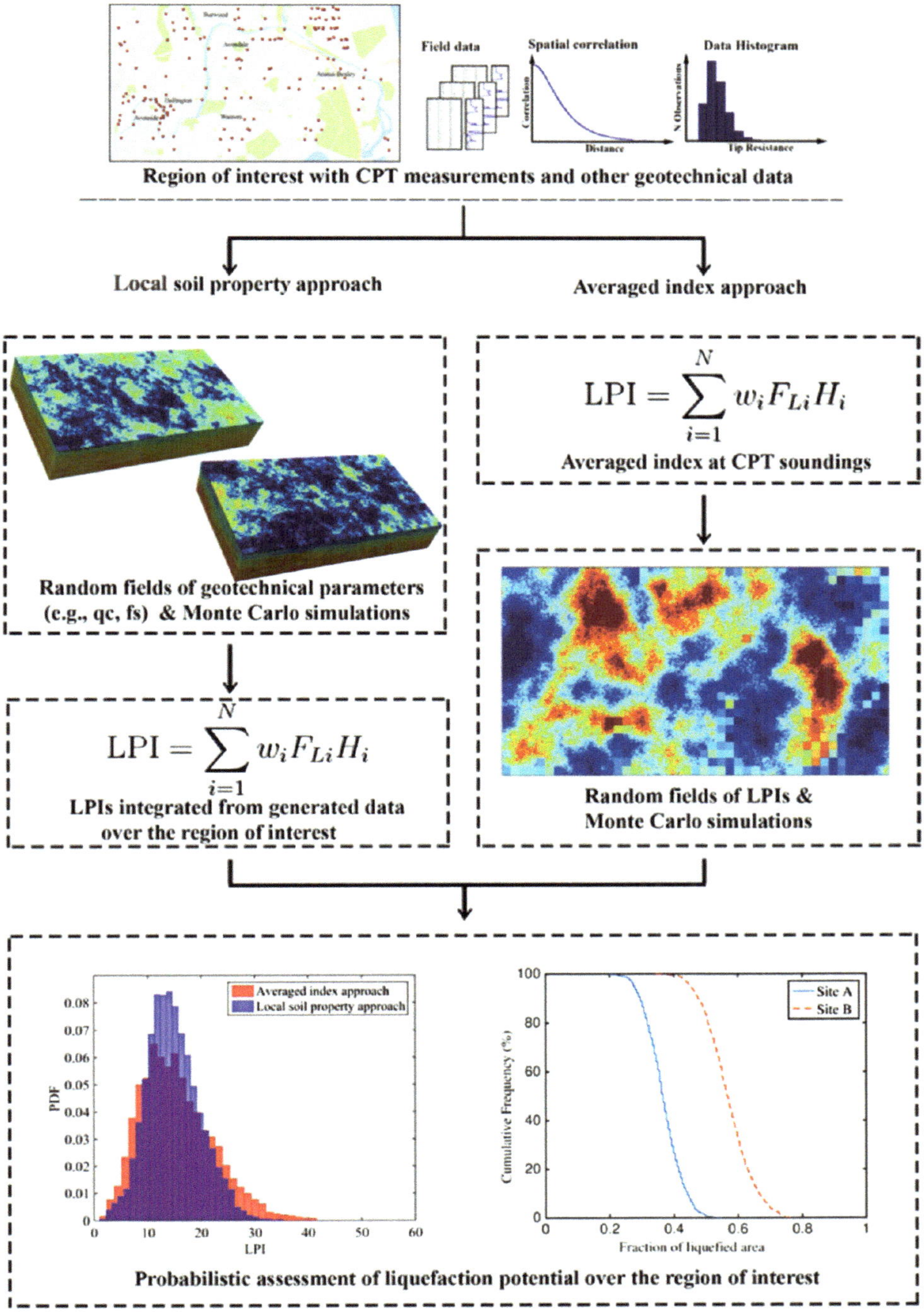

Figure 14.1 General framework of the CPT-based liquefaction potential evaluation over the extended area: the local soil property approach versus the averaged index approach.

14.5.2 Site-specific liquefaction evaluation

Site-specific liquefaction evaluation can be performed using the approach and tool described in Sections 14.3 and 14.4. In this work, we adopt the classical procedure proposed by Robertson and Wride [46] and subsequently updated by Robertson [47] and Ku et al. [32] to evaluate the liquefaction resistance of sandy soils based on CPT data. Herein, the liquefaction potential of a soil layer is evaluated using two variables – the cyclic stress ratio (CSR) and the cyclic resistance ratio (CRR). The detailed equations for CSR and CRR can be found in Wang et al. [61].

Once CSR and CRR are obtained, the factor of safety against liquefaction triggering at a particular depth z can be calculated as

$$FS = \frac{CRR}{CSR} \tag{14.1}$$

which is used to calculate the liquefaction potential index detailed in the following section. In this example, the potential of liquefaction damage will be quantified by an averaged index property, i.e., the liquefaction potential index (LPI), which was originally proposed by Iwasaki et al. [25, 26] and has been subsequently used and calibrated by many investigators, e.g., [12, 20, 21, 27, 33, 52, 53, 55, 57].

Following the definition given in References [25, 26], LPI is usually evaluated for the top 20 m of soil profile as

$$LPI = \int_0^{20} F_L w(z) dz \tag{14.2}$$

where z denotes the depth in meters and $w(z) = 10 - 0.5z$; F_L is a function of FS, the definition of it can be found in work by Sonmez [53].

14.5.3 Multiscale random field characterization

The CPT-based empirical liquefaction model described in Section 14.5.2 evaluates the liquefaction potential at individual locations at the small scale, e.g., the borehole scale. To estimate the extent of liquefaction risk over the entire region of interest, multiscale random field models are introduced and implemented in this section.

14.5.3.1 Spatial correlation

In this study, the spatial correlation of geotechnical parameters is described using the semivariogram, $\gamma(\vec{h})$, which is equal to half the variance of the difference of two random variables separated by a vector distance $\vec{h}$

$$\gamma(\vec{h}) = \frac{1}{2} Var\left[Z(\vec{u}) - Z(\vec{u} + \vec{h}) \right] \tag{14.3}$$

where $Z(\vec{u})$ is a Gaussian random variable at location $\vec{u}$. The vector $\vec{h}$ accounts for both separation distance and orientation and therefore can be used to simulate anisotropic random fields.

In previous studies of liquefaction evaluation, e.g., [33, 34, 58], the correlation $\rho(h)$ is used to describe the spatial dependence of two parameters separated by h and can be related to the semivariogram as

$$\rho(h) = 1 - \gamma(h) \tag{14.4}$$

where $\gamma(h)$ can be a model who's parameter is fit by data. As an example, an exponential model is adopted in this work

$$\gamma(h) = 1 - \exp\left(-\frac{h}{a}\right) \tag{14.5}$$

where a is the range parameter and $3a$ is the practical range, i.e., the distance at which the exponential semivariogram levels off [18].

14.5.3.2 Sequential simulation process

Given a specified probability density function and spatial correlation model, a sequential simulation process [14, 18] is adopted in this work to generate realizations of the random field across the region. Each value is simulated individually conditional upon known information as well as any previously simulated data points.

Denote $\vec{Z}_p$ as a vector of all known and previously simulated points in the random field and Z_n as the next point to be simulated, the sequential simulation process can be illustrated by

$$\begin{bmatrix} Z_n \\ Z_p \end{bmatrix} \sim N\left(\vec{\mu}, \begin{bmatrix} \sigma_n^2 & \vec{\Sigma}_{np} \\ \vec{\Sigma}_{pn} & \vec{\Sigma}_{pp} \end{bmatrix}\right) \tag{14.6}$$

where $\sim N(\vec{\mu}, \vec{\Sigma})$ denotes the vector of random variables following a joint normal distribution with mean vector $\vec{\mu v}$ and covariance matrix $\vec{\Sigma}$; σ_n^2 is the prior variance of the next simulated point; $\vec{\Sigma}_{np}$, $\vec{\Sigma}_{pn}$, and $\vec{\Sigma}_{pp}$ are the covariance matrices, where the subscripts "n" and "p" represent "next" (as in next point to be simulated) and "previous" (as in all previously simulated points), respectively. The covariance matrices are obtained by

$$COV[Z_i, Z_j] = \rho_{ij} \cdot \sigma_i \cdot \sigma_j \tag{14.7}$$

where ρ_{ij} is the correlation between random variables Z_i and Z_j with standard deviations of σ_i and σ_j , respectively.

Using the above model for joint distribution, the distribution of Z_n conditional upon all previously simulated data is given by a univariate normal distribution with updated mean and variance

$$\left(Z_n \mid \vec{Z}_p = \vec{z}\right) \sim N\left(\vec{\Sigma}_{np} \cdot \vec{\Sigma}_{pp}^{-1} \cdot \vec{z}, \sigma_n^2 - \vec{\Sigma}_{np} \cdot \vec{\Sigma}_{pp}^{-1} \cdot \vec{\Sigma}_{pn}\right) \tag{14.8}$$

It is noted that $\vec{\Sigma}_{np} \cdot \vec{\Sigma}_{pp}^{-1}$ are essentially the weights assigned in the simple Kriging process [18]. For the realization, one value of Z_n is drawn at random from the posterior univariate normal distribution.

Once simulated, Z_n becomes a data point in the vector $\vec{Z}_p$ to be conditioned upon by all subsequent data locations. This process is repeated by following a random path to each unknown location until all the values in the field have been simulated. This sequential simulation process is beneficial for the proposed work because (1) it preserves known information (e.g., field data) precisely at their locations in the simulated random field; and (2) it allows to first simulate the field at only the coarse scale, then add simulation points at the fine-scale probabilistically consistent with the previous coarse-scale realizations, a process detailed in Section 14.5.3.3.

14.5.3.3 Multiscale spatial correlation

The multiscale spatial correlation is based on the notion that material properties at the coarser scales are the arithmetically averaged values of the properties over corresponding areas at the finer scales. This relation allows for the explicit derivation of variances and spatial correlation of quantities of interest between different scales and is visually represented in Figure 14.2. For two scales of interest, we denote the coarser scale as scale "c" and the finer scale as scale "f." The cross-scale spatial correlations can then be calculated as [12]

$$\rho_{Z_I^c, Z_{II}^c} = \frac{\sum_{i=1}^{N}\sum_{k=1}^{N} \rho_{z_{i(I)}^f, z_{k(II)}^f}}{\sqrt{\sum_{i=1}^{N}\sum_{j=1}^{N} \rho_{z_{i(I)}^f, z_{j(I)}^f}}\sqrt{\sum_{i=1}^{N}\sum_{j=1}^{N} \rho_{z_{i(II)}^f, z_{j(II)}^f}}} \tag{14.9}$$

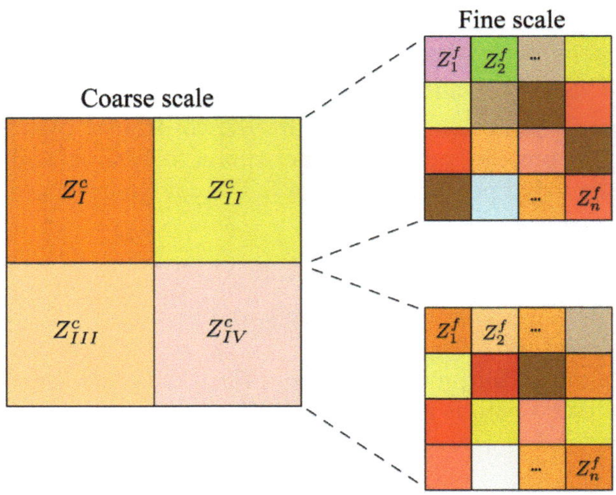

Figure 14.2 Graphic representation of material properties at two scales. The superscripts "c" and "f" refer to "coarse" and "fine" scales, respectively. The subscripts refer to the element number. Roman letters I, II, ... are used for coarse-scale element and Arabic numbers 1, 2, 3, ... are used for fine-scale element.

$$\rho_{z^f, z_I^c} = \frac{\sum_{i=1}^{N} \rho_{z^f, z_{i(I)}^f}}{\sqrt{\sum_{i=1}^{N} \sum_{j=1}^{N} \rho_{z_{i(I)}^f, z_{j(I)}^f}}} \tag{14.10}$$

where $\rho_{z_I^c, z_{II}^c}$ = correlation between two coarse-scale elements I and II; ρ_{z^f, z_I^c} = correlation between a fine-scale element and a coarse-scale element I.

The geotechnical properties will be simulated using random field models at the coarse scale (e.g., the regional scale), and then adaptively refined into smaller scales (e.g., the borehole or structure scales), conditional upon the coarse-scale random field simulations. It is worth noting that the cross-scale correlations (14.9) and (14.10) are applicable to general non-uniform arbitrarily shaped grids, and actual field CPT-based geotechnical measurements will be incorporated into the sequential simulation process.

14.5.4 Numerical examples

In this section, the developed multiscale random field model is applied to evaluate liquefaction potential in the city of Christchurch, New Zealand. The purpose of the numerical examples is twofold: (1) to demonstrate the applicability of multiscale random field models in liquefaction evaluation over extended areas; (2) to assess *the averaged index approach* and *the local soil property approach* in accounting for the spatial variability of CPT-based geotechnical parameters and their implications on the liquefaction evaluation. We assume the stationary of the random field model and focus mainly on the spatial variability of tip resistance, side friction, and the liquefaction potential index. Spatial variability of other geotechnical parameters, e.g., water table and unit weight of soil, are not included in the current study.

14.5.4.1 Analysis region and field data

Christchurch, New Zealand, is a city founded on the boundary of the alluvial Springston Formation and the marine Christchurch Formation [8]. During the period between September 2010 and December 2011, the city of Christchurch was strongly shaken by a sequence of four strong earthquake events known as the Canterbury earthquakes. For the following liquefaction analysis and liquefaction potential mapping, we pick one of the great events, the February 22, 2011, earthquake event. The moment magnitude of the earthquake M_w = 6.2. Typical peak ground surface acceleration $a_{\max}$ in the area of study ranging from 0.34 g to 0.50 g [5]. A median value of $a_{\max} = 0.42$ g is used in the following liquefaction potential calculations. A total of 155 CPT profiles with measured water tables are collected from the New Zealand Geotechnical Database (NZGD). An averaged moist unit weight of $\gamma_m = 18.5$ kN/m^3 is used for soils above water table and an saturated unit weight of $\gamma_{sat} = 19.5$ kN/m^3 is used for soils below the water table following information presented in Reference [31] Figure 14.3.

14.5.4.2 Probabilistic and spatial assessment of liquefaction potentials

In the previous two sections, we have demonstrated typical random field realizations of LPIs by the averaged index approach and the local soil property approach. In this

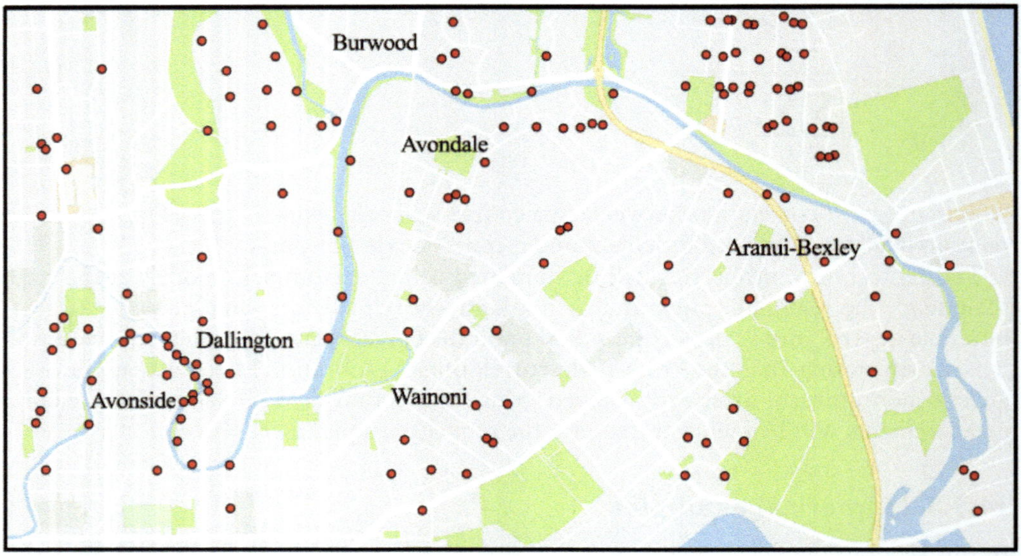

Figure 14.3 Map of the area of study in Christchurch, New Zealand. Red dots represent the locations of the 155 CPT soundings.

section, we will perform Monte Carlo simulations and assess the implications of two approaches on the probabilistic and spatial characteristics of liquefaction potentials in the Christchurch site. In addition, utilizing multiscale random field realizations, liquefaction evaluation will be performed for selected local sites to demonstrate the advantage and applicability of the proposed methodology.

A total of 1000 Monte Carlo simulations are performed to generate realizations of LPIs across the Christchurch site. Figure 14.4a shows the histogram of the averaged LPIs from 1000 Monte Carlo simulations and Figure 14.5 shows the corresponding maps of the averaged LPIs. To attenuate the effect of ergodic fluctuations, empirical semivariograms

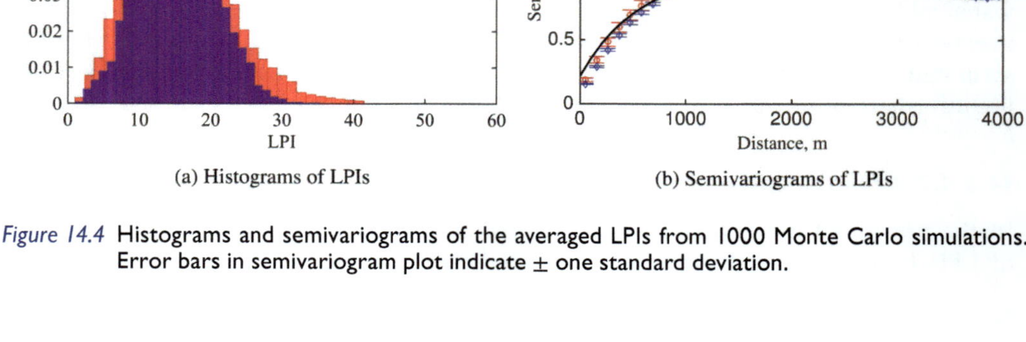

Figure 14.4 Histograms and semivariograms of the averaged LPIs from 1000 Monte Carlo simulations. Error bars in semivariogram plot indicate ± one standard deviation.

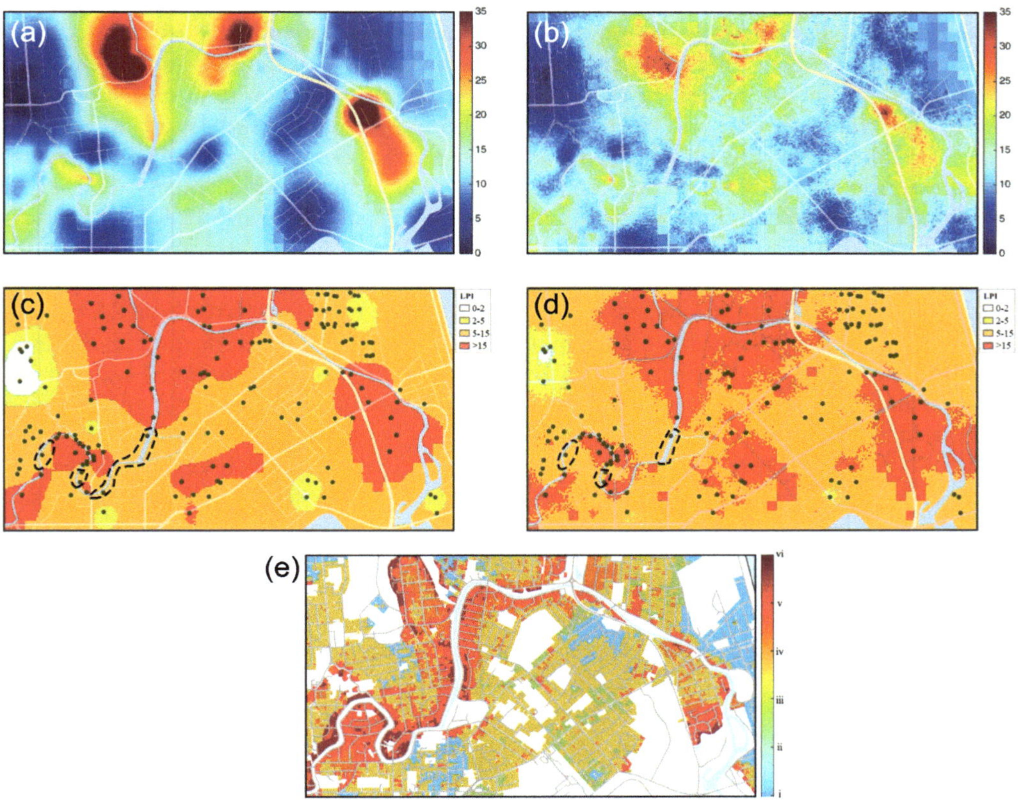

Figure 14.5 Mapping of LPI by two approaches compared with the observed liquefaction phenomena.

are calculated 1000 times, averaged, and plotted in Figure 14.4b. The solid curve is the specified exponential model with parameters fitted using LPIs at CPT soundings. The red dots and blue squares are the empirical semivariograms calculated using Equation (14.3) with LPI values generated by the averaged index approach and the local soil property approach, respectively. The error bars ($\pm$ one standard deviation) show the variations of 1000 Monte Carlo simulations. Fluctuating semivariogram results from weak and unsteady correlation of properties between separated places. This is verified in Figure 14.4b, i.e., great fluctuations (in both the average and standard deviation) of the empirical semivariograms are observed at distances greater than 1000 m. The local soil property approach results in slightly greater fluctuation of semivariogram possibly due to the nonlinear transformation from q_c and f_s to LPI.

The generated LPI maps by two different approaches are compared with the map of observed liquefaction phenomena after the February 22, 2011, earthquake event [2], as shown in Figure 14.5. Comparing the maps of predicted LPIs against the observations, it can be seen that both the averaged index approach and the local soil property approach are able to capture the varying severity levels of liquefaction at most locations across the area of study. For instance, the high LPI value area corresponds well with areas with observed moderate to several types of liquefaction. The results of the local soil property approach show more small-scale fluctuations or "noises." This is due to the high nugget effects in the specified semivariograms of q_c and f_s.

Discrepancies between simulated maps of liquefaction potential and observations are also identified. For example, in the dot–circled areas, as shown in Figure 14.5(c, d), the predicted LPI is less than 15, which is not the severest level of liquefaction. In the observation map as shown in Figure 14.5e, however, moderate to severe lateral spreading is observed in these regions and ejected material is also often observed, which usually indicates the highest level of liquefaction. Lateral spreading is a unique form of liquefaction manifestation associated with large lateral permanent ground displacements. The severity of lateral spreading could not be classified by LPI. There are separate criteria for assessing its severity. Therefore, while site profiles with thin liquefiable layers may have low LPI values, these sites may be susceptible to lateral spreading if located near rivers or on sloping ground. Actually, take Christchurch City for an example, there are about 25% of sites with lateral spreading and have LPI values less than 4.5 [38]. This means that while LPI is a useful tool for large-scale hazard assessments, it does not eliminate the need for site-specific analyses or consideration of the influence of local conditions on the manifestation of liquefaction [38].

Results from Monte Carlo simulations can also be used for the probabilistic assessment of liquefaction potentials. Various quantities of interest can be defined. As an example, we evaluate the cumulative frequency distributions of LPIs following the methodology proposed by Holzer et al. [20].

Figure 14.6 plots the cumulative frequencies of LPIs in the area of study obtained with the averaged LPI map from 1000 Monte Carlo simulations. The error bars (± one standard deviation) are also included in the cumulative frequency plots. It can be seen that the cumulative frequency curve by the averaged index approach is very close to that of the known LPI obtained from 155 CPT soundings, while a sharp change in the slope of the cumulative frequency curve for the local soil property approach is observed, which can be explained by its histogram in Figure 14.4a. Most of the LPIs simulated by this approach fall between 10 and 20, resulting in a sharp change in the cumulative frequency curve in this range. For the averaged index approach, the percentages of different LPI values are more "spread-out," which results in smoother cumulative frequency curves.

Comparisons of Figure 14.6(a, b) indicate that multiscale random fields yield consistent results as the single-scale counterparts for cumulative frequencies of LPIs. This is to be expected given the notion that properties of a coarse element are the averaged values of the properties over the corresponding areas at the fine scale.

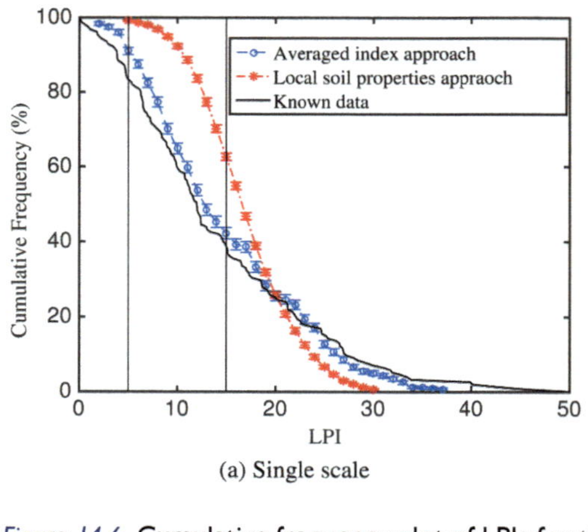

(a) Single scale

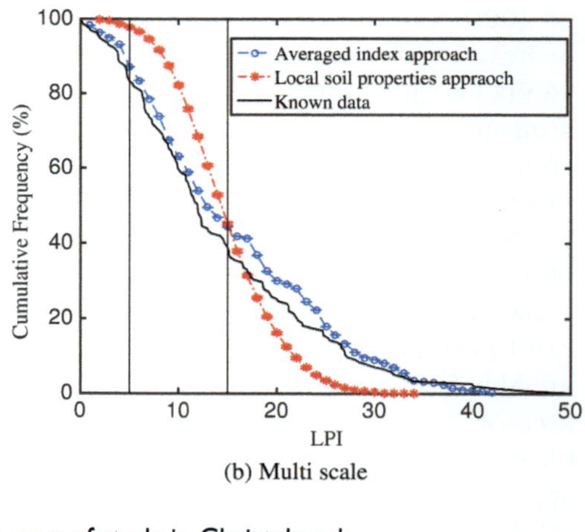

(b) Multi scale

Figure 14.6 Cumulative frequency plot of LPIs for the area of study in Christchurch.

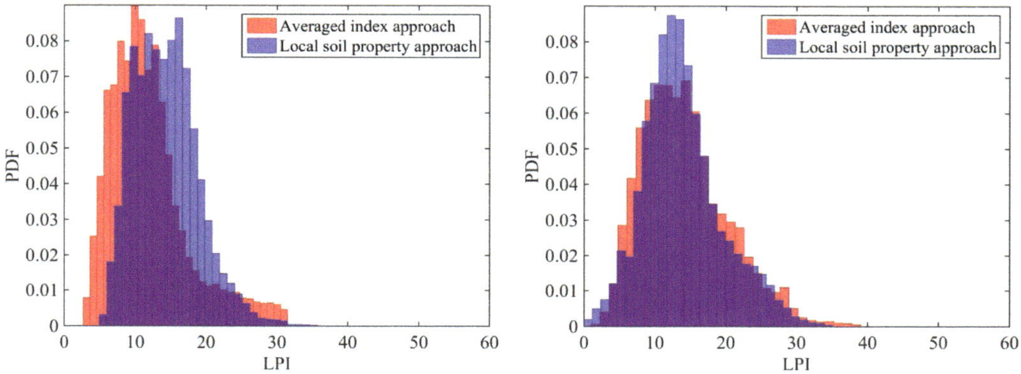

(a) N=9 (b) N=69

Figure 14.7 Histograms of predicted LPIs in the Christchurch site when a different number of CPT soundings (*N*) is used to infer model parameters.

It should be noted that all previous analyses and comparisons are based on relatively sufficient field data to infer random field model parameters. It is expected that the amount of the field will impact the results of the random field-based liquefaction mapping. While this subject itself deserves a future study, for this Christchurch site, we investigated the effect of data availability on the distribution of predicted LPIs in the area of study. The number of available CPT soundings varied from 9 to 243. It is found that when the number of CPT soundings is small, the local soil property approach yields higher mean LPI values. As the number of CPT soundings increases, the distributions obtained from the averaged index approach and the local soil approach converge. Figure 14.7 plots the histogram of LPIs obtained with 9 CPT and 69 CPT soundings.

14.5.4.3 Fine-scale site-specific liquefaction assessment

The final analysis involves the fine-scale site-specific liquefaction assessment. Compared to single-scale random field, multiscale random field provides higher resolution at local or site-specific scale, e.g., near a critical infrastructure or a particular building of interest. To illustrate this point, we evaluate the liquefaction potential at two local sites in Christchurch. The site locations are shown in Figure 14.8. Site A is a small region that consists of two schools, i.e., the Chisnallwood Intermediate School and the Avondale Primary School. Site B is the location of the Aranui High School. The inset of random fields in Figure 14.8 shows that the multiscale random field could provide much more detailed information than the single-scale counterpart, which makes it possible to perform site-specific liquefaction evaluation while consistently maintaining predictions of liquefaction for the entire region over much larger scales.

Based on the cumulative frequency distribution, Holzer et al. [20] proposed that for a given geologic unit, the percent area predicted to have liquefaction during a given earthquake-shaking scenario that could be estimated from the cumulative frequency distributions at LPI $\geq$ 5. This criterion can be further categorized to indicate the percentage of area to have minor liquefaction ($5 \leq$ LPI < 15) and major liquefaction (LPI ≥ 15).

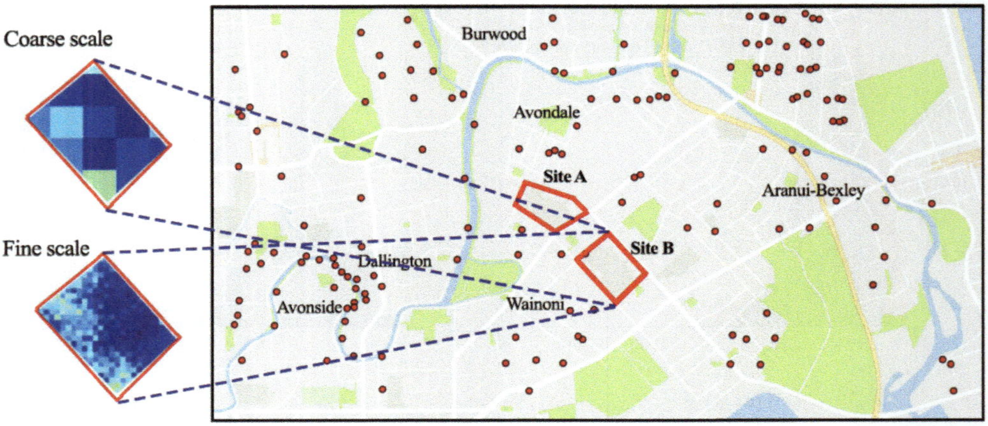

Figure 14.8 Locations of two selected sites for fine-scale liquefaction assessment.

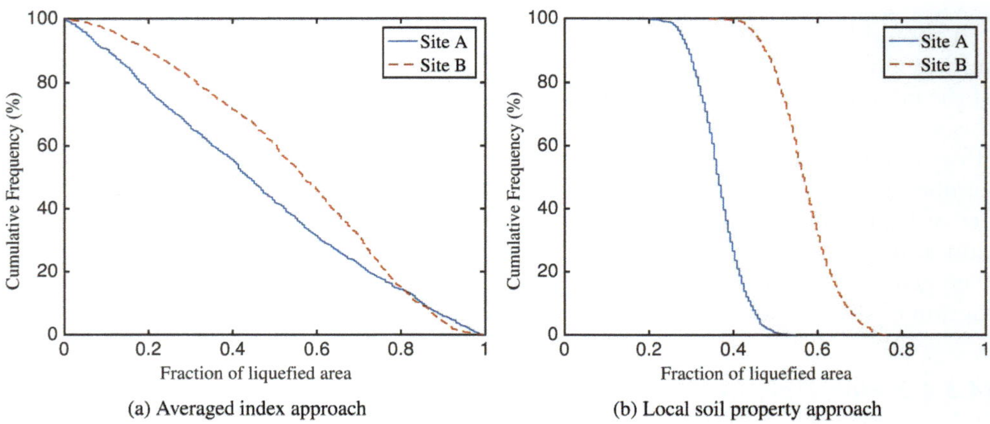

Figure 14.9 Cumulative frequency versus the fraction of liquefied area with LPI ≥ 15 for two local sites A and B selected in the area of study.

Herein, the percentage of areas in the selected sites with LPI ≥ 15 under a given earthquake-shaking scenario ($M_w = 6.2$, $a_{\max} = 0.42$ g) is calculated and plotted against the corresponding cumulative frequency in Figure 14.9.

Figure 14.9 shows cumulative frequencies of LPIs with respect to the fraction of liquefied area. The cumulative frequency curve for site B is consistently "higher" than site A for both the averaged index approach and the local soil property approach, indicating that site B is more likely to liquefy than site A in a given earthquake. Comparisons between two approaches show that the cumulative frequency for the average index approach is more "spread-out" than the local soil property approach. This is consistent with the predictions within the whole site of Christchurch, as shown in Figure 14.6.

The above analysis on the percentage of the liquefiable area underneath a particular site gives an indication of the implied reliability of the site with respect to liquefaction resistance during a given earthquake. These site-specific curves can only be obtained if a higher resolution multiscale random field is generated. Moreover, multiscale random fields are able to obtain high-resolution information at local sites while consistently maintaining low-resolution information at much larger scales, leading to a computationally efficient process.

Finally, it should be noted that the methodologies discussed in this study are sufficiently general to be applied to other earthquake-prone areas and to evaluate quantities of interest such as liquefaction-induced settlement and lateral spreads. This will be explored in future studies.

14.6 RESEARCH GAPS AND OPPORTUNITIES

While significant strides have been made in liquefaction risk analysis, primarily focusing on the prediction and mapping of liquefaction susceptibility, a comprehensive understanding of the consequent ground damage remains elusive. Current methodologies largely concentrate on assessing the likelihood of liquefaction triggering but fall short in quantitatively evaluating the aftermath – specifically, the extent of damage inflicted on the ground and underlying structures. The development of universally recognized and applied techniques for estimating liquefaction-induced damage is an ongoing challenge within the geotechnical community. The HAZUS methodology, despite its widespread application, underscores this gap. It offers a procedure rooted in historical engineering observations for estimating liquefaction-induced permanent ground deformation (PGD) based on peak ground acceleration (PGA) and site-specific susceptibility. Nonetheless, this framework is predicated on analyses of comparatively older seismic events, highlighting the need for updated and more comprehensive approaches.

Recent advancements have illuminated this research path. An increasing number of observed liquefaction instances [7, 13], coupled with the establishment of extensive databases [6], has catalyzed the development of new regression-based ground damage models [30, 54]. These models, which are being rigorously tested across various geographical contexts [11], represent a significant leap forward in our ability to predict and understand the ground damage resulting from liquefaction. Concurrently, emerging research on the vulnerability of overlying structures to liquefaction offers fresh insights into the fragility of built environments in the face of such phenomena.

Moreover, the traditional methodologies for regional liquefaction assessment – primarily reliant on geological datasets – have been superseded by more sophisticated approaches [19]. Recent propositions advocate for the use of random field models for regional-scale liquefaction hazard modeling. These models, by incorporating both geotechnical and geological data, offer a nuanced perspective that accommodates spatial uncertainties, representing a critical advancement in our quest to refine liquefaction risk assessment [59, 62, 65].

As we move forward, the geotechnical engineering field is poised at the cusp of a new era in liquefaction research. The integration of advanced modeling techniques, extensive empirical databases, and a deeper understanding of structural vulnerabilities opens up unprecedented opportunities. The challenge now lies in harnessing these insights to develop methodologies that not only predict liquefaction susceptibility but also quantify the potential damage, thereby enhancing our resilience to future seismic events.

14.7 CONCLUSION

In this chapter, the authors have delved into the complex, multifaceted nature of seismic soil liquefaction and underlined the pressing necessity for improved assessment methodologies to better forecast this catastrophic geological occurrence. Through an examination of traditional empirical assessment methods and their inherent limitations, we underscored the urgent need for a more holistic, data-centric approach.

Our exploration encompassed familiar practices like the standard penetration test (SPT), cone penetration test (CPT), and shear-wave velocity (V_s), acknowledging their contributions while highlighting their challenges due to the intricacies of soil and seismic dynamics. These techniques, while beneficial to a degree, often oversimplify the complexity of soil behavior under seismic loading, leading to less accurate assessments of liquefaction risk.

This chapter emphasized the transformative potential of contemporary approaches that harness large-scale, diverse datasets and cutting-edge computational capabilities to enhance our understanding of soil liquefaction. By leveraging advanced analytical methodologies, such as artificial intelligence, we can discern patterns and relationships within data, facilitating more precise predictions of soil liquefaction. These strides constitute significant advancements over traditional methods, revolutionizing our capacity to predict and mitigate the destructive impacts of seismic soil liquefaction.

Additionally, we touched upon the multiscale applicability of the data-centric approach, demonstrating its potential from microscale laboratory studies to macroscale and regional assessments. This ability to scale seamlessly enhances its potential as a robust assessment tool. A practical illustration of this innovative approach's effectiveness was the city-scale case study, which utilized a large dataset and sophisticated AI tools, such as the random field, to estimate liquefaction hazards. This case study was a testament to the efficacy of data-centric methodologies in understanding and predicting seismic soil liquefaction.

This chapter stands as a step in the ongoing evolution of our understanding of seismic soil liquefaction and our capabilities in predicting and mitigating its ruinous impacts. By persistently pushing the boundaries of traditional methodologies and embracing the potential of data-centric approaches, we are better equipped to address the challenges posed by soil liquefaction. As we continue to deepen our knowledge and refine our tools, we move closer to a future where we can effectively manage and mitigate the risks of seismic soil liquefaction.

REFERENCES

1. R.D. Andrus and K.H. Stokoe II. Liquefaction resistance of soils from shear-wave velocity. *Journal of Geotechnical and Geoenvironmental Engineering,* 126(11):1015–1025.
2. J. Beavan, S. Levick, J. Lee, and K. Jones. Ground displacements and dilatational strains caused by the 2010–2011 Canterbury earthquakes. *GNS Science Consultancy Report*, 67:59, 2012.
3. R.W. Boulanger and K. Ziotopoulou. Pm4sand (version 3): A sand plasticity model for earthquake engineering applications. In *Center for Geotechnical Modeling Report No. UCD/CGM-15/01, Department of Civil and Environmental Engineering*, University of California, Davis, California, 2015.
4. R.W. Boulanger and I.M. Idriss. CPT and SPT based liquefaction triggering procedures. Report No. UCD/CGM.-14, 1, 134.
5. B.A. Bradley and M. Hughes. Conditional peak ground accelerations in the Canterbury earthquakes for conventional liquefaction assessment. In *Technical Report for the Ministry of Business, Innovation and Employment*, New Zealand, 2012.

6. S.J. Brandenberg, P. Zimmaro, J.P. Stewart, D.Y. Kwak, K.W. Franke, R.E.S. Moss, K.Ö. Çetin, G. Can, M. Ilgac, J. Stamatakos, *et al.* Next-generation liquefaction database. *Earthquake Spectra*, 36(2), 939–959, 2020.
7. J.D. Bray, R.W. Boulanger, M. Cubrinovski, K. Tokimatsu, S.L. Kramer, T. O'Rourke, E. Rathje, R.A. Green, P.K. Robertson, and C.Z. Beyzaei. U.S. – New Zealand – Japan International Workshop on Liquefaction-Induced Ground Movements Effects, University of California, Berkeley, California 2–4 November 2016. Report No. 2017/02.
8. L.J. Brown and J.H. Weeber. Geology of the Christchurch urban area, Institute of Geological and Nuclear Sciences 1:25,000 Geological Map 1, 1992. https://natlib.govt.nz/records/21363240.
9. K. Onder Cetin, R.B. Seed, R.E. Kayen, R.E.S. Moss, H.T. Bilge, M. Ilgac, and K. Chowdhury. SPT-based probabilistic and deterministic assessment of seismic soil liquefaction triggering hazard. 115:698–709.
10. K. Onder Cetin and R.B. Seed. Nonlinear shear mass participation factor (rd) for cyclic shear stress ratio evaluation. 24(2):103–113.
11. Qiushi Chen, Chaofeng Wang, and C Hsein Juang. Probabilistic and spatial assessment of liquefaction-induced settlements through multiscale random field models. 211:135–149.
12. Q. Chen, C. Wang, and C.H. Juang. CPT-based evaluation of liquefaction potential accounting for soil spatial variability at multiple scales. *Journal of Geotechnical and Geoenvironmental Engineering*, 142, 04015077, 2015.
13. M. Cubrinovski, J. Bray, C. de la Torre, M. Olsen, B. Bradley, G. Chiaro, E. Stock, L. and Wotherspoon. Liquefaction effects and associated damages observed at the Wellington Centreport from the 2016 Kaikoura earthquake. 50(2):152–173.
14. C.V. Deutsch and A.G. Journel. *GSLIB: Geostatistical Software Library and User's Guide*. Oxford University Press, New York, 1992.
15. A. Elgamal, Z. Yang, and E. Parra. Computational modeling of cyclic mobility and post-liquefaction site response. *Soil Dynamics and Earthquake Engineering*, 22(4):259–271, 2002.
16. U. El Shamy and S. Farzi Sizkow. Coupled sph-dem simulations of liquefaction-induced flow failure. *Soil Dynamics and Earthquake Engineering*, 144:106683, 2021.
17. FLAC.
18. P. Goovaerts. *Geostatistics for Natural Resources Evaluation*. New York, Oxford University Press, 1997.
19. T.L. Holzer, M.J. Bennett, T.E. Noce, A.C. Padovani, and J.C. Tinsley III. Liquefaction hazard mapping with LPI in the greater Oakland, California area. 22(3):693–708.
20. T.L. Holzer, M.J. Bennett, T.E. Noce, A.C. Padovani, and J.C. Tinsley III. Liquefaction hazard mapping with LPI in the greater Oakland, California, area. *Earthquake Spectra*, 22(3):693–708, 2006.
21. T.L. Holzer, J.L. Blair, T.E. Noce, and M.J. Bennett. Predicted liquefaction of East Bay fills during a repeat of the 1906 San Francisco earthquake. *Earthquake Spectra*, 22(S2):261–277, 2006.
22. I.M. Idriss and R.W. Boulanger. *Soil Liquefaction during Earthquakes*. Berkeley, CA, Earthquake Engineering Research Institute.
23. I. Idriss. An update to the Seed-Idriss simplified procedure for evaluating liquefaction potential. In *Proceedings of the TRB Workshop on New Approaches to Liquefaction*, pp. 99–165, Federal Highway Administration.
24. K. Ishihara. Liquefaction and flow failure during earthquakes. *Geotechnique*, 43(3):351–451, 1993.
25. T. Iwasaki, F. Tatsuoka, K. Tokida, and S. Yasuda. A practical method for assessing soil liquefaction potential based on case studies at various sites in Japan. In *Proceedings 2nd International Conference on Microzonation*, pp. 885–896, 1978.
26. T. Iwasaki, K. Tokida, F. Tatsuoka, S. Watanabe, S. Yasuda, and H. Sato. Microzonation for soil liquefaction potential using simplified methods. In *Proceedings of the 3rd International Conference on Microzonation*, Seattle, WA, volume 3, pp. 1310–1330, 1982.

27. C.H. Juang, C.N. Liu, C.H. Chen, J.H. Hwang, and C.C. Lu. Calibration of liquefaction potential index: A re-visit focusing on a new CPTU model. *Engineering Geology*, 102(1):19–30, 2008.
28. E. Kavazanjian, J. E. Andrade, K. Arulmoli, B. F. Atwater, J. T. Christian, E. Green, S.L. Kramer, L. Mejia, J.K. Mitchell, E. Rathje, J.R. Rice, and Y. Wang. *State of the Art and Practice in the Assessment of Earthquake-Induced Soil Liquefaction and Its Consequences*. Washington, National Academies Press.
29. R. Kayen, R. Moss, E. Thompson, R. Seed, K. Cetin, A.D. Kiureghian, Y. Tanaka, and K. Tokimatsu. Shear-wave velocity-based probabilistic and deterministic assessment of seismic soil liquefaction potential. 139(3):407–419.
30. S. Khoshnevisan, H. Juang, Y.-G. Zhou, and W. Gong. Probabilistic assessment of liquefaction-induced lateral spreads using CPT – focusing on the 2010–2011 Canterbury earthquake sequence. 192:113–128.
31. S. Khoshnevisan, H. Juang, Y.G. Zhou, and W. Gong. Probabilistic assessment of liquefaction-induced lateral spreads using CPT – focusing on the 2010–2011 Canterbury earthquake sequence. *Engineering Geology*, 192:113–128, 2015.
32. C.S. Ku, C.H. Juang, C.W. Chang, and J. Ching. Probabilistic version of the Robertson and Wride method for liquefaction evaluation: development and application. *Canadian Geotechnical Journal*, 49(1):27–44, 2011.
33. J.A. Lenz and L.G. Baise. Spatial variability of liquefaction potential in regional mapping using CPT and SPT data. *Soil Dynamics and Earthquake Engineering*, 27(7):690–702, 2007.
34. C.N. Liu and C.H. Chen. Mapping liquefaction potential considering spatial correlations of CPT measurements. *Journal of Geotechnical and Geoenvironmental Engineering*, 132(9):1178–1187, 2006.
35. R. Luque and J.D. Bray. Dynamic analyses of two buildings founded on liquefiable soils during the Canterbury earthquake sequence. *Journal of Geotechnical and Geoenvironmental Engineering*, 143(9):04017067, 2017.
36. J. Lu, A. Elgamal, L. Yan, K.H. Law, and J.P. Conte. Large-scale numerical modeling in geotechnical earthquake engineering. *International Journal of Geomechanics*, 11(6):490–503, 2011.
37. M.T. Manzari, M. El Ghoraiby, M. Zeghal, B.L. Kutter, P. Arduino, A.R. Barrero, E. Bilotta, L. Chen, R. Chen, A. Chiaradonna, *et al.* Leap-2017 simulation exercise: Calibration of constitutive models and simulation of the element tests. In *Model Tests and Numerical Simulations of Liquefaction and Lateral Spreading: LEAP-UCD-2017*, pp. 165–185. Springer, 2020.
38. B.W. Maurer, R.A. Green, M. Cubrinovski, and B.A. Bradley. Evaluation of the liquefaction potential index for assessing liquefaction hazard in Christchurch, New Zealand. *Journal of Geotechnical and Geoenvironmental Engineering*, 140(7):04014032, 2014.
39. C.R. McGann, P. Arduino, and P. Mackenzie-Helnwein. A stabilized single-point finite element formulation for three-dimensional dynamic analysis of saturated soils. *Computers and Geotechnics*, 66:126–141, 2015.
40. R.E. Moss, R.B. Seed, R.E. Kayen, J.P. Stewart, A.Der Kiureghian, and K. Onder Cetin. CPT-based probabilistic and deterministic assessment of in situ seismic soil liquefaction potential. 132(8):1032–1051.
41. F. Oka, A. Yashima, T. Shibata, M. Kato, and R. Uzuoka. FEM-FDM coupled liquefaction analysis of a porous soil using an elasto-plastic model. *Applied Scientific Research*, 52:209–245, 1994.
42. PLAXIS.
43. R. Popescu. Finite element assessment of the effects of seismic loading rate on soil liquefaction. *Canadian Geotechnical Journal*, 39(2):331–344, 2002.
44. P.K. Robertson. Interpretation of cone penetration tests—a unified approach. 46(11):1337–1355.
45. P.K. Robertson and C.E. Wride. Evaluating cyclic liquefaction potential using the cone penetration test. 35:442–459.

46. P.K. Robertson and C.E. Wride. Evaluating cyclic liquefaction potential using the cone penetration test. *Canadian Geotechnical Journal*, 35(3):442–459, 1998.
47. P.K. Robertson. Performance based earthquake design using the CPT. In *Proceedings of the IS-Tokyo*, pp. 3–20, 2009.
48. H.B. Seed and I.M. Idriss. Simplified procedure for evaluating soil liquefaction potential.
49. H.B. Seed, I. Arango, and C.K. Chan. Evaluation of soil liquefaction potential for level ground during earthquakes. A summary report. Technical report, Shannon and Wilson, Inc., Seattle, WA; Agbabian Associates, El Segundo, CA, 1975.
50. R.B. Seed, K. Onder Cetin, R.E.S. Moss, A.M. Kammerer, J. Wu, J.M. Pestana, M.F. Riemer, R.B. Sancio, J.D. Bray, R.E. Kayen, *et al.* Recent advances in soil liquefaction engineering: a unified and consistent framework. In *Proceedings of the 26th Annual ASCE Los Angeles Geotechnical Spring Seminar*, Long Beach, CA, 2003.
51. S.F. Sizkow and U. El Shamy. Sph-dem simulations of saturated granular soils liquefaction incorporating particles of irregular shape. *Computers and Geotechnics*, 134:104060, 2021.
52. H. Sonmez and C. Gokceoglu. A liquefaction severity index suggested for engineering practice. *Environmental Geology*, 48(1):81–91, 2005.
53. H. Sonmez. Modification of the liquefaction potential index and liquefaction susceptibility mapping for a liquefaction-prone area (Inegol, Turkey). *Environmental Geology*, 44(7):862–871, 2003.
54. J.P. Stewart, S.L. Kramer, D.Y. Kwak, M.W. Greenfield, R.E. Kayen, K. Tokimatsu, J.D. Bray, C.Z. Beyzaei, M. Cubrinovski, T. Sekiguchi, *et al.* PEER-NGL project: Open source global database and model development for the next-generation of liquefaction assessment procedures. 91:317–328.
55. S. Toprak and T.L. Holzer. Liquefaction potential index: field assessment. *Journal of Geotechnical and Geoenvironmental Engineering*, 129(4):315–322, 2003.
56. S. Van Ballegooy, V. Lacrosse, M. Jacka, and P. Malan. LSN – a new methodology for characterising the effects of liquefaction in terms of relative land damage severity. In *Proceedings of the 19th NZGS Geotechnical Symposium*. Ed. CY Chin, Queenstown, 2013.
57. S. van Ballegooy, F. Wentz, and R. W. Boulanger. Evaluation of CPT-based liquefaction procedures at regional scale. *Soil Dynamics and Earthquake Engineering*, 79:315–334, 2015.
58. B. Vivek and P. Raychowdhury. Probabilistic and spatial liquefaction analysis using CPT data: a case study for Alameda County site. *Natural Hazards*, 71(3):1715–1732, 2014.
59. C. Wang, Q. Chen, M Shen, and C. Hsein Juang. On the spatial variability of cpt-based geotechnical parameters for regional liquefaction evaluation. *Soil Dynamics and Earthquake Engineering*, 95:153–166, 2017.
60. C. Wang, F. McKenna, P. Mackenzie-Helnwein, W. Elhaddad, A. Zsarnoczay, and M. Gardner. Nheri-simcenter/s3hark: Release v1.5, July 2020.
61. C. Wang, Q. Chen, M. Shen, and C.H. Juang. On the spatial variability of cpt-based geotechnical parameters for liquefaction potential evaluation. *Soil Dynamics and Earthquake Engineering*, 95:153–166, 2017.
62. C. Wang and Q. Chen. A hybrid geotechnical and geological data-based framework for multiscale regional liquefaction hazard mapping. *Géotechnique*, 68(7):614–625, 2018.
63. T.L. Youd and I.M. Idriss. Liquefaction resistance of soils: Summary report from the 1996 NCEER and 1998 NCEER/NSF workshops on evaluation of liquefaction resistance of soils. 127(4):297–313.
64. T. Leslie Youd and D.M. Perkins. Mapping of liquefaction severity index. 113(11):1374–1392.
65. J. Zhu, L.G. Baise, and E.M. Thompson. An updated geospatial liquefaction model for global application. 107(3):1365–1385.
66. O.C. Zienkiewicz, A.H.C. Chan, M. Pastor, B.A. Schrefler, and T. Shiomi. *Computational Geomechanics*. Chichester, Wiley Chichester, 1999.
67. Cliq.
68. LiqIT.
69. Liquefy-Pro.
70. NovoLIQ.

Chapter 15

Prediction for soil design properties based on a multivariate database for Shanghai soft clay

Yelu Zhou, Dongming Zhang, and Hongwei Huang

This study compiles a database for Shanghai soft clay labeled as SH-CLAY/12/4191 based on data collected from 51 investigation reports of geotechnical engineering in Shanghai. The database contains 4191 data points, each of which is composed of multivariate parameters measured simultaneously in close proximity at the same depth. There are five physical parameters and seven mechanical parameters. Then, a 12-dimensional multivariate probability distribution is constructed based on the compiled database. To make it more useful for engineering practice, the transformation models between the design parameters (i.e., K_0, E_s, c, and φ) and the index parameters as well as the in-situ test parameters are derived. The transformation models are explicit nonlinear formulas of multiple variables without cross-term. The transformation models are validated by the data from the new site, showing the generally reliable performance in predicting the design parameters.

15.1 INTRODUCTION

Determination of soil parameters for design is critical for geotechnical engineering. For a desired project, the number of boreholes and soil tests are very limited, which leads to the fact that the mechanical parameters provided in the geotechnical investigation report cannot fully reflect the characteristics of the mechanical properties of the soil at the site, making them an unreliable basis for geotechnical design. On the other hand, the geotechnical investigation report provides a relatively rich set of physic index parameters and field measurements for the soil. In view of this fashion of geotechnical site investigation, transformation models are typically adopted to estimate the value or reduce the uncertainty of the design parameters, i.e., mechanical properties, via these abundant index or field properties (Phoon and Kulhawy, 1999).

In the literature, many transformation models have been constructed based on soil databases (Kulhawy and Mayne 1990; Mayne et al. 2001). However, most of the transformation models are pairwise correlations. It is generally known that design parameters are correlated with multiple index parameters and field measurements. Annex D of ISO2394 (2015) illustrated the advantage of the multivariate probability distribution model considering the correlations between different soil parameters so that more accurate probabilistic characterization of soil design parameters could be obtained based on the multiple information of other soil parameters. Before the construction of the multivariate probability distribution, the compilation of the target multivariate soil database consisting of laboratory and field data is required. It should be noted that incomplete multivariate databases such as those consisting of multiple sets of bivariate data obtained from different locations can be used to construct the multivariate distribution, that is to say, not all parameters are measured simultaneously in close proximity. This is more

DOI: 10.1201/9781003441946-15

practical because simultaneous measurements of two parameters are more commonly encountered than those of three or more parameters, which is precisely the reason why traditional multivariate regression assumptions are often challenging to meet in geotechnical exploration.

So far, several global or regional multivariate soil databases have been compiled, such as CLAY/10/7490 (Ching and Phoon, 2014a), SAND/7/2794 (Ching et al., 2017), ROCK/10/4025 (Muzamhindo and Ferentinou, 2023), and ROCKMass/9/5876 (Ching et al., 2021) at the global scale and F-CLAY/7/216 (D'Ignazio et al., 2016), S-CLAY/7/168 (D'Ignazio et al., 2016), J-CLAY/5/124 (Liu et al., 2016), FI-CLAY/14/856 (Löfman and Korkiala-Tanttu, 2021), and SH-CLAY/11/4051 (Zhang et al., 2020) at the regional scale, and some multivariate probability distribution models have been derived successfully from these databases. The practical applications of the multivariate probability distribution may include the Bayesian updating and the establishment of the transformation models involving multivariate correlations. However, whether the database is regional or global, the transformation model developed based on it is generic for a specific site. When a generic model is applied to a specific site, the transformation uncertainty can be significantly high. This is because the generic model is designed to encompass a broad range of soil types and site conditions. Meanwhile, data collected from one single site may be too sparse to develop an imprecise site-specific model due to the statistical uncertainty. Hybridization methods for constructing a site-specific multivariate distribution by considering the generic multivariate distribution model and the multivariate site-specific data together have been proposed to reduce the transformation uncertainty (Ching and Phoon, 2019; Ching et al., 2022). In situations where ample data is available at a specific site, the methodologies will predominantly draw upon local data to formulate the transformation model. Conversely, when site-specific data are scant and local experience is limited, the transformation model should place a greater emphasis on the generic data. Nevertheless, the greater the congruity between the soil properties in the database employed to develop a generic model and those of the soil at the specific site, the more advantageous it becomes in the establishment of a site-specific model (Ching et al., 2020). The database for Shanghai's soft clay is thus compiled to facilitate the development of the transformation models for the region, because it is more relevant to Shanghai than the other global databases that cover sites from all around the world.

This chapter introduces the municipal database for the properties of Shanghai Clay labeled SH-CLAY/12/4191. Note this database is an extension of SH-CLAY/11/4051 (Zhang et al., 2020) with additional data points collected from a new site as well as differences in parameters. Transformation models to predict the soil design parameters are derived based on this database. The derived transformation models might still be generic but are practical for the cases where engineers are compelled to use the generic models. These models are also readily adaptable for constructing the site-specific models by incorporating the site-specific data, which, however, is beyond the scope of this study.

The rest of this study is organized as follows. Firstly, some basic information on the location, test condition, and type of clay parameters as well as the basic statistics and correlations pertaining to the database SH-CLAY/12/4191 is presented. Secondly, a multivariate probability distribution is constructed based on this database to characterize the multivariate correlations among the 12 parameters. The transformation models between the design soil parameters and the index parameters as well as the field measurements are then derived and subsequently validated using data from a new site that is independent of the database.

15.2 SHANGHAI CLAY DATABASE

15.2.1 Description of SH-CLAY/12/4191 database

Shanghai is situated in the eastern part of China at the mouth of the Yangtze River. It is a fairly typical soft soil area, characterized by flat terrain formed by the deposition of sediment carried by the Yangtze River over millions of years.

This study compiles a database for Shanghai's soft clay, including soil parameter information for gray muddy silty clay, gray muddy clay, gray clay, and gray silty clay, which are the primary constituents of Shanghai's soft soils. The maximum depth from the ground surface of the collected soil layers is 30 m.

The data points in SH-CLAY/12/4191 are collected from 51 site investigation reports for Shanghai. These sites are distributed in nine administrative areas of Shanghai, including Pudong, Yangpu, Huangpu, Jing'an, Putuo, Changning, Xuhui, Jiading, and Qingpu, covering an approximate area of 145 km^2, about 1/43 the size of Shanghai. There are 574 boreholes, 49 vane shear tests, and 201 cone penetration tests in total. Figure 15.1(a, b), respectively, illustrates the distribution of the 51 sites and the layout for the three types of tests at a specific site.

The collected soil parameters consist of 12 variables, which include: (1) five index parameters – the water content (ω), liquid limit (LL), plastic index (PI), liquid index (LI), and void ratio (e); and (2) seven mechanical property parameters – vertical effective stress (σ'_v), lateral pressure coefficient (K_0), modulus of compressibility (E_s), consolidated quick shear cohesion (c), consolidated quick shear angle of internal friction (φ), undrained shear strength ($S_{u(VST)}$) obtained from the vane shear test (VST), and cone-tip resistance (p_s)

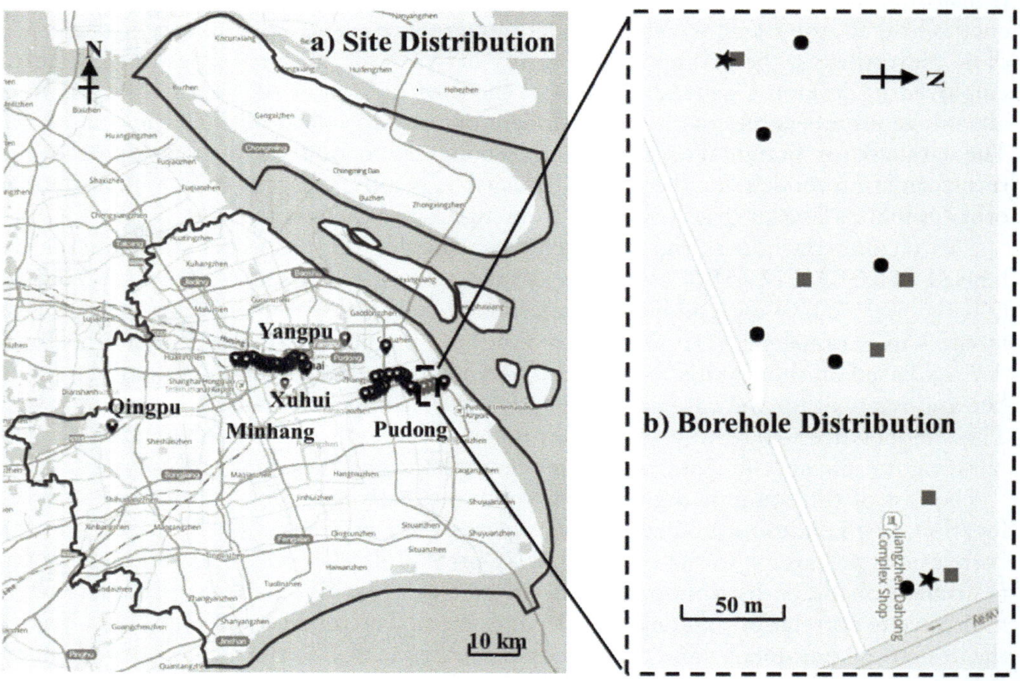

Figure 15.1 Schematic graph of site and borehole distribution.

from the cone penetration test (CPT). Note that p_s is different from the cone-tip resistance q_t adopted in international practice, due to the difference in the CPT test equipment and procedure between China and other countries or regions

There are 4191 records in the database, and each record represents the values of the 12 parameters at the same depth, which are obtained from the different types of boreholes located within a specific proximity to each other. This study constrains this distance to 35 m, within which the parameters are strongly correlated referring to the average value of the correlation distance of Shanghai (Xiao, 2018). However, it is common to find empty cells in each record because some indoor mechanical property tests are not conducted for soil samples at all depths, and data from the boreholes is not readily paired with that from the VSTs due to the less frequent execution of the VST in the proximity of boreholes.

15.2.2 Basic statistics of SH-CLAY/12/4191 database

Table 15.1 provides the statistical values of the 12 parameters in the database, including the sample size, mean value, coefficient of variation (COV), maximum value (Max), and minimum value (Min). It shows the characteristics of Shanghai soft clay with high water content, high compressibility, and low strength.

Gao and Wei (1983) collected the geotechnical test data from 17 sites in Shanghai and calculated the mean value and COV of the index and mechanical parameters for different soil layers at each site. It was found that the mean values of the collected parameters did not vary significantly among different sites, suggesting that Shanghai soft clay is relatively homogeneous. Therefore, statistical analysis of soil parameters at the municipal level is feasible. In addition, the Shanghai local code for geotechnical site investigation conducted statistical analyses on over 100 geotechnical investigation reports in Shanghai, encompassing data from tens of thousands of soil samples. It provided ranges and COV of 22 parameters for different soil layers. By comparing the statistics of parameters shared by the SH-CLAY/12/4191 database with those derived from the investigation code, it is evident that the mean values of all the parameters in SH-CLAY/12/4191 fall within the ranges determined by the investigation code, verifying the rationality of the data collected in this study. Details pertaining to statistics from Gao and Wei (1983) and the investigation code can be found in Appendix A.

Table 15.1 Basic statistics of the soil parameters of Shanghai soft clay

Parameters	*n*	*Mean*	*COV*	*Min*	*Max*
ω (%)	4011	43.29	0.15	23.30	63.60
LL (%)	2368	40.29	0.12	22.80	58.70
PI	4183	18.24	0.18	10.30	30.90
LI	2201	1.14	0.22	0.49	2.19
e	4014	1.23	0.14	0.67	1.79
σ'_v (kPa)	2801	132.61	0.36	28.69	274.38
K_0	300	0.51	0.10	0.43	0.65
E_s (Mpa)	2066	3.08	0.32	1.46	12.91
c (kPa)	1026	14.35	0.22	7.00	28.00
φ (°)	1026	13.66	0.33	5.50	28.50
S_u (kPa)	410	40.07	0.24	18.60	76.60
p_s (MPa)	1357	0.75	0.39	0.30	4.20

15.2.3 Correlation of the parameters

Pearson product–moment correlation coefficient ρ is adopted to evaluate the degree of dependency between the parameter pairs. In total, there are $12 \times (12-1)/2 = 66$ parameter pairs in SH-CLAY/12/4191. Table 15.2 presents the correlation coefficients for the 66 parameter pairs. Several inferences can be drawn from the table as follows. Firstly, there are distinct correlations between the index parameters, with the exception being that the correlation between LI and other index parameters is relatively weak. Secondly, σ'_v exhibits a substantial negative correlation with LI and marked positive correlations with c, S_u, and p_s. Since σ'_v approximates a linear increase with depth, it can be deduced that LI diminishes gradually with the increase of the soil depth, whereas c, S_u, and p_s increase with the soil depth. Other parameters, however, appear to remain unaffected by variations in soil depth. Thirdly, mechanical parameters K_0, E_s, and φ display relatively strong correlations with the index parameters, rendering the estimation of these mechanical properties using the index parameters feasible. Lastly, p_s demonstrates a certain degree of correlation with the strength index c, φ, and S_u. Hence, it is legitimate to estimate the strength index through the more readily measured in-situ parameter, p_s.

15.3 MULTIVARIATE PROBABILITY DISTRIBUTION FOR SH-CLAY/12/4191

The multivariate normal distribution is employed in this chapter due to its computational simplicity and practical applicability. In accordance with the established methodology (Ching et al., 2016), this study constructs the multivariate normal distribution using the following steps: (1) transforming non-normally distributed soil parameters into variables following a standard normal distribution and (2) calculating the correlation coefficients between the transformed variables to constitute a correlation matrix in the normal space, with which the multivariate normal distribution can be defined. Details of the two steps are described in the subsequent subsections.

15.3.1 Normal transformation for soil parameters

Soil parameters typically do not follow the normal distribution in the marginal sense. In addition to its successful applications in many soil databases, the Johnson transformation method was suggested by Zhou et al. (2021) as the optimal approach for constructing the multivariate normal distribution for real soil databases. The recommendation was based on their comparative assessment, which found the Johnson transformation method to be more robust and superior in terms of normal transformation compared to the other two methods.

The Johnson distribution system (Johnson, 1949) comprises three types of distributions: SU (for the unbounded system), SB (for the bounded system), and SL (for the lognormal system), with the commonly known lognormal distribution falling under the SL category. The formula that transforms the non-normal variable Y_i to its corresponding standard normal variable X_i is as follows:

Table 15.2 Correlation coefficients between soil parameters

	ω	*LL*	*PI*	*LI*	**e**	σ'_v	K_0	E_s	**c**	φ	S_u	p_s
ω	1.00	0.75	0.72	0.58	0.99	−0.28	0.70	−0.78	−0.39	−0.63	−0.27	−0.31
LL		1.00	0.95	−0.09	0.77	0.05	0.73	−0.68	0.03	−0.70	0.08	−0.27
PI			1.00	−0.11	0.74	0.17	0.69	−0.66	0.11	−0.72	0.13	−0.15
LI				1.00	0.54	−0.70	0.15	−0.41	−0.64	−0.06	−0.48	−0.39
e					1.00	−0.25	0.72	−0.79	−0.36	−0.64	−0.23	−0.30
σ'_v						1.00	−0.06	0.23	0.61	0.11	0.73	0.44
K_0							1.00	−0.66	−0.08	−0.58	−0.36	−0.36
E_s		Symmetry							1.00	0.23	0.72	0.19
c								1.00	−0.11	0.52	0.33	0.35
φ										1.00	0.06	0.30
S_u											1.00	0.62
p_s												1.00

Table 15.3 Distribution type and distribution parameters for $(Y_1, Y_2, \ldots, Y_{12})$

Random variable	Soil parameters	Distribution type	Distribution parameters			
			a_X	b_X	a_Y	b_Y
Y_1	$\ln(\omega)$	SB	1.60	−0.93	1.11	3.06
Y_2	ln(LL)	SB	1.33	−0.75	0.77	3.21
Y_3	ln(PI)	SB	1.22	−0.63	1.04	2.25
Y_4	ln(LI)	SB	2.11	−0.72	2.01	−1.06
Y_5	ln(e)	SB	1.62	−0.92	1.07	−0.47
Y_6	$\ln(\sigma'_v/Pa)$	SB	1.96	−3.80	6.89	−5.77
Y_7	$\ln(K_0)$	SB	1.09	0.72	0.49	−0.86
Y_8	$\ln(E_s)$	SB	1.23	0.16	1.56	0.34
Y_9	ln(c)	SU	6.05	39.77	0.003	3.72
Y_{10}	$\ln(\varphi)$	SB	1.67	0.50	2.29	1.57
Y_{11}	$\ln(S_u/\sigma'_v)$	SU	1.10	−1.24	0.11	−1.35
Y_{12}	$\ln(p_s/\sigma'_v)$	SU	1.01	−0.82	0.16	1.52

$$X = \begin{cases} b_X + a_X \sinh^{-1}\left(\dfrac{Y - b_Y}{a_Y}\right) & SU \\ b_X + a_X \ln\left(\dfrac{Y - b_Y}{a_Y + b_Y - Y}\right) & SB\ (b_Y \le Y \le a_Y + b_Y) \\ b_X + a_X \ln\left(\dfrac{Y - b_Y}{a_Y}\right) & SL\ (Y \ge b_Y) \end{cases} \tag{15.1}$$

It is a closed-form equation, where the distribution type and the corresponding distribution model parameters (a_X, b_X, a_Y, b_Y) can be determined by using a non-parametric method based on the four percentiles of Y_i proposed by Slifker and Shapiro (1980).

The natural logarithm of the 12 parameters included in SH-CLAY/12/4191 are denoted as $(Y_1, Y_2, \ldots, Y_{12})$, with some of the parameters having undergone dimensionless standardization (as detailed in Table 15.3). Following the abovementioned approach, the distribution types and corresponding parameters for $(Y_1, Y_2, \ldots, Y_{12})$ are derived and presented in Table 15.3. Standard normal variables $(X_1, X_2, \ldots, X_{12})$ can subsequently be generated using the parameters provided in Table 15.3 and Equation (15.1).

Even if each transformed variable X_i follows a normal distribution, the multivariate set $(X_1, X_2, \ldots, X_{12})$ may not necessarily follow a multivariate normal distribution, unless the linear combination of any of the components follows a normal distribution. Nevertheless, for several soil databases, the assumption of a multivariate normal distribution continues to yield satisfactory practical outcomes, despite the transformed parameters not strictly following the multivariate normal distributions. Hence, this study forgoes testing for multivariate normality.

15.3.2 Correlation matrix of multivariate distribution

The probability density function (PDF) of the multivariate joint distribution is given below:

$$f(\boldsymbol{x}) = |\mathrm{C}|^{-1/2}(2\pi)^{-m/2}\exp\left\{-\frac{1}{2}(\boldsymbol{x}-\boldsymbol{\mu})^T \times (\mathrm{C})^{-1} \times (\boldsymbol{x}-\boldsymbol{\mu})\right\} \tag{15.2}$$

where $\boldsymbol{x} = (x_1, x_2, \ldots, x_m)^{\mathrm{T}}$ is a vector composed of m random variables following standard normal distributions, with m being the number of variables (m is taken as 12 in this study), "T" refers to the transpose of the matrix, $\boldsymbol{\mu}$ is the mean vector of $\boldsymbol{x}$, C denotes the covariance matrix, and $|C|$ stands for the determinant of the covariance matrix. As each component of $\boldsymbol{x}$ is a standard normal variable, $\boldsymbol{\mu}$ simplifies to a zero vector. C corresponds to the correlation matrix, which is composed of the Pearson product–moment correlation coefficients (denoted as δ_{ij}) between X_i and X_j.

Figure 15.2 shows the scatter plots for each parameter pair (X_i, X_j). To make the best use of the collected data, δ_{ij} is estimated using independent bivariate datasets, i.e., data points in the corresponding scatter plot in Figure 15.2, which involve two parameters measured in proximity, without the requirement for simultaneous knowledge of other parameters. However, this method has the potential to generate a non-positive definite correlation matrix, rendering it mathematically invalid.

This study follows the approach proposed by Ching and Phoon (2014b) to tackle the problem. To ensure statistical robustness, the process involves obtaining 1000 samples for

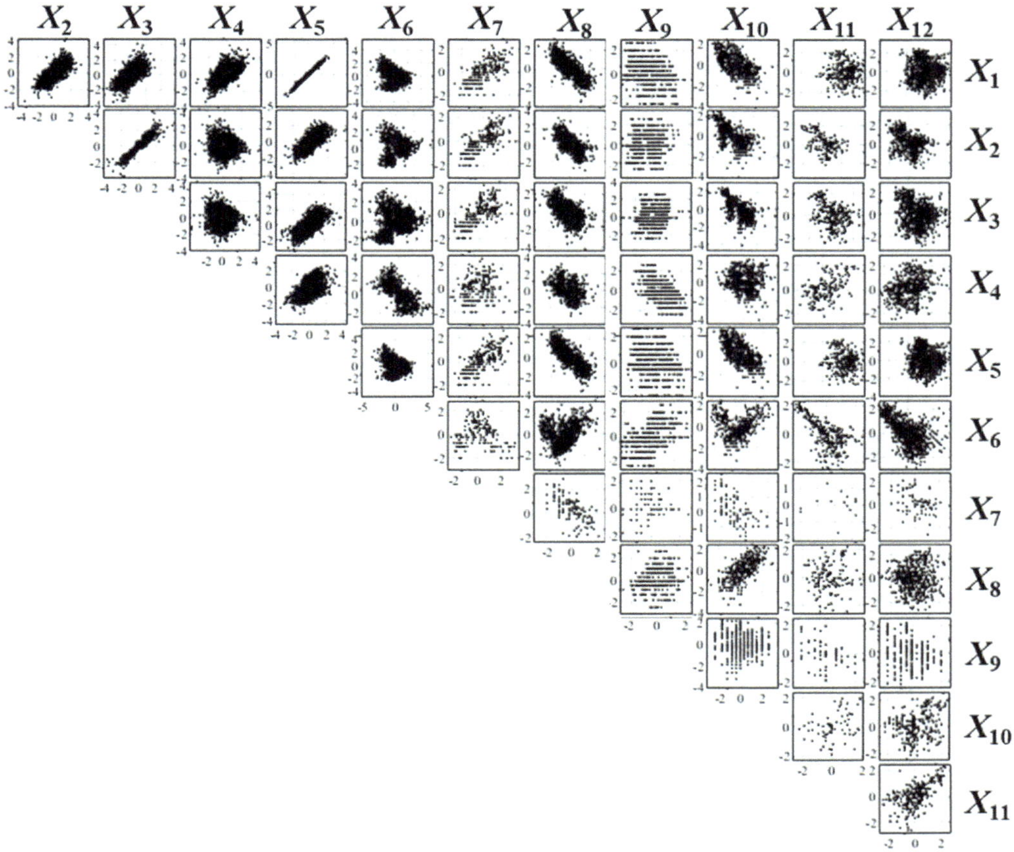

Figure 15.2 Scatter plots between X_i and X_j.

each δ_{ij} through bootstrapping technique. From this pool of samples, one δ_{ij} value is randomly selected from the 90% confidence interval of the 1000 samples, which is repeated 66 times for each parameter pair to form a new correlation matrix **C**. While only retaining the positive definite matrices, 1000 valid correlation matrices **C** are generated by this approach. The final correlation matrix **C** is derived by calculating the average value of the 1000 accepted matrices, as presented in Table 15.4.

15.4 TRANSFORMATION MODEL FOR DESIGN PARAMETERS

The established multivariate normal distribution can serve as a prior distribution for Bayesian updating, and the posterior distribution can be obtained based on the new measurements. The median and 95% confidence interval of the posterior distribution of the target parameter can be considered as predictions for the parameter given other observed data. Consequently, this allows the development of the multivariate transformation model between the target parameter and the other parameters.

15.4.1 Bayesian updating

Let the target parameters to be updated be represented as vector $\boldsymbol{X}^{u}$, which is an $m\times1$ vector. The observed parameter is denoted as vector $\boldsymbol{X}^{o}$, an $n\times1$ vector. Their corresponding mean vectors are designated as $\boldsymbol{\mu}^{u}$ (an $m\times1$ zero vector) and $\boldsymbol{\mu}^{o}$ (an $n\times1$ zero vector), and the covariance matrices are referred to as $\mathbf{C}^{u}$ (an $m\times m$ matrix) and $\mathbf{C}^{o}$ (an $n\times n$ matrix), respectively.

$\mathbf{C}^{uo}$ (an $m\times n$ matrix) and $\mathbf{C}^{ou}$ (an $n\times m$ matrix) denote the covariance matrices between the target parameters and the observed parameters, with each being the transpose of the other. The updated multivariate distribution continues to follow a multivariate normal distribution due to the conjugacy of the prior distribution, with its mean vector and covariance matrix denoted as $\boldsymbol{\mu}'^{u}$ and $\mathbf{C}'^{u}$, which can be computed using the following equation:

$$\begin{aligned} \boldsymbol{\mu}'^{u} &= \boldsymbol{\mu}^{u} + \mathbf{C}^{uo}(\mathbf{C}^{o})^{-1}(\boldsymbol{x}^{o} - \boldsymbol{\mu}^{o}) \\ \mathbf{C}'^{u} &= \mathbf{C}^{u} - \mathbf{C}^{uo}(\mathbf{C}^{o})^{-1}\mathbf{C}^{ou} \end{aligned} \tag{15.3}$$

One attractive feature of the Johnson distribution is that the functional form of the marginal distribution for Y_i remains unaltered after conditioning. Additionally, the type of the Johnson distribution, as well as its corresponding a_Y and b_Y parameters, remains constant. Only the parameters a_X and b_X are updated. The expressions for the updated parameters a'_X and b'_X are

$$\begin{aligned} a'_X &= a_X / \sigma'_X \\ b'_X &= (b_X - \mu'_X) / \sigma'_X \end{aligned} \tag{15.4}$$

where $\boldsymbol{\mu}'_x$ is the updated mean value of X_i and σ'_x is the updated standard deviation, which corresponds to the values in $\boldsymbol{\mu}'^{u}$ and $\mathbf{C}'^{u}$ for the component X_i, respectively.

In this study, the design parameters K_0, E_s, c, and φ are selected as the target parameters. Tables 15.5–15.8 present the updated Johnson parameters (a'_X, b'_X) for these four parameters conditioned on other parameters.

Table 15.4 C matrix composed of the average of the 1000 accepted δ_{ij}

X	**X_1**	**X_2**	**X_3**	**X_4**	**X_5**	**X_6**	**X_7**	**X_8**	**X_9**	**X_{10}**	**X_{11}**	**X_{12}**
X_1	1.00	0.74	0.70	0.60	0.99	−0.28	0.69	−0.81	−0.29	−0.63	−0.07	−0.04
X_2		1.00	0.94	−0.03	0.76	0.10	0.72	−0.66	0.09	−0.67	−0.45	−0.32
X_3			1.00	−0.07	0.73	0.18	0.67	−0.64	0.19	−0.65	−0.40	−0.26
X_4				1.00	0.56	−0.69	0.17	−0.44	−0.61	−0.17	0.49	0.33
X_5					1.00	−0.24	0.70	−0.82	−0.26	−0.65	−0.10	−0.07
X_6						1.00	−0.03	0.27	0.55	0.11	−0.64	−0.57
X_7							1.00	−0.63	0.03	−0.64	−0.26	−0.36
X_8		Symmetry						1.00	0.22	0.67	0.08	0.02
X_9									1.00	−0.07	−0.31	−0.17
X_{10}										1.00	0.27	0.26
X_{11}											1.00	0.62
X_{12}												1.00

Table 15.5 Updated Johnson parameters for K_0 under various combinations of information

	Updated Johnson parameter	
Information	a'_X	b'_X
ω	1.50	$-0.95X_1+1.00$
LL	1.57	$-1.04X_2+1.04$
PI	1.47	$-0.90X_3+0.97$
LI	1.11	$-0.17X_4+1.73$
e	1.53	$-0.98X_5+1.01$
σ'_v/P_a	1.09	$0.03X_6+0.72$
S_u/σ'_v	1.13	$0.27X_{12}+0.75$
p_s/σ'_v	1.17	$0.39X_{13}+0.77$
ω, LL, PI, LI, e	1.68	$-0.60X_1-0.60X_2+0.30X_3+0.35X_4-0.44X_5+1.11$
ω, LL, PI, LI, e, σ'_v/P_a	1.68	$-0.62X_1-0.56X_2+0.29X_3+0.41X_4-0.47X_5+0.04X_6+1.12$
ω, LL, PI, LI, e, σ'_v/P_a, S_u/σ'_v	1.68	$-0.62X_1-0.55X_2+0.28X_3+0.40X_4-0.46X_5+0.05X_6+0.01X_{11}+1.11$
ω, LL, PI, LI, e, σ'_v/P_a, p_s/σ'_v	1.84	$-1.30X_1+0.19X_2-0.16X_3+0.72X_4-0.17X_5+0.51X_6+0.61X_{12}+1.22$

Table 15.6 Updated Johnson parameters for E_S under various combinations of information

	Updated Johnson parameter	
Information	a'_X	b'_X
ω	2.09	$1.38X_1+0.28$
LL	1.63	$0.88X_2+0.22$
PI	1.59	$0.83X_3+0.21$
LI	1.36	$0.49X_4+0.18$
e	2.14	$1.43X_5+0.28$
σ'_v/P_a	1.27	$-0.28X_6+0.17$
S_u/σ'_v	1.23	$-0.08X_{12}+0.16$
p_s/σ'_v	1.23	$-0.02X_{13}+0.16$
ω, LL, PI, LI, e	2.17	$-0.49X_1+0.28X_2+0.26X_3+0.35X_4+1.34X_5+0.29$
ω, LL, PI, LI, e, σ'_v/P_a	2.21	$-0.34X_1+0.01X_2+0.32X_3-0.04X_4+1.52X_5-0.30X_6+0.29$
ω, LL, PI, LI, e, σ'_v/P_a, S_u/σ'_v	2.22	$-0.35X_1-0.12X_2+0.43X_3-0.005X_4+1.50X_5-0.39X_6-0.15X_{11}+0.29$
ω, LL, PI, LI, e, σ'_v/P_a, p_s/σ'_v	2.21	$-0.31X_1-0.03X_2+0.35X_3-0.06X_4+1.51X_5-0.33X_6-0.03X_{12}+0.29$

15.4.2 Transformation formula

The transformations from X to Y are as follows:

$$Y = \begin{cases} b_Y + a_Y \sinh\left(\dfrac{X - b_X}{a_X}\right) & SU \\ b_Y + a_Y\left[1 + \exp\left(-\dfrac{X - b_X}{a_X}\right)\right]^{-1} & SB \\ b_Y + a_Y \ln\left(\dfrac{X - b_X}{a_X}\right) & SL \end{cases} \tag{15.5}$$

Table 15.7 Updated Johnson parameters for *c* under various combinations of information

	Updated Johnson parameter	
Information	a'_X	b'_X
ω	6.32	$0.30X_1+41.55$
LL	6.07	$-0.09X_2+39.93$
PI	6.16	$-0.19X_3+40.50$
LI	7.63	$0.76X_4+50.18$
e	6.26	$0.27X_5+41.18$
σ'_v/P_a	7.24	$-0.66X_6+47.61$
S_u/σ'_v	6.36	$0.33X_{12}+41.83$
p_s/σ'_v	6.14	$0.17X_{13}+40.35$
ω, LL, PI, LI, e	8.03	$-0.18X_1+0.78X_2-0.94X_3+0.77X_4+0.18X_5+52.83$
ω, LL, PI, LI, e, σ'_v/P_a	8.16	$-0.04X_1+0.56X_2-0.91X_3+0.43X_4+0.33X_5-0.27X_6+53.63$
ω, LL, PI, LI, e, σ'_v/P_a, S_u/σ'_v	8.18	$-0.05X_1+0.44X_2-0.82X_3+0.47X_4+0.31X_5-0.34X_6-0.13X_{11}+53.80$
ω, LL, PI, LI, e, σ'_v/P_a, p_s/σ'_v	8.25	$0.16X_1+0.29X_2-0.76X_3+0.34X_4+0.22X_5-0.43X_6-0.21X_{12}+54.20$

Table 15.8 Updated Johnson parameters for φ under various combinations of information

	Updated Johnson parameter	
Information	a'_X	b'_X
ω	2.15	$0.81X_1+0.64$
LL	2.25	$0.90X_2+0.67$
PI	2.19	$0.85X_3+0.65$
LI	1.69	$0.17X_4+0.50$
e	2.19	$0.85X_5+0.65$
σ'_v/P_a	1.68	$-0.11X_6+0.50$
S_u/σ'_v	1.73	$-0.28X_{12}+0.52$
p_s/σ'_v	1.73	$-0.27X_{13}+0.52$
ω, LL, PI, LI, e	2.36	$-0.75X_1+0.56X_2+0.19X_3+0.16X_4+1.01X_5+0.71$
ω, LL, PI, LI, e, σ'_v/P_a	2.39	$-0.62X_1+0.34X_2+0.24X_3-0.18X_4+1.15X_5-0.26X_6+0.71$
ω, LL, PI, LI, e, σ'_v/P_a, S_u/σ'_v	2.46	$-0.66X_1-0.001X_2+0.53X_3-0.08X_4+1.12X_5-0.48X_6-0.39X_{11}+0.74$
ω, LL, PI, LI, e, σ'_v/P_a, p_s/σ'_v	2.53	$-0.17X_1-0.28X_2+0.63X_3-0.41X_4+0.95X_5-0.63X_6-0.48X_{12}+0.76$

The conditioned Y' can be obtained by substituting the original (a_X, b_X) with the updated (a'_X, b'_X) in Equation (15.5). It is thus easily derived that the p-quantile for the distribution of Y', denoted as y'_p, can be obtained from the corresponding p-quantile for the standard normal distribution, denoted as x_p. Using the Johnson SB distribution as an illustration, the median $(y'_{0.5})$ and the upper $(y'_{0.025})$ and lower $(y'_{0.975})$ boundaries of the 95% confidence interval (CI) for the updated Johnson SB distribution are as follows:

$$y'_{0.5} = b_Y + a_Y\left[1+\exp\left(\frac{b'_X}{a'_X}\right)\right]^{-1}$$

$$y'_{0.025} = b_Y + a_Y\left[1+\exp\left(-\frac{-1.96-b'_X}{a'_X}\right)\right]^{-1} \tag{15.6}$$

$$y'_{0.975} = b_Y + a_Y\left[1+\exp\left(-\frac{1.96-b'_X}{a'_X}\right)\right]^{-1}$$

The median of the conditioned distribution can serve as the point estimate for the target parameter based on other observed information, with the 95% CI indicating the precision of the estimate.

From Tables 15.5–15.8, it is evident that b'_X can be expressed as a linear combination of X_i. Therefore, Equation (15.6) is also an expression involving X_i and can be regarded as a transformation formula between the target parameter and the observed parameters.

Let's consider parameter pairs e (X_5) and K_0 (X_7) as an example to elaborate on the process of establishing the transformation formula between parameters. According to Table 15.5, the updated Johnson parameters for X_7 conditioned on X_5 are $a'_X = 1.53$ and $b'_X = 1.01–0.98\ X_5$. Notably, the parameters a_Y and b_Y remain invariant at 0.49 and –0.86, as recorded in Table 15.3. By substituting these values into Equation (15.6), the expression for Y_7 with respect to X_5 can be derived. Using Equation (15.1) and the corresponding Johnson parameters from Table 15.3, X_5 can be transformed to Y_5. Finally, after taking the exponential transformation, the functional relationship between e and K_0 can be established:

$$K_{0_{0.5}} = \exp\left\{-0.86+0.49\times\left[1+\exp\left(\frac{1.91-1.58\times\ln\left(\frac{\ln(e)+0.47}{0.60-\ln(e)}\right)}{1.53}\right)\right]^{-1}\right\}$$

$$K_{0_{0.025}} = \exp\left\{-0.86+0.49\times\left[1+\exp\left(\frac{3.87-1.58\times\ln\left(\frac{\ln(e)+0.47}{0.60-\ln(e)}\right)}{1.53}\right)\right]^{-1}\right\} \tag{15.7}$$

$$K_{0_{0.075}} = \exp\left\{-0.86+0.49\times\left[1+\exp\left(\frac{-0.05-1.58\times\ln\left(\frac{\ln(e)+0.47}{0.60-\ln(e)}\right)}{1.53}\right)\right]^{-1}\right\}$$

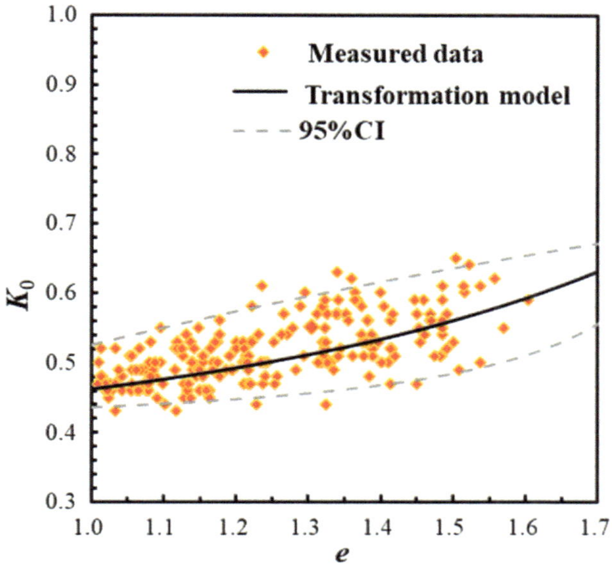

Figure 15.3 Transformation model and observed data between e and K_0.

Figure 15.3 illustrates the derived relationship between (e, K_0) based on Equation (15.7) and the corresponding data points from the database. As can be seen from the picture that the collected data align well with the transformation formula, with the majority of data points falling within the 95% confidence interval. Functional relationship between the target parameters and other univariate or multivariate parameters can also be similarly established through the same approach.

In Tables 15.5–15.8, the coefficient for X_i quantifies the sensitivity of the target parameters with respect to information of X_i. Considering the case where there is only univariate information, the coefficients for X_i exhibit good consistency with the correlation matrix presented in Table 15.4. Taking the example of the target parameter K_0: if the coefficient for X_i is negative, the target parameter increases as X_i increases, while conversely, if the coefficient is positive, the target parameter decreases with increasing X_i. For X_1 to X_5, they are positively correlated to X_7, as evidenced by their negative coefficients in b'_X. Conversely, the other parameters are negatively related to X_7, corresponding to the positive coefficients in b'_X. Additionally, the absolute ratio of the coefficient for X_i to a'_X reflects the degree of influence that X_i has on the target parameter. The ranking of the ratio values for each X_i corresponds to the ranking of the absolute value of the correlation coefficient between each X_i and X_7. The same patterns can be observed for E_s, c, and φ.

Furthermore, according to Equation (15.4), a'_X is inversely proportional to σ'_x. A higher value of a'_X indicates a smaller standard deviation for the updated variable, implying its decreased uncertainty. As can be seen from Tables 15.5–15.8, as more information is incorporated, the value of a'_X gradually increases, signifying a reduction in the uncertainty of the target parameters. This observation is consistent with our common understanding.

However, it is worth noting that for the target parameter K_0, the updated Johnson parameters obtained based on (ω, LL, PI, LI, e), (ω, LL, PI, LI, e, σ'_v/P_a), and (ω, LL, PI, LI, e, σ'_v/P_a, S_u/σ'_v) yield very similar results, with the coefficients for X_6 and X_{11} close to zero. This implies that the inclusion of σ'_v and S_u information has minimal impact on the update and prediction for the parameter K_0. Similarly, the information related to p_s has little effect on the update and prediction for the target parameter E_s.

Moreover, it is noticeable from Tables 15.5–15.8 that when multisources of information are simultaneously known, the coefficients for X_i in b'_X no longer present the same pattern as they did with univariate information. For instance, in the prediction of K_0 using (ω, LL, PI, LI, e), the coefficients for X_3 and X_4 become positive, which contradicts their positive correlation with X_7. Therefore, it is imperative to assess the validity of the transformation model.

15.4.3 Validation of transformation models

Validation of the transformation models is conducted using 156 data points from a new site in Shanghai, located in the Xuhui distinct. The actual values of the target parameters are assumed to be unknown during the prediction process. After the prediction, they are compared with the predicted medians and 95% CIs to verify the performance of the transformation model.

Figure 15.4 illustrates the median values (represented by open dots) and the 95% CI (indicated by black vertical dashed lines) for K_0 conditioning on various combinations of information. Parameters σ'_v/P_a and S_u are not included considering their little impact on the prediction of K_0. The 1:1 line represents the scenario where actual values match the predicted values. If the vertical line representing the 95% CI intersects with the 1:1 line, it indicates that the actual value falls within the predicted 95% CI. The red dashed line is used to annotate the absence of a match if there is no intersection. Note that Figure 15.4(a) signifies the statistical results of the raw data from this new site, without incorporating any information for the parameter updates.

From Figure 15.4, it is clear that the median points gradually converge toward the 1:1 line when K_0 is conditioned on other available information, indicating a closer alignment between the predicted values and actual values. However, it can be observed that the conditioned 95% CIs do not cover the 1:1 line when K_0 equals 0.4, 0.43, 0.44, 0.62, or 0.65. This is because for K_0 = 0.4, 0.43, and 0.65, these data points inherently fall outside the 95% CI of the data used to construct the model, with 0.4 even being lower than the minimum value (=0.43) in the database. Furthermore, the two data points (K_0 = 0.44 and 0.66) are very close to the 2.5% and 97.5% percentiles of the K_0 values in the database.

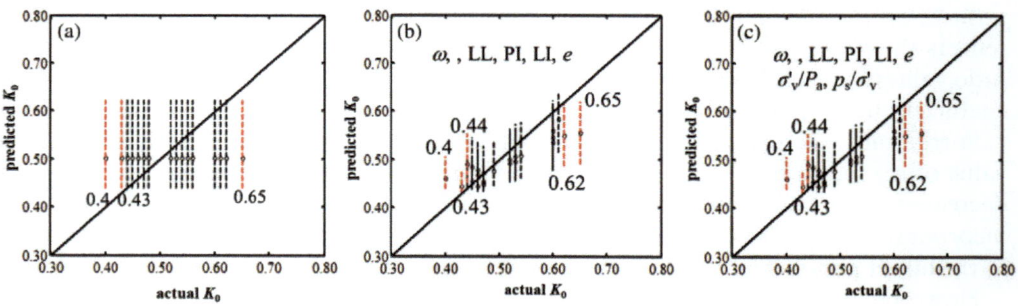

Figure 15.4 Updating results for K_0 by conditioning on: (a) no information; (b) ω, LL, PI, LI, e; (c) ω, LL, PI, LI, e, σ'_v/P_a, p_s/σ'_v.

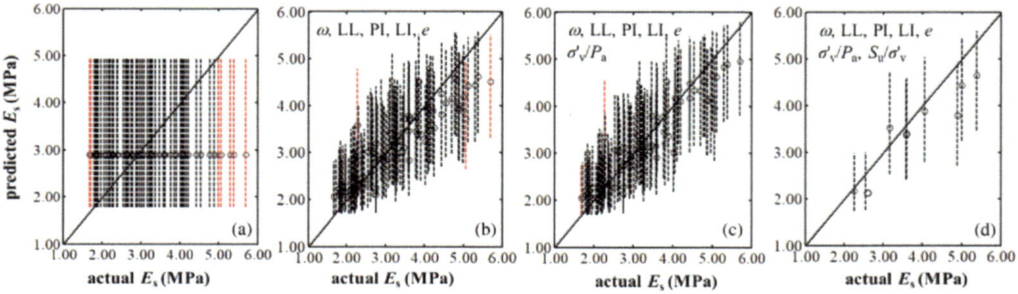

Figure 15.5 Updating results for E_s by conditioning on: (a) no information;(b) ω , LL, PI, LI, e;(c) ω , LL, PI, LI, σ'_v/P_a; (d) ω, LL, PI, LI, e, σ'_v/P_a, S_u/σ'_v.

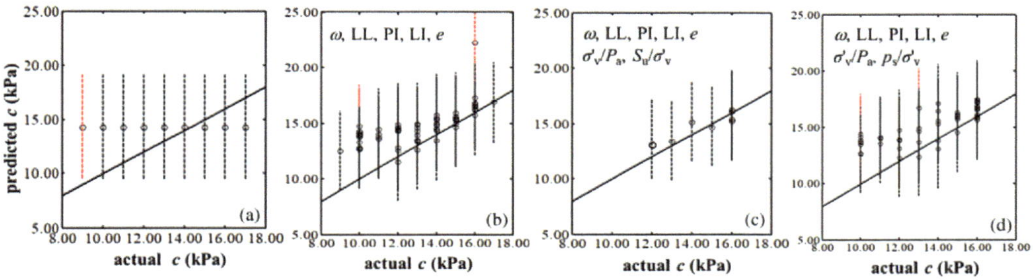

Figure 15.6 Updating results for c by conditioning on: (a) no information;(b) ω , LL, PI, LI, e;(c) ω , LL, PI, LI, σ'_v/P_a, S_u/σ'_v; (d) ω, LL, PI, LI, e, σ'_v/P_a, p_s/σ'_v

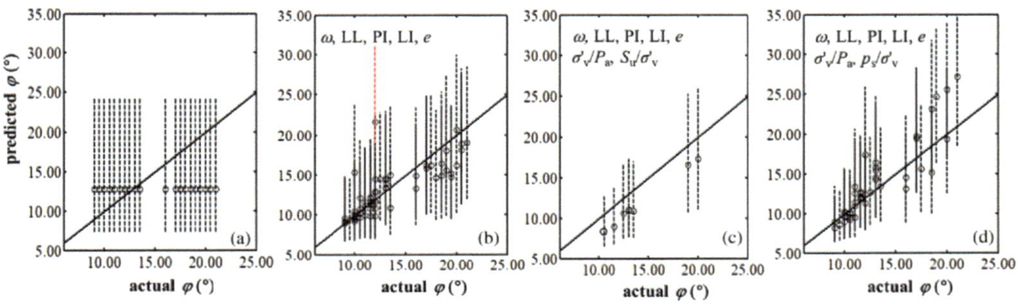

Figure 15.7 Updating results for φ by conditioning on: (a) no information; (b) ω, LL, PI, LI, e; (c) ω, LL, PI, LI, σ'_v/P_a, S_u/σ'_v; (d) ω, LL, PI, LI, e, σ'_v/P_a, p_s/σ'_v.

This implies that if any data point lies outside or close to the 95% CI of the database, extrapolation issues might arise.

Figures 15.5–15.7 show the updating results for E_s, c, and φ conditioning on various combinations of information. The results indicate that the majority of the actual values fall within the 95% CIs of the predicted values. Moreover, there is a noticeable decrease of the width of the confidence interval with the increasing information as well as the closer alignment between the predicted values and actual values.

15.5 CONCLUSIONS

This chapter presents a database for Shanghai soft clay labeled as SH-CLAY/12/4191, whose data points are collected from 51 site investigation reports in Shanghai covering approximately 145 km^2 in area, about 1/43 the size of Shanghai. General information including the geology of Shanghai, the borehole locations, the types of clay parameters, and the basic statistics of these parameters is given. By comparing the marginal statistics of SH-CLAY/12/4191 with those presented in the related lecture, it can be concluded that the quality of the data in SH-CLAY/12/4191 is generally satisfactory. A multivariate probability distribution for the 12 clay parameters is then constructed based on the SH-CLAY/12/4191 database. Finally, transformation models for soil design parameters K_0, E_s, c, and φ in relation to various index and in-situ test parameters are established. The established models are verified using data from a new site, and the results demonstrate the generally reliable performance of the established transformation models.

REFERENCES

Ching, J. & Phoon, K. K. (2014a). Transformations and correlations among some clay parameters—the global database. *Canadian Geotechnical Journal*, 51(6), 663–685.

Ching, J. & Phoon, K. K. (2014b). Correlations among some clay parameters–the multivariate distribution. *Canadian Geotechnical Journal*, 51(6), 686–704.

Ching, J., Li, D. Q., & Phoon, K. K. (2016). *Statistical Characterization of Multivariate Geotechnical Data*. Chapter 4, Reliability of Geotechnical Structures in ISO2394, CRC Press, Balkema, 89–126.

Ching, J., Lin, G. H., Chen, J. R., & Phoon, K. K. (2017). Transformation models for effective friction angle and relative density calibrated based on generic database of coarse-grained soils. *Canadian Geotechnical Journal*, 54(4), 481–501.

Ching, J. & Phoon, K. K. (2019). Constructing site-specific multivariate probability distribution model using Bayesian machine learning. *Journal of Engineering Mechanics*, 145(1), 04018126.

Ching, J., Phoon, K. K., Khan, Z., Zhang, D., & Huang, H. (2020). Role of municipal database in constructing site-specific multivariate probability distribution. *Computers and Geotechnics*, 124, 103623.

Ching, J., Phoon, K. K., Ho, Y. H., & Weng, M. C. (2021). Quasi-site-specific prediction for deformation modulus of rock mass. *Canadian Geotechnical Journal*, 58(7), 936–951.

Ching, J., Phoon, K. K., Yang, Z., & Stuedlein, A. W. (2022). Quasi-site-specific multivariate probability distribution model for sparse, incomplete, and three-dimensional spatially varying soil data. *Georisk: Assessment and Management of Risk for Engineered Systems and Geohazards*, 16(1), 53–76.

D'Ignazio, M., Phoon, K. K., Tan, S. A., & Lansivaara, T. (2016). Correlations for undrained shear strength of Finnish soft clays. *Canadian Geotechnical Journal*, 53(10), 1628–1645.

Gao, D. Z & Wei, D. D. (1983). Probability and statistical characteristics of soil engineering properties in shanghai. In *Proceedings of the 4th Conference of the Chinese Civil Engineering Society on Soil Mechanics and Foundation Engineering, Wuhan, China.* (in Chinese)

ISO2394 (2015). *General Principles on Reliability for Structures*. International Organization for Standardization, Geneva, Switzerland.

Johnson, N. L. (1949). Systems of frequency curves generated by methods of translation. *Biometrika*, 36, 149–176.

Kulhawy, F. H. & Mayne, P. W. (1990). *Manual on Estimating Soil Properties for Foundation Design*. Report EL-6800, Electric Power Research Institute, Palo Alto, CA.
Liu, S., Zou, H., Cai, G., Bheemasetti, B. V., Puppala, A. J., & Lin, J. (2016). Multivariate correlation among resilient modulus and cone penetration test parameters of cohesive subgrade soils. *Engineering Geology*, 209, 128–142.
Löfman, M. S. & Korkiala-Tanttu, L. K. (2022). Transformation models for the compressibility properties of Finnish clays using a multivariate database. *Georisk: Assessment and Management of Risk for Engineered Systems and Geohazards*, 16(2), 330–346.
Mayne, P. W., Christopher, B. R., & DeJong, J. (2001). *Manual on Subsurface Investigations*. Nat. Highway Inst. Sp. Pub. FHWA NHI-01–031. Fed. Highway Administ, Washington, DC.
Muzamhindo, H. & Ferentinou, M. (2023). Generic compressive strength prediction model applicable to multiple lithologies based on a broad global database. *Probabilistic Engineering Mechanics*, 71, 103400.
Phoon, K. K. & Kulhawy, F. H. (1999). Characterization of geotechnical variability. *Canadian Geotechnical Journal*, 36(4), 612–624.
Slifker, J. F. & Shapiro, S. S. (1980). The Johnson system: selection and parameter estimation. *Technometrics*, 22(2), 239–246.
Xiao, L. (2018). *Probabilistic Analysis of Shield Tunnel Convergence by Random Field of Multilayer Soil*. M.S. thesis, Tongji University, Shanghai. (in Chinese)
Zhang, D., Zhou, Y., Phoon, K. K., & Huang, H. (2020). Multivariate probability distribution of Shanghai clay properties. *Engineering Geology*, 273, 105675.
Zhou, Y., Zhang, D., Huang, H., & Xue, Y. (2022). Effect of normal transformation methods on performance of multivariate normal distribution. *ASCE-ASME Journal of Risk and Uncertainty in Engineering Systems, Part A: Civil Engineering*, 8(1), 04021074.

APPENDIX A

Table A1 Statistics of the muddy clay in Shanghai

Site No.	*Sample size*	*Index*	ω *(%)*	*LL (%)*	*PI*	**e**	φ (°)
1	14	Mean	49.8	45.7	21	1.43	9.5
		COV	0.16	0.20	0.54	–	–
2	16	Mean	51.0	43.9	21.8	1.44	7.4
		COV	0.25	0.23	0.97	–	–
4	30	Mean	50.5	43.8	20.6	1.45	8.0
		COV	0.22	0.23	0.72	–	–
7	9	Mean	48.0	44.1	19.9	–	8.1
		COV	0.14	0.16	0.79	–	–
8	33	Mean	49.0	45.9	22.0	1.41	8.4
		COV	0.26	0.30	0.79	–	–
9	13	Mean	51.2	44.9	20.7	1.47	7.8
		COV	0.17	0.22	0.77	–	–
10	15	Mean	50.0	40.1	18.6	1.45	–
		COV	0.30	0.25	1.04	–	–
11	90	Mean	48.0	–	20.0	1.73	–
		COV	0.26	–	0.88	–	–
Summary	220	Mean	49.1	44.3	20.5	1.6	8.2

Source: Data from Gao and Wei (1983).

Table A2 Statistics of the muddy clay, muddy silty clay, and clay in Shanghai

Soil Layer	*Index*	ω (%)	*LL (%)*	*PI*	e	E_s *(MPa)*	*c (kPa)*	φ (°)
muddy silty clay	Min–Max	36.0~49.7	29.6~40.1	10.3~17.0	1.00~1.36	2.20~5.97	8.5~14.2	12.1~28.0
	COV	0.110	0.066	0.145	0.104	0.292	0.240	0.250
muddy clay	Min–Max	40.0~59.6	34.4~50.2	17.0~25.1	1.12~1.67	1.32~3.58	11.5~15.7	8.5~16.9
	COV	0.080	0.078	0.112	0.075	0.179	0.037	0.162
clay	Min–Max	29.8~42.5	28.3~42.9	10.2~20.0	0.85~1.22	3.00~6.77	11.5~20.0	12.7~27.4
	COV	0.082	0.096	0.152	0.079	0.200	0.223	0.217
Summary	Min–Max	29.8~59.6	28.3~50.2	10.2~25.1	0.85~1.67	1.32~6.77	8.5~20.0	8.5~28.0

Source: Data from Code for Investigation of Geotechnical Engineering in Shanghai, DGJ08-37-201.

Index

"4S" (site generalizations, spatial features, sampling characteristics, and smart data), 1, 13, 38
304dB, xi, 37

A

AlexNet, 309
Algorithm implementation, 389
  likelihood function, 390–391
  posterior distribution, 392
  prior distribution, 389–390
Analysis region and field data, 441
Analysis tools and software, 435
ANI-F-CLAY/7/87 database, 137
  basic information of, 150–151
  bias factors and COV for anisotropic s_u of Finnish soft clays, 145
  multivariate ANI-F-CLAY/7/87 from Finland, 139–142
Architecture of different feature extraction modules, 254
Artificial intelligence (AI) technology, 33, 245, 305, 428, 448
Assessing approaches, 433
  mechanics-based numerical simulation, 434–435
  simplified methods, 433–434
AUT: CPTu-Geo-marine database, 186–187
Averaged index approach, 436, 441, 443–446

B

Base resistance methods, 68–70
  empirical base resistance models, 70
  relationships with rock mass property, 72–78
  semi-empirical base resistance models, 70–72
Basic quality (BQ), 305
Bayesian inverse modeling method, 388
  Bayesian framework, 389
  data classification, 388
Bayesian updating, 460–462
Benchmarking databases, 219–223
Big data, 4, 20
Big data engineering, 2
Big indirect data (BID), 5, 9–10
Bivariate database, 229, 234–235
Bothkennar site, United Kingdom, 418
  constitutive parameters calibration, 423–424
  parametric analysis, 418
  preliminary prediction of CPTU resistances, 418–423
Broader emerging trends, 25

C

Calibrated prediction models, 220–221
Calibrated transformation models, 219
Canterbury earthquakes, 441
Categorical *vs.* numerical data, 19–20
Cavity expansion theory
  of CASM model, 380
    constitutive mode, 380–382
    elastic–plastic boundary analysis, 385
    elastic zone analysis, 383
    mechanical transformation model, 385–386
    plastic zone analysis, 383–384
    spherical cavity expansion theory, 382–383
  of MCC model, 373
    CPTU mechanical model, 377–380
    definition of the problem, 373–374
    elastic analysis, 374
    elastoplastic analysis, 374–377
ChatGPT, 22, 35, 37
Christchurch, New Zealand, 441–445
Classical Bayesian model (CBM), 32–33
Classical clustering (CC), 32–33
CLAY/10/7490 database, 28–29, 32, 106–110, 333–338; *see also* Exploring challenge analysis of multivariate geotechnical property
CLAY-C_c/6/6203 database, 119
Climate change mitigation goals, 280

Coefficient of variation (COV), 10, 103, 300
Comparing regional databases, Finland and Norway, 142–143
Compression behavior of soft marine clays, 281–283
Conception module, 255–256, 262
Conception module architecture, 257
Cone penetration test (CPT), 152, 200, 345, 455
Confidence interval (CI), 29–30, 130, 460
Consolidation indexes, 182
Constitutive mode, 380–382
Constitutive parameters calibration, 423–424
Constrained modulus, 178–179
Convolutional neural network (CNN), 36, 256, 304, 309–310, 313
CPTu in overconsolidated Presumpscot clay, ME, 354–355
CPTu in soft intact NC clays at Sandpoint, Idaho, 350–351
CPTU mechanical model, 377–380
CPTu-ϕ' profile interpretation for soft intact NC Chicago clays, 351–353
Critical state line (CSL), 381
Cross-correlation and correlation coefficients, 296–300
CRS oedometer tests, 231
Cumulative frequency, 444–446

D

Dark data, 38
Data, 428
  ground motion and intensity measures, 428–429
  induced ground displacement at a site, 432
  liquefaction indices at a site, 432
  liquefaction maps, 432–433
  site characteristics, 429–430
Databases, 216; *see also individual entries*
  for data-centric geotechnics, 333–338
  driven approach, 136
  evaluation, 216–219
  of Finnish and Scandinavian soft clays
    comparing regional from Finland and Norway, 142–143
    large regional databases, 137–139
    multivariate ANI-F-CLAY/7/87 from Finland, 139–142
  subdivision analyses, 215–216
  use of
    assess the variability of s_u at a particular site, 136
    develop site-specific s_u correlations, 136
    identify and quantify bias in local data, 136
    validate s_u predictions from numerical models, 137
Databases role, evaluation of soil property
  approximate variation of consistency, *N*, and undrained cohesion of clay, 175
  chart for estimating soil type and unit weight for soil based on DMT result, 164
  classification chart, 163
  common amount of friction angle for sand and silt, 170
  comparison of D_r–q_c relationships, 169
  compiled database
    scatter of clayey soils, different SBC charts, 193–194
    scatter of deltaic soils, different SBC charts, 195–196
    scatter of sand and gravel, different SBC charts, 197–198
    scatter of sensitive soils, different SBC charts, 191–192
  consistency *vs.* undrained shear strength, 158
  constrained modulus coefficient for cohesive soils, 179
  correlation of q_c and relative density, friction angle for cohesionless/mixed soils, 173
  correlations between V_s (m/s) and SPT-*N* value, 183
  databases for data-centric geotechnics, 186
  data sources in geotechnical engineering, 153–154
  delineated zones for developed triangular chart, deltaic deposit, 201
  details of AUT: CPTu-GMD, 189
  DMT, VST and SBPMT correlation with OCR, 181
  drained elastic modulus for soils using CPT records, 177
  elastic modulus parameter of in-situ tests, 178
  empirical correlations
    for C_c, 183
    compression ratio, 182
    for G_0 based on q_c, 185
    V_s *vs.* CPT data, 185
  examples of updated classification charts, 163
  friction angles
    changes and cone tip resistance, 173
    of granular soils, 170
    relative density relations, 171
  graphic estimation, 175
  in-situ tests and prediction of geo-parameters
    geo-marine database and triangular charts, 186–201
    geotechnical parameters, 165–166
    index parameters, 166–167

related parameters, 182–186
stiffness and compressibility parameters, 176–182
strength parameters, 167–176
methods in s_u interpretation of in-situ tests, 177
normalized unit weight and R_f, 167
overview, 152–153
pioneer soil profiling charts, 159
proposed correlations
for D_r based on SPT number, 168
for friction angle based CPT result, 174
for OCR, 180
relationship between N, and s_u, 176
role of in-situ tests in soil identification, 156
soil behavior classification based on in-situ tests, 157
soil classification based on CPT data, 157–164
soil classification based on DMT records, 164–165
soil classification based on FVST records, 157
soil classification based on PMT measurements, 165
soil classification based on SPT blow counts, 157
scatter of the AUT: CPTu-GMD case study, 187
soil behavior type classification system, CPT/CPTu data, 162
soil classification based on I_{SBC}, 159
soil classification based on material index, 165
soil types on developed triangular chart, 202
SPT and CPT correlation with OCR, 180
statistical assessment of SBC diagrams via radar charts, 199–200
strength from SPT on different types of soils, 158
synopsis, 201–203
typical CPTu logs, 188
typical Menard pressuremeter values, 165
typical values for bulk and saturated unit weights, 166
uncertainty in geotechnical designs, 154–156
values of I_C on the classification chart, 159
variability
from in-situ tests, 156
from laboratory tests, 155
variation of N with σ_0/p_a and D_r, 168
variation range, 174
Database SurroundingRock/8/286, 307–309
Data-centric approach, 235–237, 448
Data-centric geotechnics, xii–xiii, 2; *see also individual entries*
agenda
data centricity, 4
fit for (and transform) practice, 4
geotechnical context, 4–5
decision making, 2
elements, 1, 3–5
geotechnical engineering and machine learning, 2
machine learning in geotechnics (MLIG), 2–3
value of MLIG, 3
Data-centric seismic soil liquefaction assessment
analysis tools and software, 435
QS3HARK, 435
tools for numerical methods, 435
tools for simplified methods, 435
assessing approaches, 433
mechanics-based numerical simulation, 434–435
simplified methods, 433–434
cumulative frequency
plot of LPIs, 444
vs. the fraction of liquefied area, 446
data, 428
ground motion and intensity measures, 428–429
induced ground displacement at a site, 432
liquefaction indices at a site, 432
liquefaction maps, 432–433
site characteristics, 429–430
framework of, CPT-based liquefaction potential evaluation over extended area, 437
graphic representation of material property at two scales, 440
histograms and semivariograms, 442
histograms of predicted LPIs, 445
locations of two sites for fine-scale liquefaction assessment, 446
map of the area of study, 442
mapping of LPI, 443
multiscale regional assessment, 435–436
general framework, 436–437
numerical examples, 441–447
random field characterization, 438–441
site-specific liquefaction evaluation, 438
overview, 427–428
research gaps and opportunity, 447
synopsis, 448
Data-driven approaches, 244–245, 429
Data-driven site characterization (DDSC), 21–22
data-informed decision support index (DIDI), 23–26
methodology, 14–16
site recognition challenge, 22–23

Data encryption, 37
Data first, practice central, and privacy enhanced, 38
Data governance, 37
Data-informed decision support index (DIDI), 26
  broader emerging trends, 25
  decisions of, 25
  judgment based solutions, 23–25
Data minimization, 37
Data protection, 36–37
Data statistical analysis, 249
Deep learning (DL) model, 309
  DL model PRL-Net, 272–273
  experimental evaluation and results
    ablation study, 259–260
    feature extraction module, 261–262
    feature transformation module, 262
    input variable order, 260–261
  Seft-Net model architecture, 252–253
    proposed feature extraction module, 253–254
    proposed feature transform module, 254–257
  Seft-Net model evaluation, 262–265
  training strategy, 257–258
    model testing, 259
    model training, 259
    performance metrics, 258–259
  variable importance analysis, 265–266
Deep learning technology, 305
Deep neural network (DNN), 35–36, 265
Departments of Transportation (DOTs), 9
"Depth," 333
Depthwise separable convolution, 254–255
Design of monitoring systems, 21
Design parameters, 452–453
Detailed geological data, 38
Developed triangular chart, 199–201
Direct field measurement of s_u techniques, 135
Dirty data, 38
Discrete element method (DEM), 428, 434, 436

E

Effective cohesion intercept, 361
Effective stress strength parameters
  fissured clays from piezocone penetration test, 355
    Fissured London Clay at Brent Cross, United Kingdom, 355–356
    organic soils, sensitive clays, and structured geomaterials, 356–357
    Triaxial–Piezocone database for NC–OC clays, 356
  intact clays from flat dilatometer test, 357–358
    Route 17 Bridge, Chesapeake, Virginia, 358–360
  intact clays from piezocone penetration test, 349–350
    CPTu in overconsolidated Presumpscot clay, ME, 354–355
    CPTu in soft intact NC clays at Sandpoint, Idaho, 350–351
    CPTu-ϕ' profile interpretation for soft intact NC Chicago clays, 351–353
Elastic analysis, 374
Elastic-plastic boundary analysis, 385
Elastic zone analysis, 383
Elastoplastic analysis, 374–377
Empirical formula for predicting shear strength, 266–269
Empirical models, 70, 223, 429
  evaluation of, 73, 92
Empirical side resistance models, 59–63
Engineering geological parameter database of tunnel surrounding rock and application, 304
  binary confusion matrix, 319
  classification of
    and characteristics of discontinuity condition, 330
    rock mass structure, 329
    in situ stress, 330
    water outflow from tunnel excavation face, 331
  confusion matrix of SVM-based, RF-based and GPC-based SRC models, 320
  curves of the accuracy of training and validation sets of the VGGNet16-based and VGGNet19-based SRC models, 314
  database SurroundingRock/8/286, 307–309
  data visualization of TIMS homepage, 325
  distribution of engineering geological parameters
  field application, 320–321
  F-measure of SVM-based, RF-based and GPC-based SRC models, 321
  hyperparameter values and validation set accuracy for intelligent SRC models, 315
  intelligent methods for surrounding rock classification
    convolutional neural network (CNN), 309–310
    Gaussian process, 310–311
    model training and evaluation, 313–319
    random forest (RF), 310
    support vector machine (SVM), 311–312
  intervals in, 308
  model training and evaluation, 313
    CNN-based SRC model, 313–317
    GPC-based SRC model, 317–318

performance evaluation of intelligent SRC model, 318–319
RF-based SRC model, 317
SVM-based SRC model, 318
monthly distribution of alteration design grade, 326
overview, 304–307
precision of SVM-based, RF-based and GPC-based SRC models, 320
qualitative and quantitative classification
of rock hardness, 328
of rock weathering degree, 328
quantitative classification of rock mass integrity, 330
recall of SVM-based, RF-based, and GPC-based SRC models, 321
route diagram of the Lexi Expressway, 324
synopsis, 324–325
validation with new cases, 319–320
verification results of new tunnel cases, 322
workflow of the intelligent SRC model, 314
Establishment of geo-gravel dataset
data statistical analysis, 249
establishment process, 247–248
Establishment process, 247
external condition, 248
particle property, 247–248
soil mass property, 248
stress–strain–volume curves, 248
Estimation of model parameters, 88–95
model formulation framework, 92–93
permeability (k), 95
rock mass deformation modulus (E_m), 93–94
unconfined compressive strength (q_u), 94–95
Eurocode 7 states, 102–103
Evaluation of compressibility property, soft marine clays
approximation of *Cc* from Janbu parameters, 285
Cc method, 299
compression behavior of
soft marine clays, 281–283
soft Nordic clays, 282
cross-correlation and correlation coefficients, 296–300
database of Finnish clays FI-CLAY/14/856, 283–286
fitting of *Cc(wn)*-transformation model using the FI-CLAY/14/822 database, 294
Haarajoki clay, homogeneous soft clay layer, 291
inherent variability, 286–289
Janbu method, 298
observed variability
in C_c effect of de-trending, 290
in C_c effect of scale, 287
parameters as function of mean value, 289
overview, 280–281
Pearson's correlation coefficient ρ for certain populations, 297
pie chart of soil types, FI-CLAY/14/856 database and Finnish classification system, 284
probability density functions (PDFs), 289–291
relationship between e_0 and C_c in the global database of marine clays, 286
results of calibration study, existing models for compressibility against FI-CLAY/14/856, 293
site-specific cross-correlation between compressibility parameters, 297
statistics for the FI-CLAY/14/856-B database and the mean values, 285
suitable pdfs for geotechnical property, settlement calculations in clay subsoil, 291
synopsis, 300–301
transformation models
for the compressibility of Finnish clays, 296
database of Finnish clays FI-CLAY/14/822, 295
uncertainty in existing and new transformation models, 291–296
Evaluation of regression strength, 214–215
Evaluation of soil/rock property, databases
coefficient of variation for inherent variability, 104
generic correlations for SAND/7/2794, 112
generic statistics for
CLAY/10/7490 (expanded), 107
CLAY-Cc/6/6203, 119
ROCK/9/4069, 113
ROCKMass/9/5876, 116
SAND/7/2794, 111
SOIL-DMT/8/7186, 121
SOIL-KSAT/7/2293, 123
generic transformation models
CLAY-Cc/6/6203, 120
ROCK/9/4069, 116
ROCKMass/9/5876, 119
SAND/7/2794, 113
SOIL-DMT/8/7186, 123
SOIL-KSAT/7/2293, 124
group-specific correlations and statistics for SOIL-KSAT/7/2293, 124
illustration of the quasi-site-specific model, 130
model structure of HBM, 127
overview, 102–106
predictive medians and 95% CIs of generic and quasi-site-specific transformation models, 130

quasi-site-specific transformation model, 125–128
HBM inference stage, construction of, 128–131
HBM learning stage, extract site-specific statistics in the database, 128
ranges of site-specific SOF_h and SOF_z for various soil types, 125
scales of fluctuation for some soil parameters, 105
site investigation data for a site in Onsoy (Norway), 129
site-specific correlations for
CLAY/10/7490 (expanded), 108–109
CLAY-Cc/6/6203, 120
ROCK/9/4069, 115
ROCKMass/9/5876, 118
SOIL-DMT/8/7186, 122
site-specific for ROCK/9/4069 (sedimentary), 114
site-specific for ROCKMass/9/5876 (sedimentary), 117
site-specific statistics for
CLAY/10/7490 (expanded), 107
CLAY-Cc/6/6203, 120
ROCK/9/4069 (igneous), 114
ROCK/9/4069 (metamorphic), 114
ROCKMass/9/5876 (igneous), 117
ROCKMass/9/5876 (metamorphic), 117
SOIL-DMT/8/7186, 121
soil/rock databases, 105
statistics for site-specific SOFz from CPT cone tip resistance for soil units, 126
synopsis, 131
transformation models calibrated by CLAY/10/7490 (expanded), 111
typical values of site-specific mean and COV_{INHER}, 106
CLAY/10/7490 database, 106–110
CLAY-C_c/6/6203 database, 119
ROCK/9/4069 database, 111–116
ROCKMass/9/5876 database, 116–119
SAND/7/2794 database, 110–111
SOIL-DMT/8/7186 database, 119–121
SOIL-KSAT/7/2293 database, 121–125
typical values of site-specific SOFS, 125
variation of vertical scale of fluctuation with horizontal scale of fluctuation, 126
Y1–Y2 data points, CLAY/10/7490 database *vs.* simulated hypothetical sites, 128
Existing databases, 51–54
lack of basic rock mass mechanical property, 54
load–displacement information, 54
number of load tests, 54
range of rock compressive strength, 54
range of rock socket size, 54
Experimental evaluation and results
ablation study, 259–260
feature extraction module, 261
feature transformation module, 262
input variable order, 260–261
Explainability and interpretability, 34
Exploring challenge analysis of multivariate geotechnical property, 332
database overview, 333–338
databases for data-centric geotechnics, 333–338
display of scatter plots for indicators competing, CLAY/10/7490, 336
list of basic percentile and statistical value measures for each parameter, 335
overview, 332–333
scattering of
correlation coefficients per site, 341, 342
means per site, 339
standard deviations per site, 340
scatter plot matrix of all collected data, 334
scatter plots on plasticity diagram, 337
synopsis, 338–342
External condition, 244, 248, 249, 270

F

Factors influencing hydraulic conductivity, 211–213
chemical constituents, 213
permeant percolation mechanism, 211
saturated and unsaturated soil, 212
temperature-dependent parameters, 211
F-CLAY/7/216 database, 137, 453
F-CLAY/10/173 database, 138–140, 144
FG/KSAT-1358 database, 219, 223
Field application, 320–321
Field Vane Shear Testing (FVST), 152
Fine-scale site-specific liquefaction assessment, 445–447
Finite-difference (FD) method, 434
Finite-element method (FEM), 428, 434, 436
Fissured clays from piezocone penetration test, 355–357
Fissured London Clay at Brent Cross, United Kingdom, 355–356
FLAC3D model, 36
Flat dilatometer test (DMT), 346
Friction angle criteria, 349, 361
Frugal HBM, 34–35

G

Gaussian process, 310–311
Gaussian process classification (GPC), 311

Generic model for clay parameters, 28
Generic rock mass database, 304
Generic transformation models, 106, 110
 CLAY-C_c/6/6203, 120
 ROCK/9/4069, 116
 ROCKMass/9/5876, 119
 SAND/7/2794, 113
 SOIL-DMT/8/7186, 123
 SOIL-KSAT/7/2293, 124
Geo-compatibility, 34
Geo-everything-dB, 37–38
Geo-Gravel dataset, 244, 269
 establishment of, 247–248
 includes triaxial compression test data, 270
Geologic characteristics, 429
Geo-marine database and triangular charts
 AUT: CPTu-Geo-marine Database, 186–187
 developed triangular chart, 199–201
 results and discussions, 190–199
 Soil Behavioral Classification methods, 187–190
Geomechanical approach, 237–241
Geotechnical databases, 210
Geotechnical data science, xii, 2, 27, 33–34, 39
Geotechnical engineering, xi–xii, 2, 21–22, 33–34, 201, 452
 applications, 135
 data sources, 153–154
 decisions, 25
 parameters, 344–345
 research, 152
 uncertainty, 154–156
Geotechnical site investigation, 153, 344, 452, 455
Gradation, 249
Griffith's fracture theory, 70
Ground motion and intensity measures, 428–429

H

Haarajoki clay, 290–291, 297–300
Hierarchical Bayesian model (HBM), 29–32, 35, 102, 126–128
 inference stage, 128–131
 learning stage, 128
Hybrid physical data (HPD), 35–36
Hyperbolic model, 81

I

Index parameters, 452–454, 456
 relative density, 167
 unit weight, 166–167
Induced ground displacement site, 432
Industry 4.0, 5, 21, 25, 33
Initial void ratio, 248, 249
Input *vs.* output parameters, 19
In-site geotechnical characteristics, 430
In-situ deformation mechanism, 135–136
In situ load tests for determination, mechanical rock property
 base resistance methods, 68–78
 estimation of model parameters, 88–95
 existing databases, 51–54
 new side-resistance and base-resistance databases, 54–59
 review of side resistance methods, 59–68
 settlement methods, 78–88
In situ test-based evaluation of soil effective stress strength property and stress history
 CPTu sounding of soft Chicago clays, 352
 deep CPTu profile at Sandpoint and Idaho, 351
 discussion and applicability
  effective cohesion intercept, c′, 361
  friction angle criteria, 361
  partial drainage effects, 361–362
 effective friction angle measured by laboratory triaxial tests and NTH solution from DMT, 360
 fissured London Clay at Brent Cross, 356
 general relationship for yield stress of soils and CPT net cone resistance, 365
 lab-measured unit weight and DMT predicted unit, 348
 laboratory triaxial effective stress friction angle ϕ′ *vs.* interpreted ϕ′ from NTH using CPTu data, 352
 laboratory triaxial-measured ϕ′ *vs.* CPTu ϕ′ from different test series, 357
 laboratory triaxial tests showing ϕ′ at qmax and NTH friction angle from CPTu at NWU site, 353
 lab soil unit weight *vs.* predicted unit weight, 348
 Martin's Point Bridge, Maine, 354
 NTH method for evaluating ϕ′ from CPTu, 350
 NTH postprocessing of CPTu data for Q and B_q at Sandpoint, 351
 overview, 344–345
 postprocessing of the DMT data in clay layer in Chesapeake, Virginia, for determining, 360
 Presumpscot clay at Martin's Point Bridge, 355
 representative DMT sounding at Route 17 bridge site, Chesapeake, VA, 359
 resistance number, normalized porewater pressure and evaluated ϕ′ profile for soft Chicago clays, 353

soil effective stress strength parameters, 346–349
fissured clays from piezocone penetration test, 355–357
friction angle criteria, 349
intact clays from flat dilatometer test, 357–360
intact clays from piezocone penetration test, 349–355
soil stress history parameters in clays
intact clays from piezocone penetration test, 362–363
organic clays from piezocone penetration test, 364–365
sensitive clays from piezocone penetration test, 363–364
soil unit weight, 345
cone penetration test (CPT), 345
flat dilatometer test, 346
shear wave velocity, 345
total unit weight *vs.* slope parameter m_p0, 347
In-situ tests, 203
and prediction of geo-parameters
geo-marine database and triangular charts, 186–201
geotechnical parameters, 165–166
index parameters, 166–167
related parameters, 182–186
stiffness and compressibility parameters, 176–182
strength parameters, 167–176
In-situ tests in soil identification, 156
soil behavior classification, 157
soil classification
CPT data, 157–164
DMT records, 164–165
FVST records, 157
PMT measurements, 165
SPT blow counts, 157
Intact clays from flat dilatometer test, 357–360
Intact clays from piezocone penetration test, 349–355, 362–363
Intelligent methods for surrounding rock classification
convolutional neural network (CNN), 309–310
Gaussian process, 310–311
model training and evaluation, 313–319
random forest (RF), 310
support vector machine (SVM), 311–312
Internal friction angle, 167–175
ISSMGE 2017 SOA/SOP Survey, 37

J

Janbu method, 282–283
Johnson distribution system, 456–458
Johnson transformation method, 456

L

Laboratory geotechnical characteristics, 430
Laboratory tests
CAUC tests, 232
clay's plasticity, 232
CRS oedometer tests, 232–234
EN ISO standard, 231
interpretation method, 234
Karlsrud method, 232–233
LL and IP, 232
Sallfors method, 232–233
Large language models, xi–xii, 22, 38
Large regional databases, 137–139
Lexi Expressway project, 320–321
Linear regression analyses, 229, 241
Liquefaction, 427–428; *see also* Data-centric seismic soil liquefaction assessment
indices at a site, 432
maps, 432–433
Liquefaction potential index (LPI), 432, 442–445
Likelihood function, 390–391
Liquefaction severity number (LSN), 432
Liquid limit (LL), 137–138, 229, 333
Load-displacement response, 79, 81
Load-transfer function, 79–81
Local soil property approach, 436, 441, 443–446
Long-short term memory (LSTM), 36, 246, 272–273
Low-carbon ground improvement methods, 280
Low-density *vs.* high-density data, 19
Low-fidelity *vs.* high-fidelity data, 19

M

Machine learning (ML), xi–xii, 1–2, 33, 245
model, 249–252
Machine learning-guided observational method (MLOM), xii, 2, 22, 35–36
Machine learning in geotechnics (MLIG), xii–xiii, 2–3
Mechanical characteristics
cohesionless soil, 371
effective internal friction angle of sand, 372
relative compactness, 371–372
Young's modulus, 372

cohesive soil, 370
compression modulus, 371
natural weight-specific density of soil, 370
overconsolidation ratio (OCR), 370
undrained shear strength, 370
undrained Young's modulus, 371
Mechanical-statistical evaluation of soil property
basic geotechnical parameters of MC model, 415
Bayesian-random field theory, 386
Algorithm implementation, 389
Bayesian inverse modeling method, 388
stochastic mechanics/random field theory, 386–387
calibrated transformation models, CPT data *vs.* regular geotechnical parameters, 407
compression modulus E_s(MPa) baseline of spatial random field, 399
constitutive model and cavity expansion theory, 372–373
CASM model, 380
MCC model, 373
correlation coefficients, key constitutive parameters, 424
CPT data and laboratory data, silty clay soil, 394
data flow and algorithm schema, spatial random fields-based Bayesian method, 389
engineering application, 392
Bothkennar site, United Kingdom, 418
national convention and exhibition center, Shanghai, 392
Suzhou river drainage and storage pipeline project, 408
evaluation method based on CPTU, 369
mechanical characteristics of cohesionless soil, 371
mechanical characteristics of cohesive soil, 370–371
excavation steps of Yunling shaft pit, 416
general master driver of connecting stochastic discretization and mechanics software, 390
geotechnical parameters of Bothkennar site, 419
in-situ CPT data in layer ⑧, 409
in-situ test data of CPTU curves, 420
lateral displacements of the diaphragm wall, 417
location and soil distribution of Bothkennar site, 420
National Convention and Exhibition Center (Shanghai), China, 393
non-informative prior knowledge of transformation models, CPT data *vs.* regular geotechnical parameters, 402
overview, 369
parametric levels of constitutive parameters, 410
parametric sensitivity of constitutive parameters, 421
pit bottom heave, 417
posterior distributions
and cross-validation of transformation model, 404–406
key constitutive parameters, 423
posterior probabilistic characterizations transformation model baseline, 400–401
predicted curves of CPTU resistances, original geotechnical parameters, 422
prior and posterior distributions, key constitutive parameters θ, 414
prior information of the constitutive parameters, 412
prior knowledge of key constitutive parameters ψ, 423
Pseudo-static equilibrium model of cone-tip resistance, 385
schema of CPT mechanical model, 378
schematic diagram
cylindrical cavity expansion, 373
spherical cavity expansion, 382
surrounding environment of Yunling complex, 409
sensitivity factors of constitutive parameters, 422
spatial sampling coordinates of
compression modulus in laboratory test, 395
consolidation shear cohesion coefficient and internal friction angle in laboratory test, 396
state parameter, space ratio and reference state parameter, 380
statistical uncertainty
bayesian reverse modeling method, 401
structural parameters θ of spatial random field by 1000 times simulation, 397
statistics of parametric sensitivity analysis, 421
three-dimensional anisotropic variogram of CPT data p_s(MPa), 397
three-dimensional numerical model, 415
three-dimensional site characterization of geotechnical parameters, 408

variance of sensitivity analysis, 411
yield surfaces of CASM, OCC and MCC models, 381
Mechanical transformation model, 385–386
Mechanics-based numerical simulation, 434–435
Mechanics-based simulations, 429
Model calibration, 219
Model structure, 214
Model training and evaluation, 313
CNN-based SRC model, 313–317
GPC-based SRC model, 317–318
performance evaluation of intelligent SRC model, 318–319
RF-based SRC model, 317
SVM-based SRC model, 318
Modified empirical formula, 272
Mohr-Coulomb criterion, 347
Mohr-Coulomb model, 62–63
Monte Carlo simulations, 442–444
Multiscale random field characterization, 438
multiscale spatial correlation, 440–441
sequential simulation process, 439–440
spatial correlation, 438–439
Multiscale regional assessment, 435–436
general framework, 436–437
multiscale random field characterization, 438–441
numerical examples, 441–447
site-specific liquefaction evaluation, 438
Multivariate ANI-F-CLAY/7/87 database from Finland, 139–142
Multivariate probability distribution, SH-CLAY/12/4191, 456
correlation matrix, 458–460
normal transformation for soil parameters, 456–458
MUSIC clay parameters, tailored cluster model, 32–33
MUSIC-X, 1, 10–13, 16

N

Nanagō Layer, 333
National convention and exhibition center, Shanghai, 392–394
spatial variability of CPT data, 394–398
three-dimensional site characterization, 407–408
transformation model
baseline, 398–402
calibration, 402–407
National Geotechnical Properties Database (NGPD), 5
New databases, 54–55
base-resistance database, 57–59
base resistance, 58
initial normal stiffness, 58
load-transfer function (q–δ relationship) for socket base, 57
property of intact rock, 58
property of rock mass, 58
rock socket geometry, 58–59
side-resistance database, 55
drilled shaft geometry and material, 55
load-transfer function (f_s–δ relationship) for rock socket sidewalls, 55
property of intact rock, 55
property of rock mass, 55
side resistance, 55
New laboratory database of hydraulic conductivity measurements, fine-grained soils
benchmarking analysis result, 224
models with smaller calibration dataset range, FG/KSAT-1358, 222
calibrated prediction models, 220–221
divergency in subdivision analysis, 215
funnel pattern residuals *vs.* predicted plot, 215
overview, 210
particle size distribution-related parameters, 213–214
permeability databases, 217–218
benchmarking databases, 219–223
compilation, 223
database evaluation, 216–219
database summary, 216
model calibration, 219
permeability of soils
definitions, 210–211
factors influencing hydraulic conductivity, 211–213
uncertainty in test results, 213
variation of hydraulic conductivity, 211
permeability transformation models, 223–225
plasticity-related parameters, 214
soil fabric, shape and composition, 214
soil property variation range, global multivariate database, 212
synopsis, 223–225
transformation models
database subdivision analyses, 215–216
evaluation of regression strength, 214–215
model structure, 214
predictors of hydraulic conductivity, 213–214
void ratio/structure-related parameters, 213
New side-resistance and base-resistance databases, 54–59
New taxonomy of site data, 13–21
sampling characteristics

categorical *vs.* numerical data, 19–20
input *vs.* output parameters, 19
low-density *vs.* high-density data, 19
low-fidelity *vs.* high-fidelity data, 19
MUSIC-X, 16
single site *vs.* multiple sites, 18–19
univariate *vs.* multivariate data, 17
site generalizations, 13–14
first dimension, 13
natural variability, 13
second dimension, 13–14
third dimension, 14
smart data
definition, 20
intelligence of algorithm, 20–21
value of information (VoI), 21
spatial features, 15–16
DDSC methods, 15–16
ML, 16
spatial variability, 15
structured and unstructured uncertainty, 15
training dataset, 15–16
New Zealand geotechnical database (NZGD), 5, 441
Next-Generation Liquefaction (NGL) database, 5
NHERI SimCenter, 435
Normalised active undrained shear strengths of soft Scandinavian clays
compilation of effective stress paths, 232
data and methods
bivariate database, 234–235
laboratory tests, 231–234
data-centric approach, 235–237
data overview, 235
effective stress paths, 237
geomechanical approach, 237–239
interpretation of preconsolidation pressure, CRS oedometer tests, 233
vs. liquid limit, 236
normalised strength *vs.* plasticity, 239
overview, 229–231
simplified yield surface, 237
synopsis, 240–241
uncertainty and sample quality, 240
Normalised undrained shear strengths, 229–230
Novel deep learning-based model Seft-Net, 270
Numerical examples, 441
analysis region and field data, 441
fine-scale site-specific liquefaction assessment, 445–447
probabilistic and spatial assessment of liquefaction potentials, 441–445

O

Ohde–Janbu method, 282
One-*versus*-rest/one-*versus*-one-based training, 311, 313
Onsoy-site data, 29–30
Organic clays from piezocone penetration test, 364–365
Organic soils, 356–357
Original Cam Clay (OCC) model, 380
Overconsolidation ratio (OCR), 179–182, 344

P

Parameters
consolidation indexes, 182
constrained modulus, 178–179
over-consolidation ratio, 179–182
stiffness, 176–178
Parametric analysis, 418
Partial drainage effects, 361–362
Particle gradation, 248
Particle hardness, 247, 249
Particle shape, 247
Particle size distribution-related parameters, 213–214
Peak side resistance, 59, 61
Permeability, 186
database compilation, 223
databases, 217–218
benchmarking databases, 219–223
compilation, 223
database evaluation, 216–219
database summary, 216
model calibration, 219
databases for fine-grained soil
benchmarking databases, 219–223
database evaluation, 216–219
database summary, 216
model calibration, 219
prediction model, 214
of soils
definitions, 210–211
factors influencing hydraulic conductivity, 211–213
uncertainty in test results, 213
variation of hydraulic conductivity, 211
testing approach, 213
transformation models, 223–225
Physics-based model, 22–23
Physics-informed machine learning (PIML), 22
Piezocone Penetration Test (CPTU), 369
Piezocone test parameters, 188–189
Plasticity index (PI), 59–60
Plastic zone analysis, 383–384
Pore pressures, 349, 361

Posterior distribution, 389–390, 392, 410, 413, 460
Precision characterization, 14
Prediction for soil design property, multivariate database for Shanghai soft clay
  C matrix composed of the average accepted δ_{ij}, 461
  correlation coefficients between soil parameters, 457
  design parameters of transformation model, 460
    Bayesian updating, 460–462
    Transformation formula, 462–466
    validation of transformation models, 466–467
  distribution type and parameters, 458
  global/regional multivariate soil databases, 453
  multivariate probability distribution, 456
    correlation matrix of multivariate distribution, 458–460
    normal transformation for soil parameters, 456–458
  municipal database, 453
  and observed data between *e* and K_0, 465
  overview, 452–453
  scatter plots between X_i and X_j, 459
  schematic graph of site and borehole distribution, 454
  Shanghai clay database
    basic statistics, 455
    correlation of the parameters, 456
    description of, 454–455
  soil parameters, 452
  synopsis, 468
  transformation model, 452–453
  updated Johnson parameters
    c under combinations of information, 463
    E_s under combinations of information, 462
    K_0 under combinations of information, 462
    φ under combinations of information, 463
  updating results
    c by conditioning, 467
    E_s by conditioning, 467
    K_0 by conditioning, 466
    φ by conditioning, 467
Prediction for the mechanical response of gravels
  actual curves *vs.* predicted curves, 272
  architecture of different feature extraction modules, 254
  conception module architecture, 257
  establishment of geo-gravel dataset
    data statistical analysis, 249
    establishment process, 247–248
  hardness category, range of hardness, average hardness, 248
  importance for indicators to evaluate q_{peak}, 252
  overview, 244–247
  performance metrics of ANN, SVR and RF models, 251
  prediction of shear strength
    deep learning model, 252–266
    empirical formula, 266–269
    machine learning model, 249–252
  prediction of strain-stress behavior, 269
  PRL-Net architecture, 270
  roundness category, range of roundness and average roundness, 248
  scatter matrix, 250
  Seft-Net model architecture, 253
  synopsis, 269–273
  test materials, 271
Prediction of deep shaft pit deformation, 413–418
  lateral displacement of the diaphragm wall, 417–418
  numerical model, 413–416
  pit bottom heave, 416–417
Predictors of hydraulic conductivity, 213–214
Preliminary prediction of CPTU resistances, 418–423
Pre-posterior analysis, 21
Prior distribution, 389–390
Privacy-enhancing technologies (PETs), 37
PRL-Net architecture, 270
Probabilistic and spatial assessment of liquefaction potentials, 441–445
Probability density functions (PDFs), 289–291
Proper characterization, 102
Proposed feature extraction module, 253–254
Proposed feature transform module
  conception module, 255–257
  depthwise separable convolution, 254–255
Pseudo-static equilibrium model, 385

Q

Quasi-regional cluster model for MUSIC clay parameters, 32
Quasi-site-specific model, 28
  for MUSIC clay parameters, 29–30
Quasi-site-specific transformation model, 125–128
  HBM inference stage, construction of, 128–131
  HBM learning stage, extract site-specific statistics in database, 128

R

Random Fields: Analysis and Synthesis (Vanmarcke), 15
Random forest (RF), 310

Random forest classification (RFC), 310
Regional advantage, 32
Regularity factor, 249
Related parameters
permeability, 186
sensitivity, 186
shear wave velocity and shear modulus at small strain, 182–186
Reliability-based design (RBD), 102
Research gaps and opportunity, 447
Review of side resistance methods, 59
discussion, 65–68
empirical models, 59–63
semi-empirical models, 63–65
ROCK/9/4069 database, 111–116
Rock/concrete interface, 63
ROCKET computer program, 64
Rock hydromechanical parameters, rock foundation design
allowable foundation pressure *vs.* rock quality designation (RQD), unweathered rock masses, 75
comparison of foundations on rock masses
with vertical closed joints, 78
with vertical open joints, 77
design method of, 65
empirical base resistance models, 73
empirical method prediction of q_p for purpose of determining nonlinear load-transfer functions for tip resistance, 80
evaluation of
base resistance models using existing models, 75
design method, 66
empirical models for base resistance with measured base resistance, in situ load tests, 73
empirical models for prediction of rock mass modulus of deformation, 92
empirical side resistance models, in situ load test database, 60
predictive models for initial normal stiffness of drilled shafts in weak rock mass, 87
predictive models for initial shear stiffness of drilled shafts in weak rock mass, 83
side resistance predictive models using databases, 68
side resistance predictive models using new databases, 69
in situ drilled shaft load test database, 64
variation of bias, 67
existing empirical methods for side resistance of rock sockets, 62
existing load test databases, 52–53
failure modes for shallow foundations on rock masses, 76
load test methods, in situ test database, 56
load-transfer models, 82
method for end resistance of foundations on rock mass, 74
normalization method, 79
overview, 50–51
prediction method of base resistance, rock quality designation (RQD), 76
predictive methods for the modulus of deformation of rock mass, 89
predictive model for drilled shaft/rock interface roughness, 61
rock mass classification and rock mass deformation modulus database, 90–91
semi-empirical base resistance models, 74
side resistance and base displacement relationships, rock sockets in weak rock, 57
in situ load tests, mechanical rock property
base resistance methods, 68–78
estimation of model parameters, 88–95
existing databases, 51–54
new side-resistance and base-resistance databases, 54–59
review of side resistance methods, 59–68
settlement methods, 78–88
synopsis, 95–96
variation bias predictions
drilled shaft tip normal stiffness with property of soft rock mass and drilled shaft geometry using the tip resistance database, 89
drilled shaft tip normal stiffness with property of weak rock mass and drilled shaft geometry using tip resistance database, 88
initial shear stiffness with property of soft rock mass and drilled shaft geometry, 85
initial shear stiffness with property of weak rock mass and drilled shaft geometry, 84
variation of modulus of deformation of rock mass with unconfined compressive strength, 94
variation of the available theory
base resistance and comparison with measured fracture initiation pressure, 72
base resistance and comparison with measured yield pressure, 71
ROCKMass/9/5876 database, 116–119
Rock mass rating (RMR), 116, 304–305
Rock quality designation (RQD), 72, 74, 304

Rock sockets, 50–51, 57–58, 95–96; *see also* Rock hydromechanical parameters, rock foundation design
load tests, 59–61
Role of site characterization information
applications, 25–27
strategy I, generic model for clay parameters, 28
strategy II (with spatial variability)–Quasi-sitespecific model for MUSIC-3X clay parameters, 30–31
strategy II (without spatial variability)–Quasisite-specific model for MUSIC clay parameters, 29–30
strategy IV–Quasi-regional cluster model for MUSIC clay parameters, 32
strategy IV–Tailored cluster model for MUSIC clay parameters, 32–33
bivariate site data, 12
challenges, 33–37
data protection, 36–37
explainability and interpretability, 34
Frugal HBM, 34–35
geo-compatibility, 34
machine learning-guided observational method (MLOM), 35–36
thick data, 35
transferability, 34
data-centric geotechnics, 2–5
decision making, 2
elements, 3–4
geotechnical engineering and machine learning, 2
machine learning in geotechnics (MLIG), 2–3
value of MLIG, 3
data-driven methods for site characterization in Project DeepGeo, 11
data-driven site characterization (DDSC), 21–22
data-informed decision support index (DIDI), 23–25
site recognition challenge, 22–23
overview, 1–2
site characterization information, 5
big indirect data (BID), 5–10
MUSIC-X, 10–13
new taxonomy of site data, 13–21
site investigation programme, 24
soil/rock property databases, 6–8
synopsis, 37–40
value of machine learning in geotechnics, function of data and method, 3
Route 17 Bridge, Chesapeake, Virginia, 358–360

S

SAND/7/2794 database, 110–111
Scatter matrix, 250
SCE-CSSM solution, 362–363
S-CLAY/10/168 database, 138–140
Seft-Net model, 259, 264–266
architecture, 252–253
proposed feature extraction module, 253–254
proposed feature transform module, 254–257
evaluation, 262–265
Semi-empirical side resistance models, 63–65
Sensitive clays, 356–357
from piezocone penetration test, 363–364
Sensitivity, 186, 409–410, 418
Serviceability limit state (SLS), 280–281
Settlement methods, 78
evaluation of
initial normal stiffness in tip resistance, 86–88
initial shear stiffness in side resistance, 81–86
load-transfer methods, 78–81
Shaft/rock interface, 61–62
Shanghai soft clay, *see* SH-CLAY/12/4191 database
SH-CLAY/12/4191 database, 452, 468; *see also* Prediction for soil design property, multivariate database for Shanghai soft clay
basic statistics of, 455
correlation of parameters, 456
description of, 454–455
Shear strength and strain–stress behavior of gravels, 269
DL model PRL-Net, 272–273
Geo-Gravel dataset includes triaxial compression test data, 270
modified empirical formula, 272
novel deep learning-based model Seft-Net, 270
three machine learning regression models, 270
Shear strength of prediction
deep learning model, 252–266
empirical formula, 266–269
machine learning model, 249–252
Shear wave velocity, 345
and shear modulus at small strain, 182–186
Simplified approach for insensitive and inorganic clays, 363
Simplified methods, 433–434
Single site *vs.* multiple sites, 18–19
Site characteristics, 429
geologic characteristics, 429

in-site geotechnical characteristics, 430
laboratory geotechnical characteristics, 430
topography, 431
water, 430–431
Site generalizations, 13–14
first dimension, 13
natural variability, 13
second dimension, 13–14
third dimension, 14
Site recognition, 1, 39
challenge, 10, 22–23, 25, 39
literature survey, 34
Site-Specific Seismic Hazard Analysis and Research Kit with Uncertainty Quantification (QS3HARK), 435
Smart data
definition, 20
intelligence of algorithm, 20–21
value of information (VoI), 21
Smoothed particle hydrodynamics (SPH), 434
Soft Nordic clays, 281–282
Software FLAC^{3D} 3.0, 390
Soil Behavioral Classification (SBC), 195
basic diagrams, 187–188
direct diagrams, 190
indirect diagrams, 188–190
SOIL-DMT/8/7186 database, 119–121
Soil effective stress strength parameters, 346–349
fissured clays from piezocone penetration test, 355–357
friction angle criteria, 349
intact clays from flat dilatometer test, 357–360
intact clays from piezocone penetration test, 349–355
Soil-fluid coupling mechanics, 392
SOIL-KSAT/7/2293 database, 121–125
Soil permeability, 213
Soil profiles, 187
Soil stress history parameters in clays
intact clays from piezocone penetration test, 362–363
organic clays from piezocone penetration test, 364–365
sensitive clays from piezocone penetration test, 363–364
Soil unit weight, 345
cone penetration test (CPT), 345
flat dilatometer test, 346
shear wave velocity, 345
Spatial features, 15–16
DDSC methods, 15–16
ML, 16
spatial variability, 15
structured and unstructured uncertainty, 15
training dataset, 15–16
Spatial variability, 15
Spatial variability-Quasi-sitespecific model for MUSIC-3X clay parameters, 30–31
Spherical cavity expansion theory, 382–383
Standard Penetration Test (SPT), 152, 155, 157
Statistical attributes to geotechnical investigation database, 332
Statistical characteristics, 249
Statistical determination of transformation models for anisotropic undrained shear strength of Finnish clays, 143–146
Statistics of the muddy clay, muddy silty clay, and clay in Shanghai, 471
Statistics of the muddy clay in Shanghai, 470
Stiffness, 176–178
Stochastic mechanics/random field theory, 386–387
Monte Carlo simulation, 387
multivariate spatial random fields, 387–388
response surface method, 387
Strain-stress behavior prediction, 269
Strength parameters
internal friction angle, 167–175
undrained shear strength, 175–176
Stress history for intact clays from piezocone penetration test
SCE-CSSM solution, 362–363
simplified approach for insensitive and inorganic clays, 363
Stress paths, 135
Structural health monitoring (SHM), 36
Structured geomaterials, 356–357
Support vector machine (SVM), 311–312
SurroundingRock/8/286 database, 304–307; *see also* Engineering geological parameter database of tunnel surrounding rock and application
Surrounding rock classification model/rock mass classification, 304–305, 307
convolutional neural network (CNN), 309–310
Gaussian process, 310–311
model training and evaluation, 313–319
random forest (RF), 310
support vector machine (SVM), 311–312
Suurpelto clay, 297
Suzhou river drainage and storage pipeline project, 408–409
posterior distribution, 410–413
prediction of deep shaft pit deformation, 413–418
prior knowledge, 409–410

T

Tailored clustering (TC), 32–33
Test materials, 271

Theory of elasticity, 78–79
Thick data, 22, 35
Three machine learning regression models, 270
Tokyo-CLAY/14/67760, 332
Tokyo lowlands, 332–333
Tools for numerical methods, 435
Tools for simplified methods, 435
Topography, 431
Total uncertainty, 286, 300
Training strategy, 257–258
  model testing, 259
  model training, 259
  performance metrics, 258–259
Transferability, 34
Transfer convolutional neural network (TCNN) model, 22
Transformation formula, 462–466
Transformation models
  database subdivision analyses, 215–216
  design parameters of, 460
    Bayesian updating, 460–462
    Transformation formula, 462–466
    validation of transformation models, 466–467
  evaluation of regression strength, 214–215
  model structure, 214
  predictors of hydraulic conductivity, 213–214
Transformation uncertainty, 110, 214–215
  existing and new transformation models, 291–296
Triaxial–Piezocone database for NC–OC clays, 356
Triaxial tests, 230–231, 244, 347, 349
Tunnel excavation, 304, 331
Tunnel information management system (TIMS), 320–321
Twitch testing, 361

U

Uncertainty and sample quality, 240
Uncertainty in permeability test results, 213
Undrained shear strength, 135, 175–176, 229
Undrained shear strength of Finnish soft clays
  ANI-F-CLAY/7/87, 137
  bias factors and COV for anisotropic s_u from ANI-F-CLAY/7/87 database, 145
  database-driven approach, 136
  databases of Finnish and Scandinavian soft clays
    comparing regional databases from Finland and Norway, 142–143
    large regional databases, 137–139
    multivariate ANI-F-CLAY/7/87 database from Finland, 139–142
  databases use
    assess the variability of s_u at a particular site, 136
    develop site-specific s_u correlations, 136
    identify and quantify bias in local data, 136
    validate s_u predictions from numerical models, 137
  direct field measurement of s_u techniques, 135
  discussion, 146–147
  geotechnical engineering applications, 135
  in-situ deformation mechanism, 135–136
  statistical determination of transformation models, 143–146
  stress paths, 135
  synopsis, 147
Uniqueness, 29, 39
Univariate *vs.* multivariate data, 17
Unweathered and sound rock masses, 72, 75

V

Validation of transformation models, 466–467
Value of information (VoI), 21
Variable importance analysis, 265–266
Variation of hydraulic conductivity, 211
"Very good to excellent" category, 139, 142
VGGNet, 309–310
Virtual ground, 15–16

W

Water, 430–431

Y

You Only Look Once version X (YOLOX), 36
Yūrakuchō Layer, 333

AF848438

Nonparametric Statistics on Stratified Spaces and Their Applications in Object Data Analysis

This book, a follow-up of *Nonparametric Statistics on Manifolds and Their Applications to Object Data Analysis* where the reader was introduced to data analysis on manifolds, is expanding the subject of data analysis on object spaces, to the case of metric spaces that have a smooth structure on an open dense subset only, governed by a nice filtration structure on the singular part of this space, that is comprised of manifolds whose boundaries are made of lower dimensional manifolds in this filtration. Such object spaces are known as stratified spaces, and their structure will be detailed in the second part of the book. Key examples of complex data from which one extracts data representable as points on a stratified space, and a review on data analysis on manifolds, are provided, alongside a summary of results on nonparametric methods on manifolds. Key results on asymptotic and nonparametric bootstrap on some stratified spaces are included. Certain object spaces with a manifold stratification arising in Statistics are considered with examples of application on data analysis and on them. Various applications include RNA based analysis of the SARSCov2 virus, 3D face identification from digital camera images, English Alphabet based comparison of certain European Languages. *Nonparametric Statistics on Stratified Spaces and Their Applications in Object Data Analysis* is intended for graduate students to expand their understanding of Nonparametric Statistics, and enhance their ability to grasp and extract complex data and analysis of manifolds.

Victor Patrangenaru is a professor in the Statistics Department at Florida State University, Tallahassee, Florida, USA. He is an honored fellow of the Institute of Mathematical Statistics, the recipient of the Dean's Travel Award from Florida State University, and the David Rothrock Teaching Award from Indiana University.

Daniel E. Osborne is an associate professor in the Mathematics Department at Florida A&M University, Tallahassee, Florida, USA. As a trained Statistician and Data Science educator, he is dedicated to training student learners and promoting best practices in data literacy, statistical reasoning, data analysis, and data visualization skills among all learners, irrespective of their backgrounds, majors, or career paths.

Nonparametric Statistics on Stratified Spaces and Their Applications in Object Data Analysis

Victor Patrangenaru and Daniel E. Osborne

CRC Press
Taylor & Francis Group
Boca Raton London New York

CRC Press is an imprint of the
Taylor & Francis Group, an **informa** business
A CHAPMAN & HALL BOOK

Designed cover image: © Victor Patrangenaru and Daniel E. Osborne

First edition published 2026
by CRC Press
2385 NW Executive Center Drive, Suite 320, Boca Raton FL 33431

and by CRC Press
4 Park Square, Milton Park, Abingdon, Oxon, OX14 4RN

CRC Press is an imprint of Taylor & Francis Group, LLC

© 2026 Taylor & Francis Group, LLC

Reasonable efforts have been made to publish reliable data and information, but the author and publisher cannot assume responsibility for the validity of all materials or the consequences of their use. The authors and publishers have attempted to trace the copyright holders of all material reproduced in this publication and apologize to copyright holders if permission to publish in this form has not been obtained. If any copyright material has not been acknowledged please write and let us know so we may rectify in any future reprint.

Except as permitted under U.S. Copyright Law, no part of this book may be reprinted, reproduced, transmitted, or utilized in any form by any electronic, mechanical, or other means, now known or hereafter invented, including photocopying, microfilming, and recording, or in any information storage or retrieval system, without written permission from the publishers.

For permission to photocopy or use material electronically from this work, access www.copyright.com or contact the Copyright Clearance Center, Inc. (CCC), 222 Rosewood Drive, Danvers, MA 01923, 978-750-8400. For works that are not available on CCC please contact mpkbookspermissions@tandf.co.uk

Trademark notice: Product or corporate names may be trademarks or registered trademarks and are used only for identification and explanation without intent to infringe.

For Product Safety Concerns and Information please contact our EU representative GPSR@ taylorandfrancis.com. Taylor & Francis Verlag GmbH, Kaufingerstraße 24, 80331 München, Germany.

Library of Congress Cataloging-in-Publication Data
Names: Patrangenaru, Victor, author. | Osborne, Daniel Eugene, 1983- author.
Title: Nonparametric statistics on stratified spaces and their applications in object data analysis / Victor Patrangenaru and Daniel E. Osborne.
Description: First edition. | Boca Raton FL : CRC Press, 2026. | Includes bibliographical references.
Identifiers: LCCN 2024039132 (print) | LCCN 2024039133 (ebook) | ISBN 9781138043138 (hardback) | ISBN 9781032914558 (paperback) | ISBN 9781315173283 (ebook)
Subjects: LCSH: Spatial analysis (Statistics) | Stratified sets. | Nonparametric statistics.
Classification: LCC QA278.2 .P3644 2026 (print) | LCC QA278.2 (ebook) | DDC 001.4/22--dc23/eng/20250121
LC record available at https://lccn.loc.gov/2024039132
LC ebook record available at https://lccn.loc.gov/2024039133

ISBN: 978-1-138-04313-8 (hbk)
ISBN: 978-1-032-91455-8 (pbk)
ISBN: 978-1-315-17328-3 (ebk)

DOI: 10.1201/9781315173283

Typeset in Nimbus Roman font
by KnowledgeWorks Global Ltd.

Publisher's note: This book has been prepared from camera-ready copy provided by the authors.

Contents

Foreword **ix**

Preface **xi**

I Data and Preliminaries for Analysis on Stratified Spaces **1**

1 Data of Complex Type **3**
1.1 DNA/RNA Basics and Historical Notes 3
1.2 Protein Data 5
1.3 Graphs 6
1.4 MRA-based Artery Data 9
1.5 Alphabet: Origin, History and Spread 10
1.6 3D Landmarks Data from Digital Camera Images 12

2 Review of Nonparametric Statistics on Manifolds **13**
2.1 Introduction 13
2.2 Fréchet Total Variance and Fréchet Means 15
2.3 Elementary CLT for Extrinsic Means 16
2.4 CLT and Bootstrap for Fréchet Means 17
2.5 Large Sample Confidence Regions for the Extrinsic Mean 23
2.6 Two-Sample Test for Total Extrinsic Variances 30
2.7 Bhattacharya's Two-Sample Test for Means 31
2.8 Test for Mean Change in Matched Pairs on Lie Groups 33
2.9 Two-Sample Test for Simply Transitive Group Actions 35
2.10 Nonparametric Bootstrap for Two-Sample Tests 39

II Nonparametric Statistics on Stratified Spaces **41**

3 Spaces with a Manifold Stratification **43**
3.1 Introduction 43
3.2 Statistics on Stratified Spaces 44
3.3 Stratified Spaces and PT Spaces 45
3.3.1 PT Spaces 46

3.4 Asymptotics of Intrinsic Sample Means: Low–Dimensional Case 46

4 Extrinsic Data Analysis on Stratified Spaces 51
4.1 Introduction 51
4.2 Extrinsic Central Limit Theorem on Open Books 53
4.3 Diffusion Tensor Imaging Data Analysis 54

5 Intrinsic Sample Means on Tree Spaces 57
5.1 Introduction 57
5.2 Open Books 60
5.3 Probability Measures on the Open Book 62
5.4 Sample Means 65
5.5 Random Sampling and the Law of Large Numbers 68
5.6 Central Limit Theorems 70
5.6.1 Sticky Central Limit Theorem 70
5.6.2 Nonsticky Central Limit Theorem 75

6 Central Limit Theorem for Random Samples on a Graph 77
6.1 Introduction 77
6.2 On a 2015 Paper by Hotz and Huckemann 79
6.3 From Graph to Line 81
6.3.1 Central Limit Theorem 84

7 Analysis of Magnetic Resonance Angiography Data 87
7.1 Brain Artery Trees 87
7.1.1 Correspondence 89
7.2 Phylogenetic Trees and Topological Heterogeneity 91
7.3 Treespace Geodesics 92
7.4 Fréchet Mean of Phylogenetic Trees 95
7.4.1 Calculating Fréchet Sample Means 96
7.4.2 Fréchet Means for Artery Trees 97
7.5 Multidimensional Scaling, Minimum Spanning Trees, and Triangles 98
7.6 Age and Sex Differences in Brain Artery Trees 101

8 An Application to Phylogenies of SARS-CoV-2 Data Analysis 105
8.1 Introduction 105
8.2 RNA-Based PT Trees 105
8.2.1 SARS-CoV-2 data 106

III Asymptotic Theory and Nonparametric Bootstrap on Special Stratified Spaces **113**

9 CLT on Low-Dimensional Stratified Spaces **115**
9.1 Introduction 115
9.2 Data Analysis on Manifolds 115
9.3 Data Analysis on One-Dimensional Stratified Spaces 117
9.3.1 Phylogenetic Trees 117
9.3.2 CLTs on Trees 120
9.4 CLTs on Graphs Including Cycles 120
9.4.1 Simulations of sticky and nonsticky sample means on graphs 122
9.5 Computational Examples on 2D Stratified Spaces 122
9.5.1 Simulations on T_4 123
9.5.2 Extrinsic Sample Mean for RNA Data 125
9.5.3 Extrinsic CLTs on Open Books 126
9.6 Discussion 126

10 Investigating Two Possible Origins of SARS-CoV-2 **129**
10.1 Introduction 129
10.2 Phylogenetic Trees and Tree Spaces 130
10.2.1 Definition and Construction 130
10.2.2 Tree Space $\mathscr{T}_n$ 131
10.3 Analyzing SARS-CoV-2 RNA Sequences Using Rooted $\mathscr{T}_3$ 131
10.3.1 RNA Sequences and Trees 131
10.3.2 Rooted Trees and Rooted $\mathscr{T}_3$ 133
10.3.3 Application to SARS-CoV-2 RNA Sequences 134
10.4 Conclusion 134

11 Applications of Tree Spaces to Language Ancestry **137**
11.1 Introduction 137
11.2 Data Collection 137
11.2.1 Words 137
11.2.2 Calculate Distances 137
11.3 Language Trees Analysis on a 3-Spider 140
11.3.1 Nonsticky Sample Mean 140
11.3.2 Sticky Sample Mean 141
11.3.3 Another Sticky Sample Mean 141
11.4 Conclusion 142

IV Applications in Object Data Analysis on Stratified Spaces 145

12 3D Face Differentiation from Digital Camera Images 147

12.1 Introduction 147
12.2 Motivation for New MANOVA on Manifolds 148
12.3 MANOVA on Manifolds 149
12.3.1 Hypothesis Testing and T^2 Statistic 149
12.3.2 Nonparametric Bootstrap Confidence Regions 152
12.4 MANOVA on $(\mathbb{R}P^3)^q$ 154
12.5 Application to Face Data 160

13 Further Directions in Statistics on Stratified Spaces 163

13.1 Introduction 163
13.2 Degrees of Stickiness and Hypothesis Testing for Distributions on Stratified Spaces 163
13.3 Additional Topics 164
13.3.1 Extrinsic Principal Component Analysis 164
13.3.2 Extrinsic Principal Components for Shapes of Planar Contours 168
13.4 Networks 171

Bibliography 173

Index 185

Foreword

It is my pleasure to introduce the first book on the statistical analysis of data lying on general stratified spaces. The book answers the need for new methodologies while covering the sophisticated underlying mathematical statistics to ultimately provide the reader with a nonparametric or model-free approach to complicated data analysis. Many of the more important and interesting types of modern data can be organized and analyzed using the newly created, most inclusive area of data analysis on stratified spaces. This includes, but is not limited to, statistics on manifolds (2020 MR classification number 62R30).

Stratified space data is a very generally complex, non-Euclidean (nonlinear or non-grid-like) data type that encompasses many of the data that require statistical analyses in the present day. During the past three decades, stratified data has become increasingly more prevalent in many scientific fields. A few examples include the following. In the natural sciences, biologists very often consider data from RNA or DNA sequencing to build evolutionary trees of species. Biomedical researchers often consider medical imaging data such as MRIs. Forensic scientists may consider 3D surface reconstructions from images of bones and chemical structures of biological remains. Within the realm of computer vision, drones have been used in risk assessment and safety management by collecting data, say, for instance, on the distribution of cars on road maps. Finally interests and activity in the world are often studied via World Wide Web data that is naturally represented as a complex graph.

This book contributes very significantly to the literature on stratified data analyses by providing a thorough description of daunting mathematical statistics underlying nonparametric statistics on stratified spaces for the practitioner who ultimately needs to understand and properly interpret the results of their stratified data analyses. Enjoy!

Robert L. Paige, Professor of Statistics, Missouri S&T

Preface

This book, a follow-up to Patrangenaru and Ellingson (2015) [106] where the reader was introduced to data analysis on manifolds, is expanding the subject of data analysis on object spaces, to the case of metric spaces that have a smooth structure on an open dense subset only, governed by a nice filtration structure on the singular part of this space, that is comprised of manifolds whose boundaries are made of lower-dimensional manifolds in this filtration. Such object spaces are known as *stratified spaces*, and their structure will be detailed in the second part of the book. This approach to **viewing random objects as points on certain metric spaces**, and taking into account the effect of the geometry and topology of such *object spaces*, takes statistics to a higher level of sophistication. Classical Statistics, based on Probability Theory, Mathematical Analysis, Linear Algebra and Computer Science, is thus fundamentally challenged to find new ways for analyzing object data problems. For this reason, our top priority is to *give an appropriate notion of mean of a random object, as a point that lives on the object space*, like every other object value of such a random point. Set aside the lack of linearity, there are some properties of Euclidean spaces that should be shared by all object spaces. These general requirements include some basic topological aspects, such as connectivity and separability. Note that for inference purposes, in case of data analysis on manifolds, homogeneity is an additional preferred features to allow for solving a two-sample problem for means, also in a manifold setting (see Patrangenaru and Ellingson (2015) [106], section 2.5.2). However, stratified spaces with more than one stratum are not homogeneous as metric spaces; therefore, belonging to a certain stratum is already a classifier for the mean on such an object space. For inference purposes, the stratum of sample mean (mean of the empirical) leads to a new type of asymptotics.

The problem of defining a sample mean for data on a metric space is a century old. It was first published by Cartan (1928) [32] in his first edition of *Lectures on the Geometry of Riemannian Spaces*, where he considered the notion of the *barycenter of a finite system of points on a Riemannian space*; the notions of stratified space (see Goresky and MacPherson [53]) or even of manifolds (see de Rham [36]) were not defined at the time. Certainly, there are object spaces, which do not have a smooth structure, where the least expected value of the square distance of the random object to a given point was considered by Fréchet (1948) [52]. All the minimizers of this expected value are regarded as mean values of that random object. On the other hand, to avoid the problem of multiple barycenters of the same finite set, Cartan imposed an additional topological restriction on the object space with a Riemannian manifold structure, by assuming that it *is simply connected and has nonpositive sectional curvature*. Cartan and Fréchet were contemporaries yet working in complementary areas of Mathematics; therefore, in Patrangenaru and Ellingson (2015) [106] both of them are recognized for their priority in defining a mean value on a metric space, by naming such values *Fréchet means* in general, and *Cartan means* (in case the sample points are from a Riemannian manifold).

Computing the value of a sample Cartan mean turns out to be a time-consuming process in Riemannian manifolds of nonzero curvature. In case of data on a startified space, it is therefore preferable to select an appropriate metric on the object space that is flat on its highest-dimensional stratum.

A key result, making inference for object data possible, is the consistency of the Fréchet sample mean set as an estimator of the mean Fréchet set (see Ziezold (1977) [150]). However, for general metric spaces, there was no known asymptotic behavior of the Fréchet sample mean, and given the computational issues with the Cartan means, Patrangenaru (1998) [104] considered a nonparametric analysis of *extrinsic means* on manifolds, which are Fréchet means associated with a distance between two observations on the object space. This is equal to the chord distance between their images of the object space under an *embedding* of a smooth object space into the Euclidean space. Later, Bhattacharya and Patrangenaru (2003) [20] introduced *intrinsic means*, by considering a Fréchet analysis w.r.t. a geodesic distance on a Riemannian manifold, allowing for asymptotic testing and estimation of means. To sum it up, Object Data Analysis (ODA) requires a sort of differentiable structure on the **object space** that **has to be** either that of **a manifold** or have some *manifold built-in structure*, no matter what the nature of the objects is.

This book is organized as follows. The first chapter of Part I is dedicated to key examples of complex data from which one extracts data representable as points on a stratified space, and a review on data analysis on manifolds. A separate chapter here is dedicated to a summary of results on nonparametric methods on manifolds. In Part II, we address some key results on asymptotic and nonparametric bootstrap on some stratified spaces, where we feature certain object spaces with a manifold stratification arising in Statistics, and we analyze data on them. In Part III, we apply this methodology to concrete examples of ODA. Part IV consists in three more applied sections only, one on MANOVA on stratified spaces, one on application to linguistics, and the last one on extrinsic PCA and other future topics of data analysis on stratified spaces.

Chapter 1 provides an overview of object data on stratified spaces extracted from various sources, presenting a number of examples of such data in practical applications. The key type of data presented here are RNA phylogenetic trees, with emphasis on the SARS-CoV-2 virus, and protein data. Another important data example is that of planar graphs. Magnetic Resonance Angiography (MRA) data is also included here, with its important ramifications in brain arteries, 3D image reconstructions. Chapter 1 introduces also alphabets-based tree data that is used in the last chapter. Last, but not least, digital camera face imaging data is presented here, used in 3D-projective shape data analysis. Chapter 2 is dedicated to a review of nonparametric statistical methods on manifolds, that is nonparametrics for smooth stratified spaces.

The notion of stratified space introduced as metric space that admits a certain dimension decreasing filtration by manifolds glued to each other, such that each boundary of a manifold part of this filtration, is a union of lower-dimensional manifolds is given in Chapter 3. The median and the mean of the probability measure associated with a random point on a Riemannian manifold $(\mathscr{M}, \rho_g)$ were introduced in the case of the empirical distribution by Cartan (1928) [31] and, in the general case of a population, by Fréchet (1948) [52]. Statistics on stratified spaces are also defined here, as well as the asymptotic behavior of the sample intrinsic mean on the simplest nontrivial type of stratified space.

Chapter 4 is dedicated to an analysis of extrinsic Fréchet mean sets, for particular stratified surfaces.

Following Fréchet's original ideas, in Chapter 5 we consider a probability distribution on an *open book* and we define the concept of sticky intrinsic mean. This new phenomenon is quantified by a LLN stating that the empirical mean of a random object with a sticky mean eventually almost surely lies on the *spine* of the open book. A CLT stating that the limiting distribution is Gaussian and supported on the spine is also given here, as well as versions of the LLN and CLT for the cases where the intrinsic mean is nonsticky or partly sticky.

In Chapter 6, we consider a connected graph G with a distance function d, so that each couple of points x,y can be connected with a geodesic whose length is exactly $d(x,y)$. Given a probability distribution $\mathbb{Q} = P_X$ on G we are interested in the Fréchet function $\mathscr{F} : G \to [0,\infty), \mathscr{F}(x) = \mathbb{E}(d(X,x)^2) = \int_G d(x,\mu)^2 \mathbb{Q}(dx)$. Here we suppose that the Fréchet function assumes its minimum in a unique point $\mu \in G$, and under the additional assumption that a small neighborhood of the cut-locus of this point has Q measure zero, we derive a CLT for independent and identically distributed random objects (i.i.d.r.o.s) from Q, including cases of stickiness.

Building on the results in the previous chapter, in Chapter 6, we consider the central limit theorem for random samples on a graph.

Chapter 7 is dedicated to an analysis of brain artery trees. Such trees do not have a natural common set of leaves; therefore, they are matched based on the *cortical correspondence* using anatomical shape via spatial locations, to place landmarks on the cortical surface, where landmarks are projected from the cortical surface to the closest point on the brain artery tree, so that all trees in the sample have the same set of labeled leaves, making possible representation of these artery trees as points in a phylogenetic trees space. This representation of MRA images of brain artery trees as points in a phylogenetic tree space enables the use of tree space geodesics, to quantify and visualize their differences, and a notion of center called the Fréchet mean. High-dimensional structure in the data is explored using multidimensional scaling, minimum spanning trees based on geodesic distances, and treespace triangles. The effect of gender and age on a brain artery system is studied, noting that the distances of the closest brain arteries to the cortical surface increase with age, tending to be smaller in females than males.

Chapter 8 is dedicated to a CLT on Stratified Spaces with an Application to Phylogenies of SARS-CoV-2 Data Analysis on phylogenetic trees. Note that such trees are built after RNA sequences via a Clustal Omega alignment method, a computer program used for multiple sequence alignment in Bioinformatics.

In Chapter 9, we consider CLT on certain stratified spaces in dimensions one and two, and on open books.

In Chapter 10 we provide an investigation of the critical question of two possible origins of the COVID-19 pandemic, using functional data on rooted RNA-based phylogenetic trees, regarded as points on open books. In Chapter 11 we address a new theme: comparing human interaction in writing based on an alphabet. Here we focus on Indo-European languages, developing a historical perspective of the genesis of West European languages, via an alphabet-based clustering of these languages. Three leaf trees are built using a single linkage method for clustering based on distances between samples from languages which use the Latin script. Taking three languages at a time, the mean is determined. If the mean exhibits nonsticky properties, then one of the languages may come from a different ancestor than the other two. If the mean is considered sticky, then the languages may share a common ancestor or all languages may have different ancestry.

Chapter 12 addresses the problem of MANOVA on smooth stratified spaces, with an application to face analysis based on 3D-projective shapes of facial configurations extracted from digital camera images.

In Chapter 13 we introduce additional topics that will be explored in the future, including extrinsic PCA on manifolds and networks analysis.

Acknowledgment

We are very grateful to David Grubbs for inviting us to write this monograph, to Rob Paige for his willingness to write a Foreword, and to referees and colleagues for their constructive suggestions and major contributions to material leading to the new subject of Statistics on Stratified Spaces. Their enthusiasm tremendously helped create this manuscript and improved its quality. This book would not have come to light without research support from the National Science Foundation, for which we are extremely grateful; the first author acknowledges support from award NSF-DMS:2311059.

Many of the results in this book are from joint work with Elizabeth Bullitt, Ian Dryden, Leif Ellingson, Harrie Hendriks, Thomas Hotz, Stephan Huckemann, Huiling Le, Steve J. Marron, Ezra Miller, Ipeq Oguz, Garett Ordway, Megan Owen, Rob L. Paige, Paul San Valentin, Sean Sqwerer, Ka Chun Wong, and Jinfeng Zhang. We thank them for their wonderful collaboration, and for their willingness to bring Statistics on Stratified Spaces to new heights. The second author is very grateful to his wife LaShun and his children Christopher, Roman, Zariyah (daughter), and Shaun, for their unwavering support during the preparation of this manuscript. He wishes to acknowledge his late father Aaron Osborne Sr. for instilling in him integrity, strength, courage, and a strong work ethic at a young age. He acknowledges the steadfast support of his brothers and sisters. Lastly, he is grateful his loving mother Vivian E. Osborne for being the metaphorical 'glue' that has bound and held our family together. Above all, the first author would like to thank his wonderful wife Olga for her unabated support during the preparation of this monograph, his son Vlad and grand-children Daniel and Julia for the love and joy they bring along in his life.

Part I

Data and Preliminaries for Analysis on Stratified Spaces

Chapter 1

Data of Complex Type

Statistics, as any other mathematical science, is changing, and the change is coming from the current data types that are collected and from the need to analyze these data sets, which no longer consist in numbers, but in complex entities, such a medical devices outputs, ribonucleic acid (RNA) sequences, trees, graphs, or digital camera images, to mention just a few. Using computational techniques from such entities, one may extract mathematical elements, nowadays called *objects*, that can be represented as points on certain metric spaces known as *object spaces* which have a smooth structure on an open dense subset. In this chapter, we provide just a small number of examples of such complex data to familiarize the reader with them. For more examples, of the smooth type, see e.g., Patrangenaru and Ellingson (2015) [106], Dryden and Mardia (2016) [39] or Marron and Dryden (2022) [92].

1.1 DNA/RNA Basics and Historical Notes

Deoxyribonucleic acid (DNA) is a polymer that was experimentally discovered by Rosalind Franklin by using x-ray diffraction crystallography images. DNA is composed of two polynucleotide chains, each one inherited from one parent, that are twisted around each other to form a curved double helix that is packed within each organism cell. DNA carries genetic instructions of all known organisms and many viruses. The discovery of the structure of DNA in 1952 was made possible by **Rosalind Franklin**'s x-ray diffraction work at King's College. Her creation of the famous Photo 51 demonstrated the double-helix structure of DNA and helped James Watson and Francis Crick to come up with the double helix model. DNA is responsible for a species' passing on of the characteristic traits from parents to children. DNA and ribonucleic acid (RNA) molecules are polymers made up of long chains of nucleotides. A **nucleotide** is the basic building block of nucleic acids (RNA and DNA). A nucleotide consists of a sugar molecule (either ribose in RNA or deoxyribose in DNA) attached to a phosphate group and a nitrogen-containing base. The bases used in DNA are adenine (A), cytosine (C), guanine (G) and thymine (T). In RNA, which is single-stranded, the base uracil (U) takes the place of thymine.

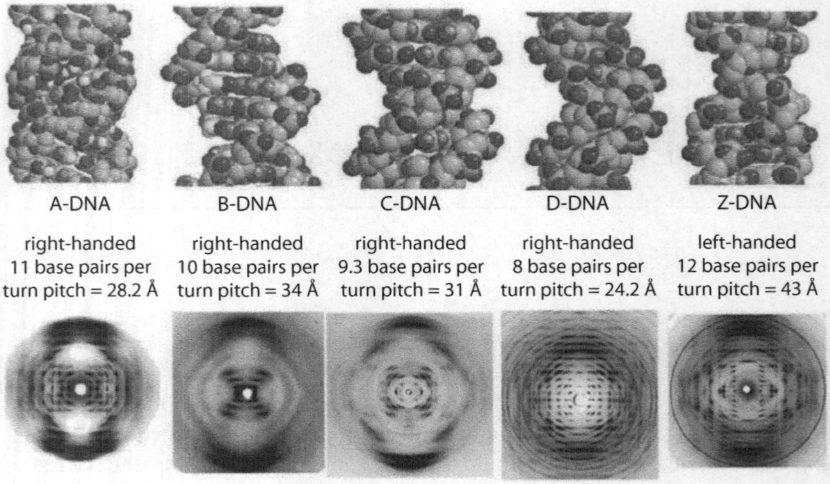

Figure 1.1: X-ray diffraction images of types of DNA.

The chemical formula for adenine is $C_5H_5N_5$, for thymine is $C_5H_6N_2O_2$, for cytosine is $C_4H_5N_3O$, for guanine is $C_5H_5N_5O$, and for uracil is $C_4H_4N_2O_2$. Nucleotides in DNA are paired in *base pairs* (see Figure 1.2); therefore, the **sequences from mother and father imply each other**. Note that *most viruses causing serious diseases in humans have their genome made entirely of RNAs* [55].

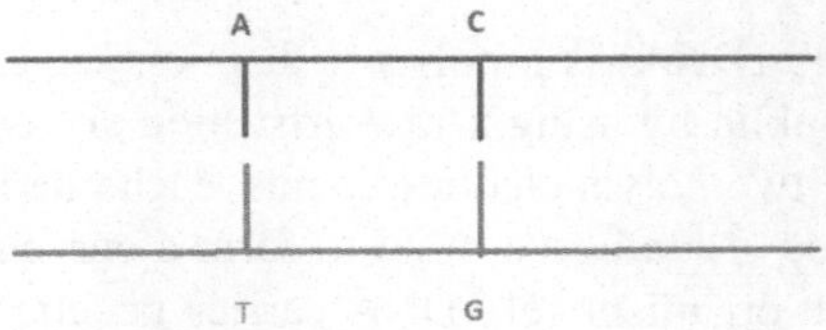

Figure 1.2: Base pairs.

DNA/RNA have two important properties: (a) they contain instructions of how organisms are put together, and (b) they replicate to successive generations. **Imperfect replications become mutations or single nucleotide polymorphisms (SNPs), which are simple changes in the sequence order of the base pairs**.

Frederick Sanger, a double Nobel Prize laureate, developed the first DNA sequencing technique. Sequence **alignment** is necessary to identify DNA regions of similarity that may be a consequence relationship between the sequences. The Sanger Institute in the UK uses **Clustal Omega**, a versatile tool for sequence alignment and phylogenetic tree building, which is used in our book. A more advanced sequence alignment method is due to the Molecular Evolutionary Genetic Analysis (**MEGA**) software, developed by Sudhir Kumar and his group from Temple University. One of the most advanced biological tools in genetics, CRISPR-Cas9 (2012-2014) is due to Nobel Prize medalists Jennifer Doudna and Emmanuelle Charpentier, and to Feng Zhang. This is a fast gene editing bioengineering technique, relevant to **RNA editing of genomic DNA** (see Jinek et al. (2013) [75]). DNA sequencing has gone from being a highly specialized job, requiring considerable labor and skill, to being routinely applied in ordinary laboratory work while also being conducted at great scale, speed and accuracy in factory-style genome centers. In the late-1970s, manually sequencing the tiny genome of a bacteriophage (a virus that infects bacteria) was a monumental task, one that earned Sanger, who led the group undertaking it, a Nobel Prize. The determination of the whole human DNA sequence (commonly referred to as the **Human Genome Project**) took more than a decade and was based the *dideoxy chain-termination method* due to Sanger as well. It started in the 1990s and concluded in 2003, expanding in speed and scale throughout. Progress since then has been so dramatic that, more recently, well over fourteen million coronaviruses have been sequenced and shared via the Global Initiative on Sharing Avian Infuenza Data. Another example that illustrates how far Genomics has come is that the cost of sequencing a whole human genome was estimated to be about 7000 British pounds in 2020, multiple orders of magnitude below the original budget of the Human Genome Project (Schwarze et al., 2020 [117]).

1.2 Protein Data

A large source of 3D size-and-shape data is the RCSB Protein Data Bank (PDB) website at http://www.rcsb.org/pdb/home/home.do, which provides a wealth of information about , including about their physical structures. As of May 27, 2024, there were $220,113$ structures posted on there. One of the many problems of interest in bioinformatics is the relationship of the physical structure and chemical sequence of a protein to its biological function, and size-and-shape analysis provides one avenue for exploring this relationship.

Here, we will consider the 3D atomic structures of two groups of binding sites, which are locations on the surface of a protein where binding activity occurs. In Figure 1.3, we show binding sites from three examples of *hydrolase (serine proteinase)* which were obtained by x-ray diffraction, as displayed using the software Rasmol. Their structure i.d.s on PDB are 1ela, 1eld and 1ele and their primary citation is Mattos et al. (1997) [93]. In Figure 1.3, the atoms are gray-level coded where hydrogen is dark gray, oxygen is medium gray, and carbon is light gray. Similarly, Figure 1.4 displays binding sites for three examples of *acid proteinase*. The coordinates and chemical types of atoms from

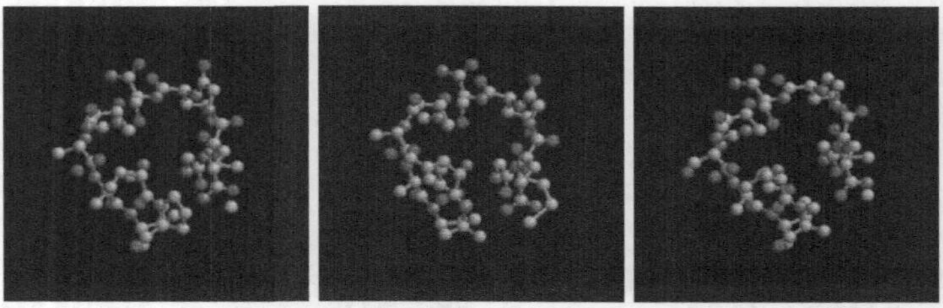

Figure 1.3: Serine proteinase protein binding site structures. (Source: Bhattacharya et al. (2012), figure 4. Reproduced by permission of John Wiley & Sons LTD.)

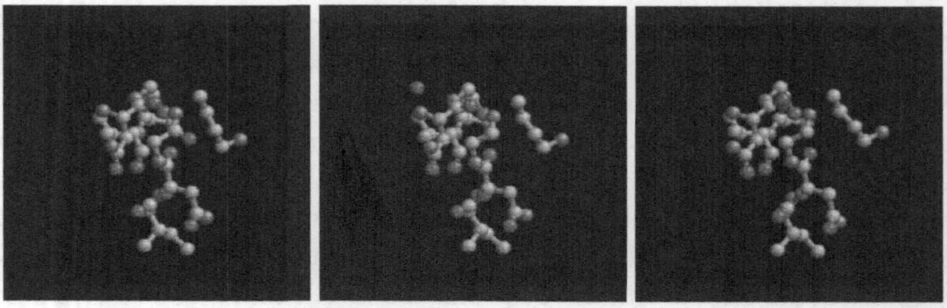

Figure 1.4: Acid proteinase protein binding site structures. (Source: Bhattacharya et al. (2012), figure 4. Reproduced by permission of John Wiley & Sons LTD.)

these acid proteinase binding sites, measured in Angstroms, are shown in tables 1.18 and 1.19 in Patrangenaru and Ellingson (2015) [106].

1.3 Graphs

Graph theory is a branch of mathematics concerned with *graphs*, which are mathematical concepts used to model pairwise relations, that are represented by connected lines, known as edges, between objects (represented by points or vertices). A simple example of graph is associated with the Seven Bridges of Königsberg problem due Leonhard Euler, dealing with the possibility of finding a path to cross over every one of seven bridges that span a forked river in Königsberg (today's Kaliningrad) flowing past an island, without crossing any bridge twice (see Figure 1.5).

Euler (1741) [45] attached to this puzzle Figure 1.6 below.

One may notice that an affirmative solution to this puzzle is equivalent to drawing the graph in Figure 1.6 from one stroke of a pen. This turns out being impossible, since the graph has all three vertices with an odd number of edges, A, B and D; therefore, the "incoming" pen should stop at two of these vertices. Here the term *graph* refers to a set of vertices (points or nodes) and of

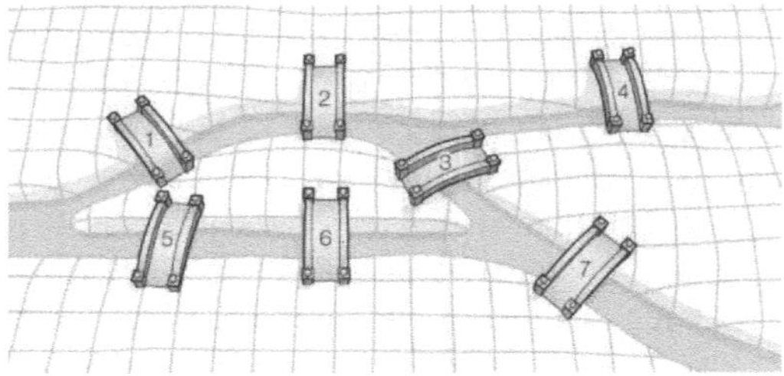

Figure 1.5: The seven Königsberg bridges puzzle.

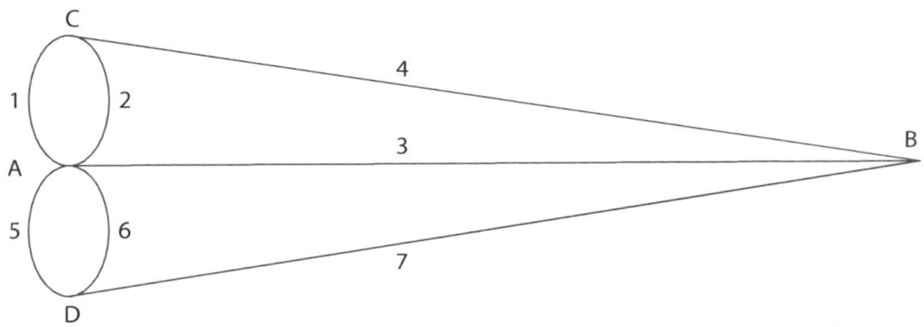

Figure 1.6: Euler's graph associated with the Köenigsberg bridges puzzle.

edges (or lines) that connect the vertices. When any two vertices are joined by more than one edge, the graph is called a multigraph. A graph without loops having at most one edge between any two vertices is called a simple graph. A tree is a connected, simply connected graph. When each vertex is connected by a unique edge to every other vertex, the graph is called a complete graph. When appropriate, a direction may be assigned to each edge, to produce what is known as a directed graph, or digraph.

An important number associated with each vertex is its degree, which is defined as the number of edges that enter or exit from it. Thus, a loop contributes 2 to the degree of its vertex. For instance, the vertices of the simple graph shown in the diagram all have a degree of 2, whereas the vertices of the complete graph shown are all of degree 3. Knowing the number of vertices in a complete graph characterizes its essential nature. For this reason, complete graphs are commonly designated as K_m, where m refers to the number of vertices, and all vertices of K_m have degree m-1. (Translated into the terminology of modern graph theory, Euler's theorem about the Königsberg bridge problem could be restated as follows: If there is a path along edges of a multigraph that traverses each edge once and only once, then there exist at most two vertices of odd degree; furthermore, if the path begins and ends at the same vertex, then no vertices will have odd degree.)

Another important concept in graph theory is the *path*, which is any continuous function from the interval $[0,1]$ starting at a vertex and ending at a (possibly the same) vertex, such that it is one to one between two consecutive vertices. If there is a path linking any two vertices in a graph, that graph is (path) connected. A path that begins and ends at the same vertex without traversing any edge more than once is called a circuit, or a closed path. A circuit that follows each edge exactly once while visiting every vertex is known as a *Eulerian circuit*, and the graph is called a *Eulerian graph*. A Eulerian graph is connected and, in addition, all its vertices have even degree.

In 1857 the Irish mathematician William Rowan Hamilton invented a puzzle (the Icosian Game). The puzzle involved finding a special type of path, later known as a Hamiltonian circuit, along the edges of a icosahedron (a Platonic solid consisting of 20 triangular faces) that begins and ends at the same corner while passing through each corner exactly once. The knight's tour (see number game: Chessboard problems) is another example of a recreational problem involving a Hamiltonian circuit. Hamiltonian graphs have been more challenging to characterize than Eulerian graphs, since the necessary and sufficient conditions for the existence of a Hamiltonian circuit in a connected graph are still unknown. For example, by removing one edge of an icosahedral graph, we obtain a planar graph that is no longer Hamiltonian.

The histories of *Graph Theory* and *Topology* are closely related, and the two areas share many common problems and techniques. Euler referred to his work on the Königsberg bridge problem as an example of geometria situs—the "geometry of position" —while the development of topological ideas during the second half of the 19th century became known as analysis situs—the "analysis of position". In 1750 Euler discovered the polyhedral formula $V - E + F = 2$ relating the number of vertices (V), edges (E), and faces (F) of a polyhedron (a solid, like the dodecahedron mentioned above, whose faces are polygons, such that the union of the faces is homeomorphic to a 2D sphere). The vertices and edges of a polyhedron form a graph on its surface, and this notion led to consideration of graphs on surfaces other than the sphere, such as a torus, and how they divide the surface into disk-like faces. Euler's formula was generalized to *orientable surfaces* as $V - E + F = 2 - 2g$, where g denotes the genus, or number of "doughnut holes" of the orientable surface (also known as Euler characteristic of the surface). Having considered a surface divided into polygons by an embedded graph, mathematicians began to study ways of constructing surfaces, and later more general spaces, by pasting polygons together. This was the beginning of the field of combinatorial topology, which later, through the work of the mathematician Henri Poincaré and others, grew into what is known today as Algebraic Topology.

The connection between Graph Theory and Topology led to a subfield called topological graph theory. An important problem in this area concerns planar graphs. These are graphs that can be drawn as dot-and-line diagrams on a plane (or, equivalently, on a sphere) without any edges crossing except at the vertices where they meet. Complete graphs with four or fewer vertices are planar, but complete graphs with five vertices (K5) or more are not. Nonplanar graphs cannot be drawn on a plane or on the surface of a sphere without edges intersecting each other between the vertices. The use of diagrams of dots and lines to represent graphs actually grew out of 19th-century Chemistry, where lettered vertices denoted individual atoms and connecting lines denoted chemical bonds (with degree corresponding to valence), in which planarity had important chemical consequences. The first use, in this context, of the word graph is attributed to the James Sylvester, a 19th-century mathematician interested in counting special types of diagrams representing molecules.

1.4 MRA-based Artery Data

The brain artery trees used in this study are reconstructed from a data set of Magnetic Resonance (MR) brain images collected by the CASILab at the University of North Carolina at Chapel Hill. This data set is publicly available and hosted at the MIDAG website Handle (2008) [60]. The database has images for various magnetic resonance modalities, including T1, T2, Magnetic Resonance Angiography (MRA), and Diffusion Tensor Imaging (DTI). The study enrolled 109 apparently healthy subjects. Each image is tagged with subject features of age, sex, handedness and self-identified race. The MRA was aquired at $0.5 \times 0.5 \times 0.8$ mm^3 accuracy. Figure 1.7 shows one MRA slice. Brain artery trees were constructed from MRA images using the tube tracking method developed in Aylward and Bullitt (2002) [4]. Each tree consists of branch segments, and each branch segment consists of a sequence of spheres fit to the bright regions in the MRA image. The sphere centers are 3D points on the center line of the artery, and the radius indicates the arterial thickness. A method for visualizing

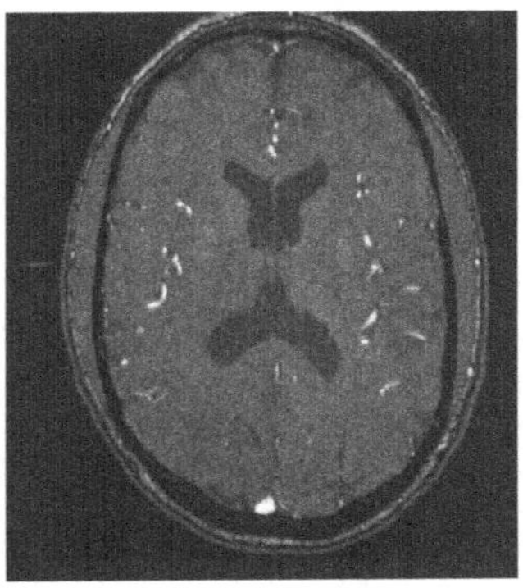

Figure 1.7: One slice of a Magnetic Resonance Angiography (MRA) image. Bright regions indicate blood flow. Bright regions are tracked through MRA slices to recover artery tubes as shown in Figure 1.8.

the structure of large trees was used to detect any remaining discrepancies (see Aydin et al. (2009) [3]).

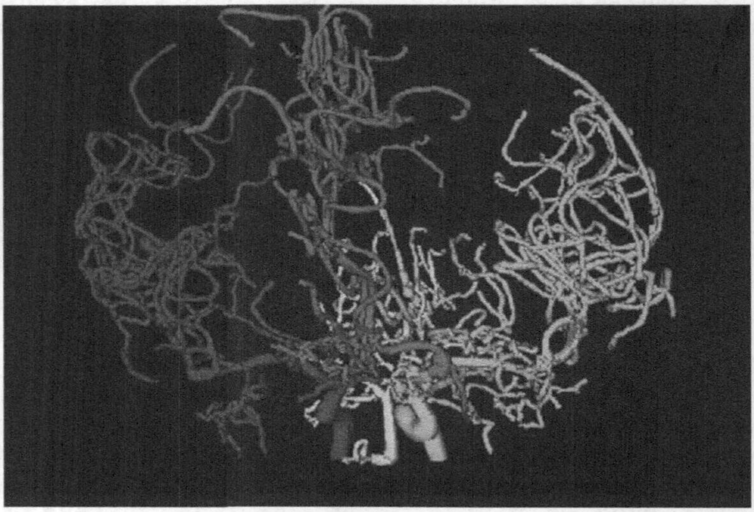

Figure 1.8: The data object is a reconstructed brain artery tree. The goal of this research is statistical analysis of a sample of such data objects.

In humans, the carotid arteries go up the neck and to the circle of Willis loop. The circle of Willis is not captured in these MRA images. Generally four brain artery trees go up from the circle of Willis to one of main brain regions: left, right, anterior, and posterior. An artificial conceptual connection of the bases of the trees together enables joint analysis of the entire artery structure above the circle of Willis.

1.5 Alphabet: Origin, History and Spread

An alphabet is a set of pictorial characters used to represent visually the phonemic structure of a language, in pursuit of faster communication and in preservation of knowledge. From this perspective, since a written communication is used to express an intended societal need, the simpler the alphabet the more effective it is to address all aspects of interest to a society: preserving life, managing health, housing, transportation, education, defense from unpredictable dangers, protecting the environment, advancing progress and democratization, etc. And, based on an archaic Greek cultural heritage, and due to their extended influence in the ancient world, the Romans helped creating such an alphabet, the Latin alphabet, which remained largely unaltered with the exception of additions (the letters J, U, W) and extensions, such as diacritics (accents), to form the Latin script that is used to write many modern European languages, including English.

Note that the Greek alphabet grew out of the consonantal writing system used for Semitic languages in the Levant, in the second millennium BCE. Most, or nearly all alphabetic scripts, including Hebrew, Arabic, and Cyrillic, used throughout the world today, ultimately are based on this Semitic alphabet, whose origins can be traced back to a Proto-Sinaitic script developed for popular language of Semitic-speaking workers and slaves in ancient Egypt, as they were unskilled to handle the hieroglyphic system used in the Egyptian language that required a large number of pictograms, as opposed to the semantic values, of the Canaanite language of slaves. This Semitic alphabet, partly influenced by the older Egyptian hieratic, related to hieroglyphs, became the ancestor of multiple writing systems across the Middle East, Europe, Northern Africa, and Pakistan, mainly through Phoenician, Paleo-Hebrew (closely related and initially virtually identical to the Phoenician alphabet or even derived from it), and later Aramaic (derived from the Phoenician alphabet), Ancient South Arabian, four closely related members of the Semitic family of scripts that were in use during the early first millennium BCE. See Figure 1.9.

In particular, the Hebrew alphabet, developed from the now defunct Phoenician alphabet, into a regional dialect of the Canaanite languages, was natively spoken by the Israelites and remained in regular use as a first language until after 200 CE (Dead Sea Scrolls) and as the liturgical language of Judaism and Samaritanism, being the only language, alongside its alphabet, to be revived as a spoken language in the 19th century, and is the only successful large-scale example of linguistic revival. It is the only Canaanite language, as well as one of only two Northwest Semitic languages, with the other being Aramaic, which is still spoken today. There is an obvious transition from the Paleo-Hebrew, א (aleph), ב (beth), ג(gimel), ד(daleth), which had important

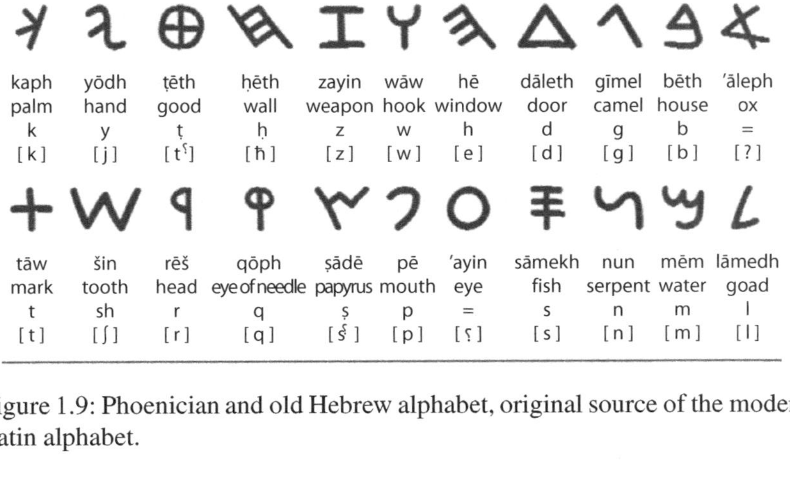

Figure 1.9: Phoenician and old Hebrew alphabet, original source of the modern Latin alphabet.

meanings (e.g., Aleph=oneness of God, beth=house, gamal=camel, daleth=door), to the abstract symbols $\alpha, \beta, \gamma, \delta$, and later to A, B, C, D, of a pragmatic use helping improve communication.

The standard modern ISO basic Latin alphabet that is used in this book for tree building purposes to analyze certain European languages is: A-B-C-D-E-F-G-H-I-J-K-L-M-N-O-P-Q-R-S-T-U-V-W-X-Y-Z. This is the alphabet we are going to use to build trees data in an effort to compare modern European languages (see Ordway and Patrangenaru (2024) [100]).

1.6 3D Landmarks Data from Digital Camera Images

The 3D surface reconstructions of faces from digital camera images, with seven labeled landmarks, are another source of object data and were obtained using the software Agisoft. These reconstructions (including texture) are displayed in Figure 1.10 (see Yao (2016) [147]).

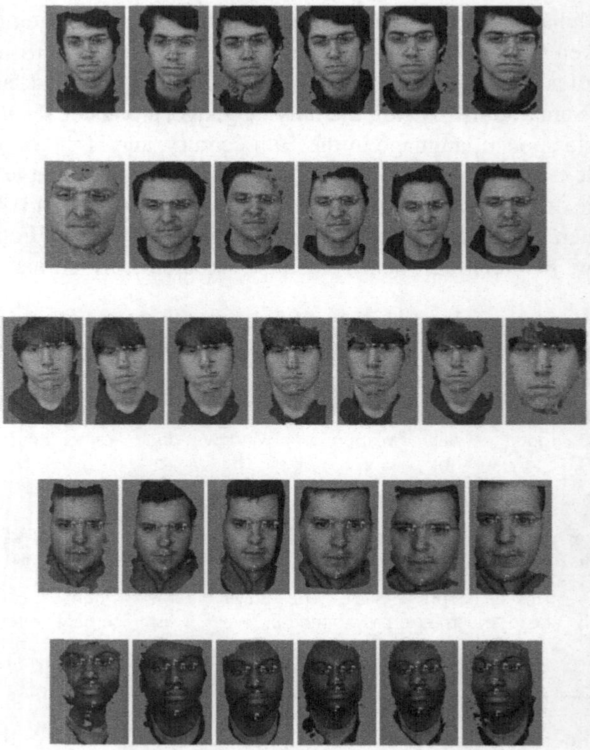

Figure 1.10: Sample of 3D facial reconstructions with marked landmarks configurations.

Chapter 2

Review of Nonparametric Statistics on Manifolds

2.1 Introduction

In preparation for data analysis on stratified spaces, in this chapter we recall certain related results on statistical analysis of data on manifolds. The main reference for this preliminary chapter is Patrangenaru and Ellingson (2015) [106]. The *Fréchet function*, $\mathscr{F}_\rho$, that is associated with a probability measure $Q = P_X$ on a complete metric space $(\mathscr{M}, \rho)$ of a random object (r.o.) X on $\mathscr{M}$. This function is defined as an expected square distance from X, that is $\mathscr{F}_\rho(x) = E(\rho^2(x,X))$ (see Fréchet (1948) [52]). The *Fréchet total variance* of X (or of Q) is the minimum value of the Fréchet function, and the set of minimizers form *Fréchet mean set* of Q. If the Fréchet mean set consists in one point only, that minimizer is the *Fréchet mean* of X (or of Q) and is labeled μ_ρ. Two families of distances are popular in case $\mathscr{M}$ has a manifold structure: (i) a *geodesic* distance $\rho =_g \rho$ is the distance associated with a Riemannian structure g on $\mathscr{M}$, and (ii) a *chord* distance $\rho = \rho_j$ induced by an embedding $j : M \to \mathbb{R}^N$. Given the random sample $x_1, \ldots, x_n$, their *Fréchet sample total variance* is the Fréchet total variance of their empirical distribution $\hat{Q}_n = \frac{1}{n}\sum_{i=1}^n \delta_{x_i}$, and their *Fréchet sample mean* is the Fréchet mean of $\hat{Q}_n$. The *intrinsic (extrinsic) sample variance, intrinsic (extrinsic) sample mean set or intrinsic (extrinsic) mean* are similarly defined respectively as intrinsic (extrinsic) variance, intrinsic (extrinsic) mean set or intrinsic (extrinsic) mean of the empirical distribution $\hat{Q}_n$. Our notations for these spread and location parameters of a random object on a metric space $(\mathscr{M}, \rho)$ are as follows:

- $t\Sigma_{\mathscr{F}}$ or $t\Sigma_{\mathscr{F},\rho}$ for the total Fréchet variance on a complete metric space $(\mathscr{M}, \rho)$
- $t\Sigma_I$ or $t\Sigma_{g,I}$ for the total intrinsic variance on a complete Riemannian manifold $(\mathscr{M}, g)$ where g is the Riemannian structure (positively definite symmetric tensor field on $\mathscr{M}$), an $\rho =_g \rho$ is the associated geodesic (Riemannian) distance on $\mathscr{M}$

- $t\Sigma_E$ or $t\Sigma_{E,j}$ for the total extrinsic variance on a closed manifold $\mathcal{M}$ associated with an embedding $j : M \to \mathbb{R}^N$
- $\mu_{\mathcal{F}}$ or $\mu_{\mathcal{F},\rho}$ for the Fréchet mean on a complete metric space $(\mathcal{M}, \rho)$
- μ_I or $\mu_{g,I}$ or μ_g for the intrinsic mean on a complete Riemannian manifold $(\mathcal{M}, g)$
- μ_E or $\mu_{E,j}$ for the extrinsic mean on a closed manifold $\mathcal{M}$ associated with an embedding $j : \mathcal{M} \to \mathbb{R}^N$.

The sample counterparts of these parameters associated with a random sample $x_1, \ldots, x_n$, are denoted as follows:

- $tS_{\mathcal{F}}$ or $tS_{\mathcal{F},\rho}$ for the total Fréchet sample variance on a complete metric space $(\mathcal{M}, \rho)$
- $tS_{I,n}$ or tS_I or $tS_{g,I}$ for the total intrinsic sample variance on a complete Riemannian manifold $(\mathcal{M}, g)$ where g is the Riemannian structure (positively definite symmetric tensor field on $\mathcal{M}$), an $\rho =_g \rho$ is the associated geodesic (Riemannian) distance on $\mathcal{M}$
- $tS_{E,n}$ or tS_E or $tS_{E,j}$ for the total extrinsic sample variance on a closed manifold $\mathcal{M}$ associated with an embedding $j : M \to \mathbb{R}^N$
- $\bar{x}_{\mathcal{F},n}$ or $\bar{x}_{\mathcal{F}}$ or $\bar{x}_{\mathcal{F},\rho}$ for the Fréchet sample mean on a complete metric space $(\mathcal{M}, \rho)$
- $\bar{x}_{I,n}$ or $\bar{x}_I$ or $\bar{x}_{g,I}$ or $\bar{x}_g$ for the intrinsic sample mean on a complete Riemannian manifold $(\mathcal{M}, g)$
- $\bar{x}_{E,n}$ or $\bar{x}_E$ or $\bar{x}_{E,j}$ for the extrinsic sample mean on a closed manifold $\mathcal{M}$ associated with an embedding $j : \mathcal{M} \to \mathbb{R}^N$.

In this chapter, it is assumed that $X_1, \ldots, X_n$ are *independent identically distributed random objects* (i.i.d.r.o.s) from a probability distribution Q on $\mathcal{M}$. The main focus of the chapter is a review of asymptotic distributions of their Fréchet total sample variance and Fréchet sample mean. Such distributions are used to derive large sample confidence regions for Fréchet means on manifolds, as well as for hypothesis testing problems. As a consequence, using standard methods, one derives pivotal as well as nonpivotal bootstrap confidence regions for Fréchet means. The asymptotic normality of the sample total intrinsic variance is the subject of Section 2.2.

Section 2.3 is devoted to asymptotics of extrinsic sample means. The first results, due to Hendriks and Landsman (1998) [66] and Patrangenaru (1998) [104] were independent of each other, using different estimators for the extrinsic mean. In this chapter, we present an extension of Patrangenaru (1998) [104]. Extrinsic means are commonly used in directional, axial, and shape data analysis. In the particular case of directional data analysis, i.e., when $\mathcal{M} = S^{m-1}$ is the unit sphere in $\mathbb{R}^m$, Fisher et al. (1996) [49] provided an approach for inference using computationally efficient bootstrapping which gets around the problem of increased dimensionality associated with the embedding of the manifold $\mathscr{M}$ in a higher-dimensional Euclidean space. In Corollary 2.5.2 confidence regions (c.r.s) are derived for the extrinsic mean $\mu_{j,E}$. Nonparametric bootstrap methods on abstract manifolds are also derived in this section (Theorem 2.5.3, Proposition 2.5.3).

If one assumes that Q has a nonzero absolutely continuous component with respect to the volume measure on $\mathscr{M}$, then from some results in section 2.13 of Patrangenaru and Ellingson (2015) [106] (see also Beran (1987) [11] and Hall (1988, 1997) [57, 58]), one derives bootstrap based confidence regions for μ_E with coverage error $O_p(n^{-2})$ (see Theorem 2.5.1) (see section 2.13 of Patrangenaru and Ellingson (2015) [106] and the references therein). One may also use the nonpivotal bootstrap to construct confidence regions based on the percentile method of Hall (1997) [58] for general Q with a coverage error no more than $O_p(n^{-m/(m+1)})$, where m is the dimension of the manifold (see Remark 2.4.4 and Proposition 2.5.3). For a definition of the coverage error, see section 2.13 of Patrangenaru and Ellingson (2015) [106]. This is particularly useful in those cases where the asymptotic dispersion matrix is difficult to compute.

2.2 Fréchet Total Variance

Since the Fréchet total variance $t\Sigma_{\mathscr{F}}(P_X)$ of a random object X on a metric space $(\mathscr{M},\rho)$ is the expected square distance from X to an arbitrary point on $\mathscr{M}$, this parameter explains the variability of X. Indeed, X is constant if and only if $t\Sigma_{\mathscr{F}} = 0$. In the particular case, when $\rho = \rho_g$ is the geodesic distance associated with a Riemannian structure g on $\mathscr{M}$, one may estimate $t\Sigma_{I,g}$ with confidence from a random sample $x_1,\dots,x_n$. Let $X_1,\dots,X_n$ be i.i.d.r.o.s with a common distribution $Q = P_{X_1}$ on $\mathscr{M}$, for which μ_g exists. Let $C(p)$ the cutlocus of a point $p \in \mathscr{M}$ (for a definition of the cutlocus see e.g., Kobayashi and Nomizu (1996) [81]). Consider the random variable

$$T_n = n^{-1}\sum_{k=1}^{n} \rho_g^2(\mu_I, X_k). \tag{2.1}$$

The intrinsic moment of order r about the intrinsic mean is $E[\rho_g^k(\mu_g, X_1)]$. Then we have:

THEOREM 2.2.1. *Assume $Q(C(\mu_g)) = 0$ and the intrinsic moments of order four or less about μ_g are finite, then $n^{1/2}(T_n - t\Sigma_g)$ converges in law to $\mathscr{N}(0, E[\rho_g^4(\mu_g, X_1)] - (t\Sigma_g)^2)$.*

REMARK 2.2.1. *Since $n^{1/2}(T_n - t\Sigma_g)$ converges weakly to $N(0, E[rho_g^4(\mu_g, X_1)] - (t\Sigma_g)^2)$, it follows that T_n converges in probability to $t\Sigma_g$, thus $t\Sigma_g(\hat{Q}_n) - t\Sigma_g$ and $T_n - t\Sigma_g$ are asymptotically equivalent. However, this falls short of saying that $n^{1/2}(t\Sigma_g(\hat{Q}_n) - t\Sigma_g)$ and $n^{1/2}(T_n - t\Sigma_g)$ are asymptotically equivalent,* *which together with Theorem 2.2.1 would suffice to yield percentile confidence regions for* $t\Sigma_g$.

A useful weaker result than Theorem 2.2.1 is the following:

THEOREM 2.2.2. *If* Q *has small flat support on* $\mathcal{M}$ *and has finite moments up to the fourth order, then* $n^{1/2}(t\hat{\Sigma}_{g,n} - t\Sigma_g)$ *converges in law to a random vector with a* $\mathcal{N}(0, Var(\rho_g{}^2(X_1, \mu_g)))$ *distribution.*

Under the assumptions of Theorem 2.2.2, if we set $S_g^2 = \frac{1}{n}\sum_{i=1}^n(\rho_g{}^2(X_i, \overline{X}_g) - \overline{\rho_g{}^2(X, \overline{X}_g)})^2$, we obtain

COROLLARY 2.2.1. *If* Q *has small flat support on* $\mathcal{M}$*, then*

$$n^{\frac{1}{2}}\left(\frac{(t\hat{\Sigma}_{g,n} - tS_{I,g})}{S_g}\right) \tag{2.2}$$

converges in distribution to a random variable with a standard normal distribution.

Further, we get:

COROLLARY 2.2.2. *A* $100(1-\alpha)\%$ *large sample symmetric confidence interval for* $t\Sigma_g$ *is given by*

$$(tS_{g,n} - z_{\frac{\alpha}{2}}\frac{S_g}{\sqrt{n}}, tS_{g,n} + z_{\frac{\alpha}{2}}\frac{S_g}{\sqrt{n}}) \tag{2.3}$$

2.3 Elementary CLT for Extrinsic Means

We consider now asymptotic distributions of Fréchet means. We start with an elementary method for constructing confidence regions for extrinsic means $\mu_E(Q)$ on closed submanifolds $\mathbb{R}^N$. The embedding j is the inclusion map of $\mathcal{M}$ in $\mathbb{R}^N$, and the projection P_j is the projection $P_{\mathcal{M}}$ on $\mathcal{M}$. Let H be the projection on the affine subspace $\mu_E(Q) + T_{\mu_E}\mathcal{M}$. We would like to determine the asymptotic distribution of $H(\overline{X})$. While $H(\overline{X})$ is not the same as $P_{\mathcal{M}}(\overline{X})$, its asymptotic distribution is easier to compute. For large samples, the extrinsic sample mean is close to the extrinsic mean and, therefore, $H(\overline{X})$ and $P_{\mathcal{M}}(\mu)$ will be close to each other. When $\mathcal{M}$ is a linear variety of $\mathbb{R}^N$, the two maps coincide. Thus for concentrated data the delta method for H gives a good estimate of the distribution of the extrinsic sample mean. Assume that around $P_{\mathcal{M}}(\mu)$ the implicit equations of $\mathcal{M}$ are $F^1(x) = ... = F^c(x) = 0$, where $F^1, ..., F^c$ are functionally independent. Then $\overline{X} - H(\overline{X})$ is in $\nu_{P_{\mathcal{M}}(\mu)}\mathcal{M}$, the orthocomplement of $T_{P_{\mathcal{M}}(\mu)}\mathcal{M}$; thus it is a linear combination of the gradients $grad_{P_{\mathcal{M}}(\mu)}F^1, ..., grad_{P_{\mathcal{M}}(\mu)}F^c$. We need to evaluate the differential of the map H at μ, in terms of $F^1, ..., F^c$. Set $\nu_\alpha = \|grad_{P_{\mathcal{M}}(\mu)}F^\alpha\|^{-1} grad_{P_{\mathcal{M}}(\mu)}F^\alpha$ $(\alpha = 1, ..., c)$, and

$$(h^{\alpha\beta}(\mu))_{\alpha,\beta=1,...,c} = ((h_{\alpha\beta}(\mu))_{\alpha,\beta=1,...,c})^{-1}, h_{\alpha\beta}(\mu) = \nu_\alpha \cdot \nu_\beta \tag{2.4}$$

Then $x - H(x) = \sum_\beta t^\beta(x,\mu)v_\beta$ where $t^\beta(x,\mu) = \sum_\alpha h^{\alpha\beta}(\mu)(x - P_{\mathcal{M}}(\mu)) \cdot v_\alpha$. Therefore,

$$H(x) = x + h^{\alpha\beta}(\mu)((P_{\mathcal{M}}(\mu) - x) \cdot v_\alpha)v_\beta, d_\mu H(v) = v - \sum_\alpha h^{\alpha\beta}(\mu)(v \cdot v_\alpha)v_\beta, \tag{2.5}$$

that is

$$G_i^j = \frac{\partial H^j}{\partial x^i}(\mu) = \delta_{ij} - \sum_{\alpha,\beta} h^{\alpha\beta}(\mu)v_\alpha^i v_\beta^j \tag{2.6}$$

where $\delta_{ij} = 1$ or 0 according as $i = j$ or $i \neq j$. By the delta method we arrive at:

THEOREM 2.3.1. *Let $\{X_k\}_{k=1,\ldots,n}$ be a random sample from a nonfocal distribution Q on the submanifold $\mathcal{M}$, given in a neighborhood of $\mu_E(Q)$ by the equations $F^1(x) = \cdots = F^c(x) = 0$. Assume Q has mean μ and covariance matrix Σ as a distribution in the ambient numerical space. If G is the matrix given by* (2.6)*, then $n^{\frac{1}{2}}(H(\overline{X}) - P_{\mathcal{M}}(\mu))$ converges weakly to $\mathcal{N}_{N-c}(0, G\Sigma G^t)$ in the tangent space of $\mathcal{M}$ at the extrinsic mean $\mu_E(Q) = P_{\mathcal{M}}(\mu)$ of Q.*

Since the sample counterpart of

$$\Gamma = G\Sigma G^t \tag{2.7}$$

may be difficult to compute and one may use Efron's percentile bootstrap (see Efron (1982) [41]) to obtain a confidence region for $\mu_E(Q)$.

COROLLARY 2.3.1. *Under the hypothesis of Theorem 2.3.1, one may construct an asymptotic $(1-\alpha)$-confidence region for $\mu_E(Q) = P_{\mathcal{M}}\mu$, using the bootstrapped statistic $n^{\frac{1}{2}}(H(\overline{X}^*) - \overline{H}(\overline{X}))$. Here $\overline{H}(\overline{X})$ is the projection of $\overline{X}$ on the affine subspace $\overline{X}_E + T_{\overline{X}_E}\mathcal{M}$, and $\overline{X}^*$ is the mean of a random sample with repetition of size n from the empirical $\hat{Q}_n$ considered as a probability measure on the Euclidean space in which $\mathcal{M}$ is embedded.*

REMARK 2.3.1. *Suppose $\mathcal{F}$ is finite and Q is nonfocal. Then there exists $\delta >$ 0 such that $\overline{X}$ is nonfocal, if $\|\overline{X} - \mu\| < \delta$. Since $P(\|\overline{X} - \mu\| \geq \delta) = O(n^{-1})$, one may define $\overline{X}_E$ to be any measurable selection from the sample extrinsic mean set, if $\overline{X}$ is focal. Theorem 2.3.1 and a corresponding version of Corollary 2.3.1 hold for this $\overline{X}_E$.*

2.4 Central Limit Theorem for Fréchet Means and Bootstrapping

Given $q \in M$, the *exponential map* $Exp_q : U \to M$ is defined on an open neighborhood U of $0 \in T_qM$ by the correspondence $v \to \gamma_v(1)$, where $\gamma_v(t)$ is the unique geodesic satisfying $\gamma(0) = q, \dot{\gamma}(0) = v$, provided $\gamma(t)$ extends at least to $t = 1$. Thus if (M, g) is geodesically complete or, equivalently, (M, ρ_g) is complete as a metric space, then Exp_q is defined on all of T_qM. In this chapter,

unless otherwise specified, all Riemannian manifolds are assumed to be *complete.*
Note that if $\gamma(0) = p$ and $\gamma(t)$ is a geodesic, it is generally not true that the geodesic distance between p and $q = \gamma(t_1)$, say, is minimized by $\gamma(t), 0 \leq t \leq t_1$ (consider e.g., the great circles on the sphere S^2 as geodesics). Let $t_0 = t_0(p)$ be the supremum of all $t_1 > 0$ for which this minimization holds. If $t_0 < \infty$, then $\gamma(t_0)$ is the *cut point of p along* γ. The *cut-locus* $C(p)$ of p is the union of all cut points of p along all geodesics γ starting at p (for example, $C(p) = \{-p\}$ on S^2).

In this section we deal with both intrinsic and extrinsic means. Hence we will often consider a general distance ρ on a complete differentiable manifold $\mathscr{M}$. In particular, (M, ρ) with the metric topology is complete as a metric space. We consider only those probability measures Q on $\mathscr{M}$ for which the Fréchet mean $\mu_{\mathscr{F}} =: \mu_{\mathscr{F}}(Q)$ exists.

REMARK 2.4.1. *(Bhattacharya and Lin (2013) [19]) The additional assumption in Bhattacharya and Patrangenaru (2005) [21] on the existence of a chart* (U, φ) *such that* $Q(U) = 1$, *or remark 5.5. in Bhattacharya and Bhattacharya (2012) [15] are in fact unnecessary, due to the consistency of the Fréchet mean (see also Bhattacharya and Patrangenaru (2014) [22]).*

Consider a chart (U, φ) on M, with $\mu_{\mathscr{F}} \in U$. If $X_i (i = 1, \ldots, n)$ are independent and identically distributed (i.i.d.) with common distribution Q and defined on a probability space $(\Omega, \mathscr{A}, P)$, let $\mu_{n,\mathscr{F}}$ be a measurable selection form the Fréchet mean set (w.r.t. ρ) of the empirical $\hat{Q}_n = \frac{1}{n}\sum_{i=1}^{n} \delta_{X_i}$. Let $\mu_n = \varphi(\mu_{n,\mathscr{F}})$ be a measurable selection from the Fréchet mean set of $\hat{Q}_n = \frac{1}{n}\sum_{i=1}^{n} \delta_{X_i}$. Assume ρ^2 is twice differentiable as a function on $M \times M$. It follows that the Fréchet function is twice differentiable on $\mathscr{M}$. We consider the local representative F^{φ} of the Fréchet function. Since μ is the point of minimum of

$$F^{\varphi}(\theta) := \int_M \rho^2(\varphi^{-1}(\theta), q) Q(dq) = E(\rho^2(\varphi^{-1}(\theta), X), \tag{2.8}$$

where X is a random object on $\mathscr{M}$ with $P_X = Q$. Similarly μ_n is a local minimum of

$$F_n^{\varphi}(\theta) := \int_M \rho^2(\varphi^{-1}(\theta), q\hat{Q}_n(dq) = \frac{1}{n}\sum_{i=1}^{n} \rho^2(\varphi^{-1}(\theta), X_i). \tag{2.9}$$

Write the Euclidean gradient of $\theta \to \frac{1}{2}\rho^2(\varphi^{-1}(\theta), q)$ as

$$\begin{aligned}\Psi(u;\theta) &= \frac{1}{2} grad_{\theta}(\rho^2(\varphi^{-1}(\theta), q)) \\ &= (\frac{1}{2}\frac{\partial}{\partial\theta^r}(\rho^2(\varphi^{-1}(\theta), q)))_{r=1}^{m} = (\Psi^r(q;\theta))_{r=1}^{m}.\end{aligned} \tag{2.10}$$

For the $\frac{1}{2}$ scalar multiple, see Schwartzman (2014) [114]. One has the Taylor expansion

$$0 = \frac{1}{\sqrt{n}} \sum_{i=1}^{n} \Psi^r(X_i;\mu_n) = \frac{1}{\sqrt{n}} \sum_{i=1}^{n} \Psi^r(X_i;\mu) +$$

$$+\frac{1}{n} \sum_{i=1}^{n} \sum_{r'=1}^{m} D_{r'} \Psi^r(X_i;\mu)) \sqrt{n}(\mu_n^{r'} - \mu^{r'}) + R_n^r, (1 \le r \le d) \tag{2.11}$$

where

$$R_n^r = \sum_{r'=1}^{m} \sqrt{n}(\mu_n^{r'} - \mu^{r'}) \frac{1}{n} \sum_{i=1}^{n} \{D_{r'} \Psi^r(X_i;\theta_n) - D_{r'} \Psi^r(X_i;\mu)\}, \tag{2.12}$$

and θ_n lies on the line segment joining μ and μ_n (for sufficiently large n). We will assume

$$\begin{aligned} E|\Psi(X_1;\mu)|^2 &< \infty, \\ E|D_{r'}\Psi^r(X_1;\mu)|^2 &< \infty (\forall r, r'). \end{aligned} \tag{2.13}$$

To show that R_n^r is negligible, write

$$u^{r,r'}(x,\varepsilon) := \sup_{\{\theta : \|\theta - \mu\| \le \varepsilon\}} |D_{r'} \Psi^r(x;\theta) - D_{r'} \Psi^r(x;\mu)|$$

and assume

$$\delta^{r,r'}(c) := E u^{r,r'}(X_i, c) \to 0, as \ c \downarrow 0, (1 \le r, r' \le m). \tag{2.14}$$

One may then rewrite (2.11) in vectorial form as

$$0 = \frac{1}{\sqrt{n}} \sum_{i=1}^{n} \Psi(X_i;\mu) + (\Lambda + \delta_n) \sqrt{n}(\mu_n - \mu) \tag{2.15}$$

where

$$\Lambda = E((D_{r'} \Psi^r(X_i;\mu)))_{r,r'=1}^{m} \tag{2.16}$$

and $\delta_n \to 0$ in probability as $n \to \infty$, if $\mu_n \to \mu$ in probability. If, finally, we assume Λ is nonsingular, then (2.15) leads to the equation

$$\sqrt{n}(\mu_n - \mu) = \Lambda^{-1}(\frac{1}{\sqrt{n}} \sum_{i=1}^{n} \Psi(X_i;\mu)) + \delta_n' \tag{2.17}$$

where δ_n' goes to zero in probability as $n \to \infty$. We have then arrived at the following theorem, due to Bhattacharya and Patrangenaru (2005) [21].

THEOREM 2.4.1. *(CLT for Fréchet sample means.) Let Q be a probability measure on a differentiable manifold $\mathcal{M}$ endowed with a metric ρ such that every closed and bounded set of (M,ρ) is compact. Assume (i) the Fréchet mean $\mu_{\mathcal{F}}$ exists, (ii) the map $\theta \to (\rho^{\varphi})^2(\theta,u)$ is twice continuously differentiable on $\varphi(U)$, (iii) the integrability conditions* (2.13) *hold as well as the relation* (2.14)*, and (iv) Λ, defined by* (2.16)*, is nonsingular. Then (a) every measurable selection μ_n from the (sample) Fréchet mean set of $\hat{Q}_n^{\varphi} = \frac{1}{n}\sum_{i=1}^{n}\delta_{\tilde{X}_i}$ is a consistent estimator of μ, and (b)*

$$\sqrt{n}(\varphi(\mu_n) - \varphi(\mu) \to_d \mathcal{N}_m(0,\Lambda^{-1}\Sigma(\Lambda^T)^{-1}), \tag{2.18}$$

where Σ is the covariance matrix of $\Psi(X_1;\mu)$.

The *proof* given here is from Bhattacharya and Patrangenaru (2005) [21]. Part (a) follows from theorem 2.3 in Bhattacharya and Patrangenaru (2005) [21]. The proof of part (b) is as outlined above, and it may also be derived from standard proofs of the CLT for M-estimators (see, e.g., Huber (1980) [70], pp. 132-134). ■

As an immediate corollary one obtains:

COROLLARY 2.4.1. *Let $(\mathcal{M},g)$ be a Riemannian manifold and let $\rho = \rho_g$ be the geodesic distance. Let Q be a probability measure on $\mathcal{M}$. Assume that (i) the intrinsic mean $\mu_g = \mu_{\mathcal{F}}$ exists, (ii) the map $\rho^2 : \mathcal{M}\times\mathcal{M} \to \mathbb{R}$ is twice continuously differentiable and Λ, defined by* (2.16)*, is nonsingular. Then the conclusions of Theorem 2.4.1 hold for the Cartan mean $\mu_{n,g}$ of $\hat{Q}_n = \frac{1}{n}\sum_{i=1}^{n}\delta_{X_i}$, with $\mu = \varphi(\mu_g)$.*

We now prove one of the main results of this section.

THEOREM 2.4.2. *(CLT for Cartan means.) Let (M,g) be a Riemannian manifold and let $\rho = \rho_g$ be the geodesic distance. Let Q be a probability measure on $\mathcal{M}$ whose support is contained in a closed geodesic ball $\overline{B}_r \equiv \overline{B}_r(x_0)$ with center x_0 and radius r which is disjoint form the cut-locus $C(x_0)$. Assume $r < \frac{\pi}{4K}$, where K^2 is the supremum of sectional curvatures in $\overline{B}_r$ if this supremum is positive, or zero if this supremum is nonpositive. Then (a) the intrinsic mean μ_g (of Q) exists, and (b) the conclusion of Theorem 2.1 holds for the image $\mu_n = \varphi(\mu_{n,g})$ of the Cartan mean $\mu_{n,g}$ of $\hat{Q}_n = \frac{1}{n}\sum_{i=1}^{n}\delta_{X_i}$, under the inverse φ of the exponential map, $\varphi = (Exp_{x_0})^{-1}$.*

REMARK 2.4.2. *If the supremum of the sectional curvatures (of a complete manifold $\mathcal{M}$) is nonpositive, and the support of Q is contained in $\overline{B}_r$ then the hypotheses of Theorem 2.4.2 are satisfied, and the conclusions (a), (b) hold. One may apply this even with $r = \infty$.*

REMARK 2.4.3. *The assumptions in Theorem 2.4.2 on the support of Q for the existence of μ_g is too restrictive for general applications. But without additional structures it cannot be entirely dispensed with, as is easily shown by letting Q be the uniform distribution on the equator of S^2. For the complex* *projective space* $\mathbb{C}P^{\frac{m}{2}}$, *m-even, necessary and sufficient conditions for the existence of the intrinsic mean* μ_g *of an absolutely continuous (w.r.t. the volume measure) Q with radially symmetric density are given in Le (1998) [87], Kendall et al. (1999) [80].*

It may be pointed out that it is the assumption of some symmetry, i.e., the invariance of Q under a group of isometries, that often causes the intrinsic mean set to contain more than one element (see, e.g., Bhattacharya and Patrangenaru (2003) [20], proposition 2.2.). The next result is therefore expected to be more generally applicable than Theorem 2.4.2.

THEOREM 2.4.3. *(CLT for Cartan means.) Let Q be absolutely continuous w.r.t. the volume measure on an m-dimensional Riemannian manifold* (M, g). *Assume that (i)* μ_g *exists, (ii) the integrability conditions (2.9) hold, (iii) the Hessian matrix* Λ *of* $F \circ \varphi^{-1}$ *at* $\mu = \varphi(\mu_g)$ *is nonsingular and (iv) the covariance matrix* Σ *of* $\Psi(\tilde{X}_i; \mu)$ *is nonsingular. Then*

$$\sqrt{n}(\mu_n - \mu) \to_d \mathcal{N}_m(0, \Gamma), \tag{2.19}$$

where $\Gamma = \Lambda^{-1}\Sigma(\Lambda^T)^{-1}$.

This theorem follows from Theorem 2.4.1 and Remark 2.4.2.
By replacing Λ and Σ with their sample counterparts $\hat{\Lambda}$ and $\hat{\Sigma}$, respectively, and assuming that Σ is nonsingular, one obtains the following:

$$n[\varphi(\mu_n) - \varphi(\mu)]^T(\hat{\Lambda}\hat{\Sigma}^{-1}\hat{\Lambda}^T)[\varphi(\mu_n) - \varphi(\mu)] \xrightarrow{\mathcal{L}} \chi^2_m, \tag{2.20}$$

where χ^2_m is the chi-square distribution with m degrees of freedom. From (2.20) one obtains a confidence region for ν (and of $\mu = \varphi^{-1}(\nu)$).

In order to obtain a confidence region for $\mu_{\mathscr{F}}$ using the CLT in Theorem 2.4.1 in the traditional manner, one needs to estimate the covariance matrix $\Gamma = \Lambda^{-1}\Sigma(\Lambda^T)^{-1}$. For this, one may use proper estimates of Λ and Σ, namely,

$$\hat{\Lambda}(\theta) := \frac{1}{n}\sum_{i=1}^{n}(Grad\Psi)(\tilde{X}_i, \mu_n), \qquad \hat{\Sigma} = Cov\hat{Q}_n^{\varphi},$$
$$\hat{\Gamma} := \hat{\Lambda}^{-1}\hat{\Sigma}(\hat{\Lambda}^t)^{-1}, \qquad \hat{\Gamma}^{-1} = \hat{\Lambda}^t\hat{\Sigma}^{-1}\hat{\Lambda} \tag{2.21}$$

The following corollary is now immediate. Let $\chi^2_{m,1-\alpha}$ denote the $(1-\alpha)$-th quantile of the chi-square distribution with m degrees of freedom.

COROLLARY 2.4.2. *Under the hypothesis of Theorem 2.4.2, if* Σ *is nonsingular, a confidence region for* $\mu_{\mathscr{F}}$ *of asymptotic level* $1-\alpha$ *is given by* $U_{n,\alpha} := \varphi^{-1}(D_{n,\alpha})$, *where* $D_{n,\alpha} = \{\nu \in \varphi(U) : n(\mu_n - \nu)^t\hat{\Gamma}^{-1}(\mu_n - \nu) \leq \chi^2_{m,1-\alpha}\}$.

We now turn to the problem of bootstrapping a confidence region for $\mu_{\mathscr{F}}$. Let $X^*_{i,n}$ be i.i.d. with common distribution $\hat{Q}_n$ (conditionally, given $\{X_i : 1 \leq i \leq n\}$). Write $\tilde{X}^*_{i,n} = \varphi(X^*_{i,n}), 1 \leq i \leq n$ and let μ^*_n be a measurable selection from the Fréchet mean set of $\hat{Q}^{*,\varphi}_n := \frac{1}{n}\sum_{i=1}^{n}\delta_{\tilde{X}^*_{i,n}}$. Let $E^*_{n,\alpha}$ be a subset of

$\varphi(U)$, such that $P^*(\mu_n^* - \mu_n \in E_{n,\alpha}^*) \to 1-\alpha$ in probability, where P^* denotes the probability under $\hat{Q}_n$.

COROLLARY 2.4.3. *In addition to the hypothesis of Theorem 2.4.1, assume C is nonsingular. Then $\varphi^{-1}(\{(\mu_n - E_{n,\alpha}^*) \cap \varphi(U)\})$ is a confidence region for $\mu_{\mathscr{F}}$ of asymptotic level $(1-\alpha)$.*

REMARK 2.4.4. *In Corollary 2.4.3, we have considered the so-called percentile bootstrap of Hall (1997) [58] (also see Efron (1982) [41]), which does not require the computation of the standard error $\hat{\Lambda}$. For this as well as for the CLT-based confidence region given by Corollary 2.4.3, one can show that the coverage error is no more than $O_p(n^{-m/(m+1)})$ or $O(n^{-m/(m+1)})$, as the case may be. One may also use the bootstrap distribution of the pivotal statistic $n(\mu_n-\mu)^T\hat{\Gamma}^{-1}(\mu_n-\mu)$ to find $c_{n,\alpha}^*$ such that*

$$P^*(n(\mu_n^*-\mu_n)^T\hat{\Gamma}^{*-1}(\mu_n^*-\mu_n) \leq c_{n,\alpha}^*) \simeq 1-\alpha, \tag{2.22}$$

to find the confidence region

$$D_{n,\alpha}^* = \{v \in \varphi(U) : n(\mu_n - v)^T\hat{\Gamma}^{-1}(\mu_n - v) \leq c_{n,\alpha}^*\}. \tag{2.23}$$

In particular, if Q has a nonzero absolutely continuous component w.r.t. the volume measure on M, then so does Q^{φ} w.r.t. the Lebesgue measure on $\varphi(U))$ (see do Carmo (1992) [38], p.44). Then assuming (a) $c_{n,\alpha}^*$ is such that the P^*-probability in (2.23) equals $1-\alpha+O_p(n^{-2})$ and (b) some additional smoothness and integrability conditions of the third derivatives of Ψ, one can show that the coverage error, i.e., the difference between $1-\alpha$ and $P(\mu \in D_{n,\alpha}^*)$ is $O_p(n^{-2})$ (see Chandra and Ghosh (1979) [33], Hall (1988) [57], (1997) [58]). It follows that the coverage error of the confidence region $\varphi^{-1}(D_{n,\alpha}^* \cap \varphi(U))$ for $\mu_{\mathscr{F}}$ is also $O(n^{-2})$. We state one such result precisely.

COROLLARY 2.4.4. *(Bootstrapping the Cartan mean) Suppose the hypothesis of Theorem 2.4.3 holds. Then*

$$\begin{aligned}\sup_{r>0} |P^*(n(\mu_n^*-\mu_n)^T\hat{\Gamma}^{*-1}(\mu_n^*-\mu_n) \leq r) - \\ -P(n(\mu_n-\mu)^T\hat{\Gamma}^{-1}(\mu_n-\mu) \leq r)| = O_p(n^{-2}),\end{aligned}$$

and the coverage error of the pivotal bootstrap confidence region is = $O_p(n^{-2})$.

REMARK 2.4.5. *The assumption of absolute continuity of the probability measure Q is reasonable for most applications. Indeed this is assumed in most parametric models in directional and shape analysis (see, e.g., Watson (1983) [141], Dryden and Mardia (2016) [39]).*

REMARK 2.4.6. *The results of this section may be extended to the two-sample problem, or to paired samples, in a fairly straightforward manner. For example, in the case of paired observations* $(X_i, Y_i), i = 1, \cdots, n$, *let* X_i *have (marginal) distribution* Q, *and intrinsic mean* μ_g, *and let* Q_2 *and* ν_g *be the corresponding quantities for* Y_i. *Let* $\varphi = Exp_{x_0}^{-1}$ *for some* x_0, *and let* μ, ν *and* μ_n, ν_n *be the images under* φ *of the intrinsic population and sample means. Then one arrives at the following*

$$\sqrt{n}(\mu_n - \mu) - \sqrt{n}(\nu_n - \nu) \to_d \mathcal{N}_m(0, \Gamma), \tag{2.24}$$

where Γ *is the covariance matrix of* $\Lambda_1^{-1}\Psi(\tilde{X}_i;\mu) - \Lambda_2^{-1}\Psi(\tilde{Y}_i;\nu)$. *Here* Λ_i *is the Hessian matrix of* $F \circ \varphi^{-1}$ *for* $Q_i (i = 1, 2)$. *Assume* Γ *is nonsingular. Then a CLT-based confidence region for* $\gamma := \mu - \nu$ *is given in terms of* $\gamma_n := \mu_n - \nu_n$ *by* $\{v \in \mathbb{R}^m : n(\gamma_n - v)\hat{\Gamma}^{-1}(\gamma_n - v) \leq \chi^2_{m,1-\alpha}\}$. *Alternatively, one may use a bootstrap estimate of the distribution of* $\sqrt{n}(\gamma_n - \gamma)$ *to derive a confidence region.*

2.5 Large Sample Confidence Regions for the Extrinsic Mean

From Theorem 2.4.1 one may derive a CLT for extrinsic sample means similar to Corollary 2.4.2. In this section, however, we use another approach which, for extrinsic means, is simpler to apply and generally less restrictive.

Recall that the extrinsic mean $\mu_{j,E}(Q)$ of a nonfocal probability measure Q on a manifold $\mathcal{M}$ w.r.t. an embedding $j : M \to \mathbb{R}^N$, when it exists, is given by $\mu_{j,E}(Q) = j^{-1}(P_j(\mu))$, where μ is the mean of $j(Q)$ and P_j is the projection on $j(\mathcal{M})$ (see Bhattacharya and Patrangenaru (2003)[20], proposition 3.1, for example). Often the extrinsic mean will be denoted by $\mu_E(Q)$, or simply μ_E, when j and Q are fixed in a particular context. To ensure the existence of the extrinsic mean set, in this section we will assume that $j(\mathcal{M})$ is closed in $\mathbb{R}^N$.

Assume $(X_1, ..., X_n)$ are i.i.d. $\mathcal{M}$-valued random objects whose common probability distribution is Q, and let $\overline{X}_E := \mu_E(\hat{Q}_n)$ be the *extrinsic sample mean*. Here $\hat{Q}_n = \frac{1}{n}\sum_{j=1}^n \delta_{X_j}$ is the empirical distribution. Two different CLTs for the extrinsic sample mean are currently available. The first one concerns a distribution on a *submanifold* $\mathcal{M}$ of $\mathbb{R}^N$ (with j the inclusion map) and was derived by Hendriks and Landsman (1996 [65], 1998 [66]). Independently, Patrangenaru (1998) [104] derived a CLT for distribution on an embedded manifold, using a different estimator. Differentiable manifolds that are not apriori submanifolds of $\mathbb{R}^N$ arise in new areas of data analysis such as in shape analysis, in high level image analysis, or in signal and image processing (see, e.g., Dryden and Mardia (1998) [39], Kendall (1984 [78], 1995 [79]), Kendall et al. (1999) [80], Srivastava and Klassen (2001) [126], Mardia and Patrangenaru (2001) [90]). These manifolds, known under the names of shape spaces and projective shape spaces, are quotient spaces of submanifolds of $\mathbb{R}^N$ (spaces of orbits of actions of Lie groups), rather than submanifolds of $\mathbb{R}^N$. Our approach is a generalization of the adapted frame method of Patrangenaru (1998) [104] to closed embeddings in $\mathbb{R}^N$. This method leads to an appropriate dimension reduction in the CLT and, thereby, reduces computational intensity. This method extends the results of Fisher et al. (1996) [49] who considered the case $M = S^m$.

Assume j is an embedding of an m-dimensional manifold $\mathscr{M}$ such that $j(\mathscr{M})$ is closed in $\mathbb{R}^N$, and Q is a j-nonfocal probability measure on $\mathscr{M}$ such that $j(Q)$ has finite moments of order two (or of sufficiently high order as needed). Let μ and Σ be, respectively, the mean and covariance matrix of $j(Q)$ regarded as a probability measure on $\mathbb{R}^N$. Let $\mathscr{F}$ be the set of focal points of $j(\mathscr{M})$, and let $P_j : \mathscr{F}^c \to j(\mathscr{M})$ be the projection on $j(\mathscr{M})$. P_j is differentiable at μ and has the differentiability class of $j(\mathscr{M})$ around any nonfocal point. In order to evaluate the differential $d_\mu P_j$ we consider a special orthonormal frame field that will ease the computations. Assume $p \to (f_1(p), \dots, f_m(p))$ is a local frame field on an open subset of $\mathscr{M}$ such that, for each $p \in M$, $(d_p j(f_1(p)), \dots, d_p j(f_m(p)))$ are orthonormal vectors in $\mathbb{R}^N$. A local frame field $(e_1(y), e_2(y), \dots, e_N(y))$ defined on an open neighborhood $U \subseteq \mathbb{R}^N$ is *adapted to the embedding* j if it is an orthonormal frame field and

$$e_r(j(p)) = d_p j(f_r(p), \forall r = 1, \dots, m, \forall p \in j^{-1}(U). \tag{2.25}$$

Let $e_1, e_2, \dots, e_N$ be the canonical basis of $\mathbb{R}^N$ and assume $(e_1(y), e_2(y), \dots, e_N(y))$ is an adapted frame field around $P_j(\mu) = j(\mu_E)$. Then $d_\mu P_j(e_b) \in T_{P_j(\mu)} j(\mathscr{M})$ is a linear combination of $e_1(P_j(\mu)), e_2(P_j(\mu)), \dots, e_m(P_j(\mu))$:

$$d_\mu P_j(e_b) = \sum_{a=1}^{m} \left(d_\mu P_j(e_b)\right) \cdot e_a\left(P_j(\mu)\right) e_a\left(P_j(\mu)\right), \forall b = 1, \dots, N. \tag{2.26}$$

By the delta method, $n^{\frac{1}{2}}(P_j(\overline{j(X)}) - P_j(\mu))$ converges weakly to a random vector V having a $\mathscr{N}_N(0, \Sigma_\mu)$ distribution. Here $\overline{j(X)} = \frac{1}{n}\sum_{i=1}^{n} j(X_i)$ and

$$\Sigma_\mu = \left[\sum_{a=1}^{m} d_\mu P_j(e_b) \cdot e_a(P_j(\mu)) e_a(P_j(\mu))\right]_{b=1,\dots,N} \Sigma \left[\sum_{a=1}^{m} d_\mu P_j(e_b) \cdot e_a(P_j(\mu)) e_a(P_j(\mu))\right]_{b=1,\dots,N}^T, \tag{2.27}$$

where Σ is the covariance matrix of $j(X_1)$ w.r.t. the canonical basis $e_1, \dots, e_N$. The asymptotic distribution $\mathscr{N}_N(0, \Sigma_\mu)$ is degenerate and can be regarded as a distribution on $T_{P_j(\mu)} j(\mathscr{M})$, since the range of $d_\mu P_j$ is $T_{P_j(\mu)} j(\mathscr{M})$. Note that

$$d_\mu P_j(e_b) \cdot e_a(P_j(\mu)) = 0, \quad \text{for} \quad a = m+1, \dots, N.$$

REMARK 2.5.1. *An asymptotic distribution of the extrinsic sample mean can be obtained as a particular case of Theorem 2.4.1. The covariance matrix in that theorem depends both on the way the manifold is embedded and on the chart used. We provide below an alternate CLT, which applies to an arbitrary embedding, leading to pivots and are independent of the chart used.*

The tangential component $tan(v)$ of $v \in \mathbb{R}^N$ w.r.t. the basis $e_a(P_j(\mu)) \in T_{P_j(\mu)}j(\mathcal{M}), a = 1,\dots,m$ is given by

$$tan(v) = (e_1(P_j(\mu))^T v \dots e_m(P_j(\mu))^T v)^T. \tag{2.28}$$

Then the random vector $(d_{\mu_E} j)^{-1}(tan(P_j(\overline{j(X)}) - P_j(\mu)))$ has the following covariance matrix w.r.t. the basis $f_1(\mu_E),\cdots,f_m(\mu_E)$:

$$\Sigma_{j,E} = (e_a(P_j(\mu))^T \Sigma_\mu e_b(P_j(\mu)))_{1\le a,b\le m} =$$
$$\left[\sum d_\mu P_j(e_b)\cdot e_a(P_j(\mu))\right]_{a=1,\dots,m} \Sigma \left[\sum d_\mu P_j(e_b)\cdot e_a(P_j(\mu))\right]^T_{a=1,\dots,m}. \tag{2.29}$$

DEFINITION 2.5.1. *The matrix $\Sigma_{j,E}$ given by (13.2) is the* extrinsic covariance matrix *of the j-nonfocal distribution Q (of X_1) w.r.t. the basis$f_1(\mu_E),\dots,f_m(\mu_E)$.*

When j is fixed in a specific context, the subscript j in $\Sigma_{j,E}$ may be omitted. If, in addition, rank $\Sigma_\mu = m$, Σ_E is invertible and we define the *j-standardized mean vector*

$$\overline{Z}_{j,n} =: n^{\frac{1}{2}} \Sigma_E^{-\frac{1}{2}} (\overline{X}^1{}_j \dots \overline{X}^m{}_j)^T. \tag{2.30}$$

PROPOSITION 2.5.1. *Assume $\{X_r\}_{r=1,\dots,n}$ are i.i.d.r.o.s from the j-nonfocal distribution Q, with finite mean $\mu = E(j(X_1))$, and assume the extrinsic covariance matrix $\Sigma_{j,E}$ of Q is finite. Let $(e_1(y),e_2(y),\dots,e_N(y))$ be an orthonormal frame field adapted to j. Then (a) the extrinsic sample mean $\overline{X}_{j,E}$ has asymptotically a normal distribution in the tangent space to $\mathcal{M}$ at $\mu_{j,E}(Q)$ with mean 0 and covariance matrix $n^{-1}\Sigma_{j,E}$, and (b) if $\Sigma_{j,E}$ is nonsingular, the j-standardized mean vector $\overline{Z}_{j,n}$ given in (2.30) converges weakly to a $\mathcal{N}_m(0_m, I_m)$-distributed random vector.*

As a particular case of Proposition 2.5.1, when j is the inclusion map of a submanifold of $\mathbb{R}^N$, we get the following result for nonfocal distributions on an arbitrary closed submanifold $\mathcal{M}$ of $\mathbb{R}^N$:

COROLLARY 2.5.1. *Assume $M \subseteq \mathbb{R}^N$ is a closed m-dimensional submanifold of $\mathbb{R}^N$. Let $\{X_r\}_{r=1,\dots,n}$ be i.i.d.r.o.'s from the nonfocal distribution Q on $\mathcal{M}$ with the finite mean vector $\mu = E(X_1)$ and covariance matrix Σ. Let $(e_1(y),e_2(y),\dots,e_N(y))$ be an orthonormal frame field adapted to $\mathcal{M}$. Let $\Sigma_E := \Sigma_{j,E}$, where $j : M \to \mathbb{R}^N$ is the inclusion map. Then (a) $n^{\frac{1}{2}} tan(j(\overline{X}_E) - j(\mu_E))$ converges weakly to $\mathcal{N}_m(0_m, \Sigma_E)$, and (b) if Σ induces a nonsingular bilinear form on $T_{j(\mu_E)}j(\mathcal{M})$, then $\|\overline{Z}_{j,n}\|^2$ in (2.30) converges weakly to the chi-square distribution χ^2_m.*

REMARK 2.5.2. *The CLT for extrinsic sample means as stated in Proposition 2.5.1 or Corollary 2.5.1 cannot be used to construct confidence regions for extrinsic means, since the population extrinsic covariance matrix is unknown. In order to find a consistent estimator of* $\Sigma_{j,E}$*, note that* $\overline{j(X)}$ *is a consistent estimator of* μ*,* $d_{\overline{j(X)}}P_j$ *converges in probability to* $d_\mu P_j$*, and* $e_a(P_j(\overline{j(X)}))$ *converges in probability to* $e_a(P_j(\mu))$ *and, further,*

$$S_{j,n} = n^{-1}\sum(j(X_r) - \overline{j(X)})(j(X_r) - \overline{j(X)})^T$$

is a consistent estimator of Σ*. It follows that*

$$\left[\sum_{a=1}^{m} d_{\overline{j(X)}}P_j(e_b)\cdot e_a(P_j(\overline{j(X)}))e_a(P_j(\overline{j(X)}))\right] S_{j,n}$$
$$\left[\sum_{a=1}^{m} d_{\overline{j(X)}}P_j(e_b)\cdot e_a(P_j(\overline{j(X)}))e_a(P_j(\overline{j(X)}))\right]^T \tag{2.31}$$

is a consistent estimator of Σ_μ*, and* $tan_{P_j(\overline{j(X)})}v$ *is a consistent estimator of* $tan(v)$.

If we take the components of the bilinear form associated with the matrix (2.31) w.r.t. $e_1(P_j(\overline{j(X)})), e_2(P_j(\overline{j(X)})), ..., e_m(P_j(\overline{j(X)}))$, we get a consistent estimator of $\Sigma_{j,E}$, called *the sample extrinsic covariance matrix*, given by

$$S_{j,E,n} = \left[\left[\sum d_{\overline{j(X)}}P_j(e_b)\cdot e_a(P_j(\overline{j(X)}))\right]_{a=1,\dots,m}\right]\cdot S_{j,n}$$
$$\left[\left[\sum d_{\overline{j(X)}}P_j(e_b)\cdot e_a(P_j(\overline{j(X)}))\right]_{a=1,\dots,m}\right]^T, \tag{2.32}$$

and obtain the following results:

THEOREM 2.5.1. *(Adapted frame CLT) Assume* $j : M \to \mathbb{R}^N$ *is a closed embedding of* $\mathscr{M}$ *in* $\mathbb{R}^N$*. Let* $\{X_r\}_{r=1,\dots,n}$ *be a random sample from the j-nonfocal distribution Q, and let* $\mu = E(j(X_1))$ *and assume* $j(X_1)$ *has finite second-order moments and the extrinsic covariance matrix* $\Sigma_{j,E}$ *of* X_1 *is nonsingular. Let* $(e_1(y), e_2(y), \dots, e_N(y))$ *be an orthonormal frame field adapted to j. If the* sample extrinsic covariance matrix $S_{j,E,n}$ *is given by* (13.4), *then for n large enough* $S_{j,E,n}$ *is nonsingular (with probability converging to one) and (a) the statistic*

$$n^{\frac{1}{2}}S_{j,E,n}^{-\frac{1}{2}}tan(P_j(\overline{j(X)}) - P_j(\mu)) \tag{2.33}$$

converges weakly to $N_m(0_m, I_m)$*, so that the following*

$$n\|S_{j,E,n}^{-\frac{1}{2}}tan(P_j(\overline{j(X)}) - P_j(\mu))\|^2 \tag{2.34}$$

converges weakly to χ^2_m, *and (b) the statistic*

$$n^{\frac{1}{2}} S^{-\frac{1}{2}}_{j,E,n} tan_{P_j(\overline{j(X)})}(P_j(\overline{j(X)}) - P_j(\mu)) \tag{2.35}$$

converges weakly to $N_m(0_m, I_m)$, *so that*

$$n \| S^{-\frac{1}{2}}_{j,E,n} tan_{P_j(\overline{j(X)})}(P_j(\overline{j(X)}) - P_j(\mu)) \|^2 \tag{2.36}$$

converges weakly to χ^2_m.

COROLLARY 2.5.2. *Under the hypothesis of Theorem 2.5.1, a confidence region for* μ_E *of asymptotic level* $1-\alpha$ *is given by (a)* $C_{n,\alpha} := j^{-1}(U_{n,\alpha})$, *where* $U_{n,\alpha} = \{\nu \in j(\mathcal{M}) : n\|S^{-\frac{1}{2}}_{j,E,n} tan_\nu(P_j(\overline{j(X)}) - \nu)\|^2 \le \chi^2_{m,1-\alpha}\}$, *or by (b)* $D_{n,\alpha} := j^{-1}(V_{n,\alpha})$, *where* $V_{n,\alpha} = \{\nu \in j(\mathcal{M}) : n\|S^{-\frac{1}{2}}_{j,E,n} tan_{P_j(\overline{j(X)})}(P_j(\overline{j(X)}) - \nu)\|^2 \le \chi^2_{m,1-\alpha}\}$.

REMARK 2.5.3. *To simplify the notation,* $\forall a = 1,\dots n$, *let* $Y_a = j(X_a)$. *Then, if* $\bar{Y}_n$ *is the sample mean vector of the i.i.d.r.vec.'s* $Y_1,\dots,Y_n$,

$$\sqrt{n}[P_j(\bar{Y}_n) - P_j(\mu)] = \sqrt{n}[(DP_j)_\mu(\bar{Y}_n - \mu)] + o_P(1) \to_d \mathcal{N}_N(0, \Sigma_\mu) \; as n \to \infty, \tag{2.37}$$

where, from (2.27), *we have*

$$\Sigma_\mu = (DP_j)_\mu \Sigma (DP_j)^T_\mu. \tag{2.38}$$

In Eq. (2.38), $(DP_j)_y$ *is the* $N \times N$ *Jacobian matrix of* P_j *at* y.

Given that $P_j : U \subset \mathbb{R}^N \to j(\mathcal{M})$ *is a differentiable function between manifolds (here* U *is an open subset of the set of nonfocal points), it follows that the differential* $d_\mu P_j$ *maps* $\mathbb{R}^N$ *to* $T_{P_j(\mu)}(j(\mathcal{M}))$, *regarded as a subspace of* $T_{P_j(\mu)}(\mathbb{R}^N) = \mathbb{R}^N$. *For this reason it was convenient to express the values of this differential on the canonical basis of* $T_{P_j(\mu)}(\mathbb{R}^N) = \mathbb{R}^N$ *as linear combinations of a basis in of* $T_{P_j(\mu)}(j(\mathcal{M}))$. *Therefore, in* (2.27), *we selected a convenient local section of the bundle of orthonormal frames of* $\mathcal{M}$, *say* $y \to (e_1(y), e_2(y), \dots, e_m(y))$ *of the subspace* $T_y(j(\mathcal{M})) \subset T_y(\mathbb{R}^N) = \mathbb{R}^N$ *(for* y *in a neighborhood of* $P_j(\mu)$ *in* $j(\mathcal{M})$*). The following proposition then follows from* (2.37).

PROPOSITION 2.5.2. *Assume* Q *is* j *nonfocal, and* $E\|Y_1\|^2 < \infty$. *Let* B *denote the* $m \times N$ *matrix whose rows* $e_i^T(P_j(\mu)), i = 1,\dots,m$, *form an orthonormal basis of* $T_{P_j(\mu)}(j(\mathcal{M}))$ *regarded as a subspace of* R^N. *Then*

$$\sqrt{n}[B(P_j(\bar{Y}_n) - P_j(\mu))] \to_d \mathcal{N}_m(0_m, \Sigma_E) as\; n \to \infty. \tag{2.39}$$

Also, writing B_n *for the* $m \times N$ *matrix whose rows are* $e_i^T(P_j(\bar{Y}_n)), i = 1,\dots,m$, *one has*

$$\sqrt{n}[B_n(P_j(\bar{Y}_n) - P_j(\mu))] \to_d \mathcal{N}_m(0_m, \Sigma_E) \; as\; n \to \infty. \tag{2.40}$$

Proof. *Equation* (2.39) *follows from Proposition 2.5.1, while* (2.40) *follows from* (2.39) *and a Slutsky type argument.*

REMARK 2.5.4. *The* extrinsic covariance matrix *of Q w.r.t. the orthogonal basis $e_1(P_j(\mu)),\dots,e_m(P_j(\mu))$ is $\Sigma_E = B\Sigma_\mu B^T$. Note that* (2.39) *expresses the asymptotic multivariate normal distribution $N_m(0_m, n^{-1}\Sigma_E)$ of the coordinates $(V_{1n},\dots,V_{mn})^T$ of the random vector $tan(P_j(\bar{Y}_n) - P_j(\mu))$ with respect to the basis $e_a^T(P_j(\mu)), a = 1,\dots,m$. Under broad conditions, the extrinsic covariance matrix Σ_E is a nonsingular $m \times m$ matrix. For this, it is sufficient to require that the distribution of $tanP_j(Y_1)$ does not have support on a $(m-1)$-dimensional subspace of $T_{P_j(\mu)}(j(\mathcal{M}))$.*

The following is a consequence of Theorem 2.5.1:

THEOREM 2.5.2. *Let the hypothesis of Theorem 2.5.1 hold. Then (a) one has the following asymptotic of the* extrinsic Hotelling statistic*:*

$$\sqrt{n}[B_n(P_j(\bar{Y}_n) - P_j(\mu))] \to_d V \sim \mathcal{N}_m(0,\Sigma_E) \text{ as } n \to \infty, \text{ and}$$
$$n[B_n(P_j(\bar{Y}_n) - P_j(\mu))]^T \hat{(}S_{j,E})^{-1}[B_n(P_j(\bar{Y}_n) - P_j(\mu))] \to_d T \sim \chi^2_m, \quad \text{as } n \to \infty. \tag{2.41}$$

Here $S_{j,E}$, is the sample extrinsic covariance matrix *given by*

$$S_{j,E} = B_n(DP)_{\bar{Y}_n}\hat{\Sigma}(DP)^T_{\bar{Y}_n}B_n^T, \tag{2.42}$$

$\hat{\Sigma}$ being the sample covariance matrix of $Y_i, i = 1,\dots,n$. Also, (b) a confidence region for $P_j(\mu)$ of asymptotic level $1-\alpha$ is given by

$$C_{1-\alpha} = = \{\nu \in j(\mathcal{M}), n[B_n(P_j(\bar{Y}_n) - \nu)]^T S_{j,E}{}^{-1}[B_n(P_j(\bar{Y}_n) - \nu)] \le \le \chi^2_m(1-\alpha). \tag{2.43}$$

Theorem 2.5.1 and Corollary 2.5.2 involve pivotal statistics. The advantages of using pivotal statistics in bootstrapping for confidence regions are well known (see, e.g., Babu and Singh (1984) [5], Beran (1987) [11], Hall (1988) [57], (1992) [58]).

At this point we recall the steps that one takes to obtain a bootstrapped statistic from a pivotal statistic. If $\{X_r\}_{r=1,\dots,n}$ is a random sample from the unknown distribution Q, and $\{X_r^*\}_{r=1,\dots,n}$ is a random sample from the empirical $\hat{Q}_n$, conditionally given $\{X_r\}_{r=1,\dots,n}$, then the statistic

$$T(X,Q) = n\|S_{j,E,n}^{-\frac{1}{2}} tan(P_j(\overline{j(X)}) - P_j(\mu))\|^2$$

given in Theorem 2.5.1.(a) has the bootstrap analog

$$T(X^*,\hat{Q}_n) = n\|S^*_{j,E,n}{}^{-\frac{1}{2}} tan_{P_j(\overline{j(X)}))}(P_j(\overline{j(X^*)}) - P_j(\overline{j(X)}))\|^2.$$

Here $S^*_{j,E,n}$ is obtained from $S_{j,E,n}$ substituting $X_1^*,.....,X_n^*$ for $X_1,.....X_n$, and $T(X^*,\hat{Q}_n)$ is obtained from $T(X,Q)$ by substituting $X_1^*,.....,X_n^*$ for $X_1,....,X_n$, $\overline{j(X)})$ for μ and $S^*_{j,E,n}$ for $S_{j,E,n}$.

The same procedure can be used for the vector valued statistic

$$V(X,Q) = n^{\frac{1}{2}} S_{j,E,n}^{-\frac{1}{2}} tan(P_j(\overline{j(X)}) - P_j(\mu)),$$

and as a result we get the bootstrapped statistic

$$V^*(X^*,\hat{Q}_n) = n^{\frac{1}{2}} {S^*_{j,E,n}}^{-\frac{1}{2}} tan_{P_j(\overline{j(X)}))}(P_j(\overline{j(X^*)}) - P_j(\overline{j(X)})).$$

For the rest of this section, we will assume that $j(Q)$, when viewed as a measure on the ambient space $\mathbb{R}^N$, *has finite moments of sufficiently high order*. If $\mathcal{M}$ is compact, then this is automatic. In the noncompact case, finiteness of moments of order twelve, along with an assumption of a nonzero absolutely continuous component, is sufficient to ensure an Edgeworth expansion up to order $O(n^{-2})$ of the pivotal statistic $V(X,Q)$ (see Hall (1988) [57], Fisher et al. (1996) [49] and Chandra and Ghosh (1979) [33]). We then obtain the following results:

THEOREM 2.5.3. *Let $\{X_r\}_{r=1,...,n}$ be a random sample from the j-nonfocal distribution Q which has a nonzero absolutely continuous component w.r.t. the volume measure on $\mathcal{M}$ induced by j. Let $\mu = E(j(X_1))$ and assume the covariance matrix Σ of $j(X_1)$ is defined and the extrinsic covariance matrix $\Sigma_{j,E}$ is nonsingular and let $p \to (e_1(p), e_2(p), \dots, e_N(p))$ be an orthonormal frame field adapted to j. Then the distribution function of*

$$n\|S_{j,E,n}^{-\frac{1}{2}} tan(P_j(\overline{j(X)}) - P_j(\mu))\|^2$$

can be approximated by the bootstrap extrinsic Hotelling distribution of

$$n\|{S^*_{j,E,n}}^{-\frac{1}{2}} tan_{P_j(\overline{j(X)})}(P_j(\overline{j(X^*)}) - P_j(\overline{j(X)}))\|^2$$

with a coverage error $0_p(n^{-2})$.

One may also use nonpivotal bootstrap confidence regions, especially when $S_{j,E,n}$ is difficult to compute. The result in this case is the following .

PROPOSITION 2.5.3. *Under the hypothesis of Proposition* 3.1, *the distribution function of $n\|tan(P_j(\overline{j(X)}) - P_j(\mu))\|^2$ can be approximated uniformly by the bootstrap distribution of*

$$n\|tan_{P_j(\overline{j(X)})}(P_j(\overline{j(X^*)}) - P_j(\overline{j(X)}))\|^2$$

to provide a confidence region for μ_E with a coverage error no more than $O_p(n^{-\frac{m}{m+1}})$.

REMARK 2.5.5. *Note that Corollary 2.5.1(b) provides a computationally simpler scheme than Corollary 2.5.1(a) for large sample confidence regions; but for bootstrap confidence regions Theorem 2.5.3, which is the bootstrap analog of Corollary 2.5.1(a), yields a simpler method. The corresponding* $100(1-\alpha)\%$ *confidence region is* $C^*_{n,\alpha} := j^{-1}(U^*_{n,\alpha})$ *with* $U^*_{n,\alpha}$ *given by*

$$U^*_{n,\alpha} = \{\mu \in j(\mathcal{M}) : n\|S_{j,E,n}^{*-\frac{1}{2}} tan(P_j(\overline{j(X)}) - P_j(\mu))\|^2 \le c^*_{1-\alpha}\}, \tag{2.44}$$

where $c^*_{1-\alpha}$ *is the upper* $100(1-\alpha)\%$ *point of the values*

$$n\|S^{*}_{j,E,n}{}^{-\frac{1}{2}} tan_{P_j(\overline{j(X)})}(P_j(\overline{j(X^*)}) - P_j(\overline{j(X)}))\|^2 \tag{2.45}$$

among the bootstrap resamples. One could also use the bootstrap analog of the confidence region given in Corollary 2.5.1(b) for which the confidence region is $D^*_{n,\alpha} := j^{-1}(V^*_{n,\alpha})$ *with* $V^*_{n,\alpha}$ *given by*

$$V^*_{n,\alpha} = \{\mu \in j(\mathcal{M}) : n\|S_{j,E,n}^{*-\frac{1}{2}} tan_{P_j(\overline{j(X)})}(P_j(\overline{j(X)}) - P_j(\mu))\|^2 \le d^*_{1-\alpha}\}, \tag{2.46}$$

where $d^*_{1-\alpha}$ *is the upper* $100(1-\alpha)\%$ *point of the values*

$$n\|S^{*}_{j,E,n}{}^{-\frac{1}{2}} tan_{P_j(\overline{j(X^*)})}(P_j(\overline{j(X^*)}) - P_j(\overline{j(X)}))\|^2 \tag{2.47}$$

among the bootstrap resamples. The region given by (2.46) *and* (2.47) *has coverage error* $O_p(n^{-2})$.

2.6 Two-Sample Test for Total Extrinsic Variances

In this short section, we present a large samples test for comparing the total intrinsic variances in two independent populations, due to A. Bhattacharya (2008) [14]. Here we will use the asymptotic distribution of the sample total extrinsic variance to construct nonparametric tests to compare two probability distributions Q_1 and Q_2 on an embedded manifold $\mathcal{M}$.

Let $X_1,\dots,X_{n_1}$ and $Y_1,\dots,Y_{n_2}$ be two iid samples from Q_1 and Q_2, respectively, that are mutually independent. Let $\mu_{i,E}$ and V_i denote the extrinsic means and total variances of $Q_i, i = 1,2$, respectively. Similarly denote by $\hat{\mu}_{i,E}$ and $\hat{V}_i$ the sample extrinsic means and total variances. We want to test the hypothesis $H_0 : V_1 = V_2 + \delta$. From the fact that the samples are independent, and the asymptotic distribution of the total sample extrinsic variance if $n = n_1 + n_2, \frac{n_a}{n} \to p_a, a = 1,2$ as $n \to \infty$, we get that, under H_0,

$$\sqrt{n}(\hat{V}_1 - \hat{V}_2) \to_d Y, Y \sim \mathcal{N}(\delta, \frac{\sigma_1^2}{p_1} + \frac{\sigma_2^2}{p_2}), \tag{2.48}$$

$$\frac{\hat{V}_1 - \hat{V}_2 - \delta}{\sqrt{\frac{s_1^2}{n_1} + \frac{s_2^2}{n_2}}} \rightarrow_d Z, Z \sim \mathcal{N}(0,1), \tag{2.49}$$

where $\sigma_1^2 = Var[d^2(X_1, \mu_{1,E})], \sigma_2^2 = Var[d^2(X_2, \mu_{2,E})]$ and s_1^2, s_2^2 are their sample estimates. Hence to test if H_0 is true, we can use the test statistic

$$T = \frac{\hat{V}_1 - \hat{V}_2 - \delta}{\sqrt{\frac{s_1^2}{n_1} + \frac{s_2^2}{n_2}}}. \tag{2.50}$$

For a test of asymptotic size α, we reject H_0 if $|T| > z_{\frac{\alpha}{2}}$.

REMARK 2.6.1. *One can also construct a bootstrap confidence interval for $V_1 - V_2 - \delta$ and use that to test $H_0 : V_1 - V_2 = \delta$, for example by taking for $a = 1,2$ bootstrap resamples $X^*_{a,i_a}, i_a = 1,\ldots,n_a$, from each of the two samples $X_{a,i_a}, i_a = 1,\ldots,n_a$, and compute the pivotal bootstrap statistics*

$$T^* = \frac{\hat{V}_1^* - \hat{V}_2^* - \delta}{\sqrt{\frac{s_1^{2*}}{n_1} + \frac{s_2^{2*}}{n_2}}}, \tag{2.51}$$

and, for a large number N of resamples, compute the upper and lower $\frac{\alpha}{2}$ the bootstrap percentiles $T^*_{1-\frac{\alpha}{2}}, T^*_{\frac{\alpha}{2}}$. The hypothesis is rejected at level α if 0 is not in the interval $(T^*_{1-\frac{\alpha}{2}}, T^*_{\frac{\alpha}{2}})$.

2.7 Bhattacharya's Two-Sample Test for Extrinsic Means of Independent Populations on a Manifold

We now turn to two-sample tests for extrinsic means of distributions on an arbitrary m-dimensional manifold M given in Bhattacharya and Patrangenaru (2014) [22]. There is a large sample test for equality of two extrinsic means, originally due to A. Bhattacharya (2008) [14] (see also Bhattacharya and Bhattacharya (2012) [15], p.42). Let $X_{ak_a} : k_a = 1,\ldots,n_a, a = 1,2$ be two independent random samples drawn from distributions $Q_a, a = 1,2$ on M, and let j be an embedding of M into $\mathbb{R}^N$. Denote by μ_a the mean of the induced probability $Q_a \circ j^{-1}$ and Σ_a its covariance matrix $(a = 1,2)$. Then the extrinsic mean of Q_a is $\mu_{a,j} = j^{-1}(P_j(\mu_a))$, assuming Q_a is nonfocal. Write $Y_{ak_a} = j(X_{ak_a}) k_a = 1,\ldots,n_a, a = 1,2$ and let $\bar{Y}_a, a = 1,2$ be the corresponding sample means. Assuming finite second moments of $Y_{a1}, a = 1,2$, which is automatic if M is compact, one has, by Theorem 2.5.2

$$\sqrt{n_a} B_a [P_j(\bar{Y}_a) - P_j(\mu_a)] \rightarrow_d \mathcal{N}_m(0, \Sigma_{a,j}), a = 1,2, \tag{2.52}$$

where $\Sigma_{a,j}$ is the extrinsic covariance matrix of Q_a, and B_a are the same as in Proposition 2.5.2 and Theorem 2.5.2, but with Q replaced by Q_a (a=1,2). That is, $B_a(y)$ is the $m \times N$ matrix of an orthonormal basis (frame) of $T_y(j(M)) \subset T_y(\mathbb{R}^N) = \mathbb{R}^N$ for y in a neighborhood of $P_j(\mu_a)$, and $B_a = B_a(P_j(\mu_a))$. Similarly, $C_a \doteq \Sigma_{\mu_a,a} = (DP_j)_{\mu_a}\Sigma_a(DP_j)^T_{\mu_a}, (a = 1,2)$. The null hypothesis $H_0 : \mu_{1,j} = \mu_{2,j}$, say, is equivalent to $H_0 : P_j(\mu_1) = P_j(\mu_2) = \pi$, say. Then, under the null hypothesis, letting $B = B(\pi)$, one has $B_1 = B_2 = B$, and

$$[B(\frac{1}{n_1}C_1 + \frac{1}{n_2}C_2)B^T]^{-1/2}B[P_j(\bar{Y}_1) - P_j(\bar{Y}_2)] \to_d \mathcal{N}_m(0_m, I_m),$$
$$\text{as } n_1 \to \infty, n_2 \to \infty, \text{and} \frac{n_1}{n_1+n_2} \to \lambda \in (0,1). \tag{2.53}$$

For statistical inference one estimates C_a by

$$\hat{C}_a = (DP_j)_{\bar{Y}_a}\hat{\Sigma}_a(DP_j)^T_{\bar{Y}_a} \tag{2.54}$$

where $\hat{\Sigma}_a$ is the sample covariance matrix of sample $a(a = 1,2)$. Also B is replaced by $\hat{B} = B(\hat{\pi})$ where $\hat{\pi}$ is a sample estimate of π. Under H_0, both $P_j(\bar{Y}_1)$ and $P_j(\bar{Y}_2)$ are consistent estimates of π, so we take a "pooled estimate"

$$\hat{\pi} = P_j(\frac{1}{n_1+n_2}(n_1P_j(\bar{Y}_1) + n_2P_j(\bar{Y}_2))). \tag{2.55}$$

We, therefore, have the following result due to Bhattacharya and Bhattacharya (2012) [15].

THEOREM 2.7.1. *Assume the extrinsic sample covariance matrix $\hat{\Sigma}_{a,j}$ is nonsingular for $a = 1,2$. Then, under $H_0 : \mu_{1,j} = \mu_{2,j}$, one has:*

$$(\hat{B}[P_j(\bar{Y}_1) - P_j(\bar{Y}_2)])^T[\hat{B}(\frac{1}{n_1}\hat{C}_1 + \frac{1}{n_2}\hat{C}_2)\hat{B}^T]^{-1}(\hat{B}[P_j(\bar{Y}_1) - P_j(\bar{Y}_2)])$$
$$\to_d \chi^2_m, \tag{2.56}$$

$$as\ n_1 \to \infty, n_2 \to \infty and \frac{n_1}{n_1+n_2} \to \lambda \in (0,1)\square$$

For the two-sample intrinsic test, given a Riemannian structure g on $\mathcal{M}$, let $\mu_{1,g}, \mu_{2,g}$ denote the intrinsic means of Q_1 and Q_2, assumed to exist, and consider $H_0 : \mu_{1,g} = \mu_{2,g}$. Denoting by $\hat{\mu}_{1,g}, \hat{\mu}_{2,g}$ the sample intrinsic means, (2.18) implies that, under H_0,

$$(\frac{1}{n_1}\Lambda_1^{-1} + \frac{1}{n_2}\Lambda_2^{-1})^{-1/2}[\varphi_p(\hat{\mu}_{1,g}) - \varphi_p(\hat{\mu}_{2,g})] \to_d \mathcal{N}_m(0, I_m),$$
$$\text{as } n_1 \to \infty, n_2 \to text and \frac{n_1}{n_1+n_2} \to \lambda \in (0,1)\infty, \tag{2.57}$$

where $\varphi_p = Exp_p^{-1}$ for some convenient p in $\mathscr{M}$, and Λ_a, Σ_a are as in Theorem 2.4.3 with Q_a in place of $Q(a = 1,2)$. One simple choice for p is the pooled estimate lying on the distance minimizing geodesic connecting $\mu_{1,g}, \mu_{2,g}$ at a distance $\frac{n_2}{n}\rho_g(\mu_{1,g}, \mu_{2,g})$ from $\mu_{1,g}$. With this choice, we write $\hat{\varphi}$ for φ_p. Let $\hat{\Sigma}_a$ and $\hat{\Lambda}_a$ be the same as $\hat{\Sigma}, \hat{\Lambda}$ in Theorem 2.4.3 and Corollary 2.4.2, but using the a-th sample and with $\hat{\varphi}$ for $\varphi(a = 1,2)$. We thus arrive at the following:

THEOREM 2.7.2. *The test to reject* $H_0 : \mu_{1,g} = \mu_{2,g}$ *iff*

$$[\hat{\varphi}(\hat{\mu}_{1,g}) - \hat{\varphi}(\hat{\mu}_{2,g})]^T \tilde{\Sigma}_p^{-1} [\hat{\varphi}(\hat{\mu}_{1,g}) - \hat{\varphi}(\hat{\mu}_{2,g})] > \chi_m^2(1-\alpha) \tag{2.58}$$

has asymptotic level of significance α, *where*

$$\tilde{\Sigma}_p = [(\frac{1}{n_1}\hat{\Lambda}_1^{-1}\hat{\Sigma}_1\Lambda_1^{-1} + \frac{1}{n_2}\hat{\Lambda}_2^{-1}\hat{\Sigma}_2\Lambda_2^{-1})]$$

REMARK 2.7.1. *In an effort to address the problem of testing for the equality of extrinsic means in matched pairs on arbitrary manifolds, in subsection 4.6.2 of Bhattacharya and Bhattacharya (2012) [15], one considers the more general case of correlated random objects (r.o.s). The matching there does not seem to play a role in the methodology developed, since the test statistic is obtained by first independently projecting the sample means* $j(\bar{X}_1)$ *and* $j(\bar{X}_2)$*. Such a choice ignores the matching, given that applying a permutation in the sample indices of the observations in the second sample, say, does not affect the value of the statistic* $\bar{Y}_2$*, and ultimately of the test statistic, so essentially only the sample means are paired, not the values* $X_{a,1}, X_{a,2}$ *of individual observations* $a = 1, \ldots, n$*. Note that while in the multivariate case, the mean of the difference is the difference of the means, for an arbitrary manifold valued data, individual differences have no meaning, and by pairing* $j(\bar{X}_1)$ *and* $j(\bar{X}_2)$*, one ignores the within samples matching. While the extension of a to sample problem for correlated r.o.s will be addressed in a series of exercises at the end of this chapter, in the next section, we will take a different approach of extending the classical methodology for hypothesis testing for matched pairs that also allows for null hypotheses when the means are not necessarily equal.*

2.8 Test for Mean Change in Matched Pairs on Lie Groups

In some particular cases when there is a Lie group $\mathscr{G}$ acting on the manifold $\mathscr{M}$, such that for each pair of points $p_1, p_2 \in \mathscr{M}$, one may find a unique element $g = g(p_1, p_2) \in \mathscr{G}$, with $g \cdot p_1 = p_2$, given matched samples $(X_i, Y_i), i = 1, \ldots, n$ from two distributions on $\mathscr{M}$, one may associate a random sample $G_i = g(X_i, Y_i), i = 1, \ldots, n$ on $\mathscr{G}$, thus moving the matched sample analysis on $\mathscr{G}$ (see Osborne et al. (2013) [102]). A general example of this kind is provided by the left action or by the adjoint action of a Lie group $\mathscr{G}$ on itself. However, this approach may be extended to the more general case, when the existence of the element unique $g = g(p_1, p_2) \in \mathscr{G}$, with $g \cdot p_1 = p_2$, is ensured only for pairs (p_1, p_2) of points that are not too far apart. This approach was used for example in Beran and Fisher (1998) [13] with $\mathscr{M} = \mathbb{R}P^2$, and $\mathscr{G} = SO(3)$, the group of rotations in $\mathbb{R}^3$, and in Mardia and Patrangenaru (2005) [91] with $\mathscr{M} = P\Sigma_2^k$, and $\mathscr{G} = SO(3)^{k-4}$, thus reducing the two-sample test for extrinsic mean multiple axes to a one-sample test for extrinsic mean multiple rotations. The drawback in Mardia and Patrangenaru (2005) [91] is that such an analysis leads to an increase in dimensionality, forcing the extrinsic sample covariance matrices to be degenerated, when the sample sizes are small, which is an impediment in computing pivotal statistics that are preferred for a lower coverage error.

Recall that the case of paired random vectors X, Y in $\mathbb{R}^m$, comparing their mean vectors, is usually done in terms of the difference vector $D = Y - X$ to eliminate much of the influence of extraneous unit to unit variation (Johnson and Wichern (2007 [76], p. 274)), without increasing the dimensionality. Similarly, for matched pairs on the unit circle $Z_1, Z_2 \in S^1$ (matched circular data), rather than testing for the equality $\mu_{j,1} = \mu_{j,2}$ of the two extrinsic means, one may test if the extrinsic mean of $\bar{Z}_1 Z_2$ is the unit of S^1. In both these cases, one deals with matched paired data on a commutative Lie group. In this section we extend these techniques to paired random objects on an arbitrary *Lie group* that is not necessarily commutative.

Assume X and Y are paired *r.o.s on a Lie group*$(\mathscr{G}, \circ)$ (see Section 4.1). The *change from X to Y* was defined by Crane and Patrangenaru (2011) [35] as the r.o. $C =: X^{-1} \circ Y$. We say that *there is no mean change* from X to Y if the mean change from X to Y is the identity of the group $\mathscr{G}$, that is, the null hypothesis is

$$H_0 : \mu_C = 1_{\mathscr{G}}, \tag{2.59}$$

where μ_C is the extrinsic mean of C with respect to an embedding of $\mathscr{G}$ in $\mathbb{R}^D$. The particular case for testing H_0 in (2.59) on the group $(\mathbb{R}P^3)^q$ is analyzed in detail in Chapter 12.

In general, to compute the p-value for the hypothesis (2.59) one may use the asymptotic χ^2_m distribution in Chapter 6, leading to the following result.

THEOREM 2.8.1. *Assume $J : \mathscr{G} \to \mathbb{R}^k$ is a closed embedding of the m-dimensional Lie group $\mathscr{G}$ in $\mathbb{R}^k$. Let $\{X_r, Y_r\}_{r=1,\dots,n}$ be a random sample of matched pairs for which the change $C_1 = X_1^{-1} \circ Y_1$ has a the j-nonfocal distribution, and let $\mu = E(j(C_1))$ and assume $j(C_1)$ has finite second-order moments and the extrinsic covariance matrix $\Sigma_{j,E}$ of C_1 is nonsingular. Let $(e_1(p), e_2(p), \dots., e_k(p))$ be an orthonormal frame field adapted to j. If $G(J,C)_n$ is the extrinsic sample mean covariance matrix, since for n large enough $G(J,C)_n$ is nonsingular (with probability converging to one) then the p-value associated with H_0 is given by*

$$p = Pr(T > n \| G(J,C)_n^{-\frac{1}{2}} tan(P_j(\overline{j(C)}) - P_j(j(1_{\mathscr{G}})))\|^2), \tag{2.60}$$

where T has a χ^2_m distribution.

Given the i.i.d.s matched pairs $(X_i, Y_i) \in \mathscr{G}^2, i = 1, \dots, n$, and the corresponding changes $C_i = X_i^{-1} Y_i \in \mathscr{G}, i = 1, \dots, n$, it is known (see Bhattacharya and Patrangenaru (2005) [21]) that

$$\sqrt{n} tan(j(\bar{C}) - j(\mu_j)) \to_d \mathscr{N}_m(0, \Sigma_j), \tag{2.61}$$

where Σ_j is the extrinsic covariance matrix of C, and $tan(v)$ is the tangential component in $T_{\mu_{a,j}} j(\mathscr{G})$ of a vector $v \in \mathbb{R}^N$ with respect to the decomposition $\mathbb{R}^N = T_{\mu_{a,j}} j(\mathscr{G}) \oplus (T_{\mu_{a,j}} j(\mathscr{G}))^\perp$
Let $S_{j,n}$ be the sample extrinsic covariance matrix, obtained from the i.i.d.r.o.s $\{X_r\}_{r=1,\dots,n}$ from the unknown distribution Q. At this point we recall from Bhattacharya and Patrangenaru (2005) [21] the steps that one takes to obtain a bootstrapped statistic from a pivotal statistic. If $\{X_r^*\}_{r=1,\dots,n}$ is a random sample from the empirical $\hat{Q}_n$, conditionally given $\{X_r\}_{r=1,\dots,n}$, then the studentized vector valued statistic

$$V(X, Q) = n^{\frac{1}{2}} S_{j,n}^{-\frac{1}{2}} tan(P(\overline{j(X)}) - P(\mu)) \tag{2.62}$$

leads to the bootstrapped statistic

$$V^*(X^*, \hat{Q}_n) = n^{\frac{1}{2}} S_{j,n}^{*\ -\frac{1}{2}} tan_{P(\overline{j(X)}))} (P_j(\overline{j(X^*)}) - P_j(\overline{j(X)})). \tag{2.63}$$

Here $S_{j,n}^*$ is obtained from $S_{j,n}$ by substituting $X_1^*, \dots\dots, X_n^*$ for $X_1, \dots., X_n$, and $T(X^*, \hat{Q}_n)$ is obtained from $T(X, Q)$ by substituting $X_1^*, \dots\dots, X_n^*$ for $X_1, \dots., X_n$, $\overline{j(X)}$ for μ and $S_{j,n}^*$ for $S_{j,n}$.
COROLLARY 2.8.1. *A $(1-\alpha)100\%$ bootstrap confidence region for μ_j is $C_{n,\alpha}^* := j^{-1}(U_{n,\alpha}^*)$ with $U_{n,\alpha}^*$ given by*

$$U_{n,\alpha}^* = \{\mu \in j(\mathscr{G}) : n \| S_{j,n}^{-\frac{1}{2}} tan(P(\overline{j(X)}) - P(\mu)) \|^2 \le c_{1-\alpha}^* \}, \tag{2.64}$$

where $c_{1-\alpha}^*$ is the upper $100(1-\alpha)\%$ point of the values

$$n \| S_{j,n}^{*\ -\frac{1}{2}} tan_{P(\overline{j(X)})} (P(\overline{j(X^*)}) - P(\overline{j(X)})) \|^2 \tag{2.65}$$

among the bootstrap resamples.

2.9 Two-Sample Tests for Extrinsic Means on a Manifold that Admit a Simply Transitive Lie Group Action

Recall that given an embedding $j : \mathscr{M} \to \mathbb{R}^N$, we consider the *chord distance* on $\mathscr{M}$, given by $d(x_1, x_2) = d_0(j(x_1), j(x_2))$, where d_0 is the Euclidean distance in $\mathbb{R}^N$. If $X_{ak_a} : k_a = 1, \dots, n_a, a = 1, 2$ are i.i.d.r.o.s drawn from distributions

$Q_a, a = 1,2$ on a $\mathscr{M}$, if we denote by μ_a the mean of the induced probability $Q_a \circ j^{-1}$ and by Σ_a its covariance matrix $(a = 1,2)$, then the extrinsic mean of Q_a is $\mu_{a,j} = j^{-1}(P(\mu_a))$, assuming Q_a is j-nonfocal, and the extrinsic sample mean is $\bar{X}_{a,j} = j^{-1}(P(\bar{Y}_a))$. Here, again, P is the projection from $\mathbb{R}^N$ to $j(\mathscr{M})$, and we write $Y_{ak_a} = j(X_{ak_a}) k_a = 1,\dots,n_a, a = 1,2$ then $\bar{Y}_a, a = 1,2$ is the corresponding sample mean. Assuming finite second moments of $Y_{a1}, a = 1,2$, which is automatic if $\mathscr{M}$ is compact, from (2.61) we have

$$\sqrt{n_a} tan(j(\bar{X}_{a,j}) - j(\mu_{a,j})) \to_d \mathscr{N}_m(0, \Sigma_{a,j}), a = 1,2, \tag{2.66}$$

where $\Sigma_{a,j}$ is the extrinsic covariance matrix of Q_a, and $tan(v)$ is the tangential component in $T_{\mu_{a,j}} j(\mathscr{M})$ of a vector $v \in \mathbb{R}^N$ with respect to the decomposition $\mathbb{R}^N = T_{\mu_{a,j}} j(\mathscr{M}) \oplus (T_{\mu_{a,j}} j(\mathscr{M}))^{\perp}$.

DEFINITION 2.9.1. *An action of a Lie group $\mathscr{G}$ on a manifold $\mathscr{M}$ is a differentiable function $\alpha : \mathscr{G} \times \mathscr{M} \to \mathscr{M}$, such that*

$$\begin{aligned} \alpha(1_{\mathscr{G}}, x) &= x, \forall x \in \mathscr{M}, \\ \alpha(g, \alpha(h,x)) &= \alpha(g \odot h, x), \forall g \in \mathscr{G}, \forall h \in \mathscr{G}, \forall x \in \mathscr{M}. \end{aligned} \tag{2.67}$$

$\mathscr{M}$ has a simply transitive Lie group of isometries $\mathscr{G}$, if there is a Lie group action $\alpha : \mathscr{G} \times \mathscr{M} \to \mathscr{M}$ by isometries with the property that given $x \in \mathscr{M}$, for any object $y \in \mathscr{M}$, there is a unique $g \in \mathscr{G}$ such that $\alpha(g,x) = y$.

A two-sample hypothesis testing problem for extrinsic means on the embedded manifold $\mathscr{M}$ that admits a simply transitive Lie group of isometries $\mathscr{G}$ can be formulated as follows:

$$\begin{aligned} H_0 &: \mu_{2,j} = \alpha(\delta, \mu_{1,j}) \\ &\text{versus} \\ H_1 &: \mu_{2,j} \neq \alpha(\delta, \mu_{1,j}). \end{aligned} \tag{2.68}$$

Given a fixed object $x \in \mathscr{M}$, the mapping $\alpha^x : \mathscr{G} \to \mathscr{M}, \alpha^x(g) = \alpha(g,x)$ is bijective; therefore, the hypothesis problem (2.68) is equivalent to the following hypothesis testing problem on the Lie group $\mathscr{G}$:

$$\begin{aligned} (1)\ H_0 &: (\alpha^{\mu_{1,j}})^{-1}(\mu_{2,j}) = \delta, \\ &\text{versus} \\ H_1 &: (\alpha^{\mu_{1,j}})^{-1}(\mu_{2,j}) \neq \delta \end{aligned} \tag{2.69}$$

Let $H : \mathscr{M}^2 \to \mathscr{G}$, defined by

$$H(x_1, x_2) = (\alpha^{x_1})^{-1}(x_2). \tag{2.70}$$

THEOREM 2.9.1. *Assume $X_{a,k_a}, k_a = 1,\ldots,n_a$ are independent identically distributed random objects (i.i.d.r.o.s) from the independent probability measures $Q_a, a = 1,2$ with finite extrinsic moments of order $s, s \leq 4$ on the m-dimensional manifold $\mathscr{M}$ on which the Lie group $\mathscr{G}$ acts simply transitively. Let $n = n_1 + n_2$ and assume $\lim_{n\to\infty} \frac{n_1}{n} \to \pi \in (0,1)$. Let $\varphi : \mathfrak{g} \to \mathscr{G}$ and L_δ be respectively, a chart with $\varphi(1_{\mathscr{G}}) = 0_{\mathfrak{g}}$, and the left translation by $\delta \in \mathscr{G}$. Then under H_0,*
(i.) The sequence of random vectors

$$\sqrt{n}(\varphi \circ L_\delta^{-1}(H(\bar{X}_{n_1,j}, \bar{X}_{n_2,j}))) \tag{2.71}$$

converges weakly to $\mathscr{N}_m(0_m, \Sigma_j)$, for some covariance matrix Σ_j that depends linearly on the extrinsic covariance matrices $\Sigma_{a,j}$ of $Q_a, a = 1,2$.
(ii.) If (i.) holds and Σ_j is positive definite, then the sequence

$$n(\varphi \circ L_\delta^{-1}(H(\bar{X}_{n_1,j}, \bar{X}_{n_2,j})))^T \Sigma_j^{-1}(\varphi \circ L_\delta^{-1}(H(\bar{X}_{n_1,j}, \bar{X}_{n_2,j}))) \tag{2.72}$$

converges weakly to χ^2_m distribution.

The following result is a direct consequence of Cramer's delta method, applied to functions between embedded manifolds.

LEMMA 2.9.1. *Assume $F : \mathscr{M}_1 \to \mathscr{M}_2$ is a differentiable function between manifolds. For $a = 1,2$, assume dim $\mathscr{M}_a = m_a$, and $j_a : \mathscr{M}_a \to \mathbb{R}^{N_a}$ is an embedding. Let X_n be a sequence of r.o.s on M_1 such that*

$$\sqrt{n} tan_{j_1(\nu)}(j_1(X_n) - j_1(\nu)) \to_d \mathscr{N}_{m_1}(0, \Sigma), \tag{2.73}$$

then

$$\sqrt{n} tan_{j_2(F(\nu))}(j_2(F(X_n)) - j_2(F(\nu))) \to_d \mathscr{N}_{m_2}(0, d_\nu F \Sigma (d_\nu F)^T), \tag{2.74}$$

where $d_\nu F : T_\nu \mathscr{M}_1 \to T_{F(\nu)} \mathscr{M}_2$ is the differential of F at μ.

By the inverse function theorem, the mapping $H : \mathscr{M} \times \mathscr{M} \to \mathscr{G}$ is continuous. Given that, according to Bhattacharya and Patrangenaru (2003) [20], for $a = 1,2$, the extrinsic sample mean $\bar{X}_{n_a,j}$ is a consistent estimator of $\mu_{a,j}$, for $a = 1,2$, by the continuity theorem (Billingsley (1995) [26], p.334) a consistent estimator for $(\alpha^{\mu_{1,j}})^{-1}(\mu_{2,j})$ is $H(\bar{X}_{n_1,j}, \bar{X}_{n_2,j})$. From Bhattacharya and Patrangenaru (2005) [21], for $a = 1,2$,

$$\sqrt{n_a} tan_{\mu_{a,j}}(j(\bar{X}_{n_a,j}) - j(\mu_{a,j})) \to_d N_m(0_m, \Sigma_{a,j}), \tag{2.75}$$

and, since $\frac{n_a}{n} \to \pi$, it follows that

$$\sqrt{n} tan_{(\mu_{1,j},\mu_{2,j})}(j^2((\bar{X}_{n_1,j}, \bar{X}_{n_2,j})) - j^2((\mu_{1,j}, \mu_{2,j}))) \to_d N_{2m}(0_{2m}, \Sigma), \tag{2.76}$$

where

$$\Sigma = \begin{pmatrix} \frac{1}{\pi}\Sigma_{1,j} & 0 \\ 0 & \frac{1}{1-\pi}\Sigma_{2,j} \end{pmatrix}. \tag{2.77}$$

We apply Lemma 2.9.1 to the function $F : \mathcal{M}^2 \to \mathcal{G}$, given by $F = L_{\delta^{-1}} \circ H$, and select a convenient chart φ, and obtain Eq. (2.71). Theorem 4.2.1(ii.) is an immediate consequence of part (i.) plus a weak continuity argument (Billingsley (1995) [26], p.334). ■

COROLLARY 2.9.1. *(Two-sample test on a homogeneous space) For $a = 1,2$, assume $X_{a,j_a}, j_a = 1,\dots,n_a$, are i.i.d.r.o.'s from independent populations on the m-dimensional embedded manifold $\mathcal{M}$ on which the Lie group $\mathcal{G}$ acts simply transitively. Let $n = n_1 + n_2$, and assume $\lim_{n\to\infty} \frac{n_1}{n} \to \pi \in (0,1)$. Assume Σ_j is positive definite and $\hat{\Sigma}_j$ is a consistent estimator for Σ_j. The asymptotic p-value for the hypothesis testing problem 2.68 is given by $p = P(T \geq T_\delta^2)$ where*

$$T_\delta^2 = n((\varphi \circ L_\delta^{-1}(H(\bar{X}_{n_1,j},\bar{X}_{n_2,j})))^T(\hat{\Sigma}_j)^{-1}(\varphi \circ L_\delta^{-1}(H(\bar{X}_{n_1,j},\bar{X}_{n_2,j}))), \tag{2.78}$$

and T has a χ^2_m distribution.

If the distributions are unknown and the samples are small, an alternative approach is to use Efron's nonparametric bootstrap (see Efron (1982) [41]). If $\max(n_1,n_2) \leq \frac{m}{2}$, the sample mean $\hat{\Sigma}_j$ in Corollary 2.9.1 does not have an inverse, and pivotal nonparametric bootstrap methodology can not be applied. In this case one may use a nonpivotal bootstrap methodology for the two-sample problem H_0 (see Hall and Hart (1990) [59], Fisher et al. (1996) [49] or Hall (1997) [58]).

REMARK 2.9.1. *Beran and Fisher (1998) [13] were the first to use group actions in hypothesis testing problems, a technique later used in Mardia and Patrangenaru (2005)[91]. The drawback in Mardia and Patrangenaru (2005)[91] was that their analysis led to an increase in dimensionality, forcing the extrinsic covariance matrices to be degenerated. In this section we consider in particular case when $\mathcal{M}$ itself has a Lie group structure $\odot$, and the group action is by left translations: $\alpha : \mathcal{M}^2 \to \mathcal{M}, \alpha(g,x) = g \odot x$. Given two objects x,y the* change from x to y is *$c = x^{-1} \odot y$. Given two random objects X,Y, from Theorem 4.2.1 we may estimate the change from the extrinsic mean of X to the extrinsic mean of Y. Note that this change of means is the mean change defined in Section 2.7 only if X,Y are matched pairs on a commutative Lie group $(\mathcal{G},\odot)$.*

REMARK 2.9.2. *The methodology of simply transitive groups on manifolds for two-sample hypothesis testing problem was first used by Osborne (2012) [101] in the context of intrinsic means. That case parallels the case of extrinsic means, with some specific adjustments.*

2.10 Nonparametric Bootstrap for Two-Sample Tests

The primary references for this section are Bhattacharya and Patrangenaru (2014) [22] and Osborne et al. (2013) [102]. The topic here concerns the use of Efron's bootstrap (Efron (1979) [40]) for the approximation of true coverage probabilities for confidence regions for the Fréchet means and for p-values of two-sample tests for equality of Fréchet means. That the bootstrap outperforms the traditional approximation of the distribution of the standardized sample mean by the CLT when the underlying distribution is absolutely continuous was first proved by Singh (1981) [124]. Extensions to studentized, or pivoted, (multivariate) sample means were obtained by Babu and Singh (1984) [5], who derived asymptotic expansions of the distribution.

Indeed, such an expansion holds for more general statistics which are smooth functionals of sample means or which admit a stochastic expansion by a polynomial of sample means (Bhattacharya (1987) [23]). This is derived from refinements of the multivariate CLT by asymptotic expansions (Bhattacharya (1977) [16], Chandra and Ghosh (1979) [33]). As a consequence, the chi-square approximation of the distributions of statistics appearing in (2.20), for example, has an error of the order $O(n^{-1})$, whereas the bootstrap approximation of these distributions has an error $O(n^{-2})$ (see, e.g., Beran (1988) [12], Hall (1987) [56], Hall (1997) [58]). Therefore, bootstrapping the statistics such as appearing in (2.20) would lead to smaller coverage errors than the classical chi-square approximations. In the two-sample case with sample sizes m and n, the error of approximation by bootstrap of the distributions of asymptotically chisquare statistics is $O(N^{-2})$ where $N = \min\{m,n\}$. In the present context of generally high-dimensional manifolds, often the bootstrap approximation of the covariance is singular, which sometimes makes the bootstrap approximation of the distributions either not feasible or subject to further errors in case of rather arbitrary augmentations.

THEOREM 2.10.1. *(Osborne et al. (2013) [102]). Under the hypotheses of Theorem 4.2.1(i), assume in addition, that for* $a = 1,2$ *the support of the distribution of* $X_{a,1}$ *and the extrinsic mean* $\mu_{a,j}$ *are included in the domain of the chart* φ *and* $\varphi(X_{a,1})$ *has an absolutely continuous component and finite moments of sufficiently high order. Then the joint distribution of*

$$V = \sqrt{n}(\varphi \circ L_{\delta}^{-1}(H(\bar{X}_{n_1,j}, \bar{X}_{n_2,j}))) \tag{2.79}$$

can be approximated by the bootstrap joint distribution of

$$V^* = \sqrt{n}(\varphi \circ L_{\delta}^{-1}(H(\bar{X}^*_{n_1,j}, \bar{X}^*_{n_2,j})) \tag{2.80}$$

with an error $O_p(n^{-\frac{1}{2}})$*, where, for* $a = 1,2$, $\bar{X}^*_{n_a,j}$ *are the extrinsic means of the bootstrap resamples* $X^*_{a,k_a}, k_a = 1,\ldots,n_a$*, given* $X_{a,k_a}, k_a = 1,\ldots,n_a$.

We briefly describe now bootstrapping procedures for two independent samples, or match pairs, testing problems on manifolds. Consider, for example, the test $H_0 : P_j(\mu_1) = P_j(\mu_2)$ for the equality of extrinsic means of two distributions on M. Let $\theta = P_j(\mu_1) - P_j(\mu_2)$, and write $T(X|\theta)$ for the statistic on the left of (2.56), but with $P_j(\bar{Y}_1) - P_j(\bar{Y}_2)$ replaced by $P_j(\bar{Y}_1) - P_j(\bar{Y}_2) - \theta$. Whatever the true θ is, the asymptotic distribution of $T(X|\theta)$ is chi-square with m degrees of freedom. Hence a bootstrap-based test of asymptotic level α is to reject H_0 iff $T(X|0) > c^*_{1-\alpha}$, where $c^*_{1-\alpha}$, is the $(1-\alpha)$th quantile of the bootstrapped values $T(X|\bar{Y}_1) - P_j(\bar{Y}_2)$. It follows that the p-value of the test is given by: p-value = proportion of bootstrapped values $T(X^*|P_j(\bar{Y}_1) - P_j(\bar{Y}_2)$ which exceed the observed value of $T(X|0)$. Similar bootstrapping procedures apply to testing the equality of sample intrinsic means, and to match pair problems.

As pointed out above, for a manifold of a high dimension m, the bootstrap version of the covariance matrix, as it appears, for example, within square brackets [ ] is often singular if the sample size is not very large. Suppose the classical chi-square approximation is not considered reliable enough, especially if the p-value that it provides is only marginally small. Then, instead, one may consider the very conservative Bonferroni procedure of looking individually at each of the d contrasts provided, for example, by a P.C.A. of the sample covariance matrix, and compute d p-values for the m tests for the m population contrasts, using the bootstrap approximations of the m distributions of sample contrasts. The estimated p-value is then set at d-times the minimum of these p-values. It should be pointed out that one should not just pick only the principal components which account for the largest variability. If the variability due to a particular contrast is small, while its true population mean value is nonzero, then the corresponding test will bring out the difference with high probability, i.e., with a small p-value (see Bhattacharya and Bhattacharya (2012) [15], p. 16).

Part II

Nonparametric Statistics on Stratified Spaces

Chapter 3

Spaces with a Manifold Stratification

3.1 Introduction

Due to a steep increase in unstructured data outputs from new technological devices, such as electronic images, the internet, RNA, DNA, and protein data, regular statistical analysis on numerical spaces or even on manifolds, where data used to be studied, became obsolete. A new way was needed of organizing objects extracted from such data, by regarding them as points on non-Euclidean metric spaces that have some singularities.
A first step in this direction was by considering, in general, random objects on metric spaces. Assume $(\mathscr{M},\rho)$ is a separable metric space with the distance function ρ and the metric topology τ_ρ generated by open balls in $\mathscr{M}$. The Borel sigma algebra $\mathscr{B}_\rho$ is the sigma algebra generated by τ_ρ. A *random object* (r.o.) in $\mathscr{M}$ is a measurable function $X : (\Omega,\mathscr{A}) \to (\mathscr{M},\mathscr{B}\rho)$, where $(\Omega,\mathscr{A},\mathbb{P})$ is a probability space.

A differentiable structure was needed on the metric spaces where the r.o.s are valued, to address inference and testing for mean objects. This has led to the study of random objects on manifolds (see, e.g., Patrangenaru and Ellingson (2015) [106]). Recall that:

DEFINITION 3.1.1. *An m-dimensional manifold of class $\mathscr{C}^r$ is a triple $(\mathscr{M},\rho,A_{\mathscr{M}})$ where $(\mathscr{M},\rho)$ is a metric space and $A_{\mathscr{M}}$ is an $\mathbb{R}^m$-valued atlas of class $\mathscr{C}^r$ on $\mathscr{M}$.*

For more on data analysis on manifolds, in addition to Patrangenaru and Ellingson (2015) [106], see Bhattacharya and Bhattacharya (2012) [15]. We will consider one such example in a subsequent chapter in this book.

DEFINITION 3.1.2. *A set $\mathscr{M}$ has a* finite filtration *$F_d, d = 0,\dots,m$, if $\forall d = 0,\dots,m$, $F_{d-1} \subseteq F_d$, and $M = F_m$. A finite dimensional stratified space $\mathscr{M}$ is a metric space having a finite filtration $\{F_d\}_{d=0,\dots,m}$, by closed subsets $F_d, d = 0,\dots,m$, such that F_0 is a discrete set, and for $d = 1,\dots,m$, the set difference $F_d \setminus F_{d-1}$ is either empty or a smooth submanifold of dimension d; this filtration* *is called stratification, and the set difference* $F_d \setminus F_{d-1}$ *is the d-dimensional stratum of the m*-dimensional stratified space $\mathcal{M}$.

Data analysis on stratified spaces is one of the most recent developments in Statistics. Examples of stratified object spaces (see Patrangenaru and Ellingson (2015) [106]) are 3D similarity shape spaces (see Kendall et al. (1999) [80]), size and shape spaces (see Dryden and Mardia (2016) [39]), and spaces of positive semi-definite matrices (see Schwartzman et al. (2008) [116]), to mention just a few. Additional important examples of object spaces with a stratified space structure, albeit recognized as such only in recent years, are binary trees (see Basrak (2010) [9]), Billera-Holmes-Vogtmann (BHV) spaces of phylogenetic trees with a fixed number of leaves (see Billera et al. (2001) [28]). In Section 3.3.1 we consider BHV phylogenetic tree spaces and their natural extension, spaces of rooted trees.

3.2 Statistics on Stratified Spaces

One of the most specific phenomena of a distribution on a stratified object space is the potential *stickiness of its intrinsic sample means*. This behavior, first recognized by Basrak (2010) [9] in the basic case of a distribution on *tripod* regarded as a binary tree, was later shown to be present for distributions on other stratified spaces, including on open books (see Hotz et al. (2013) [69]. The key property of the distance on such stratified spaces, which helps in proving stickiness, is that its restriction to the connected components of their highest-dimensional strata is Euclidean, allowing for a reduction of proofs to a combinatorial problem. In Section 3.4 we consider the large sample behavior of sample means on low-dimensional stratified spaces. The key result is on the CLT for intrinsic sample means of a distribution on a tree. Section 8.2 is dedicated to applications to RNA data analysis on phylogenetic tree (PT) spaces. Our focus is on distributions of PTs with three and four leaves, built from SARS-CoV-2 aligned RNA sequences. Given the small sample size on hand, we use an effective *PT sample trick* method: expanding the sample size, while lowering dimensionality. Our SARS-CoV-2 PT is for sample means of such PTs. Temporal and spatial representation of early SARS-CoV-2 specimens are two aspects of our analysis. Note that important additions to the MSC2020 were made with the introduction of 62R20, Statistics on metric spaces, and 62R30 Statistics on manifolds. From the perspective of nonparametric estimation of location parameters, the complete separable metric structure on an object space leads to consistency of Fréchet sample means and antimeans of distributions to their population counterparts. This general metric structure is nevertheless insufficient for proving asymptotic results for such indices. A richer metric structure allowing for some level of smoothness to prove the CLT was needed. Such a structure on an object space, first published in a 2011 CRM proceedings (see Bhattacharya et al. (2013) [17]), is that of **stratified space**.

Special cases of interest are PT spaces and rooted PT spaces (Moore et al. (2021) [97], which in case of trees with three leaves are open books (see Hotz et al. (2013) [69] for a definition).

3.3 Stratified Spaces and PT Spaces

In this chapter we consider only graphs $\mathscr{G}$, whose interior vertices have at least degree three (the degree of a vertex of a graph is the number of edges that are incident to the vertex). To be specific, each vertex V is assumed to be the center of a small enough open ball B in $\mathscr{G}$, such that $B \setminus \{V\}$ is a finite union of $k \geq 3$ open line segments. Connected one-dimensional stratified spaces are such graphs. A *tree* is a simply connected (acyclic) graph. Trees may be also characterized as contractible graphs. Any two points on a tree can be connected by a unique non-self-intersecting path; therefore, any vertex of a tree that is not interior can be regarded as a *root* of the tree (common ancestor for all other noninterior vertices) which consequently can be considered the *leaves* of the tree.

A tree with only one interior vertex, and k-edges, is also known as a **spider with $k \geq 3$ legs**. A spider with k legs may be regarded as a quotient space $S_k = \bigvee_{i=1}^{k} L_i, L_i = \{i\} \times [0,\infty)\}$ obtained by "gluing" the origin points $(i,0), i = 1,\ldots,k$ of the half-lines L_i to a common center C (see Figure 3.1).

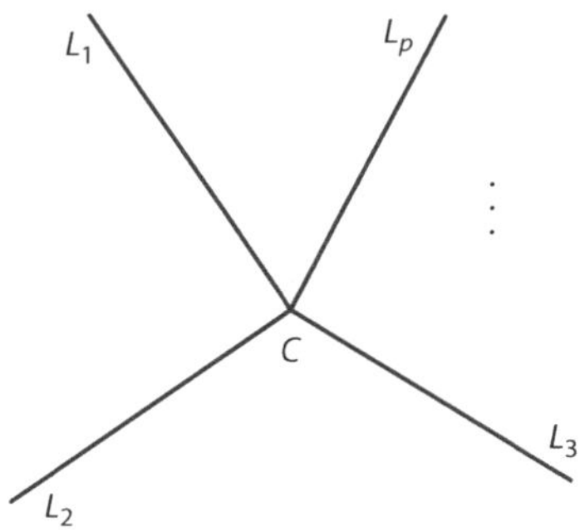

Figure 3.1: Spider, S_p.

DEFINITION 3.3.1. *A tree T is said to be centred at one of its interior vertices C, if*

$$T = \bigvee_{i=1}^{k} T_i, \tag{3.1}$$

where T_i are either half-lines originating at C or trees having C as root.

3.3.1 PT Spaces

Billera et al. (2001) [28] organized the space of trees with a p of leaves, as a metric space, T_p. A *tree with p leaves* is a tree with a distinguished **unlabeled** vertex, the *root*, and p vertices of degree 1, the *leaves*, that are labeled from 1 to p. In this chapter, such trees are also referred to *phylogenetic trees*, for a reason that will become apparent later on.
An edge of a tree with p leaves is *interior* if it is not connected to a leaf. Interior edges have positive lengths; therefore, tree T has interior edges $e_1, \ldots, e_r$ of lengths $l_1, \ldots, l_r$ respectively. If T is binary, then $r = p-2$, otherwise $r < p-2$. The vector $(l_1, \ldots, l_r)^T$ specifies a point in the positive open *orthant* $(0,1)^r$. A tree with p leaves having at most $p-2$ interior edges, the largest possible dimensional orthant is $p-2$-dimensional. The PT space T_p is obtained from $(p-2)$-dimensional closed orthants made of each of the $(2p-3)!!$ types of binary trees, by gluing them along their common faces (see Billera et al. (2001) [25]).
T_p is a $p-2$-dimensional stratified space, originating at the **star tree** (see Figure 3.2).

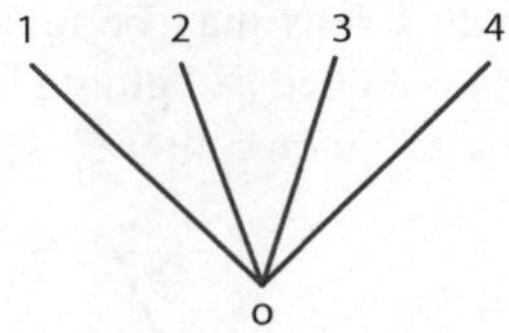

Figure 3.2: Star tree with four leaves.

In particular, it is known (see Billera et al. (2001) [28]) that the space T_3 is a spider S_3, and the space T_4 is a 2D stratified space with 15 faces meeting at the *star tree*, which is regarded as an origin of T_3. Note that the unit sphere in T_4 centered at this origin is a graph with 15 edges, known as the *Petersen graph* (see Figure 3.3).

3.4 Asymptotics of Intrinsic Sample Means: Low–Dimensional Case

Given a random object (r.o.) X on a complete separable metric space $(\mathscr{M}, \rho)$, its associated Fréchet function defined on $\mathscr{M}$ is

$$\mathscr{F}(x) = E(\rho^2(X,x)), \tag{3.2}$$

and its unique point of minimum, when it exists, is the Fréchet mean μ_ρ (Fréchet antimean $\alpha\mu_\rho$ respectively). It is known that the Fréchet sample mean is a consistent estimator of its population counterpart (see [106]); in particular this applies to Fréchet sample means on a stratified space, such as $T_p, p \geq 3$.

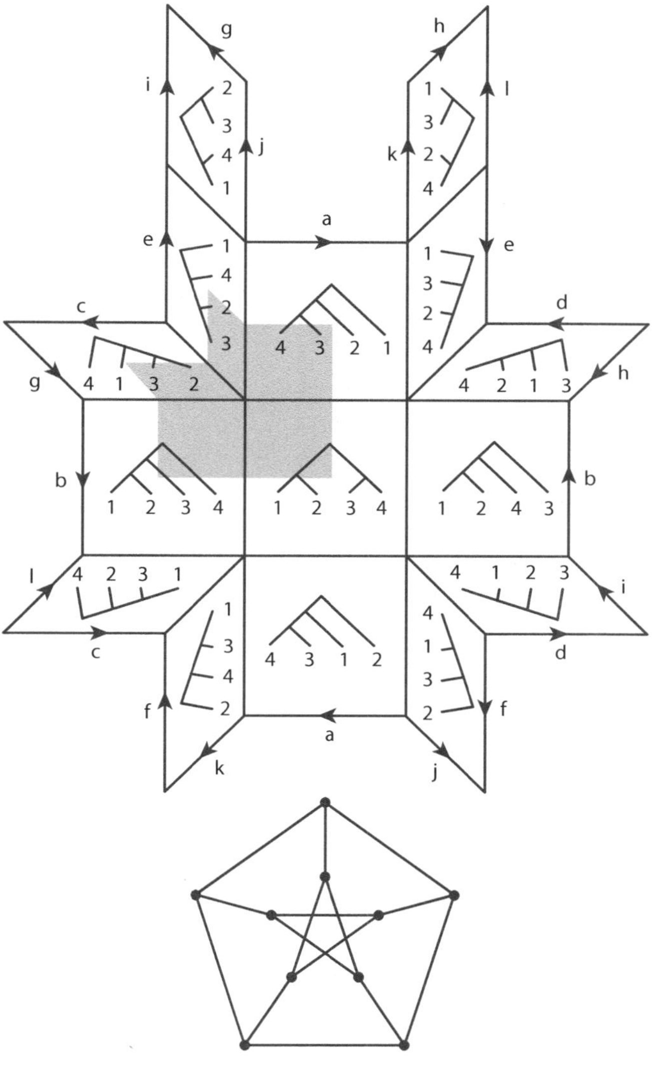

Figure 3.3: Tree space T_4 (top, from Billera et al. (2001) [28]) and its associated Petersen graph (bottom).

In this section we consider distributions on 1D and 2D stratified spaces, by extending certain asymptotic results due to Hotz et al. (2013) [69].
Assume $X_i, i = 1, \ldots, n$ are i.i.d.r.o.s on a spider S_p, having legs $L_i, i = 1, \ldots, p$ and center C. Further, assume the intrinsic mean μ_I exists and the intrinsic

variance is finite. Any probability measure Q on S_p decomposes uniquely as a weighted sum of probability measures Q_i on the leg $L_i, i = 1,\ldots,p$, and an atom Q_0 at C. There are nonnegative real numbers $\{w_k\}_{k=0}^{p}$ summing to 1 such that, for any Borel set $A \subseteq S_p$, the measure Q takes the value

$$Q(A) = w_0 Q_0(A \cap C) + \sum_{k=1}^{p} w_k Q_k(A \cap L_k).$$

Assume $w_0 = 0$ and $x \in L_a$, the Fréchet function $F_X(x) = F(x) = E(\rho^2(X,x))$. If there exists a Fréchet mean (intrinsic mean) μ_I (based on the intrinsic distance ρ), extending the Euclidean distance on each of the spider's legs, then on each leg, $F(x)$ is defined as follows:

$$\begin{aligned} F(x) = & \sum_{i=1, i\neq a}^{p} \int_0^{\infty} (x+u)^2 w_i Q_i(du) + \int_0^{\infty} (x-u)^2 w_a Q_a(du) \\ & = x^2 \sum_{i=1}^{p} \int_0^{\infty} w_i Q_i(du) + 2x[\sum_{i=1, i\neq a}^{p} \int_0^{\infty} u w_i Q_i(du) \\ & - \int_0^{\infty} u w_a Q_a(du)] + \sum_{i=1}^{p} \int_0^{\infty} u^2 w_i Q_i(du) \qquad (3.3) \\ & = x^2 + 2[\sum_{i=1, i\neq a}^{p} v_i - v_a]x + const. \qquad (3.4) \end{aligned}$$

where $v_i = \int_0^{\infty} u w_i Q_i(du), \forall i = 1,\ldots,p$.

The minimizer of the quadratic form is $x^* = v_a - \sum_{i\neq a} v_i$, where $x \in L_a = \{(a,u) : u \in [0,\infty)\}$. Thus, we have three cases to consider: (i) $v_a > \sum_{i\neq a} v_i$, (ii) $v_a = \sum_{i\neq a} v_i$ and (iii) $v_a < \sum_{i\neq a} v_i$. In case (i), μ_I is a well-defined point on L_a, then classical C.L.T. is applied. In case (ii), we can fold other legs into a half-line opposite to L_a then apply C.L.T., and since the negative part is undefined, so the result goes to a positive truncated normal distribution. In case (iii), for any $a \in \{1,\ldots,p\}$, we have $\mu_I = C$, which shows that intrinsic mean μ_I sticks to the star tree C. Taking into account the consistency of the intrinsic sample mean, one obtains the following result (see Hotz et al. (2013) [71]):

LEMMA 3.4.1. *Assume $X_i, i = 1,\ldots,n$ are i.i.d. random objects having a probability distribution Q on a spider S_p with center C, with $Q(\{C\}) = 0$, then: (i) if $v_a > \sum_{i\neq a} v_i$ for some (unique) $a \in \{1,\ldots,p\}$, then $\mu_I \in L_a$ and for n large enough $\bar{X}_{I,n} \in L_a$ and $\sqrt{n}(\bar{X}_{I,n} - \mu_I)$ has asymptotically a normal distribution, (ii) if $v_a = \sum_{i\neq a} v_i$ for some (unique) $a \in \{1,\ldots,p\}$, then after folding the legs $L_i, i \neq a$, into one half-line opposite to L_a, $\sqrt{n}(\bar{X}_{I,n} - \mu_I)$ has asymptotically a positive truncated normal distribution, and (iii) if $v_a < \sum_{i\neq a} v_i$ for all $a \in \{1,\ldots,p\}$, then $\mu_I = C$ and there is n_0 s.t. $\forall n \geq n_0$, then $\bar{X}_{I,n} = C$ a.s.*

In case (iii) of Lemma 3.4.1, we say that the intrinsic mean is *sticky*.

Lemma 3.4.1 admits two extensions: first (see Hotz et al. (2013) [71]), one considered a distribution on the product $M = \mathbb{R}^d \times S_p$, also known as *open book*. Second, one assumes that M is an arbitrary one-dimensional stratified space; simulated examples of stickiness on graphs having cycles were given in Ellingson et. al. (2014) [43]. Here we consider the asymptotic distribution of the intrinsic mean of a random point on a tree.

Assume $X_i, i = 1,\dots,n$ are i.i.d. random objects having a probability distribution Q on a tree $T = \bigvee_{a=1}^{k} T_a$ as in (3.1), where C is an arbitrary interior point of T, such that $Q(\{C\}) = 0$, and the intrinsic total variance of X_1 is finite. If $\mu_I = C$, and Q has a finite intrinsic variance σ^2, and (i) C is not a vertex, then $T \setminus \{C\} = U_1 \cup U_2$, where at least one of U_1, U_2 is a tree rooted at C. We fold U_a into a leg L_a opposite to U_b, where $\{a,b\} = \{1,2\}$, so that each tree edge is folded at its closest vertex from C and the probability measure of the edge is added to the image of the edge on L_a after being folded onto it. Using this construction, one obtains a probability measure $\tilde{Q}$ on the real line $\mathbb{R} \sim L_1 \cup \{C\} \cup L_2$, and i.i.d.r.v.s $Y_i, i = 1,\dots,n$ from $\tilde{Q}$, having the same intrinsic mean C as Q. Due to consistency, for n large enough, $\bar{X}_{I,n}$ falls inside an open interval around μ_I, the sample means $\bar{X}_{I,n}, \bar{Y}_{I,n}$ coincide a.s. and by the CLT, $\sqrt{n}(\bar{X}_{I,n} - \mu_I)$ has asymptotically a normal distribution with variance σ^2. (ii) Assume C is a vertex. (ii.1) If only two of the trees $T_a, a = 1,\dots,k$, adjacent to C have the property that $Q(T_a) > 0$, then we are in the same situation as in case (i), and after folding those two trees along two legs that form a real line with the origin at C, we notice the asymptotic distribution of $\sqrt{n}(\bar{Y}_{I,n} - \mu_I)$, where Y_i are the real i.i.d.r.v.s on the resulting real line corresponding to X_i, has a normal distribution. (ii.2) Without loss of generality we may assume that $Q(T_a) > 0, \forall a = 1,\dots p \geq 3$. Using the same folding away from C measurable map $F_a : T_a \to L_a \sim (0,\infty)$, that is an isometry from each edge of T_a onto its image on L_a, as in (i), we obtain on the legs L_a probability measures Q_a, along with weights $w_a, a = 1,p$, such that for any Borel set $A \subseteq T$, the measure Q takes the value

$$Q(A) = \sum_{k=1}^{p} w_k Q_k(A \cap L_k).$$

On each tree T_a, the Fréchet function is given by

$$\begin{aligned}
F(x) = &\sum_{i=1, i\neq a}^{p} \int_0^\infty (F_a(x)+u)^2 w_i Q_i(du) + \int_0^\infty (F_a(x)-u)^2 w_a Q_a(du) \\
&= F_a(x)^2 \sum_{i=1}^{p} \int_0^\infty w_i Q_i(du) + 2F_a(x)[\sum_{i=1,i\neq a}^{p} \int_0^\infty u w_i Q_i(du) \\
&\quad - \int_0^\infty u w_a Q_a(du)] + \sum_{i=1}^{p} \int_0^\infty u^2 w_i Q_i(du) \qquad (3.5) \\
&= F_a(x)^2 + 2[\sum_{i=1,i\neq a}^{p} v_i - v_a]F_a(x) + const. \qquad (3.6)
\end{aligned}$$

where $v_i = \int_0^\infty u w_i Q_i(du), \forall i = 1, \ldots, p$. Since $\mu_I = C$, and $\lim_{x \to C} F_a(x) = 0$, it follows that $\forall a = 1, \ldots, p$, $\sum_{i=1, i \neq a}^{p} v_i - v_a \geq 0$. We have two cases (iii.1) $\sum_{i=1, i \neq a}^{p} v_i - v_a > 0, \forall a = 1, \ldots, p$. Due to consistency of the intrinsic sample mean, the corresponding sample moments $\hat{v}_{a,n}$, for n large enough, satisfy $\sum_{i=1, i \neq a}^{p} \hat{v}_{i,n} \hat{v}_{a,n} > 0$: therefore, for n large enough, $\bar{X}_{I,n} = C$, and thus the intrinsic sample mean is sticky. This holds also if $\sum_{i=1, i \neq a}^{p} \hat{v}_{i,n} \hat{v}_{a,n} > 0$ are true for at least two indices $a \in \{1, \ldots, p\}$. If (iii.2) $v_a = \sum_{i \neq a} v_i$ for some (unique) $a \in \{1, \ldots, p\}$, then after folding the legs $L_i, i \neq a$, into one half-line opposite to L_a, $\sqrt{n}(F_a(\bar{X}_{I,n}) - \mu_I)$ has asymptotically a positive truncated normal distribution.

THEOREM 3.4.1. *If T is a tree centred at one of its interior vertices C, as in* (3.1)*, and $F_a : T_a \to L_a, Q_a, w_a, v_a$ are defined as above, then (i) if $\mu_I \neq C$ is a point on the interior of one of the edges of T, or $\mu_I = C$ is a vertex and there are only two indices $a \in \{1, \ldots, p\}$, with $Q(T_a) > 0$, then $\sqrt{n}(\bar{X}_{I,n} - \mu_I)$ has asymptotically a normal distribution with variance $\sigma^2 = F(\mu_I)$; (ii) if $Q(T_a) > 0, \forall a = 1, \ldots p \geq 3$ and $\sum_{i=1, i \neq a}^{p} v_i - v_a > 0$, for at least two indices, $a \in \{1, \ldots, p\}$, then the $\bar{X}_{I,n}$ is sticky, for n large enough; and (iii) $v_a = \sum_{i \neq a} v_i$ for some (unique) $a \in \{1, \ldots, p\}$, then after $\sqrt{n}(F_a(\bar{X}_{I,n}) - \mu_I)$ has asymptotically a positive truncated normal distribution.*

In recent years, many researchers have also become interested in studying data that arise on spaces that are not manifolds, but rather are stratified spaces, where each stratum is a manifold. The Data Analysis on Sample Spaces with a Manifold Stratification working group from the 2010-2011 SAMSI Program on Analysis of Object Data delved into problems arising in this area. Bhattacharya et al. (2011) [17] present an overview of a number of such problems. One specific area of interest that this group studied is that for a number of these spaces, Fréchet sample means exhibit a property known as *stickiness*. In short, stickiness refers to the occurrence of a Fréchet sample mean remaining at a given location when the underlying probability distribution is perturbed. Hotz et al. (2013) [69] present asymptotic results on a particular space known as an open book. Ellingson et al. (2014) [43] discussed stickiness on sample spaces that are low-dimensional stratified spaces.

A specific type of stratified space that has garnered great interest is the space of phylogenetic trees, which was introduced in Billera et al. (2001) [25] and is discussed in the papers referenced above. Barden et al. (2013) [8] present asymptotic results for Fréchet sample means on the space of phylogenetic trees. Among other topics, Feragen et al. (2013) [48] present methodology for performing principal component analysis and a permutation test for problems involving tree-valued data. Skwerer et al. (2014) [125] apply ideas developed for phylogenetic trees to the structure of brain arteries.

Chapter 4

Extrinsic Data Analysis on Stratified Spaces

4.1 Introduction

In general, given a complete separable metric space (M,ρ), the Borel σ-algebra $\mathscr{B}_M$ is the σ-algebra $\sigma(\mathscr{T}_M)$ generated by the topology of (M,ρ). Assume $(\Omega,\mathscr{A},\mathbb{P})$ is a probability space. A *random object* (r.o.) on M is function $X:\Omega\to M$, with the property that for any Borel set $B\in\mathscr{B}_M, X^{-1}(B)\in\mathscr{A}$. To any r.o. X we associate a *probability measure* Q on M, given by $Q(B)=\mathbb{P}(X^{-1}(B))$. Given a r.o. X in M, its *Fréchet mean set* is the set of all minimizers of the expected square distance of X (assuming this is finite) to an arbitrary point on M. If the Fréchet mean set consists in one minimizer, the minimizer is called *Fréchet mean* of X (see Fréchet (1948) [52], Ziezold (1977) [150]). Obviously the Fréchet mean (set) depends on the distance ρ on M. The case in point here is when the distance ρ is the *chord distance* induced by an *embedding* $j:M\to\mathbb{E}^N$, given in terms of the Euclidean norm by $\rho(x_1,x_2)=\|j(x_1)-j(x_2)\|$. In this case the Fréchet mean (set) is called *extrinsic mean (set)*. If $\mu=E(j(X))$ and we set $M(j)=\{p\in j(M),\rho(p,\mu)=\rho(j(M),\mu)\}$, then the extrinsic mean set is $j^{-1}(M(j))$, and if $M(j)$ consists in a singleton, then X has an extrinsic mean, which is labeled μ_j. Given a sample $x_1,\ldots,x_n$ of objects on M, the *extrinsic sample mean (set)* is the extrinsic mean (set) of the *empirical distribution* $\hat{Q}_n=\sum_{i=1}^{n}\frac{1}{n}\delta_{x_i}$, where δ_x assigns probability one to x. The extrinsic sample mean, when it exists, will be labeled by $\bar{x}_{j,n}$. Assume $X_1,\ldots,X_n$ are independent and identically distributed random objects (i.i.d.r.o.s) on M. For extrinsic data analysis purposes, the first objective is the understanding of the asymptotic distribution of the extrinsic sample means $\bar{X}_{j,n}$.

This chapter is a result of collaboration within the Working Group on Data Analysis on Sample Spaces with a Manifold Stratification, which ran under the Statistical and Applied Mathematical Sciences Institute (SAMSI) 2010–2011 program on Analysis of Object Data. The group, many of whose members coauthored the first paper on Statistics on stratified spaces (see Bhattacharya et al. (2013) [17]), considered practical applications from evolutionary biology to medical imaging and shape analysis (see for example Patrangenaru and Ellingson (2011) [105]). Here are some key examples of stratified spaces considered

- projective shape spacesets of equivalence classes of configurations under actions of the pseudogroup of projective, useful in analyzing 3D scenes from pinhole camera image outputs (Patrangenaru et al. (2010) [107]); affine shape, useful in analyzing aerial or satellite images (Patrangenaru and Mardia (2003) [108]); or similarity shape spaces, relevant in medical imaging (see Bandulasiri et al. (2009) [7] or Dryden and Mardia (2016) [39], chapter 6), and in other applications (see Huckemann et al. (2010) [73] for example);
- spaces of covariance matrices, arising as data points in diffusion tensor imaging (see Basser and Pierpaoli (1996) [10], Schwartzman (2008) [115] or Schwartzman et al. (2008) [116] for example); and
- tree spaces, representing spaces of metric phylogenetic trees on fixed sets of taxa (see Billera et al. (2001) [25], Aydin et al. (2009) [3], for example).

In this chapter, we briefly summarize extrinsic object data analysis, by reversing the order from the most recent to the more "classical" types of random objects. In Section 1 we are concerned with the asymptotic distribution in the case when M is a *stratified space*, that is a metric space (M,ρ) that decomposes as finite disjoint unions of manifolds (*strata*) in such a way that the *singularities* of M are constant along each stratum (see Goresky and MacPherson (1988, section 1.4) [53]. The trivial case, of a manifold of dimension d (stratified space having one d-dimensional stratum only) was settled by Hendricks and Landsman (1998) [66] and Bhattacharya and Patrangenaru (2005) [21], who showed that the asymptotic distribution of the tangential component of the extrinsic sample means $\bar{X}_{j,n}$ in the tangent space at (j image of) the extrinsic population mean, has approximately a multivariate normal distribution $\mathcal{N}_d(0_d, \frac{1}{n}\Sigma_j)$, where Σ_j is the *extrinsic population covariance matrix*. For details see Bhattacharya and Patrangenaru (2005) [21], where it is also shown that the *intrinsic sample means* have a similar asymptotic distribution. A new behavior is exhibited when the stratified space has two strata: Hotz et al. (2013) [69] considered the asymptotic distribution of a certain intrinsic sample mean on an *open book*, which a space obtained by gluing disjoint copies of a Euclidean half-space along their boundaries (*spine* of the open book). They showed that, for a natural metric on the open book, if the intrinsic mean sits on the spine, then, often the intrinsic sample means stick to the spine and have an asymptotically multivariate normal distribution there. In Section 1 we show that, for d = 1, a similar asymptotic behavior is exhibited by the extrinsic sample means on open book, with embeddings literally reminding of an infinitely thin open paper back book in 3D. In Section 2, we introduce the general concept of locally homogeneous space and show that due to consistency of Fréchet sample means, any two-sample problem can be reasonably tackled on such a sample space. Results concerning two-sample tests for extrinsic means of random objects on locally homogeneous spaces are also given in this section, the data analysis being transported on a simply locally transitive Lie group of isometries. We apply our new methodology to a concrete Diffusion Tensor Imaging example, using a small size data set previously analyzed by Schwartzman et al. (2008) [115].

4.2 Extrinsic Central Limit Theorem on Open Books

Set $S = \mathbb{R}^d$, the real vector space of dimension d with the standard Euclidean metric. If $\mathbb{R}_{\geq 0} = [0,\infty)$, then the closed half-space

$$\bar{H}_+ = \mathbb{R}_{\geq 0} \times S$$

is a metric subspace of $\mathbb{R}^{d+1} = \mathbb{R} \times S$, with boundary S, which we identify with $\{0\} \times S$, and interior $H_+ = \mathbb{R}_{>0} \times S$. We consider the following equivalence relationship

$$\mathscr{R} \subset (\bar{H}_+ \times \{1,\ldots,K\}) \times (\bar{H}_+ \times \{1,\ldots,K\}):$$

a pair $((x,k),(y,j)) = (x^{(0)},x^{(1)},\ldots,x^{(d)},k),(y^{(0)},y^{(1)},\ldots,y^{(d)},j)$ is in $\mathscr{R}$, if $x^{(0)} = y^{(0)} = 0$ and $x^{(i)} = y^{(i)}, \forall i = 1,\ldots,d$ or $x^{(0)} = y^{(0)} > 0$ and $x^{(i)} = y^{(i)}$ and $k = j$. The *open book* $\mathscr{O}$ is the quotient $(\bar{H}_+ \times \{1,\ldots,K\})/\mathscr{R}$. The following definition summarizes and introduces terminology.

DEFINITION 4.2.1 (Leaves and spine). *The open book $\mathscr{O}$ consists of $K \geq 3$* leaves L_k, *for $k = 1,\ldots,K$, each of dimension $d+1$. L_k is the set of equivalence classes of points in $\bar{H}_+ \times \{k\}$. The leaves are joined together along the* spine L_0, *set of equivalence classes of points in $\{0\} \times S \times \{1,\ldots,K\}$, i.e., L_0 can be identified with the space $S = \mathbb{R}^d$.*

EXAMPLE 4.2.1. *A piece of the open book $\mathscr{O}$ with $d = 1$ and $K = 5$ embedded in $\mathbb{R}^3$ is depicted in Figure 4.1. Each embedded leaf continues to infinity away from the spine.*

If for simplicity, we consider the case $d = 1, K$ arbitrary, and an embedding $J : \mathscr{O} \to \mathbb{R}^3$ similar to the one in Example 5.2.1, we will restrict ourselves to random objects X on $\mathscr{O}$ having a probability mass with the property (p) that the mean $E(J(X))$ is closer to the spine $J(L_0)$ than to any of the open leaves $J(L_k \backslash L_0), k = 1,\ldots,K$.

THEOREM 4.2.1. *(Extrinsic Sticky CLT on $\mathscr{O}$) Assume the independent and identically distributed random objects $X_1,\ldots,X_n$ on $\mathscr{O}$, are such that X_1 has the property (p). Let $P_{J(L_0)}$ be the projection on $J(L_0)$, and assume $J^{-1}(P_{J(L_0)}(J(X_1)))$ has finite variance $\sigma^2_{L_0}$. Then for n large enough, the*

Figure 4.1: A five-leaf open paper book.

extrinsic sample mean $\bar{X}_{J,n}$ *sticks to the spine* L_0 *and, on the spine* $\sqrt{n}\sigma_{L_0}^{-1}(\bar{X}_{J,n} - \mu_J)$ *has asymptotically a standard normal distribution.*

The proof of Theorem 4.2.1 follows from the fact that due to consistency, with probability one, for n large enough, the sample mean $\overline{J(X)}_n$ is closest to the spine than to any of the open leaves, thus the projection $P_J(X_n)$ sticks to $J(L_0)$. Since by the CLT, $\overline{J(X)}_n$ has a trivariate normal distribution, the projection $P_{J(\mathcal{O})}(\overline{J(X)}_n)$ has a normal distribution, and the result follows by standardizing this asymptotic distribution.

4.3 Diffusion Tensor Imaging Data Analysis

An example of a medical imaging data analysis problem is data analysis of spatially registered Diffusion Tensors from MRI images to track the water flow along axons in the brain. A diffusion tensor can be represented as 3×3 symmetric positive definite matrix, as shown by Basser and Pierpaoli (1996) [10]. Such matrices, when vectorized, form an open convex subset in $\mathbb{R}^6$ which has a locally homogeneous space structure under the additive group action of $\mathbb{R}^6$. Nonparametric bootstrap provides a reasonable estimation of the mean. This approach is used in a concrete example for detection from Diffusion Tensor Imaging (DTI) of a condition called dyslexia that is relatively frequent in humans. Following Schwartzman et al. (2008) [115], we consider an experiment in which DTI maps acquired for two groups of subjects, a group of healthy children and a group of children suffering from dyslexia, differ on average. In Figure 4.2 we compare the bootstrap distribution of the sample means in these

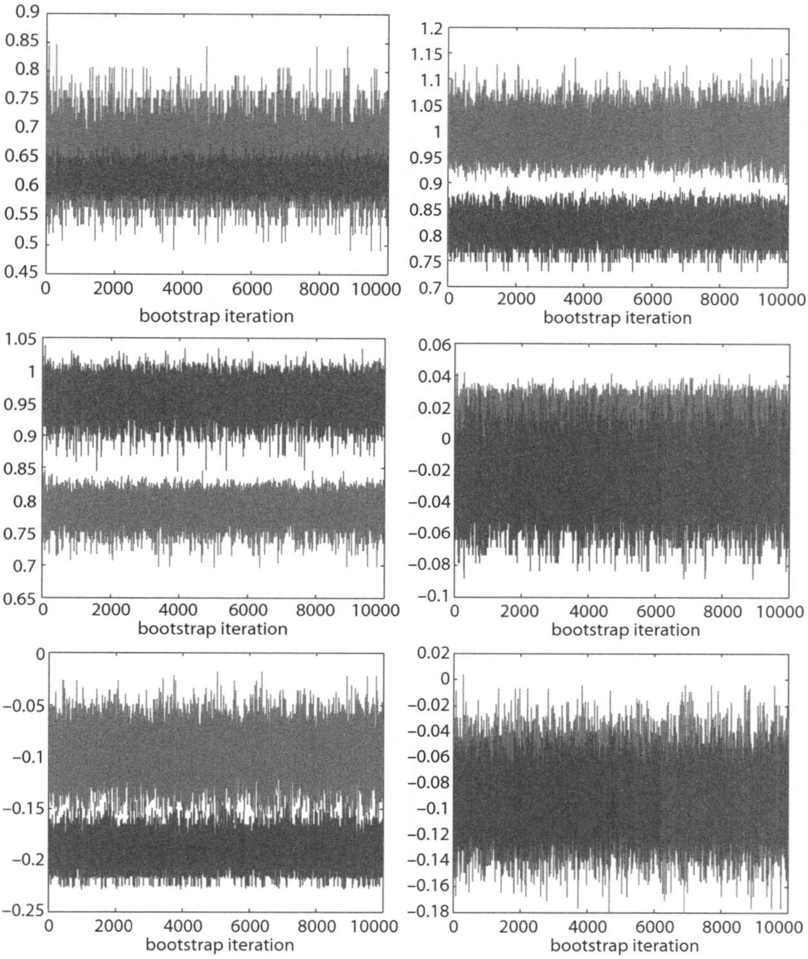

Figure 4.2: Marginals of the bootstrap distribution of the sample means: Clinically normal (red) vs. dyslexia (blue).

two populations, which are clearly showing a difference in the DTI means. Similar results were obtained by Schwartzman et al. (2008) [116] using a parametric approach. A nonparametric intrinsic approach for the same data was used in Osborne (2012) [101].

Chapter 5

Intrinsic Sample Means on Tree Spaces

5.1 Introduction

The sample mean of a sample of points in a Euclidean space moves slightly when one of the points is perturbed. This motion is pervasive in classical probabilistic and statistical situations.

In general, given a complete separable metric space (M,ρ), the Borel σ-algebra $\mathscr{B}_M$ is the σ-algebra $\sigma(\tau_M)$ generated by the metric topology of (M,ρ). Assume $(\Omega,\mathscr{A},\mathbb{P})$ is a probability space and $X:\Omega\to M$ is a *random object* (r.o.) on M, which means that for any Borel set $B\in\mathscr{B}_M, X^{-1}(B)\in\mathscr{A}$. To any r.o. we associate a probability measure Q on M, $Q(B)=\mathbb{P}(X^{-1}(B))$. Given a random object X in M, its *Fréchet mean* is the minimizer of the expected square distance of X (assuming this is finite) to an arbitrary point on M, if such a minimizer is *unique*. In complete separable metric spaces, the Fréchet sample mean (L^2-minimizer, least squares approximation), which minimizes the sum of the square distances to the given points, generalizes the notion of sample mean. Intuitively, the sample mean of a well-behaved probability distribution on a space M of dimension $d+1$ ought to avoid lying on any particular subspace of dimension d or less the covariance matrix of the distribution is singular. While this intuition has been shown to be rigorous when M is a manifold (see, e.g., Hendricks and Landman (1996) [65], Patrangenaru (1998) [104], Bhattacharya and Patrangenaru (2005) [21]), it may fail when M has certain types of singularities, as we demonstrate here for an *open book* $\mathscr{O}$: a space obtained by gluing disjoint copies of a half-space along their boundary hyperplanes; see Section 5.2 for precise definitions.

EXAMPLE 5.1.1. *The simplest singular space is the* 3-spider*: a union* $\mathscr{S}_3$ *of three rays with their endpoints glued at a point* 0 *(Figure 5.1, left). This space* $\mathscr{S}_3$ *is the open book* $\mathscr{O}$ *of dimension* 1 *with three leaves. If three points are chosen equidistant from* 0 *on the different rays, then the sample mean lies at* 0 *by symmetry (Figure 5.1, center). The unexpected phenomenon is that wiggling one or more of the points has no effect on the barycenter (Figure 5.1, right).*

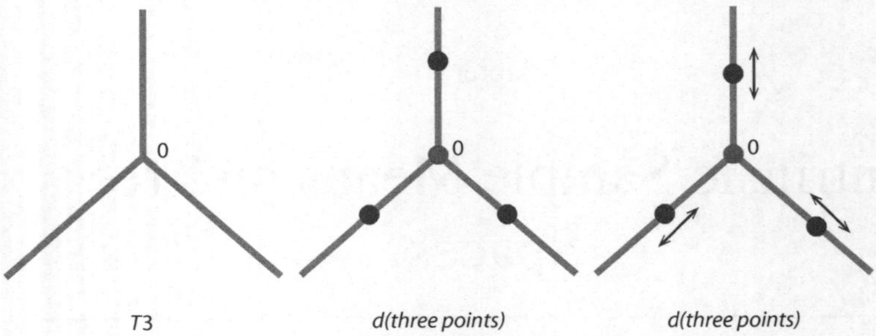

Figure 5.1: (left) The space of rooted phylogenetic trees with three leaves and fixed pendant edge lengths; (center) the probability distribution supported on three points in $\mathscr{S}_3$ equidistant from the vertex 0 has sample mean 0; (right) perturbing the distribution—and even macroscopically moving all three points a limited distance—leaves the sample mean fixed.

For instance, if the points lie at radius r from 0*, then the barycenter remains at* 0 *upon moving one of the points to radius at most* $2r$. ◊

Our main goal is to define a precise concept of when a distribution on an open book has a *sticky mean*, and to quantify this highly nonclassical condition with a Law of Large Numbers (LLN) in Theorem 5.5.1 and a Central Limit Theorem (CLT) in Theorem 5.6.1.

Roughly speaking, the sticky LLN says that, under certain generic conditions, almost surely, empirical (sample) means eventually lie on the *spine*: the hyperplane shared by all of the glued half-spaces by virtue of the gluing (in Figure 5.1, the spine is the point 0). This phenomenon contrasts with the classical LLN, where the empirical mean approaches the theoretical mean from all directions. The sticky CLT says that the limiting distribution is Gaussian and supported on the spine. Again, the nonclassical nature of this result contrasts with the classical CLT, in which the limiting distribution has full support rather than being supported on a thin (positive codimension and hence measure zero) subset of the object space. Versions of the LLN and CLT are also stated in Theorems 5.5.1, 5.6.1, and 5.6.2 for the cases where the mean is

- nonsticky—not lying on the spine—so the LLN and CLT behave classically; and
- partly sticky—lying on the spine but not sticky—so the LLN and CLT are hybrids of the sticky and nonsticky ones.

Open books are the simplest singular topologically stratified spaces. Roughly speaking, topologically stratified spaces decompose as finite disjoint unions of manifolds (*strata*) in such a way that the singularities of the total space are constant along each stratum (this is the structure described in

Goresky and MacPherson (1988, section 1.4) [53]. Every topologically stratified space that is singular along a stratum of codimension 1 is, by definition of topological stratification, locally homeomorphic to an open book along that stratum. Therefore, even if the goal is to sample from arbitrary stratified spaces possessing singularities in maximal dimension, it is first necessary to understand open books.

Sticky means on open books seem to stem from topological phenomena, rather than geometric ones. Therefore, although the topological space $\mathscr{O}$ can be endowed with many metrics, we consider only the simplest, in which each half-space has the Euclidean metric and the boundaries are glued isometrically. More general metrics on open books and more general singularities on stratified spaces form part of the wider program of probability on stratified spaces, initiated here, whose overarching goals are to

1. understand how asymptotics of sampling relates to topology and geometry of singularities, and
2. develop quantitative computational statistical methods for handling data sampled from stratified spaces.

Apropos the second goal, this chapter is output from a Working Group on sampling from stratified spaces that ran under the Statistical and Applied Mathematical Sciences Institute (SAMSI) 2010–2011 program on Analysis of Object Data. The group, in which most of the authors were members, considered practical applied questions from evolutionary biology, medical imaging, and shape analysis. Key examples of stratified spaces considered by the group therefore included:

- shape spaces, representing equivalence classes of point configurations under operations such as direct similarities, affine transformations, projective transformations, or other nonlinear transformations (see Dryden and Mardia (2016) [39], Mardia an Patrangenaru (2001) [90], Krim and Yezzi (2006) [82], for example);
- spaces of covariance matrices, arising as data points in diffusion tensor imaging (see Arsigny et al. (2006) [2], Basser and Pierpaoli (1996) [10], Schwartzman et al. (2008) [115], for example); and
- tree spaces, representing metric phylogenetic trees on fixed sets of taxa (see Billera et al. (2001) [25], Owen and Provan (2010) [103], for example).

For instance, the space $\mathscr{S}_3$ from Example parameterizes all rooted (metric) phylogenetic trees with three taxa and fixed pendant edge lengths. More generally, open books of arbitrary dimension and precisely three leaves reflect the local structure of phylogenetic tree space near any point on a stratum of codimension 1; such a point represents a tree possessing a node with nonbinary branching. Our observations of "unresolved" (that is, nonbinary) trees as barycenters of biologically meaningful samples (involving yeast phylogenies and brain arteries) constituted crucial motivation for the present study.

The relation between open books and tree spaces is that of local to global. Basrak [9] had independently and simultaneously proved a sticky CLT for certain global situations in dimension 1, namely arbitrary binary trees: connected graphs with no cycles where each node is incident to at most three edges. In contrast, our dimension 1 results are local, in that all edges meet, but there can be more than three incidents to the intersection.

It bears mentioning that unlike in open books, sample means do not stick to thin subspaces of shape spaces, or to thin subspaces of more general quotients of manifolds by isometric proper actions of Lie groups (see Huckemann (2012) [71]). The differentiating property amounts to curvature: open books are, in a precise sense, negatively curved, whereas passing to the quotient in the construction of shape spaces adds positive curvature. Basrak's binary trees (see [9]) are negatively curved in the same way that open books or spaces of trees are (see Billera et al. (2001) [25]): they are *globally nonpositively curved* (or *CAT(0) spaces*).

Finally, modern applications of statistics require knowledge of the asymptotics of distributions on singular spaces, including algebraic varieties and polyhedral complexes such as shape spaces and tree spaces. The natural generality seems to be topologically stratified spaces, or the more restrictive class of *Whitney stratified spaces* (see Schwartzman et al. (2008) [115, section 1.2], which seems to contain all of the spaces in statistical applications (including all real semi-algebraic varieties). However, for the purposes of data analysis, a central limit theorem is only a vehicle toward parameter estimation. Further steps would require the derivation of Slutsky-type theorems for random objects on open books and more general stratified spaces. Eventually, the goal lies in generalizing the rich statistical toolkit to stratified spaces. For instance, to date, there are generalizations of the powerful techniques of regression to Riemannian manifolds (see Shi et al. (2009) [122]), as well as of principal component analysis (PCA) and multivariate analysis of variance (MANOVA) to shape spaces (see Huckemann et al. (2010) [72]).

5.2 Open Books

Set $S = \mathbb{R}^d$, the real vector space of dimension d with the standard Euclidean metric. If $\mathbb{R}_{\geq 0} = [0, \infty)$ is the closed nonnegative ray in the real line, then the closed half-space

$$\bar{H}_+ = \mathbb{R}_{\geq 0} \times S$$

is a metric subspace of $\mathbb{R}^{d+1} = \mathbb{R} \times S$ with boundary S which we identify with $H = \{0\} \times S$, and interior $H_+ = \mathbb{R}_{>0} \times S$. The *open book* $\mathcal{O}$ is the quotient of the disjoint union $\bar{H}_+ \times \{1, \ldots, K\}$ of K closed half-spaces modulo the equivalence relation that identifies their boundaries. Therefore, $p = (x,k) = (x^{(0)}, x^{(1)}, \ldots, x^{(d)}, k)$ is identified with $q = (y, j) = (y^{(0)}, y^{(1)}, \ldots, y^{(d)}, j)$

whenever $x^{(0)} = 0$ and $x^{(i)} = y^{(i)}$ for all $i \in \{0, \ldots, d\}$, regardless of k and j. The following definition summarizes and introduces terminology.

DEFINITION 5.2.1 (Leaves and spine). *The open book $\mathcal{O}$ consists of $K \geq 3$* leaves *L_k, for $k = 1, \ldots, K$, each of dimension $d+1$ and defined by*

$$L_k = \bar{H}_+ \times \{k\}.$$

The leaves are joined together along the spine *L_0 which comprises the equivalence classes in $\bigcup_{k=1}^{K}(H \times \{k\})$, i.e., L_0 can be identified with the hyperplane $H = \{0\} \times S$ or with the space $S = \mathbb{R}^d$. When we speak of the spine in the following, we make clear which of these three* instances of the spine *we have in mind. The following diagram gives an overview of these instances, spaces and mappings introduced further below.*

$$\begin{array}{lccccccccc}
\mathcal{O} & \supseteq & L_k & \supseteq & L_0 & & \subseteq & L_k & \subseteq & \mathcal{O} \\
\downarrow F_k & & \updownarrow & & \updownarrow F_k|_{L_0} & & & & & \\
\mathbb{R}^{d+1} & \overset{\hat{P}}{\to} & \overline{H}_+ & \supseteq & H & & \swarrow & P_S & = & \pi_S \circ F_k \\
& & & & \updownarrow \pi_S|_H & & & & & \\
& & & & S = \mathbb{R}^d & & & & &
\end{array}$$

EXAMPLE 5.2.1. *A piece of the open book with $d = 1$ and $K = 5$ is depicted below:*

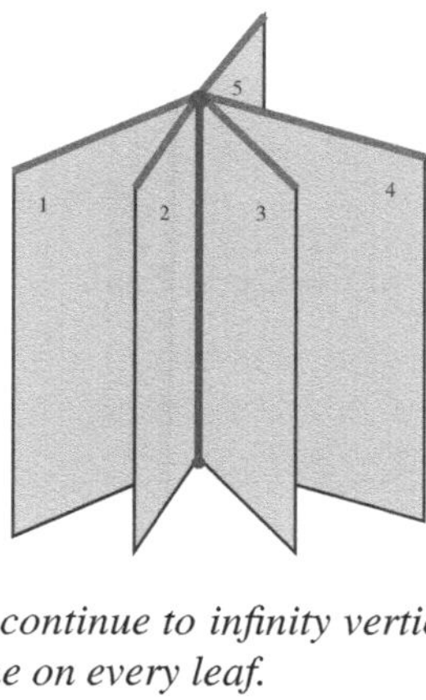

Ideally, the picture would continue to infinity vertically, both up and down as well as away from the spine on every leaf. ◇

DEFINITION 5.2.2 (Reflection). *For a given point $x \in \bar{H}_+$, let $Rx \in \bar{H}_- = \mathbb{R}_{\leq 0} \times \mathbb{R}^d = (-\infty, 0] \times \mathbb{R}^d$ denote its reflection across the hyperplane $\bar{H}_+ \cap \bar{H}_- = \{0\} \times S$.*

The metric d on $\mathcal{O}$ is expressed in terms of reflection in a natural way: given two points $p, q \in \mathcal{O}$, with $p = (x, k)$ and $q = (y, j)$,

$$\rho(p,q) = \begin{cases} |x - y| & \text{if } k = j, \\ |x - Ry| & \text{if } k \neq j, \end{cases} \tag{5.1}$$

where $|x-y|$ denotes Euclidean norm based distance on $\mathbb{R}^{d+1}$. Note that if $k \neq j$ in Eq. (5.1), then $\rho(p,q) = 0$ if and only if x and y lie on the spine and coincide. Our assumption $K \geq 3$ implies that $\mathcal{O}$ is not isometric to a subset of $\mathbb{R}^{d+1}$ (as it would be for $K \leq 2$).

LEMMA 5.2.1. *The open book $(\mathcal{O}, d)$ is a Hausdorff metric space that is globally nonpositively curved, and its spine is isometric to $\mathbb{R}^d$.*

Proof. [127, Example 3.3]. □

LEMMA 5.2.2. *The open book $\mathcal{O}$ is the disjoint union*

$$\mathcal{O} = L_0 \cup L_1^+ \cup \cdots \cup L_K^+$$

of the spine L_0 and the interiors $L_k^+ = L_k \setminus L_0$ of the leaves, $k = 1, \ldots, K$.

REMARK 5.2.1. *Although the open book $\mathcal{O}$ is not a vector space over $\mathbb{R}$, scaling by a positive constant $\lambda \in \mathbb{R}_{\geq 0}$ is defined in the natural way:*

$$\lambda p = (\lambda x, k) \text{ for all } p = (x,k) \in \mathcal{O}.$$

The open book also carries an action of the spine S, considered as an additive group, by translation, via the action of S on each leaf:

$$\mathcal{O} \ni p = (x^{(0)}, x^{(1)}, \ldots, x^{(d)}, k) \xrightarrow{z} (x^{(0)}, x^{(1)} + z^{(1)}, \ldots, x^{(d)} + z^{(d)}, k) \in \mathcal{O},$$

with $z = (z^{(1)}, \ldots, z^{(d)}) \in S$. For the above right-hand side, we write simply $z + p$.

5.3 Probability Measures on the Open Book

Our goal is to understand the statistical behavior of points sampled randomly from $\mathcal{O}$. Suppose that Q is a Borel probability measure on $\mathcal{O}$. We assume throughout the paper that $\rho(0,q)$ has bounded expectation under the measure Q:

$$\int_{\mathcal{O}} \rho(0,q)\, dQ(q) < \infty. \tag{5.2}$$

When explicitly stated, we also assume the stronger condition

$$\int_{\mathcal{O}} \rho(0,q)^2\, dQ(q) < \infty, \tag{5.3}$$

of square integrability.

LEMMA 5.3.1. *Any probability measure Q on the open book $\mathcal{O}$ decomposes uniquely as a weighted sum of probability measures Q_k on the open half-spaces (leaves) L_k^+ and a probability measure Q_0 on the spine L_0. More precisely, there* *are nonnegative real numbers* $\{w_k\}_{k=0}^K$ *summing to* 1 *such that, for any Borel set* $A \subseteq \mathscr{O}$*, the measure* Q *takes the value*

$$Q(A) = w_0 Q_0(A \cap L_0) + \sum_{k=1}^{K} w_k Q_k(A \cap L_k^+).$$

Proof. By Lemma 5.2.2 the spine and open leaves are measurable and partition $\mathscr{O}$; hence the result follows from the additivity of measures on disjoint sets. □

REMARK 5.3.1. *For* $k \geq 1$, $w_k = Q(L_k^+)$ *is the probability that a random point lies in* L_k^+*, while* $w_0 = Q(L_0)$ *is the probability that a point lies somewhere on the spine.*

CONVENTION 5.3.1. *Throughout this chapter, assume the nondegeneracy condition*

$$w_k = Q(L_k^+) > 0 \text{ for all } k \in \{1, \dots, K\}. \tag{5.4}$$

Otherwise, we would remove those leaves L_k *for which* $Q(L_k^+) = 0$ *from the open book. Nondegeneracy implies that* $w_0 < 1$ *and* $0 < w_k < 1$ *for all* $k \geq 1$ *in the decomposition from Lemma 5.3.1.*

DEFINITION 5.3.1 (Folding map). *For* $k \in \{1, \dots, K\}$ *the* k^{th} folding map $F_k : \mathscr{O} \to \mathbb{R}^{d+1}$ *sends* $p \in \mathscr{O}$ *to*

$$F_k p = \begin{cases} x & \text{if } p = (x,k) \in L_k, \\ Rx & \text{if } p = (x,j) \in L_j \text{ and } j \neq k \end{cases}$$

where the reflection operator R *was defined in Definition 5.2.2.*

REMARK 5.3.2. *In the definition of the folding map* F_k*, the leaf* L_k *is identified with the subset* $\bar{H}_+ \subset \mathbb{R}^{d+1}$*, by slight abuse of notation (again). The other leaves* L_j *are collapsed to the negative half-space* $\bar{H}_- \subset \mathbb{R}^{d+1}$ *via the reflection map. All of these identifications have the same effect on the spine* S*, which becomes the hyperplane* $H = \{0\} \times \mathbb{R}^d \subset \mathbb{R}^{d+1}$*. For example,* F_4 *takes the picture in Example 5.2.1 to* $\mathbb{R}^2$ *as follows.*

The notations H_+ *and* H_- *(with no bars) are reserved for the* strictly positive *and* strictly negative *open half-spaces that are the interiors of* $\bar{H}_+$ *and* $\bar{H}_-$*, respectively.*

LEMMA 5.3.2. *Under the folding map* F_k*, the measure* Q *pushes forward to a measure* $\tilde{Q}_k = Q \circ F_k^{-1}$ *on* $\mathbb{R}^{d+1}$ *such that, given a Borel subset* $A \subseteq \mathbb{R}^{d+1}$*,*

$$\tilde{Q}_k(A) = w_k Q_k(A \cap \bar{H}_+) + w_0 Q_0(A \cap S) + \sum_{\substack{j \geq 1 \\ j \neq k}} w_j Q_j(A \cap H_-).$$

Proof. Lemma 5.3.1. □

DEFINITION 5.3.2 (First moment on a leaf)**.** *Let $x^{(0)}, x^{(1)}, \ldots, x^{(d)}$ be the coordinate functions on $\mathbb{R}^{d+1}$. The* first moment *of the measure Q on the kth leaf L_k is the real number*

$$m_k = \int_{\mathbb{R}^{d+1}} x^{(0)}\, d\tilde{Q}_k(x) = \int_{\mathscr{O}} (\pi_0 F_k p) dQ(p),$$

where $\pi_0 : \mathbb{R}^{d+1} \to \mathbb{R}$ is the orthogonal projection with kernel $H = \{0\} \times \mathbb{R}^d$.

REMARK 5.3.3. *For any point $p \in \mathscr{O}$, the projection $\pi_0 F_k p$ is positive if $p \in L_k^+$ and negative if $p \in L_j^+$ for some $j \neq k$. Moreover, $|\pi_0 F_k p| = |x^{(0)}|$ is the distance of p from the spine. The integrability in Eq.* (5.2) *guarantees that the first moments of Q are all finite.*

THEOREM 5.3.1. *Under integrability* (5.2) *and nondegeneracy* (5.4)*, either*

1. *$m_j < 0$ for all indices $j \in \{1, \ldots, K\}$, or there is exactly one index $k \in \{1, \ldots, K\}$ such that $m_k \geq 0$, in which case, either*
2. *$m_k > 0$, or*
3. *$m_k = 0$.*

Proof. For $k = 1, \ldots, K$, let

$$v_k = \int_{H_+} x^{(0)}\, dQ_k(x).$$

The nondegeneracy (5.4) implies that $v_k > 0$. Observe that

$$m_k = w_k v_k - \sum_{\substack{j \geq 1 \\ j \neq k}} w_j v_j.$$

For any $j \neq k \in \{1, \ldots, K\}$,

$$m_j = w_j v_j - \sum_{\substack{\ell \geq 1 \\ \ell \neq j}} w_\ell v_\ell \leq w_j v_j - w_k v_k \leq \Big(\sum_{\substack{\ell \geq 1 \\ \ell \neq k}} w_\ell v_\ell \Big) - w_k v_k = -m_k,$$

since the weights w_ℓ are nonnegative. Therefore, if $m_k > 0$ for some k, then $m_j \leq -m_k < 0$ for all $j \neq k$. Also, if $m_k = 0$ for some index k, then $m_j \leq 0$ for all $j \neq k$.

Now suppose there are two indices $j, k \in \{1, \ldots, K\}$ such that $j \neq k$ and $m_j = 0$ and $m_k = 0$. Then

$$0 = m_j = w_j v_j - w_k v_k - \sum_{\substack{\ell \geq 1 \\ \ell \neq j,k}} w_\ell v_\ell$$

and

$$0 = m_k = w_k v_k - w_j v_j - \sum_{\substack{\ell \geq 1 \\ \ell \neq j,k}} w_\ell v_\ell.$$

Adding these two equalities results in

$$0 = m_j + m_k = -2 \sum_{\substack{\ell \geq 1 \\ \ell \neq j,k}} w_\ell v_\ell.$$

Since $w_\ell v_\ell \geq 0$, it follows that $w_\ell v_\ell = 0$ for all $i \neq j,k$. Consequently, $Q(L_\ell^+) = 0$ for all $\ell \neq j,k$. However, this contradicts nondegeneracy (5.4) and the fact that $K \geq 3$. Hence at most one of the numbers m_k can be nonnegative. □

Motivated by Theorem 5.5.1 and Corollary 5.5.1, we use the following terms to describe the three mutually exclusive conditions given in Theorem 5.3.1:

DEFINITION 5.3.3. *Under integrability* (5.2) *and nondegeneracy* (5.4), *we say that the mean of the measure Q is*

1. sticky *if* $m_j < 0$ *for all indices* $j \in \{1,\ldots,K\}$, *or*
2. nonsticky *if* $m_k > 0$ *for some (unique)* $k \in \{1,\ldots,K\}$, *or*
3. partly sticky *if* $m_k = 0$ *for some (unique)* $k \in \{1,\ldots,K\}$.

REMARK 5.3.4. *If square integrability (5.3) also holds, the first moment* m_k *may be identified with the partial derivative*

$$m_k = -\frac{\partial \Gamma_k}{\partial x^{(0)}}(x)\Big|_{x^{(0)}=0}$$

where $\Gamma_k : \mathbb{R}^{d+1} \to \mathbb{R}$ *is defined by*

$$\Gamma_k(x) = \frac{1}{2}\int_{\mathbb{R}^{d+1}} |x-y|^2 \, d\tilde{Q}_k(y).$$

Observe that $-\frac{\partial \Gamma_k}{\partial x^{(0)}}(x)$ *depends on* $x^{(0)}$, *but not on* $(x^{(1)},\ldots,x^{(d)})$.

5.4 Sample Means

Recall that given a sample of points $\{p_n\}_{n=1}^N \subset \mathscr{O}$, the Fréchet sample mean is a natural generalization of the arithmetic mean in Euclidean space:

DEFINITION 5.4.1. *The* Fréchet sample mean *of a sample* $\{p_n\}_{n=1}^N \subset \mathscr{O}$ *of points is*

$$\bar{p}_{\mathscr{F}} = \operatorname*{arg\,min}_{p \in \mathscr{O}} \Big(\sum_{n=1}^N \rho(p,p_n)^2 \Big).$$

By Lemma 5.2.1 and Sturm (2003) [127, proposition 4.3], the sample mean $\bar{p}_{\mathscr{F}} \in \mathscr{O}$ exists and is unique.

DEFINITION 5.4.2. *For fixed $k \in \{1,\dots,K\}$, the point $\eta_{k,N} \in \mathbb{R}^{d+1}$ defined by*

$$\eta_{k,N} = \frac{1}{N}\sum_{n=1}^{N} F_k p_n \tag{5.5}$$

is the kth folded average*: the sample mean of the pushforward under the kth folding map.*

For a sample $\{p_n\}_{n=1}^N \subset \mathscr{O}$, the condition $\bar{p}_{\mathscr{F}} \in L_0$ does not necessarily imply $\eta_{k,N} \in H$. Nevertheless, the following lemma establishes an important relationship between $\bar{p}_{\mathscr{F}}$ and $\eta_{k,N}$. Specifically, taking sample means commutes with the kth folding in two cases: if the sample mean lies off the spine in L_k^+; or if the kth folded average lies in the closure of the positive half-space.

LEMMA 5.4.1. *Let $\{p_n\}_{n=1}^N \subset \mathscr{O}$ and $b_N = \bar{p}_{\mathscr{F}}$. If $b_N \in L_k^+$, then $\eta_{k,N} \in H_+$ and $\eta_{k,N} = F_k b_N$. If $\eta_{k,N} \in \bar{H}_+$, then $b_N \in L_k$ and $F_k b_N = \eta_{k,N}$ (i.e., $b_N = (\eta_{k,N}, k)$).*

Proof. Let $k,\ell \in \{1,\dots,K\}$. If $p \in L_k$, then $\rho(p,p_n) = |(F_k p) - (F_k p_n)|$. Therefore, if $b_N \in L_k^+$ then

$$b_N = \arg\min_{p\in\mathscr{O}} \sum_{n=1}^{N} \rho(p,p_n)^2 = \arg\min_{p\in L_k^+} \sum_{n=1}^{N} |F_k p - (F_k p_n)|^2.$$

Since F_k is continuously bijective from L_k to $\bar{H}_+$, this implies that the function

$$z \mapsto \sum_{n=1}^{N} |z - (F_k p_n)|^2$$

attains a local minimum in the open set H_+. However, this functional has only one local minimizer, which must be the unique global minimizer $\eta_{k,N}$:

$$\eta_{k,N} = \arg\min_{z\in\mathbb{R}^{d+1}} \sum_{n=1}^{N} |z - (F_k p_n)|^2.$$

Consequently, $\eta_{k,N} \in H_+$ and hence $F_k b_N = \eta_{k,N}$.

If $b_N \notin L_k$, then $b_N \in L_\ell^+$ for some $\ell \neq k$. Hence $\eta_{\ell,N} = F_\ell b_N$, as we have shown. In particular, $\eta_{\ell,N} \in H_+$ and $\pi_0 \eta_{\ell,N} > 0$. Hence

$$\sum_{p_n\in L_\ell^+} \pi_0 F_\ell p_n > -\sum_{p_n\notin L_\ell^+} \pi_0 F_\ell p_n \geq -\sum_{p_n\in L_k} \pi_0 F_\ell p_n = \sum_{p_n\in L_k} \pi_0 F_k p_n. \tag{5.6}$$

Therefore,

$$\begin{aligned}\pi_0\eta_{k,N} = \frac{1}{N}\sum_{n=1}^{N}\pi_0 F_k p_n &\leq \frac{1}{N}\sum_{p_n\in L_k}\pi_0 F_k p_n + \frac{1}{N}\sum_{p_n\in L_\ell^+}\pi_0 F_k p_n \\ &= \frac{1}{N}\sum_{p_n\in L_k}\pi_0 F_k p_n - \frac{1}{N}\sum_{p_n\in L_\ell^+}\pi_0 F_\ell p_n.\end{aligned}$$

Because of Eq. (5.6), this last expression is negative. Hence, we have shown that $b_N \notin L_k$ implies $\eta_{k,N} \in H_-$. Therefore, if $\eta_{k,N} \in \bar{H}_+$ it must be that $b_N \in L_k$. Consequently, as above,

$$\begin{aligned} b_N &= \arg\min_{p\in\mathscr{O}} \sum_{n=1}^{N} \rho(p,p_n)^2 \\ &= \arg\min_{p\in L_k} \sum_{n=1}^{N} |F_k p - F_k p_n|^2 \\ &= F_k^{-1}\left(\arg\min_{z\in\bar{H}_+}\sum_{n=1}^{N}|z - F_k p_n|^2\right) \\ &= F_k^{-1}\eta_{k,N}.\end{aligned}$$

Note that $F_k^{-1}\eta_{k,N}$ is well defined, since $\eta_{k,N} \in \bar{H}_+$. □

DEFINITION 5.4.3. *Given a point* $p = (x, j) = (x^{(0)}, x^{(1)}, \ldots, x^{(d)}, j) \in \mathscr{O}$,

$$P_S\, p = (x^{(1)}, \ldots, x^{(d)}) \in S$$

is the orthogonal projection of p onto the spine S.

The following lemma shows that taking sample means commutes with projection to the spine.

LEMMA 5.4.2. *If* $\{p_n\}_{n=1}^{N} \subset \mathscr{O}$ *and*

$$\bar{y}_N = \frac{1}{N}\sum_{n=1}^{N} P_S\, p_n,$$

then $\bar{y}_N = P_S\, b(p_1, \ldots, p_N)$.

Proof. Let $\pi_S : \mathbb{R}^{d+1} \to \mathbb{R}^d$ be the orthogonal projection onto the last d coordinates. Let $b_N = \bar{p}_{\mathscr{F}}$. If $b_N \in L_k^+$ for some k, then $\eta_{k,N} = F_k b_N$ by Lemma 5.4.1. Therefore, since $P_S\, p = \pi_S F_k\, p$ for all $p \in \mathscr{O}$,

$$P_S\, b_N = \pi_S F_k b_N = \pi_S \eta_{k,N} = \frac{1}{N}\sum_{n=1}^{N}\pi_S F_k p_n = \frac{1}{N}\sum_{n=1}^{N} P_S\, p_n = \bar{y}_N.$$

On the other hand, if $b_N \in L_0$ then by definition of b_N,

$$b_N = \arg\min_{p \in L_0} \sum_{n=1}^{N} \rho(p, p_n)^2 = \arg\min_{p \in S} \sum_{n=1}^{N} \left(|\pi_0 p_n|^2 + |p - P_S p_n|^2\right).$$

Therefore, $P_S b_N = \arg\min_{y \in \mathbb{R}^d} \sum_{n=1}^{N} |y - P_S p_n|^2 = \frac{1}{N} \sum_{n=1}^{N} P_S p_n = \bar{y}_N$, as desired. □

5.5 Random Sampling and the Law of Large Numbers

We now consider random sample $\{p_n\}_{n=1}^{N}$ from the probability measure Q on $\mathscr{O}$; we wish to understand the statistical behavior of their sample mean for large N. More precisely, let $(\Omega, \mathscr{F}, \mathbb{P})$ be a probability space, and for each integer $n \geq 1$ let $p_n(\omega) : \Omega \to \mathscr{O}$ for fixed $\omega \in \Omega$ be a random point in $\mathscr{O}$. Assume that p_n and p_r are independent random variables for all $n \neq r$, and that for any Borel set $A \subseteq \mathscr{O}$ and all $n \geq 1$,

$$\mathbb{P}(p_n \in A) = \mathbb{P}\big(\{\omega \in \Omega \mid p_n(\omega) \in A\}\big) = Q(A).$$

The sample space Ω may be constructed as the set of infinite sequences $(p_1, p_2, p_3, \ldots)$ of points in $\mathscr{O}$ endowed with the product measure $\mathbb{P} = \prod_{n=1}^{\infty} Q(p_n)$ on the σ-algebra $\mathscr{F}$ generated by cylinder sets. Observe that the folded points $\{F_k p_n(\omega)\}_{n=1}^{\infty} \subset \mathbb{R}^{d+1}$ are independent, each distributed according to $\tilde{Q}_k$.

DEFINITION 5.5.1. *For any positive integer N, let $b_N(\omega) = \bar{p}_{\mathscr{F}}$ denote the sample mean of the random sample $\{p_1(\omega), \ldots, p_N(\omega)\}$. This random point in $\mathscr{O}$ is the* empirical mean *of the distribution Q. Similarly, for $k \in \{1, \ldots, K\}$, the random point $\eta_{k,N}(\omega) \in \mathbb{R}^{d+1}$ denotes the kth* folded average *of the random sample $\{p_1(\omega), \ldots, p_N(\omega)\}$, as defined by (5.5).*

The goal is to understand the statistical behavior of empirical means b_N as $N \to \infty$. A consequence of the square integrability assumption (5.3) is that the limit μ_I in the next result is the mean of a random point on $\mathscr{O}$ having probability measure Q [52]:

$$\mu_I = \arg\min_{p \in \mathscr{O}} \int_{\mathscr{O}} \rho(p, q)^2 \, dQ(q). \tag{5.7}$$

LEMMA 5.5.1 (Strong Law of Large Numbers). *There is a unique point $\mu_I \in \mathscr{O}$ such that*

$$\lim_{N \to \infty} b_N(\omega) = \mu_I$$

holds $\mathbb{P}$-almost surely. The point μ_I is the Fréchet mean relative to the intrinsic distance on $\mathscr{O}$ *(or* intrinsic mean*) of Q.*

Proof. This is a special case of [150] and can also be regarded as a special case of Sturm (2009) [127, proposition 6.6], whose generality occurs in the context of distributions on globally non-positively curved spaces. (An elementary proof from scratch is also possible, using arguments similar to the proof of Theorem 5.5.1.) □

THEOREM 5.5.1 (Sticky LLN). *Assume integrability* (5.2) *and nondegeneracy* (5.4).

1. *If the moment m_j satisfies $m_j < 0$, then there is a random integer $N^*(\omega)$ such that $b_N(\omega) \notin L_j^+$ for all $N \geq N^*(\omega)$ holds $\mathbb{P}$-almost surely. Furthermore, $\mu_I \notin L_j^+$.*
2. *If the moment m_k satisfies $m_k > 0$, then there is a random integer $N^*(\omega)$ such that $b_N(\omega) \in L_k^+$ for all $N \geq N^*(\omega)$ holds $\mathbb{P}$-almost surely. Furthermore, $\mu_I \in L_k^+$.*
3. *If the moment m_k satisfies $m_k = 0$, then there is a random integer $N^*(\omega)$ such that $b_N(\omega) \in L_k$ for all $N \geq N^*(\omega)$ holds $\mathbb{P}$-almost surely. Furthermore, $\mu_I \in L_k$.*

Proof. By the usual strong law of large numbers,

$$\lim_{N\to\infty} \eta_{k,N} = \bar{\eta}_k = \int_{\mathbb{R}^{d+1}} x d\tilde{Q}_k(x)$$

holds $\mathbb{P}$-almost surely. Observe that $m_k = \pi_0 \bar{\eta}_k$. Therefore, if $m_k > 0$, $\bar{\eta}_k \in H_+$ and $\eta_{k,N} \in H_+$ for all sufficiently large N. In that case, $b_N \in L_k^+$ for all sufficiently large N by Lemma 5.4.1. In fact, $\pi_0 b_N = \pi_0 \eta_{k,N} > m_k/2 > 0$ for N sufficiently large, so by virtue of Lemma 5.5.1, $\mu_I \in L_k^+$. The same argument starting with $m_k \geq 0$ proves the case $m_k = 0$. On the other hand, if $m_j < 0$, then $\eta_{j,N} \in H_-$ for all sufficiently large N; Lemma 5.4.1 implies that $b_N \notin L_j^+$ for all sufficiently large N, and $\mu_I \notin L_j^+$. □

As a consequence, to say that the intrinsic mean of Q is sticky implies that the empirical mean b_N sticks to the spine $L_0 \subset \mathscr{O}$ for all sufficiently large N, in the following sense.

COROLLARY 5.5.1. *If the mean of Q is sticky, then there is a random integer $N^*(\omega)$ such that $b_N(\omega) \in L_0$ for all $N \geq N^*(\omega)$ holds $\mathbb{P}$-almost surely. Moreover, $\mu_I \in L_0$. If the mean of Q is partly sticky, with $m_k = 0$, then there is a random integer $N^*(\omega)$ such that $b_N(\omega) \in L_k$ for all $N \geq N^*(\omega)$ holds $\mathbb{P}$-almost surely. Moreover, $\mu_I \in L_k$.*

Recall that P_S is the orthogonal projection onto the spine S. The measure Q pushes forward along the projection to a measure $Q_S = Q \circ P_S^{-1}$ on S:

$$Q_S(A) = Q(P_S^{-1} A)$$

for any Borel set $A \subseteq \mathbb{R}^d$. Note that $Q_0(A) \leq Q_S(A)$ for all Borel sets $A \subseteq S$, but $Q_S \neq Q_0$ by Convention 5.3.1.

COROLLARY 5.5.2. *In all cases (sticky, non-sticky, partly sticky), the limit $\mu_I \in \mathcal{O}$ satisfies*

$$P_S \mu_I = \int_S y\, dQ_S(y). \tag{5.8}$$

Proof. By Lemma 5.4.2 and Theorem 5.5.1,

$$P_S \mu_I = P_S \lim_{N \to \infty} b_N = \lim_{N \to \infty} \bar{y}_N$$

holds almost surely. By the strong law of large numbers for $\bar{y}_N \in S = \mathbb{R}^d$, the last limit is (5.8). □

5.6 Central Limit Theorems

In this section we consider fluctuations of the empirical mean $b_N(\omega)$ about the asymptotic limit μ_I, within the tangent cone at μ_I. We have shown that if the mean is either sticky or partly sticky, then $\mu_I \in S$, and the tangent cone at μ_I is an open book $\mathcal{O}$. On the other hand, if the mean is non-sticky, with $m_k > 0$, then μ_I is in the interior of the leaf L_k^+ and the tangent cone at μ_I is the vector space $\mathbb{R}^{d+1}$. We treat these two scenarios separately.

These facts essentially follow Theorem 5.5.1, which shows that in the sticky cases with probability one the fluctuations away from the mean in certain directions stop as more random variables are added to the empirical mean. In particular, this implies that the correctly normalized limit of the fluctuation from the mean cannot in the sticky case converge to a Gaussian random variable as one would have in the standard central limit theorem. Since the fluctuations in some directions are exactly zero at some point along each sequence of random variables, it is not all together surprising that limiting measure has mass concentrated on a lower-dimensional set. This is the content of Theorem 5.6.1, which is the principal result of this section.

5.6.1 *Sticky Central Limit Theorem*

Throughout this section, we assume that the mean is either sticky (with first moments $m_j < 0$ for all j) or partly sticky (with $m_k = 0$), i.e., we have that $m_j < 0$ for all $k \neq j \in \{1, \ldots, K\}$ and $m_k \leq 0$. Hence $\mu_I \in L_0$. The central limit theorem involves a centered and rescaled empirical mean.

DEFINITION 5.6.1 (Rescaled Empirical Mean ν_N). *Assume that $P_S\, \mu_I = 0$ (after the action of $-P_S\, \mu_I \in S$ on $\mathcal{O}$ as explained in Remark 5.2.1 if necessary). The* rescaled empirical mean *is the probability law ν_N on the open book $\mathcal{O}$*

induced by the random variable $\sqrt{N} b_N \in \mathscr{O}$*:*

$$\mathbb{P}\big(\{\omega \mid \sqrt{N} b_N(\omega) \in A\}\big) = \int_{\mathscr{O}\cap A} d\nu_N(p)$$

for all Borel sets $A \subseteq \mathscr{O}$.

Since in sticky settings we need to collapse fluctuations in some directions back to the spine, it is convenient to define the following projection.

DEFINITION 5.6.2. *The convex projection* $\hat{P}$ *of* $\mathbb{R}^{d+1}$ *onto* $\bar{H}_+$ *is*

$$\hat{P}x = \begin{cases} (0, x^{(1)}, \dots, x^{(d)}) & \text{if } x^{(0)} < 0, \\ (x^{(0)}, x^{(1)}, \dots, x^{(d)}) & \text{if } x^{(0)} \geq 0. \end{cases}$$

We now define measures which we will see shortly describe the limiting behaviors of ν_N as $N \to \infty$. In short, they are the limiting measures in the central limit theorem given in Theorem 5.6.1 below.

DEFINITION 5.6.3. *Assume square integrability* (5.3) *and assume that* $P_S \mu_I = 0$.

1. *The* spinal limit measure g_S *is the law of a multivariate normal random variable on the spine* $S \cong \mathbb{R}^d$ *with mean zero and covariance matrix*

$$\Sigma_S = \int_{\mathbb{R}^d} yy^T \, Q_S(dy) = \int_{\mathscr{O}} (P_S\, p)(P_S\, p)^T \, Q(dp).$$

2. *The* k^{th} costal[1] limit measure g_k *is the law of a multivariate normal random variable on* $\mathbb{R}^{d+1}$ *with mean zero and covariance matrix*

$$\Sigma_k = \int_{\mathbb{R}^{d+1}} xx^T \, \tilde{Q}_k(dx) = \int_{\mathscr{O}} (F_k\, p)(F_k\, p)^T \, Q(dp).$$

3. *The* k^{th} spinocostal[2] limit measure h_k *on the closed leaf* $L_k \cong \bar{H}_+$ *is defined by*

$$h_k(A) = h_k^0\big(F_k(A) \cap H\big) + g_k\big(F_k(A) \cap \bar{H}_+\big)$$

for Borel sets $A \subseteq L_k$*, where the* semi-spinal limit measure h_k^0 *on* L_0 *is defined by*

$$h_k^0\big((P_S|_{L_0})^{-1} B\big) = g_S(B) - g_k\big((0,\infty) \times B\big)$$

for Borel sets $B \subseteq S$*. (A possibly more natural definition of* h_k *is given in Proposition 5.6.1 below.)*

REMARK 5.6.1. *Square integrability* (5.3) *implies that the covariance matrices are finite.*

---

[1] Adjective: of or pertaining to the ribs, in anatomy.

[2] Adjective: spanning the ribs and spine, in anatomy.

REMARK 5.6.2. *The semi-spinal limit measure is generally not Gaussian. Although the orthogonal projection to $\mathbb{R}^d$ of any Gaussian measure on $\mathbb{R}^{d+1}$ is Gaussian, h_k^0 is the projection of only half of a Gaussian; this is implied by Proposition 5.6.1, an alternate direct description of h_k interpolating between the first two parts of Definition 5.6.3.*

PROPOSITION 5.6.1. *The spinocostal limit measure is the pushforward of the costal limit measure g_k under convex projection: $h_k = g_k \circ \hat{P}^{-1} \circ F_k$.*

Proof. Since the measures agree on L_k outside of L_0 by definition, it is enough to show that

$$h_k^0\big((P_S|_{L_0})^{-1}B\big) = g_k\big(\hat{P}^{-1} \circ (\pi_S|_H)^{-1}B\big) \tag{5.9}$$

for any Borel set $B \subseteq S$. For any vectors $w, w' \in \mathbb{R}^{d+1}$ that lie on the spine $H \subseteq \mathbb{R}^{d+1}$, considering them as vectors in $z = \pi_S(w), z' = \pi_S(w') \in S = \mathbb{R}^d$ results in a quantity $z^T \Sigma_S z'$, and $w^T \Sigma_k w'$. The integrals in Definition 5.6.3 directly imply that $z^T \Sigma_S z' = w^T \Sigma_k w'$. Consequently, the matrix Σ_S is a submatrix of Σ_k; the action of Σ_k on the subspace H is given by Σ_S. Thus $g_S(B) = g_k\big((-\infty,\infty) \times B\big)$, and hence by definition

$$\begin{aligned} h_k^0(B) &= g_k\big((-\infty,\infty) \times B\big) - g_k\big((0,\infty) \times B\big) = g_k\big((-\infty,0] \times B\big) \\ &= g_k\big(\hat{P}^{-1} \circ (\pi_S|_H)^{-1}B\big), \end{aligned}$$

for any Borel set $B \subseteq S$. □

Now we come to the primary result: as the sample size N becomes large, the rescaled empirical mean ν_N converges weakly to the appropriate measure from Definition 5.6.3, according to how sticky the mean is. When the mean is

1. sticky, ν_N converges weakly to the spinal limit measure g_S.
2. partly sticky, ν_N converges weakly to the spinocostal limit measure g_j supported on the (unique) leaf L_k with moment $m_k = 0$.

As discussed at the start of the section, the fact that the limiting distribution is supported on the spine S when the mean is sticky follows from Theorem 5.5.1, since then the first moments m_j are strictly negative for all j.

THEOREM 5.6.1 (Sticky CLT). *Let Q be a nondegenerate* (5.4) *probability distribution on the open book $\mathscr{O}$ with finite second moment* (5.3).

1. *If the mean of Q is sticky, then for any continuous, bounded function $\phi : \mathscr{O} \to \mathbb{R}$,*

$$\lim_{N\to\infty} \int_{\mathscr{O}} \phi(p)\, d\nu_N(p) = \int_S \phi \circ (P_S|_{L_0})^{-1}(q)\, dg_S(q).$$

2. *If the mean of Q is nonsticky, then see Theorem 5.6.2.*
3. *If the mean of Q is partly sticky, with first moment $m_k = 0$, then for any*

continuous bounded function $\phi : \mathscr{O} \to \mathbb{R}$,

$$\lim_{N\to\infty} \int_{\mathscr{O}} \phi(p)\, d\nu_N(p) = \int_{\bar{H}_+} \phi \circ F_k^{-1}(q) dh_k(q).$$

Proof. The proof works by decomposing the relevant measures—the empirical mean on the open book and its pushforward to $\mathbb{R}^{d+1}$ under folding—into pieces corresponding to the leaves and the spine.

Suppose that the mean is partly sticky with first moment $m_k = 0$. Let $\eta_N = \eta_{k,N}$ as in (5.5), and let $\nu_{\eta,N}(x)$ denote the law of $\sqrt{N}\eta_N$ on $\mathbb{R}^{d+1}$. By Lemma 5.4.1, $\nu_N(A) = \nu_{\eta,N}(F_k A)$ for any Borel set $A \subseteq L_k$, and if ϕ is a continuous and bounded function, then

$$\begin{aligned}\int_{\mathscr{O}} \phi(p)\, d\nu_N(p) &= \int_{L_k^+} \phi(p)\, d\nu_N(p) + \int_{\mathscr{O}\setminus L_k^+} \phi(p)\, d\nu_N(p)\\ &= \int_{H_+} \phi\left((F_k^{-1}|H_+)^{-1}(y)\right) d\nu_{\eta,N}(y) + \int_{\mathscr{O}\setminus L_k^+} \phi(p)\, d\nu_N(p).\end{aligned}$$

The standard CLT in $\mathbb{R}^{d+1}$ (see, e.g., Breiman (1992) [27, theorem 11.10]) implies that the random variable $\sqrt{N}\eta_N$ converges in distribution to a centered Gaussian with covariance Σ_k. Therefore,

$$\lim_{N\to\infty} \int_{H_+} \phi\left((F_k^{-1}|H_+)^{-1}(y)\right) d\nu_{\eta,N}(y) = \int_{H_+} \phi\left((F_k^{-1}|H_+)^{-1}(y)\right) dg_k(y).$$

LEMMA 5.6.1. *If the jth first moment satisfies $m_j < 0$, then $\nu_N(L_j^+) \to 0$ and*

$$\lim_{N\to\infty} \int_{L_j^+} \phi(p)\, d\nu_N(p) = 0.$$

Proof. Theorem 5.5.1.1. □

Resuming the proof of the theorem, consider the term

$$\int_{\mathscr{O}\setminus L_k^+} \phi(p)\, d\nu_N(p) = \int_{L_0} \phi(p)\, d\nu_N(p) + \int_{L_k^-} \phi(p)\, d\nu_N(p),$$

where $L_k^- = \mathscr{O} \setminus L_k = \bigcup_{j\neq k} L_j^+$, which excludes the spine L_0. With the projection $P_0 : \mathscr{O} \to L_0, (x^{(0)}, x, j) \mapsto (0, x, j)$ the function $p \mapsto \phi(P_0 p)$ is again continuous and bounded, Lemma 5.6.1 implies that

$$\lim_{N\to\infty} \int_{L_k^-} \phi(P_0 p)\, d\nu_N(p) = 0. \tag{5.10}$$

Therefore,

$$\begin{aligned}\lim_{N\to\infty}\int_{L_0}\phi(p)\,d\nu_N(p) &= \lim_{N\to\infty}\int_{L_0}\phi(P_0p)\,d\nu_N(p)\\ &= \lim_{N\to\infty}\left(\int_{L_k^-}\phi(P_0p)\,d\nu_N(p)+\int_{L_0}\phi(P_0p)\,d\nu_N(p)\right)\\ &= \lim_{N\to\infty}\left(\int_{\mathscr{O}}\phi(P_0p)\,d\nu_N(p)-\int_{L_k^+}\phi(P_0p)\,d\nu_N(p)\right).\end{aligned}$$

Observe that

$$\int_{\mathscr{O}}\phi(P_0p)\,d\nu_N(p)=\int_S\phi\circ(P_S|_{L_0})^{-1}(y)\,d\gamma_N(y),$$

where $\gamma_N=\nu_N\circ P_S^{-1}$ which is the law of $\sqrt{N}\bar{y}_N$ on S, where $\bar{y}_N$ is the projected barycenter from Lemma 5.4.2. Therefore, setting $\hat{\phi}=\phi\circ(P_S|_{L_0})^{-1}$ and applying the usual CLT to $\sqrt{N}\bar{y}_N\in\mathbb{R}^d$,

$$\lim_{N\to\infty}\int_{\mathscr{O}}\phi(P_0p)\,d\nu_N(p)=\lim_{N\to\infty}\int_S\hat{\phi}(y)\,d\gamma_N(y)=\int_S\hat{\phi}(y)\,dg_S(y).$$

We cannot apply the same argument to

$$\lim_{N\to\infty}\int_{L_k^+}\phi(P_0p)\,d\nu_N(p)=\lim_{N\to\infty}\int_{L_k^+}\hat{\phi}(y)\,d\tau_N(y)$$

with $\tau_N=\nu\circ(P_S|_{L_k^+})^{-1}$ because there is no CLT for τ_N. We have, however, above derived a CLT for $\nu_N\circ F_k^{-1}=\nu_{\eta,N}$ on $H_+=F_k(L_k^+)$:

$$\lim_{N\to\infty}\int_{L_k^+}\phi(P_0p)\,d\nu_N(p)=\lim_{N\to\infty}\int_{H_+}\widetilde{\phi}(q)\,d\nu_{\eta,N}(q)=\lim_{N\to\infty}\int_{H_+}\widetilde{\phi}(q)\,dg_k(q),$$

where $\widetilde{\phi}=\phi\circ P_0\circ F_k\circ\hat{P}^{-1}$. In summary, we have shown that

$$\begin{aligned}\lim_{N\to\infty}\int_{\mathscr{O}}\phi(p)\,d\nu_N(p) &= \int_{H_+}\phi\circ F_k^{-1}(q)dg_k(q)+\int_S\hat{\phi}(y)\,dg_S(y)\\ &\quad-\int_{H_+}\widetilde{\phi}(q)\,dg_k(q)\\ &= \int_{H_+}\phi\circ F_k^{-1}(q)dg_k(q)+\int_H\phi\circ F_k^{-1}(q)\,dh_k^0(q)\\ &= \int_{\bar{H}_+}\phi\circ F^{-1}(q)\,dh_k(q),\end{aligned}$$

where the second equality uses the fact that $\widetilde{\phi} = \phi \circ F_k^{-1}$ on H and the final equality the fact that g_k has no mass supported on the spine H, so the integral of $\phi \circ F^{-1}\, dg_k$ over H_+ can just as well be taken over $\bar{H}_+$.

The sticky case proceeds in much the same way as the partly sticky case does, except that instead of Eq. (5.10), the simpler statement

$$\lim_{N\to\infty} \int_{\mathscr{O}\setminus S} \phi(P_0 p)\, d\nu_N(p) = 0$$

holds. From that, the next step results in

$$\lim_{N\to\infty} \int_{L_0} \phi(p)\, d\nu_N(p) = \lim_{N\to\infty} \int_{\mathscr{O}} \phi(P_0 p)\, d\nu_N(p),$$

and then the usual CLT applied to $\sqrt{N}\bar{y}_N \in \mathbb{R}^d$ proves the desired result. □

REMARK 5.6.3. *The open book $\mathscr{O}$ of dimension $d+1$ admits isometric embeddings into Euclidean spaces of dimension as low as $d+2$. For this reason, extrinsic means on $\mathscr{O}$ can be also considered (see Chapter 4). A stickiness phenomenon for extrinsic means is present for some Euclidean embeddings, but not all of them. This is a topic of current research.*

5.6.2 Nonsticky Central Limit Theorem

If the mean is nonsticky with first moment $m_k > 0$, then the limit μ_I is in the interior of L_k^+. In this case, the tangent cone at μ_I is the vector space $\mathbb{R}^{d+1}$, and the fluctuations of b_N about the limit μ_I are qualitatively similar to what is described in the classical central limit theorem.

DEFINITION 5.6.4. *In this section we let $\widetilde{\nu}_N$ be the law on $\mathbb{R}^{d+1}$ of the random variable $\sqrt{N}(F_k b_N - F_k \mu_I)$:*

$$\mathbb{P}\left(\{\omega \mid \sqrt{N}(F_k b_N - F_k \mu_I) \in A\}\right) = \widetilde{\nu}_N(A)$$

for all Borel sets $A \subseteq \mathbb{R}^{d+1}$.

DEFINITION 5.6.5. *Assume $m_k > 0$. Let $\tilde{g}_k$ be the law of a multivariate normal random variable on $\mathbb{R}^{d+1}$ with mean zero and covariance matrix*

$$\tilde{\Sigma}_k = \int_{\mathbb{R}^{d+1}} (x - F_k\mu_I)(x - F_k\mu_I)^T\, d\tilde{Q}_k(x).$$

Unlike the case of the sticky and partly sticky mean, the weak limit of ν_N is that of a nondegenerate Gaussian on $\mathbb{R}^{d+1}$:

THEOREM 5.6.2 (Nonsticky CLT). *Assume $m_k > 0$. Then for any continuous bounded function $\phi : \mathbb{R}^{d+1} \to \mathbb{R}$,*

$$\lim_{N\to\infty} \int_{\mathbb{R}^{d+1}} \phi(x)\, d\widetilde{\nu}_N(x) = \int_{\mathbb{R}^{d+1}} \phi(x)\, d\tilde{g}_k(x).$$

Proof. Since $m_k > 0$, $\mu_I \in L_k^+$ and Lemma 5.4.1 implies $F_k\mu_I = \bar{\eta} = \int_{\mathbb{R}^{d+1}} x\, d\tilde{Q}_k(x)$. Also,

$$\sqrt{N}(F_k b_N(\omega) - F_k\mu_I) = \sqrt{N}(\eta_{k,N}(\omega) - \bar{\eta}), \quad \forall N \geq N^*(\omega)$$

holds with probability one. Therefore, for any Borel set

$$\left|\widetilde{\nu}_N(A) - \mathbb{P}\left(\{\omega \mid \sqrt{N}(\eta_{k,N}(\omega) - \bar{\eta}) \in A\}\right)\right| \leq R_N$$

where $R_N = \mathbb{P}(\{\omega \mid N < N^*(\omega)\})$. By the classical central limit theorem, the random variable $\sqrt{N}(\eta_{k,N}(\omega) - \bar{\eta})$ converges in law to a centered, multivariate Gaussian on $\mathbb{R}^{d+1}$ with covariance Σ_k as $N \to \infty$. Consequently,

$$\limsup_{N\to\infty} \left| \int_{\mathbb{R}^{d+1}} \phi(x)\, d\widetilde{\nu}_N(x) - \int_{\mathbb{R}^{d+1}} \phi(x)\, d\tilde{g}_k(x) \right| \leq 2 \limsup_{N\to\infty} R_N \|\phi\|_\infty = 0. \quad \square$$

Chapter 6

Central Limit Theorem for Random Samples on a Graph

6.1 Introduction

Note. This chapter is based on joint unpublished work with Harrie Hendriks; we are grateful to him for allowing us to bring this material to the reader. In the mathematical sense, we are considering an optimization problem

$$\int f(x;\theta)\mathbb{P}(dx)$$

where $f(x;\theta)$ is a bounded continuous function in both parameters. The complication is that $\theta \to f(x,\theta)$ is not differentiable. Moreover θ may vary through a stratified space. Given θ_0 there is the cut-locus

$$C_{\theta_0} = \{x \mid \theta \mapsto f(x;\theta) \text{ is not differentiable in } \theta_0\}.$$

In our setting we may assume that if a is a tangent vector at θ_0, that for all x the (one-sided) directional derivative $D_a f(x;\theta)$ exists. In particular $D_{\lambda a} f(x;\theta) = \lambda D_a f(x;\theta)$ for $\lambda > 0$.
Then $\int D_a f(x;\theta)\mathbb{P}(dx) > 0$ for all tangent vectors a at θ_0 implies that θ_0 is a local minimum. Moreover in that case θ_0 is sticky.
If $\int D_a f(x;\theta)\mathbb{P}(dx) = 0$ in some direction a at θ_0, it is possible that along this direction the value of $\int f(x;\theta)\mathbb{P}(dx)$ decreases, or that $\int D_a f(x;\theta)\mathbb{P}(dx)$ becomes negative.
We want to see how far we can go in case the integral represents the Fréchet function on a finite connected graph.

Consider a finite connected graph G with a distance function $\rho : G \times G \to [0,\infty)$. We assume that (G,ρ) is a path metric space in the sense of Gromov. In particular, each couple of points x,y can be connected with a geodesic whose length is exactly $\rho(x,y)$. Given a probability distribution $\mathbb{P}$ on G, we are interested in the Fréchet function V

$$G \ni \mu \mapsto V(\mu) = \mathbb{E}(\rho(X,\mu)^2) = \int_G \rho(x,\mu)^2 \mathbb{P}(dx)$$

where X denotes a random variable with distribution $\mathbb{P}$. We suppose that the Fréchet function assumes its minimum in a unique point $\mu \in G$. Ziezold showed then that a distribution that is close to $\mathbb{P}$ has a minimum close to μ. Say, if $\mathbb{P}_s \to_d \mathbb{P}$ and $\int_G \rho(x,\mu_s)^2 \mathbb{P}_s(dx)$ is minimal, then $\mu_s \to \mu$ for $s \to \infty$. Let S_μ be a star tree neighborhood rooted at y_0 in G, and consider its legs L_a, oriented such that each leg begins in y_0. Consider functions

$$\varepsilon_a : G \to \{+1,-1\}$$

defined as $\varepsilon_a(x) = -1$ if the distance from $y \in L_a$ to x diminishes as y runs from y_0 along L_a, and $\varepsilon_a(x) = +1$ if the distance from $y \in L_a$ to x increases as y runs from y_0 along L_a. More specifically, if L_a is parametrized by arc length, then $\varepsilon_a(x) = \frac{d}{dx}\big|_{y_0} \rho(x,y)$. Notice that y running along L_a initially either extends or shortens the geodesic from y_0 to x exactly by the distance from y to y_0. Given $x \in G$, there is at least one leg L_a for which $\varepsilon_a(x) = -1$, namely a geodesic from y_0 to x has to pass by one of the legs.

Notice that it is plausible that

$$\frac{d}{dy}\bigg|_{y_0} V(y) = \frac{d}{dy}\bigg|_{y_0} \int_G \rho(x,y)^2\, \mathbb{P}(dx) = 2\int_G \rho(x,y_0)\varepsilon_a(x)\mathbb{P}(dx).$$

In particular it follows that μ satisfies the conditions

$$\int_G \rho(x,\mu)\varepsilon_a(x)\mathbb{P}(dx) \geq 0 \text{ for all legs } L_a \text{ rooted at } \mu.$$

It is also plausible that

$$\frac{d^2}{dy^2}\bigg|_{y_0} V(y) = \frac{d^2}{dy^2}\bigg|_{y_0} \int_G \rho(x,y)^2\, \mathbb{P}(dx) = 2\int_G \varepsilon_a(x)^2\mathbb{P}(dx) = 2.$$

It is clear that if for some leg L_a it holds that $\int_G \rho(x,\mu)\varepsilon_a(x)\mathbb{P}(dx) > 0$, then for $\mathbb{P}'$ close enough to $\mathbb{P}$ it cannot be that a Fréchet mean of $\mathbb{P}'$ belongs to the interior of leg L_a. The Fréchet function of $\mathbb{P}'$ will also increase along L_a. If the previous condition is satisfied for all legs rooted at μ, then μ is *sticky*: if $\mathbb{P}'$ is close enough to $\mathbb{P}$, then its Fréchet mean is unique and equal to μ.
Notice that if the star tree at μ has only two legs, L_1 and L_2, then it cannot happen that $\varepsilon_1(y) = +1 = \varepsilon_2(y)$, in particular $\varepsilon_1(y) + \varepsilon_2(y) \leq 0$. As

a consequence μ cannot be sticky, because for $a = 1$ or $a = 2$ (or both) $\int_G \rho(x,\mu)\varepsilon_a(x)\mathbb{P}(dx) \leq 0$. Moreover, if $\int_G \rho(x,\mu)\varepsilon_a(x)\mathbb{P}(dx) \geq 0$ for $a = 1$ and $a = 2$, then $\int_G \rho(x,\mu)\varepsilon_a(x)\mathbb{P}(dx) = 0$ for $a = 1$ and $a = 2$.
Suppose μ is not sticky. Let L_a be a leg for which $\int_G \rho(x,\mu)\varepsilon_a(x)\mathbb{P}(dx) = 0$. We would like to study the behavior of an empirical Fréchet mean μ_n^* of a sample of size n from $\mathbb{P}$, where μ_n^* is chosen, such that its dependence on the sample is measurable. We are interested in the probability that μ_n^* belongs to L_a, and the conditional distribution of μ_n^* conditional on the fact that it lies in leg L_a.

Given μ there is a cut-locus $C = C^\mu$ of points x such that there are two legs from which a shortest geodesic starts from μ to x. The cut-locus of a point is a finite subset of the graph. There is an interesting covering by subsets $C_a = C_a^\mu$ of points $x \in C$ such that there is a shortest geodesic from μ to x starting off into leg L_a. Now Hotz and Huckemann (2011), theorem 3.1.(i), could generalize to

$$D_{L_a}V = -4 \sum_{x \in C_a} \rho(x,\mu)\mathbb{P}(X = x)$$

6.2 On a 2015 Paper by Hotz and Huckemann

Let us scrutinize the treatment of Hotz and Huckemann (2015) [68]. Suppose that P is a probability density on the circle. We parametrize the circle by $[-\pi,\pi) \to S^1$ given by $t \mapsto (\cos(t), \sin(t))$. The Fréchet function is given for $\theta > 0$ by

$$V(\theta) = \int_{-\pi+\theta}^{+\pi-} (t-\theta)^2 P(dt) + \int_{-\pi}^{-\pi+\theta} (-2\pi - (t-\theta))^2 P(dt)$$

and

$$V(-\theta) = \int_{-\pi}^{\pi-\theta} (t+\theta)^2 P(dt) + \int_{\pi-\theta}^{\pi-} (2\pi - (t+\theta))^2 P(dt)$$

Then

$$\begin{aligned} V(\theta) - V(0) &= \int_{-\pi+\theta}^{+\pi-} (t-\theta)^2 P(dt) + \int_{-\pi}^{-\pi+\theta} (-2\pi - (t-\theta))^2 P(dt) \\ &\quad - \int_{-\pi}^{+\pi-} t^2 P(dt) \\ &= \int_{-\pi+\theta}^{+\pi-} -2\theta t + \theta^2 P(dt) + \int_{-\pi}^{-\pi+\theta} 2(2\pi-\theta)t + (2\pi-\theta)^2 P(dt) \\ &= \theta^2 - 2\theta \int_{-\pi}^{+\pi-} tP(dt) - 4\pi\theta \int_{-\pi}^{-\pi+\theta} 1 - \frac{1}{\theta}(\pi+t) P(dt) \end{aligned}$$

and, analogously,

$$\begin{aligned}V(-\theta)-V(0) &= \int_{-\pi}^{\pi-\theta}(t+\theta)^2P(dt)+\int_{\pi-\theta}^{\pi-}(2\pi-(t+\theta))^2P(dt)\\ &\quad -\int_{-\pi}^{+\pi-}t^2P(dt)\\ &= \int_{-\pi}^{\pi-\theta}2\theta t+\theta^2P(dt)+\int_{\pi-\theta}^{\pi-}-2(2\pi-\theta)t+(2\pi-\theta)^2P(dt)\\ &= \theta^2+2\theta\int_{-\pi}^{+\pi-}tP(dt)-4\pi\theta\int_{\pi-\theta}^{\pi-}1-\frac{1}{\theta}(\pi-t)P(dt)\end{aligned}$$

Assume $V(0)$ is the minimal value of the Fréchet function. First of all, since $V(\theta)\geq V(0)$ and $V(-\theta)\geq V(0)$ and the integral terms with the $4\pi\theta$ factor being positive, we must have

$$\int_{-\pi}^{+\pi-}tP(dt)=0$$

Notice that for leg L_a of a star tree neighborhood around 0 pointing in the negative direction

$$\int_{-\pi}^{+\pi-}tP(dt)=\int_{S^1}\rho(x,\mu)\varepsilon_a(x)\mathbb{P}(dx).$$

Clearly

$$\int_{-\pi}^{-\pi+\theta}1-\frac{1}{\theta}(\pi+t)P(dt)\to\mathbb{P}(-\pi)\text{ and }\int_{\pi-\theta}^{\pi-}1-\frac{1}{\theta}(\pi-t)P(dt)\to 0\text{ as }\theta\downarrow 0$$

and it follows that $\mathbb{P}(-\pi)=0$.

Suppose that $\mathbb{P}$ has a continuous density p in the neighborhood of $-\pi$ and π such that $\lim_{t\uparrow\pi}p(t)$ exists, denoted by $p(\pi-)$. Then it seems clear that

$$\begin{aligned}&\frac{2}{\theta}\int_{-\pi}^{-\pi+\theta}1-\frac{1}{\theta}(\pi+t)P(dt)\to p(-\pi)\text{ and}\\ &\frac{2}{\theta}\int_{\pi-\theta}^{\pi-}1-\frac{1}{\theta}(\pi-t)P(dt)\to p(\pi-)\text{ as }\theta\downarrow 0.\end{aligned}$$

Notice that

$$\frac{V(\theta)-V(0)}{\theta^2}\to 1-2\pi p(-\pi)\text{ and }\frac{V(-\theta)-V(0)}{\theta^2}\to 1-2\pi p(\pi-)\text{ as }\theta\downarrow 0.$$

Then in order that 0 is a minimum of the Fréchet function, we need

$$1-2\pi p(-\pi)\geq 0 \text{ and } 1-2\pi p(\pi-)\geq 0 \text{ so that } p(-\pi)\leq \frac{1}{2\pi} \text{ and } p(\pi-)\leq \frac{1}{2\pi}.$$

6.3 From Graph to Line

Consider a finite connected graph G with intrinsic distance function ρ. Suppose given a probability measure $\mathbb{P}$ on G and a random variable X with this distribution. Then there is the Fréchet function $V: G \to \mathbb{R}$:

$$V(y)=\mathbb{E}(\rho(x,y)^2).$$

Let be given $\eta \in G$ and a (geodesic) star neighborhood of η in G, with legs L_a, $a=1,\ldots,p$. The legs are oriented in such a way that they start at η. Given leg L_a define $\varepsilon_a : G \to \{+1,-1\}$. For $x \in G$ let $\varepsilon_a(x)=+1$ if $\rho(x,y)$ increases initially while y enters leg L_a from the starting point η and let $\varepsilon_a(x)=-1$ if $\rho(x,y)$ decreases initially while y enters leg L_a. In particular there is the somewhat special case $\varepsilon_a(\eta)=+1$.
Using these functions we can define $\mathbb{P}i_a : G \to \mathbb{R}$ by

$$\mathbb{P}i_a(x)=\rho(x,\eta)\varepsilon_a(x).$$

The induced probability measure $\mathbb{P}\mathbb{P}i_a^{-1}$ on $\mathbb{R}$ will be denoted by $\mathbb{P}_a$. Notice that at some small interval $(-\rho,0]$ mapping $\mathbb{P}i_a$ is invertible, and the inverse map maps into leg L_a. Namely, on a small star neigborhood only L_a is mapped into $(-\infty,0]$, while points outside the star neighborhood are mapped in the complement of a neighborhood of 0. It is clear that for y close to η and belonging to leg L_a it holds that $\rho(x,y)\leq|\rho(x,\eta)\varepsilon_a(x)-\rho(y,\eta)\varepsilon_a(y)|$. That is why

$$V(y)\leq\int_{\mathbb{R}}(t-\mathbb{P}i_a(y))^2\mathbb{P}_a(dt), \tag{6.1}$$

with equality for $y=\eta$.

Now assume that η is a minimum of the Fréchet function and apply the above. Of course, the right-hand side of (6.1) is minimal for $\mathbb{P}i_a(y)=\int t\mathbb{P}_a(dt)$. The left-hand side of (6.1) is minimal for $y=\eta$. Since $V(\eta)=\int t^2\mathbb{P}_a(dt)$, it follows that for $y\in L_a$ close to η

$$\begin{aligned} V(\eta) &= \int_{\mathbb{R}} t^2 P_a(dt)\leq V(y)\leq\int_{\mathbb{R}}(t-\mathbb{P}i_a(y))^2\mathbb{P}_a(dt)\\ &= \int_{\mathbb{R}} t^2 P_a(dt)-2\mathbb{P}i_a(y)\int_{\mathbb{R}} tP_a(dt)+\mathbb{P}i_a(y)^2. \end{aligned}$$

Since $\mathbb{P}i_a(y) < 0$ for all $y \neq \eta$ close to η,

$$-2\int_{\mathbb{R}} tP_a(dt) + \mathbb{P}i_a(y) \leq 0$$

or $\int t\mathbb{P}_a(dt) \geq \frac{1}{2}\mathbb{P}i_a(y)$. Since $\mathbb{P}i_a(y) \uparrow 0$ as $y \to \eta$, it follows that

$$0 \leq \int_{\mathbb{R}} t\mathbb{P}_a(dt) = \int_G \rho(x,\eta)\varepsilon_a(x)\mathbb{P}(dx). \tag{6.2}$$

Summarizing we have:

THEOREM 6.3.1. *Let η be a global Fréchet mean and given a star neighborhood of η. The legs of the star neighborhood are denoted by L_a, $a = 1,\dots,p$. Then it holds for each leg L_a that*

$$0 \leq \int_{\mathbb{R}} t\mathbb{P}_a(dt) = \int_G \rho(x,\eta)\varepsilon_a(x)\mathbb{P}(dx).$$

Star neighborhood of degree 2

Notice that in general for $x \neq \eta$ one has $\varepsilon_a(x) < 1$ for at least one leg L_a. In case a star neighborhood of η has degree 2, it follows that $\rho(x,\eta)(\varepsilon_1(x) + \varepsilon_2(x)) \leq 0$. In case η is Fréchet mean, inequalities (6.2) lead to

$$0 \leq \int t\mathbb{P}_1(dt) + \int t\mathbb{P}_2(dt) = \int \rho(x,\eta)(\varepsilon_1(x) + \varepsilon_2(x))\mathbb{P}(dx) \leq 0.$$

This implies that $\mathbb{P}$-almost surely $\rho(x,\eta)(\varepsilon_1(x) + \varepsilon_2(x)) = 0$: η cannot be sticky. Moreover from inequalities (6.2)

$$0 = \int_G \rho(x,\eta)\varepsilon_1(x)\mathbb{P}(dx) = \int_G \rho(x,\eta)\varepsilon_2(x)\mathbb{P}(dx).$$

and there cannot be some $x \neq \eta$ with $\varepsilon_1(x) = \varepsilon_2(x) = -1$ and $\mathbb{P}(x) > 0$. Define for any $z \in G$ the *cut-locus* C_z of z as those points x such that there are two legs L_1 and L_2 of the star tree neighborhood of z for which $\varepsilon_1(x) = \varepsilon_2(x) = -1$. Then we have seen that $\mathbb{P}(C_\eta) = 0$ in case the star tree neighborhood of a Fréchet mean η has degree 2.

Star neighborhood of degree greater than 2

First we consider an example graph. Define a kite graph by four equilateral edges, three of them forming a triangle, and the remaining edge e_4 connected at one end point to a vertex μ of the triangle. Let x_1 be the other end point of edge e_4. Let x_2 and x_3 be the vertexes of the triangle, different from μ and let x_4 be the middle of the edge connecting x_2 and x_3. Put mass $\frac{1}{2}$ to vertex x_1, $\frac{1}{8}$ to vertices x_2 and x_3 and $\frac{1}{4}$ to x_4. Then μ is a sticky Fréchet mean, despite the fact that x_4 is in the cut-locus of μ. Suppose $\mathbb{P}^*$ is a probability distribution that is close to $\mathbb{P}$, and suppose $\mathbb{P}$ has a unique Fréchet mean η. Then $\mathbb{P}^*$ will have a Fréchet mean η^* in the neighborhood of η. The Fréchet mean of $\mathbb{P}^*$ is possibly not unique. Given that for leg L_a of a star neighborhood of Fréchet mean η it holds that

$$\int_G \rho(x,\eta)\varepsilon_a(x)\mathbb{P}(dx) \geq 0.$$

We want to see what happens for y in leg L_a, y close to η, when $\mathbb{P}$ is perturbed slightly.

If $\int_G \rho(x,\eta)\varepsilon_a(x)\mathbb{P}(dx) > 0$, then this inequality also holds for y close to η in leg L_a. Even $\int_G \rho(x,y)\varepsilon_a(x)\mathbb{P}^*(dx) > 0$ for y close to η in leg L_a. Since y has star neighborhood of degree 2, it follows that y cannot be Fréchet mean of $\mathbb{P}^*$, that is, the Fréchet mean set of $\mathbb{P}^*$ avoids L_a.

Thus assume $\int_G \rho(x,\eta)\varepsilon_a(x)\mathbb{P}(dx) = 0$. Denote by L_a^y the part of leg L_a starting at y, in the direction opposite to η. Let ε_a^y be the function such that $\varepsilon_a^y(x) = +1$ if going further into leg L_a from y increases the distance to x and $\varepsilon_a^y(x) = -1$ if not.

$$\begin{aligned}
&\int_G \rho(x,y)\varepsilon_a^y(x)\mathbb{P}(dx) - \int_G \rho(x,\eta)\varepsilon_a(x)\mathbb{P}(dx) \\
&= \int_G (\rho(x,y) - \rho(x,\eta))\varepsilon_a^y(x)\mathbb{P}(dx) + \int_G \rho(x,\eta)(\varepsilon_a^y(x) - \varepsilon_a(x))\mathbb{P}(dx)
\end{aligned}$$

I see that along segment $[\eta,y]$ it holds that $\rho(x,y)\varepsilon_a^y(x) - \rho(x,\eta)\varepsilon_a(x) = \rho(y,\eta)$.
Moreover, if $x \notin [\eta,y]$ and $\varepsilon_a^y(x) = +1$ (implying that $\varepsilon_a(x) = +1$ as well) or $\varepsilon_a^y(x) = \varepsilon_a(x) = -1$, then also $\rho(x,y)\varepsilon_a^y(x) - \rho(x,\eta)\varepsilon_a(x) = \rho(y,\eta)$.
Thus consider the region $U_a^y = \{x \in G; \varepsilon_a^y(x) = -1 \text{ and } \varepsilon_a(x) = +1\}$. Then

$$\begin{aligned}
&\int_G \rho(x,y)\varepsilon_a^y(x)\mathbb{P}(dx) - \int_G \rho(x,\eta)\varepsilon_a(x)\mathbb{P}(dx) \\
&= \rho(y,\eta) - \int_{U_a^y} \rho(y,\eta) + \rho(x,y) + \rho(x,\eta)\mathbb{P}(dx)
\end{aligned}$$

In case of a circle S^1 with unit radius we have $\rho(y,\eta) + \rho(x,y) + \rho(x,\eta) = 2\mathbb{P}i$, and U_a^y is the antipodal of the segment $(\eta,y]$ (excluding η) of length $\rho(y,\eta)$. I think for general graphs there will be many more occasions where $\rho(y,\eta) + \rho(x,y) + \rho(x,\eta)$ is locally constant. Notice that $\bigcap_{y\downarrow\eta} U_a^y = \emptyset$.
I think that for y close to η, $x \in U_a^y$ is in the interior of an edge and the shortest geodesics from x to y and from x to η start in the two different directions. For y close to η, U_a^y consists of a finite disjoint set of segments within edges, of which one of the endpoints belongs to the cut-locus of η.

More precisely:

- For leg L_a define $R_a = \{x \in G; \varepsilon_a(x) = +1\}$. Then $U_a^y \subset R_a$. Moreover, if y_1 is between η and y, then $U_a^{y_1} \subset U_a^y$.
- Let $x \in G$. If $\varepsilon_a^y(x) = -1$ for all $y \downarrow \eta$, then $\varepsilon_a(x) = -1$. This means that $U_a^y \downarrow \emptyset$ as $y \downarrow \eta$.
- In particular, one can choose y so close to η that no vertex belongs to U_a^y.
- Let $x \in U_a^y$. If a shortest geodesic from y to x and a shortest geodesic from η to x have a point $z \neq x$ in common, then let z_0 be the common point closest to η. It must be a vertex, but it also belongs to U_a^y. This contradicts the previous observation. Summarizing, the shortest geodesics from y to x and from η to x only meet in x, and x is an interior point of an edge. Moreover the (possibly nonunique) shortest geodesics from y to x, from x to η and from η to y form an embedded geodesic cycle. Along the cycle, let η^o and y^o be the points in the cut-locus of η and y, respectively.
- Along this segment $(\eta^o, y^o]$ the integrand $\rho(y,\eta) + \rho(x,y) + \rho(x,\eta) = 2\rho(\eta^o,\eta)$. The length of this interval is $\rho(y,\eta)$.
- Notice that $(\eta^o, y^o]$ is contained in the interior of an edge. That means that there is a sort of geodesic quadrangle, consisting of the edge containing x, geodesics from y and η each to an other end point of this edge, and the segment from y to η.
- There are only finitely many shortest geodesics emanating from η in the direction of L_a. Then for any such geodesic γ, consider the finitely many shortest geodesics emanating from any other leg $L_b \neq L_a$ to any point of γ. Then one has a finite set of embedded geodesic cycles. Along each of these cycles, there will be a contribution.

I consider the above as a second-order analysis of the constraints on $\mathbb{P}$, given that η is a Fréchet mean. Next we would like to see what kind of first-order constraints there are on $\mathbb{P}$, given that

$$\int_G \rho(x,\eta)\varepsilon_a(x)\mathbb{P}(dx) = 0.$$

6.3.1 *Central Limit Theorem*

Suppose given that $\int_G \rho(x,\eta)\varepsilon_a(x)\mathbb{P}(dx) = 0$ and that $\lim_{y\downarrow\eta} \int_{U_a^y} \rho(y,\eta) + \rho(x,y) + \rho(x,\eta)\mathbb{P}(dx)/\rho(y,\eta)$ exists and is equal to $\beta_a < 1$. Notice that for a random sample $X_1,\ldots,X_n$

$$D_a = \frac{1}{\sqrt{n}}\sum_{i=1}^{n} \rho(X_i,\eta)\varepsilon_a(X_i) \to_d N\left(0, \int_G \rho(x,\eta)^2\mathbb{P}(dx)\right) = N(0,V(\eta))$$

With asymptotic probability $\frac{1}{2}$, D_a will be positive, and there will be no local minimum in leg L_a. With asymptotic probability $\frac{1}{2}$, D_a will be negative, and along leg L_a the second-order Taylor expansion in L_a of the sample Fréchet function is conjectured to be

$$\begin{aligned}
\frac{1}{n}\sum_{i=1}^{n}\rho(X_i,y)^2 &= \frac{1}{n}\sum_{i=1}^{n}(\rho(X_i,\eta)+\varepsilon_a(X_i)\rho(y,\eta))^2+\\
&+\frac{1}{n}\sum_{i=1}^{n}(\rho(X_i,y)+\rho(X_i,\eta)+\varepsilon_a(X_i)\rho(y,\eta))(\rho(X_i,y)\\
&-\varepsilon_a(X_i)\rho(y,\eta)-\rho(X_i,\eta))\\
&\approx \frac{1}{n}\sum_{i=1}^{n}\rho(X_i,\eta)^2+\frac{2D_a}{\sqrt{n}}\rho(y,\eta)+(1-\beta_a)\rho(y,\eta)^2
\end{aligned}$$

The i-th term on the second line is nonzero only if $X_i \in U_a^y$, that is $\varepsilon_a(X_i) = +1$ and $\varepsilon_a^y(X_i) = -1$. Assuming that $\mathbb{P}$ has a density on U_a^y that converges on each segment in U_a^y uniformly to a uniform density for $y \to \mu$, the expected value of $(\rho(X_i,y) - \varepsilon_a(X_i)\rho(y,\eta) - \rho(X_i,\eta))/\rho(y,\eta) \to -1$, whereas $(\rho(X_i,y) + \varepsilon_a(X_i)\rho(y,\eta) + \rho(X_i,\eta))$ is constant on each segment of U_a^y.
The sample Fréchet function will assume a local minimum $y = \eta_a^*$ at $\rho(y,\eta) \approx -D_a/((1-\beta_a)\sqrt{n})$. Conditional to $D_a < 0$ the local minimum η_a^* in leg L_a behaves like $N\left(0, \frac{1}{n(1-\beta_a)^2}V(\eta)\right)$, more precisely

$$\frac{\sqrt{n}\,(1-\beta_a)}{\sqrt{V(\eta)}}\rho(\eta_a^*,\eta) \to_d N_+(0,1)$$

where $N_+(0,1)$ is the distribution of the variable $\max(Z,0)$ where Z has the standard normal distribution. In order to see which leg contains the global minimum, it may be useful to determine the covariance structure of $\{D_a\}_a$ indexed by the indices of the legs of the star neighborhood of η.

$$Cov(D_a,D_b) = \mathbb{E}(\rho(X,\eta)^2\varepsilon_a(X)\varepsilon_b(X)) \tag{6.3}$$

Chapter 7

Analysis of Magnetic Resonance Angiography Data

7.1 Brain Artery Trees

Analysis of magnetic resonance angiography (MRA) brain artery trees to uncover statistically significant trends with age or clustering by gender is performed. A data set of 3D brain artery trees was extracted from MRA images of a group of healthy subjects varying in age (see Aylward and Bullitt (2002) [4]). More details about the data set can be found in Section 1.4. This research taps into a powerful mathematical basis developed for phylogenetic inference. A phylogenetic tree can be used as a graphical representation of an evolutionary history. The space of phylogenetic trees developed in Billera et al. (2021) [25] is a geometrically interpretable sample space for phylogenetic trees. That space has the geometric property of global nonpositive curvature under which any two points are connected by a unique shortest path, called a *geodesic*, and under which any set has a unique *Fréchet mean*. The Fréchet mean was originally introduced by Fréchet (1948) [52].

In phylogenetics the leaves represent taxa. A sample of brain artery trees does not have a natural common set of leaves. This issue is addressed by accessing a technique from anatomical shape correspondence. A group-wise shape correspondence algorithm based on spatial locations is used to place landmarks on the cortical surface (see Oguz et al. (2008) [99]). The tree structure of arteries makes it inappropriate to apply that method directly to the artery system. The landmarks are projected from the cortical surface to the closest point on the brain artery tree. Under the *cortical correspondence*, all trees in the data set have the same set of labeled leaves, derived by projecting matching points on the cortical surface of the brain onto the artery trees. Thus the brain artery trees are represented as points in the space of phylogenetic trees. For more details about the steps from the brain artery trees to this new representation, see Section 1.4. Representation of MRA brain arteries as points in phylogenetic treespace enables the use of treespace geodesics to quantify and visualize their differences, and a notion of center called the Fréchet mean. High-dimensional structure in the data is explored using multidimensional scaling, minimum spanning trees based on geodesic distances, and treespace triangles (Section 7.5). Effect of sex and age on brain artery system is studied in Section 7.6. In this analysis, the cortical correspondence enabled a new strongly significant observation that the distances of the closest brain arteries to the cortical surface increase with age and tend to be smaller in females than males.

The goal of statistical analysis of populations of 3D anatomical tree structured data objects has motivated several interesting approaches to 3D anatomical tree data objects. An early paper on the topic is due to Wang (2009) [138]. That paper laid foundations by proposing a metric, several notions of center, variations, and a definition for principal components for combinatorial tree objects. The optimization problems formulated in that approach were difficult to solve on larger data sets, but effective approaches were developed later. Principal component analysis (PCA) for combinatorial trees in the same framework as Wang and Marron was developed in [3] along with fast exact algorithms to solve the associated optimization problems. A smoothing method for nonparametric regression with combinatorial tree structures as response variables and univariate Euclidean predictors was developed in [140]. Besides combinatorial tree objects, there are several other approaches that are areas of active research. Another is the Dyck path approach. The Dyck path representation of trees from Harris (1952) [61] makes every tree correspond to a curve. See [121] for details. These curves as data objects enable the use of statistical methods from *functional data analysis*. For more information about functional data analysis, see Ramsay and Silverman (2002) [111] and (2005) [112].

The problem of *shape correspondence* is determining which parts of a shape should be considered comparable across a population. One of the big ideas for tackling the shape correspondence problem is *equivalence correspondence*. That topic is covered along with other topics related to shape analysis in Dryden and Mardia [39]. Another approach, called *anatomical correspondence*, is to use other parts of the human anatomy. Anatomical correspondence of trees is explored in Dryden and Mardia by associating *cortical correspondence* with brain artery trees (see Section 1.4). Labeling leaves provides a system for determining correspondence between edges in different trees. For more information about this, see Section 7.2.

In this chapter analytical results are compared side by side using two different types of correspondence, the cortical correspondence and the *descendant correspondence*. The descendant correspondence was first used in [3]. For more information about descendant correspondence, see Section 7.1.1. These two types of correspondences yield dramatically different results. These differences are highlighted to emphasize the importance of tree correspondence and the room for improved understanding of the tree correspondence problem.

7.1.1 Correspondence

An *artery tree correspondence* is a method for determining which parts of the artery trees are comparable across a population. An ideal artery tree correspondence would combine the arteries, cortical surface and other anatomy as well. Two types of correspondence, one based on the cortical surface, called cortical correspondence, and another based only on the combinatorial information in the trees, called descendant correspondence, lead to dramatically different results in data analysis.

In addition to the brain artery trees, the data set also includes images of the cortical surface. The *cortical correspondence* is made by determining points on the cortical surface that correspond across different people. A group-wise shape correspondence algorithm based on spatial locations is used to place landmarks on the cortical surface (see Oguz et al. (2008) [99]). Each landmark is located in a corresponding spot on the cortical surface in every subject. Sixty-four landmarks on the right hemisphere and sixty-four landmarks on the left hemisphere are used in this study (see Figure 7.1a). The landmarks are combined with the artery tree by the following procedure. For each of the 128 landmarks, find the closest point on the artery tree's centerline, called the landmark connection point (see Figure 7.1b). Each landmark and the line segment to the closest point become part of the tree. Any parts of the artery tree that are not between a landmark connection and the base are trimmed (see Figure 7.1c). The base of the tree, called the *root*, and the 128 landmarks add up to a total of 129 common leaves.

The *descendant correspondence* embeds all the data trees in a single ambient plane tree according to the simple *descendant rule*: place the child with the most descendants on the left. In this study the descendant correspondence is applied separately to the four main brain artery trees. An artificial connection at their roots joins the four main trees into a single tree. The image of each data tree—the corresponding *descendance tree*—is viewed as a contraction of the *support tree*, which is the union of all descendance trees. See Ayden et al. [3, section 2.1 and figure 3] for more details about tree correspondence and support trees. Labeling the leaves of the support tree, and thus all the descendance trees, represents the trees as points in the phylogenetic tree space [25].

These two types of correspondence are based on very different information and emphasize different features of the artery trees. The descendant correspondence is a combinatorial process: it takes as input a collection of abstract trees and neither requires nor uses any geometric information. The cortical correspondence is driven by corresponding landmarks, and not any combinatorial features of the artery trees; it incorporates the proximities of the landmark points to the arteries in the analysis. Each type of correspondence offers a different view of the data, and a priori it is not certain which view will reveal statistically significant trends.

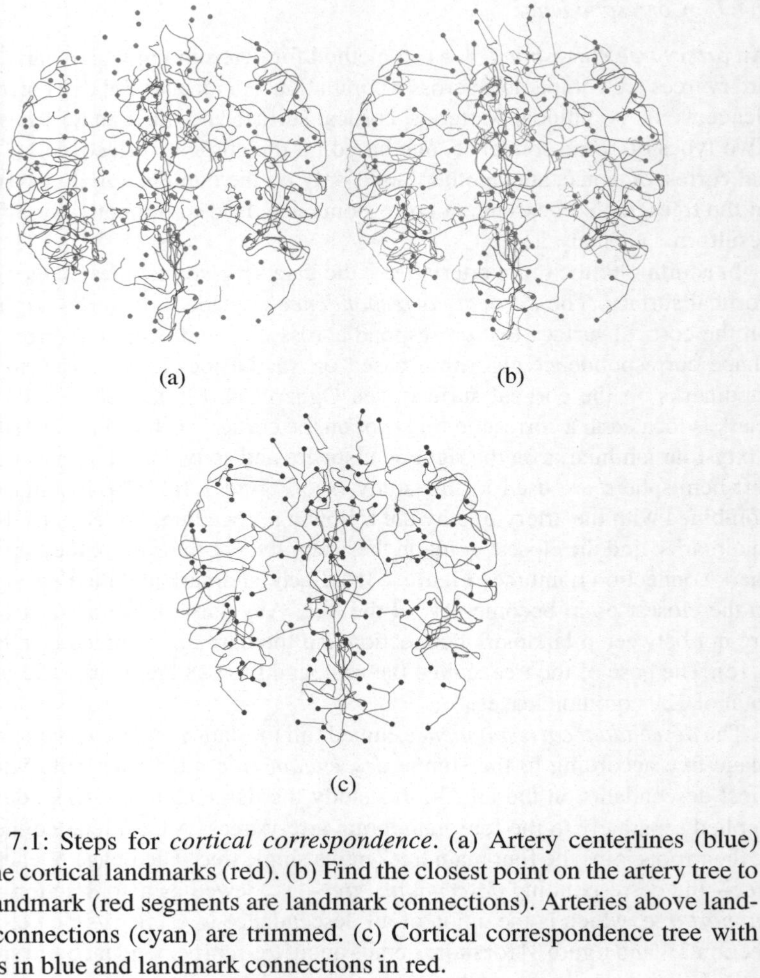

(a) (b) (c)

Figure 7.1: Steps for *cortical correspondence*. (a) Artery centerlines (blue) with the cortical landmarks (red). (b) Find the closest point on the artery tree to each landmark (red segments are landmark connections). Arteries above landmark connections (cyan) are trimmed. (c) Cortical correspondence tree with arteries in blue and landmark connections in red.

Artery trees were reconstructed satisfactorily for 98 subjects. Cortical correspondence was established for 85 subjects. Note that in the tree visualizations that follow, any cortical correspondence trees are drawn with an arbitrary planar embedding. Any descendance trees are drawn with the subtree having more descendants on the left. Also for plots of trees, the tick marks on the horizontal axes are the number of leaves to the left of that point.

7.2 Phylogenetic Trees and Topological Heterogeneity

A *graph* is a set of points, called *vertices*, and set lines connecting pairs of vertices, called *edges*. A *tree* is a graph such that every pair of vertices is connected by a sequence of edges and vertices and which has no cycles of edges and vertices. The degree of a vertex is the number of edges connected to it. The vertices of a tree with degree one are called leaves, and the edges connected to them are called *pendants*.

The fairly standard representation of phylogenetic trees outlined in this paragraph is used in Felsenstein's book (2004) [47]. Methods of analysis based on Billera et al. (2001) [25] treespace require each tree to have its leaves in bijection with a fixed set of labels. Each edge in the tree has a positive weight called its *length*. In this study, edge lengths in phylogenetic trees are the lengths of artery segments measured along their centerlines. Edgewise *correspondence* and *compatibility* are determined by how the leaves of a tree are labeled. If every leaf of a tree is labeled, then deleting any edge from the tree gives a partition of the leaf labels in two. Edges in different trees correspond if they have the same partition. Two edges are *compatible* if there exists a tree containing both edges. Two edges are compatible if and only if they partition the leaves into nested partitions. The topology of a phylogenetic tree refers to the underlying combinatorial tree divorced from the edge lengths. Thus any list of compatible splits defines a unique phylogenetic tree topology. A Euclidean orthant is a generalization of quadrant and octant to higher dimension. The space of trees is a union of many Euclidean orthants, each corresponding to a distinct phylogenetic tree topology. See figure 2 in Owen and Provan (2010) [103] for a schematic of trees embedded as points in orthants.

Full contraction of an edge in a tree means its length shrinks to zero, and it is deleted from the topology. A tree topology which is created by fully contracting some edges in a tree is called a *contraction*. A *topologically homogeneous* set of trees is a set of trees which all have the same topology or are contractions of the same topology. A *topologically heterogeneous* set of trees contains a wide variety of topologies which are also not contractions of each other. Geometrically, topologically homogeneous trees appear in the same orthant of treespace, whereas a topologically heterogeneous set is spread over multiple orthants of treespace. The topological diversity of a data set of trees has a strong relationship with the non-Euclidean structure of geodesic paths between data points and with the behavior of the Fréchet sample mean of the distribution. The characteristic attributes of geodesics and Fréchet sample means for topologically heterogeneous data sets coincides with the negatively curved geometry of tree space.

The descendant correspondence leads to a set of topologically homogeneous phylogenetic trees. Any set of topologically homogeneous phylogenetic trees can be represented as a set of points in a Euclidean orthant. In this case each axis of the orthant corresponds to an edge in the support tree. The data trees have the same topology as the support tree with some edges contracted and are embedded as points on lower-dimensional faces of the orthant. The fact that all the data trees are contractions of the support tree means that the split associated with any edge in the support tree would be compatible in any of the data tree topologies. Therefore, any edge from a tree in the data set must be compatible in every tree in the data. The support tree has 2546 internal edges, and 2529 pendants. The mean number of edges for descendant correspondence trees is 465. On the other hand, the cortical correspondence trees are topologically very heterogeneous. There are a total of 85 trees, which have 129 leaves common to all trees. The mean number of internal edges is 123.

7.3 Treespace Geodesics

Geodesics can be found quickly using the algorithm in Owen and Provan (2010) [103]. Visualization of geodesics from the artery tree of one subject to another can give insights into their differences and similarities. The two types of correspondence lead to dramatically different geodesics. Figure 7.2 shows five equally spaced points along the geodesic from one subject to another under the cortical correspondence, and Figure 7.3 shows the same except under the descendant correspondence. In both figures, common edges are drawn in green, edges that appear in the tree of the first subject and not the other are drawn in blue, and vice versa in red.

Geodesics in treespace between cortical correspondence trees tend to bow toward the center of the space. Bowing toward the center of treespace means that both the total length of edges and the number of edges shrink and then grow in moving along the geodesic. In particular, the distance between the origin and a point on the geodesic shrinks and then increases while traversing the geodesic. The bow of a treespace geodesic is not because the space is curved smoothly like a manifold, but it is rather related to the traversal of the sequence of orthants supporting the path. See figure 4 in Owen and Provan (2010) [103] for a visualization of a bowed treespace geodesic (note how the distance from the origin of the space and the number of edges in the trees shrink and then grow along the path). For artery trees, the bowing of geodesics toward the center of treespace is strongly related to how the correspondence between trees is determined.

Here is a typical example. In Figure 7.2, it is easy to see that there is little common structure between these two cortical correspondence trees. For the cortical correspondence, these subjects have five edges in common, 119 edges from the tree of the first subject are not compatible with the tree of the other subject, and 118 edges from the other subject are not compatible with the tree of the first. Notice that subsets of edges from each subject are compatible. In the tree half-way along the geodesic, blue edges from the tree of the first

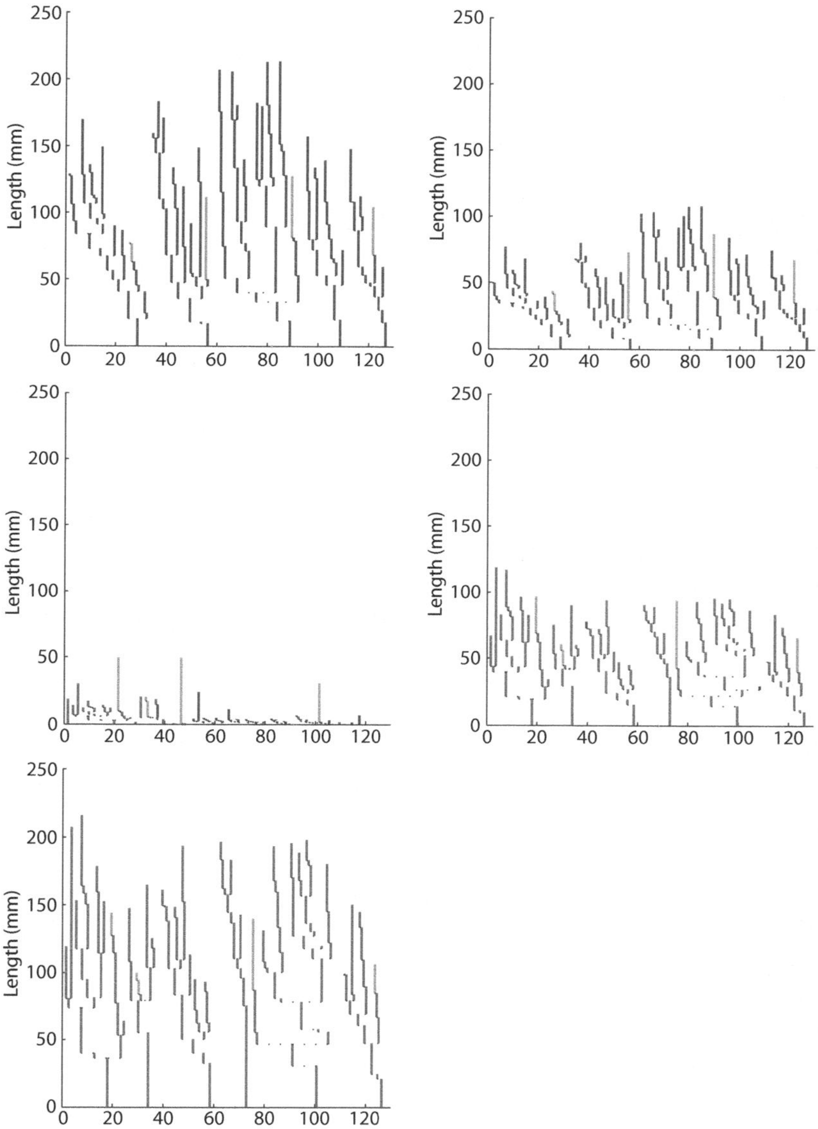

Figure 7.2: Moving along the geodesic from one cortical correspondence tree to another in four steps (same subjects as Figure 7.3). Blue edges are only in the tree of the first subject, red only in the other subject, and green are common to both (pendants are not included in this visualization). The tick marks on the horizontal axes are the number of leaves to the left of that point.

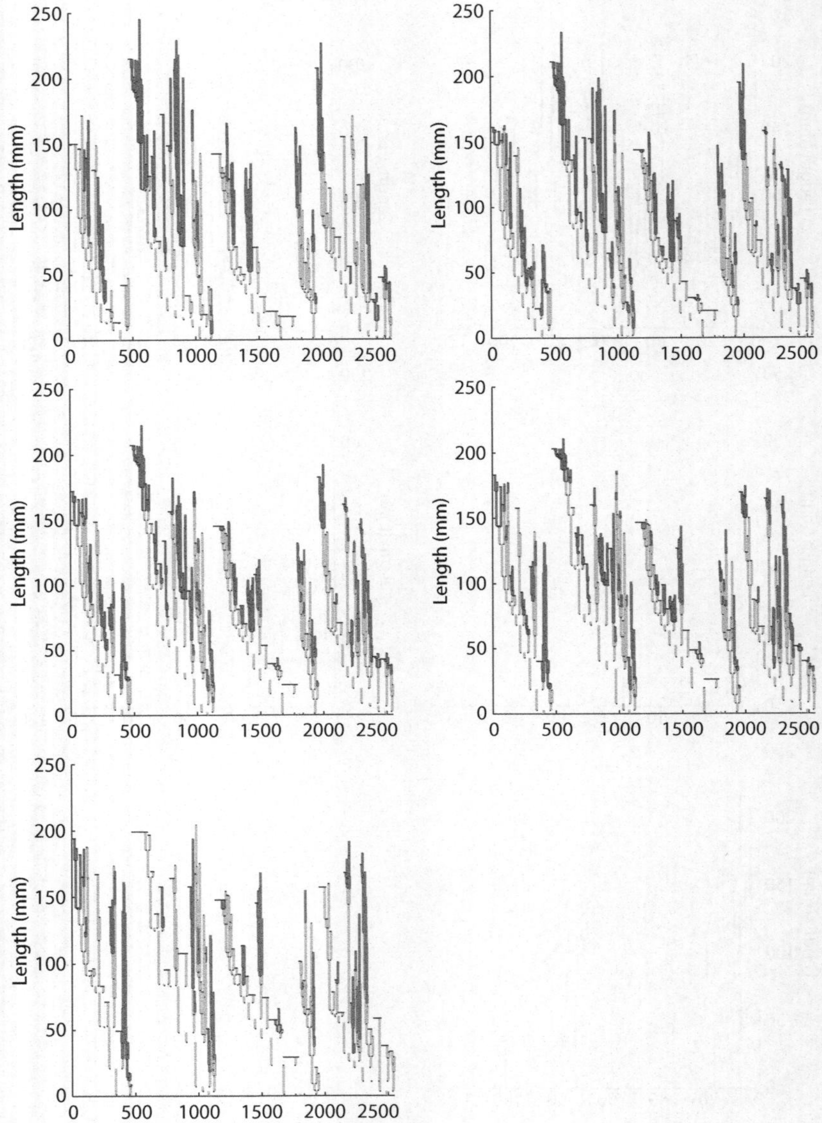

Figure 7.3: Moving along the geodesic from one descendance tree to another in four steps (same subjects and color code as Figure 7.2). The horizontal black lines appear for contracted subtrees in the support tree not present in these trees.

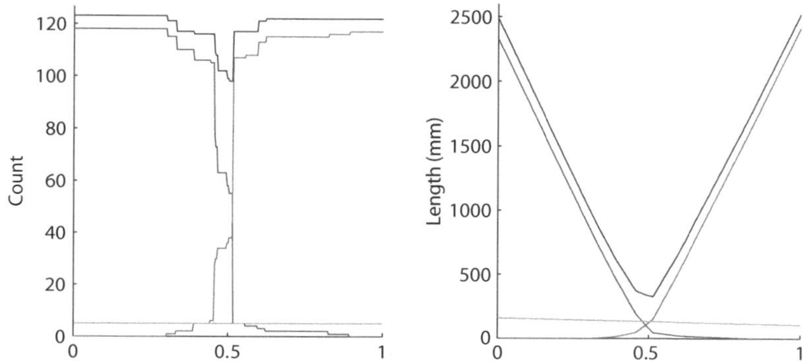

Figure 7.4: Summary of edge count and total length along geodesic between the same two cortical correspondence trees as in Figure 7.2.

subject and red edges from the tree of the second subject are present together. Figure 7.4 displays the number of edges and the total length of edges at each point along the geodesic between the cortical correspondence trees. Although the number of edges drops along the cortical correspondence geodesic, it is always at least 98. This indicates the size of compatible subsets of edges is non-negligible. The total length dips down to around 330 (mm) from a high of around 2500 (mm). A *topological transition* occurs on a treespace path when some edges leave, and/or some edges enter the tree topology (see [103] for more information about the combinatorial structure of geodesics). The geodesic between the trees in Figure 7.2 has 16 topological transitions.

The descendant correspondence was used for the trees in Figure 7.3. For the descendant correspondence, these subjects have 224 edges in common, 266 are unique to the first, 209 are unique to the other, and there are 4396 edges in the support tree that are not in either (have length zero). The total number of edges is the same in all trees along the geodesic, and the total length changes linearly from the total length in the tree of the first subject (approx. 7101.5 mm) to the total length in the other (approx. 6477.2 mm). No topological transitions occur because all edges in both trees can be present at any point on the geodesic with zero length in some cases. Here the geodesic is contained entirely in the orthant corresponding to the support tree topology. This geodesic is a straight line from a point on one lower-dimensional face to a point on some other lower-dimensional face of the same orthant.

7.4 Fréchet Mean of Phylogenetic Trees

Let $\mathcal{T}_n$ denote the space of phylogenetic trees with leaves labeled $0, 1, \ldots, n$ (see Billera et al. (2001) [25]). Let T and T' be trees in $\mathcal{T}_n$. The geodesic distance in phylogenetic treespace between T and T' will be denoted as $d(T,T')$. The Fréchet sample mean is characterized as the solution to a nonlinear optimization problem. For a set of points $\{T^1,T^2,...,T^r\}$ in $\mathscr{T}_n$ the Fréchet function $F:\mathscr{T}_n\to\mathbb{R}$ is the mean square distance

$$F(T)=\frac{1}{r}\sum_{i=1}^{r}d(T,T^i)^2$$

and the sample Fréchet sample mean is

$$\bar{T}=\underset{T\in\mathscr{T}_n}{\arg\min}\,F(T).$$

The Fréchet sample mean $\bar{T}$ is unique because $d(T,T')$ is a convex function of T for fixed T' Miller et al. (2012) [95]. The minimum $F(\bar{T})$ of the Fréchet function is the Fréchet sample variance. The Fréchet sample mean is the geometrically central tree of the data points $\{T^1,T^2,...,T^r\}$, and the Fréchet variance summarizes the variability of the data set. The purpose of the Fréchet sample mean is to give a single point summary of an entire data set.

Averaging homogeneous trees is exactly like averaging Euclidean points, and in this case, the Fréchet sample variance is the total sample variance. When a sample is from a heterogeneous distribution, a quite different behavior can result, called *stickiness*. Stickiness is the phenomenon of a sample average which does not change despite small changes in the sample. In samples from phylogenetic treespace, stickiness manifests as Fréchet sample means which have few internal branches relative to the number of leaves, and for which the tree topology remains the same under small perturbations of the data.

7.4.1 Calculating Fréchet Sample Means

A simple algorithm, referred to here as the *Inductive Mean Algorithm*, for calculating a Fréchet sample mean is based on theorem 4.7 of Sturm (2003) [127]. This algorithm was developed independently in Bakac (2012) [6] and Miller et al. (2012) [95]. Consider a sequence $(Y_i)_{i\in\mathbb{N}}$ of independent identically distributed random objects from the uniform distribution of $\{T^1,T^2,\ldots,T^r\}$. Define a new sequence of points $(S_k)_{k\in\mathbb{N}}$ in $\mathscr{T}_n$ by induction on k as follows:

$$S_1:=Y_1,$$

and

$$S_k:=\left(1-\frac{1}{k}\right)S_{k-1}+\frac{1}{k}Y_k,$$

where the right-hand side denotes the point $\frac{1}{k}$ along the geodesic from S_{k-1} to Y_k. The point S_k is called the *inductive mean value* of $Y_1,\ldots,Y_k$. The expected distance between S_k and $\bar{T}$ is less than or equal to $F(\bar{T})/k$.

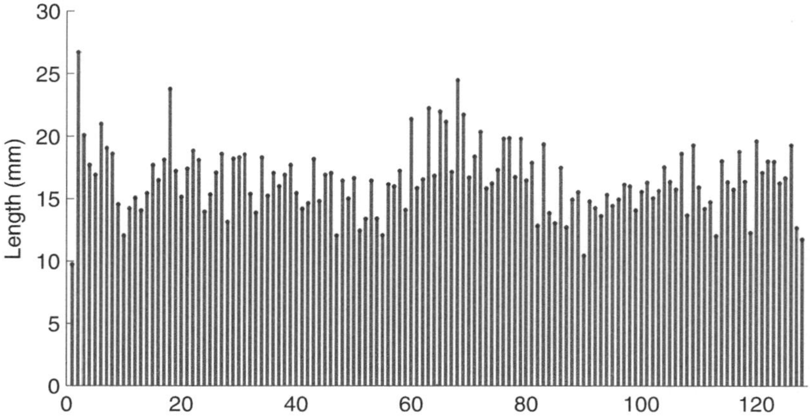

Figure 7.5: The Fréchet mean of the cortical correspondence trees is a star tree. Since all internal edges have length zero, these are just average pendant lengths.

7.4.2 Fréchet Means for Artery Trees

For the cortical correspondence trees, the Fréchet sum of squares of a random inductive mean was always higher than the one for the star tree while running the Inductive Mean Algorithm for 50,000 iterations. This suggests that the star tree is the Fréchet sample mean of the trees with cortical correspondence. See Figure 7.5 for a visualization of the star tree of the cortical correspondence trees. It is not surprising that a topologically heterogeneous set of trees has a Fréchet sample mean which is equal to, or very close to, the star tree. Intuitively, heterogeneous trees have few topological features in common, and therefore the best single tree summary of common features is empty. This is starkly contrasted by the descendance trees.

The descendance trees are topologically homogeneous and are not affected by stickiness. For the descendance trees, the Fréchet sample mean is the Euclidean mean. See Figure 7.6 for a visualization of the Fréchet sample mean of the descendance trees.

One interesting observation is that neither of these Fréchet sample means *looks like the data*! For the cortical correspondence the Fréchet mean has only pendants, whereas the data trees have on average 123 internal edges. Phylogenetic trees with 129 leaves have at most 126 internal edges, so the data trees, on average, have almost the maximum number possible. For the descendant correspondence the Fréchet mean has 2546 internal edges, and 2549 pendants, but the data trees have 465 edges on average. The striking difference

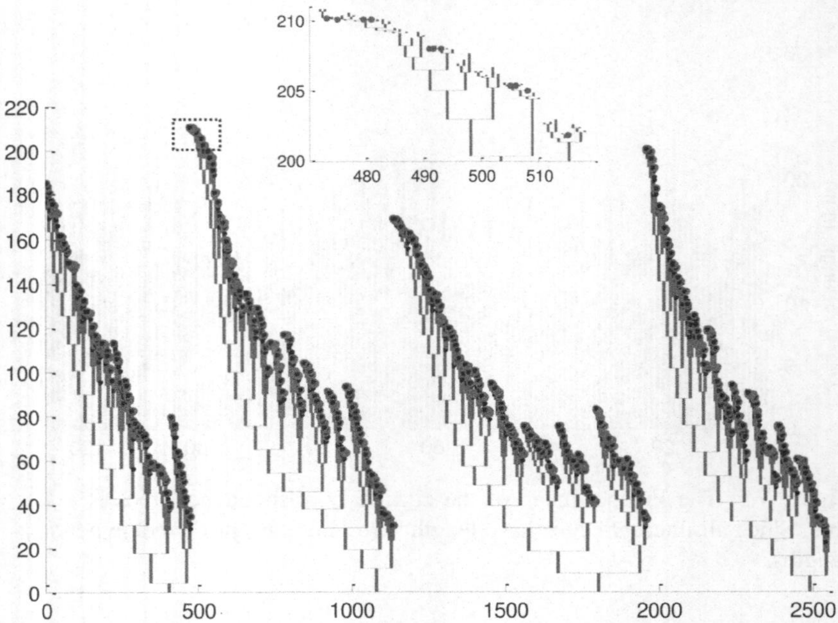

Figure 7.6: Fréchet sample mean of the descendance trees.

between these Fréchet sample means and the data trees is an indicator that there is a lot of room for improved understanding of biological tree correspondences. It may be possible to develop a method of correspondence that will result in a Fréchet sample mean which looks like an element of the data set, although this is not the most important correspondence issue. Here it is more important to establish a correspondence which facilitates averaging to cancel out *noisy* idiosyncrasies in observations. A tendency for averaging to nullify idiosyncratic features in observations is a common occurrence in data analysis. For example, consider the case of noisy curves as data objects. Small features in the observations cancel each other, and in the mean curve such details are smoothed out. In that case the average doesn't *look like the data*, but it is representative of strong trends. For the cortical correspondence trees, the idiosyncratic features are so strong that they cannot be reconciled in the same tree and they all cancel out. On the other hand, for the descendant correspondence, the fact that the Fréchet mean has about ten times more edges than the average for the data indicates that averaging is not filtering idiosyncrasies in the topological structure.

7.5 Multidimensional Scaling, Minimum Spanning Trees, and Triangles

Classical Multidimensional Scaling (MDS) is a powerful tool for studying the modes of variation of a data set, and creating low-dimensional visualizations that capture prominent structure based on the distances between data points. Essentially MDS embeds data as points in a Euclidean space with the best possible approximation of the original distances.

A *minimum spanning tree* of a set of points is a shortest possible set of geodesics (between those points) that connects all the points together. A minimum spanning tree of the mean and the data can reveal if subsets of the data points are in closer proximity to each other. For example, the minimum spanning tree might connect points within clusters.

Multidimensional scaling views of both data types are displayed in Figure 7.7. The topological structure of the minimum spanning tree is accurately depicted by the yellow lines, although the distances are distorted. Every edge in the minimum spanning tree connects the mean tree to a data point. This structure shows that the data points are all closer to the mean tree than any of the other data points for both types of correspondence. Thus both data sets are intrinsically high-dimensional in nature, as also shown by the first MDS component only explaining 2.43 and 2.49% of the variability in the cortical correspondence trees and descendance trees, respectively.

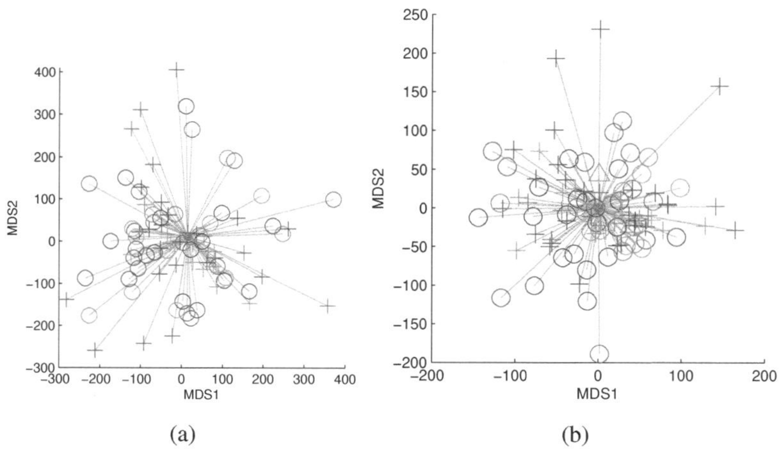

(a) (b)

Figure 7.7: Visualizations of (a) cortical correspondence trees and (b) descendance trees embedded as points in a two-dimensional Euclidean space with the minimum spanning tree (yellow) and Fréchet sample mean (red). Sex is coded in symbols (o is female, + is male, △ is transgender), and age is colored according to the bar on the right.

In non-positively curved metric spaces such as phylogenetic tree space, triangles are at least as skinny as their Euclidean comparison triangles. For nondegenerate Euclidean triangles, the angle sum is exactly 180°. For triangles in non-positively curved metric spaces, angle sums are less than 180°. As

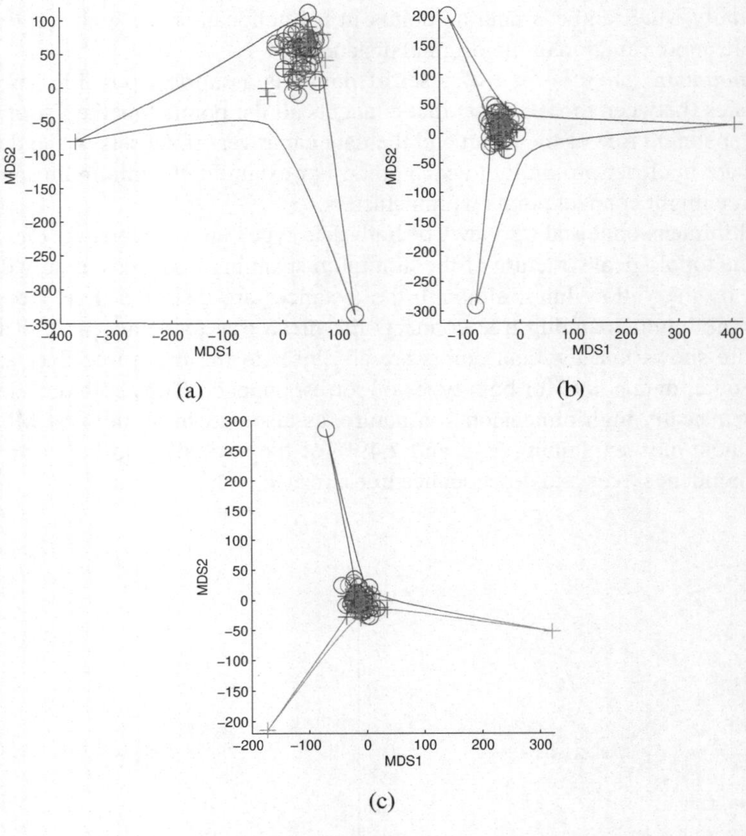

Figure 7.8: MDS embeddings of cortical correspondence data and triangles with maximum (a), median (b), and minimum angle sums. Key as in Figure 7.7.

curvature tends to negative infinity, the angle sum tends to zero. Triangles can be used to quantify the negative curvature of the space.

The shapes of triangles are very different for the two types of correspondence. Any three points from the data and the geodesics between them form a triangle. Here, angles are measured using the Alexandrov definition given on page 9 of Bridson and Haefliger (1999) [28]. Multidimensional scaling on a grid of equally spaced points on the perimeter of a triangle along with the data can be used to embed treespace triangles in two-dimensional Euclidean space. Figure 7.8 shows the triangles with maximum, median, and minimum angle sums among all triangles with vertices from the cortical correspondence trees. In these visualizations, the triangles influence how the points are embedded

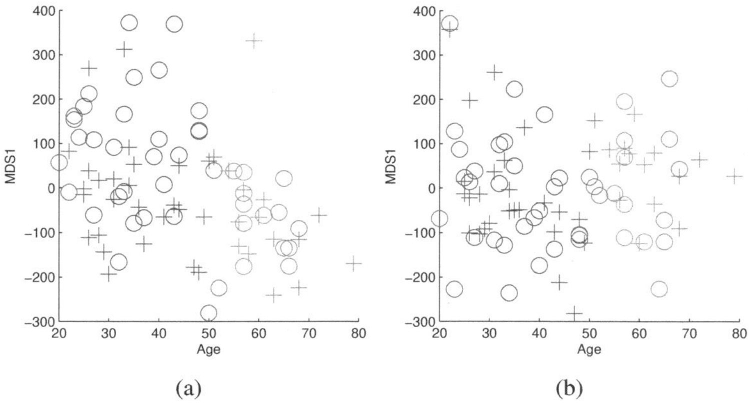

Figure 7.9: Scatterplots of cortical correspondence trees projected onto the first multidimensional scaling component plotted against age. Leaves are included in the analysis in (a) but not in (b). Key: o is female, + is male, and color is age.

in the lower-dimensional space because of the contribution from the 87 points on their perimeter. In these views driven by triangles, the distances between the other data are markedly more distorted. Data points are smashed together, suggesting that representing distances for the triangle makes it impossible to accurately represent distances for the other data. This pattern is not consistent in Figure 7.8a. The vertex near (100,100) has many data points nearby. This is because this vertex and its corner of the triangle are relatively closer to the rest of the data than the other two vertices. On the other hand, for the descendant correspondence, every data point is in the same orthant. Hence, triangles for that data are ordinary Euclidean triangles.

7.6 Age and Sex Differences in Brain Artery Trees

Multidimensional scaling can be used to embed points in a one-dimensional Euclidean space referred to as the first multidimensional scaling component. Figure 7.9a shows a scatterplot of cortical correspondence trees in the first multidimensional scaling component plotted against age. Figure 7.9b is the same plot made without including pendants in the analysis. Notice the different pattern in the two plots. When pendants are included in the analysis between age and one-dimensional MDS embeddings, the Spearman correlation is -0.4306 with a p-value 3.8962e-5 (Pearson and Kendall correlations are similar). When leaves are not included in the analysis, the correlation is not close to significant. Since taking the leaves out drastically changes the results, they are studied more closely later on. The pendant lengths are the distances from the cerebral cortex landmarks to the artery tree. That difference is investigated further by examining the connection lengths directly. The first multidimensional scaling component for descendance trees does not appear to have any relationship with age. See Figure 7.10 for a scatterplot of the first multidimensional scaling component and age for the descendance trees. Here Spearman correlation is 0.15892 with p-value 0.11805.

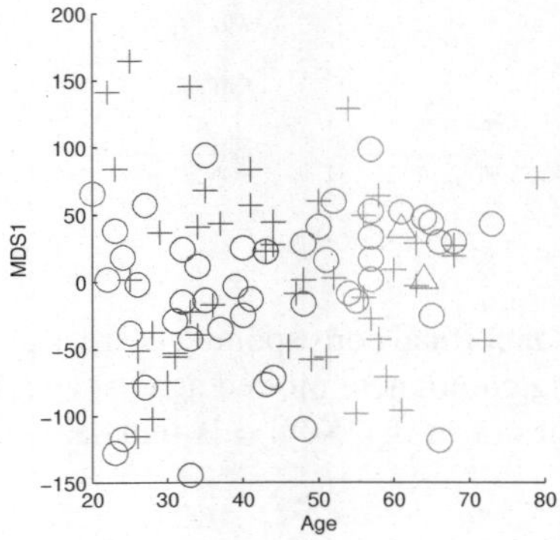

Figure 7.10: Scatterplot of descendance trees projected on the first multidimensional scaling component plotted against age. Key: o is female, + is male, △ is transgender, and color is age.

Other relationships between the trees, age, and sex can be explored. See Figure 7.11 for scatterplots of total length of arteries, the total length of the landmark subtree, and the total length of cortical landmark to artery connections are all plotted against age. Based on these scatterplots, there appears to be correlation between age and all three of these measures, and maybe even a significant sex effect. An ANCOVA study is performed to test linear regression models for each of the values against age, sex, and the interaction of age and sex. In Figure 7.11 the dashed line and solid line are the regressions for females and males, respectively. Analysis of covariance (ANCOVA) is used to test for statistical significance of sex, age, and multiplicative interaction of age and sex as predictive covariates. ANCOVA F-statistics and p-values for models in a leave-one-out fashion are listed in Table 7.1. Age and sex are statistically significant in all three measurements. The effect of age has a similar level of significance in all. But the statistical significance of sex is much stronger for the landmark connection distances. In particular, distances from the cerebral cortex to arteries that appear in MRA tend to be significantly larger for males. This last observation has not been found in previous studies. The cortical correspondence made it possible to measure the proximity of arteries to the cortical surface. These measurements were crucial to making that last observation.

Table 7.1: Summary of ANCOVA for linear models in Figure 7.11

Variable		Total Length	C.Cor. Tr. Len.	L.M. Conn.
Age	F	45.23	53.98	53.74
	p	2.274e-9	1.431e-10	1.541e-10
Sex	F	5.92	5.31	53.51
	p	0.0172	0.0237	1.654e-10
Sex* Age	F	0	0.53	0.58
	p	0.9868	0.4689	0.4492

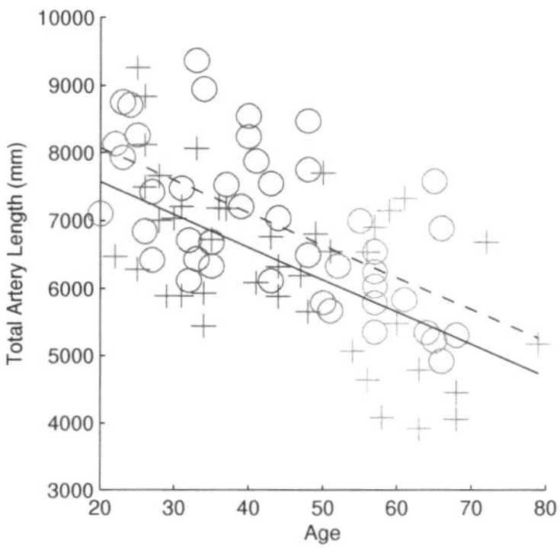

(a)

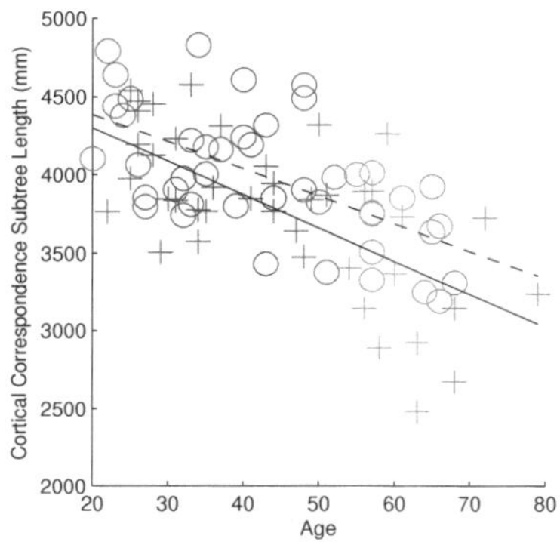

(b)

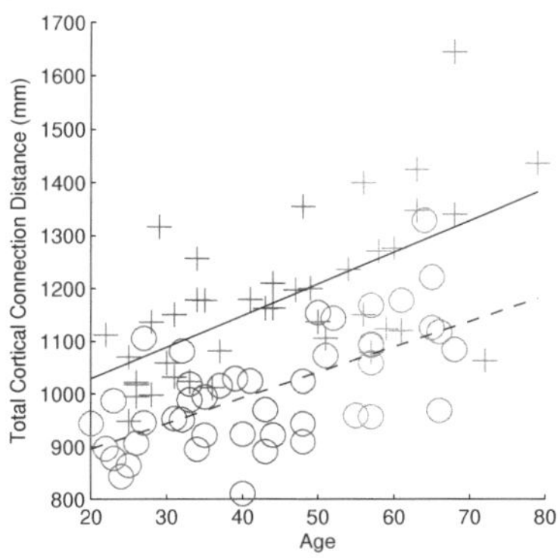

(c)

Figure 7.11: Scatterplots for (a) Total Artery Length (b) Cortical Correspondence Tree Total Length, and (c) Landmark Connection Length against age.

According to these linear models, total length in MRA decreases by 47.8 (mm) per year and females have about 511.0 (mm) more than males, cortical correspondence tree length decreases by about 19.4 (mm) per year and females have about 179.6 (mm) more than males, and total landmark connections length increases by about 5.5 (mm) per year and is about 160.3 (mm) less for females than males. The Spearman correlations for total length, cortical correspondence tree length, connection length with age for males are –0.54183 with p-value 2.5339e-4, –0.62637 with p-value 1.183e-5, and 0.6187 with p-value 1.6211e-5; and for females are –0.60169 with p-value 1.5604e-5, –0.59332 with p-value 2.1989e-5, and 0.61982 with p-value 7.2347e-6.

Chapter 8

An Application to Phylogenies of SARS-CoV-2 Data Analysis

8.1 Introduction

Covid-19 was a pandemic that ravaged the world beginning in December 2019 and led to massive geopolitical and cultural upheaval. Throughout this time, science came under an unfamiliar light as the typically behind the scenes work (that is often unsightly) became a topic of conversation for average people. While this pandemic showcased the ups-and-downs and good-and-bad of the new popularity of virology, gene editing, and Covid vaccine creation, it might have been one of the largest biological data collection undertakings in human history. Several governmental agencies around the world created (or modified or added to) databases to house the sequencing of the RNA virus that has allowed scientists unprecedented access to study evolution and mutation. As the SARS-Cov-2 virus mutated, it left a trail of phylogeny for scientists to study. Millions of observations (RNA sequences) are available for analysis. As with any data, as statisticians in this chapter we analyze the spread and sample Fréchet means for such phylogenetic trees data on tree spaces, for different continents during the early years of the pandemic caused by the SARS-Cov2 virus.

8.2 RNA-Based PT Trees

Ribonucleic acids (RNAs) are macro-molecules that look like a single-stranded helix, to fit in their tiny cellular space, and they are responsible for a virus species' passing on of the characteristics/traits from one individual to the next. The four types of chemical building blocks of RNA are the nucleotides: adenine, uracil, cytosine, and guanine, abbreviated with the letters A, U (or T), C, and G. Note that most viruses causing serious diseases in humans have a genome made entirely of RNAs (see Grubaugh et al. (2020) [55]). RNAs have two important properties: they contain instructions of how an organism is

```
>hCoV-19/Wuhan/IPBCAMS-WH-05/2020|EPI_ISL_403928
ATTAAAGGTTTATACCTTCCCAGGTAACAAACCAACCAACTTTCGATCTCTTGTAGATCTGTTCTCTAAACGAACTTTAA
AATCTGTGTGGCTGTCACTCGGCTGCATGCTTAGTGCACTCACGCAGTATAATTAATAACTAATTACTGTCGTTGACAGG
ACACGAGTAACTCGTCTATCTTCTGCAGGCTGCTTACGGTTTCGTCCGTGTTGCAGCCGATCATCAGCACATCTAGGTTT
CGTCCGGGTGTGACCGAAAGGTAAGATGGAGAGCCTTGTCCCTGGTTTCAACGAGAAAACACACGTCCAACTCAGTTTGC
CTGTTTTACAGGTTCGCGACGTGCTCGTACGTGGCTTTGGAGACTCCGTGGAGGAGGTCTTATCAGAGGCACGTCAACAT
CTTAAAGATGGCACTTGTGGCTTAGTAGAAGTTGAAAAAGGCGTTTTGCCTCAACTTGAACAGCCCTATGTGTTCATCAA
ACGTTCGGATGCTCGAACTGCACCTCATGGTCATGTTATGGTTGAGCTGGTAGCAGAACTCGAAGGCATTCAGTACGGTC
GTAGTGGTGAGACACTTGGTGTCCTTGTCCCTCATGTGGGCGAAATACCAGTGGCTTACCGCAAGGTTCTTCTTCGTAAG
...
AACTCAGGCCTAAACTCATGCAGACCACACAAGGCAGATGGGCTATATAAACGTTTTCGCTTTTCCGTTTACGATATATA
GTCTACTCTTGTGCAGAATGAATTCTCGTAACTACATAGCACAAGTAGATGTAGTTAACTTTAATCTCACATAGCAATCT
TTAATCAGTGTGTAACATTAGGGAGGACTTGAAAGAGCCACCACATTTTCACCGAGGCCACGCGGAGTACGATCGAGTGT
ACAGTGAACAATGCTAGGGAGAGCTGCCTATATGGAAGAGCCCTAATGTGTAAAATTAATTTTAGTAGTGCTATCCCCAT
GTGATTTTAATAGCTTCTTAGGAGAATGACAAAAAAAAAAAAA
```

Figure 8.1: Example of a fragment of SARS-CoV-2 RNA sequence.

put together and (b) help replication to successive generations. The replication mechanism is catalyzed by RNA-dependent RNA polymerase (RdRp), or *RNA replicase*, an enzyme that helps the replication of RNA from an RNA template. Imperfect replications become mutations or SNPs, simply a change in the sequence order of the bases. Phylogeny is the ancestral history of a species or a class of species. Organisms with **functions**, so that they can interbreed, form a *species.* There was renewed interest in bio-tech to modify the RNA of simple organisms, for *gain of function* to improve the ability of such species to multiply at an exponential rate (see Gromov (2018) [54]). An RNA phylogenetic tree (PT) is a 1-connected directed graph, where the root is presumably the common ancestor of all leaves, represented by their known RNA, but intermediate species (between leaves and the root) are unknown. A PT leaf is given by the beginning of the RNA copy of the virus hCoV-19 (SARS-CoV-2). Such a leaf is a sequence that is tens of thousands of letters long, where each letter represents one of the nucleotides that form the virus's complete genome.

8.2.1 SARS-CoV-2 data

The data we used in this chapter are from multiple resources including GenBank(see Hatcher et al. (2017) [63]) for SARS-CoV (see He et al. (2004) [64]), MERS-CoV (see Zaki et al. (2012) [149]) and SARS-CoV-bat (Li et al. (2005) [88]) and GISAID (see Elbe and Buckland-Merrett [42]) for SARS-CoV-2 [146]. The data include the virus name, host species, sampled date, and sampled locations. From the raw data, like the sample point in Figure 8.1, we first align the RNA sequences with each other via Clustal Omega method (see Sievers et al. (2011) [123]). The Clustal Omega method uses the k-tuple distance to construct the *distance matrix*. The vector distances in the distance matrix are then used to cluster the sequences into subclusters using a k-means approach. From the aligned RNA sequences, see Figure 8.2, one could compute the frac-

```
hCoV-19/Zhejiang/SX0715/2020|EPI_ISL_497950|2020-07-15       ----------TATACCTTCCCAGGTAACAAACCAACCAACTTTCGATCTCTTGTAGATCT   50
hCoV-19/Guangdong/20SF012/2020|EPI_ISL_403932|2020-01-14     -----------ataccttcccaggtaacaaaccaaccaactttcgatctcttgtagatct   49
hCoV-19/Guangdong/SYSU-IHV/2020|EPI_ISL_444969|2020-04-16    ATTAAAGGTTTATACCTTCCCAGGTAACAAACCAACCAACTTTCGATCTCTTGTAGATCT   60
                                                                        *************************************************

hCoV-19/Zhejiang/SX0715/2020|EPI_ISL_497950|2020-07-15       GTTCTCTAAACGAACTTTAAAATCTGTGTGGCTGTCACTCGGCTGCATGCTTAGTGCACT   110
hCoV-19/Guangdong/20SF012/2020|EPI_ISL_403932|2020-01-14     gttctctaaacgaactttaaaatctgtgtggctgtcactcggctgcatgcttagtgcact   109
hCoV-19/Guangdong/SYSU-IHV/2020|EPI_ISL_444969|2020-04-16    GTTCTCTAAACGAACTTTAAAATCTGTGTGGCTGTCACTCGGCTGCATGCTTAGTGCACT   120
                                                             ************************************************************

hCoV-19/Zhejiang/SX0715/2020|EPI_ISL_497950|2020-07-15       CACGCAGTATAATTAATAACTAATTACTGTCGTTGACAGGACACGAGTAACTCGTCTATC   170
hCoV-19/Guangdong/20SF012/2020|EPI_ISL_403932|2020-01-14     cacgcagtataattaataactaattactgtcgttgacaggacacgagtaactcgtctatc   169
hCoV-19/Guangdong/SYSU-IHV/2020|EPI_ISL_444969|2020-04-16    CACGCAGTATAATTAATAACTAATTACTGTCGTTGACAGGACACGAGTAACTCGTCTATC   180
                                                             ************************************************************

hCoV-19/Zhejiang/SX0715/2020|EPI_ISL_497950|2020-07-15       TTCTGCAGGCTGCTTACGGTTTCGTCCGTGTTGCAGCCGATCATCAGCACATCTAGGTTT   230
hCoV-19/Guangdong/20SF012/2020|EPI_ISL_403932|2020-01-14     ttctgcaggctgcttacggtttcgtccgtgttgcagccgatcatcagcacatctaggttt   229
hCoV-19/Guangdong/SYSU-IHV/2020|EPI_ISL_444969|2020-04-16    TTCTGCAGGCTGCTTACGGTTTCGTCCGTGTTGCAGCCGATCATCAGCACATCTAGGTTT   240
                                                             ************************************************************
...

hCoV-19/Zhejiang/SX0715/2020|EPI_ISL_497950|2020-07-15       TTTAGTAGTGCTATCCCCATGTGATTTTAATAGCTTCTTAGGAGAATGACAAAAAAAAAA   29870
hCoV-19/Guangdong/20SF012/2020|EPI_ISL_403932|2020-01-14     tttagtagtgctatccccatgtgattttaatagctt------------------------   29845
hCoV-19/Guangdong/SYSU-IHV/2020|EPI_ISL_444969|2020-04-16    TTTAGTAGTGCTATCCCCATGTGATTTTAATAGCTTCTTAGGAGAATGACAAAAAAAAAA   29880
                                                             ************************************

hCoV-19/Zhejiang/SX0715/2020|EPI_ISL_497950|2020-07-15       AAAAA    29875
hCoV-19/Guangdong/20SF012/2020|EPI_ISL_403932|2020-01-14     -----    29845
hCoV-19/Guangdong/SYSU-IHV/2020|EPI_ISL_444969|2020-04-16    AA---    29882
```

Figure 8.2: Example of fragments of aligned RNA sequence data.

tion of mismatches at aligned positions, with gaps either ignored or counted as mismatches as distances from each RNA sequence. Then one may apply a neighbor-joining method (see Saitou and Nei (1987) [113]) to build trees. The distance of the inner edge for each tree is measured by the number of equivalent aligned nongap symbols(nucleobase) dividing the smallest number of nongap positions in either of both sequences.

Shen (2021) [119] analyzed the behavior of mean SARS-CoV-2 PT distributions on T_3 and T_4 to compare the spread of this virus in different parts of the World (see also Shen et al. (2021) [120]).

Moore (2021) [96] collected a representative sample and used phylogenetic trees with only three leaves to compare SARS-CoV-2 RNA sequences samples collected within three different time periods; the first and second halves of 2020 (periods 1 and 2), and then the first four months of 2021 (period 3). He then applied the stickiness Lemma 3.4.1 and the star tree concept on the 3-spider projected generated data from these time periods.

A PT tree was built by Shen (2021) [119] from a sample of 72 aligned RNA sequences. Note that the larger the number of leaves, the smaller the sample size (in this case $n = 1$, thus impeding a statistical data analysis. For this reason, the data was subdivided into PTs with three leaves, in what we called a PT-trick. Note that such PTs have only **one** inner edge; one calculated mean and standard deviation for each of the periods mentioned above of these three leaves, PT inner edges. All the trees with three leaves built from SARS-CoV-2 RNA sequence data were represented on the space T_3 in Figure 8.4. See Shen (2021) [119] for more details.

Next a comparison with different locations and time period was performed on T_4. The virus sequences were separated into four groups to form PTs with four leaves. We used the same data as before and made seasonal groups: Jan-Mar, Apr-May, Jun-Jul, and Aug-Sep. We randomly picked one RNA sequence

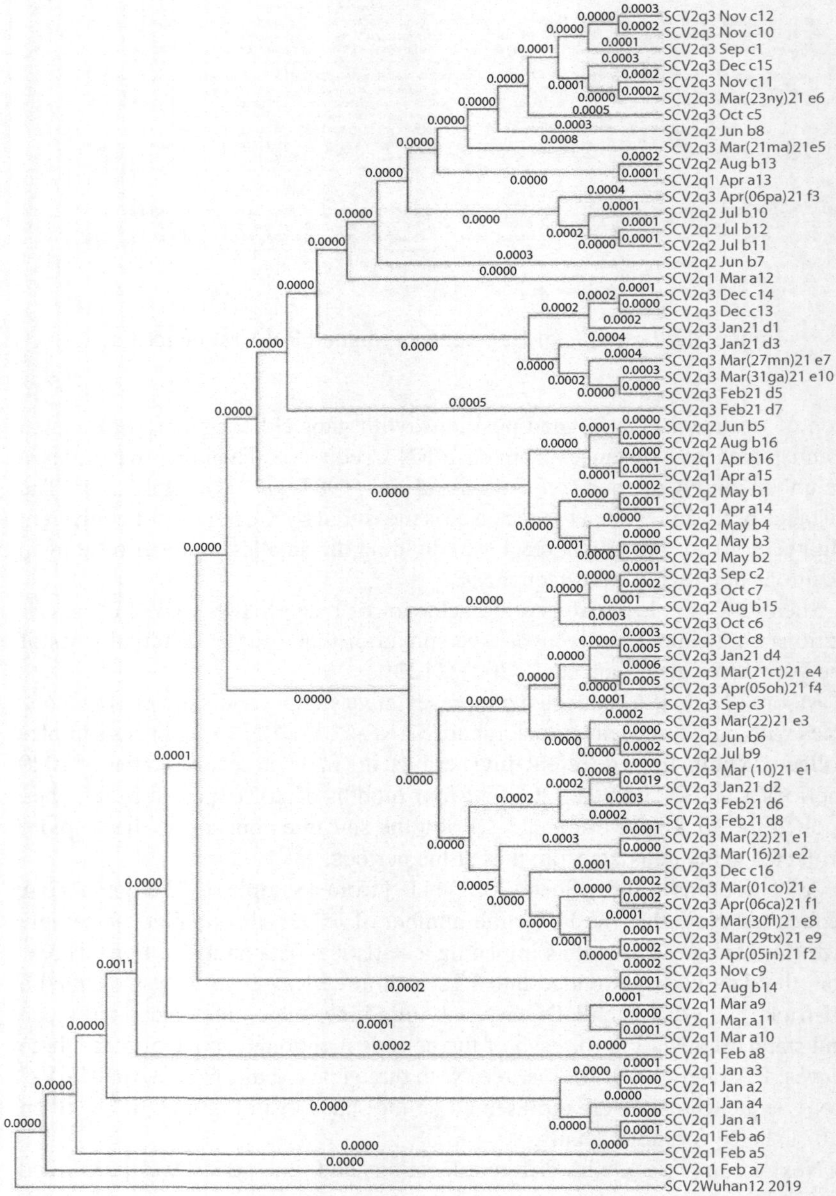

Figure 8.3: Phylogenetic tree of 72 RNA sequences of SARS-CoV-2 specimens

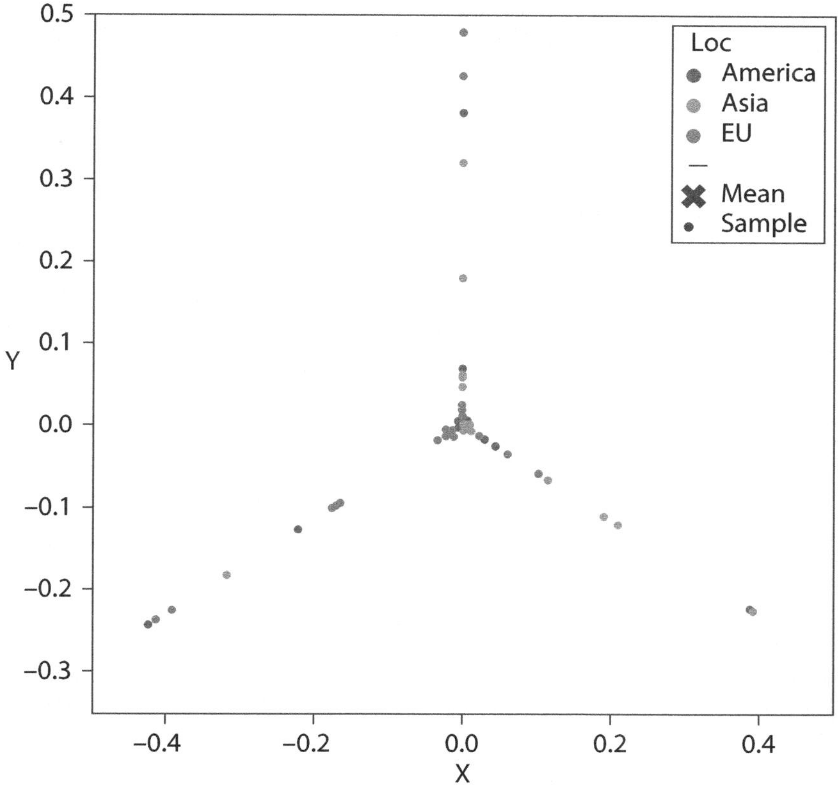

Figure 8.4: Sample trees on T_3. Orange: Asia, Green: Europe, Blue: America.

from the four groups for each location and then made an alignment. We repeated this step ten times to form aligned sequences for comparing viruses for different countries, and twenty time periods for comparing viruses on different continents. From those aligned sequences, we could build trees with four leaves.

Since the sample trees are in T_4 space, we cannot visualize the whole T_4 space in $\mathbb{R}^3$. However, we could project the points into the Petersen graph from the cone point to briefly visualize the samples, and their intrinsic means. Figure 8.5 below showed the 20 sample trees for each continent and their corresponding sample intrinsic mean trees, marked with a star.

We could find that samples are spread all over the quadrants of the T_4 space. The sample mean trees of the EU and America are close to each other, farther apart from the sample mean tree of Asia. The sample mean trees of the EU and America show a similar mutation pattern, and there is a big difference com-

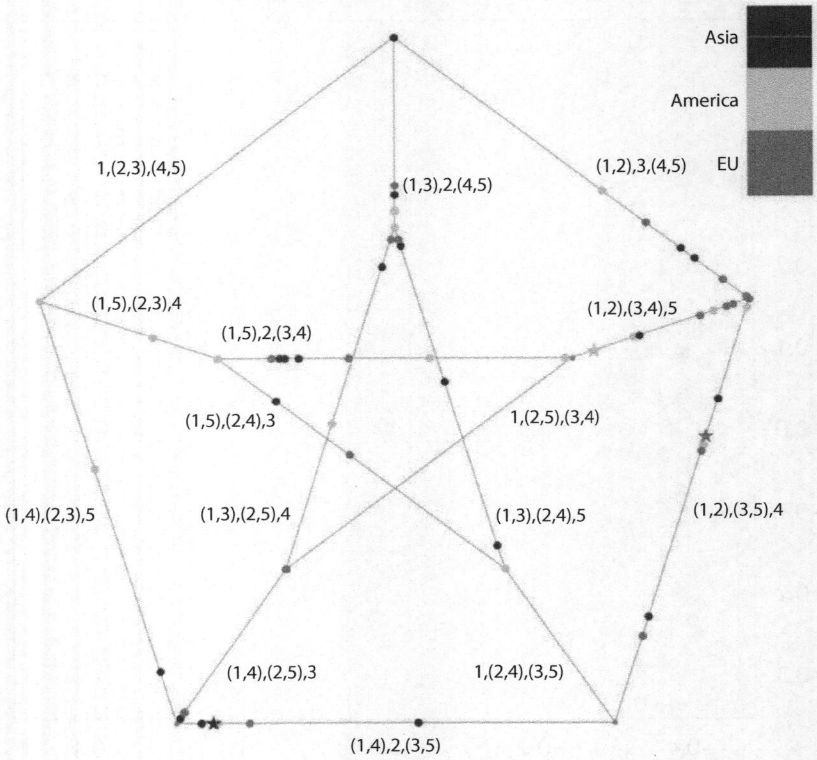

Figure 8.5: Sample trees (dots) and mean trees (stars) of different continents on T_4.

pared with Asia's. However, we cannot draw any conclusion from the figure given the standard deviation and significant low value of interior edges. The mean tree's interior edge is short; this can be explained due to the stickiness of the mean, and the small sample size.

We repeat the same steps for countries and show results as follows in Figure 8.6. The results show that most sample trees locate on several particular orthants. Some countries share the same tree type and are very close to each other. However, as we mentioned before, we probably need more samples to draw a significant conclusion. As more data were introduced, either more samples for each country or more countries are added, we found trends for the virus's evolution and mutation.

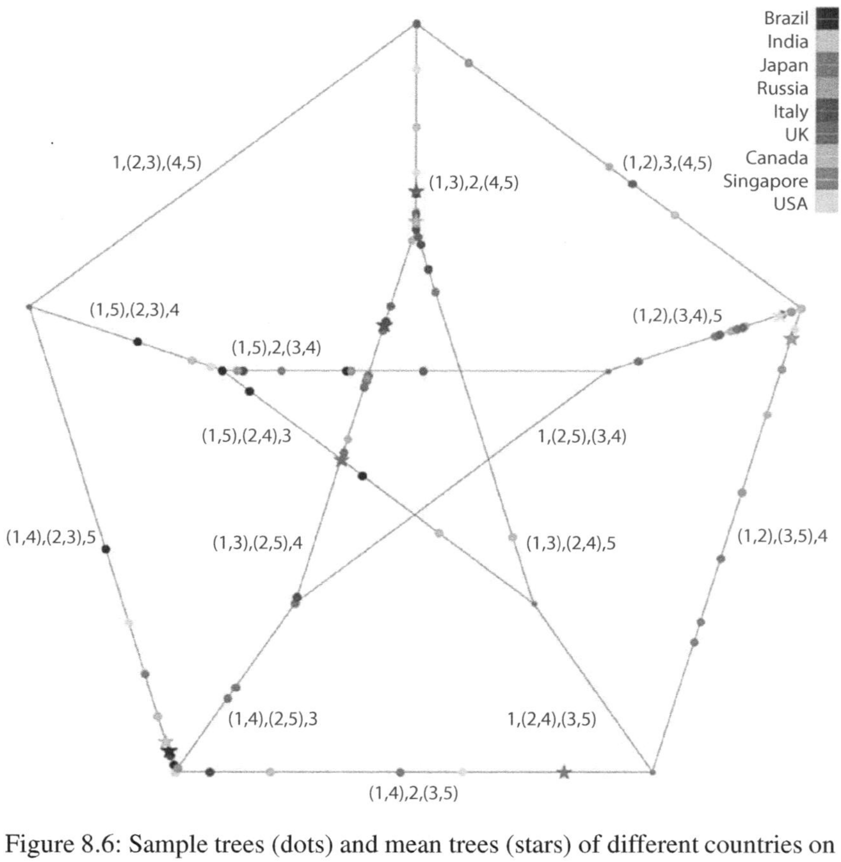

Figure 8.6: Sample trees (dots) and mean trees (stars) of different countries on T_4.

Part III

Asymptotic Theory and Nonparametric Bootstrap on Special Stratified Spaces

Chapter 9

CLT on Low-Dimensional Stratified Spaces

9.1 Introduction

Noncategorical observations, when regarded as points in a stratified space, lead to a nonparametric data analysis extending data analysis on manifolds. In particular, given a probability measure on a sample space with a manifold stratification, one may define the associated Fréchet function, Fréchet total variance and Fréchet mean set. The sample counterparts of these parameters have more nuanced asymptotic behaviors than in nonparametric data analysis on manifolds. This allows for the most inclusive data analysis known to date. Unlike the case of manifolds, Fréchet sample means on stratified spaces may stick to a lower-dimensional stratum, a new dimension reduction phenomenon. The downsize of stickiness is that it yields a less meaningful interpretation of the analysis. To compensate for this, an extrinsic data analysis that is more sensitive to input data is suggested. In this chapter one explores analysis of data on low-dimensional stratified spaces, via simulations. An example of an RNA-based extrinsic phylogenetic mean tree is also given.

9.2 Data Analysis on Manifolds

The question of studying random elements (nowadays called objects), other than random vectors, was first raised by Fréchet (1948) [52]. As an example, Fréchet suggested to analyze the shape of a contour of a closed curve, or the shape of an egg selected at random from a wire egg basket. Fréchet's approach to Analysis of Object Data (AOD) consists of identifying an object with a point in a complete metric space (M,ρ). Next, given a random object X on M, he defined what we call today the *Fréchet function* on M, given by $F_\rho(p) = E(\rho^2(X,p))$. A minimizer of F_d above is a *Fréchet mean*, and the minimum value of F_d is the *Fréchet total variance*. Ziezold (1977) [150] showed that the *Fréchet sample mean set* is a consistent estimator of Fréchet population mean set. Going beyond consistency, Hendriks and Landsman (1996) [65] and, independently, Patrangenaru (1998) [104] assumed the metric space from which the data is sampled is a submanifold of the numerical space, or forms an abstract *manifold,* thus allowing for a description of the asymptotics of the extrinsic sample mean on a manifold embedded in a numerical space, thus providing key examples of limit behavior of Fréchet sample means. In addition, Hendriks and Landsman (1996) [67], and, independently, Patrangenaru (1998) [104] used different consistent estimators of the population extrinsic covariance matrix to studentize the extrinsic sample mean vector and estimate the extrinsic population mean. The large sample behavior of the Fréchet sample mean on a manifold in general was given in Bhattacharya and Patrangenaru (2005) [21]. A manifold modeled on a Hilbert space H is locally diffeomorphic to H, which, due to consistency, allows to describe the asymptotics of the Fréchet sample means in terms of a representation in the tangent space at the Fréchet mean, if the latter exists. A 2D manifold, immersed in R^3 is shown in Figure 1. Two types of distances were considered on the sampling manifold M : a chord distance $\rho =_j d$, where $j : M \to L$ is an embedding of M into a vector space over the reals L with a Euclidean distance, or a geodesic (arc) distance $\rho = \rho_g$, associated with a Riemannian tensor g on M : The nonparametric methods for studying data of this sort has led to an *extrinsic* and an *intrinsic* data analysis.

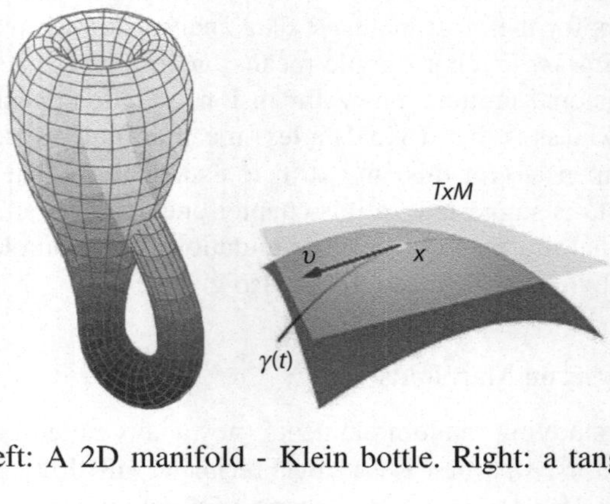

Figure 9.1: Left: A 2D manifold - Klein bottle. Right: a tangent space at a point.

A *manifold* modeled on a Hilbert space $\mathbb{H}$, is locally, but not globally, homeomorphic to $\mathbb{H}$. If $dim\mathbb{H} < \infty$, manifold data, data arising on finite dimensional Hilbert manifolds are a natural extension of multivariate data. If $dim\mathbb{H} = \infty$, data on Hilbert manifolds are an extension of functional data. Dimension reduction via geodesic PCA for HDLSS data on certain Riemannian manifolds (2010). Extrinsic analysis on nonflat manifolds is computationally faster than intrinsic analysis, see Bhattacharya et al. (2012) [18]; therefore, there is no point for manifold valued data to use an intrinsic analysis, since any manifold can be embedded in a numerical space (see Whitney (1944) [142]). Hello, extrinsic analysis!

9.3 Data Analysis on One-Dimensional Stratified Spaces

While analysis of Fréchet means on manifolds is now an established area, sample spaces in modern data analysis, including Kendall shape spaces in dimension 3 or higher, are not all manifolds. However, they do have a manifold stratification; they are stratified spaces (see, e.g., Verona(2007) [137]).

DEFINITION 9.3.1. *A filtration by closed subsets* $\emptyset = F_{-1} \subseteq F_0 \subseteq F_1 \ldots \subseteq F_n \subset \ldots$, *by closed subspaces of a metric space M, such that the difference between successive members* F_i *and* F_{i-1} *of the filtration is either empty or a smooth manifold of dimension i is called a stratification. The connected components of the difference* $F_i \setminus F_{i-1}$ *are the* strata *of dimension i.*

The *regular part* of M is the highest dimensional stratum. At each regular point, the stratified space has a tangent space. The dimension of the stratified space is m if $M = F_m \neq F_{m-1}$: otherwise $dimM = \infty$. All the other points are *singular.* The analysis of data on stratified spaces is still in a very early stage. Examples of stratified sample spaces, which are not themselves manifolds, include similarity shape spaces (see Kendall et al. (1999) [80]), affine shape spaces (see Tagare et al. (2009) [131]) and projective shape spaces (see Mardia and Patrangenaru (2005) [91]). Spaces of positive semi-definite matrices, which arise as data points in Diffusion Tensor Imaging (for example, see Schwartzman et al. (2008) [116]), and tree spaces (see Billera et al. (2001) [28], Wang and Marron (2007) [139]) are additional examples of stratified sample spaces.

9.3.1 Phylogenetic Trees

The data that biologists use usually come from one homogenous sequence, which in the biologist's language concerns the relationship between gene trees and genes that are made from one tree. The gene sequence might be about 200 base pairs long. One of the problems that have occurred in the last 40 years is that biologists believe that the way evolution works is that there would only be one *species tree.* Different genes have different histories, so you get different gene trees. Putting them together is a statistical problem that helps the study of the evolutionary process. A phylogenetic tree with p leaves is an equivalence class based on a certain equivalence, of a DNA-based connected directed graph of species with no loops, having an unobserved *root* (common ancestor) and p observed *leaves* (current observed species of a certain family of living creatures). A tree with p leaves is a simply connected graph with a distinguished vertex, labeled o, called the root, and p vertices of degree 1,

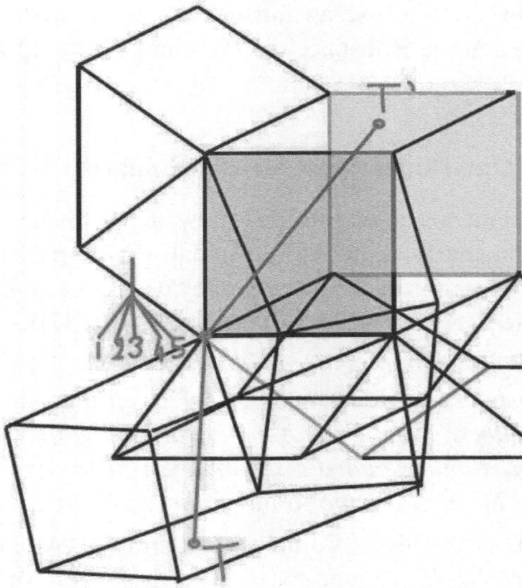

Figure 9.2: Tree space T_5.

called leaves, that are labeled from 1 to p. In addition, we assume that all interior edges have positive lengths. An edge of a p-tree is called *interior* if it is not connected to a leaf. Now consider a tree T, with interior edges $e_1, \ldots, e_r$ of lengths $l_1, \ldots, l_r$ respectively. If T is binary, then $r = p-2$; otherwise $r < n-2$. The vector $(l_1, \ldots, l_r)^T$ specifies a point in the positive open *orthant* $(0, \infty)^r$. That is to say that a binary p-tree has the maximal possible number of interior edges and thus determines the largest possible dimensional orthant; in this case the orthant is $p-2$-dimensional. The orthant corresponding to each nonbinary tree appears as a boundary face of the orthants corresponding to at least three binary trees. In particular, the origin of each orthant corresponds to the (unique) tree with no interior edges, which is known as the *star tree.* The space T_p is constructed by taking one $p-2$-dimensional orthant for each of the $(2p-3)!!$ possible binary trees and gluing them together along their common faces. Note that tree spaces are not manifolds. Singularities (points where the space does not have a tangent space) are present in the tree space structure. For further detail on phylogenetic trees and the construction of the tree space, see Billera et al. (2001) [28]. Phylogenetic trees with p leaves are points on a metric space T_p that has $p-2$-dimensional stratification. In particular, the space of trees with three leaves is $T_3 = S_3$, a 3-spider, which is the union of three line segments with a common end (see Figure 10.2).

For a probability measure on S_p, if none of the the "legs" of the p-spider has a dominant probability mass, then the Fréchet mean is the star tree. This result will be stated more formally in the following section and extended to more general spaces. T_4 is a two-dimensional stratified space obtained from $15 = (2 \times 4 - 3)!!$ 2D quadrants glued according to tree identification rules (see Billera et al.(2001) [28]. Interior points of these quadrants are combinatorial binary trees with four leaves, the coordinates of an interior point being given by the two interior edges of a binary tree in one of these combinatorial binary trees. Points on the boundaries of the quadrants correspond to combinatorial trees with four leaves, which are obtained from a combinatorial binary tree by shrinking one of the interior edges to zero length. Therefore, a representation of T_4 as a surface with singularities can be obtained from the polyhedral surface given in Figure 3.3 by identifying the edges labeled with the same letter. While it may give a 3D pictorial representation only, in fact, as mentioned in Billera et al. (2001) [25], given that $2^4 - 4 - 2 = 10$, T_4 is embedded in $\mathbb{R}^{10}$ having the *star tree* at the origin. In this representation, the intersection of a small sphere in $\mathbb{R}^{10}$ centered at the origin with the image of T_4 under this embedding is the so-called Petersen graph. An edge of a Petersen graph is the transverse intersection of one quadrant with a sphere, thus there are 15 edges (five of which are marked with a red dot below). A vertex is the point where one of the coordinate axes pierces the sphere; therefore there are ten vertices (see Figure 9.3).

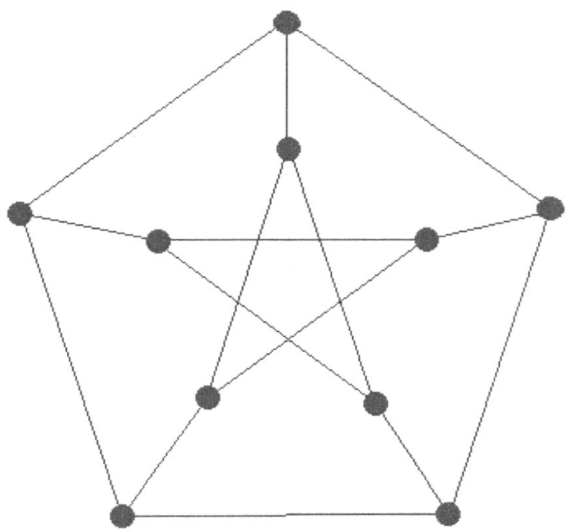

Figure 9.3: Petersen graph.

The remainder of this chapter will focus on data analysis for low-dimensional tree spaces and simple graphs.

9.3.2 CLTs on Trees

Recall from Section 3.5 that we assumed $X_i, i = 1, \dots, n$ are i.i.d.r.o.s on a spider S_p, having legs $L_a, a = 1, \dots, p$ and center C. Further, any probability measure Q on S_p decomposes uniquely as a weighted sum of probability measures Q_k on the legs L_k and an atom Q_0 at C (see Hotz et al. (2013) [69]), so that there are nonnegative real numbers $w_0, \dots, w_p$ summing to 1 such that, for any Borel set $A \subset S_p$, the measure Q takes the value

$$Q(A) = w_0 Q_0(A \cap C) + \sum_{k=1}^{p} w_k Q_k(A \cap L_k). \tag{9.1}$$

We considered there the nontrivial case when the moments $\nu_a = E(Q_a), a = 1, \dots, p$ are all positive and assumed in addition that $w_0 = 0$. We proved in Section 3.5 the following:

THEOREM 9.3.1. *(Hotz et al. (2013) [69]). Assume (i) If there exists $a = 1, \dots, p$ such that $w_a \nu_a > \sum_{b \neq a} w_b \nu_b$, then $\mu_{X_1,I} \in L_a$ and, for n large enough $\bar{X}_{n,I} \in L_a$ and $\sqrt{n}(\bar{X}_{n,I} - mu_{X_1,I})$ has asymptotically a normal distribution. (ii) If there exists $a = 1, \dots, p$ such that $w_a \nu_a = \sum_{b \neq a} w_b \nu_b$, then after folding the legs $L_b, b \neq a$ into one half-line opposite to L_a, $\sqrt{n}\bar{X}_{n,I}$ has asymptotically the distribution of the absolute value of a normal distribution. (iii) if there exists $\forall a = 1, \dots, p$ $w_a \nu_a < \sum_{b \neq a} w_b \nu_b$, then $\mu_{X_1,I} = C$ and there is n_0 s.t. $\forall n > n, \bar{X}_{n,I} = C$ a.s.*

REMARK 9.3.1. *Under the assumptions of Theorem 9.3.1 (iii), we said that the sample mean is sticky. Recall that Basrak (2010) [9] originally obtained a related result for distributions on binary metric trees. According to Basrak (2010) [12]:"Limit theorems on trees will need minor adjustments, since on a general tree, the barycenter can split the tree into more than three subtrees". Nevertheless, asymptotically, the intrinsic mean will have one of the three types of behavior described in theorem 3 in Basrak (2010) [12], meaning that the stickiness phenomenon is still present for distributions on trees.*

9.4 CLTs on Graphs Including Cycles

The above quote from Basrak (2010) [9] does not extend to arbitrary finite, connected graphs, since graphs usually have cycles (see Figure 9.4). The key property of metric acyclic graphs (trees) is that they are CAT(0) and that a random object on such a space with finite intrinsic variance has an intrinsic mean. There are random objects on a circle having an intrinsic mean set with at least two points. In fact the intrinsic mean set may contain an arbitrary number of points, or be even the entire circle. However, Basrak's condition for trees can be extended to the case of graphs in some general cases even if the graph

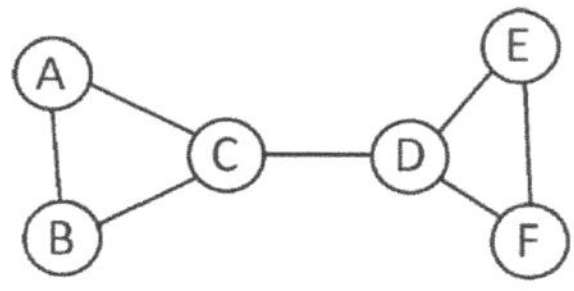

Figure 9.4: Bicyclic graph.

G has cycles with positive mass on any arc of a cycle. Consider the case that the Fréchet mean is unique, to be denoted by μ_I : In this case, by consistency of the intrinsic sample mean set (Ziezold (1977) [150]), any measurable intrinsic sample mean $\hat{\mu}_{I,n}$ selected from the sample mean set converges a.s. toμ_I; and since around μ_I the graph is homeomorphic to $\mathbb{R}$, the intrinsic sample mean behaves like in Theorem 1(i). Assume μ_I is a vertex, then there is a small neighborhood of μ_I that looks like a spider S_p with center $C = \mu_I$ and with p legs, $p \geq 3$, such that each leg belongs to a different connected component of $G \setminus \mu_I$. Then the sample Fréchet mean behaves like in Theorem 9.3.1. More generally, given a unique Fréchet mean, μ_I, there is a small neighborhood of μ_I that looks like a spider S_p with center $C = \mu_I$ and legs L_a oriented such that C is the starting point of that leg.

For each $x \in G$ and leg $L_a, \rho(x;\xi)$ is either increasing or decreasing for $\xi \in L_a$. We define $\varepsilon_a(x) = +1$ if increasing, and $\varepsilon_a(x) = -1$ if decreasing. Then μ_I is sticky if for all legs L_a

$$E(\rho(X,\mu_I)\varepsilon_a(X)) > 0. \tag{9.2}$$

This is easily identified as a condition that implies that μ_I is a local minimum of the Fréchet variance. The quantity $m(a) = E(\rho(X,\mu_I)\varepsilon_a(X))$ is called the *net moment* in the direction of L_a. As an aside, one can consider for any $x \in G$ its star neighborhood S_ξ and the "derivative" in the direction of leg L_a of S_ξ, given by

$$D_{L_a}F_\rho = E(2\rho(X,\xi)\varepsilon_a(X)). \tag{9.3}$$

Notice that the second derivative in the direction of leg L_a equals $E(2\varepsilon_a(X)^2) = 2$ is positive. In particular, at any point where the Fréchet function is differentiable, it is convex.

REMARK 9.4.1. *It would be useful to consider the case when the probability distribution of X is concentrated in a dense countable subset of the graph G, (for example, in the circle S^1 at rational angles from some base point. At every rational angle the Fréchet function cannot be minimal, since the antipodal point carries mass.)*

REMARK 9.4.2. *One may note that in the case of trees, and in particular in case of data on a spider, the condition of stickiness in this section is equivalent*

Figure 9.5: Graph exhibiting a family of distributions having non-sticky means (green dots).

with the condition in Theorem 9.3.1. Consider the situation where μ_I is a vertex, and C_a is a connected component of $G \setminus \mu_I$ that has a unique edge joined to μ_I. Then $G_a = C_a \cup \{\mu_I\}$ is a subgraph of G. For each $x \in G_a$ there is the intrinsic distance from x to μ_I, and with respect to the probability Q_a conditional to be in C_a there is an expected intrinsic distance, called the moment m_a (of Q_a or G_a with respect to μ_I). The condition for stickiness is equivalent in this case to $m_a < \sum_{a \neq b} m_b$, which in the case of a star tree (spider) is equivalent to the condition in Theorem 1 (iii).

REMARK 9.4.3. *Certain complications arise in describing stickiness phenomena, when the graph includes a cycle of positive probability mass. Consider for example the case of a simple graph such as the circle, realizable as a one vertex, one edge graph. See Hotz and Huckemann (2015) [68]. In this case, if μ_I is a Fréchet mean, then the antipodal point, $-\mu_I$, must have probability 0. Even, if the density in a neighborhood of $-\mu_I$ is continuous, then the density at $-\mu_I$ cannot exceed $2\pi^{-1}$. If the density is below $2\pi^{-1}$, Hotz and Huckemann (2015) [68] prove a central limit theorem to a normal distribution.*

In the next section we consider concrete simulations that on data on simple graphs, justifying the results in this section (see Ellingson et al. (2014) [43]).

9.4.1 Simulations of sticky and nonsticky sample means on graphs

Figure 9.5 shows a graph exhibiting nonsticky intrinsic sample mean (green dot) as we pull an observation (red dot) away. On the other hand, Figure 9.6 shows how the intrinsic sample mean sticks to its original location even after pulling an observation away from it.

A similar behavior of distributions having either sticky or regular means on bicyclic graphs is displayed in Figure 9.7.

9.5 Computational Examples on 2D Stratified Spaces

For computational algorithms for intrinsic sample means on T_p, see Owen and Provan (2010) [103]. Here we consider some computational examples to illustrate the behavior of extrinsic means for simulated data on T_4 (for the asymp-

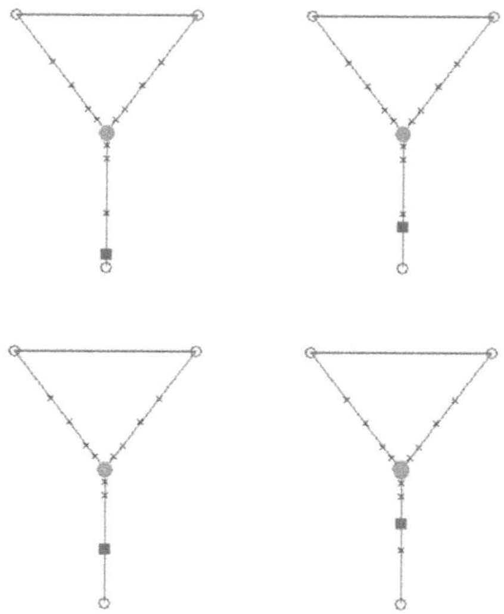

Figure 9.6: Graph exhibiting a family of distributions having sticky means (green dots).

totic distribution of the intrinsic sample means on this space, see Barden et al. (2013) [8]). For both examples, our embedding j is $j: T_4 \to \mathbb{R}^{10}$. In general T_p can be embedded in $\mathbb{R}^k, k = 2^p - p - 2$.

9.5.1 *Simulations on T_4*

Figure 9.8 shows data simulated on a portion of T_4, which is a gray region. This portion of T_4 consists of five orthants arranged radially around the origin. This portion was labeled Q_5 in Barden et al. (2013) [8]. Please note that while the region is displayed in three dimensions, this is for display purposes only. The intrinsic means, displayed in green, are computed according to the distance discussed in Billera et al. (2001) [28]. The extrinsic mean sets, displayed in red, are calculated by embedding this region into $\mathbb{R}^{10}$, as described above. For the purposes of these simulations, we assume that the data is distributed only on this region, not on the remaining portions of T_4.

In the left image, the simulated data are distributed identically over all five quadrants with equal probability of being in each quadrant, resulting in the intrinsic sample mean being the star tree and the extrinsic sample mean set consisting of five trees radially symmetric about the origin. To show the behavior of the intrinsic and extrinsic sample means, we perturbed the initial distribution by reducing the maximum possible distance in the x1 direction, adjusting the probability of an observation in the (x^1, x^2) and (x^5, x^1) quadrants accordingly.

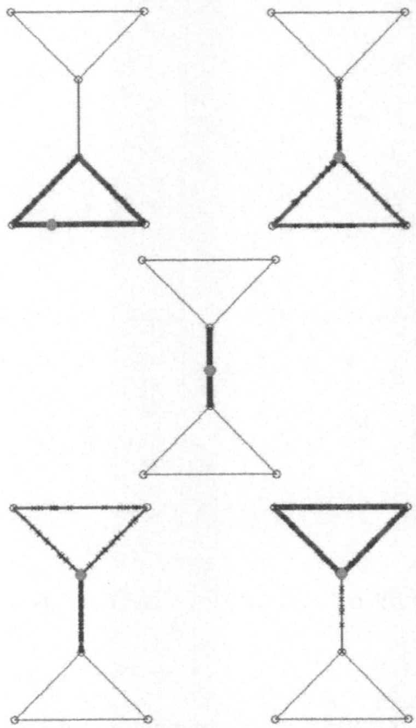

Figure 9.7: Graph exhibiting a family of distributions having both sticky and nonsticky means (green dots).

In the middle image, the distribution has been perturbed slightly in this manner. As a result, the extrinsic mean set now consists of only two trees, but the intrinsic mean is still the star tree because there is still a substantial concentration of mass on the (x^1,x^2) and (x^5,x^1) quadrants. We continue to perturb the distribution in the above manner until the intrinsic mean is no longer the star tree. The results are shown in the image on the left side of Figure 9.7. In this case, not only is the intrinsic mean no longer the star tree, but the extrinsic mean is also now unique. Please note that we had to greatly perturb the original distribution in this manner to obtain an intrinsic mean that was not the star tree. Indeed, this distribution no longer has sufficient mass on the (x^1,x^2) and (x^5,x^1) quadrants for the intrinsic mean to remain at the origin.

REMARK 9.5.1. *Note that unlike the extrinsic mean set on Q_5 which is immediate to compute, the formula for of the intrinsic mean of a random object on Q_5 is complicated, as shown in Barden et al. (2013) [8]. Moreover, the computation of the intrinsic sample mean on Q_5 is increasingly computationally intensive as the sample size grows larger. The advantage of the intrinsic analysis on Q_5 is the uniqueness of the intrinsic mean, leading to a nice asymptotic*

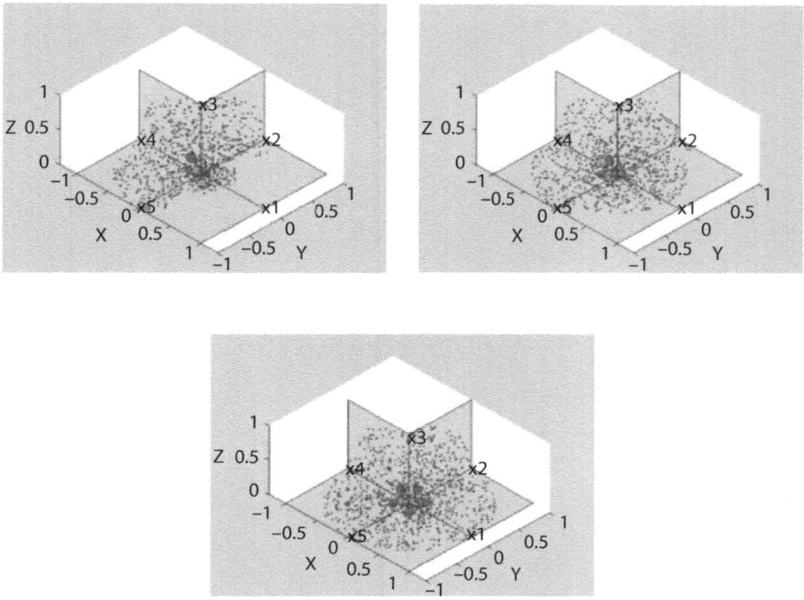

Figure 9.8: Extrinsic mean set (red) and intrinsic mean (green) on various simulations in T_4.

behavior of the intrinsic sample mean (Barden et al. (2013)[8]. The problem of the large sample behavior of the extrinsic sample mean set on Q_5 is still open and is nontrivial.

REMARK 9.5.2. *There are distributions on Q_5 having sticky intrinsic means (see Figure 9. For such a distribution the sample intrinsic mean coincides with the population intrinsic means, which is not very informative. In such cases it is preferred to use an extrinsic approach and construct confidence regions for the extrinsic mean.*

9.5.2 Extrinsic Sample Mean for RNA Data

Following Ellingson et al. (2014) [43], here we compute the extrinsic sample mean of ten simulated phylogenetic trees obtained using the Mesquite software (Maddison and Maddison (2011) [89]) and concatenated Parkinsea RNA data taken from the UT-Austin CRW website (see Cannone et al. (2002) [30]), such that there will only be four leaves in the phylogenetic tree. In Figure 9.9 we show the location of the one Parkinsea point in the tree space using its link of the origin. Figure 9.10 shows the resulting extrinsic sample mean tree and its relative location in the tree space.

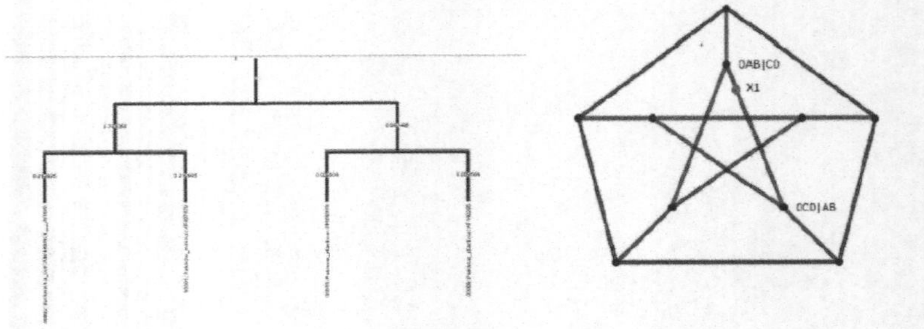

Figure 9.9: Left: A simulated phylogenetic tree of Parkinsea. Right: Location of the tree in the tree space using the link of the origin.

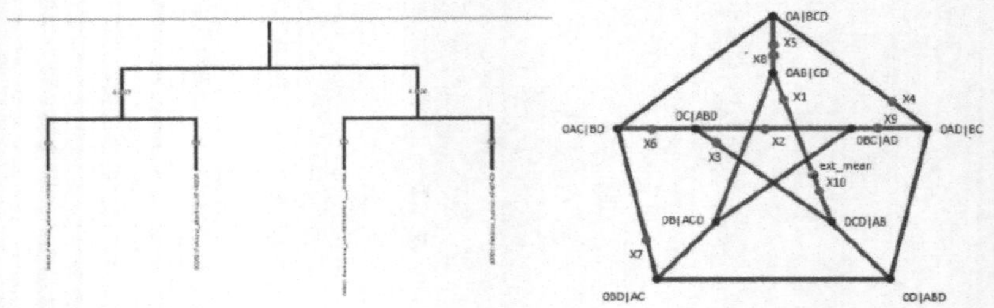

Figure 9.10: Left: The extrinsic sample mean tree. Right: Location of the extrinsic mean tree relative to the sample trees.

9.5.3 Extrinsic CLTs on Open Books

The extrinsic mean set of a distribution on a carton open book is on the spine if and only the distribution is entirely concentrated on the spine; such embeddings do not present the stickiness phenomenon. On the other hand, paperback open books present a stickiness phenomenon for a large family of distributions. In general the CLT for an embedding of stratified spaces may involve estimation of finite mean sets.

9.6 Discussion

An analysis of a population of trees based on intrinsic means may be difficult if the intrinsic sample mean sticks to a lower-dimensional stratum of the tree space. For example, the biological interpretation of the intrinsic mean of a population of phylogenetic trees with p leaves being at the star tree is that the phylogenies in the family are too diverse to offer a plausible evolutionary scenario. On the other hand, an extrinsic analysis of a population of trees based on extrinsic mean sets may be helpful, unlike the intrinsic mean tree which is unique due to the hyperbolicity of (T_p, ρ_0), if we canonically embedded T_p in

$\mathbb{R}^k$ as a $p-2$-dimensional stratified space, the extrinsic mean set reflects all average evolutionary trees from a given family of phylogenetic trees, revealing more mean evolutionary scenarios. Extrinsic data analysis should be further pursued on stratified spaces for additional reasons. The first is that computations of extrinsic means on manifolds are faster than their intrinsic counterparts (see Bhattacharya et al. (2012) [18]). Additionally, each point in the extrinsic sample mean set has, asymptotically, a multivariate normal distribution around the point of the extrinsic population mean set in the corresponding orthant, while the intrinsic sample mean might often stick to a vertex, thus making the data analysis on a graph difficult. In addition, there are no necessary and sufficient conditions for the existence of the intrinsic mean on a non-simply connected graph, while on the other hand, there are such conditions for the existence of extrinsic means on graphs and tree spaces.

Chapter 10

Investigating Two Possible Origins of SARS-CoV-2

10.1 Introduction

Late in 2019, authorities in Wuhan, China identified the first case of COVID-19 caused by the novel SARS-CoV-2 coronavirus. Shortly thereafter, in early 2020, researchers at the Wuhan Institute of Virology (WIV) identified RaTG13, a coronavirus found in *Rhinolophus affinis* bats, to be an approximately 96% match to SARS-CoV-2, which hinted at an animal origin for the disease – now, of course, the dominant theory.

From that point onward, an animal host has yet to be found, which inevitably has led to alternative hypotheses proposed to explain the pandemic, most notably the lab-accident or lab-leak origin. In this particular explanation, the SARS-CoV-2 virus escaped via negligence or malicious intent from the WIV and caused "superspreader" events in local wet markets. Although initially considered untenable, this theory has received renewed interest and is even now considered "plausible, but extremely unlikely" (see [34, 143, 94]).

Meanwhile, due to tremendous global effort, more and more SARS-CoV-2 samples have been collected and their genomes made available in databases such as GenBank or GISAID. While the prospects of either discovering an animal host or receiving increased cooperation from WIV to explore a lab-leak origin seem dim, could this ever-increasing supply of data shed light on the origins of SARS-CoV-2? That certainly is our hope and intent.

Here we introduce a method that uses SARS-CoV-2 RNA sequences and phylogenetic trees in order to define an evolutionary distance between the RaTG13 bat coronavirus and early human SARS-CoV-2 coronaviruses sampled in Wuhan to later SARS-CoV-2 samples. With this measure, we then construct a two-sample test to determine which of the candidates is on average closer in our defined distance to the later samples, the result of which suggests the Wuhan coronavirus is closer.

In Section 10.2, we cover background information related to phylogenetic trees and their construction and provide a specific construction of the space of phylogenetic trees. In Section 10.3, we extend the Billera-Holmes-Vogtmann (BHV) tree space construction, to rooted trees, define an evolutionary distance, propose a method for analyzing SARS-CoV-2 RNA sequences, and implement our method. Finally, in Section 10.4, we offer more interpretation of our results and highlight limitations and future directions for our method. Most of the figures in this chapter are due to Moore (2021) [96].

10.2 Phylogenetic Trees and Tree Spaces

10.2.1 Definition and Construction

A phylogenetic tree with m leaves or m-tree is defined as a connected, cycle-free graph consisting of m vertices of degree 1 called **leaves**, remaining vertices of degree 3 called **nodes**, $m-2$ **interior edges** that are not connected to any leaf and are paired with positive real numbers called **branch lengths**. Moreover, each tree possesses a branching pattern that determines the "shape" of that tree. Figure 10.1 displays an example of a 5-tree; recall that the space of 5-trees is a 3D stratified space depicted in Figure 9.2.

In practice, phylogenetic trees are often used to investigate evolutionary relationships or similarities between different species or even genetic sequences

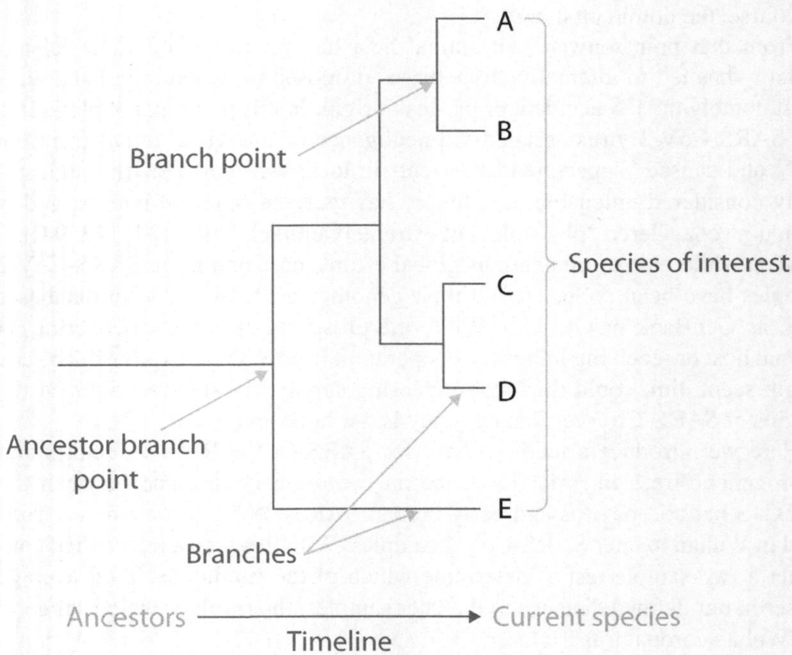

Figure 10.1: Phylogenetic tree with five leaves.

(e.g., DNA or RNA). Specifically, when viewing a collection of taxa (e.g., animal species, plant species, genetic sequences) as the tree's leaves, a **rootless** tree is one that groups the various taxa by similarity, and a **rooted** tree is one that shows an evolutionary history of bifurcations among a recent **known common ancestor**, with the branch lengths measuring the "time" until the next bifurcation, until terminating at the leaves.

Constructing phylogenetic trees remains an active area of research in phylogenetics, and current methods range from the nonparametric (parsimony), semi-parametric (neighbor-joining, minimum evolution), to the fully parametric (maximum likelihood). Nei and Kumar (2000) [98] provide a good overview of each of these construction methods.

10.2.2 Tree Space $\mathscr{T}_n$

In their 2001 article [25], Billera Holmes and Vogtmann (BHV) constructed a space $\mathscr{T}_n$ of rootless trees with n leaves according to the following. For trees with n leaves there are $(2n-3)!!$ unique maximal dimensional connected components of the highest dimensional stratum, and for each tree there are $r=n-2$ interior edges. Recall that each unrooted tree has r interior edges with positive branch lengths $l_1,\ldots,l_r$, each tree can be associated with the point $(l_1,\ldots,l_r)$ in the orthant $[0,\infty)^r$. $\mathscr{T}_n$ then is the collection of $(2n-3)!!$ orthants $[0,\infty)^r$ glued to one another at the origin and along common faces which produce the same ambiguity as one or more of the branch lengths tend toward zero. Therefore, each tree with n leaves and branch lengths $(l_1,\ldots,l_r)$ corresponds to the point $(l_1,\ldots,l_r)$ in the orthant corresponding to that tree's topology. Figure 10.2 displays $\mathscr{T}_3$, also known as a **3-spider**.

Furthermore, one can consider a metric ρ between two trees $T_1=(u_1,\ldots,v_r)$ and $T_2=(v_1,\ldots,v_r)$ given by

$$\rho(T_1,T_2)=\begin{cases}\|(u_1,\ldots,u_r)-(v_1,\ldots,v_r)\| & T_1,T_2 \text{ have the same topology}\\ \|(u_1,\ldots,u_r)-(-v_1,\ldots,v_r)\| & T_1,T_2 \text{ have different topologies}\end{cases}$$

where $\|\cdot\|$ is the Euclidean norm.

In other words, if two trees are in the same orthant, then their distance is simply the Euclidean distance between them. But, if two trees are in different orthants, then the distance between them is the length of the straight line going from the first tree to the origin and then from the origin to the second tree.

10.3 Analyzing SARS-CoV-2 RNA Sequences Using Rooted $\mathscr{T}_3$

10.3.1 RNA Sequences and Trees

Ribonucleic acid (RNA) sequences are chains of nucleotides – guanine, uracil, adenine, and cytosine – that serve a variety of roles in genetic functioning.

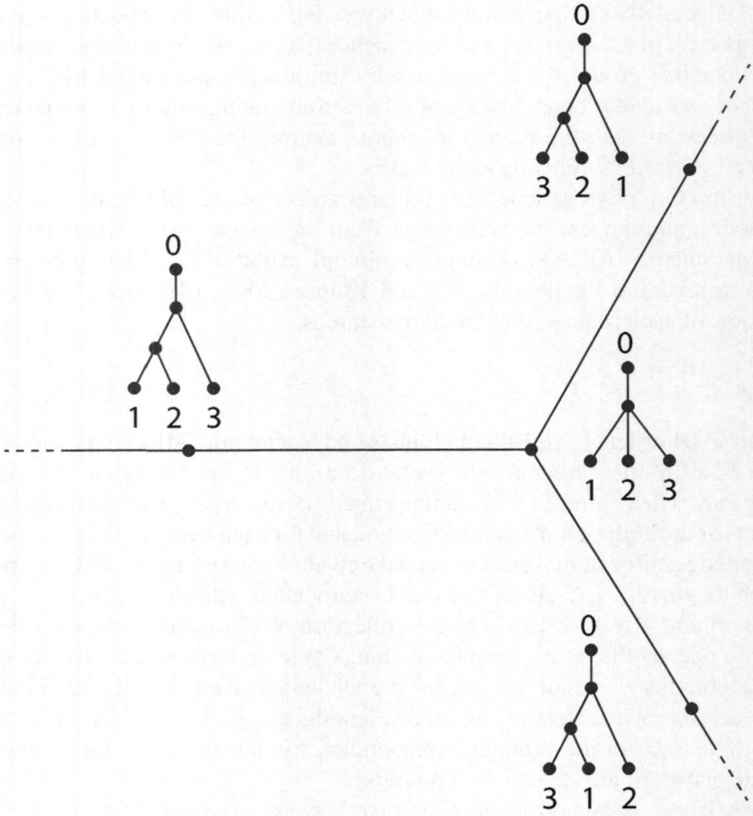

Figure 10.2: $\mathscr{T}_3$, unrooted case, from BHV [25].

In particular, coronaviruses have their entire genome encoded in RNA; they multiply through an RNA dependent RNA polymerase (RdRp) process, unlike eukaryotes that multiply from their parents' DNA.

Before genome sequencing, researchers typically constructed phylogenetic trees using morphological features of the taxa considered. However, once genome sequencing became ubiquitous, tree construction using DNA or RNA sequences became quite popular.

In essence, building phylogenetic trees using multiple RNA sequences requires comparing the sequences at specific "sites" along the chain. Crucial to this purpose then are alignment procedures, such as Clustal [86, 134], that help ensure similar sites are compared. Once aligned, these sequences can be used to build trees according to the methods already mentioned in Section 10.2.1.

10.3.2 Rooted Trees and Rooted $\mathscr{T}_3$

As mentioned in Section 10.2.1, rooted trees attempt to rebuild the evolutionary history of a set of taxa by viewing the interior nodes as most recent common ancestors and, in particular, the root serves as the most recent common ancestor (MRCA) of all the taxa considered.

In general, it is difficult to ascertain what the true root of a tree ought to be. On the one hand, if one had sufficient reason, one could assume one of the included taxa is the root and construct the tree based on that fact. On the other hand, a more common approach is to include an "outgroup" – that is, a taxa known to be distantly related to the other taxa – in order to root the tree.

Nevertheless, if one were to suppose the root were known, then one would no longer be considering trees in $\mathscr{T}_n$ but rather in $[0,\infty) \times \mathscr{T}_n$, which we call **rooted** $\mathscr{T}_n$. Figure 10.3 shows rooted $\mathscr{T}_3$, also known as an **open book** (see Hotz et al. (2013) [69]).

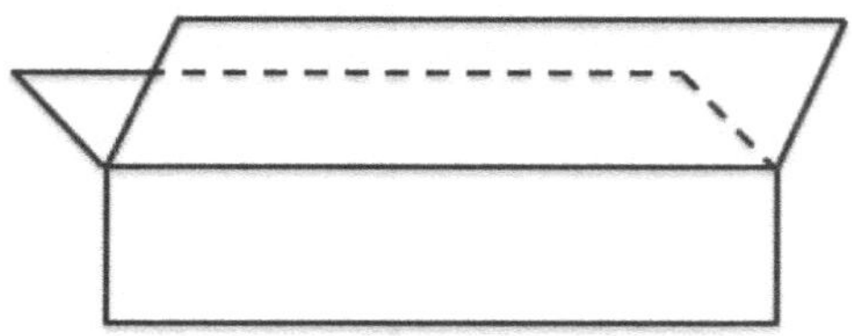

Figure 10.3: Rooted $\mathscr{T}_3$.

Rooted $\mathscr{T}_3$ proves an even more useful space because there is only one interior edge – and thus only one branch length – one can define an **evolutionary distance** from the root to the MRCA by summing the root length and the lone branch length.

Using this evolutionary distance, one then can consider a *mean evolutionary distance*, which in turn can be used to comparing two roots by comparing their means. In particular, suppose one wanted to compare the mean evolutionary distances μ_i and μ_j of roots i and j, respectively, and further suppose one had $6K$ RNA sequences for some integer $K > 0$. Then one could perform the following procedure:

1. Randomly assign $3K$ to groups for root i and j, respectively.
2. In each group, construct K rooted 3-trees and compute the evolutionary distances.
3. Test

$$H_0 : \mu_i \leq \mu_j \text{ vs } H_1 : \mu_i > \mu_j$$

using bootstrap confidence intervals on the studentized mean difference.

10.3.3 Application to SARS-CoV-2 RNA Sequences

To apply the method mentioned in the previous section, we considered 72 SARS-CoV-2 RNA sequences in Shen (2021) [119]. These sequences were initially aligned by Shen (2021) [119], using the Clustal software. For rooted trees, we used as roots: RaTG13, a coronavirus found in *Rhinolophus affinis* bats, and a SARS-CoV-2 coronavirus sequenced from a human subject in Wuhan at the beginning of the COVID-19 pandemic.

The 72 sequences were then randomly split among the two rooted 3-trees groups, after alignment, this time using the neighbor-joining method in MEGA software [132] due to Kumar et al. (2018) [83]. We used the Molecular Evolutionary Genetic Analysis (MEGA) software to construct phylogenetic trees. This method is advantageous since it is very user-friendly and has many sequence analytical tools. One does not necessarily need or have to make sequence files. However, even with sequence files, it is very easy to edit them before constructing the tree. The most popular and well-known phylogenetic tree construction tools are incorporated into MEGA, which has its own graphical interface. It automatically eliminates sequence junk (non-coding sequence [CDS]). It is able to easily ignore gaps for accurate alignment as well as compare different sequence length phylogenies. The aligned sequences are partially displayed in Figure 10.4.

For more technical details, see Moore (2021) [96] and Moore et al. (2021) [97]. The rooted trees obtained from this data yielded the evolutionary distances found in Table 10.1 where one can see the mean evolutionary distances between the bat root and the Wuhan root to the MRCA are 0.4802 and 0.1128, respectively. Table 10.1 also shows the 95% bootstrap confidence interval for the studentized mean difference when testing

$$H_0 : \mu_B \leq \mu_W \text{ vs } H_1 : \mu_B > \mu_W$$

where μ_B is the mean evolutionary distance using the bat root, and μ_W is the mean evolutionary distance using the Wuhan root. Since the interval does not contain zero, we reject H_0, which suggests the Wuhan root is on average closer in evolutionary distance to the later-sequenced SARS-CoV-2 coronaviruses than to the SARS-CoV-bat RaTG13 coronavirus root.

10.4 Conclusion

One might wonder how much further one could interpret the results in Section 10.3.3, especially concerning whether SARS-CoV-2 had an animal origin or a lab accident origin. On the one hand, these results are certainly consistent with a lab-leak hypothesis given that the Wuhan root has shown to be closer to the later samples; on the other hand, it is still possible an as-yet-discovered animal host may demonstrate otherwise. At the time this analysis was conducted, the

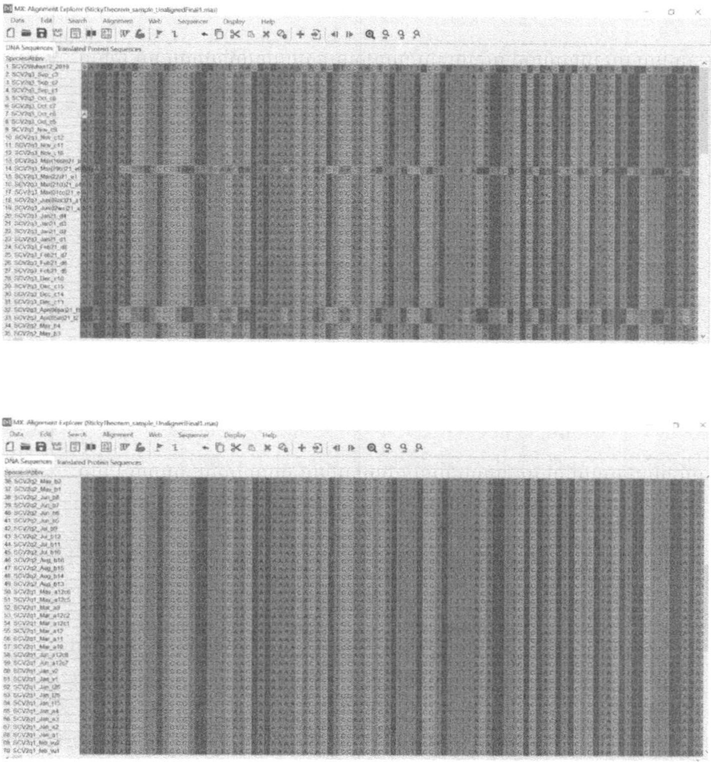

Figure 10.4: Portion of alignment of SARS-CoV-2 sequences.

RaTG13 bat coronavirus was the best candidate available, but since then the newer BANAL bat coronaviruses have proven to be even closer candidates to the SARS-CoV-2 coronavirus; it may perhaps be fruitful to repeat the procedure introduced here using these new candidates.

In general, our method has been limited by sample size, and this may not be easily ameliorated since the method requires six sequences for every additional tree one wants to include. This is also why performing the analysis on another rooted tree space – rooted $\mathscr{T}_4$, perhaps – would be costly since it would require even more sequences. However, SARS-CoV-2 data continues to abound and may only increase as time goes on, although there is no consensus about its possible evolution.

A further limitation is computation time, as the trees themselves can be costly to construct, albeit with three leaves this is not a major issue. Nevertheless, an extension of this work would be to see how the results change (or

Table 10.1: Evolutionary distances from bat and from human coronaviruses

	RaTG13			Wuhan-SARS		
Tree	Root	Interior	**Evolve Dist.**	Root	Interior	**Evolve Dist.**
1	0.4857	0.0002	**0.4859**	0.0000	0.0000	**0.0000**
2	0.3225	0.0028	**0.3253**	0.0000	0.0000	**0.0000**
3	0.2923	0.0027	**0.2950**	0.0000	0.1693	**0.1693**
4	0.2898	0.0001	**0.2899**	0.1687	0.0001	**0.1688**
5	0.4856	0.0002	**0.4858**	0.1688	0.0000	**0.1688**
6	0.4856	0.0001	**0.4857**	0.1688	0.0000	**0.1688**
7	0.4855	0.0003	**0.4858**	0.1692	0.0002	**0.1694**
8	0.4853	0.0009	**0.4862**	0.1699	0.0014	**0.1713**
9	0.4860	0.0005	**0.4865**	0.1679	0.0003	**0.1682**
10	0.4844	0.0023	**0.4867**	0.1684	0.0005	**0.1689**
11	0.2917	0.0047	**0.2964**	0.0000	0.0000	**0.0000**
12	0.2882	0.0011	**0.2893**	0.0000	0.0000	**0.0000**
		Mean	**0.4802**		**Mean**	**0.1128**
		SD	**0.0967**		**SD**	**0.0833**
Studentized Mean Diff 95% Bootstrap CI						**(6.7674, 11.3729)**

not change) by using different tree-building methods, which certainly will have larger computation costs. But, once the trees are built, the analysis itself runs quite quickly.

In conclusion, we provided a brief overview of phylogenetic trees, their uses, and their construction. Using Billera et al.'s [25] construction, we considered the space $\mathscr{T}_n$ of trees with n leaves with a main focus on $n = 3$. In particular, we further constructed the space of rooted 3-trees and used its features to define an evolutionary distance to the most recent common ancestor (MRCA). From here, we built a bootstrap two-sample test to compare the mean evolutionary distances of bat coronavirus (RaTG13) and early Wuhan SARS-CoV-2 coronavirus roots and concluded the Wuhan root is closer in evolutionary distance to later SARS-CoV-2 coronaviruses than RaTG13 is. Future directions focus mainly on increasing sample sizes, both to improve the current method and to perhaps use trees with more leaves, and evaluating the effect different tree-building methods have on our results. The current results may be fairly conclusive, as they help shed more light on new information either discovered or shared by government authorities.

Chapter 11

Applications of Tree Spaces to Language Ancestry

11.1 Introduction

This chapter is dedicated to a Latin alphabet-based analysis of a group of European language origins using a clustering-based tree building. The clustering is using as a criterion the first letter of words in various language used for the same object or concept. The R data analysis is due to Garett Ordway, to whom we are very grateful for providing his results.

11.2 Data Collection

11.2.1 Words

Of the varieties of cognate lists for language ancestry, one of the original lists developed by Morris Swadesh contains 207 words [128, 129, 130, 136]. These words range from familial relations and numbers to action verbs and anatomy. Each language has a representation of these words. The lists were accessed through the Wikipedia Module "Swadesh modules" which contains the 207 words in hundreds of languages. The languages chosen for analysis are English, German, French, Italian, and Spanish. Words were input into a data frame in R for analysis. Recently Ordway and Patrangenaru (2024) [100] imagined two populations, words were randomly sampled into groups of 102 and 105. Words were then randomly sampled three at a time to create 34 samples for population 1 and 35 for population 2. Two analyses were run involving first, English, German, and French; second, French, Italian, and Spanish.

11.2.2 Calculate Distances

Per Johnson and Wichern (2007)'s example [76], the first letter of each word is used to check for language similarity. If the letter is the same, then a distance of 0 is assigned. If the letter is different, then a distance of 1 is assigned. This is done for each of the 207 words between each pair of languages to be analyzed:

English and German, English and French, German and French, French and Italian, French and Spanish, and Italian and Spanish. Sum the distances within each of the 69 samples for two languages at a time. The minimum possible distance is 0 when all three languages begin with the same letter for all three words and the maximum possible distance is three when all three languages begin with different letters for all three words. This produces a 3×3 distance matrix.

Some basic examples of the distance calculations are given next.

EXAMPLE 11.2.1. *This is an example where all three languages are equidistant in the sample.*

English	*French*	*German*
yellow	*jaune*	*gelb*
rope	*corde*	*leine*
tooth	*dent*	*zahn*

In this example, all pairwise distances between English, French, and German are 3; therefore the point representing this triad on the sample space is the star tree; see Figure 11.1.

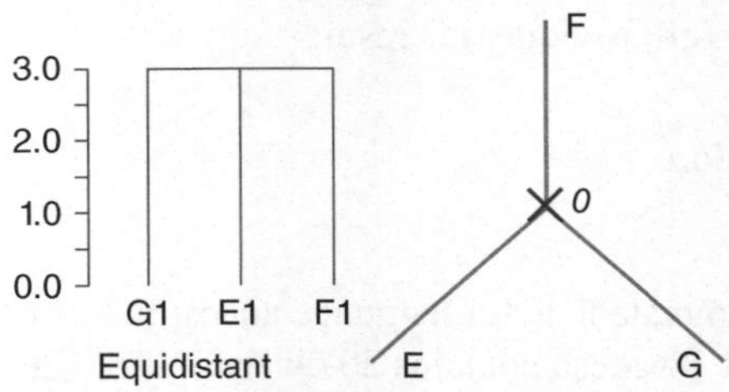

Figure 11.1: Equidistant samples (left) and corresponding location of the sample mean (right).

EXAMPLE 11.2.2. *This is an example where one language is furthest away and equidistant from the other two.*

English	*French*	*German*
child	*enfant*	*kind*
we	*nous*	*wir*
stab	*poignarder*	*stechen*

In this example, the distance between English and French is 3, between English and German is 1, and between French and German is 3; therefore, the point representing this triad is on the French leg of the tree space; see Figure 11.2.

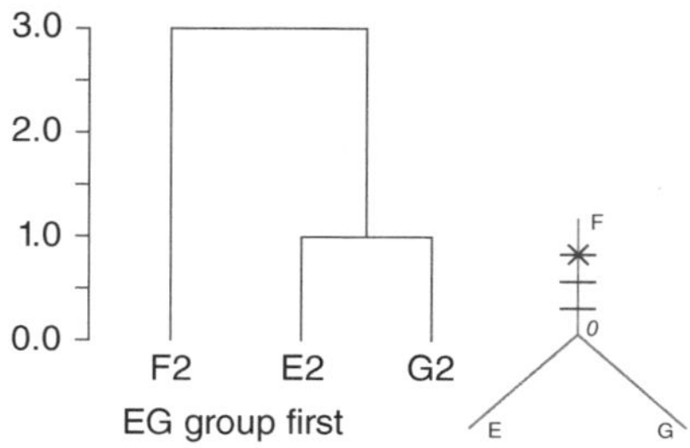

Figure 11.2: One sample equidistant from two other samples (left) and corresponding location of the sample mean (right).

EXAMPLE 11.2.3. *This is an example where one language is furthest away but is closer to one of the languages than it is to the other.*

English	*French*	*German*
blood	*sang*	*blut*
grass	*herbe*	*gras*
lake	*lac*	*see*

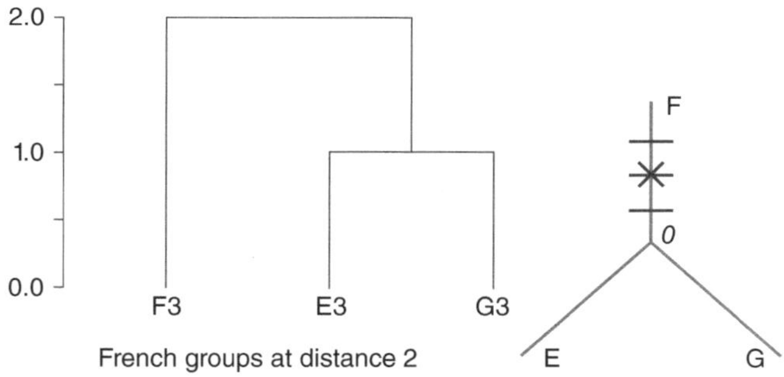

Figure 11.3: One sample distant from the other two at different distances (left) and corresponding location of the sample mean (right).

In this example, the distance between English and French is 2, between English and German is 1, and between French and German is 3. The point representing this triad is on the French leg of the tree space; see Figure 11.3.

EXAMPLE 11.2.4. *This is an example where two groups of languages tie for the minimum distance.*

English	*French*	*German*
road	*route*	*straße*
freeze	*geler*	*frieren*
because	*parce que*	*weil*

In this example, the distance between English and French is 2, between English and German is 2, and between French and German is 3; therefore, there is no unique leg representation of the triad in this case.

11.3 Language Trees Analysis on a 3-Spider

The distances used to cluster are based on single linkage, and then the language trees are represented as points on a 3-spider. Here, the three individuals are three languages with the following four possible ways to cluster and assign points on leaves.

1. All languages are equidistant, so only one cluster is needed. In the context of a 3-spider, the point is at the origin. See Section 11.2.1.
2. One language is furthest away and is equidistant from the other two. For a 3-spider, the point is on that language's leaf with a distance equal to the distance to the other languages. See Section 11.2.2. The point is at 3 on the French leaf.
3. One language is furthest away but is closer to one language than the other. For a 3-spider, the point is on that language's leaf with a distance equal to the minimum distance from that language to the other two. See 11.2.3. The point is 2 on the French leaf.
4. Two groups of languages tie for the minimum distance. For example, Language 1 is distance c_1 from Language 2, and Language 2 is also distance c_1 from Language 3, but Language 1 is a distance greater than c_1 (say c_2) from Language 3. In this case, *randomly* decide whether to cluster 1 with 2 or 2 with 3. For a 3-spider, then use the c_2 distance as the point on the Language's leaf that was not clustered originally (the point would be at c_2 on leaf 1 or leaf 3). See 11.2.4. The point is at 2 and was randomly assigned to the French leaf.

The distances are then points on a 3-spider where a point is on a particular leaf L_i where $i = E, G, F$ or $i = F, I, S$ if that language is the furthest (added to the cluster last).

11.3.1 Nonsticky Sample Mean

English, French, and German are the first three chosen languages. English is classified as a Germanic language, while French is classified as a Romance (Italic or Latin) language. As discussed in the previous sections, only one leaf may have a positive mean (although none are guaranteed). In this case, one language seems to be more different than the other two.

Language	Mean Distance 1	Mean Distance 2
English	−2.00	−2.41
German	−1.66	−1.41
French	1.49	1.41

Since French has a positive mean distance, it is considered the sample mean (barycenter) and is used to calculate the measures of variability. This is the nonsticky case for the mean. The variance of the first sample is 4.14 and the second is 4.61. Using the Nonsticky Central Limit Theorem, traditional asymptotic inference can be performed. The pooled sample variance is 4.43. These statistics yield the two sample t-statistic value of 0.16 which is far from significant. This is to be expected, since the samples had no necessary distinction; they were randomly assigned. If two groups of words were different for some reason, say, medical terms versus agricultural terms, then perhaps some significant difference may be found.

11.3.2 Sticky Sample Mean

French, Spanish, and Italian are three popular Romance languages sharing many commonalities. Latin was the language of the ancient boot-shaped peninsula and used prominently by the Roman Catholic Church for centuries. However, with no speakers today, Latin is considered a dead language. In this case, we may expect many observations at the origin and no starkly different language.

Language	Mean Distance 1	Mean Distance 2
French	−0.51	−0.76
Spanish	−0.74	−0.35
Italian	−0.11	−0.05

Since none of the means are positive, the sample mean is considered to be sticky, and the barycenter is at 0. From that mean of zero, the variances of each sample are calculated to be 2.88 and 2.30, respectively.

11.3.3 Another Sticky Sample Mean

So far, two cases have been considered where one language has been from a different ancestor and where all three languages are from the same ancestor. This case will consider all three languages from different ancestors. German, French, and Irish (Gaelic, from the Celtic ancestor) words from the Swadesh 207 list were used from the earlier procedures to find the following mean distances.

Language	Mean Distance 1	Mean Distance 2
German	−0.74	−0.03
French	−0.46	−0.97
Irish	−0.40	−0.85

In both samples, all three means are negative values indicating a sticky sample mean to be placed at the origin. From that mean of zero, the sample variances are calculated to be 4.88 and 5.45, respectively. Unlike the previous case of a sticky sample mean where the languages were considered to be of the same ancestor, here, the languages are believed to all be of different ancestors.

11.4 Conclusion

The clustering of language may be done through the use of the 3-spider. In the two applications shown, a nonsticky mean was observed when one language was from a different family, while a sticky mean was observed when all languages have the same ancestry or all languages have different ancestry. English and German are from the Germanic family, French, Italian, and Spanish are from the Latin (Romance, Italic) family, and German, French, and Irish are from different families. For these examples, the samples representing two populations were simply randomly assigned, but if two groups of words were different for some reason, say, medical terms versus agricultural terms, then perhaps some significant difference may be found in the ancestry as a different leaf may dominate in each case. While these results may not apply to all languages, they provide a start. Furthermore, inference may be done using the Central Limit Theorem for each case.

The distance calculated between each language was also exceptionally simple. An edit distance for full words may provide additional insights. While the Swadesh list of words has been the standard for linguistics, phrases or sentences introducing structure could be of interest for investigation. The languages in this study all used the Latin alphabet which is extremely common in Western civilization, but it is not the only alphabet as Cyrillic, Greek, and Arabic among others could be considered. Paradoxically, although we here compare Indo-European languages, our primary tool, the Latin alphabet sources from the Proto-Sinaitic script developed from the popular non-Indo-European language of Semitic-speaking workers and slaves in ancient Egypt, as they were unskilled to handle the hieroglyphic system used in the Egyptian language, which required a large number of pictograms, as opposed to the semantic values of the Canaanite language of slaves. This Semitic alphabet, ancestor of multiple writing systems across the Middle East, imposed itself having only about half as many letters as the Vedic Sanskrit alphabet that appeared around the same period. The written language only was considered regardless of pronunciation which needed the International Phonetic Alphabet (IPA) instead of the Latin script. With the addition of more than three languages, additional types of phylogenetic trees beyond the 3-spider would need to be considered. This grows the number of possible trees quickly following from the work by Billera et al. (2001) [28] necessitating additional computational power and theory.

Part IV

Applications in Object Data Analysis on Stratified Spaces

Chapter 12

3D Face Differentiation from Digital Camera Images

Consider g random objects (r.o.s) on a stratified space, $g \geq 2$. In order to perform a Fréchet MANOVA for the means of these r.o.s, one has to first check if there is any difference between the dimensions of the strata where these r.o.s live. A necessary condition for rejecting out of hand the equality is for a pair of these r.o.s to be on flat and convex connected components these strata of different dimensions. Once decided that the Fréchet means are inside the highest dimensional stratum, the MANOVA is reduced to a MANOVA on a manifold. In this chapter, we will therefore explore the theory necessary to run such Fréchet MANOVA, focusing on extrinsic MANOVA on a manifold.

12.1 Introduction

In this chapter we first revisit nonparametric MANOVA for comparing the mean vectors in g populations. We are extending such considerations to testing for the equality of extrinsic means from g populations on a manifold $\mathcal{M}$ embedded in a numerical space. In Section 12.2 we are introducing our new approach applied to various mean vectors. The main difference between our approach and classical parametric MANOVA is that we do not assume that all populations have a common covariance matrix Σ, and also we do not make a distributional assumption. All we assume is the existence of sufficiently high order moments of the g populations. In Section 12.3 we extend the work presented in the previous section to develop a hypothesis testing problem used to compare multiple means on smooth manifolds, and this test is performed on random samples of various sizes, collected from each of these g groups. Our newly developed MANOVA test is then applied to samples on 3D-projective shape spaces in Section 12.4. This method builds upon previous results on two-sample hypothesis testing methods, as developed in Patrangenaru et al. (2010) [107] and Crane and Patrangenaru (2011) [35]. The space $P\Sigma_3^k$ of 3D-projective shapes of k-ads, including a projective frame at given landmark indices, is isomorphic to $(\mathbb{R}P^3)^{k-5}$. Furthermore, this manifold has a Lie group structure where the multiplication operation is induced by the quaternion multiplication on S^3. Therefore, a 3D-projective shape face analysis based on landmarks can be run via recently developed nonparametric inference on Lie groups, a methodology that is presented in Section 14. Note that results by Faugeras (1992) [46] and Hartley et al. (1992) [62] show that a 3D configuration of landmarks can be obtained from pairs of noncalibrated camera images, and the configuration obtained via their algorithm is unique up to a projective transformation in 3D, thus allowing to conduct a projective shape analysis without ambiguity. Our statistical analysis is performed for sets of pictures of faces that are displayed in Section 12.5 along with conveniently selected anatomical landmarks. We make no distributional assumptions for our hypothesis testing methods.

12.2 Motivation for New MANOVA on Manifolds

For $a = 1,...,g$, suppose $X_{a,i} \sim N_p(\mu_a, \Sigma), i = 1,...,n_a$ are p-dimensional i.i.d. random vectors. To test if the mean vectors of the g groups are the same, one considers the hypothesis testing problem

$$\begin{aligned} &H_0: \mu_1 = \mu_2 = ... = \mu_g = \mu \\ &H_a: \textit{at least one equation does not hold.} \end{aligned} \tag{12.1}$$

Assuming that the covariance matrix Σ is invertible, by the Central Limit Theorem, for large sample sizes $n_a, a = 1,\dots,g$, we have

$$\sqrt{n_a}\Sigma^{-\frac{1}{2}}(\bar{X}_a - \mu) \sim N_p(0_p, I_p), \tag{12.2}$$

$$n_a(\bar{X}_a - \mu)^T \Sigma^{-1}(\bar{X}_a - \mu) \sim \chi^2_p. \tag{12.3}$$

However, Σ is always unknown, so in practice, one has to use its unbiased estimator S_a, $a = 1,...,g$.

$$n_a(\bar{X}_a - \mu)^T S_a^{-1}(\bar{X}_a - \mu) \sim \chi^2_p. \tag{12.4}$$

Let us consider the pooled sample mean $\bar{X} = \frac{1}{n}(n_1\bar{X}_1 + ... + n_g\bar{X}_g)$, $n = \sum_{a=1}^{g} n_a$.

LEMMA 12.2.1. *Under the null, $\bar{X}$ is a consistent estimator of μ, provided $\frac{n_a}{n} \to \lambda_a > 0$, as $n \to \infty$, $a = 1,...,g$.*

Proof. Indeed, for any $a \in \{1,2,...,g\}$, since $\frac{n_a}{n} \to \lambda_a > 0$, *as* $n \to \infty$, and $\bar{X}_a$ is the consistent estimator of μ; therefore,

$$\bar{X} \to_p \lambda_1\mu + \lambda_2\mu + ... + \lambda_g\mu = \mu. \tag{12.5}$$

□

THEOREM 12.2.1. *The statistic for the hypothesis in* (12.1) *is*

$$\sum_{a=1}^{g} n_a(\bar{X}_a - \bar{X})^T S_a^{-1}(\bar{X}_a - \bar{X}) \sim \chi^2_{gp}. \tag{12.6}$$

So the rejection region for the test is

$$\sum_{a=1}^{g} n_a(\bar{X}_a - \bar{X})^T S_a^{-1}(\bar{X}_a - \bar{X}) > \chi^2_{gp}(c). \tag{12.7}$$

12.3 MANOVA on Manifolds

In this section we will focus on the asymptotic behavior of statistics related to means on a manifold $\mathscr{M}$ based on samples of different sizes from different populations on $\mathscr{M}$. Now let's consider the set $X_{a,1},\ldots,X_{a,n_a}$ $(a = 1,2,...,g)$ of independent and identically distributed random objects (i.i.d.r.o.s) on $\mathscr{M}$ with common probability measure Q_a. We denote the extrinsic mean of the j-nonfocal probability measure Q_a on $\mathscr{M}$ by $\mu_{a,E}$ for ease of notation and because there is no ambiguity about the embedding used. The corresponding extrinsic sample means are written $\bar{X}_{a,E}$ for $a = 1,\cdots,g$. From this point on, we will assume that all the distributions are j-nonfocal.

12.3.1 Hypothesis Testing and T^2 Statistic

Assume $X_{a,1},\ldots,X_{a,n_a}$ are i.i.d.r.o.s on $\mathscr{M}$ a p-dimensional manifold, with probability measure Q_a with $a = 1,2,...,g$. We are interested in comparing multiple extrinsic means.

We would like to develop a test similar to (12.1) designed to test the difference between the g extrinsic means. One challenge that presents itself at the early stage is a proper definition of a pooled mean for random objects on a p-dimensional manifold $\mathscr{M}$. Linearity becomes an issue when dealing with extrinsic means. For a proper definition we will focus on the equalities tied to the assumption

$$A_0 : \mu_{1,E} = \cdots = \mu_{g,E}$$

DEFINITION 12.3.1. *Under the assumption A_0 and for any $a \in \{1,2,...,g\}$, with $\frac{n_a}{n} \to \lambda_a > 0$, as $n \to \infty$. We define*

(i) The **extrinsic pooled mean with weights** $\lambda = (\lambda_1,\ldots,\lambda_g)$, *denoted $\mu_E(\lambda)$ as the value in $\mathscr{M}$ given by*

$$j(\mu_E) = P_j(\lambda_1 j(\mu_{1,E}) + \cdots + \lambda_g j(\mu_{g,E})) \tag{12.8}$$

where $\mu_{a,E}$ is the extrinsic mean of the random object $X_{a,1}$ and $\Sigma_{a=1}^{g}\lambda_a =$ 1.

(ii) The **extrinsic pooled sample mean** *denoted* $\bar{X}_E \in \mathcal{M}$ *given by*

$$j(\bar{X}_E) = P_j\left(\frac{n_1}{n} j(\bar{X}_{1,E}) + \cdots + \frac{n_g}{n} j(\bar{X}_{g,E})\right) \tag{12.9}$$

where $\bar{X}_{a,E}$ *is the extrinsic sample mean for* $X_{a,1}$ *and* $n = \Sigma_{a=1}^{g} n_a$

Note that since A_0 implies $j(\mu_{1,E}) = \cdots = j(\mu_{g,E})$, and with our definition of the extrinsic pooled mean we get $j(\mu_E) = j(\mu_{a,E})$ for each $a = 1, \ldots, g$. Furthermore, the linear combination $\lambda_1 j(\mu_{1,E}) + \cdots + \lambda_g j(\mu_{g,E}) \in j(\mathcal{M})$. Note that for $a = 1, \cdots, g$ $\bar{X}_{a,E}$ is a consistent estimator of $\mu_{a,E}$ and therefore we get that $j(\bar{X}_E) \rightarrow_p j(\mu_E)$. Since j is a homeomorphism from $\mathcal{M}$ to $j(\mathcal{M})$ we also have that $\bar{X}_E$ is a consistent estimator of μ_E the extrinsic pooled mean. With this definition at hand, we now express the following hypothesis test, designed to test the difference between extrinsic means and is given by

$$\begin{aligned} &H_0 : \mu_{1,E} = \mu_{2,E} = \ldots = \mu_{g,E} = \mu_E, \\ &H_a : \textit{at least one equality } \mu_{a,E} = \mu_{b,E}, 1 \leq a < b \leq g \textit{ does not hold.} \end{aligned} \tag{12.10}$$

And since the embedding $j : \mathcal{M} \rightarrow \mathbb{R}^N$ is one-to-one, the hypothesis above can be interchangeably written

$$\begin{aligned} &H_0^j : j(\mu_{1,E}) = j(\mu_{2,E}) = \ldots = j(\mu_{g,E}) = j(\mu_E), \\ &H_a^j : \textit{at least one equality } \mu_{a,E} = \mu_{b,E}, 1 \leq a < b \leq g \textit{ does not hold.} \end{aligned} \tag{12.11}$$

In order to test hypothesis (12.10), we will use a T^2-like statistic. The theorem below gives us the asymptotic behavior needed to establish such a statistic. For $a = 1, \ldots, g$, we get, from Bhattacharya and Patrangenaru (2005) [21], the following:

(i) $S_{n_a} = (n_a)^{-1} \Sigma_{i=1}^{n_a} (j(X_{a,i}) - j(\bar{X}_E))(j(X_{a,i}) - j(\bar{X}_E))^T$ is a consistent estimator of Σ_a, and

(ii) $\tan_{j(\bar{X}_E)} \nu$ is a consistent estimator of $\tan_{P_j(\mu)} \nu$, where $\nu \in \mathbb{R}^N$.

It follows that $G_{\bar{X}}(j, X_a)$, given by

$$G_{\bar{X}}(j, X_a) = \left[\left[\sum_{a=1}^{m} d_{\overline{j^{(p)}(X)}} P_j(e_b) \cdot e_i(j(\bar{X}_E))\, e_i(j(\bar{X}_E))\right]_{i=1,\ldots,p}\right] \cdot S_{n_a}$$
$$\left[\left[\sum_{a=1}^{m} d_{\overline{j^{(p)}(X)}} P_j(e_b) \cdot e_i(j(\bar{X}_E)) e_i(j(\bar{X}_E))\right]_{i=1,\ldots,p}\right]^T$$

where for $\overline{j^{(p)}(X)} = \frac{n_1}{n} j(\bar{X}_{1,E}) + \cdots + \frac{n_g}{n} j(\bar{X}_{g,E})$ and is a consistent estimator of

μ such that $P_j(\mu) = j(\mu_E)$. One must note that the extrinsic sample covariance matrix $G(j,X_a)$ is expressed in terms of $d_{\overline{j^{(p)}(X)}}P_j(e_b) \in T_{j(\bar{X}_E)}j(\mathscr{M})$ and not in terms of $d_{\overline{j(Xa,1)}}P_j(e_b) \in T_{j(\bar{X}_{a,E})}j(\mathscr{M})$.

THEOREM 12.3.1. *Assume $j : \mathscr{M} \to \mathbb{R}^N$ is a closed embedding of $\mathscr{M}$. Let $\{X_{a,i}\}_{i=1}^{n_a}$ for $a = 1,...,g$ be random samples from the j-nonfocal distributions $\mathscr{Q}_a$. Let $\mu_a = E(j(X_{a,1}))$ and assume $j(X_{a,1})$'s have finite second-order moments and the extrinsic covariance matrices $\Sigma_{a,E}$ of $X_{a,1}$ are nonsingular. We also let $(e_1(p),....,e_N(p))$, for $p \in \mathscr{M}$ be an orthonormal frame field adapted to j.*
Furthermore, let $\frac{n_a}{n} \to \lambda_a > 0$, as $n \to \infty$, with $n = \Sigma_{a=1}^{g} n_a$, and $\Sigma_{a=1}^{g}\lambda_a = 1$. Then we have the following asymptotic behavior:

$$\sum_{a=1}^{g} n_a \tan_{j(\mu_E)}(j(\bar{X}_{a,E}) - j(\mu_E))^T \Sigma_{a,E}^{-1} \tan_{j(\mu_E)}(j(\bar{X}_{a,E}) - j(\mu_E)) \to_d \chi^2_{gp}.$$

It follows that the statistics for hypothesis (12.10) *have the following behaviors:*

(a) $\Sigma_{a=1}^{g} n_a \tan_{j(\mu_E)}(j(\bar{X}_{a,E}) - j(\bar{X}_E))^T G_{\bar{X}}(j,X_a)^{-1} \tan_{j(\mu_E)}(j(\bar{X}_{a,E}) - j(\bar{X}_E)) \to_d \chi^2_{gp}$.

(b) $\Sigma_{a=1}^{g} n_a \tan_{j(\bar{X}_E)}(j(\bar{X}_{a,E}) - j(\bar{X}_E))^T G_{\bar{X}}(j,X_a)^{-1} \tan_{j(\bar{X}_E)}(j(\bar{X}_{a,E}) - j(\bar{X}_E)) \to_d \chi^2_{gp}$.

Proof. Recall that from Bhattacharya and Patrangenaru (2005) [21] we have

$$\sqrt{n_a} \tan_{j(\mu_E)}(j(\bar{X}_{a,E}) - j(\mu_E)) \to_d \mathscr{N}(0_p, \Sigma_{a,E}), \quad for\ a = 1,2,...,g$$

where

$$\Sigma_{a,E} = \left[\left[\sum d_\mu P_j(e_b) \cdot e_k(P_j(\mu))\right]_{k=1,...,p}\right] \Sigma_a \times \left[\left[\sum d_\mu P_j(e_b) \cdot e_k(P_j(\mu))\right]_{k=1,...,p}^T\right]$$

where $\mu = \lambda_1 j(\mu_{1,E}) + \cdots + \lambda_g j(\mu_{g,E})$, and the Σ_a's are the covariance matrices of the $j(X_{a,1})$'s with respect to the canonical basis $e_1,...,e_N$. And under the null, from (12.10), the matrices $\Sigma_{a,E}$ are defined with respect to the basis $f_1(\mu_E),...,f_p(\mu_E)$ of local frame fields. We then have for each $a = 1,...,g$

$$n_a \tan_{j(\mu_E)}(j(\bar{X}_{a,E}) - j(\mu_E))^T \Sigma_{a,E}^{-1} \tan_{j(\mu_E)}(j(\bar{X}_{a,E}) - j(\mu_E)) \to_d \chi^2_p.$$

And since the random samples are independent, we have,

$$\sum_{a=1}^{g} n_a \tan_{j(\mu_E)}(j(\bar{X}_{a,E}) - j(\mu_E))^T \Sigma_{a,E}^{-1} \tan_{j(\mu_E)}(j(\bar{X}_{a,E}) - j(\mu_E)) \to_d \chi^2_{gp}. \tag{12.12}$$

$\bar{X}_E$ is the consistent estimator of μ_E, then the pooled sample mean

$$j(\bar{X}_E) = P_j\left(\frac{1}{n}\sum_{a=1}^{g} n_a j(\bar{X}_{a,E})\right) \to_p j(\mu_E) \text{ (by Lemma 12.2.1)} \tag{12.13}$$

And since $G_{\bar{X}}(j, X_a)$ consistently estimate Σ_a and $\tan_{j(\bar{X}_E)}$ is a consistent estimator of $\tan_{j(\mu_E)}$, we have the following

$$\sum_{a=1}^{g} n_a \tan_{j(\mu_E)}(j(\bar{X}_{a,E}) - j(\bar{X}_E))^T G_{\bar{X}}(j, X_a)^{-1} \tan_{j(\mu_E)}(j(\bar{X}_{a,E}) - j(\bar{X}_E)) \to_d \chi^2_{gp}.$$

$$\sum_{a=1}^{g} n_a \tan_{j(\bar{X}_E)}(j(\bar{X}_{a,E}) - j(\bar{X}_E))^T G_{\bar{X}}(j, X_a)^{-1} \tan_{j(\bar{X}_E)}(j(\bar{X}_{a,E}) - j(\bar{X}_E)) \to_d \chi^2_{gp}.$$

□

12.3.2 Nonparametric Bootstrap Confidence Regions

From corollary 3.2 in [21] under the hypothesis

$$\begin{cases} H_0 & : \mu_{1,E} = \mu_{2,E} = \ldots = \mu_{g,E} = \mu_E, \\ H_a & :\ni (i,j) 1 \le i < j < g, \text{s.t. } \mu_{i,E} \ne \mu_{j,E}, \end{cases}$$

we have

COROLLARY 12.3.1. *Under the assumptions of Theorem* (12.3.1), *a confidence region for* μ_E *of asymptotic level* $1 - c$ *is given by* $C_{n,c}^{(g)}$ *and* $D_{n,c}^{(g)}$ *which are defined below.*

(a) $C_{n,c}^{(g)} = j^{-1}(U_{n,c})$ *where*
$U_{n,c} = \{j(\nu) \in j(\mathcal{M}) : n \sum_{a=1}^{g} n_a \left\| G_{\bar{X}}(j, X_a)^{-1/2} \tan_{j(\nu)}(j(\overline{X}_{a,E}) - j(\nu)) \right\|^2 \le \chi^2_{gp,1-c}\}$

(b) $D_{n,c}^{(g)} = j^{-1}(V_{n,c})$ *where*
$V_{n,c} = \{j(\nu) \in j(\mathcal{M}) : \sum_{a=1}^{g} n_a \left\| G_{\bar{X}}(j, X_a)^{-1/2} \tan_{j(\bar{X}_E)}(j(\overline{X}_{a,E}) - j(\nu)) \right\|^2 \le \chi^2_{gp,1-c}\}$

where $\bar{X}_E$ *is the extrinsic pooled sample mean defined in Definition 12.3.1 (ii).*

Most of the data we will be focusing on will have value of n relatively small. We will need to use resampling, in particular bootstrap methods. For $a = 1, \ldots, g$, let $\{X_{a,i}\}_{i=1}^{n_a}$ be i.i.d.r.o.s from the j-nonfocal distributions $\mathcal{Q}_a$. Let $\{X^*_{a,r}\}_{r=1,\ldots,n_a}$ be random resamples with repetition from the empirical $\hat{Q}_{n_a}$

conditionally given $\{X_{a,i}\}_{i=1}^{n_a}$. The confidence regions $C_{n,c}^{(g)}$ and $D_{n,c}^{(g)}$ described above have the corresponding bootstrap analog $C^{*(g)}_{n,c}$ and $D^{*(g)}_{n,c}$ which are defined in the corollary below.

COROLLARY 12.3.2. *The* $(1-c)100\%$ *bootstrap confidence regions for* μ_E *with* $d = gp$ *are given by*

(a) $C^{*(g)}_{n,c} = j^{-1}(U^*_{n,c})$ *and*

$$U^*_{n,c} = \{j(\nu) \in j(\mathscr{M}) : \sum_{a=1}^{g} n_a \left\| G_{\bar{X}}(j, X_a)^{-1/2} \tan_{j(\nu)}(j(\overline{X}_{a,E}) - j(\nu)) \right\|^2 \leq c^{*(g)}_{1-c}\} \tag{12.14}$$

where $c^{*(g)}_{1-c}$ *is the upper* $100(1-c)\%$ *point of the values*

$$\sum_{a=1}^{g} n_a \left\| G_{\bar{X}^*}(j, X^*{}_a)^{-1/2} \tan_{j(\bar{X}_E)}(j(\overline{X^*}_{a,E}) - j(\bar{X}_E)) \right\|^2 \tag{12.15}$$

among the bootstrap resamples.

(b) $D^{*(g)}_{n,c} = j^{-1}(V^*{}_{n,c})$ *and*

$$V^*{}_{n,c} = \{j(\nu) \in j(\mathscr{M}) : \sum_{a=1}^{g} n_a \left\| G_{\bar{X}}(j, X_a)^{-1/2} \tan_{j(\bar{X}_E)}(j(\overline{X}_{a,E}) - j(\nu)) \right\|^2 \leq d^{*(g)}_{1-c}\} \tag{12.16}$$

where $d^{*(g)}_{1-c}$ *is the upper* $100(1-c)\%$ *point of the values*

$$\sum_{a=1}^{g} n_a \left\| G_{\bar{X}^*}(j, X^*_a)^{-1/2} \tan_{j(\bar{X}^*_E)}(j(\overline{X^*}_{a,E}) - j(\bar{X}_E)) \right\|^2 \tag{12.17}$$

where $\bar{X}^*_E$ *is the extrinsic pooled resampled mean given by*

$$j(\bar{X}^*_E) = P_j\left(\frac{n_1}{n} j(\bar{X}^*_{1,E}) + \cdots + \frac{n_g}{n} j(\bar{X}^*_{g,E})\right) \tag{12.18}$$

among the bootstrap resamples. Both of the regions given by (12.16) *and* (12.14) *have coverage erro* $O_p(n^{-2})$.

Note that $G_{\bar{X}^*}(j, X_a^*)$

$$G_{\bar{X}^*}(j,X_a^*) = \left[\left[\sum_{a=1}^{m} d_{\overline{j^{(p)}(X^*)}} P_j(e_b)\cdot e_i(j(\bar{X}_E^*))\, e_i(j(\bar{X}_E^*))\right]_{i=1,\ldots,p}\right] \cdot S_{n_a}^* \left[\left[\sum_{a=1}^{m} d_{\overline{j^{(p)}(X^*)}} P_j(e_b)\cdot e_i(j(\bar{X}_E^*)) e_i(j(\bar{X}_E^*))\right]_{i=1,\ldots,p}\right]^T$$

where $S_{n_a}^* = (n_a)^{-1}\Sigma_{i=1}^{n_a}(j(X_{a,i}^*) - j(\bar{X}_E^*))(j(X_{a,i}^*) - j(\bar{X}_E^*))^T$.
We now express the following test statistics that will be used in our analysis and are tied to the confidence regions mentioned above.

PROPOSITION 12.3.1. *Let $\{X_{a,i}\}_{i=1}^{n_a}$ for $a = 1,\ldots,g$ be random samples from the j-nonfocal distributions $\mathcal{Q}_a$. Let $\mu_a = E(j(X_{a,1}))$ and assume $j(X_{a,1})$'s have finite second-order moments, and the extrinsic covariance matrices $\Sigma_{a,E}$ of $X_{a,1}$ are nonsingular.*

(a) Then the distribution of $T_c(X^{(g)}, \hat{Q}^{(g)}) = \Sigma_{a=1}^{g} n_a \left\| G_\mu(j,X_a)^{-1/2}\ \tan_{j(\mu_E)} (j(\overline{X}_{a,E}) - j(\mu_E))\right\|^2$ can be approximated by the bootstrap distribution function of

$$T_c(X^{*(g)}, \hat{Q}^{(g)}) = \Sigma_{a=1}^{g} n_a \left\| G_{\bar{X}}(j,X_a^*)^{-1/2}\ \tan_{j(\bar{X}_E)}(j(\overline{X}_{a,E}^*) - j(\bar{X}_E))\right\|^2$$

(b) Similarly, the distribution of $T_d(X^{(g)}, \hat{Q}^{(g)}) = \Sigma_{a=1}^{g} n_a \left\| G(j,X_a)^{-1/2} \tan_{j(\bar{X}_E)} (j(\overline{X}_{a,E}) - j(\mu_E))\right\|^2$ can be approximated by the bootstrap distribution function of

$$T_d(X^{*(g)}, \hat{Q}^{*(g)}) = \Sigma_{a=1}^{g} n_a \left\| G_{\bar{X}^*}(j,X_a^*)^{-1/2}\ \tan_{j(\bar{X}_E^*)}(j(\overline{X^*}_{a,E}) - j(\bar{X}_E))\right\|^2$$

with coverage error $O_p(n^{-2})$.

Note that $T(X^{*(g)}, \hat{Q}^{(g)})$ is obtained from $T(X^{(g)}, \hat{Q}^{(g)})$ by substituting $X_1^{(g)} = (X_{1,1},\cdots,X_{g,1})^T$ with resamples $X_1^{*(g)} = (X_{1,1}^*,\cdots,X_{g,1}^*)^T$.
Using the bootstrap analog in the previous Proposition 12.3.1 yields simpler method for finding $100(1-c)\%$ confidence regions. We will utilize the test statistics expressed above to conduct our analysis with confidence regions $C_{n,c}^*$ and $D_{n,c}^*$ as shown in Corollary 12.3.2.

12.4 MANOVA on $(\mathbb{R}P^3)^q$

We start with the 3-D real projective space $\mathbb{R}P^3$, set of one-dimensional linear subspaces of $\mathbb{R}^4$. $\mathbb{R}P^3$ has a 3D manifold structure (see Patrangenaru and Ellingson (2015) [106], p.106). A point $p = [x] \in \mathbb{R}P^3$ is the equivalence class of $x = (x^1x^2x^3x^4)^T \in \mathbb{R}^4\backslash\{0\}$, where two nonzero vectors in $\mathbb{R}^4$ are equivalent, if one is a scalar multiple of the other. The point p can be represented as $p = [x^1 : x^2 : x^3 : x^4]$ (homogeneous coordinates notation). One may also represent $\mathbb{R}P^3$ as the sphere S^3 with the antipodal points identified. We will often refer to this identification as the *spherical representation* of the real projective space. $\mathbb{R}P^3$ is an embedded manifold with the Veronese-Whitney (VW)-embedding $j : \mathbb{R}P^3 \to \mathscr{S}(4,\mathbb{R})$, given by

$$j([x]) = xx^T, x^T x = 1. \tag{12.19}$$

Given a random object Y on $\mathbb{R}P^3$, $Y = [X], X^T X = 1$, such that $E(XX^T)$ has a simple largest eigenvalue, one can show that the VW (extrinsic) mean $\mu_j = [\gamma]$, where γ is a unit eigenvector of $E(XX^T)$ corresponding to this largest eigenvalue (see Bhattacharya and Patrangenaru (2005) [21]).
Our analysis will be conducted on $P\Sigma_3^k$, the projective shape space of 3D k-ads in $\mathbb{R}P^m$ for which $\pi = ([u_1], \ldots, [u_5])$ is a projective frame in $\mathbb{R}P^3$. $P\Sigma_3^k$ is homeomorphic to the manifold $\left(\mathbb{R}P^3\right)^{k-5}$ with $k-5 = q$ (see Patrangenaru et al. (2010) [107]). The embedding on this space is the VW embedding given by

$$\begin{aligned} j_k &: \left(\mathbb{R}P^3\right)^q \to (S(4,\mathbb{R}))^q \\ j_k&([x_1], \ldots, [x_q]) = (j([x_1]), \ldots, j([x_q])), \end{aligned} \tag{12.20}$$

with $j : \mathbb{R}P^3 \to S_+(4,\mathbb{R})$ the embedding given in (12.19). Additionally j_k is an equivariant embedding w.r.t. the group $(S_+(4,\mathbb{R}))^q$ and has the corresponding projection

$$\begin{aligned} P_{j_k} &: (S_+(4,\mathbb{R}))^q \backslash \mathscr{F}_q \to j_k\left(\mathbb{R}P^3\right)^q \\ P_{j_k}&(A_1, \ldots, A_q) = (j([m_1]), \ldots, j[m_q])) \end{aligned} \tag{12.21}$$

where $m_1, \ldots, m_q$ are unit eigenvectors of $A_1, \ldots, A_q$ (respectively) corresponding to the respective highest eigenvalues of those nonnegative definite symmetric matrices. Let Y be a random object from a VW distribution Q on $(\mathbb{R}P^3)^q$, where $Y = (Y^1, \ldots, Y^q)$, and $Y^s = [X^s] \in \mathbb{R}P^3$ for all $s = \overline{1,q}$. The VW mean is given by

$$\mu_{j_k} = ([\gamma_1(4)], \cdots, [\gamma_q(4)]), \tag{12.22}$$

where, for $s = \overline{1,q}$, $\lambda_s(r)$ and $\gamma_s(r), r = 1, \ldots, 4$ are the eigenvalues in increasing order and the corresponding eigenvectors of $E\left[X^s(X^s)^T\right]$.

In case of a random object $[X]$ on $\mathbb{R}^3$, let us assume that $\mu_{E,j} = [\nu_4]$, where η_r and ν_r, $r = 1,2,3,4$ are eigenvalues in increasing order and corresponding unit eigenvectors of $\mu = E[XX^T]$ corresponding to eigenvalues in their increasing order. The corresponding extrinsic sample mean, for a sample of size n, is given by $\overline{X}_{E,j} = [g(4)]$, where $d(r)$ and $g(r) \in \mathbb{R}^4$, $r = 1,2,3,4$ are eigenvalues in increasing order and corresponding unit eigenvectors of $J = \frac{1}{n}\sum_{i=1}^n X_i X_i^T$.

We now recall the result from theorem 4.1 in Bhattacharya and Patrangenaru (2005) [21] as well as represent the statistics

$$T([X],Q) = n\|S(j,X)^{-1/2}\tan_{j(\mu_{E,j})}\left(j(\overline{X}_{E,j}) - j(\mu_{E,j})\right)\|^2$$

We have for $T([X],Q) = T([X],[\nu_4])$

$$T([X],[\nu_4]) = n\, g(4)^T\ [(\nu_r)]_{r=1,2,3}S(j,X)^{-1}[(\nu_r)]_{r=1,2,3}^T\ g(4) \qquad (12.23)$$

This results extends to the statistics

$$T([X],\hat{Q}) = T([X],[g(4)]) = \|S(j,X)^{-1/2}\tan_{j(\overline{X}_{E,j})}\left(j(\overline{X}_{E,j}) - j(\mu_{E,j})\right)\|^2$$

$$T([X],[g(4)]) = n\ \nu_4^T\ [g(r)]_{r=1,2,3}S(j,X)^{-1}[g(r)]_{r=1,2,3}^T\ \nu_4, \qquad (12.24)$$

where

$$S(j,X)_{ab} = n^{-1}(d(4)-d(a))^{-1}(d(4)-d(b))^{-1} \times \sum_{i=1}^{n}(g(a)\cdot X_i)(g(b)\cdot X_i)(g(4)\cdot X_i)^2$$

and asymptotically $T([X],[\nu_4])$ and $T([X],[g(4)])$ both have a χ_3^2 distribution (see Bhattacharya and Patrangenaru (2005) [21]).
Before we express our statistics of interest, it will be important to note another result from Crane and Patrangenaru (2011) [35] concerning the statistics

$$T(Y,\mu_{E,j_k}) = n\|S_{\bar{Y}}(j_k,Y)^{-1/2}\tan_{j(\overline{Y}_{E,j_k})}\left(j(\overline{Y}_{E,j_k}) - j(\mu_{E,j_k})\right)\|^2$$

And this Hotelling T^2 type statistic is given by

$$T(Y,([\gamma_1(4)],\cdots,[\gamma_q(4)])) = n\ \left(\gamma_1(4)^T D_1 \dots \gamma_q(4)^T D_q\right)\ S_{\bar{Y}}(j_k,Y)^{-1} \times \left(\gamma_1(4)^T D_1 \dots \gamma_q(4)^T D_q\right)^T \qquad (12.25)$$

where for $s = 1,\dots,q$ we have $D_s = (g_s(1)\ g_s(2)\ g_s(3)) \in \mathcal{M}(4,3,\mathbb{R})$ and for a pair of indices $(s,a), s = 1,\dots,q$ and $a = 1,2,3$ in their lexicographic order we have

$$S_{\bar{Y}}(j_k,Y)_{(s,a),(t,b)} = n^{-1}(d_s(4)-d_s(a))^{-1}(d_t(4)-d_t(b))^{-1} \times \sum_{i=1}^{n}(g_s(a)\cdot X_i^s)(g_t(b)\cdot X_i^t)(g_s(4)\cdot X_i^s)(g_t(4)\cdot X_i^t) \qquad (12.26)$$

In the next theorem we will take advantage of these results.

$$H_0: \ \mu_{1,E} = \mu_{2,E} = \ldots = \mu_{g,E} = \mu_E, \tag{12.27}$$
$$H_a: \ \textit{at least one equality } \mu_{a,E} = \mu_{b,E}, 1 \leq a < b \leq g \textit{ does not hold.}$$

We aim to have an explicit representation of the expressions,

$$T_c\left(Y^{(g)}, \mu_E^{(p)}\right) = n_a \sum_{a=1}^{g} \left\| S_{\bar{Y}}(j_k, Y_a)^{-1/2} \tan_{j_k\left(\mu_E^{(p)}\right)} \left(j_k(\overline{Y}_{a,E}) - j_k\left(\mu_E^{(p)}\right)\right)\right\|^2 \tag{12.28}$$
$$T_d\left(Y^{(g)}, \overline{Y}_E^{(p)}\right) = n_a \sum_{a=1}^{g} \left\| S_{\bar{Y}}(j_k, Y_a)^{-1/2} \tan_{j_k\left(\overline{Y}_E^{(p)}\right)} \left(j_k(\overline{Y}_{a,E}) - j_k\left(\mu_E^{(p)}\right)\right)\right\|^2 \tag{12.29}$$

where $\mu_{a,E} = ([\nu_1^a(4)], \ldots, [\nu_q^a(4)])$ are the VW mean from distribution Q_a (of Y_{r_a}) and $(\eta_s^a(r), \nu_s^a(r))$, $r = 1, \ldots, 4$ are eigenvalues and corresponding unit eigenvectors of $E(X_{a,1}^s(X_{a,1}^s)^T]$. The corresponding VW sample mean is given by $\overline{Y}_{a,E} = ([g_1^a(4), \ldots, [g_q^a(4)])$ and for each $s = 1, \ldots, q$ we have for $r = 1, \ldots, 4$, $(d_s^a(r), g_s^a(r))$ which are eigenvalues in increasing order and corresponding unit eigenvectors of $J_s^a = \frac{1}{n_a}\sum_{i=1}^{n_a} X_{a,i}^s(X_{a,i}^s)^T$. Also $\mu_E^{(p)}$ is the VW pooled mean given by

$$j_k\left(\mu_E^{(p)}\right) = P_{j_k}\left(\sum_{a=1}^{g} \frac{\lambda_a}{\lambda} j_k(\mu_{a,E})\right) \tag{12.30}$$
$$\mu_E^{(p)} = ([\gamma_1^{(p)}(4)], \ldots, [\gamma_q^{(p)}(4)]) \tag{12.31}$$

and $\overline{Y}_E^{(p)}$ is the corresponding pooled mean given by

$$j_k\left(\overline{Y}_E^{(p)}\right) = P_{j_k}\left(\sum_{a=1}^{g} \frac{n_a}{n} j_k(\overline{Y}_{a,E})\right) \tag{12.32}$$
$$\overline{Y}_E^{(p)} = ([\mathbf{g}_1^{(p)}(4)], \ldots, [\mathbf{g}_q^{(p)}(4)]) \tag{12.33}$$

where for $s = 1, \ldots, q$, $\mathbf{d}_s^{(p)}(r)$ and $\mathbf{g}_s^{(p)}(r) \in \mathbb{R}^4$, $r = 1, 2, 3, 4$ are eigenvalues in increasing order and corresponding unit eigenvectors of the matrix $J^{(p)} = \sum_{a=1}^{g} \frac{n_a}{n} j_k(\overline{Y}_{a,E})$.

We now express the following matrices

$$\mathbf{C}_s = (\gamma_s^{(p)}(1)\ \gamma_s^{(p)}(2)\ \gamma_s^{(p)}(3)) \in \mathscr{M}(4, 3 : \mathbb{R}) \tag{12.34}$$
$$\mathbf{D}_s = (\mathbf{g}_s^{(p)}(1)\ \mathbf{g}_s^{(p)}(2)\ \mathbf{g}_s^{(p)}(3)) \in \mathscr{M}(4, 3 : \mathbb{R}) \tag{12.35}$$

COROLLARY 12.4.1. *Assume j_k is the VW embedding of $(\mathbb{R}P^3)^q$ and $\{Y_{a,r_a}\}_{r_a=1,\dots,n_a}$, $a = 1,\dots,g$ are independent random samples from j_k-nonfocal probability measures Q_a on $(\mathbb{R}P^m)^q$ that have nondegenerate j_k-extrinsic covariance matrices. Then the statistics*

(i) $T_c\left(Y^{(g)},\mu_E^{(p)}\right) = \Sigma_{a=1}^{g} n_a \ \left((g_1^a(4))^T\mathbf{C}_1\dots(g_s^a(4))^T\mathbf{C}_q\right) \ S_{\bar{Y}_a}(j_k,Y_a)^{-1} \left(g_1^a(4)^T\mathbf{C}_1\dots g_q^a(4)^T\mathbf{C}_q\right)^T$

(ii) $T_d\left(Y^{(g)},\overline{Y}_E^{(p)}\right)=\Sigma_{a=1}^{g} n_a\left[(\gamma_1^{(p)}(4)-g_1^a(4))^T\mathbf{D}_1\dots(\gamma_q^{(p)}(4)-g_q^a(4))^T\mathbf{D}_q\right] S_{\bar{Y}_a}(j_k,Y_a)^{-1} \left[(\gamma_1^{(p)}(4)-g_1^a(4))^T\mathbf{D}_1\dots(\gamma_q^{(p)}(4)-g_q^a(4))^T\mathbf{D}_q\right]^T.$

where

$$S_{\bar{Y}_a}(j_k,Y_a)_{(s,c)(t,b)} = n_a^{-1}(\mathbf{d}_s^{(p)}(4)-\mathbf{d}_s^{(p)}(c))^{-1}(\mathbf{d}_t^{(p)}(4)-\mathbf{d}_t^{(p)}(b))^{-1} \times\sum_i(\mathbf{g}_s^{(p)}(c)\cdot X_{a,i}^s)(\mathbf{g}_t^{(p)}(b)\cdot X_{a,i}^t)(\mathbf{g}_s^{(p)}(4)\cdot X_{a,i}^s)(\mathbf{g}_t^{(p)}(4)\cdot X_{a,i}^t)$$

and $s,t = 1,\dots,q$ *and* $c,b = 1,\dots,m$. *Both* $T_c\left(Y^{(g)},\mu_E^{(p)}\right)$ *and* $T_d\left(Y^{(g)},\overline{Y}_E^{(p)}\right)$ *have asymptotically, a* χ^2_{3q} *distribution.*

Proof. For part (i) we note that for each $a = 1,\dots g$ we get a natural extension of the result in theorem 4.1 of Bhattacharya and Patrangenaru (2005) [21] as shown in (12.23). For part (ii) recall that

$$T_d\left(Y^{(g)},\overline{Y}_E^{(p)}\right) = n_a\sum_{a=1}^{g}\left\|S_{\bar{Y}_a}(j_k,Y_a)^{-1/2}\tan_{j_k\left(\overline{Y}_E^{(p)}\right)}\left(j_k(\overline{Y}_{a,E})-j_k\left(\mu_E^{(p)}\right)\right)\right\|^2$$

we start by rewriting the expression above and we have

$$T_d\left(Y^{(g)},\overline{Y}_E^{(p)}\right) = n_a\sum_{a=1}^{g}\left\|S_{\bar{Y}_a}(j_k,Y_a)^{-1/2}\tan_{j_k\left(\overline{Y}_E^{(p)}\right)}\left(j_k(\overline{Y}_E^{(p)})-j_k\left(\mu_E^{(p)}\right)\right) - S_{\bar{Y}_a}(j_k,Y_a)^{-1/2}\tan_{j_k\left(\overline{Y}_E^{(p)}\right)}\left(j_k(\overline{Y}_E^{(p)})-j_k\left(\overline{Y}_{a,E}\right)\right)\right\|^2$$

$$T_d\left(Y^{(g)},\overline{Y}_E^{(p)}\right) = \sum_{a=1}^{g} n_a\left\|S_{\bar{Y}_a}(j_k,Y_a)^{-1/2}\left[(\gamma_1^{(p)}(4))^T\mathbf{D}_1\dots(\gamma_q^{(p)}(4))^T\mathbf{D}_q\right]^T - S_{\bar{Y}_a}(j_k,Y_a)^{-1/2}\left[(g_1^a(4))^T\mathbf{D}_1\dots(g_q^a(4))^T\mathbf{D}_q\right]^T\right\|^2 \quad (12.36)$$

□

If $Y_r a$ are j_k-nonfocal populations on $(\mathbb{R}P^3)^q$, we can construct an Edgeworth expansion up to order $O_p(n^{-2})$ of the pivotal statistics $T_c\left(Y^{(g)}, \mu_E^{(p)}\right)$ and $T_d\left(Y^{(g)}, \overline{Y}_E^{(p)}\right)$. Under the hypothesis

$$\begin{cases} H_0 & : \mu_{1,E} = \mu_{2,E} = \ldots = \mu_{g,E} = \mu_E^{(p)}, \\ H_a & :\ni (i,j) 1 \leq i < j < g, \text{s.t. } \mu_{i,E} \neq \mu_{j,E}. \end{cases}$$

COROLLARY 12.4.2. *The* $(1-c)100\%$ *bootstrap confidence regions for* μ_E *with* $d = gp$ *are given by*

(a) $C^{*(g)}_{n,c} = j^{-1}(U^*_{n,c})$ *and* $U^*_{n,c} = \{j_k(\nu) \in j_k((\mathbb{R}P^3)^q) : T_c\left(Y^{(g)}, \nu\right) \leq c^{*(g)}_{1-c}\}$ *where* $c^{*(g)}_{1-c}$ *is the upper* $100(1-c)\%$ *point of the values*

$$T_c\left(Y^{*(g)}, \overline{Y}_E^{(p)}\right) = \sum_{a=1}^{g} n_a \left((g^{*a}_1(4))^T \mathbf{D}_1 \ldots (g^{*a}_s(4))^T \mathbf{D}_q\right) \times S_{\bar{Y}_a^*}(j_k, Y_a^*)^{-1} \left(g^{*a}_1(4)^T \mathbf{D}_1 \ldots g^{*a}_q(4)^T \mathbf{D}_q\right)^T \quad (12.37)$$

among the bootstrap resamples.

(b) $D^{*(g)}_{n,c} = j^{-1}(V^*_{n,c})$ *and* $V^*_{n,c} = \{j_k(\nu) \in j_k((\mathbb{R}P^3)^q) : T_c\left(Y^{(g)}, \overline{Y}_E^{(p)}, \nu\right) \leq d^{*(g)}_{1-c}\}$

where

$$T_d\left(Y^{(g)}, \overline{Y}_E^{(p)}, \nu\right) = n_a \sum_{a=1}^{g} \left\| S_{\bar{Y}_a}(j_k, Y_a)^{-1/2} \tan_{j_k\left(\overline{Y}_E^{(p)}\right)} \left(j_k(\overline{Y}_{a,E}) - j_k(\nu)\right) \right\|^2$$

where $d^{*(g)}_{1-c}$ *is the upper* $100(1-c)\%$ *point of the values of*

$$T_d\left(Y^{*(g)}, \overline{Y^*}_E^{(p)}, \overline{Y}_E^{(p)}\right) = \sum_{a=1}^{g} n_a \left\| S_{\bar{Y}_a^*}(j_k, Y_a^*)^{-1/2} \tan_{j_k\left(\overline{Y}_E^{*(p)}\right)} \left(j_k(\overline{Y}^*_{a,E}) - j_k(\overline{Y}_E^{(p)})\right) \right\|^2 \quad (12.38)$$

among the bootstrap resamples. Both of the regions given by (12.16) *and* (12.14) *have coverage error* $O_p(n^{-2})$.

Figure 12.1: Faces used in the extrinsic MANOVA analysis.

Note that here

$$S_{\bar{Y}_a^*}(j_k, Y_a^*)_{(s,c)(t,b)} = n_a^{-1}(\mathbf{d}_s^{*(p)}(4) - \mathbf{d}_s^{*(p)}(c))^{-1}(\mathbf{d}_t^{*(p)}(4) - \mathbf{d}_t^{*(p)}(b))^{-1}$$
$$\sum_i (\mathbf{g}_s^{*(p)}(c) \cdot X_{a,i}^{*s})(\mathbf{g}_t^{*(p)}(b) \cdot X_{a,i}^{*t})(\mathbf{g}_s^{*(p)}(4) \cdot X_{a,i}^{*s})(\mathbf{g}_t^{*(p)}(4) \cdot X_{a,i}^{*t}), b, c = 1,2,3.$$

12.5 Application to Face Data

A digital image data set was collected using a high-resolution Panasonic-Lumix DMC-FZ200 camera. The images can be found at *ani.stat.fsu.edu/ ~ vic/E-MANOVA*. We tested for the existence of a 3D mean projective shape difference to differentiate between five faces which are represented in Figure 12.1

From digital camera face image data, we obtain 3D surface reconstructions of the faces, with seven labeled landmarks, obtained using the software Agisoft. These reconstructions (including texture) are displayed at the end of Chapter 1.

Our analysis will be conducted on $g = 5$ individuals. The 3D reconstruction was done using the Agisoft software. The images in Figure 1.10 represent 19 facial reconstructions. Each of those reconstructions was created using mostly four to five digital camera images of a given individual. We are also able to place and recover seven landmarks which are shown in Figure 12.2.

Five of those landmarks (colored in red) are selected as the projective frame, and the resulting two projective coordinates determine the 3D-projective shape of the seven landmark configurations selected. Note that we used a different projective frame than in Yao (2016) [147] to ensure that the landmarks are in general position.

We will compare these faces by conducting a MANOVA on manifold to compare $g = 5$ VW means on $P\Sigma_3^7 = (\mathbb{R}P^3)^2$. For $n = \sum_{a=1}^5 n_a = 31$ where $n_1 = n_2 = n4 = n5 = 6$ and $n_3 = 7$ our hypothesis problem will be

$$H_0: \mu_{1,E} = \mu_{2,E} = \mu_{3,E} = \mu_{4,E} = \mu_{5,E} = \mu_E,$$
$$H_a: \textit{at least one of the equalities above does not hold.}$$

Figure 12.2: Projective frame shown in red.

Since the true pulled mean is unknown and our data set is relatively small, we will reject the null hypothesis if

$$T_d\left(Y^{(3)},\overline{Y}_E^{(p)}\right) = \sum_{a=1}^{5} n_a \left\| S_{\bar{Y}_a}(j_k,Y_a)^{-1/2} \tan_{j_k\left(\overline{Y}_E^{(p)}\right)} \left(j_k(\overline{Y}_{a,E}) - j_k(\overline{Y}_E^{(p)})\right)\right\|^2$$

is greater than $d^{*(3)}_{1-\alpha}$, where $d^{*(3)}_{1-\alpha}$ is the $(1-\alpha)100\%$ cutoff of the corresponding bootstrap distribution in Eq. (12.38).

Using $\alpha = 0.05$ and 70543872 resamples, we obtain a value $T_d\left(Y^{(3)},\overline{Y}_E^{(p)}\right) = 389860$ and $d^{*(3)}_{0.95} = 60616$, and we therefore reject the null hypothesis. We conclude that there exists a statistically significant VW mean 3D-projective shape face difference between at least two of the individuals in our data set.

Chapter 13

Further Directions in Statistics on Stratified Spaces

13.1 Introduction

Throughout this manuscript, a number of stratified spaces have been considered, arising as object spaces in data analysis. In general, such an object space has an open dense smooth subset in the dimension of its highest-dimensional stratum. One should therefore expect for probability distributions on stratified spaces to have a similar behavior to those on a manifold. Along these lines, the aspects presented in the following sections are worth pursuing in future research on statistics on object spaces with a manifold stratification. Figures and computing in this chapter are due to Ka Chun Wong (2023)[144].

13.2 Degrees of Stickiness and Hypothesis Testing for Distributions on Stratified Spaces

As noticed in the previous chapters, the asymptotic distributions of the sample intrinsic means behave differently, in case the population mean lies on a lower-dimensional stratum of T_4. This behavior was studied by Huckemann et al. (2015) [74], and in general, more recently by Lammers et al. (2023) [85] and Ulmer et al. (2023) [135]. It would be worth following the degree of stickiness of a random object on an arbitrary stratified space. Note that efforts in considering the general situation, when different strata have nonzero curvature, are stifled by the increasingly time-consuming computational issues that are plaguing intrinsic data analysis in general. For this reason, since the user has the choice of a metric, Euclidean metrics on various strata are referred, as the lead only to combinatorial issues, in finding the shortest path between points on different Euclidean convex components of the stratified space. New approaches for addressing the CLT on an arbitrary stratified space are published yet not mentioned here, as such preprints are not yet refereed.

13.3 Additional Topics

As has been discussed throughout this text, nonparametric methods for statistical analysis on data with manifold sample spaces have tended to develop in at least one of two manners. First, many of these methods serve to generalize standard methods for Euclidean data analysis for sample spaces that have been previously studied. Secondly, the methods are often developed to extend nonparametric statistics on manifolds to new types of data. In this section, we will first discuss two developments that fall into the first category: principal component analysis (PCA) and spatial statistics. We will then consider shape analysis of 3D surfaces, which fits into the second category. Finally, we will discuss a further generalization of statistical methods to spaces with a manifold stratification.

13.3.1 Extrinsic Principal Component Analysis

For multivariate data, PCA greatly aids in understanding the primary directions of variation. These can, in turn, be used to perform dimension reduction by considering only those principal components that contribute greatly to the variability. For data on manifolds, there have been numerous attempts of extending PCA, most of them assuming the manifold has a high degree of symmetry, or it has a Riemannian structure on it. One early approach due to Fletcher et al. (2004) [50] was to perform Euclidean PCA onto the tangent space at the intrinsic mean by using the inverse Riemannian exponential map, on the complement of the cut-locus of the intrinsic mean. Later, Huckemann et al. (2010) [73] introduced the notion of geodesic PCA for a r.o. on a Riemannian manifold. Jung et al. (2012) [77] introduced the idea of principal nested spheres as a decomposition of data on a round sphere S^p to capture nongeodesic variation on a unit sphere using nested submanifolds of decreasing dimension. Recently, Akhoj et al. (2023) [1] are working to develop more generally the notion of *principal subbundle*, in an effort to find lower-dimensional manifolds of the numerical space where data is concentrated. Such efforts, however, are in the direction of *manifold learning*, as they do not seek a step-wise dimension reduction, capturing an increasing percentage of data variability, as done in PCA analysis.

In this chapter we follow the approach to PCA on manifolds due to Wong et al. (2024) [145]. From Patrangenaru and Ellingson (2015) [106], recall that the tangential component $tan(v)$ of $v \in \mathbb{R}^N$ w.r.t. the basis $e_a(P_j(\mu)) \in T_{P_j(\mu)}j(\mathscr{M}), a = 1, \ldots, m$ is given by

$$tan(v) = (e_1(P_j(\mu))^T v \ldots e_m(P_j(\mu))^T v)^T. \tag{13.1}$$

Then the random vector $(d_{\mu_E} j)^{-1}(tan(P_j(\overline{j(X)}) - P_j(\mu)))$ has the following covariance matrix w.r.t. the basis $f_1(\mu_E), \cdots, f_m(\mu_E)$:

$$\Sigma_{j,E} = (e_a(P_j(\mu))^T \Sigma_\mu e_b(P_j(\mu)))_{1 \le a,b \le m} =$$
$$\left[\sum d_\mu P_j(e_b) \cdot e_a(P_j(\mu))\right]_{a=1,\dots,m} \Sigma \left[\sum d_\mu P_j(e_b) \cdot e_a(P_j(\mu))\right]^T_{a=1,\dots,m}. \quad (13.2)$$

DEFINITION 13.3.1. *The matrix $\Sigma_{j,E}$ given by (13.2) is the extrinsic covariance matrix of the j-nonfocal distribution Q (of X_1) w.r.t. the basis $f_1(\mu_E), \dots, f_m(\mu_E)$.*

When j is fixed in a specific context, the subscript j in $\Sigma_{j,E}$ may be omitted.

REMARK 13.3.1. *In order to find a consistent estimator of $\Sigma_{j,E}$, note that $\overline{j(X)}$ is a consistent estimator of μ, $d_{\overline{j(X)}} P_j$ converges in probability to $d_\mu P_j$, and $e_a(P_j(\overline{j(X)}))$ converges in probability to $e_a(P_j(\mu))$ and, further,*

$$S_{j,n} = n^{-1} \sum (j(X_r) - \overline{j(X)})(j(X_r) - \overline{j(X)})^T$$

is a consistent estimator of Σ. It follows that

$$\left[\sum_{a=1}^{m} d_{\overline{j(X)}} P_j(e_b) \cdot e_a(P_j(\overline{j(X)})) e_a(P_j(\overline{j(X)}))\right] S_{j,n}$$
$$\left[\sum_{a=1}^{m} d_{\overline{j(X)}} P_j(e_b) \cdot e_a(P_j(\overline{j(X)})) e_a(P_j(\overline{j(X)}))\right]^T \quad (13.3)$$

is a consistent estimator of Σ_μ, and $tan_{P_j(\overline{j(X)})} v$ is a consistent estimator of $tan(v)$.

If we take the components of the bilinear form associated with the matrix (13.3) w.r.t. $e_1(P_j(\overline{j(X)})), e_2(P_j(\overline{j(X)})), \dots, e_m(P_j(\overline{j(X)}))$, we get a consistent estimator of $\Sigma_{j,E}$, called the *sample extrinsic covariance matrix*, given by

$$S_{j,E,n} = \left[\left[\sum d_{\overline{j(X)}} P_j(e_b) \cdot e_a(P_j(\overline{j(X)}))\right]_{a=1,\dots,m}\right] \cdot S_{j,n}$$
$$\left[\left[\sum d_{\overline{j(X)}} P_j(e_b) \cdot e_a(P_j(\overline{j(X)}))\right]_{a=1,\dots,m}\right]^T \quad (13.4)$$

Therefore, given the j-nonfocal random object X on $\mathscr{M}$, we introduce an *extrinsic principal components (p.c.s)* methodology, w.r.t. the basis $f_1(\mu_E), \dots, f_m(\mu_E)$ of the tangent space $T_{\mu_E}\mathscr{M}$. The extrinsic p.c.s are 1D stratified spaces of $\mathscr{M}$ going through the extrinsic mean that are obtained by taking the j-preimage of the intersection of the affine subspace generated by the eigenvectors v_i, $i = 1, \dots, m$ of the extrinsic covariance matrix $\Sigma_{j,E}$ corresponding to its eigenvalues λ_i, $i = 1,\ldots,m$ where λ_i are listed in their descending order, and by the orthocomplement of $T_{\mu_E}\mathcal{M}$, with $\mathcal{M}$. Here we assume the eigenvalues are simple.

Recalling the traditional principal component analysis, which was first proposed by Pearson (1901) [109], is a classical dimension reduction method for high-dimensional data. Principal component analysis seeks a space of lower dimensionality, known as the principal subspace, such that the orthogonal projection of the data points onto the subspace maximize the variance of the projected points. To achieve this goal, we are looking for the eigenvectors of the covariance matrix as the principal components. There we select the largest eigenvalues, since they contribute to most of the variability in the data. Similarly, the extrinsic principal components are working in a similar way. But instead of using the covariance matrix of the data set, we use the sample extrinsic covariance matrix.

DEFINITION 13.3.2. *The extrinsic principal components of the r.o. X on $\mathcal{M}$, with a j-nonfocal distribution Q w.r.t. the basis $f_1(\mu_E),\ldots,f_m(\mu_E)$ are 1D submanifolds going through the extrinsic mean that are obtained by intersecting the affine subspace generated by the eigenvectors v_i, $i = 1,\ldots,m$ of the matrix $\Sigma_{j,E}$ corresponding to the eigenvalues λ_i, $i = 1,\ldots,m$ where λ_i are listed in their descending order, and by the orthocomplement of the tangent space at the extrinsic mean to $\mathcal{M}$, with $\mathcal{M}$.*

The extrinsic principal components present on the normal space (or tangent space) of the extrinsic mean. The principal components represent by a (eigen-) vector on such space. These vectors can form a hyperplane with the extrinsic mean. The submanifold that these hyperplanes intercept with the embedded manifold can also represent as principal components.

We can convert the principal components from the tangent space (with basis $f_1(\mu_E),\ldots,f_m(\mu_E)$) to the projected manifolds (with standard $\mathbb{R}^N$ basis). To find the projected points, we can first look for a projection $P_T : j(\mathcal{M}) \to T_{P_j(\mu)} j(\mathcal{M})$, a projection from manifold $j(\mathcal{M})$ to the tangent space on the extrinsic mean. Transfer the projected points form standard basis to $f_1(\mu_E),\ldots,f_m(\mu_E)$ (or transfer principal components to inverse way) and project to the principal subspace in result the principal component projection on the tangent space. We will give some examples with different types of data.

EXAMPLE 13.3.1. *The following example shows the extrinsic principal components of a set of spherical data. We will illustrate the extrinsic principal components in a graphical way by a simple example. We randomly generate 300 points on a 3D unit sphere. To show the result easily, the generated data points concentrate in a single direction. The projection $P_j(x) = \frac{x}{||x||}$ which project any point in $\mathbb{R}^3$ onto S^2 through the origin. Results are shown in Figure 13.1 and Table 13.1. The red great circle in Figure 13.1 is the first principal component, and the green one is the second. The intercept of the two great circles is the*

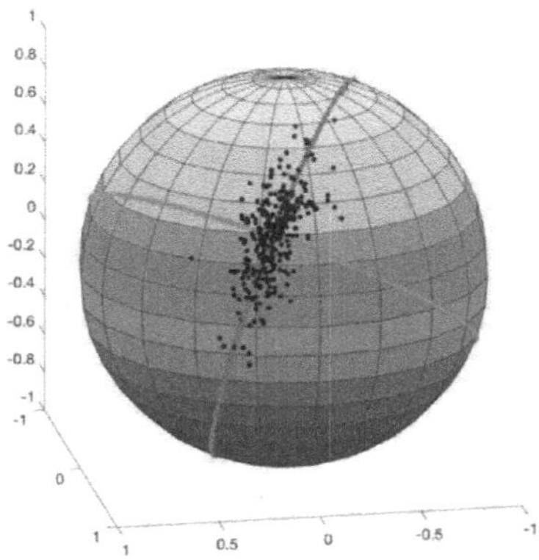

Figure 13.1: Two extrinsic principal components in Example 13.3.1.

extrinsic sample mean. As we see, the data are mainly distributed along the first principal component. The first principal component explains over 87% of the data in this example. To look for the data projected onto the first principal component, we first project the data onto the tangent space of the extrinsic sample mean, $T_{P_{j(\bar{X})}} j(\mathscr{M})$*, as shown in Figure 13.2. We project the data along the tangent space to the first principal component (the vector in red color). And we re-project back those data onto the sphere through the origin. Result are shown in Figure 13.3. The projected data (red points) stick on the first principal components on the sphere (yellow line) along the shortest distance.*

Table 13.1: Statistics for Example 13.3.1

Extrinsic sample mean	0.2153	0.8692	0.4461
Extrinsic sample covariance	0.0045 -0.0010	-0.0010 0.0305	
$f_1(\mu_E)$	-0.8692	0.3775	-0.3195
$f_2(\mu_E)$	-0.4461	-0.3195	0.8360
1st eigenvalue	0.0305		
1st p.c. under standard basis	0.4128	0.3336	-0.8475
2nd eigenvalue	0.0044		
2nd p.c. under standard basis	0.8854	-0.3651	0.2876

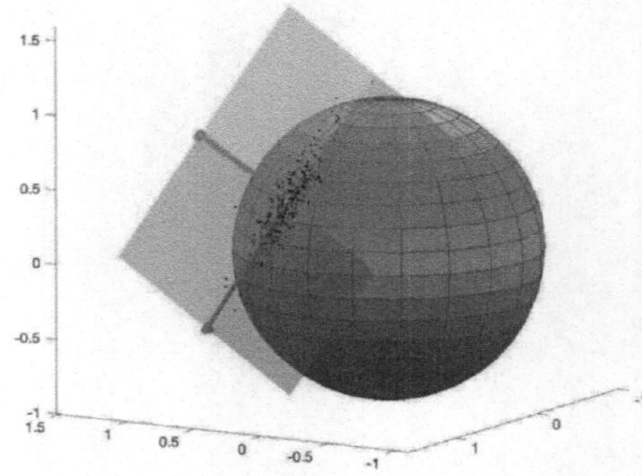

Figure 13.2: Projected data on $T_{P_{j(\bar{X})}} j(\mathscr{M})$ and the two extrinsic principal components.

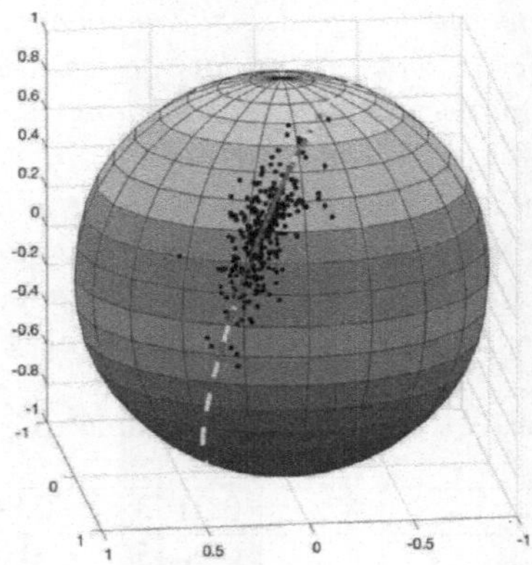

Figure 13.3: Data projected on the first extrinsic principal component.

13.3.2 Extrinsic Principal Components for Shapes of Planar Contours

In this subsection we will focus on shape analysis of planar contour. We will simply introduce shape space and some statistics on shape analysis. We will also produce extrinsic principal component analysis on some example. The general reference for this section is Patrangenaru and Ellingson (2015) [106].

We focus on contours, boundaries of 2D topological disks in the plane. To keep the data analysis stable, and to assign a *unique* labeling, we make the *generic* assumption that across the population there is a unique anatomical or geometrical landmark starting point p_0 on such a contour of perimeter 1, so that the label of any other point p on the contour is the "counterclockwise" travel time at constant speed from p_0 to p. A *regular contour* $\tilde{\gamma}$ is regarded as the range of a piecewise differentiable *regular* arclength parameterized function $\gamma : [0,L] \to \mathbb{C}, \gamma(0) = \gamma(L)$, that is one-to-one on $[0,L)$. Two contours $\tilde{\gamma}_1, \tilde{\gamma}_2$ *have the same direct similarity shape* if there is a direct similarity $S : \mathbb{C} \to \mathbb{C}$, such that $S(\tilde{\gamma}_1) = \tilde{\gamma}_2$. Two regular contours $\tilde{\gamma}_1, \tilde{\gamma}_2$ have the same similarity shape if their centered counterparts satisfy to $\tilde{\gamma}_{2,0} = \lambda \tilde{\gamma}_{1,0}$, for some $\lambda \in \mathbb{C}\backslash 0$. Therefore Σ_2^{reg}, *set of all direct similarity shapes of regular contours,* is a dense and open subset of $P(\mathbf{H})$, the projective space corresponding to the Hilbert space $\mathbf{H}$ of all square integrable centered functions from S^1 to $\mathbb{C}$ (see Ellingson et al. (2013) [44]).

The space $P(\mathbf{H})$ is a Hilbert manifold. We here introduce the Veronese-Whitney (VW) embedding $j : P(\mathbf{H}) \to \mathscr{L}_{HS} = \mathbf{H} \otimes \mathbf{H}$ given by

$$j([\gamma]) = \frac{1}{\|\gamma\|^2} \gamma \otimes \gamma^*, [\gamma] \in P(\mathbf{H}). \tag{13.5}$$

The *Veronese-Whitney mean (VW mean)* is the extrinsic mean for a random object $X = [\Gamma]$ on $P(\mathbf{H})$ with respect to the VW embedding. The VW extrinsic mean is $[e_1]$, where e_1 is the eigenvector corresponding to the largest eigenvalue of $E(\frac{1}{\|\Gamma\|^2}\Gamma \otimes \Gamma^*)$. The VW extrinsic sample mean can be computed in a similar way.

PROPOSITION 13.3.1. *Given any VW-nonfocal probability measure Q on $P(\mathbf{H})$, if $X_1, \ldots, X_n$ is a random sample from Γ, then the VW sample mean $\hat{\mu}_{E,n}$ is the projective point of the eigenvector corresponding to the largest eigenvalue of $\frac{1}{n}\sum_{i=1}^n \frac{1}{\|X_i\|^2} X_i \otimes X_i^*$.*

Once we have computed the extrinsic mean, the next step is the extrinsic covariance matrix. The following result of Prentice (1984) [110] is also needed in the sequel.

PROPOSITION 13.3.2. *(Prentice (1984) [110]) Assume $[X_i]$, $\|X_i\| = 1$, $i = 1,...,n$ is a random sample from a j-nonfocal, probability measure Q on $\mathbb{R}P^{N-1}$. Then the sample (VW-)extrinsic covariance matrix $S_{j,E}$ is given by*

$$S_{j,E_{ab}} = n^{-1}(\eta_N - \eta_a)^{-1}(\eta_N - \eta_b)^{-1} \sum_i (m_a \cdot X_i)(m_b \cdot X_i)(m \cdot X_i)^2, \tag{13.6}$$

where $\eta_a, a = 1,...,N$, are eigenvalues of $K := n^{-1}\sum_{i=1}^n X_i X_i^t$ in increasing order and $m_a, a = 1,...,N$, are corresponding linearly independent unit eigenvectors.

Here is a proof of formula (13.6). Since the map j is equivariant, without loss of generality one may assume that $j(\overline{X}_E) = P_j(\overline{j(X)})$ is a diagonal matrix,

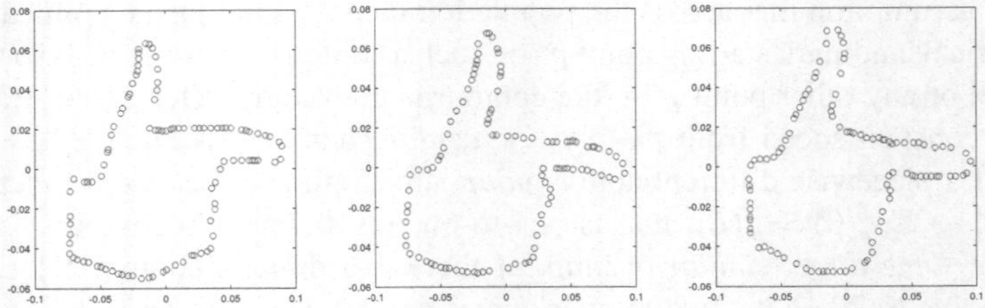

Figure 13.4: Contours of hand gestures.

$\overline{X}_E = [m_N] = [e_N]$ and the other unit eigenvectors of $\overline{j(X)} = D$ are $m_a = e_a, \forall a = 1,...,N-1$. We evaluate $d_D P_j$. Based on this description of $T_{[x]}\mathbb{R}P^{N-1}$, one can select in $T_{P_j(D)} j(\mathbb{R}P^{N-1})$ the orthonormal frame $e_a(P_j(D)) = d_{[e_N]} j(e_a)$. Note that $S(N,\mathbb{R})$ has the orthobasis $F_a^b, b \leq a$ where, for $a < b$, the matrix F_a^b has all entries 0 except for those in the positions $(a,b), (b,a)$ that are equal to $2^{-\frac{1}{2}}$; also $F_a^a = j([e_a])$. A straightforward computation shows that if $\eta_a, a = 1,...,N$, are the eigenvalues of D in their increasing order, then $d_D P_j(F_a^b) = 0, \forall b \leq a < N$ and $d_D P_j(F_a^N) = (\eta_N - \eta_a)^{-1} e_a(P_j(D))$; from this equation it follows that, if $\overline{j(X)}$ is a diagonal matrix D then the entry $S_{j,E_{ab}}$ is given by

$$S_{j,E_{ab}} = n^{-1}(\eta_N - \eta_a)^{-1}(\eta_N - \eta_b)^{-1} \sum_i X_i^a X_i^b (X_i^N)^2. \tag{13.7}$$

Taking $\overline{j(X)}$ to be a diagonal matrix and $m_a = e_a$ formula (13.6) follows.

EXAMPLE 13.3.2. *Here we illustrate an example for* VW principal component analysis *for planar contour. Consider the samples of contours of hand gestures from Sharvit et al. (1998) [118]. Some samples are shown in Figure 13.4. There are 120 contours, each has 100 sampling points. Each sample contour is a* 2×100 *real matrix; each column represents a point of the contour. We transfer the sample into a* 1×100 *complex matrix; as a result the whole sample data will be a* 120×100 *complex matrix. We compute the extrinsic sample mean by using Proposition 13.3.1. The result mean shape of contours shows in Figure 13.5. It is less sharp than those original samples. And then we compute the extrinsic sample covariance matrix by using Eq. (13.6). By applying eigenvalue decomposition on the extrinsic sample covariance matrix, we can extract the extrinsic principal components. Figure 13.6 shows the variance explained by the first ten principal components. We can see the first four principal components explained almost 95% of the total sample variance.*

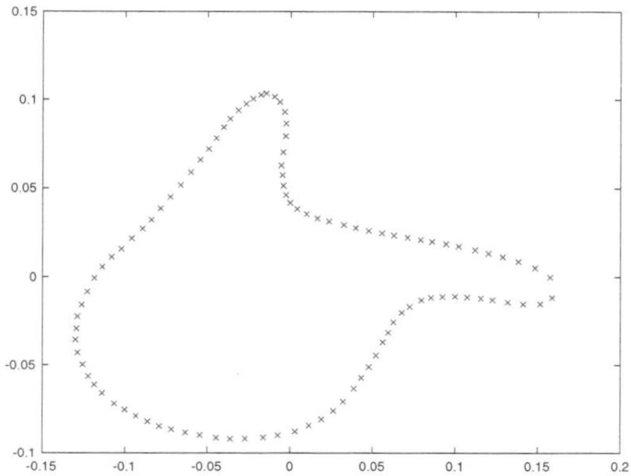

Figure 13.5: Mean shape of contours.

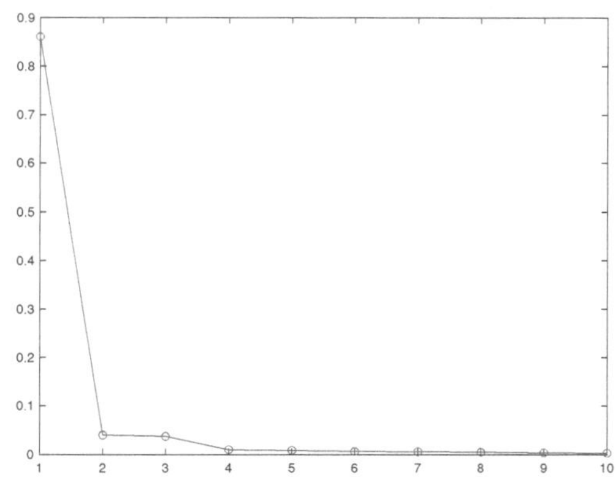

Figure 13.6: Variance rate for the first ten extrinsic principal components.

13.4 Networks

An influential paper on the statistics of the web is due to Broder et al. (2000) [29]. Statisticians followed suit and became interested in studying data arising from graphs or networks (see e.g., Hoff et al. (2002) [67]). The initial approaches were parametric in nature, using various *latent space* Bayesian models. One example of a latent space model is the *random dot product graph* (RDPG) model (see Young and Scheinerman (2007) [148]). Under the RDPG model, the probability that an edge between two nodes is present is given by the dot product of their respective latent vectors of norm 1. For example, in a social network with edges indicating friendships, the components of the vector may be interpreted as the relative interest of the individual in various topics. Recently Tang et al. (2023) [133] established asymptotic normality results for estimation of the block probability matrix B in block-model graphs. Recall that the *adjacency matrix* of a random graph with q nodes is the $q \times q$ symmetric matrix of probabilities of existence of an edge between pairs of its nodes. Another symmetric matrix of interest associated with such a random graph is the *graph Laplacian*, which in case of a *(labeled) simple graph (two nodes are connected by one edge only)*, whose diagonal entry l_{ii} are the number of nodes connected to the node having label i, and off-diagonal entry l_{ij} is -1 or 0, depending on the existence of an edge between the nodes having labels i, j. There is a vast literature on the subject of random graphs and networks, from which we randomly include Bickel and Sarkar (2016) [24], Sussman et al. (2012) [84], and Sussman (2013) [37].

While a statistical study of random graphs and networks had an explosive development in the last couple of decades, nonparametric techniques for the statistical analysis of sequences of networks were less developed. Fraiman et al. (2017) [51] focused on network sequences with a fixed number of labeled nodes and studied some statistical related aspects in a nonparametric setting. They introduced natural notions of center and a depth function for networks that evolve over time. They developed several statistical techniques including testing, supervised and unsupervised classification, and some notions of principal component sets in the space of networks. They also gave examples and asymptotic results, as well as real data examples. No data analysis for probability distributions on graphs (not just for probabilities of the existence of edges) is known to date. Moreover, unlike the simple case of object spaces for rooted or rootless trees, there are no known characterizations of spaces of graphs with nontrivial homology, which is a subject of great importance given its practical applications to engineering and medicine.

An interesting extension of this line of research is in the direction of *object spaces of simple graphs* with a given number of labeled nodes, and a fixed number of cycles, where random graphs are regarded as random objects on such spaces.

Bibliography

[1] M. Akhoj, J. Benn, E. Grong, S. Sommer, and X. Pennec. Principal subbundles for dimension reduction. *Journal of Statistical Planning and Inference*, 140:3026–3034, 2010.

[2] V. Arsigny, P. Fillard, X. Pennec, and N. Ayache. Geometric means in a novel vector space structure on symmetric positive-definite matrices. *SIAM Matrix Anal. Appl.*, 29:328–347, 2006.

[3] B. Aydin, G. Pataki, H. Wang, E. Bullitt, and J. S. Marron. A principal component analysis for trees. *The Annals of Applied Statistics*, 3:1597–1615, 2009.

[4] S. Aylward and E. Bullitt. Initialization, noise, singularities, and scale in height ridge traversal for tubular object centerline extraction. *IEEE Transactions on Medical Imaging*, 21:61–75, 2002.

[5] G. J. Babu and K. Singh. On one term Edgeworth correction by Efron's bootstrap. *Sankhya Ser. A.*, 46:219–232, 1984.

[6] M. Bacak. Computing medians and means in hadamard spaces. *arXiv.org*, arXiv:1210.2145v3, 2012.

[7] A. Bandulasiri, R. N. Bhattacharya, and V. Patrangenaru. Nonparametric inference for extrinsic means on size-and-(reflection)-shape manifolds with applications in medical imaging. *Journal of Multivariate Analysis*, 100:1867–1882, 2009.

[8] Dennis Barden, Huiling Le, and Megan Owen. Central limit theorems for Fréchet means in the space of phylogenetic trees. *Electron. J. Probab.*, 18:no. 25, 1–25, 2013.

[9] B. Basrak. Limit theorems for the inductive mean on metric trees. *J. Appl. Prob.*, 47:1136–1149, 2010.

[10] P. J. Basser and C. Pierpaoli. Microstructural and physiological features of tissues elucidated by quantitative-diffusion-tensor MRI. *J. Magn. Reson.*, 111:209–219, 1996.

[11] R. Beran. Prepivoting to reduce level error of confidence sets. *Biometrika*, 74:457–468, 1987.

[12] R. Beran. Prepivoting test statistics: a bootstrap view of asymptotic refinements. *J. Amer. Statist. Assoc.*, 83:687–697, 1988.

[13] R. Beran and N.I. Fisher. Nonparametric comparison of mean axes. *Ann. Statist.*, 26:472–493, 1998.

[14] A. Bhattacharya. Statistical analysis on manifolds: A nonparametric approach for inference on shape spaces. *Sankhya*, 70-A:223–266, 2008.

[15] A. Bhattacharya and R. N. Bhattacharya. *Nonparametric inference on manifolds. With applications to shape spaces,* volume 2. Institute of Mathematical Statistics (IMS) Monographs. Cambridge University Press, Cambridge, 2012.

[16] R. N. Bhattacharya. Refinements of the multidimensional central limit theorem and applications. *Ann. Probability*, 5:1–27, 1977.

[17] R. N. Bhattacharya, M. Buibas, I. L. Dryden, L. A. Ellingson, D. Groisser, H. Hendriks, S. Huckemann, H. Le, X. Liu, J. S. Marron, D. E. Osborne, V. Patrangenaru, A. Schwartzman, H. W. Thompson, and A. T. A. Wood. Extrinsic data analysis on sample spaces with a manifold stratification. *Advances in Mathematics, Invited Contributions at the Seventh Congress of Romanian Mathematicians, Brasov, 2011*, pages 241–252, 2013.

[18] R. N. Bhattacharya, L. Ellingson, X. Liu, V. Patrangenaru, and M. Crane. Extrinsic analysis on manifolds is computationally faster than intrinsic analysis with applications to quality control by machine vision. *Appl. Stochastic Models Bus. Ind.*, 28:222–235, 2012.

[19] R. N. Bhattacharya and L. Lin. A central limit theorem for Fréchet means. *arXiv.org*, arXiv:1306.5806:1–8, 2011.

[20] R. N. Bhattacharya and V. Patrangenaru. Large sample theory of intrinsic and extrinsic sample means on manifolds - part i. *Annals of Statistics*, 31(1):1–29, 2003.

[21] R. N. Bhattacharya and V. Patrangenaru. Large sample theory of intrinsic and extrinsic sample means on manifolds-part ii. *Annals of Statistics*, 33(3):1211–1245, 2005.

[22] R. N. Bhattacharya and V. Patrangenaru. Statistics on manifolds and landmarks based image analysis: A nonparametric theory with applications. *J. of Statist. Planning and Inference*, 145:1–22, 2014.

[23] R.N. Bhattacharya. *New perspectives in Theoretical and Applied Statistics* (M.L. Puri and J.P. Vilaplana and W. Wertz Eds), Some aspects of Edgeworth expansions in statistics and probability, pages 157–171. Wiley, NY, 1987.

[24] P. J. Bickel and P. Sarkar. Hypothesis testing for automated community detection in networks. *J. R. Statist. Soc. B*, 78:253–273, 2016.

[25] L. Billera, S. Holmes, and K. Vogtmann. Geometry of the space of phylogenetic trees. *Adv. in Appl. Math.*, 27:733–767, 2001.

[26] P. Billingsley. *Probability and measure. Third edition.* Wiley Series in Probability and Mathematical Statistics, 1995.

[27] L. Breiman. *Probability, corrected reprint of the 1968 original.* Classics in Applied Mathematics, Vol 7, Society for Industrial and Applied Mathematics, Philadelphia, PA, 1992.

[28] M. Bridson and A. Haefliger. *Metric Spaces of Non-positive Curvature.* Springer-Verlag, Berlin, 1999.

[29] A. Broder, R. Kumar, F. Maghoul, P. Raghavan, S. Rajagopalan, R. Stata, A. Tomkins, and J. Wiener. Graph structure in the web. *Computer Network*, 33:309–320, 2000.

[30] J.J. Cannone, S. Subramanian, M.N. Schnare, J.R. Collett, L.M. D'Souza, Y. Du, B. Feng, N. Lin, L.V. Madabusi, K.M. Muller, N. Pande, Z. Shang, N. Yu, and R.R. Gutell. Comparative RNA web (CRW) site: an online database of comparative sequence and structure information for ribosomal, intron, and other RNAs. *BMC Bioinforma.*, 3:2, 2002.

[31] É. Cartan. Sur certaines formes Riemanniennes remarquables des géométries à groupe fondamental simple (in French). *Ann. Sci. École Norm. Sup.*, 3(44):345–467, 1927.

[32] É. Cartan. *Léçons sur la Géométrie des Espaces de Riemann (in French).* Gauthier-Villars, Paris, 1928.

[33] T. K. Chandra and J. K. Ghosh. Valid asymptotic expansion for likelihood ratio statistic and other perturbed chisquare variables. *Sankhya*, 41:22–47, 1979.

[34] National Intelligence Council. Updated assessment on covid-19 origins. *https://www.dni.gov/files/ODNI/documents/assessments/Declassified-Assessment-on-COVID-19-Origins.pdf*, 2021.

[35] M. Crane and V. Patrangenaru. Random change on a lie group and mean glaucomatous projective shape change detection from stereo pair images. *Journal of Multivariate Analysis*, 102:225–237, 2011.

[36] G. de Rham. Sur la théorie des formes différentielles harmoniques. *Ann. Univ. Grenoble. Sect. Sci. Math. Phys.*, 22:135–152, 1946.

[37] Sussman D.L. *Foundations of Adjacency Spectral Embedding.* A dissertation submitted to Johns Hopkins University in conformity with the requirements for the degree of Doctor of Philosophy, 2013.

[38] M. P. do Carmo. *Riemannian Geometry.* Birkhäuser, Boston, 1992.

[39] I. L. Dryden and K. V. Mardia. *Statistical Shape Analysis: With Applications in R*, volume 995. John Wiley & Sons, Chichester, 2016.

[40] B. Efron. Bootstrap methods: Another look at the jackknife. *Annals of Statistics*, 7:1–26, 1979.

[41] B. Efron. *The Jackknife, the Bootstrap and Other Resampling Plans*. CBMS-NSF Regional Conference Series in Applied Mathematics, 38, SIAM, 1982.

[42] S. Elbe and G. Buckland-Merrett. Data, disease and diplomacy: Gisaid's innovative contribution to global health. *Global Challenges*, 1:33–46. DOI:10.1002/gch2.1018 PMCID: 31565258, 2017.

[43] L. Ellingson, V. Patrangenaru, H. Hendriks, and P. S. Valentin. *Topics in Nonparametric Statistics. Editors: M.G. Akritas, S.N. Lahiri and D. N. Politis*, CLT on Low Dimensional Stratified Spaces, pages 227–240. Springer, 2014.

[44] L. Ellingson, V. Patrangenaru, and F. Ruymgaart. Nonparametric estimation of means on Hilbert manifolds and extrinsic analysis of mean shapes of contours. *Journal of Multivariate Analysis*, 122:317–333, 2013.

[45] L. Euler. Solutio problematis ad geometriam situs pertinentis. *Commentarii Academiae Scientiarum Petropolitanae*, 8:128–140, 1741.

[46] O. Faugeras. What can be seen in three dimensions with an uncalibrated stereo rig? *Proc. European Conference on Computer Vision, LNCS 588*, pages 563–578, 1992.

[47] J. Felsenstein. *Inferring Phylogenies*. Sinauer Associates, 2004.

[48] A. Feragen, M. Owen, J. Petersen, M. M. W. Wille, L. H. Thomsen, A. Dirksen, and M. de Bruijne. Tree-space statistics and approximations for large-scale analysis of anatomical trees. *Information Processing in Medical Imaging, Lecture Notes in Computer Science*, 7917:74–85, 2013.

[49] N. I. Fisher, P. Hall, B. Y. Jing, and A. T. A. Wood. Properties of principal component methods for functional and longitudinal data analysis. *J. Amer. Statist. Assoc.*, 91:1062–1070, 1996.

[50] P. T. Fletcher, C. Lu, S. M. Pizer, and S. C. Joshi. Principal geodesic analysis for the study of nonlinear statistics of shape. *IEEE Trans. Med. Imaging*, 23(8):995–1005, 2004.

[51] D. Fraiman, N. Fraiman, and R. Fraiman. Nonparametric statistics of dynamic networks with distinguishable nodes. *Test*, 26(3):546–573, 2017.

[52] M. Fréchet. Les élements aléatoires de nature quelconque dans un espace distancié (in French). *Ann. Inst. H. Poincaré*, 10:215–310, 1948.

[53] M. Goresky and R. MacPherson. *Stratified Morse theory*. Ergebnisse der Mathematik und ihrer Grenzgebiete (3) [Results in Mathematics and Related Areas (3)], 14, Springer-Verlag, Berlin, 1988.

[54] M. Gromov. Mathematics behind massive artificial evolution/selection processes. *Talk at IHES, Paris, France*, https://www.youtube.com/watch?v=g4Wl3Ggho6k, 2018.

[55] N. D. Grubaugh, M. E. Petrone, and E. C. Holmes. We shouldn't worry when a virus mutates during disease outbreaks. *Nature microbiology*, 4:529–530, 2020.

[56] P. Hall. On the bootstrap and likelihood-based confidence regions. *Biometrika*, 74:481–493, 1987.

[57] P. Hall. Theoretical comparison of bootstrap confidence intervals. *Ann. Statist.*, 16:927–985, 1988.

[58] P. Hall. *The Bootstrap and Edgeworth Expansion*. Springer Series in Statistics, Springer-Verlag, NewYork, 1997.

[59] P. Hall and J. D. Hart. Bootstrap test for difference between means in nonparametric regression. *J. Amer. Stat. Asso.*, 85:1039–1049, 1990.

[60] Handle. http://hdl.handle.net/1926/594. *Biometrika*, 2008.

[61] T. Harris. First passage and recurrence distributions. *Transactions of the American Mathematical Society*, 73:1064–1070, 1952.

[62] R. I. Hartley, R. Gupta, and T. Chang. Stereo from uncalibrated cameras. *Proceedings IEEE Conference on Computer Vision and Pattern Recognition*, pages 761–764, 1992.

[63] E.L. Hatcher, S.A. Zhdanov, Y. Bao, O. Blinkova, E.P. Nawrocki, Y. Ostapchuck, A.A. Schäffer, and J.R. Brister. Virus variation resource–improved response to emergent viral outbreaks. *Nucleic Acids Research*, 45:D482–D490. DOI: 10.1093/nar/gkw1065, 2017.

[64] R. He, F. Dobie, M. Ballantine, A. Leeson, Y. Li, N. Bastien, T. Cutts, A. Andonov, J. Cao, T. F. Booth, F. A. Plummer, S. Tyler, L. Baker, and X. Lic. Analysis of multimerization of the SARS coronavirus nucleocapsid protein. *Biochem. Biophys. Res. Commun.*, 2:476–483, 2004.

[65] H. Hendriks and Z. Landsman. Asymptotic behavior of sample mean location for manifolds. *Statist. Probab. Lett.*, 26:169–178, 1996.

[66] H. Hendriks and Z. Landsman. Mean location and sample mean location on manifolds: asymptotics, tests, confidence regions. *J. Multivariate Anal.*, 67:227–243, 1998.

[67] P.D. Hoff, A. E. Raftery, and M. S. Handcock. Latent space approaches to social network analysis. *J. Amer. Statist. Assoc.*, 97:1090–1098, 2002.

[68] T. Hotz and S. Huckemann. Intrinsic means on the circle: uniqueness, locus and asymptotics. *Annals of the Institute of Statistical Mathematics*, 67:177–193, 2015/2.

[69] T. Hotz, S. Huckemann, H. Le, J. S. Marron, J. Mattingly, E. Miller, J. Nolen, M. Owen, V. Patrangenaru, and S. Skwerer. Sticky central limit theorems on open books. *Annals of Applied Probability*, 23:2238–2258, 2013.

[70] P. J. Huber. *Robust statistics*. Wiley Series in Probability and Mathematical Statistics, 1981.

[71] S. Huckemann. On the meaning of mean shape: manifold stability, locus and the two sample test. *Ann. Inst. Math. Statist.*, 64:1227–1259, 2012.

[72] S. Huckemann, T. Hotz, and A. Munk. Intrinsic MANOVA for Riemannian manifolds with an application to kendall's spaces of planar shapes. *IEEE Trans. Patt. Anal. Mach. Intell.*, 32:593–603, 2010.

[73] S. Huckemann, T. Hotz, and A. Munk. Intrinsic shape analysis: geodesic pca for Riemannian manifolds modulo isometric lie group actions. *Statist. Sinica*, 20(1):1–58, 2010.

[74] S. Huckemann, J.C. Mattingly, E. Miller, and J. Nolen. Sticky central limit theorems at isolated hyperbolic planar singularities. *Electron. J. Probab*. 20(78):34, 2015.

[75] M. Jinek, A. East, A. Cheng, S. Lin, E. Ma, and J. Doudna. RNA-programmed genome editing in human cells. *Genetics and Genomics*, https://doi.org/10.7554/eLife.00471, 2013.

[76] R. A. Johnson and D. W. Wichern. *Applied Multivariate Statistical Analysis, 6th Edition*. Pearson Prentice Hall, Upper Saddle River, New Jersey, 2007.

[77] S. Jung, I. L. Dryden, and J. S. Marron. Analysis of principal nested spheres. *Biometrika*, 99(3):551–568, 2012.

[78] D. G. Kendall. Shape manifolds, procrustean metrics, and complex projective spaces. *Bull. London Math. Soc.*, 16:81–121, 1984.

[79] D. G. Kendall. How to look at objects in a five-dimensional shape space: looking at geodesics. *Adv. in Appl. Probab.*, 27:35–43, 1995.

[80] D. G. Kendall, D. Barden, T. K. Carne, and H. Le. *Shape and Shape theory*. Wiley, New York, 1999.

[81] S. Kobayashi and K. Nomizu. *Foundations of Differential Geometry. Vol.* **I, II**. *Reprint of the 1969 original.* Wiley Classics Library, 1996.

[82] H. Krim and A. Yezzi (Eds.). *Statistics and Analysis of Shapes*. Modeling and Simulation in Science, Engineering and Technology, Birkhäuser, Boston, 2006.

[83] S. Kumar, G. Stecher, M. Li, and K. Tamura. Mega x: Molecular evolutionary genetics analysis across computing platforms. *Molecular Biology and Evolution*, 35:1547–1549, 2018.

[84] Sussman D. L., Tang M., Fishkind D. E., and Priebe C. E. A consistent adjacency spectral embedding for stochastic blockmodel graphs. *J. Am. Statist. Assoc.*, 107:1119–1128, 2012.

[85] L. Lammers, D.T. Van, T.M.W. Nye, and S. F. Huckemann. Types of stickiness in BHV phylogenetic tree spaces and their degree. *Lecture Notes in Comput. Sci., 14071*, Geometric science of information. Part I:357–365, 2023.

[86] M.A. Larkin, G. Blackshields, N.P. Brown, R. Chenna, P.A. McGettigan, H. McWilliam, F. Valentin, I.M. Wallace, A. Wilm, R. Lopez, J.D. Thompson, T.J. Gibson, and D.G. Higgins. Clustal w and clustal x version 2.0. *Bioinformatics*, 23:2947–2948, 2007.

[87] H. Le. On the consistency of procrustean mean shapes. *Adv. in Appl. Probab.*, 30:53–63, 1998.

[88] W. Li, Z. Shi, M. Yu, W. Ren, C. Smith, J. H. Epstein, H. Wang, G. Crameri, Z. Hu, H. Zhang, J. Zhang, J. McEachern, H. Field, P. Daszak, B. T. Eaton, S. Zhang, and L.-F. Wang. Bats are natural reservoirs of SARS-like coronaviruses. *Science*, DOI: 10.1126/science.1118391, 2005.

[89] W.P. Maddison and D.R. Maddison. Mesquite: a modular system for evolutionary analysis. *http://mesquiteproject.org*, Version 2.75, 2011.

[90] K. V. Mardia and V. Patrangenaru. *Proceedings of the 20th LASR Workshop*, On Affine and Projective Shape Data Analysis, pages 39–45. Leeds University Press, 2001.

[91] K. V. Mardia and V. Patrangenaru. Directions and projective shapes. *Ann. Statist.*, 33:1666–1699, 2005.

[92] J. S. Marron and I. L. Dryden. *Object-Oriented Data Analysis*. CRC Press/ Chapman & Hall, Boca Raton, 2021.

[93] C. Mattos, D. Rasmussen, X. Ding, G. A. Petsko, and D. Ringe. Analogous inhibitors of elastase do not always bind analogously. *Nat. Struct. Biol.*, 1:55–58, 1994.

[94] Amy Maxmen and Smriti Mallapaty. The COVID lab-leak hypothesis: what scientists do and don't know. *Nature*, 594(7863):313–315, 2021.

[95] E. Miller, M. Owen, and J. Provan. Averaging metric phylogenetic trees. *arXiv.org*, arXiv:1211.7046, 2012.

[96] R. Moore. *Investigating Significant Mutations of US SARS-CoV-2 RNA Sequences Using Stratified Spaces, and Genetic Connection to Drug-induced Liver Injury (DILI) Through Statistical Learning Methods*. Ph.D. Thesis, Department of Statistics, Florida State University, 2021.

[97] R. Moore, V. Patrangenaru, and A. Dixon. Investigating two possible origins of SARS-CoV-2 an RNA analysis on tree spaces. *https://stat.fsu.edu/technical-reports-page*, M1015:1–10, 2021.

[98] M. Nei and Kumar S. *Molecular Evolution and Phylogenetics*. Oxford University Press, New York, 2000.

[99] I. Oguz, J. Cates, P.T. Fletcher, R. Whitaker, D. D. Cool, S. Aylward, and M. Styner. *In: Biomedical Imaging: From Nano to Macro, ISBI 2008, 5th IEEE International Symposium*, chapter Cortical correspondence using entropy-based particle systems and local features, pages 1637–1640. IEEE, New York, 2008.

[100] G. Ordway and V. Patrangenaru. Sampling the Swadesh list to identify similar languages with tree spaces. *arXiv:2405.06549v1*, 2024.

[101] D. Osborne. Nonparametric data analysis on manifolds with applications in medical imaging. *PhD Dissertation, Florida State University*, 2012.

[102] D. Osborne, V. Patrangenaru, L. Ellingson, D. Groisser, and A. Schwartzman. Nonparametric two-sample tests on homogeneous Riemannian manifolds, Cholesky decompositions and diffusion tensor image analysis. *Journal of Multivariate Analysis*, 119:163–175, 2013.

[103] M. Owen and S.Provan. A fast algorithm for computing geodesic distance in tree space. *IEEE/ACM Transactions on Computational Biology and Bioinformatics*, 8:2–13, 2010.

[104] V. Patrangenaru. Asymptotic Statistics on Manifolds. *PhD Dissertation, Indiana University*, 1998.

[105] V. Patrangenaru and L. A. Ellingson. Statistical analysis of object data is data analysis on sample spaces with a manifold stratification. *Transition Workshop of the SAMSI AOD Program 2010-2011, Working Group of Data Analysis on Sample spaces with a Manifold Stratification*, http://www.samsi.info/working–groups/data–analysis–sample–spaces–manifold–stratification, June 13, 2011.

[106] V. Patrangenaru and L. A. Ellingson. *Nonparametric statistics on manifolds and their applications to object data analysis.* CRC Press, 2015.

[107] V. Patrangenaru, X. Liu, and S. Sugathadasa. Nonparametric 3D projective shape estimation from pairs of 2D images - i, in memory of W. P. Dayawansa. *Journal of Multivariate Analysis*, 101:11–31, 2010.

[108] V. Patrangenaru and K. V. Mardia. *Proceedings of the Leeds Annual Statistics Research Workshop*, chapter Affine Shape Analysis and Image Analysis, pages 57–62. Leeds University Press, 2003.

[109] K. Pearson. On lines and planes of closest fit to systems of points in space. *Philosophy Magazine*, 2:559–72, 1901.

[110] M. J. Prentice. A distribution-free method of interval estimation for unsigned directional data. *Biometrika*, 71:147–154, 1984.

[111] J. Ramsay and B. W. Silverman. *Applied Functional Data Analysis: Methods and Case Studies*. Springer Verlag, New York, 2002.

[112] J. O. Ramsay and B. W. Silverman. *Functional Data Analysis*. Springer Series in Statistics, 2nd edition, 2005.

[113] N. Saitou and M. Nei. The neighbor-joining method: a new method for reconstructing phylogenetic trees. *Mol Biol Evol.*, 4(4):406–425, 1987.

[114] A. Schwartzman. Lognormal distributions and geometric averages of positive definite matrices. *arXiv.org*, arXiv:1407.6383, 2014.

[115] A. Schwartzman, R. F. Dougherty, and J. E. Taylor. False discovery rate analysis of brain diffusion direction maps. *Ann. Statist.*, 36:153–175, 2008.

[116] A. Schwartzman, W. F. Mascarenhas, and J. E. Taylor. Inference for eigenvalues and eigenvectors of Gaussian symmetric matrices. *Ann. Statist.*, 36:2886–2919, 2008.

[117] K. Schwarze, J. Buchanan, J. M. Fermont, H. Dreau, M. W. Tilley, J. M. Taylor, P. Antoniou, S. J. L. Knight, C. Camps, M. M. Pentony, E. M. Kvikstad, S. Harris, N. Popitsch, A. T. Pagnamenta, A. Schuh, J. C. Taylor, and S. Wordsworth. The complete costs of genome sequencing: a microcosting study in cancer and rare diseases from a single center in the United Kingdom. *Genetics in Medicine*, 22:85–94, 2020.

[118] D. Sharvit, J. Chan, H. Tek, and B. B. Kimia. Symmetry-based indexing of image databases. *Journal of Visual Communication and Image Representation*, pages 366–380, 1998.

[119] C. Shen. *Topological Data Analysis for Medical Imaging and RNA Data Analysis on Tree Spaces*. Ph.D. Thesis, Department of Statistics, Florida State University, 2021.

[120] C. Shen, V. Patrangenaru, and R. Moore. A phylogenetic trees analysis of SARS-CoV-2. *arXiv*, page arXiv:2106.06918, 2021.

[121] D. Shen. *parse PCA Asymptotics and Analysis of Tree Data*. Ph.D. Thesis, Department of Statistics and Operations Research, University of North Carolina at Chapel Hill, North Carolina, 2012.

[122] X. Shi, M. Styner, J. Lieberman, J. G. Ibrahim, W. Lin, and Hongtu Zhu. *Medical Image Computing and Computer-Assisted Intervention—MICCAI 2009*, Intrinsic regression models for manifold-valued data, pages 192–199. Springer, 2009.

[123] F. Sievers, A. Wilm, D. Dineen, T.J. Gibson, K. Karplus, W. Li, R. Lopez, H. McWilliam, M. Remmert, J. Söding, J.D. Thompson J.D., and D.G. Higgins. Fast, scalable generation of high-quality protein multiple sequence alignments using clustal omega. *Mol. Syst. Biol.*, 7:539, DOI: 10.1038/msb.2011.75, 2011.

[124] K. Singh. On the asymptotic accuracy of Efron's bootstrap. *Ann. Statist.*, 9:1187–1195, 1981.

[125] S. Skwerer, E. Bullitt, S. Huckemann, E. Miller, I. Oguz, M. Owen, and V. Patrangenaru. Tree-oriented analysis of brain artery structure. *J. Math. Imaging Vision*, 50:126–143, 2014.

[126] A. Srivastava and E. Klassen. Monte-Carlo extrinsic estimators of manifold-values parameters. *IEEE Trans. Signal Processing*, 20:100–111, 2001.

[127] K-T. Sturm. Probability measures on metric spaces of nonpositive curvature. *Heat kernels and analysis on manifolds, graphs, and metric spaces: lecture notes from a quarter program on heat kernels, random walks, and analysis on manifolds and graphs, Contemporary Mathematics*, 338:357–390, 2003.

[128] Morris Swadesh. Lexico-statistic dating of prehistoric ethnic contacts: With special reference to north american indians and eskimos. *Proceedings of the American Philosophical Society*, 96(4):452–463, 1952.

[129] Morris Swadesh. Towards greater accuracy in lexicostatistic dating. *International Journal of American Linguistics*, 21(2):121–137, 1955.

[130] Morris Swadesh, Dell H. Hymes, and Joel Sherzer. *The origin and diversification of language*. Aldine-Atherton, 1971.

[131] H.D. Tagare, D. Groisser, and O. Skrinjar. Symmetric non-rigid registration: A geometric theory and some numerical techniques. *Journal of Mathematical Imaging and Vision*, 34:61–88, 2009.

[132] K. Tamura, G. Stecher, and S. Kumar. Mega11: Molecular evolutionary genetics analysis version 11. *Molecular Biology and Evolution*, 38:3022–3027, 2021.

[133] M. Tang, J. Cape, and C.E. Priebe. Asymptotically efficient estimators for stochastic blockmodels: The naive MLE, the rank-constrained MLE, and the spectral estimator. *Bernoulli*, 34:1049–1073, 2022.

[134] J.D. Thompson, D.G. Higgins, and T.J. Gibson. Clustal W: improving the sensitivity of progressive multiple sequence alignment through sequence weighting, position-specific gap penalties and weight matrix choice. *Nucleic Acids Research*, 22:4673–4680, 1994.

[135] S. Ulmer, D.T. Van, and S. F. Huckemann. Exploring uniform finite sample stickiness. *Lecture Notes in Comput. Sci., 14071*, Geometric science of information. Part I:349–356, 2023.

[136] Unknown. Category:swadesh modules, 2021. [Online; accessed October 10, 2023].

[137] A. Verona. *Stratified mappings-structure and triangulability*. Springer, 2007.

[138] H. Wang. Measures of agreement for vectorcardiography data. *Stat. Med.*, 28:1093–1107, 2009.

[139] H. Wang and J. S. Marron. Object oriented data analysis: sets of trees. *Ann. Statist.*, 35:1849–1873, 2007.

[140] Y. Wang, J. S. Marron, B. Aydin, A.Ladha, E. Bullitt, and H. Wang. A nonparametric regression model with tree-structured response. *J. Amer. Statist. Assoc.*, 107:1272–1285, 2012.

[141] G. S. Watson. *Statistics on Spheres*. Lecture Notes in the Mathematical Sciences, John Wiley & Sons, 1983.

[142] H. Whitney. The self-intersections of a smooth n-manifold in 2n-space. *Ann. of Math.*, 45:220–246, 1944.

[143] WHO. Preliminary report for the scientific advisory group for the origins of novel pathogens (sago). *https://www.who.int/publications/m/item/scientific-advisory-group-on-the-origins-of-novel-pathogens-report*, 2022.

[144] K. C. Wong. *Extrinsic Principal Component Analysis*. Ph.D Thesis, Department of Statistics, The Florida State University, 2024.

[145] K. C. Wong, V. Patrangenaru, R. L. Paige, and M. Pricop Jeckstadt. Extrinsic principal component analysis. *arXiv:2409.03572v2*, 2024.

[146] F. Wu, S. Zhao, B. Yu, Y.-M. Chen, W. Wang, Z.-G. Song, Y. Hu, Z.-W. Tao, J.-H. Tian, Y.-Y. Pei, M.-L. Yuan, Y.-L. Zhang, F.-H. Dai, Y. Liu, Q.-M. Wang, J.-J. Zheng, Lin Xu, E. C Holmes, and Y.-Z. Zhang. A new coronavirus associated with human respiratory disease in China. *Case Reports Nature*, Mar; 579(7798):265–269, PMID: 32015508 PMCID: PMC7094943. DOI: 10.1038/s41586–020–2008–3. Erratum doi: 10.1038/s41586–020–2202–3, 2020.

[147] K. D. Yao. *Data Analysis on Object Spaces and Applications*. Ph.D. Thesis, Department of Statistics, Florida State University, 2016.

[148] S. Young and E. Scheinerman. Random dot product models for social networks. In *Proceedings of the 5th International Conference on Algorithms and Models for the Web-Graph*, pages 138–149, 2007.

[149] A. M. Zaki, S. van Boheemen, T.M. Bestebroer, A.D. Osterhaus, and R.A. Fouchier. Isolation of a novel coronavirus from a man with pneumonia in Saudi Arabia. *N. Engl. J. Med.*, 367(19):1814–1820, 2012.

[150] H. Ziezold. On expected figures and a strong law of large numbers for random elements in quasi-metric spaces. *Transactions of the Seventh Prague Conference on Information Theory, Statistical Decision Functions, Random Processes and of the Eighth European Meeting of Statisticians*, A:591–602, 1977.

Index

j-nonfocal probability measure, 23
(BHV) phylogenetic tree spaces, 44
3D projective shape space, 155
3D surface recosntruction, 160

a Fréchet function, 13
adapted frame CLT, 26
adapted frame field, 24
adenine, 3, 131
affine shape space, 52
alphabet, 10
asymptotics of extrinsic sample means, 14
asymptotics of sample total intrinsic variance, 14

bicyclic graph, 122
binding sites, 5
Bonferroni procedure, 40
boostrap resamples, 159
bootstrap
  for two independent samples, 39
bootstrap confidence regions, 30
bootstrap extrinsic Hotelling distribution, 29
Bootstrapping the Cartan mean, 22
brain artery, 9

change on a Lie group, 34
chord distance, 51
circle of Willis, 10
CLT for Cartan means, 21
CLT for extrinsic sample means, 17, 23
CLT for Fréchet sample means, 20
CLT on a startified space, 163
cortical correspondence trees, 97
coverage error, 15
CRISPR-Cas9, 5
cytosine, 3, 131

d1, 138
d2, 138
d3, 139
d4, 139
deoxiribonucleic acid, 3
diacritic, 10
diffraction crystallography images, 3
Diffusion Tensor Imaging, 9
digital camera images, 12, 160
directional data analysis, 14
distance matrix, 106

embedding, 14
European languages origin, 137
evolutionary distance, 133
extrinsic covariance, 169
extrinsic covariance matrix, 25, 52, 154, 165
extrinsic Hotelling statistic, 28
extrinsic mean, 51
extrinsic mean set, 51
extrinsic PCA, 166
extrinsic sticky CLT, 53

Fréchet MANOVA, 147

Fréchet mean, 68
Fréchet mean set, 51
Fréchet sample variance, 96

gain of function, 106
gene editing, 5
genetic instructions, 3
genome, 106
genomics, 5
genus of a surface, 8
geodesic PCA, 164
graph, 6
graph Laplacian, 172
guanine, 3, 131

Hamiltonian circuit, 8
homogeneous space, 53
human genome, 5
hypothesis test
  for mean change, 34
  two-sample
    extrinsic means, 31
    intrinsic means, 32
    on a homogeneous space, 38
    total variance, 31
hypothesis testing MANOVA, 148

i.i.d.r.o.s, 14
icosahedron, 8
inductive mean algorithm, 96
intrinsic mean, 68
intrinsic PCA, 164

k-means clustering, 106
Kendall shape space, 44

large sample c.r. for extrinsic mean, 15
Latin alphabet, 10

m-dimensional manifold, 43
Magenetic Resonance Angiography, 9

manifold, 13
mean evolutionary distance, 133
Mega software, 134
MERS-Cov, 106
metric space, 13
metric topology, 51
minimum spanning tree, 99
multidimensional scaling, 101
multigraph, 7

negative curvature test, 100
nested spheres PCA for directional data, 164
networks, 171
nonparametric bootstrap c.r. for extrinsic mean, 15
nonplanar graph, 9
nonsticky mean, 70
nucleotide, 3, 106

object spaces, 3
open book, 44, 60
orthant, 131

parent, 3
percentile method for nonparametric bootstrap c.r., 15
Petersen graph, 46
phylogenetic tree space, 59
polynucleotide chain, 3
principal components, 164
projective shape space, 52
projective transformation, 52
proteins, 5
PT trick, 44, 107

r.o.'s on a Lie group, 34
random graphs, 172
random networks, 172
random object, 13
RDPG model, 171
RdRp, 106
real projective space, 154
ribonucleic acid, 3

Riemannian manifold, 13
Riemannian structure, 13
rooted tree, 131
rootless tree, 131

sample extrinsic covariance matrix, 26, 28
SARS-Cov, 106
SARS-Cov-2, 134
SARS-CoV-2 RNA, 44
SARS-Cov-bat, 106, 134
Semitic alphabet, 11
sequence alignment, 4
similarity shape space, 52
single linkage clustering, 140
singular points, 52
Slutsky type theorems, 60
space of rooted trees, 44
spaces of positive semi-definite matrices, 52
spherical representation, 155
spider, 45
spine of an open book, 61
spread of SARS-Cov-2 mutation, 109
startified object space, 44
startified space, 44
sticky mean, 48
sticky sample mean, 50, 120
stratified space, 43
stratified spaces, 50
stratum, 44
submanifold, 23
surface reconstruction, 12

thymine, 3
tree, 7, 45
two sample means problem, 53

uracil, 131

VW embedding, 155
VW PCA, 170
VW pooled sample mean, 157

AF804224

CIVIL LIABILITY AND FINANCIAL SECURITY FOR OFFSHORE OIL AND GAS ACTIVITIES

Civil Liability and Financial Security for Offshore Oil and Gas Activities provides insights into the liability and compensation regime for offshore-related damage. This book analyzes the legal regime in a number of countries (including the United States and the United Kingdom) as well as the EU regime. In addition, the various compensation mechanisms and amounts available today to compensate offshore-related damage are described and critically analyzed. Moreover, based on in-depth interviews with a wide variety of relevant stakeholders, including insurers, representatives from supervisory authorities, and oil and gas producers, this volume also provides a variety of policy recommendations formulated to provide an optimal compensation regime for offshore-related damage.

MICHAEL FAURE is Professor of Comparative Private Law and Economics at Erasmus University Rotterdam and Professor of Comparative and International Environmental Law at Maastricht University. He serves also as Academic Director of the Maastricht European Institute for Transnational Legal Research (METRO).

CIVIL LIABILITY AND FINANCIAL SECURITY FOR OFFSHORE OIL AND GAS ACTIVITIES

Edited by

MICHAEL FAURE

Universiteit Maastricht, Netherlands

CAMBRIDGE
UNIVERSITY PRESS

CAMBRIDGE
UNIVERSITY PRESS

University Printing House, Cambridge CB2 8BS, United Kingdom

Cambridge University Press is part of the University of Cambridge.

It furthers the University's mission by disseminating knowledge in the pursuit of education, learning, and research at the highest international levels of excellence.

www.cambridge.org
Information on this title: www.cambridge.org/9781107167162

© Michael Faure 2017

This publication is in copyright. Subject to statutory exception and to the provisions of relevant collective licensing agreements, no reproduction of any part may take place without the written permission of Cambridge University Press.

First published 2017

A catalogue record for this publication is available from the British Library.

Library of Congress Cataloging-in-Publication Data
Faure, Michael (Michael G.) editor.
Civil liability and financial security for offshore oil and gas activities / edited by Michael Faure.
New York : Cambridge University Press, 2016.
LCCN 2016023670 | ISBN 9781107167162 (hardback)
LCSH: Liability for oil pollution damages. | Oil pollution of the sea – Law and legislation. | Offshore oil well drilling – Law and legislation. | Offshore gas well drilling – Law and legislation. | Oil and gas leases. | Liability (Law). | Compensation (Law). | BISAC: LAW / Civil Law.
LCC K956 .C58 2016 | DDC 343.07/72–dc23
LC record available at https://lccn.loc.gov/2016023670

ISBN 978-1-107-16716-2 Hardback

Cambridge University Press has no responsibility for the persistence or accuracy of URLs for external or third-party Internet websites referred to in this publication and does not guarantee that any content on such websites is, or will remain, accurate or appropriate.

SCANNED

CONTENTS

List of Figures and Tables *page* xvi
List of Contributors xviii
Preface and Acknowledgements xix
List of Abbreviations xxi

1 Introduction 1
MICHAEL FAURE, NIELS PHILIPSEN AND HUI WANG

1.1 Background for This Book 1

1.2 Objective of This Book 5

1.3 Methodology 5
1.3.1 Legal Analysis 6
1.3.2 Law and Economics 8
1.3.3 Empirics 9
1.3.4 Integration 10

1.4 Structure 10

2 Offshore-Related Damage: Facts and Figures 12
KRISTEL DE SMEDT AND HUI WANG

2.1 General Issues Concerning Offshore Activities 12
2.1.1 Terminology in Offshore Activities 12
2.1.2 Industry Structure of Offshore Activities 13
2.1.3 Stakeholders 15
2.1.4 Importance of Offshore Activities 18

2.2 Location of Installations and Facilities in the European Union 18
2.2.1 General Information 18
2.2.2 Offshore Interests in the North Sea 20
2.2.3 Offshore Interests in the Mediterranean Sea 21

2.3 Overview of Major Offshore Accidents 22
2.3.1 Sources of Information 22
2.3.1.1 Worldwide Database 22
2.3.1.2 National Regulatory Authorities 24
2.3.1.3 Regional Database 31
2.3.1.4 Industry Database 33
2.3.1.5 Observation from Available Data 36
2.3.2 Overview of Serious Accidents Involving Offshore Facilities 37

2.4 Analysis of Recent Offshore Accidents in Europe: Case Studies 41
2.4.1 Gulfaks C Incident (19 May 2010): A Near Miss 41
2.4.2 Valhall PCP Production Platform Fire (13 July 2011) 41
2.4.3 Escape of Hydrocarbons at Ula Field Norway (12 September 2012) 42
2.4.4 Gannet Alpha Oil Spill (10 August 2011) 43
2.4.5 Elgin Platform Blowout (26 March 2012) 44

2.5 Analysis of Recent Offshore Accidents Outside EU Waters 45
2.5.1 Montara 45
2.5.1.1 Facts of the Incident 45
2.5.1.2 Trans-Boundary Complaints from Indonesia 48
2.5.2 Deepwater Horizon 49
2.5.2.1 Facts of the Incident and Legal Issues 49
2.5.2.2 Overview of Legal Proceedings 53
2.5.2.3 Post-Macondo Responses 56
2.5.2.4 Summary 57
2.5.3 More Accidents in 2015 58
2.5.3.1 Petrobras Brazil Offshore Explosion 58
2.5.3.2 Pemex Mexico Offshore Accidents 59
2.5.4 Risk of a Deepwater Horizon Oil Spill in the European Union? 59
2.5.4.1 Different Depths and Pressures 59
2.5.4.2 Concerns after Macondo 60
2.5.4.3 Expert Evaluation 61
2.5.4.4 Limits of Expert Opinion 64
2.5.4.5 Summary 66

3 Analysis of Existing Legal Regimes 68
MICHAEL FAURE, JING LIU AND HUI WANG

3.1 Introduction 68

3.2 International Legal Framework 69
3.2.1 The Civil Liability Convention and the Fund Convention 70
3.2.1.1 The Civil Liability Convention 71
3.2.1.2 Compensation Funds in the International Regime 77
3.2.1.3 Voluntary Mechanisms 79
3.2.2 UNCLOS 81
3.2.3 MARPOL 73/78 83
3.2.4 The OPRC Convention 84
3.2.5 Ongoing Discussion at the IMO 85

3.3 Offshore Liability Regimes 86
3.3.1 Regional Arrangements 86
3.3.2 North-East Atlantic Ocean (North Sea) 87
3.3.2.1 The OSPAR Convention 87
3.3.2.2 The Nordic Convention 88
3.3.3 The Mediterranean Sea 88
3.3.4 The Baltic Sea 90
3.3.5 The Black Sea 91
3.3.6 Summary of the Regional Seas Arrangements 91

3.4 Current EU Regime on Offshore Activities 91
3.4.1 EU Interest in Offshore Activities 91
3.4.2 Council Directive 92/91 EEC 92
3.4.3 Hydrocarbons Licensing Directive 1994 (94/22/EC) 92
3.4.4 The Marine Strategy Framework Directive 93
3.4.5 Communication of 2010 93
3.4.6 Directive on Safety of Offshore Oil and Gas Operations 94

3.5 Offshore Liability Regimes: Country Studies 97
3.5.1 Introduction 97
3.5.2 United Kingdom 98
3.5.2.1 National Interest in Offshore Activities 98
3.5.2.2 Legal Framework 98
3.5.2.3 Basis of Liability 103

3.5.2.4 Relationship with Regulation 104
3.5.2.5 Causation 104
3.5.2.6 Attribution of Liability 105
3.5.2.7 Damages and Remedies 105
3.5.2.8 Amount of Compensation 107
3.5.2.9 Claim Settlement 108
3.5.2.10 Compensation Mechanisms 109
3.5.3 Norway 115
3.5.3.1 National Interest in Offshore Activities 115
3.5.3.2 Legal Framework 116
3.5.3.3 Basis of Liability 119
3.5.3.4 Relationship with Regulation 120
3.5.3.5 Causation 120
3.5.3.6 Attribution of Liability 121
3.5.3.7 Damages and Remedies 122
3.5.3.8 Amount of Compensation 123
3.5.3.9 Claim Settlement 123
3.5.3.10 Compensation Mechanisms 124
3.5.4 Denmark 126
3.5.4.1 National Interest in Offshore Activities 126
3.5.4.2 Legal Framework 126
3.5.4.3 Basis of Liability 128
3.5.4.4 *Relationship* with Regulation 129
3.5.4.5 Attribution of Liability 129
3.5.4.6 Damages and Remedies 129
3.5.4.7 Compensation Mechanisms 130
3.5.5 United States 131
3.5.5.1 National Interest in Offshore Activities 131
3.5.5.2 Legal Framework 131
3.5.5.3 Basis of Liability 136
3.5.5.4 Relationship with Regulation 137
3.5.5.5 Causation 137
3.5.5.6 Attribution of Liability 138
3.5.5.7 Damages and Remedies 139
3.5.5.8 Amount of Compensation 141
3.5.5.9 Claim Settlement 144
3.5.5.10 Compensation Mechanisms 146
3.5.5.11 Jurisdictional Issues 153
3.5.5.12 Criminal Liability 153

3.5.6 Australia 154
3.5.6.1 National Interest in Offshore Activities 154
3.5.6.2 Legal Framework 155
3.5.6.3 Basis of Liability 158
3.5.6.4 Relationship with Regulation 160
3.5.6.5 Causation 160
3.5.6.6 Attribution of Liability 160
3.5.6.7 Damages and Remedies 161
3.5.6.8 Amount of Compensation 161
3.5.6.9 Compensation Mechanisms 161
3.5.7 Canada 163
3.5.7.1 National Interest in Offshore Activities 163
3.5.7.2 Legal Framework 163
3.5.7.3 Basis of Liability 164
3.5.7.4 Attribution of Liability 164
3.5.7.5 Damages and Remedies 164
3.5.7.6 Amount of Compensation 165
3.5.7.7 Applicability in Time 165
3.5.7.8 Compensation Mechanisms 166
3.5.8 Comparative Analysis 166
3.5.8.1 Best Practice? 166
3.5.8.2 Comparison of Legislation 167
3.5.8.3 Comparative Comments 167

3.6 Another High-Risk Sector: Nuclear 171
3.6.1 International Compensation System for Nuclear Damage 172
3.6.1.1 Origin of the International Regime 172
3.6.1.2 First-Generation Nuclear Liability Conventions 173
3.6.1.3 Second-Generation Nuclear Liability Conventions 178
3.6.2 The Compensation System for Nuclear Damage in the United States 182
3.6.2.1 Liability for Nuclear Damage 183
3.6.2.2 The Requirement of Financial Protection 187
3.6.2.3 Compensation under the Convention on Supplementary Compensation 190
3.6.3 Critical Comparison 191

3.7 Concluding Observations 195

4 Pooling Mechanisms for Offshore Liability 197
MICHAEL FAURE AND JING LIU

4.1 Theory: Pooling versus Insurance 198

4.2 OPOL 201
4.2.1 Origins of OPOL 201
4.2.2 Main Features 202
4.2.3 Rules and Membership 203
4.2.4 Financial Responsibility 204
4.2.5 Claims Handling 205
4.2.6 Enforcement 206
4.2.7 Practice 207
4.2.8 Evaluation 208

4.3 OIL and OCIL 210
4.3.1 OIL 211
4.3.2 OCIL 213

4.4 Protection and Indemnity Clubs 214
4.4.1 Origins 214
4.4.2 Coverage 214
4.4.3 Entry and Premiums 215
4.4.4 Available Amount 217

4.5 Risk Pooling in the Nuclear Sector 217
4.5.1 Risk Pooling in the Price-Anderson Act 218
4.5.1.1 A Second Tier of Compensation 218
4.5.1.2 ... Collectively Financed Through Retrospective Premiums 219
4.5.2 The Mutual Pool for Property Damage in the United States 220
4.5.3 Mutual Nuclear Pools in Europe 221
4.5.4 Nuclear Liability and Pooling System in Germany 223
4.5.5 Towards a European Pool for Nuclear Liability? 227

4.6 Concluding Observations 229
4.6.1 Advantages of Pooling 229
4.6.2 OPOL 232
4.6.3 P&I Clubs 232
4.6.4 Risk Sharing for Vessel-Based Pollution 233
4.6.5 Summary 234

5 The Use of Financial Market Instruments to Cover Liability Following a Major Offshore Accident 236
MICHAEL FAURE AND HUI WANG

5.1 Self-Insurance 237
5.1.1 Theory 237
5.1.2 Practice 238
5.1.3 Analysis 240

5.2 Capital Market 242
5.2.1 Theory 242
5.2.2 Practice 243
5.2.3 Analysis 244

5.3 Guarantees 244
5.3.1 Theory 244
5.3.2 Practice 245
5.3.3 Analysis 245

5.4 (Re-)Insurance 246
5.4.1 Theory 246
5.4.2 Practice 248
5.4.2.1 Stakeholders in the Insurance Market (Major Insurance/Re-Insurance Providers) 248
5.4.2.2 Structure of Coverage of Offshore Insurance 250
5.4.2.3 Available Insurance Amount 252
5.4.2.4 Cost Estimation 255
5.4.2.5 Calculation of Premiums 257
5.4.2.6 Influence of the Deepwater Horizon Incident on the Insurance Industry 258
5.4.3 Analysis 260

5.5 Risk-Pooling Schemes 261

5.6 OPOL 263

5.7 Combinations 264

6 Potential of Financial and Insurance Instruments to Cover Liability Following a Major Offshore Accident 266
MICHAEL FAURE AND HUI WANG

6.1 Potential Costs of an Offshore Incident 267

6.1.1 Potential Costs and Verifications 268
6.1.1.1 Safety Regulation 269
6.1.1.2 Technical Differences in Different Waters 269
6.1.1.3 Differences in Operators 271
6.1.1.4 Different Liability Standards 271
6.1.2 Insolvency Risk and Different Scenarios 273

6.2 Self-Guarantee Through Tax 276

6.3 Expanding Possibility of Insurance? 277
6.3.1 Mandatory Insurance and Increasing Coverage? 277
6.3.2 Proposal of Munich Re-Insurance 280
6.3.2.1 An SOS Cover 280
6.3.2.2 Stakeholders Assessment 283
6.3.2.3 Reaction by Munich Re 284
6.3.2.4 Analysis 285
6.3.3 Cover for Vessel-Based Pollution 286
6.3.4 Developments in the Coverage of Nuclear Risk 287
6.3.5 Summary 289

6.4 Potential of Expanding Risk-Pooling Schemes 289
6.4.1 Expanding OIL and OCIL 290
6.4.2 Expanding OPOL? 291
6.4.3 Noble Energy Proposal 293
6.4.3.1 Background 293
6.4.3.2 Contents of the Proposal 294
6.4.3.3 Implementation? 296
6.4.3.4 Reactions from Stakeholders 297
6.4.3.5 Evaluation 298
6.4.4 Challenges 299

6.5 Summary: Flexibility 301

7 Towards Optimal Liability and Compensation for Offshore Oil and Gas Activities 303
KRISTEL DE SMEDT, HUI WANG AND MICHAEL FAURE

7.1 Principles of Efficient Compensation and Liability Rules 304
7.1.1 Why Compensate? 305
7.1.2 Principles of Efficient and Fair Compensation 306

7.2 Efficient Liability Rules 308

7.2.1 Strict liability or Negligence? 308
7.2.1.1 Economic Theory 308
7.2.1.2 Application to Offshore-Related Risks 312
7.2.2 Attribution of Liability 313
7.2.2.1 Channelling of Liability 313
7.2.2.2 Joint and Several Liability 316
7.2.3 Financial Cap? 317

7.3 Liability versus Regulation 319
7.3.1 Criteria for Safety Regulation 319
7.3.1.1 Information Asymmetry as a Criterion for Regulatory Intervention 320
7.3.1.2 Insolvency Risk 320
7.3.1.3 The Threat of a Liability Suit 321
7.3.2 The Need to Regulate Offshore-Related Risks 322
7.3.3 Private or Public Regulation? 324
7.3.4 Liability and Regulation Combined 326
7.3.5 Stakeholder Assessment 328
7.3.6 Policy Conclusions 330

7.4 Mandatory Financial Security 331
7.4.1 Criteria for Mandatory Financial Security 331
7.4.2 Application to Offshore-Related Risks 332
7.4.3 Stakeholder Assessment in the European Union 333
7.4.4 (European) Policy Issues 335
7.4.5 Summary 338

7.5 Compensation Instrument 339
7.5.1 Self-Guarantee Through Taxes 339
7.5.2 Mandatory Insurance and the Munich Re Proposal 340
7.5.3 Risk-Sharing Pools 341
7.5.4 A Role for Government in Providing Compensation? 343
7.5.4.1 Direct Compensation by Government? 343
7.5.4.2 Arguments in Favour 343
7.5.4.3 Arguments Against 344
7.5.4.4 Summary 345
7.5.5 Re-Insurer of Last Resort? 346
7.5.5.1 Arguments in Favour 346
7.5.5.2 Arguments Against 347
7.5.5.3 Summary 348

7.5.6 A Compensation Fund? 349
7.5.6.1 Funds versus Insurance 349
7.5.6.2 Economic Principles to Shape a Fund 351
7.5.6.3 Experiences with Environmental Funds 352
7.5.6.4 A Fund for Offshore-Related Risks? Stakeholder Opinions 356

7.6 Rapid Claims Management 359
7.6.1 Existing Schemes 359
7.6.1.1 OPOL 360
7.6.1.2 Norway 360
7.6.1.3 The Civil Liability and Fund Conventions 361
7.6.1.4 Claims Settlement in the Deepwater Horizon Case 363
7.6.2 Normative Analysis 369
7.6.2.1 Need for a Mechanism 369
7.6.2.2 Compensation Funds 371
7.6.2.3 Stimulate Rapid Payment within Liability Law 372
7.6.2.4 Combinations 372
7.6.2.5 Evaluation 374

7.7 Trans-Boundary Harm 376

7.8 Scenarios 379
7.8.1 Scenario 1: Damage Maximum US$250 Million 379
7.8.2 Scenario 2: Damage between US$250 Million and US$750 Million 380
7.8.3 Scenario 3: Damage above US$750 Million 381

8 Concluding Remarks 383
MICHAEL FAURE, NIELS PHILIPSEN AND HUI WANG

8.1 General 383

8.2 Efficient Liability Rules 384

8.3 Mandatory Financial Security 385

8.4 A Role for Government? 386

8.5 Rapid Claims Mechanism 387

References 388
Appendix 1: Overview of Interviews with Stakeholders 406
Appendix 2: Checklist for Country Studies 408
Appendix 3: Oil Pollution Act Liability Limits, 2012 412
Index 431

FIGURES AND TABLES

Figures

1 Oil companies with exploration/production licences in Europe (operators) *page* 17
2 Distribution of accidents according to type for accidents in the WOAD 23
3 Frequency/cumulative frequency of damage costs 24
4 Consequences according to WOAD 24
5 Damage costs according to WOAD 25
6 Accident category according to WOAD 25
7 Injuries by severity, 2004–2014 27
8 Major/specified injuries 2004–2014 28
9 Total amount of oil spilled from installations (United Kingdom) 29
10 Personal injury accident frequency in the exploration and production industry 32
11 Serious and fatal personal injury accidents in the exploration and production industry 32
12 Fatalities from offshore oil and gas operations 35
13 Global offshore energy premiums by market 250
14 Proposed offshore liability towers – with "no caps" 296

Tables

1 Overview of the Ten Most Expensive Operators' Extra Expense (OEE) Losses in History 36
2 High-Profile Oil Spills from Offshore Blowouts 38
3 Upstream Losses in Excess of US$50 million, 2013 39
4 Upstream Losses in Excess of US$50 million, 2014 40
5 Compensation for Pollution Damage under the International Regime 80
6 Comparison of Liability Limits under OPA 90 and the Coast Guard and Maritime Transportation Act of 2006 142

7 Financial Responsibility for Offshore Facilities Located Partially or Wholly on the Outer Continental Shelf 149
8 Financial Responsibility for Offshore Facilities Not Located on the Outer Continental Shelf 149
9 Well Blowouts That Have Occurred in Offshore Australia Since 1965 155
10 Comparison of Legislation 168
11 The Different Coverage Caps Before and After the Conventions' 2004 Amendments (in millions of euros) 181
12 Average Per-Unit Marine Oil Spill Clean-Up Costs by Nation (in 1999 US$) 256
13 Payments 366

CONTRIBUTORS

KRISTEL DE SMEDT
Maastricht University, Netherlands

MICHAEL FAURE
Maastricht University and Erasmus School of Law Rotterdam, Netherlands

JING LIU
Wuhan University, China, and Erasmus School of Law Rotterdam, Netherlands

NIELS PHILIPSEN
Maastricht University and Erasmus School of Law Rotterdam, Netherlands

HUI WANG
Maastricht University, Netherlands

PREFACE AND ACKNOWLEDGEMENTS

On April 20, 2010, the world was shocked when an accident occurred at the mobile deep-water offshore rig Deepwater Horizon in the Gulf of Mexico, leading to an unprecedented spill that continued for several months with a total loss of several million barrels of oil. Immediately after the incident, the obvious question was asked whether the operator (BP) could be held liable to compensate for the damage that resulted from the incident. In the immediate aftermath of the disaster, steps were taken to force BP to make funds available for compensation of the victims in the light of its obligations under the US Oil Pollution Act. However, in other jurisdictions the question also arose as to how liability and compensation for the consequences of such an incident would be regulated. It appeared that whereas oil spills resulting from tankers had been largely regulated, to some extent to the surprise of many, there was no international convention with respect to civil liability and compensation for damage resulting from offshore oil and gas activities. This also led to great concern within the European Union and resulted, *inter alia*, in legislative action by the European Commission leading to a Directive (2013/30/EU of June 12, 2013) on the safety of offshore oil and gas operations. At the same time, the European Commission also commissioned an in-depth study into the liability and compensation for damage resulting from offshore oil and gas activities, where these activities occur in European waters. This study started in the fall of 2012, and the results were delivered in October 2013. Subsequently, the team of writers has substantially revised and updated the policy report into an academic study, of which this book is the result.

The aim of this book is not only to provide insights, *inter alia*, by focusing on past incidents, on the risks related to offshore incidents but also to provide an in-depth analysis of existing legal regimes at both the international and EU levels, as well as by analyzing various domestic legal regimes. This book equally analyzes how currently (through a variety of financial and insurance instruments) the liability for damage resulting from a major offshore accident is covered. In addition to describing the status quo, this book also analyzes, taking both an empirical and a theoretical (law and economics) approach, how the compensation of damage following a major offshore accident could be arranged. The analysis in this book focuses on the question of how (using a variety of compensation mechanisms in combination), on the one hand, a high amount of compensation could be provided and, on the other hand, how various compensation mechanisms could equally provide effective incentives to prevent major offshore accidents.

Because the questions addressed in this book go beyond a simple legal analysis, the team of authors constituted lawyers (Faure, Liu, and Wang) and economists (Philipsen and De Smedt). Because our aim was not only to provide an academic study but also to obtain detailed insights into the scope of offshore-related risks and the compensation mechanisms available, more than twenty interviews were held with relevant stakeholders (listed in Appendix 1). We are grateful to the stakeholders we interviewed for the valuable information they provided, which substantially contributed to the quality of the final result. During the earlier phase of the collaboration with the European Commission, we benefited from interesting insights and comments provided by collaborators at the Commission. We are grateful to all of them and, in particular, to Eero Ailio, Arthur van Daelen, Xavier Goulay, Jorg Koehli, and Taf Powell.

Obviously, the usual disclaimer applies, implying that the contents of this book obviously do not bind the European Commission in any way and that all errors are ours alone.

We are equally grateful to Marjo Mullers and Marina Jodogne (METRO, Maastricht University) for editorial assistance, as well as to the team at Cambridge University Press for a pleasant collaboration in the preparation of this manuscript for publication.

We truly hope that with this book we can provide a modest contribution to improving the safety of offshore oil and gas activities and to an adequate compensation of victims.

ABBREVIATIONS

ACV	actual cash value
AEC	Atomic Energy Commission (US)
AIPN	Association of International Petroleum Negotiators
AMSA	Australian Maritime Safety Authority
ANI	American Nuclear Insurers
ANP	National Petroleum Agency (Brazil)
AtG	Atomgesetz
AUD	Australian dollar
Barcelona Convention	Convention for the Protection of the Marine Environment and the Coastal Region of the Mediterranean
BAT	best available techniques
bio.	billion
BOE	barrels of oil equivalent
BOEM	Bureau of Ocean Energy Management
BOEMRE	Bureau of Ocean Energy Management, Regulation and Enforcement
BOP	blowout preventer
BP	British Petroleum
BSEE	Bureau of Safety and Environmental Enforcement (US)
Bucharest Convention	Convention for Cooperation in the Protection of the Black Sea Against Pollution 1992
CBT	contingent business interruption
CCC	care, custody, or control
CERCLA	Comprehensive Environmental Response, Compensation, and Liability Act (US)
CESE	Conseil Économique, Social et Environnemental
CFR	Code of Federal Regulations (US)
CGAA	Coast Guard Authorization Act (US)
CGMTA	Coast Guard and Maritime Transportation Act (US)
CLC	International Convention on Civil Liability for Oil Pollution Damage

CLEE	The Convention of Civil Liability for Oil Pollution Damage Resulting from Exploration for and the Exploitation of Seabed Mineral Resources
COFR	certificate of financial responsibility
COPE	Compensation for Oil Pollution in European Waters
CRISTAL	Contract Regarding an Interim Supplement to Tanker Liability for Oil Pollution
CSC	Convention on Supplementary Compensation for Nuclear Damage
CSSP	court-supervised settlement program
CWA	Clean Water Act (US)
DEA	Danish Energy Agency
DECC	Department of Energy and Climate Change (UK)
DEFRA	Department for Environment, Food and Rural Affairs (UK)
DEM	Deutsche mark
DEPA	Danish Environmental Protection Agency
DMITRE	Department of Manufacturing, Innovation, Trade, Resources and Energy
DNV	Det Norske Veritas
DOE	Department of Energy (US)
DOI	Department of the Interior (US)
DOJ	Department of Justice (US)
DUC	Dansk Undergrunds Consortium
EED	Energy Exploration and Development (Insurance)
EEZ	exclusive economic zone
EIA	Energy Information Administration (US)
ELD	Environmental Liability Directive
ELINI	European Liability Insurance for the Nuclear Industry
EMANI	European Mutual Association for Nuclear Insurance
EMSA	European Maritime Safety Agency
ENI	Italian Hydrocarbons Agency
ENO	extraordinary nuclear occurrence
EPA	Environmental Protection Agency (US)
EU	European Union
EUR	euro
FERC	Federal Energy and Regulatory Commission (US)
FR	Financial Responsibility Form
Fund Convention	International Convention on the Establishment of an International Fund for Compensation for Oil Pollution Damage
GCCF	Gulf Coast Claims Facility (US)

GT	tonnage of ship
HCRs	hydrocarbon releases
Helsinki Convention	Convention for the Protection of the Marine Environment of the Baltic Sea Area 1992
HP	high-pressure wells
HSE	Health and Safety Executive (UK)
HT	high-temperature wells
IADC	International Association of Drilling Contractors
IAEA	International Atomic Energy Agency
IAT	incident analysis team
IEA	International Energy Agency
IMO	International Maritime Organization
IMR	Institute of Maritime Research
INPO	Institute of Nuclear Power Operations
INTERTANKO	International Association of Independent Tanker Owners
IOPC Fund	International Oil Pollution Compensation Fund
IRF	International Regulators' Forum
IRMI	International Risk Management Institute
ITRE	European Parliament's Committee on Energy
IUMI	International Union of Marine Insurance
J&S	joint and several liability
JNCC	Joint Nature Conservation Committee
JOA	joint operating agreement
JRC	Joint Research Centre
Klif	Climate and Pollution Agency (Norway)
KPI	key performance indicator
LNG	liquefied natural gas
LOGIC	Leading Oil Government's Oil and Gas Industry Task Force
LTA	lost-time accident
MAERP	Mutual Atomic Energy Re-Insurance Pool
MARPOL 73/783	1973 International Convention for the Prevention of Pollution from Ships as Amended by the Protocol 1978
MCA	Maritime and Coastguard Agency (UK)
MDL	multidistrict litigation
MEP	Member of the European Parliament
MIA	Mutual Indemnity Agreement
mio.	million
MIT	Massachusetts Institute of Technology
MMS	Minerals Management Service (US)
MODU	mobile offshore drilling unit
MOEX	Mitsui Oil Exploration Company

MOPU	mobile offshore production units
MOU	memorandum of understanding
MPE	Ministry of Petroleum and Energy (Norway)
MRF	mutual response fund
NatCats	natural catastrophes
NCA	Norwegian Coastal Administration
NEA	Nuclear Energy Agency
NEB	National Energy Board (Canada)
NEIL	Nuclear Electric Insurance Limited
NELIA	Nuclear Energy Liability Insurance Association
NFWF	National Fish and Wildlife Foundation
NGO	nongovernmental organization
NINA	Norwegian Institute for Nature Research
NM	nautical mile(s)
NOAA	National Oceanic and Atmospheric Administration (US)
NOGEPA	Netherlands Oil and Gas Exploration and Production Association
NOK	kroner (Norway)
NOPSEMA	National Offshore Petroleum Safety and Environmental Management Authority (Australia)
Nordic Convention	Nordic Environmental Protection Convention
NPCA	Norwegian Pollution Control Authority
NPD	Norwegian Petroleum Directorate
NPDES	National Pollutant Discharge Elimination System (US)
NPFC	National Pollution Fund Centre (US)
NRC	Nuclear Regulatory Commission
NSOAF	North Sea Offshore Authorities Forum
OCES	Operators' Co-operative Emergency Services
OCIL	Oil Casualty Insurance Limited
OCS	outer continental shelf
OCSLA	Outer Continental Shelf Lands Act
OECD	Organization for Economic Cooperation and Development
OEE	operators extra expense (insurance policy)
Offshore Protocol	Protocol for the Protection of the Mediterranean Sea Against Pollution Resulting from Exploration and Exploitation of the Continental Shelf and the Seabed and Its Subsoil, 1994
OGP	International Association of Oil and Gas Producers
OIL	Oil Insurance Limited
OOG	Overlegorgaan Olie en Gas (Netherlands)
OPA	Oil Pollution Act 1990 (US)

OPEC	Organization of the Petroleum Exporting Countries
OPEP	Oil Pollution Emergency Plan (UK)
OPGGSSA	Offshore Petroleum and Greenhouse Gas Storage Act (Australia)
OPOL	Offshore Pollution Liability Agreement
OPRC Convention	International Convention on Oil Pollution Preparedness, Response, and Cooperation 1990
OSCAR	Oil Spill Contingency and Response
OSFR	Oil Spill Financial Responsibility for Offshore Facilities
OSIS	Oil Spill Information System
OSLTF	Oil Spill Liability Trust Fund (US)
OSPAR Convention	Convention for the Protection of the Marine Environment of the North-East Atlantic of 1992
OSPRAG	Oil Spill Prevention and Response Advisory Group
PAA	Price-Anderson Act (US)
P&I Clubs	Protection and Indemnity Clubs
PLA	public liability action
PSA	Petroleum Safety Authority (Norway)
PTTEP	PTT exploration and production
PTTEP AA	PTTEP Australasia
RCV	replacement-cost value
REG	regulation
RWC	restricted work cases
RWS NZ	Rijkswaterstaat Noordzee (Netherlands)
SDFI	State's Direct Financial Interest (Norway)
SDR	Special Drawing Rights
SEC	Securities and Exchange Commission
SEMS	Safety and Environmental Management System
SFS 1968:45	Swedish Nuclear Liability Act
SIR	self-insured retention
SL	strict liability
SLA	Submerged Lands Act 1953
SOS	sudden oil spill
SPV	special-purpose vehicle
SSM	State Supervision of Mines (Netherlands)
STOPIA	Small Tanker Oil Pollution Indemnification Agreement
TOPIA	Tanker Oil Pollution Indemnification Agreement
TOVALOP	Tanker Owners Voluntary Agreement Concerning Liability for Oil Pollution
UK	United Kingdom
UN	United Nations
UNCLOS	United Nations Convention on the Law of the Sea

UNEP	United Nations Environment Program
US	United States
US$	US dollars
VoOs	vessels of opportunity
WANO	World Association of Nuclear Operators
WELD	Willis Energy-Loss Database
WOAD	Worldwide Offshore Accident Databank

Introduction

MICHAEL FAURE, NIELS PHILIPSEN AND HUI WANG

1.1 Background for This Book

The background for this research on civil liability and financial security for offshore oil and gas activities is the explosion of the mobile deepwater offshore rig Deepwater Horizon on 20 April 2010, in the Gulf of Mexico that spilled 3.19 million barrels of oil in the sea as a result.[1] Luckily, at the place where the Deepwater Horizon incident occurred, US law applies, in this particular case the US Oil Pollution Act of 1990 (OPA 90). OPA 90 does have a liability regime for offshore facilities. However, at the time, the international community realized that the international regime for oil spills focused largely on vessel-source pollution. Famous incidents with e.g. the *Torrey Canyon* (1976), *Amoco Cadiz* (1978), *Exxon Valdez* (1989) and *Erika* (1999) led to the development of an impressive international liability regime.[2] Indeed, at the international level, a compensation regime for vessel-source oil pollution was already established in 1969–1971 by the adoption of two international conventions, the International Convention on Civil Liability for Oil Pollution Damage of 1969 (also called the CLC of 1969) and the International Convention on the Establishment of an International Fund for Compensation for Oil

[1] The original estimate was that around 4.9 million barrels of oil was released to sea, with estimated damages ranging between US$1 billion and US$3.5 billion. Hearing, House of Representatives of the US, Committee on Transportation and Infrastructure, 8 June 2010, 15–16. Later cost estimates were increased and amounted to US$30 billion. It was considered the largest marine oil spill in American history. However, the court ruled in January 2015 that 4.0 million barrels of oil was released to the sea, and because 810,000 barrels of oil was recovered through the cleanup actions, the final amount was determined to be 3.19 million barrels discharged into the Gulf of Mexico. See judgment of the US District Court for the Eastern District of Louisiana, filed on 15 January 2015: *In re: Oil Spill by the Oil Rig 'Deepwater Horizon' in the Gulf of Mexico, on April 20, 2010.*

[2] For a discussion of this international liability regime, see, *inter alia*, Liu, *Compensating Ecological Damage*, pp. 195–207; Verheij, 'Shifts in governance'; and Wang, *Civil Liability*, pp. 53–187.

Pollution Damage of 1971 (the Fund Convention of 1971).[3] These conventions went through many evolutions (in particular, the 1992 Protocols and the 2000 Resolutions) as a result of which, most important, the amounts were increased after every incident that had again challenged the financial limits on the liability of the tanker owner. In principle, the European Union relied on its Member States to ratify various international maritime conventions, but given its dissatisfaction with the measures taken at the international level through the International Maritime Organization (IMO), the European Commission also started to take its own initiatives for legislation at the European level.[4] The European Commission subsequently adopted the so-called Erika I and Erika II packages in which it, *inter alia*, proposed to set up a European fund (referred to as the *Cope Fund*) with an updated ceiling of €1 billion (instead of the €200 million that was then applicable under international conventions).[5] Interestingly, this European activism led the IMO to increase the limits of the 1992 CLC and the Fund Convention by 50%, with effect from November 2003. This led to adoption of a supplementary fund for oil pollution damage, leading to a total amount of compensation (again, only in the case of vessel-source pollution) of 750 million special drawing rights (SDRs), which at the time of adoption corresponded to approximately US$ 1 billion. Hence, one could note that European activism led to action at the international level whereby the IMO basically took over the initiative proposed by the European Commission; this obviously made the European initiative in that domain no longer necessary.[6]

Although EU activism thus led to a widely satisfying liability and compensation regime in the case of vessel-source pollution, the incident with the Deepwater Horizon[7] again led to a shock and the realization that a huge amount of damage also could be caused by offshore facilities for which the liability and financial security largely was left to Member State law. More particularly, given the often trans-boundary character of spills and accidents taking place from offshore facilities, there was a strong

[3] The Civil Liability and Fund Conventions will be discussed in further detail in Chapter 3.

[4] See in this respect more particularly the publication on 24 February 1993 of the long-awaited Communication on Safe Seas, COM(1993) 66 final.

[5] See the amended proposal for a regulation of the European Parliament and of the council on the establishment of a fund for compensation of oil pollution damage in European waters and related measures, *Official Journal* C227 E/487 of 24 September 2002.

[6] For an outline of these developments, see Wang, 'Shifts in governments'.

[7] For a detailed analysis of the Deepwater Horizon case, see also Perry, 'The Deepwater Horizon oil spill'.

argument for transnational regulation of civil liability and financial security, at least for EU action and (perhaps following the example of vessel-source pollution) eventually for IMO action as well.

There are also, as far as Europe is concerned, a number of reasons for concern about damage which may be caused by offshore incidents. As this chapter will show, there is in fact, internationally, quite an impressive record of offshore incidents. In addition, the North Sea has experienced quite a few offshore platform accidents, including Alexander Kielland (1980), Piper Alpha (1988), Forties Alpha (2003), Gullfaks C (2010) and most recently Gannet Alpha (2011), all incidents that occurred in the North Sea and have increased concern about the consequences in Europe. Although until very recently (June 2013) there has been no formal European regime dealing specifically with offshore pollution issues, there has been relevant legislation that plays an important role, namely, the Marine Strategic Framework Directive.[8] This Directive 'requires addressing the cumulative impacts from all activities on the marine environment' and 'is relevant to offshore oil and gas operations as it requires linking the particular concerns from each economic sector with the general aim of a comprehensive understanding of the oceans, seas and coastal areas, with the objective to develop a coherent approach to the seas taking into account all economic, environmental and social aspects through the use of Maritime spatial planning and Marine knowledge'.[9]

When the Deepwater Horizon accident occurred in the Gulf of Mexico on 20 April 2010, the European Commission immediately launched a communication to assess the risks in the offshore oil and gas industry in European waters.[10] The Commission has explored a wide range of problems, including licensing, controls by public authorities and spill response. In particular, when it addressed the liability issue, the Commission considered the possibility of extending the Environmental Liability Directive (ELD)[11] to cover environmental damage to all marine waters as defined in the Marine Strategy Framework Directive[12] and the applicability of the Waste Framework Directive.[13] The Commission also realized that the possibility of a financial cap on liability and mandatory

[8] Directive 2008/56/EC. [9] COM(2011) 688 final, 14.

[10] COM(2010) 560 final, Communication from the Commission to the European Parliament and the Council Facing the Challenge of the Safety of Offshore Oil and Gas Activities, 12 October 2010.

[11] Directive 2004/35/EC. [12] Directive 2008/56/EC. [13] COM(2010) 560 final, 8.

financial security was worth further analysis.[14] The Commission was concerned not only with offshore activities in EU waters but also with the EU-based offshore industry operating in other parts of the world, and it called on the industry's international operators to be responsible.[15]

On 27 October 2011, the Commission initiated two proposals, one for a regulation on the safety of offshore activities[16] and the other for the accession of the EU to the Offshore Protocol of the Barcelona Convention.[17] The proposed regulation (COM[2011] 688 final) follows on the principal issues raised in the 2010 Communication and imposes stricter safety standards for offshore activities in Europe and gives national regulators more power to inspect such operations. It also extends sixteen-fold the zone in which companies will be held liable for environmental damage. It specifies that the licensee shall be held 'liable for the prevention and remediation of environmental damage' pursuant to the Environmental Liability Directive.[18] On 21 February 2013, MEPs and Member States reached an agreement on a Directive to improve the safety of offshore oil and gas activities in the European Union. The Directive was signed by the Council on 12 June 2013 and published in the Official Journal on 28 June 2013.[19]

The outline of the factual and legal evolutions so far clearly justifies a thorough examination of civil liability and financial security for offshore oil and gas activities, particularly in European waters. In summary:

- Many incidents with offshore facilities have taken place in European waters.
- The Deepwater Horizon incident of April 2010 showed the potentially enormous amount of damage which could result from such an incident.
- Many offshore incidents can have a trans-boundary character, thus justifying the need for European action.

[14] COM(2010) 560 final, 8. [15] COM(2010) 560 final, 13.

[16] COM(2011) 688 final, Proposal for a Regulation of the European Parliament and of the Council on Safety of Offshore Oil and Gas Prospection, Exploration and Production Activities, 27 October 2011.

[17] COM(2011) 690 final, Proposal for a Council Decision on the Accession of the European Union to the Protocol for the Protection of the Mediterranean Sea Against Pollution Resulting from Exploration and Exploitation of the Continental Shelf and the Seabed and Its Subsoil, 27 October 2011.

[18] COM(2011) 688 final, art. 7.

[19] Directive 2013/30/EU of the European Parliament and of the Council of 12 June 2013 on Safety of Offshore Oil and Gas Operations and Amending Directive 2004/35/EC, *Official Journal* L 178/66–101, 28 June 2013. This Directive will be discussed in further detail in Section 3.4.6.

- Whereas an elaborate international (IMO founded) regime exists for vessel-source marine pollution, such a regime is absent for damage resulting from offshore facilities.
- This justifies the need to examine how a potential European liability and financial security regime for damage caused by offshore activities could be shaped.

Examining how a potential European liability and financial security regime for damage caused by offshore installations could be shaped is precisely the goal of this book.

1.2 Objective of This Book

Directive 2013/30/EU of 12 June 2013 holds that an operator of an offshore installation will have to show financial capabilities, including any financial security to cover liabilities potentially deriving from the offshore oil and gas operations.[20] The Directive, however, does not explain how these liabilities should be covered by financial security, nor what specific type or form this liability regime should take. Hence, this leads to a number of important questions which will be central throughout this book:

- To what extent can liability play a role in both preventing of accidents and compensating victims of accidental offshore oil and gas activities?
- How should such a civil liability regime be shaped to make civil liability efficient, effective and insurable?
- What options currently exist for liable operators to cover the costs resulting from such liabilities and which mechanisms (e.g. compensation funds, solvency guarantees, insurance or risk-sharing agreements) either exist or can be developed to reach that goal?
- How can mechanisms be developed to facilitate the early compensation of victims after an accident?

1.3 Methodology

Three different approaches will be used in this book: a traditional legal analysis (1.3.1), an economic analysis of law (1.3.2) and an empirical approach (1.3.3).

[20] Article 4(2)(c) of Directive 2013/30/EU.

1.3.1 Legal Analysis

The legal analysis will, in turn, consist of two approaches. The first and obvious one is legal desk research. This research requires a detailed analysis of primary legal sources that provides information on the goals and workings of the legal regimes that will be addressed in the chapters that follow. For every liability issue, a detailed literature study will take place. Many of the legal rules to be analyzed, for example, as far as environmental or nuclear liability is concerned, originate either from European rules (such as the Environmental Liability Directive) or from international conventions (such as the oil pollution conventions concluded within the framework of the IMO). In this respect, a vertical comparative legal approach is necessary to consider how some of the conventions are implemented in the national laws of some of the EU Member States, and this in particular will be analyzed.[21]

Within the legal analysis, particular attention will be paid to developments at the European level (which were briefly touched on in Section 1.1). In this respect, we, of course, refer to the Commission communication on the risks in the offshore oil and gas industry in European waters as well as the Directive on Safety of Offshore Oil and Gas Operations and in particular the impact assessment that took place in that respect. These European policy developments will be taken into account as background.

When various (national and regional) liability regimes are analyzed, this comparative analysis will take place on the basis of a particular framework. A tentative checklist will be developed so as to move towards a harmonized framework that can be used to analyze the liability and financial security mechanisms in the system under study. Questions that will be addressed include

- What is the basis of liability? Is it a negligence or strict liability rule and is the operator also liable for *force majeure* or are excuses allowed?
- How is the relationship with regulation handled? Will compliance with regulations (safety standards) excuse the operator from liability?
- Are specific rules in place relating to causation and causal uncertainty?
- How is liability attributed when multiple parties have contributed to the risk? Is a joint and several liability rule in place, or is the liability rather channelled to one particular operator?

[21] These legal regimes will be presented in Chapter 3.

- What types of damages can be compensated? Is compensation also awarded for pure economic loss, for ecological damage and for non-pecuniary losses? Is there a particular financial cap on liability, or is liability unlimited?
- Is the liability of the operator (compulsory) insured, or are other measures in place to provide financial coverage for liability?
- Are particular claims-management mechanisms in place, such as forcing the operator or the insurer to make an early offer to victims or advance payments?

These and other questions are worked out within a checklist that has been used as a common framework for the comparative analysis in order to guarantee the comparability of the results.[22]

A major offshore incident can give rise to a variety of different types of losses. To some extent, this will depend, as will be explained later, on the nature of the product involved (oil or gas) and obviously also on the location (deep or shallow water and close to the coast or not). Attention in the literature and at the policy level often has been paid mostly to pollution damage. However, in addition to pollution (usually resulting from an incident with oil), there can be equally large economic losses to the fishing and tourism industries. Given the primary attention that has often been paid to pollution damage, it is also important to focus on losses to businesses as a result of an offshore incident. From a legal perspective, it is often hard to separate those losses. Liability in most legal systems de facto will arise not only for pollution damage but also for personal injury, property damage and economic loss (to the extent that the legal system awards it). Hence, in the legal descriptions (mostly in Chapter 3), the focus will be on the compensation of damage generally without distinguishing between pollution damage and this more traditional damage. However, from the perspective of businesses that are affected by an offshore pollution incident (e.g. the fishing and tourism industries), it may be important to have mechanisms of rapid compensation available to prevent the failure of local businesses that are affected by the pollution during the civil litigation process. Hence, especially for this traditional damage, the question will be asked separately whether mechanisms of rapid claims management are in place or (to the extent that this is not the case) could be installed in order to guarantee advance payments to victims.[23]

[22] This checklist for the country studies is provided in Appendix 2.

[23] This will be addressed in Section 7.6.

As was already mentioned, there is now an EU-level Directive on Safety of Offshore Oil and Gas Operations (12 June 2013). For the purposes of this book, it is important to know that, as a result of this Directive, operators must show sufficient financial guarantees[24] and that offshore installations will be brought under the scope of application of the Environmental Liability Directive (ELD).[25] These are, however, only first steps, and as a result, many issues still need to be addressed. For example, the compulsory guarantees to be provided by operators have been stipulated in principle. But the question of how to create a mechanism that can better guarantee that operators can cover any liabilities arising from a major accident in the offshore oil and gas industry still has to be addressed.[26] Moreover, the fact that offshore installations have been brought under the ELD is perhaps more of a starting point than the final stage. First, the ELD applies only to a particular type of (pollution) damage,[27] which hence leaves unresolved the question of how more traditional damage, such as losses to the fishing and tourism industry, can be compensated and how a mechanism of rapid claims management could be installed to prevent the failure of local businesses affected by the accident while waiting for the civil litigation to be concluded.[28] Second, from a political/policy perspective, the decision to bring offshore installations under the ELD may be understandable, but from an academic/research perspective, serious questions could be asked as to whether that is the optimal solution. It could indeed lead to separate treatment of pollution damage, on the one hand (which would be, at least partially, covered under the ELD), and remaining losses, on the other. The question could be asked whether such a division of treatment is desirable.[29]

1.3.2 Law and Economics

The second approach to be followed, in addition to the more traditional legal (comparative) approach, is a law and economics methodology.

[24] See Art. 4(2)(c) of Directive 2013/30/EU of 12 June 2013 on offshore oil and gas operations.

[25] Article 7 of Directive 2013/30/EU of 12 June 2013 on offshore oil and gas operations.

[26] This will therefore be addressed especially in Chapter 6.

[27] For the scope of application of the ELD, see Bergkamp and Van Bergeijk, 'Scope of the ELD regime', pp. 51–79; and De Smedt, *Environmental Liability in a Federal System*.

[28] This will be addressed in Section 7.6.

[29] This is an aspect that will be discussed in Chapter 7.

The law and economics methodology (also referred to as *economic analysis of law*) allows insights into the working of liability rules and the effects of particular elements of liability rules on the stakeholders involved and, more particularly, the operators exposed to liability. Hence, the law and economics approach is an extremely useful way to address the core issue of this study – how an effective and efficient liability regime for operators in the offshore industry can be designed.

An important part of the law and economics methodology, especially the economic analysis of accident law,[30] also addresses the scope of liability in light of the insurability of a liability regime. Precisely because the insurability of liability may be crucial for operators active in the offshore industry, this insurability aspect will be taken into account when addressing the scope of liability.

The law and economics methodology furthermore has the advantage of not only pointing to the potential dangers of certain activities but also to their benefits, the so-called positive externalities for society. These may play an important role in the case of the offshore industry. The positive externalities created by the safety industry may be a reason to limit liability, which is a crucial aspect of the research in this book.[31]

The law and economics methodology has the major benefit that it pays attention to the goals of liability and also addresses the potentially perverse effects of liability. It also addresses the effects of liability rules on stakeholders, especially the economic effects on industrial operators, and is thus well able to indicate to what extent a limitation on the scope of liability may be warranted. In addition, law and economics analysis has indicated the extent to which a liability regime can be shaped in such a way that liability can be kept to an insurable level. Hence, the economic analysis of law is not only useful to the development of best practices but also to the critical evaluation of the potentially wide range of policy options that could be implemented with respect to the liability of the offshore industry.

1.3.3 Empirics

One of the questions that is central to this book – how an efficient and effective civil liability and financial security system for offshore oil

[30] As developed, *inter alia*, by Shavell, *Economic Analysis of Accident Law*.
[31] See Gilead, 'Tort law and internalization'.

and gas activities can be shaped – cannot be answered by theoretical research alone. Hence, an empirical approach also will be followed. This approach consists of semi-structured interviews. These interviews have been held with selected stakeholders who are experts in specific domains. An overview of these interviews is provided in Appendix I of this book.

1.3.4 Integration

The three methodological approaches have a clear relationship that will be followed. Together these approaches guarantee in an integrated manner that the goal of formulating recommendations concerning a liability and financial security regime for the offshore industry can be reached. The comparative legal approach also allows a careful analysis of existing liability regimes both in Member States and also in other offshore states and in other high-risk sectors. The law and economics approach can analyze the efficiency and effectiveness of various liability regimes. Hence, the economic analysis of law provides a benchmark to test whether particular features of the liability regimes studied would be useful in a possible future EU liability regime for the offshore industry. The empirical approach is crucial for various reasons. First, it provides good insight into the scope of the problem posed by offshore oil and gas activities; second, empirics will allow us insight into existing compensation mechanisms and financial market instruments, and third, the empirics also allow insights into the feasibility of various potential compensation mechanisms that could be developed in the future.

1.4 Structure

This book will be structured as follows. First, Chapter 2 will provide an overview of the damage resulting from major offshore incidents; next, Chapter 3 will analyze the existing legal regimes arising from major offshore and oil and gas accidents. Then Chapter 4 will list and analyze the existing risk-pooling mechanisms for damage compensation following offshore accidents in place within the European Union and elsewhere. Chapter 5 will discuss the potential of financial market instruments to cover liabilities following a major offshore accident by looking at the status quo as far as liability cover is concerned. Following this, Chapter 6 will take a forward-looking approach to the potential of financial and insurance instruments to cover liability. Chapter 7 will execute an analysis of potential civil liability regimes and financial security mechanisms. This chapter will therefore provide a critical overview of the various scenarios and options for EU action and will weigh the advantages and disadvantages of different options. On the basis of this critical analysis, finally, Chapter 8 will present concluding remarks with policy recommendations.

To some extent, it may be unavoidable that different topics will be discussed in various places. This is undoubtedly the case for the important question of whether a Deepwater Horizon–type incident could occur in Europe as well and what the potential amount of the damage in such a case could be, as well as for the functioning of the Offshore Pollution Liability Agreement (OPOL). The OPOL plays an important role in the UK legal regime (and therefore will be discussed there), but it also will be discussed when pooling mechanisms are discussed. To some extent, such overlap is therefore unavoidable. However, cross-references will indicate to readers where issues have already been discussed in order to improve the readability of this book.

The evolutions in legislation, policy documents, legal doctrine and case law in this book have, as far as possible, been incorporated up to 2 July 2015. Developments after that could not be taken into account.

2

Offshore-Related Damage: Facts and Figures

KRISTEL DE SMEDT AND HUI WANG

This chapter will first provide a broad overview of offshore activities in Europe as well as of the accidents that have occurred with offshore oil and gas installations. This will serve as a general background and is necessary for us to be able to assess the specific problem to which the legal regime should react. First, a number of general issues concerning offshore activities will be discussed, including the stakeholders involved and the importance of the offshore industry for Europe (2.1); then the location of installations and facilities in the European Union will be sketched (2.2), and an overview of major offshore accidents will be provided (2.3). A case-study approach will be used to discuss some recent offshore accidents in Europe (2.4), and specific attention will be paid to the Montara and Deepwater Horizon cases (2.5).

2.1 General Issues Concerning Offshore Activities

2.1.1 Terminology in Offshore Activities

Offshore oil and/or gas activities refer to the activities that are employed in offshore facilities (in contrast with onshore facilities) for the purpose of oil and/or gas exploration, exploitation and production. These offshore exploration activities often take place in the exclusive economic zone and on the continental shelf, where exclusive jurisdiction is granted to the coastal state under the United Nations Convention on the Law of the Sea (UNCLOS).[1]

[1] These are embodied in the following articles of UNCLOS: Art. 56, 'Rights, jurisdiction and duties of the coastal States in the exclusive economic zone'; Art. 60, 'Artificial islands, installations and structures in the exclusive economic zone'; Art. 77, 'Rights of the coastal State over the continental shelf'; and Art. 81, 'Drilling on the continental shelf.' The exclusive economic zone stretches from the seaward edge of the state's territorial sea out to 200 nautical miles from its coast. In colloquial usage, the term may include the territorial sea and even the continental shelf beyond the 200-mile limit.

Offshore facilities employed during such activities may include offshore platforms (fixed or floating),[2] offshore storage/loading systems, subsea facilities, wells, offshore pipelines, offshore drilling units and other associated offshore equipment, constructions and installations. *Oil exploitation* involves the following mining processes[3]:

- Seismic exploration,
- Exploratory drilling,
- Field development,
- Construction and installation,
- Oil/gas drilling of production wells,
- Production/maintenance and
- Field abandonment/decommissioning.

These processes often involve different types of expertise. Because of this, many players are involved in oil exploitation operations. It is beyond the scope of this book to list all the potential actors involved, but in order to comprehend the laws and regulations that will be outlined in the coming chapters, it is important to take note of the difference between licensees, operators and contractors. *Licensees* are actors that hold the rights over a particular geographic area that contains one or more oil fields.[4] Licensees may or may not be operators themselves. Moreover, to spread (financial) risks, it is common for licences to be shared among multiple parties. These parties will not all participate in oil and gas activities. An *operator* is a company actually performing oil and gas exploitation operations. Operators may hire *contractors* to perform activities on their behalf.[5]

2.1.2 Industry Structure of Offshore Activities

In most countries, offshore oil and gas exploration and exploitation activities are carried out in a similar structure. In general, the resources belong to the state, and the government has the exclusive right to grant

[2] Fixed platforms can only be used in shallow waters, where depth is no more than 400 metres. 'Deepwater' refers to a depth between 400 and 1,800 metres, and 'ultra-deepwater' refers to a depth between 1,800 and 3,000 metres and more.

[3] Janssen, *Preventing a Next Deepwater Horizon?*, 15.

[4] However, in case of exploration, the area might not contain any field. Moreover, in different legal regimes, different terms may be used; e.g. in the United States, there is the 'leasee' and the 'permittee'; in Australian law, the term 'title holder' is used. This will be further explained and analyzed in Chapter 3.

[5] Janssen, *Preventing a Next Deepwater Horizon?*, 15.

licences to carry out activities in the offshore area. The licence typically specifies the terms and conditions under which the offshore activities shall be carried out. In some countries, there even is a requirement that the state keeps a certain share in the joint operating agreement with the licensee, e.g. in Norway, Denmark, and the Netherlands. The interested party has to pay a royalty to the state in exchange for the licence.

The offshore activities are technical and complicated. Therefore, they are carried out in the form of a joint operation by parties with different expertise. The offshore industry in different jurisdictions also develops standard forms of *joint operating agreements* (JOAs). A JOA usually specifies the share of each party in the joint venture, demarcates the obligations and defines the liability of the operators during the joint operation, administers and allocates the work programmes and budgets, manages the relationship with the government and deals with the requirements in relevant regulations on health, safety and environmental protection.[6] Typically, an individual operator's liability is proportionate to its respective percentage interest in the licensed operations.[7] The standard form of the JOA that is most prevalent in offshore activities is the Association of International Petroleum Negotiators' Model International Operating Agreement (the AIPN Standard JOA).[8] In the United Kingdom, it is the Oil and Gas Standard Form Joint Operating Agreement (the UK Standard JOA).[9] Some big oil firms have their own standard or model JOAs.

In addition to the JOAs, there are standard contracts for both drilling and well services, which are also activities. The most prevalent standard forms of contracts are the International Association of Drilling Contractors' (IADC) Standard Form of International Offshore Daywork Drilling Contract and the AIPN International Model Well Services Agreement. In the United Kingdom, the generally adopted standard forms of contracts are promulgated by LOGIC, the Leading Oil Government's Oil and Gas Industry Task Force.[10]

[6] Easso, 'Managing the Risk', 19. [7] Bosma, 'The Regulation of Marine Pollution', 102.

[8] The most recent version was published in 2012 by the Association of International Petroleum Negotiators; see www.aipn.org/mcvisitors.aspx.

[9] The most recent version was published in 2008 by the industry organization in the United Kingdom, Oil and Gas UK; see www.oilandgasuk.co.uk/publications/viewpub.cfm?frmpubid=250.

[10] For a detailed discussion on these standard contracts, see Bosma, 'The Regulation of Marine Pollution', 103–4.

The standard form of a JOA is often used as a starting point for negotiations between parties.[11] However, in some jurisdictions, the change to model JOAs has to be approved by the authorities. It is important to acknowledge the use of standard forms of contracts in offshore activities because they concern not only the allocation of liability among various parties but also the insurance or financial guarantee to be taken by the related parties. This will be further analyzed in Chapter 4.

Following an incident, various types of damage and loss may occur. These may include personal injury and fatalities, property damage, damage to the marine environment, and losses suffered by related sectors, such as the fishery and tourism industries. Moreover, liability may also arise from the violation of licences and related regulations under national laws in respect of health and safety and environmental protection. In addition, liability may also arise from a contract between the contracting partners for the (joint) venture of offshore activities and/or a contract with the downstream parties and related businesses, such as shipping and transportation and storage of the oil produced. In this book, we focus on liability for damages to third parties. Hence, we do not address contractual liability.

2.1.3 Stakeholders

In the whole process of offshore oil and gas exploration activities, many stakeholders may be involved. These include national regulatory authorities, the offshore oil and gas industry and some other related industries, such as those which provide services to offshore exploration and those which provide equipment to facilitate the operation, and the insurance industry. In addition, industry organizations and other international organizations also may play important roles. As stated in our methodology section (1.3.3), an empirical approach is used in this study. Hence, interviews have been conducted with selected stakeholders and are included in Appendix 1.

State regulatory authorities. In almost every country, there are one or several governmental institutions involved in offshore activities. The authorities supervising offshore activities in the various jurisdictions are the following[12]:

[11] Roberts and Derrick, 'OGUK and AIPN Joint Operating Agreements'.

[12] Very often in one country there are separate regulatory authorities governing different aspects of the offshore activities, including licensing, health and safety, and environment issues. Therefore, the list here is non-exhaustive.

- Denmark: Danish Energy Agency (DEA);
- United Kingdom: Department of Energy and Climate Change (DECC), Health and Safety Executive (HSE);
- Germany: Landesamt für Bergbau, Energie und Geologie (State Office for Mining, Energy and Geology);
- Norway: Petroleum Safety Authority (PSA), Norwegian Environment Agency;
- Sweden: Svenska Geologiska Undersøkning (Swedish Geological Survey);
- Ireland: Department of Communications, Marine and Natural Resources;
- Netherlands: State Supervision of Mines;
- Australia: National Offshore Petroleum Safety and Environmental Management Authority (NOPSEMA);
- United States: Bureau of Ocean Energy Management (BOEM) and Bureau of Safety and Environmental Enforcement (BSEE), and
- Brazil: Brazilian National Petroleum Agency (ANP).

Industry organizations. The members of these organizations are directly involved in the exploration or exploitation of offshore activities. Each country or region may have its own industry association. Just to name a few:

- Oil and Gas UK,
- International Association of Oil and Gas Producers (OGP),
- International Association for Drilling Contractors (IADC),
- US Energy Information Administration (EIA) and
- Netherlands Oil and Gas Exploration and Production Association (NOGEPA).

Related industry organizations. These organizations are not directly involved in the process of exploration but may relate to the activities in providing their specialized services, such as the Operators' Co-operative Emergency Services (OCES). OCES is the organizational framework under which oil and gas companies operating in the waters of the North Sea and adjacent waters of the North West European Continental Shelf cooperate and share resources in the event of an emergency situation. OCES is periodically revised and has undergone a detailed scrutiny after the Deepwater Horizon incident in April 2010. The latest version of OCES, the Joint Declaration and Emergency Assistance Code, was revised in December 2011.[13]

[13] See www.oilandgasuk.co.uk/knowledgecentre/operatorscooperativeemergencyservices .cfm.

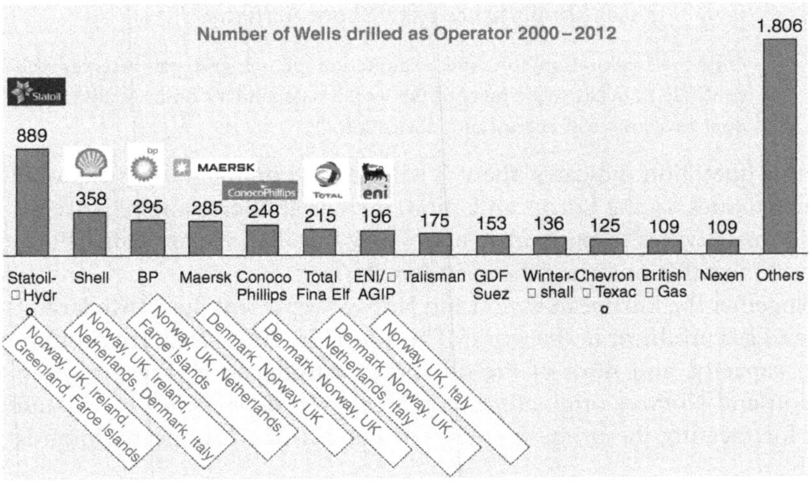

Figure 1 Oil companies with exploration/production licences in Europe (operators). (Courtesy of Dr. Philipp Wassenberg, Munich Re.)

An overview of companies active in offshore drilling can be found on various websites.[14] The overview in Figure 1 provides information on the main participants and the number of wells which have been drilled by operators in Europe in the years 2000–2012.

This figure shows that, on the one hand, a few large companies play a major role in offshore activities, the largest one being Norwegian Statoil. On the other hand, the most interesting number relates probably to the 1,806 wells that are run by other operators. Those are most likely smaller companies that may have less experience and less financial capacity than some of the major players. It is particularly those smaller companies which are especially the source of regulator attention in view of their potential insolvency.[15]

The insurance industry. Although the insurance industry is not directly involved, it plays a crucial role in the development of the whole offshore industry, e.g. the European Insurance and Re-Insurance Federation and the International Union of Marine Insurance (IUMI).

[14] See, e.g., the list of offshore drilling companies with a ranking: www.ranker.com/list/list-of-offshore-drilling-companies/business-and-company-info. A list of well-drilling service companies can be found at www.subsea.org/equipment/drilling+and+well+services/listcat.asp?cate=drillling+on.

[15] See the discussion in Section 6.1.2.

2.1.4 *Importance of Offshore Activities*

> Offshore exploration for, and exploitation of, oil and gas reserves will continue to occur while most of the world is dependent on these hydrocarbons as a source of energy and lubrication.[16]

As this quotation indicates, there is still reliance on oil and gas exploration activities. In the European Union, there is an increasing reliance on offshore activities in particular; many new installations are being built, and old installations are to be abandoned.

Together the European Union and Norway represent the fourth largest oil and gas producer in the world. The North Sea has significant production capacity, and most of the oil and gas produced by the European Union and Norway originates there. Moreover, dependence on oil and gas for meeting the primary energy demand in the European Union is high.[17]

2.2 Location of Installations and Facilities in the European Union

2.2.1 *General Information*

> Offshore oil and gas production constitutes an important indigenous energy source of hydrocarbons in Europe. In 2012, offshore oil production in the EU totalled approximately 60 million tons produced in the continental shelves of different EU member states. According to data from EUROSTAT and the petroleum national authorities, in 2012 the Offshore Crude Oil Production of EU-28 corresponded to almost 9% of the Gross Petroleum Products Consumption. Most of the crude oil in the EU Member States is produced in the North Sea, with the UK as the major contributor (75.38%), followed by Denmark (18.17%).[18]

Offshore natural gas production in 2012 in the European Union (EU-28) 'was approximately 63 million tons of oil equivalent. In this case EU-28 offshore production covered 13.8 per cent of the gross energy

[16] Speech given by Steven Rares, a judge of the Federal Court of Australia, titled, 'An International Convention on Off-Shore Hydrocarbon Leaks', presented at the International Conference on Liability and Compensation Regime for Trans-Boundary Oil Damage Resulting from Offshore Exploration and Exploitation Activities, Bali, 21–23 September 2011.

[17] Janssen, *Preventing a Next Deepwater Horizon?*, 12. See also Izundu, 'Prospects Slowly Brighten', 49–59.

[18] http://euoag.jrc.ec.europa.eu/node/63.

consumption with the UK and the Netherlands having a predominant role with 54.6% and 25.23% respectively.'[19] Norway also produces a considerable amount of natural gas. In 2012, natural gas offshore production in Norway was over 84 million tons.[20]

Europe has a large amount of offshore oil reserves in the North Sea (and hence frequent offshore exploration activities) and an increasing number of prospects in the Mediterranean Sea. Most of the offshore oil and gas activities in Europe are therefore concentrated in the North Sea, with Norway, the United Kingdom, the Netherlands and Denmark controlling most of the facilities. Offshore oil and gas installations in the Black and Baltic Seas still number fewer than ten, while there are currently over 100 installations operating in EU waters in the Mediterranean, and plans to start new explorations are reported in the Maltese and Cypriot sectors.[21] It should be noted that in the Mediterranean Sea, a large number of offshore activities also take place off the coast of North Africa, particularly off of Egypt and Libya.[22] This may have trans-boundary effects on the European Mediterranean countries, as the spill at the Jiyeh Power Station in Lebanon (2006) illustrated.

The oil and gas industry is not yet well established in the Mediterranean Sea, where deeper water and seismic activity pose particular challenges. However, there are already more than 400 wells in Spain and Italy, and extensive prospecting activities are being carried out in the south-eastern Mediterranean. Exploration and production are ongoing in the Black Sea, with some promising fields in the European waters.[23] For instance, new discoveries of gas deposits have been announced in the Black Sea, near Bulgaria and Romania.[24]

New discoveries have led to the building of new installations, whereas some old wells are not immediately abandoned but sometimes are transferred from bigger companies to smaller ones due to economic considerations. Hence, the large number of offshore facilities – not

[19] *Ibid.* [20] *Ibid.*

[21] As a consequence of the Deepwater Horizon incident, Italy issued Law 128, which, in effect, halted all new Italian offshore developments from August 2010. A cabinet decree to amend Law 128 so that offshore activities could restart was agreed on in the spring of 2012.

[22] *Inter alia*, BP is planning to start drilling operations off the coast of Libya (interview with representatives from BP on 13 March 2013).

[23] Gómez and Green, 'The Impact of Oil and Gas', 16. This is a study commissioned by the Policy Department B: Structural and Cohesion Policies (European Parliament).

[24] www.roconsulboston.com/pages/infopages/government/exxonomvgas.html.

only in the North Sea but also in the Black Sea and the Mediterranean Sea – and the fact that drilling will then take place near densely populated areas, as in the Mediterranean, are of great interest for our study.[25] What is crucial is not the exact number of offshore installations but rather the potentially significant impact of an accident involving one of Europe's offshore installations, especially one that is located close to densely populated areas. It could cause not only material losses but also damage to the environment, the economy and local onshore communities, and the lives and health of the offshore workers also might be endangered.[26]

2.2.2 *Offshore Interests in the North Sea*

The North Sea is semi-enclosed by countries, including Norway, Sweden, Denmark, Germany, the Netherlands, Belgium, France and the United Kingdom. It opens up largely to the Atlantic Ocean in the north and the English Channel in the south and connects to the Baltic Sea in the east. It is a relatively shallow sea with an average depth of about 90 metres, although in some places, e.g. off the Norwegian coast, it reaches depths of around 700 metres.[27]

Among the countries from the North Sea region, both the United Kingdom and Norway are important oil and gas producers. According to data from the Energy Information Administration, Norway is Europe's largest oil producer and the world's third largest natural gas exporter. Moreover, Norway is an important supplier of both oil and natural gas to other European countries.[28] The United Kingdom is the largest oil producer and the second largest natural gas producer in the European Union. Oil accounted for 37 per cent of the United Kingdom's total energy consumption in 2013.[29]

The Netherlands is not a major oil producer, but it plays an important role in the transportation of oil in Europe, and it acts as an important processing hub. The Netherlands is one of the largest importers and

25 Gómez and Green, 'The Impact of Oil and Gas', 16.

26 European Commission, Proposal for a Regulation of the European Parliament and of the Council on Safety of Offshore Oil and Gas Prospection, Exploration and Production Activities, COM (2011) 68 final, 2011/0309 (COD), Brussels, 27 October 2011, 'Explanatory Memorandum', 2.

27 Janssen, *Preventing a Next Deepwater Horizon?*, 11. See also Ducrotoy and Elliot, 'The Science and Management', 8–21.

28 www.eia.gov/beta/international/country.cfm?iso=nor.

29 www.eia.gov/beta/international/country.cfm?iso=gbr.

exporters of crude oil and petroleum products in the world. In addition, it is a major refining and storage centre. The Netherlands is the second largest producer and exporter of natural gas in Europe, just behind Norway.[30]

In contrast to places such as the Mediterranean Sea, offshore activities in the North Sea often have to be carried out under harsh weather conditions, which make it difficult and costly.[31] Moreover, the existence of large numbers of gas fields makes it different from offshore activities in e.g. the Gulf of Mexico in the United States.[32] The risks confronted by the offshore industry in the North Sea region are therefore unique given the climate and geological conditions (they will be further analyzed in Section 2.5.3).

2.2.3 *Offshore Interests in the Mediterranean Sea*

Offshore drilling in the Mediterranean area has increased substantially. In particular, the Eastern Mediterranean and the Levantine Basin (Israel/Lebanon) have been emerging as new provinces for oil and gas exploration in the past few years. Moreover, preliminary exploration has confirmed impressive reserves of gas and oil in the waters of Greece, Turkey, Cyprus and potentially Syria. Italy's relaxation of a ban on offshore oil and gas drilling that was introduced in the wake of the 2010 Deepwater Horizon disaster also has stimulated activity, and Malta is a new player in the offshore market as well. Finally, Spain is also attracting interest; there are plans to drill in the Valencia Basin off the eastern Spanish coast in 2014.[33]

These increased offshore oil and gas activities in the Mediterranean obviously increase the risk of offshore accidents, which in the densely populated Mediterranean area potentially could result in significant human, environmental and economic damage.

Moreover, because these new oil and gas discoveries are very recent, the legislative framework in these Mediterranean countries is also very recent, and legislation is still developing. Some of the Mediterranean countries are not EU members, and hence their legislative provisions with respect to liability for traditional and environmental damage might differ from the EU Offshore Safety Directive.

Finally, one must not forget that the Mediterranean countries are caught up in a web of regional squabbles and rivalries with respect to

[30] www.eia.gov/beta/international/country.cfm?iso=nld.

[31] www.oilgasarticles.com/articles/77/1/north-sea-oil-and-gas/page1.html. This was stressed by industry representatives from OGP; interview with OGP, 25 February 2013.

[32] Interview with OGP, 25 February 2013.

[33] Chazan, 'Mediterranean Playground for Oil Groups'.

the borders of their respective exclusive economic zones. Turkey has repeatedly told oil companies to stay away from Greek Cypriot waters, and Italy protested against Malta's publication of a tender for oil and gas exploration covering parts of the Ionian Sea claimed by Rome. Malta itself is in a dispute with Libya over offshore areas to the south, and Israel and Lebanon have sparred over Leviathan, which is close to the disputed line separating their territorial waters.[34]

2.3 Overview of Major Offshore Accidents

2.3.1 Sources of Information

2.3.1.1 Worldwide Database

One of the most comprehensive databases for offshore accidents is probably the Worldwide Offshore Accident Databank (WOAD). This databank is operated by Det Norske Veritas (DNV), and it contains more than 6,000 incidents recorded since 1975. WOAD collects data from public-domain sources such as Lloyds Casualty Reports, newspapers, and official publications. It covers various parameters of an event, including the name, type and operation mode of the unit involved in the accident, date, geographical location, main event and chain of events, causes and consequences, as well as evacuation details.[35] The latest update of the WOAD has recorded 6,451 accidents occurring to 3,795 operating units up to the Hercules 265 blowout on 23 July 2013.[36] WOAD classifies the events into four categories (see also Figure 2):

- *Insignificant events:* 'Represent hazardous situations, with very minor consequences. In most of the cases no damages were registered and repairs were not required. Small spills of crude oil and chemicals are also included in this category. Also included are very minor personnel injuries, i.e. "lost time incidents".'
- *Near misses:* 'Represent events that might have or could have developed into an accidental situation. No damage and no repairs were required in these cases.'
- *Incidents:* 'Represent hazardous situations which did not develop into an accidental situation. A low degree of damage was recorded, but

[34] *Ibid.* [35] See Christou and Konstantinidou, 'Safety of Offshore Oil', 14.

[36] DNV GL AS: www.dnvgl.com/news/new-update-of-world-offshore-accident-databank-woad-available-8318. However, this is not publicly available.

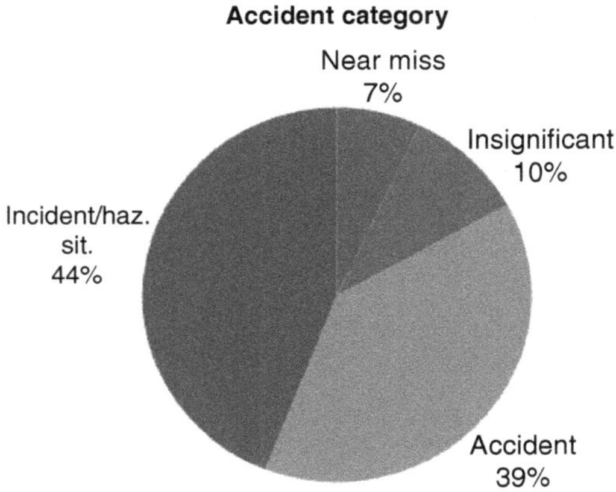

Figure 2 Distribution of accidents according to type for accidents in the WOAD.

repairs/replacements usually were required. This type includes also events causing minor injuries to personnel or health injuries.'

- *Accidents:* 'Represent hazardous situations which have developed into an accidental situation. In addition, for all situations/events causing fatalities and severe injuries, this classification of event has been used.'[37]

Of the 6,173 accidents that have been recorded in WOAD up to 2010, information on damage costs exists in only 63 cases, i.e. only in 5.83 per cent. It shows that in more than a third of the incidents (38 per cent) for which cost data exist, the cost was relatively limited: less than US$0.5 million. Forty-five per cent of the incidents had a cost of less than US$1 million. Only in 1.4 per cent of cases (with information on cost) did damage exceed US$100 million. The Joint Research Centre (JRC) provided tables summarising the data from the WOAD database, and these tables are shown in Figures 3 through 6.[38]

[37] Christou and Konstantinidou, 'Safety of Offshore Oil', 31. The WOAD does not represent complete statistics and is not even an authoritative accidents register, as suggested in the study by Christou and Konstantinidou, and can only serve as a good basis for lesson learning, not even a basis for statistical analysis. See Christou and Konstantinidou, 'Safety of Offshore Oil', 28. It is, moreover, important to realise that the number of near misses might be a lot higher than the number of accidents that have actually taken place. This is the so-called iceberg theory, where a lot of near misses lead to only one accident.

[38] We are grateful to Michalis D. Christou of the Energy Security Unit, Institute for Energy and Transport of the European Commission Joint Research Centre, for providing us with this valuable information.

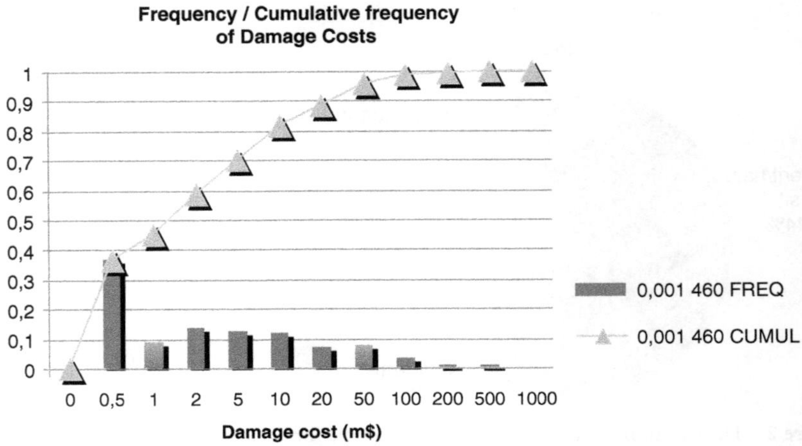

Figure 3 Frequency/cumulative frequency of damage costs.

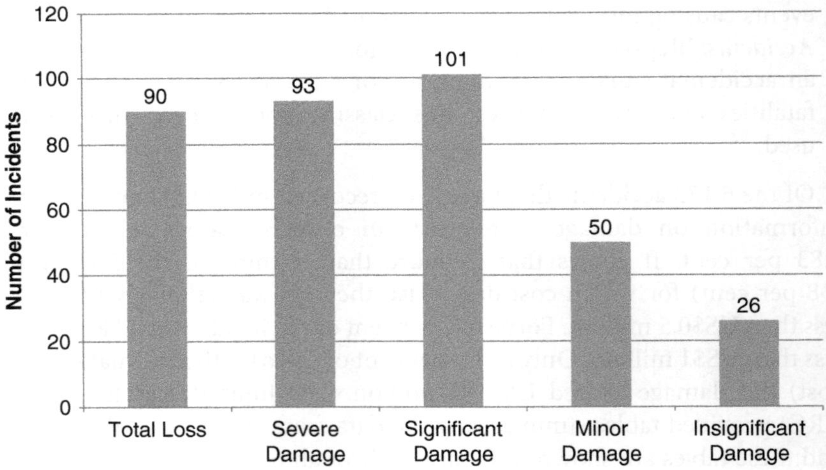

Figure 4 Consequences according to WOAD.

All costs are expressed in the year of occurrence. The database covers the period from 1970 to 2010 (Deepwater Horizon not included).[39]

2.3.1.2 National Regulatory Authorities

In various jurisdictions, there are duties to report accidents to the regulatory authorities. Hence, many national authorities have their

[39] *Ibid.*

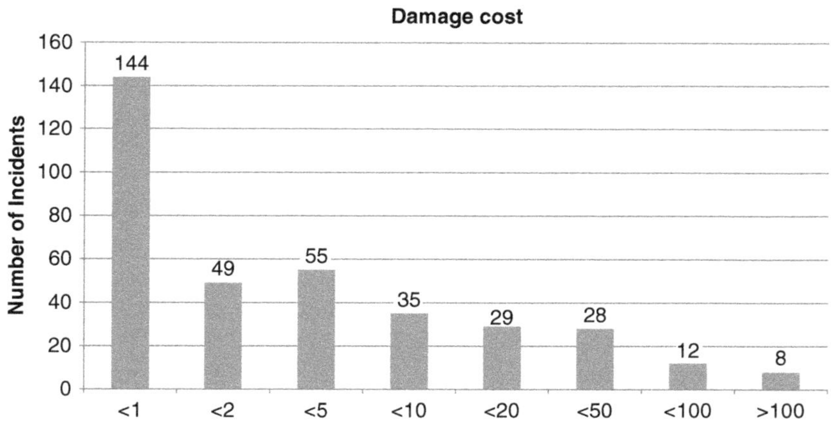

Figure 5 Damage costs according to WOAD.

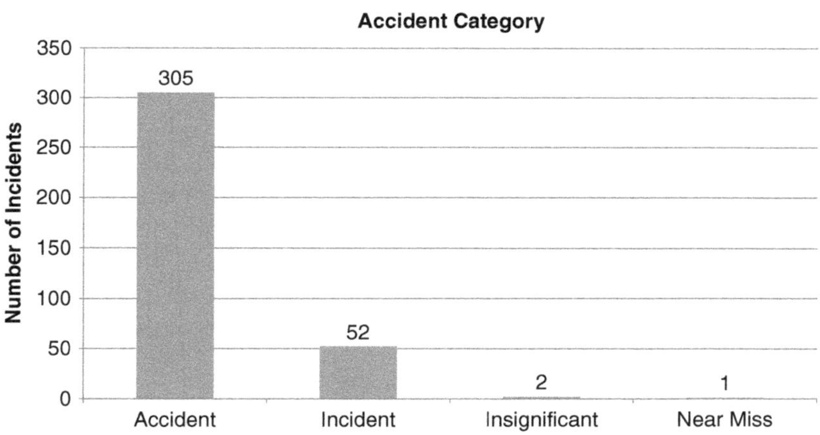

Figure 6 Accident category according to WOAD.

own databases on the accidents that occurred within their jurisdictions. In this respect, the United Kingdom (Health and Safety Executive), Norway (Petroleum Safety Authority) and Denmark (Danish Energy Agency) all keep records of accidents that occurred in the past. The United States also has a relatively developed reporting system to record past accidents.

Norway and the United Kingdom have a comprehensive system of data gathering and analysis with respect to offshore accidents; the Netherlands and Denmark also have a relatively developed reporting system. For the other EU countries with offshore facilities, information on offshore incidents is not always readily available. Especially with respect to the Mediterranean countries, very few statistics could be found. Yet, because the United Kingdom and Norway can be considered to be the most important offshore oil- and gas-producing countries, the accidents that happen on offshore facilities belonging to UK or Norwegian operators are likely to be representative. In this respect, a study of the accident statistics of the United Kingdom and Norway provides insight into the causes and consequences of offshore incidents.

In Norway, as a standard procedure, the most serious incidents are investigated by the Petroleum Safety Authority (PSA). It appears that six to nine incidents are investigated each year. Although most of these incidents did not cause significant human or environmental damage, they could have, under different circumstances, turned into disasters. For the United Kingdom, the Health and Safety Executive (HSE) is charged with overseeing the offshore industry.

The reason for choosing these data is just to give an impression of the current situation of data sources and their comparability owing to the lack of a systematic and harmonized data-collecting mechanism. Data from both the United Kingdom and Norway appear to show similarities and a similar trend. Data from the Dutch authority cover both onshore and offshore accidents, which may not be comparable with the data from the UK and Norwegian authorities, which concern only offshore accidents. However, with implementation of the Directive on Safety of Offshore Oil and Gas Operations (Directive 2013/30/EU), requiring establishment of a common reporting mechanism, the statistics in EU Member States should improve.[40]

United Kingdom The HSE, as the regulatory authority in the United Kingdom for offshore health and safety matters, produces reports on a yearly basis that provide data on offshore personal injuries and hydrocarbon releases.[41] The latest annual report[42] shows an overall downward

[40] The Directive has requirements on a common reporting format, which will be discussed further in Section 2.1.3. For further details on the Directive, see the discussion in Section 3.4.2.

[41] Available on the HSE website. These reports contain details of offshore accidents, dangerous occurrences and ill health reported to the HSE. They are based on incidents reported under the Reporting of Injuries, Diseases and Dangerous Occurrences Regulations of 1995 (RIDDOR).

[42] HSE, 'Annual Offshore Statistics and Regulatory Activity Report, 2013–2014'.

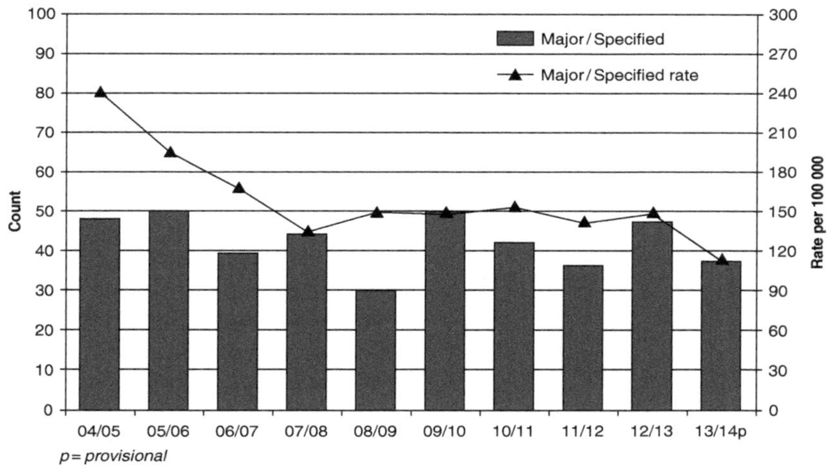

Figure 7 Injuries by severity, 2004–2014.

trend in reportable injuries over the last five years. With respect to personal injuries, the main points to be mentioned are as follows: one fatality was reported in year 2013–2014, but it occurred during routine lifeboat maintenance. There have been three fatalities in the last five years and seven in the last ten years.[43] It should be noted that in October 2013, the classification of 'major injuries' to workers was replaced with a list of 'specified injuries'. There were thirty-seven major/specified injuries, which continues the flat trend seen over the last ten years. The rate was 112 injuries per 100,000 workers. This continues the flat trend seen from 2007–2008, which was preceded by a considerable fall from 2004 to 2008 (see Figures 7 and 8).[44]

The main causes of major injuries were related to being struck by moving/flying/falling objects (forty-five accidents) and slips, trips or falls on the same level (forty-two accidents), each contributing to around 30 per cent of all the injuries. Handling, lifting or carrying (twenty-seven accidents) accounted for around 20 per cent of all injuries. Moreover, the distribution of severity by kind of accident is similar over the past three years.[45]

[43] *Ibid.* 6. See also Figure 2.7.

[44] HSE, 'Annual Offshore Statistics and Regulatory Activity Report, 2013–2014', 7. See also Figure 2.8.

[45] *Ibid.*, 11.

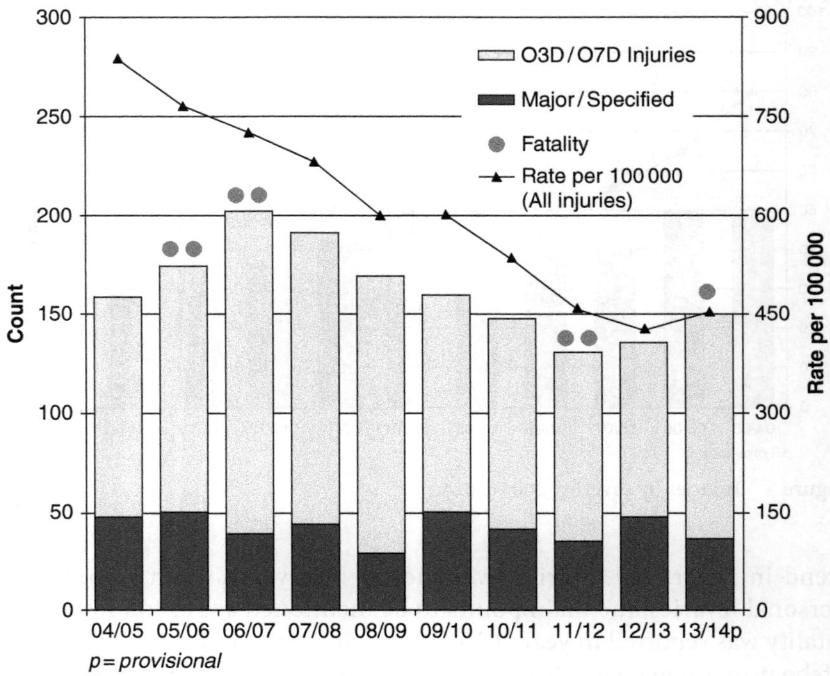

Figure 8 Major/specified injuries, 2004–2014.

With respect to hydrocarbon releases (HCRs), there were 115 HCRs in 2013–2014, an increase of about a fifth compared to 2012–2013. According to the HSE report, the increase was due to minor releases (seventy-seven releases, up over a half from 2012–2013).[46] Nevertheless, the number of significant and major HCRs together has dropped year on year for the last five years. Between 2010 and 2013, HCRs dropped by almost 50 per cent.[47]

For 2013–2014, 393 dangerous occurrences were reported, of which over 25 per cent were HCRs. There has been a gentle downward trend of dangerous occurrences over the past ten years.[48] In addition to HCRs,

[46] *Ibid.*, 14; Oil and Gas UK, 'Health and Safety Report 2014', also points out that the combined total number of HCRs is higher than in the previous year, whereas the total of major and significant releases shows a decline. Therefore, the increased number of total HCRs relates to a rise in minor releases. See Oil and Gas UK, 'Health and Safety Report 2014', 16.

[47] HSE, 'Annual Offshore Statistics and Regulatory Activity Report, 2013–2014', 14.

[48] *Ibid.*, 13.

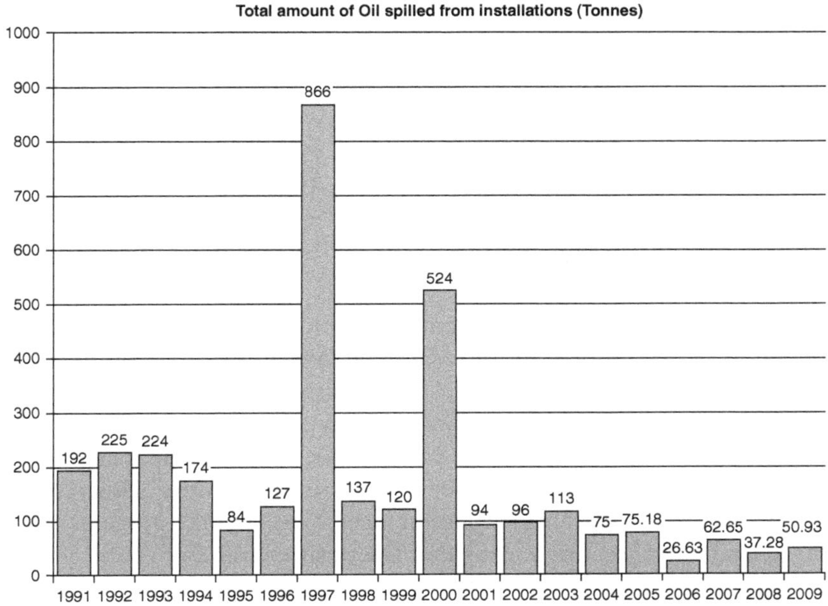

Figure 9 Total amount of oil spilled from installations (United Kingdom). (Department of Energy and Climate Change.)

other dangerous occurrences include well incidents, pipeline incidents, dropped objects, lifting machinery, collisions, subsidence or collapse of the seabed/loss of stability or buoyancy and evacuation of installations. As far as well incidents are concerned, there were forty-two reported incidents, of which twenty-five involved operation of a blowout preventer or diverter to control an unplanned flow from a well, and fifteen involved mechanical failure of a critical safety component of a development well.[49]

An overview of oil spills in tons over the last twenty years is provided by the Department of Energy and Climate Change. Note that the 2011 figure was again higher, around 216 tons, owing to the Gannet Alpha oil spill (see Figure 9).

From 2011 to 2013, almost 197 tons of oil was released from 733 accidental incidents. Sixty-five per cent of the volume of oil released resulted from a small number of releases that were greater than 10 tons.

[49] *Ibid.*, 15.

However, 17 per cent of the volume released was from a high number (719) of small-volume releases. Of these, 302 releases were less than 0.001 ton (1 kilogram). Given their small sizes, these releases should be rapidly diluted in the North Sea.[50]

Norway For Norway, the Petroleum Safety Authority provides data on offshore personal injuries and HCRs. Similar to the situation in the United Kingdom, the data show an overall downward trend in reportable injuries and accidents over the last ten years.

Personal Injuries. No fatal accidents occurred on the Norwegian continental shelf during 2013. As for serious personal injuries, the number of these accidents in the offshore industry in Norway remained fairly stable in 2013 at twenty-four, compared with twenty-three in 2012.[51] However, owing to the increase in working hours, the serious personal injury frequency fell from 0.51 to 0.48 per million working hours. This represents a statistically significant reduction compared with the previous ten-year period and therefore means that the positive trend of recent years has been maintained.[52] The Norwegian Petroleum Safety Authority maintains separate data on personal injuries for contractor personnel on production installations versus operator employees. Historically, the personal injury frequency for contractor personnel has been higher than that for operator employees. However, in 2013, the personal injury frequency for contractor employees was lower than that for operator employees.[53] The serious personal injury frequency increased slightly on mobile units in 2013 despite the general downward trend in recent years and the decrease in 2012.[54]

In 2014, there was a small increase in the serious personal injury rate per million working hours from 0.48 in 2013 to 0.53 in 2014. Nevertheless, in the long term, there has been a steady downward trend since 2005.[55]

Oil Spills. According to data from the Ministry of Petroleum and Energy, over the last forty years, offshore activities in Norway have not caused major acute discharges of oil that have led to environmental damage.[56] In 2014, seven HCRs exceeding 0.1 kilogram per second were recorded. This is the second lowest number recorded in the period

[50] Oil and Gas UK, 'Environment Report 2014', 33.
[51] Petroleum Safety Authority, 'Annual Report 2013', 5. [52] *Ibid.*, 5. [53] *Ibid.*
[54] *Ibid.* [55] Petroleum Safety Authority, 'Trends in Risk Level in Petroleum Activity', 5.
[56] Ministry of Petroleum and Energy, 'Facts: The Norwegian Petroleum Sector 2014', 51.

2000–2014.[57] Seventeen well-control incidents were recorded in 2014. This is a slight increase from 2013, when thirteen incidents were recorded, and 2012, with sixteen incidents.[58] When assessing the number of incidents against the level of activity, there were increases in both production drilling and exploration drilling. Within exploration drilling, the number of incidents in relation to the activity level varies to the greatest extent, and the level in 2014 is above the average for the period 2000–2014. In 2014, the risk indicator associated with well-control incidents was relatively low compared with previous years, which can be explained by the fact that most of the incidents in 2014 were of low severity.[59]

Yet, although there apparently is a downward trend in offshore incidents, incidents did happen, and the reasons for these incidents must be examined. The next section will examine in more detail some major incidents. However, the PSA notes in its report that the barrier indicator related to major accidents shows that a relatively large number of installations had fairly substantial nonconformities from the expected industry level. This means that the sector has a clear improvement potential with regard to ensuring that these barriers are sufficiently robust.[60]

The Netherlands In the Netherlands, the State Supervision of Mines (SSM) is the authority that supervises mineral exploration, extraction, storage and transport of minerals.[61] Its tasks are mainly to oversee compliance with statutory regulations applicable to those activities. The SSM focuses on various aspects of the activities, including, *inter alia*, health, safety and the environment.[62] It has useful statistics on past accidents (see Figures 10 and 11).

2.3.1.3 Regional Database

In addition, at the regional level, there are regional cooperation programmes which may keep regional records of incidents, e.g. the North Sea Offshore Authorities Forum (NSOAF). National regulatory authorities from North Sea countries in charge of offshore oil and gas activities have developed this forum for exchanging information concerning offshore accidents.

[57] Petroleum Safety Authority, 'Trends in Risk Level in Petroleum Activity', 3.
[58] Petroleum Safety Authority, 'Annual Report 2013', 5.
[59] Petroleum Safety Authority, 'Trends in Risk Level in Petroleum Activity', 3. [60] *Ibid.*
[61] www.sodm.nl/english/organisation. [62] *Ibid.*

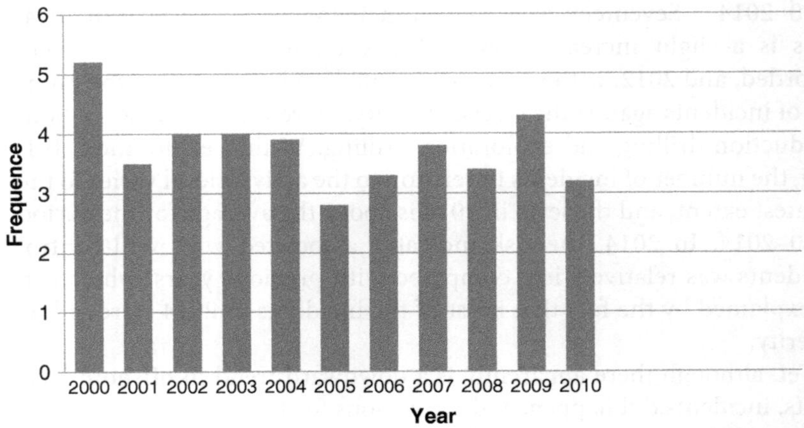

Figure 10 Personal injury accident frequency (total number of accidents per million working hours) in the exploration and production industry (on-and offshore). *Source:* State Supervision of Mines (the Netherlands), 'Staatstoezicht op de Mijnen'. The most recent data are available from Health and Safety Bulletin 05/14, 8 October 2014, available at: www.sodm.nl/sites/default/files/redactie/0514%20Analysis%20of%20the%20common%20root%20causes%20of%20the%20reported%20serious%20workplace%20accidents%20in%202013_0.pdf.
Note: Total number of accidents: lost-time accidents (LTAs) + fatal + restricted-work cases (RWCs). Lost-time accidents (LTAs), accidents that led to temporal disability to work; restricted work cases (RWCs), accidents that led to employees doing a different type of work.

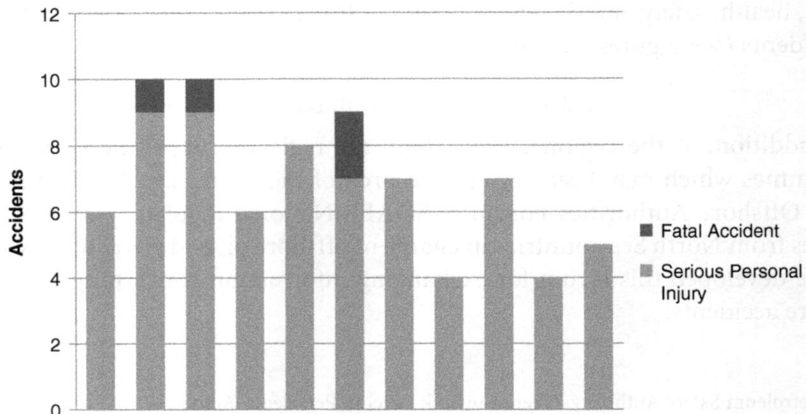

Figure 11 Serious and fatal personal injury accidents in the exploration and production industry (on- and offshore). *Source:* Same as Figure 10.

The International Regulators' Forum (IRF) has also begun a performance-measurement project to measure and compare offshore safety performance among IRF participants by collecting and comparing incident data based on a common set of criteria. Data include fatalities, injuries, gas releases, collisions, fires and losses of well control. The members of the IRF are the regulatory authorities from Australia, Brazil, Canada, Denmark, Mexico, the Netherlands, New Zealand, Norway, the United Kingdom and the United States.[63]

As mentioned earlier, Directive 2013/30/EU has some requirements on a common data-reporting format. To implement these requirements, the European Commission has adopted Regulation 1112/2014 in October 2014 specifying the common formats.[64] Regulation 1112/2014 contains specific formats and detailed information on different types of incidents related to offshore activities to be reported,[65] including

- Unintended release of hydrocarbon substances (Section A of Annex I);
- Loss of well control which requires actuation of well-control equipment or failure of a well barrier which requires its replacement or repair (Section B);
- Failure of a safety and environmentally critical element (Section C) and
- A major environmental incident (Section J).

It is reasonable to believe that with the implementation of Directive 2013/30/EU, an improved European-wide common data-reporting system can be established which will facilitate data collection for offshore-related activities.

2.3.1.4 Industry Database

Various industry organizations (as mentioned in Section 2.1.3) maintain their own records of incidents. For example, the OGP has established a Wells Committee to identify 'areas for improvement to get better as an industry across the whole cycle of well planning, construction, operation

[63] Christou and Konstantinidou, 'Safety of Offshore Oil', 13–14.

[64] Commission Implementing Regulation (EU) No. 1112/2014 of 13 October 2014 determining a common format for sharing of information on major hazard indicators by the operators and owners of offshore oil and gas installations and a common format for the publication of the information on major hazard indicators by the Member States, *Official Journal*, 22 October 2014, L302/1–25. This regulation has entered into force as of 11 November 2014, twenty days after its publication in the *Official Journal*.

[65] The events are categorised according to Annex IX of Directive 2013/30/EU.

and decommissioning' and to contribute to 'continuous safety improvements in well construction, well intervention and well integrity'.[66] Its activities include, *inter alia*, analysing incidents and disseminating lessons learned and good practices based on shared experience. For this purpose, a database has been developed. OGP members report incidents related to well control and near misses into the OGP Well Control Incident Database. All data submitted is anonymous and confidential. The data are not available to the public, only to members of the organization.[67]

However, these organizations collect data with different criteria and compile their data with different approaches, which may lead to difficulties when comparing these various data directly. For instance, a study carried out by some members of the National Commission on the BP Deepwater Horizon Oil Spill and Offshore Drilling compared the fatalities and injuries from offshore facilities in Europe with the United States. Various data sources suggest slightly different results.[68] These different data can be illustrated as follows by putting accident data together on incidents in the United States and Europe that come from the International Association of Drilling Contractors (IADC), the International Association of Oil and Gas Producers (OGP) and the International Regulators Forum (see Figure 12).

In addition, Willis, a global insurance broker, has also established a comprehensive database of energy losses in the world which also includes offshore accidents. This database, known as the Willis Energy Loss Database (WELD), has included thousands of incidents since 1972 that have led to losses in excess of US$1 million.[69] This database is available only to subscribers and is often used by insureds and insurers to analyze current and historical loss information, with which parties can 'make informed decisions on insurance strategy'.[70] The types of losses captured in the WELD mainly focus on 'property-related losses rather than individual death or injury claims'. 'Difficulties in obtaining accurate information on isolated death and injury claims have been encountered

66 www.iogp.org/wells-committee#2594267-objectives.

67 Chistrou and Konstantinidou, 'Safety of Offshore Oil', 13.

68 National Commission on the BP Deepwater Horizon Oil Spill and Offshore Drilling, 'A Competent and Nimble Regulator'.

69 Willis Energy Loss Database, August 2004; available at: www.willis.com/documents/publications/industries/energy/2155_energy%20_loss_database.pdf.

70 *Ibid.*

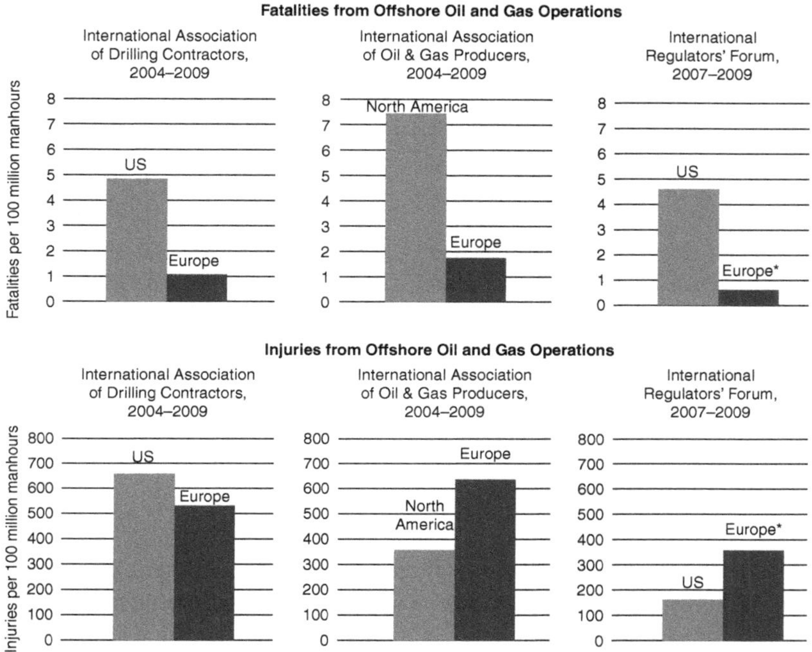

Figure 12 Fatalities from offshore oil and gas operations.
Source: National Commission on the BP Deepwater Horizon Oil Spill and Offshore Drilling, 'A competent and nimble regulator'.
Note: It is worth mentioning that data from the International Association of Oil & Gas Producers includes helicopter-related incidents. Data from the International Association of Drilling Contractors and the International Regulators' Forum include helicopter-related incidents only if it is at or near an offshore installation.

due to non-disclosure agreements, appeals, etc., and the relevance of many of the figures that included substantial punitive damage awards could be questioned. Third-party losses and sudden and accidental pollution claims are included, as are death and injury settlements that formed part of recognized figures for major property claims (e.g. Piper Alpha).'[71] The WELD also contains, *inter alia*, an overview of the ten most expensive operators' extra expense (OEE) losses in history.[72] The Willis database contains, *inter alia*, the information listed in Table 1. This table shows that

[71] *Ibid.*
[72] Willis Energy Loss Database, 1 April 2011; available at: 9781107167162c02.docx#www.willis.com/documents/publications/industries/energy/emr_april_2011.pdf.

Table 1 *Overview of the Ten Most Expensive Operators' Extra Expense (OEE) Losses in History*

Year	Type of facility	Location	Country	OEE indexed (US$)
2010	Rig	Gulf of Mexico	USA	2,000,000,000
2005	Platform	Gulf of Mexico	USA	636,047,629
1989	Well	North Sea	Norway	396,419,527
2008	Platform	Gulf of Mexico	USA	384,080,640
2005	Platform	Gulf of Mexico	USA	341,560,173
1984	Well	Nova Scotia	Canada	320,593,818
1988	Platform	North Sea	UK	308,109,489
1987	Platform	Gulf of Mexico	USA	264,476,529
1975	Well	Dubai	UAE	246,250,219
2004	Rig	Mediterranean	Egypt	230,104,683

Source: Willis Energy Loss Database, 1 April 2011.

the Macondo loss of April 2010 was huge compared to all other OEE losses in the Willis Energy Loss Database, even on an inflation-adjusted basis. Moreover, note that this only includes the amounts as they were known on 1 April 2011. As we will show when discussing the Deepwater Horizon in more detail,[73] the amounts meanwhile have already become substantially higher.

2.3.1.5 Observation from Available Data

Data for offshore incidents hence are available from various sources. The operators or licence holders even have the obligation to report to national regulatory authorities; this makes data collection possible at least at a national level. However, the offshore incidents are not always listed separately, as in the Dutch SSM system, where only total accident records are available. Offshore and onshore accidents are not recorded separately. Another difficulty in offshore incident data collection is that many resources are not publicly available, e.g. data for the OGP are only available to its participating members. Also with regard to the most comprehensive database currently available, the WOAD, the information is acquired through publicly available sources, and it is restricted in the sense that countries with a fully state-owned offshore industry do not disclose the information on offshore

[73] See Section 2.5.2.

accidents to the public.[74] A third difficulty may have to do with the fact that not all databases apply the same criteria when collecting data. Lacking harmonized criteria makes it difficult to compare the available data.[75]

Realizing these difficulties, the recent Directive on Safety of Offshore Oil and Gas Operations (Directive 2013/30/EU) has specific provisions on a 'common data reporting format' across all member states of the European Union.[76] The idea of the Directive on such a format is that operators and owners should provide the competent authority with major hazard indicators as described in Annex IX of the Directive, whereby it would make the information comparable and hence facilitate the sharing of information.[77]

2.3.2 Overview of Serious Accidents Involving Offshore Facilities

An interesting overview of high-profile oil spills from offshore blowouts is also provided in a recent report by Lloyds (see Table 2).

Overviews of upstream losses in the energy sector are also provided in the WELD. This database provides Tables 2 and 4 for 2013 and 2014.

These tables show the major loss records for the upstream energy industry in 2013 and 2014. However, the reported losses do not relate solely to offshore incidents (although many do), and in many cases, the losses reported did not cause damage to third parties but, for example, relate to the costs of reinstating a platform. The charts, however, provide an indication of the fact that in the upstream energy industry on a yearly basis substantial losses still occur all over the world.[78]

Reinstating the platform is the obligation of the oil company. However, for example, the first incident mentioned in Table 3 for 2011 shows total damage of more than US$1 billion. This is relevant to this study to the extent that insurers provide a total package of coverage for the oil industry. In other words, if large losses occur, e.g. because of business interruption, that may also reduce the available capacity for the cover for third-party liability.

[74] OGP, 'Risk Assessment Data Directory', 30.
[75] Christou and Konstantinidou, 'Safety of Offshore Oil', 15.
[76] See e.g. Preamble (42) and (43), Art. 23 and Annex IX of Directive 2013/30/EU.
[77] For further discussion of the Directive, see Section 3.4.2.
[78] Interview with Tim Taylor (Clyde & Co.) from Lloyds in London, 1 May 2013.

Table 2 *High-Profile Oil Spills from Offshore Blowouts*

Date of incident	Location	Incident and spillage details (estimated figures)	Insured loss (US$)
28.1.1969–12.2.1969	Santa Barbara, California	80,000–100,000 barrels	Not available
3.6.1979–23.3.1980	Ixtoc well, Mexico	3.3 million barrels	22,000,000
22.4.1977–30.4.1977	Ekofisk Norwegian sector, North Sea	202,381 barrels	6,887,000
1980	Funiwa Niger Delta, Nigeria	200,000 barrels	53,554,000
2.10.1980–10.10.1980	Arabian Gulf	100,000 barrels	1,300,000
21.8.2009–3.11.2009	Timor Sea, Australia/Indonesia	28,800 barrels of condensate oil	425,000,000
20.4.2010–15.7.2010	Gulf of Mexico	4.9 million barrels plus 11 fatalities and 17 injuries	2,560,000,000

Source: Rees and Sharp, *Drilling in Extreme Environments*, 17. The information is also adapted from the WELD.

Table 3 *Upstream Losses in Excess of US$50 million, 2013*

Type	Cause	Country	PD, US$	OEE, US$	BI, US$	Total actual US$
Platform	Mechanical failure	Europe	80,280,000		300,000,000	380,280,000
Rig	Leg punch-through	Africa	285,000,000			420,000,000
Platform	Faulty work/operating error	Asia	237,000,000			237,000,000
Rig	Blowout	North America	64,000,000	81,844,000		145,844,000
Well	Blowout	Asia		140,000,000		140,000,000
Platform	Impact	Africa	120,000,000			120,000,000
Rig	Faulty work/operating error	Asia	100,000,000			100,000,000
Well	Blowout	Eurasia	57,000,000	40,000,000		97,000,000
Well	Heavy weather	Europe		76,580,000		76,580,000
Rig	Mechanical failure	Latin America	4,500,000		70,700,000	75,200,000
Pipeline	Faulty design	Europe	65,400,000			65,400,000
Rig	Blowout	Asia	65,000,000			65,000,000
Rig	Mechanical failure	Latin America	60,000,000			60,000,000
MOPU	Corrosion	Africa	50,000,000			50,000,000
Pipeline	Anchor/jacking/trawl	China	50,000,000			50,000,000
Total						1,947,304,000

BI, bodily injury; MOPU, mobile offshore production units; OEE, operators' extra expense; PD, property damage. See *Willis Natural Resources*, 53.

Source: WELD, 9 March 2015 (figures include both insured and uninsured losses).

Table 4 *Upstream Losses in Excess of US$50 million, 2014*

Type	Cause	Country	PD, US$	OEE, US$	BI, US$	Total US$
Rig	Blowout	Central America	65,000,000	30,000,000	16,500,000	111,500,000
Platform	Piling operations	Central America	95,147,421			95,147,421
Platform	Subsidence/landslide	Asia	89,000,000			89,000,000
Platform	Unknown	North America	68,000,000			68,000,000
Pipeline	Anchor/jacking/trawl	Europe	63,000,000			63,000,000
Total						426,647,421

2.4 Analysis of Recent Offshore Accidents in Europe: Case Studies

Catastrophic accidents in the European Union are rare. However, accident data for offshore facilities of both the UK and Norwegian authorities reveal that incidents do happen. This section will therefore analyze the reasons for and consequences of recent offshore accidents in Europe in slightly more depth by focusing on a few specific incidents in more detail. The case-study accidents that follow constitute some of the most serious accidents that have happened in the North Sea in the last five years.

2.4.1 Gulfaks C Incident (19 May 2010): A Near Miss

On 19 May 2010, a serious situation occurred in the C06A well on the Gullfaks C platform in the North Sea. Statoil struggled for more than two months to regain control of the well.

The Petroleum Safety Authority's (PSA's) conclusion following the incident was that the planning of the drilling and completion operation in well C06A was executed with serious and extensive inadequacies. This failure was related to key factors such as risk management and change control, experience transfer and use of expertise, and familiarity with and compliance with governing documents and documenting decisions.

In slightly different circumstances, the situation could have resulted in a major accident on the Norwegian shelf. The PSA and others also pointed out similarities between the Gullfaks C incident and the dramatic incident on Snorre A in 2004, when small margins prevented the near miss from developing into a major accident. Only chance averted a subsurface blowout and/or explosion and prevented the incident from developing into a major accident.[79]

2.4.2 Valhall PCP Production Platform Fire (13 July 2011)

The incident on the Valhall PCP production platform in the North Sea (Norwegian continental shelf) on 13 July 2011 involved the breakdown of a crane engine by overheating, which led to a fire in the vent stack for the installation's compressors. A minor blaze was caused by the engine

[79] www.psa.no/news/gullfaks-c-order-fulfilled-article8675-878.html. See also Petroleum Safety Authority, 'Safety Status & Signals 2012–2013', 25.

breakdown in the crane machinery space. Overheating combined with a defective spark arrester and silencer meant that red-hot particles leaving the exhaust pipe blew across and ignited flammable gases from the vent stack.

Extinguishing the flames proved difficult, and it was decided in the end to use the water monitors on the standby ship for this purpose. One hour and thirty-seven minutes passed from detection of the fire until it had been extinguished.

No people were physically injured by the incident. In slightly different circumstances, however, it could have escalated and created a serious situation on the installation, with personal injuries or loss of life. Production on the Valhall PCP was shut down for nine and a half weeks as a result of the incident.

Norway's PSA notified the operator BP of orders following its investigation of the Valhall PCP fire on 13 July 2011. Serious non-conformities related to BP's management system for activities on the Norwegian continental shelf were identified. These relate to lack of maintenance, deficient maintenance management, inadequacies in risk identification, and deficient barrier management.

The investigation identified non-conformities in relation to

- Machinery protection of the crane engine,
- Classification of and maintenance programme for the crane engine's spark arrester and silencer,
- Updating of maintenance programme for crane engines,
- Fire and gas detection systems,
- Management of work processes and follow-up of identified risks,
- Information on hazards,
- Barrier management and handling of nonconformities,
- Risk assessment of ignition in vent lines,
- Turbine exhaust leading to a hazardous area and
- Handling of hazards and accidents.[80]

2.4.3 Escape of Hydrocarbons at Ula Field Norway (12 September 2012)

A substantial escape of hydrocarbons occurred at the Ula field in the Norwegian North Sea on 12 September 2012. The PSA decided to

[80] www.ptil.no/news/notification-of-orders-to-bp-after-investigation-of-valhall-pcp-fire-article8233-79.html.

investigate this incident. No people were injured and no damage was caused to the installation beyond the equipment directly involved, but the PSA considers the incident to have had a substantial potential.

The Ula oil field lies in Norway's North Sea sector and has three conventional steel platforms for production, drilling, and living quarters. These are linked by bridges. BP is the operator.

The leak arose in the separator module on Ula's production platform. Nobody was in the module when the incident occurred. While the facility was automatically shut down, all personnel on the installation were evacuated to the drilling platform. Production on Ula was suspended for a while.

One reason why the PSA decided to conduct an investigation is the substantial potential involved in the incident. Objectives include establishing the course of events and identifying the direct and underlying causes.[81]

2.4.4 Gannet Alpha Oil Spill (10 August 2011)

Around 216 tonnes of oil leaked from Shell's Gannett A platform into the North Sea in August 2011. It can be considered to be the single largest discharge in UK waters in the last ten years. The Department of Energy and Climate Change called it a substantial spill in the context of annual amounts of oil spilled into the North Sea, but it anticipated that the spill would disperse naturally.[82] The high winds and waves led to a substantial reduction in the size of the oil sheen on the water.

Yet, while the spill did not approach the devastating impact of the Deepwater Horizon disaster, its significance lies in the fact that it took place under the much-vaunted 'gold standard' regulatory regime of the United Kingdom and by a company that has been trading on its reputation as a responsible corporate citizen.

The Guardian claimed that Gannet Alpha should serve as a wake-up call to the government on the risk of offshore oil drilling. Indeed, Gannet Alpha and its satellite projects are among the majority of offshore installations that are approaching or have exceeded their original design life (typically twenty to twenty-five years) and are

[81] The resulting report is published on the PSA's website at www.ptil.no.
[82] www.decc.gov.uk/en/content/cms/news/statement_gfs/statement_gfs.aspx.

posing an extra danger the longer they operate.[83] According to the Health and Safety Executive, most hydrocarbon releases happen at facilities that are older than twenty years,[84] and more than 50 per cent of existing platforms fall in this category. Gannet Alpha has celebrated its twentieth birthday.

According to *The Guardian*, after the oil spill, 'the response by the government and regulators has been to downplay concerns about ageing infrastructure and oversell the regulatory regime's ability to cope with a ticking time bomb. *The Guardian* holds that there is a high probability that we will see more major oil spills and worker injuries in the coming years due to lack of regulatory capacity, a general drive towards light touch regulation and an apparent reluctance on the part of a government obsessed with "energy security" to challenge Big Oil.'[85]

2.4.5 Elgin Platform Blowout (26 March 2012)

A blowout in early 2012 at Total's Elgin platform in the North Sea, which led to a two-month crisis and the loss of a massive amount of gas, fuelled fresh scrutiny of offshore safety standards. On March 26, Total reported a gas leak that forced it to evacuate more than 200 workers from a production platform in the Elgin field of the central North Sea, about 150 miles east of Aberdeen, Scotland. It soon became clear that it was an uncontrolled blowout of natural gas and liquid gas condensate, a potentially explosive situation that caused other companies to evacuate and shut down operations at neighbouring facilities even miles away from Totals' Elgin platform.

The report by the Health and Safety Executive (HSE) raised concerns about backlogs in the maintenance on equipment critical to safeguarding life. With rising operating costs and lower revenues, companies have put pressure on facilities to produce more oil in order to break even, which means reducing the number of safety checks that could interrupt production. HSE has previously said that declines in safety checks started after a period of low oil prices in the last decade.[86]

83 www.guardian.co.uk/commentisfree/2011/aug/17/oil-spill-shell-timebomb.
84 Health and Safety Executive, 'Offshore Hydrocarbon Releases, 2001–2008'.
85 www.guardian.co.uk/commentisfree/2011/aug/17/oil-spill-shell-timebomb.
86 http://maritime-connector.com/news/offshore-oil-gas-news/uk-north-sea-oil-and-gas-leaks-at-lowest-number-in-the-last-ten-years.

2.5 Analysis of Recent Offshore Accidents Outside EU Waters

2.5.1 Montara

2.5.1.1 Facts of the Incident

The Montara incident was an oil and gas leak that took place in the Timor Sea off the northern coast of Western Australia. [87] On 21 August 2009, during drilling operations at the Montara wellhead platform, an uncontrolled release of oil and gas occurred from the H1 well. All sixty-nine workers at the platform were evacuated safely. On 1 November 2009, the leaking well was intercepted (capped). However, during operations to complete the 'well kill', fire broke out on the West Atlas rig and the Montara wellhead platform. On 3 November 2009, the fire was extinguished. For a period of seventy-four days, oil and gas continued to flow unabated into the Timor Sea, approximately 250 kilometres off the northwest coast of Australia. Patches of sheen or weathered oil could have affected at various times an area as large as 90,000 square kilometres.[88]

The Montara Development Project is owned and operated by PTTEP Australasia (PTTEP AA), a subsidiary of PTT Exploration and Production (PTTEP). PTTEP is Thailand's national petroleum exploration and production company and is one of the nation's largest publicly listed companies.[89] PTTEPAA was granted a permit and license around November 2008 to drill three development wells in the Montara oil field, one of them being the H1 well which caused the incident.[90] The West Atlas rig (owned and operated by Atlas) was positioned over the Montara wellhead platform to drill the wells (as contractor) for PTTEP AA.[91]

Under Australian law, the registered title holder of a permit, lease or license bears the obligation to prevent the occurrence of an oil spill.[92] Therefore, in the case of Montara, it is PTTEP AA that was held liable.

As to the amount of oil that was discharged into the sea, the estimates given by PTTEP AA were that the oil flowed into the sea at a rate of

[87] www.au.pttep.com/news/2012.

[88] Borthwick, 'Report of the Montara Commission', 5.

[89] www.au.pttep.com/; see also Borthwick, 'Report of the Montara Commission', 36.

[90] Later, PTTEP AA sought approval to batch drill two additional wells. Therefore, there were five wells at the Montara oil field, four production wells, H1, H2, H3 and H4, and a gas injection well, GI. See Borthwick, 'Report of the Montara Commission', 49.

[91] *Ibid.*, 49–50. [92] See the discussion in Section 3.5.6.

400 barrels per day.[93] This was questioned by the Montara Commission of Inquiry, which noticed, in particular, the testimony from PTTEPAA's chief operating officer, who even suggested that 'the initial flow may have been as high as 1,000 to 1,500 barrels per day before dropping to around 400 barrels and possibly less.'[94] The Commission believed that there are 'methods that could and should have been applied to get a more informed estimate of the amount of oil that was released, with a view to informing the public'.[95] In the final report, the Commission stated that the Well leaked 'possibly between 400 and 1,500 barrels of oil per day, and unknown amounts of gas, condensate and water, until the Relief Well operations were successful in "killing" the well over ten weeks later'.[96]

The Commission estimated the surface coverage of the released hydrocarbons to have ranged from 6,000 to 25,000 square kilometres. The evidence before the Commission indicated that hydrocarbons did enter Indonesian and Timor Leste waters to a significant degree. The estimate from the Australian Maritime Safety Authority (AMSA) of the total surface area within which oil or sheen was observed at one time or another during the spill was around 90,000 square kilometres. However, as indicated by AMSA in its submission to the Commission, most of the hydrocarbons remained 'within 35 kilometres of the platform with patches of sheen and weathered oil reported at various distances in different directions from the platform as currents, wind and temperature varied over the three month period'.[97]

Even when based on the estimate of 400 barrels per day, the volume of oil spilled from the Montara wellhead platform makes it Australia's third-largest oil spill. Only two oil spills from the tanker *Kirki* in 1991 and the *Princess Anne Marie* in 1975 were larger:

> The Blowout caused the worst oil spill in Australia's offshore petroleum industry history. Previously, there had been six offshore blowouts in Australian waters between 1965 and 1984. These involved either no oil spill or spills of only negligible amounts. It is still fortunate, in view of the highly flammable nature of the material released, that the impact of the blowout was not more severe and did not include the loss of human lives.[98]

The reason for the incident, according to the findings of the Commission of Inquiry, was that PTTEP AA 'did not observe sensible

[93] www.au.pttep.com/media/20774/montara%20oil%20spill%20incident%20response.pdf.
[94] Borthwick, 'Report of the Montara Commission', 26. [95] *Ibid.* [96] *Ibid*, 38.
[97] *Ibid*, 26. [98] *Ibid*, 38.

oilfield practices at the Montara Oilfield. Major shortcomings in the company's procedures were widespread and systemic, directly leading to the Blowout.' 'Well-control practices approved by the delegate of the Designated Authority, the Northern Territory Department of Resources, would most likely have been sufficient to prevent the Blowout if PTTEP AA had adhered to them and to its own Well Construction Standards.'[99]

The Commission expressed concerns because environmental impacts from the blowout are difficult to determine and are unlikely to be ever known. This is so because the location of the Montara wellhead platform is remote, and there are few baseline data about species and habitats. There was also a delay before the commencement of scientific monitoring of the environmental impacts of the oil spill.[100]

'PTTEP AA volunteered to the Australian government within the first few days of the incident that the company would pay for all response and clean-up costs. These have so far totalled more than AUD 320 million.' 'The company is also paying for a long term environmental monitoring programme under a separate agreement with the Australian government.'[101]

PTTEP received a fine of AUD 510,000 (approximately US$526,320) on 31 August 2012, from the Darwin Magistrates' Court for its actions in relation to the Montara spill. The court found that PTTEP was guilty on four charges: '[T]he first three charges, each of which carries a maximum penalty of $567,000 (AUD 550,000), fall under Clause 9(4) of the Offshore Petroleum and Greenhouse Gas Storage Act of 2006. The fourth charge, which carries a maximum penalty of $56,760 (AUD 55,000), lies under Section 569(1) and it relates to the failure to ensure that operations were conducted "in a proper and workman-like manner and in accordance with good oilfield practice".'[102]

PTTEP could have been confronted with the maximum fine of $1.8 million (AUD 1.7 million). The magistrate, however, recognized that PTTEP took on full responsibility for the incident and that the company had made every effort to transform its operations and culture. In addition, the magistrate also acknowledged that the company had no prior accident record.[103] The PTTEP launched a coordinated incident response, and it has spent around $41 million to $52 million (AUD 40 to 50 million) in its direct environmental clean-up operations.[104]

[99] *Ibid*, 6. [100] *Ibid*, 39.
[101] www.au.pttep.com/media/20774/montara%20oil%20spill%20incident%20response.pdf.
[102] www.rigzone.com/news/article.asp?a_id=120382. [103] *Ibid*. [104] *Ibid*.

2.5.1.2 Trans-Boundary Complaints from Indonesia

The Montara wellhead platform is located in the Timor Sea 254 kilometres northwest of the Western Australian coast and 685 kilometres from Darwin. The Ashmore Reef National Nature Reserve and Cartier Island Marine Reserve are approximately 157 kilometres away from Montara.[105] Montara is situated 250 kilometres from Australia's Kimberly coastline and 300 kilometres from the nearest Indonesian landfall. Some of the oil from the spill reached Indonesian waters. Following the incident, the government of Indonesia sought US$2.4 billion in compensation from PTTEP AA. The Indonesian government claimed that fishing grounds in the Timor Sea were polluted, that catches fell significantly as a result of the spill and inshore seaweed fishermen (particularly on the island of Rote) had their livelihoods destroyed.[106] PTTEP did not accept the claim on the basis that 'no verifiable scientific evidence' had been presented to the company to support the Indonesian government's claim for compensation.

After initially rejecting the demand for compensation, PTTEP was reportedly willing to partly compensate the fishermen.[107] PTTEP agreed to sign a memorandum of understanding (MOU) on the compensation estimated to reach around $3 million (instead of the US$2.4 billion as initially demanded by the Indonesian government).[108] However, the signing of the MOU has been postponed several times.

In general, under customary international law, a state shall exercise due diligence to ensure that activities carried out within its jurisdiction do not cause extra-territorial harm to the rights or interests of other states.[109] Such a state responsibility shall also be applicable in the case of the marine environment.[110] This was confirmed in UNCLOS, which stipulates that a coastal state has the exclusive sovereign rights over offshore exploration and exploitation activities that take place in the exclusive economic zone (EEZ) and the continental shelf; meanwhile, the coastal state also has a duty to protect and preserve the marine environment.[111] However, when an incident causes damage to other states beyond the jurisdiction of one particular coastal state, there is no

[105] Christou and Konstantinidou, 'Safety of Offshore Oil', 19.
[106] www.au.pttep.com/media/20778/government%20of%20indonesia%20compensation%20claim.pdf.
[107] www.antaranews.com/en/news/78539/montara-oil-spill-victims-still-awaiting-promised-compensation.
[108] *Ibid.* [109] Brownie, *Principles of International Law*, 455.
[110] Smith B. D., *State Responsibility*, 22. [111] This will be discussed in Chapter 3.

international regime to solve the trans-boundary damage. Very often the states rely on regional or bilateral agreements.

However, as the Montara case illustrates, in the absence of such an agreement, many hurdles exist as to the claims for compensation. One obvious difficulty relates to the evidence required in the trans-boundary claim.[112] Because the activities take place in a state other than where the damage is incurred, it is inevitably difficult for the suffering state to collect sufficient evidence to prove the causal link between the damage it has sustained and the activity in another jurisdiction. It is for this reason that the claims from the Indonesian government were not accepted by the PTTEP.

According to Article 198 of UNCLOS, when a state is aware of imminent or actual damage to the marine environment, 'it shall immediately notify other States it deems likely to be affected by such damage, as well as the competent international organizations.' On this point, the Australian government was criticized for not informing Indonesia, which could be affected by the damage caused by the Montara incident.[113]

2.5.2 *Deepwater Horizon*

2.5.2.1 Facts of the Incident and Legal Issues

On 20 April 2010, the ultra-deepwater, semi-submersible mobile offshore oil rig Deepwater Horizon exploded, caught fire, and sank in the Gulf of Mexico off the shores of Louisiana.[114] The rig was owned and operated by Transocean, a Switzerland-based offshore drilling contractor, and leased to British Petroleum (BP), Plc, one of the world's largest oil companies. The explosion and fire, which resulted in eleven fatalities and several injuries, occurred despite specialized oil spill prevention equipment called a 'blowout preventer' (BOP) designed to avert this type of disaster.[115] The failure of the BOP left

112 Bosma, 'The Regulation of Marine Pollution', 90.

113 Crawford, 'Australia's Lack of Compliance'.

114 Because the rig was named Deepwater Horizon, the case is mostly referred to as the Deepwater Horizon; however, the particular well where the blowout occurred was referred to as Macondo. Hence, one can find references to this incident as either Deepwater Horizon or Macondo.

115 Blowouts occur during offshore drilling operations when pressure exceeds the weight of the drilling fluid in the well, which results in an uncontrolled flow of oil. The oil flow could result in loss of property at the drill site. See King, 'Deepwater Horizon Oil Spill Disaster', 3.

the well unsecured and leaking from the marine riser. According to a BP report of September 2010, the incident started with a 'well integrity failure' and was then followed by a loss of control of the pressure of the fluid in the well.[116] The well was capped on 15 July 2010 and permanently sealed on 19 September 2010.

With respect to the parties involved, there was a joint venture between the following parties:

1. BP owned 65 per cent of the well;
2. Anadarko Petroleum Corporation owned 25 per cent of the well;
3. Mitsui Oil Exploration Company of Japan (MOEX Offshore LLC) owned 10 per cent of the lease for the Macondo well at the time of the incident (MOEX is a fully owned subsidiary of MOEX USA Corporation, and Mitsui Oil Exploration Co., Ltd., is the corporate parent of MOEX USA, which, in turn, is owned by Mitsui & Co., Ltd., of Japan); and
4. Transocean Ltd. is the owner of the oil rig Deepwater Horizon and is registered in the Marshall Islands.

Other parties involved include

1. Cameron International, manufacturer of the BOP, and
2. Halliburton, the drilling contractor that cemented the well.

The legal issues at stake can be summarized as follows[117]: first, as to the right of limitation of liability, according to the liability limit provision under the US Oil Pollution Act (OPA), BP's liability would be capped at $75 million if the parties were not guilty of gross negligence. The Deepwater Horizon was a semi-submersible drilling rig, thus a so-called MODU,[118] 'being used as an offshore facility, which is deemed to be a tank vessel with respect to the discharge, or the substantial threat of a discharge, of oil on or above the surface of the water'.[119] If the removal costs and damages resulting from an incident involving a MODU exceed the amount for which the responsible party is liable, then the MODU is deemed an offshore facility.[120] Thus, the responsible party for the offshore facility, the lessee or permittee, becomes liable for any excess

[116] *Ibid.*

[117] We will now merely summarise the main legal issues; the way in which the Gulf Coast Claim Facility (GCCF) functioned will be elaborated in Section 7.6.1.4.

[118] A MODU is a mobile offshore drilling unit.

[119] 33 USC §2704(b)(1).

[120] 33 USC §2704(b)(2).

liability with respect to the oil discharged on or above the surface of the water.[121]

BP announced that it would waive its right to limit under OPA[122] and that it had already paid claims in excess of the statutory limit.[123] Nevertheless, BP reserved its right to seek reimbursement or contributions from other responsible parties and from third parties for claims, costs, expenses and liabilities arising from the incident.[124]

Second, OPA has a requirement of providing a compulsory financial guarantee.[125] In the case of the Deepwater Horizon, the insurance arrangements were as follows. On the one hand, on the part of the licence holders, the insurance arrangements include

- BP (65 per cent interest holder) self-insures through its own captive insurance company, Jupiter Insurance, Ltd. Jupiter is reported to have an underwriting limit of US$700 million and does not purchase re-insurance. In other words, any insurance cover that might be needed is taken on BP's own balance sheet.
- Anadarko (25 per cent interest holder) has operator's extra expenses (OEEs) insurance with a limit of US$250 million for 100 per cent, equating to US$62.5 million for its 25 per cent interest. In addition, Anadarko has a third-party liability insurance with a limit of US$150 million.
- MOEX (10 per cent interest holder) has OEE insurance with a limit of US$300 million for 100 per cent, equating to US$30 million for its 10 per cent interest. It also has a third-party liability insurance with a limit of US$150 million.[126]

On the other hand, the contractors involved also have certain insurance arrangements. Transocean (the drill rig owner and operator) has several insurance arrangements:

[121] Force, Davies and Force, 'Deepwater Horizon', 945; see also De la Rue and Anderson, *Shipping and the Environment*, 249. The lessee or permittee also remains liable for a discharge of oil from the MODU that occurs below the surface of the water.

[122] The legal structure of the compensation under OPA will be discussed in Section 7.6.1.4.1.

[123] Statement of BP Exploration and Production, Inc. re Applicability of Limit of Liability under Oil Pollution Act of 1990, *in re Oil Spill by the Oil Rig 'Deepwater Horizon' in the Gulf of Mexico*, MDL No. 2179, 2010 WL 4151003, 18 October 2010. BP also urged Transocean, Anadarko and MOEX to waive their rights to limitation 'unequivocally' in its court filing.

[124] Force, Davies and Force, 'Deepwater Horizon', 948.

[125] For further details, see Section 4.5.

[126] Bosma, 'The Regulation of Marine Pollution', 108–9.

- Hull insurance for the Deepwater Horizon drilling rig with an agreed insurance value of US$560 million. This covers loss of the rig itself and does not cover any pollution liability or liability for third-party claims.
- A casualty insurance programme covering up to US$1 billion, among which US$50 million is self-insured by Transocean. This cover is placed with insurers in London, the United States and Bermuda.
- Separate US workers' compensation insurance for statutory limits.
- Contingent OEE insurance with US$150 million limit in the event Transocean becomes contractually liable to pay for the costs or claims arising from the pollution.

Halliburton (well service contractor) has third-party liability insurance up to US$600 million, and Cameron Iron Works (BOP supplier) has third-party liability insurance up to US$500 million.[127]

In addition, according to the Clean Water Act (CWA), there are relevant provisions concerning environmental crimes and administrative and civil penalties. Section 1319 of the CWA, titled 'Enforcement', sets forth civil and administrative penalties [§1319(d)(g)] and criminal penalties [§1319(c)] for illegal discharges. Criminal penalties are imposed for the following:

- Negligent violations §1319(c)(1)(A) and (B);
- Knowing violations §1319(c)(2)(A) and (B);
- Knowing endangerment §1319(c)(A) and (B) and
- False statements (§1319(c)(4).

As to negligent violations, the CWA provides in §1319(c)(1)[128] that anyone who 'negligently' violates specified sections of the CWA has committed an offense under the act. Such a person will be punished by a fine of between $2,500 and $25,000 per day of violation or by imprisonment for not more than one year or both. The sanctions for knowing violations are a fine of not more than $50,000 per day of violation, imprisonment for not more than two years, or both.

The core issue is whether the spill has to be considered to be due to gross negligence or not. Under the OPA, if the responsible party is not guilty of gross negligence, it may be entitled to a financial cap on its liability of $75 million in economic damages, although it may still need to pay the full amount of clean-up costs.[129] If the party were grossly negligent, it not only would lose the right of limiting its liability but

[127] *Ibid.* [128] 33 USC §1319(c)(1). [129] See further in Section 4.5.2.

also would be confronted with a higher amount of fine pursuant to the CWA.

2.5.2.2 Overview of Legal Proceedings

Within months of the incident, more than a hundred lawsuits were filed against one or more of the parties just mentioned.[130] These include civil litigation from private parties for compensation for their damages and civil and criminal charges from the US federal government,[131] and the coastal states in the Gulf area affected by the incident have all filed lawsuits seeking damages. Moreover, among the parties involved in the incident with contractual relationships, complaints have been filed against each other seeking contribution of fault by the other entities.

On 10 August 2010, many of the cases against BP and other parties were consolidated in the US District Court for the Eastern District of Louisiana under the Multi-District Litigation Docket MDL No. 2179, entitled Re: oil spill by the oil rig Deepwater Horizon in the Gulf of Mexico on April 20, 2010, presided by Judge Barbier.[132]

The litigation has gone through several phases to determine issues such as source-control efforts and the amount of oil that was discharged into the sea as a result of the incident. The 'source control' issues concern 'the conduct or inaction of BP, Transocean or other relevant parties regarding stopping the release of hydrocarbons stemming from the Incident'. The 'quantification of discharge' pertains to 'the amount of oil actually released into the Gulf of Mexico as a result of the Incident from the time when these releases began until the Macondo well was capped (on approximately 15 July 2010) and then permanently cemented shut on approximately 19 September 2010'.[133] These are crucial issues because the amount of penalty under the US federal law will depend on the amount of oil discharged into the sea and the severity of the penalty relates to the degree of negligence.

[130] This was reported by Transocean to the Congress on 27 May 2010.

[131] The US Department of Justice (DOJ) announced on 1 June 2010 that it would conduct an investigation into the incident encompassing possible violations of US civil or criminal law. On 15 December 2010, the DOJ filed a lawsuit against BP and other parties for violations under the Clean Water Act in the US District Court for the Eastern District of Louisiana, which was consolidated with the other cases, as will be discussed later.

[132] The timeline of the litigation and the court rulings are available at www.laed.uscourts.gov/oilspill/oilspill.htm.

[133] *Ibid.*

On 15 January 2015, the court issued the findings of fact and conclusions of law for the phase two trial. The court ruled that for the purpose of calculating civil penalty under the CWA, 3.19 million barrels of oil should be sustained. This refers to the 4.0 million barrels of oil discharged into the sea, deducting the amount recovered by the parties (81,000 barrels of oil). Moreover, the court found that 'BP was not grossly negligent, reckless, wilful, or wanton in its source control planning and preparation.'[134]

On 2 July 2015, BP and the five Gulf Coast states announced that they had reached a settlement of US$18.7 billion. This included

- $5.5. billion in CWA penalty,
- $7.1 billion in OPA fines in addition to $1 billion already paid under the law and
- $5.9 billion to resolve outstanding economic claims from the Gulf States and local entities.[135]

Realizing that litigation would be costly and lengthy, BP gave a high priority to negotiations with the plaintiffs' steering committee in order to reach some agreement on settlements.[136] Therefore, after a year of continuous negotiations, BP reached agreements on the economic and property-damage settlements and the medical benefits settlement. These will be discussed further in Chapter 7.

In addition to the litigation process and negotiations with various parties, BP also established some specially dedicated compensation facilities to fund compensation for victims of the incident. These private funding mechanisms will be discussed further in Section 7.6.1.4.

Prior to the settlement on 2 July 2015, several agreements had been reached related to civil penalties and criminal charges which are worth mentioning: first, among the contractual parties involved in the incident, BP has made settlements with respect to all claims related to the incident with MOEX on 20 May 2011, with Anadarko on 17 October 2011, and with Cameron on 16 December 2011.[137]

Second, as to the criminal charges against BP, it reached an agreement with the US government on 15 November 2012 mainly to resolve all

[134] Judgments and legal documents are available at www.laed.uscourts.gov/oilspill/oilspill.htm.

[135] www.eenews.net/greenwire/sotries/1060021204.

[136] See the official website of Deepwater Horizon Claims Center, https://cert.gardencitygroup.com/dwh/fs/faq?.dellogintype=faqs.

[137] For further details of these settlement agreements, see *ibid.*

federal criminal charges. BP settled for $4.5 billion in fines and penalty, including a $4 billion criminal fine to the US government over five years and a $525 million civil penalty to the Securities and Exchange Commission (SEC) paid over three years.[138] The criminal penalty settlement amounts to approximately $4 billion, which was later approved by the US District Court in the Eastern District of Louisiana on 29 January 2013. The settlement is distributed as follows:

- $2.394 million to the National Fish and Wildlife Foundation (NFWF) to support restoration efforts in the Gulf States;
- $1.15 billion to the Oil Spill Liability Trust Fund (OSLTF);
- $350 million to the National Academy of Sciences for oil spill prevention and response research;
- $100 million to the North America Wetland Conservation Fund and
- $6 million to the General Treasury.[139]

In accordance with the settlement, BP agreed to plead guilty to eleven felony counts of misconduct or neglect of ships officers related to the deaths of eleven people, as well as misdemeanor counts under the CWA and the Migratory Bird Treaty Act, and a felony count of obstruction of Congress.[140]

Third, on 3 January 2013, Transocean pleaded guilty to breach of the CWA, and it agreed to pay $1.4 billion in civil and criminal fines and penalties.[141] On 19 February 2013, Transocean reached an agreement with the Department of Justice (DOJ) for a settlement of $1 billion in civil fines and, on 14 February 2013, for $400 million in criminal fines.[142]

Fourth, on 17 February 2012, MOEX reached an agreement with DOJ, and this settlement was approved by the court on 8 June 2012. MOEX agreed to pay a $70 million civil penalty with an additional $20 million in supplemental environmental projects in the Gulf

[138] See SEC Press Release of 15 November 2012; available at: www.sec.gov/news/press/2012/2012-231.htm; see also the website of BP: www.bp.com/sectiongenericarticle800.do?categoryid=9048917&contentid=7082602.

[139] Ramseur and Hagerty, 'Deepwater Horizon Oil Spill', 5.

[140] *Ibid.* The settlement agreement is available at the website of the US Department of Justice: www.justice.gov/iso/opa/resources/43320121115143613990027.pdf.

[141] 'Transocean agrees to plead guilty to environmental crime and enter civil settlement to resolve US Clean Water Act Penalty Claims from Deepwater Horizon Incident', Department of Justice Office of Public Affairs, 3 January 2013.

[142] The Cooperation Guilty Plea Agreement is available at: www.justice.gov/opa/documents/transocean-plea-agreement.pdf.

States.[143] Among the $70 million penalty, $45 million goes to the OSLTF, and $25 million is to be distributed among the five Gulf States: $6.75 to Louisiana; $5 million each to Alabama, Florida and Mississippi; and $3.25 million to Texas.[144]

Fifth, on 20 March 2013, Judge Barbier ruled out punitive damages against Cameron (manufacturer of the BOP), and on 3 April 2013, Cameron was dismissed from the trial by Judge Barbier.

2.5.2.3 Post-Macondo Responses

The Deepwater Horizon incident served as an alert for both the industry and regulators. They all took immediate action following the Macondo incident.

First, let's look at the response by industry. In July 2010, oil majors Chevron, ConocoPhilips, ExxonMobil, and Shell reached an agreement to invest US$1 billion to build and deploy a flexible and adaptable containment system capable of capturing and containing oil from any underwater well blowout in the Gulf of Mexico. The new system, which was completed in 2012, should be able to mobilize within twenty-four hours and have an initial capacity of 100,000 barrels per day working at depths of up to 10,000 feet. The four companies also agreed to form a non-profit organization, the Marine Well Containment Company, to operate and maintain this system.[145] In September 2010, BP announced its intention to join the proposed project by providing the equipment and experienced BP technical personnel needed to respond to a deepwater well control incident.

Second, let's look at the reaction of regulators. After the incident, the US government tightened the regulatory mechanism concerning offshore activities. It restructured the federal regulatory agency responsible for monitoring offshore drilling (Minerals Management Service, known as MMS, was renamed Bureau of Ocean Energy Management, Regulation and Enforcement [BOEMRE] and divided into two divisions: Bureau of Ocean Energy Management [BOEM] and Bureau of Safety and Environmental Enforcement [BSEE]).[146] BOEMRE issued two regulations to enhance the safety of offshore

143 www.epa.gov/enforcement/air/cases/moex.html.

144 Consent Decree between the United States and MOEX Offshore 2007 LLC, 10; available at: www.epa.gov/compliance/resources/decrees/civil/caa/moex-cd.pdf.

145 New Oil Spill Containment System, Shell, 22 July 2010. See also KPMG Global Energy Institute, 'After the Gulf of Mexico Oil Spill', 7.

146 For a further discussion of these federal institutions, see Section 3.5.5.2.

activities on 30 September 2010, the Drilling Safety Rule and the Workplace Safety Rule. The first regulation sets standards to improve the safety of offshore activities on the US outer continental shelf. The second makes mandatory the safety and environmental management system (SEMS).

A ban on offshore drilling was imposed again after the incident, but it was lifted on 12 October 2010. In December 2010, the Obama administration announced that it would not allow drilling off the Atlantic coast and in the eastern Gulf of Mexico near Florida.

Royal Dutch Shell urged the US regulator to adopt the stricter drilling regulations of the North Sea and impose them on offshore drilling companies operating in the Gulf area.[147] It was estimated that the tighter regulations could cost the industry up to US$183 million per year. The new regulations were expected to add US$1.4 million to the cost of each new deepwater well and US$90,000 to that of each shallow-water well. Overall, the number of rigs in operation was expected to fall down from forty-five to thirty-five before the accident due to increased liability, tighter operating standards and higher costs.[148]

The US Environmental Protection Agency (EPA) announced in November 2012 that it had temporarily suspended BP, Plc, and other BP companies from participating in or receiving new federal contracts or renewing an expiring one. However, the suspension does not affect existing contracts BP has with the US government.[149]

2.5.2.4 Summary

The case of the Macando/Deepwater Horizon incident shows that, at least in the United States, an operator, such as in this particular case BP, but also other contractors involved in the operation of the rig, can be confronted with a large variety of claims. Five years have passed since the Deepwater Horizon incident, and the legal proceedings are not finished yet. It is important to stress that payments by BP take place (some already paid) at three different levels:

- An amount of $4 billion was paid as a record criminal penalty settlement;

[147] 'Shell Wants US to Adopt Tougher Drilling Standards', *National Post's Financial Post & FP Investing* (Canada). Quoted from Booz Allen Hamilton, 'The Offshore Oil and Gas Industry – Part one', 14.

[148] KPMG Global Energy Institute, 'After the Gulf of Mexico Oil Spill', 9.

[149] www.bp.com/sectiongenericarticle800.do?categoryid=9048917&contentid=7082602.

- An amount of $20 billion was paid to the Gulf Coast Claims Facility (GCCF) by BP (which will be discussed further in Section 7.6.1.4) and
- Substantial amounts had already been paid by BP in civil penalties, but the precise amount of these civil penalties depends on two crucial issues:
  - The total amount of oil released and
  - Whether or not there was gross negligence. Although BP waived its right to call on the limit under OPA, the question of whether there was gross negligence or not is still relevant to determine the civil fine under the CWA because that is four times as high ($4,300) in the case of gross negligence than when there is no gross negligence ($1,100).[150]

To decide on these two issues is important but remains difficult. It was only in January 2015 that the court could rule on the amount of oil discharged into the gulf as 3.19 million barrels. As to whether BP is at 'gross negligence' or not, the court ruled at the end of the first phase of the trial in September 2014 that BP was at gross negligence and wilful misconduct under the Clean Water Act.[151] However, in the judgment of January 2015, the court decided that BP was not at gross negligence.[152]

Although there was finally an agreement on 2 July 2015, this still needs to be approved by the court. Therefore, uncertainty still exists as to the conclusion of the case.

2.5.3 *More Accidents in 2015*

2.5.3.1 Petrobras Brazil Offshore Explosion

On 11 February 2015, an explosion occurred on an offshore oil and gas platform operated by Brazilian state-run Petrobras. The explosion was caused by a gas leak on board a floating oil production, storage and offloading ship (FPSO) owned by Norway-listed ship leaser BW Offshore, Ltd., which also processes natural gas that is sent to shore by an undersea pipeline.[153]

150 See Section 2.5.2.2, and see 'Daily Report for Executives', 20 February 2013; available at: http//dailyreport.bna.com/drpt/display/batch_print_display.adp.

151 'Findings of Fact and Conclusion of Laws', in *In re: Oil Spill by the Oil Rig 'Deepwater Horizon' in the Gulf of Mexico, on April 20, 2010*, filed 9 September 2014.

152 *Ibid.*, filed 15 January 2015.

153 Reuters, 11 February 2015.

2.5.3.2 Pemex Mexico Offshore Accidents

On 1 April 2015, a platform from the Mexican oil company Pemex caught fire, and at least four people died in the accident and more than 100 workers were evacuated. On 5 May 2015, Pemex had an offshore rig accident in the southern Bay of Campeche which killed two workers. 'Pemex said the Troll Solution rig, which was contracted to operate in Pemex's Abkatunpol-Chuc shallow water oil field, was positioning itself to carry out maintenance on wells linked to the Caan Alf platform.'[154] It was probably thanks to the fact that the rig was used for maintenance instead of production that no large amount of oil or gas leak into the sea.

It seems that accidents related to offshore activities do not end, although more and more attention is paid to improving safety measures. In the case of an accident, compensation and civil liability definitely play an important role, which again points to the need for a study focusing on the civil liability for offshore activities.

2.5.4 Risk of a Deepwater Horizon Oil Spill in the European Union?

2.5.4.1 Different Depths and Pressures

As discussed earlier, on 20 April 2010, an explosion occurred to the Deepwater Horizon drilling rig following a blowout of the exploratory Macondo oil well in the Gulf of Mexico. The disaster killed eleven platform workers, injured seventeen, and caused one of the greatest environmental disasters in US history. Over a period of several months, an unprecedented amount of oil flowed into the gulf. By the time the well was sealed, over 4 million barrels of oil had been released into the gulf – one barrel of oil roughly equalling 160 litres. The spill severely damaged the local oceanic environment, polluted many coastal areas and threatened the livelihoods of those living in the area. Apart from causing environmental damage, the oil spill affected both the local economy (fisheries, tourism, etc.) and the global economy – which has become more and more dependent on a steady supply of natural resources of this kind.[155]

[154] Reuters, 5 May 2015.

[155] Janssen, *Preventing a Next Deepwater Horizon?*, 11; and Kerr, Kintisch and Stokstad, 'Will Deepwater Horizon Set a New Standard?', 674–5.

In the North Sea, oil drilling conditions are somewhat different from those in the Gulf of Mexico. Most of the oil wells are located at depths of less than 100 metres. On the Dutch continental shelf, the maximum drilling depth is 50 metres. In the Norwegian part of the North Sea, the maximum drilling depth is 180 metres, off the coast of Bergen, whereas on the UK continental shelf, the deepest wells are located near the Shetland Islands. Depending on where the boundary of the North Sea is drawn, drilling depths extend to about 1,100 metres. Moreover, nowhere in the North Sea does deepwater drilling occur in high-pressure or high-temperature wells.[156]

2.5.4.2 Concerns after Macondo

Nevertheless, the Deepwater Horizon disaster aroused concern in the European Union as well. Shortly after the full extent of the disaster in the Gulf of Mexico became apparent, Norway postponed new deepwater drilling pending an investigation of what had happened. Furthermore, the United Kingdom doubled the number of environmental oil rig inspections, and the Netherlands developed new 'best practices' and equipment for oil spill mitigation.[157]

On 22 May 2010, US President Barack Obama announced the creation of the National Commission on the BP Deepwater Horizon Oil Spill and Offshore Drilling, also called the 'Oil Spill Commission'. It was to investigate the blowout that occurred when the Deepwater Horizon rig was drilling the Macondo well at a depth of more than 1,500 metres and about 4,000 metres below the sea floor. The Oil Spill Commission – which was to be independent and non-partisan – was charged with providing the president, policy makers, industry and, not in the least, the American people with a clear and fair elucidation of what happened, of the immediate and root causes and of consequent opportunities to change offshore energy production into a more safe and sound enterprise. The Commission strove to avoid singling out just one bad actor or crucial misstep, and the report took an expansive view.[158]

The investigation yielded four central findings. First, the oil spill could be seen as the embodiment of a long-lasting culture of complacency. Over the course of years, the oil industry had been moving into ever more challenging environments. However, as profits expanded and accidents

[156] Janssen, *Preventing a next Deepwater Horizon?*, 16. [157] *Ibid.*, 11.

[158] Oil Spill Commission: National Commission on the BP Deepwater Horizon Oil Spill and Offshore Drilling, 2011. See also Janssen, *Preventing a Next Deepwater Horizon?*, 59–60.

remained scarce, investments to keep up with the increased risks strongly lagged behind their actual need. Second, failures on the drilling rig itself can be traced back to a lack of consistent commitment to safety by the industry, from the highest management levels downward. The organizational culture allowed for missed warning signals, failure to share information and poor perception of the risks involved. Third, the government itself failed as well: federal regulation and oversight proved to be clearly insufficient. Legal authority, regulations, available expertise and management all fell short of providing a proper framework for deepwater drilling to occur. Fourth, the preparedness to respond to an oil spill of the magnitude of Deepwater Horizon was greatly inadequate. For over twenty years, no real improvements in response to new technologies had been made, coordination of the response effort – to integrate local, state and federal actors – exhibited many weaknesses, and the joint public–private response revealed the need for public and private investment. Consequently, the response by no means led to the desired result.[159]

2.5.4.3 Expert Evaluation

As mentioned in Section 2.2.2, differences exist between the offshore activities in, e.g., the North Sea and the Gulf of Mexico. One major difference is that the water depth in the Gulf of Mexico for oil and gas activities is much deeper than in the North Sea. In the Gulf of Mexico, offshore activities often take place at depths of thousands of metres, whereas in the North Sea, they mainly take place at a depth of less than 100 metres.[160] Then the crucial question is to what extent does the depth of offshore activities influence the accident risks and pollution costs. Incidents such as loss of well control, BOP failure or the need for relief wells also can occur in shallow water or onshore.[161]

It was pointed out by some experts in the post-Macondo studies that both the velocity and irregularity of underwater currents, as well as extreme pressures and temperatures, put extra stress on subsea equipment in the deep. Pressure control becomes more difficult as the drill bit descends because of the greater likelihood of encountering abnormal geopressures.[162] A study by Lloyd's also revealed that the cost of drilling

[159] *Ibid.* [160] Interview with OGP, 25 February 2013.

[161] National Commission on the BP Deepwater Horizon Oil Spill and Offshore Drilling, 'A Brief History of Offshore Oil Drilling', 17.

[162] *Ibid.*

in deep water is not linear with depth but increases exponentially.[163] Therefore, the risks increase significantly with depth.

However, this does not necessarily mean that the risks in shallow-water offshore activities (e.g. in the North Sea) are smaller than those in deep water (e.g. in the Gulf of Mexico). It is recognized in many studies that water depth is not the only factor that influences the risks of offshore activities.[164] Moreover, for deepwater activities, once a spill occurs, it may take longer for the spilled oil to reach shore, so in this respect the deepwater activities may have the advantage that more time is available for intervention to protect the coastline.[165] However, due to restraints on technology, the effectiveness of currently available techniques for emergency responses remains to be tested with time. Take, for instance, the use of a dispersant: when an incident occurs to facilities in deep water, using a dispersant can prevent spilled oil from coming to the surface. Therefore, the most visible pollution seems to be avoidable. However, the effect of such chemical dispersant on organisms in the sea remains to be seen.[166]

Yet, the question of how oil reacts if it is forcefully released in deep water remains unanswered.[167] Take the example of the Deepwater Horizon incident, where an estimated 4.9 million barrels of oil did not rise to the surface. 'Estimates of what happened to the spilled oil suggest that 23 per cent of the oil naturally evaporated or dissolved, 29 per cent was naturally or chemically dispersed and 25 per cent was directly recovered from the wellhead, burned or skimmed.'[168] Therefore, a balance of 23 per cent is considered to be 'residual . . . which includes oil that is on or just below the surfaces as light sheen, or as tar balls, and sand oil that has been washed ashore. The residual and dispersed oil is eventually likely to degrade naturally.'[169]

'Post Macondo studies have confirmed the presence of a deepwater plume of highly dispersed oil droplets and dissolved gases at between 3,200 and 4,200 feet deep extending for many miles, primarily to the southwest of the wellhead.'[170]

163 Rees and Sharp, *Drilling in Extreme Environments*, 10.

164 For example, the study carried out by Lloyd's, *ibid.*

165 National Commission on the BP Deepwater Horizon Oil Spill and Offshore Drilling, 'A Brief History', 16.

166 *Ibid.* 167 Rees and Sharp, *Drilling in Extreme Environments*, 17–18.

168 National Oceanic and Atmospheric Administration, 'Oil Budget Technical Report', 23 November 2010.

169 Rees and Sharp, *Drilling in Extreme Environments*, 18.

170 National Commission on the BP Deepwater Horizon Oil Spill and Offshore Drilling, 'Final Report'.

> The long-term effect of these substances on the deepwater environment remains uncertain, with depletion of the oxygen supply and levels of toxicity to exposed organisms being two areas of concern. Although studies have shown that a deepwater plume would have formed naturally (as around 15 per cent of the oil escaping the wellhead would have been physically dispersed by the fluid turbulence), the use of 18,379 barrels of dispersant injected into the oil and gas steam may have doubled its size. Injecting dispersant reduced the amount of oil rising to the surface and the risk of highly visible damage to shorelines and surface wildlife, but at the cost of more oil remaining within the water column and the risk of longer term impact on deepwater ecosystems. Further studies and monitoring may show whether the approach used at Macondo was correct and this will clearly influence future remedial operations in the event of a similar deepwater blowout.[171]

Therefore, the influence of water depth on the risks is not fully projected due to technical restraints.

In addition to water depth, the climate conditions also may influence the risks. The oil and gas industry is aware of the inhospitable climate in the North Sea,[172] and such weather conditions require the facilities to be built in such a way that they can survive the strong currents and pressure.[173] However, the advantage of strong wind and currents might be that spilled oil can be quickly dispersed.[174] However, almost any argument in favour of the 'Europe is less risky' scenario almost inevitably generates a counter-argument. One expert, for example, held that especially when addressing the risks in the West Shetlands, it should also be taken into account that the waters there are considerably colder than in the Gulf of Mexico, as a result of which natural dispersion of the oil may in fact be more difficult than in the Gulf.[175]

Another risk factor lies in the fact that in the North Sea some offshore fields, *inter alia*, in the Netherlands are mainly gas fields and not oil fields.[176] The different natures of oil and gas mean that the damage caused by an offshore incident would be different when it involves gas than when it involves oil. The major damage caused by a gas facility is mainly personal injury and probably less on the environment,[177] whereas an oil incident will have a larger influence on the environment than gas.

171 Rees and Sharp, *Drilling in Extreme Environments*, 18.
172 Interview with OGP, 25 February 2013.
173 www.oilgasarticles.com/articles/77/1/north-sea-oil-and-gas/page1.html.
174 Interview with OGP, 25 February 2013.
175 Philipp Wassenberg of Munich Re-Insurance Company, interview on 6 May 2013.
176 Interview with OGP, 25 February 2013.
177 Although methane is a strong greenhouse gas, which is damaging to the environment.

All the factors just mentioned are objective factors that may influence the accident risks for offshore activities. In addition, the subjective factors also may have some influence (e.g. the perspective of people on risks). The Deepwater Horizon incident has changed people's perspectives on offshore incidents and the risks related to offshore activities. Prior to this incident, the estimated amount of deepwater wells drilled worldwide was 14,000, and there were no known major incidents.[178] The industry was aware of the potential risks, and it had developed plans for pollution incidents based on other types of offshore spills, deepwater field tests and modelling of the likely flow dynamics.[179]

Some industry commentators argue that 50,000 wells have been drilled since 1947 with only one occurrence of a Macondo-scale event. However, others counter that these numbers may be masking reality because the absence of more incidents does not necessarily mean that the risk is a 1:50,000 event. The study completed by the Deepwater Horizon Study Group found that of the 5,000 wells drilled since 1993, only forty-three were as complex as the Macondo well. Is the probability of another Macondo-type event therefore 1:43, or even higher, given that the rate of wellbore instability in these forty-three wells is unusually high?[180]

2.5.4.4 Limits of Expert Opinion

Behavioral science tells us that people can misjudge risks due to bias. People may misjudge risk due to an inability to conceive possible negative events because desired scenarios or outcomes are easier to imagine; this is known as 'scenario bias'. Furthermore, a failure resulting from a very unlikely chain of events is considered to have a negligible probability of occurrence.[181] The Deepwater Horizon Study Group suggests that a scenario involving total blowout, total loss of the rig and complete loss of well control for eighty-seven days, as happened with Macondo, was not even considered to be possible.[182] Behavioural law and economics have shown that experts are subject to all kind of biases. The most general bias from which experts suffer is an over-reliance on their own expertise. This is referred to as 'over-confidence'. Various studies have provided evidence of this over-confidence. It was, for example, shown

178 'Society of Petroleum Engineer Notes 2010'; available at: www.spe.org/notes/2010/07/faqs-on-deepwater-drilling-gulf-spill/.

179 Rees and Sharp, *Drilling in Extreme Environments*, 17.

180 *Ibid.*, 28.

181 Lloyd's, 'Emerging Risks Report'.

182 Rees and Sharp, *Drilling in Extreme Environments*, 28.

that experts have largely overestimated the precision with which one could predict the likelihood of a meltdown of the nuclear core in a nuclear installation. Experts also showed an irresponsible large trust in the stability of the Teton Dam, notwithstanding several problems that had occurred during construction of the dam. The Teton Dam eventually collapsed in 1976.[183] Slovic and others point to several factors which cause experts to grossly underestimate specific risks and catastrophic events. These concern, *inter alia*,

- Failure to recognize that human failure can substantially influence the effectiveness of technological systems. This became clear after the nuclear incident at Three Mile Island, where operators repeatedly wrongly assessed problems with the nuclear reactor and failed to intervene on time.
- Over-estimating the current level of scientific development and knowledge.
- Under-estimating the integral functioning of technological systems. Engineers, for example, failed to discover that the reason for the failure of the DC-10 after its first flights was decompression in the cargo part of the plane which led to destruction of vital control systems.
- A well-known problem with experts is also that they make mistakes in assessing probabilities. The latter is referred to as 'calibration'. Physicians, for example, systematically over-estimate the likelihood of survival of a patient with cancer.[184]

These examples show that one has therefore to be extremely careful with expert evidence concerning the likelihood of a Deepwater Horizon–type incident in Europe given the over-confidence and other biases which may even influence expert judgments.

There is also a risk that both the energy and insurance industries do not learn from their experience with Macondo due to 'hindsight bias', by which people conclude that their ability to handle a past event was higher than it really was. It is also possible that the more time that passes without any incident, the more easily organizations lower their estimates of the probability of an accident occurring.[185] A study carried out by Lloyd's in 2011 also suggests that the absence of events in recent

[183] For a summary of these studies, see Slovic, Fischoff and Lichtenstein, 'Rating the Risks', 104ff.

[184] Meadow and Sunstein, 'Statistics, Not Experts', 634.

[185] Nancy Leveson, MIT World 2010, 'Moving Ahead: Engineering Challenges of Deep Water Drilling and Future Oil Resource Recovery', 28 September 2010.

history and a good safety record should not be taken as evidence that the risk is low.[186]

2.5.4.5 Summary

This overview of the various opinions shows that there is a lot of debate on the likelihood of a Deepwater Horizon–type accident in the European Union. Many point to the fact that there are differences between the Gulf of Mexico and European waters, but others point to the fact that these differences should not be overstated and that, hence, also in Europe, when a large-scale disaster would happen, the potential damage can be substantial. The estimate of the potential damage caused by a disaster-scale incident is, of course, very important, especially when it comes to the question how much financial capacity should be available to cover the risks from a major offshore accident. Hence, we will come back to this point when it is addressed in Chapter 6.[187]

Based on the research conducted in this chapter, the following conclusions can be drawn:

- Reliable data on the actual number of offshore facilities in the European Union are not readily available, and existing data are in some cases not comparable. In this respect, one may hope that with the implementation of Directive 2013/30/EU, the situation can be improved.
- The most important fact is that the number of offshore facilities is likely to rise in the (near) future, with oil and gas discoveries offshore in Norway, in the Mediterranean and in the Black Sea. Furthermore, although not definitive yet, there are ideas to build offshore facilities in the Arctic, which is a much more difficult and risky environment to work in.
- Although many point at differences in the drilling conditions between the Gulf of Mexico (where the Deepwater Horizon incident happened) and the European Union, data show that incidents do happen at various scales in the European Union and Norway.
- Data of Norway and the United Kingdom show that personal injuries and hydrocarbon releases decreased over the last decade. Nevertheless, a few significant releases did happen well in the last decade.

[186] Rees and Sharp, *Drilling in Extreme Environments*, 28. This was also held by Mr. De Jong, inspector-general of the State Supervision of Mines in the Netherlands.

[187] See Section 6.1.2.

- Because of fortunate circumstances (e.g. Statfjord A or Gannet Alpha), these releases did not cause severe personal or environmental damage.
- An analysis of recent incidents (last five years) shows that the reasons for these incidents have similar explanations in Norway and the United Kingdom and that these explanations might be worrisome. Both PSA and HSE investigation reports identified backlogs of maintenance, deficient maintenance management, inadequacies in risk identification and deficient barrier management as causes of the incidents.

3

Analysis of Existing Legal Regimes

MICHAEL FAURE, JING LIU AND HUI WANG

3.1 Introduction

The goal to be achieved in this chapter is to analyze existing legal arrangements to cover civil liability for offshore incidents. In addition, a number of related regimes will be discussed equally dealing with catastrophic risk because they may constitute interesting examples for a liability regime for offshore-related risks. It is of course impossible to discuss all existing legal arrangements. We will therefore focus on the legal regime in some EU Member States, from which we have obviously selected Member States that have many offshore installations and thus a larger exposure to potential liability. However, a highly interesting non-EU country is Norway. As Chapter 2 indicated, Norway has an impressive number of offshore installations, and therefore, an analysis of its legal regime is important as well. In addition to the European Union, the civil liability regime of some other major offshore nations such as the United States and Australia will be briefly addressed. Moreover, it also may be interesting to look at regional, multi-lateral and bilateral arrangements, as well as at the way civil liability is dealt with in other high-risk sectors (e.g. in particular, the nuclear industry). Looking at the nuclear industry may constitute an interesting example of how a liability regime for offshore installations could be shaped as well, or at least the study of the nuclear liability regime may show some problematic aspects from which a potential EU offshore liability regime could learn. Therefore, the nuclear liability regime will be analyzed in a rather detailed way. The reason is especially that the international nuclear liability regime has been the subject of a lot of criticism because of low limits on the liability of nuclear power plant operators and because of public funding of the compensation. In this respect, a comparison with the US Price-Anderson Act is quite interesting because public funding is absent, and moreover, a system of retrospective pooling is introduced which does not necessitate money to be paid ex ante but is based rather on ex post retrospective premiums. This may be an interesting model for the compensation for damage caused by offshore installations as well.

The remainder of this chapter is structured as follows: first, the international legal framework will be presented (3.2); then some regional arrangements concerning offshore liability regimes will be discussed (3.3). Next will come the current EU regime on offshore activities (3.4), followed by country studies concerning offshore liability regimes (3.5). Finally, consideration will be given to a liability regime in another high-risk sector – the nuclear industry (3.6). Some concluding remarks from the analysis in this chapter are presented in Section 3.7.

3.2 International Legal Framework

In order to better understand the role of civil liability for pollution damage caused by offshore activities, it may be useful first of all to have an overview of the entire legal structure on offshore oil and gas exploration activities. At the international level, the legal framework for offshore activities is established via the United Nations Convention on the Law of the Sea of 1982 (UNCLOS 1982). It contains some general provisions on the rights and obligations of states in regulating offshore related issues. UNCLOS 1982 confers on coastal states the sovereign right to explore and exploit the natural resources in the exclusive economic zone and continental shelf.[1] In addition, many international conventions may be relevant to offshore activities, but they mainly regulate the safety of the offshore industry by setting technical standards (1973 International Convention for the Prevention of Pollution from Ships, as amended by Protocol 1978, known as MARPOL 73/78) or address the emergency response and preparation for offshore pollution incidents (International Convention on Oil Pollution Preparedness, Response and Co-operation of 1990, known as the OPRC Convention). When it comes to compensation and civil liability for offshore oil pollution damage, until now, there has been no internationally accepted convention dealing specifically with liability issues.

In various (sea) regions, there are often regional sea programs under the auspices of the United Nations Environment Program (UNEP). However, the liability issue is not the focus of these programs either.

[1] Articles 56 and 77 of UNCLOS 1982.

When it comes to national legislation, the one that comes into the spotlight is US law because it has particular provisions in the Oil Pollution Act of 1990 (OPA) dealing with offshore pollution liability and financial security. However, in Europe, it wasn't until 2013 that a Directive (2013/30/EU) was adopted to address the issue of offshore-related activities. The Directive also has some provisions on the compensation and liability issue for offshore activities. However, at the level of each Member State, the domestic laws are not developed at the same level.

In the following sections, the relevant international conventions will be analyzed, focusing on the way these conventions tackle offshore issues. Before starting the overview of the specific conventions such as UNCLOS and MARPOL, the Civil Liability Convention and Fund Convention will be discussed. Although these conventions do not apply to offshore installations and only to vessel-based pollution, they provide an interesting point of comparison, especially in light of the fact that the just-mentioned US OPA does apply both to vessel-based and offshore-related damage. A study of the Civil Liability Convention and the Fund Convention is also interesting in light of the question of whether these conventions could constitute an interesting basis for a future (international or European) liability regime dealing with offshore-related risks.

3.2.1 The Civil Liability Convention and the Fund Convention

As mentioned in the Chapter 1, for vessel-based pollution, there has been a convention regulating civil liability for oil pollution damage since 1969, as well as an international fund since 1971. Moreover, voluntary agreements also have existed. As in the nuclear sector,[2] the United States did not join the international conventions but installed its own OPA with a separate but interesting regime. Hence, these compensation mechanisms are definitely interesting to look at from the perspective of comparing them with the potential liability and compensation regime for offshore installations.

This section will first describe the Civil Liability Convention (CLC) (3.2.1.1). After that, it will focus on the International Oil Pollution Compensation Fund (IOPC) (3.2.1.2) and on the related voluntary mechanisms (3.2.1.3). A discussion of the compensation and liability

[2] To be discussed in Section 3.6.

regime for vessel-based pollution is interesting not only to show the differences from the regime for offshore-related damage but also to examine how (the financing of) the compensation awarded for vessel-based pollution takes place, again from the perspective of whether that could constitute an example for offshore-related risks.[3]

3.2.1.1 The Civil Liability Convention

Historical Background and Evolution of the International Regime The international regime concerning civil liability for vessel-based marine oil pollution damage was originally established by two international conventions, the International Convention on Civil Liability for Oil Pollution Damage of 1969 (known as the Civil Liability Convention of 1969 or CLC 1969) and the International Convention on the Establishment of an International Fund for Compensation for Oil Pollution Damage of 1971 (Fund Convention of 1971). These two conventions came as a reaction to the incident caused by the tanker *Torrey Canyon* in 1967.

CLC 1969 imposes strict liability on the tanker owner and exclusively on the tanker owner (channelling of liability). Moreover, it requires a compulsory financial guarantee and limits the liability of the ship owner up to a certain amount (financial cap). The Fund Convention of 1971 established an International Oil Pollution Compensation Fund (IOPC Fund) to provide a second tier of compensation.

The CLC 1969/Fund Convention of 1971 regime came into effect in 1975–1978. However, several serious oil spills that happened afterwards, for example, the *Amoco Cadiz* in 1978 and the *Tanio* in 1980, showed the inadequacy of this regime and triggered revisions to the original conventions. Therefore, in 1992, two protocols were adopted to revise the existing regime, mainly to expand the scope of application (geographically and also with respect to recoverable damages) and increase the amount of compensation.[4]

The CLC 1969/Fund Convention of 1971 regime, after revision via the 1992 Protocols, is referred to as the CLC 1992/Fund Convention of 1992.

[3] See generally on the compensation regime for vessel-based pollution also Liu, *Compensating Ecological Damage*, 195–207; Wang, *Civil Liability*, 53–187; and Wilde, *Civil Liability*, 196–8.

[4] The first Protocols to revise the conventions were reached in 1984. Since the United States did not ratify the 1984 Protocols, it could not come into force. Nevertheless, the changes in the 1984 Protocols were needed, as illustrated by some major tanker oil spills; they are hence largely incorporated into the 1992 Protocols, which also eliminated the need for US participation.

It came into force on 30 May 1996. However, again, the adequacy of the 1992 regime was questioned when more tankers caused serious oil spills, e.g. *Nokhodka* in 1997 (Japan) and *Erika* in 1999 (France). Therefore, the compensation limits under the 1992 regime were increased by around 50 per cent through the 2000 resolutions. These increased the amounts of compensation and came into effect in November 2003.

Before the 2000 resolutions became effective, though, another tanker, *Prestige*, caused a catastrophic spill off the coast of Galicia. The states around the coastal areas, Spain, France, and Portugal, all suffered heavy losses. The sufficiency of the international regime dealing with tanker oil spills was once again highlighted.

The European Union, witnessing catastrophic oil pollution incidents caused by tankers, decided to take its own initiative instead of simply relying on the International Maritime Organization (IMO). A few months after the *Erika* incident, in March 2000, the European Commission launched a set of legislative proposals, the 'Erika I package'.[5] In these documents, the Commission developed three criteria to critically examine the existing international liability and compensation regime.[6] Subsequently, in December 2000, the Commission published a second set of proposals in the 'Erika II package',[7] in which it proposed to set up a European fund (the Fund for Compensation for Oil Pollution in European Waters, referred to as the 'COPE Fund'), with a ceiling of €1 billion,[8] to provide compensation to the pollution victims in Europe. The Commission preferred to refer the discussion to the competent international body, namely, the IMO, to obtain a similar agreement, but one that can be applied worldwide.[9] After the occurrence of the Prestige spill in

[5] COM(2000) 142 final, Communication from the Commission to the European Parliament and to the Council on the Safety of the Seaborne Oil Trading, 21 March 2000.

[6] COM(2000) 142 final, 21 March 2000, 34–7.

[7] COM(2000) 802 final, Communication from the Commission to the European Parliament and the Council on a second set of community measures on maritime safety following the sinking of the oil tanker *Erika*, 6 December 2000.

[8] In comparison, the then-existing international regime provided compensation up to approximately €200 million at the time. See in this respect the amended proposal for a regulation of the European Parliament and of the Council on the establishment of a fund for the compensation of oil pollution damage in European waters and related measures, *Official Journal* C227 E/487 of 24 September 2002.

[9] The Council had some hesitations against the idea of a regional compensation fund, and this idea was also strongly opposed by industry. In December 2000, the Transport Council reached conclusions on the need to achieve improvements to the existing international regime, including 'a substantial increase in liability and compensation ceiling'. For further discussion, see Wang, 'Shifts in Governments', 207–9.

European waters, the European Union took an even firmer position on the need for an improved compensation system.[10]

Indeed, it is due to the activism of the European Union that the IMO adopted in May 2003 the Protocol on the Establishment of a Supplementary Fund for Oil Pollution Damage (the 2003 Protocol or the Supplementary Fund Protocol). This 2003 Protocol introduced a supplementary fund of 750 million special drawing rights (SDRs; at the time of adoption, this corresponded to approximately €920 million, which is close to the amount of the COPE Fund proposed by the Commission). After adoption of the Supplementary Fund Protocol, EU Member States reacted actively and cooperatively by ratifying the 2003 Protocol. It is again thanks to the quick reaction of the European Union that the 2003 Protocol could enter into force on 3 March 2005.[11] As a result, the current international regime concerning civil liability and compensation for vessel-based oil pollutions consists of the CLC, imposing strict liability on the ship owner, an IOPC Fund contributed by the oil cargo owners, and an additional Supplementary Fund contributed by the oil industry from the countries that ratify it.

It should be mentioned that after the revision in 1992, parties have to choose to be party only to one of the regimes, either the 1969–1971 regime or the 1992 regime. A large number of parties have opted for the 1992 regime, which provides a much higher amount and more extensive compensation. However, the coexistence of the 1971 and 1992 Funds seemed problematic. The fact that the contribution from the oil industry was divided between two funds to a large extent weakened the financial viability of the fund and hence lowered the chance of full compensation in the case of a major oil spill. Therefore, it was decided in 2000 to terminate the 1971 Fund with effect from 24 May 2002. But still two CLCs coexist. As of June 2015, 133 states are parties to CLC 1992, whereas only thirty-four states remain members of CLC 1969; thirty-one states are parties to the Supplementary Fund Protocol.[12]

10 See in this respect, the European Commission, Communication on Improving Safety at Sea in Response to the Prestige Accident, 3 December 2002; see also the 'Presidency Conclusions', European Council, Brussels, 20 and 21 March 2003, 26–7.

11 Among the countries that first ratified the 2003 Protocol to make it effective, most are EU Member States: Denmark, Norway, Finland, France, Germany, Ireland and Spain.

12 Information available at the website of the IOPC Fund: www.iopcfunds.org/about-us/membership/a-z-listing/#fund-0.

Principle of Liability The CLC adopts strict liability for the ship owner. At the diplomatic conference leading to adoption of CLC 1969, delegations spent a lot of time debating where the liability should be directed (either to the ship owner or to the cargo owner) and whether it should be a strict or negligence liability rule.[13] However, no agreement could be reached until the Belgian delegation made the proposal that another layer of compensation should be contributed by the oil industry. Envisioning the prospect of a compensation fund contributed by the oil industry, a compromise was quickly reached to impose strict liability on the ship owner. It may be interesting to realize that the decision to impose strict liability on the ship owner came about as a compromise between various states with different interests (e.g. those with shipping interests vis-à-vis those with coastal interests). The strict liability on the ship owner could be decided only when the oil industry promised to contribute to the liability regime as well.[14]

CLC 1969 imposes strict liability exclusively on the ship owner without further discussion on whether such a channelling mechanism would be suitable.[15] This is probably due to the existence of the nuclear conventions that have introduced a similar channelling mechanism.[16] CLC 1969 makes clear that no other claims are eligible other than those under the convention. It mentions explicitly that no claims can be made against the servants or agents of the owner.[17]

The ship owner is defined as 'the person or persons registered as the owner of the ship or, in the absence of registration, the person or persons owning the ship'.[18] Recourse against a third party is allowed under CLC 1969.[19]

In CLC 1992, liability is still imposed on ship owners and it remains strict liability. However, there are some changes with respect to the channelling provision. According to CLC 1992, not only servants or agents of the owner but also some other parties, such as the pilot or any other person who is not a member of the crew and performs services for the ship, any charterer, any person performing salvage with the consent of the owner or on the instructions of a competent public

13 Wang, 'Shifts in Governments', 212–19.
14 For further discussion on the legal history, see Wang, *Civil Liability*, 53–97.
15 See the summary of the discussions at leg/conf/c.2/sr.5, leg/conf/c.2/sr.12, Official Records of the International Legal Conference on Marine Pollution Damage 1969, IMO 1969.
16 For further analysis of the nuclear conventions, see the discussion in Section 4.4.
17 CLC 1969, Art. III.4. 18 *Ibid.*, Art. I.3. 19 *Ibid.*, Art. III.5.

authority, any person taking preventive measures and their agents or servants are also exonerated from oil pollution liability. The broadened scope of the list of persons who can be protected from the liability strengthens the effect of channelling.

Amount of Compensation The liability established under CLC 1969 is capped at 210 million francs or 2,000 francs for each tonne of the ship's tonnage. The amount was higher than that under the 1957 Convention on the Limitation of Ship Owners' Liability but was still far from covering the whole potential damage that can be caused by oil pollution. The ship owner's right of limitation cannot be used if the incident occurred as a result of the owner's actual fault.[20]

CLC 1992 has increased the limit of liability to 4.51 million SDRs or 89.77 million SDRs depending on the size of the ship. As a compromise on the increase in the liability limit, the criteria when the ship owners lose their right to limit liability are further constricted: damage should result from their wilful misconduct.[21]

Compensable Damage CLC 1992 applies to pollution damage which happened on the territory (including territorial sea) and exclusive economic zone of a contracting party and applies to the preventive measures.[22] 'Pollution damage' is defined as

> loss or damage caused outside the ship by contamination resulting from the escape or discharge of oil from the ship, wherever such escape or discharge may occur, provided that compensation for impairment of the environment other than loss of profit from such impairment shall be limited to costs of reasonable measures of reinstatement actually undertaken or to be undertaken; the costs of preventive measures and further loss or damage caused by preventive measures.[23]

This definition is constricted by the word 'contamination', which means that damage caused by fire or explosion following a discharge is not covered.[24] It is very likely that complex issues as to causation would arise where contamination by oil is followed by contamination by fire.[25] The definition is still vague, and its concrete scope only became clear over the years. Personal injury is eligible for compensation, but not including exposure to health risks, anxiety and upset. Damage to property should

[20] *Ibid.*, Art. V.2. [21] CLC 1992, Art. V.1.2. [22] *Ibid.*, Art. II. [23] *Ibid.*, Art. I.6.
[24] Abecassis, *The Law and Practice*, 208.
[25] As illustrated by the incident involving the tanker *Aegean Sea*.

be 'real' and not speculative. Pure economic loss is eligible if the loss is quantifiable in economic terms.[26] 'Preventive measures' contain clean-up and restoration. They are compensable if they are reasonable and the loss is quantifiable in economic terms. Claims based on abstract methods of calculation are not admissible.[27]

This wording further defines the scope of compensable environmental damage. The loss of profit from impairment of the environment is compensable, including both consequential loss and pure economic loss. In addition to that, the term 'impairment of the environment' is limited to the costs of reasonable measures of reinstatement actually taken or to be undertaken. Even with this definition, the clear implication has to be understood in practice. Disputes have been based on issues of damage quantification with the state as environmental trustee and ecological restorer.[28] It is often up to national courts, where the lawsuits are brought, to interpret the meaning of vague terms such as 'reasonable measures'.

Compulsory Financial Guarantee Article VII.1 of the CLC provides that the ship owner

> registered in a Contracting State and carrying more than 2,000 tons of oil in bulk as cargo shall be required to maintain insurance or other financial guarantee, such as the guarantee of a bank or a certificate delivered by an international compensation fund, in the sums of applying the limits of liability prescribed in Article V, paragraph 1 to cover his liability for pollution damage under this Convention.

Critiques have been formulated on the fact that there is a minimum tonnage requirement, which means that tankers carrying fewer than 2,000 tonnes of oil do not need to take any financial guarantee, whereas they can cause equally large pollution damage and thus can face an insolvency risk. Without any financial guarantee, compensation to pollution victims in such an event may be rather difficult.

The CLC holds that the form of financial guarantee is not restricted to insurance[29] but also can be in other forms, which gives a certain flexibility in using other market financial instruments. In practice, this is mainly done via pooling arrangements among the 'Protection and Indemnity Clubs' (P&I Clubs), which will be discussed later.

26 Verheij, 'Shifts in Governance', 138–9. 27 *Ibid.*, 139.
28 Mason, 'Civil Liability', 3–5. 29 CLC 1992, Art. VII.2.

When it comes to enforcement of the compulsory financial guarantee requirement, Article VII.11 of the CLC provides that each contracting state 'shall ensure, under its national legislation, that insurance or other security ... is in force in respect of any ship, wherever registered, entering or leaving a port in its territory, or arriving at or leaving an off-shore terminal in its territorial sea, if the ship actually carries more than 2,000 tons of oil in bulk as cargo.' Article VII.10 provides that 'a Contracting State shall not permit a ship under its flag to which this Article applies to trade unless a certificate has been issued' with respect to the financial guarantee. Therefore, given these provisions, a state, as a flag state, should not permit its ships to trade without providing the financial guarantee; as a port state, it should ensure that all ships entering or leaving its port or offshore terminal have the required financial guarantee available.

Despite the seemingly weak enforcement provisions, the financial guarantee with respect to tankers has been available in almost all the jurisdictions that are party to the CLC.

3.2.1.2 Compensation Funds in the International Regime

The Fund Convention 1971/1992 As mentioned earlier, the prospect of a compensation fund was already decided at the 1969 conference leading to the conclusion of CLC 1969. Two years after adoption of CLC 1969, the Fund Convention of 1971 was passed.

The Fund Convention of 1971 has two aims: to provide compensation when the protection available from CLC 1969 is inadequate and to relieve ship owners from additional financial burdens.[30] Firstly, it complements the compensation from CLC 1969 in these situations: no liability can be established under CLC 1969, owners and their financial guarantors are financially incapable of compensation, and the damage exceeds the owners' liability. To encourage preventive measures, their costs raised from the voluntary activities of the owners are also treated as pollution damage.[31] The Fund has no obligation to pay if it can prove that the damage resulted from an act of war, hostilities, civil war or insurrection or oil from a warship or a state-owned/operated ship, or if the claimant cannot prove that damage resulted from a ship-related incident. Contributory negligence also can be used as a defence for the Fund.[32] The compensation available from the 1971 Fund is not unlimited: the total sum of the amount from CLC 1969 and the Fund Convention of

[30] The Fund Convention of 1971, Art. 2(1). [31] *Ibid.*, Art. 4(1). [32] *Ibid.*, Art. 4(2)(3).

1971 is capped at 450 million francs; for damage caused by a natural disaster of an exceptional, inevitable and irresistible character, the amount payable from the Fund is capped at 450 million francs. The Assembly of the Fund has the right to increase the amount up to 900 million francs. When claims exceed the amount payable from the Fund, payment should be reduced proportionally for each claimant.[33]

In addition to providing complementary compensation to victims, the 1971 Fund also indemnifies the ship owners. Indemnification is available for the payment between (1) the amount in excess of 1,500 francs for each tonne of the ship's tonnage or 125 million francs, whichever is the less, and (2) the amount not in excess of 2,000 francs for each tonne of the said tonnage or 210 million francs, whichever is less. But the indemnification is not available if the damage is caused from the wilful misconduct of the ship owner.[34] Claims for compensation or indemnification should be made within three years of the occurrence of the damage and six years of the occurrence of the incident.[35]

The Fund is financed by big oil importers in contracting states. The eligible importers need to make initial contributions to the working capital of the Fund and annual contributions to cover the administrative expenses and claims.[36] Calculation of contributions is based on a fixed sum for each tonne of oil received.[37] The contracting state has the obligation to ensure that the eligible contributing importers appear on a list and to communicate it to the Fund.[38] The contracting state also can declare that it assumes the obligation that lies on the importers within its territory to make contributions.[39]

The original CLC and Fund Convention proved to be insufficient to cover the potential catastrophic oil pollution damage. In 1992 a new compensation fund was established. The Fund Convention of 1992 removed the function of the Fund to alleviate the burden of liability of ship owners. Therefore, the only function of the 1992 Fund is to provide additional protection to the victims of oil pollution. The conditions of the application of the 1992 Fund to compensate victims are the same as those of the 1971 Fund.[40] The available compensation from CLC 1992 and the Fund Convention of 1992 is increased to 203 million SDRs.[41]

[33] *Ibid.*, Art. 4(4)(5)(6). [34] *Ibid.*, Art. 5(1). [35] *Ibid.*, Art. 6(1).
[36] Wu, *Pollution from the Carriage*, 98.
[37] The Fund Convention of 1971, Arts. 11(1), 12(2). [38] *Ibid.*, Art. 15(1).
[39] *Ibid.*, Art. 14(1). [40] The Fund Convention of 1992, Art. IV.1. [41] *Ibid.*, Art. IV.4.

The 1992 Fund is also financed by the oil industry. To remove an obstacle for ratification, the requirement of the considerable initial contributions was removed. The 1992 Fund is solely financed by annual contributions. Calculation of the contributions is also based on the amount of oil received.

The 2003 Supplementary Fund Although the limitation of compensation under the Fund Convention of 1992 had been considerably increased, shortly after its adoption, the amount was dwarfed again by yet another catastrophic oil pollution case.[42] Against this background, a Supplementary Fund was established in a 2003 Protocol.[43] The Supplementary Fund provides an additional layer of compensation for oil pollution victims under CLC 1992 and the Fund Convention of 1992. In other words, a condition for payment from the Supplementary Fund is that the victim is entitled to compensation under CLC 1992 and the Fund Convention of 1992 and is unable to obtain full and adequate compensation from them.[44] The Supplementary Fund Convention increases the aggregated amount of compensation up to 750 million SDRs.[45]

The amount of compensation available under the international regime can be summarized as in Table 5.

3.2.1.3 Voluntary Mechanisms

In fact, the international liability regime for oil pollution started to develop in the 1960s. The *Torrey Canyon* oil spill made the wider public aware of oil pollution risks. Before international conventions came into being, a voluntary scheme was adopted: the Tanker Owners Voluntary Agreement concerning Liability for Oil Pollution (the TOVALOP), under which fault liability with a reversal of the burden of proof was imposed on the tanker owner or the bareboat charter. The TOVALOP is complemented by another voluntary agreement by the oil industry: the Contract Regarding an Interim Supplement to Tanker Liability for Oil Pollution (the CRISTAL). The CRISTAL established a fund to cover oil pollution damage in addition to the TOVALOP with contributions from the oil industry.

[42] The *Nakhodka* accident near Japan in 1997 and the *Erika* disaster in France in 1999 are two examples.

[43] Protocol of 2003 to the International Convention on the Establishment of an International Fund for Compensation for Oil Pollution Damage, 1992 (The Supplementary Fund Protocol), RMC II, 7.115.

[44] The Supplementary Fund Convention, Art. IV.1.

[45] *Ibid.*, Art. IV.2.

Table 5 *Compensation for Pollution Damage under the International Regime*

Tonnage of ship (GT)	CLC 1969 (euros)	CLC 1992 (euros)	2000 Protocol (euros)	
Ships ≤ 5,000	153 per tonne	3.45 million	5.19 million	
5,000 < Ships < 140,000	153 per tonne	3.45 million + 483/ additional tonne	5.19 million + 726/ additional tonne	
Ships ≥ 140,000	16.1 million	68.6 million	103.2 million	
	1971 Fund (euros)	**1992 Fund (euros)**	**2000 Protocol (euros)**	**2003 Supplementary Fund (euros)**
Overall limit[a]	69 million	155.2 million	233 million	862 million

Note: The original unit of calculation in the Conventions is the SDR, as defined by the International Monetary Fund (IMF). For the convenience of comparison, all the units are converted to euros, based on an exchange rate on 22 February 2013 of 1 SDR=1.1498 euros.

[a] The overall limit is the maximum compensation payable by the Fund for any incident, and it includes the compensation made by the ship owner or his insurer under the CLC.

However, after the changing of the regime in 1992, those original voluntary agreements – CRISTAL and TOVALOP – were no longer necessary. But after introduction of the Supplementary Fund Convention, which imposes an additional burden on oil importers to compensate for pollution damage, two new voluntary agreements were introduced. They were introduced to alleviate the additional burden imposed on the oil importers and to balance the interests between ship owners and the oil industry. Under the Small Tanker Oil Pollution Indemnification Agreement (STOPIA) and the Tanker Oil Pollution Indemnification Agreement (TOPIA), tanker owners indemnify parts of the additional burden under the Supplementary Fund Convention.[46]

3.2.2 UNCLOS

UNCLOS 1982 provides a general legal framework for the adoption of legal regimes dealing with sea-related activities. It is therefore often called the 'constitution of the seas'. The most important provisions in the UNCLOS for offshore exploration activities are probably those related to the jurisdiction of a state relating to offshore activities. Offshore exploration activities often take place in the exclusive economic zone (EEZ) and on the continental shelf, and the jurisdiction of these areas is exclusively granted to the coastal state under the UNCLOS.[47] It grants exclusive sovereign rights to coastal states for the purpose of exploring and exploiting the natural resources of their EEZs and continental shelves, including, *inter alia*, the non-living resources with respect to the seabed and subsoil.[48]

As far as pollution from offshore activities is concerned, coastal states with the sovereign rights to exploit their natural resources shall have the duty to protect and preserve the marine environment.[49] They shall, according to the UNCLOS, either individually in national laws or jointly as appropriate, adopt laws and take measures to prevent, reduce and control pollution of the marine environment, especially take those measures designed to minimize 'pollution from installations and devices used in exploration or exploitation of the natural resources of the seabed and subsoil'.[50] Article 235 of UNCLOS further requires

[46] Wang, 'Shifts in Governments', 210.
[47] These are embodied in Arts. 56, 60, 77 and 81.
[48] See Arts. 56 and 77 of UNCLOS 1982, respectively.
[49] Article 193 of UNCLOS 1982.
[50] Articles 194 and 208 of UNCLOS 1982.

states to 'ensure that recourse is available in accordance with their legal systems for prompt and adequate compensation or other relief in respect of damage caused by pollution of the marine environment by natural or juridical persons under jurisdiction'. In order to fulfil such an objective, states shall 'cooperate in the implementation of existing international law and the further development of international law relating to responsibility and liability for the assessment of and compensation for damage and the settlement of related disputes, as well as, where appropriate, development of criteria and procedures for the payment of adequate compensation, such as compulsory insurance or compensation funds'.

Annex III of UNCLOS 1982 sets up 'basic conditions of prospecting, exploration and exploitation'. It provides for general requirements for the qualifications of applicants, transfer of technology, approval of plans of work and financial terms of contract. Some provisions may be worth special attention as they concern the liability issue.

First, Article 11 of Annex III on joint arrangements provides that

1. Contracts may provide for joint arrangements between the contractor and the Authority through the Enterprise, in the form of joint ventures or production sharing, as well as any other form of joint arrangement, which shall have the same protection against revision, suspension or termination as contracts with the Authority.
2. Contractors entering into such joint arrangements with the Enterprise may receive financial incentives as provided for in article 13 of this Annex.
3. Partners in joint ventures with the Enterprise shall be liable for the payments required by article 13 of this Annex to the extent of their share in the joint venture, subject to financial incentives as provided in that article.

The 'joint arrangements' with a certain extent of state participation are often a requirement for a legal entity to acquire a licence for offshore activities (e.g. in Norway), since the state wants to remain to a certain extent in control of its natural resources. The legal entities that have obtained the licence to carry out offshore activities may, of course, benefit from the profits of such activities, but they are at the same time required to pay in the form of taxes or royalties (e.g. the production fees in the Norwegian system) to the state. The legal entity's liability shall be proportionate to its share of the enterprise of offshore activities as provided in Article 11 of UNCLOS.

Article 22 of Annex III provides that

> The contractor shall have responsibility or liability for any damage arising out of wrongful acts in the conduct of its operations, account being taken of contributory acts or omissions by the Authority. Similarly, the Authority shall have responsibility or liability for any damage arising out of wrongful acts in the exercise of its powers and functions, including violations under article 168, paragraph 2, account being taken of contributory acts or omissions by the contractor. Liability in every case shall be for the actual amount of damage.

This provision establishes that the contractor and the state authority shall be liable for their wrongful acts during operations or when exercising their functions. The liability shall, in theory, be to the full and actual amount of the damage.

3.2.3 MARPOL 73/78

Article 2 of MARPOL defines a ship as 'a vessel of any type whatsoever operating in the marine environment and includes hydrofoil boats, air-cushion vehicles, submersibles, floating craft and fixed or floating platforms'.[51] Therefore, a ship under MARPOL may cover offshore facilities only when it is a 'fixed or floating platform'.

However, Article 2, containing the definition of 'discharge', states that 'discharge' specifically excludes a 'release of harmful substances directly arising from the exploration, exploitation and associated offshore processing of sea-bed mineral resources'.[52] Nevertheless, Annex I of MARPOL which is the Regulations for the Prevention of Pollution by Oil, provides that fixed or floating offshore oil rigs must comply with the requirements 'applicable to ships of 400 tons gross tonnage and above other than oil tankers' and may not discharge oil or oily mixtures into the sea 'except when the oil content of the discharge without dilution does not exceed 15 parts per million'.[53] This seems confusing as to the applicability of MARPOL to offshore facilities.

The Unified Interpretation of MARPOL clarifies this issue. Of the five categories of discharge from offshore installations, the Annex I requirements apply only to two categories: the discharge of oil from 'contaminated ballast' and from 'machinery space drainage', which is produced by generators, fuel tanks and pumps of offshore

[51] Article 2(4) of MARPOL. [52] Article 2(3)(b)(ii) of MARPOL.
[53] Annex I of Regulation 21 of MARPOL.

installations.[54] MARPOL Annex I does not regulate the discharge of oil from 'offshore processing drainage', 'production water discharge' or 'displacement water discharge', which are the forms of pollution 'directly related to offshore petroleum production'.[55]

3.2.4 The OPRC Convention

The International Convention on Oil Pollution Preparedness Response and Co-Operation of 1990 (OPRC Convention) specifically applies to ships and offshore units. An 'offshore unit' is defined as 'any fixed or floating offshore installation or structure engaged in gas or oil exploration, exploitation or production activities, or loading or unloading of oil'. It requires the operator of offshore units to have oil pollution emergency plans which are approved by the competent national authority. In the event of a discharge or probable discharge of oil (whether on the offshore unit or only observed at sea), the persons in charge of offshore units shall report to the coastal state to whose jurisdiction the unit is subject. The Convention requires state parties to establish a national or regional system 'for responding promptly and effectively to oil pollution incidents', and it further encourages international co-operation in pollution response.

From the discussion in preceding sections, it becomes clear that at the international level there is in fact no convention dealing directly with offshore-related damage. However, when a well is discovered and drilling activities have to take place, usually so-called mobile drilling units (MODUs) will be used. These qualify as ships (to the extent that they are mobile) and hence the so-called Bunker Convention[56] or other conventions regulating damage resulting from maritime transport could apply. This study will concentrate mostly on pure offshore-related damage (and hence not on MODUs). One should realize, however, that insofar as a unit is mobile (and hence can be considered to be a ship), other maritime international conventions (such as the Bunker Convention) or national legislation (such as the US OPA) that applies to damage caused by maritime transport could apply.[57] At the international level, as was explained, marine pollution damage caused

[54] Unified Interpretation of MARPOL 73/78, Annex I, 56.2.
[55] See Fowler, 'Offshore Oil'; see also Gao, *Environmental Regulation*, 103–4.
[56] See Shan, 'The Era after the Bunker Convention'.
[57] See Alan Spackman, Vice-President IADC, interview on 5 June 2013.

by vessels is regulated by the civil liability and fund conventions discussed earlier.

3.2.5 *Ongoing Discussion at the IMO*

It is striking that an elaborate international regime does exist for civil liability and compensation related to vessel-based marine pollution[58] but that such an international legal framework is missing as far as offshore-related damage is concerned. The international regimes discussed (mostly UNCLOS and MARPOL) do not specifically address liability or compensation issues. This shortcoming therefore raises the question of why the competent international organization, more particularly the IMO, would not issue a convention to deal with liability and compensation for offshore-related risks. This issue has in fact been discussed within the IMO. At the 99th session of the IMO Legal Committee in April 2012, the possibility of a global liability and compensation regime for offshore oil and gas activities was examined. This option was first proposed by the Indonesian delegation in September 2010 and followed the Montara offshore oil pollution incident that occurred in August 2009, where damage was caused in the Timor Sea.[59]

The Legal Committee believed that bilateral and regional arrangements are the most appropriate means to address the matter and hence decided not to pursue an international regime because there was no compelling need to do so.[60]

Later, on 22 February 2013, the Legal Committee of the IMO issued another statement concerning the liability and compensation issues connected with trans-boundary pollution damage from offshore oil exploration and exploitation activities. According to this report, the IMO was still considered the most appropriate forum to deal with the issue of trans-boundary damage caused by offshore pollution due to its extensive experience and skill in various maritime and marine environmental issues. Hence, it was agreed that Indonesia should continue the process of developing principles in order to assist states to enter into bilateral or regional arrangements. The principles to which the Legal Committee explicitly referred are the (1) precautionary approach,

[58] More particularly, the Civil Liability and Fund Conventions discussed in Section 3.2.1.

[59] For a further discussion of the Montara incident, see Section 2.5.1.

[60] www.imo.org/en/mediacentre/meetingsummaries/legal/pages/leg-99th-session.aspx.

(2) strict liability, and (3) the polluter-pays principle. The issue would hence be further discussed within the Legal Committee.[61]

However, an expert to the IMO declared that there is in fact generally little appetite in the IMO to take up the issue properly. The reason is that many members of the IMO are of the view that offshore liability is outside the scope of the IMO, which is primarily focused on shipping.[62] So the IMO has started to work on the issue, but with a very low-key approach. For those reasons, regional or bilateral arrangements are preferred to a truly IMO initiative.[63] The problem is that it is held currently that the IMO does not have the competence to, for example, make a convention for offshore-related risks. The IMO Council decided that it is not willing to take up the issue. Of course, oil companies may, especially when they are state owned, have a strong influence on the position that will formally be taken by the state which is a member of IMO.[64]

3.3 Offshore Liability Regimes

3.3.1 Regional Arrangements

In 1972, the UN Conference on the Human Environment was held in Stockholm and led to the Declaration of Principles. These Principles provide general requirements for the protection of the marine environment from toxic and other wastes, for the development of liability arrangements and for the conservation of flora and fauna. The conference further established an action plan and instituted the United Nations Environment Programme (UNEP), part of which resulted in the development of a framework convention for thirteen regional sea programmes. Today, more than 143 countries participate in thirteen regional sea programmes established under the auspices of UNEP.[65] In addition, there are five partner programmes for the Antarctic, Arctic,

[61] This statement followed an international conference that took place in Bali (21–23 November 2012) concerning liability and compensation for trans-boundary oil pollution damage resulting from offshore exploration and exploitation activities.

[62] Opinion of Richard Mason, accredited representative of the European Commission to the International Maritime Organization, e-mail of 20 March 2013 to Michael Faure.

[63] *Ibid.*

[64] Interview with Alan Spackman, Vice-President Offshore Division of IADC on 5 June 2013.

[65] These regional seas programmes cover Black Sea, Wider Carribbean, East Asian Seas, Eastern Africa, South Asian Seas, POPME Sea Area, Mediterranean, North-East Pacific, North-West Pacific, Red Sea and Gulf of Aden, South-East Pacific, Pacific and Western Africa. See the website of UNEP: www.unep.org/regionalseas/about/default.asp.

Baltic Sea, Caspian Sea and North–East Atlantic regions. Given the geographical location of the current twenty-eight EU Member States, the Member States mainly border the North Sea, the Mediterranean Sea, the Baltic Sea and the Black Sea (Bulgaria and Romania). There is at least one regional arrangement covering each sea area with respect to marine environmental protection, which also addresses the issue of offshore pollution.

3.3.2 *North-East Atlantic Ocean (North Sea)*

3.3.2.1 The OSPAR Convention

The North Sea countries (United Kingdom, Denmark, Netherlands and Norway) produce 90 per cent of all the offshore oil and gas in Europe.[66] In the North Sea, there is a Convention for the Protection of the Marine Environment of the North–East Atlantic of 1992 (the OSPAR Convention), which was adopted on 22 September 1992 and entered into force on 25 March 1998.[67] Sixteen states are parties to the OSPAR Convention: Belgium, Denmark, Finland, France, Germany, Iceland, Ireland, Luxemburg, Netherlands, Norway, Portugal, Spain, Sweden, Switzerland and the United Kingdom, as well as the European Union.

The principles enshrined in the OSPAR Convention are the precautionary principle, the polluter-pays principle, the use of best available techniques and best environmental practice and, where appropriate, clean technologies. One important provision in the OSPAR Convention concerning the liability issue is Article 2(2)(b). It requires the application of 'the polluter pays principle, by virtue of which the costs of pollution prevention, control and reduction measures are to be borne by the polluter'. The OSPAR Convention specifically states that for 'pollution from offshore sources', all contracting states shall take 'individually and jointly, all possible steps to prevent and eliminate' such pollution.[68]

Annex III of the OSPAR Convention deals with 'the prevention and elimination of pollution from offshore sources' and was last amended in 2007. It requires the use of 'best available techniques' and 'best

[66] www.oilandgasuk.co.uk/proposedeuregulation.cfm.

[67] The OSPAR Convention replaced two predecessor conventions, the Convention for the Prevention of Marine Pollution by Dumping from Ships and Aircraft of 1974 (Oslo Convention) and the Convention for the Prevention of Marine Pollution from Land-Based Sources of 1978 (Paris Convention).

[68] Article 5 of the OSPAR Convention.

environmental practice' when adopting measures to prevent and eliminate pollution from offshore sources.[69] Annex III also stresses the prohibition of dumping disused offshore facilities and appropriate measures to be taken for state authorities to deal with the abandonment of offshore facilities.[70]

3.3.2.2 The Nordic Convention

The Scandinavian countries have their own regional arrangement in addition to participating in the OSPAR Convention. In 1974, Denmark, Finland, Norway and Sweden adopted the Nordic Environmental Protection Convention (the Nordic Convention)[71] to safeguard their environmental interests.[72]

The Nordic Convention is designed to protect the Nordic countries from environmentally harmful activities, which may include discharge from (offshore) installations into the sea. The Nordic Convention applies to the continental shelves of the parties.[73] As far as compensation for environmental damage is concerned, Article 3 establishes that anyone who suffers damage caused by environmentally harmful activities in another contracting state shall have the right to institute proceedings for compensation before a court or administrative authority of that state. It further provides that '[t]he question of compensation shall not be judged by rules which are less favourable to the injured party than the rules of compensation of the State in which the activities are being carried out.'

3.3.3 The Mediterranean Sea

The Mediterranean Sea borders twenty-one countries and three continents. All these countries participate in the Barcelona Convention, a Convention for the Protection of the Marine Environment and the Coastal Region of the Mediterranean. It was signed on 16 February 1976 and entered into force on 12 February 1978 (the original name of the convention was the Convention for the Protection of the Mediterranean Sea against Pollution). The Barcelona Convention was later amended on 10 June 1995 and entered into force on 9 July 2004. The current version of the Convention refers to the amended Convention of

[69] Article 2 of Annex III of the OSPAR Convention.
[70] Articles 5–8 of Annex III of the OSPAR Convention.
[71] Stockholm, 19 February 1974.
[72] For a further discussion of the Nordic Convention, see Koivurova, 'The Future', 505–26.
[73] Article 13 of the Nordic Convention.

1995. The Barcelona Convention was concluded mainly because the Mediterranean states believe that the existing relevant international conventions could not entirely meet the special requirements of the Mediterranean Sea area.[74]

As of June 2015, all eight EU Member States in the Mediterranean Sea region are parties to the Barcelona Convention: Croatia, Cyprus, France, Greece, Italy, Spain, Malta and Slovenia. The European Union is also a party to the Convention.[75]

Article 7 provides that parties to the Barcelona Convention shall combat pollution resulting from offshore operations. It states that contracting parties shall 'take all appropriate measures to prevent, abate, combat and to the fullest possible extent eliminate pollution of the Mediterranean Sea Area resulting from exploration and exploitation of the continental shelf and the seabed and its subsoil'. Article 12 further provides that the parties shall 'cooperate as soon as possible in the formulation and adoption of appropriate procedures for the determination of liability and compensation for damage resulting from the pollution of the marine environment'.[76]

One important Protocol to the Barcelona Convention that is relevant to the offshore activities is the so-called Offshore Protocol (Protocol for the Protection of the Mediterranean Sea against Pollution Resulting from Exploration and Exploitation of the Continental Shelf and the Seabed and Its Subsoil of 1994). It was adopted on 14 October 1994 and entered into force on 24 March 2011 (parties that have ratified it are Albania, Tunisia, Morocco, Libya, Cyprus and Syria). The European Union has acceded to the Offshore Protocol through Council Decision 2013/5/EU of 17 December 2012.[77]

The Offshore Protocol covers the entire process of offshore exploration and exploitation activities from the permit requirements to the removal of abandoned or disused installations. It covers a wide range of issues including use and removal of harmful substances, liability and compensation requirements, co-ordination with other parties of the Convention and

[74] Preamble of the Barcelona Convention.

[75] For further information on the Barcelona Convention, see the website of UNEP: www.unepmap.org/index.php?module=content2&catid=001001004.

[76] This was reiterated in Art. 16 of the Barcelona Convention.

[77] Council Decision of 17 December 2012 on the accession of the European Union to the Protocol for the Protection of the Mediterranean Sea against pollution resulting from exploration and exploitation of the continental shelf and the seabed and its subsoil, *Official Journal* L4/13–14, 9 January 2013.

provisions on safety, contingency planning and monitoring.[78] The Offshore Protocol requires the use of 'best available techniques' which are 'environmentally effective and economically appropriate' to 'prevent, abate, combat and control pollution' from offshore activities.[79]

The Offshore Protocol requires that parties shall 'cooperate as soon as possible in formulating and adopting appropriate rules and procedures for the determination of liability and compensation for damage resulting from the (offshore) activities'.[80] Pending the development of such procedures, the Offshore Protocol further requires that each party shall 'take all measures necessary to ensure that liability for damage caused by activities is imposed on operators, and they shall be required to pay prompt and adequate compensation' and shall 'take all measures necessary to ensure that operators shall have and maintain insurance cover or other financial security of such type and under such terms as the Contracting Party shall specify in order to ensure compensation for damages caused by the activities covered by this Protocol'.[81]

3.3.4 The Baltic Sea

In the Baltic Sea, the Helsinki Convention applies, a Convention on the Protection of the Marine Environment of the Baltic Sea Area of 1992. It entered into force on 17 January 2000.[82] The contracting states include Denmark, Estonia, the European Community, Finland, Germany, Latvia, Lithuania, Poland, Russia and Sweden.

The Helsinki Convention seeks to assure the ecological restoration of the Baltic Sea by ensuring the possibility of self-regulation of the marine environment.[83] It has some general requirements on the prevention of pollution from offshore activities.[84] In Annex VI, the use of best available technology and best environmental practice is reinforced. The Annex also contains provisions on environmental impact assessment, discharge standards and contingency planning. As far as liability for pollution damage is concerned, the Helsinki Convention requires only application of the polluter-pays principle.[85] Article 25 of the Convention is on the responsibility for damage, and it urges states to jointly develop rules concerning damage resulting from acts or omissions in contravention of the Convention.

[78] COM(2011) 690 final, 2. [79] Article 3 of the Offshore Protocol.
[80] Article 27 of the Offshore Protocol. [81] Article 27 of the Offshore Protocol.
[82] www.helcom.fi/about-us/convention. [83] Preamble of the Helsinki Convention.
[84] Article 12 of the Helsinki Convention. [85] Article 3 of the Helsinki Convention.

3.3.5 The Black Sea

In the Black Sea, there is a regional sea program known as the Bucharest Convention, a Convention for Cooperation in the Protection of the Black Sea against Pollution of 1992.[86] The two EU member states that are in this region, Bulgaria and Romania, are both parties to the Bucharest Convention.

When it comes to liability for pollution damage caused by offshore activities, the Bucharest Convention has only some general provisions.[87] It requires parties to adopt rules and regulations on liability for pollution damage from offshore operations, with the aim of ensuring 'the highest degree of deterrence and protection for the Black Sea as a whole'. The Bucharest Convention states that the compensation should be prompt and adequate with recourse for redress in accordance with the legal systems of the parties.[88]

3.3.6 Summary of the Regional Seas Arrangements

The regional seas programmes mainly address the particularity of each individual sea area and focus mainly on general principles such as the use of best available technology/techniques and the polluter-pays principle. The major goal of these regional programmes is to establish cooperation among parties in the interest of the particular sea areas. They contain only very general provisions on the establishment of liability and a compensation system for offshore-related pollution. The conventions and regional arrangements are not sufficient for evaluation of the function of various legal regimes on offshore activities. Therefore, examination of individual states' national laws on offshore activities remains necessary.

3.4 Current EU Regime on Offshore Activities

3.4.1 EU Interest in Offshore Activities

There are, as far as Europe is concerned, a number of reasons for concern with respect to damages which may be caused by offshore incidents. There is in fact internationally a quite impressive record of offshore

[86] For further details, see www.blacksea-commission.org/convention.asp.
[87] Mainly Arts. XI and XVI of the Bucharest Convention.
[88] Article XVI of the Bucharest Convention.

incidents, but also the North Sea has witnessed many offshore accidents. Just to name a few: Alexander Kielland (1980), Piper Alpha (1988), Forties Alpha (2003), Gullfaks C (2010) and most recently Gannet Alpha (2011) are incidents that occurred in the North Sea and have increased concern about the consequences of such incidents in Europe as well.[89]

Prior to the occurrence of Deepwater Horizon incident, there was no separate European regime dealing particularly with offshore pollution issues. Since 2010, various legislative initiatives concerning different aspects of offshore activities have been proposed, and the subject has been debated by all stakeholders. Finally, in June 2013 a separate European regime on offshore safety was established.

3.4.2 Council Directive 92/91 EEC

This Directive concerns the minimum requirements for improving the safety and health protection of workers in the mineral-extracting industries through drilling.[90]

3.4.3 Hydrocarbons Licensing Directive 1994 (94/22/EC)

Directive 94/22/EC[91] is often referred to as the 'Hydrocarbons Licensing Directive'. It establishes common rules that member states must follow when issuing petroleum licences, with the goal of ensuring a competitive, non-discriminatory and transparent European market and improving the security of supply. Under Directive 94/22/EC, the sovereign rights of Member States over hydrocarbon resources are recognized.[92] Therefore, it is up to the national governments to determine the geographical areas for prospecting, exploring and producing hydrocarbons. The Member States are required to provide annual reports on the geographical areas which have been opened, the authorizations granted, the entities holding those authorizations and the reserves available in their territory.

[89] For details of offshore pollution incidents and offshore activities in Europe, see the discussion in Section 2.4.

[90] *Official Journal* L 348, 28 November 1992, 9–24.

[91] Directive 94/22/EC of the European Parliament and of the Council of 30 May 1994 on the conditions for granting and using authorization for the prospection, exploration and production of hydrocarbons, *Official Journal* 164/3–8, 30 June 1994.

[92] Preamble of Directive 94/22/EC.

On 29 July 1998, the European Commission submitted a report on the implementation of Directive 94/22/EC.[93] The report mainly centres on question of reciprocity, i.e. non-discriminatory access for EC firms to the activities of prospection, exploration and production in third countries. The conclusions from the Commission are that all the Member States have transposed Directive 94/22/EC into national laws (except for Luxembourg and Finland, which are exempt from transposition because they do not have any potential hydrocarbon resources). The Commission also states that 'no reciprocity problems have been reported, as is borne out by the increased opening-up of exploration and production activity worldwide'.[94]

3.4.4 *The Marine Strategy Framework Directive*

The Marine Strategy Framework Directive plays an important role as well.[95] This Directive 'requires addressing the cumulative impacts from all activities on the marine environment' and 'is relevant to offshore oil and gas operations as it requires linking the particular concerns from each economic sector with the general aim of a comprehensive understanding of the oceans, seas and coastal areas, with the objective to develop a coherent approach to the seas taking into account all economic, environmental and social aspects through the use of Maritime spatial planning and Marine knowledge'.[96]

3.4.5 *Communication of 2010*

When the Deepwater Horizon incident occurred on 20 April 2010 in the Gulf of Mexico, in October of the same year, the European Commission launched a Communication to assess the risks in the offshore oil and gas industry in European waters.[97] The Commission explored a wide range of problems, including the licensing, controls by public authorities and spill response. In particular, when it addressed the liability issue, the Commission considered the possibility of extending the Environmental

[93] COM(1998) 447 final, Report from the Commission to the Council of 29 July 1998 on Directive 94/22/EC on the conditions for granting and using authorizations for the prospection, exploration and production of hydrocarbons.

[94] *Ibid.* [95] Directive 2008/56/EC. [96] COM(2011) 688 final, 14.

[97] COM(2010) 560 final, Communication from the Commission to the European Parliament and the Council Facing the Challenge of the Safety of Offshore Oil and Gas Activities, 12 October 2010.

Liability Directive[98] to cover environmental damage to all marine waters as defined in the Marine Strategy Framework Directive[99] and the applicability of the Waste Framework Directive.[100] The Commission also realized that the issues of a financial cap on the liability and the possibility of mandatory financial security are worth further analysis.[101] The Commission was not only concerned with offshore activities in the EU waters but also showed concern about the EU-based offshore industry operating in other parts of the world, and it called on the industry's international obligations as responsible operators.[102]

3.4.6 Directive on Safety of Offshore Oil and Gas Operations

Following the Communication of 2010, on 27 October 2011, the Commission initiated two proposals, one for a regulation of the safety of offshore activities and the other for accession of the European Union to the Offshore Protocol of the Barcelona Convention. The proposed Regulation (COM[2011] 688 final) follows the principal issues raised in the Communication of 2010, imposes stricter safety standards for offshore activities in Europe and gives national regulators more power to inspect their operations. It also extends sixteen-fold the zone in which companies will be held liable for environmental damage. It specifies that the licensee shall be held 'liable for the prevention and remediation of environmental damage' pursuant to the Environmental Liability Directive.[103]

During the political debate, it was decided that a directive rather than the proposed regulation would be the best legal instrument to address the risks posed by offshore installations. On 12 June 2013, the directive was signed jointly by the presidents of the Council and the European Parliament. Finally, on 28 June 2013, Directive 2013/30/EU on the Safety of Offshore Oil and Gas Operations was published in the *Official Journal*. The Directive entered into force twenty days after publication in the *Official Journal* of the EU.

The objective of Directive 2013/30/EU is 'to reduce as far as possible the occurrence of major accidents relating to offshore oil and gas operations and to limit their consequences ... and to improve the response mechanisms in case of an accident'.[104] The offshore regime established in

[98] Directive 2004/35/EC. [99] Directive 2008/56/EC. [100] COM(2010) 560 final, 8.
[101] *Ibid.* [102] COM(2010) 560 final, 13. [103] COM(2011) 688 final, Art. 7.
[104] Preamble (2) of Directive 2013/30/EU, *Official Journal* L178/66.

accordance with the Directive applies to offshore operations carried out on both fixed and mobile installations and to the whole life cycle of exploration and production activities from design to decommissioning.[105]

It reiterates the importance of the provision in Directive 94/22/EC that the licensing authority is required to consider the technical and financial capability of the licensee. It stresses that Member States shall verify that the applicants have provided 'appropriate evidence and adequate provisions have been or are to be made to cover liabilities deriving from major accidents'.[106] Moreover, Member States are required to ensure effective controls through 'continuous expert regulatory oversight' by the states.[107] Member States are also required to provide periodical reports on offshore activities and incidents to the Commission.[108]

Directive 2013/30/EU makes a clear distinction between the licensee and operator,[109] and offshore operations should be conducted only by 'operators' that are to be 'appointed by licensees or licensing authorities'.[110] The operator, together with the licensee and owner, should always bear the primary responsibility for safe operations, and they are not allowed to delegate their liability through contract; even when the accident was caused by actions or omissions by contractors, the liability of operators should not be relieved.[111] The operators have an obligation to report to Member States about the occurrence of a major accident.[112] In order to ensure effective emergency response, operators are required to prepare internal response plans.[113]

The Directive includes in Annex I the information to be submitted by the operator and owner for carrying out offshore operations and the information to be included in the report on major hazards.[114] However, as to the issue of financial security by operators, Directive 2013/30/EC only has a general requirement because 'no existing financial security instruments, including risk pooling arrangements, can accommodate all possible consequences of major accidents'. Therefore, 'the Commission

[105] Preamble (24) of Directive 2013/30/EU, *Official Journal* L178/69.
[106] Preamble (10) of Directive 2013/30/EU, *Official Journal* L178/67.
[107] Preamble (12) of Directive 2013/30/EU, *Official Journal* L178/67.
[108] Article 25 of Directive 2013/30/EU. [109] Article 2 of Directive 2013/30/EU.
[110] Preamble (13) of Directive 2013/30/EU.
[111] Preamble (11), (13), (36) and Art. 3.2 of Directive 2013/30/EU.
[112] Preamble (34) of Directive 2013/30/EU, *Official Journal* L178/70.
[113] Article 28 of Directive 2013/30/EU.
[114] As to the documentation requirements, see also Arts. 11–14 of Directive 2013/30/EU.

should undertake further analysis and studies of the appropriate measures for damages relating to offshore oil and gas operations, requirements on financial capacity including availability of appropriate financial security instruments or other arrangements. This may include an examination of the feasibility of a mutual compensation scheme.'[115]

As to the application of the Environmental Liability Directive (ELD) in the offshore activities, Directive 2013/30/EU clarifies that 'holders of authorizations for offshore oil and gas operations' are also the 'operators' within the meaning of ELD that will be held liable for the prevention and remediation of environmental damage.[116] Moreover, the Directive stresses that the holders of authorizations are not allowed to delegate their liability for environmental damage to third parties through contract.[117] In addition, Directive 2013/30/EU requires amendment to be made with respect to the definition of water damage in the ELD so that the liability of the licensees under the ELD shall also apply to marine waters of Member States as defined under the Marine Framework Directive (Directive 2008/56/EC).[118]

Realizing that offshore activities may have trans-boundary effects, Directive 2013/30/EU requires Member States to share information on the operations that may have trans-boundary effects,[119] and in the event of a major accident, a Member State should inform immediately the Commission and the Member States that may be affected by the accident.[120] When a Member State considers that offshore activities under its jurisdiction are likely to have 'significant effects on the environment in another Member State', it shall forward the relevant information to the Member State that may be potentially affected. When a Member State considers that it may be potentially affected by offshore activities in the jurisdiction of another Member State, it shall require relevant information from the other Member State.[121] The goal of these provisions in the Directive is to promote joint actions between Member States in preventing damage and responses.

As to the extra-territory effect of the Directive, Preamble (38) provides that Member States 'should ensure that operators and owners include their offshore oil and gas operations outside the Union in their corporate major accident prevention policy documents'. Article 20 of the Directive

[115] Preamble (63) of Directive 2013/30/EU.

[116] Preamble (11) and Art. 7 of Directive 2013/30/EU. [117] *Ibid.*

[118] Preamble (58) of Directive 2013/30/EU. [119] Articles 31–33 of Directive 2013/30/EU.

[120] Preamble (45) of Directive 2013/30/EU, *Official Journal* L178/71.

[121] Article 31 of Directive 2013/30/EU.

requires Member States to ensure that companies registered in their territory conducting offshore activities outside the Union as licence holders or operators report to them on major accidents in which they have been involved.

The Directive requires the establishment of a common data-reporting format across all Member States by operators and owners.[122] The idea of such a format is to facilitate sharing of comparable data between Member States so that it would 'provide transparency of the safety and environmental performance of operators and owners' and 'would facilitate dissemination of lessons learned from major accidents and near misses'.[123] As we showed in Section 2.3.1, the current information on offshore accidents from various sources is difficult to compare due to a lack of harmonized criteria. A common data-reporting format as envisaged in the Directive may be useful in improving the current situation. However, the information in the format is to be provided by operators and owners in various Member States, and the transposition of the Directive into national laws may still take two years. Therefore, it will take some time before the effect of a common data-reporting format can be seen.

3.5 Offshore Liability Regimes: Country Studies

3.5.1 Introduction

In this section, various countries will be selected for a detailed study. These countries are mostly countries where offshore oil and gas activities frequently take place or where catastrophic incidents involving offshore facilities have occurred. The United Kingdom and Denmark are EU Member States with strong offshore interests because their offshore exploration and exploitation activities in the North Sea play an important role. Norway, although not an EU Member State, has offshore activities in the North Sea that clearly will have a trans-boundary influence on EU Member States in the same sea region. Furthermore, Australia and the United States have been selected for the country study because they both have witnessed some major offshore incidents, which triggered some changes in their legal regimes. Finally, the national legislation of Canada also will be addressed because offshore oil and gas activities are of large importance in Canada. This results in an examination of three

[122] Preamble (42), Art. 23 and Annex IX of Directive 2013/30/EU.
[123] Preamble (42) of Directive 2013/30/EU.

European (United Kingdom, Denmark and Norway) and three non-European (United States, Australia and Canada) countries.

The legal regimes in these six countries on liability and compensation for offshore pollution damages will be discussed. In order to carry out the research in an integrated manner, a checklist is used to examine individual legal systems in a systematic way. The checklist will facilitate the comparative study of different legal regimes. The countries will therefore be analyzed on the basis of specific similar criteria that can be found in the checklist. The checklist is included in Appendix 2.

3.5.2 United Kingdom

3.5.2.1 National Interest in Offshore Activities

The United Kingdom has strong interests in offshore oil and gas activities. Oil and gas provides more than 70 per cent of the United Kingdom's total primary energy, with oil for transport and gas for heating being dominant in these markets. Offshore oil and gas remains the largest investing sector and the largest contributor to national gross value added among the industrial sectors of the economy.[124] The United Kingdom remains in the top twenty-five global producers of both oil (twenty-fifth) and gas (twenty-second) despite the sharp decline in production in recent years.

3.5.2.2 Legal Framework

Statutory Laws and Regulations The legislation in the United Kingdom, as far as it concerns offshore oil and gas activities, may include a large number of acts and regulations. They cover various aspects of offshore exploration and exploitation activities, mainly licensing requirements, emergency response, clean-up operations and counter-pollution or remediation measures, performance/operational standards and liability/compensation issues. The offshore legal regime in the United Kingdom went through some major changes after the Piper Alpha incident in 1988.[125] Following the Deepwater Horizon incident, the UK legal regime concerning offshore activities has again been critically reviewed.[126]

124 Oil and Gas UK, 'Economic Report 2014', 11.

125 In this respect, Lord Cullen's report plays a crucial role. The changes include, *inter alia*, a separate health and safety executive, goal-setting regime and safety case regime.

126 Following the Deepwater Horizon incident, the UK government has e.g. increased the number of inspectors and the number of annual environmental inspections by the Department of Energy and Climate Change. It launched a new joint industry and

Relevant laws and regulations on offshore oil and gas exploration and exploitation in the United Kingdom include various laws covering different aspects of the offshore activity. These may include, first, for licensing

- Hydrocarbons Licensing Directive Regulations of 1995: to implement Hydrocarbons Licensing Directive of 1994 (94/22/EC);
- Petroleum Act of 1998: vests all rights to petroleum[127] in the Crown, including the rights to search for, bore for and obtain. It empowers the Secretary of State to grant licences to search for and bore for and obtain petroleum to such persons as they see fit;
- Petroleum Licensing (Exploration and Production) (Seaward and Landward Areas) Regulations of 2004 (S.I. 2004/352): this has three sets of model clauses for different types of licences known as traditional, frontier and promote licences;
- Energy Act of 2008: improves licensing to respond to changes in the commercial environment and enables the Department of Energy and Climate Change to carry out its regulatory functions more effectively.[128] This Energy Act of 2008 introduces two new licensing regimes for the storage (Chapter 2) and permanent storage of carbon dioxide (Chapter 3);
- Petroleum Licensing (Production) (Seaward Areas) Regulations of 2008 (2008 No. 225): mainly incorporate changes reflected in the Energy Act of 2008 and replace the 1988 version of the Regulations. It has a single set of model clauses that apply in the case of all three types of licences, traditional, frontier and promote;
- Offshore Gas Storage and Unloading (Licensing) Regulations of 2009: these Regulations are made under Part 1, Chapter 2 of the Energy Act of 2008, which establishes a new licensing regime for the offshore storage and unloading of combustible gas; and
- Offshore Exploration (Petroleum and Gas Storage and Unloading) (Model Clauses) Regulations of 2009: these Regulations prescribe model clauses for offshore exploration licences granted under

government group called the Oil Spill Prevention and Response Advisory Group (OSPRAG) to review the United Kingdom's ability to prevent and respond to oil spills.

127 Meaning the petroleum in the strata of Great Britain and in the territorial sea of the United Kingdom and UK continental shelf. Separate legislation applies to Northern Ireland and waters adjacent to Northern Ireland up to the landward boundary of the territorial sea.

128 www.gov.uk/oil-and-gas-offshore-environmental-legislation.

Section 3 of the Petroleum Act of 1998 and Section 4 of the Energy Act of 2008.

The Petroleum Act of 1998, which consolidated the Petroleum (Production) Act of 1934,[129] provides that the exploration for and production of oil and gas in the United Kingdom and the UK sector of the North Sea can only be undertaken in accordance with the conditions of a licence issued by the UK government.[130] The Secretary of State, on behalf of the UK government, issues licences through annual licensing rounds.[131] The Department of Energy and Climate Change (DECC) is the department responsible for the regulation of all oil and gas licensing within the United Kingdom, its territorial sea and on the UK continental shelf. The conditions to issue licences are prescribed in a series of model clauses set out in Regulations in force when the licence is granted, subject to later amendment by Act of Parliament.

Licensees are required to demonstrate that they have funds available to meet any liability for oil spills in the course of licensed activities, and the DECC requires all operators to have signed up to the Offshore Pollution Liability Agreement (OPOL).

Because Directive 2013/30/EU must be implemented in national laws by 19 July 2015, many of the requirements in the Directive have been transposed in secondary legislation in the United Kingdom already well before that date. These mainly include the Offshore Installations (Offshore Safety Directive, Safety Case, etc.) Regulations of 2015 (SI 2015/398), the Merchant Shipping (Oil Pollution Preparedness, Response and Co-operation Convention, Amendment) Regulations of 2015, the Environmental Damage (Prevention and Remediation, England) Regulations of 2015 (SI 2015/810) and the Offshore Petroleum Licensing (Offshore Safety Directive) Regulations of 2015 (SI 2015/385). However, these changes mainly concern safety requirements and pollution response and do not touch upon the liability system for offshore pollution.

Liability for Offshore Pollution As far as liability for offshore oil pollution damage is concerned, there is no specific legislation dealing

[129] The Petroleum (Production) Act of 1934 grants the rights to all oil and gas in the United Kingdom and its territorial waters to the Crown and gives the government the exclusive right to award licences to explore for and exploit these resources.

[130] For a detailed discussion, see Easo, 'Managing the Risk', 18–22.

[131] Some licensing is 'made out of round'.

specifically with this issue. There has existed since the 1970s an industry scheme known as the Offshore Pollution Liability Agreement (OPOL)[132] which provides for compensation for oil pollution caused by offshore activities. When the damages are not covered by the OPOL system or the coverage by the OPOL scheme is insufficient, the claimant still may resort to legal remedies.

The liability for offshore incidents in the United Kingdom can be threefold:

1. *Liability based on OPOL:* Strict liability;
2. *Liability based on statute and model clauses:* Environmental Damage (Prevention and Remediation) Regulations of 2009: the model clauses require all licensees to operate in accordance with the methods customarily used in good oilfield practice and to take all steps practicable to prevent the escape of oil in the licensed area;
3. *Liability based on tort.*

The industry scheme OPOL came into being because of concern at the increasing oil exploration activities taking place at the seabed off North-West Europe since the 1970s.[133] People were conscious of the potential risks of pollution caused by offshore activities. Therefore, some countries in the region decided at the end of 1976 to establish a legal regime to cover the damage caused by offshore pollution incidents.[134] They signed the Convention of Civil Liability for Oil Pollution Damage Resulting from Exploration for and the Exploitation of Seabed Mineral Resources (CLEE) on 1 May 1977.[135] However, the ratification of an international convention is a long process. OPOL thus came into effect as an interim measure while waiting for ratification of the CLEE. In the end, the CLEE could not come into effect, and the UK government decided its interest might be better protected by the continuing function of OPOL.

OPOL covers escapes or discharges of oil from offshore facilities of any 'designated state' specified in the OPOL agreement.[136] However, it

[132] It was adopted on 4 September 1974 and entered into force on 1 May 1975.

[133] OPOL will be further discussed in detail when discussing pooling arrangements in Section 4.2. However, since membership of OPOL is mandatory in the United Kingdom already at this stage, we discuss the importance of OPOL briefly.

[134] These countries include the United Kingdom, Germany, Ireland, the Netherlands, Norway and Sweden.

[135] http://sedac.ciesin.org/entri/register/reg-092.rrr.html.

[136] Nine states: the United Kingdom, Denmark, Germany, France, Ireland, the Netherlands, Norway, the Isle of Man and the Faroe Islands.

excludes offshore facilities located in the Baltic and Mediterranean Seas. (Problems in countries such as France would be that different regimes may apply to different sea areas.) Parties to OPOL are 'Operators of or intend to be the Operators of Offshore Facilities' used for oil and gas exploration and production.[137] In the United Kingdom, all offshore operators active in exploration and production on the UK continental shelf are party to the OPOL.[138] To be a member of OPOL is a pre-condition for a licence for the UK offshore operators. There is no obligation to take out financial security under statutory laws, but effectively, it is implemented through the DECC practices in the licensing process: prior to consenting to any exploration or appraisal drilling operations, DECC requires the operator of the offshore facility to be a member of OPOL. In addition, evidence of financial responsibility also is required.

OPOL does not exclude the claimant's right to pursue full compensation under statutory law of the United Kingdom in the event that the compensation provided through OPOL is inadequate. In such a case, several laws and regulations may play a role.

The Environmental Damage (Prevention and Remediation) Regulations of 2009 introduce two types of liability: strict liability in respect of environmental damage caused by a specified range of 'occupational activities' and fault liability in respect of environmental damage to protected species and natural habitats from all other occupational activities.

The 'occupational activities' listed in Schedule 2 of the Regulations that are subject to strict liability without the need of showing fault include

- Activities requiring environmental permits,
- Discharges to water,
- Groundwater discharges,
- Water abstraction or impoundment,
- Use of pesticides, biocides or dangerous substances,
- Use and release of genetically modified organisms and
- Transporting dangerous goods.

In the DECC's view, offshore oil and gas activities shall not be simply considered 'activities requiring environmental permits'. As to liability for environmental damage concerning offshore oil and gas activities, it is

137 See Preamble of Offshore Pollution Liability Agreement of 1 April 2015.
138 www.opol.org.uk/.

rather complicated to decide under the Environmental Damage (Prevention and Remediation) Regulations of 2009.

First, strict liability requires an activity listed in Schedule 2. The only paragraph which seems relevant is paragraph 2 of Schedule 2, which itself refers to activities listed in Annex I of Directive 2008/1/EC. In respect of that list, the one which seems relevant is '1.1. Combustions installations with a rated thermal input exceeding 50 MW'. For such an installation, the Regulations of 2009 also require the damage to be in an area specified in the table to Regulation 6(1). Damage to water applies only to the first nautical mile of waters around England. Otherwise, the seabed to the limit of the continental shelf or EEZ is an area of damage but only in respect of protected species or natural habitats. Given that the Regulations of 2009 apply to England and reserved matters, 'seabed' does not include areas of the sea adjacent to Scotland or Northern Ireland, which are part of those territories.

Second, as far as liability with fault (i.e. negligence or with intent) is concerned, there is no limit to the type of activity; hence, all offshore petroleum activities may be included. In terms of the area of damage, it is the seabed to the limit of the continental shelf or EEZ. However, there must be damage to a protected species or natural habitat within that area.

3.5.2.3 Basis of Liability

Under the OPOL regime, the liable parties are operators of the offshore facilities, and the liability is strict for pollution damage and cost of remedial measures. The parties are exempted from liability when the incident is caused by one of the following situations:

1. An act of war or *force majeure*,
2. Wholly by an act or omission done with intent to cause damage by a third party,
3. Wholly by the negligence or other wrongful act of a government authority or as a result of compliance with conditions imposed or instructions given by the government or
4. Wholly or partially either from an act or omission done with intent to cause damage by a claimant or from the negligence of that claimant. In such a case, the operator shall be exonerated wholly or partially from his obligations to the claimant.[139] This is known as a 'comparative negligence defence'.

[139] Clause IV.B of OPOL Agreement (of 1 April 2015). See also Budiman, 'On Liability for Offshore Oil Spillage'.

As mentioned earlier, the statutory regime also may be relevant in the case of insufficient compensation through OPOL. Under the Environmental Damage (Prevention and Remediation) Regulations of 2009, the environmental damage caused by a specified range of 'occupational activities' (e.g. activities requiring environmental permits, which is the case for offshore activities) is based on strict liability.

In the event of a fatality or personal injury, a claimant is likely to have rights to compensation under the UK statutes and regulations and under the common law of tort. In some cases strict liability will apply, and in others a claimant may have to establish fault.

3.5.2.4 Relationship with Regulation

Normally, if an incident occurs, the operator is subject under the OPOL scheme to a no-fault liability regime, and compliance with regulations does not exempt the operator from liability. OPOL provides that if the incident resulted from compliance with conditions imposed or instructions given by the government, the operators may be exempted from liability.[140] Clause IV(B) reads

> No obligation shall arise hereunder with respect to Remedial Measures and/or Pollution Damage arising from an Incident if the Incident: ... 3. was wholly caused by the negligence or other wrongful act of any Government or other authority or resulted from compliance with conditions imposed or instructions given by the Government of the State which issued the Licence as to the Offshore Facility involved.

Compliance with regulatory requirements does not absolve a party from its obligations under UK laws to pay unlimited compensation.

3.5.2.5 Causation

Because under the OPOL scheme the operators shall bear strict liability, it does not require proof of fault. The burden of proof is on the operators in the sense that the operators shall be held strictly liable unless they can establish that the incident is caused by one of the exemptions (e.g. an act of war or *force majeure*) mentioned earlier.

The claimant has to prove that there is a causal link between the incident and the damage he has suffered. The implication might be that the chances of recovering the damage might be higher than with a negligence rule. In turn, it impels the operator to operate with utmost due care.

140 *Ibid.*

3.5.2.6 Attribution of Liability

Under OPOL, the operator of an offshore facility is primarily responsible for payment of the costs of remedial measures and pollution damage. In addition, the parties to OPOL have agreed that in the event a party fails to satisfy its obligations to claimants under the Agreement, the parties will contribute sums to enable those claims to be met. This is referred to as the 'OPOL guarantee'. Contributions by the remaining parties would be proportionate to the number of offshore facilities operated by the parties.[141] To this extent, offshore operators in the United Kingdom take part in a scheme which would provide compensation for certain environmental and economic consequences of an offshore oil spill.[142]

3.5.2.7 Damages and Remedies

The claims to be covered via OPOL concern two categories, 'pollution damage' and 'remedial measures'. All claims must be 'reasonable, quantifiable and justifiable'.

'Pollution damage' to be compensated by OPOL is defined as 'direct loss or damage (other than loss of or damage to any Offshore Facility involved) by contamination which results from a Discharge of Oil'.[143] Therefore, only 'direct' damage is covered under OPOL. The meaning of 'direct loss or damage' would be determined by applying the laws of England. Those laws provide that each case will be considered on its merits. Any claims for economic loss would be decided by reference to the facts of each case. Some elements of economic loss may be recoverable by a claimant under the OPOL scheme, whilst other elements may not be recoverable. The question may arise as to whether damage to the environment itself shall be considered as such a 'direct' damage.[144]

'Remedial measures' are 'reasonable measures taken by any Party from any of whose Offshore Facilities a Discharge of Oil occurs, and of which such Party is the Operator, and by any Public Authority to prevent, mitigate or eliminate Pollution Damage following such Discharge of Oil or to remove or neutralize the Oil involved in such discharge, excluding, however, well control measures and measures taken to

[141] OPOL website: www.opol.org.uk/about-2.htm.

[142] 'Offshore Oil and Gas in the UK', 5, 76, 153. [143] Clause I.13 of OPOL Agreement.

[144] This concern was raised e.g. in Energy and Climate Change Committee, 'UK Deepwater Drilling'.

protect, repair or replace any such Offshore Facility'.[145] Claims admissible for remedial measures or pollution damage thus fall into the following categories: clean-up operations on shore or at sea, property damage, disposal costs of collected material and other losses which must be quantifiable and which must result directly from the contamination itself.[146] As explained earlier, whether the element of loss or damage is considered directly from contamination would be decided on the basis of the facts applicable to each case. Recovery of the costs of remedial measures can be made by government and public authorities (not by individuals or companies).

Recovery of the costs of pollution damage can be made by any claimant, including an individual or company, in accordance with OPOL.

Personal injury does not fall under the scope of OPOL. Since the damage compensable under OPOL must be 'direct', it is debatable whether economic losses as a result of an offshore pollution incident are compensable. For this reason, the definition of 'direct damage' under OPOL was criticized for being unclear.[147]

The judicial demarcation between 'direct' losses and indirect or consequential losses has been widely debated in the United Kingdom. According to some, current English law regards loss of profits as a 'direct' loss.[148] In practice, the contracts often define the term 'consequential loss' specifically, deeming to be consequential certain types of loss that would be regarded as direct as a matter of law. Under the joint operating agreement, 'indirect or consequential loss' is defined to include loss in relation to delay or loss of petroleum production; loss of use, profit, revenue, bargain or opportunity; and loss in relation to reservoir damage.[149]

In the Environmental Damage (Prevention and Remediation) Regulations of 2009, environmental damage is damage to '(a) protected species or natural habitats, or a site of special scientific interest, (b) surface water or groundwater, or (c) land'.[150] The Regulations apply to 'the prevention and remediation of environmental damage'.[151]

145 Clause I.15 of OPOL Agreement. 146 Guidelines for Claimants, OPOL.
147 Energy and Climate Change Committee, 'UK Deepwater Drilling', 26.
148 Easo, 'Managing the Risk', 21. 149 *Ibid.*
150 Regulation 4 of the Environmental Damage (Prevention and Remediation) Regulations of 2009.
151 Regulation 4(1) of the Environmental Damage (Prevention and Remediation) Regulations of 2009.

3.5.2.8 Amount of Compensation

The OPOL liability limit has been reviewed on a number of occasions. In August 2010, as a consequence of the Deepwater Horizon incident, the liability limit was increased to US$250 million, and this entered into effect on 1 October 2010.[152]

The overall amount per incident is US$250 million, which includes a maximum US$125 million for 'remedial measures' incurred by public authorities and another maximum US$125 million for 'pollution damage'. When all the claims in one category have been met, any surplus may be used to meet unsatisfied claims in the other category. If the total claims in a category exceed the sums available in that category, after any surplus from the other category has been fully used, the compensation is shared pro rata among the claimants.

OPOL does not take away a claimant's right to seek redress through the courts for losses which exceed the maximum recoverable under the OPOL Agreement or those beyond the scope of the OPOL Agreement.[153]

The general principles in the UK law do not impose a statutory limitation on the liability of an offshore operator. However, the Environmental Damage (Prevention and Remediation) Regulations of 2009 provide in Regulation 7 that the provisions in the Regulations 'are without prejudice to the right of an operator to limit liability in accordance with the Convention on Limitation of Liability for Maritime Claims of 1976'.[154] The Convention on Limitation of Liability for Maritime Claims of 1976 only covers liability for vessels and not for offshore facilities. Therefore, the operators of facilities under UK jurisdiction are subject to unlimited liability beyond their US$250 million limit under OPOL.

The US$250 million limit was questioned by a UK House of Commons Select Committee Report in December 2010. It raised the question of whether US$250 million would be sufficient to cover the costs of a blowout on the UK continental shelf.[155] Interestingly, another study commissioned by OPOL and Oil and Gas UK in 2012 concluded that the current limit of US$250 million was considered

[152] www.oilandgasuk.co.uk/knowledgecentre/key_issues.cfm.

[153] OPOL Guidelines for Claimants; see www.opol.org.uk/downloads/opol-guidelines-oct10.pdf.

[154] Regulation 7(2) of Environmental Damage (Prevention and Remediation) Regulations of 2009.

[155] Energy and Climate Change Committee, 'UK Deepwater Drilling', 24–6.

adequate for the vast majority of UK well operations; only in certain limited cases could the clean-up costs and compensation result in claims for amounts above that limit.[156] The process introduced by DECC on 1 January 2013, requiring evidence of financial responsibility, captures those drilling operations where the cost of pollution damage and remedial measures would exceed US$250 million. Such operations have to be 'topped up' to the financial level identified via the accepted industry assessment process to cover the potential pollution damage and remedial measures that may have to be undertaken for that particular operation.

3.5.2.9 Claim Settlement

The goal of OPOL is said to facilitate rapid payment because it does not need a lengthy legal action.[157] Clause VI of the OPOL Agreement (amended 4 July 2012) provides that 'No Party shall have any obligation under this Contract with respect to any Claim filed over one year from the date of the Incident which resulted in the Pollution Damage or the taking of the Remedial Measures in question.' Hence, the liability of the operators under OPOL will cease unless the claim is submitted within one year of the time the cause of action arises.

In the event of a dispute arising between a claimant and an operator as a party to OPOL, it should be settled under the rules of arbitration of the International Chamber of Commerce by one or more arbitrators.[158] Arbitration is generally considered faster and speedier than litigation. The compensation mechanism of OPOL is designed to provide a quick settlement of claims without the expense of court proceedings.[159] 'In arbitration settlements, the parties are able to choose their decision maker, unlike the court, who they think has specific expertise on the nature of the case. Added to a comparatively informal procedure, the dispute would be settled much faster than through litigation. Arbitration brings certainty faster, even more so as its award is final and binding, with no consequential delay as a result

[156] Oil and Gas UK and OPOL, 'Oil Spill Cost Study', 46.

[157] See OPOL website. OPOL is discussed in further detail in Section 4.2.

[158] Clause IX of OPOL Agreement.

[159] Niall Scott, Offshore Industry Compensation Regime, speech at Interspill Conference and Exhibition, London, 2012. Available at: www.interspill.com/previous-events/2012/14-march/pdfs/offshore%20forum%20panel%20-%20uk%20regime/niall%20scott%20(opol)%20-%20offshore%20industry%20compensation%20regime.pdf.

of an appeal. It also means that the process of arbitration may be cheaper than with a court settlement.'[160]

The OPOL arrangement may indeed be faster than litigation. The scheme calls on the claimant to notify an operator as soon as possible after an oil spill incident that damage has been suffered and/or costs incurred. OPOL states that the scheme is intended to ensure that all claims associated with an oil spill are settled in an orderly and swift manner. However, since OPOL is not a fund, it does not have a mechanism that would provide immediate compensation to victims. The mutual guarantee under OPOL is to assure the capability of paying sums due to claimants under OPOL.

3.5.2.10 Compensation Mechanisms

In the regulations on licensing in the United Kingdom, financial capacity and membership in OPOL are a precondition for a licence to be granted for an offshore operator.[161] It is further reinforced by the inclusion of a standard OPOL clause in all joint operating agreements. The operator must establish and maintain its financial capability to meet claims that arise under OPOL by producing evidence of insurance, self-insurance or other satisfactory measures, with the financial responsibility to meet such capability allocated contractually between the operator and non-operators under a joint operating agreement.[162]

Therefore, in the UK legal regime, the financial capacity of the (potential) offshore operators is a condition for granting licences, and compliance with regulatory requirements does not exclude operators from assuming liability for pollution damage.

The Petroleum Act of 1998 provides in the model clauses that all licensees shall take 'all steps practicable' to prevent the escape of oil in

[160] Budiman, 'On Liability for Offshore Oil Spillage'.

[161] Petroleum (Production) (Seaward Areas) Regulations of 2008, Clause 23(9). It provides that '[t]he Licensee shall comply with any reasonable instructions from time to time given by the Minister with a view to ensuring that funds are available to discharge any liability for damage attributable to the release or escape of Petroleum in the course of activities connected with the exercise of rights granted by this licence; but where the Minister proposes to give such instructions he shall before giving them –

(a) give the Licensee particulars of the proposal and an opportunity to make representations to the Minister about the proposal; and

(b) consider any representations then made to him by the Licensee about the proposal.

[162] Offshore Pollution Liability Limited, Offshore Pollution Liability Association Limited Rules, www.opol.org.uk/rules-2.htm.

the licensed area, and the licensees are required to have sufficient funds available to discharge any liability that they may accrue for damage in connection with any oil pollution. There is no statutory obligation to take out pollution liability insurance under the Petroleum Act for the licensing requirements, but OPOL and effectively all operators have some financial security through OPOL. In addition, under the requirements of the Merchant Shipping (Oil Pollution Preparedness, Response and Co-operation Convention) Regulations of 1998, there is a legal obligation upon operators to have adequate financial security in place in order to assure DECC that the provisions for implementation of counter-pollution measures, as detailed within the OPEP, can be carried out.

According to the Rules of the Offshore Pollution Liability Associated Limited (as of 1 January 2012), the financial responsibility can be established through insurance, guarantee or self-insurance. In practice, most members of OPOL use insurance as their preferred method of satisfying financial responsibility.[163]

First, for evidence of insurance, verification of insurance from an insurance company or insurance broker or agent is acceptable to OPOL for an amount of not less than US$250 million per incident and US$500 million in the annual aggregate. Second, for an operator that wishes to rely upon a guarantee, the verification of a guarantor must be acceptable to OPOL. Third, in order to qualify as a self-insurer, the operator must have one or more of the following credit or financial strength ratings from an internationally recognized credit-rating agency: Standard & Poor's: A– or higher; A. M. Best: A– or higher; Moody's: A3 or higher; and Fitch: A or higher. If the operator elects to use insurance or a guarantee to meet its obligations to evidence financial responsibility under OPOL, the insurer or guarantor must also meet one or more of the ratings as just specified.

Most joint operating agreements require the operator to take out insurance for joint operations, and partners can join in the operator's insurance or take out their own. Some large oil companies do not insure with the market.

Typically, licensees maintain various insurances relevant to blowout, including property, control of well (referred to 'operator's extra expense' [OEE] insurance), sudden and accidental pollution, and third-party legal liability.[164]

163 OSPRAG, 'Strengthening UK', 33. 164 Aldersey-Williams, 'Liability for Oil Spills'.

The UK licensing regulations require demonstration of financial competence.[165] There are two aspects of financial competence. The first is to carry out the anticipated operations under the licence, namely, the day-to-day operations (which is a matter covered primarily by the model clauses), and the second is the financial competence to deal with the adverse environmental consequences of drilling activity, namely, unplanned/accidental events, dealt with as part of the approval of OPEPs under the OPRC Regulations. According to DECC, the granting of a licence and/or drilling consent is conditional on a company showing that it has sufficient financial resources to cover the costs of a spill. However, details of the ability to pay are not publicly available and hence may be difficult to verify.[166] Whilst the ability to pay is not made public, the process by which an operator has to provide evidence to DECC that sufficient money is in place does provide assurance of the operator's ability to cover both first- and third-party costs in relation to the proposed drilling operations should a spill incident occur.

Due to concerns raised by the Deepwater Horizon incident, the Department of Energy and Climate Change announced on 8 June 2010 that it would review the indemnity and insurance requirements for operating on the UK continental shelf.[167] Following the study carried out by the Oil Spill Prevention and Response Advisory Group (OSPRAG), DECC indicated 'its need to be assured that appropriate contingency finances are available to respond to and compensate for any incident arising from licensees carrying out the work programme on their petroleum licences'.[168] DECC normally carries out checks on the finances of a company before granting a licence. However, this process, according to the OSPRAG study, focuses 'more on a company's ability to carry out the agreed work programme than its ability to pay for unforeseen events'.[169] DECC issued a letter to the operators in the United Kingdom in December 2010 stating that it now requires explicit confirmation that sufficient finance or insurance/indemnity provision is available to drill a relief well in a timely manner.[170] DECC published in November 2012 the 'Guidance Note to UK Offshore Oil and Gas

[165] 'Offshore Oil and Gas in the UK', 76.
[166] *Ibid.*, 76–7; see the opinion of Greenpeace.
[167] DECC, 'UK Increases North Sea Rig Inspections', press release 1–/067, 8 June 2010; see the website of DECC: www.gov.uk/government/news/uk-increases-north-sea-rig-inspections.
[168] OSPRAG, 'Strengthening UK', 34. [169] *Ibid.* [170] *Ibid.*

Operators on the Demonstration of Financial Responsibility before Consent May Be Granted for Exploration and Appraisal Wells on the UKCS' (DECC Guidance Note).[171] This document outlines DECC's expectations for demonstrating that 'the risks of the operation have been appropriately estimated and that the financial mechanisms are in place to meet those risks, should they materialize'.[172]

The DECC Guidance Note specifies that the financial mechanisms should be accompanied with oil pollution emergency plans (OPEPs) during the licensing process. 'OPEPs contain worst case scenario information and details are provided of the incident response actions to be implemented should such a scenario occur, including loss of well containment/well blowout.' 'All the response mechanisms to be used by operators during any such incidents will be expensive (first party costs), as may the compensation costs for pollution damage (third party costs).'[173] For this purpose, the level of financial responsibility should cover the two aspects of costs:

- Costs of well control and
- Costs of financial remediation and compensation from pollution.[174]

The DECC Guidance Note also specifies that financial responsibility can be verified by means of the following:

> reliance on the credit/financial strength rating of the operator or co-venturer;
> insurance;
> parent company guarantee/affiliate undertaking; and
> any combination of the above.[175]

DECC is aware of the work of Oil and Gas UK and OSPRAG, and it specifically mentions in the DECC Guidance Note that the Guidelines as developed by Oil and Gas UK 'are not DECC Guidance, but DECC intends to give considerable weight in each case to an operator who can show that the guidelines have been met'.[176]

[171] The DECC Guidance Note became effective on 1 January 2013.
[172] DECC Guidance Note. [173] *Ibid.* [174] *Ibid.* [175] *Ibid.*
[176] *Ibid.* The DECC strongly collaborated with Oil and Gas UK on the drafting of the Guidelines but explicitly held that the Guidelines are formally from Oil and Gas UK and not from DECC. This has the advantage that if an operator were to show financial responsibility via another means than those mentioned in the Guidelines, DECC would still have the possibility to accept that (interview with Wendy Kennedy, Head of Offshore Oil and Gas at the Department of Trade and Industry [Department of Energy and Climate Change] on 29 April 2013).

OSPRAG recommended that the industry should be issued with good-practice guidance on the requirements for operators to demonstrate adequate financial resources to DECC with respect to the costs arising from a potential incident. A working group was tasked to implement this recommendation, and as a result of the work carried out by this working group, some guidelines were published in November 2012.[177] These guidelines propose methods to assess the two elements of costs identified in the DECC Guidance Note: the cost of bringing a well under control following a blowout and the cost of remedial measures and payments of compensation to third parties for pollution damage. The figures resulting from these two estimates are then aggregated to produce the total amount of financial resources which is recommended to be demonstrated to DECC.[178] However, as explained in the Guidelines, these guidelines are to be considered as good industry practice, and following these guidelines may not necessarily fulfil all DECC's requirements in every case.[179]

According to the Guidelines, the largest part of the cost of bringing an exploration or appraisal well under control is the cost of drilling a relief well.[180] In this respect, Oil and Gas UK has also developed guidelines, namely, Guidelines for Relief Well Planning – Subsea Wells. Depending on the complexity of the relief well, different procedures should be followed by the operators. The primary factors that influence the cost of drilling a well include location, water depth, well depth and geo-environment.[181]

In addition to the cost of drilling a relief well, other elements that may have influence on the cost of well control, according to the Guidelines, include e.g. the cost of deploying a capping device, where appropriate.[182] The capping device is a further means of capping a subsea well in case the blowout preventer fails.[183] This capping device was recommended by the OSPRAG as good industry practice. These costs are estimated to be US $40 million for 30 days of operation.

As to the second aspect of the cost factor, remedial measures and compensation, because membership in OPOL is obligatory for operators in the United Kingdom, for most cases, the limit of US$250 million per incident under OPOL is considered sufficient for the cost of compensation.[184] The Guidelines further list a number of categories of wells for which the likelihood of pollution is considered 'extremely remote and

[177] Oil and Gas UK, *Guidelines to Assist Licensees*, 2. [178] *Ibid.* [179] *Ibid.*, 'Foreword'.
[180] *Ibid.*, 8. [181] *Ibid.*, 9. [182] *Ibid.*, 12. [183] *Ibid.*, 11.
[184] These are also the findings of the joint study carried out by OPOL and Oil and Gas UK in February 2012.

costs are highly unlikely to exceed' the OPOL limit of US$250 million.[185] This category mainly includes gas wells, gas condensate wells (other than HP/HT wells)[186] and wells which require artificial lift to flow.[187] For the other wells, the assessment of costs depends on the band which they fall in

(1) BAND 1: Costs unlikely to exceed US$250 m (FR demonstrated to OPOL deemed to be sufficient);
(2) BAND 2: Costs over US$250 but under US$275 m;
(3) BAND 3: Costs over US$375 but under US$500 m;
(4) BAND 4: Costs over US$500 m.[188]

The Guidelines specify that only recognized modelling methods should be used to determine the costs, including the Oil Spill Information System (OSIS), Oil Spill Contingency and Response (OSCAR) and Oilmap.[189] The results of the modelling should be analyzed to determine coverage in the following four categories:

Marine fishing areas;
Aquaculture areas;
Amount of oil on the coastline; [and]
Length of coastline becoming oiled.

185 Oil and Gas UK, *Guidelines to Assist Licensees*, 13.

186 The Guidelines define these wells as follows:

(1) Gas condensate: a portion of natural gas of such composition that it is in the gaseous phase at the temperature and pressure of the reservoirs, but that, when produced, it is in the liquid phase at surface pressure and temperature. Normally condensate has a composition of C5 and C8 and an API gravity of greater than 40.
(2) HT: any well where the anticipated undisturbed bottom hole temperature is greater than 300 degrees F or 150 degrees C.
(3) HP: any well where the maximum anticipated pore pressure of the porous formation to be drilled exceeds a hydrostatic gradient of 0.8 psi/ft, or the well requires pressure control equipment with a rated working pressure in excess of 10,000 psi or 69MPa. (*Ibid*., 14)

187 *Ibid*., 13–4. However, it is also noted in the Guidelines that for gas and gas condensate wells it will still be appropriate to demonstrate financial responsibility in relation to the costs of control of the well.

188 *Ibid*., 14.

189 *Ibid*., 14. In the United Kingdom, two models are often used to assess and plan for oil industry risks: OSIS developed by BMT ARGOSS and OSCAR developed by SINTEF. This was explained in the study in February 2012 by OPOL and Oil and Gas UK. For the purpose of the Guidelines, it is assumed that the release is a point surface release. Moreover, the modelling assumes that the well flows for thirty days before being effectively capped and the oil released during that time continues to travel until reaching a shoreline or until such oil has become insignificant at sea.

These categories are chosen because they have the most influence on cost according to the OPOL/Oil and Gas Oil Spill Cost Study – OPOL Financial Limits. Each category just mentioned should be graded as high, medium or low based on certain criteria, and the grading is translated into a numerical value: high category scores three points, medium two points and low one point. Corresponding with the scores, different amounts of financial responsibility are recommended to be demonstrated.[190]

3.5.3 Norway

3.5.3.1 National Interest in Offshore Activities

The petroleum industry is Norway's largest industry.[191] In 2012, the petroleum sector represented more than 23 per cent of the country's total value creation. The revenues from the petroleum sector constitute 30 per cent of state revenues.[192] Oil production on the Norwegian continental shelf started on 15 June 1971 at the Ekofisk oil field. According to statistics from the government, 42 per cent of the total expected resources on the Norwegian continental shelf have been produced, and there is still a huge potential for value creation on the Norwegian continental shelf. In 2012, fifty-three companies are licensees on the Norwegian continental shelf, and forty-two exploration wells have been drilled.[193]

It should be mentioned that the Norwegian government plays an important role in the oil sector. Statoil, which is 67 per cent state owned,[194] controls over 60 per cent of Norway's oil and gas production. In the Petroleum Activities Act, there are specific provisions on the state's participation in petroleum activities.[195] Initially, the Norwegian state's participation in petroleum activities was largely organized through Statoil. In 1985, the Norwegian state established the State's Direct Financial Interest (SDFI), through which the Norwegian state has direct participating interests in the licences and petroleum activities on the

[190] *Ibid.*, 15–7.

[191] From the website of the Norwegian government, Ministry of Petroleum and Energy: www.regjeringen.no/en/dep/oed/subject/oil-and-gas.html?id=1003.

[192] www.regjeringen.no/en/topics/energy/oil-and-gas/norways-oil-history-in-5-minutes/id440538/.

[193] *Ibid.*

[194] Norwegian Ministry of Trade and Industry, 'The State's Ownership Report 2013', 6.

[195] Chapter 11 of the Petroleum Activities Act, Management of the State Direct Financial Interest.

Norwegian continental shelf. Petrol AS (a wholly state-owned company) was established in 2001 to manage SDFI assets.[196]

3.5.3.2 Legal Framework

The Norwegian legal regime concerning offshore activities started to develop after the discovery of the Ekofisk oil field in December 1969. Following the incident of Alexander Kielland in 1980, Norwegian regulators recognized that the inspection-based approach was wholly inadequate. In the years subsequent to the incident, the legal regime changed to a risk-based approach, with regulations to describe what must be achieved, not how it must be achieved.[197]

Currently, the legal regime in Norway concerning offshore activities is mainly established via several statutes and related regulations. The statutes include the Petroleum Activities Act, supplemented by other statutes, including mainly the Pollution Control Act, the Petroleum Taxation Act and the Working Environment Act. On the basis of the Petroleum Activities Act, various regulations have been adopted.[198] There are, in addition, regulations focusing on the health, safety and environmental aspect of offshore activities.[199]

Norway's Petroleum Activities Act was adopted in November 1996 and later amended in June 2009.[200] The most recent amendment was made in June 2011.[201] The Petroleum Activities Act lays down the general framework for offshore oil exploration and exploitation on the Norwegian continental shelf. It establishes requirements for safe and environmentally sound oil exploitation. The Act lays down the procedure for licensing and the considerations that will be taken into account when making decisions on the allocation of licences.[202] It further requires licensees to ensure that they have a competent organization, to submit

[196] For further details, see www.regjeringen.no/en/topics/energy/oil-and-gas/norways-oil-history-in-5-minutes/id440538/.

[197] Dagg *et al.*, 'Comparing the Offshore Drilling', 37–9.

[198] All these regulations are available at the website of the Norwegian Petroleum Directorate: www.npd.no/en/regulations/regulations/.

[199] Available at the website of Petroleum Safety Authority: www.ptil.no/regulations/category216.html.

[200] Petroleum Activities Act, 29 November 1996, No. 72, amended by Act of 19 June 2009, No.104.

[201] The last changes were made by Act of 24 June 2011, No. 38. Accessible at the website of Norwegian Petroleum Directorate, which has an unofficial English translation of the Act: www.npd.no/en/regulations/acts/petroleum-activities-act/#1-1.

[202] These provisions are supplemented and implemented in the Petroleum Activities Regulations.

a development plan for approval, and to maintain efficient emergency preparedness. The Act has a separate chapter dealing with the liability of the licensee. It also has provisions on the establishment of safety zones around offshore installations and sanctions.

The liability for pollution damage is generally governed by rules in Chapter 7 of the Petroleum Activities Act under the title 'Liability for pollution damage'. Chapter 8 of the Act further provides '[s]pecial rules relating to compensation of Norwegian fisherman'. Detailed rules and methods on fishermen's compensation are implemented in two regulations, the Fishermen Compensation Regulations and the Regulations on Compensation to Fishermen for Lost Fishing Time at the location, recording and bringing ashore of scrap that is not derived from petroleum activities.

The Pollution Control Act was adopted in 1981[203] and last amended in 2003. The Pollution Control Act contains provision on the protection of the outdoor environment against pollution. Under the act, pollution is forbidden unless it is specifically permitted by law, regulations and individual permits. The Pollution Control Act is an enabling act. This means that the details in each case are outlined in discharge permits and regulations issued by the pollution control authorities.

The regulatory authorities in Norway for offshore activities are mainly as follows:

- The Norwegian Petroleum Directorate (NPD), established in 1972, is a governmental specialist directorate and administrative body, and it reports to the Ministry of Petroleum and Energy (MPE). The NPD has acquired since its establishment great authority in matters related to exploration and exploitation of oil and gas. It is a key advisor to the MPE in recommending which licences the government should award in these matters.[204] The NPD has a national responsibility for data from the Norwegian continental shelf. The NPD's data, overview and analyses constitute a crucial factual basis on which the activities are founded. The NPD sets frameworks, stipulates regulations and makes

[203] Act of 13 March 1981, No. 6, Concerning Protection against Pollution and Concerning Waste. An English version of the Act is found at: www.regjeringen.no/en/doc/laws/acts/pollution-control-act.html?id=171893.

[204] Website of the Norwegian Oil and Gas Association: www.norskoljeoggass.no/en/about/. As for the licensing procedure, application for exploration licences shall be submitted to the NPD (according to Section 3 of Petroleum Activities Regulations) and production licences shall be submitted to the Ministry of Petroleum and Energy with a copy to the NPD (Section 8 of Petroleum Activities Regulations).

decisions in areas where it has been delegated authority. It is also responsible for conducting metering audits and collecting fees from the petroleum industry.[205]

- The Norwegian Environment Agency has since 2013 taken over the original Klif (Climate and Pollution Agency), subordinate to the Ministry of Climate and Environment. It exercises regulatory authority over matters prescribed in the Pollution Act, the Product Control Act and the Climate Quota Act.
- The Petroleum Safety Authority (PSA), established in 2004,[206] subordinate to the Ministry of Labour of Social Affairs, acts as the regulatory authority for the technical and operational safety aspects of the petroleum sector in Norway.[207] This mainly covers safety, emergency preparedness and the working environment throughout all phases of petroleum activities (planning, design, construction, operation and possible ultimate removal).[208] The PSA performs inspections and provides advice and information to the Ministry of Labour and Social Affairs. Through audits (planned and systematic review of the management system, including a review of documents and interviews with selected people) and verification (random sampling to investigate whether the management system works in practice), and in cooperation with other health, safety and environmental regulators, the PSA carries out supervision of the established systems and ensures that the activities are supervised in a coherent manner.[209] Acts under the PSA's area of authority include the Petroleum Activities Act, Working Environment Act[210] and the Fire and Explosion Prevention Act, Pollution Control Act.[211]
- Norwegian Coastal Administration (NCA) is an agency of the Ministry of Fisheries and Coastal Affairs. It is responsible for maritime safety and infrastructure, transport efficiency and emergency response to acute pollution.[212]

[205] www.npd.no/en/about-us/.

[206] Before PSA was established in 2004 as an independent agency, its responsibilities were carried out by NPD.

[207] www.psa.no/about-us/category877.html.

[208] www.psa.no/role-and-area-of-responsibility/category916.html. [209] *Ibid.*

[210] Act of 17 June 2005, No. 62, Relating to Working Environment, Working Hours and Employment Protection, etc., as subsequently amended, last by the Act of 14 December 2012; available at: www.arbeidstilsynet.no/binfildownload2.php?tid=92156.

[211] An English version of the act is available at: www.regjeringen.no/en/dokumenter/pollution-control-act/id171893/.

[212] Janssen, *Preventing a Next Deepwater Horizon?*, 162.

3.5.3.3 Basis of Liability

The liability for offshore pollution damage under Norwegian law is imposed on the 'licensee' or 'operator' and is based on strict liability. Section 7-3 of the Petroleum Activities Act specifically provides that 'The licensee is liable for pollution damage without regard to fault. The provisions relating to the liability of licensees apply correspondingly to an operator who is not a licensee when the Ministry has so decided in connection with the approval of operator status.'

Section 8-3, titled 'Pollution and Waste' (which concerns compensation to Norwegian fishermen), also contains a similar provision on the strict liability of the licensee: 'The licensee is liable, regardless of fault, in respect of financial losses incurred as a result of pollution and waste from the petroleum activities, and the cost of reasonable measures to avert or limit such damage or such loss, including damage or loss as a result of such measures.'

The exemptions to the licensee's liability are mainly *force majeure*, act of war or exercise of public authority.[213] However, an exemption does not remove the full responsibility of the licensee. Under Norwegian law, his liability may 'be reduced to the extent it is reasonable, with particular consideration to the scope of the activity, the situation of the party that has sustained damage and the opportunity for taking out insurance on both sides', when the exemption event 'has contributed to a considerable degree to the damage or its extent under circumstances which are beyond the control of the liable party'. 'In the event of pollution damage from a facility located in an area outside the Norwegian continental shelf, the party who has approval from the competent authority to conduct the activities to which the facility is connected is regarded to be a licensee.'[214]

The model joint operating agreement (JOA) states in Article 7 that for 'liabilities and payments', 'unless otherwise specified in this Agreement, the Parties shall be primarily liable to each other on a pro rata basis, secondarily jointly and severally liable for all obligations arising by virtue of the joint venture's activities. This applies irrespective of a liability towards third parties.'[215]

[213] Section 7-3 of the Petroleum Activities Act. [214] *Ibid.*
[215] Available at: www.regjeringen.no/upload/oed/vedlegg/konsesjonsverk/k-verk-vedlegg-1-2-eng.pdf.

3.5.3.4 Relationship with Regulation

There are two categories of licences under Norwegian law, the exploration licence and the production licence. The exploration licence is granted by the Ministry of Petroleum and Energy,[216] and the production licence is granted by the King in Council and has to be registered in the Norwegian Register of Business Enterprises.[217] As for the production licence, 'the technical competence and financial capacity of the applicant' are the criteria for granting such a licence.[218] Therefore, in the Norwegian regime, financial capacity to pay for compensation is only a requirement for the production licence, not for the exploration licence. The Petroleum Activities Regulations only mention 'financial capacity' as a precondition for a production licence but do not specify how the financial capacity of each individual potential licensee is to be decided and what is considered sufficient capacity.

The Petroleum Activities Regulations contain provisions on the application for various types of licences, required information and documents, and access to upstream pipeline networks. In addition, none of the regulations concerning environmental aspects and health and safety issues of the offshore activities, as mentioned earlier, state that compliance with relevant regulations would exempt the licensee from liability for offshore incidents. Moreover, the specific provision in the Petroleum Activities Act makes it clear that the liability is strict and is not conditional on the breach of regulations.

3.5.3.5 Causation

As far as causation is concerned, because it is strict liability on the licensee, the claimant normally has to prove the causal link between the damage he has suffered and the incident. Where there is causal uncertainty, i.e. when it is impossible to identify who has actually caused the damage, the licensees shall be jointly and severally liable under Section 8-4 of the Petroleum Activities Act 'insofar as the damage may

[216] Section 2-1 of Petroleum Activities Act. The application for licence is submitted to the NPD; thus the NPD carries out the licensing duty on behalf of the MPE. See Petroleum Activities Regulations, Section 3.

[217] Section 3-3 of Petroleum Activities Act. The application for a production licence is submitted to the Ministry of Petroleum and Energy with a copy to NPD, as stated in Section 8 of the Petroleum Activities Regulations.

[218] In addition, another criterion is 'the applicant's plan for exploration and production in the area for which a production licence is sought'. This is provided in Section 10 of the Petroleum Activities Regulations.

be believed to have been caused by petroleum activities in connection with the licence in question'.

There are in addition particular rules when it concerns the 'financial losses incurred by Norwegian fishermen'. Section 8-1 of the Petroleum Activities Act specifies that the compensable damage has to come 'as a result of the petroleum activities occupying fishing fields or resulting in pollution and waste, or as a result of damage caused by a facility or actions in connection with the placing of a facility'.[219] Therefore, the fisherman has to prove the causal link between the damage he has suffered and the offshore incident, i.e. the occupation of fishing fields, placing of a facility or a pollution incident. Hence, the occurrence of a pollution incident is not the sole cause for compensation of the fishermen's losses, but other aspects of the offshore activities which may also hinder the fishery and cause financial losses for the fishermen are also covered by the Act and compensable.

3.5.3.6 Attribution of Liability

The Norwegian Petroleum Activities Act imposes liability on the licensee, and exclusively on the licensee. This is known as 'channelling of liability'. Section 7-4 of the Act provides that '[t]he liability of a licensee for pollution damage may only be claimed pursuant to the rules of this Act.' The Petroleum Activities Act further has a list of parties whose liability shall be excluded:

> Liability for pollution damage cannot be claimed against:
> a. anyone who by agreement with a licensee or his contractors has performed tasks or work in connection with the petroleum activities;
> b. anyone who has manufactured or delivered equipment to be used in the petroleum activities;
> c. anyone who undertakes measures to avert or limit pollution damage, or to save life or rescue values which have been endangered in connection with the petroleum activities, unless the measures are performed in conflict with prohibitions imposed by public authorities or are performed by someone other than public authorities in spite of express prohibition by the operator or the owner of the values threatened;
> d. anyone employed by a licensee or by someone mentioned under items a, b or c.[220]

[219] Section 8-1, 'Scope of Application and Definitions'.
[220] Section 7-4 of the Petroleum Activities Act.

According to the Petroleum Activities Act in Section 10–9, titled, 'Liability for Damage Caused', if 'liability in respect of a third party' (which obviously includes pollution damage) is 'incurred by anyone undertaking tasks for a licensee', 'the licensee shall be liable for damages to the same extent as, and jointly and severally with, the perpetrator and, if applicable, his employer'. Therefore, the Petroleum Activities Act provides that the liability is imposed on the licensee, and it cannot be channelled to parties down the production chain. Norwegian regulations require operators to take responsibility for their contractors. This reduces the chances for oil companies to point at others to blame.[221]

However, in the event of multiple licensees, the liability is joint and several among the licensees. When there are multiple licensees under one licence, they are 'jointly and severally responsible to the state for financial obligations arising out of petroleum activities pursuant to the licence'.[222] 'If there are several licensees under the licence and one of them is the operator, or if the Ministry has made a decision according to the first paragraph, claims for compensation shall initially be directed to the operator. If any part of the compensation is left unpaid on the due date by the operator, this part shall be covered by the licensees in accordance with their participating interest in the licence. If someone fails to cover his share, this shall be allocated proportionately between the others.'[223]

When the licensee 'fails to pay within the time limit stipulated by the judgement, the party that has sustained damage may bring action against the party that has caused the damage to the same extent as the licensee may bring action for recourse against the party causing the damage'.[224] Nevertheless, the 'licensee may claim compensation from the party causing pollution damage to him to the same extent as the licensee may bring action for recourse against the party causing the damage'.[225]

3.5.3.7 Damages and Remedies

Pollution damage is defined as 'damage or loss caused by pollution as a consequence of effluence or discharge of petroleum from a facility, including a well, and the costs of reasonable measures to avert or limit such damage or such loss, as well as damage or loss as a consequence of such

221 Janssen, *Preventing a Next Deepwater Horizon?*, 173.
222 Section 10-8 of the Petroleum Activities Act.
223 Section 7-3 of the Petroleum Activities Act.
224 Section 7-4 of the Petroleum Activities Act. 225 *Ibid.*

measures. Damage or loss incurred by fishermen as a consequence of reduced possibilities for fishing is also included in pollution damage.'[226]

The Act also specifically states that 'financial losses incurred by Norwegian fishermen as a result of the petroleum activities' are compensable.[227] It further provides that the financial losses of the fishermen are compensable when 'incurred as a result of pollution and waste from the petroleum activities, and the cost of reasonable measures to avert or limit such damage or such loss, including damage or loss as a result of such measures'.[228] They shall include 'damage and inconvenience due to pollution and waste as a result of supply vessel and support vessel traffic, as well as during relocation of the facility to or from the field concerned'.[229]

3.5.3.8 Amount of Compensation

There is no financial cap on the amount of compensation in the case of a pollution incident. Operating companies have unlimited liability for offshore pollution incidents.[230]

3.5.3.9 Claim Settlement

As far as compensation to Norwegian fishermen is concerned, there are some provisions on the time limit for the settlement of claims. Section 8-6 of the Petroleum Activities Act reads as follows:

> Claims made pursuant to this chapter shall be dealt with by a commission . . . Decisions made by the administrative appeal body may be brought straight before the district court, within two months of the party in question having been notified of the decision by summons. After expiry of the time limit for lodging a complaint or expiry of the time limit stipulated in the second paragraph, claims upheld by the commission or by the administrative appeal body shall be grounds for enforcement of distraint. When the deadline in the second paragraph is exceeded, the administrative appeal body may, pursuant to the rules in Section 31 of the Public Administration Act, decide that the district court shall hear the matter. Decisions regarding failure to comply with deadlines may be appealed to the district court.

This two-month time limit may be used to restrict the time for claims settlement. However, this applies only to the compensation of fishermen, not to other types of claims arising out of an offshore incident.

[226] Section 7-1, 'Definition of the Petroleum Activities Act', and Section 8-3 has similar provision.
[227] Section 8-1 of the Petroleum Activities Act.
[228] Section 8-3 of the Petroleum Activities Act. [229] *Ibid.*
[230] Janssen, *Preventing a Next Deepwater Horizon?*, 173.

Section 10-10 of the Petroleum Activities Act provides that '[i]f a serious accident has occurred in connection with petroleum activities comprised by this Act, the Ministry may appoint a special commission of inquiry. The same applies to incidents in the activities which have led to serious danger of loss of life or major damage to property or pollution of the marine environment. The members of the commission shall represent sufficient legal, nautical and technical expertise . . . The rules of the Maritime Act regarding an investigation of maritime accidents and regulations laid down pursuant to the Maritime Act Chapter 18 II shall apply correspondingly, as appropriate.'

3.5.3.10 Compensation Mechanisms

Both the Petroleum Activities Act and the Petroleum Activities Regulations have some provisions on financial security. The Petroleum Activities Act provides in Section 10-7 that '[u]pon granting a licence and subsequently, the Ministry may decide that the licensee shall provide such security as approved by the Ministry for fulfilment of the obligations, which the licensee has undertaken, as well as for possible liability in connection with the petroleum activities'. Therefore, the licensee must prove financial security upon the licence being granted. However, it does not specify the type of financial security and other details, such as how to decide on the amount of financial security.

The Petroleum Activities Regulations contain in Section 73 more detailed (compared with the provisions in the Petroleum Activities Act) provisions on insurance. It requires that the offshore activities carried out under Chapters 3 and 4 of the Petroleum Activities Act 'shall be insured at all times'. These activities refer to production of petroleum under a production licence. The Regulations do not mention whether activities conducted under Chapter 2 of the Petroleum Activities Act, namely, exploration activities, should also be covered by the insurance requirement.

The Petroleum Activities Regulations further provide that

> The insurance must at least cover:
>
> a. damage to facilities;
> b. pollution damage and other liability towards third parties;
> c. wreck removal and clean-up as a result of accidents;
> d. insurance of the licensee's own employees who are engaged in the activities.

The licensee shall ensure that contractors and subcontractors engaged in the activities take out insurance for their employees to the same extent as the operator insures his own employees. When taking out insurance, as mentioned in the first paragraph literas (a) to (c), the licensee shall provide reasonable insurance cover, taking into consideration risk exposure and premium costs. Insurance, as mentioned under litera (d), shall be taken out as further agreed with the organizations of the employees. At the end of each calendar year, the licensee shall inform the Ministry (of Petroleum and Energy) about existing insurance agreements, with an indication of the main terms. The Ministry may require further insurance to be taken out.[231]

Statoil, for instance, purchases insurance for, among other things, physical loss of or damage to oil and gas properties, liability to third parties, workers' compensation and employers' liability, general liability, pollution and well control.[232] The well-control policy covers costs relating to well-control incidents (including pollution and clean-up costs). The well-control policy is subject to a gross limit per incident: for the Norwegian continental shelf, it is 2,500 million NOK per incident for exploration wells and 2,000 million NOK per incident for production wells; for the Gulf of Mexico, it is US $300 million (approximately 1,800 million NOK) per incident.[233] The insurance of Statoil is subject to deductibles, excesses and self-insured retentions (SIRs) that must be borne prior to recovery, as well as exclusions and limitations. The SIR varies between approximately 16 million and 581 million NOK per loss on the Norwegian continental shelf depending on the ownership percentage interest in the well and certain other factors. The SIR in the Gulf of Mexico would be approximately 150 million NOK per incident assuming 100 per cent ownership.[234]

In addition to well-control insurance, Statoil has in place a third-party liability insurance program with a gross limit of 4,800 million NOK per incident. The SIR is insignificant (maximum 6 million NOK).[235] Statoil also acknowledges that the insurance may not adequately protect the company 'against liability for all potential consequences and damages'.[236]

[231] Section 73 of the Petroleum Activities Regulations.

[232] www.statoil.com/annualreport2012/en/abouttherereport/pages/abouttherereport.aspx.

[233] *Ibid.* [234] *Ibid.* [235] *Ibid.* [236] *Ibid.*

3.5.4 *Denmark*

3.5.4.1 National Interest in Offshore Activities

In Denmark, all the oil- and gas-producing fields are located in the North Sea.[237] The first hydrocarbon discovery in the Danish offshore was made in 1966 by A. P. Møller, and it was also the first find in the North Sea. The first oil was produced from the Dan field in July 1972.[238] Since then, the search for new oil and gas fields has brought about a multitude of activities as well as major investments in the establishment of North Sea production facilities.

As of 2013, according to the Danish Energy Agency, there were a total of nineteen producing fields of various sizes, and three operators are responsible for production from these fields: DONG E&P A/S, Hess Denmark ApS and Mærsk Olie og Gas A/S.[239] The production in the Danish part of the North Sea in 2013 derived from a total of 375 active wells, of which 196 were oil-producing wells and 72 were gas-producing wells.[240]

In order to explore, companies are required to have a licence to explore for hydrocarbons. Since 1983, areas in the North Sea have been offered to interested oil companies in a system of licensing rounds.[241] Six licensing rounds have been held, and the seventh round opened on 24 April 2014.

3.5.4.2 Legal Framework

Offshore oil and gas activities in Denmark are controlled by the following acts, orders and guidelines:

- Subsoil Act.[242] This Act lays down the basic framework for oil and gas exploration and production in Denmark. It regulates general licences for exploration and extraction activities in the Danish subsoil and on the Danish continental shelf concerning raw materials and specifically hydrocarbons.[243] The Subsoil Act covers certain safety and environmental aspects of installations used for exploration and extraction activities.[244] The Subsoil Act states that the raw materials shall belong

[237] Danish Energy Agency, 'Oil and Gas Production in Denmark', 5.
[238] See the website of Danish Energy Agency: www.ens.dk/en/oil-gas/licences.
[239] Danish Energy Agency, 'Oil and Gas Production in Denmark', 7. [240] *Ibid.*
[241] See the website of Danish Energy Agency: www.ens.dk/en/oil-gas/licences.
[242] Act No. 293 of 10 June 1981 as revised by Consolidated Act No. 960 of 13 September 2011 on the Use of the Danish Subsoil.
[243] Danish Energy Agency, 'A Guide to Hydrocarbon', 10.
[244] http://engelsk.arbejdstilsynet.dk/en/offshore/rules-guidelines.aspx.

to the Danish state and that the exploration for and production of the raw materials shall be subject to a licence granted by the Minister for Climate and Energy.[245]

- Offshore Safety Act.[246] The health and safety aspect of the offshore activities on the Danish continental shelf is regulated by this Act and regulations issued under this Act.
- Act No. 291 of 10 June 1981, as revised by Consolidated Act No. 957 of 9 September 2011 on the Establishment and Use of a Pipeline for Transport of Crude Oil and Condensate.
- Continental Shelf Act.[247] This Act regulates the exploitation of the continental shelf, including safety and environmental conditions on transit pipelines.
- Marine Environment Protection Act.[248]

The environmental aspects of the offshore activities are regulated via the Subsoil Act and the Marine Environment Protection Act.[249] The health and safety aspects are mainly regulated under the Offshore Safety Act, the Subsoil Act and the Continental Shelf Act.[250]

In addition to these laws, general regulations and guidelines are set out in model licences and joint model operating agreements. The use of model JOAs in Denmark is to secure the rights of the state participants, and their terms are negotiable. However, the JOA is subject to the approval of the Danish Energy Agency.[251]

As far as environmental supervision is concerned, the regulatory authorities for offshore activities in Denmark are the Danish Energy Agency (DEA) and the Danish Environmental Protection Agency (DEPA). The DEPA 'lays down requirements regarding the quantity of substances and materials that may be discharged into the atmosphere and the sea (the external environment)', whereas the DEA 'draws up design requirements

[245] Part 1.2 of the Subsoil Act.

[246] Act No. 1424 of 21 December 2005 on Safety, etc. on Offshore Installations for Exploration, Extraction and Transport of Hydrocarbons (Offshore Safety Act). The Offshore Safety Act was later amended by Act No. 107 of February 2007, Section 32 in Act No. 512 of 6 June 2007 and Section 1 in Act No. 1400 of 27 December 2008. Subsequently, the Offshore Safety Act was amended by Act No. 287 of 15 April 2009 and Act No. 467 of 18 June 2011.

[247] Consolidated Act No. 1101 of 18 November 2005 on the Continental Shelf as amended by Act No. 548 of 6 June 2007, Section 2, and Act No. 1400 of 27 December 2008, Section 3.

[248] Consolidated Act No. 963 of 3 July 2013 on the Protection of the Marine Environment.

[249] www.ens.dk/en/oil-gas/environment-offshore.

[250] http://engelsk.arbejdstilsynet.dk/en/offshore/rules-guidelines.aspx.

[251] Danish Energy Agency, 'A Guide to Hydrocarbon', 11.

that offshore installations must meet in order to observe the principle of using the Best Available Techniques (BAT) to reduce harmful environmental impacts'.[252] The DEA thus has the authority to decide on the granting of licences (development plans for oil and gas fields must be pursuant to the Subsoil Act, and the establishment of pipelines must be pursuant to the Continental Shelf Act) and environmental impact assessment, if necessary,[253] and the DEPA regulates issues such as emissions to the environment.

As of 1 January 2015, the health and safety aspect of the offshore activities (i.e. those relating to the Offshore Safety Act) is transferred from the Danish Energy Agency (the Ministry of Climate, Energy and Buildings) to the Working Environment Authority (the Ministry of Employment).[254] The safety aspect includes the inherent safety of the installations and equipment (process safety) and safety at the workplace and at work (occupational safety); the health aspect includes health conditions at the workplace and at work and other health conditions.[255] The tasks of the Danish Working Environment Authority mainly include: approval and permits in connection with the design, operation and modification of offshore installations; supervision of health and safety on offshore installations; and preparation of regulations on safety and health, including implementation of EU directives.[256]

3.5.4.3 Basis of Liability

According to Section 35 of the Subsoil Act, the 'licensee' is strictly liable 'to pay damages for any loss, damage or injury caused by the activities carried out under the licence'.[257] The comparative negligence defence of the victim is also taken into account.[258]

[252] www.ens.dk/en/oil-gas/environment-offshore. [253] *Ibid.*
[254] www.ens.dk/en/node/141/health-safety.
[255] http://engelsk.arbejdstilsynet.dk/en/offshore.aspx. [256] *Ibid.*
[257] Section 35 of the Subsoil Act reads that

> (1) A licensee shall be liable to pay damages for any loss, damage or injury caused by the activities carried on under the licence, even though such loss, damage or injury was caused accidentally.
> (2) In the event that the injured party contributed to the loss, damage or injury, either intentionally or through negligence, the damages payable may be reduced or lapse.

[258] Section 35(2) of the Subsoil Act provides: 'In the event that the injured party contributed to the loss, damage or injury, either intentionally or through gross negligence, the damages payable may be reduced or lapse.'

3.5.4.4 *Relationship* with Regulation

Several provisions in the Subsoil Act state that the necessary financial capacity is one of the conditions/criteria for a licence to be granted to the applicant.[259] Moreover, during the term of the licence, 'the licensees shall at all times be able to substantiate that they are financially sound.'[260] Therefore, the financial capacity is not only a requirement for obtaining a licence but also during the whole operation period of the licence. However, there is no provision in the Subsoil Act concerning what is the specific amount of financial resources to be considered 'necessary' or sufficient for the granting of a licence. Neither is there any provision on how the financial capacity is secured during the whole period of the licence.

According to Section 35 of the Subsoil Act, a licensee shall be liable for damages or injuries caused by his offshore activities under the licence, even when it is caused accidentally. Therefore, compliance with regulations does not exempt the licensee from assuming liability for compensation. In case of non-compliance with the provisions in the Subsoil Act, the licence may be revoked by the Minister for Climate and Energy.[261]

3.5.4.5 Attribution of Liability

Section 35 of the Subsoil Act apparently channels the liability to the licensee. In addition, the model licence provides in Section 31 that 'if the licence is granted to several parties jointly, they shall be jointly and severally liable for damages claimed pursuant to Section 35 of the Subsoil Act for the satisfaction of any obligations to the State under the Licence.'

3.5.4.6 Damages and Remedies

The liability for damages includes personal injury as well as property damage.[262] Section 38 of the model licence provides that the licensee 'shall indemnify the State against all claims whatsoever which may be made by any third party against the State as a consequence of the Licensee's activities. Therefore, the liability of the licensee should cover the actual amount arising from 'all claims' and should thus be unlimited.

However, when it comes to a mobile offshore installation, the liability of the ship owner under the Merchant Shipping Act shall apply and is hence limited. This is provided in Section 69 of the Offshore Safety Act

[259] Sections 5, 12 and 18b. [260] Section 6(6) of the Subsoil Act.
[261] Section 30 of the Subsoil Act. [262] *Ibid.*

on 'civil law rules for offshore installations'. It reads: 'By applying the rules in Part 9 of the Merchant Shipping Act on mobile offshore installations during exploration and extraction activities, the shipowner shall be liable when the liability is limited, regardless of the size of the installation by up to 50 million special drawing rights (SDR). For injury to a person this amount is increased by 30 million SDRs.'

3.5.4.7 Compensation Mechanisms

Section 28a of the Subsoil Act provides that '[w]hen licences or approvals are granted pursuant to subsection (3)(ii) or subsection (4), the Minister for Climate and Energy shall lay down terms and conditions regarding appropriate compensatory measures. The expenses for such compensatory measures shall be covered by the project applicant. The Minister shall notify the European Commission of the compensatory measures.'[263]

Section 29(2) contains a provision stating that '[a] licence granted under the provisions of this Act shall not be available to the recourse of creditors'. It is provided in Section 30(1) of the model licence that the 'licensee's liability for damages under the Subsoil Act shall be covered by insurance'.[264] As for the amount of insurance, the model licence provides that '[t]he insurance shall provide reasonable coverage, in the light of the risks involved in the operation of the business and the premiums to be paid.'[265] Moreover, the DEA 'may require additional insurance to be taken out' if it does not consider the insurance coverage to be sufficiently extensive.[266] At the end of each calendar year, the DEA shall be informed of the insurance then in force and of the principal terms thereof.[267]

The licensee's insurance shall cover the following issues:

- Injury to the licensee's own employees,
- Injury to any third party (including damage to property of a third party) and
- Pollution damages caused by the activities.[268]

The coverage for damages shall include all insurances which the licensee is obliged to take out under Danish law, i.e. work injury, motor vehicles,

[263] Section 28a(6) of the Subsoil Act.
[264] Section 30(1) of the (model) Licence for Exploration for and Production of Hydrocarbons.
[265] Section 30(2) of the (model) Licence for Exploration for and Production of Hydrocarbons.
[266] Section 30(3) of the (model) Licence for Exploration for and Production of Hydrocarbons.
[267] *Ibid.* [268] Danish Energy Agency, 'A Guide to Hydrocarbon', 12.

aircraft and insurance of ships' oil pollution liability. In addition the coverage for damages shall include insurance otherwise required in accordance with Section 30 of the model licence in order to cover the general third-party liability.

3.5.5 *United States*

3.5.5.1 National Interest in Offshore Activities

The United States is the world's largest oil-consuming country.[269] To secure the national supply of oil, in addition to imports, domestic oil production also plays an important role. In recent years, the domestic production of crude oil has been increasing, which reduces the net petroleum import.[270] Most offshore oil and gas activities in the United States take place in the Gulf of Mexico, which accounts for 17 per cent of US crude oil production and 5 per cent of natural gas.[271]

3.5.5.2 Legal Framework

The US legal regime on offshore oil activities has developed over decades in response to a variety of concerns and disputes raised by competing interests. Moreover, due to the so-called non-preemption of state laws, these exist in the United States in addition to federal laws, also state laws in respective states on the issue of offshore activities.

Several federal laws and regulations are of relevance. A selection will be mentioned here:

- Submerged Lands Act of 1953 (SLA). For the first time, this Act established the federal government's title to and ownership of submerged lands located on a majority of the continental margin.[272]
- Outer Continental Shelf Lands Act (OCSLA) of 1953.[273] This Act defined the outer continental shelf, separate from geologic definitions, as any submerged land outside state jurisdiction and reaffirmed federal jurisdiction over these waters and all resources they contain. Moreover, the OSCLA outlined federal responsibilities for managing and maintaining offshore lands subject to environmental constraints and safety concerns. It authorizes the Department of the Interior

[269] www.eia.gov/countries/index.cfm?view=consumption. The oil consumption of the United States alone is more than the total of all the EU Member States.
[270] www.eia.gov/todayinenergy/detail.cfm?id=20892.
[271] www.eia.gov/special/gulf_of_mexico/. [272] 43 USC 1301 *et seq.*
[273] 43 USC 1333 *et seq.*

(DOI) to lease the defined areas for development and to formulate regulations pertaining thereto as necessary. In doing so, the DOI must ensure that the US government receives fair market value for acreage made available for leasing, and it must enact regulations that guarantee resource conservation, environmental protection and operational safety for anyone involved in the activities. The OCSLA has been amended six times between 1978 and 1998 to account for changing issues. It remains the cornerstone of offshore legislation today.[274] It also authorizes regulation of workplace safety on the outer continental shelf by the Coast Guard.

Some important provisions in the OCSLA may be worth mentioning. First, Section 1332[275] provides that

> (1) the subsoil and seabed of the outer Continental Shelf appertain to the United States and are subject to its jurisdiction, control, and power of disposition as provided in this subchapter;
> (2) this subchapter shall be construed in such a manner that the character of the waters above the outer Continental Shelf as high seas and the right to navigation and fishing therein shall not be affected;
> (3) the outer Continental Shelf is a vital national resource reserve held by the Federal Government for the public, which should be made available for expeditious and orderly development, subject to environmental safeguards, in a manner which is consistent with the maintenance of competition and other national needs;
> (4) since exploration, development, and production of the minerals of the outer Continental Shelf will have significant impacts on coastal and non-coastal areas of the coastal States, and on other affected States, and, in recognition of the national interest in the effective management of the marine, coastal, and human environments –
> (A) such States and their affected local governments may require assistance in protecting their coastal zones and other affected areas from any temporary or permanent adverse effects of such impacts;
> (B) the distribution of a portion of the receipts from the leasing of mineral resources of the outer Continental Shelf adjacent to State lands, as provided under Section 1337(g) of this title, will provide affected coastal States and localities with funds which may be used for the mitigation of adverse economic and environmental effects related to the development of such resources; and
> (C) such States, and through such States, affected local governments, are entitled to an opportunity to participate, to the extent

[274] EIA, 'Overview of US Legislation', 5–6.
[275] This section is titled 'Congressional Declaration of Policy'.

> consistent with the national interest, in the policy and planning decisions made by the Federal Government relating to exploration for, and development and production of, minerals of the outer Continental Shelf . . .

Other federal relevant laws and regulations include

- Section 1333 of the OCSLA provides that with respect to disability or death of an employee resulting from offshore activities, compensation shall be payable under the provisions of the Longshore and Harbor Workers' Compensation Act.[276]
- Clean Air Act 1970. Proposed and existing natural gas and oil facilities must prepare, as part of their development plans and reporting procedures, detailed emissions data to prove compliance with this act.[277]
- Coastal Zone Management Act of 1972. This Act stresses the need to preserve, protect, develop and restore or enhance the resources of US coastal zones.[278]
- Endangered Species Act of 1973. This Act protects and promotes the conservation of all species listed as endangered by restricting federal actions that are likely to harm, harass or pursue them.[279]
- Clean Water Act of 1977 (CWA). This is the primary law governing the discharge of pollutants into all US surface waters. Under the CWA, the Environmental Protection Agency (EPA) requires that a National Pollutant Discharge Elimination System (NPDES) permit shall be obtained before any pollutant is released. The CWA holds certain industries, including oil and gas activities, to strict standards regarding direct pollution discharges into waterways. These standards are outlined in the NPDES permits and may be based on the age of a facility. The permits are issued on a five-year basis.[280]
- Oil Pollution Act 1990. This Act has provisions on the civil liability and compensation for offshore related incidents. The United States sent delegations to the negotiation of the CLC 1969 and the Fund Convention 1971.[281] However, due to some characteristics of the conventions, such as the pre-emption of state laws, low liability limits and channelling liability to ship owners, the United States failed to ratify the conventions.[282] In response to the *Exxon Valdez* accident in 1989, the US Congress quickly passed OPA. OPA has some similarities

[276] 33 USC 901 *et seq.* [277] *Ibid.* [278] EIA, 'Overview of US Legislation', 10.
[279] *Ibid.* [280] *Ibid.*
[281] See, for a detailed discussion of the Civil Liability and Fund Convention, Section 3.2.1.
[282] Wang and Faure, 'Civil Liability and Compensation', 3.

to the international regime, such as strict liability, limited liability and a compulsory financial guarantee. However, it has some substantial differences as well: the scope of compensable damage is much wider, liability is not channelled, and higher liability limits with more possibilities for the potential responsible parties to lose their right to limit. OPA does not pre-empt state laws, which means that states can still impose additional liability or financial responsibility.[283] Not only does OPA adopt a liability regime for oil pollution, but it also requires the establishment of response mechanisms for oil spills. The federal responders are required to prepare a national contingency plan outlining federal and state responsibilities, and responsible parties are also required to prepare private response plans.[284]

- OPA provides that 'each responsible party for a vessel or a facility from which oil is discharged, or which poses the substantial threat of a discharge of oil, into or upon navigable waters or adjoining shorelines or the exclusive economic zone is liable for the removal costs and damages . . . that result from such incident.'[285] A 'facility' is defined under OPA as 'any structure, group of structures, equipment, or device (other than a vessel) which is used for one or more of the following purposes: exploring for, drilling for, producing, storing, handling, transferring, processing or transporting oil. This term includes any motor vehicle, rolling stock or pipeline used for one or more of these purposes.'[286] Therefore, the offshore facilities as discussed in this book are covered under OPA.
- Energy Policy Act of 2005. This Act requires a comprehensive inventory and analysis of the estimated oil and gas resources on the outer continental shelf, establishes a new coastal impact assistance program, clarifies the exclusive jurisdiction of the Federal Energy and Regulatory Commission (FERC) under the Natural Gas Act and provides increased incentives for natural gas and oil development in offshore areas in order to maintain or stimulate production.[287]
- Code of Federal Regulations (CFR). In these regulations, there are detailed rules on natural resource damage assessment,[288] the financial responsibility for offshore facilities[289] and specific provisions on the

[283] 33 USC § 2718. [284] Sump, 'The Oil Pollution Act of 1990', 1102.
[285] 33 USC 2702(a). [286] 33 USC 2701(9). [287] *Ibid.*, 16–17.
[288] 15 CFR 990: Title 15, Commerce and Foreign Trade, Part 990, Natural Resource Damage Assessments.
[289] 30 CFR 553: Title 30, Mineral Resources, Part 553, Oil Spill Financial Responsibility for Offshore Facilities.

Oil Spill Liability Trust Fund.[290] In addition, there are also regulations authorizing the use of the outer continental shelf.[291]

The structure of the regulatory institutions has changed since 1 October 2011. The original authority, the Bureau of Ocean Energy Management, Regulation and Enforcement (BOEMRE),[292] now has been replaced by two federal institutions, the Bureau of Ocean Energy Management (BOEM) and the Bureau of Safety and Environmental Enforcement (BSEE).[293] BOEM is responsible for managing the development of US offshore resources in an environmentally and economically responsible manner. Its functions include offshore leasing, resource evaluation, review and administration of oil and gas exploration and development plans, renewable energy development, environmental studies and National Environmental Policy Act analysis.[294] BSEE is responsible for the oversight and enforcement of safety and environmental regulations. Its functions include, among other things, development and enforcement of safety and environmental regulations; issuing of permits for offshore exploration, development and production; inspections; offshore regulatory programs and oil spill response.[295] Specifically, the separation of BOEM and BSEE was designed to separate the resource management from safety oversight, thereby permitting engineers and inspectors greater independence, more budgetary autonomy and clearer senior leadership focus.[296]

[290] These are provided in 33 CFR 133 (Title 33, Navigation and Navigable Waters, Part 133, Oil Spill Liability Trust Fund; State Access) and 33 CFR 136 (Title 33, Navigation and Navigable Waters, Part 136, Oil Spill Liability Trust Fund; Claims Procedures; Designation of Source; and Advertisement).

[291] 30 CFR 550: Title 30, Mineral Resources, Part 550, Oil and Gas and Sulphur Operations in the Outer Continental Shelf.

[292] The Minerals Management Service (MMS) had been the authority until 2010, but it was renamed BOEMRE in June 2010.

[293] As explained earlier, this regulatory change was a result of the Deepwater Horizon accident. See Section 2.5.2.

[294] www.boem.gov/about-boem/reforms/reforms.aspx.

[295] The OCSLA authorises and requires the BSEE to provide for both an annual scheduled inspection and a periodic unscheduled (unannounced) inspection of all oil and gas operations on the US continental shelf. The annual inspection examines all safety equipment designed to prevent blowouts, fires, spills or other major accidents. The BSEE inspectors perform these inspections using a national checklist, called the 'potential incident of noncompliance list'. This list is a compilation of compliance elements derived from all regulated safety and environmental requirements. See Office of the Inspector of Transport Security, 'Offshore Oil and Gas', 31–2.

[296] www.boem.gov/about-boem/reorganization/reorganization.aspx.

In addition to these government institutions, there is also an industry organization in the United States: the American Petroleum Institute. It has established more than 500 industry standards covering all aspects of exploration and production, including safety, environmental and security issues.

3.5.5.3 Basis of Liability

The OPA imposes strict liability on the so-called responsible party for offshore activities. It provides that the responsible party for an offshore facility from which oil is discharged or which poses the substantial threat of a discharge of oil shall be held strictly liable for the removal costs and damages.[297] The 'responsible party' for an offshore facility is defined to be 'the lessee or permittee of the area in which the facility is located or the holder of a right of use and easement granted under applicable State law or the Outer Continental Shelf Lands Act (43 USC 1301–56) for the area in which the facility is located (if the holder is a different person than the lessee or permittee), except a Federal agency, State, municipality, commission, or political subdivision of a State, or any interstate body, which as owner transfers possession and right to use the property to another person by lease, assignment, or permit'.[298] Basically, the one who is permitted to explore the area and thus benefits from the facility is the responsible party.

When it comes to the liable party for mobile offshore drilling units, such a facility is treated as a tank vessel, as provided in Section 2704(b)(1) of OPA.[299] Therefore, the responsible party for mobile offshore drilling units will be the person 'owning, operating, or demise chartering' the mobile offshore drilling units. The liability rule remains strict liability. As in the international regime, strict liability is imposed on the responsible parties. Liability will only be excluded if the accident is caused by an act of God, an act of war, or an act or omission of a third party with no contractual or employment relationship to the responsible party.[300]

[297] 33 USC §2702(a). [298] 33 USC § 2701(32)(C).

[299] Section 2704(b)(1) of OPA provides that '[f]or purposes of determining the responsible party an applying this chapter and except as provided in paragraph (2), a mobile offshore drilling unit which is being used as an offshore facility is deemed to be a tank vessel with respect to the discharge, or the substantial threat of a discharge, of oil on or above the surface of the water.'

[300] 33 USC §2703(a).

3.5.5.4 Relationship with Regulation

According to the OPA provision on 'elements of liability', the fact of oil discharge to the sea or substantial threat of discharge of oil may trigger the strict liability of a responsible party for offshore facilities.[301] Compliance with regulations does not lead to exemption of liability of the responsible party. Moreover, if the responsible party wants to use a certain legal defence (i.e. act or omission of a third party)[302] or to benefit from the limitation of liability,[303] he has to comply with relevant federal regulations on safety, construction and operation. The violation of regulations will lead to the loss of legal defence and right of limitation.

3.5.5.5 Causation

Section 2702 of OPA provides that a responsible party is liable for removal costs and damages when the facility discharges oil or poses the substantial threat of a discharge of oil. Therefore, the actual discharge of oil is not necessary for a responsible party to be held liable; 'the substantial threat of a discharge of oil' is sufficient cause.

Moreover, Section 2703 of OPA contains provisions on defences to liability. It provides that

> [a] responsible party is not liable for removal costs or damages under Section 2702 of this title if the responsible party establishes, by a preponderance of the evidence, that the discharge or substantial threat of a discharge of oil and the resulting damages or removal costs were caused solely by
> (1) an act of God;
> (2) an act of war;
> (3) an act or omission of a third party, other than an employee or agent of the responsible party or a third party whose act or omission occurs in connection with any contractual relationship with the responsible party (except where the sole contractual arrangements arises in connection with carriage by a common carrier by rail), if the responsible party establishes, by a preponderance of the evidence, that the responsible party
> (A) exercised due care with respect to the oil concerned, taking into consideration the characteristics of the oil and in light of all relevant facts and circumstances; and
> (B) took precautions against foreseeable acts or omissions of any such third party and the foreseeable consequences of those acts or omissions; or
> (4) any combination of paragraphs (1), (2), and (3).

[301] 33 USC §2702(a). [302] 33 USC §2703(a)(3). [303] 33 USC §2704(c)(1).

Basically, the defences available to the responsible party are act of God, act of war and third party's omission. These can be used as defences only when they are the sole cause of discharge of oil. The burden of proof lies on the responsible party. Comparative negligence is also a defence available to the responsible party.[304] In particular, when the responsible party wants to use the third party's omission as a defence, he has to prove that he has exercised due care and has taken foreseeable precautions. Moreover, this section further provides that the responsible party may avail himself of these defences on condition that he has fulfilled his obligation to report the incident to the authority, to provide all reasonable cooperation and assistance requested by the authority and to comply with an order issued by the authority.[305] Therefore, these defences are conditional and require proof by responsible parties.

3.5.5.6 Attribution of Liability

Joint and several liability applies when there are multiple responsible parties. This means that each responsible party is liable for the entire amount of removal cost and damages resulting from a spill.

One significant difference between OPA and the international regime is that liability is not channelled to ship owners. OPA identifies responsible parties based on the types of facilities involved. For vessels, any person owning, operating or chartering the vessel can be the responsible party. OPA also defines the responsible party for onshore facilities, offshore facilities, deepwater ports and pipelines, respectively.[306] Thus, unlike in the international regime, as far as vessels are concerned, not only ship owners but also operators and charterers who may actually contribute to the oil pollution risks can be held liable.

However, remarkably, cargo owners can escape liability under OPA, although they benefit from and contribute to the risks of oil transportation as well. During the promulgation of OPA, there were proposals to establish a secondary liability on oil cargo owners. However, those proposals were not adopted.[307] This arrangement led to some criticism because OPA provides the cargo owners with no incentive to choose safer ships. There were also some recommendations to impose liability on

[304] Section 2703(b) of OPA. [305] Section 2703(c) of OPA. [306] 33 USC §2701(32).
[307] See Kiern, 'Liability, Compensation'.

cargo owners after the promulgation of OPA.[308] Triggered by the Deepwater Horizon accident in 2010, liability and financial responsibility under OPA became a hotly debated topic once again. The Coast Guard Authorization Act (CGAA) was enacted in 2010. Although focusing on authorizations to the Coast Guard, it contains some general provisions influencing OPA liability. One of those provisions extends liability to the cargo owners of single-hull tankers.[309] Aware of the large risk of oil spill posed by single-hull tankers, OPA 1990 requires the phasing out of single-hull tankers by 2015.[310] To further give cargo owners incentives to choose safer tankers, the CGAA extends liability to cargo owners if they choose to use a single-hull tanker for transportation.[311]

3.5.5.7 Damages and Remedies

The damages covered under OPA include two headings, 'Removal Costs' and 'Damages'. Damages include

- Natural resources,
- Real or personal property damage,
- Subsistence use,
- Revenues,
- Profits and earning capacity and
- Public services.[312]

Under OPA, 'removal' is defined as containment and removal of oil or a hazardous substance from water and shorelines or the taking of other actions to minimize or mitigate damage.[313] This definition shows that removal is the first step taken after oil pollution to clean up. Broadly speaking, removal cost is also one part of natural resource damage. But the United States treats removal costs differently from other damages.

OPA covers liability for natural resource damage, which is considered by some scholars to be an important innovation that addresses a global need – restoration of damaged ecological services and acknowledgement

[308] For example, 'Comments of the International Association of Independent Tanker Owners (INTERTANKO) on Preliminary Regulatory Impact Analysis for Financial Responsibility for Water Pollution (Vessels) (20 September 1993)'. See De la Rue and Anderson, *Shipping and the Environment*, 61.

[309] 124 Stat 2905, §713. [310] 46 USC §3703(a).

[311] Papayizas and Kiern, '2009–2010 Maritime Legislative Developments'.

[312] Section 2702 of OPA. [313] 33 USC §2701(30).

that natural resources have significant real economic value that needs to be included in the calculation of damages.[314]

However, how to determine the scope and quantify of the damage has long been a problem perplexing people when compensating damage to the environment. OPA gives a definition to the term 'natural resource damage', which includes

> (A) the costs of restoring, rehabilitating, replacing, or acquiring the equivalent of, the damage natural resources;
> (B) the diminution in value of those natural resources pending restoration; plus
> (C) the reasonable cost of assessing those damages. [315]

This definition is quite broad. Restoration is used as the primary method to evaluate the loss of natural resources and the diminution of services.[316] Not only restoration costs but also other alternatives and interim losses pending the restoration are also compensable. To quantify the damage, OPA authorizes federal and state authorities and Indian tribes as trustees of natural resources to assess the damage. It also requires the National Oceanic and Atmospheric Administration (NOAA) to promulgate regulations for the assessment of natural resources damage.[317] The assessment made based on these regulations has the force and effect of a rebuttable presumption.[318] The final rules concerning natural resources damage assessment were promulgated in 1996 by NOAA.[319] Under this regulation, a concrete procedure and available assessment methods are prescribed. Under the NOAA assessment rule, 'the costs of restoring, rehabilitating, replacing, or acquiring the equivalent of the damage natural resources' are defined as 'primary restoration'.[320] To compensate for interim losses, the NOAA rule also adopts a restoration-based approach: compensatory restoration should be taken to compensate for the lost natural resources and service during primary restoration.[321] When determining compensatory restoration, trustees should use a resource-to-resource or service-to-service approach to compensate for the lost natural resources service or value.[322] If these approaches are not possible, trustees can use other evaluation techniques to estimate the

[314] Boyd, 'Compensation for Oil Pollution Damages', 138.
[315] Lee, Bridgen and Environment International, Ltd., *The Natural resource damage*, 16.
[316] *Ibid.* [317] 33 USC §2706(e)(1). [318] 33 USC §2706(e)(2). [319] 60 Fed. Rec. 39804.
[320] 15 CFR §990.30. [321] *Ibid.*
[322] 15 CFR §990.53(d)(2). 'Resource-to-resource restoration actions are designed to restore the damaged natural resource to an equivalent quantity of the same or comparable resource.' 'A service-to-service restoration action is one designed to restore lost services

dollar value of the lost services and select the scale of the restoration action that has a cost equivalent to the lost value.[323] A number of valuation techniques are allowed to calculate the monetary value, including the disputed contingent-valuation technique.[324]

According to OPA and the assessment regulation, an assessment can either be made by the trustees and the trustees can claim against the responsible parties for the costs afterwards, or responsible parties can conduct the assessment themselves with the approval of the government. The most common approach would be a cooperative assessment involving both the trustees and responsible parties in the assessment procedures.[325] A cooperative approach can be desirable because it can promote a fast restoration and avoid costly and time-consuming litigation. An early involvement of the responsible parties and even insurers also can make the risks more predictable for insurers and thus can enhance the insurability of natural resources damage. Though it has been subject to many challenges since its promulgation, the NOAA assessment regulation provides practicable guidance in tackling the debatable and difficult task of natural resources damage assessment.[326]

3.5.5.8 Amount of Compensation

OPA also establishes limits on pollution liability according to different types of facilities. The liability of the responsible party for an offshore facility is capped at 'all removal costs plus $75 million'.[327] In comparison, the liability of a responsible party for onshore facilities and deepwater

to an equivalent quantity.' See Letourneau and Welmaker, 'The Oil Pollution Act of 1990', 188.

323 15 CFR §990.53(d)(3)(ii).

324 Letourneau and Welmaker, 'The Oil Pollution Act of 1990', 188. A 'contingent' valuation method is a method to value natural resources by doing surveys and asking people what monetary value they would like to place on certain resources. It is not based on actual market transactions but is hypothetical by assuming that people will respond to the surveys as in an actual transaction. See Cross, 'Natural Resource Damage Valuation', 315–19.

325 Allen, 'Proving Natural Resource Damage under OPA 90', 1039.

326 For a more detailed analysis of the NOAA assessment regulation, see Letourneau and Welmaker, 'The Oil Pollution Act of 1990', 148; Anderson, 'Damage to Natural Resources', 418; Seevers, 'NOAA's New Natural Resource', 1513.

327 33 USC §2704(a)(3). As far as mobile offshore drilling units are concerned, as mentioned earlier, for the 'purposes of determining the responsible party', they shall be treated as 'a tank vessel'. However, when it comes to the liability limit for such a facility, it remains subject to the same amount of liability as an offshore facility. See Section 2704(b)(2) of OPA.

Table 6 *Comparison of Liability Limits under OPA 90 and the Coast Guard and Maritime Transportation Act of 2006*

Vessel	OPA 90 liability limits (euros)	2006 (2009) liability limits (euros)
Single-hull tanker > 3000 GT	910/GT or 7.6 million	2,275 (2,427)/GT or 16.7 million (17.82 million)
Single-hull tanker ≤ 3000 GT	910/GT or 1.5 million	2,275 (2,427)/GT or 4.55 million (4.86 million)
Double-hull tanker > 3000 GT	910/GT or 7.6 million	1,441 (1592.6)/GT or 12.1 million (12.96 million)
Double-hull tanker ≤ 3000 GT	910/GT or 1.5 million	1,441 (1592.6)/GT or 3.03 million (3.24 million)
Any vessel other than a tanker	455/GT or 0.38 million	720.5 (758.4)/GT or 0.61 million (647,960)

The original currency is US$, but it is converted to euros here for the convenience of comparison. As per exchange rate IMF on 22 February 2013, 1 euro = 1.3186 US$.

ports under OPA is limited to the total amount of $350 million.[328] With the exception of an offshore facility (non-deepwater port), a cap is established for the total sum of removal costs and damages. The cap is the greater of a per-incident cap and a per-gross-ton cap.[329] The cap was increased in 2006 by the Coast Guard and Maritime Transportation Act (CGMTA). In addition, the CGMTA also established different caps for single- and double-hull tankers. A further increase was made by the Coast Guard according to the Consumer Price Index in 2009.[330] The existing caps for different facilities are listed in Table 6.

Despite these caps, a responsible party can lose his right to limitation if the incident was caused by 'gross negligence or wilful misconduct of, or the violation of an applicable Federal safety, construction, or operating regulation'.[331] A responsible party may also face unlimited liability if it fails to report an incident, fails to provide requested cooperation in connection with removal activities, or fails to comply with an order of the president.[332] Even when a responsible party can revoke the limitation

[328] 33 USC §2704(a)(4). [329] 33 USC §2704(a).

[330] 'US Coast Guard, Consumer Price Index Adjustments of Oil Pollution Act of 1990 Limits of Liability – Vessels and Deepwater Ports', *Federal Register*, Vol. 74, No. 125, 1 July 2009, 31357–69.

[331] 33 USC §2704(c)(1). [332] 33 USC §2704(c)(2).

under the OPA, it may still face unlimited liability if the applicable state statutes so require.

The OPA does not pre-empt state legislation. States can still have their own liability legislation concerning oil pollution. In the United States, thirty of the fifty states have a coastline. All but six of the coastal states have legislation on vessels liability.[333]

States may have diverging liability rules from the OPA by allowing a broader definition of the responsible parties, higher liability limits or even unlimited liability.[334] For example, cargo owners are admitted as responsible parties explicitly in some states, such as Alaska, California, Maryland and New Jersey. In some other states, cargo owners are also potentially liable. Though they are not explicitly listed as responsible parties, the broad definition of responsible parties does not exclude cargo owners either.[335]

Some states impose strict and unlimited civil liability for clean-up costs, natural resources damage and private losses caused by oil pollution, including pure economic losses, such as Alaska, California, North Carolina and Rhode Island. In some other states, unlimited liability is only established for certain categories of damage, such as in Washington (for clean-up costs and damage to persons or property), Maryland (for clean-up costs, damage to real and personal property and natural resources damage), Massachusetts (for natural resources damage) and Florida (for natural resources damage, damage to real and personal property and losses consequential upon property damage).[336]

A comprehensive discussion of the state oil pollution statutes is out of scope here, but it is worth noting that the state statutes in addition to the OPA make vessels coming into American ports face potential unlimited liability. This is quite different from the international regime. Besides, the stricter state statutes often require a higher financial responsibility. Whether and to what extent the combination of unlimited liability and financial responsibility in states can contribute to better cost internalization and better safety records also deserves more attention.

333 Force, Davies and Force, 'Deepwater Horizon'.

334 For a short summary of the states' legislations, see De La Rue and Anderson, *Shipping and the Environment*, app. 3, 1163–81; Force, Davies and Force, 'Deepwater Horizon'.

335 See De La Rue and Anderson, *Shipping and the Environment*, app. 3, 1163–81.

336 Force, Davies and Force, 'Deepwater Horizon'.

After the Deepwater Horizon incident, proposals were made to increase the liability limit or to remove it as a whole.[337] However, so far, there has been no change in legislation in this respect. For the amount of damage beyond the liability limit of the responsible parties, the claimant may seek coverage from the OSLTF[338] up to the limit of $ 1 billion per incident. If there is still damage uncompensated, the affected party may then seek redress through state laws. Due to the so-called non-pre-emption of state laws, each state has its own legislation which may provide extra compensation in addition to that under the federal law, *inter alia*, OPA.

3.5.5.9 Claim Settlement

In the United States, the claims procedure for damage caused by offshore incidents goes as follows: 'all claims for removal costs or damages shall be presented first to the responsible party or guarantor of the source designated under Section 2714(a) of this title.'[339] The designation of source shall be done by the president, after receiving information on an incident. If the designated source is an offshore facility, the president shall 'immediately' notify the responsible party and the guarantor of that designation.[340] Being informed of such a designation, the responsible party has five days to inform the president of its denial of the designation (hence, denial of responsibility). If not, an advertisement shall begin no later than 15 days after the date of the designation, and the advertisement shall continue for a period of no less than thirty days. The function of such an advertisement is to state that 'a claimant may present a claim for interim, short-term damages representing less than the full amount of damages to which the claimant ultimately may be entitled and that payment of such a claim shall not preclude recovery for damages not reflected in the paid or settled partial claim.'[341] OPA specifies these limits for parties to take action and allows compensation for interim damages. This may be useful in contributing to a rapid compensation mechanism.

[337] See e.g. HR 3534, the Consolidated Land, Energy, and Aquatic Resources Act, passed by the House of Representatives on 30 July 2010. It proposed to eliminate the liability limit for offshore facilities.

[338] This Oil Spill Liability Trust Fund (OSLTF) will be discussed in Section 3.5.5.10.

[339] 33 USC §2713. State removal costs claims are an exception in the sense that they may be presented directly to the NPFC without first presenting them to the responsible party.

[340] 33 USC §2714(a). [341] 33 USC §2714(b).

If the responsible party denies the claims or does not settle with the claimant within ninety days, claims can be submitted to the National Pollution Funds Center (which administers the Oil Spill Liability Trust Fund in the United States, as will be discussed in the next section) or file an action in court against the responsible party.[342] The ninety-day time limit also may be useful because it can avoid the victims being left in financial difficulty while the responsible party pays nothing by delaying the whole process. However, in practice and typically, claims for compensation are presented after removal is complete.[343] In the event of a serious oil spill incident, such as the Deepwater Horizon, clean-up measures and removal operations may take months. Therefore, the victims still may need to wait a while before they can file a claim with the OSLTF or an action in court.

In addition, OPA includes detailed provisions on the 'period of limitation'[344]:

> (1) Damages
> Except as provided in paragraphs (3) and (4), an action for damages under this chapter shall be barred unless the action is brought within 3 years after –
> the date on which the loss and the connection of the loss with the discharge in question are reasonably discoverable with the exercise of due care, or
> in the case of natural resource damages under Section 2702(b)(2)(A) of this title, the date of completion of the natural resources damage assessment under Section 2706(c) of this title.
> (2) Removal costs
> An action for recovery of removal costs referred to in Section 2702(b)(1) of this title must be commenced within 3 years after completion of the removal action. In any such action described in this subsection, the court shall enter a declaratory judgment on liability for removal costs or damages that will be binding on any subsequent action or actions to recover further removal costs or damages. Except as otherwise provided in this paragraph, an action may be commenced under this subchapter for recovery of removal costs at any time after such costs have been incurred.
> (3) Contribution
> No action for contribution for any removal costs or damages may be commenced more than 3 years after –
> the date of judgment in any action under this chapter for recovery of such costs or damages, or

[342] 'Oil Spill Liability Trust Fund (OSLTF) Funding for Oil Spills', January 2006, prepared by National Pollution Funds Center, Doc. NPFCPUB 16465.2, 7.
[343] *Ibid.* [344] 33 USC §2717(f).

the date of entry of a judicially approved settlement with respect to such costs or damages.

(4) Subrogation

No action based on rights subrogated pursuant to this chapter by reason of payment of a claim may be commenced under this chapter more than 3 years after the date of payment of such claim.

(5) Commencement

The time limitations contained herein shall not begin to run –

(A) against a minor until the earlier of the date when such minor reaches 18 years of age or the date on which a legal representative is duly appointed for such minor, or

(B) against an incompetent person until the earlier of the date on which such incompetent's incompetency ends or the date on which a legal representative is duly appointed for such incompetent.

Claimants have the right to submit a number of smaller claims rather than waiting to submit a single, comprehensive claim after all known damages have been analyzed and tabulated.[345] This may facilitate the claims-settlement procedure in the interest of the victim for speedy compensation.

It also may be interesting to note that under the Oil Spill Liability Trust Fund (which will be discussed later), an emergency fund is always available with an annual apportionment of $50 million for each fiscal year. Use of the emergency fund is to ensure that finances are always available to federal responses to oil spills (not for victim compensation).

3.5.5.10 Compensation Mechanisms

Financial Security Section 1016 of OPA provides that oil and gas exploration and production leases issued by the authority must establish and maintain oil spill financial responsibility (OSFR). BOEM has regulations concerning financial responsibility for offshore facilities.[346] The Code of Federal Regulations in Section 553 contains detailed provisions on the amount of financial responsibility and various methods for demonstrating that financial responsibility. This will be discussed in further detail in the section 'Compensation Mechanisms'. However, it does not identify at exactly what stage of

[345] Oil Spill Liability Trust Fund (OSLTF) Funding for Oil Spills, January 2006, presented by National Pollution Funds Center, Doc. NPFCPUB 16465.2, 9.

[346] 30 CFR 553.

the leasing and permitting process the authority would apply these regulations to check the financial responsibility of the responsible party. Although lacking specific provision, some related articles may be worth special attention.

Section 1334(b) of OCSLA reads as follows: 'Compliance with regulations as conditions for issuance, continuation, assignment, or other transfer of leases: The issuance and continuation in effect of any lease, or of any assignment or other transfer of any lease, under the provisions of this subchapter shall be conditioned upon compliance with regulations issued under this subchapter.' However, regulations on financial responsibility as in 30 CFR 553 are established under the requirements of OPA, not of OCSLA. Hence, it cannot justify that proving financial responsibility is a condition for issuance of leases.

Section 1337 contains further details on 'leases, easements, and rights-of-way on the outer Continental Shelf'. It requires that after an oil and gas lease is granted, 'the lessee, at its option, shall deliver to the Secretary upon issuance of the lease either

> (i) a cash deposit for the full amount of the exploration work commitment, or
> (ii) a performance bond in form and substance and with a surety satisfactory to the Secretary, in the principal amount of such exploration work commitment assuring the Secretary that such commitment shall be faithfully discharged in accordance with this section, regulations, and the lease; and for purposes of this subparagraph, the principal amount of such cash deposit or bond may, in accordance with regulations, be periodically reduced upon proof, satisfactory to the Secretary, that a part of the exploration work commitment has been satisfied.[347]

However, such a 'cash deposit' or 'performance bond' is meant to cover the amount of exploration work commitment and not the removal costs and cost of damages caused by an oil spill in relation to the offshore activities.

'Except as otherwise provided in this subchapter, prior to commencing exploration pursuant to any oil and gas lease issued or maintained under this subchapter, the holder thereof shall submit an exploration plan to the Secretary for approval' (Section 1340[c][1]).

[347] Section 1337(a)(7)(A).

'The Secretary may, by regulation, require any lessee operating under an approved exploration plan to obtain a permit prior to drilling any well in accordance with such plan' (Section 1340[d], 'Drilling permit').

Before receiving drilling permits, applicants must prove that they have sufficient oil spill containment equipment to respond to a worst case scenario spill. The responsible party for an offshore facility is required to establish and maintain evidence of financial responsibility when the facility 'is used for exploring for, drilling for, producing, or transporting oil from facilities engaged in oil exploration, drilling, or production'[348] and when the facility 'has a worst-case oil spill discharge potential of more than 1,000 barrels of oil'.[349] (In contrast, OPA does not have an analogous requirement for onshore facilities.[350]) This financial responsibility is to ensure that responsible parties can provide funding for oil spill responses and compensation to affected parties.

OPA provides that the financial responsibility for offshore facilities is US$35 million (for an offshore facility located seaward of the state's territorial sea) and US$10 million (for an offshore facility located landward).[351] This amount can be set higher when necessary and justified, but it should not exceed US$150 million.[352] Thus, the amounts of financial responsibility for offshore facilities are not related to the liability limits, i.e. US$75 million plus removal costs. Therefore, one can well imagine a situation where the amount between the compulsory financial evidence and the liability limit is not financially secured.

The regulations implementing the OPA financial responsibility requirement further provide that the amount of financial responsibility shall be decided on a worst-case scenario oil spill discharge volume.[353] As to the methods of financial responsibility, OPA provides that it can be in the form of evidence of insurance, surety bond, guarantee, letter of credit, qualification as a self-insurer or other evidence of financial responsibility.[354] The CFR contains further details in Section 5.5.3. (see Tables 7 and 8).

[348] 33 USC §2716(c)(1)(ii).
[349] It can be a lesser amount if the president determines that the risks posed by such facility justify it. See 33 USC §2716(c)(1)(iii).
[350] Ramseur, 'Liability and Compensation Issues', 6. [351] 33 USC §2716(c)(1)(B).
[352] 33 USC §2716(c)(1)(C). The president may determine a higher amount when it is justified by the relative operational, environmental, human health and other risks posed by the quantity and quality of oil.
[353] 30 CFR §253.13. [354] Section 2716(e).

Table 7 *Financial Responsibility for Offshore Facilities Located Partially or Wholly on the Outer Continental Shelf*

Worst-case oil spill discharge volume (barrels)	Applicable amount of financial responsibility (US$ millions)
1,000 < volume ≤ 35,000	35
35,000 < volume ≤ 70,000	70
70,000 < volume ≤ 105,000	105
Volume > 105,000	150

Table 8 *Financial Responsibility for Offshore Facilities Not Located on the Outer Continental Shelf*

Worst-case oil spill discharge volume (barrels)	Applicable amount of financial responsibility (US$ millions)
1,000 < volume ≤ 10,000	10
10,000 < volume ≤ 35,000	35
35,000 < volume ≤ 70,000	70
70,000 < volume ≤ 105,000	105

First, concerning self-insurance, the party has to annually pass either a net worth test or an unencumbered net asset test.[355] The financial test procedures for the net worth test are provided in 30 CFR 553.25:

(a) Divide the total amount of the stockholders'/owners' equity listed on the balance sheet by ten.
(b) Divide the net amount of the identifiable US assets by the net amount of the identifiable total assets.
(c) Multiply the net amount of plant, property, and equipment shown on the balance sheet by the number calculated under paragraph (b) of this section and divide the resultant product by ten.
(d) The smaller of the numbers calculated under paragraphs (a) or (c) of this section is the maximum allowable amount you may use to demonstrate OSFR under this method.

The financial test procedures for unencumbered assets are provided in 30 CFR 553.28:

[355] 30 CFR §553.21(a).

(a) Divide the total amount of the stockholders'/owners' equity listed on the balance sheet by 4.
(b) Divide the value of the unencumbered US assets by 2.
(c) The smaller number calculated under paragraphs (a) or (b) of this section is the maximum allowable amount you may use to demonstrate OSFR under this method.

Second, as for evidence of insurance, 30 CFR 553.29 provides that 'you may use only insurance certificates issued by insurers that have achieved a "Secure" rating for claims paying ability in their latest review by A. M. Best's Insurance Reports, Standard & Poor's Insurance Rating Services, or other equivalent rating made by a rating service acceptable to BOEM.'

Third, as for indemnity, 30 CFR 553.30 states that the indemnity as evidence must be 'issued by only one indemnitor to satisfy all or part' of the responsible party's obligation to demonstrate financial responsibility. Moreover, the indemnitor must be the party's corporate parent or affiliate.

Fourth, when using a surety bond as financial evidence, each bonding company that issues a surety bond must comply with certain requirements: the company must

(1) Be licensed to do business in the State in which the surety bond is executed;
(2) Be certified by the US Treasury Department as an acceptable surety for Federal obligations and listed in the current Treasury Circular No. 570;
(3) Provide the surety bond on Form BOEM-1020; and
(4) Be in compliance with applicable statutes regulating surety company participation in insurance-type risks.[356]

Other alternative methods may include 'pooling, letters of credit, pledges of treasury notes or other comparable methods'.[357]

Compensation Fund There has existed in the United States since 1986 an Oil Spill Liability Trust Fund (OSLTF), but there was no legislation to authorize the use of this OSLTF until the occurrence of the *Exxon Valdez* incident in 1989 and enactment of OPA in 1990.[358] The OSLTF is

[356] 30 CFR §553.31. [357] 30 CFR §553.32.

[358] Prior to enactment of OPA, there were debates on the compensation mechanisms and on the hurdles faced by private parties when recovering damages resulting from an oil spill. To address this and some other concerns (including e.g. the immediate accessible funds for federal agencies to conduct oil spill response activities), Congress established the

available to pay for the removal costs incurred by federal or state governments; the costs for the government in assessing natural resources damages and developing and implementing restoration plans; uncompensated removal costs and uncompensated damages and administrative costs related to the oil spill.[359]

OSLTF has two components, the 'emergency fund' and the 'principal fund'. The emergency fund has an annual apportionment of $50 million for each fiscal year. In 2003, the Coast Guard was given authority to advance up to $100 million from the OSLTF principal fund to supplement emergency fund shortfalls.[360] The goal of the emergency fund is to ensure that funds are available for federal response when needed.[361] Unless under some limited situations, the claims should be presented to the responsible parties or their guarantors first.[362] OPA limits payments of the OSLTF for removal costs and damages in each incident to US$1 billion. A sub-limit for natural resources damages payments is set at US$500 million per incident.[363] The per-incident limit for removal costs and damages was increased to US$2.7 billion under the Energy Policy Act of 2005.[364]

There are several financial sources for the principal fund of OSLTF:

> The first and largest source of financing is a 5-cent-per-barrel tax on crude oil imported to or produced in the United States.[365] The collection of this tax stops when the OSLTF has reached a certain amount, and it recommences when it drops below that amount. Thus, this tax has been suspended and reinstated several times to maintain the required fund balance.[366] Most recently, in the Emergency Economic Stabilization Act of 2008 (P.L.110-343), this tax was increased to 8 cents through the period 2009–2016. In 2017, the rate will increase to 9 cents. The tax is scheduled to terminate at the end of 2017 (Section 405 of P.L. 110–343), and the fund balance is projected to reach US$3.5 billion by 2016.[367]

OSLTF. See e.g. US Congress Committee on Merchant Marine and Fisheries, Report accompanying HR 1465, Oil Pollution Prevention, Removal, Liability, and Compensation Act of 1989, 1989, House of Representatives Report 101-242, Part 2, 101st Congress, first session, 35.

359 33 USC §2712(a).

360 'Oil Spill Liability Trust Fund (OSLTF) Funding for Oil Spills', January 2006, prepared by National Pollution Funds Center, Doc. NPFCPUB 16465.2, 5.

361 *Ibid.*, 5. 362 33 USC §2713. 363 26 USC §9509(c)(2)(A).

364 119 Stat. 1058–9, 1361.

365 The tax is addressed at Section 4611 of the Internal Revenue Code. See 26 USC §4611.

366 The 5-cent-per-barrel tax expired at the end of 1994; the Energy Policy Act of 2005 reinstated the tax with effect from April 2006.

367 In a report from the Senate Committee on Finance, Senate Report 110-228 accompanied S 2345 (110th Congress).

The tax applies to crude oil received at a US refinery and to petroleum products brought into the United States for consumption, use or warehousing. The tax also applies to any domestic crude oil used in or exported from the United States. The tax on crude oil received at a US refinery is paid by the refinery operator or, in the case of imported petroleum products, by 'the person entering the product for consumption, use or warehousing'. The tax on other crude oil is paid by 'the person using or exporting the crude oil'.[368]

A second major source of funding comes from transfers from the previously existing pollution funds, including the revolving fund under the Clean Water Act, the Deepwater Port Liability Fund, the Trans-Alaska Pipeline Liability Fund and the Offshore Oil Pollution Compensation Fund. Total transfers into the OSLTF since 1990 have exceeded US$550 million. No additional funds remain to be transferred to the OSLTF.[369]

The third (and recurring) source of funding comes from interest on the fund principal from US Treasury investments. The Department of Treasury serves as the investment manager for OSLTF.[370]

The fourth source of funding is recoveries of costs and damages from responsible parties and guarantors. The general policy of OPA is the 'polluter-pays principle'. Hence, although OPA has established a mechanism to facilitate the clean-up and response to oil spills and to encourage and expedite the actions of related parties, those responsible for oil spill incidents are liable for the costs and damages. The National Pollution Fund Center (NPFC, the competent authority established to administer the OSLTF) has a billing and collection program to recover costs expended by the OSLTF. In recent years, the NPFC has been able to collect between US$7 and US$14 million per year of the removal costs and damages it pays from the OSLTF; however, due to difficulties associated with cost recovery, the NPFC has only collected 26.5 per cent of the removal and claims expenditures incurred during the period between 1995 and 2004.[371] The major barrier to recovery, according to the US Coast Guard, is the difficulty in identifying the

[368] 26 USC §4611, Internal Revenue Code.

[369] www.uscg.mil/ccs/npfc/about_npfc/osltf.asp. [370] *Ibid.*

[371] During the period between 1995 and 2004, the Fund expended $492.3 million for removal and claims, and only $130.6 million has been recovered. See 'Report on Implementation of the Oil Pollution Act of 1990', 2005, 7.

source of the spill or the responsible party of an oil spill.[372] Moreover, costs expended in excess of a responsible party's liability limit are generally unrecoverable.

The fifth source is penalties. Responsible parties may incur fines and civil penalties under OPA, as well as under other related federal legislation concerning water pollution control, which will be deposited in the OSLTF. Penalty deposits are generally between US$4 million and US$7 million per year.[373]

3.5.5.11 Jurisdictional Issues

The Submerged Lands Act of 1953 gives states jurisdiction over any natural resources within 3 nautical miles of the coastline except Texas and the west coast of Florida, where the jurisdiction of the states extends to 9 nautical miles.[374] The Outer Continental Shelf Act grants the federal government jurisdiction over mineral resources on and under the seabed beyond the state waters, which is normally 3 nautical miles from the low tide mark. Exceptions are Texas and the west coast of Florida, whose state territory extends to 9 nautical miles. Louisiana adopts the Admiralty nautical mile, whereas other states use the international nautical mile.

3.5.5.12 Criminal Liability

When it comes to criminal liability for offshore activities, the CWA contains relevant provisions. When a person negligently or knowingly discharges oil into US waters or on adjoining shorelines without a permit or in violation of permit conditions, his act shall be punished by a fine and/or imprisonment:

- Negligent violation is subject to imprisonment of one year and/or a fine of US$2,500 to US$25,000 per day (subsequent conviction is subject to imprisonment of two years and/or a fine of US$50,000 per day) and
- Knowing violation is subject to imprisonment of three years and/or a fine of US$5,000 to US$50,000 per day (subsequent conviction is subject to imprisonment of six years and/or a fine of US$100,000 per day).[375]

[372] See the website of the US Coast Guard: www.uscg.mil/ccs/npfc/about_npfc/opa_faqs.asp.
[373] *Ibid.* [374] EIA, 'Overview of US Legislation', 4.
[375] 33 USC 1319; 40 CFR 110, 116, 117.

In addition, failure to report a discharge of oil to the federal agency also constitutes a crime and is subject to imprisonment up to five years and/or a fine.[376]

3.5.6 *Australia*

3.5.6.1 National Interest in Offshore Activities

Australia is the world's ninth largest energy producer, accounting for around 2.4 per cent of world energy production.[377] Oil and gas accounted for 17 per cent of Australia's total energy production in 2008–2009.[378] According to the official figures from the Australian government, 95 per cent of Australia's petroleum production comes from offshore sedimentary basins.[379] In the year 2009, Australia produced 559,000 barrels per day of crude oil, representing 0.6 per cent of world's crude oil production, and 42.3 billion cubic metres of natural gas, representing 1.4 per cent of the world's natural gas production, thus ranking as the world's eighteenth largest natural gas producer.[380]

Since 1965, around 3,000 wells have been drilled in Australian offshore waters, with around 390 of them located within the coastal waters of the states and the Northern Territories. About half the wells are classified as exploration wells. Since the mid-1990s, there has been an average of around 110 wells drilled offshore per year. Currently, there are approximately 410 active petroleum titles in offshore Commonwealth waters, which include 220 exploration permits, 39 retention leases, 80 production licences, 63 pipeline licences and 1 infrastructure licence.[381]

376 33 USC 1321(b)(5); 40 CFR 110, 116, 117. The amount of fine is provided in 18 USC 3571.

377 Office of the Inspector of Transport Security, 'Offshore Oil and Gas Resources Sector', 20.

378 *Ibid.*

379 www.ret.gov.au/department/documents/mir/fs_1_australia's-offshore-petroleum-industry.pdf.

380 *Ibid.* Approximately half the natural gas production was exported as liquid natural gas (LNG), which makes Australia the world's fourth largest LNG exporter.

381 Information from the website of the government of South Australia, Department for Manufacturing, Innovation, Trade, Resources and Energy (DMITRE); available at: www.pir.sa.gov.au/petroleum/prospectivity/basin_and_province_information/prospectivity_bight/grant_of_petroleum_exploration_permits_to_bp_exploration_alpha_ltd_faq.

Table 9 *Well Blowouts That Have Occurred in Offshore Australia Since 1965*

Date of incident	Name of facility	Place of incident
18 February 1965	Barracouta No. 1	Bass Strait
24 August 1966	Marlin B-1	Bass Strait
2 December 1968	Marlin A7	Bass Strait
6 August 1969	Petrel No. 1	Timor Sea
19 May 1971	Marlin A4	Bass Strait
17 December 1984	Flounder A1	Bass Strait
21 August 2009	Montara H1	Timor Sea

As of 10 July 2010, there were twelve offshore drilling rigs in operation in Australian offshore waters.[382] The wells are currently drilled in water depths ranging from 30 to 1,400 metres. As of 31 March 2010, 2902 offshore wells had been drilled in Australian offshore waters since offshore drilling commenced in offshore Bass Strait over forty years ago.[383]

The Australian Petroleum Production and Exploration Association submitted to the Montara Commission of Inquiry a list of well blowouts that have occurred in offshore Australia since 1965 (see Table 9). The incident analysis team (IAT) noted that 'of these seven incidents, only the Montara Wellhead Platform incident involved a significant oil release as the others were gas wells, except Flounder A1, which was a shallow gas release from an oil well during drilling operations'.[384]

3.5.6.2 Legal Framework

Similar to the United States, there is no uniform regime applying to offshore activities across all state jurisdictions in Australia. There has always been debate between the national government and the states on the jurisdiction over offshore issues.[385]

382 www.appea.com.au/news-a-publications/news/2010/703-montara-a-gulf-of-mexico-information.html. Source: Pex, *Who's Drilling*.
383 Source: APPEA, GA.
384 Australian Maritime Safety Authority, 'Response to the Montara', 18.
385 Daintith, *Discretion in the Administration*.

The main law that is applicable is the Offshore Petroleum and Greenhouse Gas Storage Act of 2006 (OPGGSA) (referred to as the 'Offshore Petroleum Act of 2006'). This Act applies from the 3-nautical-mile limit from the baselines out to the limits of the EEZ and the declared outer continental shelf areas, subject to exceptions. State and Northern Territory laws apply in coastal waters (the first 3 nautical miles from the baselines).

In addition to the Offshore Petroleum Act, there are several other laws and regulations in Australia that are of relevance but that remain undiscussed at this point.[386] The legal regime in Australia has gone through intense scrutiny since the Montara incident.[387] The inquiries that followed the incident have highlighted shortcomings in operational practices and how the industry is regulated, and legislative changes were recommended in order to improve safety.[388] In particular, the Montara Commission of Inquiry found that in order for the regime 'to embody the "polluter pays" principle, regulatory authorities needed to be satisfied that companies had adequate insurance to meet their obligations and "no fault" penalties needed to be in place for pollution'.[389] The Commission recommended that the obligation of companies involved in an incident to meet the full costs of monitoring and remediation should be made a condition of approval of proposals under OPGGSA 2006 and that suitable arrangements (insurance or otherwise) should be in place to ensure that companies have this capacity.[390]

On 25 May 2011, the Minister for Resources and Energy introduced five bills into Parliament that modified OPGGSA 2006 and related laws.[391] After the Montara incident of 2009, many regulatory changes took place, including the merger of various regulatory authorities into one. Prior to the Montara incident, the Australian Commonwealth government jointly administered the regulatory regime and supervised offshore activities with the state and Northern Territory governments through a Joint Authority/Designated Authority arrangement. On the advice of the Montara Commission, a single national regulator was established for all offshore activities beyond 3 nautical miles from

386 See 'Australia's Offshore Petroleum Regulatory Framework', prepared by Australian Government, Department of Resources, Energy and Tourism.

387 Bosma, 'The Regulation of Marine Pollution', 98. 388 *Ibid.* 389 *Ibid.*, 99.

390 Montara Commission of Inquiry, 'Report of the Montara Commission of Inquiry', 370–3.

391 Bosma, 'The Regulation of Marine Pollution', 99.

the territorial sea baseline, the National Offshore Petroleum Safety and Environmental Management Authority (NOPSEMA).[392]

The principal authority to carry out clean-up operations is the Australian Maritime Safety Authority.[393] Important provisions concerning civil liability for offshore activities are mainly included in Chapter of the Offshore Petroleum Act of 2006, although it is titled 'Administration'. 'Schedule 3 of the Offshore Petroleum Act of 2006 imposes duties relating to health and safety on a number of parties. The operator of the facility bears the principal duty in the regime. This duty is for the operator to take all reasonably practicable steps to ensure [that] the facility and its activities are safe and without risk to health.'[394]

Section 576(B)(1) provides that '[i]f a significant offshore petroleum incident has occurred in a title area that has caused, or that might cause, an escape of petroleum, NOPSEMA may, in accordance with this section, give a direction by written notice to the registered holder of the title.'[395] It further provides in Section 576(B)(2) that

> The direction may require the registered holder of the title to do any (or all) of the following within a reasonable period stated in the direction:
> (a) to take any action stated in the direction for the purpose of any (or all) of the following:
> (i) preventing the escape of petroleum;
> (ii) eliminating the escape of petroleum;

[392] Australian Government, 'Final Government Response to the Report of the Montara Commission of Inquiry', 2.

[393] Australian Maritime Safety Authority Act of 1990 (Cth), s6(1)(a).

[394] 'Offshore OHS Legislative Framework Information Paper', NOPSEMA.

[395] Section 576(A) has provisions on the definition of 'a significant offshore petroleum incident'. It reads

> A significant offshore petroleum incident is a significant incident or occurrence that relates to any or all of the following operations in an offshore area:
> - petroleum exploration operations;
> - petroleum recovery operations;
> - operations relating to the processing or storage of petroleum;
> - operations relating to the preparation of petroleum for transport;
> - operations connected with the construction or operation of a pipeline;
> - operations relating to the decommissioning or removal of structures, equipment or other items of property that have been brought into an offshore area for or in connection with any of the operations mentioned in paragraph (a), (b), (c), (d) or (e).

(iii) mitigating the effects of the escape of petroleum;
(iv) managing the effects of the escape of petroleum;
(v) remediating the effects of the escape of petroleum;

(b) to take any other action stated in the direction in relation to the escape of petroleum and its effects;
(c) not to take an action stated in the direction in relation to the escape of petroleum and its effects.

The direction may require the registered titleholder to take an action or not to take an action, and the direction may prohibit the doing of an action unconditionally or subject to conditions.[396] The required action (or not to take an action) can be anywhere in the offshore area, whether within or outside the title area.[397] The remedial directions may include removal of property, plugging or closing off of wells, conservation and protection of natural resources and making good of damage to seabed and subsoil.[398]

3.5.6.3 Basis of Liability

The Offshore Petroleum Act of 2006 contains no specific provision on the civil liability arising from offshore activities. However, it has some provisions which might be relevant for deciding on liability for offshore activities in Chapter 6, titled 'Administration'.

First, as to the liable party for offshore incidents under the Australian regime, Section 569(1) provides that the titleholder shall prevent the escape of petroleum, water and any mixture of water or drilling fluid. An offence against such liability is an offence of strict liability.[399] The titleholder is, as mentioned in Section 569(1), 'the registered holder of a petroleum exploration permit, petroleum retention lease or petroleum production licence', 'the registered holder of an infrastructure licence' and 'the registered holder of a pipeline licence', the registered holder of a petroleum special prospecting authority or petroleum access authority'.[400] Therefore, the registered titleholder of a permit, lease and licence

[396] Sections 576(B)(3) and (4). [397] Section 576(B)(5).

[398] Sections 585, 586 and 586(A) of the Offshore Petroleum Act of 2006.

[399] Section 569(6A) of the Offshore Petroleum Act of 2006. An offence is defined in Section 569(6). It reads:

A person commits an offence if:
- the person is subject to a requirement under subsection (1); and
- the person engages in conduct; and
- the person's conduct breaches the requirement.

[400] Section 569(1) of the Offshore Petroleum Act of 2006.

has the obligation to prevent the occurrence of an oil spill, but it seems unclear whether the civil liability for offshore incidents will be imposed on this titleholder as well.

Nevertheless, Section 571 on insurance (which will be discussed further later) imposes the liability to take out insurance on the registered holder of permits, leases and licences. Hence, the Australian legal regime may lack certain clarity in the sense that the obligation to take out insurance is with the titleholder, whereas the party liable for offshore incidents is not clear.

In addition, in Part 6.2, 'Directions Relating to Petroleum', there are also some relevant provisions that may shed light on this issue. It provides that in case of 'significant offshore petroleum incidents', the authorities may give directions to the registered titleholder of a permit/lease/licence to take action for the purpose of preventing/eliminating the escape of petroleum, mitigating/managing/remediating the effects of the escape of petroleum.[401] Hence, the registered titleholder is subject to the directions from the authority to remove the discharged oil and mitigate the damage caused by the discharge of oil. When the authority gives direction to the titleholder to pay for compensation, the titleholder shall be held liable. In the event of a breach of a direction, the party subject to the direction has to indemnify the NOPSEMA or responsible minister for its costs and expenses incurred in executing the direction.[402] Therefore, it is the registered titleholder that has to indemnify the authorities for their costs incurred. However, there is no specific provision concerning the compensation for private victims. Moreover, all these provisions relate to 'significant offshore petroleum incidents'. If there are pure threats of pollution or near-miss situations, the law remains silent.

Second, as to the question of whether the liability rule is strict or negligence, the provision in Section 569(1) states that an offence against the legal requirement is subject to strict liability. However, such a liability is based on breach of the legal obligation to prevent oil discharge. Such a liability established on the basis of a breach of legal duty is not strict liability in the traditional sense. Moreover, the legal defence for the liable party is that he can prove he 'took all reasonable steps to comply with' his obligations in Section 569(1). Hence, this is not a traditional strict

[401] Section 576(B) of the Offshore Petroleum Act of 2006.
[402] Sections 577(3) and 577(A2).

liability rule because the breach of duty is essential for establishing the titleholder's liability.

A note in the Offshore Petroleum Act of 2006 refers to the Criminal Code for an explanation of strict liability.[403] However, the concept of strict liability under criminal law may be different from the civil-law concept. Therefore, the principle of liability for offshore incidents under the Australian Offshore Petroleum Act of 2006 is not a clear-cut strict liability but based on a breach of legal requirements.

3.5.6.4 Relationship with Regulation

Section 569 of the Offshore Petroleum Act of 2006 provides that the titleholder shall 'carry out all petroleum exploration operations ... in a proper and workmanlike manner and in accordance with good oilfield practice'; he shall 'control the flow, and prevent the waste or escape ... of petroleum or water' and 'prevent the escape ... of any mixture of water or drilling fluid with petroleum or any other matter'.

The Offshore Petroleum Act of 2006 provides that a person is considered to commit an offence if his conduct breaches the regulatory requirements.[404] Therefore, compliance with regulations may be used by the titleholder as a defence against assuming liability.

3.5.6.5 Causation

A person is considered to have committed an offence if he is subject to certain legal requirements, when he engages in conduct and his conduct breached the requirement.[405] Therefore, the cause of the liability is the violation of legal requirements.

3.5.6.6 Attribution of Liability

In the case of multiple titleholders, Section 775(D) provides that where the Offshore Petroleum Act of 2006 imposes an obligation on a titleholder and there are multiple titleholders in relation to a single title, the obligation is imposed on each of them; however, the obligation may be discharged by any one of them. Therefore, they

403 Section 569(6A) of the Offshore Petroleum Act of 2006.
404 Section 569(6) of the Offshore Petroleum Act of 2006.
405 Sections 569(6) and 576(D).

are jointly and severally liable in relation to their title-related legal obligations under the Offshore Petroleum Act of 2006, but once one of the titleholders has discharged the legal obligation, the remaining titleholders are discharged from that duty.[406]

3.5.6.7 Damages and Remedies

As mentioned earlier, a national authority (NOPSEMA or the responsible minister) may take action if the registered titleholder breaches its direction.[407] Costs or expenses incurred by the authority in relation to a direction are 'a debt due to the Commonwealth by the person subject to the direction' and are 'recoverable in a court of competent jurisdiction'.[408] Therefore, the costs incurred by the public authority are compensable under the Offshore Petroleum Act of 2006. However, there may be private parties who also suffer from an offshore incident. There is no provision on the compensation for private claims.

3.5.6.8 Amount of Compensation

There is no statutory provision to impose a financial cap on the compensation. Therefore, the liability is considered unlimited.

3.5.6.9 Compensation Mechanisms

All petroleum titleholders in offshore Australia are required to hold appropriate levels of insurance prior to the commencement of any seismic, drilling or other exploration operations. This insurance must be sufficient to cover the cost of any potential remediation.[409]

Pursuant to Section 571 of the Offshore Petroleum Act of 2006, the 'registered holder' of a petroleum exploration permit, petroleum retention lease, petroleum production licence, infrastructure licence or pipeline licence 'must maintain, as directed by the responsible Commonwealth Minister from time to time, insurance against'

[406] Commonwealth Department of Resources, Energy and Tourism, Multiple Titleholders: Service of Documents, Taking Voluntary Actions and Discharging Legal Obligations Guideline (A Guideline in Relation to the Offshore Petroleum and Greenhouse Gas Storage Act 2006), effective 1 January 2012.

[407] Sections 577 and 577(A). [408] Sections 577(3) and 577(A2).

[409] Information from the website of the government of South Australia, DMITRE, at: www.pir.sa.gov.au/petroleum/prospectivity/basin_and_province_information/prospectivity_bight/grant_of_petroleum_exploration_permits_to_bp_exploration_alpha_ltd_faq.

- Expenses, or
- Liabilities, or
- Specified things arising in connection with or as a result of the carrying out of work or the doing of any other thing under the permit, lease or licence,

'including insurance against the expenses of complying with directions relating to the clean-up or other remediation of the effects of the escape of petroleum'.[410] Subsection 2 of the same article provides that '[t]he conditions of a petroleum special prospecting authority or a petroleum access authority may include a condition that the registered holder maintain, as directed by the responsible Commonwealth Minister from time to time, insurance.' Therefore, the requirement of insurance is compulsory and is imposed on the 'registered holder' of petroleum permits, leases and licences.

As to the amount of financial security, the Australian Maritime Safety Authority (AMSA) stated in a report following the Montara incident that

> Generally insurance amounts of between $100 and $300 million (US) dollars are considered to be standard practice in the offshore petroleum industry (not including third party claims). The amount of coverage for specific activities is set by the operator in consultation with the insurer and its underwriter and is based on an expert assessment of all potential liabilities.[411]
>
> Currently the level of insurance is determined by the operator and its insurer to directly reflect the potential liability associated with the type and extent of the activities undertaken on the title at that time. While higher risk activities such as well drilling, production and the decommissioning of a facility require substantial insurance, activities such as geophysical surveying are considered lower risk, unless they are undertaken in environmentally sensitive areas. The Designated Authority and/or the Joint Authority may challenge the set insurance amounts if it believes the insurance does not meet the stakeholder expectations or is considered too low based on industry best practice. The IAT [incident analysis team] has been advised that some Designated Authority's may not internally possess the necessary expertise to fully assess or challenge the insurance amounts set by the insurer.[412]

[410] Section 571(1) of the Offshore Petroleum Act.
[411] Australian Maritime Safety Authority, 'Response to the Montara', 15. [412] *Ibid.*

3.5.7 Canada

3.5.7.1 National Interest in Offshore Activities

Canada has a large amount of oil and gas resources in sedimentary basins beneath the ocean's floor offshore. The offshore oil and gas industry started exploration and development activities in Canada in 1964.[413] The offshore oil and gas activities make a great contribution to the nation's economy.[414]

3.5.7.2 Legal Framework

The main legislation concerning offshore activities includes

- Canada Oil and Gas Operations Act. This Act governs liability 'in respect of the exploration and drilling for and the production, conservation, processing and transportation of oil and gas' in the areas under the federal jurisdiction of Canada.[415]
- Oil and Gas Drilling and Production Regulations.
- Oil and Gas Certificate of Fitness Regulations.
- Oil and Gas Installation Regulations.
- Oil and Gas Diving Regulations.
- Oil and Gas Geophysical Operations Regulations.
- Oil and Gas Spills and Debris Liability Regulations (Regulations Respecting Limits of Liability for Spills, Authorized Discharges and Debris Emanating or Originating from Work or Activity Related to the Exploration for or Production of Oil and Gas). These regulations contain specific amounts of liability limits for the liable parties.

In Canada, the authority to regulate offshore drilling lies primarily with the National Energy Board (NEB). The NEB administers the pertinent acts and regulations for oil and gas exploration and production activities.[416]

Recently, in 2014–2015, Canada's House of Commons passed a new Energy Safety and Security Act (Bill C-22) that increases the absolute

[413] www.nrcan.gc.ca/energy/sources/offshore-oil-gas/1361.
[414] Canadian Centre for Energy Information, 'Canada's Evolving Offshore Oil and Gas Industry'.
[415] Section 3 of the Canada Oil and Gas Operations Act. The areas under the Canadian federal jurisdiction consist of the Northwest Territories, Nunavut and Sable Islands and submarine areas not within a province in Canada's internal waters, territorial sea and continental shelf. See Section 3 of the Canada Oil and Gas Operations Act.
[416] Dagg *et al.*, 'Comparing the Offshore', 23.

liability for offshore and nuclear companies to C$1 billion (from its current level of C$75 million). Bill C-22 amends the Canada–Nova Scotia Offshore Petroleum Resources Accord Implementation Act, the Canada–Newfoundland Atlantic Accord Implementation Act, the Canada Oil and Gas Operations Act and the Canada Petroleum Resources Act. The government of Canada worked collaboratively with the provinces of Nova Scotia, Newfoundland and Labrador to develop the offshore portion of Bill C-22, and mirror legislation is being developed and considered by those provinces' legislatures. The Energy Safety and Security Act has been referred to the Senate for consideration.[417]

3.5.7.3 Basis of Liability

'Where any discharge, emission or escape of oil or gas that is authorized by regulation, or any spill, occurs in any area to which this Act applies', the Canada Oil and Gas Operations Act provides for strict liability on the 'the person who is required to obtain an authorization' 'without proof of fault or negligence'.[418]

3.5.7.4 Attribution of Liability

The Canada Oil and Gas Operations Act provides that

> all persons to whose fault or negligence the spill or the authorized discharge, emission or escape of oil or gas is attributable or who are by law responsible for others to whose fault or negligence the authorized discharge, emission or escape of oil or gas is attributable are jointly and severally liable, to the extent determined according to the degree of the fault or negligence proved against them, for all actual loss or damage incurred by any person as a result of the spill or the authorized discharge, emission or escape of oil or gas.[419]

Therefore, in case of multiple tortfeasors, it is clear under Canadian law that joint and several liability will be imposed.

3.5.7.5 Damages and Remedies

The Canada Oil and Gas Operations Act provides that the liable party, 'the person who is required to obtain an authorization', is liable for

> (i) all actual loss or damage incurred by any person as a result of the spill or the authorized discharge, emission or escape of oil or gas, and

[417] Lloyds and Partners, *Energy Quarterly Newsletter*, January 2015, 12.
[418] Section 26(1) of the Canada Oil and Gas Operations Act.
[419] Section 26(1)(b) of Canada Oil and Gas Operations Act.

(ii) the costs and expenses reasonably incurred by Her Majesty in right of Canada or any other person in taking any action or measure in relation to the spill or the authorized discharge, emission or escape of oil or gas.[420]

3.5.7.6 Amount of Compensation

The Canada Oil and Gas Operations Act provides that the person liable under this Act is liable up to any prescribed limit of liability.[421] The Oil and Gas Spills and Debris Liability Regulations are thus adapted to spell out the specific amount of limitation. The liability limits vary based on the geographic area:

1. Land or submarine area adjacent to Arctic waters south of the sixtieth parallel of north latitude[422] – C$40 million;
2. Submarine area north of the sixtieth parallel of north latitude and to which (1) does not apply – C$40 million;
3. Any area of the Yukon Territory or Northwest Territories covered by or located at a distance of 200 metres or less from any body of inland water and to which (1) does not apply – C$25 million;
4. Any area within the Yukon Territory or Northwest Territories to which neither (1) nor (3) applies – C$10 million and
5. Any other area to which the Act applies for which no other limit is prescribed by these Regulations – C$30 million.[423]

3.5.7.7 Applicability in Time

The Canada Oil and Gas Operations Act provides that

> [a]ll claims under this section may be sued for and recovered in any court of competent jurisdiction in Canada and shall rank firstly in favour of persons incurring actual loss or damage, without preference, and secondly to meet any costs and expenses described in subsection (1) or (2).[424]

Under such a provision, the persons who incur actual loss or damage shall be ranked first to receive compensation. This may be in the interest of the victims for timely compensation. However, the time

[420] Section 26(1)(a) of Canada Oil and Gas Operations Act. [421] *Ibid.*
[422] As defined in the Arctic Waters Pollution Prevention Act.
[423] Section 3 of the Oil and Gas Spills and Debris Liability Regulations.
[424] Section 26(3) of the Canada Oil and Gas Operations Act.

limit under the Canada Oil and Gas Operations Act to bring a lawsuit is three years from the day when the loss or damage occurred and a maximum of six years from the day the spill or discharge occurred.[425] Therefore, the victims may still need to wait years before they can receive compensation.

According to Section 72.1 of the Canada Oil and Gas Operations Act, the act 'applies to every interest or right in oil or gas acquired or vested before the coming into force of this section'. Hence, it seems this Act has retrospective effect.

3.5.7.8 Compensation Mechanisms

The Canada Oil and Gas Operations Act provides that 'an applicant for an authorization' for oil and gas activities shall provide proof of financial responsibility.[426] The financial responsibility can be 'in the form of a letter of credit, a guarantee or indemnity bond or any other form satisfactory to the National Energy Board, in an amount satisfactory to the Board'.[427] It further adds that the 'holder of an authorization shall ensure that the proof of financial responsibility remains in force for the duration of the work or activity in respect of which the authorization is issued'.[428] Therefore, it is clear under the Canadian regime that financial responsibility is a compulsory requirement for the applicant to be granted an authorization, and it remains a continuing obligation during the whole period of work under authorization.

3.5.8 Comparative Analysis

3.5.8.1 Best Practice?

This overview of the legal regimes in six states shows that in most countries with a strong offshore petroleum interest, there is at least a national legal regime on civil liability, although it may consist of various pieces of legislation, some less developed than others. In some countries, such as the United Kingdom, the civil liability for offshore activities consists of different layers from the industry arrangement OPOL to statutory liability, and in the United States, the liability for offshore incidents may arise from federal and state laws. In other countries, the liability derives from rather easily identifiable primary and secondary

[425] Section 26(5) of the Canada Oil and Gas Operations Act.
[426] Section 27(1) of the Canada Oil and Gas Operations Act. [427] *Ibid.*
[428] Section 27(1.1) of the Canada Oil and Gas Operations Act.

legislation. Nevertheless, given that the offshore oil and gas activities involve many complications (technology development, various stakeholders involved, various contracts and subcontracts) and that the damages of an offshore incident may result in personal injury/fatalities, property damage and/or environmental damage, it is at least difficult and perhaps virtually impossible for any jurisdiction to cover all these aspects in one single piece of legislation. Using a dedicated checklist, an attempt is made to identify the legal parameters of each national regime (of the selected countries).

It is, however, clear from the overview provided that as opposed to the case of marine pollution resulting from vessels (where most countries have implemented the international conventions[429]), there is relatively little regulation as far as liability for damage resulting from offshore installations is concerned. As Table 10 shows, in most countries there is at best a brief mention of a liability of the operator based, for example, on a petroleum act and a provision on financial responsibility, but a detailed regulation of liability for damage resulting from offshore-related activities is in fact only present in the United States and to some extent in the United Kingdom, which relies on OPOL. However, given the large differences between the legal systems that were discussed, it is not possible to make a statement claiming that, for example, one particular legal system would constitute a 'best practice'.

A brief overview of the legal systems will be provided, followed by some comments comparing the particular elements that were used for the description of the countries. Given the fact that there are only a few legal systems where liability resulting from offshore related activities is explicitly addressed, it is not possible to draw strong normative conclusions based on this comparative analysis. Moreover, doing so would require a normative framework addressing how an optimal liability regime should be shaped in order to have a benchmark. Providing such a normative framework is too early at this stage but can be provided later using an economic framework in Chapter 7.[430]

3.5.8.2 Comparison of Legislation

3.5.8.3 Comparative Comments

Table 10 allows to facilitate some generalizations on the liability regimes in the legal systems that have been discussed, taking into account the

[429] Which have been discussed in Section 3.2.1. [430] More particularly in Section 7.2.

Table 10 *Comparison of Legislation*

	United Kingdom	Norway	Denmark	United States	Australia	Canada
Legislation	No specialized legislation; rely on OPOL	Liability on Petroleum Activities Act	Subsoil Act regulates liability	OCSLA + OPA 1990	Offshore Petroleum Act – vague	Oil and Gas Operations Act
Basis liability	SL[a] exception: war/ *force majeure*	SL, excl.: *force majeure*	SL	SL	Escape of oil = SL offence	SL
Relation with regulation	Compliance with regulations; not exclusive liability	Liability is strict, not based on breach of regulations	Compliance no exception	Compliance not exclusive liability + breach of regulations leads to loss of rights to limit liability	Only offence if breach of regulations + no liability if all reasonable steps to comply are taken	—
Causation	Proof on victim	Burden on claimant + in causal uncertainty, J&S	—	—	—	—
Attribution	J&S[b] under OPOL for insolvency	Channelling to licensee or operator Joint venture: J&S toward third parties	Channelling to licensee If several parties, J&S	Liability on responsible party and lessee; if more, J&S	Liability on title holder + J&S	Licensee is liable + J&S
Damages	Pollution damage (if direct) + remedial measures. OPOL not personal injury	Also, losses to fishermen	All	Removal costs + damages	Costs of public auth	Actual loss + costs of public authority

Cap	OPOL US$250 million	No cap	Only for MODUs	$75 million + removal costs; loss of cap if gross negligence or violation of regulations + no preemption of state law	No cap	Cap
Compensation mechanism	OPOL membership mandatory	Rapid claims settlement for fishermen Mandatory security for production licence, not exploration licence + detailed regulation of what insurance should cover	Financial capacity condition for licence	Financial security max. $150 million + details and different methods can be used. Fund (OSLTF): 1 bio. financed via tax	Mandatory insurance condition for licence	Proof of financial responsibility required

[a] SL stands for 'strict liability'.
[b] J&S stands for 'joint and several liability'.

checklist that was used for the analysis. First, the basis of liability is in general strict, and the liability is imposed on the holder of licence/permit/lease or the operator.

Second, in most countries, there are regulations aimed at safety standards for offshore operations and aimed at the prevention of incidents during offshore oil and gas activities. However, the compliance with regulations in most countries studied does not constitute a defence to exempt the responsible party from assuming liability. Only the Australian law is not clear on that point, as it includes a vague provision which could lead to the situation that compliance with regulation may be used as a legal defence.

Third, as to causation, it is mostly the occurrence of an incident that triggers the liability. Most statutes examined are silent on the causation issue.

Fourth, in the case of multiple tortfeasors, most jurisdictions follow a joint and several liability. In some jurisdictions, there is no specific provision concerning this issue.

Fifth, as to the amount of compensation, in most countries examined here, there is no upper limit on the amount to be compensated by the responsible party. Hence, in theory, the liability will be unlimited. The US and Canadian regimes are the only ones with financial caps. However, in the United States, the liability is limited only in the federal law OPA, which leaves open the possibility of state laws imposing additional liability. OPA does not pre-empt state law. Moreover, such a financial cap concerns only the damages; the removal costs remain uncapped. Although the liability is in theory unlimited in most legal regimes, it is important to realize that without the security of a financial guarantee, the unlimited liability cannot provide adequate compensation.

Sixth, as far as claims settlement is concerned, the strict liability with a financial guarantee can be considered as a mechanism in the interest of rapid claims settlement because it avoids the need for victims to prove negligence as the primary test of liability, which can be rather difficult and time-consuming.[431] This is adopted in all the liability regimes examined. However, looking at the specific compensation mechanisms in each national law, it is difficult to find useful instruments in addition to

[431] King, 'Deepwater Horizon Oil Spill Disaster', 6.

the strict liability and compulsory financial guarantee to provide rapid compensation to the pollution victims.

Seventh, as to the compensation mechanisms, there is always some requirement concerning the financial capacity of the applicant. In some countries, e.g. in the United Kingdom, the financial capability proved by OPOL is a pre-condition for the granting of a licence. The amount of such a financial guarantee is in most jurisdictions (with the exception of the United States) not specified in the regulations but assessed on a case-by-case basis by the national authority responsible for issuing licences. The US system provides for detailed requirements on the financial guarantee a responsible party has to take out. This is based on the so-called worst case scenario. The forms of financial guarantee can be different.

In addition to the financial guarantee, in the US regime, there exists a compensation fund contributed by the oil industry, the OSLTF, which provides compensation up to US$1 billion per incident.

3.6 Another High-Risk Sector: Nuclear

A most interesting sector to which compensation for marine pollution damage is often compared is nuclear liability. Like damage resulting from offshore installations, damage caused by a nuclear accident also can potentially be quite large. Hence, the international liability conventions have some features that are worth studying. For example, there is a limitation (so-called financial cap) on the liability, and the liability is channelled to the licensee of the nuclear power plant. Moreover, in second and third layers, compensation is provided by government, as a result of which doctrine holds that the nuclear industry is (at least partially) subsidized. The regime is, however, quite different in the United States under the so-called Price-Anderson Act because no legal channelling applies and there is no state intervention. Hence, the international and US regimes are certainly worth further analysis. In the remainder of this section, consideration will be given to the international compensation system for nuclear damage (3.6.1); after that, the compensation system for nuclear damage in the United States will be presented (3.6.2), followed by a brief critical comparative analysis (3.6.3).

It should be mentioned that, of course, other regimes of nuclear liability could be subject to analysis as well, such as e.g. the regime in

Japan, which is now obviously fully in the picture after the Fukushima incident[432] or the upcoming nuclear liability regime in China.[433] It would be excessive, however, to discuss those other regimes in detail as well. In this respect, reference will be made to discussions in the literature.

It also should be mentioned that when discussing the nuclear liability regimes at this stage, the focus will be on the contents of the liability regime and not on the pooling mechanisms. The pooling mechanisms for nuclear liability will be discussed in further detail in Chapter 4.[434]

3.6.1 International Compensation System for Nuclear Damage

3.6.1.1 Origin of the International Regime

At the beginning of the development of the nuclear industry, the Western European market had to rely on American suppliers and technology.[435] The American nuclear industry, however, was unwilling to bear liability for possible nuclear accidents in Europe.[436] Therefore, a 'hold harmless' clause was introduced in the first bilateral agreements between the United States and Europe, which required the European nuclear operators to indemnify the American suppliers for all claims resulting from their activities.[437] Even with those 'hold harmless' clauses, the American industry was still unsure whether it could provide sufficient protection and the US Atomic Industrial Forum conducted a comprehensive study on the possibility of European victims claiming against American suppliers in case of a nuclear incident. Two reports were published later by the Atomic Industrial Forum: the Preliminary Report on Financial Protection Against Atomic Hazards (the 'Preliminary Report')[438] and the International Problems of Financial Protection Against Nuclear Risk (the 'Harvard Report').[439] These two reports developed some principles

432 For a detailed analysis of that regime, see Faure and Liu, 'Compensating for Nuclear Accident'; and Faure and Liu, 'The Tsunami of March 2011'.

433 See on that regime Liu and Faure, 'Nuclear Liability in China'.

434 More particularly in Section 4.5.

435 For a summary of the international regime, see also Liu, *Compensating Ecological Damage*, 209–29; and Wilde, *Civil Liability*, 201.

436 See Vanden Borre, 'Shifts in Governance', 261–2.

437 Vanden Borre, 'Nuclear Liability', 180.

438 Prepared by the experts of Columbia University and published in March 1956.

439 Prepared by Harvard Law School and published in 1959.

for nuclear liability, which were later incorporated into the international conventions.

The Preliminary Report identified four parties who can be affected by nuclear risks: the nuclear industry, private insurers, the government and the victims of a potential nuclear accident. In this report, the liability of the nuclear industry is linked with insurance capacity; for the part of damage that cannot be compensated by insurance, the government needs to provide additional compensation to protect the victims. The Preliminary Report introduced a system of unlimited government intervention: any damage in excess of the limitation of liable parties' liability should be covered by the government in terms of indemnity.[440]

The channelling of liability to nuclear operators was advised in the Harvard Report. According to the Harvard Report, the suppliers and contractors were exempted from liability for the following reasons: the suppliers are afraid of being held liable instead of or jointly with the operators and then being burdened by the lengthy trials; suppliers will lose control after the delivery of goods and services and operators are more capable of obtaining insurance.[441] Those two reports proposed several principles for nuclear liability, which later were adopted in some national legislations and international conventions.

3.6.1.2 First-Generation Nuclear Liability Conventions

NEA and IAEA Conventions Two compensation regimes for nuclear damage were established in the 1960s under the auspices of the Nuclear Energy Agency (NEA) and the International Atomic Energy Agency (IAEA). Since the Paris Convention and the Brussels Convention were established under the auspices of the OECD/NEA, they are regionally confined to Western Europe, Slovenia and Turkey. The Vienna Convention, under the aegis of the IAEA, is worldwide in scope. Some principles proposed in both the Preliminary Report and the Harvard Report were accepted under those two conventions. Before describing those principles, this section first clarifies what constitutes 'nuclear damage' under those conventions or, in other words, which categories of damage the nuclear industry is liable for, according to the conventions.

According to the Paris Convention, the nuclear operator is liable for 'damage to or loss of life of any person' and 'damage to or loss of any

[440] 'Preliminary', 30. [441] See Vanden Borre, 'Shifts in Governance', 265–6.

property other than' onsite damage or property used in connection with the installations.[442] The Convention does not explain the two headings of damage further. What should be considered as damage to persons or damage to property is left to the discretion of the competent court in accordance with the applicable national law.[443] Those two headings are also recognized as 'nuclear damage' under the Vienna Convention. In addition to those two headings, 'nuclear damage' under the Vienna Convention also includes 'any other loss or damage so arising or resulting if and to the extent that the law of the competent court so provides'.[444] Therefore, more discretion is given to the competent court.

As for the geographical scope, the Paris Convention does not apply to 'nuclear incidents occurring in the territory of non-Contracting States or to damage suffered in such territory unless otherwise provided by the legislation of the Contracting Party in whose territory the nuclear installation of the operator liable is situated'.[445] The geographical scope of the Vienna Convention is less clear. Some argue that because the convention itself is silent on its geographical scope, the Vienna Convention on the Law of Treaties will apply. The convention applies to damage occurring in the territory of a state party on board aircraft registered in that state and on ships flying its flag.[446] While others deduce from the article about the competent court[447] that the geographical scope depends on the competent court and thus damage resulting from non-Vienna states could be covered by the Vienna Convention.[448] The geographical scope has been broadened by the Joint Protocol.[449]

Strict Liability In Western Europe, there is a long-established tradition of a presumption of liability for hazards resulting from a dangerous activity. The nuclear industry, covered under the Paris Convention, is obviously qualified as a dangerous activity, and there is a serious difficulty in proving the negligence of nuclear operators. Therefore, a system of absolute liability is established under the Paris Convention.[450] According to the Paris Convention, the operator is liable for damage caused by a nuclear incident in a nuclear installation or involving nuclear substances

442 The Paris Convention, Art. III(a).
443 The *expose des motifs* of the Paris Convention, point 39.
444 The Vienna Convention, Art. I(k). 445 The Paris Convention, Art. II.
446 Lamm, 'The Protocol Amending'. 447 The Vienna Convention, Art. VIII.
448 Blanchard, 'Responsibility for Environmental Damage', 233, 236.
449 This will be discussed in Section 3.6.1.3.
450 The *expose des motifs* of the Paris Convention, point 14.

coming from such installations.[451] To prove the fault of nuclear operators is no longer necessary to establish liability. Liability established under the Paris Convention is quite stringent because many classical exonerations, such as *force majeure*, acts of God or intervening acts of third persons under general tort law, are no longer applicable.[452] The available exonerations are an act of armed conflict, hostilities, civil war and insurrection. The operator is not liable for damage caused by a grave natural disaster of an exceptional character unless the legislation of the contracting party in whose territory his nuclear installation is situated provides to the contrary.[453] Similar stipulations about absolute liability and exonerations also can be found under the Vienna Convention.[454] However, under the Vienna Convention, there is an additional possibility for operators to be relieved of their liability: the competent court can, according to the applicable law, relieve the operator wholly or partly from his obligation if the operator can prove that damage resulted from gross negligence or an act or the omission of the victims.[455]

Channelling of Liability Under the Paris Convention, liability is channelled to the operators. In addition to the operators, no one else is liable for damage caused by a nuclear incident.[456] The 'operator' is defined as 'the person designated or recognized by the competent public authority as the operator of that installation'.[457] These provisions can hold other parties engaged in nuclear activities liable because the Paris Convention is the only legal basis for a claim against a nuclear operator in the case of the identified incidents.[458] This concentration of liability is based on two reasons: to avoid the complicated legal procedures to identify liable parties and to allow a concentration of insurance capacity.[459] Under the Paris Convention, the operators in principle do not have a right of recourse against other parties. This is so because it is argued that allowing recourse will make it necessary for suppliers to seek insurance coverage and will lead to costly duplication of insurance.[460] However, recourse is possible if the damage results

451 The Paris Convention, Art. III(a).
452 The *exposé des motifs* of the Paris Convention, point 48.
453 The Paris Convention, Art. IX.
454 The Vienna Convention, Arts. I(1)(k), IV(1), (3). 455 *Ibid.*, Art. IV(2).
456 The Paris Convention, Art. VI(a), (b). 457 *Ibid.*, Art. I(a)(vi).
458 *Ibid.*, Art. VI(c)(ii). 459 The *exposé des motifs* of the Paris Convention, point 15.
460 *Ibid.*, point 18.

from an act or omission done with intent to cause damage or if and to the extent provided by contract.[461] The Vienna Convention also has similar provisions.[462]

Limited Liability Under both the Paris Convention and the Vienna Convention, the operator's liability is limited both in amount and in time. The Paris Convention sets the maximum liability of the operator at 15 million SDRs but allows the contracting party to establish by legislation a greater or lesser amount, considering the capacity of insurance and financial security. The contracting party also can require a lower amount according to the nature of the installation. The lower amount should be no less than 5 million SDRs.[463] The Paris Convention introduces a cap on liability, taking into account the difficulties of operators in seeking financial security.[464] The flexible expression allows the contracting parties to set the limit higher than those set by the Paris Convention. For example, in Sweden, the limit on liability is set at 300 million SDRs according to the Nuclear Liability Act (SFS 1968:45).[465] Germany even adopted a system with unlimited liability.[466] Under the Vienna Convention, the cap on liability should be no less than US$5 million.[467] The statute of limitations under both the Paris Convention and the Vienna Convention is set at ten years from the date of the nuclear accident. They both allow extension of the extinction periods by contracting parties if the operator is covered by insurance or other financial security.[468]

Financial Security To seek financial security coverage for the operator's liability is important for international regimes of nuclear liability. Both conventions require the operator to have and maintain insurance

461 The Paris Convention, Art. VI(f). 462 The Vienna Convention, Arts. II(5), X.

463 The Paris Convention, Art. VII(b).

464 The *exposé des motifs* of the Paris Convention, point 43.

465 OECD: 'Nuclear Legislation in OECD Countries, Regulatory and Institutional Framework for Nuclear Activities: Sweden', 13; available at: www.oecd-nea.org/law/legislation/sweden.pdf.

466 'Gesetz über die friedliche Verwendung der Kernenergie und den Schutz gegen ihre Gefahren (Atomgesetz – AtG)', vom 23 Dezember 1959, Neufassung vom 15. Juli 1985, letzte Änderung vom 31. Juli 2011 (Act on the peaceful utilization of nuclear energy and the protection against its hazards (Atomic Energy Act) of 23 December 1959, as amended and promulgated on 15 July 1985, last amendment of 31 July 2011), §31.

467 The Vienna Convention, Art. V.

468 The Paris Convention, Art. VIII; The Vienna Convention, Art. VI(1).

or other financial security up to the cap of its liability.[469] Insurance is the most popularly used instrument for an operator to cover its liability. In fact, the cap on liability is usually set as the maximum available amount from the insurance market. Because insurance is available per installation for a fixed period rather than in respect of a single incident, the potential resources available for compensation may be reduced after the first incident. Under these conditions, the contracting parties may need to intervene.[470] It is for the contracting parties to decide the nature, form and extent of the compensation according to applicable national law.[471]

Exclusive Jurisdiction Both the Paris Convention and the Vienna Convention have an 'exclusive jurisdiction' clause: the jurisdiction lies only with the courts of the contracting party within whose territory the nuclear incident occurred. If the nuclear accident happens outside the contracting parties' territory or the place cannot be determined, the jurisdiction lies with the courts of the installation state of the liable operator.[472] Because nuclear damage may have a trans-boundary effect, the exclusive jurisdiction can ensure that the cap of liability is not exceeded and that a fair distribution of compensation takes place if the claims lead to insoluble problems.[473]

Public Funding The liability limits established under the Paris Convention and the Vienna Convention are quite low compared to the potential catastrophic damage that a nuclear incident can cause. The liability is limited to promote the development of the peaceful use of nuclear liability. To provide the potential victims with better protection, the Brussels Complementary Convention was established under the auspices of the NEA in 1963. Under the Brussels Complementary Convention, two additional layers of compensation are added in terms of public funds. Therefore, the total amount of money available for compensation is increased to 300 million SDRs, and a third-layer compensation system is established: firstly, the operator is liable up to an amount of at least 5 million SDRs via its financial security; between this amount and 175 million SDRs, the installation state

[469] The Paris Convention, Art. X; The Vienna Convention, Art. VII.
[470] The *exposé des motifs* of the Paris Convention, point 49.
[471] The Paris Convention, Art. XI; The Vienna Convention, Art. VIII.
[472] The Paris Convention, Art. VIII; The Vienna Convention, Art. XI.
[473] The *exposé des motifs* of the Paris Convention, point 54.

needs to make public funds available; and for the amount between 175 and 300 million SDRs, compensation is made out of public funds by all contracting parties according to a specific formula.[474] The installation state can escape its obligation under the second layer by setting the liability limit at no less than 175 million SDRs. In that case, the whole amount up to 175 million SDRs needs to be made available by the liable operators.

3.6.1.3 Second-Generation Nuclear Liability Conventions

General Issues The nuclear accident in Chernobyl in 1986 triggered people to thinking about the insufficiency of the existing nuclear liability conventions in terms of the ability to protect the victims. The Chernobyl accident led to serious damage to human health, property and the environment with a broad trans-boundary impact. However, at that time, the Soviet Union had no national legislation on nuclear liability. Nor was it a member of any international nuclear liability regime. Even if the Soviet Union had joined the Vienna Convention, serious problems still prevented effective compensation: the authorities in the Soviet Union questioned the necessity of the preventive measures[475]; the cap on liability under the Vienna Convention was dwarfed by the serious damage it caused; and Western European countries joined another regime which is independent of the Vienna Convention.

The lack of acceptance of the international regimes and the insufficient resources available in case of nuclear damage triggered an effort to revise the conventions. Another five conventions have been passed after the Chernobyl accident. The first effort was made to link the geographical scope of the existing regimes. In 1988, the Joint Protocol was adopted, under which the operator of a nuclear installation situated in the territory of a party to one convention shall be liable for the nuclear damage suffered in the territory of a party to either the Paris Convention or the Vienna Convention.[476] It was believed that the Joint Protocol would give incentives for the Central and Eastern Europe countries to join the Vienna Convention and would broaden the application of the two regimes. However, only twelve of eighteen countries in these areas which have joined the Vienna Convention became members of the Joint Protocol. Globally, of the fifty-two states

474 The Brussels Complementary Convention, Art. III(a)(b).
475 See Vanden Borre, 'Nuclear Liability'. 476 The Joint Protocol, Art. II.

party to the Paris and Vienna Conventions, only twenty-seven parties have joined the Joint Protocol.

In addition to the geographic links, some substantive changes were also adopted in the following years. Under the auspices of the International Atomic Energy Agency (IAEA) regime, two nuclear conventions were open for signature in 1997: the Protocol to the Vienna Convention and a stand-alone convention – the Convention on Supplementary Compensation for Nuclear Damage (CSC). The CSC was made open to more countries: it maintains the essential principles established under the Paris Convention and tries to establish two tiers of public funds in case of nuclear damage. It is open to any state if it is a party to either the Paris Convention or the Vienna Convention or if it has national legislation containing the same principles. An effort was also made to revise the NEA conventions. In 2004, the Protocol to the Paris Convention and the Protocol to the Brussels Supplementary Convention were open for signature.

Broadened Scope of Damage Under the regimes of both NEA and IAEA, some important changes were made, especially the broadened scope of nuclear damage and the increased amount of the limitation of liability and public funds. Under the Protocol to the Paris Convention, in addition to personal injury and property damage, four new titles have been added to the concept of 'nuclear damage': the economic loss arising from personal injury and property damage, the costs of measures of reinstatement of the impaired environment, the loss of income deriving from a direct economic interest in any use or enjoyment of the environment, and the costs of preventive measures.[477] Similar provisions can also be found in the Protocol to the Vienna Convention. The difference is that under the latter, the definition of nuclear damage is open: it admits any other economic loss which is permitted by the general civil liability law in the competent court.[478]

Because the health impact of nuclear radiation may not fully manifest for decades, the revised conventions also extend the statute of limitation for claims for personal injury and death. These kinds of claims need to be brought within thirty years from the date of the accident.[479] In addition, the available defences for operators are further limited: natural disasters

477 The Protocol to the Paris Convention, Art. I(vii).
478 The Protocol to the Vienna Convention, Art. I(k).
479 The Protocol to the Paris Convention, Art. VIII(a)(i); The Protocol to the Vienna Convention, Art. VI(1)(a)(i).

are no longer an applicable defence.[480] The Protocols are also designed to be attractive by broadening the applicable geographical scope. The Protocol to the Vienna Convention applies to nuclear damage wherever suffered, with a permitted exclusion if a non-contract party has nuclear installations but does not offer reciprocal benefits.[481] The Protocol to the Paris Convention also covers damage suffered by some non-contracting parties which satisfies some specific requirements.[482]

Increased Limits and Public Funds Another important change made under the NEA and IAEA regimes is the increased liability limitation and public funds. Under the Paris Convention, a three-tier compensation system was established in 1960. The Protocol to the Paris Convention increased the limit for nuclear operators to be no less than €700 million. The Contracting party can reduce the liability to be no less than €70 million for an incident originating from a nuclear installation or to be no less than €80 million for the carriage of nuclear substances according to the reduced risks.[483] The Protocol even allows the adoption of unlimited liability by the contracting parties as long as the financial security required is no less than the amount mentioned earlier.[484] In addition to the liability of operators, the Protocol to the Brussels Complementary Convention provides another two tiers of compensation. From the maximum amount of operators' liability set by the Protocol to the Paris Convention up to €1,200 million the installation state needs to pay. Public funds need to be made available by all contracting parties according to a set formula of compensation between the amounts of €1,200 million and €1,500 million.[485] Thus, the available resources for compensation in case of a nuclear accident are increased significantly: from 300 million SDRs (approximately €327 million) to €1,500 million.

The compensation capacity also increased under the IAEA regime. The Protocol to the Vienna Convention increases the liability limitation to no less than 300 million SDRs or no less than 150 million SDRs provided that the installation state will make public funds available to

[480] The Protocol to the Paris Convention, Art. IX; The Protocol to the Vienna Convention, Art. IV(3).

[481] The Protocol to the Vienna Convention, Art. I(A).

[482] The Protocol to the Paris Convention, Art. II(a).

[483] The Protocol to the Paris Convention, Art. VII(a)(b).

[484] The Protocol to the Paris Convention, Art. X(b).

[485] The Protocol to the Brussels Complementary Convention, Art. III(b).

cover the amount between the limitation and 300 million SDRs.[486] Thus, the revision has introduced a second layer of compensation. Though the amount is also significantly increased, it is modest compared to the possible significant damage, and it is just set at the same level as the original Paris Convention and the Brussels Supplementary Convention. The CSC also provides two tiers of compensation: a first 300 million SDRs is paid by public funds from the installation state and another 300 million SDRs from the collective funds of the contracting parties.[487]

Labyrinth of Conventions The different coverage caps before and after the 2004 amendments to the conventions provide the result shown in Table 11 as far as the total amount of compensation is concerned. These new conventions and protocols are designed to overcome the deficiencies of the first generation of nuclear liability conventions. However, the revisions have made eight international conventions available for nuclear liability, which leads to the so-called labyrinth of international conventions dealing with nuclear liability issues.[488] Moreover, among the five new conventions, only two have come into force: the Joint Protocol entered into force in 1992 and the Protocol to the Vienna Convention came into force in 2003. Until now, only five countries

Table 11 *The Different Coverage Caps Before and After the Conventions' 2004 Amendments (in millions of euros)*

	Paris (1960) and Brussels Conventions (1963) (French example)	Modification Protocol of the Paris and Brussels Conventions (2004)
Operator's liability cap	91	700
State's intervention	140	500
Contracting parties coverage	150	300
TOTAL	381	1500

486 The Protocol to the Vienna Convention, Art. V(1).
487 The CSC Convention, Arts. III(1), IV(1).
488 For the patchy and complicated approach under international nuclear liability conventions, see Currie, 'The Problems and Gaps in the Nuclear Liability Conventions'.

have ratified the protocol to the Vienna Convention: Argentina, Belarus, Latvia, Morocco and Romania, and none of them has a significant nuclear generating capacity.[489] There are still many large nuclear generating countries which are not members of any of those conventions, such as the United States, Canada, Japan, China, Korea, Russia, South Africa and Switzerland.

There is no comprehensive study on why some countries are reluctant to join nuclear liability conventions, but some hypotheses have been proposed. Firstly, some countries are geographically remote from the contracting parties to the international conventions. Thus, they can benefit less from the available compensation in case of trans-boundary damage. This may be a concern for Australia, Japan and Canada. Another reason may be that unlimited liability has been adopted in the domestic legislation in Switzerland, Austria and Japan. However, the Protocols to the Paris and Vienna Conventions have allowed the participation of states with unlimited liability. Some countries may hesitate because a discriminatory treatment between domestic and foreign victims is not allowed under the conventions, and they may be reluctant to pay foreign victims out of public funds.[490]

3.6.2 The Compensation System for Nuclear Damage in the United States

The Price-Anderson Act (PAA) and Nuclear Regulatory Commission (NRC) regulations require the licensees of nuclear installations to seek financial protection for both the potential public liability caused by the installations and for property damage of the installations.[491] The compensation instruments used in the United States can provide compensation for nuclear incidents of a catastrophic nature.

489 Of these five countries, only Argentina and Romania have nuclear capacity. The net nuclear power generating capacity of Argentina is 935 MWe and that of Romania is 1,300 MWe in 2009. See Schwartz, 'Great Expectations'.

490 *Ibid.*

491 For an overview of the US nuclear liability regime, also see Liu, *Compensating Ecological Damage*, 232–40. The Price Anderson Act was originally adopted in 1957: Act to Amend the Atomic Energy Act of 1954, Pub. L. No. 85-256, 71 Stat. 576 (1957). It was revised by the Energy Policy Act of 2005, Pub. L. No. 109-58,119 Stat. 779 (2005). The NRC also adopted many regulations about the nuclear industry. The parts relevant to financial protection and compensation are compiled under Title 10, Part 140 of Code of Federal Regulations. 10 CFR §§140.1–140.109.

3.6.2.1 Liability for Nuclear Damage

Public-Law Liability Both regulation and liability rules can be used to prevent nuclear damage. After World War II, the United States established the Atomic Energy Commission (AEC) to foster and control the peacetime development of the atomic industry. The AEC was abolished by the Energy Reorganization Act of 1974,[492] which assigned its functions to the Energy Research and Development Administration and the Nuclear Regulatory Commission (NRC). The latter is the agency in the United States to regulate the nuclear industry: it oversees reactor safety, reactor licensing and renewal, radioactive material safety and spent-fuel management. In addition, there is also a self-regulation institution, the Institute of Nuclear Power Operations (INPO), that monitors and controls nuclear risks. A detailed examination of how the nuclear industry is regulated in the United States is out of scope here. This section thus introduces briefly the liability rules for nuclear damage in the United States.

The nuclear sector may create damage during the operation of nuclear facilities, during the decommissioning and after the decommissioning of facilities. The nuclear waste, if not disposed of properly, also can lead to severe damage. Besides, cautious decommissioning and monitoring after the decommissioning period are also necessary to prevent nuclear damage.

Liability rules for nuclear damage during the operation of nuclear facilities are mainly prescribed in the PAA and in NRC regulations. When the Congress passed the Atomic Energy Act in 1954, it decided to allow private participation in nuclear power development. The thinking toward peaceful uses of nuclear energy brought with it the new problem of assigning liability caused by the industry. Without adequate insurance or financial coverage, utilities are reluctant to accept the risks.[493] The PAA was passed in 1957 to provide for a system of financial responsibility. As mentioned in the PAA, it has a dual-purpose: 'to protect the public and to encourage the development of the atomic energy industry'.[494] The dual purpose can be seen in the provisions about limitation of liability and financial protection provided by the government.

[492] Energy Reorganization Act of 1974, Pub. L. No. 92-438, 42 USC §5801.
[493] GAO Report EMD-80-80 (1980), 'Analysis of the Price Anderson Act', 3.
[494] Publ. L. No. 85-256, 68 Stat. 921.

A sole and exclusive federal cause of action for any property damage or personal injury from radiation exposure, the Public Liability Action (PLA) is established in the 1988 Amendment Act.[495] The term 'public liability action' is defined broadly to contain 'any suit asserting public liability'. The PAA also defines the related concepts 'public liability' and 'nuclear incident' broadly.[496] Through those broad definitions, Congress pre-empted all state causes of action. There are two types of PLA causes of action: the cause of action for personal injury and that for property damage.[497] Hence the PAA does not mean to cover damage to the environment.[498]

Extraordinary Nuclear Occurrence As far as substantive rules are concerned, they 'shall be derived from the law of the State in which the nuclear incident involved occurs, unless such law is inconsistent with' the PAA.[499] Thus, usually, the state tort law applies with the exception that the act has different prescriptions. The PAA provides one such exception: if the nuclear accident comprises an 'extraordinary nuclear occurrence' (ENO), it requires the nuclear plant owner to waive certain defences they might otherwise have under local tort law.[500] Hence, the operators are strictly liable if damage arises from an ENO. Whether an accident comprises an ENO is determined by the NRC. To be qualified as an ENO, a nuclear incident needs to satisfy the following criteria: the discharge or dispersal constitutes a substantial amount of source, special nuclear or by-product material or has caused substantial radiation levels offsite, and there have in fact been or will probably be substantial damages to persons offsite or property offsite.[501] The NRC regulations give detailed criteria in determining whether these conditions come due.[502] The legislation restricts the waivers to ENOs for two reasons: to avoid 'nuisance' suits and to protect the public from the consequences of catastrophic nuclear accidents.[503] However, the substantial criteria are very difficult to satisfy. Even the Three Mile Island accident was not

[495] Jose and Garza, 'The Complete Federal Preemption'. [496] 42 USC §2014(q)(w).
[497] Jose and Garza, 'The Complete Federal Preemption', 16.
[498] However, this does not mean the compensation instruments established under Price Anderson Act have no relevance to environmental damage at all. The following discussion shows that case law has successfully explained the coverage clause of the insurance policy broadly to contain clean-up costs.
[499] 42 USC §2014(hh). [500] 42 USC §2210(n). [501] 10 CFR §140.81(b)
[502] 10 CFR §140.84–140.85.
[503] Tousley, 'Abolishing the "Extraordinary Nuclear Occurrence" Threshold'.

classified as an ENO.[504] After the Three Mile Island accident, a petition was proposed to the NRC which requires the revision of the criteria of an ENO.[505] The NRC also identified the difficulties in determining an ENO[506] and published three proposed amendments to the criteria in 1985 and solicited public comment on these amendments.[507] Some scholars even advised abolishing the threshold of ENO to allow better protection to victims.[508] However, after the procedure of public comments on the proposed amendments, the NRC decided that the current criteria for determining an ENO are adequate and are consistent with the intent of Congress and thus should be sustained.[509] In this situation, most claims for nuclear damage may still not arise out of an ENO.[510] When an incident is not identified as an ENO, the claims are still subject to state tort laws. During legislating the rules of waivers for an ENO, the Joint Committee on Atomic Energy considered state law. It was expressed that by then there was no assurance that all state courts would impose a rule of strict liability in the event of a nuclear incident.[511] At that time, there were some jurisdictions which purported to reject the doctrine of strict liability and a sizable number of others where the law relative to strict liability was unsettled.[512] The approach of minimal interference with state law was kept in the legislation on waivers under an ENO. Thus, many other issues are still subject to state tort law. Though the application of state tort law does not mean strict liability could not be used, there are at least some cases where the negligence rule or fault-based approach is still followed.[513]

Another feature of liability rules for nuclear damage arising from ENOs is that other statutes of limitations are waived as long as a suit is instituted within three years of the date on which the claimant first knew or reasonably could have known of his injury or damage and the cause

504 Metropolitan Edison Co., 11 NRC 519 (1980).
505 65 FR 61283-01, 2000 WL 1577628 (Fed. Rep.)
506 NRC (1983), 'Report to Congress, The Price Anderson Act - The Third Decade', E1–E5.
507 50 Fed. Rep. 13978.
508 Tousley, 'Abolishing the "Extraordinary Nuclear Occurrence" Threshold'.
509 65 Fed. Rep. 61283-01, 2000 WL 1577628 (Fed. Rep.).
510 Roman, Fernando and Salsman, 'Canada and International Nuclear Liability'.
511 Senate Rep. No. 89-1605, 89th Cong., 2nd sess.1966. 512 *Ibid.*
513 Though under those cases, the courts disagree on the extent to which state tort law should be used or whether the federal law can pre-empt state tort law, they require the establishment of fault or identification of the fault-based standard of care. For example, *O'Connor* v. *Commonwealth Edison Co.*, 13 F.3d 1090 (7th Cir. 1994); *re TMI Gen. Pub. Utils. Corp.*, 67 F.3d 1103 (3d Cir. 1995); *Merilyn Cook et al.* v. *Rockwell International Corporation and the Dow Chemical Company*, 273 F. Supp. 2d 1175.

thereof.[514] The requirement of within twenty years of the nuclear accident was abolished by the 1988 Amendment Act.[515] As far as nuclear damage is concerned, the damage may not manifest decades after the accident. Thus, the relaxation of the statute of limitations can provide better protection to victims.

No Channelling, but Limit Unlike the situation under the international regime, there is no legal channelling of liability to nuclear operators. Instead, the PAA creates a system of economic channelling.[516] The victims can still claim from other parties besides the operators, but the financial protection clause has shifted the burden of compensation mainly to operators. For damage caused by qualified operators of a nuclear power plant, the total available amount of compensation is $12.2 billion since the most recent change.[517] This amount is far higher than the amount of financial protection needed from contractors and other licensees determined in the PAA.[518] Moreover, the PAA provides an 'omnibus' coverage. It covers 'anyone liable' for 'public liability'. Thus, victims may choose to claim from operators, the most identifiable and capable parties for their damages.

To encourage the development of the nuclear industry, the PAA exempts the licensees and contractors from unlimited liability. It adopts a limitation on aggregate public liability for a single nuclear incident. The amount of limitation is different among nuclear power plant owners with a capacity of 100,000 electrical kilowatts or more, contractors with the Department of Energy (DOE) and other licensees. This amount is closely linked to the amount of financial protection the PAA requires from those parties.[519] For a nuclear incident with damages in excess of the amount of aggregate public liability limitations, Congress will 'take whatever action is determined to be necessary (including approval of appropriate compensation plans and appropriation of funds) to provide full and prompt compensation to the public'.[520]

To prevent nuclear damage does not only require actors to take care during the operation of nuclear facilities but also asks for cautious behaviour during the decommissioning of nuclear facilities and disposal and monitoring of nuclear waste. The licensees of by-product material, production and utilization facilities as well as special nuclear material

[514] 42 USC §2210(n). [515] Pub. L. 100-408, §§10.
[516] Vanden Borre, 'Shifts in Governance', 300. [517] 10 CFR §140.11.
[518] 10 CFR §140.11–12. [519] 42 USC §2210(e)(1). [520] 42 USC §2210(e)(2).

have the obligation to decommission their facilities for unrestricted use or for use under restricted conditions.[521] As far as nuclear fuel manufacturing facilities are concerned, because the federal government has been responsible for nuclear fuel production for several decades, the decontamination and decommissioning are done through the Uranium Enrichment Decontamination and Decommissioning Fund.[522] For low-level radioactive waste, the states are responsible for disposal.[523] The federal government is responsible for disposal of high-level radioactive waste and spent nuclear fuel underground in a deep geological repository. A Nuclear Waste Fund is established to bear the costs of carrying out the disposal activities.[524]

3.6.2.2 The Requirement of Financial Protection

Under the PAA, licensees and contractors are required to maintain certain levels of financial protection for the potential public liability. Besides those third-party liabilities, the NRC regulation also requires power reactor licensees to provide financial assurance for their obligations, in the event of an accident, to stabilize and decontaminate the reactor and the reactor station site.[525] This section examines the financial requirements for third-party damage and property damage, respectively.

Financial Requirement for Third-Party Liability Both nuclear reactor operators and contractors who conduct activities according to a contract with nuclear facilities may create risks of public liability. The PAA imposes different financial protection obligations on operators and contractors.

The PAA of 1957 required that the licensees of facilities having a rated capacity of 100,000 electrical kilowatts or more should provide financial protection up to the maximum amount available from private sources. The financial protection could be private insurance, private contractual indemnities, self-insurance, other proof of financial responsibility or a combination of such measures.[526] The most popularly used protection is private insurance, which was provided by a stock company, the Nuclear Energy Liability Insurance Association (NELIA), and a mutual pool, the Mutual Atomic Energy Reinsurance Pool (MAERP). These two items

[521] 10 CFR §§30.35, 50.75, 70.38. [522] 42 USC §2297. [523] 42 USC §2021(c).
[524] 42 USC §10222. [525] 10 CFR §50.54(w)
[526] Act to Amend the Atomic Energy Act of 1954 (Price Anderson), Pub. L. No. 85-256, 71 Stat. 576, 577 (1957).

combined to provide coverage for nuclear liability up to $60 million.[527] Under the PAA of 1957, the government should agree to indemnify the licensees in case of an accident in excess of the required financial protection. The aggregated indemnity from each incident should not exceed $500 million.[528] This is so because in the 1950s, the nuclear industry was still in its infancy, and government wanted to encourage development of the nuclear industry by exonerating the operators from the burden brought by potential catastrophes.

An important step which shifted the burden from government to operators was made in 1975. By then, the Joint Committee on Atomic Energy argued that the nuclear industry should have been able to bear the financial burden after years of development.[529] Though the total amount of compensation was kept the same as that in 1957, a new tier of compensation, the 'retrospective premium', was introduced in place of the indemnity provided by the government. All operators should pay the retrospective premiums in the case of an incident in excess of the primary liability coverage up to the amount of $5 million.[530] The amount of primary financial protection and retrospective premiums has been increased over the years. Now the primary financial protection is set as $375 million, and the retrospective premium for each operator from each incident should not exceed $121.255 million from 2013.[531] As of 2013, given the current number of operating reactors, the maximum compensation for a nuclear incident is $13,616,046,000.[532]

Besides the power plant operators, the PAA and the NRC regulations also require other licensees and contractors with the Department of Energy (DOE) to provide financial protection. For example, the NRC requires power plant operators at a lower thermal power level to provide a lower amount of financial protection from $1 million to $2.5 million based on their size. For other licensees, the regulation provides a formula to calculate the necessary amount of protection, which would be in no

[527] McClure, 'A Review of Nuclear Energy Insurance', 256.

[528] Act to Amend the Atomic Energy Act of 1954 (Price Anderson), Pub. L. No. 85-256, 71 Stat. 576, 577 (1957).

[529] Report No. 94-648, House of Representatives (1975); Senate Rep. No. 454, 99th Cong., 1st Sess. (1975), 10.

[530] Pub. L. No. 94-197, 1975 HR 8631, 89 Stat 1111.

[531] 10 CFR §140.11(4); 75 Fed Rec. 16646, Apr. 2, 2010. The amounts have recently again been updated. See the website of ANI: www.nuclearinsurance.com.

[532] This comprises $375 million in the first layer (operator liability) plus an individual contribution of $121,255,000 per reactor per accident plus 5 per cent for legal expenses ($6,062,750), providing a total of $13,616,046,000.

event in excess of $4.5 million or $74 million. In addition, certain holders of construction permits and combined licenses, plutonium processing and fuel fabrication plants and uranium enrichment facilities are also imposed an obligation to seek coverage of financial protection.[533] For an incident in excess of the required financial requirement, the licensees or other relevant persons can be indemnified by the NRC via the indemnification agreement with an amount up to $500 million.[534] As far as contractors are concerned, the Secretary of Energy will enter into an indemnification agreement with them. Under this agreement, the Secretary of Energy may require the contractor to provide and maintain financial protection. For damage in excess of this amount, the Secretary will provide indemnification up to $10 billion.[535] As a price of the indemnification agreement, the NRC or the Secretary of Energy can collect a fee from the licensees or contractors. The fee is set at $30 per year per thousand kilowatts of thermal energy capacity, subject to deduction by the NRC or the Secretary.[536]

For nuclear damage in excess of the aggregated liability discussed earlier, Congress shall take 'whatever action is determined to be necessary' to provide full and prompt compensation.[537] The Secretary of Energy or the NRC shall make a survey of the causes and extent of damage and submit a report to Congress. The president shall submit an estimate of the aggregate dollar value of personal injuries and property damage, recommendations for additional sources of funds and compensation plans. The additional sources of funds include 'possible revenue measures on the sector of the economy, or on any other class, to which such revenue measures might be applied'.[538]

Financial Requirement for Property Damage In addition to damage to third parties, nuclear incidents can also create serious onsite damage. The Three Mile Island accident shows that the onsite damage to the nuclear power plant can be more important than the third-party liability. The NRC thus requires the operator to maintain a minimum coverage for onsite property damage and gives a priority to decontamination.[539] The operators are asked to seek a 'coverage limit for each reactor station site

[533] 10 CFR §140.12–140.13b. [534] 42 USC §2210(c). [535] 42 USC §2210(d).
[536] 42 USC §2210(f). [537] 42 USC §2210(e)(2). [538] 42 USC §2210(i).
[539] Changes in Property Insurance Requirements of NRC Licensed Nuclear Power Plans, 49 Fed. Rec. 44645, 44649-50 (8 November 1984).

of either $1.06 billion or whatever amount of insurance is generally available from private sources'. If the estimated costs of stabilization and decontamination exceed $ 100 million, the proceeds must be payable first for stabilization of the reactor and next for decontamination of the reactor and the reactor station site.[540] Accordingly, this rule requires a first-party insurance coverage for the property damage of the nuclear operators. The guarantee of funds for decontamination also provides protection to the potential victims and the environment.

3.6.2.3 Compensation under the Convention on Supplementary Compensation

In 2007, the Energy Independence and Security Act was passed.[541] It provides an instrument to implement the Convention on Supplementary Compensation for Nuclear Damage (CSC).[542] The United States deposited this instrument of ratification of the CSC at the Vienna headquarters of the International Atomic Energy Agency in 2008.

Under the Energy Independence and Security Act, the US Treasury will pay into the fund provided for by the CSC. But the costs can then be reimbursed. For a non-PAA incident, costs can be reimbursed from a retrospective premium scheme the suppliers have to set up: 'each nuclear supplier shall participate in a retrospective risk pooling program . . . to cover the contingent cost resulting from a covered incident outside the United States that is not a Price Anderson incident.'[543] The contribution made by each supplier into the retrospective risk pool will be fixed according to a risk-informed assessment formula determined by the Secretary of Energy.[544] In other words, in determining the contribution of each nuclear supplier, the particular risk it contributes to will be taken into account.[545] For a PAA incident, funds already available under the PAA should be used.[546] In this case, despite the fact that the country has an obligation to make public funds available under the CSC, it does not mean that taxpayers have to pay for it. This burden is actually shifted to nuclear suppliers (non-PAA incident) and operators (PAA incident). However, it is worth noting that, as already mentioned, the CSC has not come into force yet.

540 10 CFR §50.54(w)(1),(4).

541 Energy Independence and Security Act of 2007, Pub. L. No. 110-140, 934.

542 See on this CSC also in Section 3.6.1.3. 543 *Id.* 934(e).

544 Energy Independence and Security Act of 2007, Pub. L. No. 110-140, §934(e)(1), 121; Stat. 1492, 1744 (2007).

545 Faure and Vanden Borre, 'Compensating Nuclear Damage', 273–4. 546 *Id.* 934(a)(J).

3.6.3 Critical Comparison

This process of comparison began by establishing the complicated legal history of the international nuclear compensation regimes (3.6.1) and compared this to the US compensation regime (3.6.2), which, to some extent, has different features than the international regime. Early literature has already criticized the international regime from an economic perspective. The criticism was rather straightforward. The legal channelling of liability in the international conventions has the major disadvantage that many parties, other than the nuclear operator, who could equally influence the risk of a nuclear accident are not exposed to liability.[547] Also, the financial limit on the liability of the licensee of the nuclear plant remains too low, which, in combination with the large public funds made available in the international regime, leads to a substantial subsidization of nuclear energy and thus to an insufficient cost internalization.[548]

Even though it has been indicated that it is hard to make a final, positive judgment on the US compensation regime given the fact that the real costs of nuclear damage can still be higher than the compensation available, the US regime seems in many respects to be more in line with the law and economics literature with respect to nuclear liability.

A first advantage of the US regime is that it seems far more dynamic than the international regime. The PAA started in 1957 with a relatively low financial limit – $60 million – on the liability of the operator but a large amount of government intervention – $500 million. But by 1975, the PAA already provided for a dynamic system whereby the relationship between private and public funding could change, taking into account, *inter alia*, developments in the insurance market. Given that the PAA organized insurers at the federal level and not at the state level, as most US insurance markets,[549] the US nuclear insurance market could create substantially higher amounts of compensation. Today, the coverage of the nuclear risk in Europe still takes place via nuclear insurance pools, which are organized at a national, member-state level and therefore, not surprisingly, have generated amounts of insurance coverage that are too low. The US federal government has systematically removed itself from

547 See Trebilcock and Winter, 'The Economics of Nuclear Accident Law'.

548 See e.g. Faure and Fiore, 'An Economic Analysis'.

549 43 AM JUR. 2D Insurance §24 (2008) ('Insurance company regulation is matter which is traditionally left to states').

covering the nuclear risk such that by 1982, the $560 million of required compensation was entirely financed by private funds.

It is striking that in the beginning, the international regime and the American nuclear compensation scheme were very similar, but today the differences between the two systems are quite spectacular. Today, in the United States, the total amount of compensation available is $13.6 billion, of which $375 million is financed by the individual liability of the nuclear operator and the remainder via the collective responsibility of all operators financed through retrospective premiums. Today, the NEA regime requires a total amount of available compensation of 300 million SDRs (roughly representing €310.35 million or US$493.08 million). Once the Protocols to the Paris and Brussels Supplementary Conventions enter into force, the total amount of compensation available will be €1.5 billion (US$2.383 billion), of which €700 million (US$1.112 billion) will be financed by the nuclear operators and €800 million (US$1.271 billion) by public funds. Of course, the contracting parties have the freedom to charge the cost of their obligation to the nuclear operators, thus indirectly contributing to more internalization. But even if the contracting parties were to do so, thereby imposing a liability limit of €1.2 billion (US$1.907 billion), a part of the damage would still be paid for by public funds. Unless all contracting parties opt for unlimited liability of the nuclear operator, no one will be liable for damage in excess of €1.5 billion (US $2.383 billion).

The conclusion, therefore, is rather straightforward. The economic goal of cost internalization can hardly be reached in the international regime for two main reasons. In the NEA regime, the individual liability of the nuclear operator seems at first high – €700 million[550] (US$1.112 billion) compared to $375 million in the US PAA – but is only a small fraction of the potential costs of a nuclear accident, estimating the damage to be between US$10 billion and US$100 billion.[551] Second, the second layer of compensation in the international regime is entirely provided through public funds, whereby no risk-related financing takes

[550] Note, however, that the €700 million will only be applicable if the new conventions enter into force, which is not the case yet.

[551] See US General Accounting Office, 'Report to Congressional Committees'; Dubin and Rothwell, 'Subsidy to Nuclear Power', 73–9; Heyes and Liston-Heyes, 'Subsidy to Nuclear Power', 122–4. This is also the case in the IAEA regime, where the revised liability amount is 300 million SDRs ($493.083 million). See International Atomic Energy Agency, 'Status: Convention on Supplementary Compensation'.

place whatsoever. The second and third layers of public funds in the NEA regime and the second layer under CSC are a pure subsidy to the nuclear industry and contribute nothing to cost internalization. This criticism can be partially addressed if the contracting parties charge the operators for the costs of making public money available. However, these costs should be market reflective and should take into account risk differentiation, as discussed earlier. It is far from certain that any governmental institution is well equipped enough to assume this difficult task, let alone in a more efficient manner than an insurance company or mutual insurance scheme.

On the other hand, in the United States, the second layer is not only considerably higher than in the international regime (US$13.6 billion compared to €800 million[552] [US$1.271 billion] in the NEA regime), but it is also financed through the collectivity of the nuclear operators and hence contributes to a cost internalization. The situation is, moreover, only worse if one compares the PAA with the regime under the Vienna Convention, where the amounts are even dramatically lower than in the NEA regime. An important feature of the US regime is that, indeed, a system has been developed whereby the second layer of compensation does not merely consist of public funding but is the collective responsibility of industry. The task of the government in this respect is limited to pre-financing the compensation to the victim and collecting the retrospective premiums from the operators. Moreover, in order to limit the risk exposure of the operators, the annual retrospective premiums are determined by law. However, in the end, it is the nuclear operators that contribute to finance the second layer of US$13.6 billion through these retrospective premiums.

The retrospective nature of the premium does create a potentially important insolvency risk. This can, to some extent, be mitigated through other controls on the solvency of operators and via the mutual monitoring inherent in financing the second layer through an industry-run mutual or, as ANI does, by asking guarantees of affiliated companies. A potential danger of relying on retrospective premiums is indeed the insolvency risk. The advantage is, however, that no ex ante assessment of probabilities is necessary and no capital needs to be immobilized ex

[552] This €800 million consists of the state's intervention (€500 million) as well as an intervention by all contracting parties (€300 million). Again, the conventions in which these new amounts are incorporated have not entered into force yet.

ante.[553] A similar system has been worked out for financing the US contribution to the CSC through risk dependent retrospective premiums to be paid by nuclear suppliers.

The last important difference between the US and international regimes is that the US regime has no legal channelling of liability to operators. The US implementation of the CSC furthers this trend by explicitly involving nuclear suppliers in the financing of nuclear risk, as far as nuclear accidents outside the United States are concerned.[554] The international regime, on the contrary, inefficiently excludes liability for all others than the licensee who could have contributed to the risk.

The lesson seems, therefore, to be rather clear: the US Price Anderson Act and its recent amendments seem to have understood and incorporated the lessons from economic analysis. The various parties who contribute to the nuclear risk are exposed to substantial amounts of liability which may provide incentives for prevention and cost internalization. The economic literature had already often suggested that the international regime should be changed to expose more fully those creating the nuclear risk to the costs of their activity. Within a regime where insurance is only provided through nationally operating insurance pools within member states, forcing nuclear suppliers to internalize costs remains difficult. The US model demonstrates that if a compensation regime were to be organized as a collective responsibility of the nuclear industry, thereby excluding public funding, much higher amounts of compensation can be provided to victims and a better internalization of the nuclear risk can be promoted.[555] This, however, pre-supposes cooperation between nuclear operators which currently fall under the international regime and also assumes that there are possibilities of a mutual monitoring which is essential in a system of collective responsibility, such as under the Price Anderson Act. Perhaps socio-economic or institutional impediments have thus far prevented the creation of a similar risk-sharing agreement among nuclear operators in Europe. Pelzer noted that during the negotiations

[553] See Pelzer, 'International Pooling', 48. A system of pooling with retrospective premiums can therefore also be more advantageous than the payment of insurance premiums which is basically considered 'lost money'.

[554] See Vanden Borre, 'Shifts in Governance', 27.

[555] Pelzer concluded that the Price Anderson compensation system 'perfectly complements the capacity of private insurance industry in a most cost-effective way' (Pelzer, 'International Pooling', 43).

to revise the Vienna Convention, experts discussed the international pooling of operators' funds, but such suggestions did not find support and eventually failed. Pelzer argued that using private operators' money in one country to meet the obligations of operators in other countries was not feasible because '[t]here is no universal risk community of operators'.[556]

3.7 Concluding Observations

At different places, the main conclusions resulting from the separate parts in this section have already been provided. Hence, the overall concluding observations from this chapter can be relatively short, referring to the earlier conclusions, and can be presented in the following bullet points:

- There is no international legal framework dealing with liability for offshore-related incidents; the IMO will presumably not take action in this respect, at least not create an international convention.
- There is an international legal framework dealing with vessel-based pollution incorporated into the civil liability and fund conventions; the amounts of the financial limits on the compensation due by the tanker owner have increased over time. Moreover, compensation is also provided through the International Oil Pollution Compensation Fund, which is financed by oil receivers.
- There are various regional arrangements related to offshore safety. Most of those contain general principles and do not usually contain provisions with respect to liability or compensation.[557]
- At the EU level, the recent Directive 2013/30/EU on Safety of Offshore Oil and Gas Operations was adopted in June 2013. This Directive clearly states that Member States must ensure that the licensing authorities only grant a licence when there is sufficient evidence of financial security for the liability potentially deriving from the offshore oil and gas operations.[558]
- The specific countries that were examined usually do not have specific legislation aiming at damage resulting from offshore activities. If it is the case, usually a strict liability rule applies. Many legal systems do have a requirement that financial security is proven as a precondition for granting a licence.

[556] Pelzer, 'International Pooling'.
[557] See the summary in Section 3.3.6.
[558] See Section 3.4.2.

- The US Oil Pollution Act (OPA) constitutes an interesting example not only because it applies to the Deepwater Horizon case but also because the financial limits are relatively easily breakable and OPA does not pre-empt state law.
- The regime concerning liability for nuclear accidents in international conventions may not constitute an example for a potential future liability regime for offshore-related accidents in the European Union. The financial cap on liability, legal channelling of liability to the nuclear operator (thus excluding the liability of others) and compensation via public funds lead to insufficient cost internalization.
- Many of those negative features of the international regime are absent in the US Price Anderson Act. Although there is a liability limit, the amounts are substantially higher, and the second layer will not be paid through public funds but is a collective responsibility of industry, financed via retrospective premiums charged to the nuclear industry. The US Price Anderson Act hence shows how high amounts of compensation can be generated without ex ante immobilization of capital and without public funding.[559]

[559] See Section 3.3.3. The financing of the second layer in the US Price-Anderson Act via a retrospective pooling scheme will be discussed in Section 4.5.1.

4

Pooling Mechanisms for Offshore Liability

MICHAEL FAURE AND JING LIU

Earlier discussion has shown that offshore accidents have the potential to cause substantial damage and corresponding liability claims. How liability of a catastrophic scale could be covered creates a big challenge for the offshore facilities. This chapter focuses on risk-pooling mechanisms for damage compensation following an offshore accident. Various approaches are taken to pooling. First, the theoretical basis for pooling will be sketched as well as the differences with insurance (4.1). Next, attention will be paid to the Offshore Pollution Liability Agreement (OPOL), which came into effect on 1 May 1975 and is specifically focused on offshore pollution (4.2). Then attention will be paid to two pooling mechanisms for offshore-related damage, Oil Insurance Limited. (OIL) and Oil Casualty Insurance, Ltd. (OCIL) (4.3). We have already discussed civil liability and fund conventions.[1] Compensation for vessel-based pollution is largely realized via the Protection and Indemnity (P&I) Clubs, which are equally based on risk pooling and therefore deserve a discussion (4.4). Next, attention will be paid to risk-pooling schemes in other high-risk sectors. In this respect, the focus will be especially on risk pooling in the nuclear sector (4.5). The failure of large pooling mechanisms in the nuclear sector also provides important lessons for the offshore industry. Some concluding observations finish the chapter (4.6).

In this chapter, existing pooling mechanisms for offshore liability will be discussed. However, this chapter does not yet address the extent to which these pooling mechanisms are used effectively to cover liability following a major offshore accident. The coverage of liability in practice will be dealt with in Chapter 5. Overlap will, however, as far as possible be avoided by sketching how the various pooling mechanisms work in this chapter; Chapter 5 will address the extent to which the particular

[1] See Section 3.2.1.

mechanisms are used in practise. In this respect, the results of interviews with stakeholders also will be taken into account.

From the outset, it is worth mentioning that a large number of insurance and re-insurance pools do exist. Recently, a study has been commissioned by the European Commission concerning co-insurance/re-insurance pools. The study identified fifty-one pools, many of which focus on catastrophic risk (e.g. nuclear, environmental, terrorism).[2]

4.1 Theory: Pooling versus Insurance

A 'risk-sharing agreement' or 'pool' is a system whereby operators mutually agree to share each other's losses. It resembles insurance, but there are, as will be explained in more detail, a few fundamental differences. The basic difference is that insurance involves a third party (the insurance company), whereas in a risk-sharing scheme the operators are both insured and insurer; there is hence no involvement of a third party.

Nuclear pools[3] and pools under Protection and Indemnity Clubs (P&I Clubs)[4] are examples of pools or risk-sharing agreements. It is the potential injurers themselves that finance this risk pool: they can either make an advance payment or constitute an ex ante risk-sharing agreement (which can be referred to as 'risk-sharing', 'risk-sharing institutions'[5] or 'mutuals'[6] in the literature). A significant feature of this

[2] See Ernst & Young, 'Study on Co(re)insurance Pools'.

[3] In the nuclear industry, insurance is often supplied by a monopolist in a jurisdiction, which leads to high premiums and limited coverage. Thus, an effort has been made by nuclear operators to establish risk pools to cover their third-party liability or their own property damage (onsite damage). The retrospective premium system under the Price-Anderson Act and the pool established by NEIL are some examples. The Price-Anderson Act was enacted by Congress in 1957 to amend the Atomic Energy Act of 1954 (Pub. L. 85-256). A retrospective premium system was introduced in its 1975 amendment (Pub. L. 94-197). NEIL is a nuclear insurer called Nuclear Electric Insurance Limited, which insures nuclear plants and their generating units for some first-party damage.

[4] A P&I Club is a non-profit-making mutual insurance association, which is established by ship owners and charterers to cover their third-party liabilities related to the use and operation of ships. Now thirteen separate and independent principal clubs co-operate together to comprise the International Group of P&I Clubs (hereinafter the Group), accounting for approximately 90 per cent of the world's ocean-going tonnage. See the website of the Group: www.igpandi.org/home.

[5] Faure and Skogh discussed a risk-sharing agreement as an alternative compensation mechanism for environmental damage and nuclear damage. See Faure, 'Alternative Compensation Mechanisms'; Faure and Skogh, 'Compensation for Damages'; Skogh, 'Risk-Sharing Institutions'; Skogh, 'A European Nuclear Accident Pool'.

[6] See Bocken, 'Alternative Financial Guarantees'.

mechanism is that the members of this pool are both insurers and insured. Commercial insurance can play an important social function to remedy the risk aversion of the injurer without increasing the size of the expected losses.[7] Both the insurers and insured can influence the likelihood of environmental accidents and of their losses. But interests of insurers and insured are not fully congruent.[8] Rather than minimizing the risk, the insured tend to externalize their costs and get lower premiums; the insurer, a profit-maximizing firm itself, may limit and deny coverage and may, depending upon the market situation, overcharge premiums. In a risk-sharing agreement, potential injurers are both the providers and recipients of insurance. Thus, the conflicts of interest between insurers and insured are avoided in a risk-sharing agreement.

In a risk-sharing agreement, mutuality is created, whereby the contribution paid by one member depends on the claims made by all other members.[9] It is in the interest of everyone that other members' claims are as low as possible, and thus, a mutual interest in risk minimization is created.[10] To reduce risks, the members of such a group have incentives to differentiate risks to align a member's contribution to the risk each member poses and to monitor each other. Mutuality is established usually when the members are subject to similar safety rules and exposed to a comparable level of liability, and this usually occurs in highly technical industries. The members are faced with the same types of risk and have often more expertise and precise knowledge than a third-party insurer.[11] Therefore, they can evaluate the risk each member creates and can better monitor each other's behaviour.

Unlike in the case of commercial insurance, where ex ante information about the probability of a certain risk and its magnitude should be available to allow the calculation of an ex ante charged premium, in a risk-sharing agreement, each member's contribution can be agreed upon beforehand and only actually paid ex post. This characteristic makes it possible for a risk-sharing agreement to deal with uncertain risk, for which the statistical data about the occurrence are rare or the probability and size are less predictable. As long as a risk differentiation can be made among the members, a risk-sharing agreement can be feasible because an ex ante charging of premiums is no longer necessary.

[7] Van den Bergh and Faure, 'Competition on the European Market', 279.
[8] Boyer and Porrini, 'The Efficient Liability Sharing Factor'.
[9] Bennet, 'Mutual Risk', 15. [10] *Ibid.*
[11] Faure and Fiore, 'The Coverage of the Nuclear Risk', 302.

Only the relative contribution of each member to the risk has to be known.

Another difference between a risk-sharing agreement and insurance concerns the costs. In an insurance policy, the risk is shifted to the insurer at the price of a premium. The premium is not recoverable by the insured no matter whether the insured risk materialized or not. In a risk-sharing agreement, a member only contributes if an accident happens; the duty to contribute can either be postponed or the contribution can be carried over for the following year if there is no accident. A member also can recover his contribution by stopping to create the risk and leaving the pool.[12]

Summarizing, a risk-sharing agreement has some important theoretical advantages (and differences) compared to insurance:

- It creates strong incentives for mutual monitoring because the members are dependent on each other; i.e. a bad risk can create the likelihood that the pool will have to intervene.
- For highly technical and complicated (often new) risks, operators themselves may have better information (compared to insurers) on optimal preventive technologies which they can reflect in a differentiation of the contributions to the pool (or excluding membership for bad risks).
- A risk-sharing agreement does not require actuarial information ex ante on the probability of an accident and the scope of the damage. Therefore, no ex ante premium has to be fixed. Information is needed only on the relative contribution of each member to the risk, but this does not necessarily have to be translated into a premium. Ex ante costs to administer a risk pool hence can be lower, especially in cases where actuarial information (e.g. because the risk is new and statistical information is lacking) may not be available.
- Because ex ante premiums do not have to be paid, risk sharing creates fewer liquidity problems. It can be based on an agreement of the members to share in case the risk emerges.
- In complete contrast to insurance, if the risk for which the risk-sharing agreement is concluded does not emerge, the members of the risk-pooling scheme simply do not have to contribute.
- This also points at the relative flexibility of a risk-pooling mechanism: when during a particular period many accidents happen, the risk pool

[12] *Ibid.*

can ex post ask additional contributions from the member on an ad hoc basis.

- However, such a risk-pooling mechanism may have all these advantages if the number of members in the pool is relatively restricted; the comparative benefit (compared to insurance) mostly applies to highly technical (new) risks. When, however, the members of the pool might be very numerous (e.g. all car drivers in a particular area), the administrative costs of running the pool would become huge, and the comparative benefits vis-à-vis insurance would disappear.

We will now turn to a mechanism that does play an important role concerning offshore pollution liability, i.e. OPOL, although, as we will explain, this is not a risk-sharing agreement in the sense that we have described it here.

4.2 OPOL

4.2.1 Origins of OPOL

The Offshore Pollution Liability Association Limited (the Association) is an oil industry body which has been set up as a company limited by guarantee which administers a voluntary strict liability compensation scheme which is referred to as the Offshore Pollution Liability Agreement (OPOL). [13] It originated in the United Kingdom and came into effect on 1 May 1975 as an agreement between all UK offshore operators. The original founders of OPOL were

Amoco (UK) Exploration Company,
Burmach Oil (North Sea), Ltd.,
BP Petroleum Development, Ltd.,
Total Oil Marine, Ltd.,
Conoco North Sea Incorporated,
Esso Exploration and Production UK, Inc.,
Gulf Oil Production Company,
Hamilton Brothers Oil Company (GB), Ltd.,
Mobil North Sea, Ltd.,

[13] The discussion about OPOL here provides a different perspective than that in Section 3.5.2.2 in the sense that the previous section discusses OPOL as a mechanism for providing civil liability and compensation for damage caused by offshore incidents in the United Kingdom, whereas in this section, the focus is on the role of OPOL as an existing pooling mechanism for offshore liability.

Shell UK, Ltd.,
Phillips Petroleum Company,
Signal Oil and Gas Company, Ltd.,
Siebens Oil and Gas (UK), Ltd.,
Texaco North Sea UK Company,
North Sea Sun Oil Company, Ltd. and
Cluff Oil, Ltd.

In the United Kingdom, membership in OPOL is a condition for the granting of a licence by the government. The reason for creating OPOL was that at the time operators anticipated that a regional convention would regulate offshore activities, the Convention of Civil Liability for Oil Pollution Damage resulting from Exploration for and Exploitation of Seabed Mineral Resources (CLEE),[14] and there might be a convention or a regulatory duty. In the end, the convention never came, but OPOL lasted.[15] OPOL currently has 122 members.[16]

Meanwhile, OPOL has been extended to cover offshore facilities within the jurisdiction of any state that is specified in the agreement. The states currently covered include the coastal states of the European Community, Norway, the Isle of Man and the Faroe Islands (a total of ten). It is de facto, however, only mandatory in the United Kingdom, Greenland and the Faroe Islands.[17] OPOL membership outside of the United Kingdom has declined. The reason is that for operators in countries where there is no regulatory duty to be a member of OPOL (such as the UK), there may not be an incentive to be a member. OPOL is therefore rarely used, for example, in France, Germany or Denmark.[18]

4.2.2 Main Features

OPOL provides a mechanism for the settlement of claims expeditiously and without disputes.[19] It establishes a duty to compensate pollution damage based on a strict liability rule.

[14] 16 ILM 1451 (1977); available at: www.ecolex.org/ecolex/ledge/view/recorddetails?id=tre-000434&index=treaties.

[15] Interview with representatives of OPOL, 27 March 2013, in Brussels. See also Rochette *et al., Seeing Beyond the Horizon.*

[16] According to the latest annual report of OPOL, 'The Offshore Pollution'. [17] *Ibid.*

[18] *Ibid.*

[19] Preamble of Offshore Pollution Liability Agreement (OPOL Agreement), effective as of 1 April 2015.

A second feature is that OPOL guarantees that funds are available to meet claims because members in OPOL need to provide proof of financial responsibility. The overall limit on compensation is US$250 million per incident.[20] In the case of insolvency of one of the members or if a member for other reasons fails to meet its obligations, all remaining members of OPOL agree to contribute, in proportion to the number of relevant offshore facilities operated by it on the date of the incident in question, towards payment of claims due from a party that fails to meet its obligations.[21]

In principle, OPOL thus is based on a strict liability of the operator, a member of OPOL, for pollution damage and on maintenance of financial responsibility.[22] OPOL itself is not a compensation fund but only guarantees payment in case the member of OPOL fails to meet its obligations.

4.2.3 Rules and Membership

OPOL is structured in such a way that operators or intending operators of offshore facilities can become members of OPOL. In the most recent version of the OPOL, of 1 April 2015, an operator is defined as

> A person which by agreement with other persons has been authorised to, and does, manage, conduct, and control the operation of an offshore facility, subject to the terms and conditions of said agreement, or which manages, conducts and controls the operation of an Offshore Facility in which only it has an interest.[23]

The OPOL Agreement is a contract between the parties to OPOL. The rights and obligations of the parties in relation to the OPOL guarantee are specifically laid down in the founding documents applicable to the Offshore Pollution Liability Association, Ltd., namely, the Memorandum of Association and the Articles of Association and in the OPOL Agreement. Hence, an operator is both a member of OPOL, the Company, and a party to the OPOL Agreement.[24] The key to OPOL is that it obliges its members on the basis of agreement to pay compensation for remedial measures and pollution costs on a strict liability basis.[25]

[20] Clause IV(A) of OPOL Agreement, effective as of 1 April 2015. [21] *Ibid.*

[22] Clause II(C) of OPOL Agreement, effective as of 1 April 2015.

[23] Clause I(10) of OPOL Agreement, effective as of 1 April 2015.

[24] Interview with representatives of OPOL, 27 March 2013, in Brussels. [25] *Ibid.*

Clause II(B) of OPOL makes clear that any person who is or intends to be the operator of an offshore facility may become a party upon acceptance by the association of an application from the said person in a form prescribed by the association. De facto this means that the operator who wishes to join OPOL has to file a form A from the OPOL website and provide evidence of financial responsibility.[26] By applying for membership, the member agrees to abide by its memorandum and articles of association and its rules.

Indeed, in addition to the OPOL Agreement, an Offshore Pollution Liability Association has also been created which has as its main goal to administer OPOL. Operators hence become members both of OPOL and of the OPOL Association. Clause III(2) makes clear that the association administers any sums which the association receives from the parties in the event that the party fails to satisfy its obligations to claimants.

4.2.4 *Financial Responsibility*

A key element of OPOL is that every party to OPOL agrees under clause II(C)(1) to establish and maintain its financial responsibility to fulfil its obligations under clause IV of OPOL with respect to offshore facilities of which it is the operator in accordance with criteria prescribed by the rules.

According to clause IV, if discharge of oil occurs from one or more offshore facilities, the party who was the operator of that offshore facility shall reimburse the costs of remedial measures and pay compensation for pollution damage up to an overall maximum of US$250 million. Of this US$250 million, US$125 million per incident will be reimbursed for the costs of remedial measures incurred by the public authority; another sum of US$125 million per incident will be available to pay compensation for pollution damage to claimants.[27]

To cover these obligations, the operator has to comply with the rules for establishment of financial responsibility.[28] De facto, the operator may use different ways of proving financial responsibility:

1. Insurance – by having an insurance company, agent or broker, which is acceptable to the association and completing a form referred to as 'FR1' or 'FR2' which provides evidence of insurance;

[26] OPOL Continuous Certification Programme; available at: www.opol.org.uk/downloads/opol-continuous-certification-programme-flowchart.pdf.

[27] Clause IV(A) of OPOL Agreement, effective as of 1 April 2015.

[28] Form B of OPOL.

2. A guarantee on form FR3 issued by a guarantor acceptable to the association;
3. A qualification as a self-insurer, by providing a financial statement meeting certain specified criteria. The consolidated financial statement of a parent company may be used, in which even an acknowledgement of commitment is required on form FR4.[29]

Thus, the idea of OPOL is that it requires from each operator which is a party to OPOL that it provide satisfactory evidence of its ability to meet any liability under OPOL. Whether this is satisfactory will be verified by the association. OPOL itself makes clear in its information for prospective members that evidence of financial responsibility is necessary as a protection to the party and claimants to ensure that the party has made adequate provision to meet the obligations assumed under OPOL.[30] In addition, it protects the other parties because they have also agreed jointly to contribute to the payment of claims due from an operator that fails to meet its obligations. This makes clear that operators have a strong incentive to control and monitor the financial guarantees provided by other operators because they could otherwise be liable if an operator, a member of OPOL, failed to meet its obligations.

4.2.5 Claims Handling

As mentioned earlier, OPOL basically covers pollution damage. This is defined in the OPOL Agreement as 'direct loss or damage (other than loss of or damage to any offshore facility involved) by contamination which results from a discharge of oil'.[31] In addition, coverage is also provided for remedial measures executed by public authorities. Remedial measures are defined in OPOL[32] as

> [r]easonable measures taken by any party from any of whose offshore facilities a discharge of oil occurs, and of which such party is the operator, and by any public authority to prevent, mitigate or eliminate pollution damage following such discharge of oil or to remove or neutralize the oil involved in such discharge, excluding however, well control measures and measures taken to protect, repair or replace any such offshore facility.

29 'Information for Prospective Members', 2; available at: www.opol.org.uk/downloads/opol-memberinfo-jan14.pdf. All these forms are available at the website of OPOL: www.opol.org.uk/forms.htm.

30 'Information for Prospective Members', 2; available at: www.opol.org.uk/downloads/opol-memberinfo-jan14.pdf.

31 Clause I(13) of OPOL Agreement, effective as of 1 April 2015.

32 See Clause I(15).

Claimants, either public authorities or individual victims, should file a claim with the operator. Clause V of OPOL makes clear that the claimant shall 'furnish the party against whom a claim has been filed with such information, documents and testimony as are reasonably required by the said party in connection with the investigation of the claim and permit the making of such visual or other inspections or surveys as the party may request'.

Clause VI of OPOL makes clear that the claim has to be filed within one year of the date of the incident which resulted in the pollution damage or the taking of the remedial measures in question. According to clause VII, any payment to a claimant by or on behalf of a party shall be in full settlement of all said claimant's claims against that party and against any other persons associated with that party in connection with the offshore facility or facilities involved with respect to the pollution damage or the costs of remedial measures.

The claim is hence made directly against the operator. Clause VIII(C) clearly states that the association shall have no liability to any claimant other than its obligation to administer claims in the event that a party fails to satisfy its obligations to claimants.

In that case, each party is obliged to make a payment to the association of the party's share. The share to be borne by each party is in proportion to the number of units operated by that party to the total number of units of all parties, except that of the defaulting party, at the time of the incident. The brochure of OPOL for prospective member states: 'Thus, the amount involved is likely to be only a small proportion of the total sum involved in any claim.'[33]

In addition, parties have an obligation to pay administration costs for the functioning of the association via an annual subscription fee.[34]

4.2.6 Enforcement

Once a claim has been filed with an operator, only the operator remains responsible for the acceptance and payment of such claims. Payment to a claimant is required to be in full and final settlement of the claim against the operator. However, because OPOL is merely a private agreement between operators, OPOL cannot, of course, exclude the possibility that

[33] 'Information for Prospective Members', 3; available at: www.opol.org.uk/downloads/opol-memberinfo-jan14.pdf.

[34] *Ibid.*

a claimant would not use OPOL but rather file a lawsuit in ordinary courts.

However, in the event of a dispute between a claimant and an operator as to the application or interpretation of the agreement, either the claimant or the operator can submit the matter to arbitration in London.[35] According to the OPOL Agreement, arbitration shall be the exclusive means of disposing of disputes. Arbitrators are bound by the provisions of OPOL and do not have the power or authority to vary or increase the provisions of the contract or any rights or obligations thereunder. This therefore means that arbitrators could not e.g. increase the amounts provided for in OPOL.

In principle, claimants can therefore direct themselves to the operator who will, in principle, given the strict liability formula in OPOL, compensate. In case of dispute, however, arbitration in London is the only remedy for the claimant. English law will apply according to clause XII. Clause XIII, moreover, holds that no person has any right under the Contracts Act of 1999 to enforce any term of this contract, but this clause shall not affect any third-party rights which might exist apart from that Act.

This clause seems to imply, for example, that victims suffering pollution damage could not execute a direct right of action other than as provided in OPOL before ordinary courts and still call on OPOL. This clause probably has a goal to exclude victims as third-party beneficiaries.

4.2.7 Practice

OPOL publishes yearly reports on its website. The most recent report is for the year that ended 31 December 2013.[36] These annual reports, which are relatively brief, merely provide financial statements which largely relate to the administrative costs of operating the association.

Importantly, the reports systematically mention 'no incident requiring the operation of clause III(2) of the agreement occurred during the year'.[37] This means that until now, no default of the operator has ever taken place as a result of which the guarantee clause in the OPOL agreement would come into effect.

[35] Clause IX of OPOL.

[36] Available at: www.opol.org.uk/downloads/report%20and%20financial%20statements%20-%20december%202013.pdf.

[37] For example, OPOL, 'The Offshore Pollution Liability Association Limited Report', 2.

De facto, in OPOL, no member currently relies on self-insurance.[38] Of approximately eighty companies in the United Kingdom that have provided OPOL with details of financial responsibility currently in place, four companies use a company guarantee and hence rely on the credit rating of the guarantor. All but four members of OPOL show financial responsibility for operations in the United Kingdom via insurance.[39]

4.2.8 Evaluation

OPOL undoubtedly has, just on the basis of a preliminary analysis based on a review of the available documents, much strength:

1. It provides, as OPOL mentions itself, a strict liability obligation of operators to compensate pollution damage. Clause IV(A) makes clear that in the case of discharge of oil from an offshore facility, the operator 'shall reimburse' the costs. Claimants hence do not need to provide proof of negligence or fault.
2. Because claims must be lodged against the operator who has caused the pollution, making the operator solely responsible for meeting those claims, a kind of channelling of liability to the operator has taken place. For the claimant, this may have the major advantage that no discussion takes place on attribution of liability.[40]
3. The members of OPOL, moreover, agree to pay up to a total amount of US$250 million, which seems substantial.
4. The fact that a strict liability for pollution damage is incorporated into OPOL does not exclude the claimants' right to sue e.g. for damage of other types (such as personal injury or other types of losses not covered under OPOL). OPOL, of course, cannot limit the victims' rights on compensation as awarded by law and does not do so.
5. Another strength of OPOL is undoubtedly also the financial guarantee, albeit that questions can, of course, be asked with respect to the strength of the various guarantees required by OPOL.
6. Finally, one of the strengths of OPOL is the element of mutual risk sharing as far as the insolvency of one of the members would be. This is important because it guarantees that the obligations mentioned

[38] Interview with representatives of OPOL, 27 March 2013, in Brussels. [39] *Ibid.*

[40] Because no claim has been made against OPOL yet, there exists no experience to clarify the meaning of 'direct loss or damage'. See Bio by Deloitte, 'Civil Liability', 154.

earlier under OPOL will be met via a proportional obligation to intervene in the case of failure of one of the members.

Notwithstanding these strengths and advantages, initial examination also reveals that a few potential weaknesses could be mentioned:

1. The role of OPOL is in a sense relatively limited. OPOL is not, as it also stresses itself, a compensation fund. It merely is an agreement between operators to compensate losses and to provide financial security. Its role is therefore relatively limited.
2. The coverage is limited to pollution damage defined as direct loss or damage by contamination resulting from a discharge of oil. This would mean that e.g. property damage and clean-up operations would be compensated. It is, however, less clear whether other types of damage such as personal injury or economic losses would be covered as well.
3. The guarantee to provide coverage is merely based on a private agreement between operators. Only in the United Kingdom is the obligation to be a member of OPOL also a condition for obtaining a licence. However, for member states where this is not the case, the voluntary character of OPOL would result in there being no guarantee that financial coverage is available.
4. Although the amount of compensation of US$250 million seems quite substantial, recent cases such as the Deepwater Horizon seem to indicate that in some cases even those limits could be broken.
5. The structure of OPOL is such that the usual advantages of a pooling arrangement,[41] such as risk distribution and mutual monitoring, will not be attained. OPOL is indeed not a risk-sharing scheme. Risk sharing and risk distribution take place via the financial security that the operator has opted for, such as insurance or other types of security. The guarantee provided by the other operators via clause III(2) is merely to intervene in the event that the party fails to satisfy its obligations. This means that OPOL does not provide incentives for mutual monitoring of preventive measures but merely for monitoring of the solvency of the other members, which is obviously relatively limited.
6. Questions also can be asked about the importance of OPOL in practice. On the one hand, one could argue that its importance is relatively

[41] These advantages were discussed in Section 4.1.

limited because the guarantee has never been applied in practice, and in this sense, claimants never have had to call on OPOL. However, one could also hold that OPOL has worked well in the sense that apparently the financial guarantee to which OPOL members agree has been able to satisfy all claims without a need to call on the guarantee provided via clause III(2).

7. The application of OPOL is limited to damage resulting from offshore facilities within the North Sea. Offshore facilities located in the Baltic and Mediterranean Seas are hence explicitly excluded.[42] Hence, OPOL can only play a role for damage occurring in the North Sea, whereas there are substantial offshore installations in other European waters as well.
8. OPOL relies on the information provided by the members about the level of financial responsibility which the operator has agreed with the Department of Energy and Climate Change (DECC). It is, according to OPOL, up to DECC to verify the level of financial responsibility required of an operator in relation to each well: 'OPOL does not track financial viability, but it is part of the job of DECC to check this.'[43] If, for example, the financial responsibility (in case of insolvency) of an operator were to be diminished, the regulator should recognize this, so OPOL holds.[44] OPOL waits for information in that respect, provided by operators. This will then be forwarded to the regulator.[45] So this means that even as far as the (limited) pooling aspect is concerned (solvency guarantee), OPOL itself does not exercise any active control over the solvency but relies on the regulator and operators.
9. A final potential limit of OPOL is that it does not at all intervene in case of an incident as long as there is no insolvency. Because there has never been a default of an operator, OPOL has no experience and does not intervene in claims handling.[46]

4.3 OIL and OCIL

In addition to OPOL, which mainly covers activity in the North Sea, other types of pooling systems, such as the mutual insurance arrangement, are also functioning in other areas. Two of the most well-known

[42] Interview with representatives of OPOL, 27 March 2013, in Brussels. [43] *Ibid.*
[44] *Ibid.* [45] *Ibid.* [46] *Ibid.*

mutual insurers are Oil Insurance Limited (OIL) and Oil Casualty Insurance, Ltd. (OCIL).[47] As discussed later, although these insurers provide coverage on a global scale, their main market is still the United States.

4.3.1 OIL

OIL was formed in 1972 by sixteen energy companies in response to two large industry accidents that occurred in the 1960s. One was the well blowout off the coast of Santa Barbara in Southern California in 1969, and the other was a refinery explosion in 1967 in Lake Charles, Louisiana.[48] At the beginning of OIL's formation, the insurance industry showed little confidence in the organization. However, after more than forty years of operation, OIL's membership has grown from the original sixteen members in 1972 to fifty-six members in 2014.[49] Its members are spread widely over all continents, including Europe, America and Australia. Its globally insured assets have grown from US$48 billion in 1972 to over US$2 trillion in 2015.[50] OIL is owned and operated for its shareholders, all of whom are engaged in energy operations.[51]

OIL claims that it offers shareholders several principal advantages: 'significant property capacity provided at a lower cost structure than the commercial market; broad coverage, including terrorism and pollution; greater contract certainty with standardized coverage provided through the shareholders' agreement; [and] a perpetual coverage term'.[52]

OIL is rated A– (stable) by Standard & Poor's and A2 by Moody's.[53] Most of its members (twenty-seven) come from the United States,[54] seventeen come from Europe[55] and nine from Canada.[56] OIL provides coverage for the following:

47 www.businessinsurance.com/article/20100912/news/100919977.
48 www.oil.bm/about-oil/at-a-glance. 49 www.oil.bm/memberships/current-members.
50 www.oil.bm/about-oil/at-a-glance.
51 www.oil.bm/membership/reasons-to-become-a-member.
52 www.netlinkbermuda.com/articles/oil.html. 53 www.oil.bm/about-oil/at-a-glance.
54 Members include, *inter alia*, Chevron Corporation, ConocoPhillips Company, Marathon Oil Corporation and Murphy Oil Corporation, as well as Noble Energy, Inc.
55 Members include, *inter alia*, BASF SE, Electricité de France, Royal Vopak NV and Statoil ASA.
56 See www.oil.bm/membership/current-members.

- *Physical damage.* The OIL policy provides coverage 'for physical damage to, or loss of, property owned by the assured and non-owned property insured under the policy (subject to the terms, conditions and exclusions of the policy). The basis of recovery is replacement cost value (RCV). But if the property is not repaired or replaced, the recovery is on an actual cash value (ACV) basis.' The capacity for this kind of coverage is US$400 million, covering risks including, among others, cargo, construction, terrorism and windstorm.[57]
- *Control of well.* The control of well coverage is provided 'under Insuring Agreement #2b of the OIL policy'. It includes coverage for 'sue and labor' expenses, removal of debris and the costs and expenses associated with restoration and re-drilling (subject to the terms, conditions and exclusions of the policy). For each of these coverages, the limit is US$400 million with no separate premium charge.[58]
- *Third-party pollution liability.* The coverage is provided for 'legal liability (including punitive damages) or contractual liability of members for personal injury, loss of, or damage to, property arising from a seepage, pollution or contamination incident (subject to the terms, conditions and exclusions of the policy). Coverage is provided on a non-gradual basis subject to a 40 day discovery/120 day reporting requirement.' The block capacity is US$400 million.[59]

According to the premium indication request form, operators desiring coverage under OIL need to fill in a detailed form that provides, *inter alia*, information on loss history, annual report, loss control measures, assets, etc. They can also indicate a deductible sum, but the deductible sum is, as the form indicates, 'not less than $10 million' in 2014.[60]

A lengthy shareholders agreement (145 pages) provides the details of the structure of OIL, explaining, *inter alia*, illegibility requirements, participation rights, etc.[61] This shareholders' agreement also provides the condition for the insurance policy. The 2014 annual report of OIL, *inter alia*, states that US$495,932,000 premiums were earned and that a net income was made of US$731,011,000.[62] Total assets in 2014 were US$7,336,865,000, and shareholders' equity was US$4,606,088,000.[63]

57 www.oil.bm/products/all-risk-physical-damage.
58 www.oil.bm/products/control-of-well.
59 www.oil.bm/products/3rd-party-pollution-liability.
60 Marsh, 'A Guide to Oil Insurance', 9.
61 Oil Insurance Limited, 'Shareholders' Agreement'.
62 Oil Insurance Limited, 'Annual Report 2014', 18.
63 *Ibid.*

4.3.2 OCIL

OCIL was founded in 1986 as an 'industry-owned vehicle specializing in liability insurance' for energy companies.[64] By 2014, OCIL provided insurance policies to 146 companies, among which 110 are based in the United States, 17 in Canada, 14 in Europe, 3 in Australia, and 1 in South America and Africa.[65] OCIL provides excess liability policies (up limit US$100 million), excess direct and facultative property policies (up limit US$50 million) and re-insurance.[66]

The excess general liability insurance provided by OCIL is exclusively dedicated to servicing energy companies. In 1986, 'the commercial markets had ceased to provide adequate insurance coverage for liability risks.'[67] Recognizing the need to create a special industry-owned vehicle specializing in liability insurance, the member of OIL created OCIL to fill the gap. Therefore, it is not surprising that some energy companies participate in both pooling mechanisms (e.g. Chevron Corporation, ConocoPhilips Company, Marathon Oil Corporation and Murphy Oil Corporation).

Different from OIL, for which the energy companies are both shareholder and insured, for OCIL, it is not necessary to become a shareholder. Energy companies could also simply become a policyholder without being a shareholder of OCIL.[68] OCIL is rated A– excellent by A. M. Best and BBB+ stable by Standard & Poor's.[69]

From the 2014 Annual Report of OCIL, it appears that OCIL insured 146 companies and that total assets were US$1,322,813,000. Premiums written in 2014 were US$174,234,000, and a net income of US$3,697,000 was realized.[70]

OIL and OCIL hence provide an alternative model to OPOL in pooling the risks for offshore facilities. Under OPOL, companies pay only annual subscriptions to cover the administrative costs, and the obligation to pay for damage and remedial measures is only due ex post and when the liable member is insolvent. OIL and OCIL work in a more similar way to traditional insurance. The policyholders need to pay premiums ex ante. But OIL does not operate on a profit basis but 'on a fee/cost basis for its service'.[71] OCIL is not strictly a mutual in the sense that it provides not only coverage to its own shareholders.[72]

[64] www.ocil.bm/. [65] OCIL, 'Annual Report 2014', 12.
[66] www.ocil.bm/faqs/fact-sheet. [67] www.netlinkbermuda.com/articles/oil.html.
[68] www.ocil.bm/faqs/shareholder-vs-policyholder. [69] www.ocil.bm/about-ocil/overview.
[70] OCIL, 'Annual Report 2014', 18. [71] Bio by Deloitte, 'Civil Liability', 136.
[72] *Ibid.*, 138.

4.4 Protection and Indemnity Clubs[73]

4.4.1 *Origins*

In the area of maritime transportation, the technical uncertainties with regard to the occurrences of oil spills combined with the legal uncertainties about establishing liability make it difficult to cover marine oil pollution via a traditional insurance policy. P&I Clubs appeared as a response to commercial insurers' reluctance to underwrite marine risks.[74] The first English Protection Associations were founded in the mid-nineteenth century under the background of the lack of capacity and unwillingness to underwrite marine risks by private insurers. This lack of capacity and interest is the consequence of the increasing risk of English commercial vessels being seized or destroyed by the enemy navies of France, Spain and the United States during the preceding two centuries.[75] Following the development of these associations in Britain, P&I Clubs began to grow in the late nineteenth and early twentieth century in Scandinavia and America.[76] The International Groups of P&I Clubs (the P&I Group) has been formed on the basis of individual P&I Clubs. Its thirteen member clubs provide liability cover for approximately 90 per cent of the world's ocean-going tonnage.

4.4.2 *Coverage*

The insurance policies provided by traditional insurers and P&I Clubs are different. Under a traditional insurance policy, the reasonably clear existence of liability is usually sufficient for the purpose of recovery. The P&I policies are indemnity insurance, under which the indemnity obligation arises only after the insured has paid for damage.[77] The P&I policies cover the liabilities specifically enumerated in the agreement – the Club's rulebook. A typical rulebook usually provides coverage to

73 See further on the P&I Clubs Liu, 133–42.

74 Ronnerberg, 'An Introduction to Protection'. 75 *Ibid.*, 3.

76 Libby, 'Some Aspects of Protection', 171.

77 Under the P&I policies, the insured must have suffered actual monetary losses before they can seek reimbursement from the insurers. This means that a member is only entitled to seek compensation for the amount he has in fact lost due to the occurrence of a covered incident. This is called the 'pay to be paid' rule, which is usually incorporated in the Club's rule book. Noting that though the P&I Clubs follow the 'pay to be paid' principle, it has been diluted by the introduction of international conventions which allow for direct action and the common practice for P&I Clubs to intervene in case a member is insolvent. Hazelwood and Semark, *P&I Clubs*, 123–4.

collision and non-contact damage, damage to fixed and floating objects, cargo claims, property on board, loss of life, personal injury and illness, passengers, crew liabilities, supernumeraries and others on board, fines, inquires and criminal proceedings, quarantine expenses, stowaways, diversion expenses, life salvage, unrecoverable general average, ship's proportion of general average, liabilities relating to wreck of the entered vessel, pollution, towage contracts, expenses incurred pursuant to directions of the club and any other costs which the club's board of directors deem proper to cover.[78] The scope of 'oil pollution' is determined by international conventions and national legislation applicable in each case. Usually it covers the 'legal damages and compensation payable by their members', prevention and cleanup expenses and government orders and directions.[79]

A P&I Club provides more services than a pure insurer and operates as a mixture of an insurance company, a law firm and a loss adjuster. Besides offering insurance coverage, a P&I Club can also provide a worldwide network of correspondents and representatives to give on-the-spot assistance to a ship owner when required, to give letters of undertaking to offer a security when a member's vessels are arrested and to assist in claims handling and settlement.[80] These benefits to some extent also apply when P&I Clubs engage in providing offshore cover:

- They provide service to the offshore operator,
- They advise on preventive measures, and
- The handle claims.

However, because, in the offshore case, the Club acts more like a traditional insurer, profits in that case normally will not return to the 'insured'.[81]

4.4.3 *Entry and Premiums*

The P&I Clubs impose some safety-rated conditions on vessels applying for membership. Usually the Clubs require entered vessels to remain 'classed with a classification society approved by the managers'.[82] The Clubs also require members to notify the

[78] *Ibid.*, 124. [79] *Ibid.*, 179–90. [80] Ronnerberg, 'An Introduction to Protection', 25–9.
[81] Interview with representatives of the International Group of P&I Clubs, 1 May 2013, in London.
[82] Hazelwood and Semark, *P&I Clubs*, 37.

classification society of any incident with the potential to give rise to damage and to comply with the rules, recommendations and requirements imposed by their classification societies. In addition to the requirement concerning classification societies, many Clubs also 'demand more stringent condition survey reports on new entries'.[83] Different from classification societies, which focus on the seaworthiness of vessels, the Club surveys are more interested in the 'condition of the vessel as a potential source of liabilities; for example, its ability to carry cargo or passengers or crew carefully'.[84]

As a typical risk-sharing arrangement, in a P&I Club, the ship owners make contributions to a risk pool according to an ex ante agreement. Different from traditional insurance, under which premiums need to be paid ex ante, the contribution/premium for P&I Clubs is 'an agreement that each member should bear his aliquot share of the losses of the year covered by the policy'.[85] The premium is levied through calls. Before the beginning of each year, Club managers estimate the claims and expenses for the year and decide a proportion of the total to be called up as an 'advance call'. The remaining proportion will be collected via supplementary calls.[86] The Clubs have made it clear that estimation of supplementary calls does not prevent them from levying a higher or lower amount.[87] The members are 'obliged to pay the supplementary call to the extent the advance call is insufficient to cover the claims'.[88] In a catastrophic year, if the total claims exceed the amount that is 'recoverable in respect of that claim under the Group Excess Reinsurance Policies', payment can be paid from the Catastrophe Reserve Fund. In the case that this fund is insufficient, an overspill call will be levied.[89] In this way, the P&I Clubs combine an ex ante payment with a retrospective payment. The advance call is announced at the beginning of each policy year, but the payment can be made by instalments during the year. Supplemental and overspill calls are collected ex post.

The essence of the P&I Club is that the ship owners are members of the club. Hence, they, in principle, are not considered clients (as within an insurance company) but rather as members, thus having a very close relationship to the club. Different from traditional insurers, P&I Clubs

[83] *Ibid.* [84] *Ibid.*
[85] *Williams* v. *The British Marine Mutual Insurance Company* (1886), 6 Asp MLC 134 at, 136.
[86] Hazelwood and Semark, *P&I Clubs*, 104. [87] *Ibid.*, 104.
[88] Zhu, 'International Convention', 177. [89] *Ibid.*, 105.

underwrite on a cost but not profit basis, which means that it follows the equation that 'calls plus investment income should equal claims plus expenses plus reinsurance premiums.'[90] The amount of each advance is based on the ship owners' claims history, the size of the fleet, the Club's anticipated needs and the strength or profitability of the insurance market.[91]

4.4.4 *Available Amount*

Potential catastrophes may cause damage in excess of the capacity of individual Clubs. To cover such risks, the P&I Clubs pooled their claims together through a pooling agreement and formed the International Group of P&I Clubs (hereinafter the Group). The 'pooling agreement' defines the pooled risks and the allocation of losses between participating Clubs.[92] Under the current pooling agreement, each individual club retains the first US$9 million as its retention. The group pool reinsures the amount from US$9 million to US$30 million. The amount in excess of US$30 million and up to US$80 million is reinsured by a captive of the Group-Hydra Insurance Company. In addition to that, the Group collectively re-insures the excess losses up to US$3.1 billion through three general excess layers involving the international insurance market.[93] Among this amount, the compensation for oil pollution is limited to $1 billion.[94]

4.5 Risk Pooling in the Nuclear Sector

The first focus in this section will be on the risk-pooling arrangements in the Price-Anderson Act (4.5.1), where a second layer of liability is provided through retroactive premiums; then the mutual pool for property damage in the United States is discussed (4.5.2), and then attention is paid to the mutual nuclear pools in Europe (4.5.3). Finally, the nuclear liability and pooling regime in Germany is discussed (4.5.4) as well as the possibility of a European pool for nuclear liability (4.5.5).

[90] *Ibid.*, 98. [91] Ronnerberg, 'An Introduction to Protection', 29.

[92] www.igpandi.org/group+agreements/the+pooling+agreement.

[93] www.igpandi.org/downloadables/2015-reinsurance_diagram_v2.pdf. See also Hazelwood and Semark, *P&I Clubs*, 365–71.

[94] *Ibid.*

4.5.1 Risk Pooling in the Price-Anderson Act

4.5.1.1 A Second Tier of Compensation

The nuclear risk is a typical catastrophic risk with a low probability of occurring but immense losses once materialized. Considering the low occurrence, reliable data on the probability and magnitude of an accident are often lacking. The nuclear mutual pools established in the United States provide examples to cover such potential catastrophes. In the United States, nuclear liability is governed by the Price-Anderson Act of 1957,[95] which has been revised approximately every decade.[96] According to the Price-Anderson Act, each license issued should have and maintain financial guarantee to cover public liability claims.[97] The term 'public liability' is defined as 'any liability arising out of or resulting from a nuclear incident', with the exceptions of claims arising out of an act of war, workmen's compensation claims, and claims for damage to on-site property at a licensed nuclear facility.[98] The existing capacity of the insurance market (provided by a monopolistic national pool, American Nuclear Insurers [ANI]) is to provide US$375 million.[99] If an accident creates damages in excess of US$375 million, a retrospective premium needs to be called upon from all American nuclear operators licensed by the Nuclear Regulatory Commission (NRC). This premium is payable in annual instalments up to a certain maximum amount per incident per power plant and is determined according to the size and number of reactors each plant has.[100] The amount of this second tier is now set at US$122.255 million, with a maximum annual retrospective premium of US$18.963 million per reactor per calendar year.[101] This makes a total of US$13.2 billion retrospective premium available for a potential catastrophic nuclear accident.[102] The NRC guarantees those retrospective premiums or, in other words, advances the compensation in the second tier and later collects this from the operators.[103]

[95] 42 USCA §2210.

[96] The Price-Anderson Act has been revised in 1967, 1975, 1988 and 2005. See further on the Price-Anderson Act in Section 3.6.2.

[97] Pub. L. No. 85-256, 71 Stat. 576, 577 (1957).

[98] Pub. L. No. 85-256, 71 Stat. 576, 576 (1957).

[99] www.amnucins.com/?page_id=173.

[100] Faure and Vanden Borre, 'Compensating Nuclear Damage', 243.

[101] 10 CFR §140.11(4); 75 Fed. Rec. 16646, 2 April 2010. As mentioned earlier, in 2013, the contributions in the second tier have again been increased.

[102] www.amnucins.com/?page_id=171.

[103] Faure and Vanden Borre, 'Compensating Nuclear Damage', 260.

4.5.1.2 ... Collectively Financed Through Retrospective Premiums

The premium under the second layer of the Price-Anderson Act is financed through a so-called retrospective premium scheme. Hence, premiums do not have to be paid ex ante but only ex post after the nuclear accident materializes. This retrospective premium scheme was introduced in 1975. Initially, the financial requirement was satisfied with the coverage from private insurance and a government indemnity agreement. This was so because it was considered that the nuclear industry was not capable of bearing all the burdens in its infancy period, and the Price-Anderson Act intended to encourage the development of the nuclear industry. However, after years of development, it was believed that the industry should take on its own responsibilities.[104] This was achieved by phasing out federal indemnity and establishing a system of retrospective premiums paid by nuclear operators.

Under the retrospective premiums scheme, if an accident leads to damage in excess of US$375 million, all qualified nuclear operators are obliged to pay the retrospective premiums up to US$121.255 million. This arrangement is a form of a risk-sharing agreement. However, different from the usual understanding about risk-sharing agreements, this arrangement is a mandatory system. It is effectively imposed by statute. Rather than the voluntary pooling of operators, the retrospective premiums scheme is established according to the legislative requirement of the amended Price-Anderson Act. This arrangement ensures that all nuclear power plant operators participate in the system and provide strong capacity. Besides, this system is used only when the primary instrument – the insurance market – fails to cover the full damage. In other words, the retrospective premiums scheme provides an upper layer of compensation for victims.

Because the obligation of payment comes due only after the occurrence of damage, a special arrangement is needed to ensure that the operators can pay their retrospective premiums in the case of damage. According to Nuclear Regulatory Commission (NRC) regulations, the operators need to provide one of the following guarantees: surety bond, letter of credit, revolving-credit/term-loan arrangement, maintenance of escrow deposits of government securities, annual certified financial statement and other types of guarantee approved by the NRC.[105]

[104] H.R. Rep. No. 94-648, 10. [105] 10 CFR §140.21.

4.5.2 The Mutual Pool for Property Damage in the United States

The Price-Anderson Act established only a risk-sharing pool for third-party liability.[106] However, a nuclear accident also can create serious on-site damage. The Three Mile Island accident has shown that a clean-up of the on-site damage would be far more important than the third-party liability coverage of the nuclear operator.[107] This has led to the introduction of the so-called property rule by the NRC, [108] which requires each power reactor licensee to obtain a financial guarantee to stabilize and decontaminate the reactor and the reactor station site.[109]

The operators satisfy this requirement via a mutual insurance system, Nuclear Electric Insurance Limited (NEIL). NEIL covers both nuclear and non-nuclear events.[110] For the nuclear events, it traditionally provided three types of policies: on-site accidental property damage and decontamination coverage (including primary and excess programs), accidental outage coverage (covering costs associated with interruptions of electricity generation due to accidental physical damage to insured sites) and a builder's risk program (covering risks associated with the construction of new power plants).[111]

The on-site accidental property damage policy contains two programs: the primary program with a limit of US$1.5 billion and the excess program with a limit of US$1.25 billion in excess of US$1.5 billion.[112] NEIL provides coverage different from American Nuclear Insurers (ANI) and is re-insured by the latter. The primary program provides coverage of expenses incurred in connection with stabilization and decontamination of the reactor and direct physical damage to or destruction of the insured property as a result of an accident.[113] The excess program provides coverage of both property damage and decommissioning liability. The payment of expenses under the excess program is made in the following order: losses under nuclear liability coverage, losses under decommissioning liability coverage, losses under debris-removal and decontamination coverage, property damage coverage and functional total loss coverage.[114]

[106] For further details, see Liu, *Compensating Ecological Damage*, 244–6.
[107] Faure and Vanden Borre, 'Compensating Nuclear Damage', 245.
[108] See Liu, *Compensating Ecological Damage*, 246–7.
[109] Conditions of Licenses, 10 CFR §50.54(W) (2008).
[110] NEIL, 'Annual Report 2014', editorial section, 10.
[111] NEIL, 'Annual Report 2014', financial section, 6. [112] *Ibid.*
[113] Gittus, 'Introducing Nuclear Power to Australia', 226. [114] *Ibid.*, 233–4.

The premium is based on the amount of the coverage, the deductible and various rating criteria.[115] The NEIL policy is linked with the self-regulation under the Institute of Nuclear Power Operations (INPO). INPO monitors and controls nuclear risks. In the event that the site is determined by INPO as not meeting the industry standard of acceptable performance, the premium can be increased by up to 25 per cent.[116] NEIL can make a call to the member insured for payment of a retrospective premium up to ten times the annual premium to cover losses incurred by the company during the policy year.[117]

4.5.3 *Mutual Nuclear Pools in Europe*

Two mutual pools were established in Europe: the European Mutual Association for Nuclear Insurance (EMANI) and the European Liability Insurance for the Nuclear Industry (ELINI). They are both mutual insurance associations, and both provide not only coverage to nuclear power stations but also to other nuclear facilities. EMANI was established in 1978 and provides coverage for material damage, business interruption, machinery breakdown, terror and erection all risks.[118] The insurance capacity for material damage in 2013 was approximately €594 million, and the capacity for terrorism was €625 million.[119] The decontamination costs for installations in case of an accident are also covered in the damage policy.[120] However, because there is no legal obligation for nuclear operators to seek financial coverage for their onsite decontamination, as is the case in the United States, the coverage for decontamination costs by EMANI may be not significant. EMANI can act as either the leading underwriter or a co-insurer to provide insurance directly or as a re-insurer. Depending on the leading underwriter, two different types of policies exist: the national pool policy and the EMANI policy.[121] They may have different coverage and prices.

TELINI was created at the end of 2002 to provide insurance capacity for nuclear liability risks. It is the first worldwide nuclear pooling system aimed at providing nuclear third-party liability coverage.[122] ELINI was not designed as a competitor to the national pools but instead offers excess liability coverage: it provides coverage to all the

[115] *Ibid.*, 231. [116] *Ibid.* [117] *Ibid.* [118] EMANI, 'Annual Report 2013', 14.
[119] *Ibid.*, 16.
[120] Interview with Danny Van Welkenhuyzen, managing director of ELINI and Ann Geivaerts, legal manager of ELINI, 28 June 2011.
[121] *Ibid.* [122] Faure and Vanden Borre, 'Compensating Nuclear Damage', 219, 257.

headings under the revised Paris and Brussels Supplementary Conventions and also coverage for terrorism and the thirty-year prescription period.[123] The capacity to provide coverage for the less predictable risk is because of the special structure of ELINI. There are three types of memberships for ELINI: the insured members who have an insurance policy and have contributed to the constitution of the guarantee fund, non-insured members who take part in the constitution of the guarantee fund (supportive members) and non-insured members who do not take part in the constitution of the guarantee fund but pay an administration fee.[124] Both insured members and supportive members make annual contributions to the guarantee fund according to the agreed ratios. If the total losses exceed the contributions, an additional call can be made up to twenty times the annual contributions.[125] The availability of additional calls makes an ex ante calculation of premiums for the unpredictable risks less necessary. It is worth noting that ELINI is different from national mutual pools established in the United States and Germany. Under the national regimes, the insurers and insured are the same; the operators make contributions according to a preset ratio when a major accident happens to one of the operators. However, under the system of ELINI, supportive members who are not insured can also make contributions to the guarantee fund. ELINI can act as either a leading underwriter or a co-insurer. Thus, the individual insurance policies can be varied from one to another; the extent of involvement of ELINI and the percentage of each contributor in different policies also may be different. The insureds are also free to choose the amount of coverage. Through cooperation with national pools, ELINI works more like a combination of a traditional insurer and a mutual pool.

To guarantee sufficient compensation for future damage, ELINI makes reserves. The contributions made by members are put into a special bank account and can only be withdrawn after the dissolution of the pool or thirty years after the expiration of its policies. In 2013, the maximum insurance capacity per policy is €89 million, of which €50.5 million is provided through retention.[126]

ELINI can provide insurance by itself as a leading underwriter or co-insurer with national pools. The content of the contract and insured amount depend on national laws because different countries have divergent liability limits and liability and insurance requirements. When it acts

[123] ELINI, 'Annual Report 2010', 16. [124] *Ibid.*, 17.

[125] Interview with Danny Van Welkenhuyzen, managing director of ELINI and Ann Geivaerts, legal manager of ELINI, 28 June 2011.

[126] ELINI, 'Annual Report 2010', 18.

as a co-insurer to national pools, ELINI relies on the national pools to decide the content of the insurance contract but can decide its own premium. Usually, the basic premium among electricity producers is the same, and they are subject to increases or decreases according to the grading made by ELINI. When accepting one as a member of ELINI, ELINI has to make a risk review and grades the risks according to technical, insurance and financial features. The grading can be used to adjust the premium later.[127]

ELINI has attempted to create an international pooling system for nuclear liability. If the revised Paris Convention were to come into effect and would be accepted by many countries, the additional capacity provided by ELINI may become more attractive. However, given the divergent liability requirements among different countries today, and given the lack of consistent and harmonized standards on nuclear safety, as explained later, in Europe, nuclear operators may not have a strong interest in establishing a mutual pool. This may be one of the reasons why the mutual pools have developed so slowly since their creation. After several decades, the capacity of EMANI increased from only €150 million in the first decade of existence to €594 million in 2013,[128] while the capacity of ELINI was still only €89 million in 2013.

4.5.4 *Nuclear Liability and Pooling System in Germany*

In Germany, the nuclear liability clauses are mainly stipulated under the Act on the Peaceful Utilization of Atomic Energy and the Protection Against Its Hazards (Atomic Energy Act). [129] The Atomic Energy Act was

[127] For the pool in the United States, NEIL can rely on some self-regulatory organizations such as WANO and INPO to control the safety of members. WANO is the World Association of Nuclear Operators; INPO is Institute of Nuclear Power Operations in the United States. They play important roles in checking nuclear safety in nuclear power plants. In the United States, NEIL cooperates with INPO. However, to be a member of WANO or INPO, one should be the operator of a nuclear power plant. But the members of ELINI vary: they can be either operators of nuclear power plants or other nuclear installations. Thus, unlike NEIL, which can rely on INPO's information in setting premiums, ELINI is not affiliated with those institutions. Hence, ELINI cannot affiliate with a nuclear self-regulatory organization such as WANO or INPO to control the nuclear risks.

[128] Faure and Vanden Borre, 'Compensating Nuclear Damage', 219, 257; EMANI, 'Annual Report 2013', 16.

[129] See Liu, *Compensating Ecological Damage*, 299–302. Also, Atomgesetz (AtG) – Gesetz über die friedliche Verwendung der Kernenergie und den Schutz gegen ihre Gefahren.

passed in 1959, and the last revision was made in 2013. In addition to this Act, Germany is also a party to the Paris Convention and to the Brussels Supplementary Convention, as well as to the Joint Protocol. According to the Atomic Energy Act, 'the Paris Convention and the Joint Protocol shall apply as national law in the Federal Republic of Germany, unless its provisions depend on reciprocity as effected by the entry into force of the Convention.'[130]

The 1959 Atomic Energy Act aims both at promoting the use of nuclear energy and preventing damage.[131] The pro-nuclear policies continued until the 1990s, when the government started to make an effort to phase out nuclear energy. An agreement was reached on the gradual phasing out of nuclear energy in 2000 between the government and the industry,[132] and the Act on the Structural Phase-Out of Nuclear Power for the Commercial Production of Electricity was passed in 2002. According to that act, there will be no further licences issued for new nuclear power plants or reprocessing facilities, and the existing nuclear power plants are supposed to be closed before 2020.[133] In 2010, the German government decided to extend the operation of the existing nuclear reactors by an average of twelve years. However, the catastrophic Fukushima accident in 2011 in Japan triggered a change of nuclear policy in Germany again. The government stopped the extension plan and came back to the phase-out policy again. Nuclear power will only serve as a bridging tool before the replacement with renewable energy. An amendment to the Atomic Energy Act was passed in the same year.[134]

According to the existing act, nuclear liability in Germany has the following characteristics: as in the international regime, the liability is channelled to the nuclear operators, and the operators are strictly liable for damage caused by a nuclear incident.[135] The liability is even stricter in Germany because the defences under the international regimes are no longer available, such as an armed conflict, hostilities, civil war, insurrection or a grave natural disaster of an exceptional character.[136] But if the damage occurs abroad, the defences are not applicable only if that country provides the same benefits to Germany.[137] One significant difference between the

An English version can be found at: www.bfs.de/de/bfs/recht/rsh/volltext/a1_englisch/a1_08_13_atg_0115.pdf.

130 Atomic Energy Act, §25. 131 Carrol, 'Perspective on the Pros and Cons', 90.

132 Mez and Piening, 'Phasing-Out Nuclear Power', 161.

133 For the phase-out of nuclear power in Germany, see Vorwerk, 'The 2002 Amendment', 7.

134 Bernasconi-Osterwalder and Hoffmann, 'The German Nuclear Phase-Out', 2–3.

135 Atomic Energy Act, §25(1). 136 *Ibid.*, §25(3). 137 *Ibid.*

German and international systems is that an unlimited liability system is established in Germany. Only if the damage is caused by an armed conflict, hostilities, civil war, insurrection or a grave natural disaster of an exceptional character, the liability is limited to the maximum amount of the government indemnification.[138] To provide coverage for the potential liability, the operators are required to seek financial security. The administrative authority shall determine the type, terms and amount of the financial security. 'The amount of financial security shall be fixed subject to a maximum of €2.5 billion.'[139] Financial security can be either third-party liability insurance or other forms. If the liability is not covered by or cannot be satisfied by financial security, the federation shall indemnify the operator. The maximum amount of indemnification is set at €2.5 billion, and the obligation of payment is the maximum amount minus the amount that is covered by financial security.[140] After payment of the indemnification, recourse is possible if the operator disobeys specific obligations, or the operator caused the damage wilfully or by gross negligence, or if the operator did not seek financial security to the required extent.[141]

At that time, in the beginning of the 1970s, the pooling system in Germany emerged. An increase in financial security up to 1 billion DEM was on the legislative agenda. The first 500 million DEM should be covered by private means, while the government should indemnify the remaining half. The insurers and nuclear operators negotiated to cover liability up to 500 million DEM fully by insurance. The first 200 million DEM were covered by insurers, while for the remaining 300 million DEM the insurer only fronted contract. The remaining 300 million DEM were re-insured by the operators of nuclear power plants as a whole. This arrangement remained valid until 2002.[142]

In 2002, the amendment to the Atomic Energy Act increased the amount of financial security up to €2.5 billion and allowed financial security in other forms rather than through liability insurance.[143] At that time, insurers only covered €256 million. The insurers agreed that they could not provide full coverage.[144] The operators started to find alternatives: in 2001, the four parent companies of the nineteen nuclear power plants negotiated and concluded a 'Solidarity Agreement' to provide the remaining coverage mutually.[145]

138 *Ibid.*, §31(1). 139 *Ibid.*, §13(3). 140 *Ibid.*, §34(1). 141 *Ibid.*, §37.
142 Pelzer, 'International Pooling', 43.
143 Atomic Energy Act, §§13 and 14; Carroll, 'Perspective on the Pros and Cons', 91.
144 Carroll, 'Perspective on the Pros and Cons', 91.
145 Pelzer, 'International Pooling', 44, n. 24.

Under this agreement, the companies pool together to cover the second tier (in addition to the first €246 million covered by the insurers) up to €2.24 billion. Each party has an obligation to contribute a percentage of the total amount in case of any damage attributed to one of the parties. The percentage for each nuclear power station is calculated according to the square root of the thermal reactor output. The percentage of power plants is then attributed to the parent companies on the basis of their participation (Clause 1[3]). This allocation of liability is different from that in the United States, where each operator bears the same quota. In Germany, the allocation of contribution is based on the generating capacity. As in the United States, the obligation to make the contribution only comes due after damage in excess of the insurance capacity happens. However, the risk that the operators have to contribute is even smaller in Germany: the partners only have to pay if neither the operator nor the parent company is in a position to pay up to €2.5 billion (Clause 1[5]). The solidarity agreement is hence only a guarantee for the payment by the liable parties. After the payment, the partners have a right of claim for repayment on the liable parties. But the liability for third parties prevails over the claims for recourse (Clause 1[8]). In addition to mutually guaranteeing the coverage of liability, the partners also need to provide help in handling the claims, such as to make available legal and commercial staff capacity and infrastructure. For this kind of support, the partners cannot ask for repayment. The partners also provide help for the use of independent contractors, up to the amount of €122, 218 million (Clause 2). To ensure the availability of assets in the case of damage, the partners need to submit an auditor's certification each year (Clause 3).

Both the United States and Germany established a retrospective pooling system to cover nuclear liability. Both pooling systems provide a second tier of compensation: the first tier is made available through commercial insurance. However, some differences do exist: the US system is based on the statutory obligations arising from the Price-Anderson Act, while the German arrangement is achieved through a private contract. The contents in Germany are hence subject to changes in the contract. In addition, in the United States, each party's contribution is the same and is pre-determined in the statute. Hence, the total amount of compensation depends on the number of operators. In Germany, the maximum amount of compensation is set in the contract, and the percentage of each partner depends on its involvement in the power plants and on the generating capacity. In the United States, the individual liability of the operator is limited to the capacity of financial security, while in Germany, the operators are still liable if the capacity of the pooling is depleted. Another major difference lies in the conditions of payment. In the United States, the obligation to pay the retrospective premiums comes due when damage in excess of the insurance capacity happens, and the operators cannot get recourse from the liable parties. In Germany, the pooling is only a guarantee: it only pays in case the liable party fails to do so. The partners who have paid can claim repayment.

4.5.5 Towards a European Pool for Nuclear Liability?

The analysis so far shows, as was also seen when risk pooling mechanisms were introduced theoretically, that pooling systems have many advantages. Ex ante information about the possibility and extent of risks is no longer necessary. One needs only to know the proportion of each member's contribution to the risk. As a low-frequency, high-magnitude risk, nuclear liability is less predictable. A retrospective pooling system can relieve the heavy information burden. In addition, a pooling system also has financial advantages. A pooling system is not for profit and can offer a cost-effective way to cover nuclear liability, especially the less predictable new headings under the second generation of conventions. The pooling system also can be beneficial for insurers because they can be relieved of the burden of the new headings they do not feel able to cover.[146]

For these reasons, many have been pushing for risk-sharing agreements as a solution for nuclear liability at the international level.[147] It has also been argued that when the Paris and Vienna Conventions were revised, risk pooling between operators would be able to generate substantially larger amounts than are currently available for the repair of nuclear damage under the international conventions.[148] More recently, the possibility of a risk sharing by operators in Europe has also been worked out in further detail, whereby the financial consequences were calculated.[149] Notwithstanding the theoretical advantages of also using pooling by operators in the nuclear sector on a larger scale, a European-wide nuclear liability pool has so far not emerged. To some extent, this

146 Carroll, 'Perspective on the Pros and Cons', 96.
147 See e.g. Faure and Skogh, 'Compensation for Damages', 499–513.
148 Faure, 'Economic Models of Compensation', 21–43.
149 Faure and Fiore, 'The Coverage of the Nuclear Risk'.

can also be understood. Despite the promising features just mentioned, several requirements need to be satisfied for operators to pool effectively, especially at the international level. The states whose operators wish to participate in the pool should have comparable political, legal and economic backgrounds, and the pooling members should have equal or comparable levels of nuclear safety and security.[150] The countries with nuclear installations have varying political, legal and economic backgrounds, which creates a challenge for international pooling. This may be less of a problem in the European Union, where the members share more trans-boundary risks. However, with enlargement of the European Union, the development status of the members is more divergent. It is even more problematic as far as nuclear liability legislation is concerned. The EU states have a so-called patchwork nuclear liability regime.[151] They are subject to different nuclear liability conventions or are not members of any convention at all. Member State domestic legislation has different attitudes towards many important issues such as limited/unlimited liability, amounts required for financial security, and so on.[152] As for levels of nuclear safety and security, it is still problematic. Since pooling is based on trust and confidence, it should be possible for the operators to exercise mutual monitoring, or an organization should exist to control the risks of its members. However, an EU-wide approach to safety regulation and standards and a single regulatory body are still missing in Europe. Despite these weaknesses, it is worth noting that the European Union is now making efforts towards a more harmonized system of nuclear liability, and many scholars are supporting this idea.[153]

One of the reasons why risk pooling emerged successfully in the United States is the role which is played by the NRC, which guarantees US-wide nuclear safety standards. This obviously facilitates risk pooling by operators because they know that all operators within the United States at least have to comply with the minimum requirements imposed by the NRC. A similar harmonization of safety standards is largely

[150] Pelzer, 'International Pooling'. [151] See Reyners, 'Liability "problems"'.

[152] For example, in Austria and Germany, unlimited liability was adopted. But in most member states, the liability is capped. Within those countries with limited liability, the required limitations also vary. For an overview of the liability limit in EU Member States, see Gomez-Acebo and Pombo, Abogados SCP, 'Legal Study for the Accession of EURATOM', 22–3.

[153] For example, Ameye, 'Legal Study on Nuclear Third Party Liability'; Hanrilica, 'Harmonization of Nuclear Liability'.

lacking in Europe, which explains the reluctance of some operators in relatively 'safe' Member States to pool with operators in Member States where lower safety requirements would apply. This comparison between the United States and Europe teaches at least one important lesson: if one wishes pooling between operators to be successful, at least a minimum safety requirement needs to be imposed by government (and seriously enforced) to facilitate mutual monitoring between operators.

4.6 Concluding Observations

4.6.1 Advantages of Pooling

This analysis of pooling mechanisms sheds an interesting light not only on the pooling mechanisms that are currently in use in the offshore sector but also on theoretical advantages of pooling mechanisms in the nuclear sector. Even though this other sector may seem slightly remote from the offshore liability incidents, the analysis of pooling mechanisms in those other high-risk sectors is quite revealing, on the one hand showing the strengths and weaknesses of pooling mechanisms in general but on the other hand also showing some basic conditions for pooling mechanisms to develop in a successful manner. Moreover, an analysis of the compensation mechanism already in existence at the international level for vessel-based pollution, discussed in Chapter 3,[154] was equally interesting. On the one hand, it showed the possibilities to raise relatively high amounts of compensation via a combination of different techniques (coverage via P&I Clubs combined with a compensation fund). On the other hand, the international regime also showed some specific features (e.g. financial limits on liability and channelling) that were heavily criticised in the literature and hence might provide interesting warnings in the development of a liability and compensation regime for the offshore sector.

When addressing first risk pooling at a general level, the literature indicates (and the examples examined confirm) that risk pooling between operators has some attractions and advantages, especially when risks are complicated, highly technical and relatively little information is available with third parties (more particularly insurers). Under these circumstances, insurance coverage may be problematic because insurance always requires ex ante information on the probability of an accident

[154] See Section 3.2.1.

and the potential magnitude of the damage in order to be able to calculate actuarially fair premiums. In this respect, a pooling arrangement has the major advantage that coverage is possible without necessarily having correct actuarial information on risk. Even if ex ante probabilities of the accident are not known, pooling is possible. The only requirement is that the relative contribution of the different parties to the pool is known, as a result of which risks can be differentiated. Hence, as long as one knows, for example, that offshore installation A poses a higher risk than offshore installation B, pooling is possible, whereby A will have to contribute more to the pool than B, even if the precise probability of the risk for either A or B is unknown. Second, pooling is especially attractive because it will provide excellent incentives for mutual monitoring. A pool is after all a risk-sharing or risk-distribution agreement whereby parties mutually share each other's losses. Hence, if installation A did not take safety seriously, the risk for all members in the pool would increase. Mutual monitoring therefore will be executed precisely to prevent such a situation.

However, the different examples discussed, especially in the nuclear area, show that a crucial condition for a pooling arrangement to work effectively is that at least a minimum level of regulation exists. In other words, even though risks will be controlled via mutual monitoring, operators apparently rely on minimum levels of safety set by the regulator. Hence, this shows that the regulator can play an important role in facilitating pooling between operators by guaranteeing a minimum level of safety. The simple logic is that this guarantees to operators, members of the pool, that all members will at least have to comply with minimum safety requirements. It reduces the risk from 'rogue' operators and also reduces the need for the operators themselves to exercise mutual monitoring. This explains why a pool emerged for property damage in the United States and why there are mutual nuclear pools in Europe, again for property damage, but why so far a European pool for nuclear liability has failed. As opposed to the United States (where minimum safety standards are enforced via the NRC), minimum safety standards in the nuclear area are largely absent in Europe. An important lesson from these (also nuclear) examples is thus that the regulator can play an important role in facilitating pooling mechanisms by introducing safety regulations that guarantee a minimum level of safety. This may therefore be an important condition for supporting a pooling mechanism in the offshore sector.

The final and most important advantage of risk-pooling mechanisms, not only theoretically but also as was shown in practice, is first of all that money should not always be paid up front (as with premiums due to an insurance company) and that if contributions were made up front (as with P&I Clubs), they are not 'lost' if no incident occurs (as in the case of insurance) but could, in principle, lead to lower contributions in the future or eventually even be paid back to members. Moreover, because a pooling mechanism is based on mutual trust, as long as financial guarantees to back up the solvency of the members exist, contributions should not necessarily be paid ex ante. This may be quite important because contributions to be paid ex ante may lead to a substantial immobilization of capital and hence restrict the economic development of the sector. We discussed a few examples, again from the nuclear sector, where ex post pooling systems are in place based on retrospective premiums. In other words, there is no need to pay contributions ex ante (and thus immobilize capital which otherwise could be used to increase social welfare). Contributions only need to be paid ex post via retrospective premiums. The model of risk pooling in the Price-Anderson Act for nuclear risk constitutes an interesting example in this respect and allows raising considerable amounts of compensation. A similar model also exists in Germany, where the pooling is also based on ex post obligations, and a total amount of €2.24 billion has been raised which, when compared to the total amount available for vessel-based pollution in the international regime,[155] is quite substantial.

The crucial question with such an ex post pooling system is, of course, how one can guarantee the solvency of the members. In Germany, this apparently is not a major problem because the scale is relatively limited (only Germany). The pooling in Germany is also merely a guarantee that is only invoked if other liable parties fail to pay. In the US system under the Price-Anderson Act, each party's contribution is the same and predetermined in the statute. Hence, the pooling is mandatory and arranged by statute. Moreover, in the United States, it is not private parties but the NRC that will guarantee payment, compensate the victims and then claim back-payment from the operators. So the public authority (NRC) in this case guarantees payment and claims back compensation from operators on an ex post basis. Again, this can constitute an interesting example for possible models of compensation in the offshore sector

155 See Table 5 in Section 3.2.1.2, indicating that in the international regime for vessel-based pollution approximately US$1 billion is now available in compensation.

because it allows the generation of substantial compensation without necessarily immobilizing capital.

4.6.2 OPOL

The discussion of one pooling mechanism in the offshore sector today, OPOL, showed that OPOL is not a risk-sharing agreement in the sense discussed in this chapter. It is a pooling mechanism, but in fact, only the solvency risk is pooled. OPOL undoubtedly has strengths in the sense that it forces all its members to provide financial security up to the amount of compensation guaranteed by OPOL. In some member states, such as the United Kingdom, membership in OPOL is a condition for obtaining a license. In those member states, via the membership in OPOL, there is therefore a guarantee of financial security. However, the sufficiency of the guarantee is controlled by OPOL and not by regulatory authorities (at least in those countries where membership in OPOL is not a condition for obtaining a license). Moreover, because OPOL is not a risk-sharing agreement, the benefits of mutual monitoring leading to increased prevention will not exist. Members of OPOL only have an incentive to monitor the solvency of the other members. But given mandatory guarantees which have to be proven ex ante, they should not necessarily constitute a large problem. This also may explain why in practice OPOL never had to intervene and in fact only played its 'silent' role of forcing its members to provide guarantees up to the limits of OPOL. One could, with a view to the future, of course consider the possibility of reconstructing OPOL to a true risk-sharing agreement, but that would fundamentally change the nature of OPOL in its current structure.

4.6.3 P&I Clubs

Today, compensation for damage caused by offshore facilities is guaranteed via a variety of financial and market mechanisms.[156] One risk-sharing agreement which in that respect plays a (modest) role is the so-called P&I Club, also discussed in this chapter. The P&I Club is a true risk-sharing agreement that consists of ship owners that mutually cover each other's losses. Hence, this arrangement does provide incentives for mutual monitoring and in the area of vessel-based pollution in

[156] To be further discussed in detail in Chapter 5.

fact functions as insurance. The P&I Clubs play a much more important role in vessel-based pollution and only a relatively modest role as far as the coverage of pollution coming from offshore installations is concerned. However, a discussion of their structure and functioning was of interest because they are a pooling mechanism in an area closely related to the offshore sector.

4.6.4 Risk Sharing for Vessel-Based Pollution

The civil liability and fund conventions discussed in Chapter 3 show various interesting aspects. One interesting aspect is that the liability of the operator (in the case of vessel-based pollution, the tanker owner) is capped. This, of course, contrasts with the liability of operators of offshore facilities, which is, as the overview in Chapter 3 showed, largely unlimited. However, an important evolution has taken place in the sense that the amounts of the financial limit have increased over time, mostly as a result of new incidents which time after time showed that existing limits in the international conventions were too low. The international conventions are for this reason seriously criticized in the literature, arguing that limits on liability do not provide correct incentives to potential injurers (in this case, the tanker owners).

Another interesting feature of the compensation regime for a vessel-based solution is its multi-layered aspect. Indeed, a brief look at the legal history showed that the International Convention on Civil Liability for Oil Pollution Damage (CLC) of 1969, which introduced the capped strict liability of the tanker owner, was combined with the Fund Convention of 1971, to which the oil industry contributes. As a result of this, compensation awarded for vessel-based pollution is partly awarded by the shipping industry (via the strict liability under CLC 1969) and partially by the oil industry (via the Fund Convention). This has led to relatively large amounts now being available after adoption of the latest convention on supplementary funding. Currently, approximately €1 billion is available for compensating damage resulting from vessel-based pollution. This, of course, constitutes an important difference with the offshore sector because, unlike the area of vessel-based pollution, there are not two different industry sectors (shipping industry and oil) that could contribute to the compensation and thus share the burden.

Finally, it should be mentioned that, similar to OPOL, also in the area of vessel-based pollution, voluntary mechanisms (previously, CRISTAL and TOVALOP and, currently, TOPIA and STOPIA) exist. However, their role is totally different and relatively limited. These voluntary agreements in fact only provide an additional layer of compensation by the tanker owner or oil industry to supplement the amounts available under the international conventions. Their role is hence relatively limited.

4.6.5 Summary

In sum, this analysis of pooling mechanisms in other sectors shows that there are a number of features, both of the liability regime and of the available financial security, that could provide inspiration for a liability and compensation mechanism for the offshore sector. In this respect, lessons can be learned (positive as well as negative) from compensation mechanisms in the nuclear and vessel-based pollution sectors, as well as from risk-pooling mechanisms in other sectors. These lessons, summarized in this section, will of course be further developed in the chapters that follow when discussing various scenarios and possibilities of compensation for damage resulting from offshore installations. Thus, the question will, of course, again be asked to what extent risk-pooling mechanisms could play an important role in compensating damage resulting from offshore incidents. The lessons learned from this chapter have provided some insights into conditions that will have to be fulfilled to make risk-pooling mechanisms work.

In bullet points, the main lessons could be summarized as follows:

- Pooling (in the sense of risk-sharing between operators) has many theoretical advantages over insurance, especially for highly technical risks such as offshore-related damage.[157]
- OPOL has many advantages but is not a risk-sharing agreement; only the risk of insolvency is shared. Until now, the OPOL guarantee has never been used.
- The only real, existing pooling mechanisms for offshore-related damage are OIL and OCIL.
- The CLC of 1969 and the Fund Convention of 1971 provide an interesting combination of financing of compensation by tanker owner (CLC) and oil industry (Fund) but have disadvantages as well, such as the limitation of liability and the financial cap.

[157] These advantages are summarized in Section 4.1.

- The US OPA has a limit on liability, but this can be set aside in case of gross negligence or violation of regulations.[158] OPA, moreover, does not pre-empt state law.
- P&I Clubs are an interesting example of a risk-sharing agreement for marine-related risks.
- The US Price-Anderson Act provides an interesting example of an ex post risk pooling via a retrospective premium scheme.
- The emergence of a European-wide pooling system for nuclear risks is not likely given the absence of EU-wide harmonized safety standards and highly different risks created by various operators.

158 See Section 3.5.5.8.

5

The Use of Financial Market Instruments to Cover Liability Following a Major Offshore Accident

MICHAEL FAURE AND HUI WANG

After having discussed the existing pooling mechanisms available in the offshore sector (as well as more generally) in Chapter 4, this chapter now provides an inventory of the financial market instruments currently used to cover liability following a major offshore accident and describes the extent to which various mechanisms (including the pooling discussed in Chapter 4) are currently used to cover liability risks. In fact, this chapter discusses not only the coverage of liability risks but also the coverage of property damage suffered by operators and the costs of well control because de facto many of the instruments to be discussed in this chapter are also used for these purposes. From a policy perspective, the policy maker may be more worried about the coverage of third-party liability. However, the way in which first-party (property) damage and the costs of well control are covered is quite important. In practice, the available financial and insurance instruments often cover both first-party damage, including well control, and liability. Moreover, the extent to which the instruments are used to cover items such as well control can have an important influence on the available cover for liability. Thus, the more capacity that is used for first-party damage, the less may be available for liability and vice versa.

This chapter first sketches the theoretical possibilities of various mechanisms and then illustrates their current use in the offshore sector. Finally, the advantages and disadvantages will be discussed, with a view towards providing suggestions for policy makers in choosing and designing regulatory regimes for different financial security mechanisms. The latter are, of course, important in light of Chapter 6, where the potential of the various instruments to provide enlarged coverage in the future is discussed. Whereas this chapter discusses the status quo, Chapter 6 discusses the ability of various instruments to provide wider coverage than today.

Subsequently, self-insurance will be dealt with (5.1), as will use of the capital market (5.2), bank guarantees (5.3), insurance (5.4), risk-pooling schemes (5.5), the Offshore Pollution Liability Agreement (OPOL) (5.6) and various combinations (5.7). To the extent possible, a similar format will be used when describing every system. First, the theoretical advantages and limits of a particular instrument will be sketched, then the practice of that particular financial instrument for covering offshore-related risks will be discussed, and finally, a brief analysis will be provided on the use of that particular instrument.

5.1 Self-Insurance

5.1.1 Theory

Self-insurance basically is a mechanism whereby larger players in the market (often referred to as the 'majors') do not take insurance coverage at all but run the risk themselves. In fact, self-insurance can take two different forms. One is pure self-insurance, which, in fact, is nothing other than major companies holding a reserve for future losses. In a technical sense, this cannot be considered as 'insurance' for the simple reason that there is no risk spreading, no risk distribution, and hence no loss spreading after an accident happens.[1] Self-insurance in this sense is merely a reserve for potential losses whereby operators use their balance sheet to guarantee payment in case a major accident happens. Reserves do allow for risk spreading in time but not between various parties exposed to a risk.

Another possibility is the creation of a so-called captive.[2] A 'captive' could be created by a major offshore operator and would function de facto as an insurance company. In this case, the captive takes the form of an insurance company. Still, there is no loss spreading because, for example, an operator would create its own captive and would not share risks with others. The reason for a firm such as a major oil company to create a captive is that it is in some cases obliged as a result of regulation to take out insurance coverage. Instead of shifting the risk to

[1] See Faure and Hartlief (eds.), *Insurance and Expanding Systemic Risks*, 144.

[2] See on those captives, *inter alia*, Bawcut, *Captive Insurance Companies*; and Dowding, *Global Developments*.

a commercial insurer (and hence paying premiums), the oil company could create a captive, which basically is an insurance company that is wholly owned by the particular operator. From the operator's perspective, the advantage of using a captive is that profits and losses are kept within the captive (and hence within the oil company). Captives can often operate at lower costs than an insurance company and thus may be attractive to hedge risks for major offshore companies. Often companies that create such a captive still use an insurer (e.g. AIG) or a broker to provide various services. The broker could, for example, be involved in managing the captive; the insurance company could assist in providing information on policy conditions and premiums to be set by the captive.

5.1.2 Practice

Stakeholders report that major oil and gas companies in particular largely use self-insurance and captives to hedge offshore-related risks. This is obviously the case not only for liability following a major offshore accident but also, as mentioned earlier, for first-party damage (e.g. well-control costs).

During interviews, all the majors[3] confirmed that they use self-insurance. The argument that is often advanced by the majors is that for them it would make little sense to shift their risks to insurance companies, which would force them to pay higher costs with little added value as far as financial security is concerned. This is especially the case with some majors that have very high credit ratings (e.g. A– or higher). A major with such a credit rating would obviously not take insurance coverage with an insurance company that has a lower rating. This would only lead to additional costs and have no added value. The number of insurance companies that are rated as high as the major oil and gas companies is relatively limited. This also explains why the majors largely use self-insurance.[4] A second point is that stakeholders mentioned that insurance for them is a relatively costly alternative to hedge risks. BP, for example, refers to a loss experience of 30 per cent. This effectively means that for every $100 BP would pay in insurance premiums, it would also receive $30 back in compensation for losses; the

[3] More particularly, Shell, BP and ExxonMobil interviews.

[4] It was a point strongly made by representatives of BP during an interview on 13 March 2013.

remainder would go either to transaction costs or to profits for the insurance company. Because the majors consider insurance relatively costly, they prefer self-insurance. However, it was equally reported that self-insurance obviously is a solution only for the majors and not for potentially smaller or middle-sized enterprises.

Majors in some cases also make use of captives. For example, BP created a captive, Jupiter Insurance, Ltd. The captive has no staff but is managed by Willis, a large insurance broker. The captive is 100 per cent owned by BP. In its underwriting policy, BP is assisted by AIG, which consults the captive on insurance practice but carries no risk. Legally, the captive is an insurance company that can underwrite risks.[5]

The reason for creating the captive is that some countries (as in the case of BP Angola) require operators to take out insurance coverage. This explains why the captive is formally set up as an insurer and why the captive also formally underwrites risks and fixes premiums and policy conditions. Some captives do have re-insurance; Jupiter (of BP) does not.

BP's captive was, *inter alia*, involved in compensating the losses due to the Macondo incident but up to a limit of US$700 million:

- It paid out US$300 million, the limit in pollution costs, and
- US$400 million was provided for well-control costs, but this amount did not have to be used completely.

The capacity of a captive is, of course, not unlimited, but limited to the assets brought into the captive by the operator, which owns the shares of the company. In the case of Jupiter, capacity can be provided of US$1.5 billion.

The captive functions completely as a commercial insurance company. This means that commercial-rate premiums are also asked, that deductibles are required, and that risk-dependent premiums are charged, as well in order to provide incentives for prevention. Risks are also hedged on the basis of particular underwriting guidelines. The captive hence only differs from an ordinary insurance company in its ownership (it is owned by a major oil company).

In addition to relying solely on self-insurance to cover potential liability, in some cases operators may still use self-insurance to cover a first layer. The first layer would be a so-called retention and insurance

[5] BP's captive, Jupiter, therefore intervened in compensating (part of the) losses caused by the Deepwater Horizon incident (see Section 2.5.2.1).

coverage and would only be asked for higher amounts that the particular operator cannot carry itself.[6] Larger players in some cases would use, for example, €400 million in self-insured retention (SIR). But even smaller to medium-sized companies could take amounts such as €5 million to €10 million in SIR. Also, representatives of (re-)insurance companies argued that the use of self-insurance may (at least partially via retentions) even be increasing. One reason is the more limited capacity available in the market after the Deepwater Horizon accident. A second reason is due to the previously mentioned fact that (re-)insurance requires higher premiums.[7] It is important to stress that almost all insurance solutions will include a retention (or deductible). There will normally never be a so-called first-dollar coverage for the simple reason that it would be too expensive.[8]

Finally, it should also be mentioned that many representatives pointed to the fact that oil and gas companies de facto have large reserves (of oil and gas), which they carry on their balance sheet and which can de facto also constitute an important reserve to guarantee payment in the case of future losses.[9]

5.1.3 *Analysis*

Self-insurance (either via reserves, captives, or use of the capital market) obviously has several strengths and weaknesses. The advantage from the industry's perspective is that it is a relatively low-cost solution; operators can themselves take care of providing guarantees for future losses and do not have to transfer risks to an insurance company, which may create additional transaction costs. Moreover, especially for major oil and gas producers, forcing them to shift risk to an insurance company may make little sense, especially in cases where the credit rating of the oil and gas producer is in fact higher than that of the insurance company.

From the perspective of policy makers, in addition to providing a low-cost solution, self-insurance also has, at least in part, an advantage in curing the so-called moral-hazard risk that will always emerge in the case of insurance. By taking a substantial retention, the operator will still be exposed to risk as a result of which moral hazard (created through

[6] *Ibid.*
[7] Interview with representatives of Swiss Re-Insurance Company, 11 February 2013.
[8] *Ibid.* This was also mentioned in an interview with Phil Bell of RSA, 26 February 2013.
[9] This was especially mentioned by representatives of OGP.

insurance) can be controlled.[10] Some commentators think that self-insurance can dilute the deterrent effect because its low cost 'leads to a lesser sense of responsibility from the company that self-insures'.[11]

The disadvantages of self-insurance (no matter what form it takes) may also be obvious: self-insurance is not necessarily a waterproof guarantee against insolvency. This would be the case only if regulation could guarantee that the money set aside for covering the offshore-related losses would be used only for that specific goal. Moreover, self-insurance thus could lead to an externalization of risk in the case of insolvency. Especially smaller operators without strong balance sheets or credit ratings could simply run the risk of liability and, if the risk materialized, would pass on the costs to taxpayers. Self-insurance thus can only be considered effective financial security if guarantees can be provided that the reserves set aside actually will be used for the potential losses for which they were earmarked. Otherwise the risk would also exist that, for example, in case of insolvency, the trustee in bankruptcy could simply collect the assets, and money may not be available to compensate victims.

The issue in the case of offshore-related damage is obviously that self-insurance may be a valuable hedging strategy for major oil companies but not for others. For smaller and medium-sized players, self-insurance retention can only play a role as a deductible in addition to other strategies. This is why in most regulatory documents concerning financial security there are often doubts about self-insurance based on a financial test. For example, in the EU Commission guidance document concerning the geological storage of carbon dioxide,[12] self-insurance is considered the most risky option because no protection is provided from claims of creditors. Therefore, if self-insurance is offered as a financial guarantee, policy makers need to impose serious controls to verify the viability of the self-insurance as guarantee. In this respect, experience exists within the offshore sector. Both in the "Guidelines on Financial Responsibility" from Oil and Gas UK and in the rules concerning the establishment of financial responsibility under the Offshore Pollution Liability Agreement (OPOL), stringent criteria apply in order for a company to be allowed to qualify as a self-insurer. For example, in the Oil and Gas UK Guidelines, the credit or financial-strength rating

[10] See Shavell, 'On Moral Hazard'. [11] Bio by Deloitte, 'Civil Liability', 129.

[12] Implementation of Directive 2009/31/EC on the geological storage of carbon dioxide, Guidance Document 4, Brussels, European Commission, 2011.

that is considered sufficient for a company to be qualified as a self-insurer by the Department of Energy and Climate Change (UK DECC) is BBB– or above from Standard & Poor's, B+/bbb or higher from A. M. Best, Baa3 or higher from Moody's, and BBB– or higher from Fitch (investment grade). The standards under OPOL are even more stringent: an operator can only qualify as self-insurer under OPOL if it has an A or higher rating from Standard & Poor's, A– or higher from A. M. Best, or A3 or higher from Moody's.[13] These examples therefore show that it may be necessary to apply flexibility as far as the use of self-insurance is concerned: for some (major) operators, self-insurance could perhaps even be the only financial mechanism; for others, it could at least be used in part in addition to other hedging strategies. The experience with OPOL shows that stringent rules must apply to control whether a particular operator can qualify as a self-insurer.

5.2 Capital Market

5.2.1 Theory

Capital markets can also theoretically be used to hedge offshore-related liability risks. For a few centuries now, forms of so-called alternative risk transfer or securitization have been used to transfer environmental risks.[14] The idea is to buy catastrophe bonds (often abbreviated as 'cat bonds') at the stock exchange.[15] Initially, these alternative mechanisms to hedge environmental risks were mostly developed in the United States and not used to a large extent in Europe.[16] Although it is still held that the capital market will never totally replace traditional insurance products,[17] the capital markets and, more particularly, alternative risk transfer are increasingly used for environmental risks.[18]

The principle is that particular bonds are issued whereby the interest rate on the bond would reflect the accident rate. Investors thus have the opportunity to buy a bond creating a warrant in favor of the operator of their choice. If, for example, during the period of the bond (say, one year), no accident occurred, the amount of the guarantee provided

[13] See Form B and Form Fr-4 on the OPOL website; in addition, other conditions to qualify as self-insurer apply as well.

[14] See, *inter alia*, Wagner, 'Risk Securitization'.

[15] See e.g. Smith, Canelo and Tidio, 'Reinventing, Reinsurance'.

[16] *Ibid.*

[17] See Zech, 'Will the International Capital Market'.

[18] For an overview, see Bruggeman, 'Capital Market Instruments'.

by the bond would be paid with interest. If, however, the risk materialized, the bond posted would be used to cover the damage.[19]

5.2.2 *Practice*

As already mentioned, capital markets and, more particularly, bonds are used to cover certain catastrophic risks.[20] For example, in the March 11, 2011, tsunami and the subsequent nuclear incident at Fukushima, Japan, cat bonds were used. It was estimated that there were more than US$1.7 billion in cat bonds that were potentially affected by the Japanese earthquake.[21]

However, as far as offshore (liability) risks are concerned, most stakeholders reported that capital markets and, more particularly, bonds are not often used. Insurance experts hold that the problem with a bond is that it normally expires very quickly, whereas damage caused by an offshore incident often has a so-called long-tail character. Moreover, it is equally mentioned that since the 2008 financial crisis, financial institutions may be less willing to issue bonds.[22]

Also, representatives from the International Association of Oil and Gas Producers (OGP) held that operators themselves do not use bonds; only insurers may.[23] Of course, (re-)insurance companies may use the capital market to transfer particular risks, which is, in fact, nothing new. If capital markets could provide better capacity at a good price, they would do so.[24] However, they equally hold that cat bonds may be attractive, for example, for natural catastrophes such as a hurricane but not for long-tail risks. The essence of such a cat bond is that an investor takes a bond for one year, for example, and then either takes the profit or accepts a loss (if the risk materialized). This is not possible with long-tail risks such as offshore-related damage.[25] On the whole, it was held that

[19] For an explanation of the working of these cat bonds, see also Tyran and Zweifel, 'Environmental Risk'.

[20] For examples, see Bruggeman, 'Capital Market Instruments'.

[21] See Faure and Liu, 'The Tsunami'. See also Finken and Laux, 'Catastrophe Bonds and Reinsurance'; and Mortimer, 'Cat Bond Investors'.

[22] Interview with Phil Bell of RSA, 26 February 2013. This was also held by representatives of the Swiss Re-Insurance Company: financial instruments such as bonds are not suited for potentially long-tail risks related to offshore installations.

[23] Discussion with OGP representatives in Brussels, 25 February 2013.

[24] Interview with representatives of Munich Re-Insurance Company in Munich, 12 March 2013.

[25] *Ibid.*

compared to insurance for environmental risks generally, the capital market (in particular, via cat bonds) still plays a relatively minor role.

5.2.3 *Analysis*

All stakeholders, as well as the available literature, agree that capital markets and cat bonds are only well suited for sudden events and less so for long-tail risks. This is why, for a typical long-tail risk, such as the storage of carbon dioxide, the literature also holds that bonds may be difficult given the long-tail character of the risk.[26] Given the remarks of stakeholders and the fact that bonds are apparently little used today for hedging offshore-related risks, this may not be the primary instrument to look at.

5.3 Guarantees

5.3.1 *Theory*

Obviously, many parties could provide a financial guarantee to an operator. Such a financial guarantee could be provided by a mother company or a third party that would presumably have a stronger financial capacity than the operator. It could also be provided by a financial institution such as a bank. In particular cases, the guarantee could take the form of a letter of credit.

All these guarantees have a simple theoretical basis: a presumably stronger third party (either a corporate entity or a financial institution) basically puts its balance sheet at risk by guaranteeing that it will cover the liabilities of the operator in case a particular risk materializes.

For the policy maker or regulator, the advantage may be that a stronger guarantee is provided than when only the operator's assets are at stake. Operators themselves would presumably use these types of guarantees when regulatory authorities find it necessary (and hence hold that merely relying on self-insurance is not sufficient). However, in certain cases, the costs of those guarantees can be quite high. If the guarantee is provided by a related corporate entity, this higher cost would not necessarily be the case, but this might be different if guarantee has to be provided by a bank. Compared to insurance, one can hold that if insurers specialize in offshore operations, they may have more information for an appropriate

[26] See e.g. Klass and Wilson, 'Climate Change and Carbon', 160–2.

risk differentiation and premium setting, as a result of which the cost of insurance could be lower than the cost of a bank guarantee.

5.3.2 *Practice*

Stakeholders reported that alternatives to insurance or pooling, such as guarantees by banks and letters of credit, are essentially more expensive than insurance. Moreover, the financial crisis also taught us that banks may collapse if they are too heavily exposed to risk.[27] Also, operators hold that bank guarantees or letters of credit can be extremely costly.[28] In addition, they again stress that a bank guarantee may only be attractive if it is from a financial institution with a good credit rating and then again with a higher rating than the operator concerned; otherwise, the guarantee would not have a lot of added value. It was held that letters of credit are de facto not used in the European Union but only for so-called fronting arrangements in other parts of the world; de facto they are only used when commercial insurance is not available (e.g. in developing countries).[29]

5.3.3 *Analysis*

So bank guarantees and letters of credit are seldom used to cover offshore liabilities. The simple reason seems to be that they are relatively expensive compared to available alternatives in the market. Corporate guarantees are also not always considered to be good security because there is no protection from potential claims of the operator's creditors. Moreover, there may be a high risk of financial connection between the operator and the affiliated guarantor. This is why it is often held in related areas[30] that the certainty of such a corporate guarantee depends on the stringency of the required annual financial test.[31] OPOL does accept guarantees from specific guarantors that are deemed acceptable to the association and that

[27] Interview with Phil Bell, RSA Insurance, 26 February 2013.

[28] Discussion with representatives of OGP in Brussels, 25 February 2013. Letters of credit and bank guarantees are therefore de facto never used according to representatives of Oil and Gas UK, interview, 1 May 2013, in London.

[29] *Ibid.*

[30] Like the financial security for damage caused by geological storage of carbon dioxide.

[31] Implementation of Directive 2009/31/EC on the geological storage of carbon dioxide, Guidance Document 4, 27.

have to fill out particular forms by which their financial viability is screened.[32]

5.4 (Re-)Insurance

5.4.1 Theory

The theoretical advantages of insurance are well known. Economists have used the concept of risk aversion to explain that many people will be wary of risks with a relatively low probability of occurring but with a potentially large magnitude when they occur.

The utilitarian approach with respect to insurance has demonstrated that risk creates a disutility for people with risk aversion. Their utility can be increased in the case of loss by spreading the risk or if the small probability of a large loss is taken away from the insurer in exchange for the certainty of a small loss.[33] The latter is, of course, exactly the phenomenon of insurance. The risk-averse insurer has a demand for insurance; he prefers the certainty of a small loss (the payment of the insurance premium) by which the probability of a larger loss is shifted to the insurance company, thereby increasing the utility of the insurer.[34] It is remarkable that in this utilitarian approach to liability, insurance is in the first place regarded as a means to increase the utility of a risk-averse insurer and not so much as a means to protect victims, as is sometimes argued by lawyers.

The reason an insurance company can take over the risk of the insurer is well known: because of the large number of participants, the risk can be spread over a larger group of people. The insurer only has to build relatively small risk groups in which the premium is as far as possible aligned to the risk of the members of that group.

In addition to this utility-based theory of insurance, which sees insurance as an instrument to increase the expected utility of risk-averse persons through a system of risk spreading, Skogh has powerfully argued that insurance may also be used as a device to reduce transaction costs.[35]

32 See form FR-3 on verification of guarantee, which shows that the guarantor again needs to have strong proofs of credit or financial strength such as A– or higher from Standard & Poor's, A– or higher from A. M. Best or A3 or higher from Moody's.

33 See Arrow, *Aspects of the Theory*, 245–55.

34 See Shavell, *Economic Analysis of Accident Law*, 190.

35 This argument has been made by Skogh, 'The Transactions Cost Theory', 726–32.

The basic principle of insurance is therefore relatively simple: because individuals have an aversion to risk, they will seek insurance coverage. The insurer is able to aggregate many risks of risk-averse individuals facing the same risk profile. Using the law of large numbers, the insurer is able to spread the risk over a larger risk group. However, this supposes that the insurer is able to calculate an adequate premium based, on the one hand, on the probability p that a certain accident will occur and, on the other hand, on the potential magnitude of the damage D. The latter is obviously important because the insurer has to make sure that it has enough capacity at the moment when a certain systemic risk with a potentially large magnitude emerges.

These principles can be used to explain the demand in the case of offshore-related risks. Because the attitude toward risk is strongly related to the wealth of an individual, the degree of risk aversion will hence also depend on the available assets of the offshore operator. For relatively small expected losses, a wealthy operator could be risk neutral and thus not have a demand for coverage. In that case, the demand for insurance would only emerge if insurers could manage, for example, claims handling more effectively. It would then mostly be for the reduction of transaction costs and not for risk aversion that insurance would be taken out.

The degree to which offshore operators have a demand for insurance therefore may strongly depend on their own risk attitude, which, in turn, is related to their assets. Normally, it is fair to state that, especially for smaller operators, risk aversion is higher and hence a demand for some type of coverage (insurance or alternatives) may emerge.

Generally, a distinction is made between two types of insurances. There are, on the one hand, insurances that individuals take to protect themselves against the future losses that they may suffer themselves, either in their income or in their property. Fire insurance is a typical example. These types of insurance are referred to as 'first-party insurance'. Insurances also can be taken for the risk that one has to compensate damage suffered by a third party. These are therefore referred to as 'third-party insurance'. Liability insurance is a typical example of a third-party insurance. In that case, the potential injurer takes insurance against the risk it runs of having to compensate a potential victim.

In the case of offshore-related risks, both first- and third-party insurance can be relevant. First-party insurance will be taken by the operator for the property damage it could suffer to its installations. However, most likely, liability insurance will play the most important role in the case of offshore-related risks. In that case, coverage is demanded for the risk that damage is suffered by a third party.

5.4.2 *Practice*

5.4.2.1 Stakeholders in the Insurance Market (Major Insurance/Re-Insurance Providers)

Given that offshore oil and gas activities may involve many parties (including the operator or licensee, the service company, contractors, and equipment providers),[36] each party may take insurance to cover its potential risks. Therefore, the structure for offshore insurance is also complicated due to the involvement of various stakeholders. Moreover, as illustrated by the country studies in Chapter 3, many jurisdictions (both in Europe and in other places such as the United States and Australia) require the establishment and maintenance of financial security. This financial security is often provided via insurance.

The insurance for offshore facilities is mainly provided by three different actors:

- The casualty insurance market,
- The specific offshore insurance market covering platforms, and
- The marine insurance market, usually covering tankers but in some cases also offshore and even onshore installations.[37]

The offshore energy insurance market traditionally covers a large amount, sometimes even more than US$1 billion, so the insurance companies found it necessary to syndicate themselves in order to cover the entire risk of exposure. The main syndicates in this domain are the London and Bermuda insurance markets.[38]

According to a study carried out by Booz Allen Hamilton, the insurers that provide offshore oil and gas exploration and production (E&P) insurance and reinsurance (at least in the US market) mainly include, but are not limited to

- Munich Reinsurance Co.,
- Swiss Reinsurance Co., Ltd.,
- Hannover Rueckversicherung AG,

[36] Emmerson, *Arctic Opening*, 41. [37] Interview with Swiss Re, 11 February 2013.
[38] King, 'Deepwater Horizon Oil Spill Disaster'.

- Chartis (a subsidiary of American International Group, Inc.), and
- W.R. Berkley Corporation.[39]

A few protection and indemnity (P&I) clubs are also involved[40] in the cover of offshore-related risks. However, it should be noted that some of the P&I clubs that might provide coverage for offshore-related risks (e.g. GARD, SKALD, and UK) would do so more as commercial insurers than as P&I clubs. This means that for those non-maritime risks, there is no mutualization (pooling) with the maritime (traditional P&I) risk.[41]

In practice, an important role is played by large brokers (e.g. Willis or Marsh) in arranging cover for large offshore-related risks. The broker will contact underwriters and try to get the best deals for its client, the insured offshore operator. Usually there is not just one insurance company that covers the risk but maybe ten or more actors. In practice, there will be one so-called leader and many followers.[42] It is thus never just one underwriter who takes the entire risk, but the broker contacts a leader and then gets followers (this is referred to as a 'subscription market').[43]

De facto an offshore operator will present a file showing the broker that it is a good risk (and hence demand low premiums). The insurers will do a risk analysis using its own experts and relying on the information provided by the operator.[44] The figures shown in Figure 13, representing the premiums paid for global offshore energy coverage, indicate the importance of the UK market (and more particularly Lloyds) in this domain.

Many stakeholders in the insurance industry hold that there are quite a few difficulties in insuring offshore-related risks, as a result of which the number of layers may be relatively limited. One problem for insurers is the volatility of the offshore-related risk. This means that the probability of an accident may be very low, but once an accident happens, the damage could be catastrophic and could amount to €1.5 billion.[45] The reason why insurers generally dislike these low-probability, high-damage risks is that solvency regulations (so-called Basel I and Basel II regulations) may force insurance companies to immobilize large sums of

[39] Booz Allen Hamilton, 'The Offshore Oil and Gas Industry', part 1, 13.
[40] See on these P&I Clubs, Section 4.4.2.
[41] Interview with representatives of the International Group of P&I Clubs, 1 May 2013, in London.
[42] Representatives of Lloyds, interview, 1 May 2013, in London. [43] *Ibid.* [44] *Ibid.*
[45] Interview with Phil Bell, 26 February 2013.

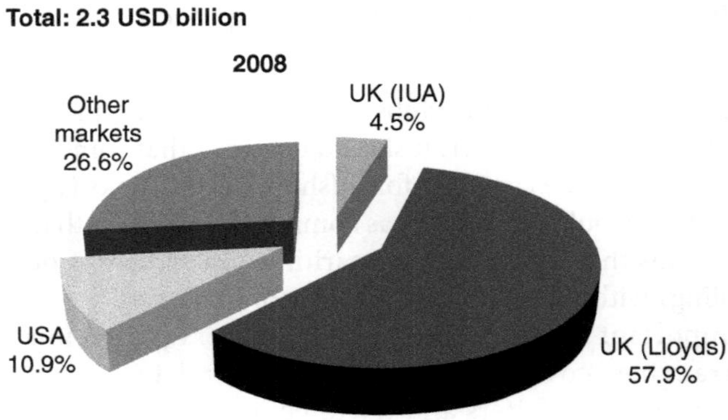

Figure 13 Global offshore energy premiums by market.
Source: Table taken from Schmitt, "Offshore Energy"; data from International Union of Marine Insurance.

capital for those future risks that cannot therefore be used in another, potentially more profitable way.[46]

Another difference between insuring against such things as hurricane and flooding losses, on the one hand, and offshore installation losses, on the other hand, is that with natural catastrophes, much better data are available on the probability and the potential damage. Moreover, there are potentially millions of people who are interested in purchasing flood insurance. Hence, actuarially fair premiums can be calculated and adequate risk spreading is possible so that coverage can be provided at realistic prices.[47] The opposite may be the case with offshore-related risks:

- There are a small number of players in the market (a small number of offshore companies requiring insurance cover),
- Limited data are available, and
- There are difficulties in predicting losses in individual cases.[48]

5.4.2.2 Structure of Coverage of Offshore Insurance

Existing insurance coverage for offshore activities mainly covers physical damage and liability exposures. The risks may involve construction, physical damage, removal of wreckage, control of the well, and liability.[49] The policies involved include

[46] *Ibid.* [47] *Ibid.* [48] *Ibid.* [49] *Ibid.*

- *Offshore physical damage coverage.* This indemnifies the insured for all risks associated with physical loss or damage to fixed offshore drilling, production and accommodation facilities.
- *Operator's extra expense (OEE).* This is a policy offered to oil and gas companies that provides coverage for expenses associated with regaining control of a well, typically covering the cost to control operations, re-drilling of the well to the depth it was previously, and removing or cleaning up seepage/pollution.[50]
- *Environmental/pollution liability.* This provides coverage for bodily injury, property damage and cleanup costs as a result of a pollution incident.
- *Business interruption/loss of production income.* This provides coverage for energy businesses against loss due to temporary interruption in oil and gas supply from an offshore facility.
- *Comprehensive general liability.* This provides coverage for claims an energy business is legally obligated to pay as a result of bodily injury or property damage to a third party.
- *Worker's compensation/employer's liability.* This provides coverage for claims arising out of employee injuries or deaths incurred while employees are in the line of duty.[51]

Usually property damage coverage is taken for offshore facilities. A casualty program covering clean-up and third-party liability can be added to the property damage insurance. In practice, usually one limit is used for the whole insurance policy. This may create a challenge for the compensation of third-party liability. For example, if a major blowout were to occur with pollution, the primary attention would be focused on reinstating the oil rig. This may take a large part of the insurance proceeds away.[52]

A distinction should be made between the insurance provided to the operator, on the one hand, and insurance coverage that may be provided to subcontractors, such as the drilling company, on the other hand. The drilling company often will work with a mobile offshore drilling

[50] The International Risk Management Institute (IRMI) provides a definition of OEE insurance, which states that OEE insurance is a 'specialized policy available to oil or gas well operators that covers the cost of regaining control of a wild well. Coverage for pollution, stuck drill stem, evacuation expense, and care, custody, or control (CCC) exposures can be added by endorsement'. See the IRMI website: www.irmi.com/online/insruance-glossary/terms/operators-extra-expense-oee.aspx.

[51] Booz Allen Hamilton, 'The Offshore Oil and Gas', part 1.

[52] Interview with Phil Bell, 26 February 2013.

unit (MODU), a floating device. In many jurisdictions, the MODU is often considered to be a ship when in transit and an offshore facility when stationed. The liability for MODUs may not be clear under various national laws, and when it is considered to be a ship, the Bunker Convention may apply, in which relatively low limits of liability are provided. Under the US Oil Pollution Act of 1990 (OPA), a MODU is treated as a tanker when determining the responsible party and as an offshore facility when deciding on excess liability.[53] Representatives of the drilling contractor hold that some of the larger drillers will self-insure, but most drillers will take insurance coverage. They experience no problem in obtaining cover on the commercial market.[54]

The policy wording that is referred to as 'energy exploration and development (EED) insurance' or 'operators' extra expense (OEE) insurance' has particular general provisions such as excluding any loss or expense caused by or attributable to earthquakes or volcanic eruptions. However, this exclusion does not apply to the Gulf of Mexico or the North Sea. The policy also contains a due-diligence clause prescribing that the insured shall exercise due care and diligence in the conduct of all operations and use safe practices and equipment generally considered prudent for such operations. Interestingly, it is also held that where the insured is the operator or joint operator of any insured well being drilled, a blowout preventer has to be set on the surface casing or the well head and installed and tested in accordance with usual practice. Moreover, the insured also should warrant that in the case of a well becoming out of control, it will use every endeavour to control the well or stop the escape or flow.

5.4.2.3 Available Insurance Amount

According to a study carried out by Bosma, information regarding the amount of cover available for pollution damage is not openly available and is difficult to access. Cover is typically based on a per-project basis and on terms that are confidential to both the insurer and the insured parties. As a result, it is difficult to ascertain what or how much insurance actually may be available to respond to any pollution incident involving offshore oil facilities.[55]

53 33 USC §2704(b). For further details on the status of MODUs, there is a so-called MODU Code, the Code for the Construction and Equipment of Mobile Offshore Drilling Units. The most recent version of this code is MODU Code 2009, which revises the 1989 version.

54 See Alan Spackman, vice-president of IADC, interview, 5 June 2013.

55 Bosma, 'The Regulation of Marine Pollution', 106.

Bosma's work establishes, after discussions with experienced offshore oilfield insurance industry specialist brokers, that the amount of pollution cover placed by oil industry participants generally is as follows:

1. *Oil companies:* US$500 million to US$1 billion (although this can often be self-insured (i.e. no insurance is actually placed with a third-party insurer).[56]
2. *Drill rig contractors:* US$300 million to US$1 billion.
3. *Other oilfield service providers:* US$50 million to US$1 billion.

However, there are significant regional variances in the levels of insurance cover placed for pollution damage, particularly within the jurisdictions of developing countries.[57]

Interviews with insurance experts suggest that as far as offshore pollution is concerned (i.e. platforms), in addition to coverage provided, *inter alia*, by maritime insurers, the so-called casualty insurers can provide coverage (for pollution and liability) in excess of amounts varying between €500 million and €750 million. This amount of coverage for casualty would then come in addition to the amount for property damage and in excess of the casualty sub-limits.[58] Other insurance experts also hold that the total amount of coverage available on the market would now vary between US$1 billion and US$1.5 billion.[59] Also, operators hold that US$1 billion is easily available for third-party liability, but when asked, they confirmed that amounts of US$1.5 billion to US$2 billion would be coverable as well.[60]

Traditional P&I cover for some disastrous scenarios can be provided with very high cover. Under certain circumstances, coverage even up to US$7.6 billion could be provided. However, these high amounts would be provided for non-oil-related risks and only under a number of very specific conditions:

[56] This also corresponds with information provided by the International Association of Oil and Gas Producers (OGP) holding that "[s]tand alone OEE (operators extra expense) insurance capacity is available up to a limit of US$500 million per event but that OEE limits under insurance packages can be procured up to US$ 1 billion per event" (e-mail from the International Association of Oil and Gas Producers, 13 May 2013, to Michael Faure).

[57] *Ibid.* [58] Interview with Swiss Re, 11 February 2013.

[59] Interview with Phil Bell, 26 February 2013.

[60] This was mentioned by a representative of the OGP (discussion on 25 February 2013), who held that 'for large companies it is in fact fairly easy to obtain coverage'. This was confirmed in an e-mail from OGP to Michael Faure on 13 May 2013.

- High retentions,
- Pooling (mutualization of the risk),
- Re-insurance, and
- A complex system consisting of different layers of cover depending on the amount of cover provided.

Note, however, that this would be exceptionally high coverage and would not be available for offshore-related risks.[61] For oil pollution, cover is de facto limited to US$1 billion.[62]

Insurance experts from Lloyds also hold that underwriters decide on whether to provide a large amount of cover for offshore-related risks based on the situation in the market. The insurance market is considered to be quite volatile. Hence, it may be possible to arrange coverage of €1 billion for a very large player in one year but not for the next. Insurance availability is often decided on a yearly basis.[63] Another issue strongly stressed by representatives of the insurance industry is that insurers always consider the total supply of cover for the various types of insurance. As a result, losses that would, for example, only affect the well itself and not create any damage for third parties could still limit insurance capacity on the market, as well as third-party liability coverage.[64] The experts reported on a large incident with a rig that drifted but where no pollution was caused because the valves apparently worked correctly. However, the cost to put the unit back in place and business interruption costs amounted almost to US$1 billion.[65] Hence, although there was no blowout or damage to third parties, such an incident still can substantially limit available insurance cover.[66]

A recent post-Macondo report of Lloyds holds that pre-Macondo insurance industry commentators estimated that the commercial insurance market was theoretically able to put a total of between US$3.5 billion and US$4 billion towards offshore energy risks. However, the capacity varies in relation to the type of risk and the pricing.[67] The commentators hold that before Macondo, there was an approximate US$1 billion of capacity for liability insurance with respect of offshore drilling, including pollution-related liabilities.[68]

61 Interview with representatives of the International Group of P&I Clubs, 1 May 2013, in London.

62 *Ibid.* 63 Interview with representatives of Lloyds, 1 May 2013, in London. 64 *Ibid.*

65 The incident is mentioned in Table 2 in Section 2.3.2, indicating a loss of more than US$1 billion due to heavy weather.

66 Interview with representatives from Lloyds, 1 May 2013, in London.

67 Rees and Sharp, *Drilling in Extreme Environments*, 26. 68 *Ibid.*, 35.

It is the mix of various stakeholders providing coverage and the mix of actors that make the structure of the coverage rather complicated. It should also be stressed that the casualty coverage does not yet include the potential extension of liability under the Environmental Liability Directive (ELD). More particularly, the requirement under the ELD to restore the environment back to baseline may be difficult to quantify, especially in the case of pollution of marine waters.

5.4.2.4 Cost Estimation

How to estimate the costs of a major offshore accident is a crucial issue because it may influence the insurer's decision on the insurance cover and the premium to be charged to the insured. In this respect, some major insurers/re-insurers (e.g. Swiss Re) may possess their own databases, but a publicly available database is lacking.

In the United States, especially in the Gulf of Mexico, all losses, even of small releases, have to be declared to the American authorities (Environmental Protection Agency and the Bureau of Ocean Energy Management). As a result of that, in the United States, a large amount of data on past loss experience is available. A similar database does not exist in Europe. However, as we indicated in Chapter 2, some countries, such as the United Kingdom, Norway and Denmark, have a reporting system, as a result of which oil releases are recorded by the respective national authorities. Those data thus also can be used by insurers as risk-management tools.

The European Commission at the Impact Assessment for the Draft Regulation on the Safety of Offshore Oil and Gas Activities estimated the average annual cost of all major offshore accidents to be between €205 million and €915 million, which is made up of the following:

- The average annual cost of blowout incidents, estimated to be between €140 million and €850 million and
- The average annual cost of other major accidents, estimated to be €65 million.

This was rebutted in the study carried out by GL Noble Denton, which gave much lower figures, with €62 million for total blowouts plus €65 million for other major accident costs, which totalled €127 million, in contrast to the €205 million to €915 million of the European Commission.[69] Note,

[69] GL Noble Denton, 'UK Responses to EC Impact', 7.

however, that even in an estimate by Oil and Gas UK in a worst-case scenario, the total damage, in the case of a major incident affecting the shoreline in the West Shetland, would be US$450 million.[70] Note, moreover, that as we will discuss in more detail later when discussing the insolvency risk,[71] some stakeholders, especially re-insurers, argue that a major incident with an offshore installation in European waters will almost inevitably lead to damage of more than €1 billion.[72]

Estimating both the probability and the cost of an offshore-related accident remains difficult but is essential for insurers in the framework of premium calculation. As far as clean-up costs are concerned, a study in 2000 which calculated the costs to clean up each unit of oil shows regional differences in per-unit clean-up costs. To clean up the same amount of oil

Table 12 *Average Per-Unit Marine Oil Spill Clean-Up Costs by Nation (in 1999 US$)*[73]

Region	Cost
Canada	$6,508.14
Denmark	$11,180.41
Estonia	$6,820.62
Finland	$2,115.29
France	$2,301.58
Germany	$10,702.67
Greece	$8,530.29
Ireland	$4,807.49
Italy	$6,541.19
Latvia	$9,212.35
Lithuania	$78.12
Netherlands	$6,655.37
Norway	$23,118.08
Spain	$438.68
Sweden	$15,642.36
United Kingdom	$3,082.80
United States	$25,614.63

Source: Schmidt Etkin, 'Worldwide Analysis'.

[70] Oil and Gas UK and OPOL, 'Oil Spill Cost Study', 39. [71] See Section 6.1.2.
[72] Interview with Philipp Wassenberg, representative of Munich Re, 6 May 2013.
[73] Schmidt Etkin, 'Worldwide Analysis'.

in the United States would cost a lot more than in a EU country (e.g. Denmark) (see Table 12).[74]

5.4.2.5 Calculation of Premiums

Insurers and re-insurers usually calculate premiums on the basis of a past loss experience and exposure basis. On that basis, a risk analysis of the specific offshore installation is made. Major re-insurers have databases of accidents that have occurred in the energy sector which they employ to calculate premiums. The overall risk is modelled with the support of experienced actuaries.[75]

Some, however, warn that it may be dangerous to assess the risk only on the basis of the past accident history. De facto, many oil companies are now searching for oil at greater depths and in places that are more difficult to reach. This may also increase the risk as a result of which merely looking at historical data should not be the only proxy to fix premiums.[76]

The premium will de facto be calculated on the basis of a variety of criteria, including the role played by the insured (as an operator or a contractor), the amount of revenues made by the insured, risks related to issues such as the oil involved in the installation and the location, which determines the applicable legal regime.[77]

For the amount of €500 million in coverage in casualty mentioned earlier, yearly premiums could amount to between €2 million and €50 million depending upon the size of the company, the importance of the activities and the risk differentiation criteria.

The premiums obviously should reflect the differing risks of each installation whereby, as mentioned, location is an important criterion. Insurers and major re-insurers employ many highly skilled engineers with the capacity to do risk assessment. In addition to the amount of revenues, type of well and location (and resulting legal regime), specific criteria for risk assessment are also taken into account. One important element is the quality and type of installation (whether it is old or new), but another is the risk exposure. For risk exposure, public awareness and

74 GL Noble Denton, 'UK Responses to EC Impact', 12.

75 Interview with Swiss Re, 11 February 2013.

76 See Phil Bell interview, 26 February 2013.

77 For example, the costs in the Gulf of Mexico of an average offshore incident would be ten times higher than in Malaysia simply because of the application of the US legal system (based on the jury, punitive damages and more generally the way in the US the liability system is used for compensation). Interview with Swiss Re, 11 February 2013.

public opinion could be crucial.[78] An additional criterion of importance is whether the offshore drilling only involves oil or gas. Gas is more dangerous in the sense that it may lead to personal injury in the case of accidents, but oil may lead to larger pollution damage.

Operators often argue that premiums are relatively high. It has already been mentioned that some of the majors hold that for every hundred dollars they would pay in insurance premiums, they would only receive thirty dollars back in compensation for losses. Insurance premiums thus would insufficiently reflect actuarial fair values.[79] This may, of course, be due to transaction costs as well as to profit margins charged by insurers. This may be the reason why, especially for major operators, alternatives such as captives may be more attractive.[80]

As far as the risks covered by P&I Clubs are concerned, it is mentioned that for very large vessels (e.g. large cruise ships or very large tankers), total premiums paid by the vessel could be substantial and even amount to more than US$1 million per year. Those are, however, extreme cases because premiums on a yearly basis would more likely vary between US$10,000 and US$100,000.[81]

5.4.2.6 Influence of the Deepwater Horizon Incident on the Insurance Industry

Following the Deepwater Horizon incident, it was claimed that there was an upward pressure on premiums from the re-insurance market.[82] John Lloyd, chief executive of Lloyd & Partners, in his written testimony to the US Senate on 11 May 2010 said that insurance coverage might decline as much as 30 per cent after the Deepwater Horizon incident.

It was noted by the insurance industry that after Deepwater Horizon, the available capacity was reduced for offshore installations. The reason was that many insurers suddenly were confronted with a huge accumulation problem. That accumulation problem is caused by the fact that especially in the offshore market a variety of different contractual relationships exist and different players whereby it is (e.g. for a re-insurer) not always visible who is exactly providing coverage for what. As a result, after the Deepwater Horizon incident, offshore installations

[78] Interview with Swiss Re, 11 February 2013.

[79] Interview with representatives of BP, 13 March 2013. [80] *Ibid.*

[81] Interview with representatives of the International Group of P&I Clubs, 1 May 2013, in London.

[82] O'Connor, 'Lloyd's Offshore Energy Market'.

were considered to be risky business, leading to a reduction in capacity. Capacity went generally down and, obviously, prices for coverage went up.[83]

The Deepwater Horizon incident was a shock to many (re-)insurers because initially it seemed as if the losses would be totally covered by the self-insurance of BP. However, in many cases BP apparently turned to redress actions against other parties involved in the operation or recovery operations surrounding the Deepwater Horizon. With most of those third parties, so-called hold harmless clauses had been concluded, as a result of which BP could not claim against those parties. However, BP apparently exercised pressure upon those parties to pay nevertheless substantial amounts of compensation. Given the commercial importance of BP, many gave in and hence compensated (although the legal basis was weak).[84] In turn, those parties that paid BP (often in settlements, precisely to avoid costly litigation before US courts) then turned to their insurers, forcing them to follow the settlement and hence to pay as well. Again, for the same commercial reasons, insurers often felt forced not to let their clients down and hence provided coverage.[85] This therefore meant that, in contrast to what was thought when the Deepwater Horizon incident occurred in 2010, (re-)insurers still largely intervened in paying part of the Deepwater Horizon–related losses, not as insurers of BP, but as insurers of some of the subcontractors of BP who, as a result of pressure exercised by BP, were forced to pay BP, notwithstanding the 'hold harmless' clauses.[86] These experiences hence have led to a much more cautious approach by insurers and to a general loss of appetite in covering offshore-related risks.

These negative effects of Deepwater Horizon on the available capacity and premiums have also been confirmed by practice. Prior to the Deepwater Horizon incident, energy underwriting rates in the offshore oil and gas industry were down 10 to 15 per cent. Shortly afterwards, a study carried out in October 2010 showed that premiums for insuring deepwater operations have increased by 25 to 30 per cent and deepwater drilling by 100 per cent or more.[87] However, the full impact of the increased insurance premiums was not felt until January 2011

[83] Interview with Swiss Re, 11 February 2013.
[84] See Philipp Wassenberg, representative of Munich Re Insurance Company, interview, 6 May 2013.
[85] *Ibid.* [86] *Ibid.*
[87] Booz Allen Hamilton, 'The Offshore Oil and Gas Industry', part 3, 15. See also Kollewe, 'Oil Industry Set for Surge'.

when re-insurance was bought.[88] Some oil and gas companies immediately purchased more insurance than they had done previously and, in some cases, even doubled it.[89] This led to a rather awkward situation that oil and gas companies improved their safety measures, but the insurance premiums did not decline. This is probably largely due to the increased demand for insurance coverage as a reaction to the Deepwater Horizon incident.

Some argued that the increased insurance costs for oil and gas companies in the Gulf of Mexico would not be passed on to end users. The purchasers for crude oil and gas may have different sources (e.g. from OPEC countries); therefore, the oil and gas companies would not increase the price to drive out the buyers.[90] If the company wanted to absorb the increasing operating costs, the larger firms may have had more assets to do so via the economies of scale.[91] In turn, increased insurance premiums make self-insurance a more attractive option for oil and gas firms. But this is feasible only for a few big firms with sufficient assets to do so.

5.4.3 *Analysis*

From the perspective of policy makers, they can rely on insurance together with self-insurance, as the most important instrument to cover offshore related risks.[92] Insurance provides a cheaper coverage than bank guarantees or letters of credit and is more easily applied as well for risks that may spread over a longer period than capital market instruments. The risk differentiation exercised by insurers can contribute to the prevention of risks.

However, while relying on insurance, the policy makers need to realize its limitations. Firstly, insurance cannot provide unlimited coverage. The earlier discussion showed that most insurers confirm that the available offshore insurance coverage in the market is around US$500 million to US$1.5 billion. It is worth noting, though, that the estimation of current coverage by insurers may not be an ideal indicator of the future available coverage. The negotiations for the Athens Convention relating to the carriage of passengers and their luggage by sea (PAL) and the OPA

88 Kollewe, 'Oil Industry Set for Surge'. 89 *Ibid.*

90 Booz Allen Hamilton, 'The Offshore Oil and Gas Industry', part 3, 17–18.

91 *Ibid.*, 18.

92 Representatives of OPOL held that most operators meet their obligations under OPOL via insurance (interview with representatives of OPOL, 27 March 2013).

of 1990 have shown that insurers have fiercely opposed the stringent liability and increased liability limits on the basis of limited insurance capacity.[93] It was even argued that the high liability limit and direct-action cause under the OPA would lead to a 'train wreck' phenomenon and crushing of the shipping industry in the United States.[94] However, after OPA went into force, the insurance market managed to mobilize resources to cover the oil pollution liability.[95] The potential to provide higher insurance coverage is confirmed by some insurers.[96] However, the provision of higher coverage may need strict conditions, such as high retentions and pooling of risks. The insurance market is volatile, which makes a stable high coverage difficult.[97] Moreover, though insurance is usually cheaper than bank guarantees or letters of credit, it is still costly compared to self-insurance. The running of an insurance company creates unavoidable transactions costs. Insurance companies are, moreover, profit-maximizing corporate entities and thus may add a profit margin to premiums. The premium charged by insurers thus may be higher than the actuarially fair (objective) value of the risk. This explains why operators, to the extent they can, use self-insurance.

5.5 Risk-Pooling Schemes

The risk-pooling schemes OIL and OCIL were discussed extensively in Chapter 4,[98] where the theoretical advantages of risk-pooling schemes also were sketched.[99] Here it will suffice to discuss the way in which those risk-pooling schemes are used in practice according to stakeholders.

Certain major oil companies argue that the standard rates charged by OIL and OCIL do not sufficiently reflect differing risks.[100] They argue that risk differentiation is better with insurance companies. The current pools mutualize (based on solidarity), but premiums do not sufficiently reflect risk and hence do not reward the good risks. For this reason, the pools are in fact only attractive for middle-size companies: because the majors usually pose better risks, they do not find the pools attractive and do not need the pools. For smaller companies, the high deductibles (of

93 Røsæg, 'The Impact of Insurance Practices'.
94 See US Coast Guard, 'Preliminary Regulatory Impact', 4–5.
95 Liu, *Compensating Ecological Damage*, 283.
96 See estimation of P&I Clubs and Lloyds in Sections 5.4.2.3 and 6.2.
97 See Section 5.4.2.3. 98 See Section 4.3. 99 See Section 4.1.
100 Interview with representatives of BP, 26 March 2013.

US$10 million) do not make membership attractive. This is why only medium-size players will join.[101] Representatives from insurers also hold that in the case of a large claim, OIL will try to recoup some of the money it had to pay to the third party. Via retrospective premiums, part of the money would have to be paid back, which makes such a model less attractive.[102]

Moreover, operators also argue that the risk pools do not have full solidarity because, depending upon the contractual arrangements, in some cases the liable operator will be compensated by OIL or OCIL but will have to repay (a part of) the damage over a specific (usually five years) period.[103] Also, other majors held that OIL and OCIL are not attractive for major players. The mutualization in OIL and OCIL could lead to the danger for major players of smaller operators taking advantage of the majors, in which case the majors would de facto become the guarantors of small players.[104] They argue that currently within these pools the risk differentiation is too low.

However, the problem here is that damage potentially can be very high, but the probability is very low. Hence, given low probabilities, the difference between a good risk and a high risk may be that the good risk pays e.g. US$30,000 in contributions and a high risk pays US$60,000. The difference is not large enough to provide adequate incentives for prevention. The high-risk operator could simply pay the contribution and still benefit from good risks that have to contribute after an accident. Pools hence provide some kind of safety net for smaller players with limited balance sheets, and risk differentiation is simply not sufficient.[105]

In sum, OIL and OCIL are apparently attractive for some middle-size players. As the discussion of OIL and OCIL earlier made clear, for those players, they do provide substantial coverage. However, for smaller players, they may not be attractive (given large retention), and for major players, they may not be attractive either, given the danger of a negative redistribution (low-risk majors contributing to high-risk members in the pool, thus creating adverse selection).

[101] *Ibid.*

[102] According to representatives of Lloyds, interview, 1 May 2013, in London. They argue that these pools were created at a time when there was less cover available in the market but that they may be less needed today.

[103] Discussion with representatives of OGP, 25 February 2013.

[104] Interview with representatives of Shell International BV. 14 March 2013, in Rotterdam.

[105] *Ibid.*

5.6 OPOL

Again, OPOL was discussed extensively earlier,[106] where also, on the basis of a preliminary analysis, the advantages and limits of OPOL were sketched.[107] In order to avoid repetition, it will suffice here to discuss the use of OPOL in practice according to stakeholders.

The function and goal of OPOL are relatively limited, and in that sense not comparable to insurance. The worldwide casualty energy market including Swiss Re excludes OPOL from the casualty coverage and policy. The reason is that not all damage is covered. An important condition is, *inter alia*, that pollution should be sudden and accidental; gradual pollution is excluded. The OPOL coverage is much larger and therefore excluded.[108]

OPOL is important in the United Kingdom, where membership in OPOL is mandatory to obtain a licence. Representatives of a large reinsurer held that some risk sharing between operators (thus enlarging OPOL) may be a good idea. This would mean that operators would take a larger part of the risk via, for instance, co-insurance. If OPOL was thus constructed as a full pooling mechanism, operators would take a higher amount of the risk, and insurers would only intervene as a second layer.[109]

Representatives from insurance companies also mention that OPOL only provides coverage of up to US$250 million and only guarantees insolvency. US$250 million is in fact not considered a high amount compared to what the total damage resulting from a catastrophic blowout could be.[110] Moreover, operators also mention that the importance of OPOL is to an important extent limited to the United Kingdom and de facto only applies to the North Sea.[111] Since OPOL in fact only provides a guarantee, it is also considered that its role is relatively limited. The duty to compensate is, after all, still on the operator, not on OPOL.[112] In this sense, it could be held that the cover provided through OPOL worked because it has never had to be used.[113]

[106] See Section 4.2. [107] See Section 4.2.8.
[108] Interview with Swiss Re, 11 February 2013. [109] *Ibid.*
[110] See Phil Bell, interview, 26 February 2013.
[111] Discussion with representatives of OGP in Brussels, 25 February 2013. In the words of the representative of DECC, 'OPOL is our thing' (referring to the fact that it is UK based) (Wendy Kennedy, Head of Offshore Oil and Gas, Department of Energy and Climate Change [DECC], 29 April 2013).
[112] *Ibid.* [113] See representatives of Oil and Gas UK, interview, 1 May 2013, in London.

The way OPOL works in relation to insurance is that operators in the North Sea who are parties to OPOL can buy an endorsement to their operators' extra expense (OEE) cover that will meet the obligations insured under OPOL. This then provides cover on a strict liability basis for compensation to third parties, including local authorities, for pollution damage and for clean-up expenses.[114] The liability under OPOL goes further than traditional insurance wordings because OPOL does not require a pollution event to be on a sudden and accidental basis, and OPOL guarantees the payment of sums to claimants in the unlikely event that a party to OPOL fails to meet its obligations. The market has addressed these issues to ensure that the coverage provided under the policies was (via OPOL) not more than intended by the insurers. This is the reason why the OPOL endorsement has been adapted to clarify that insurers will not respond to the OPOL (solvency) guarantee.[115]

5.7 Combinations

In practice, it is rare that only one type of instrument would be used. In fact, this may only be the case for the majors, who effectively only use self-insurance or captives. Others de facto often use a combination of different hedging strategies whereby, logically, the comparative benefits of the various instruments are used for an optimal combination.

As an example, a middle-sized operator may choose retention (self-insurance) of e.g. €5 million and could choose insurance or a risk-pooling scheme to cover the excess risk. Moreover, the operator could (and in the case of the UK must) also be a member of OPOL, in which case he would use the self-insurance and insurance in combination as proof of financial security.

The type of combinations that will be chosen by operators in practice may, of course, to an important extent depend upon their size and hence their assets but also on the types of risks to which they are exposed and the relative costs of various strategies to transfer risk, such as the relative costs of insurance versus risk pooling.

In summary, the following instruments are currently used to cover liability following a major offshore accident:

[114] Rees and Sharp, *Drilling in Extreme Environments*, p. 31. [115] *Ibid.*, 36.

- Self-insurance, largely by the majors, who in some cases create captives as well.
- The capital market, although actually today only to a very limited extent.
- Guarantees are possible in theory but not that often used in practice. Bank guarantees or letters of credit are simply considered too costly and are hence not used.
- (Re-)Insurance is undoubtedly the most often used mechanism of financial security for offshore related risks.
- Risk-pooling schemes such as OIL and OCIL are mostly used by middle-sized players.
- OPOL is not as such a system of financial security, but OPOL is important in the United Kingdom, where membership in OPOL is mandatory for offshore operators in order to obtain a licence.
- In practice, depending upon their size, balance sheet, assets and risks to which they are exposed, operators may use a combination of any of the financial instruments just mentioned.

These instruments provide different alternatives to cover offshore liability. However, each of them has its own limits. Capital market and guarantees are rarely used in practice due to their high costs. Some instruments are only suitable for big or medium-sized players.

Moreover, current coverage under these instruments is limited. The risk-sharing pools, including OIL, OCIL and OPOL, provide only coverage up to a few hundreds of millions of US dollars. The insurance market is currently providing higher amounts, usually up to US$0.5 billion to US$1.5 billion. These amounts, however, could still be substantially lower than the damage caused by a catastrophic accident. The experience of PAL and OPA shows that the stricter liability rules and higher financial requirements may incentivize higher amounts of insurance coverage. However, a careful study needs to be conducted to make the coverage possible.

In addition, close monitoring will be necessary to ensure the security of these instruments and to incentivize the preventive actions of offshore operators. For example, regulatory agencies need to conduct serious controls to verify the viability of the self-insurance as guarantee. Safety regulation and comparable risks are also essential for the creation of mutuality among offshore facilities and to incentivize them to pool risks together.

6

Potential of Financial and Insurance Instruments to Cover Liability Following a Major Offshore Accident

MICHAEL FAURE AND HUI WANG

This chapter will take a forward-looking approach by examining what the potential is of various insurance- and financial market–based instruments to provide coverage for damage caused by offshore installations. Different from Chapter 5, which discussed the status quo of the available financial cover, this chapter is rather forward looking and hence analyzes the potential of insurance and other financial instruments to provide cover for a potential future liability regime. To some extent, this chapter is therefore unavoidably speculative. It is based on opinions of experts on what amounts would be insurable in a European market and the literature on financial security instruments in other markets. Hence, the question that will mainly be asked is to what extent a liability and compensation scheme could be covered by insurance and financial instruments. In order to make such an assessment, it may be interesting not only to focus on the offshore liability sector but also to compare with other sectors. An obvious option is to look at amounts available for vessel-based pollution.[1] Also, developments as far as nuclear liability is concerned are interesting to consider.

This chapter will first provide an estimate of the potential costs of a major incident with an offshore installation. These estimates will be based not only on an analysis of past events but also on evidence provided by experts from both the oil and gas and insurance industries.

[1] Representatives of re-insurers (in particular Swiss Re) explicitly indicated during an interview (on 14 February 2013) that amounts of damage and compensation available for vessel-based pollution provide an interesting indication for the offshore sector as well. Even though there may be technical differences in both cases, the basis of liability is constituted by a release of oil, and a large part of compensation can be related to clean-up costs.

The experience with similar accidents in the vessel-based pollution area will be included as well.

If indeed one wishes to examine the adequacy of particular financial or insurance instruments to cover liabilities, one should at least have some idea of the potential costs of an offshore incident (6.1). Next, a proposal to cover potential catastrophes by self-guarantee is discussed (6.2). Another approach is to broaden the capacity of insurance and (re-) insurance and to cover liability caused by offshore incidents. In this section, a comparison with the coverage in other high-risk sectors will also be provided, in particular, vessel-induced oil pollution and nuclear damage. Attention will also be paid to the potential of regional or international risk pooling (6.3). The most important question that will arise in that respect is whether the current Offshore Pollution Liability Agreement (OPOL) mechanism[2] could be expanded or a different, more elaborate risk-sharing agreement for damage resulting from offshore installations could be developed. This will be based on expert opinions from OPOL, insurers and the stakeholders within the oil and gas industry. The potential to resort to government to cover major offshore accidents will also be discussed (6.4). Finally, the question arises as to what extent a combination of different insurance and financial instruments (such as operator guarantees, insurance, and international risk pooling) would be feasible in a type of multi-layered compensation scheme (6.5).

6.1 Potential Costs of an Offshore Incident

The attention currently paid to the potential for damage resulting from offshore installations is, of course, partially the result of the 'hype' after the Deepwater Horizon incident in 2010. The question, however, arises as what the potential cost is of a major offshore accident and to what extent the risk of an accident such as the Deepwater Horizon oil spill also exists in waters beyond the Gulf of Mexico. If this were the case, it would be an important justification for regulatory action. Answering a question such as that is obviously highly speculative. The goal of this section is therefore not to speculate on precise amounts of the potential costs of a (major) offshore incident but rather to attempt, based on interviews with stakeholders, to assess whether the current system of financial coverage (as sketched in Chapter 5) would be sufficient. If the answer is that this is not

[2] This was discussed in more detail in Section 4.2.

the case, this chapter will subsequently address possibilities of increasing the amounts under the current or alternative schemes. In this respect, it will first address the potential costs of an offshore accident by looking at past loss experience and the regional differences (6.1.1). Then the insolvency risks will be considered (6.1.2).

6.1.1 Potential Costs and Verifications

As discussed earlier, the accurate estimation of the potential costs of offshore accidents is very difficult. One possible way of approaching the potential costs of an offshore incident is to look, as was done in Chapter 2, at the available data on the costs of oil spills, though looking backwards may not provide an ideal indication for the future. However, even data on costs of past accidents are limited. In addition to the amounts of products spilled, they do not always provide clear indications in monetary terms of the costs of oil spills. In general, the data seem to indicate that although there have been quite a number of incidents, many of those were not very significant. However, high-magnitude damage also has occurred both in the United States and in other parts of the world. As shown in Table 1, a major accident can lead to hundreds of millions US dollar losses in terms of operators' extra expense (OEE) insurance alone.

Different regions also present different risks. The past loss experience shows that the Gulf of Mexico presented a high risk of major accidents. In addition to the Deepwater Horizon oil spill, the most expensive offshore accident so far, a few other major accidents also caused damage up to hundreds of millions of US dollars.[3] The loss experience in European waters shows a much better record.[4] Lots of

[3] See Table 1.

[4] Noting, however, that this by no means guarantees that a major accident will not occur in European waters. Various stakeholders have reported on those specific differences as well as on the likelihood of a major (catastrophic) blowout in European waters. The results show, as often, a nuanced picture. Obviously, the assessment on the likelihood of a catastrophic incident depends very much on the particular position of the stakeholders interviewed. Generalizing, not surprisingly, oil and gas producers tended to stress that the occurrence of a Deepwater Horizon incident in the European Union is highly unlikely, and if it were to occur, the damage would be substantially lower than in the Gulf; representatives from government and especially some re-insurers to the contrary hold that a Deepwater Horizon incident could equally occur in European waters and that there is no reason to assume that if this were so, the total damage would be substantially lower than in the Gulf. Obviously, the particular position taken by the stakeholders may also be reflected to some extent by the different (also economic) interests at stake.

factors can influence the potential risks, including safety regulation technical differences in various waters, differences in operators, different liability rules and so on.

6.1.1.1 Safety Regulation

Safety regulation by public and private actors is an important issue to ensure the safe functioning of offshore facilities and to prevent potential accidents. Many stakeholders, understandably mostly representatives of industry but also insurers, mentioned that safety levels and standards have strongly increased over the years. It is, for example, stressed that especially 'the majors' have an extensive knowledge and expertise in preventive technology and that their primary interest is the prevention of a blowout.[5] This is mentioned not only for reasons of reputational risks but also for the simple reason that even if there were no damage to the environment surrounding a well, the mere costs of well control can be substantial. Operators therefore will have strong incentives to invest in safety to prevent those risks.[6] 'Risks have hence substantially reduced, and as a result of this it is not so clear what still could be done additionally.'[7]

6.1.1.2 Technical Differences in Different Waters

The technical differences in different waters, such as meteorological and hydrological conditions and types of drilling facilities, can lead to different risks of offshore operations. For example, many stakeholders point at substantial differences between the technical situation in the Gulf of Mexico and in European waters.[8] In European waters, especially in the North Sea, there would be more gas drilling than oil, and in the Gulf of Mexico, there is much more oil drilling than gas. Gas can lead to personal injury but less to environmental harm, whereas the reverse is true in the

5 Interview with Phil Bell, 26 February 2013; interview with Wendy Kennedy, head of Offshore Oil and Gas, Department of Energy and Climate Change (DECC), 29 April 2013.

6 This was also stressed by a representative of ExxonMobil, interview, 19 February 2013.

7 See Wendy Kennedy, head of Offshore Oil and Gas, Department of Energy and Climate Change (DECC), 29 April 2013. See also the e-mail from the International Association of Oil and Gas Producers (OGP) to Michael Faure, 13 May 2013, stating: 'Moreover, the global upstream oil and gas industry has dedicated significant time and resources to accident prevention, intervention and oil spill response ... These cover well design and process safety procedures, and competence assurance for well operations personnel to lessen the risk of further accidents occurring offshore.'

8 Rochette *et al.*, *Seeing Beyond the Horizon*, 6.

case of oil spills: less risk of personal injuries but substantial environmental damage.[9] The argument is also made that drilling in the Gulf of Mexico is not merely deepwater drilling but even ultradeep,[10] 3 kilometres and more. Drilling would not be as deep in the North Sea, which would be less risky.[11] Moreover, it is stressed that large losses suffered in the Gulf of Mexico are often related to hurricanes. Most of the largest losses outside of Europe were all caused by hurricanes.[12] Hurricanes may be a serious problem in the Gulf of Mexico but not in European waters.

Macondo was, moreover, a high-risk, high-temperature well and a new reservoir with a lot of pressure. Some hold that many wells in the North Sea are not under high pressure, which would seriously reduce the risk of blowouts.[13] The argument is also made that the Gulf of Mexico is more risky as far as hurricanes are concerned, whereas Western Europe could have substantially high waves. For example, west of the Shetlands, waves could go as high as 20 to 25 metres. It is also held that in case of a spill, oil may disperse more easily in water as a result of high waves. At the same time, it is held that the high waves could also increase risk which is normally taken into account in the construction of oil rigs.[14] However, others stress that perhaps risks in the North Sea, especially in those areas where wells are old (and hence the pressure is low) and there is no deepwater drilling, may be relatively limited. However, increasingly, drilling also occurs in other European waters than the North Sea, where risks may be substantial.[15]

It also is held that in case of vessel-based pollution, damage would be substantially larger because the amount of oil that comes into the sea when e.g. a vessel breaks down is normally much larger than in case of an incident with an offshore installation, where through a variety of technical devices (capping devices or a so-called blowout preventer) it may be possible to better control the amount of oil leaking into the environment.[16] However,

[9] Interview with Jan de Jong, inspector general of state mines, 22 February 2013.

[10] See on this notion Section 2.1.1 and footnote 4.

[11] Interview with representatives from Shell, 26 March 2013.

[12] E-mail from the International Association of Oil and Gas Producers (OGP) to Michael Faure, 13 May 2013.

[13] Interviews with representatives from Shell, 26 March 2013; interview with Wendy Kennedy, head of Offshore Oil and Gas, Department of Energy and Climate Change (DECC), 29 April 2013.

[14] Discussion with representatives of OGP, 25 February 2013.

[15] See Philipp Wassenberg, representative of Munich Re, interview, 6 May 2013.

[16] Interview with Jan de Jong, inspector general of the state mines, 22 February 2013; interview with representatives of the International Group of P&I Clubs, 1 May 2013, in London.

there again, the counterargument was made that with a tanker the total amount of oil that can be spilled is known ex ante, whereas with a blowout from an offshore installation, the potential amount of product that can come out of an uncontrolled well can be much larger.[17]

6.1.1.3 Differences in Operators

It should also be remembered that even in the waters belonging to the Netherlands in the North Sea, on a yearly basis, still significant and large gas releases take place.[18] They are not yet disasters but could potentially give rise to disasters. Moreover, even after the Deepwater Horizon incident in 2010, some major incidents with offshore installations on a worldwide basis still took place.[19]

It is also stressed, especially by the majors, that there may be many different players in the market. The majors claim that they have a large reputation at stake which incentivizes them to strive for very high safety standards. Some smaller and medium-sized operators may engage in a 'hit and run' strategy, like venture capital players. In theory, they could borrow money to finance the drilling operations. Once they have the license, the oil and gas reservoir would basically be their balance sheet, but the way in which they operate may potentially be very risky.[20] The risk would be further amplified by the fact that there is not enough insurance coverage today for offshore installations on the global market, especially if a disaster like the Deepwater Horizon were to occur again in the Gulf of Mexico.[21]

6.1.1.4 Different Liability Standards

Liability rules also have an important influence on the potential risks to which the offshore facilities are exposed. For example, there are substantial differences between Europe and the United States both as far as procedures are concerned (very active role of the plaintiff bar, allowing to use

[17] Interview with Philipp Wassenberg, representative of Munich Re, 6 May 2013; interview with representatives of the International Group of P&I Clubs, 1 May 2013, in London (they argued that with an offshore well [contrary to with a tanker], the amount of oil that can be spilled is in principle infinite).

[18] See the overview in Chapter 2.

[19] It concerns more particularly incidents in Mexico (2011), the United Kingdom (2012), Nigeria (2012) and Brazil (2012).

[20] Interview with representatives of Shell International BV, 26 March 2013. It was a concern equally expressed by Jan de Jong, inspector general state mines, on 22 February 2013 and by representatives of BP on 26 March 2013.

[21] Interview with representatives of Noble Energy, 6 March 2013.

discovery mechanisms and trial by jury) and as far as differences in the substance of tort law (much higher amounts of pain and suffering and punitive damages) in addition to a different remuneration system for lawyers (the contingency pay system for lawyers in the United States would attract more claims). These factors may make the American system of litigation substantially more costly than the European system.[22] Recall that BP not only allocated US$20 billion to the fund but also had to pay criminal fines and was still in litigation concerning additional civil fines for violation of Oil Pollution Act (OPA) and the Clean Water Act (CWA).[23] As a result of this, the total amount due by BP amounted to almost US$40 billion. Many stakeholders hold that the costs probably would have been substantially lower if the incident had occurred in Europe.[24] However, a representative of another re-insurer holds that the costs would be comparable.[25] In this respect, it is (correctly) pointed out that the Gulf Coast Claims Facility (which handled many of the claims resulting from the Deepwater Horizon incident) is not paying out punitive damages but heads of damages, such as property losses and compensation for business interruption, which would also be largely recoverable under the law of most European legal systems.[26]

Despite the regional differences, this does not suggest that a major accident such as the Deepwater Horizon oil spill will not hit other waters beyond the Gulf of Mexico. With increasing oil prices, operators may now also move to fields that were previously not used because they were too difficult to reach (and hence too costly). This may e.g. explain why operators are now also increasingly moving to the Mediterranean or the Black Sea. However, moving to areas at greater depth and in more difficult circumstances could increase risks as well.[27] Also, regulators warn that even the Piper Alpha incident only started with a so-called significant release of gas. One has to realize, however, that a significant release of gas can potentially turn to disaster if there is a spark source present that could cause an explosion to take place.[28] Moreover, the previously mentioned increasing safety standards may have reduced considerably the

[22] Interview with Wendy Kennedy, head of Offshore Oil and Gas, Department of Energy and Climate Change (DECC), 29 April 2013.

[23] See Section 2.5.2.

[24] See interview with representatives of Swiss Re Insurance Company, 11 February 2013.

[25] Interview with Philipp Wassenberg, representative of Munich Re, 6 May 2013.

[26] *Ibid.* On the types of damages paid as civil compensation in the Deepwater Horizon case, see Section 4.9.4.

[27] See Phil Bell, interview, 26 February 2013.

[28] Interview with Jan de Jong, inspector general of the state mines, 22 February 2013.

probability of a large incident. However, the potential consequences (and hence the damage) in the case of a catastrophic blowout have not changed. The essence of an offshore incident is that even though the probability is low, it can potentially still lead to high damage when it occurs.[29]

Also, a report by Lloyds after Macondo deals explicitly with the question of whether a Macondo is likely to happen again. The report criticizes industry commentators who argue that 50,000 wells have been drilled since 1947 with only one occurrence of a Macondo-style event. It is countered that these numbers may be masking reality, as the absence of more incidents does not mean that the risk is only a 1 in 50,000 event. A study found that of the 5,000 wells drilled since 1993, only forty-three were as complex as the Macondo well. Hence, the Lloyds report raised the question of whether the probability of another Macondo-type event is rather one in forty-three or even higher given that the rate of wellbore instability in these wells is unusually high.[30] Interestingly, the Lloyds report also points to behavioural biases as a result of which experts were unable to conceive of the possible negative consequences of scenarios that were difficult to imagine. As a result, a loss of well control for eighty-seven days, as happened with Macondo, was not even considered possible. The Lloyds report also points at the well-known hindsight bias,[31] by which people conclude that their ability to handle a past event was higher than it really was. Hence, it should be warned that the more time that passes without any incidents, the more easily organizations lower their estimates of the probability of an accident occurring.[32]

6.1.2 Insolvency Risk and Different Scenarios

The preceding analysis shows that though the probability is low, the potential magnitude of damage caused by offshore accidents can be large. The question thus would be whether the potential damage will lead to insolvency risks of operators and hence will call for improvement of the current financial security system. This will depend on the size of the accident, the available assets of the operator and the financial security instruments available to them.

It is held that within Europe, since 1972 (the date of the beginning of the Willis Database), there have been no incidents that would exceed the

[29] *Ibid.* [30] Rees and Sharp, *Drilling in Extreme Environments*, 28.
[31] On this hindsight bias, see also Rachlinski, 'A Positive Psychological Theory', 571–625.
[32] Rees and Sharp, *Drilling in Extreme Environments*, 28.

market capacity. Moreover, even considered globally, with the exception of Macondo, all other historical losses did fall well within the insurance market capacity available.[33] Even some representatives from insurers argue that, especially as far as the majors are concerned, one can wonder whether there is today a serious problem that needs solving.[34] They especially warn against ideas such as regulatory solutions that would force large oil and gas producers (OGPs) to transfer risks to insurance companies that have de facto a lower rating than the major OGPs. The latter would simply amount to a mandatory transfer of wealth from stronger to weaker parties, which makes little sense economically.[35] There is, in other words, a large consensus that (as also the Deepwater Horizon case showed) the insolvency risk may not be present with the majors, and even if it were, forcing them to transfer risk to lower-rated insurance companies would only create additional costs without additional benefits.

However, representatives, especially from regulators but also from some re-insurers,[36] maintain that it would be wrong to argue that there is no problem as far as the insolvency risk is concerned. First, it is argued that as a result of increased safety standards, the probability of an incident may indeed have been reduced, but that does not mean that when a major blowout occurs, the damage could not be catastrophic.[37] Second, they also stress that the mere fact that nothing has happened (as far as a major incident is concerned) in Europe during the past twenty-five years is, of course, no guarantee that there could be no major accident. 'This could easily change tomorrow.'[38] Third, the current guarantee from the market is limited. For example, the OPOL guarantee is limited to US$250 million. The commercial insurance market can provide coverage up to US$500 million to US$1.5 billion, varying according to the estimation of different insurers. The Deepwater Horizon oil spill shows that bigger magnitudes of damage are possible. Though the majors can resort to their deep pockets and cover such damage via self-insurance, such a major accident could easily put smaller and middle-sized operators out of business.[39]

To avoid the insolvency risk of small operators, stakeholders mentioned two approaches: some argue that this particular risk, of small

[33] E-mail from International Association of Oil and Gas Producers (OGP), 13 May 2013, to Michael Faure.
[34] Interview with Phil Bell, 26 February 2013. [35] *Ibid.*
[36] Interview with Philipp Wassenberg, representative of Munich Re, 6 May 2013.
[37] Interview with Jan de Jong, inspector general of the state mines, 22 February 2013.
[38] *Ibid.* [39] Interview with representatives of Noble Energy, 6 March 2013.

stakeholders (with smaller assets) potentially creating large damage, could be taken care of by regulators by requiring smaller stakeholders to engage in joint ventures with majors as a result of which the financial viability would increase. In such cases, the majors would de facto take some of the risk of insolvency from their smaller counterparts.[40]

Another possibility is to make an accurate risk assessment (along the lines of the Oil and Gas UK Financial Responsibility Guidelines) as a result of which smaller operators would only be allowed to operate in less risky areas,[41] where the probability of accidents and the magnitude if they were to occur are lower.[42] Some representatives from middle-sized OGPs, however, argue against this, saying that it may not be desirable from the perspective of competition because it would de facto lead to an oligopoly of major oil companies which would self-insure and might drive smaller and independent actors out of the market. This may be an undesirable market development.[43]

Because both approaches have serious limitations, it is helpful to evaluate the insolvency risks seriously, considering different sizes of accidents. For smaller incidents, mandatory financial security is required in many regimes. For example, in the United States, the required financial coverage of offshore facilities varies from US$10 million to US $150 million.[44] Another private scheme, OPOL, mandates and complements financial security up to US$250 million. However, OPOL is limited to the North Sea and, moreover, membership of OPOL is mandatory only in the United Kingdom. It therefore de facto only covers UK operators.[45] Hence, OPOL cannot provide a guarantee against insolvency outside the North Sea area (such as in the Mediterranean, where offshore operations are increasing). Moreover, for non-UK operators in the North Sea, it depends on whether the national regulators require membership in

[40] Interview with representatives of BP, 13 March 2013; interview with representatives of Shell International BV, 14 March 2013.

[41] For example, more remote from the coast, shallow water instead of deepwater, less pressurized wells, no high-temperature wells and so on.

[42] Interview with representatives from Shell International BV, 14 March 2013.

[43] Interview with representatives from Noble Energy, 6 March 2013.

[44] See Section 3.5.5.10.

[45] Interview with representatives from OPOL, 27 March 2013. Also, DECC will initially rely on OPOL. However, in many cases, financial security will be required above the minimum of US$250 million provided by OPOL. This is why in practice DECC often requires a combination of different kinds of financial securities (interview with Wendy Kennedy, head of Offshore Oil and Gas, Department of Energy and Climate Change [DECC], 29 April 2013).

OPOL or another type of financial responsibility as guarantee. Therefore, in countries without mandatory financial security, it is unclear whether the cost of a smaller incident (e.g. up to US$250 million) is always recoverable.

For middle-sized accidents (defined as having a magnitude between US$250 million and the maximum insurance coverage available on the market), to the extent that operators take insurance coverage, there should be no problem. Estimates of the available amount on the insurance market varies from US$500 million to US$1.5 billion. Considering the volatility of insurance market for offshore facilities, for moderate damage estimation, US$750 million is used here to signify the limit of middle-sized accidents. A problem from a policy perspective is that there may not be a uniform regulation across Europe. Hence, this only works to the extent that regulators de facto force operators ex ante to take financial coverage (such as insurance or membership on OIL or OCIL) to provide coverage up to the maximum amount available.

For large accidents, it follows that only the majors could provide coverage beyond the limits of commercial insurance coverage available on the market via self-insurance. In such a case, financial security instruments should be further improved to ensure that the offshore actors will internalize the costs of a potential major accident.

The remaining part of this chapter discusses a few proposals by scholars and stakeholders to provide financial coverage for a potentially large accident.

6.2 Self-Guarantee Through Tax

Viscusi and Zeckhauser proposed a two-tiered compensation scheme to cover offshore liability. For the first tier, the responsible party should be able to pay through the financial security and its own assets. The amount of the first layer should be decided according to the past loss experience or an estimation that there would only be a less than 1/100 chance that it would be exceeded in the next decade.[46]

The second tier would be the low-probability catastrophe that might happen. This will be covered by the firm's annual payment to a federal fund.[47] 'For example, if government regulators thought that a deepwater well in the Gulf had an annual one in one thousand chance of imposing losses beyond $20 billion, and if the expected magnitude of total losses in

[46] Viscusi and Zeckhauser, 'Deterring and Compensating Oil-Spills', 1738. [47] *Ibid.*

such a case were $50 billion, then the well operator would pay $30 million annually into the fund, which is (1/1000) × ($50 billion – $20 billion).'[48]

Hence, no risk spreading exists with this second-layer compensation: it will be solely the responsible party himself who pays for the potential damage though the accumulation of his taxes. The payment would be affordable because the chance that a large accident beyond his capacity happens is very low.

6.3 Expanding Possibility of Insurance?

The question of the extent to which higher amounts of insurance coverage than available today will rise[49] will be approached from various angles: first, the opinions of stakeholders concerning the possibility of increasing coverage will be sketched (6.3.1); next, a particular proposal launched by the Munich Re-Insurance Company to cover catastrophic losses especially developed for the Gulf of Mexico will be discussed (6.3.2). Then the coverage available today for vessel-based pollution will be presented (6.3.3) as well as some developments concerning the coverage of nuclear risk (6.3.4).

6.3.1 Mandatory Insurance and Increasing Coverage?

One approach to alleviate insolvency risk is to make the insurance mandatory. Cohen and colleagues proposed to use mandatory insurance in place of the financial responsibility requirement under the US Oil Pollution Act of 1990 (OPA).[50] They prefer insurance to other instruments under the OPA, arguing that insurers can conduct effective monitoring on the behaviour of offshore companies and impose risk-based premiums.[51] This analysis shows that the existing insurance market provides only around US$1 billion in coverage for offshore liability. Cohen and colleagues' proposal requires the insurance market to raise adequate capital to provide a substantially higher amount of coverage. This constitutes a major challenge to the insurance market.[52]

48 *Ibid.*, 1738–9. 49 That was sketched in Section 5.4.2.3.

50 Cohen *et al.*, "Deepwater Drilling', 1899. 51 *Ibid.*

52 Some argue that it is impossible for energy insurers to raise coverage of US$10 billion given the unpredictability of offshore accidents. Hearing on the Liability and Financial Responsibility for Oil Spills Under the Oil Pollution Act of 1990 and Related Statute before the H. Comm. on Transportation and Infrastructure, 111th Cong. 22 (2010) (statement of Robert P. Hartwig, President and Economist, Insurance Information Institute).

First, as far as the question as to whether higher amounts of insurance coverage could be made available for offshore-related risks is concerned, the general principles of insurance as they were sketched earlier[53] need to be recalled. An insurance company is only able to provide coverage if a sufficiently large number of insured parties join the risk pool and sufficient premiums can be generated to actually create capacity to cover the risk. The problem of insurance capacity is one that is often mentioned as a problem when discussing catastrophic losses.[54] One problem of supplying catastrophe insurance is the so-called problem of correlation. For example when flooding occurs, it usually affects one specified area, and the risks are highly correlative, which may create an insurability problem.[55] The problem of correlation is not directly one that may affect offshore-related risks. However, there is another problem affecting the supply of catastrophe insurance as well: the absence of historical data and imperfect scientific knowledge to predict the risk. This may be a problem in the case of offshore-related risks, where many argue that objective and reliable information on probabilities and potential damage is often lacking. The third problem may be the most serious one – that insurance companies simply need sufficient financial reserves to cover the particular catastrophic risks. The capital needed by insurance firms to be able to cope with catastrophic losses must be high enough to cover, first, the expected claim costs and other expenses and, second, the costs of allocating risk capital to underwrite this risk.[56] Losses due to a catastrophe can be of such a magnitude that they can endanger the financial viability of insurance companies. Not only can the capacity of individual insurers be too limited for the large amount of the losses caused by a catastrophe, but the low-probability/high-loss character also makes the catastrophe inherently difficult to insure. The basic problem is indeed that the level of capital for insuring extreme events is often higher than what can be collected via insurance premiums.[57] This may be a serious problem in the case of offshore-related risks. It is easy to understand that for an insurance company, even in combination with other insurers (co-insurance) or with re-insurance, providing capital of, say, €2 billion

[53] See Section 5.4.1.

[54] See in this respect Gollier, 'Some Aspects'; and the contributions in Froot, *The Financing of Catastrophe Risk*.

[55] See Faure and Bruggeman, 'Catastrophic Risks', 28.

[56] See Doherty *et al.*, *Managing Large-Scale Risks*, 149.

[57] See Courbage and Stahel, *Extreme Events and Insurance*, 24.

in coverage requires not only high premiums but also efforts of many partners to put those large amounts at risk. This is why many stakeholders doubt that it is possible to generate substantially higher amounts than what is currently available on the market. It is, for example, considered that policy makers often do not correctly understand the precise functioning of insurance and the problems that arise in offshore-related risks. One problem is that the pool of insured (who have a demand for coverage) is often relatively small and consists of small players. This also means that they can only generate a relatively limited amount of assets in premium and thus in capital.[58] Again, this problem is exacerbated by the fact that usually only the small and middle-sized OGPs ask for insurance coverage because the majors self-insure.[59] This is sometimes misunderstood by policy makers, so stakeholders report. Thus, they would argue that hurricane Sandy led to US$13 billion in damage, which was largely insured, as a result of which they would question why similar amounts could not be raised for damage caused by an offshore installation. However, for weather-related damage and natural catastrophes, not only are there a lot of models and data which make the predictability very high, but the numbered of insured parties having a demand for coverage may be very large as well, thus creating potentially a large pool of policyholders that pay the premiums and thus generate sufficient capacity. Such features are precisely lacking in the offshore sector, which explains the reluctance of insurers to provide higher coverage than is available in the market today.[60]

In this respect, it should also be recalled what was mentioned earlier when discussing the effects of the Deepwater Horizon incident on the insurance market – that de facto after Deepwater Horizon the amounts of coverage decreased.[61] The reason was that insurers suddenly were unexpectedly confronted with an accumulation problem resulting from the fact that in the offshore market through different contractual relationships it is not always clear to (re-)insurers what risks they are actually covering. As a result, after Deepwater Horizon, offshore installations were considered to be an exposed business leading to a reduction of the capacity. Capacity for offshore-related risks hence generally went down, and obviously, prices for coverage went up.[62]

[58] Interview with Phil Bell, 26 February 2013. [59] *Ibid.* [60] *Ibid.*
[61] See Section 5.4.2.3. [62] Interview with representatives of Swiss Re, 11 February 2013.

6.3.2 Proposal of Munich Re-Insurance

6.3.2.1 An SOS Cover

Against all the stakeholders just mentioned, all arguing that it is not realistic to increase capacity, Munich Re-Insurance Company presented shortly after the Deepwater Horizon accident a proposal to create a facility that would be able to generate substantial capacity for offshore-related risks. The facility is referred to as 'SOS' (for 'Sudden Oil Spill'). The model presented by Munich Re would call for annual aggregate limits of US$10 billion to US$20 billion for companies engaged in offshore oil exploration. It would require the participation of multiple insurers and re-insurers and would strongly rely upon improved risk management. There would be three different models under which this facility could work: (1) a consortium of insurers and re-insurers, each providing uniform prices and conditions and fixed capacity, (2) traditional insurance or re-insurance on a subscription basis, with flexible pricing, conditions and limits and (3) a pool for oil drilling companies with contributions reflecting market share.[63] It was added that this would require the current limits under the US OPA to be raised above the US$75 million currently stipulated in the OPA.[64] Coverage, moreover, would only be provided above a retention of US$1 billion to US $1.5 billion.[65] Munich Re itself would commit as much as $2 billion to provide the total cover of $10 billion to $20 billion.[66] The concept was developed by Munich Re to cover drilling operations in the United States and, in particular, the Gulf of Mexico.[67]

So far Munich Re has developed two projects. One is for the Gulf of Mexico (the SOS cover), which could, in principle, be operational immediately; the second is the EU Oil Spill Project, showing that a similar model could be developed for the European Union as well. The basic structure of the model developed for the Gulf works as follows:

Capacity MR	US$2.00 billion
Another insurer	US$0.25 billion
Another large insurer	US$7.75 billion
TOTAL	US$10.00 billion

63 See www.businessinsurance.com/article/20100912/news/100919977.

64 See Section 4.5.2. 65 *Ibid.*

66 www.bloomberg.com/new/2010-12-21/munich-re-brokers-work-on-drilling-cover-after-gu.

67 http://reinsurancemagazine.com/article/munich-re-first-create-potential-pollution-liabi lity-solution.

In addition to a retention by the oil industry, this would already create an amount of US$12 billion.

Forty other insurers have shown interest in this model and have argued that they will participate if the model works. This would hence lead to an additional US$6 billion, and with retentions, the total coverage available under the new model would be US$20 billion.[68]

The basic reason why a much higher amount can be generated is that unlike the case today, separate coverage would be provided in liability for separate installations. The high amounts can only be provided if the cover is constructed as a natural catastrophe ('natcat') cover. Although de facto third-party liability would still be covered, the construction of the facility would be different. The cover would still follow the liability, but only to a modest extent.

The essence of the new facility would be that it provides so-called parametric trigger coverage. This means that coverage is triggered by particular events that have been specified beforehand. This is also the case with natcats, where it is, for example, determined that when a hurricane with scale five hits and/or winds of 150 kilometre/hour take place, coverage will be provided. These types of triggers are binary: either there is trigger or not – it is all or nothing.[69]

With a normal liability cover for offshore, there is indeed, on average, only US$1 billion in liability coverage available. The reason is that those liability covers are potentially long-tail liability risks. After ten years, claims still could be handed in. This creates a lot of uncertainty for shareholders, which is considered highly undesirable. Hence, the key issue in the new facility is to transform a long-term risk into a short-term risk, just as in the case of the natcats. Today, natcats are insured on a worldwide basis for several billions in many countries. These types of natcat coverage allow the generation of much higher sums than traditional liability cover.[70]

A second reason why the new facility can generate much higher amounts is that a general liability cover is not provided. Traditional liability cover would e.g. cover all installations of BP generally for €1 billion. The new facility works differently. It does not provide general liability cover for a firm but rather liability cover based on different scenarios. These scenarios are ex ante calculable and predictable. A scenario analysis would e.g. limit the coverage to all BP installations on the North Sea rather than all liability risks of BP worldwide.

[68] Interview with Philipp Wassenberg, representative of Munich Re, 6 May 2013.
[69] *Ibid.* [70] *Ibid.*

Such a scenario-based casualty approach is so far unknown. However, it leads to a better protection of the oil industry, of the consumers and potential victims, and hence of society.[71] Moreover, a large advantage of the scenario cover is that it can on the one hand provide balance-sheet protection to the industry and on the other hand (via a claims facility) rapid and certain payment to potential victims. Through such scenario coverage, the insured industry also knows in advance how much it will get, and it is also more attractive for the third-party victims. It is similar to the model followed in the Deepwater Horizon fund by Ken Feinberg.[72] Rapid payment takes place, but on the basis of theoretical conditions for liability.[73]

In the new facility, a detailed control of moral hazard during operation of the drilling activities will not take place. Munich Re does not specialize in oil drilling and cannot e.g. hire 200 engineers to verify safety on the 1,000 oil rigs in the European Union or the 5,000 in the Gulf of Mexico. The most important incentives for the prevention of harm should come from two elements: (1) self-interest of the industry to prevent accidents and (2) safety regulation imposed (*inter alia* via licensing) by government and strict enforcement.[74]

The way the Munich Re SOS cover works is that the insured are lessees and operators of federal oil and gas installations. Only operators and lessees are liable under OPA. They will bear the primary responsibility to provide financial security. In this sense, SOS is based on economic channelling. The SOS cover would be available for liability arising out of oil spills of insured activities and for defence costs up to US$10 billion plus the capacity which is provided by an insurance consortium.

In principle, the model as it has been developed for the United States could be implemented in a similar manner in the European Union as well. The main difference is that for the United States, detailed data and models are available on the basis of which a calculation could take place and the model could be implemented. This is currently still missing in Europe. However, Munich Re claims that a similar model could be developed for Europe as well and that substantial amounts then could be generated.

71 *Ibid.*

72 The Gulf Coast Claim Facility (GCCF) that was managed by Ken Feinberg will be discussed in further detail in Section 7.6.1.4.

73 *Ibid.* 74 *Ibid.*

6.3.2.2 Stakeholders Assessment

Although this sounds like a miracle solution, many stakeholders provide serious criticism. The product, launched by Munich Re, admittedly would be able to generate very high amounts of coverage for offshore pollution. This was considered by some insurance experts as balance-sheet protection product rather than traditional insurance. The current problem is that in the market there is no willingness to provide capacity; moreover, the price for providing coverage may be high, which is also a factor that will be taken into account by buyers.[75]

Swiss Re obviously does not want to comment on a product developed by a colleague but mentions that this has been put on the market as a reaction after Deepwater Horizon but that the real possibilities of providing adequate coverage do not seem very promising. The product offered by Munich Re would have a lot of prerequisites and would, moreover, not be based on traditional insurance. This also seems that the product would be limited in applicability to the Gulf of Mexico. Also, it would only cover short-tail losses, whereas damage caused by an offshore installation typically leads to long-term losses as well (long-tail risk). There also may not be a willingness of insurers and other carriers to provide capacity for this instrument. Moreover, the price may be so high that there may be no willingness to pay by the industry. Due to the severity and volatility of the energy offshore sector, the risk appetite of insurers/re-insurers is limited.[76]

The idea that (as mentioned in some marketing devices) €20 billion could be made available for losses caused by offshore installations is simply not realistic.[77] Listening to stakeholders in the London market (who are crucial in this respect), US$750 million is the realistic estimate of coverage that could be provided on the casualty market. However, it should be stressed that this US$750 million comes in addition to other coverage (e.g. provided by OPOL or on the maritime insurance market). The total available amount thus could be slightly higher.[78]

[75] Interview with Swiss Re, 11 February 2013. Although Munich Re holds that this is merely speculative because the premiums have not been made public and would, moreover, be low for smaller operators (interview with Philipp Wassenberg, Munich Re, 3 July 2013).

[76] Interview with Swiss Re, 11 February 2013.

[77] See representatives of Lloyds, interview, 1 May 2013, in London. They stress that no one has shown an interest in the Munich Re proposal, and it has not been implemented in practice.

[78] Interview with Swiss Re, 11 February 2013.

Similar comments are also heard from other stakeholders, more particularly from the major OGPs. For example, representatives from BP held that one has to take the credit ratings of companies as a starting point. The number of A-rated insurers in the energy market is relatively limited. This explains why an A-rated company such as BP would have no advantage at all in shifting risk to a B-rated insurance company, which obviously would have no added value. This is mentioned because it may explain why the number of insurers that would be willing to join the Munich Re facility may simply be too limited. Hence, it is argued that it is unclear how Munich Re would ever get the money in for its facility. It could only work if everyone helps, either insurers or operators, but that is not very likely.[79] Moreover, BP itself currently has better protection than what the Munich RE facility would offer, and the premiums to be paid would be very high. For BP, this would therefore not be an attractive scheme.[80]

Also, an American medium-sized OGP held that this Munich Re proposal is simply not realistic. It will only work if all individual operators also participate and are willing to put in assets. Moreover, the costs would be high. Presumably premiums would be between US$15 million and US$20 million for one deepwater well. Thus, if an operator has, for example, three production wells, premiums for this type of coverage would amount to US$50 million to US$60 million on a yearly basis. Such a system is obviously not economically viable. The system, moreover, has never worked in practice. For the majors, this system would, of course, not be interesting (because they can self-insure).[81]

6.3.2.3 Reaction by Munich Re

Representatives from Munich Re argue that it may be important to make such a system mandatory for Europe; otherwise, the incentives for the industry to join such a facility would be lacking.[82]

Munich Re also holds that according to them, the oil industry is systematically underestimating risks and damages in EU waters. They point to the fact that high-risk wells also exist in the European Union, e.g. in North-West Shetland but also East of Cyprus in the Mediterranean. Moreover, unlike in the United States, many coasts could potentially be affected in the European Union when a spill would occur e.g. resulting

[79] Interview with representatives of BP, 13 March 2013 [80] *Ibid.*

[81] Interview with representatives of Noble Energy, 6 March 2013.

[82] Interview with Philipp Wassenberg, representative of Munich Re, 6 May 2013.

from the West Shetlands. Even if one assumed that the damage would perhaps be 30 per cent lower in the European Union than in the United States, the damage still would be substantial. There are, for example, also wells in Greenland, and in the future, drilling in West Africa will take place as well.[83] As a strategy for Europe, Munich Re believes that making coverage mandatory would automatically create appetite for coverage. Munich Re holds that damage in the European Union could easily amount to €2 billion to €3 billion and in a worst-case scenario probably even €5 billion to €10 billion. However, the precise figures in Europe are not known yet. Unlike in the United States, where detailed studies could be done, data for the European Union are still lacking.[84]

The industry is also much too optimistic as far as the possibilities of capping devices are concerned. First of all, capping is a theoretical possibility that has never been tested. Of course, in many cases, it may help to reduce the damage, but one cannot be sure. Moreover, capping devices have to be brought in place by helicopter or by ship. The helicopter cannot fly in bad weather.

6.3.2.4 Analysis

On paper, the proposal by Munich Re seems to correspond with essential principles of risk distribution. So far, a detailed proposal has been worked out only for the Gulf of Mexico. Munich Re holds that a similar proposal can also be developed for European waters if additional information is provided. Whether this can constitute a realistic option to cover offshore-related risks is in the end, of course, not a theoretical question but will depend on the reaction of the market. The amount of retention to be held is still considerable, and the majors therefore probably still may prefer (as they apparently argue) to self-insure or look for alternative solutions. For this reason, the proposal has, although it had already been developed in 2010, never worked in practice. This can, of course, hardly be blamed on Munich Re but is due to the fact that operators (for a variety of reasons) have apparently no longer an interest in investing in this facility. Munich Re therefore holds that developing such a mechanism will de facto only be possible if there is a regulatory solution (i.e. a duty) to join such a mechanism.[85] This opinion does correspond with the analysis of the risk-sharing regimes in the nuclear sector, discussed earlier.[86] A strong aspect of the Price-Anderson Act in the United States is that risk sharing is in fact mandated by the statute and not voluntary. A lesson from the

[83] *Ibid.* [84] *Ibid.* [85] *Ibid.* [86] See Section 4.5.

nuclear sector is consequently that if the very high amounts of compensation (as proposed by Munich Re) have to be generated, this will require a regulatory intervention in the absence of which there will be no willingness of the industry to engage in such a compensation scheme.

6.3.3 Cover for Vessel-Based Pollution

Within the analysis of whether higher amounts of insurance coverage would be available for offshore-related damage, it may be interesting as well to compare the amounts of insurance cover currently available for vessel-based pollution.[87] In vessel-based pollution, the amounts of coverage available are relatively easy to identify. The reason, which also constitutes the main difference with the case of offshore-related risks, is that international conventions (or in the United States the Oil Pollution Act) mandate not only compulsory insurance but a limitation on the liability of the tanker owner as well. The amounts currently available in the international regime for a single pollution case[88] under the (now applicable) CLC of 1992 are, for the biggest ship, €103.2 million. In addition to this amount, the compensation from the 1992 Fund makes the maximum compensation available (including the compensation made under the CLC) €233 million. On top of this, the 2003 Supplementary Fund makes the total amount available €862 million if the state has ratified the Supplementary Fund Protocol.

This therefore means that under international conventions, €103.2 million is generated through insurance and almost €130 million and €629 million through the two international funds, respectively. But this does not provide much information as far as the offshore-related risks are concerned. The amounts of coverage for tanker owners are substantially lower than the amounts of coverage available today for offshore-related risks, though higher amounts may be available under the international fund and the convention on supplementary compensation. However, this is based on a limit on compensation in the CLC (which is absent for offshore-related risks) and on the combination with a fund (equally absent for offshore-related risks).

The story is basically the same for the US Oil Pollution Act (OPA): the Act caps liability at a maximum of US$75 million, which is then also

87 Earlier, we already made a comparison with the potential damage resulting from vessel-based pollution. See Section 6.1.2.

88 See Section 3.2.1.2.

insured. If the damage is higher, the Oil Spill Liability Trust Fund (OSLTF) can intervene to a maximum of US$1 billion.[89] However, the same caveats apply as to the comparison of these amounts with offshore-related risks as for the international regime.

6.3.4 *Developments in the Coverage of Nuclear Risk*

In Chapter 3, there was extensive discussion of the compensation for nuclear risk. Recall that today under the international conventions, €391 million in coverage is available, of which only €91 million is paid through insurance and the remainder through a state subsidy. After implementation of the modification protocol to the Paris and Brussels Conventions, the total amount would rise to €1.5 billion, but still only €700 million would be based on the operator's liability and coverage through insurance.[90] Hence, the international regime for nuclear liability today, even after implementation of the international conventions, would not generate higher amounts in insurance coverage than what is currently available for offshore-related risks.

The conclusion is basically the same for the US Price-Anderson Act: there is nuclear liability coverage up to US$375 million, covered by insurance.[91] The remainder is provided through a retrospective premium model that only comes into action ex post, after an accident occurs.

As mentioned earlier, the international nuclear liability regime has been subject to a lot of criticism, especially because of the low limits of compensation and because of the large state intervention. For this reason, there have been voices in Europe in favour of a European nuclear liability solution that generates much higher amounts.[92]

Within the European Commission, an expert group has been working on a proposal to generate substantially higher amounts than are currently available in nuclear liability. Again, an important role is played there by the largest re-insurer in the world, Munich Re. That company presented a proposal within the working group whereby the nuclear pools (the nuclear insurers) would provide coverage up to €2 billion. Munich Re could create a second layer providing coverage for between €2 billion and €10 billion. The coverage type would be the same as the one for natural catastrophes (natcats). This would mean that others could participate in

[89] See Section 3.5.5.8. [90] See Section 3.6.1.3. [91] See Section 3.6.2.

[92] For those suggestions, *inter alia*, see Faure and Skogh, 'Compensation for Damages'; and more recently Faure and Fiore, 'The Coverage of the Nuclear Risk'. See also Handrlica, 'Harmonization of Nuclear Liability'.

the facility and provide coverage for perhaps €10 billion for one year. Long-tail risks thus would be excluded. It would be more like a capital-market solution. If during the year nothing happened, a participant could take his benefit and leave the facility.[93] A condition for intervening through such a facility therefore would be that the damage is of a sudden and accidental nature, which excludes long-tail risks. Moreover, Munich Re itself, of course, would not generate the total capacity but would invite others to participate in the facility. The risk differentiation via the facility would be minimal. The assumption therefore would be that there would be an EU-wide regulation of the nuclear risk, which would result in stringent mandatory standards for all operators.[94]

Again, one can also see that this example may not be very useful in deciding whether additional coverage would be possible for offshore-related risks. It is still merely a proposal, so it is not known whether other commercial entities would be willing to participate in this facility. Moreover, the assumption of harmonized safety standards (to be discussed later as well) is currently not met. Finally, the facility typically works like natcat coverage (via the capital market), thus excluding long-tail risks. This may be a problem for the pollution-related risks in offshore liability, which are often of a long-tail character.

Also, other stakeholders consider that it is dangerous to compare the situation in the nuclear sector with offshore polluted risks because the two risks are completely different.[95] There is one initiative that should be mentioned in this respect as well. It is an initiative developed by the European Liability Insurance for the Nuclear Industry (ELINI), a pool of nuclear operators.[96] This group, in order to provide an alternative for the nuclear insurance pools (which constitute a national monopoly and hence ask relatively high premiums according to the nuclear industry), has developed an alternative risk-pooling scheme, very much along the lines of the Munich Re proposal. The new alternative is referred to as 'Blue Ray' and would consist of a pooling by nuclear operators as a result of which an amount of €1.2 billion could be generated.[97] However, it was equally held that this initiative will be difficult to implement as long as

93 Interview with Hermann Kramer, Munich Re Insurance Company, 12 March 2013.
94 *Ibid.*
95 Interview with Phil Bell, 26 February 2013; interview with representatives of Swiss Re, 11 February 2013.
96 Discussed in Section 4.5.3.
97 Interview with Ludo Veuchelen, company lawyer of SCK CEN, 4 June 2013.

there is no legal obligation (via increased amounts in the international conventions) to provide for those high amounts.[98]

6.3.5 *Summary*

From this brief overview, it seems as if it is not realistic to assume that the commercial insurance market would be able and willing to provide amounts of coverage for offshore-related risks higher than the amounts already available today. In fact, after the Deepwater Horizon incident, one has noticed the contrary – that available amounts of coverage have even reduced instead of increased. The only indication of increased amounts of insurance coverage was the facility proposed by Munich Re after the Deepwater Horizon incident. Originally, the proposal was especially geared towards the specific situation of the Gulf of Mexico. So far, that proposal has largely remained on paper, and many stakeholders doubt that it will ever be brought into practice. For Europe, a similar facility could be developed as well, but stakeholders will (without a regulatory duty to participate in it) undoubtedly have the same reservations.

An important issue that should be recalled in this respect is that full insurance coverage probably will never be available. Insurance, so it is argued, can never provide full coverage for all liability. First of all, there are risks which are simply uninsurable (e.g. damage which is intentionally caused). Second, insurance coverage will be more limited than liability because, for example, it is limited to sudden and accidental incidents and based on particular exclusions. Those exclusions are also necessary to avoid entrepreneurial risks and to reduce the risk of accumulation. Hence, the coverage provided by the insurance market will necessarily not be the same as full coverage for all potential liability.[99] Thus, the question arises whether it would be possible to expand the potential of regional or international risk pooling.

6.4 Potential of Expanding Risk-Pooling Schemes

The entire Chapter 4 was devoted to the various pooling mechanisms, some of which currently are already in place for offshore-related risks; others are applicable in other high-risk sectors (such as nuclear). In this section, these pooling mechanisms will again be examined, but only with

[98] *Ibid.* [99] Interview with representatives of Swiss Re, 11 February 2013.

respect to their ability to expand those mechanisms, thus providing higher amounts of coverage. This section will address the possibilities of making more use of OIL and OCIL (6.4.1), then discuss possibilities to expand OPOL (6.4.2) as well as a particular proposal that has been launched by Noble Energy towards a risk-sharing agreement (6.4.3). Finally, the challenges and problems surrounding international risk pooling in the offshore sector will be explained (6.4.4).

6.4.1 Expanding OIL and OCIL

Earlier we described the only real risk-pooling arrangements currently existing in the offshore sector, Oil Insurance Limited (OIL) and Oil Casualty Insurance, Ltd. (OCIL). It was explained that these two models are currently mainly used by middle-sized operators, smaller operators being confronted with the too high retention and larger operators preferring self-insurance. Especially when the opinions of the majors are taken into account, there does not seem to be much scope for expanding the current risk-sharing agreements in this area. It was already mentioned that a major (such as BP) criticizes the pools for being based on solidarity and mutualisation. For them, joining those pools is not attractive out of the fear that they would de facto subsidize the presumably higher risk[100] posed by smaller and medium-sized operators.[101] Also, others hold, as was mentioned earlier, that the risk differentiation in the pools is too small. To some extent, this is unavoidable: because the probability of an event occurring is relatively low, the marginal differences between good and bad risks may be small, as a result of which a differentiation of the contributions may not sufficiently incentivize.[102] This constitutes a major difference with the P&I Clubs: in P&I Clubs, ship owners, which constitute all more or less similar risks, are gathered together. In the offshore market, the players are of a different nature. This makes the pooling arrangement very difficult, given different risk profiles.[103]

In sum, although OIL and OCIL undoubtedly play a useful role today, there seems little scope for broadening them by such means as having major OGPs joining them and thus increasing capacity.

[100] In the sense that they may more easily suffer from an insolvency risk.
[101] Interview with representatives of BP, 26 March 2013.
[102] Interview with representatives of Shell International BV, 14 March 2013. [103] *Ibid.*

6.4.2 *Expanding OPOL?*

The functioning of OPOL was discussed earlier,[104] as well as the way in which OPOL is currently used in practice.[105] Of course, there are theoretical possibilities to enlarge the scope of OPOL in various ways:

1. One possibility would be to make OPOL membership mandatory in more legal systems. This would mean that the OPOL solvency guarantee would expand and might also extend to Norway, Denmark or the Netherlands.[106] Ultimately, this would be something for the local regulators to decide and is beyond the decision of OPOL. Still, one would have to take into account the limits of OPOL: because it is not a true risk-sharing agreement, it only provides a solvency guarantee and moreover only up to US$250 million, which is in our definition only for the smallest incidents.
2. A second possibility is to expand the amount of coverage of OPOL. This would mean that the current solvency guarantee would be increased from US$250 million to, say, US$500 million. Again, the majors are opposed to such a proposal for the same reason as they would not like to join risk pools like OIL or OCIL: it increases the mutualisation and hence increases the risk, which they would not desire.[107] Also, representatives of OGP were not strongly in favour of increasing the current limit on OPOL.[108]
3. A third possibility would be to expand the scope of the current OPOL beyond the North Sea (to which its application is currently limited). Not surprisingly, many are opposed to such an idea for the same reasons that they do not want to increase the amount: increasing the current OPOL to an EU-wide model, such as including the Mediterranean or the Black Sea, would mean that for current members (who may not at all be active in the Mediterranean), risks would increase, whereas the members may not have sufficient possibilities to monitor the solvency of operators in those other areas. For this reason, they would be opposed to such a territorial expansion.[109]

[104] See Section 4.2. [105] See Section 5.6.
[106] Where OPOL membership is currently not mandatory.
[107] Interview with representatives of BP, 26 March 2013; interview with representatives of Shell International BV, 26 March 2013.
[108] Discussion with OGP, 25 February 2013.
[109] Interview with representatives of BP, 26 March 2013; interview with representatives of Shell International BV, 26 March 2013. This was also confirmed in an interview with representatives of OPOL, 27 March 2013, in Brussels.

The industry is opposed to making OPOL multinational because it could lead to one country having to pay for the risks created by other countries.[110]

4. A fourth possibility is that different regional agreements, like OPOL, would be created for the Baltic, the Black Sea and the Mediterranean. In this case, there would be no risk sharing (as far as insolvency is concerned) between the operators active in the current OPOL (in the North Sea) and those active in the Mediterranean, for example. A new OPOL would then be created specifically for Mediterranean risks. This is a model that all stakeholders seem to subscribe to for the simple reason that the risks of mutualisation and cross-subsidization are thus limited. Moreover, mutual monitoring (of the insolvency risk) would be easier if new regional risk pools are created.
5. A fifth possibility would be to transform OPOL altogether from the current model (whereby it merely guarantees the solvency of its members) to a truly risk-sharing agreement like OIL and OCIL. Again, it may not be surprising that, for the same reasons the majors did not want to join OIL and OCIL, they would also not be in favour of such a transformation of OPOL from merely guaranteeing the solvency of its members to a true risk-sharing agreement. Again, the fear of mutualisation and cross-subsidization would inhibit such a model.

It should be added that some operators also point at the limits of OPOL in its current form: it is currently only mandatory in the United Kingdom, and the amount is too limited. Moreover, the mere fact that OPOL never had to intervene could (wrongly, according to some regulators) be interpreted as proof of the fact that it functioned well and that hence there would be no risk for the future. Even with the current OPOL, there is a serious risk that there may be some players that could not meet their obligations in the future.[111] Moreover, contributions to OPOL are not risk related, and so once more good risks pay for the bad risks, as a result of which a negative redistribution takes place. Since OPOL also only intervenes when the operator defaults, OPOL in its current form lacks the benefits of mutual monitoring of risk-sharing schemes.[112]

In sum, the only option to expand OPOL which was positively received by stakeholders was the fourth option mentioned earlier, that is, to create other regional risk pools for other sea areas than the North Sea along the

[110] See representatives of Oil and Gas UK, interview, 1 May 2013, in London.
[111] Interview with Jan de Jong, inspector general of the state mines, 22 February 2013.
[112] *Ibid.*

lines of OPOL – a pooling agreement where members share the insolvency risk of their members. Still, it would have to be recalled that this (1) does not have the benefits of mutual monitoring, (2) would only provide limited amounts of coverage and (3) would only intervene to guarantee solvency up to the limited OPOL amounts. Still other arrangements would have to be developed to cover for medium and large accidents.

6.4.3 *Noble Energy Proposal*

There is, however, one proposal that has been advanced by Noble Energy to come to a truly international risk-pooling mechanism.

6.4.3.1 Background

Noble Energy considered that after the Macondo incident in the summer of 2010, there was in the United States a strong reaction and almost daily media coverage of offshore pollution. At that time on the political level, a number of different proposals were introduced on how to reform the liability regime.[113] Some proposed unlimited liability. However, none of those proposals was finally accepted in both the Senate and the House. Nobel Energy, however, argues that there is in this sense still unfinished business for the simple reason that there is not enough coverage today on the global market, especially for damage that could occur in the Gulf of Mexico. If the Deepwater Horizon incident had not occurred to a major operator like BP, the consequences would have been devastating, potentially crippling for the particular enterprise.

Noble Energy agrees with a 'polluter-pays principle' as a starting point but argues that it needs a practical application. The liability cap under the Oil Pollution Act of 1990 (OPA) was US$75 million.[114] BP bypassed its right to call on the financial cap. BP could easily handle that as one of the major oil companies. However, if the same incident had happened to other (smaller) companies, they would be out of business.

Noble Energy argues that US$75 million is in fact too low a cap. For this reason, it proposed an alternative mechanism. This may still be necessary to cover the largest losses. However, as a starting point, it should be noticed that, so the company argues, in the past ten years, safety has strongly increased. The sense of corporate social responsibility has improved, and it has led to increased safety measures, such as the use

[113] On the post-Macondo responses in the United States, see also Section 2.5.2.
[114] See Section 3.5.5.8.

of so-called dual stacks. The probability of an incident therefore may have been reduced, and moreover, given investments in clean-up devices, any spill cost could now be lower. However, major financial damage is still possible. On the normal commercial market between US$1 billion and US$2 billion is now available in coverage.[115] In fact, before the Macondo incident, de facto more insurance coverage was available. What often happened was that various players each bought, for example, US$500 million in coverage. If four partners were involved in a drilling operation together, the coverage would be US$2 billion. However, (re-) insurers realized after Macondo the dangers of multiple aggregate insurance coverage. Now available insurance cover has substantially reduced.[116] However, the premiums charged are still largely the same.

Noble Energy therefore has launched a proposal to generate higher amounts of compensation via a system of joint risk sharing between operators, referred to as 'mutualisation'.

6.4.3.2 Contents of the Proposal

So far in the United States, beyond the US$75 million cap, the OSLTF is available, but this is (to a large extent) financed by taxpayers.[117] Moreover, given the financial capacity of BP, it was de facto not used in the Macondo case. Noble Energy suggests increasing the fund to a larger amount of US$5 billion. Contributions to the fund should, however, be risk dependent.

The structure would be such that the first layer is compensated by the liable operator. That first layer could vary (depending upon risk) between US$150 million and US$750 million. Criteria to determine risk could, for example, be whether drilling takes place in deep or shallow water. In addition to the US$750 million, the OSLTF would intervene with a substantially higher amount of US$5 billion.

In addition, Noble Energy proposes a second vehicle for clean-up and response costs. Again, a first layer would be paid by the responsible parties of between US$150 and US$750 million, for which operators would seek insurance coverage. The amount of exposure in the first layer again would depend upon risk.

Importantly, however, for the second layer, a model would be used whereby a fund, run by the US Treasury, intervenes. The Treasury would pre-finance (via loans to the fund) compensation if the damage is higher than the first layer. This second fund (of potentially US$5 billion and

[115] See Section 5.4.2.3. [116] See Section 5.4.2.6. [117] See Section 3.5.5.10.

US$10 billion) would be based on mutualisation. The specific amount is not known yet. Noble Energy, understandably, does not want to mention a specific amount up front. It could upset public opinion if e.g. the signal was given that drilling operations could lead to damages of, say, US$5 billion to US$10 billion.[118]

The second pool that Noble Energy proposed would have the major advantage that beyond the US$150 million to US$750 million for which individual operators would seek insurance coverage, a large industry-based mutual would be able to generate substantially higher amounts for clean-up and restoration. The advantage of this solution is that it would only be financed via retrospective premiums. Hence, ex ante immobilization of capital is not necessary. Moreover, the proposal would only affect, on average, 1 per cent of the revenues that oil companies currently make in the Gulf and can in this sense be considered reasonable. The retrospective premiums would, moreover, be risk dependent. Hence, bad risks would pay higher contributions than good risks. Of course, some additional details need to be worked out, such as how such a pool would be managed. That should be done by an organization that has sufficient information to be able to apply risk differentiation (e.g. taking into account differences in safety culture). The International Association of Oil and Gas Producers (OGP) might be a good candidate, but there may be others as well.

The example that Noble Energy has in mind was that of the Price-Anderson Act, which equally has a system of retrospective premiums.[119] The advantage is that there is on the one hand no unnecessary immobilization of capital ex ante, and on the other hand the risk differentiation allows the exclusion of free riders from the pool. Graphically, the Noble Energy proposal could be summarized as in Figure 14.

As Figure 14 shows, the Noble Energy proposal consists of two separate proposals. One proposal is, as far as third-party claims and damages are concerned, to increase the money available in the second layer of the OSLTF. However, Nobel equally proposes to create a second layer (referred to as a 'mutual response fund' [MRF]) for clean-up and response costs. Only for the latter types of costs would the industry risk-sharing agreement be created. Discussion on whether the financial limit under OPA of US$75 million could be set aside thus would disappear.

[118] Interview with representatives of Noble Energy, 6 March 2013. [119] See Section 4.5.1.

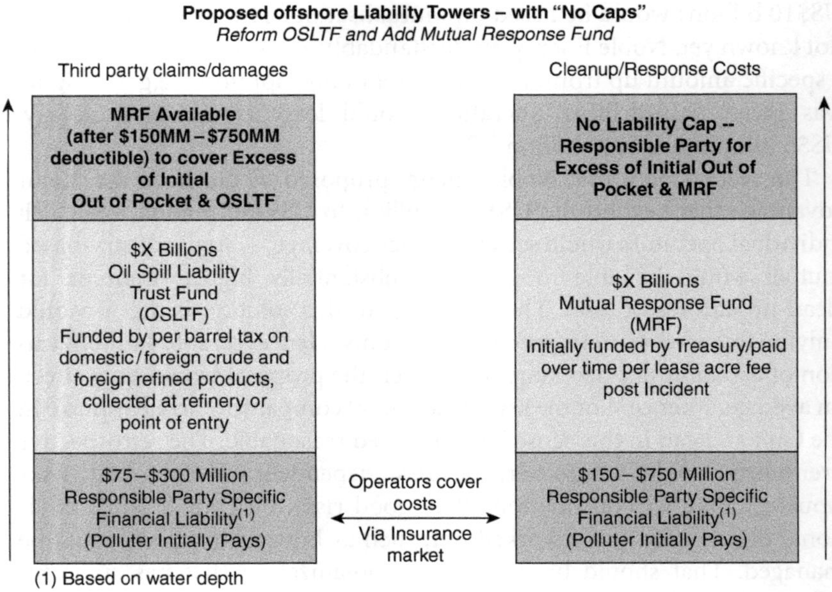

Figure 14 Proposed offshore liability towers – with "no caps".

6.4.3.3 Implementation?

Noble Energy has presented its plans, *inter alia*, to OGP. The OGP is interested in the concept of mutualisation but believes that the numbers presented by Noble Energy are very high.

As far as political feasibility is concerned, it is important to mention that no large operator will ever be able (politically) to call on the current cap of US$75 million. If another large event were to happen and no sufficient compensation were available, there would be a serious risk that there would be political calls to shut down offshore drilling in the Gulf of Mexico altogether. This is why Noble Energy is in favour of a pro-active approach and hence insists on a regulation of potential future losses, even though the political turmoil has now calmed down.

One political issue is that the interest may be different between smaller operators, on the one hand, and the majors (such as ExxonMobil, Shell, BP, and Total), on the other hand. For the majors, the current lack of coverage (for disastrous amounts) on the international (re-)insurance markets is not necessarily a problem because they can self-insure. However, for smaller operators, this is, of course, not an option. This is why alternatives need to be worked out.

Noble Energy considers the Deepwater Horizon incident an important wake-up call for the industry. It has already had the effect of substantial investments in preventive measures, thus significantly reducing the probability of an incident. Responsible care has greatly increased, and this sends an important message to the public. However, the problem remains that there simply is not enough capital available in the market to cover large losses. This is why Noble Energy is strongly in favour of an industry-based solution such as the risk pool.[120]

6.4.3.4 Reactions from Stakeholders

As the representatives from Noble Energy made clear, there is some criticism of this proposal, especially from the majors. For example, BP argues that for the same reason that it resists participation in OIL and OCIL, it is not interested in the Noble Energy proposal because it is equally based on a pooling by industry. BP (like many of the other majors) does not have a specific demand for this, and moreover, it considers itself as having a better risk profile. Hence, such a pool is not attractive for good risks. BP, moreover, argues that such a pool will be especially attractive for bad risks, and thus a serious adverse-selection problem could be created.[121] BP understands that smaller players may not have sufficient balance sheets and for that reason prefer to mutualise, as is done in P&I Clubs. However, with P&I clubs that face similar risks and are comparable (shipowners), this may make sense. The differences between offshore operators are such that for offshore installations, pooling may make less sense.[122] Operators generally are opposed to mandatory 'mutualisation of liabilities' schemes amongst all industry players, arguing that this could lead to complacency and could have a detrimental effect on applied safety standards.[123] Also, representatives from insurers hold that this proposal would amount to a levy for the risk. If operators were forced to participate in such a pool, they could subsequently consider insurance to cover their contribution in the pool.[124]

[120] Interview with representatives of Noble Energy, 6 March 2013.

[121] This was also the opinion of representatives of Oil and Gas UK (interview, 1 May 2013, in London), who were equally arguing that the danger of the Noble Energy proposal is that it can lead to cross-subsidization and reduce the incentives for prevention.

[122] Interview with representatives of BP, 13 March 2013; similar arguments were presented by representatives of Shell International BV, 14 March 2013.

[123] Letter from the International Association of Oil and Gas Producers (OGP) to Michael Faure, 13 May 2013.

[124] Representative of Lloyds, interview, 1 May 2013, in London.

Munich Re also holds that this Noble Energy proposal, which amounts to a mutual pooling between operators, will not work because there is a large amount of mutual suspicion between the operators. Every operator will claim that it has the highest safety standards, and hence, debate will arise as to how much anyone must contribute to the pool. Munich Re therefore holds that a neutral third party should differentiate risks (and hence financial contributions) based on objective criteria. This may not be possible in the case of distrust between pool members.[125]

6.4.3.5 Evaluation

Evaluating this proposal of Noble Energy, one can argue that it complies largely with the benefits of mutual monitoring that would be inherent in a risk-pooling scheme, as was explained earlier.[126] However, understandably, major operators may feel that the arguments for such a comprehensive risk-pooling scheme may be stronger in the United States, where the American plaintiff bar leads to much higher amounts of compensation than generally in Europe. Still, Noble Energy rightly mentions that such a risk-pooling model could, in theory, also be attractive for EU operators, especially for small and medium-sized operators who may be exposed to large risks as well.[127] However, this is to some extent countered by the argument of the majors that (e.g. unlike P&I Clubs) in the offshore business operators and risks are of a totally different nature, which makes risk differentiation very hard and the danger of cross-subsidization and negative redistribution very realistic. Hence, from the perspective of the majors, one can understand that they fear that such a mechanism could be used as an instrument of externalization by (potentially higher-risk) smaller and medium-sized operators who could then (in the case of mutualisation via a pool) enjoy the benefit of the balance sheets of the majors. From a social policy perspective, this would be undesirable because it could reduce incentives for care-taking by higher-risk operators.

One potential weakness to be addressed is (as in the Price-Anderson Act) how one monitors the solvency risk with individual operators. There needs to be serious monitoring not only of safety but also of solvency of individual operators because otherwise they could externalize their risk to the group and simply go out of business. This may not necessarily be a huge problem, however, and could be accounted for.

[125] Interview with Philipp Wassenberg, Munich Re, 6 May 2013. [126] See Section 4.6.1.
[127] Interview with representatives of Noble Energy, 6 March 2013.

6.4.4 *Challenges*

To some extent, one recognizes that the arguments presented by the majors against risk pooling and mutualisation are similar to the arguments presented, for example, by French and UK nuclear power plant operators against an EU-wide nuclear liability pool, including Eastern European operators in Bulgaria.[128] It was indicated earlier that the theoretical advantage of risk-sharing pools is that they provide incentives for mutual monitoring and thus could, to the extent that this monitoring can be exercised, also provide incentives for improved risk-preventive investment by all members in the pool. Theoretically, membership in the pool thus could (via the mutual monitoring) increase safety levels. However, this only works under two important conditions: (1) that risk differentiation is possible to such an extent that real differences in contributions can be made which do provide substantial incentives to operators to invest in risk-prevention measures and (2) that the risks posed by the different operators are at least of a similar nature. It may be difficult to meet both conditions in the case of risk pooling for offshore-related liabilities.

It was indeed mentioned earlier that the probability of an offshore incident is so low that the difference between the relative contributions of a high risk compared to a low risk may be too small to provide appropriate incentives for investments in risk-prevention technologies. If this were the case, risk differentiation would not work, and high-risk operators would simply prefer to pay the higher contribution to the pool, of which the cost may be lower than the investment in risk-prevention technologies. Also, the second condition (similar nature of the risks) may not be met in the case of offshore-related risks given the largely different nature not only of the operators but also of the risks related to different types of wells. Theoretically, technological information is available (as can e.g. be seen from the Oil and Gas Guidelines on Financial Responsibility) to distinguish risks on objective grounds. This could then also constitute the basis for differing contributions to a pool. However, the problem would still exist that there could be substantial differences between operators as far as their assets and credit rating are concerned. If this is the case, smaller (and potentially insolvent) operators could still obtain benefits from the pool by shifting the risk to other members with higher credit ratings. This obviously could have negative consequences for the incentive to invest in prevention.

[128] See Section 4.5.5.

Thus, one understands that either an international or regional risk-pooling scheme for substantial amounts (beyond what is currently done via OIL or OCIL) is not very likely to emerge, at least on a voluntary basis and when it should include the collaboration of the majors. Their objections to participation in those pools can be understood given the above-mentioned problems. One could then argue that learning from the Price-Anderson Act, a retrospective risk-pooling scheme could be mandatorily imposed. The advantages of such a scheme were explained earlier when the Price-Anderson Act for nuclear liability was discussed:[129] premiums could, in principle, be risk related, and the contributions would not have to be paid up front but only retrospectively. The advantage thus would be that there is no large immobilization of capital ex ante because the scheme only intervenes ex post, when an accident has happened. However, a major difference between nuclear operators in the United States and offshore installations in Europe is that nationwide safety regulation in the United States guarantees similar minimum safety standards for the entire territory. This therefore reduces the task of mutual monitoring of the members in the pool and, moreover, reduces the risk of free riding by individual members. However, such an EU-wide safety regulation for offshore installations is still largely lacking, as a result of which the risk of free riding would persist. Moreover, it would even be dangerous to argue that policy makers would have to force unwilling operators (presumably majors) into such a European-wide risk pool. Given the substantial difference between operators and risks and the absence of EU-wide high safety standards (those constitute major differences with the nuclear risk in the United States), a mandatory pooling could lead to perverse incentives for high-risk operators who could externalize their risk to the lower-risk members in the pool. In this sense, a mandatory pooling scheme could, if adequate risk differentiation and safety regulation could not sufficiently distinguish good risks from bad risks, even lead to decreased safety levels.

An important lesson therefore is that policy makers should under these circumstances refrain from mandating pooling between operators of offshore installations. The only thing a policy maker could do is to encourage the creation of industry-wide pooling by providing high standards of safety regulation EU-wide. Thus, safety regulation could facilitate and assist mutual monitoring by operators and encourage pooling arrangements.

[129] See Section 4.5.1.

6.5 Summary: Flexibility

The analysis in this chapter shows that there are no easy solutions to increase the coverage available for offshore-related risks in European waters compared to the status quo we described in Chapter 5. Various proposals do exist, using either insurance or industry pooling, but all have their disadvantages as well and are therefore understandably opposed by industry. The opposition can, moreover, also be understood taking into account economic principles because a forced mutualisation could even lead to increased safety risks, which obviously should be avoided at all costs. When addressing the question of whether there is a need for some regulatory action compared with the status quo, the starting point should be whether there is a problem in the current situation. The answer is that this strongly depends on the type of accident and the solutions that could be envisaged:

1. For smaller accidents (defined as a maximum of US$250 million damage), the current OPOL mechanism may work. However, one has to recall the limits of OPOL:
   - Only applicable to the North Sea,
   - Only mandatory in the United Kingdom,
   - Only providing solvency guarantees,
   - Never applied in practice and hence no practical experience, and
   - No risk differentiation and hence no incentives for prevention.

   OPOL relies on a variety of instruments (such as self-insurance, insurance or guarantees), but a solution will be necessary even for this lower category of accidents for the cases and territories where OPOL does not apply.
2. This is certainly also the case for medium-sized accidents (between US$250 and US$750 million). Even though they go beyond the limit provided by OPOL, insurance solutions available on the commercial market can still be used as well as pooling arrangements such as OIL and OCIL.
3. Only for the category of large accidents (damage above US$750 million), insurance may be either not or only partially (with large retentions) available. In such cases, only the majors would be able to provide cover based on their balance sheets via either self-insurance or captives.

As mentioned earlier, this could lead to a regulatory recommendation (such as e.g. already applied by DECC and following from the UK Oil and

Gas Guidelines on Financial Responsibility) to use risk assessment to determine the potential damage resulting from particular operations. This could result in only allowing majors to engage in activities that could lead to large damage or suggesting that smaller and medium-sized operators engage in joint ventures with the majors.

There only seems scope for developing other regional pools like OPOL (solvency-guarantee pools) for the Mediterranean. However, it should be clear that such pooling schemes will only develop under a regulatory duty to show financial responsibility; otherwise, operators may lack any incentives to develop such a scheme.[130] This would therefore be a strong argument in favour of a regulatory duty to show financial coverage, as we will also argue in Chapter 7.[131] Moreover, if the government were to stimulate further risk-pooling arrangements between operators (going beyond the solvency guarantees provided in OPOL), an important condition would be to impose high safety standards for offshore installations through regulation in order to facilitate the mutual monitoring inherent in risk-pooling schemes.

[130] Interview with representatives from OPOL, 27 March 2013. [131] See Section 7.4.

7

Towards Optimal Liability and Compensation for Offshore Oil and Gas Activities

KRISTEL DE SMEDT, HUI WANG AND MICHAEL FAURE

In this chapter, we will take into account the findings from previous chapters: not only the potential of financial and insurance instruments to cover a future liability regime (Chapter 6) but also an analysis of the existing legal regimes (Chapter 3) and an analysis of existing risk-pooling mechanisms (Chapter 4). The ultimate goal is obviously to improve decision making and to inform the recommendations which will be formulated as concluding remarks in the Chapter 8. In order to construct these scenarios in an informed manner, insights from economic analysis of law will be used to critically review potential options that will later be analyzed in detail. These will be based, as mentioned, on economic theory. However, they will take into account best practices (and problems) because they follow from examples such as in the compensation regime for vessel-based pollution but also from the liability and compensation regime in other high-risk sectors such as nuclear energy.

Hence, this chapter will start by sketching some principles of optimal liability rules and efficient compensation mechanisms, which will constitute the methodological basis (7.1). Next, principles of efficient liability rules will be examined and applied to the offshore sector (7.2). In this respect, the question will be asked what the optimal liability rule (negligence or strict liability) would be in the case of damage resulting from an offshore incident and how liability should be attributed, more particularly in the case where multiple tortfeasors are involved (channelling of liability, joint and several liability, etc.).

An important issue in the determination of efficient liability rules also concerns the damages and more particularly the question of whether a financial cap or limit on the amount of damages would be indicated. Attention equally will be paid to the important question of the relationship between liability and regulation. To an important extent, this may limit or expand the potential liability and, taking into account regulatory requirements, also could enlighten the task of the judge in a liability case.

Hence, the relationship between safety regulation and liability in the case of offshore installations should be carefully reviewed (7.3).

Next, similar questions will be asked concerning the construction of efficient compensation mechanisms. A first question that will be asked is whether the provision of financial security should be made mandatory, and if so, how this should be constructed (7.4). The proposed compensation mechanisms, such as a self-guarantee though tax, mandatory insurance and risk-sharing pool, are evaluated in Section 7.5. The question also arises as to how government policy can stimulate the insurability of the offshore liability risk and whether government could take measures e.g. to stimulate the emergence of industry-based solutions such as a pooling mechanism. The question could equally be asked whether, to the extent that the market would fail, government should intervene in providing compensation e.g. via a compensation fund or via state intervention of any form (direct financing of compensation or as re-insurer of last resort) (7.5).

The question also arises as to whether it is possible to develop a rapid claims management system that can provide rapid compensation especially to fishermen and hotels/restaurants in coastal areas to cover their losses (7.6). Offshore-related damage often can create harm of a trans-boundary nature. This justifies the question of whether action is needed to deal with damage of such a trans-boundary nature. At the European level, an important question relates to the liability for environmental damage in light of the Environmental Liability Directive. The question that can be asked is whether it would suffice to bring environmental damage resulting from offshore installations under the scope of the Environmental Liability Directive (7.7). Finally, a number of scenarios will be sketched indicating how an efficient liability and compensation regime for offshore-related damage could be structured (7.8). The analysis in this chapter will constitute the basis for the recommendations that will be formulated as concluding remarks in Chapter 8.

7.1 Principles of Efficient Compensation and Liability Rules

As a starting point, before looking at efficient liability rules and a compensation system in a more normative manner, it is necessary to sketch the principles on which the normative analysis will be based. As made clear in the Introduction,[1] the normative framework comes

[1] See Section 1.3.2.

from the economic analysis of law. Based on the economic analysis of accident law, the goal of liability rules is to minimize the total sum of accident costs.[2] The economic analysis of accident law has already been briefly discussed when various risk-pooling schemes were assessed, and it was stressed that according to Calabresi's framework, the goal of a liability regime is to minimize the total sum of (primary, secondary and tertiary) accident costs. For current purposes, it is therefore important to remember that from an economic perspective, the main and primary goal of a compensation regime (whether this is constituted through liability rules, insurance or a fund solution) should be to provide ex ante incentives for prevention. Moreover, compensation of victims is from this perspective not seen as the primary goal of accident law but rather as a means to reach the desirable effect of prevention.

Taking this as a starting point, the question of why there should be compensation at all, at least from an economic perspective, will be analyzed (7.1.1), and next, the consequences of these economic starting points for the shape of a compensation regime will be examined (7.1.2).

7.1.1 Why Compensate?

Using an economic perspective, one could argue that compensation of victims is, as such, not the goal of the legal system. Tort law should rather aim at welfare maximization through the minimization of accident cost. The question of how victims of offshore-related damage can be provided with appropriate compensation is, in this strict sense, not interesting from an economic perspective.[3] However, some have argued with respect to the financial compensation for victims of catastrophes that providing disaster relief is one of the principal functions of government.[4] Moreover, leaving victims of a large catastrophe such as an offshore installation blowout without any relief probably would be incompatible with the concept of the welfare state, at least as this is conceived in most EU Member States.[5]

A disaster such as the Deepwater Horizon in the United States or the Piper Alpha in Europe showed that major offshore accidents can lead to serious political and economic instability, especially for victims and local

[2] This has been made clear in the pioneering work of Calabresi, *The Costs of Accidents*; and in the works of Shavell, 'Strict Liability versus Negligence' and *Economic Analysis of Accident Law*.

[3] See Posner, *Economic Analysis of Law*, 167–77. [4] Priest, 'The Government', 235.

[5] See Schwarze and Wagner, 'In the Aftermath of Dresden'.

authorities suffering major damage in coastal areas. This therefore justifies thinking about how to structure an efficient compensation system for victims, including which role government should play in that respect. Economics can still be a useful instrument in indicating how a form of compensation can be provided at the lowest possible (administrative) cost and without unnecessary side effects (as far as affecting incentives for prevention is concerned) or negative distributional consequences.[6] Providing specific compensation for victims of catastrophes is indeed problematic from a distributional perspective for the simple reason that some victims (those hit by a catastrophe) may receive preferential treatment compared to other victims (those hit by an individual accident). This different treatment is difficult to explain from the perspective of the equality principle. Some have indeed argued that it may be problematic if compensation were only guaranteed for victims of catastrophes, whereas for victims of other accidents this generosity would not apply.[7]

From this it can be concluded that providing specific attention to victims of offshore-related damage (unlike that to other victims) may indeed constitute discrimination but may, however, not only be unavoidable (given the pressure on politicians to act) but also desirable given the potentially devastating effects of disasters and other damage resulting from offshore incidents on society. From a policy perspective, there are thus good reasons to analyze how adequate compensation to victims could be provided.

7.1.2 Principles of Efficient and Fair Compensation

In subsequently analyzing the various possible instruments of compensation, some general rules of thumb could be used to analyze the strengths and weaknesses of the various instruments. A first important principle is that no matter how compensation is organized, the incentives for preventing damage caused by offshore incidents should always remain intact. This means that, in principle, the duty to compensate should rest upon the party who actually contributed to the risk but equally that incentives are also provided to potential victims to mitigate the damage.

A second, related principle is that a duty to contribute also should be related to the amount to which a specific activity or entrepreneur

[6] See Posner, *Catastrophe*.
[7] Magnus, 'Germany'; Zeckhauser, 'The Economics of Catastrophes', 114.

contributed to the risk. This remains important if a collectivization of the compensation takes place. If such a risk differentiation is respected, the compensation will give incentives for prevention to contributors. This is not only important from the perspective of potential risk creators but also for potential victims. This also means that, in principle, the individual seeking a particular protection will pay for this protection to the extent that he or she is exposed to the risk. Higher risks from this perspective pay a higher contribution to a financing solution than lower risks. According to this principle, there should be no shifting of risk and cost to a collectivity.

The reason behind this principle is the idea that making individuals pay according to the risk they pose will make them aware of aspects such as their exposure to hazard and thus may have a positive impact on their behaviour (i.e. provide incentives for prevention). Hence, the efficiency reason behind risk differentiation is to provide incentives for prevention, risk reduction and/or mitigation of damage.

Moreover, these principles are important not only from an efficiency point of view (providing optimal incentives for prevention to all stakeholders in the risk) but also include a fairness element. Indeed, if the principles were not followed, it would mean that good risks would have to pay for the bad risks as well and would therefore in fact subsidize bad risks. This negative redistribution should be avoided, and therefore, a compensation mechanism (no matter how it is organized) should, in principle, be financed by the parties actually contributing to the damage. But again, risk differentiation is also important from the victim's perspective. A solidarity on the basis of which all (taxpayers) pay for those exposed to risk could imply a redistribution whereby those who accepted risk are rewarded by those who faced no risk.

A third general principle is that, to the extent possible, a solution should be introduced at the lowest administrative cost possible. Fourth and finally, where possible, a competitive market solution may (if it is less costly) be preferred to a bureaucratic intervention by government. The market usually will be able to provide coverage at lower cost.[8] However, there may be some (perhaps exceptional) cases where a government monopoly in the provision of disaster insurance could provide better results than competitive markets.[9]

[8] See Bruggeman, Faure and Fiore, 'The Government as Reinsurer', 381.

[9] This would be the case specifically in Switzerland. For a discussion, see Emons, 'Imperfect Tests'; Von Ungern-Sternberg, 'The Limits of Competition' and *Efficient Monopolies*.

7.2 Efficient Liability Rules

It is of course not possible to provide a detailed analysis of an efficient liability regime for offshore-related risks. However, using the tools offered by economic analysis of law, it is possible to discuss some essential features from a liability system and to suggest how they work out in the design of a liability regime for offshore-related risks. In this respect, the following aspects will be discussed: whether offshore-related risks should correspond to a strict liability or a negligence regime (7.2.1), how liability for offshore-related risks should be attributed (7.2.2), and whether there should be a financial limit on the liability for offshore-related risks (7.2.3). These issues correspond to a large extent with the checklist for country studies[10] and are hence also issues that received attention in the description of the legal system in particular countries.[11] To some extent, the solutions adopted in some of the countries therefore will also be recalled when discussing these efficient liability regimes. In Section 7.3, the relationship between liability and regulation will be discussed. Attention also will be paid there to the inter-dependencies between liability rules and regulation.[12]

7.2.1 Strict Liability or Negligence?

7.2.1.1 Economic Theory

A crucial question to be addressed in any liability setting is whether offshore-related risks should be based on strict liability or on negligence. Assume that the goal of accident law is the minimization of the total sum of accident costs C where

A = the victim;
B = the injurer;
p = probability of an accident;
L = total amount of the loss;
y = care taken by the injurer; and
x = care taken by the victim.

Then the social optimum can be found, where $C = A(x) + B(y) + p(x, y)L$ is minimized. This level can be found at efficient care levels x^* (for A) and y^* (for B).[13]

[10] Included in Appendix 2. [11] In Section 3.5. [12] See Section 7.3.5.
[13] See Shavell, 'Strict Liability versus Negligence'.

One rule which may give incentives to the injurer to follow the efficient level of care is the negligence rule. The 'negligence rule' is defined as a rule according to which the injurer will only have to bear the loss if he uses less than a legally required level of care, referred to as the 'due-care level'. The negligence rule as defined here means that the injurer will be held liable if he spends less than the due-care level required by the legal system – in other words, if he acted wrongfully. Assuming that the due-care level required by the legal system is equal to the optimal level of care y^*, the injurer will always follow the optimal-care level. This is indeed the cheapest solution for him. If the injurer spent less than y^* on care, his total costs would be equal to

$$p(o, y^\circ)L + B(y^\circ)$$

If, however, he spends the efficient level of care, he will not have to bear the expected loss. Hence, the expected costs of the injurer are in this case only his costs of taking efficient care: $B(y^*)$. The question of whether B will take efficient care or not will, therefore, depend upon

$$p(o, y^\circ)L + B(y^\circ) \lesseqgtr \quad B(y^*)$$

If the sum of expected loss and the actual costs of care are higher than the costs of efficient care y^*, the injurer will take efficient care. If, however, these costs could be lower than the costs of taking efficient care, it would be cheaper for the injurer not to take efficient care. However, because we defined y^* as the point where the social costs are as low as possible, $p(o, y^\circ)$ $L + B(y^\circ)$ will always be higher than the costs of taking y^*. Hence, the simple conclusion is that under a negligence rule, the injurer will always have an incentive to spend y^* on care, and an efficient outcome will be reached. This is true as long as the legal system defines y^* as due care.[14]

According to this economic model, an injurer will be found liable under a negligence rule if he spends less than the due care the legal system required from him ($y < y^*$). The crucial question therefore is, from an economic point of view, how the judge should establish this efficient level of care. Spending more on care will reduce the probability that an accident may occur. Requiring the injurer to spend more on care is efficient as long as the marginal costs of care are lower than the additional benefit in reduction of the expected loss. The efficient level of care y^* is found where marginal costs of care equal the marginal benefits in reduction of expected loss.[15]

[14] *Ibid.*, 14. [15] *Ibid.*; see also Landes and Posner, 'Tort Law as a Regulatory Regime'.

Economic analysis does not, however, assume that judges make explicit mathematical calculations of marginal costs and marginal benefits to establish the efficient care level. Often such a marginal cost/marginal benefit weighing takes place in a rather implicit manner. Hence, economic scholars often argue that judges act *as if* the goal of their actions was to achieve wealth maximization, even though their decisions are mostly not formulated in economic terms.[16] From the preceding, it follows that the negligence rule is in principle able to give a polluter an incentive to spend on care to reach the optimal standard. However, the negligence rule only works optimally if the legal system defines the due-care level as the optimal standard.

Also, a strict liability rule will lead to optimal incentives for care taking for the polluter because taking efficient care will minimize the expected accident costs which the potential polluter has to bear under a strict liability system[17]. Therefore, the literature generally accepts that both a negligence rule and a strict liability rule will provide a potential polluter with incentives to take the efficient care level. However, this is only valid in a unilateral accident setting, in an accident whereby only the injurer can influence the accident risk. If victims were also to be given incentives for accident reduction, a contributory-negligence defence should be added to the strict liability rule. Under negligence, victims will always have an incentive to take efficient care as well because they will, in principle, not be compensated by the injurer who, under a negligence rule, will take efficient care to avoid liability.

In the joint-care (bilateral) case, strict liability with contributory negligence and a negligence rule (with or without contributory negligence) will incite parties to adopt efficient levels of care. However, the accident risk cannot totally be minimized by increasing the levels of care. Accident losses also depend on the extent to which parties participate in the activity that might cause the damage (e.g. miles driven). Therefore, reducing the activity level will also reduce the accident risk[18]. The activity level can be interpreted as any control variable not taken into account in setting the optimal level of care. Under a negligence rule, an injurer has no incentive to adopt an optimal level of activity. This cannot be remedied because judges cannot

[16] This claim has especially been made by Landes and Posner, 'The Positive Economic Theory', 851.

[17] Polinsky, *Introduction to Law and Economics*, 11; Shavell, *Economic Analysis of Accident Law*, 8.

[18] Diamond, 'Single Activities Accidents'; Polinsky, *Introduction to Law and Economics*, 44; Shavell, 'Strict Liability versus Negligence'.

easily calculate the optimal activity level into the due-care standard[19]. A strict liability rule has the advantage that the injurer will automatically adopt an optimal activity level. This is also a means to minimize his costs. If the victim's activity has no influence on the accident risk, strict liability might have a slight advantage because it might also lead to an optimal activity level of the injurer. However, in a joint-care case, this advantage is cancelled out by the fact that the victim will not adopt an optimal activity level. This is due to the impossibility of calculating the activity level into the due-care standard when considering contributory negligence.[20]

What is the importance of the activity level for the choice between negligence and strict liability? In a unilateral accident model, whereby only the behaviour of the injurer influences the accident risk, strict liability seems to be the preferred rule because it is the only liability rule which will lead to both efficient care and an optimal activity level. In a bilateral case, the answer is more balanced.

Because activity-level changes are not calculated into the due-care standard, strict liability (with a defence of contributory negligence) will encourage activity-level changes by the injurer. However, a negligence rule will encourage activity-level changes by the victim. Therefore, several authors suggest that in bilateral cases strict liability will be a superior device if it is more important to give injurers an incentive to change their activity level than that victims be given a similar incentive[21]. This implies that if the injurer's activity is very dangerous and creates a high accident risk, even if optimal care is taken, it will be more desirable to control the injurer's activity than it is to control the victim's activity. For instance, if an injurer creates the collapse of an old house with explosives in the middle of a densely populated area, it is more desirable to control his activity than it is to control the victim's activity. However, the advantage of a strict liability rule disappears if the injurer's activity is not important enough to be controlled. If the activity of the injurer is not very dangerous, if exercised with reasonable care, it is desirable to give the victim an incentive to take an optimal activity level as well.

[19] It is difficult to determine both the optimal and the actual activity level. For examples, see Polinsky, *Introduction to Law and Economics*, 47; Shavell, 'Strict Liability versus Negligence', 2.

[20] Diamond, 'Single Activities Accidents'; Landes and Posner, 'The Positive Economic Theory'; Polinsky, *Introduction to Law and Economics*; Shavell, 'Strict Liability versus Negligence'.

[21] Landes and Posner, 'The Positive Economic Theory', 877; Shavell, 'Strict Liability versus Negligence', 7, 19.

Even though a clear-cut test is therefore difficult to give, Landes and Posner describe several factors which may lead to a preference for a strict liability rule.[22] These elements are (1) high expected accident costs, (2) the impossibility that more care by the injurer would reduce the accident risk, (3) the impracticability of constraining the victim's activity in favour of the injurer's, and (4) the desirability to reduce the risk by an activity-level change of the injurer.

7.2.1.2 Application to Offshore-Related Risks

How do these economic arguments in favour of strict liability apply to the case of offshore-related risks? To some extent, the analysis of liability for offshore-related risks resembles the analysis of liability for environmental risks and, more particularly, for vessel-based marine pollution. In that literature, it is held that risks created by (vessel-based) pollution can in most cases be considered to be a unilateral accident (i.e. an accident whereby only the injurer can influence the accident risk). In such a case, economic analysis predicts that the advantage of the strict liability rule is that it will give the injurer an incentive both to adopt an optimal activity level and to take efficient care. Because the victim in most cases cannot influence the accident risk, strict liability seems to be the best solution to give the potential injurer optimal incentives for accident reduction. It is, however, important to add that in bilateral cases, even when strict liability remains preferred because the injurer has the largest influence on the accident risk, it remains important to add a defence to take into account the victim's behaviour as well.

These arguments in favour of a strict liability rule correspond largely with the solution for marine pollution resulting from vessels as well. As was made clear earlier, the International Convention on Civil Liability for Oil Pollution Damage (CLC) adopts a principle of strict liability,[23] and the same is true for the US Oil Pollution Act (OPA).[24] The arguments that have been advanced in the economic literature in favour of strict liability for vessel-based marine pollution[25] apply, *mutatis mutandis*, to offshore-related liability risks as well. Moreover, as the comparative analysis made clear as well, the countries that have a specific liability rule for offshore-related risks all have taken strict liability as the legal basis.[26] Thus there are good reasons in favour of a strict liability for offshore-related risks.

[22] Landes and Posner, 'The Positive Economic Theory', 907. [23] See Section 3.2.1.1.
[24] See Section 3.5.5.3.
[25] See, *inter alia*, Faure and Wang, 'Economic Analysis of Compensation'.
[26] See Table 10 in Section 3.5.2.

However, it should be taken into account that although the operator undoubtedly has the most important influence on the accident risk, the damage resulting from an offshore-related incident certainly should be considered as bilateral in the sense that victims can also take optimal preventive measures to reduce the extent of the damage. Victims cannot influence the occurrence of the incident itself, but they can take measures to mitigate the damage. In order to provide optimal incentives to victims to take preventive measures as well, a comparative negligence rule, taking into account the victim's behaviour, should be added to the strict liability rule. This will allow a reduction in the compensation due to the victim to the extent that the victim's behaviour (e.g. taking insufficient measures to mitigate the damage) has contributed to the extent of the loss. To this extent, the victim's claim on compensation can then be proportionally reduced. This would give incentives to potential victims as well to take optimal measures aimed at mitigation of the loss.

7.2.2 *Attribution of Liability*

The question of attribution of liability amounts to the question of who, among the many parties are involved, should be held liable for risks related to an offshore incident. As the Deepwater Horizon incident has shown,[27] there can be a large number of parties all contributing to a different extent to the risk. The question then arises as to how liability should be allocated. This question in fact consists of two different sub-questions. On the one hand, the question arises as to whether, following the example in the nuclear liability conventions and the conventions with respect to marine oil pollution, liability should be channelled to the licensee. Another question is whether, in principle, when more parties contributed to the loss, they should be held jointly and severally liable towards third parties.

7.2.2.1 Channelling of Liability

'Channelling of liability' basically means that only one party can be held liable, thus excluding the liability of other parties involved. This type of channelling has been defended as facilitating the lives of the victims who only had to address the licensee (also in cases where through complicated contractual arrangements more parties potentially could be involved). It is also held that for operators, channelling can facilitate insurability because

[27] See the discussion in Section 2.5.2.

only one party would have to carry insurance coverage. However, in the law and economics literature, channelling is usually considered to be inefficient. It negatively affects the incentives to take care more particularly by all other parties who could have equally influenced the accident risk.[28] This has been shown by the Fukushima nuclear accident that hit Japan in 2011. According to Japanese law, the operator, TEPCO, is the only party liable for the nuclear damage. However, the flawed design of the Fukushima Daiichi power plant by GE also was an important factor leading to the accident.[29] Although GE was reported to be aware of the problematic design a few decades ago, it had no incentive to remedy it because it would not be held liable anyway.[30] The problematic nature of the channelling of liability is especially true in the nuclear sector. Although, theoretically, the operators can take recourse against the other parties (e.g. the designers and nuclear suppliers) if it is provided in the contract,[31] the bargaining power of the operators may be weak compared to the designers or nuclear suppliers, who often dominate the market.[32] In this sense, the channelling of liability may be less problematic with offshore activities. In many cases, the operators/licensees of the offshore facilities are major oil companies, which have better financial capability than the contractors. One may expect a more balanced liability arrangement in the contract between them.[33] The oil companies usually have technical expertise and the field knowledge to control the operation.[34] The oil companies also select and approve various contractors, control operations of the contractors and verify the results of the contractors' work.[35]

An alternative to legal channelling would be to impose the responsibility to seek financial security coverage on one party (e.g. the operator), referred to as 'economic channelling', instead of relieving other risk contributors from liability.[36] Economic channelling basically means

28 See Faure and Wang, 'Economic Analysis of Compensation', 188; Vanden Borre, 'Channelling of Liability', 13–39; Faure, 'Environmental Liability', 260.

29 See Faure and Liu, 'The Tsunami of March 2011', 65–7.

30 For a more detailed analysis of the Fukushima accident and the problem of channelling liability, see Faure and Liu, 'Compensating for Nuclear Accidents', 79; Ramseyer, 'Why Power Companies?', 457.

31 Article VI(f) of the Paris Convention; Arts. II(5) and X of the Vienna Convention.

32 Vanden Borre, 'Channelling of Liability'.

33 For the risk allocation between operators and the contractors in the contracts, see Cameron, 'Liability for Catastrophic Risk', 207–11.

34 Viscusi and Zeckhauser, 'Deterring and Compensating Oil-Spill Catastrophes', 1748–9.

35 Cameron, 'Liability for Catastrophic Risk', 207.

36 Vanden Borre, 'Shifts in Governance'.

that one party (e.g. the operator) takes insurance coverage whereby the potential liability of the subcontractors of the operator is also covered. This is a type of coverage that is provided by American Nuclear Insurers (ANI) under the US Price-Anderson Act. Because the insurance provides a so-called omnibus coverage to the operator, it de facto also covers the operator's contractors.[37] The advantage of such economic channelling from the perspective of subcontractors is obviously that they do not (like the operator) have to take (expensive) insurance coverage themselves as well but can rely on the umbrella coverage of the operator. The advantage of such a construction is that a larger total insurance capacity can be generated than when each subcontractor would equally have to take full liability insurance coverage for the same total amount as the operator. Such an economic channelling is an issue which most likely cannot be mandated through regulation but can be agreed upon contractually by the parties. This was, moreover a part of the proposal for the facility prepared by Munich Re.[38] The facility would provide coverage to be financed by the operator, but the benefits of coverage would extend equally to subcontractors, thus removing the need for them to buy coverage as well. This is precisely the notion of economic channelling.

Currently, apart from the operator/licensee of the offshore installation, others are liable as well to the extent that damage could be caused by their negligence. This may apply to drillers, for example. They can be held contractually liable towards the oil company that contracted them. In this case, the conditions for liability will be determined in the contract. If the driller works from a mobile unit, liability could be covered under the Bunker Convention or, at a national level in the United States, for example, under the OPA. Representatives from drillers, however, stress that they could/should only be held liable when damage occurs as a result of their drilling activity and not for all damage that could result from the well. They argue that most decisions, such as those concerning the planning of the well, the construction, and so on, are taken by the oil company, as a result of which the driller can only control the activities for which he is in charge (i.e. drilling).[39] Extending liability for all damage

[37] For details, see Faure and Vanden Borre, 'Compensating Nuclear Damage', 251–4; Vanden Borre, 'Channelling of Liability', 27: under the omnibus coverage, suppliers can be held legally liable (hence there is no legal channelling), but their liability is covered by the omnibus coverage of the nuclear operator.

[38] Interview with Philipp Wassenberg, representative of Munich Re, 6 May 2013.

[39] See Alan Spackman, vice-president IADC, interview, 5 June 2013.

resulting from the well to the driller would make the driller also liable for damage which may result from activities that he did not control.[40]

Also, other stakeholders mention that in the offshore business many partners often participate. If all those partners, each having only a small stake, were forced to buy full insurance coverage, this would lead to too high amounts of financial responsibility. Preferably one lead operator should be identified on whom a duty to provide financial security should be imposed. Again, the other parties would still remain liable, but the financial responsibility (to show financial security) would only fall to one party, who would obtain security to the benefit of the other parties involved in the risk.[41] This once more underscores the importance of so-called economic channelling (i.e. imposing the obligation to seek financial security on only one party, such as the licensee or operator).

In sum, taking into account the serious criticism that has been promulgated in the literature on legal channelling concerning both the international regime for nuclear liability and the regime for vessel-based marine pollution, legal channelling for offshore-related risks should be avoided; however, there may be an argument in favour of economic channelling which can be arranged through contractual solutions.

7.2.2.2 Joint and Several Liability

Under joint and several liability, joint tortfeasors are held liable for all the damage to which their behaviour might have contributed. Joint and several liability regimes often have been introduced to relieve the burden of proof for victims. Victims could collect the entire damage from one of the contributing tortfeasors, whereas the tortfeasor who compensated could, in turn, exercise an action in recourse against the other liable tortfeasors. An argument in favour of joint and several liability is that it provides incentives for mutual monitoring by potential injurers.[42] However, in the case of insolvency of one of the actors, inefficiencies may arise because recourse may become impossible.[43] Joint and several liability is debatable because an injurer could, in principle, also be held

[40] *Ibid.* [41] Representatives of Lloyds, interview, 1 May 2013, in London.

[42] See Tietenberg, 'Indivisable Toxic Torts'.

[43] For an analysis of joint and several liability under full solvency, see Kornhauser and Revesz, 'Sharing Damages'; and for the case of insolvency, see Kornhauser and Revesz, 'Apportioning Damages'.

liable for a part of the damage not caused by his activity and thus could potentially increase liability exposure.[44]

The application of joint and several liability is in fact a general rule in all cases where more than one tortfeasor is involved. It was therefore no surprise to notice that in the country studies, many countries have joint and several liability for the case where more than one party (e.g. an operator and a subcontractor) have contributed to the accident risk.[45] Joint and several liability, however, has been criticized from an insurance perspective under the argument that it increases the necessity to purchase insurance coverage by all parties involved.[46] However, these negative effects of joint and several liability on the insurability of the liability risk could be mitigated if economic channelling, as discussed earlier, were to be introduced.

7.2.3 Financial Cap?

A question that could be asked is whether it would be necessary to put a financial limit (a so-called cap) on the amount of compensation to be awarded to victims. The reason to discuss this is that, as was discussed in detail earlier, in the international conventions with respect to vessel-based marine pollution and nuclear liability, financial caps are included. However, it was also mentioned that it is precisely these financial caps which are heavily criticized in the literature.

Economic analysis strongly supports arguments against financial caps. From an economic perspective, it is important for the potential injurer to be fully exposed to the social costs of his activities. Otherwise, the desirable internalization of the negative externality would not take place.

In the literature, it has been indicated that there may be good reasons to favour a strict liability rule for hazardous activities, the main reason being that only a strict liability rule would lead to a full internalization of those highly risky activities.[47] This strict liability rule is especially put forward in so-called unilateral accident situations, which is where only one party influences the accident risk. Only with strict liability would the potential injurer also have an incentive to adopt an optimal activity level. This full internalization is obviously only possible if the injurer is

[44] For this reason, joint and several liability in the case of environmental harm is, for example, opposed by Bergkamp, 'The Proper Scope'.

[45] See the overview in Table 10 in Section 3.5.8.2.

[46] See Faure and Hartlief, *Insurance and Expanding Systemic Risks*, 127.

[47] Shavell, 'Strict Liability versus Negligence', 8, 11.

effectively exposed to the full costs of the activity he engages in and is therefore in principle held to provide full compensation to a victim. An obvious disadvantage of a system of financial caps is that this will seriously impair the victim's rights to full compensation. But if the cap is indeed set at a much lower amount than the expected damage, not only would this violate the victim's right on compensation, but the above-mentioned full internalization of the externality would not take place either. From an economic point of view, a limitation of compensation therefore poses a serious problem because there will be no internalization of the risky activity.

Indeed, if one believes that exposure to liability has a deterrent effect, then limitation of the amount of compensation due to victims poses another problem. There is a direct linear relationship between the magnitude of the accident risk and the amount spent on care by the potential wrongdoer. If the liability therefore is limited to a certain amount, the potential injurer will consider the accident as one with a magnitude of that limited amount. Hence, he will spend on care to avoid causing an accident with a magnitude equal to the limited amount, and he will not spend the care necessary to reduce the total accident costs. Obviously, the amount of care spent by the potential injurer will be lower, and a problem of under-deterrence arises. The amount of optimal care, reflected in the optimal standard, which is the level of care necessary to reduce the total accident costs efficiently, will be higher than the amount the potential injurer will spend to avoid an accident equal to the statutory limited amount.[48] Thus, as a result of the cap, too little care is taken.[49]

Moreover, another effect of a financial limit on liability (in addition to providing under-compensation to victims and under-deterrence to operators) is that it would constitute an indirect subsidization of the industry enjoying a particular limit on liability.[50]

Interestingly, during our many interviews with stakeholders, there has not been one, neither from industry nor from insurers, who has argued in favour of a financial cap on the liability related to offshore-related risks. Therefore, this does not seem to be an issue worth discussing at the policy level. It was also argued by a representative of a large re-insurance

48 See Faure, 'Economic Models of Compensation', 21–43.

49 The reason for the under-deterrence is obviously the same as for the under-deterrence which results from the insolvency of the injurer. Under-deterrence arises because the injurer is not exposed to full liability either as a result of his insolvency or as a result of a cap.

50 See also Radetzki and Radetzki, 'Private Arrangements'.

company that financial caps should not be imposed because they would allow a 'privatization of profits and socialization of losses'.[51] This should be avoided also as a matter of principle because the industry should in principle be fully liable for the damage it causes.[52]

7.3 Liability versus Regulation

The core of this study deals with liability rules. Economists have stressed that many of the goals of liability law, more particularly prevention and deterrence, can also be achieved through regulation. There are obviously differences in the way in which tort law and regulation work in providing incentives to parties in a potential accident setting to take optimal care.[53] However, notwithstanding those differences, both tort law and regulation, in principle, could be employed to provide incentives to operators of offshore installations to take optimal preventive measures to reduce the accident risk.[54] Hence, it is important not to discuss liability rules completely in isolation, realizing that there are other instruments that can aim at the prevention of offshore-related incidents as well, in particular, regulation.

It is therefore important to briefly address the economic criteria for safety regulation and to show why, in particular, in the offshore area, regulation may be the primary instrument to control offshore-related risks. It is important to mention this because the economic criteria that favour regulation also coincide with expert opinion from stakeholders. This implies that liability rules still may play a role, but probably a more modest role in supplementing regulation where, for a variety of reasons, the deterrent effect of regulation may fail. In the end, an optimal combination of liability and regulation thus will have to aim at preventing offshore-related incidents.

7.3.1 Criteria for Safety Regulation

Let us now examine under what kinds of circumstances liability rules may not suffice to deter environmental harm and a regulatory intervention may be necessary. The choice between regulation and liability rules was thoroughly examined by Steven Shavell in 1984 in a paper in which

[51] Interview with Philipp Wassenberg, representative of Munich Re Insurance Company, 6 May 2013.

[52] *Ibid.* [53] See also Cane, 'Tort Law as Regulation'.

[54] See Hilton, 'When Should We Prefer Tort Law'.

he advanced several criteria that influence the choice between safety regulation and liability rules.[55]

7.3.1.1 Information Asymmetry as a Criterion for Regulatory Intervention

Information deficiencies have often been advanced as a cause of market failure and as the justification for government intervention through regulation.[56] Also, for the proper operation of a liability system, information on the existing legal rules, the accident risk, and efficient measures to prevent accidents is a precondition for efficient deterrence. According to Shavell, the parties in an accident setting generally have much better information on the accident risk than that possessed by the regulatory body.[57] The parties themselves have, in principle, the best information on the costs and benefits of the activity they undertake and the optimal way to prevent accidents. This 'assumption of information' will, however, be reversed if it becomes clear that some risks are not readily appreciated by the parties in an accident setting. This may be a more significant problem if costs are external. External costs cannot always be easily quantified by the parties involved. Therefore, for every activity, the question that will have to be asked is whether either the government or the parties involved can acquire the needed information at the least cost.

7.3.1.2 Insolvency Risk

If the potential damages can be so high that they will exceed the wealth of the individual injurer, liability rules will not provide optimal incentives. The reason is that the costs of care are directly related to the magnitude of the expected damages. If the expected damages are much greater than the individual wealth of the injurer, the injurer will only consider the accident as having a magnitude equal to his wealth. He will therefore take only the care necessary to avoid an accident equal to his wealth, which can be lower than the care required to avoid the total accident risk.[58] This is a simple application of the principle that the deterrent effect of tort liability works only if the injurer has assets to pay for the damage he causes. If an injurer is protected against such liability, the problem of under-deterrence arises.[59]

[55] Shavell, 'Strict Liability versus Negligence'; Shavel, 'Liability for Harm' and 'A Model of the Optimal Use'.

[56] See the basic article by Stigler, 'The Economics of Information'; see also Schwartz and Wilde, "Intervening in Markets'; and Mackaay, *Economics of Information*.

[57] Shavell, 'Liability for Harm'. [58] *Ibid.* [59] Shavell, 'The Judgment Proof Problem'.

Safety regulation can overcome this problem of under-deterrence caused by insolvency.[60] In this case, the efficient care will be determined ex ante by regulation and will be affected by enforcement instruments which induce the potential injurer to comply with the regulatory standard, irrespective of his wealth. Therefore, a problem might still arise if the regulation also were enforced by means of monetary sanctions. Again, if these were to exceed the injurer's wealth, the insolvency problem would remain. Hence, if a safety regulation is introduced because of a potential insolvency problem, the regulation itself should be enforced by non-monetary sanctions.[61]

7.3.1.3 The Threat of a Liability Suit

Some activities can cause considerable damage, but even so, a lawsuit to recover these damages may never be brought. If this were the case, there would, of course, be no deterrent effect of liability rules. Therefore, the absence of a liability suit would again be an argument to enforce the duty of efficient care by means of safety regulations rather than through liability rules.[62] There can be a number of reasons why a lawsuit is not brought, even though considerable damage has been caused.

Sometimes an injurer can escape liability because the harm is thinly spread among a number of victims. As a consequence, the damage incurred by every individual victim is so small that he has no incentive to bring a suit. In particular, this problem will arise if the damage is not caused to an individual but to common property, such as the surface waters in which each member of the population has a minor interest. In addition, a long time might have elapsed before the damage becomes apparent; in this case, much of the necessary evidence may be either lost or not obtained. Another problem is that if the damage only manifests years after the activity, the injurer might have gone out of business.

A related problem is that it is often hard to prove that a causal link exists between an activity and a type of damage.[63] The burden of proof of a causal relationship becomes more difficult with the increasing passage of time after the damaging incident took place. Often a victim will not recognize that the harm had been caused by a tort but might think that his particular ailment (e.g. cancer) had a 'natural cause' associated with

[60] If insurance came into the picture, it could overcome the problems of under-deterrence, provided that the moral hazard problem, caused by insurance, can be cured.

[61] Shavell, 'Criminal Law and the Optimal Use'. [62] Shavell, 'Liability for Harm'.

[63] Landes and Posner, 'Tort Law as a Regulatory Regime'; Kunreuther and Freeman, 'Insurability, Environmental Risks'.

general ill health. For all these reasons, a liability suit might not be brought, and hence safety regulation is necessary to ensure that the potential polluter takes efficient care.[64]

7.3.2 The Need to Regulate Offshore-Related Risks

After having discussed these criteria for regulation,[65] it is now possible to discuss the question of how these criteria relate to offshore-related risks. If one takes the criteria for safety regulation discussed earlier and applies them to the potential risk caused by offshore installations, there is no doubt that liability rules alone cannot suffice to provide sufficient incentives for care to operators. If one looks at the first criterion, that of information costs, it must be stressed that an assessment of the risks of a certain activity often requires expert knowledge and judgement. Small organizations might lack the incentive or resources to invest in research to find out what the optimal care level would be. Also, there would be little incentive to carry out intensive research if the results were automatically available to competitors in the market: this is the well-known free-rider problem. This problem can be countered in part by legal instruments granting an intellectual property right to the results of the research. However, the problem remains that it may not be possible for small companies to undertake studies on the optimal technology for preventing environmental damage. Therefore, it is often more efficient to allow the government itself to do the research on the optimal technology (e.g. in a governmental environmental research institute). The results of this research can then be passed on to the parties in the market through the regulation. Hence, the setting of safety standards in regulation can be seen as a means of passing on information on the minimal technology required. Obviously, it is more efficient for the government to acquire information on the optimal safety standard for offshore operations than it would be for an individual firm, for instance, to find out what additional reduction in pollution would produce an optimal reduction in the expected damage from the incident. There are undeniable 'economy-of-

64 For alternatives to liability suits, see Bocken, 'Alternatives to Liability' and 'Alternatives to Liability and Liability Insurance'.

65 These are often referred to as 'public interest' criteria for regulation to contrast them with 'private interest' explanations for regulation, as advanced by public-choice scholars. In the latter perspective, regulation is often made, not in the public interest but to serve particular needs of interest groups that seek economic advantages (referred to as 'rents') via regulation. Here it is assumed that government makes regulation in the public interest.

scale' advantages in regulation. In this respect, it is of importance, as was already stressed, that there are many different operators in the offshore market. Lacking information may not necessarily be a problem for the majors, but because small and medium-sized operators are also active in this market, for them, the optimal information concerning preventive technologies may indeed be lacking.

Also, the insolvency argument points in the direction of regulation. Pollution can be caused by individuals or firms with assets which are generally lower than the damage they can cause by the pollution. In this respect, it should not be forgotten that even a small offshore operator can cause harm to a large number of individuals or to entire ecosystems. The amount of damage caused by this operation, of course, can largely exceed its individual assets.[66] Moreover, most firms have been incorporated as a legal entity and therefore benefit from limited liability. Hence, the individual shareholders are not liable to the extent of their personal assets, but a creditor of the firm can only lay claim to part or all of the total assets purchased by the firm's shareholders. As far as insolvency risk is concerned, we can also refer to the discussion in Chapter 6,[67] where it was again stressed that all operators admit (especially small and medium-sized operators) that an insolvency risk is very realistic (i.e. the likelihood that the amount of damage caused by an offshore-related incident could be higher than the company's assets). Again, this points to favouring regulation.

As far as the chance of a liability suit being brought for damage caused by an offshore incident is concerned, a balanced analysis is required. Obviously, the damage often can be spread over a large number of victims, e.g. fishermen and local businesses suffering harm as a result of pollution. These entities in some cases may have difficulties organizing themselves to bring a lawsuit. Moreover, pollution damage especially can cause long-lasting consequences and thus latency problems. However, the problem of bringing a lawsuit in the case of an offshore-related incident may not be as large as with other pollution problems such as air pollution or soil pollution, for example. In those cases, latency problems may even be more serious in the sense that one will often discover only tens of years after an emission that a particular harm has occurred, and in some cases, it may be difficult to find out who the liable injurer was.

[66] This is a point that has been discussed in detail in Section 6.1.2, which was devoted to the insolvency risk.

[67] More particularly in Section 6.1.2.

These types of causal uncertainty and multiple injurers problems are not very likely in the case of an offshore-related risk. Usually it will be well known who the operator was causing the offshore-related incident, and victims in many cases may suffer a substantial financial loss which should provide them with sufficient incentive to bring a lawsuit. Of course, victims of offshore-related damage will, like any victim in a tort suit, always face substantial hurdles in bringing lawsuits having to do with the costs of funding civil litigation,[68] but the problems in this respect are not necessarily different with respect to offshore-related liability risks.

Summarizing: the information asymmetry, especially for small and medium-sized operators, may be an important argument in favour of regulation as well as the insolvency risk, which is especially important in the case of small and medium-sized operators. In such cases, liability rules alone cannot suffice to provide appropriate incentives to operators to prevent offshore-related risks. This provides a strong argument not only to rely on the liability system for preventing those types of risks but also to rely on regulatory solutions as well.

7.3.3 *Private or Public Regulation?*

The economic criteria just discussed point strongly in the direction of at least having some regulation of offshore-related risks, especially for smaller and medium-sized operators, because liability rules alone may not provide efficient incentives. In this case, there would be serious risks of externalization of the social costs to society. Ex ante standard setting via regulation can prevent this externalization for the simple reason that the standards would have to be complied with irrespective of harm and irrespective of insolvency.

Traditionally, as well as in the economic analysis on which we rely in this study, regulation has almost implicitly been equated with public regulation (i.e. public standards set by government). However, increasingly, economic scholars have also pointed at the fact that in certain circumstances there may be strong arguments in favour of private standard setting by industry or some form of private standard setting supervised by government. The most important argument in favour of self-regulation[69] relates again to the often-mentioned problem of information asymmetry.[70]

68 See the contributions to Tuil and Visscher (eds.), *New Trends*, for a further analysis of this problem.

69 In some literature it is also referred to as 'private regulation'.

70 See Ogus, 'Self-Regulation'; Philipsen, *Regulation of and by Pharmacists*, 35–41.

In this case, it refers to the information asymmetry between the industry and the government. In many cases, it is the operators that have much better information than the government e.g. on optimal preventive technologies in the offshore sector. For this reason, self-regulation is advanced in the law and economics literature, given the larger expertise and technical knowledge of industry. As a result, information costs to fix the standards are supposed to be lower than when government would have to set standards.[71] Two more arguments in favour of self-regulation are presented by Miller.[72] Industry would be able to organize in a less bureaucratic way than government and therefore would show more flexibility. This would be especially important in an area where regulatory standards (as a result of evolutions in research and development) are very volatile and hence change quickly. Government or administrative agencies are often not able to react quickly due to bureaucratic problems and slow decision-making processes. The costs of adapting norms to rapidly changing societal circumstances and technical evolutions thus can be relatively low compared to the situation where government intervention would be necessary.[73] Second, industry would be better able to minimize the costs of regulation because in drafting the standards industry would also take into account compliance costs and enforcement costs. Moreover, industry organizations bear their own costs and hence would have incentives for cost minimization. For the government, self-regulation therefore would have the advantage that it could save on enforcement costs.

However, contrary to these advantages of safety regulation, many disadvantages are also mentioned in the literature. One problem, according to Shover,[74] is that industry does not often show that it really deserves the trust that society puts in it, for example, by reacting effectively against violations of standards. Also, the private-interest theory has warned that interest groups often have an incentive to abuse regulation to create barriers to entry and hence to limit market entry for newcomers. This would increase the income for the existing industry. Moreover, it is not always sure that the industry group will use the information advantage via self-regulation effectively to promote the public cause.[75] With

[71] Ogus, 'Rethinking Self-Regulation', 97–8; Van den Bergh, 'De maatschappelijke wenselijkheid', 793.

[72] Miller, 'The FTC and Voluntary Standards'.

[73] Ogus, 'Rethinking Self-Regulation', 98; Van den Bergh, 'De maatschappelijke wenselijkheid', 794.

[74] Shover, 'Zelfregulering door ondernemingen'.

[75] So Campbell, 'Self-Regulation and the Media', 717.

self-regulation in some cases industry could serve its own interests and not the public interest.[76] Enforcement of self-regulation is also often weak, and deterrent sanctions may be lacking.[77] Also, more recently, questions have been asked with respect to self-regulation and private standards as far as the legitimacy and accountability of self-regulation are concerned.[78] Increasingly one can see that in some situations government will make use of self-regulation, but government will delineate the circumstances under which it will rely on self-regulation. Hence, in such cases, it is not pure self-regulation by industry but rather a 'conditional self-regulation' whereby government still monitors self-regulation by industry.[79]

These arguments may apply to the case of offshore-related risks as well: many (especially large) OGPs may have much better information than government concerning offshore-related risks. This would comply with the economic criteria (information asymmetry) in favour of self-regulation. This is what one can see, for example, in the United Kingdom, where, in fact, the industry organization (Oil and Gas UK) not only drafts safety standards[80] but also rules on financial responsibility. Still the regulator (in the United Kingdom, the Department of Energy and Climate Change [DECC]) monitors the self-regulation in the sense that it will not automatically accept the analysis by Oil and Gas UK but will critically monitor it.[81] Hence, to some extent, the practice of 'conditional self-regulation' seems to be applied in the offshore area as well.

7.3.4 Liability and Regulation Combined

It was just mentioned that based on the economic criteria, there is a strong argument to control offshore-related risks through ex ante regulations. However, in individual cases, there can again still be damage

[76] Campbell, 'Self-Regulation and the Media', 717: 'Self-regulators often combine – and sometimes confuse – self-regulation with self-service.'

[77] See Reader, 'Is Self-Regulation the Best Option?', 182, 210.

[78] See, *inter alia*, Rhodes, 'The New Governance'; Coglianese and Lazer, 'Management-Based Regulation'. On the concept of trans-national private regulation, see Scott, Cafaggi and Senden, 'The Conceptual and Constitutional Challenge'; also Cafaggi, 'New Foundations'.

[79] See, *inter alia*, Baatsma *et al., Zelf doen?*, 8.

[80] See Oil and Gas UK, 'Guidelines on Relief'.

[81] Interview with Wendy Kennedy, head of Offshore Oil and Gas, Department of Energy and Climate Change (DECC), 29 April 2013.

resulting from an offshore-related incident. For compensation reasons, liability under tort therefore may come into the picture again. Moreover, although many arguments point in the direction of a preference for regulation, regulation may not be perfect in controlling offshore-related risks. Regulation (especially government regulation) cannot ex ante regulate all types of risks; it also may be outdated fast and not show sufficient flexibility. Regulation also depends upon enforcement, which may never be absolute. Given these weaknesses inherent in any regulatory regime, tort rules will still play an important role to back up failures in regulation.[82] This means that in practice there is still an important role for liability rules (to step in to back up regulatory failure). As a consequence, most legal systems hold that following a regulatory standard does not automatically exclude an operator from liability in tort. This would amount to a so-called regulatory compliance defence. Because compliance with regulation is usually merely a minimum, a judge in a tort case should still have the possibility to verify whether the optimal care levels were in fact higher than the standard laid down in regulation. If this is the case, an operator can still be held liable notwithstanding that it followed a regulatory standard.[83]

However, regulation can play an important role in liability cases. Indeed, in most legal systems, violation of a regulatory standard will automatically be considered a fault, in some cases referred to as negligence per se.[84] The economic logic is that when the safety standard was set, because the regulator possessed better information than the judge, regulation passes on information to the parties and to the judge on optimal safety standards. Hence, if the operator violates a safety standard, this facilitates the task of the judge because it will normally lead to an automatic finding of wrongfulness in a liability case. To the extent that this wrongful act stood in a causal relationship with the damage suffered by the victim, compensation is possible. In this sense, victims will also have an incentive to prove that a regulatory standard was breached. This makes the victim in fact the enforcer of safety regulation.[85]

This shows that although regulation may be the primary instrument to control offshore-related risks, liability rules still play an important role not only in compensation but also in providing incentives to operators to take optimal preventive measures, especially in cases where regulation

[82] See Schmitz, 'On the Joint Use'. [83] See Schwartz, A., 'Statutory Interpretation'.
[84] See Cane, 'Tort Law as Regulation', 427–67.
[85] See Faure and Van den Bergh, 'Negligence', 110–11.

would be sub-optimal. This is why compliance with regulation will not excuse a tortfeasor from liability in tort in most legal systems. However, because a breach of regulatory standard will constitute wrongfulness in a tort case, regulation can assist the judge (and the victim) in liability cases. In this sense, a 'smart' combination of both instruments (regulation and liability) is necessary to provide optimal incentives for operators to adopt preventive measures.

7.3.5 Stakeholder Assessment

It is striking, although perhaps not surprising, that many stakeholders also report that the safety in the offshore sector is to an important extent dependent upon compliance with regulation. Many stress that over the years and especially after the Deepwater Horizon incident in 2010, the safety standards with which offshore operators comply have increased to very high levels. Safety is, so it is especially reported by the majors, their primary concern. Irrespective of legislative duties, they would strive for high standards and state-of-the-art technology.[86] Representatives from a major such as Shell also argue that as a result of corporate social responsibility, the safety standards they apply in Europe will also be applied on a worldwide basis. Even stronger, majors are in fact strongly in favour of high standards for safety regulation on the condition that these EU standards would also be globally compatible.[87] The advantage of stringent and clear regulation is that it provides precision and avoids the discomfort that would accompany instability. Moreover, stringent rules also would have the advantage of creating a level playing field.[88] These representatives especially argue that safety standards for offshore installations may be very high, as in the Netherlands, the United Kingdom and Norway, but increasingly offshore operations also take place in the Mediterranean, in which case regulations would be drafted in countries such as Italy, Turkey and Greece where standards may not be the same.[89] Of course, increased safety levels may drive some (smaller) players out of the market, but if this is the result of increased safety, that may be an unavoidable price to pay.[90]

[86] Interview with representative from ExxonMobil, 19 February 2013; Interview with representatives from Shell International, 26 March 2013.

[87] They presented the example of the application of the European Emission Trading Scheme (ETS) to aviation, in which case EU regulation was globally out of context.

[88] Interview with representatives of Shell, 14 March 2013. [89] *Ibid.* [90] *Ibid.*

In this sense, the interests of major OGPs seems to coincide with the interests of regulators who (obviously) also plead in favour of stringent regulation of safety.[91] It is also stressed that the nature of safety regulation has changed over time. Legislation is now (e.g. in the United Kingdom and the Netherlands) focusing more on targets to be reached and less on particular details. Also, monitoring and control do not take place at a very detailed level but consist rather of controlling whether an environmental management system has been put in place that works adequately. According to regulators, this target-based regulatory system has led to an increasing sense of responsibility among industry actors and to increasing safety levels, especially in places like the Netherlands and Norway.[92] Moreover, in the Netherlands, there is equal experience with voluntary agreements. In 1995, a so-called industry-wide agreement was concluded whereby industry bound itself to voluntarily reduce pollution levels. This led to spectacular results for all parameters on which there was agreement. Since 1995, pollution has indeed decreased. Representatives of regulators hence hold that the covenant worked very well and proved to be able to reduce pollution levels. The only reason it stopped functioning was that the environmental ministry did not want to be engaged in a covenant any longer. The simple reason was that there had been a general policy change as a result of which the environmental ministry in the Netherlands no longer concluded environmental agreements.[93] At that time, the covenant was signed by all actors and hence not only by the majors. The covenant thus covered the entire market. Although the covenant no longer exists, there still is room for discussions with industry within an '*Overlegorgaan Olie en Gas*' (OOG) which provides a framework for industry and the regulator to meet and discuss in an open and transparent manner.[94]

A representative from the regulator in the United Kingdom, however, expressed a small concern about the role of the European Maritime Safety Agency (EMSA). It was feared that if EMSA were to intervene, as in controlling offshore installations within Member States' territories, a lot of expertise within the Member States would be lost, whereas there would be no guarantee that inspectors from EMSA would have similar expertise.[95] The UK regulator therefore warned against a loss of the experience and expertise within the inspection agencies of the Member

91 See e.g. the opinion of Jan de Jong, inspector-general of the state mines, interview, 22 February 2013.

92 *Ibid.* 93 *Ibid.* 94 *Ibid.*

95 Interview with Wendy Kennedy, head of Offshore Oil and Gas, DECC, 29 April 2013.

States and moreover argued that it is much more important to also invest in collaboration and mutual learning (via networking) between the various inspection agencies at the EU level. This is already taking place to a large extent.[96]

7.3.6 Policy Conclusions

- Based on theoretical starting points, safety regulation should play an important role in the prevention of offshore-related risks.
- Liability rules remain important to fulfil a complementary role where safety regulation remains sub-optimal or is inadequately enforced.
- Given informational advantages (especially of the majors in the offshore sector), safety regulation also could take the form of self-regulation or private regulation but should be supervised by government in a kind of 'conditional self-regulation'. The covenant that was concluded in the Netherlands between regulators and industry, as well as the collaboration in the United Kingdom between Oil and Gas UK and DECC, may constitute examples of such a public–private partnership in standard setting.
- It is in the interest of industry (especially those willing to comply with high safety standards) and regulators to have high and stringent safety standards; there is still room for improvement in that respect in the European Union.
- The question, however, arises whether setting those standards should be a task for the EU Commission; it seems preferable that agreements on targets and safety standards are promoted via EU guidance notes to be concluded between industry (given higher technical knowledge) and (a conglomerate of) national regulators. This recommendation, of course, complies to a large extent with the creation of the European Maritime Safety Agency,[97] which was established specifically for the purpose of ensuring a high, uniform and effective level of maritime safety and prevention of pollution by ships within the European Community as well as ensuring a response to marine pollution caused by oil and gas installations. In addition, mutual learning and collaboration (via networking) between national inspection agencies in the Member States should be encouraged as well.[98]

[96] *Ibid.* [97] Under Regulation (EC) No. 1406/2002.

[98] This was stressed by various regulators, e.g. for the Netherlands (Jan de Jong, inspector-general of the state mines, interview, 22 February 2013) and the United Kingdom (Wendy Kennedy, head of Offshore Oil and Gas, DECC, interview, 29 April 2013).

7.4 Mandatory Financial Security

7.4.1 Criteria for Mandatory Financial Security

In law and economics research, several criteria have been advanced to indicate where mandatory financial security may be indicated. The most important reason for introducing compulsory insurance is insolvency. Insolvency may pose a problem of under-deterrence. If the expected damage largely exceeds the injurer's assets, the injurer will only have incentives to purchase insurance up to the amount of his own assets. He is only exposed to the risk of losing his own assets in a liability suit. The judgement proof problem therefore may lead to under-insurance and thus to under-deterrence. Jost has rightly pointed out that in these circumstances of insolvency, compulsory insurance might provide an optimal outcome.[99] By introducing a duty to purchase insurance coverage for the amount of the expected loss, better results will be obtained than with insolvency, whereby the magnitude of the loss exceeds the injurer's assets.[100] In the latter case, the injurer will only consider the risk as one where he could at most lose his own assets and will set his standard of care accordingly. When he is under a duty to insure and exposed to full liability, the insurer obviously will have incentives to control his behaviour. Via the traditional instruments for the control of moral hazard, the insurer can make sure that the injurer will take the necessary care to avoid an accident with the real magnitude of the loss. Thus, Jost and Skogh argue that compulsory insurance can, provided that the moral hazard problem can be remedied adequately, yield better results than under the judgement proof scenario.

Indeed, this economic argument shows that insolvency may cause injurers to externalize harm: they may be engaged in activities that may cause harm which largely exceeds their assets. Without financial provisions, these costs would be transferred to society and thus would be externalized instead of internalized. Such internalization can be achieved if the insurer is able to control the behaviour of the insured. The insurer could set appropriate policy conditions and an adequate premium. This shows that if the moral hazard problem can be dealt with adequately, insurance even leads to a higher deterrence than a situation without liability insurance and with insolvency. Hence, since insolvency can

99 Jost, 'Limited Liability'. A similar argument has recently been formulated by Polborn, 'Mandatory Insurance'; and by Skogh, 'Mandatory Insurance'. Skogh also has pointed out that compulsory insurance may save on transaction costs.

100 See also Kunreuther and Freeman, 'Insurability, Environmental Risks'.

lead operators to externalize risk and social costs, mandating financial security could force them again to internalize these social costs by paying a contribution or insurance premium.

However, the literature has equally formulated a number of conditions and warnings when introducing compulsory financial guarantees.[101] One warning is that the moral hazard problem should be controlled. If moral hazard cannot be controlled, the regulator should even consider a prohibition of liability insurance.[102]

Second, if one were only to introduce compulsory insurance (as compared to mandatory financial security, which is obviously broader), this should only be done when no restrictions on the insurance market exist. If a high concentration existed, premiums would be too high, and this could equally reduce the incentives of insurers to control the moral hazard risk.

Third, from a policy perspective, it may not be wise to limit the duty to provide financial security to insurance. If the policy maker were to introduce compulsory insurance, it would become totally dependent on insurers to fulfil the duty to insure. This could create an undesirable situation whereby insurers would become de facto licensors of the industry, which could be questionable from a policy perspective.[103] This may therefore be a strong argument for a flexible approach, i.e. not to limit the provision of mandatory security necessarily to insurance but to allow the market itself to suggest a wide variety of financial and insurance instruments as long as they can guarantee compensation when an accident happens.

7.4.2 Application to Offshore-Related Risks

Earlier, it was discussed in detail[104] that there may be a serious insolvency risk, especially for small and medium-sized operators, not only for larger accidents for which no insurance coverage is generally available on the market but also for medium-sized risks. It was equally stated that whereas this insolvency risk may emerge for smaller and medium-sized operators, this is not necessarily the case for the larger operators (the so-called majors).

Thus, this calls for a balanced approach at the policy level. Obviously, it is not possible to impose a duty to show mandatory security only on small

[101] For a summary of those warnings, see Faure, 'Economic Criteria'.
[102] See Shavell, 'The Judgment Proof Problem', 43–58.
[103] See Monti, 'Environmental Risk'. [104] See Section 6.1.2.

and medium-sized operators and not on the majors (where it may not be indicated). However, this flexibility could be awarded to regulators in EU Member States when assessing whether the obligation to show financial coverage has been met. Moreover, especially as far as the instruments are concerned, it is necessary to adopt a flexible approach as well, i.e. not to limit the duty to seek financial cover to one particular instrument such as insurance.

7.4.3 Stakeholder Assessment in the European Union

Although, as we showed earlier,[105] almost all stakeholders agree that there may (depending upon the type of operator and the risks involved) be a potential insolvency risk related to offshore liability, the opinions concerning mandatory coverage are, to say the least, balanced. Major operators are very conscious of the risk that smaller operators may pose an insolvency risk and hence recognize the regulatory need for them to show financial coverage.[106] They stress, however, that it also remains important in a model of mandatory coverage to apply an effective system of risk differentiation to provide incentives for prevention.[107] As far as the amount for which coverage has to be provided, a distinction should be made according to a variety of elements, such as whether the reservoir is old or new (in which case there may be more or less pressure), whether it is (as in Macondo) a high-temperature well (which may increase the risk) and whether it is close to the shore (which may increase the pollution risk).[108] On this basis, regulators could make an assessment of the potential damage and hence of the amount that should be constituted as security by the operator. To a large extent, the Oil and Gas UK Guidelines on financial responsibility, on which DECC relies to a large extent, recognize this.[109]

Also, representatives of regulators argue that it is important that a future EU regulation should definitely guarantee the ability of operators in EU waters to compensate the damage that could emerge from their offshore installations.[110] It is equally stressed that in this respect there should be flexibility, so coverage should not necessarily be insurance.[111]

[105] *Ibid.* [106] See interview with representatives of BP, 13 March 2013.

[107] Interview with representatives of Shell International, 14 March 2013. [108] *Ibid.*

[109] *Ibid.*

[110] Interview with Jan de Jong, inspector-general of the state mines, 22 February 2013.

[111] OGP expressed it as follows: 'The key issue for the oil and gas industry is the flexibility for each company to choose how best to demonstrate financial security. A one-size-fits-all

For example, if majors decide to self-insure, this should not be a problem as long as the provided guarantee is adequately controlled. Mandatory solvency guarantees are, however, very important because the insolvency risk related to offshore liability is quite real.[112]

However, it is well known that, as was also noticed in the preparation to the Environmental Liability Directive (ELD), (a part of) the insurance sector is strongly opposed to making coverage mandatory. One important reason is that it is argued that mandatory security can negatively affect moral hazard.[113] The opposition of insurers to mandatory financial security was especially clear at the time of drafting the ELD, when only a small number of players were in the market. Since 2004, some insurers have developed products aimed at coverage of environmental damage. However, mostly this involves US insurance companies providing cover to Western European enterprises. In Central and Eastern European Member States, there is de facto no (substantial) insurance coverage for environmental damage available. It is argued that these differences would constitute a serious problem because a level playing field would be lacking.[114] Some cases (especially in Central and Eastern Europe) also showed serious insolvency risks. For example, after the Hungarian toxic sludge event, it appeared that the company had environmental impairment insurance with a very low limit. Obviously, the amount of personal injury claims in the East may be substantially lower than in the West, but it also shows important differences and the lack of a comparable operating environment.

Strikingly, a major international re-insurer argued in fact that it is not opposed to mandatory insurance whatsoever. This company argues that the traditional opposition of insurers to compulsory insurance is to some extent strange. If one is the producer of toothpaste, one should normally be happy if the government were to make the use of toothpaste mandatory. However, insurers often have the fear that compulsory insurance may lead to problems of the duty to accept bad risks. The advantage of compulsory insurance is that it excludes adverse selection because all have to be included in the pool.[115] Also, Lloyds does not seem opposed to mandatory security on the condition that the amount imposed as security

approach for meeting financial responsibility does not work for such a diverse industry as the oil and gas sector'; letter of the International Association of Oil and Gas Producers (OGP) to Michael Faure, 13 May 2013.

112 *Ibid.* 113 Interview with Swiss Re, 11 February 2013.

114 Interview with Phil Bell, 26 February 2013.

115 Interview with representatives of a major re-insurance company, 12 March 2013.

by the regulator can be available in the long term. Regulators often forget that the market is very volatile. For example, the fact that through a broker in one year US$500 million could be made available in cover does not guarantee that the same amount would still be available the next year.[116] Unexpected events that seemingly are not related to maritime or offshore risks can still influence the available cover. For example, even the Boston Marathon bombing affected the maritime insurance market because there was apparently maritime cargo in some warehouses that were hit by the bombing, as a result of which maritime insurers had to provide cover. This may lead to a reduction of capacity and a reduction of appetite by insurers to provide cover.[117]

7.4.4 (European) Policy Issues

The question, of course, arises as to whether one should at the policy level introduce mandatory insurance and also when the market would not be able to provide it. There are different experiences in this respect. When the German Environmental Liability Act of 1990 (*Umwelthaftungsgesetz*) required the owner of an installation that could cause significant damage to take out liability insurance or to have sufficient financial guarantees, this led to substantial problems because the German insurance market apparently refused to cover the risk.[118] This is thus an example of where policy makers become dependent upon the insurance market that subsequently refuses to provide coverage. There is, however, also an opposite experience in the United States with the introduction of mandatory security in the Oil Pollution Act of 1990. The literature holds that this legislative duty had an important influence on the market, providing incentives to develop financial products that could meet the regulatory duty.[119] This was also reported by several stakeholders: if there were no regulatory duty mandating financial security, it is not very likely that this market would strongly develop.[120] Moreover, it was also held that a mechanism such as OPOL could only emerge in the wake of

[116] Interview with representatives from Lloyds, 1 May 2013, in London. [117] *Ibid.*

[118] See Wagner, 'Umwelthaftung' and 'Die Zukunft'.

[119] See Boyd, 'A Market-Based Analysis', 276–85. However, representatives of the International Group of P&I Clubs (interview in London, 1 May 2013) also held that in the United States, P&I Clubs refuse to issue financial guarantees under OPA because they resist the potentially wide liability under OPA. P&I Clubs hence only provide cover in an indirect way, thus avoiding direct actions by plaintiffs against the P&I Club and still having the possibility to call on particular exceptions in the policy.

[120] Interview with a representative of Marsh, 8 February 2013.

a regulatory duty to provide mandatory security.[121] Otherwise, membership in OPOL would have little added value for operators.

Thus, these observations warrant a careful approach: mandatory security for offshore-related risks is important to internalize the externalities that could emerge in the case of insolvency, and such a duty also may stimulate the development of more financial and insurance products. Moreover, Chapter 5 showed that today OGPs make use of a large variety of existing systems of coverage that are available in the market. Hence, there is no risk (as in the case of the German Environmental Liability Act) that a duty to provide financial cover would remain a dead letter because industry would not be able to meet the duty.

There are, as far as mandating financial security is concerned, some more interesting observations that were formulated by stakeholders. A representative of a regulator held that it is crucial that the European Union provide a framework for the mandatory security, for instance, sketching which particular elements (location, size of the installation, etc.) create a necessity for requesting a particular amount of cover. Providing at least such a framework at the EU level seems crucial. Otherwise, there would be a risk that some regulators in some Member States would merely ask for very small amounts, whereas other regulators would ask for very high amounts. This could clearly endanger the level-playing-field concept for operators in Europe.[122] At the same time, it should not be the European Union that controls whether a particular operator has sufficient guarantees.[123] After the European Union has provided a framework with guidelines, the regulatory authorities in the Member States can verify what type of cover is needed for a particular offshore installation and whether the cover offered by a particular operator is adequate.[124] Representatives of a major operator held that an EU-wide framework, very much like the Oil and Gas UK Guidelines, on which DECC largely relies, could be a good example of an approach to be followed in Europe. This would mean that at the EU level, particular

[121] Interview with a representative of OPOL, 27 February 2013.

[122] Interview with Jan de Jong, inspector-general of the state mines, 22 February 2013.

[123] Interview with Wendy Kennedy, head of Offshore Oil and Gas, Department of Energy and Climate Change (DECC), 29 April 2013. Since there are huge differences between the Member States and between the situations of particular offshore installations, it is better to let regulatory authorities in the Member States take those varying location-specific circumstances into account.

[124] *Ibid.*

criteria are developed (e.g. in guidelines) indicating what type of cover needs to be required in which circumstances.[125]

The recent Directive on Safety of Offshore Oil and Gas Operations (Directive 2013/30/EU) also recognizes that the financial capabilities of the licence holders, including financial security, are crucial elements to be considered by licensing authorities. The financial capabilities shall be sufficient to cover potential liabilities arising from offshore activities. However, it is realized in the Directive that the existing financial security instruments cannot accommodate all possible consequences of major accidents.[126]

These observations are strongly in line with economic reasoning. Taking as a starting point that insolvency is a serious risk, creating the market failure of externalization of social costs, some regulatory action mandating financial security is needed. The EU policy maker has clearly already taken up this advice by prescribing in the draft Directive on safety of offshore oil and gas operations that Member States shall require the licensee to ensure that any operator has the capacity to meet the requirements for specific operations within the framework of the license. A first step towards mandating financial security therefore has already been taken. However, from the Commission's perspective, this may not be sufficient. The problem could indeed still arise that regulators in some Member States would require very low amounts of financial cover and would be very flexible and easy-going in accepting proof of self-insurance by the operator. This could lead to a race-to-the-bottom and again endanger internalization of externalities and the level playing field for operators. Hence, the EU level probably should go one step further by also mandating under what type of circumstances particular amounts should be required in security and what types of securities should be deemed sufficient. However, it may be clear that, on the one hand, the European Union wants to mandate this, but on the other hand, it would be unnecessary (and probably not within the competence of the European Union) to mandate specific amounts on an EU-wide basis, nor to prescribe which security would be sufficient in which circumstances. This may very much depend on location-specific circumstances. It is quite possible that a major incident in the North Sea with major damage in the Netherlands would amount to a different type of damage than if the same incident occurred in the Mediterranean with major

[125] Interview with representatives of Shell International, 14 March 2013.
[126] Preamble (63) and Art. 4.2 of Directive 2013/30/EU.

damage in Cyprus. This calls for some regulatory framework at the EU level, still allowing sufficient flexibility to regulators at the Member States. An example of this could be a guidance note. A similar guidance note on financial responsibility issued within the framework of the Directive on Carbon Capture and Storage was already mentioned here.[127] This guidance note provides information to regulators in Member States on the type of financial security that would be acceptable. In the case of offshore-related risks, the policy maker could go one step further by following the example of the Oil and Gas UK Guidelines on financial responsibility or any other system of a risk-based assessment of the potential damage. Such a risk assessment could then guide the regulator in Member States on the amounts to be required in financial security. Such guidance would have the advantage that it does what it says, which is to guide local regulators in Member States and at the same time still leaves sufficient flexibility with those Member States to assess the amount and form of financial security in a particular case, taking into account differing location-specific circumstances.

7.4.5 Summary

- It seems indicated to mandate financial coverage for offshore-related risks, especially for cases where smaller and medium-sized operators may create a risk of major damage, and hence, an insolvency risk would emerge.
- As far as possible at the licensing level, it should be avoided that operators would engage in offshore operations of which the risks, in the case of an accident, would outweigh their personal assets.[128] In such a case, a joint venture with OGPs with larger financial capacity may be indicated.
- It seems indicated to issue a guidance note at the EU level guiding local licensing authorities in Member States on the required amount and form of financial security for offshore related risks.
- As far as the amount is concerned, this guidance note should be based on an objective assessment of the risk, taking into account technical criteria that relate a specific operation and operator to particular amounts of potential damage.

[127] European Commission, implementation of Directive 2009/31/EC on the geological storage of carbon dioxide, Guidance Document 4, 2011.

[128] Note that, for example, in Greenland the licensing government specifies that companies must at least have US$10 billion of equity to qualify. Emmerson, *Arctic Opening*, 41.

- The guidance should allow sufficient flexibility as far as the forms of financial security are concerned and not necessarily limit those to insurance. The only condition would be that local regulators accurately verify whether the form and amount of the financial security offered by the operator would be adequate to cover the potential damage emerging from that particular offshore installation.
- The guidance thus could take into account location-specific circumstances on which local regulators in Member States can base their assessment of the amount and form of financial security.

Such an approach allows sufficient flexibility, avoids unnecessary costs (e.g. forcing majors to transfer risks to lower-rated insurance companies), encourages a level playing field for operators and avoids an externalization of social costs (and thus a market failure) in the case of insolvency.

7.5 Compensation Instrument

Now the various proposals just presented will be reviewed and analyzed in light of economic theory. First, a proposal made in the literature to finance accidents via a self-guarantee paid by taxes will be discussed (7.5.1). Next, mandatory insurance and the proposal formulated by Munich Re will be critically reviewed (7.5.2). Then we will point at the comparative advantages of risk-sharing pools between operators (7.5.3), and finally, it will be highlighted that the most effective way for government to intervene in compensation is as re-insurer of last resort rather than by providing direct compensation or creating a compensation fund (7.5.4).

7.5.1 Self-Guarantee Through Taxes

Viscusi and Zeckhauser proposed to cover a potential catastrophe by an annual tax tailored to the risk that a disaster would happen beyond the capacity of the responsible party. Hence, no risk sharing between offshore companies takes place. They have to internalize the complete social costs themselves, which can lead to efficient precaution.

However, this imposes a high requirement on the regulators to determine the potential risk of each offshore company. The tax should correspond to the past loss experience and an assessment of the company's conduct quality, which is an extremely difficult task for regulators. The administrative costs of implementing this proposal thus may be huge. If the tax cannot reflect the risk accurately, it will affect 'only the decision of whether to drill, not the degree of care with which drilling actually occurs'.[129]

Moreover, this proposal accumulates the amount through an annual tax. If, however, a catastrophe happens in the early years of such an accumulation, a full compensation will not be guaranteed.[130]

7.5.2 Mandatory Insurance and the Munich Re Proposal

Insurance policies, when designed carefully, via risk differentiation, retention and co-insurance, can provide coverage for risk-averse individuals and at the same time maintain sufficient precaution incentives. Mandating insurance, by forcing the majors to remain in the pool, can alleviate the adverse-selection problem and provide substantial coverage. Although many doubt whether high amounts of coverage could be made available, Munich Re proposed an arrangement with the ambition to provide cover up to US$10 billion to US$20 billion.

Parametric trigger coverage has an advantage in dealing with unpredictable high-magnitude risks. In this case, the trigger is no longer liability, but the occurrence of pre-determined situations. The uncertainties created by liability rules and the judges are to some extent precluded. The problem caused by long-term risk is no longer relevant. Moreover, the new facility assesses the risks on a rig-by-rig basis, and hence premiums can be more risk tailored.

However, there exist many problems as well. So far the proposal has largely remained theoretical, and many stakeholders doubt that it will ever be brought into practice. It is especially designed for the United States, where more detailed data are available. Although Munich Re argues that a similar arrangement also could be made for the European Union, it would only be possible if more data and models are available on the basis of which a calculation could take place and the model could be implemented.[131] Moreover, a substantial amount can only be raised if sufficient insurers and operators would like to join the scheme. It then depends on the willingness of the insurers to join. They may not like to have their resources frozen for the unpredictable risks, which could have been invested in more profitable areas.

[129] Abraham, 'Catastrophic Oil Spills', 1789–90. [130] *Ibid.*
[131] Interview with Philipp Wassenberg, representative of Munich Re, 6 May 2013.

To mandate all offshore operators to seek insurance coverage can substantially raise the available amount. However, it deprives the operators of the choice to cover their liability through other instruments. For example, for the majors, who often have better credit ratings than the insurance companies, little added value will be created by forcing them to join a commercial insurance pool. Mandatory insurance also involves the risk of creating a highly concentrated market, which will lead to too high premiums and insufficient risk differentiation.[132] Moreover, insurance alone cannot provide full coverage for all liability. The new facility proposed by Munich Re adopts a parametric trigger, which is not linked directly with liability. Traditional liability insurance also has limitations, such as providing coverage only for sudden and accidental incidents, and much exclusion is employed. Other instruments will be necessary to fill in the gaps created by those exclusions.

Despite this limitation, the Munich Re proposal at least provides a possibility to cover substantially higher amounts of offshore-related damage. Although introducing mandatory coverage creates many downsides as well, to require operators to provide substantially higher amounts of financial security will increase the need to seek insurance coverage. Such a regulatory duty thus could stimulate the development of market solutions.

7.5.3 Risk-Sharing Pools

Theoretically speaking, risk-sharing pools have the advantage of providing incentives for mutual monitoring and to promote risk-preventive investments by all members in the pool. Therefore, membership in a pool thus could (via mutual monitoring) increase safety levels.[133] However, this only works under two important conditions: (1) that risk differentiation is possible to such an extent that real differences in contributions can be made which do provide substantial incentives to operators to invest in risk-prevention measures and (2) that the risks posed by the different operators are at least of a similar nature.[134] It may be difficult to meet both conditions in the case of risk pooling for offshore-related liabilities.

132 Faure and Grimeaud, 'Financial Assurance Issues', 186.

133 For a discussion on the advantages of risk-sharing pools, see Faure, 'Alternative Compensation Mechanisms'; Faure and Skogh, 'Compensation for Damages'; Skogh, 'Risk-Sharing Institutions'; Skogh, 'A European Nuclear Accident Pool'.

134 *Ibid.*

On the one hand, the probability of an offshore incident is so low that the difference between the relative contribution of a high-risk compared to a low-risk operator may be too small to provide appropriate incentives for investments in risk-prevention technologies. If this were the case, risk differentiation would not work, and high-risk operators simply would prefer to pay the higher contribution to the pool, which might be lower than the investment in risk-prevention technologies. On the other hand, the second condition (similar nature of the risks) may not be met in the case of offshore-related risks given the largely different natures not only of operators but also the risks related to different types of wells. Theoretically, technological information is available to distinguish risks on objective grounds.[135] This could then also constitute the basis for differing contributions to a pool. However, the problem would still exist that there could be substantial differences between operators as far as their assets and credit ratings are concerned. If this is the case, smaller (and potentially insolvent) operators could still free ride on the pool by shifting the risk to other members with higher credit ratings. This obviously may have negative consequences for the incentives to invest in prevention.

Therefore, it is not surprising that the majors show little interest in joining pooling arrangements. Noble Energy proposed a mandatory retrospective risk-pooling scheme similar to the requirement from nuclear operators under the Price-Anderson Act. Because contributions will only be made retrospectively, there is no large immobilization of capital ex ante. However, a major difference between nuclear operators in the United States and offshore installations in Europe is that nationwide safety regulations in the United States guarantee similar minimum nuclear safety standards for the entire territory.[136] This reduces the task of mutual monitoring of the members in the pool and moreover reduces the risk of free riding of individual members. However, it would be dangerous to argue that policy makers would have to force unwilling operators (presumably majors) into such a European-wide risk pool. Given the substantial difference between operators and risks and the absence of EU-wide high safety standards (which constitutes a major difference with nuclear risk in the United States), a mandatory pooling could lead to perverse incentives for high-risk operators who could

135 Interview with representatives of Shell International, 14 March 2013.

136 For a discussion on the possibility of introducing a nuclear pooling system in the European Union, see Pelzer, 'International Pooling of Operators' Funds'; Carroll, 'Perspective on the Pros and Cons'.

externalize their risk to the lower-risk members in the pool. In this sense, a mandatory pooling scheme could even lead to decreased safety levels, especially if adequate risk differentiation and safety regulation could not sufficiently distinguish good from bad risks.

An important lesson therefore is that policy makers under such circumstances should refrain from mandating pooling between operators of offshore installations. The only thing a policy maker could do is to encourage the creation of industry-wide pooling by providing high standards of safety regulation. Thus, safety regulation could facilitate and assist mutual monitoring by operators and encourage pooling arrangements.

7.5.4 *A Role for Government in Providing Compensation?*

One important conclusion so far is that both liability rules and regulation have an important (complementary) role to play in preventing offshore-related risks (7.3) and that a balanced system of mandatory financial security can be worked out for offshore-related risks (7.5). However, the question can still be asked whether, to the extent that the market would fail, government should intervene in providing compensation. This could (at least theoretically) take a variety of forms. One possibility, often applied in practice in case of catastrophes, is outright compensation of victims by government on an ex post basis (7.5.1). Another possibility, becoming increasingly popular, again in the area of (natural) disasters and terrorism, is a role for government as re-insurer of last resort (7.5.2). A role that is most often mentioned in environmental liability is the creation of a compensation fund (7.5.3).

7.5.4.1 Direct Compensation by Government?

An option that should briefly be discussed as far as government intervention is concerned is to argue that government should provide (part of the) compensation for damage caused by offshore-related risks.

7.5.4.2 Arguments in Favour

Although there are obvious (economic) criticisms that can be formulated (and have been in the literature) on this type of government charity, there are some positive aspects as well. A positive aspect of government intervention is that the prospects of large-scale payments in the aftermath of an offshore-related disaster might encourage the government to take cost-benefit-justified precautions long before disasters strike.[137] This argument could be made in the case of damage resulting from an offshore-related disaster. The Deepwater Horizon case showed that offshore incidents may cause substantial damage. Often this can lead to a total disruption of society. Providing ex post relief may then help to restore public trust. Moreover, government may be better able to determine adaptation strategies necessary to react against the disaster, at least when compared to potential victims. One could think of the example of large infrastructure works which would be necessary to prevent damage resulting from offshore risks (if it were already possible, for example, to ex ante invest in prevention). These are typically the public goods that would not be provided through private action and thus may require government intervention.

A related argument in favour of government intervention would be that government has the capacity to diversify the risks over the entire population and to spread past losses to future generations, thus creating a form of cross-time diversification which the market could not achieve. The argument can be made that the government can be in a better position to adapt to disasters than individuals.[138] Hence, in some cases, the government may be in the best position to prevent disasters. Government intervention from this perspective would provide incentives to politicians to invest in preventive measures.

Moreover, one could also make the (too easy) argument that government can (in principle) compensate without limits. If the damage could exceed the current budgetary possibilities of the government, the already mentioned cross-diversification over time and future generations could, in principle, take place.

7.5.4.3 Arguments Against

Notwithstanding those theoretical possibilities of government compensation, the arguments against such compensation are compelling as well. The most convincing argument is a very simple one and is precisely the same argument as has often been advanced in the literature related to the international conventions on nuclear liability against government compensation (via the second and third layers).[139] If government de facto provides compensation to victims, it effectively provides a subsidy to industry, as a result of which industry will be able to externalize the social

[137] See Levmore and Logue, 'Insuring against Terrorism', 310.

[138] This argument has been strongly made by Kunreuther and Michel-Kerjan, 'Challenges for Terrorism'.

[139] On that compensation scheme, see Section 3.6.1.2.

costs of its activity. The negative consequences of such a subsidization effect are well known:

- It is no longer the one who causes the accident and creates the risk (i.e. the offshore operator) who has the burden of compensation, but society at large and ultimately the taxpayer; this is undoubtedly an undesirable distributional consequence.
- Since operators are not confronted with the total costs related to their activity, their level of prevention will be too low, and their activity (in this case production) level will be too high. Liability rules can, if government takes part of the compensation for its account, not have their desirable deterrent effect. A too-high accident risk and lower levels of safety can be the result.
- As a consequence, operators will invest less in safety measures (not being confronted with the total costs of their activity) and will not bear the total damage that could result from their activities. The consequence will be that relative prices are too low, which can be considered a market failure. In this hypothetical case, relative prices of oil and gas (especially compared to renewable energy, for example) would be relatively low because prices would not be reflected through social costs. Again, from a social welfare perspective, this would be an undesirable result.

7.5.4.4 Summary

On the basis of this reasoning, it is clear that direct compensation by government should never be a preferred option to deal with offshore-related damage. This is not to say that there should be no role whatsoever for the government in the aftermath of a disaster caused by an offshore-related incident. Relief measures and coordinating disaster management in the immediate aftermath of the disaster are undoubtedly tasks where the government can play an important role. However, an important condition would be that if steps need to be taken, either in the clean-up of polluted beaches or in providing immediate relief to victims, via liability rules, the price for those interventions is ultimately allocated to the liable operator. This corresponds to sound economic principles of cost internalization and to the 'polluter-pays-principle'.

It also should be stressed that none of the stakeholders we interviewed held that a (partial) compensation by government would be a desirable option to deal with the compensation of offshore-related damage.

7.5.5 *Re-Insurer of Last Resort?*

In various legal systems, an interesting model has been developed whereby the government acts as re-insurer of last resort. Under this approach of government involvement, the state assumes at least part of the risk for losses from catastrophes. Even though government intervention is required (because the private insurance market cannot provide adequate catastrophe insurance coverage), the underlying philosophy of this approach is that private insurance should keep on playing a significant role in allocating compensation for victims of catastrophes. This option then usually takes the form of a multi-layered insurance program. Such a program is normally administered by private insurance companies, meaning that they sell insurance, collect premiums and pay claims.

7.5.5.1 Arguments in Favour

There can indeed be arguments in favour of such a re-insurance by government, and assuming that capacity on the private insurance market is indeed severely falling behind, it can be held that without state intervention, insurance coverage for disasters would simply not have developed.[140] Re-insurance by the state thus can be considered an adequate method to resolve the uninsurability problem.[141] A condition is of course that the government charges an actuarially fair premium for its intervention.[142] This type of government intervention, moreover, has the advantage that ex post relief sponsored by the public purse can be avoided. Where the government acts as re-insurer, this at least has the advantage that a premium can be paid by those who actually cause or run the risk. It thus can facilitate market solutions, still provide incentives for prevention to potential victims and avoid the negative redistribution discussed earlier.[143] Thus, a state intervention as re-insurer may avoid the 'catastrophic responses to catastrophic risks'.[144]

Kunreuther and Michel-Kerjan also argue in favour of this type of government-provided re-insurance. They argue that one advantage is

[140] This is a point strongly made by Kunreuther, 'Mitigating Disaster Losses', 180–3; Harrington, 'Rethinking Disaster Policy'; Schwarze and Wagner, 'In the Aftermath of Dresden'.

[141] These public–private initiatives to cover extreme risks are also supported by OECD recommendations. See Monti, 'Public–private Initiatives'.

[142] Faure, 'Financial Compensation', 358. [143] See Section 7.1.2.

[144] Epstein, 'Catastrophic Responses'. In this respect, see also Kunreuther and Pauly, 'Rules Rather than Discretion', 113 (arguing that this government's role in assisting the supply side allows avoiding the inefficiencies and inequities associated with disaster assistance).

that the government has the capacity to diversify the risks over the entire population and to spread past losses to future generations, thus creating a form of cross-time diversification which the private market could not achieve.[145] However, they also argue that, especially as far as terrorism is concerned, government participation in insurance programs is crucial because the risk of terrorist attacks is partly in the government's control, and the government can have more information on ongoing terrorist groups' activities through intelligence services.[146]

7.5.5.2 Arguments Against

Both in law and economics as well as in legal scholarship, criticisms have been formulated on the facilitative role of a government stimulating insurance markets. For example, Gron and Sykes argue in several papers that it would be unjust for the government to provide (re-)insurance at a lower price than the market price.[147] This would give a wrong signal to the market as far as stimulating insurability is concerned. The authors would prefer ad hoc solutions, whereby compensation is provided to accident victims on an ex post basis. This would avoid the situation that market participants are aware that the government will guarantee compensation.

Most of this criticism is not addressed against the intervention of the government as such but is based on the assumption that the government will not ask premiums that reflect market prices. It is a criticism shared by Levmore and Logue, who argue that such a regime (of acting as re-insurer of last resort) only has its desired effect of encouraging the purchase of commercially provided terrorism coverage when it involves a substantial subsidy.[148] They are sceptical of these types of interventions in the market (for terrorism insurance), arguing that, also without government intervention, 'the market would probably have been able to provide the necessary coverage'.[149]

Apparently, the arguments against government intervention are based on the assumption that the government will not ask competitive re-insurance premiums, thereby subsidizing catastrophe insurance.

[145] See Kunreuther and Michel-Kerjan, 'Challenges for Terrorism', 210.

[146] Kunreuther and Michel-Kerjan, 'Insurability of (Mega-)Terrorism Risk'.

[147] See Gron and Sykes, 'A Role for Government'; Gron and Sykes, 'Terrorism and Insurance Markets'.

[148] Levmore and Logue, 'Insuring against Terrorism', 304 (arguing that otherwise disaster insurance would still not be 'available').

[149] *Ibid.*, 311.

Moreover, without this government support, insurance coverage probably could have developed anyway (at least for terrorism events). These points, of course, can only be validated on the basis of empirical research.

A role of government as re-insurer of last resort can only be observed in situations where normal third-party (liability) insurance is not available. Usually it concerns cases where there is no liable injurer at all (typically the case of natural hazards) or where an injurer typically has no insurance or will be judgement-proof (the case of terrorism). Models of government acting as re-insurer of last resort can therefore mostly be found in the domains of terrorism and insurance of natural hazards.

The current schemes for terrorism insurance in France (GAREAT), Germany (Extremus), the United Kingdom (Pool Re), the Netherlands (Dutch Terrorism Risk Re-insurance Company [NHT]) and the United States (the Terrorism Risk Insurance Act [TRIA]) all have features of a government acting as a re-insurer of last resort, at least in general terms.[150] For natural hazards, a well-known example is the Caisse Centrale de Réassurance (CCR) as it functions in France. Insurers can re-insure the risk of natural hazards with this CCR, which benefits from a state guarantee in the event that the CCR exhausts its resources.

7.5.5.3 Summary

However, for the case of offshore-related damage, the arguments in favour of such an intervention by government as re-insurer of last resort do not seem very compelling. One important condition for such an intervention would be that a market solution is largely failing. This may be the case for terrorism and natural hazards,[151] but it is doubtful that this is the case for third-party liability risks created by industry generally and related to offshore activities more specifically. Uninsurability on traditional insurance markets may only arise for the third category of incidents (with damage above €750 million), which may not be difficult to insure on the traditional commercial insurance market. However, as we explained earlier, various proposals have been formulated by commercial entities to create market solutions which would enable coverage for these disastrous types of offshore-related incidents. In this respect, we can refer both to the proposal formulated by Noble Energy[152] and to the proposal formulated by Munich Re.[153] Even though

150 See Jaffee, 'Report on the Role of Government'.
151 For further details, see Bruggeman, Faure and Heldt, 'Insurance against Catastrophe'.
152 See Section 6.3.3. 153 See Section 6.2.

these proposals may not have materialized yet, the regulatory answer to this obviously would not be an intervention of government as re-insurer of last resort but rather a duty imposed on industry to provide adequate coverage as a result of which industry would develop market solutions to provide appropriate coverage. There is also criticism of the fact that these public–private partnerships in practice do not always work in the way they should, more particularly because government may have a tendency to shift too much of the costs to private insurers.[154] Hence, notwithstanding the advantages of this model in the area of terrorism and natural hazards, for offshore-related risks there is no reason yet to discuss the option of an intervention of government as re-insurer of last resort because it has not been shown that market solutions providing adequate coverage could not be developed.

7.5.6 A Compensation Fund?

7.5.6.1 Funds versus Insurance

If one applies the general principles of efficient and fair compensation mentioned earlier,[155] the two main criteria to distinguish whether a fund would do a better job than insurance relate to the capacity to apply risk differentiation and to costs. Applying the principles discussed earlier, there are not many reasons why a compensation fund would provide better protection against insolvency than the private insurance markets if both are, in theory, available. One can assume that an insurer is better able to differentiate risks because it is specialized in risk differentiation and risk spreading. Insurers therefore possess techniques for determining in what way insured parties contribute to risk. Obviously, this assumes that insurance markets are competitive. In the absence of competition in insurance markets, either the supply of insurance coverage could be too limited or premiums could be excessively high, which could justify a preference for a compensation fund.[156] But if insurance markets are competitive, insurers can be assumed to be better able to deal with classic insurance problems such as moral hazard and adverse selection than the administrators of a compensation fund. One cannot see as a matter of principle why a government agency running a compensation fund would have better information on risks than an insurer. This might, however, be

[154] See Anselm Smolka of Munich Re at a conference on regulating disasters through private and public law, Haifa, 16 June 2013.

[155] See Section 7.1. [156] Faure and Van den Bergh, 'Restrictions of Competition'.

different if highly technical risks are involved where operators of certain facilities are in a much better position than insurance companies to monitor each other. This point has been made, for instance, concerning the compensation for nuclear damage and for offshore-related risks as well. One could argue that a risk-sharing agreement between offshore operators could lead to optimal monitoring between the operators because they would possess much better information on prevention and good and bad risks than an insurance company would.[157] In maritime insurance, too, the Protection and Indemnity (P&I) Clubs (already discussed), which are based on mutual risk sharing between tanker owners, play a crucial role.[158] With respect to these highly specialized matters, one could therefore argue that the operators themselves might in some cases be better equipped than an insurance company to control moral hazard because they are better able to process information on the particular risk. However, the examples given show that these risk-sharing agreements do not involve the use of a government-run compensation fund.

In sum, if both insurance and compensation funds are available, there are no clear reasons why a fund would be the preferred solution. However, there may be reasons why insurance may not provide coverage for certain risks. In such a case, funds cannot be compared with insurance because insurance is not an alternative. However, as we saw in Chapter 5, one should not conclude too quickly that offshore-related risks are uninsurable. In fact, it was shown that they are insurable to relatively high amounts.[159]

Comparing insurance with compensation funds, one should also address the comparative costs of both instruments. Insurance generally will be cheaper because liability insurance policies are not concluded for one activity but for a whole set of risks. There is thus one insurance policy with transaction costs that are incurred once and an administrative structure within an insurance company that will be forced to an adequate cost reduction by competitive pressures. The costs of risk spreading might also be lower with an insurance company than with a compensation fund. Insurers are specialized in acquiring information on the differentiation of risks. In addition, it has been argued in the literature that insurance makes a reduction in transaction costs between

157 See Faure and Skogh, 'Compensation for Damages'; Faure, 'Economic Models of Compensation'. See also the arguments formulated in Section 4.1.

158 See Coghlin, 'Protection and Indemnity Clubs'. See also Section 4.4.

159 See Section 5.4.2.3.

contracting parties possible because they can ex ante agree on a distribution of risks and losses in case of an incident.[160] The comparison obviously will also depend upon the type of compensation fund under discussion. In most cases, one immediately thinks of a compensation fund run by a regulatory authority. But then one can, of course, refer to the literature on the negative effects of bureaucracies to argue that such a publicly operated compensation fund would not necessarily provide compensation at lower costs than the private insurance market. This can be reduced if the fund is administered privately, but in that case, competition with other funds has to be organized to provide incentives for cost reduction.

Hence, if a fund were to replace the liability and insurance system, it is not so clear that it would have cost advantages. It could only play a role in an upper layer, such as when insurance would not be available. In this case, it would be for the third layer, with expected damage above the insurable amount of US$750 million. Moreover, even in this case, a fund should be used only if risk-dependent premiums can be charged in order to provide optimal incentives for prevention.

7.5.6.2 Economic Principles to Shape a Fund

If one were to create a fund, there are a few general principles, strongly in line with the principles mentioned earlier,[161] that should be followed if the fund is to create efficient incentives for prevention. *First* of all, it seems important that a fund should have a subsidiary character, meaning that whenever other solutions are available (tort law or insurance), victims should be forced to use those first. Thus, at least some guarantee can be given that incentives for victims to look for other solutions will be maintained and that they will not merely take advantage of the fund.

Second, if, nevertheless, a payment is made through a compensation fund, that fund should be subrogated in the rights of the victim against potentially liable third parties. Indeed, it may be that for political reasons (or out of sympathy for the victim), the fund manager does not consider it reasonable to expect the victim first to use the liability system.[162] If in those cases compensation is paid notwithstanding the potential of a claim in tort, the fund should be subrogated in the rights of the victim in order to provide adequate incentives for prevention for the operator.

160 This argument has been made by Skogh, "The Transactions Cost Theory'.
161 See Section 7.1.
162 Using tort law to obtain compensation can indeed be very costly and take a long time.

Third, as far as the financing of the fund is concerned, this ideally should be organized in such a way that those who contributed to the risk also finance the compensation fund. If one were to organize a fund for offshore-related damage, the way it would be realized, for example, would be through a tax on the oil and gas produced via offshore platforms. The problem with a tax, however, is that this is often merely related to the amount of oil or gas extracted from a particular well and not related to the investments in preventive technologies taken by particular operators. A flat tax thus would not include any risk differentiation and hence not provide sufficient incentives for prevention of risks. For a tax to provide accurate incentives, the tax normally should be differentiated in such a way that operators adopting high safety standards contribute less to the fund than operators creating higher risks. However, for the manager of a fund to differentiate risks in such a way may be very complicated. In practice, fund solutions are usually financed on a flat-rate basis, as a result of which they lack any positive incentive effect.

Finally, in order to keep at least some incentives with potential victims of an offshore-related incident, it would be important that the compensation fund does not provide full coverage so that victims themselves are still partially exposed to the risks. This is – as well known – one of the classic remedies against moral hazard.[163] This could be reached either by applying a deductible or by applying standardized forms of compensation (which are usually lower than full compensation).

7.5.6.3 Experiences with Environmental Funds

The preceding shows that there are many conditions that need to be fulfilled to construct a fund in accordance with economic principles. Insurance is often better able to apply principles of risk differentiation and cost minimization, and moreover, it may be difficult in practice to ask for risk-related contributions to a fund. This may explain why in practice in Europe there is not much experience with compensation funds for environmental harm. However, recently the Environmental Fund has again been put on the political agenda.

The Hungarian Proposal

Indeed, in 2011, Hungary proposed the establishment of a European Union Industrial Disaster Risk-Sharing Facility to be funded by an annual contribution from targeted industries and companies. This proposal was partly

[163] Shavell, 'On Moral Hazard and Insurance'.

in response to a major industrial accident near Kolontár, Hungary, in 2010. Based on the proposal of the Hungarian government, the European Commission published a consultation paper through Bio Intelligence Services at the end of October 2012.[164] The paper seeks to examine the feasibility of establishing a fund to address liabilities resulting from major industrial accidents involving pollution.

The main purposes of this fund would be to respond quickly in the event of a major industrial accident to relieve the suffering of persons harmed by the disaster, to remediate environmental and/or traditional damage and to prevent further damage. In the Hungarian proposal, the facility would respond if traditional and environmental damage from a major industrial accident exceeds €100 million, that is, above a level for which funding is not otherwise available.[165] A secondary purpose of the facility would be to limit the financial exposure of each company in any of the targeted industrial sectors to €100 million. It is argued that this limit of liability not only would benefit the operator/company that caused the accident but also would benefit employees who might otherwise lose their jobs due to the company becoming insolvent because of its inability to pay all the costs of the traditional and environmental damage. Member States also would benefit because they would not be called on to bear costs that the company could not pay due to insolvency.[166] In the Hungarian proposal, the fund would integrate three functions:

1. Pre-financing tool to give immediate access to funding and relief to communities.
2. Second tier of insurance (private insurance up to a point, and then the facility would intervene).
3. Unspent resources to support companies, particularly Safety and Environmental Management System (SEMs), to invest in safety and prevention.

As such, proponents anticipate a positive economic impact on national budget and citizens, a provision for improved safety and enhanced prevention measures in companies, a wide range of damages covered (besides economic activity, also human health and property) and a complement to private insurance, which could reduce costs for industry.[167]

The idea was inspired by Sweden's and Finland's Environmental Damage Insurance, which aims to guarantee full compensation for

[164] Bio Intelligence Service *et al.*, 'Study to Explore the Feasibility'.
[165] *Ibid.*, Executive Summary. [166] *Ibid.* [167] *Ibid.*

environmental damage, including the costs of measures taken to prevent or limit the damage and to restore the environment to its previous state in cases where those liable for compensation are insolvent or the liable party cannot be identified.

Environmental Funds in Sweden and Finland

Both Sweden and Finland have established pollution clean-up funds based on compulsory insurance payments for high-risk activities to finance the restoration of orphaned sites and compensate personal and property damage where the liable party is unknown or insolvent.[168] The scheme is financed by special premiums which are compulsory for operators of high-risk activities subject to environmental permits. In practice, the schemes amount to a levy, partially calibrated to reflect the risk of the activities engaged in (e.g. type and scale of the hazardous activity). The Finnish and the Swedish examples thus are considered a safety net to address circumstances where ordinary liability and insurance mechanisms will not work. In practice, these funds have not played an important role. In its first ten years of operation, no claims were indemnified by the Swedish insurance fund.[169]

Stakeholders Opinions on an Environmental Fund

Stakeholders were consulted to determine whether and how operators and the financial/insurance sector could be engaged in the development and implementation of a potential environmental fund. Yet it appears that most stakeholders and certainly the insurance industry are not keen on the idea.[170]

Major objections and concerns with respect to the creation of a fund include[171]

- Concerns that the fund or scheme would not include risk assessment and mitigation,
- Potential to hinder the development of the environmental insurance market,

[168] For Sweden, see Swedish Environmental Code, chapter 33, 'Environmental damage insurance and environmental clean-up insurance'. For Finland, see Environmental Damage Insurance Act 81/1998.

[169] Richardson, *Environmental Regulation*, 370–1.

[170] Barrett, 'European Environmental Liability Options'.

[171] Bio Intelligence Service *et al.*, 'Study to Explore the Feasibility', Executive Summary.

- Implications of the interaction between existing national funds and a new EU fund,
- Legal clarity on the definition of the baseline condition,
- Ensuring that a fund or scheme does not hinder the 'polluter-pays principle',
- Moral hazard concerns connected with the creation of an EU fund,
- Operation in Member States with widely differing economies, industries, legal liability regimes, history of environmental claims and approaches to such risks,
- Lack of a suitable impact assessment or case having been made, and
- General concerns about possible over-regulation of industry.

The insurance industry fears that the creation of a fund would be a backward step in terms of risk management. The insurance industry fears that such funds do not encourage operators to take measures to reduce environmental risks beyond what is legally required.

A fund could reduce the incentives for business to take responsibility for the consequences of its activities because the mechanism proposed for the fund would give no credit for prevention efforts or a good claims history. As such, it is contrary to the 'polluter pays principle' established in the ELD.[172]

Next, logistic issues in the organization of such a fund may arise. Not only the administration of the fund and how it will operate might cause problems, but also, with respect to the claims, problems might be expected. What if there are multiple events in a year? Or several close together? In such a case, there might be not enough money.[173]

According to the insurance industry, it is better to raise awareness of insurance. A fund should only exist where there is no insurance market. The insurance market for environmental damage is developing in the European Union, and the creation of a fund risks undoing the progress in this field. A fund would interfere with the development of the insurance market.[174]

The Bios Report therefore concludes that the purpose of the fund or scheme must be clearly identified and agreed on. It must be clarified whether its purpose is to compensate victims of a major accident for bodily injury, property damage or economic loss suffered by them; to provide emergency funding to respond to an industrial accident; and/or to remediate environmental damage. The fund or scheme is only likely to gain acceptance subject to agreement on whether a fund or scheme needs to be established for such purposes and, if so, industrial sectors to be

172 Nyman, 'Ferma Opposes'. 173 Barrett, 'European Environmental Liability Options'.
174 *Ibid.*

covered by it, the level of the threshold for the fund or scheme and many other details concerning its design.[175]

Stakeholders' reactions indicate that the creation of an environmental fund or scheme to limit the financial exposure of a liable operator is not very likely to be accepted by stakeholders for reasons such as the potential for moral hazard, conflict with the 'polluter-pays principle' or practical reasons such as feasibility of a (€100 million or other) threshold. However, using the fund to pay for orphaned sites is more likely to be acceptable. The ELD does not require Member States to remediate environmental damage if the liable operator does not do so. Depending on the nature or location of the damage, however, a Member State may consider that it has no option but to remediate the damage. The fund could provide useful financial help in such circumstances.[176]

Therefore, the necessity and desirability of such an environmental fund and its purposes would have to be carefully examined. If the purposes and functions of the fund are not well defined, the fund could have negative effects rather than contributing to protection of the environment. If there were political agreements to create an environmental compensation fund, the design, management and implementation of the fund therefore would also need to be given careful consideration in order to ensure that the fund operates effectively and efficiently. Hence, also as far as the creation of an environmental fund is concerned, many stakeholders raise important objections as a result of which the environmental fund does not really constitute an example on which one could draw as far as offshore-related damage is concerned.

7.5.6.4 A Fund for Offshore-Related Risks? Stakeholder Opinions

Not surprisingly most stakeholders, whether they are operators, regulators or insurers, are strongly opposed to any type of fund solution. Operators hold that all problems that are currently experienced in OIL/OCIL and OPOL (insufficient risk differentiation and too much mutualisation according to the majors) would only be exacerbated. Probably the contributions to the fund would be based on a flat rate tax and hence would miss any element of risk differentiation and thus of incentives for prevention.[177] Not surprisingly, this is also the opinion of representatives

[175] Bio Intelligence Service *et al.*, 'Study to Explore the Feasibility', Executive Summary.
[176] *Ibid.* [177] Interview with representatives of BP, 13 March 2013.

of OGPs.[178] Also, representatives of regulators are opposed to a compensation fund for the simple reason that it de facto creates negative redistribution: bad risks benefit more and good risks contribute more relative to the risk they pose.[179] Where available, regulators would be in favour of insurance because insurance allows risk-differentiated premiums, thus providing incentives for prevention and avoiding negative redistribution. A fund solution without risk differentiation would not provide correct psychological incentives.[180]

Similar opinions are heard from insurers. When referring to an environmental fund, they quickly recall the example of Superfund in the United States, where 40 per cent of the Superfund money went to lawyers involved in litigation.[181] Moreover, an environmental fund in the European Union would undermine the 'polluter-pays principle'. Politically, it is therefore not a wise idea. Moreover, insurers may, if a fund were to exist, be inclined to refer insured or victims to the fund and argue that they first have to try to obtain compensation from the fund. The result may be (as in the United States) that the fund would give raise to a huge amount of litigation. Moreover, a fund usually will be financed by a levy on all OGPs, for example. In such a case, it would amount to a tax on the good risks that are compensating bad risks because risk differentiation would be absent. Also, the examples in the European Union of environmental funds, in particular, Sweden and Finland, were, as just discussed, not considered a success. They were both extremely costly and only led to substantial transfers of money without any claims. As a result, the Swedish fund even closed down.[182]

With respect to environmental funds for offshore liability, it is also mentioned by stakeholders that one cannot compare with the compensation funds that do exist, both at the international level[183] and in the United States[184] for vessel-induced pollution. A major difference with the fund solution for vessel-based pollution is that in that case the fund intervenes only on top of the limited liability of the tanker owner under the CLC.[185] Moreover, the way the compensation is structured (both in the international regime and under the US OPA) is that two parties contribute: the shipping industry (tanker owners) contributes a limited amount according to its liability, and oil receivers contribute (via a flat

[178] Discussion with OGP, 25 February 2013.
[179] Interview with Jan de Jong, inspector-general state mines, 22 February 2013.
[180] *Ibid.* [181] Interview with Phil Bell, 26 February 2013. [182] *Ibid.*
[183] Discussed in Section 3.2.1.2. [184] See Section 3.5.5.10.
[185] Interview with representatives of BP, 13 March 2013.

tax) based on the oil received to the compensation fund. The compensation regime for vessel-based solution thus is typically the result of a compromise whereby two parties (shipping industry and oil industry) contribute.[186] Such a compromise is impossible in the case of offshore-related risks because there is only one party involved, i.e. oil and gas producers.[187]

It is also feared that unlike in the case of vessel-based pollution, the number of companies engaged in offshore installations is relatively small, which means that the number of contributors would be small as well. Moreover, with a generalized offshore pollution fund, the moral hazard that would be created through such a de facto tax would be huge.[188]

These arguments seem rather compelling, and as a result, there is not much scope for introducing a fund solution for offshore-related risks. It seems preferable, given the necessity of a compensation mechanism to provide incentives for prevention, to rely on insurance and other mechanisms where contributions can reflect risk. The only role one could imagine for a compensation fund would be as an upper layer (e.g. beyond US$750 million or any limit on the insurance amount available in the market). However, not only would this create a very complicated system to administer, as in the second layer, there would be no risk-related contributions and hence no positive effect on incentives. In addition, it also would de facto mean that operators would have to pay twice: first for insurance or contributions to a guarantee or pooling system and second a tax for the fund that would constitute the second layer. Moreover (this constitutes again a major difference with vessel-based pollution), given the large differences in the offshore market, for some operators, contributing to such a (costly) fund would be meaningless because they could easily take charge of the costs above the insurance limit via self-insurance. For them, a duty to contribute to a fund therefore would only create additional costs without compensating benefit. This would then amount to a situation whereby the duty to compensate to the fund would only be imposed on some (presumably smaller) operators (because they may be exposed to an insolvency risk) and not to others. This obviously would not be politically feasible. For

[186] For the history of the international regime of marine oil pollution compensation, see Wang, 'Shifts in Governments'; for a discussion of the US regime, see Wang, *Civil Liability*, 189–224.

[187] Interview with representatives of the International Group of P&I Clubs, 1 May 2013, in London.

[188] Interview with representatives of BP, 13 March 2013.

these reasons, we argue that the difficulties that prevent the creation of a compensation fund in the area of environmental liability also may inhibit the creation of a compensation fund for offshore-related risks.

7.6 Rapid Claims Management

An important question is to what extent a compensation mechanism is able to guarantee rapid claims settlement. This may be of more importance, for example, for restaurants and hotels in coastal areas affected by offshore-related pollution but also for the fishing sector. Often those industries have financed their activities based on credit. Thus, when, for example, a fisherman or a hotel owner loses income as a result of business interruption following an offshore-related incident, this can have devastating consequences and potentially lead to bankruptcy because their income may be lacking, and loans still need to be paid back. Hence, the question arises as to what extent within pooling mechanisms a rapid claims settlement can be arranged such that payments (e.g. to hotel/restaurant owners and fishermen) could be guaranteed in order to prevent further damage resulting from their insolvency. A look at the solutions in this respect in pooling arrangements is particularly interesting because the traditional liability mechanism (via civil procedure) often has the disadvantage that court procedures in order to establish liability may take very long with potentially devastating consequences for the financial situation of the victims.

First, the existing rapid claims management schemes within a few of the compensation schemes that have been discussed so far will be examined (7.6.1). Next, there will be a normative analysis addressing the need for the development of a rapid claims management scheme and the possible forms that this could take (7.6.2).

7.6.1 Existing Schemes

Rapid claims management schemes to some extent can be found within OPOL, which is said to be set up as a rapid claims management scheme, as well as in Norway as far as losses suffered by fishermen are concerned. Also, within the civil liability and fund conventions dealing with vessel-based pollution, attention has been paid to claims management. However, probably the most interesting is the claims management scheme for oil pollution in the United States, in particular, the recent experience with the Gulf Coast Claims Facility (GCCF), a form of alternative dispute resolution set up to settle the claims arising from the Deepwater Horizon incident. These examples will be reviewed in turn.

7.6.1.1 OPOL

The essence of OPOL is that the operator, a member of OPOL, agrees to settle a claim quickly on a strict liability basis.[189] The claim has to be filed within twelve months, and the operator needs to pay expeditiously.

As long as there is no default of an operator, OPOL does not intervene in claim settlement.[190] Hence, in normal cases, OPOL has no reason to intervene, and the operator will settle the claim. However, the strict liability rule always applies if there is no default of an operator. Representatives of OPOL hold that this will guarantee a rapid payment by those who have provided financial guarantees under the OPOL arrangement.[191]

As such, OPOL does not provide a direct right of action to a victim. However, if a victim is dissatisfied with the way in which one of the OPOL members has handled the claim, the victim has the possibility to call on arbitration under the rules of the International Chamber of Commerce. Therefore, a lengthy jurisdictional process is avoided.[192]

Also, others hold that the strict liability to which OPOL members agree guarantees that there will be payment. Hence, normally, there should be no problem with the rapid compensation of victims.[193]

7.6.1.2 Norway

As was made clear when discussing its compensation regime,[194] Norway has in the Petroleum Activities Act a specific regime for the compensation of Norwegian fishermen. In case of damage, the operator causing the pollution 'shall make weekly pubic announcements' concerning the information on claims submission in the '*Norwegian Gazette* and in newspapers and other publications which are generally read in those places where damage is caused or is presumed to occur'.[195] 'Decisions made by the administrative appeal body (the Commission) may be

[189] Interview with Niall Scott, managing director of OPOL and Collin Wannell, chairman of the board of OPOL, 27 March 2013.

[190] Although Wendy Kennedy, representative of DECC, held that OPOL will verify that the members comply with the strict liability duties under OPOL.

[191] Interview with Niall Scott, managing director of OPOL and Collin Wannell, chairman of the board of OPOL, 27 March 2013.

[192] 'OPOL Guidelines for Claimants'; available at: www.opol.org.uk/guidelines.htm.

[193] Interview with representatives of Oil and Gas UK, in London, 1 May 2013.

[194] See Section 3.5.3.9. [195] Bio by Deloitte, 'Civil Liability', 115.

brought straight before the district court',[196] and claims are handled by the court within strict time limits (of two months).[197] Specifically for fishermen's losses, Norwegian legislation thus has a rapid claims settlement mechanism whereby compensation for fishermen is provided within short, strict delays and via a special administrative commission.

7.6.1.3 The Civil Liability and Fund Conventions

Also, the regime of the CLC and Fund Convention, applicable to vessel-based pollution,[198] provides specific rules on claims management. Claims management is dealt with in Article 4.7 of the Fund Convention as follows:

> The Fund shall, at the request of a Contracting State, use its good offices as necessary to assist that State to secure promptly such personnel, material and services as are necessary to enable the State to take measures to prevent or mitigate pollution damage arising from an incident in respect of which the Fund may be called upon to pay compensation under this Convention.

Article 6, furthermore, provides

> The time limit to bring an action for compensation is within three years from the date of the incident which caused the damage. However, in no case shall an action be brought after six years from the date of the incident which caused the damage.

Moreover, claims handling by the International Oil Pollution Compensation (IOPC) Fund takes place on the basis of a 'Claims Manual'. The fact that there is an annual contribution and the existing finances available may facilitate payment to the victims. In this respect, the advantage of the IOPC Fund is that it always has a certain amount of funds available. Thus, when it comes to compensation for the victims, the compensation can be provided quickly (if all the conditions are complied with).

The Claims Manual of the IOPC Fund specifies that

> The 1992 Fund and the P&I Clubs try to reach agreement with claimants and pay compensation as promptly as possible. They may make provisional payments before a final agreement can be reached if a claimant would otherwise suffer undue financial hardship. However, provisional payments are subject to special conditions and limits, particularly if the total amount of claims exceeds the total amount of compensation available under the two 1992 Conventions.[199]

[196] Petroleum Activities Act, Section 8–6. [197] See Section 3.5.3.9.
[198] And discussed in detail in Section 3.2.1.
[199] International Oil Pollution Compensation Fund 1992, 'Claims Manual'.

However, as the Claims Manual states, the 'speed with which claims are agreed and paid depends largely on how long it takes for claimants to provide the required information'.[200] It is also for the reason of facilitating claims settlement that the Fund has developed a Claims Manual to assist the claimants in pursuing their claims. The Fund even has special guidelines for claimants from the fisheries, mari-culture and fish-processing sector in presenting their claims.[201]

Representatives from the International Group of P&I Clubs explain that the way the P&I and Fund mechanism works is that victims have to hand in a claim and provide detailed documentation supporting the claim, thus providing evidence of, for example, the specific amounts of damage.[202] This documentation will then be carefully reviewed, and normally, if the documentation indeed substantiates the claim, within six months a payment will be made. In principle, the P&I Club will immediately investigate a claim after the documentation substantiating the claim has been received. When it concerns a large incident, pay-out obviously may be later because assessments may be more complex. In cases that exceed the CLC limit, usually a joint claims handling office by the involved P&I Club and the Fund will be set up. Joint experts will be appointed, and the entire case will be jointly handled. In principle, the club pays first up to the amount of the limit in the CLC; the Fund pays once the CLC limit has been reached.[203]

In principle, if claimants provide their claims rapidly and provide proof to substantiate their claims, the claims will be assessed rapidly as far as their admissibility is concerned as well. However, it may require expert knowledge to analyze whether a claim is admissible. The pool of experts to assess this is relatively small.

It should be stressed that de facto claims handling by P&I Clubs and the Fund (hence for marine vessel-based pollution) has not led to a serious problem in the European Union. Evidence coming from the United States concerning long litigation (as in the case of the *Exxon Valdez*) is typically linked to the specific US situation (e.g. including debates concerning punitive damages), which does not apply to other jurisdictions.[204]

[200] *Ibid.*

[201] International Oil Pollution Compensation Fund 1992, 'Guidelines for Presenting Claims'.

[202] Interview with representatives of the International Group of P&I Clubs, 1 May 2013, in London.

[203] *Ibid.* [204] *Ibid.*

7.6.1.4 Claims Settlement in the Deepwater Horizon Case

General Claims Procedure under OPA[205] Section 2713 of the OPA sets forth the procedures for claims handling after the occurrence of an incident. Following a spill, the president, acting through the Coast Guard, identifies the responsible party, which is required to advertise that it has been designated as the responsible party and to provide information about how claims can be pursued against the responsible party.[206]

> A claimant must first submit his claims for removal costs or damages to the responsible party for payment before either making a claim against the OSLTF or filing a lawsuit under OPA against the responsible party. If the responsible party denies all liability for the claim or does not settle the claim by making payment within ninety days after the day the claim was presented or advertising was begun, whichever is later, then the claimant has an option. The claimant may either file a lawsuit against the responsible party, or it may make a claim against the OSLTF. A claim may be presented directly to the Fund, without first presenting it to the responsible party, only if
> (1) the President has advertised or otherwise advised claimants,
> (2) a responsible party has a complete defense or has exceeded its limit of liability, entitling it to recovery pursuant to §2708,
> (3) a State seeks recovery for removal costs it has incurred, or
> (4) a US claimant asserts a claim for damage for which the Fund is liable under §2712 (a) caused by a discharge from a foreign offshore unit.[207]

[205] Force, Davies and Force, 'Deepwater Horizon', 949–56.

[206] 33 USC §2714. The responsible party or guarantor has five days to deny its designation as a responsible party and, upon its failure to do so, must advertise the designation and procedures by which claims may be presented to it within fifteen days of the designation. If the responsible party does not advertise, the president is responsible, at the expense of the responsible party or the guarantor, for advertising the designation and procedures by which claims may be presented to the responsible party. The advertisement must continue for no fewer than thirty days. 33 USC §2714(b)(1). Section 2714(b)(2) provides that the advertisement 'shall state that a claimant may present a claim for interim, short-term damages representing less than the full amount of damages to which the claimant ultimately may be entitled and that payment of such a claim shall not preclude recovery for damages not reflected in the paid or settled partial claim'. Under Section 2714(c), if (1) during the five-day period both the designated responsible party and the guarantor deny the designation, (2) the source of the discharge or threat was a public vessel, or (3) the president cannot designate the source, then 'the President shall advertise or otherwise notify potential claimants of the procedures by which claims may be presented to the Fund.' It should be noted that OPA does not impose a mandatory duty on the president to designate a responsible party but instead only requires the president to designate a responsible party 'where possible and appropriate'.

[207] For a detailed discussion on the claims procedure, see Force, Davies and Force, 'Deepwater Horizon', 949–56.

If a claimant files a lawsuit, he has temporarily waived his right to make a claim against the OSLTF. If a claim is not compensated or is not compensated in full through the claimant's lawsuit, then the claimant may subsequently make a claim against the OSLTF. The present requirement under OPA is not a mere procedural technicality but is a mandatory condition precedent to filing lawsuits against a responsible party. If a claimant does not comply with this requirement, his claim will be dismissed.[208]

Following the OPA procedure, the president via the Coast Guard designated BP and Transocean as responsible parties and also provided notice to a BP-related entity and an insurer of Transocean as these parties' guarantors.

Claims Settlement in Deepwater Horizon Within weeks after the Deepwater Horizon incident, BP had already begun to pay compensation for claims.[209] On 16 June 2010, after negotiation with the Obama administration, BP established a fund of $20 billion to compensate those affected by the incident.[210] Political pressure undoubtedly was exerted on BP by President Obama to create such a fund. If BP had refused to create such a fund, it was at risk of losing its right to drill in the Gulf of Mexico, which could potentially have had even more devastating financial consequences for BP and potentially even could have led to its bankruptcy.[211] The claims would be processed through an independent claims facility, administered by Kenneth Feinberg. This facility is known as the Gulf Coast Claims Facility (GCCF). Two independent trustees, Kent Syverud and John Martin, were appointed as independent trustees of the fund.[212] The idea was to create a trust for compensation which is independent and which can manage the available money. For both the defendant (BP) and plaintiffs (the victims), it was important that someone other than the defendant (BP) held the funds. This is why a separate facility was created.[213] However, the independence of the GCCF and its claims administrator was challenged by the Plaintiffs' Steering Committee in December 2010.[214]

208 Force, Davies and Force, 'Deepwater Horizon', 950.

209 www.bp.com/sectiongenericarticle800.do?categoryId=9048917&contentId=7082602.

210 The funding of the trust was completed in the fourth quarter of 2012. See BP, 'BP Annual Report and Form 2013', 111–2.

211 See Philipp Wassenberg, representative of Munch Re, interview, 6 May 2013.

212 www.bizjournals.com/stlouis/stories/2010/10/18/focus3.html?page=all.

213 Interview with Kent Syverud, in St. Louis, 5 March 2013.

214 *Motion to Supervise Ex Parte Communications between BP Defendants and Putative Class Members*, in MDL 2179, 21 December 2010. Later on 2 February 2011, the court granted

The GCCF began to accept claims as of 23 August 2010. It is important to note that prior to establishment of the GCCF, BP had already processed and awarded claims. According to BP, the company awarded $399 million in claims from 3 May until 23 August 2010 in transition to the GCCF.[215] This GCCF was later closed in June 2012 as the court took over supervision of the claims settlement process.

Between 23 August and 23 November 2010, claimants could apply for emergency advance payments to recover damages incurred in the first six months after the incident. These mainly include compensation for the loss of earnings or profits, removal and clean-up costs, real or personal property damage, loss of subsistence use of natural resources and physical injury or death caused by the spill. These emergency advance payments fall within the category of 'interim' payments under OPA. Therefore, the claimants did not have to sign a release of liability for any future damages to receive compensation.[216]

Since 13 December 2010, the GCCF offered the option of three different types of claims: interim payment claims, quick-payment final claims and full-review final payment claims.[217] The interim payments are designed to compensate only for past losses and do not require a release of liability signed by the claimants, and the claimants still may apply for additional damages. Quick-payment final claims require a release of liability and only apply to claimants who have previously received an emergency advance payment or been found eligible for an interim payment. Full-review final payments are designed to compensate claimants for all documented past and future losses. Claimants are not obliged to accept a final payment offer, but if they do so, they must sign a release of liability and covenant not to sue.[218]

According to a study commissioned by the US Department of Justice (DOJ) during its one-and-one-half-year tenure, the GCCF processed over 1 million claims and paid a total of more than $6.2 billion to over 220,000 individual and business claimants. Approximately 99.8 per cent

in part the motion. For a detailed discussion on the GCCF, see McDonell, 'The Gulf Coast', 765–95.

215 Ramseur, 'Liability and Compensation Issues', 15.

216 GCCF, 'Gulf Coast Claims Facility Protocol for Emergency Advance Payments'.

217 On 22 November 2010, the GCCF issued a Protocol for Interim and Final Claims, which was subsequently revised on 8 February 2011. Available at: http://s3.amazonaws.com/nytdocs/docs/521/521.pdf. For an analysis of the function of GCCF, see Force, Davies and Force, 'Deepwater Horizon', 936–7.

218 McDonell, 'The Gulf Coast', 770–1.

Table 13 *Payments*

Payments	Amount paid (US$)
Individual and business claims	12,515,893,626
Government	1,474,008,457
Other	332,152,575
Total payments	14,332,054,659

of the claims paid and 96.8 per cent of the amounts paid related to claims for lost earnings or profits.[219]

The GCCF ceased to function and was taken over by the Court Supervised Settlement Program (CSSP) in June 2012. The claims administrator of CSSP is Patrick Juneau. As of 31 May 2015, BP has paid approximately US$14 billion to government, individual and business claims (see Table 13).[220]

Civil Compensation In April 2012, BP reached agreements with the Plaintiffs' Steering Committee with regard to the substantial majority of eligible private economic losses and medical claims stemming from the incident. The final agreements are embodied in the Economic and Property Damages Settlement Agreement as amended on 2 May 2012 and the Medical Benefits Class Action Settlement Agreement as amended on 1 May 2012. These agreements were approved by the court in December 2012 and January 2013.[221] After the approval of the district court, members of the Economic and Property Damages Settlement Class appealed the approval to the Fifth Circuit. A panel of the Fifth Circuit affirmed the approval of the Economic and Property Damages Settlement on 10 January 2014 but left interpretation of the settlement to another panel.[222]

[219] BDO Consulting, 'Independent Evaluation', 62. This study defines the compensation process of GCCF into two phases: Phase I starts from August 2010, and phase II starts from February 2011. The findings of the study show that the claims filed during phase II were subject to more stringent documentation requirements than those filed during phase I, whereas meanwhile it expanded the types of businesses that would be potentially eligible for compensation and granted automatic eligibility to claimants located on the Gulf Shore who were involved in businesses that were particularly reliant on Gulf resources.

[220] www.bp.com/content/dam/bp/pdf/gulf-of-mexico/public-report-may-2015.pdf.

[221] BP, 'BP Annual Report and Form 2014', 232. [222] *Ibid.*

With respect to the agreement on economic and property damages, six categories of damage are recognized:

- Specified types of economic loss for businesses and individuals,
- Specified real property damage,
- Vessel of opportunity charter payments,
- Vessel physical damage,
- Subsistence damage, and
- The seafood compensation program.

Therefore, separate mechanisms within the framework of the trust fund are established to make compensation to related parties. The Economic and Property Damages Settlement does not include claims made against BP by the DOJ or other federal agencies or by the states and local governments.

The Medical Benefits Class Action Settlement involves payments to qualifying members based on a matrix for certain specified physical conditions, as well as a twenty-one-year periodic medical consultation program for qualifying class members.[223]

The settlements are payable under the terms of the trust fund. These settlements are not capped except for economic loss claims related to the Gulf seafood industry and $105 million to be provided to the Gulf Region Health Outreach program under the Medical Benefits Class Action Settlement.[224]

As of 31 December 2014, the aggregate cash balances in the trust fund and the qualified settlement funds amounted to US$5.1 billion, including US$1.1 billion remaining in the seafood compensation fund which has yet to be distributed and US$0.4 billion held for natural resource damage early restoration. According to BP, should the cash balances in the trust fund not be sufficient, payments in respect of legitimate claims and other costs will be made by BP.[225]

Critical Review of the Deepwater Horizon Trust Fund The Deepwater Horizon trust fund is a private funding to compensate victims of mass torts which cause damage over a wide range (e.g. an environmental disaster).[226] The approach adopted by the trust fund is not unique. Such an alternative claims resolution program was also used after the September 11 attack.

[223] *Ibid.*, p. 231. [224] *Ibid.* [225] *Ibid.*, p. 112.

[226] For a discussion on various mechanisms used in mass tort claims, see Greenspan and Neuburger, 'Settle or Sue?', 97–136; Conk, 'BP and Kenneth Feinberg's Gulf Coast Gambit', 137–83.

One advantage of such a trust fund is that it can 'forestall the potentially devastating effects on families and local areas directly impacted'.[227] Compared with lengthy and costly litigation, such private funding may have the advantage of 'distributing funds to affected claimants more quickly and at less cost to the claimants'.[228]

An important advantage of an alternative dispute resolution of this type is that the costs of its functioning are considerably lower than the costs of the tort system and that hence much more funds were available for the victims.[229] In sixteen months, 220,000 real claims were settled. During this period, 3,000 employees were working for the GCCF in thirty-five offices at an approximate cost of US$40 million per month.[230] The fund could function as a confidence builder, and substantial amounts could be paid out rapidly (most victims were paid in less than eighteen months of the accident). For a catastrophe of this size, such a rapid management of claims has not been heard of before. The alternative would have been a litigation of hundreds of thousands of claims via different courts. Not only would this have taken many years and have led to a lot of uncertainties, many victims would probably still today not have been compensated. Moreover, generally in the United States in mass-damage class actions, often a third of the available compensation goes to plaintiffs' lawyers. Compared to that, the expenses for the running of the GCCF are very modest.[231] Of course, there were some problematic aspects in the functioning of the GCCF as well. For example, the costs of a consultancy firm (DBO) to verify payments (US$17 million) were extremely high.[232] Also, the mere fact that initially it was signalled to the victims that US$20 billion was available undoubtedly had an attractive effect on the amount and number of claims. This may have created a few unjustified (and perhaps even fraudulent) claims. However, on the whole, the advantages of such a fund solution (via the GCCF) are to be considered overwhelming, especially when compared to the court system. The important function was to reassure the victims that payment would be available, thus avoiding hundreds of thousands of claims initiated by victims, supported by a strong US plaintiff bar. The latter could potentially have led to the insolvency of BP as a result of which victims never would have been compensated at all with potentially

[227] Greenspan and Neuburger, 'Settle or Sue?', 99. [228] *Ibid.*
[229] Interview with Kent Syverud, in St. Louis, 5 March 2013.
[230] Interview with Kenneth Feinberg, special administrator of the GCCF, 16 June 2013, in Haifa.
[231] *Ibid.* [232] Interview with Kent Syverud, in St. Louis, 5 March 2013.

devastating consequences for British pension holders as well (because BP is largely owned by British pension funds). This is why there were many interests in avoiding the tort system and providing rapid reassurance to the victims that compensation would be rapidly available.[233]

7.6.2 Normative Analysis

7.6.2.1 Need for a Mechanism

A rapid payment can be of extreme importance especially for fishermen and hotel/restaurant owners in coastal areas whose income may be interrupted as a consequence of a disaster. Late payment could lead to their bankruptcy (especially in cases where their facilities are externally financed), and thus the losses could even become larger.

The problem is that compensation via the normal liability rules may (unless a case is settled or the insurer decides to pay on the claim of the victim) lead to litigation which could potentially take many years. This should, moreover, not always necessarily be viewed as merely negative. After all, the fact that a civil procedure related to a liability case may take a substantial amount of time can also be important in order to review whether the goals of tort law (adequate deterrence and compensation) are actually served in this particular case. Hence, to an important extent, the procedure will need to verify whether the amounts claimed by the victim indeed correctly represent the loss suffered by the victim (in order to avoid moral hazard on the victim's side) and whether it is indeed the tortfeasor (in this particular case, the liable offshore operator) who caused the loss (in order to obtain a correct risk allocation to that operator).

The alternative obviously could be that payment would not take place via liability law but rapidly e.g. through a government commission, facility or fund which would not apply the conditions of liability law in the same way as would be done under the court procedure. The advantage is that in that particular case, payment can take place rapidly; the disadvantage may be that moral hazard on the side of the victim could play (if fraudulent claims can still lead to payment) or that risks are not correctly allocated to the offshore operator (if damages are either assessed too high or too low). Representatives of the International Group of P&I Clubs[234] pointed at various problems in this respect. For example, in the case of the *Hebei*

[233] Interview with Kenneth Feinberg, special administrator of the GCCF, 16 June 2013, in Haifa.

[234] Interview in London, 1 May 2013.

Spirit,[235] some 120,000 claims were launched according to the CLC/Fund Convention, but 80,000 were deemed inadmissible, which could show that many claims are brought which do not necessarily meet the requirements of proof to be effectively compensated. A similar problem occurred in the case of the *Prestige*: the Spanish government felt guilty because it had refused a port of rescue to the *Prestige*.[236] When the spill occurred, the Spanish government generously offered compensation to every victim, at least to everyone who claimed to have suffered damage related to the *Prestige* incident. Later, these advanced payments were claimed back from CLC (hence from the P&I Clubs) and from the Fund. The problem is that in such a case, disaster relief payments may have been made by governments (in some cases for political reasons) which often do not meet the admissibility criteria of the CLC (or the Fund) under liability law.[237]

These examples show that there are particular policy considerations to be taken into account when attempting to develop a rapid claims management system for particular victims (more particularly fishermen and hotel/restaurant owners in coastal areas) such as the following:

- The offshore operators (and others who contributed to the risk) should still be exposed to the costs they created,
- Only victims who really suffered losses due to the offshore incident should be compensated and only for the damages resulting from the incident,
- A fully correct appraisal via civil procedure may take too long,
- Hence, an alternative claims settlement (via alternative dispute resolution or an administrative procedure) may have the advantages of speed and can avoid disastrous consequences for this particular category of victims.

In this respect, we can also recall the experience with the Deepwater Horizon GCCF, where it was also held that the main advantage of the creation of this GCCF was that it removed distrust from the victims and provided a clear signal that money was available within a fund (thus reducing insolvency risk) and that compensation would be paid speedily on the basis of clear criteria.

235 For further on this case, see also Hu, 'Legal Issues'.

236 On the right of entry in maritime ports for ships in distress, see Somers, 'Marine Pollution'.

237 See Interview with representatives of the International Group of P&I Clubs, in London, 1 May 2013.

7.6.2.2 Compensation Funds

One alternative, providing rapid settlements of claims, is obviously to decide for this particular category of victims to create a fund and to (particularly) bypass liability law. The disadvantages of such a radical solution have already been discussed when referring to the option of a compensation fund for offshore-related damage.[238] There is some experience with compensation funds for disasters. They usually follow an administrative procedure of claims handling whereby an administrative authority or commission verifies claims and arranges payment. Compensation, however, in the cases where a compensation fund completely replaces liability law (as, for example, in Belgium[239] and Austria[240]), is not necessarily speedier than under liability law. Moreover, the major disadvantage of such a construction would be that the operator who created the risk would no longer be confronted with the costs of the incident, and hence a correct risk allocation would fail (unless the fund would later take recourse against the liable operator).

There are some alternatives whereby no full compensation is provided to victims, but rapid disaster relief is provided e.g. to support victims in the immediate aftermath of a disaster. These payments (often by a government) are often aimed at emergency measures that government should carry out after a disaster occurs. Many countries have these systems. For example, in Japan, a Disaster Relief Act provides 'for government emergency measures to protect victims of disaster and maintain social order by causing the central government to provide needed relief services on an emergency basis in cooperation with local public entities and the Red Cross, other entities, and the people of Japan'.[241] These types of relief measures after a disaster are, however, not the type of compensation needed to avoid bankruptcy of fishermen and local restaurant/hotel owners in coastal areas. The reason is that relief measures are often just aimed at providing minimal help to restore social order e.g. by providing shelter to victims; the goal of immediate relief is not to provide recovery or compensation of all losses to victims.

To some extent, such a fund also was created in the United States with the 9/11 Victim Compensation Fund. However, there has been quite a bit of criticism of that particular fund, arguing that the authorities failed to consider the psychology of justice and compensation for harm when

[238] See Section 7.5.3. [239] See Durant, 'Belgium', 72–3.
[240] See Hinghofer-Szalkay and Koch, 'Austria', 12–26.
[241] See Faure and Liu, 'The Tsunami', 154.

establishing the fund. The fund would have violated people's perceptions of procedural fairness and violated the victims' ideas of satisfaction.[242] The 9/11 Fund therefore has not been seen as a model that should be followed in the future.[243]

7.6.2.3 Stimulate Rapid Payment within Liability Law

Another alternative is not to replace liability law with a fund solution but rather to provide within liability law a duty to provide rapid payment to victims. An interesting example of this is provided in Directive 2009/103/EC of 19 September 2009 relating to insurance against civil liability in respect of the use of motor vehicles and the enforcement of the obligation to insure against such liability.[244] Article 22 of this Directive provides in a compensation procedure forcing the insurance company of the person who caused the accident or its claims representative to make a reasoned offer of compensation in cases where liability is not contested and the damages have been qualified within three months of the date when the victim presented his claim for compensation to the insurer. This duty is typically inserted in the Directive in relation to the mandatory motor vehicle insurance. Obviously, the assessment of claims after a major disaster caused by an offshore-related incident may be much more complicated than in case of a traffic accident. But the Directive provides an interesting example of a model where, in combination with mandatory security,[245] the regulator could also oblige the institution providing the financial security to make an offer of payment within a reasonable time to the victim. Whether three months is a realistic period in case of an offshore-related incident seems doubtful. The precise time could be subject to further study but could at least allow rapid payment to this particular category of victims.

7.6.2.4 Combinations

There are also examples of interesting combinations of liability law and administrative claims handling. An obvious example is the GCCF, which handled compensation after the Deepwater Horizon incident. But that fund was basically created on the initiative of the operator concerned (BP) and therefore has no precedent value or specific legal basis because it cannot be assumed that operators in a similar situation would equally move to the creation of a similar facility in the absence of a regulatory framework.

242 See Tyler and Thorisdottir, 'A Psychological Perspective', 355–92.
243 See Bruggeman, 'Capital Market Instruments', 474.
244 *Official Journal* L263/11, 7 October 2009. 245 See Section 7.4.

There are some examples as well of a regulatory framework. For example, in response to nuclear damage, Japan chose an administrative system rather than a judicial system as the primary compensation instrument. Compensation for nuclear damage in Japan is awarded according to categories of geographical areas and government orders. The standards to identify compensable losses are also set by the administrative authority. It is held that this approach can avoid the substantial hurdles in the tort system in awarding compensation for nuclear damage.[246] Although an administrative compensation system is followed, the financing of the compensation awarded to the victims of the Fukushima incident still took place (at least partially) via the liable operator.[247]

Another model that is worth examining was created by a recent Belgian act of 13 November 2011 concerning the compensation for victims of technological accidents.[248] The reason for this new act was related to the disaster with an exploding gas pipeline that happened on 30 July 2004 in Ghislenghien. As a result of this accident, twenty-four people died and more than 130 were injured. Because in Belgium the civil procedure is linked to the criminal procedure,[249] most of the victims were only compensated seven years after the incident. This created the need for a new act which has precisely the aim of accelerating the compensation to victims. The act applies to technological disasters of great extent. It has to concern an incident affecting at least five persons (through death or hospitalization), and a specific committee has to declare the incident as an exceptional disaster. In the case that the specific procedure applies, victims can ask for compensation from the Belgian Motor Insurance Guarantee Fund. The fund will be subrogated in the rights of the victim who thus can later claim against the liable tortfeasor. If the victim disagrees with the decision of the Fund, he or she can appeal to the court. Private insurers active in Belgium are supposed to finance the fund on the basis of their market share.

Although many criticisms could be formulated on the structure of this particular Belgian solution,[250] it shows (if one disregards a few

[246] See Faure and Liu, 'The Tsunami', 195–6. [247] *Ibid.*, 199–200.

[248] *Moniteur Belge*, 24 February 2012. For commentaries, see, *inter alia*, Coune, 'Wet van 13 november 2011'; Verjans, 'Nieuwe wettelijke regeling'.

[249] Meaning that the victims cannot be compensated via civil liability as long as the criminal procedure is not finished.

[250] For example, concerning the fact that it is not clear on which basis the fund will compensate the victims; whether contributory negligence on the side of the victim will be taken into account and why unrelated insurers would have to compensate to the fund, etc.

imperfections) an interesting example of a model where an administrative agency provides (in principle, speedy) compensation to victims and subsequently still recovers from the tortfeasors. Such a model could in principle serve (1) the goal of rapid compensation of victims and (2) the goal of a correct risk allocation to the injurer. If the system worked perfectly, it would mean that the victims would be compensated as if liability were applicable (which would imply the application of a contributory negligence defence), and the fund would in that particular case only pre-finance compensation and later recover this from the liable injurer and his insurer.

7.6.2.5 Evaluation

The starting point should be that there is a strong argument in favour of designing some mechanism guaranteeing rapid payment of the claims of a particularly sensitive category of victims on the basis that without rapid payment, the damage could be disproportionally larger.[251] Thus, the nightmare of lengthy procedures (either *Exxon Valdez* in the United States or Ghislenghien in Belgium) via civil procedure in tort law could be avoided. The importance of such a mechanism is also recognized at the EU policy level, where in the Directive 2013/30 the need to establish procedures for ensuring prompt and adequate handling of compensation claims is equally stressed.[252]

The subsequent trade-off in the policy issues[253] is that there is some time necessary to evaluate claims and that there is hence always an inherent danger in being too speedy, which could lead to either moral hazard on the side of the victims or an insufficient allocation of risk to the operator. Still it seems possible to follow three different routes to realize the best of both worlds (obtaining speedy compensation for a special category of victims and still having a correct allocation of risk):

- A first possibility would be simply to follow the model of Article 22 of the Motor Vehicle Insurance Directive and to provide that those constituting financial security for operators should make an offer of

[251] This argument was also made by Philipp Wassenberg, representative of Munich Re, interview, 6 May 2013.

[252] Article 4(3) holds that Member States shall, as a minimum, establish procedures for ensuring prompt and adequate handling of compensation claims including in respect of compensation payments for trans-boundary incidents.

[253] Already referred to in Section 7.6.2.1.

compensation to the victim within several months (to be specified) of the date when the victim presented his or her claim for compensation.

- A second possibility is that one would develop, following the example of the CLC–Fund regime a Claims Manual also guaranteeing rapid payment within the compensation to victims for offshore-related risks. A formal legal basis for this currently may be lacking. For example, in the United Kingdom it could be linked to OPOL, where OPOL could possibly develop a procedure for (speedy) claims handling. Representatives of Oil and Gas UK have already indicated that this may be an interesting possibility to explore,[254] and this was confirmed by representatives of Lloyds as well.[255] Although this undoubtedly is an attractive solution as well, we have to keep in mind the limits of OPOL, since it is essentially limited to the North Sea and within the North Sea mostly to the United Kingdom. This would only be a European-wide solution if other regional pooling arrangements[256] were to emerge which could adopt a similar Claims Manual.
- The third possibility (and the most far-reaching) would be to create a possible construct (also learning from Belgium's mistakes and taking economic principles into account) whereby a facility would be created that rapidly pre-pays the damage of this specific category of victims, taking into account a few principles:
  - A claims verification takes place, verifying that under normal rules of tort law those claims would be compensable (in order to avoid moral hazard), and
  - A subrogation of the facility in the rights of the victim whereby the facility subsequently claims back the compensation from the person(s) liable for the offshore-related damage in order to create a correct allocation of the risk.

254 Interview with representatives of Oil and Gas UK, in London, 1 May 2013.

255 Interview with representatives of Lloyds, 1 May 2013, in London. They moreover added that it would make sense to arrange compensation via an independent third-party administrator who would pay out the available funds. The advantage of such a system would be that all victims would be forced to present their claim within a particular period to cut off the time bar. It would then become rapidly clear how much the total claims are, after which the available funds could be paid out.

256 Which would then also have to be created.

7.7 Trans-Boundary Harm

A point we can unfortunately not develop in detail within the scope of this research[257] relates to the fact that many offshore-related incidents could have and do have a trans-boundary character. The nature of offshore exploration is such that several different legal systems and jurisdictions could be affected. This was already made clear when we discussed the Montara case.[258] Although the incident occurred formally in an Australian exclusive economic zone, Indonesia was affected as well. It led to complaints under international law (and more particularly UNCLOS) that Australia did not inform the Indonesian government of the Montara incident.

Stakeholders also report that it is very likely that a major offshore incident in Europe would have trans-boundary effects as well. These effects could still be confined to Europe if a major incident were to occur in the northern North Sea, which could potentially affect the coasts of England, Norway and potentially the Netherlands and Belgium.[259] However, in the Mediterranean, trans-boundary effects i.e. offshore exploration in European waters affecting other states or the other way around (exploration outside of European waters affecting European territory) is very likely as well. Offshore exploration is increasingly taking place in the Eastern Mediterranean as well as, *inter alia*, near the coast of Lebanon.

The sovereign rights of states to explore their exclusive economic zone and the continental shelf have been regulated in UNCLOS, and UNCLOS equally regulates a duty of the state to notify another state that would be likely to be affected by damage to the marine environment.[260] However, there is no international regulation of the liability in case of trans-boundary incidents because, in contrast to that for vessel-based pollution, there is as yet no international convention regulating liability in case of trans-boundary damage resulting from an offshore installation. From an economic perspective, it is not difficult to make the argument that especially as far as trans-boundary damage is concerned, it would make sense to shift powers to a higher legal order and hence to create an

[257] Because it effectively may need a totally separate treatment in a separate study.
[258] See Section 2.5.1.2.
[259] Interview with Philipp Wassenberg, representative of Munich Re, 6 May 2013; Interview with Wendy Kennedy, head of the Offshore Oil and Gas Department at DECC, 29 April 2013.
[260] See Art. 198 of UNCLOS.

international convention dealing with trans-boundary pollution resulting from offshore-related risks.[261] However, the problem is that there currently is no international organization that is considered to have competencies with respect to regulating damage resulting from offshore exploration. As was made clear earlier,[262] the International Maritime Organization (IMO), which would be the logical institution to address, has considered that it is in fact not the most ideal institution to create such an international convention. Stakeholders also hold that this can be understood for the simple reason that the IMO was created especially to regulate maritime issues. Offshore installations are not primarily concerned with maritime issues, although damage resulting from an offshore installation can, of course, affect the marine ecosystem. It is therefore held that the IMO rightly refrained from taking action in this respect.[263] The legal committee of the IMO, however, recommended making bilateral or regional arrangements between various states to deal with trans-boundary damage. This was announced after an initiative from Indonesia following the Montara incident.

In this respect, again, the example of the CLC–Fund Convention can be illuminating. The CLC has explicitly provided for an exclusive liability of the courts within the jurisdiction where the incident occurred. In this respect, Article 9 of the CLC provides

> 1. Where an incident has caused pollution damage in the territory including the territorial sea or an area referred to in Article II, of one or more Contracting States, or preventive measures have been taken to prevent or minimize pollution damage in such territory including the territorial sea or area, actions for compensation may only be brought in the Courts of any such Contracting State or States. Reasonable notice of any such action shall be given to the defendant.
> 2. Each Contracting State shall ensure that its Courts possess the necessary jurisdiction to entertain such actions for compensation.
> 3. After the fund has been constituted in accordance with Article V the Courts of the State in which the fund is constituted shall be exclusively competent to determine all matters relating to the appointment and distribution of the fund.

The fund referred to as established under Article V is the Limitation Fund.

[261] Cf. Faure and Betlem, 'Applying National Liability Law', 129–32.
[262] See Section 3.2.5.
[263] See the opinion of the representatives from the International Group of P&I Clubs, interview, 1 May 2013.

Article V(3) states that in order for a ship owner to benefit from the limitation right, 'the owner shall constitute a fund for the total sum representing the limit of his liability with the Court or other competent authority in any one of the Contracting States in which an action can be brought under Article IX.'

The advantage of such an arrangement is that it creates clarity for all victims, who immediately know where their claim can be brought. Moreover, also for claims handling by insurers or other parties that provided financial security, the concentration of litigation before one court is obviously necessary to effectively distribute the available insurance proceeds. Also, representatives from insurers held that in order to effectively deal with mass claims resulting from an offshore incident, it would be important to have rules indicating that victims have to hand in their claim within specified time limits and that, moreover, claims have to be brought before a single court in one particular jurisdiction to avoid an accumulation of procedures.[264] However, one has to keep in mind that what may be an advantage from the perspective of the offshore operator and his financial guarantors (concentrating the claims in one jurisdiction and one court) may not always be beneficial for victims because it could force them to bring a case in a court outside their own jurisdiction. The desirability of focusing claims in one particular jurisdiction hence definitely needs further attention.

In sum, although this issue undoubtedly deserves further research (as far as the specific technical implementation is concerned), it may be interesting to examine whether one could introduce, also for offshore-related damage, a similar procedural rule as in the CLC–Fund Convention aiming at the competence of one single jurisdiction and court.[265] However, attention also should be paid to the question of the extent to which this corresponds with the legitimate interests of the victims as well.

[264] That was held by representatives of Lloyds, interview, in London, 1 May 2013; Interview with representatives of the International Group of P&I Clubs, 1 May 2013, in London (holding that a large spill in European waters could easily affect many Member States as a result of which it is important to have clear rules on jurisdiction).

[265] It should be noted also that Directive 2013/30 on offshore safety shares this concern for incidents with a trans-boundary element, prescribing, *inter alia*, that a procedure for adequate handling of compensation claims should include compensation payments for trans-boundary incidents (Art. 4[3]).

7.8 Scenarios

As explained in Chapter 6, three possible scenarios could be worked out with different solutions depending on (1) whether to apply OPOL or a similar regime developed for other areas than the North Sea, (2) for situations where the damage is higher than the OPOL limit (US$250 million) but still insurable on the commercial market, and (3) finally the situation where no financial cover via the regular commercial solutions can be obtained.

The starting point for each scenario is that mandatory financial security should be provided in each case but that the instruments used can differ. It will suffice here to sketch the scenarios in a table; the conditions for the specific instruments to work have been discussed in detail and therefore will be simply referred to in footnotes.

7.8.1 Scenario 1: Damage Maximum US$250 Million

UK/North Sea OPOL	Other areas: other regional arrangements (to be developed)[a]
Mandatory financial security via:	
Self-insurance[b]	
Insurance	
Industry pooling (like OIL/OCIL or comparable mechanisms[c])	
(Guarantees)[d]	
Other	

Note: The OPOL limits refer to compensation in US dollars; therefore, we have, for constructing these scenarios, also used dollar values. These proposals could, however, obviously easily be transposed to euros or other monetary units.

[a] See Section 5.6.

[b] Under the conditions mentioned in Section 5.1.3.

[c] On the condition that risk-dependent contribution schemes can be developed; see Section 5.5.

[d] Stakeholders held that these are seldom used in practice and therefore will not play a major role. See Section 5.3.3.

For small accidents, many instruments can provide financial security coverage. The majors may prefer self-insurance given the potential costs-saving possibilities. Commercial insurance can be used by SMEs, usually together with a lower amount of retention. Industry pooling provides another choice, either by existing pools such as OIL and OCIL or by establishing other comparable pooling mechanisms.

OPOL provides coverage in the North Sea. Similar schemes also could be developed in other regions by pooling similar risks together. It is worth noting, however, that OPOL does not provide direct coverage but requires its members to cover their liability up to US$250 million themselves. It intervenes only when the member or its guarantors are not able to provide compensation.

7.8.2 *Scenario 2: Damage between US$250 Million and US$750 million*

UK/North Sea OPOL	Other areas: other regional arrangements (to be developed)[a]
Mandatory financial security via:	
Self-insurance	
Insurance	
Industry pooling (like OIL/OCIL or comparable mechanisms)	
(Guarantees)	
Other	

Note: As can be seen, this scenario is basically the same as scenario 1 as far as the instruments to be used are concerned, with the difference that the OPOL limits do not apply.

[a] See Section 5.6.

For these medium-sized accidents, the available compensation instruments are similar to scenario 1. However, in this case, the existing limit of OPOL or a potentially similar arrangement in other regions is (may be) exceeded. In this case, the operators can still resort to the commercial insurance market (for SMEs) or self-insurance (the majors). The other possibility is to expand the amount available from OPOL or other pooling systems. However, this approach is strongly opposed by the majors, which argue that the different risks between operators are large and that mandatory pooling would lead to cross-subsidization from good risks to bad risks. Although a mandatory pooling is thus not desirable, an increased financial security requirement can enhance the capacity of the pooling system as a solution for SMEs.

7.8.3 Scenario 3: Damage above US$750 Million

Mandatory financial security until risk-related amount[a]

First best:
  Majors: self-insurance/captives[b]
  Others: less risky activities[c]

Second best[d]:
  Possibly:
    Munich Re facility[e]
    Noble Energy proposal (generally: industry pooling)[f]
    Fund with risk-related contributions[g]
    A retrospective pooling scheme[h]
    Government as re-insurer of last resort[i]

[a] As argued earlier, in principle, mandatory security should be provided to the amounts of the expected damage. The damage expectation can depend upon the nature and type of the offshore exploration and hence should be based on objective risk-related criteria to be decided by the licensing authority.

[b] Majors (i.e. the oil and gas producers that can in principle cover disastrous losses through their balance sheets) should retain the flexibility to choose their own hedging strategies to finance future losses but, of course, to be supervised by licensing authorities.

[c] To the extent possible, smaller OGPs should be licensed for less risky activities (i.e. drilling activities with less potential damage such as wells with less pressure, no deepwater drilling, not close to the coast, etc.). Thus, they cannot create risks of causing damage above €750 million and hence externalize a risk to society.

[d] All these possibilities are in fact 'second best' because they all have (considerable) disadvantages and are also largely rejected by the stakeholders involved, which may make implementation (politically) impossible.

[e] See Section 6.3.2.

[f] See Section 6.4.3.

[g] To the extent risk-related contributions can be charged, a fund should not be problematic. However, in practice, such a fund is often financed through a flat tax, which leads to negative effects for prevention and to negative redistribution. See Section 7.5.6.2.

[h] The advantages of such a (government-mandated) retrospective pooling scheme have been discussed at many places (following the model of the US Price-Anderson Act), *inter alia*, in Section 4.5.1.

[i] Also this is generally not a viable solution because it may (if government does not charge market-conforming premiums) lead to a subsidization. For man-made (technological) disasters, this does not seem like an ideal solution. See Section 7.5.5.2.

In case of a catastrophic accident, the existing instruments on the market are not able to provide sufficient coverage. The majors may be able to cover the potential damage out of their deep pockets through self-insurance. However, the SMEs will be insolvent and hence will have no incentive to take sufficient care. The approach to limit SMEs to less risky offshore activities is not politically possible. Another solution is to rely on the market for the new emerging arrangement. After the Deepwater Horizon accident, a few proposals have been made to provide coverage of up to US$10 billion to US$20 billion, including insurance, a retrospective pooling system, a fund and government as re-insurer of last resort. The initiators of the insurance and pooling systems argue that a legal obligation to join them will be necessary to secure enough capacity. However, the preceding analysis shows that offshore facilities possess very divergent risks. Forcing all parties to join the same risk pool will create insufficient incentives. Instead of making them mandatory, the government should take measures to promote their development. A compensation fund is also proposed as a solution, but risk-related contributions are very important to ensure sufficient precautions from the potentially liable parties. Government can act as re-insurer of last resort in covering many catastrophic risks, but this solution may be less suitable for offshore-related risks.

8

Concluding Remarks

MICHAEL FAURE, NIELS PHILIPSEN AND HUI WANG

These concluding remarks will provide a summary of recommendations that have been made concerning the formulation of an optimal liability and compensation scheme for offshore-related risks at the end of the preceding chapters. Hence, in order to avoid repetition, we will now summarize the earlier recommendations and refer in footnotes to the specific places where the motivation for those recommendations has been provided. The advantage is that all recommendations are brought together, while the reader can consult the text referred to in the footnotes to obtain the detailed motivation for the specific recommendations. The recommendations we will formulate to an important extent follow the order of the chapters in this book.

8.1 General

- Data on incidents related to damage resulting from offshore oil and gas activities are either difficult to obtain or not publically disclosed. It is recommendable that an institution at the European Union level should centrally collect those data, also in order to increase the insurability of offshore-related damage.
- It is recommendable to urge Member States to invite the offshore oil and gas producers within their jurisdictions to collaborate in the provision of those data to the central European institution.
- It is recommendable that the European Union take the initiative (eventually via a specialized UN agency or other institutions) to come to an international agreement especially focusing on offshore-related incidents with a trans-boundary character.
- In order to promote (international) risk pooling by industry, mandatory safety standards should be implemented guaranteeing a minimum

level of offshore safety in the European Union.[1] Safety regulation should play a more important role than liability rules in the prevention of offshore-related risks.
- Given the higher technical knowledge of industry on optimal safety standards, the European Union could promote (*inter alia*, via guidance notes) industry agreements (eventually with national regulators) on targets and safety standards but striving for highly harmonized EU-wide safety standards.

8.2 Efficient Liability Rules

An EU-wide regime for damage caused by offshore-related risks could be shaped along the following lines[2]:

- Liability for damage caused by offshore-related risks should be strict.[3]
- Liability should take into account the behaviour of the victim as well, meaning that the claim on compensation should in principle be reduced to the extent that the victim contributed to the loss.
- A legal channelling of liability should be avoided.[4]
- A joint and several liability of various parties who contributed to the offshore-related risk should be installed.
- Systems of so-called economic channelling, implying that the financial security of a licensee or operator also covers the liability of subcontractors, should be promoted.
- Financial caps on liability should be avoided in order to expose operators and others who contribute to offshore-related risks fully to the social costs created by their activity.
- Compliance with a regulatory standard should not automatically exclude liability for damage resulting from offshore-related risks.
- It is recommendable to provide an integrated liability regime for damage caused by offshore-related risks.

[1] See Section 7.5.3.

[2] We did not discuss whether there is any reason for action at the European Union level from an economic perspective (although it is relatively easy to argue that there is, given the trans-boundary character of the damage; see Faure, 'How Law and Economics', 40–2); nor did we discuss the question of formal EU competence in this area. Since various initiatives have already been taken in this domain, we assumed EU competence in this domain to be a given.

[3] See Section 7.2.1. [4] See Section 7.2.2.

8.3 Mandatory Financial Security

- It seems indicated to mandate financial coverage for offshore-related risks, especially for the cases where smaller and medium-sized operators may create a risk of major damage and hence an insolvency risk.[5]
- As far as possible, also at the licensing level, a situation where operators would engage in offshore operations whose risks in the case of an accident would outweigh their personal assets should be avoided. In such a case, a joint venture with OGPs with larger financial capacity may be indicated.
- It seems indicated to issue a guidance note at the EU level guiding local licensing authorities in Member States on the required amount and form of financial security for offshore-related risks.[6]
- As far as the amount is concerned, this guidance note should be based on an objective assessment of the risk, taking into account technical criteria that relate to a specific operation and operator to particular amounts of potential damage.
- The guidance should allow sufficient flexibility as far as the forms of financial security are concerned and not necessarily limit those to insurance. The only condition would be that local regulators accurately verify whether the form and amount of the financial security offered by the operator would be adequate to cover the potential damage emerging from that particular offshore installation.
- Hence, the guidance could take into account location-specific circumstances on which local regulators in Member States can base their assessments of the amount and form of financial security.[7]
- Such an approach allows sufficient flexibility, avoids unnecessary costs (e.g. forcing majors to transfer risks to lower-rated insurance companies), encourages a level playing field for operators and avoids an externalization of social costs (and thus a market failure) in the case of insolvency.
- The European Union could promote the development of regional pools like OPOL (solvency-guarantee pools) for other areas than the North Sea (e.g. the Mediterranean). The imposition of a regulatory duty to show financial responsibility, as recommended here, also should promote the development of those regional pools.

[5] See Section 7.4.2. [6] See Section 7.4.5. [7] *Ibid.*

8.4 A Role for Government?

The European Union could consider either initiating itself or suggesting to Member States particular initiatives to facilitate the provision of compensation for offshore-related risks where the damage would be higher than an amount which is insurable on the commercial market.[8] However, such a role for government should correspond with a number of fundamental principles:

- Government intervention would only be indicated for those risks for which none of the financial market solutions[9] are available.
- To the extent that market solutions (such as self-insurance or captives) are available, government intervention should be absent.
- Government should therefore only intervene as a last resort in the hypothetical situation that catastrophic losses[10] could not be covered by any mechanism available on the market.
- Even then the government intervention should be based on premiums or contributions that reflect the actual risk in order to provide adequate incentives for prevention.

When these (exceptional) conditions are met, the following types of government intervention could be envisaged:

- The creation of a compensation fund for offshore-related risk, only providing an upper layer for catastrophic damage and financed with risk-related contributions.[11]
- A retrospective pooling scheme whereby a government institution (agency) pre-finances the loss and ex post (after the accident) claims back the money paid to the victims on the basis of retrospective contributions to be paid by operators.[12] The retrospective contributions should be risk dependent.
- A re-insurance by government as a last resort.[13] The risk-dependent (re-)insurance premiums should be charged by government.

[8] It would consider scenario 3 of Section 7.8.3. [9] Described in Chapters 5 and 6.
[10] Scenario 3 from Section 7.8.3. [11] See Section 7.5.3.
[12] Based on the example of the retrospective pooling under the US Price-Anderson Act. See Section 4.5.1.
[13] See Section 7.5.2.

8.5 Rapid Claims Mechanism

The European Union should invite Member States to develop a mechanism facilitating early compensation payments to specific vulnerable groups of victims (especially fishermen and hotel/restaurant owners in coastal areas) that may be negatively affected by an offshore-related incident. The following mechanisms could be envisaged:

- An obligation imposed on the provider of the financial security or the liable person to formulate an offer for payment to the victim within a fixed period of time after the victim has presented his or her claim.
- The development of a (rapid) claims settlement mechanism via OPOL or similar regional pooling schemes.
- The potential construction of a facility allowing the pre-payment to the specific vulnerable group of victims on the basis of a rapid evaluation of the validity of their claims and subsequent recourse of the facility against the person liable for the damage caused by the offshore-related incident.

REFERENCES

Abecassis, D., *The Law and Practice Relating to Oil Pollution from Ships* (London: Butterworth, 1978).

Abraham, K., 'Catastrophic oil spills and the problem of insurance'. *Vanderbilt Law Review* 64 (2011), 1769–91.

Aldersey-Williams, J., 'Liability for oil spills – the UKCS legal regime', presented at the OSPRAG Summit, UK Oil Spill Prevention and Response, September 21, 2011.

Allen, C. H., 'Proving natural resource damage under OPA 90: out with the rebuttable presumption, in with APA-style judicial review?', *Tulane Law Review* 85 (2011), 1039–74.

Ameye, E., 'Legal study on nuclear third party liability for DG TREN of the European Commission', in N. Pelzer (ed.), *European Nuclear Liability Law in a Process of Change* (Berlin, Tagungsbericht der AIDN/INLA-Regionaltagung, 2009; Baden-Baden: Nomos, 2010), pp. 147–56.

Anderson, C. B., 'Damage to natural resources and the costs of restoration', *Tulane Law Review* 72 (1997), 417–92.

Arrow, K., *Aspects of the Theory of Risk-Bearing* (Helsinki: Yrjö Jahnssonin Säätiö, 1965).

Australian Government, 'Final government response to the Report of the Montara Commission of Inquiry', Melbourne, 2011.

Australian Maritime Safety Authority, 'Response to the Montara wellhead platform incident', Report of the Incident Analysis Team, March 2010; available at: www.amsa.gov.au/marine_environment_protection/national_plan/incident_and_exercise_reports/documents/montara_iat_report.pdf.

Baatsma, B., Felsö, F., Van Geffen, S., Mulder, J. & Oostdijk, A., *Zelf doen? Inventarisatiestudie van zelfreguleringsinstrumenten*, Onderzoek in opdracht van het Ministerie van Economische Zaken (Amsterdam, 2003).

Barrett, S., 'European environmental liability options' (January 10, 2013); available at: www.postonline.co.uk/post/feature/2313317/european-environmental-liability-options.

Bawcut, P. A., *Captive Insurance Companies: Establishment, Operation and Management* (New York: Woodhead-Faulkner, 1991).

BDO Consulting, 'Independent evaluation of the Gulf Coast claims facility: report of findings and observations', research commissioned by the US Department of Justice (June 5, 2012); available at: www.justice.gov/opa/documents/gccf-rpt-find-obs.pdf.

Bennet, P., 'Mutual risk: P&I insurance clubs and maritime safety and environmental performance', *Marine Policy* 25 (2001), 13–21.

Bergkamp, L., 'The proper scope of joint and civil liability', *Tijdschrift voor Milieuaansprakelijkheid* [*Environmental Liability Review*] (2000), 153–6.

& Van Bergeijk, A., 'Scope of the ELD regime', in L. Bergkamp and B. J. Goldsmith (eds.), *The EU Environmental Liability Directive: A Commentary* (Oxford University Press, 2013), pp. 51–79.

Bernasconi-Osterwalder, N. & Hoffmann, R. T., 'The German nuclear phase-out put to the test in international investment arbitration' (Berlin/Amsterdam 2013); available at: www.tni.org/files/download/vattenfall-icsid-case_oct2013.pdf.

Bio by Deloitte, 'Civil liability, financial security and compensation claims for offshore oil and gas activities in the European Economic Area', Final report prepared for the European Commission, DG Energy (Brussels, August 14, 2014).

Bio Intelligence Service *et al.*, 'Study to explore the feasibility of creating a fund to cover environmental liability and losses occurring from industrial accidents', Final report prepared for the European Commission, DG ENV (March 25, 2013).

Blanchard, P., 'Responsibility for environmental damage under nuclear and environmental instruments: a legal benchmarking', *Journal of Energy and Natural Resources Law* 18 (2000), 233–53.

Bocken, H., 'Alternatives to liability and liability insurance for the compensation of pollution damages', part 1, *Tijdschrift voor Milieuaansprakelijkheid* 1 (1987), 83–7.

'Alternatives to liability and liability insurance for the compensation of pollution damages', part 2, *Tijdschrift voor Milieuaansprakelijkheid* 1 (1988), 3–10.

'Alternative financial guarantees for environmental liabilities under the ELD', *European Energy and Environmental Law Review* 18 (2009), 146–70.

Booz Allen Hamilton, 'The offshore oil and gas industry report on insurance', part 1, October 5, 2010; available at: www.eoearth.org/files/172301_172400/172373/insurance_report_part-one_oct_5_4-pm_r1.pdf.

'The offshore oil and gas industry report on insurance', part 3, October 29, 2010; available at: www.eoearth.org/files/172301_172400/172376/insurance-report-part-three_oct_21_5-pm_r1.pdf.

Borthwick, D., 'Report of the Montara Commission of Inquiry', June 2010.

Bosma, S., 'The regulation of marine pollution arising from offshore oil and gas facilities: an evaluation of the adequacy of current regulatory regimes and the responsibility of states to implement a new liability regime', *Australia and New Zealand Maritime Law Journal* 26 (2012), 89–117.

Boyd, J., 'A market-based analysis of financial insurance issues associated with US national resource damage liability', in M. Faure (ed.), *Deterrence, Insurability, and Compensation in Environmental Liability: Future Developments in the European Union* (Vienna: Springer, 2003), pp. 258–302.

'Compensation for oil pollution damages: the American Oil Pollution Act as an example for global solutions?', in M. Faure and J. Hu (eds.), *Prevention and Compensation of Marine Pollution Damage* (New York: Kluwer Law International, 2006), pp. 137–63.

Boyer, M. & Porrini, D., 'The efficient liability sharing factor for environmental disasters: lessons for optimal insurance regulation', *Geneva Papers on Risk and Insurance-Issues and Practices* 33 (2008), 337–62.

BP, 'BP Annual Report and Form 20-F 2013'; available at: www.bp.com/content/dam/bp/pdf/investors/bp_annual_report_and_form_20f_2013.pdf.

BP, 'BP Annual Report and Form 20-F 2014'; available at: www.bp.com/content/dam/bp/pdf/investors/bp_annual_report_and_form_20f_2014.pdf.

Brownie, I., *Principles of International Law*, 7th edn. (Oxford University Press, 2008).

Bruggeman, V., 'Capital market instruments for natural catastrophe and terrorism risks: a bright future?', *Environmental Law Reporter* 40 (2010), 10136–53.

Bruggeman, V., Faure, M. G. & Fiore, K., 'The government as reinsurer of catastrophic risks?', *Geneva Papers on Risk and Insurance Theory* 35 (2010), 369–90.

Faure, M. G. & Heldt, T., 'Insurance against catastrophe: government stimulation of insurance markets for catastrophic events', *Duke Environmental Law & Policy Forum* 23 (2012), 185–241.

Budiman, A., 'On liability for offshore oil spillage: strictly settled', International Bar Association Oil and Gas Law Publications; available at: www.ibanet.org/document/default.aspx?documentuid=fcebab31-12B0-4f19-b5f6-1259fd824fbc.

Cafaggi, F., 'New foundations of transnational private regulation', *Journal of Law and Society* 38 (2011), 20–49.

Calabresi, G., *The Costs of Accidents: A Legal and Economic Analysis* (New Haven, CT: Yale University Press, 1970).

Cameron, P. 'Liability for catastrophic risk in the oil and gas industry', *International Energy Law Review* 6 (2012), 207–19.

Campbell, A.J., 'Self-regulation and the media', *Federal Communications Law Journal* 51 (1999), 711–72.

Canadian Centre for Energy Information, 'Canada's evolving offshore oil and gas industry 2004'; available at: www.centreforenergy.com/shopping/uploads/111.pdf.

Cane, P., 'Tort law as regulation', *Common Law World Review* 31 (2002), 305–31.

Carrol, S., 'Perspective on the pros and cons of a pooling type approach to nuclear third party liability', *Nuclear Bulletin* (2008), 75–97.

Chazan, G., 'Mediterranean playground for oil groups', *Financial Times* (2 September 2012); available at: www.ft.com/cms/s/0/ee151f26-f38a-11e1-b3a2-00144feabdc0.html#axzz3clP1naf2.

Christou, M. & Konstantinidou, M., 'Safety of offshore oil and gas operations: lessons from past accident analysis – ensuring EU hydrocarbon supply through better control of major hazards', *JRC Scientific and Policy Reports* (2012); available at: http://publications.jrc.ec.europa.eu/repository/bitstream/jrc77767/offshore-accident-analysis-draft-final-report-dec-2012-rev6-online.pdf.

Coghlin, T. G., 'Protection and indemnity clubs', *Lloyds Maritime and Commercial Law Quarterly* (1984), 403–16.

Coglianese, C. & Lazer, D., 'Management-based regulation: prescribing private management to achieve public goals', *Law & Society Review* 37 (2003), 691–730.

Cohen, M., Gottlieb, M. Linn, J. & Richardson, N., 'Deepwater drilling: law, policy, and economics of firm organization and safety', *Vanderbilt Law Review* 64 (2011), 1853–1916.

Conk, G., 'BP and Kenneth Feinberg's Gulf Coast gambit', *Roger Williams University Law Review* 17 (2012), 137–83.

Coune, C., 'Wet van 13 november 2011 betreffende de vergoeding van de lichamelijke en morele schade ingevolge een technologisch ongeval', *Tijdschrift voor Belgisch Handelsrecht* 5 (2012), 5–39.

Courbage, Chr. & Stahel, W. R. (eds.), *Extreme Events and Insurance: 2011 Annus Horribilis* (The Geneva Reports No. 5) (The Geneva Association, 2012).

Crawford, E., 'Australia's lack of compliance with international legal principles of trans-boundary harm: a case study of the Montara incident', presentation at a seminar organized by the Australian National Center for Ocean Resources and Security (ANCORS) and University of Wollongong, NSW, 12 May 2011; available at: ancors.uow.edu.au/content/groups/public/@web/@law/@ancors/documents/doc/uow103158.pdf.

Cross, F., 'Natural resource damage valuation', *Vanderbilt Law Review* 42 (1989), 269–341.

Currie, D. E. J., 'The problems and gaps in the nuclear liability conventions and an analysis of how an actual claim would be brought under the current existing treaty regime in the event of a nuclear accident', *Denver Journal of International Law and Policy* 35 (2008), 85–128.

Dagg, J., Holroyd, P., Lemphers, N., Lucas, R. & Thibault, B., 'Comparing the offshore drilling regulatory regimes of the Canadian Arctic, the US, the UK, Greenland and Norway', The Pembina Institute, June 2011.

Daintith, T., *Discretion in the Administration of Offshore Oil and Gas: A Comparative Study* (Melbourne: AMPLA, Ltd., 2006).

Danish Energy Agency, 'A guide to hydrocarbon licenses in Denmark: exploration and drilling activities' (September 2011).

'Oil and gas production in Denmark and subsoil use 2013' (June 2014).

De La Rue, C. & Anderson, C. B., *Shipping and the Environment: Law and Practice* (London: LLP, 2009).

De Smedt, K., *Environmental Liability in a Federal System: A Law and Economic Analysis* (Antwerp: Intersentia, 2007).

DECC (Department of Energy and Climate Change), 'Guidance note to UK offshore oil and gas operators on the demonstration of financial responsibility before consent may be granted for exploration and appraisal wells on the UKCS' (2012).

Diamond, P., 'Single activities accidents', *Journal of Legal Studies* (1974), 107–64.

Doherty, N. A., Grace, M. F., Klein, R. W., *et al.*, 'Managing large-scale risks in a new era of catastrophes. Insuring, mitigating and financing recovery from natural disasters in the United States' (Wharton Risk Management and Decision Processing Centre in conjunction with the Georgia State University and the Insurance Information Institute, March 2008); available at: http://opim.wharton.upenn.edu/risk/library/wharton_largescalerisks_fullreport_2008.pdf.

Dowding, T., *Global Developments in Captive Insurance* (London: Pearson, 1997).

Dubin, J. & Rothwell, G., 'Subsidy to nuclear power through Price Anderson liability limit', *Contemporary Economic Policy* 8 (1990), 73–9.

Ducrotoy, J. P. & Elliot, M., 'The science and management of the North Sea and the Baltic Sea: natural history, present threats and future challenges', *Marine Pollution Bulletin* 57 (2008), 8–21.

Durant, I. C., 'Belgium', in M. Faure and T. Hartlief (eds.), *Financial Compensation for Victims Of Catastrophes: A Comparative Legal Approach* (Vienna: Springer, 2006), pp. 37–79.

Easo, J., 'Managing the risk of catastrophic events on the UK continental shelf', *Ashurst Energy Source* 7 (2010), 18–22.

EIA (Energy Information Administration) and Office of Oil and Gas, 'Overview of US legislation and regulations affecting offshore natural gas and oil activity' (September 2005).

ELINI, 'Annual Report 2010', (Brussels 2010); available at: www.elini.net/documents/.

EMANI (European Mutual Association For Nuclear Insurance), 'Annual Report 2013' (Brussels 2013); available at: www.emani.be/files/emani_ann_rep_2013_lq.pdf.

Emmerson, C., *Arctic Opening: Opportunity and Risk in the High North* (London: Lloyd's and Chatham House, 2012).

Emons, W., 'Imperfect tests and natural insurance monopolies', *Journal of Industrial Economics* 49 (2001), 247–68.

Energy and Climate Change Committee, 'UK deepwater drilling – implications of the Gulf of Mexico oil spill', Second report of Session 2010–11, Volume 1 (House of Commons, 6 January 2011); available at: www.publications.parliament.uk/pa/cm201011/cmselect/cmenergy/450/45002.htm.

Epstein, R., 'Catastrophic responses to catastrophic risks', *Journal of Risk and Uncertainty* 12 (1996), 287–308.

Ernst & Young, 'Study on co(re)insurance pools and on ad-hoc co(re)insurance agreements on the subscription market' (Luxembourg: European Commission, February 2013); available at: ec.europa.eu/competition/sectors/financial_services/insurance.html.

Faure, M., 'Economic models of compensation for damage caused by nuclear accidents: some lessons for the revision of the Paris and Vienna Conventions', *European Journal of Law and Economics* (1995), 21–43.

'How law and economics may contribute to the harmonization of tort law in Europe?', in R. Zimmerman (ed.), *Grundstrukturen des europäischen Deliktsrechts* (Baden-Baden: Nomos, 2002), pp. 31–82.

'Alternative compensation mechanisms as remedy for uninsurability of liability', *Geneva Papers on Risk and Insurance* 29 (2004), 455–89.

'Economic criteria for compulsory insurance', *Geneva Papers on Risk and Insurance* 31 (2006), 149–68.

'Financial compensation for victims of catastrophes: a law and economics perspective', *Law and Policy* 29 (2007), 339–67.

'Environmental liability', in M. G. Faure (ed.), *Tort Law and Economics*, Vol. 1 (Cheltenham: Edward Elgar, 2009), pp. 247–86.

& Betlem, G., 'Applying national liability law to transboundary pollution: some lessons from Europe and the United States', in M. Faure and Y. Song (eds.), *China and international environmental liability: Legal remedies for Trans-Boundary Pollution* (Cheltenham: Edward Elgar, 2008), pp. 129–91.

& Bruggeman, V., 'Catastrophic risks and first-party insurance', *Connecticut Insurance Law Journal* 15 (2008), 1–52.

& Fiore, K., 'The coverage of the nuclear risk in Europe: which alternative?', *Geneva Papers on Risk and Insurance – Issues and Practices* 33 (2008), 288–322.

& Fiore, K., 'An economic analysis of the nuclear liability subsidy', *Pace Environmental Law Review* 26 (2009), 419–47.

& Grimeaud, D., 'Financial assurance issues of environmental liability', in M. Faure (ed.), *Deterrence, Inusrability, and Compensation in Environmental Liability: Future Developments in the European Union* (Wien: Springer, 2003), pp. 7–255.

& Hartlief, T. (eds.), *Insurance and Expanding Systemic Risks, Policy Issues in Insurance*, No. 5 (Paris: OECD, 2003).

& Liu, J., 'Compensating for nuclear accidents in Japan: the Fukushima case', *TGMA* 2 (2012), 74–84.

& Liu, J., 'The Tsunami of March 2011 and the subsequent nuclear incident at Fukushima: who compensates the victims?', *William & Mary Environmental Law & Policy Review* 37 (2012), 129–218.

& Skogh, G., 'Compensation for damages caused by nuclear accidents: a convention as insurance', *Geneva Papers on Risk and Insurance – Issues and Practice* 65 (1992), 499–513.

& Van den Bergh, R., 'Negligence, strict liability and regulation of safety under Belgian law: an introductory economic analysis', *Geneva Papers on Risk and Insurance* (1987), 95–114.

& Van den Bergh, R., 'Restrictions of competition on insurance markets and the applicability of EC anti-trust law', *Kyklos* (1995), 65–85.

& Vanden Borre, T., 'Compensating nuclear damage: a comparative economic analysis of the US and international liability schemes', *William & Mary Environmental Law and Policy Review* 32 (2008), 219–87.

& Wang, H., 'Economic analysis of compensation for oil pollution damage', *Journal of Maritime Law and Commerce* 37 (2006), 179–217.

Finken, S. & Laux, C., 'Catastrophe bonds and reinsurance: the competitive effect of information-insensitive triggers', *Journal of Risk and Insurance* 76 (2009), 579–605.

Force, R., Davies, M. & Force, J. S., 'Deepwater Horizon: removal costs, civil damages, crimes, civil penalties, and state remedies in oil spill cases', *Tulane Law Review* 85 (2011), 889–982.

Fowler, D., 'Offshore oil: a frontier for international lawmaking', *Chicago-Kent Journal of International and Comparative Law* (Spring 2012); available at: www.kentlaw.edu/jicl/articles/spring2012/fowler_note.pdf.

Froot, K. A., *The Financing of Catastrophe Risk* (University of Chicago Press, 1999).

Gao, Z. (ed.), *Environmental Regulation of Oil and Gas* (Alphen aan den Rijn: Kluwer Law International, 1998).

GCCF (Gulf Coast Claims Facility), 'Gulf Coast Claims Facility protocol for emergency advance payments' (23 August 2010); available at: www.restorethegulf.gov/sites/default/files/imported_pdfs/library/assets/gccf-emergency-advance-payments.pdf.

GCCF, 'Protocol for interim and final claims'; available at: http://s3.amazonaws.com/nytdocs/docs/521/521.pdf.

Gilead, I., 'Tort law and internalization: the gap between private loss and social cost', *International Review of Law and Economics* (1997), 589–608.

Gittus, J., 'Introducing nuclear power to Australia: an economic comparison', Report prepared for the Australian Nuclear Science and Technology

Organization (2006); available at: www.ansto.gov.au/_data/assets/pdf_file/0016/12445/nuclear_options_paper_Gittus_complete.pdf.

GL Noble Denton, 'UK responses to EC impact assessment on offshore regulation' (November 2011).

Gollier, C., 'Some aspects of the economics of catastrophe risk insurance', in OECD, (eds.), *Catastrophic Risks and Insurance*, Vol. 8 (Paris: OECD Publishing, 2005), pp. 13–30.

Gómez, C. & Green D. R., 'The impact of oil and gas drilling accidents on EU fisheries', Directorate-General for Internal Policies, Policy Department Structural and Cohesion Policies (Brussels: European Parliament, 2013).

Gomez-Acebo & Pombo Abogados SCP, 'Legal study for the accession of EURATOM to the Paris Convention on Third Party Liability in the Field of Nuclear Energy', Final Report, TREN/CC/01-2005 (Brussels: European Commission); available at: www.mng.org.uk/gh/private/2009_12_accession_euratom.pdf.

Greenspan, D. E. & Neuburger, M. A., 'Settle or sue? The use and structure of alternative compensation programs in the mass claims context', *Roger Williams University Law Review* 17 (2012), 97–136.

Gron, A. & Sykes, A., 'A role for government', *Regulation* 25 (2002), 44–51.

& Sykes, A., 'Terrorism and insurance markets: a role for the government as insurer?', *Indiana Law Review* 36 (2003), 447–63.

Handrlica, J., 'Harmonization of nuclear liability in the European Union: challenges, options and limits', *Nuclear Law Bulletin* 84 (2009), 37–60.

Harrington, S. E., 'Rethinking disaster policy', *Regulation – Decato Review of Business and Government* 23 (2000), 40–6.

Hazelwood, S. & Semark, D., *P &I Clubs: Law and Practice*, 4th edn. (London: Informa Law from Routledge, 2010).

Health and Safety Executive, 'Offshore hydrocarbon releases 2001–2008', RR672 Research Report (2008); available at: www.hse.gov.uk/research/rrpdf/rr672.pdf.

Heyes, A. & Liston-Heyes, C., 'Subsidy to nuclear power through Price-Anderson liability limit: comment', *Contemporary Economic Policy* 16 (1998), 122–4.

Hilton, K. N., 'When should we prefer tort law to environmental regulation?', *Washburn Law Journal* 41 (2002), 515–43.

Hinghofer-Szalskay, D. & Koch, B. A., 'Austria', in M. Faure and T. Hartlief (eds.), *Financial Compensation for Victims of Catastrophes: A Comparative Legal Approach* (Vienna: Springer, 2006), pp. 7–36.

HSE, 'Annual offshore statistics and regulatory activity report 2013–2014' (December 2014).

Hu, J. Z., 'Legal issues from the *Hebei Spirit* oil spill incident', in M. G. Faure, L. Han & H. Shan (eds.), *Marine Pollution Liability and Policy: China,*

Europe and the US (Alphen a/d Rijn: Kluwer Law International, 2010), pp. 399–412.

International Atomic Energy Agency, 'Status: Convention on Supplementary Compensation For Nuclear Damage 1 (21 May 2008); available at: www.iaea.org/publications/documents/conventions/supcomp_status.pdf.

International Oil Pollution Compensation Fund, 'Guidelines for presenting claims in the fisheries, mariculture and fish processing sector' (December 2008).

International Oil Pollution Compensation Fund, 'Claims Manual' (December 2013); available at: www.iopcfunds.org/uploads/tx_iopcpublications/claims_manual_e.pdf.

Izundu, U., 'Prospects slowly brighten for operations off Europe', *Oil and Gas Journal* 108 (2010), 49–59.

Jaffee, D. M., 'Report on the role of government in the coverage of terrorism risks', Presented at the Conference on Catastrophic Risks and Insurance, Paris. OECD, 22–23 November 2004.

Janssen, J. J. G., 'Preventing a next Deepwater Horizon? Evaluation of the regime for the prevention of and response to accidental oil spills', MSc Thesis, Utrecht University, 2012.

Jose, D. & Garza, M., 'The complete federal pre-emption of nuclear safety should prevent scientifically irrational jury verdicts in radiation litigation', *Temple Journal of Science, Technology and Environmental Law* 26 (2007), 1–25.

Jost, P. J., 'Limited liability and the requirement to purchase insurance', *International Review of Law and Economics* (1996), 259–76.

Kerr, R., Kintisch, E. & Stokstad, E., 'Will Deepwater Horizon set a new standard for catastrophe?', *Science* 328 (2010), 674–5.

Kiern, L. I., 'Liability, Compensation, and financial responsibility under the Oil Pollution Act of 1990: a review of the first decade', *Tulane Maritime Law Journal* 24 (2000), 481–590.

King, R., 'Deepwater Horizon oil spill disaster: risk, recovery, and insurance implications', Congressional Research Service Report for Congress (12 July 2010).

Klass, A. B. & Wilson, E. J., 'Climate change and carbon sequestration: assessing a liability regime for long-term storage of carbon dioxide', *Emory Law Journal* 58 (2008), 103–80.

Koivurova T., 'The future of the Nordic Environment Protection Convention', *Nordic Journal of International Law* 66 (1997), 505–26.

Kollewe, J., 'Oil industry set for surge in insurance premiums after Deepwater disaster', *The Guardian* (20 September 2010); available at: www.guardian.co.uk/business/2010/sep/deep%20water-oil-rigs-insurance-costs.

Kornhauser, L. A. & Revesz, R. L., 'Sharing damages upon multiple tortfeasors', *Yale Law Journal* 98 (1989), 831–84.

Revesz, R. L., 'Apportioning damages among potentially insolvent actors', *Journal of Legal Studies* 19 (1990), 617–51.

Koziol, H., 'Die Arzhaftung im geltenden und künftigen Recht', in H. Barta (ed.), *Haftungsrechtliche Perspektiven der ärzlichen Behandlung* (Trauner, 1997), pp. 21–35.

KPMG Global Energy Institute, 'After the Gulf of Mexico oil spill: recent developments in the oil and gas industry'; (2011); available at: www.kpmginstitutes.com /global-energy-institute/insights/2011/pdf/gei-gulf-of-mexico-oil-spillv2.pdf.

Kunreuther, H., 'Mitigating disaster losses through insurance', *Journal of Risk and Incertainty* 12 (1996), 171–87.

Freeman, P., 'Insurability, environmental risks and the law', in A. Heyes (ed.), *The Law and Economics of the Environment*, (Cheltenham, UK: Edward Elgar, 2001), pp. 304–16.

& Michel-Kerjan, E., 'Challenges for terrorism risk insurance in the United States', *Journal of Economic Perspectives* 18 (2004), 201–14.

& Michel-Kerjan, E., 'Insurability of (mega-)terrorism risk: challenges and perspectives', in *Terrorism Risk Insurance in OECD Countries* (Paris: OECD Publishing 2005), pp. 107–48.

& Pauly, M., 'Rules rather than discretion: lessons from Hurricane Katrina', *Journal of Risk and Incertainty* 33 (2006), 101–16.

Lamm, V., 'The protocol amending the 1963 Vienna Convention', in OECD Nuclear Energy Agency, International Atomic Energy Agency, *International Nuclear Law in the Post-Chernobyl Period* (Paris: OECD, 2006), pp. 169–88.

Landes, W. & Posner, R., 'The positive economic theory of tort law', *Georgia Law Review* (1981), 851–924.

& Posner, R., 'Tort law as a regulatory regime for catastrophic personal injuries', *Journal of Legal Studies* (1984), 417–34.

Lee, A. V., Bridgen, P. J. & Environment International, Ltd., *The Natural Resource Damage Assessment Deskbook: A Legal and Technical Analysis* (Washington, DC: Environmental Law Institute, 2002).

Letourneau, K. & Welmaker, W., 'The Oil Pollution Act of 1990: federal judicial interpretation through the end of the millennium', *University of San Francisco Maritime Law Journal* 12 (2000), 147–226.

Levmore, S. & Logue, K. D., 'Insuring against terrorism – and crime', *Michigan Law Review* 102 (2003), 268–327.

Libby, G. E., 'Some aspects of protection and indemnity insurance', *Marine and Inland Marine Insurance Law* (1952), 684–709.

Liu, J., *Compensating Ecological Damage: Comparative and Economic Observations* (Antwerp: Intersentia, 2013).

& Faure, M., 'Nuclear liability in China', *Washington Global Studies Law Review* (2013), 781–816.

Lloyd's, 'Emerging risks report:, behaviour: bear, bull or lemming?'; available at: www.lloyds.com/~/media/lloyds/reports/emerging%20risk%20reports/emergingriskreportbehaviour_march2010_v2.pdf.

'Drilling in extreme environments: challenges and implications for the energy insurance industry', (2011); available at: www.lloyds.com/~/media/lloyds/reports/emerging%20risk%20reports/lloyds%20drilling%20in%20extreme%20environments%20final3.pdf.

Energy Quarterly Newsletter (January 2015).

Mackaay, E., *Economics of Information and the Law* (Boston: Kluwer, 1982).

Magnus, U., 'Germany', in M. Faure and T. Hartlief (eds.), *Compensation for Victims of Catastrophes: A Comparative Legal Approach* (Vienna: Springer, 2006), pp. 119–44.

Marsh, 'A guide to oil insurance limited 2014'; available at: http://uk.marsh.com/Portals/18/documents/a%20guide%20to%20oil%20insurance%20limited%20(oil)%202014.pdf.

Mason, M., 'Civil liability for oil pollution damage: examining the evolving scope for environmental compensation in the international regime', *Marine Policy* 27 (2003), 3–5.

McClure, R., 'A review of nuclear energy insurance', in *Proceedings of the Casualty Actuarial Society*, Volume LV (Boston: Sperry Rand Corporation, 1968), pp. 255–94.

McDonell, C., 'The Gulf Coast claims facility and the Deepwater Horizon litigation: judicial regulation of private compensation schemes', *Stanford Law Review* 64 (2012), 765–95.

Meadow, W. & Sunstein, C., 'Statistics, Not Experts', *Duke Law Journal* 51 (2001).

Mez, L. & Piening, A., 'Phasing-out nuclear power generation in Germany: policies, actors, issues and non-issues', *Energy & Environment* 13 (2002), 161–81.

Miller, J. C., 'The FTC and voluntary standards: maximizing the net benefits of self-regulation', *The Cato Journal*, 4 (1985), 897–903.

Ministry of Petroleum and Energy (Norway), 'Facts: the Norwegian petroleum sector 2014' (2014).

Montara Commission of Inquiry, 'Report of the Montara Commission of Inquiry' (2010).

Monti, A., 'Environmental risk: a comparative law and economics approach to liability and insurance', *European Review of Private Law*, (2001), 51–79.

Monti, A., 'Public – private initiatives to cover extreme events' in C. Courbage and W.R. Stahel (eds.), *Extreme events and insurance: 2011 annus horribilis* (Geneva: the Geneva Association, 2012), 27–38.

Mortimer, S., 'Cat bond investors not scared off by Japan quake, *Insurance Journal*, (16 May 2011), available at: www.insurancejournal.com/news/international/2011/05/16/198576.htm.

National Commission on the BP Deepwater Horizon Oil Spill and Offshore Drilling, 'A brief history of offshore oil drilling', Staff Working Paper No. 1, (2010), available at: http://oscaction.org/resource-center/commission-reports-papers/.

'A competent and nimble regulator: a new approach to risk assessment and management', staff working paper no. 21, (2010); available at: http://oscaction.org/resource-center/staff-papers/.

'Final report' (January 2011).

National Post's Financial Post & FP Investing (Canada), 'Shell wants US to adopt tougher drilling standards used in North Sea' (15 October 2010).

NEIL (Nuclear Electric Insurance Limited), 'Annual Report 2014', Editorial Section (Wilmington/Dublin); available at: www.nmlneil.com/media/10101/neil-2014-ar-editorial-final2.pdf.

'Annual Report 2014', Financial Section (Wilmington/Dublin); available at: www.nmlneil.com/media/10102/neil-2014-ar-financials-final.pdf.

Norwegian Ministry of Trade and Industry, 'The state's ownership report 2013'; available at: www.regjeringen.no/globalassets/upload/nhd/statenseierberetning/pdf/engelsk/the_state_ownership_report_2013.pdf.

Nyman, F., 'Ferma opposes environmental liability fund' (30 November 2012); available at: www.postonline.co.uk/post/news/2312294/ferma-opposes-environmental-liability-fund.

O'Connor, R., 'Update: Lloyd's offshore energy market draws lessons from Deepwater Horizon' (16 March 2012); available at: http://insurancenewsnet.com/article.aspx?id=334831&type=newswire.

OCIL (Oil Casualty Insurance, Ltd.), 'Annual Report 2014'; available at: www.ocil.bm/images/pdfs/financials/ocil_annual%20report%202014.pdf.

Office of the Inspector of Transport Security, 'Offshore oil and gas resources sector security inquiry' (Department of Infrastructure and Transport, Australian Government, June 2012).

'Offshore oil and gas in the UK: an independent review of the regulatory regime', (December 2011); available at: www.gov.uk/government/uploads/system/uploads/attachment_data/file/48252/3875-offshore-oil-gas-uk-ind-rev.pdf.

OGP (International Association of Oil & Gas Producers), 'Risk assessment data directory', report no. 434–17 (March 2010).

Ogus, A. I., 'Rethinking self-regulation', *Oxford Journal of Legal Studies* 15 (1995), 97–108.

'Self-regulation' in B. Bouckaert and G. De Geest (eds.), *Encyclopedia of Law and Economics* (Cheltenham, UK: Edward Elgar, 2000), pp. 587–602.

Oil and Gas UK, 'Guidelines on relief well planning subsea wells' (London, January 2012).

'Guidelines to assist licensees in demonstrating financial responsibility to DECC for the consent of exploration and appraisal wells in the UKCS', Issue 1 (November 2012).

'Health and safety report 2014' (2014).
'Environment report 2014' (2014).
'Economic report 2014' (2014).
& OPOL, 'Oil spill cost study: OPOL financial limits' (February 2012); available at: http://oilandgasuk.co.uk/wp-content/uploads/2015/04/oil-spill-cost-study-120531.pdf.
Oil Insurance Limited, 'Annual Report 2014' (Hamilton); available at: http://eproxymaterials. com/interactive/oilinsurancelimited2014/.
'Shareholders' agreement' (Hamilton, March 2014); available at: www.oil.bm/images/stories/agreements/2014%20-%20march%20-%20shareholders%20agreement.pdf.
OPOL, 'The Offshore Pollution Liability Association limited report and financial statements for the year ended 31 December 2013' (London, 2013); available at: www.opol.org.uk/downloads/report%20and%20rinancial%20statements%20-%20december%202013.pdf.
OSPRAG (UK Oil Spill Prevention and Response Advisory Group), 'Strengthening UK prevention and response', Final report (September 2011).
Papavizas, C. G. & Kiern, L. I., '2009–2010 maritime legislative developments', *Journal of Maritime Law and Commerce* 41 (2011), 291–453.
Pelzer, N., 'International pooling of operators' funds: an option to increase the amount of financial security to cover nuclear liability', *Nuclear Law Bulletin* 79 (2007), 37–55.
Perry, R., 'The Deepwater Horizon oil spill and the limits of civil liability', *Washington Law Review* 86 (2011), 1–68.
Petroleum Safety Authority (Norway), 'Annual Report 2013' (2013).
'Safety status and signals 2012–2013' (2013).
'Trends in risk level in the petroleum activity: summary report 2014 – Norwegian continental shelf' (2014).
Philipsen, N. J., *Regulation of and by Pharmacists in the Netherlands and Belgium: An Economic Approach* (Antwerp: Intersentia, 2003).
Polborn, M., 'Mandatory insurance and the judgement-proof problem', *International Review of Law and Economics* (1998), 141–46.
Polinsky, A. M., *Introduction to Law and Economics* (Boston: Little, Brown, 1983).
Posner, R. A., *Economic Analysis of Law*, 6th edn. (New York: Aspen, 2003).
Catastrophe. Risk and Response (Oxford University Press, 2004).
Priest, G., 'The current insurance crisis and modern tort law', *Yale Law Journal* (1987), 1521–90.
'The government, the market and the problem of catastrophic loss', *Journal of Risk and Uncertainty* 12 (1996), 219–37.
Rachlinski, J. J., 'A positive psychological theory of judging in hindsight', *University of Chicago Law Review* 65 (1998), 571–625.

Radetzki, M. & Radetzki, M., 'Private arrangements to cover large-scale liabilities caused by nuclear and other industrial catastrophes', *Geneva Papers on Risk and Insurance* (2000) 180–95.

Ramseur, J., 'Liability and compensation issues raised by the 2010 Gulf oil spill', Congressional Research Service Report for Congress, R41679 (11 March 2011).

& Hagerty, C. L., 'Deepwater Horizon oil spill: recent activities and ongoing developments', Congressional Research Service Report for Congress (31 January 2013).

Ramseyer, M., 'Why power companies build nuclear reactors on fault lines: the case of Japan', *Theoretical Inquiries in Law* 31 (2011), 457–86.

Reader, Th. W., 'Is self-regulation the best option for the advertising industry in the European Union? An argument for the harmonization of advertising laws through the continued use of directors', *University of Pennsylvania Journal of International Business Law* 16 (1995), 181–215.

Rees, A. & Sharp, D., *Drilling in Extreme Environments: Challenges and Implications for the Energy Insurance Industry* (London: Lloyds, 2011).

Reyners, P., 'Liability problems associated with the current patchwork nuclear liability regime within the EU states', in N. Pelzer (ed.), *European Nuclear Liability Law in a Process of Change* (Baden Baden: Nomos 2009), pp. 93–104.

Rhodes, R. A. W., 'The new governance: governing without government', *Political Studies* XLIV (1996), 652–67.

Richardson, B. J., *Environmental Regulation Through Financial Organizations: Comparative Perspectives on the Industrialised Nations* (The Hague: Kluwer Law International, 2002).

Roberts, P. & Derrick, J., 'OGUK and AIPN joint operating agreements: differences on operator liability' (6 June 2011); available at: www.internationallawoffice.com/newsletters/detail.aspx?g=c3db9ed1-f9a0-47ed-9916-c1c38536e7f4.

Rochette, J., Wemaëre, M., Chabason, L. & Callet, S., 'Seeing beyond the horizon for deepwater oil and gas: strengthening the international regulation of offshore exploration and exploitation,' Studies No. 01/14 (Paris: IDDRI); available at: www.iddri.org/publications/collections/analyses/st0114_jr%20et%20al._offshore%20en.pdf.

Roman, A., Fernando, M. & Salsman, K., 'Canada and international nuclear liability'; available at: www.millerthomson.com/assets/files/article_attachments/canada_and_international_nuclear_liability.pdf.

Ronnerberg, N. J., 'An introduction to protection and indemnity clubs and the marine insurance they provide', *University of San Francisco Maritime Law Journal* (1990), 1–36.

Røsæg, Erik. 'The impact of insurance practices on liability conventions', in *Legislative Approaches in Maritime Law: Proceedings from the European*

Colloquium on Maritime Law: Lysebu, Olso (7–8 December 2000); available at: http://folk.uio.no/erikro/www/corrgr/insurance/lysebu.pdf.

Schmidt Etkin, D., 'Worldwide analysis of marine oil spill clean-up cost factors', presented at Arctic and Marine Oil Spill Program Technical Seminar (2000); available at: www.environmenttal-research.com/erc_papers/erc_paper_2.pdf.

Schmitt, M., 'Offshore energy: a reinsurance actuary's view', presented at CARe, Munich Re (6 and 7 May 2010); available at: www.casact.org/education/reinsure/2010/handouts/cs13-schmitt.pdf.

Schmitz, P. W., 'On the joint use of liability and safety regulation', *International Review of Law and Economics* 20 (2003), 371–82.

Schwartz, A., 'Statutory interpretation, capture and tort law: the regulatory compliance defense', *American Law and Economics Review* (2000), 1–57.

& Wilde, L., 'Intervening in markets on the basis of imperfect information: a legal and economic analysis', *University of Pittsburgh Law Review* (1979), 630–82.

Schwartz, J. A., 'Great expectations: where do we stand with the international nuclear liability conventions?', in N. Pelzer (ed.), *European Nuclear Liability Law in a Process of Change* (Baden Baden: Nomos, 2009), pp. 43–67.

Schwarze, R. & Wagner, G., 'In the aftermath of Dresden: new directions in German flood insurance', *Geneva Papers on Risk and Insurance* 29 (2004), 154–68.

Scott, C., Cafaggi, F. & Senden, L., 'The conceptual and constitutional challenge of transnational private regulation', *Journal of Law and Society* 38 (2011), 1–19.

Seevers, J. S., 'NOAA's new natural resource damage assessment scheme: it's not about collecting money', *Washington and Lee Law Review* (1996), 1513–70.

Shan, H., 'The era after the Bunker Convention: Is the gap in China's regime for compensating victims of vessel-source oil spills failed', in M. Faure, L. Han & H. Shan (eds.), *Marine Pollution Liability and Policy: China, Europe and the US* (Alphen a/d Rijn: Kluwer Law International, 2010), pp. 123–43.

Shavell, S., 'On moral hazard and insurance', *Quarterly Journal of Economics* (1979), 541–62.

'Strict liability versus negligence', *Journal of Legal Studies* (1980), 1–25.

'Liability for harm versus regulation of safety', *Journal of Legal Studies* (1984), 357–74.

'A model of the optimal use of liability and safety regulation', *Rand Journal of Economics* (1984), 271–80.

'Criminal law and the optimal use of non-monetary sanctions as a deterrent', *Colombia Law Review* (1985), 1232–62.

'The judgment-proof problem', *International Review of Law and Economics* (1986), 43–58.

Economic Analysis of Accident Law (Cambridge, MA: Harvard University Press, 1987).

Shover, N., 'Zelfregulering door ondernemingen; ontwikkeling, beoordeling en bange voorgevoelens', *Tijdschrift voor Criminologie* 50 (2008), 169–81.

Skogh, G., 'The transactions cost theory of insurance: contracting impediments and costs', *Journal of Risk and Insurance* (1989), 726–32.

'Risk-sharing institutions for unpredictable losses', *Journal of Theoretical and Institutional Economics* 55 (1999), 505–15.

'Mandatory insurance: transaction costs analysis of insurance', in B. Bouckaert & G. De Geest (eds.), *Encyclopedia of Law and Economics,* vol. II: *Civil Law and Economics* (Cheltenham, UK: Edward Elgar, 2000), pp. 521–37.

'A European nuclear accident pool', *Geneva Papers on Risk and Insurance – Issues and Practice* 33 (2008), 274–87.

Slovic, P., Fischoff, B. & Lichtenstein, S., 'Rating the risks', in P. Slovic (ed.), *The Perception of Risk* (London: Earthscan Publications, 2000).

Smith, B. D., *State Responsibility and the Marine Environment* (Oxford, UK: Clarendon Press, 1988).

Smith, R. E., Canelo, E. A. & Tidio, A., 'Reinventing, reinsurance using the capital markets', *Geneva Papers on Risk and Insurance* (1997), 26–37.

Somers, E., 'Marine pollution and the right of entry in maritime ports for ships in distress', in M. G. Faure, L. Han & H. Shan (eds.), *Marine Pollution Liability and Policy: China, Europe and the US* (Alphen a/d Rijn: Kluwer Law International, 2010), pp. 41–53.

State Supervision of Mines (the Netherlands), 'Staatstoezicht op de Mijnen. Strategie & programma 2012–2016'; available at: www.rijksoverheid.nl/documenten-en-publicaties/rap porten/2012/06/19/strategie-programma-2012-2016.html.

Stigler, G., 'The economics of information', *Journal of Political Economics* (1961), 213–25.

Sump, D. H., 'The Oil Pollution Act of 1990: a glance in the rearview mirror', *Tulane Law Review* (2011), 1101–20.

Tietenberg, T. H. 'Indivisable toxic torts: the economics of joint and several liability', *Land Economics* 65 (1989) 305–19.

Tousley, D., 'Abolishing the "extraordinary nuclear occurrence" threshold of the Price-Anderson Act', *University of Michigan Journal of Law Reform* (1981), 609–30.

Trebilcock, M. J., 'The social insurance-deterrence dilemma of modern North-American tort law: a Canadian perspective on the liability insurance crisis', *San Diego Law Review* 24 (1987), 929–1002.

& Winter, R. A., 'The economics of nuclear accident law', *International Review of Law and Economics* 17 (1997), 215–43.

Tuil, M. & Visscher, L. (eds.), *New Trends in Financing Civil Litigation in Europe: A Legal, Empirical and Economic Analysis* (Cheltenham, UK: Edward Elgar, 2010).

Tyler, T. R. & Thorisdottir, H., 'A psychological perspective on compensation for harm: examining the 9/11 Victim Compensation Fund', *DePaul Law Review* 53 (2003), 355–92.

Tyran, J. & Zweifel, P., 'Environmental risk internalization through capital markets (ERICAM): the case of nuclear power', *International Review of Law and Economics* (1993), 431–44.

United States Coast Guard, 'Preliminary regulatory impact analysis: financial responsibility for water pollution vessels (1993).

United States General Accounting Office (GAO), 'Report to congressional committees, nuclear regulation: a perspective on liability protection for nuclear accident', (1987); available at: http://archive.gao.gov/d28t5/133093.pdf.

Van den Bergh, R., 'De maatschappelijke wenselijkheid van gedragscodes vanuit rechtseconomisch perspectief', *WPNR* 6772 (2008), 792–8.

& Faure, M. G., 'Competition on the European market for liability insurance and efficient accident law', *Maastricht Journal of European and Comparative Law* 9 (2002), 279–306.

Vanden Borre, T., 'Channelling of liability: a few juridical and economic views on an inadequate legal construction', in N. L. J. T. Horbach (ed.), *Contemporary Developments in Nuclear Energy Law: Harmonizing Legislation in CEEC/NIS* (Alphen a/d Rijn: Kluwer Law International, 1999), 13–28.

'Shifts in governance in compensation for nuclear damage: 20 years after Chernobyl' in M. Faure, and A. Verheij (eds.), *Shifts in Compensation for Environmental Damage* (Vienna: Springer, 2007), pp. 261–311.

'Nuclear liability: an anachronism in EU energy policy?' in M. M. Roggenkamp & U. Hammer (eds.), *European Energy Law Report VII* (Antwerp: Intersentia, 2010), pp. 177–218.

Verheij, A., 'Shifts in governance: oil pollution' in M. Faure & A. Verheij (eds.), *Shifts in Compensation for Environmental Damage* (Vienna: Springer, 2007), pp. 73–102.

Verjans, E., 'Nieuwe wettelijke regeling voor de vergoeding van slachtoffers van grote technologische rampen', *Rechtskundig Weekblad* 27 (2012–2013), 1076–9.

Viscusi, K. & Zeckhauser, R. 'Deterring and compensating oil-spill catastrophes: the need for strict and two-tier liability', *Vanderbilt Law Review* 64 (2011), 1717–66.

Von Ungern-Sternberg, Th. 'The limits of competition: housing insurance in Switzerland', *European Economic Review* 40 (1996), 1111–21.

Efficient Monopolies: The Limits of Competition in the European Property Insurance Market (Oxford University Press, 2004).

Vorwerk, A., 'The 2002 Amendment to the German Atomic Energy Act concerning the phase-out of nuclear power', *Nuclear Law Bulletin* 69 (2002), 7–14.

Wagner, F., 'Risk securitization. An alternative risk transfer of insurance companies', *Geneva Papers on Risk and Insurance* (1998), 574–607.

Wagner, G., 'Umwelthaftung und Versicherung', *Versicherungsrecht* (1991), 249–60.

'Die Zukunft der Umwelthaftpflichtversicherung', *Versicherungsrecht* (1992), 261–72.

Wang, H., 'Shifts in governments in the international regime of marine oil pollution compensation: a legal history perspective', in M. Faure & A. Verheij (eds.), *Shifts in Compensation for Environmental Damage* (Vienna: Springer, 2007), pp. 197–241.

Civil Liability for Marine Oil Pollution Damage: A Comparative and Economic Study of the International, US and Chinese Compensation Regime (Alphen a/d Rijn: Kluwer Law International, 2011).

& Faure, M., 'Civil liability and compensation for marine pollution: lessons to be learned for offshore oil spills', *Oil, Gas, Energy Law Intelligence* 8 (2010), 1–27.

Wilde, M., *Civil Liability for Environmental Damage: Comparative Analysis of law and policy in Europe and the US*, 2nd edn. (Alphen a/d Rijn: Kluwer Law International, 2013).

Willis Natural Resources Market Review (April 2015); available at: www.willis.com/natural resources/pdf/naturalresourcesmarketreview2015.pdf.

Wu, C., *Pollution from the Carriage of Oil by Sea: Liability and Compensation* (Alphen a/d Rijn: Kluwer Law International, 1996).

Zech, J., 'Will the international capital market replace traditional insurance products?', *Geneva Papers on Risk and Insurance* (1998), 490–5.

Zeckhauser, R., 'The economics of catastrophes', *Journal of Risk and Uncertainty* 12 (1996), 113–40.

Zhu, L., 'International Convention on Civil Liability for Bunker Oil Pollution Damage, 2001: liability and insurance aspects', in J. Basedow & U. Magnus (eds.), *Pollution of the Sea: Prevention and Compensation* (Hamburg Studies on Maritime Affairs 10) (Berlin: Springer-Verlag 2007), pp. 171–80.

Appendix 1

Overview of Interviews with Stakeholders

1. 8 February 2013, Marsh, in Antwerp, Belgium. Karl Moens, Vice President Marsh Belgium, Claims Liability Manager Marine Practice.
2. 11 February 2013, Swiss Re. Bernard Tettamanti, Leader of Environmental Expert Group concerning ELD for Insurance Europe. Pierre Henri Francfort, expert on energy issues within Swiss Re
3. 19 February 2013, company lawyer of ExxonMobil, in Brussels, Belgium.
4. 22 February 2013, Staatstoezicht op de Mijnen (State Supervision of Mines), The Netherlands (in The Hague, Netherlands). Jan de Jong, Inspector-General.
5. 25 February 2013, International Association of Oil and Gas Producers (OGP), in Brussels, Belgium. Industry representatives from OGP: Rachel Bonfante (OGP), Interim Director of EU Affairs; Ben Dillon (Noble Energy); Barry Heagin (BG Group); Alexander Kerst (BP); Barry King (Conoco Phillips); Paul Lawson (BP); Steve Regulski (Nexen); Daniel Riesen (Shell); Marijke Schurmans (ExxonMobil); Arjiane Waayer (Shell).
6. 26 February 2013, RSA Insurance Group, London (Skype meeting). Phil Bell, Group Casualty Director.
7. 5 March 2013, Deep Water Horizon Fund (Gulf Coast Claims Facility [GCCF]), St. Louis, MO. Kent D. Syverud, independent trustee.
8. 6 March 2013, Noble Energy, Houston, TX (Skype meeting). Gary Maddock, Director of Risk and Insurance; Gerald Stevenson, Vice President and Treasurer.
9. 12 March 2013, Munich Re, in Munich, Germany. Ina Ebert; Christian Lahnstein; Hermann Kramer, Executive Manager Corporate Underwriting Casualty Munich Re; Rechtsanwältin Kristina Meyer-Rutkowski, Governmental Affairs Munich Re; Nicolas Rönneberg.
10. 13 March 2013, BP, in Brussels, Belgium. Alexander Kerst, Advisor European Government Affairs; Paul Lawson, BP Exploration Operating Company (Aberdeen); Patrick Daniel; Collin Wannell (Board Member of OPOL).
11. 14 March 2013, Shell, in Rotterdam, the Netherlands. Daniel Riesen; Gerald W. Kok, Corporate Treasury.

12. 20 March 2013, IMO. E-mail exchange with Richard Mason, accredited representative of the European Commission (EC) to the International Maritime Organization (IMO).
13. 27 March 2013, OPOL, Brussels, Belgium. Niall Scott, Managing Director of OPOL; Collin Wannell, Chairman of the Board of OPOL.
14. 29 April 2013, Department of Energy and Climate Change (DECC), UK. Wendy Kennedy, Head of Offshore Oil and Gas at the Department of Trade and Industry.
15. 1 May 2013, International Group of P&I Clubs, London. David Baker.
16. 1 May 2013, Oil and Gas UK, London. Paul Dymond, consultant; David Petrie, External Affairs Officer; John Rintoul, Environmental Issues Manager.
17. 1 May 2013, Lloyds, London. Lindsey Donnithorn, Government Policy and Affairs General Council's Division, Lloyds; Philip Sandle, Beazley Group, Head of Liability, Marine; Tim Taylor, Partner, Clyde & Co.; Neil Roberts, Senior Executive Underwriting Lloyds Market Association; James Walmsley, Senior Manager Government Policy and Affairs, Lloyds.
18. 6 May 2013, Munich Re-Insurance Company, Brussels, Belgium. Philipp Wassenberg, Munich Re-Insurance Company.
19. 4 June 2013, Study Centre for Nuclear Energy (SCK/CEN), Mol, Belgium. Ludo Veuchelen, company lawyer.
20. 5 June 2013, International Association of Drilling Contractors (IADC). Alan Spackman, Vice-President Offshore Division, Texas.
21. 16 June 2013, Gulf Coast Claims Facility (GCCF), interview in Haifa, Israel. Kenneth Feinberg, special administrator.

Appendix 2

Checklist for Country Studies

1 Basis of Liability

A first issue to be addressed is obviously whether the liability is based on either fault (negligence) or strict liability. A strict liability can potentially reach much further than a liability based on negligence. Moreover, there can be many grey zones between fault and strict liability (i.e. a tightened version of fault liability). This is applied in many countries in the case of liability for environmental damage and also corresponds to an international tendency towards introducing stricter liabilities for high-risk activities. In relation to this, the question how the behaviours of the victim and third parties are taken into account and other defences (such as *force majeure*) also will have to be addressed.

2 Relationship with Regulation

A second crucial issue determining the scope of liability is the relationship with a regulatory framework. This could be authorizations, licenses or permits or specific conditions contained in those authorizations or simply regulatory prescriptions. Currently, there is no international or EU regime dealing particularly with the liability arising from offshore oil and gas exploration activities, but a large amount of (detailed) regulations already exists on safety standards for the offshore industry. In this respect, the question arises as to whether compliance with a regulatory norm would lead to an exclusion of liability or, vice versa, whether a breach of regulation would lead to liability. For example, is insurance a precondition for issuing a license for an offshore facility? As the experience with the Environmental Liability Directive has shown, the relationship with regulation can have an important bearing on the scope of liability. If compliance with a regulatory standard excludes liability, this could seriously limit the scope of liability of the offshore industry, whereas this scope potentially would be much larger to the extent that liability also would remain possible when regulatory standards are complied with. Hence, attention will have to be paid to the question of how the regulation of the offshore industry fits into a consistent integrated system of liability.

3 Causation

In addition, the way in which the law deals with causation can determine to an important extent the scope of liability. Again, this had been made clear in relation to environmental liability as well as in cases of liability for pharmaceutical products. A major problem, especially with new technologies, is that in many cases there is uncertainty concerning the extent to which a particular activity caused or causes harm. The way in which the law deals with this issue of so-called causal uncertainty can have an important bearing on the scope of liability. If the law (as is the case in some EU Member States with respect to employers' liability for occupational diseases) shifts the burden of proving causation to operators or allows liability for exposure to risk in itself, this could potentially enlarge the scope of liability substantially because an industrial operator could in this case (if he is not able to show that his activity did not cause the harm) lead to situations where the operator is held liable also for damage that he may not have caused. Therefore, a careful determination of the criteria for causation and more particularly of the way the law deals with causal uncertainty is crucial to understanding the scope of liability.

4 Attribution of Liability

Related to the problem of causal uncertainty is how the law deals with situations where multiple parties may have contributed to the risk. One approach, in the case of multiple tortfeasors, is to adopt a joint and several liability (sometimes also referred to as 'solidarity') instead of an attribution according to each party's contribution to the risk. A joint and several liability could potentially (especially to the extent that the other tortfeasors are found insolvent) seriously expand the scope of liability of tortfeasors. The literature has indicated that the tort/insurance crisis which played a large role in the United States in the 1980s[1] may have been caused not so much by the mere shift from negligence to strict liability but rather by accompanying features of the strict liability regime, such as the way in which tort law dealt with causal uncertainty and the introduction of joint and several liability.[2]

A related important issue of attribution is the so-called channelling of liability. 'Channelling' means that liability is channelled to only one party, who then becomes fully liable for the damage. Such a channelling of liability, for example, was the solution introduced in the Nuclear Liability Convention as well as the conventions dealing with vessel-source marine oil pollution. In the first case, liability is channelled to the licensee of a nuclear power plant; in the second case, liability is channelled to the registered tanker owner. Again,

[1] See Priest, 'The Current Insurance Crisis', 1521–90.

[2] See in this respect especially Trebilcock, 'The Social Insurance-Deterrence', 929–1002.

channelling may have an important bearing on the scope of liability because it could facilitate the insurability of the liability. Moreover, it makes it clear ex ante to potential victims (but also to other stakeholders in the offshore industry) that only the party to whom liability is channelled will be held liable, thus excluding the liability of all others involved.

5 Damages and Remedies

Obviously, the scope of the damages awarded can have an important consequence on the scope of liability of operators in the offshore industry. The question thus will have to be addressed as to what heads of damages are to be compensated; a number of issues may have an important consequence for the scope of liability, such as the following:

- The highly debated issue of whether pure economic loss also should be compensated (which, again, could potentially enlarge the scope of liability of an operator). What is of particular concern here is calculation of losses suffered by fishermen and the tourism industry.
- Whether and how fatalities/personal injuries and property damage are compensated.
- How to calculate the damage to the environment itself or the ecological damage.

6 Amount of Compensation

The exploration and exploitation of offshore hydrocarbon resources often take place in a hostile environment and pose risks to humans and the environment.[3] Then the question arises as to whether the parties who carry out such activities should be held fully liable for the damage they cause (unlimited liability) or whether their liability should be limited to a certain amount (limitation of liability or financial caps).

7 Claims Settlement

Victims may suffer dramatically due to an offshore pollution incident. Their means of living may be threatened for those who completely depend on the sea area. Therefore, it is crucial for a liability system to envisage a rapid claims handling mechanism that would provide compensation to victims in a timely manner to help them survive the financial difficulty after the pollution incident.

[3] 'Offshore Oil and Gas in the UK', 3.

8 Compensation Mechanisms

It is of crucial importance not only to focus on liability but also to address compensation mechanisms. Indeed, if alternative compensation mechanisms (meaning other than liability) were available, this could reduce the need for victims to call on the tort system to seek compensation for their losses. Again, some argue that the fact that tort liability has apparently been much more popular in the United States than in Europe is to some extent due to the fact that social security may be covering less in the United States than in Europe, which thus would increase the need for victims to call on liability law as well as compensation of their primary needs.[4] Therefore, in addition to addressing merely the liability, when analyzing the various liability schemes addressed in the different work packages, attention also should be paid to the existence of other compensation mechanisms, such as (for example, but not limited to)

- The existence of compulsory liability insurance or other duties to seek financial coverage for liability,
- The existence of (mandatory or voluntary) first-party or direct insurance schemes,
- The availability of risk-sharing agreements or pooling schemes between operators,
- The use of ex ante guarantees and deposits (of a voluntary or mandatory nature), and
- The level of compensation awarded via social security mechanisms.

[4] See Koziol, 'Die Arzhaftung', 21–35.

Appendix 3

Oil Pollution Act Liability Limits, 2012

A: Incidents Exceeding Liability Limits by Vessel Type

Vessel Type: Tank Ship (Single Hull)

Project name	Incident year	Incident location	Gross tonnage	Total incident cost	Inflation factor	Total incident cost (2012 dollars)	Limits of liability	Fund exposure	Actual OSLTF costs incurred
T/V Julie N	1996	ME	18,500	$52,601,200	1.48	$77,850,000	$59,126,000	$18,723,000	$28,376,000
T/V Athos I	2004	NJ	37,900	$252,010,800	1.23	$309,973,000	$121,264,000	$188,709,000	$210,278,000
TOTAL	—	—	—	—	—	$387,823,000	$180,390,000	$207,432,000	$238,654,000

Vessel Type: Tank Barge (Single Hull)

T/B Vistabella	1991	PR	1,100	$6,105,500	1.71	$10,440,000	$6,408,000	$4,032,000	$4,782,000
T/B (Tampa Bay collision)	1993	FL	9,300	$68,900,000	1.61	$110,929,000	$29,638,000	$81,291,000	$2,397,000
T/B Morris J. Berman	1994	PR	5,400	$95,488,300	1.57	$149,917,000	$23,496,000	$126,421,000	$95,488,000
M/V Scandia & T/B North Cape	1996	RI	5,500	$49,000,000	1.48	$72,520,000	$23,496,000	$49,024,000	$9,046,000
T/B Buffalo #292	1996	TX	1,500	$21,517,000	1.48	$31,845,000	$6,408,000	$25,437,000	$16,810,000
T/B B No. 120	2003	MA	6,900	$61,237,500	1.26	$77,159,000	$23,496,000	$53,663,000	$1,753,000
T/B EMC 423	2005	IL	1,400	$12,778,500	1.19	$15,206,000	$6,408,000	$8,798,000	$5,839,000
TOTAL	—	—	—	—	—	$468,016,000	$119,350,000	$348,666,000	$136,115,000

Vessel Type: Tank Barge (Double Hull)

T/B DBL 152	2005	LA	9,700	$55,445,400	1.19	$65,980,000	$19,482,000	$46,498,000	$19,756,000
T/B DM932	2008	LA	800	$102,980,700	1.08	$111,219,000	$4,272,000	$106,947,000	$21,956,000
TOTAL						$177,199,000	$23,754,000	$153,445,000	$41,712,000

Vessel Type: Cargo/Other Self-Propelled Vessel

Project name	Incident year	Incident location	Gross tonnage	Total incident cost	Inflation factor	Total incident cost (2012 dollars)	Limits of liability	Fund exposure	Actual OSLTF costs incurred
M/V Kuroshima	1997	AK	4,200	$19,702,600	1.45	$28,569,000	$4,160,000	$24,409,000	$17,540,000
M/V Kure	1997	CA	36,000	$47,218,900	1.45	$68,467,000	$36,009,000	$32,458,000	$29,000
M/V New Carissa	1999	OR	36,600	$50,501,400	1.40	$70,702,000	$36,571,000	$34,131,000	$32,914,000
M/V Stuyvesant	1999	CA	7,100	$11,700,000	1.40	$16,380,000	$7,111,000	$9,269,000	$379,000
M/V Sergo Zakariadze	1999	PR	16,500	$15,966,700	1.40	$22,353,000	$16,502,000	$5,851,000	$6,065,000
SS J Luckenbach	2001	CA	7,900	$40,910,300	1.31	$53,593,000	$7,869,000	$45,724,000	$44,051,000
M/V Kimton	2001	PR	200	$713,700	1.31	$935,000	$854,000	$81,000	$714,000

Victoria Rose Hunt	2003	MA	100	$1,085,700	1.26	$1,368,000	$854,000	$514,000	$94,000
M/V Red Diamond	2003	FL	200	$2,595,200	1.26	$3,270,000	$854,000	$2,416,000	$2,595,000
Crane Barge Monarch	2003	CA	200	$2,481,700	1.26	$3,127,000	$854,000	$2,272,000	$2,482,000
M/V Bowstring	2003	FL	300	$1,606,500	1.26	$2,024,000	$854,000	$1,170,000	$1,606,000
M/V Selendang Ayu	2004	AK	39,800	$153,430,100	1.23	$188,719,000	$39,755,000	$148,964,000	$81,286,000
M/V Oriental I	2004	FL	200	$727,400	1.23	$895,000	$854,000	$40,000	$727,000
Albion	2005	CA	200	$1,207,100	1.19	$1,436,000	$854,000	$582,000	$1,207,000
M/V Casitas	2005	HI	300	$1,710,700	1.19	$2,036,000	$854,000	$1,181,000	$1,711,000
Mama Lere	2006	TX	400	$1,217,300	1.15	$1,400,000	$854,000	$545,000	$1,217,000
M/V Cosco Busan	2007	CA	65,100	$110,557,900	1.12	$123,825,000	$65,131,000	$58,694,000	$4,208,000
M/V Seneca	2007	MI	200	$1,211,000	1.12	$1,356,000	$854,000	$502,000	$1,211,000
LST-1166	2007	OR	2,400	$5,151,000	1.12	$5,769,000	$2,418,000	$3,351,000	$5,151,000
Catala	2007	WA	5,700	$6,138,500	1.12	$6,875,000	$5,700,000	$1,175,000	$6,138,000
C/V Sea Witch	2008	MD	17,900	$20,629,900	1.08	$22,280,000	$17,902,000	$4,378,000	$20,630,000
Big Boy & Scooby Doo	2008	PA	200	$1,010,800	1.08	$1,092,000	$854,000	$237,000	$1,011,000
Wenonah	2009	CA	300	$947,800	1.08	$1,024,000	$854,000	$169,000	$948,000
Sound Developer	2009	AK	200	$1,657,100	1.08	$1,790,000	$854,000	$935,000	$1,657,000
Monarch	2009	AK	300	$2,698,200	1.08	$2,914,000	$854,000	$2,060,000	$24,000

Vessel Type: Cargo/Other Self-Propelled Vessel (Cont.)

Project name	Incident year	Incident location	Gross tonnage	Total incident cost	Inflation factor	Total incident cost (2012 dollars)	Limits of liability	Fund exposure	Actual OSLTF costs incurred
M/V Princess Kathleen	2010	AK	5,900	$14,185,900	1.07	$15,179,000	$5,875,000	$9,304,000	$14,186,000
Davy Crockett	2011	WA	4,600	$22,454,700	1.03	$23,128,000	$4,643,000	$18,485,000	$22,455,000
Tug Tiger	2011	CA	200	$2,981,500	1.03	$3,071,000	$854,000	$2,217,000	$2,981,000
Jireh	2012	PR	1,000	$13,809,500	1.01	$13,948,000	$979,000	$12,969,000	$13,809,000
TOTAL	—	—	—	—	—	$687,525,000	$263,435,000	$424,083,000	$289,026,000

Vessel Type: Fishing Vessel

Project name	Incident year	Incident location	Gross tonnage	Total incident cost	Inflation factor	Total incident cost (2012 dollars)	Limits of liability	Fund exposure	Actual OSLTF costs incurred
F/V Tenyo Maru	1991	WA	4,200	$6,062,900	1.71	$10,368,000	$4,167,000	$6,201,000	$6,063,000
F/V Jin Shiang Fa	1993	AS	400	$2,013,000	1.61	$3,241,000	$854,000	$2,387,000	$2,420,000
F/V Yu Te No. 1	1999	AS	200	$1,164,600	1.40	$1,630,000	$854,000	$776,000	$5,296,000
F/V Amiga No. 5	1999	As	200	$3,355,700	1.40	$4,698,000	$854,000	$3,844,000	$2,766,000
F/V Kwang Myong	1999	AS	200	$1,554,800	1.40	$2,177,000	$854,000	$1,322,000	$965,000
F/V Koram No. 3	1999	AS	200	$1,403,100	1.40	$1,964,000	$854,000	$1,110,000	$813,000
F/V Kwang Myong No. 72	1999	As	200	$2,182,900	1.40	$3,056,000	$854,000	$2,202,000	$1,593,000

F/V Kwang Myong No. 58	1999	AS	200	$1,557,600	1.40	$2,181,000	$854,000	$1,326,000	$967,000
F/V Koram No. 1	1999	AS	200	$1,378,400	1.40	$1,930,000	$854,000	$1,075,000	$788,000
F/V Kwang Myong No. 51	1999	AS	200	$1,249,200	1.40	$1,749,000	$854,000	$894,000	$659,000
F/V Jessica Ann	2000	ME	200	$947,000	1.35	$1,278,000	$854,000	$424,000	$947,000
F/V Swordman I	2000	HI	100	$1,528,600	1.35	$2,064,000	$854,000	$1,209,000	$1,529,000
F/V Windy Bay	2001	AK	400	$3,396,400	1.31	$4,449,000	$854,000	$3,595,000	$3,396,000
F/V Vanguard	2001	AK	200	$699,800	1.31	$917,000	$854,000	$62,000	$700,000
F/V Genei Maru No. 7	2002	AK	100	$869,800	1.29	$1,122,000	$854,000	$268,000	$870,000
F/V Teresa Lynn	2002	FL	200	$690,800	1.29	$891,000	$854,000	$37,000	$691,000
F/V New Horizon	2004	CA	100	$805,300	1.23	$990,000	$854,000	$136,000	$305,000
F/V Mwalil Saat	2004	GU	200	$3,413,500	1.23	$4,199,000	$854,000	$3,344,000	$3,414,000
F/V The Boss	2004	OR	200	$926,100	1.23	$1,139,000	$854,000	$285,000	$926,000
F/V Milky Way	2005	WA	200	$1,039,600	1.19	$1,237,000	$854,000	$383,000	$539,000
F/V Capt Mike	2009	LA	100	$2,413,400	1.08	$2,606,000	$854,000	$1,752,000	$2,413,000
F/V Deep Sea	2012	WA	200	$2,231,000	1.01	$2,253,000	$854,000	$1,399,000	$2,231,000
TOTAL						$56,139,000	$22,101,000	$34,031,000	$40,291,000
GRAND TOTAL						$1,776,702,000	$609,045,000	$1,167,657,000	$745,799,000

Note: This listing includes all incidents regardless of vessel size or type and regardless of whether a claim to the Fund by a responsible party for amounts in excess of liability limits was received or is anticipated. Costs include federal removal costs and claims paid that have been verified. Other costs are estimated from best available information but cannot otherwise be verified. Fund exposure amounts are estimated and do not imply that the responsible parties will be able to limit their liability under the statute where the issue has not yet been determined.

Source: www.uscg.mil/npfc/docs/pdfs/reports/liability_limits_report_2012.pdf.

B: Incidents Exceeding Liability Limits by Incident Date

Incident Year: 1991

Vessel type	Project name	Incident location	Gross tonnage	Total incident cost	Inflation factor	Total incident cost (2012 dollars)	Limits of liability	Fund exposure	Actual OSLTF costs incurred
Fishing vessel	F/V Tenyo Maru	WA	4,200	$6,062,900	1.71	$10,368,000	$4,167,000	$6,201,000	$6,063,000
Tank barge (single hull)	T/B Vistabella	Pr	1,100	6,105,500	1.71	$10,440,000	$6,408,000	$4,032,000	$4,782,000
TOTAL						$20,808,000	$10,575,000	$10,233,000	$10,845,000

Incident Year: 1993

Fishing vessel	F/V Jin Shiang FA1991	AS	400	$2,013,000	1.61	$3,241,000	$854,000	$2,387,000	$2,420,000
Tank barge (single hull)	T/B (Tampa Bay Collision)	FL	9,300	$68,900,000	1.61	$110,929,000	$29,638,000	$81,291,000	$2,397,000
TOTAL	—	—	—	—	—	$114,170,000	$30,492,000	$83,678,000	$4,817,000

Incident Year: 1994

Tank barge (single hull)	T/B Morris J. Berman	PR	5,400	$95,488,300	1.57	$149,917,000	$23,496,000	$126,421,000	$95,488,000
TOTAL	—	—	—	—	—	$149,917,000	$23,496,000	$126,421,000	$95,488,000

Incident Year: 1996

Tank barge (single hull)	M/V Scandia & T/B North Cape	RIK	5,500	49,000,000	1.48	$72,520,000	$23,496,000	$49,024,000	$9,046,000
Tank barge (single hull)	T/B Buffalo No. 292	TX	1,500	$21,517,000	1.48	$31,845,000	$6,408,000	$25,437,000	$16,810,000
Tank barge (single hull)	T/V Julie N	ME	18,500	$52,601,200	1.48	$77,850,000	$59,126,000	$18,723,000	$28,376,000
TOTAL	—	—	—	—	—	$182,215,000	$89,030,000	$93,184,000	$54,232,000

Incident Year: 1999

Vessel type	Project name	Incident location	Gross tonnage	Total incident cost	Inflation factor	Total incident cost (2012 dollars)	Limits of liability	Fund exposure	Actual OSLTF costs incurred
Cargo/other SPVC	M/V New Carissa	OR	36,600	$50,501,400	1.40	$70,702,000	$36,571,000	$34,131,000	$32,914,000
Cargo/other SPVC	M/V Stuyvesant	CA	7,100	$11,700,000	1.40	$16,380,000	$7,111,000	$9,269,000	$379,000
Cargo/other SPVC	M/V Sergo Zakariadze	PR	16,500	$15,966,700	1.40	$22,353,000	$16,502,000	$5,851,000	$6,065,000
Fishing vessel	F/V Y Te No. 1	AS	200	$1,164,600	1.40	$1,630,000	$854,000	$776,000	$5,296,000
Fishing vessel	F/V Amiga No. 5	AS	200	$3,355,700	1.40	$4,698,000	$854,000	$3,844,000	$2,766,000
Fishing vessel	F/V Kwang Myong	AS	200	$1,554,800	1.40	$2,177,000	$854,000	$1,322,000	$965,000

cont.

Vessel type	Project name	Incident location	Gross tonnage	Total incident cost	Inflation factor	Total incident cost (2012 dollars)	Limits of liability	Fund exposure	Actual OSLTF costs incurred
Fishing vessel	F/V Koram No. 3	AS	200	$1,403,100	1.40	$1,964,000	$854,000	$1,110,000	$813,000
Fishing vessel	F/V Kwang Myong No. 72	AS	200	$2,182,900	1.40	$3,056,000	$854,000	$2,202,000	$1,593,000
Fishing vessel	F/V Kwang Myong No. 58	AS	200	$1,557,600	1.40	$2,181,000	$854,000	$1,326,000	$967,000
Fishing vessel	F/V Koram No. 1	AS	200	$1,378,400	1.40	$1,930,000	$854,000	$1,075,000	$788,000
Fishing vessel	F/V Kwang Myong No. 51	AS	200	$1,249,200	1.40	$1,749,000	$854,000	$894,000	$659,000
TOTAL	—	—	—	—	—	$128,820,000	$67,016,000	$61,800,000	$53,205,000

Incident Year: 2000

Fishing vessel	F/V Jessica Ann	ME	200	$947,000	1.35	$1,278,000	$854,000	$424,000	$947,000
Fishing vessel	F/V Swordman 1	HI	100	$1,528,600	1.35	$2,064,000	$854,000	$1,209,000	$1,529,000
TOTAL	—	—	—	—	—	$3,342,000	$1,708,000	$1,633,000	$2,476,000

Incident Year: 2001

Vessel type	Project name	Incident location	Gross tonnage	Total incident cost	Inflation factor	Total incident cost (2012 dollars)	Limits of liability	Fund exposure	Actual OSLTF costs incurred
Cargo/other SPV	SS J Luckenbach	CA	7,900	$40,910,300	1.31	$53,593,000	$7,869,000	$45,724,000	$44,051,000
Cargo/other SPV	M/V Kimton	PR	200	$713,700	1.31	$935,000	$854,000	$81,000	$714,000
Fishing vessel	F/V Windy Bay	AK	400	$3,396,400	1.31	$4,449,000	$854,000	$3,595,000	$3,396,000
Fishing vessel	F/V Vanguard	AK	200	$699,800	1.31	$917,000	$854,000	$62,000	$700,000
TOTAL	—	—	—	—	—	$59,894,000	$10,431,000	$49,462,000	$48,861,000

Incident Year: 2002

Fishing vessel	F/V Genei Maru No. 7	AK	100	$869,800	1.29	$1,122,000	$854,000	$268,000	$870,000
Fishing vessel	F/V Teresa Lynn	FL	200	$690,800	1.29	$891,000	$854,000	$37,000	$691,000
TOTAL	—	—	—	—	—	$2,013,000	$1,708,000	$305,000	$1,561,000

Incident Year: 2003

Cargo/other SPV	Victoria Rose Hunt	MA	100	$1,085,700	1.26	$1,368,000	$854,000	$514,000	$94,000
Cargo/other SPV	M/V Red Diamond	FL	200	$2,595,200	1.26	$3,270,000	$854,000	$2,416,000	$2,595,000
Cargo/other SPV	Crane Barge Monarch	CA	200	$2,481,700	1.26	$3,127,000	$854,000	$2,272,000	$2,482,000
Cargo/other SPV	M/V Bowstring	FL	300	$1,606,500	1.26	$2,024,000	$854,000	$1,170,000	$1,606,000
Tank barge (single hull)	T/B B No. 120	MA	6,900	$61,237,500	1.26	$77,159,000	$23,496,000	$53,663,000	$1,753,000
TOTAL	—	—	—	—	—	$86,948,000	$26,912,000	$60,035,000	$8,530,000

Incident Year: 2004

Vessel type	Project name	Incident location	Gross tonnage	Total incident cost	Inflation factor	Total incident cost (2012 dollars)	Limits of liability	Fund exposure	Actual OSLTF costs incurred
Fishing vessel	F/V New Horizon	CA	100	$805,300	1.23	$990,000	$854,000	$136,000	$305,000
Cargo/other SPV	M/V Selendang Ayu	AK	39,800	$153,430,100	1.23	$188,719,000	$39,755,000	$148,964,000	$81,286,000
Fishing vessel	F/V Mwalil Saat	GU	200	$3,413,500	1.23	$4,199,000	$854,000	$3,344,000	$3,414,000
Fishing vessel	F/V The Boss	OR	200	$926,100	1.23	$1,139,000	$854,000	$285,000	$926,000
Tanks ship (single hull)	T/V Ahtos 1	NJ	37,900	$252,010,800	1.23	$309,973,000	$121,264,000	$188,709,000	$210,278,000
Cargo/other SPV	M/V Oriental 1	FL	200	$727,400	1.23	$895,000	$854,000	$40,000	$727,000
TOTAL	—	—	—	—	—	$505,915,000	$164,435,000	$341,478,000	$296,936,000

Incident Year: 2005

Tank barge (double hull)	T/B DBL 152	LA	9,700	$55,445,400	1.19	$65,980,000	$19,482,000	$46,498,000	$19,756,000
Cargo/other SPV	Albion	CA	200	$1,207,100	1.19	$1,436,000	$854,000	$582,000	$1,207,000
Cargo/other SPV	M/V Casitas	HI	300	$1,710,700	1.19	$2,036,000	$854,000	$1,181,000	$1,711,000
Tank barge (single hull)	T/B EMC 423	IL	1,400	$12,778,500	1.19	$15,206,000	$6,408,000	$8,798,000	$5,839,000
Fishing vessel	F/V Milky Way	WA	200	$1,039,600	1.19	$1,237,000	$854,000	$383,000	$539,000
TOTAL	—	—	—	—	—	$85,895,000	$28,452,000	$57,442,000	$29,052,000

Incident Year: 2006

Cargo/other SPV	Mama Lere	TX	400	$1,217,300	1.15	$1,400,000	$854,000	$545,000	$1,217,000
TOTAL	—	—	—	—	—	$1,400,000	$854,000	$545,000	$1,217,000

Incident Year: 2007

Vessel type	Project name	Incident location	Gross tonnage	Total incident cost	Inflation factor	Total incident cost (2012 dollars)	Limits of liability	Fund exposure	Actual OSLTF costs incurred
Cargo/other SPV	M/V Cosco Busan	CA	65,100	$110,557,900	1.12	$123,825,000	$65,131,000	$58,694,000	$4,208,000
Cargo/other SPV	M/V Seneca	MI	200	$1,211,000	1.12	$1,356,000	$854,000	$502,000	$1,211,000
Cargo/other SPV	LST-1166	OR	2,400	$5,151,000	1.12	$5,769,000	$2,418,000	$3,351,000	$5,151,000
Cargo/other SPV	Catala	WA	5,700	$6,138,500	1.12	$6,875,000	$5,700,000	$1,175,000	$6,138,000
TOTAL	—	—	—	—	—	$137,825,000	$74,103,000	$63,722,000	$16,708,000

Incident Year: 2008

Tank barge (double hull)	T/B DM932	LA	800	$102,980,700	1.08	$111,219,000	$4,272,000	$106,947,000	$21,956,000
Cargo/other SPV	C/V Sea Witch	MD	17,900	$20,629,900	1.08	$22,280,000	$17,902,000	$4,378,000	$20,630,000
Cargo/other SPV	Big Boy & Scooby Doo	PA	200	$11,010,800	1.08	$1,092,000	$854,000	$237,000	$1,011,000
TOTAL	—	—	—	—	—	$134,591,000	$23,028,000	$111,562,000	$43,597,000

Incident Year: 2009

Fishing vessel	Capt Mike	LA	100	$2,413,400	1.08	$2,606,000	$854,000	$1,752,000	$2,413,000
Cargo/other SPV	Wenonah	CA	300	$947,800	1.08	$1,024,000	$854,000	$169,000	$948,000
Cargo/other SPV	Sound Developer	AK	200	$1,657,100	1.08	$1,790,000	$854,000	$935,000	$1,657,000
Cargo/other SPV	Monarch	AK	300	$2,698,200	1.08	$2,914,000	$854,000	$2,060,000	$24,000
TOTAL	—	—	—	—	—	$8,334,000	$3,416,000	$4,916,000	$5,042,000

Incident Year: 2010

Cargo/other SPV	M/V Princess Kathleen	AK	5,900	$14,185,000	1.07	$15,179,000	$5,875,000	$9,304,000	$14,186,000
TOTAL	—	—	—	—	—	$15,179,000	$5,875,000	$9,304,000	$14,186,000

Incident Year: 2011

Vessel type	Project name	Incident location	Gross tonnage	Total incident cost	Inflation factor	Total incident cost (2012 dollars)	Limits of liability	Fund exposure	Actual OSLTF costs incurred
Cargo/other SPV	Davy Crockett	WA	4,600	$22,454,700	1.03	$23,128,000	$4,643,000	$18,485,000	$22,455,000
Cargo/other SPV	Tug Tiger	CA	200	$2,981,500	1.03	$3,071,000	$854,000	$2,217,000	$2,981,000
TOTAL	—	—	—	—	—	$26,199,000	$5,497,000	$20,702,000	$25,436,000

Incident Year: 2012

Cargo/other SPV	Jireh	PR	1,000	$13,809,500	1.01	$13,948,000	$979,000	$12,969,000	$13,809,000
Fishing vessel	Deep Sea	WA	200	$62,231,000	1.01	$2,253,000	$854,000	$1,399,000	$2,231,000
TOTAL	—	—	—	—	—	$16,201,000	$1,833,000	$14,368,000	$16,040,000
TOTAL 1991–2000	—	—	—	—	—	$696,307,000	$262,491,000	$433,816,000	$238,633,000
TOTAL 2001–2013	—	—	—	—	—	$1,080,395,000	$346,554,000	$733,841,000	$507,166,000
GRAND TOTAL	—	—	—	—	—	$1,776,702,000	$1,184,323,000	$2,183,044,000	$1,362,190,000

SPV, self-propelled vessel.

Note: This listing includes all incidents regardless of vessel size or type and regardless of whether a claim to the Fund by a responsible party for amounts in excess of liability limits was received or is anticipated. Costs include federal removal costs and claims paid that have been verified. Other costs are estimated from best available information but cannot otherwise be verified. Fund exposure amounts are estimated and do not imply that the responsible parties will be able to limit their liability under the statute where the issue has not yet been determined.

C: Incidents Exceeding Liability Limits with Limits to Achieve 50 Per Cent Cost Share

Vessel Type: Tank Ship (Single Hull)

Project name	Incident year	Incident location	Gross tonnage	Total incident cost	Inflation factor	Total incident cost (2012 dollars)	Limits of liability	Fund exposure	Actual OSLTF costs incurred	Gross ton liability limits for a 50% cost share	Minimum liability for a 50% cost share
T/V Julie N	1996	ME	18,500	$52,601,200	1.48	$77,850,000	$59,126,000	$18,723,000	$28,376,000	$62,900,000	$97,000,000
T/V Athos I	2004	NJ	37,900	$252,010,800	1.23	$309,973,000	$121,264,000	$188,709,000	$210,278,000	$128,860,000	$97,000,000
TOTAL	—	—	—	—	—	$387,823,000	$180,390,000	$207,432,000	$238,654,000		

Vessel Type: Tank Barge (Single Hull)

T/B Vistabella	1991	PR	1,100	$6,105,500	1.71	$10,440,000	$6,408,000	$4,032,000	$4,782,000	$8,250,000	$33,400,000
T/B (Tampa Bay Collision)	1993	FL	9,300	$68,900,000	1.61	$110,929,000	$29,638,000	$81,291,000	$2,397,000	$69,750,000	$33,400,000
T/B Morris J. Berman	1994	PR	5,400	$95,488,300	1.57	$149,917,000	$23,496,000	$126,421,000	$95,488,000	$40,500,000	$33,400,000
M/V Scandia & T/B North Cape	1996	RI	5,500	$49,000,000	1.48	$72,520,000	$23,496,000	$49,024,000	$9,046,000	$41,250,000	$33,400,000
T/B Buffalo No. 292	1996	TX	1,500	$21,517,000	1.48	$31,845,000	$6,408,000	$25,437,000	$16,810,000	$11,250,000	$33,400,000
T/B B No. 120	2003	MA	6,900	$61,237,500	1.26	$77,159,000	$23,496,000	$53,663,000	$1,753,000	$51,750,000	$33,400,000
T/B EMC 423	2005	IL	1,400	$12,778,500	1.19	$15,206,000	$6,408,000	$8,798,000	$5,839,000	$10,500,000	$33,400,000
TOTAL	—	—	—	—	—	$468,016,000	$119,350,000	$348,666,000	$136,115,000	—	—

Vessel Type: Tank Barge (Double Hull)

T/B DBL 152	2005	LA	9,700	$55,445,400	1.19	$65,980,000	$19,482,000	$46,498,000	$19,756,000	$81,480,000	$44,300,000
T/B DM932	2008	LA	800	$102,980,700	1.08	$111,219,000	$4,272,000	$106,947,000	$21,956,000	$6,720,000	$44,300,000
TOTAL	—	—	—	—	—	$177,199,000	$23,754,000	$153,445,000	$41,712,000	—	—

Vessel Type: NTV > 300 GT

Project name	Incident year	Incident location	Gross tonnage	Total incident cost	Inflation factor	Total incident cost (2012 dollars)	Limits of liability	Fund exposure	Actual OSLTF costs incurred	Gross ton liability limits for a 50% cost share	Minimum liability for a 50% cost share
F/V Tenyo Maru	1991	WA	4,200	$6,062,900	1.71	$10,368,000	$4,167,000	$6,201,000	$6,063,000	$5,460,000	$18,900,000
F/V Jin Shiang Fa	1993	AS	400	$2,013,000	1.61	$3,241,000	$854,000	$2,387,000	$2,420,000	$520,000	$18,900,000
F/V Windy Bay	2001	AK	400	$3,396,400	1.31	$4,449,000	$854,000	$3,595,000	$3,396,000	$520,000	$18,900,000
M/V Kuroshima	1997	AK	4,200	$19,702,600	1.45	$28,569,000	$4,160,000	$24,409,000	$17,540,000	$5,460,000	$18,900,000
M/V Kure	1997	CA	36,000	$47,218,900	1.45	$68,467,000	$36,009,000	$32,458,000	$29,000	$46,800,000	$18,900,000
M/V New Carissa	1999	OR	36,600	$50,501,400	1.40	$70,702,000	$36,571,000	$34,131,000	$32,914,000	$47,580,000	$18,900,000
M/V Stuyvesant	1999	CA	7,100	$11,700,000	1.40	$16,380,000	$7,111,000	$9,269,000	$379,000	$9,230,000	$18,900,000
M/V Sergo Zakariadze	1999	PR	16,500	$15,966,700	1.40	$22,353,000	$16,502,000	$5,851,000	$6,065,000	$21,450,000	$18,900,000
SS J Luckenbach	2001	CA	7,900	$40,910,300	1.31	$53,593,000	$7,869,000	$45,724,000	$44,051,000	$10,270,000	$18,900,000
M/V Selendang Ayu	2004	AK	39,800	$153,430,100	1.23	$188,719,000	$39,755,000	$148,964,000	$81,286,000	$51,740,000	$18,900,000
Mama Lere	2006	TX	400	$1,217,300	1.15	$1,400,000	$854,000	$545,000	$1,217,000	$520,000	$18,900,000
M/V Cosco Busan	2007	CA	65,100	$110,557,900	1.12	$123,825,000	$65,131,000	$58,694,000	$4,208,000	$84,630,000	$18,900,000
LST-1166	2007	OR	2,400	$5,151,000	1.12	$5,769,000	$2,418,000	$3,351,000	$5,151,000	$3,120,000	$18,900,000
Catala	2007	WA	5,700	$6,138,500	1.12	$6,875,000	$5,700,000	$1,175,000	$6,138,000	$7,410,000	$18,900,000

Vessel Type: NTV > 300 GT (Cont.)

Project name	Incident year	Incident location	Gross tonnage	Total incident cost	Inflation factor	Total incident cost (2012 dollars)	Limits of liability	Fund exposure	Actual OSLTF costs incurred	Gross ton liability limits for a 50% cost share	Minimum liability for a 50% cost share
C/V Sea Witch	2008	MD	17,900	$20,629,900	1.08	$22,280,000	$17,902,000	$4,378,000	$20,630,000	$23,270,000	$18,900,000
M/V Princess Kathleen	2010	AK	5,900	$14,185,900	1.07	$15,179,000	$5,875,000	$9,304,000	$14,186,000	$7,670,000	$18,900,000
Davy Crockett	2011	WA	4,600	$22,454,700	1.03	$23,128,000	$4,643,000	$18,485,000	$22,455,000	$5,980,000	$18,900,000
Jireh	2012	PR	1,000	$13,809,500	1.01	$13,948,000	$979,000	$12,969,000	$13,809,000	$1,300,000	$18,900,000
TOTAL	—	—	—	—	—	$679,245,000	$257,354,000	$421,890,000	$281,937,000	—	—

Vessel Type: NTV ≤ 300 GT

Project name	Incident year	Incident location	Gross tonnage	Total incident cost	Inflation factor	Total incident cost (2012 dollars)	Limits of liability	Fund exposure	Actual OSLTF costs incurred	Gross ton liability limits for a 50% cost share	Minimum liability for a 50% cost share
F/V Yu Te No. 1	1999	AS	200	$1,164,600	1.40	$1,630,000	$854,000	$776,000	$5,296,000	$1,000,000	$976,000
F/V Amiga No. 5	1999	As	200	$3,355,700	1.40	$4,698,000	$854,000	$3,844,000	$2,766,000	$1,000,000	$976,000
F/V Kwang Myong	1999	AS	200	$1,554,800	1.40	$2,177,000	$854,000	$1,322,000	$965,000	$1,000,000	$976,000
F/V Koram No. 3	1999	AS	200	$1,403,100	1.40	$1,964,000	$854,000	$1,110,000	$813,000	$1,000,000	$976,000
F/V Kwang Myong No. 72	1999	As	200	$2,182,900	1.40	$3,056,000	$854,000	$2,202,000	$1,593,000	$1,000,000	$976,000

F/V Kwang Myong No. 58	1999	AS	200	$1,557,600	1.40	$2,181,000	$854,000	$1,326,000	$967,000	$1,000,000	$976,000
F/V Koram No. 1	1999	AS	200	$1,378,400	1.40	$1,930,000	$854,000	$1,075,000	$788,000	$1,000,000	$976,000
F/V Kwang Myong No. 51	1999	AS	200	$1,249,200	1.40	$1,749,000	$854,000	$894,000	$659,000	$1,000,000	$976,000
F/V Jessica Ann	2000	ME	200	$947,000	1.35	$1,278,000	$854,000	$424,000	$947,000	$1,000,000	$976,000
F/V Swordman I	2000	HI	100	$1,528,600	1.35	$2,064,000	$854,000	$1,209,000	$1,529,000	$500,000	$976,000
F/V Vanguard	2001	AK	200	$699,800	1.31	$917,000	$854,000	$62,000	$700,000	$1,000,000	$976,000
F/V Genei Maru No. 7	2002	AK	100	$869,800	1.29	$1,122,000	$854,000	$268,000	$870,000	$500,000	$976,000
F/V Teresa Lynn	2002	FL	200	$690,800	1.29	$891,000	$854,000	$37,000	$691,000	$1,000,000	$976,000
F/V New Horizon	2004	CA	100	$805,300	1.23	$990,000	$854,000	$136,000	$305,000	$500,000	$976,000
F/V Mwalil Saat	2004	GU	200	$3,413,500	1.23	$4,199,000	$854,000	$3,344,000	$3,414,000	$1,000,000	$976,000
F/V The Boss	2004	OR	200	$926,100	1.23	$1,139,000	$854,000	$285,000	$926,000	$1,000,000	$976,000
F/V Milky Way	2005	WA	200	$1,039,600	1.19	$1,237,000	$854,000	$383,000	$539,000	$1,000,000	$976,000
F/V Capt Mike	2009	LA	100	$2,413,400	1.08	$2,606,000	$854,000	$1,752,000	$2,413,000	$500,000	$976,000
F/V Deep Sea	2012	WA	200	$2,231,000	1.01	$2,253,000	$854,000	$1,399,000	$2,231,000	$1,000,000	$976,000
M/V Kimton	2001	PR	200	$713,700	1.31	$935,000	$854,000	$81,000	$714,000	$1,000,000	$976,000
Victoria Rose Hunt	2003	MA	100	$1,085,700	1.26	$1,368,000	$854,000	$514,000	$94,000	$500,000	$976,000
M/V Red Diamond	2003	FL	200	$2,595,200	1.26	$3,270,000	$854,000	$2,416,000	$2,595,000	$1,000,000	$976,000

Vessel Type: NTV ≤ 300 GT (Cont.)

Project name	Incident year	Incident location	Gross tonnage	Total incident cost	Inflation factor	Total incident cost (2012 dollars)	Limits of liability	Fund exposure	Actual OSLTF costs incurred	Gross ton liability limits for a 50% cost share	Minimum liability for a 50% cost share
Crane Barge Monarch	2003	CA	200	$2,481,700	1.26	$3,127,000	$854,000	$2,272,000	$2,482,000	$1,000,000	$976,000
M/V Bowstring	2003	FL	300	$1,606,500	1.26	$2,024,000	$854,000	$1,170,000	$1,606,000	$1,500,000	$976,000
M/V Oriental I	2004	FL	200	$727,400	1.23	$895,000	$854,000	$40,000	$727,000	$1,000,000	$976,000
Albion	2005	CA	200	$1,207,100	1.19	$1,436,000	$854,000	$582,000	$1,207,000	$1,000,000	$976,000
M/V Casitas	2005	HI	300	$1,710,700	1.19	$2,036,000	$854,000	$1,181,000	$1,711,000	$1,500,000	$976,000
M/V Seneca	2007	MI	200	$1,211,000	1.12	$1,356,000	$854,000	$502,000	$1,211,000	$1,000,000	$976,000
Big Boy & Scooby Doo	2008	PA	200	$1,010,800	1.08	$1,092,000	$854,000	$237,000	$1,011,000	$1,000,000	$976,000
Wenonah	2009	CA	300	$947,800	1.08	$1,024,000	$854,000	$169,000	$948,000	$1,500,000	$976,000
Sound Developer	2009	AK	200	$1,657,100	1.08	$1,790,000	$854,000	$935,000	$1,657,000	$1,000,000	$976,000
Monarch	2009	AK	300	$2,698,200	1.08	$2,914,000	$854,000	$2,060,000	$24,000	$1,500,000	$976,000
Tug Tiger	2011	CA	200	$2,981,500	1.03	$3,071,000	$854,000	$2,217,000	$2,981,000	$1,000,000	$976,000
TOTAL	—	—	—	—	—	$64,419,000	$28,182,000	$36,224,000	$47,380,000	—	—
GRAND TOTAL						$1,776,702,000	$609,045,000	$1,167,657,000	$745,799,000		

NTV, non-tank vessel.

Note: This listing includes all incidents regardless of vessel size or type and regardless of whether a claim to the Fund by a responsible party for amounts in excess of liability limits was received or is anticipated. Costs include federal removal costs and claims paid that have been verified. Other costs are estimated from best available information but cannot otherwise be verified. Fund exposure amounts are estimated and do not imply that the responsible parties will be able to limit their liability under the statute where the issue has not yet been determined.

INDEX

A. M. Best, 110, 242–3
accident law, 304–5
actual cash value (ACV), 211–12
actuarial information, 200
ACV. *See* actual cash value
AIG, 239
AIPN. *See* Association of International Petroleum Negotiators'
Alexander Kielland, 91–2, 116
Amendment Act 1988, 185–6
American Nuclear Insurers (ANI), 217
  coverage under, 314–15
American Petroleum Institute, 135
Amoco, 201–2
Amoco Cadiz, 1, 71
AMSA. *See* Australian Maritime Safety Authority
Anadarko Petroleum Corporation, 50
ANI. *See* American Nuclear Insurers
ANP. *See* Brazilian National Petroleum Agency
Applicability of Limit of Liability, 51
Articles of Association, OPOL, 203
artificial islands, 12
Ashmore Reef National Nature Reserve, 48
Association of International Petroleum Negotiators' (AIPN)
  International Model Well Services Agreement, 14
  Model International Operating Agreement, 14
Athens Convention, 260–1
Atomic Energy Act 1959, 223–7
  amendments to, 225
  Clause 2, 226
  Clause I(3), 226
  Clause I(8), 226
  goals of, 224
  Solidarity Agreement in, 225
Atomic Energy Commission, 183–4
Atomic Industrial Forum, 171–2
Australia
  basis of liability, 158–60
  energy production, 154
  legal framework, 155–8
  legislation comparisons, 167
  national interest in, 154–5
  offshore liability regimes, 158–60
  well blowouts, 155
Australian Maritime Safety Authority (AMSA), 46, 157, 162
Australian Petroleum Production and Exploration Associations, 155
Austria, 228

Baltic Sea, 19
  offshore liability regimes, 90
Barcelona Convention, 88–90, 94
  Art. 7, 89
  Offshore Protocol, 89–90
  participants in, 89
Basel I, 249–50
Basel II, 249–50
Belgian Motor Insurance Guarantee Fund, 373
Belgium, 376
Bell, Phil, 243
Bergen, 60
Best Available Techniques, 127–8
best practice, 166–7
Bios report, 355–6
Black Sea, 19, 66, 272–3
  exploration in, 19

blue ray, 288–9
BMT ARGOSS, 114
BOEMRE. *See* Bureau of Ocean Energy Management, Regulation and Enforcement
Booz Allen Hamilton, 248–9
BP, 49–50, 201–2, 365
  captive of, 239
  hold-harmless clauses and, 258–9
  Munich Reinsurance and, 283–4
  self-insurance, 239
Brazilian National Petroleum Agency (ANP), 15–16
Brussels Convention, 173–4
  coverage caps, 181–2
  modification protocol, 287–9
  public funding in, 177–8, 180–1
  Supplementary, 192
BSEE. *See* Bureau of Safety and Environmental Enforcement
Bucharest Convention, 90
Bulgaria, 19, 90, 299–300
Bunker Convention, 84–5, 251–2
Bureau of Ocean Energy Management, Regulation and Enforcement (BOEMRE), 56–7, 135, 146–7, 255–6
Bureau of Safety and Environmental Enforcement (BSEE), 15–16, 135
Burmach Oil, 201–2
business interruption insurance, 251

Cameron International, 50, 56
Cameron Iron Works, 52
Canada, 164–5
  attribution of liability, 164
  basis of liability, 164
  best practice, 166–7
  legal framework, 163–4
  legislation comparisons, 167
  national interest in, 163
Canada Oil and Gas Operations Act, 163–4
  applicability in time, 165–6
  attribution of liability, 164
  basis of liability, 164
  best practice, 166–7
  comparative analysis, 166–7
  compensation, 165
  compensation mechanisms, 166
  Section 72.1, 166
capital markets, 265
  analysis, 244
  practice, 243–4
  reinsurance and, 243
  stakeholders on, 243–4
  theory, 242–3
capping devices, 113, 285
captives
  of BP, 239
  costs of, 237–8
  creation of, 237–8
  functioning of, 239
carbon dioxide, 241–2
cargo, 414–16
Cartier Island Marine Reserve, 48
casualty insurance, 51–2
cat bonds, 242–3
  in Fukushima, 243–4
  stakeholders on, 243
catastrophe insurance, 278–9
Catastrophe Reserve Fund, 216
causation, 409
  in Norwegian offshore activities, 120–1
  offshore liability regimes and, 170
  under Offshore Petroleum and Greenhouse Gas Storage Act, 160
  in Oil Pollution Act of 1990, 137–8
  in OPOL, 104
CFR. *See* Code of Federal Regulations
CGAA. *See* Coast Guard Authorization Act
CGMTA. *See* Coast Guard and Maritime Transportation Act
Chartis, 248–9
Chernobyl, 178–9
Chevron, 56
Civil Liability Convention (CLC), 70, 234
  1969, 71
  1971, 71
  1992, 74–5
  analysis of, 70–1

Article VII.11, 77
Claims Manual, 375
compensation in, 75–6, 286
compulsory financial guarantees, 76–7
evolution of, 71–3
historical background, 71–3
liability in, 71, 74–5
rapid claims management and, 361–2
on transboundary harm, 377
vessel-based pollution and, 233–4
claims settlement, 410. *See also* rapid claims management
in Deepwater Horizon, 364–6
IOPC, 361–2
offshore liability regimes, 170–1
OPOL, 205–6
CLC. *See* Civil Liability Convention
Clean Air Act, 133
Clean Water Act (CWA), 52, 54, 58, 133, 152
CLEE. *See* Convention of Civil Liability for Oil Pollution Damage resulting from Exploration for and the Exploitation of Seabed Mineral Resources
Climate and Pollution Agency, 15–16
Cluff Oil Limited, 201–2
Coast Guard and Maritime Transportation Act (CGMTA), 141–2
OPA and, 142
Coast Guard Authorization Act (CGAA), 138–9
Coastal Zone Management Act, 133
Code of Federal Regulations (CFR), 134–5
collectivization, 306–7
COM(2011) 688 final, 4
Communication 2010, 93–4
compensation, 7. *See also* international compensation, nuclear sector
amount of, 410
Canada Oil and Gas Operations Act, 165
civil, 366–7
in CLC, 75–6, 286
efficient, 304–5
fair, 306–7
funds
economic principles, 351–2
insurance compared with, 349–51
in OPA, 150–1
OSLTF, 150–3
rapid claims management, 371–2
US offshore activities and, 150–3
government role in, 343
arguments against, 344–5
arguments for, 343–4
direct, 343
instruments, 339
market solutions, 307
mechanisms, 411
Canada Oil and Gas Operations Act, 166
OPOL, 109–15
Subsoil Act, 130–1
under NRC, 231–2
nuclear sector, 189–90
offshore liability regimes and, 170–1
under Offshore Petroleum and Greenhouse Gas Storage Act, 161–2
under OPA, 141–2
under OPOL, 107–8, 109–15, 209
limits, 203
under Petroleum Activities Act, 123
principles of, 304–5
reasons for, 305–6
workers, 251
compulsory financial guarantees, 76–7
compulsory insurance, 332, 334
Conoco North Sea Incorporated, 201–2
ConocoPhilips, 56
contamination, 75–6
continental shelf, 12
Continental Shelf Act, 127
Contract Regarding an Interim Supplement to Tanker Liability for Oil Pollution (CRISTAL), 79–81, 233–4
Contracting State, 78
Contracts Act 1999, 207
Convention for the Protection of the Marine Environment of the North-East Atlantic of 1992. *See* OSPAR Convention

Convention of Civil Liability for Oil Pollution Damage resulting from Exploration for and the Exploitation of Seabed Mineral Resources (CLEE), 101, 202
Convention on Limitation of Liability for Maritime Claims, 107
Convention on Supplementary Compensation for Nuclear Damage (CSC), 179, 190
Council Directive 92/91 EEC, 92
Court Supervised Settlement Program (CSSP), 366
criminal liability, 153–4
CRISTAL. *See* Contract Regarding an Interim Supplement to Tanker Liability for Oil Pollution
CSC. *See* Convention on Supplementary Compensation for Nuclear Damage
CSSP. *See* Court Supervised Settlement Program
CWA. *See* Clean Water Act
Cypriot, 19

damages, 15
- categories of, 367
- under EMANI, 221
- insurance, 251
- international compensation, nuclear sector, 179–80, 189–90
- in nuclear international compensation, 179–80
- under Offshore Petroleum and Greenhouse Gas Storage Act, 160
- OIL on, 211–12
- under Oil Pollution Act of 1990, 139–41
- in OPOL, 105–6, 209
- under Petroleum Activities Act, 122–3
- property, 189–90, 220–1, 230
- remedies, 410
- self-insurance and, 241–2
- under Subsoil Act, 129–30
- transboundary harm and, 379, 380, 381–2
- in US, 220–1
- WOAD cost estimates, 24

Danish Energy Agency (DEA), 15–16, 126, 127–8
Danish Environmental Protection Agency (DEPA), 127–8
Darwin Magistrates' Court, 47
DEA. *See* Danish Energy Agency
DECC. *See* Department of Energy and Climate Change
Declaration of Principles, 86–7
deductibles, 352
Deepwater Horizon, 1, 2–4, 21, 59–60, 62, 66, 98–9, 239–40, 274, 279
- casualty insurance, 51–2
- civil compensation, 366–7
- claims settlement in, 364–6
- costs of, 267–8
- critical review of trust fund, 367–9
- hold-harmless clauses, 259
- Hull insurance, 52
- insurance industry and, 258–60
- legal issues, 49–53
- limits of expert opinion on, 64–6
- Lloyds on, 258–60
- Noble Energy on, 296–7
- overview of legal proceedings, 53–6
- political instability from, 305–6
- premiums and, 259–60
- reinsurance and, 258–60

Deepwater Port Liability Fund, 152
Denmark
- basis of liability in, 128
- legal framework, 126–8
- legislation comparisons, 167
- national interest in, 126
- offshore liability regimes, 128
- regulation in, 129

DEPA. *See* Danish Environmental Protection Agency
Department of Communications, Marine and Natural Resources, 15–16
Department of Energy and Climate Change (DECC), 15–16, 29, 43–4, 98–9, 100, 111
- Guidance Note, 111–12
- OPOL, 210

Department of Interior (DOI), 131
Department of Justice (DOJ), 53, 365–6
Det Norske Veritas (DNV), 22–4
Directive 2009/103/EC, 372
Directive 2013/30/EU, 5
  on mandatory financial security, 337
Directive on Carbon Capture and Storage, 338
Directive on Safety of Offshore Oil and Gas Operations, 37, 94–7
  Annex I, 95
  Art. 20, 96–7
  data reporting format, 97
  Directive 94/22/EC, 95
  Directive 2013/30/EU, 95–6
  extra-territory effects, 96–7
  Preamble (38), 96–7
  Regulation (COM(2011) 688 final), 94–7
  transboundary effects, 96
Disaster Relief Act, 371
DNV. *See* Det Norske Veritas
DOI. *See* Department of Interior
DOJ. *See* Department of Justice
Drilling Safety Rule, 56–7
due diligence clauses, 252
Dutch Terrorism Risk Re-insurance Company, 348

Eastern District of Louisiana, 53
Eastern Mediterranean Basin, 21
Economic and Property Damages Settlement Agreement, 366
economic channelling, 314–15
EED. *See* Energy Exploration and Development Insurance
EIA. *See* Energy Information Administration
Ekofisk, 115–16
Elgin platform blowout, 44
  HSE on, 44
ELINI. *See* European Liability Insurance for the Nuclear Industry
EMANI. *See* European Mutual Association for Nuclear Insurance
Emission Trading Scheme (ETS), 328
empirics, 9–10
EMSA. *See* European Maritime Safety Agency
Endangered Species Act, 133
Energy Act 2008, 99
Energy Exploration and Development Insurance (EED), 252
Energy Independence and Security, 190
Energy Information Administration (EIA), 16, 20
energy losses, 34–5
Energy Policy Act of 2005, 134
Energy Safety and Security Act, 163–4
ENO. *See* extraordinary nuclear occurrence
Environmental Damage Insurance, 353–4
Environmental Damage Regulations 2009, 101, 102–3
  Regulation 7, 107
environmental funds, 352–6
  in Finland, 354
  Hungary, 352–4
  stakeholders on, 354–6, 357–8
  in Sweden, 354
Environmental Liability Directive, 93–4, 304
environmental liability insurance, 251
Environmental Protection Agency (EPA), 57, 133
Erika, 1
Erika I, 2
Erika II, 2, 71
Esso, 201–2
ETS. *See* Emission Trading Scheme
European Commission, 2, 20, 71
  on offshore accident costs, 255–6
European Insurance and Reinsurance Federation, 17
European Liability Insurance for the Nuclear Industry (ELINI)
  coverage under, 221–2
  establishment of, 221–2
  members of, 223
  Paris Convention and, 223
European Maritime Safety Agency (EMSA), 329–30

European Mutual Association for Nuclear Insurance (EMANI)
coverage under, 221
damages under, 221
establishment of, 221
European Union (EU)
high risk wells in, 284–5
mandatory financial security in, 332–3, 335–8
nuclear sector pooling mechanisms in, 221–3, 227–9
offshore accidents, 91–2
offshore activities, 91–2
stakeholders on, 268
EUROSTAT, 18–20
exclusive jurisdiction, 177
exploitation
mining processes, 13
Offshore Protocol on, 89–90
exploration, 13
in Black Sea, 19
licenses, 17
Offshore Protocol on, 89–90
extraordinary nuclear occurrence (ENO), 184–6
NRC on, 184–5
extra-territory effects, 96–7
Extremus, 348
Exxon Valdez, 1, 133–4, 150–3
ExxonMobil, 56

facilities, 18–20
fair compensation, 306–7
Federal Energy and Regulatory Commission, 134
Feinberg, Ken, 282
financial caps
liability, 317–19
Noble Energy on, 293–4
nuclear sector, international compensation, 181–2
Oil Pollution Act of 1990, 293
Paris Convention, 181–2
Finland, 354
fire insurance, 247
first layer coverage, 239–40
first party insurance, 247
Fishermen Compensation Regulations, 117
fishing vessel, 416–17
Fitch, 110, 242–3
fixed platforms, 13
force majeure, 6
Forties Alpha, 91–2
Fukushima, 171–2, 224, 373
cat bonds in, 243–4
channelling of liability, 314
Fund Convention, 2, 70
analysis of, 70–1
Article 4.7, 361
Article 6, 361
international legal framework, 77–9
rapid claims management and, 361–2
on transboundary harm, 377
vessel-based pollution and, 233–4

Gannet Alpha, 43–4, 91–2
GAREAT, 348
gas drilling, 257–8
GCCF. *See* Gulf Coast Claim Facility
general liability insurance, 251
geopressures, 61–2
German Environmental Liability Act of 1990, 335
Germany
nuclear pooling mechanisms in, 223–7
development of, 225
financial security in, 223–7
nuclear sector in, 224–5
unlimited liability in, 228
GL Noble Denton, 255–6
government, 386
as reinsurer, 346, 348–9
arguments against, 347–8
arguments for, 346–7
role in compensation, 343
arguments against, 344–5
arguments for, 343–4
direct, 343
Greenland, 284–5
Group Excess Reinsurance Policies, 216
Group-Hydra Insurance Company, 217

guarantees, 265
analysis, 245–6
insurance compared with, 245
OPOL on, 245–6
practice, 245
stakeholders on, 245
theory, 244–5
verification, 246
Guidelines for Relief Well Planning, 113
Guidelines on Financial Responsibility from Oil and Gas UK, 241–2
Gulf Coast Claim Facility (GCCF), 50, 63, 271–3, 359–60, 364–6, 368
declaration of losses in, 255–6
Gulf of Mexico, 53, 61, 66, 252, 259–60
drilling in, 269–71
Munich Reinsurance on, 280–1
risks in, 268–9
stakeholders on, 269–71
Gulf Region Health Outreach Program, 367
GulfOil Production Company, 201–2
Gullfaks C incident, 41, 91–2

Halliburton, 50, 52
Hamilton Brothers, 201–2
Hannover Ruekversicherung AG, 248–9
Harvard Report, 173
HCRs. *See* hydrocarbon releases
Health and Safety Executive (HSE), 15–16, 26–30
in Elgin platform blowout, 44
Hebei Spirit, 369–70
hit-and-run strategies, 271
hold-harmless clauses, 171–2
BP, 258–9
Deepwater Horizon, 259
HSE. *See* Health and Safety Executive
hull insurance, 52
Hungary, 352–4
Hurricane Sandy, 279
hurricanes, 270
hydrocarbon releases (HCRs), 28
hydrocarbons, 33
Ula field, 42–3
Hydrocarbons Licensing Directive 1994 (94/22/EC), 92–3
Hydrocarbons Licensing Directive Regulations, 99

IADC. *See* International Association of Drilling Contractors'
IAEA. *See* International Atomic Energy Agency
IAT. *See* Incident Analysis Team
iceberg theory, 23
IMO. *See* International Maritime Organization
Incident Analysis Team (IAT), 155
Indonesia, 85–6
transboundary complaints, 48–9
information asymmetry, 320
INPO. *See* Institute of Nuclear Power Operations
insolvency risk, 273–6, 337–8
liability and, 320–1
offshore accidents, 273–6
reinsurance and, 274
safety regulation and, 320–1
stakeholders on, 274–5
installations, 18–20
Institute of Nuclear Power Operations (INPO), 183–4, 221
NEIL and, 223
insurance, 17, 265. *See also specific types*
actors in, 248
analysis, 260–1
brokers, 249
business interruption, 251
capacity, 252–5, 258–9
catastrophe, 278–9
combinations, 264
compensation fund compared with, 349–51
compulsory, 332, 334
costs estimation, 255–6
Deepwater Horizon and, 258–60
demand for, 247
environmental liability, 251
expansion of, 277
fire, 247
first party, 247

insurance (cont.)
  general liability, 251
  guarantees compared with, 245
  hull, 52
  liability, 247
  limitations, 260–1
  Lloyds on, 254
  mandatory, 277–9, 340–1
  MODU, 251–2
  in nuclear sector, 198
  offshore accidents, 277
  operators extra expense, 251
  OPOL and, 263–4
  physical damage, 251
  pollution and, 253
  pooling mechanisms compared with, 198–201
  premiums, 249–50
    calculation of, 257–8
    Deepwater Horizon and, 259–60
    operators on, 258
    P&I, 215–17, 258
    retrospective, 219, 295
  principles of, 247
  risk analysis, 255–6
  risk aversion and, 246
  risk calculation, 250
  self-insurance compared with, 260–1
  stakeholders, 248–50
  structure of coverage, 250–2
  terrorism, 348
  theory, 246–8
  third party, 247
  types of, 247
  utility-based theory of, 246
  workers compensation, 251
integration of methodologies, 10
International Association of Drilling Contractors' (IADC), 16, 34
  Standard Form of International Offshore Daywork Drilling Contract, 14
International Association of Oil and Gas Producers (OGP), 34
International Atomic Energy Agency (IAEA), 173–4
  public funding and, 180–1
international compensation, nuclear sector
  channelling of liability, 175–6
  conventions, 181–2
  coverage caps, 181–2
  critical comparisons, 191–5
  exclusive jurisdiction, 177
  extraordinary nuclear occurrence, 184–6
  financial protection, 187
  financial security, 176–7
  first generation, 172–3
  origin of, 172–3
  property damage and, 189–90
  public funding in, 177–8, 180–1
  scope of damage in, 179–80
  second generation, 178–9
  strict liability, 172–3
  third party liability, 187–9
  in US, 182, 183–4, 186–7, 189–90
International Convention on Civil Liability for Oil Pollution Damage, 1–2
International Fund for Compensation for Oil Pollution Damage, 1–2
international legal framework, 69–70
  Fund Convention 1971/1992, 77–9
  voluntary mechanisms, 79–81
International Maritime Organization (IMO), 2, 72–3, 85–6, 376–7
  Legal Committee, 85–6
International Model Well Services Agreement, AIPN, 14
International Oil Pollution Compensation Fund (IOPC), 70–1, 361
  claims management, 361–2
International Regulators' Forum (IRF), 33, 34
International Union of Marine Insurance (IUMI), 17
IOPC. *See* International Oil Pollution Compensation Fund
IRF. *See* International Regulators' Forum
IUMI. *See* International Union of Marine Insurance

Japan, 373
Jiyeh, 19
JOA. *See* joint operating agreements
joint and several liability, 316–17
joint arrangements, 82
joint care case
  negligence rule in, 310–11
  strict liability rule in, 310–11
Joint Committee on Atomic Energy, 188
Joint Committee on Atomic Law, 185
joint operating agreements (JOA), 14
  Article 7, 119
  standard form of, 15
Joint Protocol, 179
joint research centre (JRC), 24
Juneau, Patrick, 366
Jupiter Insurance Ltd., 51, 239

Kimberly coastline, 48
Kirki, 46
Klif, 118

Law 128, 19
Leading Oil Government's Oil and Gas Industry Task Force (LOGIC), 14
Lebanon, 19, 376
legal analysis, 6
Letters of Undertaking, 215
Levantine Basin, 21
liability, 7, 8. *See also* offshore liability regimes
  attribution of, 313, 409–10
  basis for, 408
  cargo, 414–16
  causal relationships in, 321–2
  channelling of, 313–16
    Fukushima, 314
    nuclear sector, international compensation, 175–6
    in nuclear sector, 314
    in Paris Convention, 175–6
    in Price Anderson Act, 186–7
    TEPCO, 314
    in Vienna Convention, 175–6
  in Civil Liability Convention, 71, 74–5
  criminal
    knowing violation, 153–4
    negligent violation, 153–4
    United States offshore activities, 153–4
  as deterrent, 318
  economic theory, 308–12
  efficient, 303–5, 308, 384
  environmental, insurance, 251
  financial caps, 317–19
  fishing vessel, 416–17
  general, insurance, 251
  insolvency and, 320–1
  insurance, 247
  joint and severable, 316–17
  limited, 176
  limits exceeded by incident date, 418–30
  limits exceeded by vessel type, 412–17
  under negligence rule, 309
  in NRC, 183
  under Offshore Petroleum and Greenhouse Gas Storage Act, 158–60
  in PAA, 183
  in Paris Convention, 174–6
  under Petroleum Activities Act, 121–2
  principles of, 304–5
  public law, 183–4
  rapid claims management and liability law, 372
  regulation compared with, 319, 326–8
  relieving, 314–15
  stakeholder assessment of, 328–30
  standards, 271–3
  strict, 172–3, 174–5, 208
  strict liability rule, 310
    in joint care case, 310–11
    negligence rule compared with, 311
    in offshore accidents, 312–13
  Subsoil Act, 129
  suits, 321–2
  tank barge, 413, 414
  tank ship, 413

liability (cont.)
unlimited, 228
in Vienna Convention, 174–6
licensees, 13
operators distinguished from, 95
Lloyds
Casualty Reports, 22–4
on Deepwater Horizon, 258–60
on insurance, 254
Macondo and, 254, 273
on mandatory financial security, 334–5
LOGIC. *See* Leading Oil Government's Oil and Gas Industry Task Force
losses, 15
lost time accidents (LTA), 32

Macondo, 35–6, 53, 59–60, 64–6, 270, 273–4
Lloyds and, 254, 273
Noble Energy on, 293–4
responses after, 56–7
MAERP. *See* Mutual Atomic Energy Reinsurance Pool
Malta, 21
mandatory financial security
criteria for, 331–2
Directive 2013/30/EU on, , 337
in EU, 332–3, 335–8
Lloyds on, 334–5
offshore risks and, 332–3
stakeholder assessment, 332–3
mandatory insurance, 277–9, 340–1
Marine Environment Protection Act, 127
Marine Strategic Framework Directive, 3, 93
Directive 2013/30/EU, 96
Marine Well Containment Company, 56
MARPOL, 70
Annex I, 83
Art. 2, 83–4
definitions in, 83–4
on discharge, 83–4
Unified Interpretation of, 83
Martin, John, 364–6
Medical Benefits Class Action Settlement Agreement, 366, 367
Mediterranean Sea, 19, 21, 292
offshore activities, 21–2
offshore liability regimes, 88–90
Member State Law, 2–3
Memorandum of Association, 203
Merchant Shipping Act, 129–30
Merchant Shipping Regulations 1998, 109–10
Migratory Bird Treaty Act, 55
Minerals Management Service (MMS), 56–7
mining processes, 13
Ministry of Petroleum and Energy, 30–1, 117–18
Mitsui Oil Exploration Company of Japan (MOEX), 50, 51, 54
MMS. *See* Minerals Management Service
Mobil North Sea Limited, 201–2
mobile drilling units (MODU), 84–5
in insurance, 251–2
MOEX. *See* Mitsui Oil Exploration Company of Japan
Montara, 45–7, 49, 376
Montara Commission of Inquiry, 45–6, 155, 156–7
Montara Development Project, 45
Moody's, 110, 242–3
moral hazards, 332
control of, 331
Munich Reinsurance and, 282
remedies, 352
of self-insurance, 240–1
Motor Vehicle Insurance Directive, 374–5
Multi-District Litigation docket, 53
Munich Reinsurance, 248–9, 287–8, 314–15
analysis, 285–6
BP and, 283–4
on Gulf of Mexico, 280–1
moral hazard and, 282
on Noble Energy, 298
proposal, 340–1
reaction by, 284–5
on risk differentiation, 298–300

SOS model, 280–2
stakeholder's assessment, 283–4
mutatis mutandis, 312
Mutual Atomic Energy Reinsurance Pool (MAERP), 187–8
mutual monitoring, 200–1, 298, 342–3
in P&I Clubs, 232–3
pooling mechanisms, 200–1, 229–30

NatCat, 288
National Energy Board, 163
National Environmental Policy Act, 135
National Fish and Wildlife Foundation (NFWF), 55
National Oceanic and Atmospheric Administration (NOAA), 140–1
National Offshore Petroleum Safety and Environmental Management Authority (NOPSEMA), 15–16, 156–7, 159
National Pollutant Discharge Elimination System (NPDES), 133
National Pollution Funds Center, 145, 152–3
national regulatory authorities
in Norway, 30
offshore activities, 24–6
in UK, 26–30
natural gas, 20
NEA. *See* Nuclear Energy Agency
negative redistribution, 307
negligence rule, 309
in joint care case, 310–11
liability under, 309
strict liability rule compared with, 311
NEIL. *See* Nuclear Electric Insurance Limited
NELIA. *See* Nuclear Energy Liability Insurance Association
Netherlands, 20–1, 376
offshore activities in, 31
personal injuries in, 32
Netherlands Oil and Gas Exploration and Production Association (NOGEPA), 16
NFWF. *See* National Fish and Wildlife Foundation
9/11 Fund, 371–2
1984 Protocols, 71
NOAA. *See* National Oceanic and Atmospheric Administration
Noble Energy, 342–3
background, 293–4
on caps, 293–4
challenges in, 299–300
on Deepwater Horizon, 296–7
evaluation of, 298–300
on Macondo, 293–4
on polluter pays principle, 293
proposal contents, 294–5
stakeholders on, 297–8
weaknesses, 298
NOGEPA. *See* Netherlands Oil and Gas Exploration and Production Association
NOPSEMA. *See* National Offshore Petroleum Safety and Environmental Management Authority
Nordic Convention, 88
North Africa, 19, 252
North Sea, 3, 63, 375, 376
drilling in, 269–71
offshore activities in, 19, 20–1
offshore liability regimes, 87–8
Valhall PCP production platform fire in, 41–2
North Sea Offshore Authorities Forum (NSOAF), 31–3
North Sea Sun Oil Company Limited, 201–2
Northern Territory Department of Resources, 46–7
Norway, 20, 26, 376
national regulatory authorities in, 30
offshore activities
basis of liability in, 119
causation in, 120–1
legal framework, 116–18
national interest in, 115–16
regulation, 120
offshore liability regimes, 119
oil spills in, 30–1

Norway (cont.)
personal injuries in, 30
rapid claims management in, 360–2
Norwegian Environment Agency, 118
Norwegian Petroleum Directorate (NPD), 117–18
Norwegian Statoil, 17
NPD. *See* Norwegian Petroleum Directorate
NPDES. *See* National Pollutant Discharge Elimination System
NRC. *See* Nuclear Regulatory Commission
NSOAF. *See* North Sea Offshore Authorities Forum
Nuclear Electric Insurance Limited (NEIL), 220
INPO and, 223
self regulatory agencies and, 223
WANO and, 223
Nuclear Energy Agency (NEA), 173–4, 192–3
public funding in, 180–1
Nuclear Energy Liability Insurance Association (NELIA), 187–8
Nuclear Liability Act, 176
nuclear radiation, 179–80
Nuclear Regulatory Commission (NRC), 182, 183–4, 218
compensation under, 231–2
on extraordinary nuclear occurrences, 184–5
liability in, 183
pooling mechanisms of, 228–9
safety standards, 230
nuclear sector
channelling of liability in, 314
coverage in, 287–9
development of, 171–2
in Germany, 224–5
insurance in, 198
international compensation
channelling of liability, 175–6
conventions, 181–2
coverage caps, 181–2
critical comparisons, 191–5
exclusive jurisdiction, 177
extraordinary nuclear occurrence, 184–6
financial protection, 187
financial security, 176–7
first generation, 172–3
origin of, 172–3
property damage and, 189–90
public funding in, 177–8, 180–1
scope of damage in, 179–80
second generation, 178–9
strict liability, 172–3
third party liability, 187–9
in US, 182, 183–4, 186–7, 189–90
offshore liability regimes and, 171–2
pooling mechanisms in, 217
development of, 225
in EU, 221–3, 227–9
financial security in, 223–7
Germany, 223–7
Price Anderson Act, 217
public liability in, 217
retrospective premiums, 219
Nuclear Waste Fund, 186–7

Obama administration, 57, 60
OCIL. *See* Oil Casualty Insurance Ltd.
OEE. *See* operators extra expenses
offshore accidents
analysis of recent, 41
costs of, 255–6
distribution of, 23
energy losses from, 34–5
in EU, 91–2
EU Commission on, 255–6
funds for, 356–9
insolvency risk, 273–6
insurance, 277
large, 276
liability standards, 271–3
major, 268–9
mandatory insurance, 277–9
market capacity and, 273–4
middle-size, 276, 301
observation from available data, 36–7
operators in, 271
overview of, 37

political instability from, 305–6
potential costs of, 267–8
verification, 268–9
probability of, 272–3
safety regulation, 269
self-guarantee, 276–7
small, 275–6, 301
strict liability rule in, 312–13
technical differences in, 269–71
in US, 275–6
vessel-based pollution in, 270–1
victims, 305–6
WOAD definition, 23, 24
offshore activities
Australia
legal framework, 155–8
legislation comparisons, 167
national interest in, 154–5
well blowouts, 155
Canada
best practice, 166–7
damages under, 164–5
legal framework, 163–4
legislation comparisons, 167
national interest in, 163
Denmark
legal framework, 126–8
legislation comparisons, 167
national interest in, 126
regulation in, 129
EU interest in, 91–2
expert evaluation, 61–4
financial responsibility, 149
importance of, 18
industry database, 33–6
industry structure of, 13–15
limits of expert opinion on, 64–6
mandatory financial security and, 332–3
Mediterranean Sea, 21–2
national regulatory authorities, 24–6
in Netherlands, 31
in North Sea, 19, 20–1
Norway
causation in, 120–1
legal framework, 116–18
national interest in, 115–16
regulation, 120
pollution from, 81–2
regional database, 31–3
regulation of, 322–4
risk calculation, 250
terminology in, 12–13
United Kingdom
legislation comparisons, 167
national interest, 98
statutory laws, 98–100
United States
compensation funds, 150–3
on jurisdictional issues, 153–4
legal framework, 131–6
legislation comparisons, 167
national interest in, 131
regulation in, 137
Offshore Crude Oil Production of EU-28, 18–20
Offshore Exploration Regulations 2009, 99–100
Offshore Gas Storage and Unloading Regulations 2009, 99–100
offshore liability regimes, 86–7, 91
Australia, 158–60
Baltic Sea, 90
basis of liability in, 103–4, 170
Canada, 164
causation and, 170
claims settlement, 170–1
comparative comments, 167–71
compensation and, 170–1
country studies, 97–8
Denmark, 128
Mediterranean, 88–90
North Sea, 87–8
Norway, 119
nuclear sector and, 171–2
regulation and, 170
United Kingdom, 100–3
United States, 136, 153–4
Offshore Oil Pollution Compensation Fund, 152
Offshore Petroleum and Greenhouse Gas Storage Act, 47, 156
attribution of liability under, 160
basis of liability in, 158–60
causation under, 160
compensation under, 161–2

Offshore Petroleum (cont.)
damages under, 160
regulation under, 160
Section 569, 160
Section 569(1), 158–9
Section 571, 159, 161–2
Section 775D, 160
Offshore Pollution Liability Agreement (OPOL), 100, 234, 265, 267, 301, 335–6, 379
application, 210
arbitrators, 207
Articles of Association, 203
attribution of liability in, 105
basis of liability in, 103–4
causation in, 104
claim settlement, 108–9
claims handling, 205–6
Clause II.B of, 204
Clause III.2 of, 204
Clause IV, 204–5
Clause IV (A), 208
Clause VI, 206
Clause VII, 206
Clause XIII, 207
combinations, 264
compensation in, 107–8, 209
compensation limits in, 203
compensation mechanisms, 109–15
coverage of, 101–2, 263, 274
damages in, 105–6, 209
DECC and, 210
enforcement, 206–7
evaluation, 208–10
expansion of, 291–3
extension of, 202
financial responsibility, 204–5
FR1, 204–5
FR2, 204–5
on guarantees, 245–6
insurance and, 263–4
limitations of, 209, 210, 292
main features, 202–3
mandatory membership of, 291
member states, 232
membership of, 202
Memorandum of Association, 203
obligations in, 204–5
OEE and, 264
operators in, 203–4
origins of, 201–2
personal injuries under, 106
pooling mechanisms in, 209, 232
practice, 207–8
rapid claims management, 360
remedial measures in, 105–6, 205
risk-sharing in, 208–9
rules, 203–4
self-insurance in, 204–5, 208, 242–3
strengths, 208–10
strict liability in, 208
structure, 203–4, 209
transboundary harm and, 380
transformation of, 292
in UK, 232, 263
weaknesses of, 209–10
Offshore Pollution Liability Association, Ltd., 203
membership in, 204
Offshore Protocol, Barcelona Convention, 89
on exploitation, 89–90
on exploration, 89–90
Offshore Safety Act, 127, 129–30
offshore units, 84
OGP. *See* International Association of Oil and Gas Producers
OIL. *See* Oil Insurance Ltd.
Oil and Gas Standard Form Joint Operating Agreement, 14
regulation in, 137
Oil and Gas UK, 255–6, 375
Oil Casualty Insurance Ltd. (OCIL), 213, 265
expansion of, 290
risk sharing, 261–2
shareholders, 213
Oil Insurance Ltd. (OIL), 211–12, 265
coverage under, 211–12
expansion of, 290
formation of, 211
on physical damage, 211–12
ratings, 211
risk sharing, 261–2
shareholders, 211, 212

Oil Pollution Act of 1990 (OPA), 1, 50–1, 54, 70, 133–4, 141, 196, 335
Applicability of Limit of Liability, 51
basis of liability, 136
causation in, 137–8
CGMTA and, 142
claim settlement, 144–6
compensation funds and, 150–1
compensation under, 141–2
damages under, 139–41
financial security and, 146–53
liability cap, 293
limits under, 280–2
rapid claims management, 363–4
Section 1016, 146–7
Section 2702, 137–8
Section 2703, 137–8
section 2713, 363
state legislation and, 143
oil pollution emergency plans (OPEPs), 112
Oil Pollution Preparedness Responses and Co-Operation 1990 (OPRC Convention), 84–5
on offshore units, 84
Oil Spill Contingency and Response (OSCAR), 113–15
Oil Spill Information System (OSIS), 113–15
Oil Spill Liability Trust Fund (OSLTF), 145
compensation funds, 150–3
components of, 151
Emergency Fund, 151
financial sources, 151
Principle Fund, 151
Oil Spill Prevention and Response Advisory (OSPRAG), 111–12, 113
oil spills, 43–4
clean-up costs, 256
high profile, 37
in Norway, 30–1
from offshore blowouts, 38
Oilmap, 113–15
OOG. *See* Overlegorgaan Olie en Gas
OPA. *See* Oil Pollution Act of 1990
OPEPs. *See* oil pollution emergency plans
operators, 13
licensee distinguished from, 95
in offshore accidents, 271
in OPOL, 203–4
on premiums, 258
Operator's Cooperative Emergency Services (OCES)
Joint Declaration and Emergency Assistance Code, 16
role of, 16
operators extra expenses (OEE), 35–6, 51, 251, 252
OPOL and, 264
OPOL. *See* Offshore Pollution Liability Agreement
OPRC Convention. *See* Oil Pollution Preparedness Responses and Co-Operation 1990
OSCAR. *See* Oil Spill Contingency and Response
OSIS. *See* Oil Spill Information System
OSLA. *See* Outer Continental Shelf Lands Act
OSLTF. *See* Oil Spill Liability Trust Fund
OSPAR Convention, 87–8
Annex III, 87–8
Art. 2(2)(b), 87
on pollution, 87–8
principles of, 87
OSPRAG. *See* Operator's Cooperative Emergency Services
Outer Continental Shelf Lands Act (OCSLA), 131, 136
on jurisdictional issues, 153
Section 1334, 147
Section 1337, 147
Overlegorgaan Olie en Gas (OOG), 329

PAA. *See* Price Anderson Act
PAL. *See* Passengers and their Luggage by Sea
parametric trigger-coverage, 281, 340
Paris Convention, 173–4, 179
channelling of liability, 175–6
coverage caps, 181–2

Paris Convention (cont.)
ELINI and, 223
exclusive jurisdiction, 177
financial security, 176–7
limited liability in, 176
modification protocol, 287–9
pooling mechanisms and, 227–8
public funding in, 177–8, 180–1
strict liability in, 174–5
Supplementary, 192
Passengers and their Luggage by Sea (PAL), 260–1
Pemex Mexico, 59
personal injuries
fatal, 32
in Netherlands, 32
in Norway, 30
under OPOL, 106
PSA on, 30
Petrobras Brazil explosion, 58
Petroleum Act 1998, 99, 100, 109–10
Petroleum Activities Act, 115–16
adoption of, 116–17
attribution of liability under, 121–2
claim settlement, 123–4
compensation under, 123
damages under, 122–3
regulation under, 120, 124
Petroleum Licensing Regulations 2004, 99
Petroleum Licensing Regulations 2008, 99
Petroleum Regulations 2008 109
Petroleum Safety Authority (PSA), 15–16, 26, 42, 118
on Gullfaks C, 41
on personal injuries, 30
Section 8-3, 119
Petroleum Taxation Act, 116
Phillips Petroleum Company, 201–2
P&I Clubs. *See* Protection and Indemnity Clubs
Piper Alpha, 91–2, 98, 272–3
political instability from, 305–6
PLA. *See* Public Liability Action
Plaintiffs' Steering Committee, 54, 366
polluter pays principle, 293
pollution
insurance and, 253
from offshore activities, 81–2
OSPAR Convention on, 87–8
Pollution Control Act, 117
Pool Re, 348
pooling mechanisms, 234, 383. *See also* risk-sharing
advantages of, 200–1, 229–32
contributions to, 231
insurance compared with, 198–201
literature on, 229–30
mutual monitoring, 200–1, 229–30
mutual trust in, 231
of NRC, 228–9
nuclear, in EU, 221–3
in nuclear sector, 217
development of, 225
in EU, 221–3, 227–9
financial security in, 223–7
Germany, 223–7
Price Anderson Act, 217
public liability in, 217
retrospective premiums, 219
in OPOL, 209, 232
Paris Convention and, 227–8
premiums, 229–30
for property damage, 230
in US, 220–1
under Protection and Indemnity Clubs, 198–9
restrictions of, 201
Vienna Convention and, 227–8
Preliminary Report on Financial Protection Against Atomic Hazards, 171–2
premiums, 249–50
calculation of, 257–8
Deepwater Horizon and, 259–60
operators on, 258
P&I, 215–17, 258
retrospective, 219, 295
Prestige, 370
Price Anderson Act (PAA), 68–9, 171–2, 182, 190, 194–5, 235, 285–6, 300, 314–15, 342–3
advantages of, 191–2
channelling in, 186–7

contributions under, 231–2
coverage, 287
liability in, 183
passage of, 183
pooling mechanisms in, 217
premium system, 198
public law liability in, 183
retrospective premiums, 219, 295
primary restoration, 140–1
Princess Anne Marie, 46
private interest criteria, 322
private regulation, 324–6
property damage
nuclear sector compensation and, 189–90
pooling mechanisms for, 230
in United States, 220–1
Protection and Indemnity Clubs (P&I Clubs), 76, 235, 249, 297, 350, 369–70
available amount in, 217
benefits, 215
Catastrophe Reserve Fund, 216
contributions to, 216
coverage under, 214–15
entry, 215–17
fund mechanisms of, 362
Group Excess Reinsurance Policies, 216
Letters of Undertaking, 215
mutual monitoring in, 232–3
origins, 214
pooling under, 198–9
premiums, 215–17, 258
rapid claims management, 362
reimbursement in, 214
risk coverage, 258
risk-sharing, 216
rulebook, 214–15
services of, 215
supplementary calls, 216
Protocol on the Establishment of a Supplementary Fund for Oil Pollution Damage, 73
PSA. *See* Petroleum Safety Authority
PTTEP Australasia, 45, 47, 48
public interest criteria, 322
Public Liability Action (PLA), 184
public regulation, 324–6

rapid claims management, 387, 359. *See also* claims settlement
combinations, 372–4
compensation funds, 371–2
evaluation of, 374–5
existing schemes, 359–60
liability law and, 372
mechanisms, 369–70
in Norway, 360–2
under OPA, 363–4
OPOL, 360
Rares, Steven, 18
RCV. *See* replacement cost value
Red Cross, 371
regulation
in Denmark, 129
information asymmetry and, 320
liability compared with, 319, 326–8
in Norway, 120
offshore liability regimes and, 170
under Offshore Petroleum and Greenhouse Gas Storage Act, 160
of offshore risk, 322–4
under Oil and Gas Standard Form Joint Operating Agreement, 137
under Petroleum Activities Act, 120, 124
private, 324–6
public, 324–6
relationship with, 408
safety, 269, 319–21, 330
self-regulation, 324–6
in US, 137
Regulation 1112/2014, 33
reinsurance, 265, 266
brokers, 249
capital markets and, 243
Deepwater Horizon and, 258–60
on insolvency risk, 274
practice, 248–50
stakeholders, 248–50
structure of coverage, 250–2
theory, 246–8

reinsurer of last resort, 346, 348–9
  arguments against, 347–8
  arguments for, 346–7
remedial measures, 105–6, 205
replacement cost value (RCV), 211–12
restricted work cases (RWC), 32
retrospective premiums, 219, 295
risk aversion, 246
risk differentiation, 262, 299, 307
  Munich Reinsurance on, 298–300
risk-sharing, 199, 341. *See also* pooling mechanisms
  expansion of, 289–90
  OCIL, 261–2
  OIL, 261–2
  in OPOL, 208–9
  P&I Clubs, 216
  for vessel-based pollution, 233–4
Romania, 19, 90
Rote, 48
Royal Dutch Shell, 57
RWC. *See* restricted work cases

safety regulation
  criteria, 319–20
  insolvency risk and, 320–1
  offshore accidents, 269
  policy conclusions, 330
Santa Barbara well blowout, 211
SDR, 2
Section 1337, 132
Securities and Exchange Commission (SEC), 54–5
self-guarantee through taxes, 339–40
self-insurance, 149–50, 237, 265
  analysis, 240–2
  BP, 239
  captives
    costs of, 237–8
    creation of, 237–8
  combinations, 264
  credit ratings and, 238–9
  damages and, 241–2
  disadvantages of, 241
  first layer coverage, 239–40
  insurance compared with, 260–1
  of majors, 238–9
  moral hazards, 240–1
  in OPOL, 204–5, 208, 242–3
  practice, 238–40
  reserves, 237–8
  stakeholders on, 238–9
  theory, 237–8
self-insured retentions (SIR), 125, 239–40
self-regulation
  arguments for, 324–5
  enforcement of, 325–6
Shavell, Steven, 319–20
Shell, 43–4, 56, 328
Shell UK Limited, 201–2
Shetland Islands, 60
Siebens Oil and Gas, 201–2
Signal Oil and Gas Company Limited, 201–2
SINTEF, 114
SIR. *See* self-insured retentions
SLA. *See* Submerged Lands Act
Small Tanker Oil Pollution Indemnification Agreement (STOPIA), 79–81, 233–4
SOS model. *See* Sudden Oil Spill model
Soviet Union, 178–9
stakeholders, 15–17
  on capital markets, 243–4
  on cat bonds, 243
  on environmental funds, 354–6, 357–8
  on EU, 268
  on guarantees, 245
  on Gulf of Mexico, 269–71
  on insolvency risk, 274–5
  insurance, 248–50
  interviews with, 406–7
  on liability, 328–30
  on mandatory financial security, 332–3
  on Munich Reinsurance, 283–4
  on Noble Energy, 297–8
  on offshore-related risk funds, 356–9
  reinsurance, 248–50
  on self-insurance, 238–9
  on well blowouts, 268
Standard Form of International Offshore Daywork Drilling Contract, IADC, 14
Standard & Poor's, 110, 242–3
state regulatory authorities, 15–16

State Supervision of Mines, 15–16, 31, 36–7
Statoil, 115–16, 125
stock exchange, 242–3
STOPIA. *See* Small Tanker Oil Pollution Indemnification Agreement
strict liability rule, 310
  in joint care case, 310–11
  negligence rule compared with, 311
  in offshore accidents, 312–13
Submerged Lands Act (SLA), 131
  on jurisdictional issues, 153–4
subsidization, 378
Subsoil Act, 126–7, 132
  attribution of liability in, 129
  compensation mechanisms, 130–1
  damages under, 129–30
Sudden Oil Spill (SOS) model, 280–2
Superfund, 357
Supplementary Fund of 2003, 79–80
Supplementary Fund Protocol, 286
Svenska Geologiska Undersøkning, 15–16
Sweden, 354
Swiss Re-insurance Company, 243, 248–9, 283
Syverud, Kent, 364–6

Tanio, 71
tank barge, 413, 414
tank ship, 413
Tanker Oil Pollution Indemnification Agreement (TOPIA), 79–81, 233–4
Tanker Owners Voluntary Agreement concerning Liability for Oil Pollution (TOVALOP), 79–81, 233–4
TEPCO, 314
terrorism insurance, 348
Terrorism Risk Insurance Act (TRIA), 348
Teton Dam, 64–5
Texaco North Sea UK Company, 201–2
third party insurance, 247
Three Mile Island, 184–5, 189–90
Timor Sea, 45, 48
TOPIA. *See* Tanker Oil Pollution Indemnification Agreement
Torrey Canyon, 1, 79–81
tort law, 319
Total Oil Marine Limited, 201–2
TOVALOP. *See* Tanker Owners Voluntary Agreement concerning Liability for Oil Pollution
Trans-Alaska Pipeline Liability Fund, 152
transboundary effects, 96
transboundary harm, 376–8
  CLC on, 377
  damage above 750 million dollars, 381–2
  damage between 250 million and 750 million dollars, 380
  Fund Convention on, 377
  max damage of 250 million dollars, 379
  OPOL and, 380
Transocean, 49–50, 51–2, 53, 55
Transport Council, 72
Treasury, US, 294–5
TRIA. *See* Terrorism Risk Insurance Act

Ula field, 42–3
UN Conference on Human Environment, 86–7
UNCLOS. *See* United Nations Convention on the Law of the Sea
UNEP. *See* United Nations Environment Programme
Unified Interpretation of MARPOL, 83
United Kingdom (UK)
  legislation comparisons, 167
  liability, 100–3
  national interest in, 98
  national regulatory authorities in, 26–30
  offshore liability regimes, 100–3
  OPOL in, 232, 263
  statutory laws, 98–100
  worker injuries in, 26–7

United Nations Convention on the Law of the Sea (UNCLOS), 12, 48–9, 69, 70, 376–7
  Annex III of, 81–2
    Art. 11, 82
    Art. 23, 83
    Art. 168, 83
  Art. 56, 12
  Art. 60, 12
  Art. 77, 12
  Art. 81, 12
  Art. 198, 49–53
  Art. 235, 81–2
  international legal framework, 79–81
  on joint arrangements, 82
United Nations Environment Programme (UNEP), 86–7
United States (US)
  basis of liability, 136
  compensation funds, 150–3
  criminal liability, 153–4
  declaration of losses in, 255–6
  on jurisdictional issues, 153–4
  legal framework, 131–6
  legislation comparisons, 167
  national interest in, 131
  nuclear compensation, 182
    compensation and, 189–90
    public law liability, 183–4
  offshore accidents, 275–6
  offshore liability regimes, 136, 153–4
  property damage in, 220–1
  regulation in, 137
  Treasury, 294–5
upstream losses, 39, 40

Valencia Basin, 21
Valhall PCP production platform fire, 41–2
  investigation of, 42
vessel-based pollution
  Civil Liability Convention and, 233–4
  cover for, 286–7
  Fund Convention and, 233–4
  in offshore accidents, 270–1
  risk-sharing agreements for, 233–4
Vienna Convention, 173–4, 179
  channelling of liability, 175–6
  exclusive jurisdiction, 177
  financial security, 176–7
  limited liability in, 176
  pooling mechanisms and, 227–8
  public funding in, 177–8
  revisions of, 194–5
  Soviet Union in, 178–9
  strict liability in, 174–5
voluntary mechanisms, 79–81

W. B. Berkley Corporation, 248–9
WANO. *See* World Association of Nuclear Operators
Waste Framework Directive, 93–4
WELD. *See* Willis Energy Loss Database
well blowouts
  in Australia, 155
  catastrophic, 272–3
  Santa Barbara, 211
  stakeholders on, 268
well control equipment, 33
Well Control Incident Database, 33–4
Wells Committee, 33–4
West Africa, 284–5
West Shetlands, 63, 255–6, 270
Willis, 34–5, 239, 273
Willis Energy Loss Database (WELD), 34–5
WOAD. *See* Worldwide Offshore Accident Databank
worker injuries
  causes of, 27
  in UK, 26–7
workers compensation insurance, 251
Working Environment Act, 116
Working Environment Authority, 128
Workplace Safety Rule, 56–7
World Association of Nuclear Operators (WANO), 223
Worldwide Offshore Accident Databank (WOAD), 22–4, 36–7
  accidents, 23, 24
  consequences according to, 24
  damage costs according to, 24
  incidents, 22–3
  insignificant events, 22
  near-misses, 22

Lightning Source UK Ltd.
Milton Keynes UK
UKHW02n0936060518
322170UK00018B/433/P

9 781107 167162

AF804225

Economics, Management and Optimization in Sports

Springer
Berlin
Heidelberg
New York
Hong Kong
London
Milan
Paris
Tokyo

Sergiy Butenko
Jaime Gil-Lafuente · Panos M. Pardalos
Editors

Economics, Management and Optimization in Sports

Foreword by
Juan Antonio Samaranch

With 33 Figures
and 70 Tables

Springer

Dr. Sergiy Butenko
Department of Industrial Engineering
Texas A&M University
236E Zachry Engineering Center
College Station, TX 77843-3131, USA
butenko@tamu.edu

Dr. Jaime Gil-Lafuente
University of Barcelona
Department of Economics
Av. Diagonal, 690
E-08034 Barcelona, Spain
jaumeg@eco.ub.es

Dr. Panos M. Pardalos
Industrial and Systems Engineering Department
303 Weil Hall, University of Florida
PO Box 116595
Gainesville, FL 32611-6595, USA
pardalos@ufl.edu

ISBN 3-540-20712-0 Springer-Verlag Berlin Heidelberg New York

Cataloging-in-Publication Data applied for

A catalog record for this book is available from the Library of Congress.

Bibliographic information published by Die Deutsche Bibliothek
Die Deutsche Bibliothek lists this publication in the Deutsche Nationalbibliografie; detailed bibliographic data available in the internet at *http.//dnb.ddb.de*

This work is subject to copyright. All rights are reserved, whether the whole or part of the material is concerned, specifically the rights of translation, reprinting, reuse of illustrations, recitation, broadcasting, reproduction on microfilm or in any other way, and storage in data banks. Duplication of this publication or parts thereof is permitted only under the provisions of the German Copyright Law of September 9, 1965, in its current version, and permission for use must always be obtained from Springer-Verlag. Violations are liable for prosecution under the German Copyright Law.

Springer-Verlag is a part of Springer Science+Business Media
springeronline.com

© Springer-Verlag Berlin Heidelberg 2004
Printed in Germany

The use of general descriptive names, registered names, trademarks, etc. in this publication does not imply, even in the absence of a specific statement, that such names are exempt from the relevant protective laws and regulations and therefore free for general use.

Cover design: Erich Kirchner, Heidelberg

SPIN 10976623 42/3130 – 5 4 3 2 1 0 – Printed on acid-free paper

Foreword

Ever since the first Olympic Games in Ancient Greece, sports have become an integral part of human civilization. The last decade has been commemorated by the centennial celebration of the modern Olympic movement. With great anticipation, the Olympics return to Athens, Greece, and we are once again reminded that we live in one of the most exciting periods in the history of sports.

Reflecting back on my years of service as the International Olympic Committee president, I cannot overlook the remarkable changes that have taken place in the world of sports during these two decades. The technological development and consequent globalization of the world economy opened up a window of new opportunities for the sports industry. As a result, management, economics, and other sciences have become a significant part of modern sports.

It is my pleasure to introduce this volume comprising an interesting collection of papers dealing with various aspects of management, economics and optimization applied to sports. May this book serve as a valuable source of information to researchers and practitioners as well as to casual readers looking for a deeper insight into the magnificent world of sports.

Barcelona, Spain
October, 2003

Juan Antonio Samaranch
IOC President (1980-2001)

Preface

Modern culture is unimaginable without sports. Baron Pier de Coubertin (1863-1937) wrote in his "Ode to Sport":

> O Sport, you are Progress! To serve you, a man must improve himself both physically and spiritually. You force him to abide by a greater discipline; you demand that he avoid all excess. You teach him wise rules which allow him to exert himself with the maximum of intensity without compromising his good health.

Nowadays, not only sports is one of the most popular means of leisure and maintenance of a healthy life style, it is also an area of exciting business opportunities and an attractive object of scientific research.

Pervasive intellectualization is being displayed in sports as in any other sphere of human activities. Even though science and technology have been widely used in sports for years now, the opportunities for application of scientific methods to sports seem to be inexhaustible.

The aim of this publication is to present the up-to-date research developments in economics, management and optimization applied to sports, which would be of interest to researchers and practitioners in sports industry, and could be used as supplementary reading in related courses and seminars.

The 18 chapters gathered in this book cover a wide range of topics associated with various aspects of economics, management and optimization in sports. In particular, the volume contains research papers and reviews addressing the following issues:

- methods for ranking teams and evaluating players' performance;
- techniques for predicting outcomes of sport competitions;
- economics of professional leagues;
- optimal strategies in sports;

- optimal referee assignment techniques;
- scheduling and managing sport tournaments.

The specific sports discussed in the book include American football, soccer, baseball, cricket, basketball, track and field, and hockey.

We would like to thank the authors of chapters for providing excellent contributions, anonymous referees for their helpful comments and suggestions, and Springer-Verlag staff for their support and assistance.

Sergiy Butenko
Jaime Gil Lafuente
October 2003 *Panos M. Pardalos*

Contents

Matrix-based Methods for College Football Rankings
Vladimir Boginski, Sergiy Butenko, Panos M. Pardalos 1

Who's the Best? Data Envelopment Analysis and Ranking Players in the National Football League
Erickson H. DeOliveira, Robert Callum 15

Dynamic Programming and Time Related Strategies in Sports
Harold Sackrowitz 31

Using Statistics to Predict Scores in English Premier League Soccer
John S. Croucher 43

Modelling and Forecasting Match Results in the English Premier League and Football League
Stephen Dobson, John Goddard 59

Restructuring the Belgian Professional Football League: A Location-Allocation Solution
Trudo Dejonghe 79

The Best Systems for Appointing Referees
Jaime Gil Lafuente 101

Fashion Accessory, Social Identity or Tribal Uniform?
Nick Wilde 121

Modeling a Baseball Game to Optimize Pitcher Substitution Strategies Using Dynamic Programming
Nobuyoshi Hirotsu, Mike Wright 131

Mega Deal? A Relative Performance Analysis for Major League Baseball Players
Chii-Dean Lin, Richard A. Levine 163

Market Inefficiencies in Player Head to Head Betting on the 2003 Cricket World Cup
Michael J Bailey, Stephen R Clarke 185

Searching for Patterns in Sports Contests
Tim McGarry 203

Track and Field Performance Data and Prediction Models: Promises and Fallacies
Yuanlong Liu 225

Using Response Surface Models for Evolutionary Estimation of Optimum Running Times
Wei-Han Tan 235

Solving Sports Scheduling and Timetabling Problems with Constraint Programming
Abderrahmane Aggoun, Alkis Vazacopoulos 243

Collaboration Networks in Sports
Vladimir Boginski, Sergiy Butenko, Panos M. Pardalos, Oleg Prokopyev 265

The Holdup System for Grouping Minor Hockey Participants Into Age Divisions
William Hurley 279

Promotion and Relegation in Rent Seeking Contests
Stefan Szymanski 289

Matrix-based Methods for College Football Rankings

Vladimir Boginski[1], Sergiy Butenko[2] and Panos M. Pardalos[1]

[1] University of Florida, USA
{vb,pardalos}@ufl.edu
[2] Texas A&M University, USA
butenko@tamu.edu

1 Introduction

College football season is one of the most popular and anticipated sports competitions in the United States. Many of the National Collegiate Athletic Association (NCAA) Division I-A football games are surrounded by enormous fan interest and receive extensive media coverage. They are attended by tens of thousands of spectators and are followed by millions through the media. As a result, success of a team on the football field brings increased student applications and substantial financial profits to the institution it represents.

Due to these facts, it is especially important that ranking college football teams is as fair and unbiased as possible. However, the format of the NCAA football championship does not allow one to apply traditional ranking methods that are commonly used in professional leagues, where each team plays all other teams during the regular season, and the champion is determined in playoff series. NCAA division I-A includes more than 100 teams, and the number of games played by each team is no more than 15. Clearly, under these conditions, the "quality" of opponents is not the same for different teams, and standard ranking schemes may lead to "unfair" results. Moreover, there are no playoffs in college football, and the national champion is determined in a single game between the #1 and #2 teams in the rankings.

Until several years ago, the rankings were decided purely based on collective opinion of press writers and coaches. Clearly, these ranking principles are not acceptable, since people's opinions are in many cases "biased". For instance, a sports analyst might be impressed by the playing style of a certain team which would affect his decision, moreover, many of those whose votes are considered in the ranking polls (especially, football coaches) cannot see all games of every team during the season and rely on their personal perception or other specialists' judgements. Therefore, this ranking approach can produce "unfair" results. A major controversy took place several times, for example, in 1990, 1991 and 1997 two major polls selected different national champions. In 1998, the Bowl Championship Series (BCS) was introduced as a more trustworthy way of determining who is who in college football. The major components of the current BCS selection scheme are coaches/sports writers polls and computer-based rankings. The BCS system managed to produce an undisputed champion each year since its implementation. However, it is clearly not perfect: it was a general opinion that had Nebraska beaten Miami in 2001 Rose Bowl, the national championship would have to be split between Nebraska and Oregon. Moreover, some of the computer-based rankings included in the BCS scheme use unpublicized methodologies and have been criticized for their poor performance (Kirlin 2002, Martinich 2002).

These facts served as a motivation for many researchers to introduce their own computer-based ranking systems utilizing various mathematical techniques. The proposed approaches include models based on least-squares estimation, linear programming, maximum likelihood estimation, and neural networks (Bassett 1997, Harville 1977, Martinich 2002, Massey 2002, Wilson 1995). These methods take into account various factors and parameters, and they are often too complicated to be understood by people without an appropriate mathematical background. Moreover, in many cases the implementation of these methods is not an easy procedure. The website (Massey 2002) maintains weekly rankings produced by more than 70 different methods.

Plethora of sophisticated ranking systems made the life of ordinary football fans hard, since the rankings produced by different methods may significantly deviate, which means that the performance of their favorite teams may be underestimated or overestimated. Obviously, most of the fans cannot check if a certain ranking system is fair. One can argue that the main goal of any sports tournament (and the ranking system as one of its most important parts) is the fans' satisfaction, therefore, the ranking principles must be *consistent*, but at the same time *explicitly known* and *simple enough* to be understood and reproduced by non-specialists.

As it was pointed out above, the main difficulty one accounters in developing a college football ranking system is the fact that in the NCAA college football tournament the number of games played by every team is very small, and, obviously, one cannot expect the *quality of the opponents* of different teams to be the same. If one tries to rank teams using regular performance measures such as winning percentage, which are suitable for other competitions (for example, NBA, NHL, and MLB, where all teams play each other several times during the season), the results may be inconsistent. Therefore, one of the crucial issues that must be addressed in developing an efficient college football ranking system is taking into account the strength of the opponents of each team.

Another important subject that has been widely discussed and caused controversial opinions is whether the *margin of victory* should be taken into account in the rankings. At the first glance, one can claim that a team that outscores the opponent in a blowout game should stand higher in the rankings than a team who managed to win a close game, and considering score differentials in head-to-head games would provide more accurate rankings. However, several forcible arguments indicate that ranking systems should eliminate the motivation for teams to increase the margin of victory in blowout games, since otherwise it would lead to poor sportsmanship and greatly increase the risk of injuries. One should emphasize that the victory itself, but not the score differential, is the ultimate goal of any sports competition, therefore, the margin of victory should be either not taken into account at all, or limited by a certain (small) amount. Although Martinich (2002) claims that ignoring the margin of victory makes rankings less accurate, in this chapter we will see that it is possible to develop ranking systems that utilize relatively simple principles, take only win–loss information as the input and provide very reasonable results.

Summarizing the above arguments, a "fair" ranking system should

- utilize simple mathematical techniques;
- be available for verifying by non-specialists;
- use win–loss information only (or limit score margins);
- produce reasonable and unbiased results.

In this chapter, we describe two mathematical models for college football rankings that satisfy these criteria to a certain extent. One of these techniques is so-called Colley Matrix Method, which has been recently used as a part of the BCS system. Although the idea of this method is rather simple, it automatically takes into account the schedule strength of each team (while ignoring the margin of victory). This method is presently used as one of the official computer-based rankings in Bowl Championship Series.

Another approach presented here utilizes the *Analytical Hierarchy Process* (AHP), a universal analytic decision making tool used to rank alternatives of various types. This methodology proved to be very efficient in many practical applications, however, it remained unemployed in college football rankings, which can also be treated as ranking the alternatives (*i.e.*, football teams). The AHP method is believed to be a promising college football ranking technique.

Both of these models utilize *matrices* as their main attributes. In particular, the idea of the AHP method is to construct the *comparison matrix* whose elements have certain values determined by the comparison of different pair of alternatives (teams) based on the game outcomes. The principles of constructing this matrix are specifically designed for situations where not all pairs of alternatives can be directly compared, which is exactly the case for a college football tournament.

Numerical experiments presented in the chapter show that despite their simplicity and minimum input information, these approaches yield very reasonable results.

The remainder of this chapter is organized as follows. Section 2 provides the description of the Colley Matrix method for college football rankings. In Section 3 we briefly summarize the main ideas of the Analytic Hierarchy Process methodology, which is then used to develop a college football ranking system. Section 4 presents the results of numerical testing of the described approaches using scores from the last 2 college football seasons (2001-2002). Finally, Section 5 concludes the discussion.

2 Colley Matrix Method for College Football Rankings

One of the well-known mathematical approaches to college football rankings is the *Colley Matrix Method* (Colley 2003), which was recently developed in attempt to produce relatively "fair" and unbiased rankings and is now used as a part of BCS. Among the advantages of this approach one should mention that its main idea is rather simple, which makes this technique easy to understand and implement. Moreover, wins and losses (regardless of score differentials) are the only input information used in the model, which is reasonable due to the arguments presented above. As we will see in this section, the Colley Matrix Method can efficiently take into account the schedule strength of each team, which leads to rather realistic results. Mathematical techniques underlying this ranking system are briefly described below.

Let $n_{w,i}$ be the number of games won by a given team i, and $n_{total,i}$ be the total number of games played by this team. Instead of the winning ratio (defined simply as $n_{w,i}/n_{total,i}$) which is commonly used in practice, a modified quantitative measure of the team's performance is introduced. For any team i, the rating of this team r_i is defined as

$$r_i = \frac{1 + n_{w,i}}{2 + n_{total,i}}. \tag{1}$$

The motivation for this definition is to avoid the values of winning ratios equal to 0 (for the teams with no wins) or 1 (for the teams with no losses), which makes the comparison of such teams inconsistent: for instance, after the opening game of the season the winning team (1 win, 0 losses) is "infinitely better" than the losing team (0 wins, 1 loss). According to Formula 1, the winning team ($r = 2/3$) in this case would have a twice better score than the losing team ($r = 1/3$), which is more reasonable from the practical perspective. Also, note that the default rating of any team with no games played is equal to 1/2, which is the median value between 0 and 1. A win increases the value of r, making it closer to 1, and a loss decreases r towards 0.

After introducing this quantitative performance measure, one needs to adjust it according to the strength of the corresponding opponents. For this purpose the following transformation of the values of n_w is applied. Instead of considering the actual number of wins

$$n_w = \frac{(n_w - n_l)}{2} + \frac{n_{total}}{2} = \frac{(n_w - n_l)}{2} + \sum_{j=1}^{n_{total}} \frac{1}{2},$$

the effective number of wins n_w^{eff} is calculated by adjusting the second term of the above expression, which represents the summation of n_{total} terms equal to 1/2 (index j stands for j-th opponent) corresponding to the default rating of a team with 0 games played. In order to take into account the strength of the opponents, these terms are substituted by actual ratings of the opponent teams r_j, which yields the following formula for the effective number of wins for a given team i:

$$n_{w,i}^{eff} = \frac{(n_{w,i} - n_{l,i})}{2} + \sum_{k=1}^{n_{total,i}} \chi_{ijk} r_j, \tag{2}$$

where $\chi_{ijk} = \begin{cases} 1, \text{if team}\, i's\ k^{\text{th}} \text{ game was against team} j \\ 0, \text{otherwise.} \end{cases}$

Now, using Formulas (1) and (2), for every team i one can write the following linear equation relating the ratings of this team and its opponents:

$$(2 + n_{total,i}) r_i - \sum_{j=1}^{n_{total,i}} \chi_{ijk} r_j = 1 + \frac{(n_{w,i} - n_{l,i})}{2}. \tag{3}$$

If the total number of teams playing in the NCAA Division I-A tournament is equal to N, then the equations of this form will be written for all N teams, which results in the *linear system* with N equations and N variables. One can rewrite this system in a standard matrix form:

$$C\mathbf{r} = \mathbf{b}, \tag{4}$$

where

$$\mathbf{r} = \begin{bmatrix} r_1 \\ r_2 \\ \cdots \\ r_N \end{bmatrix}$$

represents the vector of variables,

$$\mathbf{b} = \begin{bmatrix} 1 + (n_{w,1} - n_{l,1})/2 \\ 1 + (n_{w,2} - n_{l,2})/2 \\ \cdots \\ 1 + (n_{w,N} - n_{l,N})/2 \end{bmatrix}$$

is the right-hand side vector, and

$$C = [c_{ij}]_{i,j=1\ldots n}$$

is the "Colley matrix", whose elements are defined as follows:

$$c_{ii} = 2 + n_{total,i},$$

$$c_{ij} = -n_{j,i},$$

where $n_{j,i}$ is the number of times the teams i and j played with each other during the season (most commonly equal to 0 or 1).

It turns out that the matrix C has nice mathematical properties, more specifically, it can be proved that it is positive semidefinite (Colley 2003), which enables one to efficiently solve the linear system 4 using standard techniques.

The solution of this system would represent the vector of numbers corresponding to the ratings of all N teams, and the resulting rankings are determined by sorting the elements of the solution vector $\mathbf{r}$ in a decreasing order of their values (*i.e.*, the highest-ranked team corresponds to the largest element in the solution vector, *etc.*).

3 Analytic Hierarchy Process (AHP) Method for College Football Rankings

In this section, we describe the Analytical Hierarchy Process – a powerful decision making technique for ranking alternatives. We first give a brief overview of the AHP methodology, and then apply it to college football rankings.

3.1 Analytic Hierarchy Process: General methodology

The *Analytic Hierarchy Process (AHP)* is a methodology for analytic decision making. It was introduced by Saaty in the late 1970's (Saaty 1977, Saaty 1980), and has been developed into one of the most powerful decision making tools ever since. Golden et al. (1989) describe the AHP as "a method of breaking down a complex, unstructured situation into its component parts; arranging these parts, or variables, into a hierarchic order; assigning numerical values to subjective judgments on the relative importance of each variable; and synthesizing the judgments to determine which variables have the highest priority and should be acted upon to influence the outcome of the situation".

The AHP is applicable to situations involving the comparison of elements which are difficult to quantify. It allows to structure the problem into a hierarchy[3] of simple components. For each of these components, the decision

[3] The word *hierarchy* is from Greek $\iota\epsilon\rho\alpha\ \alpha\rho\chi\eta$, meaning holy origin or holy rule.

maker performs pairwise comparisons of the alternatives which are then used to compute overall priorities for ranking the elements. In the simplest form, the hierarchy used in the AHP consists of three levels (see Figure 1). The goal of the decision is at the highest, first level. Alternatives to be compared are located at the lowest, third level. Finally, the criteria used to evaluate the alternatives are placed at the middle, second level.

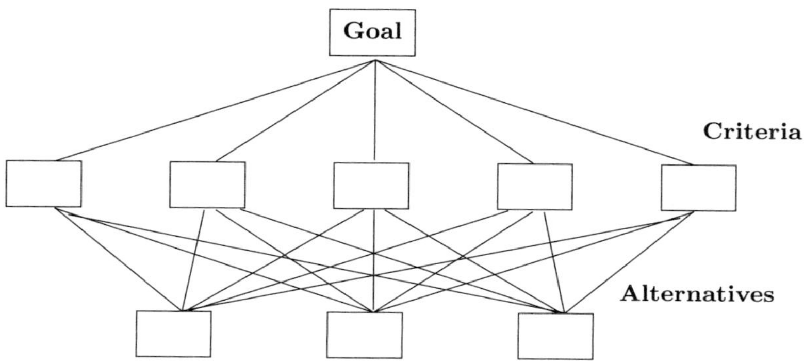

Fig. 1. A three-level hierarchy in AHP.

After defining a hierarchy, the decision maker compares pairs of alternatives using the available criteria and for each compared pair provides a ratio measure which characterizes the relative level of preference of one alternative over the other under the given criterion.

Assume that there are n elements (alternatives, options) to be ranked. As a result of performing pairwise comparisons, a matrix P is created, which is called the *dominance* or *preference matrix*, and whose elements are

$$p_{ij} = \frac{w_i}{w_j},\ i,j = 1,\dots,n.$$

Here numbers w_i and w_j are used to compare the alternatives i and j. To compare two options, a 10-point scale is often used, in which $w_i,\ i = 1,\dots,n$ are assigned values from $\{0,1,2,\dots,9\}$ as follows. If alternatives i and j cannot be compared then $w_i = w_j = 0$. If $i = j$, or i and j are equal alternatives, then $w_i = w_j = 1$. Otherwise,

$$w_i = \begin{cases} 3 & \\ 5 & \text{if } i \text{ is} \\ 7 & \\ 9 & \end{cases} \begin{matrix} \text{moderatly} \\ \text{strongly} \\ \text{very strongly} \\ \text{extremely} \end{matrix} \quad \text{preferable over } j.$$

The numbers $2,4,6,8$ are used for levels of preference compromising between two of the specified above. In all of these cases, w_j is set equal to 1. For example, if element i is strongly preferable over element j, we have $p_{ij} = 5$ and $p_{ji} = 1/5$. Zeroes are used when there is no enough information to compare two elements, in which case the diagonal element in each row is increased by the number of zeroes in that row. The above scale is used as an example, however, in general the comparisons could be made using a scale consisting of any set of positive numbers.

The constructed preference matrix is used to derive the n-vector of *priorities* which characterize values of the corresponding alternatives. The larger is the priority value, the higher corresponding alternative is ranked. Given the matrix P, one of the techniques used to derive the vector of priorities and subsequently rank the elements is the following *eigenvector solution.*

Suppose that the vector of priorities $w = [w_i]_{i=1}^n$ is known. Then if we construct the preference matrix and multiply it by w, we obtain

$$Pw = \begin{bmatrix} w_1/w_1 & w_1/w_2 & \cdots & w_1/w_n \\ w_2/w_1 & w_2/w_2 & \cdots & w_2/w_n \\ \vdots & \vdots & \ddots & \vdots \\ w_n/w_1 & w_n/w_2 & \cdots & w_n/w_n \end{bmatrix} \begin{bmatrix} w_1 \\ w_2 \\ \vdots \\ w_n \end{bmatrix} = n \begin{bmatrix} w_1 \\ w_2 \\ \vdots \\ w_n \end{bmatrix}.$$

Therefore, n is an eigenvalue of P with corresponding eigenvector w.

For the comparisons to be consistent, we need to have $p_{ij}p_{jk} = p_{ik}$ for any three alternatives i, j and k. However, in many cases, we can give only estimates of the ratios w_i/w_j, so there may be inconsistencies. In fact, in football the inconsistency and even intransitivity in scores happens quite often when, say team i beats team j, team j beats team k, who in its turn beats team i.

To find an approximation of w, we solve the problem $Pw = \lambda_{max}w$, where λ_{max} is the largest (*principal*) eigenvalue of P, and P is now an estimate of the true preference matrix with $p_{ij} = 1/p_{ji}$ forced (however, this matrix need not be consistent). The solution w is then used as the vector of priorities and the ranking of alternatives is performed as follows. Element i is assigned the value of $w(i)$, and the elements are ranked accordingly to the nonincreasing order of the absolute values of the components of vector w.

A natural question is, how good the obtained ranking is, or how to measure the error appearing as a result of inconsistency? To answer this question, a certain consistency criterion is introduced. It appears that $\lambda_{max} \geq n$ always, and P is consistent if and only if $\lambda_{max} = n$. The *consistency index* (C.I.) of a matrix of comparisons of size $n \times n$ is defined as

$$\text{C.I.} = \frac{\lambda_{max} - n}{n - 1}.$$

The *consistency ratio* (C.R.) is given by C.R. = C.I./R.I., where R.I. is an average *random consistency index* obtained from a sample of randomly generated reciprocal matrices using the corresponding scale. For example, for the aforementioned 0–9 scale, the values of R.I. for $n = 1, \ldots, 11$ are given below:

n	1	2	3	4	5	6	7	8	9	10	11	$\cdots$
R.I.	0	0	0.52	0.89	1.11	1.25	1.35	1.40	1.45	1.49	1.51	$\cdots$

The consistency ratio of up to 0.10 is considered acceptable.

Variations of the Analytic Hierarchy Process have been successfully applied to solve complex decision-making problems arising in economics, politics, technology and many other spheres. For more detail on the AHP methodology and its applications the reader is referred to (Golden et al. 1989, Saaty 1980, Saaty and Vargas 1994).

3.2 Application of AHP method to ranking football teams

In this section, we present an approach which can be considered a simple version of AHP for college football rankings. The alternatives in this model are represented by the football teams, and the only criterion used to compare them is outcomes of the games played. Our goal is to rank all the teams based solely on this criterion. To compare two teams, we use a simple three-point scale consisting of 0, 1 and 1.2. Namely, the comparison matrix $P = [p_{ij}]_{i,j=1}^{n}$ is constructed as follows.

- If teams i and j did not play each other, they cannot be compared directly and we assign $p_{ij} = p_{ji} = 0$.
- If $i = j$, or there was a tie between teams i and j, or teams i and j played each other twice within a season with alternative outcomes, then $p_{ij} = p_{ji} = 1$.
- If team i beats team j (once or twice), we set $p_{ij} = 1.2$ and $p_{ji} = 1/1.2$ (*i.e.*, we assume that the winning team is on average 1.2 times better than the loosing team).

Note, that this scale does not take into account the margin of victory, which is in agreement with the arguments presented above.

The next question that arises now is, how realistic are the results obtained by applying the ranking systems described in this chapter? It turns out that the rankings generated by both Colley Matrix method and AHP method are rather reasonable. The next section discusses the results of numerical experiments and compares them with the rankings produced by major press polls.

Table 1. Final rankings comparison: season of 2001.

AP	Team	Record	ESPN	COL	AHP	ARO
1.	Miami Fla	12-0	1	1	1	44.42
2.	Oregon	11-1	2	3	3	46.75
3.	Florida	10-2	3	4	4	38.42
4.	Tennessee	11-2	4	2	2	35.75
5.	Texas	11-2	5	6	9	53.00
6.	Oklahoma	11-2	6	10	10	53.46
7.	LSU	10-3	8	8	6	43.42
8.	Nebraska	11-2	7	5	5	47.85
9.	Colorado	10-3	9	7	8	44.00
10.	Washington St	10-2	11	11	11	49.27
11.	Maryland	10-2	10	13	13	50.42
12.	Illinois	10-2	12	9	7	48.25
13.	South Carolina	9-3	13	15	12	39.09
14.	Syracuse	10-3	14	12	15	47.38
15.	Florida St	8-4	15	16	16	38.25
16.	Stanford	9-3	17	14	14	43.33
17.	Louisville	11-2	16	17	25	69.92
18.	Virginia Tech	8-4	18	31	35	59.08
19.	Washington	8-4	19	18	18	42.00
20.	Michigan	8-4	20	20	17	40.50
21.	Boston College	8-4	23	30	31	55.83
22.	Georgia	8-4	25	24	20	45.50
23.	Toledo	10-2	22	28	43	83.33
24.	Georgia Tech	8-5	26	33	26	41.17
25.	BYU	12-2	24	19	29	76.50
27.	Fresno State	11-3	28	22	24	66.21
30.	North Carolina	8-5	27	25	22	40.77
33.	Arkansas	7-5	33	27	23	33.27
39	UCLA	7-4	41	23	19	40.36
NR	Auburn	7-5	37	26	21	37.75

4 Numerical Experiments

We now present the results of testing the approaches described above using scores from the last two college football seasons (2001–2002). These and other historical scores are available from James Howell's College Football Scores webpage (Howell 2003).

For both methods, only outcomes of the games played within NCAA Division I-A teams are taken into account (*i.e.*, all the games with lower division teams excluded). The results of the experiments are summarized in Tables 4–5. We compare the Colley rankings and AHP rankings that we obtained with Associated Press (AP) writers poll and ESPN/USA Today coaches poll. Only top 25 teams for each of the ranking systems are mentioned. The first column

Table 2. Final rankings comparison: season of 2002.

AP	Team	Record	ESPN	COL	AHP	ARO
1.	Ohio St.	14-0	1	1	1	46.36
2.	Miami	12-1	2	3	4	42.42
3.	Georgia	13-1	3	2	3	40.69
4.	Southern Cal	11-2	4	4	2	28.23
5.	Oklahoma	12-2	5	4	5	47.31
6.	Texas	11-2	7	6	6	48.31
7.	Kansas St.	11-2	6	11	11	52.09
8.	Iowa	11-2	8	8	9	52.77
9.	Michigan	10-3	9	7	7	39.77
10.	Washington St.	10-3	10	12	10	43.42
11.	Alabama	10-3	NR	10	15	48.23
12.	N.C. State	11-3	11	14	12	44.17
13.	Maryland	11-3	13	13	14	46.31
14.	Auburn	9-4	16	19	18	44.42
15.	Boise St.	12-1	12	17	30	81.46
16.	Penn St.	9-4	15	18	19	46.46
17.	Notre Dame	10-3	17	9	8	43.31
18.	Virginia Tech	10-4	14	20	22	52.21
19.	Pittsburgh	9-4	18	24	25	51.31
20.	Colorado	9-5	21	21	16	44.23
21.	Florida St.	9-5	23	15	13	33.79
22.	Virginia	9-5	25	25	17	40.71
23.	TCU	10-2	22	28	32	75.92
24.	Marshall	11-2	19	32	45	83.58
25.	West Virginia	9-4	20	23	21	44.67
26.	Florida	8-5	24	26	23	42.23
27.	Texas Tech	9-5	28	22	20	42.71
32.	South Florida	9-2	29	16	24	59.11

of each table contains a team's AP poll rank, followed by the team's name and season record in columns 2 and 3, respectively. Column 4 contains ranks due to ESPN/USA Today coaches poll. Column 5 presents ranks generated by the Colley Matrix method. Finally, the last two columns contain the ranks obtained using the AHP approach and the average final rank of the opponents (according to the AHP ranking) a team played during the season (ARO).

The first general observation is that both of the considered computer-based rankings are quite consistent with those by the press and coaches.

Another important observation is that both methods find a good balance in reflecting win–loss record and strength of the schedule in the final ranking. It should be noted that both methods adjust the rankings to the schedule strength "internally", *i.e.*, the quality of opponents is measured in terms of ranks of the opponent teams produced by the same ranking system, which excludes a possibility of overestimating or underestimating the schedule strength of any particular team by using some external parameters.

As one can see from the tables, the strength of the schedule is a significant factor in the rankings. For example, Marshall had a good win–loss record in 2002, however, they are only 32^{ns} in the Colley ranking and 45th in the AHP ranking due to the low "quality" of their opponents. On the other hand, despite the 9-5 record, Florida State is significantly higher in the AHP and Colley rankings than in the ESPN ranking because of their tough schedule.

5 Conclusion

In this chapter, we have discussed two matrix-based decision making models for ranking college football teams – Colley Matrix method and Analytical Hierarchy Process method. In both approaches, the information about all games played by every team in the season is summarized in a matrix, however, different principles are used in constructing this matrix.

Clearly, one cannot compare the obtained results with a "true" ranking, because the exact comparison measure between every pair of teams cannot be determined due to obvious reasons. However, as we have seen in this chapter, both methods produce reasonable results and exhibit a high level of agreement with respect to each other, as well as major press polls, which is the best measure of consistency of the rankings. Moreover, both methods are easy to understand and implement, which makes them attractive for practical use.

The Colley Matrix method is already being implemented as a part of the BCS. The AHP ranking method is a promising technique as well. Although in the simplest setup described in this chapter this method uses only wins and losses as input (which is motivated by the arguments presented above), the AHP method is very flexible, and it can be easily adjusted to include other criteria, such as margin of victory, home-field advantage, and experts opinion. However, even in its simplest form, the method produces very reasonable results, as evidenced by our experiments using the scores from the last two seasons. Moreover, variations of the AHP methodology could also be applied to other decision-making problems arising in sports, and in particular in college football. Therefore, we strongly believe that the AHP ranking method proposed in this chapter is practical and can be used in various sports applications to produce fair and unbiased results.

References

1. G. W. Bassett. Robust sports ratings based on least absolute errors. American Statistician, 51:17, 1997.

2. W. N. Colley. Colley's Bias Free College Football Ranking Method: The Colley Matrix Explained. http://www.colleyrankings.com/matrate.pdf. Retrieved September 26, 2003.
3. B. L. Golden, E. A. Wasil, and P. T. Harker. *The Analytic Hierarchy Process.* Springer-Verlag, 1989.
4. D. Harville. The use of linear-model technology to rate high school or college football teams. *J. Amer. Statist. Association*, 72:278-289, 1977.
5. J. Howell. James Howell's college football scores website. http://www.cae.wisc.edu/~dwilson/rsfc/history/howell/. Retrieved September 7, 2003.
6. R. Kirlin. How to fake having your own math formula rating system to rank college football teams. http://www.cae.wisc.edu/~dwilson/rsfc/history/kirlin/fake.html. Retrieved September 7, 2003.
7. J. Martinich. College football rankings: Do the computers know best? *Interfaces*, 32:85-94, 2002.
8. K. Massey. The college football ranking comparison website. http://www.mratings.com/cf/compare.htm. Retrieved September 7, 2003.
9. T. L. Saaty. A scaling method for priorities in hierarchical structures. *Journal of Mathematical Psychology*, 15:234-281, 1977.
10. T. L. Saaty. *The Analytic Hierarchy Process.* McGraw Hill Company, 1980.
11. T. L. Saaty and L. G. Vargas. *Decision Making in Economic, Political, Social and Technological Environments with the Analytic Hierarchy Process.* RWS Publications, 1994.
12. R. L. Wilson. Ranking college football teams: A neural network approach. *Interfaces*, 25:44-59, 1995.

Who's the Best? Data Envelopment Analysis and Ranking Players in the National Football League

Erickson H. DeOliveira and Robert Callum[1]

Dartmouth College, USA
edeoliveira@alum.dartmouth.org
robert.l.callum.tu03@alum.dartmouth.org

1 Introduction

The question of which players are best at any position, at any given time, is much more than a philosophical saw. Coaches must know who to play in what situations. In the face of imposed constraints (like the National Football League "salary cap") and the operational realities of running a sports franchise, personnel directors must build teams that maximize the potential winning percentage of their squads while minimizing expenditures on overall salary. Player agents and the players themselves need to know the relative strength of individuals in order to gauge what the market will bear for their services.

If the rankings are important, they are also riddled with subjectivity. There is rarely one "bottom-line" metric that can summarize all of a player's contributions. When comparing (American) football quarterbacks (QBs), which is more critical: completion percentage or the number of touchdowns thrown? And how should each be adjusted for interceptions? Further, as the game has evolved, the QB has become a threat to run with his feet, not just throw with his arm.[2] The current standard metric for QB evaluation ignores rushing yardage altogether.

[1] We would like to thank our colleagues Dan Richards and John McCairens, Professor Kenneth Baker, Dartmouth Coach Patrick O'Leary and Lloyd Lee of the Tampa Bay Buccaneers Personnel Department for their help on this paper.

[2] Donovan McNabb, Michael Vick and Daunte Culpepper personify the rushing QB, but they are not an entirely new breed. Long-time fans will remember that Fran Tarkenton and the (young) Steve Grogan were also serious running threats in their time.

We devise a methodology using data envelopment analysis (DEA) that takes disparate inputs and outputs and computes which players are "efficient." Further, taking each player's efficiency and cross-weighting it against the rest of his position can yield one measure of relative strength.

We begin with a brief overview of DEA, before explaining our ranking philosophy and methodology in detail. Decisions on which input and output measures to use, and how they are measured, are critical to achieving accurate rankings with DEA. Compared to the current standard, we believe that our methodology yields a better measure of the "best" QBs in the league. Having developed and justified the methodology, we then apply it to running backs where there are currently few, if any, competing systems. We conclude by explaining how the methodology could be expanded to defensive players.

2 Overview of Data Envelopment Analysis [3]

Data Envelopment Analysis (DEA) is a linear programming (LP) technique for measuring the efficiency of "decision-making units" (Charnes et al. 1978). Each decision-making unit (DMU) is defined by a series of input and output measures. The LP places certain weights on the inputs, such that the sum of the inputs must equal one. The LP then computes the highest possible output measure for the DMU in question, with the constraint that the weights chosen must yield an output for that DMU, and all the other DMUs under analysis, of no greater than one. DMUs that have an output of one are efficient under one possible output weighting scheme. DMUs whose outputs are less than one are inefficient, meaning that another DMU in the reference set could have produced the same output with fewer inputs.

For the literary-minded, DEA is a "Panglossian" method. Like the ever-optimistic Dr. Pangloss in Voltaire's *Candide*, the output measure for a DMU is its score in "this best of all possible worlds" (Voltaire, 1991). DEA was first developed for use in analyzing non-profit entities, where there might be several different output measures and no obvious method for ranking outcomes. If one were to rank equivalent trauma centers,[4] for example, which would be the "best": the one with the highest patient to doctor ratio, the lowest mortality rate, the least expensive cost per admittance, or the shortest time to discharge? Given these four output measures, the LP would distribute weights among the outputs endeavoring to show the trauma center under study (the DMU) in the best possible light. An expensive center with a low mortality rate could earn an efficiency of one, with a weighting scheme skewed towards

[3] The formal definitions of the linear programs that we used can be found in the methodology section.

[4] In this case, we assume that the trauma centers are drawing on the same population of patients.

the mortality output. Similarly, a relatively inexpensive trauma center with a low doctor to patient ratio could also garner a perfect efficiency score based on a different weight selection. Since, a priori, there was no clear "best" weighting scheme, the analyst has two key findings courtesy of DEA: the output weightings that result in efficient trauma centers, and those centers which, even when viewed in their best possible light, are still "inefficient." Further, by comparing the inefficient DMUs with their efficient peers, the pathway towards improvement may be illuminated.

2.1 Issues in DEA

The illustrative trauma centers and National Football League (NFL) QBs share many traits. A priori, the analyst cannot discern which of the many different QB output measures is most important, and therefore DEA is appropriate. After the analysis is run, however, the ranking may beg more questions than it answers. If a trauma center is efficient because it weighs low cost to the exclusion of every other factor (like mortality rate), the ranking risks being theoretically unassailable and practically useless. If there are too many efficient DMUs, some type of differentiation mechanism must be used to rank those units on the efficiency frontier. Finally, there is the possibility that a particular DMU, while not the best performer on any one particular output has high scores across the board. It is possible that this DMU could be viewed as "inefficient," even though it has the best overall performance.

These serious methodological issues can to a great extent be overcome by using an aggressive cross weighting scheme (Doyle and Green 1995). We discuss the implementation formally in the methodology section, but in summary each DMU is subjected to two LPs. In the first, the goal is simply to make oneself as efficient as possible. In the second, the DMU's efficiency is constrained to be greater than or equal to the efficiency of the first LP, while maximizing the difference between one's own efficiency and the average efficiency of the entire data set. This two-step process ensures that each DMU chooses the unique set of weights that not only maximizes its efficiency but also casts the other DMUs in the worst possible light. Doyle and Green show that this methodology effectively breaks ties, rewards strong overall performers whether or not they are perfectly efficient, and also penalizes DMUs that choose weighting schemes (like our low cost trauma center) that are not realistic.[5]

[5] It does so by effectively penalizing those DMUs that choose a weighting scheme far from the average of the data set. Adler et al. (2002) refer to these DMUs as "mavericks."

3 Philosophy of QB Rankings

The same caveats apply to DEA as with any optimization scheme: the results of the model are only as good as the data and the LP design. We downloaded our data from publicly available statistics pages found on www.nfl.com (the main National Football League site) and www.cnnsi.com (the web site of the magazine Sports Illustrated).[6] We cross-verified the data against each site (the data on the two sites were tabulated independently) and also performed a common-sense sanity check. We found no obvious errors.

3.1 Design of the linear programs

As is explained in greater detail in the methodology section, we designed the LPs to require no pre-defined ceilings or floors on the output weights: the aggressive cross-weightings effectively penalize those DMUs who choose weights that are not practically valid (in the language of sport, weights that are out of bounds). Rendering ceilings and floors unnecessary does not free the analyst from the taint of arbitrary judgment, however. Implicitly, analysts are zero-weighting all the outputs they choose not to use (Allen et al. 1997). Therefore, a great deal of thought needs to accompany the design of any DEA LP.

We chose six output measures on which to judge NFL QBs during a particular season: total passing yards, passing yards per game, total completions, passing attempts per interception, total rushing yards, and total touchdowns (both passing touchdowns and rushing touchdowns). Our two input measures were total passing attempts and total rushes. As explained further below, the current QB ranking system ignores a QB's rushing yards. A QB who is a threat to run adds another dimension to his team's offense, and presents greater challenges to the other team's defense. We believe that a methodology that does not include QB rushing yardage is biased in favor of pure pocket passers,[7] and ignores the flexibility a rushing QB represents.

While rushing yards are important, a QB's main job is still to throw the ball, which explains why four of the outputs measure different facets of passing skill. Accuracy is measured both by the number of completions and by the number of attempts per interception.[8] Better QBs have a strong tendency to throw for more yards, hence total passing yards in a season and the number of passing yards per game are two additional outputs.[9]

---

6 The data set as analyzed is available from the authors.

7 Those QBs that stay in the "pocket" of protection afforded by the offensive linesman (from the onrushing defense) and look solely to pass the ball. If pressured by the defense, a pocket passer will seldom use his legs to scramble away.

8 For the challenges of capturing a negative output measure in DEA, see the methodology section.

9 Different QBs in the data set played a different number of games, and so these two outputs are capturing slightly different passing dimensions. Yards per game

Our last output measure combines both rushing and passing touchdowns (TDs). The ultimate goal of an NFL QB is to win games, and that is accomplished by putting points on the board. Whether the QBs scored their touchdowns through the air or on the ground was immaterial to winning games.[10] We felt that the DEA would be more accurate by combining total touchdowns in one category, rather than placing separate output weights on rushing TDs and passing TDs.

We experimented but ultimately rejected two other potential output measures. Conversations with football coaches dissuaded us from using sacks[11] as an adverse output measure. The coaches commented that a high number of sacks could be a reflection on the weakness of the offensive line, rather than the lack of rushing ability in a QB. Further, there was little correlation between a high number of rushing yards for a QB and a low number of sacks. The measure yards per attempt was rejected since it implicitly valued the QB who threw longer passes over those whose offensive schemes favored the short pass. Whether completing a high number of short passes or a low number of long passes, the total yards should be equivalent, and as long as the QB is producing touchdowns, whether those touchdowns came on one long play or several shorter plays is irrelevant.[12]

Following the same logic, the argument can be made that the completion category is biased in favor of QBs whose offensive schemes utilized more short passing plays (with higher completion percentages). This is probably true, but ultimately we decided that completions were a valid output measure since completions are generally recognized as an indicator of QB accuracy. Further, a higher number of completions usually leads to longer drives and more points scored.

---

is obviously a per capita metric, while total yards for a season rewards QBs who were hardy enough, and successful enough, to play in the vast majority of their team's games.

[10] The question of whether a rushing TD is equivalent to a passing TD is an area of future study. Coaches and fans generally acknowledge that a passing play is riskier than a running play (the old saw is when you pass the ball, three things can happen and two of them are bad. The ball can be intercepted, fall incomplete, or be caught by the intended receiver). Yardage gained per passing play is almost always greater than yardage gained per rushing play, but the risk of the pass is greater as well. The risk/reward ratio of passing plays may call for a higher "discount rate" per yard than an identical yard gained on the ground. Following such logic, one may value a rushing TD more than a passing TD since while the points awarded for each are identical, the discount rate for the rushing play is much lower. While we find this argument compelling, for the purposes of this paper we treat all touchdowns equally.

[11] A sack occurs when the QB is tackled behind the line of scrimmage (the yard line where the play begins) for a loss.

[12] Again, assuming that the discount rate for each type of play is the same.

3.2 Improvements on the QB passing formula

The above criteria improves on the traditional means of evaluating QBs in two key respects: it combines overall performance at the QB position, and it directly compares quarterback production "outputs" against the "raw material" (the number of rushes and passes attempted) "inputs." In contrast, the current QB rating is a rigid formula indexed to the average performance of a quarterback in 1970.

The NFL.com website describes the calculation by which a score is assigned to all NFL QBs on the basis of their passing performance.
Let

C be the number of completed passes thrown
A be the number of passes attempted
Y be total passing yardage accumulated
T be the number of passes which resulted in TDs
I be the number of passing attempts which were intercepted

The passing rating is then calculated as follows:

$$\text{QBRating} = \left\{ \begin{array}{l} (C/A - 0.3)/0.2 + (Y/A - 3)/4 + \\ (0.095 - I/A)/0.04 \end{array} \right\} \times (100/6) \quad (1)$$

Some features in this formula can be linked to the DEA methodology: completions-per-attempt, yards-per-attempt and interceptions-per-attempt are each productivity metrics which link output measures to input measures. Unfortunately, the scaling multiples applied are purely arbitrary, and no attempt is made to account for a QB's rushing performance.

The DEA methodology we describe below is superior in our opinion for three reasons: the method is an objective comparison instead of an arbitrary calculation, the resulting difference in efficiency scores between two players is not subject to the arbitrary choice of scaling constants in the denominators of equation (1) (DEA results are robust to scaling), and it can be extended to other player positions for which there has been no previously accepted ranking formula.

4 Methodology

We establish a linear program by first determining the output measures against which players will be evaluated, for example, yards gained, TDs scored. Because DEA is a productivity-based tool, we also require one or more input measures. Note that the number of input measures need not equal the number of output measures: from a single input, Carries, we can have several output measures of a running back's performance – yards gained, first downs gained, TDs scored, etc.,

Let

$I_1, I_2, I_3, \ldots$ be the values for the input measures, and
$v_1, v_2, v_3 \ldots$ be the corresponding weights for each input

Similarly, let

$O_1, O_2, O_3, \ldots$ be the values for the output measures, and
$u_1, u_2, u_3 \ldots$ be the corresponding weights for each output

We can now write the efficiency for player K as follows,

$$E_k = \sum_j (O_{kj} u_{kj}) / \sum_i (I_{ki} v_{ki}) \tag{2}$$

4.1 The linear program

We construct a linear program to maximize E_k by selecting weights for both input and output measures (the v_i's and u_j's) against which every player is evaluated.

Constraints

The LP is run N times, once for each player, such that the following three constraints are met:

$$\sum_i (I_{ki} v_{ki}) = 1 \tag{3}$$

$$E_{pk} = \sum_j (O_{pj} u_{kj}) / \sum_i (I_{pi} v_{ki}) \leq 1, \forall p \tag{4}$$

$$v_{ki}, u_{kj} \geq 0, \forall i, j \tag{5}$$

The first constraint, equation (3), allows us to simplify the objective function, and maximize the weighted sum of the outputs. We then rearrange equation (4) into the inequality shown in equation (6), so that the all constraints are linearized. This allows us to implement the Simplex algorithm and locate global optima.

$$\sum_j (O_{pj} u_{kj}) - \sum_i (I_{pi} v_{ki}) \leq 0, \forall p \tag{6}$$

This is the original model developed by Charnes, Cooper and Rhodes (1978). With as many as 69 QBs to evaluate in the 2002 NFL season, however, the model exhibits a familiar problem: DEA analysis will often find a large number of the players to be 100% efficient. To overcome this, and improve discrimination among player rankings, we adopt two additional measures.

Cross-Evaluation

Firstly, we conduct a cross-evaluation of all N players by assessing each player using every other player's $v_i$and u_j weights. Each player then earns an efficiency score calculated by using his mean efficiency under the round-robin format.

That is,

$$E_{kp} = \sum_j (O_{kj} u_{pj}) / \sum_i (I_{ki} v_{pi}) \tag{7}$$

$$H_k = (\sum_p E_{kp})/N \tag{8}$$

where E_{kp} is player K's efficiency using player P's weights, and $H_k$is the average cross-evaluation for player K under all schemes. We choose to rank players according to H_k, reflecting the result of the round-robin format. In other words, the "best players" are those who have the highest average score when compared against the weighting schemes of the entire population.

Secondly, we create another LP in which we implement Doyle and Green's Type I aggressive option for further discriminating among DMUs (Doyle and Green 1995). This approach encourages the selection of weights for a particular player that not only make his E_{kk} score as strong as possible, but also makes the collective mean score for the rest of the population as weak as possible relative to his.

This can best be explained by noting that a player may have several weighting schemes that maximize his efficiency score. Because the solution that maximizes a player's efficiency is not necessarily unique, that player has the option of selecting the weighting scheme that accomplishes the dual goal of maximizing his own efficiency, while simultaneously punishing everyone else.

Doyle and Green express these two objective functions as:

For player K

$$\text{LP\#1}: \quad \text{maximize } E_{kk} \tag{9}$$

$$\text{LP\#2}: \quad \text{minimize } (\sum_p (T_{kp} - B_{kp}))/(N-1), \forall p \neq k \tag{10}$$

where T and B are the numerator and denominator of the cross-efficiency calculation. Respectively, the sum of the weighted outputs using K's weights and P's scores, and the sum of the weighted inputs using K's weights and P's scores:

$$T_{kp} = \sum_j O_{pj} u_{kj} \tag{11}$$

$$B_{kp} = \sum_i I_{pi} v_{ki} \tag{12}$$

4.2 The adverse output measure

Adverse output measures represent a challenge for DEA. QBs, for example, want to minimize the number of interceptions thrown. The LPs as constructed, however, view all of the output measures as "positives." We found the best way to account for interceptions was to create a positive measure that QBs could attempt to maximize. Hence, instead of counting interceptions as an output, we use attempts per interception (similarly, when ranking the running backs, we use carries per fumble). This positive measure yields better results than other potential solutions, including using the reciprocal of interceptions or the touchdown to interception ratio.[13]

4.3 Indexed or scaled data and tolerance

As a practical matter, one does not want more than a three or four order of magnitude difference in the raw input and output data. Scaling the data can obviate the tendency for an LP to produce linearity errors because of these magnitude differences. A more serious concern occurs because output scores are constrained to be less than or equal to one. It is possible that when using output measures of relatively large magnitude (such as the number of yards thrown by a QB during a season), the weights assigned to those measures can approach zero. Depending on the tolerance settings of the LP software used, these very small weights can be interpreted as zero. To obviate this problem, we indexed all of the input and output data on a zero to one scale, with the best score in a given category coded as one, and the other scores represented as a percentage of one.

[13] With the reciprocal, we had certain QBs with many interceptions still placing a great deal of weight on that output measure, because it distanced them from the other DMUs in the data set. By using the touchdown to interception ratio, the analyst is implicitly saying that an interception is as equally negative as a touchdown is positive. We did not want to make this somewhat questionable assertion.

4.4 Selected output measures

As we have discussed, the benefit of DEA is the ability to exercise judgment in the selection of input and output measures. Our choices of measures for both QBs and running backs appear in Table 1.

Table 1. Selected statistics for DEA analysis of QBs and RBs.

		QBs	Running backs
Inputs			
	I_1	Passing Attempts	Rushing Attempts
	I_2	Rushing Attempts	Receptions
Outputs			
	O_1	Passing Yards	Rushing Yards
	O_2	Rushing Yards	Receiving Yards
	O_3	Total TDs	Total TDs
	O_4	Attempts per Interception	Carries per Fumble
	O_5	Completions	
	O_6	Passing Yards per Game	

5 Results

5.1 Validation of existing rankings

We restricted the DEA to purely passing statistics for QBs, and purely rushing statistics for running backs in an effort to validate our results against conventional wisdom. The results for the top 10 QBs and running backs in the 2002 NFL season appear in Tables 2 and 3. We repeated this validation exercise for the 2001 NFL season and arrived at similarly encouraging results.

The results for QBs highlight the arbitrary nature of the QB passing rating formula. QBs with high passing ratings are also ranked highly under DEA. However, the DEA methodology appears to better capture the productivity of the player taking the snaps by directly linking the number of plays for which a QB is responsible, with the outcomes from a QB's throws.

In the case of running backs, there is some correlation between the league-leaders in rushing and the productivity-based DEA rankings (five of the ten rushed for more than 1000 yards). This appears to be counter-intuitive at first, in the sense that absolute rushing yardage over the course of a season would seem to have less to do with productivity than Yards per Carry, and more to do with a back's durability. Two factors contribute to this. Firstly,

Table 2. QB passing ratings and "pure-passing" DEA rankings.

Ranking	Player	DEA Efficiency	NFL QB Rating
1	Chad Pennington	99%	104.2
2	Brad Johnson	91%	92.9
3	Donovan McNabb	90%	86.0
4	Jay Fiedler	89%	85.2
5	Rich Gannon	89%	97.3
6	Trent Green	87%	92.6
7	Tommy Maddox	85%	85.2
8	Brian Griese	85%	85.6
9	Matt Hasselbeck	84%	87.8
10	Jeff Blake	83%	77.3

Table 3. League rushing leaders and "pure-rushing" DEA rankings.

Ranking	Player	Efficiency	2002 rushing yards
1	Priest Holmes	99%	1615
2	Charlie Garner	98%	962
3	Kevan Barlow	93%	675
4	LaDainian Tomlinson	92%	1683
5	Kenny Watson	91%	534
6	Jamel White	91%	470
7	James Stewart	90%	1021
8	Curtis Martin	89%	1094
9	Shaun Alexander	89%	1175
10	T.J. Duckett	87%	507

the league rushing leaders achieve their success by virtue of strong per-Carry performance, which allows a team to run the ball consistently and play possession football. Secondly, because we restricted our analysis to those backs with more than 100 carries in the 2002 season in an effort to minimize statistical fluctuations associated with small samples, we have explicitly biased the evaluation towards backs with more yards gained on the season.

Ricky Williams is conspicuous by his absence on this list. The 2002 NFL rushing leader earns a cross-evaluated efficiency score of 51% and places 36th, and an examination of the aggressive cross-weightings selected by other players offers a hint to explain his poor ranking. In the second LP, a significant number of players choose to weight fumble performance heavily. In fact, running backs with more than 100 carries in the 2002 NFL season fumbled on average once every 157 carries. Ricky Williams fumbled the ball five times on 383 carries, or twice as often as his peer backs on a per-carry basis. If we genuinely believe that running backs should be evaluated on their ball-handling as strictly as QBs are scrutinized for their interceptions, then DEA accordingly penalizes Ricky Williams for his fumbles. Note that (were the statistics readily available), we could change the LP to evaluate running backs on the basis of fumbles that resulted in a turnover, which would then be more analogous to our QB evaluation on the basis of interceptions as an adverse output measure.

5.2 Global performance ranking

Tables 4 and 5 present the results of a DEA evaluation using the input and output characteristics shown in Table 1. The results for both QBs and running backs agree with intuition. When compared against the QB passing rating, the QB DEA that includes rushing performance features many of the strongest passers, but also recognizes serious rushing threats like Michael Vick. Similarly, when running backs are evaluated on their rushing and receiving performance, more versatile players like Kevan Barlow become top performers.

5.3 Stability of results

Because we expect statistical noise to arise from rushers with few carries, or QBs who infrequently take snaps, we tested the stability or our rankings by running the LPs for larger and smaller fields of players. Our conclusion is that the QB rankings exhibit very little fluctuation for players with more than 250 throws in the 2002 season. For running backs, the rankings were stable when we limited the field to backs with more than 100 carries in the 2002 season. These thresholds correspond roughly to the average number of passing attempts and carries for players at those positions. This implies that the mean of a key input statistic may be a useful cutoff to eliminate players who lack enough playing time to be fairly evaluated (thus avoiding spurious results arising from statistical noise); this heuristic appeared equally valid when we back-tested our approach on 2001 NFL data.

6 Conclusion

6.1 Managerial applications

Our DEA offers the possibility of better informing managerial decision-making in three respects: rigorously evaluating overall player performance, identifying undervalued properties, and customizing the analysis for a given franchise's playing system.

Rigorous player evaluation

Whereas the stars in any sport are easily identified, the relative performance of one player versus another is less obvious, and DEA provides a head-to-head comparison of players on the basis of their on-field production.

Table 4. Pass ratings, DEA rankings for overall QB pass/run performance (for players with more than 250 passing attempts in the 2002 NFL regular season).

Ranking	Player	DEA Efficiency	NFL QB Rating
1	Brad Johnson	91%	92.9
2	Chad Pennington	90%	104.2
3	Trent Green	86%	92.6
4	Donovan McNabb	86%	86.0
5	Rich Gannon	86%	97.3
6	Peyton Manning	84%	88.8
7	Matt Hasselbeck	84%	87.8
8	Michael Vick	83%	81.6
9	Brian Griese	82%	85.6
10	Tommy Maddox	81%	85.2
11	Brett Favre	81%	85.6
12	Drew Bledsoe	81%	86.0
13	Mark Brunell	80%	85.7
14	Jay Fiedler	80%	85.2
15	Jeff Garcia	79%	85.6
16	Jon Kitna	79%	79.1
17	Steve McNair	79%	84.0
18	Tim Couch	79%	76.8
19	Daunte Culpepper	79%	75.3
20	Tom Brady	77%	85.7

Identification of undervalued properties

In a salary-cap world such as the NFL, DEA presents the intriguing proposition of using player salary as an input statistic for valuing player contracts. This gives General Managers (GMs) a technique for identifying undervalued players in the draft or free agency, who represent good playing value for the salary cap space they occupy.

Customized analysis

While coaches and GMs will never get away from watching tape on players they hope to acquire, DEA is a sufficiently flexible tool that franchises can adapt to their systems. For example, a team that always runs between the tackles may not want to evaluate running backs on the basis of their receiving performance; whereas another team may want to expand the scope of their evaluation to include a running back's performance on special teams.

The analysis can also accommodate a coach who wishes to exercise judgment in crafting the linear program. For example, if a running back's rushing must be given more importance than his receiving performance, Talluri (2000) proposes adding additional weight restrictions of the form $u_2 > u_1$ to the first LP.

Table 5. DEA rankings for overall RB run/receiving performance (for players with more than 100 carries in the 2002 NFL regular season).

Ranking	Player	Efficiency
1	Kevan Barlow	89%
2	Priest Holmes	89%
3	Charlie Garner	87%
4	Jerome Bettis	84%
5	James Stewart	82%
6	Shaun Alexander	82%
7	LaDainian Tomlinson	81%
8	Emmitt Smith	81%
9	William Green	79%
10	Duce Staley	77%
11	T.J. Duckett	76%
12	Curtis Martin	76%
13	Antowain Smith	76%
14	Kenny Watson	75%
15	Amos Zereoue	75%
16	Tyrone Wheatley	72%
17	Jamel White	72%
18	Clinton Portis	72%
19	Jonathan Wells	71%
20	Eddie George	71%

6.2 Professional and franchise applications

In a sports entertainment industry that is always asking who the top individual is, DEA rankings of football players are objective. This objective focus on efficiency may not satisfy those who favor total production (and longevity) in voting for MVPs and other accolades, but for the first time it may provide a means of recognizing top players performing in specialized roles. For example, in the 2002 season Oakland Raiders' running back Zack Crockett was used as a short-yardage specialist, recording 8 TDs on only 40 carries in the whole season.[14] Valuable players in this mold would finally be recognized under a DEA evaluation scheme.

6.3 Scope of DEA – defensive players

We have presented an application of DEA that has been confined to offensive players, in positions with a rich statistical basis for evaluating performance. There is reason to believe, however, on the basis of agreement with traditional

[14] Because we limited the field to backs with more than 100 carries, Zack Crockett was regrettably excluded from the rankings in our formulation of the DEA LP.

rating systems and conventional wisdom, that DEA is a robust tool that can be extended to defensive positions as well.

Given that there is no widely accepted "power ranking" for linebackers, defensive backs or defensive linesmen, DEA shows promise as a rigorous tool to directly compare player performance across the league on the other side of the ball. Table 6 summarizes possible input and output measures for evaluating linebackers, on the basis of the arguments outlined above.

Table 6. Proposed statistics for DEA evaluation of linebackers.

		Linebackers
Inputs		
	I_1	Downs played
Outputs		
	O_1	Solo tackles
	O_2	Tackles for loss
	O_3	Interceptions

6.4 Final considerations

We believe that DEA is a compelling tool for evaluating professional football players. The objectivity, rigor, and flexibility of the analysis make DEA a viable candidate for general managers and coaches looking for a means to rank players in draft or free agency situations. Similarly, a player's DEA ranking can stand as evidence of overall merit and performance. While franchise management will still have to view tape to see how well a linebacker sheds a block, and players will still have to perform on big plays (and not merely perform well on average), DEA holds great potential for settling once and for all the answer to the question: Who's the best?

References

1. Adler, N., Friedman, L., and Sinuany-Stern, Z. Review of ranking methods in the data envelopment analysis context. *European Journal of Operational Research*, 140: 249-265, 2002.
2. Allen R., Athanassopoulos A., Dyson R.G., and Thanassoulis E., Weights restrictions and value judgements in Data Envelopment Analysis: Evolution, development and future directions. *Annals of Operations Research*, 73: 13-34, 1997.

3. Charnes, A., Cooper, W.W., and Rhodes, E. Measuring the efficiency of decision-making units. *European Journal of Operational Research*, 2: 429-444, 1978.
4. Doyle, J.R., and Green, R.H. Cross-evaluation in DEA: Improving Discrimination Among DMUs. *INFOR*, 33: 205-222, 1995.
5. Talluri, S., Data Envelopment Analysis: Models and Extensions, *Decision Line – Production/Operations Management*, May 2000.
6. Voltaire, *Candide*. 2nd edition translated and edited by Robert Adams. W.W Norton & Company, 1991.

Dynamic Programming and Time Related Strategies in Sports

With an Application to Onside Kicks

Harold Sackrowitz

Rutgers University, USA
sackrowi@rci.rutgers.edu

1 Introduction

Since their introduction by Bellman (beginning with Bellman 1952) in the 1950's the ideas of dynamic programming have become among the most fundamental methods of operations research. A good source for the methodology and extensive literature is Sniedovich 1992. In Sackrowitz and Sackrowitz 1996 and Sackrowitz 2000 dynamic programming methodology was applied, respectively, to the notion of ball control in sports and two-point conversion strategy decisions in football. The goal of this paper is twofold. We will investigate strategy considerations related to the onside kick option in football. More importantly we hope, during this process, to demonstrate dynamic programming's ideal suitability for dealing with certain types of strategy issues that are almost unavoidable in most sports. By studying the various aspects of this problem we will be able to see what can and cannot be accomplished. In these settings a complete determination of an optimal strategy is usually not possible due to vagaries in the model. However, this approach can greatly improve a coach's understanding of game situations and improve decision making. In particular, we will demonstrate a strong indication that onside kicks should be used much more frequently than in current practice.

The type of situations we refer to arise in the following natural way. The rules of any sport contain provisions indicating when a game has ended and who has won. Most games are designed to be played for a fixed amount of time. Included, for such games, are rules for stopping the clock for reasons such as penalties and time outs. For example, football and hockey are played for 60 minutes while professional basketball takes 48 minutes. Of course time is not always used. Baseball is supposed to be a nine inning game while golf is to be played for 18 holes. In the most typical scenario, teams alternate playing "offense" and "defense". When a team gains possession of the ball (or puck, etc.) it has an opportunity to score points (or runs, etc.). The time consumed during any one possession can be very important in time constrained games.

The rules also explain how to proceed if the game is tied at its intended end and how a winner can be determined. It is awareness of the approaching end of play that forces time to enter strategy conditions. We often see players and teams behaving differently near the end of a game than they do at the beginning or middle of the game.

The setting we have chosen for this exposition is the game of football. The attention paid to time strategy is more pronounced in football than any other major sport. Perhaps because each team, typically, gets only 10 – 15 possessions of the ball during a game. In many ways football is ideally suited for a dynamic programming approach. There is a stoppage of action (but not necessarily the clock) between plays so that the teams can decide what play to use next. The clock is stopped after a score again giving teams an opportunity to review its options before action is resumed.

The aspect of football we have chosen to study here is the onside kick. Strategy relating to the onside kick option is not well understood. After a team has scored either a field goal or a touchdown it "kicks off" to the other team. By rule, once the kicked ball has traveled 10 yards it is a free ball. That is, the kicking team and the receiving team are equally entitled to recover the ball. The term "onside" kick describes a situation in which the kicking team makes an obvious attempt to recover its own kickoff. Typically the ball is kicked in such a way that it is difficult to catch and that it does not travel much further than 10 yards. The advantage of such a strategy to the kicking team is that they would have, at least, a modest chance to regain possession of the ball while with the standard kick the likelihood of regaining possession is near zero. However, failure by the kicking team to gain possession usually results in the receiving team gaining possession of the ball relatively close (perhaps 40 – 45 yards) to the goal line. This can dramatically improve the receiving team's chances of scoring.

Use of the onside kick is usually thought of as an act of desperation. However, without resorting to the onside kick a team that is behind with very little time remaining might not even get the opportunity to score more points. Because of the game situation and positioning of the players on the field most onside kick attempts are obvious to the receiving team. Yet, surprisingly, during the 2000 and 2001 NFL seasons, approximately 26% of onside kicks were actually recovered by the kicking team. This number suggests that, perhaps, onside kicks can and should be used more often. This is, of course, the coach's decision and most coaches do realize that score and time remaining are the key ingredients. Unfortunately, it would be impossible for even the most experienced coach to have accumulated enough empirical data to make an informed decision.

How can a team that is about to kick off, properly decide whether or not to attempt an onside kick? The answer depends on many parameters. In addition to the current score and the amount of time remaining in the game, the optimal decision depends on the probabilities of teams scoring touchdowns and field goals, as well as the probability of actually recovering an onside kick. These quantities are also unknown. The first step is to see if we can solve the problem if all these quantities were known.

2 Setting Up the Model

To fix ideas suppose that we have two teams called Team A and Team B. There are two conflicting optimization problems going on simultaneously. Team A is trying to maximize the probability that Team A wins, while Team B is trying to minimize the probability that Team A wins. In attacking the problem we soon notice that proper strategy is easily determined with "little" time remaining. Dynamic programming methods suggest that we begin at the end of the game and work back towards the beginning. To do this we must identify moments and states when choices are to be made. In football, time remaining is given in minutes and seconds. Although time remaining is always known precisely, we quickly realize that what is really important is the number of possessions remaining in a game. This is, of course, a random variable with an unknown distribution. Furthermore, its distribution depends on the manner in which both teams decide to play. Note that an experienced coach could have, based on experience, a reliable sense of how many possessions remain.

To begin we will measure the time remaining in number of possessions rather than actual time (minutes and seconds). This will allow us to identify important states (or situations) during the game needed to implement the dynamic programming process. We will focus on those moments in the game when one team ends a possession and also those moments when that team will turn the ball over to the other team. These could coincide. There are three cases.

1) If a possession ends without scoring no onside kick decision is to be made.

2) If a possession ends with the scoring of a field goal a team must next decide whether or not to attempt an onside kick.

3) If a possession ends with the scoring of a touchdown the clock is stopped and, before the clock is restarted with the ensuing kickoff, a secondary optimization problem must be resolved. The team having scored the touchdown must first decide between trying for one or two extra points (see Sackrowitz 2000 for a full discussion). Following the extra point(s) attempt the team then must decide whether or not to attempt an onside kick.

3 Notation and Solution

For ease of presentation we will study the simplest model that will still allow for useful practical results and also exhibit all the necessary elements of the dynamic programming approach. Next we describe a number of (easily relaxed in terms of programming the solution) assumptions we will make. Note that even if we relax our assumptions the solution does not become more difficult but the notation is far more cumbersome.

We will assume that both teams have the same scoring characteristics (i.e., the same probabilities of scoring touchdowns, field goals, extra points, etc.) and that the probability of winning a game that goes into overtime is one half. It can be argued that the equal scoring characteristics case is actually the most sensible. Onside kick and two point conversion strategies can only make a difference in close games. But in close games, both teams are scoring at about the same rate.

We take the probability that the kicking team will gain possession of a standard kick to be zero. Furthermore we assume the scoring probabilities for a team gaining possession of the ball by (i) receiving a standard kickoff, (ii) following a non-scoring possession by the other team, or (iii) a successful onside kick all to be P_{td}, P_{fg}, P_{ns}. We have let P_{td} and P_{fg} denote the probability of a team scoring a touchdown and field goal, respectively, during a possession while $P_{ns} = 1 - P_{td} - P_{fg}$ denotes the probability of no score. Similarly define R_{td}, R_{fg}, R_{ns} to be these same scoring probabilities but for a receiving team following a failed onside kick. Typically R_{td}, R_{fg} are considerably larger than P_{td}, P_{fg} . At any moment in the game define the score differential to be equal to the points for Team A minus the points for Team B. Next let

Wa(r ; d,n) = the probability that Team A wins given that Team A has just ended a possession by scoring r points (r = 0, 3, or 6) making the score differential = d, there are n possessions remaining in the game, and assuming both teams will behave optimally for the remainder of the game.

Wb(r ; d,n) = the probability that Team A wins given that Team B has just ended a possession by scoring r points (r = 0, 3, or 6) making the score differential = d, there are n possessions remaining in the game, and assuming both teams will behave optimally for the remainder of the game.

Ua(d,n) = the probability that Team A wins given that Team A is about to use a standard kickoff with n possessions remaining in the game, the score differential is d and both teams will behave optimally for the remainder of the game.

Va(d,n) = the probability that Team A wins given that Team A is about to use an onside kickoff with n possessions remaining in the game, the score differential is d and both teams will behave optimally for the remainder of the game.

Ub(d,n) = the probability that Team A wins given that Team B is about to use a standard kickoff with n possessions remaining in the game, the score differential is d and both teams will behave optimally for the remainder of the game.

Vb(d,n) = the probability that Team A wins given that Team B is about to use an onside kickoff with n possessions remaining in the game, the score differential is d and both teams will behave optimally for the remainder of the game.

We will adopt the convention that if the probability of winning is the same using the onside kick as for the standard kick a team will use the standard kick. Also if the probability of winning is the same for both a one-point and two-point conversion attempt a team will try the one-point conversion. Thus if Team A is about to kick off in situation (d, n), that is, the score differential is d and n possessions remain, it must use an onside kick if and only if Va(d, n) $>$ Ua(d,n). Otherwise it will use a standard kick. But if Team B is about to kick off in situation (d, n) it must use an onside kick if and only if Vb(d, n) $<$ Ub(d,n). Otherwise it will use a standard kick. Therefore the probability that Team A wins using the optimal strategy when it is about to kick off in situation (d, n) is

$$W^*a(d,n) = \max(\, Ua(d,n),\, Va(d,n)\,).$$

Similarly the probability that Team A wins when Team B, using its optimal strategy, is about to kick off in situation (d, n) is

$$W^*b(d,n) = \min(\, Ub(d,n),\, Vb(d,n)\,).$$

Next we need to relate the probabilities of winning with n possessions remaining to those of winning with n-1 possessions remaining. Suppose Team A uses a standard kick in situation (d,n). Then Team B will have the ball and score 0, 3, or 6 points (prior to possibly gaining extra points). Thus the probability that Team A would win is

$$Ua(d,n) = Wb(0;\, d,n\text{-}1)\, P_{ns} + Wb(3;\, d\text{-}3,n\text{-}1)\, P_{fg} + Wb(6;\, d\text{-}6,n\text{-}1)\, P_{td}.$$

Alternatively, suppose Team A attempts an onside kick having probability Q of being successful. Recall that if successful Team A would again be in possession of the ball but Team B would have an increased probability of scoring if it fails. Thus the probability that Team A would win in situation (d,n) if it uses an onside kick is

$$\begin{aligned} Va(d,n) = (1-Q)\,\{&Wb(0; d,n\text{-}1)\, R_{ns} + Wb(3; d\text{-}3,n\text{-}1)\, R_{fg} \\ &+ Wb(6; d\text{-}6,n\text{-}1)\, R_{td}\} + Q\,\{Wa(0; d,n\text{-}1)\, P_{ns} \\ &+ Wa(3; d\text{+}3,n\text{-}1)\, P_{fg} + Wa(6; d\text{+}6,n\text{-}1)\, P_{td}\}. \end{aligned}$$

The optimal strategy for Team B in situation (d,n) as well as Ub(d,n), Vb(d,n), and W*b(d,n) can be determined similarly. Note that all these terms with n possessions remaining can be determined from knowledge of the functions Wa(. ; . , n-1) and Wb(.; . , n-1). Next we study Wa(r ; d,n). The only difficult case will be Wa(6 ; d,n) since, when a possession ends with the scoring of a touchdown, there is an opportunity to score extra points before the ensuing kick off. When Team A ends its possession without scoring (i.e., r = 0) there is no onside kick option and Team B's ensuing scoring probabilities will be approximately as if Team A used a standard kickoff. Thus Wa(0 ; d,n) = Ua(d,n). For r = 3 we have Wa(3 ; d,n) = W*a(d,n).

Last is the case of a possession ending in a touchdown. For the last few years in the NFL the probability of successful one and two point conversion attempts has been approximately 0.987 and 0.40, respectively. For simplicity we will use those numbers. If, in situation (d,n), Team A attempts a one point conversion then the probability that Team A wins is

$$(0.987)W^*a(d+1,n) + (0.013)W^*a(d,n). \tag{1}$$

If, in situation (d,n), Team A attempts a two point conversion then the probability that Team A wins is

$$(0.4)W^*a(d+2,n) + (0.6)W^*a(d,n). \tag{2}$$

Therefore, a two-point conversion attempt is preferred to a one-point attempt if and only if the quantity in (2) is strictly greater than the quantity in (1). Furthermore, Wa(6 ; d,n) is equal to the maximum of quantities (1) and (2). We can obtain Wb(r ; d,n) in a similar fashion.

All that is still needed to begin the backwards induction process is knowledge of Wa(r; d,0) and Wb(r; d,0) for all r and d. These are the initial conditions and are simple to determine. If the last possession ends with a field goal or no score (i.e., r = 3 or 0) the game is over unless d = 0 (in which case the game goes into overtime). Thus, for r = 0 or 3,

$$\begin{aligned} &Wa(r; d,0) = Wb(r; d,0) = 1 \text{ if } d > 0, \\ &Wa(r; d,0) = Wb(r; d,0) = 0 \text{ if } d < 0, \text{ and} \\ &Wa(r; d,0) = Wb(r; d,0) = .5 \text{ if } d = 0. \end{aligned}$$

When the last possession ends with a touchdown (i.e., r = 6) there is still an opportunity for extra points. It is not hard to see that

$$\begin{aligned} &Wa(6; d,0) = 1 \text{ if } d > 0, \\ &Wa(6; d,0) = 0 \text{ if } d < \text{-}2, \end{aligned}$$

Wb(6; d,0) = 1 if d > 2,
Wb(6; d,0) = 0 if d < 0,
Wa(6; 0,0) = .987 + (.013)(.5) = .9935,
Wa(6; -1,0) = (.987)(.5) = .4935,
Wa(6; -2,0) = (.4)(.5) = .2,
Wb(6; 0,0) = (.013)(.5) = .0065,
Wb(6; 1,0) = .013 + (.987)(.5) = .565,
Wb(6; 2,0) = .6 + (.4)(.5) = .8.

Using the recursion formulas and the above initial conditions it is now a simple matter to program the solution. Thus the optimal strategy can be determined in situation (d,n) for any d and n. Such results are presented, for three choices of (R_{td}, R_{fg}, R_{ns}), in the tables. The choices reflect varying amounts on increased scoring ability following a failed onside kick attempt.

3.1 Strategy tables

In Tables 1, 2, and 3 we present optimal strategy tables for three choices of the scoring probabilities (R_{td}, R_{fg}, R_{ns}) for the receiving team following a failed onside kick. In all tables we take the usual scoring probabilities to be at the 1999, 2000 NFL rate. That is, (P_{td}, P_{fg}, P_{ns}) = (.21, .13, .66), the probability of a successful onside kick to be Q = .26, and the probabilities of successful 1-point and 2-point conversions to be .987 and .4 respectively. To use the tables in situation (d,n) we find the row corresponding to a score differential of d and the column corresponding to n possessions remaining. If the entry is "0" a standard kick is used. An entry of "1" means that an onside kick should be used. A blank indicates that the team must lose regardless of strategy as there are not enough possessions remaining to catch up.

4 Discussion and Conclusions

We have gone through the process of solving the problem of finding optimal strategies, based on the number of possessions remaining in a game, for on-side kicks when all necessary parameters are known. The approach would be similar for time related questions in most other games. Unfortunately these parameters are not known and often cannot even be estimated very reliably based on available data. Football teams only play a 16 game regular season schedule and teams can change dramatically from one year to the next. Also scoring probabilities can vary depending on opponent and teams rarely face the same opponent more than twice in a season. To make matters worse, in practice, it is difficult to know with certainty how many possessions remain. So what is a coach to do? Well, all is not lost.

When we compare Tables 1 and 2 and 3 we notice that large portions of these tables are the same despite the differences in the parameters R_{td}, R_{fg},

Table 1. (R_{td}, R_{fg}, R_{ns}) = (.30, .42, .28).

d	possessions remaining in the game (n)														
	1	2	3	4	5	6	7	8	9	10	11	12	13	14	15
-18			1	1	1	1	1	0	1	0	1	0	1	0	1
-17			1	1	1	1	1	0	1	0	1	0	1	0	1
-16		1	1	1	1	0	1	0	1	0	1	0	1	0	1
-15		1	1	0	1	0	1	0	1	0	1	0	1	0	1
-14		1	1	0	1	0	1	0	1	0	1	0	1	0	1
-13		1	1	0	1	0	1	0	1	0	1	0	0	0	0
-12		1	1	0	1	0	1	0	1	0	1	0	0	0	0
-11		1	1	0	1	0	1	0	1	0	1	0	0	0	0
-10		1	1	0	1	0	1	0	1	0	1	0	0	0	0
-9		1	1	0	1	0	1	0	1	0	1	0	0	0	0
-8	1	0	1	0	1	0	1	0	1	0	1	0	0	0	0
-7	1	0	1	0	1	0	1	0	1	0	1	0	0	0	0
-6	1	0	1	0	0	0	0	0	0	0	0	0	0	0	0
-5	1	0	1	0	1	0	0	0	0	0	1	0	0	0	0
-4	1	0	1	0	1	0	0	0	0	0	1	0	0	0	0
-3	1	0	1	0	1	0	1	0	0	0	1	0	0	0	0
-2	1	0	1	0	1	0	0	0	0	0	0	0	0	0	0
-1	1	0	1	0	1	0	0	0	0	0	0	0	0	0	0
0	0	0	0	0	0	0	0	0	0	0	0	0	0	0	0
1	0	0	0	0	0	0	0	0	0	0	0	0	0	0	0

R_{ns}. This is typical when parameters are varied (even if we had not made our notation simplifying assumptions). Although most model parameters can never be known precisely a reasonable range of possible values can usually be given. By constructing a variety of tables for parameters within this range a coach can generate considerable intuition. Watching if and how the tables change as the parameters change can be very informative. Furthermore, for many situations the optimal decision is the same for all choices of parameter values in the range. Thus for these situations we would know the proper strategy even though the precise parameter values remain unknown. As far as time goes, there are 24 possessions in a typical game. A useful intuitive device is to think of an average possession as 2.5 minutes or 6 possessions per quarter.

To get an appreciation for how much we can learn by generating a variety of tables let us just focus on that portion of the tables for which d = -15, -16, −17, and n = 3, 4, 5. All three tables agree in this region except for situation (-15, 4). It seems clear that if a team is behind by 17 or more points with no more than 7 possessions remaining they should use an onside kick. In particular that suggests using an onside kick any time in the entire fourth quarter. That is certainly not current policy and coaches should seriously consider it.

Table 2. (R_{td},R_{fg},R_{ns}) = (.30, .35,.35).

	possessions remaining in the game (n)														
d	1	2	3	4	5	6	7	8	9	10	11	12	13	14	15
-18			1	1	1	1	1	0	1	0	1	0	1	0	1
-17			1	1	1	1	1	0	1	0	1	0	1	0	1
-16		1	1	1	1	0	1	0	1	0	1	0	1	0	1
-15		1	1	1	1	0	1	0	1	0	1	0	1	0	1
-14		1	1	0	1	0	1	0	1	0	1	0	1	0	1
-13		1	1	0	1	0	1	0	1	0	1	0	1	0	1
-12		1	1	0	1	0	1	0	1	0	1	0	1	0	1
-11		1	1	0	1	0	1	0	1	0	1	0	1	0	1
-10		1	1	0	1	0	1	0	1	0	1	0	1	0	1
-9		1	1	0	1	0	1	0	1	0	1	0	1	0	1
-8	1	0	1	0	1	0	1	0	1	0	1	0	1	0	1
-7	1	0	1	0	1	0	1	0	1	0	1	0	1	0	1
-6	1	0	1	0	1	0	1	0	1	0	0	0	0	0	0
-5	1	0	1	0	1	0	1	0	1	0	1	0	0	0	0
-4	1	0	1	0	1	0	1	0	1	0	1	0	1	0	0
-3	1	0	1	0	1	0	1	0	1	0	1	0	0	0	0
-2	1	0	1	0	1	0	1	0	1	0	0	0	0	0	0
-1	1	0	1	0	1	0	1	0	1	0	0	0	0	0	0
0	0	0	0	0	0	0	0	0	0	0	0	0	0	0	0
1	0	0	0	0	0	0	0	0	0	0	0	0	0	0	0

We also notice that, throughout much of the table, onside kick strategy depends heavily on whether there are an odd or even number of possessions remaining. It is common for ones and zeros to alternate across a row of the table. This is not really surprising as onside kicks are closely tied to opportunities to score. This does appear to reduce the usefulness of the tables in their current form. For example, suppose we are behind by 5 points (d = -5) and we are not sure whether there is still time for two or three possessions. Fortunately, in the process of generating these tables we have computed (through the recursion equations) the probability of winning in situations (-5, 2) and (-5, 3) for both an onside and standard kick. This permits us to give weights (which may be subjective) to the different possibilities to help decide what to do.

An issue worth (perhaps only theoretical) consideration is suggested by Table 3. We notice that there are some entries of "1" even when the team is ahead. Initially surprising it becomes plausible with some thought. If we are ahead by 4 points near the end of the game (i. e., one possession remaining) then we can lose only if our opponent scores a touchdown. Our opponent scoring a field goal doesn't matter. If we attempt an onside kick and we successfully recover it (with probability = .26) then we win. Thus all that is to be resolved is the probability of our opponent scoring a touchdown with a standard kick (= P_{td}) versus an onside kick (= (.74) R_{td}). Of course it is

Table 3. $(R_{td}, R_{fg}, R_{ns}) = (.28, .36, .36)$.

	possessions remaining in the game (n)														
d	1	2	3	4	5	6	7	8	9	10	11	12	13	14	15
-18			1	1	1	1	1	0	1	0	1	0	1	0	1
-17			1	1	1	1	1	0	1	0	1	0	1	0	1
-16		1	1	1	1	1	1	0	1	0	1	0	1	0	1
-15		1	1	1	1	0	1	0	1	0	1	0	1	0	1
-14		1	1	0	1	0	1	0	1	0	1	0	1	0	1
-13		1	1	0	1	0	1	0	1	0	1	0	1	0	1
-12		1	1	0	1	0	1	0	1	0	1	0	1	0	1
-11		1	1	0	1	0	1	0	1	0	1	0	1	0	1
-10		1	1	0	1	0	1	0	1	0	1	0	1	0	1
-9		1	1	0	1	0	1	0	1	0	1	0	1	0	1
-8	1	0	1	0	1	0	1	0	1	0	1	0	1	0	1
-7	1	0	1	0	1	0	1	0	1	0	1	0	1	0	1
-6	1	0	1	0	1	0	1	0	1	0	1	0	1	0	1
-5	1	0	1	0	1	0	1	0	1	0	1	0	1	0	1
-4	1	0	1	0	1	0	1	0	1	0	1	0	1	0	1
-3	1	0	1	0	1	0	1	0	1	0	1	0	1	0	1
-2	1	0	1	0	1	0	1	0	1	0	1	0	1	0	0
-1	1	0	1	0	1	0	1	0	1	0	1	0	1	0	0
0	0	0	0	0	0	0	1	0	1	0	1	0	0	0	0
1	0	0	0	0	0	0	0	0	0	0	0	0	0	0	0
2	0	0	0	0	0	0	0	0	0	0	0	0	0	0	0
3	0	0	0	0	0	0	0	0	0	0	0	0	0	0	0
4	1	0	0	0	0	0	0	0	0	0	0	0	0	0	0
5	1	0	0	0	0	0	0	0	0	0	0	0	0	0	0
6	1	0	1	0	0	0	0	0	0	0	0	0	0	0	0
7	1	0	0	0	0	0	0	0	0	0	0	0	0	0	0
8	1	0	0	0	0	0	0	0	0	0	0	0	0	0	0

not likely that a coach could ever be convinced to use an onside kick when ahead.

Currently coaches rely on little more than a "sixth sense" to make most time related decisions. Dynamic programming cannot totally replace reliance on a coach's experience. However, a coach willing invest in a broad dynamic programming analysis can achieve a much better understanding of the value of his or her choices, make more correct decisions and feel more confident about those decisions.

References

1. Bellman, R., On the theory of dynamic programming, *Proceedings of the National Academy of Sciences*, 38, 716 – 719, 1952.

2. Sackrowitz, H. and Sackrowitz, D., Time management in sport: Ball control and other myths. *Chance*, 9, 41 – 49, 1996.
3. Sackrowitz, H., Refining the point(s) after touchdown decision. *Chance*, 13, 29 – 34, 2000.
4. Sniedovich, M., *Dynamic Programming*. Marcel Dekker, Inc., New York, 1992.

Using Statistics to Predict Scores in English Premier League Soccer

John S. Croucher

Macquarie University, Australia
john.croucher@mq.edu.au

1 Introduction

One of the most popular and played sports worldwide is soccer, also known as 'Association Football', with intense interest every four years in the World Cup between the 32 qualifying teams. The game is played between two teams each of 11 players including a goalkeeper who is the only player allowed to handle the ball.

There are a great many sources of information of soccer statistics, particularly those matches played in the UK, Europe and South America, where examples include the yearbooks (Sky Sports Football Yearbooks). Nowadays the internet is a rich source of information regarding current and past matches with sites such as *www.soccernet.com* being widely accessed. Apart from the intense interest of the fans that follow the fortunes of individual teams, there is another important reason why many thousands of people are interested in the final scores. This involves gambling on the outcome in an activity known as *Soccer Pools* in which punters can win vast amounts of money by using their skills to try and predict the outcome of individual matches.

In particular, gamblers in the Soccer Pools try to predict those matches that will end in a draw. Moreover, the idea is to attempt to identify those drawn matches where there will be the most goals scored in the draw since these will be the first numbers selected in the winning combination. An example of the 'Pools' is where the gambler tries to predict eight drawn matches among a selection of matches played in the English Premiership League and lower Divisions. Although it is certainly true that many winners of the Pools know nothing about soccer, this does not stop keen students of the game trying to use some science to make a profit. This involves using the data available to determine those matches that are most likely to end in a draw.

From a statistical point of view, one method of achieving this aim is to fit an appropriate model to the number of goals scored by each team and use this information to select the most likely draws. Along with predicting the final score, such a model would assist a bookmaker in framing an appropriate betting market. All these aspects are discussed in the following sections.

2 Fitting Statistical Distributions

The idea of modelling sporting data is not new with some of the earlier published research (Wood 1945) claiming that batsmen's scores in cricket follow a geometric distribution and others (Moroney 1951) attempting to fit distributions to soccer goal scoring. In soccer, goals are relatively rare with around 57% of matches having a total of two goals or less scored and initial thinking might be that they occur at random times. For this reason it is tempting to use the Poisson distribution to fit the number of goals scored.

However, studies (Moroney 1951, Pollard et al 1986 and Norman 1998) suggest that, within a match, a team's goal scoring rate may change as the match develops. That is, a team that is losing may concentrate on attack in the hope of trying to level up the score while running a greater risk of conceding even more goals. This has led to using models based on the negative binomial distribution which has proven in some cases to yield a superior fit of actual data. In others it has not been appropriate. Other research in this area include modelling of the arrival rate of goals in a match (Dixon and Robinson 1998) and the total number of goals scored (Dixon and Coles 1997).

2.1 An early example of fitting the negative binomial distribution

One example of fitting the negative binomial distribution to goal scoring in soccer is illustrated in Table 1. This shows a summary (Pollard et al 1977) of the number of goals scored in 924 matches in the English First Division Football League during 1967–68. The actual formula for calculating the expected number of goals using a negative binomial formula is given in Equation (1).

$$\text{Pr(Number of goals scored} = r) = {}^{k+r-1}C_{k-1}p^k(1-p)^r,\ r = 0, 1, 2, 3, \ldots \quad (1)$$

where $k > 0$ and $0 < p < 1$

If the method of moments is used where m is the sample mean and s^2 is the sample variance, then:

$$p = m/s^2 \text{ and } k = m^2/(s^2 - m)$$

In the case of these data, the mean number of goals scored per match was $m = 1.51$ with a variance of $s^2 = 1.75$.

Table 1. Observed and expected number of goals scored in 924 English soccer matches in 1967– 68 using a negative binomial distribution.

Number of goals	Observed	Expected
0	225	226.6
1	293	296.4
2	224	213.9
3	114	112.6
4	41	48.3
5	15	17.9
6	9	5.9
7+	3	2.5
Total	924	924.1

Table 1 demonstrates that the negative model fitted the data extremely well, as supported by a Chi-square value with a p-value of 0.57. Other attempts have been made to fit a variety of statistical distributions to soccer scores (Reep and Benjamin 1968, Hill 1974, Maher 1982 and Croucher 1984) while other variables that may contribute to the score have also been considered. For example, the evidence of a home ground advantage has been well explored (Pollard 1986, Courneya and Carron 1992) with some (Clarke and Norman 1995 and Clarke 1996) providing models based on one (Stefani 1983 and 1987) that looks at winning margins taking into account both home and away performance.

It is interesting to note in theory that if a team obtained a draw in *every* match then it would be in grave danger of relegation to a lower Division, since the three bottom teams in any Division are those relegated. This aspect is easily seen from Table 1 where a team who had drawn all their 38 matches would have been second last and relegated to Division 1.

On the other hand, a team that won all their home games but lost all away games would be on 57 competition points or in the top one-third of the table. If a team set itself a target of trying to win all their home games and be content to draw all their away games, this would yield 76 competition points and would have placed such a team in third position this season. The eventual champions, Manchester United, along with the second top team, Arsenal, each lost only four away matches while the bottom two teams West Bromwich and Sunderland each lost thirteen.

3 Fitting Statistical Models to Modern Data

Any model that attempts to predict the outcome of any sporting event must properly take account of the home ground advantage since it is well known that, in the vast majority of cases, teams will exhibit a better performance when playing in front of their home crowd. This and the following sections look at models that analyse matches played in the English Premier League during 2002/03 season along with the FA Cup Final.

3.1 The Poisson distribution

The Poisson distribution typically represents events that occur randomly in a fixed time period. If this average rate is represented by the parameter λ, the event is the scoring of a goal and the time period is the duration of a match, then we have:

$$\Pr(\text{No. of goals scored} = r) = e^{-\lambda}\lambda^r/r!, \quad r = 0, 1, 2, 3, \ldots \tag{2}$$

Data were collected from all 380 matches played in the 2002/03 English Premier League where 3 competition points are awarded for a win, 1 point for a draw and 0 points for a loss. The top three teams at the finish were Manchester United, Arsenal and Newcastle, respectively. To fit a Poisson distribution to the data while taking into account the home and away aspect discussed earlier, two models were constructed. The first of these was based on the overall average scoring rate of teams playing at home. In these 380 matches there was a total of 570 goals scored by the home team for an average rate of 1.500 goals per game. At the same time, the away teams scored a total of 430 goals at an average rate of 1.132 goals per game.

The individual team average scoring rates at home (λ_1) and away (λ_2) are shown in Table 2. The correlation coefficient between home and away scoring averages has a significant value of 0.644 (p-value $= 0.002$), showing that there is a very good degree of association between the way teams score at home and away, with the majority as expected having a larger scoring rate for home games. A notable exception was Leeds United who had scored only 25 goals at home but 33 goals away. Middlesborough was at the other extreme with 36 home goals but only 12 away goals while Aston Villa won 11 matches at home but only 1 match away.

Not surprisingly, there is a significant correlation between home scoring average and competition points ($r = 0.890$, $p < 0.001$) and only a slightly weaker one between away scoring average and competition points ($r = 0.796$, $p < 0.001$). To test whether any team finished significantly higher or lower on the ladder than their record might indicate, a multiple regression was used with the response variable being competition points and the two independent variables being the ratio of (goals scored for)/(goals scored against) at home and away, respectively. The correlation coefficient between these two variables was a significant $r = 0.733$ ($p < 0.001$) indicating that a team's performance relative to the opposition when playing at home is in proportion to what it is when they play away. These two ratios are also shown in Table 2.

Table 2. Individual team average scoring rates at home (λ_1) and away (λ_2), along with relative goal scoring rates at home and away after all 380 matches in the English Premier League 2002/03 season.

Team	Home scoring average λ_1	Home for/Home against	Away scoring average λ_2	Away for/Away against	Competition points
Manchester United	2.21	3.50	1.68	1.45	83
Arsenal	2.47	2.35	2.00	1.73	78
Newcastle	1.89	2.12	1.42	0.87	69
Chelsea	2.16	2.73	1.42	1.17	67
Liverpool	1.58	1.88	1.63	1.24	64
Blackburn	1.26	1.60	1.47	1.00	60
Everton	1.47	1.47	1.05	0.67	59
Southampton	1.32	1.56	0.95	0.60	52
Manchester City	1.47	1.08	1.00	0.68	51
Tottenham	1.58	1.03	1.11	0.64	50
Middlesbrough	1.89	1.71	0.63	0.52	49
Charlton	1.37	0.87	1.00	0.73	49
Birmingham	1.32	1.09	0.84	0.62	48
Fulham	1.37	1.44	0.79	0.47	48
Leeds	1.32	0.96	1.74	1.06	47
Aston Villa	1.32	1.79	0.89	0.52	45
Bolton	1.42	1.13	0.74	0.52	44
West Ham	1.11	0.88	1.11	0.60	42
West Bromwich	0.89	0.50	0.63	0.39	26
Sunderland	0.58	0.35	0.53	0.29	19

Using the data in Table 2, the multiple regression equation used to predict competition points from the two goal-scoring ratios at home and away was:

$$\begin{aligned} Competition\,pts = 21.4 &+ 11.4(Home\,for/Home\,against) \\ &+ 17.8(Away\,for/Away\,against) \end{aligned} \tag{3}$$

The coefficient of determination value of $r^2 = 87.4\%$ ($p < 0.001$) indicates that the equation is an excellent fit. Equation (3) also suggests that relative goal scoring performance when away may be more important than when at home since the coefficient for the former variable is over fifty percent higher. The only outlying team in this regression was the team Sunderland for whom the model suggested they should have finished with 31 competition points rather than the 19 they actually obtained. However, even if this were the case they still would have been relegated to Division 1.

Separate Poisson models for goals scored at home and away goals are shown in Table 3 using the combined scoring rate for all teams. To avoid distorting the statistical analysis, there is a combined category of '4 or more' goals to ensure that each one has an expected frequency of at least 5. With a Chi–

square value of 5.17 ($p = 0.160$) for the home data, the Poisson gives a very good fit, with the major discrepancy being that it expected far less matches where the home scored exactly two goals than actually occurred. The Poisson model was extremely accurate for the away data with a Chi–square value of only 0.53 ($p = 0.912$) with almost uncanny accuracy in every category.

Table 3. Observed and Expected number of goals scored at home and away using a Poisson model.

Goals scored	Home ($\lambda = 1.500$)		Away ($\lambda = 1.132$)	
	Observed	Expected	Observed	Expected
0	81	84.8	126	122.6
1	121	127.2	139	138.7
2	114	95.4	74	78.5
3	43	47.7	29	29.5
4 or more	21	24.9	12	10.7
Total	380	380.0	380	380.0
	$\chi^2 = 5.17$	$p = 0.160$	$\chi^2 = 0.53$	$p = 0.912$

3.2 The negative binomial distribution

To make a comparison with the Poisson distributions, an attempt was made to model the same data using a negative binomial distribution since it been successfully applied in the past (Cain et al. 2000). For these data, however, it was found to be unsatisfactory for modelling both home goals and away goals. In the case of home goals, parameter values of $m = 1.50$ and $s^2 = 1.37$ were obtained, yielding $p = 1.095$. Since the value of p must be less than 1, this renders the distribution invalid.

For the away goals, the parameters values of $m = 1.13$ and $s^2 = 1.27$ yielded a value of $p = 0.897$. Although valid, this gives an expected number of matches in which the away team scored no goals of 0.897 x 380 = 282, such a far cry from the observed value of 126 as to make it an extremely poor fit without further investigation.

It could be fairly concluded that for the matches played in the English Premier League during the 2002/3 season a Poisson distribution provides an excellent fit and this will be used in further considerations. It may well be, however, that in future seasons a negative binomial distribution could give a reasonable fit of the data and so it should not be dismissed as out of hand before trying it on other data sets.

4 Scoring Combinations

To make a reasonable prediction of the outcome of a match it is of course necessary to estimate the score of both teams. The problem of interdependence between home and away scores may be considered by using the bivariate Poisson distribution (Holgate 1964) but the approach here will use that based on two independent Poisson distributions. In this case Equation (1) will be used along with the combined parameters for all twenty teams. This means that the probability that the home team will score x goals and the away team will score y goals is given by:

$$\begin{aligned}\Pr(\text{Home goals} = x, \text{Away goals} = y) &= (e^{-\lambda_1}\lambda_1^x/x!) \times (e^{-\lambda_2}\lambda_2^y/y!) \\ &= \lambda_1^x\lambda_2^y e^{-(\lambda_1+\lambda_2)}/x!y!\end{aligned} \tag{4}$$

where

$\lambda_1 =$ the average goals per match for home teams combined
$\lambda_2 =$ the average goals per match for away teams combined

A special case of Equation (4) is the probability that a match will end in a scoreless draw (where $x = y = 0$), this simply being $e^{-(\lambda 1+\lambda 2)}$. Using the data for all 380 matches played in the 2002/03 season, the parameter values for all twenty teams combined are:

$$\lambda_1 = 1.500$$

$$\lambda_2 = 1.132$$

The observed and expected frequencies (shown in parentheses) using these parameters and Equation (4) are shown in Table 4. Although the individual score *totals* for both home and away are relatively close (as seen in Table 3), there are significant differences for individual combinations. In particular, there are about seven *less* 0 – 0 scorelines but nine *more* 1 – 0 scorelines than expected. Combined, these could be seen to 'cancel each other out' since they are adjacent. It is interesting to speculate as to why home teams might find it easier to score that single goal to break a 0 – 0 deadlock.

5 Drawn Matches

Of the 380 matches in the data set there were a total of 88 (23%) drawn games, a figure that is consistent with the overall figure of 25% that has been suggested in previous research (Haigh 2003). Table 5 shows the actual number of draws for each score along with the theoretical probability and expected values given by the Poisson distribution using the overall values of lambda.

Table 4. Observed and expected number of times given scores have appeared in 380 matches based on overall Poisson parameter values.

Home \ Away	0	1	2	3	4 or more	Total
0	20	32	15	9	5	81
	(27.3)	(30.9)	(17.5)	(6.6)	(2.4)	(84.8)
1	50	42	19	7	3	121
	(41.0)	(46.2)	(26.3)	(9.9)	(3.6)	(127.2)
2	35	40	26	10	3	114
	(30.8)	(34.8)	(19.7)	(7.4)	(2.7)	(95.4)
3	17	16	9	0	1	43
	(15.4)	(17.4)	(9.8)	(3.7)	(1.3)	(47.6)
4 or more	4	9	5	3	0	21
	(8.1)	(9.1)	(5,2)	(1.9)	(0.7)	(25.0)
Total	126	139	74	29	12	380
	(122.6)	(138.7)	(78.5)	(29.5)	(10.7)	(380.0)

Table 5. Observed and expected number of draws in 380 matches using a Poisson model.

Scoreline	Observed frequency	Theoretical probability	Overall expected frequency	Expected frequency based on 88 draws
0 – 0	20	0.0719	27.3	24.7
1 – 1	42	0.1221	46.4	42.0
2 – 2	26	0.0518	19.7	17.8
3 – 3	0	0.0098	3.7	3.4
4 – 4 or more	0	0.0004	0.1	0.1
Total	88	0.2560	97.2	88.0

The expected frequencies are based on the 380 matches. There were no drawn games higher than 2-2.

The observed number of 88 draws is somewhat short of the expected number of about 97 draws, the figures in Table 5 revealing a significantly lower number of 0–0 draws but a higher number of 2–2 draws than expected. In fact, twenty (5.3%) of all matches had no goals scored at all while 194 (51%) had a total of two or fewer goals scored by full time. This aspect has led to some soccer enthusiasts calling for techniques to increase the rate of goal scoring with the introduction of wider and/or higher goalposts.

The final column in Table 5 shows the expected frequency of the scoreline in the draw given that there were 88 draws in the season. With the last two categories combined due to the low expected frequencies, the Chi-square goodness-of-fit value of 8.07 had a significant p-value of 0.018. Interestingly, the expected number of 1–1 draws was exactly what occurred.

To reward attacking play, the Football Association decided in the early 1980s to award one competition point for a draw and three points for a win. Previously there were only two points awarded for a win. The idea was to promote the scoring of goals and make the game more exciting for spectators. The result of this change was investigated (Croucher 1984) and little evidence was found of it having the desired effect, at least in the initial stages of operation. Perhaps this is because if the scores are locked at, say, 0 – 0 late in a match that both teams will be satisfied with a draw and play defensively rather than risk their opponents gaining three valuable competition points.

6 The Chance of Avoiding Relegation

An excellent example of wanting to predict the result of particular matches using a Poisson model came at the end of the 2002/3 season. There was much excitement in the final round of matches played on 11th May 2003 since it would finally be determined which of the two teams West Ham or Bolton would be relegated to Division 1 in the next season. Bolton were in the box seat, since although both teams had 41 competition points at that stage, Bolton had the better 'for and against' goal scoring record which was a tie-breaker if both teams finished the season with the same number of competition points. This meant that West Ham had to achieve a better result than Bolton in their respective matches. In other words, if Bolton drew their match, then West Ham had to win theirs while if Bolton lost then West Ham could either draw or win. If Bolton won it would not matter what West Ham did since they would be relegated anyway. The two matches in question were Bolton at home playing Middlesbrough and Birmingham playing at home to West Ham.

In terms of outcomes, West Ham would *not* have been relegated to Division 1 if any of the following outcomes occurred:

Middlesbrough beat Bolton *and* West Ham drew with Birmingham
Middlesbrough beat Bolton *and* West Ham beat Birmingham
Middlesbrough drew with Bolton *and* West Ham beat Birmingham

Although it is quite likely that the teams involved were aware of the progressive scores in the other match, since both were played at the same time on the same afternoon, they were treated as independent for our purpose. In doing so, the probability of West Ham escaping relegation could be estimated before the matches took place.

6.1 Bolton versus Middlesbrough

Using the parameters for each team shown up to *but not including* that match, the values for Bolton at home ($\lambda_1 = 1.389$) and Middlesbrough away ($\lambda_2 = 0.611$) were used in Equation (3). The figures in Table 6 show the probability of individual scorelines occurring.

Table 6. Probabilities of various scorelines when Bolton played Middlesborough.

Bolton \ Middles	0	1	2	3	4 or more	Total
0	0.135	0.083	0.025	0.005	0.001	0.249
1	0.188	0.115	0.035	0.007	0.001	0.346
2	0.131	0.080	0.025	0.005	0.001	0.242
3	0.060	0.037	0.011	0.002	0.000	0.110
4 or more	0.029	0.018	0.005	0.001	0.000	0.053
Total	0.543	0.333	0.100	0.020	0.003	1.000

The most likely result (the one with the highest probability) in Table 6 is a 1 – 0 win to Bolton, with a 73% chance that Bolton would score between 0 and 2 goals and Middlesbrough would score either 0 or 1 goal. This indeed captured the actual final scoreline of Bolton 2 Middlesbrough 1, a therefore not terribly surprising result and, of course, not very encouraging news for West Ham. The overall probability of a draw was 0.277, slightly higher than the overall probability of 0.256 for all 380 matches. Note that 90% of the chance of drawn game came in the Bolton–Middlesbrough match from either a 0 – 0 or 1 – 1 final score.

6.2 Birmingham versus West Ham

Once again the parameters for each team shown up to *but not including* the match were used, the values for Birmingham (at home) being $\lambda_1 = 1.210$ and West Ham (away) being $\lambda_2 = 1.000$. The figures in Table 7 show the probability of individual scorelines occurring.

The most likely result (the one with the highest probability) in Table 7 was either a 1 – 0 win to Birmingham of a 1-1 draw although neither of these two outcomes actually occurred. However, there was an 81% chance that both teams would score two goals or less and this included the actual final scoreline of Birmingham 2 West Ham 2. From Table 7, the probability of a draw in this match was 0.288, again slightly higher than the overall chance of 0.256 for all 380 matches. Note that 84% of the chance of a drawn game came from either a 0 – 0 or 1 – 1 final score.

Table 7. Probabilities of various scorelines when Birmingham played West Ham.

Birmingham\West Ham.	0	1	2	3	4 or more	Total
0	0.110	0.110	0.055	0.018	0.006	0.299
1	0.133	0.133	0.066	0.022	0.077	0.361
2	0.081	0.081	0.040	0.013	0.004	0.219
3	0.032	0.032	0.016	0.005	0.002	0.087
4 or more	0.013	0.013	0.006	0.002	0.000	0.034
Total	0.369	0.369	0.183	0.060	0.019	1.000

6.3 The chance of West Ham not being regelated

The results of Tables 6 and 7 provide us with an estimate of the chance, prior to the two relevant matches being played, of West Ham being able to avoid relegation. A summary of results is given in Table 8.

Table 8. The combined probabilities of Tables 6 and 7.

Bolton versus Middlesbrough		Birmingham versus West Ham	
Bolton win	0.560	Birmingham win	0.409
Middlesbrough win	0.163	West Ham win	0.303
Draw	0.277	Draw	0.288
Total	1.000	Total	1.000

For West Ham not to be relegated they had to achieve a better result than Bolton in their match. Using the assumption of independence, Table 9 shows a summary of the possibilities for this to happen.

Table 9. Outcomes that would save West Ham from relegation to Division 1 and their probabilities.

Outcome for West Ham not to be relegated	Probability
Middlesbrough win, West Ham draw	0.163 x 0.288 = 0.047
Middlesbrough win, West Ham win	0.163 x 0.303 = 0.049
Middlesbrough draw, West Ham win	0.277 x 0.303 = 0.084
Total	0.180

According to the figures in Table 9, before that fateful afternoon the best estimate of West Ham avoiding relegation was about 18% – a rather bleak prospect. In turn, there was an 82% chance that Bolton would avoid relegation. In fact, as has been mentioned previously, although West Ham drew their match, Middlesbrough lost theirs. Consequently West Ham was indeed relegated to Division 1 for the 2003/2004 season.

7 The FA Cup Final

The FA Cup is a knockout competition played from among all teams in the English League with the highlight being the final game between the two remaining teams. This particular match, the last in the domestic season, was played between Arsenal and Southampton on 17th May 2003 at the Millennium Stadium in Cardiff. A Poisson distribution was again used to predict the result along with calculating the odds that each outcome would occur. The method used is similar to that used in Section 1.6 except that this match was played on a neutral ground. In this sense it was an 'away' game for both teams, but since neither one was actually playing at home, it seems reasonable to use a combination of both home and away goals scored in the analysis. In this case, during the regular season Arsenal scored an overall 85 goals in their 38 games for $\lambda_1 = 2.24$ while Southampton's 43 overall goals yielded $\lambda_2 = 1.13$. At first glance it seems that, based simply on these averages, Arsenal would win 2–1. However, it is necessary to gain the full picture of the situation and the likelihood of either team winning in order to calculate the true chance of success. The probability of various scorelines, calculated using Equation 3 is shown in Table 10.

Table 10. The probability of various scorelines in the FA Cup Final in 2003.

Arsenal \ Southampton	0	1	2	3	*4 or more*	*Total*
0	0.034	0.039	0.022	0.008	0.003	0.106
1	0.077	0.087	0.049	0.019	0.007	0.239
2	0.086	0.097	0.055	0.021	0.008	0.267
3	0.064	0.072	0.041	0.016	0.006	0.199
4 or more	0.061	0.069	0.039	0.015	0.005	0.189
Total	0. 322	0.364	0.206	0.079	0.029	1.000

The figures in Table 11 summarise those in Table 10 and show the probabilities for the three possible outcomes of the match. (The probability of a 4–4 draw was 0.002 while the probability of a higher score draw is negligible.) It is interesting to note that all scorelines had an individual chance of less than 10% of occurring.

The odds given in the final column of Table 11 are a reflection of the probabilities. Note that 'odds-on' occurs whenever the probability exceeds 0.50.

Table 11. Probabilities for the outcome of the FA Cup Final in 2003.

Result	Probabilities	Approximate odds
Arsenal Win	0.623	5 to 3 *on*
Southampton win	0.183	9 to 2
Draw	0.194	4 to 1
Total	1.000	

The model turned out to be a good predictor when Arsenal won the game 1–0.

8 Framing a Betting Market

Suppose a bookmaker wishes to frame the opening betting market of the match based on the probabilities in Table 11. Since the modern trend is to set a 'price' rather than odds, this will be the aim here, although of course the two are readily interchangeable. A price of $\$y$, for example, means that the successful punter will receive $\$y$ from the bookmaker for each \$1 wagered. It is clear that a price will never be less than \$1. If the bookmaker sets their market percentage at $x\%$ (where x must always be larger than 100 or else the bookmaker can be guaranteed to lose), then the price they would set for an outcome that occurs with probability p is:

$$\text{Price} = 1/px \tag{5}$$

A typical value in practice of x is 115 and so using Equation (5) together with the probabilities in Table 11, the opening market would be (to the nearest multiple of five cents) for a \$1 wager:

Arsenal win: 1/(0.623 x 1.15) = \$1.40
Southampton win: 1/(0.183 x 1.15) = \$4.75
Draw: 1/(0.194 x 1.15) = \$4.50

To compare what actually happened in practice, the prices set by eight betting agencies on the morning of the match were recorded along with a summary of their highest, lowest and mean prices which were all then compared with the prices suggested by our model. These are shown in Table 12.

The figures in Table 12 indicate that the model thought it much more likely that Arsenal would win than did any of the bookmakers and, as such, suggested a price lower than any of them. It also considered a draw far less likely and was prepared to offer a far more generous price while its price for a Southampton win, while larger than for any bookmaker, was not outrageously so. The bookmakers saw the match as less evenly balanced than the model

Table 12. A summary of the highest, lowest, mean and model prices for each outcome of the FA Cup Final 2003.

Result	Highest price	Lowest price	Mean price	Model price
Arsenal Win	1.65	1.55	1.60	1.40
Southampton win	6.00	5.00	5.40	4.75
Draw	3.40	2.20	3.20	4.50

and could well be related to the fact that the cup must produce a winner on the day and there is little incentive for a team to play for a draw at the end of ninety minutes. As it turned out, with an Arsenal win, the price structures for all models were quite reasonable.

9 Remarks

As far as this analysis went at least, the Poisson distribution yielded a more than reasonable fit of goal scoring in soccer. It certainly proved more appropriate than the negative binomial distribution that in any case requires much more data collection and calculation to use. For example, to construct a table such as Table 1 using the negative binomial distribution requires the individual scores of each match played by the two teams involved so that the appropriate values of m and s^2 can be calculated to use in the formula. For the average Pools punter, who most likely does not have the time or ability to use a computer to assist in undertaking such copious calculations, trying to find the most likely candidates for a drawn match in this way would be extremely tedious. On the other hand, the Poisson simply requires the parameter value of the overall scoring rate and is far easier to use.

There is also the natural question of independence when dealing with the separate scores of two teams playing in the same match. Although their scores in practice would not be strictly independent, for our purposes the assumption seems to have nevertheless yielded useful results.

There are many prospects for further statistical research using soccer, and indeed any football, data. One is to analyse the data separately for teams who are in, say, the top half of the competition table and those who are in the bottom half. It may well be that while one group produces significant results the other does not. The other research opportunities are left to the imagination of the statistician.

References

1. Cain, M., Law, D. and Peel, D. The favourite-longshot bias and market efficiency in UK football betting. *Scottish Journal of Political Economy*, 47: 25-36, 2000.

2. Clarke, S.R. and J.M. Norman. Home ground advantage if individual clubs in English soccer. *The Statistician*, 44: 509-521, 1995.
3. Clarke, S.R. Home advantage in balanced competitions – English soccer 1990-1996. In de Mestre, N. (editor) *Mathematics and Computers in Sport*, Bond University, Gold Coast, Queensland: 111-116, 1996.
4. Courneya, K.S. and A.V. Carron. The home advantage in sport competitions: a literature review. *J. of Sport and Exercise Psychology*, 14: 13-27, 1992.
5. Croucher, J.S. The effect of changing competition points in the English Football League. *Teaching Statistics*, 2: 39-42, 1984.
6. Dixon, M.J. and Coles, S.C. Modelling association football scores and inefficiencies in the football betting market. *Applied Statistics*, 46: 265-280, 1997.
7. Dixon, M.J. and Robinson, M.E. A birth process model for association football matches. *The Statistician*, 47: 523-538, 1998.
8. Haigh, J. *Taking Chances*, 2nd edition. Oxford University Press, 247-248, 2003.
9. Hill, I.D. Association football and statistical inference. *Applied Statistics*, 23: 203-208, 1974.
10. Holgate, P. Estimation for the bivariate Poisson distribution. *Biometrika*, 51: 241-245, 1964.
11. Maher, M.J. Modelling association football scores. *Statistica Neerlandica*, 36: 109-118, 1982.
12. Moroney, M.J. *Facts from Figures*, Harmondsworth: Penguin, 1951.
13. Norman, J.M. "Soccer". In Jay Bennett (editor) *Strategies in Sports.* Arnold: New York, 105–120, 1999.
14. Pollard, R. Home ground advantage in soccer: a retrospective analysis. *Journal of Sports Sciences*, 4: 237-248, 1986.
15. Pollard, R., P. Benjamin and C Reep. *Sport and the negative binomial distribution.* In S.P. Ladany and R.E. Machol (eds) *Optimal Strategies in Sports.* New York: North Holland, 188-195, 1977.
16. Reep, C. and P. Benjamin. Skill and chance in association football. *J. R. Statist. Soc. A*, 131: 581–585, 1968.
17. *Sky Sports Football Yearbook*, G. Rollin and J. Rollin (eds.), Pub. Headline, London, Annually.
18. Stefani, R.T. Observed betting tendencies and suggested betting strategies for European football pools", *The Statistician*, 32: 319-329, 1983.
19. Stefani, R.T. Applications of statistical methods to American football. *J. of Applied Statistics*, 14: 61-73, 1987.
20. Wood, G.H. Cricket scores and geometrical progression. *J.R. Statist. Soc. A*, 108: 12-22, 1945.

Modelling and Forecasting Match Results in the English Premier League and Football League

Stephen Dobson[1] and John Goddard[2]

[1] University of Otago, New Zealand
sdobson@business.otago.ac.nz
[2] University of Wales Swansea, UK
j.a.goddard@swansea.ac.uk

1 Introduction

This chapter discusses the statistical analysis, modelling and forecasting of match results in (association) football. Section 2 uses some of the basic tools of probability theory to summarise and describe a data set comprising 30 years of English Premier League (PL) and Football League (FL) match results. Section 3 provides a brief review of previous academic literature on modelling football match results and goal scoring. Section 4 describes the construction of a statistical model that can be used to generate match result forecasts in the form of probabilities for the three possible outcomes of home win, draw or away win. Section 5 describes the estimation of the model. A number of factors that have a statistically significant influence on match results are identified. These include the importance of the match for championship, promotion or relegation issues for either team, the involvement of either team in the (contemporaneous) FA Cup competition, the geographical distance between the home towns of the two teams, and the two teams' market sizes, measured by average home attendances relative to league positions. Section 6 assesses the forecasting model's predictive capability by comparing the model's estimated match results probabilities for the 2001-2 season with the fixed odds quoted by a leading firm of high street bookmakers. The contents of this chapter draw on, update and extend results that have been reported previously by Dobson and Goddard (2001), Audas, Dobson and Goddard (2002) and Goddard and Asimakopoulos (2003).

Table 1. Distribution of match results by home and away team scores, all matches, 1972-3 to 2001-2.

		Goals scored by away team										
		0	1	2	3	4	5	6	7	8	9	Total
	0	5210	4420	2264	856	232	51	16	3	-	-	13052
	1	6860	7807	3547	1328	364	91	16	8	4	1	20026
	2	5165	5760	3121	979	250	61	16	3	1	-	15356
Goals	3	2757	2823	1526	610	147	22	7	2	-	-	7894
scored	4	1122	1087	559	260	64	7	5	1	-	-	3105
by	5	400	387	218	76	25	4	-	-	-	-	1110
home	6	111	106	60	19	3	-	-	-	-	-	299
team	7	31	44	12	2	3	1	-	-	-	-	93
	8	9	3	3	-	-	-	-	-	-	-	15
	9	4	-	-	-	-	-	-	-	-	-	4
	10	1	1	-	-	-	-	-	-	-	-	2
	Total	21670	22438	11310	4130	1088	237	60	17	5	1	60956

2 Joint, Marginal and Conditional Probability Distributions for Goal Scoring in English League Football

Table 1 shows the distribution of the numbers of goals scored in each match by the home and away teams based on all PL and FL match results between the 1972-3 and 2001-2 seasons (inclusive). The match results data are obtained from various issues of *Rothmans Football Yearbook*. In Table 2, these raw data are converted into an empirical joint probability distribution for the home and away team scores, calculated by dividing each cell in Table 1 by the total number of matches played, 60,956. If X and Y are the random variables for the number of goals scored by the home and away teams respectively, the joint probability function $P(X = x, Y = y)$ or $P(x, y)$ specifies the probability that X is exactly equal to x, and Y is exactly equal to y, for all possible combinations of x and y. For example, 5,210 out of 60,956 or 8.55% of all matches finished as 0-0 draws, so $P(X=0, Y=0) = 0.0855$; 4,420 or 7.25% of all matches finished as 0-1 away team wins, so $P(X=0, Y=1) = 0.0725$; and so on.

The marginal probability function for X, denoted $P_x(X = x)$ or $P_x(x)$ expresses the probability that $X = x$, regardless of the value of Y. It is calculated by summing $P(X = x, Y = y)$ over all possible values y, for each x:

$$P_x(X = x) = \sum_y P(X = x, Y = y) \tag{1}$$

Similarly the marginal probability function for Y is:

Table 2. Joint and marginal probability distributions of home and away team scores.

		Goals scored by away team							
		0	1	2	3	4	5	6+	Marginal
	0	.0855	.0725	.0371	.0140	.0038	.0008	.0003	.2141
Goals	1	.1125	.1281	.0582	.0218	.0060	.0015	.0005	.3285
scored	2	.0847	.0945	.0512	.0161	.0041	.0010	.0003	.2519
by	3	.0452	.0463	.0250	.0100	.0024	.0004	.0001	.1295
home	4	.0184	.0178	.0092	.0043	.0010	.0001	.0001	.0509
team	5	.0066	.0063	.0036	.0012	.0004	.0001	.0000	.0182
	6+	.0026	.0025	.0012	.0003	.0001	.0000	.0000	.0068
	Marginal	.3555	.3681	.1855	.0678	.0178	.0039	.0014	1.0000

$$P_y(Y = y) = \sum_x P(X = x, Y = y) \tag{2}$$

In Table 2 the marginal probabilities for X are calculated by summing across the rows, and the marginal probabilities for Y are calculated by summing down the columns. For example, the home team failed to score in 21.41% of all matches, and the home team scored once in 32.85% of all matches. The corresponding percentages for away teams were 35.55% and 36.81%.

The marginal probability functions can be used to define the unconditional means and variances for X, denoted E(X) or μ_x and var(X) or σ_x^2, respectively.

$$\begin{aligned} E(X) \text{ or } \mu_x &= \sum_x x P_x(X = x) \text{ and} \\ \text{var}(X) \text{ or } \sigma_x^2 &= \sum_x (x - \mu_x)^2 P_x(X = x) \end{aligned} \tag{3}$$

Formulae similar to (3) are used to obtain E(Y) and var(Y). Applying these formulae to the data in Tables 1 and 2, the average numbers of goals scored by the home and away teams per match are E(X) = 1.5587 and E(Y) = 1.0419. The corresponding variances are var(X) = 1.6182 and var(Y) = 1.0729.

The conditional probability function for X given Y = y, denoted $P(X = x|Y = y)$, expresses the probability that X = x, given that we already know the value which Y takes is y. It is calculated as follows:

$$P(X = x|Y = y) = \frac{P(X = x, Y = y)}{P_y(Y = y)} \tag{4}$$

Similarly the conditional probability function for Y, given X =x, is:

$$P(Y = y|X = x) = \frac{P(X = x, Y = y)}{P_x(X = x)} \tag{5}$$

Tables 3 and 4 show the conditional probability functions for the numbers of goals scored by the home and away teams, respectively. The conditioning is on the number of goals scored by the opposing team. The conditional

Table 3. Conditional probability distribution of home team scores, conditioning on away team score.

		Goals scored by away team						
		0	1	2	3	4	5	6+
	0	.2404	.1970	.2002	.2073	.2132	.2152	.2289
Goals	1	.3166	.3479	.3136	.3215	.3346	.3840	.3494
scored	2	.2383	.2567	.2760	.2370	.2298	.2574	.2410
by	3	.1272	.1258	.1349	.1477	.1351	.0928	.1084
home	4	.0518	.0484	.0494	.0630	.0588	.0295	.0723
team	5	.0185	.0172	.0193	.0184	.0230	.0169	.0000
	6+	.0072	.0069	.0066	.0051	.0055	.0042	.0000

Table 4. Conditional probability distribution of away team scores, conditioning on home team score.

		Goals scored by away team						
		0	1	2	3	4	5	6+
	0	.3992	.3386	.1735	.0656	.0178	.0039	.0015
Goals	1	.3426	.3898	.1771	.0663	.0182	.0045	.0014
scored	2	.3364	.3751	.2032	.0638	.0163	.0040	.0013
by	3	.3493	.3576	.1933	.0773	.0186	.0028	.0011
home	4	.3614	.3501	.1800	.0837	.0206	.0023	.0019
team	5	.3604	.3486	.1964	.0685	.0225	.0036	.0000
	6+	.3777	.3729	.1816	.0508	.0145	.0024	.0000

distributions provide an indication as to whether there are interdependencies between the numbers of goals scored by the home and away teams, or whether the numbers of goals scored by the two teams are independent of each other.

For example, the first column of Table 3 shows, for matches in which the away team failed to score, the proportions of occasions on which the home team scored 0, 1, 2, 3 etc. goals. In 24.04% of matches in which the away team failed to score, the home team also failed to score; in 31.66% of matches in which the away team failed to score, the home team scored once; and so on. The second column shows the same for matches in which the away team scored once. In Table 4, the conditioning is reversed. The first row shows, for matches in which the home team failed to score, the proportions of occasions on which the away team scored 0, 1, 2, 3 goals, and so on. The second row shows the same for matches in which the home team scored once.

The conditional probability functions can be used to define conditional means and variances for X and Y, with conditioning (as before) on the number of goals scored by the opposing team.

$$E(X|Y = y) \text{ or } \mu_{x|y} = \sum_{x} xP(X = x|Y = y) \tag{6}$$

$$\text{var}(X|Y = y) \text{ or } \sigma^2_{x|y} = \sum_x (x - \mu_{x|y})^2 P(X = x|Y = y) \qquad (7)$$

Formulae similar to (6) and (7) are used to obtain $E(Y \mid X = x)$ and $\text{var}(Y \mid X = x)$. Applying these formulae to the data in Tables 1 to 4, the mean and variance of number of goals scored by the home team, conditional on the away team scoring once, are $E(X \mid Y{=}1) = 1.5624$ and $\text{var}(X| Y{=}1) = 1.5515$. The mean and variance of the number of goals scored by the home team, conditional on the away team scoring three times, are $E(X \mid Y{=}3) = 1.6136$ and $\text{var}(X| Y{=}3) = 1.6521$.

If the number of goals scored by the home team and the away team are independent of each other, then the number of away goals should not affect the probabilities for the number of home goals. This means the conditional probability distributions in each column of Table 3 should look similar to one another (and similar to the marginal probabilities for the number of home goals in Table 2). Similarly the number of home goals should not affect the probabilities for the number of away goals, so the conditional probability distributions in each row of Table 4 should also look similar to one another (and similar to the marginal probabilities for the number of away goals in Table 2).

In fact, however, the probabilities do not appear to conform to this condition for independence. In Table 4, for example, if the home team fails to score, then 0 is also the most likely score for the away team, with $P(Y{=}0| X{=}0) = 0.3992$. $P(Y{=}1| X{=}0) = 0.3426$ is smaller. In contrast, if the home team scores once, then 1 is the most likely score for the away team, with $P(Y{=}1| X{=}1) = 0.3898$. $P(Y{=}0| X{=}1) = 0.3386$ is smaller. The shapes of the distribution of Y conditional on X=0 and the distribution of Y conditional on X=1 are therefore quite different in the region of Y=0 and Y=1.

A more formal definition of independence, based on comparisons between the joint and marginal probability functions, is as follows. The random variables X and Y are independent if $P(X = x, Y = y) = P_x(X = x)P_y(Y = y)$ for all values of x and y. Conversely, X and Y are not independent if $P(X = x, Y = y) \neq P_x(X = x)P_y(Y = y)$ for at least one pair of values of x and y.

According to Table 2 and using X=0 and Y=0 as an example, $P(X{=}0, Y{=}0) = 0.0855$, but $P_x(X = 0)P_y(Y = 0) = 0.2141 \times 0.3555 = 0.0761$. The actual probability of a 0-0 draw is therefore higher than the product of the (unconditional) marginal probabilities that the home team and away team each fail to score. Strictly speaking, this condition for independence should be checked using the 'true' probability distributions for X and Y, rather than estimated probabilities based on a sample of match results data. Nevertheless, in this case the sample size is very large, and the discrepancy between $P(X = x, Y = y)$ and $P_x(X = x)P_y(Y = y)$ is sufficiently big, that the notion of independence between the home and away team scores can confidently be dismissed.

A useful measure of the association between two random variables is the covariance, defined as follows:

$$\begin{aligned}\mathrm{cov}(X, Y) \text{ or } \sigma_{xy} &= E[(X - \mu_x)(Y - \mu_y)] \\ &= \sum_x \sum_y (x - \mu_x)(y - \mu_y)P(X = x, Y = y) \end{aligned} \tag{8}$$

If X and Y are positively related, cov(X, Y)> 0; if X and Y are unrelated, cov(X, Y) = 0; and if X and Y are negatively related, cov(X, Y)< 0. Therefore the covariance provides an indication of the direction of the relationship (if any) between X and Y. Unfortunately, however, the covariance does not provide any information about the strength of the relationship between X and Y, because its value depends on the scaling (units of measurement) of X and Y. It is possible to standardise the covariance in such a way as to obtain the correlation coefficient defined as follows:

$$\mathrm{corr}(X, Y) \text{ or } \rho_{xy} = \frac{\mathrm{cov}(X, Y)}{\sqrt{\mathrm{var}(X)\mathrm{var}(Y)}} \tag{9}$$

corr(X,Y) always lies between −1 and +1. Its magnitude within this range conveys information about both the direction and the strength of the relationship between the random variables X and Y. If there is an exact linear positive relationship between X and Y, corr(X,Y) = 1; if X and Y are positively related, 0 < corr(X,Y)< 1; if X and Y are unrelated, corr(X,Y) =0; if X and Y are negatively related, –1 < corr(X,Y)< 0; and if there is an exact negative linear relationship between X and Y, corr(X,Y)=−1.

Using the data in Tables 1 to 4 produces cov(X,Y) = 0.0279 and corr(X,Y) = 0.0212 (p-value=0.000). These statistics confirm that there is a positive relationship between the numbers of goals scored by the home and away teams. The correlation coefficient is quite small, because the relationship between the home and away team scores is by no means exact or deterministic. Nevertheless, as the p-value indicates, the coefficient is significantly different from zero.

In order to construct Tables 1 to 4, the match results data for seasons 1972-3 to 2001-2 (inclusive) have been pooled. However, Table 5 suggests that there have been some changes in patterns of goal scoring and match results within this period. For each season Table 5 shows the proportions of matches that resulted in home wins, draws and away wins, and the average numbers of goals scored per match by the home and away teams, and in total. Over this period as a whole, there was a marked improvement in the average performance of away teams, and a corresponding deterioration in the performance of home teams. The proportion of matches resulting in away wins increased from around 20% to 26%, and the proportion of home wins fell by a similar amount, from around 52% to 46%. The proportion of draws held constant, at around 28% throughout the period. Perhaps the most striking feature of Table 5 is the dramatic jump in the average number of goals scored by away teams, which coincided with the introduction of the award of three league points (rather than two) for a win from season 1981-2 onwards.

Table 5. Percentages of home wins, draws and away wins and average goals scored, PL and FL, 1972-3 to 2001-2 inclusive.

	% home wins	% draws	% away wins	Average home goals per match	Average away goals per match	Average total goals per match
1972-3	51.7	28.4	19.9	1.565	0.917	2.482
1973-4	48.9	30.2	21.0	1.514	0.921	2.435
1974-5	52.4	27.9	19.7	1.585	0.921	2.506
1975-6	50.2	27.7	22.1	1.587	0.986	2.573
1976-7	51.9	27.9	20.2	1.656	0.986	2.642
1977-8	50.1	30.9	19.0	1.616	0.981	2.597
1978-9	48.0	29.6	22.4	1.548	1.010	2.558
1979-80	49.9	27.9	22.2	1.584	0.982	2.566
1980-1	49.6	28.0	22.4	1.525	0.941	2.466
1981-2	47.1	27.4	25.5	1.527	1.075	2.602
1982-3	52.0	26.3	21.6	1.716	1.059	2.775
1983-4	50.8	26.0	23.2	1.693	1.055	2.749
1984-5	50.1	24.3	25.6	1.623	1.101	2.724
1985-6	50.1	24.6	25.3	1.694	1.106	2.800
1986-7	48.9	27.4	23.7	1.554	1.046	2.600
1987-8	46.1	27.3	26.6	1.546	1.078	2.624
1988-9	46.8	28.6	24.6	1.583	1.079	2.662
1989-90	46.8	27.3	25.9	1.540	1.088	2.629
1990-1	48.4	27.6	24.1	1.556	1.060	2.616
1991-2	47.4	26.9	25.7	1.519	1.079	2.598
1992-3	46.5	26.3	27.1	1.575	1.117	2.692
1993-4	45.6	27.4	27.0	1.534	1.138	2.672
1994-5	46.4	27.5	26.1	1.511	1.080	2.591
1995-6	45.0	29.6	25.3	1.476	1.059	2.535
1996-7	46.4	28.1	25.5	1.473	1.032	2.505
1997-8	47.9	27.8	24.3	1.506	1.047	2.554
1998-9	45.0	28.0	27.0	1.458	1.079	2.537
1999-2000	45.5	27.1	27.4	1.460	1.074	2.533
2000-1	46.5	27.4	26.2	1.500	1.068	2.568
2001-2	46.7	26.5	26.8	1.538	1.092	2.630

3 Previous Academic Literature on Modelling Football Match Results and Goal Scoring

There are perhaps surprisingly few studies in the academic literature that seek to model match results data for football. Early contributions by Moroney (1956) and Reep, Pollard and Benjamin (1971) use the Poisson and negative binomial distributions to model the distributions of the numbers of goals scored per game. The aggregated approach adopted, however, precludes the generation of specific forecasts for individual matches based on information about the respective strengths of the two teams. By comparing final league placings with experts' pre-season forecasts, Hill (1974) demonstrates that individual match results do nevertheless have a predictable element, and are not determined solely by chance.

Maher (1982) develops a model in which the home and away team scores follow independent Poisson distributions, with means which are the product of parameters reflecting the attacking and defensive capabilities of the two teams. These parameters are estimated after the full set of match results has been observed, and the model cannot be used to obtain probabilities for scores or results in advance. Although goodness-of-fit tests show that the model provides a reasonably accurate approximation to the data, separate examination of the observed and expected distributions of the difference between the scores of the two teams reveals a tendency to underestimate the proportion of drawn matches, attributed to interdependence between the scores of the home and away teams. This tendency is corrected by modelling scores using a bivariate Poisson distribution, which allows for interdependence between the home and away scores.

Dixon and Coles (1997) adapt Maher's (1982) modelling approach for different ends: they seek to develop a forecasting model capable of estimating match outcome probabilities before matches are played. Instead of using the bivariate Poisson distribution, the marginal Poisson probabilities for the scores of both teams in low-scoring games are adjusted directly to allow for interdependence. Estimation is based on historical data only, permitting forecasts to be generated. The attacking and defensive parameters are estimated from past scores, weighted by a factor that declines exponentially over time. Each team's parameters are updated from match to match, as the scores from the most recently completed matches enter the estimation. Using a similar framework, Rue and Salvesen (2000) assume that the time-varying attacking and defensive parameters of all teams vary randomly over time. The prior estimates of these parameters are updated as new match results information is obtained. Markov chain Monte Carlo iterative simulation techniques are used for inference. Crowder et al. (2002) propose a procedure for updating the team strength parameters that is computationally less demanding.

Several researchers have investigated the impact of specific factors on match results. Barnett and Hilditch (1993), for example, investigate whether artificial playing surfaces, introduced and subsequently abandoned by several clubs during the 1980s and early 1990s, conferred (additional) home-team advantages. Ridder, Cramer and Hopstaken (1994) show that player dismissals have a negative effect on the match result from the viewpoint of the team completing the match with fewer than eleven players. Clarke and Norman (1995) use a range of non-parametric techniques to identify systematically the effect of home advantage on match results. Dixon and Robinson (1998) investigate variations in the scoring rates of the home and away teams during the course of a match. The scoring rates at any time are partly dependent on the time elapsed, and partly on which (if either) team is leading at the time.

Recently, several applied econometricians have used discrete choice regression models (known as ordered probit or ordered logit) to model match results directly, rather than indirectly through goals scored and conceded. Forrest and Simmons (2000a,b) investigate the predictive quality of newspaper tipsters' match results forecasts, and the performance of the pools panel in providing hypothetical results for postponed matches. Koning (2000) estimates a model to describe a set of match results, as part of a broader analysis of changes in competitive balance in Dutch football. Audas, Dobson and Goddard (2002) examine whether managerial change has any short-term impact on subsequent team performance measured by match results. Kuypers (2000) and Goddard and Asimakopoulos (2003) estimate forecasting models to investigate the efficiency of prices quoted by high-street bookmakers for fixed-odds betting on match results, while Cain, Law and Peel (2000) investigate similar issues for fixed-odds betting on goals scored.

4 Specification of a Match Results Forecasting Model

This section describes the specification of an ordered probit regression model, which can be used to identify factors that have a statistically significant influence on match results, and generate match result forecasts in the form of probabilities for the three possible outcomes of home win, draw or away win.

Table 1 indicates that of the 60,956 league matches played during the 30 seasons between 1972-3 and 2001-2 (inclusive), 29,438 (48.29%) finished as home wins, 16,816 (27.59%) were draws, and 14,702 (24.12%) were away wins. This implies that if any match is selected randomly from the data set, the probabilities that the match selected was a home win, a draw or an away win would be 0.4829, 0.2759 and 0.2412 respectively. The exercise of selecting a match from the data set at random and observing either a home win, a draw or an away win, can be replicated in a statistical model using the standard normal probability distribution. Below, the cumulative distribution function for the standard normal distribution (tabulated towards the back of most statsitics and econometrics textbooks) is denoted $\Phi(\ )$. Imagine drawing a value ε at random from the standard normal distribution:

the probability of obtaining ε greater than $\mu_2 = +0.0429$ is $P(\varepsilon > 0.0429) = 1 - \Phi(0.0429) = 0.4829$;

the probability of obtaining ε between $\mu_1 = -0.7025$ and $\mu_2 = +0.0429$ is $P(-0.7025 < \varepsilon < 0.0429) = \Phi(0.0429) - \Phi(-0.7025) = 0.2759$;

the probability of obtaining ε less than $\mu_1 = -0.7025$ is $P(\varepsilon < -0.7025) = \Phi(-0.7025) = 0.2412$.

These examples suggest a solution to the task of mimicking in a statistical model the process of selecting a match randomly and observing the result. Obtaining ε greater than $\mu_2 = 0.0429$ is equivalent to (or has the same probability as) observing a home win. Similarly, obtaining ε between $\mu_1 = -0.7025$ and $\mu_2 = +0.0429$ is equivalent to observing a draw. Obtaining ε below $\mu_2 = -0.7025$ is equivalent to observing an away win.

As described so far, the statistical model treats all matches that could be chosen from the data set as having identical home win, draw and away win probabilities. This corresponds to the situation when there is no information relevant to an assessment of the probable outcome before the match is played. Suppose, however, that when the match is chosen, information is available in advance, which is relevant to the assessment of the home win, draw and away win probabilities.

In the statistical model, probabilities that are specific to each match can be obtained by allowing variation in the mean of the normal distribution from which the random variable is drawn. Rather than drawing a value of ε at random from the standard normal distribution (with zero mean), the randomly chosen value, now denoted $\varepsilon_{i,j}$ can be drawn from a normal distribution with mean $y^*_{i,j}$ (where $y^*_{i,j}$ can be above or below zero), and a standard deviation of one. Therefore $y^*_{i,j}$ denotes the mean which is specific to the match between team i and team j.

Suppose for example $y^*_{i,j}$ is +0.1:

the probability of obtaining $\varepsilon_{i,j}$ greater than $\mu_2 = +0.0429$ (still equivalent to the probability of observing a home win) is $P(\varepsilon_{i,j} > 0.0429) = 1 - \Phi(0.0429 - 0.1) = 0.5228$;

the probability of obtaining $\varepsilon_{i,j}$ between $\mu_1 = -0.7025$ and $\mu_2 = +0.0429$ (observing a draw) is $P(-0.7025 < \varepsilon_{i,j} < 0.0429) = \Phi(0.0429 - 0.1) - \Phi(-0.7025 - 0.1) = 0.2661$;

the probability of obtaining $\varepsilon_{i,j}$ less than $\mu_1 = -0.7025$ (observing an away win) is $P(\varepsilon_{i,j} < -0.7025) = \Phi(-0.7025 - 0.1) = 0.2111$.

Therefore when $y^*_{i,j} > 0$ the probability of a home win is assessed to be above average, and the probability of an away win is below average. Similarly when $y^*_{i,j} < 0$, the probability of a home win is below average and the probability of an away win is above average.

In the forecasting model, $y^*_{i,j}$ is calculated as a linear function of a collection of relevant covariates. If the total number of covariates is K, and the value of covariate k for the match between teams i and j is denoted $x_{k,i,j}$, the general expression for $y^*_{i,j}$ is as follows:

$$y^*_{i,j} = \sum_{k=1}^{K} \beta_k x_{k,i,j} \tag{10}$$

Covariates that are relevant to the assessment of the match result probabilities include the win ratios of the two teams in matches played over a 24-month period prior to the match in question; the individual results of both teams' few most recent matches; dummy variables reflecting the significance of the match for championship, promotion or relegation issues for either team, and the involvement of either team in the (contemporaneous) FA Cup competition; and the two teams' market sizes, measured by average home attendances relative to league positions. Given a data set comprising covariates and match results, regression techniques can be used to obtain a set of numerical estimates of the coefficients β_k, for k=1...K.

The definitions of the covariates included in $x_{k,i,j}$ in (10) are as follows.

$P^d_{i,y,s}$ = $p^d_{i,y,s}/n_{i,y}$, where $p^d_{i,y,s}$= home team i's total 'points' score, on a scale of 1=win, 0.5=draw, 0=loss in matches played 0-12 months (y=0) or 12-24 months (y=1) before current match; within the current season (s=0) or previous season (s=1) or two seasons ago (s=2); in the team's current division (d=0) or one (d=±1) or two (d=±2) divisions above or below the current division; and $n_{i,y}$ = i's total matches played 0-12 months (y=0) or 12-24 months (y=1) before current match.

$R^H_{i,m}$ = Result (1=win, 0.5=draw, 0=loss) of i's m^{th} most recent home match.

$R^A_{i,n}$ = Result of i's n^{th} most recent away match.

$SIGH_{i,j}$ = 1 if match has championship, promotion or relegation significance for i but not for away team j; 0 otherwise.

$SIGA_{i,j}$ = 1 if match has significance for j but not for i; 0 otherwise.

CUP_i = 1 if i is eliminated from the FA Cup; 0 otherwise.

$TRSAT_{i,j}$ = interaction between linear time trend (1972-3 season = 1; 1973-4 = 2 etc.) and Saturday dummy (1 for matches played on Saturdays; 0 otherwise).

$DIST_{i,j}$ = natural logarithm of the geographical distance between the grounds of i and j.

$AP_{i,h}$ = residual for i from a cross-sectional regression of the log of average home attendance on final league position, defined on a scale of 92 for the PL winner to 1 for the bottom team in Division 3 of the Football League (FLD3), h seasons before the present season, for h=1, 2.

$P^d_{j,y,s}$, $R^H_{j,n}$, $R^A_{j,m}$, CUP_j, $AP_{j,h}$ = as above, for away team j.

The specification of the model is completed by specifying the link between $y^*_{i,j}$ and the match result, denoted $y_{i,j}$, as follows:

$$
\begin{array}{lll}
\text{Home win} & \Rightarrow \; y_{i,j} = 1 \;\; \text{if} & \mu_{2s} < y^*_{i,j} + \varepsilon_{i,j} \\
\text{Draw} & \Rightarrow \; y_{i,j} = 0.5 \; \text{if} & \mu_{1s} < y^*_{i,j} + \varepsilon_{i,j} < \mu_{2s} \\
\text{Away win} & \Rightarrow \; y_{i,j} = 0 \;\; \text{if} & y^*_{i,j} + \varepsilon_{i,j} < \mu_{1s}
\end{array}
\qquad (11)
$$

where $\varepsilon_{i,j}$ allows for the random element in the result of the match between teams i and j, and is assumed to follow the standard normal distribution. μ_{1s} and μ_{2s} are the values of the cut-off parameters in season s. These parameters control the overall proportions of home wins, draws and away wins. Table 5 indicates that there has been quite significant variation over time in these proportions; for example, following the introduction of the award of three rather than two league points for a win from the 1981-2 season onwards, there was a sustained increase in the overall proportion of away wins. Therefore a different pair of values for μ_{1s} and μ_{2s} is estimated for each season.

In order to generate either in-sample or out-of-sample estimated match result probabilities, (11) can be rearranged, as follows:

$$
\begin{array}{ll}
\text{Home win probability} & = p^H_{i,j} = \text{prob}(\varepsilon_{i,j} > \mu_{2s} - y^*_{i,j}) \\
 & = 1 - \Phi(\mu_{2s} - y^*_{i,j}) \\
\text{Draw probability} & = p^D_{i,j} = \text{prob}(\mu_{1s} - y^*_{i,j} < \varepsilon_{i,j} < \mu_{2s} - y^*_{i,j}) \\
 & = \Phi(\mu_{2s} - y^*_{i,j}) - \Phi(\mu_{1s} - y^*_{i,j}) \\
\text{Away win probability} & = p^A_{i,j} = \text{prob}(\varepsilon_{i,j} < \mu_{1s} - y^*_{i,j}) = \Phi(\mu_{1s} - y^*_{i,j})
\end{array}
\qquad (12)
$$

(12) takes the same form as the expressions used to calculate the illustrative home win, draw and away win probabilities at the start of this section.

5 Match Results Forecasting Model: Estimation Results

In order to investigate the predictive capability of the match results forecasting model specified in Section 4, the 30-year match results data set is split into a 29-year estimation period, incorporating seasons 1972-3 to 2000-1 (inclusive), and a forecast period, which is the 2001-2 season. In order to determine how the model would have performed had it been used to forecast results during the course of the 2001-2 season, it is of course essential that the 2001-2 data are not used as part of the estimation. This section describes the estimation results, by commenting on the estimated coefficients (β_k) for each set of covariates (as defined in the previous section) in turn. The estimation results are reported in Table 6.

Table 6. Ordered probit estimation results.

1. WIN RATIOS OVER PREVIOUS 24 MONTHS ($P^d_{i,y,s}$, $P^d_{j,y,s}$)								
	Home team (i)				Away team (j)			
	0-12 months (y=0)		12-24 months (y=1)		0-12 months (y=0)		12-24 months (y=1)	
Matches played:	Current season (s=0)	Last season (s=1)	Last season (s=1)	Two seasons ago (s=2)	Current season (s=0)	Last season (s=1)	Last season (s=1)	Two seasons ago (s=2)
Two divisions higher (d=2)				-0.025 (0.355)				-0.235 (0.360)
One division higher (d=1)		1.726*** (0.180)	0.589*** (0.150)	0.576*** (0.146)		-1.539*** (0.178)	-0.825** (0.150)	-0.525*** (0.145)
Current division (d=0)	1.606*** (0.107)	1.211*** (0.099)	0.513*** (0.093)	0.420*** (0.094)	-1.221*** (0.106)	-1.002*** (0.098)	-0.570*** (0.093)	-0.386*** (0.093)
One division lower (d=−1)		0.833*** (0.089)	0.374*** (0.083)	0.303*** (0.079)		-0.624*** (0.089)	-0.356*** (0.084)	-0.234*** (0.078)
Two divisions lower (d=−2)				-0.048 (0.155)				-0.375** (0.158)

2. MOST RECENT MATCH RESULTS ($R^H_{i,m}$, $R^A_{i,n}$, $R^H_{j,n}$, $R^A_{j,m}$)									
Number of matches ago (m,n)	1	2	3	4	5	6	7	8	9
Home team (i) Home matches (HH)	0.026*** (0.006)	0.010* (0.006)	0.032*** (0.006)	-0.001 (0.006)	0.004 (0.006)	-0.007 (0.006)	0.006 (0.006)	0.008 (0.006)	0.013** (0.006)
Away matches (HA)	0.011* (0.006)	0.021*** (0.006)	0.023*** (0.006)	0.008 (0.006)					
Away team (j) Home matches (AH)	-0.018*** (0.006)	-0.021*** (0.006)	-0.018*** (0.006)	-0.019*** (0.006)					
Away matches (AA)	-0.019*** (0.006)	-0.019*** (0.006)	-0.019*** (0.006)	-0.019*** (0.006)	-0.012* (0.006)	-0.010* (0.006)	-0.012** (0.006)	-0.012* (0.006)	-0.008 (0.006)

3. OTHER EXPLANATORY VARIABLES									
$SIGH_{i,j}$	$SIGA_{i,j}$	CUP_i	CUP_j	$DIST_{i,j}$	$TRSAT_{i,j}$	$\Delta AP_{i,1}$	$AP_{i,2}$	$\Delta AP_{j,1}$	$AP_{j,1}$
0.158*** (0.024)	-0.090*** (0.024)	-0.095*** (0.018)	0.066*** (0.018)	0.061*** (0.006)	-0.0012* (0.0006)	0.161*** (0.026)	0.149*** (0.016)	-0.156*** (0.026)	-0.151*** (0.016)

*** denotes coefficient significant at 1% level, ** denotes 5% level, * denotes 10% level.

Win ratios over the previous 24 months

The win ratio variables $\{P^{d}_{i,y,s}, P^{d}_{j,y,s}\}$ are the main team quality indicators. Positive home team and negative away team coefficients on these variables are expected. The coefficients on $P^{0}_{i,0,0}$, $P^{0}_{i,0,1}$ and $P^{0}_{i,1,1}$ show that for the home team, the points ratio from matches played during the present season is a stronger indicator of current team quality than the points ratio from matches played in the same division within the same 12-month period but during the previous season. In turn, the latter is a stronger indicator than the points ratio from matches played in the same division during the previous 12-month period and during the previous season. The estimated coefficients on $P^{0}_{j,0,0}$, $P^{0}_{j,0,1}$ and $P^{0}_{j,1,1}$ for the away team follow a similar pattern.

Recent match results

The recent match results variables $\{R^{H}_{i,m}, R^{A}_{i,n}\}$ for team i, and their counterparts for team j, incorporate information about each team's few most recent home and away results in the calculation of $y^{*}_{i,j}$. The estimated coefficients for the home team are predominantly positive and those for the away team predominantly negative as expected. However, these coefficients are generally more erratic than those for the win ratio variables, both in terms of their numerical magnitudes and statistical significance.

Significance of match for championship, promotion and relegation issues

If a match is significant for championship, promotion or relegation issues for one team but insignificant for the other, the match result is likely to be influenced by the difference between the incentives for the two teams. For the purposes of the estimation, a match is deemed to be significant if it is possible (before the match is played) for the team in question to win the championship or be promoted or relegated, if all other teams currently in contention for the same outcome take one point on average from each of their remaining fixtures. As expected the coefficients on $SIGH_{i,j}$ and $SIGA_{i,j}$ are positive and negative, respectively, and both are significant.

FA Cup involvement

Early elimination from the FA Cup may have implications for a team's results in subsequent league matches, although on theoretical criteria the direction of the effect is uncertain. A team eliminated from the cup may be able to concentrate efforts on the league, suggesting an improvement in league results; or cup elimination may cause a loss of confidence, suggesting a deterioration in league results. In Table 6 the estimated coefficient on CUP_i is negative and that on CUP_j is positive and both are significant. It appears that the second of the two effects described above dominates.

Geographical distance

The geographical distance between the hometowns of the teams contesting a match is a significant influence on the result. Accordingly the estimated coefficient on $DIST_{i,j}$ in Table 6 is positive and significant. The greater intensity of competition in local derbies may have some effect in offsetting home advantage, while the psychological or practical difficulties of long-distance travel for teams and supporters may increase home advantage in matches between teams from opposite ends of the country.

Saturday effect

Inspection of the match results data indicates set that at the start of the observation period, there was a tendency for home advantage to be less important for matches played on Saturdays than on other days of the week (mainly weekday evenings). By the end of the period, however, this 'Saturday effect' had virtually disappeared. Accordingly, a negative coefficient on $TRSAT_{i,j}$, a Saturdays dummy interacted with a linear time trend, is obtained.

Attendance relative to league position

The covariates $AP_{i,k}$ and $AP_{j,k}$ are positive for teams that tend to attract higher-than-average home attendances after controlling for league position, and negative in the opposite case. These covariates allow for a 'big team' effect on match results: regardless of the values of other controls, large-market teams are more likely (and small-market teams less likely) to win, either through the direct influence of the crowd on the match result, or because teams with a larger revenue base have more resources to spend on players. To reduce the effect of temporary variation in the attendance-performance relationship, the values of these variables for the two seasons prior to the match in question are included in the model. Since the values over successive seasons tend to be highly correlated, $\Delta AP_{i,1} = AP_{i,1} - AP_{i,2}$ (and its counterpart for team j) is used in place of $AP_{i,1}$ (and $AP_{j,1}$).

6 Predictive Capability

To provide a benchmark for assessing the forecasting model's estimated probabilities, the fixed odds published by one of the leading high-street bookmakers were used to compile an alternative set of match result probabilities. The prices for bets on the results of English football matches are fixed by the bookmakers several days before the match and are not adjusted as bets are placed, even if new information is received. Bookmakers' odds appear in the form: a-to-b home win; c-to-d draw; and e-to-f away win. If b is staked on a home win, the overall payoffs to the bettor are +a (the bookmaker pays the winnings and returns the stake) if the bet wins, and –b (the bookmaker keeps the stake) if the bet loses. These quoted odds can be converted to home win, draw and away win 'probabilities': $\theta^{H}_{i,j}$= b/(a+b); $\theta^{D}_{i,j}$= d/(c+d); $\theta^{A}_{i,j}$= f/(e+f). The sum of these expressions invariably exceeds one, however, because the odds contain a margin to cover the bookmaker's costs and profits. Implicit home win, draw and away win probabilities which sum to one are $\phi^{H}_{i,j} = \theta^{H}_{i,j}/(\theta^{H}_{i,j} + \theta^{D}_{i,j} + \theta^{A}_{i,j})$, and likewise for $\phi^{D}_{i,j}$ and $\phi^{A}_{i,j}$.

From a total of 2,036 league matches played in the 2001-2 season, the forecasting model is capable of generating predictions for 1,946. No predictions are available for the two teams that entered FLD3 at the start of the 2000-1 and 2001-2 seasons (Kidderminster and Rushden), because two previous years of league match results data are required to create all of the forecasting model's covariates. The bookmaker's coupons provided odds for all but one of these matches (the date of the Exeter-Plymouth match was rearranged at short notice, and this match did not appear on the coupons). The model's probabilities for the 1,945 matches are obtained by substituting the full set of covariate values for the home and away teams for each match into the model reported in Table 6, generating a fitted value of $y^{*}_{i,j}$ for each match denoted $\hat{y}^{*}_{i,j}$ for the match between home team i and away team j, and fitted values of the cut-off parameters for the last season in the data set (2000-1) denoted $\hat{\mu}_1$ and $\hat{\mu}_2$. As before, the estimated home win, draw and away win probabilities are $p^{H}_{i,j}=1-\Phi(\hat{\mu}_2-\hat{y}^{*}_{i,j})$, $p^{D}_{i,j} = \Phi(\hat{\mu}_2-\hat{y}^{*}_{i,j})-\Phi(\hat{\mu}_1-\hat{y}^{*}_{i,j})$ and $p^{A}_{i,j} = \Phi(\hat{\mu}_1-\hat{y}^{*}_{i,j})$.

Table 7 summarises features of the implicit bookmaker's probabilities, the model's estimated match result probabilities, and the match outcomes. When the bookmaker's and the model's probabilities are disaggregated by month and by division, there is little systematic variation in the mean probabilities for each outcome, but the cross-sectional standard deviations of both the bookmaker's and the model's probabilities increase systematically during the course of the season. This is because it is possible to make a more specific assessment of the probabilities for any individual match (based mainly on the current season's results) towards the end of the season than at the start (when results from previous seasons provide the only guidance). The cross-sectional standard deviations are highest in the PL and lowest in FLD3. This appears to reflect a greater degree of competitive balance in the lower than in the higher divisions of the league.

Overall the forecasting model is effective in replicating the main features of the bookmaker's probabilities. As an overall indicator of forecasting accuracy, Rue and Salvesen (2000) suggest the pseudo-likelihood (PsL) measure calculated as the geometric mean of the probabilities for the observed results. PsL can be calculated using both the bookmaker's and the model's probabilities. As Table 7 shows, it is almost impossible to separate the bookmaker's and the model's performance on this criterion: the respective values of PsL for the 2001-2 season as a whole are 0.3604 and 0.3601. According to the PsL statsitics, the bookmaker marginally out-performed the model for the

Table 7. Bookmaker's implicit probabilities, and forecast probabilities: descriptive statistics.

	Bookmaker				Model				Actual		
	$\phi^{H}_{i,j}$	$\phi^{D}_{i,j}$	$\phi^{A}_{i,j}$	PsL	$p^{H}_{i,j}$	$p^{D}_{i,j}$	$p^{A}_{i,j}$	PsL	H	D	A
PL	.4534	.2685	.2780	.3696	.4707	.2662	.2630	.3665	.4342	.2658	.3000
	.1423	*.0268*	*.1278*		*.1397*	*.0320*	*.1163*				
FLD1	.4531	.2692	.2777	.3585	.4610	.2738	.2652	.3582	.4692	.2536	.2772
	.1084	*.0173*	*.0982*		*.1102*	*.0235*	*.0931*				
FLD2	.4575	.2690	.2735	.3590	.4650	.2722	.2628	.3601	.4728	.2754	.2518
	.0994	*.0145*	*.0897*		*.1163*	*.0254*	*.0962*				
FLD3	.4561	.2683	.2756	.3571	.4678	.2737	.2585	.3571	.4707	.2690	.2603
	.1038	*.0151*	*.0949*		*.1051*	*.0230*	*.0855*				
Aug-Oct	.4516	.2715	.2769	.3529	.4609	.2762	.2628	.3538	.4560	.2610	.2830
	.0968	*.0136*	*.0873*		*.0966*	*.0196*	*.0801*				
Nov-Dec	.4575	.2709	.2715	.3562	.4671	.2730	.2599	.3564	.4632	.2779	.2589
	.1040	*.0153*	*.0936*		*.1098*	*.0235*	*.0903*				
Jan-Feb	.4586	.2693	.2721	.3615	.4678	.2712	.2610	.3596	.4519	.2716	.2765
	.1121	*.0170*	*.1024*		*.1185*	*.0256*	*.0986*				
Mar-May	.4549	.2629	.2822	.3736	.4687	.2657	.2656	.3723	.4845	.2567	.2588
	.1364	*.0250*	*.1232*		*.1437*	*.0333*	*.1199*				
All	.4551	.2688	.2761	.3604	.4656	.2718	.2625	.3601	.4638	.2658	.2704
	.1124	*.0184*	*.1017*		*.1171*	*.0259*	*.0972*				

Note: For $\phi^{H}_{i,j}$, $p^{H}_{i,j}$ etc., Table 7 reports the cross-sectional mean, together with the standard deviation in italics.

PL and FLD1, while the model marginally out-performed the bookmaker for FLD2. For FLD3 the performance was identical. The model marginally out-performed the bookmaker over the first half of the season, while the reverse was true over the second half.

Of course, the similarity between the forecasting capabilities of the bookmaker's and the forecasting model's probabilities does not mean the two sets of probabilities are identical, although they are highly correlated. The correlation coefficients for the bookmaker's and model's home win, draw and away win probabilities across all 1,945 matches for the 2001-2 season are 0.899, 0.752 and 0.892, respectively. With zero p-values, all of these correlations are significantly different from zero, and significantly different from one. Table 8 allows direct comparisons between a small set of these probabilities, which have been computed for the final round of PL matches in the 2001-2 season, played on 11 May 2002.

Table 8. Bookmaker's and model's probabilities for PL matches played on 11 May 2002.

Home team	Away team	Bookmaker			Model			Calculation of PsL		
		$\phi_{i,j}^{H}$	$\phi_{i,j}^{D}$	$\phi_{i,j}^{A}$	$p_{i,j}^{H}$	$p_{i,j}^{D}$	$p_{i,j}^{A}$	res	$\ln(\phi_{i,j}^{res})$	$\ln(p_{i,j}^{res})$
Arsenal	Everton	.6839	.1976	.1185	.8239	.1302	.0459	H	-0.380	-0.194
Blackburn	Fulham	.4844	.2690	.2466	.5221	.2692	.2087	H	-0.725	-0.650
Chelsea	Ast. Villa	.5672	.2546	.1782	.6004	.2433	.1563	A	-1.725	-1.856
Leeds	Middlesbro.	.5672	.2546	.1782	.5871	.2482	.1647	H	-0.567	-0.533
Leicester	Tottenham	.3177	.2780	.4043	.3556	.2943	.3501	H	-1.147	-1.034
Liverpool	Ipswich	.6661	.1973	.1366	.7416	.1782	.0802	H	-0.406	-0.299
Manch. Utd	Charlton	.7412	.1779	.0809	.7686	.1633	.0682	D	-1.727	-1.812
Southampton	Newcastle	.3391	.2739	.3870	.2549	.2834	.4617	H	-1.081	-1.367
Sunderland	Derby	.6135	.2684	.1181	.5339	.2658	.2003	D	-1.315	-1.325
West Ham	Bolton	.5782	.2607	.1612	.5485	.2613	.1901	H	-0.548	-0.601

7 Conclusion

This chapter has discussed the statistical analysis, modelling and forecasting of match results in (association) football. After summarising and describing a data set comprising 30 years of English Premier League and Football League match results, a statistical model has been developed that can be used to generate match result forecasts in the form of probabilities for the three possible outcomes of home win, draw or away win. Covariates used in the model include variables measuring the importance of the match for championship, promotion or relegation issues for either team, the involvement of either team in the (contemporaneous) FA Cup competition, the geographical distance between the hometowns of the two teams, and the two teams' market sizes, measured by average home attendances relative to league positions. The forecasting model's predictive capability has been assessed by comparing the model's estimated match results probabilities for the 2001-2 season with the fixed odds quoted by a leading firm of high street bookmakers.

References

1. Audas, R., Dobson, S. and Goddard, J. (2002) The impact of managerial change on team performance in professional sports. *Journal of Economics and Business*, 54, 633-650.
2. Barnett, V. and Hilditch, S. (1993) The effect of an artificial pitch surface on home team performance in football (soccer). *Journal of the Royal Statistical Society Series A*, 156, 39-50.
3. Cain, M., Law, D. and Peel, D. (2000) The favourite-longshot bias and market efficiency in UK football betting. *Scottish Journal of Political Economy*, 47, 25-36.

4. Clarke, S.R. and Norman, J.M. (1995) Home ground advantage of individual clubs in English soccer. *The Statistician*, 44, 509-521.
5. Crowder, M., Dixon, M., Ledford, A. and Robinson, M. (2002) Dynamic modelling and prediction of English Football League matches for betting. *The Statistician*, 51, 157-168.
6. Dixon, M.J. and Coles, S.C. (1997) Modelling association football scores and inefficiencies in the football betting market. *Applied Statistics*, 46, 265-280.
7. Dixon, M.J. and Robinson, M.E. (1998) A birth process model for association football matches. *The Statistician*, 47, 523-538.
8. Dobson, S. and Goddard, J. (2001) *The economics of football.* Cambridge: Cambridge University Press.
9. Forrest, D. and Simmons, R. (2000a) Forecasting sport: the behaviour and performance of football tipsters. *International Journal of Forecasting*, 16, 317-331.
10. Forrest, D. and Simmons, R. (2000b) Making up the results: the work of the football pools panel, 1963-1997. *The Statistician*, 49, 253-260.
11. Goddard, J. and Asimakopoulos, I. (2003) Forecasting football match results and the efficiency of fixed-odds betting. *Journal of Forecasting*, forthcoming.
12. Hill, I.D. (1974). Association football and statistical inference, *Applied Statistics* 23, 203-208.
13. Koning, R.H. (2000) Balance in competition in Dutch soccer. *The Statistician*, 49, 419-431.
14. Kuypers, T. (2000) Information and efficiency: an empirical study of a fixed odds betting market, *Applied Economics*, 32, 1353-1363.
15. Maher, M.J. (1982) Modelling association football scores. *Statistica Neerlandica*, 36, 109-118.
16. Moroney, M.J. (1956) *Facts from figures,* 3rd ed. London: Penguin.
17. Reep, C., Pollard, R. and Benjamin, B. (1971) Skill and chance in ball games, *Journal of the Royal Statistical Society Series A* 131, 581-585.
18. Ridder, G., Cramer, J.S. and Hopstaken, P. (1994) Estimating the effect of a red card in soccer. *Journal of the American Statistical Association*, 89, 1124-1127.
19. Rue, H. and Salvesen, O. (2000) Prediction and retrospective analysis of soccer matches in a league. *The Statistician*, 49, 399-418.

Restructuring the Belgian Professional Football League: A Location-Allocation Solution

Trudo Dejonghe

Lessius Hogeschool Antwerpen, Belgium
trudo.dejonghe@pandora.be

The aim of the article is to introduce an economic geographical solution based on empirical data for the economic problems of professional football in Belgium. The raising impact of economic rationality was accelerated by the Bosman case. The result was the introduction of economic principles in a historically protected structure. Recent evolutions are showing us the necessity of a market-oriented approach in professional football in Europe and more specifically in the smallest countries. The article shows that the use of location and a location-allocation model on an existing historically based situation is useful. It can locate the economically based and market oriented demand clubs which are not profitable in the long term. The result will be the construction of a market-oriented league with an exogenous number of clubs that will or can survive over a longer period of time.

1 The Transition of Professional Football

In Europe, prior to the 1990s, the objective of the professional football club was to provide entertainment by achieving playing success while remaining solvent and not to maximize profit. Sloane (1971) indicated that profit-making clubs in Europe has been the exception and not the rule. Chairmen, directors and patronages with a controlling function in the football clubs first reached success in other business fields. Their motives to invest in sport were more a desire for prestige, power or sporting enthusiasm and can be seen as a form of philanthropy. They get the best seats and a chance to mix with people in similar positions as themselves. Profit seems unlikely to be a motivation factor. The pursuit of playing success is the main objective and the clubs can be defined as utility maximization oriented subject to some financial constraints.

From an economic viewpoint the development of football in Europe in the 1990s was enormous. Global professionalism and the increasing importance of revenues from merchandising, sport sponsorship, broadcasting rights and in some cases the flotation of some clubs on the Stock Market resulted in a rising commercialization of professional football. A decision by the Court of Europe in December 1995, known as the "Bosman rule"[1], ruled that teams and the organising leagues could no longer place restrictions on the free movement of players across the national boundaries in the EU. The Bosman rule did not only eliminated quotas on the number of foreign players, but called into question the transfer rules, which was a set of rules and practices that inhibit competition between clubs for players who seek to sign their first contract and those whose contracts have expired. The result of the Bosman rule in combination with the diversification of the revenues is an increasing concentration and flow of capital toward the richest clubs and biggest markets. A substantial proportion of the best players from all over the world will migrate from poor countries, from rich countries with no football tradition (like the USA and Japan) or from the less lucrative professional leagues like Belgium and Holland to the rich European competitions of Spain, Italy, England, Germany and to a lesser degree France.

The growing impact of economics on professional football leads to a transition from utility maximization to a more profit maximizing approach. The natural tendency for the professional leagues is to seek exclusivity. The most important forms of exclusivity are:

- league exclusivity: a single monopoly league that controls league entrance
- team exclusivity: a limited number of teams in the highest division
- territorial exclusivity; a territorial monopoly, a region wherein a team has no competition from other professional teams combined with a franchise[2] system
- restriction of the freedom of players to transfer to or choose a new club, the "reserve clause"
- salary cap[3]

The first and in some way the second form of exclusivity exist in European football. Only one league has been recognized by the international body, the

[1] The Bosman case was based upon the Treaty of Maastricht. That treaty guarantees free movement of labour and questioned the rules for player's transfers among teams. Consequently, free trade agreements implied a policy that prevents a domestic monopoly.

[2] In a professional competition league franchise means that the national association gives the right to a club to implement a professional structure if the economic and market conditions are sufficient and if there is not already another professional club in that market segment

[3] Salary cap implies a maximum amount that clubs can spend on players. The National Basket Association (NBA) introduced it in 1980, the National Football League (NFL) in 1993 and the MLB in the mid 90s (Downward & Dawson 2000:p.47) Negotiations between the top clubs in Europe resulted in the introduction of a salary cap.

UEFA/FIFA, and the number of teams on the highest level is limited but the existence of relegation can be a distortion of the market. The system of relegation undermines territorial exclusivity. Any team with a rich investor can promote to the first division.

2 Professional Football as an Economic Geographical Item

Top-class sports, and in this case, professional football, can be seen as a person-related immaterial consumer-oriented service with a short-term utility and an economic function. (Dejonghe 2001: p.289-293). This definition implies that only densely populated locations combined with a rate of centrality can sustain a professional club. The consequence is that a "professional football club" can be described as a central place function with a high threshold, a minimum absolute or real number of potential consumers. The result in the long run is a very strong market-orientated location of professional football clubs. The revenues generated by TV, merchandising, stock market and European competitions are rising exponentially and are related to the market potentials of the clubs. Bale (1988:509) noticed already that the spatial patterns of the professional football industry had to reflect those of the growing tertiary orientation of the economy. The evolution of professional football since the 90s into a market-oriented service that has to be sold to its potential consumers emphasized his viewpoint. This transition implies that clubs in countries and cities with a small population have a competitive disadvantage. Market orientation, as a main causation in the extent of an economic product, leads to a situation of cumulative cycles. High attendance leads to more income from sponsors and media that can be invested in players which in turn leads to better results and higher attendance, ... The long-term success of professional football can be seen as an endogenous demand-oriented growth model based on Myrdalls' cumulative causation process.

These tendencies could make a change in policy of the clubs necessary and transform them to businesses with a future policy perspective in a free market. They have to transform their "service" to maximize the attraction, their total economic value. The economic objective is that total gate receipts, income from publicity, commercial inflow, television and other media earnings have to be maximized for the economic product "the league". The attractiveness of professional team sports is the fact that the total monetary inflow depends crucially on the uncertainty of outcome of the games played within the league. When the probability of any team winning approaches 100%, income from gate receipts or media fall substantially. The most important difference between the product "sport competition" and other products is that the competitive firms, namely the clubs, are obliged to co-operate to create a sellable product. This economic impact and reality has already been introduced in the

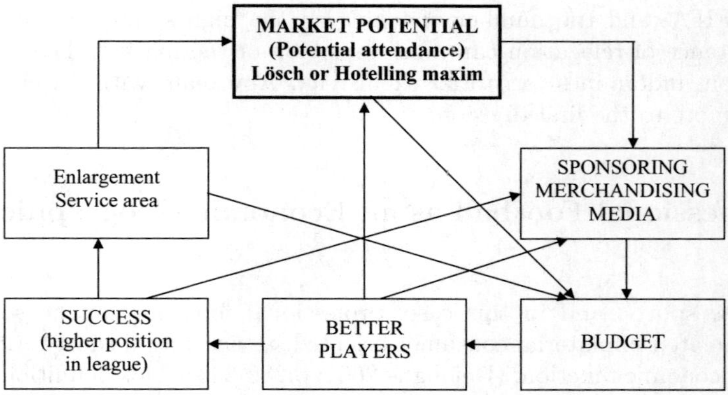

Fig. 1. The endogenous growth model (Dejonghe 2001:301).

US where National Football League (NFL), Major League Baseball (MLB), National Basketball Association (NBA) and National Hockey League (NHL) are organized according to these economic principles of rationalization and optimization.

The market of a professional football club and its league has to be located and defined in the tertiary sector. The clubs can be seen as heterogeneous, which refers to the potential product differentiation between the clubs with their individual club-specific characteristics and the necessity of a long-term budget to sustain the club results in the creation of a monopolistic competitive market structure, which is coordinated by one internationally recognized national football association. The number of clubs playing in the first league is determined by the central bureaucratic association and is not connected to the demand in the national market. The endogenous limit of the number of football clubs in the first division and the necessity of co-operation result in the creation of a producer's cartel. The purpose of such a cartel is the striving for profit-maximization by limiting mutual competition combined with supply optimization. According to De Clercq (1995:p.281), the formalization of a cartel is common in a sector with a restricted number of firms, which are creating analogous products where the total costs made in the production process are common knowledge. The substitution of the term firm by professional football club and total costs by budget is a rather good description of a top-competition. Dejonghe (2001:p.296) described the market of the product "professional football league" as a coalition in a near-cartel competition of heterogeneous interdependent oligopolic football clubs, which is organized by the international recognized football association.

The increasing impact of the economic principles on football results in a dichotomic classification of the clubs in (inter)national and local ones where for a survival on the long term the economic spatial potentials have to be analyzed. The central topic or the necessary minimal conditions are in this case the presence of potential fans and the willingness of the local firms, industry and government to invest in the club and its infrastructure. This concept results in a perception where professional football and the urban hierarchical and spatial structure are interdependent.

3 The Economic Consequences for Belgian Football Competition

The increasing importance of revenues from sponsoring, merchandising and broadcasting rights, as shown in Table 1, imposes the necessity of an organization that supports the game as an economic product that has to be sold through marketing to the potential consumers.

Table 1. Revenues first division clubs in Europe (2000-2001) (in %). Source: Deloitte & Touche(2002) Annual review of Football Finance and www.xs4all.nl(15/12/02).

UEFA ranking 2001/02	Country	Ticketing	Broadcasting Rights	Sponsoring catering and renting stadium	Merchandising,
3	England	31%	39%	30%	-
2	Italy	16%	54%	13%	17%
1	Spain	25%	51%	9%	15%
4	Germany	18%	45%	22%	14%
5	France	16%	51%	18%	15%
	Average	21%	48%	18%	12%
9	Portugal	42%	20%	18%	20%
7	Netherlands	34%	12%	41%	12%
12	Scotland	42%	23%	35%	-
21	Denmark	17%	4%	43%	3 6%
17	Norway	40%	4%	43%	13%
22	Sweden	33%	14%	28%	25%
	Average	35%	13%	35%	18%

* Data for Portugal and Scotland(1999/2000), Spain(1997/98); in England are revenues from merchandising, catering and renting the stadium with sponsoring are entered as *commercial income.*

The consequence of this trend is the domination of the above-mentioned "market agreements", that cannot be guaranteed in the contemporary structure of football in Belgium. Thinking about the economic rationalization of professional football has been accelerated through the Bosman judgement.

The combination of the loss of players as "club property", the increasing impact of the media and economic principles in professional football and "traditionalism" in the Belgian structures resulted in a sudden spectacular decline of Belgium from a 3rd place in the UEFA-ranking in1993 to a 20th place in 2000 and a recovery to a more "normal" 14th place in 2002.

The question that can be asked is "Is there a need for a professional football league in Belgium?" If the answer to this question is affirmative then an economic analysis is necessary. Because of the absence of transparency in the bookkeeping system the sources of revenue for the Belgium competition are not known. The situation today in Belgium is:

- A average attendances of 7,500 – 9,500 spectators in the last few years[4]
- Broadcasting rights of € 15 million for season 2002-2003[5]
- Lack of international impact and the associated income loss from demand oriented broadcasting rights
- Total budget of the first division (18 teams) is only € 133 million[6][7]
- No barriers between professional and amateur football
- Teams on the edge of bankruptcy and heavy debts

These facts show us that in an economic environment a restructuring of the Belgian professional football league is necessary for some reasons:

- To enter competition with other European countries
- The introduction of the licence system by de Belgian Football Association in 2000-01 and in the near future (2004-05) by the UEFA, where a club is not allowed anymore to have debts to the social security, tax-collector, players, ...
- To stop the cases of bankruptcy that have been arising since the introduction of the licence system

The competition in space, in this case the Belgium territory, will be approached as a competition not between homogenous clubs but between the different clubs with their historical and spatial characteristics. The aim is to show that some of these clubs have no economic basis to exist in a demand oriented system in the long term. The Belgian first division will through the use

[4] For the season 2001-2002: Germany (33,032), Italy (30,747), England (29,619), Spain (24,691), France (22,688) and the Netherlands (15,564). Belgium has an attendance of 9 people per every 1000 inhabitants a higher relative rate than Germany (4.0), England (6.1), France (3.8) and Italy (5.3).

[5] England € 941 million, Italy € 486 million, France 397 million, Germany € 384 million, Spain € 237 million and Holland € 57 million (UEFA, 2001,16-17)

[6] 2000-2001 season : England € 1.5 billion, Italy 850 million, Spain and Germany € 700 million (Dejonghe 2001:414)

[7] This means that the broadcasting rights are 11.2% (€ 15 million) of the total budget

of a location-allocation model be transformed in a league with an exogenous defined number of clubs with a higher spatial penetration.

4 Competition in Space

" A good location provides the firm with strategic advantages that competition may find difficult to overcome. While other marketing mix elements may be easily changed in response to a changing environment, locations represent long-term investment that can be changed only at a considerable cost." (Ghosh & Craig 1983: p.56)

4.1 The search for a location model

The Hotelling and Lösch maxim are examples of economic geographical location forms of two competitive suppliers in a uniform market area.

Lösch' classical locational equilibrium model assumed a market with rational consumers and homogeneous suppliers, operating in the simplest possible uniform spatial context. Lösch (1936) argued that firms had to establish themselves not in the location with the minimum cost or maximum turnover but in the place where profit maximization occurs. The demand for a product is a distance-decay function. The distance where the demand is equal to 0, the Christaller's range, is after rotation of the distance-decay function around the Q-axis, a circle that will be transformed into a hexagon, by the presence of spatial competition. The volume or demand cone of that market is the sum of the potential consumers in that specific area.

Hotelling (1929), on the other hand analyzed the location patterns of a linear duopolistic market for a homogeneous product with no price-competition in an area with a spatially uniform distributed population. The result was that the location of the two suppliers would be next to each other in the centre of the spatial market otherwise the market would evolve to a Löschian equilibrium. The long-term survival of more than one supplier is only possible if the total demand is sufficient to support both suppliers.

Devletoglou (1965) added the principle of the "doubtful area". He argued that the range of the market is not an arbitrary border, but a transitional zone caused by the heterogeneity in the quality of the sold products. The location-allocation models, where the location of several competitive facilities in a geographical market is analyzed, are extensions of classical equilibrium models, and can be used to locate the first division football clubs in an area. The models determine the location and the spatial interdependence of several competitive facilities in a bounded geographical market. In other words, they detect simultaneous localization of several suppliers combined with the allocation of the potential consumers. Location-allocation refers to two vital basic assumptions necessary in location decisions:

- the optimal location of the supplier or service
- the allocation to the respective suppliers or services of the potential consumers with their specific consumer behaviour and mobility constraints

A number of important questions, which need to be solved in location models, are (Daskin 1995: p.3):

- How many facilities should be planned?
- Where should each facility be located?
- How large should each facility be?
- How should the demand for the facilities' services be allocated to the facilities?

The number and size of facilities to be located is often a function of the service-cost trade-offs. In most cases, the quality of total service improves as the number of located facilities increases, but the marginal cost of providing the service can also increase. In the private sector, the decision to create or relocate an existing service has to be related to the profit maximization intention and the objectives measured in monetary units. The maximum number of services implanted in a bounded geographical area can be analyzed by estimating the maximum total profit for the sector. A service has to be closed down or relocated when the long-term perspectives show that profit will never be realised/actualised or that profit in another location will possibly be higher. In the public sector, with a non-monetary and social utility maximization principle, the aim is to maximize coverage of the area. In these models the demand is, just like in the models of Lösch and Hotelling, often assigned to the nearest facility. The main objectives of location-allocation models are the simultaneous and optimal location, considering (Dejonghe 2001: p.424):

- The company objectives (profit maximization, utility maximization, ...)
- Consumer characteristics (market potential, income, profile, ...)
- Distance and regional accessibility (public transport, highways, traffic congestion, parking facilities, ...)
- Competitive environment (competitors, spatial interaction, ...)

The importance of location strategy is very significant. According to Daskin (1995), the success or failure of both private and public sector facilities depends in part on the locations chosen for the facilities.

The combination of spatial location and optimizing the demand is known as the covering problem in the location-allocation models (Church et al. 1974; 1979, Thomas 1993, Daskin 1995, Daskin et al. 1999, Golledge et al. 1997). The objective of these models is to identify the locations that provide potential consumers with coverage within a specified distance or time constraint. One of the important characteristics of the covering models of homogeneous services is the existence of a lack of competition.

The purpose of the model is to find those locations that will be successful in the long run. The long-term equilibrium is an interdependence matrix of potential locations searching for a combination that covers the market in an optimal mode. The covering models have to maximize the utility (public sector) or the profit (private sector) of the potentials of the market and to optimize the location of the facilities in the economic and geographical space.

One of the main problems of the set of covering models is that the number of facilities that are necessary to cover the total number of potential consumers exceed the number that can, for budgetary reasons, be built. This problem resulted in the creation of some relaxation in the models. This relaxation of the purposes can be divided in:

- relaxation of the total coverage, a maximum covering problem used in the private sector
- relaxation of the coverage distance, a mini-max or centre problem used in the public sector

Professional football is typified as a market-oriented service, which means that the location-allocation problem can be analyzed as a modified maximum covering location problem of the private sector. This means that a relaxation or modification of the total coverage is necessary to ensure that every club is profitable.

In the search for a solution the term "market" can be used in two different ways:

- **Market or market area** is the nominal aggregation of the potential consumers.
- **Service area** is the areal extent of a market area and refers to the closest territory surrounding the location. It can be expressed in area measures and it is the region where the probability of the consumption of a central function located in a central place is higher than zero.

In the case of a consumer-oriented service the market is the determining factor. The range of the service area is only an indicator of the spatial competition and in a densely populated area it is not the main factor.

4.2 Professional football: A modified maximum covering model

The model

The relaxation in these models is that not ALL-potential consumers have to be covered but as many as possible in terms of not jeopardizing the economic profitability or cost-effectiveness of the facilities. The basic objective of this relaxation is to introduce or consider a fixed number of facilities, which are to be located in service areas combined with maximizing the demands covered. The used modified maximum covering first implies the presence of a status quo situation with 18 clubs. In a second stage an exogenous defined number, in this case 14 heterogeneous clubs[8], is located in a potential Hotelling maxim in the most densely populated urban centre or in Lösch maxim in the other centre. This approach leads to the results that, in contrast with the deductive location-allocation models, consumers from the same location do not necessarily prefer the nearest service. The consumer is not an "optimizer" but a "satisfier", with personal motives, preferences, and behaviour. They locate subjective quality differences and characteristics and make an individual classification of the different clubs. This subjective approach and the introduction of time distance assumes that the "top football club" has to sell its product to the market. The opening up of an area and the attraction of other clubs lead to the creation of spatial competition and the existence of unequal service areas. The heterogeneity of the clubs is projected in their differences in local consumption and impact on the region, the centrality, and has an influence on the market and/or the service area. These subjective markets will be used as the basis for the location-allocation model.

The objective of this new empirical model is - given a fixed number of existing or newly created heterogeneous clubs - to maximize the covering of potential consumers in Belgium combined with certain absolute or relative thresholds for the long term survival of the individual club.

The data

The empirical model with its heterogeneity relaxation is based on the 45,791 season ticket-holders[9] of the different clubs in the first division for the season 1997-98 in Belgium and the data of the promoted clubs in the other seasons. These data are in their own way a reflection of the subjective cultural-historical topophilia or place pride related to each club. The changes caused by the promotion and the relegation for the seasons 1998-99, 1999-2000, 2000-01 and 2001-02 are placed in the model. The total number of clubs that have

[8] The model takes 14 clubs as a fixed exogenous number. This number of first division clubs is always mentioned by the Belgian FA in the media. The purpose of this article is partly to answer the question: supposing there are 14 clubs in Belgium where should they be located?

[9] The dataset are the season ticket holders of all the Belgian first division clubs. They were allocated to the respective communities. In a forthcoming research the location-allocation model is based on the fan card holders of every team. In Belgium every visitor of a first division match has to buy a fancard of his team. Approximately 300.000 cards have been sold but the results are comparative with those here.

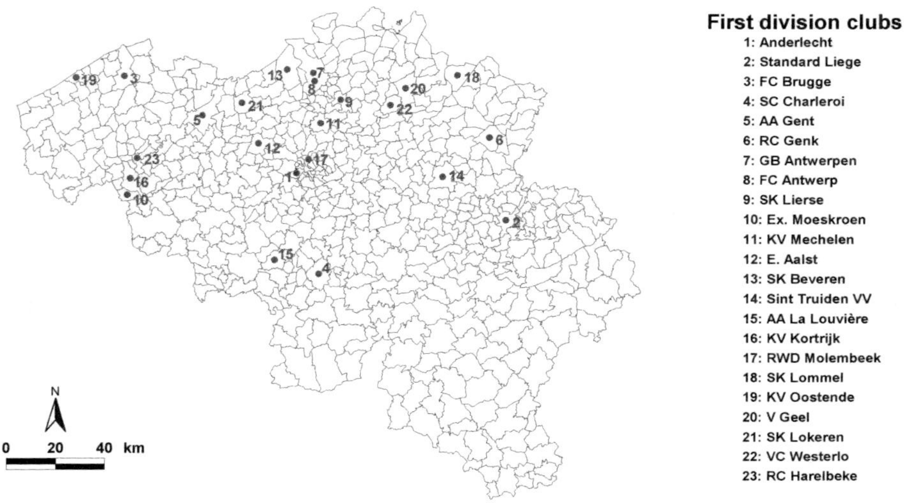

Fig. 2. The location of the 23 teams in Belgium.

played in the first division in that period was 23. Beside these 23 were some "theoretically" regional clubs created in the other regional centres such as Namur, Leuven, Tournai and Verviers.

The season ticket-holders are the consumers through which the consumption of the service "top-football" is assured, they can be considered as the "loyal" or "hard-core"-supporters. The nominal number of season ticket-holders is not a reflection of the real consumption of the product. The nominal values have to be corrected to a relative value that indicates the real attendance. Otherwise, a club with a relatively high number of season ticket-holders and a relatively minor number of floating attendances should have a larger potential market than a club with not as many season ticket-holders but a large potential market of floating attendance. The correction is based on the real number of attendants who support the home teams. The Belgian Football Association (KBVB) provided these latest data. More formally,

$$\frac{S_j^n}{S_t^n}\frac{W_j^n}{W_t^n}x100 = S_j^c,$$

where

S_j^n = the nominal number of season ticket-holders for club j
S_t^n = the sum of the nominal number of season ticket-holders for all clubs in a season
W_j^n = the nominal number of home attendances for club j
W_t^n = the sum of the nominal number of home attendances for all clubs in one season
S_j^c = the corrected number of season ticket-holders for club j

The indexes

The empirical determination of the markets and service areas implied the formulation of a model through which the 589 Belgian communities can possibly be assigned to one or more clubs. To do so, a reference-index is introduced to analyze the coverage of these communities. The **endogenous reference-index, Ri** is the average corrected number of season ticket holders for every 10,000 inhabitants.

More formally,

$$\frac{S_t^c}{B_t} \times 10,000 = R_i,$$

where

$S_t^c = S_t^n$
B_t =total population of the total market, in this case Belgium[10].
R_i =endogenous reference-index

The empirical determination of the service area and market of the clubs implies the creation of a model that should make it possible to assign the communities to one of the18 clubs. The model is based on the **community consumption index, Ci, the club specific consumption index, Cij and the club specific penetration index Pij** of " professional football" in those communities.

The **community consumption index,** Ci measures the rate of consumption of the service professional football in the communities, and is calculated as follows:

$$C_i = \frac{\sum S_{ij}^c}{B_i} \times 10,000,$$

where

C_i =local consumption index in location i
S_{ij}^c =corrected number of season ticket holders of club j in location i
B_i =total population in location i[11]

The club specific consumption index, Cij is:

$$C_{ij} = \frac{S^c_{ij}}{B_i} \times 10,000$$

The club specific penetration index, Pij measures the relative orientation of a community to one or more clubs. More formally:

$$P_{ij} = \frac{S^c_{ij}}{\sum S^c_{ij}} \times 10.000$$

An important parameter **is the local consumption index Ljj**. This is the rate of consumption of professional football in the centre where the club is located. A high rate means a strong relative attachment of the inhabitants but can also be an indicator of local saturation. This means that a growth of local consumption is not the potential solution for attracting more people. The only possible growth has to be a rise in centrality or with other words attracting attendances from outside the centre.

The **local consumption index, Ljj** is:

$$L_{jj} = \frac{S^c_{jj}}{B_j} \times 10,000,$$

where

S^c_{jj} = corrected number of season ticket holders of club j in location j, the location of the club

B_j = total population in location j, the location of club j

The measurement of the market

The determination of the markets implies the creation of a classification for the allocation of the communities and their potential consumers. This allocation is translated in an allocation rate Aij based on a combination Cij and Pij in the 589 communities in Belgium. The minimal rate of 10 for Pij is used to eliminate the sporadic consumer from the allocation model.

Pij divides the total market in:

- monopoly area: the area dominated by one club
- monopolistic competition area: the spatial competition is strong but one club still has more than 50% market share
- rest area: the area with a low rate of number of potential attendance for a club
- non-covered area

The level of Cij results, as shown in the model, in a further subdivision in an a, b or c. This is an indication for the strength of the penetration. The subdivision b or c indicates that a "new club" has the possibility of spatial competition or that the existing club(s) still has the potential to gain attendance from this area if they improve the quality of their services. The Ab and Ac areas are non-saturated monopolistic areas and indicate the direction of growth potential for a club.

The monopolistic competition area is the potential spatial growth zone. These areas can, depending on the quality of the services, be transformed in a monopoly or rest/non covered area.

Summarizing these findings the following allocation model is used:

	Pij	Cij	Aij
MONOPOLY AREA	≥ 80	≥Ri	Bi
Aa		2/3Ri ≤ Cij< Ri	
Ab		1/3Ri ≤ Cij < 2/3Ri	
Ac			
MONOPOLISTIC COMPETITION AREA	50 –80	≥Ri	Pij x Bi
Ba		2/3Ri ≤ Cij < Ri	
Bb		1/3Ri ≤ Cij < 2/3Ri	
Bc			
RESTAREA	25-50	1/5Ri ≤ Cij	Pij x Bi
Ca	or	1/5Ri ≤ Cij < 1/3Ri	1/4Bi
	≥ 50with		
RESTAREA	10-25	1/10Ri ≤ Cij	Pij x Bi
Cb	or	1/10Ri ≤ Cij < 1/5Ri	1/10 Bi
	≥ 25with		
NON COVERED AREA		Cij ≤ 1/10Ri	0

4.3 The results

The model locates the communities in one or several areas of the clubs and it indicates the non-covered areas in Belgium. The potential consumers for each area and each club is the sum of Aij. The non-covered population is therefore the sum of the population of the non-covered areas and the relative parts of the Bi that are not assigned to the monopolic competition and rest areas.

The results for the 23 heterogeneous clubs in the first division for the 1997/98 to 2001/02 seasons are shown in Table 2. The combination of spatial competition, cultural (language) barriers, country-province borders, time-distance, inertia and tradition can be put forward as an explanation for the contemporary structure and impact of the clubs. This empirical outcome shows that the total coverage equals 7,364,354 or 72.5% of the Belgian population and 37.4% of the population belongs to the monopolistic area of a club.

Table 2. The A-B-C zones of the 23 clubs.

	Total	A	Aa	Ab	Ac	B	Ba	Bb	Bc	C	Ca	Cb
SC Anderlecht	1.350.055	517.629	197.272	75.062	245.295	368.831	56.819	55.553	256.259	463.595	240.034	223.561
Standard Luik	1.080.395	872.705	34.477	102.424	735.804	27565	8.930	0	18.635	180.125	81.638	98.487
FC Brugge	768.074	386.535	207.748	127.547	51.240	198.769	79.384	44.989	74.396	182.770	97.194	85.576
SC Charleroi	462.634	365.511	0	274.052	91.459	62.718	0	30.723	31.995	34.405	26.616	7.789
AA Gent	406.902	252.919	252.919	0	0	104298	21.896	70.164	12.238	49685	30.812	18.873
RC Genk	403.848	212.194	212.194	0	0	96.512	37.679	14.645	44.185	95.142	79.649	15.093
GBAntwerpen	319.678	112.351	42.338	0	70.013	0	0	0	0	207327	203.864	3.463
SK Lierse	262.822	119.928	111.643	0	8.285	85.972	33.619	16.846	35.307	56.922	35.490	21.432
Ex Moeskroen	248.835	231.810	131.791	94.448	5.571	5.057	0	0	5.057	11.968	6.262	5.706
FC Antwerp	234.380	0	0	0	0	4.972	0	0	4.972	229.408	216.443	12.965
KV Mechelen	199.094	119.424	119.424	0	0	39.610	18.522	9.902	11.186	40.060	25.066	14.994
E Aalst	185.843	76.197	76.197	0	0	51.350	33.555	17.795	0	58.296	33.419	24.877
SK Beveren	179.252	75.877	44.533	31.344	0	89.594	0	9.363	80.231	13.781	7.951	5.830
St Truiden VV	178.971	90.497	90.497	0	0	30.717	18.929	7.547	4.241	57.757	39.495	18.262
La Louvière	173.955	76.809	76.809	0	0	86 .412	52.919	11.474	22.019	10.734	0	10.734
KV Kortrijk	170.442	75.639	75.639	0	0	52.138	44.171	0	7.967	42.665	21.062	21.603
RWDMolenbeek	163.411	0	0	0	0	74.919	57.410	0	17.509	88.492	38.052	50.440
SK Lommel	126.071	38.067	29.644	0	8.423	53.961	29.643	14.282	10.036	34.043	30.379	3.664
KV Oostende	106.458	58.245	58.245	0	0	13.742	13.742	0	0	34.471	14.302	20.169
V Geel	98.138	17.379	17.379	0	0	54.730	54.730	0	0	26.029	21.442	4.587
SC Lokeren	93.904	36.123	36.123	0	0	9.954	0	0	9.954	47.827	23.600	24.227
VC Westerlo	82.722	34.884	34.884	0	0	9.057	9.057	0	0	38.781	28.345	10.436
RC Harelbeke	68.470	26.418	26.418	0	0	23.105	0	23.105	0	18.947	17.939	1.008
	7.364.354	3.797.141										

An important parameter is the local consumption index Ljj. A high rate means a strong relative attachment of the inhabitants but can also be an indicator of local saturation. The only possible growth has to be a rise in centrality, this means attracting attendances from outside the centre.

The **local consumption index, Ljj** is:

$$L_{jj} = \frac{S^c_{jj}}{B_j} \times 10,000$$

Table 3. The local consumption indexes.

Club	Ljj	Club	Ljj
SC Anderlecht	33,37	SK Beveren	157,41
St Liège	20,15	St Truiden VV	261,80
FC Brugge	74,32	AA La Louvière	133,54
SC Charleroi	32,26	KV Kortrijk	136,60
AA Gent	49,04	RWDMolenbeek	8,92
RC Genk	208,70	SK Lommel	412,60
GBAntwerpen*	17,79	KV Oostende	103.91
SK Lierse	176,31	V Geel	271.69
Exc Moeskroen	214,20	SC Lokeren	203,47
FC Antwerp	16,51	VC Westerlo	395,31
KV Mechelen	156,20	RC Harelbeke	234,59
E Aalst	168,80		

* The value for GB Antwerpen was calculated on the data from G Ekeren, the first division club that merged with the third division club Beerschot Antwerpen

The combination of the market and Ljj make it possible to divide the heterogeneous clubs in some more homogeneous subgroups.

Table 4. The more homogeneous subgroups.

Group	Club	Location
Major clubs in major or regional cities with a low Ljj	SC Anderlecht FC Brugge Standard Liège	Brussel Brugge Liège
Regional clubs in major cities with a low Ljj	AA Gent SC Charleroi	Gent Charleroi
Regional clubs in regional cities with a high Ljj	RC Genk AA La Louvière KV Mechelen E. Aalst KV Oostende KV Kortrijk	Genk La Louvière Mechelen Aalst KV Oostende KV Kortrijk
Smaller clubs in smaller cities with a high Ljj	STVV Exc Moeskroen SK Lierse SC Lokeren SK Beveren SK Lommel V Geel	Sint-Truiden Moeskroen Lier Lokeren Beveren Lommel V Geel
Local clubs in communities with a high Ljj	RC Harelbeke VC Westerlo	Harelbeke Westerlo
Local club in major cities with very low Ljj	GBAntwerpen FC Antwerp RWDMolenbeek	Antwerpen Antwerpen Brussel

Saey & Lietaer (1980) determined that the theoretical range of a regional service area in Flanders or the accessibility class was set to 15km. According to the "traffic principle" 15km was the regional impact zone of a regional centre. A football club playing in the first division can be seen as a hierarchical marginal good that is, because of the exogenous limit of 18 clubs, an indicator of an urban hierarchical level[12]. The combination of the number of first division clubs, the Belgian urban hierarchical centre and a professional football club as a hierarchical marginal good lead us to a theoretical range of a regional club of 15km.

[12] The urban centres in Belgium are, according to their centrality, classified in a hierarchical urban system. There is one urban centre in level 0, the capital city, 3 in level 1, the major cities and 17 centres in level 2, the regional cities. (Van Der Haeghen et al. 1982: p.281).

The national clubs are monoliths and an obstacle for the extension of other clubs in the nearby region. Their real range extends to a large degree the threshold of "professional football" in Belgium. The service area exceeds the theoretical 15km. They dominate their region and determine endogenously their spatial market zone. FC Brugge is located in a regional city and can be seen as a functional substitution of Gent which has only a regional impact. These clubs have a low Ljj so that even today a potential endogenous growth of attendances is possible.

The result for the "regional" clubs in regional and major cities shows a spatial monopolistic Lösch maxim for SC Charleroi and RC Genk. The situation in E.Aalst shows the existence of spatial competition with AA Gent and SC Anderlecht/Brussel. KV Mechelen has a spatial competition with Lier and the monolith Anderlecht and to a lesser degree with GBAntwerp, FC Antwerp and VC Westerlo. KV Oostende has strong spatial competition for its service area and market with FC Brugge. The club relegated after one year (1998/99) back to the second division. KV Kortrijk relegated also after one year (1998/99) and now plays in the third division. AA La Louvière promoted after an absence of 21 years to the first division in 2000 and has a minor service area. AA Gent and SC Charleroi have a low Ljj so when the attendances can rise substantially in periods that the clubs will be successful.

The "small" clubs in the small cities which have, as seen in the values of their B and C zones, a limited centrality depend for a great part on local consumption. The lack of centrality is combined with the psychological barrier consisting of a refusal of descending in the urban hierarchy. To illustrate, the population of the regional centre Sint-Niklaas is not prepared to consume the service "professional football" offered by nearby small cities Beveren and Lokeren. (Dejonghe 2001:442-448) Harelbeke is part of the agglomeration Kortrijk and Westerlo, is in the classification of Van Der Haegen et al. (1982) a "main village". The club has a very high relative consumption in the local community, but lacks centrality and a refusal to descend in the urban hierarchy. The combination of these characteristics results in a small number of potential consumers and indicates that the attendances are at their maximum. The average attendances reach a maximum of 8,000 – 10,000 in very successful periods but drop to 2,000-3,000 in less successful periods.

The local clubs in major cities have a very low Ljj and probably an enormous growth potential. RWDMolenbeek is located in Brussel in a theoretical Hottelling maxim with SC Anderlecht. This monolith transformed Brussel into a Lösch maxim and a second club in Brussel is thereby no option. The Antwerp situation shows a Hotelling maxim of two clubs which are both not using the economic potentials of the city.

The transformation of professional football to a market/consumer-oriented service implies the presence of a market that reaches a certain "threshold". 13 of the 23 Belgian clubs have a market of less than 200,000 potential consumers, a number that in the long term is not enough to support a professional football club that can compete with Belgian top clubs or with the European clubs. The empirical verification of the markets shows us that the allocation-location solution can be applied to the existing historical-based situation.

4.4 The empirical allocation-location of 14 clubs

The transformation to 14 clubs

The problem with professional football in Belgium is not only a location-allocation problem, but for international competition it is also a matter of too many clubs in a limited market and the high international thresholds in the most successful football countries. In comparison with the dominating countries in the Champions League and the UEFA-cup, the average market in Belgium is with 550,000 for every first division club a lot smaller than the averages of the most successful countries. In Germany their is one first division club for every 4.6 million inhabitants, in England, France and Italy one for every 3 million, in Spain one for every 2 million and even in The Netherlands there is only one for every 830,000 inhabitants.

The restructuring of the first division to 14 clubs implies the elimination, relocation or merging of several clubs and the creation of totally new clubs in selected locations. The method that was used to create "new" clubs in certain regional centres was to give the potential club the characteristics of the different "regional clubs in regional cities". The result was a minimum and a maximum potential market taking into account the potential spatial competition of the existing clubs. The spatial markets and the number of potential consumers as a result of a theoretical merger of two or more clubs was determined by and deducted form the data of the individual clubs. The result is shown in Table 5 and Figure 3.

The table shows us that despite a reduction by 4 clubs:

- the coverage increased to 77.9% of the total population
- an expansion of the monopolistic area to 51.0% of the total population is seen
- an individual consumer potential that rose to at least 288,916 combined with a high Ljj for the three clubs with the smallest market.

The elimination of E Aalst, a club located in a regional city on the highway E40, a corridor for the monoliths Anderlecht and Brugge.

The potential new location in the regional city Leuven (between Brussel and Liège), has a monopolistic zone of 123,748 and a B and C zone of 59,686-95,498 was not selected. Other potential locations such as the regional city and capital of Walloon Namur (255.764) and the regional city Verviers (205,524) were not selected.

Table 5. The empirical first division with 14 teams.

Club	Merge	Location	Potential consumers	Monopolistic Area
SC Anderlecht	-	Brussel	1,619,491	743,926
Standard Liège	-	Liège	1,064,743	874,955
FC Brugge	-	Brugge	844,542	575,828
Metropool Antwerpen	FC Antwerp – GB Antwerpen	Antwerpen	634,864	453,030
SC Charleroi	-	Charleroi	469,110	361,002
Lier-Mechelen	Lierse – Mechelen	Intercept	459,844	277,033
FC Borinage-Centre	La Louvière – Mons[1]	Intercept	446,577	413,024
RC Genk	-	Genk	444,840	269,813
AA Gent	-	Gent	406,985	252,919
FC Southwest	Harelbeke-Kortrijk-Waregem	Intercept	342,892	152,014
FC Waasland	Beveren – Lokeren	Sint-Niklaas	310,840	185,222
FC Kempen	Westerlo – Geel –Turnhout	Intercept	296,174	201,033
FC Haspengouw	Sint-Truiden	Sint Truiden	291,743	169,454
Exc Moeskroen	-	Moeskroen	288,916	255,916
TOTAL			7,921,561	5,185,169

[1]AEC Mons promoted to the first division in 2001/02

Sint-Truiden is a well-equipped small town of level 3 with a strong polarisation in the urban hierarchy. Van Hecke (1997:106) determined a service area of 425km^2, a zone that is bigger that those of the regional cities Oostende (361km^2), Aalst (296km^2), Kortrijk (284km^2), St Niklaas (258km^2) and Mechelen (238km^2). The club is located in a monopolistic Lösch maxim with a high centrality in the 15km zone. The club can be seen as the historical regional relocation or functional substitution of Leuven.

Confrontation with reality

The academic study started with the data of 1997/98 and the first results were published in 2001 and created a storm of protest in the football world. Today, September 2003, we can confront the results of our research with the existing situation. RWDMolenbeek, RC Harelbeke, and SK Lommel, three clubs without any future in the model, claimed bankruptcy and were expulsed from the football competition. KV Kortrijk, according to the model had to merge with Harelbeke and the second division club Waregem, and E Aalst, a club that on empirical based data had no future in the first division, went bankrupt and was relegated to the third division. KV Mechelen, which according the model had to merge with Lierse, went bankrupt and will be placed in the third division in 2003-04. Clubs as SK Lierse, GB Antwerpen, La Louvière, SC Charleroi and the current second division club V. Geel have because of

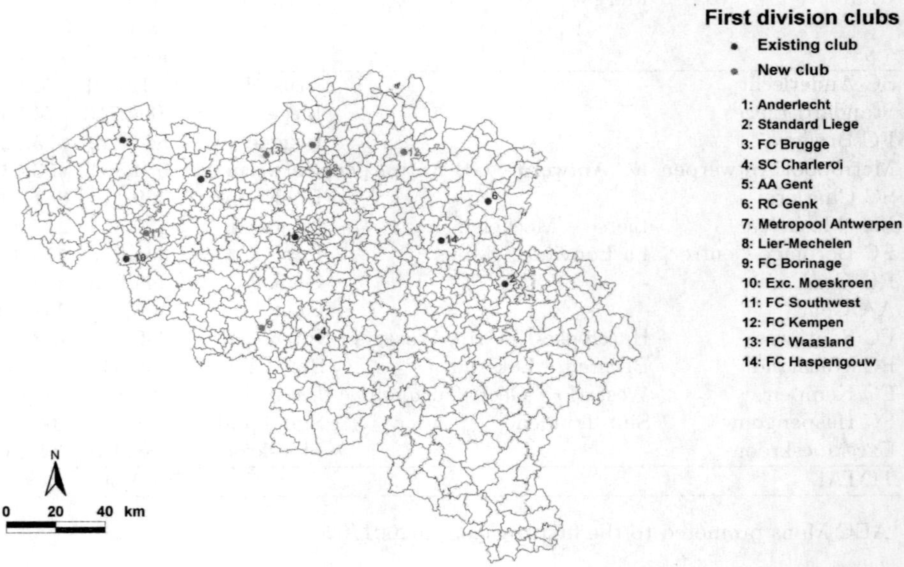

Fig. 3. The empirical rationalized first division with 14 clubs.

their debts to Social Security, taxes and others, every year problems with their licence. The potential revenues from transferring players dried up because of the implosion of the Belgian and European football market. This result shows us that a theoretical model based on empirical data can be of some use to the real world but in the short term subjective arguments can be used to postpone the reality.

5 Conclusion

The divergence between the American profit maximization leagues and the historically based utility maximization in Europe was accelerated through the Bosman Case. The necessity of revenues from broadcasting, merchandising and sponsoring resulted in a strong commercialization of professional football. Professional football was transformed into a consumer-oriented service that could be linked to an endogenous demand-oriented growth model. This transformation showed that the factor "market" became very important and had economic consequences on the Belgian professional football competition. The introduction of professional football based on economic principles resulted in a decline of Belgian football. The Belgian competition reached a relative maximum in attendances and the market is too small to compete with other European competitions. The Belgian league has too many clubs in the first division and many of them have great financial problems and a very small market. The restructuring of this market of heterogeneous clubs can be seen as an allocation-location problem. The solution of an allocation-location problem in an existing market has to take into account the historical and cultural situation. The transformation of football from a utility maximizing to a more profit maximizing consumers-oriented service implies the creation of a model based on empirical data. Tabula rasa is not an option because of the heterogeneous characteristics of the different football clubs. This means that a theoretical maximum covering solution can be used as a guide for the restructuring of an existing spatial distribution.

The problem with Belgian football competition is that in the long run, and as reality shows us in the short term, the contemporary situation will lead to market failures and the elimination of some of the clubs. The solution proposed in this paper is the creation of a new professional competition with fewer teams combined with territorial exclusivity in a centre with a potential number of consumers that reach a certain absolute or relative threshold. The empirical allocation-location model combine the maximizing of the total coverage and the individual market of the clubs and accounts for the existing situation. The result is the creation of a first division with 14 clubs combined with a larger coverage of the population in Belgium. The 14 allocated and located clubs have a greater coverage and monopolistic area than the 23 existing clubs. The transformation, based on empirical data, implies a merging of several clubs, sometimes combined with relocation or in the case of E. Aalst, RWDMolenbeek, SK Lommel and KV Oostende the elimination of the existing club in the professional league.

References

1. Bale J. (1988). The place of "place" in cultural studies of sport, *Progress in Human geography*, 12(4): 507-524
2. Church R. & Meadows M. (1979). Location modelling utilising maximum service criteria, *Geographical Analysis*, 11(1): 358-379
3. Church R. & ReVelle C. (1974). The maximal covering location problem, *The Regional Science Association Papers*, 32: 101-108
4. Daskin M. (1995). *Network and discrete location models, algorithms and applications*, New York, Wiley
5. Daskin M. & OWEN S. (1999). 2 new location covering problems: the partial p-centre problem and the partial set covering problem, *Geographical Analysis*, 31(3): 217-235
6. De Clercq M. (1995). *Economie toegelicht*, Leuven, Garant
7. Dejonghe T. (2001). *De noodzaak aan exogeen oplegde economisch-geografische principes bij het professionaliseren van het door endogene factoren ontstane*

*wereldsportstelsel. Gevalstudie: het lokalisatievraagstuk van het topvoetbal in België,*doctoraatsverhandeling, Universiteit Gent

8. Deloitte & Touche (2002). *Annual review of football and finance,* Manchester, Deloitte & Touche Tohmatsu
9. Devletoglou N.(1965). A dissenting view of duopoly and spatial competition, *Economica*, London School of economies and political science, 32: 146-160
10. Downward P. & Dawson A. (2000). *The economics of professional team sports*, Londen, Routledge
11. Ghosh A. & Craig S. (1983). Formulating retail location strategy in a changing environment, *Journal of Marketing*, 47(19): 56-68
12. Golledge R. & Stimson R. (1997). *Spatial behaviour: A geographic perspective*, New York, The Guilford Press
13. Gratton C. & Taylor P. (2000). *Economics of sport and recreation*, London, Spon Press
14. Hoehn T. & Szymanski S. (1999). The Americanization of European football, *Economic Policy*, 28: 205-240
15. Hotelling H. (1929). Stability in competition, The Economic Journal, XXXIX(153): 41-57
16. Huff D. & Batsell R. (1977). Delimiting the areal extent of a market area, *Journal of Marketing Research*, 14: 581-585
17. Lösch A. (1962). *Die raumliche ordnung der wirtschaft*, Stuttgart, G.Fischer
18. NIS (1997). Bevolking van het rijksregister 01/01/97 tabel 70.01 deelgemeenten
19. Noll R.(1999). Competition policy in European sports after the Bosman Case, in Jeanrenaud C. & Késenne S. (editors) "*Competition Policy in Professional Sports*, Antwerpen, Standaard Ed Ltd, pp. 17-44
20. Saey P. & Lietaer M. (1980). Consumer profiles and central place theory, *Tijdschrift voor Economische en Sociale Geografie*, 71(2): 180-186
21. Sheppard E. (2000). Competition between places, in Sheppard & Barnes (editors) *A companion to economic geography,* Malden, Blackwell, pp. 169-186
22. Sloane J. (1971). The economics of professional football: the football club as utility maximizer, *Scottish Journal of Political Economy*, 70: 65-94
23. Thomas I. (1993). Sensibilité du découpage spatial optimal des services d'urgence à la définition de la demande, *L'espace Géographique*, 22(4): 318-332
24. UEFA (2001). *Report Football and television 2001: c omparative analysis of major European countries,* Paris, 40p
25. Van der Haegen H., Pattyn M. & Cardyn C. (1982). The Belgian settlement system, in Van Der Haegen (editors), *West European Settlement Systems*, Leuven, Acta geographica Lovaniensia 22: 252-359
26. Van Hecke E. (1997). *Hiërarchie van de stedelijke kernen in Vlaanderen*, Heverlee, instituut voor Sociale en Economische geografie
27. www.allesoverstadions.nl, 15/12/02
28. www.xs4all.nl, 15/12/02

The Best Systems for Appointing Referees

Jaime Gil Lafuente

Universitat de Barcelona, Spain
jaumeg@eco.ub.es

1 Introduction

For many years, presidents and general managers of major football (soccer) clubs in Spain have been debating over how to assign referees for each match. When the responsible committee subjectively selected a referee, many people responsible for the "affected" teams had the right to complain about the always latent possibility that a certain amount of favoritism could occur for certain matches. While now, by using the famous "machine"[1], the whims of randomness mean that certain matches that are considered important are not refereed by the most suitable referees.

The two proposals that we are offering stem from multivalent logic which focuses their efficiency on conveniently treating the subjective opinion of experts. This is not a "magic formula", but one of the many solutions that are sufficiently good to be considered optimal.

2 Basic Requirements

A large part of the interested parties are inclined towards the well known "machine", which resolves the problem in a random manner, while others prefer to appoint the referees subjectively by a responsible committee. Therefore, a fundamental condition for the perfect operation of the proposed scheme can be found in the aggregation of said opinions taking place without any external manipulation. For this, it will be necessary to follow certain simple mathematical calculations which can be implemented in a computer program.

Intuitively, it should not be totally incorrect to assign a referee based on "human" (and therefore, subjective) opinions. Neither, in principal, is it

[1] A nickname for the computer program currently used to appoint referees in the Spanish league.

incorrect to use the "machine". But the adequate co-operation between these two methods can, we hope, lead to an optimal system for referee appointment.

It is necessary to resort to a set of experts both in sporting events such as football, and in external factors which can have an effect on the same.

The experts should possess an up-to-date knowledge on the matches that are to take place, as well as of the factors that may affect them.

The number of experts may vary between 5 and 20. The experience in other fields suggests that a part of these should be members of the Technical Committee of Referees. Let us not forget that an excessive number of specialists may bias the information, while an excessively reduced number may lead to undesired dispersion.

The only conditions "sine qua non" are that they should be *true experts* with a good knowledge of each of the elements that have an effect on the sports contest and that they be capable of maintaining total honesty in their opinions at all times.

3 Definition of the Characteristics of the Referees

In the first instance, each of these experts will be asked to establish each of the characteristics, qualities or peculiarities (from the most unusual to the most common) which are desirable and inherent to the referees so that they can adequately carry out their function.

We should not forget under this heading, the medical, psychological, technical, physical or emotional characteristics.

As an example we could list:

Physical resistance
Rationality
Top speed
General technical level
Resistance to conflict
Level of authority
Overall vision of play
Geographical origin
Fatigue level
Propensity to tiredness
Bias towards some contestants or direct rivals
General physical condition
Positioning on the field
State of mind
Tolerance level
Etc.

Once all of the various traits are grouped together, each referee is evaluated by experts in every category. Here we can utilize the tools available from the mathematics of uncertainty, which are based on valuing the proposals by means of numbers between 0 – null for the quality – and 1 – the maximum level of the characteristic.

Therefore each referee will be described by a fuzzy subset X_j, taking into account that the values are susceptible to modification, week by week, in accordance with the physical, technical or psychological tests to which they will be subjected. We arrive then at:

$$X_j \quad \begin{array}{c|c|c|c|c|} \multicolumn{1}{c}{a} & \multicolumn{1}{c}{b} & \multicolumn{1}{c}{c} & \multicolumn{1}{c}{} & \multicolumn{1}{c}{n} \\ \hline \mu_j^{(a)} & \mu_j^{(b)} & \mu_j^{(c)} & \dots & \mu_j^{(n)} \\ \hline \end{array}\ , \; j = 1, 2, ..., m,$$

where

$\{a, b, ..., n\}$ - the characteristics;

$\mu_j^{(a)}, \mu_j^{(b)}, \mu_j^{(c)}, \dots \mu_j^{(n)}$- the values assigned to the referee j with respect to each of the characteristics;

Card $\{a, b, ..., n\}$ is the number of characteristics and m is the number of referees.

As an example, we will describe 10 referees, abstaining from using their names. Likewise, in order to simplify the process, we will reduce the number of qualities, peculiarities or characteristics to 5:

Technical level, speed, personality, tolerance, rationality.

	Technical level	**Speed**	**Personality**	**Tolerance**	**Rationality**
A =	.5	.9	1	.4	.7
B =	.6	.3	.2	.8	.9
C =	.5	.4	.9	.5	.7
D =	.7	.8	.1	.1	.5
E =	.8	.5	.5	.5	.7
F =	.5	.6	.7	1	.3
G =	.4	.8	.8	.9	.2
H =	.7	.9	.3	.2	.5
I =	1	.4	.1	.3	.8
J =	.2	.3	.9	.4	.9

4 Definition of the Conditions for Each Match

It is obvious that all football matches or those of any other sport for that matter require the best judges. Nevertheless, there are matches and teams that are more conflict prone and others that are more calm, depending on the players involved and importance of the match.

Taking into account these and many other factors, we will ask our team of experts to, weekly and individually, proceed to:

a) Define, by means of a fuzzy subset, the requirements for each match relative to the characteristics of the referees by means of assigning values.

	a	b	c		n	
Y_i	$\mu_i^{(a)}$	$\mu_i^{(b)}$	$\mu_i^{(c)}$	...	$\mu_i^{(n)}$	, $i = 1, 2, \ldots, p$,

p being the number of matches to be refereed.

Following on with our example we will list 7 hypothetical league matches between the 14 clubs of the Spanish First Division:

	Technical level	**Speed**	**Personality**	**Tolerance**	**Rationality**
R. Madrid- F.C. Barcelona	.7	.4	.6	.8	.5
R. Sociedad – At. Bilbao	.5	.5	.8	.7	.4
Betis – Sevilla	.9	.8	.7	.6	.3
Deportivo - Celta	.4	.7	.7	.8	.5
Valencia – Vilarreal	.6	.4	.7	.7	.4
Español – At. Madrid	.8	.5	.8	.8	.8
R. Valladolid – Rayo V.	.5	.4	.5	.9	1

b) Indicate the weight of each of the characteristics relative to the remainder, for each match to be played. In this sense we will apply convex weighting.

$$V_i^{(k)} = \frac{w_i^{(k)}}{\sum_k w_i^{(k)}},$$

where

$k = a, b, \ldots, n$,

$i = 1, 2, \ldots, p$,

$w_i^{(k)}$ is the weight of characteristic k in match Y_i.

In our example, the table of weights for the match R. Madrid–F.C. Barcelona is

Technical level	.8
Speed	.5
Personality	.6
Tolerance	.7
Rationality	.8

so

$$\sum_k w_1^{(k)} = .8 + .5 + .6 + .7 + .8 = 3.4$$

$$V^{(Technical\ level)}_{R.Madrid--F.C.Barcelona} = \frac{.8}{3.4} = .2353;$$

$$V^{(Speed)}_{R.Madrid--F.C.Barcelona} = .1471;$$

$$V^{(Personality)}_{R.Madrid--F.C.Barcelona} = .1764;$$

$$V^{(Tolerance)}_{R.Madrid--F.C.Barcelona} = .2059;$$

$$V^{(Rationality)}_{R.Madrid--F.C.Barcelona} = .2353;$$

Followed the same steps we will have for:

– R. Sociedad – At. Bilbao:

Characteristics	Weihts	Relative weigths
Technical Level	.7	.1892
Speed	.9	.2432
Personality	.6	.1622
Tolerance	.6	.1622
Rationality	.9	.2432

– Betis – Sevilla:

Characteristics	Weihts	Relative weigths
Technical Level	.8	.2286
Speed	.7	.2000
Personality	.6	.1714
Tolerance	.7	.2000
Rationality	.7	.2000

– Deportivo – Celta:

Characteristics	Weihts	Relative weigths
Technical Level	.8	.2424
Speed	.9	.2727
Personality	.6	.1818
Tolerance	.5	.1515
Rationality	.5	.1515

– Valencia – Vilarreal:

Characteristics	Weihts	Relative weigths
Technical Level	.5	.1613
Speed	.5	.1613
Personality	.6	.1935
Tolerance	.8	.2581
Rationality	.7	.2258

– Español – At. Madrid:

Characteristics	Weihts	Relative weigths
Technical Level	.6	.2000
Speed	.7	.2333
Personality	.6	.2000
Tolerance	.7	.2333
Rationality	.4	.1334

– Valladolid – Rayo V.:

Characteristics	Weihts	Relative weigths
Technical Level	.9	.2812
Speed	.8	.2500
Personality	.6	.1875
Tolerance	.4	.1250
Rationality	.5	.1563

c) Decide overall for which characteristics of the matches, exceeding the level by the referee is considered negative and for which of them it is considered as ideal. For this we will apply the Index of Maximum and Minimum Level (Gil Lafuente 2002).

We will apply each one of these steps to our example.

Characteristics

- Speed (b)
- Tolerance (d)
- Rationality (e)

will be penalized if the required level is exceeded:

$$\delta(\mathrm{Y}_i, \mathrm{X}_j) = \sum_{k \in \{b,d,e\}} \left| \mu_i^{(k)} - \mu_j^{(k)} \right| \cdot V_i^{(k)}$$

For characteristics

- Technical Level (a)
- Personality (c)

there will be no penalization and the following will be used:

$$\zeta(\mathrm{Y}_i, \mathrm{X}_j) = \sum_{k \in \{a,c\}} \max\left\{0, \left(\mu_i^{(k)} - \mu_j^{(k)}\right)\right\} \cdot V_i^{(k)}$$

Then the index between match Y_i and referee X_j is computed as

$$\sigma(\mathrm{Y}_i, \mathrm{X}_j) = \delta(\mathrm{Y}_i, \mathrm{X}_j) + \zeta(\mathrm{Y}_i, \mathrm{X}_j).$$

5 Setting Up the Matrix for Properties

Once the requirements for each match and the weighting of the characteristics are determined, and the decision as to whether they can exceed the required level is made, we will move on to find the Maximum and Minimum Level Indices between all the matches of the day, and all the available referees.

Index between match *R. Madrid – F.C. Barcelona* and Referee ***A***:

	Technical Level	**Speed**	**Personality**	**Tolerance**	**Rationality**
R. Madrid– F.C. Barcelona =	.7	.4	.6	.8	.5
Referee **A** =	.5	.9	1	.4	.7

Technical Level	.2353
Speed	.1471
Personality	.1764
Tolerance	.2059
Rationality	.2353

σ(MB, A) = .2353 [0 $\vee$ (.7 - .5)] + .147 |.4 - .9| + .1764 [0 $\vee$ (.6 - 1)] + .2059 |.8 - .4| +.2353 |.5 - .7| = .25

$$\boxed{\sigma(\text{MB, A}) = .25}$$

Similarly, we can find all other indices between all the matches and all the referees. The results can be ordered in the Maximum and Minimum Level Matrix which will provide us with all the information for this day.

	Ref. **A**	Ref. **B**	Ref. **C**	Ref. **D**	Ref. **E**	Ref. **F**	Ref. **G**	Ref. **H**	Ref. **I**	Ref. **J**
Madrid – Barcelona	.25	.2029	.1559	.2911	.1411	.1647	.2205	.25	.2617	.3088
R. Sociedad – At. Bilbao	.1216	.2837	.1297	.308	.154	.1134	.1729	.2837	.3	.2756
R. Betis - Sevilla	.2314	.4	.2714	.2885	.2171	.2114	.1943	.2542	.3428	.42
Deportivo C. - Celta	.1454	.2605	.1575	.2424	.1666	.0878	.0878	.2181	.3121	.2787
Valencia - Villarreal	.2419	.2515	.1354	.358	.1741	.1484	.1935	.3097	.3096	.271
Español – At. Madrid	.26	.22	.1666	.4333	.1433	.2166	.2533	.3933	.28	.2733
Valladolid – Rayo V.	.2344	.1094	.0969	.3531	.1844	.1719	.2531	.3281	.1812	.1875

6 Assignment of Referees by the Kaufmann-Gil Aluja Algorithm

By means of the Kaufmann-Gil Aluja algorithm (Gil Aluja 1999) we can find which of the referees is the most suitable for the requirements of each match. The steps to follow are (Gil Lafuente 2002, p. 165):

1. Look for the element with the lowest value.

2. This element we have found establishes, by the corresponding row and column, the assignment of a referee to one of the matches of the day, that is to say, referee-match.
3. We now eliminate the relation that was initially chosen from the row and column corresponding to the assigned referee and the match he is to direct. What remains is a fuzzy relation of a lower order.
4. With this new matrix of a lower order we commence the process again by looking for the element with the least value.
5. We then repeat the process until the fuzzy relation is exhausted. In the normal event that the number of referees being greater than the number of matches, the "substitute" referees will not be assigned.

We continue our example,

	Ref. **A**	Ref. **B**	Ref. **C**	Ref. **D**	Ref. **E**	Ref. **F**	Ref. **G**	Ref. **H**	Ref. **I**	Ref. **J**
Madrid – Barcelona	.25	.2029	.1559	.2911	.1411	.1647	.2205	.25	.2617	.3088
R. Sociedad – At. Bilbao	.1216	.2837	.1297	.308	.154	.1134	.1729	.2837	.3	.2756
R. Betis - Sevilla	.2314	.4	.2714	.2885	.2171	.2114	.1943	.2542	.3428	.42
Deportivo C. - Celta	.1454	.2605	.1575	.2424	.1666	***.0878***	.0878	.2181	.3121	.2787
Valencia - Vilarreal	.2419	.2515	.1354	.358	.1741	.1484	.1935	.3097	.3096	.271
Español – At. Madrid	.26	.22	.1666	.4333	.1433	.2166	.2533	.3933	.28	.2733
Valladolid – Rayo V.	.2344	.1094	.0969	.3531	.1844	.1719	.2531	.3281	.1812	.1875

We will assign referee ***F*** to match ***Deportivo – Celta***.

By eliminating a row and column we arrive at the following assignment:

	Ref. A	Ref. B	Ref. C	Ref. D	Ref. E	Ref. G	Ref. H	Ref. I	Ref. J
Madrid – Barcelona	.25	.2029	.1559	.2911	.1411	.2205	.25	.2617	.3088
R. Sociedad – At. Bilbao	.1216	.2837	.1297	.308	.154	.1729	.2837	.3	.2756
R. Betis - Sevilla	.2314	.4	.2714	.2885	.2171	.1943	.2542	.3428	.42
Valencia - Vilarreal	.2419	.2515	.1354	.358	.1741	.1935	.3097	.3096	.271
Español – At. Madrid	.26	.22	.1666	.4333	.1433	.2533	.3933	.28	.2733
Valladolid – Rayo V.	.2344	.1094	***.0969***	.3531	.1844	.2531	.3281	.1812	.1875

We now assign referee ***C*** to match ***Valladolid – Rayo V***.

	Ref. A	Ref. B	Ref. D	Ref. E	Ref. G	Ref. H	Ref. I	Ref. J
Madrid – Barcelona	.25	.2029	.2911	.1411	.2205	.25	.2617	.3088
R. Sociedad – At. Bilbao	***.1216***	.2837	.308	.154	.1729	.2837	.3	.2756
R. Betis - Sevilla	.2314	.4	.2885	.2171	.1943	.2542	.3428	.42
Valencia - Vilarreal	.2419	.2515	.358	.1741	.1935	.3097	.3096	.271
Español – At. Madrid	.26	.22	.4333	.1433	.2533	.3933	.28	.2733

We assign referee ***A*** to match ***R. Sociedad – At. Bilbao***.

	Ref. **B**	Ref. **D**	Ref. **E**	Ref. **G**	Ref. **H**	Ref. **I**	Ref. **J**
Madrid – Barcelona	.2029	.2911	***.1411***	.2205	.25	.2617	.3088
R. Betis - Sevilla	.4	.2885	.2171	.1943	.2542	.3428	.42
Valencia - Vilarreal	.2515	.358	.1741	.1935	.3097	.3096	.271
Español – At. Madrid	.22	.4333	.1433	.2533	.3933	.28	.2733

We now assign referee ***E*** to match ***R. Madrid – F.C. Barcelona***.

We continue with the process:

	Ref. **B**	Ref. **D**	Ref. **G**	Ref. **H**	Ref. **I**	Ref. **J**
R. Betis - Sevilla	.4	.2885	.1943	.2542	.3428	.42
Valencia - Vilarreal	.2515	.358	***.1935***	.3097	.3096	.271
Español – At. Madrid	.22	.4333	.2533	.3933	.28	.2733

We assign referee ***G*** to match ***Valencia – Vilarreal***.

	Ref. **B**	Ref. **D**	Ref. **H**	Ref. **I**	Ref. **J**
R. Betis - Sevilla	.4	.2885	.2542	.3428	.42
Español – At. Madrid	***.22***	.4333	.3933	.28	.2733

Ref. ***B*** will direct match ***Español – At. Madrid***.

And finally,

	Ref. **D**	Ref. **H**	Ref. **I**	Ref. **J**
R. Betis - Sevilla	.2885	.2542	.3428	.42

The match ***Betis – Sevilla*** will be directed by Ref. ***H***; while referees D, I and J will remain in reserve.

We should point out that in the event that there were some very important matches compared to matches without any importance, we will attempt to select in the matrix those indices for the most important matches in the first place and leave the remaining less important matches for later.

If in our example, the matches to be played on that day were to be the last matches of the championship and the league was to be decided with the matches R. Madrid – F.C. Barcelona and Deportivo de la Coruña – Celta de Vigo the process to be followed would be to find the lowest indices for these two matches in the fuzzy relation.

	Ref. **A**	Ref. **B**	Ref. **C**	Ref. **D**	Ref. **E**	Ref. **F**	Ref. **G**	Ref. **H**	Ref. **I**	Ref. **J**
Madrid – Barcelona	.25	.2029	.1559	.2911	.1411	.1647	.2205	.25	.2617	.3088
R. Sociedad – At. Bilbao	.1216	.2837	.1297	.308	.154	.1134	.1729	.2837	.3	.2756
R. Betis - Sevilla	.2314	.4	.2714	.2885	.2171	.2114	.1943	.2542	.3428	.42
Deportivo C. - Celta	.1454	.2605	.1575	.2424	.1666	.0878	.0878	.2181	.3121	.2787
Valencia - Villarreal	.2419	.2515	.1354	.358	.1741	.1484	.1935	.3097	.3096	.271
Español – At. Madrid	.26	.22	.1666	.4333	.1433	.2166	.2533	.3933	.28	.2733
Valladolid – Rayo V.	.2344	.1094	.0969	.3531	.1844	.1719	.2531	.3281	.1812	.1875

We can see here that there are two minimum indices. The solution here then is to choose the index for which the value is higher than for the other match.

	Ref. **A**	Ref. **B**	Ref. **C**	Ref. **D**	Ref. **E**	Ref. **F**	Ref. **G**	Ref. **H**	Ref. **I**	Ref. **J**
Madrid – Barcelona	.25	.2029	.1559	.2911	.1411	.1647	.2205	.25	.2617	.3088
R. Sociedad – At. Bilbao	.1216	.2837	.1297	.308	.154	.1134	.1729	.2837	.3	.2756
R. Betis - Sevilla	.2314	.4	.2714	.2885	.2171	.2114	.1943	.2542	.3428	.42
Deportivo C. - Celta	.1454	.2605	.1575	.2424	.1666	.0878	.0878	.2181	.3121	.2787
Valencia - Villarreal	.2419	.2515	.1354	.358	.1741	.1484	.1935	.3097	.3096	.271
Español – At. Madrid	.26	.22	.1666	.4333	.1433	.2166	.2533	.3933	.28	.2733
Valladolid – Rayo V.	.2344	.1094	.0969	.3531	.1844	.1719	.2531	.3281	.1812	.1875

With which we will assign to match ***Deportivo - Celta*** Referee ***G***.

	Ref. **A**	Ref. **B**	Ref. **C**	Ref. **D**	Ref. **E**	Ref. **F**	Ref. **H**	Ref. **I**	Ref. **J**
Madrid – Barcelona	.25	.2029	.1559	.2911	.1411	.1647	.25	.2617	.3088
R. Sociedad – At. Bilbao	.1216	.2837	.1297	.308	.154	.1134	.2837	.3	.2756
R. Betis - Sevilla	.2314	.4	.2714	.2885	.2171	.2114	.2542	.3428	.42
Valencia - Villarreal	.2419	.2515	.1354	.358	.1741	.1484	.3097	.3096	.271
Español – At. Madrid	.26	.22	.1666	.4333	.1433	.2166	.3933	.28	.2733
Valladolid – Rayo V.	.2344	.1094	.0969	.3531	.1844	.1719	.3281	.1812	.1875

While for match ***R. Madrid – F.C. Barcelona*** Referee ***E*** will be the most suitable.

The process will now continue normally for the remaining matches and referees.

7 Assignment of Referees by Means of the Hungarian Algorithm

We will now proceed to assign the referees by means of an algorithm which goes by the name of "Hungarian" from the nationality of König (König 1916).

Starting from the matrix of dissimilarity we found before, we can obtain optimal assignments by following the steps below (Gil Aluja 1996).

1. If the matrix is not square, we add some extra rows with all elements equal to 1 (the maximum index of dissimilarity) to make the matrix square (in other words, any of the referees are not qualified to referee any of the matches corresponding to the added rows). Note that addition of these rows will not obstruct the calculation process.
2. Subtracting the lowest value from each column, and then doing the same with every row. In this way, we will obtain at least one zero in each row and each column.
3. We will look to see if it is possible to make an assignment in which all the values of the solution are null. In the event this is so, we have arrived at an optimum. Otherwise, we must continue the process. For this:

    a) Consider the rows containing the least number of zeros one by one.
    b) Draw a square around one of the zeros of each row and cross out the remaining zeros in the row and column containing the squared zero.
    c) Repeat this process with the remaining rows until there are no more zeros left.

4. We then arrive at the minimum set of rows and columns that contain all zeros. For this we must:

    a) Mark the rows in which there are no zeros in squares with an arrow $\leftarrow$.
    b) Likewise, mark with an arrow $\uparrow$ the columns in which there exists a crossed-out zero in a row marked with an arrow.
    c) Mark with an arrow $\leftarrow$ those rows in which a zero in a square exists in a column marked with an arrow.
    d) Repeat b) and c) until there are no more rows or columns to be marked.

e) We now shade in the rows that are not marked by arrows and the columns that are marked by arrows. These rows and columns constitute the minimum set of rows and columns that contain all squared or crossed-out zeros.

5. Select the lowest value among the elements of the matrix that have not been shaded. This number is then subtracted from the columns that are not shaded and added to the elements of the rows that are shaded, yielding a new matrix.
6. With this new matrix we go back to step 2, following the same process used but now with the elements of the new matrix. If we find an optimal solution, we stop (the algorithm terminates). Otherwise, we continue with steps 3 and 4 and if necessary return to 2.

We now continue with our example by converting the matrix of dissimilarity into a 10 $\times$ 10 matrix:

	Ref. **A**	Ref. **B**	Ref. **C**	Ref. **D**	Ref. **E**	Ref. **F**	Ref. **G**	Ref. **H**	Ref. **I**	Ref. **J**
Madrid – Barcelona	.25	.2029	.1559	.2911	.1411	.1647	.2205	.25	.2617	.3088
R. Soc. – At. Bilbao	.1216	.2837	.1297	.308	.154	.1134	.1729	.2837	.3	.2756
R. Betis - Sevilla	.2314	.4	.2714	.2885	.2171	.2114	.1943	.2542	.3428	.42
Deportivo - Celta	.1454	.2605	.1575	.2424	.1666	.0878	.0878	.2181	.3121	.2787
Valencia - Vilarreal	.2419	.2515	.1354	.358	.1741	.1484	.1935	.3097	.3096	.271
Español – At. Madr.	.26	.22	.1666	.4333	.1433	.2166	.2533	.3933	.28	.2733
Valladolid – Rayo V.	.2344	.1094	.0969	.3531	.1844	.1719	.2531	.3281	.1812	.1875
Reserve 1	1	1	1	1	1	1	1	1	1	1
Reserve 2	1	1	1	1	1	1	1	1	1	1
Reserve 3	1	1	1	1	1	1	1	1	1	1

Now we place under each column and after each row, the value subtracted from the corresponding elements.

	Ref. **A**	Ref. **B**	Ref. **C**	Ref. **D**	Ref. **E**	Ref. **F**	Ref. **G**	Ref. **H**	Ref. **I**	Ref. **J**	
Madrid – Barcelona	.1284	.0935	.059	.0487	0	.0769	.1327	.0319	.0805	.1213	**0**
R. Soc. – At. Bilbao	0	.1743	.0328	.0656	.0129	.0256	.0851	.0656	.1188	.0881	**0**
R. Betis - Sevilla	.0737	.2545	.1384	.001	.0399	.0875	.0704	0	.1255	.1964	**.0361**
Deportivo C. - Celta	.0238	.1511	.0606	0	.0255	0	0	0	. 13	.0912	**0**
Valencia - Vilarreal	.0873	.1091	.0055	.0826	0	.0276	.0727	.0586	.0945	.0505	**.033**
Español – At. Madr.	.1362	.1038	.0675	.1887	0	.1266	.1633	.173	.0966	.0836	**.0022**
Valladolid – Rayo V.	.1128	0	0	.1107	.0433	.0841	.1653	.11	0	0	**0**
Reserve 1	.1208	.133	.1455	0	.1013	.1546	.1546	.0243	.0612	.0549	**.7576**
Reserve 2	.1208	.133	.1455	0	.1013	.1546	.1546	.0243	.0612	.0549	**.7576**
Reserve 3	.1208	.133	.1455	0	.1013	.1546	.1546	.0243	.0612	.0549	**.7576**
	.1216	**.1094**	**.0969**	**.2424**	**.1411**	**.0878**	**.0878**	**.2181**	**.1812**	**.1875**	

We put the solitary zeros from each row in a square, and cross out those corresponding to its row and column.

	Ref. **A**	Ref. **B**	Ref. **C**	Ref. **D**	Ref. **E**	Ref. **F**	Ref. **G**	Ref. **H**	Ref. **I**	Ref. **J**
Madrid – Barcelona	.1284	.0935	.059	.0487	[0]	.0769	.1327	.0319	.0805	.1213
R. Soc. – At. Bilbao	[0]	.1743	.0328	.0656	.0129	.0256	.0851	.0656	.1188	.0881
R. Betis - Sevilla	.0737	.2545	.1384	.001	.0399	.0875	.0704	[0]	.1255	.1964
Deportivo C. - Celta	.0238	.1511	.0606	~~0~~	.0255	[0]	~~0~~	~~0~~	. 13	.0912
Valencia - Vilarreal	.0873	.1091	.0055	.0826	~~0~~	.0276	.0727	.0586	.0945	.0505
Español – At. Madr.	.1362	.1038	.0675	.1887	~~0~~	.1266	.1633	.173	.0966	.0836
Valladolid – Rayo V.	.1128	[0]	~~0~~	.1107	.0433	.0841	.1653	.11	~~0~~	~~0~~
Reserve 1	.1208	.133	.1455	[0]	.1013	.1546	.1546	.0243	.0612	.0549
Reserve 2	.1208	.133	.1455	~~0~~	.1013	.1546	.1546	.0243	.0612	.0549
Reserve 3	.1208	.133	.1455	~~0~~	.1013	.1546	.1546	.0243	.0612	.0549

Now we continue by marking with an arrow those rows in which there is no zero in a square. Then we indicate the columns in which there are crossed out zeros, continuing with the process until there are no further arrows to be placed. We will then shade the rows not marked with an arrow and the columns that are marked with an arrow.

Ref.	Ref.	Ref.	Ref.	Ref.	Ref.	Ref.	Ref.	Ref.	Ref.
.1284	.0935	.059	.0487	0	.0769	.1327	.0319	.0805	.1213
0	.1743	.0328	.0656	.0129	.0256	.0851	.0656	.1188	.0881
.0737	.2545	.1384	.001	.0399	.0875	.0704	0	.1255	.1964
.0238	.1511	.0606	0	.0255	0	0	0	.13	.0912
.0873	.1091	.0055	.0826	0	.0276	.0727	.0586	.0945	.0505
.1362	.1038	.0675	.1887	0	.1266	.1633	.173	.0966	.0836
.1128	0	0	.1107	.0433	.0841	.1653	.11	0	0
.1208	.133	.1455	0	.1013	.1546	.1546	.0243	.0612	.0549
.1208	.133	.1455	0	.1013	.1546	.1546	.0243	.0612	.0549
.1208	.133	.1455	0	.1013	.1546	.1546	.0243	.0612	.0549

We now build a matrix with the remaining elements after eliminating the shaded rows and columns and look for the element with the lowest value.

	Ref. **A**	Ref. **B**	Ref. **C**	Ref. **F**	Ref. **G**	Ref. **H**	Ref. **I**	Ref. **J**
Madrid – Barcelona	.1284	.0935	.059	.0769	.1327	.0319	.0805	.1213
Valencia - Vilarreal	.0873	.1091	***.0055***	.0276	.0727	.0586	.0945	.0505
Español – At. Madr.	.1362	.1038	.0675	.1266	.1633	.173	.0966	.0836
Reserve 1	.1208	.133	.1455	.1546	.1546	.0243	.0612	.0549
Reserve 2	.1208	.133	.1455	.1546	.1546	.0243	.0612	.0549
Reserve 3	.1208	.133	.1455	.1546	.1546	.0243	.0612	.0549

We deduct the value **.0055** from the values in the columns that are not shaded, adding them to the shaded rows, arriving at this new matrix:

	Ref. A	Ref. B	Ref. C	Ref. D	Ref. E	Ref. F	Ref. G	Ref. H	Ref. I	Ref. J
Madrid – Barcelona	.1229	.088	.0535	.0487	0	.0714	.1272	.0264	.075	.1158
R. Socie. – At. Bilbao	0	.1743	.0328	.0711	.0184	.0256	.0851	.0656	.1188	.0881
R. Betis - Sevilla	.0737	.2545	.1384	.0065	.0454	.0875	.0704	0	.1255	.1964
Deportivo - Celta	.0238	.1511	.0606	.0055	.031	0	0	0	. 13	.0912
Valencia - Vilarreal	.0818	.1036	0	.0826	0	.0221	.0672	.0531	.089	.045
Español – At. Madr.	.1307	.0983	.062	.1887	0	.1211	.1578	.1675	.0911	.0781
Valladolid – Rayo V.	.1128	0	0	.1162	.0488	.0841	.1653	.11	0	0
Reserve 1	. 1153	.1275	.14	0	.1013	.1491	.1491	.0188	.0557	.0494
Reserve 2	. 1153	.1275	.14	0	.1013	.1491	.1491	.0188	.0557	.0494
Reserve 3	.1153	.1275	.14	0	.1013	.1491	.1491	.0188	.0557	.0494

This matrix will now be the starting point for the process we have been carrying out up to this point. We will not show here all of the calculations, but will only mention the values to be subtracted from the elements of the columns that are not shaded and added to the rows that are shaded: **.0188**, **.0306**, **.0041**, **.0022**, **.0157**, **.0178**, and lastly **.0178**. We get the following matrix:

	Ref. A	Ref. B	Ref. C	Ref. D	Ref. E	Ref. F	Ref. G	Ref. H	Ref. I	Ref. J
Madrid – Barcelona	.0515	.0323	0	.0487	0	0	.0568	.0074	.0193	.0664
R. Socie. – At. Bilbao	0	.19	.0485	.1097	.057	.0256	.0851	.0854	.1345	.1079
R. Betis - Sevilla	.0211	.2176	.1037	.0253	.0642	.0349	.0178	0	.0886	.1658
Deportivo C. - Celta	.0238	.1668	.0763	.0441	.0696	0	0	.0198	.1457	.111
Valencia - Vilarreal	.0639	.1014	0	.1055	.0041	.0042	.0493	.0572	.0868	.0491
Español – At. Madr.	.0593	.0426	.0085	.1887	0	.0497	.0864	.1487	.0354	.0287
Valladolid – Rayo V.	.0971	0	0	.1391	.0717	.0684	.1496	.1141	0	.0041
Reserve 1	.0439	.0718	.0865	0	.1013	.0777	.0777	0	0	0
Reserve 2	.0439	.0718	.0865	0	.1013	.0777	.0777	0	0	0
Reserve 3	.0439	.0718	.0865	0	.1013	.0777	.0777	0	0	0

And finally we obtain the optimum assignments:

R. Madrid –Barcelona	→	Referee ***F***
R. Sociedad – At. Bilbao	→	Referee ***A***
R. Betis – Sevilla	→	Referee ***H***
Deportivo C. – Celta	→	Referee ***G***
Valencia - Villarreal	→	Referee ***C***
Español – At. Madrid	→	Referee ***E***
Valladolid – Rayo V.	→	Referee ***B***

Referees ***D***, ***I*** and ***J***. are reserve referees.

8 Conclusion

The goal of this chapter was to show that randomness or simple methods with few characteristics do not always lead to satisfactory results. It is also very clear that to take such an important decision by means of a formal meeting of a group of specialists, however neutral they may appear to be, can always arouse controversy or suspicions. However, a reliable mathematical support, working with a high number of varied parameters, suitably valued and weighted (but always respecting the subjective opinions of the experts) will lead us to a greater coherence and reduce the likelihood of fraud.

We do not want to conclude this article without stating that this work, as with many other tools developed in previous works, can be adapted to decisions in other aspects of the sporting world.

In summary, our experience allows us to state that the application of both techniques will allow to obtain a solution of the problem that has been controversial for some decades.

In the first system we attain our objective by assigning the referees by order of adaptability to the specific requirements of each one of the matches. In the second system, we arrive at the maximum balance in the assignments, in this way avoiding excessive imbalances.

In conclusion, we want to stress that the problem of assigning referees can be efficiently solved. We believe that the methods presented in this chapter should be chosen over the currently used practices of assigning referees. The result of this research work has been presented to the football federation and proper committees, it is up to them to decide if it will be taken advantage of.

References

1. De Luca, A.and Termini, S.: "A definition nonprobabilistic entropy in the setting fuzzy sets theory". Information and Control, 1972, 20, p.301-312.
2. Gil Aluja, J.: "Elements for a theory of decision in uncertainty". Kluwer Academic Publishers. Dortrech, 1999.
3. Gil Aluja, J.: "La Gestión Interactiva de los Recursos Humanos en la Incertidumbre", CEURA, Madrid 1996, p. 170-172.
4. Gil Aluja, J.: "Selección de Personal. El problema de la polivalencia y el de la uniformida". Note-books, CEURA. Madrid, 1987.
5. Gil Aluja, J.: "The interactive management of human resources in uncertainty". Kluwer Academic Publishers. Dortrech, 1998.
6. Gil Aluja, J.: Elements for a theory of decision in uncertainty. Kluwer Academic Publishers. Dortrech, 1999.
7. Gil Lafuente, J.: "Algoritmos de la Excelencia. Claves para el Éxito en la Gestión Deportiva". Publ.. Milladoiro, Publications of the Barcelona F.C., Vigo, 2002.

8. Gil Lafuente, J.: "El Mejor Sistema de Designación Arbitral: el Algoritmo Húngaro". Actas del Congreso: "International Congress A.C.S.E.G.'2002". Boulogne-Sur-Mer (France), 21and 22 , November, 2002, p. 167-180.
9. Gil Lafuente, J.: "La asignación de deportistas a los puestos de un equipo". Proceedings of the: "VI S.I.G.E.F" Congress. Morelia, Michoacán (México), 15, 16 and 17, November de 1999, p. 284-299.
10. Gil Lafuente, J.: "Les Universitats en el Centenari del Futbol Club Barcelona. Estudis en l'àmbit de l'esport", (J. Gil Aluja Ed.): "L'optimització del fitxatge d'un esportista en l'ambit de l'esport". Publications of the Barcelona F.C.. Publ. Milladoiro, Santiago de Compostela, 1999, p. 3-55.
11. Gil Lafuente, J.: "The adquisition of having inmobilized inmaterial in the sports societies". Proceedings: "International MS'99 Congress". Santiago de Compostela, 17, 18 and 19, May 1999, p. 185-196.
12. Gil Lafuente, J.: "Una Herramienta Óptima de Designación Arbitral basada en las Lógicas Multivalentes: Hombre y Máquina". Proceedings: "XI International Congress A.E.D.E.M.". París (France), 2, 3 and 4, September 2002, p. 491-498.
13. Gil Lafuente, J: "El 'Índice del Máximo y Mínimo Nivel' en la Optimización del Fichaje de un Deportista". Proceedings: "X International Congress A.E.D.E.M.". Reggio Calabria (Italia), 4, 5 and 6.
14. Gil Lafuente, J: "Marketing para el nuevo Milenio. Nuevas técnicas para la gestión comercial en la incertidumbre". Publ Pirámide. Madrid, 1996.
15. Kaufmann, A. and Gil Aluja, J.: "Introducción a la teoría de los subconjuntos borrosos a la gestión de las empresas". Publ. Milladoiro (3rd edition). Santiago de Compostela, 1993.
16. Kaufmann, A. and Gil Aluja, J.: "Técnicas especiales para la gestión de expertos". Publ. Milladoiro. Santiago de Compostela, 1993.
17. König, D.: "Theorie der endlichen und unendlichen graphen (1916), later reprinted by Chelsea Publ. C^o, New York, 1950. This work was made known by Kuhn, H.W. in his article "The Hungarian Method for the Assignment Problem", Naval Research Quaterly, vol. 2, n° 1-2 March-June 1955.
18. Zadeh, L.: "Fuzzy Sets. Information and control". 8 July 1965, p. 338-353.
19. Zimmermann, H. J.: "Results of empirical studies in fuzzy sets theory". En Klir, G. J.: "Applied General Systems Research. Plenum. New York, 1978, p.303-312.

Fashion Accessory, Social Identity or Tribal Uniform?

Exploratory research into the reasons why football consumers purchase replica football kits in Spain and England.

Nick Wilde

University of Greenwich, London, UK
n.p.wilde@gre.ac.uk

1 Introduction

The principle aims of this research are:

- To gain an understanding of the sociological, psychological or anthropological factors that make football supporters in Spain and England purchase replica football kits.
- To identify the factors that might affect the future purchasing behaviour of replica kits for football supporters in Spain and England, and help football clubs to better understand their customer base.
- To consider how to segment football supporters of clubs in Spain and England based on their purchasing behaviour of replica kits and attendance at football matches.

Merchandise is the term used for products that are made to promote football teams, which are usually characterised by the inclusion of a club name or brand logo. In Spain and the United Kingdom the main item of merchandise which generates the most revenue is the team shirt or replica shirt.

Manchester United Football Club have generated global sales of merchandise and in particular replica kits, which have made them one of highest income generating football clubs in the World, to the envy of teams like Milan, Juventus, Bayern Munich, Barcelona and Real Madrid. Surprisingly this has not been a major source of revenue generation for football clubs until the late eighties and early nineties. Some of the earliest reference to merchandising in England was by Davies in 1971, which looked at the role of the football spectator in the development of club merchandise. Edward Freedman, a leading consultant on sports merchandising however, is considered to be the pioneer of merchandising through his work with leading English football clubs, and consultancy with football clubs across Europe. As a result of Freedman's early work on developing merchandising, English clubs appear to lead the way in terms of sales of merchandise. All of the aforementioned top teams in Europe have very high average attendances, and are recognised globally as a result of their frequent appearances in the world media and through their involvement in the high profile European Champions League. However, these clubs have made no secret of their desire to emulate the commercial success of Manchester United. In 1996-1997 Manchester United sold an estimated 850,000 replica shirts, with Liverpool 600,000, with each club earning about € 2.05[1] per shirt sold, and an annual royalty payment of about € 4.1 to € 4.8 million from their kit manufacturer (Szymanski 2000: 71).

Given the introduction and growth of the European Champions Football League, the impact of its intense media coverage could potentially open up the European market for the teams taking part, with an opportunity to increase sales of merchandise on a national and international basis. According to Deloitte and Touche, in 1998 merchandising accounted for 10.3% of the turnover of Spanish clubs, compared to 9.1% in Italy and 17.2% in England. Giulianotti (2000:89) talked about the impact of merchandising on the finances of leading football clubs. He cites the example of Ronaldo, who when he signed for Internazionale in Milan, helped the club sell 35,000 shirts within ten days, thus recouping part of the transfer fee paid for him. In January 2003, the Spanish football newspaper Marca, reported that Real Madrid's recent signings of the Brazilian Ronaldo, coupled with the signing of Figo had seen sales of their replica shirts rise to around 900,000 per annum, compared to 80,000 in the season 1999 to 2000, a staggering rise in both sales and income. The report states that around 300,000 of these shirts have been made directly as a result of the signing of Ronaldo. This level of sales compares with the levels reported by Manchester United, and suggest that the impact of world media attention on football is helping to sell more merchandise. The recent signing of David Beckham from Manchester United is likely to have a significant impact on the sales of Real Madrid shirts and could potentially lead to a decline in the sales of Manchester United shirts.

There is very little research data available in the public domain which helps to explain the reasons for these apparent differences between British and other European clubs, nor which gives us a better understanding of exactly why fans purchase merchandise. There are however figures available to confirm the volume sales. The market research company Keynote report that in England around 60% of all merchandise is purchased by people under 35, and that a consumer in the north of England is twice as likely to purchase merchandise as somebody in the south of England. While this research gives insight into who buys and what they buy, it does not deal with the reasons why people buy. The

[1] All figures have been calculated in Euros with an exchange rate of 0.685 euros to the pound

sales of club merchandise for English football clubs have increased significantly during the past decade, and represent a considerable source of income for football clubs. Keynote's report also highlights the growing importance of club merchandise to sales figures. In 2001 58% of all season ticket holders of football clubs in the United Kingdom stated that they tend to buy their club's new replica football shirt, and 21% of all adults who attend football matches had purchased a replica football shirt within the past 12 months, a substantial increase of 19% from the year 2000. This equates to an average of € 89.75 per season ticket holder.

While success in European competitions is likely to have an impact on global sales, there are some exceptions to this rule. In England, Newcastle United achieved sales of replica kits of 600,000 in 1996 to 1997 without the level of media coverage experienced by top European clubs. Given the considerable revenues available, and the importance of sales to current revenue, football clubs need to know if the current level of sales will continue. A substantial fall in sales would have severe financial implications for leading clubs, who now count on this revenue as a way of financing their development. Other European clubs, on the other hand, need to be able to identify their potential markets and set realistic targets for their merchandising income. The recent increase in the sales of merchandise by clubs like Real Madrid suggest that by following a coordinated merchandising strategy, there might be potential benefits for other clubs.

There is considerable publicity concerning the manner in which football clubs in England make frequent changes to their playing kits, which encourages fans to update their own replica kits. Giulianotti (2000), talks about the exploitation of football supporters by football clubs, who frequently change the design of their kits in order to generate new sales. This merits further investigation, as a company who is seen to be exploiting fans, runs the risk of the consumer choosing not to continue buying from them, which in turn might lead to a dramatic fall in revenue. My preliminary investigations in Spain have shown that the marketing managers of football clubs are exploring merchandising as a potential generator of additional income, encouraged by the success of teams like Barcelona and Real Madrid. Another issue of concern to clubs is that of counterfeit goods, which are produced by non-licensed manufacturers, and sold on the black market with no revenue gain for the clubs. A recent interview with football clubs in the north of Spain suggested that as much as 20% of all merchandise sold is in fact counterfeit, and that in some cases the fans of a club are not aware that they are buying counterfeit goods.

There have been several studies carried out in the United States of America (USA), which deal specifically with sports supporters, and the reasons why they attend sporting events. Research findings to date suggest that there are psychological, sociological and anthropological reasons to explain the behaviour of sports fans at sports events. This research has covered predominantly American sports such as baseball, basketball, ice hockey and American football.

2 Sociological Factors

Sociological factors are classified as the social institutions and relationships that might affect the development and behaviour of any organised human groups. Cialdini (1976) looked at the behaviour patterns of university students who described themselves as supporters of a particular college football team. His research found that during the course of a sports season, students tended to wear clothing which featured the college logo more if the team had been victorious and less when they had lost. He referred to this as BIRGing, or Basking In Reflected Glory. Cialdini also discovered that when the team had won the students were more likely to refer to the team as we, but referred to the team as them when they had lost. Where a student was seen to be distancing themselves from failure, Cialdini described this as Cutting Off from Reflected Failure or CORFing.

Wann (1990) used Cialdini's framework to test the reaction of fans to the performance of their team and discovered that there was a clear tendency to BIRG. Mahony et al (2000), using the same variables, discovered that spectators are more likely to identify with teams that are winning than with teams that lose. Sutton et al. (1997) developed a conceptual framework to show how fans identified with teams, and if they received any benefits for this identification. They highlighted three levels of fan identification, namely low or social fans, medium or focused fans and high or vested fans They found that fans with low levels of identification were motivated more by the entertainment provided by a match. Medium fans were more likely to wear the team clothing and high level fans were more likely to commit themselves to the club in terms of time and finances.

My own preliminary findings suggest that there would appear to be some correlation between team success and the sales of merchandise. Alaves, a team from the Basque Country in Northern Spain, experienced a rapid increase in the sales of their replica shirts during the season in which they reached the final of a European cup competition. Teams in England also see increases in sales when they reach the final of domestic cup competitions. My future research will consider BIRGing and its impact on the sales of merchandise.

Kerstetter and Kovich (1997) developed the Consumer Involvement Profile (CIP), which was used to try and establish relationships between a spectator's sociodemographic and behavioural characteristics and their involvement with a sport. They found that spectators use sport as a way of seeking eustress, or pleasant stress, by observing or becoming involved emotionally with a crowd or by cheering a team. They also considered the perceived importance of the product. If the product is a piece of clothing with the team emblem, it might be important in so much as it is part of a fan's thoughts and interests, but it isn't necessarily an important part of their life. Kates (1998) suggests that sports marketers need to gain more of an understanding of the sub-cultures found among supporter groups. He relates this to fashion merchandising, whereby companies who understand their subcultures, are able to develop products that are more appealing to larger groups of customers. Research by Kleine (1995) supports the view that individuals use attachments such as club merchandise to define and maintain their identities. Wann et al (1999), Murrell and Dietz (1992) looked at the role of social identity and how fans reacted to the outcome of games, taking into consideration their expectations and the levels of media attention. They found that fans who identified strongly with their university team rated the team more favourably during the course of the season, even after the team had lost. The results were discussed in the context of social identity theory. People, who are members of specific social groups or categories, gain much of their self-esteem from these groups, and when their identity is threatened are more likely to move to a group with a higher perceived status. These findings have considerable implications for my own research, and suggest the possibility of fans changing their allegiance to teams, based on their success. To what extent are fans changing their allegiance, and their purchasing of club merchandise from one year to the next? Arnett et al (2000) developed this research area further and investigated the four factors of fan identity, salience, satisfaction, attachment and enduring involvement, to see if there was any variance across three types of fans, namely frequent, moderately frequent and infrequent. The research found that if a supporter owns team-related products, then this can lead to a more salient fan identity. The findings also suggest that social identity theory is a basis from which to research the area of leisure behaviour. Wann et al (2000) investigated to what extent supporters identified with a team and attempted to measure their levels of self-esteem. Their results showed that both factors determined a fans level of self-presentation, and how they might show their support to fans of another team.

Wann et al. (1999) developed the Sport Fan Motivation Scale (SFMS), to measure the different motives of fans attending events, including, eustress, self-esteem, escape, entertainment, economic, aesthetic, group affiliation and family. The research found that males experienced high levels of eustress and self-esteem motivation, whereas females who attended were more motivated by their families to attend.

3 Anthropological Factors

Anthropological factors can be classified as the study of human beings with reference to their social relations and culture. The anthropologist Desmond

Morris, in his published work in 1981, looked at the rituals of football in the United Kingdom. He suggested that supporter groups wore club colours in order to show their loyalty to the club. He further suggested that football supporters, rather than appearing to be a chaotic group of people not involved with the game, subject themselves to dress codes and conventions in the same way that any other strata of society does. Morris (1981:250) concludes that the term "Soccer Tribe" reflects the similarity between the soccer community and any other tribal society. Perhaps more significantly for my own research, he makes no reference to replica shirts as a means of showing allegiance, and there is little evidence of any replica shirts in the photographs of supporters taken during the sixties and seventies. The only significant pictorial evidence appears to be in Holland and Argentina where some fans wore replica shirts when supporting their national teams. This suggests that the phenomenon of fans wearing replica shirts to matches is probably one that began in the eighties.

We also need to understand how relationships with other supporters might affect the behaviour of supporters at sporting events. Melnick (1993) suggested that attendance at a sporting arena acted as a venue where people can enrich their social relationships. As a result of forces such as urbanisation, individualism and geographical mobility, it has become increasingly more difficult for people to develop social relationships. Not only is a spectator able to identify with similar people with a common interest, but able to share a similar experience. The replica team shirt worn by a fan may also be a way for fans to identify with other supporters at the venue. Given the rapid increase in the number of fans wearing replica football shirts to English football matches, my research will focus on the possibility that the replica shirt is now considered to be a tribal uniform among the supporters of certain clubs, or that they have, as a group, developed their own dress code. To what extent do supporters wear replica kit as a means of identifying more closely with other fans, or as a way of feeling accepted by other fans? This is an area that will be explored in my research.

Wakefield (1995) considered the interaction between, team identification, social influence (reference group and community acceptance) and perceived ticket value, and tried to find out if these factors influenced future purchase intentions of fans. The conclusion of the research was that club management need to increase the levels of social acceptance of going to a game. If individual supporters felt that their family or friends approved of them going to a game, then all of the other factors related to the experience of going to a game were also rated more favourably. Kleine et al (1995) suggested that individuals use material possessions as a way of defining or maintaining their identities, which might include the wearing of a replica kit to show allegiance. The recent experiences of Athletic Club Bilbao in Spain, who have launched their own brand, Athletic Club, suggest that owning their own brand has given their supporters more of a reason to purchase products bearing the club's brand. Recent sales growth has extended from replica shirts to sales of club overcoats and other leisurewear, and in such great numbers that other clubs will certainly be monitoring this initiative.

4 Psychological Factors

Psychological factors can be classified as the mental or behavioural characteristics of an individual or group. Ferrand (1999) researched the impact of a club's image in French football using a five step smiling face scale. He concluded that football clubs needed to be more aware of the value of their image and suggested that not only can it contribute positively to improving their brand equity, but that it also can bring about changes in the behaviour of its consumers, or supporters. Ferrand also concluded that strong images contributed towards the sale of merchandise. Supporters were more likely to associate themselves with a club with a strong image. The impact of key celebrities on sales of merchandise needs to be investigated. It is clear that key players such as David Beckham, Luis Figo and Ronaldo have an impact on sales of replica shirts and we need to understand the influence that such figures have on young people. Much has been written about celebrity endorsements in the United States of America (USA), and it is an area that merits further investigation in relation to European clubs and their key players.

The vast majority of the research on sports marketing has been undertaken in the USA, and covers predominantly American sports such as baseball, basketball, ice hockey and American football. Moreover it deals with the 'purchasing of the experience' (attendance at matches) rather than the purchasing of merchandise. The market for club merchandise clearly covers not only the fans who attend matches, but also those who are not regular attenders, whether they live in the same city or country as the club concerned or overseas. In this respect there is a growing identification of fans with the leading clubs in their country rather than local ones. This is another reason that a club like Manchester United has been able to increase its sales. The same can be said about Real Madrid and Barcelona who are increasing their fan base across the whole of Spain and not just in their respective home cities. There is also a growing Europeanisation and internationalisation of fandom with clubs gaining supporters overseas. Some clubs are now actively seeking to promote their brand in overseas markets. Everton and Manchester City have increased awareness of their clubs in China, after signing two Chinese International footballers, Li Tie and Sun Jihai respectively. Everton have also increased their links with the Chinese market by signing up Kejian, the Chinese mobile phone supplier as their main shirt sponsors. There is an estimated audience of 350 million people in China for live English Premier League games which is radically altering the markets for club merchandise, and also exposing them to the increased likelihood of counterfeit merchandise.

My research draws upon the findings of key writers, and will be of considerable interest not only to Spanish and English clubs, but also to football clubs across the world. The project as a whole focuses on an understanding of football supporters and their purchasing behaviour when consuming club merchandise in England and Spain as well as their attendance at sporting events, while taking into consideration cross-cultural issues. This paper draws on preliminary investigations that have been carried out in the two countries, and is essentially concerned with issues of methodology, given the rapidly changing market for sports-related merchandise.

5 Conclusion

I have identified a gap in the knowledge about why people buy merchandise for any particular club. While there is possibly some research commissioned privately by individual clubs, there remains a lack of published research into the purchasing behaviour of the supporters of European football clubs. My search for market research reports carried out privately continues and will inform my research further. There would appear to be several reasons why supporters purchase replica kits, and these theories will be tested through my own research. More preliminary investigations are to be undertaken through visits, and later through focus groups, before I can build up a more complete picture of purchasing behaviour of replica kits and other merchandise. In trying to understand why consumers purchase kits I will attempt to look at whether it is possible to predict future sales. In trying to understand the reasons for purchase I will consider the contributions that psychology, sociology and anthropology make, drawing on the research of people like Cialdini and Morris. Judging by my preliminary findings it would appear that each discipline will help me to understand the purchasing behaviour of football supporters. It is clear that the sales of club merchandise have become an increasing source of revenue for clubs, and that my research might contribute to help clubs better understand their relationships with their fans, and enable them to develop more effective merchandising strategies. My research will also include an attitudinal survey to be carried out in England and Spain, which will be informed by the findings of focus groups, and through my literature review. My research will hopefully help clubs to develop future merchandising strategies, and help in the setting of realistic targets.

References

1. Arnett, D.B, and Laverie, D.A. (2000) Fan Characteristics and Sporting Event Attendance: Examining Variance in Attendance. *International Journal of Sports Marketing and Sponsorship*. Vol. 2, Number 3, pp 219-238

2. Branscombe, N.R, and Wann, D.L, (1991), The Positive Social and Self-Concept Consequences of Sports Team Identification, *Journal of Sport and Social Issues*, Vol. 15, Number 2, pp 115-127
3. Brown, A, (1998), *Fanatics: Power, Identity and Fandom in Football*, Routledge, London
4. Cialdini, R.B. and Borden, R.J.(1976)Basking in Reflected Glory: Three (Football) Field Studies. *Journal of Personality and Social Psychology*. Vol. 34, Number 3, pp 366-375
5. Davies, H. (2000),*The Glory Game. The New Edition of the British Football Classic*, Edinburgh, Scotland, Mainstream Publishing.
6. Deloitte & Touche (2000) Informe Anual: Las Finanzas en la Primera Division del Futbol en Espana, Madrid, October 2000.
7. Ferrand, A. and Pages, M. (1999), Image Management in Sport Organisations: The Creation of Value, *European Journal of Marketing*, Vol. 33, Number 3/4, pp 387-401
8. Giulianotti, R, (2000), *Football, a Sociology of the Game*, Blackwell Publishers, England.
9. Kates, S.M, (1998), Consumer Research and Sport Marketing: Starting the Conversation Between Two Different Academic Discourses, *Sport Marketing Quarterly* , Vol.7, No. 2, pp 24-33
10. Kerstetter, D.L, Kovich, G.M, (1997), An Involvement Profile of Division 1 Women's Basketball Spectators. *Journal of Sport Management*, Vol. 1, No. 3, pp 234-249
11. Keynote (2002) The Football Industry, BMRB Publications Ltd, London.
12. Kleine, S.S, Kleine, R.E, Allen, C.T, (1995), How is a Possession "Me" or "Not me"? Characterizing Types and an Antecedent of Material Possession Attachment, *Journal of Consumer Research*, Vol. 22, No. 3, pp 327-343
13. Marca (January 25, 2003) www.buscador.recoletos.es/advice/marca/documento
14. Morris, D (1981), *The Soccer Tribe*, London, Jonathan Cape Ltd.
15. Madrigal, R. (1995), Cognitive and Affective Determinants of Fan Satisfaction, *Journal of Leisure Research*, Vol. 27, Number 3, pp 205-227
16. Mahony, D.F. Howard, D.R and Madrigal, R. (2000) BIRGing and CORFing Behaviors by Sports Spectators: High Self-Monitors Versus Low Self-Monitors. *International Sports Journal* Vol. 4, Number 1, pp 87-106
17. Melnick, M.J, (1993) Searching for Sociability in the Stands: A Theory of Sports Spectating, *Journal of Sport Management*, Vol.7, Number1, pp 44-60
18. Murrell, A.J, and Dietz, B (1992) Fan Support of Sport Teams: The Effect of Common Group Identity, *Journal of Sport and Exercise Psychology*, Vol. 14, Number 1, pp 28-39
19. Sutton, W.A, McDonald, M.A, and Milne, G.R, (1997), Creating and Fostering Fan Identification in Professional Sports, *Sport Marketing Quarterly*, Vol. 6, No. 1, pp15-22
20. Szymanski, S. and Kuypers, T, (2000), *Winners and Losers: The Business Strategy of Football.* Second Edition, London, Penguin Books
21. Wakefield, K.L,(1995) The Pervasive Effects of Social Influence on Sporting Event Attendance, *Journal of Sport and Social Issues*,Vol.19, Number 4 pp 335-351
22. Wann, D.L, and Branscombe, N.R, (1990), Die-Hard and Fair-Weather Fans: Effects of Identification on BIRGing and CORFing Tendencies, *Journal of Sport and Social Issues*, Vol. 14, Number 2, pp 103-117.

23. Wann, D.L, Schrader, M.P, Wilson, A.M, (1999), Sport Fan Motivation: Questionnaire Validation, Comparisons by Sport, and Relationship to Athletic Motivation, *Journal of Sport Behavior*, Vol. 22, No. 1, pp 114-139
24. Wann, D.L, Royalty, J, Roberts, A, (2000), The Self-Presentation of Sport Fans: Investigating the Importance of Team Identification and Self-Esteem, *Journal of Sport Behavior*, Vol. 23, No. 2, pp 198-206

Modeling a Baseball Game to Optimize Pitcher Substitution Strategies Using Dynamic Programming

Nobuyoshi Hirotsu[1] and Mike Wright[2]

[1] Japan Institute of Sports Sciences, Japan
hirotsu.nobuyoshi@jiss.ntgk.go.jp
[2] Lancaster University, UK
m.wright@lancaster.ac.uk

1 Introduction

The modeling of baseball games has been a topic of interest to several OR workers (Bennett 1998,Bukiet et al. 1997, D'Esopo and Lefkowitz 1977, Hirotsu and Wright 2003, Trueman 1976, Trueman 1977), and the task of finding optimal tactics can be assisted by the application of OR techniques. In a previous paper (Hirotsu and Wright 2003), we developed a Markov chain model of a baseball game between teams consisting of non-identical players, taking into account the number of runs by which the home team leads. In the paper we proposed the formulation for obtaining the optimal pinch-hitting strategy under the *designated hitter* (DH) rule, for which the substitution of pitchers is not a very complex issue, since under the DH rule pitchers do not have to bat.

In this paper, however, we consider games played *not* under the DH rule, where pitchers do have to bat unless substituted. Pitcher substitution is here a complex issue, and thus the formulation of the problem and the ensuing results are quite complex.

There are a variety of reasons why a pitcher might be substituted, including injury, tiredness and poor performance. Alternatively, late in the match, a manager may introduce a "relief" pitcher who specializes in holding on to a lead. Another reason may be because some batters prefer to bat against a right-handed (or left-handed) pitcher, and therefore it may be sensible to switch from one type of pitcher to the other when that batter is about to take his turn.

However, we are going to consider a specific tactical reason for substituting a pitcher. Since the pitcher is one of the nine batters in the line-up and is usually comparatively poor at batting, especially at professional level, a manager often faces the situation of deciding whether to bring in a "pinch-hitter" when the pitcher is otherwise about to come up to bat. This is frequently done when a side is behind late in a game and there is a good scoring chance, e.g. the bases are loaded. Of course, the team will subsequently need a pitcher when the other side is batting, and so another substitution must be made to bring in another pitcher then – not the same one as previously, since a player once substituted may not return to the side.

In this paper, we firstly make a brief explanation how to calculate the expected number of runs scored (ENRS) by a team using a Markov chain model of a baseball game. Using the Markov chain model we incorporate the effect of the defensive ability of pitchers by means of Earned Run Average (ERA), which is one of the most common indexes measuring defensive ability. In order to quantify the defensive ability of a pitcher, we introduce a parameter which calibrates the batting probabilities of players depending on ERA. After a brief explanation how to extend the method for ENRS to calculate the probability of winning, we develop a dynamic programming (DP) formulation including the effect of substitution not only of field players but also of pitchers. The problem to be solved here is, taking into account not only the batting ability of players (including pitchers), but also the defensive ability of pitchers, how the manager should optimize the use of players to maximize the probability of winning a game.

As an example, we demonstrate the method developed in this paper using a game based on the line-ups of the Colorado Rockies and the San Francisco Giants in the National League of Major League Baseball. We show how this approach may help to determine the choice of substitute and the timing of the substitution, and how much the probability of winning increases if the optimal strategy is followed.

2 The Formulation for the Expected Number of Runs Scored

In this section we make a brief explanation how to calculate the ENRS by a team using the Markov chain model of a baseball game. We will see the relationship between the ERA and the ENRS in the next section. First we define the 25 states in a half-inning corresponding to the combination of the number of outs (3 possibilities) and the occupation of the bases (8 possibilities), and the end of the half-inning ($3 \times 8 + 1$). We illustrate the 25 states in Table 1.

The definition of the 25 states is used by other researchers (e.g. Bukiet et al. 1997, Trueman 1977) for the analysis of a baseball game or other purposes. However, we further extend the number of the states to develop a formulation for obtaining the ENRS. Firstly the number of the states are extended from

Table 1. The 25-states in a half-inning (No.25 is 'three outs' state which is the end of a half-inning).

		Base runner condition							
		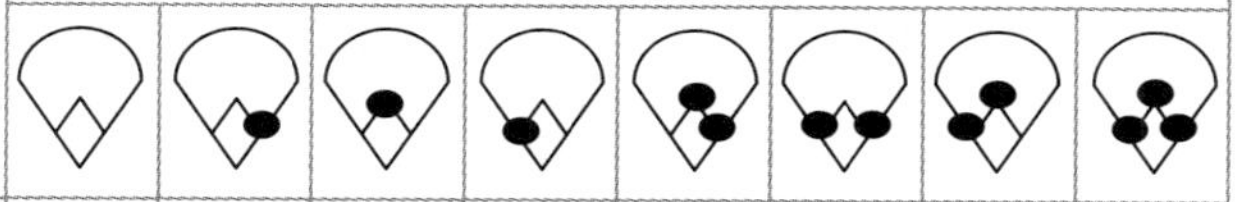							
Out count	0	1	2	3	4	5	6	7	8
	1	9	10	11	12	13	14	15	16
	2	17	18	19	20	21	22	23	24

●Runner

24, excluding the state for the end of a half-inning, to 216 by introducing the nine half-innings (9 possibilities).

Thus we now define the 216 × 216 transition matrix $\mathbf{P}_n$ for the performance of player x_n (n =1...9). That is, if player x_n comes up to bat in the current state, this state will transit to the next state following the transition matrix $\mathbf{P}_n$.

$$
\mathbf{P}_{\mathrm{n}} = \begin{array}{c} \\ 1 \\ 2 \\ \vdots \\ 8 \\ 9 \end{array}
\begin{array}{c}
\begin{array}{ccccc} 1 & 2 & & \cdots & 9 \end{array} \\
\begin{pmatrix}
\mathbf{Q}_{\mathrm{n}} & \mathbf{Q}_{0\mathrm{n}} & & & \\
 & \mathbf{Q}_{\mathrm{n}} & \mathbf{Q}_{0\mathrm{n}} & & \\
 & & \ddots & \ddots & \\
 & & & \mathbf{Q}_{\mathrm{n}} & \mathbf{Q}_{0\mathrm{n}} \\
 & & & & \mathbf{Q}_{\mathrm{n}}
\end{pmatrix}
\end{array}
\qquad (1)
$$

where the entities in $\mathbf{Q}_n$ and $\mathbf{Q}_{0n}$ consist of the probabilities P_{Sn}, P_{Dn}, P_{Tn}, P_{Hn}, P_{Bn} and $P_{out\,n}$, which are the probabilities of player x_n (n =1...9) getting a single, double, triple, home run, base on balls (walk) and out, respectively. In expression (1), the numbers outside the matrix represent the innings. The $\mathbf{Q}_n$ and $\mathbf{Q}_{0n}$ blocks are 24×24 matrices, which represent any transitions inside a half-inning and any transitions to the next half-inning, respectively.

We note that the transition matrix $\mathbf{P}_n$ can represent any transition in the nine half-innings by a performance of player x_n in a single plate appearance by setting up the entries in $\mathbf{Q}_n$ and $\mathbf{Q}_{0n}$. However, the D'Esopo and Lefkowitz runner advancement model (D'Esopo and Lefkowitz 1977) described in Table 2 is used for simplicity in this paper, though we can create any models including any possible transitions in a game by setting up these entries.

Following this runner advancement model, the 24 × 24 matrices $\mathbf{Q}_n$ and $\mathbf{Q}_{0n}$ are expressed as follows:

Table 2. D'Esopo and Lefkowitz runner advancement model (D'Esopo and Lefkowitz 1977).

Play	Probability	Rule of runner advancement
Single	P_{Sn}	Batter to first base. Runner on first reaches second. All other runners score.
Double	P_{Dn}	Batter to second base. Runner on first to third base. All other runners score.
Triple	P_{Tn}	Batter to third base. All base runners score.
Home Run	P_{Hn}	Batter scores. All base runners score.
Walk	P_{Bn}	Batter to first base. All base runners advance one base only if forced to do so.
Out	$P_{out\,n}$	Base runners do not advance.

$$\mathbf{Q}_n = \begin{pmatrix} \mathbf{A}_\mathrm{n} & \mathbf{B}_\mathrm{n} & \mathbf{0} \\ \mathbf{0} & \mathbf{A}_\mathrm{n} & \mathbf{B}_\mathrm{n} \\ \mathbf{0} & \mathbf{0} & \mathbf{A}_\mathrm{n} \end{pmatrix} \quad \mathbf{Q}_{0\mathrm{n}} = \begin{pmatrix} \mathbf{0} & \mathbf{0} & \mathbf{0} \\ \mathbf{0} & \mathbf{0} & \mathbf{0} \\ \mathbf{F}_\mathrm{n} & \mathbf{0} & \mathbf{0} \end{pmatrix}, \tag{2}$$

where $\mathbf{A}_n$, $\mathbf{B}_n$ and $\mathbf{F}_n$ are 8 × 8 matrices represented by the following expressions.

$$\mathbf{A}_\mathrm{n} = \begin{pmatrix} P_{Hn} & P_{Sn}+P_{Bn} & P_{Dn} & P_{Tn} & 0 & 0 & 0 & 0 \\ P_{Hn} & 0 & 0 & P_{Tn} & P_{Sn}+P_{Bn} & 0 & P_D & 0 \\ P_{Hn} & P_{Sn} & P_{Dn} & P_{Tn} & P_{Bn} & 0 & 0 & 0 \\ P_{Hn} & P_{Sn} & P_{Dn} & P_{Tn} & 0 & P_{Bn} & 0 & 0 \\ P_{Hn} & 0 & 0 & P_{Tn} & P_{Sn} & 0 & P_{Dn} & P_{Bn} \\ P_{Hn} & 0 & 0 & P_{Tn} & P_{Sn} & 0 & P_{Dn} & P_{Bn} \\ P_{Hn} & P_{Sn} & P_{Dn} & P_{Tn} & 0 & 0 & 0 & P_{Bn} \\ P_{Hn} & 0 & 0 & P_{Tn} & P_{Sn} & 0 & P_{Dn} & P_{Bn} \end{pmatrix}, \tag{3}$$

$$\mathbf{B}_\mathrm{n} = \begin{pmatrix} \mathbf{P}_{out\,n} & & & \\ & \cdot & & \\ & & \cdot & \\ & & & \mathbf{P}_{out\,n} \end{pmatrix} \tag{4}$$

$$\mathbf{F}_\mathrm{n} = \begin{pmatrix} \mathbf{P}_{out\,n} & 0 & \cdots & 0 \\ \cdot & \cdot & & \cdot \\ \cdot & \cdot & & \cdot \\ \cdot & \cdot & & \cdot \\ \mathbf{P}_{out\,n} & 0 & \cdots & 0 \end{pmatrix} \tag{5}$$

Here we note that off-diagonal elements of the $\mathbf{B}_n$ blocks are zeroes because runners can not advance on an out in this model.

We now describe how to obtain the ENRS in the remainder of the nine-inning game using the above transition matrix. Let v_{ni} be the ENRS in the remainder of the game starting from state i when player x_n (n =1...9) is coming up. A batting of player x_n (n =1...9) in state i results directly in an ENRS of r_{ni} and, with probability P_{nij}, a transition into state j. Then the next player x_{n+1}(n + 1 = 1 if n = 9) will come up to bat. Thus the ENRS in the remainder of the game is given by the expression

$$v_{ni} = \sum_j P_{nij} \cdot v_{n+1j} + r_{ni}. \tag{6}$$

Let $\mathbf{V}_n$ be the 216 × 1 vector where each element gives v_{ni} (i = 1...216). Then, we get the following equation:

$$\mathbf{V}_n = \mathbf{P}_n \mathbf{V}_{n+1} + \mathbf{R}_n \tag{7}$$

where $\mathbf{R}_n$ is a 216×1 vector, each element giving the expected number of runs directly scored by the batting of player x_n, represented in the following expression.

$$\mathbf{R}_n =^t (\overbrace{\mathbf{r}_n\mathbf{r}_n\ \mathbf{r}_n}^{\mathbf{1}}\overbrace{\mathbf{r}_n\mathbf{r}_n\ \mathbf{r}_n}^{\mathbf{2}}\overbrace{\mathbf{r}_n \ldots \mathbf{r}_n}^{\ldots}\overbrace{\mathbf{r}_n\mathbf{r}_n\ \mathbf{r}_n}^{\mathbf{9}}) \tag{8}$$

Here, the number above the elements of $\mathbf{R}_n$ indicate the innings. In this expression, $\mathbf{r}_n$is an 8 × 1 vector, each element corresponding to one of the eight base-runner conditions.

$$\mathbf{r}_n = \begin{pmatrix} P_{Hn} \\ 2P_{Hn} + P_{Tn} \\ 2P_{Hn} + P_{Tn} + P_{Dn} + P_{Sn} \\ 2P_{Hn} + P_{Tn} + P_{Dn} + P_{Sn} \\ 3P_{Hn} + 2P_{Tn} + P_{Dn} + P_{Sn} \\ 3P_{Hn} + 2P_{Tn} + P_{Dn} + P_{Sn} \\ 3P_{Hn} + 2P_{Tn} + 2P_{Dn} + 2P_{Sn} \\ 4P_{Hn} + 3P_{Tn} + 2P_{Dn} + 2P_{Sn} + P_{Bn} \end{pmatrix} \tag{9}$$

The first row in vector $\mathbf{r}_n$ in expression (9), for example, represents the expected number of runs directly scored in the state which is with no outs and no runners. In this state, the only chance to score is to hit a home run. Thus, the expected number of runs directly scored by the batting is P_{Hn} (= 1 run $\times P_{Hn}$) as shown in the first row in vector $\boldsymbol{r}_n$.

Here, looking at the property of a baseball game such that the nine non-identical players bat in order in cycles, we can arrange the above expression following the order of n and get

$$\begin{pmatrix} \mathbf{V}_1 \\ \mathbf{V}_2 \\ \mathbf{V}_3 \\ \vdots \\ \mathbf{V}_8 \\ \mathbf{V}_9 \end{pmatrix} = \begin{pmatrix} \mathbf{0} & \mathbf{P}_1 & & & & \\ & \mathbf{0} & \mathbf{P}_2 & & & \\ & & \mathbf{0} & \mathbf{P}_3 & & \\ & & & \ddots & \ddots & \\ & & & & \mathbf{0} & \mathbf{P}_8 \\ \mathbf{P}_9 & & & & & \mathbf{0} \end{pmatrix} \begin{pmatrix} \mathbf{V}_1 \\ \mathbf{V}_2 \\ \mathbf{V}_3 \\ \vdots \\ \mathbf{V}_8 \\ \mathbf{V}_9 \end{pmatrix} + \begin{pmatrix} \mathbf{R}_1 \\ \mathbf{R}_2 \\ \mathbf{R}_3 \\ \vdots \\ \mathbf{R}_8 \\ \mathbf{R}_9 \end{pmatrix} . \qquad (10)$$

Note that expression (10) is equivalent to a set of 1,944 simultaneous equations with 1,944 unknown variables v_{ni} in $\mathbf{V}_n$(n = 1...9, i = 1...216). We can solve it given that the probabilities of $P_{Sn}, P_{Dn}, P_{Tn}, P_{Hn}, P_{Bn}$ and $P_{out\,n}$ for each player x_n (n = 1...9) are known. These probabilities can be estimated using the batting statistics of individual players. Thus we can obtain the ENRS by solving the simultaneous equations using numerical analysis or other means. Here, the ENRS in the whole game corresponds to v_{11}, the ENRS when the first batter comes up to bat in the first inning.

3 An Evaluation of Defensive Ability by ERA

In our previous paper (Hirotsu and Wright 2003), we developed the formulation considering just the offensive aspect of teams. That is, we assumed that each batter has his own batting probabilities to result in a single, double, triple, home run, walk or out, and that these probabilities are not affected by the defensive ability of the opposing pitcher. However, we now incorporate the defensive ability of the opposing pitcher into the determination of these probabilities.

If we had sufficient data to estimate the batting probabilities of individual players against individual pitchers, we could simply use their record of "batter vs. pitcher" to set up these probabilities in the formulation. However, as a player rarely faces an opposing pitcher at the plate more than ten times in a season, it is not practical to set up robust batting probabilities in this way. Thus, we introduce a parameter w to calibrate the batting probabilities of players against any opposing pitcher reflecting their defensive ability as follows.

Suppose that a typical average pitcher is playing as the opposing pitcher in a game. Then, it is reasonable to assume that the performance of each batter in a single plate appearance will be the same as can be obtained from his overall batting statistics, obtained against a wide variety of pitchers. However, if the opposing pitcher has a better than average pitching ability, it is reasonable to assume that each batter's probability of a single, double, triple, home run or walk lower and that the probability of being *out* would be higher. On the contrary, if the opposing pitcher had a worse than average pitching ability, the batter's probability of a single, double, triple, home run or walk would be higher and the probability of being *out* would be lower.

Following the above inference, we define the defensive ability of a pitcher by the parameter w. Each batter's probabilities are obtained as follows:

$$\begin{aligned}
P'_{Sn} &= wP_{Sn}, \\
P'_{Dn} &= wP_{Dn}, \\
P'_{Tn} &= wP_{Tn}, \\
P'_{Hn} &= wP_{Hn}, \\
P'_{Bn} &= wP_{Bn}, \\
P'_{out\,n} &= 1\text{-}(P'_{Sn} + P'_{Dn} + P'_{Tn} + P'_{Hn} + P'_{Bn}),
\end{aligned} \tag{11}$$

where $P_{Sn}, P_{Dn}, P_{Tn}, P_{Hn}, P_{Bn}$ and $P_{out\,n}$ are now interpreted as the overall probabilities (against all pitchers) of player x_n(n=1...9) getting a single, double, triple, home run, walk and out respectively, and $P'_{Sn}, P'_{Dn}, P'_{Tn}, P'_{Hn}, P'_{Bn}$ and $P'_{out\,n}$ are the probabilities of player x_n getting a single, double, triple, home run, walk and out respectively against this specific pitcher.

We assume that the average pitcher's w is 1. Thus the w-value of a better pitcher than average is less than 1 and that of a worse than average is greater than 1.

There may be various ways of evaluating the w-value for each pitcher. Here, we evaluate w based on ERA, which is derived from the following formula.

$$\text{ERA} = [\text{Earned Runs}] \times 9/[\text{Innings Pitched}] \tag{12}$$

where Earned Runs (ER) is the number of runs scored against the pitcher and Innings Pitched (IP) are the number of innings the pitcher played. ERA can be roughly interpreted as the expected number of runs allowed if the pitcher were to play nine innings.

In general, a pitcher who has a higher ERA is expected to allow more runs than a pitcher who has a lower ERA. The overall ERA for the 2000 National League was 4.63. Table 3 shows the top 5 pitchers in terms of ERA in that season (STATS Inc 2000).

Table 3. Top 5 players in ERA in the 2000 National League (Minimum 162 IP).

Pitcher	Team	Innings Pitched	Earned Runs	ERA
K. Brown	Los Angeles Dodgers	230.0	66	2.58
R. Johnson	Arizona Diamondbacks	248.2	73	2.64
J D'Amico	Milwaukee Brewers	162.1	48	2.66
G. Maddux	Atlanta Braves	249.1	83	3.00
M. Hampton	New York Mets	217.2	76	3.14

Thus Brown of the Dodgers was expected to allow roughly 2.58 runs if he pitched for all nine innings of a typical game.

We can calculate the ENRS by nine non-identical players following the method developed in the previous section. Then we can adjust this using the w-value as follows.

Suppose that the Rockies play against the typical average pitcher in the National League. In this case, it seems reasonable to infer that the Rockies would score around the ENRS calculated using the overall batting probabilities of regular players of the Rockies shown in Table 4 (a), because these batting probabilities are derived by averaging out the batting results throughout the season of the National League. Looking at a specific pitcher, pitching ability measured in ERA would be also averaged out throughout the season in the League. Thus, it can be assumed that we can obtain reasonable batting probabilities against a specific pitcher by calibrating them depending on his ERA. To make clear this calibration, we will explain how to calibrate the batting probabilities using the w-value in the remainder of this section.

Figure 1 shows the relationships between the ENRS and w-value. Here, we calculate the ENRS by the team using its players' batting probabilities which are modified by various w-values. For example, if the Rockies face the pitcher whose w–value is w_1, we calculate the ENRS by the Rockies using the batting probabilities which are modified by w_1 following expressions (11). In the figure, w= 1.0 corresponds to the case that a typical average opposing pitcher plays in the game – in this case the Rockies are expected to score 5.53 runs on average. The Giants are expected to score 5.93 runs on average in the same situation. As the overall ERA of the 2000 National League is 4.63, this average pitcher has an ERA of 4.63 and w-value 1.0.

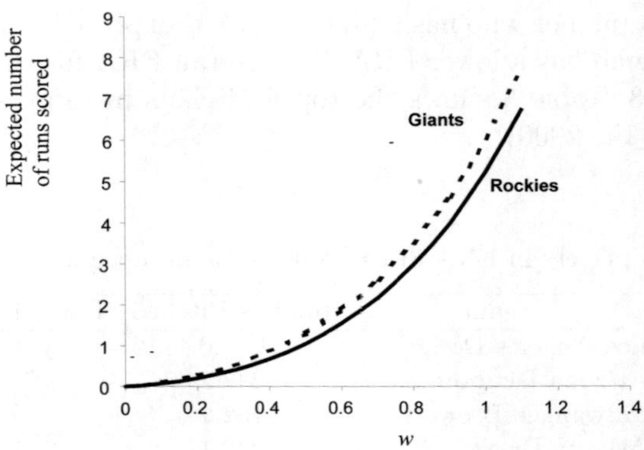

Fig. 1. The relationship between the ENRS and w-value.

Table 4. The line-ups of the Colorado Rockies and the San Francisco Giants.

(a) Colorado Rockies

Order	Players	Single	Double	Triple	Home Run	Walk	Out	Positions
1	Goodwin	0.1771	0.0218	0.0218	0.0136	0.1362	0.6294	CF
2	Lansing	0.1591	0.0354	0.0152	0.0278	0.0783	0.6843	2B
3	Walker	0.1667	0.0583	0.0194	0.0250	0.1278	0.6028	LF (RF)
4	Cirillo	0.1940	0.0797	0.0030	0.0165	0.1008	0.6060	3B
5	Helton	0.1654	0.0864	0.0029	0.0615	0.1508	0.5329	1B
6	Hammond	0.2129	0.0482	0.0040	0.0402	0.0884	0.6064	RF (LF)
7	Mayne	0.1937	0.0550	0.0000	0.0157	0.1230	0.6126	C
8	Perez	0.1865	0.0573	0.0162	0.0147	0.0441	0.6814	SS
9	Astacio	0.1124	0.0169	0.0000	0.0112	0.0253	0.8343	P
Sub	Shumpert	0.1409	0.0378	0.0241	0.0309	0.0962	0.6701	LF (2B,3B)
Sub	Jimenez	0.1124	0.0169	0.0000	0.0112	0.0253	0.8343	P
Sub	Myers	0.1124	0.0169	0.0000	0.0112	0.0253	0.8343	P

Remark: The positions shown inside ( ) represent the alternative positions.

(b) San Francisco Giants

Order	Players	Single	Double	Triple	Home Run	Walk	Out	Remarks
1	Benard	0.1637	0.0433	0.0096	0.0193	0.1011	0.6629	
2	Mueller	0.1748	0.0474	0.0065	0.0163	0.0850	0.6699	
3	Bonds	0.1106	0.0469	0.0067	0.0821	0.1960	0.5578	
4	Kent	0.1699	0.0606	0.0103	0.0487	0.1329	0.5775	
5	Snow	0.1628	0.0548	0.0033	0.0316	0.1096	0.6379	
6	Burks	0.1893	0.0468	0.0111	0.0535	0.1247	0.5746	
7	Aurilia	0.1634	0.0426	0.0036	0.0355	0.0959	0.6590	
8	Estalella	0.0871	0.0618	0.0084	0.0393	0.1601	0.6433	
9	Hernandez	0.1421	0.0368	0.0000	0.0105	0.0342	0.7763	P

We then make the reasonable assumption that the ratio of the ENRS against a particular pitcher and the ENRS against the typical average pitcher is equal to the ratio of the ERA of the particular pitcher and the overall ERA (i.e. the average ERA in the League).

$$\frac{ENRS(w(\text{particular}))}{ENRS(w=1)} = \frac{ERA(\text{particular})}{ERA(\text{overall}),} \tag{13}$$

where $ENRS(w(\text{particular}))$ is the ENRS against a particular pitcher and $ENRS(w = 1)$ is the ENRS against an average pitcher.

Using this assumption, we can obtain the w-value of a particular pitcher when he faces a particular team. For example, suppose that Hernandez, whose ERA is 3.75, pitches against the Rockies, whose ENRS against an average pitcher is 5.53. Now the overall ERA (i.e. the average pitcher's ERA) of the

2000 National League is 4.63. Under assumption (13), when Hernandez plays against the Rockies, the Rockies will be expected to score 5.53 × 3.75 / 4.63 = 4.48 runs, which corresponds to w= 0.922 in Figure 1. Thus we use 0.922 as Hernandez's w-value when playing against the Rockies to calibrate the batting probabilities of the Rockies' players. Similarly, we have obtained the w-value of the other pitchers against the Giants as shown in Table 5.

Table 5. ERA and w-value of pitchers.

Team	Name	ERA	w (Against)	Type
San Francisco Giants	Hernandez	3.75	0.922 (Rockies)	Starting pitcher
Colorado Rockies	Astacio	5.27	1.051 (Giants)	Starting pitcher
	Jimenez	3.18	0.859 (Giants)	Relief pitcher
	Myers	1.99	0.699 (Giants)	Relief pitcher

Here we have to note that ERA is not necessarily the best possible measurement of the defensive ability of pitchers. For example, relief pitchers may have an advantage in terms of ERA, because they often start pitching from one or two outs, and so it is easier to finish the inning without conceding any runs. Thus other measurements have been used such as *Component ERA* (James et al. 1998). We could easily use a different measure in our formulation, but here we are using ERA because it is the most commonly used measure.

4 The Formulation for the Probability of Winning

In this section, we make an explanation how to extend the method for the ENRS in order to calculate the probability of winning the game. Here, we brief the explanation based on (Hirotsu and Wright 2003) to show the essence of the formulation. First, for this extension, we take into account the opposing team in the formulation since winning a game depends on the performance of the opposing team, that is, a team wins a game by scoring more runs than the opposing team. Thus, we now begin with defining 432 states which are extended from the 216 states of nine half-innings by separating each inning into the top and the bottom (2 possibilities). That is, there are 432 (=216×2) different possible states which may occur in the course of a nine-inning game.

We define the transition matrix $\mathbf{P}_n^H$ for the batting of player x_n (n =1. . . 9) of the home team. Here, if player x_n comes up to bat in the current state in a bottom of an inning, this state will transit to the next state following the transition matrix $\mathbf{P}_n^H$ as shown in expression (14).

In this expression, the numbers outside the matrix, $\mathbf{Q}_n$ and $\mathbf{Q}_{0n}$ are the same as those in expression (1), though we add T representing the top of the inning and B representing the bottom of the inning in expression (14) in order to distinguish the offensive aspect of both teams. That is, the top half-inning and the bottom half-inning are for a visiting team and a home team offensive half-inning, respectively. Since a home team starts batting from the bottom of the 1st inning, we put zeroes in the first 24 rows and non-zero entries from the second 24 rows. We represent this as the $\mathbf{Q}_n$ block in the second 24 columns and as the $\mathbf{Q}_{0n}$ block in the third 24 columns.

$$
\mathbf{P}^{H}_{\mathrm{n}} = \begin{array}{cc} & \begin{array}{cccccccc} \multicolumn{2}{c}{1} & \multicolumn{2}{c}{2} & & \cdots & \multicolumn{2}{c}{9} \\ \mathrm{T} & \mathrm{B} & \mathrm{T} & \mathrm{B} & & \cdots & \mathrm{T} & \mathrm{B} \end{array} \\ \begin{array}{cc} 1 & \mathrm{T} \\ & \mathrm{B} \\ 2 & \mathrm{T} \\ & \mathrm{B} \\ \vdots & \vdots \\ 8 & \mathrm{T} \\ & \mathrm{B} \\ 9 & \mathrm{T} \\ & \mathrm{B} \end{array} & \left(\begin{array}{cccccccc} \mathbf{0} & \mathbf{0} & & & & & & \\ & \mathbf{Q}_{\mathrm{n}} & \mathbf{Q}_{0\mathrm{n}} & & & & & \\ & & \mathbf{0} & \mathbf{0} & & & & \\ & & & \mathbf{Q}_{\mathrm{n}} & \mathbf{Q}_{0\mathrm{n}} & & & \\ & & & & \ddots & & & \\ & & & & & \ddots & & \\ & & & & & & \mathbf{Q}_{\mathrm{n}} & \mathbf{Q}_{0\mathrm{n}} \\ & & & & & & \mathbf{0} & \mathbf{0} \\ & & & & & & & \mathbf{Q}_{\mathrm{n}} \end{array}\right) \end{array} \tag{14}
$$

The transition matrix $\mathbf{P}^{V}_{m}$ for the batting of a visiting team player is similarly defined as follows:

$$
\mathbf{P}^{V}_{m} = \begin{array}{cc} & \begin{array}{cccccccc} \multicolumn{2}{c}{1} & \multicolumn{2}{c}{2} & & \cdots & \multicolumn{2}{c}{9} \\ \mathrm{T} & \mathrm{B} & \mathrm{T} & \mathrm{B} & & \cdots & \mathrm{T} & \mathrm{B} \end{array} \\ \begin{array}{cc} 1 & \mathrm{T} \\ & \mathrm{B} \\ 2 & \mathrm{T} \\ & \mathrm{B} \\ \vdots & \vdots \\ 8 & \mathrm{T} \\ & \mathrm{B} \\ 9 & \mathrm{T} \\ & \mathrm{B} \end{array} & \left(\begin{array}{cccccccc} \mathbf{Q}_{m} & \mathbf{Q}_{0m} & & & & & & \\ & \mathbf{0} & \mathbf{0} & & & & & \\ & & \mathbf{Q}_{m} & \mathbf{Q}_{0m} & & & & \\ & & & \mathbf{0} & \mathbf{0} & & & \\ & & & & \ddots & & & \\ & & & & & \ddots & & \\ & & & & & \mathbf{0} & \mathbf{0} & \\ & & & & & & \mathbf{Q}_{m} & \mathbf{Q}_{0m} \\ & & & & & & & \mathbf{0} \end{array}\right) \end{array} \tag{15}
$$

Since a visiting team begins to bat at the top of the 1st inning, non-zero entries appear in the $\mathbf{Q}_m$ block at the top-left corner of the above matrix. When three members of the visiting team are out, the state transits from the top of the 1st inning to the beginning of the bottom of the inning. This transition occurs in the probabilities expressed by the entries of the $\mathbf{Q}_{0m}$ block in the first 24 rows and the second 24 columns, and the home team begins to bat. Here, we also set up the entries in $\mathbf{Q}_m$ and $\mathbf{Q}_{0m}$ following the D'Esopo and Lefkowitz model. We note that $\mathbf{P}^{H}_{n}$ and $\mathbf{P}^{V}_{m}$ can be decomposed into five matrices which correspond to the portions leading to zero runs scored, one run, two runs, three runs and four runs, following the results of the batting of player x_n and y_m as follows:

$$\mathbf{P}_n^H = \mathbf{P0}_n^H + \mathbf{P1}_n^H + \mathbf{P2}_n^H + \mathbf{P3}_n^H + \mathbf{P4}_n^H$$
$$\mathbf{P}_m^V = \mathbf{P0}_m^V + \mathbf{P1}_m^V + \mathbf{P2}_m^V + \mathbf{P3}_m^V + \mathbf{P4}_m^V$$

Let $\mathbf{W}_{nm}(r)$= $\mathbf{W}_{nm}(r)(x_1,x_2,\ldots,x_9,\ y_1,y_2,\ldots,y_9)$ be the 432 × 1 vector representing the probabilities of the home team winning in the remainder of the game from a current state where the home team leads by r runs, the nth batter x_n is coming up in the bottom of the inning and mth batter y_m is coming up in the top of the inning. Since the batting of a home and a visiting team are mutually exclusive, we get the equation for $\mathbf{W}_{nm}(r)$ as follows:

$$\begin{aligned}\mathbf{W}_{nm}(r) = {} & \mathbf{P0}_m^V \mathbf{W}_{nm+1}(r) + \mathbf{P1}_m^V \mathbf{W}_{nm+1}(r-1) + \mathbf{P2}_m^V \mathbf{W}_{nm+1}(r-2) \\ & + \mathbf{P3}_m^V \mathbf{W}_{nm+1}(r-3) + \mathbf{P4}_m^V \mathbf{W}_{nm+1}(r-4) \\ & + \mathbf{P0}_n^H \mathbf{W}_{n+1\,m}(r) + \mathbf{P1}_n^H \mathbf{W}_{n+1\,m}(r+1) + \mathbf{P2}_n^H \mathbf{W}_{n+1\,m}(r+2) \\ & + \mathbf{P3}_n^H \mathbf{W}_{n+1\,m}(r+3) + \mathbf{P4}_n^H \mathbf{W}_{n+1\,m}(r+4) \\ & + I(r)\cdot \mathbf{P}_{out\,n}^H \end{aligned} \tag{16}$$

where $\mathbf{P}_{outn}^H$ is a 432 × 1 vector leading to the end of the game from the batting of player x_n in the bottom of the 9th inning. This is represented as follows.

$$\mathbf{P}_{out\,n}^H = {}^t(0 \;\; 0 \;\; \cdots \;\; 0 \;\; P_{out\,n} \;\; \cdots \;\; P_{out\,n}) \tag{17}$$

Here, only the last 8 rows corresponding to the states in the bottom of the 9th inning with two outs have non-zero entries of P_{outn}, which is the probability of player x_n being out. $I(r)$ in expression (16) is the boundary condition at the end of the game. The value of $I(r)$ is 1 if $r > 0$ (i.e. home team wins), the probability of the home team winning in an extra inning if $r = 0$, and 0 if $r < 0$ (i.e. home team loses).

We simplify expression (16) by arranging it into a matrix representation. We firstly put the 432 × 1 vector $\mathbf{W}_{nm}(r)$ into a 17,712 (= 432 × 41) × 1 vector $\mathbf{W}_{nm}$ following the order of the number of runs by which the home team leads (41 possibilities, assuming that the number of runs by which either team may lead will never exceed 20) as follows:

$$\begin{aligned}\mathbf{W}_{nm} &= (\mathbf{W}_{nm}(r)) \\ &= {}^t(\mathbf{W}_{nm}(20)\mathbf{W}_{nm}(19)\ldots \\ &\qquad \mathbf{W}_{nm}(2)\mathbf{W}_{nm}(1)\mathbf{W}_{nm}(0)\mathbf{W}_{nm}(-1)\ldots \\ &\qquad\qquad \mathbf{W}_{nm}(-19)\mathbf{W}_{nm}(-20))\end{aligned}$$

We also put the 432 × 1 vector $\mathbf{P}_{out\,n}^H$ into a 17,712 (= 432 × 41) × 1 vector $\mathbf{P}_{out\,n}$ as follows:

$$\mathbf{P}_{out\,n} = {}^t(\mathbf{P}^H_{out\,n}\mathbf{P}^H_{out\,n}\cdots\cdots\mathbf{P}^H_{out\,n},\mathbf{0}\,\mathbf{0}\ldots\mathbf{0})$$

Note that $\mathbf{P}_{out\,n}$ consists of a series of 20 $\mathbf{P}^H_{outn}$ and a series of 21 $\mathbf{0}$ in this expression.

We then define the 1,434,672 (=17,712 × 9 × 9) × 1 vector **W**, representing the probabilities of the home team winning in the remainder of the game from a current state where the nth batter of the home team ($n = 1\ldots9$) is coming up in the bottom of the inning and mth batter of visiting team ($m = 1\ldots9$) is coming up in the top of the inning as follows:

$$\begin{aligned}\mathbf{W} &= (\mathbf{W}_{nm})\\ &={}^t(\mathbf{W}_{11}\mathbf{W}_{12}\ldots\mathbf{W}_{19}\mathbf{W}_{21}\ldots\mathbf{W}_{29}\mathbf{W}_{31}\ldots\ldots\mathbf{W}_{89}\mathbf{W}_{91}\ldots\mathbf{W}_{99})\end{aligned} \tag{18}$$

We also put the 17,712 × 1 vector $\mathbf{P}_{outn}$ into a 1,434,672 × 1 vector $\mathbf{P}_{out}$ as follows:

$$\begin{aligned}\mathbf{P}_{out} &= (\mathbf{P}_{out\,n})\\ &= {}^t(\mathbf{P}_{out1}\mathbf{P}_{out1}\ldots\mathbf{P}_{out1}\mathbf{P}_{out2}\ldots\mathbf{P}_{out2}\mathbf{P}_{out3}\cdots\\ &\qquad\ldots\mathbf{P}_{out8}\mathbf{P}_{out9}\ldots\mathbf{P}_{out9})\end{aligned} \tag{19}$$

We finally get a simple equation expressing the relation between the probability of winning as follows:

$$\mathbf{W} = \mathbf{P}^{(ns)}\mathbf{W} + \mathbf{P}_{out}, \tag{20}$$

where $\mathbf{P}^{(ns)}$ is a 1,434,672 × 1,434,672 matrix representing the transition probabilities by a batting of players. $\mathbf{P}^{(ns)}$ has the structure in expression (22). In (22), $\mathbf{P}_{Hn}$ and $\mathbf{P}_{Vm}$ are 17,712 × 17,712 matrices which have the transition probabilities for the batting of the nth batter of the home team (n = 1...9) and for the batting of the mth batter of the visiting team (m=1...9), respectively. We can see that $\mathbf{P}_{Hn}$ and $\mathbf{P}_{Vm}$ satisfy the following expression.

$$\mathbf{W}_{nm} = \mathbf{P}_{Vm}\mathbf{W}_{nm+1} + \mathbf{P}_{Hn}\mathbf{W}_{n+1\,m} + \mathbf{P}_{out\,n} \tag{21}$$

Technically, using the entries of $\mathbf{P}_{Hn}$, we have introduced the case of entering extra innings by considering an extra inning as a repetition of the 9th inning.

We note that we have known the entries of $\mathbf{P}^{(ns)}$ and $\mathbf{P}_{out}$, which are calibrated by means of the w-value of the opposing pitcher. Thus we can obtain the probability of the home team winning by solving equation (20).

As we simultaneously obtain the probability of winning in every state in a game, we are able to extend this method to develop a DP formulation for identifying the optimal substitution strategy, to be shown in the next section.

$$\mathbf{P}^{(ns)} =$$

$$\left[\begin{array}{c|c|c|c|c}
\begin{matrix} \mathbf{0} & \mathbf{P}_{V1} & & \\ & \mathbf{0} & \mathbf{P}_{V2} & \\ & & \ddots & \\ & & \mathbf{0} & \mathbf{P}_{V8} \\ \mathbf{P}_{V9} & & & \mathbf{0} \end{matrix} & \begin{matrix} \mathbf{P}_{H1} & & & \\ & \mathbf{P}_{H1} & & \\ & & \ddots & \\ & & \mathbf{P}_{H1} & \\ & & & \mathbf{P}_{H1} \end{matrix} & & & \\
\hline
& \begin{matrix} \mathbf{0} & \mathbf{P}_{V1} & \\ & \ddots & \ddots \\ & \ddots & \mathbf{P}_{V8} \\ \mathbf{P}_{V9} & & \mathbf{0} \end{matrix} & \begin{matrix} \mathbf{P}_{H2} & & \\ & \ddots & \\ & & \ddots \\ & & \mathbf{P}_{H2} \end{matrix} & & \\
\hline
& & \begin{matrix} \mathbf{0} & \mathbf{P}_{V1} & \\ & \ddots & \ddots \\ & \ddots & \ddots \\ & & \ddots \end{matrix} & \begin{matrix} \mathbf{P}_{H3} & & \\ & \ddots & \\ & & \ddots \\ \ddots & & \mathbf{P}_{H7} \end{matrix} & \\
\hline
& & & \begin{matrix} \ddots & \\ \ddots & \\ \ddots & \\ & \ddots \end{matrix} & \begin{matrix} \mathbf{P}_{H8} & & \\ & \ddots & \\ & & \ddots \\ & & \mathbf{P}_{H8} \end{matrix} \\
\hline
\begin{matrix} \mathbf{P}_{H9} & & \\ & \ddots & \\ & & \mathbf{P}_{H9} \end{matrix} & & & & \begin{matrix} \mathbf{0} & \mathbf{P}_{V1} & \\ & \ddots & \mathbf{P}_{V8} \\ \mathbf{P}_{V9} & & \mathbf{0} \end{matrix}
\end{array}\right] \qquad (22)$$

In practice, the home team has won the game and the game is actually over at the end of the top of the 9th inning if the home team leads at that time. Even though the winner of the game has been decided, the bottom of the 9th inning is included in our model because otherwise our model would be more complex than necessary.

5 A Dynamic Programming Formulation Incorporating the Substitution of Pitchers

We now describe how to apply the above method to develop a DP formulation for identifying the optimal pitcher substitution strategies. We explain the case when there is just one relief pitcher available but this formulation is easily extended to handle the case of more than one relief pitcher.

Let $\mathbf{W}[x_{rp}] = \mathbf{W}(x_1,x_2,\dots,x_9,\ y_1,y_2,\dots,y_9|x_{rp})$ be the 1,434,672 × 1 vector representing the probabilities of the home team winning with a substitute (relief pitcher x_{rp}) available. As a pitcher usually bats as the 9th batter, we assume here that x_9 is the starting pitcher.

Suppose that a manager does not make a substitution in the current state. In this case, the state transits to the next state following transition matrix $\mathbf{P}^{(ns)}$ shown in expression (22). After this transition, the probability of the home team winning is expressed by one of the elements of the same vector $\mathbf{W}[x_{rp}]$, because the home team still has a substitute x_{rp} available.

However, if the manager substitutes x_{rp} for x_9 in the current state, the state transits to the next state following the transition matrix $\mathbf{P}^{(9\to rp)}$, which represents the matrix modified not only by substituting the $\mathbf{P}_{Hrp}$ blocks for x_{rp} instead of the $\mathbf{P}_{H9}$ blocks, but also by substituting the $\mathbf{P}_{Vi}^{(9\to rp)}$ blocks for y_i instead of the $\mathbf{P}_{Vi}$ blocks (i = 1...9) in the matrix $\mathbf{P}_{ns}$, as shown in the following expression.

$$\mathbf{P}^{(9\to rp)} =$$

$$\left[\begin{array}{c|c|c|c|c}
\begin{matrix} \mathbf{0} & \mathbf{P}_{V1}^{(9\to rp)} & & \\ & \mathbf{0} & \mathbf{P}_{V2}^{(9\to rp)} & \\ & & \ddots & \\ & & \mathbf{0} & \mathbf{P}_{V8}^{(9\to rp)} \\ \mathbf{P}_{V9}^{(9\to rp)} & \mathbf{0} & & \end{matrix} & \begin{matrix} \mathbf{P}_{H1} & & & \\ & \mathbf{P}_{H1} & & \\ & & \ddots & \\ & & \mathbf{P}_{H1} & \\ & & & \mathbf{P}_{H1} \end{matrix} & & & \\ \hline
 & \begin{matrix} \mathbf{0} & \mathbf{P}_{V1}^{(9\to rp)} & \\ & \ddots & \\ & \ddots & \mathbf{P}_{V8}^{(9\to rp)} \\ \mathbf{P}_{V9}^{(9\to rp)} & \mathbf{0} & \end{matrix} & \begin{matrix} \mathbf{P}_{H2} & & \\ & \ddots & \\ & & \mathbf{P}_{H2} \end{matrix} & & \\ \hline
 & & \begin{matrix} \mathbf{0} & \mathbf{P}_{V1}^{(9\to rp)} \\ \ddots & \ddots \\ \ddots & \ddots \end{matrix} & \begin{matrix} \mathbf{P}_{H3} & & \\ & \ddots & \\ \ddots & & \mathbf{P}_{H7} \end{matrix} & \\ \hline
 & & & \begin{matrix} \ddots \\ \ddots \\ \ddots \end{matrix} & \begin{matrix} \mathbf{P}_{H8} & & \\ & \ddots & \\ & & \mathbf{P}_{H8} \end{matrix} \\ \hline
\begin{matrix} \mathbf{P}_{Hrp} & & \\ & \ddots & \\ & & \mathbf{P}_{Hrp} \end{matrix} & & & & \begin{matrix} \mathbf{0} & \mathbf{P}_{V1}^{(9\to rp)} & \\ & \ddots & \\ & \ddots & \mathbf{P}_{V8}^{(9\to rp)} \\ \mathbf{P}_{V9}^{(9\to rp)} & \mathbf{0} & \end{matrix}
\end{array}\right] \tag{23}$$

This is because, after the substitution, x_9 cannot play any more in the game and x_{rp} plays instead of x_9. Thus not only the entries in relation to the batting ability of x_{rp} but also the w-value w_{rp} should be substituted for w_9 in any $\mathbf{P}_{Vi}$ block (i = 1...9). Here, w_9 and w_{rp} represent the w-values of x_9 and x_{rp}, respectively.

After this transition, the probability of the home team winning is expressed by one of the elements of a 1,434,672 × 1 vector

$$\mathbf{W}^{(9\to rp)} = \mathbf{W}^{(9\to p)}(x_1, x_2, \ldots, x_{rp}, y_1, y_2, \ldots, y_9).$$

This represents the probability of the home team winning without any substitutes in the remainder of the game after the substitution of x_{rp} for x_9.

We compare the probabilities of the home team winning between the case of non-substitution and the case of the substitution of x_{rp}, and take the maximum of them in each state. The selection leads us to identify the optimal substitution strategy in each state. That is, following the strategy derived as optimum in each state, we maximize the probability of the home team winning with a substitute x_{rp} in the remainder of the game. We can formulate this procedure as follows:

$$\mathbf{W}[x_{\mathrm{rp}}] = \max \begin{cases} \mathbf{P}^{(ns)}\mathbf{W}[x_{\mathrm{rp}}] + \mathbf{P}_{out} & : \text{Non} - \text{substitution} \\ \mathbf{P}^{(9\to rp)}\mathbf{W}^{(9\to rp)} + \mathbf{P}_{\mathrm{out}}^{(9\to rp)} & : \text{Substitution of } x_{rp} \text{ for } x_9 \end{cases} \tag{24}$$

where $\mathbf{P}_{out}^{(9\to rp)}$represents the 1,434,672 $\times$ 1 vector modified by substituting the $\mathbf{P}_{outrp}$ blocks for the $\mathbf{P}_{out9}$ blocks in the vector $\mathbf{P}_{out}$ in expression (19).

So far, we explained the substitution between pitchers. However, a relief pitcher can be substituted not only for a pitcher but also for a field player. Therefore, in general, there are 9 possible substitutions of the relief pitcher. We can write down these situations as follows: (Here, we just show the formulation in the case when there is one relief pitcher available.)

$$\mathbf{W}[x_{\mathrm{rp}}] = \max \begin{cases} \mathbf{P}^{(\mathrm{ns})}\mathbf{W}[x_{\mathrm{rp}}] + \mathbf{P}_{out} & : \text{Non} - \text{substitution} \\ \mathbf{P}^{(1\to \mathrm{rp})}\mathbf{W}^{(1\to \mathrm{rp})} + \mathbf{P}_{out}^{(1\to \mathrm{rp})} & : \text{Substitution of } x_{rp} \text{ for } x_1 \\ \mathbf{P}^{(2\to \mathrm{rp})}\mathbf{W}^{(2\to \mathrm{rp})} + \mathbf{P}_{out}^{(2\to \mathrm{rp})} & : \text{Substitution of } x_{rp} \text{ for } x_2 \\ \vdots & \vdots \\ \mathbf{P}^{(9\to \mathrm{rp})}\mathbf{W}^{(9\to \mathrm{rp})} + \mathbf{P}_{out}^{(9\to \mathrm{rp})} & : \text{Substitution of } x_{rp} \; for \; x_9 \end{cases} \tag{25}$$

where the transition matrix $\mathbf{P}^{(1\to rp)}$,$\mathbf{P}^{(2\to rp)}$,...,$\mathbf{P}^{(8\to rp)}$ can also be expressed by (23) excepting the number of the superscript, which corresponds to the each player. $\mathbf{P}_{out}^{(i\to rp)}$ represent the 1,434,672 $\times$ 1 vector similarly modified by substituting the $\mathbf{P}_{outrp}$ blocks for x_{rp} instead of the $\mathbf{P}_{outi}$ block for x_i in the vector $\mathbf{P}_{out}$(i=1...9).

Here, when the relief pitcher is substituted for one of the field players, there are two pitchers playing in the game at the same time. In this case, we have to put one of the pitchers into a fielding position to avoid breaking the fielding condition. Here, if the fielding position can be filled by using one of the two pitchers as a field player, we move the pitcher to the field position. (If both pitchers can be, we choose the one who has smaller w-value as the pitcher). If the fielding position can not be filled after the substitution, the batting probabilities of the opposing players resulting in a single, double, triple, home run or walk are more than doubled, making the probability of being *out* close to zero, as a penalty against breaking the fielding condition, as we did in (Hirotsu and Wright 2003).

In order to cater for the case of two different types of substitute available, we show the DP formulation for the case in Appendix. In the next section we demonstrate the optimal substitution strategies where one substitute for pinch hitting (or fielding) and two relief pitchers are available. This is partly because the computing time takes about half a CPU day on an HP VISUALIZE C3600 in case of 3 substitutes available in total. The computing time in the case of 4 substitutes will take more than half a CPU month, which is too long not only to occupy our computer, but also to reflect the information from the latest game on the next game in a real baseball league. Although we do not show the result of the other combinations of the types of substitutes in this paper, the example will cover the basic patterns of substitution between a pitcher and a field player and between pitchers.

6 An Example of the Procedure for Identifying the Optimal Substitution Strategy

6.1 Sample data

We now present an example of our procedure to find the optimal pitcher substitution strategy. We have chosen the Colorado Rockies in this example because they are an average team as shown by their 82-80 won-lost record in the National League in the 2000 season, and demonstrate the game of the Rockies against the San Francisco Giants, the winners of the West Division. The players of both teams in this example have been listed in Table 4. We selected three Rockies players, Shumpert (field player), Jimenez (relief pitcher) and Myers (relief pitcher) as substitutes to see the effect of the inclusion of pitchers. We chose them because they played in a high number of games in the season.

Using the statistics taken from (Neft et al. 2001) and (STATS Inc 2000), we calculated the probability of each player in both teams achieving the following results - a single, double, triple, home run, walk or out (see Table 4) and set up the players' transition matrices following the D'Esopo and Lefkowitz runner advancement model (D'Esopo and Lefkowitz 1977). As the number of plate appearances of a pitcher to bat is not large, we aggregated all the batting statistics of pitchers for each team and used the same transition matrix for every pitcher in the same team. This highlights the effect of their own pitching ability (not batting ability) on the optimal substitution strategies, although it is easy to put each pitchers' own batting probabilities into this calculation. We allowed the Rockies players to be allocated to their fielding positions if they played these positions in more than 10 games in the season.

If the Rockies fill all the 9 fielding positions (pitcher (P), catcher (C), first baseman (1B), second baseman (2B), third baseman (3B), shortstop (SS), left fielder (LF), center fielder (CF) and right fielder (RF)), then we used the value listed in Table 4 (b) to calculate the probability of the Rockies winning. However, if the Rockies failed to fill all of these fielding positions, the probability of all the Giants players getting a single, double, triple, home run or walk was more than doubled as a penalty.

We calculated the w-value of each pitcher using his ERA, which has been shown in Table 5. In this example, we apply the w-values of relief pitchers, Jimenez and Myers only after the 7th inning. Before and including the 7th inning we set a large value for w (here $w=2.1$, a large value but not to make $P_{out\,n}$ negative) to enforce this restriction. That is, we assume that they can use a relief pitcher only after the 7th inning for the sake of simplicity. In fact they played after the 7th inning in more than 80% of their 150 appearances during the season.

Comparing the defensive ability of Astacio, Jimenez and Myers in Table 5, we can easily see that Myers is the best among them. Thus we might expect that only one substitution scenario would make any sense, i.e. that Astacio will be substituted by Myers at the beginning of the 8th inning and there will be no more substitutions in the game. However, as readers will see later in this paper, this guess is not correct.

6.2 The optimal substitution strategies

A part of the calculated results for finding the optimal substitution strategies is shown in Table 6. This table shows the results where the Rockies are the home team and the scores are level.

In this table, T and B represent the top and bottom of the innings, respectively. The three digits $d_3d_2d_1$ are 1 or 0 corresponding to whether there is or is not a runner on third base, second base and first base, respectively. The entries such as '9→M' indicate that the optimal strategy is to substitute 'M' (Myers) for '9' (here Astacio).

The inclusion of the substitution of pitchers makes the calculated results quite complicated. Thus, even though we show only the results in the situation where the scores are level, there are seven different sets of optimal strategies depending on the next batter of the Rockies, as shown in the corresponding sub-tables (i.e. (a) –(g)) in Table 6.

Looking through this table, we can state the following four main points - the result is more complicated than we guessed in the previous subsection.

Firstly, the optimal strategies in the top of the inning depend on the Rockies' next batter coming up to bat in the bottom of the inning. For example, in the situation where the 1st batter (Goodwin) is coming up in the bottom of the inning, the Rockies should substitute Myers (M) for Astacio (9) in the top of the inning as shown in Table 6 (a). However, in the situation where the 5th batter (Helton) is coming up in the bottom of the inning, the Rockies should substitute Myers (M) for Lansing (2) and Shumpert (S) for Astacio (9) in the top of the inning in some situations as shown in Table 6 (c). Therefore, the optimal strategy is not just making a substitution of Myers for Astacio.

Secondly, the first substitution chance occurs when the 9th batter (Astacio) comes up in the bottom of the 7th inning as shown in Table 6 (g). In

Table 6. The optimal substitution strategies in the case where the scores are level (2: Lansing, 9: Astacio, S: Shumpert, J: Jimenez, M: Myers).

(a) Case of the 1st, the 3rd or the 4th batter (Goodwin, Walker or Cirillo) coming up

			Base-runner conditions							
Inning	T/B	Out	000	001	010	100	011	101	110	111
8	T	0	9→M	9→M	9→M	9→M	9→M	9→M	9→M	9→M
8	T	1	9→M	9→M	9→M	9→M	9→M	9→M	9→M	9→M
8	T	2	9→M	9→M	9→M	9→M	9→M	9→M	9→M	9→M
8	B	0								
8	B	1								
8	B	2								
9	T	0	9→M	9→M	9→M	9→M	9→M	9→M	9→M	9→M
9	T	1	9→M	9→M	9→M	9→M	9→M	9→M	9→M	9→M
9	T	2	9→M	9→M	9→M	9→M	9→M	9→M	9→M	9→M
9	B	0								
9	B	1								
9	B	2								

(b) Case of the 2nd batter (Lansing) coming up

Inning	T/B	Out	000	001	010	100	011	101	110	111
8	T	0	9→M	9→M	9→M	9→M	9→M	9→M	9→M	9→M
8	T	1	9→M	9→M	9→M	9→M	9→M	9→M	9→M	9→M
8	T	2	9→M	9→M	9→M	9→M	9→M	9→M	9→M	9→M
8	B	0								
8	B	1								2→S
8	B	2		2→S						2→S
9	T	0	9→M	9→M	9→M	9→M	9→M	9→M	9→M	9→M
9	T	1	9→M	9→M	9→M	9→M	9→M	9→M	9→M	9→M
9	T	2	9→M	9→M	9→M	9→M	9→M	9→M	9→M	9→M
9	B	0								
9	B	1								2→S
9	B	2		2→S						2→S

Table 6 (cont.)

(c) Case of the 5th batter (Helton) coming up

Inning	T/B	Out	000	001	010	100	011	101	110	111
8	T	0	2→M 9→S	9→M	9→M	9→M	9→M	9→M	9→M	9→M
8	T	1	2→M 9→S	2→M 9→S	9→M	9→M	9→M	9→M	9→M	9→M
8	T	2	9→J	2→M 9→S	2→M 9→S	2→M 9→S	2→M 9→S	2→M 9→S	2→M 9→S	9→M
8	B	0								
8	B	1								
8	B	2								
9	T	0	2→M 9→S	2→M 9→S	2→M 9→S	2→M 9→S	2→M 9→S	2→M 9→S	2→M 9→S	2→M 9→S
9	T	1	2→M 9→S	2→M 9→S	2→M 9→S	2→M 9→S	2→M 9→S	2→M 9→S	2→M 9→S	2→M 9→S
9	T	2	2→M 9→S	2→M 9→S	2→M 9→S	2→M 9→S	2→M 9→S	2→M 9→S	2→M 9→S	2→M 9→S
9	B	0								
9	B	1								
9	B	2								

(d) Case of the 6th batter (Hammond) coming up

Inning	T/B	Out	000	001	010	100	011	101	110	111
8	T	0	9→M	9→M	9→M	9→M	9→M	9→M	9→M	9→M
8	T	1	2→M 9→S	2→M 9→S	9→M	9→M	9→M	9→M	9→M	9→M
8	T	2	2→J 9→S	2→J 9→S	2→M 9→S	2→M 9→S	2→M 9→S	2→M 9→S	2→M 9→S	9→M
8	B	0								
8	B	1								
8	B	2								
9	T	0	2→M 9→S	2→M 9→S	2→M 9→S	2→M 9→S	2→M 9→S	2→M 9→S	9→M	9→M
9	T	1	2→M 9→S	2→M 9→S	2→M 9→S	2→M 9→S	2→M 9→S	2→M 9→S	2→M 9→S	2→M 9→S
9	T	2	9→J	2→M 9→S	2→M 9→S	2→M 9→S	2→M 9→S	2→M 9→S	2→M 9→S	2→M 9→S
9	B	0								
9	B	1								
9	B	2								

Table 6 (cont.)

(e) Case of the 7th batter (Mayne) coming up

Inning	T/B	Out	000	001	010	100	011	101	110	111
8	T	0	9→M	9→M	9→M	9→M	9→M	9→M	9→M	9→M
8	T	1	9→J	9→M	9→M	9→M	9→M	9→M	9→M	9→M
8	T	2	9→J	9→J	9→M	9→M	9→M	9→M	9→M	9→M
8	B	0								
8	B	1								
8	B	2								
9	T	0	9→M	9→M	2→M 9→S	2→M 9→S	9→M	9→M	9→M	2→M 9→S
9	T	1	9→M	9→M	9→M	9→M	9→M	9→M	9→M	9→M
9	T	2	9→J	9→J	9→M	9→M	9→M	9→M	9→M	9→M
9	B	0								
9	B	1								
9	B	2								

(f) Case of the 8th batter (Perez) coming up

Inning	T/B	Out	000	001	010	100	011	101	110	111
8	T	0	9→M	9→M	9→M	9→M	9→M	9→M	9→M	9→M
8	T	1	9→J	9→M	9→M	9→M	9→M	9→M	9→M	9→M
8	T	2	9→J	9→J	9→M	9→M	9→M	9→M	9→M	9→M
8	B	0								
8	B	1								
8	B	2								
9	T	0	9→M	9→M	9→M	9→M	9→M	9→M	9→M	9→M
9	T	1	9→M	9→M	9→M	9→M	9→M	9→M	9→M	9→M
9	T	2	9→J	9→J	9→M	9→M	9→M	9→M	9→M	9→M
9	B	0								
9	B	1								
9	B	2								

Table 6 (cont.)

(g) Case of the 9th batter (Astacio) coming up

Inning	T/B	Out	000	001	010	100	011	101	110	111
7	B	0	9→S	9→S	9→S	9→S	9→S	9→S	9→S	9→S
7	B	1	9→S	9→S	9→S	9→S	9→S	9→S	9→S	9→S
7	B	2	9→S	9→S	9→S	9→S	9→S	9→S	9→S	9→S
8	T	0	2→M 9→S	2→M 9→S	2→M 9→S	2→M 9→S	2→M 9→S	2→M 9→S	2→M 9→S	2→M 9→S
8	T	1	9→J	2→M 9→S	2→M 9→S	2→M 9→S	2→M 9→S	2→M 9→S	2→M 9→S	2→M 9→S
8	T	2	9→J	9→J	2→M 9→S	2→M 9→S	2→M 9→S	2→M 9→S	2→M 9→S	2→M 9→S
8	B	0	9→S	9→S	9→S	9→S	9→S	9→S	9→S	9→S
8	B	1	9→S	9→S	9→S	9→S	9→S	9→S	9→S	9→S
8	B	2	9→S	9→S	9→S	9→S	9→S	9→S	9→S	9→S
9	T	0	9→M	9→M	9→M	9→M	9→M	9→M	9→M	9→M
9	T	1	9→J	9→J	9→M	9→M	9→M	9→M	9→M	9→M
9	T	2	9→J	9→J	9→M	9→M	9→M	9→M	9→M	9→M
9	B	0	9→S	9→S	9→S	9→S	9→S	9→S	9→S	9→S
9	B	1	9→S	9→S	9→S	9→S	9→S	9→S	9→S	9→S
9	B	2	9→S	9→S	9→S	9→S	9→S	9→S	9→S	9→S

any situation in the bottom of the 7th inning, when Astacio comes up, the manager should substitute Shumpert as a pinch hitter for Astacio. However, if other batters come up, the manager should not make a substitution until the 8th inning.

Thirdly, in any situation in the top of the 8th inning, Astacio should be substituted by a relief pitcher, Jimenez or Myers. This is reasonable because Astacio's ERA (5.27) is greater than Jimenez's (3.18) or Myers's (1.99). However, a little surprisingly not Myers but Jimenez may be substituted for Astacio in some situations, even though Jimenez appears worse than Myers. Which relief pitcher, Jimenez or Myers, should be chosen depends on the situation.

Finally, when the 2nd batter (Lansing) comes up, there are other pinch-hitting chances for Shumpert in the bottom of the 8th and 9th innings as shown in Table 6 (b). This is possible from the fielding point of view because Shumpert can play as a 2B, which is Lansing's fielding position.

As Shumpert should be substituted for another field player (here Lansing) in some situations, it may be reasonable to infer that it would be better to start off with Shumpert instead of Lansing, and use Lansing as a substitute. In fact, if the Rockies start off with Shumpert as the 2nd batter instead of Lansing, the probability of the Rockies winning is 0.4169. This is greater than 0.4125, which is the case when starting off with Lansing. (Of course, in practice, it is possible that Lansing may be a better fielder than Shumpert, which may affect the optimality of this decision, but such considerations are not included in our model, though they could if there were such common indexes as ERA for quantifying the fielding ability).

Thus the remainder of our analysis assumes that Shumpert is in the starting line-up and Lansing is a potential substitute. Under this line-up, here we just show the whole picture of the optimal substitution strategies as a style of a flow chart in Figure 2. We note that in the figure extra innings is considered as a repetition of the 9th inning.

In the figure there are three main scenarios. Firstly, if the 9th batter (Astacio) is coming up to bat in the bottom of the 7th inning, he should be substituted by Lansing for pinch hitting in most of the situations. After this substitution Myers starts playing as a relief pitcher and continues to play until the end of the game (Scenario 1). Otherwise, either Jimenez or Myers is first substituted at the beginning of the top of the 8th inning. Here, very roughly speaking, when the Rockies lead or tie, the manager should substitute Myers, and then Myers starts playing until the next substitution chance (Scenario 2). However, when being behind at the top of the 8th inning, the manager should substitute Jimenez first (Scenario 3).

Here we note that in terms of a pitcher's place in the batting order, there is a tendency that the pitcher should be substituted for the player who is not coming up to bat soon at the situation. For example, in the case where the 3rd batter (Walker) is coming up, Myers should be substituted for the 2nd batter (here Shumpert) who is the latest batter coming up in some situations, if it is possible to fill the fielding condition.

6.3 An example of numerical results of the optimal substitution

As there is not enough space to list all the numerical results, we demonstrate here the situations where adoption of the optimal substitution policy leads to an improvement of at least 0.01 in the probability of the Rockies winning. This probability varies slightly depending on who the Giants' next batter is, but the overall pattern is very similar. Thus, in Table 7 we represent the probability of the Rockies winning in the situation where the Giants' next batter is the 1st batter (Benard). Further, we show it in the scenario that Myers is substituted for Astacio at the beginning of the top of the 8th inning, and Lansing is substituted for Myers as a pinch hitter in the bottom of the

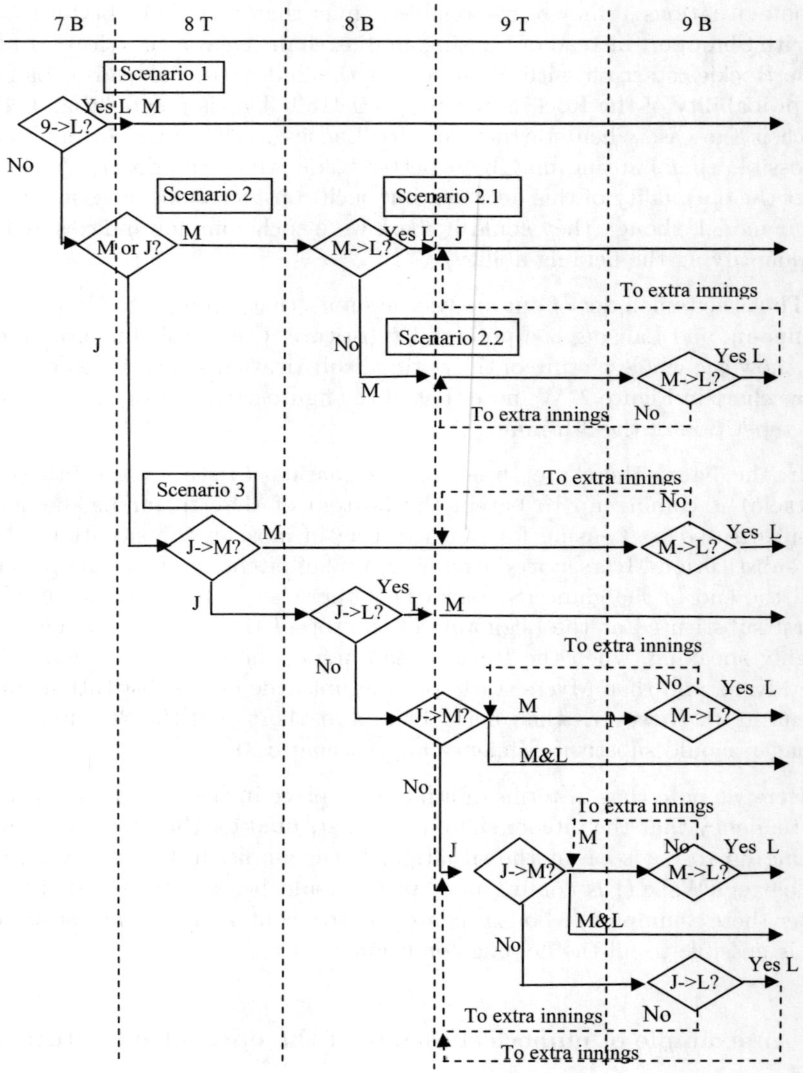

Fig. 2. Flow chart of the optimal substitution strategies (9: Astacio, L:Lansing, J:Jimenez, M:Myers).

Table 7. The situations leading to more than a 0.01 improvement by substitution in the case where the Rockies lead by from -4 to 4 in Scenario 2.1. (9: Astacio, L: Lansing, J: Jimenez, M: Myers).

				Number of runs the Rockies leading by								
Inning	T/B	Out	Runner	−4	−3	−2	−1	0	1	2	3	4
8	T	0	000					9→M (0.632→0.645)	9→M (0.846→0.860)			
8	B	0	011		M→L (0.235→0.245)							
8	B	0	101		M→L (0.235→0.245)							
8	B	0	111	M→ L (0.222→0.236)	M→L (0.342→0.359)							
8	B	1	011		M→L (0.152→0.165)	M→L (0.262→0.277)						
8	B	1	101		M→L (0.152→0.165)	M→L (0.262→0.277)						
8	B	1	110		M→L (0.174→0.183)		M→L (0.477→0.494)					
8	B	1	111	M→L (0.133→0.150)	M→L (0.225→0.249)	M→L (0.370→0.390)	M→L (0.519→0.540)					
8	B	2	010				M→L (0.277→0.289)					
8	B	2	100				M→L (0.277→0.289)					
8	B	2	011		M→L (0.089→0.101)	M→L (0.161→0.182)	M→L (0.284→0.308)					
8	B	2	101		M→L (0.089→0.101)	M→L (0.161→0.182)	M→L (0.284→0.308)					
8	B	2	110		M→L (0.098→0.114)	M→L (0.197→0.226)	M→L (0.308→0.350)					
8	B	2	111	M→L (0.060→0.080)	M→L (0.113→0.146)	M→L (0.206→0.250)	M→L (0.316→0.373)					
9	T	0	000			L→J (0.092→0.103)	L→J (0.167→0.186)	L→J (0.464→0.526)	L→J (0.755→0.839)	L→J (0.866→0.920)	L→J (0.937→0.965)	

Remark : The improvement of probability of the Rockies winning is represented in ( ).

8th innings. Then Jimenez is substituted for Lansing at the beginning of the top of the 9th inning (Scenario 2.1 in Figure 2).

In the table, we can see that the chances leading to more than a 0.01 improvement tend to occur when the Rockies are behind in the bottom of the 8th inning with runners on second and/or third base, and when Myers is coming up to bat. That is, a chance to make a tie or turn the tables directly by the substitution of Lansing for Myers.

6.4 The effect of the number of substitutes

Table 8 shows the improvement in the probability of the Rockies winning depending on the number of substitutes. As this table shows, the more substitutes, the higher the probability of winning, but it depends on the type of substitutes.

Table 8. The probability of the Rockies winning depending on the number of substitutes.

Number	Substitutes	Probability of winning
0	None	0.3258
1	Lansing (field player)	0.3266
	Jimenez (pitcher)	0.3743
	Myers (pitcher)	0.4081
2	Lansing (field player), Jimenez (pitcher)	0.3815
	Lansing (field player), Myers (pitcher)	0.4148
	Jimenez (pitcher), Myers (pitcher)	0.4081
3	Lansing (field player), Jimenez (pitcher), Myers (pitcher)	0.4169

We see that having only Lansing (field player) available as a substitute does not greatly increase the probability of the Rockies winning – the increase is just 0.0008. However, having only Jimenez (relief pitcher available from the 8th inning) increases this probability by 0.0485, and having only Myers increases it by 0.0823. This appears quantitatively to support the importance of relief pitchers in modern baseball.

The probability of winning with two substitutes (Jimenez and Myers) being available is exactly the same as the probability with only Myers being available. This is because without any pinch hitting opportunities, there are no chances for Jimenez to play. That is, there are no reasons to use Jimenez just for pitching to improve the probability of the Rockies winning, as we expected in Section 6.1. However, if Lansing is also available, then there is positive benefit in having both relief pitchers available.

6.5 The effect of home and visiting

If we do not consider the substitution of pitchers, a home team (which always bats at the bottom of each inning) has an advantage in relation to pinch hitting, since the home team can choose its strategy after knowing the result of the top of every inning (Hirotsu and Wright 2003). However, the away team derives an advantage when considering pitcher substitution because it can decide whether or not to substitute its pitcher in the bottom of the inning after knowing the result of the top of the inning.

Table 9 shows the probability of the Rockies winning both as a home and as a visiting team at the beginning of the top of each inning. This is under the condition where the Rockies are still withholding their three substitutes at the beginning of the top of each inning, the next batters of both teams are their 1st batters (i.e. Benard for the Giants and Goodwin for the Rockies), and the scores are level. There is no longer any consistent advantage for the home team.

Table 9. The probability of the Rockies winning as a home and as a visiting team at the beginning of the top of each inning withholding their three substitutes in the case where the scores are level.

Inning	Probability of winning	
	As home team	As visiting team
1	0.4169	0.4172
2	0.4333	0.4337
3	0.4502	0.4507
4	0.4726	0.4731
5	0.4955	0.4964
6	0.5323	0.5316
7	0.5652	0.5706
8	0.6449	0.6433
9	0.6349	0.6335

7 Conclusions and Further Work

In this paper we have developed the DP formulation for obtaining the optimal substitution strategy involving pitchers using ERA as a measure of pitching quality. We have applied this method to a game based on the line-ups of the Rockies and the Giants, and have shown the optimal substitution strategies for the Rockies and the increase in the probability of winning the match because of the availability of the substitutes.

As shown in the example, the optimal substitution strategies are quite complicated. This is because we have to consider pinch-hitting not only just as a matter of swapping the players, but also considering the place in the batting order with regard to the next batter, etc.

In this paper, for simplicity we do not include the tiredness of pitchers assuming that they keep their ERA at all times during a game. However, we could technically include this effect by putting the different w-values inning by inning, or out by out, in our program code. For example, we could have a gradually increasing w-value during a game.

In the example, there are no substitution chances for Jimenez to take the place of Myers when Myers is pitching, just because Myers has a lower ERA than Jimenez's. This shows the limitation of this model in which we evaluate the defensive ability of pitchers just by using the ERA. In a real game, when a slugger comes up to bat with runners on second and/or third base, a manager can introduce a relief pitcher who throws from the opposite handedness of the slugger just for facing him, taking the advantage of handedness in that situation. To make the model more useful, we should take into account the handedness, which is another important factor in a substitution strategy. The formulation developed in this paper will be extended to include handedness as a future study.

Moreover, by incorporating both the tiredness of pitchers and the handedness of players we could develop a more complicated model reflecting the real situation. Using the developed model, we could evaluate the real decision of substitution by managers and very interesting insights of a baseball game would be provided.

Appendix

In Section 5, we explained how to develop the DP formulation where there is just one relief pitcher available. This method can be extended to the case of more than one substitute being available. In this appendix, we describe the case when one substitute x_{ph} for pinch hitting (or fielding) and one relief pitcherx_{rp} are available.

Let $\mathbf{W}[x_{ph}, x_{rp}] = \mathbf{W}(x_1,x_2,\ldots,x_9,\ y_1,y_2,\ldots,y_9|x_{ph},x_{rp})$ be the 1,434,672 $\times$ 1 vector representing the probabilities of the home team winning with x_{ph} and x_{rp} available.

Here, suppose that the manager substitutes x_{ph} first. There are 9 possible substitutions for x_1,$x_{2,\ldots,}$or x_9 in the current state. If x_{ph} is substituted for x_1, the state transits to the next state by the transition matrix $\mathbf{P}^{(1\to ph)}$, which represents the matrix modified by substituting the $\mathbf{P}_{Hph}$ blocks for the $\mathbf{P}_{H1}$. After this transition, the probability of winning in the next state is expressed by one of the elements of the 1,434,672 × 1 vector $\mathbf{W}^{(1\to ph)}$ $[x_{rp}] = \mathbf{W}^{(1\to ph)}$ $(x_{ph},x_2,\ldots,x_9,\ y_1,y_2,\ldots,y_9|x_{rp})$. These elements represent the probabilities of the home team winning with a substitute x_{rp} in the remainder of the game after the substitution x_{ph} for x_1. In the same way, we define the transition matrices $\mathbf{P}^{(2\to ph)},\ldots,\ \mathbf{P}^{(9\to ph)}$ and the vectors $\mathbf{W}^{(2\to ph)}$ $[x_{rp}],\ldots,\ \mathbf{W}^{(9\to ph)}$ $[x_{rp}]$, when x_{ph} is substituted for $x_2,\ldots,x_9$.

On the other hand, suppose the manager substitutes x_{rp} first. There are also 9 possible substitutions for $x_1,x_2,\ldots,$or x_9. If x_{rp} is substituted for x_1, the state transits to the next state by the transition matrix $\mathbf{P}^{(1\to rp)}$, which represents the matrix modified by substituting the $\mathbf{P}_{Hrp}$ blocks for the $\mathbf{P}_{H1}$. Further, in this case, as x_{rp} is a pitcher so the transition matrix $\mathbf{P}^{(1\to rp)}$ should be also modified by substituting the $\mathbf{P}_{Vi}^{(1\to rp)}$ blocks for the $\mathbf{P}_{Vi}$ blocks (i = 1...9) as shown in expression (23) excepting the number of the superscript. After this transition, the probability of winning in the next state is expressed by one of the elements of the 1,434,672 × 1 vector $\mathbf{W}^{(1\to rp)}$ $[x_{ph}] = \mathbf{W}^{(1\to rp)}$ $(x_{rp},x_2,\ldots,x_9,\ y_1,y_2,\ldots,y_9|x_{ph})$. These elements represent the probability of the home team winning with substitute x_{ph} in the remainder of the game after the substitution x_{rp} for x_1. In the same way, we define the transition matrices $\mathbf{P}^{(2\to rp)},\ldots,\ \mathbf{P}^{(9\to rp)}$ and the vectors $\mathbf{W}^{(2\to rp)}$ $[x_{ph}],\ldots,\ \mathbf{W}^{(9\to rp)}$ $[x_{ph}]$.

We compare the probabilities of the home team winning between the cases of non-substitution and the above 18 possible cases of substitutions. We then take the maximum of them in each state. The selection leads us to identify the optimal substitution strategy in each state. That is, by following the optimal strategy chosen in each state, we can maximize the probability of the home team winning with two substitutes, x_{ph} and x_{rp}, in the remainder of the game. We formulate this procedure by DP as follows:

$$\mathbf{W}[x_{\text{ph}},x_{\text{rp}}] = \max\begin{cases} \mathbf{P}^{(ns)}\mathbf{W}[x_{\text{ph}},x_{\text{rp}}] + \mathbf{P}_{out} & : \text{Non}-\text{substitution} \\ \mathbf{P}^{(1\to\text{ph})}\mathbf{W}^{(1\to\text{ph})}[x_{\text{rp}}] + \mathbf{P}_{\text{out}}^{(1\to\text{ph})} & : \text{Substitution of } x_{\text{ph}} \text{ for } x_1 \\ \mathbf{P}^{(2\to\text{ph})}\mathbf{W}^{(2\to\text{ph})}[x_{\text{rp}}] + \mathbf{P}_{\text{out}}^{(2\to\text{ph})} & : \text{Substitution of } x_{\text{ph}} \text{ for } x_2 \\ \vdots & \vdots \\ \mathbf{P}^{(9\to\text{ph})}\mathbf{W}^{(9\to\text{ph})}[x_{\text{rp}}] + \mathbf{P}_{\text{out}}^{(9\to\text{ph})} & : \text{Substitution of } x_{\text{ph}} \text{ for } x_9 \\ \mathbf{P}^{(1\to\text{rp})}\mathbf{W}^{(1\to\text{rp})}[x_{\text{ph}}] + \mathbf{P}_{\text{out}}^{(1\to\text{rp})} & : \text{Substitution of } x_{\text{rp}} \text{ for } x_1 \\ \mathbf{P}^{(2\to\text{rp})}\mathbf{W}^{(2\to\text{rp})}[x_{\text{ph}}] + \mathbf{P}_{\text{out}}^{(2\to\text{rp})} & : \text{Substitution of } x_{\text{rp}} \text{ for } x_2 \\ \vdots & \vdots \\ \mathbf{P}^{(9\to\text{rp})}\mathbf{W}^{(9\to\text{rp})}[x_{\text{ph}}] + \mathbf{P}_{\text{out}}^{(9\to\text{rp})} & : \text{Substitution of } x_{\text{rp}} \text{ for } x_9 \end{cases} \quad \text{(A1)}$$

where $\mathbf{P}_{out}^{(1\to ph)},\ldots,\ \mathbf{P}_{out}^{(9\to rp)}$ represent the 1,434,672 × 1 vector modified by substituting the corresponding blocks of $x_1,\ldots,\ x_9$ in the vector $\mathbf{P}_{out}$, respectively. Expression (A1) is a DP formulation of a set of recursive simultaneous equations with 1,434,672 × 19 unknown variables of the probability of the home team winning considering any possible substitutions.

We note that we can further express $\mathbf{W}^{(i \to ph)}\,[x_{rp}]$ and $\mathbf{W}^{(i \to rp)}[x_{ph}]$ (i=1…9) in the above expression. As in this example it would be likely that x_{ph} is substituted first for x_9 as a pinch hitter and then the relief pitcher x_{rp} is substituted for x_{ph}, we just show the expression of $\mathbf{W}^{(9 \to ph)}\,[x_{rp}]$ as follows:

$$\mathbf{W}^{(9\to ph)}[x_{rp}] = \max \begin{cases} \mathbf{P}^{(9\to \mathrm{ph})}\mathbf{W}^{(9\to \mathrm{ph})}[x_{\mathrm{rp}}] + \mathbf{P}_{out} & : \text{Non} - \text{substitution} \\ \mathbf{P}^{(9\to \mathrm{ph})^{(1\to \mathrm{rp})}}\mathbf{W}^{(9\to \mathrm{ph})^{(1\to \mathrm{rp})}} + \mathbf{P}_{\mathrm{out}}^{(9\to \mathrm{ph})^{(1\to \mathrm{rp})}} & : \text{Substitution of } x_{\mathrm{rp}} \text{ for } x_1 \\ \mathbf{P}^{(9\to \mathrm{ph})^{(2\to \mathrm{rp})}}\mathbf{W}^{(9\to \mathrm{ph})^{(2\to \mathrm{rp})}} + \mathbf{P}_{\mathrm{out}}^{(9\to \mathrm{ph})^{(2\to \mathrm{rp})}} & : \text{Substitution of } x_{\mathrm{rp}} \text{ for } x_2 \\ \quad\vdots & \quad\vdots \\ \mathbf{P}^{(9\to \mathrm{ph})^{(ph\to \mathrm{rp})}}\mathbf{W}^{(9\to \mathrm{ph})^{(ph\to \mathrm{rp})}} + \mathbf{P}_{\mathrm{out}}^{(9\to \mathrm{ph})^{(ph\to \mathrm{rp})}} & : \text{Substitution of } x_{\mathrm{rp}} \text{ for } x_{\mathrm{ph}} \end{cases} \tag{A2}$$

where, for example, the transition matrix $\mathbf{P}^{(9\to ph)^{(1\to rp)}}$ is expressed in the following matrix.

$$\mathbf{P}^{(9\to ph)^{(1\to rp)}} =$$

$$\left[\begin{array}{c|c|c|c} \begin{matrix} \mathbf{0} & \mathbf{P}_{V1}^{(1\to \mathrm{rp})} & & \\ & \mathbf{0} & \mathbf{P}_{V2}^{(1\to \mathrm{rp})} & \\ & & \ddots & \ddots \\ & & & \mathbf{P}_{V8}^{(1\to \mathrm{rp})} \\ \mathbf{P}_{V9}^{(1\to \mathrm{rp})} & & & \mathbf{0} \end{matrix} & \begin{matrix} \mathbf{P}_{Hrp} & & \\ & \mathbf{P}_{Hrp} & \\ & & \ddots \\ & & & \mathbf{P}_{Hrp} \end{matrix} & & \\ \hline & \begin{matrix} \mathbf{0} & \mathbf{P}_{V1}^{(1\to \mathrm{rp})} & & \\ & \mathbf{0} & \mathbf{P}_{V2}^{(1\to \mathrm{rp})} & \\ & & \ddots & \ddots \\ & & & \mathbf{P}_{V8}^{(1\to \mathrm{rp})} \\ \mathbf{P}_{V9}^{(1\to \mathrm{rp})} & & & \mathbf{0} \end{matrix} & \begin{matrix} \mathbf{P}_{H2} & & \\ & \mathbf{P}_{H2} & \\ & & \ddots \\ & & & \mathbf{P}_{H2} \end{matrix} & \\ \hline & & \ddots & \begin{matrix} \mathbf{P}_{H3} & & \\ & \ddots & \\ & & \mathbf{P}_{H8} \end{matrix} \\ \hline \begin{matrix} \mathbf{P}_{Hph} & & \\ & \mathbf{P}_{Hph} & \\ & & \ddots \\ & & & \mathbf{P}_{Hph} \end{matrix} & & & \begin{matrix} \mathbf{0} & \mathbf{P}_{V1}^{(1\to \mathrm{rp})} & & \\ & \mathbf{0} & \mathbf{P}_{V2}^{(1\to \mathrm{rp})} & \\ & & \ddots & \ddots \\ & & & \mathbf{P}_{V8}^{(1\to \mathrm{rp})} \\ \mathbf{P}_{V9}^{(1\to \mathrm{rp})} & & & \mathbf{0} \end{matrix} \end{array}\right] \tag{A3}$$

Similarly, we can develop a DP formulation in the case of more than two substitutes being available.

References

1. Bennett JM(1998) Baseball. In: Bennett JM (ed.) Statistics in sport. Arnold, London, pp 25–64
2. Bukiet B, Harold ER, Palacios JL (1997) A Markov chain approach to baseball. *Operations Research* 45: 14–23
3. D'Esopo DA, Lefkowitz B (1977) The distribution of runs in the game of baseball. In: Ladany SP, Machol RE (eds) Optimal Strategies in Sports. North-Holland, Amsterdam, pp 55–62
4. Hirotsu N, Wright M (2003) A Markov chain approach to optimal pinch hitting strategies in a designated hitter rule baseball game. *Journal of the Operations Research Society of Japan* 46: 353-371
5. James B, Dewan J, Munro N, Zminda D (1998) STATS All-time baseball sourcebook. STATS Publishing, Morton Grove, Illinois
6. Neft DS, Cohen RM, Neft ML (2001) The sports encyclopedia: Baseball (Twenty first edition). St. Martin's Griffin, New York
7. STATS Inc (2000) Stats player profiles 2001. STATS Publishing, Morton Grove, Illinois
8. Trueman RE (1976) A computer simulation model of baseball: With particular application to strategy analysis. In: Machol RE, Ladany SP, Morrison DG (eds.) Management science in sports. North-Holland, Amsterdam pp 1–14
9. Trueman RE (1977) Analysis of baseball as a Markov process. In: Ladany SP, Machol RE (eds.) Optimal Strategies in Sports. North-Holland, Amsterdam pp 68–76

Mega Deal? A Relative Performance Analysis for Major League Baseball Players

Chii-Dean Lin and Richard A. Levine

San Diego State University, USA
cdlin@sciences.sdsu.edu
ralevine@sciences.sdsu.edu

1 Introduction

Over the past decade, baseball has entered a period of the mega-deal: multi-year, high-priced contracts that lock in superstar players to a particular team for an extended period of time. General managers and fans alike stress over these long term contracts: will the player continue at a high level of performance, and hopefully even improve? What will the long term cost be towards putting together a team capable of contending for a championship? Recent disasters have compounded these fears as the following few paragraphs of anecdotal evidence account.

Kevin Brown signed a 7 year/$105 million (M) contract before the 1999 season but has sustained debilitating injuries in each of the last three years. During each season from 1996 to 2000, he consistently started 32 plus games. In 2001 and 2002 he started only 19 and 10 games respectively.

Darren Dreifort signed a 5 year/$55M contract before the 2001 season, but has hardly pitched due to injuries in 2001, 2002, and 2003, in fact missing the entire 2002 season.

Ken Griffey Jr. signed a 9 year/$116.5M contract before the 2000 season, but was plagued by injuries in the 2001, 2002, and 2003 seasons, missing most of the 2002 season. His average year from 1996 to 1999 consisted of 52 homeruns, 142 runs batted in, and a 0.299 batting average. In the 2000 season he hit 40 homeruns and drove in 118 runs with a 0.271 batting average, in the 2001 season he hit 22 homeruns and drove in 65 runs, batting 0.286, and in an injury shortened 2002 he hit 8 homeruns, drove in 23 runs and batted 0.264. Respectable statistics, but well below the expectations he set in the latter half of the 1990s.

Mike Hampton signed an 8 year/$121M contract before the 2001 season only to fall well short of performance expectations in the 2001 and 2002 seasons. His average year from 1997, his first year with over 200 innings pitched, to 2000 consisted of 16 wins, a 3.30 earned run average (ERA), 1.35 WHIP (walks plus hits divided by innings pitched), and 151 strikeouts. In the 2000 season, he lead the New York Mets to the World Series winning 15 games, compiling a 3.14 ERA and a 1.35 WHIP, and striking out 151 batters. During the 2001 and 2002 seasons he averaged 10 wins, a 5.70 ERA, a 1.68 WHIP, and 98 strikeouts.

Chan Ho Park signed a 5 year/$65M contract before the 2002 season, but has yet to continue his pitching dominance demonstrated before 2002. His average year from 1997, his first year with over 200 innings pitched, to 2001 consisted of 15 wins, a 3.70 ERA, a 1.31 WHIP, and 193 strikeouts. In the 2002 season, he started 25 games, 8 games below his average from 1997 to 2001, wining 9 games and compiling an ERA of 5.75 and a WHIP of 1.59, and striking out 121 batters.

A mega-deal flop, whether due to a drop in player productivity/performance or injury after a contract signing, can have long run devastating effects on a team. By locking up a significant portion of payroll over an extended period of time, the team is unable to sign significant talent to offset the flop. Perhaps even worse, the team lacks the resources to extend enticing contracts to promising prospects in order to build a young talent-base towards future success. The Baltimore Orioles and Los Angeles Dodgers of the 1990s and New York Mets of the early 21st century are examples of this phenomenon. Furthermore, it is difficult to trade a player with a mega-deal as the hefty contract goes with the player.

We should not limit our anecdotes to mega-deal flops. Even if a mega-deal player performs well, the team may feel the repercussions of the long term significant dent in payroll by lacking the funds to build a consistently competitive team around the mega-deal player(s). For example, Alex Rodriguez has been a huge success for the Texas Rangers after signing a ten year/$252M contract prior to the 2001 season. His exorbitant salary has arguably prevented the team from putting together a competitive pitching staff.

Consistent, good performers may be considered mega-deal disappoints under the expectation of bigger and better productivity following the signing of a large contract. For example, Gary Sheffield has been a consistently strong performer since signing a 6 year/$61M contract before the 1997 season. His average year from 1997 to 2002 consisted of 30 homeruns, 87 runs batted in, and a 0.300 batting average. However, he has been deemed a disappointment in many circles, especially given his salary level and his pre-contract year of 1996 when he hit 42 homeruns, drove in 120 runs, and batted 0.314 (playing in all but one of his team's games, a feat he did not come close to repeating since then).

The success of a mega-deal lies primarily in a general manager's ability to predict the performance of a player over the long run. Such predictions of course are difficult to make given the large number of variables that affect a player's productivity. But the problem is compounded by the "contract year" phenomenon: players step their game up a level prior to the end of their contract in the hope of landing a mega-deal in subsequent years. For example, Richard Hidalgo signed a 4 year/$32M contract before the 2001 season. He had a career year during the 2000 season, hitting 44 homeruns, driving in 122 runs, and compiling a batting average of 0.314. In the 2001 and 2002 seasons he hit 19 and 15 homeruns, drove in 80 and 48 runs and compiled a batting average of 0.275 and 0.235 respectively. Charles Johnson signed a 5 year/$35M contract following a 2000 campaign in which he hit 31 homeruns, drove in 91 runs, and batted 0.304, by far the best year hitting, statistically, in his career. In the 2001 season he hit 18 homeruns, drove in 75 runs, and compiled a 0.259 batting average. In the 2002 season, due to injury, he played in only half of his team's games, hitting 6 homeruns, driving in 36 runs, and compiling a 0.217 batting average. In April 2001, Bobby Higginson signed a 4 year/$35.4M contract to start in the 2002 season. This offer followed a year in which he hit 30 homeruns, drove in 102 runs, each career highs, and compiled a 0.300 batting average. In 2002 he hit 10 homeruns, drove in 63 runs, and compiled a 0.282 batting average.

These examples have surrounding special circumstances of course and are just a small sample from a set of mega-deal successes and failures. The point of this discussion is that the long term deal is beset with uncertainty. Our take on this issue is to statistically evaluate whether the mega-deal is really a mega deal or if teams should be leaning towards short term contracts. General managers seem to be slowly catching on to the dangers of long term deals and potential benefits of short term contracts. Ivan Rodriguez could not obtain a mega-deal before the 2003 season, finally settling for a one year contract. Top-line free agents Jeff Kent and Kevin Millwood signed only two and one year contracts respectively before the 2003 season. Of course, players are also realizing the benefit of short term contracts as they can more easily negotiate increases to their salary following productive years, especially given the escalation and subsequent steady rise in player yearly salaries post 1990.

In this paper we consider the problem faced by general managers: are mega-deals worth it in terms of player productivity or should short term renewable contracts be the norm even for the superstars? The problem is made difficult by free agency however. The Major League Baseball players association (MLBPA) collective bargaining agreement (bigleaguers.yahoo.com/mlbpa/history) allows a player with six years experience to shop the market for the mega-deal, pitting teams against one another for his services. Consequently, players at their peak or reaching their prime sign long term deals whereas, as we will see, short term contracts are signed generally by utility players or players at the start or in the twilight of their careers. We thus need to compare player performance between contract length groups relative to or adjusting for salary and years of service in our statistical analysis.

The literature contains a number of studies on salary, contract length, and performance of MLB players. The work dates back to the econometric models developed by Scully (1974), advanced further by Zimbalist (1992), and salary evaluations by Lackritz (1990) and the references therein. More recently, Maxcy (1997), Watnik (1998), Depken (1999), and Krautmann and Oppenheimer (2002) have considered MLB player performance as a function of contract length and/or salary. However, these papers do not directly consider the effect of the mega-deal on player performance. This chapter in some sense extends this body of literature as well by introducing easily applied statistical measures to evaluate player performance relative to salary.

In Section 2, we describe the data set we constructed for analyzing the mega-deal problem. It is a unique data set derived from a number of sources, which we have made publicly available. In Section 3, we begin our comparison of player performance as a function of contract length by studying the "contract year" phenomenon. Through this initial descriptive investigation of the performance of players directly before and after a contract signing, we gain insights into the intricacies of the data and motivate the analysis of player performance and contract length relative to salary and service years. In Section 4, we introduce our relative player performance analysis for fairly comparing player performance and evaluating the mega-deal. In particular, we develop a statistic to compare the expected performance of a player during a given year, relative to salary and years of service, with the observed performance of that player. An application of this statistic to our data shows that the short term contract is in fact a better option for general managers in terms of getting the most production out of their players or at least consistent yearly performances. In Section 5, we discuss the ramifications of our analysis and other avenues open for further research about the mega-deal.

2 Data Set

A major component of this study was the compilation of a database containing player performance variables and contract information. Such a data set is most useful not only for the analysis presented here, but for the plethora of pedagogical and sabrmetric queries concerning the economic ramifications of player productivity for the interested analyst, student, or sports fan. We have thus made the data publicly available at the web site http://rohan.sdsu.edu/~cdlin/mlb_data.html with a variable coding/description document.

The data set was compiled from the following web sites, each containing different pieces of information about performance, salary, contract, and service for Major League Baseball (MLB) players:

http://roadsidephotos.com/baseball/data.htm
http://baseball1.com/statistics/

http://www.bluemanc.demon.co.uk/baseball/mlbcontracts.htm
http://www.baseball-reference.com/players.shtml
http://baseballguru.com/bbdata1.html
http://asp.usatoday.com/sports/baseball/salaries/default.aspx
http://oakland.athletics.mlb.com/NASApp/mlb/mlb/events/freeagent2001/mlb_free_agent_scorecard.jsp

The data set contains information on 255 MLB players, 149 hitters and 106 pitchers from the 1997 - 2002 seasons. The 255 MLB players encompass all recent contract signing we can find on the internet, in particular the above web resources. Here we detail only the variables used in this analysis for brevity, though we use all variables including fielding data for model selections. We refer the reader unfamiliar with these aspects of baseball to the web site mlb.com, which contains detailed descriptions of the rules of MLB and the tabulation of these summary measures. The hitter variables we consider are batting average (BAVG, number of hits divided by number of at-bats), runs batted in (RBI), homeruns (HR), games played (G), intentional walks (IBB, intentional base on balls), and stolen bases (SB). The pitcher variables we consider are earned run average (ERA, the number of earned runs allowed per nine innings), WHIP (walks plus hits divided by innings pitched), inning outs pitched (IPO), walks allowed (BB, base on balls), strikeouts (K), games started (GS), saves (SV), and homeruns allowed (HRA). We consider yearly salary, contract length, and years of service for hitters and pitchers in our study of relative player performance.

Table 1 contains a summary of the contract lengths for hitters and pitchers in the data set. We note that we consider only the most recent contract signed for each player in the database. The contract lengths include 3 contracts signed in 1998, 9 in 1999, 17 in 2000, 42 in 2001, and 78 in 2002 for hitters and include 1 contract signed in 1998, 3 in 1999, 12 in 2000, 31 in 2001, and 59 in 2002 for pitchers.

Table 1. Number of hitters and pitchers with given contract lengths during the 1998 - 2002 MLB seasons in the data set.

Contract length (in years)	Hitters	Pitchers
1	19	14
2	24	24
3	39	31
4	38	26
5	12	7
6 or more	17	4

3 Contract Year Performance Comparisons

We begin the study of the mega-deal by considering the "contract year" effect. We descriptively compare player performance before and after a contract year signing for contracts of differing lengths. The goal of this initial analysis is to motivate the development of a relative performance measure; namely a statistic which evaluates player productivity relative to yearly salary and years of service in the league.

3.1 Statistical methods

In this descriptive summary of player performance before and after a contract signing we consider some of the more commonly studied player performance statistics. Consideration of other summary measures from our data set leads to similar conclusions and thus for brevity are not included here. We note that in our relative player performance analysis in Section 4, we apply objective, quantitative statistical methods for choosing these variables, being more subjective and qualitative in the initial analysis of this section.

We consider BAVG, RBI, HR, and G for comparing performance of hitters. These measures are representative of a player's ability to reach base, to produce runs, to hit with power, and to avoid a debilitating injury over a season. For comparing performance of pitchers we use ERA, WHIP, GS + SV, and IPO. These measures are representative of a pitcher's ability to minimize a team's run production, to minimize a team's on-base percentage, and to avoid a debilitating injury, whether a starter or reliever, over a season.

We further breakdown player performance by contract length. For this discussion, we consider hitter short term contracts as those of two years or less, mid-length contracts as those of three or four years, and long term contracts as those of five or more years. For pitchers, we consider short term contracts as those of three years or less and long term contracts as those of four or more years. These break points are chosen to ensure reasonable sample sizes in each subgroup, and consequently sufficient statistical power in the analyses performed. We note though that finer or coarser subgroupings yield analogous conclusions.

For each player, we difference a summary of performance before and after the most recent contract signing. When multiple years occur between contract signings, we summarize the measures through a weighted average of each statistic across the past three seasons with weights 1/6, 1/3, and 1/2. For example, for a player signing contracts in 1997 and 2001, say, the summary of BAVG prior to a contract signing for the 2001 season is a weighted sum of the batting averages of that player during the 1998, 1999, and 2000 seasons with weights 1/6, 1/3, and 1/2 respectively. The further a season is from the contract-signing season, the less weight that season's performance has on the summary measure. We considered alternative weight functions, increasing and decreasing the weights through linear and nonlinear relationships on the length of time from contract signing to the season of interest, with similar findings.

Of course, the four statistics considered for hitters, as well as the four pitchers' statistics, are correlated; e.g., a player hitting many homeruns most likely drives in a large number of runs. We thus compare player performance before and after contract signing jointly across all four summary measures using a multivariate (Hotelling's) t-test to adjust for these potential correlations (Johnson and Wichern 2001, Chapter 6).

3.2 Hitters

Figure 1 presents the weighted average of BAVG, RBI, HR, and G in addition to salary and year of services across the hitters in each of the three contract length groups. Notice that the players perform better before a contract signing than after ($p = 0.006$ for short term contracts, $p = 0.013$ for mid-length contracts, and $p = 0.526$ for long term contracts which shows no difference before versus after signing the contract). However, contrary to our original hypothesis of the "contract year" phenomenon, the long term contract players have a smaller average performance difference than the short term contract players while the mid range contract length group has the smallest average difference performance before versus after the contract signing.

Note that this analysis is independent of yearly salary and years of service. Figure 1 presents yearly salaries before and after contract signings for each of the three contract length groups. The graphic indicates that players on long term contracts, on average, received larger salary increases (average increase of about $4.9M, 95% confidence interval $3.4M to $6.3M) than those on mid-length contracts (average increase of about $2.3M, 95% confidence interval $1.8M to $2.7M). Players on short term contracts actually showed a decrease in salary, on average (average decrease of $0.4M, 95% confidence interval - $1.2M to $0.3M). Furthermore, players on long term and mid-term contracts on average put in less service time (about 6.9 years and 5.7 years with 95% confidence intervals 5.9 to 7.9 years and 5.0 to 6.4 years respectively) than players on short term contracts (about 8.5 years, 95% confidence interval 7.3 to 9.6 years).

The effect of contract year signing on player performance is thus confounded by salary level and years of services. In particular, players on longer term contracts are typically nearer the peak of their careers and represent the superstars of the league. The group of players on short term contracts are more likely to be at the end of their careers, veteran hitters playing perhaps utility (part-time) roles, or "young" players trying to make their mark. It is thus not surprising that players on long term contracts will, on average, show

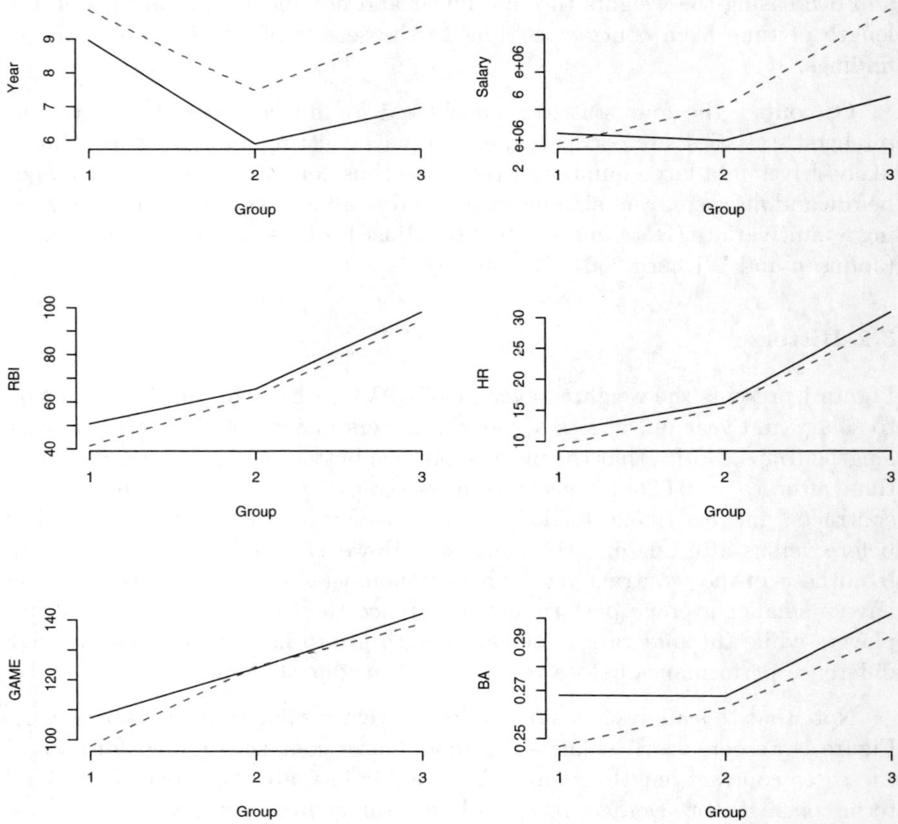

Fig. 1. The weighted average of salary, years of service, RBI, HR, G, and BAVG before signing the contract (solid lines) and after signing the contract (dash lines) for hitters in three contract length groups: less than or equal to two years (Group 1); three or four years (Group 2); and greater than four years (Group 3).

less of a drop off in performance, unadjusted for salary nor years of service, than players on shorter term contracts.

3.3 Pitchers

Figure 2 presents the weighted average of ERA, WHIP, GS+SV, and IPO across the pitchers in each of the two contract length groups. The picture for pitchers is a little less clear cut than that for hitters. Pitchers on short term contracts do not show significantly better performance before as compared to after a contract signing ($p = 0.32$). Pitchers on long term contracts show a significant difference between performance before and after contract signings ($p = 0.051$). However, contrary to our original hypothesis of the "contract year" phenomenon, the longer the contract length, the smaller the average performance difference before versus after the contract signing. Yearly salary and years of service again play a role.

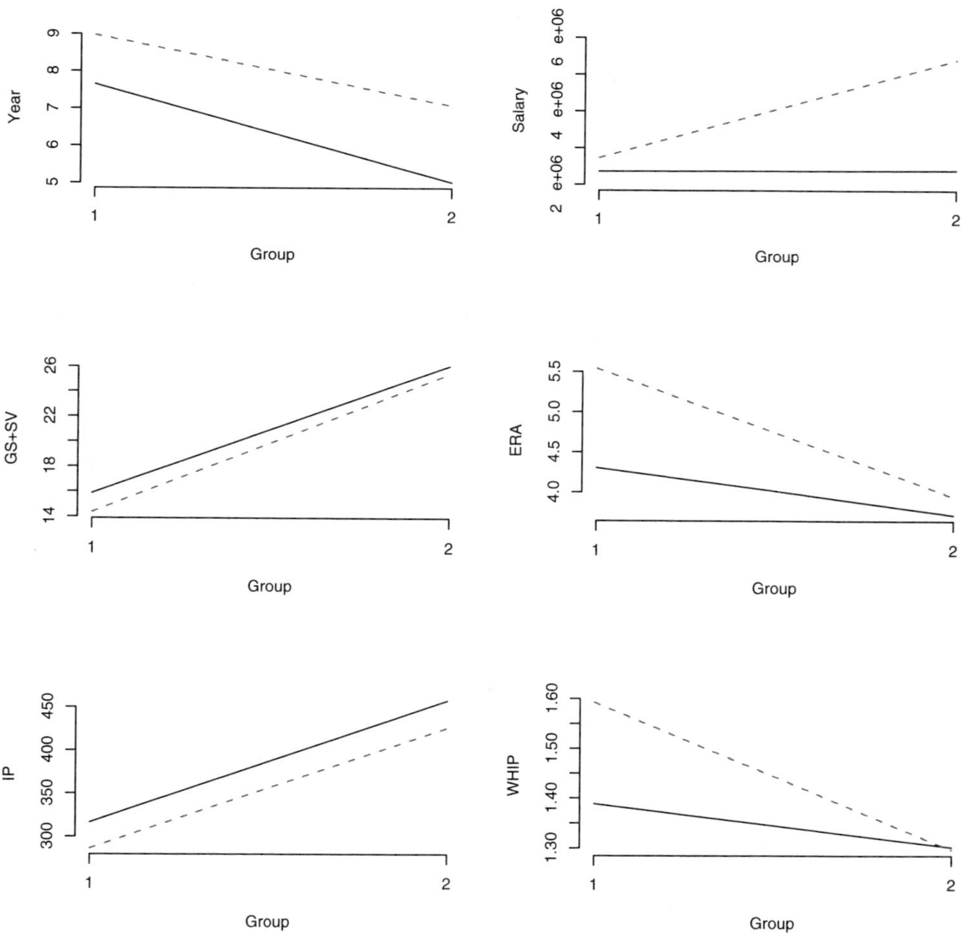

Fig. 2. The weighted average of salary, years of service, GS+SV, ERA, IPO, and WHIP before signing the contract (solid lines) and after signing the contract (dash lines) for pitchers in two contract length groups: less than or equal to three years (Group 1) and greater than three years (Group 2).

Figure 2 presents yearly salaries before and after contract signings for each of the three contract length groups. The graphic indicates that players on long term contracts on average received larger salary increases (average increase of about \$2.8M, 95% confidence interval \$2.1M to \$4.2M) than those on short term contracts (average increase of about \$0.6M, 95% confidence interval \$0.3M to \$1.2M). Furthermore, players on long term contracts on average put in less service time (about 5.0 years, 95% confidence interval 4.0 to 6.0 years) than players on short term contracts (about 7.2 years, 95% confidence interval 6.4 to 8.0 years). Analogous to our analysis of hitters' performance, we may conclude that the seemingly larger drop-off in performance for pitchers on short term contracts may be an artifact of low salary players in this group, many of which are past their prime.

4 Relative Player Performance Analysis

The analysis of Section 3 suggests that a study of the mega-deal, comparing player performance between short and long term contracts, must account for the disparity in players between these contract length groups. Of particular note is the difference in player "age" and quality, as measured by years of service and yearly salary, between contract length groups. Free agency plays a significant role in delineating these groups as players with at least six years of experience may play the market and are best positioned to obtain the mega-deal. The players on short term contracts are thus receiving smaller salaries and show greater variability in service years being either at the twilight of their careers or still wet under the collar.

In this section, we develop a relative player performance measure which allows us to evaluate player performance relative to expected performance at a given salary level, adjusted for years of service. Consequently, players are compared on an equal footing based on their potential with respect to players within the same salary cohort. We show that the statistic may be interpreted in terms of dollars earned and deserved for a given yearly performance, allowing for an objective evaluation of the mega-deal and study of the "contract year" phenomenon against contract length.

4.1 Statistical methods

Our goal in comparing relative player performance with respect to contract length is to quantify the expected performance of a player given the salary received, relative to salaries and player performance for all MLB players. To this end, we use general linear models and regression analyses to relate salary and player performance summaries.

We first outline the procedure used and then discuss each step in detail. Our relative player performance analysis is as follows. Assume we are considering p performance variables for each of n MLB players, denoted X_{ji} for the

jth variable, $j = 1, \ldots, p$, and ith player, $i = 1, \ldots, n$. We denote the p-vector storing the performance variables for player i by $\boldsymbol{X}_i$.

1. Regress before contract signing yearly salaries on the p performance variables obtaining a p-vector of regression coefficients $\hat{\beta}$.
2. Obtain an observed relative player performance score $P_{obs;i} = \hat{\beta}'\boldsymbol{X}_i$.
3. Regress contract salary on the player performance score obtaining regression coefficients $\hat{\alpha}_0$(intercept) and $\hat{\alpha}_1$ (slope).
4. Obtain an expected relative player performance score as a function of yearly salary $P_{exp;i} = (salary_i - \hat{\alpha}_0)/\hat{\alpha}_1$.
5. Obtain a relative player performance score difference $\Delta_i = P_{obs;i} - P_{exp;i}$.
6. Statistically compare relative player performance score differences, Δ_i, between contract length groups.

The first step in our relative player performance analysis is to regress the player yearly salaries on the player performance variables. The performance variables, denoted X_{ji}, are each standardized by subtracting the mean and dividing by the standard deviation of the variable across all players in the data set, both before and after contract signing. We use a combination of statistical model selection routines (Cook and Weisberg 1999, Chapter 11) and our own expertise in studying baseball data to develop a reasonable, yet strongly predictive model of salary against player performance. By "expertise," we consider only those regression models that infer the correct direction between performance and salary. For example, we exclude all models that indicate that a player's batting average is inversely related to salary. By a "reasonable, yet strongly predictive model," we choose a model that explains a large amount of the variation (R-squared value) and minimizes the information lost in using the statistical model as an approximation to the truth (Akaike information criteria, AIC). Regression diagnostics, for this data set, indicated no serious violations of underlying regression model assumptions and identified no outlying players that may affect the parameter estimates.

The regression model in step one is used to obtain an observed relative player performance score in step two. The modifier "relative" is applied to signify the role of salary in the computation of the performance score in that the score is relative to the salary earned by the player. We may think of this step as a principal components regression analysis where we first summarize a host of variables into a single (principal component) score and then regress the response of interest (salary) on this summary score. Our approach differs with principal components regression in that by regressing salary on the variables in step one to obtain a player performance score, our method ensures that the direction of the association between each variable and salary is correct. A principal components analysis ignores such a relationship leading to meaningless scores, with respect to the signs of the coefficients.

Steps three and four involve predicting the performance score for a given average salary. Step three provides a further quantification of the relationship between the observed player performance score and salary. In particular, a unit increase in the observed player performance score relates to an $\hat{\alpha}_1$ increase in the player's salary above the average across all players in the data set, $\hat{\alpha}_0$, for a given yearly performance.

Step four implements an inverse regression estimation or calibration procedure to predict the expected player performance at a given yearly salary. We note that the calibration process presented in the statistics and engineering literature provides standard error estimates for the predicted performance variables (Neter, et al., Chapter 4). For our purposes however, we need only an expected performance score, calibrated to the given salary level, for comparison with the observed performance of the player, and thus forgo the computation of the standard error. Nonetheless, the score is interpreted analogously to that of the observed relative player performance score. An expected relative player performance score $P_{exp;i}$ of 1 suggests that the player is expected to play at a level commensurate with a player earning $\hat{\alpha}_1$ above the average salary across all players in the data set, $\hat{\alpha}_0$.

The observed and expected relative player performance scores provide a means of comparing player performance after adjusting for salary and service time. In particular, the relative player performance score difference Δ gauges the observed performance of a player above or below his potential in a given year. For example, a player with a $P_{obs;i}$ of 2 and $P_{exp;i}$ of 1 is performing better than expected ($\Delta_i = 1$), and in fact should be receiving a salary $\hat{\alpha}_1$dollars above the salary he is in fact receiving, relative to all player salaries and performances. In general the quantity $\hat{\alpha}_1\Delta_i$, which we will term the relative player performance level, defines the salary level at which a player is performing, above or below expectation.

We finally, in step seven, compare player performance across contract lengths by regressing Δ_i against the contract length groups defined previously. We study the "contract year" phenomenon by taking the difference in Δ_i computed with performance data from both the season before and the season after a contract signing and regressing these values against the contract length groups. In each analysis, we perform all pairwise comparisons of contract length groups for hitters using Tukey's studentized range procedure (typically labeled the honestly significant difference, HSD, method) to control the experiment-wise type I error rate across the three tests (Lentner and Bishop 1993, Chapter 3).

4.2 Hitters

The relationship between salary and hitter performance variables is best modeled by a linear regression fit with years of service of the players, IBB, HR, and

SB. Intuitively, these variables represent a player's experience (years of service), power and productivity (HR), speed (SB), and fear factor or superstar status (IBB), a reasonable summary of hitter performance. To eliminate any scaling issues, all variables are standardized as described earlier. The resulting model for estimating the observed relative performance score is $P_{obs;i} =$ 6*years of service + 3*HR + 2*IBB + SB. This model achieves an R-squared of 58% compared to 62% for the best model (based on purely statistical measures, ignoring model realism).

The base average salary for hitters in this data set is $\hat{\alpha}_0 = 2{,}270{,}000$. The slope term defining the relative player performance level is $\hat{\alpha}_1 = 271{,}000$. Table 2 presents the average observed and expected relative player performance scores across all hitters in each contract length group in the season following a contract signing. These scores are related linearly with salary so, for example, players on short term contracts are, on average, performing at $\hat{\alpha}_1 P_{obs} = \$1.3$M above the average salary for hitters, though they are expected to perform at $\hat{\alpha}_1 P_{exp} = \$0.2$M below the average salary for hitters. Thus the positive sign on the difference in observed and expected relative player performance scores indicates that players on short term contracts are on average performing above expectations.

The relative player performance level $\hat{\alpha}_1 \Delta$ quantifies observed versus expected average player performance. Hitters on short term contracts are on average playing at a level of \$1.6M above expectation or above their salary potential. Hitters on medium range contracts are playing at \$0.5M below their salary potential. Hitters on long term contracts are playing at \$3.1 million dollars below their salary potential. These three relative player performance level values are significantly different in that hitters on short term and medium range contracts are significantly undervalued as compared to players on long term contracts and players on short term contracts are significantly more undervalued than those on medium range contracts. These statistically significant findings are achieved under the HSD test at the five percent level. All in all, players on short term contracts reach and exceed their potential, on average, significantly more than players on either medium range or long term contracts.

To study the "contract year" phenomenon across contract length groups, we compute Δ for each contract length group in the season immediately before a contract signing. Table 3 displays these values for the short term, mid-range, and long term contract length groups. Notice that players on short term contracts show a larger relative player performance level after a contract signing than before a contract signing (larger positive Δ value, though difference not significant at the five percent level as indicated by the overlapping confidence intervals). Players on mid-range contracts perform above expectation before a contract signing, but significantly below expected performance after a contract signing (difference significant at the five percent level). Players on long term contracts show relative performance levels below expected performance as indicated by the negative Δ values. Interestingly, the relative performance levels are further below expected performance after a contract signing as compared to before a contract signing (difference significant at the five percent level). Overall, we find that players on longer term contracts show a significantly greater performance drop-off, in the season following a contract signing as compared to the season before a contract signing, than players on short term contracts (HSD test significant at the five percent level). Thus the "contract year" phenomenon is present and more prevalent for hitters signing mega-deals.

Table 2. Average observed and expected relative player performance scores (P_{obs} and P_{exp}), average relative player performance level (Δ), and 95% confidence intervals around Δ for *hitters* under short term, medium range, and long term contracts, before (B) and after (A) signing a contract. Short term contracts are two years in length or less, medium term contracts are three or four years in length, and long term contracts are five years in length or more.

Contract length	Be-Af	P_{obs}	P_{exp}	$\Delta = P_{obs} - P_{exp}$	95% CI
Short	B	4.66	1.59	3.06	(0.53, 5.59)
	A	4.98	-0.82	5.79	(2.82, 8.77)
Medium	B	2.37	0.42	1.95	(0.75, 3.15)
	A	3.06	4.93	-1.87	(-3.68, -0.06)
Long	B	9.93	11.34	-1.41	(-4.62, 1.81)
	A	11.79	23.39	-11.60	(-16.96, -6.23)

4.3 Pitchers

The relationship between salary and pitcher performance variables is best modeled by a linear regression fit with years of service of the players, BB, K, GS+SV, HRA. Intuitively, these variables represent a starter and relief pitcher's yearly longevity (GS+SV) and experience (years of service), control and power (BB, K), and ability to limit the productivity of the opposing team (HRA), a reasonable summary of pitcher performance. The resulting model for estimating the observed relative performance scoreis $P_{obs;i}$= 6*years of services - 2*BB + 3*K +2.5*(GS+SV) - 0.5*HR. This model achieves an R-squared of 47% compared to 49% for the best model (based on purely statistical measures, ignoring model realism).

The base average salary for pitchers in this data set is $\hat{\alpha}_0 =$ 1,780,000. The slope term defining relative player performance level is $\hat{\alpha}_1 =$ 250,000. Table 3 presents the average observed and expected relative player performance scores across all pitchers in each contract length group in the season following a contract signing. The relative player performance levels $\hat{\alpha}_1 \Delta$ indicate that pitchers on short term contracts are performing on average at a level of $0.5M above expectation or above their salary potential. Pitchers on long term contracts on the other hand are playing on average at $1.7M below their salary potential. These two relative player performance levels, short term vs. long term contracts, are significantly different ($p < 0.0001$). Therefore, as with hitters, pitchers on short term contracts reach and exceed their potential on average significantly more than players on long term contracts.

To study the "contract year" phenomenon across contract length groups, we, as with the hitters analysis, again compute Δ for each contract length group in the season immediately before a contract. Table 3 displays these values for the short term, mid-range, and long term contract length groups. Notice that players on short term contracts show a slight decrease in relative player performance level, though this difference is not significant at the five percent level. Players on long-term contracts show relative performance levels below expected performance both before and after contract signings. The relative performance levels are further below expected performance after a contract signing as compared to before a contract signing (difference significant at the five percent level). Overall, we find that players on longer term contracts show a significantly greater performance drop-off, in the season following a contract signing as compared to the season before a contract signing, than players on short term contracts (significant at the five percent level). Thus the "contract year" phenomenon is present and more prevalent for pitchers signing mega-deals.

Table 3. Average observed and expected relative player performance scores (P_{obs} and P_{exp}), relative player performance level (Δ), and 95% confidence intervals around Δ for *pitchers* under short term and long term contracts, before (B) and after (A) signing a contract. Short term contracts are three years in length or less and long term contracts are four years in length or more.

Contract length	Be-Af	P_{obs}	P_{exp}	$\Delta = P_{obs} - P_{exp}$	95% CI
Short	B	4.94	3.70	1.24	(-0.27, 2.74)
	A	5.30	3.49	1.81	(0.18, 3.43)
Long	B	4.06	5.55	-1.49	(-4.15, 1.17)
	A	5.26	12.02	-6.76	(-10.26,-3.26)

4.4 Relative performance analysis of specific players

Our relative player performance analysis may be applied to study individual player performances relative to salary. The relevant statistics are identical to the study of average player performance across the data set as discussed in the previous two subsections. For example, consider relief pitcher Mike Williams.

Table 7 shows his observed and expected relative player performance scores and the difference between these two values, Δ. Before signing a contract, his observed relative player performance score indicates that his performance is deserving of a salary of \$3.7M ($P_{obs;i} * \hat{\alpha}_1 + \hat{\alpha}_0$ =7.82*\$0.25M + 1.78M), \$2.5M (Δ*$\hat{\alpha}_1$= 10.14*\$0.25M or \$3.7M – \$1.2M) above his yearly salary. He received a 67% salary increase under the contract, but his performance was significantly better as indicated by the statistics in Table 6. Consequently, his relative performance level after signing the contract was \$3.4M (Δ*$\hat{\alpha}_1$ = 13.43*\$0.25M) above his expected performance level indicating that he deserved a salary of \$5.4M ($P_{exp;i} * \hat{\alpha}_1 + \hat{\alpha}_0$ =14.31*\$0.25M + \$1.78M or \$2.0M + \$3.4M). Analogous computations and relative performance level interpretations may be made for hitters, though using the regression coefficients $\hat{\alpha}_0$ and $\hat{\alpha}_1$ from the hitter analyses.

The analyses of the previous two subsections indicate that, on average, players on short term contracts appear more productive than players on longer term contracts. In this subsection, we study the relative player performance level of four hitters and four pitchers representative of players from each of the contract length groups. Our goal is to further demonstrate the relative player performance analysis on individual players and illustrate differences, at the individual level, between players on short term and long term contracts.

Specific hitters

Table 4 presents batting statistics for the four hitters we consider. Representing the hitters short term contract group is Reggie Sanders, a veteran, who signed a one year/\$1.0M contract in 2002. He played in more games the season following his contract signing, but was less productive in all categories. He is playing the role of "bat-for-hire," switching teams in each of the 1999-2003 seasons. Representing the mid-length contract group is Troy Glaus, a young player with great upside potential, signing a 4 year/\$22M contract in 2002. He showed consistent performance before and after the contract signing. Representing the long term contract group are Jeff Bagwell and Chipper Jones. Bagwell is a veteran who signed a 5 year/S85M contract in 2002. He showed a marked production down swing the season following his contract signing. Jones is a superstar just off a mega-deal free-agency signing of 6 years/\$90M in 2001.

Table 5 presents the results from our relative player performance analysis on each of these four hitters before and after the contract signings, as well as yearly salary and years of service. We briefly discuss each player in turn.

Reggie Sanders showed marked decrease in production following his contract signing, however, his salary increased by 17%. Nonetheless, his productivity is strong compared to the average short term contract hitter as displayed in Figure 1. His decline in performance after the contract signing is indicated in a drop in the Δ values, though given his low salary relative to performance, he is still performing well above expectation after the contract signing. In particular, Sanders played at a relative \$4.8M ($\Delta^*\hat{\alpha}_1$ = 17.85*\$0.271M) and \$4.1M (15.21*\$0.271M) performance level above expected performance before and after contract signing respectively.

Troy Glaus showed a slight decrease in productivity, but his salary increased by 450%. Of course, given Glaus's already prolific homerun numbers, one cannot expect his power numbers to increase at a similar rate to his salary. In fact, the relative salary performance measure adjusts for this fact, indicating a relative performance level at about \$3.1M above expected performance, due to his low yearly salary before signing the contract. His salary hike was not sufficient for his level of productivity, showing a relative player performance level of \$2.2M above expected performance, though obviously this level is closer to his true value than his before contract relative performance level.

Chipper Jones showed consistent, excellent performance before and after a contract signing which saw his already large yearly salary increase by over 200%. Again, we cannot expect his performance to increase at the same rate as his salary. However, unlike Troy Glaus, Jones was making the big bucks before and after his contract signing. His relative performance level is thus \$0.2M above his expected performance level prior to the contract signing and \$3.9M below his expected performance level after the contract signing. Based on his observed relative player performance scores of 9.93 and 15.33, before and after contract signing respectively, his performances are deserving of yearly salaries of \$5.0M ($P_{obs;i} * \hat{\alpha}_1 + \hat{\alpha}_0$ =9.93*0.271M+2.3M) and \$6.5M (15.33*\$0.271M+2.3M).

Jeff Bagwell illustrates a mega-deal player who showed marked decrease in productivity, similar to that of Reggie Sanders, though following a contract signing which awarded him a 169% salary increase above an already large yearly salary. Nonetheless, his performance is strong compared to the average player and in fact similar to Chipper Jones. Thus his relative player performance levels are almost identical to that of Jones: relative performance level of \$50,000 below expected performance before his contract signing and, given the huge salary, \$4.4M below expected performance after his contract signing.

Specific pitchers

Table 6 presents batting statistics for the four hitters we consider. Representing the pitchers short term contract group are Curtis Leskanic and Jeff Suppan. Leskanic is a veteran relief pitcher who has moved between closer and middle relief roles throughout his career. He signed a 3 year/\$7.2M contract in 2001. He showed improved productivity after his contract signing as compared to before. Suppan is an average starting pitcher, a third to fifth

Table 4. Observed data (BAVG, RBI, HR, G, IBB, SB) for representative hitters under short term, medium range, and long term contracts, before (B) and after (A) contract signing.

Hitters	Be-Af	BA	RBI	HR	G	IBB	SB
R. Sanders	B	.26	90	33	119	7	14
	A	.25	85	23	137	3	18
T. Glaus	B	.28	102	47	166	6	14
	A	.25	108	41	163	7	10
J. Bagwell	B	.29	130	39	160	5	11
	A	.29	98	31	157	8	7
C. Jones	B	.31	111	36	158	10	14
	A	.33	102	38	158	20	9

Table 5. Yearly salary, years of service (frequency), observed and expected relative player performance scores (P_{obs} and P_{exp}), and average relative player performance level (Δ) for representative hitters under short term, medium range, and long term contracts, before (B) and after (A) contract signing.

Hitters	Be-Af	Salary	Freqency	P_{ob}	P_{exp}	Δ
R. Sanders	B	1,500,000	11	15.01	-2.84	17.85
	A	1,750,000	12	13.29	-1.92	15.21
T. Glaus	B	275,000	3	4.16	-7.36	11.52
	A	1,250,000	4	4.52	-3.76	8.28
J. Bagwell	B	6,500,000	11	15.41	15.61	-0.20
	A	11,000,000	12	16.04	32.21	-16.17
C. Jones	B	4,750,000	7	9.93	9.15	0.78
	A	10,330,000	8	15.33	29.75	-14.42

starter in five-man pitching rotations, who signed a 2 year/6.5M contract in 2001. He showed consistent productivity before and after his contract signing. Representing the long term contract group are Rob Nen and Denny Neagle. Rob Nen is a veteran closer who signed a 4 year/$32.5M contract in 2001. After his contract signing, his productivity dropped slightly. Denny Neagle is a veteran starting pitcher who signed a 5 year/$51M contract in 2001. After his contract signing, his productivity was significantly worse.

Table 6 presents the results from our relative player performance analysis on each of these four pitchers before and after the contract signings, as well as yearly salary and years of service. We briefly discuss each player in turn.

Curtis Leskanic illustrates a case where improved performance is balanced by salary compensation after the contract signing. Unfortunately for Leskanic though, he still was performing above his salary level, with a relative performance level of about $1.1M above expected performance both before and after contract signing. He thus deserved salaries of \$2.5M and \$3.2M before and after contract signing respectively.

Jeff Suppan also received compensation for his improved performance following the contract signing, but by no means enough to account for his performance level. His Δ value indicates that his salary, relative to performance, before contract signing was about right, being only 0.4M below expected. However, after contract signing, his relative performance level was \$2.3M above expected performance.

Denny Neagle signed a mega-deal and then performed poorly in the high altituted of Coors Field in Denver, Colorado. His relative performance level shows this decline, indicating that his relative performance level was \$0.7M below expected performance prior to the contract signing, but \$2.5M below expected performance after the contract signing, almost a four-fold decrease, despite a salary jump of 70%.

Rob Nen pitched more innings and picked up more saves after his contract signing. However, his ERA and WHIP, which were, before the contract signing, at superior levels for closers, moved to average levels after the contract signing with his ERA doubling and WHIP reaching one hit or walk per inning. Given that his before contract salary was too high even for his excellent performance, with relative performance level of \$1.0M below expected performance, his relative performance level after contract signing was \$1.5M below expected performance, indicating that he deserved a salary of \$5.1M.

Table 6. Observed data (ERA, WHIP, IPO, K, GS, SV) for representative pitchers under short term and long term contracts, before (B) and after (A) contract signing.

Pitchers	Be-Af	ERA	WHIP	IPO	K	GS	SV
M. Williams	B	3.80	1.48	192	59	0	22
	A	2.94	1.22	184	43	0	46
C. Leskanic	B	2.56	1.41	232	75	0	12
	A	1.36	2.08	208	64	0	17
J. Suppan	B	4.94	1.49	651	128	33	0
	A	4.37	1.38	655	120	34	0
D. Neagle	B	4.52	1.39	627	146	33	0
	A	5.38	1.48	512	139	30	0
R. Nen	B	1.50	.85	198	92	0	41
	A	3.01	1.03	233	93	0	45

Table 7. Yearly salary, years of service (frequency), average observed and expected relative player performance scores (P_{obs} and P_{exp}), and average relative player performance level (Δ) for representative pitchers under short term and long term contracts, before (B) and after (A) contract signing.

Pitchers	Be-Af	Salary	Frequency	P_{obs}	P_{exp}	Δ
M. Williams	B	1,200,000	10	7.82	-2.32	10.14
	A	2,000,000	11	14.31	0.88	13.43
C. Leskanic	B	1,450,000	8	2.24	-1.32	3.56
	A	2,200,000	9	5.67	1.68	4.00
J. Suppan	B	1,450,000	6	0.96	2.28	1.52
	A	2,350,000	7	3.80	3.01	9.24
D. Neagle	B	4,250,000	10	9.24	11.88	-2.64
	A	7,200,000	11	11.86	21.66	-9.82
R. Nen	B	5,500,000	8	11	14.88	-3.88
	A	6,600,000	9	13.12	19.28	-6.16

5 Conclusions and Discussion

Our relative performance analysis evens the playing field, so-to-speak, for comparison of player performance under varying contract lengths relative to salary and years of service. We find that on average, players under a mega-deal contract are earning more than their performance warrants whereas players under short term contracts are showing performance above their salary levels. Furthermore, the "contract-year" phenomenon is more prevalent after a mega-deal signing than a short term contract signing. Players on longer-term contracts show a greater drop in performance after signing the contract than players on shorter-term contracts, relative to yearly salary.

Our recommendation for MLB general managers is to shy away from the mega-deal, prefering a short term, renewable contract. Not only will the team receive more motivated players on average, but the general manager will have more opportunities to evaluate a player and adjust salary given recent performances, rather than locking in to long term, high cost deals. Of course, this recommendation benefits players too as consistently good performers will earn more frequent salary adjustments, to compensate them for a high level of performance and to keep their salary at market level.

Our analyses focus solely on player performance relative to salary and years of service. Other variables play an important role for general managers in their player management decisions: injuries, both individual player frailty and prevalence of injuries with respect to contract length, location of the team, both city and stadium environment, revenue impact, the revenue gain by having a player on the team, clubhouse chemistry, the team is better just by having the player in the lockerroom, and team payroll, both available dollars and growth potential all play. These variables are not as easily defined and formulated as performance measures, requiring analyses of economic impact at a team revenue level and a study of social well-being. We thus leave these issues to future work.

The mega-deal is observed in many team sports, including football, basketball, hockey, and soccer, and, with the increasing international popularity of team sports other than soccer, prevalent not only in the United States. Furthermore, the mega-deal exists in the advertising world where high profile sports figures receive long term, high priced contracts to market specific products. Manufacturers offering such contracts hope that the player will remain in the spotlight, which of course is directly related to performance level. The decisions made by these market excutives are then similar to those of the MLB general manager. Our relative player performance analysis may be used to analyze the mega-deal and the impact of contract length in any of these scenarios. The user merely chooses the performance variables to consider in the initial regression analysis and then proceeds through the algorithm we put forth in Section 4.1. The interpretation of the relative player performance scores and level are independent of sport or industry, being measures of salary level, in dollars (or whichever currency of interest) above or below generic expected performance.

Acknowledgement

This work was supported in part by U.S. National Science Foundation grant INT-0114738.

References

1. Cook, R.D. and S. Weisberg. *Applied* Regression *Including Computing and Graphics.* Wiley, 1999.
2. Depken, C. A. Free-agency and the competitiveness of Major League Baseball. *Review of Industrial Organization,* 14: 205-217, 1999.
3. Johnson, R.A. and D.W. Wichern. *Applied Multivariate Statistical Analysis, Fifth Edition.* Prentice-Hall, 2001.
4. Krautmann, A. C. and Oppenheimer, M. Contract length and the return to performance in Major League Baseball. *Journal of Sports Economics*, 3: 6-17, 2002.
5. Lackritz, J. Salary evaluation for professional baseball players. *The American Statistician*, 44: 4-9, 1990.
6. Lentner, M. and T. Bishop. *Experimental Design and Analysis, 2nd Edition.* Valley Book Company, 1993.
7. Maxcy, J. G. Do long-term contracts influence performance in major league baseball? *Advances in the Economicsof Sport,* 2: 157-176, 1997.
8. Neter, J., Kutner, M., Nachtsheim, C., and Wasserman, W. *Applied Linear Statistical Models, 4th Edition.* Irwin, 1996.

9. Scully, G. Pay and performance in Major League Baseball. *American Economic Review*, 64: 915-930, 1974.
10. Watnik, M. R. Pay for play: are baseball salaries based on performance? *Journal of Statistics Education,* 6, 1998.
11. Zimbalist, A. Salaries and performance: beyond the Scully model. In *Diamonds are forever: the business of baseball,* pp.109-133. Brookings Institution, 1992.

Market Inefficiencies in Player Head to Head Betting on the 2003 Cricket World Cup

Michael J Bailey and Stephen R Clarke

Swinburne University, Australia
mbailey@swin.edu.au
sclarke@groupwise.swin.edu.au

1 Background

The first official one-day International cricket match (ODI) was played at the Melbourne Cricket Ground on January 5 1971 between Australia and England. Since the inception of limited over cricket, there have been various rule changes, although general principles have remained the same. Both sides bat once for a limited time (maximum 50 overs) with the aim in the first innings to score as many runs as possible, and in the second innings to score more than the target set in the first innings. Prior to the 2003 World Cup, 16 countries had played 1936 official ODIs, although 95% of all matches have been played by 9 main cricketing nations (Australia, England, India, Pakistan, West Indies, Sri Lanka, New Zealand, South Africa and Zimbabwe). Just over 1200 cricketers have represented their country in ODIs.

Traditional betting markets have been in sports such as horse and dog racing where the sport survives primarily because of the associated betting opportunities. In recent years, betting has seen an expansion into sports such as tennis and golf in which there has been wide public interest prior to and independent of any betting activity. While initial books were set on the winner of matches, the growth of internet betting and increased competition between bookmakers has resulted in many exotic bets, such as scorer of the first goal in soccer, leader at half time etc. Many head to head bets can be created within one event by betting on a specific player to outperform another.

Although a bookmaker's task of setting prices would appear far easier in a two-outcome event, the margin of error is significantly reduced. In fixed price bookmaking, the sum of the probabilities that a bookmaker will offer on a given event will exceed one. The amount by which the summed probabilities exceed one is often referred to as the bookmaker overround, and represents the profit the bookmaker stands to win should he balance his books perfectly. The size of the overround increases with the number of competitors in the event, with most multinomial outcomes such as horse racing having an overround in excess of 15 %. For competitions in which there are only two outcomes, the overround is generally only about 7-8%, effectively meaning that the margin of error is halved.

The potentially large proportion of uninformed punters involved in sports betting, along with the smaller bookmaker's percentage taken from head to head bets, make this type of betting a prime target for exploring inefficiencies in betting markets.

Multivariate models have in the past been successfully used to identify inefficiencies in wagering markets. Dixon and Coles (1997) used Markov chain models on UK soccer matches, Dana and Knetter (1994) used Kalman filter for estimation of NFL football games, whilst Bailey (2000) has shown that market inefficiencies can be found in Australian Rules football. Perhaps the most successful work on mathematically modelling market inefficiencies has been done by Benter (1994). Following up on the work of Bolton and Chapman (1994) and Chapman (1994), Benter successfully used multinomial logit models to identify market inefficiencies in Hong Kong horse racing. Parallels can be drawn between deriving predicting probabilities for multinomial horse racing and deriving probabilities for binomial head to head match-ups in cricket. Unlike the complex multinomial approach, a logistic regression could be used to derive the probability of each player's success, with model parameters constructed from the past histories of both players. Rather than construct models with information on both batsmen, a more intuitive approach is to construct a prediction model for all batsmen in general. Because batsmen performances are essentially independent of each other, predicted scores can be directly compared and winning probabilities determined.

Clarke (1998) gives a summary of the work done on the distribution of scores in cricket. The distribution of batsman scores in test cricket was first looked at by Wood (1941; 1945) and Elderton (1945), who investigated if the scores were geometrically distributed. Kimber and Hansford (1993) also investigated the hazard function of top batsmen. In general these papers found that while the geometric is a reasonable approximation, there are slightly more ducks and very small scores than suggested by the geometric.

In one-day cricket, the time constraints on the batting team ensure that there are fewer very large scores, whilst additional constraints upon the bowler and fielders ensure that slightly fewer scores less than 5 are also recorded in comparison to test cricket. As a legacy of these constraints, an alternative to modelling batsmen scores is to use a log-linear approach. By log-transforming the number of runs that each batsman scores, it is possible to normalise the outcome of interest (runs scored) sufficiently enough to enable a parametric approach towards prediction. When constructing mulitvariate models, a parametric approach is particularly advantageous due to the relative ease with which statistical significance and thus relative importance of contributing variables can be determined. As the differences in player scores are approximately normally distributed, predicted scores can be converted into the probability of outscoring another by comparing the ratio of the difference between the two players with a combined standard deviation for both players.

2 Method

All ODI batting performances prior to the 2003 World Cup were used to identify features thought to affect the number of runs that a batsman will score. Prediction models were constructed to predict scores using a variety of approaches. Rolling averages, exponential smoothing and multiple linear regression models were all explored, with goodness of fit determined by comparison of average absolute errors (AAE). Hypothetical match-ups between leading players for past matches were created to explore the best way in which to covert predicted scores into the probability that one player will outscore another. Two alternative methods for converting predicted scores into probabilities were compared by summing the likelihood of the predicted probabilities of the winning players for each match-up.

Prediction models were applied to 2003 World Cup data and specific bookmaker comparisons calculated. Predicted probabilities were compared with bookmaker prices and potential imbalances determined. Bet sizes were calculated in accordance with the size of the perceived advantage and the probability of winning. Profitability was determined by calculating the Return on Investment (ROI) offered by each model.

The use of a range of prediction models allowed for exploration into the nature of the relationship between ROI and the quality of the prediction process as measured by the AAE for each model. All analysis was performed using SAS version 8.2. (SAS Institute Inc., Cary, NC, USA). Univariate analysis was conducted using student t-tests and Pearson correlation coefficients, whilst multivariate analysis was performed using generalized linear models. A two-sided p-value of 0.05 was considered to be statistically significant.

3 Factors Affecting Batting Performance

From the 1936 ODIs, there have been 33,606 individual batting performances. These batting performances were initially used to identify factors having a statistically significant effect on the number of runs scored by each batsman.

3.1 Position

Given the finite nature of batting time available in one-day cricket, a player's position in the batting line-up is vitally important in determining runs scored. Although batting strategies have evolved over the past 30 years, it is now generally accepted that each team's better players will bat at the top of the order.

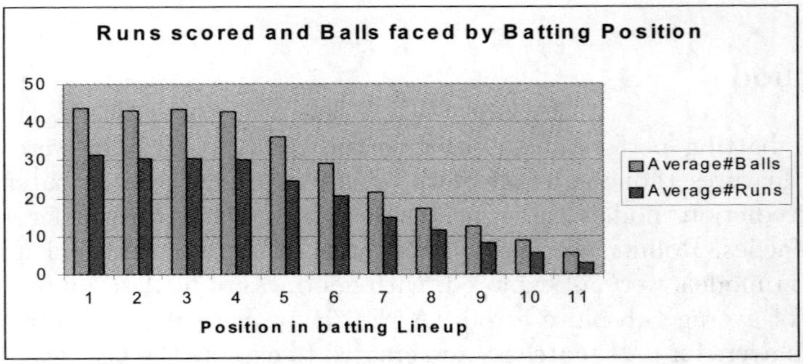

Fig. 1. Runs scored and balls faced for each given position in the batting lineup.

Analysis of batting position revealed that whilst position was a highly significant predictor of runs scored ($p<.0001$), there were no significant differences for runs scored or balls faced for the first four batsmen from each team. This result is clearly reflected in Fig. 1, where there appears to be a strong linear effect for runs scored and balls faced from position 4 onwards. By constructing a linear regression it was found that batting order could explain 17% of the variation associated with the number of runs scored. Given that resource availability plays such an integral role in determining the number of runs scored and the majority of bookmaker prices on offer were for the leading players for each team, further analysis will be restricted to only the first four batsmen from each team.

3.2 Experience

As with most sports, experience (number of matches played) plays an important role in determining performance. As can be seen from Figure 2, batsmen performance improves at a diminishing rate until a point where the increase in age and subsequent decrease in hand eye coordination outweighs the positive benefits of experience, and performance plateaus out.

By dividing the number of matches each batsmen has played into blocks of 10 and then taking the average number of runs scored for each block, it is possible to gain further insight into the effect of experience. From Fig. 3 we can see that the greatest improvement in performance generally occurs over the first 100 innings, with the average number of runs scored rising from the mid twenties to the low thirties before levelling out.

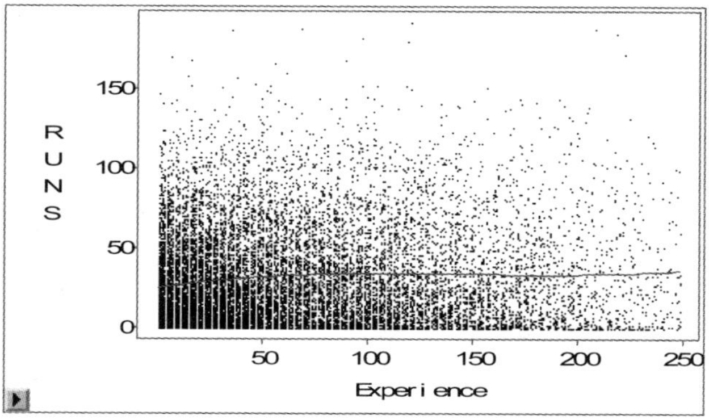

Fig. 2. The relationship between experience and runs scored.

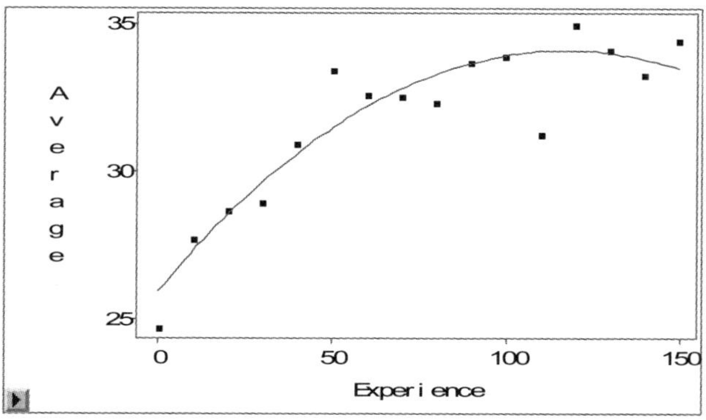

Fig. 3. The relationship between experience and runs scored.

3.3 Home country advantage

31.5% of all innings were played by batsmen on their home soil. It is not surprising that this figure is quite low, as a large percentage of ODIs are played in triangular or round robin formats, often held in neutral countries. On average, the home country advantage is worth two runs per batsman.

3.4 Match time & innings order

There are two times at which ODIs are played. Day games play the first innings prior to lunch and the second innings after lunch, with completion before nightfall. Day/night games commence mid afternoon, break at the end of the first innings for dinner and then resume with the second innings under artificial lighting. From Table 1 we can see that batsmen tend to average about one run more in day games than they do in day/night games. Batsmen that play for the team batting first will tend to score on average, about two more runs than batsmen playing for the team that bats second. Not surprisingly, there is a statistically significant interaction between Match time and Innings order, with batting under artificial lighting (batting second in day/night games) resulting in scores on average about three runs lower than if batting first.

Table 1. Average runs scored by match time and innings order.

	Bat First	Bat Second	Both
Day Games	31.2	29.6	30.4
Day/Night Games	31.2	27.9	29.5
All games	31.2	28.7	30.0

3.5 Opposition

Both the quality of the bowlers and the ability of the fielders contribute to how many runs a batsman will score. The quality of the opposition will vary, depending upon the composition of the players named in a team. Because the nucleus of cricket teams changes over time, to more accurately account for the quality of the opposition an interaction between opposition and time was explored and found to be statistically significant.

3.6 Team

Just as the opposition is a significant predictor of runs scored, the team that each batsman plays for is also a significant predictor of runs. This variable may well reflect different coaching or training techniques used across nations, or may simply be a surrogate marker for quality.

3.7 Performance and form

Batsman performance can be measured in various ways. Traditionally, a batsman's career average, defined by runs scored divided by times dismissed, is usually quoted by media outlets as a relative gauge for performance. Kimber and Hansford (1993) discuss the merits of this, and suggest alternative measures. In our case, as batting performance in head to head betting is only dependent upon runs scored and not whether a player was dismissed or not, career average is not an ideal predictor of performance. Instead, a player's arithmetic average, as defined by runs scored divided by times at bat is used as a predictor.

It is often said that a batsman can be in or out of form - how do we measure this? Does form mean a batsman's last innings, or his last 10 innings. In addition to each batsman's overall average, we use 10 moving averages (last score, last two scores, ..., last 10 scores) as measures of form. Using the AAE between the predicted and actual number of runs scored for each individual performance prior to the World Cup, it is possible to compare predictors of form, with the model producing the lowest AAEs being the best statistical predictor. From Fig. 4 we can see that the quality of the predictor is directly proportional to the numbers of innings used to create the average, suggesting that any predictor of form is not as efficient as an overall average of performance.

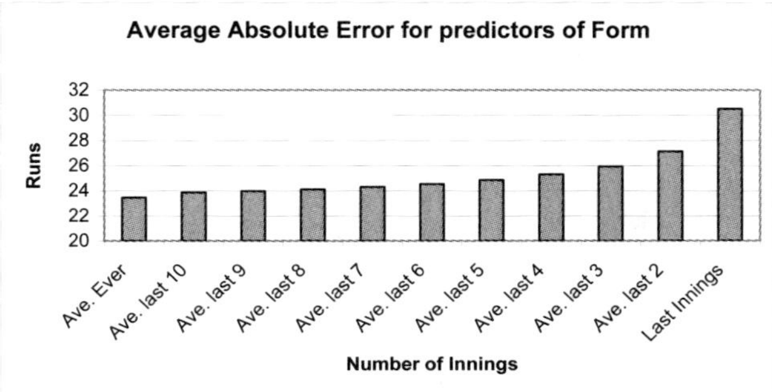

Fig. 4. Absolute errors for predictors of form.

Arithmetic averages weight each performance equally, irrespective of when it occurred. An alternative approach is to give more weight to more recent performances by exponentially smoothing past scores. It is possible to derive an unbiased prediction for future data using

$$\text{Smoothed score} = \alpha \text{ actual runs} + (1\text{-}\alpha) \text{ Previous smoothed score}$$

where α is the smoothing parameter.

In the same way that an exponentially smoothed estimate can be obtained by considering all innings played by each player, so too can we derive a smoothed estimate for performance against a specific opposition, by exponentially smoothing data from all matches played against that specific opposition. Both rolling averages and exponentially smoothed predictors were compared for each variable thought to affect the number of runs scored, with little practical difference observed between these two approaches.

4 Multivariate Models

When combined in a multivariate model, seven factors that affect batting performance remained statistically significant. The relative importance of these variables as determined by the F-statistic from the generalised linear model, can be seen in Fig. 5. A batsman's historical performance, as estimated by an exponentially smoothed predictor, was clearly the most significant factor in the model. Match time and Innings order was the second most significant predictor, whilst Home country was the third most important factor in the multivariate model. Other significant predictors include quality of the opposition, experience, the strength of the team that each player plays for, and each player's historical performance against the specific opposition.

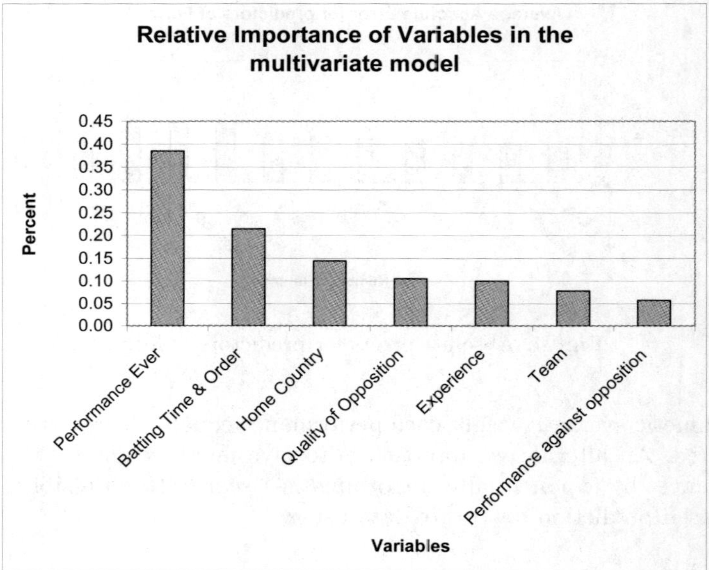

Fig. 5. Relative Importance of variables in the multivariate model.

To further account for the heterogeneity in runs scored that exists between different teams competing in the World Cup, a second approach was considered. Variables found to be significant in the multivariate model were remodelled at a team level, with different parameter estimates created depending on the relative importance of the past data with respect to each team.

About a half an hour prior to the commencement of each game, captains from opposing team toss a coin to determine which team will bat first, with the winning captain having the choice. Because it was not always possible or practical to gather this information prior to the placement of bets, a third mulitvariate model was constructed without the additional benefit of the variable Inning Order.

Although the underlying distribution of runs scored was non-normally distributed, a final linear comparison model was created without the benefit of a log-transformation prior to analysis.

5 Goodness of Fit for Predicted Score

Although past averages and exponential smoothing create unbiased estimates, to accurately measure the goodness of fit for a multivariate model, a hold-out sample separate to that from which the model was constructed, must be used to avoid the bias associated with over fitting.

From the 1936 matches played prior to the World Cup, one quarter of all matches were isolated into a hold-out sample. To avoid heterogeneity due to changes over time, a systematic sampling approach was used with every fourth match allocated into the hold-out sample. The AAE for the four multivariate models and a range of univariate predictors can be seen in Fig 6. The original log-linear multivariate model as discussed in Section 6 was the best performing prediction model, with an AAE in the hold-out sample equivalent to 23 runs. The removal of batting time had little impact on the predictive capacity of this model with the 'Early' model also having an AAE of 23 runs. Although the multivariate model constructed at a team level was significantly better performed in the training data set, the AAE in the hold-out sample was about half a run larger reflecting bias due to over fitting. The final multivariate model constructed without the benefit of a log transformation performed better than most univariate models, but was significantly worse than the log-linear approach. At a univariate level, there was little difference between using past average and exponential smoothing to create predictors. The best-performed univariate model was an exponentially smoothed average for performances based on the country in which the match was played. The two worst performing predictors were a batsman's last score, and a batsman's historical average for the venue in which he was playing.

6 Converting Predicted Scores into Head to Head Probabilities

Two methods of converting expected scores into probabilities were compared. The first exploited modeling of the distribution of scores unique to cricket. The second used the mean and standard deviation of errors of predicted scores and could be applied in other head to head situations.

When investigating the probability of one batsman outscoring another, there are advantages in assuming that scores are geometric.

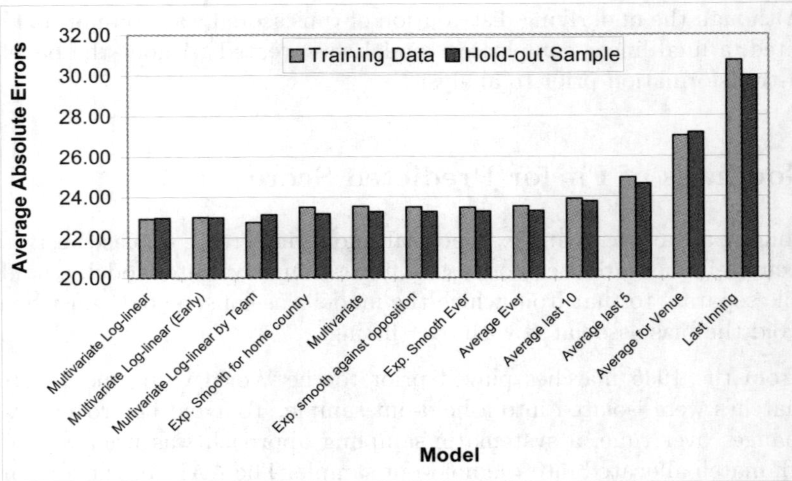

Fig. 6. Average absolute errors for prediction models.

$$\Pr\ (\text{Score} = x) = pq^x, x = 0,1,2, \text{ where } q = 1\text{-}p$$

p can be thought of as the probability of being dismissed before scoring another run, the distribution has parameters $\mu = q/p$, $\sigma^2 = q/p^2$, and is fitted using $p = 1/(1+\overline{x})$. For most top batsman, the mean is at least 30, so q is almost 1. Thus the mean and standard deviation would be approximately the same. It is easily shown the $\Pr(\text{Score} > x) = q^x$, so for two batsman we can calculate the chance one will outscore the other by summing the conditional probabilities the first will outscore the second for each possible score of the first. Thus for two batsman with parameters p_1 and p_2, we have

$$\Pr(X_1 > X_2) = \sum_{x=0}^{\infty} q_1^x p_2 q_2^x$$

$$= p_2/(1 - q_1 q_2) \quad (1)$$

$$= (n+1)/(m+n+1) \quad (2)$$

where n and m represent the predicted means for both batsmen.

This shows the odds of one batsman outscoring the other are roughly in proportion to their predicted averages. Since departures from geometric for very low and very high scores could be expected to affect both batsmen equally, we might expect this relationship to hold approximately.

Alternatively, because the difference between scores are approximately normally distributed (see Fig. 7), a player's probability of winning can readily be determined by dividing the predicted difference between players by the combined standard deviation for the two players, and comparing with a standard normal distribution.

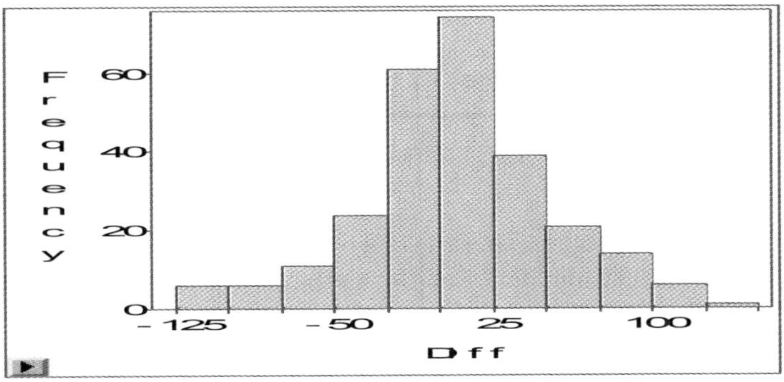

Fig. 7. Distribution of differences in scores for matched batsmen.

By creating hypothetical match-ups for the leading four batsmen for each match played prior to the World Cup and summing the log of predicted probabilities for winning players, goodness of fit could be determined. Although little practical difference existed between the approaches, a conversion based solely on the geometric distribution using (2) was statistically significantly better than a parametric approach of comparing against a normal distribution curve.

7 Head to Head Prices for the 2003 World Cup

From the 50 completed matches in the 2003 World Cup, 309 player head to head batting match-ups, were collected from three Internet sports bookmakers. On 46 occasions either one or both of the competing batsmen failed to reach the crease, leaving a database of 263 match-ups. One hundred and forty seven match-ups (56%) involved players from the same team, as opposed to match-ups from players from competing teams (44%). Prices ranged from \$1.50 to \$2.40 and were approximately normally distributed with a mean and median of \$1.90, and a standard deviation of \$0.10 (Fig. 8). The bookmaker's percentage was approximately 7%.

8 Wagering Strategy

In a fixed price market with only two outcomes, the market price becomes a reflection of supply and demand. By multiplying the predicted probability by

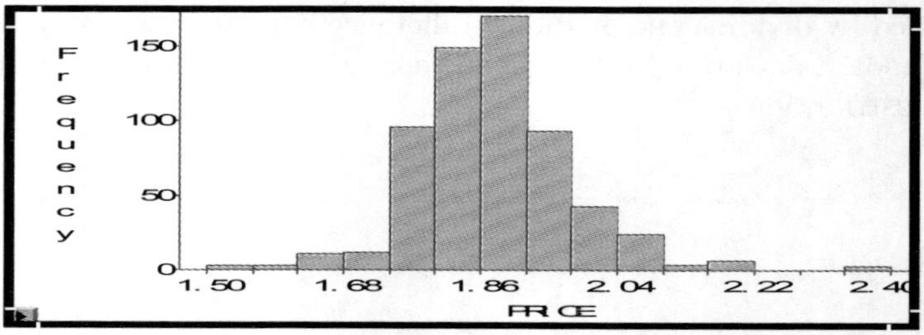

Fig. 8. Histogram of prices offered by bookmakers.

the market price and subtracting the original unit bet, it is possible to gauge the size of any market imbalance and thus the perceived advantage.

$$A = (P \times M) - 1 \tag{3}$$

where A = advantage, P = predicted probability, M = market price.

Thus for example an outcome estimated at 60% probability paying \$1.80 has an 8 % advantage.

In order to maximise growth of wealth, a betting strategy must incorporate three specific features in determining the size of the wager, namely bank size, size of advantage and probability of winning. Kelly (1956) developed a betting strategy designed to maximise the growth of wealth by maximising the expected log of wealth. This formula can be effectively simplified to

$$B = A/(M - 1) \tag{4}$$

where B = proportion of total wealth bet, A = advantage, M = market price.

Thus the above example with an 8% advantage paying \$1.80 would require a bet of 10% of total wealth to maximise expected growth rate.

In practice there are problems with using a Kelly betting strategy on a growing bank. The returns are inappropriately volatile for a short sequence of bets. In addition, in many cases bets are placed simultaneously, and the outcome of one bet (and hence total wealth) is not known before another bet is placed. The final value of wealth is also dependent upon the order in which bets occur. For this reason it is better to compare betting strategies based on a fixed bank size. It is important to note that for a wagering strategy such as the Kelly's to be successful, the modelling process must be able to produce a consistent profit whilst operating with a fixed bank size. Thus to determine the feasibility of betting on player performance in cricket, a fixed bank size has been chosen with bet size determined by only two features, size of the perceived advantage and probability of winning.

For example, in the World Cup final between India and Australia, the Indian captain Sourav Ganguly (\$1.80) was matched against fellow Indian Mohamed Kaif (\$2.00) in a head to head batting matchup. The multivariate log-linear prediction model predicted Ganguly to make 22 runs whilst Kaif was predicted to make 15 runs. Using equation (2),

$$\Pr(\text{Ganguly} > \text{Kaif}) = (22{+}1)/(22{+}15{+}1) = 0.605$$

From (3), the perceived advantage A $= 0.605 \times \$1.80 - 1 = 0.09$

From (4) the given betsize B$= 0.09/(1.8\text{-}1) = 0.11$

Based on a fixed bank size of \$1000 dollars, a betsize of .11 $\times$\$1000 = \$110 was wagered on Ganguly to outscore Kaif. Ganguly actually made 24 runs whilst Kaif made 0. The resulting profit from this wager was equal to \$110 $\times$ (1.8-1) = \$88.

9 Return on Investment

ROI is defined as profit divided by money outlaid and is expressed as a percentage. Whether a specific betting market is inefficient or not can be determined by calculating the ROI offered by specific prediction approaches. From Fig. 9 it can be seen that although most univariate models were unable to produce a positive ROI, all four multivariate models produced ROIs in excess of 20%. The most productive was the mulitvariate log-linear model with a ROI of 36%, followed by the multivariate log-linear model at a team level, which produced a ROI of 31%. The loss of information on which team would bat first reduced the ROI from 36% down to 23%, whilst the multivariate linear model also proved quite profitable with a 21% ROI.

Surprisingly, the univariate models that looked to rate player performance by using historical averages (Ave. Ever & Ave. Last 10) or exponentially smoothed predictors (Exp. Ever) all had negative ROI in excess of 10% with the average of a batsman's last 10 innings being the worst performing model with losses in excess of 30%. This most likely reflects the approach that bookmakers use to set their prices, with over compensations made for these factors.

From Table 2 it can be seen from the number of bets that the multivariate models were more selective in determining when market imbalances and thus betting opportunities were available. The model that produced the greatest raw profit was the multivariate log-linear model at a team level, although this model also had the largest degree of risk of the multivariate models, as the average bet size was larger.

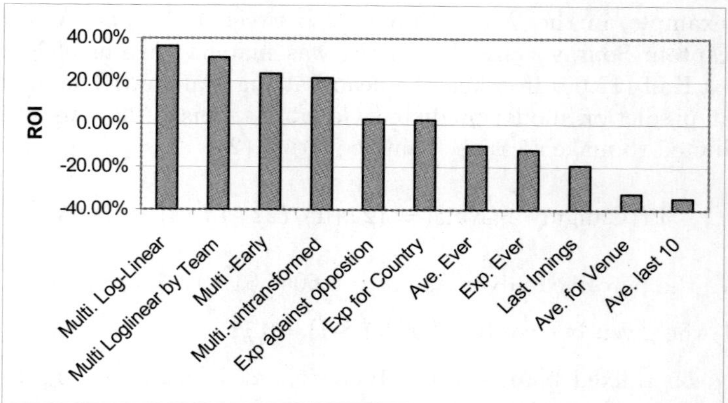

Fig. 9. Return on investment.

Table 2. Betting results from the 2003 cricket World Cup.

Model	Number of bets	Average bet size($)	Outlay fixed bank($1000)	Profit($)	ROI
Multi. Log-Linear	107	80	7480	2720	36.3%
Multi. Loglinear by Team	133	115	15250	4740	31.1%
Multi. – Early	110	63	6960	1640	23.5%
Multi. Linear	63	47	2950	630	21.6%
Exp. against oppostion	164	104	17010	410	2.4%
Exp. for Country	184	116	21290	450	2.1%
Ave. Ever	110	63	6910	-690	-10.0%
Exp. Ever	145	103	14930	-1790	-12.0%
Last Innings	246	180	99170	-19030	-19.2%
Ave. for Venue	139	90	37170	-12050	-32.4%
Ave. last 10	183	137	25060	-8660	-34.6%

10 Conclusion

Theoretically, the modelling approach that best predicts the number of runs that batsmen will score should also be the modelling approach that produces the greatest profit. Comparing the goodness of fit and the ROI for a range of models allows us to explore this relationship. From Fig. 10 it can be seen that although the relationship between goodness of fit and ROI is not perfectly linear, a statistically significant relationship does appear to exist. The negative relationship between ROI and AAE suggests that the models that produce the smallest errors for the data prior to World Cup also produce the best returns on investment for World Cup data. The implication of this relationship allows for the construction and improvement of prediction models without the necessity of large amounts of information on bookmakers' prices.

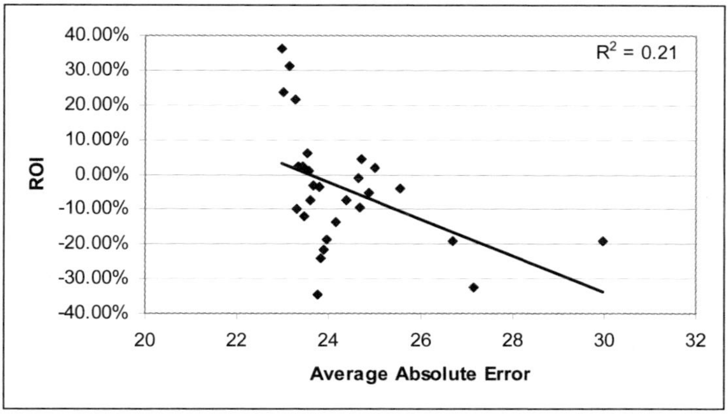

Fig. 10. Relationship between ROI and average absolute error.

Contrary to popular perception, price setting by bookmakers for sporting outcomes is not an exact science, but rather a combination of experience, knowledge and intuition. With the use of all ODIs ever played, it is possible to identify features that are known to affect batsman performance. The combination of variables in a multivariate model enables a process of comparison between batsmen that is clearly superior to that offered by bookmakers.

One of the main attributes of using a computerised modelling approach is consistency. The computer can repeatedly offer an objective price at the push of a button. Because of the infrequency in which cricket nations meet, the reduced margin of error (bookmaker overround) and the immaturity of the player performance betting market, it is of no surprise that a mathematical modelling approach for batsmen performance can be profitable.

Although ROIs in excess of 20% were achievable in theory, in practice this figure was a little less. Using a log-linear multinomial approach, 99 bets were actually placed by one of the authors during the World Cup, with a ROI of 16%. This difference between theoretical and practical profit can be primarily attributed to the time delay between recording prices and placing bets. Bookmaker prices were generally recorded about 24 hours prior to the match, whilst bets were usually placed in the final hours before the start of each game. Because of the dynamic nature of betting markets in which prices change in accordance with supply and demand, the greatest inefficiences found in the market occur when prices are initially posted. It was our experience that the most glaring inefficienceies identifed by mathematical modelling were often also identifed by the general public resulting in a reduction in price and a subsequent reduction in profit.

Having established a profitable price comparison process, addition benefits could be achieved in the future with a wagering strategy that was also dependent upon bank size as well as the probability of winning and the perceived advantage of the bet.

Mathematical models clearly have a role in the setting of prices for sports betting. Not only can they be more accurate than traditional subjective measures, but in some newer applications of betting during a sporting contest, their advantage in speed of calculation is paramount. For example, the authors have developed a model used for setting odds on the number of runs to be scored in the next over of one-day cricket. The short time frame to set odds, immediately one over is finished to allow betting before the next over begins, discounts traditional setting by a bookmaker. Ladbrokes successfully implemented this system during the cricket World Cup. With bookmakers offering an increasing range of bets on worldwide sporting events, a detailed analysis by bookmakers (or punters) with expert knowledge of those events is becoming more and more difficult. On the other hand, increasing availability of large amounts of data via the WWW is facilitating the development of computer driven mathematical models. It may be that such models will gradually replace traditional methods of setting odds by bookmakers, or searching for inefficiencies in markets by punters.

References

1. Bailey, M. Identifying arbitrage opportunities in AFL betting markets through mathematical modelling. In G. Cohen and T. Langtry (editors). *Proceedings of the Fifth Australian conference on Mathematics and Computers in Sport.* Sydney, University of Technology Sydney: 37-42. 2000.
2. Benter, W. Computer Based Horse race handicapping and wagering systems: a report. In L. Hausch (editor). *Efficiency of racetrack betting markets.*pp 183-198. Ziemba. San Diego, Academic Press Limited. 1994.
3. Bolton, R. N. and R. G. Chapman. A Multinomial Logit Model For Handicapping Horse Races. In L. Hausch (editor). *Efficiency of Racetrack betting markets.* pp 151-172. Ziemba. San Diego, Academic press limited. 1994.
4. Chapman, R. G. Still searching for positive returns at the track: Empirical results from 2,000 Hong Kong Races. In L. Hausch (editor). *Efficiency of racetrack betting markets.* pp173-183. Ziemba. San Diego, Academic Press Limited. 1994.
5. Clarke, S. R. Test Statistics. In J. Bennett (editor). *Statistics in Sport*, pp 83-101.Arnold, 1998.
6. Dana, J. D. and M. M. Knetter. Learning and efficiency in a gambling market. *Management Science* **40**(10): 1317-1328. 1994.
7. Dixon, M. J. and S. G. Coles. Modelling Association football scores and inefficiencies in the UK football betting market. *Applied Statistics* **46**: 265-280. 1997.
8. Elderton, W. E. Cricket scores and some skew correlation distributions. *Journal of the Royal Statistical Society (Series A)* **108**: 1-11. 1945.
9. Kelly, J. A New Interpretation of Information Rate. *Bell System Technical Journal* **35**: 917-926. 1956.

10. Kimber, A. C. and A. R. Hansford. A statistical analysis of batting in cricket. *Journal of the Royal Statistical Society A* **156**(3): 443-445. 1993.
11. Wood, G. H. What do we mean by consistency? *The Cricketer Annual*: 22-28. 1941.
12. Wood, G. H. Cricket scores and geometrical progression. *Journal of the Royal Statistical Society (Series A)* **108**: 12-22. 1945.

Searching for Patterns in Sports Contests

Tim McGarry

University of New Brunswick, Canada
tmcgarry@unb.ca

1 Introduction

There is structure in sports contests. Identifying the presence of such behavioral patterns in sports contests using a formal level of description, however, remains very much in its infancy. In short, the patterns that we think we see in sports contests are at present very difficult to identify in the recorded data that purport to describe them. We report on some of the challenges that have presented themselves thus far in obtaining a satisfactory description for sports contests. These challenges lead us in this article to consider evidences from other research inquiries that may well inform on our search for behavioral structures in sports contests. To this end we present some ideas for future research in our efforts to explain the behaviors that give shape to various types of sports contests.

2 Structure in Sports Contests[1]

2.1 The written rules

Structure in sports contests depends in the first instance on the game rules. On this basis sports contests may be divided into at least two separate categories. Score-dependent sports such as tennis and volleyball, for example, finish only when a given score is reached. Time-dependent sports, however, such as basketball and hockey finish when a fixed amount of time expires.[2] In addition, score-dependent sports tend to follow structured sequences, with

[1] In this article, we limit our consideration to sports that consist of two players or teams that compete against each other.

[2] Tournament organizers may select to use various methods of breaking a tie at the end of game time in these types of sports. Nonetheless, we still consider these types of sports as time-dependent.

possession being traded between opponents in equal fashion by virtue of the game rules. Then again, time-dependent sports exhibit looser structures because of the unequal exchanges that occur as possession is won and lost. For further information, the reader is referred to Hughes and Bartlett (2002) for an extended description of sports structures as based on Read and Edwards' catalogue of net-and-wall games, invasion games and striking-fielding games.

2.2 The unwritten rules

We noted briefly that the game rules influence the type of structure that various types of sports contests exhibit. Of course, sports within the same classification system exhibit unique patterns of behavior to that sport. Similarly, contestants within the same sport seemingly demonstrate unique, or signature, patterns of play, as the sports practice of scouting the opponents would suggest. The patterned (structured) behaviors of players within sports contests, and the resultant patterned features that typify sports contest, serves as the focus of interest for this article.

While the playing patterns in sports contests are produced within the constraints of the game rules, we should note that the patterns themselves are not brought about because of the game (system) constraints. Instead, the patterns are formed as a result of the interactions among the contestants, with the game rules only limiting the types of behavioral patterns that are possible. In the language of complex systems, it might be suggested that the system self-organizes within the constraints that surround it. Other than game rules, the behavior of the system might be constrained by some hierarchy of customs and habits that results in a set of strategies and tactics. For example, in squash it is customary for each player to seek control of the T-position (centre-court). Other system constraints will include the cognitive, psychological, physical and technical factors, each of which will of course influence the types of decisions and actions of the players. These considerations may go some way to explaining the patterned features that are unique to any given sport.

Visual inspections of sports contests lead us to accept the presence of structured behaviors. There would therefore seem to be a good justification (on the grounds of experience) for such an acceptance. Scientific undertakings would thus be expected to offer further justification on the presence of sports structures, however, the reliable detection of patterns in sports contests remains surprisingly equivocal to date. Unfortunately, the scientific method has yet to identify, at a formal level of description, the patterned behaviors that typify various sports contests. The unwritten rules that describe and give structure to sports contests have yet to be written, if you will.

3 Uncertainty (Chance) in Sports Contests

The behaviors and outcomes of sports contests are uncertain and may thus be described in terms of chance (or probability). The sequences of sports actions (i.e., behaviors and outcomes) can therefore be analyzed using a probability structure. In this view, the next action in the sequence is linked to the former action on the basis of probability. In this section, we present some brief examples of probability analyses as used to investigate sports behaviors.

3.1 Probability structures

The frequency distribution of passing sequences in soccer is well described using the negative binomial distribution (Reep & Benjamin, 1968). In short, the frequencies of passing sequences get fewer as their lengths get longer and they do so in a lawful way as described in the negative binomial. In addition, the shot/goal ratio showed good stability at the level of about 10 shots per goal. These findings led Reep and Benjamin to the interpretation that chance wields an important influence on the behaviors in soccer. Some soccer practitioners interpreted Reep and Benjamin's data as a template for soccer success, upon which the "long-ball" or "direct-play" strategy was formulated, whereas others rejected this interpretation. While the particulars of this interpretation for sports practice remains a point of contention, there is little serious argument with regard to Reep and Benjamin's thesis that soccer behaviors can be described (and prescribed) on the basis of chance.

If sports behaviors can be described in terms of probabilities (see above), then it follows that managing chance so as to maximize the upsides (rewards) and minimize the downsides (risks) will yield good dividends. With this in mind, we (McGarry and Franks, 1996) analyzed the shot selection patterns in squash contests as a probability structure as follows. First, the probabilities of various types of shot responses, as well as various outcomes, were recorded as a function of various types of antecedent shots for each player in a squash contest. Thus, the data from a squash contest yielded a "playing profile" for each player (see Table 1). The playing profiles were analyzed for signature patterns of behavior using statistical analysis (see Section 3.2).

Table 1 presents data from a playing profile for an individual. The playing profile contains the shot response matrix (i.e., shot x shot) as well as the outcome matrix (shot x outcome). The shots listed in the first column are those of the player whereas those in the first row (see prime symbols) are the antecedent shots of the opponent. Thus, in response to a serve from the opponent, the player responded with 0.04 (or 4%) drive shots, 0.08 (or 8%) boast shots, and 0.88 (or 88%) volley shots from the repertoire of shots available to him (or her). Similarly, the player responded to a drive from the opponent with 0.59 (or 59%) drive shots, 0.13 (or 13%) boast shots, and so on. The transition probabilities between each possible shot pair (i.e., the antecedent shot from the opponent and the shot response of the player) are therefore contained within the shot response matrix. In the same way, the transition probabilities between the shot response of the player and the outcome associated with his (or her) shot, if any, are contained in the outcome matrix. For example, a drop shot by the player in response to any type of antecedent shot from the opponent, yielded 0.30 (or 30%) winners, 0.08 (or 8%) errors and 0.03 (or 3%) lets, respectively. The opponent responded to the remaining percentage (i.e., 59%) of drop shots that failed to produce an outcome as determined from his (or her) playing profile (not shown). The next section details how the playing profiles were used to analyze sports behavior.

Table 1. Example of a playing profile recorded from a squash contest.*

	Serve′	Drive′	Boast′	Drop′	Volley′	Lob′	Winner	Error	Let
Serve	0.00	0.00	0.00	0.00	0.00	0.00	0.00	0.00	0.00
Drive	0.04	0.59	0.31	0.26	0.56	0.27	0.03	0.01	0.02
Boast	0.08	0.13	0.13	0.03	0.14	0.11	0.08	0.08	0.00
Drop	0.00	0.05	0.51	0.54	0.12	0.05	0.30	0.08	0.03
Volley	0.88	0.23	0.00	0.00	0.15	0.57	0.10	0.03	0.05
Lob	0.00	0.00	0.04	0.17	0.03	0.00	0.00	0.00	0.00

* The shot responses in the first column and the outcomes in the first row are those of the player. The shot responses in the first row (indicated using the prime) are the antecedent shots of the opponent. See text for further details.

3.2 Markov processes

The patterned behaviors of squash players were described using a probability matrix on the hypothesis that the signature features that characterize a player's behavior would be contained within that playing profile. The idea, then, is that each profile is unique to the player, and that these characteristics yield important information for future contests. To investigate this hypothesis we analyzed the behavior of squash contests as a stochastic (Markov) process (see Fig. 1) as described in brief below (see McGarry and Franks (1996) for further details). For a recent analysis of the actions of soccer contests using a similar type of probability analysis, see Mendo and Anguera (2001).[3]

The squash contest consists of an array of shot types (Serve, Drive, Boast, Drop, Volley and Lob) and outcomes (Winner, Error and Let) whose behavior is characterized in the random transitions between states (e.g., Serve → Volley → Boast → Drop → Drive → Winner). The stochastic process contains Markov properties because the transition between any pair of states depends only on the present state. Thus, the sequence of steps that the system took

[3] This article is written in Spanish.

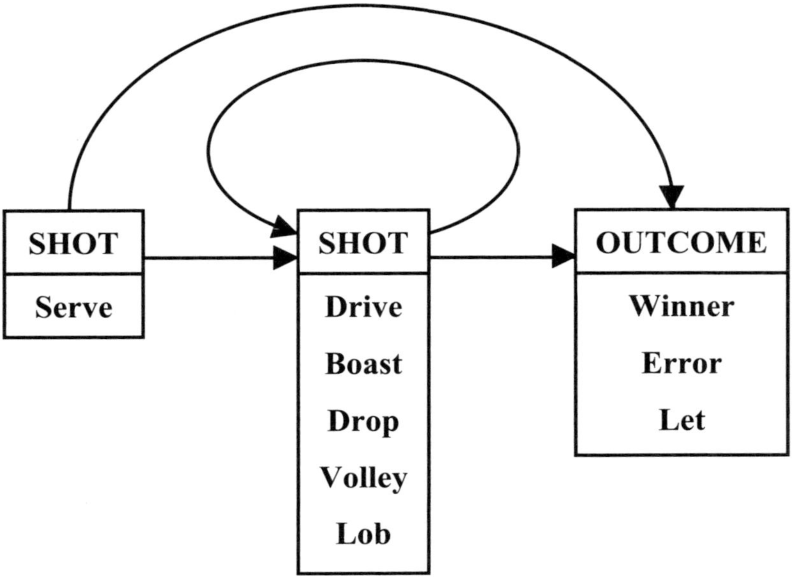

Fig. 1. Stochastic (Markov) processes for the sequence of shots and outcomes in squash.

to reach the present state bears no influence on the next step that the system will take. In the first instant the squash contest "lives" in the state of Serve, which marks the start of the squash rally. Thereafter, the system switches at random among the various states as based on the probabilities contained in the playing profiles (cf. Table 1). In the last instant the squash contest lives in one of the outcome states, which marks the end of the squash rally. The path (known as a random walk) that the system took represents the stochastic sequence of shots and outcome that describes that squash rally. The process repeats itself from squash rally to squash rally until a winning score is reached whereupon the squash contest ends.

3.3 Transition probabilities

If a playing profile contains a signature pattern of behavior, as hypothesized, then the playing matrix should not change significantly from one contest to the next. The transition probabilities contained in the playing profile should therefore be invariant. We (McGarry & Franks 1996) used χ^2 analysis to examine the playing profiles for invariant properties. The data yielded mixed findings. In general, the playing profiles of squash players changed as the opponent changed, however, the playing profiles tended to show more invariance when the context in which the antecedent shot was specified was more detailed. These results illustrate the importance (and difficulties) of finding invariant features in sports behavior (cf. structure) if the data from a sports contest are to produce useful information for future applications.

The use of behavioral probabilities as a management aid in the decision-making process is widespread in many different fields (e.g., actuarial tables, insurance statistics, stock market analyses, etc.). The behaviors in sports contests would likewise seem to present themselves as good candidates for some type of probability analysis. Indeed, some game strategies for sports contests are often made on the basis of future expectations, as for example are strategies based on "percentage play". This said, some limitations of using a probability analysis for squash contests based on Markov processes were reported (see McGarry & Franks, 1996, for further considerations on this point). These findings demonstrate that a system description of sports contests based on transition probabilities might not be as straightforward as first imagined.

Sports analysts have tended to record all of the data from a sports contest and to search those data for performance measures, a practice that assumes each datum item is of equal importance in its information value. Our consideration of squash contests led us to question this assumption (McGarry & Franks, 1996). Instead, we suggested that some actions in a sports contest might reasonably be expected to be more important than other actions. For example, the shot that creates an opening on which to mount a successful attack might be considered as being more important than a shot that simply trades possession within some type of stable exchange. If some shots in a sequence of shots are more important than other shots, then the problem for a probability analysis is that some important transitions are masked by lesser important transitions. In the next section, we present the possibility for sports contests of probability structures that exist below the surface level on which they are recorded.

3.4 Unidentified probability structures: T-patterns

The use of probabilities for analyzing the transitions between sports behaviors may take other structures than the Markov chain that we used. For example, imagine a sequence of behaviors as represented in the upper panel of Fig. 2. On closer inspection, we see that some repetitive sequences are immersed within the data sequence (Fig. 2B and 2C). These types of repeat behaviors would most likely remain undetected when using the traditional types of stochastic analysis that we reported for squash contests. This section reports in brief on how these types of buried patterns within the behavior stream, referred to as T-patterns (see Magnusson, 2000), might be subject to detection using appropriate statistical analysis.

The term T-pattern was coined by Magnusson to distinguish this type of pattern as a particular kind that is based on temporal structure. The central

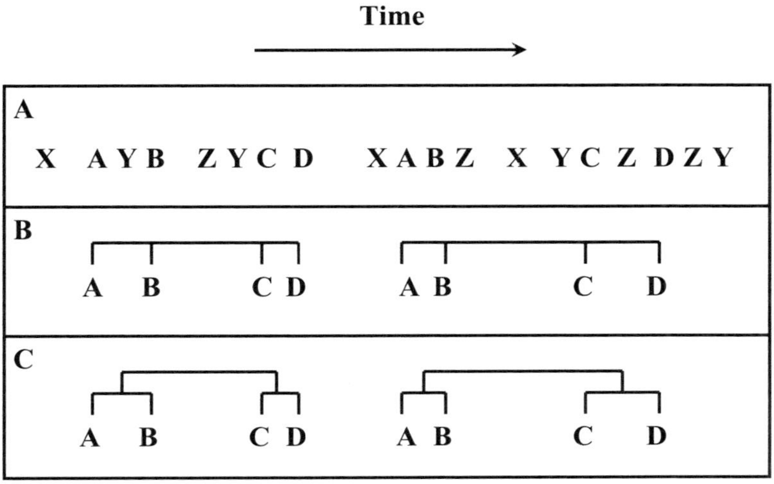

Fig. 2. Example of buried patterns within a data stream as detected using T-pattern analysis. (**A**) Sequence of behaviors on a time line. (**B**) Repeat sequences of behavior (ABCD) drawn from the data stream. (**C**) Repeat sequences of behavior (AB)(CD), albeit in a different structure from that identified in panel B. Not all possible structures for repeat sequences are presented. Adapted from (Magnusson 2000, Fig. 1).

theme of Magnusson's T-pattern analysis is to pull out repeat sequences in the data stream that might otherwise be masked with other data interspersed between their occurrences. In Fig. 2, the second panel provides an example of a repeated data structure (ABCD) that is extracted from amongst other behavioral events within the data stream (i.e., X, Y and Z's). The third panel provides another example extracted from the initial data set, where in this instance the data structures (AB and CD) are repeated in a binary tree type of relation. Other data structures are also possible for extraction (e.g., AD and BC), but such structures are not presented in Fig. 2 for reasons of expediency. In this way, T-patterns are identified within the behavioral sequences that might well otherwise be unattainable from other methods of analysis. For an application of T-pattern analysis to soccer, see Borrie et al. (2002).

The details for data extraction from a behavioral stream of unknown temporal structures are documented in Magnusson (2000). The basis of the analysis lies in a hierarchical structure in which a "... bottom-up detection strategy ..." is implemented using a "... recursive binary tree condition ...". Since there are many possible data structures that can be detected within a data set, a bottom-up strategy is used to find simpler types of patterns first, and thereafter more complex patterns as composites of these simpler patterns. One drawback of this type of analysis is the generation of data patterns that are the result of chance occurrences. This drawback notwithstanding, the detection of T-patterns within a behavioral stream, while interesting, offers little information as to their root causes. In some instances it may therefore be difficult to assign meaning to T-patterns post detection.

3.5 Situational probabilities

Some recent evidences for the use of situational (subjective) probabilities – a term used to refer to the context of the situation on which the probabilities are based – were reported from an analysis of squash behaviors (Abernethy et al., 2001, Experiment 2). These authors used simulated squash contests against the same squash player to assess the ability of experts and non-experts to predict the direction and depth of the opponent's shot on the basis of advance visual cues. Experts were better than non-experts, as expected, when the visual information was occluded using liquid crystal goggles at various times during the opponent's stroke production (up to 260 ms after ball contact). The experts likewise retained an advantage over the non-experts when the visual information was occluded at various times before the stroke was made (up to 620 ms before ball contact), with the reported judgment errors of experts being less than that expected from chance. These findings support the earlier reports of Alain and colleagues (e.g., Sarrazin et al 1983, Sarrazin et al 1986, Alain and Sarrazin 1990) as well as that of Buckolz et al. (1988).

Using computer simulations to analyze the decision-making processes of shot selections of the defending player in squash, Sarrazin et al. suggested that the defending player uses various sources of information in formulating the subjective probabilities for the next shot by the opponent. The information sources identified by these authors were the respective court positions of the players, the playing habits of the attacking player, the attacking player's ability to accurately place the ball in the intended location, the angle of the attacking player relative to the front wall, the motion of the player when striking the ball, and the flight trajectory of the ball. In addition, these authors reported that while all the information need not be used in a single prediction, the court positions of the players was used most of the time, either alone or in combination with other information sources. Other factors such as playing experience, time constraints, the expected shot responses of the opponent, the performer's confidence in his (her) prediction and the mistakes caused by incorrect predictions were also listed as possible influences on the decision-making process.

Similar findings were reported in tennis (Buckolz et al, 1988). These authors reported that the defending player in tennis uses two primary information sources on which to anticipate the passing shots of the attacking player.

The "context" cues include situational variables such as knowledge of the opponent's strengths, weaknesses and preferences, the position of the two players on court and the surrounding conditions. The "body language" cues include the stance, visual focus and racket preparation. These authors further suggested that the contextual cues are usually available first and that they may, or may not, be combined at a later time with the body language cues. The existence of body language cues for accurately predicting the passing shot types in tennis were reported, together with the finding that the more advanced players were better able to predict these outcomes than were lesser able players.

In brief, the many information sources available to sports contestants must be processed quickly if a player is to make an accurate decision in a timely fashion. Indeed, a hallmark of sports excellence, albeit anecdotal, is the seeming abundance of time afforded to the expert player, even when the temporal constraints of the situation remain tight. One possibility is that the latency costs for experts are lessened as a result of reducing the information that must be attended to, perhaps as a result of somehow prioritizing the information within the visual array. The temptation to posit a kinship between sports contests and chess games is a strong one, given Chase and Simon's (1973) observations from de Groot's work that expert chess players are "...very good at coming up with the 'right' moves for further consideration ...' (p. 55). Furthermore, the latency with which players can detect patterned information on the board depends on their chess skill. The next section continues with this line of thought.

4 Pattern Detection

Using human raters offers an interesting way forward of trying to uncover the important information that underlies the detection of patterns in sports contests. This method has been used before to good effect in various other studies of human behavior. One advantage of using human raters is that the experimenter can manipulate the information presented to the viewer in various ways. In this way, the information that is necessary for pattern detection might be teased out from superfluous information that may be contained within the visual display. This section details some studies that have used human raters as a means of getting at the information that gives rise to the processes of pattern detection. In this regard, we present some studies on pattern recognition in chess that offer useful insights on the extensive knowledge structures that expert chess players acquire. Other studies have sought to analyze sports contests using similar methods. In addition to recognizing patterns of static stimuli within the visual field (cf. chess), we detail some studies of pattern detection using dynamic stimuli. These studies used light-point displays of locomoting humans to investigate the processes of dynamic pattern recognition. We propose that a similar technique for identifying patterns in sports contests might likewise yield valuable information. This suggestion is in keeping with our recent considerations of sports contests as dynamical systems (see McGarry et al., 1999, 2002; McGarry, in press), where the spatial-temporal relations of the players are considered as a key descriptor for the system.

4.1 Static pattern detection: Structured versus unstructured patterns

Squash is sometimes described using colloquial language as high-speed chess, a reference of course to the importance attached to the decision-makings of squash players that must be made under demanding temporal constraints. In chess (and squash), the players must select an action from an array of possible actions in an alternating sequence. Since the intention behind each action is to maximize the benefits accrued to the player, and to minimize the opportunities afforded to the opponent, an analysis based on game theory might be predicted to yield good dividends for an understanding of chess plies.[4] We note with some interest that a probability analysis of chess plies would not be expected to yield good information in this regard. For example, even if the transition probabilities between chess plies were to give a good long-run approximation of a chess game, an unknown supposition, we may assert with some certainty that the transition probabilities between chess plies would depend on the game context (cf. squash). This is known because of the lawful relations that exist in a chess game. For instance, each chess piece is limited in its possible travels on the chess board by the game rules, a check on the king must be answered immediately, threats tend to be countered, when possible, and so on. Thus, the rules of the game, the strategies of the players, and the like produces patterns of chess relations that in all likelihood cannot be described in meaningful terms using transition probabilities. This same point is made with much more force by de Groot (1966, p. 39) who states that "...superior achievement of masters in perceptual experiments can not be explained by a supposed *general* knowledge of chess possibilities and probabilities. Their superior performance in perception experiments is not based on the availability of (first order) 'probability tables' in the backs of their minds." The same comment might well apply for sports contests too.

De Groot was one of the first authors to report on the recognition patterns of chess players. This author analyzed the recall ability of various master level chess players by providing them with a brief visual display of various structured chess positions taken from the mid-stages of a game. The masters demonstrated better recall than less expert chess players, as expected, suggesting an extensive knowledge (memory) base for chess patterns, presumably

[4] We use the definition of Chase and Simon (1988) who referred to a chess move as consisting of two successive plies. Thus, one ply by each player (i.e., white and black) comprises a move in chess. In keeping with our considerations of chess and squash, a ply in chess would be equal to a shot in squash.

as a result of repeated exposures. Importantly, however, the memory for patterns is specific to chess games, a feature demonstrated by Chase and Simon (1973).[5] Unlike de Groot, these authors reported a study on chess that used unstructured (random) patterns as well as structured patterns. The master demonstrated superior recall for the structured patterns but not for the unstructured patterns. Thus, the knowledge structures that support the recall of chess patterns is a function of repeated exposure, instead of the result of a better short-term memory per se. Subsequent studies of sports contests from football (e.g., Garland and Barry, 1991) and soccer (e.g., Williams et al., 1993), amongst other sports, reported finding similar differences between experts and non-experts for structured patterns only. These data led both groups of authors to suggest a specific cognitive advantage for the experts based on their prior experiences. Other authors on chess patterns suggest a perceptual advantage afforded the experts as a result of previous experiences (e.g, Reingold et al., 2001).

The general thesis for the improved recall of experts is based on the well-documented idea of chunking, whereupon bits of information get grouped together in larger chunks as learning progresses. For example, Bryan and Harter (1899) showed that telegraphers of the Morse code language first learned to recognize individual letters of the alphabet as a particular sequence of dots and dashes. In time, however, the telegraphers learned to identify longer sequences of dots and dashes as words instead of letters, and then later as phrases instead of words. These learning features were referred to as a hierarchy of habits (Bryan and Harter, 1899). Thus, the learner will recognize more complex patterns by the process of chunking, a process that changes the way that the information gets coded. This change seems necessary to get around the limits of short-term memory (approximately 5 to 9 items of information, sometimes referred to as 7 plus or minus 2). The reader is referred to Miller (1956) for further details on the limits of short-term information processing. The knowledge structures in long-term memory, which the short-term memory access, are not subjected to these processing constraints.

4.2 Structured versus unstructured patterns in sports contests: Some possible limitations

There is a key distinction in the processes of pattern recognition between the chess studies and sports studies. In chess, the patterns to be detected (and recalled) are static, whereas in sports contests they are dynamic. Some

[5] The following excerpt is taken from de Groot (1966). "... a group of Russian investigators (Djakow, Petrovsky, and Rudik, 1927) demonstrated that chess masters are *not generally* superior to control subjects in visual memory achievement. By their *specialty*, therefore, masters must have a superior coding system immediately available. What they 'see' must be something quite different from what lesser players see." (p. 35).

early studies of sports contests used schematic diagrams (e.g., Garland and Barry, 1991) presented in static form, thus yielding good correspondence with the research on chess at the possible expense of some field validity. Other studies (e.g., Williams et al., 1993) used video clips of short durations (e.g., 10 seconds) following which the final positions of the players were recalled by the observers. The unclear relation between assessing pattern recognition using recall of the last instant of a dynamic visual display requires further scrutiny. Studies of biological motions (see Section 4.5) have demonstrated that dynamic pattern recognition is strongly dependent on the relations among the moving parts, and that these relations often vanish on the instant that the dynamic visual display is stopped (see Fig. 3, for an example from gait pattern analysis). Thus, the structure in a dynamic display is oftentimes not detected as structure in a static display. This concern may be weakened for sports contests when structure in static displays is identifiable from recognized formations (cf. Garland and Barry, 1991), however, the possible confoundings of structure in dynamic versus static displays should be considered further. For example, it remains unknown whether the method of recall of final positions from dynamic patterns post presentation gets at dynamic structure or at static structure as based on the final instant (i.e., the last static representation in a dynamic series of static representations). Our instinct leads us towards the latter interpretation.

The reports that experts are better able than non-experts to identify patterns in structured but not unstructured behaviors provide good evidence for the existence of structure in sport. Unlike the lawful relations that give rise to chess patterns, however, the nature of structure in sports contests is not well understood in terms of a formal level of description. The reason for this is that the information that is used to detect patterned behaviors in sports contests remains unidentified for the most part (cf. chess). Further studies on expert and non-expert differences will be useful in this regard, particularly if the information that is presented in the structured displays can be usefully manipulated by the experimenter for further analysis (see later).

4.3 Visual search mechanisms

Treatment of the possible mechanisms that lie behind the expert knowledge structures are beyond the scope for this review. For our purposes, we simply note that the search processes that the viewer undertakes might, or might not, change as a function of expertise, with most studies tending to support the former position. Eye registration systems allow for the detection of gaze when scanning a visual scene, with the experts usually showing fewer scans and longer fixation points. In addition, the locus of the fixations seems to change with expertise, thus yielding possible hints as to the information sources that are most important to the viewer. Of lesser contention perhaps is the suggestion that the key differences between experts and non-experts are found

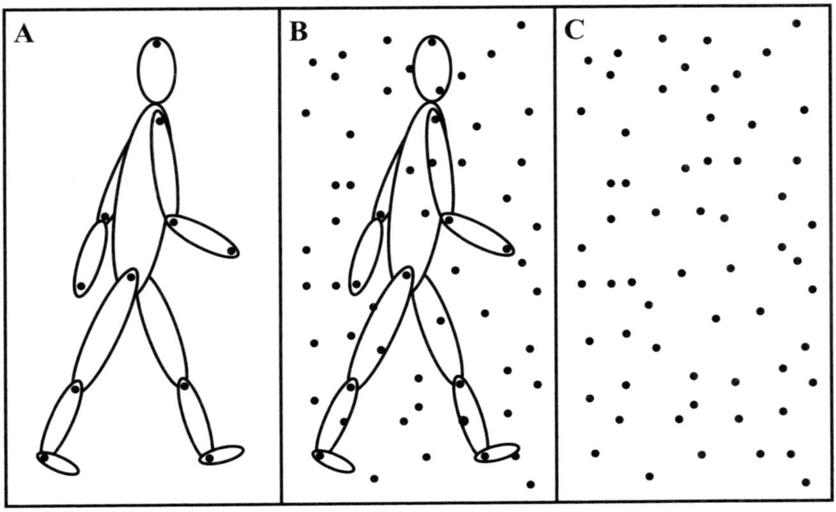

Fig. 3. Point light display of a walking human. (**A**) Point-light display depicting a walking human from markers attached at landmarks on the anatomy. While the outline of the human is presented for ease of exposition, it is not presented in the point-light display proper. (**B**) The walking human can be detected quite easily in the point-light display, even when presented against a dynamic background display of like random points. (**C**) The image of the walking human vanishes on the instant that the complete display is stopped, i.e., when the points are presented in a static display. Adapted from Bertenthal and Pinto, 1994, Fig. 1.

in the visual "software" as opposed to the visual "hardware". This is to say that experts do not detect patterned features of sports behaviors more readily because they demonstrate better eye function such as visual acuity and depth perception (e.g., Ward and Williams, 2003). Instead, experts are better able to recognize patterns by picking up information from various aspects of the visual scene and assembling the pieces of information into a coherent percept.

In their report on de Groot's findings from chess, Chase and Simon (1988) indicate that the search heuristics of experts and non-experts are similar. For example, the search through a decision tree of possible moves is usually not wider nor deeper for experts than for non-experts, and in fact might even be a little narrower and/or shallower instead. For example, the search path for a difficult position may typically consist of 30 to 50 moves and go to a depth of only 2 or 3 moves. In addition, it is unusual for a player regardless of level to search a tree of 100 plus moves, or to search deeper than 5 moves (Chase and

Simon, 1988).[6] The exception to this finding of a shallow search is through a series of forcing moves, whereupon the search path can be quite extensive.

In sum, experts do not see further ahead than non-experts, as might otherwise be supposed. Instead, it seems that the key distinction in the search processes between experts and non-experts is that somehow the former do not spend time on bad moves, rather the best moves somehow present themselves to the experts for further consideration. This type of finding reminds us of the psychological phenomenon known as "pop-out", in which the searched target seemingly jumps out from a distracter set due to parallel processing (Saarinen, 1997). Indeed, Reingold et al. (2001) suggested the presence of "... automatic and parallel encoding procedures for chess relations in experts" (p. 504) after using latency reactions to assess the processing costs of checking status on chess relations. This said, any connection of visual pop-out to the abilities of chess experts to quickly reach the most appropriate decisions remains a matter of speculation at this time.

4.4 Dynamic pattern detection: Examples from studies on biological motion

Space and time are key variables for sports contests that players try to control. In this section, we reflect on these variables in our search for patterned behaviors. Specifically, we seek to identify patterns in the spatial-temporal relations among the players on the field of play. This standpoint represents a shift in focus for our analysis of sports contests, from the actions of players in a behavioral stream to the relations in space and time that exists between the players, much like would be provided from a "birds eye" view of the playing surface. In this regard, we have begun to consider sports contests in terms of dynamical (self-organizing) systems, whereupon the behaviors emerge spontaneously from within the system as a product of constraints imposed on the system (see McGarry et al., 2002; McGarry, in press). For reasons of space, we will not pursue this line of inquiry further in this article. Instead, we will present some considerations from the extant literature on studies of biological motion with the intent of informing further on our continuing search for patterned relations in sports contests.

The point-light method was first used by Marey to investigate movement features by highlighting certain landmarks of the anatomy (e.g., head, shoulder, elbow, writs, hip, knee and ankle) while masking all other features. In this way, the human is represented only as an arrangement of individual markers

[6] The same finding of a shallow search path would likely be true for sports contests, a point made also by Ward (2003) in personal communication. We reason that a deeper search by players would yield little benefits given the changing nature of sports contests, and in fact may well incur increased costs due to the temporal demands under which decisions must frequently be taken.

(usually points), with the positions of the points (or dots) changing as the human moves around in the visual display. It is a remarkable fact that the dot pattern is not easily recognizable (if indeed it can be recognized) when the visual representation is still, but that it is quickly recognized under dynamic conditions as human (or animal) movement (see Fig. 3A). Somehow the viewer is able to form a complete percept from extractions of the dynamic relations that exists between the moving points. We turn in brief to the work of Johansson, Cutting, and others, in an attempt to further explain this impressive feature of perceptual function.

The visual system readily detects the movements of humans and animals and distinguishes them from other movements. For example, walking, running and dancing actions are each identified and distinguished from each other on the basis of the patterns that describe them (see Johansson, 1975). Similarly, minor inconsistencies in these otherwise standard patterns, as in the case of a slight limp when walking, are quickly detected. Furthermore, we seem to be able to recognize a person exclusively from his (or her) walking style, gestures and the like (a supposition forwarded by Johannson, 1973). Soon thereafter data obtained using the point-light technique demonstrated that humans could indeed distinguish the gait patterns of friends within a group (Cutting and Kozlowski, 1977), as well as identify the sex of the walker (Kozlowski and Cutting, 1977).

The gait of humans and animals might be well described in terms of combinations of pendulum-like actions, each of which show periodic movements that are highly regular (Johansson, 1973). Such regularities can be detected using vector analysis. Using examples of vector analysis, Johansson (1973) showed that "... when the common component [of the dots in a visual display] was subtracted ... the residual motion formed a translatory motion, a rotary motion, or a pendulum motion [depending on the examples that were used]" (p. 207, bracket inserts mine). These findings indicate strongly that it is the lawful spatial-temporal relations in the visual display that forms the basis of the perceptual response. Cutting and Profitt (1982) extended Johansson's work to examine the issue of whether extraction of common motions from a visual display leaves relative motions as the residual (cf. Johansson, 1973, though see Cutting and Proffitt, 1982, Footnote 2), or vice versa. We will leave this question for Cutting and Proffitt to answer, however, we will make use of their example to demonstrate these important issues.[7]

Imagine an unseen wheel dressed with one point-light at the center and one on the rim. The wheel rolls from left to right in a visual display. The

[7] The example used by Cutting and Profitt (1982) is the same example, amongst others, used by Johannson (1973). In fact, Cutting and Profitt (1982) attribute the example to Gestalt psychologists in general, and Rubin (1927) and Duncker (1938) in particular. These secondary references (see list) are taken from Cutting and Profitt (1982).

path that the inner point takes is one of translation, a line from left to right, whereas that of the outer point describes a cycloid (see Fig. 4A). The viewer detects the path of each point when it is presented on its own, however, the paths are perceived to change when the points are presented together. In this instance, the viewer forms one of three percepts – a wheel rolling on a flat surface (Fig. 4B), the outer point rotating around the inner point in a looping fashion as the latter translates on a horizontal line (Fig. 4C), or as a "tumbling stick" with a light at each end (Fig. 4D). These different percepts are due to the lawful relations that are extracted, correctly or incorrectly, from the moving points (see Cutting and Proffitt, 1982, for further details on the particulars of these lawful relations). Our thesis is that the kinematic relations among sports contestants might likewise yield percepts, or patterns, although the lawful relations that are hypothesized to give rise to them remain unidentified.

Cutting et al. (1978) analyzed gait patterns for biomechanical invariants that viewers might use to form percepts on which to identify the sex of the walker. They considered the center of moment (i.e., the point about which the moving parts go around) and the center of gravity (i.e., the point about which the mass of the parts act upon) as possible discriminators for perception, not least since both centers differ between the sexes due to differences in anatomy. From these considerations, Cutting et al. identified the center of moment as the feature on which the percept is based. This feature, of which the viewer need not have explicit knowledge, does not constitute the percept itself. Instead, it was hypothesized that it is the invariant relations of the individual parts as they move around the center of moment that forms the percept. The supposition is that since the parts share a common fate with the center of moment (i.e., each light point at each instant is a function of the same center of moment), the parts demonstrate a lawful relation to this common point (Cutting et al., 1978). The viewer thus extracts these invariant relations and the percept is formed. Such relations are further complicated in that actions are produced in three dimensions, making the ease with which the viewer detects patterned features within the visual surrounds even more remarkable.

Investigations of human actions using point-light displays are an interesting way of getting at the important information on which patterns are formed. For example, Cutting et al.'s work demonstrated that viewers could recognize the gait patterns of friends (and males and females) in the absence of familiarity cues other than those provided by the dynamic point-light sequences. Thus, the essential information for recognizing gait patterns seems to lie in the relations between the dymnamic markers. This information can be further investigated through experimental manipulations of the visual display, for example, removing points (e.g., Johansson, 1973, Demonstration 3), adding points (e.g., Bertenthal and Pinto, 1994, cf. Fig. 3), changing the hypothesized

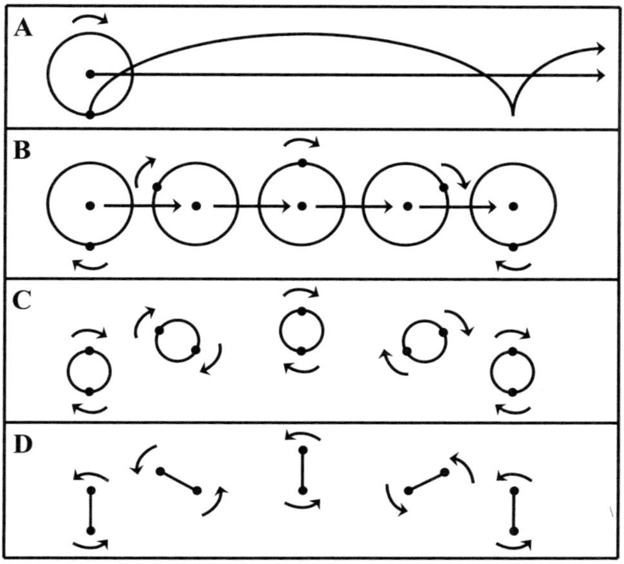

Fig. 4. Example of different percepts formed from extraction of lawful relations between two points. (**A**) Rolling wheel dressed in two point lights, one on the center of the wheel and one on the rim of the wheel. The inner point describes a linear path from left to right whereas the outer point describes a cycloid path. However, when the two lights are presented together, the viewer sees one of three percepts as follows. (**B**) Rolling wheel on flat surface. The outer point circle the inner point that translates left to right. (**C**) Looping wheel. The outer point loop around the inner point that again translates left to right. (**D**) Tumbling stick. A stick with a light at each end tumbles as each point (seemingly) rotates around the other in looping fashion. Adapted from Cutting and Proffitt (1982), Fig. 4.

invariant relation(s) (e.g., Cutting, 1978), amongst other possibilities.[8] In this

[8] The addition or removal of points in the visual display is typically undertaken before onset of the presentation. Using computer software, however, allows for the addition or removal of the points in real time as the visual display unfolds. This unique variant (to our knowledge) is interesting to us because previous research would indicate that such changes may not get noticed by some viewers (see, for example, Simons and Levin, 1997). This inability to sometimes detect change in a visual scene, static or dynamic, is known as "change blindness", with an auditory cousin known as "change deafness" being reported too (Vitevitch, 2003). In brief, the reason for change blindness is attributed to the viewer perceiving the gist of a scene, without attending to every detail in that scene, for whatever reason. Thus, changes in detail that do not affect the gist seem to go unnoticed by the viewer, at least at the level of conscious awareness (Simons and Levin, 1997). Our thinking is that the ability or inability to detect changes in the dynamic point

way, the experimenter may gather important information on the processes of pattern recognition that may otherwise be inaccessible through other means.

For reasons explained above, the point-light technique is a useful method for investigating skilled actions, including sports actions (see Abernethy et al., Experiment 1, for an example from squash). We think that the same method will likewise be useful for analyzing the patterns in sports contests. These patterns are hypothesized to be the result of the spatial-temporal relations that exists among the players. In addition, we hypothesize that the players in a sports contest share a common fate, or center of moment, to which each player is linked. We acknowledge that gait patterns consist of moving parts with fixed linkages between them, whereas sports contests contain moving parts (players) with no set linkages. Nonetheless, we suppose the existence of linkages between players because of local constraints. In recent considerations of sports contests as dynamical systems, we therefore postulated that linkages (couplings) between players are forged and broken as the game context demands (McGarry et al., 2002, McGarry in press). On this basis, the possibility of elastic linkages between players is forwarded. Our research program in sports contests will begin to investigate these supposed linkages and their lawful relations to the center of moment, as hypothesized, using point-light displays.

We suggest that players aim to control the sports contest through the control of space and time, much like the chess player tries to control the game through the control of squares (space) and tempo (time). Similar to chess, multiple battles among players occur at various locations on the playing surface at varying times, each battle taking place in pursuit of a larger objective, namely winning the sports contest. These pieces of the jigsaw must be put together if we are to understand the patterns exhibited in sports contests in a lawful way.

5 Conclusion

There is structure in sports contests. However, this structure is not well understood, at least at any formal level of system description. In this article, we reported on some research that analyzed squash contests using probability structures. Some dissatisfactions of this approach for understanding sports patterns led us to forward some new considerations as to how structure in sports contests might be tackled. Varied examples of pattern recognition from research on chess structures and gait analysis were presented. The latter research prompted us to postulate a center of moment for sports contests. The hypothesis is that the players in a sports contest move around the center of

---

light display documenting sports contests may speak to the importance of that point, or points, to the pattern that is being represented.

moment in lawful ways, and furthermore that the lawful relations are extracted by the viewer to form the percept of patterned behavior. Use of the point-light technique as a method for future research to investigate this hypothesis was recommended.

References

1. Abernethy, B., D.P. Gill, S.L. Parks, and S.T. Packer. Expertise and the perception of kinematic and situational probability information. *Perception*, 30, 233-252, 2001.
2. Alain, C., and C. Sarrazin. Study of Decision-Making in Squash Competition: A Computer Simulation Approach. *Canadian Journal of Sport Sciences*, 15, 193-200, 1990.
3. Bertenthal, B. and J. Pinto. Global processing of biological motions. *Psychological Science*, 5, 221-225, 1994.
4. Borrie, A., G.K. Jonsson, and M.S. Magnusson. Temporal pattern analysis and its applicability in sport: an explanation and exemplar data. *Journal of Sporst Sciences*, 20, 845-852, 2002.
5. Bryan, W.L. and Harter, N. Studies in the telegraphic language: The acquisition of a hierarchy of habits. *Psychological Review*, 6, 345-375, 1899.
6. Buckolz, E., H. Prapavesis, and J. Fairs. Advance cues and their use in predicting tennis passing shots. *Canadian Journal of Sport Sciences*, 13, 20-30, 1988.
7. Chase, W.G. and A. Simon. Perception in chess. *Cognitive Psychology*, 4, 55-81, 1973.
8. Chase, W.G. and H.A. Simon. The mind's eye in chess. In Collins, A.M. and E.E. Smith (editors), *Readings in cognitive science: A perspective from psychology and artificial intelligence*, pp. 461-494. M. Kaufmann Publishers, 1988.
9. Cutting. J.E. Generation of synthetic male and female walkers through manipulation of a biomechanical invariant. *Perception*, 7, 393-405, 1978.
10. Cutting. J.E. and L.T. Kozlowski. Recognizing friends by their walk: Gait perception without familiarity cues. *Bulletin of the Psychonomic Scoiety*, 9, 353-356, 1977.
11. Cutting. J.E., C. Moore, and R. Morrison. Masking the motions of human gait. *Perception and Psychophysics*, 44, 339-347, 1988.
12. Cutting. J.E., and D.R. Proffitt. The minimum principle and the perception of absolute, common and relative motions. *Cognitive Psychology*, 14, 211-246, 1982.
13. Cutting. J.E., D.R. Proffitt, and L.T. Kozlowski. A biomechanical invariant for gait perception. *Journal of Experimental Psychology: Human Perception and Performance*, 4, 357-372, 1978.
14. Duncker, K. Induced motion. In Ellis, W.D. (editor), *A sourcebook of Gestalt psychology*. pp. Routledge and Kegan Paul, 1938. (Original published in German, 1929)
15. de Groot, A.D. Perception and memory versus thought: some old ideas and recent findings. In Kleinmuntz, B. (editor), *Problem solving: research, method, and theory*, pp. 19-50. Wiley, 1966.

16. Garland, D.J. and J.R. Barry. Cognitive advantage in sport: The nature of perceptual structures. *American Journal of Psychology*, 104, 211-228, 1991.
17. Hughes, M.D., and R.M. Bartlett. The use of performance indicators in performance analysis. *Journal of Sports Sciences*, 20, 739-754, 2002.
18. Johansson, G. Visual perception of biological motion and a model for its analysis. *Perception and Psychophysics*, 44, 339-347, 1973.
19. Johansson, G. Visual motion perception. *Scientific American*, 232, 76-88, 1975.
20. Kozlowski L.T. and J.E. Cutting. Recognizing the sex of a walker from a dynamic point-light display. *Perception and Psychophysics*, 21, 575-580, 1977.
21. Magnusson, M.S. Discovering hidden time patterns in behavior: T-patterns and their detection. *Behavior Research Methods, Instruments and Computers*, 32, 93-110, 2000.
22. McGarry, T. Soccer as a dynamical system: Some theoretical considerations. In Reilly, T., J. Cabri, and D. Araujo (editors), *Science and Football. Proceedings of the Fifth World Congress of Science and Soccer*, in press
23. McGarry, T., D.I. Anderson, S.A. Wallace, M. Hughes, and I.M. Franks. Sport competition as a dynamical self-organizing system. *Journal of Sports Sciences*, 20, 771-781, 2002.
24. McGarry, T., and I.M. Franks. In search of invariant athletic behaviour in competitive sport systems: An example from championship squash match-play. *Journal of Sports Sciences*, 14, 445-456, 1996.
25. Mendo, A.H., and M.T. Anguera. Estructura conductual en deportes sociomotores: Futbol. (Behavioral structure in sociomotor sports: Soccer.) *Revista de Psicologia Social*, 16, 71-93, 2001.
26. Miller, G.A. The magical number seven, plus or minus two. Some limits on our capacity for processing information. *Psychological Review*, 63, 81-97, 1956.
27. Reep, C. and B. Benjamin Skill and chance in Association Football. *Journal of the Royal Statistical Society, Series A,* 131, 581-585, 1968.
28. Reingold, E.M., N. Charness, R.S. Schultetus, and D.M. Stampe. Perceptual automaticity in expert chess players: Parallel encoding of chess relations. 8, 505-510, 2001.
29. Rubin, E. Visuell wahrgenommene wirkliche Bewegungen. *Zeitschrift fur Psychologie*, 103, 384-392.1927.
30. Saarinen, J. Localization and discrimination of "pop-out" targets. *Vision Research*, 36, 313-316, 1996.
31. Sarrazin, C., D. Lacombe, C. Alain, and C., and J. Joly. Simulation study of a decision-making model of squash competition. Phase one: The analysis of the protocol. *Human Movement Science*. 4. 279-306. 1983
32. Sarrazin, C., Alain, C., and Lacombe, D. Simulation study of a decision-making model of squash competition. Phase two: Testing the model through the use of computer simulation. *Human Movement Science*. 5. 373-391. 1986.
33. Simons, D.J. and D.T. Levin. Change blindness. *Trends in Cognitive Sciences*, 1, 261-267, 1997.
34. Vitevitch, M.S. Change deafness: The inability to detect change between two voices. *Journal of Experimental Psychology: Human Perception and Performance*, 29, 333-342, 2003.
35. Ward, P. and A.M. Williams. Perceptual and cognitive skill development in soccer: The multidimensional nature of expert performance. *Journal of Sport and Exercise Psychology*, 25, 93-111, 2003.

36. Williams, M., K. Davids, L. Burwitz, and J. Williams. Cognitive knowledge and soccer performance. *Perceptual and Motor Skills*, 76, 579-593, 1993.

Track and Field Performance Data and Prediction Models: Promises and Fallacies

Yuanlong Liu

Western Michigan University, USA
yuanlong.liu@wmich.edu

1 Introduction

Prediction is always a fascinating obsession no matter what we predict. We predict what tomorrow's weather will be, which team will win the game, who will be the next president of the United States, how fast one will run and how high one will jump, and even how much money we will make next year. Mathematical and statistical models have been used to predict future events in different disciplines. Some prediction models can be used to predict specific magnitudes of an event in the future. For example, because of the high technology development in the last two decades, the models used for weather forecasting became more and more accurate. In some other fields, the prediction models are actually not for a specific magnitude of an event in the future but for predicting various developmental trends in the future. For example, statistical models are commonly used for predicting the developmental trends in stocks but not for predicting specific values of the stocks.

In track and field, the curiosity with respect to the limits of human athletic performances has always been of interest to athletes, coaches and scientists (Liu 2002). Since track and field performances are the oldest competitive sports, the interest in a human being's ability and accomplishment in running, jumping and throwing has a long and diversified history. Mathematicians, statisticians, physiologists and operational researchers have developed numerous models to predict future performances, world records and 'ultimate' performances in track and field, and to compare male and female performances (Liu and Schutz 1998).

There have been many publications regarding the prediction of track and field performances. Some of them are opinion based and lack scientific support. Others used track and field data and mathematical and statistical models. There has been some confusion as to what the track and field performance data provide and what the prediction models show. A review of the prediction models in literature leads us to believe that there is no prediction model available to accurately predict the future magnitude of any performance in track and field (Liu 2002, Liu and Schutz 1998). So, what are the promises and fallacies of the prediction models in track and field? The primary focus of this chapter is a discussion of the following issues in predicting track and field performances: Track and field performances data and its validity in predicting future performances and the promises and fallacies of the prediction models.

2 Track and Field Data and Prediction Models

2.1 Track and field performance data

The sources of track and field performance data

Collecting track and field performance data is different from collecting data in laboratories. The major advantage of collecting track and field performance data is that it can be obtained easily from different published sources. The International Association of Athletic Federation (IAAF, http://www.iaaf.org) is a governing authority for athletic programs, standardized technical equipment and world records. It provides the verified world records in track and field since 1912. The federations in individual countries affiliated to IAAF have the system to record and provide the track and field performance data in each countries of the world. The International Olympic Committee (http://www.olympic.org) provides all the records of the modern Olympic games. Track and field performances data can also be obtained from newspapers and magazines (e.g., Track and Field News), yearbooks and organization web-pages.

Track and field data in predicting future performances

The track and field performances data are usually accurately recorded with great precision (e.g., to the hundredth of a second in track competitions). There are many different ways to collect the track and field performance data. Men's and women's records at different competition levels (world, Olympic, national, college, and high school, etc.) provide historical documentation for human beings breaking their limits in running, jumping and throwing. However, one should notice that the discontinuity of the record data (e.g., the world records and Olympic records) may seriously affect the validity when it is used in the purpose of prediction, because there may be no records for several years or several records in one single year. The best performance per year data provide better continuous longitudinal trend without missing magnitudes. Coaches use individual athlete's longitudinal performance data to evaluate and predict his/her future performances.

Scientists use track and field performance data to develop prediction models. Historically, there are three types of data used in developing the prediction models and in comparing male and female track and field performances in literature. Some researchers used the performance percentages of improvement over certain time period in evaluating and predicting the track and field performances (Dyer 1982, Ryder et al. 1976, Stefani 1977). World records and Olympic records data are also commonly used (Whipp and Ward 1992, Gembris et al. 2002). Liu and Schutz (1998) pointed out that the annual best performances should be used in the development of the track and field prediction models and these data would overcome the drawback of discontinuity in improvement percentage and world record data. Annual best performance data have also been used in academic research publications (Liu 2002, Liu and Schutz 1998, Schutz and Liu 1998).

Validity of the track and field performance data

As shown above, all the three types of data have been employed to develop the prediction models and to compare male and female track and field performances. However, the discontinuous characteristic of the improvement percentage and world records data may seriously affect the validity of the prediction models. For example, between 1960 and 1967, there has been only one new world record established in men's 1500 meters run. The current women's 200 meters world record has stood for 21 years.

Using only world records or the improvement rates, and ignoring the fact that there has been no improvement over years would not provide a valid model for reality. Liu (1993) used the best annual performance data and world record data to fit nonlinear regression and linear models and to examine the impact of the discontinuity in the world record data on the validity of prediction models. The results showed that the parameter estimates of the prediction models were much more unstable with using the world record data than with the annual best performance data. Schutz and Liu (1993) indicated that using word records would lead to an inflation of the estimate of linearity and/or improvement rates that cannot be empirically justified. Because of the continuous property and the stable estimation of the parameters in the prediction models, the annual best performance data has been shown to be the best data in developing deterministic prediction models in track and field.

2.2 Prediction models in track and field

A brief overview of existing prediction models

Evaluating and predicting track and field performances has a long history and there have been numerous models developed to evaluate and predict future track and field performances and to compare male and female performances. Kennelly (1906) is probably the earliest researcher to examine the track and field data using a statistical modeling technique. He showed that a linear relation between velocity and running distance was stable over all the examined running events. Physiologists Hill (1925) and Peronnet and Thilbault (1989) developed metabolism based models to provide a physiological explanation of the time-distance relation and to predict record performance. Keller (1974) and Senator (1982) used biomechanics to develop mechanics-based models to predict the ultimate performances. However, models based on either physiology or biomechanics do not work well when evaluating and predicting the performance trend in track and field.

The most common approach in developing the prediction models for track and field performances are the ones used by mathematicians, statisticians, and psychometicians (e.g., Chateerjee and Chatterjee 1982, Deakin 1967, Liu 2002). These models usually examine longitudinal improvement of running, jumping and throwing through types of asymptotic mathematical function.

Mathematical models

The following section provides a brief description of the types of mathematical models that have been utilized to predict track and field records (also see Schutz and Liu 1998 for partial review).

Historical Date versus Time

In these models a time-year, velocity-year, or a distance-year relationship is developed separately for each event. An exponential model relates time, velocity, or distance and historical date for a specific event:

$$T(n) = b_0 + b_1 e^{-b_2 n}, \tag{1}$$

where $T(n)$ is the predicted performance in year n, b_0 is the estimated asymptotic value of $T(n)$ as n approaches infinity, and b_1 and b_2 are calculated parameters which govern the shape of the curves (Chatterjee and Chatterjee 1982, Lucy 1958, Schutz and McBryde 1983).

A linear model relates time, velocity, or distance and historical date for a specific event:

$$T(n) = b_0 + b_1 n, \tag{2}$$

where $T(n)$ is the predicted performance in year n, b_0, the intercept, is the world record or best performance in year zero (usually 1900), and b_1, the slope, is the estimated yearly amount of improvement (e.g., Ballerini and Resnick 1985, Ballerini 1987, Whipp and Ward 1992).

A polynomial model relates velocity and historical time for a specific event:

$$V(n) = b_0 + b_1 n + b_2 n^2 + \ldots\ldots + b_m n^m, \tag{3}$$

where $V(n)$ is velocity, n is the historical year, and b_0, b_1, b_2, $\ldots$, $b_m$are calculated parameters which determine the shape of the curve (Mognoni et al. 1982).

Time versus Distance

In these models the time or velocity versus distance relationship, for all running events, is developed separately for specific years. By comparing the different yearly relationships, future projections are made via a second level modeling of the derived parameters.

An exponential model relates mean running velocity and distance for all events in a specific year:

$$V(d) = b_0 + b_1 e^{-b_2 d}, \tag{4}$$

where $V(d)$ is the mean velocity for a specific year, d is the running time or distance, and b_0, b_1, and b_2 are calculated parameters reflecting the shape of the curve (Furusawa et al. 1927, Ward-Smith 1985).

A power function model relates running time and distance for all events in a specific year:

$$T(d) = b_1 d^{b_2}, \tag{5}$$

where $T(d)$ is running time, d is distance, and b_1 and b_2 are calculated parameters (Lloyd 1966, Riegel 1981). This model is equivalent to a linear relationship between time and distance on a logarithmic scale (e.g., Kennelly 1906).

A linear-log function model relates velocity and distance for all events in a specific year:

$$V(d) = b_1 f(\log(d)), \tag{6}$$

where $V(d)$ is velocity, d is distance, and $f()$ is a function which is varies with different researchers (Craig 1963, Francis 1943, Hill 1925).

3 The Promises and Fallacies of the Prediction Models

3.1 The expectation and reality on the rredictions in track and field

When we talk about predictions or forecasting, we usually expect to accurately know what would happen. For example, when we listen to weather forecasting, we would like to know whether it will be rainy or sunny tomorrow. Because of the development of technology and science, short term weather forecasting is becoming trustworthy. The mathematical and statistical models are also used to predict the stock market. However, the prediction for stocks for tomorrow may or may not be very accurate. The inaccuracy of the stock prediction for tomorrow is due to many external factors that are out of the researchers' control and the unsystematic effect cannot be exactly modeled. Yet the models in stock predictions usually work very well to predict the trend of the stocks over a period of time.

What do we expect from the mathematical models in track and field performance predictions? Similar to predicting tomorrow's weather, it is not uncommon for people to expect that a mathematical model would provide an accurate magnitude of the next world record and the year in which that the record may be established. Therefore, there has been so much effort to predict future world records and waiting time for establishing new world records. The journal *Nature*, in spite of its reputation around the world, had published several predictions that were far from the reality. Whipp and Ward (1992) with limited literature review on prediction modeling published an essay in *Nature*. They used discontinuous word record data with a linear regression model and proposed that the track and field performances would improve linearly. They asserted that female performances in track and field would catch up with male performances as early as 1998 in some events. Based on their assertion, one might run 100 meters at no time in some day! *Nature* published another manuscript that again ignored the academic literature and used the discontinuous record data (Gembris et al. 2002). The Figure 1 in the article already shows that the proposed model was not valid and the models could not represent the trend of the improvement of their track and field data.

Schutz and Liu (1998) summarized some published predictions of the world records literature (Peronnet and Thibault 1989, Ryder et al. 1976, Schutz and McBryde 1983, Stefani 1977). The reality differs from the model predictions. Some people say that the track and field performances would improve linearly they do not. Others say that in some events the performances have reached an asymptotic level, however new world records are still established. The dilemma is that either all these models do not demonstrate what they promise to show or what we expected is not what the models are supposed to reveal. In the next section, the promises and fallacies of the prediction models in track and field are discussed.

3.2 The promises and fallacies of the prediction models in track and field

Research has shown that the nonlinear regression modeling technique with the annual best performance data is the most appropriate method to describe track and field performances and predict the deterministic trend of improvement in track and field. When the general trend of the performances has met the asymptotic level of the deterministic models, regression models cannot assess the future performances. Other approaches such as a random sampling model suggested by Schutz and Liu may be used to predict the general trend of the world records (Schutz and Liu 1998).

Regression type models only promise to describe the past performance patterns over time and predict the trend of future performances in track and field. Although one may wish to use the prediction models to predict the specific magnitude of future world records, however it is not what the regression models promise to provide.

In the regression models shown in the previous sections, the parameters in the regression equations are derived based on the Least Squares (LS) and Maximum Likelihood (ML) estimating theory, and the regression line is fitted through the performance data (Ratkowsky 1983). When track and field performance data and historic year were used in the regression modeling, the residual variance or error variance is always nonzero magnitude. Therefore, precise prediction of the track and field performances is not what the prediction models can offer. Due to this reason, it may not be appropriate to evaluate the prediction models by comparing the predicted world records in order to see which model makes more accurate predictions. However, it is a valid comparison in which a model has a better fit to the data measured by some goodness of fit indexes (see Liu and Schutz 1998 for a detailed model comparisons).

In summary, it is a fallacy to use the prediction models to predict accurate track and field performance in the future. The prediction models worked well to show the historic trend of the track and field performances and to predict the general trend of the performances in the future.

3.3 The factors that affect goodness of fit about the prediction models

Speaking about the goodness of fit, there are many external factors that affect prediction models. Liu (2002) and Schutz and Liu (1998) compared the mathematical models in the literature and pointed out that the variability among the model predictions is large because different models have utilized different data sets and assumptions.

In fact, many factors affect accuracy and validity of prediction models. Internal factors include the validity of the data and the methods used for model development. External factors that are out of researchers' control include potentials and limitations of the human body, psychological effect in competitions, science and technology influence, performance-enhancing drugs, and political and economic situations in a specific time period. Because the asymptotic mathematical models have treated the external effects as random variables for evaluating and predicting the longitudinal trend of the track and field, it is more valid and preferable than other models in the literature. Liu (2002) summarized the external factors and discussed how these factors would affect future track and field performances and prediction models.

4 Conclusion

This chapter focused on the discussion of prediction models in track and field performances and their promises and fallacies. Although one might wish to use the prediction models to predict the specific magnitude of future world records, it is not what regression models promise to provide. The promise of all the regression type of models in track and field is to describe the past performance patterns over time and predict the trend for future performances. After the asymptotic level is reached, the regression models cannot be used for the evaluation and prediction of the track and field performances. Some new approaches such as the random sampling model (Liu 1993, Schutz and Liu 1998) should be employed for this purpose. The future holds surprising changes, in equipment, in technology, in performance enhancing drugs, in social and economic status, etc. It is still true that predicting future track and field performances is good fun, and probably about as exact as some of the uses that mathematics is put to elsewhere (Deakin 1961).

References

1. Ballerini, R. and S. I. Resnick. Records from improving populations. *Journal of Applied Probability,* 22: 487-502, 1985.
2. Ballerini, R. Records, in the presence of a linear trend. *Advances in Applied Probability,* 19: 801-828, 1987.
3. Chatterjee, S., and S. Chatterjee. New Lamps for old: An exploratory analysis of running times in Olympic Games. *Applied Statistics,* 31, 14-22, 1982.
4. Craig, A. B. Evaluation and predictions of world running and swimming records. *Journal of Sport Medicine,* 3: 14-21, 1963.
5. Deakin, B. Mathematics of athletic performance. *Matrix,* Melbourne University Mathematical Society Magazine, issue 4, 1961.
6. Deakin, B. Estimating bounds on athletic performance. *The Mathematical Gazette,* 51: 100-103, 1967.
7. Dyer, K.F. *Catching up the men: Women in sport.* Junction Books, 1982.
8. Francis, A. W. Running records. *Science,* 98: 315-316, 1943.
9. Furusawa, K., A. V. Hill, J. L. Parkinson The dynamics of sprint running. *Proceedings Royal Society,* B(102): 29-42, 1927.
10. Gembris, D., J.G. Taylor, and D. Suter. Trends and random fluctuations in athletics. *Nature,* 417: May 506, 2002.
11. Hill, A. V. The physiological basis of athletic records. *British Association for the Advancement of Science, Report of the 93rd Meeting:* 156-173, 1925.
12. Keller, J. B. Optimal velocity in a race. *The American Mathematical Monthly,* 81: 474-480, 1974.
13. Kennelly, A. An approximate law of fatigue in speeds of racing animals. *Proceedings of the American Academy of Arts of Scienc,* 42: 275-331, 1906.
14. Liu, Y. *Evaluation and prediction of world records and ultimate performance in track and field.* Unpublished Master Thesis, University of British Columbia, 1993.

15. Liu, Y. and R. W. Schutz. Prediction models for track and field performances. *Measurement in Physical Education and Exercise Science*, 2: 205-223, 1998.
16. Liu, Y. Track and field performances and competitions in the new millennium: Will it be a thematic change? *International Sports Journal*, 6: 29-35, 2002.
17. Lloyd, B. B. The energetics of running: An analysis of world records. *British Association for the Advancement of Science, Report of the 133rd Meeting:* 515-530, 1965.
18. Lucy, L. B. Future progress in the mile. *IOTA,* 1: 8-11, 1958.
19. Mognoni, P., C. Lafortuna, G. Russo, and A. Minetti. An analysis of world records in three types of locomotion. *European Journal of Applied Physiology,* 49: 287-299, 1982.
20. Peronnet, F. and G. Thibault. Mathematical analysis of running performance and world running records. *Journal of Applied Physiology,* 67: 453-465, 1989.
21. Ratkowsky, D. A. *Nonlinear Regression Modeling.* Marcel Dekker, Inc, 1983.
22. Riegel, P. S. Athletic records and human endurance. *American Scientist,* 69: 285-290, 1981.
23. Ryder, H.W., H.J. Carr, P. Herget. Future performance in footracing. Scientific American, 234: 109-119, 1976.
24. Schutz, R. and Y. Liu. Statistical Modeling in Track and Field. In Bennett, J. (editor). *Statistics in Sport,* pp. 173-197. London: Arnold. 1998.
25. Schutz, R. and Y. Liu. Track and field world records: optimal models, datasets and projections. *American Statistical Association 1993 Proceedings of the Section on Statistics in Sport:* 52-57, 1993.
26. Schutz, R. W. and J. P.McBryde. The prediction of ultimate track and field performance: Past, present, future. *Proceedings of the FISU Conference Universiade' 83 in Association with the Xth HISPA Congress*: 498-515, 1983.
27. Senator, M. Extending the theory of dash running. *Journal of Biomechanical Engineering,* 104: 209-213, 1982.
28. Stefani, R.T. Trends in Olympic winning performances. *Athletic Journal,* Dec.: 44-46, 1977.
29. Ward-Smith, A. J. A mathematical theory of running, based on the first law of thermodynamics, and its application to the performance of world-class athletes. *Journal of Biomechanics,* 18: 337-349, 1985.
30. Whipp, B.J. and S. A. Ward. Will women soon outrun men? *Nature*, 335(2): January 25, 1992.

Using Response Surface Models for Evolutionary Estimation of Optimum Running Times[1]

Wei-Han Tan

University of Texas Health Science Center, San Antonio, USA
wtan@alumni.rice.edu

1 Background

Coaches in the track world are very limited in what they can say to an athlete who does not run to the best of his or her ability in competition. Most often they can only provide four general statements:

1) You (the athlete) started your race too fast and were unable to finish the race strong.

2) You started too slow and were too far back to catch up to the rest of the runners.

3) You started to sprint too soon and lost your energy at the end.

4) You started to sprint too late, the rest of the runners were too far ahead, and the distance left in the race was too short for you to catch them.

This problem is especially true for long distance races. The statements by the coaches provide very little guidance for the athlete. Eventually, a coach might suggest an optimum combination for a runner based on the coach's informal judgment model. Until then, runners can only hope to guess the right combination of the start/pace portion of the race and the final sprint distance. Is it feasible to combine empirical data with judgments to estimate an optimum combination?

Mathematical models have been used to represent judgment policies (See, e.g., Christal, 1968, 35- 41). Response surface models have been developed to estimate combinations of independent variables that optimize the response (dependent) variable (See, e.g., Mason et al., 1989, 204- 233).

[1] An earlier version of this paper has appeared in *STATS*, 24: 14-18, 1999

2 Project

The objective of my project was to see whether statistical models and computers could be used to find a runner's optimum combination of sprint distance and first 400 meter time to optimize (minimize) their 800 meter time (response variable). The sprint distance is the distance (in meters) from the end of the race at which the runner begins sprinting. Previous applications of statistics and computers in running have concentrated on training schedules and procedures. To my knowledge, there has not been an application of these topics in the same manner that I performed the experiment. The 800 meter race was chosen because it was the shortest "long distance" race that contained obvious signs of needing an optimum combination of the variables of interest for best performance. By using the 800 meter race more data could be collected in a reasonable period of time as compared to a longer distance race.

3 Methodology

To approach the problem, the applications of policy modeling and response surface models were combined. Policy capturing and judgment analysis were the first techniques used. They were applied when the runner filled out their survey form shown in Figure 1. On this form, the runner expressed his or her judgments on their 800 meter time for various combinations of sprint distance and first 400 meter time. I set the general combination, but the runner filled in data that would best fit their running style. The first 10 data sets were filled by the runner's estimations. The process of estimation or judgment modeling was used so that each runner could begin the process of evolutionary estimation of their optimum immediately. The judgment model also guaranteed that the optimum combination was bounded by the judgment data, whereas if empirical data was collected for this portion, the proper spread in data was not guaranteed. For example, I asked each runner to estimate his or her total time for a combination of a fast first 400 meters and a long sprint distance. Runner #1 estimated a total time of 190 seconds for a combination of a 65 second first 400 meters and a sprint distance of 300 meters. The runners estimated a total of 10 judgments on the survey form, 9 of which bounded the optimum 400 meter time and sprint distance and the last judgment represented what they thought their best combination would be. (Refer to Table 1 for an example)

Least squares regression was then used to fit a response surface model for each individual's judgment data. The general form of the response surface model was

$$y = a_0 + a_1\mathrm{T} + a_2\mathrm{D} + a_3\mathrm{TD} + a_4\mathrm{T}^2 + a_5\mathrm{D}^2 + \mathrm{e}$$

where:

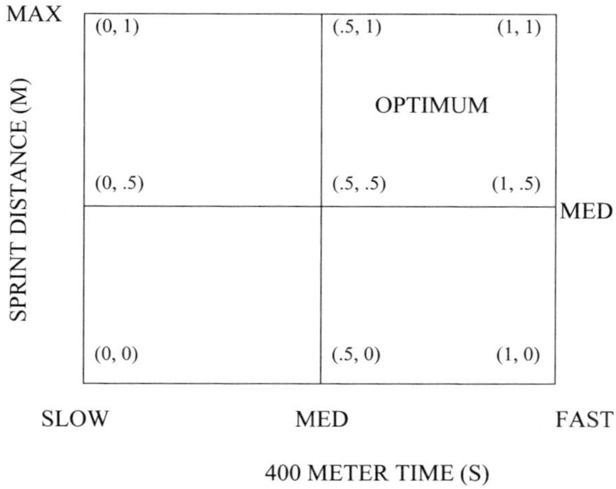

JUDGEMENTS	COMBINATIONS	400 METER TIME (S)	SPRINT DISTANCE (M)	800 METER TIME (S)
		T	D	Y
1	(0, 0)			
2	(0, .5)			
3	(0, 1)			
4	(.5, 0)			
5	(.5, .5)			
6	(.5, 1)			
7	(1, 0)			
8	(1, .5)			
9	(1, 1)			
10	OPTIMUM			

Fig. 1. Runner survey form.

Table 1. Judgments for Runner # 1.

Trial	First 400 time (s)	Sprint distance (m)	800m time (s)
1	65	25	165
2	65	150	170
3	65	300	190
4	75	25	160
5	75	150	150
6	75	300	158
7	85	25	180
8	85	150	168
9	85	300	166
10	70	120	145

y = TOTAL TIME
T = TIME FOR FIRST 400 METERS
D = SPRINT DISTANCE
TD = T*D
e = ERROR

The initial fitted regression equation for Runner #1's set of data containing the first 10 observations was:

$$y = 1132.63 - 26.9990T + .4011453D - .0072833TD + .1868641T^2 + .0004862D^2$$

In order to estimate the optimum combination of 400 meter time and sprint distance, the partial derivatives of y with respect to T and D were determined. By setting the two partial derivatives equal to zero, the optimum combination could be estimated. The coefficients from the regression model were used as the data to solve two simultaneous equations that would result in the optimum combination of final sprint distance and first 400 meter time. The simultaneous equations developed to estimate the optimum were:

$$(2a_4)T^* + (a_3)D^* = -a_1$$
$$(a_3)T^* + (2a_5)D^* = -a_2$$

where a_1, a_2, a_3, a_4, and a_5 are the coefficients from the regression model, T^* is the optimum 400 meter time, and D^* is the optimum sprint distance.

The simultaneous equations developed from the coefficients of Runner #1's response surface model were:

$$2*(.1868641)T^* + (-.0072833)D^* = (26.9990)$$
$$(-.0072833)T^* + 2*(.0004862)D^* = (-.4011453)$$

The optimum combination was

$T^* = 75.176$ (FIRST 400 METER TIME)
$D^* = 150.54$ (SPRINT DISTANCE)

After the optimum combination was produced, Runner #1 was asked to go to the track in an attempt to run this combination. Even if he or she missed the optimum combination, the data were added as another actual value to the data set. The optimum combination was then recomputed. The empirical data were added to the 10 judgment data points already included in the data set. The data point was included because it further defined the response surface. This process of inputting more data to change the curve is called evolutionary operations. As more data were collected, the model changed until no substantial improvements could be made (Refer to Table 2). The final optimum combination was T*=75 seconds and D*=145 meters. Graphical representation of the optimum combination is shown in Figures 2 and 3.

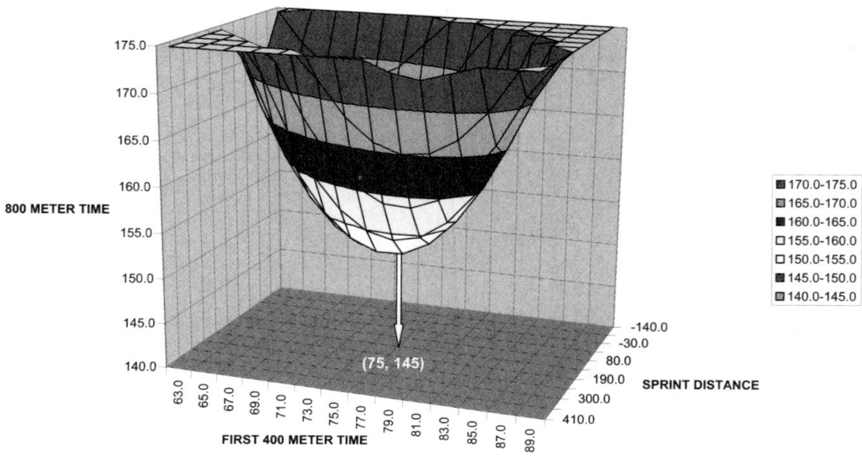

Fig. 2. Response surface model for Runner # 1: final run. The graph's equation is $\hat{y} = 1010.49 - 23.8034T + .04360634D - .007365563TD + .166084619T^2 + .000398947D^2$.

This experiment included five runners; three of which were experienced and two were inexperienced. The experienced runners had great results. In general, the standard error of estimates for the experienced runners for their response surface ranged from 3 to 5 seconds; compared to the range of the standard deviation for the 800 meter variable of 9 to 13 seconds. The lowest standard error of estimate obtained was 3.317. The t-tests and response surface analysis consistently showed that the first 400 meter variable was the most influential variable on the final 800 meter time for all runners. My methodology did not appear to benefit the two inexperienced runners. This was because the inexperienced runners gave me judgment data that contradicted the empirical data that were collected. It is important to note here that experimental conditions could have helped enhance their data, but I wanted to have the runners perform on an actual track where the data were more meaningful. Despite these results no data were excluded.

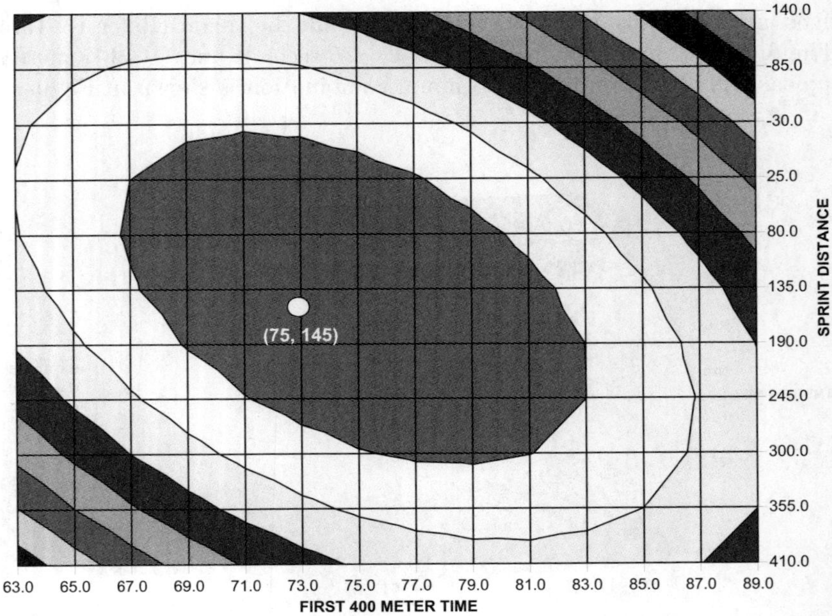

Fig. 3. Response surface model for Runner # 1: final run. The graph's equation is $\hat{y} = 1010.49 - 23.8034T + .04360634D - .007365563TD + .166084619T^2 + .000398947D^2$.

4 Conclusion and Future Work

This experiment was very successful for the experienced runners. I placed fourth grand prize at the 1997 Intel International Science and Engineering Fair in the Behavioral Sciences category and the second place special award in statistics from the American Statistical Association. The application of statistical models and computers can be used to predict a runner's best combination of first 400 meter time and sprint distance in an 800 meter race. New procedures are being tested so that this project benefits inexperienced runners as well as experienced. It helps the runners because it gives them an idea of how to run to get their best time as an end result. Now they have an idea of what to aim for instead of having to waste half of the season trying to get a rough approximation. Currently runners from a local high school have joined this project and are preparing for the upcoming season.

Future studies involve an addition of the first 100 meter time as a variable and the shifting of the response surface. Heart rate monitors are being used to monitor the runners and are being evaluated for estimating optimum running conditions. This methodology could possibly benefit longer distance races since the same basic principles of sprint and pace are applied. It would also

Table 2. Running trials and optimum combinations for Runner # 1.

FIRST 400 METERS (SECONDS)	SPRINT DISTANCE (METERS)	800 METER TIME (SECONDS)
	RECOMPUTED	
75	151	150.987
	RUNNING TRIAL	
73	150	151
	RECOMPUTED	
75	150	150.986
	RUNNING TRIAL	
70	150	152
	RECOMPUTED	
75	150	150.983
	RUNNING TRIAL	
77	150	158
	RECOMPUTED	
75	146	150.976
	RUNNING TRIAL	
75	146	155
	RECOMPUTED	
75	145	150.976
	RUNNING TRIAL	
75	145	151
	RECOMPUTED	
75	145	150.976

be interesting to study the optimum combination of type of exercise, exercise time, and other variables in a physical fitness response. This would help those who are attempting to reach a level of fitness in a reasonable time period. The effect of the inexperienced runners caused me to consider shifting the judgment data towards the empirical data in the future. This would reduce some of the error found in contradicting statements.

Acknowledgments

I would like to thank all the runners for making my research possible. I would also like to thank my family for their support. Finally I would like to thank Dr. Robert Mason for his critique during the project and my mentor Dr. Joe Ward Jr. for his countless hours of hard work, support and guidance.

References

1. Christal, Raymond E.(1968), Selecting , A Harem-And Other Applications Of The Policy-Capturing Model. *The Journal of Experimental Education,* 36: 35-41.
2. Mason, R.L., Gunst R.F., and Hess, J.L. (1989), *Statistical Design and Analysis of Experiments with Applications to Engineering and Science.* New York: John Wiley and Sons, Inc.

Solving Sports Scheduling and Timetabling Problems with Constraint Programming

Abderrahmane Aggoun[1] and Alkis Vazacopoulos[2]

[1] COSYTEC S. A., France
abderrahmane.aggoun@cosytec.com

[2] Dash Optimization, Inc., USA
av@dashoptimization.com

1 Introduction

In the last fifteen years, many successful industrial applications have been implemented using constraint programming tools. Many of those applications include specific constraints, which are difficult to model using the existing modeling and optimization tools. Industrial applications that are particularly suitable for constraint programming include production planning, scheduling and resource allocation.

In this paper we begin with a brief overview of the CHIP system. We then describe a set of successful applications in the areas of production planning and scheduling. Finally, we detail the CP models of three examples from timetabling and sports scheduling.

2 The Constraint Programming CHIP System

Constraint programming is based on the idea that many difficult problems can be expressed declaratively in terms of variables and constraints. The variables range over a finite set of values and typically denote alternative decisions to be taken. The constraints are expressed as relations over subsets of variables and restrict feasible value combinations for these variables. A solution is an assignment of values to variables, which satisfies all constraints.

CHIP is a constraint programming system designed to tackle real world "constrained search" problems efficiently and with a short development time. The CHIP system provides a number of constraint solvers over different computational domains. In this paper we focus on finite domains in CHIP (Simonis 1999).

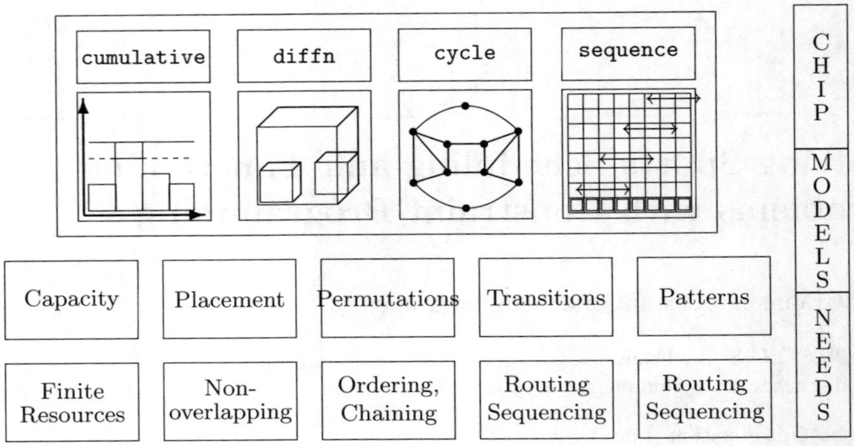

Fig. 1. Global Constraints/Problem domains.

The first versions of finite domain constraint systems used rather simple constraints (e.g. arithmetic constraints, `alldifferent` constraint). The problem with these finite domain constraint systems was that formulation of applications with simple constraints is difficult and inefficient. Global constraints were introduced in the CHIP system (Aggoun and Beldiceanu 1993, Beldiceanu and Contejean 1994) and are now a key concept of constraint programming. Global constraints also called semantic constraints and use domain specific knowledge (Discrete Mathematics, Operational Research, and Constraint Satisfaction Problems (Van Hentenryck 1989) techniques) to derive better propagation results (removing infeasible solutions). Instead of expressing a problem with many primitive constraints, we describe it with few global constraints, which work on sets of variables. The abstraction of global constraints has many advantages: problems can be expressed more easily and with fewer constraints, the constraint can be used for many different applications (not area specific), the abstraction is as general as possible and efficient methods for constraint reasoning can be constantly improved. Since global constraints were introduced in CHIP in 1993, they have been continuously studied in the constraint programming literature (Caseau and Laburthe 1994, Régin and J. C. Rueher 2000, Bockmayr et al. 2001, Beldiceanu 2000, Simonis et al. 2000). Each global constraint has its own set of algorithms (solver engines using techniques from Artificial Intelligence, Operations Research, Discrete Mathematics) and share variables with other constraints.

In CHIP, the development of global constraints is based on the constraint morphology/problem domains as shown in Figure 1. There are three basic concepts: 1) application constraints (needs), 2) mathematical formulation of these problem constraints (models) and 3) definition of the global constraint in constraint programming (CHIP) to model and solve these problem constraints. The definition of a global constraint is an answer to a set of requirements from applications, which cannot be easily expressed and modeled with the simple constraints available in a CP system. In addition, global constraints should be abstract enough to model many different applications.

`Cumulative` constraint: The `cumulative` constraint (Aggoun and Beldiceanu 1993) expresses resource constraints between sets of domain variables. It can be used for many types of problems, e.g., scheduling with resource constraints (machines, manpower, etc), packing and placement problems (one-dimensional bin packing, 2D and 3D problems), or assignment problems (time tabling). The `cumulative` constraint uses rather complex propagation techniques to analyze and propagate necessary conditions for consistency. A typical use of `cumulative` constraint is found in resource scheduling. We have to schedule a number of tasks (Figure 2a) of different durations where the tasks require certain amounts of resources (e.g. manpower) during their operation. The overall amount of manpower available during the scheduling period is fixed and the total requirements at each time point by all tasks should not exceed the available limit.

`Diffn` constraint: The `diffn` constraint (Figure 2b) (Beldiceanu and Contejean 1994) expresses non-overlapping constraints on n-dimensional rectangles. Obviously, the `diffn` constraint can be applied to many types of packing and placement problems in two or more dimensions. In two dimensions this corresponds to cutting problems of various forms. In three dimensions, typical problems are packing of palettes or containers with rectangular boxes. In four dimensions, we extend the three-dimensional problem with the choice of the container. The first three co-ordinates describe the position of a box within the container; the fourth dimension describes which of the several containers we place each box in. The `diffn` constraint is also applied to scheduling and assignment problems, where it can easily model disjunctive machine assignment for tasks. The x-dimension of the constraint represents time; each value in the y-dimension stands for a possible machine assignment. Each task is represented as a rectangle, which is given by its start time, its resource assignment, its duration, and a height 1 (Figure 2b). The non-overlapping condition states that no tasks can be scheduled at the same time on the same machine.

The basic version of cumulative can be defined as follows. Let O be the set of operations, and $n = |O|$. Operation o is characterized by three parameters, which can be either domain variables or values; the starting time s_o, the duration d_o, and the amount r_o of some resource used by the operation. We are also given the upper bound e on the resource consumption and the completion time l. The global constraint:

$$\texttt{cumulative}([s_1, s_2, \ldots, s_n], [d_1, \ldots, d_n], [r_1, \ldots, r_n], l, e)$$

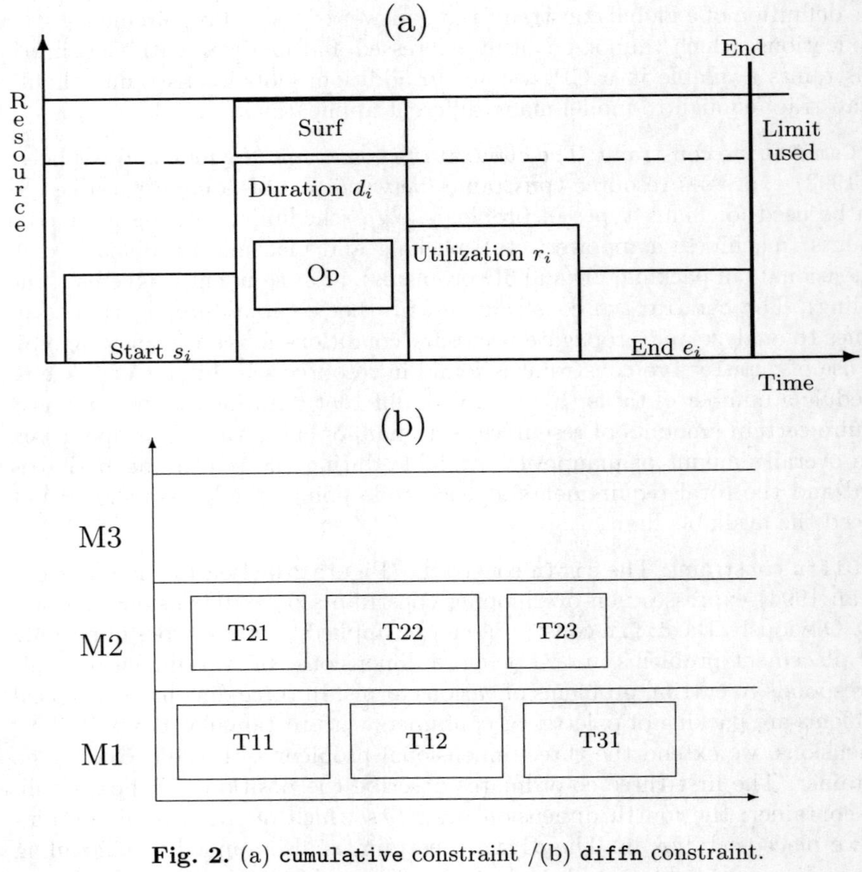

Fig. 2. (a) `cumulative` constraint /(b) `diffn` constraint.

is satisfied if the following conditions hold:

$$\sum_{i \in O: s_i \leq t \leq s_i + d_i} r_i \leq e, t = 1, \ldots, l;$$

$$\max_{i \in O}(s_i + d_i) \leq l$$

Below we give the definition of the global constraints used in the examples discussed in this paper.

`Bin packing` global constraint: A variant of the `cumulative` global constraint is the `bin packing` constraint where each operation is to be assigned to a bin. Each operation is defined only by the x_i and the r_i since all bins are equal which means all durations are equal. The value of x_i is the bin where the operation is assigned to.

`bins`$([x_1, \ldots, x_n], [r_1, \ldots, r_n], e)$, O is the set of operations and e the upper limit for all bins, e is a domain variable similarly to the `cumulative` constraint.

`binsupper`$([\ x_1, \ldots, x_n], [r_1, \ldots, r_n], [c_1, ..., c_{|B|}]$: c_i is the upper limit for bin i which is either a domain variable or a constant.

`nonoverlap`$([x_i, \ldots, x_n])$: All variables x_i take different values; this is equivalent to the well-known `alldifferent`$([x_1, \ldots, x_n])$ constraint.

`nonoverlapdisj`$([\{x_1, d_1\}, \ldots, \{x_n, d_n\}])$: It expresses a non-overlapping constraint of operations on a disjunctive machine. Each operation is defined by $\{x_i, d_i\}$ where x_i is the starting time while d_i is its duration.

`nonoverlap2D`$([\{x_1, m_1, d_1\}, \ldots, \{x_n, m_n, d_n\}])$: It expresses a 2D non-overlapping constraint of operations on disjunctive machines (where a machine can execute one operation at a time). Each operation is defined by $\{x_i, m_i, d_i\}$ where x_i is the starting time, m_i the machine and d_i its duration. This constraint is applied to scheduling problems; it expresses the assignment of operations to machines. A machine can run only one operation at a time and each operation is assigned only to one machine.

`cardinality`$\{N, [c_1, \ldots, c_n])$: It expresses the cardinality constraint, where C is a set of c_i constraints; c_i is a simple constraint like $x = v$ or $x \neq v$. N is a domain variable indicating the number of satisfied constraints. In addition it constrains the c_i constraints such that at any time at least $\min(N)$ and at most $\max(N)$ c_i constraints are satisfied. The domain variable may change during the propagation and is usually subject to constraints.

A number of other constraint systems include variants of global constraints, see (Aggoun and Beldiceanu 1993, Beldiceanu and Contejean 1994, Bockmayr et al. 2001). The constraints can be used and combined for many different problems and significantly extend the range of problems which can be handled with CP (Simonis 1999, Bockmayr et al. 2001).

3 Application Areas

There are several application areas in which CP is used. We will cite few of them developed in CHIP showing the contribution of CP in the field of decision support systems.

Configuration tool in the Telecommunications Industry: The CP application developed by ALCATEL (Rohan 1999) deals with the computation of optimized configurations of telecommunication systems. Each configuration may involve several types of telecommunications equipment (e.g. stations, cards, and bays), each of which provides different services. The aim of the configuration tool is to compute the best solution in terms of equipment and services (even in a back-up mode). The variety of constraints is quite large. Some of the constraints are similar to scheduling problems: precedence, use of resources with finite capacity, assignment of equipment (most of the items consume energy) to resources (power supply) subject to specific constraints, and a variety of 2D and 3D placement constraints of components. The flexibility and the power of global constraints in CHIP enable the telecommunications industry to model almost all the application constraints. The optimization is powered by a set of preference rules guiding the search to compute the best solution satisfying the obligatory constraints. This application is a mixture of four types of domains: planning, scheduling, placement, and AI techniques to express rules guiding the search. ALCATEL's application has been solved successfuly and used daily since 1999.

Continuous production scheduling in chemical industries: In the context of the LISCOS[3] project, BASF (Timpe 2002) solved a production scheduling problem requiring a hybrid approach combining MIP (mixed integer programming) and constraint programming. The application deals with production campaigns: While the total amount of products to be produced in the scheduling horizon is fixed, the splitting of this total amount into campaigns of different lengths (lot-sizing), the order in which the campaigns of different products are produced (sequencing), and also the machines each campaign is produced on (assignment), are to be determined by the solver. The objective of the production scheduling problem is to minimize both late and early production. The sequence-dependent changeover time of the end-product machines should also be minimized.

The problem requires combining modeling techniques from lot sizing (Wolsey 2001) and techniques from scheduling combined with machine assignment. The initial models have shown that neither MIP nor CP techniques can solve the problem. A hybrid approach combining MIP and CP formulation has been developed and implemented using Xpress-MP of Dash Optimization and CHIP of COSYTEC. In the CP part, most of the available global constraints of CHIP were used. Initial results of this technique are encouraging and show the advantage of the hybrid approach to solve planning and scheduling problems.

EVA: This application developed for EDF (Electricity of France) is used for planning the transport of wasted nuclear fuel from the French nuclear power stations to the reprocessing factory over a period of two years. The problem is highly constrained by the limited availability of expensive resources (i.e., containers) and tight time windows imposed on the reactors. The application deals with many constraints coming from different fields of expertise

[3] Project supported by the European Commission, Growth Programme, Research Project LISCOS -Large Scale Integrated Supply Chain Optimisation Software, Contract N ° G1RD-CT-1999-00034.

such as resource management and synchronization, logistics, storage management, nuclear physics and/or regulation. Due to the large number of complex constraints and the limited number of effective solutions, constraint programming appears to be the best-suited technology to tackle such problems. The system developed with CHIP has been used successfully since 1996.

OPTI-CHANNEL®: OPTI-CHANNEL® of COSYTEC is a planning tool for resources and activities in the audio-visual business. It is made up of several modules with the following capabilities: 1) Workload: Analyzes the production requirement and adapts it to the resources according to the real physical and human constraints (technical problems, holidays, training); 2) Equipment planning: Allocation of equipment according to the constraints (studios, OB units, editing suite, etc.); 3) Personnel planning: Optimized personnel timetable management to enable organization of working periods, fair distribution of working hours, compliance with legislation, control of overtime, and use of part-time working.

Another example of crew scheduling is the **GYMNASTE** tool, which creates nurses' shift schedule subject to various complex constraints. For more details, we advise the user to refer to the technical paper (Chan et al. 1998).

4 Examples of Timetabling and Sports Scheduling Problems

In this section we discuss the three selected examples modeled and solved using constraint programming.

Establishing a college timetable: A director of a college is in charge of establishing the weekly timetable for two classes in the last year of college. The objective is to make a schedule subject to various constraints. Notice that although this problem is titled and described as a college timetable scheduling problem it actually occurs more in a grade school setting (ex. 2 classes of 5th graders). The problem has not been changed in this paper because of the reference to the book (Heipcke 2002).

The progressive party example: The problem is to organize a party for 42 yacht teams at a regatta. Some of the teams act as hosts and have an available capacity ranging from 4 to 10 guests on their boat. The remaining teams act as guests and proceed to visit a number of the host boats. The size of the guest teams is known and ranges from 2 to 7 people. In all there are 13 host boats and 29 guest teams.

Sports league scheduling: The problem deals with scheduling tournaments in sports leagues such as soccer, hockey, and basketball.

In presenting the models for the above problems we may introduce redundant global constraints. Our goal in this paper is to present a set of global constraints for each specific constraint of the problem. Further study of the formulations can simplify them.

5 Establishing a College Timetable

A director of a college is in charge of establishing the weekly timetable for two classes in the last year of college. The objective is to make a schedule while respecting various constraints. This example is taken from the book (Heipcke 2002). Detailed definition and formulation in Xpress-MP is also given in that book.

Mr. Miller is in charge of establishing the weekly timetable for two classes in the last year of college. The two classes have the same teachers, except for mathematics and sports. In the college, all lessons have a duration of two hours. Furthermore, all students of the same class attend exactly the same courses. From Monday to Friday, the slots for courses are the following 8:00-10:00, 10:15-12:15, 14:00-16:00, and 16:15-18:15. The following table lists the number of two-hour lessons that every teacher has to teach the students of the two classes per week:

Table 1. Number of 2-hour lessons per teacher and class.

t	Teacher	Subject	Lessons for class 1	Lessons for class 2
1	Mr. Cheese	English	1	1
2	Mrs. Insulin	Biology	3	3
3	Mr. Map	History-Geography	2	2
4	Mr. Effofecks	Mathematics	0	4
5	Mrs. Derivate	Mathematics	4	0
6	Mrs. Electron	Physics	3	3
7	Mr. Wize	Philosophy	1	1
8	Mr. Muscle	Sports	1	0
9	Mrs. Biceps	Sports	0	1

The aim of this problem is to find a schedule, which fulfills all constraints listed below. From the CP point of view, this problem can be seen as an assignment of lessons to teachers much like the assignment of operations to machines found in scheduling problems. In our example, teachers are similar to machines and lessons similar to operations. Each day contains 4 slots and the total number of slots per week is $4 \times 5 = 20$. In this assignment, the time horizon (x-axis) is the number of slots, the number of resources (y-axis) is the number of teachers.

5.1 Constraints

Let $NP = 4$, the number of time slots per day, $ND = 5$: number of days per week, $N = NP \times ND$: the number of total slots, $COURSE_{t,c}$ denotes the number of two-hour lessons that the teacher t has to teach class c every week, $CLASS$ set of classes and $TEACHERS$ the set of teachers.

Each lesson is modeled as an operation with the following attribute $\{start, teacher, duration = 1\}$, see the definition 3 of the `nonoverlap2D` constraint. Each teacher is responsible for one specific subject. Therefore, the assignment problem is simplified, for each subject the teachers are known. The unique unknowns are the start times of the lessons, in which day and in which slot.

Below we give the constraints and a CP formulation for each constraint. We will use the following notation for the standard three attributes of each operation (lesson): $\{s_{i,c,t}, t, d\}$ where $s_{i,c,t}$ is the start time for lesson i, class j, t is the teacher and d the duration.

Constraint 1: At most one two-hour lesson per subject is taught on the same day per class. In order to avoid the same course at the end of the day i (last slot) and immediately followed by a similar course in day $i+1$ (first slot), we suggest to balance over the time the different courses on a given subject and per class. Therefore, we assume that there is at least a full day separating two consecutive lessons on the same subject.

This constraint forbids a teacher from giving two lessons on the same day for the same class. This constraint can be modeled with the `non-overlapping` constraint. Then $\forall c \in CLASS$ (for each subject and for each class):

$$\texttt{nonoverlap2D}(\{s_{1,c,t}, t, 4\}, \ldots, \{s_{i_k,c,t}, t, 4\}) \text{ where } i_k = COURSE_{t,c}$$

There are 2 `nonoverlap2D` constraints per subject and per class; since the number of subjects is 9 then the number of constraints is 2×9 (number of classes $\times$ the number of subjects). Each lesson is modeled as an operation where the duration is 4 slots (1 day) to ensure that there is at most one lesson of a given subject per day.

Example: History-Geography: There are two lessons for the first class $\{s_{1,1,3}, 3, 4\}, \{s_{2,1,3}, 3, 4\}$ and two lessons for the second class $\{s_{1,2,3}, 3, 4\}$, $\{s_{2,2,3}, 3, 4\}$: $s_{1,1,3}$ is the starting time for lesson 1, class 1 and the teacher 3 (Mr Map), 4 is the duration of the operation in the non-overlapping constraint.

$$\texttt{nonoverlap2D}\ (\{s_{1,1,3}, 3, 4\}, \{s_{2,1,3}, 3, 4\})$$

$$\texttt{nonoverlap2D}(\{s_{1,2,3}, 3, 4\}, \{s_{2,2,3}, 3, 4\})$$

Table 2. If $s_{1,1,3}$ is assigned to day 1 then the other lessons can only start in the remaining days from 2 (slot 5) to 5.

	Day 1				Day 2				Day 3				Day 4				Day 5			
	1	2	3	4	1	2	3	4	1	2	3	4	1	2	3	4	1	2	3	4
$s_{1,1,3}$	*	*	*	*																
$s_{2,1,3}$	•	•	•	•																

Constraint 2: A teacher must not teach more than one course at a time. This constraint can be modeled by a `non-overlapping` constraint between all lessons. As explained in Figure 2, the y-axis represents the teachers; the variables t (teachers) of lessons $\{s_{i,c,t}, t, d\}$ are known which makes this problem simple. For each teacher (y-axis) it constrains the lessons to not overlap and in the x-axis it constrains lessons to be in different slots.

$$\texttt{nonoverlap2D}(\{s_{1,1,t}, t, 1\}, \ldots, \{s_{i_{k1},1,t}, t, 1\}, \{s_{1,2,t}, t, 1\}, \ldots, \{s_{i_{k2},2,t}, t, 1\})$$

where $i_{k1} = COURSE_{t,1}$,$i_{k2} = COURSE_{t,2}$.

Constraint 3: The sports lessons have to take place on Thursday afternoon from 14:00 to 16:00. Slot associated with Thursday afternoon from 14:00 to 16:00 is 15. Lesson given by Mr. Muscle ($t = 8$) is $\{s_{1,1,8}, 8, 1\}$, therefore we fix the start decision variable of the lesson. The same holds for the teacher Mrs. Biceps.

$$s_{1,1,8} = 15,\ s_{1,1,9} = 15$$

Constraint 4: The first time slot on Monday morning is reserved for supervised homework. This constraint is modeled by the different constraint. Then $\forall c \in CLASS, \forall i \in COURSE_{t,c}$:

$$s_{i,c,t} \neq 1$$

Constraint 5: Mr. Effofecks is absent every Monday morning because he teaches courses at another college. This constraint means that the teacher can start teaching on Monday afternoon, which corresponds to slot 3 in the time horizon. Therefore all of Mr. Effofeck's lessons have a release date, which is slot 3. Then $\forall c \in CLASS, \forall i \in COURSE_{t,c} | t = 4$ (Mr Effofecks):

$$s_{i,c,4} \geq 3$$

Constraint 6: Mrs. Insulin does not work on Wednesday. Slots corresponding to Wednesday are 9 to 12. The handling of this constraint is to remove these slots from the domain of the start decision variables Mrs. Insulin's lessons. Then $\forall c \in CLASS, \forall i \in COURSE_{t,c} | t = 2$ (Mrs. Insulin):

$$\forall k \in \{9..12\} : s_{i,c,2} \neq k$$

Constraint 7: A class can attend only one course at a time. This constraint is formulated as a no overlapping between all lessons for each class. There are two classes, so we need two `nonoverlap2D` constraints. Then $\forall c \in CLASS$:

$$\texttt{nonoverlap2d}(\cup_{\forall t \in TEACHER}\{\{s_{1,c,t}, t, 1\}, \ldots, \{s_{i_k,c,t}, t, 1\}\})$$

where $i_k = COURSE_{t,c}$.

5.2 Search

Variable selection: The enumeration is straightforward. We suggest enumerating lessons class by class. We fix lessons of the first class and then those of the second class. We can improve the selection by selecting first subjects where the number of lessons is higher.

Value selection: We suggest fixing the starts of the lessons as early as possible (starting from the smallest value to the maximum value).

This example has many solutions. Tables 3 and 4 show examples of valid schedules.

Table 3. Example of scheduling.

Teacher	Subject	M1	M2	M3	M4	T1	T2	T3	T4
Mrs. Biceps	Sports	–							
Mr. Muscle	Sports	–							
Mr. Wize	Phil.	–							
Mrs. Electron	Phy.	–					1	2	
Mrs. Derivate	Math.	–				1			
Mr. Effofecks	Math.	–			2				2
Mr. Map	Hist.	–			1	2			1
Mrs. Insulin	Biol.	–	2	1			2	1	
Mr. Cheese	Eng.	–	1	2					

This type of problems is easy to model and solve using CP solvers. This example is solved without any backtrack.

6 The Progressive Party Example

The problem is to organize a party for 42 yacht teams at a regatta. Some of the teams act as hosts; they remain on board, and have an available capacity

Table 4. Example of scheduling.

Teacher	Subject	W1	W2	W3	W4	T1	T2	T3	T4	F1	F2	F3	F4
Mrs. Biceps	Sports							2					
Mr. Muscle	Sports							1					
Mr. Wize	Phil.		2	1									
Mrs. Electron	Phy.		1	2					1	2			
Mrs. Derivate	Math.	1					1				1		
Mr. Effofecks	Math.				2				2				
Mr. Map	Hist.	2											
Mrs. Insulin	Biol.					1	2						
Mr. Cheese	Eng.	-	-	-	-								

ranging from 4 to 10 guests on their boats. The remaining teams act as guests and proceed to visit a number of the host boats; the size of the guest teams is known and ranges from 2 to 7 people. We assume that the hosts are selected. In all there are 13 host boats and 29 guests. The complete description and the different formulations, MIP and CP, plus their comparisons are well detailed in the paper (Smith et al. 1996).

Table 5. Capacities of Hosts/Size of Guests Teams.

Host capacities (13)	10, 10, 9, 8, 8, 8, 8, 8, 8, 7, 6, 4, 4
Size of Guest Teams (29)	7, 6, 5, 5, 4, 4, 4, 4, 4, 4, 4, 4, 3, 3, 2, 2, 2, 2, 2, 2, 2, 2, 2, 2, 2, 2, 2

The guest teams cannot be split and the party must take place over 7 consecutive hours, each guest team changing boats at each hour. The constraints to be respected are the boat capacity and that any two groups meet only once over the course of the party. The goal is to minimize the number of boats needed to organize the meetings.

Let G be the set of guests ($|G| = 29$), T be the set of periods ($|T| = 7$) and H be the set of hosts ($|H| = 13$).

6.1 Model

To solve this problem we model it in the form of a `bin-packing` problem. Each bin expresses the available capacity of a host boat. In fact, at any time the total number of people, total team sizes, meeting at a host boat should not exceed the capacity of that boat.

We are able to express the available capacity of a host boat by using the `bin-packing` constraint (a variant of the `cumulative` constraint), we ensure

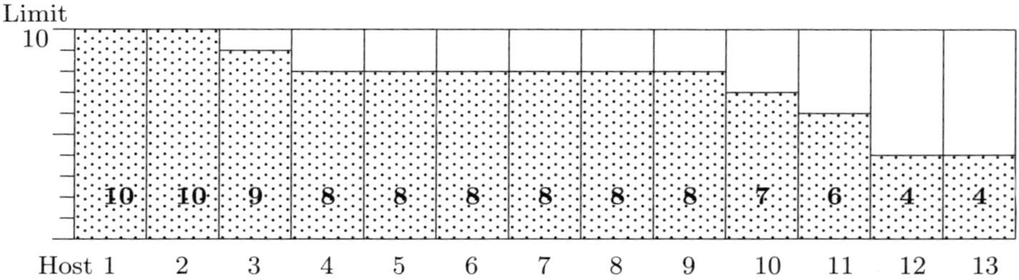

Fig. 3. Capacity of hosts modeled as bins.

that a guest team does not return to the same boat with the `alldifferent` constraint and use the `cardinality` constraint to express that the guest teams should only meet once at a party.

6.2 Decision variables

For each hour of the party, we create $|G|$ (guests) decision variables $x_{t,1}, \ldots x_{t,i}, \ldots x_{t,|G|}$. Indeed, for each hour we have 29 decision variables. The total number of domain variables is $|G| \times |T|$ which is 29×7.

$x_{t,i} = m$ means guest i at time t is visiting the host m,
then $x_{t,i} \in H$, for all $i \in G$, for all $t \in T$

6.3 Constraints

Constraint 1: A guest team cannot visit a host boat more than once over the course of the party.

For each guest there exists $|T|$ variables $x_{t,i} \in T$ indicating the visited hosts; therefore they should take all different values to ensure that a guest i cannot visit a host more than once. This constraint is expressed naturally with the well-known `alldifferent` constraint. In our model we will use the non-overlapping notation defined below.

$$\texttt{nonoverlap}([x_{1,i}, \ldots, x_{|T|,i}]) \text{ , for all } i \in G$$

Constraint 2: At any time do not exceed the capacity of the host boat.

The `bin packing` constraint, which is a variant of the `cumulative` constraint, is the appropriate global constraint in this case. Each team is modeled by the attribute $\{x_{t,i}, c_i\}$ where the value of $x_{t,i}$ indicates in which bin (host) item i (guest) go (visit). When an item i is assigned to a bin (host) it occupies a size c_i (size of the guest crew). The bin constraint ensures that at any time the size of the bin is never exceeded.

For each hour t, we create a `bin packing` constraint:

$$\texttt{binsupper}([x_{t,i},\ldots,x_{t,|G|}],[c_1,\ldots,c_{|G|}],[C_1,\ldots,C_{|H|}])$$

where $x_{t,i} = m$ indicates which host m is visited by guest i in period t; c_i is the size of the guest i, C_j is the capacicty of the host j.

The number of `bin packing` global constraints is equal to the number of periods; in our example we have 7 global constraints.

Constraint 3: The guest teams should meet only once at the party.

This constraint is complex and can not be modeled by a single global constraint. The idea is to combine two global constraints, `distance` and `cardinality` constraints, to express this type of requirement. We first sketch an example of meeting between the two teams 1 and 2 and then we give the general model.

A meeting between guests 1 and 2 at hour 1 is true if $x_{t,1} = m$ & $x_{t,2} = m, t = 1$, have the same value m. This means that the two guests are hosted by m at period $t = 1$.

Let $l_{i,j,t} = |x_{t,i} - x_{t,j}|$, this is expressed by the CHIP distance constraint where $l_{i,j,t}$, $x_{t,i}$, and $x_{t,j}$ are domain variables.

If $l_{i,j,t}$ takes the value 0 then it means $x_{t,i}$ and $x_{t,j}$ are guests of host m at period t. For example, all possible meetings between guest i and j are:

$[l_{i,j,1} = |x_{1,i} - x_{1,j}|, l_{i,j,2} = |x_{2,i} - x_{2,j}|, \ldots, l_{i,j,7} = |x_{7,i} - x_{7,j}|]$

The goal is to model the fact that the two teams i and j should meet only once at the party. This requirement can be expressed by the `cardinality` constraint:

$$\texttt{cardinality}([0..1],[l_{i,j,1},\ldots l_{i,j,|T|}])$$

constraining the number of variables $\{l_{i,j,t}\}$ for two teams i and j taking the value 0 to range between 0 and 1.

We generalize this model for all meetings between teams as follows

$$\texttt{forall } i \in \{1..|G|\}, j \in \{i+1..|G|\}\text{: }\texttt{cardinality}([0..1],[l_{i,j,1},\ldots l_{i,j,|T|}])$$

6.4 Search

Variable selection: We suggest a formulation where principal variables are structured into groups. The natural way is to group principal variables per periods. First, we schedule all meetings of period 1, then period 2, and so on until period 7. The motivation for considering this approach is that it leads to maximize the number of satisfied constraints period per period, which reduces

Table 6. Example of a solution.

Host	T1	T2	T3	T4	T5	T6	T7
1	1, 14	2, 7	3, 4	5, 6	8, 11, 20	9, 13, 22	10,16,25, 29
2	2, 6	1, 15	5, 7	3, 8	4, 9	10,18,23, 28	11, 12, 19
3	3, 7	4, 6	1, 16	2, 14	5, 10	8, 15, 21	9, 20, 28
4	4, 15	3, 14	2, 17	1	6, 7	5, 24	8, 22, 27
5	5, 16	8, 10	6, 9	4, 17	1	2, 25	3, 23
6	8, 9	5, 17	10, 12	7, 11	2, 16	1	4, 24
7	10, 11	9, 12	8, 13	21,26,27, 29	3, 15	4, 14	1
8	12, 13	11, 16, 18	14, 15, 19	9, 10	17,22,24, 28	3, 29	2, 26
9	17,18,19, 20	13, 21, 24	11, 22, 26	12, 25, 28	14, 23, 29	6, 16, 27	5, 15
10	21, 22, 23	19, 25, 27	18, 24, 29	13, 15	12, 26	7, 20	6, 14
11	24, 25, 26	20, 22, 29	21, 28	16, 19, 23	13, 27	11, 17	7, 18
12	27, 28	23, 26	20, 25	18, 22	19, 21	12	13
13	29	28	23, 27	20, 24	18, 25	19, 26	17, 21

the number of pending constraints as we get deeper in the search tree. The second suggestion consists of considering the principal variables in each group in decreasing order of crew size and in case of ties select the variable appearing in more constraints. This criterion will lead to a more active propagation and will avoid failures for the last variables. The idea is that, at the end of the enumeration of a group, it is much easier to find an assignment for a crew with a small size than for a larger one.

Value selection: Once a decision variable is selected, values are tried in an increasing order. The first boat has the largest capacity, and so on.

This problem has the advantage of being a well-studied time tabling problem with a variety of constraints. The problem is hard to solve with a mathematical approach using MIP models as reported in the article (Smith et al. 1996). However they have shown that a CP formulation is able to solve the problem in less than 30 minutes. The formulation presented in this paper allows taking advantage of the power of the global constraints of CHIP and solves the problem in a few ms.

7 Sports League Scheduling

Scheduling tournaments in sports leagues (e.g. soccer, hockey, basketball) contains various complex constraints and different objectives to optimize. The scheduling problem presented is detailed in the paper (Hamiez and Hao 2001). Below we describe the league scheduling problem:

7.1 Definitions

1. There are $|T|$ teams, T is even and denotes the set of all teams.
2. The season lasts $|T|$ - 1 weeks
3. There are $|T|/2$ periods.

7.2 Constraints

1. One game is scheduled in every period.
2. Every team plays one game in every week of the season.
3. No team plays more than twice in the same period over the course of the season.
4. All teams play each other exactly once; this is a half competition.

The problem consists of finding a schedule for the tournament, which respects all the constraints of the problem.

7.3 Notations

1. $N = |T|$ is the number of teams. We label the teams from 1 to N.
2. $W = |T| - 1$ is the number of weeks. We label the weeks from 1 to W.
3. $P = |T|/2$ is the number of periods. We label the periods from 1 to P.
4. C is the set of all constraints of the problem.

7.4 CP formulation

We first present a CP formulation of the problem based on CHIP constraints. Following the problem description in (Hamiez and Hao 2001), a configuration (solution) is represented as a two-dimensional array with weeks in columns and periods in rows. Each cell in the array is a tuple $\{x_{(w,p)}, y_{(w,p)}\}$ where $x_{(w,p)}$ and $y_{(w,p)}$ are the teams and which means team $x_{(w,p)}$ plays $y_{(w,p)}$ in week w at period p. The problem to solve is: For $w = 1 \dots W$ and for $p = 1 \dots P$, find an assignment to $x_{(w,p)}$ and $y_{(w,p)}$ such that all the constraints C are satisfied. The domain of all decision variables $x_{(w,p)}$ and $y_{(w,p)}$ range from 1 to N.

7.5 Decision variables

The problem to solve is a half competition where each team meets other teams only once. Consequently, the total number of matches is:

$M = W \times P, M$ is the number of matches where a match $m = \{x, y\}$ is competition between x and y.

For example if $|T| = 6$ then $W = (6 - 1) = 5$, $P = 3$ and $M = 5 \times 3 = 15$ matches.

The following table shows an example of a valid schedule with $|T|=6$ teams, $W = 5$, $P = 3$.

Table 7. Example of scheduling.

P / W	w:1	w:2	w:3	w:4	w:5
p:1	{1, 6}	{1, 4}	{2, 6}	{2, 4}	{3, 5}
p:2	{2, 5}	{2, 3}	{1, 3}	{1, 5}	{4, 6}
p:3	{3, 4}	{5, 6}	{4, 5}	{3, 6}	{1, 2}

Column 2, labeled w:1, shows the schedule for the first week. Line 2, labeled p:1, shows the schedule for the first period. The tuple {1,6} means team 1 meets team 6 in week 1 during the period 1.

Each match is a competition of two teams x and y, the total number of variables is: $NM = 2M$, where M is exactly the size of the two-dimensional array presented in Table 7.

The model described in this paper contains M matches, tuples, and NM decisions variables. Let M^* be the set of all $m = \{x_{(w,p)}, y_{(w,p)}\}, w = 1 \ldots W, p = 1 \ldots P$. Then we have that $\forall m = \{x_{(w,p)}, y_{(w,p)}\} \in M^*$ then $x_{(w,p)}, y_{(w,p)} \in N$, the domain of decision variables of a match m.

In the formulation presented in this paper there are no binary variables.

7.6 Constraints

This section deals with constraints of the sports league scheduling problem. A CP formulation of the constraints requires their decomposition regarding the structure proposed in terms of matches (tuples) in the above section. The idea is to use global constraints of CHIP and especially the non-overlapping global constraint.

Constraint 1: A match m is a competition between two teams $x_{(w,p)}$ and $y_{(w,p)}$, therefore $x_{(w,p)}$ and $y_{(w,p)}$ must take different values. Then we have that $\forall m = \{x_{(w,p)}, y_{(w,p)}\} \in M^*$,

$$\texttt{nonoverlap}(\{x_{(w,p)}, y_{(w,p)}\})$$

This constraint ensures that $x_{(w,p)}$ and $y_{(w,p)}$ always take different values.

Constraint 2: In each $w \in W$ all teams are playing. In fact, each column satisfies the `cardinality` constraint: each team appears exactly once, which is modeled, as all teams are different. In order to take advantage of the power of global constraints, each team is modeled as an operation where its duration is equal to one and the start is the value of decision variable $x_{(w,p)}$ or $y_{(w,p)}$. For example two operations are associated with each match $\{x_{(w,p)}, y_{(w,p)}\}$, the first one for the variable $x_{(w,p)}$ and the second for the variable $y_{(w,p)}$. These operations are defined as follows. We have that $\forall w \in W$ there are P matches and then there are $2P$ teams playing in each week; then we have for $w = 1 \ldots W$:

$$\texttt{nonoverlapdisj}(\ \{x_{(w,1)}, 1\}, \ldots, \{x_{(w,P)}, 1\}, \{y_{(w,1)}, 1\}, \ldots, \{y_{(w,P)}, 1\}).$$

The modeling of this constraint consists in stating a `non-overlapping` constraint between all teams of a week (column), the number of variables is $2P$.

For the first column in Table 7, the `non-overlapping` constraint is stated as follows:

$\texttt{nonoverlapdisj}(\{x_{(1,1)}, 1\}, \{x_{(1,2)}, 1\}, \{x_{(1,3)}, 1\}, \{y_{(1,1)}, 1\}, \{y_{(1,2)}, 1\}, \{y_{(1,3)}, 1\}\}$

Constraint 3: In each row, no team appears more than twice.

This constraint can be modeled as a bin packing problem where each team $x_{(w,p)}$ and $y_{(w,p)}$ is modeled as operations to assign it to bins; the starting time, assignment to bins, are the variables $x_{(w,p)}$ and $y_{(w,p)}$, the resource consumption is 1 and the total amount of resource H is a domain variable where the upper bound is 2. Indeed, for clarity, the bins correspond to the teams. Then we have For $p = 1 \ldots P$:

$$\texttt{bins}(\{x_{(1,p)}, \ldots, x_{(W,p)}, y_{(1,p)}, \ldots, y_{(W,p)}\}, \{1, \ldots, 1\}, e)\ ,\ e \text{ in } 0..2$$

The modeling of this constraint consists in stating a bin packing problem where for each period, there are $2 \times W$ operations.

Constraint 4: The last and the most difficult constraint is the global constraint on all matches: each match appears only once which means that all matches are different.

This constraint is easily formulated in CP as a 2D non-overlapping constraint. We extend the previous formulation where each team is defined as an operation as follows $\{start : x_{(w,p)}, machine : y_{(w,p)}, duration : 1\}$. Each match can be seen as square where its length is 1. In order to ease the modeling we sketch the solution given in Figure 4.

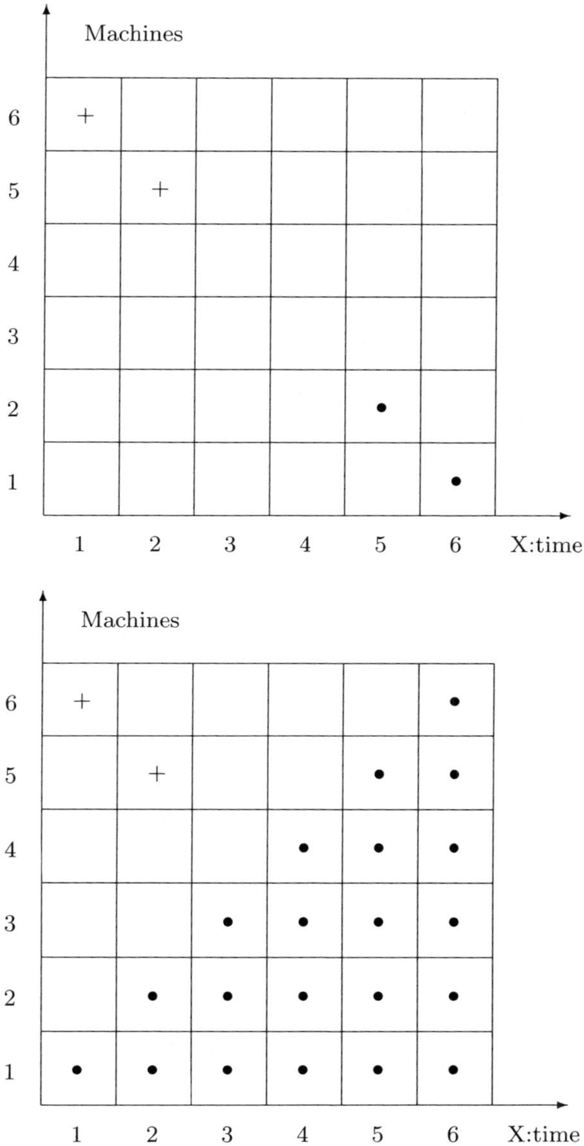

Fig. 4. All pairs are different.

The solution $\{x_{(1,1)}, y_{(1,1)}\} = \{1, 6\}$ means team 1 meets team 6 in week 1 at period 1. This is seen as as task which occupies 1 slot in Figure 4 (x is time, y is the machine, the duration of slot 1}, the number of machines is N and the time horizon is also equal to N. In this model, we are also interested in modeling the constraint that if x meets y then we should forbid that y meets x because it is a half competition. Therefore in our formulation we forbid the assignment {6, 1}. The solution $\{x_{(1,2)}, y_{(1,2)}\} = \{2, 5\}$ means team 2 meets team 5 in week 1 at period 2, therefore we can forbid the assignment $\{5, 2\}$. In fact, it is possible to remove symmetrical matches by forbidding all the cells below the diagonal (cells marked with the symbol •). After forbidding this region, the total remaining cells is exactly the number of matches to schedule (half competition). The handling of this constraint is similar to a perfect placement where we would like to place cells (1x1) in a big square of dimension $N \times N$.

Let F contain all cells marked with the symbol • then:

$$\texttt{nonoverlap2D}(F \cup \{\{x_{(w,t)}, y_{(w,t)}, 1\}, w = 1, \ldots W, t = 1, \ldots, P\})$$

Let $|T| = 4$ then the above constraint will be:

$\texttt{nonoverlap2D}(\{1, 1, 1\}, \{1, 2, 1\}, \{1, 3, 1\}, \{1, 4, 1\}, \{2, 2, 1\}, \{2, 3, 1\}, \{2, 4, 1\}, \{3, 3, 1\}, \{3, 4, 1\}, \{4, 4, 1\}, \{x_{(1,1)}, y_{(1,1)},\}, \{x_{(1,2)}, y_{(1,2)},\}, \{x_{(2,1)}, y_{(2,1)},\}, \{x_{(2,2)}, y_{(2,2)},\}, \{x_{(3,1)}, y_{(3,1)},\}, \{x_{(3,2)}, y_{(3,2)},\})$

7.7 Search

The search uses the high-level primitives, allowing the user to easily define various enumeration strategies by setting the right parameters. In fact, these primitives work on semantic objects (tuples) instead of on basic domain variables. As defined above, each match is a tuple $m = \{x_{(w,p)}, y_{(w,p)}\}$ where $x_{(w,p)}$ and $y_{(w,p)}$ are the decision variables.

Tuple selection: first criterion: $x_{(w,p)}$ has the smallest number of values in its domain, second criterion: $y_{(w,p)}$ has the smallest number of values, third criterion: p is the smallest value, fourth criterion: w is the smallest.

This strategy tries to select tuples where $x_{(w,p)}$ and $y_{(w,p)}$ have smallest domains. In case of ties it tries to schedule columns first, i.e., weeks.

Value selection: assign values first to variables $x_{(w,t)}$ and $y_{(w,t)}$. In order to avoid exploring the entire search tree, we suggest limiting the number of fails per variable to 1.

Optimization: There is no objective function to minimize or maximize. For this problem we are seeking feasibility, a solution that satisfies all constraints.

Example: Table 8 shows an example of a valid schedule with $|T|$=12 teams, $W = 11$, $P = 6$.

Table 8. Example of scheduling for 12 teams.

P \ W	w:1	w:2	w:3	w:4	w:5	w:6	w:7	w:8	w:9	w:10	w:11
p:1	{5,11}	{5,12}	{11,12}	{2,6}	{7,9}	{1,10}	{1,4}	{8,9}	{8,10}	{2,3}	{3,4}
p:2	{3,9}	{1,8}	{4,7}	{10,12}	{3,6}	{7,12}	{2,11}	{6,10}	{4,5}	{5,9}	{1,2}
p:3	{1,7}	{3,10}	{2,9}	{4,11}	{2,10}	{4,8}	{6,8}	{5,7}	{9,11}	{1,12}	{5,6}
p:4	{2,8}	{4,9}	{1,6}	{5,8}	{1,5}	{3,11}	{7,10}	{3,12}	{2,12}	{4,6}	{7,11}
p:5	{4,10}	{2,7}	{3,8}	{1,9}	{4,12}	{6,9}	{3,5}	{1,11}	{6,7}	{10,11}	{8,12}
p:6	{6,12}	{6,11}	{5,10}	{3,7}	{8,11}	{2,5}	{9,12}	{2,4}	{1,3}	{7,8}	{9,10}

Column 2, labeled w:1, shows the schedule of the first week. Line 2, labeled p:1, shows the schedule of the first period.

This problem is still difficult for MIP and CP as the number of matches increases; the best results are obtained by using sophisticated models based on Tabu Search (Hamiez and Hao 2001).

8 Conclusion

A significant number of real-life applications have been developed with CHIP. The different applications, as described above, show the applicability of constraint programming to a wide range of combinatorial search and optimization problems. Among these applications are timetabling such as nurse scheduling, personnel planning and scheduling in production. Timetabling and sports scheduling are well-studied problems with a variety of constraints. The important aspect is that CP has the capability to model each requirement based on the concept of global constraints. These formulations are comprehensive for many users. The future trends in CP are the development of hybrid approaches (Bockmayr and Pisaruk 2003) (e.g. combination of Mixed Integer Programming and CP, meta-heuristics and CP) and the development of online optimization techniques (Cabon et al. 1998).

Acknowledgments

We would like to thank I. Gouachi, X. Savalle and N. Boucher for interesting discussions and for their remarks on the earlier drafts of this paper.

References

1. Aggoun, A. Beldiceanu, N. Extending CHIP in Order to Solve Complex Scheduling Problems, Journal of Mathematical and Computer Modelling, Vol. 17, No. 7, pages 57-73, Pergamon Press, 1993.
2. Beldiceanu, N. Contejean, E., Introducing Global Constraints in CHIP, *Journal of Mathematical and Computer Modelling*, Vol. 20, No 12, pp 97-123, 1994.

3. Beldiceanu, N. Global Constraints as graph properties on a structured network of elementary constraints of the same type, *Constraint Programming 2000*, Singapore, pages 384-395, volume 1894 of Lecture Notes in Computer Science, Springer-Verlag, 2000.
4. Bockmayr, A. Pisaruk, N. Aggoun, A. Network flow problems in constraint programming, *Constraint Programming 2001*, Cyprus, 2001.
5. Bockmayr, A. Pisaruk, N. Detecting Infeasibility and Generating Cuts for MIP using CP, *CP-AI-OR 2003*, Montréal, Canada, Ma 8-10, 2003.
6. Cabon, B., de Givry, S. Verfaillie, G. Anytime Lower Bounds for Constraint Optimization Problems. *CP'98*, pp. 117-131, Pisa, Italy.
7. Chan, P. Heus, K. Weil, G. 1998. Nurse Scheduling with Global Constraints. In *CHIP : GYMNASTE, PAPPACT 98*, London.
8. Caseau, Y. Laburthe, F. 1994. Improved CLP Scheduling with Task Intervals, *Proc 11th ICLP 1994*, Italy, June 1994. MIT Press.
9. Hamiez, J.P. Hao, J.K. Solving the sports league scheduling problem with Tabu search. *Lecture Notes in Artificial Intelligence* 2148: 24-36, Springer 2001.
10. Heipcke, S. *Applications of optimization with Xpress-MP*. Published by Dash Optimization Ltd., www.dashoptimization.com, 2002.
11. Régin, J. C. Rueher, M. A. Global Constraint combining a sum constraint and difference constraint. In *Principles and practice of Constraint Programming, CP'2000*, Singapore, pages 384-395, Springer, LNCS 1894, 2000.
12. Rohan, Y. Configuration Tool In Telecom Industry. *CHIP Users' Club 99*, COSYTEC SA, 1999.
13. Simonis, H. The CHIP System and its Applications. *Compulog Americas First Summer School on (Constraint) Logic Programming*, New Mexico State University, Las Cruces, N.M. August 1-6, 1999.
14. Simonis, H. Aggoun, A. Beldiceanu, N. Bourreau, E. Global Constraint Visualization in P. Deransart. In J. Maluszynski, M. Hermenegildo, editors, *Analysis and Visualisation Tools for Constraint Programming*, volume 1870 of Lecture Notes in Computer Science, Springer-Verlag, 2000.
15. Smith, B. Brailsford, S. Hubbard, P. Williams, H. The progressive party problem: Integer linear programming and constraint programming compared. *Constraints* 1: 119-138, 1996.
16. Timpe, C. Solving Planning and Scheduling Problems with Combined Integer and Constraint Programming. *OR Spectrum*, 24, 2002.
17. Van Hentenryck, P. *Constraint Satisfaction in Logic Programming*. The MIT Press, 1989.
18. Wolsey, L.A. MIP Modeling for Production Planning. *CP-AI-OR 2001*, Imperial College, London, April 8-12, 2001.

Collaboration Networks in Sports

Vladimir Boginski[1], Sergiy Butenko[2], Panos M. Pardalos[1], and Oleg Prokopyev[1]

[1] University of Florida, USA
{vb,pardalos,oap4ripe}@ufl.edu
[2] Texas A&M University, USA
butenko@tamu.edu

1 Introduction

We live in the world of information, where huge amounts of data of diverse nature and origin arise in various spheres of life, including sports. To get useful information from this data, one should apply special techniques of summarizing and visualizing the information contained in a certain dataset. In many practical situations, a real-life dataset can be represented as a large *graph* (*network*) (Boginski et al. 2003). A graph is a set of vertices (dots) and edges (links) connecting them. In a graph representation of a dataset, certain attributes are usually associated with vertices and edges of the corresponding graph. These attributes may contain specific information characterizing the given application, which often provides a new insight into the internal structure and patterns of the data.

The examples of representing real-life datasets as graphs come from diverse areas. Among these examples, one can mention the Web graph that naturally represents the World-Wide Web (Broder et al. 2000), the Call graph arising in the telecommunications traffic data (Abello et al. 1999, Aiello et al. 2001, Hayes 2000); the Market graph reflecting the structure of the stock market (Boginski et al. 2003a), and metabolic networks arising in biology (Jeong et al. 2000).

In this chapter, we will discuss one of the most interesting real-life graph applications – so-called "social networks" where the vertices are real people (Hayes 2000, Watts 1999). The main idea of this approach is to consider the "acquaintanceship graph" connecting the entire human population. In this graph, an edge connects two given vertices if the corresponding two persons know each other.

Social networks are associated with a famous "small-world" hypothesis, which claims that despite the large number of vertices, the distance between any two vertices (or, the *diameter* of the graph) is small. More specifically, the idea of "six degrees of separation" has been introduced. It states that any two persons in the world are linked with each other through a sequence of at most six people (Hayes 2000, Watts 1999, Watts and Strogatz 1998).

Clearly, one cannot verify this hypothesis for the graph incorporating more than 6 billion people living on the Earth, however, smaller subgraphs of the acquaintanceship graph connecting certain groups of people can be investigated in detail. One of the most well-known graphs of this type is the *scientific collaboration graph* reflecting the information about the joint works between all scientists. Two vertices are connected by an edge if the corresponding two scientists have a joint research paper. Another graph of this type is known as the "*Hollywood graph*": it links all the movie actors, and an edge connects two actors if they ever appeared in the same movie. Well-known concepts associated with these graphs are so-called "Erdös number" (in the scientific collaboration graph) and "Bacon number" (in the Hollywood graph), which are assigned to every vertex and characterize the distance from this vertex to the vertex denoting the "center" of the graph. In the collaboration graph, the central vertex corresponds to the famous graph theoretician Paul Erdös, whereas in the Hollywood graph the same position is assigned to Kevin Bacon.

In this chapter, we discuss graphs of a similar type arising in sports, that represent the players' "collaboration". In these graphs, the players are the vertices, and an edge is added to the graph if the corresponding two players ever played together in the same team. One of the examples of this type of graphs is the graph representing baseball players (see http://www.baseball-reference.com/oracle/). For any two baseball players who ever played in the Major League Baseball(MLB), a path connecting them can be found in this graph.

As another instance of collaboration networks in sports, we study the "NBA graph" where the vertices represent all the basketball players who are currently playing in the NBA. We apply standard graph-theoretical algorithms for investigating the properties of this graph, such as its connectivity and diameter (*i.e.*, the maximum distance between all pairs of vertices in the graph). As we will see later in the chapter, this study also confirms the "small-world hypothesis". Moreover, we introduce a distance measure in the NBA graph similar to the Erdös number and the Bacon number. The central role in this graph is given to Michael Jordan, arguably the greatest basketball player of all times, and we refer to this measure as the *Jordan number*.

The rest of the chapter is organized as follows. In Section 2, we provide several basic definitions from graph theory, in order to familiarize the reader with standard properties of graphs. Section 3 describes the examples of collaboration networks (scientific collaboration graph, Hollywood graph and baseball graph) in more detail. Section 4 presents various properties of the NBA graph. Finally, Section 5 concludes the discussion.

2 Graph Theory Basics

To give a brief introduction to graph theory, we mention several basic definitions and notations. Let $G = (V, E)$ be an undirected graph with the set of n vertices V and the set of edges E.

The graph $G = (V, E)$ is *connected* if there is a path from any vertex to any vertex in the set V. If the graph is disconnected, it can be decomposed into several connected subgraphs, which are referred to as the *connected components* of G.

The *distance* between two vertices is the number of edges in the shortest path between them (it is equal to infinity for vertices representing different connected components). The *diameter* of a graph G is usually defined as the maximal distance between pairs of vertices of G. Note, that in the case of a disconnected graph the usual definition of the diameter would result in the infinite diameter, therefore the following definition is in order. By the diameter of a disconnected graph we mean the maximum finite shortest path length in the graph (which is the same as the largest of diameters of the graph's connected components).

The *degree* of a vertex is the number of edges emanating from it.

The *edge density* the graph is defined as a ratio of its number of edges to the maximum possible number of edges. One can easily check that the maximum possible number of edges is given by $n(n-1)/2$ (where n is the number of vertices).

It is also useful to define certain special formations in a graph, which are referred to as *cliques* and *independent sets.* A *clique* in a graph is a set of completely interconnected vertices (*a complete subgraph*), and an *independent set* is a set of vertices without connections.

3 Examples of Collaboration Networks

In this section, we give a more detailed description of the examples of collaboration networks mentioned in the introduction – the scientific collaboration graph, the Hollywood graph, and the baseball graph.

3.1 Scientific collaboration graph and Erdös number

As it was mentioned above, the vertices of the scientific collaboration graph are scientists, and the edges in this graph connect the scientists who have ever collaborated with each other (*i.e.*, had a joint paper). In order to measure the distances in this graph, the "central vertex" is introduced. This vertex corresponds to Paul Erdös, the father of the theory of random graphs. This vertex is assigned *Erdös number* equal to 0. For all other vertices in the graph, the Erdös number is defined as the distance (i.e., the shortest path length) from the central vertex. For example, those scientists who had a joint paper with Erdös have Erdös number 1, those who did not collaborate with Erdös, but collaborated with Erdös' collaborators have Erdös number 2, *etc.*

Following this logic, one can construct the *connected component* of the collaboration graph with "concentric circles", which would incorporate almost all scientists in the world, except those who never collaborate with anybody. This connected component is expected to have a relatively small diameter.

The idea of constructing collaboration graphs encompassing people in different areas gave a rise to several other applications. Next, we discuss the Hollywood graph and the baseball graph, where the number of vertices is significantly smaller than in the scientific collaboration graph, which allows one to study their structure in more detail.

3.2 Hollywood graph and Bacon number

The Hollywood graph is constructed using the same principles as the scientific collaboration graph, however, the number of Hollywood actors is much smaller than the number of scientists, therefore, one can investigate the characteristics of every vertex in this graph. This information is maintained at the "Oracle of Bacon" website (http://www.cs.virginia.edu/oracle/). The most recent Hollywood graph contains 595,578 vertices (actors). The central vertex in this graph represents the famous actor Kevin Bacon, and this vertex obviously has Bacon number 0. Since the number of vertices in this graph is small enough, one can explicitly calculate the Bacon number for every actor. It turns out that most of the actors have Bacon numbers equal to 2 or 3, and the maximum possible Bacon number is equal to 8, which is the case for only 3 vertices.

The distribution of Bacon numbers in the Hollywood graph is shown in Figure 1. The average Bacon number (i.e., the average path length from a given actor to Bacon) is equal to 2.946. As one can see, both the average and the maximum Bacon numbers of the Hollywood graph are very small, which provides an argument in favor of the "small world hypothesis" mentioned earlier.

3.3 Baseball graph and Wynn number

Collaboration networks similar to the ones mentioned above can also be constructed in sports. One example of such a network is the "baseball graph" representing all baseball players who ever played in the MLB. In this graph, two players are connected if they ever were teammates. The most recent

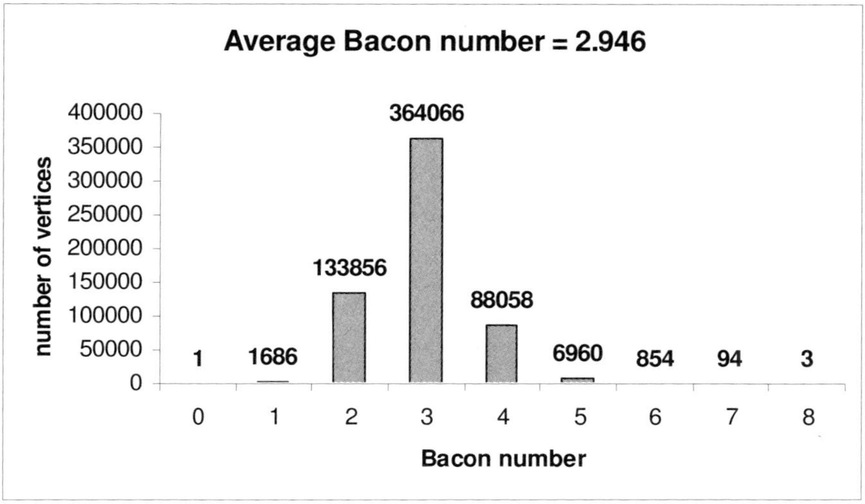

Fig. 1. Number of vertices in the Hollywood graph with different values of Bacon number. Average Bacon number = 2.946.

baseball graph has 15817 vertices. Links between any pair of baseball players can be found at the "Oracle of Baseball" website (http://www.baseball-reference.com/oracle/).

One can assign the central role in this graph to Early Wynn, a member of the Hall of Fame who spent 23 seasons in the MLB. Figure 2 shows the distribution of Wynn numbers in the baseball graph. The maximum Wynn number is 6, which is smaller than the maximum Bacon number since total number of baseball players is less than the number of Hollywood actors.

3.4 Diameter of collaboration networks

Another aspect that should be mentioned here is that the maximum from the central vertex in the collaboration graphs certainly depends on the choice of this central vertex. The reason for choosing Kevin Bacon as the center of the Hollywood graph, and Early Wynn as the center of the baseball graph is the fact that it is reasonable to expect them to be connected to many vertices: Bacon appeared in many movies, and Wynn played in several baseball teams had quite a number of teammates during his long career. However, one can choose less "connected" centers of these graphs, and in this case the maximum distance from the new center of the graph may significantly increase. For example, if one chooses Barry Bonds as the center of the baseball graph, the maximum Bonds number will be 9 instead of 6. Moreover, in the Hollywood graph, it is possible to choose the center so that the maximum distance from

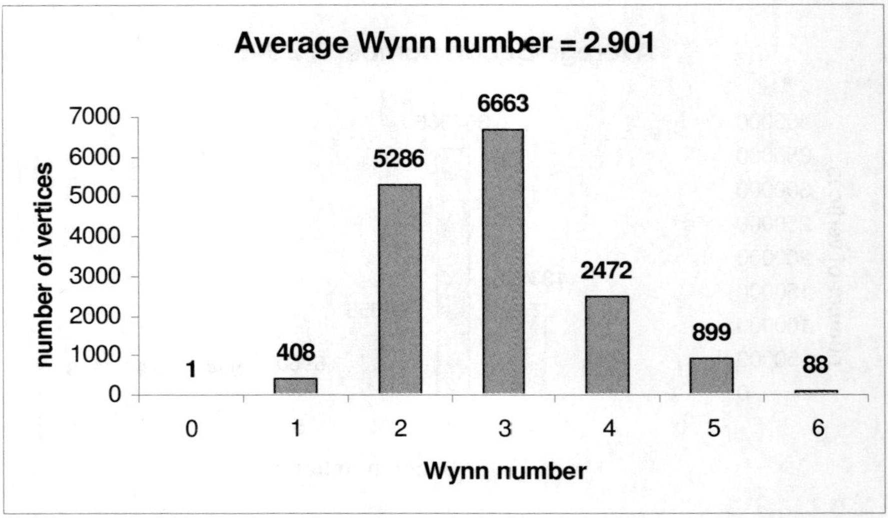

Fig. 2. Number of vertices in the baseball graph with different vaues of Wynn number. Average Wynn number = 2.901.

it is equal to 14, and the average distance is greater than 6 (instead of 2.946). Therefore, in order to have a more complete information about the structure of these graphs, one should calculate the maximum possible distance among *all* pairs of vertices in the graph. Recall that this quantity is referred to as the *diameter* of the graph. Clearly, the diameter can be found by considering each vertex as the center of the graph, calculating corresponding maximal distances, and then choosing the maximum among them.

In the next section, we study the properties of the NBA graph incorporating basketball players playing in the world's best basketball league. In a similar fashion, we introduce the *Jordan number*, investigate its values corresponding to different vertices, and calculate the diameter of this graph.

4 NBA Graph

The NBA graph considered in this section is constructed using the same idea as the graphs described above. Here we provide a detailed description of the structural properties of this graph. As we will see, its properties are rather similar to the properties of other social networks, which confirms the small-world hypothesis.

4.1 General properties of the NBA graph

The instance of the NBA graph that we consider in this section is relatively small and contains only those players who are *currently* playing in the NBA (as of the season of 2002-2003). However, this information is sufficient to reveal that the NBA graph follows similar patterns as other social networks. As of May 2003, the total number of players in the rosters of all the NBA teams is equal to 404 (players picked in the 2003 NBA draft and transfers that occurred after the end of the 2002-2003 season are not taken into account). An edge connects two given players if they ever played in the same team. Consequently, the constructed NBA graph has 404 vertices, and 5492 edges connecting them. Note that the maximum possible number of edges is equal to $404 \times (404 - 1)/2 = 81406$, therefore, the *edge density* of this graph (i.e., the ratio of the number of edges to the maximum possible number of edges) is rather small: $5492/81406 = 6.75\%$.

As one can easily see, this graph has a highly specific structure: the players of every team form a *clique* in the graph (i.e., the set of completely interconnected vertices), because all the vertices corresponding to the players of the same team must be interconnected. Since many players change teams during or between the seasons, there are edges connecting the vertices from different cliques (teams). Note that this type of structure is common for all "collaboration networks" (see Figure 3).

It should be pointed out that the number of players in a basketball team is relatively small, and the players' transfers between different teams occur rather often, therefore, it would be logical to expect that the NBA graph should be *connected*, i.e., there is a path from every vertex to every vertex, moreover, the length of this path must be small enough. As we will see below, calculations confirm these assumptions.

First, we used a standard breadth-first search technique for checking the connectivity of the considered graph. Starting from an arbitrary vertex, we were able to locate all other vertices in the graph, which means that every vertex is reachable from another, therefore, the graph is connected. In the next subsection, we will also see that every pair of vertices in this graph are connected by a short path, which is in agreement with the "small-world hypothesis".

4.2 Diameter of the NBA graph and Jordan number

The next subject of our interest is verifying if the NBA graph follows the small-world hypothesis. We need to answer the question, what is the distance between any two vertices in this graph?

Similarly to the collaboration graphs mentioned above, we define the "central vertex" in the NBA graph corresponding to Michael Jordan, who played

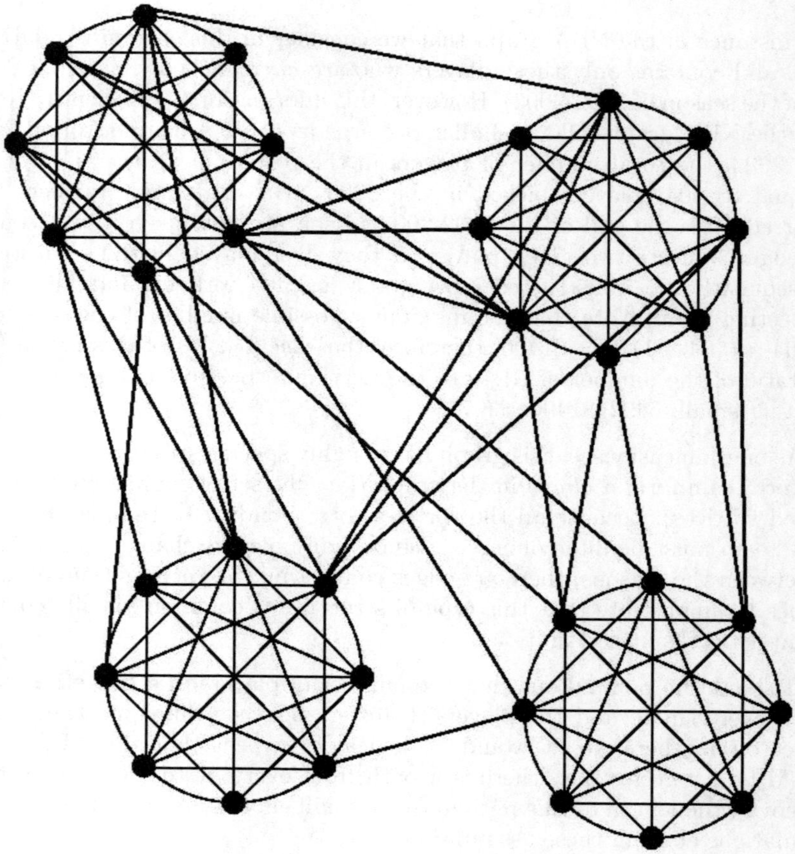

Fig. 3. General structure of the NBA graph and other collaboration networks.

for Washington Wizards during his final NBA season. Obviously, all other players in the Wizards' roster for 2002-2003, as well as all the players who have played with Jordan during at least one season in the past, have Jordan number 1. It should be noted that Michael Jordan played only for two teams (Chicago Bulls and Washington Wizards) through his entire career, therefore, one can expect that the number of players with Jordan number 1 is rather small. In fact, only 24 players currently playing in the NBA have Jordan number 1.

Following similar logic, the players who have played with Jordan's "collaborators" have Jordan number 2, and so on. However, it turns out that the maximum Jordan number in this instance of the NBA graph is only 3, *i.e.*, all the players are linked with Jordan through at most two vertices, which is

Table 1. Jordan numbers of some NBA stars (end of the 2002-2003 season).

Player	Team	Jordan Number
Kobe Bryant	Los Angeles Lakers	2
Vince Carter	Toronto Raptors	2
Vlade Divac	Sacramento Kings	2
Tim Duncan	San Antonio Spurs	2
Michael Finley	Dallas Mavericks	2
Steve Francis	Houston Rockets	3
Kevin Garnett	Minnesota Timberwolves	3
Pau Gasol	Memphis Grizzlies	3
Richard Hamilton	Detroit Pistons	1
Allen Iverson	Philadelphia 76ers	2
Jason Kidd	New Jersey Nets	2
Toni Kukoc	Milwaukee Bucks	1
Karl Malone	Utah Jazz	2
Stephon Marbury	Phoenix Suns	2
Shawn Marion	Phoenix Suns	2
Kenyon Martin	New Jersey Nets	3
Jamal Mashburn	New Orleans Hornets	2
Tracy McGrady	Orlando Magic	2
Reggie Miller	Indiana Pacers	3
Yao Ming	Houston Rockets	3
Dikembe Mutombo	New Jersey Nets	2
Steve Nash	Dallas Mavericks	2
Dirk Nowitzki	Dallas Mavericks	2
Jermaine O'Neal	Indiana Pacers	2
Shaquille O'Neal	Los Angeles Lakers	2
Gary Payton	Milwaukee Bucks	2
Paul Pierce	Boston Celtics	2
Scottie Pippen	Portland Trail Blazers	1
David Robinson	San Antonio Spurs	2
Arvydas Sabonis	Portland Trail Blazers	2
Jerry Stackhouse	Washington Wizards	1
Predrag Stojakovic	Sacramento Kings	2
Antoine Walker	Boston Celtics	2
Ben Wallace	Detroit Pistons	2
Chris Webber	Sacramento Kings	2

certainly not surprising: with 29 teams and only around 15 players in each team, NBA is really a "small world". Figure 4 shows the distribution of Jordan numbers in the NBA graph. The average Jordan number is equal to 2.27, which is smaller than the average Bacon number in the Hollywood graph, and the average Wynn number in the baseball graph, due to smaller number of vertices.

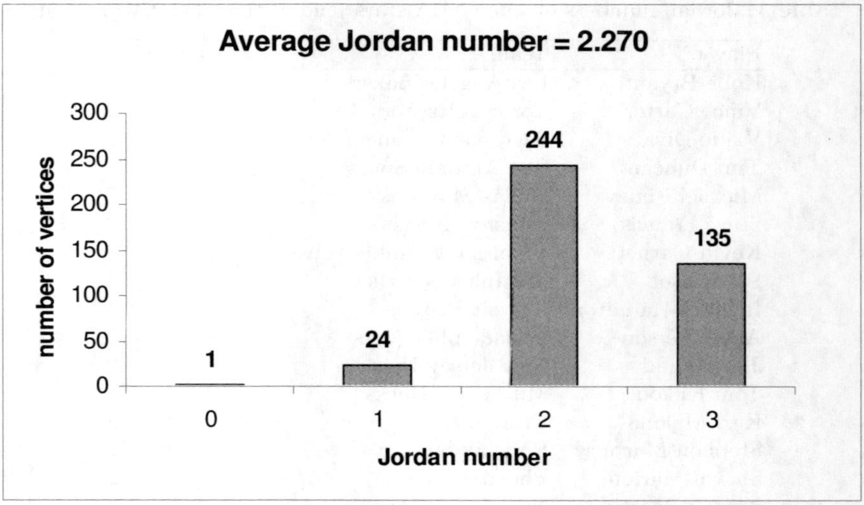

Fig. 4. Number of vertices in the NBA graph with different values of Jordan number. Average Jordan number = 2.270.

Table 1 presents Jordan numbers corresponding to some well-known NBA players. Not surprisingly, most of them have Jordan number 2, except for several players with Jordan number 3: those who joined this league recently, and therefore did not have many teammates through their career, as well as Reggie Miller who spent 16 seasons in the same team (Indiana Pacers), and Kevin Garnett who played in Minnesota for 8 years. Scottie Pippen, Toni Kukoc, and Jerry Stackhouse were Jordan's teammates at different times, therefore, they have Jordan number 1.

Furthermore, we calculated the *diameter* of the NBA graph, i.e., the maximum possible distance between any two vertices in the graph. Since the maximum Jordan number in the NBA graph is equal to 3, one would expect the value of the diameter to be of the same order of magnitude. As it was mentioned in the previous section, the diameter of the NBA graph can be found as follows: for every given vertex, we calculate the distances between this vertex and all others. In this approach, we need to repeat this procedure 404 times, and every time a different vertex is considered to be the "center" of the graph. Our calculations show that the diameter of the NBA graph (the maximum distance between all pairs of vertices) is equal to 4. Therefore, one can claim that the NBA graph actually follows the small-world hypothesis, since its diameter is small enough.

Table 2. Degrees of the vertices in the NBA graph.

degree interval	number of vertices
11-20	134
21-30	116
31-40	103
41-50	42
51-60	8
61+	2

Table 3. The most "connected" players in the NBA graph.

Player	Team	Degree	Av. Distance
Corie Blount	Chicago Bulls	63	1.906
Jim Jackson	Sacramento Kings	68	1.923
Robert Pack	New Orleans Hornets	57	1.936
Grant Long	Boston Celtics	50	1.946
Bimbo Coles	Boston Celtics	54	1.958

4.3 Degrees and "connectedness" of the vertices in the NBA graph

As it was pointed out above, the maximum and the average distance from the center of the graph actually depend on the choice of this center. One can easily guess that Michael Jordan is not the most "connected" central vertex of the NBA graph, since he played only for two teams and the number of his former teammates among currently active players is rather small. In fact, the *degree* of the vertex (i.e., the number of edged starting from it, or, the number of teammates) corresponding to Jordan is only 24. Table 2 presents the number of vertices in the NBA graph corresponding to different intervals of the degree values.

It would be reasonable to assume that if one picks a vertex with a high degree as the center of the NBA graph, the average distance in the graph corresponding to this vertex would be smaller than the average Jordan number. We have found the most "connected" players in the NBA graph with the smallest corresponding average distances. Table 3 presents five players who could be the most "connected" centers of the NBA graph. As one can notice, all of them are "bench players" who have changed many teams during their career, therefore, they have high degrees in the NBA graph. Also, an interesting observation is that although Corie Blount's vertex is degree smaller than Jim Jackson's, the average connectivity is higher for Corie Blount, which could be explained by the fact that his teammates were highly "connected" themselves.

5 Conclusion

In this chapter, we have discussed different properties of social collaboration networks in sports. As one could predict, the properties of these graphs confirm the small-world hypothesis. Investigating these graphs is also interesting from the perspective of "linking" different players, and in many cases it turns out that two players who never played together can be connected by a short chain.

Although the instance of the NBA graph considered in this chapter contains only currently active basketball players, it can be easily extended to reflect all players in the history of the NBA. Moreover, since a lot of foreign players from different countries and continents have come to the NBA in recent years, one would expect that the graph covering all basketball players playing in major foreign championships is also connected and has a small diameter.

Moreover, besides baseball and basketball, similar networks can be constructed for other team games, such as soccer, American football, *etc.* This approach is universal, and it allows one to obtain non-trivial information from player transfers data.

References

1. J. Abello, P.M. Pardalos, M.G.C. Resende, On maximum clique problems in very large graphs, DIMACS Series, 50, American Mathematical Society, 1999, pp. 119-130.
2. W. Aiello, F. Chung, L. Lu, A random graph model for power law graphs, Experimental Math, 10, 2001, pp. 53-66.
3. V. Boginski, S. Butenko, P.M. Pardalos, Modeling and Optimization in Massive Graphs, In: Novel Approaches to Hard Discrete Optimization (P. M. Pardalos and H. Wolkowicz, eds.), American Mathematical Society, 2003, 17-39.
4. V. Boginski, S. Butenko, P.M. Pardalos, On Structural Properties of the Market Graph, In: Innovations in Financial networks (A. Nagurney, ed.), Edward Elgar Publishers, 2003, in press.
5. A. Broder, R. Kumar, F. Maghoul, P. Raghavan, S. Rajagopalan, R. Stata, A. Tomkins, J. Wiener, Graph structure in the Web, Computer Networks, 33, 2000, pp. 309-320.
6. B. Hayes, Graph Theory in Practice, American Scientist, 88: 9-13 (Part I), 104-109 (Part II), 2000.
7. H. Jeong, B. Tomber, R. Albert, Z.N. Oltvai, A.-L. Barabasi, The large-scale organization of metabolic networks, Nature, 407, 2000, pp. 651-654.
8. Oracle of Baseball. http://www.baseball-reference.com/oracle/ Retrieved September 26, 2003.
9. The Oracle of Bacon at Virginia. http://www.cs.virginia.edu/oracle/ Retrieved September 26, 2003.
10. D. Watts, Small Worlds: The Dynamics of Networks Between Order and Randomness, Princeton University Press, 1999.

11. D. Watts, S. Strogatz, Collective dynamics of 'small-world' networks, Nature, 393, 1998, pp. 440-442.

The Holdup System for Grouping Minor Hockey Participants Into Age Divisions

William Hurley

Royal Military College of Canada
hurley_w@rmc.ca

1 Introduction

In almost all minor sports programs where there is early competitive streaming, including hockey, soccer, and baseball, players born early in the year are more likely to reach elite levels than those born late in the year. To give some idea of the extent of the problem in Canadian minor hockey, consider the following data. For the 1999-2000 season, over 100,000 participants registered with the Ontario Minor Hockey Association (OMHA). These players are grouped into competitive divisions depending on their age and skill. The most highly skilled players play at the AAA level. I classified each AAA registrant as either a First Half birth (meaning the player was born in one of the first six months of the year) or a Second Half birth (the last six months of the year). The resulting distribution is as follows:

	Age	**First Half**	**Second Half**
Major Atom	11	266	117
Minor Peewee	12	258	128
Major Peewee	13	253	119
Minor Bantam	14	286	107
Major Bantam	15	280	118
Midget	16,17	268	161
Totals		1611	750

Note that the number of births in the First Half is over twice the number in the Second Half. This distribution stands in contrast to the near uniform distribution of male births in Canada.

This disparity is present at all levels of hockey including Major Junior (ages 16-21) and the National Hockey League (NHL). I examined NHL rosters for the 1996-97 season. 349 of 1091 players had birthdays in the first quarter of the year. One way to demonstrate that this proportion is not consistent with normal statistical variation is to first assume that players born in a particular quarter are no more likely to play in the NHL than players born in any other quarter, and then ask what the chance is that we would observe at least 349 of 1091 NHL players with birthdays in the first quarter. This probability, obtained using the binomial distribution with 1091 trials and a probability of success equal to .25, is approximately 1 in 10,000,000! Clearly, we need something other than chance to explain the distribution of NHL birthdays.

The academic literature terms this discrimination the *Relative Age Effect (RAE)*. A very good review can be found in Musch and Grondin (2001). Notable contributions include Barnsley, Thompson and Barnsley (1985), Grondin, Deshaies and Nault (1984), Daniel and Janssen (1987), Barnsley and Thompson (1988), Thompson, Barnsley, and Stebelsky (1991) and Hurley, Lior and Tracze (2001). The impetus for this research in minor sports was motivated by similar research in education. Among others, Davis et al (1980) have found that younger students entering Grade 1 do not do as well as their older classmates in early scholastic achievement. However, unlike minor sports, it is not clear that this advantage persists as children progress through the system. See Angrist and Kreuger (1992) on this last point.

The explanation for the Relative Age Effect is tied to the *Calendar Year System* (*CYS*) that Canadian minor hockey organizations use to group players into competitive groups. Those players born early in the calendar year have, on average, significant size and motor skill advantages over their younger competitors and consequently a higher proportion of them play on elite teams. There is significant evidence that this also happens in a number of other sports (soccer and baseball) that rely on birthdates to group players for the purposes of competition.

Hockey Canada (HC), the organization that governs minor hockey in Canada, is aware of the problem. In the summer of 1999, they organized and sponsored a conference, the Open Ice Hockey Summit, to examine the state of hockey in Canada. Participants included players, coaches, and managers from all levels of Canadian hockey including the NHL. The conference produced eleven recommendations to improve hockey in Canada. Recommendation 3 is this:

> Examine the date of age determination. (The cut-off date is currently December 31st, but some thought has been given to rotating it throughout the calendar year. The objective is for a player to not always be the youngest or oldest in a given division.)[1]

To date, however, there has not been any consensus within HC or its branch organizations on just what to do. There have been some sensible suggestions. But HC has not acted on any of them. The interested reader is referred to

[1] See *http://www.canadianhockey.ca/openice/e/recommendations/3.html*

Hurley, Lior and Tracze (2001) for a discussion of the advantages and disadvantages of these alternative systems. This paper presents yet another.

The most popular criteria for grouping minor sports participants into competitive divisions are age, size (weight and/or height), and skill. In this paper I consider *Birthdate Slotting Systems (BSS),* systems which group players into age divisions *using only player birthdates.* Boucher and Halliwell (1991) were the first to offer a *BSS* different than the *CYS.* Their *Novem System* is a calender rotation system for a limited number of age groupings. In the example Boucher and Halliwell use, only a limited number of age groupings are affected, and even within these groupings, there is no sense in which the system is fair. In this paper, I offer a new *BSS* which I term the *Holdup System.* It has two important characteristics. First, it is easy for parents and minor hockey administrators to understand. And second, it is more equitable than the *CYS* and the *Novem System.*

Effectively, the *Holdup System* gives players an option to do an additional year in a particular age division. The key is that younger players, those born late in a calendar year, get to exercise this option earlier in their minor hockey careers than do players born earlier in the year. When they exercise this option, they become the oldest players in a particular age division. In the end, I will show that this system has the characteristic that all players will face the same *relative age competition,* a notion I will define more fully in Section 3.

2 Age Division Assignment Tables and Relative Age

I conceive a minor hockey career in three parts: *pre-CORE* seasons, *CORE* seasons, and *post-CORE* seasons. What I want to do is find an age categorization system which is more equitable than the *CYS over a player's CORE seasons.* For instance, in Canadian minor hockey, I could define the *CORE* seasons to be Minor Novice (age 8) through Major Bantam (age 15). However I make no specific assumption. The definition of the CORE seasons is left for those who might implement this system.

To compare the effects of various *BSS*, I employ an *Age Division Assignment Table* (*ADAT*) which specifies who plays with whom from season to season. An example is shown in Table 1. In this table, birthdate segments are listed in the first column. The segment **y0(1)** refers to players born in quarter **1** (i.e. January, February and March) of year **y0**. Here I have broken the year into quarters. However I could have used any other fraction of a year just as easily. Elements of the table, $AD1, AD2, ...$, refer to the *age divisions* a player will play in over his minor sports career. The age divisions are numbered in a natural way. Players in $AD1$ are younger that those in $AD2$, those in $AD2$ are younger than those in $AD3$, and so on. Columns 2 through 9 refer to specific seasons of play. Rather than list specific calendar years, I have simply numbered the seasons 1 through 8.

To see how this table works, consider a player born on September 16 in year **y1**. This player is a third quarter birth in year **y1**, hence we check row **y1(3)** to determine what age divisions he will be assigned to in particular seasons. This player begins in *AD*1 in Season **1** and finishes in *AD*4 in Season **8**. In fact this table is just the *Age Division Assignment Table* for the *CYS* with two-year age divisions.

It should be understood that, in general, *Age Division Assignment Tables* will extend down and to the right as more birthdate segments enter the system and successive seasons are played. In addition it is also important to understand that an *Age Division Assignment Table* specifies "floor" age divisions in the sense that exceptional players may play "up" a division if an organization allows such movement.

Now consider the *Holdup System*. It is assumed to have the following rules:

1. a 4th quarter birthdate segment would be given an additional year in *AD*1;
2. a 3rd quarter birthdate segment would be given an additional year in *AD*2;
3. a 2nd quarter birthdate segment would be given an additional year in *AD*3; and
4. a 1st quarter birthdate segment would have an additional year in *AD*4;

That is, the younger the birthdate segment, the earlier it gets to repeat an age division. What, then, are the effects of this rule?

The *Age Division Assignment Table* for this system is shown in Table 2. It assumes that all players will exercise their option to holdup as soon as it becomes available to them. Note that in Season 3, the fourth quarter birthdays of year **y1** are the oldest players in the *AD*1 division. This would never happen under the *CYS* since 4th quarter birthdays are always the youngest in an any age division under the *CYS*. The key question is how this relative advantage is conferred on the various quarters. To answer, I need to define the concept of *relative age*.

Consider the birthdate segment **y2(4)** under the *Holdup System* (Table 2). This segment begins *CORE* play in Season 2 in *AD*1 which is highlighted in the table. Note that this age division is made up of 9 birthquarters ranging from the oldest (**y0(4)**) to the youngest (**y2(4)**). I associate the relative age 1 with the oldest birthquarter (**y0(4)**), a relative age of 2 with the second oldest birthquarter (**y1(1)**) and so on up to a relative age 9 with the youngest (**y2(4)**).

Now consider what happens to each fourth quarter birthdate segment in the table. The sequence of relative ages in successive seasons is 9, 5, 1, 6, 2,

Table 1. The age division assignment table for the CYS.

Birthdate	**Season**							
Segments	**1**	**2**	**3**	**4**	**5**	**6**	**7**	**8**
y0(1)	*AD*1	*AD*2	*AD*2	*AD*3	*AD*3	*AD*4	*AD*4	*AD*5
y0(2)	*AD*1	*AD*2	*AD*2	*AD*3	*AD*3	*AD*4	*AD*4	*AD*5
y0(3)	*AD*1	*AD*2	*AD*2	*AD*3	*AD*3	*AD*4	*AD*4	*AD*5
y0(4)	*AD*1	*AD*2	*AD*2	*AD*3	*AD*3	*AD*4	*AD*4	*AD*5
y1(1)	*AD*1	*AD*1	*AD*2	*AD*2	*AD*3	*AD*3	*AD*4	*AD*4
y1(2)	*AD*1	*AD*1	*AD*2	*AD*2	*AD*3	*AD*3	*AD*4	*AD*4
y1(3)	*AD*1	*AD*1	*AD*2	*AD*2	*AD*3	*AD*3	*AD*4	*AD*4
y1(4)	*AD*1	*AD*1	*AD*2	*AD*2	*AD*3	*AD*3	*AD*4	*AD*4
y2(1)		*AD*1	*AD*1	*AD*2	*AD*2	*AD*3	*AD*3	*AD*4
y2(2)		*AD*1	*AD*1	*AD*2	*AD*2	*AD*3	*AD*3	*AD*4
y2(3)		*AD*1	*AD*1	*AD*2	*AD*2	*AD*3	*AD*3	*AD*4
y2(4)		*AD*1	*AD*1	*AD*2	*AD*2	*AD*3	*AD*3	*AD*4
y3(1)			*AD*1	*AD*1	*AD*2	*AD*2	*AD*3	*AD*3
y3(2)			*AD*1	*AD*1	*AD*2	*AD*2	*AD*3	*AD*3
y3(3)			*AD*1	*AD*1	*AD*2	*AD*2	*AD*3	*AD*3
y3(4)			*AD*1	*AD*1	*AD*2	*AD*2	*AD*3	*AD*3
y4(1)				*AD*1	*AD*1	*AD*2	*AD*2	*AD*3
y4(2)				*AD*1	*AD*1	*AD*2	*AD*2	*AD*3
y4(3)				*AD*1	*AD*1	*AD*2	*AD*2	*AD*3
y4(4)				*AD*1	*AD*1	*AD*2	*AD*2	*AD*3
y5(1)					*AD*1	*AD*1	*AD*2	*AD*2
y5(2)					*AD*1	*AD*1	*AD*2	*AD*2
y5(3)					*AD*1	*AD*1	*AD*2	*AD*2
y5(4)					*AD*1	*AD*1	*AD*2	*AD*2
y6(1)						*AD*1	*AD*1	*AD*2
y6(2)						*AD*1	*AD*1	*AD*2
y6(3)						*AD*1	*AD*1	*AD*2
y6(4)						*AD*1	*AD*1	*AD*2
y7(1)							*AD*1	*AD*1
y7(2)							*AD*1	*AD*1
y7(3)							*AD*1	*AD*1
y7(4)							*AD*1	*AD*1

7, 3, 8, and 4. This sequence contains each of the integers 1 through 9. That is, over the *CORE* seasons, all fourth quarter birthdates get to be the oldest in one season, the second oldest in one season, all the way to the youngest in one season. This is clearly an improvement over the *CYS* for fourth quarter birthdates since, under the *CYS*, their relative age sequence is 8, 4, 8, 4, 8, 4, 8, 4, 8. But there is more. I can also show that all other birthdate segments (including first, second, and third quarter segments) get this same list of relative ages, 1-9, although not necessarily in the same order as the fourth

Table 2. The age division assignment table for the Holdup System with quarters.

Birthdate Segments	Season 1	2	3	4	5	6	7	8	9
y0(1)	AD1	AD2	AD2	AD3	AD3	AD4	AD4	AD4	AD5
y0(2)	AD1	AD2	AD2	AD3	AD3	AD3	AD4	AD4	AD5
y0(3)	AD1	AD2	AD2	AD2	AD3	AD3	AD4	AD4	AD5
y0(4)	AD1	AD1	AD2	AD2	AD3	AD3	AD4	AD4	AD5
y1(1)	AD1	AD1	AD2	AD2	AD3	AD3	AD4	AD4	AD4
y1(2)	AD1	AD1	AD2	AD2	AD3	AD3	AD3	AD4	AD4
y1(3)	AD1	AD1	AD2	AD2	AD2	AD3	AD3	AD4	AD4
y1(4)	AD1	AD1	AD1	AD2	AD2	AD3	AD3	AD4	AD4
y2(1)		AD1	AD1	AD2	AD2	AD3	AD3	AD4	AD4
y2(2)		AD1	AD1	AD2	AD2	AD3	AD3	AD3	AD4
y2(3)		AD1	AD1	AD2	AD2	AD2	AD3	AD3	AD4
y2(4)		AD1	AD1	AD1	AD2	AD2	AD3	AD3	AD4
y3(1)			AD1	AD1	AD2	AD2	AD3	AD3	AD4
y3(2)			AD1	AD1	AD2	AD2	AD3	AD3	AD3
y3(3)			AD1	AD1	AD2	AD2	AD2	AD3	AD3
y3(4)			AD1	AD1	AD1	AD2	AD2	AD3	AD3
y4(1)				AD1	AD1	AD2	AD2	AD3	AD3
y4(2)				AD1	AD1	AD2	AD2	AD3	AD3
y4(3)				AD1	AD1	AD2	AD2	AD2	AD3
y4(4)				AD1	AD1	AD1	AD2	AD2	AD3
y5(1)					AD1	AD1	AD2	AD2	AD3
y5(2)					AD1	AD1	AD2	AD2	AD3
y5(3)					AD1	AD1	AD2	AD2	AD2
y5(4)					AD1	AD1	AD2	AD2	AD2
y6(1)						AD1	AD1	AD2	AD2
y6(2)						AD1	AD1	AD2	AD2
y6(3)						AD1	AD1	AD2	AD2
y6(4)						AD1	AD1	AD1	AD2
y7(1)							AD1	AD1	AD2
y7(2)							AD1	AD1	AD2
y7(3)							AD1	AD1	AD2
y7(4)							AD1	AD1	AD1

quarter segments. There is a sense, then, in which all segments face the same relative age competition. This argument is formalized in the next section.

3 Relative Age Fairness

Suppose that each calendar year of births is divided into m equal *birthdate segments,* each of length $1/m$ of a year. I label each of these segments with the integers $1, 2, ..., m$ where 1 is assigned to the segment beginning January 1 and m is assigned to the segment ending December 31. I define $\mathbf{y(j)}$ to be birthdate segment $\mathbf{j}$ within calendar year $\mathbf{y}$ where $\mathbf{j} \in \{1, 2, ..., m\}$. I label age divisions $AD1, AD2, ...$ The *CORE* period is assumed to comprise $2m + 1$ seasons.

I define a *Birthdate Slotting System* to be an assignment of birthdate segments to age divisions $(AD1, AD2, ...)$ over the *CORE* seasons. Any *BSS* can be summarized in a *Age Division Assignment Table.*

Let $r_t(\mathbf{y(j)})$ denote the *relative age* of birthdate segment $\mathbf{y(j)}$ in season t. All age divisions are assumed to be made up of exactly $2m + 1$ consecutive birthdate segments. Hence $r_t(\mathbf{y(j)}) \in \{1, 2, ..., 2m, 2m + 1\}$. To define my notion of fairness, I will employ the concept of a *Relative Age List*:

Definition 1: The *Relative Age List* for birthdate segment $\mathbf{y(j)}$, $L(\mathbf{y(j)})$, is a list of the *relative ages* that birthdate segment $\mathbf{y(j)}$ experiences over $2m+1$ *CORE* seasons ordered from smallest to largest. Denote these ordered *relative ages* $r_{(1)}, r_{(2)}, ..., r_{(n)}$ where $r_{(1)} \leq r_{(2)} \leq ... \leq r_{(n)}$ and therefore $L = [r_{(1)}, r_{(2)}, ..., r_{(n)}]$.

Definition 2: A *Birthdate Slotting System* is said to be *RA-Fair* ("Relative Age Fair") if every birthdate segment experiences the same *Relative Age List*, L_0, i.e.

$$L(\mathbf{y(j)}) = L_0 \qquad \text{for all } \mathbf{y(j)}. \tag{1}$$

I make no attempt to measure the value of different *relative age* vectors. By way of example, consider two birthdate segments, $\mathbf{y(j)}$ and $\mathbf{y(j{+}1)}$, and suppose their respective *relative age* vectors for a particular *BSS* different than the *CYS* are as follows:

	Season								
	1	**2**	**3**	**4**	**5**	**6**	**7**	**8**	**9**
$R(\mathbf{y(j)})$	1	2	3	4	5	6	7	8	9
$R(\mathbf{y(j+1)})$	9	8	7	6	5	4	3	2	1

The temptation is to compare these two vectors, perhaps with some sort of weighted average. However, given that there is no consensus among hockey experts on when it is best to have an age advantage, I make no such attempt. Hence, my notion of fairness is the simple requirement that all segments have the same *Relative Age List*. If it is true that it is best to have a relative age advantage at a particular point in time, then the *Holdup System* will not be completely fair, *but it will be more equitable than the CYS*.

Suppose each age division is $2m + 1$ periods rather than $2m$ periods. For instance, if m were 4, age divisions would be 2.25 years rather than 2 years. To characterize what happens, I need to develop a simple variation of standard modulo arithmetic. Consider the integers p and n. $p \bmod n$ is defined as the integer remainder when p is divided by n. For example, $10 \bmod 4 = 2$ and $15 \bmod 3 = 0$. For my purposes, I need to make a simple change. Normally if n divides p evenly, $p \bmod n$ is equal to zero. However, I am going to set $p \bmod n$ equal to n. Hence $24 \bmod 6 = 6$, not 0. I employ the notation

$$p \bmod n = \langle p \rangle_n \,. \tag{2}$$

Consider the assignment in Table 2. It is easy to show that, for birthdate segment $\mathbf{y(j)}$, the relative age in a particular season depends only on the relative age in the previous season in the following way.

$$r_t(\mathbf{y(j)}) = \langle r_{t-1}(\mathbf{y(j)}) - m \rangle_{2m+1} \,. \tag{3}$$

Using this recursion, it is easy to derive the relative ages for each birthdate segment. For instance consider segment $\mathbf{y2(4)}$:

$$
\begin{aligned}
r_2(\mathbf{y2(4)}) &= 9 \\
r_3(\mathbf{y2(4)}) &= \langle r_2(\mathbf{y2(4)}) - 4 \rangle_9 = \langle 9 - 4 \rangle_9 = 5 \\
r_4(\mathbf{y2(4)}) &= \langle r_3(\mathbf{y2(4)}) - 4 \rangle_9 = \langle 5 - 4 \rangle_9 = 1 \\
r_5(\mathbf{y2(4)}) &= \langle r_4(\mathbf{y2(4)}) - 4 \rangle_9 = \langle 1 - 4 \rangle_9 = 6 \\
r_6(\mathbf{y2(4)}) &= \langle r_5(\mathbf{y2(4)}) - 4 \rangle_9 = \langle 6 - 4 \rangle_9 = 2 \\
r_7(\mathbf{y2(4)}) &= \langle r_6(\mathbf{y2(4)}) - 4 \rangle_9 = \langle 2 - 4 \rangle_9 = 7 \\
r_8(\mathbf{y2(4)}) &= \langle r_7(\mathbf{y2(4)}) - 4 \rangle_9 = \langle 7 - 4 \rangle_9 = 3 \\
r_9(\mathbf{y2(4)}) &= \langle r_8(\mathbf{y2(4)}) - 4 \rangle_9 = \langle 3 - 4 \rangle_9 = 8 \\
r_{10}(\mathbf{y2(4)}) &= \langle r_9(\mathbf{y2(4)}) - 4 \rangle_9 = \langle 8 - 4 \rangle_9 = 4
\end{aligned}
$$

Note that this system returns the relative age list $[1, 2, 3, 4, 5, 6, 7, 8, 9]$ for segment $\mathbf{y2(4)}$. In fact, one can show that the assignment in Table 2 produces the same relative age list for each birthdate segment. The relative ages for each birthquarter are shown in the following table:

	Season								
Quarter	**1**	**2**	**3**	**4**	**5**	**6**	**7**	**8**	**9**
1	6	2	7	3	8	4	9	5	1
2	7	3	8	4	9	5	1	6	2
3	8	4	9	5	1	6	2	7	3
4	9	5	1	6	2	7	3	8	4

Consequently we conclude that the *Holdup System* is *RA-Fair*.

Note that the *Holdup System* produces an additional season of play for all quarters. For instance, under the *CYS*, suppose that players play 8 seasons, 2 in each of 4 age divisions. Then, under the *Holdup System,* players will play 9 seasons since each player repeats an age division.

4 Variations

There is nothing special about using quarters ($m = 4$). I could just as easily have chosen any other integer. The logical integers to use in Canadian minor hockey would be $m = 3$ and 4. For $m = 3$, the *CORE* would be 7 seasons long. For $m = 4$, it would be 9 seasons long.

In Canadian minor hockey, the elite levels (AAA, AA, and A) are played with one-year age divisions. It is straightforward to construct a *Holdup System* for one-year age divisions. In this case the various teams will have to be drawn from, on average, an age division slightly longer than a year.

The *Holdup System* is not equivalent to a calendar rotation system in the sense of Recommendation 3 of the Open Ice Summit where the date of age determination is changed from season to season. To see this, examine the *Holdup System* example above with $m = 4$ (quarters). Note that each age division comprises nine contiguous birthquarters. There is no way to specify a date of age determination and have the outcome be age divisions with nine birthquarters.

Finally, there is nothing special in the argument above that makes the *Holdup System* only applicable to minor hockey. It could be used in any minor sports program.

5 Conclusion

I have proposed a new system to group minor sports participants into age divisions. This new system has two important properties:

1. it is more equitable than the *CYS*; and
2. it is relatively easy for parents and administrators to understand.

I believe it is superior to the *CYS*. Whether this system or other such systems are ever implemented will depend on whether a minor sports organization has the political will to do so.

References

1. Angrist, J.D. and Krueger, A.B. (1992). "The Effect of Age at School Entry on Educational Attainment: An Application of Instrumental Variables With Moments from Two Samples," *Journal of the American Statistical Association*, 87(418), 328-336.
2. Barnsley, R.H., Thompson, A.H., and Barnsley, P.E. (1985). "Hockey success and birthdate: the relative age effect." *Canadian Association for Health, Physical Education, and Recreation*, 51, 23-28.

3. Barnsley, R.H., and Thompson, A.H. (1988). "Birthdate and success in minor hockey: the key to the NHL." *Canadian Journal of Behavioral Science*, 20(2), 167-176.
4. Boucher, J. and Halliwell, W. (January/February, 1991). "The Novem System: A Practical Solution to Age Grouping," *CAPHER Journal*, 16-20.
5. Daniel, T.E., and Janssen, C.T.L. (1987). "More on the Relative Age Effect." *Canadian Association for Health, Physical Education, and Recreation*, 53, 21-24.
6. Davis, B.D., Trimble, C.S., and Vincent, D.R. (1980). "Does Age of Entrance Affect School Achievements," *The Elementary School Journal*, 80, 133-143.
7. Grondin, S., Deshaies, P., and Nault, L.P. (1984). "Trimestres de naissance et participation au hockey et au volleyball." *La Revue Quebecoise de l'Activite Physique*, 2, 97-103.
8. Hurley, W.J., Lior, Dan, and Tracze, S.J. (2001a), "A Proposal to Reduce the Age Discrimination in Canadian Minor Hockey," *Canadian Public Policy*, 27, 65-75.
9. Musch, Jochen, and Grondin, Simon, (2001), "Unequal Competition As An Impediment to Personal Development: A Review of the Relative Age Effect in Sport," *Developmental Review*, 21, 147-167.
10. Thompson, A.H., Barnsley, R.H., and Stebelsky, G. (1991). ""Born to Play Ball" The Relative Age Effect and Major League Baseball," *Sociology of Sport Journal*, 8, 146-151.

Promotion and Relegation in Rent Seeking Contests

Stefan Szymanski*

Imperial College London, UK
szy@imperial.ac.uk

1 Introduction

The National League, founded in 1876, was the world's first professional sports league (for a detailed discussion see Seymour (1960)). The model developed by the National League and by organized baseball became the standard model of a North American sports league. One of the key characteristics of such leagues is that entry is controlled by the incumbent teams through the occasional sale of expansion franchises. In the case of the major leagues, the ability to restrict entry at the level of the club is matched by a limited threat of entry at the level of the league. The sunk costs are such that entry is rare, that when it occurs it seldom succeeds, and that the first inclination of the entrant is to collude with the incumbents (for an account of various attempts to enter the professional sports league market see Quirk and Fort (1992)). A consequence of the restricted entry opportunities is that incumbent major leagues are able to extract monopoly rents, most notably through the relocation threat used to obtain large subsidies from local taxpayers for municipal facilities (see e.g. Noll and Zimbalist (1997) or Siegfried and Zimbalist (2000)).

The world's second professional sports league was the English Football League, founded in 1888. While similar in some ways to the National League's formula, the British chose to differ in one significant dimension- they chose to make the league open to entry by new teams (see e.g. Inglis (1988) for historical details). After a few years the organizers decided upon a very specific mechanism to facilitate entry- promotion and relegation. This system works by ranking each team at the end of the season according to its performance (e.g. winning percentage). A number of teams with the worst performance are then demoted to play in an immediately junior division in the following

* I am indebted to Steve Ross, who contributed many ideas that have borne fruit with this analysis. Errors, of course, are my own.

season, to be replaced by the best performing teams in the lower division[1]. As football (soccer) spread around the world, most countries outside the US adopted the promotion and relegation system. This paper develops a simple model to show how this system provides greater incentive to compete than the closed North American system. The reason is simple enough- teams in leagues with promotion and relegation must supply effort to win matches even when they are out of contention for the league title, otherwise they risk demotion to an immediately junior division. Promotion and relegation also has the effect of expanding the opportunity to play at the level of the major league to a much larger number of teams and cities. Because of this it undermines the ability of major league teams to extract subsidies from local government (the relocation threat has little or no relevance).

This is one of a number of papers that have recently attempted to analyze the promotion and relegation issue. Noll (2002) discusses the effect on team mobility of promotion and relegation in the English leagues, mostly from the perspective of an analyst accustomed to the American model. Ross and Szymanski (2002) consider whether promotion and relegation might be a suitable mechanism in antitrust law for dealing with the monopoly power of sports leagues in the US. Szymanski and Valletti (2003) examine not only the effect on effort contribution of promotion and relegation but also the impact of revenue sharing incentives. Finally, Szymanski (2003) discusses promotion and relegation in the context of the broad issue of the optimal design of sporting contests.

The paper is set out as follows. In the next section I examine the issue of optimal league size, section 3 develops a model of a professional sports league as a rent seeking contest, section 4 considers the impact on fan utility and section 5 discusses the threat of franchise relocation. Section 6 concludes.

2 Optimal League Size

In general, a larger league will be more attractive than a smaller one. A World Championship title is more prestigious than a national title, which is more prestigious than a regional competition. The more inclusive the competition, the more gratifying the victory. However, there are limits to the optimal size of a league. If each club is to play each other at least once during a season, there are limits imposed by the physical ability of players to perform in a sequence of matches and by the total supply of talent. Given the skewed distribution of talent, a larger league tends to create more unbalanced contests. If expansion

[1] For example, in the English Premier League, the three worst performing teams are relegated to the Football League Division One and replaced by the three best performing teams from that division.

leads to more unbalanced contests it can be argued that very large leagues will sacrifice quality for quantity. Finally, even if talent were evenly spread across clubs, the talent-level for each team will be diluted; although in many cases the value-added for fans of additional clubs outweighs the marginal decline in attractiveness for fans of existing teams, at some point this ceases to be the case. These reasons explain why members of a league would want to control access in order to maintain an optimally sized league, taking into account the underlying distribution of talent, and why the law could find such control to be legitimate.

However, there are reasons to suppose that league members will be have a tendency to restrict access below the socially optimal level. The major North American sports leagues are organized as joint ventures, where entry and other major business decisions are made jointly by the clubs seeking to maximize their own profits, as opposed to a "single entity" where entry and other decisions would be made by an executive or Board seeking to maximize overall league profits. Thus, while a single entity league would ordinarily be expected to expand franchises until the point where marginal revenue equals zero, the objective of teams in the league will be to choose the number of franchises so as to maximize average revenue per club.[2] This is analogous to the well known observation that a labor-managed firm will not expand employment as far as a profit maximizing firm (see e.g. Ward (1958) and Meade (1974)).

Even though expansion franchises can be assessed a fee to compensate the existing teams for the loss of expected income (reduced probability of winning a championship, reduced percentage of revenue from league-wide ventures), for a conventional league revenue function this will imply under-expansion. Moreover, as noted above, league members have an incentive to expand sub-optimally in order to provide clubs with a credible threat to move to economically viable open markets unless local taxpayers provide generous tax subsidies. Any expansion can be expected to remove the most viable markets as threatened relocation sites, thus reducing the ability to acquire subsidies. As Fort and Quirk (1995) put it, "monopoly profits earned by leagues invite entry, so that one critical aspect of league decision making is acting to inhibit

[2] The point is easily made with a simple model. Suppose each team in a league has a profit function of the form where π_i = V + w_i (t_i)- t_i where $w_i = \frac{t_i}{\sum_{j=1}^{n} t_j}$ is the success of the team (e.g. the probability of winning the championship), t is the playing talent that produces success (at a constant marginal cost normalized to unity) and V is some fixed utility associated with the presence of a sports team in a particular location (which the team is assumed able to extract through stadium subsidies or other means). Social welfare in this model is simply nπ where n is the number of teams. The symmetric profit maximizing talent investment for each team $t^* = (n-1)/n^2$, and hence $\pi^* = V + 1/n^2$. Clearly profits are everywhere decreasing in n. However, total welfare is simply nV + 1/n and so social welfare is increasing in n up to $n = 1/\sqrt{V}$. This is the basis of the under-expansion result.

entry." (p.1292.) Creating new franchises in locations most likely to be attractive to an entrant league is one strategy a league can adopt. Moreover, when entry does occur, it is open the incumbent to absorb all or part of the new league: "If a rival league is successful the inevitable outcome is merger with the existing league in order to exploit the resulting market power over players, TV networks and stations, and local governments." (p1294). This suggests that while strategic entry deterrence considerations may lead to expansion beyond the point that maximizes average franchise revenue, it is likely that significant opportunities will remain to obtain monopoly rents.[3]

We now develop a model to illustrate the point that an open league system, with promotion and relegation, will significantly dissipate rents, as well as inefficient rent-seeking activity such creation of suboptimal entry restrictions, without any of the problems that might be associated with overexpansion.

3 A Model of Promotion and Relegation in a Rent Seeking Contest

There is a close analogy between league competition and a rent seeking contest (see e.g. Tullock (1970), Gradstein and Konrad (1999)). In a rent seeking contest, the competitors expend effort in pursuit of a fixed prize; the higher each contestant's effort, the greater the probability of winning. In a league competition teams also expend effort in pursuit of a prize. Although generally the explicit money value of the Championship Title is small, more successful teams tend to attract a higher degree of support and thus generate a higher revenue through the sales of tickets, merchandising, etc. In the rent seeking literature considerable attention has been given to the notion of efficient rent seeking – whereby all rents are spent in the contest. Here we are interested in a similar issue – to what extent will the possibility of promotion or relegation (a change in the structure of expected rewards) lead to increased investment in talent and therefore a reduction in economic rents accruing to the club?

[3] This observation is perfectly illustrated by Major League Baseball's experience in the late period from 1959-62. Despite a huge population growth and major population shift from the northeastern United States throughout the continent, the American and National Leagues remained constant at eight teams each for the entire century. In 1959, legendary baseball executive Branch Rickey sought to create a new "Continental League" with franchises in New York (which at the time had only one club, the Yankees) and Los Angeles (which had only the Dodgers), Houston and Minneapolis-St. Paul (which had no major league club at the time) and four other locations without any established major league team. Many felt this league would have been viable, but the two established leagues responded by adding additional teams in New York (the Mets) and Los Angeles (the Angles) as well as new teams in Houston and Minneapolis-St. Paul. This deprived Rickey of sufficient markets to operate at minimum viable scale, and forestalled entry.

(a) A (closed) monopoly league

Suppose there existed four feasible locations for a sports team, based on the drawing power of those teams. I assume that each location can support only one team, while two locations possess a greater drawing power than the others. In such a universe a number of league configurations are possible. I focus on the following plausible candidates:

(i) A single symmetric two-team league (based on the larger drawing areas)
(ii) Two symmetric two-team leagues
(iii) A three team league (with one location unserved)
(iv) A four team league

It is natural to suppose that a two-team league in the two largest drawing locations would emerge initially. Each team (call them 1 and 2) generates income from its franchise and I ignore for the time-being the possibility of revenue sharing. Revenue is thus a function of the drawing power of each team in its franchise area and the success of the team (which could be measured by indicators such as win percent or championship victories). Success in turn depends on the investment made in playing talent relative to the investments made by the rival team. The relative investment in talent can be thought of as the *ex ante* probability of winning. I assume that units of talent can be purchased a constant marginal cost in the market. I normalize this marginal cost to unity. I analyse per period profits, assuming that all investment in talent relates to a single year (i.e. clubs sign only one year contracts). In a closed league, absent any shifts in exogenous parameters, the equilibrium choices of the clubs will not change from year to year.

The drawing power of the teams 1 and 2 is assumed to be to $\mu > 1$ which is the drawing power of any teams that might choose one of the two smaller locations (potentially teams 3 and 4). Thus the expected profit of a club in the two-team league in any year is

$$E(\pi_1) = \frac{\mu t_1}{t_1 + t_2} - t_1 \tag{1}$$

In reality the two largest drawing locations might differ in size but this problem has been assumed away for the closed league. Assuming profit maximization the equilibrium for the two-team league is thus

$$t_1^* = t_2^* = \frac{\mu}{4}, \quad \pi_1 = \pi_2 = \frac{\mu}{2} \tag{2}$$

Admission of a team from a third location would not be profitable for the two incumbents. Given the lesser drawing power of a team from the third location (team 3) we can compute profits and talent investments of each team:

$$\begin{aligned} t_1^* = t_2^* &= \tfrac{2}{(4-\mu)^2}, & \pi_1 = \pi_2 &= \tfrac{\mu(4-\mu)-2}{(4-\mu)^2} \\ t_3^* &= \tfrac{2(2-\mu)}{(4-\mu)^2}, & \pi_3 &= \tfrac{(2-\mu)^2}{(4-\mu)^2} \end{aligned} \tag{3}$$

Aggregate profits have now fallen by an amount equal to $\mu(3\text{-}\mu)/(4\text{-}\mu)$. Expansion is not profit enhancing in aggregate because (a) expected revenues fall and (b) investment in talent increases.[4] Expected revenues fall because revenues are equal to the probability of success multiplied by drawing power. The probabilities are unchanged in aggregate, but the average drawing power has fallen with the addition of the third team. The impact of expansion from the point of view of the league members is to dissipate rents through competition for talent. However, the addition of the third team is likely to be welfare enhancing, because the citizens of the third location would receive the benefit of having a team, thus offsetting any reduction in welfare of fans caused by the reduced winning probabilities of the two large drawing teams. The same arguments extend naturally to expansion of the league to a fourth team. We can compute the investments in talent and profits as follows:

$$\begin{array}{ll} t_1^* = t_2^* = \frac{3\mu(2\mu-1)}{4(1+\mu)^2}, & \pi_1 = \pi_2 = \frac{\mu(2\mu-1)^2}{4(1+\mu)^2} \\ t_3^* = t_4^* = \frac{3\mu(2-\mu)}{4(1+\mu)^2}, & \pi_3 = \pi_4 = \frac{(2-\mu)^2}{4(1+\mu)^2} \end{array} \tag{4}$$

Aggregate profits are even lower and investment in talent is even higher in the four team model.

(b) Competing closed leagues

Given that a monopoly league would not voluntarily expand, it is natural to consider the possibility of two two-team leagues. Thus a second two-team league could be established in the two lesser drawing locations. I assume that this league would be intrinsically less interesting than the league based on teams from the large drawing locations. Their revenues would be smaller and their profit maximizing investment in talent would be smaller. Thus I suppose that participation in the second tier league would confer a lower drawing power, which I represent by a factor $\lambda < 1$. Thus the equilibrium investments in talent and profits in closed two league system are:

$$\begin{array}{ll} t_1^* = t_2^* = \frac{\mu}{4}, & \pi_1 = \pi_2 = \frac{\mu}{2} \\ t_3^* = t_4^* = \frac{\lambda}{4}, & \pi_3 = \pi_4 = \frac{\lambda}{2} \end{array} \tag{5}$$

Note that these assumptions imply that the second-tier league has a negligible effect on profits in the top tier league, but the existence of the top-tier league reduces the profitability of the second-tier league.

(c) Open leagues

We now consider the possibility that instead of expanding the top tier league, a system of promotion and relegation is instituted. This means that at the

[4] Note from (3) that the third team would want to invest in a negative quantity of talent for all $\mu > 2$. Assuming that in such cases the team is restricted to zero units of talent, admitting a third team never increases total profits.

end of each season the worst performing team from the top-tier (e.g. in terms of win percent) is relegated to the second-tier league and replaced by the best performing team from the second-tier league. I maintain the assumption that competition in the second-tier league is less attractive than participation in the top-tier league. I assume that the starting point for the system is one where the two larger drawing teams are in the top-tier league.

Once the threat of relegation is introduced, the time dimension of the problem becomes significant, since current investment decisions will be based on expectations about future profits as well as current profits. I develop a two period model, so that investments in the first period will be affected by the prospect of promotion or relegation. In the top tier of an open league the first period expected profit of team 1 can be written as

$$E(\pi_1) = \frac{t_1}{t_1 + t_2}(\mu + \delta D_{\mu 1}) + \frac{t_2}{t_1 + t_2}\delta D_{\mu 2} - t_1 \tag{6}$$

where $D_{\mu 1}$ is the expected profit of a large drawing team from retaining a place in the top-tier league and $D_{\mu 2}$ the expected profit following from demotion to the lower-tier league and δ is the discount factor. Similarly, the expected profit of a team in the second tier is

$$E(\pi_3) = \frac{t_3}{t_3 + t_4}(\lambda + \delta D_1) + \frac{t_4}{t_3 + t_4}\delta D_2 - t_3 \tag{7}$$

where D_1 is the expected profit of a small drawing team from promotion to the top-tier league and D_2 the expected profit from remaining the lower-tier league. The symmetric profit maximizing investment in talent for each club in the top tier the first period is

$$\begin{aligned} t_1^* = t_2^* &= \tfrac{1}{4}[\mu + \delta(D_{\mu 1} - D_{\mu 2})] \\ t_3^* = t_4^* &= \tfrac{1}{4}[\lambda + \delta(D_1 - D_2)] \end{aligned} \tag{8}$$

The prospect of demotion for teams in the top tier and the prospect of promotion for teams in the second tier both induce an increase in investment in the first period. The size of these extra investments depend on the expected profits form competition in the second period.

It is clear that in the second period there will be competition between teams of unequal drawing power. Given that both teams have the same probability of relegation from the top tier, I assume it is team 2 that is relegated. Likewise I assume team 3 is promoted. It is straightforward to compute the investments in talent and levels of profit of the teams in each league. These are

$$\begin{aligned} t_1^{2*} &= \frac{\mu^2}{(1+\mu)^2}, \quad t_3^{2*} = \frac{\mu}{(1+\mu)^2} \\ t_2^{2*} &= \frac{\lambda\mu^2}{(1+\mu)^2}, \quad t_4^{2*} = \frac{\lambda\mu}{(1+\mu)^2} \\ \pi_1^2 &= \frac{\mu^3}{(1+\mu)^2}, \quad \pi_3^2 = \frac{1}{(1+\mu)^2} \\ \pi_2^2 &= \frac{\lambda\mu^3}{(1+\mu)^2}, \quad \pi_4^2 = \frac{\lambda}{(1+\mu)^2} \end{aligned} \tag{9}$$

(superscripts refer to time period). Given the expected profit in period 2 it is straightforward to derive the profit maximizing investment in talent and expected profits in period 1

$$\begin{aligned} t_1^{1*} &= t_2^{1*} = \tfrac{1}{4}\left[\mu + \frac{\delta(1-\lambda)\mu^3}{(1+\mu)^2}\right] \\ t_3^{1*} &= t_4^{1*} = \tfrac{1}{4}\left[\lambda + \frac{\delta(1-\lambda)}{(1+\mu)^2}\right] \\ \pi_1^1 &= \pi_2^1 = \tfrac{1}{4}\left[\mu - \frac{\delta(1-\lambda)\mu^3}{(1+\mu)^2}\right] \\ \pi_3^1 &= \pi_4^1 = \tfrac{1}{4}\left[\lambda - \frac{\delta(1-\lambda)\mu}{(1+\mu)^2}\right] \end{aligned} \tag{10}$$

Note also that, as is to be expected, the value promotion is positive ($D_1 > D_2$) and the value of relegation is negative ($D_{\mu 2} < D_{\mu 1}$). These are the spurs to competition that dissipate the rents of a closed monopoly league. Rent dissipation occurs through increased investment in playing talent. Higher levels of investment in playing talent means higher quality of competition on the field.

Of the three main configurations considered (a two-team monopoly league, a closed two-team two league system and an open two-league system), the open system delivers the highest level of investment in talent. Investment in talent is also higher than a three-team league or a four-team league as long as μ is not too close to unity. For example, assuming a discount factor of 0.9, and a discount for second-tier competition of 50% (= 0.5), the total investment in talent of the teams in the top-tier league in the first period will be greater than talent investment in a three team league for $\mu > 1.3$. Thus even if it were decided somehow to mandate an optimal expansion of a closed league (which seems unlikely), the open system would in fact deliver a better outcome in terms of talent investment. The open two-league system not only leads to greater aggregate investment, but also greater investment at the highest level of competition. Note that the difference in revenue generating potential between the divisions (λ) drives the result- if there were no difference then the threat of relegation would be an empty one. If a closed lower-tier league exists, it will clearly have a lower level of investment in talent than an open system.

4 Fan Utility

By modeling leagues as contests, we abstracted to a degree from the utility of the fans who value these contests. In fact, I would argue that fan utility is a complex notion with several dimensions. Although most economic models have focused only on the issue of competitive balance - - arguing that fans derive greater utility from more balanced contests – in fact one might imagine four ways in which the fans derive utility. Assuming that success is proportional to a team's share of playing talent these are:

(i) team success- utility is increasing in $\frac{t_i}{\sum_{j=1}^{n} t_j}$

(ii) total talent- utility is increasing in the quality of the players $\sum_{j=1}^{n} t_j$

(iii) average talent per team- utility is increasing in the quality of players at each match played

$$\frac{1}{n}\sum_{j=1}^{n} t_j$$

(iv) competitive balance- utility is increasing for $\frac{t_i}{\sum_{j=1}^{n} t_j} < 1/n$ and decreasing for $\frac{t_i}{\sum_{j=1}^{n} t_j} > 1/n$

This does not take into account fans' expenditure on tickets, merchandise and so on. If teams were capable of perfect price discrimination they would simply extract all the consumer surplus of the fans and fan utility would be identical to revenue. While this is perhaps extreme, it is not obvious that this issue is critical to the evaluation of the promotion and relegation system. For example, one might suppose that the clubs can extract a constant fraction of fan utility, thus obviating the need to analyze issues such as ticket pricing.

The model suggests not only that total talent is increased with open competition but also average talent per match. Even if all the best potential baseball, basketball, football and hockey players in the entire world already appear in the North American major leagues, there is likely to be some scope for improvement through increased expenditure on methods training, and coaching. These improvements would be driven by the fact the failure would now involve a heavier price than simply waiting for the next season. Players would try harder, and coaches would work harder to as a natural consequence of openness so that that "effective" talent on display at matches would increase. Moreover, if there is any elasticity in the supply of talent, investment in new talent would increase. Note that if the number of the teams in the top tier were expanded, and promotion involved the admission of significantly inferior teams, then average talent per match could fall. The significance of promotion and relegation is that it does not require potentially inefficient expansion to produce the efficient increase in competition.

In our model, closed two team leagues are always perfectly balanced while an open system with promotion and relegation will lead to unbalanced competition in some years. This effect should not be exaggerated. If we extended our model one period further forward the most likely outcome would be for the larger drawing team relegated in period to win promotion back to the top tier and the smaller drawing team that was promoted in period 1 to be relegated. Thus competition would once again be evenly balanced. In a league with many more teams (say 18 in each tier, as is usually the case in European soccer), the larger drawing teams are highly unlikely to be relegated. Moreover, it is open to leagues to share revenues or promote competitive balance in other ways, which is, after all, the basic rationale for the permissive antitrust regime they currently enjoy.

Because leagues can extract much of the surplus that consumers derive from increased utility in competitive balance, leagues have an incentive to promote such balance. However, the problem with schemes like revenue sharing is that while it may improve balance, it reduces the incentives each team has to make its own product as attractive as possible for its fans, through better players, improved facilities and stadium, effective marketing, etc. Revenue sharing allows teams, to some degree, to free-ride on the ancillary revenue generating activities of other teams. For this reason, it may be preferable to allow teams to retain some fraction of locally generated revenues, thus preserving some competitive imbalance (as indeed happens even in the NFL, the most egalitarian of all leagues). Compared to a closed league structure, open leagues mitigate the dilemma leagues face of promoting balance and preserving incentives to compete, to the ultimate benefit of fans. Although any sharing of revenues derived from local promotion may lessen the incentives for that promotion, open leagues significantly limit the ability of clubs to free-ride with respect to player talent; a club that seeks to pocket shared revenue rather than spend it on player talent will find itself relegated. Moreover, because more clubs will participate in the top-tier league over time, the diversity of fans is better served. In sum, within-season competitive balance may well increase as clubs in an open league are more willing to share revenue (knowing that the recipients will have to spend the money on improved balance), while between-season competitive balance is clearly enhanced. Thus openness might be thought to spread the benefits of league competition more evenly and thus generate a net increase in welfare.

5 Reduced Opportunity for Rents Through Franchise Scarcity

The prior discussion demonstrates how, with four possible locations for teams, the members of a two-team closed league would resist expansion. If it were argued that a rival league in the remaining two locations would significantly reduce the rents of the incumbents (perhaps by driving up the marginal cost of talent), then expansion to a three team league might still occur to pre-empt competition. Leaving a single market unserved has an additional benefit for the incumbents, in that they can threaten relocate and use this threat to extract rents from their current location. The repeated use of this threat is well documented in Noll and Zimbalist. I think of this in the following way. Suppose there are rents that teams cannot extract through conventional means of ticket pricing, broadcasting rights sales and merchandising- at its most simple this is the consumer surplus associated with reading about the local team in the newspaper or talking about it with friends. It is socially efficient to extract the quasi-rents associated with the maintenance of a team of high quality through some form of public subsidy (like the fixed fee of a two-part tariff). However, it is socially wasteful to extract pure economic rents by threatening to relocate unless a heavily subsidized facility is provided. This possibility is a consequence of franchise scarcity and the amount of the rent extracted is the willingness to pay of the unserved location (and if this is larger than the willingness to pay of taxpayers/voters in the current location the team moves). The existence of an unserved location enables all of the teams to extract economic rents from their existing location, whether or not any of them actually move.

Once a system of promotion and relegation is instituted, all credible locations will be served, even if only by teams competing in the lower tier. The threat of relocation is now of limited value. The only credible relocation offer in the open system is when a team in small drawing location currently in the top tier offers to relocate to a large drawing area. However, the amount of rent is much smaller because (a) second tier competition has a value (b) a second tier team can be promoted in the future (and will be likely to if it is form a large drawing area) and (c) a team currently in the top tier might end up getting relegated at some point in the future. All of this suggests that relocation is quite unlikely with a system of promotion and relegation.

6 Conclusions

This paper has explored the implications of the institution of promotion and relegation for rent seeking in professional sports leagues. It has shown that teams will typically supply more effort and dissipate more rent in an open system with promotion and relegation. It also shows that promotion and relegation undermines the relocation threat that is so widely used in America to extract subsidies but is unknown, for example, in European soccer.

Much research remains to be done to analyze in full the impact of promotion and relegation on league performance, but this paper may at least point to the direction in which such research should tend.

References

1. Fort, Rodney and James Quirk. 1995. "Cross Subsidization, Incentives and Outcomes in Professional Team Sports Leagues" *Journal of Economic Literature*, XXXIII, 3, 1265-1299
2. Gradstein, Mark and Kai Konrad. 1999. "Orchestrating Rent-Seeking Contests" *Economic Journal*, 109, 536-545.
3. Inglis, Simon. 1988. League Football and the men who made it. Willow Books. London.
4. Meade, James. 1974. "Labor-Managed Firms in Conditions of Imperfect Competition", *Economic Journal*, 84, 817-824
5. Noll R. (2002) "The economics of promotion and relegation in sports leagues: the case of English football" *Journal of Sports Economics*, 3, 2, 169-203.
6. Noll R. and A. Zimbalist (1997) Sports, Jobs and Taxes. Brookings Institution Press.
7. Quirk, James and Rodney Fort. 1992. Pay Dirt: The Business of Professional Team Sports, Princeton N.J.: Princeton University Press.
8. Ross S. and Szymanski S. (2002) "Open competition in league sports" *Wisconsin Law Review*, vol 2002, 3, 625-656.
9. Seymour, Harold. 1960. Baseball: The Early Years. Oxford University Press, New York
10. Siegfried, John and Andrew Zimbalist. 2000. "The economics of sports facilities and their communities" *Journal of Economic Perspectives*, 14, 95-114.
11. Szymanski S. (2003) "The Economic Design of Sporting Contests" *Journal of Economic Literature*, forthcoming
12. Szymanski, Stefan. and Tommaso Valletti . 2003. " Promotion and Relegation in Sporting Contests." Mimeo.
13. Tullock, Gordon. 1980. "Efficient Rent Seeking" in J. Buchanan R Tollison and G. Tullock, eds. *Toward a Theory of Rent Seeking Society.* Texas A&M University Press, 97-112.
14. Ward, Benjamin. 1958. "The Firm in Illyria: Market Syndicalism" *American Economic Review*, 48, 4, 566-589

Druck: betz-druck GmbH, D-64291 Darmstadt
Verarbeitung: Buchbinderei Schäffer, D-67269 Grünstadt

AF848344

Platonism & Positivism in Psychology

Platonism & Positivism in Psychology

MORTIMER J. ADLER

with a new introduction by
James R. Weiss

Routledge
Taylor & Francis Group
LONDON AND NEW YORK

Originally published in 1937 by Longmans, Green and Co.

Published 1995 by Transaction Publishers

Published 2017 by Routledge
2 Park Square, Milton Park, Abingdon, Oxon OX14 4RN
711 Third Avenue, New York, NY 10017, USA

First issued in hardback 2018

Routledge is an imprint of the Taylor & Francis Group, an informa business

New material this edition copyright © 1995 Taylor & Francis

All rights reserved. No part of this book may be reprinted or reproduced or utilised in any form or by any electronic, mechanical, or other means, now known or hereafter invented, including photocopying and recording, or in any information storage or retrieval system, without permission in writing from the publishers.

Notice:
Product or corporate names may be trademarks or registered trademarks, and are used only for identification and explanation without intent to infringe.

Library of Congress Catalog Number: 94-27446

Library of Congress Cataloging-in-Publication Data

Adler, Mortimer Jerome, 1902-
[What man has made of man]
Platonism and positivism in psychology / Mortimer J. Adler ; with a new introduction by James R. Weiss.
p. cm.
Originally published: New York : Longmans, Green and Co., 1937.
Includes bibliographical references.
ISBN 1-56000-772-9 (pbk.) : $21.95
1. Psychology. 2. Psychology—Philosophy. 3. Psychology and philosophy. 4. Psychoanalysis. I. Title.
BF38.A34 1994
150'.1—dc20 94-27446
CIP

ISBN 13: 978-1-138-52994-6 (hbk)
ISBN 13: 978-1-56000-772-2 (pbk)

TO

MY PARENTS

CONTENTS

INTRODUCTION TO THE TRANSACTION EDITION ix

INTRODUCTION xvii

AUTHOR'S PREFACE xxv

LECTURE 1. THE CONCEPTION OF SCIENCE IN THE MODERN WORLD 3

LECTURE 2. THE POSITION OF PSYCHOLOGY: IN PHILOSOPHY AND AMONG THE NATURAL SCIENCES 31

LECTURE 3. THE HISTORY OF PSYCHOLOGY 61

LECTURE 4. PSYCHOANALYSIS AS PSYCHOLOGY 94

SUPPLEMENTARY NOTES 124

EPILOGUE 235

LIST OF PRINCIPAL NOTES 245

INTRODUCTION TO THE TRANSACTION EDITION

BY DR. JAMES R. WEISS

SEVERAL summers ago I attended a week-long seminar led by Mortimer Adler at the Aspen Institute's campus in Colorado. As a psychoanalyst and a psychiatrist engaged in a busy clinical practice, I am continually struggling to understand my patients and their conflicts in all their complexity. I was becoming alarmed by a trend within psychiatry to dismiss as outmoded the insights that psychoanalysis has to offer about the nature of man and his inner conflicts, replacing them with simplistic approaches to understanding emotional difficulties by virtue of their being "more scientific." Recent advances in medicine, psychopharmacology, neurobiology, cognitive science, and genetics have resulted in the 1990s being dubbed "the decade of the brain." The mind has been reduced to a mere by-product of neurochemical events. Some psychiatrists listen more to Prozac than to their patients. Authors of one recent book in the field of cognitive neuroscience purport to "paint a particular type of picture of how the mind is produced by the brain" (Kosslyn and Koenig 1992). The uniqueness of the individual is being ignored, while distinctions between brain (body) and mind (soul, psyche) are being blurred or denied. Man is seen as "merely an animal," albeit one of great complexity, differing from apes only in degree, but not in kind. As an alternative to this materialistic monism there is a growing popular "New Age" literature, which, in its denial of man's physical and biological characteristics, is equally reductionistic. Illness is attributed to negative thoughts or "bad karma." Psychic health is promised via the use of crystals, séances, mysticism or faith healing.

I feel strongly that psychoanalysis is a discipline specially positioned and disposed to investigate man in a way that does justice to both brain and mind, body and soul, psyche and soma, and their complex interrelatedness, without reducing him to one or the other, or to some ill-conceived hybrid. Wishing to learn more about the place of psychoanalysis in the history of Western thought, I engaged in several conversations about the topic with Mr. Adler that summer in Aspen. This led to my inviting him to come to Chapel Hill in the fall of 1992 as guest speaker at the monthly scientific meeting of the North Carolina Psychoanalytic Society. His presentation was entitled "Aristotle and Freud," and he told me it would be based on the series of lectures he had given in 1936 at the Institute for Psychoanalysis in Chicago. In preparation for his visit I was disappointed to discover that the outline of those 1936 lectures was out of print. When Mr. Adler provided me with a photocopy of *Platonism and Positivism in Psychology*, I eagerly set out to read it. I became concerned almost immediately, however, as I read the introduction to that first edition, which was written by Dr. Franz Alexander, who was then director of the Chicago Institute and who had invited Mr. Adler there to speak about the philosophy of science. Alexander declared Adler's views "an anachronism" which illustrated "the insurmountable gap between what could be called a scientific Weltanschauung and the dogmatic attitude of a Thirteenth Century scholastic." Claiming the higher ground of scientific certitude for psychoanalytic investigative method, Alexander was dismissive of Adler's philosophical approach to the knowledge of man as mere speculation. He saw the difference between man and animal as only a quantitative one. He saw moral phenomena as only "the results of parental influences, the nature of which are determined by the structure of the civilization in which the parents live," and seemed unwilling to consider that one set of moral precepts and injunctions might be superior to another, or that judgments of such matters could be validly approached philosophically. Here was no less a spokesperson for psychoanalysis than Franz Alexander displaying the very problems I had hoped psychoanalysis could best address: closed-mindedness and biological reductionism.

Thus, it was with some trepidation that I anticipated Mr. Adler's visit to our psychoanalytic society. As it turned out, his message was far more warmly received by the psychoanalytic audience of 1992 than it had been in 1936. I wondered what could account for this change in attitude among groups of analysts.

Early analysts, Freud included, were extremely concerned about establishing in the public eye the validity and legitimacy of their new discipline. To this end, they emphasized that their methods and theories were scientific, and they took pains to distance themselves and their theories from philosophy, which was deemed mere opinion or superstition. Those were the days of drive theory. Powerful biologically based forces were seen to be at work in man, although out of his awareness. These unconscious forces largely motivated and determined an individual's behavior, even though that individual may have been of the conceit that he was rationally choosing and determining his own actions. It was this valuable Freudian insight, the idea of psychic determinism, that allowed those early analysts to make claims that psychoanalysis was a pure, biologically based science of man.

I would hope that today's analysts are more secure in their sense of the value and legitimacy of their unique point of view, finding it less necessary to hitch the wagon of their professional self-esteem to a narrowly scientific approach that conceives of human nature in only materialistic terms.

In fact, our 1992 audience was far more receptive of Adler's assertions than Alexander had been. However, I think that today's analysts still struggle with a professional ambiguity about the nature of man and the nature of their theories. Is psychoanalysis a branch of natural science or of the humanities and philosophical psychology? Is the mind ultimately reducible to brain, or is it of a different order of reality? Does man differ from the brute animals only in degree, or also in kind? To what extent is man a free agent, and to what extent is he a pawn of his own biological urges and passions? Sensing that they may become bogged down in fruitless debate, some analysts turn their backs on these questions. They content themselves with immersion in their clinical work, leaving the work of solving these complex philosophical problems to others of a more theoretical bent.

Adler does not flinch from the task. He addresses these difficult issues head-on in this volume and in his subsequent writings on the subject. He strongly asserts that human beings differ from other species not merely in degree, but in kind. Only the human intellect is capable of conceptual, abstract thinking based on sensory data. Animals are incapable of conceptual thought. Man can make judgments and refinements of his abstractions that are impossible for an animal.

If this is not true, that is, if we humans differ only in degree from the animals, then a number of consequences follow. Most important for psychoanalysts is that human freedom of choice or "free will" would be impossible. The demands of our animal passions and instinctive life can only be channeled, compromised with, and ordered by an ego (psyche) capable of attaining some degree of autonomy and distance from those passions. Psychic determinism cannot be absolute.

In *Intellect: Mind Over Matter,* Adler says,

> "Anyone who wishes to think rationally should have the habit of thinking coolly, with all affective feelings or sentiments and all emotions parked outside. The heat of the passions, especially if they are strong and violent bodily commotions, cannot help but cause a disturbance or even a distortion of all intellectual work.
>
> William Wordsworth, in the preface to the 1800 edition of *Lyrical Ballads*, said that 'poetry is emotion recollected in tranquility.' No statement could be more significant about the role of the emotions in the work of the intellect. Emotion has its place in poetry, as well as in music and the visual arts, but that place is in the past, to be remembered, not in the present while the artist is engaged in the production of a poem, a musical composition, or a work of visual art.
>
> Emotional thinking is, to use Freud's phrase, 'wishful thinking'—controlled by the drive of subjective desires and passions rather than by the objective realities to which dispassionate thinking should respond. Paradoxically, and obviously not recognized by him, Freud is caught in self-contradiction at this point.
>
> If psychoanalytic theory claims that its hypotheses can be empirically verified or falsified by the data obtained by scientifically conducted, clinical research, then psychoanalytic theorizing is not wishful thinking. Yet Freudian psychology also claims that the passions control all human thinking, which is therefore wishful thinking throughout. Both claims cannot be true." (Adler 1990b)

Most contemporary analysts would agree that a goal of the treatment process is to help the individual attain greater freedom from the inner constraints of anxiety, defenses, symptoms, and inhibitions, as well as from the "tyranny of the passions." This freedom would be impossible if humans were not endowed with a faculty or dimension capable of transcending those passions (c.f. Waelder's [1976] concept of ego transcendence). That dimension of man, Adler asserts, is immaterial, and it cannot be ignored or swept aside in favor of simpler, materialistic approaches to understanding human nature, intellect, or will.

Mr. Adler contends that man is both body and soul, soma and psyche, material and immaterial, brain and mind. A proper study of his nature requires contributions from both the physical sciences and from philosophy. His work has been influenced by

Aristotle, Aquinas, and Jacques Maritain. Philosophy and science must cooperate in seeking answers to questions that require both of their methods of investigation.

It was partially in reaction to the overemphasis of psychic determinism and its implicit denial of freedom of choice that Heinz Kohut (1977) rejected classical ego psychology. He founded psychoanalytic self-psychology, which sees the self as a whole as agent, a source of initiative, ambition, and action. Schafer (1976), Klein (1976), and Gill (1976) have also addressed what they felt to be the failure of psychoanalytic structural theory to satisfactorily explain the individual's choice of action. Other writers such as Leo Rangell have followed more closely in the footsteps of Freud, his daughter, Anna Freud, Hartman, Kris, Lowenstein, Rapaport and the other ego psychologists. They also have disagreed with the trend in psychoanalysis toward the biological reductionism of drive theory taken as an absolute. In fact, beginning with Anna Freud's (1988) "The Ego and the Mechanisms of Defense," originally published in 1936, numerous analysts over the decades have greatly expanded the implications of Freud's original structural theory. Modern ego psychologists, for example, are more able to understand how such ego functions as memory, cognition, or judgment may become able to function autonomously, free from the demands of the drives or the tensions of intersystemic conflict. Rangell says, "it is precisely man's ego autonomy, and the degree of freedom this bespeaks, which defines and makes room for his subsequent assumption of responsibility." He later states, "I cannot visualize practicing psychoanalysis without having within my therapeutic armamentarium, the interpretation that 'you have a choice,' or 'you had a choice, and you chose this or that' or at least that 'you set up the conditions which made such-and-such a decision necessary or inevitable'"(Rangell 1990).

Of course, the point behind helping an individual achieve a greater degree of autonomy of rational thought and free choice is to enable him to act and direct himself toward the pursuit of real goods of life for himself and his fellow men. Mr. Adler has dedicated his life to this pursuit. He was born in New York City in 1902, and later taught at Columbia University for seven years (1923-1930). He moved to the University of Chicago (1930-1952) where he became professor of the philosophy of law. In close association with Robert M. Hutchins, he led seminars on the great books and ideas of Western civilization for a number of years.

Eventually he left formal academia to concentrate on the broad dissemination of philosophical ideas to the general public rather than to the narrower community of professional philosophers. He became editor of the Great Books of the Western World, for which he wrote the masterful Syntopicon, with its 102 essays on the great ideas. He has been chairman of the board of editors of the Encyclopaedia Britannica, director of the Institute for Philosophical Research, and editor of *The Great Ideas Today*. Since 1953, he has conducted annual seminars at the Aspen Institute for Humanistic Studies, of which he is an honorary trustee. The 1981 seminars on his book *Six Great Ideas* were televised by Bill Moyers as a series on the Public Broadcasting System.

Mr. Adler has authored or co-authored more than fifty books, many of them written expressly for the thoughtful lay population. They include: *How to Think About War and Peace*, 1944; *The Idea of Freedom*, two volumes, 1958 and 1981; a revision of his best-selling *How to Read a Book* (with Charles Van Doren), 1972; *Aristotle for Everybody*, 1978; *How to Think About God*, 1980; *How to Speak/How to Listen*, 1983; and *A Vision of the Future: Twelve Ideas for a Better Life and a Better Society*, 1984. He has written two autobiographies: *Philosopher at Large*, 1977, and *A Second Look in the Rearview Mirror*, 1992. His works have at times been wrongly dismissed by university professors of philosophy as "popularizations," which they are not.

Adler has written:

"Philosophy is the only academic subject listed in college and university catalogues that, in varying degrees, should be the vocation of everyone. Everyone should not aim to be a mathematician, a physicist, a molecular biologist, an economist, or an historian. These and most other subjects listed in college and university catalogues are fields of academic specialization as philosophy has, unfortunately, also become in the twentieth century. Unlike all these fields of specialization, philosophy, properly understood, is the vocation of all thinking human beings who confront fundamental problems and issues about the world in which they live, about human society, and about themselves and their place in nature. If confronting these problems, they are inclined to think about them, they are involved in philosophical thought.

It is this fact that distinguishes the seven or eight major philosophical books I have written since *Aristotle for Everybody* in 1978 from books written by professors of philosophy in the same period. The books they have written, like the articles they write for the technical journals of philosophy, are written for their peers (i.e., other professors of philosophy). In contrast, the books I have written attempt to restore philosophy to its proper place in our society and culture.

In the twenty-five centuries of Western philosophical thought, it is only recently, with Kant and other German philosophers after him, that philosophical books were written by men who held university professorships of philosophy. I do not think that I would, or could, have written the books I have produced since 1978 had I been a professorial philosopher, concerned with esteem of other professors of the subject. Instead, philosophy has been my lifelong vocation and I have at last learned how to write, not for professors of philosophy, but for other human beings who have the same vocation in some degree." (Adler 1990a)

Other books by Mr. Adler that deal with subject matter closely related to *Platonism and Positivism in Psychology* include: *The Difference of Man and the Difference It Makes*, 1967; *Some Questions About Language: A Theory of Human Discourse and Its Objects*, 1976; *Intellect: Mind Over Matter*, 1990; and *Desires: Right and Wrong: The Ethics of Enough*, 1991.

Adler has been a lifelong proponent of liberal education, believing that it is a cornerstone of a free society. In 1981 his Paideia Project called for a major reform of public school education and provided a blueprint for its structure. On the behalf of the Paideia Group, he authored *The Paideia Proposal: An Educational Manifesto* (1982). Later, he wrote *The Paideia Program: An Educational Syllabus* (1984). He is the honorary chairman of the Paideia Group, Inc., which was founded by his colleagues and which continues to pursue his goal of educational reform.

Platonism and Positivism in Psychology is an intellectual tour-de-force dealing with the relationships of science and philosophy, the nature of man, and the place of psychoanalysis in the history of psychological and philosophical thought. Few individuals are better equipped to discuss such matters than Mortimer Adler. Teaching, learning, and the quest for philosophical truth have been his lifelong occupations. The extensive notes that the author has appended to these lecture outlines make them come alive for the general reader, and stimulate him to think philosophically, which, the author suggests, is a proper vocation for everyone.

REFERENCES

Adler, M. J. *A Second Look in the Rearview Mirror* (New York: Macmillan Publishing Co., 1990a).

———*Intellect: Mind Over Matter* (New York: Macmillan Publishing Co., 1990b).

Freud, A. "The Ego and the Mechanisms of Defense." In *The Writings of Anna Freud*, vol. 1 (Madison, Conn.: International Universities Press, 1988).

Gill, M. M. "Metapsychology is Not Psychology." In *Psychology vs. Metapsychology. Psychological Issues*, Monograph 36 (New York: International Universities Press, 1976).

Klein, G. S. *Psychoanalytic Theory* (New York: International Universities Press, 1976).

Kohut, H. *The Restoration of the Self* (New York: International Universities Press, 1977).

Kosslyn, S., and Koenig, O. *Wet Mind. The New Cognitive Neuroscience* (New York/Toronto: The Free Press, 1992).

Rangell, L. *The Human Core* (Madison, Conn.: International Universities Press, 1990).

Schafer, R. *A New Language for Psychoanalysis* (New Haven: Yale University Press, 1976).

Waelder, R. "The Problem of Freedom in Psychoanalysis and the Problem of Reality Testing." In *Psychoanalysis*, ed. S. A. Guttman (New York: International Universities Press, 1976), pp. 101-20.

INTRODUCTION

BY DR. FRANZ ALEXANDER

It is unusual to write an introduction to a book of an author whose conclusions, approach to his problems and whole outlook are diametrically opposite to those of the author of the introduction. Why did I then accept Mr. Adler's suggestion to write an introduction to his book and why did Mr. Adler ask me to do so, are both questions which require an explanation. The circumstances under which these four lectures originated will elucidate this paradox.

Engaged in psychoanalytic teaching and clinical studies for a long period of time, I gradually came to the conviction that in this field as in others where students are using a highly standardized technical procedure and are mainly absorbed in minute observation of facts,—briefly in all preeminently empirical fields,—the students are apt to lose perspective towards their own work. This conviction goes back to those early days that I spent as a research worker in physiology in an experimental laboratory. There, I became first acquainted with the characteristic mentality of modern scientific research. There I learned the *mores* and virtues of modern research and first recognized the danger which confronts the scientific worker of the present day. This danger is not restricted to scientific laboratories, it is a general problem of the present age. Man, the inventor of the machine, has become the slave of the machine, and the scientist, in developing highly refined methods of investigation, has become not the master but the slave of his laboratory equipment. An extreme amount of specialization of interest and mechanization of activity has taken place and a scotoma for essentials has developed; a naive belief in the magic omnipotence of specific technical procedures leads to a routine, often sterile submersion in details without interest in or understanding of larger connections.

It is no exaggeration to say that in many scientific centers not the interest in certain fundamental problems but the fortuitous possession of some new apparatus directs the research work: a new laboratory technique is introduced which spreads like a fad to all laboratories; then everywhere problems are selected which can be approached by this new technique or apparatus. Scientific interest in the fundamentals is lost, research is dictated more or less at random by the technical facilities at the worker's disposal.

This attitude necessarily must lead to that caricature of scientific ethics which regards suspiciously everything that entails reasoning and not merely observation and is contemptuous about theories, not to say hypotheses that are not as yet proven. There is a naive adoration of "pure facts" which are collected without any leading ideas.

Psychoanalysis is a highly empirical field in which the student is exposed to an extreme variety of observations and—in a certain sense—unique facts, as every patient presents a unique combination of common elements. Today the psychoanalytic clinician is undergoing a healthy reaction against the present abundance of theory and generalizations. He is in the process of accepting the mentality of the natural scientist and is assuming all the virtues and weaknesses of our era of laboratory research. Like his other clinical colleagues also he uses a highly standardized and refined technique but pays a high price for his technical skill: he is gradually losing perspective and correct judgment regarding the validity and limitations of his technique and of his scientific work in general.

To expose the students and the members of the staff of the Psychoanalytic Institute of Chicago to a lecture course dealing with epistemology and methodology seemed to me therefore highly indicated. I hoped that such a course would make them more conscious of and critical toward the methods, achievements and significance of their own field. Although one could not anticipate full agreement with all of Mr. Adler's conclusions, it was to be expected that the discussion of problems of psychological methodology by an expert such as Mr. Adler would be a real challenge and a healthy stimulus for all of us, which would stop us for a moment in our daily work and induce us to give ourselves an account of what we are really doing.

This experiment was successful in one respect. Although Mr. Adler was not able to convince his audience of the validity of any of his major statements, he aroused interest in this type of reflection, so direly needed in any group of specialists.

During the discussions that followed each of his lectures Mr. Adler has realized the basic disagreement of his audience with his theses and was therefore hesitant to publish his lectures without giving me an opportunity to state these fundamental controversies in the form of an introduction to his book.

During these discussions it became evident that our disagreement represents two diametrically opposite points of view, and illustrates the unsurmountable gap between what could be called a scientific *Weltanschauung* and the dogmatic attitude of a Thirteenth Century scholastic. As Mr. Adler himself very correctly put it, his lectures represent something which might have been the attitude of a contemporary pupil of Thomas Aquinas, or perhaps of Thomas himself if he could come to life again and be confronted with the new discoveries of Sigmund Freud. If for nothing else, then as such an anachronism, Mr. Adler's lectures may have the interest and value of a curiosity. It became obvious that according to Mr. Adler since the times of Aristotle and his belated scholastic pupil, Thomas Aquinas, man really did not contribute anything much of value to the inventory of his knowledge about the surrounding world and himself. Such radical new accomplishments as the recognition of the fact of biological evolution, the technical achievements of physics, chemistry and medicine, the better understanding of mental diseases, which have resulted in an ever-growing capacity of modern psychiatry to cure psychopathological processes, are of no great importance for him. This contempt for "practical" accomplishments explains why he considers the futile speculations and meditations about human nature of ancient and medieval philosophers like Aristotle and Thomas Aquinas to be superior to our detailed and precise knowledge of normal and pathological mental processes which enable us to influence these processes in a desired direction and thus cure mental ailments. Mr. Adler does not evaluate scientific concepts according to the degree to which they increase our ability to control and influence natural phenomena but according to their ability to be fitted into some preconceived rigid and abstract logical construction. The gap between these two attitudes cannot be bridged. In the following I shall state the main points of our controversy.

1. Mr. Adler defines philosophy as a body of logical conclusions drawn from common sense observations, and science as a body of conclusions drawn from specific observations obtained by specific investigative methods. I agree with Mr. Adler's definition of science but not with his definition of philosophy. Mr. Adler

reduces philosophy to reasoning about inadequate (common sense) observations, science representing at the same time reasoning about more adequate observations obtained by refined and improved methods of investigation. And yet, in order to save the medieval hegemony of philosophy, with a peculiar twist of reasoning Mr. Adler tries *to subordinate science*—that is to say conclusions drawn from improved observations—to philosophy, which according to his own definition consists of conclusions from inadequate observations. If Adler's definition of philosophy is correct, philosophy should be discarded in the proportion to which scientific knowledge progresses by the use of steadily improving special techniques of investigation. With this definition, Adler himself speaks the death sentence of philosophy. However I am much more inclined to accept the view of those philosophers (a view which was most convincingly expressed by Bernard Alexander) who consider philosophic reflection as the manifestation of a deep and universal need in human beings to orient themselves towards the universe with which they are confronted. Science must necessarily leave this deep need for orientation to a high degree unfulfilled because due to its nature its answers are disconnected and restricted to isolated portions of reality, namely to those fields that have already been scientifically explored, and also because scientific answers are necessarily incomplete due to the high claims regarding validity which science sets to itself. In view of this incomplete nature of science, only two attitudes remain for a human being in handling his desire to orient himself in the world in which he lives:

a. A kind of agnosticism, contenting himself with the answers which science can offer at any given time, abandoning the wish for an integrated concept of the universe.

b. An attempt to integrate the isolated and disconnected answers of science into a *Weltbild*. This is philosophy. Such a philosophical picture of the world however must be consistent with what is known by science. It tries to integrate the results of scientific investigations into the best possible synthesis of current knowledge. Therefore philosophy never can disprove science, on the contrary it must always be ready to be corrected by new scientific discoveries. In this sense philosophy never can become superfluous as long as human beings continue to ponder over the world in which they live.

2. One of Mr. Adler's fundamental convictions is that the act of reasoning itself cannot be studied scientifically: that in this regard it has an exceptional position among all psychological

processes. Since the reasoning faculties have been subjected successfully both to psychological and physiological investigation, this statement does not need further discussion. Freud's studies of rational thinking, his understanding of it as motor innervations with small quantities of energy on a trial and error basis, his comparative studies of rational thinking in contrast to thinking in dreams, his classical study of denial, and finally his demonstration of the influence of emotional factors upon rational thought have shown that the reasoning process can be psychologically analysed and genetically understood as the result of a process of adjustment of the biological individual to its environment. Furthermore the first promising steps have been made by the method of brain operation to establish the relation of rational processes such as the faculty for abstraction, to brain functions.

3. Perhaps the greatest gap between Adler's concepts and the viewpoint assumed by psychoanalysts consists in his insistence upon the non-animal nature of human beings, in his contention that there is no evolutionary continuity between men and brutes. Adler belongs to that rapidly diminishing, in fact today almost extinct species of thinkers who still try desperately to disregard the solid facts of evolution. The type of argument which is used can be best characterised by quoting the distinguished American thinker, James Harvey Robinson.[1]

"Having myself given much time to the comings and goings of beliefs in the past, I see how great a part mere ignorance and confusion always play in blocking the ready acceptance of new knowledge. . . . It is true the biologists have, many of them, given up what *they* call 'Darwinism'; they have surrendered Spencer's notion of the hereditary transmission of acquired characters, and they even use the word 'evolution' timidly and with many reservations. *But this does not mean that they have any doubts that mankind is a species of animal, sprung in some mysterious and as yet unexplained manner from extinct wild creatures of the forests and plains.* This they simply take for granted; for, unlike the public at large, they distinguish carefully between the varied and impressive evidence which appears to confirm man's animalhood and the several theories which have been advanced from time to time by Lamarck, Darwin, Spencer, Haeckel, and others, to account for the process by which organic life, including man, has developed. The first confusion of which we must relieve ourselves is that between the *facts*, on the one hand, revealed by geology, biology and comparative anatomy, and, on the other hand, the *conjectures* suggested to

[1] Robinson, James Harvey: The Human Comedy, New York, Harper & Bros., 1937, pp. 23–4.

explain the history of life. As time has gone on the facts which compel anyone acquainted with them to accept man's essentially animal nature have become more abundant and unmistakable, while many of the older theories of evolution have, as a result of further study and increasing knowledge, shown themselves to a great extent untenable. Much light has been cast of late on the history of life, but in some respects it seems more mysterious than ever before."

Adler, like some theologists, in order to refute the unassailable fact of evolution, takes advantage of recent controversies regarding those detailed mechanisms by which man has developed into his present form from more primitive beings.

His methodological discussion about the discontinuity of the different realms of knowledge, his insistence upon unsurmountable gaps between the inanimate world, animals and man, are based on a refusal to draw the correct conclusion from the factual evidence which we possess regarding the relationships between these different sections of the universe. Even though scientists are not able to reduce the processes of life to physics and chemistry there is no doubt that the process of life consists in a specific combination of exactly the same processes as are known in the physics and chemistry of nonliving substances. The specificity of life consists *in the specificity of the combination* of those partial processes which constitute life and which are identical with the chemical processes of nonliving substances. It is true that the specific principles of this coördination are not known as yet. Nevertheless it is quite possible that some time in the future when we shall possess the knowledge of the characteristic combination of physico-chemical processes in the living cell we shall be able to create living substance from nonliving substance. It is needless to say that the difference between the higher animals and man is far less incisive than that between animals and inanimate nature. The psychoanalyst considers the difference between man and animals as definitely of quantitative nature because he can observe that the new-born infant shows no human but merely animal characteristics. The difference between men and animals, which is one of Adler's fundamental theses, comes gradually about during the individual's development from an ovum to an adult.

4. Adler's efforts to demonstrate that the Freudian concepts of personality have been anticipated if not more completely elaborated by Aristotle are based upon entirely formalistic and verbal comparisons. Aristotle's theory of personality has the same relation to the Freudian as Democritus' concept of the atom bears

to the scientific concepts of Dalton or as Edward von Hartmann's "unconscious" has to the Freudian concept of the unconscious. The similarity is based mainly on the use of the same word. Democritus' "atom" was the result of deductive speculation and did not contribute to a better knowledge of matter, whereas Dalton's atomic concept was concluded from specific observations regarding certain permanent quantitative relationships in chemical compounds and has led to the body of knowledge represented by modern chemistry. Similarly Aristotle's speculations about human nature did not enable any contemporary physician to understand better the nature of paranoia or hysterical blindness, whereas Freud's concepts were derived from painstaking minute observations made by the help of a refined technique of psychological inquiry and have made possible a better understanding and methodical influencing of psychological processes.

5. Mr. Adler's statement that the Aristotelian concepts about personality in a certain respect were more complete than Freud's is obviously true. But this completeness is anything but an asset as compared with the incompleteness of the Freudian concepts. The Freudian concepts are necessarily incomplete; they are, like every scientific concept, in a continuous process of change and improvement because, being of a scientific nature, they must remain in accordance with new observational material whereas the systematic philosophic concepts of Aristotle are "eternal," since they had to be only consistent in themselves, had to fit not observations but vague generalizations and an abstract system of thought.

6. The gap between the scholastic attitude of Adler and the scientific views of Freud becomes widest when Adler takes a stand regarding the psychoanalytic concepts of moral phenomena. Adler evaluates the Aristotelian concepts of morals as superior because they are absolute whereas the Freudian concepts are relative. Freudians consider moral concepts to be the results of parental influences, the nature of which are determined by the structure of the civilization in which the parents live. Aristotle has an anthropomorphic concept and declares the moral ideas of the Greek Peninsula to be absolute values. Freud's concept of morality compares to the Aristotelian as the concept of present cosmology is related to the Aristotelian concept of the earth as the center of the universe. The facts of modern anthropology and ethnology fully corroborate the Freudian view and disprove the Aristotelian, showing that the concepts of good and bad radically differ in different cultural settings depending upon the political, economic and cultural history and situation of a group.

If there is such a thing as turning back the clock of history and science, here we see a classical example of it. Scholasticism, a sterile form of deductive thinking, developed as a harmless outlet for the reasoning powers of man in a period of intellectual servitude when man could not observe the world around himself lest any observation come in contradiction with prevailing dogmas. He had to content himself with flawless deductions from incorrect premises. Free observation of facts was forbidden; rigid acceptance of preconceived ideas was the highest requisite of these medieval centuries. What else remained for the human mind to do but indulge in playful deductive meditations starting from the accepted and prescribed dogmas as premises? Our present freedom of inquiry is nothing but the continuation of that great period of the liberation of the mind which started in the Renaissance, with Galileo, Giordano Bruno, and Bacon. It took centuries before this freedom of investigation was fully achieved. At first the realm of the celestial bodies, the stars, then the human body, was claimed as a legitimate subject for scientific study. The last step in this process of emancipation of the human mind was the accomplishment of Sigmund Freud, the objective unbiased study of the human personality, including its reasoning and moral powers.

In spite of these fundamental disagreements, I am very much indebted to Mr. Adler for his courage in presenting his scholastic philosophy to a group which was so thoroughly biased by what is a scientific credo. The consistency of his presentation, the astuteness and skill of his logical deductions, aroused in us the same admiration as those of some medieval scholasticists. The internal consistency of his lectures representing literally a medieval viewpoint, through the effect of contrast, afforded unprecedented opportunity for his audience to understand better the nature of psychoanalytic methods and the position which this science takes in the present realm of knowledge.

Chicago, January 25, 1937

PREFACE

THESE lectures were delivered before the Institute of Psychoanalysis in Chicago in the Spring of 1936. I was asked to discuss methodological problems, especially in relation to the scientific procedures of psychoanalysis. But it seemed important, almost unavoidable, to shift the discussion to more fundamental considerations about the nature of any science of psychology, its limited subject-matter, the scope of its problems. Only in such light could questions about method be properly answered. This, furthermore, necessitated a distinction, in general, between the scope of philosophy and of science; and, in particular, a determination of the parts which philosophy and science contribute to the knowledge of man.

As Dr. Alexander says in his Introduction, it became evident during the course of these lectures that my analysis provoked sharp differences of opinion on the part of the audience. I must confess to having been naively unsuspecting. I was surprised to discover that psychoanalysts were just as ready to tell me what philosophy was about and how to solve its problems, as they were willing to instruct me concerning the scientific facts they had learned from clinical experience. Furthermore, what I had to say seemed to me high praise of the contribution of psychoanalysis as a body of scientific knowledge. I did not expect that the praise would seem so faint because it did not glorify psychoanalysis, or even the whole of natural science, by abolishing all other knowledge, by condemning philosophy as empty talk or futile speculation. But the chief issue of the contemporary world, whether science is enough, once agitated, obscures all other questions, or determines the acceptance or rejection of the answers which are proposed. That was the case in the forum of the Institute last Spring.

I, therefore, felt it would be desirable to have Dr. Alexander write the Introduction to this book, in order to bring the line of these disagreements more sharply into focus. The dialectic would benefit from the expression of opposite voices on the same points. Dr. Alexander could not, of course, do more than indicate his position briefly. In fairness to him, the reader must be referred to the works of Freud and to Dr. Alexander's own writings for a fuller exposition of the view which psychoanalysts take of these basic questions. Since Dr. Alexander covers in ten pages the same points that I discuss in two hundred, it would hardly be fair to his position for the reader to make a final judgment on the issues involved. At the same time, it would not be fair to me for the reader to make a judgment about my position by *merely* consulting what Dr. Alexander says I say.

The lectures are here printed in the form in which they were given. The structure of an outline has the virtue of making explicit the progress of an analysis. It gives the leading propositions due prominence, and exhibits the order and relation of the parts of a complicated argument. But an outline is, strictly, not an argument. It is much more like an analytical index to one. This is its chief disadvantage. For those who know the full substance of the arguments, an outline can serve excellently the purpose of summarizing; it may even have the additional merit of putting a multitude of relevant discussions into the perspective of a single frame. But for those who do not know intimately what these arguments are, an outline must necessarily seem dogmatic, almost simple-minded in its brevity. On the surface it is dogmatic because, as a teaching instrument, it proceeds by the sequence of assertions, few of which it can stop fully to support. Yet whatever is affirmed is either self-evident or demonstrable, and in philosophy demonstrations can be given. I have attempted to show this, and also to compensate for the brevity and condensation of the outline form, by adding to the lectures a series of notes, appended at the rear and referred to by the bracketed numbers in the text. These notes are not only expansions of points made in the lectures; they are argumentative discussions of what the lectures may appear to say dogmatically because too simply or too briefly. The reader is asked to regard the lectures and the notes as integral parts of a unity, just as he would be obligated to examine the evidence on which the premises of a demonstration rested before judging the truth of the conclusion.

The issue which is central to this book, the relation of science and philosophy, is the crucial stumbling block of modern times. As Dr. Alexander recognizes, the issue is not merely between psychoanalysis and other psychologies, or between Freud and Aristotle. It involves all members of the household of human knowledge in the modern triangle of science, philosophy and theology. This, unfortunately, is not material for a domestic comedy. The issue is not a remote, academic one. The practical consequences of one or another solution are momentous. Science seems to be the modern hero. The *hybris* of science, the pride in its ability to answer all questions or to dismiss those it cannot answer as unanswerable,—a pride which is comic only when scientists pretend to tell philosophers what philosophy is about,—may make science the pivot of a tragic turn in modern affairs.

I would like to say, finally, that although what follows is written for the most part in the language of the philosophy of Aristotle and St. Thomas Aquinas, it is not "scholasticism," as that word is currently used to suggest necromancy with buried syllogisms. Even when not used opprobriously, scholasticism is an unfortunate designation. It confuses the fact that medieval philosophers were Christians with the fact that in the middle ages thinkers worked in a continuous tradition of philosophical thought, and did not seek to invent "systems" which would bear their separate names to posterity. Lack of education, combined with the burning desire to be original, prevents most modern philosophers from being scholastic in this latter sense. But what is more important to say to a contemporary audience is that although the truths of traditional philosophy are compatible with the dogmas of a revealed religion,—and therefore a man can be a Christian without ceasing to be a philosopher,—nevertheless, philosophy is not religion. Nor is it sacred theology. St. Thomas is a philosopher in the same sense that Aristotle and Santayana are. The truths of reason are formally established by reason independently of faith, even when they are truths in which faith concurs.

The passages I quote from the *Summa Theologica*,—the mere name of which may be misleading to some,—are entirely philosophical, as, indeed, all the points which I have argued in this book are philosophical. They rest neither on theological dogma nor on scientific fact. The citation of Aristotle and St. Thomas in the notes and footnotes is not an appeal to authorities on my part. It does not justify the dismissal of what I have to say as Aristotelian or Thomistic in the utterly absurd sense of a nostalgic return to the dead past. One can be an Aristotelian or a Thomist only by being a philosopher facing contemporary issues in the light of reason and experience, as they did in their day; and, as they were, through being respectful of the tradition of human knowledge wherever it bears witness to the truth.

MORTIMER J. ADLER

Chicago, 1937

PLATONISM AND POSITIVISM IN PSYCHOLOGY

LECTURE ONE

THE CONCEPTION OF SCIENCE IN THE MODERN WORLD

I. Introduction

1. Why the modern world? Because it is only in the modern world that science is self-conscious, that it considers itself distinct from other disciplines, as having a method and a mission. Because it is only in the modern world that science and philosophy are sharply separated and that there are supposed to be conflicts between science and philosophy and between science and religion. Because it is only in the modern world that the scientist wonders about himself and his methods. Because it is only in the modern world that positivism has rightly conceived science as different from philosophy, adding thereto a denial that philosophy is knowledge, if it claims to be different from science.
2. This does not mean that science is entirely a modern invention. It occurs and appears in the ancient and medieval worlds, but differently. It is only in the modern world that science has revolted from religion, revolted from philosophy. Thus, in the 17th century, physics revolted and gained autonomy; in the 18th, biology; in the 19th, psychology and the so-called social sciences. As a result of these revolts, it is in the modern world, and particularly in the 19th and 20th centuries, that it is supposed that any question worth answering is capable of being answered only by the method of the scientist; or, on the other hand, that any question which is not capable of being thus answered is not a genuinely answerable question.
3. There are, therefore, in modern times a number of peculiar problems which confront and perplex the scientist. (By the sciences I shall henceforth mean the natural sciences, including psychology. I shall, for the most part, ignore the social sciences.)

a. The difference between science and philosophy, and the relation between science and philosophy.

b. In considering this first problem, the scientist usually differentiates himself by his method, only to be further perplexed by the plurality of methods (procedures, techniques) which claim to be scientific. He must, therefore, determine what is common to all these different procedures.

c. He is, furthermore, aware of the plurality of the sciences as they are distinguished according to subject-matter as well as by method, and he is often embarrassed by their several claims to priority or independence. He faces here two sorts of questions:

(1) Questions of order: priority and dependence.

(2) Questions of composition: cooperating disciplines, hybrid subject-matters.

4. All of these problems are more intensified and, in a sense, more embarrassing for the psychologist than for any other natural scientist.

a. Thus, for example, his relation to the philosopher, because of the recency of his revolt. Boring's insight: "Psychology has never succeeded in taking philosophy to itself or in leaving it alone."[1]

b. Similarly, his attempt to define his subject-matter and his methods. Consider here the existence of "schools" of psychology; the fact that there are rival pretenders to the title of "scientific psychology." Consider the controversial character of these plural psychologies in contrast to the situation in physics or chemistry, where there is neither this plurality nor such polemic [1].

c. Finally, his relation to other scientists: particularly the relation of the psychologist to the physiologist, on the one hand, and to the social scientist, on the other.

5. And even more so are these problems exasperating for the psychoanalyst; and here the following should be noted:

a. His relation,—in both subject-matter and method,—to the rest of the psychologists. Is psychoanalysis a *part* of or the *whole* of psychology? Is it just another sect or school?

[1] *A History of Experimental Psychology*, New York, 1929: p. 660. Professor Boring goes on to say: "Such a diagnosis is not founded upon the necessary relation of science to philosophy, for this conflict is not to be found in the natural sciences. In psychology it lies very near the surface."

b. The character of his methods and of his research in relation to the better recognized and more certainly accepted canons of scientific procedure.

c. His acceptance of a philosophy,—the "philosophy of evolution",—and, at the same time, his opposition to philosophy.[2]

6. I propose, therefore, in this opening lecture to approach the problems of the psychologist and the psychoanalyst by considering, first, the two most general problems concerning science in modern times:

(1) The distinction between philosophy and science, and their relation.

(2) What is common to the plurality of scientific methods, procedures, techniques.

Then, subsequently, I shall consider the position of psychology in the order and relation of the natural sciences and in relation to philosophy.

II. Two Further Introductory Observations

1. We must understand a distinction within the field of logic, made according to differences in the relation of logic, as an organon, to knowledge and the types of cognition. (This is the distinction between the prior and posterior analytics in Aristotle's *Organon,* and between *petite logique* and *grande logique,* or formal and material logic, in the modern tradition.)

a. Formal logic: the analysis of terms, propositions, syllogisms, but not induction.

b. Material logic (which necessarily presupposes formal logic): the analysis of bodies of knowledge according to principles of order and system; the distinction of the kinds of knowledge; the formulation of the methods of knowing (evidence, proof, probability, induction, measurement, etc.)

[2] Freud proclaims himself a positivist. He regards philosophy, along with religion, as outside the realm of established knowledge,—as superstition, speculation, or "mere poetry." He speaks of his own "philosophy of life" as the rigorous *weltanschauung* of science. Unfortunately, this is not true. It would be better if Freud contented himself with being a scientist; but, like so many scientists who are positivists, Freud denies philosophers the right to their "opinions," but does not restrict himself to the "facts" of science. Vd. *New Introductory Lectures on Psychoanalysis,* New York, 1933: Ch. 7; also *The Future of an Illusion,* New York, 1928. Cf. Fritz Wittels, *Freud and His Time,* New York, 1931: Ch. II, "Freud, the Anti-Philosopher."

(1) Material logic is, in part, a sort of "theoretical rhetoric": the methods and types of exposition appropriate to the communication of different kinds of knowledge.

(2) Material logic is necessarily related to psychology: the analysis of human powers and acts of cognition.

c. In these lectures, I shall presuppose formal logic in order to discuss the larger material problems [2].

2. But, as we have seen, material logic involves psychological analysis. We cannot distinguish the kinds of knowledge, we cannot formulate the methods of knowing, without an account of the objects of human knowledge and of the cognitive processes of man. This fact has critical consequences:

a. Any scientist who entertains certain convictions about scientific method, about abstraction and induction, about observation and conception, about proof and probability, is necessarily, implicitly or explicitly, asserting propositions in psychology about the cognitive operations of man. Thus, for example, consider the ancient paradox of the strict atomic materialist who tried to explain himself as a scientist in terms of his own theory while at the same time appealing for the truth of that theory in terms of the rational canons of scientific method [3].

b. This paradox is even more embarrassing for the psychologist:

(1) It has been exposed by Lovejoy in the case of the "thinking behaviorist," and by Bradley in the case of the British associationists [4].

(2) In short, the psychologist in his capacity as a scientist appealing for the truth of his conclusions and their status as scientific knowledge, must assert or admit

(a) That man is in some sense rational, that he is bound by evidence and proof, seeks truth and avoids error, knows the difference between the better and the worse reason, etc. (In short, the man who claims to be a scientist cannot as a psychologist so conceive man that there is no difference between a scientist and a sophist).

(b) That man is intellectual as well as sensitive, that he knows by operations other than sensing, that he has abstract concepts, which are not sensory images, that, in short, he can generalize and thus go beyond all sense-data.

c. This point is particularly important for the Freudian, and particularly important for me to make clear at the beginning of this course of lectures.

(1) If psychoanalysis is a body of scientific knowledge which employs abstract concepts,—and it most certainly does employ such abstract concepts as *complex*, *regression*, *sexuality*, *cathexis*, *abreaction*, etc., —then the Freudian cannot without contradiction at one time be proud of his scientific knowledge and at another deny the intellect.[3] His position as a scientist considering his own knowledge must agree with his position as a psychologist considering man's cognitive powers.

(2) Furthermore and similarly, if psychoanalysis is scientific in its methods and appeals to the binding canons of reason in evidence and proof, then the Freudian as a psychologist cannot deny the independence of human reason from passion and prejudice. He cannot assert that everything which appears to be *rational* is merely *rationalization* without making psychoanalysis itself a mere rationalization instead of a body of objectively true scientific knowledge.

3. Therefore, as I proceed I shall not hesitate to present a psychological analysis,—an account of man as a scientific knower,—which I shall expect the Freudian to accept, not only because he is a scientist, but in his capacity as a psychologist as well.

III. The Kinds of Knowledge [4]

1. There are two basic ways in which one body of knowledge can be distinguished from another:

[3] The intellect is denied whenever it is held to be indistinguishable in operation from sense and imagination, or is treated as completely reducible thereto. The modern attempt to identify idea, the product of intellectual abstraction, with image, the product of sensitive memory, always fails. The "abstract image" is as self-contradictory as the "round square." Vd. Locke, *An Essay Concerning Human Understanding*, Bk. II, Ch. 2, ¶ 9, 10; Bk. III, Ch. 3; Bk. IV, Ch. 6, 7, 17; Berkeley, *The Principles of Human Knowledge*, Introduction, ¶ 6–27; Hume, *An Enquiry Concerning Human Understanding*, Sections II, XII.

[4] For the fullest discussion of the distinctions indicated in this section of the lec-

a. By *subject-matter*, and here
   (1) Either in terms of *that which* is known: the knowable *thing* or material object of knowledge.
   (2) Or in terms of *what* is known of it: the thing *as known* or formal object of knowledge; the cognitive selection, the content of the knowledge.

b. By *method*.

c. All other distinctions,—such, for instance, as that made in terms of the modality of the knowledge (its certainty or probability),—can be ignored for the time being as subordinate.

2. But first we must consider what is common to *almost* all human knowledge.

a. Almost all human knowledge is a product of the cooperative activity of the senses and intellect, as these are involved in processes of observation and reflection.
   (1) By sense, or the powers of observation, I mean all those perceptual abilities which are exercised through the activity of bodily parts called sense-organs and brain.
   (2) By intellect, or the powers of reflection, I mean the following abilities: abstraction (the ability to conceive, define, distinguish, generalize, make inductions and analyses); judgment (the ability to formulate propositions and to assert them); reasoning (the ability to infer and demonstrate, to systematize knowledge deductively.)

b. The one exception is important: the kind of knowledge we call immediate experience,—and perhaps, also, the memory of such experiences,—*seems* to be purely sensitive.
   (1) No new reflective process is involved here in the act of knowing.
   (2) But even here there is an involvement of past reflections (the results of past intellectual activity) as evidenced by the fact that we cannot observe or remember without some interpretation. (This is

ture, see J. Maritain, *Les Degrés du Savoir*, Paris, 1932; vd. also my mimeographed *Analysis of the Kinds of Knowledge*, at the University Press Bookstore, Chicago, 1935; and *Art and Prudence*, New York, 1937: Ch. 9 on knowledge and opinion. This chapter is a refinement as well as a condensation of the earlier analysis.

what Kant meant by saying that percepts without concepts are blind.)

(3) This exception need not concern us now. We shall return to it in a later analysis of the observational process itself, in order to determine what are the "pure data" of scientific observation.

c. What is crucially important here is the basic proposition: *All human knowledge arises from the operation of the senses, and most human knowledge goes beyond sense, i.e., has a reflective development.*

(1) The first clause denies that ideas are primary objects of knowledge, capable of being directly or intuitively known by the intellect (supposedly the error of the Platonists); it must also be interpreted as denying innate ideas (the Platonic error made by Descartes and repeated in a somewhat disguised form by Kant), and the *a priori* in any of the absurd constructions which have been straw-men attacked by modern empiricism.

(2) The second clause asserts that although *all* human knowledge depends initially upon the activity of the senses, it does not depend exclusively thereon. It depends also upon intellectual activity, abstracting ideas from the presentations of sense. (Ideas are neither innate nor intuited, but the products of abstraction.)

d. The fact that most human knowledge goes beyond sense can be made clear simply by the limitations of the senses themselves.

(1) The senses apprehend only this particular thing or its qualities. This is true also of sensitive memory and imagination.

(2) But man is able to apprehend kinds of things, of which the particulars are recognizable instances; and man is able to conceive or to construct conceptually what is not observable at all by sense, even though the construction depends upon sense-data of some sort.

(3) Therefore we can say that understanding goes beyond sensation in grasping universals, knowing kinds, conceiving and constructing entities not observable. All of this depends upon the first intellectual act, the act of abstraction.

(4) If this were not the case, if man could not abstract as well as sense,—and the intellect is *primarily* the power of abstracting,—then induction and generalization would not be possible, and we could not distinguish between science and history, or even between history and aesthetic experience.

3. *The distinction between science and history.*

a. By historical knowledge is here meant the conclusions of historical research, as by scientific knowledge we mean the conclusions of scientific research. It is necessary to say this in order to exclude the interpretation of historical fact which is made in the course of historical (i.e., narrative) writing.

b. The historical proposition is singular: it is always about this particular thing, this change, this event, this agent, this action, etc.

(1) This does not mean that historical knowledge is purely sensitive. Historical research involves inferences and hence general propositions. Moreover, the historical proposition usually reports knowledge of things not immediately experienced. Finally, the particulars we know historically are known as instances of kinds.

(2) It means only that the historical proposition is like sensitive knowledge in that it refers to particular things.

c. Scientific knowledge, considering now only the conclusions of scientific research, consists of general propositions: either propositions which are first generalizations, that is, inductions from the data of sense-experience, or derived generalizations, propositions which have been either inferred (i.e., deduced) from the primary inductions, or otherwise elaborated.

d. In both history and science, observation is primary. In history, the basic evidences are discovered by observation, and from these the inferences to unobserved facts are made. In science, also, the basic evidences are matters of observation, and from these the primary inductions and further proofs are made.

e. In both kinds of knowledge, *that which* is known are the things of the physical world: physical things in change. Both kinds of knowledge have the same material object. The two kinds of knowledge can, however, be distinguished by reference to *what* is known of this common object:

(1) The formal object of historical knowledge is the thing in its particularity.

(2) The formal object of scientific knowledge is the thing in its universality.

There is a further distinction between history and science. The historian is primarily concerned with questions of existence,—whether or not his material object, a certain particular, exists, or existed. But the scientist is never concerned with the existence of his material object, physical things in general. He does not question the existence of the phenomenal order. (A full understanding of these distinctions depends not only upon a grasp of the distinction between sense and intellect, but also upon the comprehension of basic metaphysical distinctions between the form and matter, the essence and accidents, of *entia naturae*, as existing and as objects of knowledge.)

4. *The distinction between science and philosophy.*

a. They are alike in one essential respect as knowledge; and this differentiates them both from history. They are both knowledge of things in their universality.

(1) Their propositions, their conclusions, are general propositions.

(2) In both, the primary inductions,—on which everything else rests,—are inductions from sense-experience.

b. To the extent that they have the same material object, —the things of the physical world,—which is the proper or cognate object of all human knowledge, they cannot be distinguished in subject-matter. (Nor can we, at this point, completely distinguish them in terms of their formal objects. We shall return to this principle of distinction later, at 5. *infra.*)

c. They must be differentiated, then, by method. This differentiation can be made only with respect to the activity of the senses and not with respect to the activity of the intellect. In other words, philosophy and science do not differ methodologically in their reflective dimensions. In both induction is the same, judgment is the same, analysis is the same, proof is the same, etc. Hence, their difference in method must be in the observational dimension. This difference can be stated in terms of the difference between common and special experience [5].

(1) *Common experience:* perceptions not produced by, or in any way resulting from, processes of research or investigation, but rather arising from the normal operation of the sense-organs in the ordinary life of the animal. Common experience is the sense-experience of the ordinary man. It is constituted by undirected, unmethodic observations.

(2) *Special experience:* sense-data resulting from the special operations which we shall call "investigation," i.e., all the motions and manipulations, with or without apparatus and instruments, of men *seeking* and *doing something* for the sake of observing. The experience is special in the sense that the scientist differs from other men in what he does in order to observe, and therefore what he observes is different. Thus, we have the experience of the laboratory (usually with apparatus), of field research, of the clinic, etc.

(3) It is because investigation is a special kind of doing, a kind of physical activity or operation, that we speak of the *data* of science, the special observations resulting from such doing. We do not speak of the data of philosophy since the experience that philosophy reflects about is the common experience of all men. But the scientist reflects about his special data. This must not be interpreted to mean that he can ignore the common experience of all men; it means only that that is insufficient for him; and, therefore, he investigates.

(4) *Digression:* This difference is made clear by the analogous difference between investigative and non-investigative history. The knowledge which one has of one's own life as a matter of personal memory is non-investigative history; almost all other history is investigative.

(5) *Terminology:* We shall use the words "science" and "philosophy," without qualification, to name investigative and non-investigative knowledge, which

has generality. We can also speak of the investigative and the philosophical sciences.[5]

d. This distinction between philosophy and science explains a number of things that are generally recognized about their difference [6]:

(1) Their different relations to common sense.

(2) Their different degrees of change and progress, and the conditions of progress in the two fields.

(3) The difference between the locus of contingency and the possibility of error in the two fields.

(4) The utility of science and the inutility of philosophy, *in one meaning of utility*, namely, the use of knowledge to direct operations for the alteration or change of things, which is commonly called prediction and control. There is another meaning of utility: i.e., the use of knowledge to determine means and ends in human conduct, in which sense philosophy is practical, as in ethics and politics. But in the first sense philosophy is not useful. Metaphysics and mathematics, both of which are philosophical knowledge, illustrate this negative point. There is no metaphysical or mathematical engineering. Science, on the other hand, can become practical or applied in the technological sense because the knowledge it has achieved results from doing with respect to things and hence can again lead to doing with respect to things [6a].

e. Finally, we can understand why there *should* be no conflict between philosophy and science, no rivalry, but rather good order and cooperation.

(1) The reason is that special experience does not *alter* common experience; it merely *supplements* common experience with respect to details. But philosophical truths are the highest generalities of common experience. In respect to the sort of data phi-

[5] In the 17th century, the word "philosophy" was the commonly accepted name for knowledge which had generality, in contradistinction to history. Newton distinguished himself by taking the name of "experimental philosopher." The word "science" today, as it is commonly used, thus denotes what the phrase "experimental philosophy" did in the 17th century. The basic distinction between philosophy and science remains the same whether we name the kinds of knowledge distinguished as "abstract philosophy" and "experimental philosophy," or as "philosophical science" and "investigative science." The criterion of the distinction is method: the employment of investigative procedures.

losophy finds in common experience it is adequate. Hence, inductions properly made from common experience are not affected by the findings of investigation, as the latter, in turn, cannot be forecast by philosophy from its limited data. Nor can they be judged by philosophy so long as they are properly restricted. If the reflective processes are properly disciplined in both fields, scientific and philosophical knowledge must be compatible and complementary [6b].

(2) This does not mean that actually there have been or are no conflicts. There are always bad philosophers and bad scientists who go beyond their fields, who transgress upon domains not their own, and when this happens there is a bad order and antagonism instead of cooperation and harmony.

(3) The occurrence of such conflicts in modern times is a significant comment on the lack of intellectual discipline in this period. Whenever there is a conflict between science and philosophy, one or the other is out of order [7].

5. We can now differentiate the formal objects of philosophical and scientific knowledge, as before we distinguished the formal objects of historical and scientific knowledge.

a. The proper or cognate object of all human knowledge is the physical world: the order of physical things as sensible and intelligible. This is the case because only corporeal things are sensible, directly or indirectly, and all human knowledge involves the senses.

b. But, as we have seen, most human knowledge involves the intellect as well, and therefore we can distinguish, analytically, the proper objects of sense and intellect:

(1) The proper object of sense: the thing as particular; as *this*

(2) The proper object of intellect: the thing as universal; as *such and such*.

c. Within the sphere which is common to both science and philosophy, distinction can be made between two cognitive aims: the aim of science is to know what is universally observable about the nature of things; the aim of philosophy is to know what is universally intelligible about their nature, their substance and causes.

(1) The formal object of the natural sciences is the phenomenal order. Things are observable in their operations and changes. The phenomenal order is the changing surface of nature. Science studies this surface: it classifies, correlates, systematizes, measures the given phenomenal diversity.

(2) The formal object of philosophy is the noumenal order. Things are intelligible in the principles and causes of their being. The noumenal order is, in part, the substance of sensible nature. (We shall presently see that the noumenal includes objects which are purely intelligible.) Philosophy penetrates the phenomenal surface to the intelligible constitution of sensible things.

(3) This distinction in the formal objects of science and philosophy does not abrogate the point that both are intellectual knowledge, i.e., knowledge of things in their universality; nor does it violate the basic principle of human cognition, that intellectual knowledge is based upon the data of sense-experience. (The accidents which constitute the phenomenal order are both sensible and intelligible. Although the substance of things is, strictly, not sensible, the process of inference by which it is intellectually known originates from a knowledge of the accidents.)

d. Since what is characteristically human knowledge involves both sense and intellect, we must note

(1) The infra-intelligible object (below the intellect): the unique individual thing, known as *individual* and not as a particular instance of such and such a kind. (This object is always physical.)

(2) The supra-sensible object (above the senses): nonphysical things, such as souls separately existing, angels, God. (Such objects always transcend the physical order.)

(3) Since neither of these two objects are proper objects of human knowledge,—although both, let it be added, are intrinsically knowable, the latter more so than the former,—neither are known adequately. The infra-intelligible object is known

probably. The supra-sensible object is known *analogically*, by *a posteriori* modes of remotion from experience [8].

e. What further distinguishes science from philosophy is that it never seeks to know supra-sensible objects; in fact, it is prevented from doing so by its method, which requires that its objects be subject to human investigation. Philosophy, not bound by the limitations of the investigative method, can at least *seek* knowledge of transcendent objects, but in so doing it is, like science, bound by conditions common to all human knowledge: it must do so *a posteriori*, that is, it must learn about such objects from sense-experience, both with respect to their existence and their natures [8a].

f. Neither science nor philosophy, nor even history in the usual sense of that term, seeks to know the infra-intelligible object. That is the field of aesthetic or immediate experience.

6. In order to complete the distinction of philosophy and science, and to state their relation, we must indicate the formal divisions of philosophy.[5a] It should be noted that the fields of philosophy are distinguished formally in terms of grades of abstraction; whereas, we shall see, the different sciences, all of the same grade of abstraction, can be distinguished only materially. There are three grades of abstraction. Thus:

a. *Physics:* knowledge of physical things as such: i.e., with all their accidents and as changing. The range of physics includes everything from astronomy through chemistry and biology to human physiology and even psychology, in so far as man is a physical thing. It is this range of subject-matters which requires a careful formulation of the relation between philosophical physics and the special natural sciences.

Mathematics: knowledge of physical things not as such,

[5a] We are here concerned only with the divisions within speculative or theoretic philosophy. Theoretic philosophy as a whole must be divided against practical philosophy (ethics, politics, poetics), and these two together must be distinguished from the organon of knowledge (the liberal or speculative arts of grammar, logic and rhetoric) which it is the task of the philosopher to formulate. The Stoic tripartite classification of philosophical disciplines (logic, physics, ethics) is accurate if logic be taken as representing all the liberal arts, physics the whole of theoretic philosophy, and ethics all practical philosophy.

but as determined in quantity and quantitative order. In other words, mathematics is an abstraction from physics. It considers only one aspect of physical things, their quantity due to their matter, and apart from their changes, motions and other accidents.

Metaphysics: knowledge of physical things merely as being, abstracted from all special determinations due to matter and motion. In short, metaphysics is concerned with whatever is *as it is*, and thus extends from the knowledge of the being of physical things to the knowledge of the being of anything, both as possible and as actually existent. Metaphysics considers both physical and transcendent objects. The conditions of becoming presuppose the conditions of being, and the order of becoming presupposes an order of being both in and apart from the conditions of becoming.

b. Another way of saying this is: metaphysics is about being abstracted from any special determinations; mathematics is about being as determined by quantity and order but abstracted from the material conditions which determine change, and from change itself; and physics is about being as materially determined in the order of becoming [9].

c. Of these three philosophical subject-matters, distinguished not by that which is known but by what is known of that which is known, only one can be shared in common by philosophy and science, because only one is capable of supporting investigative procedures. That one obviously is physics.

(1) First, because only physical things *as such* are sensible and operable, and investigation can yield knowledge only of sensible, operable things.[6] Thus, the mathematical and metaphysical objects,—being abstractions from the physical thing *as such* and hence neither sensible nor operable,—are not capable of investigation in the sense defined. There is, in short, no empirical, experimental or investiga-

[6] A science can be experimental, in the strict sense of that term, only if its objects are both sensible and physically operable. If its objects are sensible but not physically operable, as in the case of astronomy, the science can be investigative or empirical, but not experimental.

tively observational mathematics or metaphysics [10].

(2) Second, because common experience is sufficient for both mathematics and metaphysics. Neither need to have more experience, as a source of reflective development, than that which the least inquiring ordinary man possesses as the result of his ordinary observations.

d. It must be added at once that although physics *can be investigative, it need not be*. Physics names a subject-matter that is common to both philosophical and scientific knowledge. Whereas physics may be either philosophical or investigative knowledge, all the sciences, all bodies of investigative knowledge, must be branches of physics.

(1) It is proper to identify science in its modern meaning with *natural* science, because science is investigative knowledge of natural, that is, changing things. Whether social phenomena are in a strict sense natural and can be proper objects of science need not be discussed here. If they are, then the social studies may some day become natural sciences. If they are not, then the social studies will never become scientific.[7]

(2) I shall refer to philosophical physics,—the philosophy of nature, the knowledge man has of the physical world by reflection about common experience,—as general physics (or as philosophy of nature); and to scientific physics, which includes the entire range of the particular sciences of nature resulting from the investigation of particular kinds of bodies, particular kinds of change, etc., as the particular physics. There is only one general phys-

[7] Societies are not natural substances, but as systems of phenomena they are not different from other phenomenal entities. If science were knowledge of the substantial order, social science would be impossible. But science is knowledge of the phenomenal order, and social science of either a non-mathematical or a mathematical sort can be constructed in the same way as any other natural science. The possibility of social science seems indisputable. Whether such science exists in any adequate development is another question. Vd. Notes 9 and 47 *infra*. The problems of social science must not be confused with those of social psychology; the latter tries to solve the philosophical problem of the causes of human behavior. Vd. my discussion of this point in *Art and Prudence*, New York, 1937: Ch. 9.

ics, or natural philosophy, and many particular physics, or natural sciences.

e. One further point is important. Since physics deals with changing things and with change itself, it treats of causes, the conditions of change; whereas mathematics, having abstract objects, i.e., quantities and orders apart from change, has nothing to do with causes.[8] It should be added here that the philosophy of nature treats of causes theoretically, while the natural sciences are concerned with causes only practically.[8a]

7. *Summary:* Physics, as the name for a domain of knowledge, a subject-matter determined by the lowest grade of abstraction, is both philosophical and scientific [10a].

a. *As philosophical,* it is what Aristotle developed in the eight books of the *Physics,* and in some part also the books of the *De Caelo* and the *De Generatione et Corruptione.* These books contain discussions of the causes, principles and elements of all change, the types of change, the nature of matter, space, time, infinity, and the types of physical existence. Such discourse is a philosophy of nature and constitutes an introduction to all of Aristotle's particular scientific works, his astronomy, his mechanics, his biology, etc.

b. *As scientific,* it consists of the whole range of investigative natural sciences, differentiated from each other materially by the kind of change or the kind of body (i.e., the selected phenomenal area) being investigated. (This differentiation is primarily in subject-matter; subordinate distinctions can be made in terms of method.)

(1) Thus, according to the kind of change: local mo-

[8] This is perfectly clear in Galileo's *Two New Sciences* (trans. by De Salvio and Favaro), New York, 1914. He conceives his study of acceleration as the kind of science "where mathematical demonstrations are applied to natural phenomena, as is seen in the case of perspective, astronomy, mechanics, music, and others, where the principles once established by well chosen experiments, become the foundation of the entire superstructure" (p. 178). Such a science, i.e. any mathematical physics or applied mathematics, is not concerned with causes, except in so far as causes are involved in the experimental operations. Galileo writes: "The present does not seem to be the proper time to investigate the cause of the acceleration of natural motion concerning which various *opinions* have been expressed by various *philosophers.* . . . Now, all these *fantasies,* and others too, ought to be examined; but it is not really worth while. At present it is the purpose of our Author merely to *investigate* and to *demonstrate* some of the properties of accelerated motion" (pp. 166–167). Italics mine.

[8a] Vd. Note 47 *infra.*

tion (mechanics); generation, growth, and decay (biology).

(2) Or, according to the kind of body: celestial or earthly, inanimate or animate (astronomy or geology, chemistry or botany).

c. The history of natural science is a history of increasing particularization of subject-matters and the separation of the particular physics. For our present purposes, we shall consider only the main grouping of these particular sciences into the physical (inanimate bodies) and the biological (animate bodies). We shall return to this point later.

d. It is clear from this analysis that general physics is illuminated by metaphysics because whatever holds for being in general must hold for beings which change. The particular natural sciences may, however, be informed by mathematical principles, thus giving rise to what the ancients treated as the mixed (or hybrid) science of mathematical physics, or mathematics applied to nature by measurement [11]. But, whether mathematical or not, these sciences presuppose the principles of philosophical physics, because whatever holds for any change and any body, must hold for particular types of change and particular kinds of bodies.[9]

e. In order to make this last point clear, in order to understand the relation of the natural sciences to philosophy (to general physics, to mathematics and to metaphysics) we must now consider the history of the natural sciences.

IV. THE HISTORY OF THE NATURAL SCIENCES

1. In the ancient world there was no clear separation of science and philosophy according to method.

[9] Vd. J. Maritain, *La Philosophie De La Nature*, Paris, 1935. This dependence is clearly illustrated in the works of Newton and Galileo. Thus, in the opening discussion of the Third Day of the *Two New Sciences*, Galileo presupposes the distinction between local motion and other types of changes, and the distinction between natural and violent motion (*op. cit.*, pp. 153–54). These distinctions he learned and accepted from the traditional philosophy of nature, i.e., general physics. They are to be found, of course, in Aristotle's *Physics*, Bk. I, II, and *De Generatione et Corruptione*, Bk. I. The presupposition of an analysis of the principles, causes and elements of change, and of the distinction of the kinds of change, is even more obvious in works of empirical biology.

a. Aristotle's *Physics* is a work in general physics and is purely philosophical.

b. Aristotle's astronomical and biological works are works in particular physics. Being non-mathematical, they depend for their first principles upon general physics, and to this extent they are philosophical. But to apply the principles to particular types of physical phenomena, there is a large addition of investigative knowledge based upon special observations. To this extent they are scientific works. (*Note:* This composite character of ancient physical writings,—partly philosophical in method and partly scientific,—requires great care in the modern reader, who must distinguish philosophical principles from scientific knowledge of particular matters of fact. In the latter sphere it is to be expected that the ancients would make many errors; but these errors do not affect the validity of the principles.) Galen is another good example of such a composite work. In his treatise *Of Natural Faculties*, the principles of general physics are applied by investigation to human physiology.

c. Mixed or intermediate sciences: applied mathematics, such as astronomy of the Alexandrian sort, mechanics (Archimedes), harmonics (Aristoxenus), etc. These sciences are mixed in two senses:

   (1) In subject-matter: They are mathematics *formally* and physics *materially*.

   (2) In method: They are philosophical (in their mathematical dimension) and scientific or investigative (in their physical dimension).

d. The mixture to be found in mathematical physics is different from the mixture of philosophy and science to be found in Aristotle's non-mathematical physical works in particular fields. In the latter there is no mixture of subject-matter, but only a composition of two sorts of knowledge, of the same grade of abstraction, but obtained by different methods.[9a]

2. In the modern world there has been an ever sharper separa-

[9a] Aristotelian science was bad as science for a more fundamental reason than imperfect data or wrong conclusions. Innocent of the distinction between the formal objects of philosophical and scientific inquiry, Aristotle employed a substantive logic throughout and failed to see that data of investigation could yield purely relational generalizations. Vd. Notes 2 and 47 *infra.*

tion of the sciences from philosophy in terms of their investigative methods. (In fact, the great achievement of the modern world has been this separation of science from philosophy, and the understanding of its autonomy.)

a. Mixed science in the modern world.

(1) Galileo: *Two New Sciences.* (This work is of the same type as Newton's *Principia Mathematica*, the work of Clerk-Maxwell, Einstein, etc.) Analysis:

(a) Philosophical physics is presupposed.[10]

(b) The application of mathematics: the requirement of investigation in order to make metrical observations, and the need of measurement for the sake of applying mathematics.[11]

(c) By means of experiments: the construction of an approximation to the ideal to show that the mathematically possible is physically actual; also for the sake of precision in measurement. But there is a limited use of experiments. They are used to establish the definitions, from which all else follows deductively.[12]

[10] Vd. fn. 9 *supra.*

[11] "First of all it seems desirable to find and explain a definition best fitting natural phenomena. For anyone may invent an arbitrary type of motion and discuss its properties; thus, for instance, some have imagined helices and conchoids as described by certain motions which are not met with in nature, and have very commendably established the properties which these curves possess in virtue of their definitions; but we have decided to consider the phenomena of bodies falling with an acceleration such as actually occurs in nature and to make this definition of accelerated motion exhibit the essential features of observed accelerated motions. And this, at last, after repeated efforts we trust we have succeeded in doing. In this belief, we are confirmed mainly by the consideration that the experimental results are seen to agree with and exactly correspond with those properties which have been, one after another, demonstrated by us" (*op. cit.*, p. 160). It should be noted that the demonstration here referred to is not experimental but mathematical. Experiment is used not to demonstrate the truth of theorems but to test the agreement of a mathematically coherent analysis with observed nature.

[12] Vd. fn. 8 *supra.* This limited use of experiments,—to find observable exemplifications of an analysis,–is indicated by a number of passages. Thus, at one point Galileo says: "Without depending upon the above experiment, which is doubtless very conclusive, it seems to me that it ought not to be difficult to establish such a fact by reasoning alone" (*op. cit.*, p. 164). And at another: "Whether this acceleration is that which one meets in nature in the case of falling bodies, I am still doubtful; and it seems to me, not only for my own sake, but also for those who think as I do, that this would be the proper moment to introduce one of those experiments–and there are many of them, I understand–which *illustrate* in several ways the *conclusions reached*" (*op. cit.*, p. 178). Simplicio is the

(d) *Digression:* such mathematical physics is not inductive. But the relation between the conclusions and the data is nevertheless the same. This can be shown by supposing apparatus which Galileo did not have [12].

(2) The works of Kepler and Copernicus are examples of applied mathematics without experiments but with a tremendous amount of metrical data.

b. Inductive, non-statistical and non-mathematical physics. Newton's *Optics*. (The work is of the same type as that of Boyle, Rumford, Faraday, Kelvin, etc.) It is typical *experimental*, as opposed to *mathematical*, physics. Analysis:

(1) Difference in the use of definitions and in the meaning and use of axioms.[13]

(2) The experiment is used as the source of the induction, not as a means by which to apply mathematical formulae. It is a construction of the ideal case in order to be able to see the universal in the particular by abstraction of essence from accidents, i.e., the separation of relevant conditions from the irrelevant.

(3) There is both measurement and mathematics in the *Optics*, even though it is not applied mathematics, but pure physics. The mathematics is used for the sake of precision of formulation.

(4) There is "speculation" about causes,—the real na-

speaker at this point in the dialogue. The italics are mine to indicate that Simplicio has admitted the demonstration of the theorem, by mathematical reasoning, and asks to have its applicability to nature shown, by experiment. Cf. fn. 8 *supra.*

[13] Newton's *Principia Mathematica* is a work of the same sort as Galileo's *Two New Sciences*. It is applied mathematics. This can be seen in the way in which the mathematical demonstrations come first, and the tests of applicability to nature by reference to selected phenomena follow. In such works, the axioms and definitions are used probatively to demonstrate the theorems. But in the *Optics*, the definitions and axioms are not the premises whereby Newton demonstrates his theorems. Each of the theorems is an experimental induction. Of the axioms Newton says that they are "the sum of what hath hitherto been treated of in Opticks. For what hath been generally agreed on I content myself to assume under the notion of Principles, in order to what I have farther to write. And this may suffice for an Introduction to Readers of quick Wit and good Understanding not yet versed in Opticks: Although those who are already acquainted with this Science, and have handled Glasses, will more readily apprehend what followeth."

ture of light, the cause of its properties, etc.,—whereas Galileo, being a mathematical physicist, excluded causes.[14] (Here note the dependence of a particular physics on general physics. Newton unfortunately made the error of accepting Greek atomism as his "philosophy of nature.")

c. Mendel's work in genetics, and subsequent work by Morgan and Bateson. These works are examples of inductive science, which is non-mathematical, and yet may be either metrical or non-metrical, statistical or non-statistical. Even when statistical methods are employed, the conclusions are not statistical generalizations but genuine inductions. The experiments in this field are clearly different from the experiments in Newton's *Optics* and are, for the most part, merely devices for controlling the phenomena for the sake of special observation.

d. Lyell's *Principles of Geology* is non-experimental, and inductive without being statistical. The inductions state secular trends which are typical of all life-history patterns.

e. We can briefly summarize the foregoing by the following classification of all investigative natural sciences:

(1) Mixed sciences: applied mathematics. Either experimental or not. Thus, astronomy is not experimental; mechanics is.

[14] It should be noted, however, that the speculation about causes does not occur in the main body of the treatise, but rather at the end of the work in a series of queries, to stimulate, as Newton says, "a farther search to be made by others." See particularly, Book III, Part 1, Query 31, in which Newton propounds the atomic hypothesis for the causal explanation of natural phenomena. It is here also that Newton discusses scientific method: "As in Mathematics, so in Natural Philosophy, the Investigation of difficult Things by the Method of Analysis, ought ever to precede the Method of Composition. This Analysis consists in making Experiments and Observations, and in drawing general Conclusions from them by Induction, and admitting of no Objections against the Conclusions, but such as are taken from Experiments, or other certain Truths. For Hypotheses are not to be regarded in experimental Philosophy. And although the arguing from Experiments and Observations by Induction be no Demonstration of general Conclusions; yet it is the best way of arguing which the Nature of Things admits of, and may be looked upon as so much the stronger, by how much the induction is more general. . . . This is the Method of Analysis: And the Synthesis consists in assuming the Causes discovered, and established as Principles, and by explaining the Phaenomena proceeding from them, and proving the Explanations." Cf. Query 28.

(2) Purely physical sciences: essentially non-mathematical, although they may involve mathematics, measurements, and statistics in some subordinate use.

(a) They may be experimental or not. Thus, geology is not experimental; genetics is.

(b) They may be statistical or not. Either the generalizations are primary inductions or they are only statistical generalizations. The statistical sciences are quasi-mathematical.

(c) With respect to the way in which they describe the phenomena they study, they are:

(1) Taxonomic (merely classificatory)

(2) Physionomic (secular trends; life histories

(3) Analytic (quasi-aetiological).

Note: Mathematical physics is analytic, but not in the same way as the non-mathematical sciences. In either case, however, a practical knowledge of causes is involved [13].

f. In the light of this summary it can be seen that there is a rivalry between two philosophical fields for the control of the particular natural sciences: mathematics (which does not consider causes), on the one hand, and general physics (which is primarily aetiological), on the other. In fact, the story of modern science can be told in terms of this rivalry. The particular natural sciences must get their principles somewhere, since they have a limited scope of investigation. They borrow their principles either from mathematics or from the philosophy of nature. There are, as a result, two sorts of sciences and accordingly two different philosophies of science, i.e., two different conceptions of the nature, aims and scope of science. I shall now attempt briefly to summarize this.

V. The Relation of the Several Natural Sciences *Inter Se* and to Philosophy

1. The view one takes of the relation *inter se* of the particular natural sciences depends upon one's metaphysics and philosophy of nature (general physics).

a. Thus, in terms of an atomistic or Cartesian philosophy, all the physical sciences are reducible to a basic mathematical physics like mechanics, because the only properties of matter are quantitative and these are the properties to which mathematics is applicable. The mixed kind of physics (mathematical physics) looks forward to the day when all the other natural sciences will be reduced to it.

b. In terms of an Aristotelian philosophy,—and this is the only other alternative,[15]—there is an acknowledged and irreducible heterogeneity of the physical sciences, and the non-mathematical sciences are not regarded as inferior. In fact, they are more purely physical than the mixed sciences of applied mathematics. The systems of phenomena which constitute the subject-matters of the several sciences are based on deeper distinctions in the nature of physical things. The particular natural sciences can, therefore, be arranged according to the kinds of bodies studied.

(1) The primary distinction would be between the animate and the inanimate.

(2) Subordinate distinctions would be made within these two realms: thus, within the inanimate, mechanical and chemical systems; within the animate, biological, psychological and social systems. This denies continuity and reducibility in nature in one sense and affirms it in another, in so far as the philosophy of nature and metaphysics provide principles which are common to all the particular sciences of nature.

2. The view one takes of the relation between the sciences and philosophy depends, of course, upon one's conception of philosophy.

a. In terms of many contemporary conceptions of philosophy,—the positivist's, the pragmatist's, the sociologist's,—philosophy is at best the handmaiden of the sciences.

(1) It is not a body of knowledge having a content and validity independent of the results of scientific research.

(2) It performs one or another literary or rhetorical function, which is ancillary to the acquisition of knowledge by the scientific method [14].

(3) Like religion, it satisfies emotional needs, which a scientist is able to conquer or deprive. The pursuit

[15] Vd. Lecture II *infra*.

of philosophy is the expression of a persistent human desire to know the world more fully and more profoundly than it can be known by science. But this desire must be fruitless if the only genuine knowledge is scientific. Dissatisfaction and impatience with the achievements of science is the regrettable emotional source from which philosophy,—the illusory quest for ultimates,—perpetually arises.[16]

b. In terms of a conception of philosophy which views it as the tradition of wisdom, prior logically, as well as historically, to the accomplishments of science, the sciences are at best the servants of philosophy.

(1) Philosophy rules the sciences negatively.

(a) It restricts them to their proper spheres, and governs them justly in that it recognizes their autonomy within their own domains.

(b) In other words, it does not dictate the conclusions of scientific research, but it does require scientists to confine their "interpretations" to whatever can be concluded from their limited data. This negative rule is unfortunately often violated by modern scientists, particularly when they give Gifford lectures or win Nobel prizes [15]. It must be added that many modern philosophers have also exceeded their powers and have attempted to answer questions which should properly be reserved for scientific investigation. Such dogmatic and speculative excesses on the part of modern philosophers, both in the 17th and the 19th centuries, are in large part responsible for the bad reputation which philosophy has in the modern world [16].

(c) That the sciences cannot rule themselves is simply indicated by the fact that any discussion of science, its methods, its scope, the nature

[16] This is the psychoanalyst's conception of philosophy. Vd. S. Freud, *op. cit.* in fn. 2 *supra.* It is shared, of course, by sociologists and anthropologists. Vd. K. Mannheim, *Ideology and Utopia,* New York, 1936; A. Herzberg, *The Psychology of Philosophers,* New York, 1929; P. Masson-Oursel, *Comparative Philosophy,* New York, 1926. That the pragmatist is a positivist in his denial of philosophy as knowledge of the noumenal, and that he is a sociologist in his conception of philosophy as nothing more than the highest rationalization of local cultural values is clearly seen in Dewey's *Reconstruction in Philosophy* (New York, 1920: pp. 22–27). Vd. also *Experience and Nature,* Chicago, 1929.

and validity of its knowledge, raises questions which are not answerable by scientific method, but belong to the province of the logician and the epistemologist, i.e., the philosopher. In short, the question, What is the relation of science to philosophy? is a philosophical and not a scientific question.

(2) The sciences serve philosophy positively by affording it additional materials with which to exemplify its principles.

VI. Conclusion

1. Summary of the basic distinctions:
   a. Science and philosophy, together, are distinguished from history in terms of generality and particularity: different formal objects.
   b. Science and philosophy are distinguished from each other *negatively* in terms of method.
      (1) Both are observational; both are reflective;
      (2) But philosophy is *not* investigatively observational, whereas science is impossible without investigation.
   c. Science and philosophy are distinguished from each other *positively* in terms of their different formal objects, i.e., their different cognitive aims. They are essentially different types of knowledge.
      (1) The positivist is right in his conception of science as concerned exclusively with the phenomenal order, as descriptive rather than explanatory, as systematizing and correlating phenomena rather than as seeking to know the substance and causes of things.
      (2) The positivist is wrong only in his denial of another type of knowledge, namely, philosophy which is concerned exclusively with the noumenal order, which explains rather than describes, and which succeeds in knowing the substance and causes of things. The goodness of positivism can be separated from its denials and exclusions, its *negativism* in short.
      (3) For a modern audience, philosophy can perhaps be most easily defined by saying that it is precisely the kind of knowledge which the positivist declares to be impossible. It is an unfortunate,

though an understandable, circumstance that the achievement of a sound conception of science, largely due to modern philosophers, should be accompanied by a misunderstanding and even a denial of philosophy [16a].

d. The fields of philosophy,—the philosophy of nature, mathematics, metaphysics,—are formally distinguished *inter se* by their different grades of abstraction.

e. The particular natural sciences are all determined by the same grade of abstraction which defines the province of the philosophy of nature, but they are materially distinguished *inter se* according to the phenomenal area they select for investigation. These material distinctions divide the sciences according to subject-matters. But there are also quasi-formal distinctions among the sciences according to their principles, their methods and their aims. Thus:

(1) They are mixed or pure: consisting of mathematical principles applied to natural things, or of physical principles applied to natural things.

(2) They are experimental or not; and if experimental, the experiments are either for the sake of applying mathematics by measurement or for the sake of induction.

(3) They are statistical or not; and if statistical, statistics is employed either for the sake of testing or qualifying generalizations made otherwise, or for the sake of making generalizations (statistical inductions).

(4) They are analytic, physionomic or taxonomic.

Note: These various traits combine in many different ways.

2. Summary of the basic propositions:

a. All human knowledge arises with sense-experience, but most human knowledge goes beyond sense by means of intellectual processes (abstraction, judgment, inference).

b. All scientific knowledge and all philosophical knowledge, consisting of generalities, must involve both sense and intellect as cognitive faculties.

c. Because they are different types of knowledge, employing different methods, there can be no conflict between the evidence and proper conclusions of scientific research and the principles and proper conclusions of philosophical analysis.

(1) "Proper" here does not mean "true in fact" but rather "achieved by the proper method and within the proper province." Thus, there are conflicts between trespassing scientists and dogmatic philosophers in the modern world.

(2) If a question cannot be answered by a given method, answers given to it cannot be refuted by knowledge achieved through the given method. Thus, mathematics and metaphysics are concerned with questions which cannot be answered by any process of investigation; hence no knowledge achieved by investigative procedures can be used to refute philosophical answers to such questions.

3. The problem which remains for us is formulated by the question: What is the position of psychology as a branch of philosophy, on the one hand, and as a particular natural science, on the other?

   a. It is obviously a peculiar subject-matter. No other particular sphere of natural phenomena is the subject-matter of philosophical knowledge. In every other case, mechanics, chemistry, botany, etc., only investigative knowledge is possible.

   b. Yet it can be shown that psychology is, in one sense, a particular branch of natural science, and, in another sense, a very special philosophical discipline. The positivistic denial of philosophy as knowledge of the real is, therefore, crucially important to psychology. If, as I shall try to show, the basic questions in psychology are entirely philosophical, then the attempt to construct a science of psychology on positivistic principles is doomed to failure and to perpetuate the misguided polemic of the "schools."

4. The themes of the lectures to follow are determined by the foregoing analysis, the insights it yields, the questions it raises. They are in order:

Lecture Two: THE POSITION OF PSYCHOLOGY AMONG THE SCIENCES AND IN PHILOSOPHY
Lecture Three: THE HISTORY OF PSYCHOLOGY
Lecture Four: THE POSITION OF PSYCHOANALYSIS

LECTURE TWO

THE POSITION OF PSYCHOLOGY: IN PHILOSOPHY AND AMONG THE NATURAL SCIENCES

I. Introduction

1. However one conceives psychology, it is clear that the psychologist has a double status: on the one hand, he is a scientist or philosopher who is a special student of man (human nature); on the other hand, he is a psychologist who must give an account of man which accounts for himself as a man who is either a scientist or a philosopher. In short, the psychologist is himself the subject-matter of psychology as a field of knowledge.
   a. This double status of the psychologist as *knower* and as *object known* is related to the reflexivity of psychological knowledge. And this reflexivity imposes a peculiar burden on the psychologist. He cannot, as scientist or philosopher, claim to have knowledge of a certain sort, and then as a psychologist deny that men are capable of such knowledge.
   b. It is important to note here that the analysis of the kinds of knowledge, made in the first lecture, was philosophical, i.e., not the result of scientific research or investigation; that that philosophical analysis involved psychological considerations about man's cognitive powers and operations; and hence that there is psychological knowledge which is not scientific but philosophical. In other words, there is a philosophical psychology.
   c. The philosophical analysis of the kinds of knowledge and the nature of science included, at least, the two following propositions in psychology:

(1) There is a distinction between sense and understanding as powers of knowing, and between sensitive and intellectual knowledge as kinds of knowledge.

(2) Reason is autonomous. In both philosophy and science men assent to conclusions only because they are bound by evidence or premises to do so, and not as a result of their desires or passions. Men can reason as well as rationalize. Otherwise there is neither science nor philosophy. (Freud, in a recent lecture, recognized this point.)

(3) These two psychological points are crucial to the psychoanalyst in his double role as a scientist (or a philosopher sometimes) and as a psychologist.

2. At the conclusion of the first lecture, Dr. Alexander raised a number of questions which went immediately to the heart of the matter. His questions focus the chief issues for our present discussion. They follow:

a. Why all the reflective processes, attributed by me to a distinct and separate intellectual faculty, could not be reduced to, or treated as extensions of, man's sensitive powers and operations?

b. Whether such a reduction was not required by the physiology of cognition which showed that man's reflective processes were disturbed by brain pathology? (It was admitted, of course, that the brain is nothing but an elaborated sensorium plus complicated connections with a primary motorium. The essential structure of the brain is the same as that of the simplest spinal segment.)

c. Whether we must not consider reflective processes as serving biological functions, just as we consider the operations of sense in relation to the life of the animal?

d. Whether there is not perfect biological continuity, that is, whether man differs from brute animals essentially in *kind* or only accidentally in *degree?*

3. I must answer all of these questions in the course of this lecture because the answers are indispensable to a proper definition of the subject-matter of psychology and its place among the natural sciences.

a. But, first, I must make an inescapable phenomenological point. By phenomenology I mean a description of the facts, regardless of how they are explained. Thus, the answer to the first question phenomenologically must be made independently of the answer to the second question. In other words, physiology or neurology cannot determine the answer to the first question. The distinction between sense and intellect as an essential distinction, a distinction in kind, a distinction incapable of reduction, is made in terms of common experience.

(1) Sense and understanding are essentially different. The difference is that of the two intentions of the mind: the particular and the universal.

(2) This is best illustrated by mathematics, since mathematics is impossible if all knowledge is merely sensitive.

(a) First, geometry. There is a distinction between the defined object (a universal) and the drawn figure (a particular). No geometrical knowledge is about the drawn figure; nor can any sensitive knowledge of the drawn figure establish the simplest geometrical truth. For example, the truth that the number of diagonals in an n-sided figure is equal to $n(n-3)/2$, cannot be learned by counting the diagonals in any number of particular figures. It can be learned only by a process of demonstration employing universals. The examination of particular figures may, of course, be *suggestive* of this truth.

(b) Second, arithmetic. The obvious truth that any positive number over 100 multiplied by any positive number over 100 yields a product which is a positive number over 100, can be established only by intellectual processes. No sensitive evidence here is more than remotely relevant.

(3) The distinction is one which every one knows. It is the distinction between recognizing John Smith in a crowd (by sense) and understanding what a man is as distinct from a mouse (by intellect). We commonly distinguish the following items of knowledge: (1) that John Smith is; (2) what John Smith is, i.e., a man; (3) that men are in this room; (4) what men, in this room or elsewhere, are, i.e., rational animals [17].

b. I turn next to the physiology of cognition (because of Dr. Alexander's second question). This cannot possibly alter the foregoing distinction, although we shall have to understand the foregoing distinction in terms of our physiology.

(1) Physiology teaches us the distinction between the peripheral and central nervous system, between the organs of sense and movement and the organs of connection between senses, muscles and glands.

(2) The brain, in addition to being an elaborate system of connections, is the organ of imagination, because it contains the primary and secondary sensory projection areas. It is imagination which is an extension of sense, and it is imagination which is reducible to sense. Dr. Alexander's point applies properly to imagination but not to understanding.

(3) But the distinction between imagination and understanding is the same as the distinction between sensation and understanding. What is imagined is always a particular, even though what have been called "generic" or "abstract" images indicate the way in which the imagination contains the universal implicitly and potentially [18].

(4) I shall show, subsequently, that there is no bodily organ of understanding (no physiology of the intellect), but that, since intellect depends upon sensation and imagination (as providing the materials from which the intellect abstracts universal intentions), there is a physiological account that is relevant to the intellect. Thus, fatigue and brain lesions hinder understanding by hindering its indispensable condition, the activity of the imagination [19]. This does not reduce understanding to imagination any more than the fact that a fire needs oxygen reduces a fire to oxygen.

(5) The psychiatrist is right who refuses to surrender his subject-matter to the neurologist. The psychology of the organic psychoses shows the *essential*, not merely the *present*, inadequacy of brain physiology to account for psychological phenomena. (Neuropathology deals only with the material cause of insanity. It does not, and cannot, account for the individual content of the same psychosis in different individuals. This is a matter of psychogenesis.)

c. The remaining two of Dr. Alexander's questions I shall

answer later, since both depend upon the full demonstration of the answer I have given to the second, namely, that understanding is not directly physiological, but only indirectly in so far as understanding depends upon sense and imagination.

(1) I shall answer the third question by showing that the biological functions of sense and intellect are different; and

(2) I shall answer the fourth question by showing that there is a specific and essential difference between men and brute animals, animals lacking an intellect, though having intelligence of a sensitive and imaginative sort.

d. My answers to the first and second questions, with their consequent answers to the third and fourth, are fundamental propositions in philosophical, not scientific, psychology. I must, therefore, first consider the position of psychology in philosophy and among the natural sciences.

4. That I must do this at once is also indicated by the analysis made in the first lecture. That analysis showed a basic paradox about psychology as subject-matter, partly indicated by the reflexivity of psychological knowledge, and partly indicated by the following questions, which can be asked about psychology and only about psychology:

(1) Is it, as a body of knowledge, a branch of philosophy?

(2) Is it, as a body of knowledge, a particular natural science?

(3) Is there a unity of these two bodies of knowledge?

a. These three questions cannot be asked about metaphysics and mathematics, on the one hand, since they are entirely philosophical in method; nor, on the other hand, can they be asked about mechanics, geology, electricity, genetics, qualitative chemistry, etc., since they are entirely scientific in method.

b. If the affirmative answer is given to these three questions, this fact will then be true only of psychology, and this will indicate something unique about its subject-matter, a uniqueness already somewhat indicated by the fact of reflexivity.

c. The affirmative answer *must* be given to these questions. To show this,—and particularly to answer the second of these questions,—we must understand the relation *inter se* of the particular natural sciences. To do this, we must first analyze the prevailing issue in what is currently called the "philosophy of science." We turn to this at once.

II. The Philosophy of Science

1. The "philosophy of science," i.e., the sort of discussion to be found in the writings of Poincaré, Cassirer, Meyerson, Mach, Whitehead, Russell,[1] deals with four leading topics:
   a. The character of scientific knowledge. (This raises psychological questions.)
   b. The principles of scientific method. (This raises logical questions, viz., the issue about predicational vs. relational logic.[1a])
   c. The aims of science: description or explanation. (This is the issue raised by 19th century positivism.[1b])
   d. The order and relation of the particular sciences in terms of their subject-matters, their objects of knowledge. (This raises metaphysical questions about the different kinds of *knowable objects.*)
2. It is only the fourth question with which we are here concerned, and we must, therefore, enter the realm of metaphysics in order to answer it. (It is important to remember here that metaphysics is knowledge based upon observation, and developed by reflection from common experience. It is open to argument in the reflective dimension only.)
3. There are two "philosophies of science" in modern times:
   a. The first position is that all the natural sciences are homogeneous with respect to their object, hence continuous, and hence ultimately all reducible to the basic science of that one common object.
      (1) This is the position taken by some physicists with

[1] Vd. T. Merz, *History of European Thought in the 19th Century*, London, 1896; W. Whewell, *Philosophy of the Inductive Sciences*, London, 1840; J. B. Stallo, *Concepts and Theories of Modern Physics*, London, 1890; H. Poincaré, *Foundations of Science*, New York, 1921; E. Cassirer, *Substance and Function*, Chicago, 1923; E. W. Hobson, *Domain of Science*, Cambridge, 1926; J. S. Haldane, *The Sciences and Philosophy*, New York, 1929; A. S. Eddington, *Nature of the Physical World*, New York, 1928; E. Meyerson, *Identity and Reality*, New York, 1930; G. N. Lewis, *The Anatomy of Science*, New Haven, 1926; F. S. C. Northrop, *Science and First Principles*, New York, 1931; B. Russell, *The Scientific Outlook*, New York, 1931; F. R. Tennant, *Philosophy of the Sciences*, Cambridge, 1932.

[1a] Vd. Notes 2 and 47 *infra.*

[1b] Vd. Notes 16a and 47 *infra.*

regard to the biological sciences. Mechanics is the basic science of matter in motion, and all the other natural sciences are reducible thereto.

(2) This is the Cartesian philosophy of science, and it is based on the metaphysical point that matter exists as such, that its substantial nature is extension, and its essential operation is local motion. The Cartesian cosmology is a 17th century Christianized version of Plato's *Timaeus*. (The study of living matter must, therefore, be only a special case of the study of matter; the study of vital functions, of the kinds of change involved in digestion, sensation, reproduction, etc., must be reducible to the study of the proper change of matter in place, which is mechanics. It also involves the reduction of all causes to one type: the external, efficient cause, i.e., a local motion caused by a local motion).

(3) The Cartesian philosophy of science, furthermore, makes mathematical physics the supreme achievement of scientific endeavor, and presents it as the ideal toward which all scientific work must be directed. (The opposite position criticizes mathematical physics as a misleading ideal, a mixed science rather than real physics. But the Cartesian position idealizes it because if the essence of matter is quantity or extension or dimensionality, then all of the sciences of matter must be formally mathematical.)

(4) The Cartesian position with respect to the sciences is also taken by materialists, differing from Descartes in metaphysics in that they are monists whereas he is a dualist.[1c] In either case the properties

[1c] This sort of materialism is called by Marxists "natural-science materialism." Dialectical materialism goes beyond this by being "natural-science materialism" *plus* "historical materialism." Vd. A. Thalheimer, *An Introduction to Dialectical Materialism*, New York, 1936: Ch. VII, VIII; also Note 28a *infra*. It should be added here that the Cartesian position was taken by Kant; it was because of his Cartesian insistence upon dimensionality in science, that he despaired of "empirical psychology" as a science. The existence of psychometrics as an application of mathematics to psychical phenomena disposes of Kant as a prophet,—or, perhaps, Kant would have called this science "anthropometrics." Finally, it should be added here that the Cartesian position on the unity of science is taken by the logical positivists. Vd. R. Carnap, *The Unity of Science*, London, 1935. That they take this position is not due to their logic or, more properly, their grammar, but rather to their unacknowledged metaphysics which is simple-minded materialism.

of matter are the same, and hence the sciences of matter are homogeneous.

(5) The issue in the field of biology between mechanism and vitalism is to be understood in terms of the opposition between the first of these two philosophies of science and the second.

b. The second position is that the objects of natural science are heterogeneous, that there is only a partial continuity among the natural sciences, in which one may include the findings of another, but that the different sciences are ultimately irreducible to one another.

(1) This is the position taken by some biologists with respect to physics. The name "vitalist" is an ambiguous way in which to designate all those who resist the attempt to reduce all the natural sciences to mechanics and who insist that the heterogeneity of the animate and the inanimate is absolute and ultimate and, who, therefore, also insist that biology, though it may depend on knowledge gained by physicists and chemists, cannot be reduced thereto.[2]

(2) This is the Aristotelian philosophy of science, and it is based on the metaphysical point that matter does not exist as such, but only as formed; and that the forms under which units of matter can exist are never exclusively quantitative or dimensional, because quantity is at best an accidental and not a substantial determination of whatever exists materially. As a consequence of this basic point, it follows that matter existing under different substantial forms is radically different in nature. Hence the sciences of dead and of living matter belong to a different order because their objects belong to a different order of natural existence. (This philosophy of science insists not only upon different kinds of natural things, but upon different types of change, of which local

[2] Vd. J. S. Haldane, *Mechanism, Life and Personality*, London, 1913; A. N. Whitehead, *Science and the Modern World*, New York, 1925; J. Needham, *The Skeptical Biologist*, New York, 1930; van Uexküll, *Theoretical Biology*, New York, 1926; C. Lloyd-Morgan, *Emergent Evolution*, New York, 1923; and *Life, Mind and Spirit*, London, 1926; W. McDougall, *Modern Materialism and Emergent Evolution*, New York, 1929; R. F. Hoernle, *Matter, Life, Mind and God*, New York, 1923.

motion is only one, and upon different causes, of which the external or moving cause is only one.)

(3) The Aristotelian philosophy of science, furthermore, denies that mathematical physics is the supreme achievement of scientific endeavor. On the contrary, it is not even good as natural science,—except for practical purposes in engineering applications,—because its basic principles are not physical but mathematical. The biological sciences, and among these physiology, are much nearer to the ideal of investigative natural science, because the principles of physiology, for instance, are derived from general physics (the philosophy of nature). The tradition of physiology from Galen to Haldane is predominantly Aristotelian. (In other words, although all the empirical sciences deal only with the phenomenal surface of nature, those of the mathematical type are much more superficial than those which Maritain calls the empiriological type. It is their greater superficiality,—their remoteness from ontological significance,—which makes them appear "more abstract" and also enables them to be more exact.)

(4) The Aristotelian position is as much opposed to materialism of the simple-minded, atomistic-mechanistic sort as it is opposed to the dualism of Descartes, which separates matter and form as if they were independent substances.

c. This issue in the philosophy of science cannot be solved "empirically" by consulting the facts of the history of science. It is erroneous to try to use the history of science, stressing practical advances made in accordance with one or the other position, to solve the philosophical point.

d. This issue in the philosophy of science is of paramount importance for scientific psychology.[8]

(1) Either as a part of biology: reducible or not reducible to mechanics.

(2) Or as independent of biology in the same way as

[8] Cf. A. G. A. Balz, *The Metaphysical Infidelities of Modern Psychology* in the Journal of Philosophy, XXXIII, 13.

biology is independent of, and irreducible to, physics (mechanics or mathematical physics).

4. The two philosophies of science arise from radically different metaphysics. The issue between them can, therefore, be resolved only by recourse to metaphysics. It is necessary, therefore, to consider the metaphysical alternatives upon which the resolution turns. The appropriate formulation of these alternatives is in terms of cosmological propositions. We are here concerned with metaphysics only in so far as it is involved in the philosophy of nature [20].
   a. There are only four possible positions in metaphysics on the nature of existing things in change (the objects of investigative knowledge). And only one of these four positions can be right.
      (1) *Absolute materialism:*[4] best illustrated by the Greek atomists and their modern followers. (We have already seen how this metaphysical position insists upon the reducibility of all natural science to mechanics and mathematical physics, because matter has only dimensional properties.)
      (2) *Dualism:*[5] the position sometimes attributed to Plato, sometimes found in the neo-Platonic tradition, but explicitly stated by Descartes. The dualists assert that there are two independent substances: matter, the essence of which is dimensionality (extension); and mind, the essence of which is rationality (thought).
         (a) It follows, then, that there are only two sciences: physics, as mechanical and mathematical, on the one hand, and psychology, on the other. And these two sciences are independent of each other [20a].
         (b) Furthermore, just as all sciences of matter are ultimately homogeneous and mathematical, so psychology cannot be scientific at all, in the sense of being investigative knowledge, since investigative knowledge involves sense-observation. Psychology is purely philosophical knowledge and its method is entirely introspec-

[4] Vd. G. Plekhanov, *Essays in the History of Materialism*, London, 1934; A. Thalheimer, *op. cit.*; T. A. Jackson, *Dialectics*, New York, 1936: Ch. II; F. Engels, *Ludwig Feuerbach and the Outcome of Classical German Philosophy* (1888).
[5] Vd. A. O. Lovejoy, *The Revolt Against Dualism*, New York, 1930.

tive. It is the soul's knowledge of itself; it is the "science of consciousness." It was Kant's point that an *empirical science* of the "ego" could never be developed because it could not be experimental or mathematical. This further reveals Kant as Cartesian. But, curiously enough, Kant did concede an empirical science of man,—the phenomenal being whose behavior is observable,—and called this anthropology.

(c) As a result of this conception of psychology, modern psychology, which is for the most part Cartesian, is burdened with *insolubilia*. The various problems involved in the understanding of the relation of body and soul can be intelligently solved only when they are intelligibly stated. But the Cartesian misconception of form and matter as *res cogitans* and *res extensa* made insoluble riddles out of the relation of the mind and body, the problem of introspection, the problem of sensation and the passions, which seem to be connected both with the body and with the mind.

(3) *Absolute idealism:* [6] best illustrated by the philosophy of Hegel, and by other post-Kantian forms of German romantic and English epistemological idealism. This metaphysics does not occur in the ancient or mediaeval world at all, unless it can be attributed to Parmenides. (The monism of the Eleatics makes both change and plurality illusory. Only the motionless one exists.)

(a) Absolute idealism either denies matter and the plurality of individuals existing independently of Mind, or it treats them as self-contradictory appearances, as illusions due to an incomplete experience of reality. This is the familiar doctrine of Bradley's *Appearance and Reality* [20b].

(b) It makes psychology the primary, if not the only, science of reality. All the other sciences are about illusions or appearances.

(c) It makes psychology as a science non-investigative and purely dialectical, because all exist-

[6] Vd. J. Royce, *Lectures on Modern Idealism*, New Haven, 1917; R. F. Hoernle, *Idealism as a Philosophy*, New York, 1927.

ences are thoughts in an Absolute Mind which operates dialectically. In a sense, this dialectical character preserves a hierarchical structure within reality, and provides for the ordering of all the sciences, even those that deal with illusions, as subordinate to what Hegel called the pure phenomenology of spirit, which is psychology.[7]

(4) *Formal materialism:*[8] illustrated by the metaphysics of Aristotle and St. Thomas Aquinas [21].

(a) This position can best be stated by restating all the others. Absolute materialism denies substantial and qualitative forms, and faces the impossible task of having to reduce all differences and determinations to those of quantity. Absolute idealism denies matter and individuality except as illusions. Dualism insists upon both matter and form (extension and thought), but separates them as independent substances not involved in each other's existence. Formal materialism insists upon both matter and form as involved together in the nature of all existing changing things. It denies that matter exists apart from forms, and that the forms of matter exist except in matter. It denies that the only forms of matter are quantitative. It denies that matter is illusory.

(b) In terms of the analysis of matter and form, we can give an account of change as the transition from potentiality to actuality, and we can distinguish between the essential and the accidental forms by which matter is determined. (We can also give a better account than Kant

[7] Vd. Hegel, *The Phenomenology of Mind*, trans. by Baillie, New York, 1931. Cf. W. T. Stace, *The Philosophy of Hegel*, London, 1924: Part III on the Philosophy of Nature and Part IV on the Philosophy of Spirit.

[8] Vd. J. Maritain, *An Introduction to Philosophy*, London, 1930: Ch. V, VI, VII; I. Gredt, *Elementa Philosophia Aristotelico-Thomisticae*, Fribourg, 1932; R. P. Philips, *Modern Thomistic Philosophy*, London, 1934; E. Gilson, *The Spirit of Mediaeval Philosophy*, New York, 1936. For a discussion of controversial points within the doctrine of hylomorphism itself, see the account of the neo-Platonic objections in D. E. Sharp, *Franciscan Philosophy at Oxford in the Thirteenth Century*, London, 1930; also Pedro Descoqs, *Essai Critique sur l'Hylémorphisme*, Paris, 1924.

did of the distinction between the noumenal and the phenomenal orders in terms of the relation of being and operation, the properly intelligible and the properly sensible.)

(c) It clearly follows from this metaphysical position that the natural sciences (investigative knowledge of changing things) are as radically heterogeneous as there are essentially different types of formed-matter. In other words, the objects of science are different because matter is formed by irreducibly different genera of substantial forms.

(d) The picture of the sciences can be given as follows:

1.) The physical group: deals with natural things that are dead: inorganic forms. (Elemental and mixed bodies.)

2.) The biological group: deals with natural things that are alive, involving organic forms, which are called *souls*, not *minds*. (Vegetables, or plants, and animals.)

3.) With respect to psychology there are two ways in which its independent status can be conceived: in Aristotelian terms, as a separate body of knowledge dealing with the nature of the soul *as such*, considered in abstraction from plants and animals in which souls exist as material forms; or with the soul in its self-subsistence; in Thomistic terms, as the science of man, distinguished from biological science by the fact that man is rational. (In Thomistic terms, anthropology is a better name than psychology for knowledge about man, about human nature as integrally composite of body and soul, matter and form. Human psychology is in one sense a part of anthropology, dealing with the human soul as an integral part of man, and in another it belongs to natural theology since it is knowledge of a transcendent, extra-physical sort of being.) The position taken by St. Thomas is more Aristotelian than that taken by Aristotle in the *De Anima*, which shows Platonic influence in treating of soul primarily rather than of man. Vd. Note 36c *infra*.

b. The issue made by these four metaphysics is resolvable as follows. (I can merely indicate here the dialectic of the resolution. It is rationally binding. It is not a matter of opinion. There is only one right metaphysics.)

(1) Two of the four positions are at once seen to be impossible in terms of the evidence and in terms of internal contradictions. Thus:

(a) Absolute idealism arises from the simple epistemological error of identifying the *id quo* and the *id quod* of knowledge [22].

(b) Dualism has a number of inescapable internal difficulties. Its premises convert difficult but soluble problems into baffling paradoxes and impossible riddles. Thus, the problem of the unity of man throughout the neo-Platonic tradition and in modern psychology; the problem of the interaction of independent substances (body and mind), in the case of sensation, passion and volition, badly solved by recourse to the Divine providence of a pre-established harmony or left unsolved by a psychophysical parallelism which ignores the facts of vital action; the problem of change itself in a view of world which makes matter as completely actual as form; and most important of all, the impossibility of giving an account of the physical sciences, which are possessed by *minds*, but are knowledge of *bodies*.

(2) Only two of the four positions are seriously worth considering as possible, and both are materialistic, differing in that one is simple-minded and absolutistic, and the other is dialectical and relativistic.

(a) Absolute materialism ultimately reaches insuperable difficulties.[8a] On the one hand, it can-

[8a] For refutations of absolute ("natural-science") materialism, vd. Hook's account of Feuerbach's "critique of absolute materialism," and his account of Marx's criticism of "*all* materialisms from Democritus to Feuerbach" in *From Hegel to Marx*, New York, 1936: pp. 237 ff. and 274 ff.; also W. McDougall, *Modern Materialism and Emergent Evolution*, New York, 1929. The errors of "natural science materialism," —common to the atomists and the Cartesian mechanists,—are not avoided by the

not give an account of the nature of things, since it cannot reduce all forms to quantities or combinations of quantities, and it cannot give an account of all the phenomena of change, since it cannot reduce biological generation to local motion. On the other hand, it cannot give an account of knowledge itself, neither of the process by which matter knows matter sensitively, nor of the sciences of matter which seem to involve ideas as well as sensations. In short, just as absolute idealism is ultimately forced to call matter an illusion of the senses, so absolute materialism is forced to call all of our sensitive experience and all our intellectual knowledge illusory, however practical and useful it may seem to be in the world of matter.

(b) Hence, negatively, by elimination, hylomorphism or formal materialism is the only tenable metaphysics. Further, positively, it can be shown to be the only metaphysics which, without shifting its ground, without inconsistency, and without talking about fictions or illusions, can give a coherent account of physical nature, of our process of knowing it, and of the sciences which are knowledge of it. It is the only metaphysics which agrees at all points with common sense and common experience, because it is the only one which arises without error from the analysis of common experience.

c. The consequences of this metaphysical resolution are critical for the philosophy of science.

(1) It indicates the rightness of the Aristotelian account of the sciences as heterogeneous and irreducible, and the wrongness of the Cartesian or materialistic account.

(2) It leads, as we shall see, to a tripartite division of the sciences: physics, biology and psychology (or anthropology in the Thomistic sense).

(3) But, most important of all, it imposes the burden of metaphysics upon every scientist who is intelli-

addition of "historical materialism" on the level of human or social phenomena. In so far as dialectical materialism accepts atomism and mechanism as a philosophy of nature, it makes an error which such an addition does not expunge.

gent enough to consider the order and relation of the sciences. The opposition between the philosophies of science is so strictly correlated with metaphysical oppositions that any choice in the one case necessarily dictates a choice in the other, and conversely. Thus:

(a) If any one holds to homogeneity and reducibility, he must be either an absolute materialist, in which case psychology has no subject-matter; or a Cartesian dualist, in which case psychology is not an investigative science, if it is a science at all.

(b) If any one holds to heterogeneity and irreducibility, he must be either an absolute idealist or a formal materialist. In either case, psychology is separate from biology; but in the former case, it is in no sense an observational or investigative science.

(4) *Digression:* For the sake of brevity, I shall proceed upon the assumption that few scientific psychologists (certainly no experimentalists, and I doubt if any psychoanalysts) are absolute idealists.

(a) Those who are absolute materialists (such as the behaviorists) are not only wrong in metaphysics, but are involved in destroying themselves, since psychology must ultimately become mechanics and mathematical physics.[8b]

(b) And those who are dualists (as were most of the old-fashioned introspectionists or structuralists) are not only wrong in metaphysics, but are involved with such impossible problems as the relation of mind and body, the relation of psychology to physiology, etc.

(c) The only psychologists who are right in metaphysics, who can give an account of their subject-matter which properly relates it to physiology and yet preserves its independence, and

[8b] Vd. K. N. Kornilov, *Psychology in the Light of Dialectic Materialism*, in *Psychologies of 1930*, Worcester, 1930. Kornilov rejects the simple mechanism and materialism of American behaviorism and also the subjectivism of German introspective psychology. He does not see, however, as Marxists usually cannot or will not, that these denials necessarily lead to formal materialism, to the position of Aristotle and St. Thomas. Cf. Freud's comment on the dogmatic blindness of Marxism in *New Introductory Lectures on Psychoanalysis*, New York, 1933: pp. 246–7.

who have no ultimate *insolubilia* in psychology itself, are the formal materialists, of whom there are very few, if any [22a]. There are two possibilities:

1.) The *gestalters:* but they waver between dualism and hylomorphism, though they should choose the latter.
2.) The *psychoanalysts:* but they waver between materialism (complete reduction to physiology), dualism (Freud in his more neo-Platonic moments), and formal materialism, though they also should choose the last.

d. Without further demonstration, I am going to take two positions:
   (1) formal materialism in metaphysics
   (2) heterogeneity and irreducibility in the philosophy of science.

   And in terms of these two positions I shall now discuss the subject-matter of psychology and its position among other bodies of knowledge.

III. The Position of Psychology Among the Investigative Sciences

1. Let us first consider psychology in the domain of biology.
   a. Matter has various forms or perfections. Its potentialities are actualized in the existence of different kinds of bodies: elemental, mixed and living bodies. As living, matter not only has the perfection of existence but also the perfection of vital operations, such as, vegetation, sensation, locomotion and understanding [23].
   b. There are no vital actions (operations) where there is no life.
      (1) Soul is the actuality of an organic body having the potentiality of life. (Body besouled is body alive; the soul is the form of matter as living.)
      (2) Soul is to body as form to matter. Soul and body are the *analytically* separable components in the composite nature of living things.
      (3) Soul is the first act, the act of being, of a living thing. Vital operation is the second act. A living thing is not always vitally active in all the ways in which it is able to operate. It is impossible for Cartesians to make this distinction. For them, the prob-

lem of sleep is insoluble. The soul which is not thinking simply ceases to be.

(4) Biological study is of vital operations. It is concerned with the powers and operations of living things, powers and operations which are their properties because they are besouled.

(5) The different powers of vital action are related to the different organs of the body. The parts of the soul, which are these powers, are related to the parts of the body, which are endowed with these powers of vital operation because they are parts of a living (besouled) body. Thus, digestion is related to the stomach, vision to the eye, so long as these are parts of a living body.

c. The foregoing indicates the field of problems which are those of animal and human physiology.

(1) Physiology is the study of the acts of bodily organs.

(2) It is an investigation, not of the powers and operations of the body as mere matter, but of the operations of the composite: matter formed or besouled and hence alive and having powers of vital action.

(3) The question is how to distinguish the study made by the psychologist in the domain of biological phenomena, from that of the physiologist.

(a) Galen tried to make physiology the study of the operations of the vegetative organs (digestion, excretion, etc.) and of some of the organs of local motion; and to exclude, as belonging to psychology, the study of cognitive operations, such as sensation.[9] But this is untenable, because sensation is just as much the act of a bodily organ as is reproduction or local motion.

(b) The only possible distinction is one based on the distinction between the parts and the whole, the separate organs and the organism as a whole. (This must not be understood to mean that the physiologist studies organs in isolation from the organism as a whole which

[9] "Since feeling and voluntary motion are peculiar to animals, whilst growth and nutrition are common to plants as well, we may look on the former as effects of the soul (*psyche*) and the latter as effects of the nature (*physis*)" (*On the Natural Faculties*, I, 1).

is their environment; moreover he sometimes studies the organism as a whole in its physical relation to the environment.)

1.) Thus, the physiologist studies the action of the circulatory system by itself, or of the eye, or of the bladder, etc.

2.) Whereas the animal psychologist studies the behavior of the whole organism, based, of course, upon the vital action of the parts as integrated. He studies animal learning, the instincts, etc.

3.) There is no possible reduction here of animal psychology to animal physiology because the whole is not just a sum of the parts, not a collection but an integrative organization which functions integratively. Those physiologists who study the organs in relation to the organic totality recognize that the totality is not a mere summation. (It would be the same error to suppose that the whole was reducible to the sum of the organic parts, as to suppose that any of the organic parts were reducible to a mere sum or collection of atoms in their mechanical play. This is the error of the materialist and the Cartesian.)

(c) The one remaining problem is that of human psychology and human physiology.

1.) This problem depends for its solution upon the question whether man is essentially or only accidentally different from other animals.

2.) If the answer is that man differs from other animals only in degree, then psychology is a biological science, related to organ physiology, as that in turn is related to cell-physiology or cytology. In other words, the picture of the sciences of life would then be as follows:

Cytology: the life of the cell, *considered* separately, whether or not it *exists* separately.

Physiology: the life of a particular organ, *considered* separately, though it does not *exist* separately.

Psychology: the life of the organism as a whole.

3.) If this answer is given, that man does not differ radically or essentially in kind, then

a.) Psychology is only a particular natural science, and not a branch of philosophy (and it is hard to account for what appears to be a body of philosophical and non-investigative knowledge about man's nature); and furthermore

b.) Psychoanalysis is queer. There is no animal psychoanalysis, so far as the *method* of psychoanalysis is concerned. Man is at least capable of being distinguished from all other animals as able to be psychoanalyzed.

c.) Finally, though psychoanalysis may be, as speculative, only a study of human behavior, it is as practical and therapeutic more than that, and that more is intelligible only in terms of the difference between man and other animals.

2. Summary: The position of psychology seems to be peculiar among the investigative sciences. As animal psychology, it is properly located in the biological domain. But human psychology raises some difficulties.[10]

a. On the one hand, it is like animal psychology, and *in so far* is properly located in the biological domain [24].

b. But, on the other hand, it seems to be *something else*, such as psychoanalysis or philosophical psychology or introspective psychology, etc.

c. This *something else*, which human psychology seems to be, can be properly understood only *if* man is understood as essentially and radically different from other animals, and *if*, to this extent, human psychology ex-

[10] Vd. R. P. McKeon, *De Anima: Psychology and Science* in the Journal of Philosophy, XXVII, 25.

ceeds animal psychology and cannot be properly located in the biological domain.

(1) Furthermore, we have the problem whether what is peculiarly human psychology is entirely philosophical in method or whether as a body of knowledge it can be both philosophical and scientific, that is, investigative in method. (This problem does not occur with respect to animal psychology. It properly includes investigative knowledge.)

(2) We can now construct another possible picture of the allocation of psychology.

Animal psychology:
a particular natural science concerned with the behavior of animal organisms as wholes; a science in the biological domain.

Human psychology:
(a) As if it were the same as animal psychology, by ignoring what is characteristically and differently human.[11]
(b) Of man as such: but investigative in method, such as psychoanalysis, and not in the biological domain.
(c) Of man as such: but philosophical in method.

(3) The chief problem which this picture raises concerns the existence and nature of a body of psychological knowledge, relevant only to man and yet investigative in method. Psychoanalysis *seems* to be such knowledge. It is investigative in method; it is relevant only to man. If there is such a body of knowledge it is not reducible to biology, and it is not even, strictly speaking, in the biological domain.

(4) In order to solve this problem—in order to determine whether psychoanalysis is philosophical or investigative, or *both*,—we must now consider the nature of psychology as a body of philosophical knowledge.

IV. The Position of Psychology in Philosophy

1. We must, first, consider the philosophical demonstration that the human soul differs from the vegetative soul, which

[11] Vd. Note 24 *infra.*

is not cognitive, and from the animal soul, which is sensitive, by being intellective, i.e., by the power and operations of understanding (abstraction, judgment, inference).

a. It should be reiterated here that the distinction between sensation and understanding is phenomenologically unquestionable.

b. It should be emphasized, furthermore, that any account of science must acknowledge this distinction, and hence every psychologist considering himself as a scientist must admit that he has a human soul.

c. The radical difference of the human soul, that it is intellective, can be understood only in terms of a metaphysical analysis. And we must remember that whatever we say here follows from the choice we have already made in the philosophy of science. We cannot at this point give up our position that the objects of science are heterogeneous and hence that all the sciences are not reducible to one basic science, such as mechanics; and if we hold to it, we must hold to the metaphysical position upon which it is based: formal materialism or hylomorphism [25]. In terms of this position, we must give the following account of understanding.

(1) All existing physical things are composite of form and matter.

(2) But all existing physical things are individual, i.e., unique particulars, though particulars of a certain sort.

(3) Hence, the forms of individual things are individuated. The form, being in matter, is subject to material conditions which individuate it.

(4) But understanding is grasping the universal, which means separating the form from the conditions of individuation. In other words, John Smith cannot be understood as *hoc aliquid*, but only in so far as he is a man. John Smith is understood as a man when the form which in him exists individually, in *his* matter, is separated from *that* matter.

(a) The separation of the form from the matter requires two cognitive steps: first, the acts of sensation and imagination in which the form of the individual thing is separated from its matter but is apprehended under the conditions of individuality because sensation is the act of a bodily organ, and hence sensible forms as apprehended by sense are in matter (that is, sensitive apprehension is always of the particular); and second, the intellectual act of abstraction which separates the form, grasped by sense under individuating conditions, from those conditions; and thus converts *intelligibility* into something *actually understood.*

(b) Things are intelligible (potentially understood) in so far as they are composite of matter and form. They are actually understood by the separation of the form from the matter by sense, and of the form from material conditions by the intellect.

(5) Whatever is received is in the recipient according to the mode of being of the recipient.

(6) In terms of the foregoing, it can be shown now that the primary intellectual operation (the acts of abstraction and of receiving abstracted universals) cannot be the act of a bodily organ as the sensitive operation (the act of sensation) is the act of a bodily organ. For, if it were, the forms received in the intellect would be received into matter, since the act of a bodily organ is an act of matter, and forms which are acts of matter exist in matter as individuated. Hence, they would still be potentially intelligible and not actually understood. Hence, if men are able to understand,–to apprehend things in their universality (their intelligibility) as well as in their individuality (their sensibility),—intellection must be an operation that is not the act of a bodily organ. This truth is not inconsistent with the fact that intellectual operation depends upon prior sensitive operation, since the forms abstracted are abstracted from the informed matter of sensation and imagination [26].

d. This analysis of intellectual power and operation explains three things characteristic of man and no other animal.

(1) Science and philosophy, involving knowledge of universals. This does not mean that brute animals

lack knowledge, but that they have only sensitive knowledge.

(2) The practical and productive arts, involving the practical and intellectual operations of free choice, deliberate doing and intentional making [27]. Thus:

(a) In the practical sphere: men freely determine their moral characters, their social institutions, their governments, their laws. Other animals are social, but only instinctively. Animal societies are, therefore, capable of being studied entirely in the domain of biology; but not human society, which can be properly studied only in terms of a psychology that recognizes the intellect, practical reason and voluntary conduct.

(b) In the productive sphere: men make things, fine and useful productions, by conceiving and intending them. Other animals are productive, but only instinctively. Whereas man can make many different kinds of things, and by manufacturing can multiply individual instances of each kind, productive animals can build or make only one sort of thing, for the creation of which they are instinctively endowed. Birds of a given species make nests of a given kind; so in the case of ants, bees, termites. Animals cannot vary their works or multiply them by manufacturing. Animal productions are, therefore, capable of being studied entirely in the domain of biology, but not the productions of human art, fine or useful.

2. That man differs essentially and specifically from all other animals is capable of being demonstrated *a posteriori* from the observable operations of men and brutes.

a. Thus, only man manufactures. Manufacturing involves the grasp of a species of thing and its duplication by imprinting the same form upon many units of matter. To manufacture means to make according to specifications, choosing material capable of receiving such and such determinations. Only an intellectual animal could manufacture or, as Marxians say, machinofacture.

Marxian materialists, in defining man as a productive and machinofacturing animal, acknowledge his intellectuality. Unless man can comprehend universals, can separate forms from the material conditions of individuality, he cannot make things according to specifications; he cannot, in machinofacturing, impose the *same* artificial form on multiple, *different* units of matter [28].

b. Similarly, arguments could be drawn from a comparison of human and animal societies, human and animal knowledge, human and animal language [28a]. In all these cases, there is an analogy between the human and the animal mode, but the analogy also involves a specific or essential difference. The difference is not merely a matter of degree. (It should be noted here that the right use of analogy *always* includes the marking of *essential* differences. Otherwise, analogical reasoning is deceptive and illicit. Only things which are essentially different can be properly analogized, e.g., God and creatures, man and animals, art and nature.)

(1) Thus, bees live socially but not according to a political theory or in terms of moral principles; thus, birds make nests but cannot manufacture them; thus, monkeys jabber and communicate in a rudimentary way, but have no language which has syntactical structure, and without syntax there cannot be propositions or science.[12]

(2) That animals seem to think is not surprising when one remembers that most men for the most part think, not intellectually, but in terms of sense and imagination. If men can do a sort of thinking,—making comparisons, remembering, projecting, estimating, etc.,—through imagination, so can animals who have imagination because they, like men, are sensitive [29].

c. Psychoanalysis itself provides us with strong *a posteriori* evidence of an essential difference between men and brutes. The latter cannot be psychoanalyzed because

(1) They lack reason; lacking it, they have no grammatical language; and without such language, in-

[12] Even logical positivists would admit this. Vd. Carnap, *Philosophy and Logical Syntax*, London, 1935.

tellectual communication (the communication of judgments, of opinions, etc.) is impossible;

(2) They lack reason (or intellect); without intellect, free choice is impossible; without free choice and voluntary conduct, the therapeutics of psychoanalysis is impossible.

3. I have now answered all of Dr. Alexander's questions [30]. I have shown in what sense understanding has a physiological basis and in what sense it cannot have. I have shown that the difference between men and brutes is not a matter of degree. I have shown, therefore, that the type of "biological continuity" which is supposed by those who find no *essential* difference between men and brutes, does not exist. There is, of course, the type of continuity which obtains in the series of integral numbers. Such a continuity exists because *man is an animal*,–a *rational* but not a *brute* animal [31].

4. In the light of the foregoing, the nature of psychology as a body of philosophical knowledge can be understood.

   a. *In terms of its object:* the intellect is able to know reflexively as sense is not. Thus, the senses only know their proper sensible objects, but not each other or themselves (i.e., the eye cannot see, the ear cannot hear, the *eye as seeing*); but the intellect is able to understand all intelligible objects, including the senses and, most important of all, it is able to understand itself. This means that the intellect is able to know, through sensitive experience, the various organs of the human body and their several powers of operation, and such knowledge is the investigative science of physiology. But the intellect can also know directly (i.e., without investigation) its own operations, *in se* and in relation to all other human operations, and such knowledge is the branch of philosophy called psychology.[12a]

[12a] Cf. C. Alibert, *La Psychologie Thomiste et les Théories Modernes*, Paris, 1903; pp. 1–22. Alibert defines the province of philosophical psychology in terms of reflexive knowledge. It extends to all the powers of man, even to the vegetative, of which he says: "Son existence m'est révélée seulement par des inductions tirées des données physiologiques" (p. 9). It does not include, however, fields of phenomena which require investigation. "L'observation externe est encore necéssaire pour prendre connaissance des états anormaux, tels que le rêve, le somnambulisme, l'hallucination, la folie, et pour decrire la vie psychique de l'animal, délimiter la partie des facultés humaines qu'il convient de lui attribuer, et celles que nous devons tenir pour incommunicables" (p. 13).

(1) We have now solved the fundamental paradox of psychology as knowledge, which is due to its reflexivity: namely, that the psychologist as a man, as a scientist or philosopher, knows himself as a man and as a scientist or philosopher.

(2) Philosophical psychology arises from man's direct knowledge of himself as operating in a distinctively human way [31a].

b. *In term of its method:* the reflexivity of the understanding (that we know ourselves as knowing, understand ourselves as understanding) is the root of philosophical psychology and explains how psychological knowledge is possible *without investigation.* We know the nature of man *a posteriori* in terms of his operations, and these operations, in so far as they are typically human, we can know directly because the intellect is able to understand its own operations and the operations of sense and appetite as these are immediately experienced. Thus, the concept of the understanding is an abstraction from particular operations of understanding; similarly, in the case of sense, desire, passion, will, etc.

(1) This does not mean that human behavior *cannot* be investigated. It does not mean that rational operations *cannot* be investigated. The clinical studies, made by psychiatrists and psychoanalysts of the pathology of reason and will, are investigations of intellectual activity. Even behavioristic studies of human "problem-solving," devised through the analogy of rats in mazes and monkeys in boxes, are investigations of rational behavior. But all such investigations must accept the principles of philosophical psychology, or be involved in self-contradictions. When the data of such studies are properly interpreted, it is found that scientific knowledge is about the *accidents* of human, i.e., rational, behavior [32].

(2) Common experience is inadequate as a basis for knowledge of *kinds* of physical things, with the one exception of ourselves as an object of knowledge.[12b]

[12b] This statement is insufficient as an explanation. In so far as it emphasizes quantity it is wrong. The root of the matter is, as Maritain points out, that man does not know the essential natures of physical things in their ultimate specific diversity. Vd. *Degrés*

We must investigate, and these things, therefore, are exclusively the objects of the natural sciences. But, in the case of ourselves, common experience is *sufficient*, and investigation, while *possible*, is not *necessary*. Philosophical psychology is nothing more than the analysis of the immediate experience of our own operations. In short, the self is a unique object for man to know.

5. *Now, it follows from all that has preceded:*
   a. That psychology is the only branch of philosophical knowledge which is about a particular type of changing thing: namely, man in his operations [32a]. Human change consists of all sorts of learning, the formation of moral and intellectual habits, etc.
   b. That psychology, in a qualified sense, as animal psychology, is not a branch of philosophy but a particular natural science in the domain of biology, concerned with the behavior of organisms in their totality.
   c. That psychology as philosophical is concerned with the powers and operations of man and from these it seeks to learn what it can of the essence of man, the nature of the human soul, etc. It is not concerned with *human biography as such:* how a man develops. Since human development is always for better or for worse and is the result of choices freely made, the ancient and mediaeval tradition relegated developmental problems to the field of ethics [33]. This last point is, as we shall see, crucially important for the understanding of psychoanalysis.
   d. That psychology as scientific is, in large part, concerned with the behavior of animals, and with human behavior *in so far as it is animal.*
   e. One important problem remains: Can psychology as scientific be concerned with human behavior *in so far as it is distinctively human?* This is what psychoanalysis seems to be doing. Psychoanalysis seems to have the same object of knowledge as philosophical psychology,

au Savoir, Paris, 1932: p. 77. The only *infima species* man knows is human nature: man is the only physical object man can adequately define. It is in this sense that man is a unique object of human knowledge on the physical level, and this explains the distinction of psychology among all subject-matters constituted by the first-grade of abstraction.

but it seems to be proceeding by the investigative method and clearly hopes to be a body of scientific knowledge. Is this possible? If so, how? And how is it related to philosophical psychology? This last question directs us to what is, perhaps, the most difficult problem, namely, the unity of psychology as including both philosophical and scientific knowledge. We have already solved this problem in general (vd. Note 32a *infra*), but the precise nature of the scientific contribution must be determined before the problem can be more fully discussed. We shall return to it, therefore, in Lecture IV.

(1) The same questions can be asked of other types of psychological research by investigative methods, which seem to be directed to the study of man as man and not merely as an animal.

(2) Such studies as those made by psychometrics seem to be concerned with accidental features of human nature. Thus, the measurement of human intelligence, or other human abilities, is concerned with quantitative accidents; and the experimental study of human memory or perception or imagination is concerned with such accidents as are involved in time, place, number, the material, etc. We are excluding, of course, all research which is physiological in method or interpretation.

(3) It is possible for there to be a mixed science of the same sort as mathematical physics: thus, Edgeworth's *Mathematical Psychics*.[13] The two ideals of natural science,—the mathematical and the empiriological,—are to be found in psychology. In psychology, more than in physics or physiology, the mathematical type of science is superficial.

(4) Psychoanalysis is much nearer to the consideration of man as man than any of these other scientific attempts at human psychology; and it is not a mixed or mathematical science.

(5) Hence, the problem of psychoanalysis as a scientific study of man, in his peculiar character as man, is most acute.

[13] London, 1881. The subtitle is: An Essay on the Application of Mathematics to the Moral Sciences. Cf. C. Spearman, *The Abilities of Man*, New York, 1928; L. L. Thurstone, *Vectors of Mind*, Chicago, 1935.

V. Conclusion

1. We have now accounted for the peculiarities of psychology among all bodies of knowledge, philosophical and scientific. This account will also explain, as we shall see, the peculiar history of psychology.
2. What remains to be understood is the nature of knowledge which claims both to be about *man as man* and to be *scientific*. This is the claim of psychoanalysis. And it must be said here that it appears more rightfully to make this claim than any other pretender to the title of scientific psychology.[14]
3. As we have seen, these questions are most acute for psychoanalysis as a body of knowledge about man, though they can be asked about psychometrics, gestalt psychology, functional psychology, behaviorism, etc.
4. The best way to approach this general problem is through the history of psychology as the study of man by any method, philosophical or scientific.[15]

[14] Psychometrics also has ground for making this claim. It is significant that Galton who, with Fechner, was a founder of psychometrics, rightly conceived the subject-matter of psychology in his *Inquiries into Human Faculty*. Vd. Note 46 *infra* for a comparative evaluation of psychoanalysis and psychometrics.

[15] The distinction between philosophical and scientific psychology, which has been adumbrated throughout this lecture, must not be confused with the distinction, in vogue since the days of Christian Wolff, between rational and empirical psychology. Vd. Note 39c *infra.*

THE HISTORY OF PSYCHOLOGY

I. Introduction

1. I shall first summarize the points upon which we seem to agree. I shall then restate the issues raised in the last discussion, and I shall remind you of the problem with which the last lecture ended. It is for the solution of this problem that we turn today to the history of psychology.
2. The points of agreement.
   a. We are Aristotelians rather than Cartesians in our philosophy of science. That means that we view the objects of natural science as intrinsically and irreducibly heterogeneous.
      (1) There is a hierarchy of the sciences of natural things.
      (2) The biological sciences may, for instance, depend upon knowledge obtained by physical and chemical researches, but they have an independent field of inquiry. In the advance of knowledge, they will never be replaced by mechanics [33a].
      (3) Mathematical physics should not be the exclusive ideal of investigative science.
   b. We are Aristotelians in metaphysics. Negatively, this means that we are not simple-minded, absolute materialists of the atomistic sort, or dualists of the Cartesian sort, or absolute, but not quite so simple-minded, idealists of the Hegelian sort. Our unanimous negation of these positions is correlated with our rejection of what follows from each of them with respect to the philosophy of science. Positively, this means that we hold the following propositions:
      (1) The realm of natural things in change,—the subject-matter of all investigative science,—consists of things formed of matter. All natural things are material or corporeal.

(2) But the forms in which matter exists, constituting the variety of natural things, are not merely or primarily quantitative. These forms are the different perfections of matter. Thus, we find matter existing but not living, or living but not sensing, or living and sensing but not thinking. As existing but not living, matter is either elemental or mixed in form. As living, matter either merely vegetates, or senses and moves as well as vegetates.

(3) This hierarchy of perfections in existing things is the basis for the hierarchy of the natural sciences investigating them [34].

(4) The superior perfection necessarily includes inferior perfections. Thus, what is alive must also exist as a body exists; that is, living things have all the properties of inanimate bodies, and some other properties in addition; similarly, whatever is able to sense and move must also vegetate, but not conversely [35].

(5) This order of the perfections of matter indicates both the continuity in nature (the inclusion of inferior perfections) and the discontinuity (the addition of new properties).

c. The premises we have thus accepted and asserted bind us to certain conclusions.

Digression: We must avoid the error of "wishful thinking," sometimes called "rationalization."

(a) There are only two kinds of thinking: logical and illogical. "Wishful thinking" may be either. When the psychoanalyst uses the phrase "logical thinking" in a derogatory sense, he means a wishful use of logic. Thus, it is so often said that you can prove anything by logic. That is absolutely true. No one knows it better than a logician. Given a set of false premises, you can prove a conclusion which logically follows from them, and that conclusion may be either true or false in fact. If you desire to prove that conclusion, and invent the premises to that end, without any honest regard for their intrinsic truth or falsity, your thinking is only formally logical, but essentially wishful. It is rationalization. Such rationalization is the more treacherous because the wishful thinking is *formally* good by *logical* criteria.

(b) Whereas you can prove anything by logic, you cannot prove anything without it. There is, therefore, another kind of wishful thinking. The first kind is rationalization which is good logically but is based on false premises chosen wishfully. The second kind is irrational thinking. Here we assert a conclusion despite the fact that it cannot be proved logically by sound premises, or deny a conclusion even though it follows necessarily from premises which we accept as sound. This is obviously also wishful thinking, differing from the first sort, which uses logic for its purposes, by trying to avoid logic. Thus, there are two kinds of *wishful* or bad thinking: (a) logical: rationalization; (b) illogical: irrationalization.

(c) Good thinking, the thinking of a competent scientist or philosopher, must not be wishful in either sense. The premises used must be chosen and accepted, not because they prove a conclusion we desire to believe, but because by the criteria of evidence they are sound and true in fact. If they yield a conclusion, they must do so logically; and if they yield a conclusion logically, that conclusion must not be avoided even though it is contrary to our desires. Finally, no conclusion can be accepted which does not follow logically from sound premises. The two criteria of good thinking are (1) sound premises and (2) correct inference. The only *wish* must be to know the truth.

(1) In the light of this digression, let me say, first, that the analysis I presented last time was not an instance of wishful thinking of the first sort. In this connection, let me remind you that you accepted the premises,—the Aristotelian philosophy of science and formal materialism in metaphysics,[1]—*before* you saw the conclusions which followed from them. I presented these premises,

[1] Vd. Lecture II *supra*. It should be added here that the argument there explicitly determined the choice of formal materialism as against the alternatives of absolute idealism, Cartesian dualism and absolute materialism. It also implicitly determined a choice between the sound and corrupt forms of that doctrine,—the issue between Aristotle and all sorts of Platonism. Vd. Note 21 *infra*. Leibnitz, Spinoza, Whitehead are neo-Platonists, i.e. corrupt Aristotelians. For an example of neo-Platonic plural-

not because they were able to yield the conclusion I desired, but because they seemed to me to be the only sound premises and were, therefore, unavoidable.

(2) I must say, therefore, that any refusal to accept the conclusions which follow from these premises looks to me like wishful thinking of the second sort, not rationalization but irrational thinking. The remark, *that you can prove anything by logic*, is a typical defense mechanism on the part of those who wish to avoid the conclusions which do follow logically from sound premises.

(3) The only way in which to avoid a conclusion which follows logically from premises is to deny the premises as unsound and wrong in fact.

(4) Therefore, unless it can be shown that the Aristotelian philosophy of science is wrong and the Cartesian right, unless it can be shown that absolute materialism or dualism or absolute idealism is right in metaphysics, and formal materialism is wrong, the conclusions which follow logically from an Aristotelian philosophy of science and from formal materialism, must be accepted by anyone who does not do the irrational sort of wishful thinking.

3. The conclusion which some members of this group tried to reject, *without* denying the premises and *without* finding any flaw in the reasoning, was: that man is essentially different from other animals, as animals are from vegetables, and as all living things are from inanimate bodies.

a. This essential difference resides in man's intellectual power, his power of understanding as distinctly different from his power of sensing. The latter he shares in common with brute animals.

b. I shall not repeat the various proofs of this essential difference. I shall merely remind you of

(1) The phenomenological distinction between sensing and understanding discoverable by any consideration of scientific knowledge;

(2) The *a posteriori* evidence of an essential difference between men and brutes in terms of language, cul-

ism, see Whitehead's discussion of the hierarchy of "societies" as a graded continuum, in *Adventures of Ideas*, New York, 1933: pp. 260–267.

tural history, social institutions, the arts and sciences.

(3) The metaphysical proof that understanding is not the act of a corporeal organ, such as the brain; but that understanding is an act which depends upon the act of a corporeal organ such as the brain. The latter activity is imagination. If I were to use Dr. Alexander's language, I would make this point as follows: Understanding is a function of imagination; imagination is a function of the brain. The word "function" is ambiguous. It has two meanings in the foregoing statement. A secondary function, a function of a function, is not the same as a primary function; only the latter is the act of a corporeal organ. Furthermore, this point holds if brute animals understand as well as men. If they do, understanding in them, as in man, must be a secondary function [36]. It should be added that when it is said here that understanding depends on, or is a function of, imagination, imagination is being regarded as a *necessary* but not a *sufficient* condition of intellection.

c. I shall accept Dr. Alexander's formulation as adequate. The only point of disagreement which remains, then, is whether men are essentially different from brute animals with respect to their exclusive possession of this so-called secondary function, understanding.

(1) As against the strong *a posteriori* evidence in favor of the difference, Dr. Alexander offered the case of the cretin idiot. The Mongolian idiot can be added, and also the dementia praecox patient in the later stages of dementia and catatonic stupor.

(2) The objection raises only a question of verbal usage. If by the word "man" we mean an offspring of human parents, then aments and dements are men. In which case, the class of men would include individuals who perform only as animals do, differing from animals only to the extent that they are curable, as the cretin idiot is known to be by thyroid.

(3) But what does curable here mean? It indicates at once another meaning of the word "man." We cure a cretin in order to make him able to perform as a man can. This second meaning of "man" does not define man as an offspring of human parents, but as a creature able to be an artist or a scientist or a statesman,

a creature able to be psychoanalyzed by verbal methods.

(4) When I say that man is essentially different from brutes, I mean a class of individuals defined by the second meaning of the word "man." I will now give you three criteria for telling whether a creature is *actually* a man, or actually but *incompletely* a man as the cretin is, or *merely* a brute.

(a) The ancient criterion: *rationality*, which means the ability to learn or be taught science and philosophy, the ability to govern other men by laws or to be governed by them, the ability to communicate by grammatical language, etc.

(b) The Marxian criterion: *productivity*, which means the ability to manufacture things according to specifications and plans.

(c) The Freudian criterion: *psychoanalyzability*, which means the ability to be changed in moral character by a technique dependent on the use of language and grammatical communication; or, more generally, *psychological curability*, i.e., the potentiality for becoming more fully a man, whether the efficient cause of therapy be the administration of thyroid, psychiatric devices, or psychoanalysis. These three criteria exclude all brutes absolutely. They also exclude the Mongolian idiot and the dementia praecox patient *for the time being*, that is, until we learn how to cure them, where curing them means making them actually men [36a].

d. I wish to add here only one other point, with which you have not so far, and I think cannot ever, disagree. If the statement, man is a rational animal, means that man is able to be a scientist, and if being a scientist means asserting as true only those propositions which are based upon sound evidence in a logical manner (i.e., doing non-wishful thinking), then the psychoanalyst, who claims to be a scientist, cannot say that *all* human thinking or knowing is wishful, whether the wishful thinking be rationalization or the sort we have called irrationalization. In other words, a psychoanalyst cannot claim to be a scientist and at the same time deny that man is a rational animal.

4. I now turn to the problem of to-day's lecture, which is to

discover the subject-matter and scope of psychology as a body of scientific or investigative knowledge.

a. There is no problem about psychology as a body of philosophical or non-investigative knowledge. I have already indicated the subject-matter of psychology, as developed by philosophical analysis, namely, the powers, operations and essence of man.[1a] This analysis man is able to make through self-knowledge, without any research of the investigative sort. I shall make this clearer in the course of reporting the history of psychology.

b. With respect to scientific psychology, we have seen

(1) That it must be different from physiology, which studies the powers and operations of particular bodily organs;

(2) That it achieves this difference by studying the operations of the organism as a whole.

(3) But the problem still remains whether there is a specifically human psychology, different from animal psychology, and related to human physiology as animal psychology is related to animal physiology.

c. In short, we are here trying to understand the subject-matter of psychoanalysis. It is not, at least in its present formulation, an animal psychology. It deals specifically with man. Its method seems to be primarily investigative rather than philosophical. The primary question is whether psychoanalysis is *the* science of human psychology. If so, there are three further questions:

(1) What are its investigative methods and problems?

(2) Is it aetiological, physionomic or taxonomic? (or any two or three?)

(3) How is it related to philosophical psychology?

5. I suggested that these questions could best be answered by a brief survey of the history of psychology. I proceed now to that history. It divides, first, into two periods: from the beginning until the 19th century, and from the 19th century to the present day; and the first of these two periods divides further into the period from the beginning until the 17th century, and from the 17th century to the 19th century. Thus:

a. The first division into periods is based on the distinction between philosophical and scientific method. Prior to the 19th century, psychology was non-investigative. The science of psychology has a fairly recent history [36b].

[1a] Vd. Note 32a *infra*.

b. The second division, of the first period, is based on the discrimination of what is sound from what is unsound in philosophy. From the 17th to the 19th centuries, psychology as a branch of philosophy is for the most part unsound, filled with errors, confused as to subject-matter, etc.

c. Psychology is the most recent of the natural sciences. It became autonomous as a branch of investigative knowledge by revolting from philosophy. To the extent that there can be a scientific psychology, the revolt was fortunate and justified; it was only extremely unfortunate that scientific psychology arose at a time when its basic concepts were derived from the bad philosophical psychology of the 17th and 18th centuries, and when the prevalent positivism made it difficult, if not impossible, for psychologists to understand the relation of their science to the philosophy of man. If it had been born earlier, it would have had a less tainted heritage and we would have less trouble getting it over its present feeble-mindedness.

d. In general, it is a misfortune to be born in modern times, a period in which each thinker commits the sin of pride, acknowledges no ancestors, rests in no tradition, but insists with consummate egotism, upon his complete originality and his complete independence of all prior human learning. (This is peculiarly true of psychoanalysis, and particularly of its great founder, Freud, who is curiously unaware of the tradition to which he belongs as a thinker and a student.)

II. From the Beginning to the 17th Century: Philosophical Psychology

1. So far as the literary tradition goes, the beginning of psychology, as of all other bodies of knowledge except mathematics, is in the work of Aristotle. His method required him to define separate subject-matters and to develop separate bodies of knowledge, both philosophical and investigative. Thus, his *Metaphysics* and *Physics*, unlike his treatises in astronomy and zoology, were entirely non-investigative, entirely philosophical. His psychology (*De Anima* and *Parva Naturalia*) was partly investigative and partly philosophical. Thus, the double status of psychology is indicated at the beginning of its history. It would have been clearer had Aristotle written a treatise on man [36c].

II.

2. But, first, as an initial digression, we must consider Plato. (Freud is most indebted to Plato and Schopenhauer among philosophers, even though he seldom cites Schopenhauer, and his references to Plato are peculiarly trivial, considering his debt to Plato.)
   a. We shall ignore the metaphysical error which Plato seems to make in separating soul and body as if they were independently existing things. We shall also ignore the errors which follow in Platonic psychology. All these are repeated by Descartes [37].
   b. Here we are only concerned to state Plato's true insights about the nature of man:
      (1) That he is a rational animal, and is, therefore, subject to the fundamental conflicts between animality and rationality. (Thus, the myth of the charioteer in the *Phaedrus*, which is devoted to speeches about love.[2]) Here Plato anticipates Freud in great detail concerning sexuality, and love as its proper sublimation. I am referring to the three papers by Freud on the psychology of love in 1910, 1912 and 1918.[3] Here, furthermore, Plato describes in detail the conflicts in man between his lower and higher nature, between his animality and his rationality, between his id and his ego and super-ego.[4]
      (2) That he is a loving animal. This means that in all his activities, whether they be sexual or artistic or philosophical, man acts through love. The Platonic insistence upon Eros as a fundamental trait of the psyche is the Freudian insistence that all human activities are an expression of libido. Libido, desire, love—all name the same fundamental root of activity. In the *Symposium*, and elsewhere, Plato enumerates and orders the various objects of love. His distinction between sensual and intellectual love is the Freudian distinction between sexuality and its various sublimations.
      (3) That man can be rational without ceasing to love, and love without ceasing to be rational. Here the

[2] Vd. especially the second speech which Socrates makes: *Phaedrus*, 244–257.

[3] In *Collected Papers*, London, 1925: Vol. IV, Papers XI–XIII, three contributions to the psychology of love.

[4] Vd. *Republic*, IX, 588 C–590 B, in which Plato employs the image of a three-headed beast to describe the soul.

Platonic points are that reason is not reason unless it rules the irrational parts of man, his animality; that love is not human love unless it is ruled by reason (*sexuality* sublimated into *eros.*)

(4) Finally, that health is a harmony in the body and that happiness is a harmony in the soul. The important principle here is the insight that human well-being requires integration to resolve conflicts and dissociations; that dissociation is basic to all psychopathology, and that dissociation can be cured or prevented only by the harmony or integration which results from an ordering of the parts. This is achieved only when reason rules, when man is a rational lover.

c. It is important to note, in passing, that the Platonic discussion does not separate psychological analysis from medical and moral precepts. The close relation in the practical application of psychology, between the medical and the moral aspects, should also be noted.

3. Aristotle sharply separates his psychological analysis from its various practical applications.

a. In the first place, Aristotle corrects the metaphysical error of Plato with regard to the relation of soul to body. They are not related as two separate existences, but as the form and matter of the same existence. In short, man exists and operates as he does, by virtue of the composition in him of form and matter. (The form of any living matter is called "soul"; hence, the form of man is the human soul. All of the powers and operations of man are functions of the composite of form and matter, except one, understanding, which is only secondarily a function of the composite.)

b. The *De Anima*[4a] is concerned with the powers and operations of all living things (all things having souls).

(1) Here we learn the distinction of three grades of soul in terms of distinctive operations: *vegetable, animal, human.*

(2) Here also we learn the hierarchy of living powers, the kinds of vital action: the vegetative (nutritive, augmentative, generative), the sensitive (the various

[4a] To this should be added, of course, the short treatises of the *Parva Naturalia*, especially the *De Sensu et Sensibili, De Memoria et Reminiscentia, De Somno et Vigilia* and *De Somniis*. The last should be of particular interest to psychoanalysts.

senses), the locomotive, the appetitive, and the intellectual or rational.

c. In the case of man, the following points are made:

(1) The distinction between the cognitive and non-cognitive powers. Only men and animals are cognitive, and hence only they have an appetitive power (i.e., can be said to desire) because desire depends upon objects known, either by sense or by intellect.

(2) The distinction between the rational and the non-rational parts of man, the rational part being both cognitive and appetitive. Will is rational desire (intellectual appetite), what Plato would have called rational as opposed to animal love.

(3) The various dependencies: of sense and locomotion upon vegetative operations; of intellect upon sense; of appetite upon both sense and intellect; and of appetite also upon locomotion. Yet appetite, in a sense, rules all the other powers, in that whatever we do, we do from desire, either sensual or rational.[5]

(4) The analysis of the senses: the division of exterior and interior senses; the enumeration and description of the various exterior senses; the interior sense as perception, imagination, memory. (Exterior and interior sensitivity can be regarded as functions of the sense-organs and brain respectively.)

(5) The analysis of the intellect: the distinction of the active and the possible intellect; and the separation of the three intellectual acts of abstraction or understanding, judgment or forming propositions, and reasoning or inference. (Also, the order of abstraction, the grades of abstraction, and the distinction between the speculative and practical dimensions of intellectual operation.)

(6) The analysis of desire (appetite): first into the sensual and the rational (or will); then, the sensual into the concupiscible and irascible (with the dominance

[5] There are two aspects of appetite, the affective or emotional side, the conative or impulsive side. The states of the appetite are, therefore, both passions and sources of activity. This, of course, must be qualified by the distinction between the sensitive and the intellectual appetite. The same names are often given to the passions and the states of will, but they are used analogically. Thus, sensual and intellectual *love* are not the same, any more than brutal *anger* and righteous *indignation.*

of love as the root of all the passions). The nature of the passions: as sensual appetites or desires able to control action *without* or *in spite of* reason, though also to obey reason.

d. The basic terms of this psychological analysis are power and act: the capacity to operate or perform actions, and the actions which express those capacities.

(1) Powers are either active or passive; thus, vegetation is active, and sense is passive.

(2) To the basic terms must be added the notion of habit or disposition, as a mean between power and act; [6] thus, habits of locomotion; sensitive habit, known as memory; habits of action and passion (now spoken of as sentiments, attitudes or complexes); habits of intellect, the knowledge we possess but are not at the moment considering.

e. Psychology is concerned with acts, habits and powers. It is concerned with acts in order to learn about habits and powers. It studies acts in relation to their objects. It is *not* concerned with human behavior as such, or with the development of human character as the result of habit-formation of different sorts.

4. The fuller account of human life (of behavior, of character, of development and biography and the types of men) is not, according to the Aristotelian mode of treatment, the province of psychology, but the application of psychology in the practical order in which we are interested in controlling other men, or in directing our own lives well [38].

[6] A habit is a developed power, a determinate potentiality. It is a mean between the power as pure operative capacity and actual operation. Habits thus participate in both potentiality and actuality; with respect to the powers they determine, they are as acts; with respect to the operations they incline toward, they are as powers. Habits are both the results of operation and the causes of operation. A habit is formed by specific acts within a given power, and the habit of that power thus formed enables us better to perform further acts of the same sort. What we are able to do at first with difficulty, by virtue of an undeveloped power, we are able to do after repeated acts with ease and celerity, because a habit has been formed. Habitual operation is pleasant because unimpeded. It approaches the perfection of instinct and it is for this reason that habit is rightly regarded as second nature. Traditionally, the word "habit" is restricted to those formations in which rational power is directly the subject or indirectly involved, and the word "disposition" is used for what we would call "bodily habits." The concept of habit and disposition is indispensable in psychological analysis. It is of crucial importance for understanding what Freudians mean by an "active unconscious." Vd. Note 55 *infra*.

a. The control of others: the practical field of politics, with its subordinate fields of education and rhetoric. Here we have a fuller discussion of the emotions (appetites, desires, passions) and of types of men; the techniques to be used by the orator, the educator and the statesman to control behavior.[6a]

b. The direction of one's own life: the practical field of ethics.

(1) The practical aim of ethics is human well-being, happiness.

(2) But human well-being depends upon the nature of man, and hence ethics as practical is applied psychology.[6b]

(3) The basic points of theoretical ethics are psychological:

(a) The distinction between voluntary, involuntary, and non-voluntary conduct by reference to the way in which reason is involved, and knowledge is used.

(b) The analysis of the virtues as habits of action, passion and cognition (moral and intellectual virtues). The moral virtues are good habits of man's animal nature, good in the sense of habits formed rationally, and good because means to happiness. (The intellectual virtues are good habits of the rational powers; the moral virtues are good habits of those powers which are rational *only by participation*, through submitting to the regulation of reason.)

(c) The fundamental conflict between reason and the passions (what, in Freudian terms, is the conflict between the reality principle and the pleasure principle).

(d) The types of human life: sensual and practical lives (life-instinct dominance) and contemplative activity (death-instinct dominance).

[6a] Vd. *Rhetoric*, Bk. II, Ch. 2–11.

[6b] "The student of politics,"—or ethics,—"must know somehow the facts about the soul, as the man who is to heal the eyes or the body as a whole must know about the eyes or the body" (*Ethics*, I, 13, 1102^{a}18–22). In other words, psychology is to ethics and politics, as physiology is to medicine. The *Ethics* introduces psychological discussions at various relevant points; thus, for instance, the analysis of wish and volition in Bk. III, Ch. 1–4; and the analysis of intellectual power and operation in Bk. VI devoted to the intellectual virtues.

(4) The various practical treatises (*Ethics*, *Politics*, *Rhetoric*) also contain classifications of types of men, descriptions of types of characters, and accounts of human biography in terms of the aetiology of character. None of this material is scientific in the sense of being the result of deliberate, special investigations or researches.

5. This Aristotelian distinction between theoretical and practical psychology (the application of psychology in ethics and politics) is significant in the history of psychology.

a. In the Hellenistic period and in the Renaissance, men wrote books about the classification of characters and about the types of lives (biographies). Theophrastus's *Book of Characters* and Plutarch's *Comparative Lives* are examples of such taxonomic and physionomic knowledge of men.[7] This knowledge was not thought to be part of theoretical psychology. The subject-matter of human psychology was rather, according to the *De Anima*, the powers and acts of man. Furthermore, such knowledge was not the result of what we would call scientific research.

b. This interest in the types of men and the types of lives was predominantly practical, that is, moral, even when the author was merely a biographer or a classifier of characters.

c. This separation of the theoretical and the practical in psychology did not deny the dependence of the practical upon the theoretical.

6. The mediaeval development in psychology was primarily Platonic from St. Augustine to the beginning of the 13th century.[8] In the 13th century, the works of Aristotle were in the possession of Jews, Moors and Christians, and the issue between Plato and Aristotle was finally resolved by St. Thomas, who purified the Jewish and Arabic, as well as earlier Christian, Aristotelians of their Platonic misreading of the Aristotelian texts. Aquinas's work was an enlargement and a perfection of Aristotle's analysis [38a.]

[7] Vd. La Bruyère, *Les Caractères*, 1688; and also the contemporaneous work of Sir Thomas Overbury and Bishop Earle.

[8] Cf. A. Pegis, *St. Thomas and the Problem of the Soul in the Thirteenth Century*, Toronto, 1934; E. Gilson, *The Spirit of Mediaeval Philosophy*, New York, 1936; Ch. IX, XII–XV; D. E. Sharp, *Franciscan Philosophy at Oxford in the Thirteenth Century*, London, 1930.

a. In the *Treatise on Man*,[9] the *De Anima* was amplified. Here we have theoretical psychology: the analysis of human nature, its powers and operations. (It should be noted that St. Thomas points out that the acts of the appetitive part of the soul are reserved for later consideration in the subsequent moral treatises.[10] It should be noted also that although this treatise pays great attention to cognitive operations, there is no "epistemology" in the modern sense of that word, no analysis of the "content and validity of knowledge.")

b. In the *Moral Treatises*,[11] the *Ethics* and *Politics* were amplified. Here, in addition to an analysis of the will, the passions and desires, we have strictly practical psychology: the analysis of habits or virtues and vices; the analysis of the conditions of human well-being, of the ends and means of a good life.

c. But there is little or no taxonomy or physionomy, such as was started by Theophrastus and Plutarch. Their work is not continued in the mediaeval tradition, and when it is taken up again in the later renaissance, it is the work of literary men primarily and not moralists.[12]

III. From the 17th to the 19th Centuries: Philosophical Psychology

1. Philosophical psychology during this period can be summarized under a number of heads: (a) the resurrection of old errors and confusions, (b) the misunderstanding of, and attack upon, the doctrine of faculties, and (c) the misconception of the subject-matter of psychology. There is a sequence and relation of these points which will become apparent.

[9] *Summa Theologica*, I, QQ. 75–88. For other parts of psychological doctrine, one must go to the moral treatises: *Summa Theologica*, I-II, QQ. 6–17 (on will); QQ. 22–48 (on acts and passions); QQ. 49–54 (on habit). Cf. *Summa Contra Gentiles*, II, Ch. 56–78; also *De Veritate* and *De Anima* among the Disputed Questions; and *De Unitate Intellectus* among the Philosophical Opuscula.

[10] Vd. Notes 33 and 38 *infra*.

[11] Vd. fn. 9 *supra*. The following references may be added here: *Summa Theologica*, I-II, QQ. 1–5 (on the ends of life); II-II, QQ. 171–189 (on types and states of life).

[12] That this was the case can be explained by the great influence of Aristotle's *Poetics* during these centuries, an influence complicated unfortunately by misunderstandings of the text.

2. *The resurrection of old errors and confusions: materialism and Platonism.*
   a. Hobbes[13] revived once more the ancient error of the atomists: that there is only matter in motion, the matter being reducible to small particles differing only in quantity, the motion being entirely local. Although Hobbes used an Aristotelian vocabulary, even to the extent of distinguishing faculties, his treatment of psychological phenomena as results of atomic motions in sense-organs and brain, had the familiar consequences of materialism: sensationalism in the analysis of cognition, and the denial of intellect; the dominance of the passions in the analysis of motivation, and the denial of free-will. Hobbes was the first behaviorist. (To his credit it must be said that he was inconsistent with his principles. Despite his analysis, Hobbes recognized that man is rational both in his account of knowledge and of political behavior.)
   b. Descartes[14] revived once more the Platonic error of dualism, and exaggerated that error, making a sharper separation between body and soul by identifying soul with mind or intellect [39]. As a result:
      (1) The substantial unity of man was destroyed, and psychology became burdened with the impossible riddles of the mind-body relation, of sensation and free-will as interactions between incommunicable substances, etc. (Where Hobbes had written *Of Man*, Descartes wrote of the soul.)
      (2) The basic terms of psychological analysis (power, habit, act) were obscured or lost because the soul was viewed as a complete substance rather than as the formal principle of the composite and the proximate source of its powers. Furthermore, since its essence was identified with the operation of thought, the soul could have no variety of parts or powers. (It became impossible, therefore, for Descartes properly to distinguish cognition and appetite, the spheres of intellect and will. Psychology became almost exclusively an analysis of cognition and ignored or distorted the non-cognitive and non-rational

[13] *Human Nature* (1650); *Leviathan* (1651), Pt. I, Ch. I–II.

[14] *Rules for the Direction of the Mind* (1628); *Meditations* (1641), especially I and VI; *Principles of Philosophy* (1644), especially Pt. I; *The Passions of the Soul* (1645–6) in *Philosophical Works*, ed. by Haldane and Ross, Cambridge, 1931: v. I.

aspects of human nature. This ultimately led to the confusion of psychology with epistemology.)

(3) As sensationalism was the error of Hobbes' materialism, so intellectualism or intuitionalism was the characteristic error of Cartesian Platonism. Ideas are both innate and directly known, rather than acquired by abstraction from sense, the instruments not the objects of knowledge. (A materialism consistently carried out reduces man to the grade of brutes; the Platonic error goes to the opposite extreme of elevating the soul to the substantial and cognitive level of angels. The Aristotelian conception of man as a rational animal is the mean between these errors; as St. Thomas so frequently says, man is on the boundary between the material and the spiritual worlds.)

(4) Physiology, animal and human, became the science of mechanical automata.[14a]

c. Locke, Berkeley and Hume [15] followed Hobbes in his sensationalism, but also made the Platonic error of treating "ideas," i.e., sensations or impressions, as the objects of knowledge.

(1) In these three writers, the confusion of psychology with epistemology reached its height. They tried to give a genetic account, an inventory, and a critical evaluation of the contents of the mind. Thus, Locke stated his problem to be "the original, certainty and extent of human knowledge." [16] As a result of the simple error of treating sensations or ideas as *that which* (*id quod*) is known instead of as *that by which* (*id quo*) man knows *what* (*quid*) he knows of *that which* he knows (physical things), the development of psychological doctrine from Locke to Hume ended in subjectivism, phenomenalism and positivism.

(2) In all three, it is not man but the mind, cognition, or

[14a] La Mettrie in *L'Homme Machine* (1748) showed how readily Cartesianism became mechanistic materialism. Vd. also Holbach and Helvetius.

[15] Locke, *An Essay Concerning Human Understanding* (1690); Berkeley, *A Treatise Concerning the Principles of Human Knowledge* (1710); Hume, *Treatise of Human Nature* (1739–40); *Enquiry Concerning Human Understanding* (1748).

[16] *Op. cit.*, I, 1, ¶ 2.

experience, which is made the subject-matter of psychology.

(3) Apart from its obvious errors, Locke's *Essay* is the best work in the British tradition, largely because its author, like Hobbes, was not always consistent with his principles. Thus, Locke was a realist despite his definition of an idea as the *id quod* of knowledge. He is to be praised for his basic distinction between ideas of sensation and reflection, the latter being the understanding's reflexive knowledge of its own operations. His analysis of sensation is sound although it made no advance on the Aristotelian tradition. Furthermore, both in Book II and Book IV, Locke retains the distinction of power and act, and that between active and passive powers.[16a] Locke is a faculty psychologist despite the fact that he objected to the doctrine of faculties because he misconceived faculties as separate agents rather than diverse channels through which the vital energy of the soul expresses itself. (The best that can be said for Locke is that his *Essay* was in the best tradition of philosophical psychology, although neither adequate nor clear analytically. The worst that must be said of him is that his many errors and inconsistencies permitted Berkeley and Hume completely to nullify all the sound insights of his analysis.)

(4) Hume, following Hobbes, emphasized the passive association of ideas as the only principle of mental life, or experience.[17] Hume's positivism, itself the result of psychological errors, discarding the concepts of substance and cause, not only denied soul, either as a substance or as a principle, but also made nonsense of such basic terms as power and operation. Thus, we see how psychological error (of a Platonic variety) is a cause of positivism, and how positivism is in turn the source of even more egregious psychological error (pure phenomenalism). Hume is the

[16a] Locke's chief defect is with respect to the act of abstraction. As a result of his failure to distinguish sensation and intellection, the having of "simple ideas" is entirely a matter of passive reception, though the mind actively compounds and divides, judges and reasons. Cf. Books II and IV of the *Essay*. Vd. Lecture I, fn. 3, *supra*.

[17] *Enquiry Concerning Human Understanding*: Section III.

fountain-head of British associationism from Hartley, Brown and Reid to James Mill, Bain and Spencer.[18]

(5) The French psychologists of the 18th century, such as Condillac,[19] were followers of the English school, and were, therefore, associationists and sensationalists.

3. *The misunderstanding of, and attack upon, the doctrine of faculties.*

a. The traditional analysis of Aristotle and St. Thomas can be called "faculty psychology" only if the concept of faculty is rightly formulated and is not the ridiculous notion it has come to be in modern usage.[19a] Briefly stated, it involves the following points:

(1) That the faculties are powers of operation, i.e., abilities, capacities. (The Wolffian distinction between faculty and power, related to the distinction between habit and disposition, is a misleading confusion of the distinction between the intellect as an incorporeal power and the powers of bodily organs, with the distinction between those powers which are rational or submissive to reason and those which are not. I shall ignore Wolffian terminology in what follows; faculty and power are synonyms.)

(2) That the faculties or powers are neither parts of the essence of soul nor separate from the soul as independent agents, but rather distinctions within the total power of the soul. They are its potestative parts. The faculties are related to the soul as the properties of a substance to its essence.

(3) That the distinction of the faculties from each other and from the soul, which is their principle, is real rather than formal. It is a real distinction of a modal sort. Faculties, being accidents, cannot exist inde-

[18] Hartley, *Observations on Man* (1749); Thomas Reid, *Inquiry into the Human Mind* (1764); Thomas Brown, *Lectures on the Philosophy of Mind* (1820); James Mill, *Analysis of the Phenomena of the Human Mind* (1829); Alexander Bain, *The Senses and the Intellect* (1855), *The Emotions and the Will* (1859); Herbert Spencer, *The Principles of Psychology* (1870–2).

[19] Condillac, *Treatise on the Sensations* (1754), trans. by H. W. Carr, London, 1930. Cf. N. Malebranche, *Recherche de la Verité* (1674); vd. also Maine de Biran, *Essai Sur Les Fondements de la Psychologie* (1812). Malebranche and Maine de Biran are, in contrast to Condillac, the best examples of French Augustinism.

[19a] Vd. C. A. Hart, *The Thomistic Concept of Mental Faculty*, Washington, 1930.

pendently of their subject, man, but they are really distinct from their principle, the soul, as is clear from their independent variability. Such modal distinction of the faculties from each other and from the soul does not violate the unity of the soul or of human behavior, in which the powers *cooperate*.

(4) That the faculties are not *that which* act but *that in virtue of which* man acts. The soul is the principle but man is the subject of operation, and hence of the operative powers. In other words, the powers are of the composite, of man, and not of the soul. (It is important to remember this since it is almost impossible for any writer, even St. Thomas, to avoid such inaccurate phrasing as "the intellect knows" or "the will commands" when what is meant is that man by his intellectual power knows, etc. St. Thomas would agree with Locke that it is not will which is free, but man through the will. The misreading of St. Thomas on this point is extraordinary, since knowledge of his philosophy would make it impossible to suppose that he could be treating a power as if it were a substance, a subject of operation.)

b. We have already seen how Descartes mistakenly supposed the distinction of the faculties to be inconsistent with the soul's unity. In this he was followed by Spinoza. We have also seen how Locke employed the concept of powers, but attacked the false notion of faculties as independent agents. Unfortunately, Locke made the understanding (i.e., soul) itself the subject of the powers he distinguished rather than man, so that Hume denying the soul as substance made an end of faculty analysis in the tradition of British psychology.

c. Leibnitz [20] reduced the faculties to "mere empty dispositions which are the remains or traces of earlier impressions." This was tantamount to denying their existence entirely. Leibnitz, like Descartes, was a neo-Platonist; following the Augustinian concept of the seminal reasons, he accounted for psychological phenomena in terms of the awakening of slumbering, innate ideas rather than as the activation of native powers.

[20] *Monadology* (1714); vd. also *New Essays on the Human Understanding* (1704), an answer to Locke.

d. Kant, influenced by Locke and Hume, on the one hand, and by Wolff,[21] on the other, presents a confused picture. Kant is a faculty psychologist; his philosophy, in all three critiques, but especially in the *Critique of Pure Reason*,[22] depends largely on his analysis of the soul, its powers and operations. Viewed positively, Kant is important as an antidote to British associationism. Unfortunately there is the negative side: his erroneous analysis of the relations of sense, understanding and reason, his Rousseauian addition of feeling as a separate faculty, his failure to see that man and not the transcendental ego is the subject of the powers,—these are blemishes which give Kant, in psychology as in everything else, the pathetic status of being almost right [39a].

e. Herbart [23] took the position of Leibnitz, rejecting the faculties as mere fictions, and returning to a modified form of associationism,—a dynamic sort of "mental chemistry," —in which Wundt followed him,[24] though not always consistently.

f. Brentano [25] is the only outstanding exception in modern times to the denial or misconception of faculty psychology. His "act-psychology" is clearly in the Aristotelian tradition. He is followed by Husserl and others [26] who expanded his analysis in some respects but, because of characteristically modern deviations concerning metaphysics, muddled its clarity and simplicity.

[21] The word "psychology" as the proper name for a subject-matter is probably due to the influence of Christian Wolff (1679–1754), though its invention is attributed to Melancthon (1497–1560). Thus, Wolff, in this as in many other things, obscured the clarity of the scholastic tradition which had made man rather than the soul the central object of analysis.

[22] First ed., 1781.

[23] Vd. *Lehrbuch zur Psychologie* (1816), trans. by M. K. Smith (1891); *Psychologische Untersuchungen* (1839–40); *Psychologie als Wissenschaft* (1824–25). For a succinct comment on Herbart's relation to Kant, on the one hand, and to experimental psychology, on the other, vd. Boring, *A History of Experimental Psychology*, New York, 1929: pp. 237–250.

[24] Vd. Note 39a *infra.*

[25] *Psychologie von empirischen Standpunkte* (1874).

[26] Husserl, *Logische Untersuchungen* (1900–1); Meinong, *Untersuchungen zur Gegenstandtheorie und Psychologie* (1904); Witasek, *Grundlinien der Psychologie* (1908); Lipps, *Die Grundtatsachen des Seelenlebens* (1883); Stumpf, *Erscheinungen und psychische Functionen* (1907). Vd. Titchener, *Experimental Psychology of the Thought Processes*, New York, 1909.

4. *The misconception of the subject-matter of psychology.*

a. The proper subject-matter of psychology, as we have seen, is man *qua* man, his powers, operations and nature. It is neither the soul *per se*, nor the body, nor immediate experience (consciousness or psychological phenomena, i.e., the content of cognitive operation). But as a consequence of the errors of materialism and Platonism, and of the denial or misunderstanding of the doctrine of faculties, modern philosophical psychology has chosen one or another inappropriate subject-matter.

(1) Thus, Hobbesian materialism, consistently carried out, must lead to behaviorism and the reduction of psychology to physiology or animal psychology.

(2) Cartesian dualism made the soul the subject-matter of psychology, and since the soul in operation is thinking or knowing, psychology became the study of consciousness (cognitive content).

(3) The Cartesian influence upon more recent developments must be understood as modified by the positivism of Hume and Kant, for it is in terms of their phenomenalism that the experimental enterprise, beginning in the 19th century, discarded soul and mind, as well as the distinction of power and act, and conceived itself as the investigation of consciousness, or the phenomena of mental life [39b].

b. The influence of Hume and Kant on the subject-matter of modern psychology is twofold:

(1) Locke's error of making ideas the object of knowledge ultimately led to Hume's denial of substance and cause, and hence of the soul as a substantial principle and as a cause of powers and operations. Psychology necessarily became the report of the content of immediate experience. All psychological phenomena, even those of the emotions, desire and will, were reduced to elements in the stream of consciousness, governed by the laws of association. William James' chapter on the Stream of Thought is representative of this tradition in philosophical psychology.[26a]

(2) Following Wolff's distinction between rational and empirical psychology, Kant's denial, on positivistic grounds, of rational psychology, led to the errone-

[26a] *The Principles of Psychology* (1890): I, 9.

ous supposition that psychology must be restricted to the study of phenomena. This is shared by post-Kantian idealists,[26b] as well as by those who began experimental psychology in the 19th century. The experimentalists also misinterpreted Kant's attack on rational psychology as a denial of the possibility of philosophical psychology. The absurdity of this is obvious when it is realized that Kant's own philosophy was in large part psychological [39c].

c. Thus we can see how psychological errors of Platonic lineage led to positivism and how positivism in turn led to the circumscription of psychological subject-matter, on the one hand, and to the attempt to develop a scientific psychology apart from, or in lieu of, a philosophical analysis of man's nature, on the other. Yet, paradoxically, it is only with the development of scientific psychology that we return to a proper definition of subject-matter. Both psychoanalysis and psychometrics study man *qua* man.

5. *Summary:*
   a. There is no investigative or scientific psychology in this period (from the beginning through the middle of the 19th century).
   b. ***The works in philosophical psychology during this period add nothing except error and confusion to Aristotelian psychology as that was perfected by St. Thomas Aquinas.*** Even most of the errors are ancient ones, to be found in the atomistic or Platonic traditions. Only positivism, with its consequent denial of philosophical psychology itself, is new in modern times.
   c. If one were writing the history of psychology as a branch of philosophy and were concerned only to report genuine advances in analysis, everything from the 17th to the 19th centuries could be ignored, *without exception.* Even the best writers need not be mentioned because at their best they are only poor statements of what is better found in Aristotle or St. Thomas.[26c]
   d. But in writing the history of psychology, as both philosophical and scientific, it is necessary to mention these

[26b] Vd. F. H. Bradley, *A Defense of Phenomenalism in Psychology*, Mind, N. S. IX (1900).

[26c] Vd. C. Alibert, *La Psychologie Thomiste et les Théories Modernes*, Paris, 1903.

writers because of their influence upon those who tried to be scientific psychologists in the 19th and 20th centuries.[27] Thus, the influence of Locke and Kant on Wundt; the influence of Hume and British associationism on William James and American behaviorists; the influence of Brentano and Wundt on gestalt psychology, etc.

e. The classification of types of character (types of personality) and the interest in human development, in biography, as an account of the ways in which men develop, is largely lost during this period, except what is left of it in the hands of literary men.

IV. From the 19th Century to the Present Day: Scientific Psychology

1. The rise of scientific psychology cannot be understood apart from the development of the biological sciences in the 19th century. Three different fields of biological research exerted an influence: (1) physiology and neurology, particularly of the sense-organs and brain; (2) natural history, particularly accounts of animal behavior and the mythological part of the doctrine of evolution; (3) medical researches, and particularly the development of general pathology. I shall reserve the third of these until the end.

2. The influence of physiology is to be seen in the development of the Wundtian type of experimental psychology and in the development of psycho-physics.

a. Wundt, the father of experimental psychology, was influenced by such physiologists as Weber, Müller, Helmholtz, who were primarily concerned with physiology of sense-organs and nervous system.[28]

(1) But Wundt's work was a curious mixture of philosophical and scientific psychology. So far as his great works were "systematic," they were philosophical psychology.[29] His basic analysis did not

[27] Cf. G. Murphy, *An Historical Introduction to Modern Psychology*, New York, 1929; E. G. Boring, *A History of Experimental Psychology*, New York, 1929; G. S. Brett, *A History of Psychology*, New York, 1921. Vd. Note 36b *infra*.

[28] J. Müller, *Handbuch der Physiologie des Menschen* (1833–40); E. H. Weber, *Der Tastsinn und das Gemeingefühl* (1846); H. von Helmholtz, *Handbuch der physiologischen Optik* (1856–66), *Tonempfindungen* (1863).

[29] *Physiological Psychology*, 6th ed., 1911; *Outlines of Psychology* (trans. by Judd) 3rd ed., 1907; *System der Philosophie* (1889).

depend at all upon his experimental findings [40].

(2) As a philosophical psychologist, he did not have a consistent doctrine. He followed Descartes in the substantial dualism of mind and body.[29a] (He is not, however, an interactionist, but a parallelist. In his *Physiological Psychology*, he tried to treat psychic phenomena as mere correlates of brain processes.) He also followed Locke and Hume in being a sensationalist and an associationist; yet he followed Kant and Herbart in retaining some mental activity, e.g., the act of apperception.

(3) His experimental work on reaction time, on association, on attention, had no bearing whatsoever on his analysis.

(4) It must be said in his favor that he broadened psychological subject-matter to include the affects and will, but these he treated as if they were *mental phenomena*, i.e., elements of consciousness.

(5) Psychology, instead of being an analysis of man's powers and acts, became in Wundt's hands an analysis of mental elements and their combinations. This is characteristic of all of Wundt's followers, whether they are structuralists or functionalists.[30] It should be noted that this was not the result of research. It was not an investigative finding. It was the result of bad philosophical analysis.[30a] (The differences between Wundt and Kulpe are of little importance, except, perhaps, in so far as the Wurzburg school helped to further the beginnings of *gestalt* psychology.)

[29a] Father Brennan (*General Psychology*, New York, 1937: p. 14) quotes Wundt as saying in his *Physiological Psychology*: "The products of my labors do not square with the materialistic hypothesis nor with the dualistic theories of Plato and Descartes. It is only the animism of Aristotle, in which psychology is combined with biology, that issues as a plausible metaphysical conclusion from experimental psychology" (Bk. II, Ch. 23). However gratifying this declaration by the great experimentalist may be, his work, as well as this very statement itself, betrays little or no understanding of the principles of hylomorphism as a philosophy of nature or in psychology. He is as much an Aristotelian as Descartes is.

[30] Thus, for example, H. Ebbinghaus, *Grundzüge der Psychologie* (1897–1908); Külpe, *Outlines of Psychology* (trans. by Titchener), 1909; E. B. Titchener, *A Text-Book of Psychology* (1909), *Systematic Psychology*, 1929.

[30a] Vd. Note 39a *infra*.

b. Psycho-physics (the work of Weber, Fechner,[31] etc.) also resulted from researches in sense-physiology. It was the beginning of a mathematical psychology, imitating mathematical physics in applying mathematics to the measurement of "psychical" phenomena (e.g., acts of the power of sense). This developed into what has come to be called psychometrics, including not only the measurement of judgments in the field of sensation but all kinds of mental tests and measurements (the work of Spearman, Thorndike, Thurstone,[32] etc.)

(1) This type of research contributed nothing to general psychological theory.

(2) It was, nevertheless, a body of investigative knowledge, much more precise in its methods and findings than any other part of scientific psychology.

(3) But it was neither an analysis of man's powers and operations, nor a science of human behavior. It was at best a science of the quantitative accidents of human operations, particularly the operations of perception and judgment; just as mathematical physics is a science of the quantitative accidents of physical things, particularly their local motions.

3. The influence of natural history, 19th century studies of animal behavior and, in general, the doctrine of evolution, is best seen in the contributions of William James to a science of psychology.[33]

a. James was not himself an experimenter; nor, for the most part, even an investigator except in such fields as telepathy, psychical research, etc.

b. Like Wundt, he was primarily a philosophical psychologist, but one who tried to incorporate into his exposition of psychology everything that seemed relevant to the study of man. As a result, his *Principles* are not in any sense a system; but rather a collection of chapters that do not hang together.

(1) On the one hand, his chapters on the stream of

[31] G. T. Fechner, *Elemente der Psychophysik* (1860); also *Vorschule der Aesthetik* (1876).

[32] C. Spearman, *Nature of Intelligence and the Principles of Cognition* (1923), *The Abilities of Man* (1927); E. L. Thorndike, *The Measurement of Intelligence* (1927); L. L. Thurstone, *Vectors of Mind* (1935). Cf. R. J. Slavin, *The Philosophical Basis for Individual Differences, According to St. Thomas Aquinas*, Washington, 1936: esp. pp. 115–140.

[33] *The Principles of Psychology* (1890).

consciousness, conception, imagination, reasoning, are a confused mixture of Aristotle and Hume. They are bad philosophical psychology.

(2) On the other hand, his chapters on instinct, habit, emotion, hypnosis, personality, draw largely from the results of investigative work in physiology, biology and medicine. They expanded the subject-matter of psychology to include much more than cognition and immediate experience. They created an interest in animal psychology and pointed to its significance for human psychology. (The influence of evolution is apparent here. Man like any other animal lives by adjusting himself to his environment.)

(3) James lacked an analysis whereby to organize all this knowledge about man, and failed to distinguish psychology and ethics. Thus, his chapter on habit is a confusion of psychology and morals [41].

4. We can now trace the way these various influences mix and converge to constitute the currently existing schools of psychology.[33a] These schools differ from one another either philosophically or in terms of some limitation of field of research [42]. The amazing thing is the way in which they combine a positivistic attitude toward philosophy with the espousal of some form of Platonism or materialism in philosophical psychology. The results of materialism and Platonism are errors in philosophical psychology. The result of positivism is the attempt to construct psychology as purely scientific and as independent of philosophy. This is impossible. But, in addition, the two results are inconsistent with each other. Either by itself would produce bad psychology. Together, they make the scandal of this academic subject in the 20th century.

a. *The introspective or structural school: followers of Wundt.*[34] As philosophical psychologists, they are concerned with the elements of consciousness and their

[33a] This enumeration omits psychometrics and psychoanalysis because they can be treated as bodies of scientific knowledge rather than "schools of thought." Vd. Note 46 *infra*. Except for verbal differences and negligible shades of meaning, this enumeration is otherwise as exhaustive as those made by Murchison, Woodworth, and Heidbreder. Vd. Note 1 *infra*.

[34] *E.g.*, Titchener, Judd, Warren, Witmer, Weld, to mention only some Americans.

combinations. As investigators, they do work in sensation, perception, memory, association, imagination, attention, feeling, reaction time, etc.

b. *The functionalists: followers of James.*[85] They are broadly interested in anything having to do with man, his mind, his knowledge, his personality, his life, etc. Primarily, however, they are interested in the facts of human behavior and in its physiological basis. Man can be studied both in the way animals are studied and by introspective techniques. They do any kind of research. Theirs is the broadest conception of psychological subject-matter and the one nearest to the entire ancient tradition.

c. *The behaviorists: radical functionalists who ceased compromising with structuralism.* (Thus, Angell, the functionalist, is followed by Watson, the behaviorist.[86]) They are concerned only with behavior, human or animal. Their methods are those of the physiologist and the animal psychologist. Their only concepts are physiological. They deny any real distinction between man and other animals, and as "philosophers" they are ultimately materialists and sensationalists, following Hobbes and Hume.

d. *The gestalt school: originally followers of Brentano and Husserl, von Ehrenfels, Meinong and Witasek.*[87] They started as critics of both structuralism in human psychology and behaviorism in animal psychology.[88] In one sense, their philosophical analysis is in the tradition of Aristotle; it is an analysis in terms of powers and acts and in terms of form and matter.[88a] But, in another

[85] *E.g.*, Dewey, Angell, Woodworth, Thorndike, Carr. Vd. J. R. Angell, *The Relations of Structural and Functional Psychology to Philosophy*, Dec. Public of Univ. of Chicago, 1903. Cf. T. J. Ragusa, *The Substance Theory of Mind and Contemporary Functionalism*, Washington, 1937.

[86] J. B. Watson studied under J. R. Angell at the University of Chicago. *Psychology from the Standpoint of a Behaviorist* appeared in 1919. Other behaviorists are Weiss, Hunter, Tolman, Lashley.

[87] M. Wertheimer, *Ueber Gestalttheorie* (1925); W. Köhler, *Gestalt Psychology* (1929); K. Koffka, *Principles of Gestalt Psychology* (1935); K. Lewin, *A Dynamic Theory of Personality* (1935).

[88] Vd. Köhler, *Gestalt Psychology*, New York, 1929: Ch. 1–3.

[88a] Vd. R. Allers, *The New Psychologies*, New York, 1933; T. V. Moore, *Gestalt Psychology and Scholastic Philosophy* in The New Scholasticism, VII, 1.

sense, they are sensationalists and physiologists of the same sort as the behaviorists, denying any essential difference between man and brute animals. Their experimental work is narrow, being mainly concerned with cognition and animal learning. In so far as they are concerned with the behavior of the organism as a whole, they cannot be distinguished from the behaviorists, although they use a different language and a more sophisticated neurology [43].

5. *Summary: Where is the science of psychology?*

a. In the first place, the continuing existence of "schools" of psychology during the investigative period indicates that psychology did not become purely scientific when it started to experiment at the end of the 19th century. From its beginning until the present day, experimental or investigative psychology has neither mastered philosophy nor left it alone.[89] In the philosophical dimension, the variety of doctrines in the post-Wundtian schools is constituted by new combinations of earlier errors and confusions in analysis. They add as little to philosophical psychology as did their predecessors from the 17th to the 19th century. If we ignore the differences among the schools as nothing more than bad philosophical polemic, consisting of ungenuine issues and wrong solutions, what have we left? We do have, it must be admitted, a body of knowledge resulting from research and investigation.

(1) This can be summarized under the following heads: sensation, perception, imagination, problem-solving, emotion, feeling, learning, habit, memory, reaction time, association, fatigue, movement, attention, distraction, drug-effects, psychophysics, mental tests and measurements, genetics, individual differences, etc. [44].

(2) This body of investigative findings cannot be denied by any "school of psychology." It is a tremendous mass of scientific data, data discovered by research. But, for the most part, there are no outstanding generalizations, no clear formulae, nothing that resembles the scientific facts of physiology or chemistry. Psychometrics and, as we shall see,

[89] Vd. E. G. Boring, *History of Experimental Psychology*, New York, 1929: p. 660. Cf. pp. 249–50. Vd. also Note 42 *infra*.

psychoanalysis are the only exceptions to the inductive sterility of psychological research [45].

b. In the second place, if one examines all the results of research and investigation done by experimental psychologists, one finds that it can be classified as follows: [40]
   (1) As physiology: all the materials on sensation, reaction time, memory, emotion, fatigue, etc.
   (2) As animal psychology (including studies of human beings *as if* they were animals essentially): all the material on instinct, learning, behavior, development, etc.
   (3) As psychometrics: all the results of tests, measurements, psychophysics, etc. (This can be made to include all experimental data which are metrical.)

c. There is no question that we have acquired much knowledge as a result of the attempt to make psychology a science, but these additions to knowledge:
   (1) *Negatively:* do not alter or expand the traditional philosophical psychology, which is the analysis of human powers and acts.
   (2) *Positively:* are contributions to physiology, animal psychology, or psychometrics (mathematics applied to psychological measurements).

d. Our problem was to find a subject-matter, studied investigatively, which was neither physiology nor animal psychology. The history of so-called experimental psychology, with one exception, reveals none. That exception is psychometrics. Otherwise, experimental psychologists do physiological research or study the behavior of animals and the behavior of men as if they were animals [45a].

e. The only other exception, psychoanalysis, is not in the tradition of experimental psychology. It arose as a result of the third way in which biology influenced psychology in the 19th century.[41]
   (1) Medical researches and the science of pathology are responsible for the rise of psychiatry and psychopathology.[42] The early psychiatrists were not

[40] Vd. *The Foundations of Experimental Psychology*, ed. by C. Murchison, Worcester, 1929. Cf. *Psychology, a Factual Textbook* (New York, 1935) by Boring, Langfeld, Weld and others.

[41] Vd. Section IV, 1 of this lecture, *supra.*

[42] Vd. B. Hart, *Psychopathology, Its Development and Its Place in Medicine,*

experimenters or testers. As physicians, they employed a clinical method, a method which involved the classification of types of cases (the diagnostic determination of syndromes) and the study of case histories for the sake of diagnosis and prognosis [45b].

(2) Psychiatry, from the beginning, was a taxonomic and physionomic study of men as men, and not in the manner of the behaviorists, a study of men as brutes.

(3) A basic divergence arose with respect to aetiology. On the one hand, right-wing psychiatry looked to neurology for explanations of pathology; on the other hand, psychoanalysis attempted to give a psychogenesis of the disorders of men.

(4) Ordinary psychiatry is in unstable position between neurology, on the one hand, and psychoanalysis, on the other. It is psychoanalysis which is truly a science of psychology; that is, an investigative study of man as man, and not of man as an animal.

f. The historical significance of this fact can easily be shown. We saw that in the ancient and mediaeval period, psychology had both a theoretical and practical dimension. Its theoretical dimension was the analysis of human powers and acts. Its practical dimension was ethics, concerned with the problem of the formation of habits and the development of character, and, in general, with the conditions of human well-being. We must also remember the tradition of Theophrastus and Plutarch: the classification of the types of character, and human biography.

(1) Psychoanalysis, originating in medicine as practical, as concerned with the health of man in the fullest sense, his complete well-being,—happiness is equivalent to health in this sense,—is naturally in the tradition of ethics. We shall find that the doctrine of psychoanalysis is in part ethical. This is quite proper, to the extent that psychoanalysis is applied psychology. (To the extent that the psy-

New York, 1927; R. S. Woodworth, *Dynamic Psychology*, New York, 1922: Ch. I; S. Freud, *The History of the Psychoanalytic Movement* in *Collected Papers*, London, 1924: I, pp. 287–359; also in the same volume a paper on Charcot: pp. 9–23.

choanalyst is merely a therapist, his practical interest is much narrower than that of the moralist, who is primarily concerned with prevention rather than cure.)

(2) But psychoanalysis is not merely practical. It does not merely apply knowledge. It seeks knowledge; it is investigative. As investigative, it is a body of research that is primarily taxonomic and physionomic, achieved by the appropriate clinical methods of diagnosis, case-history, etc.

(3) In other words, the classification of the types of men and the formulation of the types of development in human biography is the proper subject-matter for a science of psychology. We had knowledge of these matters before psychoanalysis, but until psychoanalysis we did not have scientific or investigative knowledge in this field. Furthermore, the investigative method of psychoanalysis, being clinical, is properly adapted to such subject-matter.

g. In the next lecture, I shall more fully substantiate the point that I have made in this one, namely, that psychoanalysis is the only *non-mathematical* science of psychology, the only body of investigative knowledge about man as man,—as essentially different from other animals,—knowledge achieved by a method of investigation that could not be used in studying animal life [46]. In the course of the final lecture, I shall attempt to establish this point.

h. In so far as I shall find it necessary to criticize psychoanalysis and to suggest corrections of its errors, the faults and errors will all be due to its misfortune of having been born in the 19th century.[48] But I shall try to

[48] As is so often the case, Freud is both a positivist and given to "speculations" which are his unacknowledged philosophy. He is a positivist in so far as he holds that, other than science, "there are no sources of knowledge or methods of research" (*New Introductory Lectures on Psychoanalysis*, New York, 1933: p. 218). He rejects the notion that science is only one field of inquiry and that philosophy and religion are others with equal claims to truth. "The bare fact is that truth cannot be tolerant and cannot admit compromise or limitations, that scientific research looks on the whole field of human activity as its own" (*op. cit.*, p. 219). Yet Freudian metapsychology cannot be purified of "philosophy." Vd. Note 51 *infra*. The task is not one of freeing psychoanalysis from all philosophical presuppositions, but rather of correcting the mixture of materialism, dualism and evolutionism which is the Freudian philosophy. Cf. R. Allers, *The New Psychologies*, New York, 1933: pp. 5–24.

show that psychoanalysis can be cured of its heritage of bad philosophy,—the currents of error which dominated its 19th century origin. *Upon such corrections, its future depends.*

LECTURE FOUR

PSYCHOANALYSIS AS PSYCHOLOGY

I. Introduction

1. This lecture is devoted to the status of psychoanalysis as psychology, both theoretically as a body of knowledge and practically as a moral doctrine. To state the problem of this lecture, it is necessary here briefly to reiterate:
   a. The distinction between philosophical and scientific knowledge. (Generalizations which are neither are *opinion*. The opinions of scientists are often confused with philosophical knowledge, because scientists think that philosophy is "speculation," in the sense of guessing about what they do not know.)
   b. The distinction in kind between man and brute animals, established *a posteriori* in terms of the absolute difference between what man can do and brutes can do.
   c. The thesis that the proper subject-matter of human psychology is man as man, and not merely as an animal.
      (1) Human psychology must, therefore, be differentiated from animal psychology.
      (2) Furthermore, man is a unity, and psychology is the study of that unity, both in being and in operation, whereas anatomy and physiology are studies of the nature and operation of the parts, considered separately. (Psychology is distinguished from physiology in two respects: first, in studying the cooperation of the parts in the activity of the whole; and second, in studying certain operations which can be best known reflexively because they are not acts of bodily organs.)
   d. The thesis that psychology is the only branch of philosophical knowledge about a particular type of physical being. Thus, with respect to nature in general, there is philosophical knowledge, which we have called general

physics, or the philosophy of nature; with respect to different types of natural things and different types of change, there are the particular natural sciences, all of them bodies of investigative knowledge. *Psychology is distinguished from all these by the fact that, in part, it is philosophical in method, though particular in subject-matter.*[1] The reason for this exception is that psychology is man's knowledge of himself, and because of this reflexivity investigation is not necessary, although it is possible; not only possible, but desirable as providing supplementation and detail. In the case of other particular subject-matters, investigation is necessary.

e. The thesis that a sound philosophical psychology exists.[1a] It is the analysis of the unity of man as body and soul, and the analysis of man's powers and operations in relation to each other and as cooperating in human life.

f. The thesis that psychology unites both philosophical analysis and scientific findings in one continuous doctrine, in which philosophy, answering the basic questions, necessarily subordinates, as well as interprets and regulates, scientific research.[1b]

2. As a result of these distinctions and theses, we then faced the following problem: what is the nature and scope of scientific psychology, that is, a body of knowledge about man as such, which is neither physiology nor animal psychology, and yet is obtained by investigative methods? To answer this question, we surveyed the history of psychology, and discovered:

a. That until the 19th century, there was no attempt at a scientific (investigative) study of man;

b. That the history of psychology from the beginning until the 19th century is a history of philosophical psychology, in which can be found:

(1) The correct analysis, made by Aristotle and St.

[1] In other words, there is both a philosophical and a scientific psychology, but not both a philosophical and a scientific chemistry, botany, hydrodynamics, histology, etc. The philosopher must not give psychology up to the scientist, nor can the scientist ever dispossess him from this field. Vd. Note 32a *infra.*

[1a] It has existed in a continuous tradition since the Greeks, obscured somewhat by the interruptions of modern philosophy. Those who ignore or neglect this tradition are led to suppose that knowledge of man is a modern or even a recent achievement, or to place it in the future, still to be achieved.

[1b] Vd. Notes 32a and 39c *infra.*

Thomas, of the union of soul and body as a composition of form and matter. It follows that the basic terms of psychological analysis are power, habit and act. This analysis distinguishes the various powers, habits and operations, formulates the order and relation of these powers and operations, etc.

(2) The three errors of dualism (or Platonism), materialism and idealism (another form of Platonism). Each of these errors is in metaphysics, but each metaphysical error has psychological consequences. Thus, dualism in separating soul and body as existing independently, assigns to the soul powers and operations it does not have, and improperly orders and relates the various powers of man. This error is the source of the overemphasis on cognition, the confusion of psychology and epistemology, and, worst of all, the misconception of the subject-matter of psychology as immediate experience or consciousness. Similarly, materialism, denying the immateriality of the soul, is unable to distinguish sense and intellect and reduces psychology to physiology and animal behaviorism; and idealism, denying matter, is forced to the absurd extremes of phenomenalism, solipsism and skepticism, or, what is worse, pantheism (objective idealism).

(3) The way in which these errors entailed a misunderstanding of the doctrine of faculties and hence the loss of the basic principles and concepts of philosophical psychology. It is interesting to observe that both positivism and idealism converge upon, and reinforce each other in making, a purely phenomenalistic construction of the subject-matter of psychology.

Note: The four metaphysics thus have four psychologies correlated with them. The three metaphysical errors generate errors in psychology. The history of modern philosophical psychology (usually without an explicit metaphysics) is a history of these errors, two of which are the consequences of Platonism. If we are Aristotelians in the philosophy of science and in metaphysics, we must be Aristotelians in psychology.[2] These issues cannot be avoided

[2] Vd. Lecture II *supra.*

by psychologists, because scientific work in psychology depends upon philosophical analysis. Other natural scientists can be positivists without impairing their research, but positivism either blinds scientific psychologists to the scope and nature of their field of investigation, or encourages them to continue unwitting and incompetent "philosophizing."

c. That the various "schools" of psychology since the 19th century do not differ "scientifically," but only as one bad philosophy from another,[2a] and this polemic is perpetuated by the fact that most scientific psychologists are at once positivists and also philosophically doctrinaire.

d. That an examination of all the *experimental* work done in the name of psychology shows it to fall under the following heads:

(1) *Animal psychology:* the work of the behaviorists and gestalters.

(2) *Physiology:* the work of the behaviorists and gestalters, as well as work done by introspective psychologists in the field of sensation.

(3) *Psychometrics:* measurement of the quantitative accidents of human performances.

e. That psychometrics,—the analogue of mathematical physics,—satisfies all the requirements of a science of psychology. (Psychometrics can be extended to include all the experimental work on perception, judgment, memory, association, etc., in so far as it is metrical. In so far as such work is not metrical,—i.e., qualitative introspective research, —it is a description of the field of consciousness or the flow of immediate experience, and belongs to what Santayana called "literary psychology.") Furthermore, psychometrics as a scientific psychology employs, as it should, the basic conceptual schema of philosophical psychology,—the analysis of the faculties.[3]

[2a] Vd. Note 42 *infra*. All those who have recently surveyed "schools of psychology" concur in Boring's judgment that the varieties of doctrine present only philosophical issues. This is the meaning of Woodworth's statement that "the contentions of each school seem to me not of the nature of hypotheses that are susceptible to proof or disproof by the weight of evidence" (*Contemporary Schools of Psychology*, New York, 1931: p. 217). Cf. E. Heidbreder, *Seven Psychologies*, New York, 1935: Ch. XI.

[3] The relation between psychometrics and faculty analysis as Spearman formulates it, nicely exemplifies the proper ordering of scientific and philosophical knowledge in psychology. Vd. Notes 39a and 45 *infra*.

f. That psychometrics as the only science of man in the experimental tradition does not exclude psychoanalysis, which is also a science of man, but non-metrical and non-experimental. In the sense that any non-mathematical science of a subject-matter is more obviously an account of that subject-matter than a science applying mathematics, psychoanalysis is more obviously knowledge of man than psychometrics.[3a] But, unlike psychometrics, psychoanalysis obscures, or even denies, its dependence upon a philosophical analysis of human nature. As a consequence, psychoanalytic doctrine is in large part a confusion of science and philosophy.[3b]

3. In order to clarify the scope and content of psychoanalysis, we must answer the following questions:
   a. What are the methods and problems of psychoanalysis as a non-mathematical science of psychology?
      (1) With respect to method: the distinction between clinical and experimental procedures.
      (2) With respect to problems: the distinctions of taxonomy,—including morphology,—physionomy, and aetiology [47].
   b. What is the relation between psychoanalysis and philosophical psychology? What, in general, is its relation to philosophy?
   c. What is the relation between psychoanalysis as theoretical (as a body of knowledge) and as practical (as medicine and morals)?

II. The Strictly Scientific Contribution of Psychoanalysis

1. The traits of scientific knowledge can be briefly summarized as follows: [4]
   a. Scientific propositions, describing the phenomenal order by characterizations and correlations, are inductive gen-

[3a] Vd. Notes 46 and 47 *infra.*

[3b] Freud would hardly say of traditional philosophical psychology what Professor Spearman does when he points to the fact that modern efforts in this field never render "the older fine works in any way obsolete. Thus we moderns (in physics) can afford to look at Archimedes patronizingly, as down at a brilliant child; but (in psychology) Plato, Aristotle, Aquinas, not to mention many others, even yet oblige us to look upward as at our masters" (*The Nature of Intelligence*, New York, 1923: pp. 27–28). *Parentheses mine.*

[4] Vd. Lecture I; also my *Analysis of the Kinds of Knowledge* (privately issued, Chicago, 1935): pp. 21–32.

eralizations from observations made under special circumstances and by special methods.

b. Such generalizations are a contribution to knowledge only if men are not able to make them from common experience, or if men are not able to make them as well, that is, with as much precision or in as much detail. We must remember, of course, that precision and detail are in some matters gratuitous. Thus, many of the generalizations of social research are in no sense a contribution, because they are common knowledge. The precision supposed to be added by investigation is, for the most part, pretentious.[5]

c. A scientific fact is a general proposition inductively achieved from scientific evidence. Scientific evidence consists of the data obtained by special investigation.

(1) Scientific facts are the basic inductions. What are called scientific laws are usually inductions from these inductions; i.e., higher generalizations. Both scientific facts and scientific laws are descriptions of what is the case in the phenomenal order. The law does not "explain" the facts, except in the sense in which the less general is made intelligible by subsumption under the more general.

(2) The words "hypothesis," "theory" and "explanation" have two quite distinct meanings in relation to science. Thus, an hypothesis means either a supposed entity (a fictive hypostasis), or a proposition to be proved (a problem); a theory means either an imaginative description of the properties and behavior of hypothetical entities or a systematic organization of the facts and laws which constitute a body of scientific knowledge; an explanation means either a statement of the causes or the demonstration of a conclusion by premises.

(3) A scientist should not make "hypotheses" in the first sense, and should not have a "theory" in the first sense. If he does, the result is merely guessing; it is neither scientific knowledge nor philosophy [48].

(4) A scientist may attempt explanation by turning to descriptions of phenomenal systems of lower order, as the biologist turns to chemistry;[5a] but as a scientist he

[5] Cf. *Art and Prudence*, New York, 1937: Ch. 9 on Knowledge and Opinion.
[5a] Vd. Notes 33a and 47 *infra*.

must not employ fictitious entities whose behavior he imagines. (If he seek explanations in the sense of genuine aetiology, he is in danger of becoming either a guesser or a philosopher.) A scientist need not explain; explanation is not a necessary part of scientific knowledge. Thus, taxonomies and physionomies are bodies of scientific knowledge and do not treat of causes; so also mathematical physics is descriptive and not explanatory [49].

2. The scientific contribution of psychoanalysis must be limited to taxonomic and physionomic problems. (These, with the addition of morphology, are the only problems proper to a non-mathematical natural science and capable of being solved by investigation. But, as we shall see, "psycho-morphology" is primarily philosophical rather than scientific.)

a. *Human taxonomy:* the classification of the types of men.

(1) At this point, psychoanalysis is in the tradition of the study of man, following Theophrastus, la Bruyère, and the characterologists, who classified types of characters.[5b] This kind of knowledge has always been the possession of moralists and novelists, and of men in general, but psychoanalysis has made a contribution in precision and detail, by means of scientific method.

(2) Psychoanalysis has made a number of basic classifications. Thus:

(a) The infantile, the neurotic, the adult.

(b) The abnormal types: the various neuroses.

(c) The normal types: the neurotic characters: the analerotic, the narcissistic, the anacilitic, etc.

(3) The classifications are, of course, developed by detailed characterizations of each type.

Note: It was Freud's brilliant insight that the normal and the abnormal are continuous; that every character is a neurosis, although we call a character a neurosis only when its typifying characteristics become unduly predominant.

b. *Human physionomy:* the formulation of the pattern of human development as a uniform secular trend. There is a universal invariancy here, as in the functional statements of covariation in mathematical physics; the invariancy differs with regard to the time dimension. Human physi-

[5b] For the history of these writings, vd. A. A. Roback, *The Psychology of Character*, New York, 1927: Ch. II.

onomy is analogous to the embryological series, the growth curves and life-histories in biology.

(1) Here, too, psychoanalysis is in the tradition of the study of man, following Plutarch and other biographers, whether moralists or novelists. This kind of knowledge has always been the possession of moralists and novelists, but here even more than in taxonomy, psychoanalysis has made a contribution to knowledge of the pattern of human development, by means of scientific research.

(2) This contribution is of two sorts:

(a) *The general physionomy:* the life-history of any man, the elements of which are such episodes as the intra-uterine period, birth trauma, infancy, weaning, the Oedipus crisis, latency, adolescence, adulthood, either neurotic or complete.

(b) *The special physionomies:* typically different life-histories for different types of characters. (Here taxonomy and physionomy are related.)

c. These two sorts of knowledge about man are scientific. They consist of generalizations based on scientific evidence. This evidence is gathered by a method which I shall call "clinical" to distinguish it from experimental procedures.

(1) The clinical method in the study of human beings is merely a special case of a method commonly employed in the biological sciences, the method of the natural historian. That we call this special case of the method of natural history "clinical" is due to the understandable accident that a clinic affords excellent opportunities for taxonomic and physionomic research with regard to human beings.

(2) The biological sciences are in good part taxonomic and physionomic, concerned with the classification of plants and animals and the formulation of life histories. As such, they are not experimental and could not be. In distinction, such biological sciences as pathology, physiology, genetics, are experimental.

(3) The method of psychoanalysis extends the clinical procedures of medicine.

(a) Diagnosis is rudimentary classification; clinical pictures are typologies, characteristic sets of symptoms. Psychoanalytic diagnosis is much more elaborate; its syndromes are detailed formulations of character types.

(b) Ordinary medical procedure is satisfied with a case-history reported by the patient, whereas psychoanalysis has invented a technique for getting a case-history in detail far beyond what is obtainable by the ordinary exercise of memory. *The psychoanalytic couch is an instrument of case-history research which has discovered many details of human biography hitherto unknown. It stands to biography as the microscope to anatomy or the telescope to astronomy.*

d. To the extent that psychoanalysis is purely taxonomic and physionomic, its original method,—an extraordinarily ingenious modification of clinical procedure—is the only proper method of investigation. (This explains why the condemnation of psychoanalysis as not scientific because not experimental, or mathematical, or statistical, is so absurd and irrelevant.)

e. But for the most part, psychoanalysts will not be satisfied with this appraisal of their scientific accomplishment. They claim that by the use of scientific methods they have contributed two other sorts of knowledge: (1) a morphology, or analysis of the structure of the psyche, which they call "topography"; and (2) an aetiology, or an account of the causes of psychic action and development, and this includes what they call "dynamics" and "genetics." It remains to show, therefore, that this claim is not justified. I do not mean that topography, dynamics and genetics are not a part of psychoanalytical doctrine; in fact, they comprise a large part of the theoretical literature written by psychoanalysts. I mean, rather, that to the extent that this group of problems is capable of solution, they belong properly to philosophical psychology and, for the most part, have been already solved. If psychoanalysis has made any contribution here, it has been made by psychoanalysts, not in their capacity as scientists drawing inductions from clinical data, but as philosophers. Even a psychoanalyst cannot repress his reflective and analytical powers. In spite of his emotional reaction against the name and tradition of philosophy, Freud has had a number of insights which genuinely enrich philosophical psychology.

f. I shall devote the next section of this lecture to Freudian topography, dynamics and genetics, in an attempt to separate that part of it which is sound philosophical analysis from that part which, unfortunately, is the kind of bad "philosophizing", of the sort current in the late 19th century, which consists of *guesses* (hypotheses, theories, explanations, all in the bad sense) *beyond the evidence*. In that part which is good analysis, I shall try to distinguish what is an original contribution from what is merely a translation of the Aristotelian tradition into the conceptual vocabulary of psychoanalysis. But, before turning to that discussion, there are a number of preliminary observations.

(1) The morphological problem may belong either to scientific investigation or to philosophical analysis. Thus, anatomy is a natural science dealing with this problem. But the parts of the soul are not observable as the parts of the body are. The parts of the soul are the powers of man. They are not phenomenal entities, but belong rather to the ontological order, since they are founded in the essence or specific nature of man. They are known only by inference from operation or activity. The data, on which such inferences rest, are facts of immediate experience. They fall within man's reflexive knowledge of himself. The analysis of the powers of the soul is the best established part of philosophical psychology throughout its long ancient and mediaeval tradition. It remains only to show that the scientific evidence, assembled by psychoanalysis, can do no more than exemplify distinctions already and better known; and that the topographical analysis which psychoanalysts have made of the structural parts, the layers or levels, of the psyche, is at best a repetition of traditional points, and at worst a confused or inadequate statement.

(2) The aetiological problem, strictly, does not belong at all to theoretic science. In so far as psychoanalysis is intimately related to therapeutic medicine, it must of course be concerned with causes. To the extent that the reading of psychoanalytical doctrine is colored by practical considerations, it may appear to be aetiological. That is not inconsistent with what is here being said, namely, that as a body of theoretic knowledge, obtained by the methods of science and

therefore restricted to the phenomenal order, psychoanalysis does not treat of the causes of human behavior or psychic development or, in the broadest sense, the changes of which man specifically is the subject [47].

(a) Psychoanalysts may attempt to explain the phenomena they describe by reference to physiology, which is the study of phenomenal systems of lower order. But this sort of explanation must always be incomplete, because human reason, which is the essential and distinguishing cause of human activity, is not the act of a bodily organ. Hence, the reduction to physiological terms can at best explain man as an animal, but not man as man.

(b) Psychoanalysts may unwittingly convert their physionomies into aetiologies. This type of interpretation is justifiable in the practical dimensions of diagnosis and therapy, but it is not, in any sense, an addition of new theoretic knowledge [50]. Furthermore, if the interpretation be regarded as theoretic rather than practical, it transgresses the positivistic restrictions of science, as well as commits the fallacy of *post hoc ergo propter hoc*.

(c) This is not to deny that an aetiology of human activity is possible, but only to assert that it belongs properly to philosophical psychology which, being knowledge of a distinctly different type from science, can achieve real causal insights. (As the principle and determinations of causality in general belong to the philosophy of nature and not to natural science, so human causation belongs to philosophical and not scientific psychology.) Even so, there is a limitation of our knowledge of human causation, because reason is a principal and essential cause, and it is different from every other natural cause in that it is free.[6] With this qualification, philosophical psychology treats of the causes of human operation by formulating the dynamic and cooperative relation of the various powers, the tendencies

[6] Vd. *Art and Prudence*, New York, 1937: Ch. 8 and 9.

of their habits, and the determination of their acts by objects. Psychoanalytical theory, in that part of it which is called "dynamics", is aetiological in the same way; but here it is philosophy rather than science, and furthermore it is here, for the most part, a repetition of the traditional analysis. There are two exceptions to this statement: on the one hand, psychoanalysis makes the crucial error of misconceiving the dynamic and special role of reason as a cause; on the other hand, psychoanalysis makes a number of genuine philosophical contributions to dynamics, especially the concept of repression. We shall discuss this later [55].

(d) The remaining part of what the psychoanalyst thinks is an aetiology,—his genetics,—is neither scientific nor philosophical knowledge. We shall deal with this after the dynamics.

III. Psychoanalysis as Philosophical Psychology and as Guessing:

1. I repeat: only the taxonomic and physionomic knowledge which psychoanalysts have contributed is scientific psychology. Everything else in psychoanalytic literature is either *physiology* or *metapsychology*. I shall use Freud's word "metapsychology" to name two quite different sorts of thing: on the one hand, philosophical analysis; on the other hand, opinions of a speculative sort (neither scientific nor philosophical knowledge, but guesses, conjectures,—"theories" in the sense in which a scientist should not have theories).

   a. In using the word "metapsychology" Freud recognized that the doctrine advanced under that head was not strictly scientific knowledge [51].

   b. In so far as the metapsychology is philosophical psychology, its chief defect arises from the fact that it is based too much on special, or clinical, experience. (This may explain why the originality of the metapsychology, to the extent that it is philosophical, is largely its peculiar terminology.)

   c. It should be remembered that the criteria for judging the goodness or badness of the metapsychology are not the same as those before used to judge the contribution of psychoanalysis to science. *The part of the metapsychol-*

ogy that is philosophy must be judged by the criteria of philosophical knowledge. The part that is opinion or wild speculation can, of course, be dismissed as such. Our task, therefore, is to separate the metapsychological doctrine into these two parts.

2. But first we must distinguish three sorts of problems dealt with in the metapsychology.

   a. *The topographical problem:* Here the concern is with the structure of the psyche. Psycho-topography is the analogue of anatomy, as psycho-physionomy is the analogue of embryology [52].

      (1) The topographical problem is the basic philosophical problem of distinguishing the various powers of the soul.

      (2) Topography is related to taxonomy (the parts of the psyche in different types of characters) and to physionomy (the parts of the psyche at different stages of development).

   b. *The dynamic problem:* Here the concern is with the cooperation, dependence and interdependence of the various parts of the psyche.[6a]

      (1) The dynamic problem is also a basic philosophical problem, that of ordering the various powers of the soul in terms of their dependence and cooperation with each other.

      (2) To the extent that the dynamic problem involves the consideration of the conditions under which the powers become operative, it leads to an aetiology: either the sort of scientific explanation accomplished by turning to physiology, or a philosophical analysis of causes.

      (3) Dynamics is related to taxonomy (the order of the parts in different types of characters) and to physionomy (the order of the parts at different stages of development).

   c. *The genetic problem:* This is concerned with the origins of the psyche itself: how the individual man comes to have the psyche that he has: or how man as a kind of animal comes to have his psychological nature.

[6a] It is interesting to note that Alibert conceives the empirical province of Thomistic psychology as *la dynamologie.* This is the province both of philosophical analysis and scientific inquiry. Vd. Note 39c *infra.*

(1) This is neither a scientific nor a philosophical problem [53].

(2) It is at this point that the metapsychology becomes speculative in the bad sense of unfounded "theory" about unobservable events.

3. I shall survey the metapsychology, first by considering its topography and dynamics in the light of the traditional philosophical psychology [53a]; and then by showing why the genetics has no status either as scientific or philosophical knowledge.

4. The Freudian topography is for the most part sound as philosophical psychology. It is, however, analytically inadequate. This can be shown by indicating briefly the translation of the Freudian and Aristotelian analyses.[7]

a. The structures of the psyche ≡ the parts or powers of the soul. (In neither case is the psyche or soul supposed to exist apart from the body. The Freudian correlates psychic structures with bodily parts; the Aristotelian correlates psychic powers with bodily organs, except in the case of understanding. This difference is, of course, crucial and is the source of all the Freudian errors.[7a])

b. The basic distinction of ego and id ≡ the basic distinction between the rational and the non-rational parts of the soul.

(1) Here the Aristotelian analysis is better because more complete. It distinguishes between the cognitive and the appetitive powers; in the case of the cognitive power, between sense and intellect; in the case of the appetitive power, between the natural or vegeta-

[7] The translation which follows does not aim to be exhaustive but illustrative. Moreover, it will not be understood unless it is read as a statement of formal equivalence between the conceptual schema of two doctrines which, though the same, are expressed in different vocabularies. I have used the sign ≡ to signify such formal equivalence. In making this translation between Freudian and Aristotelian doctrines, I have not distinguished between the contributions to the latter which were made by St. Thomas and by Aristotle.

[7a] The problem in natural theology of the self-subsistence of the soul turns on the proposition that understanding is not the act of a bodily organ. Vd. Note 26 *infra*. There is no inconsistency between the fact that the soul is the substantial form of the composite and the fact that its one incorporeal power (intellect) is a mark of its capacity to subsist apart from the body upon the corruption of the composite. From the point of view of psychology rather than natural theology, it is important to stress here that the soul is a formal principle, the proximate source of all the vital powers of which man is the subject, only one of which is a power of the soul *per se*.

tive appetite,[7b] the sensitive appetite and the intellectual appetite. The rational part of man includes intellect as cognitive and as appetitive; the non-rational part of man includes the senses as cognitive and as appetitive, as well as the vegetative and the locomotive powers.

(2) The incompleteness of the Freudian analysis here is confusing because the id is non-rational *both* as non-cognitive *and* as appetitive only in the vegetative or sensitive way. In other words, the Freudian distinction of ego and id is partly in terms of the cognitive vs. the appetitive powers (knowledge and desire), and partly in terms of that which is rational and that which is not (reality vs. pleasure principle).

(3) Furthermore, the Freudian analysis of the ego in terms of its cognitive functions and in terms of its rationality is very inadequate. The Freudian analysis is adequate only with respect to the id.

c. The distinction between the pure id and the organized id (complexes, erotic organizations, etc.) ≡ the natural appetites (non-cognitive, vegetative) and the cognitive appetites, which are specific desires based on the determination of the appetite by objects known sensitively or intellectually [54].

(1) The Freudian analysis fails to make a clear distinction between need and desire in terms of the cognitive determination of appetite.

(2) The analysis of object-fixations, erotic organizations, complex formation, can be improved by a better understanding of the relation of cognition to appetite.

(3) The Freudian theory of instincts is the analysis of the vegetative powers (natural appetites, needs).

d. The differentiations of the conscious, the preconscious, the unconscious and the repressed unconscious. Here the translation is as follows:

(1) Conscious ≡ any knowledge or desire in act or operation as such.

(2) Preconscious ≡ any knowledge or desire in habit.

[7b] The natural appetite is not identical with the vegetative appetite. What is here meant is only that the appetitive inclinations of the vegetative powers are natural in the sense of not being cognitively determined to a particular object. Vd. Note 54 *infra.*

The habit may be either sensitive or intellectual. But the Freudian analysis fails to distinguish between the sensitive and the intellectual preconscious (sensitive and intellectual habit).

(3) Unconscious ≡ the vegetative needs which have not yet become cognitively determined. (The unconscious is the pure id. The organized id may consist either of preconscious desires or repressed complexes, etc.) In a broader conception of the instinctive, the unconscious includes all the powers of the soul as undetermined by habit and as naturally appetitive, i.e., as tending to their appropriate acts. Habits belong either to the preconscious or the repressed unconscious.

(4) Repressed unconscious ≡ habitual desires which achieve motility without ideational representation and issue in all forms of overt conduct without the control of reason. The repressed tendencies may be temporarily barred from motility by the opposition of reason, but such interference is never permanent, nor does it abolish the tendency [55].

e. The ego and the super-ego ≡ the distinction between the speculative and the practical reason, between reason as knowing the truth and reason as using the truth for the purposes of action.

(1) Here, as we shall later see, the Freudian analysis is poorest, because it fails to understand the identity of reason as knowing and reason as prescribing, or directing or commanding, action. As a result of its questionable genetics, the Freudian account of the super-ego separates it from the ego, and makes it a secondary function of ego and id.

(2) In one important point here the Freudian analysis is correct. The super-ego is related to the id as the practical reason (the reason as moralistic) is to the appetitive part of the soul. The Freudian, however, fails to distinguish the rational appetite from sensuality, the will from the passions.

(3) And the Freudian analysis is bad in so far as it makes the super-ego a function of the id in relation to prevailing social conventions as these are learned by the ego. This reduces morality to mere conformity to the tribal *mores*. From the point of view of Freud-

ian therapy and moral doctrine, this is the worst error in psychological analysis.

f. The various complexes and erotic organizations of the id ≡ the various passions and determinations of the sensitive appetite.[8]

g. *Summary and criticism* [56].

(1) In regard to the ego: the Aristotelian analysis is much more adequate, not only because it distinguishes between sense and intellect and between intellect and will, but because it analyzes their operations and relationship.

(2) In regard to the ego and super-ego: the Aristotelian analysis is much more adequate with respect to the distinction between speculative and practical intellect, and between the practical intellect and the intellectual appetite or will.

(3) In regard to the id: here the Aristotelian and Freudian analyses are about the same, but the Aristotelian is clearer because it more sharply distinguishes needs (vegetative appetite) from desires (appetite cognitively determined to an object). There are minor differences in the account of the various passions (or complexes, erotic organizations, object-fixations).

(4) In general, the recognition of the bifurcation of the human psyche into ego and id is the recognition, however unclear, of man's dual nature, i.e., his rationality and animality. Furthermore, both Freud and Aristotle agree on the physionomic point: man is born with an actual id (at least with regard to instincts or vegetative needs), but not with an actual ego. The ego develops in the course of human life as man learns from his environment. So the Aristotelian would say with the Freudian that the ego (or the rational part of the soul) is potential, or merely a power, in the infant and only becomes actual, i.e., operative and habituated, in the course

[8] This translation can be carried out in great detail for the various passions, which are determined as different according to object or tendential mode. The analysis of the passions is scattered throughout Freud's writings. The Thomistic analysis can be found in the *Summa Theologica*, I–II, QQ. 22–48. The Aristotelian analysis is in the *Rhetoric*, II, 2–12. Cf. H. I. Smith *The Classification of Desires in St. Thomas*, Washington, 1915.

of development. This is the difference between an infant and a man.

5. The Freudian dynamics is, to a lesser degree, sound philosophical psychology. Here the Freudian analysis makes one original contribution (the concept of repression), but in other respects, it is inadequate and even in error. As before, this can again be shown by a brief translation of the Freudian and Aristotelian analyses.
   a. The relation of ego and id. Freud recognizes, as Aristotle recognizes *more fully*, (1) the reciprocity of intellect and will, (2) the subordination of the will to both reason and the passions, and (3) the subordination of the sensitive appetite to both reason and vegetative needs.
      (1) According to Freud, the id is able to influence the ego, to determine its processes, and the ego is able to influence the id, in repression, censorship, etc. So in Aristotelian psychology, desire is the source of all activity, the acts of all the powers of the soul, on the one hand; and on the other hand, the acts of one of these powers, the cognitive power, determines desires. Not only does the intellect move the will and the will move the intellect, but the passions move the will and the will is able to control the passions. In short, there is a very complicated reciprocal interdependence of the rational and non-rational parts of the soul, as well as between the cognitive and the appetitive functions. This is because reason rules the non-rational part not as an absolute despot but only as a ruler suffered by free subjects, who can disobey and even usurp his power and subject him to their will [57].
      (2) Reason is able to operate independently of the passions, and the passions independently of reason. Freudian psychology must also recognize some independence of ego and id; that is, the ego can function to some extent apart from the id, and conversely. If this were not the case, their reciprocal subordination would be impossible [58].
   b. Types of dominance by the ego of the id (conflict and harmony, repression and discipline) = reason ruling the passions by counteracting their motions, or ruling them

by moderating or perfecting them through prudence (the moral virtues).[8a]

(1) Here, although the two types of dominance are recognized by both Aristotle and Freud, the Aristotelian analysis is better because it more explicitly recognizes that the non-rational part of the soul is not anti-rational, but is rational to the extent that it is able to participate in reason, to be governed by reason.

(2) The Aristotelian analysis of the way in which reason positively rules the passions (by prudence in the formation of the moral virtues) is much more adequate. But the Freudian account of the way in which reason negatively rules the passions,—the account of the conflict between ego and id, in terms of censorship and repression,—is a notable addition to the Aristotelian analysis at this point.[9]

c. Types of dominance by the id of the ego (either determining its processes as in dreams, wishful thinking, etc. or in completely submerging it) ≡ the influence which the passions exert upon reason and the complete obstruction of reason by temporary excesses of passion.[9a] The Freudian analysis here is much more detailed, and adds positively and definitely to the Aristotelian account.

d. The sources of conflict in the soul, i.e., causes of schizophrenia.[10]

(1) The basic conflict is due to the fact that desire is subject to two determinations, by the pure id (or instincts) on the one hand, and by the ego, on the other. The ego conflicts with the instincts (or pure id) in the control or determination of desires. So the Aristotelian analysis speaks of the appetitive power as subject to determination by the vegetative power in so far as sensitive operations serve vegetative needs, and also as subject to determination by the cognitive power of reason.[10a] The source of the conflict is the same in both cases.

8a Vd. Note 57 *infra.*

9 This is the Freudian's chief claim to originality. Vd. Note 55 *infra.*

9a Vd. Note 57 *infra.*

10 I am using the word "schizophrenia" very generally in its etymological meaning of "split soul," and not as a synonym for the psychosis *dementia praecox.*

10a Cf. St. Thomas Aquinas, *Summa Theologica,* I–II, Q. 9, AA. 1, 2; Q. 10, A. 3; Q. 17, AA. 7, 8, 9.

(2) Subordinate conflicts are described in the two analyses: between the ego and the fixated or organized id ≡ between reason and the particular passions (between will and sensitive appetite); between the various fixations or libidinous organizations themselves ≡ the conflict of the passions *inter se*.

e. The focal point of conflict: in the Freudian analysis, the focal point may either be in exterior actions or in thought; so in the Aristotelian analysis, the conflict may focus either upon the control of exterior motility or upon the interior actions of the imagination or intellect (as in dreaming or wishful thinking).

f. Sublimation ≡ the determination of the passions by intellectual objects. Thus, for Aristotle, the transformation of sensuality into intellectual love; for Freud, the transformation of sexuality into love for any idealized (which is the same as an intellectualized) object.

g. Degradation ≡ the determination of the will by sensible objects. (The transformation of love into sexuality.)

h. Abreaction ≡ the catharsis or purgation of the passions.

i. The basic principles of human behavior, as they occur in the two analyses:

(1) Instincts ≡ vegetative needs and natural appetite.

(2) Pleasure principle ≡ sensitive appetite, i.e., the passions as moving toward sensual gratifications.

(3) Reality principle ≡ intellectual appetite, i.e., the will as moving according to reason commanding through knowledge.

(I have omitted the repetition compulsion and the death instinct. Vd. Note 60 *infra*.)

k. *Summary and Criticism* [56]

(1) The chief inadequacy of Freudian dynamics arises from its incomplete analysis of intellect and appetite (ego and id).

(2) But, as we have seen, the Freudian dynamics in a number of important respects is a contribution to the philosophical analysis of the relation of the powers of the soul.[11]

(3) The chief error is due to the supposition that the ego is a function of the id in relation to physical reality, and that the super-ego is a function of the id

[11] *E.g.*, repression, censorship, sublimation, degradation.

and the ego in relation to social reality. The ego cannot be merely a function of the id, if it is able to control the id or to oppose it in conflict. In short, to be consistent with itself Freudian dynamics cannot assert that the ego is a function of the id, and also admit that the ego can function independently of the id and can subordinate the id [58].

(4) This error is really not an error in dynamics. It occurs often in Freudian dynamic accounts because the Freudian permits his genetics to color all his other thinking. This, as I shall now show, is his great misfortune. Freudian topography and dynamics are, to some extent, sound; against the background of modern academic psychology, these parts of the Freudian analysis are a great original contribution, although in fact they add only a few significant points to the older tradition of philosophical psychology. (The genius of Freud must be differently estimated if it is measured against the modern or the ancient achievement.) They are erroneous in part because of the Freudian genetics they involve. We must now turn to this third part of the metapsychology.

6. The Freudian genetics is of two sorts: (a) the origin of the different parts of the psyche in the development of the individual (ontogenetics); and (b) the origin of the whole psychic structure in the development of animal life (phylogenetics).

a. In both of these genetic accounts, the leading points made are:

(1) That all psychic organizations or functions are determined by the needs of animals in their adjustment to environment;

(2) That man is merely a stage in the evolutionary development of animal life;

(3) That there is continuity throughout the animal series, the basic functions of life being present in the single-cell organism;

Note: This proposition fails to take account of the obvious discontinuities. Those who hold it are forced to say not only that human sensitivity is essentially the same as unicellular irritability, but that reasoning, if it is a vital function, is also present in the unicellular life.

(4) That in the life of the individual, the phylogeny is repeated: the ego (or theoretical reason) is a function of the id in its adjustment to the *physical* environment; the super-ego (or practical, moral reason) is a function of the id in its adjustment to the *social* environment.[11a]

b. These propositions are not philosophical truths. They are not analyses of common experience. They are not propositions in metaphysics, mathematics or general physics. They are not propositions in philosophical psychology. In fact, the Freudian ontogenetics is directly inconsistent with the philosophical truth that one generic power is not derived from another in any way. Now, either sense and understanding (the ego) are not powers at all or they cannot be developmentally derived from the vegetative powers (the id).

c. These propositions are not scientific truths. They are not inductions from scientific evidence. They are not propositions in any biological science, not in anatomy, physiology, zoology, genetics, etc.

d. They are propositions in what is called "the theory of evolution." This "theory" is not a theory in the sense of a systematic organization of scientific facts and laws, in the sense in which Newton's *Principia* is a theory. It is a theory in the sense in which there is an attempt to explain certain facts, which have been scientifically established in the biological sciences, by making *hypotheses* which are not propositions to be proved, but are merely imaginative guesses about unobservable processes or events. This is the sense of hypothesis in which Newton said no scientist should make them.

e. This "theory" is the kind of wild speculation which, in modern times and particularly in the 19th century, is confused with philosophy. Darwin himself is partly responsible for much of this speculation. *The Origin of Species* is full of guesses which are clearly unsupported by the evidence. (To the extent that *The Origin of Species* contains scientifically established facts, these facts are not organized into any coherent system.) Fur-

[11a] Vd. Note 53 *infra*.

thermore, these guesses, which constitute the theory of evolution, are not in the field of scientific knowledge anyway. They are historical. This conjectural history, begun by Darwin, was even more fancifully elaborated by the 19th century evolutionary "philosophers" [59].

(1) On the one hand, the post-Darwinian "scientific cosmologies": Spencer, Haeckel.[12]

(2) On the other hand, the post-Darwinian "evolutionary philosophies": e.g., Bergson.[13]

Note: The genetics to be found in Freud's metapsychology has its best expression in Bergson's *Creative Evolution*, in which all the forms of life are the result of the life-force,–the *élan vital* or cosmic id,–struggling against the inorganic environment.[13a]

f. The criticism can be briefly made.[14]

(1) In the first place, "evolution" is not a scientific fact, but at best a probable history, a history for which the evidence is insufficient and conflicting.

(2) The relevant facts are facts in embryology, genetics, palaeontology, comparative anatomy. These facts establish only one historical probability: that types of animals which once existed no longer exist, and that types of animals now existing at one time did not exist. They do not establish the elaborate story

[12] H. Spencer, *First Principles*, London, 1862; 4th American ed., 1880; E. Haeckel, *The Riddle of the Universe at the Close of the Nineteenth Century* (trans. by McCabe), New York, 1900.

[13] H. Bergson, *Creative Evolution*, London, 1911. Although Schopenhauer antedates Darwin, his influence upon "evolutionary philosophy" is tremendous. It is clearly reflected in Bergson. Vd. *The World as Will and Idea*, 1st ed. in 1818; trans. by Haldane and Kemp.

[13a] The theories of Schopenhauer and Bergson have a bearing on the form which the doctrine of the unconscious takes in Jung. Vd. *Contributions to Analytical Psychology*, New York, 1928. For a general discussion of Jung and other variants of psychoanalysis, vd. Woodworth, *Contemporary Schools of Psychology*, New York, 1931: Ch. 5; also Allers, *The New Psychologies*, New York, 1933: pp. 25–51; Crichton-Miller, *Psycho-Analysis and its Derivatives*, New York, 1933.

[14] Vd. L. T. More, *The Dogma of Evolution*, Princeton, 1925; T. H. Morgan, *Evolution and Genetics*, Princeton, 1925. The latter is a revision of Professor Morgan's Vanuxem Lectures for 1915–16, entitled *A Critique of the Theory of Evolution*. Vd. also *Evolution in Modern Thought* by Weismann, Bateson, Lloyd-Morgan and others.

which is the myth of evolution; nor do they establish any of the aetiological guesses about the way in which species originated or became extinct, such as natural selection, adaptation to environment, struggle for existence, transmission of acquired characteristics, etc.

(3) If the grand myth of evolution, as a history of the development of the forms of life, and the grand theory of evolution, as an explanation of how it all happened, are not scientific knowledge, how much less are they philosophical knowledge. (This type of speculation, peculiar to the 19th century, did much to bring discredit upon the name of philosophy which it so wrongly arrogated to itself.)

(4) For the most part, the wild speculations of Spencer, Haeckel, Schopenhauer and Bergson are now generally discredited both by scientists and philosophers.[15] Yet this is all that Freudian genetics amounts to,—a repetition of these speculations in a form even wilder, if that is possible, than in Spencer, Schopenhauer, Bergson [60].

7. *Conclusion so far*

a. Psychoanalysis as a scientific psychology is taxonomic and physionomic but not aetiological. Here we have a contribution by psychoanalysis to the knowledge of man.

b. Psychoanalysis is aetiological in that part of the metapsychology which we have called dynamics. Here it contributes the concept of repression. Otherwise it is a reworking of the traditional philosophical analysis of the order and interdependence of the powers of the soul.

c. Where the dynamics *appears* to be most original, in attempting to account genetically for the origin of the ego, it is most in error. The error is demonstrable. The genetic account is inconsistent with the analysis of the relation of ego and id. This error occurs in the metapsychology as a result of the uncritical acceptance or espousal of the "evolutionary speculations" of the 19th

[15] Cf. H. Höffding, *The Influence of the Conception of Evolution on Modern Philosophy* in *Evolution and Modern Thought;* John Dewey, *The Influence of Darwin on Philosophy and Other Essays in Contemporary Thought*, New York, 1910.

century, which are neither scientific nor philosophical knowledge, but have only the status of unsupported opinion.

d. Psychoanalysis as scientific knowledge would be totally unaffected by the complete elimination of this genetics. Psychoanalysis as philosophical knowledge (the topography and dynamics of the metapsychology) would be greatly improved by its elimination.

e. In any case, it is perfectly clear that the genetic fictions have absolutely nothing to do with psychoanalysis as scientific psychology.

f. The argument that I have with psychoanalysts at this point is not with them as psychologists or even as scientists, but with them as uncritical speculators, as bad philosophers. This is the misfortune of their 19th century heritage. To cure them of their superstitions would, of course, take more time than is proper at this place. It could be done by showing them how little there is in the way of strictly scientific evidence to support the popular myth of evolution.

IV. Psychoanalysis as Practical Psychology: Medicine and Morals

1. I wish to consider psychoanalysis as practical for two reasons: first, to make the point that the genetics has no practical significance, whereas the topography and dynamics are significant in psychoanalytic therapy and morals; and second, to show the correlation between Freudian and Aristotelian ethics, a correlation which could occur only through their sharing the same analysis of man's nature. To the extent that the Freudian analysis is inadequate, the ethics will be so also.

2. The end of ordinary medical therapy is the health of the body, considered with respect to the functioning of its various organs. The end of psychoanalytic therapy is health in a more general sense, i.e., the health of the whole man, of body and soul together. Health in this sense is what Aristotle means by happiness: an activity of man in accordance with the perfection of all parts of his nature. The end of the psychoanalyst as a practitioner is the same as the end of the moralist as a teacher. Both are practical psychologists. They use psychological knowledge for the sake

of making men happy, so far as one man can help another to become happy.[16]

3. In general we can make a translation between Aristotelian and Freudian ethics. I shall briefly indicate it as follows:
   a. The moral virtues ≡ the socialized id, i.e., the id as it is governed by the ego and super-ego.[16a] (Here there is an important difference. The moral virtues are the passions and social actions moderated by reason in the light of knowledge. The id is socialized by the ego and super-ego merely in the direction of conformity to the prevailing conventions or customs of the tribe.)
   b. The intellectual virtues ≡ insight or self-understanding. (Here also there is a difference. The Freudian does not understand the intellectual virtues because his psychological analysis is inadequate with respect to the intellect and the cognitive process. He does not know what it means to say that the good of the intellect is the truth. He does not fully admit,[16b] because of his genetic superstitions, that the ego (reason) is the measure of goodness in the id, as reality in turn is the measure of goodness in the ego.)
   c. The aim of psychoanalysis: to cure mental disorder, that is, to reduce the conflict in the psyche between ego and id; not to destroy the id, but by relieving repression to make a man understand himself, and through understanding his desires to adjust them to reality, which is another way of saying, to make a man reasonable; to sublimate the libido so far as this is conformable to a proper satisfaction of the vegetative needs.
   d. The aim of morality: to reduce the conflict between reason and the passions; not to destroy the passions, but to make them participate in reason through submitting

[16] The essential difference in aim between the moralist and the psychotherapist is that the former is primarily concerned with positive training, the latter with the elimination of defects. Their difference is analogous to that between preventive medicine and surgery.

[16a] This translation could be carried out in great detail by comparing the Freudian account of the genesis of neurosis with the Thomistic account of unhappiness in terms of sin and vice, especially the relation of capital to superficial sins which corresponds to the relation between complex and symptom in the Freudian analysis.

[16b] Yet, on the question of truth, Freud is opposed to skepticism and relativism, as on the question of the contribution of the intellect, he is opposed to materialism. Vd. *New Introductory Lectures on Psychoanalysis*, New York, 1933: pp. 240 ff.

to prudent government; to subordinate the sensitive appetite to the intellectual appetite; so to order all the goods desired by man that he is able to achieve all of them in a proper measure and in a proper subordination of lesser to greater goods.

e. Both the psychoanalyst and the moralist recognize the same basic difficulty in human life: the conflict or disorder in the soul due to the imperfect rule of the passions by reason, either because of the weakness of reason or because of the strength of the passions. The difference between them is with respect to what they do about this difficulty. The moralist appeals directly to reason; he hopes by giving the reason the knowledge it needs, to strengthen it; but he knows that this is not enough, that the virtues are moral habits, and that the virtues cannot be simply taught as geometry is. The sound moralist knows that his analysis of what a good man is does not enable him to make men good. Aristotle said that it was almost impossible to teach ethics to young men because of their subjection to the passions. The psychoanalyst seems to recognize this difficulty, and so approaches the problem, not from the side of reason but from the side of the passions. He tries to relieve the pressure of the passions; he tries to help reason without using it directly. And he knows, when he is honest, that he cannot succeed any more than the moralist in making men good; and ultimately for the same reason. His knowledge is not sufficient. He does not know how.

f. Yet the moralist succeeds better than the psychoanalyst *to the extent that* the psychoanalyst proceeds as if the goodness of a man did not depend essentially upon the cultivation of his reason.

(1) Freud's criticism of hypnosis as therapy was that it only cured the symptoms of hysteria and did not cure the disease. It removed the symptoms by a trick without attacking their causes.

(2) My criticism of psychoanalysis as therapy is similar. It only relieves or alters one of the conditions of the moral problem. By an extraordinary trick,—much more ingenious and extraordinary than hypnosis,—it influences the passions which are involved in the disorder of the soul. But this is negative. So is the cure of repression. The problem is fully

solved only if the individual acquires sound moral principles and then is able to form habits and direct his life according to these principles by the rule of reason [61].

(3) To the extent, then, that the psychoanalyst is not competent as a moralist,—to the extent that his philosophy is incomplete or erroneous,—he cannot achieve the end he sets himself: to make men happy [62]. At best, he can give them a little help by a kind of purgation of the passions through understanding them objectively. (In other words, psychoanalysis works in the same way that art, particularly the drama, does. It makes a man a spectator of his own passions by a process of identification, transference, etc.)

(4) What a good friend who is a wise man and a competent moralist can do in some cases, the psychoanalyst who is a wise man and a competent moralist can do in other cases. The difference in the cases is the difference between the normal and the neurotic, a difference in the degree to which the passions are disordered and reason is weak.

g. Unfortunately, the psychoanalyst is seldom a wise man and a competent moralist, because he is seldom if ever a philosopher. The Freudian conception of a good man as a complete adult is inadequate because all that this ideal involves is normal biological functioning, primarily on the vegetative and social level. The crucial error in his moral insight arises from his crucial error in psychology: his failure to understand the nature of intellect and will. (As a result of his analysis of the super-ego the Freudian thinks that morality is nothing more than conformity to the prevailing *mores*. He does not recognize that moral principles are based upon speculative truths, that they hold for all men, that they are not relative and changing.)

V. Conclusions

1. I have tried to show positively that psychoanalysis has a place in the European tradition, both among the sciences and in relation to philosophy. It is due to the great genius of Freud in rediscovering man as the subject-matter of psychology that the vitality and significance of the ancient tradition

has been at last infused into psychology as a scientific enterprise.

2. But psychoanalysts do not understand their place in this tradition, and as a result they do not understand their own doctrine. This may account for their failure to make an intelligible presentation of it to those who are not psychoanalysts and who employ critical standards to judge what is presented.
   a. Psychoanalysts fail to understand what part of their contribution is scientific, and what part of their doctrine is philosophical.
   b. They do not understand the relation between the theoretical and the practical aspects of their undertaking.
   c. These failures of understanding may account for the failure so far of any psychoanalyst to make a clear, systematic statement of psychoanalysis in terms of its basic concepts, its principles, its evidences, its facts.[17]
3. To what are these failures of understanding due? To lack of philosophical training, on the one hand; and to the influence of the 19th century,—its bad philosophy, its prejudices and superstitions,—on the other. The only cure that I know of for the influence of the 19th century is education in the European tradition which was almost completely obscured in that century.
4. But education is not enough. To suppose that it is is to make the error of assuming that men, psychoanalysts among them, are completely rational and are capable of being moved by the truth. Prejudices are like passions. May I borrow the technique of the psychoanalyst for a moment, and analyze his prejudices in the hope that he may get the insight that will work as a cure.
   a. If you were to psychoanalyze me,—and you will probably start as soon as I leave the room,—you would say that I have the prejudices and the passions of an anal-erotic. I am trying to be completely submissive to objective truth and order. I am masochistic toward reason.
   b. If I were to psychoanalyze psychoanalysts, I would say that as a group they have the prejudices and passions of the narcissist. They are trying to be original; they are trying to swallow everything in psychoanalysis. To be orig-

[17] Vd. the not quite successful attempt made for psychoanalysis by W. Healy, A. F. Bronner and A. M. Bowers, *The Structure and Meaning of Psychoanalysis*, New York, 1930.

V. 4. b.

inal in this way, they must ignore their sources in the European tradition, and they must be sadistic toward order and reason. In other words, psychoanalysts commit the sin of pride. They are unwilling to recognize their place in the great intellectual tradition of western Europe, and to make what contribution they can to its science and wisdom, however slight that may be.

c. I have added this little psychoanalytical ending for the sake of warning you that you cannot dismiss what I have said in these lectures by psychoanalyzing me. That defense can be used against you as readily as against me. Psychoanalysis is irrelevant to the merits of any intellectual position. The truths of science and philosophy, resting on evidence and demonstration, are not challenged by the *ad hominem* of calling their proponents narcissists or anal-erotics. A man becomes an authority by speaking the truth; the truth does not rest on the authority of its human source and is thus unaffected by states of personality.

NOTES

1. In the history of physics and chemistry there have been, of course, many instances of theoretical controversy. The opposition between the corpuscular and the wave theory of light, the issue concerning phlogiston in the science of heat, the war between the Neptunists and the Vulcanists in geology, the argument about the hypostatical or veridical character of molecules in the kinetic theory of gases,—these and many others constitute interesting chapters in the modern history of the natural sciences. But in every case, the polemic is between alternative theories within the area of a science, the fundamental data and principles of which the controversialists commonly share. The controversy has usually, if not always, been as to which of two competing theories gives a more adequate account of all the facts. The acknowledged facts, the observational data and basic inductions, constitute the unity of a science within which there may be two or more opposing theories. The situation in psychology, from the beginning of its history as an empirical science to the present day, is different. The "schools of psychology" are not rival theories within a single science, defined by common principles and given unity by common data. The schools of psychology do not even agree about the subject-matter of their "science." The phenomenal area which behaviorists investigate is not the same as that explored by introspective structuralists; structuralists and psychoanalysts do not seek their data in the same quarter. Not only do the different schools of psychology,—however they be named and enumerated,—not have the same subject-matter, but they do not acknowledge each other's methods as "scientific"; they share neither common principles nor common data. This is not an opposition of theories within a science. It is rather an opposition of rival pretenders to the title of being *the* science of psychology. So long as this situation persists, may one not wonder justifiably whether there is *any* science of psychology? The nearest analogue to psychological controversy is the polemic between vitalists and mechanists in biology, but even this is much more like an opposition of theories than like the civil war which makes the field of psychology anarchic.

   Vd. *Psychologies of 1930*, Worcester, 1930; R. S. Woodworth, *Contemporary Schools of Psychology*, New York, 1931; and E. M.

Heidbreder, *Seven Psychologies*, New York, 1935. In what other field of natural science could such books be published? Consider the varieties assembled by Professor Murchison in his 1930 circus of psychologies: hormic psychology, associationism and act psychology, functionalism, motor psychology, structural psychology, configurational psychology, Russian (i.e., Marxist) psychology, behaviorism, reaction psychology, dynamic psychology, factor school of psychology, analytical psychologies. Professor Woodworth devotes his attention only to five: introspective psychology, behaviorism, gestalt psychology, psychoanalysis and related schools, purposivism or hormic psychology. In his introductory chapter Professor Woodworth writes: "With all these divergent movements going on, contemporary psychology might be expected to show no coherence at all, but curiously enough the psychologists of the day enjoy meeting and discussing the problems of psychology, and still treat one another with respect. . . . The mutual respect of psychologists reflects the obvious zeal shown by members of all schools for the continued progress of the science. Some schools are more concerned to make psychology scientific and others to make it more human and adequate. One would like to see it both scientific and adequate, and the psychologists keep together in the hope that such results can be achieved" (*Op cit.*, pp. 15–16; by permission of The Ronald Press Company). That divergent psychologists still respect each other in their common aim to establish a science which will be adequate may be regarded as a hopeful sign; but it can hardly be regarded as indicating that a science of psychology now exists.

Professor Woodworth admits that "schools such as we have been considering seem almost peculiar to psychology. It may be there are some in the social sciences, but at least in the natural sciences at the present time they are hard to find" (*Op. cit.*, pp. 214–215). And he is forced to conclude that this looks "like an indication that all is not well and that psychology does not yet know where it stands or what it has to do" (*Op. cit.*, p. 215). In the face of this situation the forced optimism which current commentators, such as Woodworth and Heidbreder, try to summon up in their concluding chapters seems ever so much like whistling in the dark. Cf. E. G. Boring, *A History of Experimental Psychology*, New York, 1929: Ch. 24. Vd. Note 42 *infra*.

2. The way in which formal logic is presupposed by material logic is clearly illustrated by the relation between the prior and posterior analytics in Aristotle's *Organon*. In the *Prior Analytics* we find the formal principles of inference. The account of syllogisms considers only the formal characteristics of the nexus of terms and propositions which constitute reasoning. But all syllogisms are not scientific, nor is the formal structure of a science fully stated by the analysis of syllogisms. The account of science depends upon this prior analysis,

but requires also a consideration of the relation between definition and demonstration, of both to causal analysis, and of the ordering of the syllogisms of a science by its ultimate principles, those primary definitions and indemonstrable or assumed propositions which generate the science. To say that these considerations are material is not to deny the formality which everywhere rules the posterior analysis; it is merely to say that the account of science is an application of the forms to particular types of matter. The difference between the *Prior* and the *Posterior Analytics* can be seen in another way. There is little or no psychology needed for the analysis of syllogisms. But the analysis of science requires an account of induction, an account of the origin of its first principles. Any attempt to give a formal account of induction reduces it to syllogisms, and the search for the grounds of demonstration, its first premises, must continue. That is why the exposition of induction with which the *Posterior Analytics* concludes (Bk. II, Ch. 19) is necessarily and unavoidably psychological.

The relation between these two parts of Aristotle's *Organon* is found in modern treatises which try to be as comprehensive analytically. Thus, Book I of Jevons' *Principles of Science* (1874) is devoted to what he calls "formal logic." The concluding chapter of this part deals with induction, but attempts to reduce it to syllogisms. The rest of the treatise is concerned with induction genuinely; Book II considers number, variety and probability, Book III, the methods of measurement, Book IV, inductive investigation, and Book V, generalization, analogy and classification. In these books we find the exposition of scientific method and it is inextricably interwoven with material considerations, on the one hand, and psychological analysis, on the other. The same relation obtains between Books I and II of J. S. Mill's *System of Logic* (1843), and Books III to VI. The three parts of Johnson's monumental work on *Logic* (1921–1924) are similarly divided: the first and second parts are formal, even in their consideration of induction; the third part is concerned with "the logical foundations of science." In some instances, modern logicians have devoted special treatises to the application of formal logic to the matter and method of scientific knowledge, as in the case of Venn's *Empirical Logic* (1889). But, generally, the structure of Aristotle's *Organon* is imitated both by original treatises and by text-books. Vd. Cohen and Nagel, *An Introduction to Logic and Scientific Method* (1934); Stebbing, *A Modern Introduction to Logic* (1930). In their material portions, the modern treatises differ from Aristotle in the extent to which they are methodological. This is due obviously to the crucial difference in the meaning of "science" in ancient and modern times. Aristotle was not concerned with the procedures of research: measurement, experimentation, etc.

There is one further point of difference between Aristotelian logic and those modern treatments which are sensitive to the nature and methods of science. The formal logic of Aristotle is predicational.

The distinctions of subject and predicate, the analysis of the predicaments, the post-predicaments, and the predicables, the division of the kinds of predication, the concern with real definitions and the role they play in syllogisms,—these are essential to Aristotelian logic. In modern times, what is called a non-Aristotelian logic has been developed. It is relational rather than predicational. It reduces all subjects and predicates to classes and distinguishes classes as belonging to different orders according to the way in which they are related. Its basic distinctions are among types of relation and relational order. It is not concerned with real definitions, and treats the syllogism as a special case of substitutional operation in linear inference. The point that logic and mathematics are continuous is true for this relational logic; it could not be maintained for predicational logic.

It has been held by some that relational logic is the general case, and by others that relational logic can be translated into the logic of predication. Whether such subsumptions or translations can be achieved seems to me only a matter of grammatical subtlety. There is a real difference between these two logics. Aristotelian logic is the logic of philosophical knowledge in so far as it is concerned with ontological problems. It is the logic appropriate to an inquiry into substance and causes. Relational logic is the logic of modern mathematics and investigative science. Science does not treat of substance and causes; it investigates the phenomena with a view to classification, order, correlation, measurement, systematization. The operational character of relational (or mathematical) logic is appropriate to the operational principles of scientific method. Aristotelian "science" is bad as science because it attempts to apply a predicational logic to the study of phenomena. Modern "philosophy" is bad as philosophy to the extent that it makes the opposite error. Vd. Notes 9, 10a, 16a and 47 *infra.*

3. In Book IV of the *De Rerum Natura,* Lucretius expounds the psychology of cognition. In this context, he states the basic canon of scientific method in terms of the relation of sensation and reasoning. "You will find that the concept of the true is begotten first from the senses, and that the senses cannot be gainsaid. For something must be found with a greater surety, which can of its own authority refute the false by the true. Next then, what must be held of greater surety than sense? Will reason, sprung from false sensation, avail to speak against the senses, when it is wholly sprung from the senses? For unless they are true, all reason too becomes false" (trans. by C. Bailey, Oxford, 1921: p. 159). "And if reason is unable to unravel the cause, why those things which close at hand were square, are seen round from a distance, still it is better through lack of reasoning to be at fault in accounting for the causes of either shape, rather than to let things clear seen slip abroad from your grasp, and to assail the grounds of belief and to pluck up the whole foundations on which life and

existence rest" (*Ibid.*, p. 160). Yet the atomic philosophy which Lucretius espouses is an attempt to explain away by reasoning the immediate sensitive experiences we have of the qualitative variety which exists among physical things. Sensible qualities are not really *in* things, though they are apparently there, because the atomistic materialist claims to know, not by sense but through reasoning, that the atoms have only quantitative properties and that composite bodies are arrangements of atoms, differing *inter se* only in geometrical properties. Lucretius' account of scientific method is thus inconsistent with what he claims to know as a scientist. The inconsistency is even more devastating when Lucretius attempts, at the end of Book II and in Books III and IV, to give an analysis of the soul and of the psychological processes of sensation and reasoning in terms of the size, shape and motion of atoms. Atomism, or what is the same, simple-minded absolute materialism, is always thus self-refuting. It claims to be knowledge gained in a certain way, and this "knowledge" makes it impossible to give a tenable account of knowledge itself and violates the procedure by which it is said valid knowledge must be obtained. The *reductio ad absurdum* which can easily be made of Lucretius applies as readily to the materialism of Hobbes in the 17th century and to that of Büchner, Moleschott, Vogt and Feuerbach, in the 19th century. (Vd. The account of Feuerbach's refutation of "absolute materialism" in Hook, *From Hegel to Marx*, New York, 1936: pp. 237–242. This refutation, made in 1838 when Feuerbach was still an idealist, applies to the materialism he himself later adopted as well as to that of Marx.) Materialism does not escape its insoluble dilemmas by becoming dialectical in the post-Hegelian sense of that word, but only by becoming dialectical in the Greek manner. It then ceases to be an absolute materialism and becomes the materialism of hylomorphism, or formal materialism, which sees that being, becoming and knowing require a formal as well as a material principle for a consistent and adequate account both of the nature of things and of our knowledge of them. Vd. Notes 21 and 28 *infra*.

4. The ancient error of the materialistic philosophy is nicely illustrated in the *reductio ad absurdum* which Professor Lovejoy makes of behaviorism, which is distinguished as a contemporary "school of psychology" not only by its restricted methods but by its bad metaphysics. Vd. A. O. Lovejoy, *The Paradox of the Thinking Behaviorist*, in the Philosophical Review, XXXI, pp. 135–147 (1922). It is not only that the behaviorist is a philosopher while at the same time denying the validity of any knowledge not obtained by scientific methods (vd. J. B. Watson, *Psychology From the Standpoint of a Behaviorist*, Philadelphia, 1919: Ch. I). That is an error which most scientific psychologists share. In addition, the behaviorist presents the conclusions of his research *as if* they were rationally founded in his data, yet at the same time reduces all "reasoning" to conditioned reflexes in the

sphere of the laryngeal musculature. His scientific work indicates plainly *that* he thinks, though not always well, as is clear from the fact that *what* he thinks makes it impossible for him to explain, without contradiction, *how* he thinks. Cf. Scott Buchanan, *An Aeolian Theory* in Psyche, April, 1927: pp. 100–101.

What Lovejoy and Buchanan have done humorously to the behaviorist, F. H. Bradley did to the British associationist psychologists with dialectical severity. The error is strictly the same. The attempt to reduce all rational operations to the mechanical association of ideas, —an "idea" being nothing more than a sensation or a sense-image,— is the prototype of the more recent attempt to reduce thinking to the mechanical conditioning of reflexes. Bradley not only shows the impossibility of all such reductions, but exhibits the crucial error of sensationalism. If ideas are purely the particulars of sense, if they are not actually universals in the mind, thinking of any sort is impossible, even the thinking which comes to the conclusions of sensationalistic psychology. Any psychology which denies the distinction between sense and intellect, between perceiving and understanding, is thus self-annihilating. Vd. *The Principles of Logic*, London, 1883: Bk. II, Pt. II, Ch. I on the theory of the association of ideas.

The point here being made is not that the mind cannot know itself as it knows other things. Quite the contrary! The fact of reflexivity is not itself the stumbling block in the path of a scientific psychology. All that is here being insisted is that the mind which is the object of psychological science be described in a way that is consistent with the characteristics of the mind which the psychologist reveals himself as having in his operations as a scientist.

5. The distinction between common and special experience can be illuminated by an understanding of the nature of experience. Experience is usually identified with sense-experience, although it can be extended to include the immediate but non-sensitive awareness we have of our rational operations. But, taking experience in the more restricted meaning of the word, it must be further observed that sense-experience is not identical with, and hence not reducible to, sensations. While it is true that all sense-experience arises from the activity of the senses,—the sensations which result from their activation by sensible objects,—experience goes beyond sensation in that it does not refer exclusively to particular objects here and now present, as all sensations do. Experience involves memory and imagination and this explains the kind of generality it has. It is a product of sensation, not simply but in consequence of the summation and composition of sensations, which is the work of memory and imagination. Aristotle makes this point in his account of induction. "Out of sense-perception comes to be what we call memory, and out of frequently repeated memories of the same thing develops experience; for a number of memories constitute a single experience" (*Posterior Analytics*, II,

100ª4–5). An experience thus contains the "rudimentary universals" to be made explicit and genuine universals by the act of abstraction which is essential to the process of induction or generalization. Cf. *Metaphysics*, I, 1. What Aristotle called "a single experience" is treated by modern psychologists as a "generic image," an image which has quasi-universality because it lacks the individual intention of the perceptual image. Vd. Wundt, *Outlines of Psychology*, III, 16; Kulpe, *Outlines of Psychology*, IV, 27; Galton, *Inquiries into Human Faculty*, Appendix A.

The notion of common experience must not be interpreted to mean that all men have the same experiences. That is plainly not the case. I have used the phrase "common experience" to denote the whole set of experiences which men have naturally through the ordinary operation of their senses, their memories and imaginations. Though the experiences of two men are not in all details the same, they both have common experience to the extent that they share experiences which have thus arisen naturally in the course of their lives. That men share experiences follows from the fact that they live, for the most part, in the same world of objects and from the fact that as sensitive animals they have the same specific powers and operations. The negative definition of common experience is that it does not result from investigation. This does not mean that ordinary men, men who are not scientists, never investigate. It means only that their common experience is not the result of any deliberate research or inquisitive effort on their part. (Ordinary men investigate whenever they make a deliberate use of their powers of observation in order to answer definite questions. Such investigation differs from scientific inquiry, however, in that the latter is directed by problems formulated in terms of technical concepts. It is only scientific investigation which creates a new order of experience.)

Special experience is positively defined as being the result of deliberate research and specially contrived inquisitive efforts. An experiment can be defined as an artistically devised experience. Instead of waiting for the multiplication of sensations and memories to do their work toward the production of an experience, the scientist performs operations and manipulations which create for the senses at once what the imagination might never produce naturally. The experiment as an artificial experience has the same quasi-universality that a natural experience has. It is this quasi-universality which makes both the natural experience and the experiment the proximate matter from which an induction can be made. The experiment is the ideal of scientific method, not only because it so readily yields an induction, but because it differs so radically from the natural manner of experiencing. Scientific procedures which consist in the collection of huge amounts of data which must be statistically refined before they are anything like an experience that is pregnant with inductions, are not only usually infertile but lack the artistry which distinguishes scientific from ordinary experiencing.

The distinction between common and special experience is thus seen to be a distinction between natural and artificial experience. Scientific method, from its best experimental to its crudest empirical forms, is the art of creating experiences which men do not naturally have. The value of experience, whether common or special, whether natural or artificial, is the same: it is the immediate source of inductions, of generalized knowledge of the world of experienceable objects. The maxims of common sense, the axioms or principles of philosophy, the general facts and laws of science, all arise from experience, but not the same kind of experience, because the method of ordinary and philosophical knowing is thoroughly natural, whereas the method of scientific knowing involves operations of art in the production of special experiences.

6. The conception of scientific knowledge as based on special experience is generally accepted in modern times, i.e., it is generally said that the data of science result from investigation. But the conception of philosophical knowledge as based on common experience seems scarcely to be understood at all in modern times. The history of the modern degradation of philosophy to the status of being handmaiden to the sciences is a long and complicated one. But it is clear that, in the course of the last three centuries, philosophy has gradually lost the status of a body of knowledge having independent validity in its own right. Nineteenth and twentieth century positivism was the inevitable result of a development which at first looked upon scientific method as a way of correcting the "unchecked speculations" of the philosopher and then came to regard scientific knowledge as the type of all valid knowledge of a generalized sort. Positivism expresses itself in many ways, but all of them reduce to a denial of philosophy as knowledge of the world of experienceable objects. The conception of the philosopher as working on the periphery of the scientific domain, trying but failing to solve problems which the scientist will later succeed in solving as science advances (vd. William James, *Some Problems of Philosophy*); the conception of the philosopher as endlessly trying but failing to answer unanswerable questions (vd. M. W. Calkins, *The Persistent Problems of Philosophy* and John Dewey, *The Quest for Certainty*); the conception of the philosopher as "fusing religion and science into one rational scheme of thought," as tentatively formulating "ultimate generalities" in the light of science, and therefore compelled to change his philosophy from time to time as science advances (vd. A. N. Whitehead, *Process and Reality*); the reduction of the philosopher to the role of a logician, nay a grammarian, whose chief task is to keep scientific language in order and undefiled (vd. R. Carnap, *Philosophy and Logical Syntax* and *The Unity of Science*)—these are all expressions of positivism which holds that the

only questions genuinely answerable by knowledge are those which can be answered by the investigative procedures of science. What is most amazing is that so few recent philosophers have seen the absurdity, if they have not resented the indignity, of a role which makes their work entirely subordinate to and dependent on the contingencies of changing scientific knowledge. No self-respecting man should want to be a philosopher according to this conception of his task.

There are a few exceptions, of course. I am not thinking here of those who work in the tradition of the perennial philosophy, such as Jacques Maritain and Étienne Gilson; but rather of those others among contemporary thinkers who, though somewhat more divergent from the ancient and mediaeval tradition, nevertheless share in its conception of philosophy as knowledge; thus, E. Husserl, who as a philosopher refers the truth of his analysis directly to common experience in utter disregard of the content of changing science; and similarly George Santayana, from whom I quote the following passage. "There is one point, indeed, in which I am truly sorry not to be able to profit by the guidance of my contemporaries. There is now a great ferment in natural and mathematical philosophy, and the times seem ripe for a new system of nature, at once ingenious and comprehensive, such as has not appeared since the earlier days of Greece. . . . But what exists today is so tentative, obscure and confused by bad philosophy, that there is no knowing what parts may be sound and what parts merely personal and scatter-brained. If I were a mathematician I should no doubt regale myself, if not the reader, with an electric or logistic system of the universe expressed in algebraic symbols. For good or ill, I am an ignorant man, almost a poet, and I can only spread a feast of what everybody knows. Fortunately exact science and the books of the learned are not necessary to establish my essential doctrine, nor can any of them claim a higher warrant than it has in itself: for it rests on public experience. It needs, to prove it, only the stars, the seasons, the swarm of animals, the spectacle of birth and death, of cities and wars. My philosophy is justified, and has been justified in all ages and countries, by the facts before every man's eyes; and no great wit is required to discover it, only (what is rarer than wit) candour and courage. Learning does not liberate men from superstition when their souls are cowed or perplexed; and, without learning, clear eyes and honest reflection can discern the hang of the world and distinguish the edge of truth from the might of imagination. In the past or in the future, my language and my borrowed knowledge would have been different, but under whatever sky I had been born, since it is the same sky, I should have had the same philosophy" (*Skepticism and Animal Faith*, New York, 1923: pp. ix–x; by permission of Charles Scribner's Sons).

This statement by Santayana epitomizes the points involved in the methodological distinction of philosophy from science. The distinction turns, in the first place, on the fact that philosophy appeals only to what Santayana here calls "public experience",—the common experience of men, essentially the same at any time because under the same sky. From this fact it follows, in the second place, that philosophical doctrine must ultimately agree with the well-tested opinions of common sense because both arise by induction from the same common experience. Philosophical knowledge differs from common sense in the subtlety of reflection by which it discovers the implications of the obvious first truths,—"the facts before every man's eyes." Common sense is the beginning of wisdom and philosophy is its fruit. The relation of scientific knowledge to common sense is different. Philosophy deepens common sense into wisdom by performing the dialectical task, which Socrates rightly called intellectual midwifery, of making common sense deliver the truths it implicitly contains. Science merely makes up for the inadequacy of common sense by obtaining knowledge of the detailed operations of particular sorts of things, knowledge which cannot be gained from common experience. Investigation is required to know such details. But the knowledge of such details is not necessary for wisdom. As Santayana said, "exact science and the books of the learned are not necessary to establish" a philosophical doctrine. The changing content of the positive sciences can never alter a single philosophical truth. The works of Aristotle and St. Thomas Aquinas contain many statements which science has taught us are wrong as a matter of fact; but those statements were not in the first instance philosophical truths. The modern reader of such ancient books must remember that they contain both philosophical and scientific knowledge, without distinguishing labels attached to the different sorts of propositions.

This leads to the next point. We can find scientific errors in the works of Aristotle because scientific knowledge is essentially contingent. Its conclusions are only hypothetically necessary. The contingency of their truth is due to the fact that their truth is always relative to the data of special experience, and these data change from time to time, as scientists work ingeniously at contriving new and better special experiences. The essential contingency of scientific knowledge is based upon the same fact which properly leads us to expect interminable progress in the accomplishment of more and better scientific knowledge. But the data of philosophy are the facts of common experience. Errors in philosophy are due to failures of analysis, inconsistencies, inadequacies, etc.; they are not due to the inadequacy of the data. When a philosophical work is well-done, it is finished, though the work of philosophy to refine analyses and follow implications may itself never be finished. We should not hope for the same sort of progress in philosophy that we expect in science; in fact, we can do very little to improve upon ancient wisdom, because the limits of wisdom are not set by the limits of our ingenuity or industry in doing research, but by the natural limits of our reflective powers, our powers of understanding and judgment. For the same reason that progress in philosophy is accidental, whereas it is essential in science, philosophical truths are necessary and not contingent. I would go further than Santayana's statement that scientific knowledge cannot claim a higher warrant than philosophy has in itself; I would say that the warrant which philosophy has in common experience establishes it with a necessity and a permanence which makes it superior to science. I do not mean that philosophy can replace science or make science gratuitous. It is not as if in the same sphere philosophy were better knowledge than science. Their spheres are different, complementary rather than competitive, as science is complementary to common sense, as special experience completes common experience. (Vd. Dr. G. Phelan's discussion of the difference between progress in the sciences and in philosophy,—"progress by substitution and progress by deepening insight,"—in his essay on *Jacques Maritain*, New York, 1937. The first intellectual duty is "to defend thought against false progress;" the second is "to defend perpetual novelty against immobilists.")

The superiority of philosophical knowledge is not merely the necessity and permanence due to its source and method. The practical value of philosophy is greater. The modern world does not understand this because it thinks of utility narrowly. Knowledge is useful to the extent that it enables us to control and direct the operations of things. In this sense of utility, scientific knowledge is highly useful and philosophy is useless. But man operates on things as means to ends. The determination of the ultimate ends of human conduct and the ordination thereto of all means, cannot be accomplished by scientific knowledge. This is the work of philosophy in the practical sphere. Wisdom in the speculative sphere becomes practical when it is used for solving the problems of action. In the same sense that philosophy is superior to science in the speculative sphere because the principles which constitute its knowledge have greater generality and belong to higher orders of abstraction, so ethics and politics, which are practical philosophy, are superior to engineering and medicine, which are applied science, because they are concerned with the ends and the ordering of means, whereas the applied sciences give us at best a mastery of the means. But the means can be directed to the right or the wrong ends. It is, therefore, clear that scientific knowledge has the kind of utility which enables it to be used or misused, used for good or evil; whereas philosophy as practical wisdom cannot be misused. At worst, we can fail to use it. This is the unfortunate plight of the modern world which, denying philosophy in the speculative order, fails to use it practically. The danger of vast scientific knowledge unaccompanied and undirected by wisdom is obviously enormous. As Santayana said, learning, —of the scientific sort,—does not liberate men from superstition which, practically, is the confusion of means and ends.

The denial that philosophy is a body of knowledge, utterly independent of the results of scientific investigation, is currently due, as we have seen, to the curious fallacy of supposing that common experience, the existence of which cannot be denied, is unable to yield inductions and to validate generalizations in exactly the same way that special experience does. (Historically, of course, it is due to the psychological errors which underlie the positivism of Hume and Kant. Vd. Note 16a *infra.*) Those who hold this position must deny that, prior to science, men knew general truths, the significance and validity of which have remained unchanged despite all the results of research. They can do this only at the expense of denying truths which lie at the foundation of science, which science did not itself achieve and without which science cannot interpret its own findings. Those who claim that all valid knowledge, having generality, is the product of scientific research must, furthermore, be made to consider the nature of mathematics, which certainly has generality and which is certainly not the result of special investigations or based on special experience. (When we talk about "mathematical research," we do not mean any sort of observational inquiry, but a purely reflective process. Mathematics is arm-chair knowledge. Scientific research creates new experiences; mathematical research discovers new implications, new conclusions, new orders of propositions.)

Confronted by mathematics, the positivists are forced to admit that it is not the same sort of knowledge that science is; but instead of recognizing in mathematics knowledge that is typically philosophical, they take the characteristic modern alternative of treating mathematics as not knowledge at all, but rather as a sort of fictional logistic. All mathematical truths are *merely* conventional; they are not knowledge of the real world; they are nothing but the deceitful rules of a complicated grammar of symbolic operations. I say "deceitful" because mathematics appears to be knowledge of the real world of physical things, among the properties of which quantity and order are primary. The difference between mathematics and physics in subject-matter is that the former abstracts from the matter and motion of physical things, but it deals nevertheless with the properties of physical things. The ultimate criterion of truth is the same in mathematics and in physics. Unless mathematics is not knowledge of reality *in any sense* (i.e. not even from the point of view that its object is a certain grade of abstraction from reality), then the existence of mathematics establishes the existence of knowledge which does not depend upon the investigative methods of science. That mathematics is philosophical knowledge can be tested by applying to it all the criteria employed in differentiating philosophy from science. But mathematics is not *the type* of philosophical knowledge, particularly not modern mathematics which tries to be faithful to the positivistic ideal of being conventional and employing assumptions. Metaphysics is the type of philosophical knowledge; it proceeds from first principles and makes no assumptions; its concepts clearly refer to the real world and not to fictive entities. In the sense in which the word "metaphysics" is currently used by positivists, this statement is, of course, unintelligible or ridiculous. Vd. *Art and Prudence*, New York, 1937: Ch. 9; cf. P. Descoqs, *Essai Critique sur l'Hylémorphisme*, Paris, 1924: Appendice I.

I conclude this discussion by stating one dilemma: *either* philosophy is not genuinely valid knowledge, which rests directly on experience and is independent of scientific research, *or* philosophy is knowledge which is superior, both speculatively and practically, to science. The modern world, in its violent repugnance to the second horn of this dilemma, has gradually torn itself free from wisdom by embracing the first. Cf. J. Maritain, *Les Degrés du Savoir*, Paris, 1932: Ch. 2.

6a. In his introduction to this book Dr. Alexander speaks of my contempt for "practical" accomplishments and my failure to evaluate scientific concepts "according to the degree to which they increase our ability to control and influence natural phenomena." But, as we have seen, there are two meanings of utility applicable to knowledge: (1) the use of knowledge to control and influence natural phenomena, and in this sense science is magnificently useful whereas philosophy is totally useless; (2) the use of knowledge to determine the ultimate ends of human conduct and the ordination thereto of all proximate and remote means, and in this sense philosophy is practically indispensable whereas science can be disregarded. Scientific knowledge, at its best, provides us with a mastery over natural means or enables us to devise new means where nature fails us. The simplest rule of wisdom in the practical order of human life requires us to choose means according to the ends they serve, and if these ends be not ultimate they must, in turn, be ordered to a final end. The practical utility of science is, therefore, dependent upon a determination of the order of all human goods according as they are means and ends. It is clear that science itself does not make this determination; in fact, since such an ordering of goods is normative and involves what is currently called "value judgments", the scientist would be the first to admit, nay to insist, that the question of the order of goods is beyond him. If he dare to say that there is no objective order of goods, no ultimate end, no right determination of means,—as the psychoanalyst does in his claim that all moral standards are conventional or relative,—then he speaks as a philosopher and not as a scientist. And if he were right in his moral relativism, then how could he consistently defend his own dogmatic assertion of the unquestionable *value* of scientific knowledge because it enables us to control nature? Is such control good really or is it only a matter of our western European opinion? And if really, why?

I do not deny the utility of science, but I hold that the utility of philosophy is paramount. Ethics is architectonic in the practical order; all the applied sciences and the practical arts, all forms of engineering and medicine, are subordinate. All the scientific knowledge which has been accumulated since the Renaissance does not change, in one essential detail, the analysis of man's moral problem, the problem of virtue and happiness, as that is set forth in Aristotle's *Ethics* or the moral treatises of the *Summa Theologica*. Nor will any further accretions of scientific knowledge make any difference. If this can be called contempt for the scientific achievement, then the opposite position is an idolatry of science. Both contempt and idolatry are avoided by a just appraisal of the goodness of science in relation to the goodness of wisdom. Science unquestionably increases human power to achieve the ends of human life. But the goodness of this gift of power depends upon whether or not it is properly used; and since the power is entirely a mastery of means, that, in turn, depends upon wisdom about means and ends. Science without wisdom is dangerously ambivalent; the greater the power, the more ominous is its improper use. The hazards of human welfare in modern times can be understood in the light of the ever increasing disproportion between science and wisdom, between the operative power of men in an age of science and that clarity of direction which guides men only in so far as philosophy rules in the practical order. Even if philosophy were not attenuated in modern times, its directive task would be made more difficult by the augmented utilities which scientific advances confer.

This analysis of the practical importance of philosophy and science applies to the relation of ethics and psychotherapeutics. Psychoanalysis may be useful medically when it is subordinated to sound morals; it is full of danger to human well-being when it is substituted for morality. Vd. Notes 61 and 62 *infra*.

6b. Fully to understand that philosophy is independent of the changing content of scientific knowledge requires an understanding of the difference of their formal objects. It is not sufficient to define philosophy *negatively*, as we have so far done, by saying that it is non-investigative. For contemporary readers, the negative approach has pedagogic advantages. Science is knowledge and it is based upon empirical research. To those who hold this it is important to say philosophy is knowledge, though it is *not* based on empirical research. This negative point is essential to the distinction between science and philosophy, but it does not adequately define philosophical knowledge. We turn, therefore, to a positive distinction in terms of different formal objects of knowledge.

The material object,—the thing which is known,—is the same for science and philosophy. (For the time being, we shall disregard natural theology which has a different material object. Vd. Notes 8 and 9 *infra*.) Let us call that object the physical order: that which exists sensibly and under conditions of change. But of *that which* is known, *what* is known may be different, according to distinctions in the manner of knowing. Thus, sensitive and intellectual modes of apprehension are different, and we speak of the physical thing in its particularity and in its universality as different formal objects of knowledge according to this difference in the mode of considering it. Thus, the formal object of history is distinguished from the formal object of science and philosophy, but both sorts of knowledge are relevant to the same reality, the same material object.

This first distinction in formal objects depends upon the primary intellectual act of total abstraction. Further distinctions depend upon the difference between total, or extensive, abstraction, and formal, or intensive, abstraction. (Vd. St. Thomas Aquinas, *In Librum Boetii de Trinitate*, Q.V., A.3; J. Maritain, *Les Degrés du Savoir*, Paris, 1932: pp. 74–75; J. T. Casey, *The Primacy of Metaphysics*, Washington, 1936: pp. 81–82.) Within the sphere constituted by an abstract mode of consideration, the scientist is concerned with the phenomenal order, i.e., the sensible accidents or properties and their correlation, and the philosopher is concerned with the ontological order, the intelligible conditions of being and the essential nature of physical things. As Maritain says, the scientist goes from the observable to the observable, while the philosopher goes from the observable to the purely intelligible. The proper domains of science and philosophy do not overlap. Though they may have the same material object,–the world of bodies, –they do not have the same formal object. A scientific account of the phenomena can never displace or replace a philosophical understanding of the existences involved, or vice versa. Scientific descriptions, or even explanations, do not give us the intimate being of things. Philosophy is thus formally independent of science. The sciences by their proper means "will never suffice adequately to decide a philosophical question, for these questions depend on principles and on a light which are not within the range of science." (Vd. Maritain, *op. cit.*, pp. 93–95, 100–102.)

Further distinctions within the sphere of intensive abstraction,–what are called the grades of abstraction,–make the difference between science and philosophy clearer. Thus, all the natural sciences are of the same grade of abstraction, whereas within philosophy, there are mathematics and metaphysics, which differ from the philosophy of nature by being higher grades of abstraction. No one can doubt that the formal objects of mathematics and metaphysics are different from the formal object of the natural sciences. It is only the philosophy of nature which has a formal object constituted by the same grade of abstraction as that of the natural sciences. The problem of the independence of philosophy is, therefore, difficult, if at all, only with regard to the philosophy of nature. (For a discussion of this problem, vd. Note 10a *infra;* for a discussion of the three grades of abstraction, vd. Note 9 *infra.*)

That a difference in the formal objects of two bodies of knowledge is the sufficient formal condition of their independence can, perhaps, be most readily seen in the case of science and history. The general truths of natural science are unaffected by the changing content of historical knowledge; and scientific advances do not modify the facts established as probable by historical research. Here, then, are two bodies of knowledge, having the same material object, which are formally independent of each other because they are different modes of considering that identical object. (The exceptions to this independence are accidental; thus, the scientific generalization employed in an historical inference may be subject to alteration in such a way that the probability of the historical conclusion is affected; or an historical fact may be used as a singular datum in scientific research, e.g., the historical fact of an eclipse in astronomy.) The formal independence of mathematics and natural science will also be readily acknowledged. (Vd. Note 10 *infra.*) All that is here being said is that philosophy, comprising metaphysics and the philosophy of nature as well as mathematics, is related to investigative science as mathematics is, or as the whole field of science is to history.

The independence of philosophy and science can finally be understood in terms of the difference between common and special experience. Special experience does not alter common experience, though it adds a multitude of particular details in regard to which common experience is silent. Since philosophical facts are derived from common or vulgar experience, as scientific facts are inductions from experiment, the inadequacy of common experience with respect to details is indifferent, and the birth of science through special experience does not displace philosophy in the household of human knowledge. Two misunderstandings must be prevented here. (1) The facts of philosophy, its general truths, are not the facts of vulgar experience, any more than the facts of science, its inductive generalizations, are the same as its experimental data. The facts of ordinary experience,—what can be called the facts of common-sense knowledge,—play the same role in philosophical thought that experimental data do in scientific thought. They are the starting point for reflection and analysis. (2) Ordinary experience is more, rather than less, certain than scientific experience. Maritain quotes Duhem as saying: "The layman believes that the result of a scientific experiment is distinguished from vulgar observation by a higher degree of certainty; he is deceived, for the description of a physical experiment has not the immediate and relatively easy to control certainty of the common and non-scientific account. Less sure than the latter, it is of greater value by the number and precision of the details it makes known to us; there is its veritable and essential superiority." But, as Maritain says, the addition of these details is irrevelant to such properly philosophical facts as "there exists something, there exists a multiplicity, there exists change and becoming, there exists knowledge and thought, there exists desire" (*Op cit.*, p. 114).

7. The conflict between science and philosophy in modern times is strictly a conflict between scientists and philosophers. The fault is on both sides. Philosophers who have exceeded their domain, who have "speculated" in the bad sense of that word, who have guessed at

conclusions which could properly result only from painstaking research, have been justly rebuked by scientists as "arm-chair thinkers". Philosophers have been guilty of such disorder almost to the extent that they have not recognized their true vocation, i.e., reflection about the data of common experience. In the 17th century Descartes is the example of a philosopher attempting to answer scientific questions without recourse to scientific method; in the 19th century, Herbert Spencer is the much more distressing example of a thinker who did little else but "speculate" in scientific fields. He is one of the causes of philosophy's bad name at present. Hegel and Schopenhauer are, for different reasons, equally subject to the charge of fomenting a conflict between philosophy and science by transgressing the line which distinguishes scientific from philosophical questions. The charge, in other words, is that they are bad philosophers; only when they are bad in this sense of lacking their proper discipline, of failing to recognize the limits of their data and the character of the questions they can answer, do philosophers come into conflict with science.

Modern scientists, on the other hand, have been equally guilty of such transgressions, moving, of course, in the opposite direction. Newton, for instance, great scientist that he was, did not seem to recognize the difference between questions which could be answered by experiment or research and those which could not. He pretended to speak about philosophical matters with the same assurance that characterizes his well-founded scientific conclusions, although he exhibited little or no competence in the learning and discipline of philosophy. The concluding sections of the *Principia* and the *Optics* illustrate the unwarranted and incompetent dogmatism of a scientist trying to do arm-chair thinking without realizing that it has a technique just as rigorous as, and much less obvious than, the manipulations of the laboratory and the observatory. It is true that philosophical questions can be answered in an arm-chair, but success in scientific work is neither preparation for, nor a mark of ability to perform, the philosophical task. Darwin is another example of a scientist who concerned himself with questions his evidence could not possibly answer. The concluding chapter of *The Origin of Species* so confused scientific with philosophical and theological questions that the 19th century never fully recovered from the vertigo it suffered in trying to separate them. As a result we are all heirs to the myth, the religion, of evolution. If Darwin had had any competence in philosophy or theology or if, lacking it, he had contented himself with reporting the data and conclusions of his research, the conflicts about "evolution" which embroiled science with religion and generated the elaborate guess-work of 19th century thought, could never have occurred. In more recent years, the Gifford lectureship is the source of similar confusions and transgressions. Successful scientists are called to the platform to discourse on the problems of natural theology, which is a branch of metaphysics, in the light of scientific findings. They suppose, in the first place, that the scientific findings are relevant; they proceed, in the second place, as if it required years of training to become a scientist but only moments of leisure to indulge in the "idle speculation" which they take to be the understanding and solution of philosophical problems.

Formally, the chief sources of conflict between science and philosophy are two: (1) both general physics, i.e., the philosophy of nature, and the particular natural sciences, are knowledge of natural things, from the point of view of the same grade of abstraction; (2) metaphysics, though it is of a different grade of abstraction than the philosophy of nature, supplies the latter with its fundamental principles and concepts. (Vd. Note 9 *infra* for a discussion of the grades of abstraction.) If, in the first instance, the difference in method between natural philosophy and natural science is not carefully observed, the kind of question which can be answered by philosophical thought will not be distinguished from the kind of question which can be answered about the same subject-matter only by scientific method. The difference provides a clear, formal criterion. Philosophy can answer questions only about the mode of existence and the kinds of change to which all physical things are subject; it cannot answer questions about particular types of physical existence or particular causes of change. Such questions, as Aristotle pointed out, require investigation involving "intimate association with nature" (*De Generatione et Corruptione*, I, 316^a6). By this criterion it must be said that when Aristotle concludes that the heavenly bodies are imperishable and suffer only local motion, he is doing so as a scientist because the object of his knowledge is a particular kind of physical body, and not natural things or change in general. He may have been right in terms of his observations, but we have other and undoubtedly better observations. In his day, the observations of an astronomer were not radically different from the common experience of men. Under such circumstances the distinction between philosophy and science is not so readily made. But in our day, the experience of an astronomer, implemented by telescopes and spectroscopes, is radically different from "public experience." We can, therefore, easily make the distinction. But the formal principle of it is the same in the 4th century B.C. and today. To understand it is to read the works of those ancient and mediaeval authors who were at once philosophers and scientists, more intelligently. As Maritain points out, "on the side of the human subject, one must recognize that too great a confidence in the intelligibility of things and in the procedure of reason, in a domain which precisely is not that of philosophy but of experience, and where essences are not revealed to us, had its part (and perhaps a predominant part) in the errors of ancient science." From this point of view, he would declare that "modern science has rendered philosophy the immense service of delivering it from the foreign care, which has oppressed it so long, of describing phenomena." But, on the side of the object, "there is no necessary connexion between ancient mechanics and physics and astronomy, on the one hand, and the metaphysics or natural philosophy of the scholastic tradition, on the other. The whole edifice of the experimental sciences of the ancients could fall, and this immense ruin could appear to hurried minds as the ruin of all that the ancients had thought, whereas in reality their metaphysics and their philosophy of nature, in the essential principles" would not have been touched (*Les Degrés du Savoir*, Paris, 1932: pp. 118–119).

In the second instance, if the difference between metaphysics and natural philosophy in grade of abstraction, is ignored, the metaphysical principles which are involved in natural philosophy will be subjected to the same type of exemplification to which the conclusions of natural philosophy are susceptible. But this is not possible without confusion or distortion. All the details of scientific knowledge about nature can be interpreted so as to exemplify the general propositions of the philosophy of nature because both are knowledge having the same grade of abstraction. But to require that metaphysical propositions be similarly exemplified is to mistake their grade of abstraction and hence their true signification. The process of interpreting the results of scientific research in terms of metaphysical propositions can be accomplished only through the mediation of the philosophy of nature, and not directly. It follows, therefore, that when the formal difference between metaphysics and the philosophy of nature is ignored, or when the relation of these two is not understood, *apparent conflicts* are likely to occur between metaphysical truths and scientific facts. This has happened many times in the history of modern philosophy, from Descartes to the present day.

8. The individual thing, in its individuality, is infra-intelligible in the sense that it cannot be defined or comprehended by abstractions of any sort. This does not mean that the individual thing is unintelligible in the strict and absolute sense. Only prime matter, matter totally without form and hence purely potential, is strictly unintelligible. The individual thing, as an object of aesthetic intuition or of personal acquaintance, is not grasped by the senses alone, but by the senses and the intellect in that mode of cooperation which constitutes what the middle ages called "the knowledge of vision". God's knowledge is vision; the beatific vision of the blessed approaches it in type; the aesthetic intuition is a remote analogue in natural human knowledge. Although the intellect cooperates in the aesthetic intuition, it does so defectively because of its inadequate comprehension of individuals. As Kant said, it is knowledge without concept, and yet it is not subject to the blindness of pure perception. The Thomistic statement of the beautiful object as *id quod visum placet*, suggests the relation between aesthetic intuition and knowledge of vision. Vd. *Summa Theologica*, I, Q.5, A.4, ad.1. *Visum* does not refer to the operation of the sense of sight. Cf. *Summa Theologica*, I, Q.12, A.13; Q.14, A.7; Q.58, A.7, ad.3. Professor A. E. Taylor's essay on Knowing and Believing

provides an interesting commentary on these points. Vd. *Philosophical Studies*, London, 1934: pp. 386–394.

A word must be added about supra-sensible objects of knowledge. Just as the intellect cooperates in the knowledge of infra-intelligible objects, so the senses cooperate in the knowledge of the supra-sensible. In the former case, the cooperation is by conjunction; in the latter, sensitive apprehension is ordered instrumentally to the knowledge of transcendent objects, since all natural knowledge arises from sense-experience. Supra-sensible objects are the concern of that branch of metaphysics which is natural theology. Within the scope of natural knowledge, it is possible for men to know only that such objects are, but not what such objects are, except negatively and analogically. In a sense, such objects, whose mode of being transcends the physical order, are not only supra-sensible but also eminently intelligible. They are more intelligible in themselves than they are for us, because they transcend the physical order and thus exceed our intellectual capacity, rooted as it is in sense and imagination. The natural knowledge of transcendent beings must not be confused with that supernatural knowledge of them, which is proposed by the dogmatic theology of a revealed religion. The greater adequacy of the latter is due to the supplementation of reason by faith; the former is inadequate because it is merely rational and rational in a way that is limited by its dependence on sense.

8a. We have seen that the historian, unlike the scientist, is frequently concerned with questions of existence. In fact that is his primary interest; it is only after he has established the existence of some historical particular, a person, an event, etc., that he proceeds to questions of characterization and relationship. The scientist is concerned primarily with the characteristics and relational patterns of the phenomenal world. He never asks, because he need not, whether that world exists. If he sometimes seems to be asking the existential question, viz., whether atoms exist, he has unwittingly turned philosopher. The proper philosophical question is, of course, about the divisibility of matter, and not about the existence of real beings corresponding to the constructions of mathematical physics. The physical scientist usually becomes a philosopher only after he has committed the fallacy of reification.

Nor is the philosopher concerned with questions about whether this or that exists, except in the domain of natural theology. With the exception of natural theology, philosophy has the same material object as natural science. No more than the scientist, need the philosopher ask whether the world of nature exists. But as a metaphysician considering the being of whatever is, the philosopher is inevitably led to inquire about the existence of beings *other than* the physical. Such beings are possible. Are they actual? At this point, the inquiry of the metaphysician is like that of the historian and, in general way, so is his procedure. As the historian passes inferentially from present sensible evidences to the determination of a past existence,—an existence which is, of course, homogeneous in ontological type with the present evidences,—so the natural theologian passes inferentially from the entire order of physical existence to the existence of a transcendent being. Both inferences are *a posteriori*, both involve the principle of causality. The essential differences are: (1) that basic metaphysical truths are needed to pass from one ontological order to another, and (2) our knowledge of the nature of that which is proved to exist is, in the theological case, only negative and analogical. Precisely because it is negative and analogical, it can be said to be derived from the realm of experience.

9. The differentiation of the three spheres of philosophical knowledge, —physics (the philosophy of nature), mathematics and metaphysics,— is first clearly made by Aristotle. Thus, in the *Physics*, he inquires how the mathematician differs from the physicist. "Obviously physical bodies contain surfaces and volumes, lines and points, and these are the subject-matter of mathematics. . . . Now the mathematician, though he treats of these things, nevertheless does not treat of them as the limits of a physical body; nor does he consider the attributes indicated as attributes of such bodies. That is why he separates them; for in thought they are separable from motion, and it makes no difference, nor does any falsity result, if they are separated" (II, 2, $193^{b}22$–35). The physicist is concerned with things whose forms, though analytically separable from matter, do not exist separately. It is the business of the primary type of philosophy, metaphysics, to consider "the mode of existence and essence of the separable" (II, 2, $194^{b}15$). And in the *Metaphysics* we find: "Physics deals with things which are inseparable from matter but not immovable, and some parts of mathematics deal with things which are immovable but probably not separable, but embodied in matter; while the first science deals with things which are both separable and immovable. . . . There must, then, be three theoretical philosophies, mathematics, physics and what we may call theology, since it is obvious that if the divine is present anywhere, it is present in things of this sort" (VI, 1, 1026^{a} 13–21). Cf. *Metaphysics*, XI 3, 4.

The Aristotelian distinction of the three divisions of philosophy is thus made in terms of the mode of being of their proper objects: the object of physics is material and changing; the object of mathematics consists of abstractions from the physical object, forms which are analytically separable from matter in motion, but do not exist apart from matter; the objects of metaphysics are forms which are actually separated from matter and self-subsistent. This method of making the distinctions defines the subject-matter of metaphysics too narrowly, identifying it with knowledge of transcendent objects, making it natural theology. Similarly, it overemphasizes the formal object of mathematics and fails to indicate that the material object of mathematics is the same as that of physics. The capital text for confusing the two parts of metaphysics,—ontology and natural theology,—has long been the famous passage which says that "the first philosophy considers being *qua* being, both what it is and the attributes which belong to it *qua* being" (*Metaphysics*, VI, 1, 1026^a33). To avoid such confusion, two interpretations must be made of this text: first, that metaphysics considers the being of anything which is, the attributes of any being according to its mode of being (ontology); and second, that metaphysics considers *that which is* simply and absolutely, both that such pure being is and what its attributes are (natural theology). The material object of ontology can be the same as that of mathematics and the philosophy of nature; it is only in so far as metaphysics is natural theology that its material object is not physical existence.

It is true to say that metaphysics is knowledge of being. But this must be understood, first, in the sense of the formal object of metaphysical knowledge, namely, anything *considered* as being; and only second, in the sense of the material object of that part of metaphysics which is natural theology, namely, being as such *existing* simply. The central term of metaphysics is being, and this carries with it all the transcendentals which are convertible with being. But transcendental terms are essentially analogical in their universal predicability. The subject-matter of metaphysics is, therefore, everything which is, according as it is. (The existential questions about transcendent beings are a very special part of metaphysics.) The marks of a metaphysical proposition can be simply enumerated: it is about (1) being, or (2) any term which is convertible with being, such as the transcendentals, or (3) any division of being, such as act and potentiality, or (4) any term which is convertible with these divisions, such as form and matter. It is in this sense that it can be said that metaphysics is about anything *qua* being and the attributes which belong to it *qua* being. The opusculum *De Ente et Essentia* of St. Thomas Aquinas is a much better example of a metaphysical treatise than is the *Metaphysics* of Aristotle. Cf. J. Maritain, *Sept Leçons sur L'Être*, Paris, 1932–33; and N. Balthasar, *L'Abstraction Métaphysique et l'Analogie des Êtres dans l'Être*, Louvain, 1935.

Philosophy is knowledge of existing things, and not of ideas. Mathematics, in so far as it is philosophical, is not knowledge of abstractions as such, but of physical things considered formally at a certain level of abstraction. Metaphysics is also knowledge of physical things,—though not exclusively,—considered at an even higher level of abstraction. It is this maximum grade of abstraction which enables metaphysics to include within its scope beings which transcend the physical order.

The distinction of the three departments of philosophy is better made by Aquinas than by Aristotle, because it is made explicitly in terms of the grades of abstraction. The Thomistic analysis includes all that is sound in the Aristotelian account, and avoids its difficulties and ambiguities. The theory of abstraction enables us to understand how physics (the philosophy of nature), mathematics and metaphysics,–except for natural theology,–have the same object materially, i.e., physical existences, but different objects formally. Vd. St. Thomas Aquinas, *In Librum Boetii De Trinitate*, Q. 5, Q. 6, A. 1. (This elaborate and subtle analysis is summarized in a single passage in the *Summa Theologica*: I, Q. 85, A. 1, ad. 1, 2.) Nor is this inconsistent with an ontological distinction between the objects of physics and metaphysics, in so far as the latter is natural theology. As St. Thomas says, "the proportionate object of the human intellect, which is united to a body, is a quiddity or nature existing in corporeal matter; and through such natures of visible things it rises to a kind of knowledge of things invisible" (*Summa Theologica*, I, Q. 84, A. 7). Physical things are the *proper* (proportionate and appropriate) object of human knowledge, but only being,–infinite being,–is its *adequate* object.

The three grades of abstraction can be understood as degrees of remotion from matter. The first grade, like total or extensive abstraction, departs from matter only in so far as matter is the principle of individual or numerical diversity. Thus, the physical thing is seized in its universality, but as fully subject to the material conditions of change. This formal object constitutes the domain shared by the natural sciences and the philosophy of nature. The second grade departs further from matter, disregarding the matter of physical things in so far as it is the principle of change and motion, but regarding it in so far as it is the principle of dimensionality and quantity in things. This formal object constitutes the domain of mathematics. The third grade departs furthest from matter, disregarding every accident of the thing which is due to matter except the radical accident of contingency in being itself; positively it regards matter only as having being through form and it is, therefore, able to regard form as being apart from matter. This formal object constitutes the domain of metaphysics, both ontology and natural theology. Cf. Maritain, *Les Degrés du Savoir*, Paris, 1932: pp. 72–74.

This analysis raises three problems. (1) The relation between the philosophy of nature and the natural sciences. They would appear to have the same formal object. In a sense this is true. They are both concerned with the physical thing as subject to motion and change. But the philosophy of nature has an ontological interest, whereas the sciences of nature consider only the phenomenal surface. If the formal object, constituted by the first grade of abstraction, be called *ens mobile*, the difference can be stated as follows: the philosopher is concerned with the intelligible *being* of the changing thing; the scientist with its sensible *motions* and *changes*. Limited to the data of common experience, limited by the restrictions of ontological analysis to essential traits, the method of philosophy is incapable of achieving demonstrable knowledge of nature in its phenomenal variety and contingency. This is the task of science, which, by experiment and measurement, is able to determine observable properties from observable properties. The philosophy of nature seeks only to demonstrate the *ratio* of the being of changing things; or, in other words, the ontological principles which underlie the phenomena of change. The sciences of nature seek to discover the interconnexion of the phenomena of change, the correlation of the accidental variables which, in their covariancy, present to us the uniform face of a changing world. To the extent that the specific natures of physical substances escape man's power of definition, the philosopher must be content with generic distinctions among the types of bodies and their operations. But the scientist, not concerned with definitions that grasp essential natures, is able to make multifarious accidental classifications that introduce relative order into the maze of phenomenal variety. The philosophy of nature is one, but the sciences of nature are many. Their multiplicity arises from the different cuts which can be made in the phenomenal order, separating this sort of thing from that, this type of change from that. Generically, all the natural sciences have the same material object; but they differ, as particular divisions of natural science in general, by having different material objects, different segments of the phenomenal world of changing things. Cf. Maritain, *op. cit.*, pp. 76–79; also *La Philosophie de la Nature*, Paris, 1935.

(2) The philosophy of nature must not be confused with metaphysics, although it is continuous with metaphysics in its ontological interest. Both, in a sense, are concerned with being. But the philosophy of nature considers the being of the changing thing as changing, whereas metaphysics considers the being of the changing thing only as being. The philosophy of nature may, thus, be thought of as an application of metaphysics. The ontological principles it employs are derived from metaphysics, but they are used by the philosopher of nature to understand change, and the thing which changes in so far as it changes. In other words, the philosophy of nature has some independent principles. It is not merely applied metaphysics. Cf. J. T. Casey, *The Primacy of Metaphysics*, Washington, 1936: pp. 93–94.

A philosophy of nature is impossible unless it is founded upon metaphysics. Lacking metaphysical principles, it would be, what it is so frequently in modern times, nothing more than a conjectural embroidery on the fringe of natural science. It would not be philosophy in short, but the guess-work of persons who are dissatisfied with the superficiality of science. The other error in modern times is worse: the identification of metaphysics with the philosophy of nature. It is implicit in the speculations of Descartes and Spinoza. It is made explicit for the modern tradition in the bad scholasticism of Christian Wolff, who divided metaphysics into ontology (general metaphysics), on the one hand, and cosmology, psychology and natural theology (special metaphysics) on the other. (Vd. I. Gredt, *Elementa Philosophiae*, Fribourg, 1932: II, p. 2.) Kant learned these distinctions out of Baumgarten's *Metaphysica*, the text-book which he made the basis of his lectures. His failure to understand the difference between metaphysical problems, on the one hand, and the problems of cosmology (the philosophy of nature) and psychology, on the other, is largely responsible for all the nonsense that is written about metaphysics since the 18th century, as well as for the excessively dogmatic *naturphilosophie*, Hegelian absolutism and its *sequelae*. Vd. A. D. Lindsay, *Kant*, London, 1934: Ch. II, VI.

(3) Mathematics is not in the same line with metaphysics and the philosophy of nature. It has no ontological interest. This does not mean that the formal object of mathematics has no place in the order of actual existences, but rather that the mathematician is not concerned with that order, as are the metaphysician and the philosopher of nature. When the mathematician discusses existence, when he postulates or demonstrates existence, he means only to assume the *possibility* of his object or to show, by construction, that it is possible, i.e., not self-contradictory within the rules laid down by his postulates. The possible, in this sense, may include both what can be real or actual and what is *merely* ideal or fictive, a construction of the intellectual imagination, an *ens rationis*. As Maritain points out, since the order of real things includes both the actual and the possible and the order of ideal beings includes both the possible and the fictive there are inherent in mathematics, as knowledge of ideal beings, two opposed tendencies. On the one hand, in so far as it is concerned with the possible as a division of the real, it is knowledge relevant to the physical order and is properly associated with the natural sciences and the philosophy of nature. On the other hand, in so far as it is primarily concerned with its own constructions and with the possible only as a division of the ideal, it is something apart, something tangential to the whole hierarchy of human knowledge, since the natural sciences, like the philosophy of nature, are concerned with the real as the actual. (The natural sciences are in line with metaphysics and the philosophy of nature only in so far as they are concerned with the actual rather than the possible, but they are in line with mathematics in so far as they are concerned with the surface of substantial being, with the relation and order of accidents. This explains how mathematics can be aligned with science against philosophy, despite the fact that, being non-investigative, mathematics *as knowledge* is philosophical in type. Cf. P. Descoqs, *loc. cit.* in Note 6, on p. 136 *supra*.)

The problem of the classification of mathematics is a difficult one. There can be no question that in method and as a type of knowledge, it is philosophy rather than investigative science. Yet only to the extent that it is construed as knowledge of the real does it belong properly with philosophy. As so construed, I have used it to exemplify the difference between philosophical and scientific knowledge; otherwise, the exemplification would have been false. But it need not be so construed. In the modern world it is for the most part regarded either as knowledge of ideal constructions or as a purely conventional logistic. Yet this prevailing interpretation of the nature of mathematics does not prevent it from being of immense fertility in physical applications or from providing regulative principles to all those natural sciences which try to be metrical. The inner duplicity of mathematics is thus exhibited. Mathematics is like an incarnate angel suffering from schizophrenia. It seems to have affinities for two worlds, the strain between which it can endure only through a fundamental ambiguity. Yet the paradox of its nature is that almost in proportion as it more and more madly explores the ideal realm of its construction, does it become more and more useful in physical applications to the real. Cf. Maritain, *op. cit.*, pp. 78–83.

It is doubtful whether there is any cure for the madness of mathematics. But it can at least be kept from infecting the rest of philosophy and science. The history of both philosophy and science, from Descartes and Leibnitz on, records the spread of the mania for a universal mathesis. The most recent example of the derangement of both metaphysics and the philosophy of nature under the influence of mathematics is in the writings of A. N. Whitehead. (Whitehead's "scientific cosmology" is strictly not a philosophy of nature, which must employ independently established metaphysical principles. Rather it is an ontological projection of natural science, a reification of relational or mathematical logic. Vd. Notes 20b and 47 *infra.*) The dominance of mathematical physics over the other empirical, but non-mathematical, sciences is another example. What protection is there to ensure the rectification of the intellect in the line of the actual? In the sphere of philosophy, it is metaphysics, superior to mathematics in its grade of abstraction and therefore regulative of it as well as of all other bodies of knowledge. The philosophy of mathematics which is today largely the principles of logistic must become a metaphysical consideration of the objects of mathematical thought. In the sphere of the sciences, it is the philosophy of nature which, sharing the ontological light of metaphysics, must exert a counter attraction upon the empirical disciplines. But mathematics, because it is of a higher grade of abstraction than the philosophy of nature, is regulative of the natural sciences in its own right, whereas the philosophy of nature can be regulative only by virtue of the metaphysical principles it incorporates. It is an unequal struggle which, in the modern world, is becoming more and more a victory for mathematics in the sphere of science. The amazing technological utility of mathematical physics is an important factor in this trend. But this must not occur in the sphere of philosophy, where mathematics is of little or no utility and metaphysics is paramount.

One further point must be made. In an earlier discussion, the independence of philosophy and science as bodies of knowledge was argued. (Vd. Note 6b *supra.*) Such independence is, of course, reciprocal. It means that scientific knowledge is autonomous in its own sphere as philosophical knowledge is in its. Two qualifications must now be added. (1) Knowledge of a higher grade of abstraction can be regulative of knowledge of an inferior grade. In this sense, metaphysics and mathematics are regulative of natural science. This does not mean that metaphysics or mathematics can determine the conclusions of scientific research,—so far science is autonomous,—but rather that the interpretation and formulation of these conclusions may employ principles which science must take from philosophy because they exceed its powers. (2) Knowledge of an inferior grade can exert a material influence upon knowledge of a superior grade by supplying it with new exemplifications of its principles or by requiring it to extend its principles to new fields. In this way, the progress of the natural sciences may have an effect upon mathematics and the philosophy of nature and, indirectly, even upon metaphysics. But this does not abrogate the formal independence of philosophy, any more than mathematical physics ceases to be an empirical science because it employs mathematical formulae. Vd. Note 10a *infra.*

10. As we have already seen (Note 6b *supra*) the different grades of abstraction indicate why there can be no conflict between metaphysical and mathematical philosophy, on the one hand, and natural science, on the other. They are concerned with formally distinct objects. Yet the fact that there exists a mathematical physics,—in which mathematics is regulative with respect to physical formulations, and physical research is provocative of further mathematical analysis,—may lead some to suppose that philosophical and scientific knowledge are not reciprocally independent in such a way that conflict between them is impossible. This must be shown, and it can be most readily shown to a contemporary audience by taking the case of mathematics. Everyone knows that a mathematical question cannot be properly answered by any sort of empirical research. We may not be able to answer it at all, but if we can, we can do so without the benefit of investigation of an observational sort. It follows that if empirical methods are totally ineffective in the answering of mathematical questions, knowledge which is obtained by these methods cannot be inconsistent with such answers as are achieved by the appropriate method of mathematics. The principle, in short, is that if a cognitive procedure is incompetent to answer a question, it is equally incompetent to propose anything as a refutation to answers obtained by a method which is adapted to that type of question. If there can be no empirical mathematics, there can be no empirical or investigative refutations of mathematical propositions.

The same rule applies to metaphysics. The metaphysical object is even more incapable of empirical investigation than the mathematical, being of an even higher grade of abstraction. (Strictly, investigation, which always involves physical operations, is possible only within the sphere of knowledge constituted by abstractions of the lowest grade.) There can, therefore, be no empirical or experimental metaphysics, and equally no empirical or investigative refutations of metaphysical propositions.

The case of the philosophy of nature is, of course, the only difficult one, because what is, in general, the same formal object determine its sphere of knowing and that of all the natural sciences. But here, as we have seen, there is a subordinate formal distinction with regard to the aspect of the object being studied. The *being* of the changing thing is incapable of investigation; only the *phenomena of change* itself are open to empirical research. Investigation is, therefore, gratuitous in the philosophy of nature, but indispensable for the advancement of the natural sciences. Here the emphasis should be placed upon the necessity of research in the particular fields of natural science. The problems of optics, hydrodynamics, organic chemistry and genetics cannot be solved philosophically, any more than the problems of the philosophy of nature can be solved investigatively.

It should be said here that psychology constitutes the peculiar case which it was the aim of these lectures to examine. Psychology seems to be classifiable along with optics and hydrodynamics as a special field of physics: its object is a particular physical thing, man, subject to particular types of changes, such as learning and deciding. But psychology also seems to be capable of development in the light of common experience, and to violate the rule that in the particular fields of physics, investigation is indispensable. As we have seen, both indications are correct. Psychology is unique in that it is both a special natural science and a branch of philosophy, the only one which deals with a particularized physical object. It does not have this special status because of any distinctive formal constitution of its object of knowledge. Rather its uniqueness is due to the reflexivity which obtains in man's knowledge of himself.

10a. For an adequate discussion of the special problem of the philosophy of nature in relation to the changing content of natural science, vd. J. Maritain, *La Philosophie de la Nature*, Paris, 1935; *Les Degrés du Savoir*, Paris, 1932: pp. 83–144; and W. O'Meara, *Science and a Philosophy of Nature*, Proceedings of the American Catholic Philosophical Association, 1936: pp. 123–128.

The general solution of this problem can be indicated briefly here. As Maritain points out, "scientific data are like illustrations which it is normal for the philosopher to use to embody his ideas." This is particularly true for the philosopher of nature. Here the advance of scientific knowledge renews and enriches the matter which the philosopher can use for exemplification. Thus, for example, modern cytology and embryology have a descriptive precision which requires the philosopher to refine his formulation of substantial change, i.e. the process of generation. This dependence of philosophy upon natural science is material throughout. "The changes it brings in its train affect especially that imagery whose importance is after all so great with regard to the vocabulary and the halo of associations which envelop the didactic terms themselves. To imagine, however, that the philosophical doctrines must be transformed as the result of scientific changes would be as absurd as to imagine that our soul is transformed according to the diversity of the foods we assimilate" (*Les Degrés du Savoir*: pp. 100–102). In short, scientific facts can at best bring new matter to the consideration of the philosopher. They are never the original data upon which his inductions depend, but merely the occasion for new exemplifications and, perhaps, refinements in formulation of principles already known. "It is an illusion to believe that in appealing to scientific facts without considering them in a superior light, one can decide a philosophical debate, that of hylomorphism for example. For themselves they say nothing about it. Don't torture them to get false confessions! Never solicit them! Of course, let us question them; and that supposes that we are already provided with some information. They must be brought into court, as numerous as possible, and one must refer to the learned scientist for everything which concerns their civil status, their experimental significance and the way in which they have been established. But it is only by bringing them in touch with philosophical knowledge already acquired elsewhere, and with philosophical principles, that one can draw from them an intelligible content that can be handled by philosophy, that one can discern and judge what is their ontological value" (Maritain, *op. cit.*, pp. 115–116).

This can be explained by the analogous relation of mathematics and experimental physics. Experimental data are never the logical ground of mathematical discoveries; but new physical data may nevertheless provoke or occasion mathematical research, with the consequence that mathematical formulations are provided which subsume the data. It must be noted here that the empirical data do not test, verify or refute in any way, the demonstrable truths of mathematical analysis; nor are they the matter which generates those truths inductively. The data only discriminate between mathematical formulae which are or are not applicable to the physical world. If two formulae are applicable so far as the data are concerned, it is for the mathematician to choose between them by criteria of elegance and simplicity. The principles of the philosophy of nature are *similarly* related to the inductively established facts of natural science. The latter may provoke more precise formulations or even require an extended interpretation of old principles. To this extent the progress of scientific knowledge may cause development in the philosophy of nature. But whereas scientific progress is a transition to new and better knowledge, definitely enlarging or replacing the old, the philosophical development is not progressive in this way; rather it is a richer understanding of principles already known. It is intensive, not extensive. And it must be reiterated,—so widespread and entrenched is the popular illusion to the contrary,—that the propositions of the philosophy of nature are not founded upon scientific facts, nor can they, therefore, be "supported" or "refuted" by scientific knowledge. The relation of material dependence is one of exemplification.

What is here said of the relation between natural science and the philosophy of nature can be more sharply said of that between natural science and metaphysics. The philosophy of nature is, in a sense, an intermediary between metaphysical truths and scientific facts. The latter can be seen to embody the principles of metaphysics only after their ontological significance has been interpreted in the light of natural philosophy. One example will suffice to make this clear. The biologist gives us a mass of precise descriptive knowledge about the observable characteristics of embryonic development. These facts will not tell the natural philosopher what substantial change is. But they will enable him, perhaps, to delineate within this complicated phenomenal pattern the distinctions of substantial and accidental change. The facts will thus become illuminated with ontological significance. Then, and only then, will the metaphysician be able to determine the mode of being of a foetus. This is strictly not an advance in metaphysical knowledge; we are not learning what a substance is, but whether the foetus is a substance. Through such application of metaphysics we obtain a better understanding of an embryo than science can give us, though,—be it ten times repeated,—not a more useful one for the practice of obstetrics.

These points can be simply summarized. (1) The influence of the sciences upon philosophy is entirely material; it is psychological rather than logical. (2) The influence of philosophy upon the sciences is twofold: (a) it is positive or informative, in the way in which mathematics supplies empirical physics with formulae and principles; (b) it is negative or regulative, in the way in which metaphysics restricts science to the solution of phenomenal problems. (3) The philosophy of nature is regulative by virtue of its metaphysical principles; it is informative only after its substantial concepts and ontological principles have been transformed by relational logic and made applicable to the phenomenal order. Vd. Note 47 *infra*.

11. In the *Timaeus* Plato constructs the science of nature as an application of mathematics. In the *Republic*, he proposes music and astronomy, along with arithmetic and geometry, as pure mathematical studies, to be divorced as far as possible from observations (528 E–531 C). This again indicates the duplicity of mathematics, which continues to our day in the tradition of mathematical physics. Aristotle recognized this duplicity and distinguished both mathematics and physics from that mixed or intermediate kind of science which combines them, treating of problems that are formally physical in terms of principles which are formally mathematical. Thus, he writes of

"the more physical branches of mathematics, such as optics, harmonics and astronomy. These are in a way the converse of geometry. While geometry investigates physical lines but not *qua* physical, optics investigates mathematical lines, but *qua* physical and not *qua* mathematical" (*Physics*, II, 2, 194^{a}6–11). The mediaeval tradition of the Quadrivium recognized this distinction by separating the pure mathematics of arithmetic and geometry from their respective applications in music and astronomy. The latter involved the arts of measurement. St. Thomas clearly distinguished applied mathematics from physics. The former he treated as a science *intermediate* between mathematics and physics. Vd. *In Librum Boetii De Trinitate*, Q. 5, A. 3, ad. 5, 6, 7.

And at the beginning of modern physical research, the distinction is preserved both by Newton and Galileo. The title of Newton's greatest work tells the story: *The Mathematical Principles of Natural Philosophy*. Galileo, in the *Two New Sciences* enumerates perspective, astronomy, mechanics and music as sciences in which "mathematical demonstrations are applied to natural phenomena." In contrast, the tradition of physiology from Galen to Claude Bernard and J. B. Haldane is that of a purely physical science, the principles of which have the same formality as the problems they are employed to solve. Galen's work on *The Natural Faculties* plainly reveals that the principles he is applying to the analysis of the facts of vegetative functioning are derived from Aristotle's *Physics*, i.e., a philosophy of nature. For an illuminating discussion of the paradoxes which arise from the failure to distinguish mathematical from physical solutions of physical problems,—paradoxes of the sort that attended the popular, and even the technical, misunderstandings of the theory of relativity,—see Maritain, *Theonas*, New York, 1933: pp. 61–102.

12. Galileo's procedure was applicative rather than inductive in the strict sense of the word. By mathematical reasoning he first arrived at a formulation of the rate of acceleration of bodies falling freely to the earth, and then proceeded to test his equation by the measurement of a carefully constructed experimental case. The numbers which he obtained from the observation of the times and distances indicated by a ball rolling down an inclined plane were sufficiently accurate approximations to the ideal numbers required by his formula to assure him that his equation applied to the natural motion, the changing quantity of which he sought to define. It is perfectly clear that he did not first construct the experiment, make the measurements, and then from the resulting data sheet of observed times and distances *induce* the proposition that the acceleration was as the square of the times. The apparatus available to him made such a procedure impossible. Let us suppose, however, that he had such instruments as photoelectric cells and chronometers. Considering the problem of acceleration, he might then have constructed the following experiment. At various distances from the top of a metal tube, he could have placed

photo-electric cells which would be activated by the passage of an opaque body to affect the registrations of a chronometer. Dropping a ball through this tube, and reading the clock, he would then immediately obtain numbers which stated the covariation of times and distances. From these numbers the formula for the rate of acceleration could then be readily induced.

This indicates that the formal relation between the observed data and the general proposition is the same whether the procedure is applicative or inductive. The difference is psychological. In the inductive procedure, the universal is discovered in the particular. Experiments aid induction by creating a particular case which approximates ideal conditions, separating the irrelevant and accidental conditions from the relevant and essential. In the applicative procedure, the universal, already known, is located in the particular, and is thus tested or verified. Experiments are as useful for application as for induction, because the location of the universal in the particular requires the creation of an approximation to the ideal.

13. A scientific proposition is here called analytic when it is capable of an aetiological interpretation. But the causal interpretation is always something added to the content of the proposition which, strictly, is descriptive and not explanatory. The natural sciences, mathematical or otherwise, as bodies of theoretic knowledge, do not treat of causes any more than they do of substances. The physionomic statements of biology and the functional formulae of mathematical physics can, however, yield causal interpretations. In fact, in so far as they have any practical or operational significance, they must be so interpreted. Vd. Notes 16a and 47 *infra.*

A genuine causal analysis usually involves asymmetry in the temporal order and, what is more important, always an ontological asymmetry. The being of the cause does not depend on the being of its effect, as the latter does upon the former. The efficient cause is prior to the material and formal cause in the determination of a change. Functional statements, formulae of covariation, are essentially symmetrical. Any variable can be treated as independent or dependent. The equations of mathematical physics are functions and, in terms of mathematical principles, it is therefore arbitrary which variable is designated as the cause (independent) and which as the effect (dependent). The analytic statements of mathematical physics are thus seen to be descriptive rather than aetiological. From the time of Galileo, mathematical physicists have naturally disclaimed an interest in causes. Vd. Eddington, *The Nature of the Physical World*, New York, 1928: Ch. III, XIV. But mathematical physics is physics to the extent that it is experimental and observational. The investigative procedures necessary for the application of mathematics to nature involve the performance of physical operations upon physical things. In the construction of an experiment, in the performance of any complicated observation, the physicist operates by giving his descriptive formulae a causal interpretation. He does this or that in order to obtain this or that effect. The language which reports an experimental performance plainly reveals that the physicist knows the difference between the independent and the dependent variables. Yet this knowledge is practical or operational, rather than theoretic.

The same can be said for the formulations which are achieved by the non-mathematical natural sciences. These statements are, like the equations of mathematical physics, essentially descriptive, but they can also be viewed as analytic. Thus, the elements of a secular trend are asymmetrically ordered in time; this order can and, for practical purposes, must be causally interpreted. For the operations of research or for the practice of medicine, the descriptive knowledge which constitutes such sciences as physiology and pathology must be made to yield an aetiology.

The principles of aetiology are, of course, to be found in the philosophy of nature. The natural scientist as a technician, the engineer and the practitioner who apply scientific knowledge, are employing philosophical concepts when they make a causal interpretation of their descriptive knowledge for operational purposes. We must regard the detailed aetiologies which thus grow on the operational side of science as essentially practical, because they are not part of its theoretic content.

14. Vd. Note 6 *supra.*

15. Vd. Note 7 *supra.*

16. The position of Kant in the history of modern philosophy is strikingly defined by the errors of the 17th century which preceded him, and the errors of the 19th century which followed him. The whole story can be told in terms of before and after Kant. Kant used the words skepticism and dogmatism to name equally wrong extremes which diverge from the proper mean of discipline, critically self-imposed by sound philosophical procedure. Critical discipline in philosophy means two things: (1) the determination of the sphere of questions which can be answered without research or investigation; this in Kantian terms is the distinction between the pure and the empirical, the *a priori* and the *a posteriori;* (2) the delimitation of this sphere of inquiry, bounded as it is on one side by dogmatic theology, which requires faith as an auxiliary, and on the other by science, which requires investigation as an aid. The sphere of inquiry, thus determined and restricted, is the domain of philosophy. Those who deny it entirely, by error of defect, are skeptics. Those who transgress it in either direction, by error of excess, are dogmatists. Vd. *The Critique of Pure Reason*, Preface to the Second Edition (1787); *Prolegomena to Any Future Metaphysic*, Section 13, Remark III, and

Appendix C, critique of the fourth paralogism of transcendental psychology.

In the 17th century, or at least before Kant, philosophers had committed both errors. On the one hand, Descartes, Leibnitz and Spinoza were dogmatists. They tried by philosophical methods to answer questions which should have been reserved either for theology or experimental science. The badness of their philosophies, particularly their metaphysics, resulted from these confusions and transgressions. (To the extent that questions in natural theology concerning God, creation, the soul, are parts of larger questions which require sacred theology, founded on Revelation, for their adequate solution, Kant was right in regarding a purely rational approach to the whole question in each case as doomed to the frustrations of an illusory dialectic, a dialectic which did not properly recognize its own limits in philosophical mysteries. Descartes, Leibnitz and Spinoza deserved Kant's critique for violating the distinction between faith and reason, but unfortunately Kant himself misapprehended that distinction, denying to reason what is properly within its province,—metaphysics, both ontology and natural theology.) On the other hand, Hume was a skeptic. He tried to deny that there were any genuinely answerable questions about the realm of existence, other than those which required empirical research to answer. He refuted himself in doing so, because he employed the method of philosophy to answer a philosophical question, even though he answered it wrongly. Kant's critical regimen was an attempt to cure philosophy of these ills and to establish it within its own proper domain as vigorously as the sciences seemed to be flourishing in theirs.

The 19th century is the record of his failure. After Kant,—and in a sense because of him,—philosophers committed the same errors. On the one hand, the romantic and absolute idealists of the Fichtean or Hegelian sort were worse dogmatists than the metaphysicians of the 17th century, worse both because of the extent of their dogmatic excess and because they combined metaphysical dogmatism with the kind of epistemological skepticism that originated in Berkeley and Hume. On the other, in revolt against them, as Hume in a sense was in revolt against the "abstract" and "abstruse" philosophy of his 17th century predecessors, positivists and pragmatists of all sorts revived the error of skepticism,—philosophers denying that philosophy has an independent domain of inquiry and knowledge.

The state of philosophy after Kant is, in a sense, worse than its condition before Kant. The very errors of dogmatism and skepticism which Kant sought to correct not only recurred after him with greater violence, but, through commixture, became almost obscurantist in their impurity. The fault is Kant's. Had he repudiated Hume's premises instead of trying to avoid his conclusions, had he understood the tradition of European philosophy prior to the 17th century so as to have been able simply to correct the disorder of faith, reason and observation in Leibnitz, Descartes and Spinoza, he could have accomplished his critical reforms without the transcendental nonsense of the *Critique of Pure Reason.* Of course, he would have done nothing original; he would merely have regained the sound tradition of European philosophy prior to the Renaissance. That, of course, is too much to expect of almost any modern man. Discarding the burden of tradition, the heavy authority of human wisdom, and emphasizing the individual accomplishment that is marked by novelty and originality,—these are the correlative expressions of the protestantism and egoism which characterize modern thought. Kant is, indeed, the tragic hero of the modern story. He is a good man, rightly motivated and honest in execution, but with fatal defects of ignorance and pride. The catastrophe of the 19th century is the more painful as a tragic denouement because Kant came so near to being sound.

16a. The appraisal one makes of Hume and Kant must differ according as one views their work in relation to the natural sciences, on the one hand, or in relation to philosophy, on the other. In the first connection, they must be credited with having solved what can be thought of as the most critical problem of modern times. Our epoch is certainly distinguished from all others by the existence, amplification and application of empirical science. The novelty of science, its difference both in method and utility from philosophy, was recognized early in the 17th century. This new thing challenged the traditional theory of knowledge. It required a radical extension of that part of philosophy which has come to be called epistemology. In other words, just as the nature of philosophy is itself a philosophical problem, so is the nature of science. And in so far as Hume and Kant solved the latter problem, they made an epochal contribution to philosophy. Hume's radical insights were the basis for Kant's constructive formulations. Hume saw that empirical knowledge did not penetrate to the substance of things; he saw that it could not be a theoretic apprehension of causes, although the customary sequences which such knowledge described could, for practical purposes, be given a causal interpretation. He was the originator of the positivistic conception of science as restricted to observable relations among phenomena. Kant developed this positivistic conception constructively by discussing the forms and categories of empirical science, such as space, time, quantity, quality and relation; by reducing substance and causality to relational concepts; by understanding the role of mathematics and experimentation in the investigation of phenomenal reality. (Vd. *Metaphysical First Principles of Natural Science*. Kant did not realize, of course, that his own "pure physic" and transcendentally deduced categories were marks of an ill-conceived philosophy of nature.) The phenomenalistic positivism of the 19th century, as found in Comte, Mach and Mill, adds little if anything to the conception of science created by Hume and Kant. It does correct the *a priorism* of Kant by returning to Hume's view of the contingency of science. The logical positivism of the 20th century adds only the heuristic principles of relational and mathematical logic. The positivistic tradition, beginning with Hume and Kant, has done a great intellectual service in defining the traits and the limitations of scientific knowledge. It has not only solved a philosophical problem, but it has given the scientists, who gradually became self-conscious, an understanding of themselves. Vd. Eddington, *The Nature of the Physical World*, New York, 1928; Bridgman, *The Logic of Modern Physics*, New York, 1927; and similar works.

It is unfortunate, indeed, that the enlightening contribution of positivism was mixed with the darkness of denials. These also began with Hume and Kant. They solved one part of the problem (i.e., the nature of science); but their epistemological principles were not sufficiently refined or sound to solve the other part as well (i.e., the nature of philosophy). Although Hume proclaimed the principle of causality as the basis of his analysis of probability (vd. *An Enquiry Concerning Human Understanding*, Sections VI, IX); although Hume both discussed the enduring order of nature and advocated practical procedure in terms of causes and substances; and although he recognized in mathematics a body of non-empirical knowledge, treating it, of course, as "knowledge of ideas," Hume was unable to see the possibility of metaphysical knowledge. Of whatever does not contain "any abstract reasoning concerning quantity or number" or "any experimental reasoning concerning matter of fact and existence," Hume could say: "Commit it then to the flames: for it can contain nothing but sophistry and illusion" (*Op cit.*, Section XII, Part 3). Similarly, Kant was not able to see that his own critical philosophy, in so far as it was knowledge, was not empirical science and that, in consequence, non-scientific knowledge and knowledge of something other than the phenomenal order were possible. The fatal epistemological error in Kant's case, as in Hume's, arises from an inadequate psychology. Hume was a sensationalist. His failure to distinguish ideas from sense-images made it impossible for him to understand philosophical knowledge; in fact, it almost made it impossible for him to understand science. His positive contribution with regard to science was qualified by a raw empiricism and skepticism which Kant had to remedy. Unfortunately, the transcendental medicine was almost as bad as the disease. Kant's own psychological error was also on the nature of ideas. He failed to understand ideas as the products of intellectual abstraction, and hence did not see that the intuitive faculty of sense provided materials for abstractive and non-intuitive knowledge of an intelligible or noumenal order. As a consequence, Kant denied that there could be any knowledge of noumenal reality or of purely intelligible objects. As a further consequence, and one intrinsically damaging to Kant's conception of natural science, his psychological error distorted the nature of mathematics as intuitive knowledge of imaginative constructions. A fuller discussion of these psychological errors occurs elsewhere (Notes 21 and 22 *infra*). They are mentioned here only to remark that, on the one hand, the limitations of positivism as a theory of knowledge are due to bad psychology and, on the other, that much of the badness of psychology since the 18th century is due to a positivistic conception of knowledge. Alibert says in his Preface to *La Psychologie Thomiste et les Théories Modernes*, Paris, 1903: "Sans se proposer une étude spéciale du positivisme et du kantisme, l'auteur s'attache à combattre le principe de ces systèmes, celui du premier en particulier, dans les pages consacrées à la conscience intellective et à l'origine des idées de cause et de substance."

In its critical and negative aspects, contemporary logical positivism adds nothing new to the denials made by Hume and Kant, unless it be the hollow pretense that the dismissal of metaphysics as nonsense is supported by logical or grammatical considerations. Modern logistic in all its varieties,—whether relational logic or Cambridge symbolic logic or Viennese grammatics,—cannot consistently be regarded as *formal* and at the same time be used as a *criteriology*, a theory of knowledge. Neither formal logic nor universal grammar provide a basis for demonstrating (1) the impossibility of self-evident propositions, (2) the impossibility of philosophical knowledge, i.e., ontological propositions, (3) the reducibility of all propositions to atomic or elementary statements of sense-observation. The logical positivist's theory of knowledge is not *logical*; logic as logic is necessarily indifferent to the issue between positivism and realism; rather it is based on the same erroneous and inadequate psychology that generated the position of Hume and Kant. If the grammar and logistic of contemporary positivists were divorced from this bad psychology, which they seldom make explicit, the commotion which the Viennese have agitated in the shallow currents of contemporary thought would soon subside. Vd. Note 17 *infra*.

We can undo the errors of Hume, Kant and the whole positivistic tradition without in any way nullifying the intelligible rendering they made of scientific knowledge. On the contrary, if by correcting their errors in psychology, we can add a sound conception of philosophical knowledge, that by contrast will reflect new light upon the nature of science. This can be briefly done. Let us define philosophy, first, by affirming of it everything Hume denied of empirical science: it is knowledge of substance and causes; it goes behind the phenomena to a substantial reality; it has certitude where science is only probable; it is abstract rather than experimental reasoning. Then, in Kantian terms, let us say that philosophy is knowledge of the noumenal, but not transcendent, order (metaphysics as ontology, and the philosophy of nature), as well as knowledge of purely intelligible, and hence transcendent, objects (metaphysics as natural theology). With the denials of positivism thus removed, a sound and adequate epistemology enables us to understand the nature of philosophy, as the ancient tradition understood it, as well as the nature of science, according to the modern insight which most contemporary thinkers and scientists share. And how is this *rapprochement* of the two traditions to be accomplished? How is modern positivism to be retained for its vision of science and cured of its blindness to philosophy? Simply by an analysis of cognition which shows the cooperation of sense and intellect and which understands understanding itself as a faculty of abstraction rather than of intuition or of innate ideas. Such a theory of knowledge will correct the psychologism of Hume and Kant by recognizing the priority of metaphysics in philosophy.

There is one further comment on all this. One must not forget the radical difference between the modern and mediaeval epochs. The middle ages never doubted the intellect's ability to know reality and were interested in error only as pathological. In modern times, when as a matter of fact research has immensely increased man's knowledge, philosophers start by treating knowledge as a bewildering mystery, supposing that it is truth rather than error which must be explained.

17. The phenomenological question is, Whether man understands? and this is answered whenever we affirm that we do understand and mean thereby something distinctly different from what we mean when we affirm that we sense or imagine. The analytical question, What is understanding? is necessarily subsequent. The fact that we do understand cannot be altered by subsequent considerations concerning the nature of understanding, its relation to the senses and the brain, etc. *That understanding is* cannot be refuted by any analysis of *what understanding is*. It can be refuted only by a negative answer to the phenomenological question. It is impossible to give that negative answer and continue to talk intelligently. To those for whom this is not apparent, it can be shown by a logical analysis of the propositions which constitute the fabric of human discourse.

Propositions can be divided, first, according as their intention is singular or universal. The intention of a proposition is determined by the mode of signification of its subject term. On the grammatical side, the intention of a proposition is usually indicated by whether the words constituting the subject term are proper or common names. Within the singular intention, there is a further division into individual and particular intentions, according as the subject term has the minimum connotation of *this* or has the more fully determinate connotation of *this such and such*. Within the universal intention, there is the traditional distinction of first and second intention, according as the subject term has denotative reference to possible or actual members of a class or according as the subject term is shorn of all denotation except the designation of an *ens rationis*. The problem of the intention of the predicate term occurs only in propositions having universal intention. If the subject is in the first intention, the predicate must be so also. But if the subject is in the second intention reflexively, the predicate may either be in the second intention reflexively or predicatively. The propositions of mathematics are of the first sort; the propositions of logic, of the second.

The fact that human discourse cannot be interpreted without distinguishing between the singular and the universal intention is all that is needed to establish the distinction between sensitive and intellectual apprehensions.

Far from denying these distinctions, logical positivism has been unable to avoid them in its attempt to classify different orders of statement and to discriminate language about things from language about language. Unfortunately, contemporary analysts have not been able to achieve the clarity and subtlety of mediaeval logic in treating distinctions of this sort, partly because of the ill-concealed nominalism which is the core of all attacks on philosophy since the days of Hume. Nominalism is, of course, a psychological doctrine, a denial of intellect, a reduction of all knowledge to sense-knowledge. Nominalism has not only muddled the grammar and logic of contemporary positivists, but it has made it as difficult for them as for Hume to give an account of science which would be consistent with their denial of metaphysics.

18. The imagination is traditionally called the passive intellect because it is the faculty productive of experiences. An experience, as we have seen, is a sort of universal; the *quasi*-universality of the products of imagination has been recognized, in the modern tradition, under the head of generic or abstract images. Vd. Note 5 *supra*. Strictly, of course, images are not ***abstract***. The word "abstract" can be said of them only with the signification that they have the potentiality for abstraction; they are the proximate matter upon which the intellect operates in the first act of abstraction. The difference between the perceptual image and the so-called generic image is that the former has an individual intention whereas the latter has a particular intention. The marks of determinate individuation can be lost without the loss of the marks of singularity. The generic image, therefore, provides the transition to the universal intention of the idea which can be abstracted from it. It is a mean between the extremes of sensitive and intellectual apprehension.

19. "When the act of imagination is hindered by a lesion of the corporeal organ, for instance, in a case of frenzy; or when the act of memory is hindered, as in the case of lethargy, we see that a man is hindered from actually understanding things of which he had previous knowledge." This indicates that "for the intellect to understand actually, not only when it acquires fresh knowledge, but also when it applies knowledge already acquired, there is need for the act of the imagination and of the other powers." In addition to the evidence from pathology and fatigue, this can be learned by anyone from his own experience, for "when he tries to understand something, he

forms certain phantasms to serve him by way of examples, in which as it were he examines what he is desirous of understanding. For this reason it is that when we wish to help someone understand something, we lay examples before him, from which he forms phantasms for the purpose of understanding" (St. Thomas Aquinas, *Summa Theologica*, I, Q.84, A.7). In the following article, St. Thomas proceeds to argue that the operations of the intellect are hindered by the suspension, through disease, fatigue or sleep, of the sensitive powers. In light sleep the senses are not entirely inactive, and the imagination may operate as in dreaming. "Sometimes while asleep a man may judge that what he sees is a dream, discerning, as it were between things and their images. Nevertheless, the common sense remains partly suspended; and therefore, although it discriminates some images from reality, yet is it always deceived in some particular. Therefore, while man is asleep, according as sense and imagination are free, so is the judgment of his intellect unfettered, though not entirely. Consequently, if a man syllogizes while asleep, when he wakes up he invariably recognizes a flaw in some respect" (*Ibid.*, Q.84, A.7, ad.2).

According to this analysis, the act of understanding is causally dependent upon acts of the bodily organs of sense and imagination, but it is not itself the act of a bodily organ. This analysis accounts for all the experimental and clinical evidences which modern researches have been able to add to the essentially correct knowledge which St. Thomas possessed through common experience. As the eye and the ear are the bodily organs of particular sensitive operations, vision and audition, so the cerebral cortex of the brain is the bodily organ of what St. Thomas would have treated as the various operations of the interior sense and which we, with some loss of detail, call memory and imagination. The cerebral cortex contains the primary reception areas for the various senses. Although all neurological evidence is against the specific localization in brain areas of any of the higher functions, such as learning or discrimination or judgment, the same sort of evidence supports the localization of the sensory reception areas. Thus, injury to the calcarine fissure of the occipital lobe can cause blindness though the eye is healthy and there are no lesions in the optic nerve. The neurological evidence, both experimental and clinical, supports the Thomistic analysis. Specific lesions can destroy the power of a particular sense or, if the lesion be central, destroy a particular field of imagination and memory, such as the visual or the auditory. But such destruction does not destroy intellection, or what the neurologist calls the "higher mental functions". They may be hindered or impaired by the loss of sensitive or imaginative powers, but they cannot be destroyed without the total annihilation of all sensitive functions. The powers of the understanding continue to operate as long as sense-organs and brain are sufficiently intact to provide it with any matter to operate upon. The neurological evidence shows precisely what St. Thomas held: that understanding is causally dependent on sense and imagination, but is not directly the function of a corporeal organ, such as the cerebral cortex. The nervous functions are the material, but not the efficient or formal, cause of intellection.

Another sort of evidence is relevant and favorable to the Thomistic analysis. Muscular fatigue can be increased to a point at which the muscle being studied is no longer able to operate. But all the experimental studies of so-called "mental fatigue" have been unable to discover any similar phenomenon of total collapse. The weariness of the individual doing incessant mental work over long periods of time is discovered to be muscular. This physical weariness affects mental activity, as that is measured by the time taken and the errors made in performing such operations as arithmetical calculations. In other words, if we speak of a man's mind becoming weary, such predication of weariness is *per accidens*. The weariness of the body, impairing the operations of sense and imagination, hinders understanding "inasmuch as the intellect requires the operation of the sensitive powers in the production of phantasms" (*Ibid.*, Q.75, A.3, ad.2).

In conclusion, it must be pointed out that the sort of evidence which is gathered by modern physiological and neurological research does not establish the proposition that understanding is not the act of a bodily organ. (Vd. the authorities cited in Note 36 *infra.*) That proposition is established by philosophical demonstration, independently of all such evidence. The present discussion merely makes explicit the compatibility of the scientific facts with the philosophical proposition, by showing that these facts are relevant to the way in which the understanding depends on sense and imagination.

20. The word "cosmology" is sometimes used to name that part of metaphysics which is knowledge of the mode of being appropriate to physical things. It is sometimes used, even more narrowly, to name the metaphysical propositions which are involved in the philosophy of nature, and sometimes it is used as a synonym for the philosophy of nature. It is a regrettable name and carries with it the distortions of the Wolffian tradition which have infected modern philosophy, scholastic and otherwise. Vd. Note 9 *supra*. "Cosmology" should be restricted to mean the philosophy of nature. It is not a branch of metaphysics, but it necessarily presupposes metaphysical principles. I used the word in Lecture II to indicate that metaphysical principles were being discussed *only* in so far as they were involved in the philosophy of nature. I could not in the scope of that lecture discuss the deeper ontological significance of those principles or their significance for natural theology. Where in that Lecture I talk superficially of four metaphysical doctrines I mean four philosophies of nature, four "cosmologies", which are divergent because arising from different metaphysical positions. Thus, the traditional name "hylomorphism" is more appropriately applied to a philosophy of nature

than to a metaphysic. Hylomorphism names a metaphysical doctrine only in so far as an analysis of the physical order in terms of matter and form rests upon ontological insights, that being is divisible into potentiality and actuality, into essence and accident, etc.

Some clarification would be achieved if the word "cosmology" could be used, in contradistinction to "philosophy of nature," to designate two peculiarly modern aberrations: (1) the illicit extrapolation and projection of scientific knowledge into an imaginative world-view, and (2) the attempt to solve the problem of the unity of the phenomenal world. The philosophy of nature treats of the interconnexion of individual substances only in terms of causality. Any attempt to pass beyond the pluralism of primary substances causally connected must lead either (a) properly to the unity which pervades things because they are gathered and ordered in Divine Providence, and this is strictly a theological unity,—or (b) improperly to the kind of monism which denies plural substances and uses a part-whole analysis to solve the problem of the one and the many. The "block-universe" which results from the latter alternative, with its absolute determinism, makes it impossible to account for real contingency in nature, which is intelligible only in terms of Aristotelian pluralism and which is compatible with Providence. (Aristotelian pluralism is a pluralism of individual substances. It must not be confused with neo-Platonic pluralism which is a pluralism of forms hierarchically received in one pervasive matter, which has some being apart from form. Vd. Note 21 *infra.*) The point to be reiterated here is that modern "cosmologies," from Leibnitz and Spinoza to Hegel and Whitehead,—having their ancient prototype in Plotinos,—are neo-Platonic confusions, influenced by the relationalism of science, on the one hand, and by unacknowledged borrowings from theology, on the other. Vd. Notes 20b and 25 *infra.*

20a. "When Descartes established the scientific dichotomy between mind and body, he provided both the *raison d'être* of modern psychology and the mystery which it has never completely dissolved. Descartes cut the world in two, into mind and matter, just at the time when science was about to begin the course of development which has made it the dominating influence in modern civilization. We all know how successful the physical sciences have been and we can also see that biology has prospered in abandoning a vitalism and identifying itself with the physical side of the Cartesian dichotomy. If Descartes was right, if there are these two worlds, then the success of science in attacking the one forms a challenge for the creation of a science of the other. This view is common in psychology. When psychologists speak of the classification of the sciences they are usually sure of only two sciences—psychology and physics. Yet, if psychology is coordinate with physics and if the scientific method is applicable to both, then it seems strange that psychology has come such a little way when physics has ramified into many fields and has come so far" (E. G. Boring, *The Physical Dimensions of Consciousness*, New York, 1933: p. 3; by permission of D. Appleton-Century Co.).

And Freud says: "Strictly speaking, indeed, there are only two sciences—psychology, pure and applied, and natural science" (*New Introductory Lectures on Psychoanalysis*, New York, 1933: p. 245).

20b. Absolute idealism is a subtle form of dualism. This is revealed by the historical transformations which produce Hegel out of Leibnitz and Spinoza, through Kant. The point of origin is, of course, Descartes. Making Platonic errors in psychology and misusing Aristotelian terms in metaphysics,—i.e. using them as a neo-Platonist does, —Descartes gave birth to the monstrosity of mind and matter as the changing independent substances which constitute the being of the phenomenal world. Both Spinoza and Leibnitz tried to change the unintelligible face this gave the world, but neither did more than accomplish a cosmetic disguise. The mask of monism which they imposed was not the genuine substantial unity of matter and form. Add now the complicating factor of the English tradition in psychology with its multifarious errors and you get Kant, whose bewilderment takes the form of the most elaborate philosophical analysis made in the modern world. The utterly incomprehensible rift between the noumenal and the phenomenal orders takes the place of the Cartesian bifurcation. Though the language and much of the meaning is different, the essential paradoxes of the two dualisms are the same. In both cases, the world is irremediably split asunder, and the transitivity that is indispensable for coherence is either abolished or rendered inexplicable. Hegel, in attempting to avoid the *insolubilia* in Kant, does much the same that Spinoza and Leibnitz did with Descartes, and in consequence he resembles them. The monism of the Absolute is applied to heal the wound caused by the diremption between the noumenal and the phenomenal orders. Within this One there is still the dualism of Appearance and Reality,—the plurality of inadequate, even self-contradictory, beings which constitute the phenomenal world, somehow embraced in the transcendent fullness and intelligibility of the noumenal real.

If it be said that hylomorphism is also dualistic, it need only be answered that it is a dualism which renders the unity of things at once intelligible and does not require the immanence of a transcendent One, which is the last contradiction in the Hegelian series,—so great that it blurs all the others into the background. A philosopher should be able to appear wise without making the man of common sense appear to be a fool. This is the obvious merit of Aristotle and St. Thomas; and the equally obvious failure of Descartes, Leibnitz, Spinoza, Kant and Hegel. Of these, Kant's failure is the most striking because he so well recognized the errors of materialism and idealism. But he never discerned their modern source in the dualism of Descartes. If Kant had understood the union of matter and form in *being* as well as in *experience*, if he had grasped the connection of the noumenal and the phenomenal as the procession of *being* and *operation*, he might not have been followed by Hegel and Marx as he was preceded by Spinoza and Leibnitz. Cf. Note 16 *supra.*

In a sense there are only two positions, monism and dualism. The only monism which is not a dualism in disguise is absolute materialism and it is untenable on many counts. The only dualism which does not require a super-imposed one for either the unity or the intelligibility of things is hylomorphism. The face of nature that it perceives is clear and whole. The *plurality* of individual substances constitutes the *one totality* of things by their causal connections. Any deeper understanding than this of the oneness of the universe is theological and not metaphysical: the immanence of God, not substantially, but by Power and Providence. Whitehead's criticism of Aristotelian cosmology is based on his failure to understand that "the disjunction of primary substances" is qualified by their causal connection and that Plato's doctrine of the Receptacle and Leibnitz's monadology achieved a better vision of the unity of the world only at the expense of confusing metaphysics and theology. Beyond causality there is no philosophical problem of the interconnection or procession of things. Vd. *Adventures of Ideas,* New York, 1933: Ch. VIII on Cosmologies and Ch. IX on Science and Philosophy. Whitehead's *Process and Reality* is at best a current version of Plato and Leibnitz. Vd. Note 25 *infra.*

21. I have often been tempted to use the name "dialectical materialism" for the traditional metaphysics of Aristotle and St. Thomas Aquinas; for if the Marxists fully understood their own doctrine, they would be hylomorphists or formal materialists. Marx and Engels said they rejected the simple-minded, mechanistic materialism of their day, the 19th century forms of a doctrine which had its historical antecedents in the ancient Greek atomists, and in the philosophy of Hobbes. This materialism was unable to give a satisfactory account of change, physical or historical, and reduced itself to skepticism or subjectivism whenever it tried to explain the processes of cognition. The phenomena of change require the notion of contraries which inhere in matter disjunctively and lead to a conception of matter as having the potentiality for the opposite contrary whenever a given form determines it actually. The contraries are forms; they are determinations of matter. The perpetual process of change is understood when it is grasped that no single form determines matter absolutely, i.e., exhausts its appetite or potentiality for contrary forms. Efficient causality in change is the energetic reduction of the potentiality of matter for a given determination to that determinateness actually. The Hegelian dialectic appealed to Marx because of its obvious apparatus of contraries and oppositions. Marx used it mainly

in the sphere of human history; it supplemented the inadequacy of simple materialism as an account of historical change. Marx rejected the idealistic premises of Hegel, quite properly, but failed to explain why the principles of change he learned from the Hegelian dialectic did not apply to the processes of the material and infra-human world. He lacked the philosophical sophistication and technique to see that the dialectical materialism he was looking for, the materialism which could give a consistent and adequate account of all natural and historical processes, as well as of the knowledge of them, was the materialism of Aristotle and St. Thomas Aquinas. As the result of this failure, Marxian materialism today is in an unstable equilibrium between the position which it definitely rejected,—simple, absolute materialism,—and the position which its premises indicate it is trying to achieve. (In this connection, it should be noted that Marx's doctoral dissertation involved a critique of Democritus. Vd. F. Mehring, *Karl Marx*, New York, 1935: pp. 53–59. Unlike Marx, Engels wrote Hegelianism more thoroughly into his materialism and insisted, with manifold inconsistencies, upon a dialectical process in nature. Vd. his essay on *Ludwig Feuerbach and the Outcome of Classical German Philosophy* (1888). Professor Hook has recently exposed all the contradictions in Engels on this point. Vd. *Dialectic and Nature* in the Marxist Quarterly, I, pp. 253–284 (April-June, 1937). In what follows I am treating Marxism as if it were free from the philosophical ineptitudes of Engels.)

That Marxian materialism must be the formal materialism of Aristotle and Aquinas can be simply shown. The Marxian recognizes the difference between physical and historical change, i.e., the motions studied by the physicist and biologist and the unique course of development to be found in human history. In fact, it was precisely this recognition which led him to supplement "natural science materialism" by "historical materialism" and to borrow from Hegel the principles of dialectic to describe the pattern of history. It is true that historical movement is *partly* dialectical. But that it is only partly so and that the processes of infra-human nature are not dialectical, in a univocal sense of the word, the Marxian has not been able to explain because he developed his position as a too-simple negation of Hegel's idealism, itself the negation of materialism. The solution of these difficulties is to be found in the Aristotelian distinction between material and rational potencies, the potentiality of prime matter and the potentiality of the intellect. Matter is able to receive contraries only *disjunctively*. It is this fact which explains the characteristics of physical change, in which matter is the substratum. Material motions are not asymmetrically progressive but cyclical, from one contrary to another and back again; thus, the phenomena of generation and corruption, of growth and decline, even of local motion and alteration, are understood in the philosophy of nature, the basic principles of which are matter and the contraries. But the intellect, or reason, is able to receive contraries *conjunctively*, as species embraced in a genus. The opposition of species is not the simple negation of A by not-A; it is the relation of contraries, both of which are positive terms, negating each other only partially within the positive ground of the genus. The way in which the intellect receives contraries and resolves their opposition by ascending to a concept which unifies them, indicates the essentially dialectical character of its procedure. In short, as the potentiality of matter gives physical motion its distinctive character, so the power of mind is the source of dialectical motion; matter is the substratum of contraries disjunctively received, mind is the substratum of contraries synthesized.

In the physical world, only man has an intellect. Therefore, dialectical motion will occur only in human history, and there only to the extent that the intellect is operative as a cause. The asymmetrical line of progress, which is characteristic of dialectical change, is most apparent in intellectual history, what is called the history of ideas, and much less so in social, economic and political development. But man is not a pure intellect. He is an animal and many material factors are among the causes of his activity. To the extent that matter is causally effective in human history, it is not simply or purely dialectical. This is true even of intellectual history. The history of philosophy is hardly an unbroken or undeviating progress. In its alternating advances and declines, it partakes of the character of material change; yet there is dimly discernible that dialectical movement forward which indicates that the intellect is operative in intellectual history. The line of progress is even more obscured in social and economic history, probably because the efficacy of material factors is greater there. Yet to the extent that there are genuine advances, the intellect is at work, and one's optimism about future progress must depend upon how much power one attributes to the intellect in its struggle against material forces.

Hegel, denying matter and making an Absolute Mind the basic ontological principle of change, fell into two errors: (1) he failed to see that only human history is a dialectical motion; and (2) he failed to see that no change is purely dialectical because material factors are always present in some degree and, hence, he was an unqualified optimist about progress. The Marxian has tried to avoid all of these Hegelian errors: he has distinguished natural change and human history; he has admitted regressions as well as advances. He is not a naive optimist precisely because he is a materialist. Yet he cannot be simply a materialist because, in negating Hegel, he does not wish to return to the mechanical materialism which Hegel's insight about dialectical motion negated. Dialectically to negate the negation, he must become a formal materialist, a materialist who distinguishes between the potentiality of matter and the power of mind. He will then realize that his philosophy of history is nothing but a reading of the facts of historical change, in so far as these are known, in the light of psychology, an analysis of man as a rational animal, in whom rationality is limited by animality. Strictly, there is no philosophy of history, i.e., no separate branch of philosophy with principles of its own. What the Marxian propounds as a philosophy of history is merely an interpretation of known historical change in terms of the principles of the philosophy of nature and of man, i.e., philosophical psychology. (He may, of course, include prophecy in his exposition, as an extrapolation of the curve which past history obscurely reveals.) Finally, Marx's recognition of a genuine contingency in historical change requires him to reject the block-universe of any absolute monism and to take the pluralist view of many individual substances causally connected. Vd. Note 20 *supra*.

In short, no materialism can account for physical change without adding to matter the formal principles of the contraries (*sound* natural-science materialism); nor can materialism account for dialectical change without adding to matter and the contraries the principle of mind (historical materialism). Formal materialism comprises all these principles. It is the only philosophy which includes them all and is able to give an intelligible account of their relation, of the substantial union of form and matter and of intellect as the substantial form of man. If Marxian doctrine wishes to avoid the errors of atomism, on the one hand, and the errors of Hegel, on the other, it must eventually learn that, as a philosophy, it is formal materialism. Only formal materialism can understand the world studied by the natural sciences and the order of history. Only formal materialism is dialectical materialism.

The history of philosophy contains many instances of failure to achieve the perfection of dialectical materialism. There is, first of all, the Platonic error which assigns to matter apart from all forms some actual existence,—the being of the receptacle in the *Timaeus*. This leads inevitably to the error of the plurality of substantial forms which dominates all the neo-Platonic thinking of the Middle Ages, even when it is ostensibly an effort to reconstruct the Aristotelian metaphysics and philosophy of nature. The tradition of hylomorphism is, for the most part, corrupted by neo-Platonism. The perfection of this doctrine has seldom been achieved. The philosophy of St. Thomas is its consummate expression, in many respects an improvement upon the Aristotelian sources.

Corrupt hylomorphism leads to a second and worse error, the Cartesian error of explicit dualism which assigns separate substantial existence to both matter and form. This contains the seeds of a third error, the worst of all. The problems of cognition are insoluble in terms of Cartesian premises, and beginning with Locke's criticism of innate ideas, we pass in a series of psychological mistakes to the subjectivism of Berkeley and the skepticism of Hume. In the course of this episode in intellectual history, the substantial existence of matter is denied. Kant's failure to establish a sound metaphysics,—

although matter and form are the leading principles of his analysis of experience,—thus finally leads to the absolute idealism, or spiritualism, of Hegel. Facing the issue between spiritualism and materialism, Marx attempted to employ the dialectical principles of the former and to retain the honest realism of the latter. The cycle would have been completed had Marx been able to weld the two into a sound philosophy of dialectical materialism.

Neo-Platonism, Cartesian dualism and Hegelian idealism are all corruptions of, or derivations from, the doctrine of formal materialism. Of the cosmological alternatives, only atomistic materialism is an independent doctrine; in fact, it is the philosophy which Aristotle most honors as being nearest to the facts, when he considers the attempts of his predecessors to analyze the processes of change. Vd. *De Generatione et Corruptione*, I, 2. Aristotle accepted materialism in part, and corrected its inadequacies by enriching it with the Platonic theory of ideas, transforming them both by his conception of substance as a union of form and matter, and rendering the analytically separable principles of form and matter intelligible in terms of actuality and potentiality, which constitute the primary division of being. It is interesting to note that both Platonism and atomism make the error of giving matter an actuality apart from form, and that both lack the concept of substance, all forms being equally the accidental determinations of matter. The concept of substance is crucial, carrying as it does the distinctions of essence and accident, of existence and operation, as well as the distinction of actuality and potentiality. It avoids the atomist's mistake of making quantities the prime, if not the sole, determinations of matter, but unless it is rightly understood, the ever-present error of pluralism begins the process of corruption which allows formal materialism to degenerate into dualism and idealism. Cf. Note 20b *supra*. (It should be added here that the concept of substance which is indispensable to a sound metaphysics and philosophy of nature is negligible to, in fact, should be disregarded by, the natural sciences. This point expresses the basic difference between philosophy and science as types of knowledge: the former penetrating to the ontological center, the latter staying on the empirical surface. It is significant that the concept of substance is placed by Kant in the category of relation. Natural science cannot employ it in any other way. As E. Cassirer shows, science progressively *reduces* substance to function, i.e., being to operation, the noumenal to the phenomenal.)

Spinoza is almost *sui generis* in this account of the transformations of metaphysical doctrine. He is certainly not a materialist, not an idealist, nor simply a Cartesian dualist. He comes nearest to being an Aristotelian formal materialist. Substance is constituted by the composition of inseparable formal and material principles (the attributes of thought and extension); neither of these principles has being in itself; only substance is. There is furthermore the distinction between essence and accidents (the infinite attributes and the finite modes). So far the analytical terminology appears to be Aristotelian. But the absence of the distinction between potentiality and actuality, the absolute unity of substance and the denial of individual substances which are *per se* and *in se* but not *a se*, the perplexing definition of the attributes which regards each separately as expressing the whole essence of substance, combined with the infinite plurality of attributes,—these are indications that Spinoza's *Ethics* is a confusion of many strands in the philosophical tradition. Primarily, it is a confusion of the neo-Platonism of mediaeval Jewish and Arabic thought with the dualism of Descartes. It must be remembered that the neo-Platonists of the 12th and 13th centuries talked the language of Aristotle even when they defiled his principles. This accounts for much of Spinoza, who, like Leibnitz, made the mistake of trying to solve in philosophy the theological problem of the procession of creatures within the embracing totality of an immanent One. Vd. R. P. McKeon, *The Philosophy of Spinoza*, New York, 1928: Ch. I on the background of Spinoza's thought. As Professor McKeon says, "Spinoza could not know that the cartesian distinction of body and mind was to wipe out the aristotelian distinction of matter and form; if he could have thought of it in those terms he might have hoped that his treatment of extension and thought be confused with neither, or at least that it be recognized to be nearer to Aristotle than to Descartes" (*op. cit.*, p. 51). The other element which complicates the characterization of Spinoza's philosophy is the absolute monism that has its ancient expression in the doctrine of Parmenides,—or the All-One of Plotinos,—and its modern form in the absolutisms of Hegel and Bradley. The truth which Parmenides grasped is a truth in natural theology,—the absolute oneness of absolute being; the error which Parmenides made, and which Aristotle corrected, was the error of giving this principle a cosmological interpretation. The result is, of course, pantheism, with its failure to acknowledge the real existence of modes of imperfect being and modes of imperfect unity. That Spinoza could contrive a novel doctrine by the mixture of these errors, can be explained by the fact that he lived at a time when philosophy and theology were disordered with respect to each other. Vd. Note 16 *supra*. Cf. A. E. Taylor, *Some Incoherencies in Spinozism*, in Mind, N. S. 182, pp. 137–158. Professor Taylor exposes the way in which Spinoza's argument covertly admits the plurality of individual substances while at the same time Spinoza dismisses such pluralism as a "vulgar notion."

22. As we have already seen, absolute idealism in so far as it is *absolute*, arises from the Kantian distinction between the unconditioned and the conditioned, which Hegel converts into the absolute one of Reality and the relative many of Appearance. Vd. Note 20b *supra*. But in so far as it is *idealistic*, there is no difference between the so-

called objective idealism of Hegel and the subjective idealism of Berkeley. Both result from the same error in the analysis of cognition. The opposition here is not between idealism and materialism but between idealism and realism. This can be shown simply as follows.

The basic principle of realism in the theory of knowlege is that sensations and ideas are not *that which* we know primarily, but *that by which* we know *what* we know of *that which* we know. They constitute both the instrumentalities of knowledge and its content, but not its objects. (Ideas are objects only secondarily in reflexive knowledge. We can understand that we understand, or in other words we can understand the primary object of knowledge, the thing, as something understood, the idea.) As the instrumentalities of knowledge, ideas refer the mind intentionally to that which it knows,—the order of existing things. To identify the *id quo* and the *id quod* of knowledge is to identify the objects of knowledge with its content. This is the root of all subjective idealisms. Locke is the chief source of this error in modern times, for in the Introduction to his *Essay Concerning Human Understanding* he defines an idea as "whatsoever is the *object* of the understanding when a man thinks" (I, 1, ¶ 8). (Italics mine.) Locke's common sense realism saved him from the consequences of this statement, though not from the inconsistencies he had to suffer in order to avoid the consequences. But Berkeley and Hume, with perverse logic, preferred to reduce themselves to the absurdities of subjectivism and solipsism. (For a full discussion of the realistic principle which distinguishes the *id quo* and the *id quod* of knowledge, vd. Maritain, *Les Degrés du Savoir*, Paris, 1932: Ch. III, Le Realisme Critique; also *Réflexions sur L'Intelligence*, Paris, 1931. Cf. G. E. Moore, The Refutation of Idealism and related papers in *Philosophical Studies*, New York, 1922. For a magnificent statement of the crucial importance of realism in contemporary thought, vd. Professor Gilson's address at the Harvard Tercentenary in September, 1936; also *Le Réalisme Methodique*, Paris, n. d.)

The same error is the root of all objective idealisms. The difference arises from the metaphysical status of the ideas which are the objects of knowledge. If we combine the premises, (1) that the objects of knowledge must be existences and (2) that ideas are the objects of knowledge, we reach the conclusion that ideas must exist *per se*, prior to and independent of the mind which knows them. In its primitive form this is Platonic idealism, called extreme realism in the Middle Ages because it answered the question about the status of universals by assigning them real subsistence. In modern times, objective idealism assimilates the Parmenidean doctrine of the one, to produce the pantheism of the One Absolute Idea. This is Hegelianism in all its varieties. (If one describes Hegelianism as a neo-Platonic combination of Parmenides and Heraclitus,—the immanent One which is perpetually in Process, constituted by or at least involving the Ideas,—one can discern how close are the affinities between Hegel and Spinoza, on the one hand, and between Bradley and Whitehead, on the other. Plotinos is in them all.)

Further, the same error is the source of phenomenalism among Hume's followers, and it is unmistakably present in the psychological doctrine of positivism. There is some justice in the Marxian charge that positivism and pragmatism, accepting the phenomenalism of Mach or Mill, are idealism in disguise.

The critical point can be expressed in another way in terms of the distinction between *secundum se* and *quoad nos*. The thing *as known* is *quoad nos*. But knowledge is a proportion between intellect and thing, and the thing which measures the adequacy of the intellect in knowing is the thing *secundum se*. If the instrumentality of the idea as the medium of knowledge is ignored, no difference is possible between the thing as it is, *secundum se*, and the thing as it is known, *quoad nos:* and it is impossible to distinguish truth and falsity. If, on the other hand, the meaning of *quoad nos* is not qualified by the intentional reference of knowledge to its object, there is again no distinction between the thing as it is and as it is known, and we are imprisoned in the world of our ideas. Radical objectivism and complete subjectivism seem to be opposite errors but they are the same essentially and arise from the same failure to see that ideas as the instrumentality of knowledge relate its subject and object by intentional reference and separate them according to the mode of being of the thing, *secundum se* and *quoad nos*.

This is related to the problem of the way in which men have ideas. In the history of thought, there have been only three solutions: (1) ideas are intellectually intuited, as sensible things are sensed; (2) ideas are innate principles of the understanding itself; and (3) ideas are in the intellect only as a result of abstraction from sense-experience. The first and second of these positions are combined in the Platonic tradition which, through St. Augustine, influenced Descartes and Malebranche and, through Malebranche, Berkeley and Hume. (Cf. E. Gilson, *The Spirit of Modern Philosophy*, New York, 1937: pp. 15 ff.; 86 ff.) It is through the illumination of the innate ideas, the seminal reasons, by the light of ideal objects, that intellectual intuition takes place. Through light we see light. The distinction of Kant is that he denied the first position and embraced the second; in fact, he opposed the first so vigorously that he also denied the reflexive power of the understanding to know its own ideas in second intention. In the Platonic tradition ideas exist *per se* because they are the objects as well as the content of knowledge. According to Kant, purely intelligible objects, if there are any, transcend our knowledge because the ideas, the pure concepts of the understanding, serve only as the forms of judgment, the matter of which must be provided by sensuous intuition. In contrast, the third position, taken by Aristotle and St. Thomas, denies that ideas exist *per se;* agrees with Kant in the sense that such intelligible objects as separate souls, the angels and God, are not capable of being known intellectually by abstraction but only negatively and analogically by construction; but makes the intellectual knowledge of sensible things,—the apprehension of their intelligible being,—as genuine as the sensitive knowledge of their accidents and operations. Ideas are not the forms but the content of judgments. Apart from and prior to judgment, they are the simple apprehensions of the quiddities of sensible things, an apprehension which requires both an intuitive phase, on the part of sense, and an abstractive phase, on the part of intellect. The doctrine of abstraction is, thus, crucial in psychology, as the doctrine of substance is in metaphysics and the philosophy of nature. Misunderstanding or error in regard to the one is always and everywhere correlative with misunderstanding or error in regard to the other. These are the twin fallacies of Platonism in European thought productive of all the basic speculative departures from the truth, except atomistic materialism which is correlative with sensationalism in psychology and the denial of ideas. Though metaphysics is logically prior to psychology, it seems clear, in modern times at least, that the psychological error is the original and generative one.

That the Platonic error of making ideas the objects of intuitive knowledge is the challenging mistake in the psychology of cognition can be judged from the effort St. Thomas made to correct it and, more than that, to guard against the many ways in which it intrudes itself. Vd. *Summa Theologica*, I, Q. 84, AA. 1, 2, 3, 5, 6. The Platonic error about the nature of the soul as a substance, rather than as the formal principle in the composite substance of man, follows from the erroneous supposition that the intellect knows ideas directly by intuition rather than the nature of things through them, by abstraction from the materials made available by sense and imagination. The unity of man is thus destroyed and the extremity of Cartesian dualism is the inevitable consequence.

It is interesting to note that a sound cosmology is materialistic and that a sound theory of knowledge is realistic. There is not only a harmony between these two parts of philosophical doctrine, not only a felicitous union of them in a sound psychological analysis of the nature of the soul, but, what is most striking, there is the agreement of a sound philosophy in all its parts with common sense, with public experience, with the plain facts known to every man. (Even Whitehead who treats the doctrine of substance as the villain in European thought cannot, as a matter of common sense, avoid it in his concept of "a Society." Vd. *Adventures of Ideas*, New York, 1933; pp. 260–267.) Idealism and dualism in all their forms are forced to justify their departure from common sense by renouncing it as superficial and illusory or by despising it as unlearned and gross. But subtlety alone, or subtlety and novelty, are marks merely of intellectual ingenuity and not of soundness. The philosophy of Aristotle and St. Thomas is both subtle and sound.

22a. What is said in the text of Lecture II at this point is restricted to "schools" of scientific psychology. I am disregarding those writers who, directly or indirectly, discuss psychological questions in the Aristotelian and Thomistic tradition. They fall into three groups: (1) They are philosophers, such as Brentano or even Husserl, whose expositions are entirely philosophical psychology but not entirely Thomistic. (2) They are neo-Scholastics, such as Dom T. V. Moore or Father R. E. Brennan in this country, who have attempted to put together the principles of Thomistic psychology and the findings of modern research, in physiology as well as psychology. Vd. *Dynamic Psychology*, Philadelphia, 1924; and *General Psychology*, New York, 1937; also *Experimental Psychology* by Father H. Gruender (Milwaukee, 1932). (3) Such books are to be distinguished from expositions of the Aristotelian-Thomistic analysis of soul, regarded entirely as a part of philosophy, such as I. Gredt, *Elementa Philosophiae*, Fribourg, 1932: I, pp. 328–488; or C. Alibert, *La Psychologie Thomiste et les Théories Modernes*, Paris, 1903.

The chief problem raised by all these books, though in different ways, is whether the two bodies of knowledge, same in subject-matter but distinct in method and type of knowing, should be left in separation or joined in a well-ordered doctrine which is psychology, both philosophical and scientific. Since most scientific psychologists would deny that there is any valid knowledge other than that acquired by research, the problem is not for them. It is rather for the philosophers who, unlike the scientists, recognize that psychological investigation does contribute knowledge which is relevant to their principles and analyses. The solution which seeks to establish the unity of psychology by ordering the results of empirical research and philosophical analysis, seems to me the better one. But the problem is not solved until the right principle of order is used. It is at this point that current expositions seem to me regrettably defective, especially the undue importance which neo-Scholastic writers attach to physiological details and the vagaries of experimental data. Vd. Note 32a *infra*.

23. The hierarchy of substances is the hierarchy of the perfections of matter and this in turn can be expressed in terms of a hierarchy of properties or operations. The principle of order is as follows: a given perfection A is higher than B, if the presence of A entails the presence of B, but not conversely. Thus, life is a higher perfection of matter than existence, because whatever is alive exists but not everything which exists lives. Similarly, in the realm of living things, sensitive powers are higher than vegetative, because not everything that grows is sensitive, but nothing is able to sense which does not also grow. And understanding is highest in the order of vital operations, because it necessarily entails vegetative and sensitive powers, but is not entailed by them. Substantial forms are named according to their highest power or property. Those forms which confer life upon matter are

distinguished from those which merely confer existence, by being called souls. A body alive is matter besouled, as a body existing is matter informed. In the realm of living things, vegetative, sensitive and intellective souls are distinguished according to the vital operations with which the matter thus besouled is endowed. The hierarchy of the natural sciences is determined accordingly: the study of material existences, the physical sciences; the study of living things, the biological sciences. In so far as all living things are besouled, psychology is a biological science. In so far as man shares many of the powers and operations of other living things, he can be studied biologically. But in so far as man, through having an intellective soul, exceeds the animal kingdom, psychology as the study of the specific nature of man becomes a science apart from the rest of biology. That psychology is not merely a biological science turns ultimately upon the proposition that the operations of understanding are not the acts of a bodily organ. It follows from this proposition that man is hierarchically graded with respect to all other living things as they in turn are graded with respect to inanimate bodies. The order of the sciences is, therefore: the physical group, the biological group, and human psychology.

24. Human psychology is like animal psychology to the extent that men and animals share many vital operations, i.e. sensation and imagination, locomotion, the passions, not to mention all the vegetative functions. If the distinctive intellectual power of man is ignored, there is no essential difference between human and animal psychology. Vd. Note 23 *supra*. The programme of behavioristic psychology involved an attempt to reduce human to animal psychology. It failed, not because human behavior is so much more complex than animal motions, but because it is radically different. A science of animal behavior is possible because the motions of animals are either instinctively determined or the products of reflex habituation. But human behavior is voluntary and human habits are voluntarily formed. The resulting contingency in the phenomena of human behavior makes them much less susceptible to scientific formulation. Cf. *Art and Prudence*, New York, 1937: Ch. 9.

25. The "philosophy of science,"—in so far as it concerns the ordering of the sciences,—is a rendering of the philosophy of nature, in so far as it considers the hierarchy of physical things. The order of the sciences of nature must depend upon the order of their objects. What we have distinguished as the Aristotelian and the Cartesian philosophies of science are strictly correlative to Aristotelian and Cartesian cosmologies. And these, in turn, depend upon different metaphysics. If we are bound by simple consistency, we are not free to take one position in the philosophy of science and another, which is not its correlative, in the philosophy of nature or metaphysics.

Nor is one free to choose a metaphysical position according to the view of science or of man's nature which, *for other reasons*, one prefers to take. Quite the contrary! If one proceeds according to reason, the resolution of the issues in metaphysics must come first. This subsequently determines (1) the account one gives of the order and relation of the natural sciences, and (2) the position of psychology. If either of these two determinations is given temporal priority, it must be done in ignorance of metaphysical presuppositions. Such is usually the case among scientists who talk about the outlook of their science or of the sciences, unaware that they are betraying metaphysical judgments which would shock them if made explicit. Psychologists especially are given to philosophizing unwittingly, as witness the introductory chapter of any treatise or text-book of scientific psychology.

Even worse is the construction of what pretends to be a philosophy of nature out of the reigning concepts and laws of natural science. Whitehead's *Process and Reality* is such a "scientific cosmology," an attempt at a "speculative synthesis" which will be for our epoch what Plato's *Timaeus* and the Cartesian cosmology were in their day. (Vd. *op. cit.*, New York, 1929: Preface and Ch. I.) What Whitehead takes to be the central problem of cosmology is not a philosophical problem at all, namely, the problem of the inter-connection of the totality of things. There is little to support Whitehead's interpretation of the Receptacle in the *Timaeus* as Plato's principle for solving this problem. There is even less evidence to show that the Greeks ever considered the problem which Whitehead formulates. It is a typical modern problem, beginning with Descartes and occupying Leibnitz and Spinoza. It originates in the 17th century, partly because these philosophers failed to distinguish the problems of the philosophy of nature from those of metaphysics and theology, and partly because they converted natural science into a cosmology. The theologian is able to consider the totality of things in their interconnection and in process only in so far as their nexus and order is Providential. But Providence is not a principle in philosophy. It is only as a theologian, aided by revealed truth, that St. Augustine can develop a doctrine about history or that St. Thomas can discuss the procession of creatures. The neo-Platonic tradition confused theology and philosophy. This is true even of Plotinos who was not aided by Holy Scripture. There are traces of Plotinos in Descartes and Spinoza and Whitehead. These modern thinkers, all neo-Platonists, are pre-occupied with a problem which is not Greek but Christian and which, in the Christian tradition, belongs to theology and not philosophy. Their "cosmologies" are not philosophies of nature, but attempts to answer this problem by "ontologizing" science. Again, the influence of the *Timaeus* is apparent. Plato did not conceive the philosophy of nature as related to metaphysics and as having principles of its own. He told what he called a "likely story" about the realm of becoming, basing it upon the science of his day, particularly mathematics. That is what Descartes did, and what Whitehead is re-doing.

I use the phrase "scientific cosmology" to name a projection of the relational analyses of science *behind* the phenomena to which they properly apply. Such projection is a reification of mathematical relations, of the relational logic of science, but it is not a philosophy of nature, first, because the method of philosophy is not to interpret the findings of science, and second, because the conceptual limitations of science make that knowledge incapable of providing any genuine philosophical insight. One example will suffice. The concept of substance has no place in science which, as Cassirer points out, studies functions, or correlations of accidents. (Vd. E. Cassirer, *Substance and Function*, Chicago, 1923.) Whitehead, illicitly projecting natural science into a cosmology, makes the denial of substance pivotal, and naturally turns to the Platonic and Cartesian traditions to support his exclusion of what must be the central term in the philosophy of nature, which seeks to answer ontological questions properly and does not look to the sciences for help they cannot give. As I have already indicated (Note 22 *supra*), Whitehead does not succeed in avoiding the concept of substance. His enduring societies, personal or otherwise, are Aristotelian primary substances. But his analysis of their nature makes, on the one hand, the typical Platonic error of a plurality of forms in that societies are composed of societies, though ultimately they are reduced to the togetherness of actual occasions of the ingression of the ideas, and shows, on the other hand, that a society is a relational rendering of substance, which is appropriate to the sciences of phenomena but not to ontology. Vd. Note 47 *infra*.

26. Cf. Aristotle, *De Anima*, III, 3–8; St. Thomas Aquinas, *Summa Contra Gentiles*, II, 59–78, and *Summa Theologica*, I, Q.75, AA.2, 5; Q. 76, A. 1; Q. 84, AA. 1, 2, 3, 4, 6; Q. 85, AA. 1, 2.

The argument can be reduced to the following propositions. (1) Knowledge involves some *communication* of the knower and the known. (2) A thing is known in so far as its form is somehow in the knower. (3) Whatever is received is in the recipient according to the mode of being of the recipient. (4) Forms in matter are individual because that is the material mode of reception. (5) Sensitive knowledge is the apprehension of things in their individuality because sensation is the act of a bodily organ and sensible forms are therefore received materially. (6) Man understands as well as senses, i.e., man knows intellectually. (7) Intellectual knowledge is the apprehension of things in their universality. (8) Therefore, the intellect must receive universal forms. (9) But no form is universal according as it is received in matter. (10) Therefore, intellectual reception must be immaterial, which is to say that understanding is not the act of a bodily organ. This last proposition is the capital premise for the conclusion that the soul, having this immaterial mode of operation, must also to that extent have an immaterial mode of being, and hence it is not only the substantial form of the body but is capable of separate self-subsistence upon the corruption of the composite. (Cf. Whitehead's reasoning which comes to the conclusion that "in some important sense the existence of the soul may be freed from its complete dependence upon the bodily organization." *Adventures of Ideas*, New York, 1933: p. 267.)

27. Vd. my analysis of artistic production in a paper on *Creation and Imitation: An Analysis of Poiesis*, published in the Proceedings of the American Catholic Philosophical Association, 1935: pp. 153–174. The intellect is the principle, not only of philosophy and science, but of art and morality; of art, because intentional making requires that the artist order his material according to a form which he possesses in abstraction from any matter; of morality, because the moral act must be a voluntary one, and the will is appetite as intellectually determined to an object. With respect to all means, all partial goods which are the subject of choice, the will is free because the intellect, operating in the practical order, determines the appetite imperfectly. Human freedom is proportionate to this indeterminacy.

28. The Marxian conception of man as a manufacturing, or machinofacturing, animal is another indication that Marxism must become a truly dialectical, or formal, materialism. Marxian materialism in its present development cannot give a satisfactory account of the way in which men multiply material things according to specifications. In this respect it is as inadequate as the simple, mechanical materialism which it so resolutely rejected. If the Marxian seems equally resolute not to follow out the intellectual consequences of that rejection, one is led to suspect that he *dislikes* the consequences rather than that he is unable to discover them. He must then be charged with superstition and ignorance; the first because one does not make rational judgments according to one's likes, and the second because if the Marxian fully understood *all* the consequences of his implicit premises, both speculative and practical, he would see that he had no reason to dislike them.

It should be added, in commentary on the text at this point, that the material out of which man makes useful productions is not prime matter but the formed matter of substances; and that the artificial forms which he imposes upon these materials are not substantial forms, but accidental determinations of composition and order. These qualifications for the sake of metaphysical precision do not, of course, diminish the force or alter the direction of the argument.

28a. What is perhaps the most striking fact distinguishing man as radically different in kind from all other animals, is the fact of human history, the fact of cultural development. As Allers says, "the fact of human progress alone would show the impossibility" of the opposite view. "The inheritance of tradition and progress in technique are things not to be found in animals: a hundred thousand years ago the termites built their admirable houses just as they do today; but what a difference between the dwellings of primitive man and the skyscrapers of modern New York!" (*The New Psychologies*, New York, 1933: pp. 33–4).

There is a nice paradox in this for the Marxists. Man is an historical animal because man is a rational animal. This does not deny the Marxian point that material causes operate in human history,—that what men make and do depends upon the material and animal conditions of their life,—but it does deny that material causation is sufficient to explain the phenomena. The material and animal conditions of human life are essentially the same as those for any other animal; but what men are able to produce under such conditions is different because the factor of reason or free intelligence has been added. Marxism distinguishes itself from simple materialism,—what it calls "natural science materialism,"—by adding thereto a materialistic account of history. But that account must either insist that material factors are the only essential causes and everything else is epiphenomenal, or it must acknowledge that in human history reason is an *independent* cause. In the first alternative, dialectical materialism, even though it is historical, is no better than natural-science materialism, because it is using the same principles, disguised in Hegelian language; it simply cannot explain the facts of man's progressive mastery over materials. In the second alternative, dialectical materialism, recognizing the role of reason in imposing forms upon matter, is hylomorphism. The Marxian can take the second position without altering his view of the shifting struggle of the classes, without yielding in his justifiable moral condemnations of capitalism. In fact, he is strengthened in the latter if he become a Thomist. (Vd. Note 21 *supra.*)

29. Brutes and men are alike in the possession of the powers of the *interior sense,* such as comparison and discrimination, memory, imagination, and estimation. The interior sense consists of powers which follow from the operations of the proper or special senses. As sensations are the functions of particular sense-organs and primary projection areas of the brain, so the operations enumerated under the head of the interior sense are functions of the total involvement of the brain, going beyond the activity of the primary areas of sensory reception. Another name for these powers in man is the *passive intellect,* because the performance of such operations as comparison and discrimination, imagination and cogitation is a kind of thinking. Men do a sort of thinking with images alone, the kind of thinking that occurs in reverie and day-dreaming. Such thinking does not require an intellect. To the extent that animals have brain power, they are able to do thinking of this sort. The one difference here between men and animals is that men possess an "intellectual imagination" able to com-

pound images freely. Animals, lacking intellect, lack this power. They can estimate but not cogitate. Vd. *Summa Theologica*, I, Q. 78, A. 4; Q. 81, A. 3, ad. 3. Cf. H. A. Wolfson, *The Internal Senses in Latin, Arabic, and Hebrew Philosophic Texts* in the Harvard Theological Review, April, 1935.

It is this similitude in human and animal powers which enables animal psychologists to construct experiments for the investigation of animal intelligence in problem solving. When, on the supposition that human intelligence differs from animal intelligence only in degree, experiments which use analogous problems, only more complicated, are constructed to test human powers, it is not surprising that the supposition is verified. This *petitio principii* is as much involved in the experimental work of the *gestalt* psychologists as in that of the behaviorists, although the former being more subtle conceals it better. When an experimental psychologist is able to make an animal solve a *scientific* or an *artistic* problem, he will have proved that animals have intellects as well as men. It is not necessary to take men into a laboratory to discover that they are intellectual.

30. The text does not answer Dr. Alexander's third question, concerning the biological utility of understanding. The utility of sense in the economy of animal life is, of course, its implementation of vegetative functioning by aiding the animal to satisfy his basic needs. Sensitive knowledge not only determines desires but helps to gratify them. It is difficult to say whether animals have any purely speculative mode of sensitive operation, i.e., a use of the senses purely for the sake of knowing. But we do know that men employ their cognitive faculties for both speculative and practical ends. On the practical side, intellectual knowledge determines desires and helps to satisfy them in the same way that the senses do. So far there can be no question about the biological utility of the intellect. But "biological utility" usually has the narrow meaning of "survival value" or "adaptation to environment" or the "satisfaction of animal wants." The intellect, therefore, has more than biological utility because it directs men in a mode of life which involves other goods than these. It directs human desire to intelligible goods which exceed all animal wants and which have nothing to do with survival value or adaptation to environment in the struggle for existence. Beauty and truth and justice are things for which men work, even die, because their intellects enable them to understand, and hence desire, goods that no animal seems to apprehend. Biological utility is so far transcended by the gift of intellect that men are given to intellectual operation for its own sake. Since the ultimate perfection of any thing lies in the performance of its proper operation, the intellectual life is the happiness to which all human virtues and external goods must be ordered. The latter are the goods of "biological utility." They are for the sake of the intellect rather than the intellect for the sake of them. One cannot praise

the intellect by talking of its biological utility; that is fitting praise for the senses. The intellect should be honored for its biological inutility because it is that power whereby men are able to live and love differently from animals. If, through a misuse of the intellect, men become unhealthy, ill-adapted and even unhappy, that must be charged to the animality which complicates man's attempt to lead a human life.

31. Aristotle, in the *Metaphysics*, compares the species of things to integral numbers, which differ in species by the addition or subtraction of unity. And, as St. Thomas adds, in the *De Anima* "he compares the various souls to the species of figures, one of which contains another, as the pentagon contains and exceeds a tetragon" (*Summa Theologica*, I, Q. 76, A. 3). The type of continuity which obtains among natural forms is like that of the series of whole numbers. It is not the continuity of a dense series, between any two members of which there is an infinite number of intermediates. To speak of it as a continuity at all is only for the sake of emphasizing that each member of the series contains its hierarchical inferiors and is contained by its hierarchical superiors. To the extent that each member exceeds, or is exceeded by, another in terms of an indivisible unit, discontinuity predominates. This unit is the specific difference. To deny this unit difference is to deny species, since species cannot be different kinds if they differ only in degree. The radical error in *The Origin of Species* is the attempt to define species as the extremes of a series of graded intermediates, differing quantitatively. Species are said to *originate* through the extinction of the intermediate varieties. It is this error which the discovery of mutations corrects and which changes the interpretation of all of Darwin's data. Whether mutations produce accidental or essential differences is not here the question. If they are only accidental, mutations do not constitute an origin of species; but the varieties of species which result are discontinuous. In any case, therefore, the true maxim about nature's work in the order of generation is the contradictory of *natura non facit per saltum.*

31a. "The intellect knows itself not by its essence, but by its act. This happens in two ways: In the first place, singularly, as when Socrates or Plato perceives that he has an intellectual soul because he perceives that he understands. In the second place, universally, as when we consider the nature of the human mind from knowledge of the intellectual act. . . . There is, however, a difference between these two kinds of knowledge, and it consists in this, that the mere presence of the mind suffices for the first, the mind itself being the principle of action whereby it perceives itself, and hence it is said to know itself by its own presence. But as regards the second kind of knowledge, the mere presence of the mind does not suffice, and there is further required a careful and subtle inquiry. Hence many are ignorant of the soul's nature and many have erred about it" (*Summa Theologica*, I, Q. 87, A. 1). Here St. Thomas distinguishes between the facts of experience, as they are known to all men, and the analysis of them which the philosopher makes by "a careful and subtle inquiry." Psychology as a body of philosophical knowledge arises with the analysis of such apparent facts as that men understand, that men have habits, etc. The reflexive action of the intellect is the source of these facts, but the understanding of them,–the analysis of the habits, powers and ultimately the nature of man from a knowledge of his operations,–is not different in method from the philosophical penetration into the nature of other things, whose phenomenal manifestations are not known reflexively. Cf. Note 17 *supra.*

32. Any investigative study of rational processes can do no more than determine the accidents of rationality in operation. In the first place, a scientific investigation cannot proceed upon the assumption that men are not rational without self-contradiction. The fact of reason is a brute fact of experience. It underlies the existence of science itself. It does not wait for science to establish its existence. Rationality is no more discovered by psychological investigation than gravitation is discovered by physical investigation. The physicist determines certain quantitative accidents of gravitation, the rate of acceleration of bodies falling toward the earth, the relation of the mass and the distance to gravitational force, etc. Scientific psychology starts from the commonly known fact that men are rational and proceeds to investigate the circumstances and qualifying accidents of rational performances.

That this is so is clear, in the second place, from the data which the investigative psychologist has collected. The data of the experimental laboratory, of psychometric research, and of clinical studies, do not define the essence of reason. They cannot. What is discovered can be summarized as follows. (1) Behavioristic experiments have shown that under many circumstances human beings solve problems by "trial and error" rather than "rational thought." Such data have been misinterpreted to mean that human beings are not rational, or that animals are. But, as we have seen, all that the experiments show is that trial and error problems are solved by trial and error, in the case of men as in the case of animals. Vd. Note 29 *supra.* (2) Psychometric research has determined the quantitative accidents of intelligence. It has shown the way in which the different operations of man's rational ability are related to each other in terms of quantitative correlations. It has measured the relation of rational power to other abilities. It did not discover the fact of individual differences, but it has made this fact quantitatively significant and precise by manifold measurements. (3) Clinical studies, by psychiatrists and psychoanalysts, have documented what is commonly known about the relation of reason to the passions. They have not studied rationality but rather the conditions under which rational men do not operate rationally because they are also animals. They have determined, in short, the accidents of irrationality in men. They did not discover that fact. (4) Experimental and clinical studies by physiologists and neurologists have determined, but not discovered, the dependence of rational performances, and also the pathology of reason, upon various bodily conditions, such as fatigue, brain lesions, etc. In some instances, they have been able to measure this dependence by fairly precise quantitative techniques.

The scientific evidence does not tell us what reason is or what rational operations are. It does not establish the existence of reason or the equally obvious facts about its pathology, the abeyance of reason and irrationality. The data cannot be interpreted to mean that men are not rational or that human and animal intelligence differ only in degree. Their only significance is the determination of the extrinsic conditions and the qualifying accidents of reason as an ability and in operation.

Philosophical psychology rests upon the same obvious facts of common knowledge, which scientific investigations have, at best, supplemented by the sorts of determinations enumerated above. But, starting from these facts, it proceeds in a different direction because its interest is in definitions and causes rather than in the correlation or measurement of phenomena. It seeks to understand these facts; it seeks to determine the essence of rationality, not its accidents. The philosophical analysis defines the intellectual power, distinguishes its subordinate abilities, determines the different fields of its operation, divides the several acts by which the power is fulfilled, and formulates the relation of reason to those other powers of man which, though not intrinsically rational, participate in rationality either as its instruments or by its ordination. For all of this analysis, even when it becomes exceedingly subtle, only the facts of common experience are needed. The scientific evidence adds nothing that is *essential*. At best, it contributes exemplifications of what is independently known.

The foregoing discussion has singled out the study of reason, in its speculative dimension, as a cognitive power. But what has been said applies to all *human* behavior, because human behavior is specifically different from animal behavior in so far as it is rational or participates in reason. The phrase "human behavior" has become ambiguous, largely through the errors of that scientific "school" which claims this to be its subject-matter. If it means the whole field of human conduct, the development of human character and the social intercourse of men, and if the scientist aims to discover the causes of human behavior in this sense, the scientific effort is doomed to failure. Human behavior is voluntary and contingent in a way that the rest of natural motions are not. Reason is a first cause in human behavior, and the essential operations of this cause are not susceptible to scientific investigation. Vd. Note 24 *supra*. It should be added here that investigative psychologists, for the most part, have not yet learned the lesson positivism has taught physicists, namely, that science is incapable of being theoretic knowledge of causes. Vd. Note 16a *supra* and Note 47 *infra*.

But "human behavior" can be given a more restricted meaning. It can mean the operations of specifically human powers. The subject-matter of psychology is human behavior in this sense: the powers and acts of men,—what the psychometrist calls the abilities and performances. Common experience provides the basic facts concerning the existence of these powers and acts. Philosophical psychology analyzes their essential nature. Scientific psychology investigates their external conditions and their accidental traits. In short, philosophical and scientific knowledge differ in psychology according to the general difference in their type. Vd. Notes 6b and 9 *supra*.

32a. The uniqueness of psychology makes it difficult to define. It is the only body of philosophical knowledge which deals with a single species of physical thing. It is therefore the only body of philosophical knowledge which has a subject-matter capable of being investigated scientifically. It is not a part of the philosophy of nature because there are no other parts strictly coordinate with it. It is rather a special application of the principles of natural philosophy; yet it is not merely that because it has primary inductions of its own from the data of reflexive experience. Whereas the philosophy of nature considers the mode of being of any changing thing,—the types of such contingent beings, the kinds and conditions of change,—philosophical psychology studies particular phenomena of change,—the operations of human nature, the formation of habits, etc. To this extent it is like scientific knowledge, though it differs both in method and in aim: it is not investigative and it seeks definitions; its principles are in the order of substance and causality.

Another way of indicating these difficulties is to classify the questions which relate to man: (1) Questions concerning the self-subsistence of the human soul, and its mode of being, belong to metaphysics and, within metaphysics, to natural theology. (2) Questions concerning the mode of being of man as composite of form and matter, of soul and body, and, in view of the subsistence of the human soul, questions concerning man's unity belong to metaphysics in the province of ontology. (3) Questions concerning man as living, as subject to generation and corruption, and to accidental motions, belong to the philosophy of nature. (4) Questions concerning the specific nature of man, his powers, habits and operations, belong to psychology. (4a) In so far as these questions are concerning the essence of man, the essence of his powers and their essential relationships, the answers must be achieved by philosophical analysis from the data of reflexive experience. The answers constitute philosophical psychol ogy. (4b) In so far as these questions are concerning the accidental conditions of human operation and accidental determinations of man's powers, the answers must be achieved by investigative research, by special observations and measurements. The answers constitute scientific psychology.

Strictly then, psychology begins with the fourth type of question. The answers to the second and third types are necessarily presupposed and may, therefore, be properly included in the exposition of philosophical psychology, but the distinctions in type should be preserved for the sake of clarity. It makes a difference whether the locus of an argument is in metaphysics, in natural philosophy, or in psychology proper. Questions of the first type cannot be answered except *a posteriori;* in the order of knowing they presuppose psychological analysis itself (Type 4a). There is good reason to restrict the discussion of these questions to treatises on natural theology. If we turn now to what are strictly psychological questions, we can see at once the relation of Types 4a and 4b. The principles of philosophical psychology are presupposed by any sort of scientific psychology: essential distinctions are the ground for accidental determinations. As the natural sciences in general can be informed and regulated by the superior conceptual order of mathematics or the philosophy of nature, so the science of psychology is informed and regulated by the special philosophical analysis of its own subject-matter. All of the basic questions in psychology are purely philosophical. Denial or ignorance of philosophy is the cause not only of the inadequacies of scientific psychology but also of the vain controversies of the "schools." On the other hand, just as the sciences provide exemplifications of superior principles, so the empirical findings in psychology exemplify the relevant philosophical principles. This does not mean that exemplification is the sole contribution of scientific research. It also adds a wealth of detailed knowledge about all sorts of accidents, which philosophy cannot reach to. It would lose its proper posture if it tried, as history unfortunately reveals. This briefly indicates the ordering principles for the exposition of all knowledge relevant to man. Vd. Notes 36c, 39c and 46 *infra.*

Two comments remain to be added. (1) If the word "psychology" were restricted to name the consideration of the soul itself, either as the formal principle of living things or as self-subsistent, and if "anthropology" were used to name the study of man specifically, his nature, powers and operations, then the body of knowledge which is a special branch of philosophy as well as a special natural science would be called anthropology, and psychological questions would belong to the philosophy of nature and to natural theology. (2) Man is of all natural things the one with which man is most intimately acquainted, the only one whose specific nature he is able to comprehend fully. In the light of this fact, it may be thought paradoxical that scientific efforts in psychology are among the most recent and least successful. Why should not a science of psychology be better *qua* science than any other? The paradox is removed when it is realized that the greater knowability of man for man produces as its natural result in the hierarchy of human knowledge, not the *best* science, but rather the *only* body of philosophical knowledge relevant to a species of physical thing. Those who will not recognize this fact must remain perplexed about the poorness of psychology as a science although men are so well acquainted with themselves. Those who acknowledge the existence of philosophical psychology can learn from it why the obstacles to the investigation of human nature are so great, so great that psychology will never achieve the precision and plenitude of physics.

33. The ancient and mediaeval tradition not only relegated developmental problems,—the growth of character, the types of human life,—to the field of ethics, but also located the discussion of certain other psychological topics in practical treatises on ethics, rhetoric and poetics. Professor McKeon promises soon to publish a paper on the history of the treatment of the passions, showing how this topic passed from a locus in rhetoric to poetics, then to ethics, and finally in modern times to physiology and pathology. Ancient and mediaeval treatises on psychology dealt in detail with sense and intellect in their speculative dimension. In so far as sense and intellect are involved in the operations of appetite, the acts of the passions and the will, they are in the practical dimension, directed toward the end of action rather than the end of knowledge. It, therefore, seemed appropriate to place the discussion of these topics in practical treatises. Aristotle pointed out that the moralist must be a student of psychology, as the physician is a student of physiology. Vd. *Nichomachean Ethics*, I, 13, 1102^a 16–25. But the subject-matter of physiology is essentially the same, whether it is studied academically or for practical purposes in a medical school. So, too, the psychological analyses that Aristotle develops in the *Ethics* and the *Rhetoric*,—the nature of will and the conditions of voluntary action, the types of pleasure and love, the operations of practical reason, in the *Ethics;* the passions, their nature, the conditions of their arousal, and their course, in the *Rhetoric*,—are strictly psychological regardless of their practical contexts.

St. Thomas makes a similar disposition of topics. In the Treatise on Man, after he has enumerated and analyzed all the powers of man, he proceeds to the detailed analysis of the operations of one of these powers, the intellect, reserving for other treatises the consideration of the acts of the other powers. "The acts of the appetitive part of the soul come under the consideration of the science of morals; wherefore we shall treat of them in the second part of this work, to which the consideration of moral matters belongs" (*Summa Theologica*, I, Preamble to Q. 84).

Such allocations of psychological topics to various parts of a systematic philosophy must not obscure the unity of the subject-matter of psychology, which comprises all the powers and acts of man, the appetitive as well as the cognitive. The parts of Aristotelian psychology can be gathered together from the *De Anima*, the *Ethics* and the *Rhetoric*, as in the case of Aquinas, they can be collected from the Treatise on Man and from the Treatises on Human Acts, the Passions, and Habit in the *Summa Theologica*. The collection will readily form the systematic whole of philosophical psychology. That Aristotle and St. Thomas made the disposition they did of its parts is ultimately a matter of relevance and convenience in exposition,—the dependence of practical philosophy upon psychological principles, and the avoidance of repetition. To suppose otherwise would lead one to say that Spinoza had no doctrines in metaphysics, physics and psychology because all of these are contained in a book called *Ethics*. (Cf. Alexander Bain's division of his psychological doctrine into two treatises: *The Senses and the Intellect* and *The Emotions and the Will*.)

33a. When a natural scientist seeks to "explain" the phenomena which he is describing, he usually turns to another science which describes a lower order of phenomena. Thus, the psychologist turns to neurology and physiology for explanations, the biologist to the chemist, the chemist to the physicist. What is meant by "lower" here is simply that the system of phenomena which one science studies *includes* systems of phenomena studied by other sciences. The latter sciences, dealing with the included systems, are lower. But this reference of higher to lower sciences for the sake of explanation is not a reduction of the sciences because the proper task of each is the description of a special segment of phenomenal reality. The phenomena of life may be explained mechanically,—at least, in so far as explanation is "scientific,"—but they are not mechanical phenomena. Vd. Note 13 *supra* and Note 47 *infra*.

34. The word "perfection" must be understood in its metaphysical sense primarily. The ethical meaning is derivative. Forms perfect matter in so far as they actualize its potentialities. The fulfillment of anything's potentialities is its completion or perfection. The substantial form is, in this sense, the first perfection or act of a thing's being. It is the term of generation. But in so far as the thing which is substantially actual has powers of operation, there are second perfections or acts, i.e., the operations of these powers. The ethical meaning of perfection follows from the metaphysical insight that goodness and being are correlative, and that anything is good in proportion as it achieves actuality. The highest perfection of a thing is its fullest actuality. To judge a thing we must, therefore, know its specific powers of operation. Vd. Aristotle, *Nichomachean Ethics*, I, 7.

For a discussion of the hierarchy of perfections in nature, vd. St.

Thomas Aquinas, *Summa Theologica*, I, Q. 76, AA. 3, 4. The evolutionist similarly grades the lower and higher forms of life. Evolution exhibits a hierarchy of perfections in nature. Cf. Note 31 *supra.*

35. "There is no other substantial form in man besides the intellective soul; and the soul, as it *virtually* contains the sensitive and nutritive souls, so does it *virtually* contain all inferior forms, and it alone does whatever the imperfect forms do in other things. The same is to be said of the sensitive soul in brute animals, and of the nutritive soul in plants, and universally of all more perfect forms with regard to the imperfect" (*Summa Theologica*, I, Q. 76, A. 4). I have underlined the word "virtually" because its full analytical significance is essential to understanding how St. Thomas avoided the error of the plurality of substantial forms. An inferior form, virtually contained by a superior form, is not actually present as a form but only operatively present in the order of powers which the superior form involves. As habit is a mean between an operative power and its act, so a virtual form is a mean between the potentiality of substantial being and its act. A single substance has a hierarchy of powers, but not a hierarchy of forms. "Of one thing there is but one substantial being. But the substantial form gives substantial being. Therefore of one thing there is but one substantial form. But the soul is the substantial form of man. Therefore it is impossible for there to be in man another substantial form besides the intellective soul" (*Ibid.*).

A single example may make this clear. On a sheet of blank paper, a triangle is present *potentially*. It can become physically actual there through the addition of three connected straight lines. On a sheet of paper on which a square is drawn, a triangle is present *virtually*. It can become actual there first by the drawing of the diagonal and then by the removal of two sides. Whatever has the form of a square is thus said to contain the form of a triangle virtually; and although the properties of a square are not identical with the properties of a triangle, they include the latter properties. The triangular properties are, of course, not the same in the square, in which the triangle is present only virtually, and in an actual triangle. In the same way, the human soul includes an animal soul virtually, and all the properties, i.e., powers, of animal life are present in human nature, though differently because they are subordinated to a higher power. (Thus, a man and an animal, but not a vegetable, can suffer emotionally caused indigestion, and only a man can suffer thus by worrying about ideas.)

This point can be seen *analogically* in the processes of development and decay. When an infant or a cretin idiot becomes a man, a potentiality is actualized. When a man suffers *dementia praecox* and decays to a persistent catatonic stupor, a virtuality is actualized. The distinction is strictly exemplified in the processes of generation and corruption.

36. Vd. Notes 19 and 32 *supra*. Cf. C. J. Herrick, *The Brains of Rats and Men*, Chicago, 1926; K. S. Lashley, *Brain Mechanisms and Intelligence*, Chicago, 1929; G. E. Coghill, *Anatomy and the Problems of Behavior*, Cambridge, 1929; W. G. Summers, *The Psychology of Sensation* in The Proceedings of The American Catholic Philosophical Association, 1934: pp. 104–113.

36a. In his Introduction, Dr. Alexander says that human infants show only animal characteristics, and that the difference between men and animals "comes gradually about during the individual's development from an ovum to an adult" (pg. xiv *supra*). While it is true that an infant is not fully a man, it must be understood that the infant is actually a potential adult. Even though the infant at birth behaves as an animal and only that way, the infant is not even then actually an animal because it can become a man as an animal cannot. The educability of an infant, like the curability of a cretin idiot, signifies the possession of powers not shared by animals. It is this difference in powers which marks the essential difference between animals and men. Generation and development must not be confused. The infant is born a human being, and not an animal; it develops into an adult. The development is with respect to powers and not specific nature; the adult differs from the infant not in power simply, but in habits and operations, i.e., the developments of power. Vd. Note 35

36b. Those who tell the history of psychology usually recognize this division but as usually misunderstand its significance because they are positivists. Supposing as they do that the only valid and adequate knowledge in this field must be scientific,—the result of experiment or research,—they treat what precedes the middle of the 19th century as the era of "pre-scientific psychology," which means, of course, that they treat it as a period in which philosophers "speculated rationally,"—a polite way of saying "guessed,"—about problems later to be investigated when scientific method reached psychology. They may even use the phrases "philosophical psychology" and "scientific psychology," but they do not attribute to the former any status as established knowledge; it consists merely of "systems," or "hypotheses" which do not become knowledge until they are tested by scientific research. Thus it is clear that the historians of psychology, explicitly or implicitly, take a positivist view of the nature of philosophy. What is worse, they proceed for the most part as if modern philosophy were intelligible other than as an attenuation of the ancient and mediaeval tradition, confounded by the repetition of old errors. That they have little respect for philosophical psychology is pardonable considering that they start with Descartes and Locke. Thus, in a hundred pages dealing with "the preparation for experimental psychology within philosophical psychology," Professor Boring devotes three to Aristotle and none to any mediaeval writer. He admits that Aristotelian traces can be found in all modern psychology; in fact, "many modern fundamental concepts of psychology can be found in Aristotle, and the historian can show for some of them a continuous descent. For this history, however, the reader must go to other sources. It is enough, as we begin with Descartes, for us to realize that he is by no means the first systematic psychologist" (*A History of Experimental Psychology*, New York, 1929: p. 158). Vd. the seven pages devoted by G. Murphy to the "intellectual background of seventeenth-century psychology" in his *Historical Introduction to Modern Psychology* (New York, 1929); and the chapter on "prescientific psychology" in E. Heidbreder's *Seven Psychologies* (New York, 1933). Although Professor Brett gives a whole volume to ancient and patristic theories, and another volume to the mediaeval and early modern period, in his three-volume *History of Psychology* (New York, 1921), he, like the others, writes as if the transition from ancient to modern philosophy, and from philosophy to science in psychology, were unqualified progresses. Until the anti-philosophical virus of positivism is completely overcome, an adequate history of psychology will not be written. Until psychologists rid themselves of this version of the history of their subject, "schools of psychology" will persist in interminable controversy about ungenuine issues, and scientific work will be misled by the failure to recognize that the basic questions in psychology are philosophical and cannot be answered by research. This is the root of error in psychoanalysis. Freud is a positivist. Vd. *New Introductory Lectures on Psychoanalysis*, New York, 1933: Ch. 7 on A Philosophy of Life.

Furthermore, most historians of psychology who write from the scientific point of view, confuse philosophical psychology with what Wolff and Kant called "rational psychology." Vd. Note 39c *infra* and cf. C. Alibert, *La Psychologie Thomiste et les Théories Modernes*, Paris, 1903.

36c. In writing a treatise on the soul rather than a treatise on man, Aristotle betrays the influence of Plato and the pre-Socratics. His method of reaching his own doctrine through a consideration of his predecessors has unfortunate consequences in this case. The *De Anima* lacks unity of subject-matter. It not only combines philosophical and scientific questions, but is concerned with problems that must be distinguished according to their various formalities. Thus, its questions concerning soul in general and the grades of soul constitute a problem in the philosophy of nature, the problem of living things in the order of composite substances. This can be seen in another way. To the *Physics*, which contains the most general considerations of the philosophy of nature, the *De Caelo* adds a treatment of local motion, the *De Generatione et Corruptione* a treatment of substantial change and alteration, and the *De Anima* a treatment of the motions which are vital operations. The interest of the *De Anima* extends to a consideration of vegetative, locomotive and sensitive powers *per se*, as acts of bodily organs. The questions raised here are physiological and belong properly to investigative biological science. So the consideration of plants as wholes, as composites, is botanical science, and the consideration of animals as composites is zoology and, perhaps, animal psychology,–both in the biological domain and scientific in type. That part of the *De Anima* which is not a section of the philosophy of nature, on the one hand, or dispersed among the biological sciences, on the other, is the treatment of human nature. To the extent that the analysis of human powers and acts arises from man's reflexive knowledge of himself, the *De Anima* is philosophical psychology. But as such it is not knowledge of the soul *per se*, but of man who, as a composite substance, has these vital powers by virtue of having as a principle of his life a soul of human grade. Vd. Note 32a *supra*.

37. The three Platonic errors in psychology are: (1) the substantial separation of soul and body, (2) the assignment of sensitivity to the soul as a power of operation apart from bodily organs, (3) the operation of the intellect in the direct intuition of ideas, apart from sensitive knowledge. Aristotle and St. Thomas corrected these errors by painstaking analysis, but Descartes returned to two of them (1 and 3) and bequeathed the "problem of the soul," in an insoluble form, to modern psychology. For a citation of the Thomistic texts in which the refutation of Platonism is accomplished, vd. Note 22 *supra*. *Summa Contra Gentiles*, II, Ch. 56–79 should be added. For a critique of Descartes, vd. Maritain, *Three Reformers*, New York, 1932: pp. 53–89; *Le Songe de Descartes*, Paris, 1932; *Réflexions Sur L'Intelligence*, Paris, 1931: Ch. II.

38. Vd. Note 33 *supra*. Distinction must be made between two kinds of psychological knowledge which occur in the practical treatises of Aristotle. (1) Propositions about the nature of man and about human powers and operations. (2) Propositions about types of human character, the conditions which influence human development, the variables of human behavior, the contingent circumstances of action, etc. Knowledge of the first sort is strictly speculative philosophy. Such propositions have the universality and certainty that mark speculative knowledge. Knowledge of the second sort is necessarily uncertain and inexact. Such knowledge is not "scientific," to use Aristotle's words, but "practical." We desire it chiefly because it is useful in directing conduct, and not for its own sake. Psychological knowledge of the first sort provides ethics with some of its leading principles. But these principles cannot be applied without knowledge of the second sort. That part of contemporary psychology and social science which is concerned with human behavior in the large, is of the latter kind,–practically useful knowledge about the general facts of life.

38a. The magnificence with which St. Thomas perfects the Aristotelian tradition is more evident in psychology than anywhere else, except natural theology. For one thing, where Aristotle treats human nature as a part of the study of soul in general, St. Thomas makes man the central object of psychological analysis and subordinates the discussion of soul thereto. Strictly, soul as substantial form cannot be in itself either a material or formal object of knowledge. Man can be such an object, and psychology is not adequately defined in subject-matter until it treats of man rather than the soul. (We are, of course, here considering only human psychology or what would be better named anthropology. The consideration of soul as self-subsistent belongs to natural theology.) Furthermore, where Aristotle refutes Platonism and materialism, St. Thomas purifies Aristotle himself of the tendencies to these very same errors which his neo-Platonic commentators could easily read into his texts. The Thomistic rendering of Aristotle is not only an answer to Averroes, Avicenna and Ibn Gebirol, but to Duns Scotus and Roger Bacon, to Descartes and Leibnitz as well, because as an exposition of the errors of Platonism, it is so much clearer than the account which Aristotle gave.

With respect to such crucial matters as the relation of the active and the possible intellect, it may even be conceded that Averroes was nearer to a right reading of the Aristotelian text than St. Thomas. But in the issue against Averroism, Aquinas was right philosophically, and those who praise the Averroistic reading are really admitting that Aristotle himself was more of a Platonist than St. Thomas was willing to construe.

39. Vd. Note 37 *supra*. Though he followed Descartes and was much influenced by him, Spinoza was a better psychologist. Yet he could not solve the problem of the unity of man by making mind the idea of the body. This led to the identification of mind as knower with the ideas it knows. The account of understanding and knowledge given in Book II of the *Ethics* suffers from the denial of universals *in abstracto*. Book III is an excellent analysis of the passions. The separation of theoretical psychology (Books II and III) from its practical application to the problems of conduct and happiness (Books IV and V) provides a model of systematic order which unfortunately is seldom imitated by other modern writers.

39a. We have already discussed many errors of Kant's philosophical psychology, his misconception of the intellect as a faculty of judgment rather than abstraction, his distinction of understanding and reason as scientific and dialectical,—in which he curiously distorts the Platonic division of the knowledge segment of the divided line (vd. *Republic*, VI, 509 D–511 E),—his failure to understand the faculties as the powers of man rather than as the mysterious operations of the transcendental ego in constituting experience. But at least he did proceed in terms of faculties, at least he did see that the separation of powers was not inconsistent with their unified operation,—even though he spoke of this as a transcendental unity,—at least he did understand that scientific psychology was restricted to the phenomenal order,—even though he did not see how it could be constituted as an experimental and mathematical science. All of these insights were lost or nullified by Herbart who succeeded him at Königsberg. Herbart denied the genuine distinction between rational, or philosophical, and empirical psychology, conceiving psychology at once as an empirical, mathematical and metaphysical science. As Boring points out, it was on the mathematical side that Herbart influenced Fechner who, also influenced by Weber, surmounted Herbart's Kantian repudiation of experimentation in psychology. It is not Herbart's misconception of the method or status of psychology as a science which is most important, however; that was remedied by Fechner and Wundt. It is rather his misconception of its subject-matter. Herbart rejected the analysis of faculties, and with it he rejected man as the subject-matter of psychology. Instead he attempted to formulate the relations of active ideational elements,—an analysis of consciousness which was a sort of animated associationism. (Vd. Heidbreder, *Seven Psychologies*, p. 65 ff.) Thus, the British and German lines converge on Fechner and Wundt to frustrate the early efforts to found a scientific psychology by reason of an impossible definition of its subject-matter. If it were true, as Boring says, that Herbart's attack on faculty psychology rejected a "kind of analysis to which psychology has never returned" (*History of Experimental Psychology*, p. 242), there would be no psychological science today. Fortunately it is not true; both psychometrics and psychoanalysis, however much they are indebted to Herbart and Fechner, have returned to the study of the abilities, habits and performances of men; neither are concerned with the elements of consciousness as such. Thus, Spearman writes: "The attacks made long ago by the Herbartians appeared to be irresistible; no serious defense was even attempted. Yet the sole permanent effect of these attacks was only to banish the word 'faculty', leaving the doctrine represented by this word to escape scot free. As much may be said for the onslaught of Thorndike" (*The Abilities of Man*, New York, 1927: p. 39). Cf. C. Pratt, *Faculty Psychology*, Psychological Review, March, 1929: pp. 142–171.

It is often said that Herbart's discussion of the vitality of inhibited ideas,—inhibited as a result of conflict with other ideas, and active though they are below the limen of consciousness,—is not only a memory of Leibnitz's *petites perceptions*, but also an anticipation of the Freudian notion of the active, though repressed, unconscious. (Boring denies a direct influence here. Vd. *op. cit.*, p. 246. Allers, on the contrary, claims Freud to have been influenced by Herbart through Breuer. Vd. *The New Psychologies*, New York, 1933: p. 4.) But the nature of the active unconscious cannot be explained either in Leibnitzian or Herbartian terms, and it remains a mystery in Freud. The problem is solved only by the conception of habit as a mean between potentiality and act, and hence as an active tendency toward operation although not fully operative. Vd. Note 55 *infra*.

39b. Until the recent attacks, made from different angles by behaviorists and by psychoanalysts, text-books of psychology defined the subject-matter of their science as consciousness, or the phenomena of consciousness. In doing so, they betrayed the unfortunate influence of either the Cartesian or the British tradition. The content of consciousness or, what is the same thing, immediate experience, cannot be the object of any body of knowledge, scientific or philosophical. Immediate experience belongs to all knowledge as its brute datum. An adequate account of the "content of consciousness" can be given only by reporting the data and concepts of all the natural sciences, of philosophy, of poetry, of law and morality. But such an encyclopedia is not psychology. Beyond that, there is only the report of one's personal experiences and biography. The chief difficulty of most of modern psychology, in so far as it is influenced by Descartes and Hume, is that it has no subject-matter which it can properly call its own. Lacking subject-matter, it can hardly become a science, not to mention clear philosophical analysis. Santayana aptly described it as "literary psychology" and added that it is better written by poets than by empiricists or experimenters. (Scientific psychology rightly discarded Soul as subject-matter, but failed to select Man.)

It is extremely significant for the translation which Lecture IV attempts between Aristotle and Freud, to note that both the Freudian and the contemporary Thomist criticize the subject-matter of modern psychology in the same way. The Freudian rightly objects to the narrowness and superficiality of academic psychology. It is limited to the study of the ego and of consciousness,—the phenomena of immediate experience. In contrast, Freudian analysis is truly a "psychology of the deeper layers," a study of the whole man, of all the powers of the soul. So Alibert objects to the contraction of psychological subject-matter due to positivism and phenomenalism: "A la différence de la psychologie Thomiste, qui avait pour objet l'âme prise avec l'ensemble de ses puissances, conscientes ou inconscientes, la psychologie moderne n'en accepte qu'une partie: savoir, le moi, ses facultés et leurs phénomènes" (*La Psychologie Thomiste et les Théories Modernes*, Paris, 1903: p. 9.)

39c. The distinction between rational and empirical psychology, instituted by Wolff, adopted by Kant, and still flourishing among many neo-scholastic writers, has been the source of much confusion in the modern history of psychology because of ambiguity in the words "rational" and "empirical." They convey both a distinction in subject-matter and a distinction in method. Originally, the subject-matter distinction was primary. For Wolff and Kant, rational psychology was a branch of metaphysics, treating such questions as whether the soul is self-subsistent and what is its mode of being, in separation from the body, as a simple substance. As we have seen, such questions strictly belong to natural theology and in some part require the supplementation of sacred theology and revelation. Vd. Note 32a *supra.* Empirical psychology is the field of faculty analysis; it is concerned with man and not with the soul *per se.* (The object of rational psychology is a purely intelligible and simple being; the object of empirical psychology is a sensible being, a composite of form and matter; hence there is some justification for the use of the words "rational" and "empirical" to distinguish knowledge of these different objects. As so used, the words have no methodological connotation whatsoever.) As Alibert makes the distinction in subject-matter, rational psychology deals with ontological questions and questions of existence, and empirical psychology deals with dynamological and phenomenological questions. (Vd. *La Psychologie Thomiste et les Théories Modernes,* Paris 1903.)

Wolff's use of the words "rational" and "empirical" to name this distinction in type of question had extremely unhappy consequences because in post-Kantian thought the words acquired a specious methodological meaning,—"rational" naming the supposedly *a priori* or purely dialectical method of philosophy, "empirical" naming the *a posteriori* or investigative method of science. (I say *specious* because philosophical knowledge, though dialectical rather than investigative, is as much *a posteriori,* as much the result of inductions from experience, as science. There is no knowledge which is purely rational or purely empirical.) Kant's attack on rational psychology is based on the same grounds as his attack on every part of what he thought was metaphysics, namely, that ontological knowledge of noumenal and transcendental being is impossible. (It must be remembered here that Kant accepted Wolff's ill-advised classification of the parts of special metaphysics as psychology, cosmology and theology.) But Kant was not attacking rational (philosophical or, even in his sense, *a priori*) as opposed to empirical (scientific and *a posteriori*) knowledge. He distinguished between pure or rational physics and empirical natural science; he made the same methodological distinctions in psychology; in fact, Kant's own philosophy is in large part philosophical psychology, dealing with questions which belong to empirical psychology (in subject-matter) but, in his terms, rational in method. It is ironic, therefore, that Kant should be cited by historians of psychology as authority for the position that philosophical psychology (identified by them with rational psychology) is impossible.

These confusions can be clarified only if we are able to eliminate the treacherous words "rational" and "empirical." We can then properly order the related distinctions in subject-matter and in method. (To show how these eliminations can be made, I shall use the words in what follows to make substitutions for them.) 1. *Subject-matter*: (a) "psychological" questions concerning existence and mode of being of the soul: rational psychology; (b) "anthropological" questions concerning the nature of man, his powers, habits and acts: empirical psychology. 2. *Method*: (a) philosophical, by analysis of the phenomena of reflexive experience, involving *a posteriori* inferences from operations to powers and nature: rational; (b) scientific, by investigation of the phenomena of human behavior, reaching accidental determinations of powers and acts: empirical. Now, it should be noted, in the first place, that philosophical method is not rational in Kant's sense of *a priori;* like scientific method, it also depends upon the data of experience and in the same way, though the data are different. And, in the second place, it should be noted that whereas the subject-matter here called "anthropological" is capable of both philosophical analysis and scientific investigation, the subject-matter here called "psychological" is restricted to philosophy. Yet even these ontological questions are not answered *a priori*, but only by *a posteriori* inference from the facts concering human nature which are the basic facts of philosophical psychology. This *a posteriori* inference from knowledge of human nature to questions concerning the soul differs from the *a posteriori* inference from the facts of human operation to knowledge of the powers and essence of man in that the former requires special metaphysical premises. This is important because it indicates the impropriety of treating questions concerning man and questions concerning the soul as if they were parts of a single subject-matter. The relation between empirical and rational psychology,—when these words name a subject-matter distinction,—is that of premises to conclusions, but the conclusions require additional premises not provided by empirical psychology. The relation between empirical and rational psychology, —when these words name a methodological distinction,—is that of accidental to essential determinations of human nature and its mode of life. Here there is a real unity: empirical psychology as a subject-matter combines and orders two bodies of knowledge which have the same subject-matter, man, but differ in method, as philosophical and scientific. Rational psychology as a subject-matter is purely philosophical, a group of questions in metaphysics (more narrowly, in natural theology, since the primary question is one of existence of a transcendent, simple, purely intelligible being, i.e., the separated soul). These latter questions are solved in terms of the knowledge provided by empirical psychology, in so far as it is philosophical, and by strictly metaphysical principles (knowledge of substance and cause).

When the distinctions are thus understood and ordered, the errors of Kant become plain, and plainer still those who followed and misunderstood his errors. Kant not only made the positivistic error of denying the possibility of rational psychology as a branch of natural theology, but he failed to see that the field of empirical psychology included problems capable of clear philosophical solution and also problems capable of scientific determination by methods which, in so far as they are mathematical and experimental, do not differ from those employed by physicists. Only to the extent that rational psychology includes questions concerning the mode of being of the separated soul was Kant right in saying that rational psychology exceeded the bounds of reason and belonged to the province of faith (sacred theology). Such questions are not capable of being solved within the limits of philosophy. Any attempt to do so must lead to paralogisms and dialectical frustrations. But the question, Whether the soul has any self-subsistence, is as much a purely philosophical question (in natural theology), susceptible of clear affirmative answer, as the question, Whether God exists.

40. The contrast between Wundt and Brentano can be made in terms of their analytical principles, as well as in terms of their methods. In method Wundt was an experimentalist whereas Brentano was frankly a philosophical psychologist. But the issue between Wundt's principles and those of Brentano is philosophical. It is in no way affected by whatever experimental evidence was obtained in the laboratory at Leipzig. Titchener's discussion of this opposition proceeds without reference to "scientific data." Vd. his *Systematic Psychology: Prolegomena*, New York, 1929; cf. E. G. Boring, *A History of Experimental Psychology*, New York, 1929: Ch. 18 on act psychology in relation to experimental psychology.

41. It is to Alexander Bain rather than to William James that we owe the reintroduction of the concept of habit into psychological theory. Vd. G. Murphy, *Historical Introduction to Modern Psychology*, New York, 1929: p. 111. But neither of these writers fully realized its true significance, because neither placed it in the context of the distinction between power and act. Their partial and confused consideration of habit is one among many indications of how attenuated the vestiges of a sound tradition can become.

42. "All the movements in psychology that have led to self-conscious schools—Wundt's physiological psychology, introspectionism in America, functional psychology, *Gestalt* psychology, behaviorism, but not animal psychology nor the mental tests—have been philosophical movements, conducted, for the most part, by men who would eschew philosophy and rely solely upon the experimental method. The degree to which the text of this book has entered upon quasi-philosophical and systematic matters, while purporting to be an account of the history of experimental psychology, is a measure of the admixture of philosophy with experimentation within psychology. Ever since their foundation, the journals of psychology in all countries have been weighted down with 'theoretical' papers that are really expositions of psychologists, untrained to philosophy but writing on philosophical

matters. Incompetent work can be ignored, but a division of the mind within psychology is not healthy. Inevitably it must hinder work in the individual and thus the most rapid progress in the science. There is too much to psychology now for psychologists to master their own material and philosophy too" (E. G. Boring, *A History of Experimental Psychology*, New York, 1929: pp. 660–661; by permission of D. Appleton-Century Co.). Boring's complaint can be reversed for much of contemporary philosophy. These philosophers,—the pragmatists, positivists, etc.,—have given psychology up, or what is much worse, they have adopted the "theory" of one or another "school" of so-called scientific psychology. In fact, bad psychology is the villain throughout the history of modern philosophy. It works havoc enough in Descartes, Hobbes and Kant, but that is as nothing to the ruin which follows upon its being made, by Locke and Hume, the "first philosophy" in place of metaphysics.

43. Much of the experimental evidence which has been obtained by researchers who have espoused *gestalttheorie* is in the field of animal psychology. In this respect, the gestalters and the behaviorists are alike. They are alike, furthermore, in that they are both reactions against introspectionism of the structuralist sort, and in that both regard men and animals as differing only in degree. As the behaviorist uses the concepts of reflex and conditioned reflex to explain human and animal behavior without distinction, so the gestalt psychologist applies the concepts of configuration and insight to both human and animal perception and learning. They differ with respect to their main interests: the behaviorist concerns himself chiefly with sensori-motor coordinations and emotions; the gestalt psychologist primarily with visual perception and problem solving. Gestalt experiments are, for the most part, more ingenious, and the neurological hypothesis which *gestalttheorie* so largely depends upon is much more sophisticated than the earlier behavioristic neuro-mythology of the reflex arc. But Lashley's neurology, as a basis for behavioristic interpretations, has been recognized by Köhler to be essentially the same as the psychophysiology suggested by his *Die physischen Gestalten in Ruhe und im stationären Zustand* (1920). Cf. *Gestalt Psychology*, New York, 1929: Ch. IV.

Both behaviorism and gestalt psychology started out as great reform movements in experimental psychology but although both stimulated many departures in experimental work, especially of a physiological sort or concerned with animal behavior, neither has achieved the fruition it promised,—the establishment of a science of psychology. In this respect, of course, they are no worse than all the other movements generated by Wundt's laboratory. Cf. E. G. Boring, *A History of Experimental Psychology*, New York, 1929: Ch. 22 on *Gestalt* Psychology and Behaviorism.

44. The chapter headings of a recent text-book of psychology,—*Psychology, A Factual Textbook*, by Boring, Langfeld, Weld and others (New York, 1935),—reveals the structure of psychology to be nothing more than a collection of topics that have been experimentally investigated. They are as follows: the nature of psychology, the response mechanism, psychological measurement, vision, audition, taste and smell, somesthesia, intensity, the perception of spatial relations, temporal perception, the perception of movement, perceiving, learning, imagery, pleasantness and unpleasantness, emotion, action, thought and personality. The greater part of this material is physiological, and of that the greater part is sense-physiology. The same picture of psychological subject-matter can be obtained by examining *The Foundations of Experimental Psychology*, Worcester, 1929. Experimental psychology began in the physiological laboratory and seems to be returning to it, if it ever departed therefrom. Boring has attempted to make a virtue of this defect by turning psychology into a systematic interpretation of physiological data. Vd. *The Physical Dimensions of Consciousness*, New York, 1933. Cf. on this point, H. P. Weld, *Psychology as Science*, New York, 1928.

45. The inductive sterility of experimental research in psychology is plain to anyone who has examined vast amounts of data and found no significant generalizations yielded by them. What are the "laws" of learning, of reaction time, of memory, of perception, of attention, of association? These are all fields in which the pile of raw data is tremendous. At best the generalizations with which experimental reports conclude are matters of common sense knowledge, given an air of precision and elegance by a pseudo-technical language and a parade of metrical determinations. I do not mean to say that we have not learned a great many details about the functioning of sense-organs, muscular fatigue, drug-effects, the role of the glands in emotional disturbance, and so forth. But this is all physiology. The criteria by which we shall know that an experimental science of psychology exists are few and simple: (1) its generalizations must not be common sense knowledge, but genuinely novel inductions from experimental evidence; (2) its generalizations must be about specifically human operations and must be incapable of assimilation to physiology; (3) its generalizations must be compendant, i.e. they must hang together by sharing a few simple concepts, in the way in which the formulae of mechanics are rendered compendant by such concepts as time, space and mass. By this test experimental psychology is still in limbo, if it is in potentiality at all.

Dr. Crichton-Miller (*Psychoanalysis*, New York, 1933: p. 7) quotes William James as writing to James Sully in 1890 that "psychology is like physics before Galileo's time—not a single elementary law yet caught a glimpse of. A great chance for some future psychologue to make a name greater than Newton's; but who will then read the books of this generation? Not many, I trow." He adds, in comment, that forty years later William McDougall declared Freud to have done more for psychology than anyone since Aristotle. It is reasonable to suppose, he says, that James would have accepted Freud as the Newton he was looking for. (The comparability of Newton and Freud is also proposed by Dr. J. S. Van Teslaar, *An Outline of Psychoanalysis,* New York, 1924: p. 362.) Leaving comparisons aside, it is certainly true that the contribution of Freud to scientific psychology, as measured by criteria which William James seems to have shared, is genuine. Psychoanalysis is psychological science: it defines the scope of its subject-matter properly as man; its method is investigative; it is inductively fertile. Vd. Note 39b *supra.*

The only other genuine contribution is that made by psychometrics. It satisfies the same criteria. It can be viewed either as separate from what is ordinarily called experimental psychology or as including it, because all the metrical data gathered by psychological research can be used to apply mathematics to the quantitative accidents of human abilities and acts. In either view, psychometrics is an existing science of psychology. Two qualifications must be made, however: first, that psychometrics is applied mathematics and adds little conceptually to psychological analysis; second, that it presupposes the concepts of philosophical psychology,—which is concerned with the essence of human operation,—and is satisfied to make precise determinations of the quantitatively variable accidents. The formulae of psychometrics,—not only precise, but genuine contributions to knowledge,—are, like those of mathematical physics, functional statements. They do not, therefore, solve the aetiological problem. The psychometrician is not concerned with human behavior in the large or with its causes. He is in theory a faculty psychologist,—dread name for the doctrine of powers and acts, abilities and performances, —and in practice he is a mathematician who tries to make crude measurements approximate rational equations. The amazing thing is the degree to which he has succeeded in this effort. It is in the spirit of Fechner rather than of Wundt that psychological experimentation has proved fruitful,—the spirit of Fechner's method, but not of his Cartesian ideology. The latter has obscured the bloom. Perhaps, all of experimental psychology can be assimilated to psychometrics. This may be the meaning of Boring's effort in *The Physical Dimensions of Consciousness,* but it is vitiated by his inability to avoid or cure the chronic Cartesian affliction by substituting an ambiguous monism. To effect the consummation with clarity and without pretension the philosophical analysis which is faculty psychology must be explicitly recognized as providing the conceptual matrix for the mathematical formulations. I say "without pretension" because by such acknowledgment it will be conceded that psychometrics can add to the traditional account of the essence of human nature only the determination of accidental details. This, after all, is the business of empirical science in any field,—the description of phenomenal relationships. Vd. L. L. Thurstone, *Vectors of Mind*, Chicago, 1935: pp. 44–54 also C. Spearman, *The Abilities of Man*, New York, 1927: pp. 38–39. Spearman more than Thurstone recognizes that psychometrics employs the the concepts of faculty analysis, and is thus subordinated to philosophical psychology. Vd. Note 39a *supra.*

45a. The criterion of subject-matter is in some respects the best to use in searching for a science of psychology. We have already shown that a science, distinct from physiology, can be constituted only by taking human nature as its object. Philosophical psychology is similarly constituted. It approaches the essence of man in terms of an analysis of the proper powers flowing from that essence, and in the course of that analysis, it studies the relation of the powers in operation and habituation. Scientific knowledge, limited by method and in aim, to the phenomenal order, studies the accidental determinations or conditions of the powers in operation, and correlates observable relationships synchronically or diachronically. Scientific psychology may, of course, be either like mathematical physics or like the non-mathematical biological sciences.

All the errors and false starts of modern psychology can be stated in terms of misconceptions of subject-matter. Thus, philosophical psychology blocked itself as soon as it took for itself (1) the task of describing the states of the soul as a separate substance or (2) the task of describing, genetically or chemically, the contents of immediate experience or the elements of consciousness and their combination. In so far as investigators or experimenters accepted either of these definitions of their subject-matter, they too were doomed to fail in their effort to create a science of psychology. Out of the despair which was the inevitable consequence within the first half century of the experimental effort, were born the extreme reactions of behaviorism and animal psychology. But animal psychology, if it is a science, cannot be extended to man, despite the behaviorist's effort to do so. Behaviorism is thus either animal psychology or reducible to human physiology. The tremendous mass of experimental findings, common to all schools, can be assimilated to *one* psychology only by interpreting them somehow in the light of a unified subject-matter, the only possible subject-matter, man or human nature. The distinction of psychometrics and psychoanalysis is that they have taken this as their subject-matter. It should be added that human behavior is included in that subject-matter only if by "behavior" is meant the performances or acts of human abilities. Vd. 39b *supra.*

45b. The history of psychiatry should, of course, be told in relation to the history of psychology. Unfortunately, it is the least developed part of the history of medicine. The Noguchi Lectures of Dr. Gregory Zilboorg on *The Medical Man and the Witch During the Renaissance*

(Baltimore, 1935) are a substantial contribution to the subject,—intelligent scholarship to be extended, one hopes, to other periods, particularly the Greek and Hellenistic. Dr. Zilboorg points out that "psychology—that is, academic psychology—contributed little if anything to psychopathology. The process is here reversed in that medical psychology has always been the pioneer in the general field, stimulating the most important advances in the study of normal psychology" (p. 204). And thus writes Alibert: "La *psychopathologie* ou *psychologie morbide* a rendu et continue á rendre les plus grands services à la psychologie normale" (*La Psychologie Thomiste et les Théories Modernes*, Paris, 1903: p. 16). One must ask, therefore, whether the advance made by psychopathology in conceiving "mental disease" as a natural phenomenon does not impugn the adequacy of philosophical psychology.

So far as the organic psychoses are concerned, there is no problem. The pathology of the nervous system is of the same sort as other organic disorders. The mental aspect of these psychoses is to be understood in terms of the functional dependence of reason upon sense and imagination which are directly functions of bodily organs. St. Thomas recognized that the operation of reason is affected by an individual's organic condition, whether that be a matter of native temperament, fatigue or lesions. Vd. Note 19 *supra*. (This applies both to cognition and volition.) The functional psychoses,—dementia praecox, cyclothymia and paranoia,—are rapidly being assimilated to the organic group, and there is every reason to expect that medical research will discover their physical pathogenesis and conditions. It is inevitable that all mental disorders will eventually fall into two groups: organic diseases (the psychoses) and abnormalities of character (the neuroses). There is no middle ground. The so-called functional psychoses are at present anomalies due to ignorance. Cf. R. E. Brennan, *A Theory of Abnormal Cognitive Processes, According to the Principles of St. Thomas Aquinas*, Washington, 1925; and C. A. Dubray, *The Theory of Psychical Dispositions*, Washington, 1905: pp. 157 ff.

The analyses of the neuroses,—the determination of the types of character and the account of their development,—does not, of course, belong to philosophical psychology; in the ancient and mediaeval traditions, these problems belonged to ethics which depended for its theoretical principles upon psychology. Nor is there any question that we have made an advance in subjecting the neuroses to scientific study. Psychoanalysis is the great gift which practical medicine has made to psychology as a science. This advance does not alter any philosophical principles, however. It is clear that in the strict meaning of "disease" only the body can suffer lesions; the intellect as such cannot be diseased, though it can be affected by a diseased body. Abnormalities which are not strictly diseases are the results of habit formation, contrary to the *norms* of good conduct; in other words, vicious characters. Psychoanalysis has learned much by investigating the nature and development of moral character, but it has *discovered* nothing which is inconsistent with the principles of philosophical psychology, nor can it, either as knowledge or as therapy, ever be more than an aid to the moralist. Vd. Notes 61 and 62 *infra.*

That the modern history of psychiatry is a record of the struggle of scientific inquiry against the practices of the Church may be true, but it is not relevant to the point under discussion. That the Devil, being of angelic nature, can cause the movement of matter and thus be one,—but never the sole or sufficient,—cause of human activity is a theological doctrine. This doctrine is thoroughly compatible with all philosophical knowledge, but demonology is not in itself a part of philosophical psychology. Furthermore, it is as much beyond the power of scientific evidence to refute the existence of demons as to establish the existence of angels. Science, philosophy and theology view "mental disease" in terms of causes of different orders. Logically, there is no conflict between them; in fact, they are necessarily complementary. But this is not to say that actually and historically there were no conflicts between scientists and churchmen who exceeded the bounds of theological doctrine or who even converted what was merely theology into a dogma of faith. The Dominican Inquisitors who composed the *Malleus Maleficarum* were far from being good Thomists, and the ecclesiastical and temporal authorities who applied it in practice were equally far from being good Christians. It is not only psychiatry which has progressed since the days of Johann Weyer, but theology and the Church as well. Unfortunately, the kind of naive demonology which guided the Inquisitors in their efforts to extirpate witchcraft can still be found today among those psychoanalysts who have reified such analytical terms as complex and super-ego into demonic forces with which neurotics,—nay, all of us,—are supposed to be possessed. Historic roles are reversed. Today, the Thomist must exorcise the "demons" out of psychoanalytical theory. To do this the only instrument he needs is Ockham's razor. Vd. Notes 48 and 51 *infra.*

46. In the light of Note 45 *supra*, the text at this point requires a word of explanation. Unless the word "non-mathematical" is carefully noted, the statement that psychoanalysis is the *only* science of psychology is inconsistent with the acknowledgment that psychometrics is an inductively fertile field of experimental research, employing psychological concepts.

The honors must be divided between psychometrics and psychoanalysis, the one mathematical and experimental, the other non-quantitative and clinical. Both provide genuine contributions to our knowledge of man by means of investigative techniques. Neither is reducible to physiology. The three criteria which test the accomplishment of scientific knowledge in the field of psychology are satisfied, to some degree, by both. Psychometrics and psychoanalysis are related as the mathematical and the non-mathematical branches of physical science. Neither excludes the other or competes with the other in any way.

The interesting fact, which Lecture IV elucidates, is that the concepts of psychoanalysis, like those of psychometrics, are re-phrased versions of the traditional terminology of philosophical psychology,—the analysis of the powers of man and their operations. As psychometrics, employing these concepts, has added metrical determinations of the quantitatively variable accidents of human ability and action, and has embraced these determinations in functional statements of co-variation, so psychoanalysis has used the same concepts to formulate generalizations about human development and character differentiation, inductively extracted from a vast amount of clinical data. These generalizations are entirely qualitative: they are classificatory, or taxonomic, and physionomic, i.e. the formulation of secular alterations. Finally, psychoanalysis, no less than psychometrics, is not aetiological in the strict sense of that term. This is a limitation which arises from the very nature of scientific knowledge. Vd. Note 47 *infra*. But even in the sense in which scientific knowledge can be given a causal interpretation for the sake of practical applications, psychoanalysis,—in fact any scientific psychology,—fails. The failure is not due to any incompetence in scientific method or any lack of analytical skill in the interpretation of the evidence. The failure is the inevitable consequence of the fact that the subject-matter being investigated is human activity which, being voluntary, has reason for its essential cause. As psychometrics can do no more than study the quantitative accidents of this activity, so psychoanalysis must be content to study its biographical accidents. In either case, the knowledge, given an aetiological interpretation, throws light only on accidental causes. Success in an undertaking thus limited is appropriate glory for an empirical science.

For a discussion of the unity of psychology, achieved by the proper ordering of its philosophical and scientific domains, vd. Note 32a *supra*. In terms of this unity, peculiar to psychology, it will be seen that psychology is the only particular natural science which depends upon philosophy for its proper concepts and principles.

47. To understand the different problems of natural science, we must first distinguish between the mathematical and the non-mathematical types of science. The former aims to describe the phenomena in terms of functional formulations. That it eschews causes we have already seen. Vd. Note 13 *supra*. But mathematical physics is not only non-explanatory; it does not classify; it does not do morphological analysis; it does not describe changes in which time's arrow is real, i.e. secular trends. All the problems of a mathematical science are necessarily of the same sort because mathematics is undeviating in setting as the ideal the accomplishment of statements of the equational or

functional kind. As both Eddington and Jeans have pointed out, the aim is to express everything in differential equations. It is only in the non-mathematical natural sciences that different sorts of problems are genuinely discriminated. Since psychoanalysis is non-mathematical, these discriminations are a relevant concern.

The problems of the non-mathematical natural sciences cannot be understood except in terms of the concepts of the philosophy of nature. The possibility of classification is founded on the fact that whatever exists is specifically determined, i.e., is a substance possessing determinate characteristics, both essential and accidental. The possibility of morphological analysis is founded on the fact that substances as integral wholes are determinate organizations of parts. The possibility of temporal ordering arises from the fact that changes occur in regular sequences because individuals of a given species develop and change according to their determinate nature. But, as we have seen, natural science is not ontological knowledge. It does not apprehend substances; it does not employ real definitions. It must, therefore, transform concepts which are appropriate to the substantial order into concepts applicable to the phenomenal surface which science is able to investigate. The logic of science is not the Aristotelian logic of predication. That belongs properly to metaphysics and the philosophy of nature. Cassirer and Whitehead are right in so far as they insist upon the inapplicability of a predicational logic in scientific work. The logic of science is relational, and relational logic, in its non-quantitative form, is continuous with mathematical logic. The latter is, in a sense, a special case of relational logic. The non-mathematical natural sciences employ this general relational logic. Thus, substance is relationalized by the concepts of class, order and system. The phenomena can be classified, ordered, systematized by the principles of relational logic. Where the philosopher talks of substance, the scientist considers a persistent relational system, the structural form of which endures throughout substitutions of particular matter; where the philosopher is concerned with the attributes of substance, the scientist classifies relational systems; where the philosopher determines the integral parts of a substantial whole, the scientist distinguishes subordinate systems included in the relational structure of a larger system; where the philosopher discerns the uniform career of change which a substance undergoes because of its determinate nature, the scientist apprehends the relational transformations of system structure, distinguishing this serial pattern as a diachronic order of relations from the part-whole pattern as a synchronic order. (Whitehead's term "society" is the same in meaning as what is here called a "system." It is a relationalized version of substance. Vd. Note 25 *supra*.)

Two further points must be added. (1) Russell is right that scientific notions are conceptual constructions and not abstracted concepts. (Vd. *Our Knowledge of the External World*, Chicago, 1914.)

And (2) Bridgman is right that the empirical content of these notions is their reference to scientific operations and not ordinary perceptions or imaginations. The meaning of a scientific construction is determined by the operations it calls for. (Vd. *The Logic of Modern Physics*, New York, 1927.)

In short, the scientist deals only with the relations of phenomena; his conceptual schema, constructed entirely by a logic of relations, enables him to *impose or find* systems, classes and orders *on or in* nature. But which is it, impose *or* find? If he is a radical positivist, he will say "impose," almost in the manner of Kant. But since he will not attribute the imposition to a transcendental mind, science becomes for him conventional, using logical fictions which are practically expedient. If he is not an extreme positivist, or if he is a Platonist, he will say "find," holding that the relatedness of phenomena is a real interconnectedness, and that the discoveries of science are true or false according as they agree or not with what is independently the case. Such a scientist cannot avoid having a philosophy of nature. And he will either make the error of Whitehead, reifying the relational logic of science into a Heraclitean cosmology, in which the Ideas may be substituted for the Logos, or he will be guided by what is here the crucial principle, namely, that philosophy is knowledge of a different type, informed by a different logic. He will realize that the concept of substance is indispensable in the philosophy of nature as ontological knowledge and that a philosophy of nature, whose conceptual schema is constructed by Aristotelian logic, far from being incompatible with the aims and methodology of science, provides the only intelligible basis for the scientist's classifications and orderings of phenomena, if they are viewed as knowledge of an independent reality. The relations of phenomena are real connections, rooted in the determinate substantiality of the actual. Aristotelian science is bad for the same reason that "scientific cosmologies" are bad: both attempt to extend a logic, appropriate to one type of knowledge, to fields in which that logic is inapplicable. Similarly, Aristotelian philosophy is good for the same reason that modern science is good. As the former employs the logic of definition and predication for knowing the substance of things, so the latter employs the logic of relations for investigating phenomena. Contemporary issues which call upon us to choose between these logics, which ask us to adhere to science and forsake philosophy or which in the name of philosophy require a false apologetics for science, are obviously ungenuine. And just as obviously the modern synthesis is clear. It remains to be done. It will be done only when both scientists and philosophers learn the full wisdom of Maritain's principle: *distinguer pour unir*.

To return to the problems of the non-mathematical sciences: they are taxonomic (classification of relational systems), morphological (synchronic analysis: the part-whole ordering of inclusive and included systems), and physionomic (diachronic formulation: the secular ordering of systematic transformations). The taxonomic can be made to include the morphological as a special case. The latter is a classification of systems in terms of part-whole structure. Taxonomy, furthermore, is more than the mere identification of persistent systems; it involves the characterization of these systems in terms of relatively stable or unstable traits. (Scientific descriptions do not distinguish essence, property and accident, but the discrimination between universal laws and statistical generalizations is the relational analogue of the distinction among the predicables.) The transition from taxonomy, including morphology, to physionomy is a shift from a synchronic to a diachronic mode of ordering. Synchronic problems disregard change and time as its dimension. But this indicates that the separation of taxonomy and physionomy is only analytical. According as a system is of a certain type and structural form, so does it undergo systematic transformations; and conversely. (In the substantial order, one would say that according as a thing is, so does it change; its career of modification reveals its specific character.) Although the taxonomic and physionomic problems are correlative, they can be independently solved. As a matter of fact, however, they are seldom solved in complete independence of each other. Sciences which seem to be exclusively taxonomic and morphological, such as classificatory botany, zoology and anatomy, have a kind of partial autonomy, but they need to be completed by physionomic sciences, such as physiology and genetics. The latter sciences, of course, require the former. Lecture IV shows the correlation of the taxonomic and the physionomic phases of psychoanalytical research.

One problem remains. Can scientific knowledge be aetiological? Can science give an analysis of causes? The positivist rightly says No. The concept of cause is inextricably associated with the concept of substance. The notion of causality and the distinction of the four causes depends upon the concepts of actuality and potentiality and the distinctions of form and matter. Cause and substance belong together to the philosophy of nature. The same insight which excludes substance from natural science, must exclude causality. Since the days of Galileo, sciences of the mathematical type have restricted themselves to description. Explanation, in the sense of determining real causes, was clearly beyond their province. In the non-mathematical sciences, there is a relational substitute for causality, just as persistent system takes the place of substance. Uniform or at least usual sequences become the attenuated phenomenal substitute for causal relationship. The aetiological problem in science thus appears to be the same as the physionomic. Of course, the distinction of causes is lost, unless it be thought that vestiges of material and formal causality are retained in the part-whole analysis of morphology. Of all the causes, final cause seems most abhorrent to the scientist, although in the scientific reduction of causality to sequence the efficient is to the final cause as a beginning is to an end. In any event, it is clear that scientific knowledge of any field of phenomena is either mathematical and functional or non-mathematical and taxonomic, morphologic and physionomic. There is no aetiology in addition. If the scientist appears to talk or think causally, he does so by making the substitutions above indicated, but not by doing any further research or making new discoveries.

Yet the scientist cannot avoid talking or thinking in causal terms. The reason for this is that the scientist as an experimenter or investigator is also an artist or operator. Even though scientific knowledge is purely descriptive and does not *as such* contain knowledge of causes, the scientist must have such knowledge or he could not construct experiments in which he controlled certain conditions to produce certain effects. Because it is operational, all technology, including the art of research itself as well as the so-called applied sciences, presupposes an aetiology. But where does this aetiology come from if it is not strictly a part of scientific knowledge? The answer is that the man of common sense, the ordinary practical man, makes genuine causal determinations. He need not, like the philosopher, possess an analysis of the principle of causality and its relation to substance; he does not try to rise from a knowledge of proximate and instrumental causes to ultimate and principal causes; but he must, because he is a daily practitioner of one sort or another, know the causes of change in the things (substances) with which he deals. This knowledge of causes, being operative, is not an attenuated fiction. The kind of knowledge which the practical man of experience has of cause and substance is essentially the same as that which the philosophy of nature begins with and reflectively develops. And the scientist, as an operator in engineering or research, must use the same sort of knowledge. For all practical and productive purposes, the scientist can give his purely descriptive formulations a causal interpretation; he can convert his mathematical formulations into an aetiology by determining what is really the independent variable; he can translate his physionomy into a causal diagnosis or prediction by penetrating to relations of genuine efficiency beneath his secular trends. But it must be reiterated that the scientist does not make these conversions and translations as a *knower*, i.e., as a scientist, but only as a doer or maker. If the scientist could be merely a knower, he could avoid the concepts of substance and cause entirely. That he cannot avoid these concepts is due to the fact that scientific knowing involves real operations. Nevertheless, aetiology is no part of scientific knowledge, any more than ontology is. The scientist uses the concepts of substance and cause only when he must operate as any other practical man, and at that point his knowledge is of the same order. In exceptional cases, he may possess philosophical insight as well. The purity of theoretic natural science is that it is entirely descriptive in relational terms. The applied or practical sciences, such as engineering and medicine, are impure because they are aetiological and, therefore, require admixtures of common sense or philosophy. It is only because we sometimes permit their intimate relation to medicine to color them, that such sciences as physiology and pathology appear to be aetiological. In the strictly theoretic dimension, they are as purely descriptive as mathematical physics.

There is, of course, another sense in which the theoretic natural sciences are "explanatory," namely, by reference to phenomenal systems, described by other sciences and included as parts of the whole system which the science, making this reference, is investigating. Vd. Note 33a *supra*. This point does not alter the foregoing analysis. Cf. Dr. E. G. Salmon, *Physical Sciences and Causality*, Proceedings of the American Catholic Philosophical Association, 1936, pp. 117–123; also Dr. C. deC. DeKoninck, *Thomism and Scientific Indeterminism*, ibid, pp. 58–75.

48. Newton's famous dictum *Hypotheses non fingo* is repeated at the end of the *Opticks* (Book III, Q. 31) where, in discussing scientific method he says: "Hypotheses are not to be regarded in experimental philosophy." The word "hypothesis" has two traditional meanings which have for long made Newton's remark an almost unintelligible puzzle. (1) The sense in which Euclid speaks of an hypothesis, i.e. the knowledge which is available, through assumption or prior demonstration, for the solution of a given problem. (2) The sense in which the word has come to be used in investigative science. An hypothesis is a hypothetical proposition, in the form 'if P, then Q.' Such propositions constitute problems when they challenge us to establish the connection between P and Q. The theorems in Euclid are hypothetical propositions to be demonstrated. In this sense, every well formulated scientific problem is an hypothesis, a hypothetical proposition to be established or refuted by the evidence. The first meaning of hypothesis is involved also, since what the scientist already knows, —that which is given as the condition of the problem's solution,—is indispensable to the formulation of the problem and to the determination of what evidence will be relevant to its being solved. Scientists certainly employ hypotheses, in the first sense, and make hypotheses, in the second. Only in strictly exploratory research,—which is far from the ideal of scientific method,—does the scientist proceed without hypotheses. The ideal is quite opposite to this: research is good in proportion as all its hypotheses,—its conditions and its problems,—are explicitly defined.

What, then, did Newton mean by *Hypotheses non fingo?* There are two clues, one from etymology, the other from the context of Newton's writings. The word "hypothesis" has traditionally a logical significance. The closely related word "hypostasis" has traditionally an ontological significance. An hypostasis is a substance, that which underlies accidents and change as their subject. While no one who was well instructed in the philosophical tradition could confuse these two words, their closely related Greek roots make it possible for a modern scientist, even one of Newton's stature, to do so. That in fact he did so is evidenced by his attack on Descartes for making hypotheses in the sense of postulating what Newton called "occult substances" and "occult causes" to explain the phenomena of nature, e.g., the vortices. Newton was right in principle; he was insisting upon the positivism of science. Experimental philosophy, i.e., investigative science, is not concerned with the substance of things or their causes; it is concerned with accidents, mainly quantitative variables, with phenomenal change and its covariations or secular trends. Descartes did confuse the problems of philosophy and science in his physical writings. Unfortunately so did Newton. A man who really understood the sense in which a scientist does not, should not, make hypotheses would not have postulated the "aether" to "explain" the phenomena of light. Vd. *Opticks*, Book III, QQ. 18–24.

Philosophers do not *assume* or *postulate* substances or causes. The nature of substances and the analysis of causes is properly the field of philosophical knowledge. But when a scientist goes beyond his empirical evidence to make "hypostases" or to discover "occult" causes, he is guessing and, worse than that, he is violating the principles and scope of scientific investigation. The history of science is full of such violations, from Descartes and Newton to Darwin and Freud. Ockham's razor has been used by modern scientists to decimate metaphysics entirely, but it has too often not been used to cut off the growth of fictitious entities that have reified on the face of science.

49. Vd. Notes 13 and 47 *supra*.

50. While psychoanalysts cannot study the operations of reason essentionally, they can throw some light upon the external circumstances and accidental causes of rational behavior. The essential cause of any specifically human act is reason, but it is not the sole cause. To the extent that psychoanalytical observations have discovered the way in which environmental circumstances and biographical factors determine the conflict between reason and the passions, to the extent that the interpretation of the clinical data has isolated the predisposing conditions of irrationality, psychoanalysis has made a contribution to the aetiology of human activity. Just as the kind of "practical aetiology" which is involved in psychometrics can be useful in educational administration to allocate pupils and to apply the principles of pedagogy, so the kind of "practical aetiology" which is involved in psychoanalysis can be useful to the medical practitioner and the moralist. Vd. Note 47 *supra*. It is important to remember in both these cases that the scientific knowledge is inadequate because, even in the practical dimension, it deals only with accidental causes. The principles of pedagogy and of morality must include the kind of knowledge of man which only philosophical psychology can provide. The radical

flaw in contemporary educational practice can be traced to the kind of educational psychology which dominates our teachers' colleges and our schools of education. It is entirely scientific and, therefore, pathetically inadequate. The same can be said for much of contemporary morality, and particularly the sort that has been foisted on the world by behaviorism and psychoanalysis. Vd. Notes 24, 32 *supra*, and 61, 62 *infra*.

51. "In our opinion a presentation which seeks to estimate, not only the *topographical* and *dynamic*, but also the economic element, is the most complete that we can at present imagine, and deserves to be distinguished by the term *meta-psychological*" (Freud, *Beyond the Pleasure Principle*, London, 1922: p. 1). By the economic element Freud means the quantitatively variable factor of psychic tension that is related to the avoidance of pain and the production of pleasure. Cf. F. Alexander, *The Psychoanalysis of the Total Personality*, New York, 1930.

But the use of the word "metapsychology" is not fully explained by the statement that this part of psychoanalysis is concerned with the "total personality," that it is the most complete account that can be given. The metapsychological papers are more "theoretical," in what for a scientist is the bad sense of that word, than other monographs written by Freud and his students. It may be insisted that the metapsychology consists of the most general inductions from clinical data, but a careful logical examination of the documents will not support this doctrinaire effort to make all of psychoanalysis scientific. *Beyond the Pleasure Principle*, for example, reveals Freud as making hypotheses in the sense in which Newton said no scientist should make them. At the end Freud says: "We wish to make it clear that the uncertainty of our *speculation* is enhanced by the necessity of *borrowing* from biological science. Biology is truly a realm of limitless possibilities; we have the most surprising revelations to expect from it, and cannot conjecture what answers it will offer in some decades to the questions we have put to it. Perhaps they may be such as to overthrow the whole *artificial* structure of *hypotheses*. If that is so, someone may ask why does one undertake such work as the one set out in this article, and why should it be communicated to the world? Well, I cannot deny that some of the *analogies*, relations and connections therein traced appeared to me worthy of *consideration*" (*Op. cit.*, p. 78). Italics mine. This is not the way a scientist talks about the inductions he has obtained from the data of his investigations. It is rather the way Newton talks, as a guesser and "speculator," in the Queries appended to his *Opticks*. There Newton was being a "meta-physician" in the bad sense of that word which means "guessing beyond the evidence." Here Freud is being a "meta-psychologist." Furthermore, Freud's references to biology should be noted, because they indicate that he is guessing beyond the evidences of biological science, as well as exceeding the significance of his own clinical data.

The metapsychological writings contain two quite different kinds of material. On the one hand, there is the analysis of the structure of the soul (topography), of the functional interdependence of its parts (dynamics), and of the fundamental principles of psychic operation (economics). On the other hand, there is a variety of speculations about ultimate origins and causes which are fanciful conjectures, fruits of the myth-making power of evolutionary biology. That Freud himself recognized this difference,—though not in these terms,—is clear from his opening remarks in *The Ego and the Id* (London, 1928): "In my essay, *Beyond the Pleasure Principle*, published (in German) in 1920, I began the discussion of a train of thought, my personal attitude toward which, as I mentioned there, might be described as a sort of benevolent curiosity; in the following pages this train of thought is developed further. I have taken up those *ideas* and brought them into connection with various *facts observed* in psychoanalysis and have endeavored to draw *fresh conclusions* from the combination; in the present work, however, no further contributions are *levied from biology, and it consequently stands in a closer relation to* psychoanalysis than does *Beyond the Pleasure Principle.* The thoughts contained in it are *synthetic* rather than *speculative* in character, and their aim appears to be an ambitious one. I am aware, however, that they do not go beyond the baldest outlines and I am perfectly content to recognize their limitations in this respect" (*Op. cit.*, pp. 7–8). Italics mine, to indicate, first, that neither book is simply the product of investigative science, and second, that the difference between them is a difference between evolutionary speculations and a psychological analysis which may be sound even though it is not scientific.

The two parts of the metapsychology can be distinguished as Freud's philosophical psychology and his "speculative opinions." Neither is strictly *based* on scientific evidence in the sense in which a scientific induction is a generalization that is derived from the data. That Freud writes his philosophical psychology as if he were interpreting clinical data is intelligible in the light of his positivistic prejudices against philosophy, but his language cannot deceive us if we scrutinize the analysis and examine the data. Furthermore, a sound philosophical analysis of the structure and operations of the soul should be applicable to the clinical evidence as well as to the facts of common experience. That it is so applicable indicates nothing about its mode of derivation. It may even be that Freud obtained his philosophical insights from his clinical studies, as in the great tradition of psychology, those same insights were obtained by other men in the light of common experience. If Freud had had a greater knowledge of that tradition, he might have demonstrated the truth of his claim that "priority and originality are not among the aims which psychoanalysis sets itself" (*Beyond the Pleasure Principle:* p. 2). He would have found more in that tradition which anticipated his chief tenets than the myth which Plato puts into the mouth of Aristophanes in the *Symposium*. Vd. *op. cit.*, p. 74–75. If he had understood the way in which common experience is the source of philosophical analysis, he might not have had to apologize for his speculations by saying that he could not work out these ideas "except by combining facts with pure imagination many times in succession, and thereby departing far from observation" (*Ibid.*, p. 77). He would have been able to distinguish between the certainty of philosophical knowledge and those "speculative opinions" of which he felt obliged to say: "Self criticism does not render obligatory any special tolerance of divergent opinions. One may inexorably reject theories that are contradicted by the very first steps in the analysis of observation, and yet at the same time be aware that those one holds oneself have only a tentative validity" (*Ibid.*, p. 77).

That part of the metapsychology is sound which can be understood in the light of common experience and which, for the most part, can be translated into the analytical terms of the traditional anthropology of Aristotle and St. Thomas. That part which is sound is, therefore, not novel or original, for the most part, except in vocabulary. (The outstanding exception is the concept of repression. This is a genuine contribution to philosophical psychology. Vd. Note 55 *infra.*) The rest is either a perpetuation of the errors of Descartes which Freud learned from Fechner or, what is worse, "speculations" which are neither scientific nor philosophical knowledge. They are not even errors in philosophical analysis of the Cartesian sort. They are myths and fancies, guesses about "occult substances and causes", opinions which violate Ockham's principle that a scientist should not reify his imaginations or multiply hidden causes and convert them into a demonology. Vd. Note 48 *supra.* It is not surprising that the greatest novelty of the Freudian doctrine should be in the opinionative part of the metapsychology.

In writing of the metapsychology, I have considered those articles which the editors of Freud's *Collected Papers* have assembled under that head in Volume IV; also Papers XXI–XXIV of Volume II, and the separate monographs *Beyond the Pleasure Principle*, *The Ego and the Id*, *Group Psychology and the Analysis of the Ego*.

52. The parts of the body are organs. The parts of the soul are powers. As the whole soul is related to the body as the actuality of its organic potentiality for life, so the parts of the soul are the powers of the particular organs. They are not the acts of the potentialities of these particular organs, because these organs are not always active, as the body is alive actually as long as it is besouled. The parts of the body are sensible, physical things and are, therefore, capable of being known directly by observation. But powers cannot be observed. Nothing which is not fully actual can be observed. Our knowledge

of the powers of a body is the result of inference from the operations or changes which we can observe. Men are sufficiently acquainted through common experience with their own operations or changes to make an adequate division of the powers which constitute the parts of the soul. Anatomical division and classification, especially the histological, requires investigation. Psycho-topography does not. Furthermore, what many psychoanalysts like to suppose is not the case, namely, that Freudian topography is to the traditional analysis of the powers of man as histology is to gross anatomy. (It should be noted here that the way in which Freud "discovers" the unconscious is precisely the mode of inference which Aristotle and St. Thomas employ in passing from the operations to the powers. Freud, unfortunately, speaks of "assuming" rather than inferring, and reifies what he assumes. Vd. *New Introductory Lectures on Psychoanalysis*, New York, 1933: pp. 99–100. The modern attack upon the faculties proceeds in terms of Ockham's razor. But this, as we have seen, is due to a misconception of the faculties, as substantial agents rather than powers, active or passive. Locke's attack is much more applicable to Freudian writing which does appear to reify the parts of the psyche and to speak of them as if they were agents rather than psychic properties. The Freudian reifications can be corrected by the translation we have made of these parts into powers.)

Although both topography and what I have called the psychoanalyst's taxonomy are classificatory, they can be as readily distinguished as classificatory botany, which groups varieties of organisms, and anatomy, which classifies the parts of an organism of a given species. (Morphology is a kind of classification. Vd. Note 47 *supra*.)

53. The genetic problem, as the Freudian states it, consists of a set of questions which are intelligible only in terms of the myth of evolution. I say "myth" in order to refer to the elaborate conjectural history, which vastly exceeds the scientific evidence in the fields of genetics, palaeontology, comparative anatomy, geology and geography. This myth is the story of evolution which is told to school children and which they can almost visualize as if it were a moving picture. It is the concoction of such evolutionary "philosophers" as Herbert Spencer, Ernst Haeckel and Henri Bergson, as well as the invention of popularizers of science.

Questions about the origin of life or of various psychological structures are clearly historical, and neither scientific nor philosophical. If there is any philosophical question involved in all this discussion, it is a question about God, Creation and Providence which the well-disciplined philosopher knows is, for the most part, beyond the scope of natural theology. Only those modern thinkers, such as the Freudians and the Marxians, whose faith is the religion of evolution, can answer such questions dogmatically. They know all the answers in a way that no Christian theologian ever pretended to know them. Although Darwin confused the 19th century badly by the preposterous issue he agitated between evolution and "special creation", he at least recognized some limits to scientific knowledge and historical conjecture about ultimate origins.

In the case of individual development, the Freudian also uses the evolutionary myth. He analogizes ontogeny and phyologeny in the light of Haeckel's "biogenetic law". He supposes that the soul evolves from its primitive seed in the *id*, as the forms of life have come from some undifferentiated primary plasm. It is clear that this supposition outruns both scientific evidence and common experience. The genetic problem thus formulated can, therefore, be neither scientific nor philosophical. As Freud unwittingly indicated, it is a challenge to the myth-maker of an Aristophanic sort. Vd. *Beyond the Pleasure Principle*, New York, 1922: pp. 72–76.

Dr. Van Teslaar, discussing the significance of psychoanalysis, declares the contribution of Freud to be as revolutionary as Darwinism. Freud's addition of a "genetic, developmental, evolutionistic viewpoint" to psychology has resulted in "as radical a transformation of all branches of psychology as that which Darwin has inaugurated in the biological sciences." Dr. Van Teslaar acknowledges that much scientific criticism has been directed against Haeckel's biogenetic law, but insists that the facts support recapitulation. He adds: "Freud's discovery shows that in the course of its development the individual mind repeats racial history. The details of Freud's work amounts to a restatement of the recapitulation theory applied to the biologic history of mind. For the first time there has been disclosed to us the manner in which psychic recapitulation operates and its consequences" (*An Outline of Psychoanalysis*, New York, 1934: p. 375).

53a. The most compact statement of Freudian topography and dynamics is to be found in the lecture on The Anatomy of the Mental Personality in the *New Introductory Lectures on Psychoanalysis*, New York, 1933. Here Freud distinguishes and attempts to define the id, the ego and the super-ego, and shows that these divisions must not be confused with the distinctions of unconscious, pre-conscious, repressed unconscious. Cf. Healy, Bronner and Bowers, *The Structure and Meaning of Psychoanalysis*, New York, 1930: pp. 22–57; also pp. 192–267. For an equally compact statement of the traditional philosophical account of the powers of man and their cooperative inter-dependence, vd. St. Thomas Aquinas, *Summa Theologica*, I, QQ. 77–83. Anyone who can read the Freudian and Thomistic texts with ability to discern analytical equivalence beneath seemingly incommensurable vocabularies, can make in full the kind of translation which I briefly indicated in Lecture IV.

54. Vd. St. Thomas Aquinas, *Summa Theologica*, I, QQ. 80–82; I–II, Q. 17, AA. 7–9. "Some acts proceed from the natural appetite,

others from the animal, and others from the intellectual appetite" (I–II, Q. 17, A. 8). Cf. I–II, Q. 1, A. 2; Q. 6, A. 4. The natural appetite of animals consists of the tendencies of the various powers to the performance of their proper acts. This holds for sense and intellect as well as for the vegetative powers. There is, however, some justification for thinking of natural appetite especially in connection with these latter powers, because their operation is not determined directly by cognition or subject to will. If we regard natural appetite narrowly in the sphere of the vegetative as the inclination of these powers to their fulfilling operations, the concept thus restricted is equivalent to what in modern terms is the concept of biological needs, the basic urges or drives.

The remaining appetites, i.e., animal and human, are cognitively determined, by sense and intellect respectively. The difference between need and desire can be stated clearly in terms of the difference indicated between natural and cognitively determined appetites. A need is a biological want, arising from some condition of the vegetative organs. It is an indeterminate tendency to action, having no *particular* object as its goal. Thus, hunger, fatigue, sexuality are needs. A desire is a tendency toward a particular object, known through sense or intellect. The sensitive desires are, in large part, related to animal needs, as the desire for this morsel of food is related to hunger. Sensitive desires are acts of bodily organs; intellectual desires, the acts of will, are not. Sensitive desires cannot be reduced to needs. It is obvious that we often seek to possess an object of sensuality which we do not need. The problem of the nature of appetite,—whether it is a single power of the soul embracing the three forms here distinguished, whether the natural appetite is primary and the cognitively determined appetites are developments therefrom, whether there is any power of appetite, other than the natural, which is antecedent to sensitive and intellectual determinations,—consists of questions difficult to answer. But the facts of experience are clear and they require us to distinguish between indeterminate needs and definite desires. To ask why objects have a tendential mode of being for man, i.e., are objects of appetite when known, is as impossible a question as to ask why objects have an intentional mode of being, i.e., are objects of knowledge. Cf. Freud, *Instincts and Their Vicissitudes* (1915), and *The Unconscious* (1915) in *Collected Papers,* IV, London, 1925.

55. In view of the fact that "consciousness" was not a technical term in psychological doctrine until the late 18th century, it is not surprising that Aristotle and St. Thomas did not discuss "the unconscious". The Freudian can claim to have made an advance over the associationist and introspective psychologies which examined only the contents of consciousness, but he cannot extend this claim against the traditional philosophical analysis in which the basic terms are

power, habit and operation. On the contrary, if his conception of the unconscious is to be saved from reification as a level, province or structural part of the psyche, the Freudian must be freed from the prepossessions of 19th century psychology in which consciousness was entitized. It is not to anticipations in Leibnitz and Herbart, von Hartmann and Schopenhauer, that one must turn if one seeks a thoroughly analytical understanding of unconsciousness, but rather to the traditional doctrine of faculties. A Thomistic rendering of the unconscious will not only clarify the Freudian theory, but will show that the metapsychology is a poorly expressed version of this philosophical analysis. (It should be repeated here that a contemporary Thomist, such as Alibert, claims the superiority of Thomistic analysis over modern philosophical psychology to be precisely the fact that it includes the unconscious. Vd. Note 39b *supra.*)

In the first place, it is necessary always to replace the substantive words, "consciousness" and "unconsciousness" by adjectival or adverbial forms. Modes of potentiality and operation are being distinguished, not modes of being. In order to make the Thomistic translation, it is then necessary to consider how an Aristotelian would use the words "conscious" or "consciously" if he were to introduce them into a psychological exposition. The answer is simple and clear. That operation is conscious which is the act of a cognitive power, either sense or intellect; and that operation occurs consciously which, while not an act of cognition, participates in actual cognition; thus, either an act of sensitive appetite, a passion or emotion, or an act of will. (So a Freudian would say that ideational activity is conscious or that desires occur consciously in so far as they have ideational representation.) We can now ask to what "unconscious" and "unconsciously" apply. To answer this question, we must first distinguish between the primary use of the word "conscious", exclusively applicable to acts of cognition, and a derivative use to qualify objects cognized. Thus, the object which we know or desire, i.e. of which we are conscious, is said to be a conscious object or end. In the same way, there is a primary use of the word "unconscious" to distinguish all non-cognitive operations, such as the purely vegetative and locomotive; and a derivative use to qualify whatever cannot be directly cognized. Now, neither a power nor a habit is a direct object of sense or of intellectual apprehension. All powers and habits, including those of cognition itself, are therefore said to be unconscious in this derivative sense. Vd. *Summa Theologica*, I, Q. 87, AA. 1–3.

But, the Freudian will at once object, this is superficial, even though true. The psychological significance of the unconscious is that it is dynamic. A power of the soul, called "unconscious" because not an object of immediate cognition, is no different from an atom or an angel in this respect. The Freudian is right. We must proceed, therefore, to another conception of the unconscious, namely, as a mode of operation defined without reference to cognition. Here we must distinguish power, habit and activity according to their grades of actuality. A vital power is not a pure potentiality. It exists as the property of a living substance; it is a determinate accident. As possessing some actual being, a power also possesses a proportionate degree of operation. In so far as a thing is actual it can act. The essence of a power is an actual inclination to a particular mode of activity. If the power is a passive power it must be acted upon before it is fully actualized in activity, but short of such actualization, it is still an actual power and, as such, an energetic inclination, a dynamic disposition. If we use the phrase "perfect operation" for those acts which are determined by particular objects, then we can speak of the powers, actual tendencies not thus determined, as imperfectly operative. Powers are energetic factors in behavior because they are operative, though imperfectly. They are one part of what the Freudian means by the dynamic unconscious. To the extent that its behavior is directed by its powers, an organism behaves unconsciously.

Habits are another part of the dynamic unconscious. In the ordinary meaning of the word, a habit is acquired by the actual exercise of a power. Once acquired, the habit exists as a modification of the power. It is more determinate than the power because it has been formed by acts determined to particular objects, but less determinate than these acts because it regards only the species of these objects. In other words, a habit is not a perfect operation but only a tendency to such operation; yet as a tendency it is more determinate than mere power. Thus, habits have more operative actuality than powers; though their mode of operation is imperfect, they are less imperfect. We know that habits exist in all degrees of strength, ranging from something more than the imperfection of power to something less than the complete actuality of perfect operation. In all these degrees, habits are actual; according to their degrees of actuality, they have energy and are, therefore, weaker or stronger energetic factors in behavior. This applies to habits of cognition as well as to habits of desire. We can be said to think or desire unconsciously to the extent that our thinking or appetition is affected by our habits.

An instinct is like a habit in that it is only a tendency to action rather than perfectly operative; it is not mere power because it is always relatively determinate, i.e. an inclination to a specific mode of behavior. As an energetic factor, instinct does not differ from habit. Habits of great strength so closely resemble instincts as operative tendencies that they are called "second nature". This indicates the difference between instincts and habits. An instinct is native and not acquired. Instinct can be defined as a natural habit. This is a difference in mode of origin, not mode of operation, yet it is related to the latter. The alteration of instincts is different from the alteration of habits; the latter are less stable and more modifiable. (In the lower animals, instincts remain largely unmodified; in man they are usually, if not always, overlaid with habit formations. In man, instincts are nothing more than vegetative needs.) For our present purposes, it suffices to point out that instinctive tendencies are energetic factors in behavior and are a part of what the Freudian calls the dynamic unconscious.

Perfect operations, i.e. particular, perfectly determinate acts, remain. These can be properly spoken of as operations, in contrast to powers, habits, instincts, which are only more or less indeterminate tendencies to operation. But, as we have seen, all operations are not conscious. At this point, cognition becomes the essential distinction between those operations which are conscious or occur consciously and those which do not. Operation, whether conscious or unconscious, can be a factor in behavior in two ways: (1) being fully actual, an act can communicate its energy by reducing what is in potentiality to act and, in this way, one operation can be the efficient cause of another; (2) an act can determine a power, alter the strength of a habit, modify an instinct. The only strictly conscious factors in behavior are thus seen to be those acts which are conscious (cognitions) or which occur consciously (involve cognitions). All the remaining factors in behavior,—powers, habits, instincts and acts which are not cognitive,—are unconscious. These others could not be factors in behavior unless they had some actuality, the actuality of tendency if not the actuality of perfect operation. This actuality is the energy of the unconscious. And, furthermore, it is clear that no behavior is ever determined by conscious factors alone. The Freudian is right in comparing the proportions of conscious and unconscious energy to the visible and submerged portions of an iceberg. This quantitative fact must not, however, be misinterpreted. In the order of powers, reason is most perfect; in the order of factors, operation is more perfect than habit, i.e. has more actuality, as habit is more perfect than power. Conscious operation, and particularly rational operation or operation involving reason, is the prime factor in human behavior. Since habits are formed by operations, we can further distinguish habits which have been formed by rational operation from all other habits which are mere bodily dispositions. The former are more significant unconscious factors in human behavior than the latter.

So far we have done no more than to translate the imaginative language of Freud into the clear concepts of Thomistic psychology. (Freud, at one place, defines a process as unconscious "when we have to assume that it was active *at a certain time*, although *at that time* we knew nothing about it;" but then, a little later, he adds that the unconscious is a "mental province rather than a quality which mental things have." Vd. *New Introductory Lectures on Psychoanalysis*, New York, 1933: pp. 99–102.) But there is a further step in analysis which marks an original contribution made by Freud. Although it can be translated into Thomistic terms, the psychological insight conveyed by the concept of repression seems to be an addition to the traditional understanding of the relation of reason to the passions.

Freud says that "the essence of the process of repression lies, not in abrogating or annihilating the ideational presentation of an instinct, but in withholding it from becoming conscious" (*The Unconscious* in *Collected Papers*, London, 1925: IV, p. 98; vd. also the essay on *Repression*). Repression must be sharply differentiated from such inhibitive processes as renunciation or conscious abandonment of a tendency. I am indebted to Dr. Alexander for the statement of repression as a specific form of control mainly characteristic of the infantile personality, excluding certain tendencies not only from motility but from consciousness. It is due to the fact that the infantile or the weak ego cannot stand the temptation, and therefore the tendencies, in order to be checked, must be excluded not only from motor expression but also from consciousness. Repressed tendencies do not lose their dynamic efficiency, but express themselves through substitutive vents (symptoms), and they remain conserved in the unconscious in their original form. Being repressed they cannot undergo those modifications which are called sublimation (domestication of original non-adjusted tendencies). Freud stresses this last point, "that the repressed remains unaltered by the passage of time" (*New Introductory Lectures*, p. 105).

We must separate the fact of repression from the explanation of the origin and effects of repression. The fact of repression is best understood by distinguishing the repressed unconscious from both the pure unconscious and the pre-conscious. The pure or original unconscious consists of those unconscious factors in behavior which we have called powers and instincts; to these might be added all purely bodily dispositions and unconscious acts. The pre-conscious consists of those habits which are tendencies to conscious operation, habits either of cognition or desire. As factors in overt behavior or thought, they are unconscious, but they can also become conscious, e.g., the transition of an idea from memory to actual consideration. The repressed unconscious consists of the same sort of habits as the pre-conscious. Only tendencies to conscious operation can be repressed because repression is exclusion from consciousness of what, were it not for repression, would naturally occur as conscious operation. Repression does not destroy the habit which it bars from normal operation and development. The habit retains all its energetic efficiency as an unconscious factor in behavior and is not susceptible to alteration in the same way that normally operative habits are. Repressed ideas are those which we will not consider, but they remain in the memory, nevertheless, and have operative tendency. Herbart discussed repression in terms of the incompatibility of ideas. This is insufficient. The permanence of repression can be accounted for only as a habit of the will not to consider certain ideas. Acts of understanding, i.e., actual consideration, are within the power of the will. The will, like any other power, is subject to habituation. Repression must, therefore, be defined as a habit of the will not to consider a certain idea, and the idea which is the content of this habit belongs to the repressed unconscious in distinction from the pre-conscious memory.

The repression of desires, i.e., habits of the passions, is more complicated. It depends upon ideational repression because these habits are formed by conscious operation, involving the intellect. Emotional or appetitive habits exist as tendencies toward or away from objects which can be ideationally represented. Repression here consists in the exclusion of such representation. But the conative tendency, which may have a deep instinctive origin, remains with all the energy possessed by the habit according to its strength. As unconscious factors in behavior, such repressed habits can influence bodily dispositions and operations as well as conscious activities of thought or desire. In this respect they do not differ from other habits. But because they are repressed and are prevented from normal operation, their influence is indirect and by means of all sorts of substitutions. Symptomatic actions, dreams, wishful thinking are good examples of the influence of repressed factors. What is most important of all is that repressed habits are not subject to modification by conscious operation. Though the will was involved in the conscious formation of these habits, the will is barred by its own habit of repression from further direct control or modification.

The explanation of repression requires us to understand the nature of the will as intellectual appetite determined by the rational estimate of things as good and evil. "In popular language," Freud says, "the ego stands for reason and circumspection, while the id stands for the untamed passions" (*New Introductory Lectures*, p. 107). "The id knows no values, no good and evil, no morality" (*Op. cit.*, p. 105). It is the ego which is the agency of repression. But within the ego, reason and will must be distinguished. The will is moved by the reason in so far as the rational consideration of objects is the ground for approval or disapproval. Inadequate knowledge may, of course, result in wrong estimates of good and evil. But even if the estimates be right, the will may not be able to cope with the strength of the passions tending toward what should be avoided. The traditional analysis here goes no further than the distinction between the successful operation of the will in forming habits of desire in accordance with reason (the virtues) and the failure of the will with the consequence of undisciplined habits (the vices). It is at this point that the concept of repression extends the analysis. The pain of failure may cause the will to withdraw from its conflict with the passions. This withdrawal becomes repression when the will moves the reason not to consider the objects to which the passions tend. The repressed habits are not only undisciplined but, by the habit of the will itself, become free from subjection to reason. They are, therefore, worse than ordinary vices because they are vices which cannot be directly cured by ordinary rational or educative methods. They are protected from criticism by a vicious habit in the will itself. Repression can be defined as the vice of a will lacking fortitude in its relation to the passions. Repression is not the cause of moral vice; it is, on the one hand, the result of moral vice, and, on the other, the cause of its perpetuation. The cure of neurosis which, though Freudians do not admit it, is the beginning of moral reformation, must attack repression as the primary obstacle to the rectifying work of reason in the discipline of the passions. The will must be strengthened by a removal of the fears and resistances which underlie repression. The attack upon repression must be oblique because the habit of the repressive will prevents a direct appeal to reason. Such oblique attack is the essence of psychoanalytic therapy. Though the path is through the passions, through the gradual understanding of their objects and hence the reduction of repression, the goal is the restoration of reason, by means of a cooperating will, to its proper role of regulation in human life. The inaccessible portions of the id thus become open to the ego. Freud describes psychoanalysis as a work of reclamation. Its object is to strengthen the ego, to widen its field of vision, "so to extend its organization that it can take over new portions of the id. Where id was, there shall ego be" (*Op. cit.*, p. 112). How close this is to the goal of morality may be judged from Freud's further statement of the ideal that "the intellect—the scientific spirit, reason—should in time establish a dictatorship over the human mind. The very nature of reason is a guarantee that it would not fail to concede to human emotions and to all that is determined by them, the position to which they are entitled" (*Op. cit.*, p. 234). Not a dictatorship, but the *just* rule of reason is the condition of good human life. Cf. Notes 61 and 62 *infra*.

For a discussion of the relation of reason and the passions, vd. Note 57 *infra*. For an examination of the concept of the unconscious in the light of the traditional psychology, vd. G. Dwelshauvers, *Traité de Psychologie*, Paris, 1928. Cf. *L'Inconscient*, Paris, 1919. Also, C. M. Dubray, *The Theory of Psychical Dispositions*, Washington, 1905: pp. 157 ff.

56. It has been objected that this translation of the Aristotelian and Freudian vocabularies is superficial, and purely verbal. This objection is based on the ground that the special experience achieved by investigative science is much more complete and adequate in detail than the common experience upon which philosophical knowledge rests. Thus, it can be said that although both Democritus and Dalton used the word "atom", the concept of the former is "purely speculative" whereas the concept of the latter can be referred to many careful observations. It is true that pre-scientific speculations in regard to questions which are essentially scientific are incommensurable

with answers which result from investigation. Thus, popular notions about the causes of disease, based on ordinary experience, must give way before the science of pathology, because such questions by their very nature call for research. Democritus was not a scientist in our modern sense of that word. He was a philosopher. Dalton was a chemist. The word "atom" clearly does not mean the same thing for Democritus and Dalton. The questions Dalton answered cannot be answered by common experience or by philosophical analysis. But the questions Democritus did answer are not affected in any way by scientific evidence. Atomistic materialism,—a philosophical and not a scientific doctrine,—is the same in the 19th century as it was in the 5th century B.C. So far as Dalton was a materialist, as well as a chemist, he did not improve upon Democritus.

The situation in psychology is unique, because psychology is the only field of subject-matter that is at once scientific and philosophical. In so far as Freud is answering philosophical questions about man, all the clinical data at his command are logically irrelevant. In so far as Freud is answering scientific questions,—is solving the problems of taxonomy and physionomy,—the data are extremely relevant and should certainly be the source of better answers than those which can be given by common sense or by pre-scientific speculations in regard to such questions. The issue turns, therefore, on whether the points under review in this translation are philosophical or scientific. The question cannot be begged by denying the existence of a philosophical psychology, i.e. by asserting that *all* questions about man are exclusively scientific. I have attempted carefully to distinguish those problems in regard to which psychoanalysis can proceed scientifically, and those in regard to which it cannot. The problem of the structure of the soul is clearly not a scientific question. That is the problem on which Freud and Aristotle are being compared.

What is said above applies equally to the translation of Freud and Aristotle on the problem of the functional relationship of the various powers of the soul, i.e., the dynamic problem.

57. The following texts should be noted. "As the Philosopher says (*Politics*, I, 2) the reason, in which resides the will, moves by its command, the irascible and concupiscible powers, not, indeed, by a *despotic sovereignty*, as a slave is moved by his master, but by a *royal and politic sovereignty*, as free men are ruled by their governor and can nevertheless act counter to his commands. Hence both the irascible and concupiscible powers can move counter to the will, and accordingly nothing hinders the will from being moved by them at times" (*Summa Theologica*, I–II, Q. 9, A. 3, ad. 3). Cf. *Summa Theologica*, I–II, Q. 17, A. 17. "Whence it is that we experience that the irascible and concupiscible powers do resist reason, inasmuch as we sense or imagine something pleasant, which reason forbids, or unpleasant, which reason commands. But from the fact that the irascible

and concupiscible resist reason in something, we must not conclude that they do not ever obey" (*Summa Theologica*, I, Q. 81, A. 3, ad. 2). "The passion of the sensitive appetite moves the will, in so far as the will is moved by its object, inasmuch as, to wit, man through being disposed in such and such a way by a passion, judges something to be fitting and good, which he would not judge thus were it not for the passion. Now this influence of a passion on man occurs in two ways. First, so that his reason is wholly bound, so that he has not the use of reason: as happens in those who through a violent access of anger or concupiscence become furious or insane, just as they may from some other bodily disorder. . . . Sometimes, however, the reason is not entirely engrossed by the passion, so that the judgment of reason retains, to a certain extent, its freedom; and thus the movement of the will remains in a certain degree. Accordingly, in so far as the reason remains free, and not subject to the passion, the will's movement, which also remains, does not tend of necessity to that whereto the passion inclines. Consequently, either there is no movement of the will in that man, and passion alone holds sway; or if there be a movement of the will, it does not necessarily follow the passion" (*Summa Theologica*, I–II, Q. 10, A. 3).

In these and many other passages, St. Thomas formulates the relation of passion to reason and will. The only alternatives considered seem to be the dominance of the passions by reason, or the subordination of the will to the passions. Freud adds another course: repression, the willful withdrawal of reason from the conflict. Vd. Note 55 *supra*. The nearest approach to this psychological insight is made by Aristotle in his consideration of the incontinent man who is led by his emotions to do what he *knows* to be wrong. Incontinence becomes repression when the passions further lead one to ignore what one knows, if such knowledge would require reason to oppose passions over which the will does not seem able to exercise command. Vd. *Nichomachean Ethics*, VII. Aristotle points out that children are naturally licentious and incontinent, and that, in this sense, vicious adults, not governed by their reason, are like children who do not live according to the direction of their tutors. Or, as Freud would say, the neurotic adult behaves in an infantile fashion. He is neurotic because he has never outgrown being a child.

58. The question about the independence of reason and the passions, the ego and the id, must not be interpreted genetically. Whether the Freudian account of the evolution of the ego out of the id is right or wrong is not relevant. Nor does it matter whether the Freudian speaks of the ego as a function of the id. The point is one of freedom of operation. If the ego were not, in some degree, operatively free from the id, it could not control the id, as it does both by inhibition and domestication. The same is true of the id. If the id were not, to some degree, operatively free of the ego,—if, in Aristotle's terms,

reason ruled the passions despotically instead of royally,—there could be no conflict between ego and id, no independent determination of behavior by repressed tendencies, no problems of vice and neurosis. In short, the primary fact is the fact of conflict between two parts of the soul, the rational and the non-rational. This fact,—and no other is needed,—establishes the analytical point that the forces in conflict are operatively independent of each other to some degree. What is here being said does not apply only to reason and passion, ego and id; it applies equally to the conflict of two passions, such as love and fear. The fact of this conflict requires us to understand these two motions of the sensitive appetite as operatively independent of each other. I would go further and say that just as this makes it impossible to say that fear is *simply* a function of love or love of fear, in any intelligible meaning of the word "function", so it is impossible to say that the ego is *simply* a function of the id, or reason of the passions. This is the type of error which intrudes itself into Freudian dynamic analysis as a result of the evolutionary genetics to which psychoanalysis religiously subscribes. Vd. Notes 51 and 53 *supra.* Freud supposes that all psychic energy is originally vested in the id, and that the actuality of the ego is nothing but energy borrowed from the id. This error is corrected by an analysis which shows that each of the independent powers of the soul is unconscious energy. Vd. Note 55 *supra.*

59. The important point here is not how much scientific truth there is in the propositions contributed by the biological sciences to the "theory of evolution"; nor is it a question concerning the *facts* which constitute all that is scientific knowledge about the origin of species. Whatever is scientific knowledge cannot be gainsaid, either by common sense or by philosophy. But a right understanding of philosophy and science shows that the facts ascertained by science make no difference to the solution of questions properly answerable by philosophy. An "evolutionary philosophy" is therefore a violation of both science and philosophy. It is the kind of speculation which discredits them both, because it goes beyond the scientific facts, on the one hand, and makes philosophy appear to be a sort of guessing, on the other. That the facts of evolution caused such agitation in philosophical circles in the 19th century indicates how far from their proper domain philosophers must by then have wandered, how little they must have understood their own principles and the regimen of their method. That the facts of relativity theory did the same thing in the very recent past is further proof that philosophy has not yet recovered itself.

60. Vd. *Beyond the Pleasure Principle* and cf. Bergson's *Creative Evolution.* The ideological affinities are plain. I must confess that I originally misread Freud's work as being entirely analytical. I took

the death instinct to be the desire for freedom from the passions and the attainment of the peace of the contemplative life, which is a kind of death in life in itself. I had supposed that the basic ideology was derived from Schopenhauer, that the pleasure principle was the will to live, and the death instinct its negation. But I have since learned that the book contains a cosmic biological fantasy. Cf. G. Santayana's misreading in his review under the title *A Long Way Round to Nirvana* in The Dial, Nov., 1923: pp. 435–42.

Whatever be the right interpretation of the text, one point is clear. In so far as the metapsychology is genetic, it does not rest on psychological evidence exclusively. On the contrary it seems to depend more upon biology, and even then not upon the established scientific facts of the various biological sciences, but upon the conjectural history which I have called the myth of evolution. Freud is a great storyteller, as witness the imaginative sociology of *Group Psychology and the Analysis of the Ego*. Many of his stories have a merit common to all great fiction. They convey profound insights into human nature. But poetry and science are not the same; much less poetry and philosophy. A scientific psychology must appeal to nothing but psychological evidence and must not go beyond the strict confines of its investigatively ascertained data. It must plainly exhibit each of its propositions as the inductive product of these data. It must order these propositions in such a way that their compendancy and interconnections are clear. The writings of Freud, or for that matter of any other psychoanalyst, cannot pass this simple test. Psychoanalysts would do well, first, to separate their strictly psychological doctrine from all the adventitious materials and reified fantasies with which it is at present garnished; second, to expound this doctrine in a strictly logical manner rather than in the style and language of poets; and third, to separate sharply the truths they have achieved as scientists from the philosophical analysis that is obscurely discernible in the metapsychology. It would be better if they left philosophy to philosophers, or at least sought instruction from the tradition in which they are thinking, whether they know it or not.

61. The psychoanalyst's therapeutic aim can be described, negatively, as the elimination of repression. Repression stands in the way of the control of the passions by reason; or, in psychoanalytical language, it stands in the way of the total integration of the personality. As repression is relieved, there is a consequent re-invigoration of conscious control of the separated tendencies of the id. The ego and the id assume an harmonious functional relationship. Desires which are irrational, i.e., desires which violate the reality principle, are modified by reason through sublimation and domestication, so that they can be assimilated. The result, supposedly, is a unified and adjusted personality,—unified in itself through a rational ordering of the appe-

tites and passions, adjusted to the environment because reason moderates and modifies desire according to the dictates of reality.

I do not mean to deprecate the therapeutic technique of psychoanalysis by referring to it as an extraordinary trick. Rather I wish to praise its ingenuity, its artistry. Nor do I mean to suggest that the psychoanalyst underestimates the power of reason, which must be utilized to make men happy or healthy; or that he caters to the passions, coddles them, compromises with them. I readily accept the statement of psychoanalytical technique which Dr. Alexander has made for me. What appears to be catering to the passions is quite the opposite. The therapeutic effort is to lure the passions out from their recesses, to remove the barriers of repression, not in order to give the passions freedom, but for the sake of making them submissive to reason. To suppose otherwise is to misunderstand the central concept of repression. Repressed tendencies are not merely inhibited desires and impulses. They are tendencies which have become anarchic through a kind of secession from the rule of reason; or, as the psychoanalyst would prefer to put it, through the abdication of reason from dominion over them because they have too painfully challenged its commands. In any case, adjustment, harmony, health and happiness, depend upon the reduction of such anarchy to a minimum. Therapy, therefore, must enable reason to regain its full sovereignty. It does this through a process of emotional clarification, on the one hand, and through augmenting the power of reason by giving it a fuller knowledge of the intransigent passions, on the other. A psychoanalysis is at best a course of education concerning one's emotions. I have elsewhere described psychoanalytical therapy as the process whereby the split soul speaking two languages,—the symptomatic language of the repressed id and the rational, social language of the ego,—learns to make a translation of these tongues and thus becomes able to regain unity by the communication of its parts. Vd. *Dialectic*, New York, 1927: pp. 107–125. The maxim of the Greek wise men, "Know thyself", is the guiding rule of psychotherapy.

But such knowledge is not, by itself, the power to live well, to attain happiness. The psychoanalyst's therapy, therefore, seems to me predominantly negative. He has merely cleared the ground of obstacles to moral training, which is never merely an acquirement of knowledge, but always a right ordering of loves. He has done nothing positively to institute that training. While it may be true that a moralist would fail with neurotics because of the heavy burden of repression he could not cope with, the lightening of this burden by psychoanalysis can do no more than to make the individual more susceptible to moral training. Unfortunately, it seldom does even that. Vd. Note 62 *infra*.

A simple analogy may make clear the relation of psychoanalytic therapy to moral training. Consider the case of a paralyzed man who wishes to perform well gymnastically. His trainer might be prevented from developing these skills in him, or at least impeded, by the physical handicap he suffers. A surgeon is able to remove the obstacle of paralysis. When the paralysis is cured, the patient is much better able to learn what the gymnastic trainer can teach him. But the surgical cure of the paralysis is not in and by itself the same as a course of gymnastic training. The surgery as a negative step is here instrumental to the positive training. Psychoanalytic therapy is a kind of surgery, removing repression and curing neurosis,—the obstacles to moral cultivation. But the neurotic who is seeking to live well needs the positive help of a moralist, just as the paralytic needs the gymnast as well as the surgeon. Granted that repressions have kept the neurotic from being morally mature, their removal will not magically transform the infantile character into an adult. A child must be morally habituated. So must the psychoanalyzed neurotic if he is to grow up.

Human happiness depends upon the moral virtues, the rectification of the appetites through ordination to their due end, the moderation of the passions through prudence. The passions, after they cease to be submerged by repression and clandestine in their motions, must be positively disciplined. They must be habituated in obedience to reason. And reason itself must be disciplined in the habit of applying knowledge to the affairs of action and emotion, of making choices deliberately and in the light of counsel. The virtue of prudence must be formed,—prudence which is the operative habit of practical reason tending to appoint a mean between the extremes of too much and too little, and taking from the will its power to command.

Psychoanalysis may, as I have said, make a person, who would otherwise be intractable, susceptible to moral training. But the art of moral training exceeds its powers, and this is the art which positively aims to make men happy. It is the possession of those who have learned well the science of ethics, who not only know the principles and rules of human welfare but appreciate the inexactitude of these rules in practice, their relativity to particular cases. It is one thing to know the practical science of ethics academically. It is another to use it practically, in the guidance of one's own life or in counseling and training others. The difficult art of casuistry must be mastered. It cannot be learned in a classroom, but only through much experience. Unless the psychoanalyst is a sound moralist and practically wise, unless he understands the fundamental principles of ethics and is an accomplished casuist, he can cure men negatively but he cannot make them happy. It is obvious that if the psychoanalyst has these additional virtues and accomplishments, he does not have them as a psychoanalyst but as a man of wisdom and experience. Though the combination is rare, it is not impossible, unless the psychoanalyst makes it impossible by denying morality itself. Vd. Note 62 *infra*. Two things must, therefore, be said of psychoanalytical therapy: (1) that it is far from being indispensable; it is necessary, perhaps, only in those pathological cases in which special technique is required to prepare the ground; and (2) that even when it is necessary, it never by itself achieves the end of making men virtuous and happy. That is the work of morality. Without it, the accomplishment of psychoanalysis must be an empty, or at best, a partial gain. If, through psychoanalysis, the patient renounces morality or "transcends" it, the result is worse than a frustration. It is pernicious.

62. The preceding discussion (Note 61 *supra*) indicates that the final evaluation of psychoanalytic therapy depends upon the relation of psychoanalytic theory to morality. Sound moral theory recognizes the distinction between those principles which obtain universally in the direction of human life,—the analysis of its ultimate end and proximate means, the classification of the goods and the determination of their order, the definition of the virtues and their relation to pleasure and pain,—and the conventional determinations of these principles which constitute the customary regulations and the prevailing sentiments of a given people at a given time. One set of customs may not be intrinsically better than another, but all conventions are not indifferent in moral rectitude. In the light of principles a whole culture can be judged as better or worse, according as its conventions are more enlightened by reason, according as its ruling sentiments are sounder evaluations of fundamental goods. This judgment is not, as the contemporary sociologist would say, vitiated by the ethnocentric predicament of the moralist making it. He is not making it according to the *mores*,—the customs and sentiments,—of his tribe, but in the light of principles that are universal because men are everywhere men, and human society is everywhere essentially the same.

But the psychoanalyst claims to investigate morality as the natural result of the functional relationship between the individual and his environment. This in itself is not wrong, because morality has a natural basis in man's specific nature operating under natural environmental conditions. In fact, it is the psychological understanding of the fact of morality which shows us that its first principles are not merely conventional, not purely arbitrary, but rather intrinsically natural. The Freudian, however, has adopted the point of view of sociology and anthropology since the days of William Graham Sumner, the view that morality can be completely equated with the *mores*, that it is *nothing but* the prevailing sentiments and the ruling customs of a tribe or society at a given time and place. This is utter relativism. The reduction of morality to the *mores*, in the current understanding of the ethnocentric predicament, is an absolute denial of morality. The *Ethics* of Aristotle is nothing but the rationalization of Greek *mores* in the 4th century B.C.; as morality it is neither better nor worse than the practices of Hottentots or contemporary

Germans. It is superior only in having received a more cultivated literary expression or a subtler rationalization.

I wish to emphasize the fact that the psychoanalysts did not formulate this doctrine in the light of their own clinical data. They borrowed it from anthropology, as they made levies upon evolutionary biology for their genetic theory. Nor did the anthropologist or sociologist learn from their data that morality is entirely relative, entirely a matter of arbitrary *mores*. What their data showed them, –if they interpreted them properly,–was that *mores* differ from tribe to tribe and from place to place. The *mores* are the set of practices and values which prevail. So much the data can establish and no more. The next step is to understand the *mores* in the light of the principles of morality which are common to all men. The obvious fact that the *mores* are not everywhere absolutely different, but rather that they differ in details within a frame of similarities, should teach an open mind that the *mores* are conventional determinations of principles which are natural and common to all men and all cultures. The investigative scientist could not have made this interpretation, of course, if he did not understand the scope and character of the practical science of ethics, if he did not understand the relation between nature and convention, if he did know the difference between principles and casuistically determined rules. In short, unless he were a competent philosopher he had no competence to interpret his data beyond the simple fact they obviously established, the fact of the variety of conventions. But the anthropologist and the sociologist, like their brothers in the natural sciences, were not satisfied with being investigative scientists. The urge to "philosophize" could not be resisted, and so we have the false doctrine of the *mores*, which is nothing but a misrepresentation of the facts. Just as the physicist when he turns philosopher soon denies that there is any philosophical knowledge, and especially metaphysics, so the social scientist when he turns philosopher repudiates morality, denies that there is any practical philosophy that has a truth above and apart from the acknowledgment of custom.

This is the psychoanalyst's doctrine about morality. As a scientist he need not, in fact he should not, be a moralist. As a psychotherapist, he may perform the negative task of relieving repression without recourse to moral principles. But if he aims further to make the whole man healthy, in the full Greek sense of making him happy, then his denial of morality is serious. The reality principle is insufficient if it merely proclaims the reality of the customs and sentiments which prevail in the community in which the patient lives and to which he must conform in part. The full reality of those customs and sentiments is their natural basis, their rational source in the principles of human morality. Happiness does not consist in conformity to the *mores*. Such blind submission is the docility of a slave, the pusillanimity of a domesticated animal. The *mores* themselves must be judged. They are not the standards of judgment. Rather they are measured by conformity to reason and the practical principles it dictates. The reality principle, then, must be the rule of reason in human life. Morality transcends the *mores*. If the psychoanalyst denies this, the patient he cures is, at best, a "socialized beast", whose anti-social tendencies have become domesticated, or, at worst, he is a thoroughly vicious man who subverts reason to the degraded task of *rationalizing* his appetites and passions, and who uses his intelligence, not to seek the good and avoid evil, but only to avoid the painful consequences of transgression of society's laws. The philosopher, said Aristotle, does from wisdom and virtue what other men do from fear of the law. The maxim applies not only to the philosopher but to every good man. He who acts *only* through fear of the law is a bad man aided by cunning and abetted by all the contrivances of expediency.

I have tried to show how psychoanalytic therapy might be of some instrumental value in the larger task of making men good and happy. Even if the service is only negative, it may be praiseworthy as an aid, perhaps an indispensable one in some cases. But if, as appears to be the fact, psychoanalytic therapy is practised in the light of a false doctrine about morals, it is condemnable as frustrating its own potentialities. The psychoanalyzed patient is cured by being taught the errors of the psychoanalyst. He learns to psychoanalyze himself, which means that he learns to take a purely "scientific attitude" toward all moral problems. He becomes objective. He is at all points a student of life rather than a practitioner involved in its perplexities. The error here is the supposition that to be scientific entails a denial of norms, that to be objective means that knowledge by itself can decide practical questions. The fact of norms is the cardinal fact that there is good and evil in the world, which the human appetites must be trained to seek or avert. Knowledge is not enough, even when it is "scientific". Knowledge must be converted into moral principles by the acknowledgment of an objective difference between good and evil. This is a true objectivity, beside which the "scientific attitude" of the psychoanalyst and his patient is the worse kind of moral solipsism, which converts everything into a matter of personal or tribal taste.

But the root error is deeper. It is the prevalent positivism to which psychoanalysis subscribes. Just as, on the theoretical side, the psychoanalyst stumbles because he denies philosophy and then proceeds to philosophize beyond the scientific significance of his data; so on the practical side, he blunders when he denies morality and then imposes his a-morality upon the patient. In both cases he has exceeded the task for which he is disciplined. In both cases he has proclaimed that science is enough, only to reveal the error of this dogma by self-contradiction,—by going beyond science precisely because it is not enough. Of the two mistakes, the latter is clearly more important. In the first instance, he only fools himself. In the second, he leads other men astray.

EPILOGUE

I WROTE the foregoing notes as a commentary after I had re-read the text of the lectures. I have now re-read the notes and feel impelled to add a brief comment on them. The discussion in the notes engages, at many places, in intellectual history, not for its own sake but for the purpose of discriminating the true from the false. Historical positions have not been simply reported. They have been judged. But the fact that the positions have been recognized as belonging to this or that moment of the historical sequence must mean that intellectual *history* contributes something to the understanding of philosophical issues. This does not mean that they cannot be stated entirely apart from proper names and dates. But the history of the natural sciences is much less relevant, much less useful, to the understanding of what they present today as knowledge, than the history of philosophy is to the understanding of what a contemporary philosopher holds to be the case. Thinkers of a Platonic temperament do not find the history of philosophy embarrassing. It exhibits a continuous discussion of the few basic ideas which, in every age, are implicated in human discourse. Thus, Whitehead reports intellectual history as the adventures of ideas, their incidence and transformation according to the conditioning circumstances of different epochs. And an Hegelian describes these transformations as a dialectical development which marches along the straight line of progress to culmination in his own philosophy. But an Aristotelian finds the history of philosophy embarrassing precisely because he conceives philosophy as *knowledge*, and not as progressively enlightened controversy. Far from seeing in the sequence of historical positions a closer and closer approximation of the truth, he discerns error on all sides. He must face the question, therefore, why, if philosophy is knowledge obtained by reason from common experience, should most philosophers have been unable to speak without error, error often grievous though never unmitigated by some truth. This is the question I feel called upon to answer because of the historical portions of my notes.

In the first place, the obvious difference between the history of science and the history of philosophy must be noted. The line of progress in science is undeviating. In each successive period, the state of scientific knowledge is an improvement on what went before. There is both more and better knowledge. The history of science can be reported as a correction of errors and inadequacies, but once the correction is made, the advance is consolidated. There is no back-sliding, no atavism. A scientific error always occurs at an earlier time than the knowledge which corrects it. This is not true of the history of philosophy, which moves forward in spiral rather than rectilinear fashion. Aristotle corrected the errors of Plato and the pre-Socratics only to have these same errors recur throughout the middle ages, especially the neo-Platonic reversions. Aquinas corrected the neo-Platonism which dominated most of mediaeval thought only to have modern philosophy return to these same errors, as well as to materialism and other positions taken by the pre-Socratics. In this view of the history of philosophy, atavisms are the rule and advances are exceptions. But the fact, however startling, that error and its correction are not related in the history of philosophy as they are in the history of science, does not show that philosophy is not knowledge. On the contrary, it shows that philosophy is knowledge of a different sort from science. There is not, on the whole, more error in philosophy than science, but the way in which it occurs, and recurs, is different. This difference is to be explained by the difference between science and philosophy as distinct types of knowledge employing different methods. Science, concerned with the phenomenal order, is investigative and inductive. Its regular advance is due to the progressive exploration of the sensible world and to the new fields of data discovered by experimental ingenuity, under the impetus of technology and invention. Philosophy, concerned with ontological problems, is inductive and dialectical. If philosophy were merely inductive, it would soon achieve a dead stability. The same is true of science. Without the continual increment of data achieved by investigation, induction would soon exhaust the fertility of what data there were. The role which investigation plays in science, dialectic plays in philosophy. The great philosophers do not disagree about the data of common experience or even about the primary inductions which these yield. But the movement of philosophy is from these data and inductions upward to the principles of highest generality which are the sources of intelligibility, and then from these principles downward in more and more articulated analyses to illuminate the concrete. This is the upward and the downward way of dialectic as Plato describes it in the *Republic*. This is the technique peculiar to philosophy, as investigation is to science. Unlike some modern Platonists who rely upon portions of the seventh epistle, I hold that this technique does not make philosophy merely dialectical, that Plato as well as Aristotle valued this technique as an instrument for getting valid knowledge and not merely as an instrument of controversy or the clarification of discourse. Plato used "dialectic" to name both the philosophical process and the product it achieves,— the wisdom that Aristotle called *philosophia prima*. If I stress dialectic as method, it is because it is necessary to make clear that the method of philosophy is not demonstrative in the sense in which modern mathematics is; it is not linear inference, deducing theorems from postulated premises. Finally, it should be added that the dialectical method of philosophy is a completely natural process of knowing; whereas the investigative method of science is technical, involving many auxiliary arts.

Dialectic is the source both of advances in philosophical knowledge and of the multiple repetition of errors. It is a subtle technique, mastered only by a few. Although experimental ingenuity and inductive genius may be equally rare in science, division of labor in the scientific enterprise makes it possible for a multitude of drones to perform useful tasks which are within the competence of many. Science is always productively at work, even in the absence of great scientists. This is not true of philosophy. Here the many, lacking either great or distinctive dialectical skill, do little more than repeat old errors or old truths. And the errors are easier to repeat than the truths. Being the product of inferior dialectic, they are naturally more within the competence of the many. This partly explains why the errors of Platonism and materialism, originating in the ancient world, have been so often repeated in the mediaeval and the modern epochs; it also explains why the dialectic of St. Thomas improves upon, rather than merely repeats, Aristotelian principles and analysis. If Aristotle in his day was the master of those that know, St. Thomas, mastering Aristotle, was an even greater master in his day.

Philosophical truth does not stand to philosophical error as white to black, if one considers a whole philosophy and not merely its constituent propositions. A true and a false philosophical proposition are related as contradictory, but a true philosophy must be compared with a false one in terms of dialectical criteria. The measures of truth in a whole philosophy are its intelligibility, its coherence and its adequacy, as well as the truth of its propositions. One philosophy is better than another as containing more truth and more wisdom, better articulated. It is by such criteria that I have judged the superiority of Aristotle to Plato, and the superiority of St. Thomas to the neo-Platonic philosophies of the middle ages and modern times, not to mention materialism and the various confusions of doctrine which proliferate in all ages.

The history of philosophy in modern times is complicated, first, by psychologism, and second, by the rise of positivism. When the question, How do we know? is made to precede all other philosophical analysis, metaphysics and the philosophy of nature suffer from the priority of psychology, and it in turn suffers from occupying the improper position of "first philosophy"; but worse than that, philosophers soon become occupied with questions about philosophical method as prior to all other philosophical questions. This ultimately leads to positivism and the denial that philosophy is knowledge of the real, independent of science. However sharp the contrast between the history of science in modern times and the history of philosophy in the ancient and mediaeval periods, it is even sharper if the comparison be made with the history of philosophy in the era of science. Not only is there no progress in philosophy, but modern philosophers do not seem able to agree among themselves about their problems and methods. This is rarely the case among scientists; rarer, or even unthinkable, would be the case of a scientist who denied that science was knowledge, having separate problems and an independent method. Yet that is precisely the position of those paradoxical creatures, the philosophical positivists,—philosophers denying or, at least, abandoning philosophy. There is reason, therefore, for a philosopher to feel most embarrassed about the history of modern philosophy. He is confronted not only by the return to ancient and mediaeval errors, usually in a form that is less clear or significant, but also by the apparent loss of philosophy itself, or, at least, its degradation in the cultural nexus.

This last point brings me to another aspect of the notes which calls for comment. At many points explicitly, and generally in tone, they *seem* to condemn the whole of modern times. I say "seem" because the condemnation is really restricted to what is condemnable: not the whole of modern times, but the errors and misdirections which have unfortunately obscured and marred the genuine advances made in this period. As Maritain says, "to denounce a fundamental spiritual deviation in a period of culture is not to condemn that period. . . . An error in spiritual principle bears its inevitable fruit. We must expose the principle and avow the loss. During the same period there is an evolution in human affairs, an expansion of history; there are conjoined to certain evils, gains and achievements of mankind that have an almost sacred value" (*Freedom in the Modern World*, New York, 1936: pp. 84–85). The spiritual deviation of modern times is in philosophy and it is an evil that seems to have been necessarily conjoined with the gain and achievement which is science. We have already seen that positivism made a philosophical advance in so far as, beginning with Hume and Kant, it formulated a clear conception of the nature of science. So far positivism must be praised. It is to be condemned only for its negations and exclusions which originate in the psychologism of modern philosophy. Similarly, science must be praised without stint for its magnificent contribution to human culture, even though the advent of science in the household of human knowledge almost dispossessed the older children, philosophy and theology. Again, as Maritain says, "if the loss or lowering of the metaphysical spirit is an incalculable damage for the general order of intelligence and human affairs, nevertheless, the predominance of the metaphysical spirit, when it is not accompanied by exceptionally vigorous critical rectifications, serves as an accidental hindrance to the particular interests of experimental research. And this accidental hindrance is a heavy cost, for experimental research, and the least advance in the least truth of fact, is also a thing of the spirit, and the spirit does not tolerate any hindrance" (*Les Degrés du Savoir*, Paris, 1932: pp. 118–119).

In other words, if we take a more comprehensive view of intellectual history, in which philosophy is only a part, we shall see modern times in a better light, and in the altered perspective the career of philosophy in the epoch of science may also be better understood. We may even discern the promise of a new *Aufklärung*. To be brief, I shall report this more comprehensive view in a story. The characters are Plato (and his retainers), Aristotle (and his), Philosophy, Theology (and Religion), Science (and Technology). There are three episodes.

(1) In the ancient world, Philosophy, Theology and Science were together as relatively undifferentiated parts of Human Knowledge. Yet their different personalities can be dimly seen if we note three moments in the development of Human Knowledge in the ancient world. It is as if Human Knowledge appeared upon the stage successively wearing three masks: the pre-Socratic mask is the face of Science turned towards the changing world; the Platonic mask turns the eyes of knowledge upward theologically; the Aristotelian mask bears the features of Philosophy not yet perfectly composed, because the line of its vision shifts now in the direction

of Science and now in the direction of Theology. It is in its Aristotelian personality that Human Knowledge was most mature in antiquity. Although real differentiation of characters has not yet occurred, the Aristotelian moment is the one in which the eyes of Philosophy have at least distinguished three directions in which Knowledge can turn. That is the Aristotelian achievement. In a sense, it is the birth of Philosophy.

(2) In the mediaeval world,—a world made different by the Christian revelation,—Religion enters as a separate character. Human Knowledge is deeply moved by the part it must play, a role in which hospitality to this newcomer is mixed with strange fears. At first, and almost for a thousand years, Human Knowledge plays this part wearing the Platonic mask, but there are momentary, almost furtive, shifts of expression. Philosophy, having achieved some sense of its separate character in the ancient world, can no longer completely submerge itself in the personality of Theology. If it continues for many centuries to wear the Platonic mask, it is because that is most suited to expressions of hospitality toward Religion. The fears with which this geniality is mixed betray themselves in nominalistic caricatures of the face of Science, subtle reversions to a pre-Socratic mood. It is not until Aristotle is remembered and resurrected that this conflict of emotions is resolved. The happy climax is reached in the 13th century when Aquinas effects the full separation of Philosophy from Theology. Human Knowledge is still ambiguously personified as Philosophy and Science, but it is in the now clear character of Philosophy that Human Knowledge is self-conscious of its distinction from another sort of knowledge which is the heart of Religion. Far from being tragic, this sharp differentiation enables Philosophy and Theology to understand each other as never before, and out of that understanding is born their compatibility, their harmonious union. This is the Thomistic achievement. It is more than an echo of the Aristotelian moment of antiquity. It is a genuine advance toward the greater maturity of Human Knowledge, an advance in which Philosophy passes out of its infancy.

(3) In the modern world, Technology is the innovator. Again Human Knowledge is profoundly stirred; this time by all the promises which Technology makes. The gifts with which Religion came are not forgotten entirely, but Technology offers a new world to conquer. The dowry it brings is an obvious one, as measurable and assessable as the things of this world. Human Knowledge is torn between what seems to be an assured happiness and the bright novelties which make Technology alluring. The strain

is too much. There is a gradual alienation of affection. At first Knowledge avoids a real breach by being deceptive. When it stays at home with Theology it is clearly philosophical in character, but when it goes out into the world with Technology it puts on the mask of Science. What is at first only a mask becomes through much worldliness a more and more distinct character. It is now Philosophy which is torn between Science and Theology. And extraordinary things being to happen. Philosophy finds it easier to get along with both Theology and Science by wearing the Platonic mask which twice before it had discarded. It tries to be all things at once. But the difficulties increase as Science and Theology move further apart. Human Knowledge in its Platonic personality aids in establishing the autonomy of Science, as in the early Christian period it made way for Theology. Under these circumstances, Philosophy finds it increasingly difficult to retain its own character. It tries so hard to understand Science that it almost ends in utter self-abnegation and in fitful forgetfulness of Theology. The Platonic mask, which in the ancient world reflected theological light, assumes the self-effacing features of positivism looking only at Science.

The birth of Science is the achievement of the modern world. But unless one believes that Human Knowledge *is* Science and that Theology and Philosophy are immature expressions which it has outgrown, the historical drama is not yet concluded. The third act will not come to its happy ending until it reaches another Aristotelian moment in which Philosophy will attain full maturity, separate not only from Theology but from Science. Just as in the Thomistic synthesis, Philosophy and Theology became well-related through being well-separated, so in the synthesis yet to come Science and Theology will be united through being divided by Philosophy, itself in clear distinction from them both. If I am right in the prophecy of this next turn, the story is the tragi-comedy of human progress. There has been and will be a progressive differentiation of Human Knowledge. Theology, Philosophy and Science will be better related as three distinct characters than as three aspects of one. Religion and Technology will assume their proper places, together though differently encouraging this triumvirate of Knowledge. And Philosophy, which must play the central role because of its relations to Science and Theology, will have advanced to a better understanding of itself precisely because in our times it has almost lost itself in understanding Science, as once before it almost did in serving Theology.

The intellectual history of western Europe thus told is not only the story of a general cultural advance, but also the story of the progress of philosophy, in which Aristotle is the hero and Plato the somewhat benign villain. It was through correcting pre-Socratic and Platonic errors that Aristotle won the first fruits of *purely* philosophic wisdom. It was through correcting the many times multiplied neo-Platonic errors,—much more subtle than the original ones because colored by misunderstandings of Aristotle—that St. Thomas ripened these fruits. As heretics are the indispensable causes of a sounder orthodoxy in theology, so the increasing subtlety of philosophical error is necessary for the deepening of wisdom. St. Thomas was able to improve upon Aristotle, not only because he understood the master so well, but also because his understanding was magnified by the mastery of errors which were unknown to Aristotle. The neo-Platonists of the middle ages were more subtle adversaries for St. Thomas than Plato was for Aristotle. But St. Thomas did not finish the work of philosophy. Because of the rise of science in modern times, because of another tide of neo-Platonism,—this time complicated by a vigorous ret ırn to the materialism of the pre-Socratics and by the denials of positivism,—further philosophical tasks remain to be done. The fruits of wisdom can be brought to even fuller bloom in a third *Aufklärung*, provoked by the errors of modern times,—a consummation which may be achieved when men realize that Technology has failed to bring peace on earth. Their wisdom will turn upward again to the theological virtues instead of down to science and the industrial arts. These matters are too contemporary to permit an assured judgment, but I discern in Maritain's *Les Degrés du Savoir* the outlines, at least, of a synthesis of science, philosophy and theology which will do for us what St. Thomas did for philosophy and theology in the middle ages. In that synthesis, philosophical wisdom will be enriched. The understanding of science will make for an improved Thomism, as the understanding of theology enabled St. Thomas to improve upon Aristotle. Whether or not Maritain has accomplished this or merely forecasts it, he seems to me the only contemporary philosopher who has deeply sensed the movement of history and the point at which we stand. More than that, he has divined the principle of intellectual progress. It is a maxim which can be used as the title for the whole historical drama: *distinguer pour unir*.

If one turns from prophecy for the moment to survey the contemporary scene, one is forced to realize that the whole story is a tragi-comedy and that the tragic accent dominates the present. Positivism, which reached its high point in the 19th century, is still ascendent in many quarters. Marxian materialism has not yet become truly dialectical materialism. With its affinities for mathematical physics and relational logic, on the one hand, and for Theology, on the other, neo-Platonism in many subtle varieties is still attempting to fuse science and religion. Formal materialism, though it is renewing its ancient and mediaeval vigor, has still many obstacles to overcome and victories to win if it is to produce a modern *Aufklärung*. It must assimilate Marxism, by showing itself to be the true dialectical materialism which Marx failed to derive from Hegel. It must do justice to the insight of positivism with respect to science, but avoid its blindness with respect to theology. This it can do by proper discrimination of the kinds of knowledge, converting Hume's denials about philosophy into affirmations of the limits of science, and Kant's antinomies and paralogisms into indications of the philosophical mysteries which mark the boundary between reason and faith. And, finally, formal materialism, today as ever, must strengthen itself against Platonic weaknesses,—particularly in regard to the relation of matter and form in the doctrines of substance and abstraction,—weaknesses which appear as soon as dialectical vigor is the least relaxed.

In facing all of these tasks, in all that remains to be done, the position of psychology seems to me crucial because philosophy cannot be re-established in its separate autonomy until peculiarly modern epistemological errors are corrected. We have seen how complicated is the network of reciprocal influences here,—how Platonism and psychologism have converged, how both have led to positivism, and how that in turn has prevented the science of psychology from reaching intellectual maturity. The cure of Platonism, the cure of positivism, both depend upon the rectification of psychology, as a science and in philosophy. But that also means that psychologists must be cured of Platonism and positivism. Psychology, therefore, seems to me that part of the field of science on which philosophy must win its decisive victory. Contemporary psychology,—particularly the doctrines which proclaim themselves as scientific,—is the source of much disorder in the practical world: in education, in morals, in religion, and even in economics and politics. That this is so follows necessarily from the relation of psychology to all practical problems; it is the theoretic root of the moral disciplines. The view which one takes of human nature determines how one lives as a man, alone and with others. The modern view of man, produced by the divorce of psychology from philosophy and its wedding to science, brings man to a lower state than he fell to from grace. The nobility with which man walked as a rational animal in Greece, the spirituality of man as a person in the middle ages,—these have been lost or obscured. And in their place is the bare objectivity of the man whose nature has been equated to the limited devices of the laboratory and the clinic. The position of psychology in modern culture is crucial both in the practical and the speculative dimension. I have no doubt that it will be rectified and that man will again rise to proper self-esteem, because I have no doubt that philosophy will reach a new fruition, healing thereby the disharmonies and blemishes of modern times. But this is prophecy, and if the prophet turn his eyes from the future to the present, the note of hope in his voice must give way to the tone of a Jeremiad.

> "If this belief from heaven be sent,
> If such be Nature's holy plan,
> Have I not reason to lament
> What Man has made of Man?"

LIST OF PRINCIPAL NOTES

[1] Schools of Psychology and Controversy in Other Sciences 124

[2] Formal and Material Logic 125

[3] Materialism and Cognition 127

[4] Psychology and Cognition 128

[5] Common and Special Experience 129

[6] Common Experience and the Status of Philosophy 131

[6a] The Different Utilities of Philosophy and Science 136

[6b] The Positive Distinction between Philosophy and Science 137

[7] Conflicts Between Philosophy and Science 139

[8] Infra-intelligible and Suprasensible Objects 142

[8a] Questions of Existence in History, Science and Philosophy 143

[9] The Divisions of Philosophy: Grades of Abstraction 144

[10] The Proper Sphere of Investigation 150

[10a] Relations Between Philosophy of Nature and Natural Science 151

[11] Mathematical Physics as a Mixed Science 153

[12] Applicative and Inductive Use of Experiment 154

[13] Science and Causality 155

[16] Modern Philosophy Before and After Kant 156

[16a] The Positive Contribution of Positivism, and its Negativism 158

[17] The Fact of Understanding and Logical Positivism 161

[18] Imagination and Intellect 162

[19] Dependence of Understanding on Sense and Body 162

[20] The Word "Cosmology" 164

[20b] Monism, Dualism and Modern Philosophy 166

[21] Dialectical Materialism: Aristotle, Aquinas, Marx 167

[22] The Basic Principle of Realism, and Idealistic Errors 172

[23] Hierarchy of Substances and the Order of the Sciences 176

[24] Human and Animal Psychology 177

[25] Philosophy of Science and Philosophy of Nature 177

[28] Marxian and Formal Materialism 180

[28a] The Significance of Human History for Materialism 180

[29] Interior Sense in Animals and Man 181

[30] The Utility of the Intellect 182

[31] Continuity in Nature 183

[32] Scientific and Philosophical Accounts of Rationality 184

[32a] The Uniqueness of Psychology: The Order of its Questions 186

[33] Distribution of Psychological Topics in the Ancient and Mediaeval Tradition 188

[35] Virtuality and the Plurality of Forms 190

[36b] Positivism and the Histories of Psychology 191

[36c] The Subject-Matter of the *De Anima* 192

[38a] Aristotle and Aquinas as Aristotelians 194

[39a] Kantian and post-Kantian Psychology 194

[39b] Freudian and Thomistic Critiques of Modern Psychological Subject-Matter 196

[39c] Rational and Empirical Psychology: Wolff, Kant and thereafter 196

[43] Gestalt and Behavioristic Psychology 200

[45] The Inductive Fertility of Psychometrics and Psychoanalysis 201

[45a] The Subject-Matter of Scientific Psychology 203

[45b] The Significance of the History of Psychiatry 203

[46] The Doctrine of Faculties in Psychometrics and Psychoanalysis 205

[47] The Different Logics of Science and Philosophy in Relation to their Different Problems 206

[48] The Meaning of "Hypothesis" 211

[50] Psychoanalysis and Aetiology 212

[51] The Metapsychology of Freud 213

[52] Topography of the Psyche: Reification of Powers 215

[54] Need and Desire: Kinds of Appetite 217

[55] The Conscious, the Unconscious, and Repression 218

[56] The Nature of the Translation between Freud and Aristotle 224

[57] The Relation of Reason and Will to the Passions 225

[58] Reciprocal Independence of Reason and the Passions 226

[59] "Evolutionary Philosophy" 227

[60] Freudian Genetics 227

[61] Psychoanalytic Therapy and Moral Training 228

[62] Psychoanalysis and the Theory of Morality 231

AF828800

APPLIED BIOPOLYMER TECHNOLOGY AND BIOPLASTICS

Sustainable Development by Green Engineering Materials

APPLIED BIOPOLYMER TECHNOLOGY AND BIOPLASTICS

Sustainable Development by Green Engineering Materials

Edited by

Neha Kanwar Rawat, PhD

Tatiana G. Volova, DSc

A. K. Haghi, PhD

AAP | APPLE ACADEMIC PRESS

First edition published 2021

Apple Academic Press Inc.
1265 Goldenrod Circle, NE,
Palm Bay, FL 32905 USA

4164 Lakeshore Road, Burlington,
ON, L7L 1A4 Canada

CRC Press
6000 Broken Sound Parkway NW,
Suite 300, Boca Raton, FL 33487-2742 USA

2 Park Square, Milton Park,
Abingdon, Oxon, OX14 4RN UK

© 2021 Apple Academic Press, Inc.

Apple Academic Press exclusively co-publishes with CRC Press, an imprint of Taylor & Francis Group, LLC

Reasonable efforts have been made to publish reliable data and information, but the authors, editors, and publisher cannot assume responsibility for the validity of all materials or the consequences of their use. The authors, editors, and publishers have attempted to trace the copyright holders of all material reproduced in this publication and apologize to copyright holders if permission to publish in this form has not been obtained. If any copyright material has not been acknowledged, please write and let us know so we may rectify in any future reprint.

Except as permitted under U.S. Copyright Law, no part of this book may be reprinted, reproduced, transmitted, or utilized in any form by any electronic, mechanical, or other means, now known or hereafter invented, including photocopying, microfilming, and recording, or in any information storage or retrieval system, without written permission from the publishers.

For permission to photocopy or use material electronically from this work, access www.copyright.com or contact the Copyright Clearance Center, Inc. (CCC), 222 Rosewood Drive, Danvers, MA 01923, 978-750-8400. For works that are not available on CCC please contact mpkbookspermissions@tandf.co.uk

Trademark notice: Product or corporate names may be trademarks or registered trademarks and are used only for identification and explanation without intent to infringe.

Library and Archives Canada Cataloguing in Publication

Title: Applied biopolymer technology and bioplastics : sustainable development by green engineering materials / edited by Neha Kanwar Rawat, PhD, Tatiana G. Volova, DSc, A.K. Haghi, PhD.

Names: Rawat, Neha Kanwar, editor. | Volova, Tatiana G., editor. | Haghi, A. K., editor.

Description: Includes bibliographical references and index.

Identifiers: Canadiana (print) 20200363255 | Canadiana (ebook) 20200363328 | ISBN 9781771889216 (hardcover) | ISBN 9781003045458 (ebook)

Subjects: LCSH: Biopolymers. | LCSH: Biodegradable plastics. | LCSH: Green technology. | LCSH: Sustainable engineering.

Classification: LCC TP248.65.P62 A67 2021 | DDC 572/.33—dc23

Library of Congress Cataloging-in-Publication Data

CIP data on file with US Library of Congress

ISBN: 978-1-77188-921-6 (hbk)
ISBN: 978-1-00304-545-8 (ebk)

About the Editors

Neha Kanwar Rawat, PhD
Researcher, Materials Science Division,
CSIR – National Aerospace Laboratories, Bangalore, India

Neha Kanwar Rawat, PhD, is a recipient of a prestigious DST Young Scientist Postdoctoral Fellowship and is presently a researcher in the Materials Science Division, CSIR – National Aerospace Laboratories, Bangalore, India. She received her PhD in chemistry from Jamia Millia Islamia (A Central University), India. Her main interests include nanotechnology-nanostructured materials synthesis and characterization, her main focus comprising green chemistry, novel sustainable chemical processing of nano-conducting polymers/nanocomposites, conducting films, ceramics, silicones, and matrices: epoxies, alkyds, polyurethanes, etc. She also pursues her interest in using new technology in areas that include electrochemistry, organic-inorganic hybrid nanocomposites, and protective surface coatings for corrosion inhibition and MW shielding materials. She has published numerous peer-reviewed research articles in journals of high repute. Her contributions have led to many chapters in international books published with the Royal Society of Chemistry, Wiley, Elsevier, Apple Academic Press, Nova U.S., and many others in progress. She has been working on many prestigious research and academic fellowships in her career. She is a member of many groups, including the Royal Society of Chemistry and the American Chemical Society (USA) and a life member of the Asian Polymer Association.

Tatiana G. Volova, DSc
Professor and Head, Department of Biotechnology,
Siberian Federal University, Krasnoyarsk, Russia

Tatiana G. Volova, DSc, is a Professor and Head in the Department of Biotechnology at Siberian Federal University, Krasnoyarsk, Russia. She is the creator and head of the Laboratory of Chemoautotrophic Biosynthesis at the Institute of Biophysics, Siberian Branch of the Russian Academy of

Sciences. Professor Volova is conducting research in the field of physico-chemical biology and biotechnology and is a well-known expert in the field of microbial physiology and biotechnology. Dr. Volova has created and developed a new and original branch in chemoautotrophic biosynthesis, in which the two main directions of the 21st century technologies are conjugate, hydrogen energy, and biotechnology. Under the guidance of Professor Volova, the pilot production facility of single-cell protein, utilizing hydrogen had been created and put into operation. The possibility of involvement of man-made sources of hydrogen into biotechnological processes as a substrate, including synthesis gas from brown coals and vegetable wastes, was demonstrated in her research. She had initiated and deployed in Russia comprehensive research on microbial degradable bioplastics; the results of this research cover various aspects of biosynthesis, metabolism, physiological role, structure, and properties of these biopolymers and polyhydroxyalkanoates (PHAs), and have made a scientific basis for their biomedical applications and allowed them to be used for biomedical research. Professor Volova is the author of more than 300 scientific works, including monographs, inventions, and a series of textbooks for universities.

A. K. Haghi, PhD

Professor Emeritus of Engineering Sciences, Former Editor-in-Chief, International Journal of Chemoinformatics and Chemical Engineering and Polymers Research Journal; Member, Canadian Research and Development Center of Sciences and Culture

A. K. Haghi, PhD, is the author and editor of 200 books, as well as 1000 published papers in various journals and conference proceedings. Dr. Haghi has received several grants, consulted for a number of major corporations, and is a frequent speaker to national and international audiences. Since 1983, he served as a professor at several universities. He is former Editor-in-Chief of the *International Journal of Chemoinformatics and Chemical Engineering* and *Polymers Research Journal* and is on the editorial boards of many international journals. He is also a member of the Canadian Research and Development Center of Sciences and Cultures (CRDCSC), Montreal, Quebec, Canada.

Contents

Contributors .... *ix*

Abbreviations .... *xi*

Preface .... *xv*

1. **Processing Relation to Rheological Behavior of Some Systems Containing Green Polymers** **1**
Andreea Irina Barzic

2. **Properties of the Magnetic Polymer Nanocomposites in Magnetic Fields** **23**
Jimsher Aneli and Grigor Mamniashvili

3. **Microbial Fuel Cell: A Green Microbial Approach for Production of Bioelectricity Using a Conducting Polymer** **33**
Swapnali A. Patil, Akshada A. Bakliwal, and Swati G. Talele

4. **Sustainability of the Tire Industry: Through a Material Approach** **53**
Sanjit Das, Hirak Satpathi, S. Roopa, and Saikat Das Gupta

5. **Silk: An Explorable Biopolymer in the Biomedical Arena** **99**
Athira John, Ananthu Prasad, and Neha Kanwar Rawat

6. **Influence of Nanotechnology on Prevention of Biofouling in Aquaculture Cage Nets: Green Conducting Polymers and Their Significance in Case Studies** **113**
P. Muhamed Ashraf

7. **Biodegradable PHAs: Promising "Green" Bioplastics and Possible Ways to Increase Their Availability** **127**
Tatiana Gr. Volova, Evgeniy G. Kiselev, Aleksey V. Demidenko, Svetlana V. Prudnikova, Ekaterina I. Shishatskaya, and Sabu Thomas

8. **Membrane Technology for Green Engineering** **155**
Supriya Dhume and Yogesh Chendake

9. **Green Hydrogen Energy: Storage in Carbon Nanomaterials and Polymers** **183**
Brahmananda Chakraborty and Gopal Sanyal

10. Self-Assemblies of Macromolecular Systems Containing Green Polymers 241
Andreea Irina Barzic, Raluca Marinica Albu, and Cristian Logigan

Index 259

Contributors

Raluca Marinica Albu
"Petru Poni" Institute of Macromolecular Chemistry, Laboratory of Physical Chemistry of Polymers, 41A Grigore Ghica Voda Alley, Iasi – 700487, Romania

Jimsher Aneli
Institute of Machine Mechanics, 10, Mindeli Str., Tbilisi – 0186, Georgia, E-mail: jimaneli@yahoo.com

P. Muhamed Ashraf
ICAR Central Institute of Fisheries Technology, Cochin – 682029, Kerala, India, E-mail: ashrafp2008@gmail.com

Akshada A. Bakliwal
Department of Pharmaceutics, Sandip Institute of Pharmaceutical Sciences, Nashik, Maharashtra, India

Andreea Irina Barzic
"Petru Poni" Institute of Macromolecular Chemistry, Laboratory of Physical Chemistry of Polymers, 41A Grigore Ghica Voda Alley, Iasi – 700487, Romania

Brahmananda Chakraborty
High Pressure and Synchrotron Radiation Physics Division,

Yogesh Chendake
Department of Chemical Engineering, Bharati Vidyapeeth (Deemed to be) University, College of Engineering, Pune, Maharashtra, India, E-mail: yjchendake@bvucoep.edu.in

Sanjit Das
Material Science Section, Hari Shankar Singhania Elastomer and Tyre Research Institute (HASETRI), Plot No 437, Hebbal Industrial Area, Mysore, Karnataka – 570016, India, E-mail: sanjit.das@jkmail.com

Aleksey V. Demidenko
Siberian Federal University, 79 Svobodnyi Av., Krasnoyarsk – 660041, Russia; Institute of Biophysics SB RAS, Federal Research Center "Krasnoyarsk Science Center SB RAS," 50/50 Akademgorodok, Krasnoyarsk – 660036, Russia

Supriya Dhume
Department of Chemical Engineering, Bharati Vidyapeeth (Deemed to be) University, College of Engineering, Pune, Maharashtra, India

Saikat Das Gupta
Material Science Section, Hari Shankar Singhania Elastomer and Tyre Research Institute (HASETRI), Plot No 437, Hebbal Industrial Area, Mysore, Karnataka – 570016, India, E-mail: saikat.dasgupta@jkmail.com

Athira John
Center for Biopolymer Science and Technology (CBPST), CIPET, Kochi, Kerala, India

Evgeniy G. Kiselev
Siberian Federal University, 79 Svobodnyi Av., Krasnoyarsk – 660041, Russia; Institute of Biophysics SB RAS, Federal Research Center "Krasnoyarsk Science Center SB RAS," 50/50 Akademgorodok, Krasnoyarsk – 660036, Russia

Cristian Logigan
Chemical Company SA, 14 Chemistry Bdv., Iasi – 700293, Romania

Grigor Mamniashvili
Andronikashvili Institute of Physics of Ivane Javakhishvili Tbilisi State University,
6, Tamarashvili Str., Tbilisi – 0177, Georgia

Swapnali A. Patil
Department of Pharmaceutics, Sandip Institute of Pharmaceutical Sciences, Nashik, Maharashtra, India

Ananthu Prasad
International and Inter-University Center for Nanoscience and Nanotechnology (IIUCNN),
Mahatma Gandhi University, Kottayam, Kerala, India

Neha Kanwar Rawat
CSIR – National Aerospace Laboratories Bangalore, Karnataka, India

S. Roopa
Department of Polymer Science and Technology, Sri Jayachamarajendra College of Engineering,
JSS Science and Technology University, Mysore – 570006, India, E-mail: roopasm2000@jssstuniv.in

Gopal Sanyal
Mechanical Metallurgy Division, Bhabha Atomic Research Center, Trombay, Mumbai – 400085, India

Hirak Satpathi
Material Science Section, Hari Shankar Singhania Elastomer and Tyre Research Institute (HASETRI),
Plot No 437, Hebbal Industrial Area, Mysore, Karnataka – 570016, India, E-mail: hirak@jkmail.com

Ekaterina I. Shishatskaya
Siberian Federal University, 79 Svobodnyi Av., Krasnoyarsk – 660041, Russia;
Institute of Biophysics SB RAS, Federal Research Center "Krasnoyarsk Science Center SB RAS,"
50/50 Akademgorodok, Krasnoyarsk – 660036, Russia

Swati G. Talele
Department of Pharmaceutics, Sandip Institute of Pharmaceutical Sciences, Nashik, Maharashtra, India,
E-mail: swatitalele77@gmail.com

Sabu Thomas
Siberian Federal University, 79 Svobodnyi Av., Krasnoyarsk – 660041, Russia;
International and Interuniversity Center for Nano Science and Nano Technology,
Mahatma Gandhi University, Kottayam, Kerala, India

Tatiana Gr. Volova
Siberian Federal University, 79 Svobodnyi Av., Krasnoyarsk – 660041, Russia;
Institute of Biophysics SB RAS, Federal Research Center "Krasnoyarsk Science Center SB RAS,"
50/50 Akademgorodok, Krasnoyarsk – 660036, Russia, E-mail: volova45@mail.ru

Abbreviations

ACL	anterior cruciate ligaments
AgNPs	silver nanoparticles
ASW	artificial/natural seawater
ATP	adenosine triphosphate
BLYP	Becke-Lee-Yang-Parr
BMIAc	1-butyl-3-methylimidazolium acetate
BmimCl	1-butyl-3-methylimidazolium chloride
BOD	biochemical oxygen demand
CC	climate change
CdO	cadmium oxide
CHT/SS	chitosan/silk sericin
CMC	critical micelle concentration
CMPs	conjugated microporous polymers
CMT	critical micelle temperature
CNCs	cellulose nanocrystals
CNF	cellulose nanofibers
CNHs	carbon nanohorns
CNTs	carbon nanotubes
CO_2	carbon dioxide
COFs	covalent organic frameworks
CS-g-PNIPAAm	chitosan-graft-poly(N-isopropyl acrylamide)
CTFs	covalent triazine frameworks
CVD	chemical vapor deposition
DADS	diallyl disulfide
DCMH	doxycycline monohydrate
DFT	density functional theory
DMF	N,N-dimethylformamide
DMSO	organic solvents-dimethyl sulfoxide
DoE	department of energy
DSC	differential scanning calorimetry
DVA	dynamically vulcanized alloy
DWCNTs	double-walled carbon nanotubes
ECM	extracellular matrix
EGF	epidermal GF

EmimAc	1-ethyl-3-methylimidazolium acetate
ENR	epoxidized natural rubber
E-SBR	emulsion styrene-butadiene rubber
EtOH	ethanol
ETRMA	European Tire and Rubber Manufacturer's Association
FFA	free fatty acids
FO	forward diffusion
FTIR	Fourier transform infrared
GGA	generalized gradient approximation
GT	gelatin
H_2	hydrogen
HCPs	hyper cross-linked polymers
HDS	highly dispersive silica
HFC	hydrogen fuel cells
HPAM	hydrolyzed polyacrylamide
HPC	hydroxypropyl cellulose
HPMS	hydroxypropylmethylcellulose
HSM	high strength Mixer
IBET	integrated biology, English, and technology
ICE	internal combustion engine
IgG	immunoglobulin
IUPAC	International Union of Pure and Applied Chemistry
IUU	illegal, unreported, and unregulated
LA	lactic acid
LDA	local density approximation
MD	membrane distillation
MeCN	acetonitrile
MF	microfiltration
MFCs	microbial fuel cells
MgO	magnesium oxide
MOFs	metal-organic frameworks
MWCNTs	multi-walled nanotubes
NF	nanofiltration
NR	natural rubber
PAFs	porous aromatic frameworks
PAH	polyaromatic hydrocarbons
PBE	Perdew-Burke-Ernzerhof
PbO	lead oxide
PCA	polycyclic aromatics

PCL	polycaprolactone
PDMS	polydimethylsiloxane
PECs	polyelectrolyte complexes
PHAs	polyhydroxyalkanoates
PI	process intensification
PIMs	polymers of intrinsic microporosity
PNC	polymer nanocomposites
POP	porous organic polymer
PP	polypropylene
PPO	poly(propylene oxide)
PS	polystyrene
PTFE	polytetrafluoroethylene
PV	pervaporation
PVA	polyvinyl alcohol
PVB	polyvinyl butyral
PVI	prevulcanization inhibitor
RCIM	Russian Collection of Industrial Microorganisms
RH	hydrodynamic radius
RO	reverse osmosis
RRI	Rubber Research Institute
SAXS	small-angle x-ray scattering
SEM	scanning electron microscope
SF	silk fibroin
S-SBR	solution styrene-butadiene rubber
STE	science-technology-engineering
SWCNT	single-walled carbon nanotube
TBR	truck bus radial
TBT	tributyltin
TBzTD	tetrabezylthiuram disulfide
TE	tissue engineering
TEO	thyme essential oil
TGA	thermogravimetric analysis
UF	ultrafiltration
USTMA	U.S. Tire Manufacturing Association
VP	vapor permeation
WPC	whey protein concentrates
WPU	waterborne polyurethane
XRD	x-ray diffraction
ZnO	zinc oxide

Preface

With growing concern for the environment and the rising price of crude oil, there is increasing demand for non-petroleum-based polymers from renewable resources. Recognizing emerging developments in biopolymer systems research with fully updated and expanded chapters, this book brings together a number of key biopolymer and bioplastic topics in one place for a broad audience of engineers and scientists, especially those designing with biopolymers and biodegradable plastics, or evaluating the options for switching from traditional plastics to biopolymers.

Bioplastics are, on the one hand, bio-based plastics (produced from renewable resources) and, on the other hand, bioplastics may well be biodegradable plastics. Bio-based plastics have a huge market and a wide range of applications and then have become an increased in economy and research. In today's world, bioplastics are becoming increasingly prominent owing mainly to scarcity of oil, an increase in the cost of petroleum-based commodities, and growing environmental concerns with the dumping of non-biodegradable plastics in landfills.

Bioplastics are more sustainable because they can break down in the environment faster than fossil-fuel plastics, which can take more than 100 years. Some, but not all, bioplastics are designed to biodegrade. Bioplastics that are designed to biodegrade can break down in either anaerobic or aerobic environments, depending on how they are manufactured.

The volume provides insight into the diversity of polymers obtained directly from, or derived from, renewable resources. The book highlights the importance and impact of eco-friendly green biopolymers and bioplastics, both environmentally and economically. The contents of this book will prove useful for students, researchers, and professionals working in the field of green technology.

Green chemistry concepts are becoming more and more important in the production of polymer materials as a result of increasing demand for green and renewable products on various markets. Chapter 1 is focused on presentation the importance of rheology as a tool for predicting the optimum conditions of processing of the materials containing green polymer as main or secondary components. Polymer solutions in less toxic solvents have specific flow behavior as a function of the composition of the analyzed system. The viscoelastic character is also a key factor in shaping green polymers in the form of the desired product. The practical importance of some macromolecular systems containing green polymers is also briefly reviewed. Current developments in this field are described from a rheological point of view.

In Chapter 2, the self-assembling processes were studied in the magnetic polymer nanocomposites (PNC) synthesized based on the carbon nanoparticles doped with cobalt nanoclusters, synthesized by a unique CVD technology developed by the authors, under the combined effect of magnetic field and heating. These processes took place with the diffusion of magnetic nanoparticles stimulated by the combined effect of the outer steady magnetic field and heating. The obtained magnetic polymer nanocomposites have good electrical and adhesive properties and are promising for potential practical applications in magneto-electronics.

In Chapter 3, it is shown that microbial fuel cell (MFC) is one of the recent advanced sciences of using microbes for the generation of electricity. Microbial fuel cell undergoes an aerobiotic ferment of organic and inorganic matter to digest fuel which is obtained in energy form. This methodology is not so applicable due to expensiveness. MFC is an electrochemical science where chemical energy is converted to electrical energy using different microbes. In other words, MFC can also be titled as cellular respiration. The basic concept in is the reduction-oxidation respiration in the microbial culture which generates energy in electrical form. The movement of electrons produces a potential difference which in turn helps to generate fuel. A path of hope for MFC conceptual is progressively studied to develop a new electrode in recent days. MFC is into a study to upgrade the concept with cost-effectiveness and helpful application of the technique in a new glance.

Tire industry is basically a slow-changing industry, however, in the present era of fast change, a number of environmental and sustainability-related issues are impacting the tire industry to a large extent. Non-pneumatic tire, tires for electric vehicles, tires with sensors is talk of the day to meet the future requirements of the automobile industries. All these tires need to have fuel efficiency, safety at high speed, and low noise. Tire designers and material scientists are jointly involved in tire design, material design to achieve targeted tire performance properties. Material scientists are focusing their research mainly on sustainable materials as an alternative to present fossil-based material. In Chapter 4 authors shared different research works that are being carried out on sustainable materials particularly for the tire industry. Reuse-regenerate-recycle is one of the research areas to reduce the consumption of virgin fossil fuel-based materials. Natural rubber (NR) is one of the commonly used rubbers in the tire industry. It is a naturally occurring material and helps the tire industry to achieve sustainability in the long run. However, the imbalance between the demand and supply of Natural rubber, worldwide scientists are doing research on Guayule, Dandelion to obtain a suitable alternative of natural rubber. With the introduction of stringent requirements of tire labeling criteria by the European Union, materials scientists are bound to use the alternative of carbon black reinforcing filler. One of the major alternatives of carbon black is silica which is also an eco-friendly sustainable material. Synthetic rubbers using plant-derived materials or biomass (agricultural waste) are came into the picture, Liquid Farnesene rubber is also another newly introduced polymer from bio origin. Research efforts in developing microbes that can grow an isoprene monomer used to make a mimic NR is also undergoing. Fillers which are the next bulk consuming material after polymer are being explored from non-petroleum-based origin like silica, clay, and from bio-origin like rice husk ash, cornflower starch, bagasse; eggshell derived nano-calcium carbonate. Coupling agents like Baker's yeast as an alternate of TESPT. The process aids from non-petroleum-based origin like soybean oil, orange oil, canola oil, sunflower oil, neem oil, etc., are also under the different stages of research for application in tire. Apart from that, many resins from a natural source are used to replace petroleum-based process oil and also to enhance tire performance. Cellulose-based synthetic fibers are of profound interest to replace synthetic fiber in tire reinforcement. Researchers are working to get anti-oxidant from flowers, fruits peels, and food wastes to replace petro-based antioxidants. Tea polyphenols as anti-oxidant have already found potential for the rubber compound. Investigations are also being done for a single compounding ingredient that would possess several functions and thus the term surfactant-accelerator-processing aid (SAPA) or termed as multi-functional additives (MFA).

The main scope of Chapter 5 is to show that the biopolymer silk as a recognizable implement in the biomedical field is manifested in nanotechnology due to its promising avenues. Silk has been esteemed and caught the attention of the human race, by virtue of its lustrous appeal since long ago. Therefore, it rules the textile market across the globe even today. But apart from its aesthetic value researchers have attributed some medicinal values to silk in the recent past and its scope is expected to shoot up shortly. It was surprising that, even sericin which was once considered the waste material in the silk industry had got special consideration due to its unique properties. Silk proteins were found to have excellent biocompatibility and slow biodegradability which provides it an unavoidable role in biomedical applications. The structure composition and current researches like wound healing, tissue engineering (TE), and the emerging biomedical applications of silk in its various forms like films, nanofiber, hydrogels, and particles were discussed. Silk by undergoing major and minor chemical modification or by combining with different synthetic or natural polymers can work wonders in the biomedical field. There is still a vast world of silk unexplored; the researchers have to research these hidden capable avenues and their applications.

Biofouling in aquaculture cages has been an important issue for the farmers since its management incurs a huge sum of money and labor. Antifouling strategies failed due to the diversity of organisms are different from region to region and increased concern about pollution in the water bodies. Chapter 6 describes the strategies to combat biofouling using nanomaterials coated over a polyaniline coated polyethylene and its efficiency towards antifouling. Green synthesis of nanomaterial incorporated hydrogel and mixed charged polymeric hydrogel over polyethylene aquaculture cage nets against biofouling also described. Brief description of the environmental impact assessment of nanomaterials treated nets in the marine environments included at the end.

In Chapter 7, mesoporous SiO_2, gender, climate change (CC), and science in parliament investigated.

In Chapter 8, two approaches have been investigated in order to reduce the cost and increase the availability of the use of biodegradable polymers of microbiological origin (polyhydroxyalkanoates, PHA): (1) expanding the raw material base and attracting glycerol as a carbon substrate (large-tonnage waste from the production of biodiesel) and (2) filling the polymer with natural materials for industrial use. An effective technology for the synthesis of destructible polymers was developed using a new and more affordable substrate compared to sugars-glycerol. The *Cupriavidus eutrophus* B-10646 strain was studied as a product of polymers on glycerin, and the kinetic and production characteristics of the strain providing polymer yields on glycerol of various cheat, comparable with the process on sugars, were studied. Replacing sugars with glycerol provides a reduction in the cost of a carbon substrate in the production of PHA. To increase the availability of PHA and their technical applications, mixtures of PHA with natural materials were formed, the structure and properties, as well as the patterns of degradation in the soil, were studied. The possibility of long-term functioning of mixed forms in the soil with a gradual release of active substances (herbicides and fungicides) and the possibility of providing plants with protective equipment during the growing season are shown. Both approaches are effective for enhancing accessibility and expanding the scope of application of perspective green plastics-destructible PHA.

Green technology is that to operate the ecological skills for the development and application of merchandise, instrumentality, and systems to conserve the natural resources and surroundings. This field is moderately innovative. It has developed rapidly as people have become aware of the detrimental effects of environmental variation including global warming and adverse effects on the living beings. It is believed that growth in green technology will lead to worldwide, sustainable, and economic powers that impact economics, societies, and way of life in the future. The application of green technology has multiple challenges while it can provide many opportunities during real-life chemical processes. The membrane technology can provide an economical pathway to many of these challenges. Here applicability of membranes during green technology is discussed. Over the past few years, the membrane separation process has become one of the major separation technologies. It is rapidly becoming an important part of green technology due to some of its benefits. In membrane processes do not require the addition of chemicals, it works on physical separation. This improves the feasibility of recycling the recovered chemicals or using them in further processes. Additionally, membrane technology has relatively low energy requirements, easy to design and integrated nature may advance, and can be integrated with current processes. This gives benefits of reduced production cost, equipment size, energy consumption, and waste generation and improving controls and process flexibility.

Chapter 9 would consider two aspects of membrane technologies for process enhancements; as process integration and effluent treatment. Process integration would be focused on for improvements in economical separation, recovery, and recycle; this would reduce the generation of waste and effluent component generation, thus enhancing the process financial system. Membrane separation technologies are very helpful in effluent treatment. It is investigated for a reduction in effluent volume and its treatment for components to recover. The work starts by presenting an overview of membrane operations. It would be followed for the application of membrane processes for process improvement and *in situ* recovery of components. This would be followed by the effluent treatment aspects. Such recovery of components harmful to the environment and their further utilization by membrane processes would increase the process economy and product purity. To get superior removal of contaminants compared to conventional technologies, the membrane technology would provide better support to the system. The conventional technologies frequently required continuous modification of conditions and the use of chemicals. This means a high level of operator involvement and the risk of contaminants in the effluent.

One of the greatest challenges of this century is to find an alternative fuel for vehicular propulsion which can replace widely used fossil fuels. Hydrogen is one of the finest choices due to its clean energy and highly plentiful in nature. And also, it has the highest energy value per unit weight (142 MJ kg^{-1}) compared to liquid hydrocarbons (47 MJ kg^{-1}). The conventional hydrogen storage technologies, namely, high-pressure tank and liquid state storage, are not applicable due to large size and higher energy cost for liquefaction. Solid-state storage may become a viable technology provided the storage medium can absorb a large amount (~6.5 wt%) of hydrogen and can release them easily as recommended by the department of energy (DoE), USA. Recently, carbon nanomaterials have drawn immense attention from researchers due to their high surface area, high electrical and thermal conductivity, and high mechanical strength. Although for pure carbon nanostructures, the physisorption of hydrogen is negligible at room temperature, the functionalized carbon nanomaterials are the promising candidates for hydrogen storage media at ambient conditions. This chapter describes the issues and challenges for hydrogen storage in functionalized carbon nanomaterials. It will highlight the bonding, charge transfer mechanism, and hydrogen storage capability of novel carbon nanostructure (carbon nanotube, graphene, and graphyne).

Polymers are also potential hydrogen storage materials due to their pure organic nature, tunable structures, and large scale synthesis. Chapter 10 discusses about different types of polymers such as conjugated micro porous polymers, porous polymers networks, polymers of intrinsic microporosity (PIM), hyper cross-linked polymers, and their hydrogen storage capabilities.

Materials developed on the basis of green polymers are of major importance not only for the keeping a clean environment, but also for creating of advanced products with architecture controlled at the micro- and nano-scale level. The last chapter reviews the progress made in the field of self-assemblies in green macromolecular systems. The first short section presents some theoretical aspects and the most used synthesis and preparation routes of macromolecular assemblies of biopolymers. Then, the commonly employed characterization techniques of such soft nanomaterials. The newest developments in green polymer self-assemblies are described emphasizing their practical importance. The presented research examples prove the range of possibilities and future trends in self-assembly processes that will significantly contribute to the creation of materials nanoarchitectonics. The green materials, here the topic under analysis, cover several application fields.

CHAPTER 1

Processing Relation to Rheological Behavior of Some Systems Containing Green Polymers

ANDREEA IRINA BARZIC

"Petru Poni" Institute of Macromolecular Chemistry, Laboratory of Physical Chemistry of Polymers, 41A Grigore Ghica Voda Alley, Iasi – 700487, Romania

ABSTRACT

Green chemistry concepts are becoming more and more important in the production of polymer materials as a result of increasing demand for green and renewable products in various markets. The chapter is focused on the presentation of the importance of rheology as a tool for predicting the optimum conditions of processing of the materials containing green polymer as main or secondary components. Polymer solutions in less toxic solvents have specific flow behavior as a function of the composition of the analyzed system. The viscoelastic character is also a key factor in shaping green polymers in the form of the desired product. The practical importance of some macromolecular systems containing green polymers is also briefly reviewed. Current developments in this field are described from a rheological point of view.

1.1 INTRODUCTION

The serious consequences of our planet the pollution have drawn the attention of the scientific community and industrial market towards green chemistry and related products [1–4]. In the 1990s, the International Union of Pure and

Applied Chemistry (IUPAC) introduced green chemistry as a field that is connected to the "design of chemical products and processes that reduce or eliminate the use or generation of substances hazardous to humans, animals, plants, and the environment" [5]. In other words, green chemistry attempts to diminish and prevent pollution at its source. Based on these concepts, the research tendencies in material science begin to change opening new perspectives concerning synthesis techniques and processing procedures [6–8]. In some cases, the material's level of performance is not as high as expected, but opens new paths for a healthier environment.

When it comes to polymers, there is a slight difference between *natural* and *green* ones [5]. As stated by their name, *natural polymers* (or biopolymers) are represented by macromolecules that can be encountered naturally or they are synthesized by living organisms. Among them, one can mention cellulose, silk, chitin, DNA, and protein. Another definition of natural polymers refers to macromolecular compounds which can be obtained from raw materials that came across in nature [5]. Each year, around 300 million tons of plastics are manufactured, among which natural polymers represent less than 1%, but their production tends to increase [5]. For instance, the food packaging, cosmetic, medical, pharmaceutical, and beverage industries require large amounts of biopolymers, supporting the development of the natural polymers market. The latter is dominated by cellulose ethers, starch-based materials, fermentation polymers, exudates, protein-derived polymers, and marine polymers [5].

On the other hand, when referring to *green polymers* one should have in mind macromolecular compounds achieved through green (or sustainable) chemistry procedures [8–10]. In order to reduce pollution during polymer preparation, one should use adapted techniques that minimize the contamination of the environment. For this reason, natural polymers are usually considered green polymers [5].

The 21st century brought again in attention renewable polymers and major thrust towards the design of various bio-materials of macromolecular nature. For a close control of the pursued properties, a careful attention of each component of the final material is performed. So, there is a gradual transition from petrochemistry to bioeconomy [11], making inevitable the development of the green polymer market. Analyzing these aspects economically, it is obvious that remaining oil resources will be achieved with higher expenses, whereas the growing worldwide energy demands cannot be fully satisfied. Such aspects might have undesired effects on the cost-effectiveness and competitiveness of plastics [5]. Switching chemical raw material manufacturing to renewable resources might secure plastics production against a potential future oil crisis.

Another important issue that could contribute in favor of green plastics are the environmental problems that arise from plastics wastes from numerous industries, like electronics, packaging, paints, and so on [12–14]. Thus, the consumers concerns related to global warming can be solved by replacing the current products with sustainable and 'green' ones. Moreover, several states formulated environmental legislation and regulations which support the thrive of eco-friendly products with a reduced carbon footprint.

In the synthesis and processing of macromolecular compounds, the green principles are referring to [5, 7, 15]:

- A large amount of raw material in the product;
- A clean (lack of waste) production process;
- No use of supplementary substances like organic solvents;
- High energy efficiency in manufacturing;
- Utilization of renewable resources and renewable energy;
- No generation of health and environmental hazards;
- Raised safety standards;
- Reduced carbon footprint;
- Controlled product life cycles with effective waste recycling.

The utilization of renewable resources for green polymer preparation should not compete with food production, but more important must not encourage farming or deforestation, while most not involve transgenic plants or genetically adapted bacteria [5]. The latter aspects are crucial for human health since biodegradable polymers should not generate inhalable spores or nanoparticles.

Having all these in view, one may distinguish three main routes to produce renewable plastics [5, 7, 15]:

- Utilization of biomass and/or carbon dioxide to prepare 'renewable oil' and green reactants for highly resource- and energy-effective polymer synthesis procedures.
- Use of living cells, which are viewed as solar-powered chemical reactors through genetic engineering and biotechnology techniques to generate bio-based polymers.
- Via activation and polymerization of carbon dioxide.

To ensure and consolidate the progress in green polymers, researchers, and manufacturers work together and make known their ideas at specific scientific events dedicated to production of green/renewable polymers and bioplastics and from various resources, such as plants and bio-refineries, algae, waste, and carbon dioxide [5, 15]. Nowadays this is viewed in the emerging biofuels, like bioethanol, which is made by fermentation of sugar achieved from cellulose. This biofuel is also utilizable in the synthesis of green polyethylene, which is entirely recyclable. The energy-efficient processing techniques allow the transformation of the biomass into renewable coal and oil. There are certain agricultural and forestry residues already which can be used to prepare renewable monomers.

Carbohydrates, proteins, and polyesters are chemically adapted for green material processing and applications. Natural fibers are considered outstanding reinforcement agents for thermosets and thermoplastics. As an example, micro-fibrillated cellulose has been used in nanocomposites with polymer matrix in many applications, including in medical implants. Lignin is considered to be a renewable energy source in paper production, and also as filler for several polymers and rubbers. Thermoplastic lignin combined with natural fibers (e.g., Arboform) preserves the advantages of wood and synthetic thermoplastic materials [5]. Bio-hybrids have been prepared by the introduction of starch in polyolefins or in compostable polyesters. Biopolymers, such as chitosan and polylactic acid display numerous medical applications. Phosphoproteins, like Casein, have been proved to be good binders or adhesives. Renewable monomers are already substituting for "oil-made" monomers. This is widely present in plastic bottles that currently are made of bio-based materials which can be recycled. Generally speaking, it is not so easy for polymer production to fulfill the demands of green chemistry, but not impossible. Thus, green polymer preparation routes must involve non-toxic solvents or even solvent-free procedures, no byproduct formation and efficient use of renewable resources [5, 15].

The processing of the materials containing green polymer as main or secondary components in the shape of the desired products especially starts from solution or melt state [16–18]. A large category of polymers degrade prior to melting so solution processing remains the only path toward finite product manufacturing. In both situations, it is of paramount importance to know the flow behavior and viscoelasticity of the system before processing. One of the most sensitive tools for analysis of polymer solutions (or melts) in the semi-dilute concentration domain is *rheology*. This technique allows extraction of a large amount of information that facilitate better understanding of the response of the polymer sample to shear deformation, which is often encountered during the processing stage [19].

This chapter is devoted to present the state-of-art of rheological characterization of solution containing green polymers and the applicability of such systems in various industries. Also, a short presentation concerning the environmental impact of these materials is reviewed.

1.2 SOLUTION RHEOLOGY CORRELATION WITH POLYMER PROCESSING

Polymer dissolution is quite distinct than that of low molecular weight materials since it involves two stages: solvent diffusion and chain disentanglement [20]. When an amorphous and glassy polymer (with no cross-links) is placed in a thermodynamically compatible solvent, the solvent begins to diffuse among the macromolecules. Owing to the plasticization of the polymer in the presence of the solvent, a gel-like swollen layer is created along with two separate interfaces: one between the glassy polymer and gel layer and the other between the solvent molecules and the polymer gel. After a period of time, there are three possible situations:

- The first case refers to fully polymer dissolution;
- The second probability is linked to a partial dissolution of the polymer; and
- In the last case, the polymer cracks and no gel layer is produced.

When macromolecules are dissolved, they tend to adopt a variety of conformations [21–23]. A large number of chain shape is enabled by many internal rotations which can form through simple carbon-carbon bonds creating several rotational isomers. As a result of energy restrictions, not all conformations display a similar probability of occurrence [22]. Therefore, the most stable conformations are the ones that prevail in solution. If the macromolecules are highly flexible they often change their conformation and instead of having a linear form in solution, the chains adopt a very typical conformation, named random coil. It is worth mentioning that the random coil would not remain the same after the polymer has been dissolved. This is because the solution behavior of polymers is strongly affected by several factors, such as [21–23]:

- Polymer structure, composition, and conformation;

- Molecular weight and polydispersity index;
- Thermodynamic quality of the solvent;
- Polymer-solvent interactions;
- Temperature.

The dissolution process can be influenced by the chain chemistry, composition, and stereochemistry [20]. The polymer dissolves either by displaying a thick swollen layer or by experiencing cracking, as a result of relieving ability of the osmotic pressure stress those results in the polymer. Hence, the nature of the polymers and dissimilarities in free volume and chain segment stiffness are ascribed to the properties variations from polymer to polymer.

It is widely known that the dissolution rate is lower as the polymer molecular weight is higher, while the polydisperse polymers dissolved about twice as fast as monodisperse ones of identical molecular weight [20, 24–26].

In a binary solution prepared from a polymer-solvent pair having similar solubility parameters, the forces acting between chain segments become smaller than those occurring between the polymer and the solvent [24]. In such a system, the random coil adopts an unfolded conformation. In a less compatible solvent, the polymer-solvent interactions are overcome by those formed between chains, so random coil follows a tight and contracted conformation. In extremely "poor" solvents, the forces between polymer-solvent are eliminated thoroughly, and the random coil become so contracted that it precipitates [27, 28].

External factors, like agitation, temperature, and radiation exposure have a great impact on the dissolution process. It was found that the velocity of dissolution is higher as the agitation and stirring frequency of the solvent are intensified. This is because of the decrease of the thickness of the surface layer, while the dissolution rate reaches a critical value when the pressure of the solvent on the polymer surface is high, regardless the temperature [24]. Other reports [29–31] showed that in the absence of agitation the solvent diffuses into the polymer forming the gel layer, while subjection the solution to agitation the gel is stripped off very fast during the stirring stage. If in the first case the gel layer is diminished with time owing to desorption of the macromolecules, in the second case agitation favor sorption of solvent and subsequently desorption of chains from the swollen gel. Radiation dose is another factor that influences the dissolution process [32]. First, the polymer swells and then the chains begin to disentangle and dissolve. The swelling rate was higher in comparison to the dissolution rate, the gel layer size increased linearly with the square root of time. Oppositely, if the dissolution rate was bigger than the swelling one, then the gel thickness shrinks with time.

The polymer concentration in the solution has a strong impact on the overall material behavior viewed in the changes recorded in the viscosity (η). The Debye criterion, $c[\eta] = 1$ is useful to delimitate the concentration ranges of dissolute macromolecules. So, when discussing polymer solutions, there are three concentration domains [33]:

- The range of *dilute* solutions at $c[\eta] < 1$, where the macromolecules are mainly surrounded by solvent molecules so they behave independently of one another;
- The range of *semi-dilute* or *semi-concentrated* solutions at $1 < c[\eta] < 7$, where concentration is high enough that facilitates the contacts between different macromolecules, but there is no continuous three-dimensional entanglement network; and
- The range of *concentrated solutions* at $c[\eta] > 7$, where the macromolecular coils begin to overlap and thus an entanglement network is formed throughout the entire volume of solution.

It is presumed that polymers in dilute solution are independently moving and they do not display permanent contacts with each other [34]. As a result, viscosity of the dilute polymer solution is decreasing as the shearing is more intense reflecting a non-Newtonian behavior. This can be explained by deformation of the individual macromolecules in the stream. Polymer chains subjected to shearing modify their shape, thus reducing their resistance to flow owing to streamlining at higher shear rates [33]. On the other hand, for long-chain and flexible polymers the viscosity is unchanged at very low shear rates. This is the so-called the zero-shear rate Newtonian viscosity or limiting viscosity at null shear rate. The viscosity might further remain constant even at intermediate shear rates and at a critical shear rate the viscosity begin to decrease and at very higher shear rates reaches a minimum value, which is known as the upper Newtonian viscosity (or limiting viscosity) [33].

As the polymer amount in solution is higher more contacts between macromolecular coils occur and thus network of entanglements is formed [33, 35]. Chains at the sites of contact are able to slide and for this reason the network is considered quasi-permanent. The intermolecular contacts responsible for the occurrence of the entanglement network are presumed to appear at a critical concentration (noted c_e) as a function of the molecular weight. Non-Newtonian flow becomes more and more obvious with the augmentation of polymer concentration in solution [33]. Below the c_e point, the solution viscosity is strongly dependent on the intramolecular, excluded-volume effects, whereas above c_e the intermolecular entanglements have a prevalent impact on the rheology of the solution [33–35].

Most polymer products are obtained from semi-diluted or concentrated solutions [19, 36]. In these concentration domains, the analysis of the shear flow behavior is essential for proper processing/shaping of the polymer material. The rheology is the most suitable methods to examine the polymer fluid response to shear deformation [33]. The viscosity curves contain data of paramount importance concerning the flow-ability of polymer solution subjected to a broad interval of shear rates and processing conditions. The experiments recorded at small shear rates are especially useful for analysis of the problems which arise from manufacturing procedure. In the low shearing domain, the experiments are framed in linear viscoelastic region, and hence the data can be connected to the polymer's molecular structure [37–40]. The latter influences the polymer processability and the performance of the final product. This is why electronic, cosmetic or paint industry often employs rheological testing for product perfecting, as well as for novel product development.

Polymer films are generally prepared from solution phase using techniques like:

- **Drop Casting [41]:** Relies on preparation of a thin solid film by dropping a polymer solution onto a support with flat surface and left to evaporate the solvent;
- **Tape Casting [42]:** It is based on a reservoir of polymer solution with a slit-shaped outlet which leaves a layer of wet film;
- **Spin Coating Deposition [43]:** Small quantity of polymer solution is applied on the center of the substrate, which is subjected to spinning at variable rotation speeds or sometimes not spinning at all. In order to spread the solution, the substrate is rotated at high speed and thus the coating material is under action of centrifugal force and covers the support. Rotation is continued while the fluid reaches the edges of the support or until the wanted film is attained. The used solvent is generally volatile and mainly evaporates during spinning. As the rotation speed is higher, the obtained film is thinner [44];
- **Spray Deposition [45]:** It is a relatively new method that involves a highly diluted solution of a polymer, which is nebulized into air and concentrated under specific evaporation conditions. The

produced aerosol is transported by a carrier gas and cast onto a solid substrate;

- **Langmuir-Blodgett [46]:** It is mainly a nanostructured system obtained when Langmuir films—are transferred from the liquid-gas interface to solid substrate during the vertical crossing of the support through the monolayers.

Among the presented techniques of polymer film preparation from solution, phase the most commonly used ones are drop-casting, tape casting, and spin coating deposition. In these cases, flow behavior of the polymer solution is very important. The uniformity of the wet and implicitly solid film thickness is strongly related to the viscosity dependence of the shear rate. It was proved that Newtonian polymer fluids are suitable for obtaining a uniform liquid surface contour [47]. This is not the case of non-Newtonian solutions for which the thickness profile is far from uniform [48]. Particularly the polymer wet layer is thinner towards the substrate edge since solution viscosity is changing from wafer center to its margins as a result of a variation of the shear rate along spinning disk radius. This is also available for tape casting where the film thickness is affected by speed of blade movement and polymer solution response to variable shearing [42, 49]. Hence, the parameters extracted from rheological properties, such as the energy of activation and flow activation entropy, show the effect of the polymer structure, and provide more precise information on the proper processing conditions adapted to each macromolecular system. Having all these in view, a close understanding of the rheological functions (dynamic viscosity, shear stress, elastic shear modulus, loss tangent, viscous shear modulus, and relaxation time) of the polymer solutions and their dependence on temperature and solvent are of paramount importance for processing them into stable films with uniform thickness. The uniform film thickness is highly desirable for achieving isotropic physical properties (e.g., transparency or electrical/thermal conductivity) and securing the functionality of the components implemented in the devices.

1.3 FLOW BEHAVIOR OF SOME SYSTEMS CONTAINING GREEN POLYMERS AND THEIR APPLICABILITY

The rheological behavior of solutions prepared from green polymers as main or secondary components has known an increasing attention. Such materials can be viewed as additives for a large product portfolio, such as plasticizers, lubricants, viscosity depressants, antistatic, and antifogging components, but also as release, agents attained from sustainable materials that are prepared in such way that they optimize production and enhance efficiency [50]. In this section of the chapter, the state-of-art concerning rheological investigation of solutions containing certain green polymers is presented. The attention was directed towards the following types of green polymers:

- cellulose and its derivatives;
- silk;
- chitin; and
- chitosan.

1.3.1 SOLUTION RHEOLOGY OF CELLULOSE AND ITS DERIVATIVES

As already known, *cellulose* as found in nature is not soluble in water or in many other organic solvents because its chain stiffness and numerous inter and intramolecular interactions. However, cellulose dissolution was proved to be optimal in ionic liquids [51–55]. Most of them, particularly those halide anions are highly viscous fluids, having a viscosity of about three times higher in comparison with that of common organic solvents [51]. It was shown that even at elevated temperatures cellulose solution in chloride-based ionic liquids present large viscosity similar to certain polyolefin melts [54]. In practice, the high viscosity of cellulose-ionic liquids solutions is not so desirable because it will make difficult the mass transfer and require considerable power consumption for blending to achieve a heterogeneous solid-liquid system. The introduction of less viscous co-solvents into such solutions is a good alternative to reduce the mixture viscosity and enhance the efficiency of polymer processing, including heat and mass transfer rate [51]. For instance, dimethylsulfoxide, dimethylformamide, and pyridine have demonstrated to display a good ability to diminish the viscosity of cellulose solution in ionic liquids without an impact on the ionic liquids dissolving capability of polymer in a specific concentration domain of these co-solvents [56–58]. The addition of organic solvent is useful for adequate processing of cellulose-ionic liquids solutions with reduced costs. However, so far, literature is not abundant in studies concerning the viscoelastic behavior and solution behavior of cellulose solved in ionic liquids in the presence of co-solvents.

Lv and co-workers [51] studied the properties in solution of cellulose in a mixture of ionic liquid/dimethylsulfoxide. The viscosity is diminished by addition of dimethylsulfoxide in the system. It is presumed that this solved reduces monomer friction coefficient in the polymer solution, but it does not significantly modify the entanglement state of cellulose, while it shrinks the relaxation time. The steady shear experiments are useful to monitor the behavior of cellulose in dilute, semidilute unentangled and semidilute entangled domains. The investigated cellulose solutions display as Newtonian properties at very low polymer concentration, whereas the solution viscosity increases with concentration and presents a shear-thinning behavior at higher shear rates. The critical concentration for macromolecular coil overlapping (c*) and the entanglement concentration (c_e) of cellulose are found to be 0.4 wt% and 1.3 wt%, respectively. The parameters extracted from the power low concentration dependencies indicate that the ionic liquid/ dimethylsulfoxide are like theta-solvents for cellulose at room temperature, while thermodynamic characteristics of the solvent mixtures were close to those of the ionic liquids for cellulose. In other words, the conformation of cellulose in ionic liquids would not be modified with the introduction of dimethylsulfoxide, not only in the dilute domain, but also in the entanglement concentration interval. These conclusions were supported by results achieved previously by Gericke et al. [52] for several types of cellulose in 1-ethyl-3-methylimidazolium acetate (EmimAc), proving that this ionic liquid is like a theta-solvent for cellulose. Sescousse and collaborators [53] performed a comprehensive investigation of viscosity concentration dependency in two distinct ionic liquids (EmimAc and 1-butyl-3-methylimidazolium chloride (BmimCl)) at several temperatures. They suggested that solvents present similar thermodynamic quality, in spite of their different state at room temperature, namely EmimAc is a liquid, and BmimCl is a solid. Chen et al. [54] studied linear viscoelastic and steady shear viscosity of cellulose solutions in BmimCl from dilute to highly entangled solutions. They obtained concentration dependencies of rheological functions consistent with scaling predictions for neutral polymers in a theta-solvent. However, failure of the Cox-Merz rule with the steady shear viscosity bigger than complex viscosity and the noticed internal mode structure of dilute and semidilute unentangled examined solutions indicate that cellulose in BmimCl is not simply a flexible polymer in a theta-solvent. Druel and co-workers [59] used JetCutting technology to prepare cellulose aerogel beads from ionic liquid solutions and dried them by supercritical carbon dioxide extraction. The rheological characteristics of cellulose solutions in 1,5-diazabicyclo[4.3.0]non-5-enium propionate ([DBNH][CO_2Et]) and in 1-ethyl-3-methylimidazolium acetate ([Emim][OAc]) were analyzed at several concentrations and temperatures. The first ionic liquid is a better solvent from thermodynamic point of view since the cellulose intrinsic viscosity is twice bigger in regard with that in [Emim][OAc]. From rheological data it was extracted the processing window of JetCutter which imposes a low solution viscosity at elevated shearing rates and also the polymer concentration above which the coils are overlapping ensuring the formation of intact aerogel beads. For their preparation, cellulose-[DBNH][CO_2Et] solutions having a concentration of 2–3 wt% were achieved and then coagulated in non-solvents like water and alcohols. Bead dimensions are varying from 0.5 to 0.7 mm when obtained from 2% solutions and up to 1.8 mm when made from 3% solution. The cellulosic beads have wide applicability in cosmetics, food, medical, sorption, filtration, and separation [59].

Owing to its insolubility, cellulose was chemically modified resulting a broad category of *cellulosic materials* that are more facile to be processed from solution phase. A recent study of Gauche and Felisberti [56] discusses the viscoelastic behavior of cellulose nanocrystals (CNCs) grafted with poly(2-alkyl-2-oxazoline)s. This procedure is a good strategy to enhance CNC dispersion in solvents (of low polarity) or polymeric matrixes. The grafting of cellulose with flexible macromolecules generates important changes in colloidal behavior. Poly(2-alkyl-2-oxazoline)s achieved by cationic ring-opening polymerization reaction was grafted onto CNC surfaces, where the coupling reaction was enabled by the presence of partially hydrolyzed polymers-attaining a reaction yield of about 64%. The modified cellulosic particles can be redispersed in water after freeze-drying, forming stable dispersions. In addition, they were not cytotoxic up to 10 wt% aqueous dispersion. The rheological properties of grafted nanoparticles indicate that there are significant differences in regard with original CNC dispersions. CNC dispersions in water present an organization typical for nematic liquid crystals and rheological behavior similar to true gel (at 5 wt%) before drying. Conversely, when in the same conditions is inserted 10 wt% of CNC-g-poly(2-ethyl-2-oxazoline-stat-ethylene imine) (P(EtOx-s-Ei)) the gel properties are less pronounced. Dispersions of CNC-g-P(PEtOx-s-Ei) particles prepared by redispersion of freeze-dried particles behaved as an isotropic fluid, lacking the nematic molecular order. For pristine CNC dispersions, the magnitudes of shear moduli were found to be almost independent of frequency having a prevalent elastic behavior. Moreover, the ratio of loss and elastic moduli has subunitary values in the range of applied angular frequency supporting the idea that the sample is a true gel. Upon addition of CNC-g-P(EtOx95-s-Ei5) to the initial CNC dispersion it is noticed that elastic modulus and viscosity decrease with 3–4 orders of magnitude, while frequency increased to 0.03 Hz, a shear-thickening zone was remarked. This could be connected to the coil-to-stretch transition in the tethered polymer chains subjected oscillatory shearing, wherein the macromolecule disentangles to an almost fully extended state. After 0.03 Hz, a shear-thinning zone appears since the effect of chain rigidity on the viscosity is no longer prevalent over the shearing, thus the alignment of the crystals along the flow direction is taking place, followed by breaking off the nematic organization [60]. The loss tangent (tanδ) values showed relevant modifications over the angular frequency interval: the transition from liquid-like to solid-like behavior was noticed at the overlap frequency, corresponding to $\tan\delta \cong 1$. This sol−gel transition appeared at the exact point that coincides with the beginning of the shear-thinning zone. In the range of low frequency, the tanδ value was 5.75, as expected for a viscous polymer solution. For higher frequencies, tanδ decreased up to a value of 0.17, and the CNC dispersion in the presence of 10 wt% CNC-g-POx might be regarded as a weak gel. These results support the hypothesis that the nanoparticle self-assembly was not present in these samples. A CNC dispersion attained from a physical blend of pristine CNCs and CNC-g-P(EtOx95-s-Ei5) followed by freeze-drying and redispersion in water (5 wt%) displayed a considerable reduction of the complex viscosity, shear moduli of several orders of magnitude. The CNC-g-P(EtOx95-s-Ei5) dispersion had a similar behavior.

Other cellulose derivatives, like hydroxypropyl cellulose (HPC) or hydroxypropylmethylcellulose (HPMS), have a rheological behavior strongly dependent on the solvent and concentration [37, 39, 61–64]. HPC forms in water, acetic acid or N,N-dimethylacetamide isotropic solutions at concentrations below 27%-value dependent on solvent and molecular weight. After that there is a small concentration interval where a biphasic (isotropic and cholesteric phases) behavior is remarked, followed by a cholesteric phase, where viscosity versus concentration has a peak. The anisotropic solutions of HPC are prevalently viscoelastic [61]. Upon heating, anisotropic HPC solutions in acetic acid endure an anisotropic to isotropic phase transition which is reflected in a maximum in the temperature variation of the first normal-stress difference and of the solution viscosity [61]. After shear cessation, HPC solutions exhibit a band texture-noticed as alternating bright and dark lines orthogonal to the shear direction caused by reorientation of the average molecular orientation. Such morphology is highly stable and is noticed when the macromolecules are strongly oriented along the shear direction [62–64]. The band texture can be maintained even after solution drying. Such surface pattern can be transferred upon blending to other polymers with isotropic orientation of macromolecular chains [37, 39, 65]. Cellulose derivative films with banded pattern on their surface can be used for cell culture purposes or for nematic, alignment in liquid crystal displays [37].

1.3.2 SOLUTION RHEOLOGY OF SILK POLYMERS

The study of Keerl [66] was focused on rheological characterization of regenerated *Bombyx mori* fibroin and silk solutions in water of recombinant spider silk protein. The introduction of kosmotropic salts and application of elevated temperatures determine a solution behavior similar to that of a native silk spinning dope. In the studied shear rate interval, the silk-based samples presented a non-Newtonian behavior. This is expected because of the increasing flow alignment of the underlying microstructure. The onset of the shear thinning domain, that is, the critical shear rate noticed in other reports on natural silk dopes, was usually around 0.5–4 s^{-1} [67–69]. Measurements performed on the silk solutions highlighted shear-thinning right from the beginning, with critical shear rates starting around 0.01 s^{-1}. Oscillatory tests below 5% w/v show a storage modulus independent of frequency viewed as a plateau between shear frequencies of 2 and 10 rad/s [66] indicating a gel-like behavior. Above 5% w/v, silk solutions in water present frequency dependence and an overlap point about 15–30 rad/s, which is in good accordance with the viscoelastic properties of native silk dope [66]. These analyses are relevant for determining proper conditions of achieving green polymer fibers of elevated mechanical resistance used in various domains, from biomedicine to military purposes [66].

The investigation of Tao and collaborators [70] was focused on flow behavior of silk fibroin (SF) reinforced waterborne polyurethane (WPU). These green materials have promising features for tissue engineering (TE) applications. In the shear rate domain of 10^{-2}–10^{3} s^{-1}, the composite solutions have a shear thinning behavior regardless the amount of WPU in the system. This is sustained by the subunitary values of the flow index extracted from Ostwald-de Waele law applied to shear stress data as a function of shear rate. Dynamic oscillation behavior of the SF-WPU solutions. All WPU dispersions were found to be highly frequency dependent. At 13 wt%, shear moduli curves were crossed at the middle of the frequency range, displaying a rheological behavior similar to a dilute solution. When the percent of WPU dispersion is varying between 15 to 25 wt%, storage modulus is lower than that of the loss one and parallels to each other at low frequency range, while the moduli approach each other at high shear frequencies. This proves that the dispersions have features analogs to Brownian suspensions. At high WPU contents, like 22 and 25 wt% one may notice a gel-like network structure or aggregates formation, which is seen in higher viscosity compared to that of dispersions containing lower WPU amounts. The formation of aggregates or gel-like networks in dispersion system with WPU contents above 17 wt% might restrict the dispersion of SF particles into the polymer matrix.

Laity et al. [71] analyzed the rheological features of native silk protein feedstock specimens at temperatures ranging from 2 to 55°C. They observed no thermally-driven phase change behavior. All solutions have flow properties commonly noted for concentrated polymer solution, with viscoelastic characteristics dominated by two main relaxation modes with time constants around 0.44 and 0.055 s at 25°C. The samples present a temperature dependence following the Arrhenius relation, consistent with the kinetics which is governed by activation energy of flow, which varied from 30.9 to 55.4 kJ mol^{-1} according to oscillatory data. The compiled master-curves for native silk feedstock samples based on the principles of time-temperature superposition, using oscillatory data prove their viscoelastic behavior peculiar for a polymer solution across a wide temperature range. The rheological data showed the processing range of natural silks and explained the molecular mechanisms governing the flow properties of these green materials.

1.3.3 SOLUTION RHEOLOGY OF CHITIN

Bochek and collaborators [72] reported the rheological behavior of cellulose/chitin mixtures in 1-butyl-3-methylimidazolium acetate (BMIAc). They noticed a non-Newtonian flow of this green polymer blend in solution. Positive and negative deviations of values of the viscosity from additive values are noted even if polymerization degree of chitin is higher than that of cellulose. This is probably because destruction of polysaccharide chains takes place during dissolution in BMIAc, but that of chitin is more pronounced [72]. Another reason for this result arises from the fact that BMIAc is a thermodynamically lower solvent for cellulose and better for chitin. Addition of more chitin in the system determines a decrease of the solution viscosity. The network of intermolecular entanglements and the hydrogen bonds often noted for cellulose are disrupted, resulting a new blend system of hydrogen bonds, concomitantly with a weaker network of intermolecular entanglements. Such aspects determine a lower energetic barrier of flow process.

1.3.4 SOLUTION RHEOLOGY OF CHITOSAN

The rheological properties of chitosan solutions in several solvents-acetic, lactic, and hydrochloric acid-and various temperatures (25–45°C) were reported [73]. At shear rates up to 1 s^{-1}, the real part of the complex viscosity is almost Newtonian after which it decreases. The analyzed chitosan solutions present a loss modulus higher than the elastic one showing that viscous effects are more significant than elastic ones. Application of the Cox-Merz rule for 2% chitosan solution in acetic acid indicates a decrease of the viscosity and thus a lower energetic barrier of flow.

Hydrophobically modified chitosan was prepared by Shedge and Badgier [74] from natural resources, like cashew nut shell liquid. Chemical modification of this green polymer was done with 3-pentadecyl cyclohexane carbaldehyde. Above the critical association concentration, the hydrophobically modified chitosan solutions present aggregation properties. As the polymer concentration is higher, the intermolecular hydrophobic associations begin to play a prevalent role and determine the formation of flower-like micelles. The frequency sweep experiments beyond the overlap concentration reveal for elastic modulus weak frequency dependence, having almost a plateau modulus at about 1 Pa, while loss modulus increases with frequency. The increase of hydrophobic content determines the increase of shear moduli and the disappearance of their overlap, showing a soft-solid gel-like behavior. For the same degree of hydrophobic modification and variable concentration, it was found that there is an evolution from liquid-like behavior to a solid-like structure, which reveals a transition to a gel phase. For the latter case, the amount of elastically active chains, that is, the bridges between flower-like micelles is higher and form an extensive network which is reflected in increased viscosity of solutions. These materials have a great potential in applications like thickeners for cosmetic lotions, paints, oil recovery, textile materials, and other industrial products.

Chitosan/glycerol-phosphate solutions present a thermo-gelling behavior as reported by Chenite et al. [75]. They prove that chitosan solutions can be neutralized and no gel-like precipitation is taking place upon using glycerol-phosphate. Subjecting the samples to increased temperatures it was noticed the formation of hydrogel. The addition of the basic salt modulates hydrophobic interactions and hydrogen bonding which are useful to keep chitosan in solution at neutral pH at 4°C, and then allow thermogelling at 37°C. Rheological data reveal endothermic gelation of samples and facilitated the formulation of a sol/gel diagram. The gel occurrence is the result of the interplay between temperature and pH. These materials are utilized in pharmaceutical and food industries.

Lewandowska [76] analyzed the rheological characteristics of chitosan blends with partially hydrolyzed polyacrylamide (HPAM) in several solvents (aqueous acetic acid, lactic acid (LA), and aqueous acetic acid/NaCl). The examined chitosan-based solutions displayed a non-Newtonian behavior with shear-thinning and/or shear-thickening zones. In other words, the flow index values are subunitary as expected for pseudoplastic behavior for HPAM, chitosan, and their blends in most solvents. When using aqueous acetic acid, the flow index is higher than 1 showing shear-thickening behavior.

The highest flow activation energy was observed for the polymer solutions in aqueous acetic acid. In addition, this parameter is higher for the polymer blends in regard with that of the individual counterpart's solutions in aqueous 0.1 M CH_3COOH/0.2 M NaCl. The result reveals that there are certain interactions among the polymer components in the blend systems.

1.4 CONCLUDING REMARKS

The market of green polymer materials is in continuous progress as the environmental requirements are demanding less polluting products. The rheological behavior of polymers is paramount for the processing stage. Polymer films of uniform thickness are achieved when the solution has a viscosity independent of the shear deformation. The role of rheology in solution processing into polymer films is discussed. Several systems of green polymers in solution are reviewed from rheological point of view showing that the solvent, temperature, or additive is strongly impacting the viscosity dependence on shear rate and viscoelastic behavior.

ACKNOWLEDGMENT

The financial support of European Social Fund for Regional Development, Competitiveness Operational Programme Axis 1-Project "Petru Poni Institute of Macromolecular Chemistry-Interdisciplinary Pol for Smart Specialization through Research and Innovation and Technology Transfer in Bio(nano)polymeric Materials and (Eco)Technology," InoMatPol (ID P_36_570, Contract 142/10.10.2016, cod MySMIS: 107464) is gratefully acknowledged.

KEYWORDS

- **cellulose nanocrystals**
- **green polymers**
- **processability**
- **viscoelasticity**
- **viscosity**
- **waterborne polyurethane**

REFERENCES

1. Ellis, B., (2000). Environmental issues in electronics manufacturing: A review. *Circuit World*, *26*, 17–21.
2. *How Industrial Applications in Green Chemistry are Changing Our World*. https://chemical.report/whitepapers/how-industrial-applications-in-green-chemistry-are-changing-our-world/2067 (accessed on 3 August 2020).
3. Doble, M., & Kruthiventi, A. K., (2007). *Catalysis and Green Chemistry, in Green Chemistry and Engineering* (pp. 53–67). Elsevier: India.
4. Anastas, P., (2011). Twenty years of green chemistry. *Chem. Eng. News*, *89*(26), 62–65.
5. *Green and Natural Polymers are on the Rise*. https://www.polymersolutions.com/blog/green-and-natural-polymers-on-the-rise/ (accessed on 3 August 2020).
6. Jin, S., Byrne, F., McElroy, C. R., Sherwood, J., Clark, J. H., & Hunt, A. J., (2017). Challenges in the development of bio-based solvents: A case study on methyl(2,2-dimethyl-1,3-dioxolan-4-yl)methyl carbonate as an alternative aprotic solvent. *Faraday Discuss*, *202*, 157–173.
7. Kobayashi, S., (2017). Green polymer chemistry: New methods of polymer synthesis using renewable starting materials. *Struct. Chem.*,*28*, 461–474.
8. Worthington, M. J. H., Kucera, R. L., & Chalker, J. M., (2017). Green chemistry and polymers made from sulfur. *Green Chem., 19*, 2748–2761.
9. Gomez, J. G. C., Méndez, B. S., Nikel, P. I., Pettinari, M. J., Prieto, M. A., & Silva, L. F., (2011). Making green polymers even greener: Towards sustainable production of polyhydroxyalkanoates from agro-industrial by-products. In: Petre, M., (ed.), *Advances in Applied Biotechnology* (pp. 41–62). InTech: Croatia.
10. Hong, M., & Chen, E. Y. X., (2019). Future directions for sustainable polymers. *Science and Society, Special Issue Part Two: Big Questions in Chemistry, 1*, 148–151.

11. Gawel, E., Pannicke, N., & Hagemann, N., (2019). A path transition towards a bio-economy-The crucial role of sustainability. *Sustainability, 11*, 3005.
12. Thompson, R. C., Moore, C. J., Vom, S. F. S., & Swan, S. H., (2009). Plastics, the environment and human health: Current consensus and future trends. *Philos Trans. R. Soc. Lond. B Biol. Sci., 364*, 2153–2166.
13. Prata, J. C., Silva, A. L. P., Da Costa, J. P., Mouneyrac, C., Walker, T. R., Duarte, A. C., & Rocha-Santos, T., (2019). Solutions and integrated strategies for the control and mitigation of plastic and microplastic pollution. *Int. J. Environ. Res. Public Health, 16*, 2411.
14. D'Ambrières, W., (2019). Plastics recycling worldwide: Current overview and desirable changes. *Field Actions Science Reports* [Online], p. 19 http://journals.openedition.org/factsreports/5102 (accessed on 3 August 2020).
15. Mathers, R. T., & Meier, M. A. R., (2011). *Green Polymerization Methods: Renewable Starting Materials, Catalysis, and Waste Reduction*. Wiley: Germany.
16. National Research Council, (1994). *Polymer Science and Engineering: The Shifting Research Frontiers*. Washington, DC: The National Academies Press. https://doi.org/10.17226/2307 (accessed on 3 August 2020).
17. Denson, C. D., (1984). Polymer processing operations other than shaping. In: Astarita, G., & Nicolais, L., (eds.), *Polymer Processing and Properties*. Springer: Boston.
18. Baird, D. G., & Collias, D. I., (2014). *Polymer Processing: Principles and Design*. Wiley: USA.
19. Han, C. D., (2007). Rheology and processing of polymeric materials. *Volume 1: Polymer Rheology*. Oxford University Press: USA.
20. Miller-Chou, B. A., & Koenig, J. L., (2003). A review of polymer dissolution. *Prog. Polym. Sci., 28*, 1223–1270.
21. Su, W. F., (2013). Polymer size and polymer solutions. In: Su, W. F., (ed.), *Principles of Polymer Design and Synthesis* (pp. 9–26). Lecture notes in chemistry. Springer: Germany.
22. Danner, R. P., & High, M. S., (1993). *Handbook of Polymer Solution Thermodynamics*. Wiley: USA.
23. Tsuchida, E., (1982). *Interactions between Macromolecules in Solution and Intermacromolecular Complexes*. Springer: USA.
24. Ueberreiter, K., (1968). The solution process. In: Crank, J., & Park, G. S., (eds.), *Diffusion in Polymers* (pp. 219–257). Academic Press: New York.
25. Cooper, W. J., Krasicky, P. D., & Rodriguez, F., (1985). Effects of molecular weight and plasticization on dissolution rates of thin polymer films. *Polymer, 26*, 1069–1072.
26. Manjkow, J., Papanu, J. S., Hess, D. W., Soane (Soong), D. S., & Bell, A. T., (1987). Influence of processing and molecular parameters on the dissolution rate of poly-(methyl methacrylate) thin films. *J. Electrochem. Soc., 134*, 2003–2007.
27. Wang, R., & Wang, Z. G., (2012). Theory of polymers in poor solvent: Phase equilibrium and nucleation behavior. *Macromolecules, 45*, 6266–6271.
28. Ouano, A. C., & Carothers, F. A., (1980). Dissolution dynamics of some polymers: Solvent-polymer boundaries. *Polym. Eng. Sci., 20*, 160–166.
29. Pekcan, O., Ugur, S., & Yilmaz, Y., (1997). Real-time monitoring of swelling and dissolution of poly(methyl methacrylate) discs using fluorescence probes. *Polymer, 38*, 2183–2189.

30. Pekcan, O., & Ugur, S., (2002). Molecular weight effect on polymer dissolution: A steady state fluorescence study. *Polymer, 43*, 1937–1941.
31. Ugur, S., & Pekcan, O., (2000). Fluorescence technique to study thickness effect on dissolution of latex films. *J. Appl. Polym. Sci., 77*, 1087–1095.
32. Drummond, R., Boydston, G. L., & Peppas, N. A., (1990). Properties of positive resists. III. The dissolution behavior of poly(methyl methacrylate-co-maleic anhydride). *J. Appl. Polym. Sci., 39*, 2267–2277.
33. Malkin, A. Y., & Isayev, A. I., (2006). *Rheology: Concepts, Methods, and Applications.* ChemTec. Publishing: Toronto.
34. *How Polymers Behave in Dilute Solutions.* https://pslc.ws/macrog/property/solpol/ps5.htm (accessed on 3 August 2020).
35. Chisca, S., Barzic, A. I., Sava, I., Olaru, N., & Bruma, M., (2012). Morphological and rheological insights on polyimide chain entanglements for electro spinning produced fibers. *J. Phys. Chem. B., 116*, 9082–9088.
36. Han, C. D., (2007). Rheology and processing of polymeric materials. *Volume 2: Polymer Processing*. Oxford University Press: USA.
37. Cosutchi, A. I., Hulubei, C., Stoica, I., & Ioan, S., (2010). Morphological and structural-rheological relationship in epiclon-based polyimide/hydroxypropylcellulose blend systems. *J. Polym. Res., 17*, 541–550.
38. Barzic, A. I., Rusu, R. D., Stoica, I., & Damaceanu, M. D., (2014). Chain flexibility versus molecular entanglement response to rubbing deformation in designing poly(oxadiazolenaphthylimide)s as liquid crystal orientation layers. *J. Mater. Sci., 49*, 3080–3098.
39. Stoica, I., Barzic, A. I., & Hulubei, C., (2016). Polyimide embedding in lyotropic polymer matrix for surface-related applications: Rheological and microscopy investigations. *Rev. Roum. Chim., 61*, 575–581.
40. Cosutchi, A. I., Hulubei, C., & Ioan, S., (2007). Rheological study of some epiclon-based polyimides. *J. Macromol. Sci. Part B., 46*, 1003–1012.
41. Zhao, K., Yu, X., Li, R., Amassian, A., & Han, Y., (2015). Solvent-dependent self-assembly and ordering in slow-drying drop-cast conjugated polymer films. *J. Mater. Chem. C., 3*, 9842–9848.
42. Albu, R. M., Hulubei, C., Stoica, I., & Barzic, A. I., (2019). Semi-alicyclic polyimides as potential membrane oxygenators: Rheological implications on film processing, morphology, and blood compatibility. *Express Polym. Lett., 13*, 349–364.
43. Barzic, A. I., Soroceanu, M., Albu, R. M., Ioanid, E. G., Sacarescu, L., & Harabagiu, V., (2019). Correlation between shear-flow rheology and solution spreading during spin coating of polysilane solutions. *Macromol. Res.* doi: 10.1007/s13233-019-7164-7.
44. Bornside, D. E., Macosko, C. W., & Scriven, L. E., (1989). Spin coating: One-dimensional model. *J. Appl. Phys., 66*, 5185–5193.
45. Fujita, K., Ishikawa, T., & Tsutsui, T., (2002). Novel method for polymer thin film preparation: Spray deposition of highly diluted polymer solutions. *Jpn. J. Appl. Phys., 41*, L70.
46. Matsui, J., Yoshida, S., Mikayama, T., Aoki, A., & Miyashita, T., (2005). Fabrication of polymer Langmuir-Blodgett films containing regioregular poly(3-hexylthiophene) for application to field-effect transistor. *Langmuir, 21*, 5343–5348.
47. Emslie, A. G., Bonnet, E. T., & Peck, L. G., (1958). Flow of a viscous liquid on a rotating disk. *J. Appl. Phys., 29*, 858–862.

48. Acrivos, A., Shah, M. J., & Petersen, E. E., (1960). On the flow of a non-Newtonian liquid on a rotating disk. *J. Appl. Phys., 31*, 963–968.
49. Berni, A., Mennig, M., & Schmidt, H., (2004). Doctor blade. In: Aegerter, M. A., & Mennig, M., (eds.), *Solgel Technologies for Glass Producers and Users* (pp. 89–92). Springer: Boston.
50. *Green Polymer Additives*. https://greenpolymeradditives.emeryoleo.com/additives/ (accessed on 3 August 2020).
51. Lv, Y., Wu, J., Zhang, J., Niu, Y., Liu, C. Y., He, J., & Zhang, J., (2012). Rheological properties of cellulose/ionic liquid/dimethylsulfoxide (DMSO) solutions. *Polymer, 53*, 2524–2531.
52. Gericke, M., Schlufter, K., Liebert, T., Heinze, T., & Budtova, T., (2009). Rheological properties of cellulose/ionic liquid solutions: From dilute to concentrated states. *Bio-Macromolecules, 10*, 1188–1194.
53. Sescousse, R., Le, K. A., Ries, M. E., & Budtova, T., (2010). Viscosity of cellulose-imidazolium-based ionic liquid solutions. *J. Phys. Chem. B., 114*, 7222–7228.
54. Chen, X., Zhang, Y. M., Wang, H. P., Wang, S. W., Liang, S. W., & Colby, R. H., (2011). Solution rheology of cellulose in 1-butyl-3-methyl imidazolium chloride. *J. Rheol., 55*, 485–494.
55. Druel, L., Niemeyer, P., Milow, B., & Budtova, T., (2018). Rheology of cellulose-[DBNH][CO_2Et] solutions and shaping into aerogel beads. *Green Chem., 20*, 3993–4002.
56. Fort, D. A., Swatloski, R. P., Moyna, P., Rogers, R. D., & Moyna, G., (2006). Use of ionic liquids in the study of fruit ripening by high-resolution ^{13}C NMR spectroscopy: 'Green' solvents meet green bananas. *Chem. Commun., 7*, 714–716.
57. Evlampieva, N. P., Vitz, J., Schubert, U. S., & Ryumtsev, E. I., (2009). Molecular solutions of cellulose in mixtures of ionic liquids with pyridine. *J. Appl. Chem., 82*, 666–672.
58. Rinaldi, R., (2011). Instantaneous dissolution of cellulose in organic electrolyte solutions. *Chem. Commun., 47*, 511–513.
59. Gauche, C., & Felisberti, M. I., (2019). Colloidal behavior of cellulose nano crystals grafted with poly(2-alkyl-2-oxazoline)s. *ACS Omega, 4*, 1189 3–11905.
60. Larson, R. G., & Magda, J. J., (1989). Coil-stretch transitions in mixed shear and extensional flows of dilute polymer solutions. *Macromolecules, 22*, 3004–3010.
61. Barzic, A. I., Albu, R. M., Gradinaru, L. M., & Buruiana, L. I., (2018). New insights on solvent implications in flow behavior and interfacial interactions of hydroxypropylmethylcellulose with cells/bacteria. *E-Polymers, 18*, 135–142.
62. Navard, P., & Haudin, J. M., (1986). Rheology of hydroxypropyl cellulose solutions. *J. Polym. Sci., 4*, 189–227.
63. Ernst, B., & Navard, P., (1989). Band textures in mesomorphic (hydroxypropyl) cellulose solutions. *Macromolecules, 22*, 1419–1422.
64. Peuvrel, E., & Navard, P., (1991). 24 Band textures of liquid crystalline polymers in elongational flows. *Macromolecules*, 5683–5686.
65. Barzic, A. I., Hulubei, C., Avadanei, M. I., Stoica, I., & Popovici, D., (2015). Polyimide precursor pattern induced by banded liquid crystal matrix: Effect of dianhydride moieties flexibility. *J. Mater. Sci., 50*, 1358–1369.
66. Keerl, D., & Scheibel, T., (2014). Rheological characterization of silk solutions. *Green Mater., 2*, 11–23.
67. Kojic, N., Bico, J., Clasen, C., & McKinley, G. H., (2006). *Ex vivo* rheology of spider silk. *J. Exp. Biol., 209*, 4355–4362.

68. Moriya, M., Roschzttardtz, F., Nakahara, Y., Saito, H., Masubuchi, Y., & Asakura, T., (2009). Rheological properties of native silk fibroins from domestic and wild silkworms, and flow analysis in each spinneret by a finite element method. *Biomacromolecules, 10*, 929–935.
69. Terry, A. E., Knight, D. P., Porter, D., & Vollrath, F., (2004). pH-induced changes in the rheology of silk fibroin solution from the middle division of *Bombyx mori* silkworm. *Biomacromolecules, 5*, 768–772.
70. Tao, Y., Hasan, A., Deeb, G., Hu, C., & Han, H., (2016). Rheological and mechanical behavior of silk fibroin reinforced waterborne polyurethane. *Polymers, 8*, 94.
71. Laity, P. R., & Holland, C., (2017). Thermo-rheological behavior of native silk feedstocks. *Eur. Polym. J., 87*, 519–534.
72. Bochek, A. M., Murav'ev, A. A., Novoselov, N. P., Zabivalova, N. M., Petrova, V. A., Yudin, V. E., Popova, E. N., & Lavrent'ev, V. K., (2013). Rheological properties of mixtures of cellulose with chitin in 1-butyl-3-methylimidazolium acetate of the obtained composite films obtained. *J. Appl. Chem., 86*, 1913–1917.
73. Martínez-Ruvalcaba, A., Chornet, E., & Rodrigue, D., (2004). Dynamic rheological properties of concentrated chitosan solutions. *Appl. Rheol., 14*, 140–147.
74. Shedge, A. S., & Badiger, M. V., (2014). Hydrophobically modified chitosan: Solution properties by rheology and light scattering. *Advancesin Planar Lipid Bilayers and Liposomes, 19*, 91–102.
75. Chenite, A., Buschmann, M., Wang, D., Chaput, C., & Kandani, N., (2001). Rheological characterization of thermo gelling chitosan/glycerol-phosphate solutions. *Carbohydr. Polym., 46*, 39–47.
76. Lewandowska, K., (2012). Rheological properties of chitosan blends with partially hydrolyzed polyacrylamide in different solvents. *Progr. Chem. Appl. Chitin and its Derivatives, 17*, 53–62.

CHAPTER 2

Properties of the Magnetic Polymer Nanocomposites in Magnetic Fields

JIMSHER ANELI[1] and GRIGOR MAMNIASHVILI[2]

[1]*Institute of Machine Mechanics, 10, Mindeli Str., Tbilisi – 0186, Georgia, E-mail: jimaneli@yahoo.com*

[2]*Andronikashvili Institute of Physics of Ivane Javakhishvili Tbilisi State University, 6, Tamarashvili Str., Tbilisi – 0177, Georgia*

ABSTRACT

The processes of self-assembly were studied in the magnetic polymer carbon nanocomposites doped with cobalt nanoclusters. These processes proceed due to the diffusion of magnetic nanoparticles stimulated by a combined effect of an outer steady magnetic fields and heating. The obtained polymer composites are promising for practical applications.

2.1 INTRODUCTION

In the last decade, the investigation of such new nanostructure forms of carbon as nanoparticles, nanotubes, and nanowires has become very topical. This is associated with the fact that, due to their sizes and peculiarities of their atomic structure, nanostructural particles reveal such unique physical-mechanical properties that the range of their promising applications covers many areas of activity from microelectronics to medicine.

In recent years, there increased interest in technologies of production of carbon-based materials oriented on the production of doped carbon nanoparticle modifications (nanotubes, nanoclusters, nanowires). This gives scientists and engineers the opportunity of aimed control of the unique properties of these materials, which are their natural properties [1].

As a matter of fact, the nanoobject control at the nanometer level using nanoparticles with the aim to arrange them in rows, signatures, and grids is the clue to the production of new functional materials. Hence, in recent years, for obtaining of constructional units of different nanometric sizes, many methods of self-assembling and synthesis were developed. In this connection, the possibility to control perfectly the self-assembling and synthesis processes of nanoparticles is a serious challenge from the point of view of both fundamental and applied investigations.

Based on the fundamental principles, the process of self-assembling requires the existence of an interaction between atoms and clusters, as well as of thermodynamic and kinetic driving forces, so that the organization of atoms and clusters for creation of nano-size domain structures should be realized. From this point of view, magnetic nanoparticles deserve a particular interest due to their unique physical-chemical properties and applicability in the new functional material technologies.

Carbon shells provide both the protection of ferromagnetic impurities against aggressive environments and new unique properties for the hydride nanostructures. The self-assembling of magnetic clusters coated with carbon shells represents just such an example that could be used in the contemporary materials, for instance, in strong rare-earth free bonded magnets, analytical instruments (nuclear magnetic resonance tomographs) and nanosensors using magnetic one-dimensional nanowires.

Moreover, currently, due to their low toxicity, magnetic carbon nanoparticles are under testing for therapeutic and diagnostic applications.

In recent years, the magnetic field was used for the creation of nanoscale materials, which resulted in significant achievements in fabrication of macro- and microstructure synthesized materials possessing unique properties.

In contrast with other existing self-assembling technologies, ordering induced by the magnetic field defines the formation of magnetic nanoparticles in ordered structures with unique properties. Therefore, the area of application of carbon magnetic nanoparticles is quite large. It is enough to name such applications as magnetic fluids, plastic scratch-resist glasses, information storage magnetic media, sensors, biomedicine, etc. It should be noted that, in spite of their broad prospects for multifunctional applications, the carbon nanoparticles doped with ferromagnetic clusters have not been well researched [2].

In work [3] the electric and magnetic properties measurements were carried out to study the gradiently anisotropic conducting and magnetic polymer composites synthesized due to self-assembly processes of nanoparticles under influence of the elastic forces developed by the stretching of polymer film composites fabricated on the basis of polyvinyl alcohol (PVA) doped with graphite powder and nickel nanoparticles.

In this work, the study of self-assembling properties of carbon nanoparticles doped with cobalt nanoclasters in magnetic polymer nanocomposites (PNC) under the combined influence of magnetic field and heating will be carried out using methods similar ones developed in work [3].

2.2 RESULTS AND DISCUSSION

To achieve this goal, we planned experiments with magnetic nanopowders, preparation of filled and unfilled polymer films, and the study of self-organization processes in them. This process was facilitated by heating the films above the glass transition temperature. The particles in conventional composites are essentially immobile in contrast to PNC particularly above the glass transition temperature T_g. The nanoparticles mobility can affect polymer dynamics resulting in changes in the viscosity modulus, kinetics of the particle-cluster formation, etc., [3].

The tensile measurements showed that, below T_g, the conventional composites and PNCs behave similarly with respect to mechanical properties. Though, above T_g, the toughness of PNC can increase by an order of magnitude with increasing temperature. It was assumed that the mechanism of toughness enhancement is the mobility of nanoparticles. The development of self-healing materials and coatings where nanoparticles migrate towards various defect sites requires better understanding of the process of nanoparticle diffusion. Heating of PNC above T_g enhances the mobility of polymer chains, which should facilitate the boundary diffusion between polymer interfaces, and this effect should be visualized using the magnetic nanoparticles introduced in the polymer. This process could be improved by applying additional stimuli, in particular a low-frequency (ac) magnetic field, a stationary magnetic field, pressure, heating separately or in combinations, etc. Such an impact stimulates self-assembling processes in the prepared films in the result of which one could produce the films "glued" to each other without using other type glues and polymer melting temperature. One of the objectives of this work was the development of a simple technology of production of carbon nanoparticles doped with ferromagnetic clusters and the study of their morphology and composition.

In particular, for production of carbon-based nanopowders and nanocoatings, the method of chemical vapor deposition (CVD) is used mainly along with application of the process of hydrogen reduction of volatile chlorides. The carbon nanoparticles doped with cobalt magnetic nanoclusters with mean sizes in the range 50–100 nm were synthesized by technology using the combination of pyrolysis of ethanol (and other hydrocarbons), vapor pyrolysis, and the CVD process in the mode of a closed recirculation de cycle with monitored technology parameters.

The developed technological process was realized in the installation, the reactor design and basic units of which provided for the possibility of monitoring of parameters such as the vapor content in reactor zones, catalytic capacity of substrates, partial oxygen pressure (over the range of $10^{-20} \div 10^{-25}$ atm.). This allows carrying out the investigations with the aim to establish the optimal technological parameters for production of finely dispersed carbon nanopowders doped with magnetic nanoclusters. Detailed description of this technology is given in work [4].

For preparation of polymer films, polyvinyl butyral (PVB) polymer with low $T_g \sim 45–55°C$ was chosen. Polymer PVB is a resin mostly used wherever strong binding, optical clarity, adhesion to many surfaces, toughness, and flexibility is required. As filler, we used the carbon nanopowder doped with magnetic (Co or Fe) nanoclusters (C/Co) of our production. For a comparative study, we used commercial Co nanopowder with average diameter of 28 nm (Sun Co., USA). The concentration of used Co-doped carbon nanopowders in the polymeric composite was in the range of 10–50 wt. %, while for the Co nanopowder filled polymer composite concentration was equal to 20 wt. %.

At first 10%, alcohol solution was prepared, than this solution was poured into Teflon press molds and, after their drying for 48 hours, the films one mm thick was obtained. The filled composites were prepared as follows: magnetic nanopowders were taken in the appropriate proportion (in terms of dry weight) and PVB was mixed with alcohol in a usual way, and then followed the ultrasonic treatment for 10–15 min for the destruction of magnetic nanopowder agglomerates. After thorough mixing for 7–10 min, magnetic polymer composite films similar to unfilled ones were obtained in Teflon press molds. From these films, the circular shape disk samples were cut.

To study the self-assembly processes in these PNC at different concentrations of carbon magnetic nanopowder, we used a simple method from work [5]. In this case, circular samples of the polymer composite (diameter-28 mm, thickness-1 mm) were exposed to the magnetic field which was provided by two attached permanent neodymium magnets and temperature of 85°C for two hours (Figure 2.1).

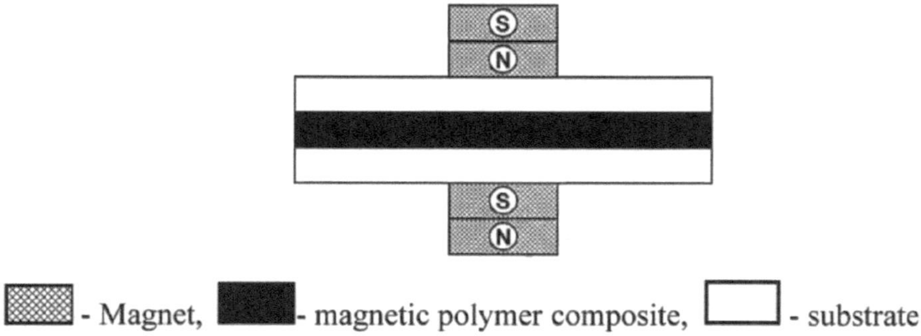

FIGURE 2.1 Geometry of the samples.

Resulting self-assembly of C/Co nanopowders caused changes in their concentration and modulation of local resistance along the radius of the sample, which was measured by a two-contact method as in our previous work [6] (Figures 2.2–2.7). The resistance was measured between the points spaced 2 mm apart along the radius in all following cases except for Figure 2.3 where the resistance was measured between the sample center and the points 2 mm.

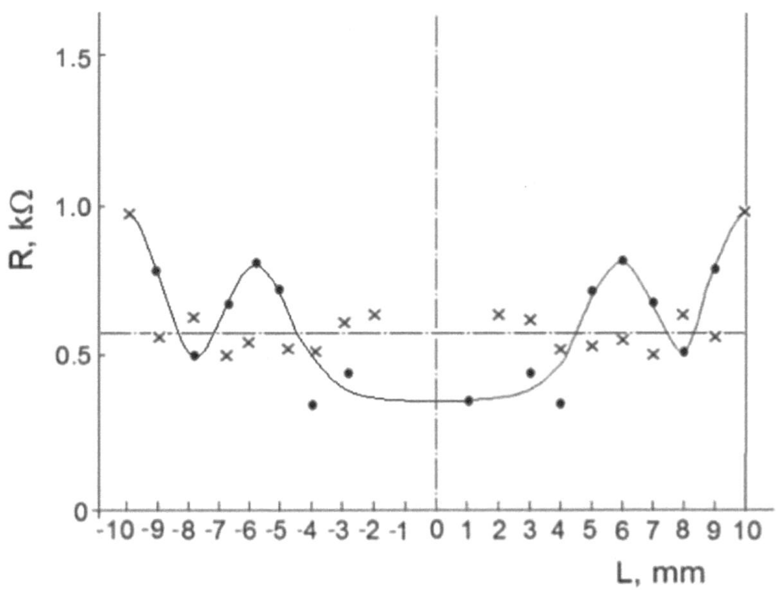

FIGURE 2.2 Polymer composite C/Co, wt. 30%. ×: Sample prepared without magnetic field treatment; •: sample prepared under combined magnetic field and heat treatment.

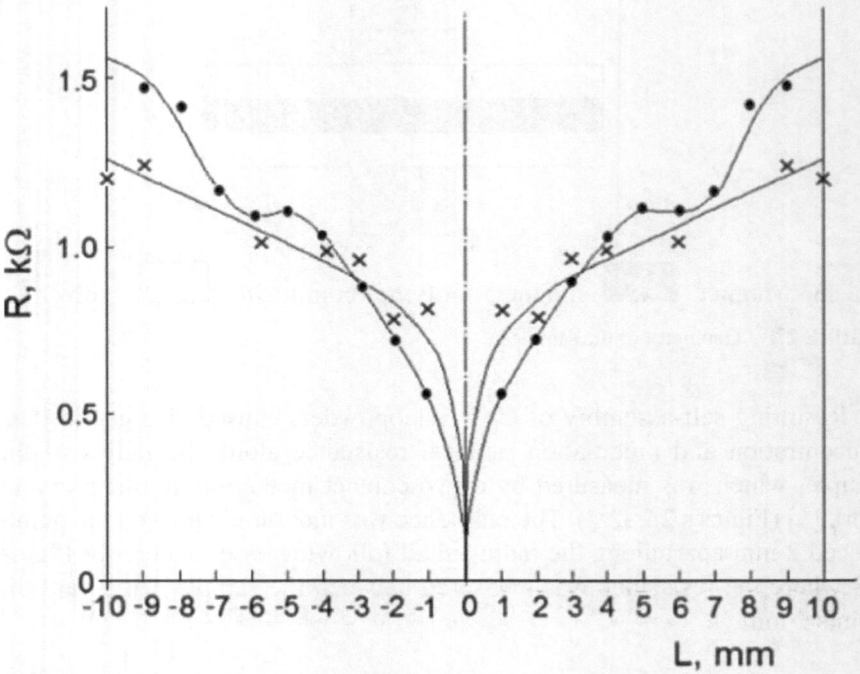

FIGURE 2.3 Polymer composite C/Co, wt. 30%.

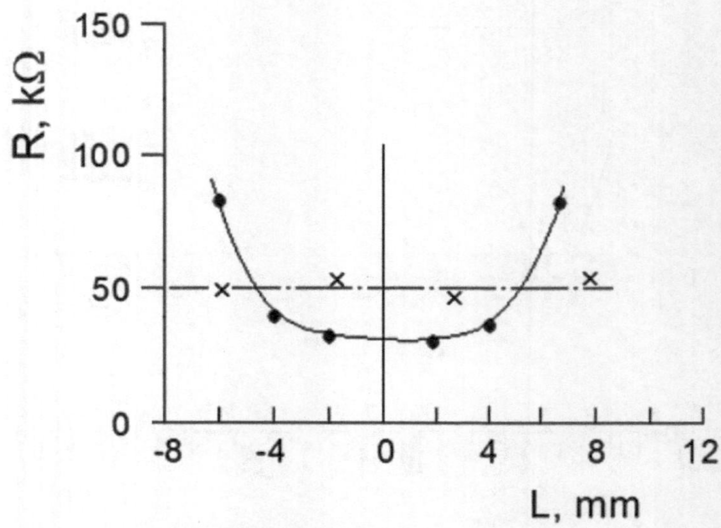

FIGURE 2.4 Polymer composite C/Co, wt. 50%

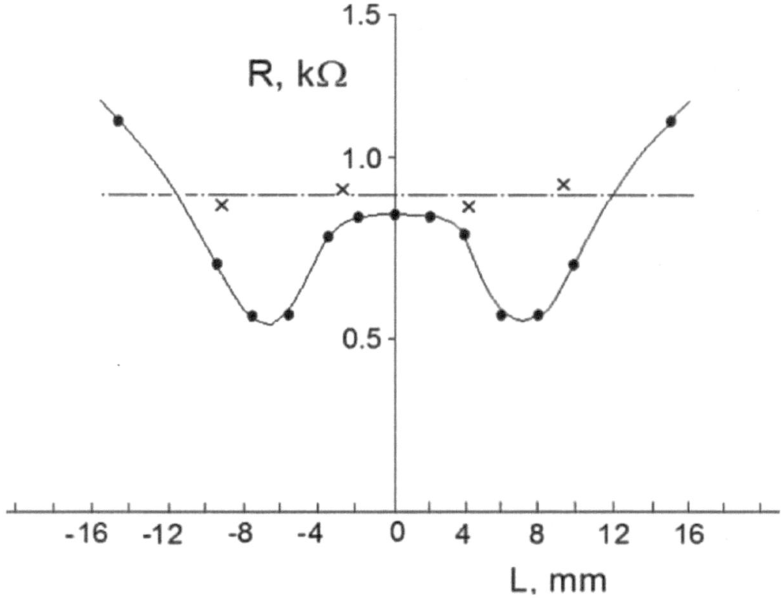

FIGURE 2.5 Polymer composite C/Co, wt. 20%.

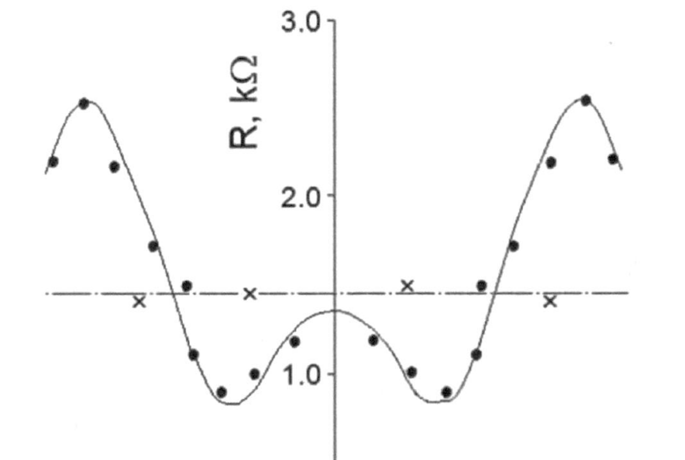

FIGURE 2.6 Polymer composite C/Co, wt. 15%.

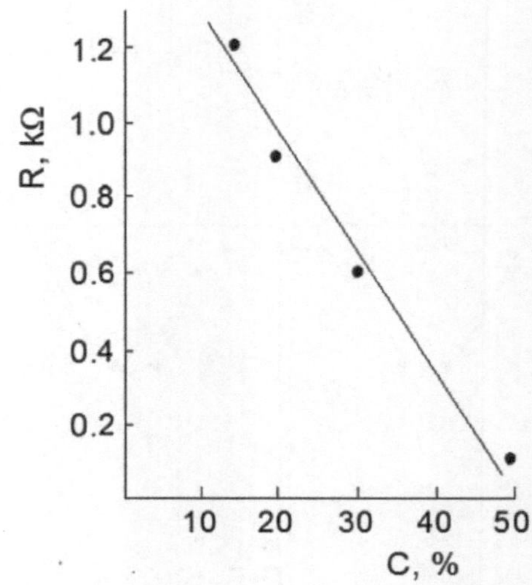

FIGURE 2.7 Dependence of resistance of initial untreated samples on the nanopowder concentration.

Because polymeric composites are magnetic carbon nanocomposites, we were able to study the processes of the self-assembling of nanoparticles using radio frequency resonance magnetic susceptibility measuring devices [5].

The experimental set-up is presented in Figure 2.8. In the inductive coil of the resonance contour of LC-generator, cylindrical tipped ferrite rod is used as a probe. The investigated rectangular shape magnetic polymer composite film is displaced relatively the immovable ferrite tip. The scanning of the film surface is realized along the previously marked disk radius.

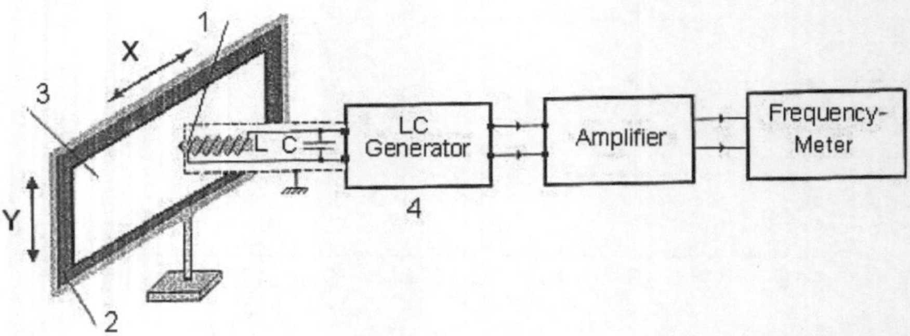

FIGURE 2.8 Scheme of the measuring of the magnetic characteristics of the polymer films. 1-ferrite probe; 2-frame; 3-magnetic polymer nanocomposite film; 4-LC generator.

The change of magnetic particle concentration causes the inductance change dL of the resonance contour of LC-generator resulting in the frequency displacement of LC-generator df related with dL by relation $df/f = \frac{1}{2}\, dL/L$. This frequency displacement could be precisely measured that stipulates the high sensitivity of the method. At the natural frequency of used LC-generator near ~ 2 MHz the observed range of the frequency change df was about ~ 1000 Hz at the precision of the frequency measurement ~1 Hz.

For example, we represent the result of self-assembly measurements of Co/C magnetic nanocomposite film in the magnetic field of neodymium magnet (Figure 2.9) one similar to the results obtained during electrical resistance measurements.

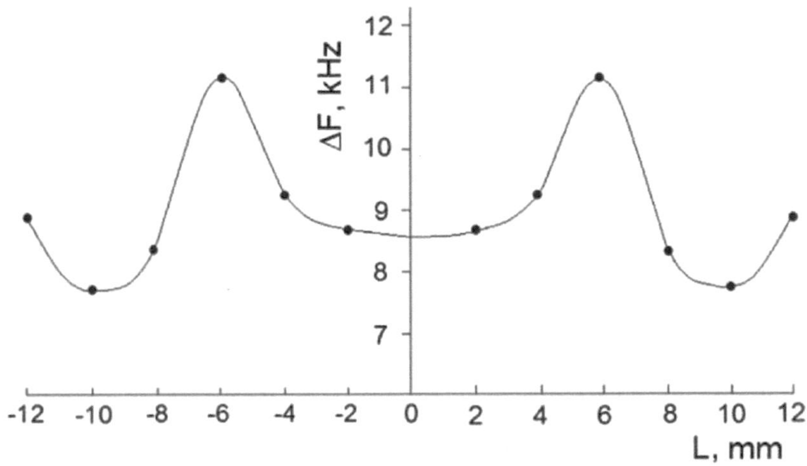

FIGURE 2.9 Magnetic susceptibility measurements of Co/C 50% polymer nanocomposite along the polymer composite film circular disk diameter.

When we move along the diameter of the disk, the frequency measurements indicate frequency change (frequency interval change of several kHz) associated with the change in the concentration of magnetic nanoparticles in the polymer composite.

Based on the obtained experimental results, it is possible to investigate the self-assembling processes in magnetic nanocomposite polymer films synthesized by the elaborated technology using the carbon magnetic nanopowders fabricated by a technology described in [4] under combined influence of magnetic field and heating.

2.3 CONCLUSION

The self-assembling processes were studied in the magnetic PNC synthesized based on the carbon nanoparticles doped with cobalt nanoclusters, synthesized by a unique CVD technology developed by the authors, under combined effect of magnetic field and heating. These processes took place with the diffusion of magnetic nanoparticles stimulated by the combined effect of outer steady magnetic field and heating. The obtained magnetic PNC have good electrical and adhesive properties and are promising for potential practical applications in magneto-electronics.

ACKNOWLEDGMENT

The work was supported by the Shota Rustaveli National Science Foundation and STCU grant # 6213.

KEYWORDS

- **chemical vapor deposition**
- **diffusion**
- **magnetic carbon nanopowder**
- **polymer nanocomposites**
- **polyvinyl butyral**
- **resistance**

REFERENCES

1. Jimenez-Contrares R., (2007). *Nanotechnology Research Developments*. Nova Science Publishers, Inc., New York.
2. King, V. R., (2007). *Nanotechnology Research Advances*. Nova Science Publishers, Inc., New York.
3. Grabowski, C. A., & Mukhopadhyay, A., (2014). *Macromolecules, 47*, 7238.
4. Gegechkori, T., Mamniashvili, G., Kutelia, E., et al., (2015). *J. Magn. Magn. Mater., 373*, 200.
5. Mette, C., Fischer, F., & Dilger, K., (2015). Proceedings of the institution of mechanical engineers. *Part L: Journal of Materials: Design and Applications, 229*(2), 166–172.
6. Aneli, J., Nadareishvili, L., & Shamanauri, L., (2016). *Journal of Electrical Engineering 4* (4), 196–202.

CHAPTER 3

Microbial Fuel Cell: A Green Microbial Approach for Production of Bioelectricity Using a Conducting Polymer

SWAPNALI A. PATIL, AKSHADA A. BAKLIWAL, and SWATI G. TALELE

Department of Pharmaceutics, Sandip Institute of Pharmaceutical Sciences, Nashik, Maharashtra, India, E-mail: swatitalele77@gmail.com (Swati G. Talele)

ABSTRACT

Microbial fuel cell (MFC) is one of the recently advanced sciences of using microbes for the generation of electricity. Microbial fuel cell undergoes an aerobiotic ferment of organic and inorganic matter to digest fuel which is obtained in energy form. This methodology is not so applicable due to expensiveness. MFC is an electrochemical science where chemical energy is converted to electrical energy using different microbes. In other words, MFC can also be titled as cellular respiration. The basic concept in is the reduction-oxidation respiration in the microbial culture which generates an energy in electrical form. The movement of electrons produces a potential difference which in turn helps to generate fuel. A path of hope for MFC conceptual is progressively studied to develop a new electrode in recent days. MFC is into a study to upgrade the concept with cost-effectiveness and helpful application of technique in new glance.

3.1 HISTORY

In 20th century, the idea of generating current from microbial cell came into highlight. M. C. Potter studied the concept in 1911 [1]. Potter produced electricity from the microbe *Saccharomyces cerevisiae*, but his work was not up to completion. Branet Cohen in the year 1931 made microbial half fuel cells which when was lined up in series was able to generate about 35 volts electric current [2].

A work by DelDuca et al. added hydrogen generated by the fermenting glucose by *Clostridium butyricum* as the reacting agent at the +ve end that is anode of a hydrogen along with air fuel cell. The cell worked out but it was not successfully carried out due to the owing to hydrogen instability by microbial production [3]. This complication was solved in 1976 by Suzuki et al. [4] who developed the MFC [5].

After in 1970 the concept of MFC was little, clear. The concept was put forward by Robin M. Allen and by H. Peter Bennetto. Population found the fuel cell as a possibility concept for the generating electric current in exploring countries. Bennetto's studies in 1980s, made a clear picture of operation of fuel cell came into focus.

The University of Queensland, Australia, in 2007 studied MFC as a collective effort with Foster's Brewing. The prototype, a 10 L design, transferred brewery wastewater into CO_2(carbon dioxide), pure water and electric current. The group planned to develop a pilot batch for presenting in an international conference on bioenergy [6].

3.2 INTRODUCTION

Due to growing population there is increase in necessitate of energy. So the new concept of deriving energy from bacterial has widely coming up. Microbial fuel cells (MFCs) is a process in which bacterium is selected as the catalyzer that does oxidation of organic and inorganic material and produce current flow [7–11]. The resources of energy are fuel fossil, nuclear energy and recycling resources depend on their sources and ability to exist. The nonrenewable resources are fossil fuel and nuclear energy. Unearthing and the carbon dioxide gas present under the earth and the fossil fuel are the reason for the global warming and environmental pollution [12]. For solving the problem of energy deficiency and problem associated with fossil fuel all countries of the world are trying to find a solution for recycling of energy sources. So MFC is one of the renewed techniques for production of electric energy by using metals as a catalyzer. MFC technique is one of the advance tech above the different energy-producing techniques to over the harmful oxides with greater effect. But the big drawback of this advanced technique is that the cost of application is too high [13]. An MFC is a process in which chemical energy is transformed formed by the oxidizing of organic and inorganic compound to produce adenosine triphosphate (ATP) by reacting with electrons to produce electric energy [14]. The common MFC consist of (+ve) Anode and (–ve) cathode part which is separated by a membrane. The microbes settle on anode and metabolism of organic compound such as glucose takes place. That acts as electron donor. The organic compound the electron and proton are produced by metabolism of organic compound. On the anode surface, the electron gets transferred.

From anode, the movement of electron to cathode takes place through the electrical circuit, while the protons move through the electrolyte and then reach the cationic membrane. To increase the speed of oxygen reduction on the cathode, Platinum is used as a catalyzer. Platinum electrodes are expensive the replacement of platinum electrode with other electrode does give proper oxygen reduction at cathode electrode [15, 16].

3.2.1 MFC DESIGNS

Different types of MFC are available. Most commonly use and less expensive is the two-chamber MFC developed on H pattern which has two bottles linked to each other by a tube which acts as a separator which is cation exchange membrane called Nafion, Ultex or plain salt bridge [17, 18]. The cation exchange membrane is also named as proton exchange membrane. In the H pattern, the proton exchange membrane is fixed in between the tubes of the connecting bottles. The two membranes can be kept together to get a large surface area. U shape tube heated and bent are used to join the bottles. The bottles are filled with agar and salt. Which acts as a cation exchange membrane. This method is less expensive. The salt bridge gives less amount of power because of high internal resistance. H shape pattern is used for initial research work for studying the generation of energy with different material or different microbial compounds. The surface area affects the amount of energy generated. H-shape type is the common parameter, such as studying power generation using recent available materials, of microbial communities that come up due to the degradation of certain compounds, but they commonly generate density of low power.

The power created relies upon the surface territory of cathode contrasted with that of anode [18] and the outside of the film [19]. The power thickness (P) produced by these systems is for the most part reliant on high inner opposition and cathode based misfortunes. When contrasting force produced by this sort, it makes to look at them based on similarly measured anodes, cathodes, and layers. By ferricyanide as an electron, acceptor at cathode chamber causes increment in the power thickness because of the great electron availability acceptor at high focuses. Due to Ferricyanide, causes increment in power by 1.5 to 1.8 occasions to a Pt-impetus cathode and broke down oxygen (H-structure reactor with a Nafion CEM). The most noteworthy thickness of intensity so far has detailed for MFC frameworks has low inside obstruction framework with ferricyanide at the cathode. While ferricyanide is a best catholyte as far as execution, it ought to be artificially recovered and its motivation of utilization isn't worth by and by. In this manner, the utilization of ferricyanide is restricted for central lab studies. It isn't important to put the cathode in water or in a different chamber while utilizing oxygen at the cathode. The cathode is in direct contact with air either in the nearness or nonappearance of a film [21]. In one sort of framework kaolin dirt, based separator and graphite cathode are joined to shape a consolidated separator-cathode structure [31]. High power densities have been gotten utilizing oxygen as the electron acceptor when watery cathodes are changed with air-cathodes. In the most straightforward framework type, the anode and cathode are put on either side of a cylinder, with the anode shut with a level plate and the cathode presented to air on one side, and water on the opposite side. At the point when a film is applied in this air-cathode framework, it conveys as fundamentally to repel water from spilling through the cathode, however, it additionally lessens oxygen dispersion at the anode chamber. The utilization of oxygen by microbes in the anode chamber can cause in a lower Coulombic productivity (which can be characterized as the portion of electrons secured as present versus the greatest conceivable recuperation) [21]. Hydrostatic weight on the cathode makes the break of water; however, that can be decreased by coatings, for example, polytetrafluoroethylene (PTFE), to the external side of the cathode which permits oxygen dispersion yet stops enormous loss of water [22]. Numerous minor departures from these structures have created in to attempt to improve control thickness or produce ceaseless course through the anode chamber (in change to the above frameworks which were altogether examined in group mode). Systems have been created with an external barrel-shaped reactor with a concentric inward cylinder that is at the cathode [23, 24], and with an internal round and hollow reactor (anode comprising of granular media) with the cathode outwardly [25]. Another change in plan the framework an up-flow fixed-bed biofilm reactor, with the liquid streaming persistently through permeable anodes toward a layer isolating the anode from the cathode chamber [26]. Frameworks have been created to look like hydrogen energy components, where a CEM is sandwiched between the anode and cathode. To expand the framework voltage,

MFCs can be contemplated with the frameworks structured in the arrangement of level plates or connected together in arrangement structure.

3.3 MANUFACTURING AND COST

The majority of the above have been exhibited from a fundamental science lab-based point of view, favorable with prior innovation arrange improvement, under controlled conditions. In any case, the best challenge for any innovation moving into this present reality is its reasonableness for assembling, which thus, drives economies of scale. The equivalent applies to the MFC improvement and this has been a piece of the test in the innovation taking off and becoming financially available. Most of the center parts and segments can be talked and along these lines costly, even at proto kind level, and there is a logical test in recognizing choices that would (i)perform similarly well and for delayed periods however, above all (ii)be economical and broadly accessible. One of the roads scientists have investigated is the elective ease materials including ceramic.

3.4 APPLICATION

One of the main applications could be the advancement of pilot-scale reactors at mechanical areas where a high caliber and solid influent are accessible. Nourishment handling wastewaters and digester effluents are great competitors. To look at the potential for power age at such a site, consider a nourishment preparing plant creating 7500 kg/d of waste organics in an emanating. This speaks to a potential for 950 kW of intensity, or 330 kW expecting 30% productivity. At an accomplished intensity of 1 kW/m^3, a reactor of 350 m^3 is required, which would generally cost 2.6 M Euros, at current costs. The created vitality, determined based on 0.1 Euros per kWh, is worth about 0.3 M Euros every year, giving ten-year restitution without different contemplations of vitality misfortunes or additions contrasted with other (high-impact) advancements. The reduction in production will reduce the payback time. In the long haul progressively weaken substrates, for example, household sewage, could be treated with MFCs, diminishing society's have to put generous measures of vitality in their treatment. A changed exhibit of elective applications could likewise rise, running from biosensor improvement and continued vitality age from the ocean bottom, to biobatteries working on different biodegradable powers. While full-scale, profoundly successful

MFCs are not yet inside our grip, the innovation holds impressive guarantee, and real obstacles will without a doubt be overwhelmed by designers and researchers. The increasing weight on our condition, and the call for sustainable power sources will further invigorate the advancement of this innovation, driving soon we would like to its effective usage.

3.4.1 GENERATION OF POWER

MFCs are alluring for power generation applications that require just low control, however where supplanting batteries might be unrealistic, for example, remote sensor networks [27–29]. Wireless sensors, controlled by microbial energy units can then for instance be utilized for remote checking (conservation) [30].

Basically, any natural material could be utilized to sustain the energy unit, including coupling cells to wastewater treatment plants. Compound procedure wastewater [31, 32] and engineered wastewater [33, 34] have been utilized to create bioelectricity in double and single-chamber mediators MFCs (uncoated graphite cathodes).

Higher power creation was seen with a biofilm-secured graphite anode [35, 36]. Fuel cell discharges are well under administrative limits [35]. MFCs use vitality more effectively than standard interior ignition motors, which are restricted by the Carnot Cycle. In principle, a MFC is fit for vitality proficiency a long way past 50% [36]. Rozendal acquired vitality change to hydrogen multiple times that of traditional hydrogen generation innovations.

MFCs can likewise work at a little scale. Anodes sometimes need just be 7 μm thick by 2 cm long [37] to such an extent that a MFC can supplant a battery. It gives an inexhaustible type of vitality and shouldn't be revived.

MFCs work well in gentle conditions, 20 °C to 40 °C and furthermore at pH of around 7 [38]. They do not require for long term medicinal applications, for example, in pacemakers. Power stations can be founded on amphibian plants, for example, green growth. Whenever sited neighboring a current power framework, the MFC framework can share its power lines.

3.4.2 EDUCATION

Soil-based microbial energy components fill in as instructive devices, as they include numerous logical orders (microbiology, geochemistry, electrical building, and so forth.) and can be made utilizing normally accessible materials, for example, soils, and things from the freeze. Packs for home science tasks and study halls are available. One case of microbial power devices being utilized in the study hall is in the IBET (Integrated Biology, English, and Technology) educational plan for Thomas Jefferson High School for Science and Technology. A few instructive recordings and articles are additionally accessible on the International Society for Microbial Electrochemistry and Technology (ISMET Society).

3.4.3 BIOSENSOR

The flow produced from a microbial power module is straightforwardly relative to the vitality substance of wastewater utilized as the fuel. MFCs can gauge the solute centralization of wastewater (i.e., as a biosensor). Wastewater is ordinarily evaluated for its biochemical oxygen demand (BOD) values. MFC BOD sensors underestimate BOD values in the presence of these electron acceptors normally actuated slop gathered from wastewater plants. A MFC-type BOD sensor can give constant BOD esteems. Oxygen and nitrate are favored electron acceptors over the terminal, lessening current age from a MFC. MFC BOD sensors think little of BOD esteem within the sight of these electron acceptors. This can be kept away from by hindering oxygen-consuming and nitrate breath in the MFC utilizing terminal oxidase inhibitors, for example, cyanide, and azide. Such BOD sensors are monetarily accessible.

The United States Navy is thinking about microbial power devices for ecological sensors. The utilization of microbial energy components to control natural sensors would almost certainly give capacity to longer periods and empower the accumulation and recovery of undersea information without a wired framework. The vitality made by these energy units is sufficient to support the sensors after an underlying startup time [34]. Due to undersea conditions (high salt focuses, fluctuating temperatures, and constrained supplement supply), the Navy may send MFCs with a blend of salt-tolerant microorganisms. A blend would consider a progressively complete usage of accessible supplements. *Shewanella oneidensis* is their essential competitor, however, may incorporate other warmth and cold-tolerant *Shewanella* spp. [35].

A first self-controlled and independent BOD/COD biosensor has been created and permits to distinguish natural contaminants in freshwater. The senor depends just on power delivered by MFCs and works constantly without upkeep. The biosensor turns on the caution to educate about sullying level: the expanded recurrence of the sign advises about higher defilement level, while low recurrence cautions about low tainting level [36].

3.4.4 BIORECOVERY

In 2010, Heijne et al. [37] Developed an instrument which was capable of generating electricity and reducing Cu (II) (ion) to copper metal. Microbial electrolysis cells have been constructed to produce hydrogen [38].

3.4.5 WASTEWATER TREATMENT

MFCs are being applicable in water treatment to harvest energy utilizing anaerobic digestion. The system can also decrease pathogens. However, it needs temperatures above of 30°C and requires an extra energy in order to convert biogas to electricity. Spiral spacers can be utilized to expand power generation by making a helical stream in the MFC. Scaling MFCs is an incredible test because of the power yield difficulties of a bigger surface region [39].

3.5 MATERIALS OF CONSTRUCTION

Anode–anodic materials must be conductive, biocompatible, and misleadingly stable in the reactor plan. Metal anodes comprising of noncorrosive tempered steelwork can be used [40], yet copper isn't valuable because of the harmfulness of even follow copper particles to microbes. The most adaptable anode material is carbon, accessible as smaller graphite plates, bars, or granules, as sinewy material (felt, fabric, paper, strands, froth), and as lustrous carbon. There are various carbon providers around the world, for instance, E_TEK and Electro combination Co. Inc. (USA), GEE Graphite Limited, Dewsbury (UK), Morgan, Grimbergen (Belgium), and Alfa-Aesar (Germany). The least complex materials for anode terminals are graphite plates or bars as they are generally economical, simple to deal with, and have a characterized surface zone. Graphite felt cathodes have large surface area [41] which can have high surface regions (0.47 m^2g^{-1}, GF series, GEE Graphite limited, Dewsbury, UK). In any case, not all the showed surface zone will fundamentally be accessible to microscopic organisms. Carbon fiber, paper, froth, and fabric (Toray) have been widely utilized as terminals.

It has been demonstrated that present increments with by and large inside surface region in the request carbon felt > carbon foam > graphite [42]. Generously higher surface region are accomplished either by utilizing a smaller material like reticulated vitreous carbon (RVC; ERG Materials and Aerospace Corp., Oakland, CA) [43] which is accessible with various pore sizes, or by utilizing layers of stuffed carbon granules (Le Carbone, Grimbergen, Belgium) or dots [48]. In the two cases keeping up high porosity is essential to avert stopping up. The long time impact of biofilm development or particles in the stream on any of the above surfaces has not been enough inspected. To expand the anode performance, diverse compound and physical systems have been pursued. Park et al. consolidated Mn (IV) and Fe(III) and utilized covalently connected unbiased red to intercede the electron move to the anode. Electrocatalytic materials, for example, polyanilins/Pt composites have additionally been appeared to improve the present age through helping the immediate oxidation of microbial metabolites.

Coordinating the watercourse through the anode material can be utilized to expand control. Cheng et al. found that stream coordinated through carbon fabric toward the anode, and diminishing terminal dividing from 2 to 1 cm, expanded power densities (standardized to the cathode anticipated surface territory) from 811 to 1540 mW/m^2 in an air-cathode MFC. The expansion was believed to be because of limited oxygen dissemination into the anode chamber, despite the fact that the advective stream could have assisted with proton transport toward the cathode also. Expanded power densities have been accomplished utilizing RVC in an up-flow UASB type MFC or in a granular anode reactor with ferricyanide cathodes. Move through an anode has likewise been utilized in reactors utilizing exogenous go-betweens cathode. Because of its great execution, ferricyanide (K3[Fe(CN)6]) is exceptionally well known as a test electron acceptor in microbial power modules. The best favorable position of ferricyanide is low over potential utilizing a plain carbon cathode, bringing about a cathode working potential near its open circuit potential. The best detriment, be that as it may, is the inadequate reoxidation by oxygen, which requires the catholyte to be consistently supplanted. Likewise, the long haul execution of the framework can be influenced by dispersion of ferricyanide over the CEM and into the anode chamber. Oxygen is the most reasonable electron acceptor for a MFC because of its high oxidation potential, accessibility, minimal effort (it is free), maintainability, and the absence of a compound waste item (water is shaped as the main end product). The decision of the cathode material incredibly influences execution, and is differed dependent on application. For residue power modules, plain graphite plate terminals drenched in the seawater over the silt have been utilized. Due to its extraordinary execution, ferricyanide (K3[Fe(CN)6]) is especially outstanding as a test electron acceptor in microbial power modules. The best good position of ferricyanide is the low over potential using a plain carbon cathode, achieving a cathode working potential close to its open circuit potential. The best impairment, in any case, is the insufficient reoxidation by oxygen, which requires the catholyte to be reliably displaced. Moreover, the whole deal execution of the system can be impacted by scattering of ferricyanide over the CEM and into the anode chamber. Oxygen is the most sensible electron acceptor for a MFC due to its high oxidation potential, openness, insignificant exertion (it is free), practicality, and the nonattendance of a compound waste thing (water is formed as the primary end product). The choice of the cathode material amazingly impacts execution, and is varied reliant on application. For buildup control modules, plain graphite plate terminals doused in the seawater over the residue have been used [51].

3.6 MANUFACTURING AND COST

The perspective, aim summary of the above study has being derived from past technology, research under unmaintained conditions. The wide scope which was the great challenge of the technique was to move into the present world and develop proper proved technique to carry with the economy of the world. Similarly, Microbial fuel cell is one of the challenging jobs in development and technology to make available in the commercial available in market. The development is a challenge to make the product available that to with cost reduction and widely and easily available. One of the option to explore and make the product develop at low cost is the material used for electrodes should be prepared from ceramic, cardboard [55, 56] or derivatives of plant. Geand Heinlong reported a long term pilot study in wastewater treatment plant which reduced the material cost by 60% of Microbial fuel cell because of cation exchanger membrane and so the expenses of the separator was reduced by using ceramic separators as a valid option for this technique. One of the important points of the parts availability like electrodes, membrane, wiring to compile a large scale. The technique developed in laboratory scale need to be expanded in pilot study to make the availability of the technique assessable to the real world under different environmental condition such as temperature change, humidity, batch, and continuous flow pattern. The institutional level knowledge would be a boon in the microbial fuel cell in real-time. Microbial fuel cell reactors and grade can produce electricity spontaneously, potentially, effectively with cost reduction, energy consumption, and treatment cycle. The energy from the vitality investigation perspective, 14 Wh achievedatthe last Pee Power field preliminary at Glastonbury-2017 would be the equivalent to 01.23 British pence of vitality putting something aside for each kWh. This depends on crude power information delivered during the field preliminary where the MFC stack was worked on perfect human urine and doesn't consider the sparing that would be picked up for each liter of wastewater treated. Further comprehension of particle transport selectivity and financial layer arrangement techniques are imperative to en-capable more extensive work of particle trade films in technical forms for manageable improvement. Further progress is expected to give field hardware that is increasingly powerful and solid additional time just as the development of novel vitality stockpiling and vitality collecting strategies. The stored energy is applied as external capacitor is new path in this field. The study complied on the present data is an attempt of new concept in the market. Surface modification of the anode materials also contribute to the boosted up performance of the MFCs. A carbon cloth was modified by coating with biologically reduced Palladium nanoparticles to develop a functional anode. With this surface-modified carbon cloth, MFC has shown improved power density and columbic efficiencies by 14% and 31% respectively. The better execution could be because of the diminished charge move obstruction by the biogenic palladium. Mink et al. have built up an anode of vertically developed multiwall carbon nanotubes (CNTs) with nickel silicide. This manufactured terminal delivered a current thickness of 197 mA/m^2 and a power thickness of 392 mW/m^3 in a micronized MFC of 1.25 μL limit. The outcomes were nearly higher than that accomplished with a plain carbon material. It was asserted that the upgraded presentation was credited to expanded surface to volume proportion for microbial bond, electron move, and low opposition by the nickel silicide reaching material. It has been reasoned that since the procedure doesn't include any costly materials to manufacture MFC, it can significantly diminishes the related expenses. Carbon paper modified with multiwall CNTs and polyethyleneimine through layer by layer assembled technique was used as anode in an MFC. The modification has brought about reduction in interfacial charge transfer through providing three-dimensional network structures for bacterial adhesion. This modified anode was capable of producing 20% higher power densities than an unmodified carbon cloth electrode. Stainless steel cross-sections covered with warmth treated and untreated goethites acquired from mining mud were utilized as anode materials in MFC applications. Higher powers and columbic efficiencies were recorded with goethite catalyst based electrode than conventional stainless steel mesh electrode. The improved performance was due to reduced mass transfer losses and accelerated electron transfer between biocatalysts and electrode. Goethite being cheaper can provide the economically feasible electrode materials with acceptable power generations in MFC. The use of carbon cloth or graphite granules as anodes in lab-scale MFC even though showed better performances, still combating the challenges of scale-up. Carbon cloth being expensive cannot be used in large scale systems. In the same way, well-establishedlab-scale configurations of carbon cloth would be difficult to use in large scales. Similarly, graphite granules are vulnerable to clogging due to low porosities. Materials to be used in biofilm-based bioreactors (MFCs) for wastewater treatment must have high structural strength and should be free from clogging since, they are to be in constant contact with the electron transfer. In addition, the material supports should be intact with the biofilm under any operational condition from open-flow to saturated flow systems. Hence, the use of such support materials is impractical. To overcome the clogging issues, Logan et al. have configured a new anode with high surface area graphite fiber brush bounded to a central core collecting material. The design has led to the development of a successful high conducting and high specific surface area anode which can be used for scale-up MFC applications. 4.2 Cathode materials: In MFC, the cathode chamber is considered as a sink of electrons, in which oxygen is reduced to water. In cathode chamber, reduction of oxygen occurs at air, liquid, and solid three-phase interface. A typical MFC cathode is composed of an electrode support, catalyst, and air diffusion layer. Electrode materials used for anode can also be used for cathode but a potential cathode should have the properties of high electric conductivity, high mechanical strength, and effective catalytic nature. Most commonly, MFCs are being operated under neutral PH and mild temperatures. At these conditions, the rate of oxygen reduction is very low, leading to increase over potentials and thereby limiting the performance of an MFC. For robust cathodic reactions in MFCs, the cathodic carbon-based support materials must be amended with additional catalysts.

Platinum is most regularly utilized cathode impetus because of its high productivity towards oxygen decrease. The utilization of costly metal impetuses as cathode materials in MFCs is obliging the viable uses of MFC innovation. Notwithstanding significant expense, Platinum impetuses are progressively at risk of fouling when utilized with low-quality water. A lot of research endeavors have been put to diminish the expense of cathodic impetuses by finding less expensive options in contrast to Platinum without relinquishing the exhibition. This audit talks about the financially savvy cathode materials and their impact on MFC execution. In one investigation, the carbon material was covered on one side with a blend of carbon powder and poly-tetra-fluoro-ethylene base layer pursued by a few coatings of poly-tetra-fluoro-ethylene as dispersion layer. The water confronting side of the carbon fabric was covered with platinum impetus. In this way, manufactured carbon material was utilized as cathode in a solitary chamber MFC. The MFC has indicated upgraded control densities (42%) and columbic efficiencies (200%) when contrasted with carbon material with base layer alone. It has been accounted for that four coatings of poly-tetra-fluoro-ethylene delivered greatest power thickness as it can diminishes the water misfortunes through cathode. Cathode impetuses made of metal porphyrins and phthalocyanines bolstered on Ketjen black carbon were read for their synergist action towards oxygen decrease in MFCs. Iron phthalocyanine based cathode has indicated more noteworthy oxygen decrease rates at unbiased PH than Platinum impetus in impartial PH. On looking at the impact of carbon substrate, Ketjen black carbon has revealed preferable action over Vulcan XC carbon because of higher surface region of Ketjen black carbon. Most extreme power thickness of 634 mW/m^2 was accomplished with Iron phthalocyanine-Ketjen black carbon at unbiased pH, which is higher than that of costly Platinum impetus (593 mW/m^2) at comparable conditions. It has been concluded from the examinations that the progress metal-based macrocyclicimpetuses are less expensive and can be effectively utilized in enormous scale MFCs for commercialization of the innovation. In a way to deal with create financially savvy interchange to Platinum impetus, Zhang et al. have created initiated carbon air cathode. The cathode was created by virus squeezing actuated carbon with poly-tetra-fluoro-ethylene around a nickel work current authority. The firm has announced the greatest power thickness of 1220 mW/m^2 with the subsequently manufactured cathode, which is relatively a lot higher than that acquired with platinum impetus (1060 mW/m^2). From the outcomes, it was finished up by the gathering that initiated carbon-metal work authority determined cathodes can be utilized as productive and monetarily possible air cathodes in MFCs. Another firm has revealed that initiated carbon-poly-tetra-fluoro-ethylene air cathodes have demonstrated upgrades in power densities with the expansion in actuated carbon stacking. This could be because of the decreases in Warburg impedance, contact obstruction, and accuse move opposition of the actuated carbon load. Regardless of cathodes debasement over 1.5–5 months of utilization, still these cathodes displayed control densities similar to platinum-based cathodes. In an alternate report business actuated carbons from various source materials were alkali gas treated to improve the exhibition as oxygen lessening impetuses in MFCs at unbiased PH. Smelling salts gas treated initiated carbon impetuses have come about preferred exhibitions over untreated enacted carbon on account of the decrease in oxygen substance and improvement in nitrogen bunches on the outside of the impetus material. To upgrade the cathode reactant action of initiated carbon, Xia et al. have built up an actuated carbon-Iron ethylene di-amine tetra-acetic corrosive cathode with tempered steel work current authority. In this way created cathode has displayed altogether improved power thickness (1580 ± 80 mW/m^2), which is 10% more noteworthy than plain actuated carbon cathode, and furthermore, the power thickness is tantamount to platinum impetus based cathode. It was accounted for that the anode is tougher than platinum cathodes. The improved exhibition of the carbon-Iron ethylene di-amine tetra-acetic corrosive cathode was seen as because of the dynamic pyridine, quaternary nitrogen, and iron gatherings. The investigation has announced that the reactant movement of financially feasible enacted carbon cathodes will be improved by pyrolyzing actuated carbon with Iron ethylene di-amine tetra-acetic corrosive. Afterward, Zhang et al. have mixed enacted carbon with carbon dark to lift up the presentation and life span of initiated carbon cathodes. It was accounted for that the most extreme power thickness was diminished by 7% for the carbon dark based cathode while, 61% decrease was accounted for with platinum impetus following five months of ceaseless activity.

Zuo and Logan have examined the power densities of carbon cloth by changing the configurations (tube/flat). Initially, use of a single carbon cloth tube resulted in lower power density than a flat carbon cloth electrode. It was noticed that with the increase in number of concentric tubes, the power density increased due to increase in the cathode surface area. The firm has obtained a power density of 83 mW/m^3 with two carbon cloth tubes as cathode. It was also reported that on wrapping the cathode around anode like a tubular MFC, power density increased to 128 mW/m^3. Finally, from the analysis of the results, it has been concluded that geometry of cathode either tube or flat will not affect the power density. The power density is a dependent factor of cathode surface area rather than geometry. Generally, major costs of MFC cathodes are ascribed by catalysts and catalyst binders. Packed bed air cathodes were developed with four carbon-based materials (granular activated carbon, granular semi-coke, carbon felt cubes, and granular graphite) and the fabrication was devoid of expensive binders and catalysts. Granular activated carbon-based packed bed air cathode has shown the highest power density of 676 ± 93 mW/m^2 and lowest was reported with carbon felt based cathode (60 ± 43 mW/m^2). It was observed that on increasing the amount of granular activated carbon and semicoke with an intention to increase the surface area, power generation was significantly dropped. Thicker layers of carbon materials limited the diffusion of oxygen, thereby reducing the oxygen reduction rate. From the reports, it was deduced that packed bed air cathodes of granular activated carbon and granular semi coke can be used as potential cost-effective alternates to platinum air cathodes. Mesoporous nitrogen-rich carbon developed at different temperatures was used as cathode material to address the issue associated with an expensive platinum catalyst. Mesoporous nitrogen-rich carbon cathode reported 14% less power production than platinum cathode but showed only a 7% reduction after a month operation when it was 11% for platinum cathode. Zhang et al. [49] have experimented with nitrogen-dopedionothermal carbon aerogel as air cathode to improve the rate of oxygen reduction. This aerogel based electrode resulted in 1.7 times higher power density than usual platinum electrode as well as most of the oxygen reducing catalyst air cathodes due to highly porous nature, high surface area and large pore volume. The firm has reported that the proposed aerogel electrode could be an efficient cost-effective air cathode catalyst for MFC applications. In recent years, tremendous studies have been carried out towards the development of design and electrode materials to boost up the performance of MFCs. However, still the technology is facing the challenges for commercialization. In sum up, stringent research is needed to further develop the hardware.

3.7 OUTLOOK

MFC plans need enhancements before an attractive item will be conceivable. Both the issues recognized above and the scale-up of the procedure stay basic issues. The greater part of the structures investigated here can't be scaled to the level required for an enormous wastewater treatment plant which requires many cubic meters of reactor volume. Either the characteristic change pace of MFCs should be expanded, or the plan should be improved so a savvy, huge scale framework can be created. Plans that can most effectively be made in stacks, to create expanded voltages, will be helpful as the voltage for a solitary cell is low. The achievement of explicit MFC applications in wastewater treatment will rely upon the fixation and biodegradability of the natural issue in the influent, the wastewater temperature, and the nonattendance of lethal synthetics. Materials costs will be a huge factor in the all-out reactor costs. For the most part, anodic materials generally utilized in MFC reactors, for example, graphite froths, reticulated vitreous carbon, graphite, and others, are very costly. Rearranged cathodes, for example, carbon strands, may reduce these anode costs. The utilization of costly impetuses for the cathode should likewise be stayed away from. Another essential perspective is the evacuation of non-carbon-based substrates from the waste streams: nitrogen, sulfur, and phosphorus-containing mixes frequently can't be released into nature at influent focuses. So also, even particulate natural mixes should be evacuated and changed over to effectively biodegradable mixes, as a feature of a successful wastewater treatment activity.

Soil-based microbial energy units fill in as instructive apparatuses, as they incorporate various logical orders (microbiology, geochemistry, electrical building, and so on.) and can be made utilizing usually accessible materials, for example, soils, and things from the cooler. Units for home science tasks and study halls are accessible. One case of microbial power devices being utilized in the homeroom is in the IBET educational program for Thomas Jefferson High School for Science and Technology. A few instructive recordings and articles are likewise accessible on the International Society for Microbial Electrochemistry and Technology (ISMET Society).

3.8 CONCLUSION

MFC is a cutting edge innovation for creation of power from digestion of microorganisms. In this survey, we have managed significant squanders and xenobiotic, for example, hexavalent chromium, agro wastes, nitrates, and azo colors. Some of them, for example, hexavalent chromium, and azo colors are exceptionally lethal to the biological system and cause demise of living beings. In MFC, they are utilized for power generation and furthermore they are changed into less harmful metabolites, which exhibits it's another potential use in waste administration and contamination control. Till now, countless microorganisms and a waste assortment of substrates (counting waste and xenobiotic) have been utilized to deliver power. Be that as it may, a noteworthy disadvantage of this innovation is that the power yield is extremely low and scaling up prompts a lessening in power yield. This is the primary motivation behind why this innovation has yet not been marketed. Thus, much more work is required with the goal that this innovation ends up effective, material, and broadly acknowledged.

KEYWORDS

- **adenosine triphosphate**
- **aerobiotic ferment**
- **anode**
- **biochemical oxygen demand**
- **cathode**
- **microbial fuel cells**

REFERENCES

1. Potter, M. C., (1911). Electrical effects accompanying the decomposition of organic compounds. *Proceedings of the Royal Society B: Biological Sciences, 84*(571), 260–276.
2. Cohen, B., (1931). The bacterial culture as an electrical half-cell. *Journal of Bacteriology, 21*, 18–19.
3. DelDuca, M. G., Friscoe, J. M., & Zurilla, R. W., (1963). Developments in industrial microbiology. *American Institute of Biological Sciences, 4*, 81–84.
4. Karube, I., Matasunga, T., Suzuki, S., & Tsuru, S., (1976). Continuous hydrogen production by immobilized whole cells of *Clostridium butyricum. Biochimica et Biophysica Acta, 24*(2), 338–343.
5. Karube, I., Matsunaga, T., Tsuru, S., & Suzuki, S., (1977). Biochemical cells utilizing immobilized cells of *Clostridium butyricum. Biotechnology and Bioengineering, 19*(11), 1727–1733.
6. Jurg Keller, & Rabaey, (2014). *Brewing a Sustainable Energy Solution.* The University of Queensland Australia.
7. Berk, R. S., & Canfield, J. H., (1964). Bioelectrochemical energy conversion. *Appl. Microbiol., 12*, 10–12.
8. Rao, J. R., Richter, G. J., Von, S. F., & Weidlich, E., (1976). The performance of glucose electrodes and the characteristics of different biofuel cell constructions. *Bioelectrochem. Bioenerg., 3*, 139–150.
9. Davis, J. B., & Yarbrough, H. F., (1962). Preliminary experiments on a microbial fuel cell. *Science, 137*, 615–616.
10. Cohen, B., (1931). The bacterial culture as an electrical half-cell. *J. Bacteriol., 21*, 18, 19.

11. Potter, M. C., (1911). Electrical effects accompanying the decomposition of organic compounds. *Proc. R. Soc. London Ser. B, 84*, 260–276. https://doi.org/10.1098/rspb.1911.0073
12. Rahimnejad, M., Adhami, A., Darvari, S., Zirepour, A., & Oh, S. E., (2015). Microbial fuel cell as new technology for bioelectricity generation: A review. *AEJ, 54*, 745–756.
13. Peighambardoust, S., Rowshanzamir, S., & Amjadi, M., (2010). Review of the proton exchange membranes for fuel cell application. *Int. J. Hydrog Energy, 35*, 9349–9384.
14. Torres, C. I., Brown, R. K., Parameswaran, P., Marcus, A. K., Wanger, G., Gorby, Y. A., & Rittmann, B. E., (2009). Selecting anode-respiring bacteria based on anode potential: Phylogenetic, electrochemical, and microscopic characterization. *Environ Sci. Technol., 43*, 9519–9524.
15. Sun, J., Bi, Z., Hou, B., Cao, Y., & Hu, Y., (2011a). Further treatment of decolorization liquid of azo dye coupled with increased power production using microbial fuel cell equipped with an aerobic biocathode. *Water Res., 45*, 283–291.
16. Sun, J., Hu, Y. Y., & Hou, B., (2011b). Electrochemical characterization of the bioanode during simultaneous azo dye decolorization and bioelectricity generation in an air-cathode single chambered microbial fuel cell. *Electrochim. Acta, 56*(19), 6874–6879.
17. Bond, D. R., Holmes, D. E., Tender, L. M., & Lovley, D. R., (2002). Electrode reducing micro organisms that harvest energy from marine sediments. *Science, 295*, 483–485.
18. Park, D. H., (1999). Utilization of electrically reduced neutral red by *Actinobacillus succino genes*: Physiological function of neutral red in membrane-driven fumarate reduction and energy conservation. *J. Bacteriol., 181*, 2403–2410.
19. Oh, S., Min, B., & Logan, B. E., (2004). Cathode performance as a factor in electricity generation in microbial fuel cells. *Environ. Sci. Technol., 38*, 4900–4904.
20. Oh, S., & Logan, B. E., (2006). Proton exchange membrane and electrode surface areas as factors that affect power generation in microbial fuel cells. *Appl. Microbiol. Biotechnol., 70*, 162–169.
21. Liu, H., & Logan, B. E., (2004). Electricity generation using an air-cathode single chamber microbial fuel cell in the presence and absence of a proton exchange membrane. *Environ. Sci. Technol., 38*, 4040–4046.
22. Cheng, S., Liu, H., & Logan, B. E., (2006). Increased performance of single chamber microbial fuel cells using an improved cathode structure. *Electrochem. Commun., 8*, 489–494.
23. Habermann, W., & Pommer, E. H., (1991). Biological fuel cells with sulfide storage capacity. *Appl. Microbiol. Biotechnol., 35*, 128–133.
24. Liu, H., Ramnarayanan, R., & Logan, B. E., (2004). Production of electricity during wastewater treatment using a single chamber microbial fuel cell. *Environ. Sci. Technol., 38*, 2281–2285.
25. Rabaey, K., Clauwaert, P., Aelterman, P., & Verstraete, W., (2005). Tubular microbial fuel cells for efficient electricity generation. *Environ. Sci. Technol., 39*, 8077–8082.
26. He, Z., Minteer, S. D., & Angenent, L. T., (2005). Electricity generation from artificial wastewater using an up flow microbial fuel cell. *Environ. Sci. Technol., 39*, 5262–5267.
27. Subhas, C. M., & Joe-Air, J., (2013). Application of microbial fuel cells to power sensor networks for ecological monitoring. *Wireless Sensor Networks and Ecological Monitoring: Smart Sensors, Measurement, and Instrumentation* (Vol. 3, pp. 151–178). Springer link.

28. Wang, V. B., Song-Lin, C., Cai, Z., Sivakumar, K., Zhang, Q., Kjelleberg, S., Cao, B., et al., (2014). A stable synergistic microbial consortium for simultaneous azo dye removal and bioelectricity generation. *Bioresource Technology, 155*, 71–76.
29. Wang, V. B., Song-Lin, C., Cao, B., Seviour, T., Nesatyy, V. J., Marsili, E., Kjelleberg, S., et al., (2013). Engineering PQS biosynthesis pathway for enhancement of bioelectricity production in Pseudomonas aeruginosa microbial fuel cells. *PLoS One, 8*(5), e63129.
30. Venkata, M. S., Mohanakrishna, G., Srikanth, S., & Sarma, P. N., (2008). Harnessing of bioelectricity in microbial fuel cell (MFC) employing aerated cathode through anaerobic treatment of chemical wastewater using selectively enriched hydrogen producing mixed consortia. *Fuel, 87*(12), 2667–2676.
31. Venkata, M. S., Mohanakrishna, G., Reddy, B. P., Saravanan, R., & Sarma, P. N., (2008). Bioelectricity generation from chemical wastewater treatment in mediator less (anode) microbial fuel cell (MFC) using selectively enriched hydrogen producing mixed culture under acidophilic microenvironment. *Biochemical Engineering Journal, 39*, 121–130.
32. Mohan, S. V., Veer, R. S., Srikanth, S., & Sarma, P. N., (2007). Bioelectricity production by mediator less microbial fuel cell under acidophilic condition using wastewater as substrate: *Influence of substrate loading rate. Current Science, 92*(12), 1720–1726. JSTOR24107621.
33. Venkata, M. S., Saravanan, R., Raghavulu, S. V., Mohanakrishna, G., & Sarma, P. N., (2008). Bioelectricity production from wastewater treatment in dual chambered microbial fuel cell (MFC) using selectively enriched mixed micro flora: Effect of catholyte. *Bioresource Technology, 99*(3), 596–603.
34. Venkata, M. S., Veer, R. S., & Sarma, P. N., (2008). Biochemical evaluation of bioelectricity production process from anaerobic wastewater treatment in a single chambered microbial fuel cell (MFC) employing glass wool membrane. *Biosensors and Bioelectronics, 23*(9), 1326–1332.
35. Venkata, M. S., Veer, R. S., & Sarma, P. N., (2008). Influence of anodic biofilm growth on bioelectricity production in single chambered mediator less microbial fuel cell using mixed anaerobic consortia. *Biosensors and Bioelectronics, 24*(1), 41–47.
36. Choi, Y., Jung, S., & Kim, S., (2000). *Development of Microbial Fuel Cells Using Proteus Vulgaris Bulletin of the Korean Chemical Society, 21*(1), 44–48.
37. Ting Chen, Scott Calabrese Barton, Gary Binyamin, Zhiqiang Gao, Yongchao Zhang, Hyug-Han Kim, & Adam Heller, (2001). A miniature biofuel cell. *J. Am. Chem. Soc., 123*(35), 8630–8631.
38. Bullen, R. A., Arnot, T. C., Lakeman, J. B., & Walsh, F. C., (2006). Biofuel cells and their development(PDF). *Biosensors and Bioelectronics, 21*(11), 2015–2045.
39. Tanisho, S., Kamiya, N., & Wakao, N., (1989). Microbial fuel cell using *Enterobacter aerogenes*. *Bioelectrochem. Bioenerg., 21*, 25–32.
40. Gil, G. C., Chang, I. S., Kim, B. H., Kim, M., Jang, J. K., Park, H. S., & Kim, H. J., (2003). Operational parameters affecting the performance of a mediator-less microbial fuel cell. *Biosens. Bioelectron., 18*, 327–334.
41. Chaudhuri, S. K., & Lovley, D. R., (2003). Electricity generation by direct oxidation of glucose in mediator less microbial fuel cells. *Nat. Biotechnol., 21*, 1229–1232.
42. Kim, N., Choi, Y., Jung, S., & Kim, S., (2000). Development of microbial fuel cell using *Proteus vulgaris*. *Bull. Korean Chem. Soc., 21*, 44–49.

43. Sell, D., Kramer, P., & Kreysa, G., (1989). Use of an oxygen gas diffusion cathode and a three-dimensional packed bed anode in a bioelectrochemical fuel cell. *Appl. Microbiol. Biotechnol., 31*, 211–213.
44. Lowy, D. A., Tender, L. M., Zeikus, J. G., Park, D. H., & Lovley, D. R., (2006). Harvesting energy from the marine sediment-water interface II-Kinetic activity of anode materials. *Biosens. Bioelectron., 21*, 2058–2063.
45. Niessen, J., Schroder, U., Rosenbaum, M., & Scholz, F., (2004). Fluorinated polyanilines as superior materials for electrocatalytic anodes in bacterial fuel cells. *Electrochem. Commun., 6*, 571–575.
46. Schroder, U., Niessen, J., & Scholz, F., (2003). A generation of microbial fuel cells with current outputs boosted by more than one order of magnitude. *Angew. Chem., Int. Ed., 42*, 2880–2883.
47. Cheng, S., Liu, H., & Logan, B. E., (2006). Increased power generation in a continuous flow MFC with advective flow through the porous anode and reduced electrode spacing. *Environ. Sci. Technol., 40*, 2426–2432.
48. Bergel, A., Feron, D., & Mollica, A., (2005). Catalysis of oxygen reduction in PEM fuel cell by seawater biofilm. *Electrochem. Commun., 7*, 900–904.
49. Cheng, S., Liu, H., & Logan, B. E., (2006). Power densities using different cathode catalysts (Pt and CoTMPP) and polymer binders (Nafion and PTFE) in single chamber microbial fuel cells. *Environ. Sci. Technol., 40*, 364–369.
50. Zhao, F., Harnisch, F., Schroder, U., Scholz, F., Bogdanoff, P., & Herrmann, I., (2005). Application of pyrolyzed iron(II) phthalocyanine and CoTMPP based oxygen reduction catalysts as cathode materials in microbial fuel cells. *Electrochem. Commun., 7*, 1405–1410.
51. Rabaey, K., & Verstraete, W., (2005). Microbial fuel cells: Novel biotechnology for energy generation. *Trends Biotechnol., 23*, 291–298.
52. Subhas, C. M., & Joe-Air, J., (2013). Application of microbial fuel cells to power sensor networks for ecological monitoring. *Wireless Sensor Networks and Ecological Monitoring: Smart Sensors, Measurement, and Instrumentation* (Vol. 3, pp. 151–178). Springer link.
53. Wang, V. B., Song-Lin, C., Cai, Z., Sivakumar, K., Zhang, Q., Kjelleberg, S., Cao, B., et al., (2014). A stable synergistic microbial consortium for simultaneous azo dye removal and bioelectricity generation. *Bioresource Technology, 155*, 71–76.
54. Wang, V. B., Song-Lin, C., Cao, B., Seviour, T., Nesatyy, V. J., Marsili, E., Kjelleberg, S., et al., (2013). Engineering PQS biosynthesis pathway for enhancement of bioelectricity production in *Pseudomonas aeruginosa* microbial fuel cells. *PLoS One, 8*(5), e63129.
55. Venkata, M. S., Mohanakrishna, G., Srikanth, S., & Sarma, P. N., (2008). Harnessing of bioelectricity in microbial fuel cell (MFC) employing aerated cathode through anaerobic treatment of chemical wastewater using selectively enriched hydrogen producing mixed consortia. *Fuel, 87*(12), 2667–2676.
56. Venkata, M. S., Mohanakrishna, G., Reddy, B. P., Saravanan, R., & Sarma, P. N., (2008). Bioelectricity generation from chemical wastewater treatment in mediator less (anode) microbial fuel cell (MFC) using selectively enriched hydrogen producing mixed culture under acidophilic microenvironment. *Biochemical Engineering Journal, 39*, 121–130.

CHAPTER 4

Sustainability of the Tire Industry: Through a Material Approach

SANJIT DAS,[1] HIRAK SATPATHI,[1] S. ROOPA,[2] and SAIKAT DAS GUPTA[1]

[1]*Hari Shankar Singhania Elastomer and Tire Research Institute (HASETRI), Plot No 437, Hebbal Industrial Area, Mysore, Karnataka – 570016, India*

[2]*Department of Polymer Science and Technology, Sri Jayachamarajendra College of Engineering, JSS Science and Technology University, Mysore – 570006, India*

ABSTRACT

The tire industry is basically a slow-changing industry, however, in the present era of fast change, a number of environmental and sustainability-related issues are impacting the tire industry to a large extent. Non-pneumatic tire, tires for electric vehicles, tires with sensors is talk of the day to meet the future requirements of the automobile industries. All these tires need to have fuel efficiency, safety at high speed, and low noise. Tire designers and material scientists are jointly involved in tire design, material design to achieve targeted tire performance properties. Material scientists are focusing their research mainly on sustainable materials as an alternate to present fossil-based material. In this chapter, authors will share different research work are being carried out on sustainable materials particularly for the tire industry. Reuse-Regenerate-Recycle is one of the research areas to reduce the consumption of virgin fossil fuel-based materials. Natural rubber (NR) is one of the commonly used rubbers in the tire industry. It is a naturally occurring material and helps the tire industry to achieve sustainability in the long run. However, the imbalance between the demand and supply of NR, worldwide scientists are doing research on Guayule, Dandelion to obtain a suitable alternative of NR. With the introduction of stringent requirements of tire labeling criteria by the European Union, materials scientists are bound to use alternative of carbon black reinforcing filler. One of the major alternatives of carbon black is silica which is also an eco-friendly sustainable material. Synthetic rubbers using plant-derived materials or biomass (agricultural waste) are came into the picture, Liquid Farnesene rubber is also another newly introduced polymer from bio-origin. Research efforts in developing microbes that can grow an isoprene monomer used to make a mimics NR is also undergoing. Fillers which are the next bulk consuming material after polymer are being explored from non-petroleum-based origin like silica, clay, and from bio-origin like rice husk ash, cornflower starch, bagasse; eggshell derived nano calcium carbonate. Coupling agents like Baker's yeast as an alternate of TESPT. Process aids from non-petroleum-based origin like soybean oil, orange oil, canola oil, sunflower oil, neem oil, etc. are also under the different stages of research for application in the tire. Apart from that, many resins from natural source are used to replace petroleum-based process oil and also to enhance tire performance. Cellulose-based synthetic fibers are of profound interest to replace synthetic fiber in tire reinforcement. Researchers are working to get anti-oxidant from flowers, fruits peels, and food wastes to replace petro-based antioxidants. Tea polyphenols as anti-oxidant has already found potential for the rubber compound. Investigations are also being done for a single compounding ingredient which would possess several functions and thus the term surfactant-accelerator-processing aid (SAPA) or termed as multi-functional additives (MFA).

4.1 INTRODUCTION

Sustainability and sustainable development are the key focus area of technology, research, and development of any industry. Sustainability implies meeting the present needs without compromising the freedom of future generations need. The main goal of sustainability is to comply with three requirements such as environment, economy, and society [1] (see Figure 4.1). New regulations from different government bodies, environmental bodies forces all the industrial sectors to adhere to the norms to make the environment more sustainable by reducing carbon footprints. Sustainability seeks an improved quality of life and embraces equality for all [2]. Sustainable development aims at assuring the ongoing productivity of exploitable natural resources and conserving all species of fauna and flora [3], Brundtland Commissions (United Nations) 1987.

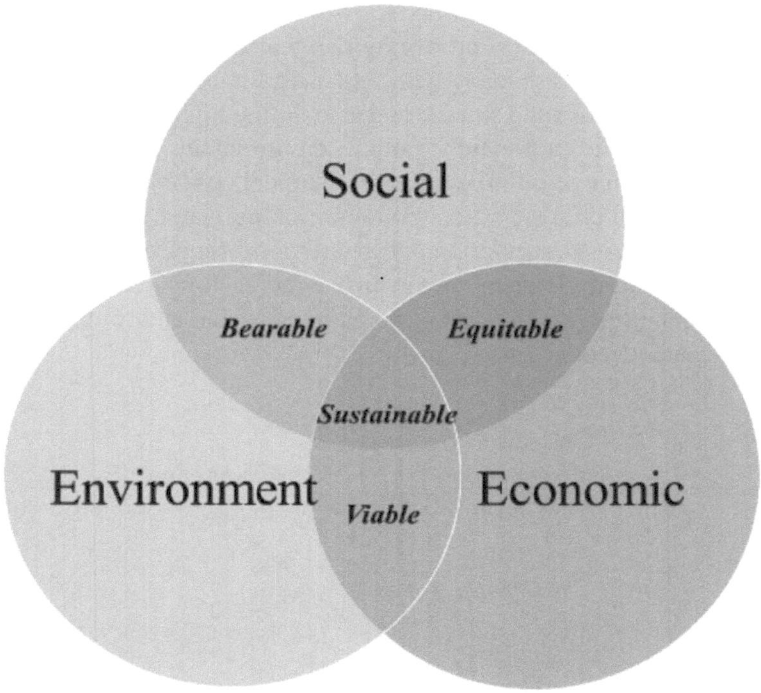

FIGURE 4.1 Graphical representation of sustainability (Venn diagram of sustainable development).

According to Philip Sutton [4], environmental sustainability can be achieved by reducing usage of physical resources and using more and more renewable rather than depletable resources. To achieve this, redesign of product and process is necessary to avoid toxic material generation. Sustainability goals and targets are universal. Reaching the goals is spread over all fonts-governments, business, civil society, and people everywhere to play the role. Currently, the way goods and services are produced; it is unsustainable and contributes significantly towards environmental problems. A new focus on sustainable developments needs new approach. Figure 4.2 illustrates the transition from business as usual to sustainable development [5].

Private sectors have to understand that sustainability may be breaking even or accommodate a loss in the short-term but if done correctly, then it can be profitable in long-term. Collaboration across sectors is inevitable to handle big issues. Breaking of silos in between sectors and collaboration between different stakeholders-suppliers, customers, partners in value chain, universities, and research institutes is one of the ways to reach the goal of sustainability. Sustainability is growing demand for tire producers. Eventually chemical industry producing raw materials has the big challenge as well. In rubber industry, many types of gases, fumes, and vapors come out of chemicals during use and storage and also during high temperature curing. These toxic materials ultimately pollute environment. Almost all rubber manufacturer discharge oil mixed water, chemical contaminated water from testing labs. These effluent waters pollute drinking water resources, land of irrigation [6]. REACH regulation and restriction brings some measure and put band on some toxic chemicals which is being followed by all reputed rubber manufacturers and ultimately some respite from environmental pollution level is achieved.

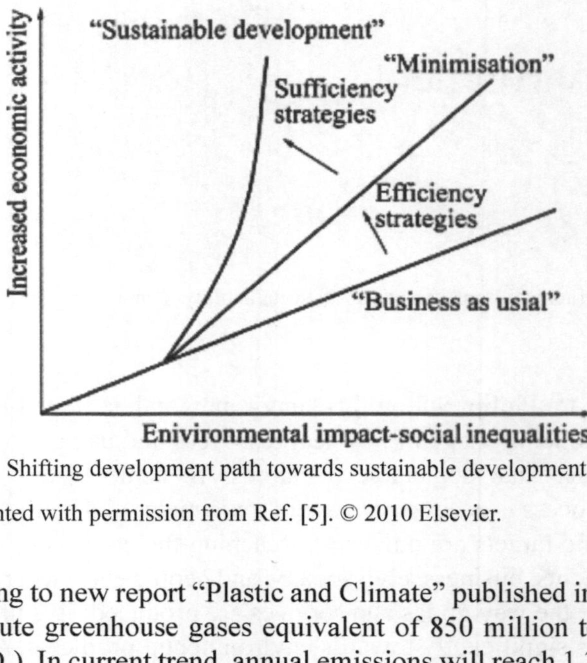

FIGURE 4.2 Shifting development path towards sustainable development.

Source: Reprinted with permission from Ref. [5]. © 2010 Elsevier.

According to new report "Plastic and Climate" published in 2019, plastic will contribute greenhouse gases equivalent of 850 million tons of carbon dioxide (CO_2). In current trend, annual emissions will reach 1.34 billion tons by 2030. By 2050, plastic could emit 56 billion tons of greenhouse gas emissions, as much as 14% of the earth's remaining carbon budget. Published research in 2019, showing that insects are destroyed by human activities like habitat destruction, pesticide poisoning, invasive species, and climate change (CC). Unless it cannot be stopped, it will result in ecological imbalance on earth in the next 50 years. Presently we are following linear material economy. Our material using fashion is characterized by the sequence take-make-use-scrap. To achieve sustainable goal, present days novel idea of circular economy emphasizes on 3R, i.e., repair-reuse-recycle. Raw Material availability and cost is going to affect business profoundly. Population growth and economic development provide spurs and opportunities for business, especially as it relates to massive developing countries China, India, and Brazil. Sustainable materials are those whose use brings benefits, unlike conventional materials. It must be naturally abundant, an ease of extraction with minimum energy spent and easy to recycle. Two general classes may be categorized. First, one is those materials which are mainly from plant origin like wood, natural fibers and polymers and the second one is mostly recycled materials, generally produced from waste products and raw materials. Automobiles uses report says by 2017 approx. 78.6 million of passenger cars are in use worldwide [7]. Considering approx. 6.5 billion people, around 87% of world population is to receive benefits of automobile [1]. The number of vehicles (excluding motorcycle) in use around the world is expected to reach 1.2 billion by 2020 and 1.6 billion in 2030 due to increase demand in developing countries [8–10]. Although modern automobile elastomers component is 8% by weight, but it is one of the key materials from engineering and design viewpoint [11]. With increase of vehicle introduction definitely will create environmental impact, traffic congestion, traffic congestion, accidents. Industries need to reduce these negative issues and to enhance aspects such as excitement, delight, and comfort (Figure 4.3).

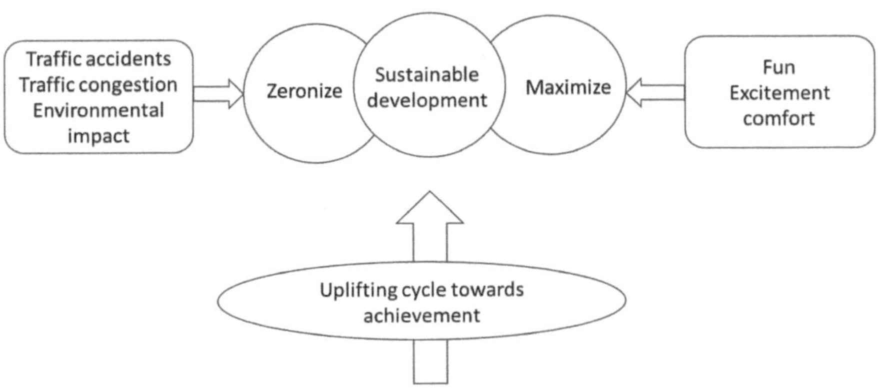

FIGURE 4.3 Zeronize and maximize (vision and philosophy) [1].

Toyota scientists have established technology revolution that brings about improved physical properties of recycled EPDM by selectively breaking the cross-links of three-dimensional EPDM network. Three factors are of more concerns of automotive industries-carbon dioxide emission, disposal of automobiles and air pollution from exhaust emissions. With introduction of Government, regulations for sustainability automotive industries are aiming for major changes in vehicle design, from Combustible Engine type vehicle to Electric Vehicle. By 2035, almost 90% of vehicle will be of electric type across the globe. To comply with future sustainability tire industries Research and Technology team goal is to develop new tire through design, manufacturing practice and introduction of sustainable material. Many new concepts are already under development like self-inflating tires, airless tires or Tweel tires, self-healing tires, 3D printed tires, tall, and narrow tires, triple tube tires, BOH_3 concept tire (to produce electricity while driving that would fill the battery) [12–16]. E-mobility trend in automotive sector means higher weight of vehicle contributed by battery. Higher weight of vehicles creates higher torque so electric engines need more efficient tire with higher wear resistance. For saving 1% fuel from fully loaded truck, only 2% to 4% rolling resistance reduction of the tire is sufficient, whereas passenger car tire requires 5–7% rolling resistance reduction to achieve same level of fuel economy [17]. The tire industry is raw material intensive which means that raw materials account for 65% of the production cost. Main raw materials for the tire are rubber (natural and synthetic), carbon black, silica, petroleum-based rubber process oil, nylon, polyester, steel cord, bead wire, and other rubber chemicals like resins, zinc oxide (ZnO), stearic acid, sulfur, etc. Raw materials used in the tire may be classified into two categories-from non-petroleum origin like NR, silica, steel cord, bead wire. Another raw material class is from petroleum origin like synthetic rubber, carbon black, rubber process oil, synthetic fibers like nylon, polyester. Table 4.1 shows the proportion of all raw materials in terms of their value and weight [18]. Most of the tire raw materials are petroleum-based. Hence, for tire companies are always getting influenced by raw material price because of the dependency of fluctuations of international prices of petroleum products.

Global tire chemical suppliers and mainly synthetic rubber manufacturers doing continuous research with non-fossil-based fuel materials to bring down the petroleum-based raw material usage by 30%. Synthetic Rubber manufacturers are trying to produce latest generation functionalized solution styrene-butadiene rubber (S-SBR) to facilitate silica mixing which will reduce carbon black usage. Industries are taking a new pathway, i.e., Styrene Circular Solution. This is nothing but recycling of polystyrene (PS) back into new styrene monomer for future usage [19].

TABLE 4.1 Proportion of Raw Materials in Terms of Value and Weight

Raw Materials	By Value (%)	By Weight (%)
Polymer	52	49
Reinforced filler	23	10
Petroleum-based tire cord	8	24
Rubber chemicals	15	12
Others	2	5
Total	**100**	**100**

4.2 POLYMER

Tire industry is raw material intensive and which accounts for 60% of the production cost. Most of the raw materials used in tire are petroleum-based except NR, bead wire, steel cord and few chemicals like stearic acid hence as we have spoken earlier that tire companies will be affected by the fluctuation of petroleum price. NR obtained from *Hevea brasiliensis* is the mostly used elastomer in rubber industries. Tire industries alone consume almost 75% of NR produced across the globe (Figure 4.4).

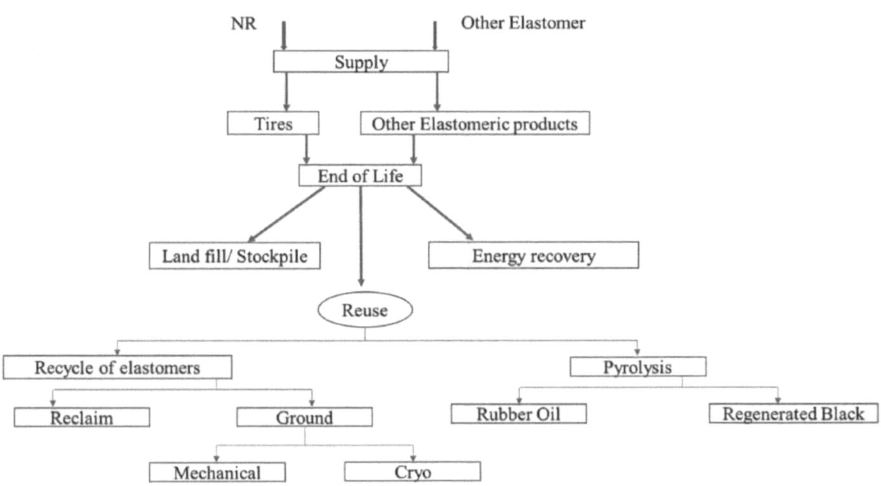

FIGURE 4.4 Flowchart showing the life cycle of elastomeric products.

To combat with the demand and price of NR, similar materials has been developed, e.g., isoprene or synthetic polyisoprene. They are well introduced and being used by all rubber industries. Because of the recent volatility in petroleum price, synthetic polyisoprene is also facing challenges to meet cost compatibility of industries. Scientists have found out other natural sources of NR, e.g., Guayule (*Parthenium argentatum*), which mostly grow in Mexico and Chihuahuan deserts of southern Texas [20], Dandelion (*Taraxacum kok-saghyz*) native to Kazakhstan of former the Soviet Union. Russian Dandelion is discovered by USSR during 1932 as a part of strategic programme of harvesting domestic source of NR [21]. It uses resources that resemble the one from a frequently used *Hevea brasiliensis* rubber tree. Different Research organizations across the Europe and the US are working to enhance the yield of Dandelion rubber approximately 2000 kg per hector [21]. The above research came into picture with the hope to create an alternative to Asia's rubber monopoly. There are other advantages for these two alternatives of NR, beside the monopoly issue. *Hevea brasiliensis* is vulnerable to several pests and its growth depends on tropical climatic conditions. Besides this, guayule and dandelion do not cause allergic reaction which is very common for *Hevea* rubber. The Neiker-Technalia center has been commissioned in Spain to research genotyping of the Guayule (a type of shrub) and Russian dandelion (a perennial plant) [22]. The former is cultivated in Mediterranean areas while the latter thrives in northern and eastern European countries. Guayule rubber is linear in structure and its crystallization rate is faster than NR, hence gum guayule rubber exhibits improved failure properties [23]. Resin content is less than 5% hence oxidative degradation properties are better than NR [24]. Apart from Hevea, Guayule, and Dandelion, other plants are also known which produce polyisoprene-Goldenrod (*Solidago altissima*), Jelutong (*Dyeraostulata*) [25] (Figure 4.5).

Sarkar et al. [25] in his review have mentioned many types of terpene based sustainable elastomers which are under development and few are already commercialized for industry uses. Elementary chemical structure of most of the terpenes are of isoprene type (2-methyl-1,4 butadiene, C5, "isoprene rule"). β-myrcene (7-methyl-3-methyl-1,6-octadiene) is one of the important monomers of the terpene family. It is available in oils from various plants like wild thyme, ylang-ylang bay, and lemongrass. Commercially β-myrcene is produced by pyrolysis of β-pinene [26]. Poly (β-farnesene), poly (ocimene), and poly (limonene) are synthesized by emulsion polymerization. Recent synthesis report of poly (styrene-co-myrcene) has found that its chemical structure is similar to SBR [27]. β-myrcene structure is similar to 1,3-polybutadiene, so this monoterpene is used for many chemical synthesis [25]. β-myrcene is usually found in liquid stage, easy for storage.

Although liquid handling in industry is a challenge but it makes the process greener and truly sustainable solution.

FIGURE 4.5 Various rubber plants: (a) Hevea, (b) Russian dandelion, and (c) Guayule

Source: Reproduced from Rubber Asia.

Newmark and Majumdar [28] have done the synthesis of poly (β-farnesene). Synthesis route that has been followed is anionic polymerization technique. The product's main constituents are of 1,4-cis microstructure. Sumitomo Rubber has introduced liquid Farnesene rubber very recently. This Farnesene rubber can result in better ice-grip properties for its new stud-less winter tire [29]. This proprietary liquid Farnesene rubber is developed by Kuraray Co. Ltd, Tokyo in collaboration with Amyris Inc.

Abdelrahman et al. [30] have reported catalytic ring-opening dehydration of tetrahydrofuran to produce 1,3-butadiene. Azeem et al. [31] have synthesized a styrene monomer from *Penicillium expansum* which is cultivated on forest waste biomass in the presence of yeast extract broth. Styrene biosynthesis from sugar over *E. coli* strains has been invented by Mckenna and Nielson [32].

Amyris and other companies have aimed for large scale production of terpenes via fermentation of plant sugars and cellulosic wastes [33–35]. Amyris has modified the common yeast *Saccharomyces cerevisial* to efficiently produce trans-β-farnesene [36]. Global energies in partnership with Synthos are developing a fermentation process to produce butadiene using modified *Escherichia coli* [37]. Cattle hair keratin is developed from tanning industry wastes and hydrolysate of bird feathers from poultry wastes [38]. Barone et al. [39–41] have explained that bird feather keratin has a 40% hydrophilic part and 60% hydrophobic part that enable it to react easily with hydrophobic polymer and hence improving its strength parameter. Prazepiorkowska [42] in his review have mentioned that elastic polymeric films containing keratin can be used as motor car parts. Keratin incorporation in NR and nitrile rubber improves thermal aging resistance of the matrix. Keratin contains sulfide bonds, which during aging are converted into groups with anti-oxidant character.

The BF Goodrich Co. has manufactured passenger car tires and the United States Rubber Co. Ltd has produced heavy-duty rayon cord truck tires from Dandelion rubber [43]. Road tests reveal that the tread wear, of dandelion, made tires are 94% as good as that of guayule rubber. Building tack of dandelion is better than guayule and similar to that of NR. Soybean oil used in Goodyear Tire can increase tread life by 10%. In 2012, Bridgestone have announced a tire made out of soybean oil [44]. Goodyear and DuPont Industrial Biosciences are working together to develop bio isoprene. Pirelli and Versalis have initiated joint research project for the use of guayule based NR for tire production [45]. In 2012, Dutch Tire Company Apollo Vredestein have produced prototype of tires using NR made from Guayule and Russian Dandelion. In 2015, Bridgestone Japan produced tires using Guayule rubber. Michelin has announced that by 2048, all of its tires will be manufactured using 80% sustainable materials and 100% of all tires will be recycled. Currently, they are using 28% sustainable materials like NR, sunflower oil, limonene, etc., and 2% recycled materials like crumb rubber [36]. The Bio Butterfly program is launched in 2012 with Axens and IFP energies Novelties, to produce synthetic elastomers from biomass like wood, straw, beat. Michelin and Axens have announced that first industrial scale prototype plant will start by end of this year 2019 for manufacturing butadiene from bioethanol [46]. Falken's clever new Enasave 100 tire has been made using natural alternatives replacing synthetic rubber, and conventional carbon black with biomass-derived carbon black. Instead of using fossil resources for rubber antioxidants, the tire is made by a unique process. They have used plant oil instead of mineral oil with NR and silica. Obviously, these tires have eco-friendly benefits. Enasave 100 tires consist of better wear resistance, better fuel efficiency, and wet grip performance. Michelin are aiming for replacing oil-based elastomers by wood chips from wood industry waste. They are hoping to have first wooden tire sometime in future in 2020 [47]. Bridgestone have displayed concept tire of 100% sustainable material at the 2012 Paris Motor show and commitment is that by 2050, 100% sustainable material usage. Bridgestone is hoping for, by the year 2020 commercial sales of sustainable tire will take place [48].

Epoxidized natural rubber (ENR) is capable of substituting the expensive solution polymerized styrene butadiene (S-SBR) rubber which is widely used for high performance car tires, including winter tires. Truck bus radial (TBR) tire compound formulations are mostly based on NR along with carbon black as reinforcement filler to cater mechanical and wear properties. Replacement of carbon black from TBR formulation is a challenge since silica incorporation as reinforcing filler in NR is difficult. So, to minimize the usage of carbon black ENR along with NR is being practiced which facilitate silica incorporation in TBR tire. Recently Apollo Tires and Rubber Research Institute of India (RRI) partnership has patented special grade of ENR.

Solution styrene butadiene (S-SBR) is one of the key elastomers used in tire apart from emulsion styrene butadiene rubber (E-SBR) and polybutadiene rubber. S-SBR plays an important role for low rolling resistance in combination of silica filler. Precise molecular design for controlled microstructure and chain end structure with special functional groups are pursued [49]. Such functional components are amino, carboxyl, and epoxy groups (Table 4.2).

TABLE 4.2 Possible Molecular Design for Controlled Microstructure and Chain End Structure [49]

	M_W	ML_{1+4} 100°C	Resilience
	35.7	56	57
	34.8	51	64
	37.7	63	62
	29.5	53	65

Materials development trends are moving in two distinct directions—Collaborative research with polymer producers on tailored synthetic elastomers for niche applications and development of sustainable alternatives to traditional petrochemical-based materials. Researcher has developed S-SBR with functional groups at one chain end, both chain end and also in backbone chain. Different degree and type of functionalization provide better silica-rubber interaction for improved snow traction, wet traction, and rolling resistance properties of ultra-high-performance passenger radial tires. Tire pressure is one of the important criteria to be maintained by the user to get the maximum benefit of fuel economy. The tire component responsible for retaining tire pressure is called inner liner. Butyl and halo butyl rubber are targeted for substitution by fossil-free material. Dynamically vulcanized rubber/resin thermoplastic elastomers alloy (DVA: dynamically vulcanized alloy) is therefore evolved. DVA is a sea-island structure, with a resin as sea phase and crosslinked rubber as island phase [49]. The resin has some barrier properties. A DVA-based inner liner can support both high barrier performance and endurance. This information is unveiled at the 2009 Tokyo motor show as next generation eco-tire technology.

Continental launches bike tire made from sustainable dandelion rubber [50]. Goodyear's Fuel Max Technology features a proprietary fuel saving tread compound. Hankook Tire Corporation has introduced its future generation electric vehicle tire which features strengthened handling and noise reduction capabilities. It provides hyper low noise environment, comfortable riding, and driving performance similar to ultrahigh performance tire. Aramid hybrid reinforcement belt features heavier load carrying capacities. This tire also contains aqua pine compound, resins extracted from conifers and vegetable oil. The new compound exhibits better wet braking and driving stability. This new generation treads compound offers improved tread abrasion resistance to combat high power of the electric vehicle and initial acceleration with no loss [51].

4.3 RECYCLING MATERIALS

Uses of rubber products in day to day life are increasing with time. Disposal of used end life rubber products is of serious concern. Tire is one of the rubber products which is contributing major stockpile up. Approximately 73% of scrap tires are disposed of in landfill [52]. Goodyear Tire has achieved their goal of not sending any waste to landfill from any of their manufacturing facilities and in 2006; they achieved zero waste to landfill [51]. Scrap tire dumping to the land continues to pose serious environmental, health, aesthetic problems as they are bulky, non-biodegradable, good breeding area for mosquitoes and rodents, shelter for animals like snake, rats, and also high risk of fire. Moreover, many toxic chemicals and process oils leaching to the soil is detrimental for fertility of the land as well as pollution of underground water supplies. Incineration of tire as a supplement of coal, wood can again multiply the environmental hazards. Reuse of scrap tire is thus a burning issue on which scientists are working to find out solution to maintain sustainability by converting it into a secondary resource of raw material for industry use. Recycling is one of the solutions which are practiced more than two decades. That we are trying to discuss in this chapter.

The U.S. Tire Manufacturing Association (USTMA) released its "2017 U.S. Scrap Tire Management Summary" report in July 18, 2018. The report reveals that in 1991 over 1 billion scrap tires are in stockpiles, now it's 60 million. So, there is 94% decrease [53]. This improvement has come due to the enforcement of stringent regulations. According to report of the European Tire and Rubber Manufacturer's Association (ETRMA), the treatment route of end-life-tire has changed from land-filling to material and energy recovery regime. Distribution report of scrap tire disposition in the U.S. in 2017 is shown in Table 4.3 [54].

TABLE 4.3 Scrap Tire Disposition in the U.S. in 2017

Disposal Category	Percent (%)
Tire derived fuel	43
Ground tire rubber	25
Land disposed	16
Civil engineering application	8
Exported	3
Electric furnace fuel	1
Reclaim project	1
Others	3

Some of the ways of using end-life-tire include; use in asphaltic concrete, for paving streets and highways, sport surfacing, production of steam through incineration, as fuel for cement kiln. In the tire industry, it is used after converting into recycled material like rubber crumb of finer mesh, chemically treated, or de-vulcanized rubber as secondary raw material. As of 1990, established tire and rubber industries are using only around 2% recycled material. However, in the past decades the tire recycling industry has a tremendous growth. Coupled with legal restrictions and environmental pressure present consumptions of recycled materials is nearly 10%. Recycling means conversion of used rubber into a re-processable and reusable form by the currently used physical and chemical process to save resource, energy, and to control the spread of hazardous/non-biodegradable materials [1]. The converted new product is treated as a secondary source of raw material for use in tire and other rubber industry. Reversing the vulcanization process is sometimes compared with reversing breadcrumb into fresh dough or like un-baking a cake and then reusing the eggs [1, 55]. In chemical term, de-vulcanization means reversing the rubber from its thermoset state to elastic state by cleaving sulfur bonds in the molecular structure [55]. The recovery of useful, uncontaminated materials from tire is very complex process. Scrap vulcanized rubber utilization is first thought in 1858 when Hiram Hall have developed heater pan process for reclaiming NR vulcanizates [52]. The major component of steel and fabric free scrap reclaim rubber is-rubber (50%), carbon black (27.5%), oil (17.5%), and ash (5–10%) [56]. The heating value of scrap tire is approximately 35×10^6 Joule/Kg [52]. Chemical analysis of whole scrap tire reveals carbon – 83%, hydrogen – 7%, oxygen – 2.5%, sulfur – 1.2%, nitrogen – 0.3% [56]. In 1990 Watson have used a novel mixer to explain that in recycling process C-S bonds can rupture rather than C-C bonds, since C-S bond strength is less than C-C bond [57]. A high strength mixer (HSM) is presented in 2009 as an efficient reclaiming machine [57]. In this process mechanical strain and chemical reaction occurs to rupture cross-linked bond and polymer backbone remain unaffected. ASTM 184A defines devulcanization as a "combination of depolymerization, oxidation, and increased plasticity. In other words, devulcanization is the process of cleaving of monosulfide, disulfide, and polysulfide crosslinks of rubber. In comparison to this process, reclaiming is different to from devulcanization since rupture of the C-C bonds also happen in this process.

The major form of recycled rubber is still ground rubber. This is produced either by ambient grinding, wet grinding, and cryogenic grinding. The product outcome of this process is called rubber powder or crumb rubber which may be of different types of coarse or finer grades depending on mesh size used for making crumb. Out of the above three techniques cryogenic grinding provides better quality tire crumb rubber and it is more economical and safer than ambient grinding [58] (Table 4.4).

Recently Michelin have acquired Lehigh, a specialized in high technology rubber powders which derived from recycled tire. Michelin proposes to develop partnership and identify new ways to recycle tires or new outlets for recycled tires [48]. Recycling of rubber can be followed by different techniques like mechanical, chemical, thermo-mechanical, cryogenic, microwave, and ultrasonic process. The ultrasonic reclaiming process has been explored in 1973 by immersing rubber articles in a liquid followed by treatment with ultrasonic radiation in the range of 20 KHz26/79. It is reported that by this process C-S and S-S bonds cleavage occurs but the C-C bonds remain intact. The properties of this recycled rubber are found to be similar those of original vulcanizates. Aoudia et al. [60] have reported microwave devulcanization of waste tire; the same has been used for making epoxy composites of improved mechanical properties.

TABLE 4.4 Pyrolysis Products of Used Tires [59]

Primary Products	Wt (%)	Content	Products Derived from Primary Products
Gas	10–30	Hydrogen, CO_2, CO, methane, ethane, propane, butane, sulfur, and other hydrocarbons	
Oil	38–55		Carbon Black
Char	33–38		Activated Carbon

Chemical process for reclaiming is more effective and faster devulcanization is selective for crosslinking only. Mercaptans, disulfide, phenol, amines, and metal chlorides are basic chemicals used for this method. Studies on Mechanochemical recycling processes of NR based scrap rubber indicate that diallyl disulfide (DADS) played an important role in the process [61, 62]. Diphenyl disulfide is found to be an effective reclaiming agent for NR based scrap rubber but it has bad odor [63–65]. Kojima et al. [66] have used supercritical carbon dioxide and diphenyl disulfide for chemical recycling of NR based scrap rubber. TMTD has been introduced as an alternative reclaiming agent of DADS by De et al. [67–69]. Reclaim prepared by using TMTD as reclaiming agent provides better tensile properties when mixed with SBR compounds. Verbrugger et al. [70] have used various types of amines as reclaiming agent for EPDM based scrape rubber compound and explained that there is no difference reactivity in different types of amines. Yehia et al. [71] studied the reclamation reaction by phenyl hydrazine system in brabender mixer along with ferrous chloride, ferric chloride and zinc chloride. The use of alkaline metals to break the crosslinks in rubber was patented in 1997 [72]. In this process, devulcanization has been done in solvent like, toluene, benzene, naphtha in the presence of sodium.

Surface modification method for making ground rubber has been done without breaking the bonds in the vulcanized material. Abrasion resistance and tear strength properties are influenced by surface modification. The most common method for chemical activation is treatment of powder by cross-linkable polymer-like latex, low molecular weight polymer and a curing system. The additives are latex, epoxidized oligomers, acryl amide. Grafting of ground tire rubber with maleic anhydride, methyl methacrylate, ethylene vinyl acetate, acrylamide has been reported [73–75]. Extensive work has been done on anaerobic desulfurization by the microorganisms *Pyrococcus furiosus*. This bacterium converts sulfur to hydrogen sulfide [76]. Another bacterium *Sulfolobus acidocaldarious* can split C-S bond. It is oxidizing sulfur but the process can be stopped at intermediate stages [77, 78]. Pyrolysis is another route for conversion of scrap tire into useful products like TDF and making regenerated carbon black [79, 80]. Pyrolysis is usually followed through different techniques like in inert atmosphere, oxidative pyrolysis, steam atmosphere, molten salt technique. Manuel and Dierkes[81] review on pyrolysis development in Europe and Japan. In general pyrolysis products are 5–20% light hydrocarbon gas, 20–50% hydrocarbon oil and aromatic liquids and 30–50% carbon char. Glenn and Ward studied the use of ground scrap rubber to prevent weed growth particularly on soybeanfields [82]. When crumb powder is used as mulch soybean plants are reduced in size thus minimizing wind damage, no adverse effect on seed quality. Mandal et al. [83] in his research has worked on regenerated CB from tire scrap, explained that regenerated CB has higher surface area (due to increased surface roughness) than virgin black which resulted better aging properties than the virgin black.

Bandyopadhyay et al. [84] have studied tire crumb use in NR based tire tread cap compound. He observed marginal deterioration in tensile strength, fatigue to failure and abrasion properties. 100 mesh crumbs have shown better properties retention compared to 40 and 80 mesh crumbs. In another study Bandyopadhyay et al. have explained that incorporation of superfine reclaims in the NR-based tire tread compound retained the properties to a greater extent compared to other recycled product like devulcanized rubber and other coarser reclaim. Regenerated carbon black from scrap tire compound is studied in NR based Lug tire tread compound and compared with virgin N330 black. Accelerated aging properties of compounds containing regenerated black have been found to be marginally better than that of compounds containing N330 black [85].

4.4 SUSTAINABLE RUBBER ADDITIVES

Elastomers are very often used as its normal form. For the manufacturing of finished product, a series of compounding ingredients are getting along with the base elastomeric material. For the ease of understanding, a typical tire tread formulation with all the possible rubber additives has been given in Table 4.5. Each and every compounding ingredient has their specific roles and responsibilities. That's make them inescapable from the product formulation and compounding. Most of these rubber chemicals are from inorganic sources or from petrochemical sources. Hence, mostly all are associated with health hazard and have a huge impact on environment. One solution to get rid of such problems can be the full substitution of these chemicals without disturbing the properties that those chemicals indulge into the finished product. But after an extensive search, a complete substitution is not at all possible. So, on a second thought, one can minimize the risk associated with those chemicals. And that can be achieved by following a sustainable pathway. Thinking about the market, scientific communities have put a very careful and slow-moving way forward towards the plausible alternatives of these rubber chemicals (Figure 4.6).

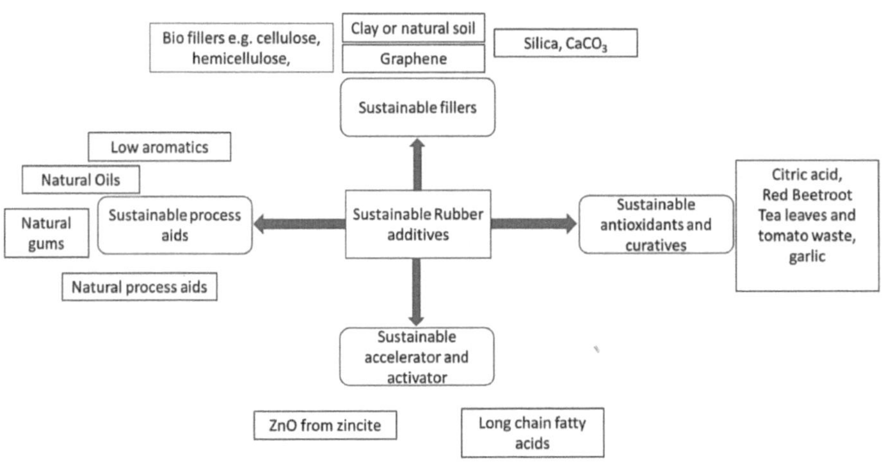

FIGURE 4.6 Schematic representation of various sustainable rubber additives.

In this section, we are trying to have a short discussion on these additives from a sustainable viewpoint. It includes sustainable rubber additives and curatives, various process aids such as vegetable oil and natural gums, sustainable fillers such as clay, calcium carbonate, silica, graphene, bio fillers from plant wastes, sustainable steel cord, etc. That also includes some possible sustainable alternatives.

TABLE 4.5 Typical Tire Tread Formulation (phr: Parts Per Hundred Parts of Rubber)

Ingredients	Amount (phr)
Natural rubber	100
Carbon black	45
Low PCA Oil	10
Stearic acid	2
ZnO	4
Antioxidant 6PPD	2
Antioxidant TMQ	1.5
Accelerator	1.5
Prevulcanization inhibitor (PVI)	0.2
Soluble sulfur	2.0
NOBS	0.5

4.4.1 ECO-FRIENDLY PROCESS OILS AND VEGETABLE OILS

Process oil is an important tool, which is being added during the processing step for the incorporation of filler materials within the rubber matrix. It affects the properties of the rubber vulcanizates from hardness and modulus point of view. With a growing concern for environmental safety, there is a global demand for adopting a green concept or a sustainable pathway. Depending on the source of the material and their chemical composition, "oil" can be classified. For composite application, use of oil is mostly as a process aid which is nothing but an extender. In case of automotive application such as for tires, high aromatic oils are most commonly used. High aromatic oil is well compatible with a wide range of synthetic and NR. But its source lies on the petroleum crude only. These oils are having a high content of polycyclic aromatics (PCA) which are potentially carcinogenic. From the wear of a tire tread PCA is getting leached into the environment since they are not chemically attached to base rubber material. This environmental effect leads to a global need to reduce the dependence of petroleum-based oils. So, there is a need to aware the user to promote the use of low PCA and low polyaromatic hydrocarbons (PAH) containing process oils in place of using old conventional aromatic and naphthenic process oils. PCA levels upper than 3% are alarming for the environment, according to REACH regulations (Figures 4.7 and 4.8).

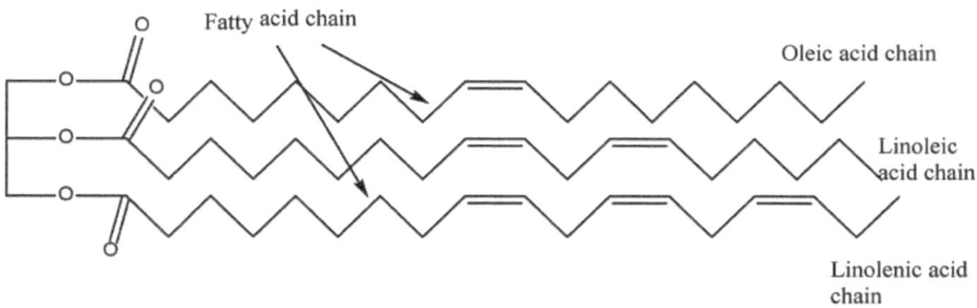

FIGURE 4.7 Chemical structure of vegetable oil (in general).

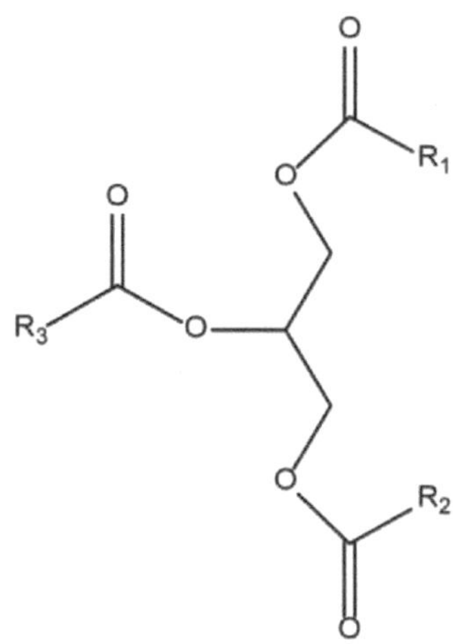

FIGURE 4.8 Chemical structure of triglycerides where R1, R2, and R3 are the fatty acid units.

The best alternative for petroleum-based resin can be vegetable oil. Vegetable oil belongs to such a platform where it is inherently biodegradable, extremely low cost and having excellent credentials towards environment. Figures 4.7 and 4.8 shows the general structure of triglycerides which is nothing but the general chemical formula for vegetable oil. Their tautomeric structure possesses unsaturation that can polymerize in the presence of atmospheric oxygen.

Dasgupta et al. [86–89] have studied the effect of natural source oil as an alternative to low PCA oils for rubber industries. They have compared the effect of Neem oil and Kurunj oil with aromatic oil on 100%NR, NR/BR blend, and SSBR/NR/BR ternary blend. And they have observed some interesting facts. For 100% NR compound, aromatic oil, Neem oil, and Kurunj oil shows similar flow characteristics. Filler dispersion rate and retention of modulus after aging is much higher in case of natural oils as compared to aromatic oil. Again in between Neem oil and Kurunj oil, Kurunj oil have better filler dispersion property whereas Neem oil containing NR compound have lower Mooney scorch. In case of NR/BR blend (mostly for bias truck tread cap), compound with natural oils have higher Mooney viscosity and higher abrasion as compared to aromatic oil containing compound. In this case, neem oil has shown better filler dispersion characteristics. In case of ternary system (mostly for PCR tread cap), compound with natural oil shows higher polymer to filler interaction. They also show higher modulus but tensile strength retention is pretty poor as compared to aromatic oil. Dasgupta's group have also studied the effect of aneco-processing aid and zinc soap-based process aid known as Zincolet PN60 with SBR/NR/BR ternary system. It shows improvement in polymer filler interaction that leads to better mixing, heat buildup, fatigue to failure properties. Vegetable oils such as linseed oil are long time used for coating application mostly as paint and varnishes. Vegetable oil can be easily dried. Castor oil and soybean oil are another typical examples of vegetable oil which has very good color retention property. Vegetable oil can be used as a resinous plasticizer as well as heat stabilizer. Typical resinous plasticizers are cottonseed oil or coconut oil. Common example of heat stabilizer oil is rubber seed oil which imposes heat stability on polyvinyl chloride application. Heat treatment of fatty acids using sulfur and accelerator lead to vulcanization of the fatty acids. Botros et al. have studied the effect of vulcanized oils on rubber compound and how it affects flow characteristics and the aging properties [90–93]. Ismail et al. [94] have used palm oil in silica filled NR compounds and studied their properties. It appears to have better vulcanizate properties. In case EPDM rubber, Wang et al. [95] have studied the effect of palm oil in comparison to paraffinic oil. The effect of both the oils is comparable if we consider processing and mechanical properties. But due to low cost and natural source, palm oil can be a useful alternative of paraffinic oil in case of EPDM rubber. In another study epoxidized, palm oil comes in direct comparison to distillate aromatic extract oil. It has been observed that both the vulcanizate possess similar properties such as hardness, modulus, tensile strength, elongation at break, etc. These types of findings have placed epoxidized palm oil as a plausible alternative for distillate aromatic extract oil [96–98]. Even for the carbon black containing butyl rubber compound, the use of norbornylized soybean oil has been reported [99]. Linseed oil have found its use as plasticizer in rubber/EG matrix. It appears to have higher thermal stability, resistance to crack formation as compared to the naphthenic oil [100].

4.4.2 SUSTAINABLE ACCELERATOR AND ACTIVATORS

ZnO is an extensively used material in rubber industries because of its role it plays in rubber vulcanization process. It acts as an accelerator activator in combination with stearic acid. ZnO is a multifunctional material based on its unique physical and chemical properties. The preparation of ZnO involves metallurgical or chemical method. Chemical method of preparing ZnO is nothing but the emulsion and microemulsion technique. And metallurgical process includes mechanochemical, sol-gel, hydrothermal, and solvothermal process directly from Zincite [101]. Because of its extensive use, Tire industries are the biggest consumer of ZnO. They consumed almost a half of the total worldwide production of ZnO [102]. Commonly ZnO is used in tire formulations in 3–8 phr (parts per hundred parts of rubber) (Table 4.5). The growing concern for ZnO is its environmental impact. ZnO always comes with a small fraction of cadmium oxide (CdO) and lead oxide (PbO) and this leads to a proper REACH regulation of controlling the amount of this trace element (heavy metals). On the other hand, stearic acid is a long chain fatty acid with a C18 chain. Since both ZnO and stearic acid originates from natural source, so from sustainability point of view one can consider both these materials as green material.

A continuous effort is ongoing to reduce the amount of ZnO to get rid of the environmental issues and also to lower the production cost of tire. Out of different approaches, one solution can be the use of nano ZnO in place of normal ZnO. Nano ZnO possesses higher surface area and that significantly affect the mixing characteristics. The dispersion of high surface area nano ZnO is better than the normal rubber grade ZnO and that allows using ZnO at a lower level with much more confidence [103]. Sahoo and Bhowmick [104] have studied the role of nano ZnO in XNBR compound and reported that ZnO nanoparticles induces higher state of curing and higher maximum torque as compared to rubber grade ZnO. In another part of the research, people have worked to find out better alternative of ZnO. Magnesium oxide (MgO) is found to be the most promising metal oxide out of the block. Being a non-heavy metal oxide, it insists the breakdown of the accelerator during the sulfur vulcanization process. And the rate appears to be faster than ZnO. The main drawback of using MgO is the lower percentage of crosslinking that actually hinders industrial application.

To overcome such issues some new technique has been introduced. Przybyszewaska et al. [105] have studied that in case of NBR, zinc chelates can be used in the vulcanization process. Guzman et al. [106] have reported the development of a new activator which consist a mixture of both ZnO and MgO. In this accelerator, Mg is incorporated inside the ZnO lattice with a very precise stoichiometry to form mixed metal oxide $Zn_{1-x}Mg_xO$ in nano scale. In the sulfur vulcanization step, it has been observed that this mixed metal oxide takes advantages from both MgO and ZnO. The reactions that generally happen during the scorch time, the chemical decay of the accelerator and the formation of the MBT happens much faster than it was expected earlier. And that happens due to the presence of Mg inside the ZnO lattice structure. We have already discussed that the low crosslink density is one of the major drawbacks of using MgO alone. That problem can be resolved by using this mixed nano oxides.

4.4.3 NATURAL GUM AND OTHER MULTIFUNCTIONAL PROCESS AIDS

The best source of natural gum is seed coatings or woody parts of plants. It is nothing but utilizing waste materials. So, it has gained potential interest from environmental aspect. A lot of advancement has been done on this field. Earlier very, less attention has been paid for utilizing such waste in rubber formulation. One of the eco-friendly approaches is by utilizing waste natural gum in virgin rubber.

In recent times Bahera gum have received a lot of attention from the rubber fraternity. It comes out from the bark of a medicinal plant, called *Terminalia bellerica*. This tree can be easily found in south East Asia, mostly in India. The extract of this tree is rich in polyphenolic compounds, which makes it useful as a chemo preventive agent. Except that Bahera gum is full of different tannins and calcium oxalate [107–109]. The gum is basically a multifunctional additive. The fatty acid/esters present in the gum can serve as accelerator activator whereas the phenolic part acts as antioxidant. Guhathakurta et al. [110] have used Bahera gum in NR and brominated iso-butylene-co-paramethyl styrene rubber. Addition of the gum in Rubber increases the tack strength. Activation energy has also been decreased with higher loading of gum. Saha et al. [111] have used Bahera gum in water-based NR formulation. Lap-shear strength and peel strength are found to be increased by a huge margin. Shanmugharaj et al. [112] have studied the influence of Asan gum in water-based NR latex system. At an optimum 5 phr of gum loading, they have observed around 400% increases in lap tear strength for NR latex system. The same system has been used for poly (vinyl acetate) and poly (vinyl alcohol) mixture. At an optimal 5 phr loading there is around 15% increase in lap tear strength for the PVA system.

Another vulcanized animal/vegetable oil (can also be called natural gum) which is never been in limelight is Factice. It can be used as a polishing ingredient for finished rubber product. Factice possess all the minimum characteristics it can have to act a processing aid or a property enhancer for rubber compounds. Two main classes of factice are heat-cured brown factice and cold cured white factice. They are introduced into the rubber field almost at the same time [113].

Rubber seed oil is another kind of natural gum that is extensively getting used as a multipurpose reagent in NR and SBR compounds. Nandanan et al. [114] have studied that. From one angle rubber seed oil can enhance the mechanical properties and from the other side it can enhance abrasion, flex resistance and aging resistance. Rubber seed oil is also helpful for decreasing cure time and blooming on the vulcanizate surface. There are other different resin, promoters, and some supporting literature also exist [115, 116].

4.4.4 ANTIOXIDANTS AND CURATIVES

Antioxidant and curatives are added into the rubber formulation in a lesser amount as compared to other ingredients. But from sustainability point of view, their replacement can certainly bring something positive for the environment. Here is this study we are trying to present some of the benign source for these antioxidants and curatives that are useful for rubber compounding formulation.

Tomato fruit is an important source of antioxidants. Pinela et al. [117] have studied that. It requires some extraction process that can generate those antioxidants from tomato. It is proven that a lot of phytochemicals and phenolics such as catechin, quercetin, cyanidin, coumarins, etc. have a huge application potential as a benign antioxidant. Shahidi et al. [118] have reported the ease availability of antioxidants from different nuts. Wu et al. [119] have done something interesting with citric acid. From citric acid, through a microwave pyrolysis technique, carbon nanodots are generated. These carbon nanodots are none other than antioxidants which are not radical friendly. Wu and group have used these nanodots while mixing with SBR. Antioxidants can easily be termed as stabilizers for several polymer systems [120].

Tran et al. [121] have worked with red beetroot which contain a specific antioxidant called betanin. They have incorporated this beetroot powder into a starch-based elastomer to obtain flexible bio-composites with tunable antioxidant properties. This smart bio elastomer composite can find its application in the field of pharmaceutical and food packaging. Most of the thiuram, those are used as accelerators for rubber industries are not environment friendly. They release carcinogenic nitrosamine during the heating. TBzTD (tetrabezylthiuram disulfide) has come to the market with the prospect of minimizing risk of using conventional thiurams. Metal salt of sorbic acid and ferulic acid has been introduced by Lin and his group [122]. Zinc salt of these acids is found to be a suitable crosslinking agent for ENR/silica composites. There has been noticeable increase of abrasion resistance in case of zinc salt of ferulic acid. Garlic and related genus allium plants are a good source of organic sulfur. This diallyl disulfide can be strong contender for replacing common sulfur in rubber compounding [123, 124].

4.4.5 SUSTAINABLE FILLERS

Virgin elastomer does not have load-bearing capability. It does not have that sufficient strength. So, it is necessary to add fillers into it. It can be carbonaceous (carbon black, CNT, and graphene) or silica, clay, or other renewable bio fibers such as cellulose, hemicellulose, or lignin. In this part of the study, we are trying to present some of the renewable fillers from sustainability point of view (Figure 4.9).

Introducing nanofillers into the tire industries is also something related to sustainability. By introducing 2–4 phr of nanofiller, someone can achieve same level of properties as it is for the conventional filler loading. That can reduce the weight of the final product, i.e., tire for automobile industries. So, it has a direct/indirect impact on fuel consumption.

4.4.5.1 SILICA

Silica can be of two types; one is fumed silica and another is precipitated silica. Precipitated silica is most common for rubber and tire industries.

Silicon dioxide, also known as silica (from the Latin silex), is a chemical compound that is a dioxide of silicon with the chemical formula SiO_2. It has been known since ancient times. Silica is most commonly found in nature as quartz.

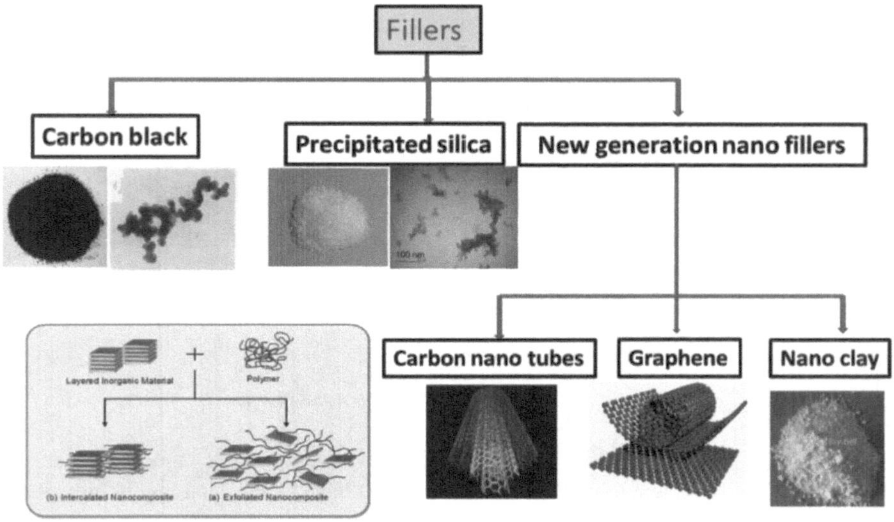

FIGURE 4.9 Schematic presentation of the available fillers/nanofillers for tire industries.

Addition of silica to a rubber compound offers a number of advantages such as improve tear strength, less heat buildup and increase in compound adhesion in multi-component products such as tires. Two fundamental properties of silica influence their use in rubber compounds: ultimate particle size and the extent of hydration. Properties such as pH, chemical composition, and oil absorption are having pretty less importance. Silica, an amorphous material consists of silicon and oxygen arranged in a tetrahedral structure of a three-dimensional lattice. Particle size ranges from 1 to 150 nm and surface area from 20 to 300 m^2/g. There is no long-range crystal order, only short-range ordered domains in a random arrangement with neighboring domains. Surface silanol concentration (silanol groups-Si-O-H) influences the degree of surface hydration. Silanol types fall into three categories; namely isolated, geminal (two -OH hydroxyl groups on the same silicon atom), and vicinal (on adjacent silicon atoms). Using new generation silane coupling agent, silica rubber interaction can be improved drastically. The formation of filler-rubber linkages via organosilanes has a major influence on the properties of rubber compounds. By lowering the specific component of surface energy of silica, the level of adsorption of hydrocarbon rubber on the filler surface can be enhanced and subsequently filler-filler interactions are reduced. Coupling agents are used as a surface modifier, either applied on the silica particle itself before the mixing or during mixing to the rubber compound [125–130] (Figure 4.10).

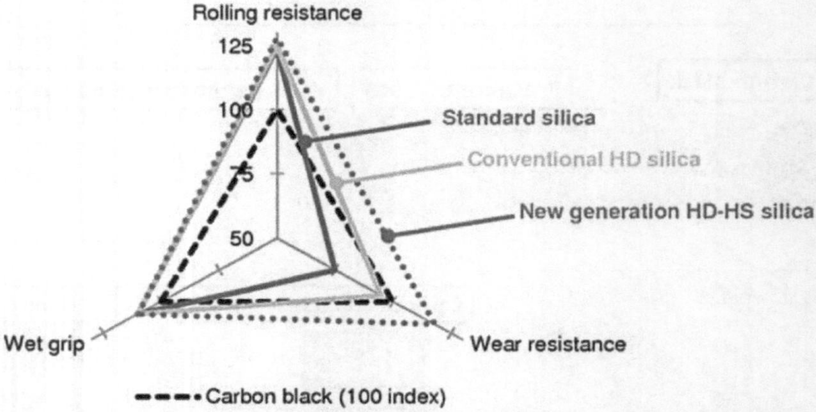

FIGURE 4.10 The magic triangle of tire with carbon black, silica, and highly dispersive silica.

Source: Reproduced from Ref. [131].

Michelin are the first company to introduce silica in a tire formulation to have their first-ever fuel-efficient green tire. Using silica as reinforcing filler, they are able to reduce tire rolling resistance by 20% and hence making the tire 3–4% more fuel economic. Also, silica tires are suitable for winter applications because of its good flexibility at low temperature [132]. Silica manufacturing giants such as PPG, Solvay, and Evonik are working to produce highly dispersive silica (HDS). This HDS can be one of the solutions of "Magic Triangle," that all tire industries are aiming for (Figure 4.10). Some typical examples of HDS are ZEOSIL, ULTRASIL, etc., [133–135]. These HDSs are leading to better dispersion that can affect the tire properties from three different angles. Higher rubber filler interaction can lead to better wear resistance. Lower filler-filler interaction can lead to lower hysteresis at high temp, hence lowering the rolling resistance [136]. Nayek et al. and Bhowmick et al. have reported the wear behavior of NR/SBR/silica composites. The idea of introducing green tire by varying silane and silica variant [137, 138].

From the earlier reviews, it has been estimated that the amount of biomass from plant resources is more or less 10^{13} tons worldwide and only 3% of it is getting renewed per annum. To maintain overall oxygen balance, the theory of sustainability restricts the usage only at the renewed percentage.

4.4.5.2 CELLULOSE

Introduction of cellulose opens a new era of polymer composite. Cellulose is a kind of renewable natural polymer which is most plentiful in the world. Plants are the source of cellulose as each and every plant cell wall is consisting of cellulose. From wood, grass to even seed fibers, fruit leaves, fungi, cellulose can be extracted [139–141].

Cellulose is basically a polymer of semicrystalline polysaccharide with a repeating unit of D-anhydroglucose. Cellulose consists of crystalline, paracrystalline, and amorphous domains [142]. Chemical structure of cellulose is illustrated in Figure 4.11 which is basically a cellobiose unit where n is the degree of polymerization. Degree of polymerization and hence the molar mass depends on the source. In case of wood fiber DP is around 300 whereas in case of plant fiber or bacterial cellulose DP is much higher, around 10000. Crystalline domain is responsible for the mechanical behavior of the cellulose fiber. Sakurada et al. calculated Young's modulus of elementary cellulose fibrils of bleached ramie fiber to be 134 GPa. Ishikawa et al. published a similar article where they measured the elastic modulus of different polymorphs sourced from ramie fiber. Their morphology and aspect ratio can vary greatly depending upon the source and also the polymorphs, so as their mechanical properties. The secondary hydroxyl groups at C2 and C3 that can form intra and inter molecular hydrogen bonds, play an important role in crystalline packing and hence physical properties of cellulose [143].

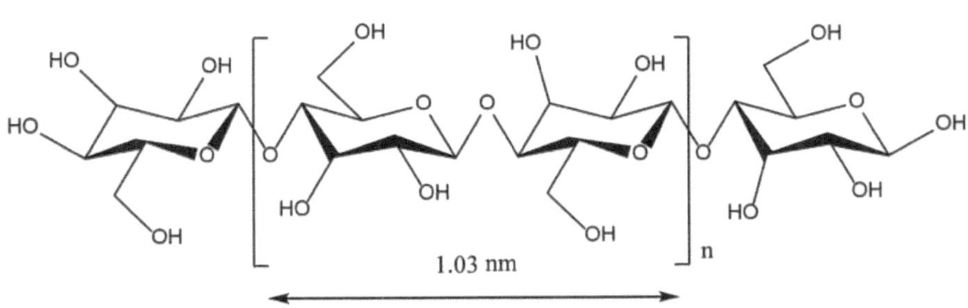

FIGURE 4.11 Chemical structure of cellulose fiber.

However, high-end application of cellulose is pretty limited due to lack of melting properties and its nature of being hygroscopic. The application of cellulose has been drastically increased only after the introduction of nanocellulose. Any cellulose microfibril, having a sideways measurable length below 50 nm can be considered as nanomaterials [144–146].

From 1970 onwards, the news of using natural fiber-reinforced composites start coming in Refs. [147, 148]. In the past, few years' nanotechnologies emerged as a unique technology with a very diverse role in the development of nanocomposites. It also reflects a constant growth in the market mostly for industrial applications [149]. Several groups are working on the development of nanocellulose fiber reinforced polymer composites. Compared to industrially made fiber and filler, nanocellulose fiber has gained much interest especially among automotive industry, in the sense of its availability and an improved CO_2 balance [150–157].

Agricultural biomass refers to the profusion of organic materials, so as to say the agricultural feed consisting of. It is the most abundant resource as well as it is renewable. Concerning it to be a fact, the society of fossil resources has been a precursor of the usage of agricultural biomass mostly in Southeast Asia and thus establishment of a sustainable production towards the usage of agricultural resources is pertinent [158]. Recent advances towards the biomass waste development and their conversion into products offer a huge opportunity for further research on improvised plant resources.

Cellulose fiber holds biorefinery functional properties towards agricultural waste and holds a good step towards the renewability of cellulose. With accordance to the assembled nanofibers are not more than a few micrometers in length. Because of the high gas barrier properties and their morphology, cellulose nanofiber stands out to be an optimistic material in various fields [158]. Some of the nanocellulose materials obtained from agricultural products or waste.

4.4.5.3 HEMICELLULOSE

Wood hemicellulose is also a kind of polymer with asymmetric polysaccharides. It has a similarity with the cellulose structure. Figure 4.12 shows one of the three characteristic hemicellulose microstructures.

Hemicellulose helps in improving some of the properties of cellulose used in paper industries. The rest of these polysaccharides remain attached with lignin and most of the processes it is getting burned with it. In the last few years new areas of using hemicellulose has been studied for several industrial sector in particular for food additive sectors.

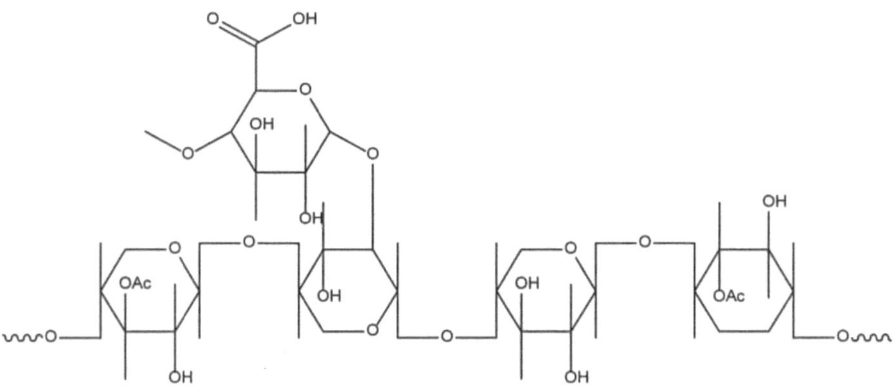

FIGURE 4.12 One of three possible typical hemicellulose structure.

4.4.5.4 LIGNIN

One of the predominant components of wood is lignin which is amorphous in nature. It comprises of asymmetric structure (Figure 4.13) having the most characteristic building blocks as compared to cellulose. That structure variates with source and involves depolymerization mechanism during the isolation process. During the pulping method for paper, making lignin fragments with sulfonate moieties is produced by the delignification mechanism based on the use of sulfites. Due to delignification cellulose, fiber isolates. Dissolved lignin can be utilized as fuel in this context. It enhances the process and also helps the inorganic catalyst to rejuvenate.

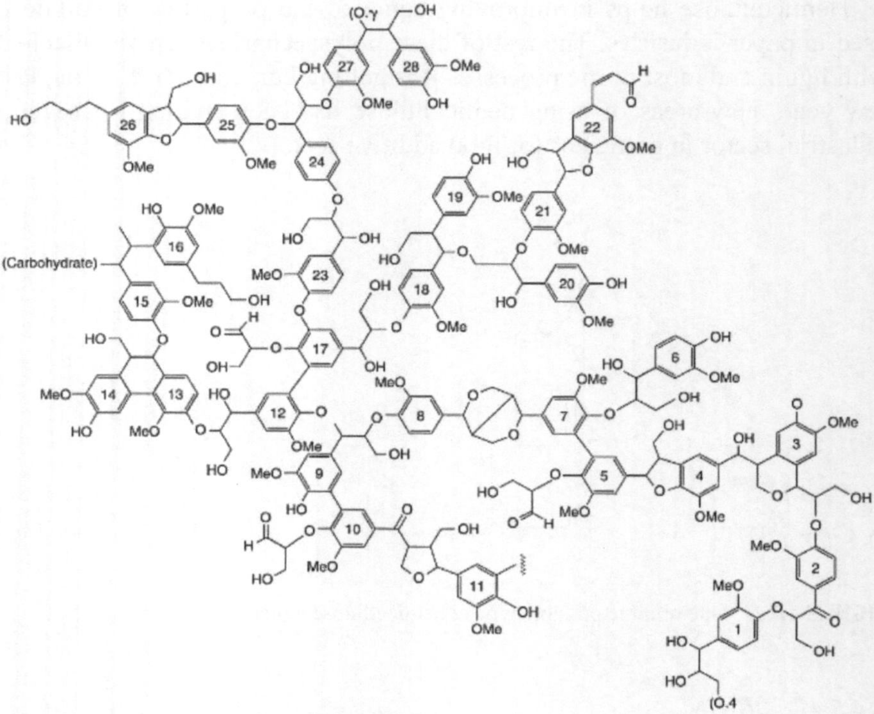

FIGURE 4.13 Polymer assembly of lignin.

Source: Reprinted with permission from Ref. [159]. © 2008 Elsevier.

Considering the mechanical properties, the exiting cellulose fiber has been a paradigm for the existing synthetic fibers [153, 160]. But still, there are existing shortcomings that ensue. There can be incompatibility issues with the polymer matrix due to hydrophobicity of the polymer chain. There is also a tendency of forming aggregates during processing. Biomaterials have a tendency of moisture absorption that can reduce the potential of these materials as reinforcing filler in case of thermoplastic system. Having said that, the design parameter for a polymer composite succumbs to the adhesion property of the fiber with the base polymer material and also the frictional coefficient of the fiber. Shalwan et al. have studied the ways to increase the adhesion between the fiber and the base polymer. According to them, one of the easier ways to achieve that is by treating with a base namely sodium hydroxide. Any other ways which affect directly on the mechanical strength of the fiber, are easily excluded [161]. In order to avoid that Gassan et al.

and Ragoubi et al. have reported strategies based on the physical treatment to improve the polymer/natural fiber compatibility [162, 163]. Li et al. and Belgacem et al. have worked on the surface structure of the fibers and modified them to reduce hydrophobicity and to a certain polarity to achieve higher polar interaction [164, 165].

In terms of electronic devices, cellulose nanofibers (CNF) find its utility. Electrical conductivity and the surface area of the cellulose nanofiber are very high compared to other fibers including some of the synthetic ones. So they find their usage in the field of electrical storage systems and other nano-electronic components [166]. Masoodi et al. has done some work on CNF. They have studied that cellulose finds its usage as a layered film in a bio-resin [167]. For the same purpose sometimes, silica is also deployed along with [168]. Singh et al. [169], De Maura et al. [170], Youssef et al. [171], and Kalia et al. [172] have gone through the cellulose nanocomposites and their utilization as an unconventional packaging material. The characterization of the composites is an essential tool and for that simple polymer characterization techniques such as Fourier transform infrared (FTIR), thermogravimetric analysis (TGA), and differential scanning calorimetry (DSC) are adopted and for the morphological study scanning electron microscope (SEM) has been used. By using a conducting polymer of different concentrations in the composite, one can vary electrical conductivity of the composite and it can have a potential use as an anti-static or anti-bacterial packaging material. Now from a different aspect nanocellulose have two different but important features. Sonic velocity of nanocellulose is high and it arrests the frictional loss. Actually, it has been observed that cellulose transparent film have similar sonic velocity as compared to aluminum and titanium [173]. Jonas et al. [174] have reported that SONY has successfully used it in its headphone diaphragms. These nanocellulose diaphragms are of a thickness of 20 μm and are prepared by dehydration followed by compression. The beauty of nanocellulose is lying in such kind of application only where its similar sound velocity as compared to metallic diaphragm comes into action [176].

When it comes to food and other non-food packaging industry, lignocellulose material finds its application [158]. It can be used for packaging of solid as well as liquid or for medical and pharmaceutical packaging [176, 177]. Nano-cellulose imparts its gas barrier and heat stability properties on polymer matrix. So, researchers have utilized this property of nanocellulose for generating new generation packaging material [178–183]. Reinforcing CNF offer high performance application in case of biofiber based bio-composites [184]. John and Thomas et al. [185] have stated that the amalgamation of nano-cellulose fiber with the renewable polymers produces bio-composites having the vortex as biofiber. These materials are compatible to human body and hence find its utilization in various artificial organs, mostly in kidney dialysis membrane and tubing, blood bags, artificial skin etc. Pineapple leaf fiber and polyurethane composites find its application in various medical implants [186]. Addition of small amount of CNF to polyurethane can increase the strength and stiffness of the product. PVA based composite with bacterial cellulose fibers are also compatible to human body and find its usage in the cardiology. There are reporting of other bionanomaterials for cosmetics, veterinary, dental, drug delivery, medical implants, wound heal dressing, tissue, and cell culture [187–200].

For meeting consumer need as well as regulatory requirements, the highly required field is performance improvement of automobiles. Nanocellulose, thermoplastic composite materials are largely used in automobile industry as it meets the performance needs and car safety. In the case of automotive sector nanocellulose, hemicellulose, lignin fiber can deliver high performance and contribute towards physical and chemical properties at a minimal cost.

Natural fiber draw attention as reinforcing fiber in polymer composites from different thermoplastic matrix such as polypropylene (PP), polyethylene, nylon for other kinds of automotive application. As a further move now, a day's automakers are using these natural fibers composite thermoset matrices for some of their car components [201–205]. According to Shinoj et al. [206], Mercedes Benz is using coconut fiber for seat parts along with NR latex. Some companies find the usage of epoxy resin reinforced with flax or sisal fiber mat for door panels. According to Suddell group, Audi, and Ford use composites from polyurethane matrix reinforced with flax/sisal/kenaf mat for door trim panels [207, 208].

4.4.5.5 CLAY

Clay is a naturally abundant material which has a huge potential in tire industries. Clay minerals are crystalline mostly layered aluminosilicates. It may contain certain amount of iron, alkaline metals, and alkaline earth metals. The crystalline lattice can be derived as mono-, bi-, and tri-dimensional types. Clays are consisting of alternate structure of silicate and alumina. Both this tetrahedral and octahedral structure is bound together by oxygen atom. When the clay is classified as 1:1 clay that means in each layer there is one tetrahedral and one octahedral unit present. When it is 1:2, that indicates one tetra sheet and two octa-sheet in each layer. Similarly, it will change in case of 2: 1 structured clay.

Smectite is in 2:1 phyllosilicate family with a silicate-gibbsite-silicate stacking. Different example of smectite is montmorillonite, saponite, bentonite, vermiculite, laponite, etc. Therefore, hydroxides groups are exposed at one side and the other side by an oxygen atom (Figure 4.14).

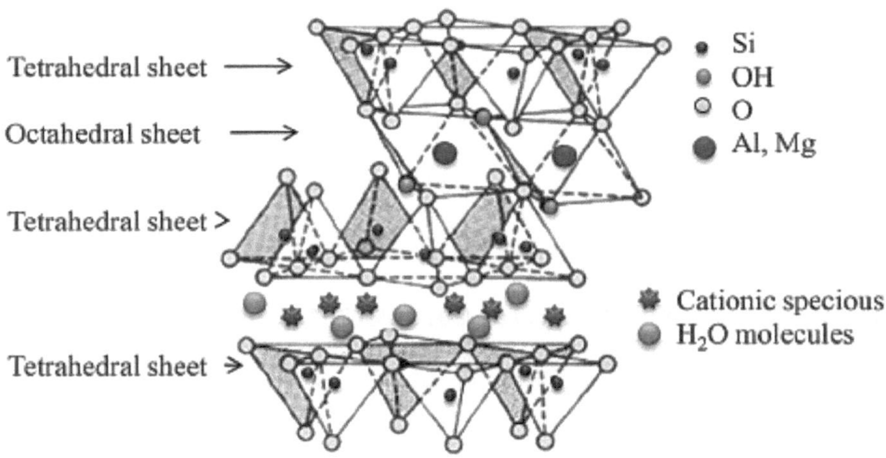

FIGURE 4.14 Crystal stacking structure of Smectite clay (2:1).

Source: Reproduced with permission from Ref. [209]. © 2011. Royal Society of Chemistry.

Kaolinite is 1:1 type clay. A very high cohesion results in kaolinite layers and that hinders intercalation. Another class of clay does exist that is known as layered double hydroxides or anion clay. Brucite is one of the examples of LDH. In brucite magnesium ions are connected with OH groups, building regular octahedron that are linked with one another, forming layers which interact with each other by very week forces. Mostly all clays are hydrophilic that makes them not suitable for mixing. Hence, it is essential to modify the clay before mixing. One of the popular techniques is by ion exchanging. Small cations can replace the big cations. For example, in some cases sodium is being replaced by ammonium ion. Elastomer/rubber is one of the most promising materials for the preparation of PCNs. There are a lot of studies have been done related to PCNs.

Varghese et al. [210] have prepared NR based nanocomposites with 10 wt% bentonite and layered silicate for the latex compounding method. The properties are being compared with the commercial clay. In mechanical, thermal or in swelling test, layered silicate performs much better than commercial clay. Wang et al. [211] have studied NR-MMT and CR-MMT by coagulating latex and clay suspension. The properties are also being compared with NR carbon black composite instead of clay. NR-MMT nanocomposites exhibit higher hardness, higher modulus, high tear strength, excellent gas barrier properties. Higher aspect ratio of clay layers is responsible for diffusing gas and hence have better anti-aging properties as compared to carbon black one. Hence, it finds it application in mostly inner liner, dumper, and inner tube compounds. Sadhu et al. [212] have studied the SBR and NBR system with a combination of copolymer and organomodifier. Unmodified clay did not disperse properly forming agglomerated structure whereas modified clay can disperse properly because of its exfoliated structure. Maji et al. [213] have studied the effect of Cloisite 30B. It is proved to be that 8 phr clay-based nanocomposite can possess two times increase in properties. Several researchers such as Khalil et al.; Ezema et al. and Aleksandra et al. have studied the effect of filler, its content, and size on the rubber matrix [214–216]. Incorporation of nano clay filler into the rubber matrix lead to an improve in mechanical properties. Mechanical properties improve mainly due to the better polymer filler interaction. Rubber interacts with the functional groups present on the clay surface. Rubber clay nanocomposites are basically having an application in the tire tread and base material because rubber clay composites help to gain certain properties like lower weight, lower energy dissemination, and higher air retaining capability. It also helps to extend the magic triangle performance (rolling resistance, traction, and abrasion) of a tire tread [216]. By reviewing several literatures it has been found that clay incorporates longer life into the compound as well as lower fuel consumption without compromising tread grip for winter application [217–219].

Steel cords are used for the radial tire application. So, there is a certain requirement for bringing this reinforcing material under the sustainability umbrella. The question is how can steel cord be sustainable. Steel cord production giant Bekaert have claimed to produce a lower weight steel cord (super tensile or ultra-tensile steel cord) with equivalent strength. That can reduce the tire weight and hence with proper air inflation, tire will have lower rolling resistance which increase fuel efficiency. Lots of work are conducting worldwide related to that. Longer tire life with improve adhesion can definitely lead to a sustainable tire. Beside that to improve adhesion, sometime inorganic cobalt salts are getting mixed in compound stage with the ply compounding formulation. To avoid that TAWI has been introduced where the brass coating contains not only Copper and Zinc but certain amount of Co as well. That certainly a more sustainable approach than earlier [220].

4.5 CONCLUSION

This book chapter is not first of its kind since several articles has been already published regarding this. In the above text, we have set our target pretty simple. We have tried to cover more and gather as much as information possible related to renewable and sustainable source of material which can act as potential replacement for tire industries. Most of the world leading tire industries has taken their goal set for complete replacement of the petroleum-based materials with sustainable eco-friendly materials to have a better future for the next generation to come.

"If it can't be reduced, reused, repaired, rebuilt, refurbished, refinished, resold, recycled, or composted, then it should be restricted or removed from production."

—Pete Seeger (Folk Singer and Social activist)

KEYWORDS

- **biomass**
- **cellulose fiber**
- **dandelion**
- **differential scanning calorimetry**
- **emulsion styrene-butadiene rubber**
- **guayule**
- **liquid Farnesene rubber**
- **multi-functional additives**
- **natural rubber**
- **recycle**
- **sustainability**

REFERENCES

1. Mukhopadhyay, R., Sengupta, R., & Das, Gupta, S., (2008). Recent advances in eco-friendly technology. In: Bhowmick, A. K., (ed.), *Current Topics in Elastomer Research* (p. 1021). CRC Press, Taylor, and Francis Group: NW.
2. ARIES (The Australian Research Institute for Environment and Sustainability). Department of Environmental Science, Macquive University North Ryde, Australia (Accessed on 3 August 2020).
3. https://www.mcgill.ca/sustainability/files/sustainability/what-is-sustainability.pdf (accessed on 10 August 2020).
4. Sutton, P., (2004). *A Perspective on Environmental Sustainability* (pp. 1–32). Paper on the Victorian Commissioner for Environmental Sustainability.
5. Koltun, P., (2010). Materials and sustainable development. *Progress in Natural Science: Materials International, 20*, 16–29.
6. Jagadale, S. C., et al., (2015). Environmental concern of pollution in rubber industry. *International Journal of Research in Engineering and Technology, 4*(11), 187–191.
7. Wagner. I., (2017). *Statistics and Facts on the Global Automotive Industry.*
8. Haraguchi, T., (2006). *Trends in Automobiles and Rubber Parts* (pp. 151–157). KGK.
9. Haraguchi, T., (2006). *Trends in Rubber Parts for Automobiles* (Vol. 79, No. 3, p. 103). Nippon GomuKyokaishi.
10. Haraguchi, T., (2010). The prospects of reducing fuel consumption with automobile basic efficiency. In*: Proceedings of FISITA, Scientific Society for Mechanical Engineering (GTE), F2010-A-013.*
11. Mukhppadhyay, R., (2007). Future perspective of Indian automotive rubber component industry. *Indian and International Rubber Journal.*
12. Aryton, B. A., (2017). *Smart Mobility*. New technology in tires: Electrical tires are the Future.
13. Wills, H., (2014). *The Future of Tires: The Latest Tech and Eco Tires*. A.E.
14. Crawfords, The Future of Tyre Technology., (2016). *In Industry News and Info.*
15. James Burbank the London Economic, The Future of Tyre Technology., (2018). *In Automotive, Tech. and Auto.*
16. David, S., (2014). Key risks for sustainable tyre business., Tire *Asia.*
17. Jones, K. P., (1997). Rubber and Environment., *UNCTD/IRSG Workshop* (pp. 8–19). Manchester.
18. http://content.indiainfoline.com/wc/archives/sect/tyre/ch03.html, India infoline tyre newsletter (2004).
19. Brown, K., https://www.rubbernews.com/suppliers/trinseo-innovation-key-sustainable-tires Rubber and Plastic news (accessed on 3 August 2020).
20. Rasutis, D., et al., (2015). A sustainability review of domestic rubber from the guayule plant. *Industrial Crops and Products, 70*, 383–394.
21. Beilen, J. V. B., & Poirer. Y., (2007). *Crit. Rev. Biotechnol.,* 27.
22. https://newatlas.com/russian-dandelion-guayule-latex/23362/ (accessed on 3 August 2020).
23. Santangelo, P., & Roland, C., (2001). The fatigue life of Hevea and guayule rubbers. *Rubber Chemistry and Technology, 74*(1), 69–78.

24. Bhowmick, A., Rampalli, S., & McIntyre, D., (1985). Effect of resin components on the degradation of guayule rubber. *Journal of Applied Polymer Science, 30*(6), 2367–2388.
25. Sarkar, P., & Bhowmick, A. K., (2018). Sustainable rubbers and rubber additives. *Journal ofApplied Polymer Science, 135*(24), 45701.
26. Behr, A., & Johnen, L., (2009). Myrcene as a natural base chemical in sustainable chemistry: A critical review. *Chem. Sus. Chem: Chemistry and Sustainability Energy and Materials, 2*(12), 1072–1095.
27. Sarkar, P., & Bhowmick, A. K., (2016). Green approach toward sustainable polymer: Synthesis and characterization of poly(myrcene-co-dibutyl itaconate). *ACS Sustain. Chem. Eng., 4*, 2129–2141.
28. Newmark, R. A., & Majumdar, R. N., (1988). 13C-NMR spectra of cis-polymyrcene and cis-polyfarnesene. *Journal of Polymer Science Part A: Polymer Chemistry, 26*(1), 71–77.
29. Liquid farnesene rubber used in car tires for first time. https://www.kuraray.com/release/2017/170220.html, (accessed on 3 August 2020).
30. Abdelrahman, O. A., et al., (2017). Biomass-derived butadiene by dehydra-decyclization of tetrahydrofuran. *ACS Sustainable Chemistry and Engineering, 5*(5), 3732–3736.
31. Azeem, M., Borg-Karlson, A. K., & Rajarao, G. K., (2013). Sustainable bio-production of styrene from forest waste. *Bioresource Technology, 144*, 684–688.
32. McKenna, R., & Nielsen, D. R., (2011). Styrene biosynthesis from glucose by engineered *E. coli. Metabolic Engineering, 13*(5), 544–554.
33. AMYRIS, Biopharma, hhtp://amyris.com/products.biopharma (Accessed on November 2019).
34. Leavell, M. D., McPhee, D. J., & Paddon, C. J., (2016). Developing fermentative terpenoid production for commercial usage. *Current Opinion in Biotechnology, 37*, 114–119.
35. Ajikumar, P. K., et al., (2008). Terpenoids: Opportunities for biosynthesis of natural product drugs using engineered microorganisms. *Molecular Pharmaceutics, 5*(2), 167–190.
36. Yoo, T., & Henning, S. K., (2017). Synthesis and characterization of farnesene-based polymers. *Rubber Chemistry and Technology, 90*(2), 308–324.
37. Buhl, T., (2014). Global bioenergies announces break-through in direct biological production of butadiene http://www.businesswire.com/news/home/20141126005551/en/Global-Bioenergies-Announces-Break-Through-Direct-Biological-Production (accessed on 3 August 2020).
38. Prochon, M., & Yves, H. T. N., (2017). *Rubber Chem. Technol., 88*(2), 258–275.
39. Barone, J. R., & Schmidt, W. F., (2006). Effect of formic acid exposure on keratin fiber derived from poultry feather biomass. *Bioresource Technology, 97*(2), 233–242.
40. Barone, J. R., & Schmidt, W. F., (2005). Polyethylene reinforced with keratin fibers obtained from chicken feathers. *Composites Science and Technology, 65*(2), 173–181.
41. Barone, J. R., Schmidt, W. F., & Liebner, C. F., (2005). Thermally processed keratin films. *Journal of Applied Polymer Science, 97*(4), 1644–1651.
42. Przepiórkowska, A., Chronska, K., & Prochon, M., (2008). Biodegradable elastomeric vulcanizates. *Rubber Chemistry and Technology, 81*(4), 723–735.
43. Whaley, W. G., & Bowen, J. S., (1947). *Russian Dandelion (Kok-Saghyaz), an Emergency Source of Natural Rubber*. U.S. Department of Agriculture: Washington D.C.

44. Rubber Journal Asia., (2017). *Green Rubber: New Bio Rubber Tires with Greener Footprint.*
45. Burgundy, M., The new tires are bio-based., *The Bio Journal.* http://thebiojournal.com/the-new-tires-are-bio-based/ (accessed on 3 August 2020).
46. *The Smithers Report, September.,* (2019)., *32*(39e).
47. Anthony, A., (2018). *Michelin Researching Eco-Friendly Wood Based Tires.*
48. *Michelin Aims High for Use of Sustainable Materials and Recycling of Tyres.* Tire News A.A-2019. https://tyrenews.com.au/michelin-aims-high-for-use-of-sustainable-materials-and-recycling-of-tyres/ (accessed on 3 August 2020).
49. Hara, Y., & Kirino, Y., (2012). Environmental compound technology for tires. *Nippon Gomu Kyokaishi, 85*(6), 187–192.
50. Simon, S., (2019). https://www.cyclingweekly.com/news/product-news/continental-launches-bike-tyre-made-dandelions-424634 (accessed on 3 August 2020).
51. Rich, A., (2016). *Eco Friendly Tires; Tire Review Magazine*. https://www.tirereview.com/157213/ (accessed on 3 August 2020).
52. Crane, Elefritz, R. A., Kay, E. L., & Laman. J. R., (1978). *Rubber Chem. Technol., 51*(3), 577–599.
53. https://www.waste360.com/recycling/look-what-s-ahead-scrap-tire-recycling (accessed on 3 August 2020).
54. http://www.statista.com/statistics/614554/scrap-tire-generation-share-by-type-in-the-united states/ (Accessed on November 2019).
55. https://waste-management-world.com/a/scrap-tyre-recycling (accessed on 3 August 2020).
56. Beckman, J., et al., (1974). Scrap tire disposal. *Rubber Chemistry and Technology, 47*(3), 597–624.
57. Brown, D. A., Watson. W. A., (2000). *Novel Concepts in Environmentally Friendly Rubber Recycling*. Presented at the International Rubber Forum 2000, Antwerp.
58. Srinivasean, A. Shanmugharaj, A. M., & Bhowmick, A. K., (2008). Waste rubber recycling. In: Bhowmick, A. K., (ed.), *Current Topics in Elastomer Research* (p. 1043). Taylor and Francis Group, Boca Rayton, FL.
59. Bredberg, K. P. J., Christiansson, M., Stenberg, B., & Holst, O., (2001). *Appl. Microbiol. Biotechnol., 55*, 43.
60. Aoudia, K., et al., (2017). Recycling of waste tire rubber: Microwave devulcanization and incorporation in a thermo set resin. *Waste Management, 60*, 471–481.
61. Jana, G. K., & Das, C. K., (2005). *Polymer Plast. Technol. Eng.*, 44.
62. Jana, G. K., & Das, C. K., (2005). *Prog. Rubber Plast. Recycling Technol.,* 21.
63. Rajan, V., et al., (2005). Comparative investigation on the reclamation of NR based latex products with amines and disulfides. *Rubber Chemistry and Technology, 78*(5), 855–867.
64. Rajan, V., et al., (2005). Model compound studies on the devulcanization of natural rubber using 2,3-dimethyl-2-butene. *Rubber Chemistry and Technology, 78*(4), 572–587.
65. Rajan, V. V., & Dierkes, W. K., & Noordermeer, J. W. M., (2006). *J. Appl. Polym. Sci.,* 102.
66. Kojima, M., Tosaka, M., & Ikeda, Y., (2004). Chemical recycling of sulfur-cured natural rubber using supercritical carbon dioxide. *Green Chemistry, 6*(2), 84–89.
67. De, D., et al., (2006). Reclaiming of ground rubber tire (GRT) by a novel reclaiming agent. *European Polymer Journal, 42*(4), 917–927.

68. De, D., & De, D., (2011). Processing and material characteristics of a reclaimed ground rubber tire reinforced styrene butadiene rubber. *Materials Sciences and Applications, 2*(05), 486.
69. De, D., De, D., & Singharoy, G., (2007). Reclaiming of ground rubber tire by a novel reclaiming agent. I. Virgin natural rubber/reclaimed GRT vulcanizates. *Polymer Engineering and Science, 47*(7), 1091–1100.
70. Verbruggen, M. A. L. Devulcanization of EPDM Rubber, (2007). PhD Thesis. University of Twente, Enschede, The Netherlands.
71. Yehia, A. A. I., M. N., Hefny, Y. A., Abdelbary, E. M., & Mull, M. A., (2004*). J. Elast. Plast.*, 33.
72. Meyers, R. D., Nicholson, P., MacLeod, J. B., & Moir, M. E., (1997). *U.S. Patent 5,602,186*. (To Exxon Research and Engineering Company).
73. Grigoryeva, O., et al., (2008). Reactive compatibilization of recycled polyethylenes and scrap rubber in thermoplastic elastomers: Chemical and radiation-chemical approach. *Rubber Chemistry and Technology, 81*(5), 737–752.
74. Amash, A., Giese, U., & Schuster. R., (2002). Interphase grafting of reclaimed rubber powder. *Kautschuk Gummi Kunststoffe, 55*(5), 218–218.
75. Rezaei, A. M., Jalali, A. A., & Nazockdast, H., (2010). Partial replacement of NR by GTR in thermoplastic elastomer based on LLDPE/NR through using reactive blending: Its effects on morphology, rheological, and mechanical properties. *Journal of Applied Polymer Science, 115*(4), 2416–2422.
76. Bredberg, K., et al., (2001). Anaerobic desulfurization of ground rubber with the thermophilic archaeon *Pyrococcus furiosus*-a new method for rubber recycling. *Applied Microbiology and Biotechnology, 55*(1), 43–48.
77. Romine, R. A., Romine, M. F., & Snowden-Swan, L., (1995). *Microbial Processing of Waste Tire Rubber*. Paper 56, presented at the 148th fall meeting of the rubber division, ACS, Cleveland.
78. http://tirebusiness.com/article/19950626/ISSUE/306269960/researchers-find-bacteria-that-eat-tiresfor-lunch, (1995). *Tire-Eating Bacteria Under Study.*
79. Beckman, J. A. C., Kay, E. L., & Laman, J. R., (1972). Akron rubber group meeting. *Rubber Age, 105*(4), p. 43.
80. IeBeau, D. S., (1967). *Rubber Chem. Technology*., (Vol. 40, p. 217).
81. Manuel, H. J., & Dierkes, W., (1995). *RAPRA Review Report*. IRSG.
82. Glenn, R. C., Ward, C. Y., (1972). *Presented at the Southern Agricultural Workers Conference*. Richmond, Virginia.
83. Mandal, N., Gupta, S. D., & Mukhopadhyaya, R., (2005). Regeneration of carbon black from waste automobile tires and its use in carcass compound. *Progress in Rubber Plastics and Recycling Technology, 21*(1), 55–72.
84. Bandyopadhyay, S., et al., (2006). Use of recycled tire material in NR/BR blend based tire tread compound: Part II (with ground crumb rubber). *Progress in Rubber, Plastics, and Recycling Technology, 22*(4), 269–284.
85. Bandyopadhyay, S., et al., (2006). Effect of regenerated carbon-black on a bias tire tread cap compound. *Progress in Rubber, Plastics, and Recycling Technology, 22*(3), 195–210.
86. Dasgupta, S., Agrawal, S., Bandyopadhyay, S., Chakraborty, S., Mukhopadhyay, R., Malkani, R. K., & Ameta, S. C., (2007). *Polymer Testing, 26*, 489.

87. Dasgupta, S., et al., (2008). Characterization of eco-friendly processing aids for rubber compound: Part II. *Polymer Testing, 27*(3), 277–283.
88. Dasgupta, S., et al., (2009). Eco-friendly processing oils: A new tool to achieve the improved mileage in tire tread. *Polymer Testing, 28*(3), 251–263.
89. Dasgupta, S., et al., (2009). Improved polymer-filler interaction with an eco-friendly processing aid: Part 1. *Progress in Rubber Plastics and Recycling Technology, 25*(3), 141–164.
90. *Encyclopedia of Polymer Science and Engineering, Cellular Materials to Composites, Second Edition, A Wiley-Interscience Publication,* (1985), vol 3, 619.
91. Joseph, R., Madhusoodhanan, K. N., Alex, R., Varghese, S., George, K. E., & Kuriakose, B., (2004). *Plastics Rubbers Compos., 33*, 217.
92. Nag, A., Halder, S. K., (2006). *Kautschuk Gummi Kunstsoffe*, 322.
93. Botros, S. H., El-Mohsen, F. F., & Meinecke, E. A., (1987). *Rubber Chem. Technol., 60,* 159.
94. Ismail, H., (2000). Effect of palm oil fatty acid additive (POFA) on curing characteristics and vulcanizate properties of silica filled natural rubber compounds. *Journal of Elastomers and Plastics, 32*(1), 33–45.
95. Wang, Z., et al., (2016). Investigation of palm oil as green plasticizer on the processing and mechanical properties of ethylene propylene diene monomer rubber. *Industrial and Engineering Chemistry Research, 55*(10), 2784–2789.
96. Pechurai, W., Chiangta, W., & Tharuen, P., (2015). Effect of vegetable oils as processing aids in SBR compounds. In: *Macromolecular Symposia*. Wiley Online Library.
97. Petrović, Z. S., et al., (2013). Soybean oil plasticizers as replacement of petroleum oil in rubber. *Rubber Chemistry and Technology, 86*(2), 233–249.
98. Sahakaro, K., & Beraheng, A., (2011). Epoxidized natural oils as the alternative safe process oils in rubber compounds. *Rubber Chemistry and Technology, 84*(2), 200–214.
99. Li, J., et al., (2018). Sustainable plasticizer for butyl rubber cured by phenolic resin. *Journal of Applied Polymer Science, 135*(24), (4550). 0.
100. Fernandez, S. S., Kunchandy, S., & Ghosh, S., (2015). Linseed oil plasticizer based natural rubber/expandable graphite vulcanizates: Synthesis and characterizations. *Journal of Polymers and the Environment, 23*(4), 526–533.
101. Nanoshel, (2019). *Application of Zinc Oxide for Industrial Rubber and Plastics.* Zinc oxide for Rubber Industries.
102. Walter, J., (2009). *Tire Technology International,* 18.
103. Chapman, A. V., & Johnson, T., (2005). *Kautschuk Gummi Kunststoffe, 58*, 358.
104. Sahoo, S. & Bhowmick, A. K., (2007). *J. Appl. Polym. Sci., 106*, (3077).
105. Przybyszewska, M., Zaborski, M., Jakubowski, B., & Zawadiak, J., (2009). *Express Polym. Lett., 3*, 256.
106. Guzmán, M., et al., (2012). Zinc oxide versus magnesium oxide revisited: Part 2. *Rubber Chemistry and Technology, 85*(1), 56–67.
107. Bhatia, K., & Ayyar, K. S., (1980). *Indian Forester, 106*, 363.
108. Bhatia, K. L., & Swaleh. M., (1980). *Indian Forester, 103*, 273.
109. Bhatia, K. L., & Swaleh, M., (1980). *Indian Forester, 107*, 519.
110. Guhathakurta, S., et al., (2006). Waste natural gum as a multifunctional additive in rubber. *Journal of Applied Polymer Science, 102*(5), 4897–4907.

111. Saha, S., et al., (2005). Studies of *Terminalia bellerica* (Bahera), a natural gum, as an additive in a water-based adhesive composition. *Journal of Adhesion Science and Technology, 19*(15), 1349–1361.
112. Shanmugharaj, A., et al., (2005). Studies on the adhesion behavior of water-based adhesives blended with as a gum. *Journal of Adhesion Science and Technology, 19*(8), 639–658.
113. Hurlston, E., (1936). The use of factice in rubber manufacture. *Rubber Chemistry and Technology, 9*(4), 621–625.
114. Nandanan, V., Joseph, R., & George, K., (1999). Rubber seed oil: A multipurpose additive in NR and SBR compounds. *Journal of Applied Polymer Science, 72*(4), 487–492.
115. Bhattacharyya, S. K., et al., (2012). Exploring microcrystalline cellulose (mcc) as a green multifunctional additive (MFA) in a typical solution-grade styrene butadiene rubber (S-SBR)-based tread compound. *Industrial and Engineering Chemistry Research, 51*(32), 10649–10658.
116. Manoharan, P., & Naskar, K., (2017). Biologically sustainable rubber resin and rubber-filler promoter: A precursor study. *Polymers for Advanced Technologies, 28*(12), 1642–1653.
117. Pinela, J., et al., (2017). Valorization of tomato wastes for development of nutrient-rich antioxidant ingredients: A sustainable approach towards the needs of the today's society. *Innovative Food Science and Emerging Technologies, 41*, 160–171.
118. Alasalvar, C., & Shahidi, F., (2009). Natural antioxidants in tree nuts. *European Journal of Lipid Science and Technology, 111*(11), 1056–1062.
119. Wu, S., Weng, P., Tang, Z., & Guo, B., (2016). *ACS Sustainable Chem. Eng., 4*, 247.
120. Zhan, K., Ejima, H., & Yoshie, N., (2016). Antioxidant and adsorption properties of bioinspired phenolic polymers: A comparative study of catechol and gallol. *ACS Sustainable Chemistry and Engineering, 4*(7), 3857–3863.
121. Tran, T. N., et al., (2017). Starch-based bio-elastomers functionalized with red beetroot natural antioxidant. *Food Chemistry, 216*, 324–333.
122. Lin, T., et al., (2015). Renewable conjugated acids as curatives for high-performance rubber/silica composites. *Green Chemistry, 17*(6), 3301–3305.
123. Ding, C., & Matharu, A. S., (2014). Recent developments on bio-based curing agents: A review of their preparation and use. *ACS Sustainable Chemistry and Engineering, 2*(10), 2217–2236.
124. Gomez, I., et al., (2016). Inverse vulcanization of sulfur using natural dienes as sustainable materials for lithium-sulfur batteries. *Chem. Sus. Chem., 9*(24), 3419–3425.
125. Ten, B. A., (2002). *Silica Reinforced Tire Rubbers: Mechanistic Aspects of the Role of Coupling Agents.* University of Twente: Enschede.
126. Kraus, G., (1978). Reinforcement of elastomers by particulate fillers. In: Eirich, F. (ed.), *Science*, and *Technology of Rubber*. Academic Press: New York.
127. Wischhusen, M., (2010). *Balancing Rolling Resistance with Other Performance Characteristics.* Tire Review.
128. *Organization for Economic Cooperation and Development (OECD)*, (2004). SIDS Initial assessment Report for SIAM 19: Synthetic Amorphous Silica and Silicates.
129. Schlomach, J., & Kind, M., (2004). Investigations on the semi-batch precipitation of silica. *Journal of Colloid and Interface Science, 277*(2), 316–326.

130. Integrated Pollution Prevention and Control, (2006). Reference document on best available techniques for the manufacture of large volume inorganic chemicals-solid and other industry, European Commission.
131. https://www.pneurama.com/en/rivista_articolo.php/Nanomaterials-a-tire-s-biggest-friends?ID=36715 (accessed on 3 August 2020).
132. Michelin Innovation. http://www.michelin.in/IN/en/whymichelin/more-on-brand-innovation.html, (accessed on 3 August 2020).
133. Tire Solutions from Solvay. http://www.solvay.com/en/binaries/TIRE % 20SOLUTIONs-177294.pdf (accessed on 3 August 2020).
134. Highly dispersible silica from EVONIK. http://ultrasil.evonik.com/product/ultrasil/en/products/highly dispersiblesilica/pages/default.aspx, (accessed on 3 August 2020).
135. PPG Silica Products. http://www.ppgsilica.com/getmedia/0de26f2f-8399-492e-9170-5a492d38c6c5/HiSiEZ200G.pdf.aspx (accessed on 3 August 2020).
136. Wang, M. J., et al., (2002). New generation carbon-silica dual phase filler part I. Characterization and application to passenger tire. *Rubber Chemistry and Technology, 75*(2), 247–263.
137. Nayek, S., et al., (2005). Wear behavior of silica filled tire tread compounds by various rock surfaces. *Rubber Chemistry and Technology, 78*(4), 705–723.
138. Lodha, V., Bhattacharyya, S., Dasgupta, S., Mukhopadhyay, R., Sarkar, P., & Bhowmick, A. K., (2017). In: *The Proceedings of the 192nd Technical Meeting of the Rubber Division*. Cleveland, OH: ACS.
139. Asim, M., Abdan, Khalina., Jawaid, M., et al., (2015). A review on pineapple leaves fiber and its composites. *nt. J. Polym. Sci.*, 16.
140. Jawaid, M., & Abdul, Khalil, H. P. S., (2011). Cellulosic/synthetic fiber reinforced polymer hybrid composites: A review. *Carbohydrate Polymers, 86*(1), 1–18.
141. Lamaming, J., et al., (2015). Isolation and characterization of cellulose nanocrystals from parenchyma and vascular bundle of oil palm trunk (*Elaeisguineensis*). *CarbohydrPolym., 134*, 534–540.
142. Morton, W. E., & Hearle, J. W. S., (1962). *Physical Properties of Textile Fibers*. Butterworth & Co. Ltd: London.
143. Abdul, K. H. P. S., Bhat, A. H., & Ireana, Y. A. F., (2012). Green composites from sustainable cellulose nanofibrils: A review. *Carbohydrate Polymers, 87*(2), 963–979.
144. Ishikawa, A., Okano. T., & Sugiyama, J., (1997). Fine structure and tensile properties of ramie fibers in the crystalline form. of cellulose I, II, III, and IV. *Polymer, 38*(2), 463–468.
145. Vanderhart, D. L., & Atalla, R. H., (1984). Studies of microstructure in native celluloses using solid-state carbon-13 NMR. *Macromolecules, 17*(8), 1465–1472.
146. Isogai, A., (2013). Wood nanocelluloses: Fundamentals and applications as new bio-based nanomaterials. *Journal of Wood Science, 59*(6), 449–459.
147. Gray, D. G., (1974). Polypropylene transcrystallization at the surface of cellulose fibers. *J. Polym. Sci. Polym. Lett. Edn., 12*(9), 509–515.
148. McAllister, D. H., Pearson, P., & Wells, H., (1982). *Proceedings of the Reinforced Plastics Congress*. Brighton, UK: British Plastics Federation.
149. Abdul Khalil, H. P. S., et al., (2017). 1-Nanofibrillated cellulose reinforcement in thermo set polymer composites. In: Jawaid, M., Boufi, S., & Abdul Khalil, H. P. S., (eds.), *Cellulose-Reinforced Nanofiber Composites* (pp. 1–24) Woodhead Publishing.

150. Bledzki, A. K., & Gassan, J., (1999). Composites reinforced with cellulose based fibers. *Progress in Polymer Science, 24*(2), 221–274.
151. Ansell, M. P. et al., (2001). Review of current international research into cellulosic fibers and composites. *Journal of Materials Science, 36*, 2107–2131.
152. Saheb, D. N., & J. N., (1999). Natural fiber reinforced composites is an emerging area in polymer science. *Adv. in Polymer Techn., 18*, 351–363.
153. Wambua, P. I., J., & Verpoest, I., (2003). Natural fibers: Can they replace glass in fiber reinforced plastics. *Compos. Sci. Technol., 63*(9), 1259–1264.
154. Anandjiwala, R. D., & Blouw. S., (2004). Bast fibrous plants for healthy life. In: *Proceedings of the FAO Global Workshop*. Banja Luka, Bosnia-Herzegovina.
155. Bodros, E., Pillin, I., Montrelay, N., & Baley, C., (2007). Could biopolymers reinforced by randomly scattered flax fiber be used in structural applications? *Compos. Sci. Technol., 67*, 462–470.
156. Carus, M., G. C., Pendarovski, C., Vogt, D., Ortmann, S., Grotenhermen, F., Breuer, T., & Schmidt, C., (2008). Studie zur markt-und konkurrenzsituation bei naturfasern und naturfaserwerkstoffen (Deutschland Und Eu). In: *Fachagentur Nachwachsende Rohstoffe (FNR)*. Guelzow.
157. Karus, M., & K. M., (2002). Natural fibers in the European automotive industry. *Article in Journal of Industrial Hemp, 7*(1), 119–131.
158. Dungani, R., et al., (2017). 3-Bionanomaterial from agricultural waste and its application. In: Jawaid, M., Boufi, S., & Abdul Khalil, H. P. S., (eds.), *Cellulose-Reinforced Nanofiber Composites* (pp. 45–88). Woodhead Publishing.
159. Gandini, A., & Belgacem, M. N., (2008). State of the art. In: Gandini, A. M. N. B., (ed.), *Monomers, Polymers, and Composites from Renewable Resources*. Elsevier: Oxford OX5 1GB, UK.
160. Faruk, O., Bledzki, A. K., Fink, H. P., & Sain, M., (2012). 2000–2010, Biocomposites reinforced with natural fibers. *Prog. Polym. Sci., 37*(11), 1552–1596.
161. Shalwan, A., & Yousif. B. F., (2013). In state of art: Mechanical and tribological behavior of polymeric composites based on natural fibers. *Mater. Des., 48*, 14–24.
162. Gassan, J., & G. V., (2000). Effects of corona discharge and UV treatment on the properties of jute-fiber expoxy composites. *Compos. Sci. Technol., 60*(15), 2857–2863.
163. Ragoubi, M., Bienaime, D., Molina, S., George, B., & Merlin, A., (2010). Impact of corona treated hemp fibers onto mechanical properties of polypropylene composites made thereof *Ind. Crops Prod., 31*(2), 344–349.
164. Li, Z., Wang. L., & Wang, X., (2007). Cement composites reinforced with surface modified coir fibers. J. Compos. Mater., 41(12), 1445–1457. Li, Z., Wang. L., & Wang, X., (2007). Cement composites reinforced with surface modified coir fibers. *J. Compos. Mater., 41*(12), 1445–1457.
165. Belgacem, M. N., & Gandini. A., (2005). The surface modification of cellulose fibers for use as reinforcing elements in composite materials. *Composite Interfaces, 12*(1/2), 41–75.
166. Liu, H., Bai, J., Wang, Q., Li, C., Wang, S., Sun, W., & Huang, Y., (2014). Preparation and characterization of silver nanoparticles/carbon nanofibers via electrospinning with research on their catalytic properties. *Nano Brief Rep. Rev., 9*(3), 145004-1-145004-7.
167. Masoodi, R., El-Hajjar. RF., Pillai, K. M., & Sabo, R., (2012). Mechanical characterization of cellulose nanofiber and bio-based epoxy composite. *Mater. Des., 36*, 570–576.

168. Tabatabaei, S., Shukohfar, A., Aghababazadeh, R., & Mirhabibi, A., (2006). Experimental study of the synthesis and characterization of silica nanoparticles via the sol-gel method. *J. Phys. Conf. Ser., 26*, 371–374.
169. Singh, A., Sharma. P. K., & Malviya, R., (2011). Eco friendly pharmaceutical packaging material. *World Appl. Sci. J., 14*(11), 1703–1716.
170. De Moura, M. R., Mattoso, L. H. C., & Zucolotto, V., (2012). Development of cellulose-based bactericidal nanocomposites containing silver nanoparticles and their use as active food packaging. *J. Food Eng., 109*, 520–524.
171. Youssef, A. M., El-Samahy, M. A., & Abdel, R. M. H., (2012). Preparation of conductive paper composites based on natural cellulosic fibers for packaging applications. *Carbohydrate Polymers, 89*(4), 1027–1032.
172. Kalia, S., et al., (2014). Nanofibrillated cellulose: Surface modification and potential applications. *Colloid and Polymer Science, 292*, 5–31.
173. Iguchi, M., Yamanaka. S., & Budhiono, A., (2000). Bacterial cellulose-a masterpiece of nature's arts. *J Mater Sci., 35*(2), 261–270.
174. Jonas, R., F. L., (1998). Production and application of microbial cellulose. *Polym. Degrad. Stab., 59*(3), 101–106.
175. Kalia, S., Dufresne, A., Cherian, B. M., Kaith, B., Avérous, L., & Njuguna, J., (2011). Cellulose-based bio-and nanocomposites: A review. *Int. J. Polym. Sci.,* 1–35.
176. Nafchi, A. M., et al., (2012). Antimicrobial, rheological, and physicochemical properties of sago starch films filled with nanorod-rich zinc oxide. *Journal of Food Engineering, 113*(4), 511–519.
177. Zadbuke, N., et al., (2013). Recent trends and future of pharmaceutical packaging technology. *Journal of Pharmacy and Bioallied Sciences, 5*(2), 98–110.
178. Lewis, H., Verghese, K., & Fitzpatrick, L., (2010). Evaluating the sustainability impacts of packaging: The plastic carry bag dilemma. *Pack. Technol. Sci., 23*, 145–160.
179. Freire, M. G., Teles, A. R. R., Ferreira, R. A. S., Carlos, L. D., Lopes-Da-Silva, J. A., & Coutinho, J. A. P., (2011). Electrospun nanosized cellulose fibers using ionic liquids at room temperature. *Green Chem., 13*, 3173–3180..
180. Zhang, X., Tu, M., & Paice, M., (2011). Routes to potential bioproducts from lignocellulosic biomass lignin and hemicelluloses. *Bio. Energy Res., 4*(4), 246–257.
181. Ul-Islam M., Khan, T., & Park, J. K., (2012). Nanoreinforced bacterial cellulose-montmorillonite composites for biomedical applications. *Carbohydr. Polym., 89*(4), 1189–1197.
182. Azeredo, H. M. C., (2013). Antimicrobial nanostructures in food packaging trends. *Food Sci. Technol., 30*(1), 56–69.
183. Sirviö, J. A., et al., (2013). Sustainable packaging materials based on wood cellulose. *RSC Advances, 3*(37), 16590–16596.
184. Chakraborty, A., Sain. M., & Kortschot, M., (2006). Cellulose microfibers as reinforcing agents for structural materials; cellulose nanocomposites: Processing, characterization, and properties. In: *ACS Symposium Series.* Washington, DC: ACS.
185. John, M. J., & Thomas. S., (2008). Biofibers and biocomposites. *Carbohydr. Polym., 71*(3), 343–364.
186. Cherian, B. M., Leao, A. L., de Souza. S. F., Thomas, S., Pothan, L. A., & Kottaisamy, M., (2010). Isolation of nanocellulose from pineapple leaf fibers by steam explosion. *Carbohydr. Polym., 81*(3), 720–725.

187. Millon, E., & Wan, W. K., (2006). The polyvinyl alcohol-bacterial cellulose system as a new nanocomposite for biomedical applications. J. Biomed. Mater. *Res. B Appl. Biomater., 79*(2), 245–253.
188. Costa, L. M. M., de Olyveira, GM., Cherian, B. M., Leão, A. L., de Souza, S. F., & Ferreira, M., (2013). Bionanocomposites from electrospun PVA/pineapple nanofibers/ *Stryphnodendron adstringens* bark extract for medical applications. *Ind. Crops Prod.,* 41, 198–202.
189. Baumann, M. D., Kang. CE., Stanwick, J. C., Wang, Y., Kim, H., Lapitsky, Y., et al., (2009). An injectable drug delivery platform for sustained combination therapy. *J. Control Release, 138*(3), 205–213.
190. Nadagouda, M. N., & Varma, R. S., (2008). Green synthesis of silver and palladium nanoparticles at room temperature using coffee and tea extract. *Green Chemistry, 10*(8), 859–862.
191. Czaja, W. K., et al., (2007). The future prospects of microbial cellulose in biomedical applications. *Biomacromolecules, 8*(1), 1–12.
192. Iamaguti, L. S., Brandao. CVS., Minto, B. W., Mamprim, M. J., Ranzani, J. J. T., & Gomes, D. C., (2008). Utilização de membrana biossintética de celulose na trocleoplastia experimental emcães. Avaliações clínica, radiográfica e macroscópica. *Vet e Zootec, 15*(1), 160–218.
193. Macedo, N. L., Matuda. F., Macedo, L. G. S., Monteiro, A. S. F., Valera, M. C., & Carvalho, Y. R., (2004). Evaluation of two membranes in guided bone tissue regeneration: Histological study in rabbits. *Braz J. Oral Sci., 3,* 395–400.
194. Klemm, D., Schumann. D., Udhardt, U., & Marsch, S., (2001). Bacterial synthesized cellulose- artificial blood vessels for microsurgery. *Prog. Polym. Sci., 26,* 1561–1603.
195. Ikada, Y., (2006). Challenges in tissue engineering. *J. R. Soc. Interface, 3*(10), 589–601.
196. Kumar, R., Roopan, S. M., Prabhakarn, A., Khanna, V. G., & Chakroborty, S., (2012). Agricultural waste *Annona squamosa* peel extract: Biosynthesis of silver nanoparticles. *Spectrochim Acta Part A., 90,* 173–176.
197. Seal, B. L., Otero, T., & Panitch, A., (2001). Polymeric biomaterials for tissue and organ regeneration. *Mater. Sci. Eng. Rep., 34,* 147–230.
198. Capes, J. S., Ando, H. Y., & Cameron, R. E. J., (2005). Fabrication of polymeric scaffolds with a controlled distribution of pores. *Mater. Sci. Mater. Med., 16,* 1069–1075.
199. Bhattacharya, M., Malinen, M., Lauren, P., Lou, Y. R., Kuisma, S. W., Kanninen, L., et al., (2012). Nanofibrillar cellulose hydrogel promotes three-dimensional liver cell culture. *J. Control Release, 164*(3), 291–298.
200. Hua, K., et al., (2014). Translational study between structure and biological response of nanocellulose from wood and green algae. *RSC Advances, 4*(6), 2892–2903.
201. Bartus, S. D., Vaidya, U., & Ulven, C. A., (2005). Design and development of a long fiber thermoplastic bus seat. *Compos, Struct., 67,* 263–277.
202. Jeyanthi, S., & Rani, J. J., (2012). Improving mechanical properties by kenaf natural long fiber reinforced composite for automotive structures. *J. Appl. Sci. Eng., 15*(3), 275–280.
203. Cicero, J. A., Dorgan, J., Dec, S. F., & Knauss, D. M., (2002). Phosphite stabilization effects on two step melt-spun fibers of polyactide. *Polym. Degrad. Stab., 78,* 95–105.
204. Mohanty, A. K., Misra, M., & Hinrichsen, B. G., (2000). Biodegradable polymers and biocomposites: An overview. *Macromol. Mater. Eng., 276, 277*(1), 1–24.

205. Brouwer, W. D., (2000). Natural fiber composites: Where flax compete with glass? *Sampe J., 36*(6), 18–23.
206. Shinoj, S., Visvanathan, R., Panigrahi, S., & Kochubabu, M., (2011). Oil palm fiber (OPF) and its composites: A review. *Ind. Crops Prod., 33*(1), 7–22.
207. Suddell, B. C., & Evans, W. J., (2005). Natural fiber composites in automotive applications in natural fibers in biopolymers and their biocomposites. In: *Natural Fibers, Biopolymers, and Biocomposites*. Boca Raton, FL: CRC Press.
208. Pickering, K., (2008). *Properties and Performance of Natural-Fiber Composites*. Cambridge: Woodhead Publishing.
209. Ghadiri, M., Chrzanowski, W., & Rohanizadeh, R., (2015). Biomedical applications of cationic clay minerals. *RSC Advances, 5*(37), 29467–29481.
210. Varghese, S., & Karger-Kocsis, J., (2003). *Polymer, 44*.
211. Wang,Y., et al., (2005). *J. Appl. Polym. Sci, 96*.
212. Sadhu, S., & Bhowmick, A. K., (2004). Preparation and properties of nanocomposites based on acrylonitrile-butadiene rubber, styrene-butadiene rubber, and polybutadiene rubber. *Journal of Polymer Science Part B: Polymer Physics, 42*(9), 1573–1585.
213. Maji, P. K., Guchhait, P. K., & Bhowmick, A. K., (2008). Effect of the microstructure of a hyperbranched polymer and nanoclay loading on the morphology and properties of novel polyurethane nanocomposites. *ACS Applied Materials and Interfaces, 1*(2), 289–300.
214. Khalil, A., Saikh, N. S., Nudrat, R. Z., & Khalid, M., (2013). Effect of micro-sized marble sludge on physical properties of natural rubber composites. *Chem. Indust. Chem. Eng. Quart., 19*(2), 281–293.
215. Ezema, I. C., Menon, A. R R., Obayi, C. S., & Omah, A. D., (2014). Effect of surface treatment and fiber orientation on the tensile and morpholoical properties of banana stem fiber reinforced natural rubber composite. *JMMCE, 2*(3), 216–222.
216. Aleksandra, I. D., Gordana, B. G., Aleksandra, B., Igor, G., & Luljeta, R., (2014). Preparation and properties of natural rubber/organo-montmorillonite: From lab samples to bulk material. *Macedonian J. Chem. Chem. Eng., 33*(2), 249–265.
217. Arroyo, M., López-Manchado, M. A., & Herrero, B., (2003). Organo-montmorillonite as substitute of carbon black in natural rubber compounds. *Polymer, 44*(8), 2447–2453.
218. Fidelis, C., Piwai, S., Benias, N. C., Upenyu, G., & Mambo, M., (2013). Maize stalk as reinforcement in natural rubber composites. *Int. J. Sci. Technol. Research, 2*(6), 263–271.
219. Ku, H., et al., (2011). A review on the tensile properties of natural fiber-reinforced polymer composites. *Composites Part B: Engineering, 42*, 856–873.
220. https://www.bekaert.com/en/products/automotive/corner/steel-cord-for-tire-reinforcement (accessed on 3 August 2020).

CHAPTER 5

Silk: An Explorable Biopolymer in the Biomedical Arena

ATHIRA JOHN,[1] ANANTHU PRASAD,[2] and NEHA KANWAR RAWAT[3]

[1]*Center for Biopolymer Science and Technology (CBPST), CIPET, Kochi, Kerala, India*

[2]*International and Inter-University Center for Nanoscience and Nanotechnology (IIUCNN), Mahatma Gandhi University, Kottayam, Kerala, India*

[3]*CSIR – National Aerospace Laboratories Bangalore, Karnataka, India*

ABSTRACT

The biopolymer silk as a recognizable implement in the biomedical field is manifested in nanotechnology due to its promising avenues. Even though silk sutures were known for centuries, yet an expansion of this research had taken place in these recent decades. The excellent biocompatibility and slow degradation make it well applicable in this field. Silk is mainly composed of two types of proteins namely fibroin and sericin, each of which have created its benchmark in the biomedical world. Wound dressing, drug delivery, implants, and tissue engineering (TE) is the major fields of applications. Silk blended with various other synthetic and natural polymers have found unique applications in these fields. The prominent techniques like electrospinning have found wide applications in incorporating silk in biomedical field. The chapter discusses the scope for this unexplorable material: silk, its prospects and limitations are also explored.

5.1 INTRODUCTION

The *Bombyx mori* silkworm silk has found its application in textile industries for centuries. However, the biomedical field remained unexplored since its conception. Fortunately, researchers found the unique properties of silk to be applicable in biomedical grounds in recent decades and are still research is under progress [1]. Silk is a novel class of proteins having high molecular weight. They possess unique features like self-assembly, robust mechanical properties, biocompatibility, and biodegradability which makes them more suitable for biomaterials application [2]. It is in 1969, that the first patent available in literature by Bloch and Messores, proposed the use of silk reconstituted by LiSCN/LiBr as are placement for the standard wax coating used on silk sutures to reduce their limpness, fraying, and unwanted capillary action [3]. After this cited, there was mass research initiated to enhance the properties of silk and its application in the biomedical arena.

5.2 STRUCTURE AND CONSTITUENTS

Biologically silk can be defined as a structural protein that has been spun into as fiber for using outside the body. The most well-studied silk is the one made of the domesticated silkworm (*Bombyx mori*) for the reason that it possesses a semi-crystalline silk core (silk fibroin: SF) which excellent load-bearing capacity and an external layer called sericin which functions as a gumming agent [4]. Typically, silk produced by *Bombyx Mori*, silkworm consists of 70–80% fibroin, 20–30% sericin, 0.2–0.4% wax, and 1.2–1.6% carbohydrates. 0.7% inorganic matter and. 0.2% pigments [6] (Figure 5.1).

5.2.1 SILK FIBROIN (SF)

Silkworm fibroin is used as biomedical sutures for ages; it is also used in textile production. It is thin, long, light, and soft. However, one of the most potent natural fibers. This was due to the chemical structure of the protein [7]. Fibroin consists of light (L) chain polypeptide, and heavy (H) chain polypeptide linked together via a single disulfide bond. The main sequence in the heavy chain is due to the alanine-glycine repeats. Fibroin has higher proportions of glycine and alanine acids. Fibroin has water-absorbing, insulating, and lustrous nature. Thermotolerance is yet another noticeable property of fibroin [6, 11] (Figure 5.2).

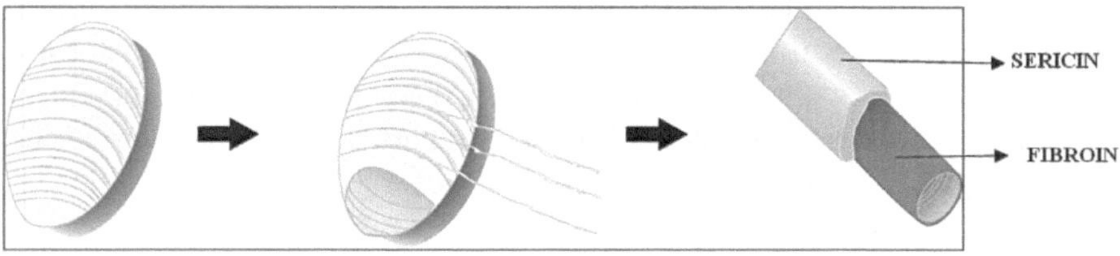

FIGURE 5.1 Components of silk.

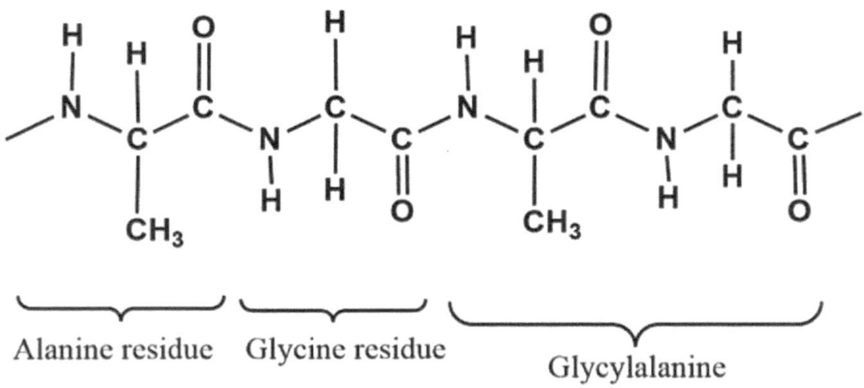

FIGURE 5.2 Structure of silk fibroin.

5.2.2 SILK SERICIN

Sericin is yet another constituent of silk which provides adhesion to the fibroin in cocoon. It was considered a waste material in the textile industry until recent past. After the properties of the sericin were closely observed, it found a lot of applications in biomedical field. It is very easily hydrolyzed. It becomes a user due to its specific properties like resistance to oxidation, antibacterial activity. They also possess UV resistance and absorb/ release moisture efficiently, inhibitory activity. It is carefully separated from fibroin during the silk manufacturing process, to make silk appearance lustrous. The sericin was typically considered waste until these properties of it were known [5]. The chemical structure and molecular weight of sericin depend on the method of separation of sericin and fibroin and process of recovering sericin from degumming. Sericin is composed of more with serine and aspartic acids

[6]. Sericin has been to have properties to become hydrogel with sustained-release property, pH-dependent degradation dynamics, and good elasticity. It is also reported in literature that it has no immunogenic responses by various reporters [11]. Zhang et al. [40] reported that sericin protein could be cross-linked, copolymerized, and blended with other macromolecular materials, to make materials with enhanced properties. They also reported its usage, as an improved coating material for natural and artificial fibers, fabrics, and articles (Figure 5.3).

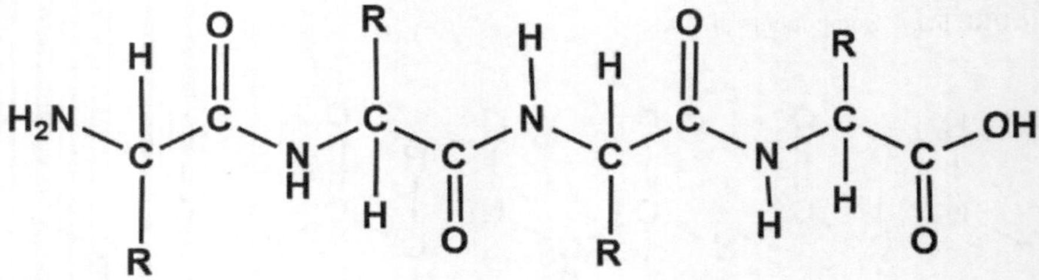

FIGURE 5.3 Structure of sericin.

5.3 SILK IN BIOMEDICAL FIELD

Silk has undergone many chemical modifications to emerge as a biomedical tool in various forms, such as silk solution, films, scaffolds, particles, nano-fibers, hydrogels, etc. Owing to its biocompatibility, slow biodegradability, controllable structure and morphology, self-assembly, and excellent mechanical properties silk has contributed widely towards the biomedical field. Silk is less inflammatory than other conventional biodegradable polymers. Among this biocompatibility and biodegradation are the major ones Silk is less inflammatory than other conventional biodegradable polymers and also found to have less or no immunogenic response unless fibroin and sericin are compared [11].

5.3.1 DRUG DELIVERY

Drug delivery is the method of administering a medicated compound to achieve a therapeutic effect in the body. The controlled release of the drugs is at the target size is a major concern in treatment. The literature reports numerous applications of silk proteins to have in delivering drugs at a controlled rate. The prominent processes like Gelation, microparticle, freeze-drying, electrospinning, electrospraying, etc. are the primary techniques explored in this area [11].

DadrasChomachayi, Masoud et al. reported the sustained drug release by SF and gelatin (GT) nanofibrous substrate thyme essential oil (TEO) and doxycycline monohydrate (DCMH), both as anti-bacterial agents, were loaded and studies were conducted. Both of them showed excellent cell-compatibility. The addition of GT has helped improving hydrophilicity, surface wettability, and a mass loss percentage of silk [8]. Suktham, Kunat et al. successfully attempted to load resveratrol to silk protein (SP) nanoparticles via a solventless precipitation method. Cell viability assay proved that Resveratrol-loaded SP nanoparticles could effectively inhibit the growth of colorectal adenocarcinoma (Caco-2) cells. Although the advantages found were that, it was non-cytotoxic to skin fibroblasts [9]. Gao Ya et al. has studied a transdermal drug delivery with SF, where silk acted as a scaffold to adsorb prepolymer solution and initiate microneedles formation of PEGDA/sucrose. It was successful in transdermal delivery and controllable release of Rhodamine B, indocyanine green, and doxorubicin by regulating the sucrose content. SF aided the hasty fabrication of polymeric microneedles for transdermal drug delivery applications [10]. Gupta et al. prepared SF-encapsulated with curcumin nanoparticles less using the devised capillary-microdot technique for drug delivery applications [12]. The drug encapsulation efficiency of various nanoparticles like pramipexole, curcumin, and propranolol hydrochloride on encapsulation with SF was studied by Gianak, Olga et al. The studies revealed among solvents propranolol showed the highest effect [13]. Zhuping Yin et al. synthesized swellable microneedles with SF using polydimethylsiloxane (PDMS). They were quickly penetrable to porcine skin with a depth of ~200 μm in vitro, and transform into semi-solid hydrogels and form 50–700 nm porous networks inside [14].

5.3.2 *WOUND DRESSING*

A wound is biologically defined as a fracture in healthy body integrity. Therefore, repairing the wound comprises the activation of intracellular and intercellular pathways for the restoration of this tissue integrity and homeostasis [16]. Literature reports SF-based dressings being used as carriers for, growth factors, delivering drugs, and bioactive agents to the wound area, along with appropriate aid for complete healing [15–20]. Antimicrobial properties imparting dressings have been traditionally used to minimalize the bacterial infection of burns and wounds. Meanwhile, Silk has been used in suturing for centuries.

Uttayarat, Pimpon et al. manufactured SF mats coated with silver nanoparticles (AgNPs) to act as a prototypic wound dressing and also evaluated for antimicrobial properties. The colloidal AgNPs were prepared by irradiating silver nitrate by gamma rays which were confirmed by transmission electron microscopy. The growth of *Staphylococcus aureus* and *Pseudomonas aeruginosa* was effectively inhibited on the application of this [15]. Karahaliloglu, Zeynep et al. reported the usage of chitosan/silk sericin (CHT/SS) scaffolds collective with lauric acid (LA) and zinc oxide (ZnO) nanoparticles for the successful wound dressing applications [16]. Mehrabani, MojtabaGhanbari et al. synthesized an interconnected microporous structured wound dressing material from chitin/SF/TiO_2 nanocomposite which has high antibacterial, blood clotting, and mechanical strength properties using the freeze-drying method. The studies proved that swelling and water uptake of the dressing was 93%, which highly contributes to wound dressing applications. Hemostatic potential studies showed that blood clotting ability of the nanocomposites was high on comparing with pure components and commercially available products.

Moreover, cell viability, attachment, and proliferation were confirmed by MTT assay and DAPI staining [17]. A lot of other natural dopant incorporated silk has also been the focus of researches these days [18–20]. Selvaraj et al. included fenugreek, a natural antioxidant to SF to form nanofiber by a co-electrospinning method. The biocompatibility and antioxidant activity were evaluated were found to be increased and were confirmed through MTT assay and DPPH scavenging assay, respectively. The experiment has proven that on incorporation of fenugreek thermal and mechanical properties of SF nanofibers has increased [20]. Sericin has also proven its wound dressing capability. Huawei He et al. reported the sericin/polyvinyl alcohol (PVA) blend film incorporated with AgNPs to have functional antibacterial activities against *E. coli* and *S. aureus* [21].

Lamboni, Lallepak et al. have reported the functionalization of bacterial cellulose using silk sericin for proper wound healing [23]. Hence, silk has become a successful biomimetic candidate for wound dressing and potential replacements for treating skin injuries [22].

5.3.3 TISSUE ENGINEERING (TE)

Tissue engineering (TE) is an interdisciplinary field in which engineering of tissues and organs by integrating science and technology of cells, materials, and biochemical factors is carried out. The most challenging part is the replication of the natural extracellular matrix (ECM). Hence, scaffold engineering is the prime concern in this field. Amongst the variety of materials tested, SF has proven as a promising material for this. SF has been reported as films, porous matrices, hydrogels, nonwoven mats, etc., for various TE applications in bone, tendon, ligament, cartilage, skin, liver, trachea, nerve, cornea, eardrum, dental, bladder [23–27]. Biman B. Mandal et al. have prepared novel 3-D sericin/GT scaffolds and 2-D films with uniform pore distribution, improved compressive strength, high swell ability, and high overall porosity all of which support TE and biomedical applications. The enhanced cell attachment, viability, and low immunogenicity were also reported which proved its potential as a future biopolymeric graft material [27]. Chao, Pen-Hsiu Grace et al. have reported the successful attempts of silk hydrogels in cartilage TE [25]. Lawrence, Brian et al. said the application of silk in cornea TE [28]. Kim, Hyeon Joo et al. reported the biomimetic growth of calcium phosphate on porous SF polymeric scaffolds to generate organic/inorganic composites as scaffolds for bone TE. Hence prepared mineralized protein-composite scaffolds were subsequently seeded with human bone marrow stem cells and found to be successful [29]. Nandana Bhardwaj et al. fabricated three-dimensional (3D) blended scaffolds of silk and keratin by freeze-drying method and characterized their physicochemical, mechanical, and degradable properties extensively. SF and human hair keratin blended scaffolds were demonstrated as the promising dermal substitute for skin TE [30]. Manchineel, Shivaprasad et al. manufactured the SF/melanin composite films and electrospun fiber mats for skeletal muscle TE. The incorporation of melanin modulated the thermal stability and electrical conductivity of scaffolds. It has also checked the alignment of fibers in electrospun mats and provided excellent antioxidant properties to the scaffolds [31]. Franck et al. observed the ability of gel spun silk scaffolds in bladder TE and also its cell responses [41]. Despite all the signs of progress made, there are still fields unexplored in the TE application for silk.

5.3.4 MEDICAL IMPLANTS

The purpose of implantable devices is to imitate an organ and to replace an injured part to retain the normal functioning of the body. The most extensively used medical implants are heart, ears, eyes, knees, breasts, hips, bones, and cardiovascular system. Conventional materials like metals, ceramics, and synthetic polymers may lead to immunological rejection by the body and hence a replacement with biopolymer is always welcomed. Silk has been reported as biomedical implants in literature these days [32–37]. Mehrjou, Babak et al. fabricated homogeneous nanocones which were undergone oxygen plasma etching on the surface of silk films. This has increased surface energy [34]. SF is also found to be successful in dental implants [35]. Elia, Roberto et al. reported formulation of Risperidone by gelation of SF as injectable implants [36]. SF tubular scaffolds are found to be ideal in vascular graft implants. Various methods like dip coating, electrospinning has been reported in literature. Blends with polyethylene oxide are also found to be effective [37]. Lorenz Meinel et al. assessed the appropriateness of silk biomaterials as implants for bone healing. It was found that novel porous SF scaffolds permitted adequate temporal control of hydroxyapatite deposition and resulted in the formation of a trabecular-like bone matrix in the studies [38]. Altman et al. have developed a wire rope matrix for the expansion of autologous tissue-engineered anterior cruciate ligaments (ACL) using a patient's adult stem cells with silk [39]. Films made out of sericin and fibroin has excellent oxygen permeability it mimics the human cornea and its functional properties hence it is explorable in the future to make corneal implants out of silk [40].

5.4 ELECTROSPUN SILK IN BIOMEDICAL APPLICATIONS

Electrospinning is an evolving technique in synthesizing scaffolds and wound dressing materials using silk in biomedical field. Affordability, very high surface-to-volume ratio, variable porosity, and versatility in materials selection are the factors that make electrospinning a favorable one in biomedical applications. The ECM of human body is composed fibrous proteins that can be mimicked by the nanofibrous mat produced by electrospinning.

Li et al. reported the electrospinning of bioactive SF/hydroxyapatite (nHAp) to improving bone formation; he was successful in developing human bone marrow-derived mesenchymal stem cells on the scaffolds. Schneider et al. spun epidermal GF (EGF)-loaded silk nanofibers which could aid the healing process, thereby decreasing the time of wound closure [43].

Our research group has been involved in understanding the role of silk in elctrospun fibers of biodegradable polymers and their vast applications. We have carried out emulsion electrospinning of SF and polycaprolactone (PCL) for TE applications. The findings revealed various combinations of emulsion electrospinning rather than the conventional electrospinning of polymer blends. PCL has been dissolved in chloroform and SF in formic acid. The concentration of PCL was continuously taken at 10% (w/v), while that of SF concentration was varied. The physicochemical, mechanical, and interfacial properties of the nanofibrous electrospun mats were studied. Different combinations of the emulsion were prepared (Figure 5.4). The structure of SF and scaffolds were confirmed using FTIR (Figure 5.5). The resultant nanofibers exhibited a distinct core-shell structure without using co-axial electrospinning (Figure 5.6) [44]. These nanofibers could be a prospective vehicle for drug-delivery, growth factors, or biomolecule encapsulation in upcoming TE applications (Table 5.1).

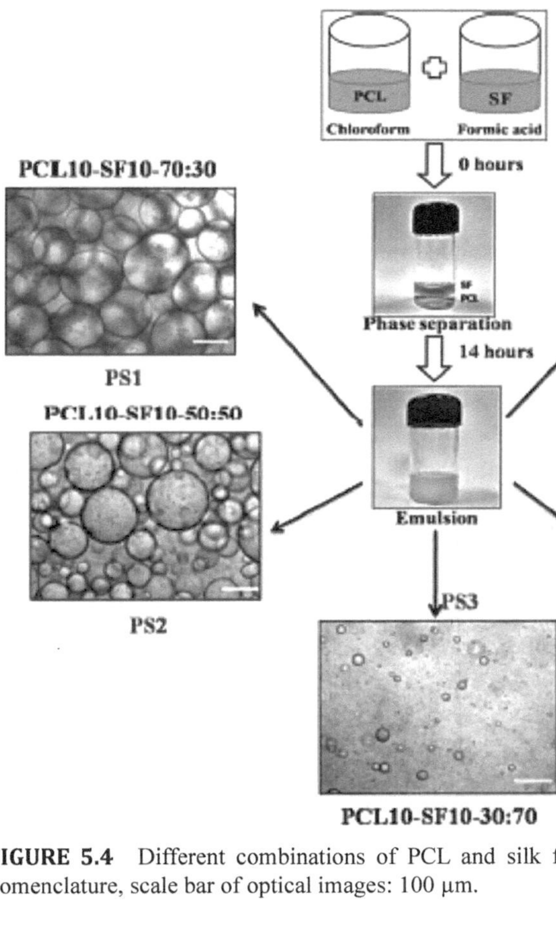

FIGURE 5.4 Different combinations of PCL and silk fibroin (SF) emulsion with their nomenclature, scale bar of optical images: 100 μm.

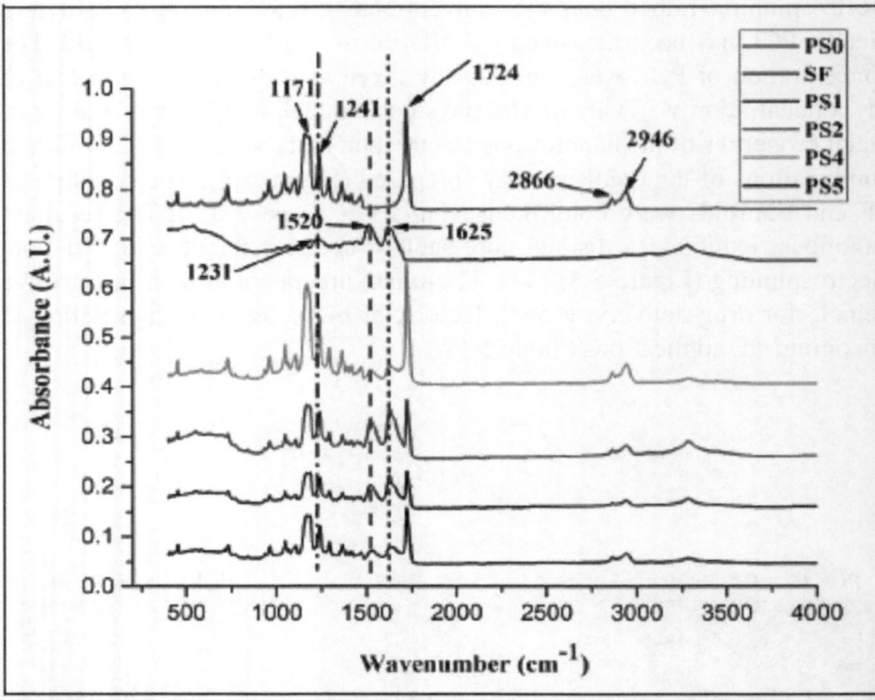

FIGURE 5.5 FTIR spectra of different nanofibrous scaffolds and SF powder.

5.5 CONCLUSION

Silk has been esteemed and caught the attention of the human race, by virtue of its lustrous appeal since long ago. Therefore, it rules the textile market across the globe even today. But apart from its aesthetic value researchers have attributed some medicinal values to silk in the recent past and its scope is expected to shoot up shortly. It was surprising that, even sericin which was once considered the waste material in silk industry had got special consideration due to its unique properties. Silk proteins were found to have excellent biocompatibility and slow biodegradability which provides it an unavoidable role in biomedical applications. The structure composition and current researches like wound healing, TE and the emerging biomedical applications of silk in its various forms like films, nanofiber, hydrogels, and particles were discussed. Silk by undergoing major and minor chemical modification or by combining with different synthetic or natural polymers can work wonders in the biomedical field. There is still a vast world of silk unexplored; the researchers have to research these hidden capable avenues and their applications.

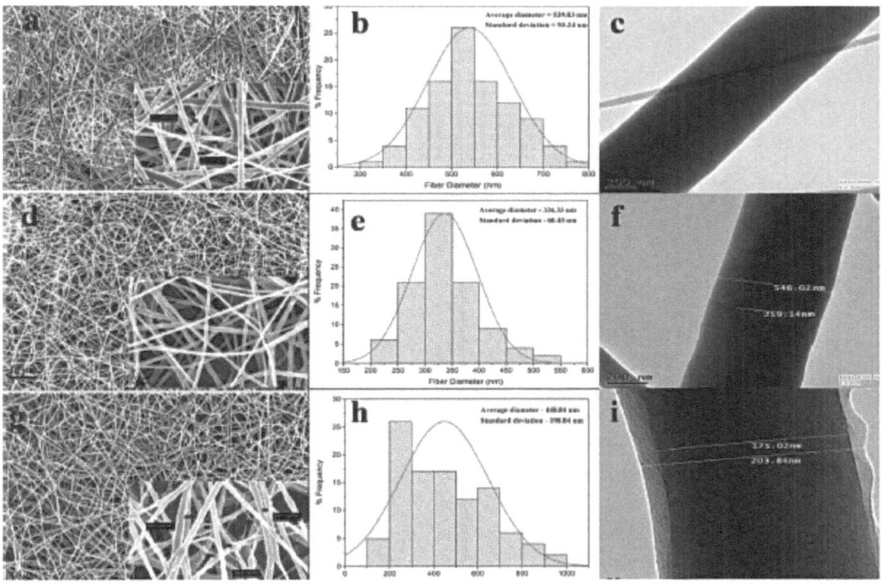

FIGURE 5.6 Different combinations of PCL and silk fibroin (SF) emulsion with their nomenclature, scale bar of optical images: 100 μM.
Scanning electron microscope (SEM) images of PS0 to PS5 nanofibers (a,d,g,j,m,p) (scale bar:10 μm) with magnified images of same (inset) (scale bar: 1 μm). Frequency distribution of nanofiber diameters of PS0 to PS5 b,e,h,k,n,q). Transmission electron microscope (TEM) images of PS0 to PS5 nanofibers (c,f,i,l,o,r).

TABLE 5.1 Properties of the Different Emulsion Combinations

Emulsion Combination	Nanofiber Name	Spinnability	Nanofiber Diameter
PCL 10%-SF 0%	PS0	Y	539.8 ± 93.2 nm
PCL 10%-SF 10%-70:30	PS1	Y	336.3 ± 60.4 nm
PCL 10%-SF 10%-50:50	PS2	¥¥	448.8 ± 198.0 nm
PCL 10%-SF 10%-30:70	PS3	N	1021.6 ± 351.9 nm
PCL 10%-SF 5%-70:30	PS4	Y	212.6 ± 43.4 nm
PCL 10%-SF 5%-50:50	PS5	Y	370.1 ± 111.43 nm

Y – Yes; N – Not spinnable; ¥ – Spannable with secondary jets of nanofibers, random fiber.*

KEYWORDS

- **anterior cruciate ligaments**
- **biocompatibility**
- **electrospinning**
- **fibroin**
- **nanofibers**
- **sericin**

REFERENCES

1. Leng-Duei, K., et al., (2015). Structures, mechanical properties, and applications of silk fibroin materials. *Progress in Polymer Science, 46*, 86–110.
2. Murphy, A. R., & David, L. K., (2009). Biomedical applications of chemically-modified silk fibroin. *Journal of Materials Chemistry,19*(36), 6443–6450.
3. Bloch, A., & Arthur, S. M., (1969). *Silk Suture*. U.S. Patent No. 3,424,164.
4. Pérez-Rigueiro, J., et al., (2000). Mechanical properties of single-brin silkworm silk. *Journal of Applied Polymer Science,75*(10), 1270–1277.
5. Mondal, M., (2007). The silk proteins, sericin, and fibroin in silkworm, *Bombyx mori* Linn.: A review. *Caspian Journal of Environmental Sciences,5*(2), 63–76.
6. Rangi, A., & Lalit, J., (2015). The biopolymer sericin: Extraction and applications. *J. Text Sci. Eng., 5*(1), 1–5.
7. Murphy, A. R., & David, L. K., (2009). Biomedical applications of chemically-modified silk fibroin. *Journal of Materials Chemistry,19*(36), 6443–6450.
8. Dadras, C. M., et al., (2018). Electrospun nanofibers comprising of silk fibroin/gelatin for drug delivery applications: Thyme essential oil and doxycycline monohydrate release study. *Journal of Biomedical Materials Research Part A,106*(4), 1092–1103.
9. Suktham, K., et al., (2018). Efficiency of resveratrol-loaded sericin nanoparticles: Promising bionanocarriers for drug delivery. *International Journal of Pharmaceutics,537*(1/2), 48–56.
10. Gao, Y., et al., (2019). Highly porous silk fibroin scaffold packed in PEGDA/sucrose micro needles for controllable transdermal drug delivery. *Biomacromolecules,20*(3), 1334–1345.
11. Dutta, D., Chowdhury, M. H., & Avijit, B., (2018). Silk proteins in drug delivery: An overview. *RPHS,4*(4), 514–518.
12. Gupta, V., et al., (2009). Fabrication and characterization of silk fibroin-derived curcumin nanoparticles for cancer therapy. *International Journal of Nanomedicine,4*, 115.
13. Gianak, O., et al., (2018). Silk fibroin nanoparticles for drug delivery: Effect of bovine serum albumin and magnetic nanoparticles addition on drug encapsulation and release. *Separations,5*(2), 25.

14. Yin, Z., et al., (2018). Swellable silk fibroin micro needles for transdermal drug delivery. *International Journal of Biological Macromolecules, 106*, 48–56.
15. Uttayarat, P., et al., (2012). Antimicrobial electro spun silk fibroin mats with silver nanoparticles for wound dressing application. *Fibers and Polymers,13*(8), 999–1006.
16. Karahaliloglu, Z., Ebru, K., & Emir, B. D., (2017). Antibacterial chitosan/silk sericin 3D porous scaffolds as a wound dressing material. *Artificial Cells, Nanomedicine, and Biotechnology,45*(6), 1172–1185.
17. Mehrabani, M. G., et al., (2018). Chitin/silk fibroin/TiO_2 bio-nanocomposite as a biocompatible wound dressing bandage with strong antimicrobial activity. *International Journal of Biological Macromolecules,116*, 966–976.
18. Elakkiya, T., et al., (2014). Curcumin loaded electro spun *Bombyx mori* silk nano fibers for drug delivery. *Polymer International,63*(1), 100–105.
19. Kheradvar, S. A., et al., (2018). Starch nanoparticle as a vitamin E-TPGS carrier loaded in silk fibroin-poly(vinyl alcohol)-Aloe vera nanofibrous dressing. *Colloids and Surfaces B: Biointerfaces,166*, 9–16.
20. Selvaraj, S., & Nishter, N. F., (2017). Fenugreek incorporated silk fibroin nanofibers a potential antioxidant scaffold for enhanced wound healing. *ACS Applied Materials and Interfaces,9*(7), 5916–5926.
21. He, H., et al., (2017). Preparation and characterization of silk sericin/PVA blend film with silver nanoparticles for potential antimicrobial application. *International Journal of Biological Macromolecules,104*, 457–464.
22. Farokhi, M., et al., (2018). Overview of silk fibroin use in wound dressings. *Trends in Biotechnology,36*(9), 907–922.
23. Lamboni, L., et al., (2016). Silk sericin-functionalized bacterial cellulose as a potential wound-healing biomaterial. *Biomacromolecules,17*(9), 3076–3084.
24. Kasoju, N., & Utpal, B., (2012). Silk fibroin in tissue engineering. *Advanced Healthcare Materials,1*(4), 393–412.
25. Pen-Hsiu, G. C., et al., (2010). Silk hydrogel for cartilage tissue engineering. *Journal of Biomedical Materials Research Part B: Applied Biomaterials,95*(1), 84–90.
26. Gupta, P., et al., (2016). Biomimetic, osteoconductive non-mulberry silk fiber reinforced tricomposite scaffolds for bone tissue engineering. *ACS Applied Materials and Interfaces8*(45), 30797–30810.
27. Mandal, B. B., Anjana, S. P., & Kundu, S. C., (2009). Novel silk sericin/gelatin 3-D scaffolds and 2-D films: Fabrication and characterization for potential tissue engineering applications. *Acta Biomaterialia,5*(8), 3007–3020.
28. Lawrence, B. D., et al., (2009). Silk film biomaterials for cornea tissue engineering. *Biomaterials,30*(7), 1299–1308.
29. Kim, H. J., et al., (2008). Bone tissue engineering with premineralized silk scaffolds. *Bone42*(6), 1226–1234.
30. Bhardwaj, N., et al., (2014). Silk fibroin-keratin based 3D scaffolds as a dermal substitute for skin tissue engineering. *Integrative Biology,7*(1), 53–63.
31. Manchineella, S., et al., (2016). Pigmented silk nanofibrous composite for skeletal muscle tissue engineering. *Advanced Healthcare Materials,5*(10), 1222–1232.
32. Rebelo, R., Margarida, F., & Raul, F., (2017). Biopolymers in medical implants: A brief review. *Procedia Engineering,200*, 236–243.
33. Elia, R., et al., (2015). Electrodeposited silk coatings for bone implants. *Journal of Biomedical Materials Research Part B: Applied Biomaterials,103*(8), 1602–1609.

34. Mehrjou, B., et al., (2019). Antibacterial and cytocompatible nanoengineered silk-based materials for orthopedic implants and tissue engineering. *ACS Applied Materials and Interfaces,11*(35), 31605–31614.
35. Ebrahimi, A., et al., (2018). Preparation and characterization of silk fibroin hydrogel as injectable implants for sustained release of risperidone. *Drug Development and Industrial Pharmacy,44*(2), 199–205.
36. Elia, R., et al., (2015). Silk electro gel coatings for titanium dental implants. *Journal of Biomaterials Applications,29*(9), 1247–1255.
37. Kundu, S., (2014). *Silk Biomaterials for Tissue Engineering and Regenerative Medicine* (p. 53). Elsevier.
38. Meinel, L., et al., (2005). Silk implants for the healing of critical size bone defects. *Bone,37*(5), 688–698.
39. Altman, G. H., et al., (2003). Silk-based biomaterials. *Biomaterials,24*(3), 401–416.
40. Yu-Qing, Z., (2002). Applications of natural silk protein sericin in biomaterials. *Biotechnology Advances,20*(2), 91–100.
41. Franck, D., et al., (2013). Evaluation of silk biomaterials in combination with extracellular matrix coatings for bladder tissue engineering with primary and pluripotent cells. *PloS One,8*(2), e56237.
42. Holland, C., et al., (2019). The biomedical use of silk: Past, present, future. *Advanced Healthcare Materials,8*(1).
43. Eatemadi, A., et al., (2016). Nanofiber: Synthesis and biomedical applications. *Artificial Cells, Nanomedicine, and Biotechnology, 44*(1), 111–121.
44. Roy, T., et al., (2018). Core-shell nanofibrous scaffold based on polycaprolactone-silk fibroin emulsion electro spinning for tissue engineering applications. *Bioengineering,5*(3), 68.

CHAPTER 6

Influence of Nanotechnology on Prevention of Biofouling in Aquaculture Cage Nets: Green Conducting Polymers and Their Significance in Case Studies

P. MUHAMED ASHRAF

ICAR Central Institute of Fisheries Technology, Cochin – 682029, Kerala, India, E-mail: ashrafp2008@gmail.com

ABSTRACT

Biofouling in aquaculture cages has been an important issue for the farmers since its management incurs a huge sum of money and labor. Antifouling strategies failed due to the diversity of organisms are different from region to region and increased concern about pollution in the water bodies. The chapter describes the strategies to combat biofouling using nanomaterials coated over a polyaniline coated polyethylene and its efficiency towards antifouling. Green synthesis of nanomaterial incorporated hydrogel and mixed charged polymeric hydrogel over polyethylene aquaculture cage nets against biofouling also described. Brief description of environmental impact assessment of nanomaterials treated nets in the marine environments included at the end.

6.1 INTRODUCTION: FISHERIES SCENARIO INDIA AND ABROAD

Growing populations in the world and demand for food security is increasing year by year. A part of the food security is met through marine and aquaculture produced fishery resources. India has 8041 km of coastline, 3 million hectares of reservoirs, and 1.2 million hectares of brackish water. India has shown incremental improvement in fish production and it provides employment to 14 million people. According to the National Fishery Development Board, report the fish production during 2017–2018 was 12.60 million tons of which 67% were from inland and culture production. This shows the contribution from the culture sector in the nutritional security is significant. Aquaculture production in India was second in the world and plays a major role in India, food security, labor, exports, and the economy [1]. Major aquaculture based fishes are carps, catfishes, prawns, tilapia, freshwater pearl culture, coldwater fisheries, and ornamental fish culture.

Dwindling fish catch, overexploitation, illegal, unreported, and unregulated (IUU) fishing, environmental issues, climate change (CC) and a labor shortage are major concerns regarding marine fishery has led to planners to concentrate on the aquaculture production [2–4]. Fish production through aquaculture is an absolute necessity to feed the growing population and fish is also considered as the cheapest protein source [5]. India produces fish 5% of the total world production and 7% of the total aquaculture production. In view of this, planners stressed the need to improve the aquaculture production by extending shallow waters to marine areas. Increased aquaculture production led to a series of related issues such as viral and bacterial diseases, deteriorated water quality, influence of climate change, unutilized nutrients, and organic wastes, increased input cost, low rate to the produce due to increased catch, escapement of exotic species in the surrounding waters, and finally the cost of maintenance of cages. Maintenance of the cages is a major problem and it cost about 25% of the project cost. The major issue in cages was biofouling and closure of lumens of meshes by fouler.

6.2 BIOFOULING IN UNDERWATER STRUCTURES

Submerged natural and artificial materials are used for the settlement of micro and macro-organisms are called biofouling. When a material is submerged in the marine or aquatic environment, initially the surface was absorbed by organic molecules like proteins, polysaccharides, and macro organic molecules. Followed by microorganisms like bacteria, diatoms, microalgae, and fungi are formed a complex nutrient-rich layer and the same layer was called biofilms [6–8]. Macro-organisms like mussels, barnacles, sponges, polychaetes, oysters, etc., are attached over this biofilm. These biofoulers form over the surfaces in parallel or overlapping mode and the attachment was strong and difficult to remove [9, 10].

6.2.1 AQUACULTURE

Aquaculture or aquafarming is the farming of fish, crustaceans, mollusks, algae, *aquatic organisms, and other aquatic organisms*. The farming implies, rearing of organisms under controlled conditions in fresh or saltwater to improve the growth, protection from predators, diseases, and enhance production. This was done open ponds, reservoirs, or farms. Cage culture implies the fishes reared in cages, exposed in the aquatic environment from fingerlings to table or marketable sizes. Cages are traditionally fabricated by different ways and it depends on the region and culture of the area. Commercial cage culture was practiced by fabricating circular or square cages comprised of polyethylene nets covered top and bottom. The cages are made upright by using floating frames, and sinkers. Intensive cage culture is coming up in India recent years and it was directly carried out marine environments (Figure 6.1).

FIGURE 6.1 Models of aquaculture cages in the coastal marine waters.

6.3 BIOFOULING AND ITS MANAGEMENT

Biofouling is the major nuisance in the aquaculture practices. Different types of marine organisms were attached over the cage nets and the lumens were clogged. This will prevent the water circulation, accumulation of wastes in the cages, nutrient accumulation, increase the weight of the cage, reduce the drag, and retard the growth of organisms in the cage. Biofouling also reduce the efficiency of the materials used for fabrication of cages by increasing weight, abrasion, and brittleness. Current practice is cleaning the cage meshes by mechanical or manual methods. The accumulation of biofoulers diversity and intensity will be varied with the seasons and geographic locations [11]. A common strategy to combat biofouling is very difficult since the species and nature of organism are different from places to places. A group of benthic organisms produce toxic metabolites which will affect the growth and life of organisms in the cages. Biofouling communities compete for resources with cultured fishes and may harm the fishes in the cages.

The cost of antifouling management in aquaculture cages are very expensive [12] and it was estimated about 25% of the total project cost. The cost involves periodic cleaning, replacement of nettings and pollution due to the cleaning process. This will make the farming become nonprofitable. The question how to mitigate the problem of biofouling in the aquaculture cages.

6.4 CURRENT APPROACHES TOWARDS ANTIFOULING

Biofouling is a serious problem in shipping industries and a large number of coating technologies were developed and employed to combat biofouling in ship hulls. Earlier days copper oxide-based antifouling paints, then tributyltin (TBT) and their analogous compounds were used to combat biofouling in ship hulls. TBT was banned due to environmental concerns. New generation antifouling coatings are mainly from low surface energy, self-polishing type paints. In fishing boats and other submerged materials such as pillars, dam shutters mainly, employed CuO-based antifouling paints. The basic principle behind the CuO-based paints are a specific amount of CuO was released to the environment at a fixed rate and this will be continued till concentration is minimum. In fishing boats, it will stay up to one year. Generally, CuO-based antifouling coating life is one year and they need to repaint. The world over aquaculture is getting prominence and the major issue raised by the farmers is biofouling and its management. Antifouling intervention in aquaculture system is very sensitive since it deals with organisms and food. It needs careful evaluation. So intervention using chemical methods needs lots of tests and safety approvals. A series of research attempts were made to combat biofouling in aquaculture systems.

The current scenario of management of aquaculture cages from fouling is mainly manual cleaning by brushing or mechanical scrubbers. Mechanical cleaning involves employing mechanical scrubbers, water jets, and scrappers. These are labor-intensive, pollute the cages and its vicinity by dispersing minute particles, disturbance to the organisms, and expensive. Other methods include air or sun drying by taking out the nets. This will not remove the fouler but kill the fouler and needs physical removal [13].

Popular antifouling strategy in aquaculture cages are by treating low-levelcopper-based chemical coating and a fixed quantity of biocide leaching the aquatic environment will deter the fouler. This treatment will last only for one season of culture. The cost of treating the net is very high. Costello et al. [14] developed a biodegradable copper oxide composite and these are not planned to use for aquaculture. Copper and tin-based molecules using against biofouling was banned by the EU and other countries [15]. Most of the copper-based antifouling coating over aquaculture cage nets are very heavily loaded with copper and some types of organisms will attach over these nets. Cleaning of these nets will create copper and sludge waste and difficult to dispose of. The other methods are:

- Biological control using grazers [16, 17];
- Silicon-based fouling-release coating [18];
- Acetic acid-based antifouling strategies over fish cage nets [19];
- Modifying the cage net design to avoid shellfish biofouling;
- Natural material based antifouling strategies [20, 21].

These are considered in laboratory-scale not much practical application under field conditions. Introduction of nanotechnology as potential tools for varying applications may be attempted for combating biofouling under submerged environments.

6.5 NANOTECHNOLOGICAL INTERVENTION

In the Japanese Society of Precision Engineering Conference, Japan, 1974, Prof. Taniguchi introduced the word nanotechnology, which means synthesis, fabrication, development of devices or materials having size 1 to 100 nm [22]. The principle of nanotechnology is those materials with known properties and functions will exhibit different behavior and functions at nano-sized state. By reducing, the size of the particles will increase enormously the surface area of the particle per unit weight. In other words per unit weight the number of nano-sized particles and surface area will be higher than the microstate. Generally, nanomaterial size will be varied between 1 to 100 nm. Nano-sized inorganic and carbon-based molecules are employed for solving the problems of drug delivery, sensors, catalysis, surface coatings, and antibacterial applications. Depends on applications different nanomaterials were employed. The advantage of nanotechnology is the use of very low amounts of materials, increased surface area, high activity, and high efficiency. Series of research undertaken to the application of nanomaterials against the use of antibacterial and biofouling applications.

Polyethylene is a non-polar molecule and its fibers used for the fabrication of aquaculture cage nettings. High-density polyethylene was extensively used for the fabrication of cage nettings (Figure 6.2). Due to its nonpolar nature difficult, to make any chemical-based interventions. As described earlier sections the antifouling strategies over polyethylene nets is difficult. ICAR CIFT is introduced as a new strategy to prepare biofouling resistant aquaculture cage nets by employing a different method [23]. Polyaniline is a conducting polymer and has good conductivity. Trials were conducted to coat the polyaniline over the net was failed. Then the polyaniline was synthesized in situ over polyethylene. The method involves the cage net was immersed in acidified aniline solution overnight and the ammonium persulfate was added slowly to the solution without disturbing the system. Dark greenish precipitates of polyaniline were formed and attached strongly over the surface. Due to immersion in acidic aniline solution, the surfaces were covered by adsorbed aniline. The slow polymerization process probably made the surface to attach a strong coating of polyaniline. The formation of polyaniline over the polyethylene examined through a scanning electron microscope (SEM) showed the nanorods of polyaniline were formed over the surface. The more nanorods of polyaniline over the surface higher will be the efficiency. Excess polyaniline is removed by washing. Now the surface becomes conductive and ready to accept biocides over the surface. As stated earlier copper oxide is a known potential biocide against marine fouler. The concern raised against its use as biocide was due to the higher amount of biocide need to prevent biofouling and its principle is a fixed percentage of copper oxide continuously leached to the environment, this will deter the larvae of fouler from the surface. Application of nano copper oxide will reduce the amount used for prevention is from gram to milligrams or lesser. The minute amount will provide a higher biocidal effect. Hence, nano biocide is a good option to combat biofouling. The polyaniline coated cage nets were immersed in nano copper oxide dispersed aqueous solution overnight. The active copper oxide nanoparticles were adsorbed over the conductive polyaniline uniformly over the surface. The nano-sized particles easily penetrated to the interior layers of the polyaniline-polyethylene composite. The nanoparticles were having increased surface area, smallest size made them to adsorb over the surface, increased activity, and exhibit the efficient toxic effects. Now the material is ready for the field use.

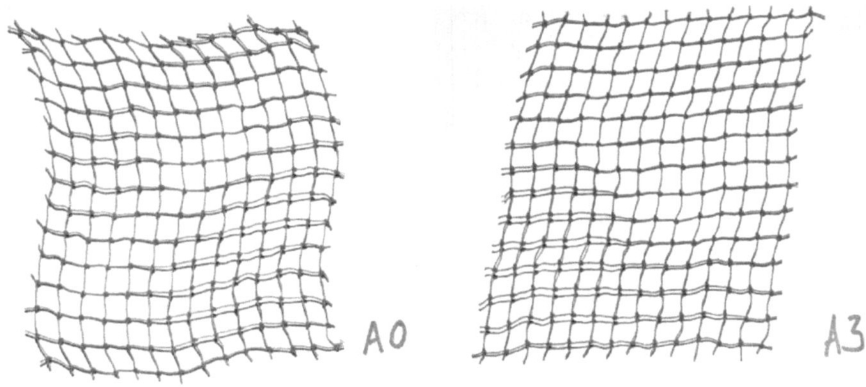

FIGURE 6.2 The aquaculture cage nets untreated and treated.

6.5.1 SURFACE CHARACTERIZATION

Surface of the cage nets were evaluated sophisticated techniques like SEM. The results clearly showed that the surface of the treated webbing was coated polyaniline and over that copper was adsorbed. Similarly, the atomic force microscope measurement exhibited uniform coating over the surface.

Fourier transform infrared (FTIR) spectroscopy measure the vibrational characteristics of the functional groups present in the molecule at it equilibrium position. The infrared light scanning from frequencies 400 to 4000 cm^{-1} will depict the characteristics of different functional group present in the molecules. The region 800 to 1800 cm^{-1} was known as fingerprint region and most of the functional groups will exhibit in this region. Few functional groups like OH, amines, CH_3, CH_2, and CH was generally identified from 2800 to 4000 cm^{-1}. The characteristic IR absorption frequencies of different functional groups will vary with the environment such as neighboring groups, or surrounding chemical bonds. Polyethylene has two types of CH groups viz., CH_3, CH_2, and their absorption was at 1460 ± 10 (symmetric stretching), 1360 ± 20 (Wagging), 2950 (symmetric str), 2850 (symmetric str), and 720 cm^{-1}. The polyethylene netting exhibited the IR absorption at frequencies 729, 1373, 1396, 1471, 1488, 2853, and 2924 cm^{-1}. The polyethylene was coated with polyaniline and IR spectra of the PE wagging doublet were shifted to 1362 and 1396 cm^{-1} indicating the polyaniline was coated over the polyethylene. The polyaniline will exhibit the specific quinonoid peak of NH_4^+/NH^+ at 1027/1141 cm^{-1}. The polyaniline formed over polyethylene exhibited peaks at 1047 /1181

cm^{-1} and nano copper oxide coated over the polyaniline showed a peak at 1070/1179 cm^{-1}. The peak shift from in 1027 to 1047 and 1070 clearly indicated the formation of polyaniline and adsorption of copper oxide over polyaniline respectively. Another functional group is benzenoid ring which was generally shows IR absorption at 1500 and 1585 cm^{-1} in polyaniline. The study showed the benzenoid ring absorption was shifted to 1508 and 1568 cm^{-1}. The characteristics of benzenoid ring were influenced due to the coating over polyethylene and adsorption of nano copper oxide. The broadening of the IR spectral peak implied that the formation of O-H-N or N---H-N bonding in the matrix. Specific IR absorption of copper oxide is 624 cm^{-1} which was shifted to 579 cm^{-1} when adsorbed over polyaniline. These results highlighted the formation of nano copper oxide coated polyaniline coating over polyethylene. The material needs to be tested its anti-biofouling efficiency in the laboratory or marine environments.

6.5.2 NANO CUO HYDROGEL-GREEN SYNTHESIS

Another approach towards the development of antifouling strategies is to use hydrogel. Hydrogels are considered as environmentally safe and its hydrophilicity will deter the adsorption of proteins. This will prevent the formation of biofilm, a prerequisite to attach microbes and fouler [24]. Hydrogels mainly used for development of antifouling strategies were from polyethylene glycol and their related compounds. These hydrogels alone under submerged condition loses its activity after some time. This leads to incorporate a biocide along with hydrogel. Ashraf [25] studied nano copper oxide incorporated PEG hydrogel was synthesized in situ over polyethylene aquaculture cage nets coated with polyaniline and the results showed that significant reduction in the accumulation of biofoulers after three months exposure in the Cochin estuary. The SEM and AFM micrograph showed the formation and uniformly coated nettings (Figure 6.3). The surface was uniformly covered with hydrogel and the treatment was 53% more efficient in controlling fouling attachment than untreated control. The nano copper oxide incorporated hydrogel showed the medium hydrophilicity than hydrogel alone coated nets. The order of hydrophilicity was hydrogel >CuO incorporated hydrogel > Untreated. The nano copper oxide incorporated hydrogel coating was semi hydrophilic, free from any defect, and more compact. The nano copper oxide in the hydrogel acted as a point source of biocide against microorganisms, which generally attacks the hydrogel.

FIGURE 6.3 SEM and AFM images of untreated and treated net samples.

6.5.3 MIXED CHARGED ZWITTERIONIC HYDROGEL

The hydrogels are not having much surface energy difference with surrounding waters and hence application slight thermodynamic force makes the fouler to stay permanently over the surface. Search for an alternative to lead to the new class of hydrogels called zwitterionic hydrogels. Zwitterionic polymers are a family of hydrogel which contain equal number of cations and anions. Generally synthesized by using sulfur, carboxy, or phosphor betaines. They are superhydrophobic in nature due to the presence of abundant ions and formation subsequent strong hydration layers. The protein and microorganisms were deterred by the betaine pendent structures. But these molecules are not very efficient in biofouling resistant. A new class of compounds called mixed charged copolymers which can resist biofouling as that of the zwitterionic hydrogel. Mohan and Ashraf [26] synthesized a nano-sized SiO_2-incorporated poly (Nisopropylacrylamide-co-2-(methacryloyloxy)ethyl]-Trimethylammonium/3-sulfopropyl methacrylate) mixed-charged zwitterionic polymeric hydrogel in situ over a polyethylene aquaculture cage netting material treated with polyaniline. The hydrogel was synthesized microwave method. The SiO_2 incorporated mixed charged polymeric hydrogel coated nets were exposed in marine environment exhibited biofouling resistance few months and it needs to modify to meet the industrial standards.

6.6 FIELD EVALUATION STUDIES

The pre-weighed treated aquaculture cage net and untreated cage nets were tied over PVC rack. The same is exposed in the sea by immersing it at 1 m from the surface through a floating platform. If it is exposing the estuarine environment, it must be immersed in such a way that 1 m down from the surface in non-tidal condition. The panels can be exposed for 1 to 9 months depending on the requirement of the researcher. After retrieval the samples were washed with running water to remove the sediments, loosely bound materials and other non-fouler. Allowed to dry in air and accumulation of biomass is noted (Figure 6.4). Analyze the diversity of fouling organisms by visual and microscopic evaluations. The diversity of fouler will be varied from season to season.

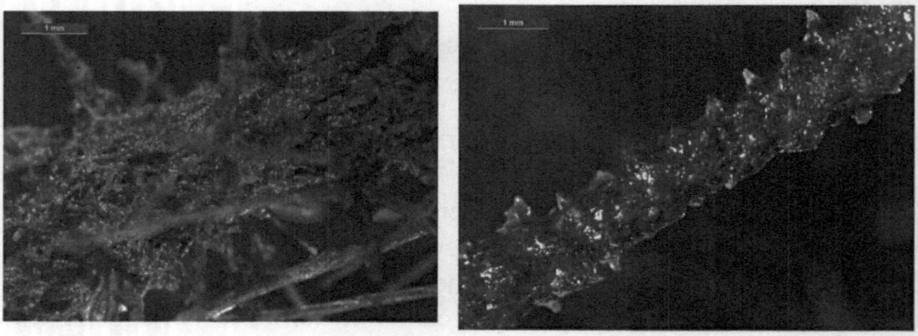

FIGURE 6.4 Field exposed samples untreated and treated.

6.7 LABORATORY EVALUATION

Ekbalad et al. [23] described in detail about the laboratory evaluation of biofoulers. The method describes the quantitative evaluation of the amount settlement of organisms and adhesion strength of the attached organisms. The testing materials were coated over an acid-washed clean glass slides and the same was used for different types of evaluations against fouler.

6.7.1 BIOFOULING ORGANISMS

One of the major foulers under marine environment is barnacles. Cyprid settlement assays were done to assess the biofouling inhibition [27]. Three days old cyprids of about 20 Nos were added in 1 ml of artificial/natural seawater (ASW) over the biocide coated glass slides (about 10 replicates). The cyprids were incubated at 28°C in dark for 48 h. After 24 and 48 h, the slides were taken out and enumerated the cyprid settlement and calculated the mean percentage settlement. Krusal-Wallis test and related statistical tests to be performed for evaluating statistical significance.

6.7.2 MICRO ORGANISM

Callow et al. [28] described the assays for the settlement of Ulva zoospore over the treated surface of glass plate in Quadriperm dish (Greiner Bio-one Ltd). 10 ml of suspension containing 1.5×10^6 mL^{-1} zoospores were added to the dish kept in dark for 45 min. Take out the slides and wash to remove unsettled swimming zoospores. The slides were fixed in 2.5% (v/v) glutaraldehyde, washed, and air-dried. The density of settled organism was measured through microscope as described by Callow et al. [29]. The adherence strength of zoospores can be estimated by exposing the slides to a wall shear stress of 52 Pa in a calibrated flow channel. The details will be available in Schulz et al. [30]. The data expressed in percentages.

Diatom assays can be carried out by using cells of *Navicula perminuta* [31]. It is resuspended in artificial seawater having chlorophyll-a concentration 0.30 $\mu g\ ml^{-1}$ and the treated surfaces were placed in Quadriperm dishes. 10 ml of diatom suspension was added into it. The slides were exposed for 2 h and removed. The unattached diatoms were eliminated by washing and the slides were examined through a microscope as described above.

Antibacterial influence of the treated biocide was tested by using two marine bacteria species viz. Cobetia marina and *Marinobacterhydrocarbonoclasticus*. The former represents hydrophilic and the latter is hydrophobic in nature. C marina was used earlier as a model organism as it was considered bacteria of initial attachment in marine environment [32, 33]. The ASW conditioned (2 h) biocide coated slides were exposed in 8 ml bacterial suspension with standardized absorption of 0.2 AU at 595 nm for 1 h. the experiment was done polystyrene (PS) quadriperm dish. The slides were removed and washed to remove unattached bacteria. Then suspend the slide again quadrperm dish containing 8 ml sterile filtered seawater with fortified growth medium. Incubate the slides for 4 h at 30°C. Removed slides were stained using fluorochrom SYTO 13 and quantified using Tecan plate reader.

6.8 CONCLUSION

Biofouling in aquaculture cage nets an important problem and its management is very expensive. Biofouling organism's nature and diversity is different from marine, estuary, lakes, and region. Currently, not many efficient technologies are available. Application of nanomaterials based intervention can offer better antifouling management since it will not affect much the materials and organisms. The concentration of nanomaterials is very meager and the materials like copper will easily undergo complexation with water-soluble salts.

KEYWORDS

- **aquaculture**
- **artificial/natural seawater**
- **biofouling**
- **nanomaterials**
- **polyaniline**

REFERENCES

1. Sam, S., Maheswaran, M., & Gunalan, B., (2015). Indian seafood industry strength weakness, opportunities, and threat in the global supply chain. *International Journal of Fisheries and Aquatic Studies, 3*(2), 199–205.
2. Salagrama, V., (2004). *Policy Research: Implications of Liberalization of Fish Trade for Developing Countries: A Case Study for India*. Unpublished report. Kakinada: ICM.
3. FAO, (2014). *The State of World Fisheries and Aquaculture-Opportunities and Challenges*.
4. Shyam, S. S., & Narayanakumar, R., (2012). *World Trade Agreement and Indian Fisheries Paradigms: A Policy Outlook*.
5. FAO, (2004). *The State of World Fisheries and Aquaculture* (p. 153). Rome FAO.
6. Dobretsov, S., Teplitski, M., & Paul, V., (2009). Mini-review: Quorum sensing in the marine environment and its relationship to biofouling. *Biofouling*, *25*(5), 413–427.
7. Jain, A., & Bhosle, N. B., (2009). Biochemical composition of the marine conditioning film: Implications for bacterial adhesion. *Biofouling*, *25*(1), 13–19.
8. Avelelas, F., Martins, R., Oliveira, T., Maia, F., Malheiro, E., Soares, A. M., Loureiro, S., & Tedim, J., (2017). Efficacy and ecotoxicity of novel anti-fouling nanomaterials in target and non-target marine species. *Marine Biotechnology*, *19*(2), 164–174.

9. Qian, P. Y., Lau, S. C., Dahms, H. U., Dobretsov, S., & Harder, T., (2007). Marine biofilms as mediators of colonization by marine macro organisms: Implications for antifouling and aquaculture. *Marine Biotechnology*, *9*(4), 399–410.
10. Yebra, D. M., Kiil, S., & Dam-Johansen, K., (2004). Antifouling technology-past, present, and future steps towards efficient and environmentally friendly antifouling coatings. *Progress in Organic Coatings*, *50*(2), 75–104.
11. Lane, A., & Willemsen, P., (2004). Collaborative effort looks into biofouling. *Fish Farming Int.*, *44*, 34–35.
12. Beaz, D., Beaz, V., Dürr, S., Icely, J., Lane, A., Thomason, D., Watson, P., & Willemsen, P. R., (2005). Sustainable solutions for mariculture biofouling in Europe. In: *ASLO Conference, Santiago Da Compostela*. Spain. Home page address: http://www.crabproject.com/index.php/57/publications (accessed on 3 August 2020).
13. Arakawa, K., (1980). *Prevention and Removal of Fouling on Cultured Oysters: A Handbook for Growers*. Maine Sea Grant Technical Report No. 56.
14. Costello, M. J., Grant, A., Davies, I. M., Cecchini, S., Papoutsoglou, S., Quigley, D., & Saroglia, M., (2001). The control of chemicals used in aquaculture in Europe. *J. Appl. Ichthyol.*, *17*, 173–180.
15. IMO, (2002). Biocides Directive EC 98/8/EC (European Commission. 1988).
16. Hidu, H., Conary, C., & Chapman, S. R., (1981). Suspended culture of oysters: Biological fouling control. *Aquaculture*, *22*, 189–192.
17. Lodeiros, C., & García, N., (2004). The use of sea urchins to control fouling during suspended culture of bivalves. *Aquaculture*, *231*(1–4), 293–298.
18. Baum, C., Meyer, W., Fleischer, L. G., & Siebers, D., (2002). *Biozidfreie Antifouling Beschichtung*. EU Patent EP1249476A2.
19. Carver, C. E., Chisholm, A., & Mallet, A. L., (2003). Strategies to mitigate the impact of *Ciona intestinalis* (L.) biofouling on shellfish production. *Journal of Shellfish Research*, *22*(3), 621–631.
20. McCloy, S., & De, N. R., (2000). Novel technologies for the reduction of biofouling in shellfish aquaculture. In: Fisheries, N., (ed.), *Flat Oyster Workshop* (pp. 19–23). Sydney.
21. De, N. P. C., Steinberg, P. D., Charlton, T. S., & Christov, V., (2004). *Antifouling of Shellfish and Aquaculture Apparatus*. U.S. Patent 6,692,557.
22. Tret'yakov, Y. D., (2007). Challenges of nanotechnological development in Russia and abroad. *Herald of the Russian Academy of Sciences*, *77*(1), 15–21.
23. Ashraf, P. M., Sasikala, K. G., Saly, N. T., & Leela, E., (2017). Biofouling resistant polyethylene cage aquaculture nettings: A new approach using polyaniline and nano copper oxide. *Arabian Journal of Chemistry*. http://dx.doi.org/10.1016/j.arabjc.2017.08.006 (accessed on 3 August 2020).
24. Ekblad, T., Bergström, G., Ederth, T., Conlan, S. L., Mutton, R., Clare, A. S., Wang, S., et al., (2008). Poly(ethylene glycol)-containing hydrogel surfaces for antifouling applications in marine and freshwater environments. *Biomacromolecules*, *9*(10), 2775–2783.
25. Ashraf, P. M., (2019). Nano CuO incorporated polyethylene glycol hydrogel coating over surface modified polyethylene aquaculture cage nets to combat biofouling. *Fishery Technology, 56*(2019), 115–124.
26. Mohan, A., & Ashraf, P. M., (2019). Biofouling control using nano silicon dioxide reinforced mixed-charged zwitterionic hydrogel in aquaculture cage nets. *Langmuir*, *35*(12), 4328–4335.

27. Hellio, C., Marechal, J. P., Veron, B., Bremer, G., Clare, A. S., & Le, G. Y., (2004). Seasonal variation of antifouling activities of marine algae from the Brittany coast (France). *Marine Biotechnology*, *6*(1), 67–82.
28. Callow, M. E., Callow, J. A., Pickett-Heaps, J. D., & Wetherbee, R., (1997). Primary adhesion of enteromorpha (chlorophyta, ulvales) propagules: Quantitative settlement studies and video microscopy 1. *Journal of Phycology*, *33*(6), 938–947.
29. Callow, M. E., Jennings, A. R., Brennan, A. B., Seegert, C. E., Gibson, A., Wilson, L., Feinberg, A., et al., (2002). Micro topographic cues for settlement of zoospores of the green fouling alga enteromorpha. *Biofouling*, *18*(3), 229–236.
30. Schultz, M. P., Finlay, J. A., Callow, M. E., & Callow, J. A., (2000). A turbulent channel flow apparatus for the determination of the adhesion strength of micro fouling organisms. *Biofouling*, *15*(4), 243–251.
31. Pettitt, M. E., Henry, S. L., Callow, M. E., Callow, J. A., & Clare, A. S., (2004). Activity of commercial enzymes on settlement and adhesion of cypris larvae of the barnacle *Balanus*amphitrite, spores of the green alga *Ulva*linza, and the diatom Navicula perminuta. *Biofouling*, *20*(6), 299–311.
32. Akesso, L., Pettitt, M. E., Callow, J. A., Callow, M. E., Stallard, J., Teer, D., Liu, C., et al., (2009). The potential of nano-structured silicon oxide type coatings deposited by PACVD for control of aquatic biofouling. *Biofouling*, *25*(1), 55–67.
33. Ista, L. K., Fan, H., Baca, O., & López, G. P., (1996). Attachment of bacteria to model solid surfaces: Oligo (ethylene glycol) surfaces inhibit bacterial attachment. *FEMS Microbiology Letters*, *142*(1), 59–63.

CHAPTER 7

Biodegradable PHAs: Promising "Green" Bioplastics and Possible Ways to Increase Their Availability

TATIANA GR. VOLOVA,[1,2] EVGENIY G. KISELEV,[1,2] ALEKSEY V. DEMIDENKO,[1,2] SVETLANA V. PRUDNIKOVA,[1] EKATERINA I. SHISHATSKAYA,[1,2] and SABU THOMAS[1,3]

[1]*Siberian Federal University, 79 Svobodnyi Av., Krasnoyarsk – 660041, Russia, E-mail: volova45@mail.ru (T. G. Volova)*

[2]*Institute of Biophysics SB RAS, Federal Research Center "Krasnoyarsk Science Center SB RAS," 50/50 Akademgorodok, Krasnoyarsk – 660036, Russia*

[3]*International and Interuniversity Center for Nano Science and Nano Technology, Mahatma Gandhi University, Kottayam, Kerala, India*

ABSTRACT

Two approaches have been investigated in order to reduce the cost and increase the availability of the use of biodegradable polymers of microbiological origin (polyhydroxyalkanoates, PHA): (1) expanding the raw material base and attracting glycerol as a carbon substrate (large-tonnage waste from the production of biodiesel); and (2) filling the polymer with natural materials for industrial use. An effective technology for the synthesis of destructible polymers was developed using a new and more affordable substrate compared to sugars-glycerol. The *Cupriavidus eutrophus* B-10646 strain was studied as a product of polymers on glycerin, and the kinetic and production characteristics of the strain providing polymer yields on glycerol of various cheat, comparable with the process on sugars, were studied. Replacing sugars with glycerol provides a reduction in the cost of a carbon substrate in the production of PHA. To increase the availability of PHA and their technical applications, mixtures of PHA with natural materials were formed, the structure and properties, as well as the patterns of degradation in the soil, were studied. The possibility of long-term functioning of mixed forms in the soil with a gradual release of active substances (herbicides and fungicides) and the possibility of providing plants with protective equipment during the growing season are shown. Both approaches are effective for enhancing accessibility and expanding the scope of application of perspective green plastics-destructible PHA.

7.1 INTRODUCTION

The concept of sustainable development is the basic idea of the 21st century. Annual production of synthetic plastics has exceeded 320 million tons, and they largely accumulate in landfills, occupying fertile arable lands. Plastic wastes ruin municipal sewage and drainage systems. Up to 10% of the annual plastic production leaks into the Global Ocean [1]. New methods of economic management should introduce new, functional, and environmentally friendly recyclable materials in industrial production of the consumption of nonrenewable fossil resources, preserving them for future generations, enable more effective use of energy resources. Therefore, development of new environmentally friendly materials, which will completely degrade without releasing toxic products, should be the priorities for technologists of the 21st century.

Industrial ecology and green chemistry are tools for creating new eco-friendly materials including those manufactured from renewable sources [2]. Hybrid materials based on synthetic polyolefins and natural materials have been increasingly used in various applications. Polymers synthesized by living systems, the so-called biopolymers, have attracted considerable attention, too. Among various natural materials, a special place is occupied by polyhydroxyalkanoates (PHAs)-microbial polymers that can be used in different areas. PHAs are degraded to harmless products (CO_2 and H_2O); they are biocompatible, mechanically strong, thermoplastic, capable of blending with other materials, and processable by conventional methods from different phase states [3–7].

To a large extent (up to 45–48%), the cost of PHAs is determined by the cost of high purity carbon substrates [8, 9]; therefore, one of the most high-priority areas of research is the development of technologies using low-cost substrates.

However, the high cost hinders the use of PHAs in technical applications such as agriculture (films and pots for greenhouses, carriers for fertilizers and agrochemicals); production of packages, containers, and household goods; construction industry; etc., which require very large amounts of material. PHA cost can be reduced by upgrading production processes and by using cheaper substrates. Another approach is to use PHA blends with cheaper materials: this can both reduce the cost of the material and modify the properties of the polymers.

There are few published data on production and properties of PHA blends with solid-phase fillers. Some relatively recent studies investigated natural materials as fillers for modifying properties of PHAs. They described results of filling PHAs with clay and its derivatives [10], plant fibers [11], lignin, and holocellulose from a lignocellulosic biowaste [12], and wood chips and powder [13, 14]. As those studies describe different methods of production and investigation of PHA-based blends, it is difficult to compare them and draw any conclusion about the consistent effects of the fillers on the properties of these polymers. So, obtaining composite materials based on PHA is an unconditional resource for increasing their accessibility and expanding the scope of application.

Great potential for reducing the cost of PHA is the use of waste as raw materials. Potential raw materials for PHA synthesis are various substrates with different oxidation states of carbon, energy content, and cost, including individual compounds (carbon dioxide and hydrogen, sugars, alcohols, organic acids) and byproducts of alcohol and sugar industries, chemical processing of plant raw materials, and production of olive, soybean, and palm oils [15–17]. PHAs can be synthesized from both individual carbon compounds and various industrial wastes, making them economically feasible materials for different applications. Thus, the cost-effectiveness of PHAs can be enhanced by using new strains and carbon sources and by improving manufacturing processes, including large-scale ones. Scaling up laboratory biotechnologies to pilot production (PP) is a necessary step towards commercial production.

One of the promising substrates for large-scale production of PHAs is glycerol, the scale of production of which is currently increasing. This is due to the growing production of biodiesel as an alternative renewable energy source [18]. Biodiesel production increased dramatically from 500,000 gallons in 1999 to 450 million gallons in 2007, and it has been growing steadily over the past few years. In 2010, the global production of biodiesel reached 19 billion L [19], rising to more than 28 billion L in 2017 [20].

Glycerol is a by-product, amounting to about 10%, of biodiesel production by transesterification of animal and vegetable fats and oils (rapeseed, mustard, soybean, and palm oils) [15]. In industrial glycerol grades, the water content varies between 5.3 and 14.2%; methanol content - 0.001-1.7%; NaCl-traces-5.5%; K_2SO_4-0.8–6.6%. The analysis of publications generally indicates that glycerol has the potential to be used as a substrate for PHA production. However, the presence of impurities in glycerol indirectly affects the synthesis of PHA and their properties. It is obvious that involving new strains and upgrading the technological stages of the process will contribute to the improvement of PHA production.

In this work, our results are presented that are aimed at increasing the availability of PHA, which requires the use of glycerol as a substrate [6, 21–23], as well as the preparation of PHA composites with available natural materials [24].

7.2 MATERIAL AND METHODS

The strains used in this study were *Cupriavidus eutrophus* B-10646, registered in the Russian Collection of Industrial Microorganisms (RCIM). Schlegel's mineral medium was used as a basic solution for growing cells: $Na_2HPO_4{\cdot}H_2O$ – 9.1; KH_2PO_4 – 1.5; $MgSO_4{\cdot}H_2O$ – 0.2; $Fe_3C_6H_5O_7{\cdot}7H_2O$ – 0.025; $CO(NH_2)_2$ – 1.0 (g/L). Nitrogen was provided in the form of urea, and, thus, no pH adjustment was needed. The pH level of the culture medium was stabilized at 7.0 ± 0.1.

To cultivate bacteria in shake flask culture, an Innova 44 constant temperature incubator shaker (New Brunswick Scientific, U.S.) was used. Inoculum was prepared by resuspending the reference bacterial culture maintained on agar medium. The reference culture was grown in 1.0–2.0 L glass flasks half-filled with liquid saline medium, with the initial concentration of glycerol from 5 to 10 g/L. Growth kinetics of bacterial cells was studied in automated laboratory fermenters (Bioengineering AG, Switzerland), with 30-L and 150 L fermentation vessels and the working volume of the culture from 18 to 100 L, under strictly aseptic conditions.

Fermenters were equipped with systems for monitoring pH level, foam level, temperature, pressure, and dissolved oxygen. The fermenters were controlled by BioScadaLab software in automatic mode. To supply the feeding substrates, the fermenters were equipped with Bioengineering Peripex peristaltic pumps. The concentration of dissolved oxygen was maintained at

DO 30%. The air supply control was carried out in a cascade mode (DO-air flow-mixer revolutions). During cultivation in a 30-L fermenter, the amount of air supplied per cultivation process varied from 0 to 5.5 L/min, the speed varied from 500 to 1000 rpm. During cultivation in a 150-L fermenter, the amount of air supplied per cultivation process varied from 10 to 5.5 L/min, the speed varied from 300 to 750 rpm.

A two-stage process was used. In the first stage, cells were grown under nitrogen deficiency: the amount of nitrogen supplied in this stage was 60 mg/g cell biomass synthesized (i.e., 50% of the cell's physiological requirements – 120 mg/g); the cells were cultured in complete mineral medium and with glycerol flux regulated in accordance with the requirements of the cells. In the second stage, cells were cultured in a nitrogen-free medium; the other parameters were the same as in the first stage. The temperature of the culture medium was 30 ± 0.5°C and pH was 7.0 ± 0.1.

During the cultivation, samples of culture medium were taken for analysis every 4–5 h (from fermenters) or every 8–10 h (from flasks); cell concentration in the culture medium was determined based on the weight of the cell samples dried at 105°C for 24 h (CDM); the total biomass (X_{total}) and catalytically active biomass (X_c) ($X_c = X_{total}$-PHA) g/L, were distinguished. Cell concentration in the culture medium was determined every hour by converting the optical density of culture broth at 440 nm to cell dry mass by using a standard curve prepared previously. The criteria for evaluating the process of PHA biosynthesis were as follows: concentration of cell biomass in culture, polymer content in cells, consumption of the main growth substrate, duration, and productivity of the process. The kinetic and production parameters of the culture were determined by conventional methods. The biomass concentration (X, g/L), the catalytic biomass concentration (X_c = X-PHA, g/L), the yield coefficient of the polymer (Y, g PHA/g substrate), the specific growth rate (m, h^{-1}), and the volumetric productivity (P, g/(L·h)) were calculated.

Intracellular polymer content at different time points was determined by analyzing samples of dry cell biomass. Intracellular PHA content and composition of extracted polymer samples were analyzed using a GC-MS (6890/5975C, Agilent Technologies, U.S.).

Three natural materials were used as fillers: peat, wood flour, and clay. High-moor peat "Agrobalt-N," state registration 0428-06-209-139-0-0-01, was produced by OOO "Akademiyatsvetovodstva," Russia). Wood floor was produced by grinding wood of birch (*Betula pendula* Roth) using an MD 250-85 woodworking machine ("StankoPremyer" Russia). Then it was dried at 60°C for 120 h until it reached constant weight, and 0.5 mm mesh was used to separate the particle size fraction. Clay was taken from the mine "Kuznetsovskoye," the Krasnoyarskii Krai, Russia. Its composition: loss on ignition – 6.92%; SiO_2 – 60.1%; Al_2O_3 – 19.17%; Fe_2O_3 – 6.72%; CaO – 2.02%; MgO – 2.12%; SO_3 – 0.65%; Na_2O – 0.88%; K_2O – 1.45%.

The polymer (poly-3-hydroxybutyrate) and fillers were pulverized by impact and shearing action in ultra-centrifugal mill ZM 200 (Retsch, Germany). To achieve high fineness of polymer grinding, the material and the mill housing with the grinding tools were preliminarily cooled at –80°C for about 30 min in an Innova U101 freezer (NEW BRUNSWICK SCIENTIFIC, U.S.). Grinding was performed using a sieve with 2-mm holes at a rotor speed of 18,000 rpm. Fillers were not precooled before grinding; a sieve with 2-mm holes was used, and the grinding was performed at 6000 rpm. Fractionation of the polymer and filler powders was carried out using vibratory sieve shaker AS 200 control (Retsch, Germany). Fractioning time was 10 min, amplitude 1.5 mm; a 100 g sample was used. Fractions of the polymer and filler powders of 200 μm and less were selected and used. Then, polymer powder was mixed with different filler powders in bench top planetary mixer SpeedMixer DAC 250 SP (Hauschild Eng., Germany); the blend time was 1 min, and the speed was 1000 rpm. The blends contained different amounts of the fillers: 10, 30, and 50%.

The homogenized blends were used to produce pellets and granules. Pellets were prepared by cold pressing at 36 bar using a laboratory semiautomatic press (Minipress, Minsk, Belarus). Granules were prepared by mixing wet polymer paste (40%) in ethanol in planetary mixer SpeedMixer DAC 250 SP (Hauschild Eng., Germany); with the mixing time 1 min at 1000 rpm, and granulating in screw granulator Fimar EAC (Italy) with a 6 mm nozzle. The granules were dried in a fume hood at room temperature for 24 h.

Poly-3-hydroxybutyrate, peat, clay, and wood powders, powdered P(3HB)/filler blends, and granules and pellets were examined by using state-of-the-art physicochemical methods (SEM microscopy, FTIR spectroscopy,differential scanning calorimetry (DSC), and x-ray diffraction (XRD).

7.3 RESULTS AND DISCUSSION

7.3.1 SYNTHESIS OF PHA ON GLYCEROL OF VARIOUS PURIFICATION

Three brands of glycerin were used: Glycerol purified (Corporate Oleon, Sweden): glycerol – 99.3; chloride – 0.0001; salts (NH_4) – 0.005; Fe – 0.0005;

Ar – 0.00004; moisture – 0.09; fatty acid and ester – 0.25 (% mass); max heavy metal – 0.00005 μg/g (Glycerol I). Glycerol refinery B.V. (Dutch glycerol refinery, *Netherlands*): glycerol – 99.7; chloride <0.001; moisture – 0.09; fatty acid and ester – 1.0; sulfate <0.002; total organic impurities – 0.5–1.0; individual organic impurities – 0.1 (% mass); heavy metal <5 (μg/g) (Glycerol II). Crude glycerol (Prisma Comercial Exportadora de Oleoquimicos LTDA, Brazil): glycerol – 82.1; chloride – 4.35; mong – 0.13; methanol – 0.13; ash – 6.59; moisture – 9.88 (% mass); pH 5.8 (Glycerol III).

7.3.1.1 THE EFFECT OF GLYCEROL CONCENTRATION ON BACTERIAL GROWTH AND PHA SYNTHESIS

C. eutrophus B-10646 cells were cultivated in the media with concentrations of purified glycerol varied within a wide range to determine the limits of physiological action of this substrate for the strain and the kinetic constants [21]. The limits of physiological action of glycerol for this strain are very wide, varying between 0.5 and 60.0 g/L. There was a wide plateau (from 1 to 30 g/L), and the zones of limitation and inhibition of bacterial growth by glycerol were 0.1–3.0 and 30–60 g/L, respectively. The dependence of the specific growth rate (μ) on the substrate concentration (S) was described by the Andrews equation, which is a modified Monod equation. Using the graphical analysis method of Lineweaver-Burke (1/μ:1/S) and Dickson's method (1/μ:S), we calculated kinetic constants for this strain (saturation constant (*Ks*) and inhibition constant (*Ki*), μ_{max}). For the strain *C. eutrophus* B-10646, the limits of the physiological action of glycerol were found to be 1–30 g/L; K_s and K_i were 0.36 g/L (0.004 mol/L) and 62.0 g/L (0.673 mol/L), respectively; $\mu_{max} = 0.085\ h^{-1}$.

7.3.1.2 A STUDY OF THE GROWTH AND SYNTHESIS OF PHA BY C. EUTROPHUS B-10646 ON PURIFIED AND CRUDE GLYCEROL

The content of the main component (glycerol) in purified glycerol is more than 95–99%. In unpurified (crude) glycerol, depending on the raw material and the technology used, the content of glycerol is 80–85%; the other part comprises impurities, including free fatty acids (FFA) and methyl esters of FFA, alcohols, as well as water and salts, which, as a rule, inhibit the microorganisms responsible for PHA production.

The productivity of culture of *C. eutrophus*B-10646on different glycerol is shown in Table 7.1. Two grades of purified glycerol, with a 99.3 (Glycerol I) and 99.7% (Glycerol II) content of the basic material, and unpurified, crude glycerol (82.1%) were used in this study. Cultivation of *C. eutrophus* B-10646 cells was carried out in a 30-L fermenter with a starting cell concentration in the inoculate of 1.0–1.5 g/L, 25 g/L and 18 L working volume of the culture. The process was performed in a two-stage culture (30 hours) on saline medium: the growth of cells was limited by nitrogen deficiency in the first stage for 30–32 h; in the second stage (24–30 h), cells were grown in the nitrogen-free medium. The concentration of glycerol in the culture was maintained at 5–10 g/L, and it was fed into the culture with a peristaltic pump dispenser.

TABLE 7.1 Production Indicators of PHA Biosynthesis in a Culture of *C. eutrophus* B-10646 on Glycerol of Various Purification

Type of Glycerol	X, g/L	PHA, %	μ*, ч$^{-1}$	Y_x	Y_p	Productivity (P), g/L*h	
						P_x	P_p
Glycerol-I	69.3 ± 3.5	72.1 ± 3.6	0.101	0.38	0.29	1.10	0.81
Glycerol-II	69.8 ± 3.5	73.3 ± 3.6	0.097	0.39	0.29	1.15	0.90
Glycerol-III	69.3 ± 2.9	73.3 ± 3.2	0.100	0.36	0.26	1.15	0.89

Source: Volova et al. [21].

Comparable cell concentrations and polymer contents were achieved in the experiments with *C. eutrophus* B-10646 cultivated on purified glycerol (Glycerol I and Glycerol II) (Table 7.1). When glycerol purified and glycerol refinery B.V. was used, the maximum yield of total biomass was 70 g/L, and the polymer content in cells was 72–75%. The analysis of the results revealed some differences in the kinetic parameters during the cultivation of bacterial cells. The specific growth rate of cells cultivated on glycerol purified was the highest (estimated from total and active biomass) in the initial period of the first stage of cultivation on complete nutrient medium with a limited supply of nitrogen (50% of the physiological requirement of bacteria), reaching 0.15 h^{-1} and 0.14 h^{-1}, respectively. This period corresponded to the most active consumption of glycerol by the culture, at an average specific rate of 4.0 ± 0.2 g/(g·h). The rate of polymer synthesis at this stage was 0.18 h^{-1}, and tended to decrease. After 30–32 hours, the total cell biomass concentration was 42.1 ± 1.7 g/L, and the polymer content in the cells reached 47.8 ± 2.3%. During the second stage, the supply of nitrogen to the culture was stopped; the controlled supply of glycerol and mineral elements precluded the deficiency of these substrates in the culture. In the second stage, the specific growth rates of bacterial cells and polymer synthesis decreased gradually, as the consumption of glycerol dropped. At the end of the fermentation period (60 h), the total cell concentration was 69.3 ± 3.5 g/L, and the polymer content in the cells was 72.4 ± 3.6%. The consumption of glycerol for the whole period was 3.5 ± 0.2 kg.

On the second type of purified glycerol (Glycerol II), the maximum values of the specific growth rate of bacteria (in terms of total and active biomass) in the first stage of the process were 0.15 h^{-1} and 0.13 h^{-1}, and they were comparable to the results for glycerol purified during this period. The consumption of glycerol in this period was between 3.8 and 4.0 g/(g·h)while the polymer synthesis rate was 0.18 ± 0.02 h^{-1} and tended to decrease. After 30 hours, the total cell biomass concentration was 45.6 ± 2.2 g/L, and the polymer content in the cells reached 56.1 ± 2.7%. At the end of the fermentation period, the total cell biomass concentration was 69.4 ± 3.5 g/L, and the polymer content in the cells was 73.3 ± 3.6%. The consumption of glycerol for the whole period was 3.4 ± 0.2 kg (Table 7.1).

On glycerol III, the maximum values of bacteria specific growth rate (in terms of total and active biomass) in the first stage of the process were 0.14 h^{-1} and 0.13 h^{-1}, respectively, when glycerol was consumed by the culture at a rate of 4.2 ± 0.2 g/g h. The average rate of polymer synthesis in this stage was 0.17 ± 0.02 h^{-1}. After 30 hours, the total cell biomass concentration was 46.2 ± 1.9 g/L, and the polymer content in the cells reached 52.2 ± 2.1%. At the end of the fermentation period, the total cell biomass concentration was 69.3 ± 2.9 g/L, and the polymer content was 78.1 ± 3.2% (Table 7.1). The consumption of raw glycerol for the whole period amounted to 3.8 ± 0.2 kg, which corresponded to the yield coefficient for the polymer – 0.26 ± 0.01, taking into account that the concentration of raw glycerol was 82.07%.

The consumption of raw glycerin for the entire period was 3.8 ± 0.2 kg, which corresponds to the economic coefficient for the polymer Yp – 0.26 ± 0.02. Considering that its concentration in crude (raw) glycerin is 82.07%, the economic coefficient in terms of absolute glycerin is comparable to the values obtained with purified glycerin (Yp 0.29 ± 0.02 g/g).

Thus, it was shown that the production characteristics of *C. eutrophus* B-10646 when using glycerol of various purifications are comparable; the use of crude glycerol did not inhibit bacterial growth and PHA synthesis.

7.3.1.3 PILOT PRODUCTION OF PHA ON GLYCEROL

The process of PHA synthesis by *C. eutrophus* B-10646 on Glycerol II was scaled-up and studied under pilot production (PP) conditions [21, 22]. Pilot production consisted of units for media and inoculum preparation; a unit for fermentation; a unit for polymer extraction and purification. The PP fermentation unit included a steam generator (Biotron, South Korea) for sterilizing fermenters and service lines, a compressor (Remeza, Belarus) for air supply, a 30-L seed culture fermenter, a 150-L production fermenter, an ultrafiltration (UF) unit (Vladisart, Russia) to concentrate the culture, and a unit for cool dehumidification of the condensed cell suspension (LP10R ILSHIN C, South Korea).

The *C. eutrophus* B-10646 seed culture was derived from a reference stock culture maintained on agar medium, by growing cells in 2.0-L flasks on a complete nutrient medium in an incubator shaker. The resulting culture was concentrated by centrifugation in compliance with the sterility rules. Inoculum (4 L) with a cell concentration of 14–17 g/L was inoculated into a 30-L fermenter containing phosphate buffer. The process of building up the seed material was carried out on complete nutrient medium. For this purpose, dosing pumps were used to feed continuously separate flows of the solutions of glycerol, urea, magnesium sulfate, and ferric citrate with trace elements into the fermenter. The residual concentration of glycerol in the culture was maintained at 5–20 g/L; urea – 0.1–0.2 g/L; magnesium sulfate – 0.05–0.1 g/L.

The inoculum produced in a seed fermenter, was pumped into a 150-L production fermenter containing a sterile buffer solution of potassium and sodium phosphates in a sterile seeding line; thus, the initial concentration of the culture was 7.0 g/L. The process was carried out in two stages with continuous supply of sterile air and feeding solutions. At the end of fermentation (45 h), the cell biomass concentration was 110 ± 5.5 g/L, and the polymer content 78 ± 3.1%. Thus, the average productivities of the polymer and cell biomass were 1.83 and 2.29 g/(L·h), respectively.

The results of the study of the chemical composition and physicochemical properties of PHA samples synthesized on glycerol of different purification degrees are presented in Table 7.2.

TABLE 7.2 Physico-Chemical Properties of Samples P(3HB) Synthesized on Glycerol of Various Purities

Substrate	M_w, кDa	D	C_x, %	T_g, °C	T_c, °C	T_{melt}, °C	T_{degr}, °C
Glucose	920	2.52	76	-	92	178	295
Glycerol I	355	3.42	50	2.9	96	174	296
Glycerol-II	416	3.63	55	-	103	176	296
Glycerol-III	304	3.49	52	2.7	99	172	295

"–" means no peak.

Source: Volova et al. [21, 22].

The properties of PHA samples synthesized by *C. eutrophus* B-10646 on three types of glycerol did not differ dramatically. The polymers synthesized by the glycerol-adapted productive culture *C. eutrophus* B-10646 on two types of purified glycerol had similar values of Mn-104 and 115 kDa, Mw-355 and 416 kDa, and polydispersity – 3.42 and 3.63, respectively. The Mn and Mw values of the PHAs produced on crude glycerol were somewhat lower, 87 and 304 kDa. These values were generally lower than those obtained earlier on other substrates. The ^{1}H-NMR spectra of three PHA samples synthesized on three glycerol grades were similar and showed the expected resonances for P(3HB) as demonstrated by the methyl group at 1.25 ppm, the methylene group between 2.45 and 2.65 ppm, and the methine group at 5.25 ppm. The ^{1}H-NMR-obtained spectra of PHA samples synthesized on glycerol were similar to those obtained by other researchers. The P(3HB) samples synthesized on glycerol had a reduced C_x (50–55%), i.e., the amorphous and crystalline regions had become nearly equal to each other. No deviations in the temperature properties were found in the samples of P(3HB). T_m and T_d values were within the previously identified value limits, 172–176, and 295–296°C, respectively.

It should be noted that the specific consumption of various substrates (sugars and glycerol) is comparable for the synthesis of PHA by the strain *C. eutrophus* B-10646. Table 7.3 presents a comparison of unit costs and the cost of various carbon substrates for the production of PHA.

It is shown that fructose accounts for the largest costs and amounts to 303.0 rubles (₽)/kg of polymer. When replacing fructose with glycerin, the cost of the carbon substrate is reduced by 41.6% when using purified glycerin and by 55% when using "raw" glycerin. When glucose is replaced with glycerin, the cost of the carbon substrate is reduced by 8.4% when using purified glycerol and by 29.4% when using "raw" glycerin. Thus, the technology for the synthesis of PHA on glycerol, developed, and implemented in a pilot version, provides an efficient process for producing polymers with a significant reduction in production costs.

TABLE 7.3 Costs of Carbon Substrates for the Production of 1 kg of PHA

Name of Substrate	Cost of Substrates ₽/kg	Substrate Consumption, kg/kg PHA	Unit Cost of Substrate, ₽/kg PHA
Fructose	101.00	3.0	303.0
Glucose	69.00	2.8	193.2
Purified Glycerol "Duth glycerol refinery," Netherlands	59.00	3.0	177.0
Crude Glycerol "Prisma comercialexportadora de oleoquimicos LTDA,"Brazil	44.00	3.1	136.4

Source: Volova et al. [22].

7.3.2 COMPOSITES BASED ON PHA AND NATURAL MATERIALS

7.3.2.1 CHARACTERIZATION OF P(3HB) AND INITIAL NATURAL MATERIALS

Poly-3-hydroxybutyrate and natural materials (clay, peat, and birch wood flour) with considerably different properties (Table 7.4) were used to produce P(3HB)/filler blends: fine powder and paste prepared with ethanol (40% moisture content).

P(3HB) is a highly crystalline material, with the prevailing crystalline phase, and its degree of crystallinity (C_x) is usually above 60–65%; the C_x of the P(3HB) sample used in this study is 75%. XRD pattern of P(3HB) is showed, that diffraction peaks in 2θ = 13.4, 16.8, 20, 22.2 and 25.5°. XRD patterns of the fillers differed from the diffraction pattern of P(3HB). Wood flour showed distinct diffraction peaks at 2θ = 14.2, 19.6, and 25.5°, in contrast to the amorphized peat. X-ray examination of clay revealed numerous narrow diffraction peaks, and the most distinct one was at 23.5°. Clay had the highest C_x (53%). In wood flour and, especially, peat, amorphous phase prevailed, and their degrees of crystallinity were low: 26 and 9%, respectively.

Thermal properties of the polymer and fillers are different too. P(3HB) is a thermoplastic material, with a considerable difference between melting point (176°C) and thermal decomposition temperature (287°C). Its peak of thermal decomposition is in the 280°C. P(3HB) crystallization temperature is 108°C and melting enthalpy is 89.3 J/g. By contrast, none of the fillers is a thermoplastic material. Initial heating of these materials (and blends) gives a small peak at a temperature between 70 and 130°C, which is associated with evaporation of moisture and volatile organic compounds. After moisture evaporation, thermal decomposition of peat begins at 130°C and thermal decomposition of wood flour at 220°C. For clay, at a temperature between 60 and 200°C, moisture evaporates and decomposition and evaporation of volatile organic compounds begin (the process is endothermal); however, at temperatures between 200 and 400°C, decomposition is accompanied by heat release.

TABLE 7.4 Physicochemical Properties of Initial Materials and Blends

Sample	**C_x,%**	**T_{melt}, °C**	**T_{cryst}, °C**	**T_{degr}, °C**	**Melting Enthalpy, (J/g)**
Initial Materials:					
P(3HB)	75	176	108	287	89.3
Clay	53			200*	
Peat	9			130*	
Wood flour	26			220*	
Blends (P(3HB)/Filler = 70/30):					
P(3HB)/Clay:					
90:10	72	177	110	286	52.4
90:30	60	175	110	283	38.5
90:50	56	176	109	276	29.8
P(3HB)/Peat:					
90:10	60	174	107	288	64.1
90:30	48	176	108	292	59.7
90:50	42	177	110	290	48.7
P(3HB)/Wood Flour:					
90:10	65	178	107	279	66.9
90:30	47	175	106	273	51.3
90:50	44	171	104	274	45.4

* – *start of* $T_{degr.}$

Possible structural differences between the blends were detected using IR spectroscopy, which is employed to study the structures of various macromolecules. Variations in the intensity of the bands in the low-frequency range are indicators of both the crystalline to amorphous phase ratio in the blend and the type of interactions of the components in the blend.

FT-IR spectra of P(3HB) and the three fillers were taken in the 400–4000 cm^{-1}range. The spectra show not only the groups of carbon compounds (C=O, C-OH, CH_2, CH_3) contained in all initial materials but also the groups and compounds characteristic of individual components. These are CH, CH_2, CH_3, C-OH, and COOH groups for P(3HB), S-S-, C=C, C=N, and N-H groups for peat, Si-O-Si(AL) and Fe-O in α-Fe_2O_3 for clay, and -C-O-;-C-O-C-for birch wood flour. The IR absorption spectra of P(3HB) contain absorption bands corresponding to vibrations of the main structural components of polymers except the absorption bands of vibrations of the terminal C-OH and COOH groups. The bands of the ordered optical densities (crystalline phase) are in the 1228 cm^{-1} range while the bands of the amorphous phase are shifted to 1182 cm^{-1}. The spectra show distinct absorption bands of asymmetric stretching of CH_3- and CH_2-groups (2978 and 2994 cm^{-1}); symmetric stretching of CH- and CH_2 groups (2994 and 2934 cm^{-1}); stretching of conjugated (1687 cm^{-1}) and unconjugated (1720 cm^{-1}) carbonyl groups C=O; skeletal CH vibrations (599 cm^{-1}); and CH bending (622 cm^{-1}). PHAs are hydrophobic compounds, in which molecules of water interact with the hydrogen atom of the methyl group or with the oxygen atom of the carboxyl group through hydrogen bonding. The narrow bands at 3436 cm^{-1} are typical of the hydrogen bonded OH group. The 1687 cm^{-1} band is characteristic of hydrogen vibration in the OH group interacting with the oxygen atom through hydrogen bonding. The 2874, 2934, 2978, and 2994 cm^{-1} absorption bands are typical of the bonded OH group and may be a component of the COOH dimer group. In the 3000–2500 cm^{-1} range, there is a group of weak bands typical of the dimers of carboxylic acids.

In contrast to P(3HB) IR spectra, the IR spectra of peat contain absorption bands of the -S-S- bond groups (521 cm^{-1}); -C-O- groups characteristic of carbohydrates (1029 cm^{-1}); deformation vibrations of -CH_2- groups (718 cm^{-1}); skeletal vibrations of the benzene ring C=C (lignin), C=N, N-H (1513 and 1597 cm^{-1}), and stretching of the -OH groups at 3300 cm^{-1}. The IR spectra of clay contain absorption bands in the 776 and 795 cm^{-1} range, corresponding to stretching and symmetrical vibrations of Si-O-Si(Al) and Fe-O in α-Fe_2O_3; stretching vibrations of (AI-0), (Si-O-Al), and -$A1_20_3$ in the 692 cm^{-1} range. In the 978 cm^{-1} range, there are absorption bands corresponding to (O-Si) deformation vibrations. The absorption frequency observed at 1635 cm^{-1} is indicative of deformation vibrations of H-O-H, which suggest water sorption on the surface of the mineral. The absorption band at 3615 and 3392 cm^{-1} corresponds to stretching and symmetric vibrations of AlO-H for clay bound water. IR spectra of birch wood flour show plane bending of the guaiacyl ring of lignin of leaves (1031 cm^{-1}), stretching of -C-O-;-C-O-C- (1235 cm^{-1}), skeletal vibrations of the syringyl ring of lignin of leaves (1324 cm^{-1}), stretching, and skeletal vibrations of benzene ring (1422 cm^{-1}, 1504 cm^{-1}, 1593 cm^{-1}), stretching of -C=O of conjugated and unconjugated groups (1652 cm^{-1}, 1734 cm^{-1}), and intramolecular and intermolecular stretching of -OH groups and water (3337–3342 cm^{-1}).

7.3.2.2 CHARACTERIZATION OF P(3HB)/FILLER BLENDS

Powdered P(3HB) and fillers were processed by cold pressing to produce pellets with a diameter of 6 mm and height 3 mm and granules with a diameter of 2 mm and length of between 2 and 4 mm (Figure 7.1).

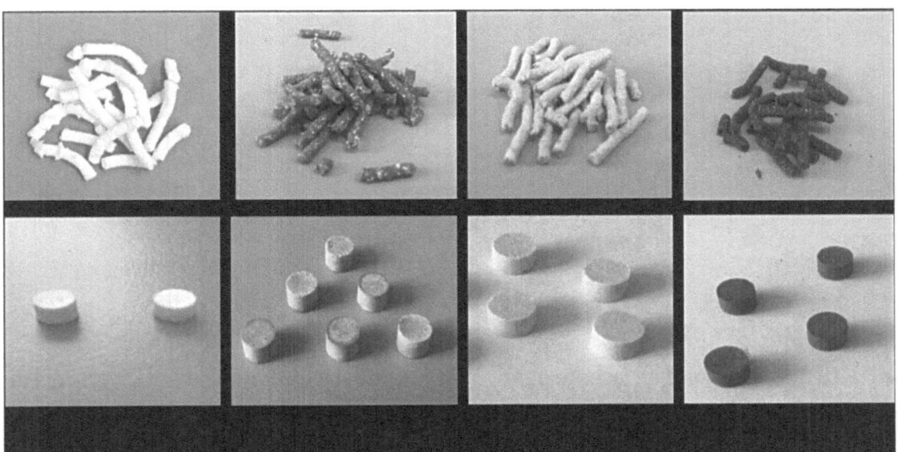

FIGURE 7.1 Granules and 3D forms obtained from PHA mixed with natural materials.

Source: T. Volova photo.

Pellets and granules prepared from blends with fillers that had different properties exhibited diverse surface structures (Figure 7.2). SEM images showed porous and rough surface microstructures. The surface of the

P(3HB)/wood flour pellets showed rectangular sawdust particles embedded in the surface layer; on the surface of the P(3HB)/peat pellets, there were asymmetrical heteraxial inclusions. On P(3HB)/clay pellets, clay inclusions were uniformly distributed over the surface.

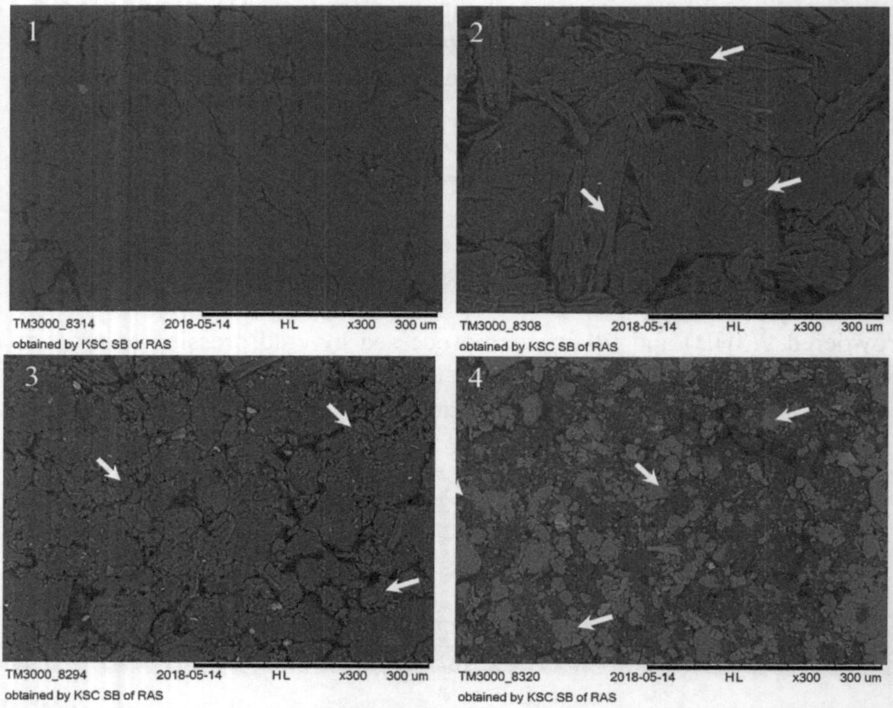

FIGURE 7.2 SEM images of pressed pellets (P(3HB)/filler = 70/30): 1-P(3HB); 2-P(3HB)/wood flour; 3-P(3HB)/peat; 4-P(3HB)/clay.

The blends were characterized using DSC, X-ray examination, and IR spectroscopy. IR spectrometry showed that characteristic absorption bands of initial components were separately retained in all blends: for P(3HB) those were CH, CH_2, CH_3, C-OH, and COOH groups; for peat-the additional and characteristic S-S-, C=C, C=N, N-H groups; for clay-Si-O-Si(AL) and Fe-O in α-Fe_2O_3; for birch wood flour: -C-O-;-C-O-C-. In the spectra of the blends, absorption bands were either separate bands or coinciding peaks, some of them transformed. An insignificant difference in numbers was within the limits of spectral error – 4 cm^{-1} (accumulation of 256 scans) and could result from increased moisture content of some of the samples. No new significant absorption bands were revealed in the blends, suggesting that no new chemical bonds were formed and that the blends prepared by the techniques employed in this study were physical mixtures.

Schematically, the process of obtaining mixtures from polymers and natural materials is presented in Figure 7.3.

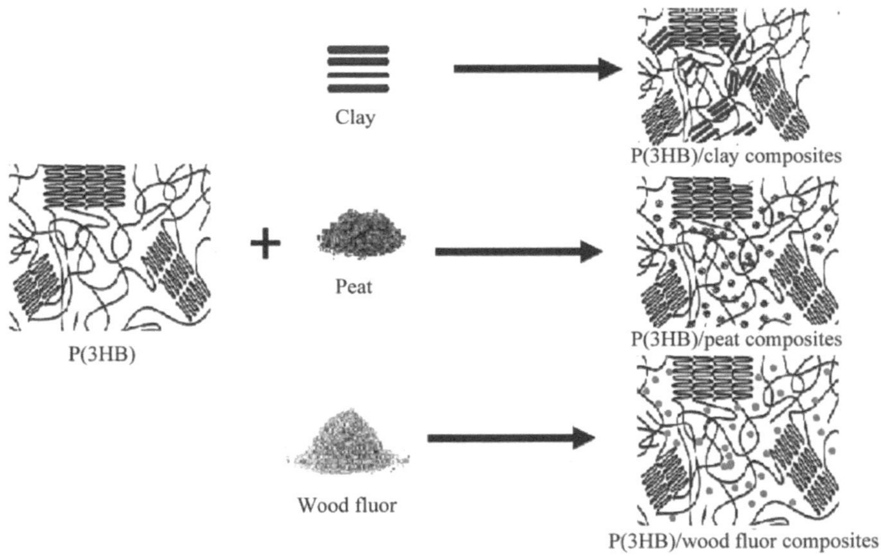

FIGURE 7.3 Schemes of initial materials and formulated mixtures (Sabu Thomas schemes).

Blending of P(3HB) with the fillers changed the properties of the polymer. The blends became progressively amorphous as the percentage of the fillers was increased (10, 30, and 50% fillings of the polymer with natural materials were tested). The degree of crystallinity of P(3HB) was decreased more noticeably in the blends with peat and wood flour (materials with low C_x). The C_x of the P(3HB)/peat and P(3HB)/wood flour blends with 30% filler decreased to 47 and 48%, respectively; the C_x of the blends with 50% of the same fillers decreased to 42 and 44%, respectively. That is, amorphous phase prevailed in these blends. In the P(3HB)/clay blends, the crystalline phase prevailed at all levels of filling of the polymer with clay; the C_x was no less than 56%. Changes in the crystallization kinetics of the polymer in the blends were associated with the rather high percentages of fillers (30 and 50%). This must have decreased the free volume necessary for nucleation of spherulites to occur, preventing the development of the crystalline phase and, thus, reducing it.

The study of thermal properties of the blends did not show any significant influence of the fillers on the temperatures of melting and thermal decomposition and the position of the T_{melt} and T_{degr} peaks in thermograms. However, the onset of thermal decomposition of the blends was shifted to the left relative to the initial P(3HB). The enthalpy of melting of the blends was significantly decreased (by a factor of 1.5–2.2). It is difficult to compare results obtained in this study with literature data because of data scarcity and different techniques used to process blends into products.

The study of physicochemical properties of P(3HB) blends with the three natural materials did not reveal any dramatic impairment of the properties of the polymer. However, Young's moduli of all blends decreased as the amount of the filler in the blend increased. The most significant decrease in Young's modulus was observed in the P(3HB)/peat and P(3HB)/clay blends containing 50% of the filler: the parameter dropped to 150 and 76 MPa, respectively, compared to Young's modulus of the initial polymer of approximately 360 MPa.

As P(3HB) blends can be considered as potential carriers for pesticides and fertilizers for soil application and as materials for fabricating degradable packaging and other goods, it is important to study their degradation behavior in soil. Degradation of P(3HB) blends with natural materials was first investigated in this study, showing that the mass loss of the blends over 35 days of incubation in soil varied between 30 and 50% of the initial mass of the products, depending on the type of the filler.

In this way, natural materials-clay, birch wood flour and peat-were used as fillers for degradable poly-3-hydroxybutyrate. The P(3HB)/filler to produce of pellets and granules. All initial materials and blends were investigated using IR spectroscopy, DSC, X-ray analysis, electron microscopy with microanalysis and X-ray spectrometry. Analysis of IR spectra showed that no chemical bonds were established between the polymer and fillers and that the blends were physical mixtures. Their temperature characteristics and degrees of crystallinity were lower than those of P(3HB), suggesting different crystallization kinetics of the blends. Degradation behavior of pellets and granules prepared from P(3HB)/filler blends in soil was similar to degradation behavior of pure polymer. The mass loss of all pellets and granules over 35 days of incubation in soil varied between 30 and 50% of the initial mass of the products, depending on the type of the filler. Thus, the blends prepared and tested in this study can be used to fabricate degradable packaging, greenhouse accessories, and delivery systems for fertilizers and pesticides.

7.3.3 A STUDY OF THE SUITABILITY OF MIXTURES OF P(3HB) AND NATURAL MATERIALS FOR THE CONSTRUCTION OF PLANT PROTECTION PRODUCTS OF A NEW GENERATION

The newest trend in research is development of new-generation slow-release pesticides, including fungicides, embedded in degradable matrices. The use of such formulations will decrease the amounts of pesticides applied to soil, ensure slow and targeted release of the active ingredients, and reduce the risk of uncontrolled distribution of the pesticides in the biosphere. The main condition for constructing such formulations is the availability of appropriate materials with the following properties: degradability and compatibility with global biosphere cycles, safety for living organisms and their nonliving environment, long-term presence in the natural environment and controlled degradation followed by formation of non-toxic products, chemical compatibility with pesticides, and processability by available methods.

Therefore, the purpose of this study was to construct and investigate slow-release fungicide formulations, with the active ingredient embedded in the degradable matrix of P(3HB) blended with readily available natural materials (peat, clay, wood flour) [23]. Slow-release fungicide formulations (azoxystrobin, epoxiconazole, and tebuconazole) shaped as pellets and granules in a matrix of biodegradable poly(3-hydroxybutyrate) and natural fillers (clay, wood flour, and peat) were constructed.

Antifungal activity of the matrix-loaded fungicides and the free fungicides is shown in Figure 7.4. The *in vitro* study showed that all formulations had a significant inhibitory effect on *F. verticillioides*: the size of the fungus colonies was smaller in the presence of the fungicides than in the control group. Photographs show *F. verticillioides* colonies on the nutrient medium in the presence of matrix-loaded fungicides and in the control group. The average radius of the fungus colonies in the Petri dishes with the experimental azoxystrobin formulations was comparable to the radius of the colonies in the dishes with pure fungicide (positive control) (1.3–1.7 cm) while in the dishes without any fungicide (negative control), the size of the fungus colonies was much larger (3.5–3.8 cm). Thus, the experimental formulations were as effective as the free fungicides. Similar results were obtained for the tebuconazole and epoxiconazole formulations. Formulations with different fillers produced similar fungicidal effects. No significant differences in the inhibitory effects on the growth of fungal colonies were found between the pellets and granules.

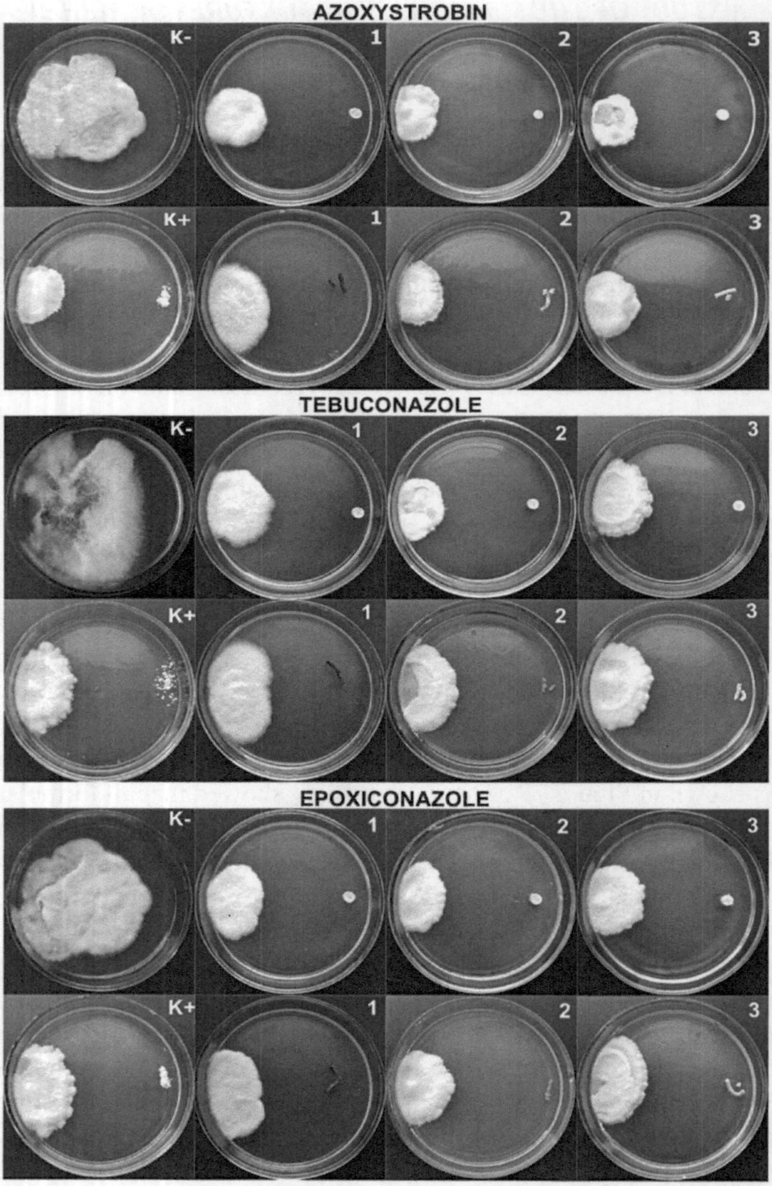

FIGURE 7.4 Sensitivity of *Fusarium verticillioides* to the fungicides in different forms: K^- – negative control (without fungicide), K^+ – positive control (active ingredient), 1 – pellets and granules of P(3HB)/peat/fungicide, 2 – pellets and granules of P(3HB)/clay/fungicide, 3 – pellets and granules of P(3HB)/wood flour/fungicide.

Source: Reprinted from Volova et al. [23].

Biodegradation of the experimental fungicide formulations was studied in laboratory soil microecosystems. Degradation behavior of the experimental formulations and the release of fungicides to soil were studied by determining the mass of the specimens during incubation of formulations in soil, taking into account residual concentrations of polymer and fungicides. These processes are certainly determined by various factors. During the incubation of the experimental granules and pellets in soil, only the polymer was degraded but the fillers (peat, clay, and wood flour) were not. Even if the formulations were degraded completely, the fillers would stay in the soil. Therefore, a study was performed to investigate degradation of poly(3-hydroxybutyrate) pellets and granules in soil. The granules were fully degraded in 126 days while degradation of pellets took longer-about 160 days. The type of the filler did not produce any major effect on degradation of the specimens, although specimens containing clay were degraded at a somewhat faster rate. Therefore, the residual mass of clay-containing specimens was the lowest – 55% of their starting mass.

All experimental formulations were capable of functioning in soil longer than the experimental period of 83 days, enabling gradual and slow delivery of fungicides to plants. Fungicide release from the polymer matrix was determined by measuring fungicide contents in the formulations before and after incubation of the specimens in soil. The active ingredients were released rather uniformly, especially from the pellets. By Day 20, a significantly higher amount of tebuconazole had been released from the granules than from the pellets. Fungicide concentrations in the experimental formulations with different fillers and fungicide concentrations in the soil were measured during incubation of the specimens in soil. Fungicide release kinetics was influenced by various factors such as kinetics of degradation of formulations in soil, shape of the specimens, composition of the matrix, solubility of the fungicides, and stability of the fungicides in soil. The fungicides accumulation in the soil released from the granules was somewhat higher than from pellets, especially in the early phase of incubation. The highest fungicide concentrations in the soil with the experimental granules were measured by the third week of the experiment. By contrast, the highest concentrations of the fungicides released from the experimental pellets were observed later, between week five and week seven of the experiment. The dynamics of accumulation of the three relatively poorly soluble fungicides were similar. Measurements of their residual concentrations in the specimens suggested that the fungicides were gradually released from both granules and pellets. After the fungicide concentrations in soil reached their maxima, no dramatic changes in this parameter occurred until the end of experiment (83 days).

The highest concentration of tebuconazole, whose solubility was higher than the solubility of azoxystrobin and epoxiconazole, varied between 20 and 26 µg/g soil, depending on the filler material (that is 20–26% of the fungicide content in the formulations before soil incubation) and the maximal concentrations of azoxystrobin and epoxiconazole in soil were 15 and 11 µg/g soil, respectively. These concentrations are corresponded to the recommended fungicide rates. The differences in their concentrations in soil were most likely caused by their dissimilar stabilities.

The studies have shown that all fungicide formulations had a pronounced inhibitory effect in vitro against the fungus *Fusarium verticillioides*. Biodegradation behavior of the experimental fungicide formulations in the soil was mainly determined by the shape of the specimens (granules or pellets) without significant influence of the filler type. The content of fungicides present in the soil due to degradation of the formulations and fungicide release was primarily determined by their solubility [23]. The use of P(3HB) and fillers as a matrix for embedding fungicides made it possible to develop slow-release formulations of fungicides, which can function in soil for a long time, enabling gradual and slow delivery of fungicides.

The next study was to develop ecofriendly herbicide formulations. Its main aim was to develop and investigate slow-release formulations of herbicides (metribuzin, tribenuron-methyl, and fenoxaprop-P-ethyl) of different structure, solubility, and specificity, which were loaded into a degradable matrix of poly-3-hydroxybutyrate (P(3HB)) blended with available natural materials (peat, clay, and wood flour) [24, 25]. Studies of biodegradation of the experimental herbicide formulations were performed in laboratory soil micro ecosystems. The specimens of herbicide loaded polymer/filler carriers were degraded gradually. The residual masses of the pellets and granules at the end of the experiment were 45–70% and 60–80%, respectively, of their starting masses. The composition of the matrix was a minor factor affecting mass loss of the slow-release herbicide formulations. Degradation kinetics of the majority of the experimental slow-release herbicide formulations was as follows: $\ln M_i = \ln M_0 + kt$, where M_0; M_i are the starting mass of the specimen and the mass of the specimen after degradation, g, k is degradation rate constant, and t is time, h. The validity of this equation for describing degradation kinetics of all the specimens was supported by the high accuracy of approximation (between 0.91 and 0.98). The highest values of degradation rate constant (0.00746 and 0.0094) were found for the specimens with metribuzin loaded into the clay- and peat-containing matrices and tribenuron loaded into the clay- and peat-containing matrices (0.0074 and 0.0077).

Thus, herbicide formulations were capable of functioning in soil longer than the experimental period of 83 day.

The effects of the experimental herbicide formulations on soil microflora were studied. After 83 days of incubation of the slow-release herbicide formulations in the soil, the total abundance of copiotrophic bacteria increased by a factor of 4–10 compared to the control, without herbicide application. The abundance of fungi in soil samples with slow-release metribuzin and tribenuron-methyl formulations increased by a factor of 1.3–3.6 on average while in the samples with the slow-release fenoxaprop-P-ethyl, the abundance of fungi decreased by a factor of 1.2–2.0. The effect of the herbicides on fungi was apparently determined by the type of the herbicide. Examination of the species composition of soil microflora at the end of the experiment (83 d) revealed certain changes in the proportions of the major taxa in soil samples with slow-release herbicide formulations compared to the control soil. *Bacillus* species, constituting up to 28.7%, prevailed in both groups of soil samples. The proportion of *Streptomyces* increased to 21–24.5% on average and *Pseudomonas* to 5.6–12.3%, while *Stenotrophomonas rhizophila* decreased to 8–9.7%. The type of the filler in formulations did not substantially influence the taxonomic composition of the dominant species in microbial communities. Thus, application of slow-release herbicide formulations did not decrease soil capacity for microbial degradation of complex organic compounds, including pollutants.

Herbicides released from the granules reached their peak concentrations in soil after 20 days of incubation while the herbicides released from most of the pellets reached their highest concentrations in soil much later, at Days 48–50. The solubility of the rapidly dissolving metribuzin was 1.2 mg/L, whereas the solubilities of the poorly soluble tribenuron-methyl and fenoxaprop-P-ethyl were 2 mg/L and 0.7 μg/L, respectively. Regardless of the type of the filler, metribuzin released from the pellets reached its peak concentrations in soil after 6–7 weeks of incubation: 60 to 80 mg/g soil, i.e., 50–80% of the loaded active ingredient, and remained steady for a long time, declining insignificantly by the end of the experiment. Residual metribuzin in the formulations was no more than 5–8% of its starting amount, i.e., the loaded herbicide was almost completely released to soil. Concentrations of the metribuzin released from the granules varied in a similar way, but the peak concentration was reached sooner: after three weeks for the specimens containing peat and after four weeks for the granules containing clay and wood flour. Metribuzin release from the pellets corresponded to the Fickian diffusion mechanism and from the granules to the non-Fickian transport, regardless of the filler type, suggesting considerable structural differences between the carriers. Thus, metribuzin was steadily and gradually released from both pellets and granules and, hence, was present in the soil over extended time periods.

So, herbicide formulations enabling slow controlled release of the active ingredients were fabricated. The degree of solubility of the herbicide considerably affected its release rate and degradation of the formulations in soil: the rapidly dissolving metribuzin and tribenuron-methyl were released at faster rates than the less readily soluble fenoxaprop-P-ethyl. Loading of the herbicides into the polymer matrix composed of the slowly degraded P(3HB) and available natural materials enabled both sustained function of the formulations in soil (lasting between one and a half and three or more months) and stable activity of the otherwise rapidly inactivated herbicides such as tribenuron-methyl and fenoxaprop-P-ethyl.

7.4 CONCLUSION

Two approaches have been investigated in order to reduce the cost and increase the availability of the use of biodegradable polymers of microbiological origin (polyhydroxyalkanoates, PHA):(1) expanding the raw material base and attracting glycerol as a carbon substrate (large-tonnage waste from the production of biodiesel); and (2) filling the polymer with natural materials for industrial use.

An effective technology for the synthesis of destructible polymers was developed using a new and more affordable substrate compared to sugars-glycerol. The *Cupriavidus eutrophus* B-10646 strain was studied as a product of polymers on glycerin, and the kinetic and production characteristics of the strain providing polymer yields on glycerol of various cheat, comparable with the process on sugars, were studied. Replacing sugars with glycerol provides a reduction in the cost of a carbon substrate in the production of PHA.

Mixtures were formed using P (3HB) and available natural materials (clay, peat, birch sawdust). From mixtures in the form of dry powders and raw paste, mixed forms in the form of granules and extruded 3D forms were obtained; their structure, physicochemical, and mechanical properties were studied. A comparative study of the influence of the type of form and method of manufacture (by pressing powders, extrusion from melts, mortar technologies) on the intensity of biodegradation and the duration of functioning in the soil substantiated the feasibility of using forms that can provide plants with preparations during the growing season (from 1.5 to 3.0 and more months) of two methods-obtaining granules from crude paste of polymer and filler mixtures and cold pressing of powders. Based on the results obtained and their analysis, a pioneer family of prolonged preparations of herbicidal and fungicidal action for pre-emergence soil application was constructed.

Regardless of the geometry of the developed formulations, the composition of the mixed base, the solubility of pesticides in water and their stability, it is shown that the achieved concentrations of active substances in the soil are within the limits corresponding to the recommended application rates. The results showed the suitability of the developed and studied mixtures and forms of them for the construction of long-term new generation plant protection products.

ACKNOWLEDGMENT

This study was financially supported Project "Agropreparations of the new generation: a strategy of construction and realization" (No 074-02-2018-328) 2018–2019 in accordance with Resolution No 220 of the Government of the Russian Federation of April 09, 2010, "On measures designed to attract leading scientists to the Russian institutions of higher learning."

KEYWORDS

- **differential scanning calorimetry**
- **free fatty acids**
- **pilot production**
- **polyhydroxyalkanoates**
- **x-ray diffraction**

REFERENCES

1. Urbanek, A. K., Rymowicz, W., & Mirończuk, A. M., (2018). Degradation of plastics and plastic-degrading bacteria in cold marine habitats. *Appl. Microbiol. Biot., 102*, 7669–7678. https://doi.org/10.1007/s00253-018-9195-y (accessed on 3 August 2020).

2. Qaiss, A., Bouhf, R., & Essabir, H., (2015). Characterization and use of coir, almond, apricot, argan, shell, and wood as reinforcement in the polymeric matrix in order to valorize these products, In: Hakeem, K., Jawaid, M., & Alothman, O. Y., (eds.), *Agricultural Biomass Based Potential Materials* (pp. 305–339). Cham: Springer,Switzerland. https://doi.org/10.1007/978-3-319-13847-3_15 (accessed on 3 August 2020).
3. Chen, G. Q., (2010). Plastics from bacteria: Natural functions and applications. *Microbiology Monographs*. Springer-Verlag, Berlin.
4. Chen, G. Q., (2012). New challenges and opportunities for industrial biotechnology. *Microb. Cell Fact., 11*, 111. doi: 10.1186/1475-859-11-111.
5. Volova, T., Vinnik, Y., Shishatskaya, E., Markelova, N., & Zaikov, G., (2017). *Natural-Based Polymers for Biomedical Applications*. Apple Acad. Press, Canada.
6. Volova, T., Shishatskaya, E., Prudnikova, S., Zhila, N., & Boyandin, A., (2019a). *New Generation Formulations of Agrochemicals: Current Trends and Future Priorities* (p. 286). Toronto-Canada: CRC/Taylor & Francis: Appl. Acad. Press. ISBN: 9781771887496.
7. Koller, M., Maršálek, L., De Sousa, D. M. M., & Braunegg, G., (2017). Producing microbial polyhydroxyalkanoate (PHA) biopolyesters in a sustainable manner. *New Biotechnol., 37*, 24–38. https://doi.org/10.1016/j.nbt.2016.05.001 (accessed on 3 August 2020).
8. Hermann-Krauss, C., Koller, M., Muhr, A., Fasl, H., Stelzer, F., & Braunegg, G., (2013). Archaeal production of polyhydroxyalkanoates (PHA) co-and terpolyesters from biodiesel industry-derived by-products. *Archaea* (p. 10). Article ID 129268. doi: 10.1155/2013/129268.
9. Kourmentza, C., Plácido, J., Venetsaneas, N., Burniol-Figols, A., Varrone, C., & Gavala, H. N., (2017). Reis MAM recent advances and challenges towards sustainable polyhydroxyalkanoate (PHA) production. *Bioengineering, 4*(2), e55. doi: 10.3390/bioengineering4020055.
10. Torres-Giner, S., Montanes, N., Boronat, T., Quiles-Carrillo, L., & Balart, R., (2016). Melt grafting of sepiolite nanoclay onto poly(3-hydroxybutyrate-co-4-hydroxybutyrate) by reactive extrusion with multi-functional epoxy-based styrene-acrylic oligomer. *Eur. Polym. J., 84*, 693–707.https://doi.org/10.1016/j.eurpolymj.2016.09.057 (accessed on 3 August 2020).
11. Gunning, M., Geever, L., Killion, J., Lyons, J., & Higginbotha, C., (2013). Mechanical and biodegradation performance of short natural fiber polyhydroxybutyrate composites. *Polym. Testing, 32*, 1603–1611. https://doi.org/10.1016/j.polymertesting.2013.10.011 (accessed on 3 August 2020).
12. Angelini, S., Cerruti, P., Immirzi, B., Scarinzi, G., & Malinconico, M., (2016). Acid-insoluble lignin and holocellulose from a lignocellulosic biowaste: Bio-fillers in poly(3-hydroxybutyrate). *Eur. Polym. J., 76*, 63–76.https://doi.org/10.1016/j.eurpolymj.2016.01.024 (accessed on 3 August 2020).
13. Khunthongkaew, P., Murugan, P., Sudesh, K., & Iewkittayakorn, J., (2018). Biosynthesis of polyhydroxyalkanoates using Cupriavidus necator H16 and its application for particleboard production. *Journal of Polymer Research, 25*, 131. https://doi.org/10.1007/s10965-018-1521-7 (accessed on 3 August 2020).
14. Luigi-Jules, V., Clement, M. C., Alan, W., Des, R., Bronwyn, L., & Steven, P., (2018). Wood-PHA composites: Mapping opportunities. *Polymers, 10*,751–766. https://doi.org/10.3390/polym10070751 (accessed on 3 August 2020).

15. Du, C., Sabirova, J., Soetaert, W., & Lin, S. K. C., (2012). Polyhydroxyalkanoates production from low-cost sustainable raw materials. *Curr. Chem. Biol., 6*, 14–25. doi: 10.2174/2212796811206010014.
16. Możejko-Ciesielska, J., & Kiewisz, R., (2016). Bacterial polyhydroxyalkanoates: Still fabulous. *Microbiol. Res., 192*, 271–282. doi: 10.1016/j.micres.2016.07.010.
17. Sabbagh, F., & Muhamad, I., (2017). Production of polyhydroxyalkanoates as secondary metabolite with main focus on sustainable energy. *Renew. Sust. Energ. Rev., 72*, 95–104. https://doi.org/10.1016/j.rser.2016.11.012 (accessed on 3 August 2020).
18. Fernández-Dacosta, C., Posada, J., Kleerebezem, R., Cuellar, M., & Ramirez, A., (2015). Microbial community-based polyhydroxyalkanoates (PHAs) production from wastewater: Techno-economic analysis and ex-ante environmental assessment. *Bioresour. Technol., 185*, 368–377. doi: 10.1016/j.biortech.2015.03.025.
19. GHG emission reductions from world biofuel production and use (2015). *Global Renewable Fuels Alliance*. https://www.epure.org/media/1298/final_report_ghg_emissions_biofuels_2015-3.pdf (accessed on 14 August 2020).
20. Statista, (2018). *Leading Biodiesel Producers Worldwide in 2017*. By country (in billion liters). https://www.statista.com/statistics/271472/biodiesel-production-in-selected-countries (accessed on 3 August 2020).
21. Volova, T., Demidenko, A., Kiselev, E., Baranovskii, S., Shishatskaya, E., & Zhila, N., (2019b). Polyhydroxyalkanoate synthesis based on glycerol and implementation of the process under conditions of pilot production. *Applied Microbiology and Biotechnology, 103*(1), 225–237 https://doi.org/10.1007/s00253-018-9460-0 (accessed on 3 August 2020).
22. Volova, T., Kiselev, E., Zhila, N., & Shishatskaaya, E., (2019c). Synthesis of PHAs by hydrogen bacteria in a pilot production process. *Biomacromol., 20*, 3261–3270. doi: 10.1021/acs.biomac.9b00295.
23. Volova, T., Prudnikova, S., Boyandin, A., Zhila, N., Kiselev, E., Shumilova, A., Baranovsky, S., et al., (2019d). Constructing slow-release fungicide formulations based on poly-3-hydroxybutyrate and natural materials as a degradable matrix. *Journal of Agricultural and Food Chemistry, 67*, 9220–9231. https://doi.org/10.1021/acs.jafc.9b01634 (accessed on 3 August 2020).
24. Kiselev, E. G., Boyandin, A. N., Zhila, N. O., Prudnikova, S. V., Shumilova, A. A., Baranovskiy, S. V., Shishatskaya, E. I., Thomas, S., & Volova T. G., (2020). Constructing sustained-release herbicide formulations based on poly-3-hydroxybutyrate and natural materials as a degradable matrix. *Pest Manag Sci. 76*, 1772–1785. https://doi.org/10.1002/ps.5702.
25. Thomas, S., Baranovsky, S., Vasiliev, A., Kiselev, E., Kuzmin, A., Nemtsev, I.,Sukovatyi, A., et al., (2019f). Thermal, mechanical, and biodegradation studies of biofiller based poly-3-hydroxybutyrate biocomposites. *International Journal of Biological Macromolecules*.

[illegible]

CHAPTER 8

Membrane Technology for Green Engineering

SUPRIYA DHUME and YOGESH CHENDAKE

Department of Chemical Engineering, Bharati Vidyapeeth (Deemed to be) University, College of Engineering, Pune, Maharashtra, India, E-mail: yjchendake@bvucoep.edu.in (Y. Chendake)

ABSTRACT

Green technology is that to operate ecological skills for the development and application of merchandise, instrumentality, and systems to conserve the natural resources and surroundings. This field is moderately innovative. It has developed rapidly as people have become aware of the detrimental effects of environmental variation including global warming and adverse effects on the living beings. It is believed that growth in green technology will lead to worldwide, sustainable, and economic powers that impact economics, societies, and way of life in the future. Application of green technology has multiple challenges while it can provide many opportunities during real-life chemical processes. The membrane technology can provide an economical pathway to many of these challenges. Here applicability of membranes during green technology is discussed. Over past few years, membrane separation process has become one of the major separation technologies. It is rapidly becoming important part of green technology due some of its benefits. In membrane processes does not require addition of chemicals, it works on physical separation. This improves the feasibility of recycling the recovered chemicals or using it in further processes. Additionally, membrane technology has relatively low energy requirements, easy to design and integrated nature may advance, and can be integrated with current processes. This gives benefits of reduced production cost, equipment size, energy consumption and waste generation and improving controls and process flexibility [1–3].

The works would consider two aspects of membrane technologies for process enhancements; as process integration and effluent treatment. Process integration would be focused on for improvements in economical separation, recovery, and recycle; this would reduce generation of waste and effluent component generation, thus enhancing process financial system. Membrane separation technologies are very helpful in effluent treatment. It is investigated for reduction in effluent volume and its treatment for components to recover. The work starts by presenting an overview of membrane operations. It would be followed for application of membrane processes for process improvement and *in situ* recovery of components. This would be followed by the effluent treatment aspects. Such recovery of components harmful to environment and their further utilization by membrane processes would increase the process economy and product purity. To get superior removal of contaminants compared to conventional technologies, the membrane technology would provide better support to system. The conventional technologies frequently required continuous modification of conditions and use of chemicals. This means a high level of operator involvement and the risk of contaminants in the effluent [4–7].

8.1 INTRODUCTION

8.1.1 GREEN TECHNOLOGY

Green Technology is nothing but modification of process parameters for the expansion and application of products, equipment, and systems to conserve the natural environment and resources. In case of unavoidable effluent and waste generation, it requires it recovery and recycle to system for improvement in conversion and yield for desired product, with reduction or elimination of hazardous effluent generation. It helps to reduce the volume of effluent generation; avoid harmful effluent stockpiling or contamination with environment. It shows the harmful impact on human activities. Green technology refers to use or synthesis of products and equipments which satisfy the following criteria:

1. It reduces the ecological degradation;
2. It has low greenhouse gas emission;
3. It is safe and provide healthy and good environment;
4. Conservation of natural resources and energy usage;
5. It assists to maximize consumption of renewable resources, than diminishing mineral and petroleum oil ones.

Overall, it can be said that, green technology is an eco-friendly clean technology contributes to sustainable development to conserve the natural resources and environment. Development of such synthesis route, processes, equipments, and systems will meet the demands of present and future generations. These developments would help humans, to eliminate the utilization or generation of hazardous substances [8–12].

The principles of green technology are relevant to chemical products including its usage, design of process, manufacturing, and final disposal. Green technology is also referred to be sustainable developments. This technology development and chemical engineering focused on such designed process which reduces or eliminates the generation of hazardous substances. The environmental engineering focuses on the effects of polluting chemicals on nature, where green technology focuses on the environmental impact of chemicals their synthesis and waste reduction. It consists of industrial approaches to prevent the pollution and reducing consumption of non-renewable resources, enhancing output within defined resources, optimal conversion, and reduction in energy consumption.

Green technology concerns are also related to production of food and its process which requires considerable usage of land. Such land-use changes and shows an incredible reduction in biological assortment, aquatic organic process with phosphorus substances which caused by over-fertilization, water shortages due to irrigation, climate changes (CCs), eco-toxicity, human effects of pesticides, etc. They are lead to contamination of surroundings with different toxic substances contains arsenic, cadmium, chromium, lead, and mercury. These are the priority metals that are having high level of toxicity. This toxicity depends upon several factors including their dose, route of exposure, and chemical species [13–15].

Another one of the important aspects of chemical processing is energy conversion. Some of the energy sources considered today is fossil fuels, hydro energy, other forms of electrical energy generation (tidal, wind, atomic, etc.). All such conventional energy sources affect the environment differently. Fossil fuels like coal, oil, and natural gas are considerably more harmful than renewable energy sources. Frequently its usage leads to air and water pollution. This intern results in damage of public health, wildlife, and habitat loss, water resource contamination, land use, and global warming emissions. This can be overcome by utilization of renewable energy resources. These renewable energy resources would replace fossil fuels, thereby reducing emissions that cause CC. Increasing renewable energy allows decreasing the costs described, and that is obviously very economically beneficial in combination with the environmental benefits. There are many more ways green energy benefits the economy, directly, and indirectly. By reducing the use of traditional energy sources, natural ecosystems can be maintained to the best conditions. Further use of Green Energy can be economically attractive. The economic benefits of green energy can be summarized as follows:

1. Increase in energy efficiency.
2. Lower energy cost from the renewable sources viz., solar, wind, and biomass.
3. Nowadays, carbon emission is being reduced and same has been mandated through various laws. Some of the laws have established laws related to carbon credits and fines for higher release of these gases. Today fuel cost has to be finalized considering, carbon emission.

To achieve these goals of green technology, there is a need for design of chemical processes and its development or application of products; equipments for conservation of natural resources are beneficial to environment. In the environmental point of view, the necessity is to avoid use or generation of hazardous substances to reduce the pollution in environmental point of view. And the process modification and propose feature must think to reduce or eliminate energy and avail the economical benefits.

8.1.2 NEED AND GOALS

As discussed until now, Green Technology possesses multiple benefits. They are achieved by following principals of green technology:

- **Prevent Waste:** Chemical production is design such a way that remains no waste to treat or clean up.
- **Atom Economy:** Design of synthesis in such a way to the final product contain highest amount of initial material. Waste a small number of or no atoms.
- **Design Less Hazardous Chemical Production:** Design synthesis process to generate the substances with minimum or no toxics to humans or environment.
- **Design Harmless Chemicals and Products:** The chemical products are designing in such a way that has little or no toxicity.
- **Use Unhazardous Solvents and Reaction Conditions:** Avoid using hazardous solvents, separation agents, or other auxiliary chemicals. If necessity to use these chemicals, use safer ones.

- **Increase Energy Efficiency:** Run chemical reactions at room temperature and pressure whenever possible.
- **Use Renewable Feedstocks:** Use of raw materials which are renewable. The source of renewable feedstocks is frequently the agricultural products or the wastes of other processes; the source of depletable feedstocks is often fossil fuels (petroleum, natural gas, or coal) or mining operations.
- **Avoid Chemical Derivatives:** Avoid using blocking or defensive groups or any temporary modifications if possible. Derivatives use additional reagents and generate waste.
- **Use Catalysts, Not Reagents:** Minimize waste by using catalytic reactions. Catalysts are efficient with small amount and can carry out a single reaction many times. They are preferable to stoichiometric reagents, which are used in excess and carry out a reaction only once.
- **Design Chemicals and Products to Degrade After Use:** Design chemical products to break down to harmless substances after use so that they do not accumulate in the environment.
- **Analyze in Real-Time to Prevent Pollution:** Include in-process, real-time observation, and control during synthesis to diminish or eliminate the development of byproducts.
- **Minimize the Potential for Adversity:** Design chemicals and their physical forms (solid, liquid, or gas) to diminish the potential for chemical accidents together with explosions, fires, and releases to the environment.

Hence there is a need to improve or modify the process so as to incorporate the Green Technology principals. At many places, economical separation of components and their composition variations can be useful to achieve the desired progress towards Green Technology. Membrane processes and their process applications are an excellent way for the same. Membrane-based systems carry out physical separations based upon component and membrane properties. Additionally, the separation conditions can be tuned as per the need. Further section deals with membrane processes and their applications towards Green Technology.

8.2 MEMBRANE PROCESSES

Membrane is nothing but semipermeable barrier, which allows selective transport of component across the same. The transport and separations are in the same phase. This reduces requirement of heat separation of components. Components are separated upon their molecular size, shape, charge, interactions, etc. with use of pressure and particularly considered semi-permeable membranes. This separation of components does not require accumulation of chemicals or any contaminations. The separation is carried out in single-stage without phase inversion hence generation of hazardous waste is eliminated.

Membrane separation in green technology is describe in such a way that, with help of membrane process reduction in byproduct/waste due to less usage of chemicals during production. It can be combined with the synthesis processes for continuous separation which gives the benefits of recovery and reuse of waste/byproduct can be usable as raw feed to the system. This reduces generation of waste, and effective recycle enhances conversion along with conservation of natural resources.

Membrane separation technology is physical separation. It provide excellent alternative for recovery of components in pure form. This helps in the waste processing and recovery of components. These recovered components can be recycled to the processing streams. A proper and careful design of membrane properties and separation parameters can be used to recover or separate the components from effluent and utilize economically. It would reduce or eliminate the generation of hazardous substances in environment. These hazardous substances are harmful to the nature/environment.

Contamination of these hazardous components can affect CC, availability of harmless and sufficient water supply, and food production. Additionally, it would require additional energy for the treatment, which in turn would affect the emission of hazardous substances in the atmosphere. This would have a vital impact over the water cycle, sterilization precipitation patterns, and poignant the supply and quality of surface and groundwater, agricultural production, and associated ecosystems. Water usages for agricultural is up to the seventieth each primary food production and food process is critically dependent upon reliable installation and adequate water quality. Each variety of activities may also have serious effects on water resources. This would affect human and surrounding to a greater extent. Recovery and removal of such components before their discharge into the atmosphere can prove highly beneficial to fight these demons. The membrane-based processes can provide answers to these issues.

As mentioned up till currently, there is have to be compelled to minimize losses in terms of waste generated through reaction selection, method style and use or recycling. This may cause effective use of obtainable resources and generate the littlest doable quantity of stuff. It might give massive economical and ecological advantages to the economic makers. Use of membrane methods will be one among the wonderful choices throughout process modifications. The membrane processes of microfiltration (MF), ultra-filtration, and reverse osmosis (RO) supply glorious potential for continuous removal of elements from effluent streams. Choice of the optimum method could be a operate of material properties and physical forms. Membrane separation technologies cannot build the contaminants disappear; but, they are extraordinarily effective at concentrating the contaminants, i.e., a pretreatment part, recovery of part for his or her applications in any processes. The membrane method has a plan of use or convalescent specific, i.e., parts for re-use through the applying of membrane separation technologies.

Due to increasing limitations placed on waste and effluent disposal, innovative ways of dangerous waste treatment are required; which can be fulfilled by membrane technologies. The benefit of these technologies is that they can be tuned as per requirement and application in hand. They can be combined with current process to enhance the output, decreases the byproduct formation and scale back or eliminate waste generation. In recent years, membrane separation has gained increasing quality for these applications. The three major varieties of membrane separation technologies in use nowadays square measure reverse diffusion, immoderate filtration, and electrochemical analysis.

The popularity and applicability of membrane-based processes in increasing due to subsequent reasons. The combination of membranes based separation system can be done in line with the synthetic streams. This helps to monitor and control the product properties and desired environment.

A strong increase within the world's population and an eternal growth of commercial productivity the global energy consumption continues to extend. On the other hand, the fossil energy resources viz. oil, gas, and coal are progressively exhausted. So as to secure the energy offer for the close to future energy-saving measures ought to be taken and also the development of energy-saving technologies ought to be aroused. During this respect membrane technology is an associate emerging technology with several prospects as associate energy-saving separation technology in a range of classical separation techniques like distillation, evaporation, refrigeration, condensation, a phase change occurs and the energy consumption is relatively high. Membrane technology is one amongst the new separation techniques with comparatively low energy consumption since no phase change takes place.

In membrane process continuous separation is, occurs though some unwanted components are clogging the membrane to avoiding this unnecessary derivatization a little modification has to done on temporary basis. Like doing some changes in membrane, preparation helps to reduce the clogging or give backwash to the system to avoiding the clogging. Also in life processing and control helps to avoid hazardous products. Moderate separation conditions and minimize potential of chemical accidents like releases, explosions, fires, and formation of hazardous components and effluents. Further benefits of membrane processes can be summarized as below.

8.3 BENEFITS OF MEMBRANE SEPARATIONS

Membrane separations are often used for very large number of separation because membrane processes separate at the molecular scale up to a scale at that particle can be seen. Membranes have intensely high property for the separation of components. There can be a big struggle of management over separation property by having larger range of polymers and inorganic media usage as membranes. Energy necessities area unit sometimes low because membrane separations typically don't need a phase transition. They sometimes supply a simple and straightforward operation and low maintenance method possibility as they are doing not have moving parts, complicated management schemes, and little subsidiary instrumentation. Membranes will have intensely high property for the separation of parts. Membrane processes area unit sometimes higher for the surroundings once there is a usage of comparatively straightforward and non-harmful materials and they will recover minor however valuable constituents from a mainstream with a minor energy value. Due to these benefits, many membrane processes have been designed and utilized in green technology.

8.4 MEMBRANE PROCESSES IN GREEN TECHNOLOGY

Membrane filtration could be a technique of separating particles in liquid solutions or gas mixtures. This system is employed in a very big selection of applications starting from dairy process to waste treatment. The four main ways of pressure-driven membrane filtration are MF, Ultrafiltration, Nanofiltration, and RO, so as of decreasing pore size. Their brief properties are summarized below:

1. **Microfiltration (MF):** These membranes have pore sizes ranging from 0.1 to 10 μm. They are mainly used for the removal of huge particles, colloids, and bacteria from feed streams. This is often

particularly well-liked within the food trade for treating waste material before discharging it to a municipal sewer.

2. **Ultra-Filtration (UF):** It is a process that is similar to MF, but with smaller pore size, ranging from 0.01 to 0.1 μm. UF membranes are used in removal of viruses and polypeptides, and are extensively used in protein concentration and wastewater treatments.
3. **Nanofiltration (NF):** These membranes are similar to RO membranes in that they contain a thin-film composite layer (<1 μm) on top of a porous layer (50 to 150 μm) for small ion selectivity. NF membranes are rejecting multivalent salts and uncharged solutes, whereas permitting some monovalent salts to accept. They operate at lower pressures than RO membranes, creating them ideal for achieving the associate best combination of flux and rejection.
4. **Reverse Osmosis (RO):** These membranes are even tighter than NF membranes, and are ready to reject all monovalent ions whereas permitting water molecules to meet up with in aqueous solutions. They additionally take away viruses and microorganisms found in feed solutions. Common applications for RO filtration embrace seawater desalination and industrial water treatment.

The other main methods of concentration driven membrane filtration are pervaporation (PV), gas separation, dialysis, and electrodialysis. Chemical analysis could be a method during which the solute, usually an electrolyte, transfers across the membrane driven by the distinction in concentration between the two sides of the membrane. Two conditions ought to be consummated for the method to be effective:

i. The concentration on the permeate aspect has got to be unbroken low so the driving force remains as high as possible; and
ii. The pressure should be low, and stay low throughout the method, so a counter flux of solvent does not end in feed dilution.

8.5 ELECTRODIALYSIS

Electrodialysis could be a method within which ion-selective membranes measure used alongside an electrical field normal to the membrane phases. An electrodialysis stack consists of parallel compartments separated alternately by cation-exchange and anion-exchange membranes. The feed is introduced into every compartment, so power-assisted by the electrical field and selected by the membranes, the cations, and also the anions are transferred in opposite directions to the neighboring compartments. Betting on the sort of membrane used, ions are often selected consistent with their valence. So this method is often used for fractionating ions of various valences. The most application of electrodialysis is in demineralization normally and within the chemical process of saltwater.

Membrane electrolysis could be a method whereby each electrode reactions, i.e., the cathodic reduction additionally because the electrode oxidization, are coupled to the transport and transfer of charged ions. In membrane electrolysis, the electrode reaction is crucial to the particular separation method. The aim of the membrane is to separate the anode loop (anolyte) from the cathode loop (catholyte) by a fluid, so as to avoid unwanted secondary reactions, thus on mix the electrode reaction with a separation step or to isolate individually the products formed on the electrode. In water electrolysis, such products is also in an exceedingly gaseous form such as oxygen and hydrogen element additionally because the acids (H+) and bases (OH–) fashioned on the electrode or the mixture of gas halogen and hydroxide answer and element as in common sodium chloride electrolysis.

8.6 PERVAPORATION (PV)

Pervaporation (PV) is that the combination of the selective separation and transfer of an element across the membrane and its evaporation on the permeate aspect. So as to attain evaporation the pressure on the permeate aspect should be specified the partial pressure of this element is not up to its saturation vapor pressure. Therefore, although the driving force for transfer is the difference in activity of the transferred species, this is the result of applying a vacuum on the permeate side.

PV is dearer than different membrane processes due to the specified offer of warmth to provide the evaporation. It is, therefore, utilized in the separation of parts from mixtures that area unit tough to treat, like azeotopic solutions and chemical compound mixtures, that standard processes would be additional pricey.

8.7 MEMBRANE DISTILLATION (MD)

Membrane distillation (MD) could be a separation method wherever a microporous hydrophobic membrane separates two liquid solutions at totally different temperatures. The property of the membrane prevents mass transfer of the liquid, whereby a gas-liquid interface is made. The gradient on the membrane leads to a vapor pressure distinction, whereby volatile elements within the offer combine evaporate through the pores (10 nm–1 μm) and, via diffusion and/or convection of the compartment with high vapor pressure, area unit transported to the compartment with low vapor pressure wherever they are condensed within the cold liquid/vapor phase. To offer solutions that solely contain non-volatile substances, like salts, water vapor are going to be transported through the membrane whereby dematerialized water is obtained on the distillation-side and an extra targeted salt flow on the supply-side.

8.8 PROCESS INTENSIFICATION (PI)

One of the essential challenges presently facing the planet is "to support sustainable industrial growth." A doable answer is obtainable by process intensification (PI), a design approach presenting existing edges in producing and process. Extra edges may be summarized as shrinking equipment size, boosting plant effectiveness, saving energy, reducing capital costs, increasing safety, minimizing environmental impact, and maximizing raw materials exploitation.

Membrane methods deal with the goals of PI; as a result of they need the potential to interchange conventional energy-intensive techniques, to accomplish the selective and economical transport of specific elements, and to enhance the performance of reactive processes. On a variety of occasions, industrial separation processes in trade were replaced or power-assisted by membrane separation processes with important reductions in cost, energy, and environmental impact.

It can be seen from the following examples that the use of membranes for continuous processing and process improvements can be useful to achieve the goals for PI. The membrane engineering contributes to the comprehension of the principles of green technology. Continuous separation and recovery of components, separation in milder conditions, recycle of separated components, improve reaction kinetics, conversions, energy conservation, energy generation at higher efficiency, and improved product quality are all the benefits of membrane technology. They help to reduce cost of production and provide an environmental-friendly system. Let go to the examples for their complete descriptions. It can be seen from the following examples, viz., membrane reactor for esterification, fermentation-based component synthesis, ethanol concentration, and dairy processing.

8.8.1 MEMBRANE REACTOR-ESTERIFICATION REACTION

Esterification of acids and alcohols could be a typical example of an equilibrium restricted reaction that generates water as a by-product. The conversion is mostly low due to restrictions forced by thermodynamic equilibrium. To urge a high organic compound yield, it is necessary to shift the position of the equilibrium to the organic compound facet by either employing a massive way over one among the reactants (usually alcohol).

One of the foremost ordinarily investigated samples of esterification could be a chemical process that happens between carboxylic acid and alcohol to get esters. The reaction takes place in acidic environments. During this method, water is additionally generated as byproduct.

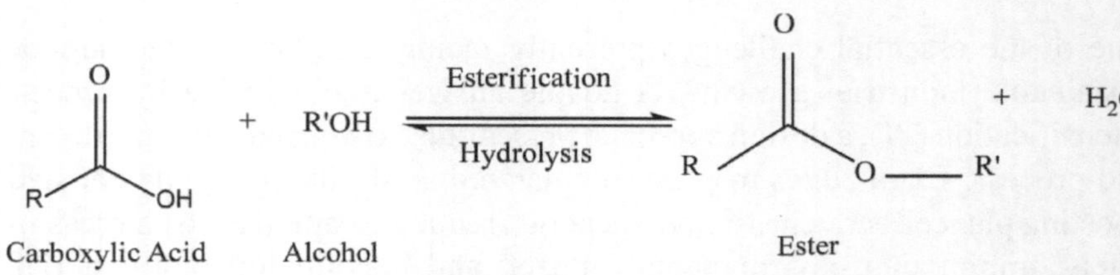

The reaction between carboxylic acid and alcohol is understood because the Fischer esterification. It is one amongst the foremost vital reactions of carboxylic acids. Most carboxylic acids are appropriate for the reaction; however, the alcohol ought to usually be a primary or secondary alkyl. Tertiary alcohols are liable to elimination. Once an acid and an alcohol are mixed along, no reaction takes place. However, upon addition of chemical action amounts of an acid, the two components are mix with equilibrium method to contribute an ester and water. The presence of the acid catalyst within the mechanism of ester formation helps in two ways: It causes the carbonyl perform (makes the carbonyl carbon a lot of electrophilic) to bear nucleophilic attack by the alcohol; and protonation of the hydroxyl group offers water, that could be a superior leaving group (i.e., weaker base) within the elimination step. Ordinarily used catalysts for a Fischer esterification sulfuric acid, p-toluenesulfonic acid, and Lewis acids.

Esterification reactions are generally reversible processes and generate water as part of production of ester. The example of acetic acid and methanol reaction at 60°C is shown these phenomena. This reaction shows an investigation into the impact of water on liquid phase sulfuric acid. In order to moderate, the effect of water on the catalysis as a result in a reversible reaction, initial reaction kinetics were measured using a low concentration of sulfuric acid and different initial water concentrations. It was found that the catalytic activity of sulfuric acid was strongly inhibited by water. The catalysts lost up to 90% activity as the amount of water present increased. The decreased activity of the catalytic protons is suggested to be caused by preferential solvation of them by water over methanol. The results indicate that, as esterification progresses and byproduct water is produced, deactivation of the sulfuric acid catalyst occurs.

As explain before esterification is an equilibrium reaction and water is one of the products of this reaction. The generalized esterification reaction is already discussed above.

In any esterification, reaction water is the byproduct of the reaction. If this water is not removed immediately after the formation of esters then it reacts to get back the reactants due to reversible reaction conditions. Therefore, as per the thermodynamic considerations, to produce more ester and shift the equilibrium in the forward direction; water should be removed immediately.

The membrane reactors are working very effectively for continuous removal of water generated during process. Applications of membrane reactors for water removal during catalytic reactions in food, pharmaceutical, cosmetics, and petrochemical industries are well investigated and established. Currently, global target in this direction is design of a compact and more efficient catalytic membrane reactor. By applying water removal, two objectives have been pursued:(i) overcoming the thermodynamic limitations of the reaction; and (ii) avoiding catalyst poisoning.

Organic solvents are extensively utilized in a range of commercial sectors. Reclaiming and reusing the solvents is also the foremost economically and environmentally useful possibility for managing spent solvents. Purifying the solvents to meet reuse specifications can be challenging. For hydrophilic solvents, water must be removed prior to reuse, yet many hydrophilic solvents form hard-to-separate azeotropic mixtures with water. Such mixtures make separation processes energy-intensive and cause economic challenges. The membrane technologies PV and vapor permeation (VP) can be less energy-intensive than distillation-based processes and have proven to be very effective in removing water from azeotropic mixtures. In PV/VP, separation relies on the solution diffusion interaction between the dense layer of the membrane and therefore the solvent/water mixture. A variability of membrane materials, such as polymeric, inorganic, mixed matrix, and hybrid, has been used for industrial applications. A small set of those commercially obtainable and highlighted here: poly (vinyl alcohol), polyimides, amorphous perfluoro polymers, NaA zeolites, chabazite zeolites, T-type zeolites, and hybrid silicas. Solvents targeted for recovery and then common solvents are chosen for analysis: acetonitrile, 1-butanol, *N,N'*-dimethyl formamide, ethanol, methanol, methyl isobutyl ketone, methyl tert-butyl ether, tetrahydrofuran, acetone, and 2-propanol.

8.8.2 DAIRY PROCESSING FOR MILK COMPONENTS

In the last thirty years, membrane processes became major tools within the food trade. This trade represents the second sector of membrane applications, once water treatment, and on equal terms with pharmaceutical and biotechnology applications. Among the food sector, the farm trade has beyond any doubt developed the foremost advanced filtration procedures for concentration and fractionation of molecules from milk and it is derivate. The primary membrane development within the separation procedures of milk elements occurred within the late 1960's with the appearance of membrane separation. The membrane equipment's are enforced throughout the farm process chains: milk reception, cheese-making, whey protein concentration, fractionation of protein, and effluents treatment (Figure 8.1).

Four different membrane operations are distinguished. Ultrafiltration (UF) is the greatest extensively used process in the dairy industry. NF has lately remained established for whey demineralization. Other membrane processes are RO and MF which develop to retain micro-organisms or casein micelles.

1. **Microfiltration (MF):** The MF of milk concerns two main industrial applications:

   i. **Bacteria Removal from Skimmed Milk:** Similar to most foods, milk provides a good media for spoilage micro-organisms, which might cause alteration of chemical composition of milk and difficulties in milk preservation and transformation. Therein medium frequency projected a noteworthy different to heat-treatment. It is significantly custom-made to the bacterium removal from skim milk; the dimensions of the micro-organisms being within the same vary as fat globules.

   For this application, ceramic membranes used with a pore size of almost 1.4 μm. By combining high cross flow velocity with a low transmembrane pressure, the membranes lead to low matter losses at high fluxes. In large equipment, a second MF stage can be added to concentrate further the first retentate. Due to high retention of bacteria and spores, such a process is used, with some version, by the cheese-makers to produce safe raw milk cheeses.

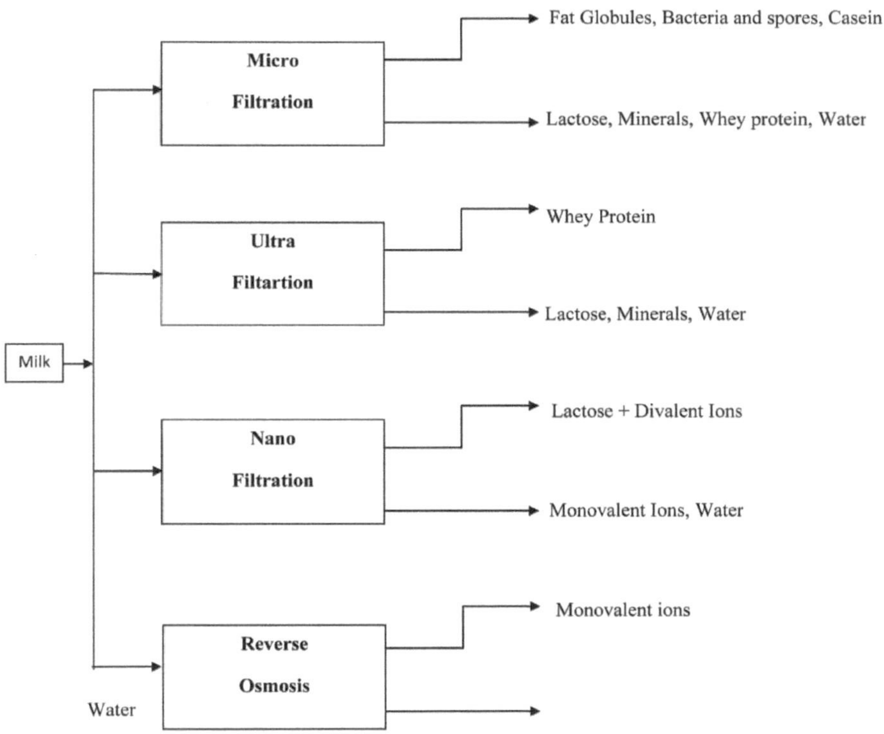

FIGURE 8.1 Milk processing with membrane technologies.

ii. **Selective Concentration:** The permeate, with a composition on the brink of that of a whey, contains whey proteins in their native state, is healthful, free from phase particles and fats, and becomes a helpful fluid to organize whey protein concentrates(WPC).

The retentate is employed for the casein enrichment of cheese milk to boost the cheese-making method, shorter coagulation time and magnified final curd firmness. Additionally, partial removal of whey proteins considerably reduces the prejudicial effects of heat treatment on retentate milk coagulation, and is employed to develop a brand new high-heat milk powder, with a cheese-making ability just like that of raw milk.

Finally, purified casein molecules, obtained by filtration against water, and WPC acquired after UF of milk MF, are exceptional starting fluids for further fractionation of individual caseins or whey proteins. Separation of small milk fat globules from large ones has recently been proposed using MF. It is claimed that the use of small globule with fraction yields makes yogurts soft and hard cheeses.

2. **Ultrafiltration (UF):**

i. **Standardization and Concentration of Milk Proteins:** UF has been used on milk process plants, for the milk standardization in total macromolecule or protein for drinking and cheese milks. Standardization of the protein content of the milk leads to a regular milk composition all year spherical, freelance of seasonal differences, and so leads to additional constant method and products quality, higher use of existing instrumentality and accrued cheese yields. Spiral wound or hollow ceramic membranes with a molecular cut-off of less than 10 kg. mol^{-1} area unit used for this purpose, at a pressure of 200–400 kPa and flux of 30–120 $L.h^{-1}.m^{-2}$.

   UF method permits removal of excess water and lactose by an initial step before coagulation, thereby eliminating the requirement to separate whey from curd. All the proteins are so focused along within the pre-cheese fluid. The acceptable enzymes and micro-organisms are then added into the protein concentrate to coagulate. The benefits of the method for cheese production are nice. The cheese yield is concerning 10–30% above within the ancient method since the whey proteins are enclosed within the pre-cheese. Accelerator usage is usually reduced. The UF method eliminates the requirement for the massive storage tanks historically used for heating and preparation the curds, leading to a saving in each capital investment and energy prices. Visible of the actual fact that the many new cheeses are made exploitation UF.

ii. **Concentration of Whey Proteins:** By far the foremost effective application of membrane processes within the dairy business is that the production of a WPC by UF. UF membranes, with associate acceptable cut-off (10–20 kg mol^{-1}) is familiar take away each the lactose and ions, yielding a retentate with a high protein concentration. The ensuing product will then be more processed by evaporation and spray drying to provide whey proteins caseins (WPC) with various interesting functions. And its purity is ranging from 35 to 85%.

In industrial processes, whey UF is performed in the main in multi-stage spiral wound systems with polysulfone membranes. Processes are presently operated at a temperature of virtually 50°C, requiring a pre-treatment so as to avoid severe fouling throughout operation. Flux is regarding doubly as high as flux at 10°C, which could be a major incentive for operation at such high temperature. However, because of the fast decrease in membrane costs, that alter new systems to work at a temperature of below 15°C, and in respect of microbiological quality of the top product, less temperature method is currently favored.

3. **Nanofiltration (NF) and Reverse Osmosis (RO):** Concentration of whey on its manufacture site is that the major application of RO attributable to its flexibility and energy consumption compared to vacuum evaporation. The utmost concentration which will be obtained by 18–27% dry out matter, restricted by high pressure, high retentate viscousness, calcium phosphate precipitation and lactose crystallization.

Because of the high salt content of whey, that generates various process difficulties and nutritionary imbalance in food, it became advantageous to remove whey before evaporation. Whey was demineralized within the vary 50–95% by electrodialysis and/or ion exchange, however these operations crystal rectifier to an outsized volume of polluting effluents, high investment and running prices. NF became thus industrial alternative, creating it doable to succeed in at the same time the concentration of dry matter and demineralization (25–50% to even 90% with diafiltration). The method is competitive to RO and electrodialysis. Demineralization is additional selective; NF property remains satisfactory even at the very best demineralization rate; losses of lactose and nitrogen are low. Additionally, the NF step considerably improves the technological characteristics of the concentrate and offers it higher worth.

So membrane operations give the dairy industry with reliable safe, clean, and sober processes. These operations have typically been to substitute, either whole or part, different normal dairy processes and so improve ancient processes. However, additional typically, they need created it doable for brand spanking new and original product to be created that were merely impossible before. Due to the number of analysis that has been targeted on the characterization of milk components, analytical methodologies and separation processes, it is seemingly that the final quality and production potency of the varied milk protein ingredients can increase within the close to future.

8.9 MEMBRANE TECHNOLOGY FOR EFFLUENT TREATMENT

Water contamination by serious metals, cyanides, and dyes is increasing globally and desires to be self-addressed as this may cause water inadequacy in addition as water quality. Totally different techniques are accustomed clean and renew water for human consumption and agricultural functions however they every have limitations. Among those techniques, membrane technology is promising to unravel the problems. Membrane processes have an excellent potential in waste matter treatment to boost treatment potency of waste matter treatment plants. This chapter reviews recent development in membrane technology for waste matter treatment. Differing types of membrane technologies, their properties, mechanisms benefits, limitations, and promising solutions are mentioned.

Several techniques are developed for treatment of wastewater; such strategies embrace RO ion exchange gravity and adsorption among others. Adsorption has been wide wont to take away water contaminants because of its low value, offered of various adsorbents and straightforward operation. Completely different adsorbents that are used embrace use of magnetic nanoparticles activated carbon, nanotubes, and polymer nano-composites; these will take away completely different contaminants together with heavy metals that are terribly harmful even at low concentrations. Although adsorption will take away most of water pollutants, it is some limitations like lack of applicable adsorbents with high adsorption capability and low use of those adsorbents commercially. Thus, there has been a necessity for additional economical techniques like membrane technology. Membrane separation or treatment method chiefly depends on three basic principles, specifically adsorption, and electrostatic phenomenon. The adsorption mechanism within the membrane separation method relies on the hydrophobic interactions of the membrane and therefore the solute. These interactions unremarkably result in additional rejection as a result of it causes a decrease within the pore size of the membrane. The separation of materials through the membrane depends on pore and molecule size. For this reason, varied membrane processes with completely different separation mechanisms are developed. These embrace MF, UF, NF, forward diffusion (FO), and RO.

Membrane processes like MF, NF, UF, and RO are presently used for water use, saltwater, and brackish water. Polymer-based most primarily based membranes are mostly used membrane material however as a result of polymers like polysulfone and polyethersulfone, are hydrophobic, polymeric membranes are cared-for fouling. These results in blockage of membrane pores and decrease membrane performance conjointly will increase operation price by the stringent additional cleanup method.

For polymeric membranes, surface modification of the polymer is important; such surface modification includes grafting, blending, and incorporation of nanomaterials such as TiO_2, ZnO, Al_2O_3, carbon nanotubes (CNTs), and graphene oxide.

From all membrane processes, MF removes very little or no organic matter; but, once pre-treatment is applied, accrued removal of organic material will occur. MF is used as a pre-treatment to RO or NF to reduce fouling potential. The most disadvantages of MF are that it cannot eliminate contaminants (dissolved solids). Also, the NF method is capable of removing ions that contribute extensively to the osmotic pressure hence permits operation pressures that are not up to RO. For NF to be effective pre-treatment is required for a few heavily contaminated waters; Membranes are sensitive to free chlorine. NF membrane for textile sewer water treatment, the ready membrane displayed smart removal of heavy metal ions, common salts, and dyes, showing high removal potency toward metal ions and cationic dyes.

RO is pressure-driven technique wont to take away dissolved solids and smaller particles; RO is simply permeable to water molecules. The applied pressure on RO should be enough so water may be able to overcome the osmotic pressure. The pore structure of RO membranes is way tighter than UF, they convert hard to soft water, and that they are much capable of removing all particles, bacteria, and organics, it needs less maintenance. Some disadvantages embrace the utilization of high pressure, RO membranes are valuable compared to alternative membrane processes and also are tend to fouling. RO has extraordinarily little pores and able to take away particles smaller than 0.1 nm.

Let us see the applicability of membranes separation with selected examples. Two examples are selected for the discussion is ammonia purge stream and acid recovery system:

1. **Ammonia Purge Stream-Prism Membranes:** The membrane processes are the feasible of separation of hydrogen from hydrogenation tail gas, enrichment of refinery gas and air separation. Currently, the growing request of hydrogen in many industrial processes moves forward to hydrogen source and production. However, separation of hydrogen is the most important issue in its production cycle. The first commercial application of membranes in gas separation was the hydrogen separation in the ammonia purge stream by using prism membranes. The gas stream is fed into a membrane gas separation unit with hollow fiber operating at 110–130 bars. The gas components include high concentration of hydrogen (66.5%) and nitrogen (22.2%). Membrane system is able to recover the stream at lower pressure (25–70 bars) with a hydrogen concentration up to 94% and hydrogen recovery up to 90%.

   The prism technology is also applied for other separations, i.e., methanol separation. The methanol/hydrogen stream is further treated in gas separation unit for downstream processing to enrich methanol production in retentate with hydrogen separated in permeate that can be recycled to reactor.

2. **Acid Recovery System:** It is one of the most important systems in effluent technology. Figure 8.2 shows acid recovery system. It's a continuous system of acid recovery in pickling industry. Feed containing acid and salts in water. Almost 7 to 10% HCl acid and 7% salts in feed for recovery system. Process steam (feed) send for pretreatment. Which is primary filtration system where from steam remove the suspended solids. 0.1–0.6% $FeCl_2$or oil sludge is sent to the roasting unit for formation of oxides from salt at temperature 845°C. This unit produces HCl gas for resorption/reuse and form mineral oxides for reuse. But this roasting unit is optional process in acid recovery system.

After pretreatment sludge free process steam is fed to the first separation unit where dissolved solid contains 10% HCl and 7% salts. In stage, one separation the process is optimized to get desired salt concentration. It can be done by stage-wise to avoid clogging and choking while getting desired salt concentration. Optimized to get desired salt concentration that is 60% is advisable salt concentration is sent back to the roasting unit for avoiding salt crystallization and help for some economical benefits to process.

Then the next salt free process steam is fed to the second separation unit. This unit working with such membrane separation techniques which help to get maximum acid concentration, i.e., up to 20%. It is also done by stage-wise by avoiding clogging and chocking in the process while getting desired acid concentration. This separated acid with concentration of 20% can be recycled to system to getting 25% acid recovery. And at the conclusion of the process, the water can be removed from the entire separation unit with may containing less than 1% acid for precleaning and rinsing before pickling. For better cleaning acidic water is one of the great options.

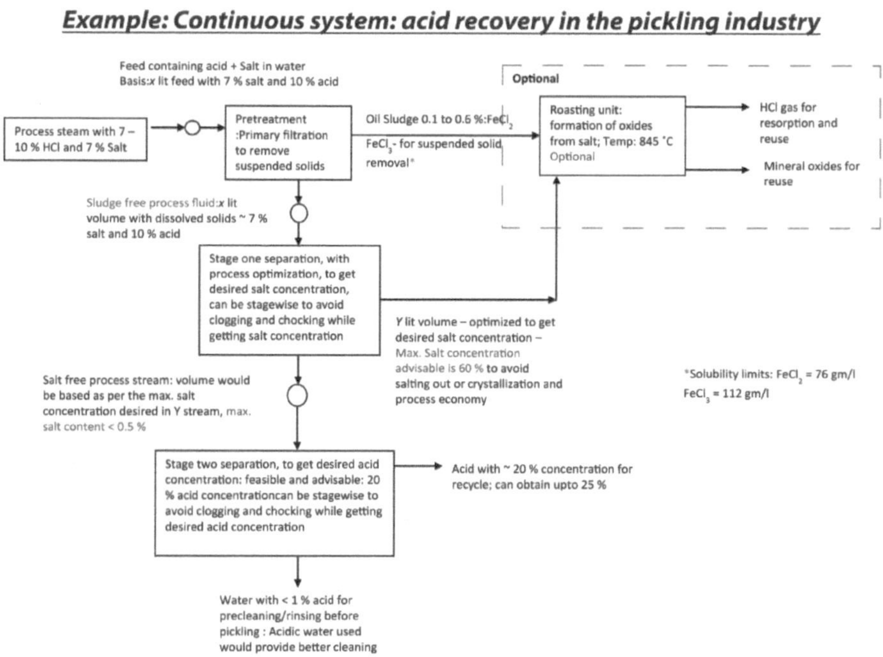

FIGURE 8.2 Acid recovery in the pickling industry.

8.10 ECONOMIC BENEFITS BY COMBINATION WITH MEMBRANE PROCESSES

As discussed in this chapter, membrane technology is very helpful to environment. It helps to reduce the waste, very less consumption of energy though it reduces operating cost, also its environmental friendly and clean technology. Membrane processes show improvement in reaction rate to the formation of desired product. Membrane processes done separations continuously and separation at milder conditions. Easy to recycle and recover the unconverted components with higher conversion and it also helps to improve the product quality. However, the membrane system helps to reduction in waste, losses, and improved component usage for further processes. Automatically these all help to reduction in economy. Some economical benefits with combination of membrane processes are explained with some examples below:

1. **Fermentation-Based Biomolecule Synthesis-Acid, Alcohol, and Biofuel:** Fermentation technology is widely used to produce various economically important compounds which have applications in the energy production, pharmaceutical, chemical, and food industry. Fermentation is highly efficient technique for synthesis of chemicals. It utilizes biological feedstock, provides highly selective functionality. But the issue with the fermentation technique is limitations for higher concentrations and its separation. Also, large waste generated during separations, e.g., separation of LA generates 1 ton of waste / ton LA, and for CA it is 2 ton.

   Lactic acid (LA) is an alpha-hydroxy acid with twin practical teams, creating it appropriate to be used during a sort of chemical transformations and products. LA is employed globally for applications in food, pharmaceutical, textiles, cosmetics, and chemical industries. There are two processes utilized in LA production: chemical synthesis and carbohydrate fermentation. Chemical synthesis produces solely a ceramic mixture of LA using acetaldehyde as a beginning material. Fermentation is most well-liked to supply LA, and concerning 90% of LA is created by fermentation. In fermentation, the substrates for LA production will be renewable and might utilize low-value raw materials like starchy materials, sugarcane, whey, and glycerin. Concerning 90% LA is created by fermentation technique however, this method has not been additionally profitable as a result of the high value in separation method. To deal with this downside, the event of additional economical and viable separation technologies is required. Therefore, that is completed with membrane separation.

   Since 1960, membrane separation processes have been recommended as a substitute for LA extraction. Advances in separation and purification based on membranes, particularly MF, UF, and electrodialysis technology allowed the manufacture of new production processes, which do not produce a salt residue, instead of the traditional precipitation process of LA.

Electrodialysis is a separation method employing selectively permeable cationic and anionic exchange membranes. These membranes are alternately arranged between the cathode and the anode; by spread over an electric potential between the electrodes, cations migrate to cathode and anions to the anode. Membranes used in electrodialysis are polymeric, non-porous, and have a thickness between 10 and 500 μm. Electrodialysis is applied to remove salts from solutions or to concentrate ionic substances. Membranes can operate at approximately 80% of theoretical thermodynamic efficiency.

So this concludes that the precipitation is that the typical technique for LA recovery, however it is unattractive for economic and environmental viewpoint as a result of it generates LA with low purity and solid waste residue. Also, solvent extraction has the advantage of not generating heaps of waste because of the precipitation technique. However, it needs high exchange space, high price with solvent recovery, and high toxicity of extractants, that limits the large-scale potential. Advances in solvent extraction, notably with the novel extractants, have given promising results. Oppositely, membrane-based separation technologies have according to high property, high levels of purification and separation. Despite its benefits, high price of membranes, polarization, and fouling issues forestall the utilization of those processes.

2. **Ethanol Concentration-Absolute Ethanol Formation:** Most of the world's ethanol is produced by fermentation of crops (93%) with artificial ethanol (7%) being produced by direct hydration of ethane. The fermentation of plant material (e.g., barley and rice) is the route by which alcoholic drinks (e.g., beer, whiskey, gin, and vodka) are produced. It is also how bioethanol for biofuels is produced. Of the uses of bioethanol, easily the greatest significant fuel for cars. However, an increasing one is in the manufacture of ethane as a route to poly (ethane). The other main uses of ethanol as a chemical intermediate are for:

- glycol ethers;
- ethanolamine/ethylamine; and
- ethyl propenoate.

It is conjointly used as a solvent within the manufacture of cosmetics, pharmaceuticals, detergents, inks, and coatings. While bioethanol is taken into account as a sustainable energy supply, it needs any purification for uses aside from fuel. The foremost common purification technique used within the fermentation alcohol business is rectification by any distillation. However, distillation has vital disadvantages as well as high value and restricted separation capability. Although associate azeotrope could be a technique for separation with constant-boiling mixture within which the composition of the vapor is that the same as that of the liquid. Thus, the two parts cannot be separated by fractional distillation process. As imagine, it is very tough to distil this kind of substance. In fact, the foremost targeted sort of fermentation alcohol, associate azeotrope, is around 95.6% ethanol by weight as a result of pure ethanol is essentially nonexistent. Azeotropes exist in resolution at a boiling purpose specific for that part.

The separation of ethanol-water mixtures is of great significance for the production of ethanol from biomass. Together UF and PV processes are used for the continuous processing of fermentation and separation, the removal of ethanol from the UF permeate can be accomplished by PV. Separation of ethanol-water mixtures by the PV process is examined. The preparation of several membrane types (homogeneous, asymmetric, and composite) are used in PV separation technology. Dissimilar types of membrane provide altered characteristics. The self-supported polymeric membranes may provide an economic attraction in membrane production, but not in performance, due to their poor mechanical properties, low chemical resistance, and swelling. The supported membranes improve mechanical strength and separation performance. The inorganic membranes provide an outstanding separation performance, but are extremely difficult in process ability, leading to a higher price in industrial usage. The mixed matrix membranes thus provide an attractive alternative that are able to overcome all barriers while achieving both permeation flux and separation factor.

8.11 WAY INTO FUTURE-SCOPE AND BENEFITS

The discussion up till now shows large growth potential and applicability of membrane technology in green technology applications. Though multiple applications are available in current date, their applicability needs to be established with the help of research and developments and its implementation in real life situations. For that purpose, some of the targets have to be established and reached as described below:

1. **Research Targets:** Technological advances towards process and energy economy: The separation parameters and separation efficiency need to be optimized to improve the material quality. Such improved quality material can be utilized in further processes and can provide techno-economical benefits. Additionally, the research needs to be progressed for improvement in energy economy. As per an estimate, separation processes are major energy-intensive systems in process design and engineering. A reduction in energy consumption for separation can be highly beneficial towards the goal of energy economy in green technology.

   Further process modifications can be targeted towards improvement in solvent systems, which would provide efficient heat and mass transfer. This would be beneficial in improving energy economy along with processing benefits. An improvement in mass transfer can affect the reaction kinetics and process conversions. Major issues with the mass transfer controlled reactions are catalytic reactions. An improved catalyst with feasibility of its easy intrinsic separation would provide large benefits. Many of the reactions are either equilibrium reactions and require conversions after product formation. The unconverted reactants are recycled back to the system. This causes loss of energy, reaction kinetics, and conversion. Additionally use of chemicals for separation and recovery unreacted component might result in waste generation. This needs *in situ* separation of desired components. It would provide solution towards many of these issues. A carefully designed membrane system can be combined with conventional processing for *in situ* separation; can be useful in drawing these technoeconomical benefits. These types of *in situ* separations are highly essential for materials with sensitive functional groups. Their separation and further utilization would avoid need of chemical processing and contamination. Thus, avoid degradation of functionality and loss of product properties.

   In the designed membrane system, development of some synthetic methodologies is also help to technoeconomical assistances. These synthetic methodology supports to make atom efficient and benign to human health/environment. According to membrane process designed for intensive manufacturing of chemicals which required less energy to process for separation of molecules. And some of these molecules are sent back to the process. This recyclability of molecules increases the product quality and reduction in large economical requirements.

Today the waste streams of energy-producing units are some of the major polluting factors. The wastes from thermal or nuclear power plants are harmful for humans and environment based upon their sheer volume and properties. Membrane-based systems like fuel cells can be an option, which are reported to produce negligible amount of waste. Though the processes are well reported a large investigation are needed for their scale up and commercial feasibility.

2. **Implementation:** Though multiple steps are being observed towards research and development, their practical applicability is still limited. Industry is still reluctant and skeptical about accepting the technological development. The steps needed to be taken by industries and enforcement agencies for implementation of the technological advances. It can be done by taking the following measures:
   a. The steps of the process can be redesigned for improved adoption, quality enhancement, and product improvements.
   b. Develop the research programs to facilitate technology transfer.
   c. Designed and manipulated the systematic recognition of hazard/toxicity as property of molecular structure.
   d. Development and utilization of practical investigations for design and manipulation with green technology principles.

Implementation of some of these steps would help us to make the systems better.

KEYWORDS

- **economical separation**
- **effluent treatment**
- **green technology**
- **membrane separation**
- **process intensification**
- **recovery and recycle**

REFERENCES

1. Angelo, B., & Fausto, G., (2011). *Membranes for Membrane Reactors: Preparation, Optimization, and Selection.* John Wiley & Sons, UK.
2. Baker, R. W., Cussler, E. L., Eykamp, W., Koros, W. J., Riley, R. L., & Strathman, H., (1991). *Membrane Separation Systems: Recent Developments and Future Directions.* Noyes Data Corporation, USA.
3. Fendler, J. H., (1994). *Membrane-Mimetic Approach to Advanced Materials.* Springer-Verlag Berlin, Heidelberg.
4. Grandison, A. S., & Lewis, M. J., (1996). *Separation Processes in Food and Biotechnology Industries: Principles and Applications.* Wood head Publishing Ltd. England.
5. Ian, D. W., Edward, R. A., Michael, C., & Colin, F. P., (2000). *Encyclopedia of Separation Science.* Elsevier Science Ltd., Academic Press, UK.
6. Marcel, M., (1996). *Basic Principals of Membrane Technology.* Kluwer Academic Publishers, Netherlands.
7. Mark, C. P., (1990). *Handbook of Industrial Membrane Technology.* Noyes Publications, USA.
8. Mel, S., (2002). *Encyclopedia of Smart Materials.* A Wiley-Interscience Publication, John Wiley & Sons, Inc. USA.
9. Michael, C. F., & Stephen, W. D., (1999). *Encyclopedia of Bioprocess Technology: Fermentation, Biocatalysis, and Bioseparation.* A Wiley-Interscience Publication, John Wiley & Sons, Inc. Canada.
10. Neogi, P., (1996). *Diffusion in Polymers.* Marcel Dekker Inc. USA.
11. Nicholas, P. C., (1996). *Biotechnology for Waste and Wastewater Treatment.* Noyes Publications, USA.
12. Nunes, S. P., & Peinemann, K. V., (2001). *Membrane Technology in Chemical Industry.* Wiley-VCH Verlag GmbH.
13. Paul, T. A., & Mary, M. K., (2002). Origins, current status, and future challenges of green chemistry. *Acc. Chem. Res.*, *35*, 686–694.
14. Richard, W. B., (2002). *Membrane Technology and Applications.* John Wiley & Sons, Ltd., USA.
15. Winston, H., & Sirkar, K. K., (1992). *Membrane Handbook.* Van Nostrand Reinhold, Springer USA.

[illegible]

REFERENCES

[illegible]

CHAPTER 9

Green Hydrogen Energy: Storage in Carbon Nanomaterials and Polymers

BRAHMANANDA CHAKRABORTY[1] and GOPAL SANYAL[2]

[1]High Pressure and Synchrotron Radiation Physics Division, Bhabha Atomic Research Center, Trombay, Mumbai – 400085, India

[2]Mechanical Metallurgy Division, Bhabha Atomic Research Center, Trombay, Mumbai – 400085, India

ABSTRACT

One of the greatest challenges of this century is to find an alternative fuel for vehicular propulsion which can replace widely used fossil fuels. Hydrogen is one of the finest choices due to its clean energy and highly plentiful in nature. And also, it has the highest energy value per unit weight (142 MJ kg^{-1}) compared to liquid hydrocarbons (47 MJ kg^{-1}). The conventional hydrogen storage technologies, namely, high-pressure tank and liquid state storage, are not applicable due to large size and higher energy cost for liquefaction. Solid-state storage may become a viable technology provided the storage medium can absorb a large amount (~6.5 wt%) of hydrogen and can release them easily as recommended by the department of energy (DoE), USA. Recently, carbon nanomaterials have drawn immense attention from researchers due to their high surface area, high electrical and thermal conductivity, and high mechanical strength. Although for pure carbon nanostructures, the physisorption of hydrogen is negligible at room temperature, the functionalized carbon nanomaterials are the promising candidates for hydrogen storage media at ambient conditions. This chapter describes the issues and challenges for hydrogen storage in functionalized carbon nanomaterials. It will highlight the bonding, charge transfer mechanism, and hydrogen storage capability of novel carbon nanostructure (carbon nanotube, graphene, and graphyne).

Polymers are also potential hydrogen storage materials due to their pure organic nature, tunable structures, and large scale synthesis. This chapter discusses about different types of polymers such as conjugated microporous polymers, porous polymers networks, hyper cross-linked polymers, polymers of intrinsic microporosity (PIM) and their hydrogen storage capabilities.

9.1 INTRODUCTION

Everything is energy and that's all there is to it. Fuel is invariably the precursor to energy. Conventional fossil fuels are at the verge of culmination through rapid depletion with progress of civilization [1, 2] with a daily global consumption of over 89 million barrels they suffer the largest drawback being rapidly exhaustive, causing huge pollutant emission and volatile in price-controlled by political factors [3]. Essentially, that thrusts people exploring alternative fuels such as bio-diesel, bio-alcohol, hydrogen, etc. for sustaining advancement [4–6]. Precisely, rise in price and negative impact on environment for depleting fossil fuels are the chief motivation behind development of alternative fuel technologies (Figure 9.1(a)). On the other hand, alternative fuels are obviously aimed at reduction in fuel cost and emission, and off course a cleaner environment. Among all alternative fuels, hydrogen is a promising energy carrier with zero-emission while burnt with oxygen. Unlike conventional liquid fuel, it can be generated from a broader spectrum of primary stationary energy resources including nuclear and solar power being cleaner than fossil fuel; and in turn, the produced hydrogen can be utilized for generating electricity (Figure 9.1(b)) [7], or fueling vehicles either through direct burning inside an internal combustion engine (ICE) or as precursor for fuel cells [3, 8, 9]. The energy released through oxidation enables hydrogen to act as a fuel. Hydrogen storage is indispensable machinery for the progress of hydrogen and fuel cell skills in applications encompassing stationery as well as portable power [10–14]. Storage of hydrogen is one of the major technical obstacles which are being addressed by researchers worldwide for the last a few decades [15–17]. The following section would discuss about importance of hydrogen energy in brief. Thereafter, issues, and challenges regarding hydrogen storage would be narrated in detail including an ephemeral insight about various hydrogen storage means. The succeeding three sections would focus on progress in research about hydrogen storage within carbon nanomaterials and organic polymers highlighting theoretical insights behind the fundamental concepts responsible for the phenomena. Finally, the chapter would be concluded elucidating future research in this direction.

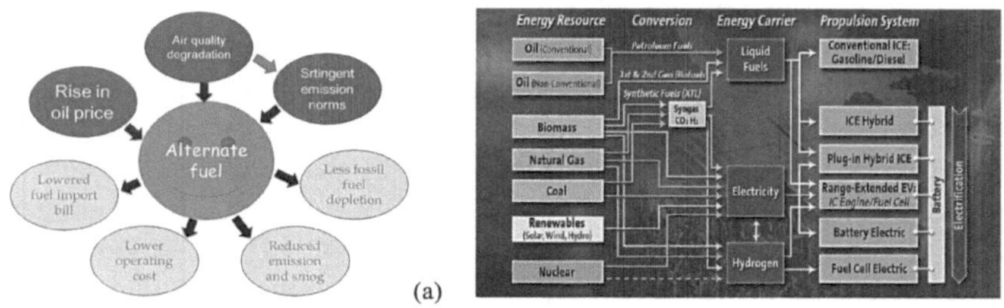

FIGURE 9.1 (a) Alternate fuels' emergence and benefits, and (b) transformation of stationary power into transportation system.

Source: b) Reprinted with permission from National Research Council [7]. © 2008 National Academies Press

9.2 HYDROGEN FUEL: IMPORTANCE, MERITS, AND DEMERITS

Hydrogen is an energy carrier which can be utilized for a broad range of applications [18–20]. Hydrogen promises huge long-term potential for utilization as fuel in usage, where presently fossil fuels are in practice-precisely for driving the vehicles, which would tender immediate benefits in terms of curbed pollution and fresher environment (Figure 9.2(a)) [21]. Among all fuels, hydrogen has the highest gravimetric energy density of 143 MJ kg^{-1} and octane number more than 130 [3]. The rock bottom flashpoint of –231°C designates hydrogen easily burnable on one hand and highly fire hazardous on the other. However, auto-ignition temperature of hydrogen is still quite high (585°C). Besides usage in automobile fuel, it has commenced to be utilized in commercial fuel cell vehicles and spacecraft propulsion [22, 23]. A fuel cell coupled with an electric motor is much more efficient than a conventional ICE operating on gasoline [24–26]. Again, hydrogen itself can serve as fuel for ICEs. Energy generated from 1 kg of hydrogen is equivalent to that generated from 2.8 kg of gasoline [27]. The world's dependence on hydrogen energy exhibits a steady rising trend since the past a few decades (Figure 9.2(b)) [28].

Molecular hydrogen is unfortunately unavailable on earth in expedient innate reservoirs [29]. Most hydrogen on globe is bonded to oxygen in water and to carbon in surviving or dead and/or fossilized biomass. Hydrogen must thus be separated [19] from any of its compound using primary energy source such as solar energy [30], biomass [31], electricity [32], or hydrocarbons [33] (Figure 9.1(b)). If it is produced by splitting water into hydrogen and oxygen [34–36], water is again obtained, when hydrogen is burnt without any formation of any harmful substance unlike conventional fuel. Alternatively, hydrogen may be obtained by breaking the chemical bonds from a host of chemical compounds [39, 40]. Conventional natural gas reforming process for hydrogen production induces significant environmental impacts for huge emission of carbon dioxide [41]. Greener processes are still not technologically that advanced to become commercially viable at large scale. Comprehensively, generation of hydrogen is yet an expensive and energy-intensive process [42]. Being gaseous and extremely lightweight, it is not easy to store enough hydrogen through any easy and convenient way for usage as per needs. In other words, containing good amount of hydrogen is a costly affair till date for its escaping tendency. Hence, researchers have been endeavoring to find suitable storage means for hydrogen especially for mobile applications since past a few decades.

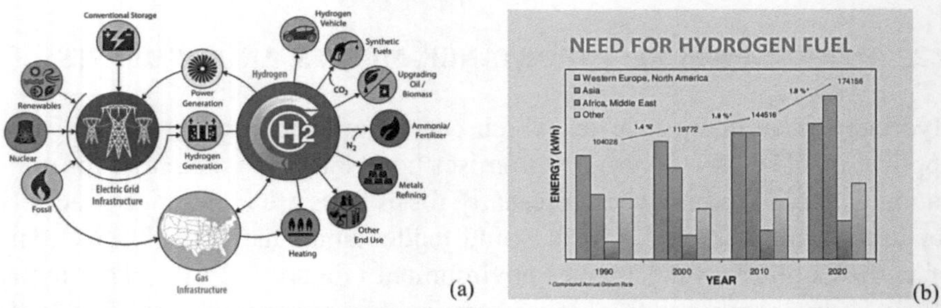

FIGURE 9.2 (a) Potential for wide-scale hydrogen production and utilization (From https://www.energy.gov/eere/fuelcells/h2scale, with permission), and (b) worldwide emerging demand of hydrogen fuel (From https://www.slideshare.net/SridharSibi/hydrogen-fuel-amp-its-sustainable-development, with permission).

9.3 HYDROGEN STORAGE: CHALLENGES AND ISSUES

Being carrier of energy, hydrogen storage technically means storing energy. Depending on functionality, hydrogen storage technologies may be classified into five major groups: (1) high-pressure gas storage, (2) liquid hydrogen, (3) physically bound hydrogen, (4) chemically bound hydrogen, and (5) hydrolytic evolution of hydrogen [11]. In view of application, all hydrogen storage technologies fall into either of these two domains: stationary storage and portable storage or on-board storage or storage for mobile applications. High pressure gas storage is regarded as the most common hydrogen storage solution till date. Liquid hydrogen although being quite mature technology, needs very low temperature for liquefaction. Among all forms of hydrogen, liquid hydrogen bestows both the highest volumetric and gravimetric storage density (Figure 9.3(a [43], b [44])) The above two solutions are very much suitable for stationary applications such as underground gas storage to provide grid energy availability for intermittent energy sources or 'power to gas' technology to translate electrical power to a gas fuel. Hydrogen energy density is not a great concern for stationary applications.

On the other hand, for on-board hydrogen storage for fueling a vehicle, energy density is very much critical. The existing options for hydrogen storage need large storage volume that turns them impractical for portable as well as stationary applications. Portability is one of the greatest challenges in the automotive engineering, where high density storage systems are infected with problems for safety concerns. Fuel cell-powered motor vehicles are required to offer driving range, more than 300 miles and this is impossible to attain with traditional storage methods. Secondly, both liquefaction and gas pressurization are too expensive being very much energy-intensive for mobile applications. Physically bound hydrogen essentially refers to adsorption of molecular hydrogen at solid surface through Van der Waals interaction. Hydrogen in that case is thought to be stored within that matter. Similarly, chemically bound hydrogen is that which is captured through chemical reaction with the carrier material. Hydrolytic evolution of hydrogen refers to any chemical or electrochemical process involving water, capable of releasing hydrogen. These technologies are collectively termed as storing hydrogen in solid state. Most current research into hydrogen storage for mobile applications is concentrated on accumulating hydrogen as a low-weight, compact energy carrier. A long-term objective set by the Fuel Cell Technology Office engrosses usage of nano-materials to perk up maximum range. The DoE has set the aims for onboard hydrogen storage for light cars (Table 9.1) [45]. The list of necessities includes parameters involving operability, volumetric, and gravimetric capacity, cost, and durability. These targets have been placed as the goal for a multiyear research plan anticipated to offer a substitute to fossil fuels. These targets or required properties are closely linked with three sets of parameters: structure (crystalline form, particle size or specific surface area, and structural defects), chemistry (components, phases, composition, catalysts, and impurities), and reaction or diffusion path (elementary reactions, transient species, etc.). These are mutually interacting and the critical issues vary for different types of hydrogen storage materials (Figure 9.3(c)) [46].

Solid-state hydrogen storage essentially requires hydrogen gas to be in contact with solid. The pressure can be markedly lower compared to gas storage but will evidently depend on the operative features of the storage technology. Particularly, service temperature ranges distinguish solid storage from the gaseous one. Control of adsorption-desorption cycles in solid storage devices is most often managed thermally, hence operating temperatures could surpass the near-ambient temperatures emblematic for hydrogen gas storage. The volumetric and gravimetric storage densities are directly influenced by hydrogen storage material properties together with physical, thermodynamic, and chemical/kinetic. Extra influences on energy densities include, but not limited to, properties of the containment material, operational environment, available energy, and required hydrogen delivery rates (Figure 9.3(d) [11]). Two more concerns are often associated with solid-state hydrogen storage devices. The first one is stress in the containment material persuaded by expansion of the solid storage media during absorption or desorption of hydrogen. Such stresses may, in principle, be rather large and should be well-thought-out in addition to the stress associated with gas pressure. The second consideration is probable chemical interaction between the solid storage media and containment material that could lead to poisoning, corrosion of the solid storage media, and aggravated hydrogen embrittlement of the structural metal.

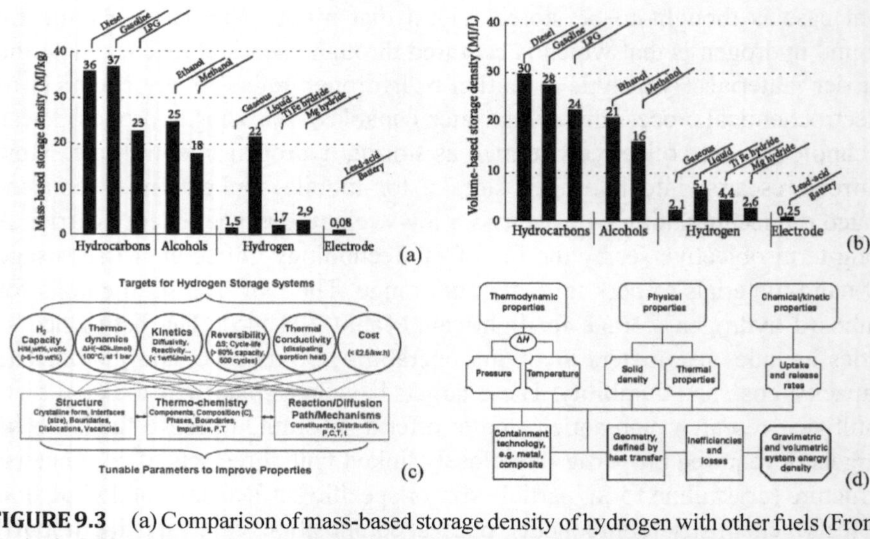

FIGURE 9.3 (a) Comparison of mass-based storage density of hydrogen with other fuels (From Zacharia [43], with permission), (b) comparison of volume-based storage density of hydrogen with other fuels (Reprinted with permission from Das [44]. © 1996 Elsevier), (c) Practical targets for on-board hydrogen storage in relation to key materials parameters (Reprinted with permission from Guo, Shang, and Aguey-Zinsou [46]. © 2008 Elsevier), and (d) the hierarchical influence of various hydrogen storage material properties on the system characteristics and efficiencies (Reprinted with permission from Walker [11]. © 2008 Elsevier.)

TABLE 9.1 Technical System Targets: Onboard Hydrogen Storage for Light-Duty Fuel Cell Vehicles

Storage parameter	Units	2020	2025	Ultimate
System Gravimetric Capacity				
Usable, specific energy from H_2 (net useful energy/max system mass)	kWh/kg (kg H_2/kg system)	1.5 (0.045)	1.8 (0.055)	2.2 (0.065)
System Volumetric Capacity				
Usable energy density from H_2 (net useful energy/max system volume)	kWh/L (kg H_2/L system)	1.0 (0.030)	1.3 (0.040)	1.7 (0.050)
Storage System Cost				
Storage system cost	\$/kWh net (\$/kg H_2)	10 (333)	9 (300)	8 (266)
Fuel cost	\$/gge at pump	4	4	4
Durability/Operability				
Operating ambient temperature	°C	–40/60 (sun)	–40/60 (sun)	–40/60 (sun)
Min/max delivery temperature	°C	–40/85	–40/85	–40/85
Operational cycle life (1/4 tank to full)	cycles	1500	1500	1500
Min delivery pressure from storage system	bar (abs)	5	5	5
Max delivery pressure from storage system	bar (abs)	12	12	12
Onboard efficiency	%	90	90	90
'Well' to power plant efficiency	%	60	60	60
Charging/Discharging Rates				
System fill time	min	3–5	3–5	3–5
Minimum full flow rate (e.g., 1.6 g/s target for 80 kW rated fuel cell power)	(g/s)/kW	0.02	0.02	0.02
Average flow rate	(g/s)/kW	0.004	0.004	0.004
Start time to full flow (20°C)	s	5	5	5
Start time to full flow (–20°C)	s	15	15	15
Transient response at operating temperature 10%–90% and 90%–0% (based on full flow rate)	s	0.75	0.75	0.75

TABLE 10.1 *(Continued)*

Fuel Quality				
Fuel quality (H_2 from storage)	$\%H_2$	Meet or exceed SAE J2719		
Dormancy				
Dormancy time target (minimum until first release from initial 95% usable capacity)	Days	7	10	14
Boil-off loss target (max reduction from initial 95% usable capacity after 30 days)	%	10	10	10
Environmental Health and Safety				
Permeation and leakage		-	Meet or exceed SAE J2579 for system safety	
Toxicity		-	Meet or exceed applicable standards	
Safety		-	Conduct and evaluate failure analysis	

Useful constants: 0.2778 kWh/MJ; Lower heating value for H2 is 33.3 kWh/kg H2; 1 kg H2≅ 1 gal gasoline equivalent (gge) on energy basis.

*Source:*https://www.energy.gov/eere/fuelcells/doe-technical-targets-onboard-hydrogen-storage-light-duty-vehicles. With permission.

9.4 CONVENTIONAL AND ADVANCED MATERIALS FOR HYDROGEN STORAGE

There is a host of materials which are in developmental stage for on-board usage (Figure 9.4(a)). For example, chemical storage could tender high storage performance for strong binding of hydrogen and high storage densities. But regeneration of storage material is still a big issue. Similarly, as an alternative to chemical storage mentioned above, there are many physical means with which hydrogen can be stored. Hydrogen remains in that case in physical forms, i.e., as gas, adsorbate, molecular inclusions, or supercritical fluid. Theoretical limitations, as well as experimental findings, are considered [47] concerning the gravimetric and volumetric capacity of the storage media, as well as safety, refilling-time demands, etc. The following discussion shall be restricted to only important solid-state gas storage materials having promising track record or potential for on-board usage.

A huge number of chemical storage systems are under study, involving hydrogenation/dehydrogenation reactions, hydrolysis reactions, ammonia,

ammonia borane and other boron hydrides, alane or aluminum hydride, etc., [48]. The most promising chemical approach is invariably the electrochemical hydrogen storage, since release of hydrogen may be controlled by applied electricity. For example, metal hydrides, such as LiH, MgH_2, $NaAlH_4$, $LiAlH_4$, $LaNi_5H_6$, $TiFeH_2$ and palladium hydride, with changeable degrees of efficiency, can be utilized as a storage method for hydrogen, often reversibly [49]. These materials have great energy density, although their specific energy is habitually worse than the foremost hydrocarbon fuels (Figure 9.4(b)) [50]. Nano-metal hydrides have a number of properties that turn them even better contenders for future hydrogen storage systems [51]. At nanoscale, chemical, and structural properties such as sorption site density and particle size demonstrate a significant improvement in sorption kinetics and temperature for hydrogen diffusion or release. However, the downsides of nanoscale materials comprise poor total sorption capacity and heat transfer. Hydrazine is a promising non-metal hydride for ease in handling and storage. Aluminum has been projected as an energy storage method by many researchers [52, 53]. Hydrogen can be obtained from aluminum by reacting it with water. Before that, aluminum must be exposed from its natural oxide layer. That process requires pulverization and chemical reactions with caustic substances. The byproduct, aluminum oxide, can be recycled back into aluminum with the well-known Hall-Héroult process, causing the process theoretically renewable. Graphene is chemically efficient to store hydrogen efficiently. Upon hydrogenation, graphene becomes graphane. It releases stored hydrogen at 450°C [54].

Physisorption of gases is happened by the interplay between surface area and variations in entropy and enthalpy. Though surface area is frequently used as a proxy for adsorption sites, determination of exactly available surface area for porous materials is difficult. The enthalpy of adsorption determines the temperatures at which reversible adsorption happens. For physisorption at supercritical temperatures, the density of stored hydrogen is predictable to be lower than that of liquid hydrogen. Although this limitation is not valid when chemisorption or a chemical reaction happens between hydrogen and a substrate, it is relevant to most of nanoporous hydrogen storage materials and may be used to define a saturation limit for hydrogen adsorption in such materials. Activated carbons are essentially amorphous carbon materials having high porosity with very high apparent surface area. Hydrogen physisorption can be raised in these substances by optimizing pore diameter to around 7 Å with a corresponding increase in apparent surface area [55]. They can be recycled from waste materials like cigarette butts, indicating economic mileage [56]. A clathrate is a chemical substance having a lattice that contains or traps molecules. They are able to cage hydrogen but need very high pressure for stability. Tetrahydrofuran has been demonstrated to act as a catalyst to reduce the pressure requirement for clathrate hydrate [57]. However, they are yet to achieve DoE's target (Figure 9.4(c)) [58]. Metal-organic frameworks (MOFs) represent another class of synthetic porous matter that store energy in terms of hydrogen and at molecular level [59, 60]. These are highly crystalline organic-inorganic hybrid structures that include organic ligands as linkers and metal clusters or ions (secondary building units or SBU) as nodes. When guest molecules or solvent dwelling in the pores are detached during the process of solvent exchange through heating under vacuum, MOFs' porous structure may be achieved without subverting the frame and hydrogen molecules will be adsorbed onto the pores' surface by physisorption. Glass capillary arrays [61] and glass microsphere [62] are other notable technologies in this direction.

Apart from cost, operability, and durability issues, the three limiting factors for the use of hydrogen fuel cells (HFC) embrace size, efficiency, and safe onboard storage of the gas. To address these issues, the use of nanomaterials has been proposed as an alternative option to the traditional hydrogen storage systems. The use of nanomaterials could endow with a higher density system which is expected to enhance the driving range limit (300 miles) set by the DoE. Carbonaceous materials, e.g., graphite, carbon nanotubes (CNTs), and metal hydrides are the major focus of researchers in this direction. Yu et al. pictorially formulated schemes of three different approaches for nanomaterials to store hydrogen (Figure 9.4(d)) [63]. Carbonaceous materials are at present being considered for on-board storage systems for their multifunctionality, robustness, versatility, suitable mechanical properties and low cost comparing other alternatives [64]. However, low hydrogen capacity of CNTs (~3.0–7.0 wt% at 77 K) and high release temperature of fullerene (600°C) does not attract pure carbon nanomaterial as promising candidate for hydrogen storage. It is pertinent to note here that incorporation of nano-materials within carbonaceous materials through chemical functionalization is proven not only to be beneficial in enhancing hydrogen storage capacity but also are helpful in enrichment of other performances such as transport and catalytic properties (e.g., Nafion membranes) through improvement in hydrogen splitting kinetics by virtue of catalytic activity of nano-particles (e.g., TiO_2/SnO_2) [64]. In essence, another such promising class of materials is the organic polymers containing nanosized pores. They have added weight-advantage being comparatively lightweight. Like carbon nanomaterials, their hydrogen storage performance is also not appreciable at ambiance; however, addition of little quantum nanomaterials causes significant improvement in their storage performance. Hence, decoration of pure carbonaceous nanomaterials and organic polymers with suitable nanoparticles has virtually unlimited potential in enhancing their hydrogen storage performance.

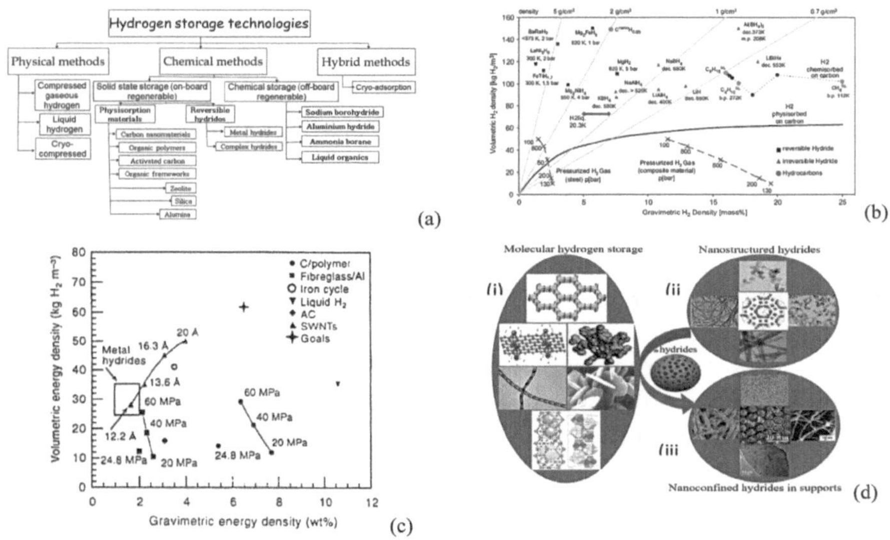

FIGURE 9.4 (a) Different hydrogen storage methods, (b) Comparison of volumetric hydrogen storage density against gravimetric for different pure storage materials (Reprinted with permission from Klell [50]. © 2010 Wiley-VCH Verlag GmbH & Co.), (c) Installed energy densities for several vehicular hydrogen storage technologies (Reprinted with permission from Dillon et al. [58]. © 1997 Springer Nature), (d) Scheme of three approaches for nanostructured materials applied for hydrogen storage. (i) Several typical nanostructures (porous carbon, graphene, BN nanotube, MOF, COF, and CMP) for molecular hydrogen storage; (ii) several nanosized hydrides (MgH2, Mg(BH4)2, Pd, Li2NH); (iii) several hydrides nanoconfined in supports (MgH2@MOF, MgH2@graphene, MgH2@carbon, NaAlH4@SBA-15, LiNH2@carbon) (Reprinted with permission from Yu et al. [63]. © 2017 Elsevier.)

9.5 HYDROGEN STORAGE IN CARBON NANOMATERIALS

Carbon nanomaterials are useful for effectual energy translation and storage. They possess unique size-/surface-dependent (e.g., morphological, mechanical, optical, and electrical) properties responsible for their remarkable performance for storing hydrogen at high pressure and low temperature in pure form by virtue of intercalation. The carbon-based materials have received outstanding consideration as potential storage materials for their low cost, good recycling characteristics, easy accessibility, low densities, and reasonably good chemical stability, wide diversities of bulk and pore structures, and amenability to synthesize variants or post-synthetically engineer traits using an extensive range of manufacturing, activation, and carbonization routes. A number of theoretical investigations confirm that proper incorporation of selective elemental nanoparticles within carbon lattice causes change in binding energy and desorption temperature resulting in great storage of hydrogen at ambient condition feasible. However, experimental efforts are still not able to tally with all theoretically ambitious predictions for practical limitations such as oxidation of decorated nano-particles, agglomeration, uneven distribution, etc. The science of the hydrogen multilayer intercalation with carbonaceous nanostructures can be understood well and further enriched in the light of so far available analytical and observational research outputs. On the other hand, many researchers do not believe in a leeway of a physical-chemical interpretation and a practice comprehension of the hydrogen multilayer intercalation with carbonaceous nanostructures at technological pressures and ambient temperature. Hence, it seems valuable to deepen and broaden the discussion in scientific publications of this genuine open question, and/or international cooperation. The following subsections essentially review progress in the research and development of carbon nanomaterials during the past two decades for hydrogen storage, along with some discussions on contests and standpoints regarding improvement in hydrogen storage performance through decoration with nanoparticles.

9.5.1 CARBON NANOMATERIALS: A BRIEF OVERVIEW

Carbon, having atomic level ability to polymerize, is one of the most appealing elements, with the capability to form an extensive range of structures, often with fundamentally different properties. Orthodox examples of carbon allotropes encompass "hard" diamond and "soft" graphite used in a wide range of products, including consumer goods in a range of human activity [65]. All nanomaterials composing of carbon atoms are termed as carbon-based or carbon nanomaterials. Classification of carbon-based nanomaterials is frequently performed as per geometrical structure. Manipulation of atomic bonding and shape of a single atomic layer of graphite may lead to formation of all different carbon nanomaterials' allotropes as shown in Figure 9.5(a). Carbon nanostructures comprise particles that can be tube-shaped, horn-shaped, spherical, or ellipsoidal. Nanoparticles having the shape of tubes are termed as CNTs whereas horn-shaped particles are nano-horns or spheres or ellipsoids belong to the group of fullerenes (Figure 9.5(b)) [66].

Fullerenes, discovered in 1985, are an allotropic alteration of carbon, often termed as a molecular form of carbon, or carbon molecules [67]. The fullerene family comprises a number of atomic C_n clusters (n> 20), composed of carbon atoms on a spherical surface. The carbon atoms are typically located on the surface of the sphere at the vertices of hexagons and pentagons. The carbon atoms are habitually present in the sp^2-hybrid form and connected together by covalent bonds. Fullerene C_{60} is the most familiar and best-investigated fullerene. The spherical molecule is highly symmetric and contains 60 carbon atoms, situated at the vertices of twelve pentagons and twenty hexagons. The diameter of fullerene C_{60} is 0.7 nm [68].

CNTs, discovered in 1991, are one of the carbon allotropes with outstanding properties suitable for technical applications. These are characterized by cylindrical structures with a diameter of numerous nanometers, consisting of rolled graphene sheets. These may vary in length, diameter, chirality (symmetry of the rolled graphite sheet) and the number of layers. As per their structure, CNTs may be classified into two major groups: single-walled nanotubes (SWCNTs) and multi-walled nanotubes (MWCNTs). Some researchers in addition isolate double-walled carbon nanotubes (DWCNTs) as a detach class of CNTs. Generally, SWCNTs have a diameter around 1 to 3 nm and a span of a few micrometers. MWCNTs have a diameter of 5 to 40 nm and a length around 10 μm. However, recently synthesis of CNTs with a length of yet 550 mm has been reported [69]. The organization of CNTs leads to excellent properties with a sole combination of elasticity, strength, and rigidity compared with other fibrous competitors. For instance, CNTs display considerably higher aspect ratios (length to diameter ratios) than other materials and superior aspect ratios for SWCNTs as compared with MWCNTs due to their smaller diameter. In addition, CNTs show high electrical and thermal conductivity compared to other conductive resources. Electrical properties of SWCNTs depend on their chirality, i.e., hexagon orientation with respect to the tube axis. Consequently, SWCNTs are classified into three sub-classes: (i) armchair (electrical conductivity > copper), (ii) zigzag (semiconductive properties), and (iii) chiral (semiconductive properties). By contrast, MWCNTs having multiple carbon layers, often with variable chirality, can exhibit astonishing mechanical properties instead of exceptional electrical characteristics.

Graphene is a two-dimensional allotropic form of carbon, shaped by single layers of carbon atoms [70, 71]. Here, carbon atoms exhibit sp^2-hybridization linked by σ- and π-bonds in a two-dimensional hexagonal crystal lattice with a separation of 0.142 nm between adjoining atoms of carbon hexagons.

Graphene also essentially represents a structural element of some other carbon allotropes, such as graphite, fullerenes, and CNTs. Theoretical studies on graphene started a long time before the real material samples were obtained. Comprehensive studies on the properties of graphene are still ongoing. Graphene has many exclusive physical properties, such as very high mechanical rigidity and an elevated thermal stability. The electric properties of this carbon allotrope are fundamentally dissimilar to the properties of three-dimensional materials. Though theoretically physisorption (0.01–0.06 eV) of molecular hydrogen on graphene sheet is much less energy-intensive than chemisorption (~1.5 eV), the achievable gravimetric density through chemisorption (8.3%) is much higher than physisorption (3.3%) (Figure 9.5(c)) [72]. Carbon nanotube can be considered as a curved form of graphene having added advantage of more hydrogen storage potential for its curvature.

Besides these three real carbon nanomaterials mentioned above, there are a number of theoretical two-dimensional carbon nanomaterials such as graphyne [73], graphdiyne [74] pentagraphene [75], etc., having superior hydrogen storage properties under study.

9.5.2 HYDROGEN STORAGE IN PURE CARBON NANOMATERIALS

The physisorption of hydrogen along high-surface carbons has clarified one focal point in view of ease of hydrogen uptake and release [76, 77]. A range of carbon materials have been studied both theoretically (e.g., fullerene, grapheme, etc.) and experimentally (e.g., carbon nanotube, carbon nanofibers, etc.). In transcending from an assortment of porous activated carbons to the intermediary class of MOFs, hydrogen uptake measurements at 77 K had recognized that up to 5 wt% and 7.5 wt% hydrogen can be stored in porous carbons and MOFs, respectively [78]. A preceding disclosure by Chambers et al. which obscured that CNFs could physisorbed up to 67.55 wt% hydrogen, prompted a remarkable renaissance of research related to potential adsorbents [79]. Amongst the diversity of investigations that pursued, some entailed the exploitation of carbon materials using metal dopants while others related to the synthesis of high purity carbons of diverse geometrical structures [58, 80]. Despite so many efforts, none of the carbon nanomaterials could be formed to store hydrogen at a level comparable to the alleged findings of Chambers or the target value specified by the DOE. Similarly, carbon materials had again elicited great expectations after the publicized results of Dillon et al. [58]. In particular, the account of Dillon enthused a significant and apparently somewhat inconsistently-concluding number of studies utilizing single-walled CNTs (SWCNTs), multi-walled CNTs (MWCNTs), CNFs, and carbon nanohorns (CNHs), which had been prepared using chemical vapor deposition (CVD), laser vaporization or arc discharge methods. Nevertheless, more reliable, self-consistent results have been published later with certain carbon nanomaterials [81–83]. A wide variation of uptake values, as found through theoretical calculation and practical experimentations ranging from insignificant to large, is depicted herein for various carbon adsorbents (Table 9.2) [78, 84–114].

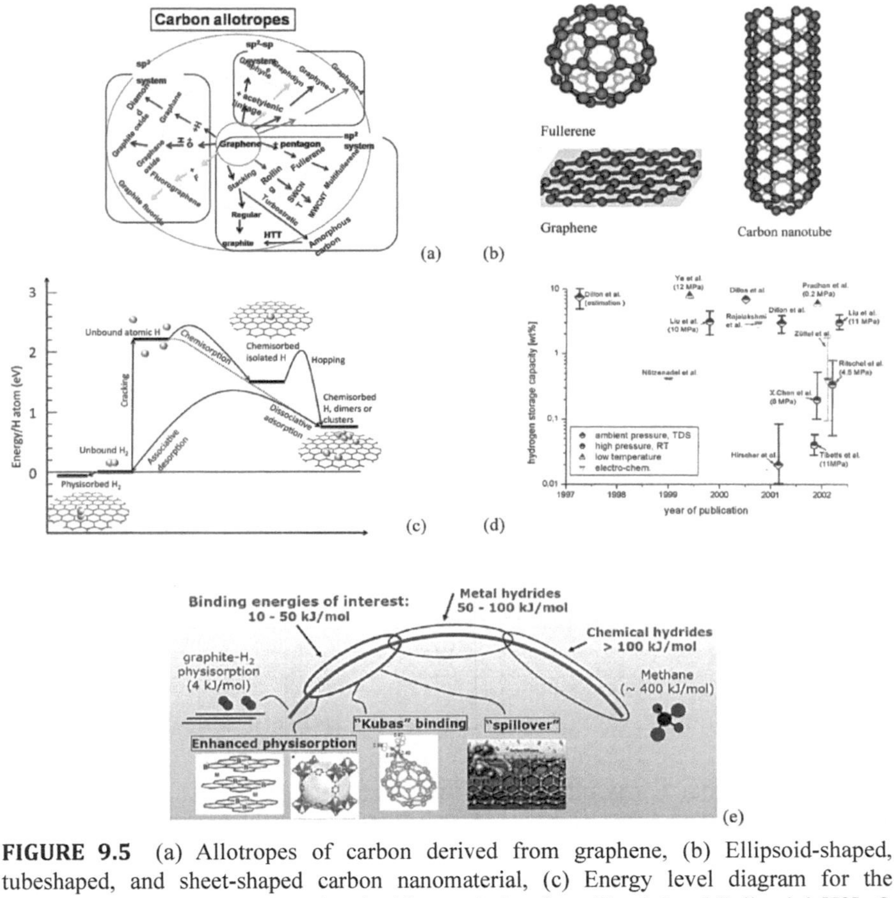

FIGURE 9.5 (a) Allotropes of carbon derived from graphene, (b) Ellipsoid-shaped, tubeshaped, and sheet-shaped carbon nanomaterial, (c) Energy level diagram for the graphenehydrogen system (Reprinted with permission from Tozzini and Pellegrini [72]. © 2013 Royal Society of Chemistry), (d) Year-on-year experimental scatter in hydrogen storage for single-walled carbon nanotubes (Reprinted with permission from Hirscher et al. [96]. © 2003 Elsevier), and (e) Favorable theoretical binding energy regime for easy adsorption and retention of hydrogen (Reprinted with permission from Ströbel et al.[116]. © 2006 Elsevier).

The experimental findings regarding hydrogen storage performance are often less reliable for lack of reproducibility, as judged by many researchers [115–118]. The noticeable fact is that most experimental efforts are visible for CNTs and nanofibers. It appears from Figures 9.5(d) [119] that since end of the last millennium, there is an increasing trend in efforts towards reproducing hydrogen storage results for various carbon nanomaterials, e.g., SWNTs, resulting in considerable scatter in experimental findings. The chief reasons behind such diverse observation are variation of sample quality in terms of purity across different laboratories and variety of processing routes for preparation and characterization of the samples. Among them, Chamber's [78] result concerning gravimetric hydrogen storage is abysmally high (~67 wt%) and that surpasses even promising theoretical estimates too by many folds. The reason behind this observation is attributed to expansion of spacing of graphite layers for excessive pressure generated by high hydrogen pressure and water vapor. However, no other research group till date could claim reproducibility of obtaining such high performance and thus doubt exists about authenticity of purity and characterization of such samples as pointed out by researchers [115, 116]. Most theoretical findings predict hydrogen uptake by carbon lattice to be around 8 wt%, however considering experimental efforts, practical achievable hydrogen adsorption is often less than 1 wt% at ambiance. With lower temperature and higher pressure however, better results could be achieved [100, 106], sometimes even exceeding the theoretical prediction. This is so as for lower temperature and higher pressure the binding energy is more favorable for hydrogen physisorption in pure carbon nanomaterials.

9.5.3 HYDROGEN STORAGE IN FUNCTIONALIZED CARBON NANOMATERIALS

Chemical functionalization of carbon nanomaterial is beneficial in enhancement of hydrogen uptake at similar condition and possibility of retention of hydrogen at higher temperature and lower pressure by bringing the binding energy into the favorable regime of 10–50 kJ mol^{-1} (Figure 9.5(e)) [116]. Fundamentally, this happens due to either extension of the interlayer distance of graphite lattice planes, and an equivalent increase in specific surface area [120, 121] through incorporation of alkali metal atoms or metal halides, or charge transfer from molecular hydrogen (the highest occupied orbital) to empty transition metal orbitals and a subsequent back-donation from the filled d orbitals into the lowest unoccupied orbital of the ligand [122–125], or dissociation of hydrogen into atomic form in presence of catalyst favoring local chemisorption, popularly termed as spillover [126–128]. Significant theoretical and experimental findings concerning hydrogen adsorption in decorated carbon nanomaterials have been shown in Table 9.3 [94, 96, 129–153].

TABLE 9.2 Hydrogen Storage Performance of Pure Carbon Nanomaterials through Physisorption Observed by Researchers Worldwide

Carbon Nanomaterial Type	Storage Condition		Hydrogen Storage Capacity (wt%)	Remarks	References
	Temperature	Pressure			
Fullerene	300 K	Ambient	>8	Theoretical; fullerene nanocage	[84]
	300 K	Ambient	~8	Theoretical; charged fullerene	[85]
Graphite	773 K	3.5 MPa	1.6	Experimental	[86]
	300 K	70 MPa	8	Theoretical; fullerene-intercalated graphite	[87]
	77 K	20 bar	1.2	Experimental; exfoliated graphite	[88]
Graphene	300 K	1 MPa	7.4	Experimental	[89]
	Ambient	Ambient	Up to 8	Theoretical; graphene multilayer	[90]
	Ambient	Ambient	7.7	Theoretical	[91]
Activated carbon	293 K	3 MPa	0.39–0.53	Experimental	[92]
Activated carbon fiber	298 K	10 MPa	2.29	Experimental	[93]
Carbon nanotube	Ambient	0.067 MPa	3.5–4.5	Experimental; High purity single-walled	[94]
	Ambient	4.8 MPa	1.2	Experimental; Purified SW	[95]

TABLE 10.2 *(Continued)*

Carbon Nanomaterial Type	Storage Condition		Hydrogen Storage Capacity (wt%)	Remarks	References
	Temperature	Pressure			
	Ambient	0.9 MPa	0	Experimental; Ball-milled SW	[96]
	80 K	~7 MPa	8.25	Experimental; High purity SW	[97]
	295 K	0.1 MPa	0.93	Experimental; Impure SW	[98]
	~300–700 K	Ambient	0.25	Experimental; Purified MW	[99]
	300 K	1.0 MPa	13.8	Experimental; Acid treated MW	[100]
	298 K	3.6 MPa	0.03	Experimental; Low purity MER MW	[101]
	298 K	10 MPa	0.68	Experimental; Random orientation MW	[102]
Nanoporous carbon material	77 K	3.8 MPa	6	Theoretical; quasi-periodic icosahedral	[103]
	Ambient	207 bar	8.13	Experimental; valid for a pore density of 1 cm^3/g and size of 0.9 Å	[104]
Carbon nanofiber	77–300 K	1.5 MPa	1–1.8	Experimental; Purified GNF herringbone	[105]
	77 K	12 MPa	12.38	Experimental; CNF	[106]
	93 K	10 MPa	1	Experimental; CNF	[107]
	298 K	11.35 MPa	53.68	Experimental; GNF platelet	[78]
	298 K	3.6 MPa	<0.1	Experimental; Vapor grown	[101]
Ordered mesoporous carbon	298 K	30 MPa	2.14	Experimental	[108]

TABLE 10.2 *(Continued)*

Carbon Nanomaterial Type	Storage Condition		Hydrogen Storage Capacity (wt%)	Remarks	References
	Temperature	Pressure			
Templated carbon	298 K	10 MPa	1.43	Experimental	[109]
Carbon xerogel	298 K	20	1.0	Experimental	[110]
Carbon sphere	298 K	9	0.25–0.43	Experimental	[111]
Carbon nanosheet	298 K	0.11	0.1–0.26	Experimental	[112]
Carbon cloth	298 K	90 bar	0.5	Experimental	[113]
Carbon nanoscroll	293 K	10 MPa	1.21	Theoretical	[114]

Comparing with Table 9.2, it can be seen that decoration with metal nanoparticle is often capable of causing the adsorption at room temperature and normal pressure feasible which is otherwise almost impossible for pure carbon nanomaterials. A good review highlighting mechanism responsible for enhancement of adsorption by heteroatoms is available [154]. As explained there, heteroatoms work as activation centers to activate hydrogen in a facile manner. To turn that process effective, heteroatoms must be incorporated both geometrically and chemically within carbon network. Efforts in early finding of potential of heteroatoms in tailoring binding energy and desorption temperature of carbon nanomaterials is quite vast, although not entirely exhaustive. A few remarkable recent developments will be highlighted here. A single zirconium atom attached on graphene surface can absorb maximum of 9 H_2 molecules with mean binding energy of 0.34 eV and typical desorption temperature of 433 K leading to a wt% of 11 (Figure 9.6(a)) [155]. Again, calcium atom attached on graphyne surface can absorb up to a maximum of 6 molecular hydrogen (H_2), with a uniform binding energy of ~0.2 eV/H_2, leading to 9.6 wt% of hydrogen (Figure 9.6(b)) [156]. Similar promising results are achieved for CNTs decorated with yttrium (Figure 9.6(c)) [157] and titanium (Figure 9.6(d)) [130]. However, in most cases, experimental observational values are often much less than theoretical estimates. This is because in practical situation it is frequently impossible to reproduce the theoretical conditions due to purity issues, oxidation problems, and atomic clustering of the decorating elements resulting in uneven distribution within the carbonaceous support.

TABLE 9.3 Hydrogen Storage Performance of Functionalized Carbon Nanomaterials Observed by Researchers Worldwide

Carbon Nanomaterial Type	Storage Condition		Decorating Element	Hydrogen Storage Capacity (wt%)	Remarks	References
	Temperature	Pressure				
Fullerene	Ambient	Ambient	Ca	>8.4	Theoretical	[129]
	Ambient	Ambient	Ti	7.5	Theoretical	[130]
Graphene	77 K	Ambient	Zirconia	1.2	Experimental; Reduced GO	[131]
	298 K	4 MPa	Pt and N	4.4	Experimental	[132]
	77 K	10 bar and above	Li	>5	Theoretical; Pillared Graphene Oxide	[133]
	Ambient	Ambient	Ca	~5	Theoretical; zigzag graphene nanoribbon	[134]
	298 K	100 bar	P	2.2	Experimental	[135]
	298 K	3 MPa	Pt	1.4	Experimental	[136]
	298 K	30 bar	Pd3Co	4.6	Experimental; boron-doped	[137]
	Ambient	20 bar	Li	6.5	Theoretical; pillared graphene sheet	[138]
	Ambient	Ambient	Li	12.8	Theoretical	[139]
	Ambient	Ambient	Al	13.79	Theoretical	[140]
Carbon nanotube	473–673 K	0.1 MPa	Li	20	Experimental	[141]
	<313 K	0.1 MPa	K	21	Experimental; Wet H_2	[142]
	473–663	0.1 MPa	Li	0.7–4.2	Experimental; 10–40% pure	[143]

	Ambient	0.067 MPa	Ti-Al-V	6	Experimental; Sonicated	[94]
	Ambient	0.08 MPa	Fe	<0.005	Experimental; Purified	[96]
	298 K	15.85 atm	Ni	0.812	Experimental; multiwalled	[144]
		12.64 atm	Cu	2.59		
		15.45 atm	Agglomerated Fe	0.909		
	298 K	32 bar	Pd, N	1.25	Experimental; Graphene Nanoplatelets	[145]
	283–335 K	Ambient	Ca	1.05	Experimental; Multiwalled	[146]
			Co	1.51		
			Fe	0.75		
			Ni	0.4		
			Pd	7		
	Ambient	80 bar	B	<2	Experimental	[147]
	Moderate	Ambient	K	4.47	Experimental	[148]
Carbon nanofiber	298	50 bar	Ni	0.56	Experimental decorated CNF grown on ACF substrate	[149]
Porous carbon	298 K	100 bar	P	1.2	Experimental	[150]
Graphyne	Ambient	-	Ca	7	Theoretical	[151]
Graphitic carbon nitride	298 K	4 MPa	Pd	2.6	Experimental	[152]
Superactived carbon	298 K	10 MPa	Pt	1.46	Experimental	[153]
Carbon nanoscroll	293 K	10 MPa	Li	3.78	Theoretical	[114]

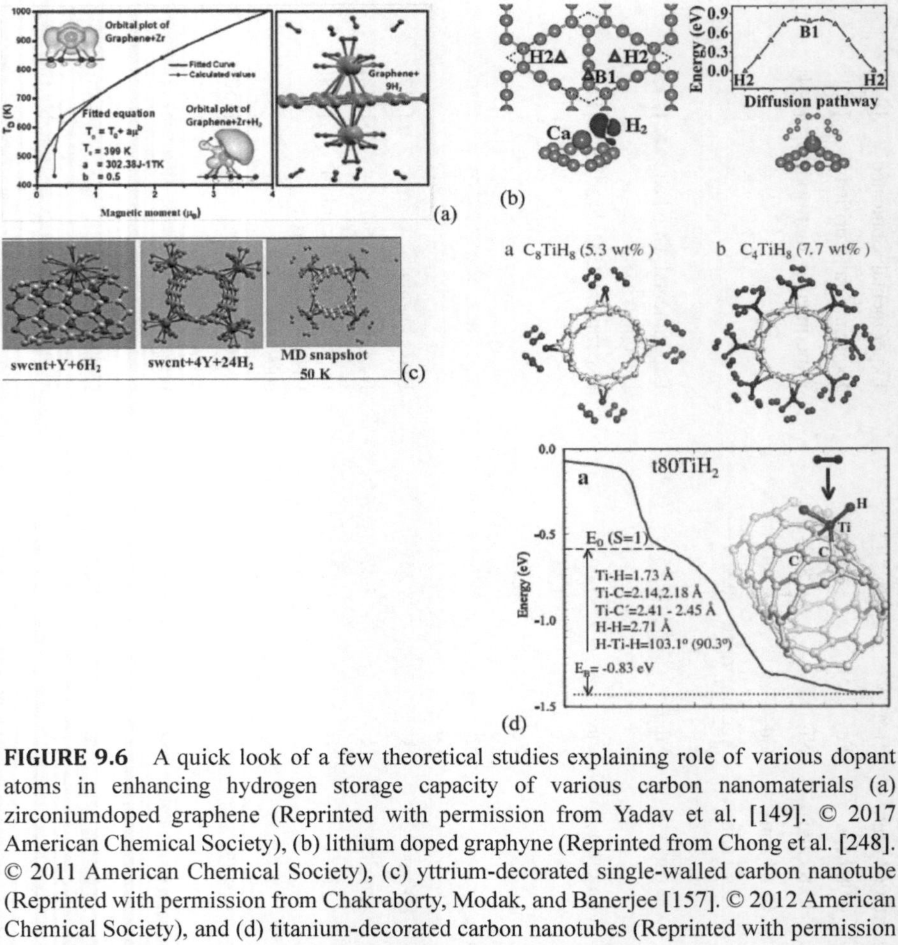

FIGURE 9.6 A quick look of a few theoretical studies explaining role of various dopant atoms in enhancing hydrogen storage capacity of various carbon nanomaterials (a) zirconiumdoped graphene (Reprinted with permission from Yadav et al. [149]. © 2017 American Chemical Society), (b) lithium doped graphyne (Reprinted from Chong et al. [248]. © 2011 American Chemical Society), (c) yttrium-decorated single-walled carbon nanotube (Reprinted with permission from Chakraborty, Modak, and Banerjee [157]. © 2012 American Chemical Society), and (d) titanium-decorated carbon nanotubes (Reprinted with permission from Yildirim and Ciraci [130]. © 2005 American Physical Society.)

9.6 HYDROGEN STORAGE IN POLYMERS

Organic hydrides or liquid organic hydrogen carriers which fix hydrogen by chemical or covalent bonds' formation have been commercialized in existing infrastructure for oil storage [158]. But they fail to fulfill the requisites for being a candidate hydrogen transport material for many safety risks like volatility, flammability, and toxicity. In search of a safer alternative, a few solid-state materials such as metal hydrides [159], porous materials [160] (e.g., activated carbon, metal-organic frameworks (MOFs)) seem to be quite preferable. However, each of them has characteristic limitations over their specific advantages. For example, recovery of hydrogen from metal hydrides is an energy-intensive process involving high temperature. Again, requirement of high-pressure conditions for storing sufficient quantum of hydrogen turns them less attractive. Keeping above in mind, solid formed organic polymers of categories like microporous polymers, hyper-cross-linked polymers, and polymers with intrinsic micro-porosity also find option for in-depth analytical and exploration research for identifying their candidature as efficient hydrogen storage material. This section will focus on efficacy of organic polymers in storing hydrogen in detail. The following subsection would cover a brief and general overview about organic polymers and the other subsequent section would describe progress in research on this frontier including perspectives of decoration with heteroatoms in nanoparticulate form.

9.6.1 ORGANIC POLYMERS: A BRIEF OVERVIEW

A polymer is a substance having a molecular structure composed of essentially from a huge number of collectively bonded similar units [161]. Polymerization is the process by which macromolecules are formed by linking together monomer molecules through chemical reactions [162]. For their organic nature, the polymeric frameworks consist of light elements providing a burden-advantage in many purposes. The long-chain nature sets polymers apart from other materials and gives rise to their characteristic properties. Depending on the nature of the skeletal structure they possess, polymers can be classified as linear, branched, and network or cross-linked polymers [163]. It is the skeletal structure's variation which is responsible for major difference in their properties. A higher crosslink density imparts superior rigidity to a polymer.

Solid organic polymers having visible pores are termed as porous polymers (PP) [164]. They simultaneously contain properties of both a porous material and polymer. They can easily be processed through chemical routes. Some of them may even be dissolved in appropriate solvent and then directly processed using solvent-based methods without destroying their porosity. Hypercross-linked polymers (HCPs) symbolize a class of nano-porous plastics with a broad range of practical and potential applications, e.g., gas sorption and separation, drug delivery, etc., [165]. These polymers manifest numerous significant advantages, including moderate synthetic situations, an enormous stock-room of cheap monomers, robust structures, and good chemical and thermal stabilities. Porous organic polymers (POPs) are those amorphous solids that contain interconnected pores with dispersed sizes and structures, resulting in possessing accessibly large surface areas [166]. Chemical structure of typical microporous polymers of all varieties is shown in Figure 9.7(a) [167]. All the three types of polymers described above are capable of absorbing gas and store within their enclosed intermolecular space at low temperature and high pressure in pure form. The introduction of suitable elements in limited quantity is found to enhance their gas retention properties by many folds at near ambient conditions. The hydrogen adsorption by porous polymer is a completely reversible process and thus it is dependent on applied pressure and temperature. In general, lower is the temperature and higher is the pressure, better is hydrogen adsorption. Measurement of absorption is done both volumetrically (e.g., accelerated surface area and porosimetry system) and gravimetrically (e.g., intelligent gravimetric analyzer). Hydrogen adsorbs only feebly to most surfaces, but adsorption is improved between two surfaces which are close together (i.e., within micropores).

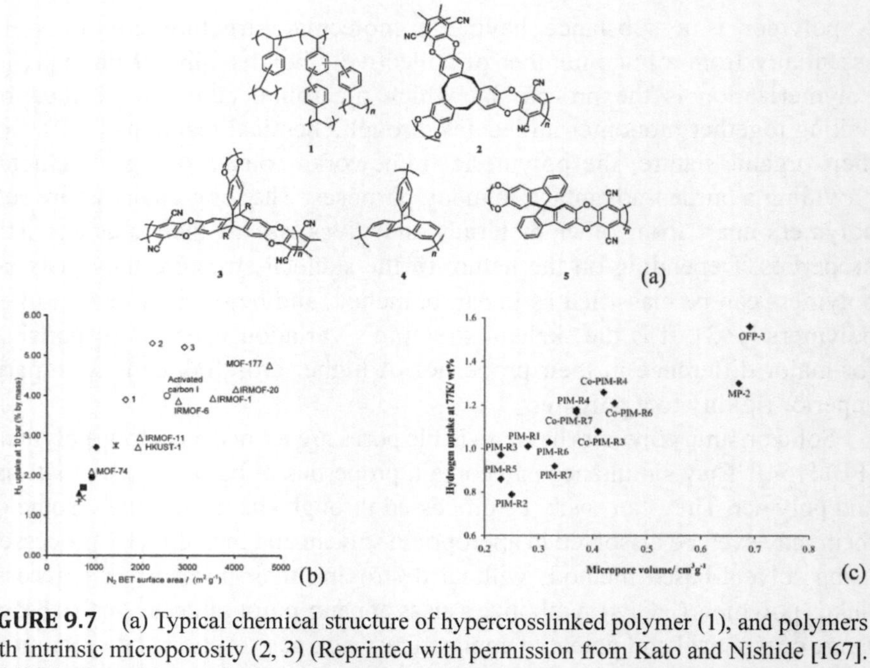

FIGURE 9.7 (a) Typical chemical structure of hypercrosslinked polymer (1), and polymers with intrinsic microporosity (2, 3) (Reprinted with permission from Kato and Nishide [167]. © 2018 Springer Nature) (b) Hydrogen uptake in PIMs in relation with BET surface area and comparison with other porous materials (Reprinted with permission from McKeown, Budd, and Book [187]. © 2007 John Wiley (c) Hydrogen uptake in PIMs in relation with pore volume and comparison with other polymers (Reprinted with permission from Ramimoghadam, Gray, and Webb [188]. © 2016 Elsevier).

9.6.2 HYDROGEN STORAGE IN PURE POLYMERS

In general, porous solid polymers adsorb hydrogen through physisorption, whereas chemisorption is the mechanism by which liquid polymers adsorb hydrogen. But there are a few exceptions. For instance, ketone/alcohol polymer cycles (Table 9.4) [168–174] are capable of storing hydrogen through chemisorption legally and conveniently for designing mobile energy storage device. The diversity of artificial routes for polymers eases the synthesis of numerous porous polymers competent to incorporate multiple chemical functionalities at the pore surface or into the porous framework. In essence, efficacy of hydrogen storage performance through physisorption is vastly dependent on the chemical processing route adopted for their synthesis. Nevertheless, it should be kept in mind that organic microporous polymers are still novices to the variety of potential hydrogen storage materials unlike carbon nanomaterials and only subtle research progress is visible towards maximizing their effectual surface area and optimizing them for efficient uptake and release of hydrogen towards meeting DoE targets (Table 9.1). It is interesting to compare the initial available data for microporous polymers with the best-reported results for other types of material having high surface area. Table 9.5 [175–185] is for comparing hydrogen storage for a number of porous polymers, developed by researchers worldwide. Upon careful observation of the obtained results, the following can be appreciated for hydrogen storage within pure porous polymers.

TABLE 9.4 Hydrogen Storage by Pure Polymers Through Chemisorption

Polymer Name	Characteristic Details	References
Quinaldine-substituted poly(acrylic acid) (PQD) and its hydrogenated 1,2,3,4-tetrahydroquinaldine derivative (PHQD)	Hydrogen release at mild condition (room temperature or ~80°C) at normal pressure	[168, 169]
Fluorenol-fluorenone cycle	Efficient release of hydrogen in presence of iridium catalyst	[170–172]
Polystyrene-poly (cyclohexyl ethylene) cycle	In presence of 10 MPa pressure and palladium catalyst at 200°C	[173, 174]

A high enthalpy of absorption ensures great hydrogen uptake capacity from chemistry point of view. Additionally, a high surface area signifies more and more capacity as adsorption is proportional to surface area [186]. Although hypercrosslinking of nanoporous polymers results in enhancement in hydrogen adsorption capacity, the maximum achievable limit, which is ~ 1.5 wt% still needs upgradation through tailoring of polymer chemistry in order to raise the heat of adsorption for enabling adsorption at higher temperatures [175]. Self-condensation of bischloromethyl monomers leads to generation of a plethora of microporous hypercrosslinked polymers [176]. Networks based on BCMBP display a gravimetric storage capacity of 3.68 wt% at 15 bar and 77.3 K, which is quite appreciable for pure polymers. The porosity in these polymers is governed not only by the level of cross-linking but also by slight changes in the design of the rigid monomer unit. PIMs effectively link the gap between conventional microporous materials and polymers as they possess properties common to both types of material. The prospective offered by the structural variety of PIMs that can be controlled just by the choice of monomer precursors remains still largely unexplored [177]. Available surface area, aggregate pore volume, and distribution of micropores are believed to be the key deciding parameters for their hydrogen storage performance. As shown in Figure 9.7(b), despite having a relatively smaller surface area, many PIMs exhibit superior hydrogen uptake comparing MOFs [187]. Again, though having less pore volume, by virtue of superior pore interconnectivity, many PIMs are better hydrogen retainer than other kinds of polymers (Figure 9.7(c)) [188]. A range of polymers of intrinsic microporosity (PIMs) have been studied by McKeown et al. [166]. The form of hydrogen isotherm is found to be influenced by the micropore distribution. A higher concentration of ultramicropores (pore size <0.7 nm), as observed in PIMs is associated with enhanced low-pressure adsorption. The extended π-electron system and incorporation of sulfur heteroatom did not appear to significantly influence the adsorption of hydrogen for nanoporous polymers with stereocontrolled cores [189]. At ambient temperature, for significantly increasing the heat of adsorptions thus adsorption capacities, new functional groups with sturdy interactions with hydrogen molecules are required. Theoretical studies show that maximum adsorption of hydrogen would happen between two surfaces with a separation of 0.7 nm [190]. This can be achieved by either pore design or inclusion of open metal sites or heteroatoms. These approaches can be easily tailored to microporous polymers, where optimization of characteristics even seems to be much easier. Again, pure organic materials have a discrete weight advantage for being composed solely of light elements. It is conceived from the above studies that improvement of molecular design in terms of rise in molecular weight might lead to rise in hydrogen storage density even beyond DoE target, as exemplified by Kato et al. [167] (Figure 9.8(a)).

TABLE 9.5 Hydrogen Storage Performance of Pure Porous Polymers Observed by Researchers Worldwide

Polymer Name	Type	Storage Condition		Hydrogen Storage Capacity (wt%)	Remarks	References
		Temperature	Pressure			
Amberlite XAD4 poly(styrene-co-divinylbenzene)	Nanoporous	77 K	0.12 MPa	0.8		[175]
Amberlite XAD16 poly(styrene-co-divinylbenzene)				0.6		
Hayesep N poly(divinylbenzene-co-ethylenedimethacrylate)				0.5		
Hayesep B polydivinylbenzene modified with polyethyleneimine				0.5		
Hayesep S poly(divinylbenzene-co-4-vinylpyridine)				0.5		
Wofatit Y77 poly(styrene-co-divinylbenzene)				1.2		
Lewatit EP63 poly(styrene-co-divinylbenzene)				1.3		
Lewatit VP OC 1064 poly(styrene-co-divinylbenzene)				0.7		
Hypersol-Macronet MN200 hypercrosslinked polystyrene	Hypercrosslinked			1.3		
Dichloroxylene (ortho-, meta-, and para-isomers, 0.33 each)	Microporous Hypercrosslinked		1.13 bar	1.3 to 1.56	Process dependent	[176]
4,4¢-Bis(chloromethyl)-1,1¢-biphenyl				1.56		
Bis(chloromethyl)anthracene				1.41		
PIM-1	Soluble non-network PIM		1 bar and 10 bar	0.95 and 1.45		[177]

TABLE 10.5 *(Continued)*

Polymer Name	Type	Storage Condition		Hydrogen Storage Capacity (wt%)	Remarks	References
		Temperature	Pressure			
PIM-7				1.00 and 1.35		[178]
HATN-PIM	Network PIM			1.12 and 1.56		[166]
CTC-PIM				1.35 and 1.70		
Porph-PIM				1.20 and 1.95		[178]
Trip-PIM				1.63 and 2.71		[179]
HCP	Hyper crosslinked			1.27 and 2.75		[180]
PS4AC1	Nanoporous Polymers Containing Stereocontorted Cores	77 K and 298 K	60 bar and 70 bar	2.8 and 0.44		[181]
PT4AC				2.2 0. and 50		
PS4AC2				3.7 and 0.43		
PS4TH				3.6 and 0.45		
poly(glycidyl methacrylate-co-ethylene	Macroporous copolymer	77 K	0.12 MPa	0.2		[182]
poly(chloromethyl styrene-co-divinylbenzene)	highly polar polymeric sorbents	77 K	0.12 MPa	0.4		[183]
poly(4-vinylpyridine-co-divinylbenzene)				0.6		
Melamine+1,4-diiodobenzene	porous polymer networks of aromatic rings	77 K	1 bar and 20 bar	0.16 and 1.17		[184]
Melamine+ 1,4-dibromobenzene				0.12 and 1.04		
Melamine+ 1,3,5-tribromobenzene				0.10 and 0.94		
diaminobenzene/tribromobenzene	Hypercrosslinked using Buchwald conditions	273 K	9 MPa	0.22		[185]

9.6.3 HYDROGEN STORAGE IN FUNCTIONALIZED POLYMERS

Functional porous polymers can be intended to demonstrate stimuli-responsive uniqueness capable of reversibly altering the pore structure or even toggling between the closed and open porous state after disclosure to environmental stimulus. Such exclusive characteristics are unavailable in other porous materials. Chemical functionalization (in other words, incorporation of heteroatoms) of POP results in improvement in gas storage properties. In essence, incorporation of heteroatoms in a polymer itself is a misnomer as a polymer itself is composed of at least two elemental atoms, i.e., carbon, and hydrogen unlike carbon nanomaterials. Additionally, there is frequent existence of atoms of various non-metals such as oxygen, nitrogen, sulfur, etc. Hence, decoration by heteroatoms typically implies incorporation of either metallic elements or with non-metals having properties more inclined towards metals as per periodic table. Conjugated microporous polymers (CMPs) [191–194], covalent organic frameworks (COFs) [195–199], covalent triazine frameworks (CTFs) [200–202], porous aromatic frameworks (PAFs) [203–205],metal-organic frameworks (MOF) are a few among these highly assorted varieties where substantial mixing of foreign elements results in change in entire nature of the mother polymer and generation of new variety of polymers typically known as coordination polymers. Though these are fairly well performer for hydrogen storing, typical weight advantage of pure polymers is somehow compromised for incorporation of heteroatoms in large quantities. Discussion within this section thus will be restricted to experimental and theoretical progress investigating hydrogen storage performance of POPs functionalized marginally with elemental nano-particles.

Most of such efforts are unfortunately found to be theoretical in nature and there is rarity of experimental instances in this direction till date. Decorating elements such as Ti, Li, Sc etc. are found to cause sufficient shift of binding energy and desorption temperature towards favorable condition to enable porous polymers to contain enough amount of hydrogen near ambient temperature and pressure. For example, Lee et al. showed through first principle calculation that Ti-decorated *cis*-polyacetylene might attain a reversibly usable gravimetric density of 7.6 wt% [206] (Figure 9.8(b)). A similar study involving ethylene by Durgun et al. concludes that Ti is capable of raising hydrogen intake capacity of ethylene complex up to 14 wt% [207]. For CMPs, Li doping is found to enhance hydrogen uptake up to 6.1 wt% at 77 K and normal pressure [208] (Figure 9.8(c)). Sc and Li functionalized [2,2] paracyclophane complexes are found to hold up to 10

H_2 molecules and 8 H_2 molecules by charge polarization mechanism with hydrogen weight percentage of 11.4 and 13.5, respectively [209]. Such instances are promising and encouraging for endeavoring future experiments for validating such theoretical findings and ruling out any oversimplifying assumption during theoretical recalculation for accomplishing more realistic observation. Boron was experimentally found to enhance hydrogen storage property of POP in the form of carborane by Yuan et al. [210]. They found that boron is efficient in raising initial isosteric heats of adsorption toward hydrogen by 3–5 kJ/mol. Pd/MIL-101 Nanocomposites have been recently found to exhibit reducing hydrogen absorption enthalpy by Zhao et al. [211]. The above findings strongly suggest that decorated organic porous polymer has huge potential for cost-effective exploitation as a future hydrogen carrier.

9.7 THEORETICAL INSIGHTS FOR HYDROGEN STORAGE

9.7.1 IMPORTANCE OF THEORETICAL SIMULATIONS FOR HYDROGEN STORAGE

Theoretical simulations play an important role in predicting the interaction mechanism of hydrogen molecules with the nanomaterials as well as with the functionalized atoms or groups in the case of functionalized Nanomaterials. Simulation techniques can give important information like preferred adsorption configuration, adsorption energy, charge transfer, etc. which are difficult to get from experiments. With the invention of high speed, large memory supercomputers, extensive computer simulations are now very much possible. Also, so many well-established computer codes, both open access and commercial are available to compute theoretical data on hydrogen storage. So, the availability of trusted software packages and supercomputers makes it possible to generate hydrogen storage database for a large number of systems and recommend the experimentalist the potential systems to try in experiments. Theoretical simulations can provide nice picture on orbital interactions and charge transfer which is difficult or may not be possible to get from experimental measurements. From the simulations data, it is straightforward to conclude whether specific H_2 is having physisorption, chemisorptions, or Kubas [122, 212] types of interactions. The stability of the structures and desorption temperature can also be obtained from theoretical simulations. One of the important issues for hydrogen storage in metal-doped carbon nanomaterial or polymer is the metal-metal cluster.

From the computation of diffusion energy barrier of metals, we can get idea on possible metal loading pattern to avoid metal-metal clustering. But there are challenges and issues for predicting hydrogen wt% from theoretical simulations. Simulations results are sensitive to the simulations methods and setting parameters. For example, LDA exchange-correlation [213] over-binds the H_2 molecules whereas GGA exchange-correlation [214] under-binds the H_2 molecules. So, we need to be careful regarding choice of the simulation methods and prediction coming from simulation details. For the adsorption of H_2 molecules on carbon nanostructures or polymers, weak Van der Waals interactions are present. So, we should carefully include those interactions while computing binding energy of H_2 molecules on the host. Keeping in mind, the role of theoretical simulations for predicting hydrogen storage parameters and the limitation of simulations methods, the following sub-sections will highlight overview of quantum simulations, important parameters obtained from theoretical data, importance of dispersion corrections and sensitivity of theoretical data.

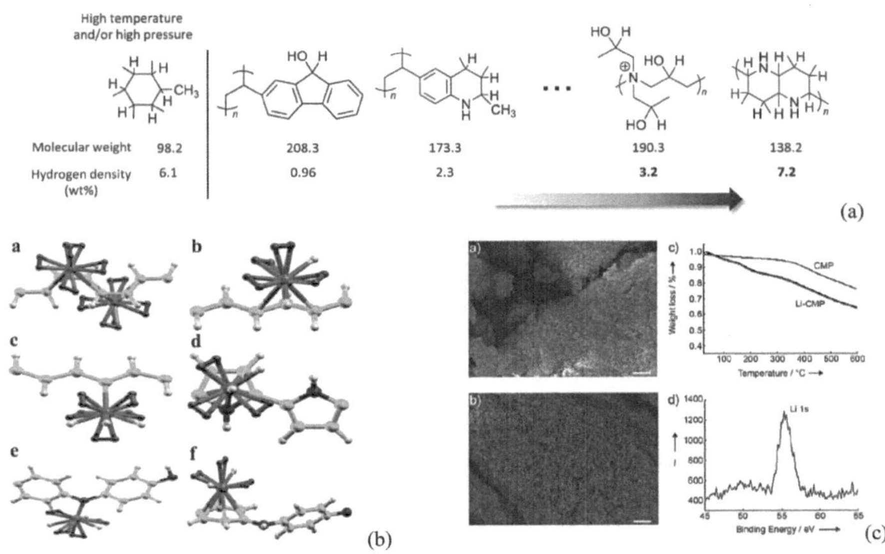

FIGURE 9.8 (a) Progressive improvement in hydrogen storage density through chemical modification in polymer (Reprinted with permission from Kato and Nishide [167. © 2018 Springer Nature), (b) Atomic structures of Ti decorated polymers with maximum no. of hydrogen molecules attached with Ti atom (Reprinted with permission Lee and Choi [180]. © 2006 American Physical Society), (c) Existence of Li in ionic state with one negative charge per Li atom within Li-doped CMP polymer (Reprinted with permission Li et al. [208]. © 2010 John Wiley).

9.7.2 OVERVIEW OF QUANTUM SIMULATIONS

Quantum simulation is very powerful in terms of providing the physical insight of the system as well as producing data which can be compared from experimental data. Quantum simulations find applications in many branches of science and engineering. In last few decades, it has gain popularity due to development of high speed, large memory supercomputer. One of the notable features of quantum simulations is that it does not require any model potential to be provided by the user. Quantum simulations involve solving the Schrodinger's equation for the concerned system containing so many numbers of electrons and ions. The major challenge for quantum simulations is that the equations like the Schrodinger's equation are difficult to solve analytically except for a few simple systems. Solving such an equation needs sophisticated numerical techniques and many approximations. These approximations and sophisticated numerical techniques have given birth to various quantum simulations methods, e.g., density functional theory (DFT) [215–217], Hartree-Fock [218, 219], Quantum Monte Carlo [220–222], coupled cluster [223, 224], multi-reference configuration interaction [225], etc. Each of the methods has their own merits and demerits.

9.7.2.1 MANY-BODY HAMILTONIAN

The central idea of quantum mechanical simulations is to find the electronic structure of the concerned system by solving the many-body Schrödinger equation. For a system containing the positively charged nuclei and negatively charged electrons, the non-relativistic many-particle Hamiltonian as:

$$\widehat{H}=\widehat{T}_{nucl}+\widehat{T}_{elec}+\widehat{V}_{n-e}+\widehat{V}_{e-e}+\widehat{V}_{n-n} \tag{9.1}$$

where the different terms are as follows:

$$\widehat{T}_{nucl}=-\frac{\hbar^2}{2}\sum_i\frac{\nabla^2_{R_i}}{M_i},\widehat{T}_{elec}=-\frac{\hbar^2}{2}\sum_i\frac{\nabla^2_{r_i}}{m_i},$$

$$\text{and}\widehat{V}_{n-e}=-\frac{1}{4\pi\varepsilon_0}\sum_{i,j}\frac{Z_ie^2}{|R_i-r_j|},\widehat{V}_{e-e}=\frac{1}{8\pi\varepsilon_0}\sum_{i\neq j}\frac{e^2}{|r_i-r_j|},\widehat{V}_{n-n}=\frac{1}{8\pi\varepsilon_0}\sum_{i\neq j}\frac{e^2Z_iZ_j}{|R_i-R_j|} \tag{9.2}$$

First-term represents the kinetic energy operators for the nuclei, second term is the kinetic energy operators for the electrons, third, fourth, and fifth terms indicate the Coulomb interactions between nuclei-electron, electron-electron, and nuclei-nuclei respectively. M_i and m_i refer to the masses of the nuclei and electrons at the positions R_i and r_i, respectively. The wave functions ($\psi\{R_n, r_e\}$) of the system can be found out from the above Hamiltonian by solving the following Schrödinger equation:

$$\hat{H}\psi\{R_n, r_e\} = E\psi\{R_n, r_e\} \tag{9.3}$$

As it is very difficult to solve it, we first use Born-Oppenheimer approximation to separate the electron and nuclei degree of motion. Since the velocities of the nuclei, are much less than that of the electrons, then nuclei can be considered to be fixed at their position. So, the nuclei act as external particles with a potential and the electrons move [226]. With this assumption, the kinetic energy part of the Hamiltonian can be neglected and the Coulomb interactions between the nuclei can be taken as constant. So, the Hamiltonian of much electron system is reduced to:

$$\hat{H} = \hat{T}_{elec} + \hat{V}_{ext} + \hat{V}_{e-e} \tag{9.4}$$

But still solving the Schrödinger equation with the reduced Hamiltonian is quite challenging and different methods have been formulated. Now we will mention some of the methods for quantum simulations.

9.7.2.2 HARTREE-FOCK METHOD

Hartree-Fock method [218, 219] is the corrected version of the Hartree method where wave function of many-electron systems is considered as a simple product of single-electron orbitals. But as the Hartree method does not satisfy the Pauli's exclusion principle, in Hartree-Fock method, the wave function is described by the anti-symmetries many-electron wave function using the slater determinant.

$$\psi_{HF}(x_1, x_2, \cdots, x_N) = \frac{1}{\sqrt{N!}}\begin{bmatrix} \varphi_1(x_1) & \cdots & \varphi_1(x_N) \\ \vdots & \ddots & \vdots \\ \varphi_N(x_1) & \cdots & \varphi_N(x_N) \end{bmatrix} \tag{9.5}$$

This leads to the Hartree-Fock equation given by:

$$\varepsilon_i \varphi_i(\vec{r}) = [\hat{T}_{elec} + \hat{V}_{ext} + \hat{V}_{e-e}]\varphi_i(\vec{r}) - \frac{1}{2}\sum_j \int \varphi_j^*(\vec{r}^{\prime})\varphi_i(\vec{r}^{\prime})\frac{1}{|\vec{r}-\vec{r}^{\prime}|}\varphi_i(\vec{r})d^3\vec{r}^{\prime} \quad (9.6)$$

The main demerit of this method is that the exchange term is non-local and it neglects the electron correlations. Hartree-Fock method gives good results for molecules but fails for metals. This scheme is backbone for various Post Hartree-Fock methods like Coupled cluster [223, 224], Multi-reference Configuration Interaction [225], etc. where correlation effects are taken care of.

9.7.2.3 DENSITY FUNCTIONAL THEORY (DFT)

Density functional theory (DFT) is one of the powerful and fast quantum simulation methods which are widely used in many branches of science and engineering. DFT is based on two theorems given by Hohenberg and Kohn [227] in 1964 and followed by few equations formulated by Kohn and Sham [228]. Two theorems of Hohenberg and Kohn are as follows:

1. **First Theorem:** There is a one-to-one correspondence between the ground-state density $\rho(\vec{r})$ of a many-electron system and the external potential $\hat{V}_{ext}$. An immediate consequence is that the ground-state expectation value of any observable $\hat{O}$ is a unique functional of the exact ground-state electron density.

$$\langle \psi | \hat{O} | \psi \rangle = O[\rho] \quad (9.7)$$

2. **Second Theorem:** For $\hat{O}$ being the Hamiltonian $\hat{H}$, the ground-state total energy functional:

$$H[\rho] \equiv E_{V_{ext}}[\rho] \quad (9.8)$$

is of the form:

$$E_{V_{ext}}[\rho]=\underbrace{\langle\psi|\hat{T}+\hat{V}|\psi\rangle}_{F_{HK[\acute{A}]}}+\langle\psi|\hat{V}_{ext}|\psi\rangle=F_{HK[\rho]}+\int\rho(\vec{r})V_{ext}(\vec{r})d\vec{r} \tag{9.9}$$

where, the Hohenberg-Kohn density functional $F_{HK[\rho]}$ is universal for any many-electron system. $E_{V_{ext}}[\rho]$ reaches its minimal value (equal to the ground-state total energy) for the ground state density corresponding to $\hat{V}_{ext}$.

The theorem tells the existence of a one-to-one correspondence between the external potential and the electron density. So, everything regarding the system can be known from the functional F_{HK}. By minimizing the functional F_{HK}, we can find out the exact energy and density of the many electron system. But the problem is that F_{HK} is not known which makes it difficult for the implementation of this formulism for real system.

9.7.2.4 THE KOHN-SHAM EQUATIONS

Kohn-Sham equations led the foundation for the practical implantation of DFT. They have replaced the real interacting system by a non-interacting system with some external potential. The Hamiltonian is now given by:

$$\hat{H}_{KS}=\hat{T}_0+\hat{V}_H+\hat{V}_{xc}+\hat{V}_{ext} \tag{9.10}$$

With $\hat{T}_0$ being the kinetic energy operator and $\hat{V}_H$, $\hat{V}_{xc}$ and $\hat{V}_{ext}$ are the Hartree, exchange-correlation, and external potentials respectively. Now the energy functional can be expressed as:

$$E_{V_{ext}}[\rho]=\underbrace{T_0[\rho]+E_H[\rho]+E_{xc}[\rho]}_{F_{HK[\rho]}}+\int\rho(\vec{r})V_{ext}(\vec{r})d\vec{r} \tag{9.11}$$

With $F_{HK[\rho]}$ denoting the sum of the first three terms and $T_0[\rho]$ being the kinetic energy functional along with $E_H[\rho]$ and $E_{xc}[\rho]$ representing the Hartree and exchange-correlation functional. Now the ground state density can be obtained from the solution of Kohn-Sham equation given by:

$$\left[-\frac{\hbar^2}{2m_e}\nabla+\widehat{V}_{eff}(\vec{r})\right]\psi_i(\vec{r})=\varepsilon_i\psi_i(\vec{r}) \tag{9.12}$$

with

$$\widehat{V}_{eff}(\vec{r})=\widehat{V}_H+\widehat{V}_{xc}+\widehat{V}_{ext} \tag{9.13}$$

and $\hat{V}_{xc}$ and $\hat{V}_H$ are given by:

$$\widehat{V}_{xc}=\frac{\delta E_{xc}[\rho]}{{}'n},\widehat{V}_H=\frac{e^2}{4\pi\varepsilon_0}\int\frac{\rho(\vec{r}')}{|\vec{r}-\vec{r}'|}d\vec{r}' \tag{9.14}$$

We can see that the only unknown term is exchange-correlation functional. The above equation can be solved using a self-consistent procedure. So the accuracy of DFT data depends on how good the exchange-correlation functional can be formulated. LDA (local density approximation) [213] and GGA (generalized gradient approximation) [214] are two widely used exchange-correlation functional. LDA is based on homogenous electron gas model and is expressed as [214]:

$$E_{xc}^{LDA}[\rho]=\int\rho(\vec{r})\varepsilon_{xc}(\rho(\vec{r}))d\vec{r} \tag{9.15}$$

where, the exchange part of the ε_{xc} can be obtained analytically and the correlation is obtained numerically using quantum Monte-Carlo simulations [213].

In GGA, the gradient of the density is also taken into account. There exist different versions of GGA exchange-correlation Functional, e.g., Perdew-Burke-Ernzerhof (PBE) [214], and Becke-Lee-Yang-Parr (BLYP) [229, 230]. The notable difference between GGA and LDA results that LDA overbinds the system whereas GGA underbinds the system.

To perform simulations with different quantum mechanical methods, various software packages are available. Softwares like Gamess [231] or Gaussian [232] are designed to deal with molecular structures using a localized basis set like the Gaussian center orbitals or atomic orbitals. Codes like VASP [233] or Quantum-espresso [234] which are designed to deal with metals and other solids aiming for band structure calculations use the plane wave basis set. Packages like WIEN2K [235] or ELK [236] are made mainly to deal with heavy elements where the pseudopotential methods are not suitable and all the electron potential should be considered.

9.7.3 PROPERTIES FROM THEORETICAL SIMULATIONS

Theoretical simulations can provide very beautiful and clear picture regarding adsorption configuration, adsorption energy, charge transfer, and orbital interactions which are difficult to get in experiments.

9.7.3.1 ADSORPTION SITE

Finding the adsorption configuration with minimum energy is an important step for hydrogen storage simulations. We need to perform geometry optimization for a longer period till the forces between the atbecomeomes very less (typically 0.01 eV/Å). For the hydrogen adsorption on carbon Nanomaterials as well as polymer due to involvement of weak Van der Waal's forces, the convergence may take longer time. Sometimes, the gas molecules may get stuck to the local minima and we may miss the actual configuration with minimum energy. To avoid this, it is suggested to make 2–3 different initial configurations and allow them relax separately and choose the one with minimum energy as the stable configuration. If the system contains transition metal with magnetic signature, the relaxation and total energy calculation should be performed with spin polarization into consideration. The stable configuration also depends on simulations setting parameters such as energy cut off for the plane wave as well as K-points for sampling the Brillouin zone. So, one needs to test the convergence with respect to energy cut off for the plane wave as well as number of K-points. The distance of the hydrogen molecule with the nearest atom of the host or the dopant can give idea regarding strength of the bonding. Lower the distance (bond length) stronger is the adsorption. Figure 9.9 displays two possible adsorption configurations of Y on graphyne [237], one on hexagonal site and other on triangular site. The adsorption is strong on triangular site with binding energy of -4.05 eV compared to hexagonal site with –3.15 eV. Figure 9.9 also depicts the configuration of 9 H_2 attached on hexagonal site and 5H_2 attached on triangular site.

9.7.3.2 ADSORPTION ENERGY

The adsorption energy of H_2 is a key parameter in finding the capability of hydrogen storage of the material. For practical hydrogen storage device, this energy must fall within a window. If the adsorption energy is too high, it will be difficult to release them, whereas if the H_2 molecules are attached weakly, then they may come out of the system before using. In this regard, DoE [45] has prescribed that the binding energy of H_2 must fall within the range 0.2–0.7 eV for practical implementation of hydrogen as fuel. The adsorption energy of H_2 on the host can be computed from the expression:

$$E_{ads} = E_{Host+H_2} - E_{Host} - E_{H_2} \tag{9.16}$$

where, E_{Host+H_2} is the energy of the host (nanostructure + dopant) plus H_2, E_{Host} is the energy of the host and E_{H_2} is the energy of an isolated H_2 molecule. As per the convention, negative value of adsorption energy indicates that binding is energetically favorable. All three energies in the above equation should be computed with same simulations setting as the computed energy may vary with the setting parameters. The adsorption energy may be different for different adsorption sites. If there is no periodic boundary condition in any direction (for molecules, clusters, etc.) sufficient vacuum should be considered to avoid interactions between periodic images. Table 9.6 presents the binding energy of the metal Y and H_2 molecules on two different sites of on graphyne [237]. We can notice that binding energy is given with GGA and GGA+ Van der Waal's corrections. With the inclusion of Van der Waal's corrections, there is slight increase in binding energy of H_2 molecules.

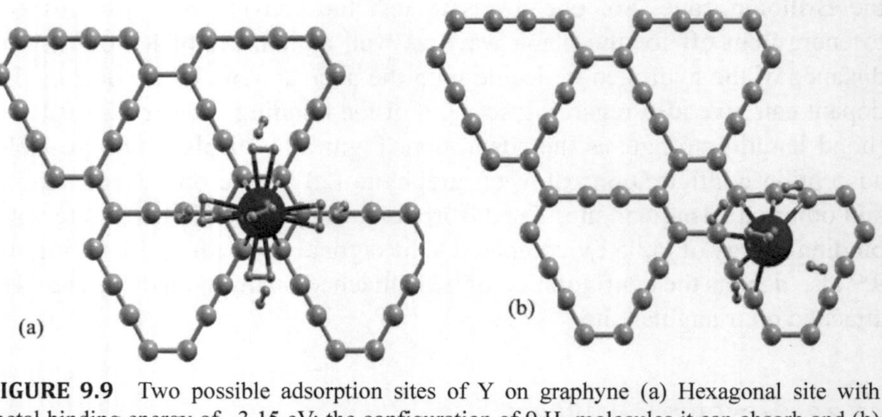

FIGURE 9.9 Two possible adsorption sites of Y on graphyne (a) Hexagonal site with metal binding energy of –3.15 eV; the configuration of 9 H_2 molecules it can absorb and (b) Triangular site with metal binding energy of –4.05 eV; the configuration of 5 H_2 molecules it can absorb (Reprinted with permission f Gangan et al. [237]. © 2019 Elsevier).

TABLE 9.6 The Computed Binding Energy per H2 and Desorption Temperature for All Configurations for Y Doped Graphyne Using GGA and GGA + Van Der Waals Corrections

Configurations	Binding Energy (eV)				Average Desorption Temperature (K)			
	Site 1		Site 2		Site 1		Site 2	
	GGA	GGA vDW	GGA	GGA vDW	GGA	GGA vDW	GGA	GGA vDW
Y	3.15	3.433	4.05	4.32	–	–	–	–
H_2	0.304	0.357	0.29	0.35	388	456	371	447
$3H_2$	0.292	0.378	0.30	0.40	373	483	383	511
$5H_2$	0.28	0.376	0.26	0.35	356	480	332	447
$9H_2$	0.261	0.3	–	–	333	383	–	–
Average BE	0.297	0.366	0.28	0.366	379	468	358	468

Source: Reprinted with permission from Gangan et al. [237], © 2019 Elsevier.

9.7.3.3 *CHARGE TRANSFER AND ORBITAL INTERACTIONS*

Theoretical insight regarding bonding and charge transfer mechanism can be obtained from simulation data. We can get quantitatively the amount of charge getting transferred between the bonded species. From the charge transfer we can get some idea regarding bonding mechanism. The H_2 molecule can be attached on the host through physisorption, chemisorptions, or Kubas types of interaction. The bonding is strong for the chemisorption, weak for physisorption and intermediate for the Kubas types of interaction. More the charge transfer, stronger is the bonding. For pure polymer or the carbon Nanomaterials, the adsorption is physisorption and the charge transfer is negligible. For chemisorptions, the hydrogen atoms get dissociated and charge transfer is large. Both physisorption and chemisorptions are not suitable for practical implementation of hydrogen storage. For metal doped carbon nanomaterials or polymer, there is charge transfer from metal d orbitals to H's orbital followed by back donation through Kubas interaction. In this case, the binding energy comes in the range 0.2–0.7 eV as prescribed by DoE and is most suitable for practical implementation of hydrogen storage. So, the charge transfer can give a picture whether the interaction is chemisorptions, physisorption, or Kubas types of interactions [122, 212]. To find the charge transfer one need to take care regarding number of valence electrons taken for the simulations. Bader charge analysis [238] and Mulliken's charge analysis [239] are two powerful techniques to find the amount of charge transfer during the orbital interactions. To visualize the spatial variation of charge density, the pictorial plot of charge density can be obtained using visualize software like XCRYSDEN [240], VESTA [241] etc. The orbital interactions and hybridization can be obtained qualitatively from plot of partial density of states which describe the contribution of Density of states from each orbital of each atom. There should consistency between the charge transfer from Bader charge analysis, analysis of partial density of states and charge density plot. Chakraborty et al. [157] have shown that when Y atom is absorbed on single-walled carbon nanotube (SWCNT), there is 1.1446e charge transfer from Y atom to SWCNT. Figure 9.10 depicts [157] the charge density plot for the charge density difference between SWCNT+Y and SWCNT for the isovalue of 0.2e and -0.2e. We can notice that there is a polarization between inner and outer surfaces due to charge transfer and charge redistribution.

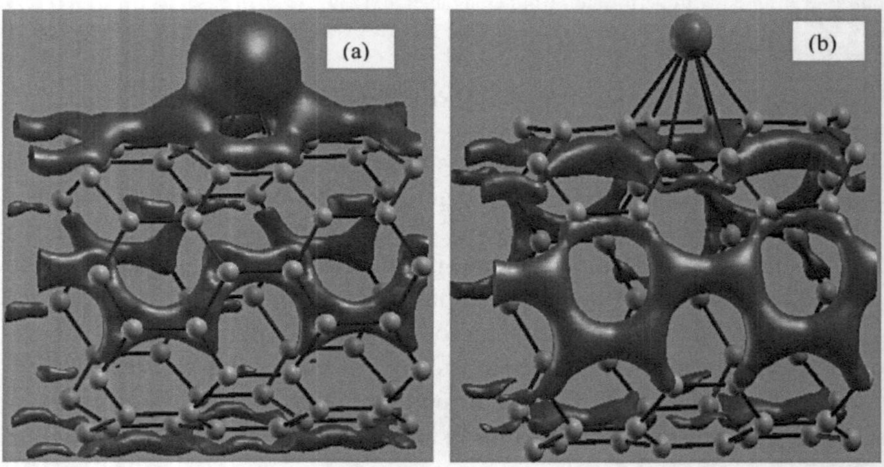

FIGURE 9.10 Charge density isosurface for charge density difference between Y-SWCNT and SWCNT systems (a) iso-value 0.2e (b) iso-value –0.2e (color online). (Reprinted with permission from Chakraborty, Modak, and Banerjee [157]. © 2012 American Chemical Society).

9.7.4 IMPORTANCE OF DISPERSION CORRECTIONS FOR HYDROGEN STORAGE

As the H_2 adsorption involves weak Van der Waals interactions, we must include dispersion corrections in our simulations as the standard DFT with GGA exchange-correlation functional can't handle long term electron correlation effects. The dispersion correction increases the binding energy of H_2 molecules slightly. There exist various schemes for incorporating the dispersion corrections in DFT and the binding energy of H_2 molecules will differ little bit for various dispersion schemes. Here we will briefly outline the different schemes employed for dispersion corrections.

1. **Grimme's DFT-D2** [242]: In this scheme, the total energy is expressed as a sum of the Kohn−Sham energy, E_{DFT}, and vdW semi-empirical pair correction, $E_{disp}^{(2)}$. The pair corrections depend on two interacting atoms and independent of environment.
2. **Grimme's DFT-D3:** Here the environment dependent dispersion coefficients (A three body component) were considered during the DFT simulations to improve the accuracy of Grimme's DFT-D2 scheme.
3. **vdW-DF** [244–246]: In this approach, dispersion interactions are described in terms of electron density and no additional inputs needed.
4. **optPBE-vdW:** It is a different version of vdW-DF.

In Table 9.7, we have presented the adsorption energy of Co molecule on graphene and defected graphene employing different schemes of dispersion corrections [247]. It is clear from Table 9.7 that various scheme yields different adsorption energy and we get a window for adsorption energy. We can notice that the PBE+vdW and DFT-D3 give almost identical binding energies, 100 meV higher than the DFT-D2 method. So, we need to select the proper dispersion correction schemes depending upon the system used for the simulations.

TABLE 9.7 Comparison of the Calculated Adsorption Energy for Adsorption of CO Molecule on Graphene and Defective Graphene for Different vdW Functional

Method	**Graphene**		**Defected Graphene**	
	Adsorption Energy (eV)	**Distance (Å)**	**Adsorption Energy (eV)**	**Distance (Å)**
PBE + vdW	−0.11	2.88	−2.15	1.33
DFT-D2	−0.09	2.87	−2.25	1.33
DFT-D3	−0.10	2.92	−2.17	1.33
vdW-DF	−0.14	3.02	−1.80	1.33
optB88-vdW	−0.11	2.77	−2.04	1.33
optPBE-vdW	−0.16	2.93	−2.01	1.33

Source: Jiang et al. [247],with permission.

9.7.5 *SENSITIVITY OF THEORETICAL DATA*

The data obtained from DFT simulations depend on the exchange-correlation functional, simulations parameters, dispersion correction schemes and simulation software. So, one need to be careful regarding choice of the simulation parameters. As the binding energy is computed from difference of energy, simulations should be considered with identical setting parameters otherwise, it may give wrong result. Here we will mention two important parameters, namely, energy cut-off and *k*-point grid.

1. **Energy Cut-Off:** It is a key parameter to get accurate value of ground state properties. Before taking the data, one must check the convergence of energy with respect to cut-off energy. For a periodic system, the solution of the Schrödinger equation can be expressed as [214]:

$$\phi_n(r) = \exp(ik \cdot r) f_n(r) \tag{9.17}$$

where, the periodic function $f_n(r)$ can be expanded in terms of finite number of plane waves with the reciprocal lattice vectors of the crystal:

$$f_n(r) = \Sigma_G C_{n,G} \exp[iG \cdot r] \tag{9.18}$$

The summation is carried out over the reciprocal lattice vector G. Now the wave function can be written as:

$$\phi_n(r) = \Sigma_G C_{n,k+G} \exp[i(k+G) \cdot r] \tag{9.19}$$

where, $C_{n,k+G}$ indicates the coefficients of the plane waves with kinetic energy:

$$E = \frac{h^2}{2m} |k+G|^2 \tag{9.20}$$

This gives a kinetic energy cut-off for the plane waves given by:

$$E_{cut} = \frac{h^2}{2m} G_{cut}^2 \tag{9.21}$$

and the infinite sum becomes:

$$\phi_i(r) = \sum_{|G+k|<G_{cut}} C_{i,k+G} \exp[i(k+G)\cdot r] \tag{9.22}$$

Higher the energy cut-off, expensive is the simulation and accurate is the result. One should take the energy cut-off such that energy convergence is achieved.

2. ***k*-Point Grid:** It is another key parameter for DFT simulations which is used for sampling the Brillouin-zone. One needs to check the convergence of energy with respect to K-points. As the K-points are in the reciprocal space, higher the cell length in any direction, lower should be the number of K-points in that direction. K-points are required only in the direction where periodicity is there. The computational cost increases with increase in K-points.

9.8 CONCLUSIVE REMARKS WITH FUTURE SCOPES

In this chapter, we have presented an overview of various materials for efficient hydrogen storage. We have also included the features and capabilities of quantum simulations techniques in predicting various hydrogen storage parameters, e.g., binding energy, hydrogen wt%, desorption temperature, etc. Functionalized carbon Nanomaterials are future for efficient hydrogen storage at ambient condition. This field has been explored quite a lot in theoretical prediction. But the experimental results are so far not very promising and experimental wt% in most cases is lower than theoretical prediction. The main issue is to synthesize functionalized carbon Nanomaterials and polymer. During the synthesis, oxygen interference, metal-metal clustering, impurity, defects, etc. hinders the hydrogen storage and experimentally obtained wt% comes much below the requirement set by DoE. So, the developing sophisticated and innovated method for synthesis of functionalized carbon Nanomaterials and polymer is the key for implementing hydrogen fuel in transport applications.

ACKNOWLEDGMENTS

BC would like to thank Dr. R. C. Rannot and Dr. A. K. Mohanty for their support and encouragement. BC would also like thank Dr. S. Banerjee for inspiration and scientific discussions. GS would like to thank Dr. R. N Singh Dr. V. Kain and Dr. Madangopal Krishnan for their support.

KEYWORDS

- **carbon nanomaterials**
- **conjugated microporous polymers**
- **density functional theory**
- **hydrogen fuel**
- **hydrogen storage**
- **organic polymers**

REFERENCES

1. Shafiee, S., & Topal, E., (2009). When will fossil fuel reserves be diminished? *Energy Policy, 37*(1), 181–189.
2. Hoel, M., & Kverndokk, S., (1996). Depletion of fossil fuels and the impacts of global warming. *Resource and Energy Economics, 18*(2), 115–136.
3. Mazloomi, K., & Gomes, C., (2012). Hydrogen as an energy carrier: Prospects and challenges. *Renewable and Sustainable Energy Reviews, 16*(5), 3024–3033.
4. Bechtold, R., (1997). *Alternative Fuels Guidebook.* SAE.
5. National Research Council, (2013). *Transitions to Alternative Vehicles and Fuels.* National Academies Press.
6. Lee, S., Speight, J. G., & Loyalka, S. K., (2014). *Handbook of Alternative Fuel Technologies*. CRC Press.
7. National Research Council, (2008). *Transitions to Alternative Transportation Technologies: A Focus on Hydrogen.* National Academies Press.
8. Singh, S., Jain, S., Venkateswaran, P. S., Tiwari, A. K., Nouni, M. R., Pandey, J. K., & Goel, S., (2015). Hydrogen: A sustainable fuel for future of the transport sector. *Renewable and Sustainable Energy Reviews, 51*, 623–633.
9. Mandal, T. K., & Gregory, D. H., (2010). Hydrogen: A future energy vector for sustainable development. *Proceedings of the Institution of Mechanical Engineers, Part C: Journal of Mechanical Engineering Science, 224*(3), 539–558.
10. Yartys, V. A., & Lototsky, M. V., (2004). An overview of hydrogen storage methods. In: *Hydrogen Materials Science and Chemistry of Carbon Nanomaterials* (pp. 75–104). Springer, Dordrecht.
11. Walker, G., (2008). *Solid-State Hydrogen Storage: Materials and Chemistry*. Wood head Publishing, Elsevier.
12. Stetson, N. T., (2012). Hydrogen storage overview. In: *DoE Annual Merit Review and Peer Evaluation Meeting.*

13. Durbin, D. J., & Malardier-Jugroot, C., (2013). Review of hydrogen storage techniques for on board vehicle applications. *International Journal of Hydrogen Energy, 38*(34), 14595–14617.
14. Ni, M., (2006). An overview of hydrogen storage technologies. *Energy Exploration and Exploitation, 24*(3), 197–209.
15. Beaudin, M., Zareipour, H., Schellenberglabe, A., & Rosehart, W., (2010). Energy storage for mitigating the variability of renewable electricity sources: An updated review. *Energy for Sustainable Development, 14*(4), 302–314.
16. Abe, J. O., Ajenifuja, E., & Popoola, O. M., (2019). Hydrogen energy, economy, and storage: Review and recommendation. *International Journal of Hydrogen Energy*.
17. Chen, H., Cong, T. N., Yang, W., Tan, C., Li, Y., & Ding, Y., (2009). Progress in electrical energy storage system: A critical review. *Progress in Natural Science, 19*(3), 291–312.
18. Dahiya, R. P., (1985). Progress in hydrogen energy. In*: Proceedings of the National Workshop on Hydrogen Energy* (pp. 1019–1036). New Delhi.
19. Züttel, A., Remhof, A., Borgschulte, A., & Friedrichs, O., (2010). Hydrogen: The future energy carrier. *Philosophical Transactions of the Royal Society A: Mathematical, Physical, and Engineering Sciences, 368*(1923), 3329–3342.
20. Winter, C. J., & Nitsch, J., (2012). *Hydrogen as an Energy Carrier: Technologies, Systems, Economy*. Springer Science and Business Media.
21. https://www.energy.gov/eere/fuelcells/h2scale (accessed on 10 August 2020).
22. Fayaz, H., Saidur, R., Razali, N., Anuar, F. S., Saleman, A. R., & Islam, M. R., (2012). An overview of hydrogen as a vehicle fuel. *Renewable and Sustainable Energy Reviews, 16*(8), 5511–5528.
23. Veziro, T. N., & Barbir, F., (1992). Hydrogen: The wonder fuel. *International Journal of Hydrogen Energy, 17*(6), 391–404.
24. Pudukudy, M., Yaakob, Z., Mohammad, M., Narayanan, B., & Sopian, K., (2014). Renewable hydrogen economy in Asia-opportunities and challenges: An overview. *Renewable and Sustainable Energy Reviews, 30*, 743–757.
25. Crabtree, G. W., Dresselhaus, M. S., & Buchanan, M. V., (2004). The hydrogen economy. *Physics Today, 57*(12), 39–44.
26. Edwards, P. P., Kuznetsov, V. L., David, W. I., & Brandon, N. P., (2008). Hydrogen and fuel cells: Towards a sustainable energy future. *Energy Policy, 36*(12), 4356–4362.
27. Prasath, B. R., Leelakrishnan, E., Lokesh, N., Suriyan, H., Prakash, E. G., & Ahmed, K. O., (2012). Hydrogen operated internal combustion engines-a new generation fuel. *International Journal of Emerging Technology and Advanced Engineering, 2*(4), 52–57.
28. https://www.slideshare.net/SridharSibi/hydrogen-fuel-amp-its-sustainable-development (accessed on 3 August 2020).
29. Williams, Q., & Hemley, R. J., (2001). Hydrogen in the deep earth. *Annual Review of Earth and Planetary Sciences, 29*(1), 365–418.
30. Ohta, T., (2013). *Solar-Hydrogen Energy Systems: An Authoritative Review of Water-Splitting Systems by Solar Beam and Solar Heat: Hydrogen Production, Storage, and Utilization*. Elsevier.
31. Balat, H., & Kırtay, E., (2010). Hydrogen from biomass-present scenario and future prospects. *International Journal of Hydrogen Energy, 35*(14), 7416–7426.
32. Iwahara, H., Esaka, T., Uchida, H., & Maeda, N., (1981). Proton conduction in sintered oxides and its application to steam electrolysis for hydrogen production. *Solid State Ionics, 3*, 359–363.

33. Cortright, R. D., Davda, R. R., & Dumesic, J. A., (2011). Hydrogen from catalytic reforming of biomass-derived hydrocarbons in liquid water. In: *Materials for Sustainable Energy: A Collection of Peer-Reviewed Research and Review Articles from Nature Publishing Group,* (pp. 289–292).
34. Turner, J. A., (2004). Sustainable hydrogen production. *Science, 305*(5686), 972–974.
35. Holladay, J. D., Hu, J., King, D. L., & Wang, Y., (2009). An overview of hydrogen production technologies. *Catalysis Today, 139*(4), 244–260.
36. Ni, M., Leung, M. K., Leung, D. Y., & Sumathy, K., (2007). A review and recent developments in photocatalytic water-splitting using TiO_2 for hydrogen production. *Renewable and Sustainable Energy Reviews, 11*(3), 401–425.
37. Kapdan, I. K., & Kargi, F., (2006). Bio-hydrogen production from waste materials. *Enzyme and Microbial Technology, 38*(5), 569–582.
38. Hallenbeck, P. C., & Benemann, J. R., (2002). Biological hydrogen production; fundamentals and limiting processes. *International Journal of Hydrogen Energy, 27*(11/12), 1185–1193.
39. Hawkes, F. R., Dinsdale, R., Hawkes, D. L., & Hussy, I., (2002). Sustainable fermentative hydrogen production: Challenges for process optimization. *International Journal of Hydrogen Energy, 27*(11/12), 1339–1347.
40. Haryanto, A., Fernando, S., Murali, N., & Adhikari, S., (2005). Current status of hydrogen production techniques by steam reforming of ethanol: A review. *Energy and Fuels, 19*(5), 2098–2106.
41. Zehner, O., (2012). *Green Illusions: The Dirty Secrets of Clean Energy and the Future of Environmentalism.* U of Nebraska Press.
42. McDowall, W., & Eames, M., (2006). Forecasts, scenarios, visions, back casts and roadmaps to the hydrogen economy: A review of the hydrogen futures literature. *Energy Policy, 34*(11), 1236–1250.
43. Zacharia, R., (2015). Review of solid state hydrogen storage methods adopting different kinds of novel materials. *Journal of Nanomaterials, 2015*, 4.
44. Das, L. M., (1996). On-board hydrogen storage systems for automotive application. *International Journal of Hydrogen Energy, 21*(9), 789–800.
45. https://www.energy.gov/eere/fuelcells/doe-technical-targets-onboard-hydrogen-storage-light-duty-vehicles (accessed on 3 August 2020).
46. Guo, Z. X., Shang, C., & Aguey-Zinsou, K. F., (2008). Materials challenges for hydrogen storage. *Journal of the European Ceramic Society, 28*(7), 1467–1473.
47. Zhevago, N. K., (2016). Other methods for the physical storage of hydrogen. In: *Compendium of Hydrogen Energy* (pp. 189–218). Wood head Publishing.
48. Satyapal, S., Petrovic, J., Read, C., Thomas, G., & Ordaz, G., (2007). The US department of energy's national hydrogen storage project: Progress towards meeting hydrogen-powered vehicle requirements. *Catalysis Today, 120*(3/4), 246–256.
49. Sakintuna, B., Lamari-Darkrim, F., & Hirscher, M., (2007). Metal hydride materials for solid hydrogen storage: A review. *International Journal of Hydrogen Energy, 32*(9), 1121–1140.
50. Klell, M., (2010). Storage of hydrogen in the pure form. *Handbook of Hydrogen Storage*, 187–214.
51. Schneemann, A., White, J. L., Kang, S., Jeong, S., Wan, L. F., Cho, E. S., Heo, T. W., et al., (2018). Nanostructured metal hydrides for hydrogen storage. *Chemical Reviews, 118*(22), 10775–10839.

52. Li, Q., & Bjerrum, N. J., (2002). Aluminum as anode for energy storage and conversion: A review. *Journal of Power Sources, 110*(1), 1–10.
53. Graetz, J., Reilly, J. J., Yartys, V. A., Maehlen, J. P., Bulychev, B. M., Antonov, V. E., Tarasov, B. P., & Gabis, I. E., (2011). Aluminum hydride as a hydrogen and energy storage material: Past, present, and future. *Journal of Alloys and Compounds, 509*, S517–528.
54. Pumera, M., & Wong, C. H., (2013). Graphane and hydrogenated graphene. *Chemical Society Reviews, 42*(14), 5987–5995.
55. Sevilla, M., & Mokaya, R., (2014). Energy storage applications of activated carbons: Super capacitors and hydrogen storage. *Energy and Environmental Science, 7*(4), 1250–1280.
56. Blankenship, II. T. S., Balahmar, N., & Mokaya, R., (2017). Oxygen-rich microporous carbons with exceptional hydrogen storage capacity. *Nature Communications, 8*(1), 1545.
57. Florusse, L. J., Peters, C. J., Schoonman, J., Hester, K. C., Koh, C. A., Dec, S. F., Marsh, K. N., & Sloan, E. D., (2004). Stable low-pressure hydrogen clusters stored in a binary clathrate hydrate. *Science, 306*(5695), 469–471.
58. Dillon, A., Jones, K. M., Bekkedahl, T. A., Kiang, C. H., Bethune, D. S., & Heben, M. J., (1997). Storage of hydrogen in single-walled carbon nanotubes. *Nature, 386*(6623), 377.
59. Zhou, H. C., Long, J. R., & Yaghi, O. M., (2012). *Introduction to Metal-Organic Frameworks*.
60. James, S. L., (2003). Metal-organic frameworks. *Chemical Society Reviews, 32*(5), 276–288.
61. Zhevago, N. K., Denisov, E. I., & Glebov, V. I., (2010). Experimental investigation of hydrogen storage in capillary arrays. *International Journal of Hydrogen Energy, 35*(1), 169–175.
62. Wicks, G. G., Heung, L. K., & Schumacher, R. F., (2008). SRNL's porous, hollow glass balls open new opportunities for hydrogen storage, drug delivery, and national defense. *American Ceramic Society Bulletin, 87*(6), 23–27.
63. Yu, X., Tang, Z., Sun, D., Ouyang, L., & Zhu, M., (2017). Recent advances and remaining challenges of nanostructured materials for hydrogen storage applications. *Progress in Materials Science, 88*, 1–48.
64. Hussein, A. K., (2015). Applications of nanotechnology in renewable energies: A comprehensive overview and understanding. *Renewable and Sustainable Energy Reviews, 42*, 460–476.
65. Roston, E., (2010). *The Carbon Age: How Life's Core Element Has Become Civilization's Greatest Threat*. Bloomsbury Publishing USA.
66. Zaytseva, O., & Neumann, G., (2016). Carbon nanomaterials: Production, impact on plant development, agricultural, and environmental applications. *Chemical and Biological Technologies in Agriculture, 3*(1), 17.
67. Kroto, H. W., Heath, J. R., O'Brien, S. C., Curl, R. F., & Smalley, R. E., (1985). *Nature, 318*(162), 1216.
68. Yadav, B. C., & Kumar, R., (2008). Structure, properties, and applications of fullerenes. *International Journal of Nanotechnology and Applications, 2*(1), 15–24.
69. Zhang, R., Zhang, Y., Zhang, Q., Xie, H., Qian, W., & Wei, F., (2013). Growth of half-meter long carbon nanotubes based on Schulz-Flory distribution. *ACS Nano, 7*(7), 6156–6161.

70. Li, D., & Kaner, R. B., (2008). Graphene-based materials. *Science, 320*(5880), 1170, 1171.
71. Ferrari, A. C., Meyer, J. C., Scardaci, V., Casiraghi, C., Lazzeri, M., Mauri, F., Piscanec, S., et al., (2006). Raman spectrum of graphene and graphene layers. *Physical Review Letters, 97*(18), 187401.
72. Tozzini, V., & Pellegrini, V., (2013). Prospects for hydrogen storage in graphene. *Physical Chemistry Chemical Physics, 15*(1), 80–89.
73. Narita, N., Nagai, S., Suzuki, S., & Nakao, K., (1998). Optimized geometries and electronic structures of graphyne and its family. *Physical Review B., 58*(16), 11009.
74. Jiao, Y., Du, A., Hankel, M., Zhu, Z., Rudolph, V., & Smith, S. C., (2011). Graphdiyne: A versatile nanomaterial for electronics and hydrogen purification. *Chemical Communications, 47*(43), 11843–11845.
75. Shunhong, Z., Jian, Z., Qian, W., Xiaoshuang, C., Yoshiyuki, K., & Puru, J., (2015). Penta-graphene: A new carbon allotrope. 7(2).
76. Lueking, A. D., Pan, L., Narayanan, D. L., & Clifford, C. E., (2005). Effect of expanded graphite lattice in exfoliated graphite nanofibers on hydrogen storage. *The Journal of Physical Chemistry B., 109*(26), 12710–12717.
77. Jin, H., Lee, Y. S., & Hong, I., (2007). Hydrogen adsorption characteristics of activated carbon. *Catalysis Today, 120*(3/4), 399–406.
78. Thomas, K. M., (2007). Hydrogen adsorption and storage on porous materials. *Catalysis Today, 120*(3/4), 389–398.
79. Chambers, A., Park, C., Baker, R. T., & Rodriguez, N. M., (1998). Hydrogen storage in graphite nanofibers. *The Journal of Physical Chemistry B., 102*(22), 4253–4256.
80. Liu, C., Fan, Y. Y., Liu, M., Cong, H. T., Cheng, H. M., & Dresselhaus, M. S., (1999). Hydrogen storage in single-walled carbon nanotubes at room temperature. *Science, 286*(5442), 1127–1129.
81. Texier-Mandoki, N., Dentzer, J., Piquero, T., Saadallah, S., David, P., & Vix-Guterl, C., (2004). Hydrogen storage in activated carbon materials: Role of the nanoporous texture. *Carbon, 12*(42), 2744–2747.
82. Gundiah, G., Govindaraj, A., Rajalakshmi, N., Dhathathreyan, K. S., & Rao, C. N., (2003). Hydrogen storage in carbon nanotubes and related materials. In: *Advances in Chemistry: A Selection of CNR Rao's Publications (1994–2003)* (pp. 305–309).
83. Gogotsi, Y., Dash, R. K., Yushin, G., Yildirim, T., Laudisio, G., & Fischer, J. E., (2005). Tailoring of nanoscale porosity in carbide-derived carbons for hydrogen storage. *Journal of the American Chemical Society, 127*(46), 16006, 16007.
84. Pupysheva, O. V., Farajian, A. A., & Yakobson, B. I., (2007). Fullerene nanocage capacity for hydrogen storage. *Nano Letters, 8*(3), 767–774.
85. Yoon, M., Yang, S., Wang, E., & Zhang, Z., (2007). Charged fullerenes as high-capacity hydrogen storage media. *Nano letters (Print), 7*(9), 2578–2583.
86. Zhong, Z. Y., Xiong, Z. T., Sun, L. F., Luo, J. Z., Chen, P., Wu, X., Lin, J., & Tan, K. L., (2002). Nanosized nickel (or cobalt)/graphite composites for hydrogen storage. *The Journal of Physical Chemistry B., 106*(37), 9507–9513.
87. Kuc, A., Zhechkov, L., Patchkovskii, S., Seifert, G., & Heine, T., (2007). Hydrogen sieving and storage in fullerene intercalated graphite. *Nano Letters, 7*(1), 1–5.
88. Lueking, A. D., Pan, L., Narayanan, D. L., & Clifford, C. E., (2005). Effect of expanded graphite lattice in exfoliated graphite nanofibers on hydrogen storage. *The Journal of Physical Chemistry B., 109*(26), 12710–12717.

89. Orimo, S., Majer, G., Fukunaga, T., Züttel, A., Schlapbach, L., & Fujii, H., (1999). Hydrogen in the mechanically prepared nanostructured graphite. *Applied Physics Letters, 75*(20), 3093–3095.
90. Tozzini, V., & Pellegrini, V., (2011). Reversible hydrogen storage by controlled buckling of graphene layers. *The Journal of Physical Chemistry C., 115*(51), 25523–25528.
91. Lin, Y., Ding, F., & Yakobson, B. I., (2008). Hydrogen storage by spillover on graphene as a phase nucleation process. *Physical Review B., 78*(4), 041402.
92. Zieliński, M., Wojcieszak, R., Monteverdi, S., Mercy, M., & Bettahar, M. M., (2005). Hydrogen storage on nickel catalysts supported on amorphous activated carbon. *Catalysis Communications, 6*(12), 777–783.
93. Im, J. S., Park, S. J., Kim, T., & Lee, Y. S., (2009). Hydrogen storage evaluation based on investigations of the catalytic properties of metal/metal oxides in electro spun carbon fibers. *International Journal of Hydrogen Energy, 34*(8), 3382–3388.
94. Dillon, A. C., Gennet, T., Alleman, J. L., Jones, K. M., Parilla, P. A., & Heben, M. J., (1999). *DOE Hydrogen Program, FY Progress Report.*
95. Smith, M. R., Bittner, E. W., Shi, W., Johnson, J. K., & Bockrath, B. C., (2003). Chemical activation of single-walled carbon nanotubes for hydrogen adsorption. *The Journal of Physical Chemistry B., 107*(16), 3752–3760.
96. Hirscher, M., Becher, M., Haluska, M., Dettlaff-Weglikowska, U., Quintel, A., Duesberg, G. S., Choi, Y. M., et al., (2001). Hydrogen storage in sonicated carbon materials. *Applied Physics A., 72*(2), 129–132.
97. Ye, Y., Ahn, C. C., Witham, C., Fultz, B., Liu, J., Rinzler, A. G., Colbert, D., et al., (1999). Hydrogen adsorption and cohesive energy of single-walled carbon nanotubes. *Applied Physics Letters, 74*(16), 2307–2309.
98. Nishimiya, N., Ishigaki, K., Takikawa, H., Ikeda, M., Hibi, Y., Sakakibara, T., Matsumoto, A., & Tsutsumi, K., (2002). Hydrogen sorption by single-walled carbon nanotubes prepared by a torch arc method. *Journal of Alloys and Compounds, 339*(1/2), 275–282.
99. Wu, X. B., Chen, P., Lin, J., & Tan, K. L., (2000). Hydrogen uptake by carbon nanotubes. *International Journal of Hydrogen Energy, 25*(3), 261–265.
100. Chen, Y., Shaw, D. T., Bai, X. D., Wang, E. G., Lund, C., Lu, W. M., & Chung, D. D., (2001). Hydrogen storage in aligned carbon nanotubes. *Applied Physics Letters, 78*(15), 2128–2130.
101. Tibbetts, G. G., Meisner, G. P., & Olk, C. H., (2001). Hydrogen storage capacity of carbon nanotubes, filaments, and vapor-grown fibers. *Carbon, 39*(15), 2291–2301.
102. Zhu, H., Cao, A., Li, X., Xu, C., Mao, Z., Ruan, D., Liang, J., & Wu, D., (2001). Hydrogen adsorption in bundles of well-aligned carbon nanotubes at room temperature. *Applied Surface Science, 178*(1–4), 50–55.
103. Kowalczyk, P., Hołyst, R., Terrones, M., & Terrones, H., (2007). Hydrogen storage in nanoporous carbon materials: Myth and facts. *Physical Chemistry Chemical Physics, 9*(15), 1786–1792.
104. Gallego, N. C., He, L., Saha, D., Contescu, C. I., & Melnichenko, Y. B., (2011). Hydrogen confinement in carbon nanopores: Extreme densification at ambient temperature. *Journal of the American Chemical Society, 133*(35), 13794–13797.
105. Ströbel, R., Jörissen, L., Schliermann, T., Trapp, V., Schütz, W., Bohmhammel, K., Wolf, G., & Garche, J., (1999). Hydrogen adsorption on carbon materials. *Journal of Power Sources, 84*(2), 221–224.

106. Fan, Y. Y., Liao, B., Liu, M., Wei, Y. L., Lu, M. Q., & Cheng, H. M., (1999). Hydrogen uptake in vapor-grown carbon nanofibers. *Carbon, 10*(37), 1649–1652.
107. De la Casa-Lillo, M. A., Lamari-Darkrim, F., Cazorla-Amoros, D., & Linares-Solano, A., (2002). Hydrogen storage in activated carbons and activated carbon fibers. *The Journal of Physical Chemistry B., 106*(42), 10930–10934.
108. Saha, D., & Deng, S., (2009). Hydrogen adsorption on ordered mesoporous carbons doped with Pd, Pt, Ni, and Ru. *Langmuir, 25*(21), 12550–12560.
109. Wang, L., & Yang, R. T., (2008). Hydrogen storage properties of carbons doped with ruthenium, platinum, and nickel nanoparticles. *The Journal of Physical Chemistry C., 112*(32), 12486–12494.
110. Zubizarreta, L., Menéndez, J. A., Job, N., Marco-Lozar, J. P., Pirard, J. P., Pis, J. J., Linares-Solano, A., et al., (2010). Ni-doped carbon xerogels for H_2 storage. *Carbon, 48*(10), 2722–2733.
111. Zubizarreta, L., Menéndez, J. A., Pis, J. J., & Arenillas, A., (2009). Improving hydrogen storage in Ni-doped carbon nanospheres. *International Journal of Hydrogen Energy, 34*(7), 3070–3076.
112. Hu, Z. L., Aizawa, M., Wang, Z. M., Yoshizawa, N., & Hatori, H., (2010). Synthesis and characteristics of graphene oxide-derived carbon nanosheet-Pd nanosized particle composites. *Langmuir, 26*(9), 6681–6688.
113. Zubizarreta, L., Arenillas, A., & Pis, J. J., (2009). Carbon materials for H_2 storage. *International Journal of Hydrogen Energy, 34*(10), 4575–4581.
114. Mpourmpakis, G., Tylianakis, E., & Froudakis, G. E., (2007). Carbon nanoscrolls: A promising material for hydrogen storage. *Nano Letters, 7*(7), 1893–1897.
115. Atkinson, K., Roth, S., Hirscher, M., & Grünwald, W., (2001). Carbon nanostructures: An efficient hydrogen storage medium for fuel cells. *Fuel Cells Bulletin, 4*(38), 9–12.
116. Ströbel, R., Garche, J., Moseley, P. T., Jörissen, L., & Wolf, G., (2006). Hydrogen storage by carbon materials. *Journal of Power Sources, 159*(2), 781–801.
117. Hirscher, M., & Becher, M., (2003). Hydrogen storage in carbon nanotubes. *Journal of Nanoscience and Nanotechnology, 3(*1, 2), 3–17.
118. Cheng, H. M., Yang, Q. H., & Liu, C., (2001). Hydrogen storage in carbon nanotubes. *Carbon, 39*(10), 1447–1454.
119. Hirscher, M., Becher, M., Haluska, M., Von, Z. F., Chen, X., Dettlaff-Weglikowska, U., & Roth, S., (2003). Are carbon nanostructures an efficient hydrogen storage medium? *Journal of Alloys and Compounds, 356*, 433–437.
120. Rzepka, M., Lamp, P., & Casa-Lillo, M. A. D. L., (1998). Physisorption of hydrogen on microporous carbon and carbon nanotubes. *The Journal of Physical Chemistry B., 102*(52), 10894–10898.
121. Wang, Q., & Johnson, J. K., (1999). Computer simulations of hydrogen adsorption on graphite nanofibers. *The Journal of Physical Chemistry B., 103*(2), 277–281.
122. Kubas, G. J., (2001). Metal-dihydrogen and σ-bond coordination: The consummate extension of the Dewar-Chatt-Duncanson model for metal-olefin π bonding. *Journal of Organometallic Chemistr., 635*(1/2), 37–68.
123. Skipper, C. V., (2013). *The Kubas Interaction in Transition Metal Based Hydrogen Storage Materials*. (Doctoral dissertation, UCL (University College London)).
124. Chung, C., Ihm, J., & Lee, H., (2015). Recent progress on Kubas-type hydrogen-storage nanomaterials: From theories to experiments. *Journal of the Korean Physical Society, 66*(11), 1649–1655.

125. Hoang, T. K., & Antonelli, D. M., (2009). Exploiting the Kubas interaction in the design of hydrogen storage materials. *Advanced Materials, 21*(18), 1787–1800.
126. Wang, L., & Yang, R. T., (2008). New sorbents for hydrogen storage by hydrogen spillover-a review. *Energy Environmental Science, 1*(2), 268–279.
127. Lueking, A. D., & Yang, R. T., (2004). Hydrogen spillover to enhance hydrogen storage-study of the effect of carbon physicochemical properties. *Applied Catalysis A: General, 265*(2), 259–268.
128. Cheng, H., Chen, L., Cooper, A. C., Sha, X., & Pez, G. P., (2008). Hydrogen spillover in the context of hydrogen storage using solid-state materials. *Energy and Environmental Science, 1*(3), 338–354.
129. Yoon, M., Yang, S., Hicke, C., Wang, E., Geohegan, D., & Zhang, Z., (2008). Calcium as the superior coating metal in functionalization of carbon fullerenes for high-capacity hydrogen storage. *Physical Review Letters, 100*(20), 206806.
130. Yildirim, T., & Ciraci, S., (2005). Titanium-decorated carbon nanotubes as a potential high-capacity hydrogen storage medium. *Physical Review Letters, 94*(17), 175501.
131. Kaur, M., & Pal, K., (2016). An investigation for hydrogen storage capability of zirconia-reduced graphene oxide nanocomposite. *International Journal of Hydrogen Energy, 41*(47), 21861–21869.
132. Parambhath, V. B., Nagar, R., & Ramaprabhu, S., (2012). Effect of nitrogen doping on hydrogen storage capacity of palladium decorated graphene. *Langmuir, 28*(20), 7826–7833.
133. Tylianakis, E., Psofogiannakis, G. M., & Froudakis, G. E., (2010). Li-doped pillared graphene oxide: A graphene-based nanostructured material for hydrogen storage. *The Journal of Physical Chemistry Letters, 1*(16), 2459–2464.
134. Lee, H., Ihm, J., Cohen, M. L., & Louie, S. G., (2010). Calcium-decorated graphene-based nanostructures for hydrogen storage. *Nano Letters, 10*(3), 793–798.
135. Ariharan, A., Viswanathan, B., & Nandhakumar, V., (2016). Heteroatom doped multi-layered graphene material for hydrogen storage application. *Graphene, 5*(02), 39.
136. Divya, P., & Ramaprabhu, S., (2014). Hydrogen storage in platinum decorated hydrogen exfoliated graphene sheets by spillover mechanism. *Physical Chemistry Chemical Physics, 16*(48), 26725–26729.
137. Samantaray, S. S., Sangeetha, V., Abinaya, S., & Ramaprabhu, S., (2018). Enhanced hydrogen storage performance in Pd3Co decorated nitrogen/boron doped graphene composites. *International Journal of Hydrogen Energy, 43*(16), 8018–8025.
138. Deng, W. Q., Xu, X., & Goddard, W. A., (2004). New alkali doped pillared carbon materials designed to achieve practical reversible hydrogen storage for transportation. *Physical Review Letters, 92*(16), 166103.
139. Ataca, C., Aktürk, E., Ciraci, S., & Ustunel, H., (2008). High-capacity hydrogen storage by metalized graphene. *Applied Physics Letters, 93*(4), 043123.
140. Ao, Z. M., & Peeters, F. M., (2010). High-capacity hydrogen storage in Al-adsorbed graphene. *Physical Review B., 81*(20), 205406.
141. Chen, P., Wu, X., Lin, J., & Tan, K. L., (1999). High H_2 uptake by alkali-doped carbon nanotubes under ambient pressure and moderate temperatures. *Science, 285*(5424), 91–93.
142. Yang, R. T., (2000). Hydrogen storage by alkali-doped carbon nanotubes-revisited. *Carbon, 38*(4), 623–626.
143. Pinkerton, F. E., Wicke, B. G., Olk, C. H., Tibbetts, G. G., Meisner, G. P., Meyer, M. S., & Herbst, J. F., (2000). Thermo gravimetric measurement of hydrogen absorption

in alkali-modified carbon materials. *The Journal of Physical Chemistry B., 104*(40), 9460–9467.

144. Prakash, J., Tripathi, B. M., Dasgupta, K., Chakravartty, J. K., Pai, M. R., Kumar, A., & Bharadwaj, S. R., (2016). Tuning of hydrogen storage property of Multi-walled carbon nanotube by decorating Ni, Cu, and Fe nanoparticles. *Current Nanomaterials, 1*(2), 124–131.
145. Vinayan, B. P., Sethupathi, K., & Ramaprabhu, S., (2012). Hydrogen storage studies of palladium decorated nitrogen doped graphene nanoplatelets. *Journal of Nanoscience and Nanotechnology, 12*(8), 6608–6614.
146. Reyhani, A., Mortazavi, S. Z., Mirershadi, S., Moshfegh, A. Z., Parvin, P., & Golikand, A. N., (2011). Hydrogen storage in decorated multi walled carbon nanotubes by Ca, Co, Fe, Ni, and Pd nanoparticles under ambient conditions. *The Journal of Physical Chemistry C., 115*(14), 6994–7001.
147. Sankaran, M., & Viswanathan, B., (2007). Hydrogen storage in boron substituted carbon nanotubes. *Carbon, 45*(8), 1628–1635.
148. Chen, C. H., & Huang, C. C., (2007). Hydrogen storage by KOH-modified multi-walled carbon nanotubes. *International Journal of Hydrogen Energy, 32*(2), 237–246.
149. Yadav, A., Faisal, M., Subramaniam, A., & Verma, N., (2017). Nickel nanoparticle-doped and steam-modified multiscale structure of carbon micro-nanofibers for hydrogen storage: Effects of metal, surface texture and operating conditions. *International Journal of Hydrogen Energy, 42*(9), 6104–6117.
150. Ariharan, A., Viswanathan, B., & Nandhakumar, V., (2016). *Phosphorous-Doped Porous Carbon Derived from Paste of Newly Growing Ficus benghalensis as Hydrogen Storage Material.*
151. Hwang, H. J., Kwon, Y., & Lee, H., (2012). Thermodynamically stable calcium-decorated graphyne as a hydrogen storage medium. *The Journal of Physical Chemistry C., 116*(38)20220–20224.
152. Nair, A. A., Sundara, R., & Anitha, N., (2015). Hydrogen storage performance of palladium nanoparticles decorated graphitic carbon nitride. *International Journal of Hydrogen Energy, 40*(8), 3259–3267.
153. Wang, Z., & Yang, R. T., (2010). Enhanced hydrogen storage on Pt-doped carbon by plasma reduction. *The Journal of Physical Chemistry C., 114*(13), 5956–5963.
154. Sankaran, M., & Viswanathan, B., (2006). The role of heteroatoms in carbon nanotubes for hydrogen storage. *Carbon, 44*(13), 2816–2821.
155. Yadav, A., Chakraborty, B., Gangan, A., Patel, N., Press, M. R., & Ramaniah, L. M., (2017). Magnetic moment controlling desorption temperature in hydrogen storage: A case of zirconium-doped graphene as a high capacity hydrogen storage medium. *The Journal of Physical Chemistry C., 121*(31), 16721–16730.
156. Li, C., Li, J., Wu, F., Li, S. S., Xia, J. B., & Wang, L. W., (2011). High capacity hydrogen storage in Ca decorated graphyne: A first-principles study. *The Journal of Physical Chemistry C., 115*(46), 23221–23225.
157. Chakraborty, B., Modak, P., & Banerjee, S., (2012). Hydrogen storage in yttrium-decorated single walled carbon nanotube. *The Journal of Physical Chemistry C., 116*(42), 22502–22508.
158. Yadav, M., & Xu, Q., (2012). Liquid-phase chemical hydrogen storage materials. *Energy and Environmental Science, 5*(12), 9698–9725.

159. Orimo, S. I., Nakamori, Y., Eliseo, J. R., Züttel, A., & Jensen, C. M., (2007). Complex hydrides for hydrogen storage. *Chemical Reviews, 107*(10), 4111–4132.
160. Thomas, K. M., (2007). Hydrogen adsorption and storage on porous materials. *Catalysis Today, 120*(3/4), 389–398.
161. Brandrup, J., Immergut, E. H., Grulke, E. A., Abe, A., & Bloch, D. R., (1999). *Polymer Handbook.* New York: Wiley.
162. Odian, G., (2004). *Principles of Polymerization.* John Wiley & Sons.
163. Billmeyer, F. W., & Billmeyer, F. W., (1984). *Textbook of Polymer Science.*
164. Wu, D., Xu, F., Sun, B., Fu, R., He, H., & Matyjaszewski, K., (2012). Design and preparation of porous polymers. *Chemical Reviews, 112*(7), 3959–4015.
165. Huang, J., & Turner, S. R., (2018). Hyper cross linked polymers: A review. *Polymer Reviews, 58*(1), 1–41.
166. McKeown, N. B., & Budd, P. M., (2006). Polymers of intrinsic microporosity (PIMs): Organic materials for membrane separations, heterogeneous catalysis, and hydrogen storage. *Chemical Society Reviews, 35*(8), 675–683.
167. Kato, R., & Nishide, H., (2018). Polymers for carrying and storing hydrogen. *Polymer Journal, 50*(1), 77.
168. Kato, R., Oya, T., Shimazaki, Y., Oyaizu, K., & Nishide, H., (2017). A hydrogen-storing quinaldine polymer: Nickel-electro deposition-assisted hydrogenation and subsequent hydrogen evolution. *Polymer International, 66*(5), 647–652.
169. Yamaguchi, R., Ikeda, C., Takahashi, Y., & Fujita, K. I., (2009). Homogeneous catalytic system for reversible dehydrogenation-hydrogenation reactions of nitrogen heterocycles with reversible interconversion of catalytic species. *Journal of the American Chemical Society, 131*(24), 8410–8412.
170. Kato, R., Yoshimasa, K., Egashira, T., Oya, T., Oyaizu, K., & Nishide, H., (2016). A ketone/alcohol polymer for cycle of electrolytic hydrogen-fixing with water and releasing under mild conditions. *Nature Communications, 7*, 13032.
171. Oyaizu, K., & Nishide, H., (2009). Radical polymers for organic electronic devices: A radical departure from conjugated polymers? *Advanced Materials, 21*(22), 2339–2344.
172. Kawahara, R., Fujita, K. I., & Yamaguchi, R., (2012). Dehydrogenative oxidation of alcohols in aqueous media using water-soluble and reusable Cp* Ir catalysts bearing a functional bipyridine ligand. *Journal of the American Chemical Society, 134*(8), 3643–3646.
173. Staudinger, H., & Wiedersheim, V., (1929). On high polymer compounds, Part 21:About the reduction of polystyrene. *Reports of the German Chemical Society (A and B Series), 62*(8), 2406–2411.
174. Mülhaupt, R., (2004). Hermann Staudinger and the origin of macromolecular chemistry. *Angewandte Chemie. International Edition, 43*(9), 1054–1063.
175. Germain, J., Hradil, J., Fréchet, J. M., & Svec, F., (2006). High surface area nanoporous polymers for reversible hydrogen storage. *Chemistry of Materials, 18*(18), 4430–4435.
176. Wood, C. D., Tan, B., Trewin, A., Niu, H., Bradshaw, D., Rosseinsky, M. J., Khimyak, Y. Z., et al., (2007). Hydrogen storage in micro porous hyper cross linked organic polymer networks. *Chemistry of Materials, 19*(8), 2034–2048.
177. McKeown, N. B., Gahnem, B., Msayib, K. J., Budd, P. M., Tattershall, C. E., Mahmood, K., Tan, S., et al., (2006). Towards polymer-based hydrogen storage materials: Engineering ultra micro porous cavities within polymers of intrinsic microporosity. *Angewandte Chemie International Edition, 45*(11), 1804–1807.

178. Budd, P. M., Butler, A., Selbie, J., Mahmood, K., McKeown, N. B., Ghanem, B., Msayib, K., et al., (2007). The potential of organic polymer-based hydrogen storage materials. *Physical Chemistry Chemical Physics, 9*(15), 1802–1808.
179. Ghanem, B. S., Msayib, K. J., McKeown, N. B., Harris, K. D., Pan, Z., Budd, P. M., Butler, A., et al., (2007). A triptycene-based polymer of intrinsic microposity that displays enhanced surface area and hydrogen adsorption. *Chemical Communications, 1*, 67–69.
180. Lee, J. Y., Wood, C. D., Bradshaw, D., Rosseinsky, M. J., & Cooper, A. I., (2006). Hydrogen adsorption in microporous hyper-crosslinked polymers. *Chemical Communications,25*, 2670–2672.
181. Yuan, S., Kirklin, S., Dorney, B., Liu, D. J., & Yu, L., (2009). Nanoporous polymers containing stereo contorted cores for hydrogen storage. *Macromolecules, 42*(5), 1554–1559.
182. Šmigol, V., Švec, F., (1993). Preparation and properties of uniform beads based on macro porous glycidyl methacrylate-ethylene dimethacrylate copolymer: Use of chain transfer agent for control of pore-size distribution. *Journal of Applied Polymer Science, 48*(11), 2033–2039.
183. Fontanals, N., Marcé, R. M., Galià, M., & Borrull, F., (2003). Preparation and characterization of highly polar polymeric sorbents from styrene-divinylbenzene and vinylpyridine-divinylbenzene for the solid-phase extraction of polar organic pollutants. *Journal of Polymer Science Part A: Polymer Chemistry, 41*(13), 1927–1933.
184. Shaffei, K. A., Atta, A. M., Gomes, C. S., Gomes, P. T., El-Ghazawy, R. A., & Mahmoud, A. G., (2014). Preparation of polymer networks for hydrogen storage using the ullmann synthetic protocol. *Journal of Polymer Research, 21*(5), 445.
185. Germain, J., Svec, F., & Fréchet, J. M., (2008). Preparation of size-selective nanoporous polymer networks of aromatic rings: Potential adsorbents for hydrogen storage. *Chemistry of Materials, 20*(22), 7069–7076.
186. Thomas, A., Kuhn, P., Weber, J., Titirici, M. M., & Antonietti, M., (2009). Porous polymers: Enabling solutions for energy applications. *Macromolecular Rapid Communications, 30*(4/5), 221–236.
187. McKeown, N. B., Budd, P. M., & Book, D., (2007). Micro porous polymers as potential hydrogen storage materials. *Macromolecular Rapid Communications, 28*(9), 995–1002.
188. Ramimoghadam, D., Gray, E. M., & Webb, C. J., (2016). Review of polymers of intrinsic micro porosity for hydrogen storage applications. *International Journal of Hydrogen Energy, 41*(38), 16944–16965.
189. Yuan, S., Kirklin, S., Dorney, B., Liu, D. J., & Yu, L., (2009). Nanoporous polymers containing stereocorcorted cores of hydrogen storage. *Macromolecules, 42*(5), 1554–1559.
190. Rzepka, M., Lamp, P., & De La Casa-Lillo, M. A., (1998). Physisorption of hydrogen on microporous carbon and carbon nanotubes. *The Journal of Physical Chemistry B., 102*(52), 10894–10898.
191. Xu, Y., Jin, S., Xu, H., Nagai, A., & Jiang, D., (2013). Conjugated micro porous polymers: Design, synthesis, and application. *Chemical Society Reviews, 42*(20), 8012–8031.
192. Cooper, A. I., (2009). Conjugated micro porous polymers. *Advanced Materials, 21*(12), 1291–1295.
193. Dawson, R., Laybourn, A., Clowes, R., Khimyak, Y. Z., Adams, D. J., & Cooper, A. I., (2009). Functionalized conjugated micro porous polymers. *Macromolecules, 42*(22), 8809–8816.

194. Jiang, J. X., Wang, C., Laybourn, A., Hasell, T., Clowes, R., Khimyak, Y. Z., Xiao, J., et al., (2011). Metal-organic conjugated micro porous polymers. *Angewandte Chemie International Edition, 50*(5), 1072–1075.
195. Ding, S. Y., & Wang, W., (2013). Covalent organic frameworks (COFs), from design to applications. *Chemical Society Reviews, 42*(2), 548–568.
196. Feng, X., Ding, X., & Jiang, D., (2012). Covalent organic frameworks. *Chemical Society Reviews, 41*(18), 6010–6122.
197. Cote, A. P., Benin, A. I., Ockwig, N. W., O'Keeffe, M., Matzger, A. J., & Yaghi, O. M., (2005). Porous, crystalline, covalent organic frameworks. *Science, 310*(5751), 1166–1170.
198. El-Kaderi, H. M., Hunt, J. R., Mendoza-Cortés, J. L., Côté, A. P., Taylor, R. E., O'Keeffe, M., & Yaghi, O. M., (2007). Designed synthesis of 3D covalent organic frameworks. *Science, 316*(5822), 268–272.
199. Waller, P. J., Gándara, F., & Yaghi, O. M., (2015). Chemistry of covalent organic frameworks. *Accounts of Chemical Research, 48*(12), 3053–3063.
200. Kuhn, P., Antonietti, M., & Thomas, A., (2008). Porous, covalent triazine-based frameworks prepared by ionothermal synthesis. *Angewandte Chemie International Edition, 47*(18), 3450–3453.
201. Hug, S., Mesch, M. B., Oh, H., Popp, N., Hirscher, M., Senker, J., & Lotsch, B. V., (2014). A fluorene based covalent triazine framework with high CO_2 and H_2 capture and storage capacities. *Journal of Materials Chemistry A.,2*(16), 5928–5936.
202. Chen, X., Yuan, F., Gu, Q., & Yu, X., (2013). Light metals decorated covalent triazine-based frameworks as a high capacity hydrogen storage medium. *Journal of Materials Chemistry A., 1*(38), 11705–11710.
203. Ben, T., & Qiu, S., (2013). Porous aromatic frameworks: Synthesis, structure, and functions. *Cryst. Eng. Comm., 15*(1), 17–26.
204. Ben, T., Pei, C., Zhang, D., Xu, J., Deng, F., Jing, X., & Qiu, S., (2011). Gas storage in porous aromatic frameworks (PAFs). *Energy and Environmental Science, 4*(10), 3991–3999.
205. Lan, J., Cao, D., Wang, W., Ben, T., & Zhu, G., (2010). High-capacity hydrogen storage in porous aromatic frameworks with diamond-like structure. *The Journal of Physical Chemistry Letters, 1*(6), 978–981.
206. Lee, H., Choi, W. I., & Ihm, J., (2006). Combinatorial search for optimal hydrogen-storage nanomaterials based on polymers. *Physical Review Letters, 97*(5), 056104.
207. Durgun, E., Ciraci, S., Zhou, W., & Yildirim, T., (2006). Transition-metal-ethylene complexes as high-capacity hydrogen-storage media. *Physical Review Letters, 97*(22), 226102.
208. Li, A., Lu, R. F., Wang, Y., Wang, X., Han, K. L., & Deng, W. Q., (2010). Lithium-doped conjugated micro porous polymers for reversible hydrogen storage. *Angewandte Chemie International Edition, 49*(19), 3330–3333.
209. Sathe, R. Y., & Kumar, T. D., (2018). Paracyclophane functionalized with Sc and Li for hydrogen storage. *Chemical Physics Letters, 692*, 253–257.
210. Yuan, S., White, D., Mason, A., & Liu, D. J., (2013). Porous organic polymers containing carborane for hydrogen storage. *International Journal of Energy Research, 37*(7), 732–740.
211. Zhao, Y., Liu, F., Tan, J., Li, P., Wang, Z., Zhu, K., Mai, X., et al., (2019). Preparation and hydrogen storage of Pd/MIL-101 nanocomposites. *Journal of Alloys and Compounds, 772*, 186–192.

212. Kubas, G. J., (2009). Hydrogen activation on organometallic complexes and H_2 production, utilization, and storage for future energy. *Journal of Organometallic Chemistry, 694*(17), 2648–1653.
213. Ceperley, D. M., & Alder, B. J., (1980). Ground state of the electron gas by a stochastic method. *Physical Review Letters, 45*(7), 566.
214. Perdew, J. P., Burke, K., & Ernzerhof, M., (1996). Generalized gradient approximation made simple. *Physical Review Letters, 77*(18), 3865.
215. Martin, R. M., & Martin, R. M., (2004). *Electronic Structure: Basic Theory and Practical Methods.* Cambridge University Press.
216. Capelle, K., (2006). A bird's-eye view of density-functional theory. *Brazilian Journal of Physics, 36*(4A), 1318–1343.
217. Argaman, N., & Makov, G., (2000). Density functional theory: An introduction. *American Journal of Physics, 68*(1), 69–79.
218. Mayer, I., (2013). *Simple Theorems, Proofs, and Derivations in Quantum Chemistry.* Springer Science & Business Media.
219. Thijssen, J., (2007). *Computational Physics*. Cambridge university press.
220. Foulkes, W. M., Mitas, L., Needs, R. J., & Rajagopal, G., (2001). Quantum Monte Carlo simulations of solids. *Reviews of Modern Physics, 73*(1), 33.
221. Hammond, B. L., Lester, W. A., & Reynolds, P. J., (1994*). Monte Carlo Methods in Ab Initio Quantum Chemistry*. World Scientific.
222. Kolorenč, J., & Mitas, L., (2011). Applications of quantum Monte Carlo methods in condensed systems. *Reports on Progress in Physics, 74*(2), 026502.
223. Shavitt, I., & Bartlett, R. J., (2009). *Many-Body Methods in Chemistry and Physics: MBPT and Coupled-Cluster Theory*. Cambridge university press.
224. Jeziorski, B., & Monkhorst, H. J., (1981). Coupled-cluster method for multi determinantal reference states. *Physical Review A., 24*(4), 1668.
225. Sherrill, C. D., & Schaefer, III. H. F., (1999). The configuration interaction method: Advances in highly correlated approaches. In: *Advances in Quantum Chemistry* (Vol. 34, pp. 143–269). Academic Press.
226. Born, M., & Oppenheimer, R., (1927). To the quantum theory of molecules. *Annals of Physics, 389*(20), 457–484.
227. Hohenberg, P., & Kohn, W., (1964). Inhomogeneous electron gas. *Physical Review, 136*(3B), B864.
228. Kohn, W., & Sham, L. J., (1965). Self-consistent equations including exchange and correlation effects. *Physical Review, 140*(4A), A1133.
229. Becke, A. D., (1988). Density-functional exchange-energy approximation with correct asymptotic behavior. *Physical Review A., 38*(6), 3098.
230. Lee, C., Yang, W., & Parr, R. G., (1988). Development of the Collesalvetti correlation-energy formula into a functional of the electron density. *Physical Review B., 37*(2), 785.
231. https://www.msg.chem.iastate.edu/ (accessed on 3 August 2020).
232. http://gaussian.com/ (accessed on 3 August 2020).
233. https://www.vasp.at/ (accessed on 3 August 2020).
234. https://www.quantum-espresso.org/ (accessed on 3 August 2020).
235. http://susi.theochem.tuwien.ac.at/ (accessed on 3 August 2020).
236. http://elk.sourceforge.net/ (accessed on 3 August 2020).

237. Gangan, A., Chakraborty, B., Ramaniah, L. M., & Banerjee, S., (2019). First principles study on hydrogen storage in yttrium doped graphyne: Role of acetylene linkage in enhancing hydrogen storage. *International Journal of Hydrogen Energy*.
238. Henkelman, G., Arnaldsson, A., & Jónsson, H., (2006). A fast and robust algorithm for Bader decomposition of charge density. *Computational Materials Science, 36*(3), 354–360.
239. Mulliken, R. S., (1955). Electronic population analysis on LCAO-MO molecular wave functions. I. *The Journal of Chemical Physics, 23*(10), 1833–1840.
240. Kokalj, A., (1999). XCrySDen: A new program for displaying crystalline structures and electron densities. *Journal of Molecular Graphics and Modeling, 17*(3/4), 176–179.
241. Momma, K., & Izumi, F., (2011). VESTA 3 for three-dimensional visualization of crystal, volumetric, and morphology data. *Journal of Applied Crystallography, 44*(6), 1272–1276.
242. Grimme, S., (2006). Semi empirical GGA-type density functional constructed with a long-range dispersion correction. *Journal of Computational Chemistry, 27*(15), 1787–1799.
243. Dion, M., Rydberg, H., Schröder, E., Langreth, D. C., & Lundqvist, B. I., (2004). Van der Waals density functional for general geometries. *Physical Review Letters, 92*(24), 246401.
244. Lee, K., Murray, É. D., Kong, L., Lundqvist, B. I., & Langreth, D. C., (2010). Higher-accuracy van der Waals density functional. *Physical Review B., 82*(8), 081101.
245. Klimeš, J., Bowler, D. R., & Michaelides, A., (2009). Chemical accuracy for the van der Waals density functional. *Journal of Physics: Condensed Matter, 22*(2), 022201.
246. Klimeš, J., Bowler, D. R., & Michaelides, A., (2011). Van der Waals density functionals applied to solids. *Physical Review B., 83*(19), 195131.
247. Jiang, Y., Yang, S., Li, S., Liu, W., & Zhao, Y., (2015). Highly sensitive CO gas sensor from defective graphene: Role of van der Waals interactions. *Journal of Nanomaterials,2015*, 5.
248. Chong, L., Li, J., Wu, F., Li, S., Xia, J., Wang, L., (2011). *J. Phys. Chem. C, 115*, 23221.

CHAPTER 10

Self-Assemblies of Macromolecular Systems Containing Green Polymers

ANDREEA IRINA BARZIC,[1] RALUCA MARINICA ALBU,[1] and CRISTIAN LOGIGAN[2]

[1]*"Petru Poni" Institute of Macromolecular Chemistry, Laboratory of Physical Chemistry of Polymers, 41A Grigore Ghica Voda Alley, Iasi – 700487, Romania*

[2]*Chemical Company SA, 14 Chemistry Bdv., Iasi – 700293, Romania*

ABSTRACT

Materials developed on the basis of green polymers are of major importance not only for the keeping a clean environment, but also for creating of advanced products with architecture controlled at micro- and nano-scale level. This chapter reviews the progress made in the field of self-assemblies in green macromolecular systems. The first short section presents some theoretical aspects and the most used synthesis and preparation routes of macromolecular assemblies of biopolymers. Then, the commonly employed characterization techniques of such soft nano-materials. The newest developments in green polymer self-assemblies are described emphasizing their practical importance. The presented research examples prove the range of possibilities and future trends in self-assembly processes that will significantly contribute to creation of materials having specific nano-architecture. The green materials, which is here under analysis, cover several application fields.

10.1 INTRODUCTION

Macromolecular self-assemblies are still a hot topic in materials and polymer science they facilitate the manufacturing of a variety of materials with extended applicability [1–7]. Preparation of soft nanomaterials in solution allows mimicking the exquisite structure and function of certain compounds found in nature, particularly biomacromolecules (proteins or DNA). Such nature inspired materials are outstanding from the point of view of their selectivity and precision, advancing applications in domains like information storage, transportation, and delivery. To achieve this purpose, it is paramount to utilize small molecules, amphiphiles, colloids, and polymers that enable fabrication of upgraded materials based on nanoscale blocks/units. To this end, multidisciplinary research must be conducted by combining skills of nanotechnology, polymer science, supramolecular chemistry, and biochemistry, but not limited to these ones.

In contrast to the macromolecules created and used by nature, synthetic polymers can be achieved from numerous monomers and combinations of them. Polymerization of these reactants enables different kinds of more or less complex homopolymers and copolymers. In this way, macromolecular engineering facilitates the access to uncommon architectures and shapes. Unfortunately, none of these macromolecular structures possess the sophistication and complexity noticed for those structures derived from the green materials (example: combination of 20 amino acids-natural monomers). The developments recorded in the past decades are mainly referring to refinements of procedures of polymer synthesis, allowing a higher control over the system's composition, molar mass, and entire architecture, however, they less address an assembly of structures of comparable complexity to proteins or other green polymers [2]. More recent reports [8, 9] provided another perspective by showing the possibility of diminishment tedious and time-consuming synthetic steps by engineering polymers of relatively small size that can self-organize via purely noncovalent forces (physical interactions). They are able to emulate the folding of peptide segments in proteins [10–14]. The most important aspect in such approaches is represented by the chemical structure of the synthetic macromolecules which must be able to transport all the information to direct the self-assembly process. Furthermore, self-assembly of polymer is a good and rapid pathway for the preparation of objects from nanometer to micrometer range, which are very hard if not impossible to be attained using the traditional chemical reactions. As a function of the resulted morphologies (size, shape, periodicity, etc.), these self-assembled materials are suitable for a huge number of applications, such as electronics [15], reusable elastomers [16], drug delivery [17], detergents [18], paints [19], and cosmetic products [20].

Having all these in view, the chapter present a short description of the theoretical aspects involved in macromolecular assemblies, followed by preparation and formation of such structures. The most employed characterization techniques of these soft materials are also described. The current approaches used in preparation of self-assemblies using green polymer are depicted in regard with the impact produced in various applicative fields.

10.2 THEORETICAL ASPECTS

The self-assembly of macromolecules in solution require some structural peculiarities. For instance this phenomenon is often encountered in block copolymers, which are containing covalently linked, and more recently non-covalently linked, macromolecular building blocks. The chemically distinct monomer units are combined in discrete blocks along the main chain. Depending on the used synthesis strategy and technique, one may obtain linear, branched, or cyclic molecular architectures. The compositional versatility of block-copolymers created innovative synthetic routes useful for producing previously unachievable levels of architectural complexity. In other words, self-assemblies of block copolymers are a good method for the attaining soft-matter-based core-shell nanoparticles (micelles) with extremely interesting properties and functions [2, 4, 5, 21–24].

A block copolymer with precise architecture is essential for controlling the solution self-assembly process by "adjusting" the interactions among the chain segments, both with each other, and the molecules of solvent [25]. Also, by tuning the molecular parameters (the degree of polymerization (N), the Flory-Huggins interaction parameter (χ), the, and the volume fraction (f)) it is possible to accomplish numerous types of morphologies, as experimentally and theoretically revealed [26, 27]. The architecture is also of paramount importance in controlling both the final morphologies and their extent of long-range order [28]. As a result of the exciting features of the block copolymers, the number of publications on this topic increased substantially since 1990s.

When placed in solvent media, block copolymers are suffering two main processes: micellization and gelation. The first one is taking place when the block copolymer is brought in contact with a large quantity of a selective solvent for one of the blocks. In such circumstances, the macromolecular chains are beginning to organize themselves in a variety of structures from micelles or vesicles to cylinders. The soluble block will be moving and orienting towards the continuous solvent medium and forms the 'corona' of the resulted micelle, while the insoluble part of the copolymer will be shielded from the solvent molecules in the 'core' of the structure. In contrast to micellization, gelation process appears in the semidilute to the high concentration domain of block copolymer solutions and results from the specific positioning of ordered micelles. The relative length of the blocks affects the micellar structure, namely when the soluble block prevails the micelles have a small core and a large corona (star micelles), while in the opposite situation the micelles have a short soluble corona (crew-cut micelles).

The analysis of a micellar system involves determining of several parameters, including the thermodynamic compatibility of the solvents, equilibrium constant, the critical micelle temperature (CMT), the critical micelle concentration (CMC), the molar mass M_w of the micelle, its aggregation number Z and the morphology. The mentioned variables influence the hydrodynamic radius (RH) R_H, the radius of gyration R_G, their ratio which impacts the micellar shape, the core radius R_C, and the thickness L of the corona. The shape and the dimensions of the aggregates are impacted by a numerous of parameters that influence the balance between three main forces acting on the system. They are related to the following aspects:

- The constraints between the blocks constituting the core, namely as a function of the type of solvent the block will be more or less stretched.
- The forces between chains forming the corona.
- The surface at the interface of the solvent/core of the micelle.

From a theoretical point of view, for the characterization of the aggregate structure is needed to account for the thermodynamic properties of self-assembly but also for the interactions among the macromolecules from the interior of aggregates. The last two factors (thermodynamics and intra-aggregate forces) together with the forces acting between different aggregates are influencing the type of self-assembled structure resulted in equilibrium conditions. In this context, it is necessary to comprehend the fundamentals that govern the relation between morphology and dimension of the aggregates acquired by self-assembly, including key factors such as concentration, composition, temperature, block length, copolymer architecture, and the nature of the solvents.

The theories developed to understand the behavior of block copolymers in solution can be categorized as follows:

- **Scaling Theory:** Proposed by de Gennes [29] estimated parameters such as the aggregation number or the radius for crew-cut micelles from specific information like the block length and interfacial tension. Another study [30] extended this theory for star micelles, while the report of Zhulina [31] based on this subject suggested that there are four types of micelles;
- **Self-Consistent Mean Field Theory:** Formulated by Noolandi and Hong [32] allow the evaluation of the dimension of spherical micelles at equilibrium, and the changes in the aggregation number as the degree of polymerization varies. Their theory is constructed by accounting for the molecular characteristics of the polymer, its concentration in solution, and prediction of the interfacial tension of the core/corona [33]. Further development of this approach enabled elucidation of other aspects, such as the evolution of the CMC with copolymer chemical structure (triblocks versus diblocks) [33], the temperature impact on the RH and aggregation number [34], the transition from spherical and cylindrical micellar systems [35], and the effects of polydispersity on CMC [36];
- **The Theory of Israelachvili and Coworkers [37]:** Employs geometrical considerations that foretell the micellization phenomenon and the corresponding formed morphologies. This model is applicable for both amphiphilic molecules of low molar and block copolymers. When discussing in the case of amphiphilic molecules one must consider the action of the forces governing the assembly into well defined structures: the hydrophobic attraction among insoluble hydrophobic segments, and the repulsion among the hydrophilic sequences due to electrostatic or steric interactions that both impose the contact with solvent of the amphiphilic molecules. When the attractive forces are prevailing, the interfacial area (a_o) per molecule will decrease, while if repulsive forces predominate, a_o will increase. The balance between such opposing forces is strongly affected by the geometry of the blocks, thus creating a variety of known morphologies. According to Israelachvili [37], the geometric properties are dictated by three parameters, namely the optimal interface a_o, the volume () taken by the hydrophobic segments, and the maximum length (l_c) of the chains. The described parameters are included in the definition of the packing parameter (p):

$$p = v/a_0 lc \tag{10.1}$$

The shape factor p is < 1/3, 1, and > 1 for spherical micelles, bicontinuous bilayers, and inverted structures, respectively.

- **The Theory of Disher and Eisenberg [38]:** Attempted unify the experimental results reported in many investigations on amphiphilic block copolymers. They stated a rule for the appearance of polymersomes (polymer-based vesicles) in water: namely a ratio (denoted f) of the mass of the hydrophilic part to the total mass 35 ± 10%. The molecules having f > 45% will lead to micelles, while those having f < 25% will be able self-assemble into inverted structures.

10.3 PREPARATION OF MACROMOLECULAR SELF-ASSEMBLIES

The literature [2] indicates two main procedures for preparation of macromolecular self-assemblies in solution (micelles):

- Introduction of the copolymer in a good solvent for both blocks. In certain situations, the wanted micellar structure can be achieved by the manipulation of some solution parameters, such as temperature or a cosolvent. For other systems, the subsequent utilization of a selective solvent, into an already prepared solution in a common solvent it enables micelle formation.
- Removal of the common solvent from the system through a dialysis procedure. The approach to accomplish micelles is based on simply of inserting the dry solid copolymer powder into a selective solvent.

The choice for one or another method is mainly dictated by the system peculiarities. For instance, the glassy blocks like those from polystyrene (PS) generate the formation of highly stable micelles that are lacking an exchange between individual macromolecular chains and micelles ('frozen micelles'). For this situation, the first approach is more adequate.

10.4 METHODS OF CHARACTERIZATION OF MACROMOLECULAR SELF-ASSEMBLIES

10.4.1 ANALYSIS OF CMC

Fluorescence spectroscopy is one of the most common methods used in determination of the CMC [38, 39]. Pyrene is often preferred fluorescent probe given its strong fluorescence in nonpolar domains and its poor radiation in polar media. The shift of the excitation peak is monitored to prove the transfer of pyrene molecules into an increasingly nonpolar micellar medium. The ratio of intensities of the excitation maxima at 339 and 333 nm is changing with concentration. This leads to a plot with where a sudden variation of the data is noted. The crossover of the lines having distinct slopes represents the CMC value.

UV-absorption spectroscopy is another powerful technique for analysis of the CMC [40, 41]. The approach relies on the tautomerism of 1-phenyl-1,3-butadione between keto and enol forms which display distinct absorption maxima: 312 nm for the enolic form and 250 nm for the keto form, the former occurring in nonpolar solvents such as cyclohexane while the latter is formed in polar solvents, where H-bonding is disrupted in favor of the keto configuration [42].

There are other methods for CMC characterization which can be extended from small surfactants to block copolymer micelles, such as static or dynamic light is scattering or small-angle X-ray scattering (SAXS). However, there are some limitations of these techniques since signal intensity is very small considering the much lower CMC of block copolymers in regard with that of low molar mass surfactants.

10.4.2 DETERMINATION OF MICELLAR SIZE AND SHAPE

Specific micelle features (shape, size, and aggregation number Z) can be examined through several techniques. Depending on the chosen characterization method, a variety of data about the investigated system can be extracted. The most relevant are the following [2]:

- **Scattering Methods:**
  - **Static Light Scattering:** Technique to determine the average molar masses of self-assembled structures, including their CMC. Moreover, if scattering from the core and the corona of the micellar material is not so distinct one may estimate the RG value.
  - **Dynamic (or Quasi-Elastic) Light Scattering:** Allows analysis RH of a block copolymer micelles from the data regarding the diffusion coefficient. The sensitivity and versatility of this approach allow modifications in the micelle equilibrium owing to changes of temperature, pH, or other parameters.

- **Small-Angle X-Ray Scattering (SAXS):** Is adequate for the study of micellar solutions to attain overall and internal dimensions from dissimilarities in electron density of the solvent and the solute.
- **Small-Angle Neutron Scattering:** Provides significant data concerning the shape, and also the cross-section.
- **Non-Scattering Methods:**
- **Transmission Electron Microscopy:** Enable visualization of the micelle morphology, size or internal structure.
- **Atomic Force Microscopy:** Gives images from which the dimension, shape or mechanical resistance of the micelles can be determined.
- Dilute solution capillary.
- **Viscometry:** The changes in the solution time of flow are related to the micellization phenomenon.
- Membrane osmometry.
- Ultracentrifugation.
- Size exclusion chromatography.
- Nuclear magnetic resonance spectroscopy.

10.5 CURRENT DEVELOPMENTS IN SELF-ASSEMBLIES OF SYSTEMS CONTAINING GREEN POLYMERS

The association phenomenon in some systems containing a green polymer is briefly presented based on the latest advances in this research field. In this section of the chapter are reviewed the developments concerning cellulose, chitosan, and DNA.

10.5.1 CELLULOSE BASED SYSTEMS

The self-assembly of cellulose-based polyelectrolyte complexes (PECs) was investigated by Zhao et al. [43]. The procedure occurs in two stages:

- A weak anionic polyelectrolyte is dissolute in water and then is blended with four types of cationic polyelectrolytes (chitosan; cationic cellulose; poly-diallyl dimethylammonium chloride, PDDA; poly2-methacryloyloxy ethyl trimethylammonium chloride, PDMC) in HCl aqueous solutions; and

- The resulted PECs and then are re-dispersed in NaOH aqueous solutions.

Infrared spectroscopy data show that the ionic complexation indeed takes places between sodium carboxymethyl cellulose and the cationic polyelectrolytes and also probes the "two-step" method for obtaining PECs aqueous dispersions. PECs interfacial self-assembly process is interesting but not reported for other PECs dispersions processed in analogous conditions. Dynamic light scattering analyses reveal that R_H values of the PEC aggregates are comprised in the interval of 100–500 nm, while field emission electron scanning microscopy data indicate smaller values. Average R_H values of the PEC aggregates are ranging as follows: sodium carboxymethyl cellulose-cationic cellulose > sodium carboxymethyl cellulose-PDMC > sodium carboxymethyl cellulose-PDDA > sodium carboxymethyl cellulose-chitosan. From this, it can be extrapolated the idea that cationic polyelectrolytes with larger molecular weights produce larger PEC aggregates [44]. The basic constructing segments of PECs are ionically cross-linked and nano-sized PEC particles, which are amorphous considering their ionic cross-linking and "scrambled egg" structures [43]. The PECs dispersions in water and their component polyelectrolyte solutions are subjected to a drying process at 30°C on silicon supports. The corresponding FESEM data show tree-shaped and fractal surface patterns. Such morphologies are useful in surface patterning and membrane separation [43].

Cellulose nanocrystals (CNCs) are able to self-assembly [45, 46] having a remarkable ability to organize into photonic structures. The colloidal stability of these materials is a mandatory to achieve self-assembly, but in the same time, insufficiently charged CNCs (such as isolated after hydrolysis) also impede the possibility to self-assemble. After preparation of the suspensions are subjected to purification through dialysis and then introduced in polar solvents [47]. If a surfactant [48] or further functionalization [49, 50] is applied, colloidal stability is extended towards extreme conditions.

Hata et al. [51] studied the oligomerization-induced self-assembly of crystalline cellulose into nanoribbon networks supported by organic solvents. The controlled macromolecular self-assembly into organized hierarchical structures using a simple method is hard, but they showed that organic solvents can serve as small-molecule additives and thus enable the control of the self-assembly of cellulose oligomers into the desired network structures. They used cellodextrin phosphorylase-catalyzed oligomerization of phosphorylated glucose monomers from D-glucose primers, which generate precipitates of nanosheet-shaped crystals in water. The introduction of water-miscible organic solvents-dimethyl sulfoxide (DMSO), N,N-dimethylformamide (DMF), acetonitrile (MeCN), and ethanol (EtOH)-favor occurrence of nanoribbon networks and avoid the irregular aggregation and the precipitation of the nanosheets. The results prove that small-molecule additives are useful for controlling the self-assembly of crystalline oligosaccharides for the fabrication of hierarchically structured materials with large robustness in a facile manner.

10.5.2 CHITOSAN BASED SYSTEMS

The self-assembly of enzymatic hydrolysates of chitosan and carboxymethyl cellulose allows formation of biocompatible particles [52]. Both polymers were hydrolyzed with chitosanase and cellulase respectively to attain fragments of lower molecular weights. Polymer nanoparticles are almost instantly formed only by blending the two hydrolysate solutions. The particle size distribution resulted to be relatively narrow, around 200 nm in mean size. The mean particle dimension diminished from 226 nm to 165 nm as the molecular weight of chitosan hydrolysate was lowered from 9.5 to 6.8 kDa. The composition of the system also impacted the particle size. The blended polymer nanoparticles are found to be stably suspended over 1 week even for small pH (pH 3.0) conditions, high ionic strength (NaCl 1 M), or temperatures around 4°C.

Brunel et al. [53] obtained nanohydrogels of pure chitosan, which are lacking any toxic solvent or chemical cross-linker. This can be achieved by an ammonia-generated physical gelation of a reverse emulsion of a chitosan/triglyceride mixture. The produced colloids are accomplished with a controlled distribution of dimensions and present a positive surface charge. Such macromolecular assemblies opened new perspectives toward novel nano-carriers for bioactive compounds. Chondroitin sulfate renders PECs having positively charged surface and thus determining negative chitosan-based colloidal hydrogels that maintain of the original average size of the dispersion. The mode of assembly of HIV-1 p24 protein with the formed colloids is mainly based on multiple interactions among the nanohydrogels and the protein, regardless of their surface charges. In any case, the amounts of introduced protein are not so high, suggesting a surface association. For comparison, the assembly of an immunoglobulin (IgG) is studied and led to different data from p24. No association is noted with the positive colloidal hydrogels while a larger loading capacity could be attained with the negative ones. The reported fully biodegradable submicrometric physical hydrogels based on green polymers could cargo a several types of biomolecules recommending them as versatile carrier's high impact in biomedicine [54].

On the other hand, the chitosan-based nanoparticles, prepared via self-assembly, can be made to become sensitive to some stimuli [55]. Chitosan-graft-poly(N-isopropylacrylamide) copolymers (CS-g-PNIPAAm) were obtained by polymerization of NIPAAm monomer in the presence of chitosan in water, where cerium ammonium nitrate has the role of initiator. The copolymer solution was further diluted using deionized water and subjected to a higher temperature in order to ensure the occurrence of the self-assembly. Based on this procedure micelles of CS-g-PNIPAAm are attained. For reinforcement of the desired micelle structure and to convert them into nanoparticles, in the system is introduced glutaraldehyde. Transmission electron microscopy of samples showed a porous or hollow structure of nanoparticles. The dimensions of the nanoparticles can be tuned thought the temperature of the environment and can be applied in hydrophilic drug delivery.

Another report on sensitive macromolecular assemblies based on chitosan is made by Chen et al. [56]. They prepared an innovative pH-sensitive PEC for insulin designed for oral administration which result after self-assembly of two oppositely charged nanoparticles of 200–300 nm (chitosan covered nanoparticles and alginate covered nanoparticles) via electrostatic interaction and optimized double emulsion procedure. The morphology of the PEC system is pH-sensitive and the reasons for the morphological variation arise from the strength of the electrostatic interaction among the green polymer and alginate at several pH values. The PEC materials display non-cytotoxicity against Caco-2 cell. The insulin-loaded systems are effective in decreasing the insulin level in rats and keep prolonged release of the active substance.

Chitosan/cellulose systems are able to self-assemble into a 3D porous structure resulting fascinating materials [57]. Cellulose extracted from plants and chitosan taken from marine sources was used by Zhang and collaborators [57] together with a clear dissolution strategy. The renewable materials are dispersed into a homogeneous solution in green alkali/urea media. Given the hydrogen-bond interaction among the green polymers, they were able to self-assemble into a unique 3D alloy structure. The high amount of interface interactions enhanced the tensile strength and breaking elongation of the alloy sample up to 69%–260%, in regard to pure chitosan film. The 3D alloy materials present a pore diameter of 920–1435 nm together with an outstanding high cell viability that exceeds 95%. Moreover, accelerated wound healing rate of 95.6% is attained in the presence of the alloy sample after 14 days using a full-thickness wound in a rat.

10.5.3 DNA BASED SYSTEMS

DNA is a widely studied biopolymer which in certain conditions leads to outstanding self-assembled structures. The self-assembly of micellar or vesicular structures in DNA-polymer conjugates was reviewed by Kedracki and co-workers [58]. When a hydrophobic polymer sequence is coupled to a nucleotide unit it is possible self-assemble in solution rendering to core-shell micelles made of a hydrophobic core covered by a hydrophilic corona of the DNA fragment. DNA-polymer conjugates that form micelles for the targeting of antisense oligonucleotide to cells are studied [59]. Another category of amphiphilic DNA-polymer conjugates which generate micellar structures in water solution are suitable for molecular recognition to hybridize them with gold nanoparticles, leading to the construction of stimuli-responsive materials [60].

The self-assembly in DNA-based systems can be used not only in drug delivery but also in imaging applications. Peterson and Heemstra [61] prepared micelles and vesicles via phase-driven self-assembly of monomer structures containing discrete hydrophilic and hydrophobic sequences. In these materials, DNA plays the role of the hydrophilic block of the amphiphilic monomers and facilitates assemblies with the advantage of the good information storage and ability of molecular recognition.

By altering monomer structure, one may tune the morphology of DNA-polymer conjugates and monitor the influence of assembly on DNA stability and hybridization.

Another investigation developed by Alemdaroglu [62] the DNA is coupled to a poly(propylene oxide) (PPO) of higher biocompatibility than other macromolecular compounds used for the synthesis of DNA-polymer conjugates. In aqueous solution, the DNA-b-PPO copolymers begin to self-assemble resulting micelles composed of a hydrophobic PPO core and a shell made from nucleotide segments.

These materials are suitable as nanoreactors to carry out organic reactions [62]. Moreover, in other investigation it was proved that DNA-based micelles can suffer a morphological transition from spherical to rod-like structures after hybridization with long DNA segments, which were selected in such a way that they codify several times the complementary part of the DNA representing the self-assembling conjugate [63].

Aside from the self-organization of DNA-polymer systems into micelles, the occurrence of vesicular structures has been also reported [64, 65]. The vesicles are prepared by the self-assembly of an amphiphilic DNA-polymer conjugate leading to the coupling among a highly hydrophobic macromolecular segment and a DNA unit. The formation of vesicular structures is of great importance for some pharmaceutical and biomedical applications because they are able to be filled with substances which can be specifically delivered to cells. Vebert-Nardin and colleagues [64, 65] showed for the first time the self-organization of vesicular structures self-assembled from DNA-polymer system via reaction of a suitable polymer segment to a 12 nucleotide sequence.

In recent trends [66], the scientists are focused on self-assembled materials with upgraded programmable features made from DNA-based systems. There are several strategies of directing the self-organization of DNA-inspired, sequence-specific polyphosphodiesters into structures with unique characteristics.

This highlights the fact that DNA-derived systems are not fully exploited in solving some issues in certain technological areas, particularly controlled polymer folding, and assembly. The accurate positioning of structurally distinct molecules within a macromolecular chain renders unmatched possibilities for encoding information on the molecular level and broadcast it to the microscopic and even macroscopic scale through the noncovalent interactions.

10.6 CONCLUDING REMARKS

The chapter aimed at presenting an up-to-date report on the developments made in the area of self-assemblies in the systems containing green polymers. The theoretical fundaments that lie at the basis of organization of macromolecules in solutions are described. The most used methods employed in preparation and characterization of self-assembled structures is presented. The most significant researches involving green polymer self-organization and the practical importance are also depicted. The control of the shape and size of such nano-objects and the construction of particular properties are no longer unreachable challenges. This research domain offers opportunities for chemists, physicists, and biochemists, to create systems that might eventually mimic in sophistication and accuracy the biological structures developed by nature.

ACKNOWLEDGMENTS

The financial support of European Social Fund for Regional Development, Competitiveness Operational Programme Axis 1-Project "Petru Poni Institute of Macromolecular Chemistry-Interdisciplinary Pol for Smart Specialization through Research and Innovation and Technology Transfer in Bio(nano)polymeric Materials and (Eco)Technology," InoMatPol (ID P_36_570, Contract 142/10.10.2016, cod MySMIS: 107464) is gratefully acknowledged.

KEYWORDS

- **cellulose nanocrystals**
- **critical micelle concentration**
- **green polymers**
- **immunoglobulin**
- **organic solvents**
- **self-assemblies**

REFERENCES

1. Zhulina, E. B., & Borisov, O. V., (2002). Self-assembly in solution of block copolymers with annealing polyelectrolyte blocks. *Macromolecules, 35*, 9191–9203.
2. Rodrıguez-Hernandez, J., Checot, F., Gnanou, Y., & Lecommandoux, S., (2005). Toward 'smart' nano-objects by self-assembly of block copolymers in solution. *Prog. Polym. Sci., 30*, 691–724.
3. Patterson, J. P., Robin, M. P., Chassenieux, C., Colombani, O., & O'Reilly, R. K., (2014). The analysis of solution self-assembled polymeric nanomaterials. *Chem. Soc. Rev., 43*, 2412–2425.
4. Huang, F., O'Reilly, R., & Zimmerman, S. C., (2014). Polymer self-assembly: A web themed issue. *Chem. Commun., 50*, 13415–13416.
5. Tritschler, U., Pearce, S., Gwyther, J., Whittell, G. R., & Manners, I., (2017). 50th Anniversary perspective: Functional nanoparticles from the solution self-assembly of block copolymers. *Macromolecules, 50*, 3439–3463.
6. Feng, H., Lu, X., Wang, W., Kang, N. G., & Mays, J. W., (2017). Block copolymers: Synthesis, self-assembly, and applications. *Polymers, 9*, 494–520.
7. Ariga, K., Nishikawa, M., Mori, T., Takeya, J., Shrestha, L. K., & Hill, J. P., (2019). Self-assembly as a key player for materials nanoarchitectonics. *Sci. Technol. Adv. Mat., 20*, 51–95.

8. Van Rijn, P., (2013). Polymer directed protein assemblies. *Polymers, 5*, 576–599.
9. Luo, Q., Dong, Z., Hou, C., & Liu, J., (2014). Protein-based supramolecular polymers: Progress and prospect. *Chem. Commun., 50*, 9997–10007.
10. Hendricks, M. P., Sato, K., Palmer, L. C., & Stupp, S. I., (2017). Supramolecular assembly of peptide. *Acc. Chem. Res., 50*, 2440–2448.
11. Cavalli, S., Albericio, F., & Kros, A., (2010). Amphiphilic peptides and their cross-disciplinary role as building blocks for nanoscience. *Chem. Soc. Rev., 39*, 241–263.
12. Rho, J. Y., Cox, H., Mansfield, E. D. H., Ellacott, S. H., Peltier, R., Brendel, J. C., Hartlieb, M., et al., (2019). Dual self-assembly of supramolecular peptide nanotubes to provide stabilization in water. *Nat. Commun., 10*, 4708.
13. Raymond, D. M., & Nilsson, B. L., (2018). Multi component peptide assemblies. *Chem. Soc. Rev., 47*, 3659–3720.
14. Ottera, R., & Besenius, P., (2019). Supramolecular assembly of functional peptide-polymer conjugates. *Org. Biomol. Chem., 17*, 6719–6734.
15. Edrington, A. C., Urbas, A. M., DeRege, P., Chen, C., Swager, T. M., Hadjichristidis, M., et al., (2001). Polymer-based photonic crystals. *Adv. Mater., 13*, 421–425.
16. Holden, G., Legge, N. R., Quirck, R., & Schroeder, H. E., (1996). *Thermoplastic Elastomers* (2nd edn.). Cincinnati: Honser/Gardner Publication.
17. Wang, D., Tong, G., Dong, R., Zhou, Y., Shen, J., & Zhu, X., (2014). Self-assembly of supramolecularly engineered polymers and their biomedical applications. *Chem. Commun., 50*, 11994–12017.
18. Sundararajan, P. R., (2017). *Physical Aspects of Polymer Self-Assembly*. Wiley: USA.
19. Baradie, B., & Schoichet, M. S., (2002). Synthesis of fluorocarbon-vinyl acetate copolymers in supercritical carbon dioxide: Insight into bulk properties. *Macromolecules, 35*, 3569–3575.
20. Nakai, S., Nakai, A., & Michida, T., (2016). Microencapsulation of ascorbic acid for cosmetic by utilizing self-assembly of phase separated polymer. *Chem. Pharm. Bull., 64*, 1514–1518.
21. Rodríguez-Hernández, J., Chécot, F., Gnanou, Y., & Lecommandoux, S., (2005). Toward 'smart' nano-objects by self-assembly of block copolymers in solution. *Prog. Polym. Sci., 30*, 691–724.
22. Ge, Z., & Liu, S., (2013). Functional block copolymer assemblies responsive to tumor and intracellular microenvironments for site specific drug delivery and enhanced imaging performance. *Chem. Soc. Rev., 42*, 7289–7325.
23. Kempe, K., Wylie, R. A., Dimitriou, M. D., Tran, H., Hoogenboom, R., Schubert, U. S., Hawker, C. J., Campos, L. M., & Connal, L. A., (2016). Preparation of non-spherical particles from amphiphilic block copolymers. *J. Polym. Sci., Part A: Polym. Chem., 54*, 750–757.
24. Kataoka, K., Harada, A., & Nagasaki, Y., (2001). Block copolymer micelles for drug delivery: Design, characterization, and biological significance. *Adv. Drug Delivery Rev., 47*, 113–131.
25. Mai, Y., & Eisenberg, A., (2012). Self-assembly of block copolymers. *Chem. Soc. Rev., 41*, 5969–5985.
26. Lennon, E. M., Katsov, K., & Fredrickson, G. H., (2008). Free energy evaluation in field-theoretic polymer simulations. *Phys. Rev. Lett., 101*, 138302.
27. Matsen, M. W., & Bates, F. S., (1996). Origins of complex self-assembly in block copolymers. *Macromolecules, 29*, 7641–7644.

28. Poelma, J. E., Ono, K., Miyajima, D., Aida, T., Satoh, K., & Hawker, C. J., (2012). Cyclic block copolymers for controlling feature sizes in block copolymer lithography. *ACS Nano, 6*, 10845–10854.
29. De Gennes, P. G., (1978). In: Libert, L., (ed.), *Solid State Physics*. Academic Press: New York.
30. Daoud, M., & Cotton, J. P., (1982). Star-shaped polymers: A model for the conformation and its concentration dependence. *J. Phys., 43*, 531–538.
31. Zhulina, E. B., & Birshtein, T. M., (1985). Conformations of molecules of block copolymers in selective solvents (micellar structures). *Vysokomol. Soed., Seriya A.,27*, 511–517.
32. Noolandi, J., & Hong, K. M., (1983). Theory of block copolymer micelles in solution. *Macromolecules, 16*, 1443–1448.
33. Linse, P., (1993). Micellization of poly(ethylene oxide)-poly(propylene oxide) block copolymers in aqueous solution. *Macromolecules, 26*, 4437–4449.
34. Linse, P., & Malmsten, M., (1992). Temperature-dependent micellization in aqueous block copolymer solutions. *Macromolecules, 25*, 5434–5439.
35. Linse, P., (1994). Micellization of poly(ethylene oxide)-poly(propylene oxide) block copolymers in aqueous solution: Effect of polymer polydispersity. *Macromolecules, 27*, 6404–6417.
36. Gao, Z., & Eisenberg, E., (1993). A model of micellization for block copolymers in solutions. *Macromolecules, 26*, 7353–7360.
37. Israelachvili, J. N., (1992). *Intermolecular and Surface Forces*. London: Harcourt Brace and Company.
38. Lupu, M., Macocinschi, D., Epure, V., Ioanid, A., Grigoras, C. V., & Ioan, S., (2006). Micellization process in poly(ether urethane) solutions. *J. Macromol. Sci. Part B, 45*, 395–405.
39. Kalyanasundaram, K., & Thomas, J. K., (1977). Environmental effects on vibronic band intensities in pyrene monomer fluorescence and their application in studies of micellar systems. *J. Am. Chem. Soc., 99*, 2039–2044.
40. Geng, J., Johnson, B. F. G., Wheatley, A. E. H., & Luo, J. K., (2014). Spectroscopic route to monitoring individual surfactant ions and micelles in aqueous solution: A case study. *Cent. Eur. J. Chem., 12*, 307–311.
41. Tanhaei, B., Saghatoleslami, N., Chenar, M. P., Ayati, A., Hesampour, M., & Mänttäri, M., (2012). Experimental study of CMC evaluation in single and mixed surfactant systems, using the UV-Vis spectroscopic method. *J. Surfactants Deterg., 16*, 357–362.
42. Dominguez, A., Fernández, A., González, N., Iglesias, E., & Montenegro, L., (1997). Determination of critical micelle concentration of some surfactants by three techniques. *J. Chem. Education, 74*, 1227–1231.
43. Zhao, Q., Qian, J., Gui, Z., An, Q., & Zhu, M., (2010). Interfacial self-assembly of cellulose-based polyelectrolyte complexes: Pattern formation of fractal "trees." *Soft Matter, 6*, 1129–1137.
44. Schatz, C., Bionaz, A., Lucas, J. M., Pichot, C., Viton, C., Domard, A., & Delair, T., (2005). Formation of polyelectrolyte complex particles from self-complexation of N-sulfated chitosan. *Biomacromolecules, 6*, 1642–1647.
45. Parker, R. M., Guidetti, G., Williams, C. A., Zhao, T., Narkevicius, A., Vignolini, S., & Frka-Petesic, B., (2018). The self-assembly of cellulose nanocrystals: Hierarchical design of visual appearance. *Adv. Mater., 30*, 1704477.

46. Rofouie, P., Alizadehgiashi, M., Mundoor, H., Smalyukh, I. I., & Kumacheva, E., (2018). Self-assembly of cellulose nanocrystals into semi-spherical photonic cholesteric films. *Adv. Funct. Mater., 28*, 1803852.
47. Bruckner, J. R., Kuhnhold, A., Honorato-Rios, C., Schilling, T., & Lagerwall, J. P. F., (2016). Enhancing self-assembly in cellulose nanocrystal suspensions using high-permittivity solvents. *Langmuir, 32*, 9854–9862.
48. Salajková, M., Berglund, L. A., & Zhou, Q., (2012). Hydrophobic cellulose nanocrystals modified with quaternary ammonium salts. *J. Mater. Chem., 22*, 19798–19805.
49. Tang, J., Sisler, J., Grishkewich, N., & Tam, K. C., (2017). Functionalization of cellulose nanocrystals for advanced applications. *J. Colloid Interface Sci., 494*, 397–409.
50. Natterodt, J. C., Petri-Fink, A., Weder, C., & Zoppe, J. C., (2017). Cellulose nanocrystals: Surface modification, applications, and opportunities at interfaces. *Int. J. Chem., 71*, 376–383.
51. Hata, Y., Fukaya, Y., Sawada, T., Nishiura, M., & Serizawa, T., (2019). Biocatalytic oligomerization-induced self-assembly of crystalline cellulose oligomers into nanoribbon networks assisted by organic solvents. *Beilstein J. Nanotechnol., 10*, 1778–1788.
52. Ichikawa, Iwamoto, S., & Watanabe, J., (2005). Formation of biocompatible nanoparticles by self-assembly of enzymatic hydrolysates of chitosan and carboxymethyl cellulose. *Biosci. Biotechnol. Biochem., 69*, 1637–1642.
53. Brunel, F., Véron, L., David, L., Domard, A., Verrier, B., & Delair, T., (2010). Self-assemblies on chitosan nanohydrogels. *Macromol. Biosci., 10*, 424–432.
54. Quiñones, J. P., Peniche, H., & Peniche, C., (2018). Chitosan based self-assembled nanoparticles in drug delivery. *Polymers, 10*, 235.
55. Chuang, C. Y., Don, T. M., & Chiu, W. Y., (2011). Preparation of environmental-responsive chitosan-based nanoparticles by self-assembly method. *Carbohydr. Polym., 84*, 765–769.
56. Chen, T., Li, S., Zhu, W., Liang, Z., & Zeng, Q., (2019). Self-assembly pH-sensitive chitosan/alginate coated polyelectrolyte complexes for oral delivery of insulin. *J. Microencapsul., 36*, 96–107.
57. Zhang, R., Xie, J., Yang, B., Fu, F., Tang, H., Zhang, J., Zhao, Y., et al., (2019). Self-assembly of chitosan and cellulose chains into a 3D porous polysaccharide alloy films: Co-dissolving, structure, and biological properties. *Appl. Surf. Sci., 493*, 1032–1041.
58. Kedracki, D., Safir, I., Gour, N., Ngo, K. X., & Vebert-Nardin, C., (2013). DNA-polymer conjugates: From synthesis, through complex formation and self-assembly to applications. *Adv. Polym. Sci., 253*, 115–150.
59. Jeong, J. H., & Park, T. G., (2001). Novel polymer-DNA hybrid polymeric micelles composed of hydrophobic poly(D,L-lactic-co-glycolic acid) and hydrophilic oligonucleotides. *Bioconjug. Chem., 12*, 917–923.
60. Li, Z., Zhang, Y., Fullhart, P., & Mirkin, C. A., (2004). Reversible and chemically programmable micelle assembly with DNA block-copolymer amphiphiles. *Nano Lett., 4*, 1055–1058.
61. Peterson, A. M., & Heemstra, J. M., (2001). Controlling self-assembly of DNA-polymer conjugates for applications in imaging and drug delivery. *WIREs Nanomed. Nanobi., 7*, 282–297.
62. Alemdaroglu, F. E., Ding, K., Berger, R., & Herrmann, A., (2006). DNA-templated synthesis in three dimensions: Introducing a micellar scaffold for organic reactions. *Angew. Chem. Int. Ed. Engl., 45*, 4206–4210.

63. Ding, K., Alemdaroglu, F. E., Borsch, M., Berger, R., & Herrmann, A., (2007). Engineering the structural properties of DNA block copolymer micelles by molecular recognition. *Angew. Chem. Int. Ed. Engl., 46*, 1172–1175.
64. Teixeira, F. Jr., Rigler, P., & Vebert-Nardin, C., (2007). Nucleo-copolymers: Oligonucleotide-based amphiphilic dib lock copolymers. *Chem. Commun.*, 1130–1132.
65. Vebert-Nardin, C., (2011). Towards biologically active self-assemblies: Model nucleotide chimeras. *Chimia., 65*, 782–786.
66. Vybornyi, M., Vyborna, Y., & Haner, R., (2019). DNA-inspired oligomers: From oligophosphates to functional materials. *Chem. Soc. Rev., 48*, 4347–4360.

Index

1

1-butyl-3-methylimidazolium
  acetate (BMIAc), 15
  chloride (BmimCl), 11
1-ethyl-3-methylimidazolium acetate (EmimAc), 11, 12

A

Acetic acid, 13, 16, 17, 166
Acetonitrile (MeCN), 168, 250
Acid recovery system, 173, 174
Acidic aniline solution, 118
Activation
  energy, 74
  entropy, 9
Adenosine triphosphate (ATP), 35, 49
Adhesion, 26, 77, 82, 86, 101, 122
Adsorption energy, 212, 219, 220, 223
Aerobiotic ferment, 33, 49
Aerogel, 11, 12, 47
Agricultural waste, 54, 80
Alanine acids, 100
Alkaline
  earth metals, 84
  metals, 67, 84
Alpha-hydroxy acid, 176
Aluminosilicates, 84
Ambient
  grinding, 66
  temperature, 189, 194, 208, 211
Ammonia, 173, 174, 190, 191, 250
  purge stream-prism membranes, 174
Ammonium persulfate, 118
Amphiphilic molecules, 245
Anaerobic
  desulfurization, 68
  digestion, 40
Andrews equation, 133
Anionic exchange membranes, 163, 177
Anodes, 34–36, 38, 40–44, 46, 48, 49, 164, 177
Anterior cruciate ligaments (ACL), 106, 110
Anti-aging properties, 86
Anti-bacterial
  agents, 103
  applications, 117
  packaging material, 83
Antifouling, 113, 116
  current approaches, 116
  paints, 116
  strategies, 117, 118, 120
Antioxidants, 54, 62, 75, 76, 104
Aquaculture, 113–122, 124
Aquafarming, 115
Aquatic organic process, 157
Aqueous
  acetic acid, 17
  dispersion, 12
Aromatic oil, 70, 72
Arrhenius relation, 15
Artificial/natural seawater (ASW), 123, 124
Aspartic acids, 101
Atomic
  force microscopy, 248
  structure, 23
Auto-ignition temperature, 185
Auxiliary chemicals, 158
Azeotropes, 178
Azo colors, 48
Azoxystrobin, 145, 148

B

Bacillus, 149
Bacteria specific growth rate, 135
Bacterial
  adhesion, 43
  cells, 130, 134, 135
  cellulose, 79, 84
Bader charge analysis, 221, 222
Bahera gum, 74, 75
Baker's yeast, 54
Becke-Lee-Yang-Parr (BLYP), 218

Benthic organisms, 116
Benzene, 68, 140, 141
Benzenoid ring, 120
Binding energy, 194, 197, 198, 201, 211, 213, 219–225
Biochemical oxygen demand (BOD), 39, 49
Biocidal effect, 118
Biocompatibility, 99, 100, 102, 104, 108, 110, 252
Bio-composites, 76, 84
Biodegradation, 102, 148, 150
Biodiesel, 127, 129, 130, 150, 184
Bioengineering, 130
Bioethanol, 4, 62, 177, 178
Biofiber, 76, 83, 84
Biofilms, 36, 38, 41, 44, 114, 120
Biofoulers, 114, 115, 120, 122
Biofouling, 113–118, 120–122, 124
Biofuels, 4, 177
Biogas, 40
Biogenic palladium, 43
Biomass, 3, 4, 54, 61, 62, 79, 80, 87, 122, 131, 134–136, 158, 178, 185, 186
Biomaterials application, 100
Biomedical
  applications, 105, 106, 108, 253
  arena, 100
  field, 99–102, 106, 109
  world, 99
Biomedicine, 14, 24, 251
Biomimetic growth, 105
Bionanomaterials, 84
Bioplastics, 4
Biopolymers, 2, 4, 99, 106, 128, 241, 252
Biorefinery functional properties, 80
BioScadaLab software, 130
Biosensor, 37, 39
Biotechnology, 3, 168
Block copolymers, 243, 244
  Disher and Eisenberg theory, 246
  Israelachvili and coworkers theory, 245
  scaling theory, 245
  self-consistent mean field theory, 245
Bombyx mori, 14, 100
Born-Oppenheimer approximation, 215
Brillouin zone, 219, 225
Brownian suspensions, 15

C

Cadmium, 73, 157
  oxide (CdO), 73
Carbohydrates, 4, 100, 140
Carbon
  allotropes, 194–196
  black, 54, 58, 62, 63, 66, 68, 76, 78, 86
  carbon bonds, 5
  cloth, 43, 44, 46
    electrode, 46
    tube, 46
  compounds, 129, 140
  dioxide, 3, 4, 11, 34, 56, 58, 67, 129, 186
  emission, 158
  felt cubes, 47
  footprints, 3, 54
  hexagons, 195
  lattice, 194, 198
  magnetic
    nanoparticles, 24
    nanopowder, 26
  materials, 47, 191, 196
  molecules, 195
  nanodots, 76
  nanohorns (CNHs), 197
  nanomaterials, 183, 184, 192–194, 196–199, 201, 202, 204, 207, 211, 212, 221, 226
  nanoparticles, 24, 25
  nanopowder, 26
  nanostructures, 183, 194, 213
  nanotubes (CNTs), 43, 173, 183, 192, 195–198, 201, 204
  network, 201
  poly-tetra-fluoro-ethylene, 45
  shells, 24
  substrates, 45, 127, 128, 137, 150
Carbonization routes, 194
Carboxyl group, 140
Carboxylic acids, 140, 166
Carcinogenic nitrosamine, 76
Cardiovascular system, 105
Casein
  micelles, 168
  molecules, 170
Catalytic
  membrane reactor, 167
  reactions, 159, 167, 179

Catalyzer, 34, 35
Cathodes, 35, 36, 38, 40–42, 44–49, 164, 177
Cathodic reactions, 44
Cationic dyes, 173
Cations, 85, 121, 164, 177
  exchange membranes, 35, 163
Cell
  biomass, 135, 136
  dry mass, 131
Cellobiose unit, 79
Cellular respiration, 33
Cellulose, 2, 4, 10–13, 15, 16, 18, 76, 79–84, 87, 104, 248–251, 254
  fiber, 79, 80, 82, 84, 87
  ionic liquids, 10
  microfibril, 80
  nanocrystals (CNCs), 12, 13, 18, 249, 254
  nanofibers (CNF), 80, 83, 84, 200, 203
Centrifugal force, 8
Ceramic membranes, 168, 170
Charge transfer mechanism, 183, 221
Chemical
  accidents, 159, 162
  bonds, 119, 143, 144, 186
  compounds, 186
  vapor deposition (CVD), 26, 32, 197
Chemisorption, 191, 196, 199, 207, 221
Chirality, 195
Chitin, 2, 10, 15, 104
Chitosan, 4, 10, 16, 17, 104, 248, 250, 251
  silk sericin (CHT/SS), 104
Climate change (CC), 56, 114, 157, 160
Clostridium butyricum, 34
Clusters, 24, 192, 195, 220
Cobalt nanoclusters, 23, 32
Colloids, 162, 242, 250
Combat biofouling, 116, 118
Conjugated microporous polymers (CMPs), 184, 211, 226
Copiotrophic bacteria, 149
Copolymers, 121, 242, 243, 245–247, 251, 252
Coulomb interactions, 215
Coumarins, 75
Covalent
  organic frameworks (COFs), 211
  triazine frameworks (CTFs), 211
Cox-Merz rule, 11, 16
Critical
  micelle
    concentration (CMC), 244–247, 254
    temperature (CMT), 244
  shear rate, 7, 14
Crumb rubber, 62, 66
Cryogenic grinding, 66
Crystalline, 79, 84, 100, 137, 138, 140, 143, 187, 192, 249, 250
  domain, 79
  lattice, 84
  packing, 79
Crystallinity, 138, 143, 144
Crystallization, 60, 139, 143, 144, 174
Cupriavidus eutrophus, 127, 130, 133, 134, 136, 137, 150
Cyclic molecular architectures, 243
Cyprid settlement assays, 122

D

Dandelion, 53, 60–62, 64, 87
D-anhydroglucose, 79
Degradation, 35, 60, 61, 83, 99, 102, 128, 144, 145, 147–150, 156, 179
Dehydrogenation, 190
Demineralization, 164, 168, 171
Density functional theory (DFT), 214, 216–218, 222–226
Department of energy (DoE), 183, 187, 192, 207, 208, 220, 221, 225
Depolymerization, 66, 81
Devulcanization, 66, 67, 68
Diafiltration, 171
Diallyl disulfide (DADS), 67, 76
Diatoms, 114, 123
Differential scanning calorimetry (DSC), 83, 87, 132, 142, 144, 151
Diffraction
  pattern, 138
  peaks, 138
Diffusion, 5, 23, 25, 32, 44, 47, 149, 161, 165, 167, 173, 187, 191, 213, 247
Dilute solution capillary, 248
Dimethylformamide (DMF), 10, 250
Dimethylsulfoxide, 10, 11
Diphenyl disulfide, 67
Distillation, 164, 165, 167, 178

Double-walled carbon nanotubes (DWCNTs), 195
Doxorubicin, 103
Doxycycline monohydrate (DCMH), 103
Drug delivery, 84, 99, 102, 103, 117, 205, 243, 251, 252
Dynamic
light scattering, 247
viscosity, 9

E

Economical separation, 156, 159, 180
Eco-tire technology, 64
Elastic
moduli, 12
modulus, 13, 16, 79
shear modulus, 9
Elastomers, 53, 57, 59, 60, 62–64, 69, 76, 85, 243
Elctrospun fibers, 106
Electric conductivity, 44
Electrical conductivity, 83, 105, 195
Electrochemical analysis, 161
Electrodes, 35, 42, 164, 177
Electrodialysis, 163, 164, 171, 176, 177
Electrolyte, 35, 163
Electron
acceptor, 36, 39, 41, 42
gas, 218
microscopy, 144, 251
potential, 218
transfer, 44
Electrospinning, 99, 103, 104, 106, 107, 110
Electrospraying, 103
Electrospun, 105, 107
Electrostatic phenomenon, 172
Elementary reactions, 187
Ellipsoids, 195
Empty transition metal orbitals, 198
Emulsion styrene butadiene rubber (E-SBR), 63, 87
Energy
cut-off, 224, 225
saving separation technology
condensation, 161
distillation, 161
evaporation, 161
refrigeration, 161
Epidermal GF (EGF), 106
Epoxiconazole, 145, 148
Epoxidized natural rubber (ENR), 63, 76
Epoxy groups, 63
Equilibrium, 119, 166, 167, 179, 244, 245, 247
Escherichia coli (*E. coli*), 61, 62, 104
Esterification, 165–167
Ethanol (EtOH), 26, 132, 138, 165, 168, 177, 178, 250
European
Tire and Rubber Manufacturer's Association (ETRMA), 65
Union (EU), 54
Extracellular matrix (ECM), 105, 106

F

Fermentation, 2, 4, 62, 130, 135, 136, 165, 176–178
Fermenters, 130, 131, 136
Ferric chloride, 67
Ferricyanide, 36, 41, 42
Ferromagnetic
clusters, 24, 25
impurities, 24
Fertility, 65
Fertilizers, 129, 144
Ferulic acid, 76
Fibroin, 14, 99–102, 106, 107, 109, 110
Filler-filler interaction, 78
Fischer esterification, 166
Flory-Huggins interaction parameter, 243
Fluorenol-fluorenone cycle, 207
Fluorescence spectroscopy, 246
Fluorochrom, 123
Food security, 113, 114
Formic acid, 107
Forward diffusion (FO), 173
Fossil fuel, 34, 53, 184
Fourier transform infrared (FTIR), 83, 107, 108, 119, 132
Fractionation, 132, 168, 170
Free fatty acids (FFA), 133, 151
Freeze-drying, 12, 13, 103–105
Fullerene, 192, 195, 196, 199
Functional groups, 63, 64, 86, 119, 179, 208
Fungicides, 128, 145–148
Fusarium verticillioides, 145, 148

G

Gasoline, 185, 190
Gelatin (GT), 103, 105
Gelation, 17, 102, 106, 243, 244, 250
Generalized gradient approximation (GGA), 213, 218, 220–222
Geochemistry, 38, 48
Glass
  microsphere, 192
  transition temperature, 25
Global biosphere cycles, 145
Glycerin, 127, 132, 135, 137, 150, 176
Glycerol, 16, 127, 129–138, 150
Glycol ethers, 177
Goodyear tire, 62, 64
Granular media, 36
Granules, 40, 41, 44, 132, 141, 144–151
Graphene, 70, 76, 173, 183, 191, 193, 195–197, 199, 201–204, 223
Graphite
  granules, 44
  lattice planes, 198
Graphyne, 183, 196, 201, 204, 219, 220
Gravimetric
  capacity, 187
  storage, 187, 188, 208
Grazers, 117
Green
  chemistry, 1, 2, 4, 128
  energy, 158
  materials, 14, 15, 241, 242
  plastics, 3, 128
  polymers, 1, 2, 4, 5, 9, 10, 17, 18, 241, 242, 248, 251, 253, 254
  synthesis, 113
  technology, 155–160, 162, 165, 178–180
  technology principals, 158
    atom economy, 158
    avoid chemical derivatives, 159
    design chemicals and products, 159
    design harmless chemicals and products, 158
    design less hazardous chemical production, 158
    increase energy efficiency, 159
    minimize the adversity potential, 159
    prevent waste, 158
    real-time analyze to prevent pollution, 159
    use catalysts, not reagents, 159
    use renewable feedstocks, 159
    use unhazardous solvents and reaction conditions, 158
Greenhouse, 56, 129, 144
  accessories, 144
  gas, 56, 156
Ground-state electron density, 216
Guaiacyl ring, 141
Guayule, 53, 60–62, 87

H

Hartree-Fock
  equation, 216
  methods, 215, 216
Hazardous
  effluent generation, 156
  solvents, 158
  substances, 157, 158, 160
  waste, 160
Headphone diaphragms, 83
Hemicellulose, 76, 80, 81, 84
Herbicides, 128, 148–150
Heteroatoms, 201, 205, 208, 211
Hevea brasiliensis, 59, 60
High strength mixer (HSM), 66
Highly dispersive silica (HDS), 78
Hohenberg-Kohn density, 217
Holocellulose, 129
Homeostasis, 103
Homopolymers, 242
Hybrid
  materials, 128
  silicas, 168
  structures, 192
Hybridization, 195, 222, 252
Hydration, 77, 121, 177
Hydride nanostructures, 24
Hydrocarbon, 26, 67, 183, 186
  fuels, 191
  gas, 68
Hydrodynamic radius (RH), 244, 245, 247
Hydrogels, 16, 102, 105, 108, 113, 120, 121, 250, 251
Hydrogen (H_2), 15–17, 26, 34, 36, 38, 40, 66–68, 79, 129, 140, 164, 174, 183–194,

196–202, 204–213, 219–221, 225, 226, 251
adsorption, 191, 198, 199, 206, 207, 219
bonding, 140
fuel, 185, 186, 192, 225, 226
cells (HFC), 192
isotherm, 208
storage, 183, 184, 186–189, 191–194, 196–199, 202, 204, 205, 207–209, 211–213, 219, 221, 225, 226
carbon nanomaterials, 193
challenges and issues, 186
conventional and advanced materials, 190
dispersion corrections importance, 222
functionalized carbon nanomaterials, 198
functionalized polymers, 211
polymers, 204
pure carbon nanomaterials, 196
pure polymers, 207
theoretical insights, 212
Hydrogenation, 174, 190, 191
Hydrolysis, 190, 249
Hydrolytic evolution, 187
Hydrolyzed polyacrylamide (HPAM), 17
Hydrophilic solvents, 167
Hydrophilicity, 103, 120
Hydrophobic compounds, 140
Hydrophobicity, 82, 83
Hydroxides groups, 85
Hydroxyl group, 166
Hydroxypropyl
cellulose (HPC), 13
methylcellulose (HPMS), 13
Hyper cross-linked polymers (HCPs), 184, 205

I

Illegal, unreported, and unregulated (IUU), 114
Immunoglobulin (IgG), 250, 254
Incubation, 144, 147–149
Infrared (IR) spectroscopy, 119, 140, 141, 142, 144
Inorganic
cobalt salts, 87
composites, 105
membranes, 178
Integrated biology, English, and technology (IBET), 39, 48
Intelligent gravimetric analyzer, 206
Internal combustion engine (ICE), 184, 185
International
Society for Microbial Electrochemistry and Technology (ISMET), 39, 48
Union of Pure and Applied Chemistry (IUPAC), 2
Intrinsic microporosity, 206
Ionic liquids, 10, 11, 12
Isoprene, 54, 59, 60, 62
Isotropic phase transition, 13

K

Kaolinite layers, 85
Ketjen black carbon, 45
Kinetic, 15, 25, 130, 143, 144, 147, 148, 165, 167, 179, 191, 192
constants, 133
driving forces, 24
energy, 215, 217, 224
Kohn-Sham
energy, 223
equation, 217
Krusal-Wallis test, 123
Kubas interaction, 221

L

Lactic acid (LA), 17, 104, 176, 177
Lactose crystallization, 171
Langmuir films, 9
Lap-shear strength, 75
Lead oxide (PbO), 73
Lewis acids, 166
Liquefaction, 183, 187
Liquid
Farnesene rubber, 61, 87
hydrogen, 186, 187, 191
polymers, 207
saline medium, 130
Local density approximation (LDA), 213, 218
Lumens, 114, 115

M

Macrocyclicimpetuses, 45
Macromolecular
chains, 14, 243, 246, 253
coil overlapping (c*), 11

materials, 102
nature, 2
self-assemblies characterization methods, 246
CMC analysis, 246
micellar size and shape determination, 247
structures, 242
systems, 1, 9, 241
Macromolecules, 2, 5–7, 12–14, 140, 170, 205, 242–244, 253
Magic triangle, 78
Magnesium oxide (MgO), 73, 74, 132
Magnetic
carbon nanopowder, 32
field, 23–27, 31, 32
nanoclusters, 26
nanocomposite film, 31
nanoparticles, 32
Magneto-electronics, 32
Marine polymers, 2
Mass transfer, 10, 44, 165, 179
Matrix, 4, 15, 62, 73, 82–84, 86, 106, 120, 145, 147, 148, 150, 167, 178
Medical implants, 4, 84, 105
Melting point, 139
Membrane
applications, 168
distillation (MD), 131, 164
electrolysis, 164
operations, 156, 168, 171
osmometry, 248
phases, 163
processes, 155, 156, 159, 161, 162, 164, 168, 170, 173–176
properties, 159, 160
reactors, 167
separation, 155, 156, 160–162, 165, 168, 172, 175, 176, 180, 249
technology, 155, 156, 161, 165, 167, 169, 172, 175, 178
Mesenchymal stem cells, 106
Metal
hydrides, 191, 192, 204, 205
metal cluster, 212
organic frameworks (MOFs), 192, 196, 204, 208, 211
Metallic diaphragm, 83
Metalorganic frameworks (MOFs), 192
Methanol separation, 174
Methine group, 137
Methyl group, 137, 140
Methylene group, 137
Metribuzin, 148–150
Microalgae, 114
Microanalysis, 144
Microbes, 33, 35, 36, 40, 54, 120
Microbial
compounds, 35
culture, 33
energy
components, 38, 39
units, 38, 48
fuel cell (MFC), 33–49
power
devices, 39, 48
modules, 41, 42
Microemulsion, 73
Microfiltration (MF), 161–163, 168–170, 173, 176
Micropores, 206, 208
Moduli, 15, 144
Molar mass, 79, 242, 244, 247
Molecular
design, 63, 208
hydrogen, 187, 193, 198
inclusions, 190
orientation, 13
scale, 162
size, 160
structure, 8, 66, 180, 205
weight, 5–7, 13, 68, 100, 101, 208, 250
Mollusks, 115
Monolayers, 9
Monomer, 4, 60, 205, 208, 242, 249, 252
Monte-Carlo simulations, 218
Mulliken's charge analysis, 221
Multi-functional additives (MFA), 54, 87
Multireference configuration interaction, 216
Multi-walled nanotubes (MWCNTs), 195, 197

N

Nafion membranes, 192
Nano clay, 86
Nano copper oxide, 118, 120
Nano oxides, 74

Nanocellulose, 80, 83, 84
Nanoclasters, 25
Nanocomposites, 4, 23, 30, 31, 80, 83, 86, 104
Nanocones, 106
Nanodots, 76
Nano-electronic components, 83
Nanofibers, 80, 83, 102, 104, 106–110, 196, 198, 200, 203
Nanofillers, 76, 77
Nanofiltration (NF), 83, 163, 168, 171, 173
Nanomaterials, 80, 113, 117, 118, 124, 173, 183, 184, 187, 192–194, 196–199, 201, 202, 204, 207, 211, 212, 219, 221, 225, 226, 241, 242
Nanoparticles, 3, 12, 23–26, 30–32, 43, 73, 103, 104, 118, 172, 193, 194, 243, 250–252
Nanopowders, 25–27, 31
Nanoscale, 24, 191, 242
Nanosensors, 24
Nanosheets, 250
Nanotechnological intervention, 117
  mixed charged zwitterionic hydrogel, 121
  nano CuO hydrogel-green synthesis, 120
  surface characterization, 119
Nanotechnology, 99, 117, 242
Nanotubes, 23, 172, 195
Nanowires, 23, 24
Natural
  gum, 74, 75
  polymers, 2
  rubber (NR), 53, 54, 58–60, 62, 63, 66–68, 70, 72, 74, 75, 78, 84, 86, 87
Navicula perminuta, 123
Nematic
  molecular order, 12
  organization, 13
Newtonian polymer fluids, 9
Nickel nanoparticles, 25
Non-Fickian transport, 149
Non-heavy metal oxide, 73
Non-Newtonian behavior, 7, 14, 17
Non-polar molecule, 118
Non-scattering methods, 248
Non-tidal condition, 122
Non-volatile substances, 165
Nuclear
  energy, 34
  magnetic resonance
    spectroscopy, 248
    tomographs, 24
  power plants, 180
Nucleotide, 252, 253
Nutritional security, 114

O

Oligomers, 68, 249
Oligosaccharides, 250
Open circuit potential, 41, 42
Organic
  acids, 129
  compound, 35, 166
  materials, 80, 208
  molecules, 114
  polymers, 184, 192, 193, 205, 226
  solvents, 3, 10, 167, 249, 250, 254
    dimethyl sulfoxide (DMSO), 250
  wastes, 114
Organosilanes, 77
Oscillatory data, 15
Oxidation, 33, 34, 41, 42, 66, 101, 129, 184, 194, 201
Oxidative pyrolysis, 68
Oxygen reduction rate, 47
Oysters, 114

P

Paracrystalline, 79
Parthenium argentatum, 60
Pathogens, 40
Pee power field, 43
Pellets, 132, 141, 142, 144–150
Penicillium expansum, 61
Perdew-Burke-Ernzerhof (PBE), 218, 223
Periodicity, 225, 242
Permeation flux, 178
Pervaporation (PV), 163, 164, 167, 178
Pesticides, 144, 145, 151, 157
Petrochemical sources, 69
pH, 17, 38, 45, 77, 102, 130, 131, 247, 250, 251
Pharmaceutical packaging, 83
Phenolics, 75
Phosphate buffer, 136
Phosphoproteins, 4

Physisorption, 183, 191, 192, 196, 198, 199, 207, 212, 221
Pilot production, 129, 136, 151
Plasticization, 5
Plasticizers, 10, 72, 73
Platinum, 35, 44–47
Polar
  interaction, 83
  solvents, 247, 249
Polarity, 12, 83
Polarization, 177, 212, 219, 222
Pollutant, 149, 172, 184
Poly(3-hydroxybutyrate), 145, 147
Poly(acrylic acid) (PQD), 207
Poly(propylene oxide) (PPO), 252
Poly(vinyl acetate), 75
Poly(vinyl alcohol), 75, 168
Poly-3-hydroxybutyrate (P(3HB)), 132, 137–145, 148, 150
Polyaniline, 113, 118–121, 124
Polyaromatic hydrocarbons (PAH), 70
Polybutadiene rubber, 63
Polycaprolactone (PCL), 106, 107, 109
Polychaetes, 114
Polycyclic aromatics (PCA), 70, 71
Polydimethylsiloxane (PDMS), 103
Polydispersity, 6, 137, 245
Polyelectrolyte complexes (PECs), 248–250
Polyesters, 4, 58
Polyethylene, 4, 84, 106, 113, 115, 118–121
  aquaculture cage nets, 113
Polyethyleneimine, 43, 209
Polyhydroxyalkanoates (PHAs), 127–131, 133, 134, 136–138, 141, 150, 151
  pilot production, 136
  synthesis, 129, 136
Polyisoprene, 59, 60
Polymer, 1, 5–7, 17, 26–29, 53, 59, 82, 173, 207, 209, 210, 250
  film preparation techniques, 8
    drop casting, 8
    Langmuir-Blodgett, 9
    spin coating deposition, 8
    spray deposition, 8
    tape casting, 8
  intrinsic microporosity (PIM), 184, 206, 208–210
  nanocomposites (PNC), 25, 26, 32
  processing
    solution rheology correlation, 5
  solution, 8, 9, 13, 15
  solvent interactions, 6
  yields, 127
Polymeric matrixes, 12
Polymerization, 3, 12, 15, 60, 61, 79, 118, 205, 242, 243, 245, 251
Polyolefins, 4, 128
Polypeptide, 100
Polyphenolic compounds, 74
Polypropylene (PP), 84, 129, 136, 205
Polysaccharide, 15, 79, 80, 114
Polystyrene (PS), 58, 123, 207, 209, 246
Polytetrafluoroethylene (PTFE), 36, 45
Polyvinyl
  alcohol (PVA), 25, 75, 84, 104
  butyral (PVB), 26, 32
  chloride, 72
Porosity, 41, 105, 106, 191, 205, 208
Porous
  aromatic frameworks (PAFs), 211
  organic polymers (POPs), 205, 211, 212
Power
  density, 43, 46, 47
  generation applications, 38
Pressure-driven membrane filtration, 162
  microfiltration, 162
  nanofiltration, 162
  reverse osmosis, 162
  ultrafiltration, 162
Prevulcanization inhibitor (PVI), 70
Process intensification (PI), 165, 180
  membrane reactor-esterification reaction, 166
  milk components, 168
Proton, 35, 41
Protonation, 166
Prototype, 34, 62
P-toluenesulfonic acid, 166
Pyridine, 10, 46
Pyrolysis, 26, 60, 67, 68, 76

Q

Quadriperm dishes, 123
Quantum simulations, 213–215, 225
  density functional theory (DFT), 216
  Hartree-Fock method, 215

Kohn-Sham equations, 217
many-body Hamiltonian, 214

R

Radial tire application, 86
Raw materials, 2, 56–59, 129, 159, 165, 176
Recycle, 53, 57, 67, 87, 156, 160, 165, 176, 180
Reduction-oxidation respiration, 33
Relaxation time, 9, 11
Renewable
energy, 3, 4, 129, 157
natural polymer, 79
resources, 3, 4, 156
sources, 128, 158
Reoxidation, 41, 42
Reverse osmosis (RO), 161, 163, 168, 171–173
Rheology, 1, 4, 8, 17
Rubber
clay nanocomposites, 86
compounds, 75, 77, 78
matrix, 70, 86
powder, 66
Research Institute (RRI), 63
Russian Collection of Industrial Microorganisms (RCIM), 130

S

Sabu Thomas schemes, 143
Saccharomyces cerevisiae, 33
Scanning electron microscope (SEM), 83, 109, 118–121, 132, 141, 142
Schrödinger equation, 214, 215, 224
Secondary
building units (SBU), 192
raw material, 65
Self-assemblies, 241–243, 246, 253, 254
Sericin, 99–102, 104–106, 108, 110
Shear
cessation, 13
deformation, 5, 8, 17
frequencies, 14, 15
moduli, 12–14, 16
rates, 7–9, 11, 14, 16, 17
stress, 9, 14, 123
thickening zone, 13
thinning zone, 13
Silane, 77, 78
Silica, 54, 58, 62–64, 70, 72, 76–78, 83
Silk
fibroin (SF), 14, 15, 100, 101, 103–109
hydrogels, 105
polymers, 14
protein (SP), 14, 15, 103, 132
sericin, 101
Silver nanoparticles (AgNPs), 104
Single-walled carbon nanotube (SWCNT), 197, 204, 222
Size exclusion chromatography, 248
Skin fibroblasts, 103
Small-angle x-ray scattering (SAXS), 247, 248
Sodium
carboxymethyl cellulosechitosan, 249
chloride electrolysis, 164
hydroxide, 82
phosphates, 136
Solid waste residue, 177
Solution
polymerized styrene butadiene (S-SBR), 58, 63, 64
rheology, 5, 10
styrene butadiene rubber (S-SBR), 58, 63
viscosity, 9, 11–13, 15
Sorption, 6, 12, 141, 191, 205
Staphylococcus aureus, 104
Static light scattering, 247
Stearic acid, 58, 59, 73
Stoichiometric reagents, 159
Storage modulus, 14, 15
Styrene circular solution, 58
Sulfuric acid, 166, 167
Supramolecular chemistry, 242
Surface
energy, 78, 106, 116, 121
hydration, 77
surface-to-volume ratio, 106
Surfactant-accelerator-processing aid (SAPA), 54
Sustainability, 53–56, 58, 65, 73, 75, 76, 79, 86, 87
Sustainable
fillers, 70, 76
cellulose, 79
clay, 84

hemicellulose, 80
lignin, 81
silica, 76
industrial growth, 165
materials, 10, 53, 57, 62
rubber additives, 69

T

Tebuconazole, 145, 147, 148
Tecan plate reader, 123
Teflon press molds, 26
Tensile strength, 68, 72, 251
Textile sewer water treatment, 173
Thermal
aging, 62
conductivity, 9, 183, 195
Thermodynamic
characteristics, 11
considerations, 167
force, 121
limitations, 167
point, 12
quality, 6, 11
Thermogelling, 17
Thermograms, 144
Thermogravimetric analysis (TGA), 83
Thermoplastic, 4
composite materials, 84
lignin, 4
materials, 4, 139
matrix, 84
system, 82
Thermoset matrices, 84
Thermotolerance, 101
Three-dimensional (3D), 43, 58, 77, 105, 196
EPDM network, 58
network structures, 43
Thyme essential oil (TEO), 103
Time-temperature superposition, 15
Tissue engineering (TE), 14, 99, 104–108
Toxicity, 24, 157, 158, 177, 180, 204
Transdermal drug delivery applications, 103
Transmission electron microscope (TEM), 104, 109, 248
Tributyltin (TBT), 116
Truck bus radial (TBR), 63

U

U.S. Tire Manufacturing Association (USTMA), 65
Ultrafiltration (UF), 136, 163, 168, 170, 171, 173, 176, 178
Ultrasonic process, 67
Ultra-tensile steel cord, 86
Unsaturation, 71

V

Vacuum evaporation, 171
Van der Waals
corrections, 220
forces, 219
interactions, 187, 213, 222
Vapor permeation (VP), 167, 209
Velocity, 6, 83, 168
Viscoelasticity, 4, 18
Viscometry, 248
Viscosity, 7–13, 15–18, 25, 72
Viscous shear modulus, 9
Volatile
chlorides, 26
organic compounds, 139
Vulcanization, 66, 72, 73, 74

W

Waste matter treatment plants, 172
Wastewater treatment, 38, 40, 42, 44, 47, 48
Water
cycle, 160
electrolysis, 164
Waterborne polyurethane (WPU), 14, 15, 18
Whey protein, 169–171
concentrates (WPC), 169–171
Wire rope matrix, 106
Wound dressing, 103, 104, 106

X

X-ray
diffraction (XRD), 132, 138, 151
spectrometry, 144

Y

Young's modulus/moduli, 79, 144

Z

Zero
emission, 184
shear rate Newtonian viscosity, 7
Zinc oxide (ZnO), 58, 70, 73, 74, 104, 173
Zoospores, 123
Zwitterionic
hydrogels, 121
polymers, 121